CHILTON'S TRUCK AND VAN MANUAL 1991-1995

Publisher Editor-In-Chief	Kerry A. Freeman, S.A.E.
Managing Editors	Peter M. Conti, Jr., W. Calvin Settle, Jr., S.A.E.
Assistant Managing Editor	Nick D'Andrea
Senior Editors	Debra Gaffney, Michael L. Grady, Richard J. Rivele, S.A.E. Richard T. Smith, Jim Taylor, Ron Webb
Project Managers	Benjamin E. Greisler, A.S.E., Martin J. Gunther Jeffrey M. Hoffman, James Steele, Michael M. Carroll, Jacques Gordon
Editorial Staff	Chris Armenti, Peter A. Bilotta, A.S.E., Lawrence C. Braun, S.A.E., A.S.C., Thomas P. Browne III, Dean G. Callahan, William C. Cottman, A.S.E., Leonard Davis, A.S.E., Paul DeGuiseppi, Joseph L. Defrancesco, A.S.E., Robert F. Dougherty, Jr., Robert E. Doughten, John J. Ferraro, A.S.E., Sam Fiorani, Andrew J. Folz, A.S.E., Matthew E. Frederick, Edward J. Giacomucci, A.S.E., Shawn Hibbert, A.S.E., Neil Leonard, Will Kessler, Kevin Maher, Robert McAnally, Raymond K. Moore, Craig P. Nangle, A.S.E., Charles Ramsey, Roy Ripple, A.S.E., Don Schnell, A.S.E., Richard Schwartz, Anthony Tortorici, A.S.E., S.A.E.
Production Manager	Andrea M. Steiger
Assistant Production Manager	Marsha Park Herman
Production Specialists	Margaret Stoner, Kim Hayes
Director of Manufacturing	Mike D'Imperio
Asst. Manufacturing Manager	Robin Norman
OFFICERS	
President, Chilton Enterprises	David S. Loewith
Senior Vice President	Ronald A. Hoxter

CHILTON BOOK COMPANY

ONE OF THE **DIVERSIFIED PUBLISHING COMPANIES,**
A PART OF **CAPITAL CITIES/ABC, INC.**

Manufactured in
© 1994 Chilton Book Company
Chilton Way, Radnor, PA 19089
ISBN 0-8019-7911-0
ISSN 0742-0315

1234567890 3210987654

HOW TO USE THIS MANUAL

HOW TO USE THIS MANUAL

Truck Section

Truck sections are grouped by manufacturer and arranged in alphabetical order. The text and illustrations that comprise the service procedures in each Truck Section are arranged in the following order of systems and components: Engine Mechanical, Engine Lubrication, Engine Cooling, Engine Electrical, Emission Controls, Fuel System, Drive Axle, Manual Transmission/Transaxle, Clutch, Automatic Transmission/Transaxle, Front Suspension, Rear Suspension, Steering, Brakes, Chassis Electrical.

Specification charts are always located at the front of each section. All illustrations are located as close as possible to the pertinent text. Procedures are for all models in the particular section unless specifically noted otherwise.

Locating Information

The Table of Contents, at the front of the book, lists the beginning of each Car Section in the manual.

To find where a particular Truck Section is located in the book, you need only look in the Table of Contents. Once you have found the proper section, you may wish to find where specific procedures are located in that section. Turn to the Index at the front of the section. At the upper left-hand side is a listing of the main topics within the section and the page number they will be found on. Following the main topics is an alphabetical listing of all the procedures within the section and their page numbers.

Safety Notice

Proper service and repair procedures are vital to the safe, reliable operation of all motor vehicles, as well as the personal safety of those performing repairs. This manual outlines procedures for servicing and repairing vehicles using safe effective methods. The procedures contain many NOTES and CAUTIONS which should be followed along with standard safety procedures to eliminate the possibility of personal injury or improper service which could damage the vehicle or compromise its safety.

It is important to note that repair procedures and techniques, tools and parts for servicing motor vehicles, as well as the skill and experience of the individual performing the work vary widely. It is not possible to anticipate all of the conceivable ways or conditions under which vehicles may be serviced, or to provide cautions as to all of the possible hazards that may result. Standard and accepted safety precautions and equipment should be used when handling toxic or flammable fluids, and safety goggles or other protection should be used during cutting, grinding, chiseling, prying, or any other process that can cause material removal or projectiles.

Some procedures require the use of tools specially designed for a specific purpose. Before substituting another tool or procedure, you must be completely satisfied that neither your personal safety, nor the performance of the vehicle will be endangered.

Part Numbers

Part numbers listed in this book are not recommendations by Chilton for any product by brand name. They are references that can be used with interchange manuals and aftermarket supplier catalogs to locate each brand supplier's discrete part number.

Although information in this manual is based on industry sources and is as complete as possible at the time of publication, the possibility exists that some car manufacturers made later changes which could not be included here. Information on very late models may not be available in some circumstances. While striving for total accuracy, Chilton Book Company cannot assume responsibility for any errors, changes, or omissions that may occur in the compilation of this data.

Copyright Notice

Truck Sections

Unit Repair Sections

CHRYSLER CORP. 1

Full Size Trucks
Pickup • Ramcharger • Van

SPECIFICATIONS

ENGINE IDENTIFICATION

Year	Model	Engine Displacement Liters (cc)	Engine Series (ID/VIN)	Fuel System	No. of Cylinders	Engine Type
1991	D/W 150 Pick-up	3.9 (3916)	X	TFI	6	OHV
	D/W 150 Pick-up	5.2 (5211)	Y	TFI	8	OHV
	D/W 150 Pick-up	5.9 (5899)	Z	TFI	8	OHV
	D 250 Pick-up	3.9 (3916)	X	TFI	6	OHV
	D/W 250 Pick-up	5.2 (5211)	Y	TFI	8	OHV
	D/W 250 Pick-up	5.9 (5882)	8	DSL	6	OHV
	D/W 250 Pick-up	5.9 (5899)	Z	TFI	8	OHV
	D/W 350 Pick-up	5.9 (5882)	8	DSL	6	OHV
	D/W 350 Pick-up	5.9 (5899)	5	TFI	8	OHV
	B150 Van	3.9 (3916)	X	TFI	6	OHV
	B150 Van	5.2 (5211)	Y	TFI	8	OHV
	B250 Van	3.9 (3916)	X	TFI	6	OHV
	B250 Van	5.2 (5211)	Y	TFI	8	OHV
	B250 Van	5.9 (5899)	W	TFI	8	OHV
	B350 Van	5.2 (5211)	Y	TFI	8	OHV
	B350 Van	5.9 (5899)	W	TFI	8	OHV
	Ramcharger	5.2 (5211)	Y	TFI	8	OHV
	Ramcharger	5.9 (5899)	Z	TFI	8	OHV
1992	D/W 150 Pick-up	3.9 (3916)	X	MFI	6	OHV
	D/W 150 Pick-up	5.2 (5211)	Y	MFI	8	OHV
	D/W 150 Pick-up	5.9 (5899)	Z	TFI	8	OHV
	D 250 Pick-up	3.9 (3916)	X	MFI	6	OHV
	D/W 250 Pick-up	5.2 (5211)	Y	MFI	8	OHV
	D/W 250 Pick-up	5.9 (5882)	8	DSL	6	OHV
	D/W 250 Pick-up	5.9 (5899)	Z	TFI	8	OHV
	D/W 350 Pick-up	5.9 (5882)	8	DSL	6	OHV
	D/W 350 Pick-up	5.9 (5899)	5	TFI	8	OHV
	B150 Van	3.9 (3916)	X	MFI	6	OHV
	B150 Van	5.2 (5211)	Y	MFI	8	OHV
	B250 Van	3.9 (3916)	X	MFI	6	OHV
	B250 Van	5.2 (5211)	Y	MFI	8	OHV
	B250 Van	5.9 (5899)	Z	TFI	8	OHV
	B350 Van	5.2 (5211)	Y	MFI	8	OHV
	B350 Van	5.9 (5899)	Z	TFI	8	OHV
	Ramcharger	5.2 (5211)	Y	MFI	8	OHV
	Ramcharger	5.9 (5899)	Z	TFI	8	OHV
1993	D/W 150 Pick-up	3.9 (3916)	X	MFI	6	OHV
	D/W 150 Pick-up	5.2 (5211)	Y	MFI	8	OHV
	D/W 150 Pick-up	5.9 (5899)	Z	MFI	8	OHV
	D 250 Pick-up	3.9 (3916)	X	MFI	6	OHV
	D/W 250 Pick-up	5.2 (5211)	Y	MFI	8	OHV
	D/W 250 Pick-up	5.9 (5882)	8	DSL	6	OHV
	D/W 250 Pick-up	5.9 (5899)	Z	MFI	8	OHV
	D/W 350 Pick-up	5.9 (5882)	8	DSL	6	OHV
	D/W 350 Pick-up	5.9 (5899)	5	MFI	8	OHV
	B150 Van	3.9 (3916)	X	MFI	6	OHV
	B150 Van	5.2 (5211)	Y	MFI	8	OHV

79111C02

VEHICLE IDENTIFICATION CHART

Code	Liters	Cu. In. (cc)	Cyl.	Fuel Sys.	Eng. Mfg.
8	5.9	359 (5882)	6	DSL	Cummins
5	5.9	360 (5899)	8	TFI	Chrysler
C	5.9	359 (5882)	6	DSL	Cummins
A	5.9	360 (5899)	8	MFI	Chrysler
W	8.0	488 (7997)	10	MFI	Chrysler

Code	Year
N	1992
R	1994

DSL - Diesel

MFI - Multiport fuel injection

TFI - Throttle body fuel injection

79111C01

GENERAL ENGINE SPECIFICATIONS

Year	Engine ID/VIN	Engine Displacement Liters (cc)	Fuel System Type	Net Horsepower @ rpm	Net Torque @ rpm (ft. lbs.)	Bore x Stroke (in.)	Compression Ratio	Oil Pressure @ rpm
1991	X	3.9 (3916)	MFI	125@4000	195@2000	3.91x3.31	9.1:1	30-80@3000
	Y	5.2 (5211)	TFI	170@4000	260@2000	3.91x3.31	9.2:1	30-80@3000
	8	5.9 (5882)	DSL	160@2500	400@1700	4.02x4.72	17.5:1	30-70@2500
	W	5.9 (5899)	TFI	190@4000	292@2400	4.00x3.58	8.1:1	30-80@3000
	Z	5.9 (5899)	TFI	193@4000	292@2400	4.00x3.58	8.1:1	30-80@3000
	5	5.9 (5899)	TFI	193@4000	292@2400	4.00x3.58	8.1:1	30-80@3000
1992	X	3.9 (3916)	MFI	175@4800	220@3200	3.91x3.31	9.1:1	30-80@3000
	Y	5.2 (5211)	MFI	170@4000	260@2000	3.91x3.31	9.2:1	30-80@3000
	8	5.9 (5882)	DSL	160@2500	400@1700	4.02x4.72	17.5:1	30-70@2500
	Z	5.9 (5889)	TFI	193@4000	292@2400	4.00x3.58	8.1:1	30-80@3000
	5	5.9 (5889)	TFI	193@4000	292@2400	4.00x3.58	8.1:1	30-80@3000
1993	X	3.9 (3916)	MFI	175@4800	220@3200	3.91x3.31	9.1:1	30-80@3000
	Y	5.2 (5211)	MFI	230@4800	280@3200	3.91x3.31	9.1:1	30-80@3000
	8	5.9 (5882)	DSL	160@2500	400@1750	4.02x4.72	17.5:1	30-70@2500
	Z	5.9 (5899)	MFI	230@4000	325@2500	4.00x3.58	8.9:1	30-80@3000
	5	5.9 (5899)	MFI	230@4000	230@2800	4.00x3.58	8.9:1	30-80@3000
1994-95	X	3.9 (3916)	MFI	175@4800	220@3200	3.91x3.31	9.1:1	30-80@3000
	Y	5.2 (5211)	MFI	220@4400	300@3200	3.91x3.31	9.1:1	30-80@3000
	C	5.9 (5882)	DSL	160@2500	400@1750	4.02x4.72	17.5:1	30-70@2500
	5	5.9 (5899)	MFI	230@4000	330@3200	4.00x3.58	8.9:1	30-80@3000
	Z	5.9 (5899)	MFI	230@4000	330@3200	4.00x3.58	8.9:1	30-80@3000
	A	5.9 (5899)	MFI	230@4000	330@3200	4.00x3.58	8.9:1	30-80@3000
	W	8.0 (7997)	MFI	300@4000	450@2400	4.00x3.58	8.6:1	50-60@3000

MFI - Multiport fuel injection

TFI - Throttle body fuel injection

DSL - Diesel

79111C03

GASOLINE ENGINE TUNE-UP SPECIFICATIONS

Year	Engine ID/VIN	Engine Displacement Liters (cc)	Spark Plugs Gap (in.)	Ignition Timing (deg.)		Fuel Pump (psi)	Idle Speed (rpm)		Valve Clearance	
				MT	AT		MT	AT	In.	Ex.
1991	X	3.9 (3916)	0.035	10B	10B	13-16	750	750	HYD	HYD
	Y	5.2 (5211)	0.035	10B	10B	13-16	700	700	HYD	HYD
	5	5.9 (5899)	0.035	10B	10B	13.5-15.5	700	700	HYD	HYD
	W	5.9 (5899)	0.035	10B	10B	13.5-15.5	700	700	HYD	HYD
	Z	5.9 (5899)	0.035	10B	10B	13.5-15.5	700	700	HYD	HYD
1992	X	3.9 (3916)	0.035	1	1	37-41	750	750	HYD	HYD
	Y	5.2 (5211)	0.035	1	1	37-41	700	700	HYD	HYD
	5	5.9 (5899)	0.035	10B	10B	13.5-15.5	700	700	HYD	HYD
	Z	5.9 (5899)	0.035	10B	10B	13.5-15.5	700	700	HYD	HYD
1993	X	3.9 (3916)	0.035	1	1	37-41	750	750	HYD	HYD
	Y	5.2 (5211)	0.035	1	1	37-41	700	700	HYD	HYD
	5	5.9 (5899)	0.035	1	1	35-45	700	700	HYD	HYD
	Z	5.9 (5899)	0.035	1	1	35-45	700	700	HYD	HYD
1994-95	X	3.9 (3916)	0.035	1	1	35-45	2	2	HYD	HYD
	Y	5.2 (5211)	0.035	1	1	35-45	2	2	HYD	HYD
	5	5.9 (5899)	0.035	1	1	35-45	2	2	HYD	HYD
	Z	5.9 (5899)	0.035	1	1	35-45	2	2	HYD	HYD
	W	8.0 (7997)	0.035	1	1	35-45	2	2	HYD	HYD

NOTE: The Vehicle Emission Control Information label reflects changes made during production. The label figures must be used if they differ from above

B - Before top dead center

HYD - Hydraulic

1 Ignition timing cannot be adjusted. Base engine timing is set at TDC during assembly.

2 Refer to the Vehicle Emission Control Information (VECI) label for correct specification

79111C04

DIESEL ENGINE TUNE-UP SPECIFICATIONS

Year	Engine ID/VIN	Engine Displacement cu. in. (cc)	Valve Clearance Intake (in.)	Valve Clearance Exhaust (in.)	Intake Valve Opens (deg.)	Injection Pump Setting (deg.)	Injection Nozzle Pressure (psi) New	Injection Nozzle Pressure (psi) Used	Idle Speed (rpm)	Cranking Compression Pressure (psi)
1991	8	5.9 (5882)	0.010	0.020	NA	1	3550	NA	2	NA
1992	8	5.9 (5882)	0.010	0.020	NA	1	3550	NA	2	NA
1993	8	5.9 (5882)	0.010	0.020	NA	1	3550	NA	2	NA
1994	C	5.9 (5882)	0.010	0.020	NA	1	3550	NA	2	NA

NOTE: The Vehicle Emission Control Information label reflects changes made during production. The label figures must be used if they differ from above

NA - Not Available

1 Align marks on pump flange and gear housing

2 Automatic transmission with A/C - 700 rpm; Manual transmission with A/C - 750 rpm

79111C05

FIRING ORDERS

NOTE: To avoid confusion, always replace spark plug wires one at a time.

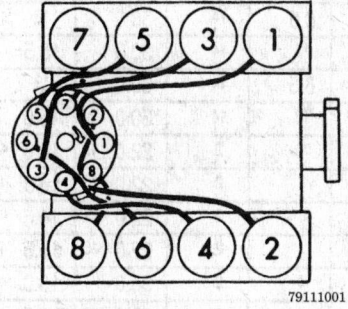

79111001

5.2L and 5.9L Engines
Engine Firing Order:1–8–4–3–6–5–7–2
Distributor Rotation: Clockwise

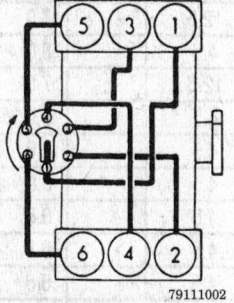

79111002

3.9L Engine
Engine Firing Order:
1–6–5–4–3–2
Distributor Rotation:
Clockwise

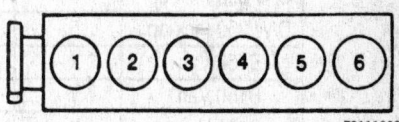

79111003

Cummins 5.9L Turbodiesel Engine
Engine Firing Order: 1–5–3–6–2–4

CAPACITIES

| Year | Model | Engine ID/VIN | Engine Displacement Liters (cc) | Engine Oil with Filter | Transmission (pts.) | | | Transfer Case (pts.) | Drive Axle | | Fuel Tank (gal.) | Cooling System (qts.) |
					4-Spd	5-Spd	Auto.		Front (pts.)	Rear (pts.)		
1991	D/W 150 Pick-up	X	3.9 (3916)	4.5	7.0	4.0	1	13	18	14	22.0 3	15.1
	D/W 150 Pick-up	Y	5.2 (5211)	4.5	7.0	4.0	1	13	18	14	22.0 3	17.0
	D/W 150 Pick-up	Z	5.9 (5899)	4.5	7.0	4.0	1	13	18	14	30.0	15.5
	D250 Pick-up	X	3.9 (3916)	4.5	7.0	4.0	1	-	-	14	22.0 3	15.1
	D/W 250 Pick-up	Y	5.2 (5211)	4.5	7.0	4.0	1	13	18	14	22.0 3	15.5
	D/W 250 Pick-up	8	5.9 (5882)	12.5	-	7.0	22.0	13	6.5	14.0	30.0	18.0
	D/W 250 Pick-up	Z	5.9 (5899)	4.5	7.0	4.0	1	13	18	14	30.0	15.5
	D/W 350 Pick-up	8	5.9 (5882)	12.5	-	7.0	22.0	13	6.5	14.0	30.0	18.0
	D/W 350 Pick-up	5	5.9 (5899)	4.5	7.0	4.0	1	13	18	14	30.0	15.5
	B150 Van	X	3.9 (3916)	4.5	7.0	4.0	1	-	-	2	22.0 15	14.6 4
	B150 Van	Y	5.2 (5211)	4.5	7.0	4.0	1	-	-	2	22.0 15	16.5 4
	B250 Van	X	3.9 (3916)	4.5	7.0	-	1	-	-	2	22.0 15	14.6 4
	B250 Van	Y	5.2 (5211)	4.5	7.0	-	1	-	-	2	22.0 15	16.5 4
	B250 Van	W	5.9 (5899)	4.5	7.0	-	1	-	-	2	35.0	15.0 4
	B350 Van	Y	5.2 (5211)	4.5	7.0	-	1	-	-	2	22.0 15	16.5 4
	B350 Van	W	5.9 (5899)	4.5	7.0	-	1	-	-	2	35.0	15.0 4
	Ramcharger	Y	5.2 (5211)	4.5	7.0	4.0	1	13	18	14	34.0	16.5
	Ramcharger	Z	5.9 (5899)	4.5	7.0	4.0	1	13	18	14	34.0	15.0
1992	D/W 150 Pick-up	X	3.9 (3916)	4.5	-	8.0	1	13	18	14	22.0 3	15.1
	D/W 150 Pick-up	Y	5.2 (5211)	4.5	-	8.0	1	13	18	14	22.0 3	17.0
	D/W 150 Pick-up	Z	5.9 (5899)	4.5	-	8.0	1	13	18	14	30.0	15.5
	D 250 Pick-up	X	3.9 (3916)	4.5	-	8.0	1	-	18	14	22.0 3	15.1
	D/W 250 Pick-up	Y	5.2 (5211)	4.5	-	8.0	1	13	18	14	22.0 3	17.0
	D/W 250 Pick-up	8	5.9 (5882)	12.5	-	7.0	22.0	13	6.5	14	30.0	19
	D/W 250 Pick-up	Z	5.9 (5899)	4.5	-	8.0	1	13	18	14	30.0	15.5
	D/W 350 Pick-up	8	5.9 (5882)	12.5	-	7.0	22.0	13	6.5	14	30.0	19
	D/W 350 Pick-up	5	5.9 (5899)	4.5	-	8.0	1	13	18	14	30.0	15.5
	B150 Van	X	3.9 (3916)	4.5	-	6.8	1	-	-	2	22.0 15	14.6 4
	B150 Van	Y	5.2 (5211)	4.5	-	6.8	1	-	-	2	22.0 15	16.5 4
	B250 Van	X	3.9 (3916)	4.5	-	6.8	1	-	-	2	22.0 15	14.6 4
	B250 Van	Y	5.2 (5211)	4.5	-	6.8	1	-	-	2	22.0 15	16.5 4
	B250 Van	Z	5.9 (5899)	4.5	-	6.8	1	-	-	2	35.0	15.0 4
	B350 Van	Y	5.2 (5211)	4.5	-	-	1	-	-	2	22.0 15	16.5 4
	B350 Van	Z	5.9 (5899)	4.5	-	-	1	-	-	2	35.0	15.0 4
	Ramcharger	Y	5.2 (5211)	4.5	-	8.0	1	13	18	14	34.0	17.0
	Ramcharger	Z	5.9 (5899)	4.5	-	8.0	1	13	18	14	34.0	15.5
1993	D/W 150 Pick-up	X	3.9 (3916)	4.5	-	8.0	1	13	18	11	22.0 15	15.1
	D/W 150 Pick-up	Y	5.2 (5211)	4.5	-	8.0	1	13	18	11	22.0 15	17.0
	D/W 150 Pick-up	Z	5.9 (5899)	4.5	-	8.0	1	13	18	11	30.0	15.5
	D250 Pick-up	X	3.9 (3916)	4.5	-	8.0	1	-	18	11	22.0 15	15.1
	D/W 250 Pick-up	Y	5.2 (5211)	4.5	-	8.0	1	13	18	11	22.0 15	17.0
	D/W 250 Pick-up	8	5.9 (5882)	12.5	-	7.0	22.0	13	6.5	11	30.0	19
	D/W 250 Pick-up	Z	5.9 (5899)	4.5	-	8.0	1	13	18	11	30.0	15.5
	D/W 350 Pick-up	8	5.9 (5882)	12.5	-	7.0	22.0	13	6.5	11	30.0	19
	D/W 350 Pick-up	5	5.9 (5899)	4.5	-	8.0	1	13	18	11	30.0	15.5
	B150 Van	X	3.9 (3916)	4.5	-	6.8	1	-	-	2	22.0 15	14.6 4

79111C06

CAPACITIES

Year	Model	Engine ID/VIN	Engine Displacement Liters (cc)	Engine Oil with Filter	Transmission (pts.) 4-Spd	5-Spd	Auto.	Transfer Case (pts.)	Drive Axle Front (pts.)	Rear (pts.)	Fuel Tank (gal.)	Cooling System (qts.)
	B150 Van	Y	5.2 (5211)	4.5	-	6.8	1	-	-	2	22.0 [15]	16.5 [4]
	B250 Van	X	3.9 (3916)	4.5	-	6.8	1	-	-	2	22.0 [15]	14.6 [4]
	B250 Van	Y	5.2 (5211)	4.5	-	6.8	1	-	-	2	22.0 [15]	16.5 [4]
	B250 Van	Z	5.9 (5899)	4.5	-	6.8	1	-	-	2	35.0	15.0 [4]
	B350 Van	Y	5.2 (5211)	4.5	-	-	1	-	-	2	22.0 [15]	11.5 [4]
	B350 Van	Z	5.9 (5899)	4.5	-	-	1	-	-	2	35.0	15.0 [4]
	Ramcharger	Y	5.2 (5211)	4.5	-	8.0	1	13	18	14	34.0	17.0
	Ramcharger	Z	5.9 (5899)	4.5	-	8.0	1	13	18	14	34.0	15.5
1994-95	D 1500 Pick-up	X	3.9 (3916)	4.5	-	21	22	-	-	25	26.0 [15]	20.0
	D/W 1500 Pick-up	Y	5.2 (5211)	5.0	-	21	22	23	24	25	26.0 [15]	20.0
	D/W 1500 Pick-up	Z	5.9 (5899)	5.0	-	21	22	23	24	25	26.0 [15]	20.0
	D/W 2500 Pick-up	Y	5.2 (5211)	5.0	-	21	22	23	24	25	26.0 [15]	20.0
	D/W 2500 Pick-up	C	5.9 (5882)	12.5	-	21	22	23	24	25	26.0 [15]	26.0
	D/W 2500 Pick-up	Z	5.9 (5899)	5.0	-	21	22	23	24	25	26.0 [15]	20.0
	D/W 2500 Pick-up	W	8.0 (7997)	8.0	-	21	22	23	24	25	26.0 [15]	24.0
	D/W 3500 Pick-up	C	5.9 (5882)	12.5	-	21	22	23	24	25	26.0 [15]	26.0
	D/W 3500 Pick-up	5	5.9 (5899)	5.0	-	21	22	23	24	25	26.0 [15]	20.0
	D/W 3500 Pick-up	W	8.0 (7997)	8.0	-	21	22	23	24	25	26.0 [15]	24.0
	B150 Van	X	3.9 (3916)	4.5	-	-	26	-	-	27	22.0 [15]	14.6
	B150 Van	Y	5.2 (5211)	4.5	-	-	26	-	-	27	22.0 [15]	15.6
	B250 Van	X	3.9 (3916)	4.5	-	-	26	-	-	27	22.0 [15]	14.6
	B250 Van	Y	5.2 (5211)	4.5	-	-	26	-	-	27	22.0 [15]	15.6
	B250 Van	A	5.9 (5899)	4.5	-	-	26	-	-	27	35.0	15.0 [28]
	B350 Van	Y	5.2 (5211)	4.5	-	-	26	-	-	27	22.0 [15]	15.6
	B350 Van	A	5.9 (5899)	4.5	-	-	26	-	-	27	35.0	15.0 [28]

1 A904/A998/A999 and A727: 17.1 pts.
   A500/A518: 20.4 pts.
2 Chrysler: 4.5 pts.
   Dana 60: 6.25 pts.
3 Optional fuel tank: 30 gals.
4 With HD cooling or A/C, add one quart
5 not used
6 Optional fuel tank: 20 gals.
7 Fleet vehicles: 19 pts.
8 not used
9 not used
10 With 7.25 in. rear: 3.0 pts.
    With 8.25 in. rear: 4.4 pts.
11 Optional fuel tank: 22 gals.
12 not used
13 NP205: 4.5 pts.; NP241: 6.0 pts.
14 Chrysler 8.25 in. and 9.25 in.: 4.5 pts.
    Spicer/Dana 60: 6.0 pts.
    Dana 70: 7.0 pts.
15 Optional fuel tank: 35 gals.
16 Overrunning clutch: 0.78 pts.
17 not used
18 Dana 44: 5.6 pts.
    Dana 60: 6.5 pts.
19 Manual transmission: 15.5 qts.
    Automatic transmission: 16.5 qts.
20 FWD: 20 gals.; AWD: 18 gals.

21 NV3500: 4.2 pts.
    NV4500: 8 pts.
    AX15: 6.6 pts.
    Getrag: 7 pts.
22 32RH: 17 pts.
    36RH: 16.6 pts.
    42RH: 20.2 pts.
    46RH: 21.8 pts.
    47RH: 30.8 pts.
23 NP231HD: 2.5 pts.
    NP241: 4.7 pts.
    NP241HD: 13 pts.
24 7.25 in.: 3 pts.
    Dana 44: 5.6 pts.
    Dana 60: 6.5 pts.
25 Chrysler 7.25 in.: 3 pts.
    Chrysler 8.25 in. and 9.25 in.: 4.8 pts.
    Spicer and Dana 60: 6.0 pts.
    Dana 70 and 80: 7.0 pts.
26 A998/A999 and A727: 17.2 pts.
    A500: 20.4 pts.
    A518: 21.4 pts.
27 Chrysler 8.25 in.: 4.4 pts.; 9.25 in.: 4.8 pts.
    Dana 60: 6.3 pts.
28 With rear heater: 16 qts.
29 Optional fuel tank: 22 gals.

CAMSHAFT SPECIFICATIONS
All measurements given in inches.

Year	Engine ID/VIN	Engine Displacement Liters (cc)	Journal Diameter					Elevation		Bearing Clearance	Camshaft End Play
			1	2	3	4	5	In.	Ex.		
1991	X	3.9 (3916)	1.998-1.999	1.967-1.968	1.951-1.952	1.561-1.562	-	0.373 [3]	0.400 [3]	0.001-0.005	0.002-0.010
	Y	5.2 (5211)	1.998-1.999	1.982-1.983	1.967-1.968	1.951-1.952	1.561-1.562	0.400 [3]	0.400 [3]	0.001-0.005	0.002-0.010
	8	5.9 (5882)	2.125	2.125	2.125	2.125	-	1.852 [2]	1.841 [2]	0.001-0.005	0.006-0.010
	W	5.9 (5899)	1.998-1.999	1.982-1.983	1.967-1.968	1.951-1.952	1.561-1.562	0.410 [3]	0.410 [3]	0.001-0.005	0.002-0.010
	Z	5.9 (5899)	1.998-1.999	1.982-1.983	1.967-1.968	1.951-1.952	1.561-1.562	0.410 [3]	0.410 [3]	0.001-0.005	0.002-0.010
1992	X	3.9 (3916)	1.982-1.983	1.967-1.968	1.951-1.952	1.561-1.562	-	0.373 [3]	0.0400 [3]	0.001-0.005	0.002-0.010
	Y	5.2 (5211)	1.998-1.999	1.982-1.983	1.967-1.968	1.951-1.952	1.561-1.562	0.373 [3]	0.400 [3]	0.001-0.005	0.002-0.010
	8	5.9 (5882)	2.125	2.125	2.125	2.125	-	1.852 [2]	1.841 [2]	0.001-0.005	0.006-0.010
	5	5.9 (5899)	1.998-1.999	1.982-1.983	1.967-1.968	1.951-1.952	1.561-1.562	0.410 [3]	0.410 [3]	0.001-0.005	0.002-0.010
	Z	5.9 (5899)	1.998-1.999	1.982-1.983	1.967-1.968	1.951-1.952	1.561-1.562	0.410 [3]	0.410 [3]	0.001-0.005	0.002-0.010
1993	X	3.9 (3916)	1.998-1.999	1.982-1.983	1.951-1.952	1.561-1.562	-	0.432 [3]	0.432 [3]	0.001-0.005	0.002-0.010
	Y	5.2 (5211)	1.998-1.999	1.982-1.983	1.967-1.968	1.951-1.952	1.561-1.562	0.432 [3]	0.432	0.001-0.005	0.002-0.010
	8	5.9 (5882)	2.125	2.125	2.125	2.125	-	1.852 [2]	1.841 [2]	0.001-0.005	0.006-0.010
	5	5.9 (5899)	1.998-1.999	1.982-1.983	1.967-1.968	1.951-1.952	1.561-1.562	0.410 [3]	0.410 [3]	0.001-0.005	0.002-0.010
	Z	5.9 (5899)	1.998-1.999	1.982-1.983	1.967-1.968	1.951-1.952	1.561-1.562	0.410 [3]	0.410 [3]	0.001-0.005	0.002-0.010
1994-95	X	3.9 (3916)	1.998-1.999	1.982-1.983	1.951-1.952	1.561-1.562	-	0.432 [4]	0.432 [4]	0.001-0.005	0.002-0.010
	Y	5.2 (5211)	1.998-1.999	1.982-1.983	1.967-1.968	1.951-1.952	1.561-1.562	0.432 [4]	0.432 [4]	0.001-0.005	0.002-0.010
	C	5.9 (5882)	2.125	2.125	2.125	2.125	-	1.852 [2]	1.841 [2]	0.001-0.005	0.006-0.010
	A	5.9 (5899)	1.998-1.999	1.982-1.983	1.967-1.968	1.951-1.952	1.561-1.562	0.410 [4]	0.410 [4]	0.001-0.005	0.002-0.010
	5	5.9 (5899)	1.998-1.999	1.982-1.983	1.967-1.968	1.951-1.952	1.561-1.562	0.410	0.410	0.001-0.005	0.002-0.010
	Z	5.9 (5899)	1.998-1.999	1.982-1.983	1.967-1.968	1.951-1.952	1.561-1.562	0.410	0.410	0.001-0.005	0.002-0.010
	W	8.0 (7997)	2.091-2.092	2.074-2.075	2.059-2.060	2.043-2.044	2.027-2.028 [3]	0.390	0.407	0.001-0.003 [5]	0.005-0.015

NA - Not Available

1 Height of cam lobe: 1.604-1.624 in.
2 Minimum diameter at peak lobe
3 Lift at valve
4 Journal No. 6: 1.917-1.918
5 Lift at valve
6 No. 2: 0.0015-0.0035

79111C07

CRANKSHAFT AND CONNECTING ROD SPECIFICATIONS

All measurements are given in inches.

Year	Engine ID/VIN	Engine Displacement Liters (cc)	Crankshaft				Connecting Rod		
			Main Brg. Journal Dia.	Main Brg. Oil Clearance	Shaft End-play	Thrust on No.	Journal Diameter	Oil Clearance	Side Clearance
1991	X	3.9 (3916)	2.500-2.501	1	0.002-0.010	2	2.124-2.125	0.0005-0.0022	0.006-0.014
	Y	5.2 (5211)	2.4995-2.5005	1	0.002-0.010	3	2.124-2.125	0.0005-0.0022	0.006-0.014
	8	5.9 (5882)	3.2662 2	0.0047	0.005-0.012	6	2.714 2	0.0035	0.004-0.012
	5	5.9 (5899)	2.8095-2.8105	1	0.002-0.010	3	2.124-2.125	0.0005-0.0022	0.006-0.014
	W	5.9 (5899)	2.8095-2.8105	1	0.002-0.010	3	2.124-2.125	0.0005-0.0022	0.006-0.014
	Z	5.9 (5899)	2.8095-2.8105	1	0.002-0.010	3	2.124-2.125	0.0005-0.0022	0.006-0.014
1992	X	3.9 (3916)	2.500-2.501	1	0.002-0.007	2	2.124-2.125	0.0005-0.0022	0.006-0.014
	Y	5.2 (5211)	2.4995-2.5005	1	0.002-0.007	3	2.124-2.125	0.0005-0.0022	0.006-0.014
	8	5.9 (5882)	3.2662 2	0.0047	0.005-0.012	6	2.714 2	0.0035	0.004-0.012
	5	5.9 (5899)	2.8095-2.8105	1	0.002-0.007	3	2.124-2.125	0.0005-0.0022	0.006-0.014
	Z	5.9 (5899)	2.8095-2.8105 1	1	0.002-0.007	3	2.124-2.125	0.0005-0.0022	0.006-0.014
1993	X	3.9 (3916)	2.500-2.501	1	0.002-0.007	2	2.124-2.125	0.0005-0.0022	0.006-0.014
	Y	5.2 (5211)	2.4995-2.5005	1	0.002-0.007	3	2.124-2.125	0.0005-0.0022	0.006-0.014
	8	5.9 (5882)	3.2662 2	0.0047	0.005-0.012	6	2.714 2	0.0035	0.004-0.012
	5	5.9 (5899)	2.8095-2.8105	1	0.002-0.007	3	2.124-2.125	0.0005-0.0022	0.006-0.014
	Z	5.9 (5899)	2.8095-2.8105	1	0.002-0.007	3	2.124-2.125	0.0005-0.0022	0.006-0.014
1994-95	X	3.9 (3916)	2.500-2.501	1	0.002-0.007	2	2.124-2.125	0.0005-0.0022	0.006-0.014
	Y	5.2 (5211)	2.4995-2.5005	1	0.002-0.007	3	2.124-2.125	0.0005-0.0022	0.006-0.014
	C	5.9 (5882)	3.2662 2	0.0047	0.005-0.012	6	2.714 2	0.0035	0.004-0.012
	A	5.9 (5899)	2.8095-2.8105	1	0.002-0.007	3	2.124-2.125	0.0005-0.0022	0.006-0.014
	5	5.9 (5899)	2.8095-2.8105	1	0.002-0.007	3	2.124-2.125	0.0005-0.0022	0.006-0.014
	Z	5.9 (5899)	2.8095-2.8105	1	0.002-0.007	3	2.124-2.125	0.0005-0.0022	0.006-0.014
	8	8.0 (7997)	2.9995-3.0005	0.0002-0.0023	0.0020-0.017	NA	2.124-2.125	0.0002-0.0029	0.010-0.018

NA - Not Available

1 0.0005-0.0015 on No. 1
   0.0005-0.0020 on Nos. 2-4
2 Maximum wear limit

79111C08

VALVE SPECIFICATIONS

Year	Engine ID/VIN	Engine Displacement Liters (cc)	Seat Angle (deg.)	Face Angle (deg.)	Spring Test Pressure (lbs. @ in.)	Spring Installed Height (in.)	Stem-to-Guide Clearance (in.) Intake	Stem-to-Guide Clearance (in.) Exhaust	Stem Diameter (in.) Intake	Stem Diameter (in.) Exhaust
1991	X	3.9 (3916)	45	45	2	3	0.001-0.003	0.002-0.004	0.0372-0.373	0.371-0.372
	Y	5.2 (5211)	45	45	2	3	0.001-0.003	0.002-0.004	0.372-0.373	0.371-0.372
	8	5.9 (5882)	1	1	65@1.94 4	2.19	0.002-0.006	0.002-0.006	0.313-0.314	0.313-0.314
	5	5.9 (5899)	45	45	2	3	0.001-0.003	0.002-0.004	0.372-0.373	0.371-0.372
	W	5.9 (5899)	45	45	2	3	0.001-0.003	0.002-0.004	0.372-0.373	0.371-0.372
	Z	5.9 (5899)	45	45	2	3	0.001-0.003	0.002-0.004	0.372-0.373	0.371-0.372
1992	X	3.9 (3916)	44.25-44.75	43.25-43.75	5	6	0.001-0.002	0.002-0.004	0.313	0.313
	Y	5.2 (5211)	44.25-44.75	43.25-43.75	2	6	0.001-0.003	0.002-0.004	0.313	0.313
	8	5.9 (5882)	1	1	65@1.94 4	2.19	0.002-0.006	0.002-0.006	0.313-0.314	0.313-0.314
	5	5.9 (5899)	45	45	8	6	0.001-0.003	0.002-0.004	0.372-0.373	0.371-0.372
	Z	5.9 (5899)	45	45	8	6	0.001-0.003	0.002-0.004	0.372-0.373	0.371-0.372
1993	X	3.9 (3916)	44.25-44.75	43.25-43.75	85@1.64	1.64	0.001-0.003	0.001-0.003	0.311-0.312	0.311-0.312
	Y	5.2 (5211)	44.25-44.75	43.25-43.75	85@1.64	1.64	0.001-0.003	0.001-0.003	0.311-0.312	0.311-0.312
	8	5.9 (5882)	1	1	65@1.94	2.19	0.002-0.006	0.002-0.006	0.313-0.314	0.313-0.314
	5	5.9 (5899)	44.25-44.75	43.25-43.75	85@1.64	1.64	0.001-0.003	0.002-0.004	0.372-0.373	0.371-0.372
	Z	5.9 (5899)	44.25-44.75	43.25-43.75	85@1.64	1.64	0.001-0.003	0.002-0.004	0.372-0.373	0.371-0.372
1994-95	X	3.9 (3916)	44.25-44.75	43.25-43.75	85@1.64	1.64	0.001-0.003	0.001-0.003	0.311-0.312	0.311-0.312
	Y	5.2 (5211)	44.25-44.75	43.25-43.75	85@1.64	1.64	0.001-0.003	0.001-0.003	0.311-0.312	0.311-0.312
	C	5.9 (5882)	1	1	65@1.94	2.19	0.002-0.006	0.002-0.006	0.313-0.314	0.313-0.314
	5	5.9 (5899)	44.25-44.75	43.25-43.75	85@1.64	1.64	0.001-0.003	0.002-0.004	0.372-0.373	0.371-0.372
	A	5.9 (5899)	44.25-44.75	43.25-43.75	85@1.64	1.64	0.001-0.003	0.002-0.004	0.372-0.373	0.371-0.372
	Z	5.9 (5899)	44.25-44.75	43.25-43.75	85@1.64	1.64	0.001-0.003	0.002-0.004	0.372-0.373	0.371-0.372
	W	8.0 (7997)	44.5	45	81-89@1.64	1.64	0.001-0.003	0.001-0.003	0.311-0.312	0.311-0.312

1 Intake: 30 degrees; Exhaust: 45 degrees
2 Intake: 78-88 at installed height Exhaust: 80-90 at installed height
3 Intake: 1.625-1.688; Exhaust: 1.453-1.516
4 Minimum acceptable specification
5 Intake: 78-88 at 1.688 in. Exhaust: 80-90 at 1.203 in.
6 Intake: 1.688; Exhaust: 1.203
7 Intake: 78-88 at 1.688 in. Exhaust: 80-90 at 1.484 in.

79111C09

PISTON AND RING SPECIFICATIONS

All measurements are given in inches.

Year	Engine ID/VIN	Engine Displacement Liters (cc)	Piston Clearance	Ring Gap			Ring Side Clearance		
				Top Compression	Bottom Compression	Oil Control	Top Compression	Bottom Compression	Oil Control
1991	X	3.9 (3916)	0.0005-0.0015	0.010-0.020	0.010-0.020	0.015-0.055	0.0015-0.0030	0.0015-0.0030	0.0002-0.0050
	Y	5.2 (5211)	0.0005-0.0015	0.010-0.020	0.010-0.020	0.015-0.055	0.0015-0.0030	0.0015-0.0030	0.0002-0.0050
	8	5.9 (5882)	NA	0.016-0.028	·0.010-0.021	0.010-0.021	0.0030-0.0060	0.0030-0.0060	0.0020-0.0050
	5	5.9 (5899)	0.0005-0.0015	0.010-0.020	0.010-0.020	0.015-0.055	0.0015-0.0030	0.0015-0.0030	0.0002-0.0050
	W	5.9 (5899)	0.0005-0.0015	0.010-0.020	0.010-0.020	0.015-0.055	0.0015-0.0030	0.0015-0.0030	0.0002-0.0050
	Z	5.9 (5899)	0.0005-0.0015	0.010-0.020	0.010-0.020	0.015-0.055	0.0015-0.0030	0.0015-0.0030	0.0002-0.0050
1992	X	3.9 (3916)	0.0005-0.0015	0.010-0.020	0.010-0.020	0.015-0.055	0.0015-0.0030	0.0015-0.0030	0.002-0.005
	Y	5.2 (5211)	0.0005-0.0015	0.010-0.020	0.010-0.020	0.015-0.055	0.0015-0.0030	0.0015-0.0030	0.002-0.005
	8	5.9 (5882)	NA	0.016-0.028	0.010-0.021	0.010-0.021	0.0030-0.0060	0.0030-0.0060	0.0020-0.0050
	5	5.9 (5899)	0.0005-0.0015	0.010-0.020	0.010-0.020	0.015-0.055	0.0015-0.0030	0.0015-0.0030	0.002-0.005
	Z	5.9 (5899)	0.0005-0.0015	0.010-0.020	0.010-0.020	0.015-0.055	0.0015-0.0030	0.0015-0.0030	0.002-0.005
1993	X	3.9 (3916)	0.0003-0.0008	0.010-0.020	0.010-0.020	0.010-0.050	0.0015-0.0030	0.0015-0.0030	0.002-0.008
	Y	5.2 (5211)	0.0003-0.0008	0.010-0.020	0.010-0.020	0.015-0.055	0.0015-0.0030	0.0015-0.0030	0.002-0.008
	8	5.9 (5882)	NA	0.016-0.028	0.010-0.021	0.010-0.021	0.0030-0.0060	0.0030-0.0060	0.0020-0.0050
	5	5.9 (5899)	0.0005-0.0015	0.010-0.020	0.010-0.020	0.015-0.055	0.0015-0.0030	0.0015-0.0030	0.002-0.008
	Z	5.9 (5899)	0.0005-0.0015	0.010-0.020	0.010-0.020	0.015-0.055	0.0015-0.0030	0.0015-0.0030	0.002-0.008
1994-95	X	3.9 (3916)	0.0003-0.0008	0.010-0.020	0.010-0.020	0.010-0.050	0.0015-0.0030	0.0015-0.0030	0.002-0.008
	Y	5.2 (5211)	0.0003-0.0008	0.010-0.020	0.010-0.020	0.015-0.055	0.0015-0.0030	0.0015-0.0030	0.002-0.008
	C	5.9 (5882)	NA	0.016-0.028	0.010-0.021	0.010-0.021	0.0030-0.0060	0.0030-0.0060	0.002-0.005
	5	5.9 (5899)	0.0005-0.0015	0.012-0.022	0.022-0.031	0.015-0.055	0.0016-0.0033	0.0016-0.0033	0.002-0.008
	A	5.9 (5899)	0.0005-0.0015	0.012-0.022	0.022-0.031	0.015-0.055	0.0016-0.0033	0.0016-0.0033	0.002-0.008
	Z	5.9 (5899)	0.0005-0.0015	0.010-0.020	0.010-0.020	0.015-0.055	0.0015-0.0030	0.0015-0.0030	0.002-0.008
	W	8.0 (7997)	0.0098-0.0218	0.010-0.020	0.010-0.020	0.015-0.055	0.0029-0.0038	0.0029-0.0038	0.0073-0.0097

NA - Not Available

79111C10

TORQUE SPECIFICATIONS
All readings in ft. lbs.

Year	Engine ID/VIN	Engine Displacement Liters (cc)	Cylinder Head Bolts	Main Bearing Bolts	Rod Bearing Bolts	Crankshaft Damper Bolts	Flywheel Bolts	Manifold Intake	Manifold Exhaust	Spark Plugs	Lug Nut
1991	X	3.9 (3916)	105	85	45	135	55	45	4	30	5
	Y	5.2 (5211)	105	85	45	135	55	40	4	30	5
	8	5.9 (5882)	6	7	8	92	101	-	32	-	5
	5	5.9 (5899)	105	85	45	135	55	40	4	30	5
	W	5.9 (5899)	105	85	45	135	55	40	4	30	5
	Z	5.9 (5899)	105	85	45	135	55	40	4	30	5
1992	X	3.9 (3916)	105	85	45	135	55	45	4	30	5
	Y	5.2 (5211)	105	85	45	135	55	40	4	30	5
	8	5.9 (5882)	1	7	8	92	101	-	32	-	5
	5	5.9 (5899)	105	85	45	135	55	40	4	30	5
	Z	5.9 (5899)	105	85	45	135	55	40	4	30	5
1993	X	3.9 (3916)	105	85	45	135	55	45	4	30	5
	Y	5.2 (5211)	105	85	45	135	55	40	4	30	5
	8	5.9 (5882)	1	7	8	92	101	-	32	-	5
	5	5.9 (5899)	105	85	45	135	55	40	4	30	5
	Z	5.9 (5899)	105	85	45	135	55	40	4	30	5
1994-95	X	3.9 (3916)	105 [12]	85	45	135	72	10	25	30	5
	Y	5.2 (5211)	105 [12]	85	45	135	55	17	25	30	5
	C	5.9 (5882)	1	7	8	135	101	-	32	-	5
	5	5.9 (5899)	105 [12]	85	45	135	55	17	25	30	5
	A	5.9 (5899)	105 [12]	85	45	135	55	17	25	30	5
	Z	5.9 (5899)	105 [12]	85	45	135	55	17	25	30	5
	W	8.0 (7997)	105 [12]	85	45	135	55	11	30	30	5

1 All bolts: 66 ft. lbs.
Long bolts: 89 ft. lbs.
All bolts and additional 1/4 turn
2 In sequence: 45, 65, 65 plus 1/4 turn
3 Plus 1/4 turn
4 Bolts: 20 ft. lbs.; Nuts: 15 ft. lbs.
5 1/2x20 stud: 85-110 ft. lbs.
5/8x18 stud with cone nut: 175-225 ft. lbs.
5/8x18 stud with flanged nut: 300-350 ft. lbs.
6 In sequence: 29, 62 and 93 ft. lbs.

7 In sequence: 45, 88 and 129 ft. lbs.
8 In sequence: 26, 51 and 73 ft. lbs.
9 In sequence: 45, 65, 65 plus 1/4 turn
Torque small bolt in rear of head to 25 ft. lbs.
10 In sequence: Tighten intake plenum bolts to 24, 48 and 84 in. lbs.
11 Lower intake: 40 ft. lbs.
Upper intake: 16 ft. lbs.
12 In sequence: 50, 105

79111C11

BRAKE SPECIFICATIONS
All measurements in inches unless noted

Year	Model	Master Cylinder Bore	Brake Disc Original Thickness	Brake Disc Minimum Thickness	Maximum Runout	Brake Drum Diameter Original Inside Diameter	Brake Drum Diameter Max. Wear Limit	Brake Drum Diameter Maximum Machine Diameter	Minimum Lining Thickness Front	Minimum Lining Thickness Rear
1991	D150 Pick-up	1.125	1.24	1.18	0.004	11.00	11.09	11.06	0.062	0.062
	W150 Pick-up	1.125	1.24	1.18	0.004	11.00	11.09	11.06	0.062	0.062
	D250 Pick-up	1.125	1.24	1.18	0.005	12.00	12.09	11.06	0.062	0.062
	W250 Pick-up	1.125	1	2	0.005	12.00	12.09	12.06	0.062	0.062
	D350 Pick-up	1.125	1.18	1.125	0.005	12.00	12.09	12.06	0.062	0.062
	W350 Pick-up	1.125	1.18	1.125	0.005	12.00	12.09	12.06	0.062	0.062
	B150 Van	1.125	1.24	1.18	0.004	11.00	11.09	11.06	0.062	0.062
	B250 Van	1.125	1.24	1.18	0.004	11.00	11.09	11.06	0.062	0.062
	B350 Van	1.125	1	2	0.004	12.00	12.09	12.06	0.062	0.062
	AD Ramcharger	1.125	1.24	1.18	0.004	11.00	11.09	11.06	0.062	0.062
	AW Ramcharger	1.125	1.24	1.18	0.004	11.00	11.09	11.06	0.062	0.062
1992	D150 Pick-up	1.125	1.24	1.18	0.004	11.00	11.09	11.06	0.062	0.062
	W150 Pick-up	1.125	1.24	1.18	0.004	11.00	11.09	11.06	0.062	0.062
	D250 Pick-up	1.125	1.24	1.18	0.005	12.00	12.09	11.06	0.062	0.062
	W250 Pick-up	1.125	1	2	0.005	12.00	12.09	12.06	0.062	0.062
	D350 Pick-up	1.125	1.18	1.125	0.005	12.00	12.09	12.06	0.062	0.062
	W350 Pick-up	1.125	1.18	1.125	0.005	12.00	12.09	12.06	0.062	0.062
	B150 Van	1.125	1.24	1.18	0.004	11.00	11.09	11.06	0.062	0.062
	B250 Van	1.125	1.24	1.18	0.004	11.00	11.09	11.06	0.062	0.062
	B350 Van	1.125	1	2	0.004	12.00	12.09	12.06	0.062	0.062
	AD Ramcharger	1.125	1.24	1.18	0.004	11.00	11.09	11.06	0.062	0.062
	AM Ramcharger	1.125	1.24	1.18	0.004	11.00	11.09	11.06	0.062	0.062
1993	D150 Pick-up	1.125	1.24	1.18	0.004	11.00	11.09	11.06	0.062	0.062
	W150 Pick-up	1.125	1.24	1.18	0.004	11.00	11.09	11.06	0.062	0.062
	D250 Pick-up	1.125	1.24	1.18	0.005	12.00	12.09	11.06	0.062	0.062
	W250 Pick-up	1.125	1	2	0.005	12.00	12.09	12.06	0.062	0.062
	D350 Pick-up	1.125	1.18	1.125	0.005	12.00	12.09	12.06	0.062	0.062
	W350 Pick-up	1.125	1.18	1.125	0.005	12.00	12.09	12.06	0.062	0.062
	B150 Van	1.125	1.24	1.18	0.004	11.00	11.09	11.06	0.062	0.062
	B250 Van	1.125	1.24	1.18	0.004	11.00	11.09	11.06	0.062	0.062
	B350 Van	1.125	1	2	0.004	12.00	12.09	12.06	0.062	0.062
	AD Ramcharger	1.125	1.24	1.180	0.004	11.00	11.09	11.06	0.062	0.062
	AW Ramcharger	1.125	1.24	1.18	0.004	11.00	11.09	11.06	0.062	0.062
1994-95	D1500 Pick-up	1.125	1.26	3	0.004	11.00	11.09	11.06	0.062	0.062
	W1500 Pick-up	1.125	1.26	3	0.004	11.00	11.09	11.06	0.062	0.062
	D2500 Pick-up	1.250	1.50	3	0.005	13.00	13.09	13.06	0.062	0.062
	W2500 Pick-up	1.250	1.50	3	0.005	13.00	13.09	13.06	0.062	0.062
	D3500 Pick-up	1.250	1.50	3	0.005	13.00	13.09	13.06	0.062	0.062
	W3500 Pick-up	1.250	1.50	3	0.005	13.00	13.09	13.06	0.062	0.062
	B150 Van	1.125	-	3	0.004	11.00	11.09	11.06	0.062	0.062
	B250 Van	1.125	-	3	0.004	11.00	11.09	11.06	0.062	0.062
	B350 Van	1.125	-	3	0.004	12.00	12.09	12.06	0.062	0.062

1 With 3300 lb. or 3600 lb. rear axle: 1.24 in.
  With 4000 lb. rear axle: 1.18 in.
2 With 3300 lb. or 3600 lb. rear axle: 1.180 in.
  With 4000 lb. rear axle: 1.125 in.
3 Minimum thickness indicated on rotor hub

79111C12

WHEEL ALIGNMENT

Year	Model		Caster Range (deg.)	Caster Preferred Setting (deg.)	Camber Range (deg.)	Camber Preferred Setting (deg.)	Toe-in (in.)	Steering Axis Inclination (deg.)
1991	D150 Pick-up		1N-2P	1/2P	0-1P	1/2P	1/8P	NA
	W150 Pick-up		1/2P-3 1/2P	2P	1/2P-1 1/2P	1P	1/8P	8 1/2 [1]
	D250 Pick-up		1N-2P	1/2P	0-1P	1/2P	1/8P	NA
	W250 Pick-up		1/2P-3 1/2P	2P	1/2P-1 1/2P	1P	1/8P	8 1/2 [1]
	D350 Pick-up		1N-2P	1/2P	0-1P	1/2P	1/8P	NA
	W350 Pick-up		1/2P-3 1/2P	2P	1/2P-1 1/2P	1P	1/8P	8 1/2 [1]
	B150 Van		1 1/4P-3 3/4P	2 1/2P	5/8N-5/8P	0	0	NA
	B250 Van		1 1/4P-3 3/4P	2 1/2P	5/8N-5/8P	0	0	NA
	B350 Van		1 1/4P-3 3/4P	2 1/2P	5/8N-5/8P	0	0	NA
	AD Ramcharger		1N-2P	1/2P	0-1P	1/2P	1/8P	NA
	AW Ramcharger		1/2P-3 1/2P	2P	1/2P-1/12P	1P	1/8P	8 1/2 [1]
1992	D150 Pick-up		1N-2P	1/2P	0-1P	1/2P	1/8P	NA
	W150 Pick-up		1/2P-3 1/2P	2P	1/2P-1 1/2P	1P	1/8P	8 1/2 [1]
	D250 Pick-up		1N-2P	1/2P	0-1P	1/2P	1/8P	NA
	W250 Pick-up		1/2P-3 1/2P	2P	1/2P-1 1/2P	1P	1/8P	8 1/2 [1]
	D350 Pick-up		1N-2P	1/2P	0-1P	1/2P	1/8P	NA
	W350 Pick-up		1/2P-3 1/2P	2P	1/2P-1 1/2P	1P	1/8P	8 1/2 [1]
	B150 Van		1 1/4P-3 3/4P	2 1/2P	5/8N-5/8P	0	0	NA
	B250 Van		1 1/4P-3 3/4P	2 1/2P	5/8N-5/8P	0	0	NA
	B350 Van		1 1/4P-3 3/4P	2 1/2P	5/8N-5/8P	0	0	NA
	AD Ramcharger		1N-2P	1/2P	0-1P	1/2P	1/8P	NA
	AW Ramcharger		1/2P-3 1/2P	2P	1/2P-1 1/2P	1P	1/8P	8 1/2 [1]
1993	D150 Pick-up		1N-2P	1/2P	0-1P	1/2P	1/4P	NA
	W150 Pick-up		1/2P-3 1/2P	2P	1N-1P	0	1/8P	8 1/2 [1]
	D250 Pick-up		1N-2P	1/2P	0-1P	1/2P	1/8P	NA
	W250 Pick-up		1/2P-3 1/2P	2P	1N-1P	0	1/8P	8 1/2 [1]
	D350 Pick-up		1N-2P	1/2P	0-1P	1/2P	1/8P	NA
	W350 Pick-up		1/2P-3 1/2P	2P	1N-1P	0	1/8P	8 1/2 [1]
	B150 Van		1 1/4P-3 3/4P	2 1/2P	5.8N-5/8P	0	0	NA
	B250 Van		1 1/4P-3 3/4P	2 1/2P	5.8N-5/8P	0	0	NA
	B350 Van		1 1/4P-3 3/4P	2 1/2P	5.8N-5/8P	0	0	NA
	AD Ramcharger		1N-2P	1/2P	0-1P	1/2P	1/8P	NA
	AW Ramcharger		1/2P-3 1/2P	2P	1/2P-1 1/2P	1P	1/8P	8 1/2 [1]
1994-95	D1500 Pickup		2 3/4P-4 3/4P	3 3/4P	0-1P	1/2P	1/4P	NA
	W1500 Pick-up		2 1/2P-4 1/2P	3 1/2P	0-1P	1/2P	1/4P	NA
	D2500 Pick-up		2 1/2P-4 1/2P	3 1/2P	0-1P	1/2P	1/4P	NA
	W2500 Pick-up		2P-4P	3P	0-1P	1/2P	1/4P	NA
	D3500 Pick-up		2 1/4P-4 3/4P	3 1/4P	0-1P	1/2P	1/4P	NA
	W3500 Pick-up		2 1/4P-4 3/4P	3 1/4P	0-1P	1/2P	1/4P	NA
	B150 Van		1 1/4P-3 3/4P	2 1/2P	5/8N-5/8P	0	0	NA
	B250 Van		1 1/4P-3 3/4P	2 1/2P	5/8N-5/8P	0	0	NA
	B350 Van		1 1/4P-3 3/4P	2 1/2P	5/8N-5/8P	0	0	NA

NA - Not Available

1 King pin inclination

79111C13

ENGINE ELECTRICAL

NOTE: Disconnecting the negative battery cable on some vehicles may interfere with the functions of the on board computer systems and may require the computer to undergo a relearning process, once the negative battery cable is reconnected.

Distributor

REMOVAL

1. Disconnect the negative battery cable.
2. Disconnect the distributor pickup lead wires and vacuum hose, if equipped.
3. Unfasten the distributor cap retaining clips and lift off the distributor cap with all ignition wires still connected. Remove the coil wire if necessary.
4. Matchmark the rotor to the distributor housing.

NOTE: Do not crank the engine during this procedure. If the engine is cranked, the matchmark must be disregarded.

5. Remove the hold-down bolt and clamp.
6. Remove the distributor from the engine.

INSTALLATION

Timing Not Disturbed

1. Install a new distributor housing O-ring.
2. Install the distributor in the engine so the rotor is aligned with the matchmark on the housing and the housing is aligned with the matchmark on the engine. Make sure the distributor is fully seated and the distributor shaft is fully engaged.
3. Install the hold-down clamp and snug the hold-down bolt.
4. Connect the distributor pickup lead wires.
5. Install the distributor cap and snap the retaining clips into place.
6. Connect the negative battery cable.
7. Adjust the ignition timing and tighten the hold-down bolt.

Timing Disturbed

1. Install a new distributor housing O-ring.
2. Position the engine so the No. 1 piston is at TDC of the compression stroke and the mark on the vibration damper is aligned with **0** on the timing indicator.
3. Install the distributor in the engine so the rotor is aligned with the position of the No. 1 ignition wire on the distributor cap and the housing is aligned with the matchmark on the engine. Make sure the distributor is fully seated and the distributor shaft is fully engaged.
4. Install the hold-down clamp and snug the hold-down bolt.
5. Connect the distributor pickup lead wires.
6. Install the distributor cap and snap the retaining clips into place.
7. Connect the negative battery cable.
8. Adjust the ignition timing and tighten the hold-down nut or bolt.

Ignition Timing

ADJUSTMENT

1. Start the engine, set the parking brake and run the engine until normal operating temperature is reached. Keep all lights and accessories **OFF**.
2. If a magnetic timing unit is available, insert the probe into the receptacle near the timing scale. The scale is located on the timing chain cover above and to the left of the vibration damper on Pick-Ups and below the damper on Vans.
3. If a magnetic timing unit is not available, connect a conventional power timing light to the No. 1 cylinder spark plug wire.
4. Connect the red lead of a tachometer to the negative primary terminal of the coil and connect the black lead to a good ground.
5. Set the idle speed according to the Vehicle Emission Control Information (VECI) label. Disconnect and plug the distributor vacuum advance hose at the distributor or computer, if equipped.
6. Connect the Diagnostic Readout Box II (DRBII) and access the Basic Timing Mode. If the DRBII is not available, disconnect the coolant sensor located near the thermostat housing and confirm that the Check Engine light on the instrument panel is **ON**.

7. Aim the timing light at the timing scale or read the magnetic timing unit.
8. If the timing is advanced (higher than the specification on the VECI label), the distributor should be turned clockwise. If the timing is retarded, (lower than the specification on the VECI label), the distributor should be turned counterclockwise.
9. Loosen the distributor hold-down bolt just enough so the distributor can be rotated. Turn the distributor in the proper direction until the specified timing is reached. Tighten the hold-down bolt and recheck the timing and idle speed.
10. Turn the engine off. Remove the jumper wire, if used. Connect the vacuum hose or coolant sensor (make sure the Check Engine light does not come on when started). Disconnect the timing apparatus and tachometer.
11. If the coolant temperature sensor was disconnected, erase the created fault code using the Erase Fault Code mode on the DRBII. If the DRBII is not available, the code can be erased by disconnecting the battery, although it is not recommended.

NOTE: If the battery is disconnected, radio memory will be lost and other fault codes that may have been stored in the computer's memory will be erased. If the coolant sensor code is not erased at this point, it will disappear after 50–100 vehicle key on/off cycles providing there is no problem with that circuit.

Alternator

PRECAUTIONS

Several precautions must be observed when working with the alternator to avoid damaging the unit.

- If the battery is removed for any reason, make sure it is reconnected with the correct polarity. Reversing the battery connections may result in damage to the one-way rectifiers.
- When utilizing a booster battery as a starting aid, always connect the positive to positive terminals and the negative terminal from the booster battery to a good engine ground on the vehicle being started.
- Never use a fast charger as a booster to start vehicles.
- Disconnect the battery cables when charging the battery with a fast charger.
- Never attempt to polarize the alternator.

• Do not use test lights of more than 12 volts when checking diode continuity.

• Do not short across or ground any of the alternator terminals.

• The polarity of the battery, alternator and regulator must be matched and considered before making any electrical connections within the system.

• Never separate the alternator on an open circuit. Make sure all connections within the circuit are clean and tight.

• Disconnect the battery ground terminal when performing any service on electrical components.

• Disconnect the battery if arc welding is to be done on the vehicle.

BELT TENSION ADJUSTMENT

NOTE: The belt tension is automatically adjusted by the tensioner on the 5.9L diesel engine. Periodic adjustment is not necessary.

1. Disconnect negative battery cable.

2. Loosen alternator pivot bolt slightly.

3. Loosen the adjuster strap nut or bolt just enough so the alternator can be moved.

4. If the alternator bracket is not equipped with an adjuster bolt, use a prybar and apply tension to the alternator until the belt(s) deflect about ¼–½ in. under a 10 lb. load. Torque the adjuster strap bolt to 200 inch lbs. (23 Nm). Torque the pivot bolt to 30 ft. lbs. (41 Nm).

5. If the bracket is equipped with an adjuster bolt, tighten it until the belts deflect about ½ in. under a 10 lb. load. Torque the adjuster strap nut to 200 inch lbs. (23 Nm). Torque the pivot bolt to 30 ft. lbs. (41 Nm).

6. Reconnect negative battery cable.

REMOVAL AND INSTALLATION

1. Disconnect the negative battery cable.

2. On the 5.9L diesel engine, use a ⅜ in. drive breaker bar to lift the belt tensioner and remove the belt. On all other engines, loosen the mounting bolts, move the alternator toward the engine and remove the drive belt(s).

NOTE: On some Vans, it may be easier to remove the alternator through the right front wheelwell since the fan shroud and possibly air conditioning and heater plumbing under the hood will prevent removal from in front of the vehicle. Lift the vehicle and safely support. Remove the right front wheel and gain access to the alternator on these applications.

3. Remove the mounting bolts, spacers and adjuster bolt. Remove the alternator from the brackets.

4. Remove the battery positive, field and ground terminals from the rear of the alternator. Remove the wire harness hold-down screw from the alternator.

To install:

5. Connect all wiring to the proper terminals on the rear of the alternator and install the wire harness hold-down screw.

6. Position the alternator in the mounting brackets.

7. Install the spacers, pivot bolt, adjuster strap bolt or nut and adjuster bolt.

8. On the 5.9L diesel engine, torque the upper bolt to 18 ft. lbs. (24 Nm), the lower bolt to 32 ft. lbs. (43 Nm) and install the belt. On all other engines, install the drive belt(s) and adjust to specification. Torque the adjuster strap nut or bolt to 200 inch lbs. (23 Nm). Torque the pivot bolt to 30 ft. lbs. (41 Nm).

9. Connect the negative battery cable.

Voltage Regulator

REMOVAL AND INSTALLATION

NOTE: The voltage regulator is integrated into the circuitry of the Single Module Engine Controller (SMEC) or Single Board Engine Controller (SBEC) on vehicles with EFI and is not serviceable.

1. Disconnect the negative battery cable.

2. Unplug the connector from the voltage regulator.

3. Remove the retaining screws and remove the regulator from the vehicle.

4. The installation is the reversal of the removal procedure. Make sure the connector retainer is properly clipped in place.

Starter

REMOVAL AND INSTALLATION

1. Disconnect the negative battery cable.

2. Raise the vehicle and support safely.

3. Remove the heat shield from the starter.

4. Disconnect the solenoid lead wires from the starter.

5. Unbolt the starter, remove the exhaust bracket and automatic transmission oil cooler tube bracket, if equipped, and remove the starter from the vehicle.

To install:

6. Install the starter to the bell housing and install the upper mounting bolt loosely.

7. Install the automatic transmission oil cooler tube bracket and exhaust bracket, if equipped. Install the lower mounting nut or bolts. Torque the mounting nut and bolt evenly to 50 ft. lbs. (68 Nm) on all engines except the 5.9L diesel engine. On 5.9L diesel engine, torque the mounting bolts to 32 ft. lbs. (43 Nm).

8. Connect the solenoid lead wires.

9. Install the heat shield.

10. Connect the negative battery cable and check the starter for proper operation.

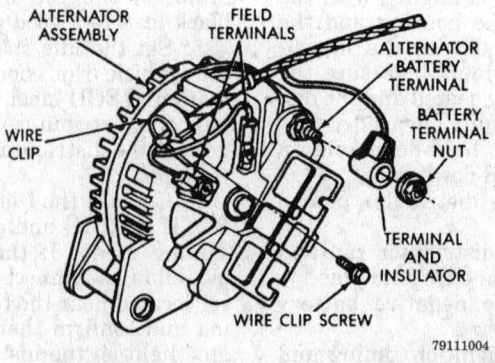

79111004

Chrysler 60 and 78 amp alternator terminals

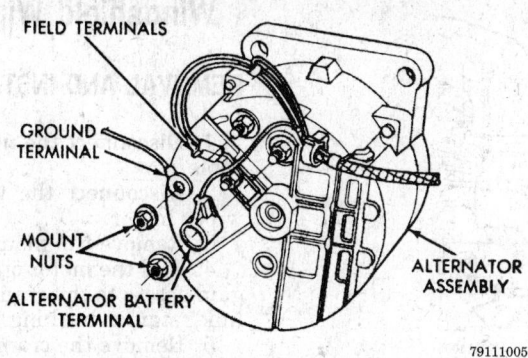

Chrysler 114 amp alternator terminals

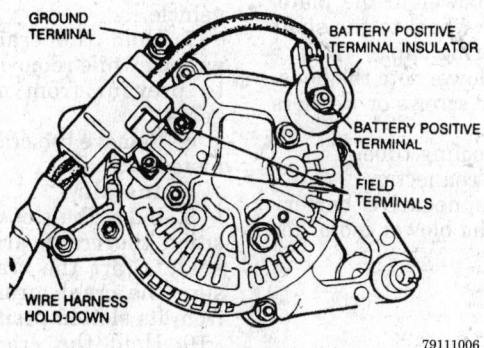

Chrysler 90HS, 120HS and Nippondenso alternator terminals

Intake Manifold Heater

REMOVAL AND INSTALLATION

5.9L Diesel Engine

1. Disconnect the negative battery cable.

2. Remove the intercooler outlet duct from air inlet tube, if equipped.

3. Remove the 4 bolts that attach the air crossover to the intake manifold. Loosen the lower throttle control bracket mounting bolt and move the bracket away from the engine.

4. Loosen the hose clamps on the turbocharger end of the crossover tube and remove the tube and gasket. Cover the turbocharger opening with a clean shop towel.

5. Disconnect the electrical wiring from the intake manifold heater, remove the heater and remove the gasket.

To install:

6. Clean the gasket mounting surface and install a new gasket.

7. Install the intake manifold heater and connect the wiring.

8. Remove the towel from the turbocharger opening and install the air crossover tube and gasket. Tighten the hose clamps.

9. Rotate the throttle control bracket back into place. Torque all mounting bolts to 18 ft. lbs. (24 Nm).

10. Attach the throttle rod to the throttle lever. Reconnect negative battery cable.

TESTING

1. Disconnect the negative battery cable.

2. Disconnect the wires from the intake manifold heater.

Rotating the throttle bracket away from the engine — 5.9L diesel engine

3. Using an ohmmeter, check the resistance from a good ground to each heater terminal.

4. If there is any resistance, or if the circuit is open, inspect the assembly for dirty or corroded connections.

5. If the resistance is 0 ohms, the heater is functioning properly.

6. If the circuit is open, the heater is defective.

CHASSIS ELECTRICAL

Heater Blower Motor

REMOVAL AND INSTALLATION

Van

WITHOUT AIR CONDITIONING

1. Disconnect the negative battery cable.

2. Remove the air intake duct and top half of the fan shroud, if necessary. Disconnect the blower connector.

3. Remove the 7 screws that fasten the back plate to the heater housing.

4. Remove the blower motor from the vehicle.

5. Remove the spring clip fastening the blower wheel to the blower shaft and pull off the wheel.

6. Remove the vent tube.

7. Remove the nuts fastening the blower motor to the back plate and remove the motor.

To install:

8. Check the seal for breaks or poor adhesion; repair as needed.

9. Install the blower motor to the back plate.

10. Install the vent tube.

11. Install the blower wheel to the shaft and secure the spring clip.

12. Install the assembly to the heater housing and install the 7 screws.

13. Install electrical connector.

14. Install the fan shroud and air duct if removed.

15. Connect the negative battery cable and check the blower motor for proper operation.

WITH AIR CONDITIONING

1. Disconnect the negative battery cable.

2. Remove the air intake duct and top half of the fan shroud.

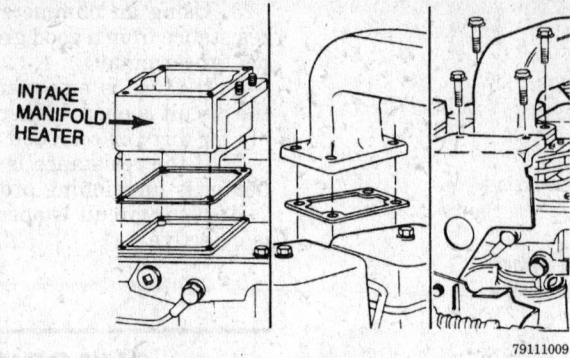

INTAKE MANIFOLD HEATER

79111009

Intake manifold heater and related parts — 5.9L diesel engine

3. Disconnect the blower connector.

4. Remove the blower motor cooling tube.

5. Remove the retaining nuts and washers from the studs holding the blower.

6. Pull the air conditioning lines inboard and upward while removing the blower assembly from the vehicle. Remove the spring clip fastening the blower wheel to the blower shaft and pull off the wheel.

To install:

7. Install the blower wheel to the shaft and install the spring clip. Inspect the blower mounting plate seal and repair as needed. Apply rubber adhesive to the seal to aid in assembly.

8. Install the blower into the housing and install the washers and nuts.

9. Install the cooling tube.

10. Reconnect the electrical connector.

11. Install the fan shroud and air intake duct.

12. Connect the negative battery cable and check the blower motor for proper operation.

Pickup and Ramcharger

1. Disconnect the negative battery cable.

2. Disconnect the blower connector.

3. Remove the blower motor cooling tube.

4. Remove the screws or retaining nuts retaining the blower plate to the housing.

5. Remove the assembly from the housing.

6. Remove the spring clip fastening the blower wheel to the blower shaft and pull off the wheel. Remove the blower from the plate.

To install:

7. Inspect the blower mounting plate seal and repair as necessary.

8. Install the blower to the plate. Install the blower wheel to the shaft and install the spring clip.

9. Install the blower into the housing and install the screws or washers and nuts.

10. Install the cooling tube.

11. Connect the connector.

12. Connect the negative battery cable and check the blower motor for proper operation.

Windshield Wiper Motor

REMOVAL AND INSTALLATION

1. Disconnect the negative battery cable.

2. Disconnect the wires from the wiper motor.

3. Remove the mounting bolts.

4. Pull the motor out far enough to gain access to the crank arm to motor link retainer bushing.

5. Remove the crank arm from the drive link by prying the retainer bushing from the crank arm.

6. Remove the motor from the vehicle.

7. Hold the crank arm with a wrench while removing the crank nut to prevent from overloading the gears.

8. Remove the crank arm from the motor.

To install:

9. Index the slot correctly and position the crank arm on the motor shaft. Start the crank nut making sure the crank arm does not move from its slotted position.

10. Hold the crank arm with a wrench and torque the nut to 95 inch lbs. (11 Nm).

11. Lubricate the drive link retainer bushing and install the crank

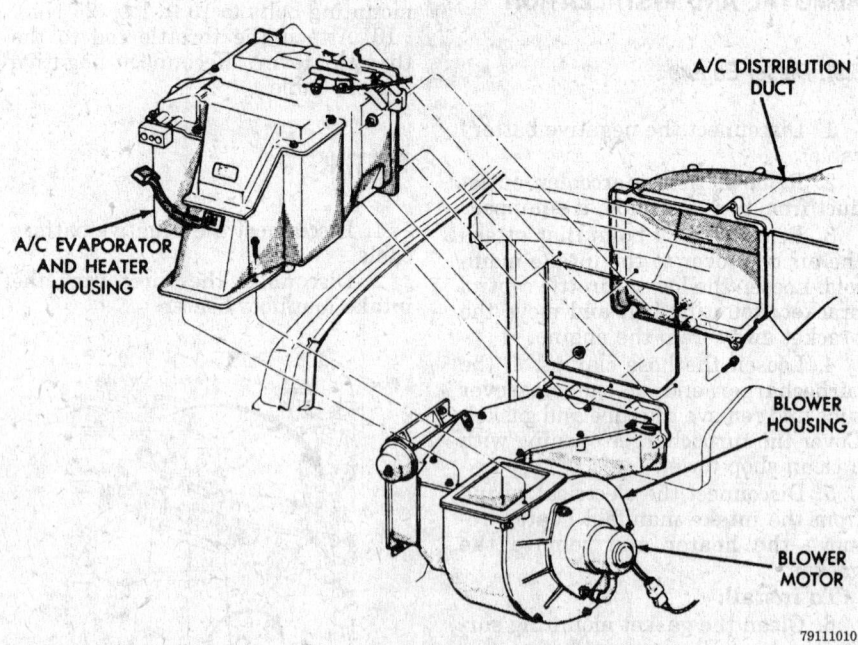

A/C DISTRIBUTION DUCT

A/C EVAPORATOR AND HEATER HOUSING

BLOWER HOUSING

BLOWER MOTOR

79111010

Underhood location of the blower motor — Van with air conditioning

arm pin to the bushing by snapping them together.

12. Install the motor to the vehicle and torque the mounting bolts to 55 inch lbs. (6 Nm).

13. Connect the wires to the motor.

14. Connect the negative battery cable and check the wiper motor for proper operation.

Instrument Cluster

REMOVAL AND INSTALLATION

Van

1. Disconnect the negative battery cable.

2. Open the glove box. Remove the screws that fasten the hood and bezel assembly. Pull the bezel off of the upper retaining clips.

3. Disconnect the gearshift pointer cable from the arm on the steering column, if equipped.

4. Remove the cluster screws. Pull the cluster out far enough to disconnect the speedometer cable by releasing the spring clip.

5. Remove all printed circuit board multiple connectors and the message center connector.

6. Remove the cluster assembly.

7. To remove the speedometer, remove the cluster lens, unplug the Emissions Maintenance Reminder (EMR) timer wires from the speedometer, if equipped, and remove the retaining screws.

To install:

8. Install the speedometer if removed and connect the EMR wiring, if equipped. Position the cluster to the panel and connect the speedometer cable, multiple connectors and message center connector.

9. Push the cluster in place and install the retaining screws.

10. Connect the gearshift pointer cable to the arm on the steering column, if equipped, and adjust if necessary.

11. Install the hood and bezel assembly and install the retaining screws.

12. Connect the negative battery cable.

Pickup and Ramcharger

1. Disconnect the negative battery cable.

2. Remove the map light.

3. Remove 6 screws which attach the faceplate to the base panel. There is a screw below the heater-air conditioning control panel which is not visible from above.

4. If equipped with automatic transmission, place the shift lever in its lowest position.

5. Remove the faceplate by pulling the top edge rearward to clear the brow and pulling the bottom out, disengaging the attaching clips. If equipped with 4WD, disconnect the indicator wires.

6. Remove the upper and lower steering column covers. Disconnect the shift indicator actuator cable from the steering column, if equipped.

7. Loosen the heater and air conditioning control and pull it out enough to clear the cluster housing.

8. Remove the screws that retain the cluster and pull the cluster out far enough to disconnect the speedometer cable by releasing the spring clip.

9. Remove all printed circuit board multiple connectors.

10. Remove the cluster assembly.

11. To remove the speedometer, remove the cluster lens, unplug the EMR timer wires from the speedometer, if equipped, and remove the retaining screws.

To install:

12. Install the speedometer, if removed, and connect the EMR wiring, if equipped. Position the cluster to the panel and connect the speedometer cable and multiple connectors.

13. Push the cluster in place and install the retaining screws.

14. Install the heater air conditioning control screws.

15. Connect the shift indicator actuator cable and check for alignment.

16. Install the upper and lower steering column covers.

17. Connect the 4WD indicator, if equipped. Install the cluster faceplate and map light.

18. Connect the negative battery cable.

Radio

REMOVAL AND INSTALLATION

Van

1. Disconnect the negative battery cable.

2. Open the glove box. Remove the screws that fasten the hood and bezel assembly. Pull the bezel off of the upper retaining clips.

3. Remove the screws that attach the radio to the dash.

4. Pull the radio out, disconnect the connectors, ground cable and antenna and remove the radio.

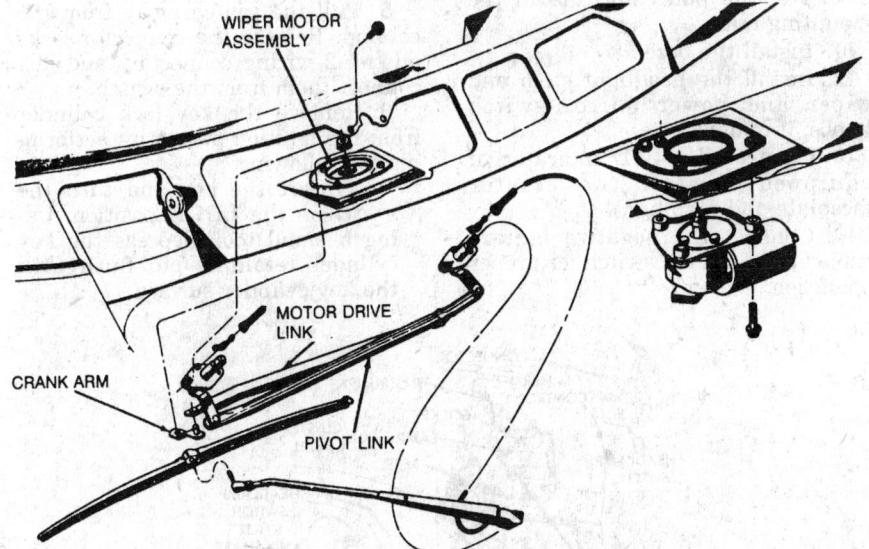

WIPER MOTOR ASSEMBLY

MOTOR DRIVE LINK

CRANK ARM

PIVOT LINK

79111012

Typical windshield wiper motor and linkage — Van

5. The installation is the reversal of the removal procedure.

Pickup and Ramcharger

1. Disconnect the negative battery cable.
2. Remove the map light.
3. Remove 6 screws which attach the faceplate to the base panel. There is a screw below the heater-air conditioning control panel which is not visible from above.
4. If equipped with automatic transmission, place the shift lever in its lowest position.
5. Remove the faceplate by pulling the top edge rearward to clear the brow and pulling the bottom out, disengaging the attaching clips. If equipped with 4WD, disconnect the indicator wires.
6. Remove the screws that attach the radio to the dash.
7. Pull the radio out, disconnect the connectors, ground cable and antenna and remove the radio.
8. The installation is the reversal of the removal procedure.

Headlight Switch

REMOVAL AND INSTALLATION

Van

1. Disconnect the negative battery cable.
2. Remove the lower steering column cover.
3. Unscrew the hood release handle and lower.
4. Working under the instrument panel, depress the spring button on the headlight switch and pull the stem out.
5. Open the glove box. Remove the screws that fasten the dash bezel assembly. Pull the bezel off of the upper retaining clips.
6. Remove the switch bezel and remove the illumination bulb socket.
7. Remove the switch mounting nut from the panel, remove the switch and disconnect the wiring.
To install:
8. Connect the switch, install the switch to the panel and install the mounting nut.
9. Install the illumination bulb socket to the switch bezel and install the bezel.
10. Install the dash bezel and headlight switch stem.
11. Connect the negative battery cable and check the switch for proper operation.

12. Install the hood release handle and lower steering column cover.

Pickup and Ramcharger

1. Disconnect the negative battery cable.
2. Remove the map light.
3. Remove 6 screws which attach the faceplate to the base panel. There is a screw below the heater-air conditioning control panel which is not visible from above.
4. If equipped with automatic transmission, place the shift lever in its lowest position.
5. Remove the faceplate by pulling the top edge rearward to clear the brow and pulling the bottom out, disengaging the attaching clips. If equipped with 4WD, disconnect the indicator wires.
6. Reach under the instrument panel, depress the spring button and remove the headlight switch knob.
7. Remove the wiper and power mirror switch knobs off their levers, if equipped.
8. Remove the bezel.
9. Remove the switch mounting nut from the panel, remove the switch and disconnect the wiring.
To install:
10. Connect the switch, install the switch to the panel and install the mounting nut.
11. Install the bezel.
12. Install the headlight stem and wiper and power mirror switch knobs, if removed.
13. Connect the 4WD indicator, if equipped. Install the cluster faceplate and map light.
14. Connect the negative battery cable and check the switch for proper operation.

Combination Switch

REMOVAL AND INSTALLATION

1. Disconnect the negative battery cable.
2. Remove the tilt lever, if equipped.
3. Remove the steering column covers.
4. Remove the combination switch tamper-proof mounting screws and pull the switch away from the steering column.
5. Loosen the connector screw; the screw will remain in the connector.
6. Disconnect the connector from the switch.
7. The installation is the reverse of the removal procedure.
8. Connect the negative battery cable and check all functions of the combination switch for proper operation.

Ignition Lock/Switch

REMOVAL AND INSTALLATION

1. Disconnect the negative battery cable.
2. Remove the tilt lever, if equipped.
3. Remove the upper and lower column covers.
4. Remove the 3 ignition switch tamper-proof Torx screws; APEX 440–TX20H or equivalent is required.
5. Pull the switch away from the column. Release the connector locks on the 2 wiring connectors and disconnect them from the switch.
6. Remove the key lock cylinder from the ignition switch by performing the following:
   a. Insert the key and turn the switch in the **LOCK** position. Using a small tool, depress the key cylinder retaining pin flush with the key cylinder surface.

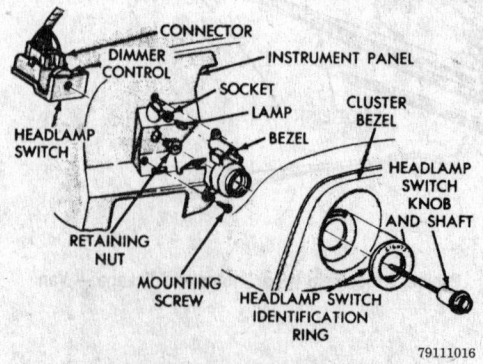

Headlight switch removal — Van

b. Rotate the key clockwise to the **OFF** position to unseat the key cylinder from the ignition switch assembly. The cylinder bezel should be about 1/8 in. above the ignition switch halo light ring. Do not attempt to remove the key cylinder at this point.

c. With the key cylinder in the unseated position, rotate the key counterclockwise to the **LOCK** position and remove the key.

d. Remove the key cylinder from the ignition switch.

To install:

7. Connect the wiring connectors.

8. Mount ignition switch to the column by performing the following:

a. Position the shifter in **P** position. The park lock dowel pin on the ignition switch assembly must engage with the column park lock slider linkage.

b. Verify that the ignition switch is in the **LOCK** position. The flag should be parallel to the ignition switch terminals. Apply a small amount of grease to the flag and pin.

c. Position the park lock link to mid-travel.

d. Align the locating pin hole and its pin and position the ignition switch against the lock housing face. Make sure the pin is inserted into the park lock link contour slot. Torque the retaining screws to 17 inch lbs.

9. With the key cylinder and ignition switch in the **LOCK** position, key not in cylinder, gently insert the key cylinder into the ignition switch until it bottoms.

10. Insert the key. Simultaneously push in on the cylinder and rotate the key to the **RUN** position. This action should fully seat the cylinder in the ignition switch.

11. Install the column covers and the tilt lever, if equipped.

12. Connect the negative battery cable and check the push-to-lock and park lock functions, halo lighting and all ignition switch positions for proper operation.

Stoplight Switch

ADJUSTMENT

1. Push the stoplight switch forward in the mounting bracket as far as it will go; the brake pedal should move forward slightly.

2. Pull back on the brake pedal bringing the striker toward the switch until the pedal will not go

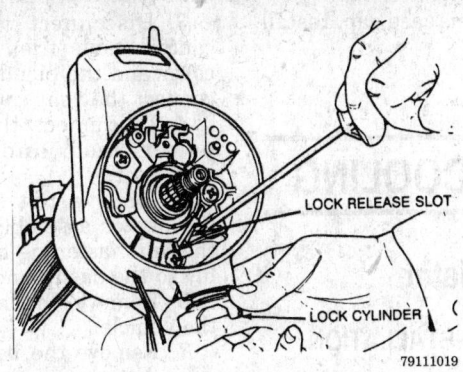

Removing the key lock cylinder

back any farther. This will cause the switch to ratchet backward into position and automatically adjust.

REMOVAL AND INSTALLATION

1. Disconnect the negative battery cable.

2. Remove the switch mounting bracket assembly from the brake pedal bracket.

3. Remove the switch from its bracket.

4. The installation is the reversal of the removal procedure.

5. Connect the negative battery cable and check the switch for proper operation.

Neutral Safety Switch

REMOVAL AND INSTALLATION

1. Disconnect negative battery cable.

2. Raise and safely support the vehicle. Position drain pan under switch.

3. Disconnect switch wiring harness.

4. Remove switch from case.

5. Verify that switch operating lever fingers are centered in the switch opening in the case when in **P** and **N**.

6. Install new seal on switch and install switch into case. Torque to 25 ft. lbs.

7. Test continuity of new switch. Connect wire harness and negative battery cable.

8. Add transmission fluid as required to correct level.

Fuses, Circuit Breakers and Relays

LOCATION

The fuse block and relay bank is located to the left of the glove box on Vans and directly under the steering column on Pick-Ups and Ramchargers. Circuit breakers are also located on the fuse block. The auto shutdown relay, starter relay, A/C cutout relay and part throttle unlock relay are located in the engine compartment near the 50-way bulkhead connector

Computers

LOCATION

The Single Module Engine Controller (SMEC) is located in engine compartment, mounted to the left or right fender depending on model. The Electronic Control Unit (ECU) is located on the righ side of the dash panel in engine compartment.

Flashers

LocationThe turn signal flasher and the hazard flasher are located in the fuse block.

Cruise Control

ADJUSTMENT

Diesel Engine

1. The clearance between the throttle stud and cable clevis should be as small as possible without moving the throttle. If gap is not correct, remove adjustment clip.

2. Push protective sleeve into housing to decrease gap or pull sleeve

out of housing to increase gap. Install adjustment clip.

ENGINE COOLING

Radiator

REMOVAL AND INSTALLATION

1. Disconnect the negative battery cable.
2. Drain cooling system. Remove heater hoses from core and cap.
3. Remove the upper hose and coolant reserve tank hose from the radiator.
4. Remove the shroud from the radiator and position aside.
5. Remove the upper radiator mounting screws. On Van, the radiator will be removed from the bottom of the vehicle. Raise and safely support vehicle.
6. Remove the lower hose from the radiator.
7. Disconnect the automatic transmission cooler lines, if equipped, position aside and cap ends.
8. Remove the mounting screws and carefully lift or lower the radiator out of the engine compartment.

To install:
9. Lift or lower the radiator into position and install the mounting screws.
10. Connect the automatic transmission cooler lines, if removed.
11. Connect the lower hose.
12. Install the upper radiator mounting bolts and shroud.
13. Reconnect negative battery cable.
14. Connect the upper radiator hose and coolant reserve tank hose.
15. Fill the cooling system.
16. Run the vehicle until normal operating temperature is reached. Recheck coolant level and automatic transmission fluid level and add as required.

Heater Core

REMOVAL AND INSTALLATION

Van

WITHOUT AIR CONDITIONING

1. Disconnect the negative battery cable.
2. Drain the cooling system. Disconnect and plug the heater hoses.

3. Disconnect the temperature control cable from the heater core cover and the blend door crank. Disconnect the vent cable.
4. Disconnect the blower motor power feed and ground wire connector.
5. Remove the screws retaining the heater assembly aside cowl and the nuts fastening the heater assembly to the dash panel.
6. Remove the heater unit from the vehicle.
7. Remove the back plate and remove the screws holding the heater core cover to the heater housing. Lift the cover from the housing.
8. Remove the retaining screws from the heater core and remove the core from the heater housing.

To install:
9. Clean out the inside of the housing. Place the heater core into the housing and fasten.
10. Position the blend air door and right vent door in the housing and fasten the heater core cover to the housing.
11. Check the dash panel and side cowl seals for breaks or lack of adhesion and repair as needed.
12. Install the heater assembly to the vehicle.
13. Connect the blower motor power feed and ground wire connector.
14. Position temperature control cable on blend air door crank and attach cable to heater core cover. Connect and adjust vent cable.
15. Connect the heater hoses. Connect negative battery cable.
16. Refill the cooling system.
17. Run the vehicle until normal operating temperature is reached. Recheck coolant level and automatic transmission fluid level and add as required.

WITH AIR CONDITIONING

1. Disconnect the negative battery cable. Discharge the air conditioning system.
2. Disconnect the freeze control connector from the wire harness at the H-valve.
3. Drain the cooling system. Place a layer of non-conductive waterproof material over the alternator to prevent coolant from spilling on it when disconnecting the heater hoses. Disconnect and cap the heater hoses.
4. Slowly disconnect the refrigerant plumbing from the H valve. Remove the 2 screws from the filter drier bracket and swing the plumbing out of the way. Cap open air conditioning lines to prevent contamination of the system.

5. Remove the temperature control cable from the cover.
6. Working from inside the vehicle, remove the glove box, spot cooler bezel and the appearance shield. Working through the glove box opening and under the instrument panel, remove the screws and nuts attaching the evaporator core housing to the dash panel.
7. Remove the 2 screws from the flange connection to the blower housing. Separate the evaporator core housing from the blower housing.
8. Carefully remove the evaporator core housing from the vehicle.
9. Remove the cover from the housing. Remove 1 screw from strap on heater core tubes and pull core out of housing.

To install:
10. Clean the inside of the housing. Place the heater core into the housing and install the retaining strap and screws.
11. Install the housing cover.
12. Attach the blower housing to the evaporator housing.
13. Inspect all air seals and mating surfaces for possible breaks and leaks and repair as needed.
14. Through the glove box opening and under the dash, install the screws and nuts attaching evaporator housing assembly to dash panel.
15. Install the appearance shield, spot cooler bezel and glove box.
16. Attach the temperature control cable to the cover.
17. Position the plumbing and install the 2 screws onto the filter drier bracket.
18. Install a new gasket and connect the refrigerant plumbing to the H-valve.
19. Connect the heater hoses and remove the waterproof material from the alternator. Connect the freeze control wire harness.
20. Evacuate and recharge the air conditioning system.
21. Connect the negative battery cable. Refill the cooling system.
22. Run the vehicle until normal operating temperature is reached. Recheck coolant level and add as required. Check the operation of the entire climate control system.

Pickup and Ramcharger

WITHOUT AIR CONDITIONING

1. Disconnect the negative battery cable.
2. Drain the cooling system. Remove and plug the heater core hoses.
3. Remove the right side cowl trim panel, if equipped. Remove the glove

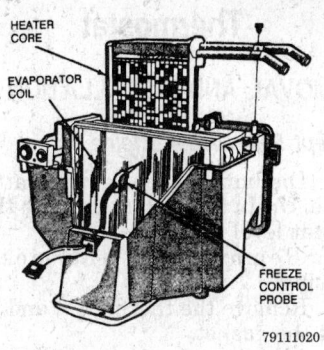

Heater core removal — Van with air conditioning

box assembly. Remove the structural brace through the glovebox opening.

4. Remove the right half of the instrument panel lower reinforcement and disconnect the radio ground strap.

5. Disconnect the control cables from the heater housing and the blower motor wires on the engine side.

6. Remove the retaining screw between the package to cowl side sheet metal.

7. Remove the 6 heater housing retaining nuts on the engine side of the heater assembly and remove the heater housing assembly.

8. Remove the heater housing cover retaining screws and the mode door crank. Separate the cover from the housing.

9. Carefully lift the heater core from the heater housing.

To install:

10. Clean the inside of the housing. Install the heater core into the housing.

11. Install the housing cover.

12. Inspect the dash panel seal for damage and repair as required.

13. Install the assembly to the dash panel and install the retaining nuts.

14. Install the cowl side retaining screws.

15. Connect the blower motor connector.

16. Connect the control cables.

17. Install the right lower instrument panel reinforcement, structural brace, glove box and cowl side trim panel, if equipped.

18. Connect the heater hoses.

19. Refill the radiator.

20. Connect the negative battery cable.

21. Run the vehicle until normal operating temperature is reached. Check coolant level and add as required.

WITH AIR CONDITIONING

1. Disconnect the negative battery cable. Discharge the air condition system. Drain the cooling system. Disconnect and plug the heater hoses and the refrigerant lines.

2. Remove the condensation tube from the housing.

3. Move the transfer case and gear shift levers away from the instrument panel.

4. Remove the right side cowl trim panel, if equipped. Remove the glove box and swing it out from the bottom.

5. Remove the structural brace from the through hole in the glove box opening. Remove the ash tray.

6. Remove the right lower half of the dash reinforcement by removing the retaining screws holding it to the instrument panel and to the cowl side trim panel.

7. Disconnect the radio ground strap. Remove the center and floor air distribution ducts.

8. Disconnect the temperature control cable from the assembly and tape it out of the way.

9. Disconnect the vacuum lines from the extension on the control unit and unclip the vacuum lines from the defroster duct.

10. Remove the wiring connector from the resistor block. Remove the blower motor electrical connector from the engine side of the assembly.

11. Disconnect the vacuum lines on the engine side and make sure the grommet is free from the dash panel.

12. Remove the retaining nuts on the engine side. Remove the screw that retains the assembly to the cowl side of the sheetmetal.

13. Remove the assembly from the vehicle.

14. Remove the vacuum actuators, door crank levers, evaporator case cover retaining nuts and screws and the heater core retaining screws. Lift the cover off of the asembly and remove the heater core from its mounting.

To install:

15. Clean the inside of the housing. Install the heater core into the housing.

16. Install the housing cover, retaining screws and nuts, levers and actuators.

17. Inspect the dash panel seals for damage and repair as required.

18. Feed the vacuum lines through the hole in the dash panel, install the assembly to the dash panel and install all retaining nuts and screws.

19. Connect the resistor block and blower motor.

20. Connect the vacuum lines to the extension on the control unit and clip the vacuum lines to the defroster duct.

21. Connect the temperature control cable to the assembly.

22. Connect the radio ground strap. Install the center and floor air distribution ducts.

23. Install the dash reinforcement, structural brace, glove box and right side cowl trim panel, if equipped. Install the ash tray.

24. Install the condensation tube.

25. Connect the heater hoses and vacuum lines.

26. Install a new gasket and connect the refrigerant lines.

27. Evacuate and recharge the air conditioning system. Refill the radiator and connect negative battery cable.

28. Run the vehicle until normal operating temperature is reached. Recheck coolant level and add as required. Check operation of the entire climate control system.

Water Pump

REMOVAL AND INSTALLATION

If the water pump is being replaced due to bearing or shaft damage, the mechanical cooling fan should be carefully inspected for fatigue cracks, loose blades or loose rivets resulting from excessive vibration. If the fan is damaged in any way, it could snap at any time, possibly causing serious personal injury or damage to the vehicle.

Gasoline Engines

1. Disconnect the negative battery cable. Drain the coolant.

2. Remove the shroud from the radiator and slide it over the fan, if 1-piece.

3. Remove the radiator and lower hose.

4. Remove the fan blade, spacer or viscous drive unit, pully, bolts and shroud (if it was not removed in Step 2) together. Remove the air pump belt and power steering pump belt.

NOTE: Do not place the viscous fan in an upright position because the silicone fluid in the drive could drain into the bearing and contaminate its lubricant.

5. Loosen the alternator mounting bolts and remove the alternator/air conditioner belts.

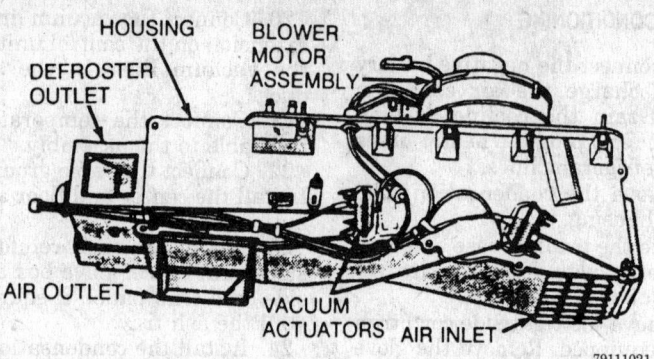

Evaporator heater assembly — Pick-Up and Ramcharger with air conditioning

6. Remove the alternator bracket. This bracket also supports the air conditioner compressor or idler pulley, which will remain supported by its rear mount.

7. Remove the air pump(s) and the air pump bracket.

8. Unbolt the power steering pump bracket with the pump still attached and position it out of the way.

9. Disconnect the heater and bypass hoses.

10. Remove the remaining pump retaining bolts and remove the pump from the engine.

To install:

11. Clean and dry the pump mating surfaces. Install a new bypass hose to the engine.

12. Install a new gasket and install the water pump to the engine.

13. Install any bolts that do not retain a bracket. Tighten the bypass hose clamps. Install the heater hose.

14. Install the water pump to compressor front mount bolts and bracket, if removed.

15. Install all remaining brackets and components that were removed during the removal procedure. Torque all water pump retaining bolts that do not go through adjusting slots to 30 ft. lbs. (41 Nm).

16. Install and adjust the alternator/air conditioner belts. Install the remaining belts.

17. Install the fan blade, spacer or viscous drive unit, pulley and bolts along with the shroud (if it is a 1-piece unit) together.

18. Install the radiator, lower hose and shroud.

19. Adjust all belts and torque the remaining water pump retaining bolts to 30 ft. lbs. (41 Nm).

20. Fill the radiator with coolant.

21. Connect the negative battery cable, run the vehicle until the thermostat opens, fill the radiator completely and check for leaks.

22. Once the vehicle has cooled, recheck the coolant level.

5.9L Diesel Engine

1. Disconnect the negative battery cable.

2. Drain the coolant.

3. Use a ⅜ in. drive breaker bar to lift the belt tensioner and remove the belt.

4. Remove the 2 water pump retaining bolts and remove the pump from the engine.

5. Remove the O-ring from the pump groove.

To install:

6. Clean the O-ring groove and install a new O-ring.

7. Clean the pump mating surfaces and install the pump to the engine.

8. Torque the mounting bolts to 18 ft. lbs. (24 Nm). Fill the radiator with coolant.

9. Install the drive belt.

10. Connect the negative battery cable, run the vehicle until the thermostat opens, fill the radiator completely and check for leaks.

11. Once the vehicle has cooled, recheck the coolant level.

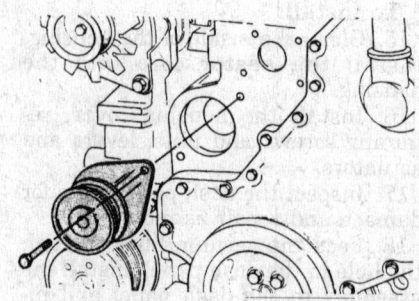

5.9L diesel engine water pump

Thermostat

REMOVAL AND INSTALLATION

Except 5.9L Diesel Engine

1. Disconnect the negative battery cable. Drain the coolant down to thermostat level or below.

2. Remove the thermostat housing.

3. Remove the thermostat and discard the gasket.

4. Clean the housing mating surfaces and use a new gasket.

5. The installation is the reversal of the removal procedure.

6. Connect the negative battery cable, run the vehicle until the thermostat opens, fill the radiator completely and check for leaks.

7. Once the vehicle has cooled, recheck the coolant level.

5.9L Diesel Engine

1. Disconnect the negative battery cable.

2. Drain the coolant.

3. Use a ⅜ in. drive breaker bar to lift the belt tensioner and remove the belt.

4. Disconnect the upper radiator hose from the thermostat housing.

5. Loosen the alternator mounting bolts and lower the alternator.

6. Unbolt the thermostat housing and remove the housing, engine lifting bracket and thermostat with seal.

To install:

7. Install the new thermostat to the housing, making sure the tang on the thermostat is aligned with the slot in the housing. This will ensure correct positioning of the jiggle pins in the housing. This is important because during cooling system filling. Air vents through the jiggle pin openings, through the upper hose and out the radiator fill neck.

8. Clean the pump mating surfaces.

9. Install the engine lifting bracket, new seal, thermostat and housing. Make sure the seal is installed with the beveled side facing out.

10. Torque the thermostat housing bolts to 18 ft. lbs. (24 Nm). Connect the upper hose to the housing.

11. Reinstall the alternator into position. Torque the upper bolt to 18 ft. lbs. (24 Nm) and the lower bolt to 32 ft. lbs. (43 Nm).

12. Install the drive belt.

13. Connect the negative battery cable, run the vehicle until the thermostat opens, fill the radiator completely and check for leaks.

14. Once the vehicle has cooled, recheck the coolant level. 5.9L diesel engine thermostat

79111023

Cooling System Bleeding

Air trapped in the cooling system gathers under the radiator cap during engine operation. The next time the engine is operated thermal ex-

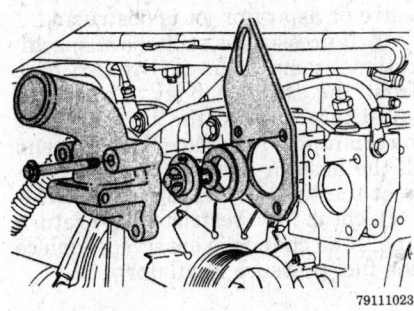

79111023

791111024 not found

791111024

Installing the thermostat in the housing — 5.9L diesel engine

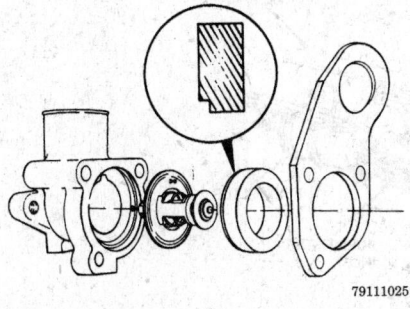

79111025

Proper positioning of the rubber seal — 5.9L diesel engine

pansion of coolant will force trapped air past the radiator cap and into the coolant reserve tank where it is vented into the atmosphere. As the engine cools down coolant will be drawn from the reserve tank into the radiator to replace the vented air.

GASOLINE FUEL SYSTEM

Fuel System Service Precautions

Safety is the most important factor when performing not only fuel system maintenance but any type of maintenance. Failure to conduct maintenance and repairs in a safe manner may result in serious personal injury or death. Maintenance and testing of the vehicle's fuel system components can be accomplished safely and effectively by adhering to the following rules and guidelines.

• To avoid the possibility of fire and personal injury, always disconnect the negative battery cable unless the repair or test procedure requires that battery voltage be applied.

• Always relieve the fuel system pressure prior to disconnecting any fuel system component (injector, fuel rail, pressure regulator, etc.), fitting or fuel line connection. Exercise extreme caution whenever relieving fuel system pressure to avoid exposing skin, face and eyes to fuel spray. Please be advised that fuel under pressure may penetrate the skin or any part of the body that it contacts.

• Always place a shop towel or cloth around the fitting or connection prior to loosening to absorb any excess fuel due to spillage. Ensure that all fuel spillage (should it occur) is quickly removed from engine surfaces. Ensure that all fuel soaked cloths or towels are deposited into a suitable waste container.

• Always keep a dry chemical (Class B) fire extinguisher near the work area.

• Do not allow fuel spray or fuel vapors to come into contact with a spark or open flame.

• Always use a backup wrench when loosening and tightening fuel line connection fittings. This will pre-

vent unnecessary stress and torsion to fuel line piping. Always follow the proper torque specifications.

• Always replace worn fuel fitting O-rings with new. Do not substitute fuel hose or equivalent where fuel pipe is installed.

RELIEVING FUEL SYSTEM PRESSURE

1. Loosen the fuel filler cap to release fuel tank pressure.
2. Disconnect the injector wiring harness from the engine harness.
3. Connect a jumper wire to ground terminal 1 of the injector harness to engine ground.
4. Being careful not to allow contact between the jumper leads, connect a jumper wire to pin 2 next to the 1 that is grounded and touch the other end of the jumper to the positive battery post for no longer than 5 seconds. This will relieve fuel pressure.
5. Remove the jumper wires and connect the connector.

Fuel Tank

REMOVAL AND INSTALLATION

1. Disconnect negative battery cable.
2. Relieve the fuel pressure.
3. Raise the vehicle and support safely.
4. Using the proper equipment, drain the fuel tank.
5. Remove vent hoses from hose routing bracket attacked to the top of the frame rail. Disconnect the electrical connection.
6. Place a transmission jack or equivalent, under the center of the tank and apply slight pressure. Remove the tank straps.
7. Lower tank slightly to provide access to the fuel supply and return hoses, and disconnect hoses. Disconnect fuel vapor line from pressure relief/rollover valve.
8. Lower tank from vehicle.
To install:
9. Raise the tank into position and connect all electrical harnesses and vacuum hoses.
10. Install the tank straps and tighten the retaining nuts.
11. Install vent hoses, if not done, and check all connections.
12. Fill fuel tank, install fuel filler cap and reconnect negative battery cable.

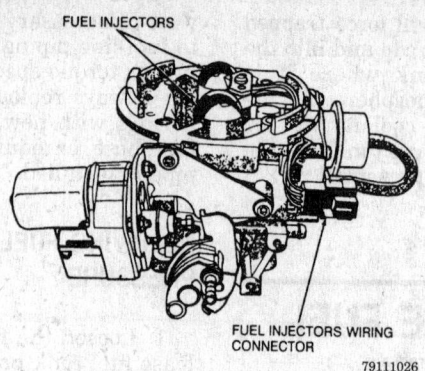

Injector wiring connector location

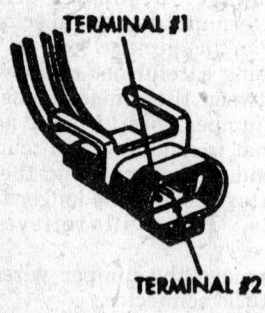

Fuel injector harness connectors

Fuel Filter

REMOVAL AND INSTALLATION

——— **CAUTION** ———
Do not use conventional fuel hoses or clamps when servicing this fuel system. They are not compatible with the injection system and could fail, causing personal injury or damage to the vehicle. Use only hoses and clamps specifically designed for fuel injection.

1. Relieve the fuel pressure.
2. Disconnect the negative battery cable.
3. Remove the filter retaining screw and remove the filter assembly from the mounting plate.
4. Loosen the outlet hose clamp on the filter and inlet hose clamp on the rear fuel tube.
5. Wrap a shop towel around the hoses to absorb fuel. Remove the hoses from the filter and fuel tube and discard the clamps and the filter.
   To install:
6. Install the inlet hose on the fuel tube and tighten the new clamp to 10 inch lbs. (1 Nm).

7. Install the outlet hose on the filter outlet fitting and tighten the new clamp to 10 inch lbs. (1 Nm).
8. Position the filter assembly on the mounting plate and tighten the mounting screw to 75 inch lbs. (8 Nm).
9. Connect the negative battery cable, start the engine and check for leaks.

Electric Fuel Pump

PRESSURE TESTING

1. Relieve the fuel pressure.
2. Disconnect the larger diameter fuel supply hose from the engine fuel line assembly.
3. Connect the fuel system pressure tester C–4799B or equivalent, between the fuel supply hose and the engine fuel line assembly.
4. With the key in the **RUN** position, put the DRB I or II in the activate auto shutdown relay mode; this will activate the fuel pump and pressurize the system.
5. The pressure specification is 13.5–15.5 psi. If the pressure is within specifications, reinstall the fuel hose.
6. If fuel pressure is below specifications, install the tester in the fuel

supply line between the tank and the filter and repeat the test.
7. If the pressure is 5 psi higher than in Step 5, replace the fuel filter. If no change is observed, squeeze the return hose. If pressure increases, replace the pressure regulator. If no change is observed, the problem is either a plugged in-tank sock filter or a defective pump.
8. If fuel pressure is above specifications, remove the fuel return line hose from the chassis line at the fuel tank and connect a 3 foot piece of fuel hose to the return line. Put the other end into a 2 gallon minimum capacity approved gasoline container. Repeat the test. If pressure is now correct, check the in-tank return hose for kinking. Replace the fuel pump assembly if the in-tank reservoir check valve or aspirator jet is obstructed.
9. If pressure is still above specifications, remove the fuel return hose from the throttle body. Connect a substitute hose to the throttle body return nipple and place the other end of the hose in a clean container. Repeat the test. If pressure is now correct, check for a restricted fuel return line. If no change is observed, replace the fuel pressure regulator.

REMOVAL AND INSTALLATION

The fuel pump module is installed in the top of the fuel tank. It contains the fuel pump, fuel pump reservoir, pressure relief/rollover valve, electrical connector for the sending unit, fuel filler vent, supply and return tube connections and the drain tube nipp
1. Disconnect negative battery cable.
2. Relieve the fuel pressure.
3. Raise the vehicle and support safely.
4. Using the proper equipment, drain the fuel tank.

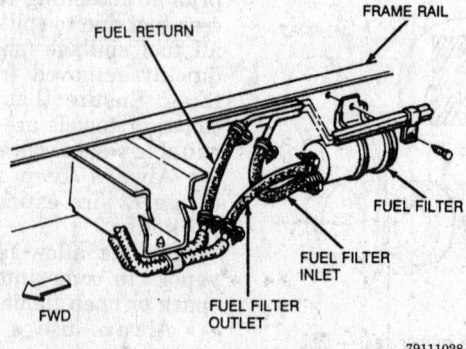

Fuel filter location — fuel injected engine

5. Remove vent hoses from hose routing bracket attacked to the top of the frame rail. Disconnect the electrical connection.

6. Place a transmission jack or equivalent, under the center of the tank and apply slight pressure. Remove the tank straps.

7. Lower tank slightly to provide access to the fuel supply and return hoses and disconnect hoses. Disconnect fuel vapor line from pressure relief/rollover valve. Lower tank from vehicle.

8. Remove the fuel pump module collar and module from tank. Mark position of sender wires and disconnect. Remove sender unit from module.

9. Replace module and pump assembly as a unit. Reinstall sender to module and reposition into tank. Install collar to original position.

10. Raise the tank into position and connect all harnesses and vacuum hoses. Install the tank straps and tighten the retaining nuts.

11. Connect fuel filler tube and vent hoses.

12. Fill fuel tank, install fuel filler cap and reconnect negative battery cable.

Fuel Injection

IDLE SPEED ADJUSTMENT

1. Start the engine and allow it to reach normal operating temperature. If engine is already hot, run for 2 minutes.

2. Turn the engine OFF and allow 1 minute for the Idle Speed Control (ISC) actuator shaft to fully extend.

3. Disconnect the ISC actuator connector and the coolant temperature sensor.

4. Connect a tachometer to the engine and start the engine.

5. Adjust the extension screw on the actuator shaft until the rpm is within specifications. 3.9L engine — 2500–2600 rpm 5.2L and 5.9L engines — 2750–2850 rpm

6. Turn the engine OFF. Reconnect the ISC actuator connector and the coolant temperature sensor. Remove the tachometer.

Idle Mixture ADJUSTMENT

There is no idle mixture adjustment provided with this fuel injection system.

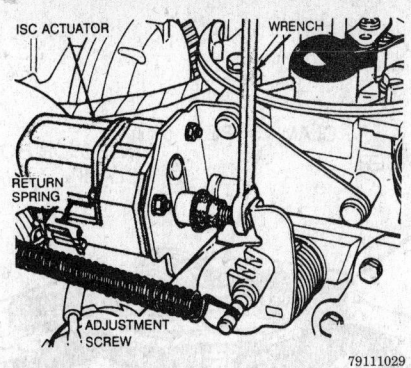

Adjusting the idle speed — fuel injected engine

Fuel Injector

REMOVAL AND INSTALLATION

1. Remove the air cleaner assembly.

2. Relieve the fuel pressure.

3. Disconnect the negative battery cable.

NOTE: There is a small spacer below the injector hold-down clamp. Take the proper precaution to prevent this spacer from falling into the throttle body.

4. Remove the injector hold-down Torx screw, the hold-down and the spacer.

5. Using a small flat-tipped tool, lift the caps off of the injector.

6. Using the same tool, gently pry the injector from its pod.

7. Remove the lower O-ring from the pod.

To install:

8. Install the new lower O-ring on the injector.

9. Align the injector terminal housing with the locating socket in the injector cap.

10. Press the injector cap so the upper O-ring flange is flush with the lower surface of the cap.

11. Spray the inner surfaces of the injector pod with a carburetor parts cleaner to remove residual varnish and gasoline.

12. Lubricate the O-rings sparingly with unmedicated petroleum jelly.

13. Place the injector and cap into the injector pod and align the cap locating pin with the locating hole in the casting. The right side locating pin in 5mm in diameter and will not fit into the other hole.

14. Press firmly on the injector cap until it is flush with the casting surface.

15. Install the spacer and align the holes in the hold-down with the pins on the caps and install.

16. Push down on the caps, install the screw and torque to 35 inch lbs. (4 Nm).

17. Connect the negative battery cable and check for leaks using the DRB I or II to activate the fuel pump.

18. Install the air cleaner.

DIESEL FUEL SYSTEM

Fuel System Service Precaution

Safety is the most important factor when performing not only fuel system maintenance but any type of maintenance. Failure to conduct maintenance and repairs in a safe manner may result in serious personal injury or death. Maintenance and testing of the vehicle's fuel system components can be accomplished safely and effectively by adhering to the following rules and guidelines.

• To avoid the possibility of fire and personal injury, always disconnect the negative battery cable unless the repair or test procedure requires that battery voltage be applied.

• Always relieve the fuel system pressure prior to disconnecting any fuel system component (injector, fuel rail, pressure regulator, etc.), fitting or fuel line connection. Exercise extreme caution whenever relieving fuel system pressure to avoid exposing skin, face and eyes to fuel spray. Please be advised that fuel under pressure may penetrate the skin or any part of the body that it contacts.

• Always place a shop towel or cloth around the fitting or connection prior to loosening to absorb any ex-

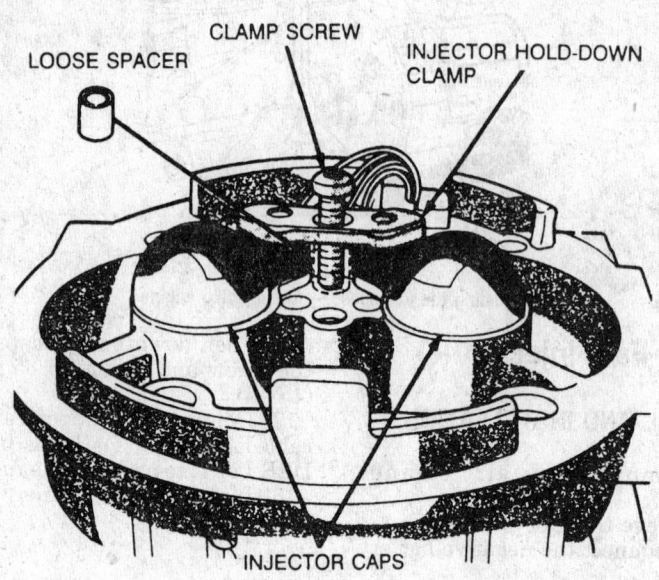

Injector hold-down and spacer — fuel injected engine

LOOSE SPACER — CLAMP SCREW — INJECTOR HOLD-DOWN CLAMP — INJECTOR CAPS

79111030

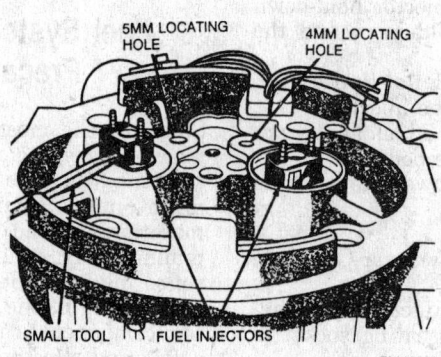

Removing the injector from its pod — fuel injected engine

5MM LOCATING HOLE — 4MM LOCATING HOLE — SMALL TOOL — FUEL INJECTORS

79111031

cess fuel due to spillage. Ensure that all fuel spillage (should it occur) is quickly removed from engine surfaces. Ensure that all fuel soaked cloths or towels are deposited into a suitable waste container.

• Always keep a dry chemical (Class B) fire extinguisher near the work area.

• Do not allow fuel spray or fuel vapors to come into contact with a spark or open flame.

• Always use a backup wrench when loosening and tightening fuel line connection fittings. This will prevent unnecessary stress and torsion

to fuel line piping. Always follow the proper torque specifications.

• Always replace worn fuel fitting O-rings with new. Do not substitute fuel hose or equivalent where fuel pipe is installed.

Fuel Filter

REPLACEMENT

1. Disconnect the negative battery cable.

2. Disconnect the Water In Filter (WIF) sensor connector.

3. Remove the separator filter assembly from the filter head with a standard oil filter wrench.

4. Remove the square cut O-ring from the filter mounting bushing.

5. Drain the fuel/water separator filter and remove the assembly from the fuel filter.

To install:

6. Install a new O-ring to the WIF assembly and install to the new separator filter.

7. Install a new square cut O-ring to the mounting bushing.

8. Fill the fuel/water separator filter with clean diesel fuel.

9. Apply a light coat of oil to the sealing surface of the separator filter.

10. Install the assembly and tighten it ½ turn after the seal contacts the filter head.

11. Reconnect the WIF sensor connector.

12. Connect the negative battery cable, start the engine and check for leaks.

DRAINING WATER FROM THE SYSTEM

Filtration and separation of water from the fuel is important for trouble free operation and long life of the fuel system. Regular maintanence, including draining moisture from the fuel/water separator filter is essential to keep water out of the fuel pump. To remove collected water place drain pan under drain tube on separator. Push up on the drain at the bottom filter separator. Hold drain open until all water and contaminants have been removed and clean fuel exits the drain.

Diesel Injection Pump

REMOVAL AND INSTALLATION

NOTE: The Bosche VE lever is indexed to the shaft during pump calibration. Do not remove it from the pump during removal.

1. Disconnect the negative battery cable.

2. Remove the throttle linkage and bracket.

3. Disconnect the fuel drain manifold.

4. Remove the injection pump supply line.

5. Remove the high pressure lines.

6. Disconnect the electrical wire to the fuel shut off valve.

7. Remove the fuel air control tube.

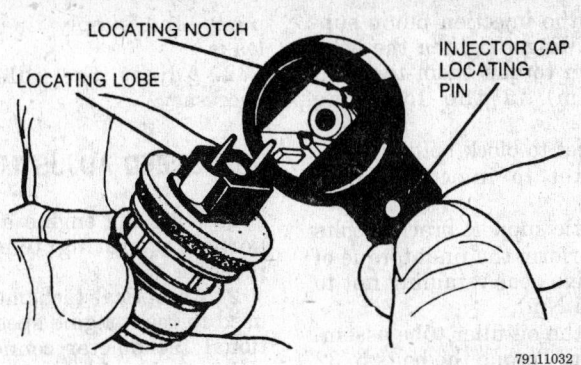

LOCATING NOTCH

LOCATING LOBE

INJECTOR CAP
LOCATING
PIN

79111032

Installing the injector to the cap — fuel injected engine

8. Remove the pump support bracket.

9. Remove the oil fill tube bracket and adapter from the front gear cover.

10. Place a shop towel in the gear cover opening in a position that will prevent the nut and washer from falling into the gear housing. Remove the gear retaining nut and washer.

11. Install the turning tool into the flywheel housing opening on the exhaust side of the engine. Place a ½ in. drive universal joint in the turning tool and attach enough extensions to the joint to make it convenient to turn the tool.

12. Using a ratchet to turn the turning tool, turn the engine until the key way on the fuel pump shaft is pointing approximately in the 6 o'clock position.

13. Locate TDC for cylinder No. 1 by turning the engine slowly while pushing in on the TDC pin. Stop turning the engine as soon as the pin engages with the gear timing hole. Disengage the pin after locating TDC and remove the turning equipment.

14. Loosen the lockscrew, remove the special washer from the injection pump and wire it to the line above it so it will not get misplaced. Retighten the lockscrew to 22 ft. lbs. (30 Nm) to lock the driveshaft.

15. Using a puller, extract the pump drive gear from the driveshaft.

NOTE: Be careful not to drop the drive gear key into the front cover when removing or installing the pump. If it does drop in, it must be removed before proceeding.

16. Remove the 3 mounting nuts and remove the injection pump from the vehicle.

17. Remove the gasket and clean the mounting surface.

To install:

18. Install a new gasket.

NOTE: The shaft of a new or reconditioned pump is locked so the key aligns with the drive gear keyway with cylinder No. 1 at TDC.

19. Install the pump and finger-tighten the mounting nuts; the pump must be free to move in the slots.

20. Install the pump drive gear, washer and nut to the driveshaft. The pump will rotate slightly because of gear helix and clearance. This is acceptable providing the pump is free to move on the flange slots and the crankshaft does not move. Torque the nut to 11–15 ft. lbs. (15–20 Nm). This is not the final torque; do not overtighten.

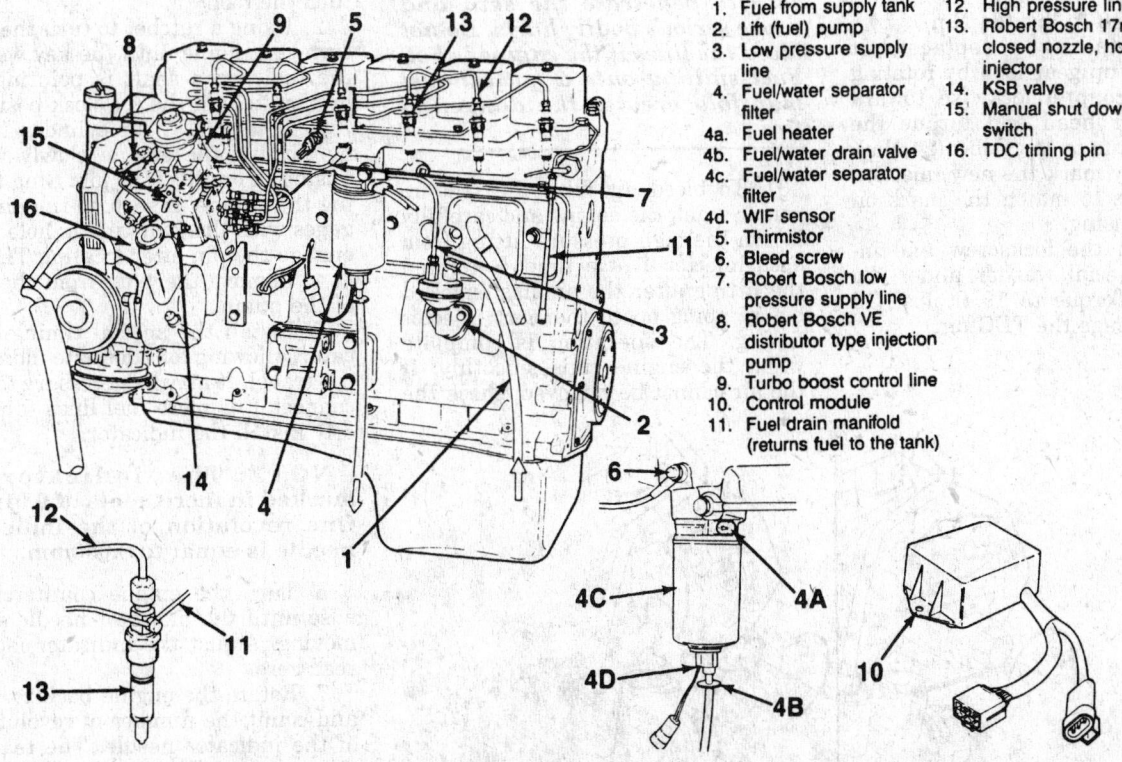

1. Fuel from supply tank
2. Lift (fuel) pump
3. Low pressure supply line
4. Fuel/water separator filter
4a. Fuel heater
4b. Fuel/water drain valve
4c. Fuel/water separator filter
4d. WIF sensor
5. Thirmistor
6. Bleed screw
7. Robert Bosch low pressure supply line
8. Robert Bosch VE distributor type injection pump
9. Turbo boost control line
10. Control module
11. Fuel drain manifold (returns fuel to the tank)
12. High pressure lines
13. Robert Bosch 17mm, closed nozzle, hole type injector
14. KSB valve
15. Manual shut down switch
16. TDC timing pin

79111033

Fuel system components — 5.9L diesel engine

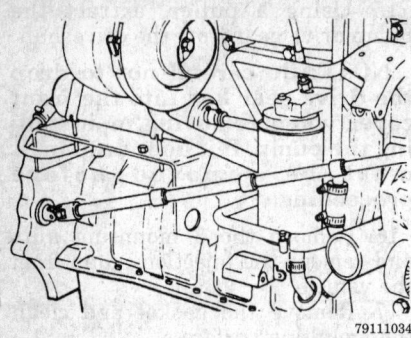

Installing the turning tool — diesel fuel injection system

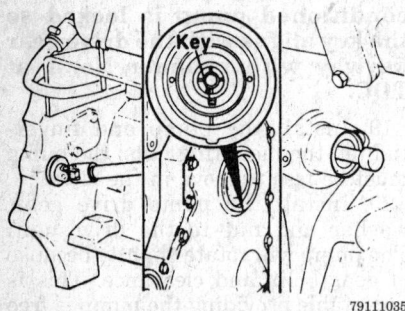

Locating TDC with the TDC pin — diesel fuel injection system

21. If installing the original pump, rotate the pump to align the original timing marks and torque the mounting nuts to 18 ft. lbs. (24 Nm).

22. If installing a replacement pump, take up gear lash by rotating the pump counterclockwise toward the cylinder head and torque the mounting nuts to 18 ft. lbs. (24 Nm). Permanently mark the new injection pump flange to match the mark on the gear housing.

23. Loosen the lockscrew and install the special washer under the lockscrew. Torque to 19 ft. lbs. (13 Nm). Disengage the TDC pin.

24. Install the injection pump support bracket. finger-tighten the bolts initially, then torque them to 18 ft. lbs. (24 Nm) in the following sequence
   a. Bracket to block bolts.
   b. Bracket to injection pump bolts.
   c. Throttle support bracket bolts.

25. Now perform the final torque of the pump drive gear retaining nut to 48 ft. lbs. (65 Nm).

26. Install the oil filler tube assembly and clamp. Torque the bolts to 32 ft. lbs. (43 Nm).

27. Install all fuel lines and the electrical connector to the fuel shut off valve. Tighten the high pressure lines to 18 ft. lbs. (24 Nm).

28. Install the fuel air control tube. Torque the banjo fitting bolt to 9 ft. lbs. (12 Nm).

29. Install the throttle bracket and linkage. When connecting the cable to the control lever, adjust the legnth so the lever has stop-to-stop movement.

30. Connect the negative battery cable.

CAUTION

Do not place any part of the hand near the base of the high pressure line. A fuel leak from a high pressure fuel line has sufficient pressure to penetrate the skin and cause serious bodily harm. Do not bleed the lines if the engine is hot. Fuel spilling onto a hot exhaust manifold creates the danger of fire.

---

31. To bleed air from the system, run or crank the engine and carefully loosen the high pressure fitting from each injector 1 at a time. Retighten the fitting after the air has expelled before going on to the next injector fitting. The operation is complete when the engine runs smoothly. If the air cannot be removed, check the

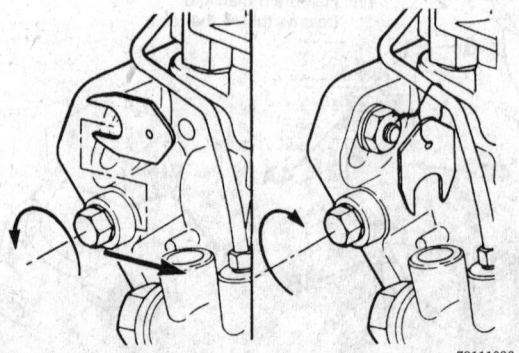

Removing the special washer from the injection pump

pump and supply line for suction leaks.

32. Adjust the idle speed, if necessary.

IDLE SPEED ADJUSTMENT

1. Start the engine and run until normal operating temperature is reached.

2. An optical tachometer must be used to read engine speed; a conventional tachometer connected to the coil is useless in this instance.

3. Turn the air conditioning **ON**, if equipped.

4. Turn the idle speed screw until the desired idle speed is obtained. The specification if equipped with automatic transmission is 700 rpm or if equipped with manual transmission is 750 rpm.

Injection Pump Timing

ADJUSTMENT

1. Install the turning tool into the flywheel housing opening on the exhaust side of the engine. Place a ½ in. drive universal joint in the turning tool and attach enough extensions to the joint to make it convenient to turn the tool.

2. Using a ratchet to turn the tool, turn the engine until the key way on the fuel pump shaft is pointing approximately in the 6 o'clock position.

3. Locate TDC for cylinder No. 1 by turning the engine slowly while pusing in on the TDC pin. Stop turning the engine as soon as the pin engages with the gear timing hole. Disengage the pin after locating TDC.

4. Remove the plug from the end of the pump.

5. Install the special timing indicator, allowing for adequate indicator pin travel. It may be necessary to disconnect 1 or more fuel lines to properly install the indicator.

NOTE: The indicator is marked in increments of 0.01mm. One revolution of the indicator needle is equal to 0.050mm.

6. Turn the engine counterclockwise until the indicator needle stops moving. Adjust the indicator face to read zero.

7. Rotate the engine back to TDC and count the number of revolutions of the indicator needle. The reading shown when the engine timing pin engages is the amount of plunger lift the pump has at that point.

8. Readjust the indicator face to read zero. Loosen the pump mounting nuts and rotate the pump clockwise toward the cylinder head until the indicator reads the correct value for plunger lift (the reading in Step 7). Torque the mounting nuts to 18 ft. lbs. (24 Nm).

9. Remove the engine turning equipment and timing indicator. Install the timing plug and torque to 7.5 ft. lbs. (10 Nm). Connect any fuel lines that were disconnected.

10. Road test the vehicle.

Fuel Injector

REMOVAL AND INSTALLATION

1. Disconnect the negative battery cable. Remove the throttle linkage and bracket, if necessary.

2. Disconnect the high pressure fuel supply line to the injector.

3. Disconnect the fuel drain manifold.

4. Clean the area around the injector.

5. Using a 24mm deepwell socket, remove the injector from the cylinder head.

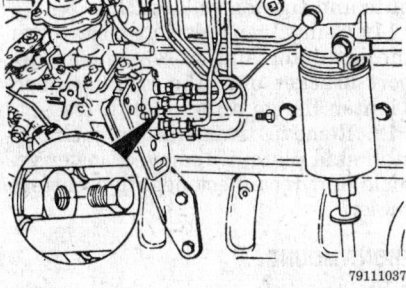

Removing the plug from the pump — diesel fuel injection system

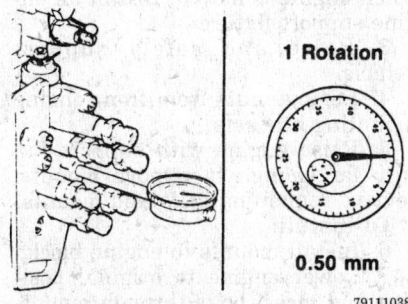

Installing the timing indicator — diesel fuel injection system

To install:

6. Clean the injector bore with a bore brush.

7. Assemble the injector and 1 new copper sealing washer. Never use more than 1 copper washer.

8. Apply a thin coat of anti-sieze compound to the threads of the injector hold-down nut and between the top of the nut and the injector body.

9. Align the protrusion in the injector with the notch in the bore and install the injector. Torque the injector retainer nut to 44 ft. lbs. (60 Nm).

10. Push the O-ring into the groove at the top of the injector.

11. Using new sealing washers, assemble the fuel drain manifold and high pressure lines. Torque the banjo fitting bolt to 6 ft. lbs. (8 Nm). Leave the high pressure line loose temporarily.

--- **CAUTION** ---

Do not place any part of the hand near the base of the high pressure line. A fuel leak from a high pressure fuel line has sufficient pressure to penetrate the skin and cause serious bodily harm. Do not bleed the lines if the engine is hot. Fuel spilling onto a hot exhaust manifold creates the danger of fire.

12. Reconnect negative battery cable. Install the throttle linkage and bracket, if removed.

13. To bleed air from the system, run or crank the engine and tighten the fitting after the air has expelled. If more than 1 injector was replaced, tighten each fitting after the air has expelled before going on to the next injector fitting. Torque the fittings to 18 ft. lbs. (24 Nm). The operation is complete when the engine runs smoothly. If the air cannot be removed, check the pump and supply line for suction leaks.

EMISSION CONTROLS

Emissions Warning Lamp

RESETTING

1. Connect the DRBII to the diagnostic connector.

2. Turn the ignition switch to the **RUN** position and access the Emissions EMR Tests on the DRBII.

3. Select EMR Memory Check.

4. Select Reset EMR Light. This will reset the EMR timing in the computer and turn the light OFF.

5. Disconnect the DRBII.

GASOLINE ENGINE MECHANICAL

NOTE: Disconnecting the negative battery cable on some vehicles may interfere with the functions of the on board computer systems and may require the computer to undergo a relearning process, once the negative battery cable is reconnected.

Engine Assembly

REMOVAL AND INSTALLATION

1. Relieve the fuel pressure, if equipped with fuel injection. Disconnect the battery cables and remove battery from the vehicle.

2. Scribe hinge position on hood and remove the hood. Remove the oil

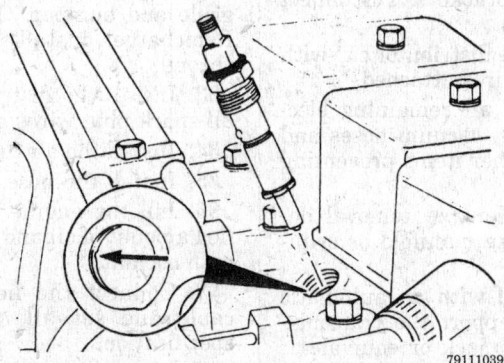

Installing the fuel injector — diesel fuel injection system

dipstick and tube. Discharge the air conditioning.

3. Raise the vehicle and support safely. Drain the engine oil and coolant. Remove the lower radiator hose.

4. Remove the starter. Remove the engine to transmission struts, if equipped.

5. Remove the exhaust pipe from the exhaust manifold(s).

6. If equipped with manual transmission, remove the transmission.

7. If equipped with an automatic transmission, remove the inspection plate. Matchmark the flexplate to the converter, remove the torque converter bolts and push the torque converter backwards as far as it will go. Remove the lower bell housing bolts.

8. Remove the engine mount lower nuts.

9. Disconnect and plug the rubber fuel inlet and return hoses from the fuel lines at the right front of the vehicle. Lower the vehicle.

10. Remove the air cleaner assembly, disconnect all linkages and cables and remove the throttle body. Stuff a clean shop towel into the intake manifold opening to prevent foreign objects from entering. Remove the left exhaust manifold and heat shield on Van.

11. Remove the discharge and suction lines from the air conditioning compressor. Cap the openings on the compressor to prevent foreign objects and moisture from entering the system.

12. On Van, remove the front bumper, grille and support brace. Remove the radiator, shroud, condenser and support as an assembly. On Pick-Up and Ramcharger, remove the radiator and shroud. Remove the fan and all related parts.

13. Unbolt the power steering pump brackets from the engine and position it aside.

14. Remove the alternator, air pump and all brackets. Disconnect the heater hoses.

15. Remove the distributor cap with all spark plug wires attached.

16. Disconnect all remaining electrical connectors, vacuum hoses and check for any other items preventing engine removal.

17. Attach an engine removal device to the intake manifold or cylinder head.

18. If equipped with an automatic transmission, support the transmission with a floor jack or equivalent. Remove the remaining bell housing bolts.

19. Remove the engine from the vehicle. Van may need to be raised slightly, depending on the size of the removal device.

To install:

20. Lower the engine into position and install the upper bell housing bolts. Remove the engine removal device. Install the left side exhaust manifold, if it was removed. Install the oil dipstick.

21. Raise the vehicle and support safely.

22. Install the engine mount nuts and the remaining bell housing bolts.

23. If equipped with a manual transmission, install the transmission and all related parts.

24. If equipped with an automatic transmission, align the torque converter and flexplate and install the bolts. Install the inspection plate, starter and the engine to transmission struts, if equipped.

25. Connect the rubber fuel inlet and return hoses to the fuel lines at the right front of the vehicle.

26. Install the exhaust pipe to the exhaust manifold(s). Install the air pump tube to the exhaust pipe, if equipped. Lower the vehicle.

27. Connect the heater hoses.

28. Install the alternator, air pump, power steering pump and all brackets.

29. Install the air conditioning compressor and connect the lines with new gaskets, if disconnected.

30. Install the throttle body using a new base gasket and connect all linkages and cables. Connect all electrical connectors and vacuum hoses that were disconnected during the engine removal procedure.

31. Install the lower radiator hose. Install the fan and all related parts. Adjust all belt tensions.

32. On Van, install the radiator, shroud, condenser and support as an assembly. Install the support brace, grille and bumper. On Pick-Up and Ramcharger, install the radiator and shroud.

33. Install the distributor cap with all spark plug wires attached.

34. Install the air cleaner assembly.

35. Install the hose.

36. Fill the engine with the specified amount of oil and fill the radiator with coolant.

37. Connect the negative battery cable and set all adjustments to specification.

38. Evacuate and recharge the air conditioning system.

Engine Mounts

REMOVAL AND INSTALLATION

3.9L Engine

REAR MOUNT

1. Disconnect negative battery cable.

2. Raise and safely support vehicle. Remove skid plate, if equipped.

3. Install transmission jack into position and raise transmission slightly.

4. Remove rear mount through-bolt and nut.

5. Remove the flange nuts from the crossmember support bracket.

6. Raise rear of transmission and remove insulator flange nuts and remove mount. If necessary, remove attaching bolts holding transmission support bracket to transmission.

To install:

7. Install attaching bolts holding transmission support bracket to transmission.

8. Position insulator in transmission support bracket and install through-bolt.

9. With insulator installed in a level position tighten through bolt nut to 50 ft. lbs.

10. Position crossmember support bracket into insulator. Install flange nuts and tighten to 30 ft. lbs.

11. Using transmission jack, lower the insulator and crossmember support bracket onto the crossmember. Tighten flange nuts to 30 ft. lbs.

12. Remove transmission jack, install skid plate, if removed, lower vehicle and reconnect negative battery cable.

FRONT MOUNT

1. Disconnect negative battery cable. Remove skid plate, if equipped.

2. Position fan assembly to insure clearance for radiator and top hose when engine is moved. Install an engine support fixture.

3. Raise and safely support vehicle.

4. Remove nuts from front engine mounting brackets.

5. Raise engine with support fixture far enough to remove mounts, remove remaining bolts and mounts.

To install:

6. Install mounts to engine block.

7. Lower engine to original position and insert bolts through mounting brackets.

8. Install nuts and tighten to 75 ft. lbs. Install skid plate, if removed.

9. Lower vehicle and remove engine support fixture. Reconnect negative battery cable.

Except 3.9L Engine

1. Disconnect negative battery cable.
2. Raise and safely support vehicle.
3. Install transmission jack into position and raise transmission slightly.
4. Remove rear mount through-bolt and nut.
5. Remove U-bracket from frame crossmember and remove insulator from U-bracket.
6. Remove rear engine mount from the bottom face of transmission extension housing.
   To install:
7. Install rear mount to bottom of transmission. Torque bolts to 50 ft. lbs.
8. Install insulator to U-bracket. Tighten nuts to 30 ft. lbs.
9. Install through-bolt through the U-bracket/insulator assembly and the rear engine support. Tighten to 50 ft. lbs.
10. Attach U-bracket to crossmember tightening bolts to 30 ft. lbs. Remove transmission jack and lower vehicle. Connect negative battery cable.

Cylinder Head

REMOVAL AND INSTALLATION

1. Relieve the fuel pressure if equipped with fuel injection. Disconnect the negative battery cable from the battery and drain the cooling system.
2. Raise the vehicle and safely support. Disconnect the exhaust pipe from the manifolds.
3. Remove the alternator, if the right head is being removed, and the air pump and battery ground cable, if the left head is being removed.
4. Remove the air cleaner assembly. Unbolt the air conditioning compressor, if equipped, and lay it aside. Remove the distributor cap with all wires attached.
5. Disconnect all wires, hoses, linkages and cables from the throttle body. Disconnect the fuel line.
6. Disconnect the ignition coil, coolant temperature sending unit wire and all other connectors along the wiring harness connected to items on the intake manifold.

7. Disconnect the heater hose, upper radiator hose and the lower bypass hose clamp.
8. Remove the valve cover(s).
9. Remove the intake manifold assembly. Remove the exhaust manifold(s).
10. Remove the rocker arm and shaft assembly from the head(s). Do not disassemble unless service is required.
11. Remove the pushrods and identify them to ensure installation in their original locations.
12. Remove the head bolts and remove the cylinder head(s).
    To install:
13. Clean and dry all gasket surfaces of the cylinder block and head. Inspect all surfaces with a straightedge. If the flatness exceeds 0.0075 times the length of the span measured (in any direction), replace or machine the head gasket surface.
14. Using no sealer, install the new head gasket(s) to the block. Clean, dry and lightly oil all head bolts threads. Install the head(s) and install the head bolts.
15. Torque the head bolts in sequence to 50 ft. lbs. (68 Nm). Repeat the sequence retightening the bolts to a final torque of 105 ft. lbs. (143 Nm)

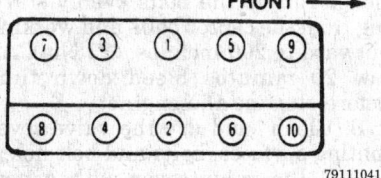

Cylinder head bolt torque sequence — 5.2L and 5.9L engines

and repeat the second step to ensure all bolts are accurately torqued.
16. Assemble the rocker shaft assembly, if it was serviced. Make sure all rocker arms with a **RH** stamp are installed to the right of those with an **LH**. Install the pushrods, rocker arms and shaft(s) with the notch on the end of the shaft pointing to the engine centerline and to the rear of the right bank or to the front of the left bank. Make sure the long stamped steel retainers are at the No. 2 and 4 positions. Torque the bolts evenly and gradually to 200 inch lbs. (23 Nm).
17. Clean and dry the intake manifold contact surfaces. Coat the intake manifold side gaskets lightly with sealer and install the gaskets to the heads. Cutouts at the front of the gaskets differentiate the right and left sides.
18. Apply a thin uniform coat of quick dry cement to the front and rear intake manifold gaskets and mounting surfaces on the block and apply a thin bead of sealer to each of 4 the corners. Install the front and rear gaskets engaging the hole in the block and the tangs from the head gaskets. Apply a second thin bead of sealer above the gaskets in the 4 corners.
19. Carefully lower the intake manifold into position engaging the bypass hose. Inspect the gaskets to make sure they have not become dislodged.
20. Install the intake manifold bolts and torque in sequence to 25 ft. lbs. (34 Nm). Repeat the sequence retightening the bolts to a final torque of 40 ft. lbs. (54 Nm). Repeat the second step to ensure all bolts are accurately torqued.
21. Install the exhaust manifold(s) and torque the bolts to 20 ft. lbs. (27 Nm). Torque the end nuts to 15 ft. lbs. (20 Nm).
22. Clean and dry the valve cover mating surfaces, bolts and bolt holes.

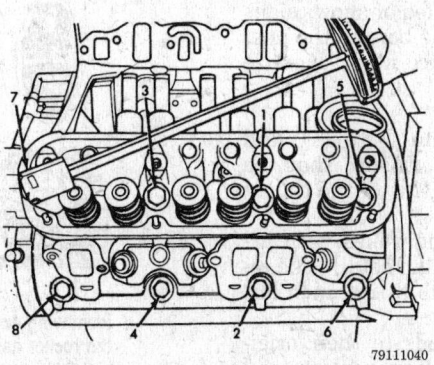

Cylinder head bolt torque sequence — 3.9L engine

Install the valve cover(s) each with a new gasket.

23. Connect the heater hose, upper radiator hose and the lower bypass hose clamp.

24. Connect the ignition coil, coolant temperature sending unit wire and all other connectors that were disconnected along the wiring harness.

25. Install the air conditioning compressor, if equipped. Install the distributor cap and all spark plug wires.

26. Install the alternator, battery ground and air pump, if removed.

27. Connect all wires, hoses, cables and the fuel line to the throttle body. Install the air cleaner assembly.

28. Raise the vehicle and safely support. Connect the exhaust pipe to the manifolds.

29. Fill the cooling system.

30. Connect the negative battery cable and set all adjustments to specification.

Valve Lifters

REMOVAL AND INSTALLATION

1. Disconnect negative battery cable. Remove air cleaner.

2. Relieve the fuel pressure if equipped with fuel injection.

3. Remove the valve cover(s) and rocker arm assembly.

4. Remove the pushrods and identify them to ensure installation in their original locations.

5. Remove the intake manifold assembly, yoke retainer and aligning yokes.

6. Use an appropriate valve lifter removal tool to remove each lifter from its bore. If reinstalling the tappets, identify each upon removal to ensure installation in the original position.

7. If the tappet or bore in cylinder block is scored, scuffed or shows signs of sticking, ream the bore to the next oversize and replace with oversize tappet.

To install:

8. Lubricate the lifter(s) and bore(s) and install. Ensure that the oil feed hole in the side of the tappet body faces up.

9. Install aligning yokes with arrow toward camshaft. Install yoke retainer. Tighten bolts to 200 inch lbs. (23 Nm).

10. Install pushrods to their original positions. Install rocker arm and shaft assembly.

11. Install valve cover(s). Connect negative battery terminal and start vehicle. Check for oil leaks.

Rocker Arms and Shaft

REMOVAL AND INSTALLATION

1. Disconnect the negative battery cable.

2. Remove the valve cover and gasket.

3. Note the positioning of the notch and remove the rocker arms and shaft assembly from the head.

4. Disassemble the assembly as required and replace all worn parts.

NOTE: If equipped with exhaust valve rotators, the exhaust rocker arm must have relief for clearance.

To install:

5. Make sure all rocker arms with an **RH** stamp are installed to the right of those with an **LH**. Install the assembly with the notch on the end of the shaft pointing to the engine centerline and to the rear of the right bank or to the front of the left bank. Make sure the long stamped steel retainers are at the No. 2 and 4 positions. Torque the bolts evenly starting from the center bolts and working outward to 200 inch lbs. (23 Nm). Allow 20 minutes bleed down time before starting the engine.

6. Clean and dry the valve cover mating surfaces, bolts and bolt holes. Install the valve cover with a new gasket. Torque the screws or nuts to 95 inch lbs. (11 Nm).

7. Connect the negative battery cable and check for leaks.

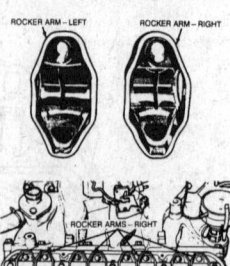

Identifying and assembling the rocker assembly — 5.2L and 5.9L engines — 3.9L engine parts are similar

Intake Manifold

REMOVAL AND INSTALLATION

1. Relieve the fuel pressure if equipped with fuel injection. Disconnect the negative battery cable from the battery and drain the cooling system.

2. Remove the air pump and bracket. Removal of the bracket will allow for easier installation of the left front corner of the intake manifold.

3. Remove the air cleaner assembly. Remove the air conditioning compressor or unbolt it and lay it aside, if equipped. Remove the distributor cap with all wires attached.

4. Disconnect all wires, hoses, linkages and cables from the throttle body. Disconnect the fuel line.

5. Disconnect the ignition coil, coolant temperature sending unit wire and all other connectors along the wiring harness connected to items on the intake manifold.

6. Disconnect the heater hose, upper radiator hose and the lower bypass hose clamp.

7. Remove the valve covers.

8. Unbolt the intake manifold from the heads and remove the intake manifold assembly. Disassemble the manifold as required and clean out the exhaust crossover passages.

To install:

9. Clean and dry the intake manifold contact surfaces. Coat the intake manifold side gaskets lightly with sealer and install the gaskets to the heads. Cutouts at the front of the gaskets differentiate the right and left sides.

10. Apply a thin uniform coat of quick dry cement to the front and rear intake manifold gaskets and mounting surfaces on the block and apply a thin bead of sealer to each of 4 the corners. Install the front and rear gaskets engaging the hole in the block and the tangs from the head gaskets. Apply a second thin bead of sealer above the gaskets in the 4 corners.

11. Carefully lower the intake manifold into position engaging the bypass hose. Inspect the gaskets to make sure they have not dislodged.

12. Install the intake manifold bolts with the aspirator tube, air pump bracket and kickdown linkage bracket in place. Torque the bolts in sequence to 25 ft. lbs. (34 Nm). Repeat the sequence retightening the bolts to a final torque of 40 ft. lbs. (54 Nm). Repeat this step to ensure bolts are accurately torqued.

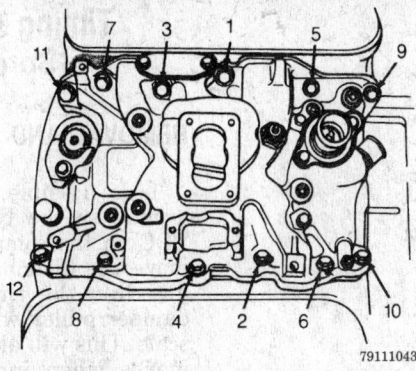

Intake manifold bolt torque sequence — 3.9L engine

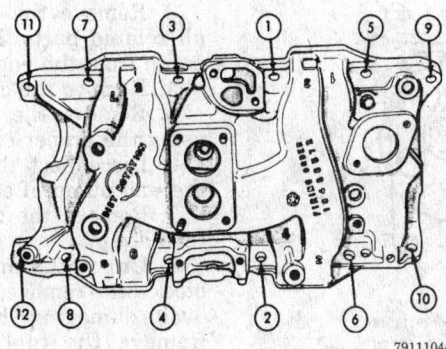

Intake manifold bolt torque sequence — 5.2L and 5.9L engine

13. Clean and dry the valve cover mating surfaces, bolts and bolt holes. Install the valve covers with a new gasket. Torque the screws or nuts to 95 inch lbs. (11 Nm).

14. Connect the heater hose, upper radiator hose and the lower bypass hose clamp.

15. Connect the ignition coil, coolant temperature sending unit wire and all other connectors that were disconnected along the wiring harness.

16. Install the air conditioning compressor, if equipped. Install the distributor cap and all spark plug wires.

17. Install air pump.

18. Connect all wires, hoses, cables and the fuel line to the throttle body. Install the air cleaner assembly.

19. Fill the cooling system.

20. Connect the negative battery cable and check for leaks.

Exhaust Manifold

REMOVAL AND INSTALLATION

1. Disconnect the negative battery cable. Remove the hot air tube and heat shield.

2. Raise the vehicle and support safely. Remove the exhaust pipe from the exhaust manifold. Lower the vehicle.

3. Take note of all conical washer locations and remove the bolts, nuts and washers attaching the manifold to the head.

4. Remove the manifold.

To install:

5. If either of the end studs came out with the nuts, install a new stud using sealer on the coarse threads.

6. Position the manifold on the end studs. Install conical washers and nuts on the studs.

7. Install the remaining bolts and washers in their proper locations. The inner bolts are not mounted with washers. Working outward from the center, torque the bolts to 20 ft. lbs. (27 Nm) and the nuts to 15 ft. lbs. (20 Nm).

8. Install the exhaust pipe to the manifold.

9. Connect the negative battery cable and check for exhaust leaks.

Timing Chain Front Cover

REMOVAL AND INSTALLATION

1. Disconnect the negative battery cable.

2. Drain the cooling system.

3. Remove the radiator, fan and all related parts. Remove the water pump from the engine.

4. Remove the crankshaft pulley.

5. Remove the vibration damper using the proper puller.

6. Disconnect the fuel lines from the fuel pump, if equipped with a mechanical pump.

7. Remove the 2 front bolts from the oil pan.

8. Unbolt the chain cover from the block and remove using caution to avoid damaging the oil pan gasket. Remove the fuel pump from the cover, if equipped.

To install:

9. Clean and dry the mating surfaces of the cover and block. Apply a thin bead of sealer to the oil pan gasket.

10. Install a new cover gasket and install the cover. Torque bolts to 30 ft. lbs. (41 Nm).

11. Install the water pump with a new gasket.

12. Install the damper with tool C-3638 or equivalent. Install the bolt and washer and torque to specification. Apply a small amount of sealer to the bolts and install the crankshaft pulley.

13. Install the mechanical fuel pump using a new gasket, if equipped, and connect the fuel lines. Install the 2 oil pan bolts.

14. Install the radiator, fan and all related parts.

15. Fill the cooling system.

16. Connect the negative battery cable and check for leaks.

Front Cover Oil Seal

REPLACEMENT

1. Disconnect the negative battery cable.

2. Remove the belts from the crankshaft pulley.

3. Remove the fan and shroud from the vehicle.

4. Remove the crankshaft pulley.

5. Remove the vibration damper using the proper puller.

6. Using a seal remover pry outward behind the lip of the oil seal. Take care not to damage the crankshaft seal surface of the cover.

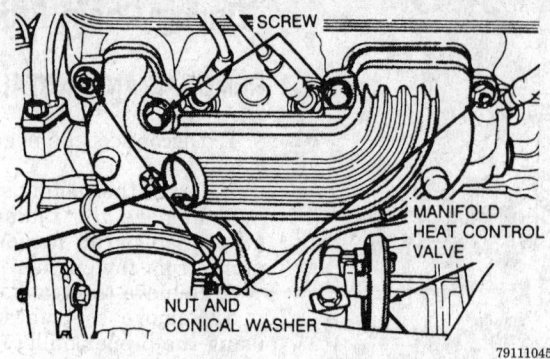

Exhaust manifold installation — 3.9L engine (right side shown)

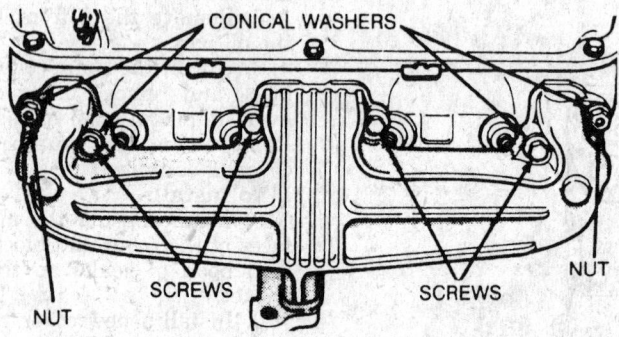

Exhaust manifold installation — 5.2L and 5.9L engines

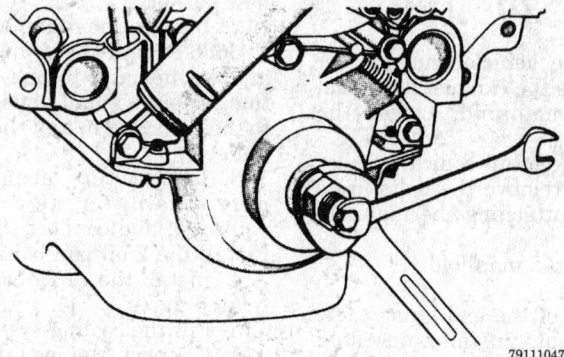

Installing the front cover oil seal — 3.9L, 5.2L and 5.9L engines

To install:

7. Install the new seal by installing the threaded shaft part of the special tool C–4251 into the threads of the crankshaft.

8. Place the seal into the opening with the spring toward the engine. Place the installing adapter C–4251–3 or equivalent, with the thrust bearing and nut on the shaft. Tighten the nut until the tool is flush with the timing chain cover. Remove the tool.

9. Install vibration damper using tool C–3638 or equivalent. Install the

bolt and washer and torque to specification.

10. Apply a small amount of sealer to the bolts and install the crankshaft pulley.

11. Install the fan and shroud.

12. Connect the negative battery cable and check for leaks.

Timing Chain and Sprockets

REMOVAL AND INSTALLATION

1. If possible, crank the engine around so the No. 1 cylinder is at TDC on the compression stroke. Remove the distributor cap to confirm and line the timing mark on the damper pulley with **0** on the timing scale. This will aid in aligning timing marks when installing the timing gears. Disconnect the negative battery cable.

2. Drain the cooling system.

3. Remove the radiator, fan and all related parts. Remove the water pump from the engine.

4. Remove the crankshaft pulley.

5. Remove the vibration damper using the proper puller.

6. Disconnect the fuel lines from the fuel pump, if equipped.

7. Remove the 2 front bolts from the oil pan.

8. Unbolt the chain cover from the block and remove, using caution to avoid damaging the oil pan gasket. Remove the fuel pump from the cover, if equipped.

9. Remove the camshaft gear retaining bolt, cup washer and fuel pump eccentric, if equipped. Remove the timing chain and gears.

To install:

10. Place both camshaft and crankshaft gears on the bench with the timing marks on the exact imaginary center line through both gear bores as they are installed on the engine. Place the timing chain around both sprockets.

11. Turn the crankshaft and camshaft so the keys line up with the keyways in the gears when the timing marks are in proper position.

12. Slide both gears over their respective shafts and use a straightedge to check timing mark alignment.

13. Install the fuel pump eccentric and cup washer, if equipped. Torque the camshaft gear retaining bolt to 35 ft. lbs. (47 Nm).

14. Clean and dry the mating surfaces of the timing chain cover and block. Apply a thin bead of sealer to the oil pan gasket.

15. Install a new cover gasket and install the cover. Torque the bolts to 30 ft. lbs. (41 Nm).

16. Install the water pump with a new gasket.

17. Install the vibration damper with tool C–3638 or equivalent. Install the bolt and washer and torque

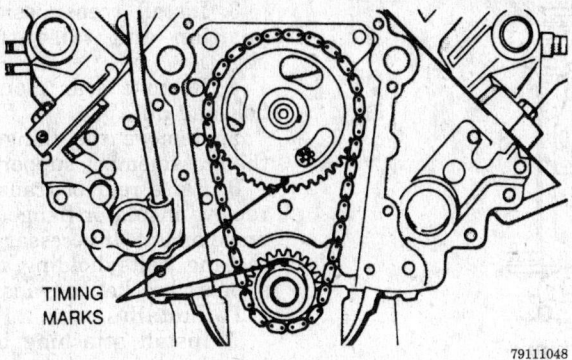

TIMING
MARKS

79111048

Alignment of timing marks — 3.9L, 5.2L and 5.9L gasoline engines

to specification. Apply a small amount of sealer to the bolts and install the crankshaft pulley.

18. Install the fuel pump using a new gasket, if equipped, and connect the fuel lines. Install the 2 oil pan bolts.

19. Install the radiator, fan and all related parts.

20. Fill the cooling system.

21. Connect the negative battery cable, set all adjustments to specifications and check for leaks.

Camshaft

REMOVAL AND INSTALLATION

1. If possible, crank the engine around so the No. 1 cylinder is at TDC on the compression stroke. Remove the distributor cap to confirm and line the timing mark on the damper pulley with **0** on the timing scale. This will aid in aligning timing marks when installing the timing gears. If equipped with fuel injection, relieve the fuel pressure. Disconnect the negative battery cable.

2. Drain the cooling system.

3. Remove the engine from the vehicle.

4. Remove the valve cover(s).

5. Remove the rocker shaft assembly(s). Identify and remove the pushrods.

6. Remove the intake manifold, identify and remove all lifters.

7. Remove the distributor. Lift out the oil pump and distributor driveshaft.

8. Remove the radiator, fan and all related parts.

9. Remove the fuel pump, if equipped. Remove the timing chain cover, timing chain and gears.

10. Note the location of the oil tab and remove the camshaft thrust plate.

11. Install a long bolt into the front of the camshaft to facilitate removal. Remove the camshaft, being careful not to damage any of the cam bearings with the cam lobes.

To install:

12. Install the camshaft to within 2 in. of its final installation position. Install the camshaft Gear Installer tool C–3509 with tongue in back of distributor drive gear. Bolt it in place with the distributor lock plate bolt. This will prevent the camshaft from being pushed in too far and knocking out the welch plug at the rear of the block. This tool should remain in place until the timing chain installation has been completed.

13. Install the camshaft thrust plate and chain oil tab with 3 bolts. Make sure the tang of the oil tab enters the lower right hole in the thrust plate. Torque the bolts to 210 inch lbs. (24 Nm). Make sure the top edge of the oil tab is flat against the thrust plate or it will not feed oil to the chain.

14. Place both camshaft and crankshaft gears on the bench with the timing marks on the exact imaginary center line through both gear bores as they are installed on the engine. Place the timing chain around both sprockets.

15. Turn the crankshaft and camshaft so the keys line up with the keyways in the gears when the timing marks are in proper position.

16. Slide both gears over their respective shafts and use a straight-edge to check timing mark alignment.

17. Install the fuel pump eccentric and cup washer, if equipped. Torque the camshaft gear retaining bolt to 35 ft. lbs. (47 Nm). Remove the camshaft blocking tool.

18. Measure camshaft endplay. Replace the thrust plate, if not within specifications.

19. Coat the oil pump and distributor driveshaft with oil. Install 3.9L

engine shaft so when the gear spirals into place and drops into the oil pump, the slot in the top of the gear is pointing directly to the left front intake manifold bolt hole. Install 5.2L and 5.9L engine shaft in a similar manner, except position it so the slot is parallel with the center line of the camshaft.

20. If the camshaft was not replaced, lubricate and install the lifters in their original locations. If the camshaft was replaced, new lifters must be used.

21. Install the pushrods and rocker shaft assembly.

22. Install the intake manifold. Install the valve cover(s).

23. Install the distributor so the rotor points to the No. 1 spark plug wire position on the cap.

24. Install the timing chain cover and all related parts, fuel pump, if equipped, and radiator.

25. Install the engine into the vehicle.

26. When everything is bolted in place, change the engine oil and replace the oil filter.

NOTE: If the camshaft or lifters have been replaced, add 1 pint of Mopar crankcase conditioner or equivalent, when replenishing the oil to aid in break-in. This mixture should be left in the engine for a minimum of 500 miles and drained at the next normal oil change.

27. Fill the radiator with coolant.

28. Connect the negative battery cable, set all adjustments to specifications and check for leaks.

DIESEL ENGINE MECHANICAL

NOTE: Disconnecting the negative battery cable on some vehicles may interfere with the functions of the on board computer systems and may require the computer to undergo a relearning process, once the negative battery cable is reconnected.

Engine

REMOVAL AND INSTALLATION

This engine has a dry weight of 880 lbs. Make sure the engine removal

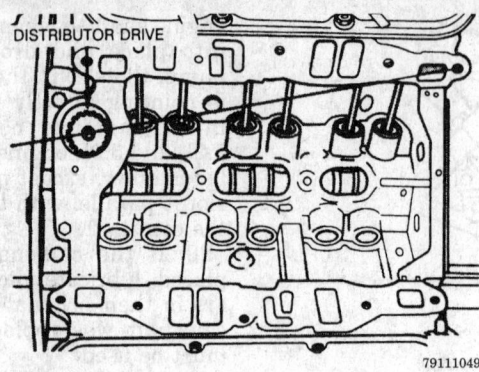

Installed distributor drive gear position — 3.9L engine

equipment is rated adequately, or personal injury may result.

1. Scribe location of hood hinge bolts for proper alignment and remove the hood.

2. Disconnect the cables from the battery and from the engine. Remove the battery from the vehicle.

3. Drain the coolant.

4. Remove the radiator, shroud, belt, fan and all related parts.

5. Remove the air inlet tube from the turbocharger and the air intake housing. Remove the exhaust pipe from the turbocharger outlet flange. Remove the inlet duct from the turbocharger and inlet duct. Remove the outlet duct.

6. Remove the air conditioner compressor and set aside.

7. Disconnect and remove the alternator and all other electrical connections to the engine.

8. Disconnect the acclerator linkage, the speed control linkage and the throttle valve linkage.

9. Raise the vehicle and support safely. Remove the starter.

10. If equipped with automatic transmission, remove the torque converter bolts and remove the lower bell housing bolts. If equipped with manual transmission, remove the transmission.

11. Drain the oil from the engine.

12. Disconnect the transmission oil cooler lines from their brackets, if equipped.

13. Lower vehicle, disconnect and plug the fuel lines. Disconnect the power steering lines and vacuum pump lines.

14. Hoist the engine slightly using the lifting eyes and support the transmission if still installed.

15. Remove the motor mounts.

16. Remove the upper bell housing bolts.

17. Remove the engine from the vehicle.

To install:

18. Position the engine in the engine compartment and install the motor mounts. Torque the nuts and bolts to 57 ft. lbs. (77 Nm).

19. Install the bell housing bolts and torque converter bolts, if equipped. Install the manual transmission, if equipped.

20. Install the starter.

21. Connect the exhaust pipe.

22. Connect the transmission oil cooler lines to their brackets, if equipped.

23. Connect the fuel lines.

24. Connect the power steering lines.

25. Connect all engine driven accessories.

26. Connect the accelerator linkage.

27. Connect the throttle linkage to the control lever.

28. Mount the air conditioner compressor.

29. Connect the alternator and all other electrical connections to the engine.

30. Install the intake and exhaust pipes and air ducts to the turbocharger and intercooler.

31. Install the fan and all related parts, shroud and radiator.

32. Fill the enine with the proper amount of diesel engine oil.

33. Fill the radiator with coolant.

34. Install and connect the battery cables. Set all adjustments to specifications and check for leaks.

Engine Mounts

REMOVAL AND INSTALLATION

Engine Rear Mount

1. Disconnect negative battery cable.

2. Raise and safely support vehicle. Remove skid plate , if equipped.

3. Install transmission jack into position and raise transmission slightly.

4. Remove rear mount through-bolt and nut.

5. Remove the flange nuts from the crossmember support bracket.

6. Raise rear of transmission and remove insulator flange nuts and remove mount. If necessary, remove attaching bolts holding transmission support bracket to transmission.

To install:

7. Install attaching bolts holding transmission support bracket to transmission.

8. Position insulator in transmission support bracket and install through-bolt.

9. With insulator installed in a level position tighten through bolt nut to 50 ft. lbs.

10. Position crossmember support bracket into insulator. Install flange nuts and tighten to 30 ft. lbs.

11. Using transmission jack, lower the insulator and crossmember support bracket onto the crossmember. Tighten flange nuts to 30 ft. lbs.

12. Remove transmission jack, install skid plate if removed, lower vehicle and reconnect negative battery cable.

Engine Front Mount

1. Disconnect negative battery cable. Remove skid plate, if equipped.

2. Position fan assembly to insure clearance for radiator and top hose when engine is moved. Install an engine support fixture.

3. Raise and safely support vehicle.

4. Remove nuts from front engine mounting brackets.

5. Raise engine with support fixture far enough to remove mounts, remove remaining bolts and mounts.

To install:

6. Install mounts to engine block.

7. Lower engine to original position and insert bolts through mounting brackets.

8. Install nuts and tighten to 75 ft. lbs. Install skid plate, if removed.

9. Lower vehicle and remove engine support fixture. Reconnect negative battery cable.

Cylinder Head

NOTE: The cylinder head design was changed in mid — 1991, and the new design is not interchangeable with earlier models.

REMOVAL AND INSTALLATION

1. Disconnect the negative battery cable.
2. Drain the coolant.
3. Disconnect the radiator hose and heater hoses.
4. Remove the turbocharger and air crossover.
5. Remove the exhaust manifold.
6. Remove all fuel lines from the injection pump and remove injector nozzles. Remove the fuel filter.
7. Remove the valve covers.
8. Remove the rocker arms and pushrods.
9. Unbolt the cylinder head from the block. Remove the cylinder head.
10. Inspect the coolant passages. A large accumulation of rust or lime will require service to the block.
11. Inspect the surface of the head for flatness. The maximum variation is 0.0004 in. (0.010mm) within any 2 in. diameter area or 0.012 in. (0.30mm) overall end-to-end or side-to-side.

To install:

12. Thoroughly clean and dry the mating surfaces of the head and block. Position the new head gasket on the dowels.
13. Install the head onto the dowels on the block.
14. Lubricate the pushrod sockets and install the pushrods and rocker arms.
15. Clean, dry and lightly lubricate the head bolts. Install and torque in sequence first to 66 ft. lbs. (90 Nm) and then to 89 ft. lbs. (120 Nm). Tighten an additional 90 degrees. Tighten the 8 mm bolts to 18 ft. lbs. (24 Nm). 16.
16. Install the rocker arm pedastal bolts. Torque to 18 ft. lbs. (24 Nm).
17. Adjust the valve clearance.
18. Install the valve covers with new gaskets. Torque the bolts to 18 ft. lbs. (24 Nm).

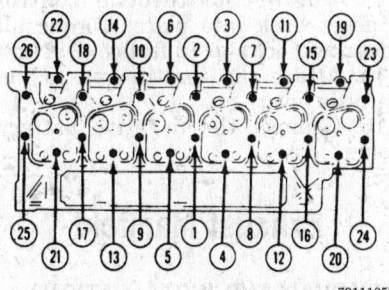

79111052

Cylinder head bolt tightening sequence — 5.9L diesel engine

19. Install all fuel lines and the fuel filter.
20. Install the exhaust manifold.
21. Install the turbocharger and air crossover.
22. Connect the radiator hose and heater hoses.
23. Fill the radiator with coolant.
24. Connect the negative battery cable, set all adjustments to specifications and check for leaks.

Valve Lifters/Tappet

REMOVAL AND INSTALLATION

1. Disconnect negative battery cable.
2. Remove the valve covers.
3. Drain the cooling system. Remove the rocker lever assemblies and push rods.
4. Remove the drive belts and fan hub assembly.
5. Remove the vibration damper.
6. Remove the gear housing cover. Remove the lift pump.
7. Insert the dowel tool through the push tube holes and into the top of each tappet. Pull the tappets up and wrap a rubber band around the top of the dowel rods. This will prevent the tappets from dropping into the engine.
8. Align the engine timing marks and remove the bolts in the thrust plate. Remove the camshaft, gear and thrust plate.
9. Insert a trough the full length of the cam bore. Make sure the trough is in the position to catch any falling tappets. Cummins tappet changing tool 3822513 or equivalent, should be used for this procedure.
10. Remove the wooden dowel and catch the tappet in trough. Move the trough so the tappet falls over in the bottom of the trough. Mark each tappet for proper assembly, each tappet must be installed into their original location.
11. Carefully pull the trough and tappet from the cam bore. Repeat as needed to remove each tappet. Inspect the tappet socket, stem and face for excess wear, cracks and other damage. Minimum tappet stem diameter is 0.627 inch (15.925mm). If out of limits, replace tappet.

To install:

12. Install the trough into the cam bore. Feed the installation tool down the tappet bore and into the trough.
13. Feed the installation tool cord through the cam bores. Carefully pull the trough and installation tool out the front.

14. Lubricate the tappets with lubriplate 105 or equivalent.
15. Insert the installation tool into the tappet. Place the tappet and tool in the trough and slide the trough back into the cam bore.
16. Align the tappet below the proper bore and lift the tappet into the bore. After the tappet is in position, rotate the trough 180 degrees to hold tappet in position while the installation tool is removed. Install the dowel and rubber band to hold tappet in the bore and repeat process as required.
17. Lubricate camshaft lobes, journals and thrust washer and install camshaft into bore. Install lift pump, gear housing cover and vibration damper.
18. Install fan hub assembly and drive belts.
19. Install pushrods, rocker lever assemblies and valve covers.
20. Reconnect negative battery cable, refill coolant and oil as required. Start and check for leaks.

Valve Lash

ADJUSTMENT

1. Perform the adjustment when the engine is below 140°F (60°C).
2. Disconnect the negative battery cable.
3. Use the timing pin to locate TDC for cylinder No. 1. Disengage the pin.
4. Remove the valve covers.
5. With the engine is in this position, the first group of valves may be adjusted: No. 1 intake: 0.10 in. (0.254mm) No. 1 exhaust: 0.20 in. (0.508mm) No. 2 intake: 0.10 in. (0.254mm) No. 3 exhaust: 0.20 in. (0.508mm) No. 4 intake: 0.10 in. (0.254mm) No. 5 exhaust: 0.20 in. (0.508mm)
6. Mark the pulley and rotate the engine 360 degrees.
7. With the engine is in this position, the second group of valves may be adjusted: No. 2 exhaust: 0.20 in. (0.508mm) No. 3 intake: 0.10 in. (0.254mm) No. 4 exhaust: 0.20 in. (0.508mm) No. 5 intake: 0.10 in. (0.254mm) No. 6 intake: 0.10 in. (0.254mm) No. 6 exhaust: 0.20 in. (0.508mm)
8. Torque all locknuts to 18 ft. lbs. (24 Nm).
9. Install the valve covers with new gaskets. Torque the bolts to 18 ft. lbs. (24 Nm).
10. Connect the negative battery cable.

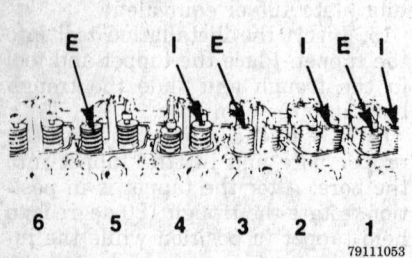

First group of valves to be adjusted — 5.9L diesel engine

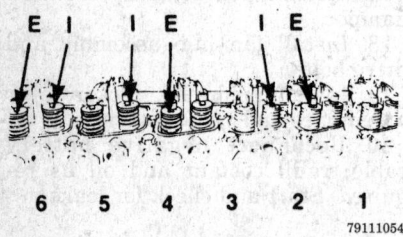

Second group of valves to be adjusted — 5.9L diesel engine

Rocker Arm and Pedestal Assembly

REMOVAL AND INSTALLATION

1. Disconnect the negative battery cable. Remove the crossover tube.
2. Remove the valve covers.
3. Loosen the adjusting screw locknuts. Loosen the screws until they stop.
4. Remove the 8mm bolt and 12mm head bolt from the pedestal.
5. Remove the pedestal and rocker arm assembly. Remove the pushrods if necessary.
6. Remove the retaining ring and thrust washer.
7. Remove the rocker arm from the pedestal.

NOTE: Do not disassemble the rocker shaft and pedestal; they must be replaced as an assembly.

8. Remove the locknut and adjusting screw from the rocker arm.
To install:
9. Install the adjusting screw and locknut.
10. Lubricate the shaft with oil and install the rocker arm to the shaft.

Install the thrust washer and snapring.
11. Install the pushrods to the engine, if removed.
12. Install the pedestal and rocker arm assembly to the head aligning the dowel in the pedestal with the dowel bore in the head. If the pushrod is holding the pedestal off the head, turn the engine until the pedestal will set on the head without interference.
13. Lubricate the threads of the bolt with oil. Install and torque first to 29 ft. lbs. (40 Nm), then to 62 ft. Lbs. (85 Nm) and finally to 93 ft. lbs. (126 Nm). If all of the pedestals were removed, follow the entire head bolt torque sequence including those head bolts that were not removed in this procedure.
14. Tighten the 8mm bolts to 18 ft. lbs. (24 Nm).
15. Adjust the valves.
16. Install the valve cover with a new gasket. Torque the bolts to 18 ft. lbs. (24 Nm).
17. Connect the negative battery cable.

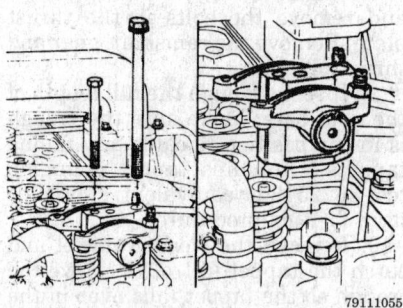

Removing the rocker arm and pedestal assembly — 5.9L diesel engine

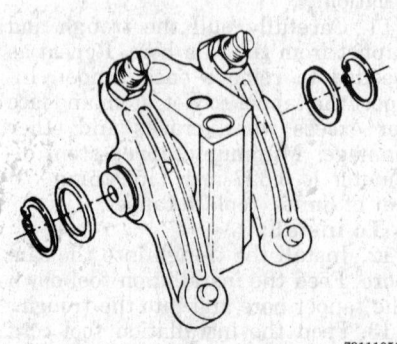

Removing or installing the rocker arm from the pedestal

Intake Manifold Cover

REMOVAL AND INSTALLATION

1. Disconnect negative battery cable.
2. Remove the intercooler outlet duct from air inlet tube, if equipped.
3. Remove the 4 bolts that attach the air crossover to the intake manifold. Loosen the lower throttle control bracket mounting bolt and move the bracket away from the engine.
4. Loosen the hose clamps on the turbocharger end of the crossover tube and remove the tube and gasket. Cover the turbocharger opening with a clean shop towel.
5. Disconnect the electrical wiring from the intake manifold heater, remove the heater and remove the gasket.
6. Disconnect the charge air temperature sensor, fuel heater ground wire and the air temperature switch from the intake manifold cover.
7. Remove the manifold intake cover and gasket. Clean all sealing surfaces.
   To install:
8. Using a new gasket install the intake manifold cover.
9. Connect the charge air temperature sensor, fuel heater ground wire and the air temperature switch to the intake manifold cover.
10. Some of the intake manifold bolt holes are drilled through and must be sealed. Apply liquid teflon sealant to the bolts. Install the intake manifold cover bolts and tighten to 18 ft. lbs. (24 Nm). 11.
11. Clean the gasket mounting surface and install a new gasket between the intake manifold cover and the air intake heater.
12. Install the intake manifold heater and connect the wiring.
13. Remove the towel from the turbocharger opening and install the air crossover tube and gasket. Tighten the hose clamps.
14. Rotate the throttle control bracket back into place. Torque all mounting bolts to 18 ft. lbs. (24 Nm).
15. Attach the throttle rod to the throttle lever.
16. Connect the negative battery cable.

Exhaust Manifold

REMOVAL AND INSTALLATION

1. Disconnect negative battery cable.

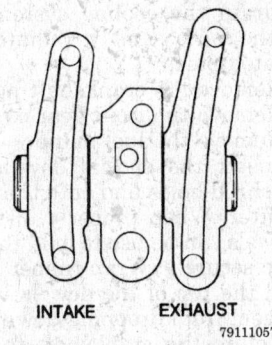

INTAKE   EXHAUST

79111057

Proper rocker arm installation — 5.9L diesel engine

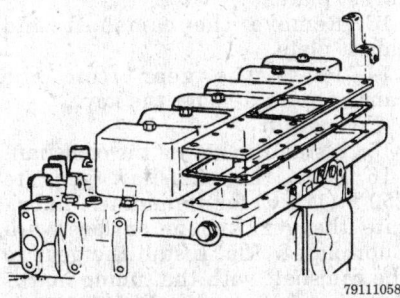

79111058

Intake manifold cover and gasket — 5.9L diesel engine

2. Disconnect air intake and exhaust pipes.

3. Disconnect the turbocharger oil supply line and the oil drain tube from the turbocharger.

4. Disconnect the intercooler inlet duct from the turbocharger. Remove the turbocharger and gasket.

5. Remove the cab heater supply and return lines. Remove the exhaust manifold and gasket. Clean all sealing surfaces.

To install:

6. Install manifold and new gasket. Tighten bolts in sequence to 32 ft. lbs. (32 Nm).

7. Install turbocharger. Tighten the turbocharger mounting nuts to 24 ft. lbs. (32 Nm).

8. Position the intercooler inlet duct to the turbocharger. Tighten the clamp to 72 inch lbs. (8 Nm).

9. Position the air intake pipe and the exhaust pipe onto the turbocharger and tighten clamps to 72 inch lbs. (8 Nm).

10. Install oil drain tube and oil supply line to the turbocharger. Tighten the drain tube bolts to 18 ft. lbs. (24 Nm). Tighten the oil supply line fitting to 11 ft. lbs. (15 Nm). 11.

11. Connect the cab heater supply and return lines. Tighten line nuts to 18 ft. lbs. (24 Nm).

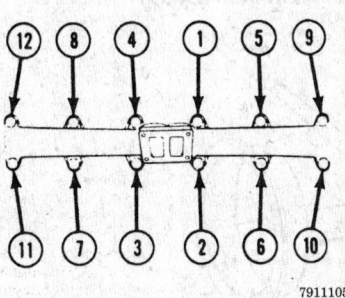

79111059

Exhaust manifold bolt torque sequence — 5.9L diesel engine

12. Reconnect the negative battery cable, start the engine and check for exhaust leaks.

Turbocharger

REMOVAL AND INSTALLATION

1. Disconnect negative battery cable.

2. Disconnect air intake and exhaust pipes.

3. Disconnect the turbocharger oil supply line and the oil drain tube from the turbocharger.

4. Disconnect the intercooler inlet duct from the turbocharger. Remove the turbocharger and gasket.

To install:

5. Inspect the mounting surface for cracks and damage. Install the turbocharger. Apply anti-sieze compound to the mounting studs and tighten the turbocharger mounting nuts to 24 ft. lbs. (32 Nm).

6. Position the intercooler inlet duct to the turbocharger. Tighten the clamp to 72 inch lbs. (8 Nm).

7. Position the air intake pipe and the exhaust pipe onto the turbocharger and tighten clamps to 72 inch lbs. (8 Nm).

8. New turbochargers must be prelubricated with fresh engine oil before operation. To do so, poor 2–3 oz. of oil into the supply fitting and rotate the turbine wheel to circulate the oil. Install oil drain tube and oil supply line to the turbocharger. Tighten the drain tube bolts to 18 ft. lbs. (24 Nm). Tighten the oil supply line fitting to 11 ft. lbs. (15 Nm).

9. Connect the cab heater supply and return lines. Tighten line nuts to 18 ft. lbs. (24 Nm).

10. Reconnect the negative battery cable, start the engine and check for leaks.

Timing Chain Front Cover

REMOVAL AND INSTALLATION

1. Disconnect the negative battery cable.

2. Remove the fan drive assembly and belt.

3. Remove the belt tensioner.

4. Remove the oil fill tube and adaptor.

5. Remove the crankshaft pulley.

6. Remove all bolts that attach the cover to the gear housing.

7. Gently pry the gear cover away from the housing and remove from the engine.

To install:

8. Clean the gasket sealing surfaces.

9. Lubricate the gear train with oil.

10. Thoroughly clean and dry the seal area of the crankshaft.

11. Install the front cover and a new gasket. Install the bolts finger-tight.

12. Using the alignment/installation tool, align the cover to the crankshaft.

13. Torque the cover bolts to 18 ft. lbs. (24 Nm). Remove the tool.

14. Install the oil fill tube and adapter.

15. Install the crankshaft pulley but do not torque the bolts at this point.

16. Install the fan, belt tensioner and belt.

17. Torque the crankshaft pulley bolts to 92 ft. lbs. (125 Nm).

18. Connect the negative battery cable and check for leaks.

Front Cover Oil Seal

REPLACEMENT

1. Disconnect the negative battery cable.

2. Remove the drive belt.

3. Remove the crankshaft pulley.

4. Drill two ⅛ in. holes into the seal face, 180 degrees apart.

5. Using a slide hammer with a No. 10 sheet metal screw, pull the seal out alternating from side to side until the seal is out.

6. Thoroughly clean and dry the crankshaft.

To install:

7. Apply a bead of Loctite®277 to the outside diameter of the seal.

8. Install the pilot from the seal kit onto the crankshaft.

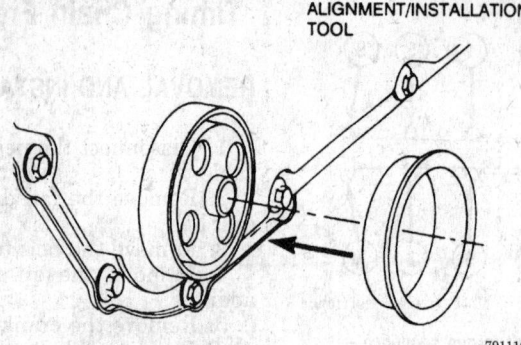

ALIGNMENT/INSTALLATION TOOL

79111060

Aligning the cover with the crankshaft — 5.9L diesel engine

9. Install the seal onto the pilot and start it into the front cover seal bore. Remove the pilot.

10. Use the alignment/installation tool and a plastic hammer to fully install the seal.

11. Install the crankshaft pulley but do not torque the bolts at this point.

12. Install the drive belt.

13. Torque the crankshaft pulley bolts to 92 ft. lbs. (125 Nm).

14. Connect the negative battery cable and check for leaks.

Timing Sprockets

REMOVAL AND INSTALLATION

1. Disconnect the negative battery cable.

2. Remove the fan drive assembly and belt.

3. Remove the belt tensioner.

4. Remove the oil fill tube and adaptor.

5. Remove the crankshaft pulley.

6. Remove all bolts that attach the cover to the gear housing.

7. Gently pry the gear cover away from the housing and remove from the engine.

8. Using a puller, remove the camshaft gear from the camshaft. Remove the key and replace.

9. Remove the crankshaft gear using heavy duty puller. Remove and install new alignment pin leaving it protrude 0.063 inch above crankshaft surface.

To install:

10. Heat the camshaft gear and crankshaft gear to 250°F (121°C) for 45 minutes. Lubricate the gear mount surface with Lubriplate® 105. Install the gears with the timing marks facing away from the shafts and aligned.

11. Clean the gasket sealing surfaces.

12. Lubricate the gear train with oil.

13. Install the front cover and a new gasket. Install the bolts finger-tight.

14. Using the alignment/installation tool, align the cover to the crankshaft.

15. Torque the cover bolts to 18 ft. lbs. (24 Nm). Remove the tool.

16. Install the oil fill tube and adapter.

17. Install the crankshaft pulley but do not torque the bolts at this point.

18. Install the fan, belt tensioner and belt.

19. Torque the crankshaft pulley bolts to 92 ft. lbs. (125 Nm).

20. Connect the negative battery cable and check for leaks.

Camshaft

REMOVAL AND INSTALLATION

1. Disconnect the negative battery cable.

2. Remove the valve covers.

3. Remove the rocker pedestal and arm assemblies.

4. Remove the pushrods.

5. Remove the drive belt.

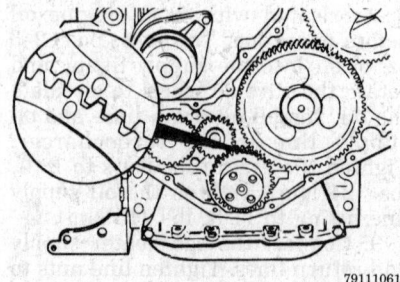

79111061

Crankshaft gear and camshaft gear timing marks — 5.9L diesel engine

6. Drain the cooling system. Remove the fan assembly, radiator and all related parts.

7. Remove the crankshaft pulley.

8. Remove the front gear cover.

9. Remove the fuel pump.

10. Insert the special dowels into the pushrod holes and onto the top of each lifter. When properly installed, the dowels can be use to hold the tappets up securely. Wrap rubber bands around the top of the dowels to prevent them from dropping down.

11. Rotate the crankshaft to align the crankshaft to camshaft timing marks.

12. Remove the bolts from the thrust plate.

13. Remove the camshaft and thrust plate.

14. Press the gear from the camshaft and remove the key.

To install:

15. Install the key on the camshaft.

16. Heat the camshaft gear to 250°F (121°C) for 45 minutes. Lubricate the gear mount surface with Lubriplate® 105. Install the gear to the camshaft with the timing marks facing away from the shaft.

17. Lubricate the camshaft bores, lobes, journals and thrust washer with Lubriplate® 105.

NOTE: Do not push the camshaft in too far or it may dislodge the plug in the rear of the camshaft bore, possibly creating a leak.

18. Install the camshaft and thrust washer so the **E** timing mark on the injection pump gear aligns with the **C** timing mark on the camshaft gear and the timing mark on the crankshaft gear align with those on the camshaft gear.

19. Install the thrust washer bolts and torque to 18 ft. lbs. (24 Nm).

20. Check the endplay of the camshaft. The specification is 0.006–0.010 in. (0.152–0.254mm).

21. Check the backlash of the camshaft gear. The specification is 0.003–0.013 in. (0.080–0.330mm).

22. Install the tappets and pushrods.

23. Install the rocker pedestal and arm assemblies.

24. Install the front cover and crankshaft pulley.

25. Install the drive belt and fan assembly.

26. Install the fuel pump.

27. Adjust the valves.

28. Install the valve covers. Refill cooling system.

29. Connect the negative battery cable and check for leaks.

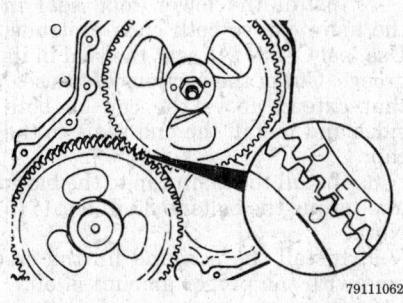

Camshaft gear and injection pump gear timing marks — 5.9L diesel engine

Piston and Connecting Rod

POSITIONING

ENGINE LUBRICATION

Oil Pan

REMOVAL AND INSTALLATION

1. Disconnect the negative battery cable.
2. Remove the engine oil dipstick and engine controller.
3. On Vans, remove the engine cover. Remove the air intake duct.
4. Raise the vehicle and support safely. Drain the engine oil.
5. Remove the transmission support braces.
6. Remove the starter and torque converter inspection cover, if equipped.
7. Remove the oxygen sensor and air injection tube, if equipped.

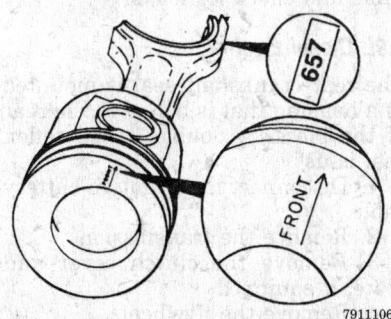

The ""FRONT" marking and the number on the rod should be oriented as shown — 5.9L diesel engine

8. Lower the exhaust crossover pipe.
9. Remove the nut on right engine mount. Loosen nut on left mount. Remove the rear transmission insulator bolts.
10. Using the proper equipment, support the transmission.
11. Support the engine with a jackstand. Raise the engine and transmission as required to allow for pan removal.
12. Remove the oil pan bolts and remove the pan from the vehicle.

To install:

13. Using a new pan gasket set, apply a thin bead of sealer to the 4 corners where the rubber seals and cork gasket meet.
14. Thoroughly clean and dry all bolts and bolt holes. Install the pan and tighten the bolts to 75 inch lbs. (4 Nm), then retighten to 200 inch lbs. (23 Nm).
15. Lower the engine and transmission and install the engine mount nuts and transmission mount bolts.
16. Install the exhaust crossover pipe. Install the oxygen sensor and air injection tube, if equipped.
17. Install the torque converter inspection cover, if equipped. Install the starter and support braces. Lower the vehicle.
18. Install the dipstick, air intake duct and engine cover, if removed.
19. Fill the engine with the proper amount of oil.
20. Connect the negative battery cable and check for leaks.

Oil Pump

REMOVAL AND INSTALLATION

Except 5.9L Diesel Engine

1. Disconnect the negative battery cable.
2. Remove the oil pan.
3. Remove the screen.
4. Unbolt the oil pump from the rear main bearing cap and remove it from the vehicle.

To install:

5. Prime the pump by pouring clean oil into the pump intake and turning the driveshaft until oil comes out the pressure port. Repeat a few times until no air bubbles are present. Install the oil pump with a rotating motion to ensure proper pump driveshaft engagement.
6. Hold the pump flush against the main cap and finger-tighten the attaching bolts.
7. Torque the bolts to 30 ft. lbs. (41 Nm).

8. Install the screen.
9. Install the oil pan with a new gasket.
10. Connect the negative battery cable and check the oil pressure.

5.9L Diesel Engine

1. Disconnect the negative battery cable.
2. Remove the drive belt.
3. Remove the radiator.
4. Remove the fan assembly.
5. Remove the oil fill tube and adaptor.
6. Remove the crankshaft pulley and damper.
7. Remove the front cover.
8. Remove the 4 pump mounting bolts and remove the pump from the block.

To install:

9. Prime the pump by pouring clean oil into the pump intake and turning the driveshaft until oil comes out the pressure port. Repeat a few times until no air bubbles are present. Align the idler gear pin with the locating bore in the block and install the pump.
10. Tighten the mounting bolts in the proper sequence to 44 inch lbs. (5 Nm), then repeat the sequence torquing to 18 ft. lbs. (24 Nm).

NOTE: When the pump is correctly installed, the flange on the pump should not touch the block; the back plate on the pump seats against the bottom of the bore.

11. Measure the backlash of the idler to pump drive gears. The specification is 0.003–0.013 in. (0.08–0.33mm).
12. Measure the backlash of the idler to crankshaft gears. The specification is 0.003–0.013 in. (0.08–0.33mm).
13. Install the front cover, crankshaft damper and pulley.
14. Install the oil fill tube and adaptor.
15. Install the fan assembly, radiator and drive belt.
16. Connect the negative battery cable and check the oil pressure.

Checking

EXCEPT 5.9L DIESEL ENGINE

1. Disassemble the pump.
2. Replace the pump assembly, if the cover is scratched, grooved or warped more than 0.0015 in. (0.038mm).
3. The minimum thickness of the outer rotor is 0.825 in. (20.96mm). The minimum diameter of the outer rotor is 2.469 in. (62.70mm). The

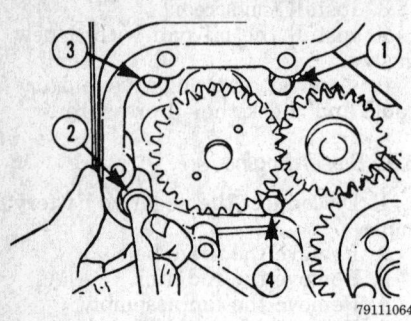

Oil pump attaching bolts tightening sequence — 5.9L diesel engine

minimum thickness of the inner rotor is 0.825 in. (20.96mm). If any of the above measurements are not within specifications, replace the shaft and both rotors.

4. The maximum clearance between the outer rotor and the pump body is 0.014 in. (0.356mm). Replace the pump assembly, if not within specifications.

5. Install the inner rotor and place a straight-edge across the bolt holes.If a feeler gauge of 0.004 in. (0.101mm) or more fits, replace the pump assembly.

6. The maximum clearance between the rotors is 0.080 in. (0.254mm). Replace the shaft and both rotors if not within specifications.

7. Inspect the relief valve plunger for scoring. Small marks may be removed with 400-grit wet of dry sandpaper.

8. The relief valve spring should have a freelength of 1.95 in. (49.5mm). Replace the spring if it fails to meet specifications.

9. Assemble the pump using new parts where necessary.

5.9L DIESEL ENGINE

1. Inspect the drive and idle gears for damage of any type. The maxi-mum backlash between the gears is 0.015 in. (0.38mm).

2. Remove the back plate.

3. The maximum tip clearance of the rotors is 0.007 in. (0.178mm).

4. Place a straight-edge across the rotors. The maximum clearance is 0.005 in. (0.127mm).

5. The maximum rotor to body clearance is 0.015 in. (0.381mm).

6. Remove the rotor and inspect all parts for visible damage.

7. Assemble the pump using new parts where necessary.

Rear Main Bearing Oil Seal.

The 3.9L and 5.2L engines rear main seal is a 2 piece, fitted rope type seal. The 5.9L engine rear main seal is a split rubber type. In all cases, the upper half can be installed with the crankshaft in place. The 2 halves should always be replaced as a set.

REMOVAL AND INSTALLATION

3.9L and 5.2L Engines

1. Raise the vehicle and support safely. Drain the engine oil. Remove the oil pan and oil pump.

2. Remove the rear main bearing cap.

3. Remove the lower seal from the cap.

4. To remove the upper rope seal, use oil seal remover and installer kit KD–492 or equivalent, following the instructions provided with the tool.

To install:

5. Wipe the crankshaft surface clean and coat it lightly with oil.

6. Use oil seal remover and installer kit KD–492 or equivalent, to install the upper rope seal. Trim the ends of the upper seal to eliminate frayed ends.

7. Install the lower rope seal in the main cap so both ends protrude. Use tool C–3511 to seat the seal in its groove. Cut of the portions of the seal that extend above the cap on both sides and install the end seals to the cap.

8. Install the main cap to the block and torque the bolts to 85 ft. lbs. (115 Nm).

9. Install the pan and fill the engine with the proper amount of oil.

10. Connect the negative battery cable and check for leaks.

5.9L Gasoline Engine

1. Raise the vehicle and support safely. Drain the engine oil. Remove the oil pan and oil pump.

2. Remove the rear main bearing cap.

3. Remove the lower seal from the cap.

4. To remove the upper seal, press on one end of the seal with a small blunt tool, rotate the crankshaft slightly and pull out the other end of the seal.

To install:

5. Wipe the crankshaft surface clean and coat it lightly with oil.

6. When installing the upper seal, hold the seal (with the paint stripe to the rear) tightly against the crankshaft and rotate the crankshaft while sliding the seal into the groove. If any rubber has peeled off of the back of the new seal, do not use it; it will leak.

7. When installing the lower rubber seal, make sure the paint stripe is positioned to the rear and place a drop of sealer next to both ends of the seal.

8. Install the main cap to the block and torque the bolts to 85 ft. lbs. (115 Nm).

9. Install the oil pump and pan and fill the engine with the proper amount of oil.

10. Connect the negative battery cable and check for leaks.

5.9L Diesel Engine

The rear crankshaft seal is mounted in a housing that is bolted to the rear of the block. A double lipped teflon seal is us

1. Disconnect the negative battery cable.

2. Remove the transmssion.

3. Remove the clutch cover and plate, if equipped.

4. Remove the flywheel.

5. Drill two 1/8 in. holes 180 degrees apart into the seals. Be extremely careful not to drill against the crankshaft.

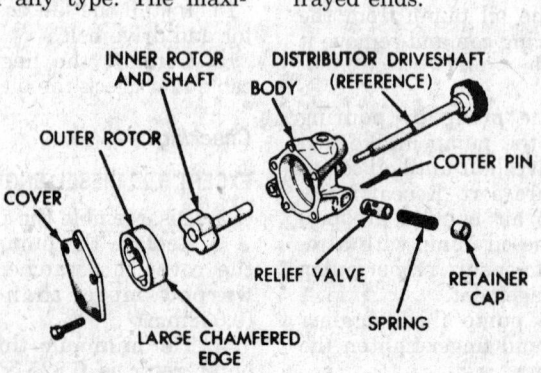

Exploded view of the oil pump — 5.2L and 5.9L gasoline engines

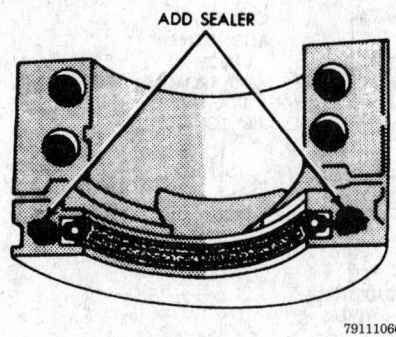

ADD SEALER

79111066

Lower rear main seal — 5.9L gasoline engine

6. Using a No. 10 sheet metal screw and a slide hammer, remove the rear seal.

To install:

7. Thoroughly clean and dry the crankshaft surface. Do not oil the crankshaft or seal prior to installation or the seal will leak.

8. Install the seal pilot included in the replacement seal kit, on the crankshaft. Push the seal on the pilot and crankshaft. Remove the pilot.

9. If the new seal has a rubber outer diameter, lubricate it with soapy water. If the seal does not have a rubber outer diameter, use Loctite®277 or equivalent on the outer diameter.

10. Use the alignment tool to install the seal to the proper depth in the housing. Drive the seal in gradually and evenly until the alignment tool stops against the housing.

11. Install flywheel and clutch and cover, if equipped.

12. Install transmission and reconnect negative battery cable. Add oil to correct level, start vehicle and check for leaks.

MANUAL TRANSMISSION

Transmission Assembly

REMOVAL AND INSTALLATION

1. Disconnect the negative battery cable.

2. Remove the shift lever. If equipped with the NP-435 4 speed, push the retainer (not the gearshift lever itself) down and rotate it counterclockwise slightly to release. If equipped with the NP-2500 5 speed,

unbolt the shifter base assembly from the transmission. If equipped with the A-833 4 speed, label and disconnect the linkages to the shifter assembly and unbolt the shifter from its support. If equipped with the G-360 5 speed, remove snapring under lower boot on shift tower.

3. Raise the vehicle and support safely. Remove the skid plates, if equipped. Drain the transmission and transfer case.

4. Disconnect the distance sensor and the speedometer cable from the transmission or transfer case.

5. Matchmark and remove the driveshaft(s). Disconnect the PTO, if equipped.

6. If equipped with 4WD, disconnect all linkage, electrical connectors and vacuum lines from the transfer case. Using a jack, support the transfer case, unbolt the transfer case from the transmission and slide backward to remove from the vehicle.

7. Disconnect the reverse light switch connector and remove all wiring from any clips on the transmission case.

8. Install an appropriate engine support fixture to hold the engine in place.

9. Support the transmission with a transmission jack.

10. Remove the transmission crossmember.

11. Remove the transmission to bell housing bolts.

12. Slide the transmission backwards until the input shaft clears the clutch disc. Remove the transmission from the vehicle.

To install:

13. Lubricate the pilot bushing and input shaft splines very lightly with high temperature lubricants.

14. Mount the transmission securely on a transmission jack and lift it in place until the input shaft is centered in the bell housing opening. Roll the transmission forward until the input shaft splines fully engage with the clutch di

15. Install the transmission to bell housing bolts. Torque the bolts to 50 ft. lbs. (68 Nm).

16. Install the transmission crossmember. Remove the transmission and engine support fixtures.

17. Install the transfer case and connect all linkage, electrical connectors and vacuum lines to the transfer case.

18. Connect the reverse light switch connector and clip all wiring to the transmission case.

19. Connect speedometer cable and distance sensor, if equipped.

20. Install the driveshaft(s) and connect the PTO, if equipped.

21. Fill the transmission and transfer case with the proper lubricant.

22. Install the shifter assembly and linkages.

23. Install the skid plates, if equipped.24.Connect the negative battery cable and check the transmission for proper operation.

LINKAGE ADJUSTMENT

A-833 4 Speed

1. Place the shifter in the neutral position.

2. Raise the vehicle and support safely.

3. Install the fabricated aligning tool to hold the levers in the neutral crossover position.

4. Disconnect the control rods from the levers in the shifter assembly. Make sure the levers are still in the neutral position.

5. Rotate the threaded ends of the shift control rods to adjust the rod length. Starting with the 1/2 rod, adjust it so the fabricated tool does not bind. Repeat with the 3/4 and the reverse rods.

6. Install the rods to their levers with the washers and clips.

7. Remove the alignment tool and check the shifting action for smoothness.

CLUTCH

Clutch Assembly

REMOVAL AND INSTALLATION

1. Disconnect the negative battery cable.

2. Raise the vehicle and support safely.

3. Remove the transmission and transfer case, if equipped.

4. Remove the clutch housing, release fork and bearing assembly, if equipped.

5. Remove the clutch cover bolts. If cover is to be reused, loosen cover bolts evenly a few threads at a time to avoid warping the cover.

6. Remove the clutch cover and disc.

To install:

7. Install the clutch cover and disc. Use a clutch aligning tool or

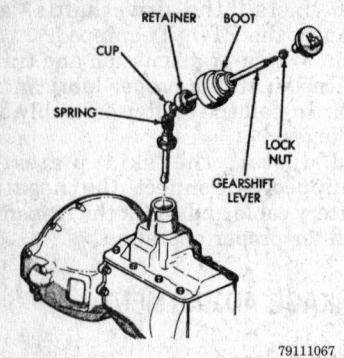

79111067

New Process 435 shifter components

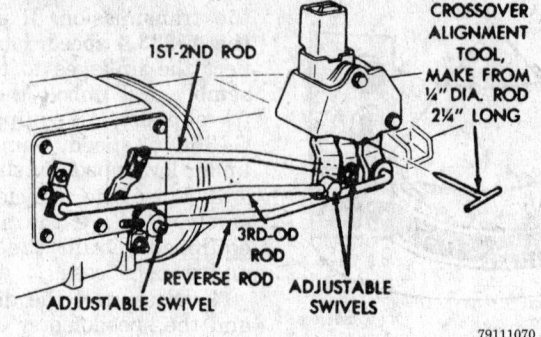

79111070

A-833 shifter and linkage

spare input shaft to center the disc. Tighten all of the bolts finger-tight.

8. The cover bolts must be turned gradually, evenly and to the proper torque to avoid distorting the cover. Torque $5/16$ in. diameter bolts in the aforementioned manner to 17 ft. lbs. (23 Nm). Torque $3/8$ in. diameter bolts similarly to 30 ft. lbs. (41 Nm).

9. Install the transmission and transfer case, if equipped.

10. Install the inspection cover.

11. Connect the negative battery cable and check the clutch for proper operation.

Pedal Height/Free-Play
ADJUSTMENT

NOTE: Chrysler uses a hydraulic clutch release system on their full size Pick-Ups and Vans. There is no adjustment for free-play on vehicles with this system.

Clutch Master Cylinder and Slave Cylinder

The clutch master cylinder, remote reservoir, slave cylinder and connect-

ing lines are all serviced as a complete assembly. The cylinders and connecting lines are sealed units. They are prefilled with fluid from the factory and cannot be disassembled or serviced separately.

REMOVAL AND INSTALLATION

1. Disconnect the negative battery cable.

2. Raise the vehicle and support safely. On diesel models, remove the slave cylinder shield from the clutch housing.

3. Remove the nuts attaching the slave cylinder to the bell housing.

4. Remove the slave cylinder and clip from the housing.

5. Lower the vehicle. On diesel models, disconnect the clutch pedal interlock switch wires.

6. Remove the locating clip from the clutch master cylinder mounting bracket.

7. Remove the retaining ring, flat washer and wave washer that attach the clutch master cylinder pushrod to the clutch pedal. Slide the pushrod off of the pedal pin. Inspect the bushing on the pedal pin and replace if it is excessively worn.

8. Verify that the cap on the clutch master cylinder reservoir is tight so fluid will not spill during removal.

9. Remove the screws attaching the reservoir and bracket, if equipped, to the dash panel and remove the reservoir.

10. Pull the clutch master cylinder rubber seal from the dash panel.

11. Rotate the clutch master cylinder counterclockwise 45 degrees to unlock it. Remove the cylinder from the dash panel.

12. Remove the clutch master cylinder, remote reservoir, slave cylinder and connecting lines from the vehicle.

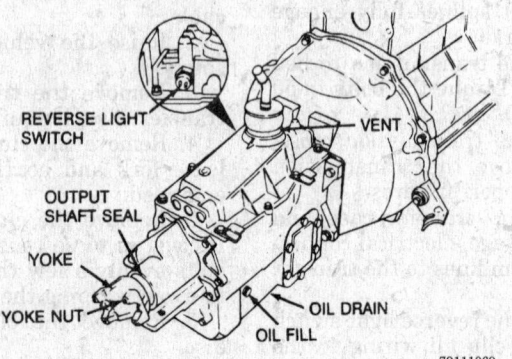

79111068

New Process 2500 transmission

79111069

Getrag G-360 transmission coupled with the 5.9L diesel engine

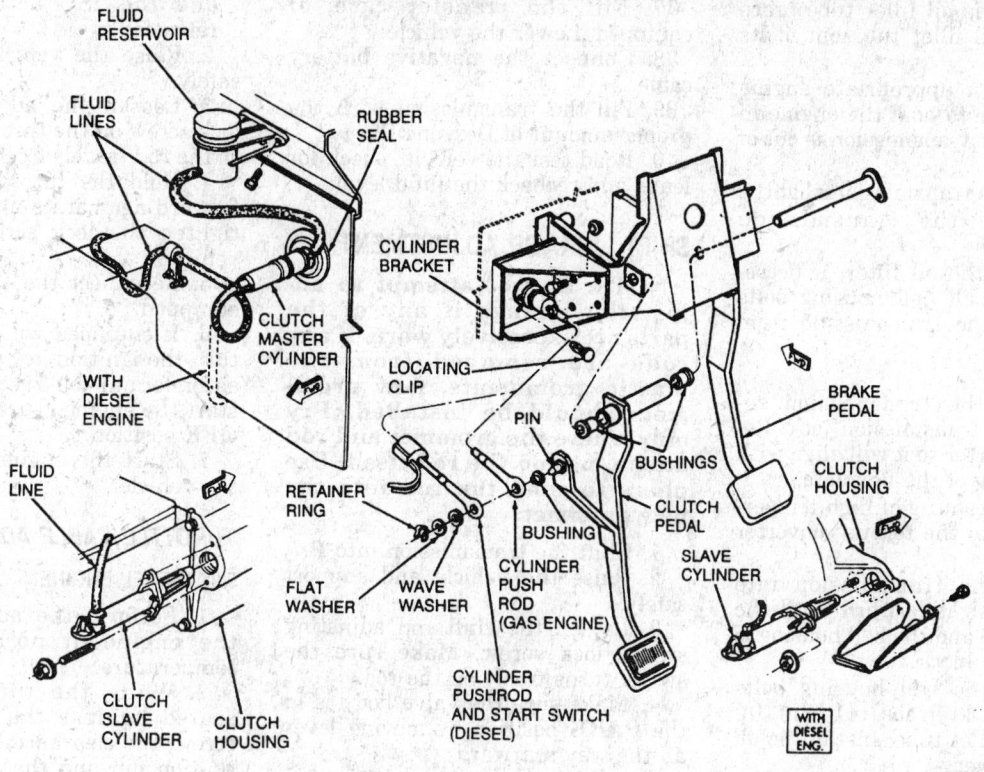

Hydraulic clutch release components — Pick-Up and Ramcharger

79111071

To install:

13. Verify that the cap on the fluid reservoir is tight so fluid will not spill during installation.

14. Position the components in the replacement kit in their places on the vehicle.

15. Insert the master cylinder in the dash. Rotate clockwise 45 degrees to lock in place.

16. Lubricate the rubber seal with a lubricant to ease installation. Seat the seal around the cylinder in the dash.

17. Install the fluid reservoir and bracket, if equipped, to the dash panel.

18. Install the master cylinder pushrod to the clutch pedal pin. Secure the rod with the wave washer, flat washer and retaining ring. Install the locating clip. Do not remove the plastic shipping stop from the pushrod until the slave cylinder has been installed.

19. Raise the vehicle and support safely.

20. Insert the slave cylinder pushrod through the opening and make sure the cap on the end of the pushrod is securely engaged in the release lever before tightening the attaching nuts. Torque the nuts to 200 inch lbs. (23 Nm).

21. Install slave cylinder cover, if equipped. Lower the vehicle. Remove the plastic shipping stop from the master cylinder pushrod. Connect clutch pedal interlock switch wires.

22. Operate the clutch pedal a few times to verify proper operation of the system. The system will self-bleed any air in the system and vent through the reservoir.

23. Connect the negative battery cable and road test the vehicle.

AUTOMATIC TRANSMISSION

Transmission Assembly

REMOVAL AND INSTALLATION

1. Disconnect the negative battery cable.

2. Raise the vehicle and support safely. Drain the transmission and transfer case, if equipped.

3. Disconnect and lower or remove any exhaust parts as required.

4. Remove the skid plates, if equipped.

5. Matchmark and remove the driveshaft(s).

6. Disconnect the distance sensor, if equipped, and the speedometer cable.

7. If equipped with 4WD, disconnect all linkage, electrical connectors and vacuum lines from the transfer case. Support the transfer case, unbolt the transfer case from the transmission and slide it backwards to remove from the vehicle.

8. Remove the engine to transmission struts.

9. Remove the starter and the fluid cooler lines bracket.

10. Remove the torque converter inspection cover.

11. Matchmark the converter to the flexplate. Remove the torque converter bolts.

12. Disconnect the wires to the neutral safety switch and lockup solenoid, if equipped.

13. Disconnect the oil cooler lines from the transmission.

14. Disconnect the gearshift rod and torque shaft assembly from the transmission.

15. Disconnect the throttle rod from the lever.

16. Unbolt the oil filler tube brace and lift the oil filler tube out of its bore.

17. Install an appropriate engine support fixture to hold the engine in place when the transmission is out of the vehicle.

18. Raise the transmission slightly.

19. Remove the transmission crossmember.

20. Remove the oil filter, if necessary. Remove all bell housing bolts and remove the transmission from the vehicle.

To install:

21. Install the transmission securely on the transmission jack. Rotate the converter so it will align with the positioning of the flexplate.

22. Apply a coating of high temperature grease to the torque converter pilot hub.

23. Raise the transmission into place and push it forward until the dowels engage and the bell housing is flush with the block.

24. Install the bell housing bolts and torque to 30 ft. lbs. (41 Nm). Install the oil filler tube. Install the oil filter, if removed.

25. Install the transmission crossmember. Remove the engine support fixture and the transmission jack.

26. Install the torque converter bolts and torque as follows:

   a. On models with 9.5 inch, 3-lug converter, tighten bolts to 40 ft. lbs. (54 Nm).

   b. On models with 9.5 inch, 4-lug converter, tighten bolts to 55 ft. lbs. (74 Nm).

   c. On models with 10.75 inch, 4-lug converter, tighten bolts to 270 inch lbs. (31 Nm).

27. Connect the oil cooler lines.

28. Connect the throttle rod to the lever and adjust, if necessary.

29. Connect the gearshift rod and torque shaft assembly to the transmission and adjust if necessary.

30. Connect the wires to the neutral safety switch, and lockup solenoid, if equipped. Make sure all wires are routed correctly and clipped in place.

31. Install the torque converter inspection cover, starter and transmission struts.

32. Install the transfer case, if equipped.

33. Connect the distance sensor, if equipped.

34. Connect the the speedometer cable.

35. Install the driveshaft(s).

36. Install exhaust parts that were removed.

37. Fill the transfer case, if equipped. Lower the vehicle.

38. Connect the negative battery cable.

39. Fill the transmission with the proper amount of Dexron®II.

40. Road test the vehicle, check for leaks and recheck the fluid level.

SHIFT LINKAGE ADJUSTMENT

NOTE: Do not attempt to adjust the linkage if any of the parts are excessively worn. If any rods are removed from the plastic grommets, new grommets should be installed. Pry only where the grommet and rod attach, not on the rod itself. Use pliers to snap the rod into the new grommet.

1. Shift the transmission into **P**.

2. Raise the vehicle and support safely.

3. Loosen the shift rod adjusting swivel lock screw. Make sure the swivel turns freely on the rod.

4. Make sure the valve body is in the **PARK** position by moving lever all the way rearward.

5. Adjust the swivel position on the shift rod to obtain a free pin fit in the toque shaft lever. Tighten the lock screw.

6. If the vehicle starts in any gear other than **P** or **N**, or does not start in both **P** and **N**, then either the adjustment is wrong or another problem exists.

THROTTLE LINKAGE ADJUSTMENT

Except 5.9L Diesel Engine

1. Retract the ISC actuator by doing one of the following:

   a. If the DRBII is available, connect its connector to the diagnostic connector. Start the engine and place the DRBII in the "Throttle Body Minimum Air Flow Test" mode. Disconnect the electrical connector on the ISC actuator. Shut off the engine and disconnect the DRBII; the actuator is now fully retracted.

   b. If the DRBII is not available, 2 jumper wires may be used. With the engine OFF, disconnect the connector to the ISC actuator. Connect a pair of jumper wires to the battery. Connect the negative jumper to the top pin of the ISC actuator and the positive jumper to the other pin. Do not leave the jumpers connected for more than 5 seconds. Disconnect the jumpers

and the ISC actuator is fully retracted.

2. Raise the vehicle and support safely.

3. Loosen the adjustable swivel lock screw on the throttle rod enough so the rod travels freely in the swivel.

4. Hold the throttle lever firmly forward against its internal stop and tighten the lock screw. Lower the vehicle.

5. Reconnect the ISC actuator, if equipped.

6. If equipped with fuel injection, turn the ignition key to the **RUN** position for at least 5 seconds but do not start the engine. Turn the key to the **OFF** position.

7. Start the engine and road test the vehicle.

THROTTLE CABLE ADJUSTMENT

5.9L DIESEL ENGINE

1. Perform the adjustment with the engine at normal operating temperature.

2. While the throttle lever is seated against the low idle stop screw, the clearance between the actuation pin and the rear end of the slotted cable should be 0.180 inch (4.57 mm).

3. If it is not at specification, lift the locking pawl and slide the cable to the proper position to obtain the specified clearance.

4. Lock the pawl back into place and road test the vehicle.

TRANSFER CASE

Transfer Case Assembly

REMOVAL AND INSTALLATION

1. Disconnect the negative battery cable.

2. Raise the vehicle and support safely.

3. Remove the skid plates, if equipped. Drain the transfer case fluid.

4. Disconnect the distance sensor, if equipped, and disconnect the speedometer cable from the transfer case.

5. Matchmark and remove front and rear driveshafts.

6. Disconnect the PTO, if equipped.

7. Disconnect the linkage, electrical connectors and vacuum lines from

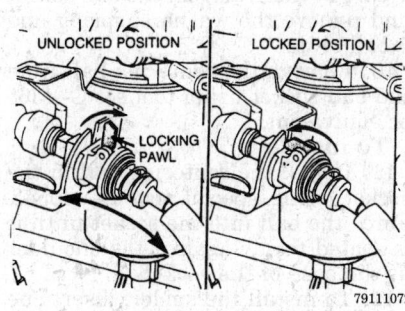

Unlocking and locking the throttle cable pawl

the transfer case. Support the transfer case, unbolt the transfer case from the transmission and slide it backwards to remove it from the vehicle.

To install:

8. Apply silicone sealer to both sides of transfer case-to-transmission gasket and position on transmission.

9. Align and seat transfer case on transmission making sure gear splines are aligned with transmission output shaft. If necessary, align splines by rotating transfer case rear output shaft yoke.

10. Install and tighten attaching nuts to 35 ft. lbs. (47 Nm).

11. Install rear crossmember. Remove transmission support.

12. Align and connect driveshafts.

13. Connect distance sensor wires. Reconnect transfer case shift lever to range lever tightening locknut to 90 inch lbs. (10 Nm).

14. Fill transfer case with lubricant. Install skid plate and lower vehicle.

15. Reconnect negative battery cable and check operation of transfer case.

LINKAGE ADJUSTMENT

New Process 205

1. Shift the transfer case into neutral.

2. Move the shift rod boot upward for access.

3. Loosen the shift bracket bolts and move the bracket as far forward as possible. Tighten the bolts.

4. Check the smoothness of operation of the transfer case.

New Process 241

1. Move the transfer case shift lever boot aside for access to the shift lever and gate.

2. Move the shift lever into the **4H** position. Make sure the lever is against the **4H** gate.

3. Raise the vehicle and support safely.

4. Loosen the shift rod clamp screw until the shift rod is free to slide in the swivel.

5. Verify that the lever is in the **4H** position and correct location if moved out of position.

6. Tighten the clamp screw.

7. Check the smoothness of operation of the transfer case.

DRIVE AXLE

Rear Driveshaft

REMOVAL AND INSTALLATION

One-Piece Rear Driveshaft

1. Raise the vehicle and support safely.

2. Matchmark the driveshaft and the rear axle drive pinion gear shaft yoke.

3. Remove the rear U-joint attaching bolts and both strap clamps from the rear axle drive pinion gear shaft yoke.

4. Fluid may run from the rear of the extension housing or transfer case when the shaft is removed, so position a drain pan under the area.

5. Remove the driveshaft from the transmission or transfer case.

6. The installation is the reversal of the removal procedure. Torque ¼–28 clamp bolts to 14 ft. lbs. (19 Nm) and ⁵⁄₁₆–24 to 25 ft. lbs. (34 Nm).

Two-Piece Rear Driveshaft

1. Raise the vehicle and support safely.

2. Matchmark the driveshaft and the rear axle drive pinion gear shaft yoke.

3. Remove the rear U-joint attaching bolts and both strap clamps from the rear axle drive pinion gear shaft yoke.

4. Detach the protective boot clamp from the front shaft splines, if equipped, and slide the rear shaft slip yoke from the front shaft at the center bearing. Remove the rear shaft from the vehicle.

5. Matchmark the yokes at the transmission or transfer case and remove the clamp retaining bolts and clamp straps.

6. Remove the center bearing retaining bolts and nuts and remove the front shaft with center bearing from the vehicle.

To install:

7. Lubricate shaft splines. Install front yoke of front shaft to transmission or transfer case and install retaining screws loosely. Loosely install retaining nuts and bolts in center support. As applicable, tighten ¼-inch clamp screws to 14 ft. lbs. (19 Nm). Torque ⁵⁄₁₆ — clamp screw to 25 ft. lbs. (34 Nm).

8. Align master spline in shafts and slide together. Position boot over front splines, if equipped.

9. Align the reference marks, position rear U-Joint in axle yoke saddles and install straps and bolts. Torque ¼–28 clamp bolts to 14 ft. lbs. (19 Nm) and ⁵⁄₁₆–24 to 25 ft. lbs. (34 Nm).

10. Raise and safely support the vehicle. Rotate the driveshaft via the engine to allow the center bearing to self-align. Torque the center bearing bolts and nuts to 50 ft. lbs. (68 Nm).

Front Driveshaft — 4WD Vehicle

REMOVAL AND INSTALLATION

1. Raise the vehicle and support safely.

2. Remove the skid plate, if equipped.

3. Matchmark the driveshaft and the front axle drive pinion gear shaft yoke.

4. Remove the joint to transfer case flange capscrews and lockwashers.

5. Remove the front U-joint attaching bolts and both strap clamps.

6. Remove the front driveshaft from the vehicle.

To install:

7. Install driveshaft and front U-joint attaching bolts and both strap clamps. Torque ¼–28 clamp bolts to 14 ft. lbs. (19 Nm) and ⁵⁄₁₆–24 to 25 ft. lbs. (34 Nm).

8. Install the joint to transfer case flange capscrews and lockwashers and torque to 25 ft. lbs. (34 Nm).

9. Install the skid plate and lower vehicle.

Rear Driveshaft Center Bearing

REMOVAL AND INSTALLATION

1. Remove the shafts and center bearing from the vehicle.

NOTE: Do not clamp the driveshaft tube in a vise. Clamp only the forged portion of the welded yoke in a vise. Do not overtighten the vise jaws.

2. Clamp the front shaft in a vise and remove the bearing support and rubber insulator from the center bearing.
3. Bend the slinger away from the center bearing to provide sufficient clearance for installing a puller.
4. Remove the bearing from the front shaft with a puller and remove the slinger. The replacement package contains the bearing, slinger and retainer.
To install:
5. Press the replacement slinger, bearing and retainer onto the front shaft until seated on shoulder.
6. Install the shafts and center support bearing on vehicle.

Single Cardan Universal Joint

REMOVAL AND INSTALLATION

1. Remove the driveshaft from the vehicle.

NOTE: Do not clamp the driveshaft tube in a vise. Clamp only the forged portion of the welded yoke or the slip yoke in a vise. Do not overtighten the vise jaws.

2. Clamp the yoke in a vise and remove the bearing cap retainers.
3. Place a socket which has an inside diameter larger than the outside diameter of the bearing cap, against the yoke around the perimeter of the first cap to be removed. Place a socket which is slightly smaller than the cap, on the cap opposite the cap to be removed. Then position the yoke in a vise.
4. Compress the jaws until the smaller socket has driven the other cap into the larger socket.
5. Release the jaws and remove the cap that is partially out of the yoke.
6. Repeat the procedure for the remaining cap(s).

To install:
7. Clean and remove any rust from the yoke bores and lubricate lightly with lithium based grease.
8. Position the spider cylinders in the yoke bores. Insert the seals into the yoke bores and against the spider cylinders. Tap the bearing caps into the yoke bores far enough to keep the spider in place.
9. Place the socket that is slightly smaller than the cap against the first cap and position the assembly in a vise.
10. Compress the jaws to force the bearing caps into the yoke bores far enough so the retainer grooves are visible.
11. Repeat the procedure for remaining caps.
12. Install the retaining clips.
13. Install the driveshaft assembly to the vehicle.

Double Cardan Joint — 4WD Vehicles

REMOVAL AND INSTALLATION

1. Remove the front driveshaft from the vehicle.
2. Matchmark the yokes before disassembling so they will be installed in their original locations to retain driveshaft balance.
3. To expedite removal, remove the bearing caps in the sequence indicated.
4. Support the driveshaft horizontally and aligned with the base plate of the press. Shear the bearing cap plastic retaining ring and position the first link yoke rear arm over a 1⅛ in. socket. Place spider press tool C–4365–1 or equivalent, on the bearing caps in the flange yoke arms. Force the bearing cap out of the yoke with a press.
5. If the bearing cap is not completely removed, insert a spacer between the spider and bearing cap and complete the removal.
6. Rotate the driveshaft 180 degrees and repeat the procedure.
7. Disengage the spider trunnions from the link yoke. Pull the flange yoke and the spider from the centering ball on the ball support tube yoke.
8. To remove the ball socket, separate the joint between the link yoke and the flange yoke by forcing the spider trunnion bushing from the link yoke. Pull the flange yoke and the spider with the ball socket from the centering ball as an assembly.

9. Pry the seal from the ball socket and remove the washers, spring and 3 ball seats.
10. Remove the centering ball from the ball socket using tool set C–4365 or equivalent.
To install:
11. Install the centering ball in the socket using special tool C–4365–3. Force the ball into the socket until it is seated firmly against the shoulder at the base of the socket.
12. To install the spider, insert one bearing cap partially into one of the yoke bores and then rotate the yoke 180 degrees. Insert the spider into the yoke bore and seat the spider trunnion in the bearing cap. Partially insert the opposite bearing cap in the remaining yoke bolt.
13. Force the bearing caps inward while pivoting the spider back and forth to provide free movement of the trunnions in the bearing.
14. When the retainer grooves become visible, install the retainer.
15. Continue to force the caps inward until the opposite retainer can be installed in its groove.
16. Lubricate the centering ball and socket with the lubricant provided in the replacement key.
17. Repeat the installation procedure with the remaining portion of the assembly.

Front Axle Shaft, Bearing and Seal

REMOVAL AND INSTALLATION

Model 44

RIGHT SIDE SHAFT

1. Raise the vehicle and support safely.
2. Remove the wheel and remove the brake caliper from the rotor. Do not allow the caliper to hang by the hose.
3. Remove the dust cap and driving hub snapring.
4. Remove the driving hub and retaining spring.
5. Remove the wheel bearing nut lock using tool C–4170–A or equivalent. Remove the retaining washer and the wheel bearing adjusting nut.
6. Remove the rotor/hub with wheel bearings and retainer spring plate. Remove the grease seal and bearing from the rotor.
7. Remove the splash shield and spindle from the steering knuckle.
8. Remove the brake caliper adaptor from the knuckle.

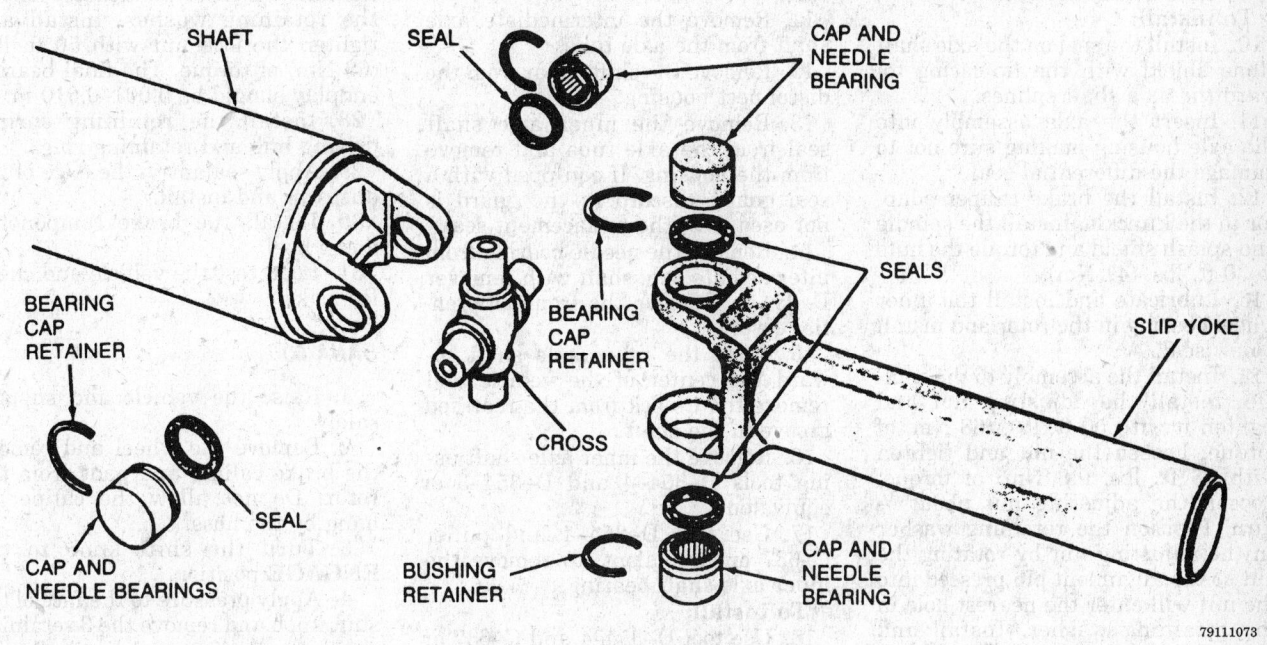

Single cardan universal joint components

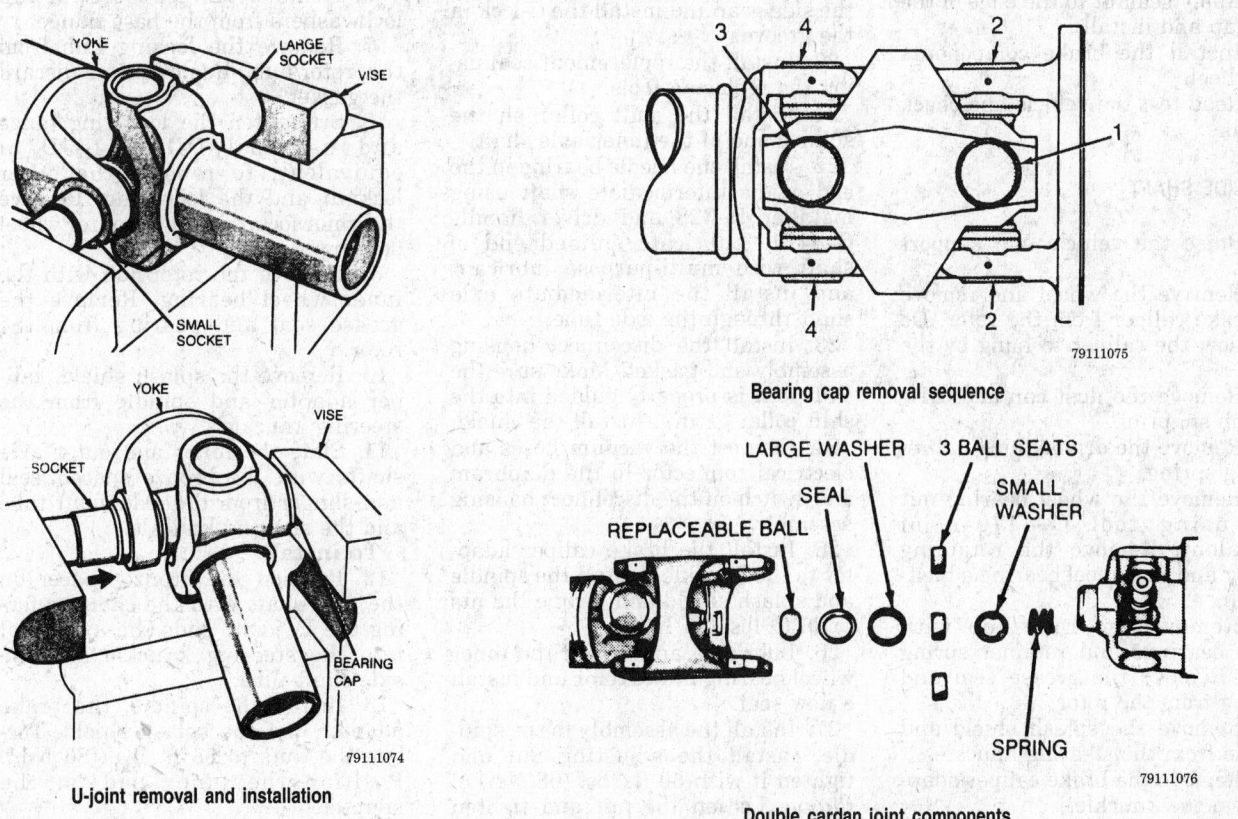

U-joint removal and installation

Bearing cap removal sequence

Double cardan joint components

9. Remove the axle shaft from the axle housing. Remove the seal and stone guard from the shaft.

To install:

10. Install the seal on the axle shaft stone shield with the lip facing toward the axle shaft splines.

11. Insert the axle assembly into the axle housing making sure not to damage the differential seal.

12. Install the brake caliper adaptor to the knuckle. Install the spindle and splash shield and torque the nuts to 30 ft. lbs. (41 Nm).

13. Lubricate and install the inner wheel bearing in the rotor and install a new seal.

14. Install the assembly to the spindle. Install the adjusting nut and tighten it with 50 ft. lbs. (68 Nm) of torque. Loosen the nut and tighten with 35 ft. lbs. (48 Nm) of torque. Loosen the adjusting nut about 3/8 turn. Position the retaining washer on the adjusting nut by rotating the nut so the alignment pin pressed into the nut will enter the nearest hole in the retaining washer. Install and tighten the nut lock with 50 ft. lbs. (68 Nm) of torque. The final bearing endplay should be 0.001–0.010 in.

15. Install the retaining spring, driving hub and retaining ring.

16. Apply sealant to the edge of the dust cap and install.

17. Install the brake components and wheel.

18. Road test the vehicle and check for leaks.

LEFT SIDE SHAFT

1. Raise the vehicle and support safely.

2. Remove the wheel and remove the brake caliper from the rotor. Do not allow the caliper to hang by the hose.

3. Remove the dust cap and driving hub snapring.

4. Remove the driving hub and retaining spring.

5. Remove the wheel bearing nut lock using tool C–4170–A or equivalent. Remove the retaining washer and the wheel bearing adjusting nut.

6. Remove the rotor/hub with wheel bearings and retainer spring plate. Remove the grease seal and bearing from the rotor.

7. Remove the splash shield and spindle from the steering knuckle.

8. Remove the brake caliper adaptor from the knuckle.

9. Disconnect the vacuum hoses and electrical connector from the disconnect housing assembly.

10. Remove the disconnect housing assembly, cover and shield from the axle.

11. Remove the intermediate axle shaft from the axle tube.

12. Remove the shift collar from the disconnect housing.

13. Remove the inner axle shaft seal from the axle tube and remove from the housing. If equipped with a seal guard, discard it; the guard is not used with the replacement seal.

14. Remove the needle bearing from intermediate axle shaft with remover D–354–1. Remove the front differential cover.

15. Force the inner axle shaft toward the center of the vehicle and remove the C–lock from the recessed groove in the shaft.

16. Remove the inner axle shaft using tools D–354–4 and D–354–3 or equivalent.

17. Use tool D–354–1 and puller C–637 or equivalent, to remove the inner axle shaft bearing.

To install:

18. Use tool D–354–4 and C–637 or equivalent, to install the inner axle shaft bearing.

19. Install the inner axle shaft using D–354–4 and D–354–2 or equivalent. Slide the axle shaft into the side gear and install the C-lock in the groove.

20. Install the replacement seal using the replacing tools.

21. Install the shift collar on the splined end of the inner axle shaft.

22. Install the needle bearing in the end of the intermediate shaft using installer D–328 and driver handle C–4171. Lubricate splined end of shaft with multi-purpose lubricant and install the intermediate axle shaft through the axle tube.

23. Install the disconnect housing assembly and gasket. Make sure the shift fork is properly guided into the shift collar groove. Install the shield.

24. Connect the vacuum hoses and electrical connector to the diaphram and switch on the disconnect housing assembly.

25. Install the brake caliper adaptor to the knuckle. Install the spindle and splash shield and torque the nts to 30 ft. lbs. (41 Nm).

26. Lubricate and install the inner wheel bearing in the rotor and install a new seal.

27. Install the assembly to the spindle. Install the adjusting nut and tighten it with 50 ft. lbs. (68 Nm) of torque. Loosen the nut and tighten with 35 ft. lbs. (48 Nm) of torque. Loosen the adjusting nut about 3/8 turn. Position the retaining washer

on the adjusting nut by rotating the nut so the alignment pin pressed into the nut will enter the nearest hole in the retaining washer. Install and tighten the lock nut with 50 ft. lbs. (68 Nm) of torque. The final bearing endplay should be 0.001–0.010 in.

28. Install the retaining spring, driving hub and retaining ring.

29. Apply sealant to the edge of the dust cap and install.

30. Install the brake components and wheel.

31. Road test the vehicle and check for leaks.

DANA 60

1. Raise the vehicle and support safely.

2. Remove the wheel and remove the brake caliper and pads from the rotor. Do not allow the caliper to hang by the hose.

3. Turn the shift knob to the **ENGAGE** position.

4. Apply pressure to the face of the shift knob and remove the 3 retaining screws located nearest to the flange. Pull outward and remove the shift knob from the base.

5. Remove the snapring from the axle shaft.

6. Remove the capscrews and lockwashers from the base flange.

7. Remove the locking hub from the rotor/hub. Remove and discard the gasket.

8. Straighten the lock ring tangs and use tool DD–1241–JD or equivalent, to remove the outer locknut and the lock ring. Remove the inner locknut and the outer wheel bearing.

9. Remove the rotor/hub with the inner wheel bearing. Remove the grease seal and bearing from the rotor.

10. Remove the splash shield, caliper adaptor and spindle from the steering knuckle.

11. Slide the inner and outer axle shafts with the bronze spacer, seal and slinger from the axle shaft tube and the steering knuckle.

To install:

12. Position the bronze spacer on the axle shaft with the chamfer facing the U–joint. Slide the axle shaft into the steering knuckle and the axle shaft tube.

13. Install the spindle, the brake adaptor and the splash shield. Torque the nuts to 65 ft. lbs. (86 Nm). Position the inner pad on the adapter.

14. Lubricate and install the inner wheel bearing in the rotor and install a new seal. Install the rotor to the

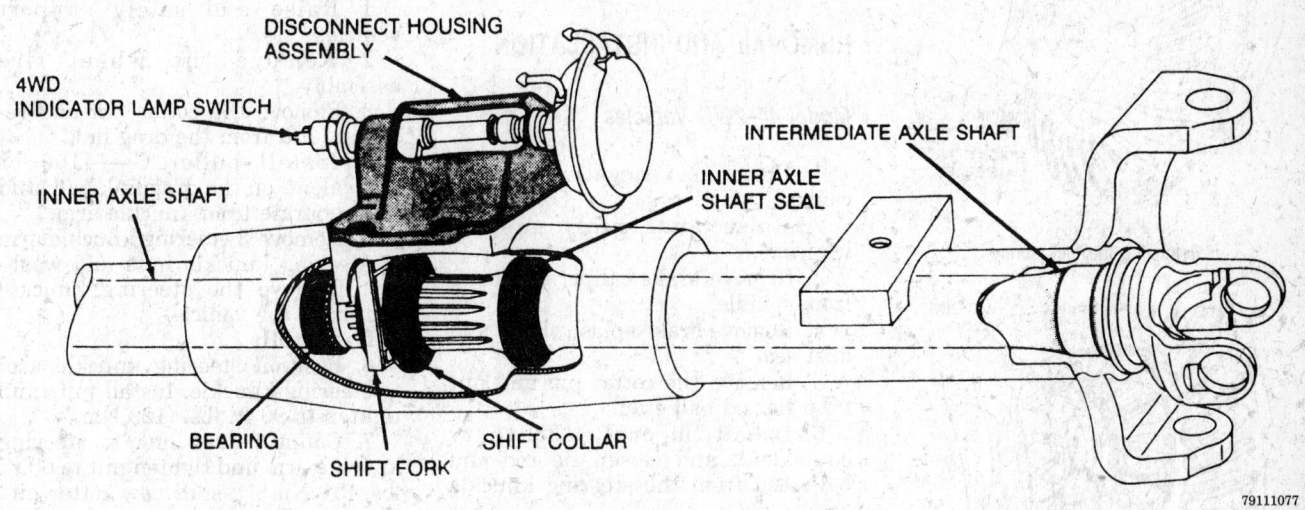

Model 44 left side front axle

4WD INDICATOR LAMP SWITCH

DISCONNECT HOUSING ASSEMBLY

INNER AXLE SHAFT

BEARING

SHIFT FORK

SHIFT COLLAR

INNER AXLE SHAFT SEAL

INTERMEDIATE AXLE SHAFT

79111077

spindle. Install the outer wheel bearing and inner locknut.

15. Install the locknut nut and tighten it with 50 ft. lbs. (68 Nm) of torque. Loosen the nut and tighten with 35 ft. lbs. (48 Nm) of torque. Loosen the adjusting nut about ⅜ turn. Install the lock ring and outer locknut. Install and tighten the locknut with 65 ft. lbs. (88 Nm) of torque. Bend one tang over each of the locknuts. The final bearing end-play should be 0.001–0.010 in.

16. Install a new gasket on the hub. Install the drive flange and torque the nuts to 35 ft. lbs. (48 Nm). Install the snapring.

17. Position the locking hub shift knob on its base. Align the splines by pushing inward on the shift knob and rotating it clockwise to lock it in place.

18. Install and tighten the 3 screws.

19. Install the brake components and wheel.

20. Road test the vehicle and check for leaks.

Rear Axle Shaft, Bearing and Seal

REMOVAL AND INSTALLATION

Chrysler 8⅜ in. and 9¼ in.

1. Raise the vehicle and support safely.

2. Remove the wheel and brake drum.

3. Remove the differential housing cover.

4. Rotate the differential case as required to expose the lock screw and remove it. Remove the pinion mate gear shaft from the case.

5. Force the axle shaft toward the center of the vehicle and remove the axle shaft C–clip lock from the recessed groove in the axle shaft.

6. Remove the axle shaft from the axle housing.

7. Pry the axle shaft seal from the end of the axle tube using a prybar.

8. To remove the bearing from an 8⅜ in. rear, use removal tool C–637 attached to a slide hammer. To remove the bearing from an 9¼ in. rear, use removal tool C–4828 attached to a slide hammer.

To install:

9. Clean the bearing bore in the axle tube.

10. Insert the new axle shaft bearing onto the pilot of tool C–4198 for 8⅜ in. rear or tool C–4826 for 9¼ in. rear. The bearing is fully installed when it is seated firmly against the shoulder in the axle tube.

11. Install the new seal to the axle tube.

12. Lubricat the bearing bore and seal lip with grease and insert the axle shaft into the axle tube engaging its splines with the differential side gear splines.

13. Install the C–clip lock in the groove at the end of the shaft. Force the shaft outward to seat the C–clip.

14. Insert the differential pinion gear mate shaft into the case and through the thrust washers and pinion gears. Align the hole in the shaft with the lock screw hole in the differential case and install the lock screw. Torque the screw to 14 ft. lbs. (19 Nm).

15. Thoroughly clean and dry the case cover, mating surface, bolts and bolt holes. Apply silicone sealer to the cover and install.

16. Install the drum and wheel.

17. Fill the differential with the proper lubricant.

18. Road test the vehicle and check for leaks.

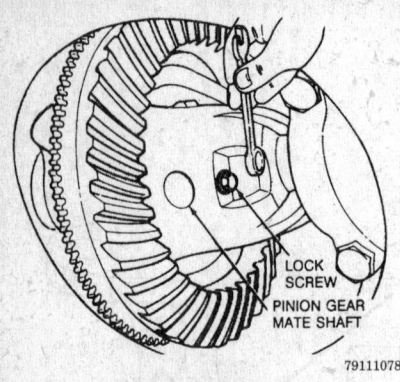

Removing the lock screw

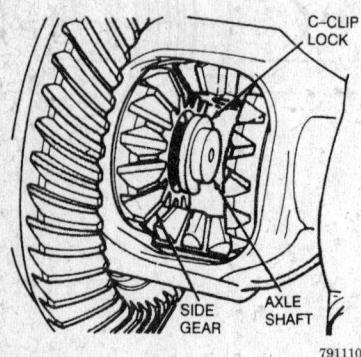

Removing the C-clip lock

Model 60, 60M and 70

1. Raise the vehicle and support safely.

2. Remove the axle flange lock bolts or nuts.

3. Remove the axle shaft.

4. Remove the locknut and remove the special adjustment nut.

5. Remove the outer bearing, brake drum and inner bearing. Remove the inner seal from the drum.

To install:

6. Lubricate and install the inner bearing to the drum and install a new seal.

7. Install the drum to the axle housing. Install the outer bearing to the drum.

8. Tighten the adjustment nut with 130 ft. lbs. (175 Nm) while rotating the wheel.

9. Loosen the adjustment nut $1/3$ — turn to provide about — 0.005 — inch of bearing endplay. Install the locknut.

10. Install the axle with a new flange gasket.

11. Install the axle flange bolts and torque to 70 ft. lbs. (95 Nm).

Front Wheel Knuckle/Spindle and Bearings

REMOVAL AND INSTALLATION

Model 44-2WD Vehicles

1. Disconnect negative battery cable.

2. Raise and safely support vehicle.

3. Remove brake caliper and rotor from spindle.

4. Remove brake splash shield and dust seal.

5. Remove the cotter pin and nut from tie rod ball stud.

6. Install puller C — 3894A or equivalent, and loosen tie rod end ball stud from the steering knuckle arm.

7. Remove the shock absorber from vehicle.

8. Install spring compressor tool DD — 1278 or equivalent, in the spring and snug nut by hand.

9. Remove cotter pins and lower and upper ball stud nuts at the steering knuckle.

10. Loosen upper and lower joint studs from the steering knuckle and remove ball studs. Slowly loosen tension on spring until spring is in a relaxed position.

11. Remove steering knuckle from vehicle. Remove the brake adapter and steering knuckle attaching bolts and separate knuckle from the steering arm.

To install:

12. Install brake adapter on steering knuckle and tighten to 100 ft. lbs. (136 Nm).

13. Install steering knuckle arm to steering knuckle bolts and torque to 215 ft. lbs. (291 Nm).

14. Mount the knuckle to the suspension arms and torque upper nut to 105 ft. lbs. (142 Nm). Torque lower nut as follows:

    a. If $11/16$–16 bolt to 135 ft. lbs. (183 Nm). Install cotter pin.

    b. If $3/4$–16 bolt to 175 ft. lbs. (237 Nm). Install cotter pin.

15. Connect the tie rod end. Install nut and tighten to 45 ft. lbs. (61 Nm). Install cotter pin.

16. Install new dist seal on steering knuckle. Install splash shield.

17. Install shock absorber. Install rotor and brake caliper.

18. Install wheel and tire assembly and reconnect negative battery cable.

19. Road test vehicle for proper operation.

Model 44 and 60–4WD Vehicles

1. Raise and safely support vehicle.

2. Remove the wheel tire assembly.

3. Remove the cotter pin and retaining nut from the drag link.

4. Install puller C — 4150 or equivalent, on the drag link ball stud and separate from knuckle arm.

5. Remove 3 steering knuckle arm to steering knuckle nuts and washers. Remove the steering knuckle arm from the vehicle.

To install:

6. Position steering knuckle arm to steering knuckle. Install nuts and tighten to 90 ft. lbs. (122 Nm).

7. Connect drag link to steering knuckle arm and tighten nut to 60 ft. lbs. (81 Nm). Install new cotter pin. Lower vehicle.

Manual Locking Hubs

REMOVAL AND INSTALLATION

1. Raise the vehicle and support safely.

2. Remove the wheel and remove the brake caliper and pads from the rotor. Do not allow the caliper to hang by the hose.

3. Turn the shift knob to the **ENGAGE** position.

4. Apply pressure to the face of the shift knob and remove the 3 retaining screws located nearest to the flange. Pull outward and remove the shift knob from the base.

5. Remove the snapring from the axle shaft.

6. Remove the capscrews and lockwashers from the base flange.

7. Remove the locking hub from the rotor/hub. Remove and discard the gasket.

To install:

8. Install a new gasket on the hub. Install the drive flange and torque the nuts to 35 ft. lbs. (48 Nm). Install the snapring.

9. Position the locking hub shift knob on its base. Align the splines by pushing inward on the shift knob and rotating it clockwise to lock it in place.

10. Install and tighten the 3 screws.

11. Install the brake components and wheel.

12. Road test the vehicle and check for leaks.

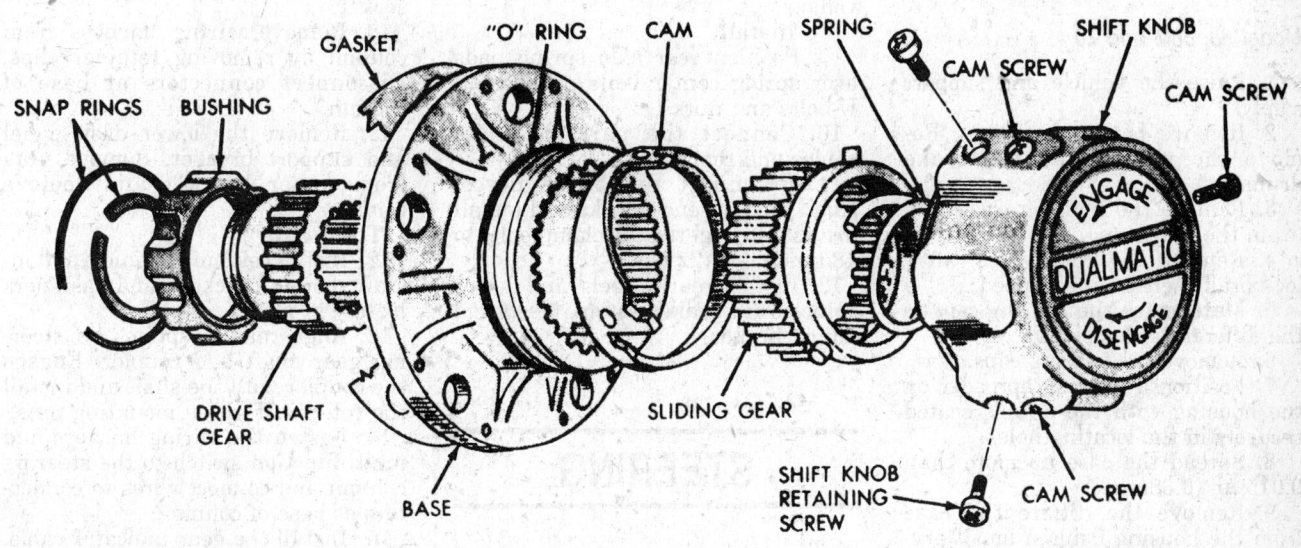

SNAP RINGS   BUSHING   GASKET   "O" RING   CAM   SPRING   CAM SCREW   SHIFT KNOB   CAM SCREW

DRIVE SHAFT GEAR   BASE   SLIDING GEAR   SHIFT KNOB RETAINING SCREW   CAM SCREW

ENGAGE   DUALMATIC   DISENGAGE

79111080

Dualmatic locking hub

Pinion Seal

REMOVAL AND INSTALLATION

1. Raise the vehicle and support safely.
2. Matchmark and remove the driveshaft.
3. Remove the rear wheel and brake drums to prevent any drag.
4. Using an inch lb. torque wrench, measure the pinion bearing preload. Read the torque while the handle of the wrench is moving through several complete revolutions.
5. Using the proper tools, hold the companion flange and remove the drive pinion nut and washer.
6. Remove the companion flange using tool C–452 or equivalent. Lower the rear of the vehicle to prevent fluid loss.
7. Using a seal remover tool, remove the seal from the carrier and clean the seal seat.
To install:
8. The outside diameter of the seal is precoated with a special sealer so no sealing compound is required for installation. The seal is properly installed when the flange contacts the housing flange face.

9. Install the companion flange and the washer with the convex side out.
10. For 8⅜ and 9¼ rears, tighten the pinion nut to 210 ft. lbs. (285 Nm) and check the pinion bearing preload. If the preload is less than the original preload measured, continue tightening the nut in very small increments until the proper preload is reached.
11. For 60, 60M and 70 rears, torque the pinion nut to 260 ft. lbs. (350 Nm).
12. Install the driveshaft, drums and rear wheels.
13. Refill the differential with the proper lubricant.
14. Road test the vehicle.

Differential Case

REMOVAL AND INSTALLATION

Chrysler 8⅜ in. and 9¼ in.

1. Raise the vehicle and support safely.
2. Remove the wheels and the brake drums.
3. Remove the rear housing cover and drain the lubricant.
4. Remove the rear wheel anti-lock brake sensor, if equipped.
5. Remove both axle shafts.

6. Matchmark the bearing caps to the differential housing.
7. Remove the differential bearing threaded adjuster lock from each cap.
8. Loosen but do not remove the bearing caps.
9. Loosen the side adjusters using tool C–4164.
10. Remove the bearing caps, the threaded adjusters and the differential case.
To install:
11. Position the assembled differential case in the housing.
12. Install the bearing caps in their original positions according to the matchmarks made during the disassembly.
13. Torque the upper bolts to 10 ft. lbs. (14 Nm) and finger-tighten the bottom bolts.
14. Tighten the side adjusters until the proper backlash specifications are reached with each adjuster tightened to 10 ft. lbs. (14 Nm).
15. Torque the bearing caps bolts to 70 ft. lbs. (95 Nm) for 8⅜ in. rears or 100 ft. lbs. (136 Nm) for 9¼ in. rears.
16. Install threaded adjuster locks. Tighten lock screw to 90 inch lbs. (10 Nm) torque. Check and adjust side clearance, if necessary.
17. Install both axle shafts.
18. Install the rear wheel anti-lock brake sensor, if equipped.

19. Install the housing cover and fill with the proper lubricant.
20. Install the drums and wheels.
21. Road test the vehicle.

Model 60, 60M and 70

1. Raise the vehicle and support safely.
2. Remove both axle shafts. Remove the wheels and the brake drums.
3. Remove the housing cover and drain the lubricant.
4. Remove the rear wheel anti-lock brake sensor, if equipped.
5. Matchmark the bearing caps to the differential housing.
6. Remove the bearing caps.
7. Position a housing spreader on the housing with the dowels seated securely in the locating holes.
8. Spread the case no more than 0.015 in. (0.38mm).
9. Remove the differential case from the housing using a small prying tool, if necessary.
To install:
10. Spread the housing and install the assembled case.
11. Install the bearing caps in their original positions according to the matchmarks made during the disassembly.
12. Torque the cap bolts to 85 ft. lbs. (115 Nm).
13. Check and adjust all measurements to specifications.
14. Install the rear wheel anti-lock brake sensor, if equipped.
15. Install the housing cover and fill with the proper lubricant. Include hypoid gear lubricant if the differential is a TraC–Lok.
16. Install the brake drums and wheels.
17. Install both axle shafts.
18. Road test the vehicle.

Axle Housing

REMOVAL AND INSTALLATION

1. Disconnect the negative battery cable. Raise vehicle and support safely.
2. Remove the rear wheel anti-lock brake sensor, if equipped.
3. Remove the rear wheels.
4. Disconnect the brake hose at the T-fitting.
5. Disconnect the parking brake cables.
6. Matchmark and remove the driveshaft.
7. Support the weight of the assembly with the proper equipment.

Disconnect the shock absorbers and remove the leaf spring nuts and U-bolts.
8. Remove the assembly from vehicle.
To install:
9. Position rear axle spring pads over spring center bolts and install U-bolts and nuts.
10. Connect the parking brake cables and rear shocks.
11. Reconnect the brake hoses. Align the reference marks and install driveshaft. Tighten the clamp bolts to 185 inch lbs. (22 Nm).
12. Install rear wheels and lower vehicle. Fill brake system, bleed and adjust brakes.

STEERING

Steering Wheel

REMOVAL AND INSTALLATION

1. Disconnect the negative battery cable.
2. Remove the horn pad.
3. Remove the steering wheel hold-down nut. Matchmark the steering wheel to the shaft.
4. Using a steering wheel puller, pull the steering wheel off the shaft.
To install:
5. Install steering wheel onto shaft and torque to 45 ft. lbs. (61 Nm).
6. Install horn pad and connect negative battery cable.

Steering Column

REMOVAL AND INSTALLATION

1. Disconnect the negative battery cable.
2. Remove the horn pad.
3. Remove the steering wheel hold-down nut. Matchmark the steering wheel to the shaft.
4. Using a steering wheel puller, pull the steering wheel off the shaft.
5. Matchmark column shaft to coupler and remove steering coupling screws.
6. Remove lower dash cover and fuse block cover.
7. If equipped with column shift, disconnect the link rod from shift lever. Position shift lever in **P** and remove indicator cable.

8. Remove tilt lever, if equipped.
9. Remove upper and lower shrouds. Remove the turn signal multi-function switch with 7 mm socket.
10. Remove wiring harness from column by removing retainer clips. Disconnect connectors at base of column.
11. Remove the lower dash panel and support bracket. Remove nuts from upper bracket and remove column.
To install:
12. Install column in vehicle and install mount brackets and fasteners loosely in place.
13. Align master splines on steering gear shaft and coupler. Engage the coupler with the shaft and install the roll pin. Tighten mounting nuts.
14. Fasten the wiring harness and multi-function switch to the steering column and connect wires to connectors at base of column.
15. Install the gear indicator cable. Install lower fixed shroud.
16. Install the lock housing shrouds and tilt lever, if equipped.
17. Install steering wheel to shaft and tighten to 45 ft. lbs. (61 Nm).
18. Install horn pad and connect negative battery cable.
19. Check operation of multi-function switch and any other related switches.

Manual Steering Gear

REMOVAL AND INSTALLATION

1. Disconnect the negative battery cable.
2. Remove the 2 bolts from the wormshaft to steering shaft coupler.
3. Raise the vehicle and support safely.
4. Matchmark and remove the pitman arm from the pitman shaft using tool C–4150 or equivalent.
5. Remove the steering gear mounting bolts and remove the gear from the vehicle.
6. The installation is the reversal of the removal procedure. Torque the pitman arm nut to 175 ft. lbs. (237 Nm).

ADJUSTMENT

Wormshaft Preload Torque

1. Raise the vehicle and support safely.
2. Remove the pitman arm from the pitman shaft.
3. Remove the horn pad.

4. Loosen the sector shaft adjusting screw locknut and back off the adjusting screw about 1½ turns.

5. Turn the steering wheel to the right stop and then back ½ turn. Measure the torque required to turn the steering while back to the straight ahead position. The specification is 4–6 inch lbs.

6. If not within specifications, loosen the large adjustment cap locknut and turn the adjustment cap until the proper preload is reached. Turning the adjuster clockwise increases the preload torque.

7. Tighten the locknut and recheck the preload.

8. Tighten the sector shaft adjuster screw locknut.

Sector Shaft

1. Perform the wormshaft preload procedure.

2. Center the steering wheel.

3. Loosen the sector shaft adjuster screw locknut and screw the adjuster screw all the way down. Tighten the locknut.

4. Rotate the steering wheel ¼ turn away from the overcenter position. Measure the torque required to rotate the wheel past the overcenter position. The specification is 14 inch lbs.

5. If not within specifications, adjust the screw accordingly and tighten the locknut.

6. Install the horn pad.

7. Install the pitman arm.

Power Steering Gear

ADJUSTMENT

1. If the vehicle wanders of the steering has too much play, the sector shaft can be adjusted.

2. Loosen the adjusting screw locknut and turn the screw all the down.

3. Back the screw off ¼–½ turn.

4. Tighten the locknut.

5. Road test the vehicle. If the steering wheel does not return easily after a turn, back the screw off until the wheel returns easily.

REMOVAL AND INSTALLATION

1. Place the wheels in the straight ahead position.

2. Remove the windshield washer solvent reservoir and the coolant overflow tank, if necessary.

3. Position a drain pan under the steering gear.

4. Disconnect the fluid hoses from the gear and plug them.

5. Disconnect the steering column shaft from the stub shaft.

6. Raise the vehicle and support safely. Matchmark and remove the pitman arm from the center link on 2WD vehicles. On 4WD vehicles, disconnect the drag link from the pitman arm. Remove the pitman arm from the pitman shaft.

7. Remove the retaining bolts and remove the steering gear from the vehicle.

To install:

8. For 2WD vehicles, position steering gear at frame rail and install bolts loosely. Align steering shaft and stub shaft and install bolts to 33 ft. lbs. (45 Nm) torque. Realign gear at frame and torque bolts to 100 ft. lbs. (136 Nm).

9. On 4WD vehicles, install steering gear to reinforcement and tighten screws to 100 ft. lbs. (136 Nm). Position steering gear at frame rail and install bolts loosely. Align steering shaft and stub shaft and install bolts to 33 ft. lbs. (45 Nm) torque. Realign gear at frame and torque bolts to 100 ft. lbs. (136 Nm).

10. Install pitman arm to steering shaft and torque nut to 175 ft. lbs. (237 Nm). Connect steering linkage to arm. Install replacement cotter pins.

Power Steering Pump

REMOVAL AND INSTALLATION

Gasoline Engine

1. Disconnect the negative battery cable.

2. Position a drain pan under the power steering pump.

3. Disconnect the fluid hoses from the pump and plug them.

4. Remove the front bracket attaching bolts and remove the belt from the pulley.

5. Remove the rear pump to bracket nut and remove the pump.

6. Remove the bracket from the pump.

7. Remove the pulley from the pump with the proper puller. Install the pulley on the new pump using the special installation tools.

To install:

8. Position pump in rear bracket and install retaining bolts.

9. Install pump to engine and install retainer bolts. Attach fluid lines.

10. Tighten all mounting bolts. Add power steering fluid to reservoir. Install drive belt to correct tension.

Diesel Engine

1. Disconnect the negative battery cable.

2. Position a drain pan under the power steering pump.

3. Disconnect and cap vacuum and steering pump hoses.

4. Disconnect oil pressure sender electrical connector and remove sender from block.

5. Disconnect and cap the oil feed from bottom of vacuum pump.

6. Remove lower vacuum/steering pump mounting bolts, gasket and pump from the engine.

7. Remove the steering pump to vacuum pump bracket attaching nuts and slide the steering pump from bracket.

To install:

8. Install body spacers to pump and install pump into bracket. The steering pump and spacer must mate completely with vacuum pump bracket.

9. Install new gasket to pump assembly using sealer to retain the gasket. Install pump assembly to engine.

10. Install oil pressure sending unit and electrical connector.

11. Install oil feed line and vacuum hoses to vacuum pump. Install fluid hoses to power steering pump.

12. Fill reservoir with power steering fluid. Connect negative battery cable, start engine and check for leaks.

Belt Adjustment

1. Loosen the bracket mounting bolts.

2. Using a ½ in. drive breaker bar in the square hole provided in the bracket, move the pump away from the engine. Do not pry against the fluid reservoir.

3. With the pump moved enough so the belt deflects about ¼–½ in. under a 10 lb. load, tighten the bolts.

System Bleeding

1. Fill the reservoir with power steering fluid.

2. Turn the wheels to the full left turn position and add fluid until the reservoir is full.

3. Start the engine and add fluid to bring the level to the correct level.

4. To purge the system of air, turn the steering wheel from side to side without contacting the stops.

5. Return the wheel to the straight ahead position and operate the engine for 2 minutes before road testing. This should bleed the system completely.

Tie Rod Ends

REMOVAL AND INSTALLATION

1. Raise the vehicle and support safely.
2. Remove the cotter pin and nut from the tie rod end.
3. Using a puller, remove the tie rod from the steering knuckle or center link.
4. Loosen the sleeve clamp nut and bolt and unscrew the tie rod end from the sleeve.
5. The installation is the reversal of the removal procedure. Torque the stud nuts to 45 ft. lbs. (61 Nm) and install a new cotter pin.
6. Perform a front end alignment as required to adjust toe-in.

BRAKES

For all brake system repair and service procedure not detailed below, please refer to "Brakes" in the Unit Repair section.

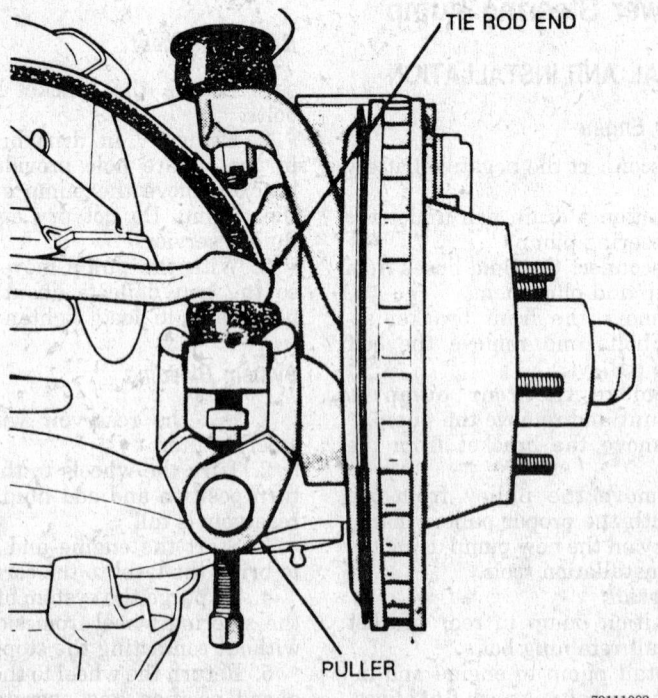

Removing the tie rod end

Master Cylinder

REMOVAL AND INSTALLATION

1. Disconnect the negative battery cable.
2. Disconnect and plug the brake lines from the master cylinder.
3. Remove the nuts attaching the master cylinder to the power booster.
4. Remove the master cylinder from the mounting studs.
To install:
5. Bench bleed the master cylinder.
6. Install to the studs and install the nuts.
7. Install the brake lines to the master cylinder loosely.
8. Slowly push brake pedal to the floor and hold in this position while tightening brake lines at master cylinder. Refill master cylinder and check for leaks and proper pedal resistance.

Proportioning/Combination Valve

REMOVAL AND INSTALLATION

1. Disconnect the negative battery cable.
2. Raise the vehicle and support safely.
3. Tag and disconnect the brake lines from the valve.
4. Disconnect the wires to the pressure switch.
5. Remove the combination valve from the frame bracket.
6. The installation is the reversal of the removal procedure.
7. Bleed the brakes in the following order:
   a. Rear Wheel Anti-Lock valve
   b. Right rear wheel cylinder
   c. Left rear wheel cylinder
   d. Right front caliper
   e. Left front caliper

Power Brake Booster

REMOVAL AND INSTALLATION

1. Disconnect the negative battery cable. Disconnect the vacuum hose(s) from the booster.
2. Remove the nuts attaching the master cylinder to the booster and move the master cylinder aside.
3. From inside of the vehicle, remove the clip that secures the booster pushrod to the brake pedal.
4. Remove the nuts that attach the booster to the dash panel and remove it from the vehicle.
5. Transfer the check valve to the new booster.
To install:
6. Position booster on dash panel and install retainer nuts.
7. Install booster pushrod on brake pedal and install new retainer clip.
8. Install brake master cylinder to booster and secure. Install vacuum hose to booster check valve.

Brake Caliper

REMOVAL AND INSTALLATION

1. Raise the vehicle and support safely. Remove the tire and wheel assembly.
2. Remove the caliper retaining clips and anti-rattle springs.
3. Lift the caliper off of the rotor. Remove the outer pad from the caliper.

4. Remove the brake hose retaining bolt from the caliper.

To install:

5. Install the brake hose to the caliper using new copper washers.

6. Adjust the ears of the outer pad to provide a tight fit in the caliper recesses.

7. Position the caliper over the rotor so the caliper engages the adaptor correctly.

8. Install the anti-rattle springs and retaining clips.

9. Fill the master cylinder and bleed the brake system.

Disc Brake Pads

REMOVAL AND INSTALLATION

1. Remove some of the fluid from the master cylinder. Raise the vehicle and support safely. Remove the tire and wheel assemblies.

2. Remove the caliper and remove the outer pad from the caliper.

3. Remove the inner pad from the adaptor.

To install:

4. Use a large C–clamp to compress the piston back into the caliper bore.

5. Adjust the ears of the outer pad to provide a tight fit in the caliper recesses.

6. Install the inner pad to the adaptor.

7. Position the caliper over the rotor so the caliper engages the adaptor correctly.

8. Install the anti-rattle springs and retaining bolt(s).

9. Install tire and wheel assembly. Refill the master cylinder. Pump brake pedal before moving the vehicle.

Brake Rotor

REMOVAL AND INSTALLATION

Chrysler Disc Brake

1. Raise the vehicle and support safely.

2. Remove the wheel.

3. Remove the caliper and disc brake pads. It is not necessary to remove the brake line from the caliper

4. Remove the dust cap.

5. Remove the cotter pin, locknut, wheel bearing nut and washer from the spindle.

6. Remove the outer wheel bearing.

7. Remove the rotor with the inner wheel bearing from the spindle. Remove the grease seal.

To install:

8. Lubricate and install the inner wheel bearing. Install a new grease seal.

9. Install the rotor to the spindle.

10. Lubricate and install the outer wheel bearing, washer and nut. When the bearing preload is properly set, install the nut lock and a new cotter pin.

11. Install the grease cap.

12. Install the brake pads and caliper.

13. Install the wheel. Pump brake pedal before moving vehicle.

Bendix Disc Brake

1. Raise and safely support vehicle.

2. Remove wheel and tire assembly.

3. Remove support key retaining screw and support key. Remove caliper from adapter.

4. Remove hub cap. Remove drive flange snapring, flange nuts and lockwashers. Remove drive flange and gasket.

5. Straighten tang on lockring. Remove locknut, lockring, inner adjusting nut and bearing.

6. Remove hub and rotor from spindle.

To install:

7. Repack wheel bearings and install in hub. Install hub and rotor on spindle.

8. Install outer bearing and adjuster nut and tighten to 50 ft. lbs. (68 Nm). Back off adjusting nut and retorque to 40 ft. lbs. (54 Nm) while rotating wheel. Back off adjuster nut 3/8 turn. Install lockring and nut. Bearing endplay should be 0.001–0.010 in.

9. Bend 1 tang of lock-ring over locknut and 1 tang over adjusting nut.

10. Install hub with new gasket, drive flange, lockwashers and nuts. Tighten nuts to 35 ft. lbs.

11. Install flange snapring and hub cap. Install caliper, brake shoes and wheel assembly. Pump brake pedal before moving vehicle.

Brake Drums

REMOVAL AND INSTALLATION

Except Dana Axle

1. Raise the vehicle and support safely.

2. Remove the wheel.

3. Remove the factory clips from the wheel studs, if equipped.

4. Remove the drum. If the drum is difficult to remove, remove the plug from the rear of the backing plate and push the self adjuster lever away from the star wheel. Rotate the star wheel to retract the shoes.

5. The installation is the reverse of the removal procedure.

Dana Axle

1. Raise the vehicle and support safely.

2. Remove the axle shafts.

3. Remove the bearing adjuster nut and the outer bearing.

4. Remove the drum. If the drum is difficult to remove, remove the plug from the rear of the backing plate and push the self adjuster lever away from the star wheel. Rotate the star wheel to retract the shoes.

5. The installation is the reversal of the removal procedure.

Brake Shoes

REMOVAL AND INSTALLATION

11 inch Brake Drum

1. Raise the vehicle and support safely. Remove the wheels and drums. Remove the primary and secondary shoe return springs from the anchor pin.

2. Lift the adjuster lever and disconnect the actuator cable.

3. Remove the shoe retainers and springs.

4. Remove the shoes (held together by the lower spring) while separating the parking brake actuating lever from the shoe with a twisting motion.

To install:

5. Thoroughly clean and dry the backing plate. To prepare the backing plate, lubricate the bosses, anchor pin and parking brake actuating lever pivot surface lightly with lithium based grease.

6. Remove, clean and dry all parts still on the old shoes. Lubricate the star wheel shaft threads with antisieze lubricant and transfer all parts to their proper locations on the new shoes.

7. Spread the shoes apart, engage the parking brake lever and position them on the backing plate so the wheel cylinder pins engage and the anchor pins hold the shoes.

8. Install the parking brake strut and hold-down spring assemblies.

9. Install the anchor plate. Lubricate the sliding surface of the actuator cable plate lightly and install the cable.

10. Install the shoe return spring opposite the cable, then install the remaining spring.

11. Adjust the star wheel.

12. Remove any grease from the linings and install the drum.

13. Complete the brake adjustment with the wheels installed.

12 inch Brake Drum

1. Raise the vehicle and support safely. Remove the axles and drums.

2. Unhook the adjuster lever return spring from the lever.

3. Remove the lever and return spring from the lever pin.

4. Unhook the adjuster cable from the lever.

5. Remove the shoe to shoe upper spring.

6. Remove the shoe hold-down springs.

7. Disconnect the parking brake cable from the parking brake lever.

8. Remove both brake shoes, the lower spring and star wheel assembly.

To install:

9. Thoroughly clean and dry the backing plate. To prepare the backing plate, lubricate the bosses, anchor pin and parking brake actuating lever pivot surface lightly with lithium based grease.

10. Remove, clean and dry all parts still on the old shoes. Lubricate the star wheel shaft threads with antisieze lubricant and transfer all parts to their proper locations on the new shoes. Install the assemblies to the backing plate.

11. Install the shoe hold-down springs and pins.

12. Connect the parking brake cable to the lever.

13. Install the upper spring.

14. Position the adjuster lever return spring on the pin. Install the adjuster lever and attach the cable.

15. Adjust the star wheel.

16. Remove any grease from the linings and install the drum.

17. Complete the brake adjustment with the wheels (but not the axles) installed.

18. Install the axles.

Wheel Cylinder

REMOVAL AND INSTALLATION

1. Raise the vehicle and support safely.

2. Remove the wheel, drum and brake shoes.

3. If equipped with a 12 in. drum, remove the anchor bolt and nut, washer, spring, parking brake lever, adjuster cable, cam plate and anchor spring bushing.

4. Remove the brake line from the wheel cylinder.

5. Remove the wheel cylinder bolts and remove the cylinder from the backing plate.

To install:

6. Install cylinder to backing plate and install bolts. Install brake line to cylinder.

7. If equipped with a 12 in. drum, install the anchor bolt and nut, washer, spring, parking brake lever, adjuster cable, cam plate and anchor spring bushing.

8. Install brake shoes, drum and wheel.

9. Bleed wheel cylinder and correct brake fluid level.

Parking Brake Cable

ADJUSTMENT

1. Release the parking brakes fully.

2. Raise the vehicle and support safely.

3. Adjust the rear brakes.

4. Loosen the nut on the front cable until there is slack in all the cables.

5. Rotate the rear wheels and tighten the cable adjusting nut until there is a slight drag at the wheels.

6. Continue to rotate the rear wheels and loosen the nut until all drag is eliminated.

7. Back off the nut an additional 2 turns.

8. Apply and release the parking brake several times. Upon the last release, verify that there is no drag at the rear wheels.

9. To check the operation, make sure the parking brake holds on an incline.

REMOVAL AND INSTALLATION

Front Brake Cable

1. Raise the vehicle and support safely.

2. Remove the front cable adjusting nut.

3. Remove the clip securing the cable to the anchor bracket and slide the cable out of the bracket.

4. Remove the retainer attaching the cable to the pedal assembly

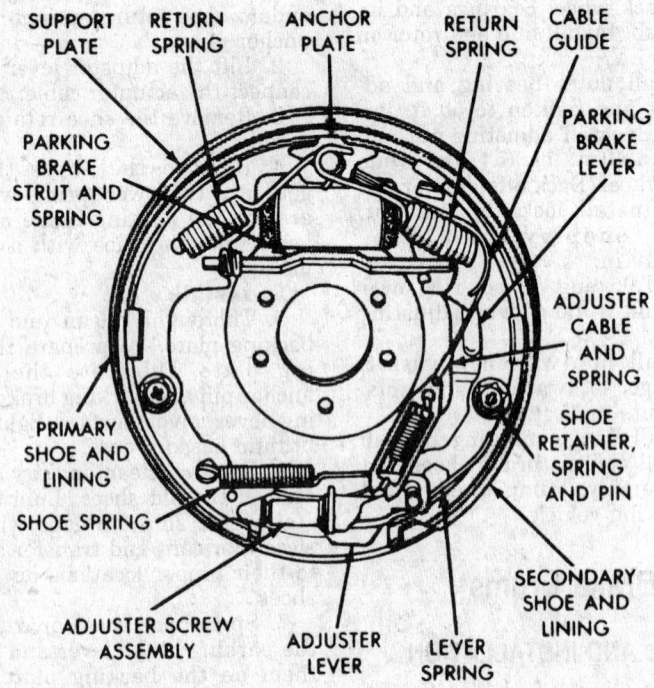

SUPPORT PLATE — RETURN SPRING — ANCHOR PLATE — RETURN SPRING — CABLE GUIDE — PARKING BRAKE LEVER — PARKING BRAKE STRUT AND SPRING — ADJUSTER CABLE AND SPRING — PRIMARY SHOE AND LINING — SHOE RETAINER, SPRING AND PIN — SHOE SPRING — SECONDARY SHOE AND LINING — ADJUSTER SCREW ASSEMBLY — ADJUSTER LEVER — LEVER SPRING

79111083

11 in. drum brakes

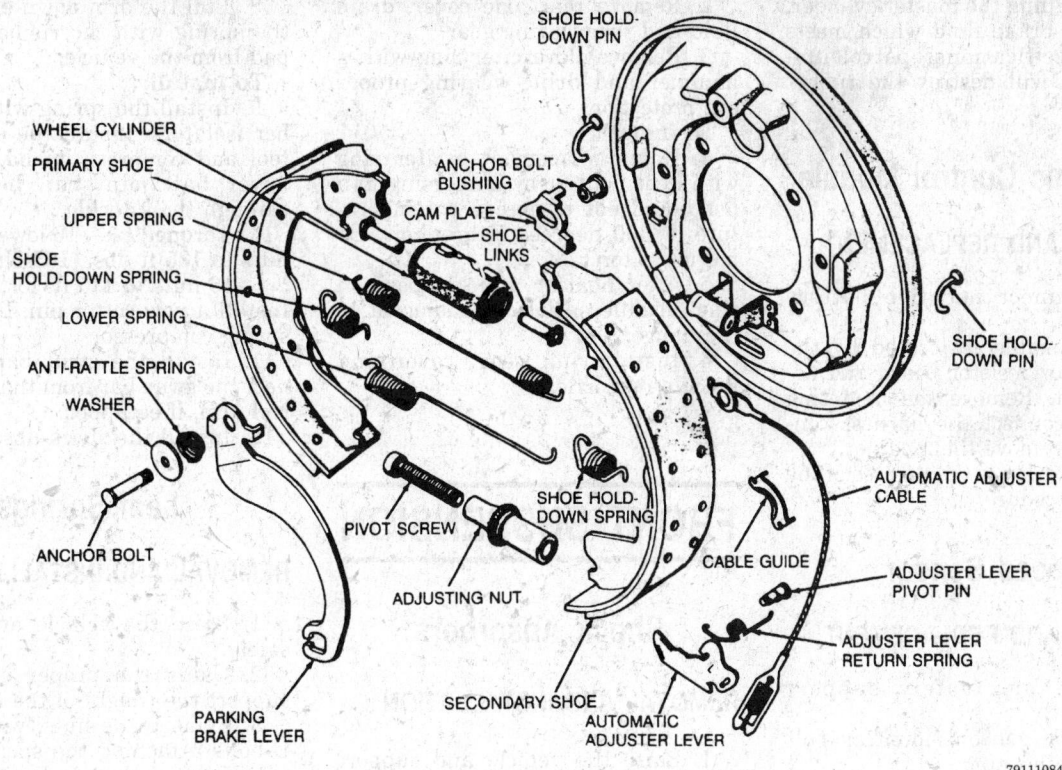

12 in. drum brakes

frame. Disengage the cable from the pedal clevis.

5. Remove the cable grommet from the floor pan and remove the cable.

6. The installation is the reversal of the removal procedure.

Rear Brake Cable

1. Release the parking brake fully.
2. Raise the vehicle and support safely.
3. Remove the adjusting nut from the front cable.
4. Remove the brake drums. Remove the shoes, if necessary. Disconnect the cable from the lever and compress the cable retainer tabs and remove the cable from the backing plate.
5. Remove the cable from the equalizer and ratio lever.
6. The installation is the reversal of the removal procedure.

Brake System Bleeding

1. If master cylinder service has been performed, bleed master cylinder on bench before installing on vehicle.

2. Clean master cylinder reservoir cap to prevent contamination. Open cap and fill with brake fluid.

3. If pressure bleeding, follow instructions with bleeder for preparation and operation, and bleed the system in the following order:
   a. Master cylinder
   b. RWAL hydraulic valve
   c. Right rear wheel
   d. Left rear wheel
   e. Right front wheel
   f. Left front wheel

4. If manual bleeding, follow the above order of bleeding.

5. Open the bleeder valve and have a helper depress brake pedal to the floor. While pedal is held on the floor, tighten bleeder and repeat procedure until no air is expelled from the bleeder. Check the fluid level in the master cylinder and repeat at next site until complete system is free of air.

6. If brake light on dash remains on or the pedal feels spongy, there is air in system. Repeat procedure as required.

Anti-Lock Brake System Service

PRECAUTIONS

Failure to observe the following precautions may result in system damage.

• Before performing electric arc welding on the vehicle, disconnect the Electronic Brake Control Module (EBCM) and the hydraulic modulator connectors.

• When performing painting work on the vehicle, do not expose the Electronic Brake Control Module (EBCM) to temperatures in excess of 185°F (85°C) for longer than 2 hrs. The system may be exposed to temperatures up to 200°F (95°C) for less than 15 min.

• Never disconnect or connect the Electronic Brake Control Module (EBCM) or hydraulic modulator connectors with the ignition switch ON.

• Never disassemble any component of the Anti-Lock Brake System (ABS) which is designated non-servicable; the component must be replaced as an assembly.

• When filling the master cylinder, always use brake fluid which meets DOT-3 specifications; petroleum-based fluid will destroy the rubber parts.

Electronic Control Module

REMOVAL AND REPLACEMENT

1. Disconnect negative battery cable.
2. The module is located by the blower motor resistor board and defroster duct. Remove the mounting screws, disconnect the harness connector and remove the module.
3. Installation is the reverse of the removal procedure.

Speed Sensor

REMOVAL AND REPLACEMENT

1. Raise and safely support vehicle.
2. Remove sensor mounting bolt on rear axle housing.
3. Remove sensor shield and sensor from housing. Disconnect sensor wire harness.
4. Installation is the reversal of the removal process.

Hydraulic Control Valve

REMOVAL AND INSTALLATION

1. Disconnect negative battery cable.
2. Raise and safely support vehicle.
3. Disconnect valve to sensor harness connector.
4. Disconnect brake lines connecting valve to rear brakes and to combination valve.
5. Remove valve attaching screws and remove valve.
To install:
6. Install brake lines into hydraulic valve and install valve to frame bracket.
7. Bleed hydraulic valve and brake system. Add brake fluid as required.

Exciter Ring

REMOVAL AND INSTALLATION

1. Raise and safely support vehicle.

2. Remove rear axle cover, drain fluid and remove ring gear.
3. Remove old exciter ring with a hammer and drift, wearing proper eye protection.
To install:
4. Heat replacement exciter ring with heat light or by immersing in a hot liquid not to exceed a temperature of 300 degrees Fahrenheit. Do not use a torch.
5. After heating, quickly position ring on differential case adjacent to the flange.
6. Install ring gear, cover and fluid to rear axle.

FRONT SUSPENSION

Shock Absorbers

REMOVAL AND INSTALLATION

1. Raise the vehicle and support safely.
2. On 2WD vehicles, remove the upper shock nut, washer and bushing. Remove the lower mounting bolts and remove the shock from the vehicle.
3. On 4WD vehicles, remove the upper mounting nut, lower mounting stud and retainers. Remove the shock from the vehicle.
4. The installation is the reversal of the removal procedure.

Coil Springs

REMOVAL AND INSTALLATION

1. Raise the vehicle and support safely.
2. Remove the shock absorber.
3. Remove the strut bar and disconnect the sway bar from the lower control arm, if equipped.
4. Install spring compressor tool DD–1278 or equivalent, to the coil spring and tighten the nut finger-tight, then back off nut ½ turn.
5. Remove the cotter pin and lower ball joint nut.
6. Release the lower ball joint taper using ball stud loosening tool C–3564–A or equivalent.
7. Remove the tool and remove the ball stud from the control arm. Release the compressor tool from the coil spring.

8. Pull the arm down and remove the spring with the rubber isolation pad from the vehicle.
To install:
9. Install the spring with the rubber isolators. Install the compressor tool and compress it enough so the lower ball joint can be inserted through the knuckle.
10. Torque $^{11}/_{16}$ — 16 lower ball joint nuts to 135 ft. lbs. (183 Nm). Torque $^{3}/_{4}$ — 16 nuts to 175 ft. lbs. (237 Nm). Install a new cotter pin. Remove the spring compressor.
11. Install the strut bar and connect the sway bar from the lower control arm, if equipped.
12. Install the shock absorber.

Leaf Springs

REMOVAL AND INSTALLATION

1. Raise the vehicle and support safely.
2. Using the proper equipment, support the weight of the front axle.
3. Remove the nuts, washers and U-bolts attaching the springs to the axle housing. Remove the spring pad.
4. Remove the spring shackle bolts, shackle and spring front bolt.
5. Remove the spring from the vehicle.
6. The installation is the reverse of the removal procedure. Torque the U-bolt nuts to 110 ft. lbs. (149 Nm).

Upper Ball Joint

INSPECTION

To inspect the ball joints, unload the suspension. Upper ball joints on 2WD vehicles and any ball joint on 4WD vehicles should be replaced if any play exists at all.

REMOVAL AND INSTALLATION

2WD Vehicles

1. Raise the vehicle and support safely.
2. Position a support at the outer end of the lower control arm and lower the vehicle so the support compresses the coil spring.
3. Remove the tire and wheel assembly.
4. Release the upper ball joint taper using ball stud loosening tool C–3564–A or equivalent.
5. Unthread the ball joint from the control arm with tool C–3561 or equivalent.

6. The installation is the reversal of the removal procedure. Torque the ball joint itself to 125 ft. lbs. (169 Nm).

7. Torque the upper ball stud nut to 135 ft. lbs. (183 Nm).

4WD Vehicles

1. Raise the vehicle and support safely.

2. Remove the front axle shaft.

3. Disconnect the tie rod end from the steering knuckle. On the left side, disconnect the drag link ball stud from the steering knuckle.

4. On the left side, remove the nuts and washers from the steering knuckle arm and remove the arm and spring, if equipped, from the knuckle.

5. If equipped with a Model 44 front axle, remove the ball joint nuts and discard the lower nut. Use a brass drift and hammer to separate the steering knuckle from the axle tube yoke. Use tool C–4169 to remove the sleeve from the upper yoke arm.

6. Remove the snapring from the ball joint. Install the knuckle in a vise and use tools D–150–1, D–150–3 and C–4212–L or equivalent, to remove the ball joint from the knuckle.

7. If equipped with a Dana 60 front axle, remove the bolts from the

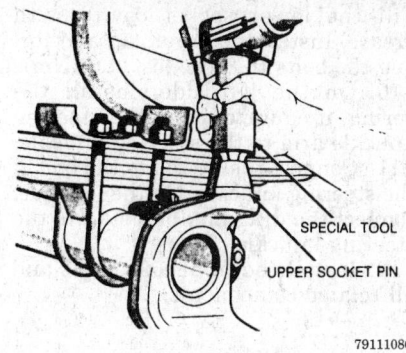

SPECIAL TOOL

UPPER SOCKET PIN

79111086

Removing or installing the upper socket pin

knuckle lower cap. Dislodge the cap from the steering knuckle and axle tube yoke. Remove the steering knuckle. Use tool D–192 or equivalent, to remove the upper socket pin from the axle tube upper arm bore. Remove the seal.

To install:

8. If equipped with a Model 44, use tools C–4212–L and C–4288 or equivalent, to force the upper ball joint into the steering knuckle. Install the snapring and install a new rubber boot. Thread the replacement sleeve into the upper yoke bore so 2 threads are exposed at the top of the yoke. Position the knuckle on the

axle tube yoke and install a new lower ball stud nut. Torque to 80 ft. lbs. (108 Nm). Using the special socket, torque the sleeve to 40 ft. lbs. (54 Nm). Install the upper ball stud nut and torque to 100 ft. lbs. (136 Nm) and install a new cotter pin.

9. If equipped with a Dana 60 front axle, use tool D–192 or equivalent, to install the upper socket pin to the axle tube upper arm bore. Install a new seal. Torque to 500–600 ft. lbs. (668–813 Nm). Position the knuckle over the socket pin. Fill the lower socket cavity with grease. Install the lower cap and torque the bolts to 80 ft. lbs. (110 Nm).

10. On the left side, install the spring, if equipped, and the steering knuckle arm to the steering knuckle.

11. Connect the tie rod to the end of the steering knuckle. On the left side, connect the drag link ball stud to the steering knuckle.

12. Install the front axle shaft and all related components.

Lower Ball Joint

INSPECTION

To inspect the ball joints, unload the suspension. Lower ball joints on 2WD vehicles should be replaced if the have more than 0.020 in. play. Any ball joint on 4WD vehicles should be replaced if any play exists.

REMOVAL AND INSTALLATION

2WD Vehicles

1. Raise the vehicle and support safely.

2. Remove the shock absorber.

3. Remove the strut bar and disconnect the sway bar from the lower control arm, if equipped.

4. Install spring compressor tool DD–1278 or equivalent, to the coil spring and tighten the nut finger-tight, then back off half a turn.

5. Remove the cotter pin and lower ball joint nut.

6. Release the lower ball joint taper using ball stud loosening tool C–3564–A or equivalent.

7. Remove the tool and remove the ball stud from the control arm. Release the compressor tool from the coil spring.

8. Pull the arm down and remove the spring with the rubber isolation pad from the vehicle. Remove the ball joint boot. Use tool C–4212 or a ball joint press, to remove the ball joint from the arm.

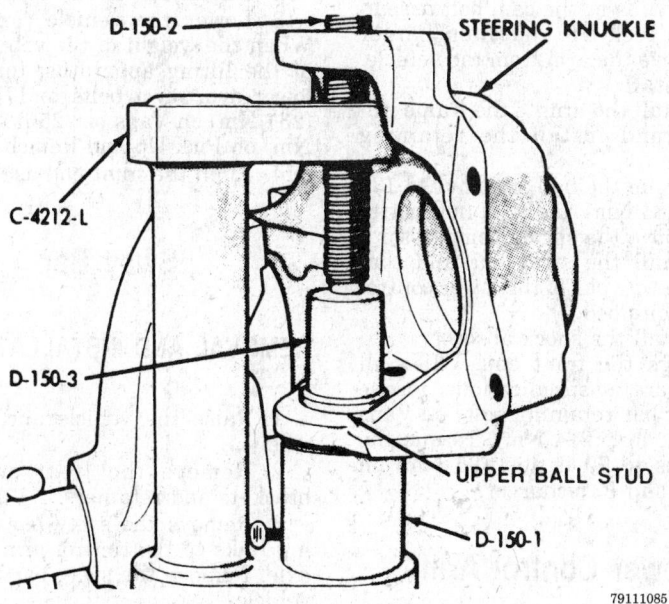

D-150-2

STEERING KNUCKLE

C-4212-L

D-150-3

UPPER BALL STUD

D-150-1

79111085

Removing the upper ball joint from the knuckle

To install:

9. Use the remover tool to press the ball joint into the arm. Install a new rubber boot. Install the spring with the rubber isolators. Install the compressor tool and compress it enough so the lower ball joint can be inserted through the knuckle.

10. Torque $^{11}/_{16}$ — 16 lower ball joint nuts to 135 ft. lbs. (183 Nm). Torque $^3/_4$ — 16 nuts to 175 ft. lbs. (237 Nm). Install a new cotter pin. Remove the spring compressor.

11. Install the strut bar and connect the sway bar from the lower control arm, if equipped.

12. Install the shock absorber.

4WD Vehicles

1. Raise the vehicle and support safely.

2. Remove the front axle shaft.

3. Disconnect the tie rod end from the steering knuckle. On the left side, disconnect the drag link ball stud from the steering knuckle.

4. On the left side, remove the nuts and washers from the steering knuckle arm and remove the arm and spring, if equipped, from the knuckle.

5. If equipped with a Model 44 front axle, remove the ball joint nuts and discard the lower nut. Use a brass drift and hammer to separate the steering knuckle from the axle tube yoke.

6. Remove the snapring from the ball joint. Install the knuckle in a vise and use tools D-150-1, D-150-3 and C-4212-L to remove the ball joint from the knuckle.

7. If equipped with a Dana 60 front axle, use tools C-4212-L, C-4366-1 and C-4366-2 to remove the lower ball joint.

To install:

8. If equipped with a Model 44, use tools C-4212-L and C-4288 or equivalent, to force the lower ball joint into the steering knuckle. Install the snapring and install a new rubber boot. Position the knuckle on the axle tube yoke and install a new lower ball stud nut. Torque to 80 ft. lbs. (108 Nm). Install the upper ball stud nut and torque to 100 ft. lbs. (136 Nm) and install a new cotter pin.

9. If equipped with a Dana 60 front axle, use tools C-4212-L, C-4366-3 and C-4366-4 to install the seal and lower bearing cup into the axle tube yoke lower bore. Reposition the tools and install the lower bearing and seal into the bore. Position the knuckle over the socket pin.

Fill the lower socket cavity with grease. Install the lower cap and torque the bolts to 80 ft. lbs. (110 Nm).

10. On the left side, install the spring, if equipped, and the steering knuckle arm to the steering knuckle.

11. Connect the tie rod to the end of the steering knuckle. On the left side, connect the drag link ball stud to the steering knuckle.

12. Install the front axle shaft and all related components.

Upper Control Arm

REMOVAL AND INSTALLATION

1. Raise the vehicle and support safely.

2. Remove the shock absorber.

3. Remove the strut bar and disconnect the sway bar from the lower control arm, if equipped.

4. Install spring compressor tool DD-1278 or equivalent, to the coil spring and tighten the nut finger-tight, then back off $^1/_2$ turn.

5. Remove the cotter pin and upper ball joint nut. Suspend the rotor assembly with a wire so there is not excessive pull on the brake hose.

6. Release the upper ball joint taper using ball stud loosening tool C-3564-A or equivalent.

7. Remove the tool and remove the ball stud from the control arm.

8. Remove the pivot bar retaining bolts on Vans or the cam bolt assemblies on Pick-Up and Ramcharger and remove the arm from the vehicle.

To install:

9. Install the arm to the frame rail bracket and install the retaining bolts.

10. Torque the ball joint nut to 135 ft. lbs. (183 Nm). Install a new cotter pin. Remove the spring compressor.

11. Install the strut bar and connect the sway bar to the lower control arm, if equipped.

12. Install the shock absorber.

13. Align the front end. When all settings are at specifications, torque the pivot bar retaining bolts on Vans to 195 ft. lbs. (264 Nm). Torque the cam bolts to 70 ft. lbs. (95 Nm) on Pick-Up and Ramcharger.

Lower Control Arm

REMOVAL AND INSTALLATION

1. Raise the vehicle and support safely.

2. Remove the shock absorber.

3. Remove the strut bar and disconnect the sway bar from the lower control arm, if equipped.

4. Install spring compressor tool DD-1278 or equivalent, to the coil spring and tighten the nut finger-tight, then back off half a turn.

5. Remove the cotter pin and lower ball joint nut.

6. Release the lower ball joint taper using ball stud loosening tool C-3564-A or equivalent.

7. Remove the tool and remove the ball stud from the control arm. Release the compressor tool from the coil spring.

8. Pull the arm down and remove the spring with the rubber isolation pad from the vehicle. Remove the lower control arm pivot bolt from the crossmember and remove the arm from the vehicle.

To install:

9. Install the arm to the cross-member finger-tight. Install the spring with the rubber isolators. Install the compressor tool and compress it enough so the lower ball joint can be inserted through the knuckle.

10. Torque $^{11}/_{16}$ — 16 lower ball joint nuts to 135 ft. lbs. (183 Nm). Torque $^3/_4$ — 16 nuts to 175 ft. lbs. (237 Nm). Install a new cotter pin. Remove the spring compressor.

11. Install the strut bar and connect the sway bar from the lower control arm, if equipped.

12. Install the shock absorber.

13. Lower the vehicle completely. When the weight of the vehicle is off of the lifting apparatus, torque the lower arm pivot bolts to 175 ft. lbs. (237 Nm) on Vans or 225 ft. lbs. (305 Nm) on Pick-Up and Ramcharger.

14. Align the front end as required.

Sway Bar

REMOVAL AND INSTALLATION

1. Raise the vehicle and support safely.

2. Remove the front sway bar brackets and retainers.

3. Remove the sway bar connecting links to the control arm or front axle. Remove the sway bar from the vehicle.

4. The installation is the reversal of the removal procedure. Tighten the nuts just enough so the bushings compress to the same outer diameter as the washer adjacent to it.

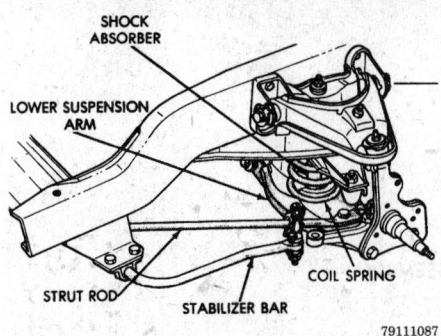

Upper control arm and related components — 2 wheel drive Pick-Up and Ramcharger

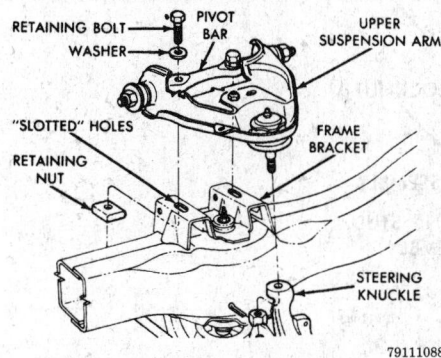

Upper control arm — Van

Front Wheel Bearings

REMOVAL AND INSTALLATION

For 4WD models, please refer to "Drive Axle" in this manu

1. Raise the vehicle and support safely.
2. Remove the tire and wheel assembly.
3. Remove the caliper and disc brake pads.
4. Remove the dust cap.
5. Remove the cotter pin, castelated locknut, wheel bearing nut and washer from the spindle.
6. Remove the outer wheel bearing.
7. Remove the rotor with the inner wheel bearing from the spindle. Remove the grease seal and inner wheel bearing.
8. If bearing replacement is required, remove bearing races with brass drift or puller. Install races with an appropriate installation tool.
To install:
9. Lubricate and install the inner wheel bearing. Install a new grease seal.

10. Install the rotor to the spindle.
11. Lubricate and install the outer wheel bearing, washer and nut. When the bearing preload is properly set, install the locknut and a new cotter pin.
12. Install the grease cap.
13. Install the brake pads and caliper.
14. Install the wheel.

ADJUSTMENT

1. Tighten the wheel bearing nut to 240–300 inch lbs. (27–34 Nm) while turning the rotor.
2. Loosen the wheel bearing adjusting nut completely.
3. Tighten the nut finger-tight.
4. Check the wheel bearing endplay. The specification is 0.0001–0.003 in.
5. Install the locknut and cotter pin.

REAR SUSPENSION

Shock Absorber

REMOVAL AND INSTALLATION

1. Raise the vehicle and support safely.
2. Remove the bolts that attach the shock to the frame or bracket.
3. Remove the shock from the vehicle.
4. The installation is the reversal of the removal procedure.

Leaf Springs

REMOVAL AND INSTALLATION

1. Raise the vehicle and support safely.
2. Using the proper equipment, support the weight of the axle.
3. Remove the nuts, washers and U-bolts attaching the springs to the axle housing. Remove the spacer.
4. Remove the spring shackle bolts, shackle and spring front bolt.
5. Remove the springs and auxiliary spring, if equipped, from the vehicle.
To install:
6. Install spring on the axle tube so the spring center bolt is inserted into the locating hole in the axle tube spring pad.
7. Align front of spring at eye and install pivot bolt and nut. Install rear eye bolt and nut. Tighten mounting bolts on front and rear of spring until all separation between metal is removed.
8. Install U-bolts with new lockwashers and retaining nuts. Align auxiliary spring with primary spring, if equipped. Lower vehicle.

Rear Wheel Bearings

For rear wheel bearing **REMOVAL AND INSTALLATION procedures, refer to Drive Axle section.**

Rear Axle Assembly

For rear axle **REMOVAL AND INSTALLATION** procedures, refer to Drive Axle section.

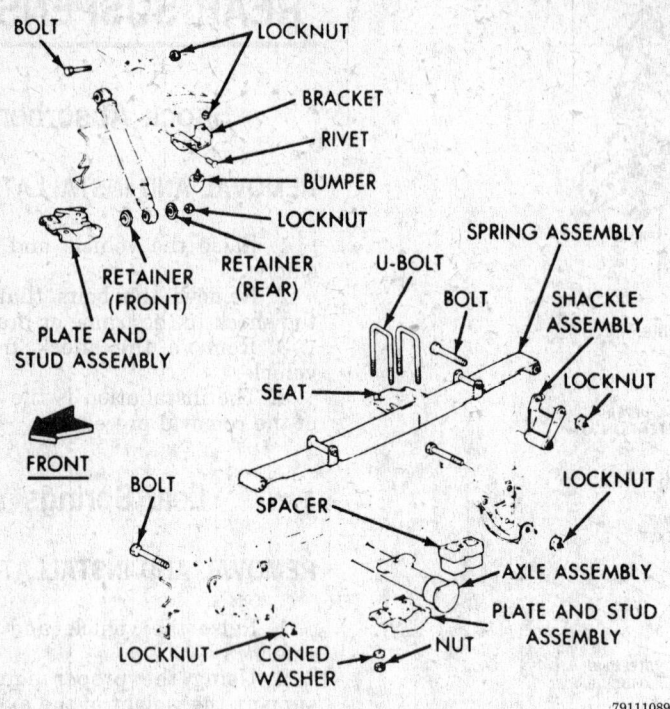

Rear suspension components — 150-models

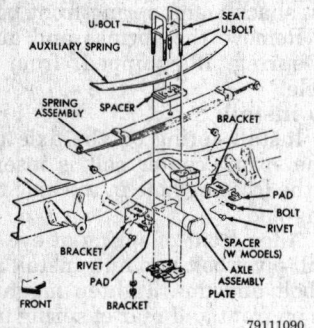

Rear suspension with auxiliary springs — D- and W250 and 350-models

SPECIFICATIONS

VEHICLE IDENTIFICATION CHART

		Engine Code					Model Year	
Code	Liters	Cu. In. (cc)	Cyl.	Fuel Sys.	Eng. Mfg.		Code	Year
K	2.5	153 (2507)	4	TFI	Chrysler		N	1992
3	3.0	181 (2972)	6	MFI	Mitsubishi		R	1994
R	3.3	201 (3300)	6	MFI	Chrysler			
L	3.8	231 (3785)	6	MFI	Chrysler			
X	3.9	238 (3916)	6	TFI	Chrysler			
Y	5.2	318 (5211)	8	TFI	Chrysler			

DSL - Diesel
MFI - Multiport fuel injection
TFI - Throttle body fuel injection

ENGINE IDENTIFICATION

Year	Model	Engine Displacement Liters (cc)	Engine Series (ID/VIN)	Fuel System	No. of Cylinders	Engine Type
1991	Caravan	2.5 (2507)	K	TFI	4	SOHC
	Caravan	3.0 (2972)	3	MFI	6	SOHC
	Caravan	3.3 (3300)	R	MFI	6	OHV
	Voyager	2.5 (2507)	K	TFI	4	SOHC
	Voyager	3.0 (2972)	3	MFI	6	SOHC
	Voyager	3.3 (3300)	R	MFI	6	OHV
	Town&Country	3.3 (3300)	R	MFI	6	OHV
	Dakota	2.5 (2507)	K	TFI	4	SOHC
	Dakota	3.9 (3916)	X	TFI	6	OHV
	Dakota	5.2 (5211)	Y	TFI	8	OHV
1992	Caravan	2.5 (2507)	K	TFI	4	SOHC
	Caravan	3.0 (2972)	3	MFI	6	SOHC
	Caravan	3.3 (3300)	R	MFI	6	OHV
	Voyager	2.5 (2507)	K	TFI	4	SOHC
	Voyager	3.0 (2972)	3	MFI	6	SOHC
	Voyager	3.3 (3300)	R	MFI	6	OHV
	Town&Country	3.3 (3300)	R	MFI	6	OHV
	Dakota	2.5 (2507)	K	TFI	4	SOHC
	Dakota	3.9 (3916)	X	MFI	6	OHV
	Dakota	5.2 (5211)	Y	MFI	8	OHV
1993	Caravan	2.5 (2507)	K	TFI	4	SOHC

ENGINE IDENTIFICATION

Year	Model	Engine Displacement Liters (cc)	Engine Series (ID/VIN)	Fuel System	No. of Cylinders	Engine Type
1993	Caravan	3.0 (2972)	3	MFI	6	SOHC
	Caravan	3.3 (3300)	R	MFI	6	OHV
	Voyager	2.5 (2507)	K	TFI	4	SOHC
	Voyager	3.0 (2972)	3	MFI	6	SOHC
	Voyager	3.3 (3300)	R	MFI	6	OHV
	Town&Country	3.3 (3300)	R	MFI	6	OHV
	Dakota	2.5 (2507)	K	TFI	4	SOHC
	Dakota	3.9 (3916)	X	MFI	6	OHV
	Dakota	5.2 (5211)	Y	MFI	8	OHV
1994-95	Caravan	2.5 (2507)	K	TFI	4	SOHC
	Caravan	3.0 (2972)	3	MFI	6	SOHC
	Caravan	3.3 (3300)	R	MFI	6	OHV
	Caravan	3.8 (3785)	L	MFI	6	OHV
	Voyager	2.5 (2507)	K	TFI	4	SOHC
	Voyager	3.0 (2972)	3	MFI	6	SOHC
	Voyager	3.3 (3300)	R	MFI	6	OHV
	Voyager	3.8 (3785)	L	MFI	6	OHV
	Town&Country	3.8 (3785)	L	MFI	6	OHV
	Dakota	2.5 (2507)	K	TFI	4	SOHC
	Dakota	3.9 (3916)	X	MFI	6	OHV
	Dakota	5.2 (5211)	Y	MFI	8	OHV

DSL - Diesel
MFI - Multiport fuel injection
TFI - Throttle body fuel injection
OHV - Overhead valve
SOHC - Single overhead camshaft

GENERAL ENGINE SPECIFICATIONS

Year	Engine ID/VIN	Engine Displacement Liters (cc)	Fuel System Type	Net Horsepower @ rpm	Net Torque @ rpm (ft. lbs.)	Bore x Stroke (in.)	Compression Ratio	Oil Pressure @ rpm
1991	K	2.5 (2507)	TFI	96@4400	133@2800	3.45x4.09	8.9:1	35-65@2000
	3	3.0 (2972)	MFI	143@5000	168@2500	3.59x2.99	8.9:1	30-80@3000
	R	3.3 (3300)	MFI	150@4800	185@3600	3.66x3.19	8.9:1	30-80@3000
	X	3.9 (3916)	MFI	125@4000	195@2000	3.91x3.31	9.1:1	30-80@3000
	Y	5.2 (5211)	TFI	170@4000	260@2000	3.91x3.31	9.2:1	30-80@3000
1992	K	2.5 (2507)	TFI	96@4400	133@2800	3.45x4.09	8.9:1	35-65@2000
	3	3.0 (2972)	MFI	143@5000	168@2500	3.59x2.99	8.9:1	30-80@3000
	R	3.3 (3300)	MFI	150@4800	185@3600	3.66x3.19	8.9:1	30-80@3000
	X	3.9 (3916)	MFI	175@4800	220@3200	3.91x3.31	9.1:1	30-80@3000
	Y	5.2 (5211)	MFI	170@4000	260@2000	3.91x3.31	9.2:1	30-80@3000
1993	K	2.5 (2507)	TFI	100@4800	135@2800	3.45x4.09	8.9:1	35-65@2000
	3	3.0 (2972)	MFI	143@5000	168@2500	3.59x2.99	8.9:1	30-80@3000
	R	3.3 (3300)	MFI	150@4800	185@3600	3.66x3.19	8.9:1	30-80@3000
	X	3.9 (3916)	MFI	175@4800	220@3200	3.91x3.31	9.1:1	30-80@3000
	Y	5.2 (5211)	MFI	230@4800	280@3200	3.91x3.31	9.1:1	30-80@3000

MFI - Multiport fuel injection
TFI - Throttle body fuel injection
DSL - Diesel

GENERAL ENGINE SPECIFICATIONS

Year	Engine ID/VIN	Engine Displacement Liters (cc)	Fuel System Type	Net Horsepower @ rpm	Net Torque @ rpm (ft. lbs.)	Bore x Stroke (in.)	Compression Ratio	Oil Pressure @ rpm
1994-95	K	2.5 (2507)	TFI	100@4800	135@2800	3.45x4.09	8.9:1	35-65@2000
	3	3.0 (2972)	MFI	143@5000	168@2500	3.59x2.99	8.9:1	30-80@3000
	R	3.3 (3300)	MFI	150@4800	185@3600	3.66x3.19	8.9:1	30-80@3000
	L	3.8 (3785)	MFI	162@4400	213@3300	3.78x3.43	9.0:1	30-80@3000
	X	3.9 (3916)	MFI	175@4800	220@3200	3.91x3.31	9.1:1	30-80@3000
	Y	5.2 (5211)	MFI	220@4400	300@3200	3.91x3.31	9.1:1	30-80@3000

MFI - Multiport fuel injection
TFI - Throttle body fuel injection
DSL - Diesel

GASOLINE ENGINE TUNE-UP SPECIFICATIONS

Year	Engine ID/VIN	Engine Displacement Liters (cc)	Spark Plugs Gap (in.)	Ignition Timing (deg.) MT	Ignition Timing (deg.) AT	Fuel Pump (psi)	Idle Speed (rpm) MT	Idle Speed (rpm) AT	Valve Clearance In.	Valve Clearance Ex.
1991	K	2.5 (2507)	0.035-0.043	12B	12B	13.5-15.5	850	850	HYD	HYD
	3	3.0 (2972)	0.039-0.043	12B	12B	46-50	800	800	HYD	HYD
	R	3.3 (3300)	0.048-0.053	1	1	46-50	750	750	HYD	HYD
	X	3.9 (3916)	0.035	10B	10B	13-16	750	750	HYD	HYD
	Y	5.2 (5211)	0.035	10B	10B	13-16	700	700	HYD	HYD
1992	K	2.5 (2507)	0.035	12B	12B	37-41	850	850	HYD	HYD
	3	3.0 (2972)	0.039-0.043	12B	12B	46-50	800	800	HYD	HYD
	R	3.3 (3300)	0.048-0.053	1	1	46-50	750	750	HYD	HYD
	X	3.9 (3916)	0.035	1	1	37-41	750	750	HYD	HYD
	Y	5.2 (5211)	0.035	1	1	37-41	700	700	HYD	HYD
1993	K	2.5 (2507)	0.035	12B	12B	37-41	850	850	HYD	HYD
	3	3.0 (2972)	0.039-0.043	12B	12B	46-50	800	800	HYD	HYD
	R	3.3 (3300)	0.048-0.053	1	1	46-50	750	750	HYD	HYD
	X	3.9 (3916)	0.035	1	1	37-41	750	750	HYD	HYD
	Y	5.2 (5211)	0.035	1	1	37-41	700	700	HYD	HYD
1994-95	K	2.5 (2507)	0.035	12B	12B	37-41	2	2	HYD	HYD
	3	3.0 (2972)	0.039-0.043	12B	12B	46-50	2	2	HYD	HYD
	R	3.3 (3300)	0.048-0.053	1	1	46-50	2	2	HYD	HYD
	L	3.8 (3785)	0.048-0.053	1	1	46-50	2	2	HYD	HYD
	X	3.9 (3916)	0.035	1	1	35-45	2	2	HYD	HYD
	Y	5.2 (5211)	0.035	1	1	35-45	2	2	HYD	HYD

NOTE: The Vehicle Emission Control Information label reflects changes made during production. The label figures must be used if they differ from above
B - Before top dead center
HYD - Hydraulic
1 Ignition timing cannot be adjusted. Base engine timing is set at TDC during assembly.
2 Refer to the Vehicle Emission Control Information (VECI) label for correct specification

FIRING ORDERS

NOTE: To avoid confusion, always replace spark plug wires one at a time.

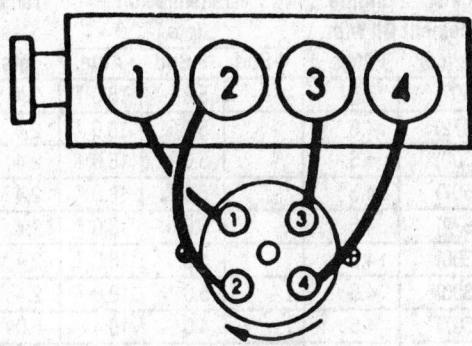

2.2L and 2.5L Engines
Engine Firing Order: 1–3–4–2
Distributor Rotation: Clockwise

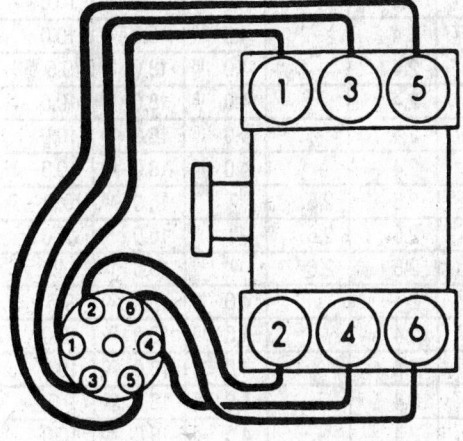

3.0L Engine (Mitsubishi)
Engine Firing Order: 1–2–3–4–5–6
Distributor Rotation: Counterclockwise

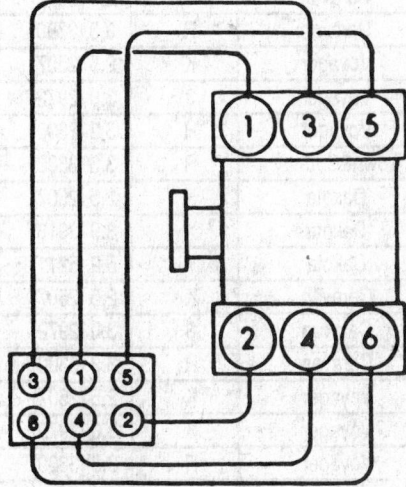

3.3L Engine
Engine Firing Order: 1–2–3–4–5–6
Distributorless Ignition System

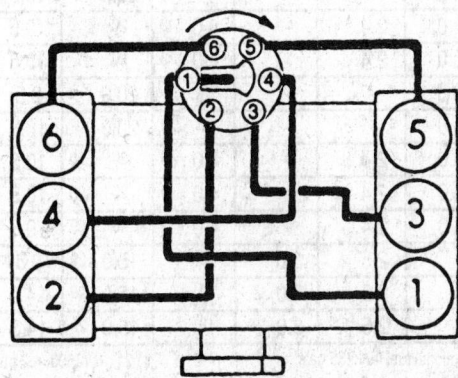

3.9L Engine
Engine Firing Order: 1–6–5–4–3–2
Distributor Rotation: Clockwise

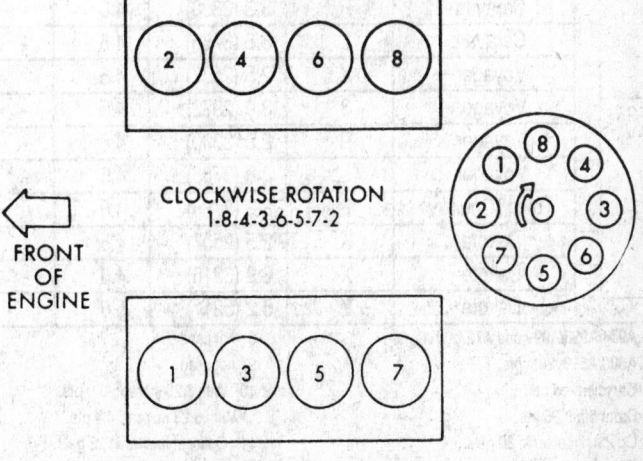

FRONT OF ENGINE

CLOCKWISE ROTATION
1-8-4-3-6-5-7-2

5.2L Engine
Engine Firing Order: 1–8–4–3–6–5–7–2
Distributor Rotation: Clockwise

CAPACITIES

Year	Model	Engine ID/VIN	Engine Displacement Liters (cc)	Engine Oil with Filter	Transmission (pts.) 4-Spd	5-Spd	Auto.	Transfer Case (pts.)	Drive Axle Front (pts.)	Rear (pts.)	Fuel Tank (gal.)	Cooling System (qts.)
1991	Caravan	K	2.5 (2507)	4.5	-	5.0	18.0 [7]	2.4	-	4.0 [16]	15.0 [6]	9.5
	Caravan	3	3.0 (2972)	4.5	-	5.0	18.0 [7]	2.4	-	4.0 [16]	15.0 [6]	10.0
	Caravan	R	3.3 (3300)	4.5	-	5.0	18.0 [7]	2.4	-	4.0 [16]	15.0 [6]	10.0
	Voyager	K	2.5 (2507)	4.5	-	5.0	18.0 [7]	2.4	-	4.0 [16]	15.0 [6]	9.5
	Voyager	3	3.0 (2972)	4.5	-	5.0	18.0 [7]	2.4	-	4.0 [16]	15.0 [6]	10.0
	Voyager	R	3.3 (3300)	4.5	-	5.0	18.0 [7]	2.4	-	4.0 [16]	15.0 [6]	10.0
	Town&Country	R	3.3 (3300)	4.5	-	5.0	18.0 [7]	2.4	-	4.0 [16]	15.0 [6]	10.0
	Dakota	K	2.5 (2507)	4.5	-	4.0	10.4	5.0	2.6	10	15.0 [11]	9.8
	Dakota	X	3.9 (3916)	4.5	-	4.0	10.4	5.0	2.6	10	15.0 [11]	14.0
	Dakota	Y	5.2 (5211)	4.5	-	4.0	10.4	5.0	2.6	10	15.0 [11]	14.3
1992	Caravan	K	2.5 (2507)	4.5	-	4.8	18.0 [7]	2.4	-	4.0 [16]	18.0 [20]	9.5
	Caravan	3	3.0 (2972)	4.5	-	4.8	18.0 [7]	2.4	-	4.0 [16]	18.0 [20]	10.0
	Caravan	R	3.3 (3300)	4.5	-	4.8	18.0 [7]	2.4	-	4.0 [16]	18.0 [20]	10.0
	Voyager	K	2.5 (2507)	4.5	-	4.8	18.0 [7]	2.4	-	4.0 [16]	18.0 [20]	9.5
	Voyager	3	3.0 (2972)	4.5	-	4.8	18.0 [7]	2.4	-	4.0 [16]	18.0 [20]	10.0
	Voyager	R	3.3 (3300)	4.5	-	4.8	18.0 [7]	2.4	-	4.0 [16]	18.0 [20]	10.0
	Town&Country	R	3.3 (3300)	4.5	-	4.8	18.0 [7]	2.4	-	4.0 [16]	18.0 [20]	10.0
	Dakota	K	2.5 (2507)	4.5	-	4.0	10.4	2.5	2.6	10	15.0 [15]	9.5
	Dakota	X	3.9 (3916)	4.5	-	6.6	10.4	2.5	2.6	10	15.0 [15]	14.0
	Dakota	Y	5.2 (5211)	4.5	-	6.6	10.4	2.5	2.6	10	15.0 [15]	14.3
1993	Caravan	K	2.5 (2507)	4.5	-	4.8	18.0 [7]	2.4	-	4.0 [16]	18.0 [20]	9.5
	Caravan	3	3.0 (2972)	4.5	-	4.8	18.0 [7]	2.4	-	4.0 [16]	18.0 [20]	10.0
	Caravan	R	3.3 (3300)	4.5	-	4.8	18.0 [7]	2.4	-	4.0 [16]	18.0 [20]	10.0
	Voyager	K	2.5 (2507)	4.5	-	4.8	18.0 [7]	2.4	-	4.0 [16]	18.0 [20]	9.5
	Voyager	3	3.0 (2972)	4.5	-	4.8	18.0 [7]	2.4	-	4.0 [16]	18.0 [20]	10.0
	Voyager	R	3.3 (3300)	4.5	-	4.8	18.0 [7]	2.4	-	4.0 [16]	18.0 [20]	10.0
	Town&Country	R	3.3 (3300)	4.5	-	4.8	18.0 [7]	2.4	-	4.0 [16]	18.0 [20]	10.0
	Dakota	K	2.5 (2507)	4.5	-	6.6	10.4	4.5	2.6	10	15.0 [15]	9.8
	Dakota	X	3.9 (3916)	4.5	-	6.6	10.4	4.5	2.6	10	15.0 [15]	14.0
	Dakota	Y	5.2 (5211)	5.0	-	6.6	10.4	4.5	2.6	10	15.0 [15]	14.3
1994-95	Caravan	K	2.5 (2507)	4.5	-	4.6	18.0	-	-	-	20.0	9.5
	Caravan	3	3.0 (2972)	4.5	-	4.6	18.0	-	-	-	20.0	10.5
	Caravan	R	3.3 (3300)	4.5	-	4.6	18.0	2.4	-	4.0 [13]	20	10.5
	Caravan	L	3.8 (3785)	4.5	-	4.6	18.0	2.4	-	4.0 [13]	20	10.5
	Voyager	K	2.5 (2507)	4.5	-	4.6	18.0	-	-	-	20.0	9.5
	Voyager	3	3.0 (2972)	4.5	-	4.6	18.0	-	-	-	20.0	10.5
	Voyager	R	3.3 (3300)	4.5	-	4.6	18.0	2.4	-	4.0 [13]	20 [20]	10.5
	Voyager	L	3.8 (3785)	4.5	-	4.6	18.0	2.4	-	4.0 [13]	20 [20]	10.5
	Town&Country	L	3.8 (3785)	4.5	-	4.6	18.0	2.4	-	4.0 [13]	20 [20]	10.5
	Dakota	K	2.5 (2507)	4.5	-	1	2	2.5	3.0	10	15.0 [29]	9.8
	Dakota	X	3.9 (3916)	4.5	-	1	2	2.5	3.0	10	15.0 [29]	14.0
	Dakota	Y	5.2 (5211)	5.0	-	1	2	2.5	3.0	10	15.0 [29]	14.3

1 A904/A998/A999 and A727: 17.1 pts.
 A500/A518: 20.4 pts.
2 Chrysler: 4.5 pts.
 Dana 60: 6.25 pts.
3 Optional fuel tank: 30 gals.
4 With HD cooling or A/C, add one quart
5 not used
6 Optional fuel tank: 20 gals.
7 Fleet vehicles: 19 pts.

8 not used
9 not used
10 With 7.25 in. rear: 3.0 pts.
 With 8.25 in. rear: 4.4 pts.
11 Optional fuel tank: 22 gals.
12 not used
13 NP205: 4.5 pts.; NP241: 6.0 pts.
14 Chrysler 8.25 in. and 9.25 in.: 4.5 pts.
 Spicer/Dana 60: 6.0 pts.
 Dana 70: 7.0 pts.

15 Optional fuel tank: 35 gals.
16 Overrunning clutch: 0.78 pts.
17 not used
18 Dana 44: 5.6 pts.
 Dana 60: 6.5 pts.
19 Manual transmission: 15.5 qts.
 Automatic transmission: 16.5 qts.
20 FWD: 20 gals.; AWD: 18 gals.

21 NV3500: 4.2 pts.
 NV4500: 8 pts.
 AX15: 6.6 pts.
 Getrag: 7 pts.
22 32RH: 17 pts.
 36RH: 16.6 pts.
 42RH: 20.2 pts.
 46RH: 21.8 pts.
 47RH: 30.8 pts.

CAPACITIES

Year	Model	Engine ID/VIN	Engine Displacement Liters (cc)	Engine Oil with Filter	Transmission (pts.)			Transfer Case (pts.)	Drive Axle		Fuel Tank (gal.)	Cooling System (qts.)
					4-Spd	5-Spd	Auto.		Front (pts.)	Rear (pts.)		

23 NP231HD: 2.5 pts.
   NP241: 4.7 pts.
   NP241HD: 13 pts.
24 7.25 in.: 3 pts.
   Dana 44: 5.6 pts.
   Dana 60: 6.5 pts.

25 Chrysler 7.25 in.: 3 pts.
   Chrysler 8.25 in. and 9.25 in.: 4.8 pts.
   Spicer and Dana 60: 6.0 pts.
   Dana 70 and 80: 7.0 pts.
26 A998/A999 and A727: 17.2 pts.
   A500: 20.4 pts.
   A518: 21.4 pts.

27 Chrysler 8.25 in.: 4.4 pts.; 9.25 in.: 4.8 pts.
   Dana 60: 6.3 pts.
28 With rear heater: 16 qts.
29 Optional fuel tank: 22 gals.

CAMSHAFT SPECIFICATIONS
All measurements given in inches.

Year	Engine ID/VIN	Engine Displacement Liters (cc)	Journal Diameter					Elevation		Bearing Clearance	Camshaft End Play
			1	2	3	4	5	In.	Ex.		
1991	K	2.5 (2507)	1.375-1.376	1.375-1.376	1.375-1.376	1.375-1.376	1.375-1.376	NA	NA	NA	0.005-0.020
	3	3.0 (2972)	NA	NA	NA	NA	NA	1	1	-	NA
	R	3.3 (3300)	1.997-1.999	1.980-1.982	1.965-1.967	1.949-1.952	-	0.400 3	0.400 3	0.001-0.005	0.005-0.012
	X	3.9 (3916)	1.998-1.999	1.967-1.968	1.951-1.952	1.561-1.562	-	0.373 3	0.400 3	0.001-0.005	0.002-0.010
	Y	5.2 (5211)	1.998-1.999	1.982-1.983	1.967-1.968	1.951-1.952	1.561-1.562	0.400 3	0.400 3	0.001-0.005	0.002-0.010
1992	K	2.5 (2507)	1.375-1.376	1.375-1.376	1.375-1.376	1.375-1.376	1.375-1.376	NA	NA	NA	0.005-0.020
	3	3.0 (2972)	NA	NA	NA	NA	NA	1	1	NA	NA
	R	3.3 (3300)	1.997-1.999	1.980-1.982	1.965-1.967	1.949-1.952	-	0.400 3	0.400 3	0.001-0.005	0.005-0.012
	X	3.9 (3916)	1.982-1.983	1.967-1.968	1.951-1.952	1.561-1.562	-	0.373 3	0.0400 3	0.001-0.005	0.002-0.010
	Y	5.2 (5211)	1.998-1.999	1.982-1.983	1.967-1.968	1.951-1.952	1.561-1.562	0.373 3	0.400 3	0.001-0.005	0.002-0.010
1993	K	2.5 (2507)	1.375-1.376	1.375-1.376	1.375-1.376	1.375-1.376	1.375-1.376	NA	NA	NA	0.005-0.013
	3	3.0 (2972)	NA	NA	NA	NA	NA	1	1	NA	NA
	R	3.3 (3300)	1.997-1.999	1.980-1.982	1.965-1.967	1.949-1.952	-	0.400 3	0.400 3	0.001-0.005	0.005-0.012
	X	3.9 (3916)	1.998-1.999	1.982-1.983	1.951-1.952	1.561-1.562	-	0.432 3	0.432 3	0.001-0.005	0.002-0.010
	Y	5.2 (5211)	1.998-1.999	1.982-1.983	1.967-1.968	1.951-1.952	1.561-1.562	0.432 3	0.432	0.001-0.005	0.002-0.010
1994-95	K	2.5 (2507)	1.375-1.376	1.375-1.376	1.375-1.376	1.375-1.376	1.375-1.376	NA	NA	NA	0.005-0.013
	3	3.0 (2972)	NA	NA	NA	NA	NA	1	1	NA	NA
	R	3.3 (3300)	1.997-1.999	1.981-1.982	1.965-1.967	1.949-1.952	-	0.400	0.400	0.001-0.005	0.005-0.012

CAMSHAFT SPECIFICATIONS
All measurements given in inches.

Year	Engine ID/VIN	Engine Displacement Liters (cc)	Journal Diameter					Elevation		Bearing Clearance	Camshaft End Play
			1	2	3	4	5	In.	Ex.		
1994-95	L	3.8 (3785)	1.997-1.999	1.980-1.982	1.965-1.967	1.949-1.952	-	0.400	0.400	0.001-0.005	0.005-0.012
	X	3.9 (3916)	1.998-1.999	1.982-1.983	1.951-1.952	1.561-1.562	-	0.432 4	0.432 4	0.001-0.005	0.002-0.010
	Y	5.2 (5211)	1.998-1.999	1.982-1.983	1.967-1.968	1.951-1.952	1.561-1.562	0.432 4	0.432 4	0.001-0.005	0.002-0.010

NA - Not Available
1 Height of cam lobe: 1.604-1.624 in.
2 Minimum diameter at peak lobe
3 Lift at valve
4 Journal No. 6: 1.917-1.918
5 Lift at valve
6 No. 2: 0.0015-0.0035

CRANKSHAFT AND CONNECTING ROD SPECIFICATIONS
All measurements are given in inches.

Year	Engine ID/VIN	Engine Displacement Liters (cc)	Crankshaft				Connecting Rod		
			Main Brg. Journal Dia.	Main Brg. Oil Clearance	Shaft End-play	Thrust on No.	Journal Diameter	Oil Clearance	Side Clearance
1991	K	2.5 (2507)	2.3620-2.3630	0.0003-0.0031	0.002-0.007	3	1.968-1.969	0.0008-0.0034	0.005-0.013
	3	3.0 (2972)	2.3610-2.3630	0.0006-0.0020	0.002-0.010	3	1.968-1.969	0.0008-0.0028	0.001-0.004
	R	3.3 (3300)	2.519	0.0007-0.0022	0.001-0.007	2	2.283	0.0008-0.0030	0.005-0.015
	X	3.9 (3916)	2.500-2.501	1	0.002-0.010	2	2.124-2.125	0.0005-0.0022	0.006-0.014
	Y	5.2 (5211)	2.4995-2.5005	1	0.002-0.010	3	2.124-2.125	0.0005-0.0022	0.006-0.014
1992	K	2.5 (2507)	2.3620-2.3630	0.0004-0.0028	0.002-0.007	3	1.968-1.969	0.0008-0.0034	0.005-0.013
	3	3.0 (2972)	2.3610-2.3620	0.0006-0.0020	0.002-0.010	3	1.968-1.969	0.0008-0.0028	0.001-0.004
	R	3.3 (3300)	2.519	0.0004-0.0028	0.003-0.009	2	2.283	0.0008-0.0030	0.005-0.015
	X	3.9 (3916)	2.500-2.501	1	0.002-0.007	2	2.124-2.125	0.0005-0.0022	0.006-0.014
	Y	5.2 (5211)	2.4995-2.5005	1	0.002-0.007	3	2.124-2.125	0.0005-0.0022	0.006-0.014
1993	K	2.5 (2507)	2.3620-2.3630	0.0004-0.0028	0.002-0.007	3	1.9680-1.9690	0.0008-0.0034	0.005-0.013
	3	3.0 (2972)	2.3610-2.3620	0.0006-0.0020	0.002-0.010	3	1.968-1.969	0.0008-0.0028	0.001-0.004
	R	3.3 (3300)	2.519	0.0004-0.0028	0.003-0.009	2	2.283	0.0008-0.0030	0.005-0.015
	X	3.9 (3916)	2.500-2.501	1	0.002-0.007	2	2.124-2.125	0.0005-0.0022	0.006-0.014
	Y	5.2 (5211)	2.4995-2.5005	1	0.002-0.007	3	2.124-2.125	0.0005-0.0022	0.006-0.014

CRANKSHAFT AND CONNECTING ROD SPECIFICATIONS

All measurements are given in inches.

Year	Engine ID/VIN	Engine Displacement Liters (cc)	Crankshaft				Connecting Rod		
			Main Brg. Journal Dia.	Main Brg. Oil Clearance	Shaft End-play	Thrust on No.	Journal Diameter	Oil Clearance	Side Clearance
1994-95	K	2.5 (2507)	2.3620-2.3630	0.0004-0.0028	0.002-0.007	3	1.9680-1.9690	0.0008-0.0034	0.005-0.013
	3	3.0 (2972)	2.3610-2.3620	0.0006-0.0020	0.002-0.010	3	1.968-1.969	0.0008-0.0028	0.001-0.004
	R	3.3 (3300)	2.519	0.0004-0.0028	0.003-0.009	2	2.283	0.0008-0.0030	0.005-0.015
	L	3.8 (3785)	2.519	0.0007-0.0030	0.004-0.012	2	2.283	0.0007-0.0030	0.005-0.015
	X	3.9 (3916)	2.500-2.501	1	0.002-0.007	2	2.124-2.125	0.0005-0.0022	0.006-0.014
	Y	5.2 (5211)	2.4995-2.5005	1	0.002-0.007	3	2.124-2.125	0.0005-0.0022	0.006-0.014

NA - Not Available

1 0.0005-0.0015 on No. 1
   0.0005-0.0020 on Nos. 2-4

2 Maximum wear limit

VALVE SPECIFICATIONS

Year	Engine ID/VIN	Engine Displacement Liters (cc)	Seat Angle (deg.)	Face Angle (deg.)	Spring Test Pressure (lbs. @ in.)	Spring Installed Height (in.)	Stem-to-Guide Clearance (in.)		Stem Diameter (in.)	
							Intake	Exhaust	Intake	Exhaust
1991	K	2.5 (2507)	45	45	115@1.65	1.65	0.0009-0.0047	0.0030-0.0047	0.3124	0.3103
	3	3.0 (2972)	44.5	45.5	73@1.59	1.59	0.001-0.002	0.002-0.003	0.313-0.314	0.312-0.313
	R	3.3 (3300)	45	44.5	60@1.56	1.56	0.001-0.003	0.002-0.003	0.311-0.312	0.311-0.312
	X	3.9 (3916)	45	45	2	3	0.001-0.003	0.002-0.004	0.0372-0.373	0.371-0.372
	Y	5.2 (5211)	45	45	2	3	0.001-0.003	0.002-0.004	0.372-0.373	0.371-0.372
1992	K	2.5 (2507)	45	45	115@1.65	1.65	0.0009-0.0047	0.0030-0.0047	0.3124	0.3103
	3	3.0 (2972)	44.5	45.5	73@1.59	1.59	0.001-0.002	0.002-0.003	0.313-0.314	0.312-0.313
	R	3.3 (3300)	45	44.5	60@1.56	1.56	0.001-0.003	0.002-0.003	0.311-0.312	0.311-0.312
	X	3.9 (3916)	44.25-44.75	43.25-43.75	5	6	0.001-0.002	0.002-0.004	0.313	0.313
	Y	5.2 (5211)	44.25-44.75	43.25-43.75	2	6	0.001-0.003	0.002-0.004	0.313	0.313

VALVE SPECIFICATIONS

Year	Engine ID/VIN	Engine Displacement Liters (cc)	Seat Angle (deg.)	Face Angle (deg.)	Spring Test Pressure (lbs. @ in.)	Spring Installed Height (in.)	Stem-to-Guide Clearance (in.)		Stem Diameter (in.)	
							Intake	Exhaust	Intake	Exhaust
1993	K	2.5 (2507)	45	45	115@1.65	1.65	0.0009-0.0047	0.0030-0.0047	0.3124	0.3103
	3	3.0 (2972)	44.5	45.5	73@1.59	1.59	0.001-0.002	0.002-0.003	0.313-0.314	0.312-0.313
	R	3.3 (3300)	45	44.5	95@1.57	1.622-1.681	0.001-0.003	0.002-0.006	0.312-0.313	0.311-0.312
	X	3.9 (3916)	44.25-44.75	43.25-43.75	85@1.64	1.64	0.001-0.003	0.001-0.003	0.311-0.312	0.311-0.312
	Y	5.2 (5211)	44.25-44.75	43.25-43.75	85@1.64	1.64	0.001-0.003	0.001-0.003	0.311-0.312	0.311-0.312
1994-95	K	2.5 (2507)	45	45	115@1.65	1.65	0.0009-0.0047	0.0030-0.0047	0.3124	0.3103
	3	3.0 (2972)	44.5	45.5	73@1.59	1.59	0.001-0.002	0.002-0.003	0.313-0.314	0.312-0.313
	R	3.3 (3300)	45	44.5	95@1.57	1.622-1.681	0.001-0.003	0.002-0.006	0.312-0.313	0.311-0.312
	L	3.8 (3785)	45	44.5	95@1.57	1.622-1.681	0.001-0.003	0.002-0.006	0.312-0.313	0.311-0.312
	X	3.9 (3916)	44.25-44.75	43.25-43.75	85@1.64	1.64	0.001-0.003	0.001-0.003	0.311-0.312	0.311-0.312
	Y	5.2 (5211)	44.25-44.75	43.25-43.75	85@1.64	1.64	0.001-0.003	0.001-0.003	0.311-0.312	0.311-0.312

1 Intake: 30 degrees; Exhaust: 45 degrees
2 Intake: 78-88 at installed height
  Exhaust: 80-90 at installed height
3 Intake: 1.625-1.688; Exhaust: 1.453-1.516
4 Minimum acceptable specification
5 Intake: 78-88 at 1.688 in.
  Exhaust: 80-90 at 1.203 in.
6 Intake: 1.688; Exhaust: 1.203
7 Intake: 78-88 at 1.688 in.
  Exhaust: 80-90 at 1.484 in.

PISTON AND RING SPECIFICATIONS

All measurements are given in inches.

Year	Engine ID/VIN	Engine Displacement Liters (cc)	Piston Clearance	Ring Gap			Ring Side Clearance		
				Top Compression	Bottom Compression	Oil Control	Top Compression	Bottom Compression	Oil Control
1991	K	2.5 (2507)	0.0005-0.0015	0.010-0.021	0.011-0.021	0.015-0.035	0.0015-0.0031	0.0022-0.0080	NA
	3	3.0 (2972)	0.0008-0.0015	0.012-0.018	0.010-0.016	0.012-0.035	0.0020-0.0035	0.0008-0.0020	NA
	R	3.3 (3300)	0.0009-0.0022	0.012-0.022	0.012-0.022	0.010-0.040	0.0012-0.0037	0.0012-0.0037	0.0005-0.0089
	X	3.9 (3916)	0.0005-0.0015	0.010-0.020	0.010-0.020	0.015-0.055	0.0015-0.0030	0.0015-0.0030	0.0002-0.0050
	Y	5.2 (5211)	0.0005-0.0015	0.010-0.020	0.010-0.020	0.015-0.055	0.0015-0.0030	0.0015-0.0030	0.0002-0.0050
1992	K	2.5 (2507)	0.0005-0.0015	0.010-0.021	0.011-0.021	0.015-0.035	0.0015-0.0031	0.0002-0.0080	NA
	S	3.0 (2972)	0.0008-0.0015	0.012-0.018	0.010-0.016	0.012-0.035	0.0020-0.0035	0.0008-0.0020	NA
	R	3.3 (3300)	0.0009-0.0022	0.012-0.022	0.012-0.022	0.010-0.040	0.0012-0.0037	0.0012-0.0037	0.0005-0.0089
	X	3.9 (3916)	0.0005-0.0015	0.010-0.020	0.010-0.020	0.015-0.055	0.0015-0.0030	0.0015-0.0030	0.002-0.005
	Y	5.2 (5211)	0.0005-0.0015	0.010-0.020	0.010-0.020	0.015-0.055	0.0015-0.0030	0.0015-0.0030	0.002-0.005
1993	K	2.5 (2507)	0.0005-0.0015	0.010-0.020	0.011-0.021	0.015-0.055	0.0015-0.0031	0.0015-0.0037	NA
	3	3.0 (2972)	0.0012-0.0020	0.012-0.018	0.010-0.016	0.012-0.035	0.0020-0.0035	0.0008-0.0020	NA
	R	3.3 (3300)	0.0009-0.0022	0.012-0.022	0.012-0.022	0.010-0.040	0.0012-0.0037	0.0012-0.0037	0.0005-0.0089
	X	3.9 (3916)	0.0003-0.0008	0.010-0.020	0.010-0.020	0.010-0.050	0.0015-0.0030	0.0015-0.0030	0.002-0.008
	Y	5.2 (5211)	0.0003-0.0008	0.010-0.020	0.010-0.020	0.015-0.055	0.0015-0.0030	0.0015-0.0030	0.002-0.008
1994-95	K	2.5 (2507)	0.0005-0.0015	0.010-0.020	0.011-0.021	0.015-0.055	0.0015-0.0031	0.0015-0.0037	NA
	3	3.0 (2972)	0.0012-0.0020	0.012-0.018	0.010-0.016	0.012-0.035	0.0020-0.0035	0.0008-0.0020	NA
	R	3.3 (3300)	0.0010-0.0022	0.012-0.022	0.012-0.022	0.010-0.040	0.0012-0.0037	0.0012-0.0037	0.0005-0.0089
	L	3.8 (3785)	0.0010-0.0022	0.012-0.022	0.012-0.022	0.010-0.040	0.0012-0.0037	0.0012-0.0037	0.0005-0.0089
	X	3.9 (3916)	0.0003-0.0008	0.010-0.020	0.010-0.020	0.010-0.050	0.0015-0.0030	0.0015-0.0030	0.002-0.008
	Y	5.2 (5211)	0.0003-0.0008	0.010-0.020	0.010-0.020	0.015-0.055	0.0015-0.0030	0.0015-0.0030	0.002-0.008

NA - Not Available

TORQUE SPECIFICATIONS
All readings in ft. lbs.

Year	Engine ID/VIN	Engine Displacement Liters (cc)	Cylinder Head Bolts	Main Bearing Bolts	Rod Bearing Bolts	Crankshaft Damper Bolts	Flywheel Bolts	Manifold Intake	Manifold Exhaust	Spark Plugs	Lug Nut
1991	K	2.5 (2507)	2	30 3	40 3	50	70	17	17	26	95
	3	3.0 (2972)	80	60	38	110	70	17	17	20	95
	R	3.3 (3300)	9	30 3	40 3	50	70	17	17	26	95
	X	3.9 (3916)	105	85	45	135	55	45	4	30	5
	Y	5.2 (5211)	105	85	45	135	55	40	4	30	5
1992	K	2.5 (2507)	2	30 3	40 3	50	70	17	17	26	95
	3	3.0 (2972)	80	60	38	110	70	17	17	20	95
	R	3.3 (3300)	9	30 3	40 3	40	70	17	17	30	95
	X	3.9 (3916)	105	85	45	135	55	45	4	30	5
	Y	5.2 (5211)	105	85	45	135	55	40	4	30	5
1993	K	2.5 (2507)	2	30 3	40 3	50	70	17	17	26	95
	3	3.0 (2972)	80	60	38	110	70	17	17	20	95
	R	3.3 (3300)	9	30 3	40 3	40	70	17	17	30	95
	X	3.9 (3916)	105	85	45	135	55	45	4	30	5
	Y	5.2 (5211)	105	85	45	135	55	40	4	30	5
1994-95	K	2.5 (2507)	2	30 2	40 2	85	70	17	17	20	95
	3	3.0 (2972)	80	60	38	112	70	17	17	20	95
	R	3.3 (3300)	2	30 2	40 2	40	70	17	17	20	95
	L	3.8 (3785)	2	30 2	40 2	40	70	17	17	20	95
	X	3.9 (3916)	105 12	85	45	135	72	10	25	30	5
	Y	5.2 (5211)	105 12	85	45	135	55	17	25	30	5

1 All bolts: 66 ft. lbs.
  Long bolts: 89 ft. lbs.
  All bolts and additional 1/4 turn
2 In sequence: 45, 65, 65 plus 1/4 turn
3 Plus 1/4 turn
4 Bolts: 20 ft. lbs.; Nuts: 15 ft. lbs.
5 1/2x20 stud: 85-110 ft. lbs.
  5/8x18 stud with cone nut: 175-225 ft. lbs.
  5/8x18 stud with flanged nut: 300-350 ft. lbs.
6 In sequence: 29, 62 and 93 ft. lbs.

7 In sequence: 45, 88 and 129 ft. lbs.
8 In sequence: 26, 51 and 73 ft. lbs.
9 In sequence: 45, 65, 65 plus 1/4 turn
  Torque small bolt in rear of head to 25 ft. lbs.
10 In sequence: Tighten intake plenum bolts to 24, 48 and 84 in. lbs.
11 Lower intake: 40 ft. lbs.
  Upper intake: 16 ft. lbs.
12 In sequence: 50, 105

BRAKE SPECIFICATIONS
All measurements in inches unless noted

Year	Model	Master Cylinder Bore	Brake Disc Original Thickness	Brake Disc Minimum Thickness	Brake Disc Maximum Runout	Brake Drum Diameter Original Inside Diameter	Brake Drum Diameter Max. Wear Limit	Brake Drum Diameter Maximum Machine Diameter	Minimum Lining Thickness Front	Minimum Lining Thickness Rear
1991	Caravan	0.940	0.861	0.80	0.005	9.00	9.09	9.06	0.06	0.06
	Voyager	0.940	0.861	0.80	0.005	9.00	9.09	9.06	0.06	0.06
	Town&Country	0.940	0.861	0.80	0.005	9.00	9.09	9.06	0.06	0.06
	Dakota	NA	0.861	0.81	0.004	10.00	10.09	10.06	0.06	0.06
	Dakota	NA	0.861	0.81	0.004	9.00	9.09	9.06	0.06	0.06
1992	Caravan	0.940	0.861	0.80	0.005	9.00	9.09	9.06	0.06	0.06
	Voyager	0.940	0.861	0.80	0.005	9.00	9.09	9.06	0.06	0.06
	Town&Country	0.940	0.861	0.80	0.005	9.00	9.09	9.06	0.06	0.06
	Dakota	NA	0.861	0.81	0.004	9.00	9.09	9.06	0.06	0.06
	Dakota	NA	0.861	0.81	0.004	10.00	10.09	11.06	0.06	0.06

BRAKE SPECIFICATIONS

All measurements in inches unless noted

Year	Model		Master Cylinder Bore	Brake Disc Original Thickness	Brake Disc Minimum Thickness	Maximum Runout	Brake Drum Diameter Original Inside Diameter	Max. Wear Limit	Maximum Machine Diameter	Minimum Lining Thickness Front	Minimum Lining Thickness Rear
1993	Caravan		0.940	0.940	0.88	0.005	9.00	9.09	9.06	0.06	0.06
	Voyager		0.940	0.940	0.88	0.005	9.00	9.09	9.06	0.06	0.06
	Town&Country		0.940	0.940	0.88	0.005	9.00	9.09	9.06	0.06	0.06
	Dakota		NA	0.861	0.81	0.004	9.00	9.09	9.06	0.06	0.06
	Dakota		NA	0.861	0.81	0.004	10.00	10.09	10.06	0.06	0.06
1994-95	Caravan		0.940	0.940	0.88	0.005	9.00	9.09	9.06	0.06	0.06
	Voyager		0.940	0.940	0.88	0.005	9.00	9.09	9.06	0.06	0.06
	Town&Country		0.940	0.940	0.88	0.005	9.00	9.09	9.06	0.06	0.06
	Dakota		NA	0.861	0.81	0.004	9.00	9.09	9.06	0.06	0.06
	Dakota		NA	0.861	0.81	0.004	10.00	10.09	10.06	0.06	0.06

1 With 3300 lb. or 3600 lb. rear axle: 1.24 in.
  With 4000 lb. rear axle: 1.18 in.
2 With 3300 lb. or 3600 lb. rear axle: 1.180 in.
  With 4000 lb. rear axle: 1.125 in.
3 Minimum thickness indicated on rotor hub

WHEEL ALIGNMENT

Year	Model		Caster Range (deg.)	Caster Preferred Setting (deg.)	Camber Range (deg.)	Camber Preferred Setting (deg.)	Toe-in (in.)	Steering Axis Inclination (deg.)
1991	Caravan	F	-	1 11/6P	1/4N-3/4P	5/16P	1/8P	12.2
	Caravan	R	-	-	1N-1/2P	0	0	-
	Voyager	F	-	1 11/6P	1/4N-3/4P	5/16P	1/8P	12.2
	Voyager	R	-	-	1N-1/2P	0	0	-
	Town&Country	F	-	1 11/6P	1/4N-3/4P	5/16P	1/8P	12.2
	Town&Country	R	-	-	1N-1/2P	0	0	-
	Dakota	F	1/2P-2 1/2P	1 1/2P	0-1P	1/2P	1/8P	NA
	Dakota	R	-	-	1/16P-3/32P	1/32P	1/32P	NA
1992	Caravan	F	-	1 51/6P	1/8N-3/4P	5/16P	1/16P	12.2
	Caravan	R	-	-	13/16N-7/16P	1/4P	0	-
	Voyager	F	-	1 51/6P	1/8N-3/4P	5/16P	1/16P	12.2
	Voyager	R	-	-	13/16N-7/16P	1/4P	0	-
	Town&Country	F	-	1 51/6P	1/8N-3/4P	5/16P	1/16P	12.2
	Town&Country	R	-	-	13/16N-7/16P	1/4P	0	-
	Dakota	F	1/2P-2 1/2P	1 1/2P	0-1P	1/2P	1/8P	NA
	Dakota	R	-	-	1/16P-3/32P	1/32P	1/32P	NA
1993	Caravan	F	-	1 5/16P	1/4N-1P	3/4P	1/16P	12.2
	Caravan	R	-	-	13/16N-7/16P	1/4P	0	-
	Voyager	F	-	1 5/16P	1/4N-1P	3/4P	1/16P	12.2
	Voyager	R	-	-	13/16N-7/16P	1/4P	0	-
	Town&Country	F	-	1 5/16P	1/4N-1P	3/4P	1/16P	12.2
	Town&Country	R	-	-	13/16N-7/16P	1/4P	0	-
	Dakota	F	1/2P-2 1/2P	1 1/2P	0-1P	1/2P	1/8P	NA
	Dakota	R	-	-	1/16P-3/32P	1/32P	1/32P	NA

WHEEL ALIGNMENT

Year	Model		Caster Range (deg.)	Caster Preferred Setting (deg.)	Camber Range (deg.)	Camber Preferred Setting (deg.)	Toe-in (in.)	Steering Axis Inclination (deg.)
1994-95	Caravan	F	-	1 5/16P	1/4N-3/4P	5/16P	1/16P	NA
	Caravan	R	-	-	13/16N-7/16P	1/4P	0	NA
	Voyager	F	-	1 5/16P	1/4N-3/4P	5/16P	1/16P	NA
	Voyager	R	-	-	13/16N-7/16P	1/4P	0	NA
	Town&Country	F	-	1 5/16P	1/4N-3/4P	5/16P	1/16P	NA
	Town&Country	R	-	-	13/16N-7/16P	1/4P	0	NA
	Dakota	F	1/2P-2 1/2P	1 1/2P	0-1P	1/2P	1/8P	NA
	Dakota	R	-	-	0-1P	1/2P	1/8P	NA

NA - Not Available

ENGINE MECHANICAL

NOTE: Disconnecting the negative battery cable on some vehicles may interfere with the functions of the on board computer systems and may require the computer to undergo a relearning process, once the negative battery cable is reconnected.

Engine Assembly

REMOVAL & INSTALLATION

Caravan, Voyager and Town & Country

2.5L ENGINES

1. If equipped with fuel injection, relieve the fuel pressure. Disconnect the negative battery cable and all engine ground straps.
2. Mark the hood hinge outline on the hood and remove the hood.
3. Drain the cooling system. Remove the radiator hoses, fan assembly, radiator shroud and radiator.
4. Remove the air cleaner, duct hoses and oil filter.
5. Unbolt the air conditioning compressor from its mount, if equipped, and position it to the side.
6. Remove the power steering pump mounting bolts and position the pump to the side, without disconnecting any fluid lines.
7. Label and disconnect all electrical connectors from the engine, alternator and carburetor or fuel injection system.

8. Disconnect the fuel line, heater hoses and accelerator linkage.
9. Disconnect the air pump lines and remove the pump, if equipped.
10. Remove the alternator.
11. Disconnect the shift linkage(s), clutch linkage, as required, and speedometer cable.
12. Raise the vehicle and support safely. Disconnect the exhaust pipe from the manifold. Remove the right inner fender shield.
13. If equipped with a manual transaxle, remove the transaxle.
14. If equipped with an automatic transaxle, perform the following procedures:
    a. Remove the lower cover from the transaxle case.
    b. Remove the exhaust pipe-to-exhaust manifold bolts. Separate the pipe from the manifold.
    c. Remove the starter and set it aside.
    d. Matchmark the flexplate to the torque converter, for installation purposes.
    e. Remove the torque converter bolts. Separate the converter from the flexplate. Remove the lower bell housing bolts.
15. Lower the vehicle and support the transaxle (if still in the vehicle) with a floor jack or equivalent. Attach an engine lifting device to the engine.
16. To lower the engine, separate the right side engine bracket from the yoke bracket. To raise the engine, remove the yoke/insulator long bolt.

NOTE: If removing the insulator-to-rail screws, first mark the position of the insulator on the side rail to insure proper alignment during reinstallation.

17. Remove the remaining bell housing bolts. Remove the front engine mount nut/bolt and the left insulator through bolt or the insulator bracket-to-transaxle bolts.
18. Lift the engine from the vehicle.
To install:
19. Lower the engine into the engine compartment. Loosely install all of the mounting bolts. With all bolts installed, torque the:
    Engine to mount bolts to 40 ft. lbs.
    Engine to transaxle bolts to 70 ft. lbs.
    Torque converter bolts to 40 ft. lbs.
20. Remove the engine hoist and transmission holding fixture.
21. Secure ground strap. Install right inner splash shield and starter assembly.
22. Connect the exhaust system to the engine manifold.
23. Reinstall the alternator, power steering pump and the air conditioner compressor.
24. Connect the fuel line, heater hoses and the accelerator cable. Reconnect all electrical connectors.
25. Install oil filter and refill the engine crankcase to the correct level.
26. Reinstall the radiator hoses, fan assembly, radiator shroud and radiator. Add coolant to the proper level.
27. Install the hood.
28. Start the engine and run until operating temperature is reached. Recheck all fluid levels and add as required.

3.0L AND 3.3L ENGINES

1. If equipped with fuel injection, relieve the fuel pressure. Disconnect the negative battery cable.

2. Matchmark the hinge-to-hood position and remove the hood.

3. Drain the cooling system. Disconnect and label all engine electrical connections.

4. Remove the coolant hoses from the radiator and engine. Remove the radiator and cooling fan assembly.

5. Remove the air cleaner assembly. Disconnect the fuel lines from the engine. Disconnect the accelerator cable from the engine.

6. Raise the vehicle and support safely. Drain the engine oil.

7. Remove the air conditioning compressor mounting bolts, the drive belt(s) and position the compressor to the side. Disconnect the exhaust pipe from the exhaust manifold.

8. Remove the transaxle inspection cover, matchmark the converter to the flexplate and remove the torque converter bolts.

9. Remove the power steering pump mounting bolts and set the pump aside, upright, with the fluid lines attached.

10. Remove the lower bell housing bolts. Disconnect and label the starter motor wiring and remove the starter motor from the engine.

11. Lower the vehicle. Disconnect and label the vacuum hoses and engine ground straps.

12. Support the transaxle with a floor jack or equivalent. Attach an engine lifting device to the engine.

13. Remove the upper transaxle-to-engine bolts.

14. To separate the engine mounts from the insulators, mark the right insulator-to-right frame support and remove the mounting bolts. Remove the front engine mount through bolt. Remove the left insulator through bolt, from inside the wheel housing. Remove the insulator bracket-to-transaxle bolts.

15. Lift and remove the engine from the vehicle.

To install:

16. Lower the engine into the engine compartment. Align the engine mounts and install the bolts; do not tighten the bolts until all bolts have been installed. Install the upper transaxle-to-engine mounting bolts and torque to 75 ft. lbs.

17. Remove the engine lifting fixture from the engine.

18. Raise the vehicle and support safely.

19. Align the converter marks and install the torque converter bolts. Install the transaxle inspection cover.

20. Connect the exhaust pipe to the exhaust manifold. Install the starter motor and connect the wiring.

21. Install the power steering pump and air conditioning compressor. Adjust the drive belt tension, if necessary.

22. Lower the vehicle. Reconnect all vacuum hoses and electrical connections to the engine.

23. Connect the fuel lines and accelerator cable.

24. Install the radiator and fan assembly. Connect the fan motor wiring. Connect the radiator hoses and refill the cooling system.

25. Refill the engine with the proper oil to the correct level.

26. Connect the engine ground straps. Install the hood and connect the battery.

27. Start and run the engine until normal operating temperature is reached. Check for leaks and correct levels.

Dakota

2.5L ENGINE

1. If equipped with fuel injection, relieve the fuel pressure. Disconnect the negative battery cable and all engine ground straps.

2. Mark the hood hinge outline on the hood and remove the hood.

3. Drain the cooling system. Remove the radiator hoses, heater hoses, radiator shroud and radiator.

4. Remove the air cleaner, duct hoses and oil filter.

5. Unbolt the air conditioning compressor from its mount, if equipped, and position it to the side.

6. Remove the power steering pump mounting bolts and position the pump to the side, without disconnecting any fluid lines.

7. Label and disconnect all electrical connectors from the engine, alternator and carburetor or fuel injection system.

8. Disconnect the fuel line, heater hoses and accelerator linkage.

9. Disconnect the air pump lines and remove the pump, if equipped.

10. Remove the alternator. Remove the charcoal canister and the horn(s).

11. Disconnect the shift linkage(s), clutch linkage, as required, and speedometer cable.

12. Raise the vehicle and support safely. Disconnect the exhaust pipe from the manifold. Remove the right inner fender shield.

13. Remove the exhaust pipe-to-exhaust manifold bolts. Separate the pipe from the manifold. Remove the starter and set it aside.

14. Matchmark the flexplate to the torque converter, for installation purposes. Remove the torque converter bolts. Separate the converter from the flexplate.

15. Remove the lower bell housing bolts.

16. Lower the vehicle and support the transmission, if still in the vehicle, with a floor jack or equivalent. Attach an engine lifting device to the engine.

17. Remove the engine mount bolts.

18. Lift the engine from the vehicle.

To install:

19. Lower the engine into the engine compartment. Loosely install all of the mounting bolts. With all bolts installed, torque the:

Transmission to clutch housing bolts to 30 ft. lbs.

Engine front mount through bolts and nuts to 50 ft. lbs.

Torque converter bolts to 40 ft. lbs.

20. Remove the engine hoist and transmission holding fixture.

21. Secure ground strap. Install right inner splash shield and starter assembly.

22. Connect the exhaust system to the engine manifold.

23. Reinstall the alternator, power steering pump and the air conditioner compressor.

24. Connect the fuel line, heater hoses and the accelerator cable. Reconnect all electrical connectors.

25. Install throttle body with new gasket to intake manifold and torque mounting nuts to 50 ft. lbs. (68 Nm).

26. Install oil filter and refill the engine crankcase to the correct level.

27. Reinstall the radiator hoses, heater hoses, radiator shroud and radiator. Add coolant to fill to the proper level.

28. Install the hood.

29. Start the engine and run until operating temperature is reached. Recheck all fluid levels and add as required.

3.9L AND 5.2L ENGINES

1. If equipped with fuel injection, relieve the fuel pressure. Disconnect the negative battery cable and all engine ground straps.

2. Mark the hood hinge outline on the hood and remove the hood.

3. Drain the cooling system. Remove the radiator hoses, heater hoses, radiator shroud and radiator.

4. Remove the air cleaner, duct hoses and oil filter. Remove the distributor cap with the wires installed.

5. Unbolt the air conditioning compressor from its mount, if equipped and position it to the side.

6. Remove the power steering pump mounting bolts and position the pump to the side, without disconnecting any fluid lines.

7. Label and disconnect all electrical connectors from the engine, alternator and carburetor or fuel injection system.

8. Disconnect the fuel line, heater hoses and accelerator linkage.

9. Disconnect the air pump lines and remove the pump, if equipped.

10. Remove the alternator. Remove the charcoal canister and the horn(s).

11. Disconnect the shift linkage(s), clutch linkage (as required) and speedometer cable.

12. Raise the vehicle and support safely. Disconnect the exhaust pipe from the manifold. Remove the right inner fender shield.

13. Remove the exhaust pipe-to-exhaust manifold bolts. Separate the pipe from the manifold. Remove the starter and set it aside.

14. Matchmark the flexplate to the torque converter, for installation purposes. Remove the torque converter bolts. Separate the converter from the flexplate.

15. Remove the lower bell housing bolts.

16. On manual transmission, support transmission with a stand and disconnect the clutch release mechanism. Remove the transmission to clutch housing bolts. Move the engine forward until the drive pinion clears the clutch disc.

17. Remove the engine mount bolts.

18. Install the engine lifting fixture. Do not lift the engine by the intake manifold.

19. If equipped with 2WD, remove the engine front mount nuts.

20. If equipped with 4WD, the engine and front driving axle are connected through insulators and support brackets. Support the axle and separate as follows:

   a. On the left side, remove the 2 bolts attaching the bracket to the transmission bell housing and 2 bracket to pinion nose adaptor bolts. Separate the engine from the insulator by removing the upper nut washer assembly and through-bolt from the engine support bracket.

   b. On the right side, remove the 2 bracket to axle bolts and 1 bracket to bellhouisng bolt. Separate the engine from the insulator by removing the upper nut washer assembly and the through-bolt from the engine support bracket.

21. Lower the vehicle and remove the engine.

To install:

22. Lower the engine into position and install the engine mount bolts. Install the upper bell housing bolts. Remove the engine removal device. Install the oil dipstick.

23. Raise the vehicle and support safely.

24. Install the engine mount and interconnect nuts and bolts. Install the remaining bell housing bolts.

25. If equipped with a manual transmission, install the transmission and all related parts.

26. If equipped with an automatic transmission, align the torque converter and flexplate and install the bolts. Install the inspection plate, starter and the engine to transmission struts, if equipped.

27. Connect the rubber fuel inlet and return hoses to the fuel lines at the front of the vehicle.

28. Install the exhaust pipe to the exhaust manifold(s). Install the lower radiator hose.

29. Lower the vehicle. Install the horn(s) and the charcoal canister.

30. Connect the heater hoses to the engine.

31. Make sure the negative battery cable is not connected to the battery. Connect the engine side of the negative cable to the engine. Install the alternator, air pump, power steering pump and all brackets.

32. Install the air conditioning compressor.

33. Install the throttle body, if equipped using a new base gasket and connect all linkages and cables. Connect all electrical connectors and vacuum hoses that were disconnected during the engine removal procedure.

34. Install accessory drive belts and adjust belt tensions as required.

35. Install the radiator and shroud.

36. Install the distributor cap with all spark plug wires attached.

37. Install the air cleaner assembly.

38. Install the hood.

39. Fill the engine with the specified amount of oil and fill the radiator with coolant.

40. Connect the negative battery cable and set all adjustments to specification.

Engine Mounts

REMOVAL & INSTALLATION

Front Mount

2.5L ENGINES

1. Disconnect negative battery cable. Remove skid plate, if equipped.

2. Position fan assembly to insure clearance for radiator and top hose when engine is moved. Install an engine support fixture.

3. Raise and safely support vehicle.

4. Remove nuts from front engine mounting brackets.

5. Remove the insulator support to crossmember bracket bolts and nuts.

6. Raise engine with support fixture far enough to remove mounts, remove remaining bolts and mount.

To install:

7. Install mounts to engine block. Tighten the bolts to 40 ft. lbs. (54 Nm).

8. If removed, install the insulator support to crossmember bracket bolts and nut and tighten to 30 ft. lbs. (41 Nm).

9. Lower engine to original position and insert bolts through mounting brackets and tighten to 50 ft. lbs. (68 Nm).

10. Install skid plate, if removed.

11. Lower vehicle and remove engine support fixture. Reconnect negative battery cable.

3.0L AND 3.3L ENGINES

1. Disconnect negative battery cable.

2. Support engine and transmission so it will not rotate. Remove the bolt from the insulator and the front crossmember mounting bracket.

3. Remove the front engine mount bracket to crossmember screws and nuts. Remove the insulator.

4. The installation is the reverse of the removal procedure.

3.9L AND 5.2L ENGINES WITHOUT 4WD

1. Disconnect negative battery cable. Remove skid plate, if equipped.

2. Position fan assembly to insure clearance for radiator and top hose when engine is moved. Install an engine support fixture. Do not lift engine by the intake manifold.

3. Raise and safely support vehicle.

4. Remove bolts and washers hold-

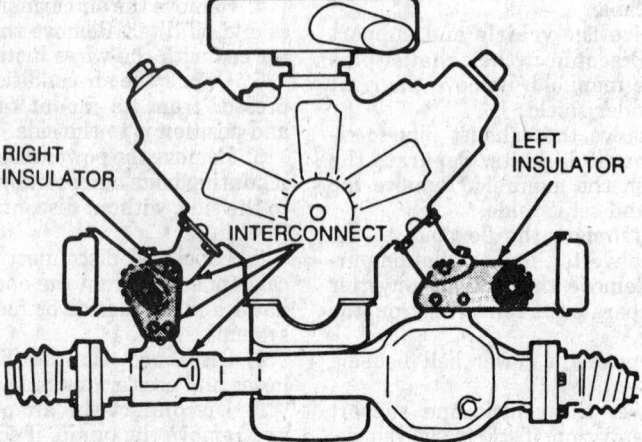

RIGHT INSULATOR   LEFT INSULATOR

INTERCONNECT

Interconnected motor mounts – Dakota with 4WD

ing insulator to the adapter and remove the insulator bolt.

5. Raise engine with the lifting fixture and remove insulator.

To install:

6. If adapter was removed, install to block and tighten bolts to 30 ft. lbs. (41 Nm).

7. If bracket was removed, install and tighten bolts to 30 ft. lbs. (41 Nm).

8. Position the insulator onto the adapter and install bolts and washers. Lower the engine.

9. Install insulator to bracket bolt and tighten nut to 50 ft. lbs. (68 Nm). Lower vehicle and remove support fixture.

3.9L AND 5.2L ENGINES WITH 4WD

The front engine mount on the 4WD vehicle attaches directly to engine adapters and the axle housing. The engine and front axle must be supported during any serrvice procedures involving these insulators.

1. Disconnect negative battery cable. Remove skid plate, if equipped.

2. Position fan assembly to insure clearance for radiator and top hose when engine is moved. Install an engine support fixture. Do not lift engine by the intake manifold.

3. Raise and safely support vehicle.

4. Install front axle lifting fixture. Remove insulator thru bolt and nut. Remove the nut and washer attaching the engine mount bracket to the insulator bracket.

5. Raise engine slightly to clear insulator bracket stud. Remove bolts attaching the insulator to the axle housing.

6. Remove insulator to frame bolt and remove insulator.

To install:

7. Position insulator in frame and align attaching points. Install bolts and tighten nuts to 75 ft. lbs. (102 Nm).

8. Lower engine while guiding engine bracket onto insulator stud and install nut and bolt. Tighten stud nut to 30 ft. lbs. (41 Nm). Tighten bolt to 75 ft. lbs. (102 Nm).

9. Remove front axle lifting fixture, lower vehicle and remove the engine supporting fixture.

Left Side Mount

3.0L AND 3.3L ENGINES

1. Disconnect negative battery cable.

2. Raise and safely support vehicle.

3. Remove the left front wheel and tire assembly.

4. Remove the inner splash shield. Support transmission and remove the insulator bolt from the mount.

5. Remove the transmission mount fasteners and remove the mount.

6. The installation is the reverse of the removal procedure.

Right Side Mount

3.0L AND 3.3L ENGINES

1. Remove the right engine mount insulator vertical fasteners from the frame rail.

2. Support the engine in a way that releases the load from the engine mounts.

3. Remove the bolt in the insulator assembly. Remove the insulator.

4. Installation is the reverse of the removal procedure.

Cylinder Head

REMOVAL & INSTALLATION

2.5L Engines

1. Relieve the fuel pressure if equipped with fuel injection. Disconnect the negative battery cable and unbolt it from the head. Drain the cooling system. Remove the dipstick bracket nut from the thermostat housing.

2. Remove the air cleaner assembly. Remove the upper radiator hose and disconnect the heater hoses.

3. Disconnect and label the vacuum lines, hoses and wiring connectors from the manifold(s), carburetor or throttle body and from the cylinder head. Remove the air pump, if equipped.

4. Disconnect all linkages and the fuel line from the carburetor or throttle body. Unbolt the cable bracket. Remove the ground strap attaching screw from the firewall.

5. If equipped with air conditioning, remove the upper compressor mounting bolts. The cylinder head can be remove with the compressor and bracket still mounted. Remove the upper part of the timing belt cover.

6. Raise the vehicle and support safely. Disconnect the converter from the exhaust manifold. Disconnect the water hose and oil drain from the turbocharger, if equipped.

7. Rotate the engine by hand, until the timing marks align (No. 1 piston at TDC). Lower the vehicle.

8. With the timing marks aligned, remove the camshaft sprocket. The camshaft sprocket can be suspended to keep the timing intact. Remove the spark plug wires from the spark plugs.

9. Remove the valve cover and curtain, if equipped. Remove the cylinder head bolts and washers, starting from the middle and working outward.

10. Remove the cylinder head from the engine.

NOTE: Before disassembling or repairing any part of the cylinder head assembly, identify factory installed oversized components. To do so, look for the tops of the bearing caps pained green and O/SJ stamped rearward of the oil gallery plug on the rear of the head. In addition, the barrel of the camshaft is painted green and O/SJ is stamped onto the rear end of the camshaft. Installing standard sized parts in an head equipped with oversized parts—or visa versa—will cause severe engine damage.

11. Clean the cylinder head gasket mating surfaces.

12. Using new gaskets and seals, install the head to the engine. Using new head bolts assembled with the old washers, torque the cylinder head bolts in sequence, to 45 ft. lbs. Repeating the sequence, torque the bolts to 65 ft. lbs. With the bolts at 65 ft. lbs., turn each bolt an additional ¼ turn.

NOTE: Head bolt diameter is 11mm. These bolts are identified with the No. 11 on the head of the bolt. The 10mm bolts used on previous vehicles will thread into an 11mm bolt hole but will permanently damage the cylinder block. Make sure the correct bolts are being used when replacing old head bolts.

13. Install the camshaft sprocket and timing belt.

14. Install valve cover using new side cover gaskets. Install all vacuum hoses, upper air conditioning compressor mounting bolts and throttle linkage.

15. Install all electrical wiring and oil fill tube retainer nut.

16. Raise vehicle and reinstall exhaust pipe to engine manifold. Install turbocharger lines, if equipped.

17. Adjust all engine fluid levels as required and reconnect negative battery cable. Start the engine and check for leaks.

3.0L Engine

1. Relieve the fuel pressure. Disconnect the negative battery cable. Drain the cooling system.

2. To remove the accessory drive belt, insert a ½ in. drive breaker bar into the square hole of the serpentine drive belt tensioner and rotate it counterclockwise to reduce the belt tension. Remove the drive belt. Remove the alternator and power steering pump from the brackets and move them aside.

3. Remove the air conditioning compressor from its mount and support it aside. Remove the adjustable drive belt tensioner from the block.

4. Raise the vehicle and support safely. Remove the right front wheel assembly and the right inner splash shield.

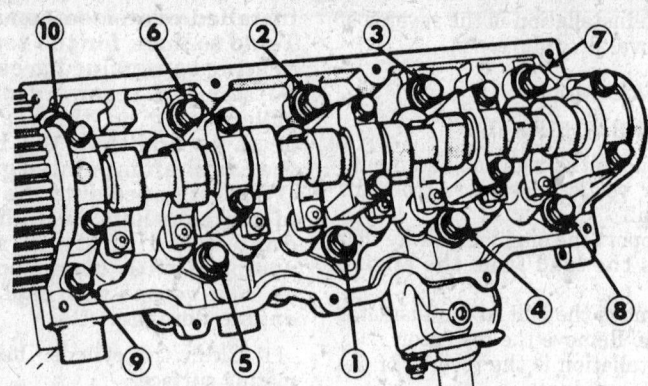

Cylinder head bolt torque sequence — 2.5L engines

4. Remove the crankshaft pulleys and the torsional damper.

5. Lower the vehicle. Using a floor jack and a block of wood positioned under the oil pan, raise the engine slightly. Remove the engine mount bracket from the timing cover end of the engine.

6. Remove the timing belt covers. If the same timing belt will be reused, mark the direction of the timing belt's rotation, for installation in the same direction. Make sure the engine is positioned so No. 1 cylinder is at the TDC of it's compression stroke and the sprockets timing marks are aligned with the engine's timing mark indicators.

7. Loosen the timing belt tensioner bolt and remove the belt.

NOTE: When removing the timing belt from the camshaft sprocket, make sure the belt does not slip off of the other camshaft sprocket. Support the belt so it can not slip off of the crankshaft sprocket and opposite side camshaft sprocket.

8. Hold camshaft sprocket with spanner wrench and remove bolt and washer. Remove camshaft sprocket. Remove valve cover(s) and install auto lash adjuster retainers tool MD998443 or equivalent, on the rocker arms.

9. If removing the right cylinder head, matchmark the distributor rotor to the distributor housing and the housing to distributor extension locations. Remove the distributor and the distributor extension.

10. Note position of rocker arms and bearing caps and loosen but do not remove the bolts from the camshaft bearing caps. Remove the rocker arm, rocker shafts and bearing caps as an assembly. Remove the camshafts from the cylinder head and inspect them for damage, if necessary.

11. Remove the intake manifold assembly.

12. Remove the exhaust manifold.

13. Remove the air cleaner assembly.

Label and disconnect the spark plug wires and the vacuum hoses.

14. Remove the cylinder head bolts starting from the outside and working inward. Remove the cylinder head from the engine.

15. Clean the gasket mounting surfaces and check the heads for warpage; the maximum warpage allowed is 0.008 in. (0.20mm).

To install:

16. Install the new cylinder head gaskets over the dowels on the engine block.

17. Install the cylinder heads on the engine and torque the cylinder head bolts in sequence using 3 even steps, to 80 ft. lbs. (108 Nm).

18. Install the air cleaner assembly. Install spark plug wires and vacuum hoses.

19. Install exhaust and intake manifold to engine.

20. Install rocker arm assembly to cylinder head making sure to torque in the correct order. Install distributor drive and adaptor assembly.

21. Install the camshaft sprocket(s).

22. Install and align timing belt as follows:

a. Position both camshafts so the marks line up with timing marks on the alternator bracket (rear bank) and inner timing cover (front bank). Rotate the crankshaft so the timing mark aligns with the mark on the oil pump.

b. Install the timing belt on the crankshaft sprocket and while keeping the belt tight on the tension side (right side), install the belt on the front camshaft sprocket.

c. Install the belt on the water pump pulley, rear camshaft sprocket and then the belt tensioner.

d. Rotate the front camshaft counterclockwise to tension the belt between the front camshaft and the crankshaft. If the timing marks came out of line, repeat the procedure.

e. Install the crankshaft sprocket flange.

f. Loosen the tensioner bolt and allow the spring to tension the belt.

g. Turn the crankshaft 2 full turns in the clockwise direction until the timing marks align again. Now that the belt is properly tensioned, torque the tensioner lock bolt to 21 ft. lbs. (29 Nm).

23. Refill the cooling system. Connect the negative battery cable. Start the engine and check for leaks. Adjust the timing as required.

3.3L Engine

1. Relieve the fuel pressure. Disconnect the negative battery cable. Drain the cooling system.

2. Remove the intake manifold with throttle body.

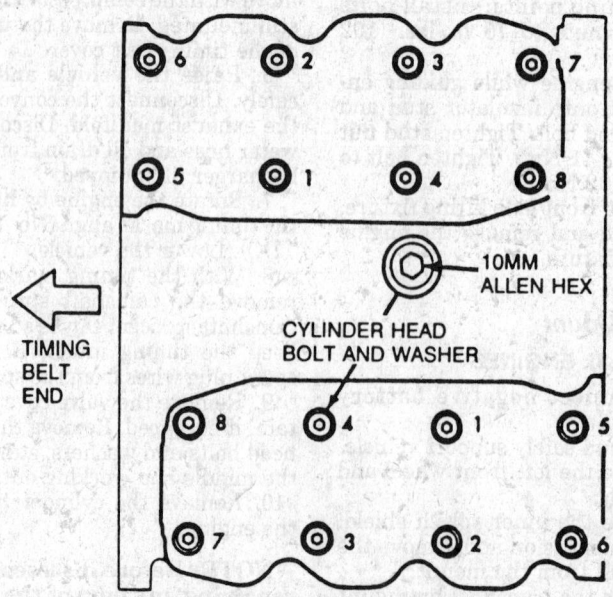

Cylinder head bolt torque sequence — 3.0L engine

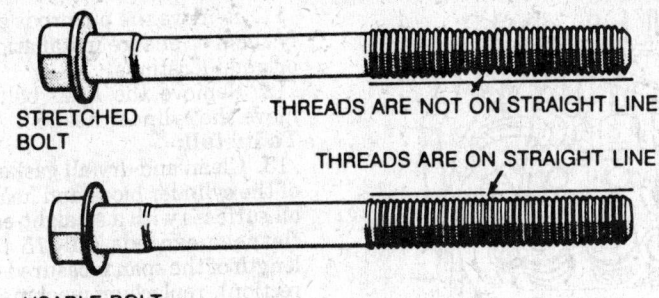

STRETCHED BOLT

THREADS ARE NOT ON STRAIGHT LINE

THREADS ARE ON STRAIGHT LINE

USABLE BOLT

Checking the head bolt for stretching

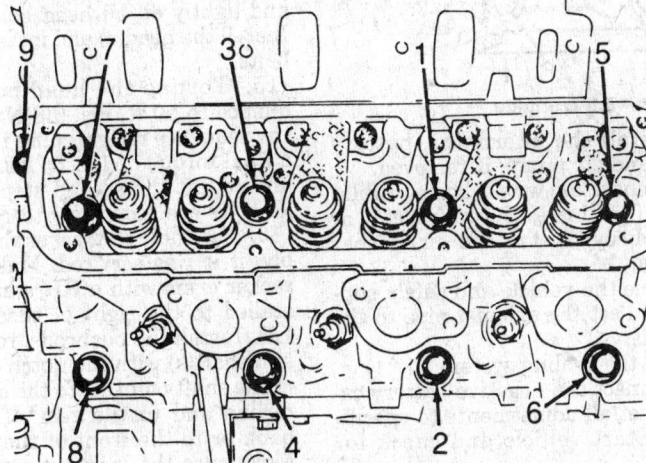

Cylinder head bolt torque sequence—3.3L engine

3. Disconnect the coil wires, sending unit wire, heater hoses and bypass hose.

4. Remove the closed ventilation system, evaporation control system and cylinder hear cover.

5. Remove the exhaust manifold.

6. Remove the rocker arm and shaft assemblies. Remove the pushrods and identify them in ensure installation in their original positions.

7. Remove the head bolts and remove the cylinder head from the block.

To install:

8. Clean the gasket mounting surfaces and install a new head gasket to the block.

9. Install the head to the block. Before installing the head bolts, inspect them for stretching. Hold a straight edge up to the threads. If the threads are not all on line, the bolt is stretched and should be replaced.

10. Torque the bolts in sequence to 45 ft. lbs. (61 Nm). Repeat the sequence and torque the bolts to 65 ft. lbs. (88 Nm). With the bolts at 65 ft. lbs., turn each bolt an additional ¼ turn.

11. Torque the lone head bolt to 25 ft. lbs. (33 Nm) after the other 8 bolts have been properly torqued.

12. Install the pushrods, rocker arms and shafts and torque the bolts to 21 ft. lbs. (12 Nm).

13. Place a drop of silicone sealer onto each of the 4 manifold to cylinder head gasket corners.

——— CAUTION ———

The intake manifold gasket is composed of very thin and sharp metal. Handle this gasket with care or damage to the gasket or personal injury could result.

14. Install the intake manifold gasket and torque the end retainers to 105 inch lbs. (12 Nm).

15. Install the intake manifold and torque the bolts in sequence to 10 inch lbs. Repeat the sequence increasing the torque to 17 ft. lbs. (23 Nm) and recheck each bolt for 17 ft. lbs. of torque. After the bolts are torqued, inspect the seals to ensure that they have not become dislodged.

16. Lubricate the injector O-rings with clean oil and position the fuel rail in place. Install the rail mounting bolts.

17. Install the valve cover with a new gasket. Install the exhaust manifold.

18. Install or connect all remaining items that were removed or disconnected during the removal procedure.

19. Refill the cooling system. Connect the negative battery cable. Start the engine and check for leaks.

3.9L Engine

1. Relieve the fuel pressure if equipped with fuel injection. Disconnect the negative battery cable and drain the cooling system.

2. Raise the vehicle and safely support. Disconnect the exhaust pipe from the manifolds.

3. Remove the alternator if the right head is being removed. Remove the air pump and battery ground cable if the left head is being removed.

4. Remove the air cleaner assembly. Unbolt the air conditioning compressor and lay it to the side, if equipped. Remove the distributor cap with all wires attached.

5. Disconnect all wires, hoses, linkages and cables from the carburetor or throttle body. Disconnect and plug the fuel line.

6. Disconnect the ignition coil and coolant temperature sending unit electrical connectors.

7. Disconnect the heater hose, upper radiator hose and the lower bypass hose clamp.

8. Remove the valve covers.

9. Remove the intake manifold assembly. Remove the exhaust manifolds.

10. Remove the rocker arm and shaft assembly from the heads. Do not disassemble unless service is required.

11. Remove the pushrods and identify them to ensure installation in their original locations.

12. Remove the head bolts and remove the cylinder head(s).

To install:

13. Clean and dry all gasket surfaces of the cylinder block and head. Inspect all surfaces with a straightedge. If the flatness exceeds 0.0075 times the length of the span measured, in any direction, replace or machine the head gasket surface.

14. Using no sealer, install the new head gasket(s) to the block. Clean, dry and lightly oil all head bolts threads. Install the heads and install the head bolts.

15. Torque the head bolts in sequence to 50 ft. lbs. (68 Nm). Repeat the sequence retightening the bolts to a final torque of 105 ft. lbs. (143 Nm) and repeat the second step to ensure all bolts are accurately torqued.

16. Assemble the rocker shaft assembly, if it was serviced. Make sure all rocker arms with an **RH** stamp are installed to the right of those with **LH**. Install the pushrods, rocker arms and shaft(s) with the notch on the end of the shaft pointing to the engine centerline and to the rear of the right bank or to the front of the left bank. Make sure the long stamped steel retainers are at the No. 2 and 4 positions. Torque the bolts evenly and gradually to 17 ft. lbs. (23 Nm).

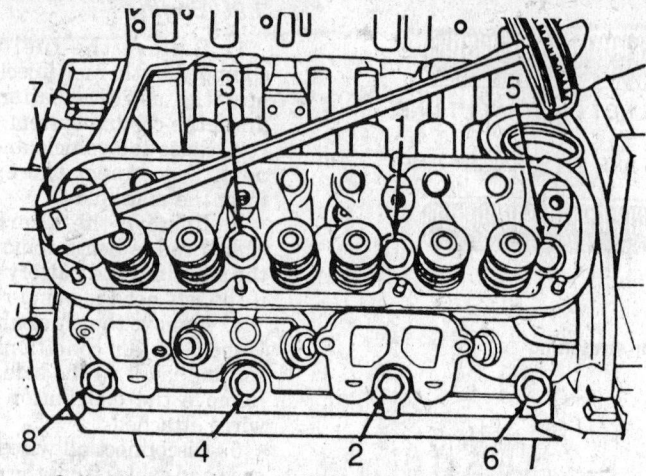

Cylinder head bolt torque sequence—3.9 L engine

17. Clean and dry the intake manifold contact surfaces. Coat the intake manifold side gaskets very lightly with sealer and install the gaskets to the heads. Cutouts at the front of the gaskets differentiate the right and left sides.

18. Apply a thin uniform coat of quick dry cement to the front and rear intake manifold gaskets and mounting surfaces on the block and apply a thin bead of sealer to each of 4 the corners. Install the front and rear gaskets engaging the hole in the block and the tangs from the head gaskets. Apply a second thin bead of sealer above the gaskets in the 4 corners.

19. Carefully lower the intake manifold into position engaging the bypass hose; after it is satisfactorily in place, inspect the gaskets to make sure they have not become dislodged.

20. Install the intake manifold bolts. On 1988–91 engine torque in sequence to 25 ft. lbs. (34 Nm). Repeat the sequence retightening the bolts to a final torque of 40 ft. lbs. (54 Nm) and repeat the second step to ensure all bolts are accurately torqued. On 1992 engine torque in steps to 6 ft. lbs. (8 Nm). Repeat torque procedure to 12 ft. lbs. (16 Nm). Repeat to assure all bolts are torqued to 12 ft. lbs. (16 Nm).

21. Install the exhaust manifold(s) and torque the bolts to 20 ft. lbs. (27 Nm). Torque the end nuts to 15 ft. lbs. (20 Nm).

22. Clean and dry the valve cover mating surfaces, bolts and bolt holes. Install the valve covers each with a new gasket.

23. Connect the heater hose, upper radiator hose and the lower bypass hose clamp.

24. Connect the ignition coil, coolant temperature sending unit wire and all other connectors that were disconnected along the wiring harness.

25. Install the air conditioning compressor, if equipped. Install the distributor cap and all spark plug wires.

26. Install the alternator, battery ground and air pump, if removed.

27. Connect all wires, hoses, cables and the fuel line to the carburetor or throttle body. Install the air cleaner assembly.

28. Raise the vehicle and safely support. Connect the exhaust pipe to the manifolds.

29. Fill the cooling system.

30. Connect the negative battery cable and set all adjustments to specification. Start vehicle and check for leaks.

5.2L Engine

1. Relieve the fuel pressure if equipped with fuel injection. Disconnect the negative battery cable from the battery and drain the cooling system.

2. Raise the vehicle and safely support. Disconnect the exhaust pipe from the manifolds.

3. Remove the alternator, if the right head is being removed, and the air pump and battery ground cable, if the left head is being removed.

4. Remove the air cleaner assembly. Unbolt the air conditioning compressor, if equipped, and lay it aside. Remove the distributor cap with all wires attached.

5. Disconnect all wires, hoses, linkages and cables from the carburetor or throttle body. Disconnect the fuel line.

6. Disconnect the ignition coil, coolant temperature sending unit wire and all other connectors along the wiring harness connected to items on the intake manifold.

7. Disconnect the heater hose, upper radiator hose and the lower bypass hose clamp.

8. Remove the valve cover(s).

9. Remove the intake manifold assembly. Remove the exhaust manifold(s).

10. Remove the rocker arm and shaft assembly from the head(s). Do not disassemble unless service is required.

11. Remove the pushrods and identify them to ensure installation in their original locations.

12. Remove the head bolts and remove the cylinder head(s).

To install:

13. Clean and dry all gasket surfaces of the cylinder block and head. Inspect all surfaces with a straight-edge. If the flatness exceeds 0.0075 times the length of the span measured (in any direction), replace or machine the head gasket surface.

14. Using no sealer, install the new head gasket(s) to the block. Clean, dry and lightly oil all head bolts threads. Install the head(s) and install the head bolts.

15. Torque the head bolts in sequence to 50 ft. lbs. (68 Nm). Repeat the sequence retightening the bolts to a final torque of 105 ft. lbs. (143 Nm) and repeat the second step to ensure all bolts are accurately torqued.

16. Assemble the rocker shaft assembly, if it was serviced. Make sure all rocker arms with a **RH** stamp are installed to the right of those with an **LH**. Install the pushrods, rocker arms and shaft(s) with the notch on the end of the shaft pointing to the engine centerline and to the rear of the right bank or to the front of the left bank. Make sure the long stamped steel retainers are at the No. 2 and 4 positions. Torque the bolts evenly and gradually to 200 inch lbs. (23 Nm).

17. Clean and dry the intake manifold contact surfaces. Coat the intake manifold side gaskets lightly with sealer and install the gaskets to the heads. Cutouts at the front of the gaskets differentiate the right and left sides.

18. Apply a thin uniform coat of quick dry cement to the front and rear intake manifold gaskets and mounting surfaces on the block and apply a thin bead of sealer to each of 4 the corners. Install the front and rear gaskets engaging the hole in the block and the tangs from the head gaskets. Apply a second thin bead of sealer above the gaskets in the 4 corners.

19. Carefully lower the intake manifold into position engaging the bypass hose. Inspect the gaskets to make sure they have not become dislodged.

20. Install the intake bolts and torque to 6 ft. lbs. (8 Nm). Repeat torque procedure to 12 ft. lbs. (16 Nm). Repeat to assure all bolts are torqued to 12 ft. lbs. (16 Nm).

21. Install the exhaust manifold(s) and torque the bolts to 20 ft. lbs. (27 Nm). Torque the end nuts to 15 ft. lbs. (20 Nm).

22. Clean and dry the valve cover mating surfaces, bolts and bolt holes. Install the valve cover(s) each with a new gasket.

23. Connect the heater hose, upper

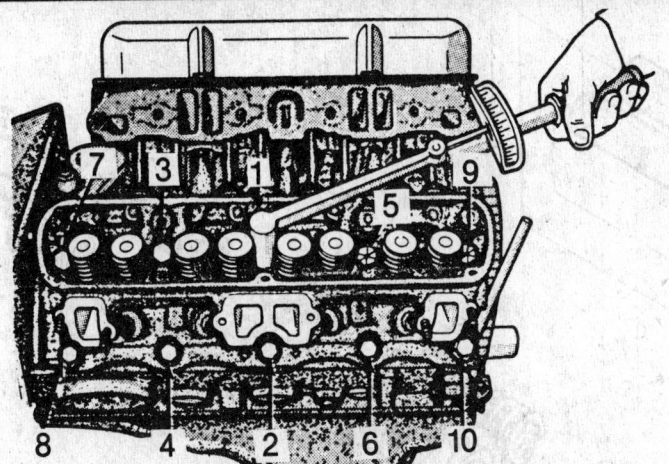

Cylinder head torque sequence—1992 5.2L engine

radiator hose and the lower bypass hose clamp.

24. Connect the ignition coil, coolant temperature sending unit wire and all other connectors that were disconnected along the wiring harness.

25. Install the air conditioning compressor, if equipped. Install the distributor cap and all spark plug wires.

26. Install the alternator, battery ground and air pump, if removed.

27. Connect all wires, hoses, cables and the fuel line to the carburetor or throttle body. Install the air cleaner assembly.

28. Raise the vehicle and safely support. Connect the exhaust pipe to the manifolds.

29. Fill the cooling system.

30. Connect the negative battery cable and set all adjustments to specification.

Valve Lifters

REMOVAL & INSTALLATION

2.5L Engines

1. Disconnect the negative battery cable.

2. Remove the valve cover and curtain. If removing all lifters, remove the camshaft and rocker arms.

3. If only removing 1 lifter, rotate the crankshaft until the low point of the desired cam lobe is contacting the rocker arm.

4. Using the special valve spring compressor tool 4682 or equivalent, depress the valve spring without dislodging the keepers and slide the rocker arm out.

5. Remove the valve lifter(s) from the bore(s).

6. Lubricate the lifter(s) and their bore(s) with clean engine oil.

7. The installation is the reverse of the removal procedure.

8. Connect the negative battery cable and check the lifters for proper operation.

3.0L Engine

1. Disconnect the negative battery cable. Remove the air cleaner assembly.

2. Remove the valve cover.

3. Using the valve lifter retainer tools MD998443 or equivalent, install them on the rocker arms to keep the lifters from falling out.

4. On the right side cylinder head, remove the distributor extension.

5. Have a helper hold the rear end of the camshaft down. If the rear of the camshaft cannot be held down, the belt will dislodge and the valve timing will be lost. Loosen the camshaft cap bolts but do not remove them from the caps. Remove the caps, arms, shafts and bolts all as an assembly.

6. Remove the lifter(s) from the rocker arm(s).

7. Lubricate the lifter(s) and their bore(s) with clean engine oil.

8. The installation is the reverse of the removal procedure.

9. Connect the negative battery cable and check the lifters for proper operation.

3.3L Engine

1. Disconnect the negative battery cable. Relieve the fuel pressure.

2. Remove the cylinder head(s) to gain access to the valve lifter(s).

3. Remove the yoke retainer and aligning yoke(s).

4. Use an appropriate valve lifter removal tool to remove each lifter from its bore. If reinstalling the tappets, identify each upon removal to ensure installation in the original position.

To install:

5. Lubricate the lifter(s) and bore(s) and install.

6. Install aligning yoke(s).

7. Install the yoke retainer and torque the bolts to 105 inch lbs. (12 Nm).

8. Install the cylinder head(s) and all related components.

9. Connect the negative battery cable and check the lifters for proper operation.

3.9L and 5.2L Engines

1. Disconnect negative battery cable. Remove air cleaner.

2. Relieve the fuel pressure if equipped with fuel injection.

3. Remove the valve cover(s) and rocker arm assembly.

4. Remove the pushrods and identify them to ensure installation in their original locations.

5. Remove the intake manifold assembly, yoke retainer and aligning yokes.

6. Use an appropriate valve lifter removal tool to remove each lifter from its bore. If reinstalling the tappets, identify each upon removal to ensure installation in the original position.

7. If the tappet or bore in cylinder block is scored, scuffed or shows signs of sticking, ream the bore to the next oversize and replace with oversize tappet.

To install:

8. Lubricate the lifter(s) and bore(s) and install. Ensure that the oil feed hole in the side of the tappet body faces up.

9. Install aligning yokes with arrow toward camshaft. Install yoke retainer. Tighten bolts to 200 inch lbs. (23 Nm).

10. Install pushrods to their original positions. Install rocker arm and shaft assembly.

11. Install valve cover(s). Connect negative battery terminal and start vehicle. Check for oil leaks.

Rocker Arms and Shafts

REMOVAL & INSTALLATION

2.5L Engines

1. Disconnect the negative battery cable.

2. Remove the valve cover.

3. Rotate the crankshaft until the low point of the desired cam lobe is contacting the rocker arm.

4. Use the special valve spring compressor tool or equivalent, depress the valve spring (without dislodging the keepers) and slide the rocker arm out.

5. The installation is the reverse of the removal procedure.

3.0L Engine

1. Disconnect the negative battery cable. Remove the air cleaner assembly.

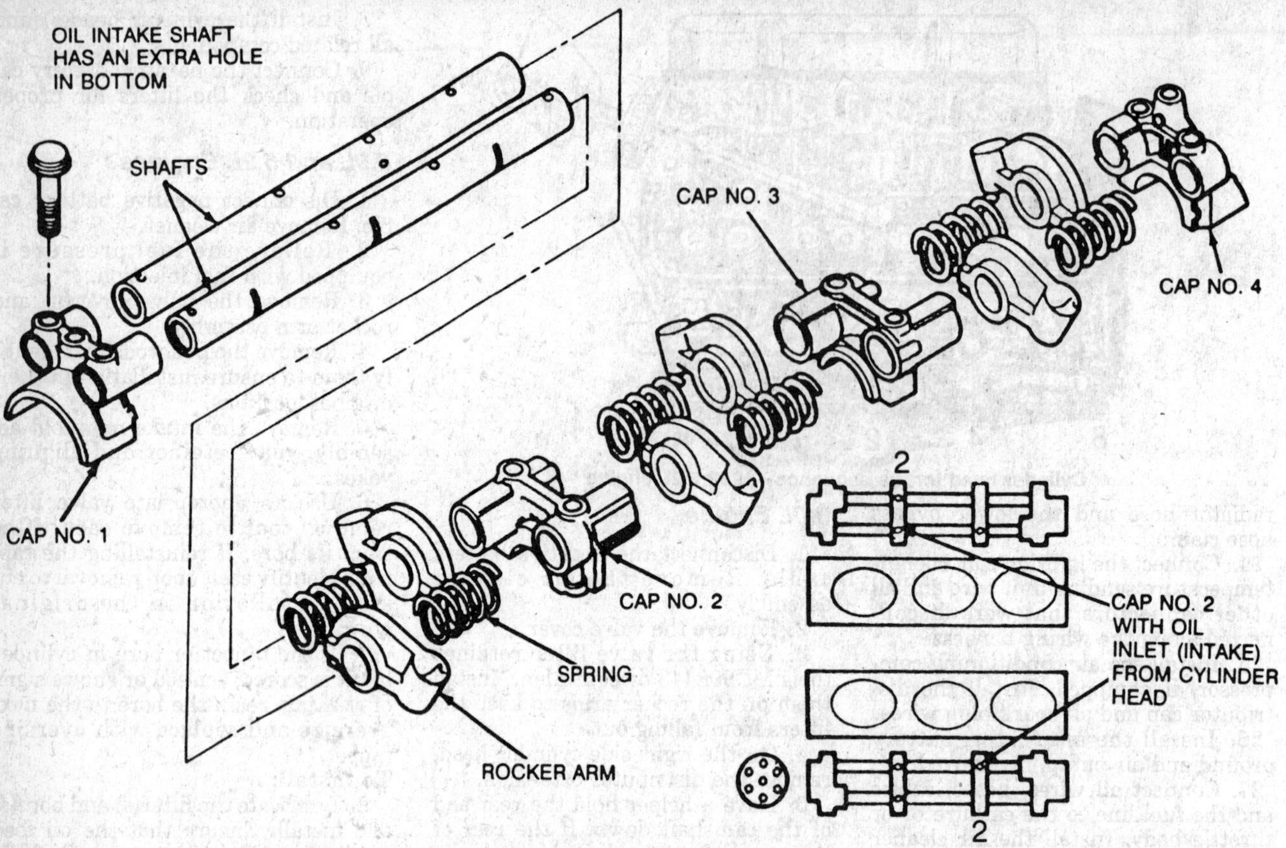

OIL INTAKE SHAFT
HAS AN EXTRA HOLE
IN BOTTOM

SHAFTS

CAP NO. 3

CAP NO. 4

CAP NO. 1

CAP NO. 2

CAP NO. 2
WITH OIL
INLET (INTAKE)
FROM CYLINDER
HEAD

SPRING

ROCKER ARM

Rocker shafts/arms assembly—3.0L engine

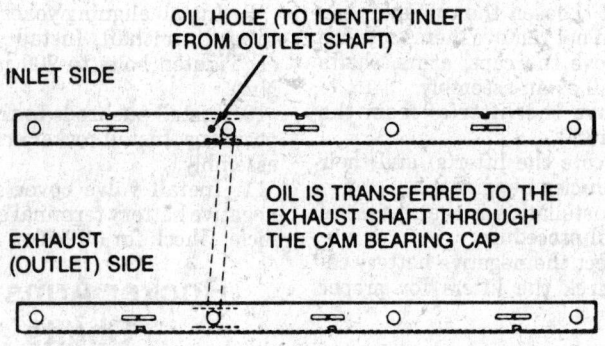

OIL HOLE (TO IDENTIFY INLET
FROM OUTLET SHAFT)

INLET SIDE

OIL IS TRANSFERRED TO THE
EXHAUST SHAFT THROUGH
THE CAM BEARING CAP

EXHAUST
(OUTLET) SIDE

Identifying the rocker shafts—3.0L engine

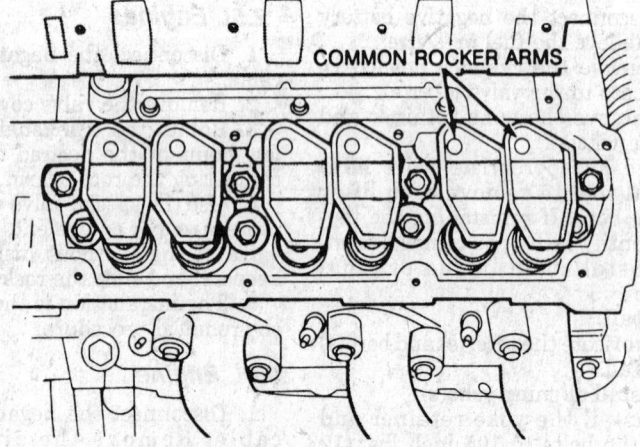

COMMON ROCKER ARMS

Rocker shaft/arms assembly—3.3L engine

2. Remove the valve cover.

3. Using the auto lash adjuster retainer tools MD998443 or equivalent, install them on the rocker arms to keep the lash adjusters from falling out.

4. On the right side cylinder head, remove the distributor extension.

5. Have a helper hold the rear end of the camshaft down. If the rear of the camshaft cannot be held down, the belt will dislodge and the valve timing will be lost. Loosen the camshaft cap bolts but do not remove them from the caps. Remove the caps, arms, shafts and bolts all as an assembly.

6. Disassemble the unit keeping all parts in order and repair as required.

7. The installation is the reverse of the removal procedure. Apply a drop of sealant to the rear edge of the rear cap. Torque the cap bolts first to 85 inch lbs. (19 Nm), then to 180 inch lbs. (19 Nm) in the following order: No. 3 cap, No. 2 cap, No. 1 cap, No. 4 cap.

3.3L Engine

1. Disconnect the negative battery cable.

2. Remove the upper intake manifold assembly.

3. Remove the valve cover.

4. Remove the rocker shaft retaining bolts and retainers.

5. Remove the rocker shaft and arm

assembly. Disassemble and repair as required.

6. The installation is the reverse of the removal procedure. Torque the retaining bolts gradually and evenly to 21 ft. lbs. (28 Nm).

7. Allow 20 minutes tappet bleed down time after rocker shaft installation before starting the engine.

3.9L and 5.2L Engines

1. Disconnect the negative battery cable.

2. Remove the valve cover and gasket.

3. Note the positioning of the oil notch and remove the rocker arms and shaft assembly from the head.

4. Disassemble the unit as required and replace all worn parts.

NOTE: On engines with exhaust valve rotators, the exhaust rocker arm must have relief for clearance.

To install:

5. Make sure all rocker arms with an **RH** stamped on them are installed to the right of those with an **LH**. Install the assembly with the notch on the end of the shaft pointing to the engine centerline and to the rear of the right bank or to the front of the left bank. Make sure the longer stamped steel retainers are at the No. 2 and 4 positions. Torque the bolts evenly and gradually to 21 ft. lbs. (28 Nm).

6. Install the valve cover.

7. Connect the negative battery cable and check for leaks.

Intake Manifold

REMOVAL & INSTALLATION

3.0L Engine

1. Relieve the fuel system pressure. Disconnect the negative battery cable.

2. Drain the cooling system.

3. Remove the throttle body to air cleaner hose.

4. Remove the throttle body and transaxle kickdown linkage.

5. Remove the AIS motor and TPS

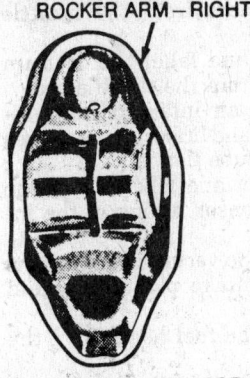

ROCKER ARM—LEFT   ROCKER ARM—RIGHT

RELIEVED FOR ROTATOR CLEARANCE

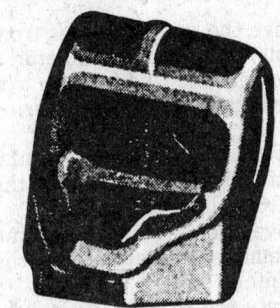

INTAKE ROCKER ARM   EXHAUST ROCKER ARM

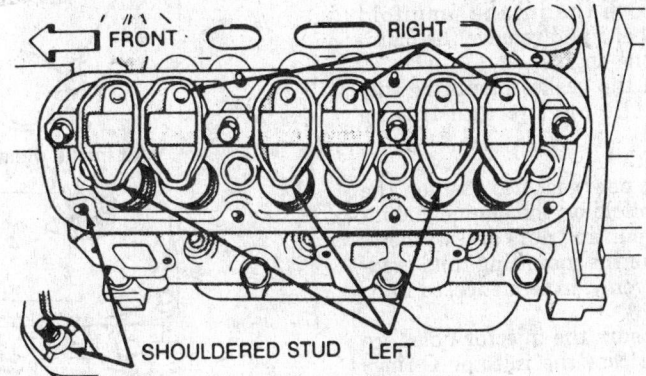

FRONT   RIGHT

SHOULDERED STUD   LEFT

SHORT STAMPED RETAINER

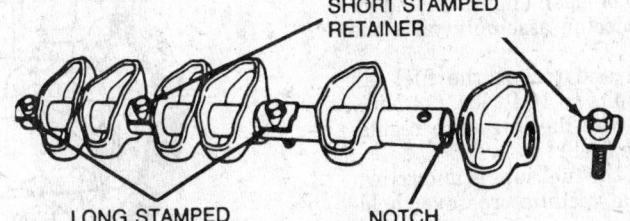

LONG STAMPED RETAINERS   NOTCH (END OF ROCKER SHAFT)

Rocker shaft/arms assembly— 3.9L and 5.2L

ROCKER ARMS   CYLINDER HEAD

Rocker arm assembly—1992 5.2L engine

wiring connectors from the throttle body.

6. Remove and label the vacuum hose harness from the throttle body.

7. From the air intake plenum, remove the PCV and brake booster hoses and the EGR tube flange.

8. Disconnect and label the charge temperature sensor wiring at the intake manifold.

9. Remove the vacuum connections from the air intake plenum vacuum connector.

10. Remove the fuel hoses from the fuel rail.

11. Remove the air intake plenum mounting bolts and remove the plenum.

12. Remove the vacuum hoses from the fuel rail and pressure regulator.

13. Disconnect the fuel injector wiring harness from the engine wiring harness.

14. Remove the fuel pressure regulator mounting bolts and remove the regulator from the fuel rail.

15. Remove the fuel rail mounting bolts and remove the fuel rail from the intake manifold.

16. Separate the radiator hose from the thermostat housing and heater hoses from the heater pipe.

17. Remove the intake manifold mounting bolts and remove the manifold from the engine.

18. Clean the gasket mounting surfaces on the engine and intake manifold.

To install:

19. Using new gaskets, position the intake manifold on the engine and install the mounting nuts and washers.

20. Torque the mounting nuts gradually and evenly, in sequence, to 15 ft. lbs. (20 Nm).

21. Make sure the injector holes are clean. Lubricate the injector O-rings with a drop of clean engine oil and install the injector assembly onto the engine.

22. Install and torque the fuel rail mounting bolts to 10 ft. lbs. (14 Nm).

23. Install the fuel pressure regulator onto the fuel rail.

24. Install the fuel supply and return tube and the vacuum crossover hold-down bolt.

25. Connect the fuel injection wiring harness to the engine wiring harness.

26. Connect the vacuum harness to the fuel pressure regulator and fuel rail assembly.

27. Remove the cover from the lower intake manifold and clean the mating surface.

28. Place the intake plenum gasket with the beaded sealant side up, on the intake manifold. Install the air intake plenum and torque the mounting bolts gradually and evenly, in sequence, to 10 ft. lbs. (14 Nm).

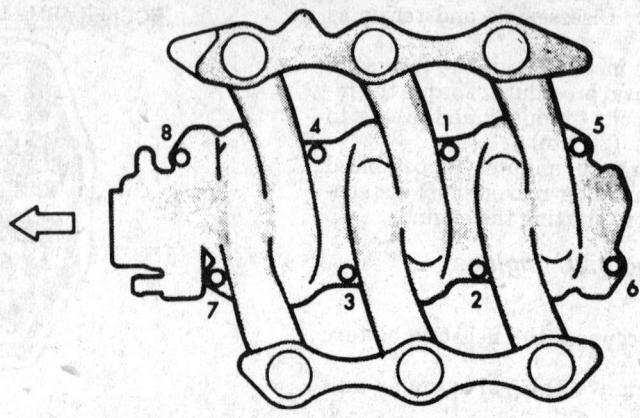

Intake manifold bolt torque sequence—3.0L engine

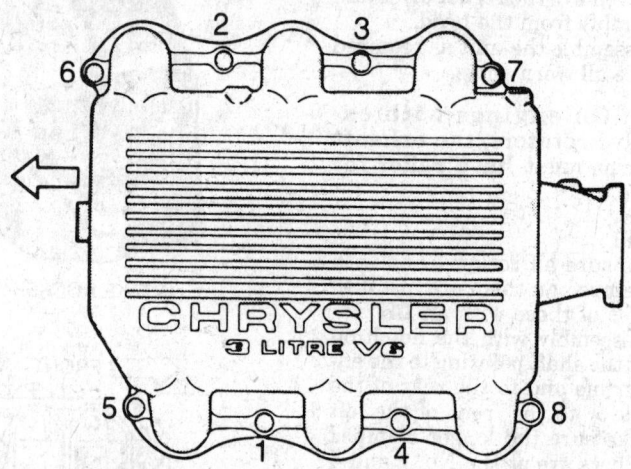

Air intake plenum bolt torque sequence—3.0L engine

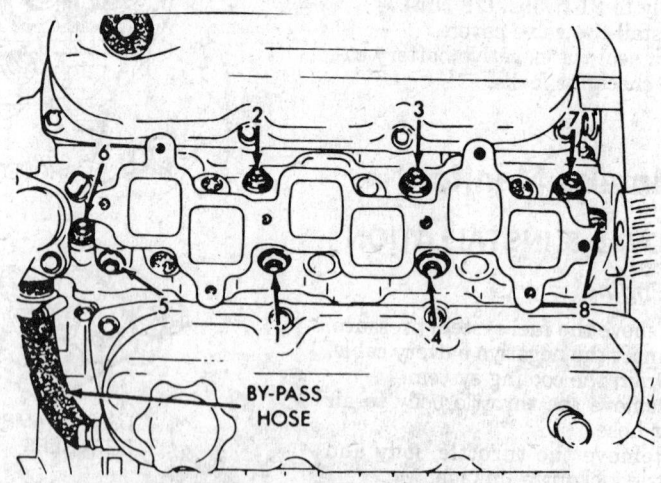

Intake manifold bolt torque sequence—3.3L engine

29. Connect or install all remaining items that were disconnected or removed during the removal procedure.

30. Refill the cooling system. Connect the negative battery cable and check for leaks using the DRB I or II to activate the fuel pump.

3.3L Engine

1. Disconnect the negative battery cable. Relieve the fuel pressure. Drain the cooling system.

2. Remove the air cleaner to throttle body hose assembly.

3. Disconnect the throttle cable and remove the wiring harness from the bracket.

4. Remove AIS motor and TPS wiring connectors from the throttle body.

5. Remove the vacuum hose harness from the throttle body.

6. Remove the PCV and brake booster hoses from the air intake plenum.

7. Disconnect the charge temperature sensor electrical connector. Remove the vacuum harness connectors from the intake plenum.

8. Remove the cylinder head to the intake plenum strut.

9. Disconnect the MAP sensor and oxygen sensor connectors. Remove the engine mounted ground strap.

10. Remove the fuel hoses from the fuel rail and plug them.

11. Remove the DIS coils and the alternator bracket to intake manifold bolt.

12. Remove the upper intake manifold attaching bolts and remove the upper manifold.

13. Remove the vacuum harness connector from the fuel pressure regulator.

14. Remove the fuel tube retainer bracket screw and fuel rail attaching bolts. Spread the retainer bracket to allow for clearance when removing the fuel tube.

15. Remove the fuel rail injector wiring clip from the alternator bracket.

16. Disconnect the cam sensor, coolant temperature sensor and engine temperature sensor.

17. Remove the fuel rail.

18. Remove the upper radiator hose, bypass hose and rear intake manifold hose.

19. Remove the intake manifold bolts and remove the manifold from the engine.

20. Remove the intake manifold seal retaining screws and remove the manifold gasket.

21. Clean out clogged end water passages and fuel runners.

To install:

22. Clean and dry all gasket mating surfaces.

23. Place a drop of silicone sealer onto each of the 4 manifold to cylinder head gasket corners.

CAUTION

The intake manifold gasket is composed of very thin and sharp metal. Handle this gasket with care or damage to the gasket or personal injury could result.

24. Install the intake manifold gasket and torque the end retainers to 10 ft. lbs. (12 Nm).

25. Install the intake manifold and torque the bolts in sequence to 10 inch lbs. Repeat the sequence increasing the torque to 17 ft. lbs. (23 Nm) and recheck each bolt for 17 ft. lbs. torque. After the bolts are torqued, inspect the seals to ensure that they have not become dislodged.

26. Lubricat the injector O-rings with clean oil and position the fuel rail

in place. Install the rail mounting bolts.

27. Connect the cam sensor, coolant temperature sensor and engine temperature sensor.

28. Install the fuel rail injector wiring clip to the alternator bracket.

29. Install the fuel rail attaching bolts and fuel tube retainer bracket screw.

30. Install the vacuum harness to the pressure regulator.

31. Install the upper intake manifold with a new gasket. Install the bolts only fingertight. Install the alternator bracket to intake manifold bolt and the cylinder head to intake manifold strut and bolts. Torque the intake manifold mounting bolts to 21 ft. lbs. (28 Nm) starting from the middle and working outward. Torque the bracket and strut bolts to 40 ft. lbs. (54 Nm).

32. Install or connect all items that were removed or disconnected from the intake manifold and throttle body.

33. Connect the fuel hoses to the rail. Push the fittings in until they click in place.

34. Install the air cleaner assembly.

35. Connect the negative battery cable and check for leaks using the DRB I or II to activate the fuel pump.

3.9L and 5.2L Engines

1991

1. Relieve the fuel pressure if equipped with fuel injection. Disconnect the negative battery cable from the battery and drain the cooling system.

2. Remove the air pump and bracket. Removal of the bracket will allow for easier installation of the left front corner of the intake manifold.

3. Remove the air cleaner assembly. Unbolt the air conditioning compressor and lay it to the side, if equipped. Remove the distributor cap with all

wires attached. Remove the alternator.

4. Disconnect all wires, hoses, linkages and cables from the carburetor or throttle body. Disconnect the fuel line.

5. Disconnect the ignition coil, coolant temperature sending unit wire and all other connectors along the wiring harness connected to items on the intake manifold.

6. Disconnect the heater hose, upper radiator hose and the lower bypass hose clamp.

7. Remove the valve covers.

8. Unbolt the intake manifold from the heads and remove the intake manifold assembly. Disassemble the manifold as required and clean out the exhaust crossover passages.

To install:

9. Clean and dry the intake manifold contact surfaces. Coat the intake manifold side gaskets very lightly with sealer and install the gaskets to the heads. Cutouts at the front of the gaskets differentiate the right and left sides.

10. Install the front and rear gaskets engaging the hole in the block and the tangs from the head gaskets. Apply a thin bead of sealer above the gaskets in the 4 corners.

11. Carefully lower the intake manifold into position engaging the bypass hose; after it is satisfactorily in place, inspect the gaskets to make sure they have not become dislodged.

12. Install the intake manifold bolts with the aspirator tube, air pump bracket and kickdown linkage bracket in place, if equipped. Torque the bolts in sequence to 25 ft. lbs. (34 Nm). Repeat the sequence retightening the bolts to a final torque of 40 ft. lbs. (54 Nm) and repeat the second step to ensure all bolts are accurately torqued.

13. Clean and dry the valve cover mating surfaces, bolts and bolt holes.

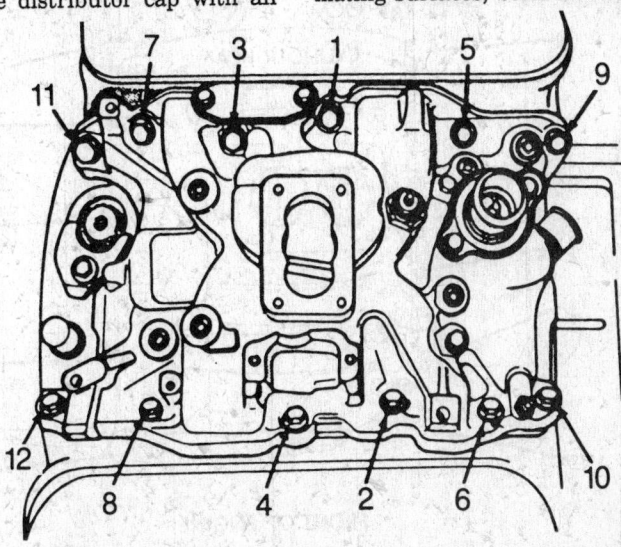

Intake manifold bolt torque sequence—3.9L engine

Install the valve covers with a new gasket. Torque the screws or nuts to 95 inch lbs. (11 Nm).

14. Connect the heater hose, upper radiator hose and the lower bypass hose clamp.

15. Connect the ignition coil, coolant temperature sending unit wire and all other connectors that were disconnected along the wiring harness.

16. Install the air conditioning compressor, if equipped. Install the distributor cap and all spark plug wires.

17. Install air pump.

18. Connect all wires, hoses, cables and the fuel line to the carburetor or throttle body. Install the air cleaner assembly.

19. Fill the cooling system.

20. Connect the negative battery cable. Use the DRB II to activate the fuel pump on fuel injected vehicles. Check for leaks.

3.9L Engine

1. Relieve the fuel pressure. Disconnect the negative battery cable from the battery and drain the cooling system.

2. Remove the alternator and fuel lines from the fuel rail.

3. Disconnect the accelerator linkage, speed control and transmission kickdown cables.

4. Remove the distributor cap and wires. Disconnect the coil wires, heat indicator sending unit, heater hoses and bypass hose.

5. Remove the closed crankcase ventilation and evaporation control systems.

6. Remove the intake manifold

bolts and lift the intake from the engine. Remove and discard the flange side gaskets and the rear cross-over gaskets.

7. Clean and dry all mating surfaces and inspect for flatness.

To install:

8. Install 4 plastic locator dowels into the block and install the new flange gaskets into position. Ensure **MANIFOLD SIDE** is visible on each flange gasket. Apply silicone sealant to the 4 corner joints. Install the front and rear cross-over gaskets onto the dowels.

9. Carefully lower the intake manifold onto the engine block and cylinder heads. Check gasket alignment.

10. Tighten the intake mounting bolts following the torque sequence as follows:

    a. Tighten bolts 1 and 2 to 72 inch lbs. (8 Nm).

    b. Tighten bolts 3 thru 12 in sequence to 72 inch lbs. (8 Nm).

    c. Check that all bolts are tightened to 72 inch lbs. (8 Nm).

    d. Tighten bolts in sequence to 12 ft. lbs. (16 Nm).

    e. Check that all bolts are tightened to 12 ft. lbs. (16 Nm).

11. Install the closed crankcase ventilation and evaporation control systems.

12. Install the distributor cap and wires. Connect the coil wires, heat indicator sending unit, heater hoses and bypass hose.

13. Install the distributor cap and wires.

14. Connect the accelerator linkage, speed control and transmission kickdown cables.

15. Install alternator and drive belt. Adjust belt tension as required. Install fuel lines.

16. Install the air cleaner and refill the cooling system. Reconnect the negative battery cable and start engine. Check for leaks and allow engine to reach normal operating temperature. Recheck fluid levels and add as required.

5.2L Engine

1. Relieve the fuel pressure. Disconnect the negative battery cable from the battery and drain the cooling system.

2. Remove the alternator and fuel lines.

3. Disconnect the accelerator linkage, speed control and transmission kickdown cables. Remove the return spring.

4. Remove the distributor cap and wires. Disconnect the coil wires, heat indicator sending unit, heater hoses and bypass hose.

5. Remove the closed crankcase ventilation and evaporation control systems.

6. Remove the intake manifold bolts and lift the intake from the engine. Remove and discard the flange side gaskets and the rear cross-over gaskets.

7. Clean and dry all mating surfaces and inspect for flatness.

To install:

8. Install 4 plastic locator dowels into the block and install the new flange gaskets into position. Ensure **MANIFOLD SIDE** is visible on each flange gasket. Apply silicone sealant to the 4 corner joints. Install the front and rear cross-over gaskets onto the dowels.

9. Carefully lower the intake manifold onto the engine block and cylinder heads. Check gasket alignment.

10. Tighten the intake following the torque sequence as follows:

    a. Tighten bolts 1 thru 4 in sequence to 72 inch lbs. (8 Nm).

    b. Tighten bolts 5 thru 12 in sequence to 72 inch lbs. (8 Nm).

    c. Check that all bolts are tightened to 72 inch lbs. (8 Nm).

    d. Tighten bolts in sequence to 12 ft. lbs. (16 Nm).

    e. Check that all bolts are tightened to 12 ft. lbs. (16 Nm).

11. Install the closed crankcase ventilation and evaporation control systems.

12. Install the distributor cap and wires. Connect the coil wires, heat indicator sending unit, heater hoses and bypass hose.

13. Install the distributor cap and wires.

14. Connect the accelerator linkage,

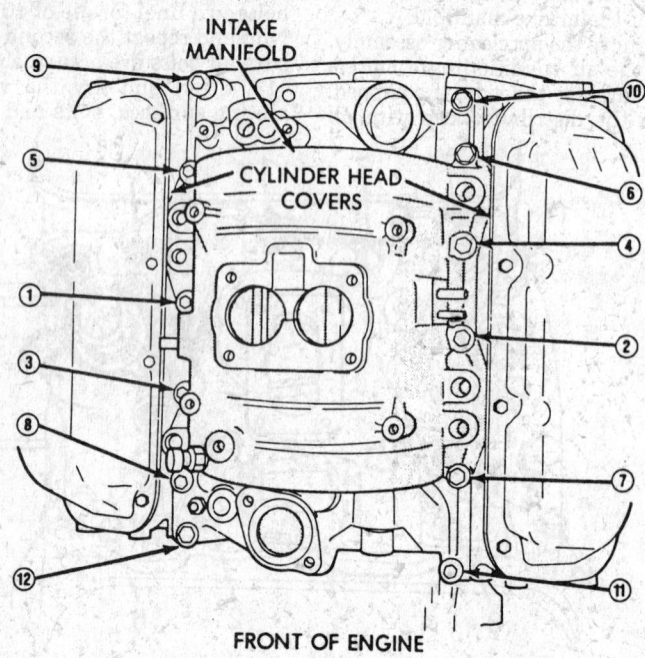

INTAKE MANIFOLD

CYLINDER HEAD COVERS

FRONT OF ENGINE

Intake manifold torque sequence—3.9L engine

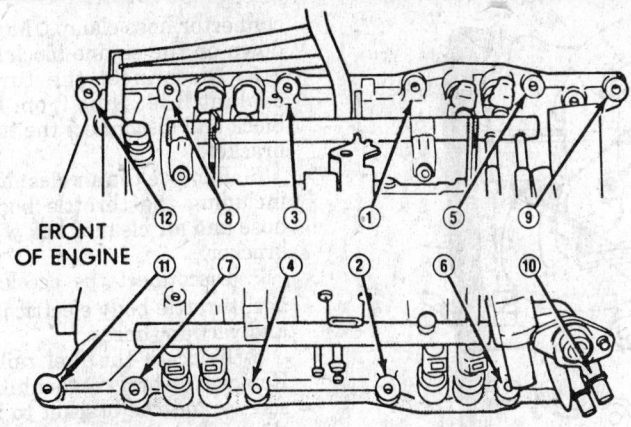

Intake manifold torque sequence – 5.2L engine

FRONT OF ENGINE

speed control and transmission kickdown cables. Install the return spring.

15. Install alternetor and drive belt. Adjust belt tension as required. Install fuel lines.

16. Install the air cleaner and refill the cooling system. Reconnect the negative battery cable and start engine. Check for leaks and allow engine to reach normal operating temperature. Recheck fluid levels and add as required.

Exhaust Manifold

REMOVAL & INSTALLATION

3.0L Engine

1. Disconnect the negative battery cable. Raise and safely support the vehicle.

2. Disconnect the exhaust pipe from the rear exhaust manifold, at the articulated joint.

3. Disconnect the EGR tube from the rear manifold and the oxygen sensor wire.

4. Remove the crossover pipe to manifold bolts.

5. Remove the rear manifold to cylinder head nuts and the manifold.

6. Lower the vehicle and remove the heat shield from the manifold.

7. Remove the front manifold to cylinder head nuts and the manifold.

8. Clean the gasket mounting surfaces. Inspect the manifolds for cracks, flatness and/or damage.

To install:

9. When installing, the numbers 1–3–5 on the gaskets are used with the rear cylinders and 2–4–6 are on the gasket for the front cylinders. Torque the manifold to cylinder head nuts to 14 ft. lbs. (19 Nm).

10. Install the crossover pipe to the manifold.

11. Connect the EGR tube and oxygen sensor wire.

12. Connect the exhaust pipe to the rear exhaust manifold, at the articulated joint.

13. Connect the negative battery cable and check the manifolds for leaks.

3.3L Engine

1. Disconnect the negative battery cable.

2. If removing the rear manifold, raise the vehicle and support safely. Disconnect the exhaust pipe from the rear exhaust manifold at the articulated joint.

3. Separate the EGR tube from the rear manifold and disconnect the oxygen sensor wire.

4. Remove the alternator/power steering support strut.

5. Remove the bolts attaching the crossover pipe to the manifold.

6. Remove the bolts attaching the manifold to the head and remove the manifold.

7. If removing the front manifold, remove the heat shield, bolts attaching the crossover pipe to the manifold and the nuts attaching the manifold to the head.

8. Remove the manifold from the engine.

9. Install rear exhaust manifold and torque all exhaust manifold attaching bolts to 17 ft. lbs. (23 Nm).

10. Attach exhaust pipe to exhaust manifold and tighten bolts to 21 ft. lbs. (28 Nm).

11. Attach cross-over to exhaust manifold and tighten to 25 ft. lbs. (33 Nm). Connect the oxygen sensor electrical connector.

12. Install EGR tube and alternator and power steering strut.

13. Install front exhaust manifold and attach the cross-over pipe. Install front manifold heat shield.

14. Connect negative battery cable. Start the engine and check for exhaust leaks.

3.9L and 5.2L Engines

1. Disconnect the negative battery cable. Remove the hot air tube and heat shield, if necessary.

2. Raise the vehicle and support safely. Remove the exhaust pipe from the exhaust manifolds. Lower the vehicle.

3. Take note of all conical washer locations and remove the bolts, nuts and washers attaching the manifold to the head.

4. Remove the manifold.

To install:

5. If either of the end studs came out with the nuts, install a new stud using sealer on the coarse threads.

6. Position the manifold on the end studs. Install conical washers and nuts on the studs.

7. Install the remaining bolts and washers in their proper locations. The inner bolts are not mounted with washers. Working outward from the center, torque the bolts to 20–25 ft. lbs. (27–34 Nm) and the nuts to 15 ft. lbs. (20 Nm).

8. Install the exhaust pipe to the manifolds.

9. Connect the negative battery cable and check for exhaust leaks.

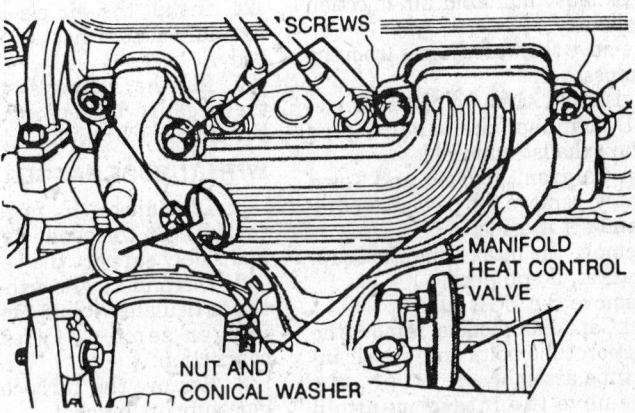

Exhaust manifold bolt installation – 3.9L engine

SCREWS

MANIFOLD HEAT CONTROL VALVE

NUT AND CONICAL WASHER

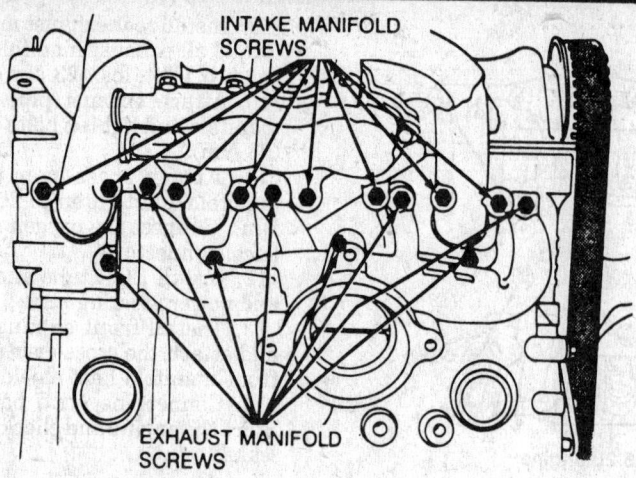

INTAKE MANIFOLD SCREWS

EXHAUST MANIFOLD SCREWS

Combination manifold attaching nuts and bolts—2.5L non-turbocharged engine

Combination Manifold

REMOVAL & INSTALLATION

2.5L Engines

WITHOUT TURBOCHARGER

NOTE: On some vehicles, some of the manifold attaching bolts are not accessible or too heavily sealed from the factory and cannot be removed on the vehicle. Head removal would be necessary in these situations.

1. Disconnect the negative battery cable.
2. Drain the cooling system.
3. Remove the air cleaner and disconnect all vacuum lines, electrical wiring and fuel lines from the carburetor or throttle body.
4. Disconnect the throttle linkage.
5. Loosen the power steering pump and remove the drive belt. On Dakota, remove the power steering and air pump support bracket.
6. Remove the power brake vacuum hose from the intake manifold.
7. On Canadian models, remove the coupling hose from the diverter valve to the exhaust manifold air injection tube assembly.
8. Remove the water hoses from the water crossover.
9. Raise and safely support the vehicle. Disconnect the exhaust pipe from the exhaust manifold.
10. On Caravan, Voyager and Town & Country, remove the power steering pump and set it aside.
11. Remove the intake manifold support bracket, if equipped.
12. Remove the EGR tube.
13. On Canadian models, remove the air injection tube bolts and the air injection tube assembly.
14. Remove the intake manifold bolts.

15. Lower the vehicle and remove the intake manifold.
16. Remove the exhaust manifold nuts.
17. Remove the exhaust manifold.

To install:

18. Install new combination intake and exhaust manifold gasket to engine. Coat lightly the manifold side with Mopar gasket sealer or equivalent. Install the exhaust manifold and torque the manifold to cylinder head nuts to 17 ft. lbs. (23 Nm).
19. Install the intake manifold. Raise vehicle and torque bolt, starting from the center and working outward, to 17 ft. lbs. (23 Nm).
20. Install the EGR tube with new gasket and tighten attaching bolts to 17 inch lbs. (23 Nm).
21. Install the exhaust pipe to manifold and tighten bolts to 20 ft. lbs. (27 Nm). Lower vehicle.
22. Connect the water hoses to the cross-over. Install diverter valve assembly and the air tube to the exhaust manifold.
23. Install the power steering and air pump support bracket. Install the power brake vacuum hose to the intake manifold.
24. Install the air cleaner and connect all vacuum hoses to the throttle body.
25. Fill the cooling system, connect the negative battery cable and start the engine. Check for exhaust leaks.

WITH TURBOCHARGER

1. Disconnect the negative battery cable. Drain the cooling system. Raise and safely support the vehicle.
2. Disconnect the exhaust pipe at the articulated joint. Disconnect the oxygen sensor at the electrical connection.
3. Remove the turbocharger to engine support bracket.
4. Loosen the oil drain back tube

connector hose clamps. Move the tube down on the engine block fitting.
5. Disconnect the turbocharger coolant inlet tube from the engine block and disconnect the tube support bracket.
6. Remove the air cleaner assembly, including the throttle body adaptor, hose and air cleaner box with support bracket.
7. Disconnect the accelerator linkage, throttle body electrical connector and vacuum hoses.
8. Relocate the fuel rail assembly. Remove the bracket to intake manifold screws and the bracket to heat shield clips. Lift and secure the fuel rail (with injectors, wiring harness and fuel lines intact) up and out of the way.
9. Disconnect the turbocharger oil feed line at the oil sending unit Tee fitting.
10. Disconnect the upper radiator hose from the thermostat housing.
11. Remove the cylinder head, manifolds and turbocharger as an assembly.
12. With the assembly on a workbench, loosen the upper turbocharger discharge hose end clamp.

NOTE: Do not disturb the center deswirler retaining clamp.

13. Remove the throttle body to intake manifold screws and throttle body assembly. Disconnect the turbocharger coolant return tube from the water box. Disconnect the retaining bracket on the cylinder head.
14. Remove the heat shield to intake manifold screws and the heat shield.
15. Remove the turbocharger to exhaust manifold nuts and the turbocharger assembly.
16. Remove the intake manifold bolts and the intake manifold.
17. Remove the exhaust manifold nuts and the exhaust manifold.

To install:

18. Place a new 2-sided Grafoil type intake/exhaust manifold gasket; do not use sealant.
19. Position the exhaust manifold on the cylinder head. Apply anti-seize compound to threads, install and torque the retaining nuts, starting at center and progressing outward in both directions, to 17 ft. lbs. (23 Nm). Repeat this procedure until all nuts are at 17 ft. lbs. (23 Nm).
20. Position the intake manifold on the cylinder head. Install and torque the retaining screws, starting at center and progressing outward in both directions, to 19 ft. lbs. (26 Nm). Repeat this procedure until all screws are at 19 ft. lbs. (26 Nm).
21. Connect the turbocharger outlet to the intake manifold inlet tube. Position the turbocharger on the exhaust manifold. Apply anti-seize compound

to threads and torque the nuts to 30 ft. lbs. (41 Nm). Torque the connector tube clamps to 30 ft. lbs. (41 Nm).

22. Install the tube support bracket to the cylinder head.

23. Install the heat shield on the intake manifold. Torque the screws to 105 inch lbs. (12 Nm).

24. Install the throttle body air horn into the turbocharger inlet tube. Install and torque the throttle body to intake manifold screws to 21 ft. lbs. (28 Nm). Torque the tube clamp to 30 inch lbs.

25. Install the cylinder head/manifolds/turbocharger assembly on the engine.

26. Reconnect the turbocharger oil feed line to the oil sending unit Tee fitting and bearing housing, if disconnected. Torque the tube nuts to 10 ft. lbs. (14 Nm).

27. Install the air cleaner assembly. Connect the vacuum lines and accelerator cables.

28. Reposition the fuel rail. Install and torque the bracket screws to 21 ft. lbs. (28 Nm). Install the air shield to bracket clips.

29. Connect the turbocharger inlet coolant tube to the engine block. Torque the tube nut to 30 ft. lbs. (41 Nm). Install the tube support bracket.

30. Install the turbocharger housing to engine block support bracket and the screws hand tight. Torque the block screw 1st to 40 ft. lbs. (54 Nm). Torque the screw to the turbocharger housing to 20 ft. lbs. (27 Nm).

31. Reposition the drain back hose connector and tighten the hose clamps. Reconnect the exhaust pipe.

32. Connect the upper radiator hose to the thermostat housing.

33. Refill the cooling system.

34. Connect the negative battery cable and check the manifolds for leaks.

Turbocharger

REMOVAL & INSTALLATION

NOTE: On some vehicles, some of the turbocharger to exhaust manifold nuts are not accessible enough to loosen and cannot be removed on the vehicle. Head removal would be necessary in these situations.

1. Disconnect the negative battery cable. Drain the cooling system.

2. Disconnect the EGR valve tube at the EGR valve.

3. Disconnect the turbocharger oil feed at the oil sending unit hex and the coolant tube at the water box. Disconnect the oil/coolant support bracket from the cylinder head.

4. Remove the right intermediate shaft, bearing support bracket and outer halfshaft assemblies.

5. Remove the turbocharger to engine block support bracket.

6. Disconnect the exhaust pipe at the articulated joint. Disconnect the oxygen sensor at the electrical connection.

7. Loosen the oil drain-back tube connector clamps and move the tube hose down on the nipple.

8. Disconnect the coolant tube nut at the block outlet (below steering pump bracket) and tube support bracket.

9. Remove the turbocharger to exhaust manifold nuts. Carefully routing the oil and coolant lines, move the assembly down and out of the vehicle.

To Install:

NOTE: Before installing the turbocharger assembly, be sure it is first charged with oil. Failure to do this may cause damage to the assembly.

10. Position the turbocharger on the exhaust manifold. Apply an anti-seize compound, Loctite® 771-64 or equivalent, to the threads and torque the retaining nuts to 40 ft. lbs. (54 Nm).

11. Connect the coolant tube to engine block fitting. Torque the tube nut to 30 ft. lbs. (41 Nm).

12. Position the oil drain-back hose and torque the clamps to 30 inch lbs.

13. Install and torque the:
Turbocharger to engine support bracket block screw to 40 ft. lbs. (54 Nm).
Turbocharger housing screw to 20 ft. lbs. (27 Nm).
Articulated joint shoulder bolts to 21 ft. lbs. (28 Nm).

14. Install the right driveshaft assembly, the starter and the oil feed line at the sending unit hex. Torque the oil feed tube nut to 10 ft. lbs. (14 Nm) and the EGR tube to EGR valve nut to 60 ft. lbs. (81 Nm).

15. Refill the cooling system. Connect the negative battery cable and check the turbocharger for proper operation.

Timing Chain Front Cover

REMOVAL & INSTALLATION

3.3L Engine

1. Disconnect the negative battery cable. Drain the cooling system.

2. Support engine and remove the right side motor mount.

3. Raise the vehicle and support safely. Drain the engine oil, remove the oil pan and oil pickup tube.

4. Remove the right wheel and tire assembly and the splash shield.

5. Remove the drive belt.

6. Unbolt the air conditioning compressor and position it to the side. Remove the compressor mounting bracket.

7. Remove the crankshaft pulley bolt and pulley.

8. Remove the idler pulley from the engine bracket and remove the bracket.

9. Remove the cam sensor from the timing chain cover.

10. Unbolt and remove the cover from the engine. Make sure the oil pump inner rotor does not fall out. Remove the 3 O-rings from the coolant passages and the oil pump outlet.

11. Thoroughly clean and dry the gasket mating surfaces. Install new O-rings to the block.

12. Remove the crankshaft oil seal from the cover. The seal must be removed from the cover when installing to ensure proper oil pump engagement.

To install:

13. Using a new gasket, install the chain case cover to the engine.

14. Make certain the oil pump is engaged onto the crankshaft before pro-

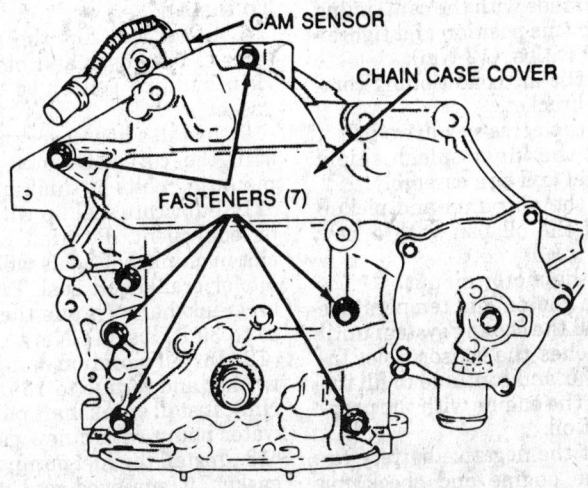

Timing chain cover—3.3L engine

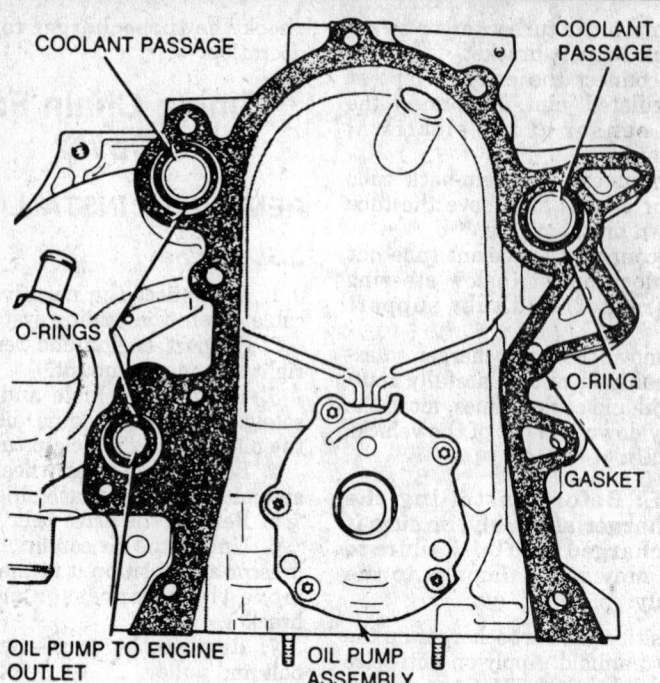

COOLANT PASSAGE

COOLANT PASSAGE

O-RINGS

O-RING

GASKET

OIL PUMP TO ENGINE OUTLET

OIL PUMP ASSEMBLY

Timing chain cover removed—3.3L engine

ceeding, or severe engine damage will result. Install the attaching bolts and torque to 20 ft. lbs. (27 Nm).

15. Install the crankshaft oil seal and pulley. Make sure the pulley bottoms out on the crankshaft seal diameter.

16. Install the engine bracket and torque the bolts to 40 ft. lbs. (54 Nm). Install the idler pulley to the engine bracket.

17. To install the cam sensor, first clean off the old spacer from the sensor face completely. Inspect the O-ring for damage and replace if necessary. A new spacer must be attached to the cam sensor prior to installation; if a new spacer is not used, engine performance will be adversely affected. Oil the O-ring lightly and push the sensor into its bore in the chain case cover until contact is made with the cam timing gear. Hold in this position and tighten the bolt to 9 ft. lbs. (12 Nm).

18. Install the air conditioning compressor and bracket.

19. Install the accessory drive belt.

20. Install the inner splash shield and the wheel and tire assembly.

21. Install the oil pump and pickup tube. Install the oil pan with a new gasket.

22. Install the motor mount.

23. Remove the engine temperature sensor and fill the cooling system until the level reaches the sensor hole. Install the sensor and continue to fill the radiator. Fill the engine with the proper amount of oil.

24. Connect the negative battery cable, start the engine and check for leaks.

3.9L and 5.2L Engines

1. Disconnect the negative battery cable.

2. Drain the cooling system.

3. Remove the radiator, fan and all related parts. Remove the water pump.

4. Remove the crankshaft pulley.

5. Remove the vibration damper using the proper puller.

6. Disconnect the fuel lines from the fuel pump, if equipped.

7. Remove the 2 front bolts from the oil pan.

8. Unbolt the chain cover from the block and remove, using caution to avoid damaging the oil pan gasket. Remove the fuel pump from the cover, if equipped.

To install:

9. Clean and dry the mating surfaces of the cover and block. Apply a thin bead of sealer to the oil pan gasket.

10. Install a new cover gasket and install the cover. Do not tighten the mounting bolts at this time.

11. Lubricate seal lip with lubriplate or equivalent. Position damper hub slot in crankshaft, this will act as a pilot for crankshaft seal. Press damper on crankshaft. Torque the cover bolts to to 30 ft. lbs. (41 Nm).

12. Install vibration damper bolt and washer and torque to 135 ft. lbs. (183 Nm). Install crankshaft pulley and the water pump with a new gasket.

13. Install the fuel pump using a new gasket, if equipped and connect the fuel lines. Install the 2 oil pan bolts.

14. Install the radiator, fan and all related parts.

15. Fill the cooling system.

16. Connect the negative battery cable and check for leaks.

Front Cover Oil Seal
REMOVAL & INSTALLATION

3.3L Engine

1. Disconnect the negative battery cable.

2. Raise the vehicle and support safely. Remove the right front tire and wheel assembly and the inner splash shield.

3. Remove the drive belt.

4. Remove the crankshaft bolt and pulley.

5. Use tool C–4991 to remove the seal.

To install:

6. Clean out the bore. Place the seal with the spring toward the engine. Install the new seal using tool C–4992 until it is flush with the cover.

7. Install the crankshaft pulley and retainer bolt.

8. Install the accessory drive belt and adjust tension as desired.

9. Install the splash shield and the tire and wheel assembly.

10. Connect the negative battery cable and check for leaks.

3.9L Engine

1. Disconnect the negative battery cable.

2. Remove the belts from the crankshaft pulley.

3. Remove the fan and shroud from the vehicle.

4. Remove the crankshaft pulley.

5. Remove the vibration damper using the proper puller.

6. Using a flat bladed tool behind the lips of the oil seal, carefully pry outward. Take care not to damage the crankshaft seal surface of the cover.

To install:

7. Install the new seal by installing the threaded shaft part of the special tool C–4251 into the threads of the crankshaft.

8. Place the seal into the opening with the spring toward the engine. Place the installing adapter C–4251-3 with the thrust bearing and nut on the shaft. Tighten the nut until the tool is flush with the timing chain cover. Remove the tool.

9. Install the damper with tool C–3638, install the bolt and washer and torque to specification.

10. Apply a small amount of sealer to the bolts and install the crankshaft pulley.

11. Install the fan and shroud.

12. Connect the negative battery cable and check for leaks.

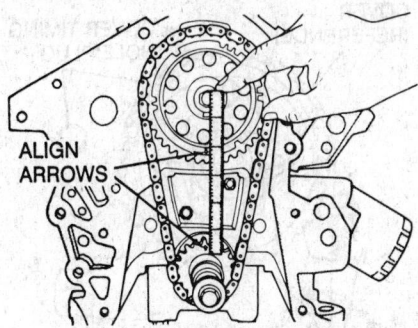

Timing mark alignment—3.3L engine

Timing Chain and Sprockets

REMOVAL & INSTALLATION

3.3L Engine

1. If possible, position the engine so No. 1 piston is at TDC on the compression stroke. Disconnect the negative battery cable. Drain the coolant.

2. Remove the timing chain case cover and chain snubber.

3. Remove the camshaft gear attaching cup washer and bolt and remove the timing chain with the camshaft gear attached. Remove the crankshaft sprocket with suitable puller taking care not to damage the crankshaft surface.

To install:

4. Install the crankshaft sprocket onto the crankshaft making sure sprocket is seated into position. Rotate sprocket to position timimg mark in the 12 o'clock position.

5. Place timing chain around camshaft sprocket and align timing mark in the 6 o'clock position. Place timing chain around the crankshaft sprocket and install camshaft sprocket in position.

6. Check alignment of gears with straight edge.

7. Install camshaft bolt and washer and tighten bolt to 35 ft. lbs. (47 Nm).

8. Check camshaft endplay. The specification with a new plate is 0.005–0.012 in. (0.0127–0.304mm) or 0.012 in. (0.31mm) maximum with a used plate. Replace the thrust plate if not within specifications.

9. Install the timing chain snubber. For 1992 engine, these bolts are 20mm long and should **NOT** be used with previous year engines.

10. Thoroughly clean and dry the gasket mating surfaces.

11. Install new O-rings to the block.

12. Remove the crankshaft oil seal from the cover. The seal must be removed from the cover when installing to ensure proper oil pump engagement.

13. Using a new gasket, install the chain case cover to the engine.

14. Make certain the oil pump is engaged onto the crankshaft before proceeding or severe engine damage will result. Install the attaching bolts and torque to 20 ft. lbs. (27 Nm).

15. Install the crankshaft oil seal, crankshaft pulley and bolt. Make sure the pulley bottoms out on the crankshaft seal diameter.

16. Install engine bracket, idler pulley, air conditioning bracket and accessory belt(s).

17. To install the cam sensor, first clean off the old spacer from the sensor face. Inspect the O-ring for damage and replace if necessary. A new spacer must be attached to the cam sensor prior to installation; if a new spacer is not used, engine performance will be adversely affected. Oil the O-ring lightly and push the sensor into its bore in the chain case cover until contact is made with the cam timing gear. Hold in this position and tighten the bolt to 10 ft. lbs. (12 Nm).

18. Install the inner splash shield and tire and wheel assembly.

19. Install the oil pump pickup tube and oil pan. Install engine mount, lower vehicle and fill crankcase to proper level.

20. Refill the cooling system.

21. Connect the negative battery cable, road test the vehicle and check for leaks.

3.9L and 5.2L Engines

1. If possible, crank the engine so No. 1 cylinder is at TDC on the compression stroke. Remove the distributor cap to confirm and line the timing mark on the damper pulley with **0** on the timing scale. This will aid in aligning timing marks when installing the timing gears. Disconnect the negative battery cable.

2. Drain the cooling system.

3. Remove the radiator, fan and all related parts. Remove the water pump.

4. Remove the crankshaft pulley.

5. Remove the vibration damper using the proper puller.

6. Disconnect the fuel lines from the fuel pump, if equipped.

7. Remove the 2 front bolts from the oil pan.

8. Unbolt the chain cover from the block and remove, using caution to avoid damaging the oil pan gasket. Remove the fuel pump from the cover, if equipped.

9. Remove the camshaft gear retaining bolt, cup washer and fuel pump eccentric, if equipped. Remove the timing chain and gears.

To install:

10. Place both camshaft and crankshaft gears on the bench with the timing marks on the exact imaginary center line through both gear bores as they are installed on the engine. Place the timing chain around both sprockets.

11. Turn the crankshaft and camshaft so the keys line up with the keyways in the gears when the timing marks are in proper position.

12. Slide both gears over their respective shafts and use a straightedge to check timing mark alignment.

13. Install the fuel pump eccentric and cup washer, if equipped. Torque the camshaft gear retaining bolt to 35 ft. lbs. (47 Nm).

14. Clean and dry the mating surfaces of the timing chain cover and block. Apply a thin bead of sealer to the oil pan gasket.

15. Install a new cover gasket and install the cover. Do not tighten the mounting bolts at this time.

16. Lubricate seal lip with lubriplate or equivalent. Position damper hub slot in crankshaft, this will act as a pilot for crankshaft seal. Press damper on crankshaft. Torque the cover bolts to to 30 ft. lbs. (41 Nm).

17. Install vibration damper bolt and washer and torque to 135 ft. lbs. (183 Nm). Install crankshaft pulley and the water pump with a new gasket.

18. Install the fuel pump using a new gasket, if equipped and connect the fuel lines. Install the 2 oil pan bolts.

19. Install the radiator, fan and all related parts.

20. Fill the cooling system.

21. Connect the negative battery cable and check for leaks.

Timing Belt Cover

REMOVAL & INSTALLATION

2.5L Engines

1. Disconnect the negative battery cable.

2. Remove the nuts that attach the upper cover to the valve cover. Remove the bolt that attaches the upper cover to the lower cover.

3. Remove the upper cover.

4. Raise the vehicle and support safely. Remove the right side splash shield.

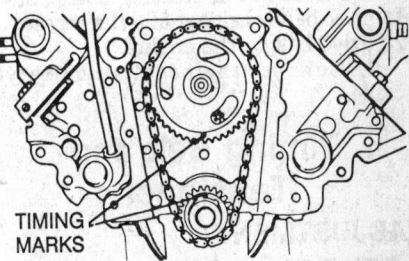

Timing mark alignment—3.9L and 5.2L engines

5. Remove the crankshaft pulley, water pump pulley and the accessory belts.

6. Remove the lower cover attaching bolts. Remove the lower cover.

7. The installation is the reverse of the removal procedure.

3.0L Engine

1. Disconnect the negative battery cable.

2. If equipped with air conditioning, loosen the adjustment pulley locknut, turn the screw counterclockwise to reduce the drive belt tension and remove the belt.

3. To remove the serpentine drive belt, insert a ½ in. breaker bar in to the square hole of the tensioner pulley, rotate it counterclockwise to reduce the drive belt tension and remove the belt.

4. Remove the air conditioning compressor and the air compressor bracket if equipped, power steering pump and alternator from the mounts; support them aside. Remove power steering pump/alternator automatic belt tensioner bolt and the tensioner.

5. Raise and safely support the vehicle. Remove the right inner fender splash shield.

6. Remove the crankshaft pulley bolt and the pulley/damper assembly from the crankshaft.

7. Lower the vehicle and place a floor jack under the engine to support it.

8. Separate the front engine mount insulator from the bracket. Raise the engine slightly and remove the mount bracket.

9. Remove the timing belt cover bolts and the upper and lower covers from the engine.

10. Install upper and lower timing belt covers and install retaining screws.

11. Install the mount bracket and connect the engine mount insulator to the mount bracket.

12. Raise and safely support vehicle. Install crankshaft pulleys and damper. Install right inner splash shield and wheel and tire assembly.

13. Install the air conditioner bracket and compressor to the engine.

14. Install power steering pump, automatic belt tensioner and the accessory drive belt to the engine.

15. Connect the negative battery cable.

Timing Belt and Tensioner

ADJUSTMENT

2.5L Engines

1. Disconnect the negative battery cable.

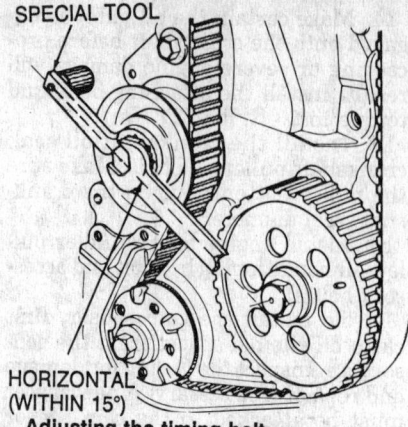

Adjusting the timing belt— 2.5L engines

2. Raise the vehicle and support safely. Remove the right front inner splash shield.

3. Remove the timing belt upper and lower covers.

4. Position camshaft so the timing marks are in the proper position and check crankshaft and intermediate shaft for proper alignment.

5. If properly aligned, turn the crankshaft ¼ turn counterclockwise. Install tensioning weight onto the hex of the tensioner and loosen the adjuster bolt. Slowly turn the crankshaft clockwise to remove any slack that might be in the timing belt. Secure tensioner adjuster nut. Rotate crankshaft clockwise for 2 engine revolutions and recheck engine timing marks.

6. If not properly aligned, inspect belt for wear and replaced as required. Adjust engine timing and belt tension.

7. Reinstall timing belt covers and install splash shield.

3.0L Engine

1. Loosen the bolt that holds the timing belt tensioner in place. The bolt is located to the left of the accessory belt tensioner mounting.

2. Allow the spring to pull the tensioner in automatically.

3. Tighten the tensioner locking bolt.

REMOVAL & INSTALLATION

2.5L Engines

1. If possible, position the engine so No. 1 piston is at TDC. Disconnect the negative battery cable.

2. Remove the timing belt covers. Remove the timing belt tensioner and allow the belt to hang free.

3. Place a floor jack under the engine and separate the right motor mount. Raise the engine from the mount slightly.

4. Remove the air conditioning compressor belt idler pulley, if equipped and remove the mounting

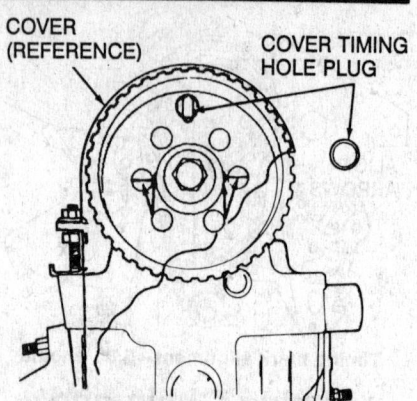

Alignment of the arrows on the camshaft sprocket with the camshaft bearing cap to cylinder head mounting line— 2.5L engines

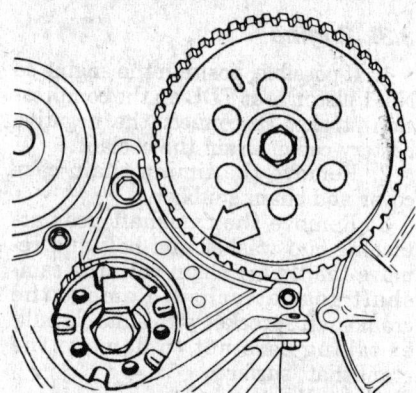

Alignment of the crankshaft sprocket and intermediate shaft sprocket— 2.5L engines

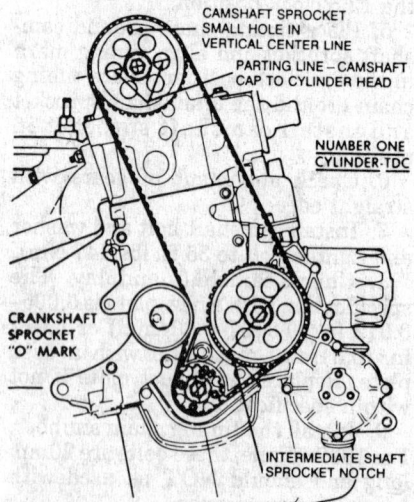

Timing belt installation— 2.5L engines

stud. Unbolt the compressor/alternator bracket and position it to the side.

5. Remove the timing belt from the vehicle.

To install:

6. Remove spark plugs from the engine. Turn the crankshaft sprocket and intermediate shaft sprocket until the marks are in line. Use a straight-

edge from bolt to bolt to confirm alignment.

7. Install timing belt onto engine and lower engine into engine mount. Tighten retainer bolts and nut and remove the floor jack.

8. Turn the camshaft until the small hole in the sprocket is at the top and rows on the hub are in line with the camshaft cap to cylinder head mounting lines. Use a mirror to see the alignment so it is viewed straight on and not at an angle from above. Install belt over camshaft pulley. Turn the idler pulley counterclockwise slightly and install belt over the idler pulley. Hold belt in position and turn idler pulley clockwise until the slack is out of the belt. The timing mark should be in aligned with the crankshaft timing mark. If not, repeat this step until the mark on the pulley is aligned with crankshaft mark.9. Repeat this step for the crankshaft pulley working back to the idler pulley. Do not rotate the belt far enough to disturb the camshaft pulley.

10. Install the weight on the hex of the tensioner and allow to hang freely. Rotate the crankshaft 2 engine revolutions clockwise and recheck engine timing. Tighten the adjuster nut while holding the tensioner in position.

NOTE: If the timing marks are in line but slack exists in the belt between either the camshaft and intermediate shaft sprockets or the intermediate and crankshaft sprockets, the timing will be incorrect when the belt is tensioned.

11. Install the air conditioning compressor/alternator bracket and the idler pulley.

12. Connect the negative battery cable and road test the vehicle.

3.0L Engine

1. If possible, position the engine so No. 1 cylinder is at TDC. Disconnect the negative battery cable. Remove the timing covers from the engine.

2. If the same timing belt will be reused, mark the direction of the timing belt's rotation, for installation in the same direction. Make sure the engine is positioned so the No. 1 cylinder is at the TDC of it's compression stroke and the sprockets timing marks are aligned with the engine's timing mark indicators.

3. Loosen the timing belt tensioner bolt and remove the belt. If not removing the tensioner, position it as far away from the center of the engine as possible and tighten the bolt.

4. If the tensioner is being removed, paint the outside of the spring to ensure that it is not installed backwards. Unbolt the tensioner and remove it along with the spring.

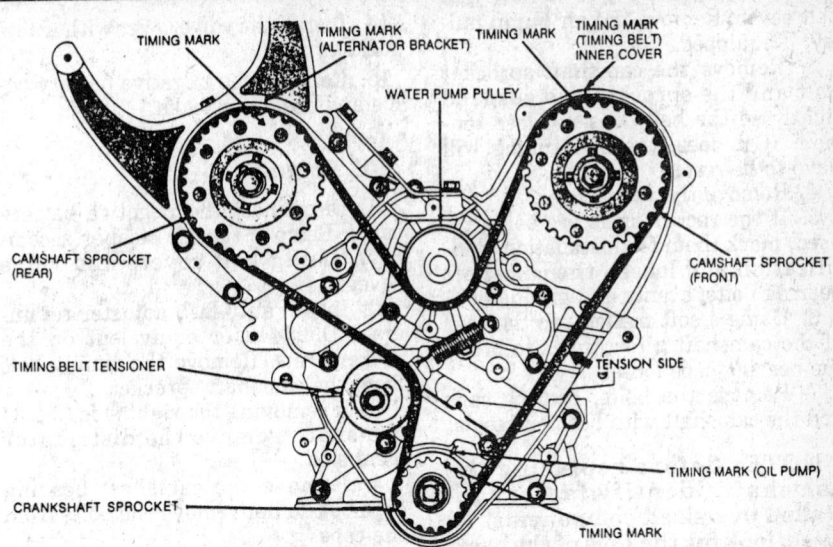

Timing belt installation–3.0L engine

To install:

5. Install the tensioner, if removed and hook the upper end of the spring to the water pump pin and the lower end to the tensioner in exactly the same position as originally installed. If not already done, position both camshafts so the marks line up with those on the alternator bracket (rear bank) and inner timing cover (front bank). Rotate the crankshaft so the timing mark aligns with the mark on the oil pump.

6. Install the timing belt on the crankshaft sprocket and while keeping the belt tight on the tension side (right side), install the belt on the front camshaft sprocket.

7. Install the belt on the water pump pulley, then the rear camshaft sprocket and the tensioner.

8. Rotate the front camshaft counterclockwise to tension the belt between the front camshaft and the crankshaft. If the timing marks came out of line, repeat the procedure.

9. Install the crankshaft sprocket flange.

10. Loosen the tensioner bolt and allow the spring to tension the belt.

11. Turn the crankshaft 2 full turns in the clockwise direction only until the timing marks align again. Now that the belt is properly tensioned, torque the tensioner lockbolt to 21 ft. lbs. (29 Nm).

12. Install the timing belt covers and all related parts.

13. Connect the negative battery cable and road test the vehicle.

Timing Sprockets

REMOVAL & INSTALLATION

2.5L Engines

1. Disconnect the negative battery cable. Remove the timing belt.

2. Remove the crankshaft sprocket bolt. Using the puller tool C–4685 or equivalent, and the button from tool L–4524 or equivalent, remove the crankshaft sprocket.

3. Using the tool C–4687 or equivalent, hold the camshaft and/or intermediate sprocket, remove the center bolt and the sprocket(s).

4. The installation is the reverse of the removal procedure. Torque the camshaft and intermediate sprocket bolts to 65 ft. lbs. (88 Nm) and the crankshaft sprocket bolt to 50 ft. lbs. (68 Nm).

3.0L Engine

1. Disconnect the negative battery cable.

2. Remove the timing belt.

3. To remove the camshaft sprocket, hold the sprocket with tool MB990775 or equivalent, and remove the retaining bolt and washer.

4. To remove the crankshaft sprocket, remove the bolt and remove the sprocket from the crankshaft.

5. The installation is the reverse of the removal procedure. Torque the camshaft sprocket bolt to 70 ft. lbs. (95 Nm) while holding the sprocket with the holding tool. Torque the crankshaft sprocket bolt. to 110 ft. lbs. (150 Nm).

Camshaft

REMOVAL & INSTALLATION

2.5L Engines

1. Disconnect the negative battery cable. Relieve the fuel pressure, if equipped with fuel injection.

2. Turn the crankshaft so the No. 1 piston is at the TDC of the compression stroke. Remove the upper timing

belt cover. Remove the air pump pulley, if equipped.

3. Remove the camshaft sprocket bolt and the sprocket and suspend tightly so the belt does not lose tension. If it does, the belt timing will have to be reset.

4. Remove the valve cover.

5. If the rocker arms are being reused, mark them for installation identification and loosen the camshaft bearing bolts, evenly and gradually.

6. Using a soft mallet, tap the rear of the camshaft a few times to break the bearing caps loose.

7. Remove the bolts, bearing caps and the camshaft with both end seals.

NOTE: Before replacing the camshaft, identify factory installed oversized components. To do so, look for the tops of the bearing caps pained green and O/SJ stamped rearward of the oil gallery plug on the rear of the head. In addition, the barrel of the camshaft is painted green and "O/SJ" is stamped onto the rear end of the camshaft. Installing standard sized parts in an head equipped with oversized parts—or visa versa—will cause severe engine damage.

Also, take note of the color of the paint stripe on the rear camshaft seal. These stripes differentiate seal sizes. If a seal with a different color stripe is installed, a severe leak will develop if the seal is too small or the cap will not be able to be fully installed if the seal is too big.

8. Check the oil passages for blockages and the parts for wear and damage and replace parts, as required. Clean the gasket mounting surfaces.

To install:

9. Transfer the sprocket key to the new camshaft. New rocker arms and a new camshaft sprocket bolt are normally included with the camshaft package. Install the rocker arms and lubricated camshaft.

10. Install the bearing caps with No. 1 at the timing belt end and No. 5 at the transaxle end. The camshaft bearing caps are numbered and have arrows facing forward. Torque the camshaft bearing bolts evenly and gradually to 18 ft. lbs. (24 Nm).

11. Install a small amount of sealer into No. 1 and No. 5 camshaft cap and install seals into place. Check for proper alignment of seals.

12. Mount a dial indicator to the front of the engine and check the camshaft endplay. Play should not exceed 0.006 in.

13. Install the camshaft sprocket using the new bolt. Install the air pump pulley, if equipped.

14. Install the valve cover with a new gaskets.

15. Connect the negative battery cable and check for leaks.

3.0L Engine

1. Disconnect the negative battery cable. Remove the air cleaner assembly, timing belt covers and valve covers.

2. Install auto lash adjuster retainers MD998443 or equivalent on the rocker arms. Remove the timing belt from the camshaft sprocket.

3. If removing the right side (front) camshaft, remove the distributor extension.

4. Remove the camshaft bearing caps but do not remove the bolts from the caps.

5. Remove the rocker arms, rocker shafts and bearing caps, as an assembly.

6. Remove the camshaft from the cylinder head.

7. Inspect the bearing journals on the camshaft, cylinder head and bearing caps.

To install:

8. Lubricate the camshaft journals and camshaft with clean engine oil and install the camshaft in the cylinder head.

9. Install the rocker arm shaft assembly making sure the arrow on the bearing cap and the arrow mark on the cylinder head are in the same direction. The direction of the arrow marks on the front and rear assemblies are opposite to each other.

10. Torque the bearing cap bolts, in the following sequence: No. 3, No. 2, No. 1 and No. 4 to 85 inch lbs. (10 Nm).

11. Repeat the sequence increasing the torque to 175 inch lbs. (18 Nm).

12. Install the distributor extension, if it was removed.

13. Install the valve cover and all related parts.

3.3L Engine

1. Relieve the fuel pressure. Disconnect the negative battery cable.

2. Remove the engine from the vehicle. Remove the intake manifold, cylinder heads, timing chain cover and timing chain from the engine.

3. Remove the rocker arm and shaft assemblies.

4. Label and remove the pusrods and lifters.

5. Remove the camshaft thrust plate.

6. Install a long bolt into the front of the camshaft to facilitate its removal. Remove the camshaft being careful not to damage the cam bearings with the cam lobes.

To install:

7. Install the camshaft to within 2 in. of its final installation position.

8. Install the camshaft thrust plate and 2 bolts and torque to 105 inch. lbs. (12 Nm).

9. Place both camshaft and crankshaft gears on the bench with the timing marks on the exact imaginary center line through both gear bores as they are installed on the engine. Place the timing chain around both sprockets.

10. Turn the crankshaft and camshaft so the keys line up with the keyways in the gears when the timing marks are in proper position.

11. Slide both gears over their respective shafts and use a straightedge to check timing mark alignment.

12. Measure camshaft endplay. If not within specifications, replace the thrust plate.

13. If the camshaft was not replaced, lubricate and install the lifters in their original locations. If the camshaft was replaced, new lifters must be used.

14. Install the pushrods and rocker shaft assemblies.

15. Install the timing chain cover, cylinder heads and intake manifold.

16. Install the engine in the vehicle.

17. Change the engine oil and replace the oil filter.

NOTE: If the camshaft or lifters have been replaced, add 1 pint of Mopar crankcase conditioner or equivalent, when replenishing the oil to aid in break in. This mixture should be left in the engine for a minimum of 500 miles and drained at the next normal oil change.

18. Fill the radiator with coolant.

19. Connect the negative battery cable, set all adjustments to specifications and check for leaks.

3.9L and 5.2L Engines

1. If possible, crank the engine around so No. 1 cylinder is at TDC on the compression stroke. Remove the distributor cap to confirm and line the timing mark on the damper pulley with 0 on the timing scale. This will aid in aligning timing marks when installing the timing gears.

2. If equipped with fuel injection, relieve the fuel pressure. Drain the cooling system.

3. Removal of the engine is required for this procedure.

4. After engine is removed, remove the valve covers and rocker shaft assemblies. Identify and remove the pushrods.

5. Remove the intake manifold. Identify and remove all lifters.

6. Remove the distributor.

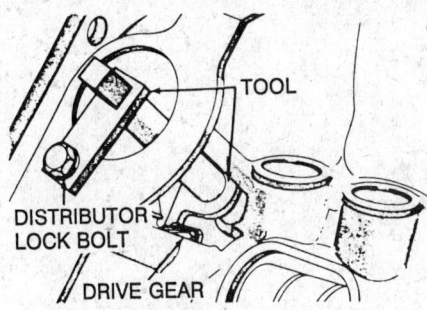

Installing the camshaft blocking tool—3.9L and 5.2L engines

7. Lift out the oil pump and distributor driveshaft.

8. Remove the fuel pump, if equipped. Remove the timing chain cover, timing chain and gears.

9. Note the location of the oil tab and remove the camshaft thrust plate.

10. Install a long bolt into the front of the camshaft to facilitate removal. Remove the camshaft, being careful not to damage any of the cam bearings with the cam lobes.

To install:

11. Install the camshaft to within 2 in. of its final installation position.

12. Install the camshaft blocking tool C–3509 and bolt it in place with the distributor hold-down bolt. This will prevent the camshaft from being pushed in too far and knocking out the welch plug at the rear of the block. This tool should remain in place until the timing chain installation has been completed.

13. Install the camshaft thrust plate and chain oil tab. Make sure the tang of the oil tab enters the hole in the thrust plate at the lower right. Torque the bolts to 18 ft. lbs. (24 Nm). Make sure the top edge of the oil tab is flat against the thrust plate or it will not feed oil to the chain.

14. Place both camshaft and crankshaft gears on the bench with the timing marks on the exact imaginary center line through both gear bores as they are installed on the engine. Place the timing chain around both sprockets.

15. Turn the crankshaft and camshaft so the keys line up with the keyways in the gears when the timing marks are in proper position.

16. Slide both gears over their respective shafts and use a straightedge to check timing mark alignment.

17. Install the fuel pump eccentric, cup washer and retainer bolt. Remove the camshaft blocking tool, if it was installed.

18. Measure camshaft endplay, if applicable. Replace the thrust plate if not within specifications.

19. Coat the oil pump and distributor driveshaft with oil. On 3.9L engine, install the shaft so when the gear spirals

Position of installed distributor drive gear caps—5.2L engine

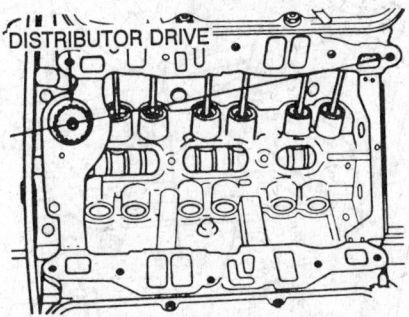

Installed distributor drive gear position—3.9L engine

into place and drops into the oil pump, the slot in the top of the gear is pointing directly to the left front intake manifold bolt hole. On 5.2L engine, install the shaft so when the gear spirals into place and drops into the oil pump, the slot in the top of the gear is pointing in a direction parallel to the center-line of the crankshaft.

20. If the camshaft was not replaced, lubricate and install the lifters in their original locations. If the camshaft was replaced, new lifters must be used.

21. Install the pushrods and rocker shaft assemblies.

22. Install the intake manifold, if it was removed. Install the valve covers.

23. Install the distributor so the rotor points to the No. 1 spark plug wire position on the cap.

24. Install the timing chain cover and all related parts.

25. Install the fuel pump.

26. Install the engine.

27. Change the engine oil and replace the oil filter.

NOTE: If the camshaft or lifters have been replaced, add 1 pint of Mopar crankcase conditioner or equivalent, when replenishing the oil to aid in break in. This mixture should be left in the engine for a minimum of 500 miles and

drained at the next normal oil change.

28. Fill the radiator with coolant.

29. Connect the negative battery cable, set all adjustments to specifications and check for leaks.

Intermediate/Balance Shaft

REMOVAL & INSTALLATION

Intermediate Shaft

2.5L ENGINES

1. Disconnect the negative battery cable.

2. Crank the engine around until the No. 1 piston is at TDC. Remove the timing belt covers to confirm that all timing marks are lined up.

3. Remove the fuel pump, if equipped. Remove the distributor. Looking down at the oil pump, the slot in the shaft must be parallel with the center line of the crankshaft. Remove the oil pan, dipstick tube and oil pump.

4. Remove the timing belt. Remove the intermediate sprocket using tool C–4685. Remove the seal retainer mounting bolts and remove the retainer from the block.

5. Remove the retainer and the intermediate shaft from the block.

6. If necessary, remove the front intermediate shaft bushing using tool C–4697–2 and the rear bushing using tool C–4686–2.

To install:

7. If replacement of the intermediate shaft bushings are required, install the front bushing using tool C–4697–1 until the tool is flush with the block. Install the rear bushing using tool C–4686–1 until the tool is flush with the block.

8. Lubricate the distributor drive

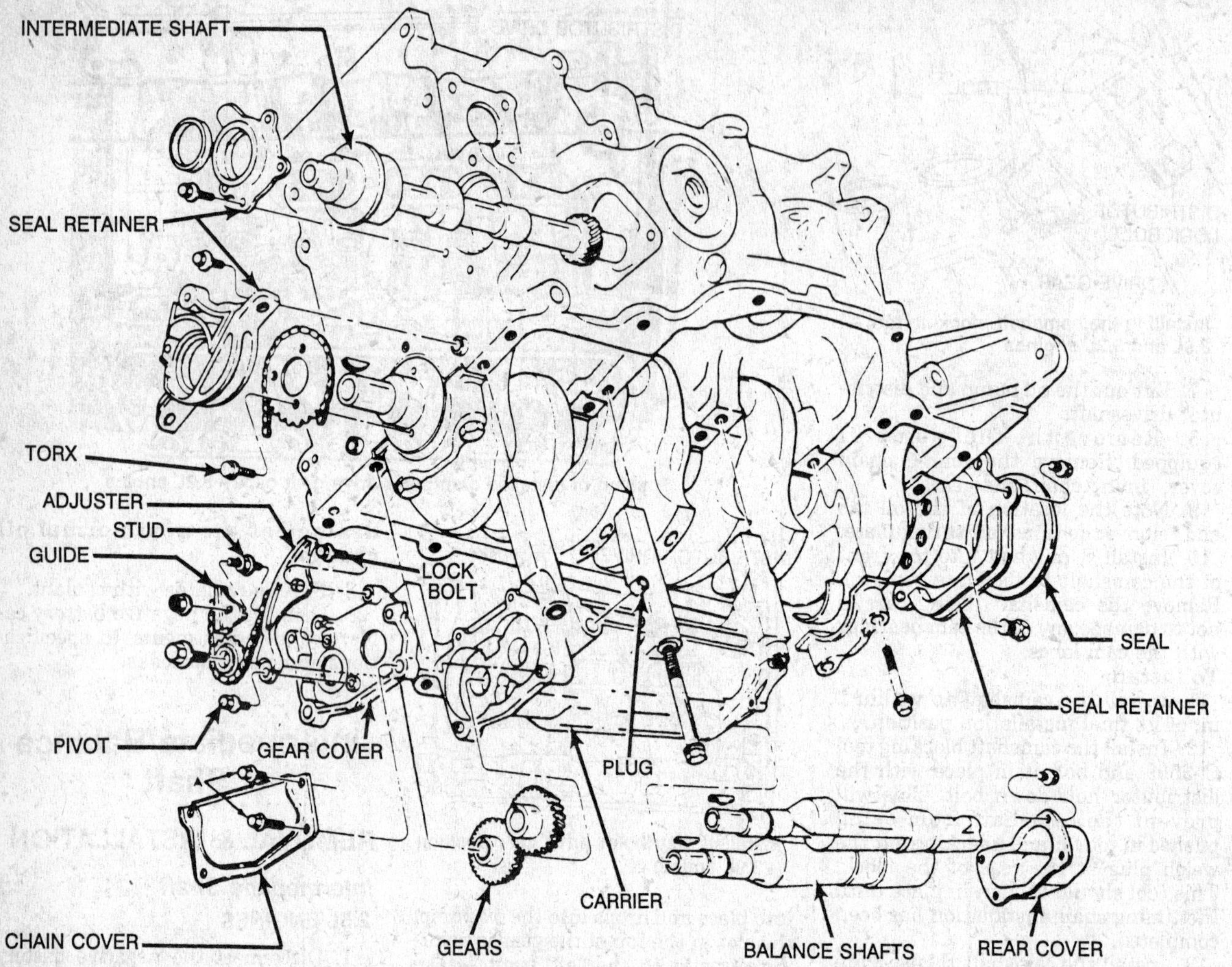

INTERMEDIATE SHAFT

SEAL RETAINER

TORX

ADJUSTER

STUD

GUIDE

LOCK BOLT

PIVOT

GEAR COVER

PLUG

CHAIN COVER

GEARS

CARRIER

BALANCE SHAFTS

REAR COVER

SEAL

SEAL RETAINER

Exploded view of the balance shafts and related parts—2.5L engine

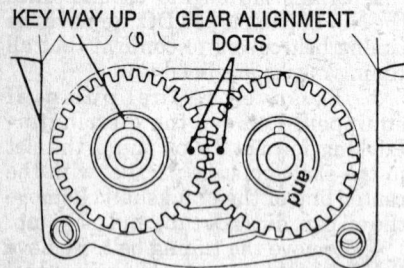

KEY WAY UP   GEAR ALIGNMENT DOTS

Alignment of balance shaft sprockets— 2.5L engine

gear and install the intermediate shaft into the block.

9. Install the intermediate shaft sprocket and the timing belt.

10. With the timing belt properly installed, install the oil pump so the slot is parallel to the center line of the crankshaft. Install the distributor so the rotor is aligned with the No. 1 spark plug wire tower on the cap.

11. Install the fuel pump, if equipped.

12. Connect the negative battery cable and road test the vehicle.

Balance Shaft

2.5L ENGINE

1. Disconnect the negative battery cable. Raise the vehicle and support safely.

2. Remove the timing belt. Remove the oil pan, the oil pickup, the crankshaft belt sprocket and the front crankshaft oil seal retainer.

3. Remove the balance shafts chain cover, guide and tensioner. Remove the balance shaft gear and chain sprocket retaining screws and crankshaft chain sprocket torx screws. Remove chain and sprocket assembly.

4. Remove the gear cover retaining stud. Remove cover and balance shaft gears.

5. Remove the carrier rear cover and balance shafts.

To install:

6. Install balance shafts and rear carrier cover to the block. Install gears and set gear timing as follows:

   a. Turn balance shafts until both shaft keyways are parallel to the vertical centerline of the engine. (facing up)

   b. Install short hub drive gear on sprocket driven shaft.

   c. Install the long hub gear on gear drivenshaft.

   d. Check that the timing marks on the gears are facing each other and that the keyways are facing up.

7. Install the gear cover and tighten double ended stud/washer to 105 inch lbs. (12 Nm).

8. Install crankshaft sprocket and tighten head torx screw to 130 inch lbs. (13 Nm).

9. Rotate the crankshaft to position the No. 1 cylinder on the TDC of the compression stroke; the timing marks on the chain sprocket should align with the parting line on the left side of the No. 1 main bearing cap.

10. Position the balance shaft sprocket into the balance chain so the sprocket (yellow dot) timing mark mates with the yellow link on the chain.

11. Install the balance chain/sprocket assembly onto the crankshaft and the balance shaft. Torque the sprocket to shaft bolts to 21 ft. lbs. (28 Nm). If necessary to secure the crankshaft

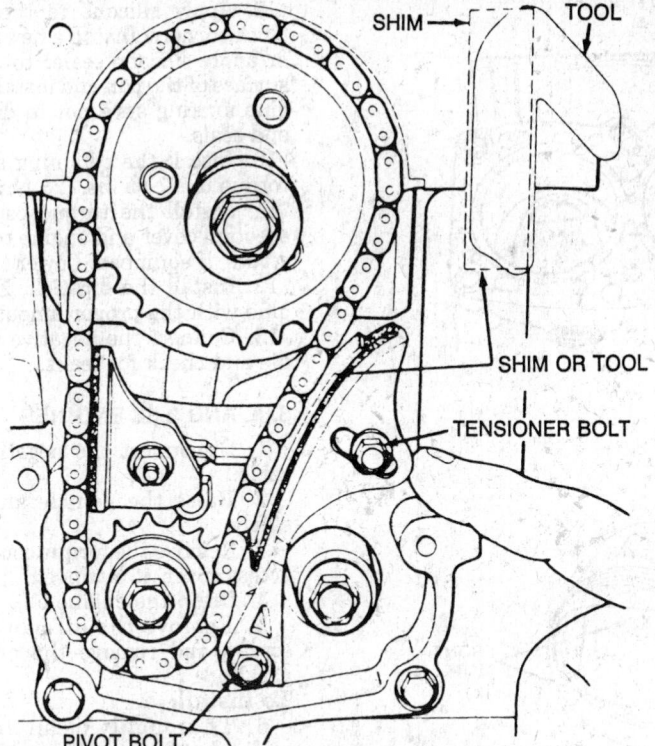

Adjusting the balance shaft chain tensioner—2.5L engine

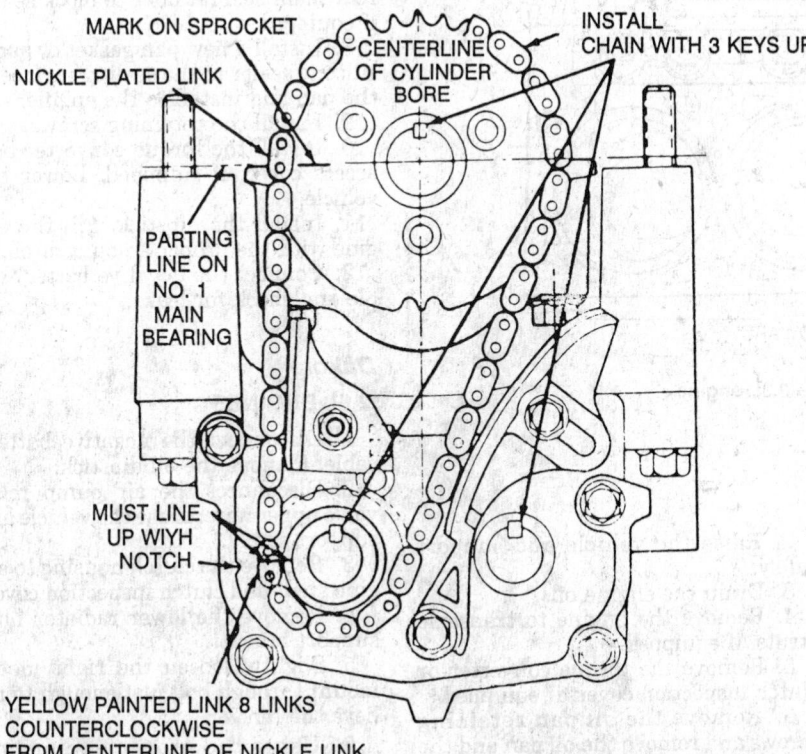

MARK ON SPROCKET
NICKLE PLATED LINK
CENTERLINE OF CYLINDER BORE
INSTALL CHAIN WITH 3 KEYS UP
PARTING LINE ON NO. 1 MAIN BEARING
MUST LINE UP WIYH NOTCH
YELLOW PAINTED LINK 8 LINKS COUNTERCLOCKWISE FROM CENTERLINE OF NICKLE LINK

Timing the balance shaft sprocket with the crankshaft sprocket—2.5L engine

while tightening the bolts, place a block of wood between the crankcase and the crankshaft counterbalance.

12. Loosely install the chain tensioners and place a shim (0.039 in. × 2.75 in.) between the chain and the tensioner. Apply firm pressure (to reduce the chain slack) to the tensioner shoe.

Torque the tensioner to front gear cover bolts to 8.5 ft. lbs. (12 Nm). Remove the shim.

13. Install the chain cover and the rear cover to the carrier housing and torque the bolts to 8.5 ft. lbs. (12 Nm).

14. Replace the crankshaft retainer seal, apply silicone sealer to the mating surface and install the retainer.

15. Install the oil pickup and oil pan.

16. Install the crankshaft sprocket and the timing belt.

17. Connect the negative battery cable and road test the vehicle.

Piston and Connecting Rod

POSITIONING

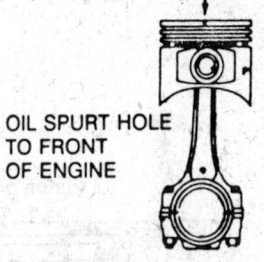

NOTCH TO FRONT OF ENGINE
OIL SPURT HOLE TO FRONT OF ENGINE

Piston positioning—2.5L engines

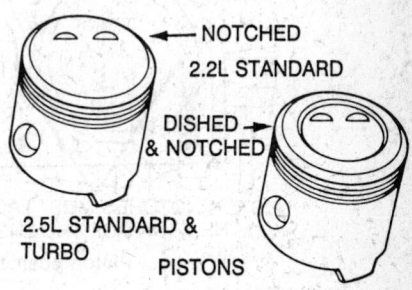

NOTCHED 2.2L STANDARD
DISHED & NOTCHED
2.5L STANDARD & TURBO
PISTONS

2.5L engines—piston

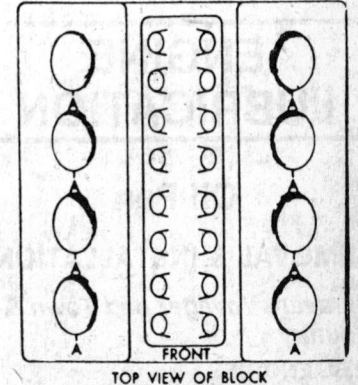

FRONT
TOP VIEW OF BLOCK

Piston positioning—3.9L and 5.2L engines

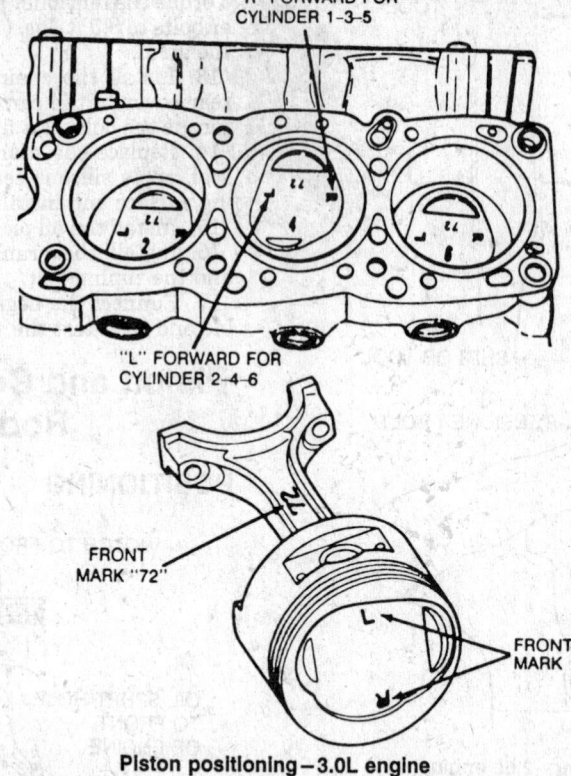

"R" FORWARD FOR
CYLINDER 1-3-5

"L" FORWARD FOR
CYLINDER 2-4-6

FRONT
MARK "72"

FRONT
MARK

Piston positioning—3.0L engine

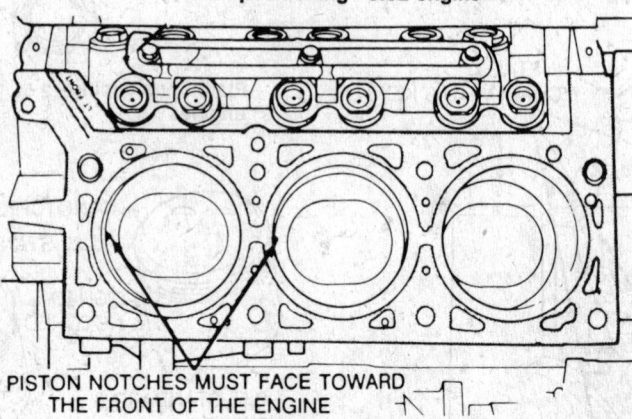

PISTON NOTCHES MUST FACE TOWARD
THE FRONT OF THE ENGINE

Piston positioning—3.3L engine

ENGINE LUBRICATION

Oil Pan

REMOVAL & INSTALLATION

Caravan, Voyager and Town & Country

2.5L ENGINES

1. Disconnect the negative battery cable. Remove the oil dipstick.

2. Raise the vehicle and support safely.

3. Drain the engine oil.

4. Remove the engine to transaxle struts, if equipped.

5. Remove the torque converter or clutch inspection cover, if equipped.

6. Remove the oil pan retaining screws and remove the oil pan and the side seals.

To install:

7. Thoroughly clean and dry all sealing surfaces, bolts and bolt holes.

8. Apply silicone sealer to the 4 end seal to block corners and install the end seals making sure the corners are not twisted.

9. Apply silicone to the 4 pan to block corners. Install a new pan gasket or apply silicone sealer to the sealing surface of the pan and install to the engine making sure not to dislodge the end seals.

10. Install the retaining screws and torque to 17 ft. lbs. (23 Nm).

11. Install the torque converter inspection cover and engine to transaxle struts, if equipped. Lower the vehicle.

12. Install the dipstick. Fill the engine with the proper amount of oil.

13. Connect the negative battery cable and check for leaks.

3.0L AND 3.3L ENGINES

1. Disconnect the negative battery cable.

2. Raise the vehicle and support safely.

3. Remove the torque converter bolt access cover, if equipped.

4. Drain the engine oil.

5. Remove the oil pan retaining screws and remove the oil pan and gasket.

To install:

6. Thoroughly clean and dry all sealing surfaces, bolts and bolt holes.

7. Apply silicone sealer to the chain cover to block mating seam and the rear main seal retainer to block seam, if equipped.

8. Install a new pan gasket or apply silicone sealer to the sealing surface of the pan and install to the engine.

9. Install the retaining screws.

10. Install the torque converter bolt access cover, if equipped. Lower the vehicle.

11. Install the dipstick. Fill the engine with the proper amount of oil.

12. Connect the negative battery cable and check for leaks.

Dakota

2.5L ENGINES

1. Disconnect the negative battery cable. Remove the oil dipstick.

2. Disconnect the air pump relief valve upper hose. Raise the vehicle and support safely.

3. Remove the clutch housing to engine strut and clutch inspection cover.

4. Remove the lower radiator hose support bracket.

5. Slightly loosen the right motor mount through bolt just enough to relieve the tension.

6. Using the proper equipment, support the weight of the engine. Loosen the left motor mount through bolt enough to clear the bracket.

7. Raise the left side of the engine about 2 inches.

8. Remove the oil pan retaining screws and remove the oil pan and gasket.

To install:

9. Thoroughly clean and dry all sealing surfaces, bolts and bolt holes.

10. Apply silicone sealer to the 4 end seal to block corners and install the end seals making sure the corners are not twisted.

11. Apply silicone to the 4 pan to block corners. Install a new pan gasket or apply silicone sealer to the sealing surface of the pan and install to the engine making sure not to dislodge the end seals.

12. Install the pan retaining screws and torque to 17 ft. lbs. (23 Nm).

13. Lower the engine. Torque the motor mount through bolts to 50 ft. lbs. (68 Nm).

14. Install the clutch inspection cover and housing to engine strut. Install the lower radiator hose support bracket. Lower the vehicle.

15. Install the dipstick and air pump hose. Fill the engine with the proper amount of oil.

16. Connect the negative battery cable and check for leaks.

3.9L ENGINE WITH 2WD

1. Disconnect the negative battery cable. Remove the oil dipstick. Disengage the distributor cap and remove it away from the firewall.

2. Raise the vehicle and support safely. Drain the engine oil.

3. Remove the exhaust crossover.

4. Loosen the motor mount bolts. Using the proper equipment, raise the engine. When the engine is high enough, install replacement bolts (similar in size to the motor mount bolts), in the engine mount attaching points on the frame brackets. Lower the engine so the bottom of the motor mounts rest on the 2 replacement bolts. Remove the torque converter inspection cover, if equipped.

5. Remove the oil pan retaining screws and remove the oil pan and gaskets.

To install:

6. Thoroughly clean and dry all sealing surfaces, bolts and bolt holes.

7. Place a drop of silicone sealer to the timing chain cover to block mating seam.

8. Install the new gaskets to the engine and add a drop of silicone sealer to the corners where the rubber and cork meet. Install the rubber seals to the pan.

9. Install the pan to the engine and torque the retaining screws to 17 ft. lbs. (23 Nm). Install the torque converter inspection cover, if equipped.

10. Reinstall the engine to the mount and install the exhaust crossover. Lower the vehicle.

11. Install the distributor cap.

12. Install the dipstick. Fill the engine with the proper amount of oil.

13. Connect the negative battery cable and check for leaks.

3.9L ENGINE WITH 4WD

1. Disconnect the negative battery cable. Remove the oil dipstick.

2. Raise the vehicle and support safely.

3. Using the proper equipment, support the weight of the engine. Remove the front driving axle.

4. Remove the exhaust crossover and the lower transmission cover.

5. Remove the oil pan retaining screws and remove the oil pan and gaskets.

To install:

6. Thoroughly clean and dry all sealing surfaces, bolts and bolt holes.

7. Place a drop of silicone sealer to the timing chain cover to block mating seam.

8. Install the new gaskets to the engine and add a drop of silicone sealer to the corners where the rubber and cork meet. Install the rubber seals to the pan.

9. Install the pan to the engine and torque the retaining screws to 17 ft. lbs. (23 Nm). Install the lower transmission cover, if equipped.

10. Install the exhaust crossover.

11. Install the front driving axle. Lower the vehicle.

12. Install the dipstick. Fill the engine with the proper amount of oil.

13. Connect the negative battery cable and check for leaks.

5.2L ENGINE

1. Disconnect the negative battery cable. Remove the engine dipstick.

2. Raise the vehicle and support safely.

3. Drain the engine oil. Remove the exhaust crossover pipe.

4. Remove left engine to transmission strut.

5. Remove oil pan retaining bolts and remove oil pan.

To install:

6. Using a new pan gasket, add drop of sealer at corners of rubber and cork. Install pan and tighten bolts to 200 inch lbs. (23 Nm).

7. Install engine to strut and exhaust crossover. Lower vehicle.

8. Install oil dipstick. Connect negative battery cable, add engine oil, start engine and check for leaks.

Oil Pump

REMOVAL & INSTALLATION

2.5L Engines

1. Crank the engine around so No. 1 piston is at TDC. Disconnect the negative battery cable.

2. Matchmark the rotor to the block and remove the distributor to confirm that the slot in the oil pump shaft is parallel to the centerline of the crankshaft. Matchmark the slot to the distributor bore, if desired.

3. Remove the dipstick. Raise the vehicle and support safely. Drain the engine oil and remove the pan.

4. Remove the oil pickup.

5. Remove the 2 mounting bolts and remove the oil pump from the engine.

To install:

6. Prime the pump by pouring fresh oil into the pump intake and turning the driveshaft until oil comes out the pressure port. Repeat a few times until no air bubbles are present.

7. Apply sealer Loctite® 515 or equivalent, to the pump body to block machined surface interface. Lubricate the oil pump and distributor driveshaft.

8. Align the slot so it will be in the same position as when it was removed. If it is not, the distributor will not be timed correctly. Install the pump fully and rotate back and forth to ensure proper positioning between the pump

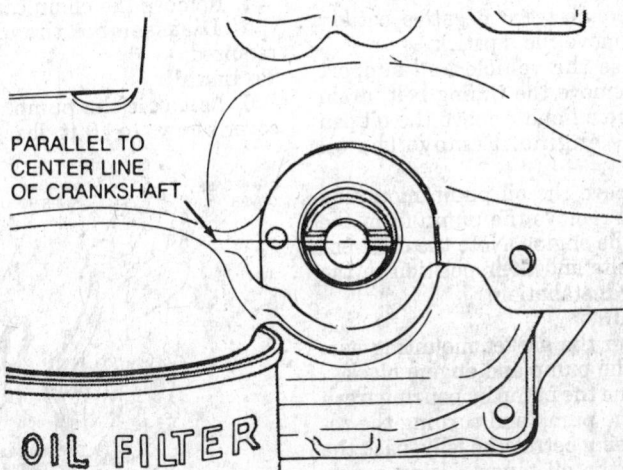

PARALLEL TO CENTER LINE OF CRANKSHAFT

OIL FILTER

Aligning the slot in the oil pump shaft– 2.5L engines

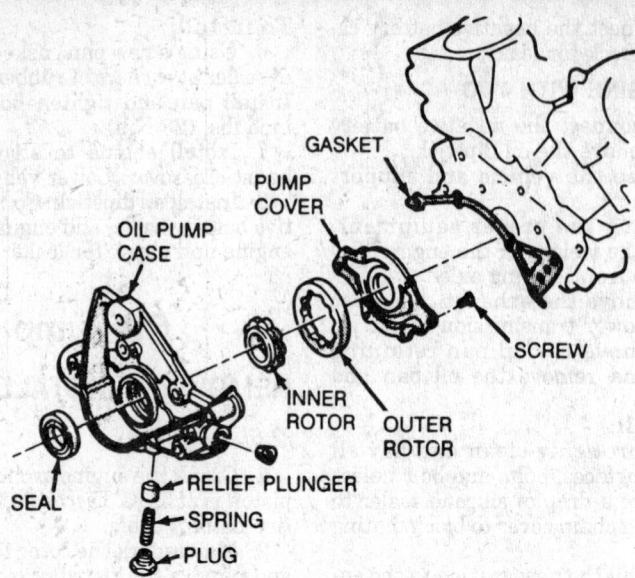

Exploded view of the oil pump—3.0L engine

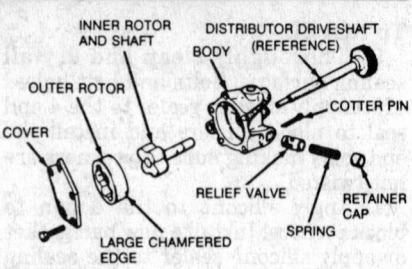

Exploded view of the oil pump—3.9L engine

mounting surface and the machined surface of the block.

9. Install the mounting bolts fingertight and lower the vehicle to confirm proper slot positioning. If the slot is not properly positioned, raise the vehicle and move the gear as required. If the slot is correct, hold the pump firmly against the block and torque the mounting bolts to 17 ft. lbs. (23 Nm).

10. Clean out the oil pickup or replace as required. Replace the oil pickup O-ring and install the pickup to the pump.

11. Install the oil pan using new gaskets. Lower the vehicle.

12. Install the distributor.

13. Install the dipstick. Fill the engine with the proper amount of oil.

14. Connect the negative battery cable, check the timing and check the oil pressure.

3.0L Engine

1. Disconnect the negative battery cable. Remove the dipstick.

2. Raise the vehicle and support safely. Remove the timing belt, drain the engine oil and remove the oil pan from the engine. Remove the oil pickup.

3. Remove the oil pump mounting bolts and remove the pump from the front of the engine. Note the different length bolts and their position in the pump for installation.

To install:

4. Clean the gasket mounting surfaces of the pump and engine block.

5. Prime the pump by pouring fresh oil into the pump and turning the rotors or, using petroleum jelly, pack the inside of the oil pump. Using a new gasket, install the oil pump on the engine and torque all bolts to 11 ft. lbs. (15 Nm).

6. Install the balancer and crankshaft sprocket to the end of the crankshaft.

7. Clean out the oil pickup or replace as required. Replace the oil pickup gasket ring and install the pickup to the pump.

8. Install the timing belt, oil pan and all related parts.

9. Install the dipstick. Fill the engine with the proper amount of oil.

10. Connect the negative battery cable, start the engine and check the oil pressure.

3.3L Engine

1. Disconnect the negative battery cable. Remove the dipstick.

2. Raise the vehicle and support safely. Drain the oil and remove the oil pan.

3. Remove the oil pickup.

4. Remove the chain case cover.

5. Disassemble the oil pump as required.

To install:

6. Assemble the pump. Torque the cover screws to 10 ft. lbs. (12 Nm).

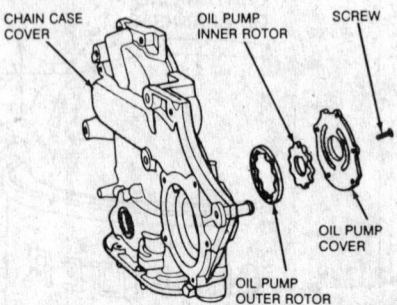

Oil pump components—3.3L engine

7. Prime the oil pump by filling the rotor cavity with fresh oil and turning the rotors until oil comes out the pressure port. Repeat a few times until no air bubbles are present.

8. Install the chain case cover.

9. Clean out the oil pickup or replace as required. Replace the oil pickup O-ring and install the pickup to the pump.

10. Install the oil pan.

11. Install the dipstick. Fill the engine with the proper amount of oil.

12. Connect the negative battery cable, start the engine and check the oil pressure.

3.9L and 5.2L Engines

1. Disconnect the negative battery cable.

2. Raise the vehicle and support safely. Drain the oil and remove the oil pan.

3. Remove the screen.

4. Unbolt the oil pump from the rear main bearing cap and remove it from the vehicle.

To install:

5. Prime the pump by pouring fresh oil into the pump intake and turning the driveshaft until oil comes out the pressure port. Repeat a few times until no air bubbles are present. Install the oil pump with a rotating motion to ensure proper pump driveshaft engagement.

6. Hold the pump flush against the main cap and finger-tighten the attaching bolts.

7. Torque the bolts to 30 ft. lbs. (41 Nm).

8. Install the screen.

9. Install the oil pan with a new gasket and lower vehicle.

10. Connect the negative battery cable and add oil to correct the level. Start the engine and check the oil pressure. Check for oil leaks.

CHECKING

2.5L Engines

1. Remove the cover from the oil pump.

2. Check endplay of the inner rotor using a feeler gauge and a straight

edge placed across the pump body. The specification is 0.001–0.004 in. (0.03–0.09mm).

3. Measure the clearance between the inner and outer rotors. The maximum clearance is 0.008 in. (0.20mm).

4. Measure the clearance between the outer rotor and the pump body. The maximum clearance is 0.014 in. (0.35mm).

5. The minimum thickness of the outer rotor is 0.944 in. (23.96mm). The minimum diameter of the outer rotor is 2.77 in. (62.70mm). The minimum thickness of the inner rotor is 0.943 in. (23.95mm).

6. Check the cover for warpage. The maximum allowable is 0.003 in. (0.076mm).

7. Check the pressure relief valve for damage. The spring's freelength specification is 1.95 in. (49.50mm).

8. Assemble the outer rotor with the larger chamfered edge in the pump body. Torque the cover screws to 10 ft. lbs. (12 Nm).

3.0L Engine

1. Remove the rear cover.

2. Remove the pump rotors and inspect the case for excessive wear.

3. Measure the diameter of the inner rotor hub that sits in the case. Measure the inside diameter of the inner rotor hub bore. Subtract the first measurement from the second; if the result is over 0.006 in. (0.15mm), replace the oil pump assembly.

4. Measure the clearance between the outer rotor and the case. The specification is 0.004–0.007 in. (0.10–0.18mm).

5. Check the side clearance of the rotors using a feeler gauge and a straightedge placed across the case. The specification is 0.0015–0.0035 in. (0.04–0.09mm).

6. Check the relief plunger and spring for damage and breakage. Replace as required.

7. Install the rear cover to the case.

3.3L Engine

1. Thoroughly clean and dry all parts. The mating surface of the chain case cover should be smooth. Replace the pump cover if it is scratched or grooved.

2. Lay a straightedge across the pump cover surface. If a 0.003 in. (0.076mm) feeler gauge can be inserted between the cover and straightedge, the cover should be replaced.

3. The minimum thickness of the outer rotor is 0.301 in. (7.63mm). The minimum diameter of the outer rotor is 3.14 in. (79.78mm). The minimum thickness of the inner rotor is 0.301 in. (7.64mm).

4. Install the outer rotor onto the chain case cover, press to one side and

measure the clearance between the rotor and case. If the measurement exceeds 0.022 in. (56mm) and the rotor is good, replace the chain case cover.

5. Install the inner rotor to the chain case cover and measure the clearance between the rotors. If the clearance exceeds 0.008 in. (0.203mm), replace both rotors.

6. Place a straightedge over the chain case cover between bolt holes. If a 0.004 in. (0.102mm) thick feeler gauge can be inserted under the straightedge, replace the pump assembly.

7. Inspect the relief valve plunger for scoring and freedom of movement. Small marks may be removed with 400 grit wet or dry sandpaper.

8. The relief valve spring should have a freelength of 1.95 in.

9. Assemble the pump using new parts where necessary.

3.9L and 5.2L Engines

1. Disassemble the pump.

2. Replace the pump assembly if the cover is scratched or grooved or if the cover is warped more than 0.0015 in. (0.038mm).

3. The minimum thickness of the outer rotor is 0.825 in. (20.96mm). The minimum diameter of the outer rotor is 2.469 in. (62.70mm). The minimum thickness of the inner rotor is 0.825 in. (20.96mm). If any of the above measurements are not within specifications, replace the shaft and both rotors.

4. The maximum clearance between the outer rotor and the pump body is 0.014 in. (0.356mm). Replace the pump assembly if not within specifications.

5. Install the inner rotor and place a straightedge across the bolt holes. If a feeler gauge of 0.004 in. (0.101mm) or more fits, replace the pump assembly.

6. The maximum clearance between the rotors is 0.010 in. (0.254mm). Replace the shaft and both rotors if not within specifications.

7. Inspect the relief valve plunger for scoring. Small marks may be removed with 400 grit wet or dry sandpaper.

8. The relief valve spring should have a freelength of 1.9 in. Replace the spring if it fails to meet specifications.

9. Assemble the pump using new parts where necessary.

Rear Main Bearing Oil Seal

REMOVAL & INSTALLATION

Except 3.9L and 5.2L Engines

1. Disconnect the negative battery cable.

2. Remove the transmission or transaxle. Remove the flywheel or flexplate.

3. If there is any leakage coming from the rear seal retainer, drain the engine oil and remove the oil pan, if necessary. Remove the rear main oil seal retainer.

4. Remove the seal from the retainer.

To install:

5. Lightly coat the seal outer diameter with Loctite® Stud N' Bearing Mount or equivalent.

6. Install the seal to the retainer.

7. If removed, thoroughly clean and dry the retainer to block sealing surfaces. Install a new gasket or apply silicone sealer and install the retainer to the block. Install the pan, if removed.

8. Install the flywheel or flexplate and the transmission or transaxle.

9. Connect the negative battery cable and check for leaks.

3.9L and 5.2L Engines

1. Raise the vehicle and support safely. Drain the engine oil. Remove the oil pan and oil pump.

2. Remove the rear main bearing cap.

3. Remove the lower seal from the cap.

4. To remove the upper rope seal, use oil seal remover and installer kit KD–492 or equivalent, following the instructions provided with the tool.

To install:

5. Wipe the crankshaft surface clean and coat it lightly with oil.

6. Use oil seal remover and installer kit KD–492 or equivalent to install the upper rope seal. Trim the ends of the upper seal to eliminate frayed ends.

7. Install the lower rope seal in the main cap so both ends protrude. Use tool C–3511 to seat the seal in its groove. Cut of the portions of the seal that extend above the cap on both sides and install the end seals to the cap.

8. Install the main cap to the block and torque the bolts to 85 ft. lbs. (115 Nm).

9. Install the pan and fill the engine with the proper amount of oil.

10. Connect the negative battery cable, start engine and check for leaks.

ENGINE COOLING

Radiator

REMOVAL & INSTALLATION

1. Disconnect the negative battery

cable. Unclip throttle cable at fan shroud, if equipped.

2. Open the radiator petcock and drain the antifreeze. Once the antifreeze has stopped draining, close the petcock.

3. Remove the upper hose and coolant reserve tank hose from the radiator.

4. Remove the fan shroud and electric cooling fan from radiator. On 5.2L and 3.9L engines, move fan shroud towards engine, it is not necessary to remove from the vehicle.

5. Disconnect the automatic transmission or transaxle cooler hoses, if equipped and plug them.

6. Remove upper mounting screws and lift radiator from lower supports.

To install:

7. Lower the radiator into position and install the mounting screws.

8. Connect the automatic transmission or transaxle cooler lines, if removed.

9. Connect the lower radiator hose.

10. Install the shroud or electric cooling fan, if removed.

11. Connect the upper hose and coolant reserve tank hose.

12. Fill the system with coolant.

13. Connect the negative battery cable. Run the vehicle until the thermostat opens, fill the radiator completely and check the coolant and transmission fluid levels and add as required.

14. Once the vehicle has cooled, recheck the coolant level.

Electric Cooling Fan

REMOVAL & INSTALLATION

1. Disconnect the negative battery cable.

2. Unplug the connector.

3. Remove the mounting screws.

4. Remove the fan assembly from the vehicle.

5. The installation is the reverse of the removal procedure.

Heater Core

REMOVAL & INSTALLATION

Without Air Conditioning

CARAVAN, VOYAGER AND TOWN & COUNTRY

1. Disconnect the negative battery cable. Drain the cooling system.

2. Remove the lower steering column cover and both side under panel silencers.

3. Remove the lower reinforcement under the steering column, right side cowl and sill trim. Remove the bolt holding the right side instrument panel to the right cowl.

4. Loosen the 2 brackets supporting the lower edge of the heater housing. Remove the instrument panel trim covering and reinforcement. Remove the retaining screws from the right side to the steering column.

5. Disconnect the vacuum line at the brake booster. Remove the accessory switch carrier and storage module.

6. Clamp off the heater hoses near the heater core and remove the hoses from the core tubes. Plug the hose ends and the core tubes to prevent spillage of coolant. Remove the heater assembly retaining nuts at the firewall.

7. Disconnect the blower motor wiring, resistor wiring and the temperature control cable. Disconnect the hanger strap from the package and rotate it out of the way.

8. Pull the right side of the instrument panel out as far as possible. Fold the carpeting and insulation back to provide a little more working room and to prevent spillage from staining the carpeting.

9. Pull the heater core assembly out from under instrument panel and remove it from the vehicle.

10. Remove vacuum harness and top cover from assembly. Remove nut and lever from the door shaft and remove the temperature door from the cover.

11. Remove the retaining screw from the heater core and remove the core from the housing assembly.

To install:

12. Clean out the inside of the housing. Wrap the heater core with foam tape and place it in position. Secure it with its retainer screw. Install temperature door and vacuum harness to heater unit.

13. Install the cover to the heater unit.

14. Install the assembly to the vehicle and install the nuts to the firewall. Fold the carpeting back into position.

15. Connect the hanger strap from the package and rotate it out of the way. Install the 2 brackets supporting the lower edge of the heater housing. Connect the blower motor wiring, resistor wiring and the temperature control cable.

16. Install the retaining screws from the right side to the steering column. Install the instrument panel trim covering and reinforcement.

17. Install the bolt holding the right side instrument panel to the right cowl. Install the lower reinforcement under the steering column, right side cowl and sill trim.

18. Install the heater hoses to the core tubes. Install the storage compartment and accessory switch carrier to original position.

19. Connect the vacuum line at the brake booster.

20. Fill the cooling system.

21. Connect the negative battery cable and check the heater for proper operation and leaks.

DAKOTA

1. Disconnect the negative battery cable. Drain the coolant.

2. Remove the steering column cover, intermittent wiper control, if equipped and the lower instrument panel module retaining screw to the right of the steering column.

3. Remove the center distribution duct retaining screws and panel support screw at the bottom of the module.

4. Remove the courtesy light at the lower right corner of the module and the screw near the ash tray.

5. Open the glovebox and remove the screws along the top edge.

6. Move the module out and down far enough to unclip the wiring harness and antenna cable and disconnect the speaker wire, if equipped with monaural radio, and glovebox light wire. Remove the module from the vehicle.

7. Remove the center air distribution duct.

8. Remove the antenna wire from retaining clip at the right end of the heater unit.

9. Disconnect the blower motor connector and remove the terminal insulator retainer from the heater unit.

10. Disconnect the demister hoses from the adaptor at the top of the heater unit.

11. Disconnect the vacuum feed line from the check valve.

12. Disconnect the temperature control cable flag retainer from the heater unit.

13. Remove the adjusting clip from the blend air door crank.

14. Disconnect the heater hoses from the core tubes and plug them.

15. Remove 4 heater unit attaching nuts from the rear engine compartment dash panel.

16. Remove the heater unit support attaching screw and rotate the brace out of the way.

17. Remove the heater unit from the vehicle.

18. To disassemble the housing assembly, vacuum diaphragm and retaining screws from the cover and remove the cover.

19. Remove the retaining screw from the heater core and remove the core from the housing assembly.

To install:

20. Remove the temperature control door from the unit and clean the unit out with solvent. Lubricate the lower pivot rod and its well and install. Wrap

the heater core with foam tape and place it in position. Secure it with its screw.

21. Assemble the unit, making sure all vacuum tubing is properly routed.

22. Install the assembly to the vehicle and connect the vacuum harness. Install the nuts to the firewall. Install the support brace to the heater unit.

23. Connect the demister hoses to the adaptor at the top of the heater unit.

24. Connect the blower motor connector and install the thermal insulator retainer to the heater unit.

25. Connect the vacuum feed line to the check valve.

26. Connect the temperature control cable flag retainer to the heater unit and install the adjusting clip from the blend air door crank.

27. Install the center air distribution duct.

28. Install the antenna wire from retaining clip at the right end of the heater unit.

29. Install the instrument panel module and all related parts.

30. Connect the heater hoses to the core tubes.

31. Fill the cooling system.

32. Connect the negative battery cable and check the heater for proper operation and leakage.

With Air Conditioning

CARAVAN, VOYAGER AND TOWN & COUNTRY

1. Disconnect the negative battery cable. Properly discharge the air conditioning system. Drain the cooling system.

2. Remove the lower steering column cover. Remove the left and right under panel silencers.

3. Remove the lower reinforcement under the steering column, right side cowl and sill trim. Remove the bolt holding the right side instrument panel to the right cowl.

4. Loosen the 2 brackets supporting the lower edge of the heater housing. Remove the instrument panel trim covering and reinforcement. Remove the retaining screws from the right side to the steering column.

5. Disconnect the vacuum lines at the brake booster and water valve.

6. Clamp off the heater hoses near the heater core and remove the hoses from the core tubes. Plug the hose ends and the core tubes to prevent spillage of coolant.

7. Disconnect the H-valve connection at the valve and remove the H-valve. Remove the retaining nuts from the package mounting studs at the firewall.

8. Disconnect the blower motor wiring, resistor wiring and the tempera-

ture control cable. Disconnect the vacuum harness at the connection at the top of the heater unit.

9. Disconnect the hanger strap from the package and rotate it out of the way. Remove the rubber condensate drain tube located in the engine compartment.

10. Pull the right side of the instrument panel out as far as possible. Fold the carpeting and insulation back to provide a little more working room and to prevent spillage from staining the carpeting.

11. Remove the entire housing assembly from the dash panel and remove it from the vehicle.

12. To disassemble the housing assembly, remove the vacuum diaphragm, vacuum harness and retaining screws from the cover and remove the cover.

13. Remove the retaining screw from the heater core and remove the core from the housing assembly.

To install:

14. Remove the temperature control door from the unit and clean the unit out with solvent. Lubricate the lower pivot rod and its well and install. Wrap the heater core with foam tape and place it in position. Secure it with its screw.

15. Assemble the unit, making sure all vacuum tubing is properly routed.

16. Install the assembly to the vehicle and connect the vacuum harness. Install the nuts to the firewall and install the condensation tube. Fold the carpeting back into position.

17. Connect the hanger strap from the package and rotate it out of the way. Install the 2 brackets supporting the lower edge of the heater housing. Connect the blower motor wiring, resistor wiring and the temperature control cable.

18. Install the retaining screws from the right side to the steering column. Install the instrument panel trim covering and reinforcement.

19. Install the bolt holding the right side instrument panel to the right cowl. Install the lower reinforcement under the steering column, right side cowl and sill trim.

20. Connect the vacuum lines at the brake booster and water valve.

21. Connect the heater hoses to the core tubes.

22. Using new gaskets, install the H-valve and connect the hose connection at the valve.

23. Evacuate and recharge the air conditioning system.

24. Fill the cooling system.

25. Connect the negative battery cable and check the entire climate control system for proper operation and leakage.

DAKOTA

1. Disconnect the negative battery cable. Properly discharge the air conditioning system. Drain the coolant.

2. Remove the steering column cover, intermittent wiper control and the lower instrument panel module retaining screw to the right of the steering column.

3. Remove the center distribution duct retaining screws and panel support screw at the bottom of the module.

4. Remove the courtesy light at the lower right corner of the module and the screw near the ash tray.

5. Open the glovebox and remove the screws along the top edge.

6. Move the module out and down far enough to unclip the wiring harness and antenna cable and disconnect the speaker wire, if equipped with monaural radio, and glovebox light wire. Remove the module from the vehicle.

7. Remove the center air distribution duct.

8. Remove the antenna wire from retaining clip at the right end of the heater unit.

9. Disconnect the blower motor connector and remove the terminal insulator retainer from the heater unit.

10. Disconnect the demister hoses from the adaptor at the top of the heater unit.

11. Disconnect the vacuum harness connector from the air conditioning control hose and vacuum feed line from the check valve.

12. Disconnect the temperature control cable flag retainer from the heater unit and remove the adjusting clip from the blend air door crank.

13. Disconnect the heater hoses from the core tubes and plug them.

14. Remove the condensation drain tube.

15. Remove 4 heater-air conditioning unit attaching nuts from the rear engine compartment dash panel.

16. Remove the heater-air conditioning unit support attaching screw and rotate the brace out of the way.

17. Remove the heater-air conditioning unit from the vehicle.

18. To disassemble the housing assembly, vacuum diaphragm and retaining screws from the cover and remove the cover.

19. Remove the retaining screw from the heater core and remove the core from the housing assembly.

To install:

20. Remove the temperature control door from the unit and clean the unit out with solvent. Lubricate the lower pivot rod and its well and install. Wrap the heater core with foam tape and place it in position. Secure it with its screw.

21. Assemble the unit, making sure all vacuum tubing is properly routed.

22. Install the assembly to the vehicle and connect the vacuum harness. Install the nuts to the firewall and install the condensation tube. Install the support brace to the heater-air conditioning unit.

23. Connect the demister hoses to the adaptor at the top of the heater unit.

24. Connect the blower motor connector and install the thermal insulator retainer to the heater unit.

25. Connect the vacuum harness connector to the air conditioning control hose and vacuum feed line to the check valve.

26. Connect the temperature control cable flag retainer to the heater unit and install the adjusting clip from the blend air door crank.

27. Install the center air distribution duct.

28. Install the antenna wire from retaining clip at the right end of the heater unit.

29. Install the instrument panel module and all related parts.

30. Connect the heater hoses to the core tubes.

31. Using new gaskets, install the H-valve and connect the hose connection at the valve.

32. Evacuate and recharge the air conditioning system.

33. Fill the cooling system.

34. Connect the negative battery cable and check the entire climate control system for proper operation and leakage.

Water Pump

REMOVAL & INSTALLATION

2.5L Engines

1. Disconnect the negative battery cable.

2. Drain the cooling system. Remove the alternator and position aside.

3. If equipped with air conditioning, remove the compressor from the bracket and position it to the side.

4. Raise the vehicle and support safely, if necessary and remove the alternator and bracket. Remove the pulley from the water pump.

5. Disconnect the lower radiator hose and heater hose from the water pump.

6. Remove the water pump housing attaching screws and remove the assembly from the vehicle. Discard the O-ring.

7. Remove the water pump from the housing.

To install:

8. Using a new gasket or silicone

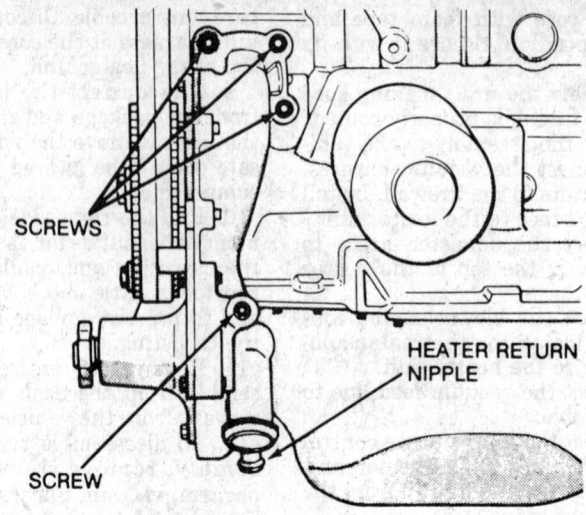

Water pump—2.5L engines

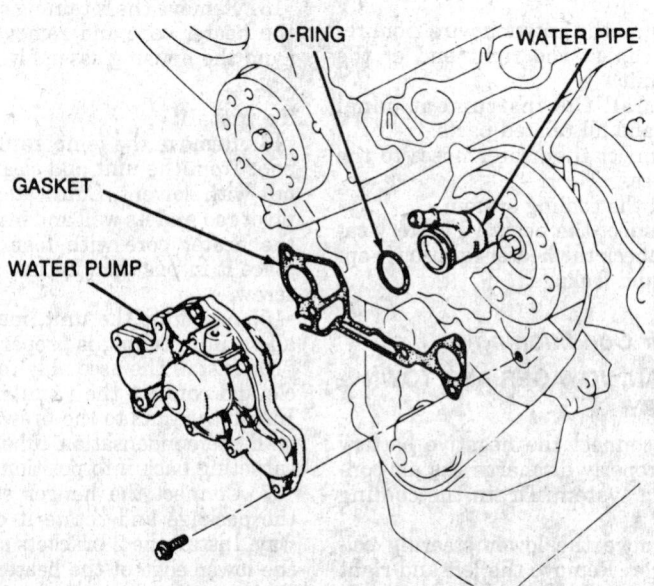

Water pump assembly—3.0L engine

sealer, install the water pump to the housing.

9. Install a new O-ring to the housing and install to the engine. Torque top 3 bolts to 21 ft. lbs. (30 Nm). Install lower screw and tighten to 50 ft. lbs. (68 Nm).

10. Install the water pump pulley and belt. Connect the radiator hose and heater hose to the water pump.

11. Install all items removed to gain access to the water pump and adjust the belts.

12. Remove the hex-head plug or vacuum switching valve on the top of the thermostat housing. Fill the radiator with coolant until the coolant comes out the plug hole. Install the plug or valve and continue to fill the radiator.

13. Connect the negative battery cable, run the vehicle until the thermostat opens, fill the radiator completely and check for leaks.

14. Once cool, recheck the coolant level.

3.0L Engine

1. Disconnect the negative battery cable.

2. Drain the cooling system.

3. Remove the upper front outer timing cover. If the same timing belt will be reused, mark the direction of the timing belt's rotation, for installation in the same direction. Make sure the engine is positioned so No. 1 cylinder is at the TDC of it's compression stroke and the sprockets timing marks are aligned with the engine's timing mark indicators.

4. Loosen the timing belt tensioner bolt and remove the belt. Position the tensioner as far away from the center of the engine as possible and tighten the bolt. Remove the water pump

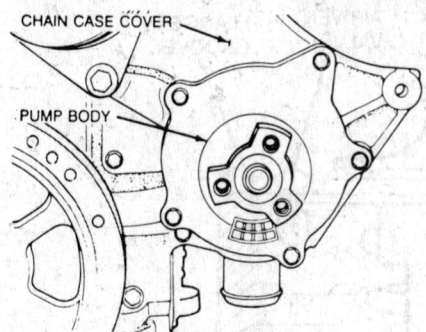

CHAIN CASE COVER

PUMP BODY

Water pump—3.3L engine

mounting bolts, separate the pump from the water inlet pipe and remove the pump from the engine.

To install:

5. Clean all gasket and O-ring surfaces on pump and install new O-ring on water inlet pipe. Dampen O-ring with water to ease in assembly. Install the pump with a new gasket to the engine. Torque the water pump mounting bolts to 20 ft. lbs. (27 Nm).

6. Position both camshafts so the marks line up with timing marks on the alternator bracket (rear bank) and inner timing cover (front bank). Rotate the crankshaft so the timing mark aligns with the mark on the oil pump.

7. Install the timing belt on the crankshaft sprocket and while keeping the belt tight on the tension side (right side), install the belt on the front camshaft sprocket.

8. Install the belt on the water pump pulley, rear camshaft sprocket and then the belt tensioner.

9. Rotate the front camshaft counterclockwise to tension the belt between the front camshaft and the crankshaft. If the timing marks came out of line, repeat the procedure.

10. Install the crankshaft sprocket flange.

11. Loosen the tensioner bolt and allow the spring to tension the belt.

12. Turn the crankshaft 2 full turns in the clockwise direction until the timing marks align again. Now that the belt is properly tensioned, torque the tensioner lock bolt to 21 ft. lbs. (29 Nm).

13. Refill the cooling system. This system uses a self-bleeding thermostat, so there is no need to bleed the system. Connect the negative battery cable and road test the vehicle.

3.3L Engine

1. Disconnect the negative battery cable.

2. Drain the cooling system.

3. Remove the serpentine belt.

4. Raise the vehicle and support safely. Remove the right front tire and wheel assembly and lower fender shield.

5. Remove the water pump pulley.

6. Remove the 5 mounting screws and remove the pump from the engine.

7. Discard the O-ring.

To install:

8. Using a new O-ring, install the pump to the engine. Torque the mounting bolts to 9 ft. lbs. (12 Nm).

9. Install the water pump pulley and torque screws to 21 ft. lbs. (30 Nm).

10. Install the fender shield and tire and wheel assembly. Lower the vehicle.

11. Install the serpentine belt.

12. Remove the engine temperature sending unit. Fill the radiator with coolant until the coolant comes out the sending unit hole. Install the sending unit and continue to fill the radiator.

13. Connect the negative battery cable, run the vehicle until the thermostat opens, fill the radiator completely and check for leaks.

14. Once cool, recheck the coolant level.

3.9L and 5.2L Engines

1. Disconnect the negative battery cable. Drain the coolant.

2. Remove the shroud from the radiator and slide it over the fan. Unclip the throttle cable and coolant overflow bottle from the fan shroud.

3. Remove the upper radiator hose at the radiator.

4. Remove the fan blade, spacer or viscous drive unit, pulley, bolts and shroud together. Remove the air pump belt and power steering pump belt.

NOTE: Do not place the viscous fan in an upright position because the silicone fluid in the drive could drain into the bearing and contaminate its lubricant.

5. Remove the accessory drive belt(s).

6. Remove the air pump. Remove the air pump bracket and unbolt the power steering pump bracket with the pump still attached and position it out of the way.

7. Remove the water pump pulley from the pump.

8. Disconnect the heater hose, bypass hose and lower hose from the pump.

9. Remove the remaining water pump mounting bolts. Remove the pump from the engine.

To install:

10. Clean and dry the pump mating surfaces. Install a new bypass hose to the engine.

11. Install a new gasket and install the water pump to the engine.

12. Install any bolts that do not retain a bracket. Tighten the bypass hose clamps. Install the heater hose.

13. Install water pump pulley and tighten bolts to 20 ft. lbs. (27 Nm).

14. Install the water pump to compressor front mount bolts and bracket, if removed.

15. Install all remaining brackets and components that were removed during the removal procedure. Torque all water pump retaining bolts that do not go through adjusting slots to 30 ft. lbs. (41 Nm).

16. Install the fan blade, spacer or viscous drive unit, pulley and bolts along with the shroud.

17. Install and adjust the accessory drive belt(s) as required.

18. Install the upper radiator hose and fan shroud. Reattach the throttle cable and overflow bottle to radiator shroud.

20. Fill the radiator with coolant. This cooling system is self-bleeding, so system bleeding is not required.

21. Connect the negative battery cable, run the vehicle until the thermostat opens, fill the radiator completely and check for leaks.

22. Once cool, recheck the coolant level.

Thermostat

REMOVAL & INSTALLATION

Except 3.9L and 5.2L Engines

1. Disconnect the negative battery cable. Drain the coolant down to thermostat level or below.

2. Remove the radiator hose from the thermostat and remove the thermostat housing.

3. Remove the thermostat and discard the gasket.

4. Clean the housing mating surfaces and use a new gasket.

To install:

5. Install the thermostat into the engine and install housing using new gasket. Install the upper radiator hose.

6. On 2.2L and 2.5L engines, remove the hex-head plug or vacvuum switching valve on the thermostat housing. Fill the radiator with coolant until the coolant comes out the plug hole. Install the plug or valve and continue to fill the radiator. On the 3.3L engine, remove the engine temperature sending unit. Fill the radiator with coolant until the coolant comes out the sending unit hole. Install the sending unit and continue to fill the radiator. All other engines are self-bleeding.

7. Connect the negative battery cable, run the vehicle until the thermostat opens, fill the radiator completely and check for leaks.

8. Once cool, recheck the coolant level.

3.9L and 5.2L Engines

1. Disconnect the negative battery cable. Drain the coolant down to thermostat level or below.

2. Remove the thermostat housing.

3. Remove the thermostat and discard the gasket.

4. Clean the housing mating surfaces and use a new gasket.

5. The installation is the reversal of the removal procedure.

6. Connect the negative battery cable, run the vehicle until the thermostat opens, fill the radiator completely and check for leaks.

7. Once the vehicle has cooled, recheck the coolant level.

COOLING SYSTEM BLEEDING

The thermostat in the 3.0L engines is equipped with a small air vent valve that allows trapped air to bleed from the system during refilling. This valve negates the need for cooling system bleeding in those engines. The 3.9L and 5.2L engines cooling system is also self-bleeding and does not require bleeding when refilling.

To bleed air from the 2.2L and 2.5L engines, remove the 8mm hex-head plug on the top of the thermostat housing (some engines have a vacuum switching valve at that location). On the 3.3L engine, remove the engine temperature sending unit. Fill the radiator with coolant until the coolant comes out the hole. Install the plug, valve or switch and continue to fill the radiator. This will vent all trapped air from the engine.

ENGINE ELECTRICAL

NOTE: Disconnecting the negative battery cable on some vehicles may interfere with the functions of the on board computer systems and may require the computer to undergo a relearning process, once the negative battery cable is reconnected.

Distributor

REMOVAL

Except 1992 3.9L and 5.2L Engines

1. Disconnect the negative battery cable.

2. Disconnect the distributor pick-

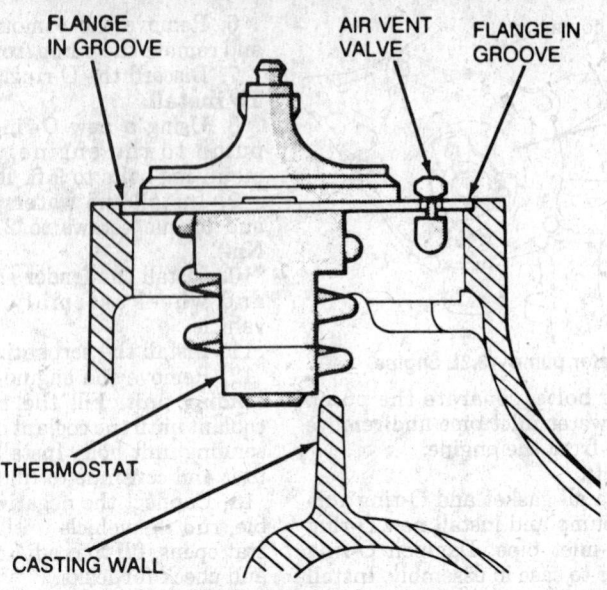

Thermostat with air vent valve—3.0L engine

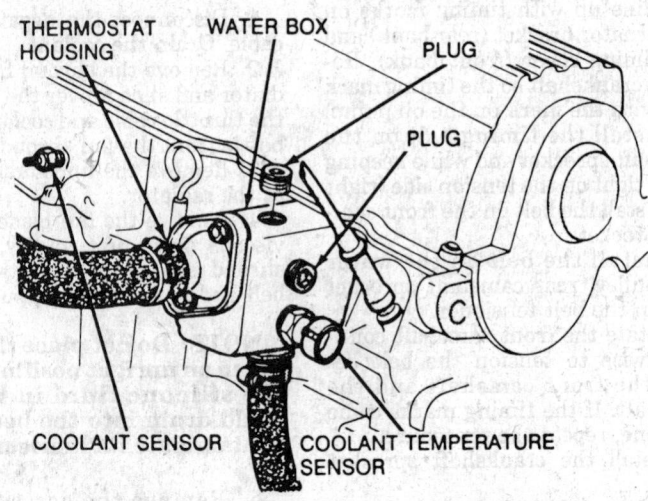

Cooling system bleed plug—2.2L and 2.5L engines

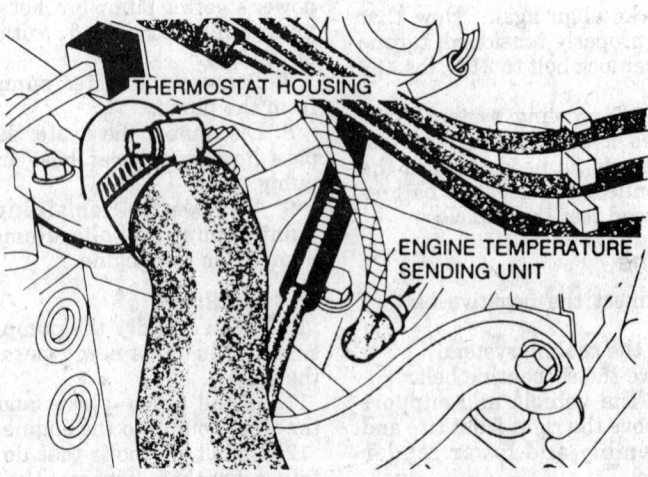

Engine temperature sending unit—3.3L engine

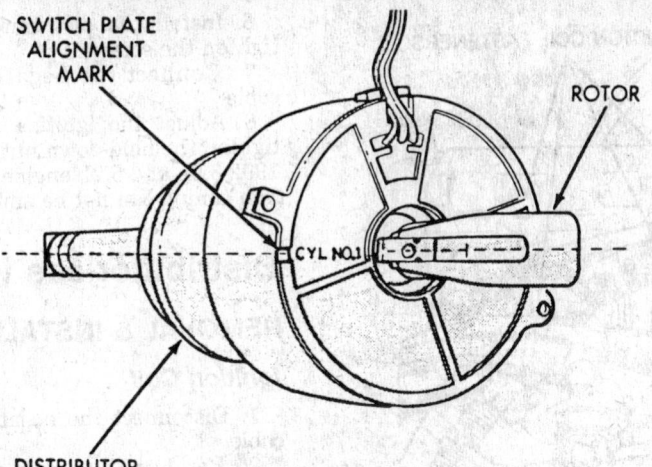

SWITCH PLATE
ALIGNMENT
MARK

ROTOR

CYL NO.1

DISTRIBUTOR

Rotor alignment mark— 3.9L and 5.2L engines

up lead wires. Remove the splash shield, if equipped.

3. Unscrew the distributor cap hold-down screws and lift off the distributor cap with all ignition wires connected. Remove the coil wire, if necessary.

4. Matchmark the rotor to the distributor housing and the distributor housing to the engine.

NOTE: Do not crank the engine during this procedure. If the engine is cranked, the matchmark must be disregarded.

5. Remove the hold-down bolt and clamp.

6. Remove the distributor from the engine.

3.9L and 5.2L Engines

1. Disconnect negative battery cable.

2. Matchmark the original position of the distributor housing to the engine.

3. Unscrew the distributor cap hold-down screws and lift off the distributor cap with all ignition wires still connected. Remove the coil wire, if necessary.

4. Before distributor is removed the No. 1 piston must be at TDC of the compression stroke and the mark on the vibration damper must be aligned with **0** on the timing indicator.

5. Check to assure alignment of the distributor rotor to the alignment mark on the distributor switch plate. If mark is not aligned, reset engine position.

6. Disconnect distributor switch plate wiring harness and remove distributor rotor from shaft.

7. Remove the hold-down bolt and clamp.

8. Remove the distributor from the engine.

INSTALLATION

Timing Not Disturbed

1. Install a new distributor housing O-ring.

2. Install the distributor in the engine so the rotor is aligned with the matchmark on the housing and the housing is aligned with the matchmark on the engine. Make sure the distributor is fully seated and the distributor shaft is fully engaged. On 1992 3.9L and 5.2L engines, install distributor in engine so the matchmark on housing is aligned and the rotor is aligned with mark on the switch plate. Turn housing to align, if necessary.

3. Install the hold-down clamp and snug the hold-down bolt or install the nut.

4. Connect the distributor pickup lead wires. Install the splash shield, if equipped.

5. Install the distributor cap and tighten the retaining screws.

6. Connect the negative battery cable.

7. Adjust the ignition timing and tighten the hold-down bolt. On 1992 3.9L and 5.2L engines, base ignition timing can not be adjusted.

Timing Disturbed

1. Install a new distributor housing O-ring.

2. Position the engine so the No. 1 piston is at TDC of the compression stroke and the mark on the vibration damper is aligned with **0** on the timing indicator.

3. Install the distributor in the engine so the rotor is aligned with the position of the No. 1 ignition wire on the distributor cap and the housing is aligned with the matchmark on the engine. On 1992 3.9L and 5.2L engines, install the distributor in engine so the matchmark on housing is aligned and the rotor is aligned with the No.1 ignition wire. Turn the distributor housing as needed to align the rotor with the mark on the switch plate. Make sure the distributor is fully seated and the distributor shaft is fully engaged.

NOTE: There are distributor cap runners inside the cap on 3.0L engine. Make sure the rotor is pointing to where the No. 1 runner originates inside the cap and not where the No. 1 ignition wire plugs into the cap.

4. Install the hold-down clamp and snug the hold-down bolt or install the nut.

5. Connect the distributor pickup lead wires. Install the splash shield, if equipped.

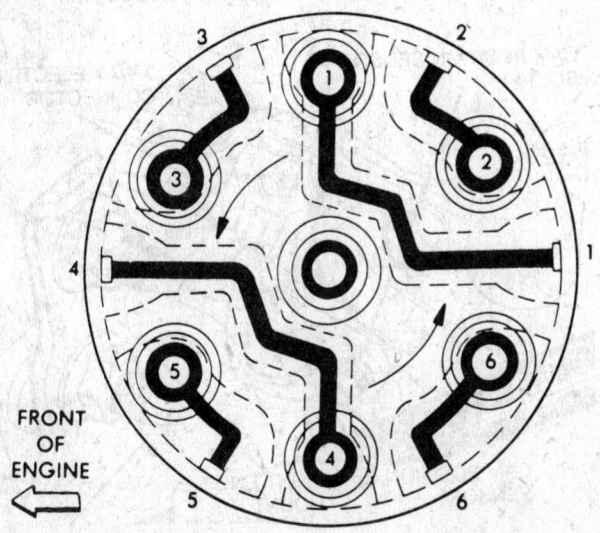

3   2

1

3   2

4   1

5   6

FRONT
OF
ENGINE

5   4   6

Distributor cap terminal routing—3.0L engine

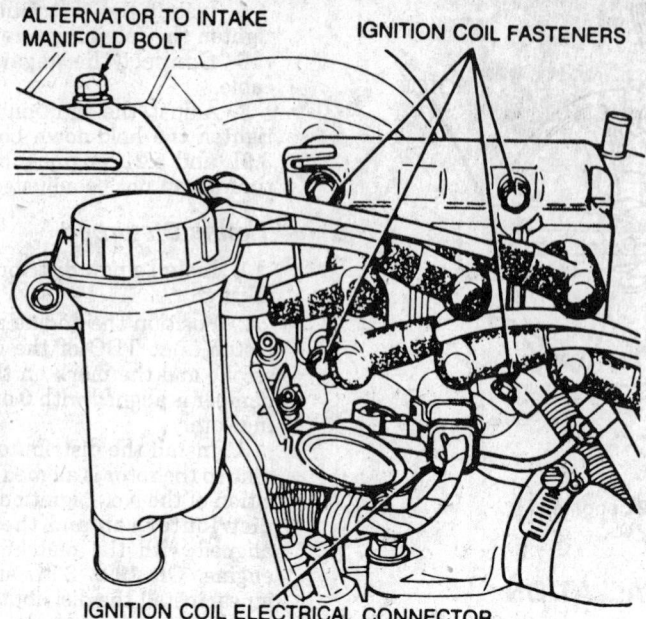

Ignition coil removal and installation—3.3L engine

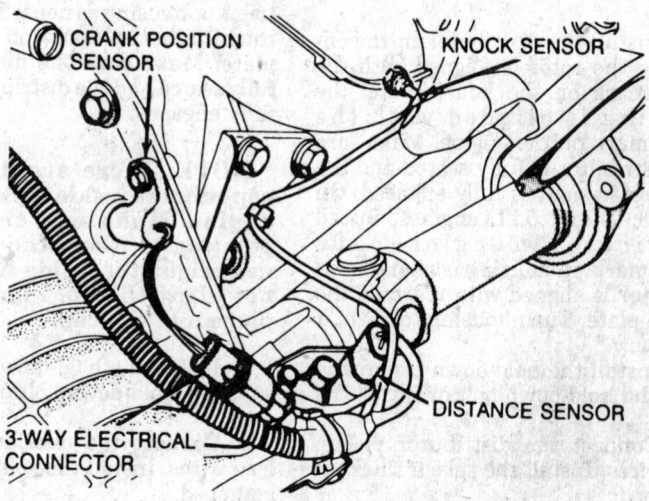

Crank position sensor location—3.3L engine

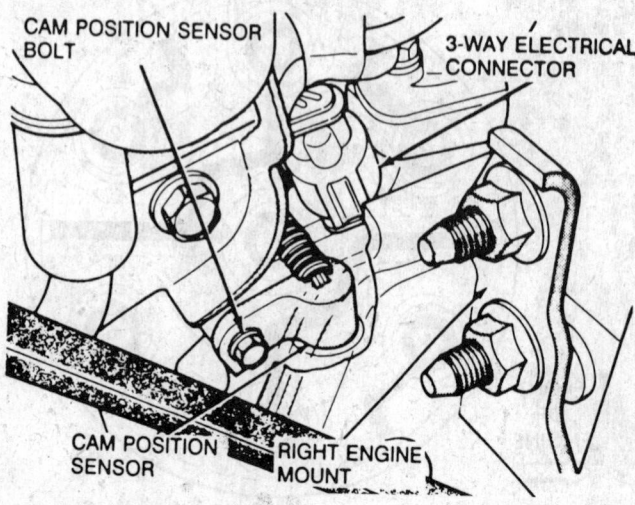

Cam position sensor location—3.3L engine

6. Install the distributor cap and tighten the screws.

7. Connect the negative battery cable.

8. Adjust the ignition timing and tighten the hold-down nut or bolt. On 1992 3.9L and 5.2L engines, base ignition timing can not be adjusted.

Distributorless Ignition

REMOVAL & INSTALLATION

Ignition Coil

1. Disconnect the negative battery cable.

2. Remove the spark plug wires from the coil.

3. Disconnect the electrical connector.

4. Remove the coil fasteners.

5. Remove the coil from the ignition module.

6. The installation is the reverse of the removal procedure.

Crankshaft Position Sensor

1. Disconnect the negative battery cable.

2. Disconnect the sensor lead at the harness connector.

3. Remove the sensor retaining bolt.

4. Pull the sensor straight up out the transaxle housing.

5. If the sensor is being reinstalled, remove any remains of the old spacer completely and attach a new spacer to the sensor. If a new spacer is not used, the sensor will not function properly. New sensors are packaged with a new spacer.

To install:

6. Install the sensor to the tranaxle housing and push the sensor down until it contacts the drive plate.

7. Hold in this position and install the retaining bolt. Torque to 9 ft. lbs. (12 Nm).

8. Connect the sensor lead wire.

Camshaft Position Sensor

1. Disconnect the negative battery cable.

2. Disconnect the sensor lead at the harness connector.

3. Loosen the sensor retaining bolt sufficiently to allow the slotted mounting surface to slide past.

4. Pull the sensor straight up and out of the chain case cover. Resistance may be high due to the rubber O-ring.

5. If the sensor is being reinstalled, remove any remains of the old spacer completely and attach a new spacer to the sensor. If a new spacer is not used, the sensor will not function properly. New sensors are packaged with a new spacer.

To install:

6. Inspect the O-ring for damage and replace if necessary.

7. Lubricate the O-ring with oil. Install the sensor to the chain case cover and push the sensor into its bore until contact is made with the cam timing gear.

8. Hold in this position and tighten the bolt to 9 ft. lbs. (12 Nm).

9. Connect the wire and position away from the accessory drive belt.

Ignition Timing

ADJUSTMENT

NOTE: The ignition timing on the distributorless 3.3L engine and 1992 3.9L and 5.2L engines cannot be checked or changed.

1. Start the engine, set the parking brake and run the engine until normal operating temperature is reached. Keep all lights and accessories **OFF**.

2. If a magnetic timing unit is available, insert the probe into the receptacle near the timing scale. The scale is located near the crankshaft pulley on 3.0L, 3.9L and 5.2L engines or on the bell housing on 2.2L and 2.5L engines.

3. If a magnetic timing unit is not available, connect a conventional power timing light to the No. 1 cylinder spark plug wire.

4. Connect the red lead of a tachometer to the negative primary terminal of the coil and connect the black lead to a good ground.

5. Disconnect and plug the distributor vacuum advance hose at the computer or distributor. If equipped with a carbureted 2.2L engine, disconnect the carburetor 6-way electrical connector and remove the violet wire from the connector and reconnect the connector. This disables the electronic spark advance. Set the idle speed according to the Vehicle Emission Control Information (VECI) label.

6. If equipped with Electronic Fuel Injection (EFI), disconnect the coolant sensor located near the thermostat housing. The Check Engine light on the instrument panel must be **ON**. On 1989–92 vehicles, connect the Diagnostic Readout Box II (DRB II) and access the Basic Timing Mode. If the DRB II is not available, disconnect the coolant sensor located near the thermostat housing. The Check Engine light on the instrument panel must be **ON**.

7. Aim the timing light at the timing scale or read the magnetic timing unit.

8. Loosen the distributor hold-down bolt just enough so the disributor can be rotated.

9. Turn the distributor in the prop-er direction until the specified timing according to the VECI label is reached. Tighten the hold-down bolt and recheck the timing and idle speed.

10. Turn the engine OFF. Remove the jumper wire, if used. Connect the vacuum hose(s) or coolant sensor (make sure the Check Engine light does not come on when restarted). Disconnect the timing apparatus and tachometer. Reinstal the violet wire into the carburetor connector, if removed.

11. If the coolant temperature sensor was disconnected, erase the created fault code using the Erase Fault Code mode on the DRB II. If the DRB II is not available, the code can be erased by disconnecting the battery, although it is not recommended.

NOTE: If the battery is disconnected, radio memory will be lost and other fault codes that may have been stored in the computer's memory will be erased. If the coolant sensor code is not erased at this point, it will disappear after 50–100 vehicle key on/off cycles providing there is no problem with that circuit.

Alternator

PRECAUTIONS

Several precautions must be observed when working with the alternator to avoid damaging the unit.

• If the battery is removed for any reason, make sure it is reconnected with the correct polarity. Reversing the battery connections may result in damage to the one-way rectifiers.

• When utilizing a booster battery as a starting aid, always connect the positive to positive terminals and the negative terminal from the booster battery to a good engine ground on the vehicle being started.

• Never use a fast charger as a booster to start vehicles.

• Disconnect the battery cables when charging the battery with a fast charger.

• Never attempt to polarize the alternator.

• Do not use test lights of more than 12 volts when checking diode continuity.

• Do not short across or ground any of the alternator terminals.

• The polarity of the battery, alternator and regulator must be matched and considered before making any electrical connections within the system.

• Never separate the alternator on an open circuit. Make sure all connections within the circuit are clean and tight.

• Disconnect the battery ground terminal when performing any service on electrical components.

• Disconnect the battery if arc welding is to be done on the vehicle.

BELT TENSION ADJUSTMENT

NOTE: The belt tension is automatically adjusted by a dynamic tensioner on the 3.0L, 3.3L, 3.9L and 5.2L engines. Periodic adjustment is not necessary.

1. Loosen the pivot bolt slightly.

2. Raise the vehicle and support safely, if necessary. Remove the splash shield, if equipped. Loosen the adjuster slot nut or bolt just enough so the alternator can be moved.

3. Apply tension to the alternator until the belt(s) deflect about ¼–½ in. under a 10 lb. load. Torque the adjuster strap bolt to 17 ft. lbs. (23 Nm). Torque the pivot bolt to 30 ft. lbs. (41 Nm).

4. If the bracket is equipped with an adjuster bolt, first loosen the locknut, if equipped and tighten the bolt until the belt(s) deflect about ¼–½ in. under a 10 lb. load. Tighten the adjuster slot bolt and pivot bolt.

REMOVAL & INSTALLATION

1. Disconnect the negative battery cable.

2. On some 2.2L and 2.5L engines, remove the air conditioning compressor and position it to the side. Remove the oil filter to allow the alternator to be removed from above, if possible.

3. On 3.0L, 3.3L, 3.9L and 5.2L engines, release the dynamic belt tensioner and remove the belt. On all other engines, loosen the mounting bolts, move the alternator toward the engine and remove the drive belt(s).

4. Remove the mounting bolts, spacers and adjuster bolt. Remove the alternator from the brackets. On 1992 3.9L and 5.2L engine the factory recommends removal of the idler pulley.

5. Remove the battery positive, field and ground terminals from the rear of the alternator. Remove the wire harness hold-down screw from the alternator, if equipped.

To install:

6. Connect all wiring to the proper terminals on the rear of the alternator and install the wire harness hold-down screw, if equipped.

7. Position the alternator in the mounting brackets.

8. Install the spacers, pivot bolt, adjuster slot bolt or nut and adjuster bolt, if equipped. Install the air conditioning compressor and oil filter, if removed.

9. Install idler pulley and bushing,

if removed and tighten bolts to 30 ft. lbs. (41 Nm).

10. Install belt(s) and adjust the belt tension, if necessary.

11. Connect the negative battery cable.

Voltage Regulator

REMOVAL & INSTALLATION

NOTE: The voltage regulator is integrated into the circuitry of the Single Module Engine Controller (SMEC) or Single Board Engine Controller (SBEC) on vehicles equipped with EFI and is not serviceable.

1. Disconnect the negative battery cable.

2. Unplug the connector from the voltage regulator.

3. Remove the retaining screws and remove the regulator from the vehicle.

4. The installation is the reverse of the removal procedure. Make sure the connector retainer is properly clipped in place.

Starter

REMOVAL & INSTALLATION

Except 1992 3.9L and 5.2L Engines With 4WD

1. Disconnect the negative battery cable.

2. Some engines are equipped with bolts that attach starter to flywheel housing. Remove as required.

3. Raise the vehicle and support safely. Remove skid plate, if equipped.

4. Remove the rear mount from the starter. Remove the heat shield from the starter, if equipped.

5. Unbolt the starter, remove the exhaust bracket and automatic transmission oil cooler tube bracket, if equipped. To aid in removal slide starter forward to clear starter gear housing nose and allow starter to come down past the exhaust pipe.

6. Disconnect the solenoid lead wires from the starter and remove starter from the vehicle.

To install:

7. Connect the solenoid lead wires and install the heat shield, if equipped.

8. Install the starter to bell housing attaching bolt, if equipped. Install the mounting bolt loosely. Install the automatic transmission oil cooler tube bracket and exhaust bracket, if equipped. Install the lower mounting nut. On 3.9L and 5.2L engines, torque starter upper mounting bolt to 50 ft. lbs. (68 Nm) and the stud nut to 20 ft. lbs. (27 Nm).

9. On all other engines, install

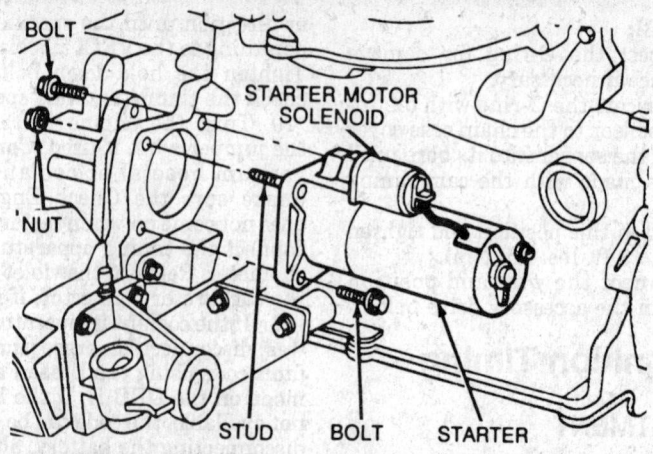

Removing or installing the starter—2.2L and 2.5L engines

mounting bolts to 40 ft. lbs. (54 Nm). Install the rear mount to the starter.

10. Connect the negative battery cable and check the starter for proper operation.

3.9L and 5.2L Engines With 4WD

1. Disconnect negative battery cable.

2. Raise and safely support vehicle.

3. Remove skid plate, if equipped.

4. Remove lower retaining nut from stud through the front of the vehicle.

5. Lower vehicle and remove 15mm steering shaft coupler bolt.

6. Rotate coupler clip on shaft and slide the upper half of the shaft toward the rear of the vehicle.

7. Remove the upper starter mounting bolt. Move starter forward to clear the stud and lift starter upward to gain access to the cable connections.

8. Remove cable from terminal adapter and remove starter motor.

To install:

9. Install cable to starter adapter and install into position. Install upper mounting bolt to 50 ft. lbs. (68 Nm). Install lower retainer nut to 20 ft. lbs. (27 Nm).

10. Install steering shaft and tighten bolt to 35 ft. lbs. (47 Nm).

11. Install negative battery cable and skid plate. Check operation of starter motor.

EMISSION CONTROLS

Emission Warning Lamps

Please refer to "Emission Con-

trols" in the Unit Repair section for system maintenance procedures. Due to the complex nature of modern electronic engine control systems, comprehensive diagnosis and testing procedures fall outside the confines of this repair manual. For complete information on diagnosis, testing and repair procedures concerning all modern engine and emission control systems, please refer to "Chilton's Guide to Fuel Injection and Electronic Engine Controls".

RESETTING

1. Connect the DRB II to the diagnostic connector.

2. Turn the ignition switch to the **RUN** position and access the Emissions EMR Tests on the DRB II.

3. Select EMR Memory Check.

4. Select Reset EMR Light. This will reset the EMR timing in the computer and turn the light off.

5. Disconnect the DRB II.

FUEL SYSTEM

Fuel System Service Precaution

Safety is the most important factor when performing not only fuel system maintenance but any type of maintenance. Failure to conduct maintenance and repairs in a safe manner may result in serious personal injury or death. Maintenance and testing of the vehicle's fuel system components can be accomplished safely and effectively by adhering to the following rules and guidelines.

• To avoid the possibility of fire and personal injury, always disconnect the negative battery cable unless the repair or test procedure requires that battery voltage be applied.

• Always relieve the fuel system pressure prior to disconnecting any fuel system component (injector, fuel rail, pressure regulator, etc.), fitting or fuel line connection. Exercise extreme caution whenever relieving fuel system pressure to avoid exposing skin, face and eyes to fuel spray. Please be advised that fuel under pressure may penetrate the skin or any part of the body that it contacts.

• Always place a shop towel or cloth around the fitting or connection prior to loosening to absorb any excess fuel due to spillage. Ensure that all fuel spillage (should it occur) is quickly removed from engine surfaces. Ensure that all fuel soaked cloths or towels are deposited into a suitable waste container.

• Always keep a dry chemical (Class

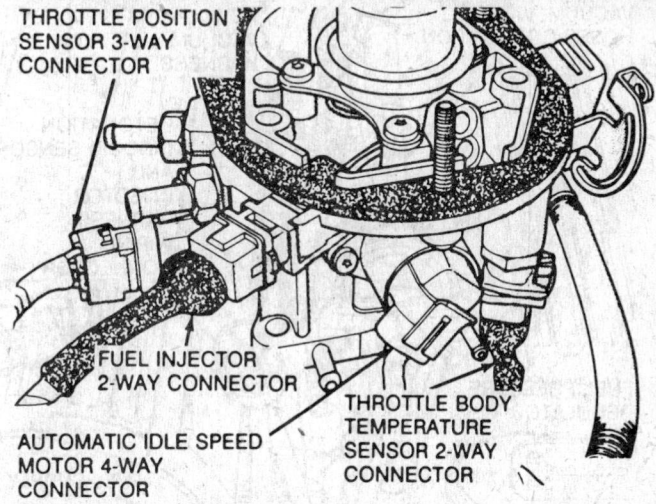

Fuel Injector harness connector location — 2.5L non-turbocharged engine

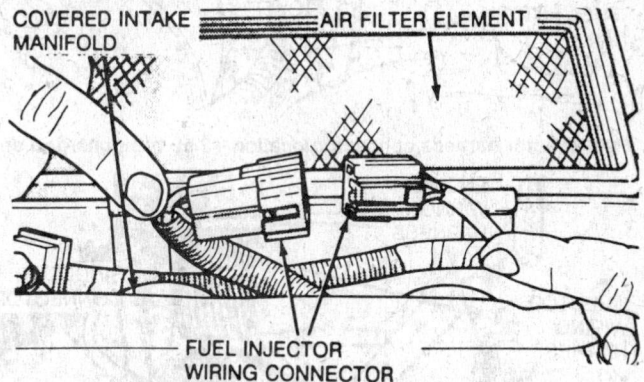

Fuel Injector harness connector location — 3.0L engine

B) fire extinguisher near the work area.

• Do not allow fuel spray or fuel vapors to come into contact with a spark or open flame.

• Always use a backup wrench when loosening and tightening fuel line connection fittings. This will prevent unnecessary stress and torsion to fuel line piping. Always follow the proper torque specifications.

• Always replace worn fuel fitting O-rings with new. Do not substitute fuel hose or equivalent where fuel pipe is installed.

RELIEVING FUEL SYSTEM PRESSURE

Except 1992 3.9L and 5.2L Engines

1. Loosen the fuel filler cap to release fuel tank pressure.
2. Locate the fuel injector harness connector.
3. Connect a jumper wire from terminal No. 1 of the appropriate connector to ground.
4. Being careful not to allow contact between the jumper leads, connect a jumper wire to terminal No. 2 of the connector and touch the other end of the jumper to the positive battery post for no longer than 5 seconds. This will relieve fuel pressure.

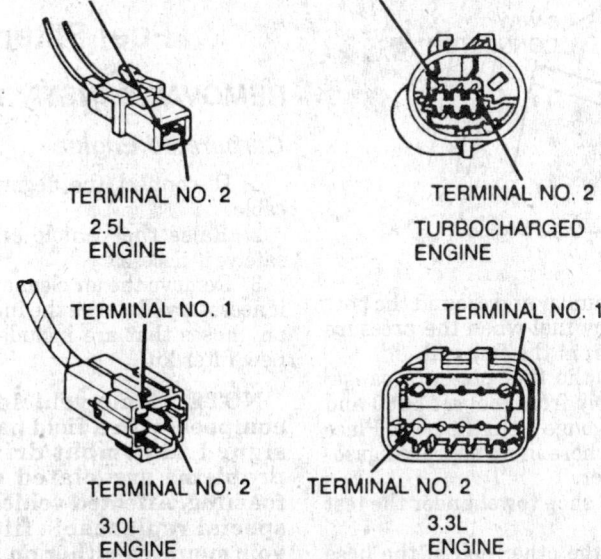

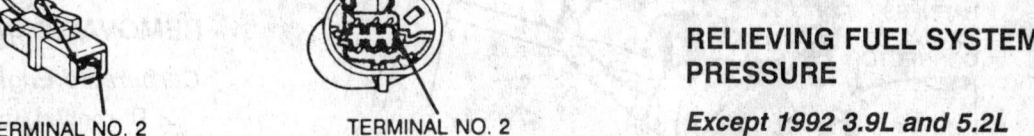

Fuel Injector harness connector terminals — except 3.9L engine

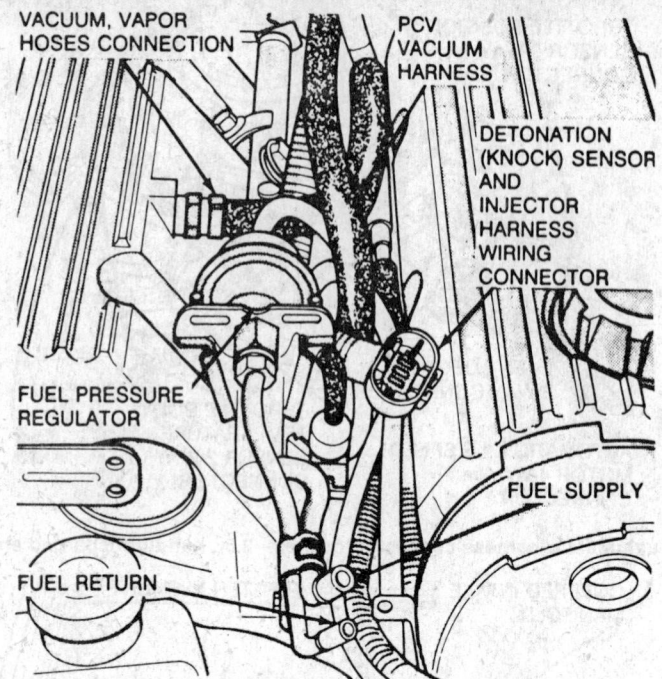

Fuel Injector harness connector location—2.5L turbocharged engine

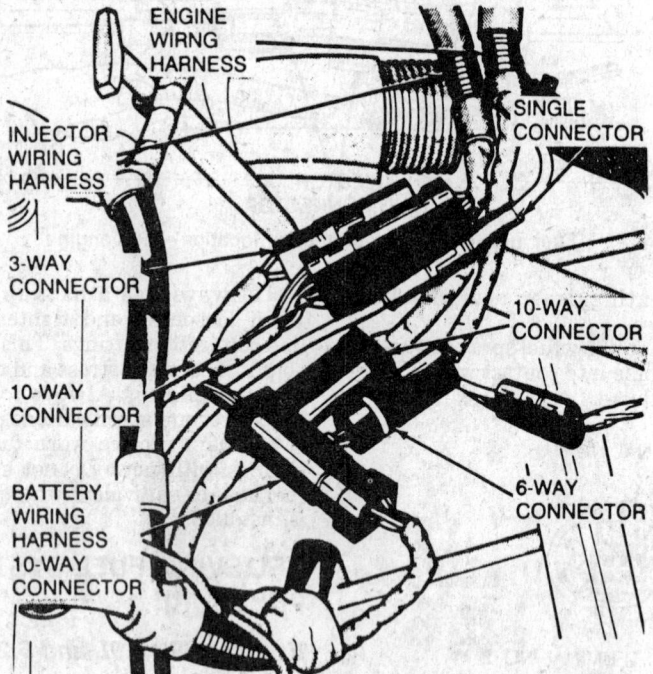

Fuel Injector harness connector location—3.3L engine

5. Remove the jumper wires and connect the connector.

3.9L and 5.2L Engines

1. Loosen the fuel filler cap to release fuel tank pressure.

2. Disconnect negative battery cable.

3. Remove the protective cap from pressure test port on the fuel rail. Do not allow fuel to spill onto the intake manifold or exhaust manifold. Place

shop towels under and around the port to absorb any fuel when the pressure is released from the fuel rail.

4. Obtain the fuel pressure gauge/hose assembly from tool set 5069 and remove the gauge from the hose. Place 1 end of the hose in an approved gasoline container.

5. Place a shop towel under the test port.

6. Screw the other end of the hose to the test pressure port. Install the

protective cap after all pressure has been released.

Fuel Tank

REMOVAL & INSTALLATION

1. Disconnect the negative battery cable.

2. Relieve the fuel pressure.

3. Raise the vehicle and support safely.

4. Using the proper equipment, drain the fuel tank.

5. On Van, remove the screws that fasten the filler tube to the inner and outer fender. On All Wheel Drive Van, remove the fasteners that hold the park brake cable support bracket in place.

6. Disconnect the electrical wiring and all hoses from the tank.

7. Place a transmission jack or equivalent, under the center of the tank and apply slight pressure.

8. Remove the nuts from the mounting straps and remove tank straps.

9. Lower the tank far enough to allow access to the fuel supply and return hoses, and disconnect hoses. Disconnect fuel vapor line from pressure relief/rollover valve.

10. Lower tank from vehicle.

To install:

11. Raise the tank into position and connect all electrical harnesses and vacuum hoses.

12. Install the tank straps and tighten the retaining nuts to 35–40 ft. lbs. (47–54 Nm).

13. Install vent hoses, if not done, and check all connections.

14. Fill fuel tank, install fuel filler cap and reconnect negative battery cable. Pressurize the fuel system and inspect for leaks.

Fuel Filter

REMOVAL & INSTALLATION

Carbureted Engine

1. Disconnect the negative battery cable.

2. Raise the vehicle and support safely, if necessary.

3. Remove the air cleaner assembly, if necessary. Remove the fuel filter and any hoses that are included with the new filter kit.

NOTE: Some vehicles may be equipped with a field package designed to combat driveability problems associated with fuel foaming. Affected vehicles have a special replaceable filter/reservoir mounted either on a plate on the carburetor or near the ther-

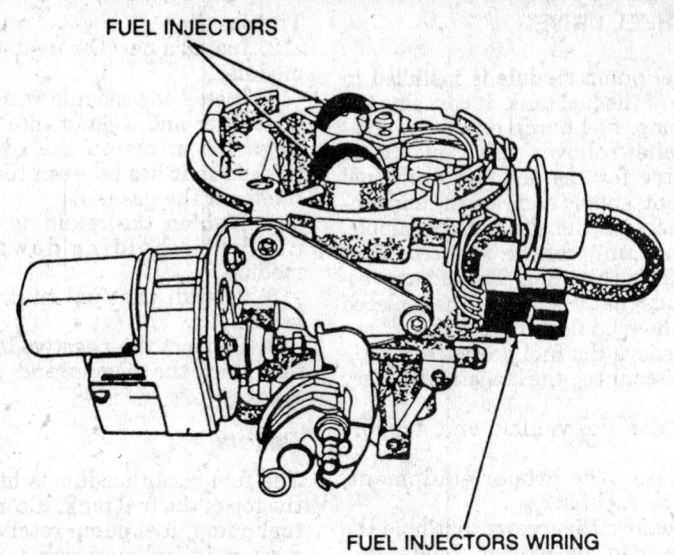

FUEL INJECTORS

FUEL INJECTORS WIRING HARNESS

Injector wiring harness connector — 3.9L engine

mostat. **When the kit is installed on a 2.2L engine, the conventional filter is discarded. Do not install a conventional filter or conventional clamps or hoses in place of the special replacement parts; they are not compatible with the electronic fuel pump installed with the kit.**

4. Install the new filter, hoses and clamps.

5. Connect the negative battery cable, start the engine and check for leaks.

Fuel Injected Engine

—————— **CAUTION** ——————

Do not use conventional fuel filters, hoses or clamps when servicing this fuel system. They are not compatible with the injection system and could fail, causing personal injury or damage to the vehicle. Use only hoses and clamps specifically designed for fuel injection.

1. Relieve the fuel pressure.
2. Disconnect the negative battery cable. On All Wheel Drive Van, it will be necessary to remove the exhaust heat shield and the converter support bracket.
3. The filter is located on the frame rail towards the rear of the vehicle. Raise the vehicle and support safely. Remove the filter retaining screw and remove the filter assembly from the mounting plate.
4. Loosen the outlet hose clamp on the filter and inlet hose clamp on the rear fuel tube.
5. Wrap a shop towel around the hoses to absorb fuel. Remove the hoses from the filter and fuel tube and discard the clamps and the filter.

To install:

6. Install the inlet hose on the fuel tube and tighten the new clamp to 10 inch lbs.
7. Install the outlet hose on the filter outlet fitting and tighten the new clamp to 10 inch lbs.
8. Position the filter assembly on the mounting plate and tighten the mounting screw to 75 inch lbs. (8 Nm).
9. Connect the negative battery cable, start the engine and check for leaks.

Mechanical Fuel Pump

PRESSURE TESTING

1. Raise the vehicle and support safely.
2. Connect a pressure gauge (0–15 psi minimum range) to the fuel pump outlet fitting.
3. Crank the engine several times while observing the gauge.
4. The 2.2L should develop 4.5–6.0 psi. If equipped with the field package noted above, fuel pressure should be 13–15 psi.

REMOVAL & INSTALLATION

1. Disconnect the negative battery cable.
2. Raise the vehicle and support safely, if necessary. Remove the oil filter.
3. Disconnect the fuel lines from the pump and plug.
4. Remove the fuel pump retaining bolts or nuts and remove the pump from the engine.
5. Clean and dry the mounting surfaces and bolt holes. Install new gasket on mating surface.

6. The installation is the reverse of the removal procedure.
7. Connect the negative battery cable, start the engine and check for leaks.

Electric Fuel Pump

PRESSURE TESTING

1. Relieve the fuel pressure.
2. Properly connect the fuel system pressure tester.
3. With the key in the **RUN** position, put the DRB I or II in the activate auto shutdown relay mode; this will activate the fuel pump and pressurize the system.
4. If the pressure is within specifications, remove pressure tester from the fuel system.
5. If fuel pressure is below specifications, install the tester in the fuel supply line between the tank and the filter and repeat the test.
6. If the pressure is 5 psi higher than in Step 5, replace the fuel filter. If no change is observed, squeeze the return hose. If pressure increases, replace the pressure regulator. If no change is observed, the problem is either a plugged in-tank sock filter or a defective pump.
7. If fuel pressure is above specifications, remove the fuel return line hose from the chassis line at the fuel tank and connect a 3 foot piece of fuel hose to the return line. Put the other end into a 2 gallon minimum capacity approved gasoline container. Repeat the test. If pressure is now correct, check the in-tank return hose for kinking. Replace the fuel pump assembly if the in-tank reservoir check valve or aspirator jet is obstructed.
8. If pressure is still above specifications, remove the fuel return hose from the throttle body. Connect hose to the throttle body return nipple and place the other end of the hose in a clean container. Repeat the test. If pressure is now correct, check for a restricted fuel return line. If no change is observed, replace the fuel pressure regulator.

REMOVAL & INSTALLATION

Caravan, Voyager and Town & Country

FRONT WHEEL DRIVE

The fuel pump/level unit assembly is installed in the top of the fuel tank. It consists of the fuel pump, fuel pump reservoir, pressure relief/rollover valve, electrical connector for the sending unit, fuel filler vent, supply and return tube connections and the drain tube nipple. The level unit can be serviced separately from the pump.

1. Relieve the fuel pressure.
2. Disconnect the negative battery cable.
3. Raise the vehicle and support safely.
4. Using the proper equipment, drain the fuel tank.
5. Remove the screws that hold the filler neck to the quarter panel.
6. Disconnect the wiring and hoses from the tank.
7. Place a transmission jack or equivalent, under the center of the tank and apply slight pressure. Remove the tank straps.
8. Lower the tank and remove the filler tube from the tank.
9. Disconnect the vapor separator rollover valve hose and remove the fuel tank from the vehicle.
10. To remove fuel pump/level unit, use a hammer and a brass drift and tap the lock ring counterclockwise to release the pump.
11. Remove the fuel pump from the tank.
12. To remove fuel inlet strainer from the pump/level unit, gently pry sock from the ferrule and pump body with broad bladed tool taking care not to damage the pump inlet. On 1990–92 pumps, gently bend locking tabs on pump reservoir body to clear locking tangs on the fuel pump strainer. Remove strainer. To install new strainer, lubricate strainer O-ring and insert strainer onto the inlet of the fuel pump reservoir body. Bend locking tabs into position.

To install:
13. Install a new O-ring to the pump.
14. Install the pump with the new inlet filter and locknut into the tank.
15. Install the lock ring with a hammer and brass punch turning the ring clockwise.
16. Install the fuel tank into the vehicle.
17. Connect the negative battery cable, start the engine and check for leaks.

ALL WHEEL DRIVE

The fuel pump module is installed in the top of the fuel tank. It contains the fuel pump, fuel pump reservoir, pressure relief/rollover valve, electrical connector for the sending unit, fuel filler vent, supply and return tube connections and the drain tube nipple. The level unit can be serviced separately from the pump. The factory recommends that the locknut be replaced when the module is removed.
1. Relieve the fuel pressure.
2. Disconnect the negative battery cable.
3. Raise the vehicle and support safely.
4. Using the proper equipment, drain the fuel tank.
5. Remove the screws that hold the filler neck to the quarter panel.
6. Disconnect the wiring and hoses from the tank. Remove the support bracket for the parking brake cable and the bracket for the converter.
7. Place a transmission jack or equivalent, under the center of the tank and apply slight pressure. Remove the tank straps.
8. Lower the tank and remove the filler tube from the tank.
9. Disconnect the vapor separator rollover valve hose and remove the fuel tank from the vehicle.
10. Remove the fuel module from the tank. It will be necessary to hold down on module while removing the retainer clamp or nut. The module is spring loaded into the tank. Mark position of sender wires and disconnect. Remove sender unit from module. Replace module and pump assembly as a unit.
11. To remove the inlet strainer, gently bend locking tabs on pump reservoir to clear locking tangs on the fuel pump filter. Remove strainer. To install new strainer, lubricate strainer O-ring and insert strainer onto the inlet of the fuel pump reservoir body. Bend locking tabs into position.

To install:
12. Install a new O-ring to the pump module.
13. Install the module with the new inlet filter and locknut into the tank. Align the arrow on the edge of the module so it lies between the lines on the lip of the gas tank.
14. Tighten the retaining clamp or nut while holding down on the module.
15. Install the fuel tank into the vehicle.
16. Connect the negative battery cable, start the engine and check for leaks.

Dakota

The fuel pump module is installed in the top of the fuel tank. It contains the fuel pump, fuel pump reservoir, pressure relief/rollover valve, electrical connector for the sending unit, fuel filler vent, supply and return tube connections and the drain tube nipple. The level unit can be serviced separately from the pump. The factory recommends that the locknut be replaced when the module is removed.
1. Relieve the fuel pressure.
2. Disconnect the negative battery cable.
3. Raise the vehicle and support safely.
4. Using the proper equipment, drain the fuel tank.
5. Disconnect the vent hoses and filler hose from the tank and remove the hoses from the bracket attached to the top of the frame rail, if equipped.
6. Place a transmission jack or equivalent, under the center of the tank and apply slight pressure.
7. Remove the inboard retaining strap J-bolt nuts and remove the straps.
8. Lower the tank enough to reach in and disconnect the electrical connector and the remaining fuel tubes from the top of the module. Lower the tank from the vehicle.
9. Remove the fuel module from the tank. It will be necessary to hold down on module while removing the retainer clamp or nut. The module is spring loaded into the tank. Mark position of sender wires and disconnect. Remove sender unit from module. Replace module and pump assembly as a unit.
10. To remove the inlet strainer, gently bend locking tabs on pump reservoir to clear locking tangs on the fuel pump filter. Remove strainer. To install new strainer, lubricate strainer O-ring and insert strainer onto the inlet of the fuel pump reservoir body. Bend locking tabs into position.
To install:
11. Install a new O-ring to the pump module.
12. Install the module with the new

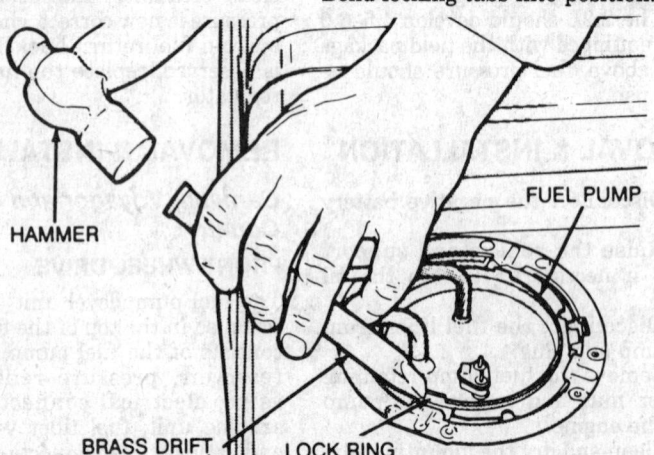

HAMMER

FUEL PUMP

BRASS DRIFT   LOCK RING

Removing the fuel pump from the tank—Caravan, Voyager and Town & Country

inlet filter and locknut into the tank. Align the arrow on the edge of the module so it lies between the lines on the lip of the gas tank.

13. Tighten the retaining clamp or nut while holding down on the module.

14. Install the fuel tank into the vehicle.

15. Connect the negative battery cable, start the engine and check for leaks.

Carburetor

REMOVAL & INSTALLATION

1. Disconnect the negative battery cable.

2. Remove the air cleaner assembly.

3. Remove and install the fuel tank cap to relieve any pressure in the tank.

4. Matchmark all vacuum hoses and electrical connectors and remove them from the carburetor.

5. Disconnect the throttle, cruise control, choke and kickdown cables and linkages, if equipped.

6. Disconnect and plug the fuel inlet line.

7. Remove the mounting bolts and/or nuts and remove the carburetor from the intake manifold.

To install:

8. Clean the mounting surface of the manifold and install a new base gasket.

9. Install the mounting bolts and/or nuts and tighten them alternately to compress the base gasket evenly.

10. Connect the fuel line.

11. Connect the throttle, cruise control, choke and kickdown cables and linkages, if equipped.

12. Install all vacuum hoses and electrical connectors in their proper locations.

13. Install the air cleaner.

14. Connect the negative battery cable, start the engine and perform all necessary adjustments.

IDLE SPEED ADJUSTMENT

2.2L Engine

1. Before checking or adjusting the idle speed, check ignition timing and adjust if necessary.

2. Disconnect and plug the vacuum connector at the coolant vacuum switch cold closed valve.

3. Unplug the connector at the radiator fan and install a jumper wire so the fan will run continuously.

4. Remove the PCV valve from the valve cover and allow it to draw under hood air.

5. Connect a tachometer to engine.

6. Ground the carburetor switch with a jumper wire.

7. Disconnect the oxygen system test connector located on left fender shield, if equipped.

8. Start and run the engine until normal operating temperature is reached.

9. If tachometer indicates idle speed is not at specifications, turn the idle speed screw until correct idle speed is obtained. The screw is located on top of the solenoid mounted on the back of the carburetor.

10. Reconnect the PCV valve, oxygen connector and vacuum connector.

11. Remove jumper wire and reconnect the radiator fan.

12. After Steps 9 and 10 are completed, the idle speed may change slightly. This is normal and engine speed should not be readjusted.

13. Make sure the solenoid kicker moves the throttle blade adequately so the idle does not change when the air conditioning is working.

3.9L Engine

1. Start the engine and run until at normal operating temperature. Check and adjust the ignition timing. Turn the engine **OFF**.

2. Disconnect and plug the EGR hose. Disconnect the oxygen sensor, if equipped.

3. Disconnect and plug the 3/16 in. hose at the canister.

4. Remove the PCV hose from the valve cover and allow it to draw under hood air.

5. Ground the carburetor switch with a jumper wire, if equipped.

6. Disconnect the vacuum hose from the computer, if equipped and connect an auxiliary vacuum supply of 16 in. Hg.

7. Make sure all accessories are off. Install a tachometer and start the engine.

8. Allow the engine to run for 2

minutes to stabilize. Turn the idle speed screw until the correct idle speed according to the Vehicle Emission Control Information label is reached.

9. Connect all wires and hoses. It is normal for the idle speed to vary after all hoses and wires are reconnected; do not readjust.

SERVICE ADJUSTMENTS

For all carburetor service adjustment procedures and Specifications, please refer to "Carburetor Service" in the Unit Repair section.

Fuel Injection

IDLE SPEED ADJUSTMENT

Single and Dual Point Injection

1. Start the engine and allow it to reach normal operating temperature. If it is already hot, run it for 2 minutes.

2. Turn the engine OFF and allow 1 minute for the Idle Speed Control (ISC) actuator shaft to fully extend.

3. Disconnect the ISC actuator connector and the coolant temperature sensor.

4. Connect a tachometer to the engine and start the engine.

5. Adjust the extension screw on the actuator shaft until the rpm is within specifications.

6. Turn the engine OFF. Reconnect the ISC actuator connector and the coolant temperature sensor. Remove the tachometer.

Except Single and Dual Point Injection

The idle speed is controlled by the automatic idle speed motor (AIS) which

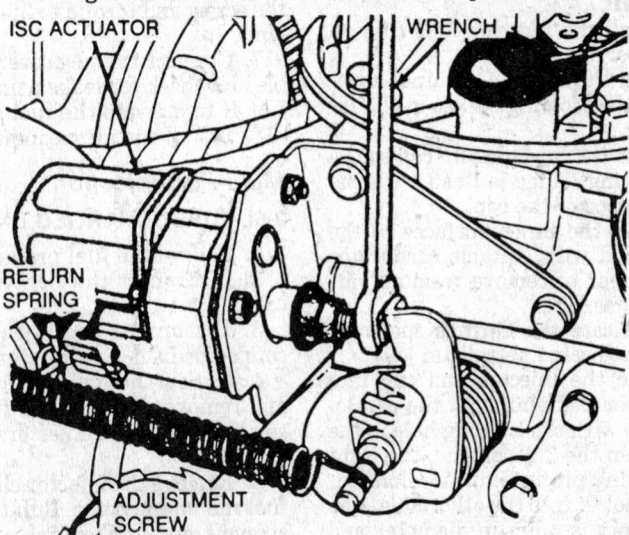

Adjusting the idle speed—Single and Dual Point Injection

is controlled by the logic module. The logic module receives data from various sensors and switches in the system and adjusts the engine idle to a predetermined speed. Idle speed specifications can be found on the Vehicle Emission Control Information (VECI) label located in the engine compartment. If the idle speed is not within specifications and there are no problems with the system, the throttle body should be replaced.

IDLE MIXTURE ADJUSTMENT

There is no idle mixture adjustment provided with any Chrysler fuel injection system.

Fuel Injector

REMOVAL & INSTALLATION

Single and Dual Point Injection

2.5L AND 3.9L ENGINES

1. Remove the air cleaner assembly.
2. Relieve the fuel pressure.
3. Disconnect the negative battery cable.

NOTE: On the 3.9L engine, there is a small spacer below the injector hold-down clamp. Take the proper precaution to prevent this spacer from falling into the throttle body.

4. Remove the injector hold-down Torx screw, the hold-down and the spacer, if equipped.
5. Using a small flat-tipped tool, lift the cap off of the injector.
6. Using the same tool, gently pry the injector from its pod.
7. Remove the lower O-ring from the pod.

To install:

8. Install the new lower O-ring on the injector.
9. Align the injector terminal housing with the locating socket in the injector cap.
10. Press the injector into cap so the upper O-ring flange is flush with the lower surface of the cap.
11. Spray the inner surfaces of the injector pod with suitable carburetor parts cleaner to remove residual varnish and gasoline.
12. Lubricate the O-rings sparingly with unmedicated petroleum jelly.
13. Place the injector and cap into the injector pod and align the cap locating pin with the locating hole in the casting. On the 3.9L engine, the right side locating pin is 5mm in diameter and will not fit into the other hole. The left side pin is 4mm in diameter and will be loose in the other hole.

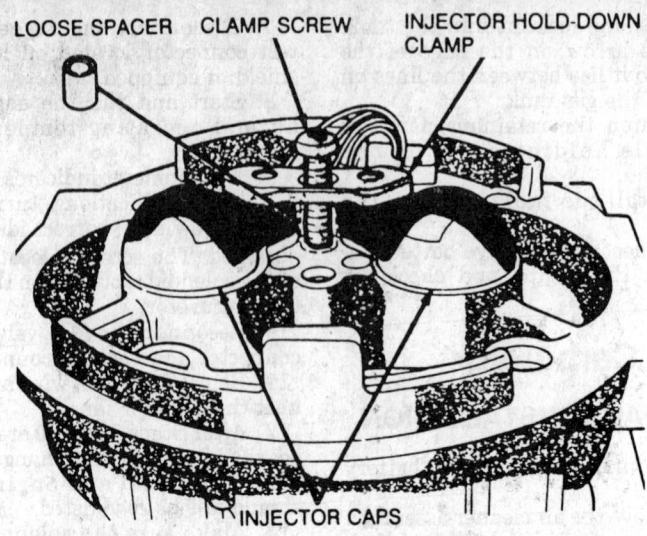

Injector hold-down and spacer—Single and Dual Point Injection

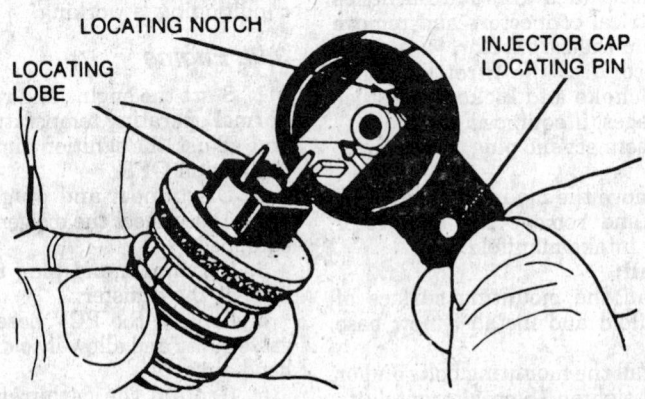

Installing the injector to the cap—Single and Dual Point Injection

14. Press firmly on the injector cap until it is flush with the casting surface.
15. Install the spacer, if equipped and align the holes in the hold-down with the pins on the caps and install.
16. Push down on the caps, install the screw and torque to 35 inch lbs. (4 Nm).
17. Connect the negative battery cable and check for leaks using the DRB I or II to activate the fuel pump.
18. Install the air cleaner.

Multi-Port Injection

2.5L TURBOCHARGED ENGINE

1. Relieve the fuel pressure.
2. Disconnect the negative battery cable.
3. Disconnect the injector wiring connector from the injector.
4. Unbolt the fuel rail from engine and remove. Position the fuel rail assembly so the fuel injectors are easily accessible.
5. Remove the injector clip from the fuel rail and injector. Pull the injector straight out of the fuel rail receiver cup.

6. Check the injector O-ring for damage. If the O-ring is damaged, replace it. If the injector is being reused, install a protective cap on the injector tip to prevent damage.
7. Repeat the procedure for the remaining injectors.

To Install:

8. Before installing an injector the rubber O-ring should be lubricated with a drop of clean engine oil to aid in installation.
9. Install injector top end into fuel rail receiver cup.
10. Install injector clip by sliding the open end into top slot of the injector and onto the receiver cup ridge.
11. Repeat the steps for the remaining injectors.
12. Install the fuel rail to the engine and secure.
13. Connect the negative battery cable and check for leaks using the DRB I or II to activate the fuel pump.

3.0L ENGINE

1. Relieve the fuel pressure.
2. Disconnect the negative battery cable.

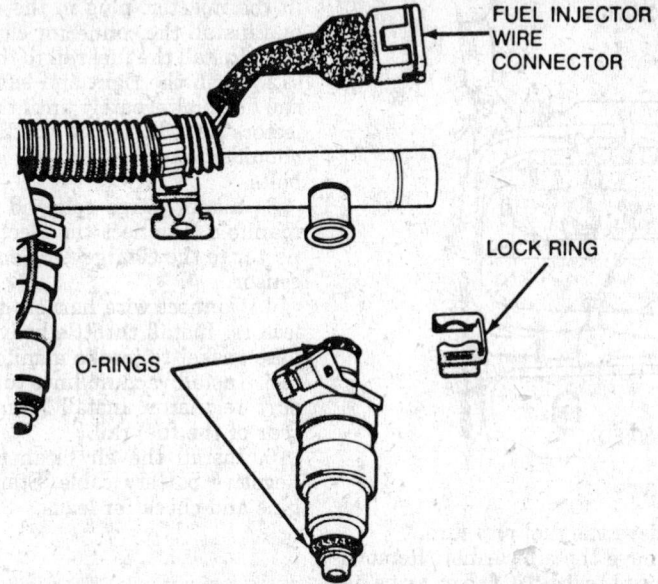

Fuel Injector removal and Installation – 2.5L turbocharged engine

3. Remove the air cleaner to throttle body hose.

4. Disconnect the throttle cable from the throttle body and disconnect the kickdown linkage. Remove the throttle cable bracket attaching bolts.

5. Disconnect the electrical connectors to the throttle body.

6. Matchmark and carefully remove the vacuum hoses from the throttle body.

7. Remove the PCV and brake booster hoses from the air intake plenum.

8. Remove the ignition coil from the intake plenum, if mounted there.

9. Remove the EGR tube flange from the intake plenum, if equipped.

10. Unplug the coolant temperature sensor and charge temperature sensor, if equipped.

11. Remove the vacuum connection from the air intake plenum vacuum connector.

12. Remove the fuel hoses from the fuel rail and plug them.

13. Remove the air intake plenum to intake manifold bolts and remove the plenum and gaskets. Cover the intake manifold openings.

14. Remove the vacuum hoses from the fuel rail.

15. Disconnect the fuel injector wiring harness.

16. Remove the fuel rail attaching bolts and remove the fuel rail from vehicle. Position the rail on the bench upside down so the injectors are easily accessible.

17. Remove the small connector retainer clip and unplug the injector. Remove the injector clip off the fuel rail and injector. Pull the injector straight out off the rail.

To install:

18. Lubricate the rubber O-ring with clean oil and install to the rail receiver cap. Install the injector clip to the **TOP** slot of the injector, plug in the connector and install the connector clip.

19. Install the fuel rail to the vehicle and plug in the injector harness. Connect the vacuum hoses to the fuel rail.

20. Install new intake plenum gaskets with the beaded sealer side up and install the intake plenum. Torque the attaching bolts and nuts to 115 inch lbs. (13 Nm).

21. Install the fuel hoses to the fuel rail.

22. Install or connect all items that were removed or disconnected from the intake plenum and throttle body.

23. Connect the negative battery cable and check for leaks using the DRB I or II to activate the fuel pump.

3.3L ENGINE

1. Relieve the fuel pressure.

2. Disconnect the negative battery cable.

3. Remove the air cleaner and hose assembly.

4. Disconnect the throttle cable. Remove the wiring harness from the throttle cable bracket and intake manifold water tube.

5. Remove the vacuum hose harness from the throttle body.

6. Remove the PCV and brake booster hoses from the air intake plenum.

7. Remove the EGR tube flange from the intake plenum, if equipped.

8. Unplug the charge temperature sensor and unplug all vacuum hoses from the intake plenum.

9. Remove the cylinder head to intake plenum strut.

10. Disconnect the MAP sensor and oxygen sensor connector. Remove the engine mounted ground strap.

11. Release the fuel hose quick disconnect fittings and remove the hoses from the fuel rail. Plug the hoses.

12. Remove the Direct Ignition System (DIS) coils and the alternator bracket to intake manifold bolt.

13. Remove the intake manifold bolts and rotate the manifold back over the rear valve cover. Cover the intake manifold.

14. Remove the vacuum harness from the pressure regulator.

15. Remove the fuel tube retainer bracket screw and fuel rail attaching bolts. Spread the retainer bracket to allow for clearance when removing the fuel tube.

16. Remove the fuel rail injector wiring clip from the alternator bracket.

17. Disconnect the cam sensor, cool-

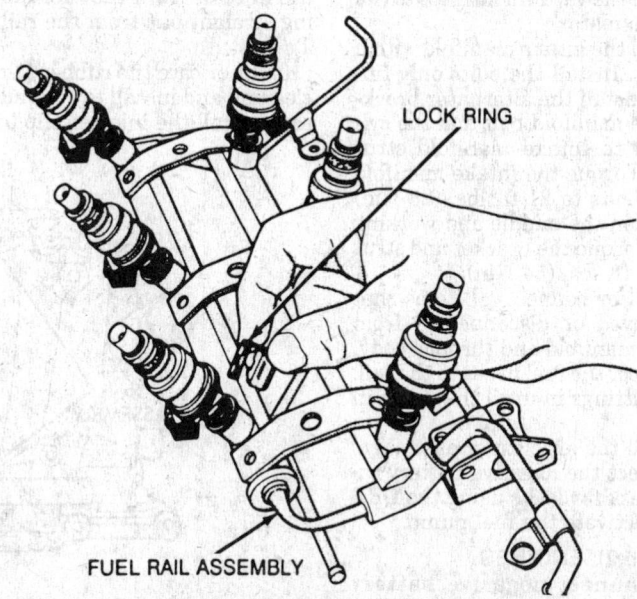

Fuel rail assembly – 3.0L engine

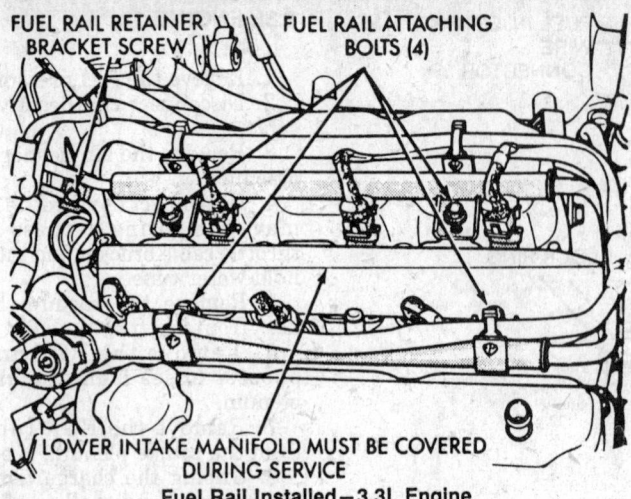

FUEL RAIL RETAINER BRACKET SCREW

FUEL RAIL ATTACHING BOLTS (4)

LOWER INTAKE MANIFOLD MUST BE COVERED DURING SERVICE

Fuel Rail Installed—3.3L Engine

ant temperature sensor and engine temperature sensor.

18. Remove the fuel rail.

19. Position the rail on the bench so the injectors are easily accessible.

20. Remove the small connector retainer clip and unplug the injector. Remove the injector clip off the fuel rail and injector. Pull the injector straight out of the rail.

To install:

21. Lubricate the rubber O-ring with clean oil and install to the rail receiver cap. Install the injector clip to the slot in the injector, plug in the connector and install the connector clip.

22. Install the fuel rail.

23. Connect the cam sensor, coolant temperature sensor and engine temperature sensor.

24. Install the fuel rail injector wiring clip to the alternator bracket.

25. Install the fuel rail attaching bolts and fuel tube retainer bracket screw.

26. Install the vacuum harness to the pressure regulator.

27. Install the intake manifold with a new gasket. Install the bolts only finger-tight. Install the alternator bracket to intake manifold bolt and the cylinder head to intake manifold strut and bolts. Torque the intake manifold mounting bolts to 21 ft. lbs. (28 Nm) starting from the middle and working outward. Torque the bracket and strut bolts to 40 ft. lbs. (54 Nm).

28. Install or connect all items that were removed or disconnected from the intake manifold and throttle body.

29. Connect the fuel hoses to the rail. Push the fittings in until they click in place.

30. Install the air cleaner assembly.

31. Connect the negative battery cable and check for leaks using the DRB I or II to activate the fuel pump.

3.9L AND 5.2L ENGINES

1. Disconnect negative battery cable.

2. Relieve the fuel pressure.

3. Remove the air cleaner. Remove the throttle body from intake manifold.

4. Disconnect the electrical connectors at fuel injectors.

5. Remove vacuum line at fuel pressure regulator.

6. On 3.9L engine, disconnect electrical connector at Charge Air Temperature sensor.

7. Remove canister purge solenoid and bracket assembly from intake manifold. Disconnect 2 fuel lines at the rear of the fuel rail and remove the remaining fuel rail mounting bolts.

8. Gently rock the left side fuel rail until the injectors start to clear the injector bores. Repeat the procedure on the right side fuel rail. Work both side fuel rails until the clearance is large enough to extract both fuel rails from the engine.

9. Remove the clips retaining the fuel injector to the fuel rail. Remove the injector from the fuel rail by pulling straight out from the rail.

To install:

10. Lubricate the rubber O-ring with clean oil and install to the rail receiver cap. Install the injector clip to the slot

in the injector, plug in the connector and install the connector clip.

11. Install the fuel rail to the engine.

12. Push the right and left side fuel rail down alternately until the fuel injectors have bottomed on the injector shoulder. Install fuel rail mounting bolts.

13. Install purge solenoid to intake manifold. Connect the electrical connector to the Charge Air Temperature sensor.

14. Connect wire harness to fuel injectors. Install throttle body and new base gasket to intake manifold.

15. Install vacuum lines to fuel pressure regulator. Install 2 fuel lines to rear of the fuel rail.

16. Install the air cleaner and the negative battery cable. Start the engine and check for leaks.

DRIVE AXLE

Halfshaft

REMOVAL & INSTALLATION

1. Disconnect the negative battery cable.

2. Raise the vehicle and support safely.

3. Remove the tire and wheel assembly.

4. Remove the cotter pin from the end of the halfshaft. Remove the nut lock, spring washer, axle nut and washer.

5. Remove the ball joint retaining bolt and pry the control arm down to release the ball stud from the steering knuckle.

6. If removing the right side drive axle, it is necessary to remove the speedometer driven gear from the transaxle.

7. Position a drain pan under the

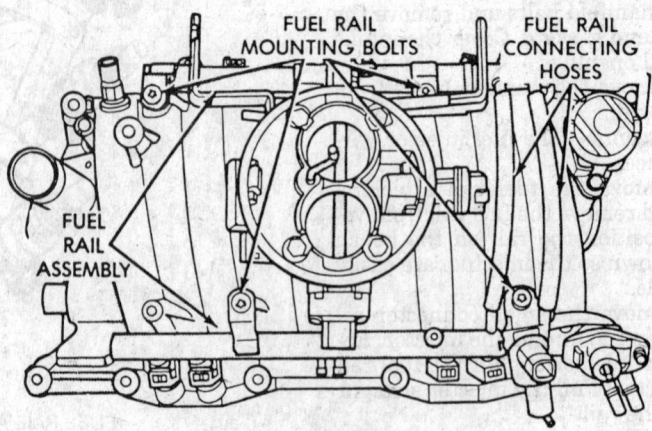

FUEL RAIL MOUNTING BOLTS

FUEL RAIL CONNECTING HOSES

FUEL RAIL ASSEMBLY

Fuel rail assembly—1992 3.9L and 5.2L engines

transaxle where the halfshaft enters the differential or extension housing. Remove the halfshaft from the transaxle or center bearing. Unbolt the center bearing from the block and remove the intermediate shaft from the transaxle, if equipped.

To install:

8. Install the halfshaft or intermediate shaft to the transaxle, being careful not to damage the side seals.

Make sure the inner joint clicks into pace inside the differential. Install the center bearing retaining bolts, if equipped. Install the outer shaft to the center bearing if equipped.

9. Pull the front strut out and insert the outer joint into the front hub.

10. Turn the ball joint stud, if necessary to position the bolt retaining indent to the inside of the vehicle. Install the ball joint stud into the steering knuckle. Install the retaining bolt and nut.

11. Install the axle nut washer and nut and torque the nut to 180 ft. lbs. (244 Nm). Install the spring washer, nut lock and a new cotter pin.

12. Install the speedometer driven gear into the transaxle and secure retaining bolt.

13. Install the tire and wheel assembly.

CV-Boot

REMOVAL & INSTALLATION

NOTE: Use only clamps provided with the replacement package when servicing. Plastic wire ties and other straps will not clamp tightly enough and grease will sling out causing costly damage to the joint.

Inner Joint

EXCEPT 4WD VEHICLE

1. Raise the vehicle and support safely. Remove the halfshaft from the vehicle.

2. If cutting the boot away, mark and note the boot positioning on the shaft relative to the raised shoulders. Remove the boot clamps to gain access to the tripod retention system.

3. Separate the housing from the tripod according to the following:

NOTE: Always hold the rollers in place when removing the housing from the tripod or the needle bearing may fall out.

a. S.S.G. — Uses a wire ring tripod retainer which expands into a groove around the top of the housing. Use a tool to pry the wire ring, without damaging it, out of the groove and slide the tripod from the housing.

b. G.K.N. — Has retaining tabs integral with the housing cover. Hold the shaft on an angle while gently pulling on the shaft until 1 of the tripod bearings is free of the retainer collar. Repeate until all rollers are free of the retainer collar.

4. Remove the snapring from the end of the shaft and remove the tripod.

5. If not already done, mark the boot positioning on the shaft, relative to the raised shoulders and remove the boot from the shaft.

6. Remove as much old grease as possible from the joint. Inspect all parts for wear or damage.

NOTE: Do not use petroleum-based solvents on the joints, shaft or boot to clean; it will ruin hidden rubber seals within the joint. Use only chlorine-based cleaner or hot soapy water to clean the joint, if necessary. Make sure the joint is completely dry before assembling.

To install:

7. If equipped, slide a new rubber

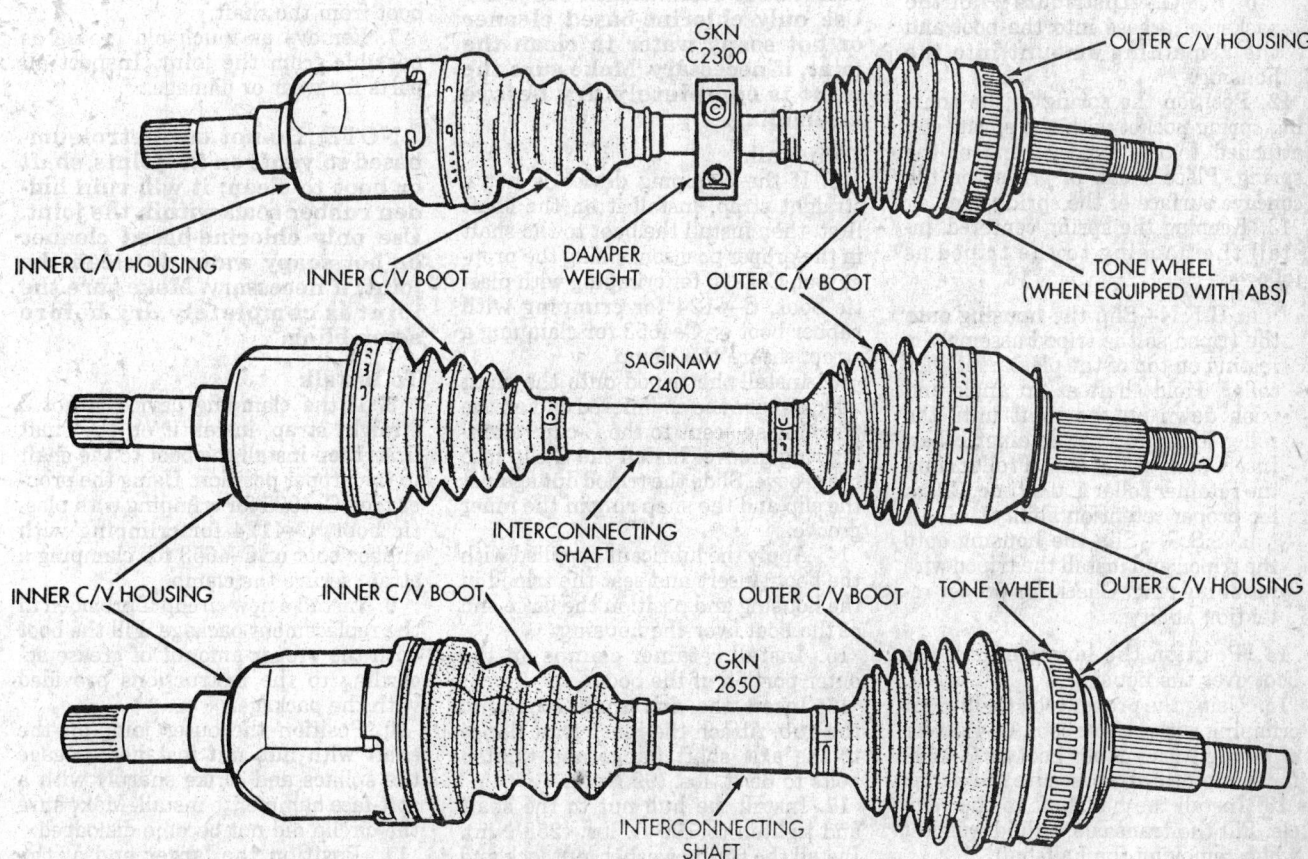

Driveshaft identification chart—axle boot identification

washer seal over the stub shaft and down into the groove provided.

8. If the clamping device is not a staight strap, install it on the shaft first, then install the boot to the shaft in the proper position. Using the proper tool, C–4975 for crimping with plastic boot, C–4124 for crimping with rubber boot or C–4653 for clamping a strap, secure the clamp.

9. Install the tripod onto the shaft as follows:

a. G.K.N—Slide the tripod on the shaft with the non-chamfered edge facing the tripod retainer ring groove.

b. S.S.G.—Place the wire ring tripod retainer over the shaft, then slide on the tripod. The tripod may be installed either way; both ends are the same.

10. Install the new snapring into its groove on the shaft to lock the tripod in position.

11. Distribute the grease provided in the grease package as follows, or according to the instructions in the package:

a. G.K.N—If equipped with 3 packets of grease, distribute 2 of the 3 packets into the boot and the remaining packet into the housing. Otherwise, distribute ½ of the packet of grease into the boot and the remaining amount into the housing.

b. S.S.G.—Distribute ½ of the packet of grease into the boot and the remaining amount into the housing.

12. Position the spring in the housing spring pocket with the spring cup attached to the exposed end of the spring. Place a dab of grease on the concave surface of the spring cup.

13. Keeping the spring centered, install the housing to the tripod as follows:

a. G.K.N—Slip the housing onto the tripod so the tripod assembly is resting on top of the plastic retainer collar. Hold shaft at an angle and push down on the shaft until the roller locks into the retainer collar. Insert each of the tripod rollers into the retainer collar 1 at a time. Check for proper retention ability.

b. S.S.G.—Slip the housing onto the tripod and install the tripod wire retaining ring. Check for proper retention ability.

14. Position the larger end of the boot over the housing.

15. Using the proper tool, C–4975 for crimping with plastic boot, C–4124 for crimping with rubber boot or C–4653 for clamping a strap, secure the clamp.

16. Install the halfshaft to the vehicle. Fill the transaxle if fluid was lost when removing the halfshaft.

17. Road test the vehicle.

4WD VEHICLE

1. Remove the cotter pin, nut lock and spring washer from the stub shaft.

2. Remove the halfshaft from the vehicle.

3. Raise and safely support the vehicle. Remove the skip plate, if equipped.

4. Remove the wheel and tire assembly.

5. Remove the bolts that attach the inner housing to the axle shaft flange.

6. Separate the joint housing from the hub. Remove the CV-driveshaft from the vehicle.

7. Remove the inner rubber boot retaining clamps and clamp protector. Pull the inner boot back onto the shaft.

8. Pull the tripod and shaft straight out from the inner CV-joint housing.

9. Move the snapring from the groove behind the tripod. Slide the tripod toward the conter of the shaft and remove the C-clip on the outer end of the shaft.

10. Remove the joint and the boot from the shaft.

11. Remove as much old grease as possible from the joint. Inspect all parts for wear or damage.

NOTE: Do not use petroleum-based solvents on the joints, shaft or boot to clean; it will ruin hidden rubber seals within the joint. Use only chlorine-based cleaner or hot soapy water to clean the joint, if necessary. Make sure the joint is completely dry before assembling.

To install:

12. If the clamping device is not a straight strap, install it on the shaft first, then install the boot to the shaft in the proper position. Using the proper tool, C–4975 for crimping with plastic boot, C–4124 for crimping with rubber boot or C–4653 for clamping a strap, secure the clamp.

13. Install the tripod onto the shaft making sure the chamfered end on the tripod is adjacent to the C-clip retaining ring groove. Install the C-clip into the groove. Slide the tripod out against the clip and the snap ring in the inner groove.

14. Apply the lubricant supplied with the boot. Insert and seat the tripod in the housing and position the large end of the boot over the housing.

15. Install retainer clamps on the outer portion of the boot.

16. Insert the driveshaft stub into the hub. Attach the inner joint flange to the axle shaft flange and tighten bolts to 65 ft. lbs. (90 Nm).

17. Install the hub nut to the shaft and tighten to 190 ft. lbs. (258 Nm). Install the spring washer, nut lock and new cotter pin.

18. Install the wheel and tire assembly.

Outer Joint

EXCEPT 4WD VEHICLE

1. Remove the halfshaft from the vehicle.

2. If cutting the boot away, mark and note the boot positioning on the shaft relative to the raised shoulders. Remove the boot clamps to gain access to the joint retention system.

3. Separate the housing from the tripod according to the following:

a. G.K.N—Using a soft-jaw vise, support the halfshaft. Strike the joint assembly sharply with a soft-face hammer to dislodge the internal circlip and remove from the shaft.

b. S.S.G.—Loosen the damper weight bolts and slide it and the boot toward the inner joint. Expand the snapring and slide the joint from the shaft. Reinstall the damper weight and torque the bolts to 21 ft. lbs. (28 Nm).

4. If damaged, remove the wear sleeve from the CV-joint machined ledge.

5. Remove the circlip from the groove.

6. If not already done, mark the boot positioning on the shaft, relative to the raised shoulders and remove the boot from the shaft.

7. Remove as much old grease as possible from the joint. Inspect all parts for wear or damage.

NOTE: Do not use petroleum-based solvents on the joints, shaft or boot to clean; it will ruin hidden rubber seals within the joint. Use only chlorine-based cleaner or hot soapy water to clean the joint, if necessary. Make sure the joint is completely dry before assembling.

To install:

8. If the clamping device is not a straight strap, install it on the shaft first, then install the boot to the shaft in the proper position. Using the proper tool, C–4975 for crimping with plastic boot, C–4124 for crimping with rubber boot or C–4653 for clamping a strap, secure the clamp.

9. Install a new circlip if provided in the replacement package. Fill the boot with the proper amount of grease according to the instructions provided with the package.

10. Position the outer joint on the shaft with hub nut installed, engage the splines and strike sharply with a soft-face hammer to install. Make sure the circlip did not become dislodged.

11. Position the larger end of the boot over the housing.

12. Using the proper tool, C–4975 for crimping with plastic boot, C–4124 for crimping with rubber boot or C–4653 for clamping a strap, secure the clamp.

13. Install the halfshaft to the vehicle. Fill the transaxle if fluid was lost when removing the halfshaft.

14. Road test the vehicle.

4WD VEHICLE

1. Remove the halfshaft from the vehicle.

2. If cutting the boot away, mark and note the boot positioning on the shaft relative to the raised shoulders. Remove the boot clamps to gain access to the joint retention system.

3. To separate the housing from the tripod, expand the snapring and slide the joint from the shaft.

4. If damaged, remove the slinger from the CV-joint .

5. If not already done, mark the boot positioning on the shaft, relative to the raised shoulders and remove the boot from the shaft.

7. Remove as much old grease as possible from the joint. Inspect all parts for wear or damage.

NOTE: Do not use petroleum-based solvents on the joints, shaft or boot to clean; it will ruin hidden rubber seals within the joint. Use only chlorine-based cleaner or hot soapy water to clean the joint, if necessary. Make sure the joint is completely dry before assembling.

To install:

8. If the clamping device is not a straight strap, install it on the shaft first, then install the boot to the shaft in the proper position. Using the proper tool, C–4975 for crimping with plastic boot, C–4124 for crimping with rubber boot or C–4653 for clamping a strap, secure the clamp.

9. Fill the boot with the proper amount of grease according to the instructions provided with the package.

10. Align the outer CV-joint splines with the axle shaft splines and push joint onto axle until the snap ring seats in the groove. Ensure the snapring is properly seated in the housing.

11. Place the large end of the boot onto the housing and secure with clamp.

12. Install the axle into the vehicle.

Driveshaft and U-Joints

The driveshaft on the All Wheel Drive (AWD) model is a hollow shaft with a plunging tripod joint on the front of the shaft and a fixed tripod joint on the rear of the shaft. The shaft is serviced as an assembly only. The flange patterns are different at each end to insure proper installation.

REMOVAL & INSTALLATION

Except All Wheel Drive Vehicles

1. Raise the vehicle and support safely.

2. Matchmark the driveshaft and the rear axle drive pinion gear shaft yoke. Unbolt the support collar from the frame, if equipped.

3. Remove the rear U-joint attaching bolts and both strap clamps from the rear axle drive pinion gear shaft yoke.

4. Fluid may run from the rear of the extension housing or transfer case when the shaft is removed, so position a drain pan under the area.

5. Remove the driveshaft from the transmission or transfer case.

6. The installation is the reverse of the removal procedure.

All Wheel Drive Vehicles

1. Place vehicle in neutral. Raise and safely secure the vehicle.

2. Remove the driveshaft rear mounting bolts.

3. Remove the driveshaft from the vehicle. Do not allow the driveshaft to hang freely while installing.

4. To install, reverse the removal procedure making sure mating surfaces are clean.

Front Axle Shaft, Seal and Bearing

REMOVAL & INSTALLATION

Left Side Shaft

1. Raise the vehicle and support safely.

2. Disconnect the left CV-driveshaft from the axle shaft flange.

3. Remove the differential housing cover. Rotate the differential case so the differential pinion mate gear shaft lock screw is accessible. Remove the lock screw and the pinion mate gear shaft from the differential case.

4. Force the left axle shaft toward the center of the vehicle and remove the shaft C-clip lock from the recessed groove in the axle shaft.

5. Remove the axle shaft from the differential housing. Inspect the axle shaft bearing contact surface for brinelling, spalling and pitting. If any of these conditions exist, replace the axle shaft and bearing.

6. Remove the axle shaft seal from the end of the housing bore with a prybar. Remove the axle shaft bearing only if it is being replaced; do not reinstall used bearings. Use removal tool

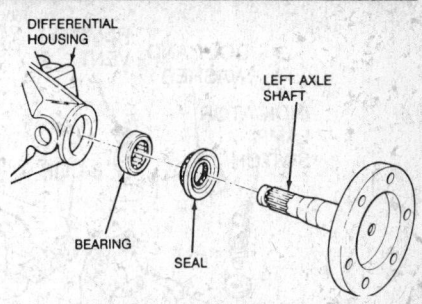

Left side axle shaft assembly

C–4167 and slide hammer tool C–637 to remove the bearing.

To install:

7. Thoroughly clean and dry the bearing bore in the differential housing.

8. Insert the new bearing into the pilot of bearing installation tool C–4198 and attach to the handle. Insert the bearing to the housing until it is seated against the bore shoulder.

9. Install the new seal using tool C–4203 with the flat side of the tool facing the seal. When the installation tool contacts the housing flange face, the seal is installed to the correct depth.

10. Lubricate the bearing bore and seal lip, insert the axle shaft into the housing and engage the splines. With the shaft in place, install the C-clip lock and push the shaft outward to seal the lock.

11. Install the mate shaft, align the hole in the shaft with the lock screw hole in the differential case and install the lock screw. Torque the lock screw to 8 ft. lbs. (11 Nm).

12. Thoroughly clean and dry the case cover, mating surface, bolts and bolt holes. Apply silicone sealer to the cover and install.

13. Connect the CV-driveshaft.

14. Level the vehicle and fill the differential with multi-purpose gear lubricant.

15. Road test the vehicle and check for leaks and correct front axle operation.

Right Side and Intermediate Shaft

1. Raise the vehicle and support safely.

2. Remove the tire and wheel assembly.

3. Remove the cotter pin, nut lock and spring washer from the stub shaft. Remove the hub nut and washer.

4. Remove the bolts that attach the inner CV-joint to the axle shaft flange.

5. Separate the stub shaft splines from the hub bearing splines and remove the CV-joint shaft.

6. Remove the differential housing cover.

7. Disconnect the 4WD indicator light switch and the vacuum tube from

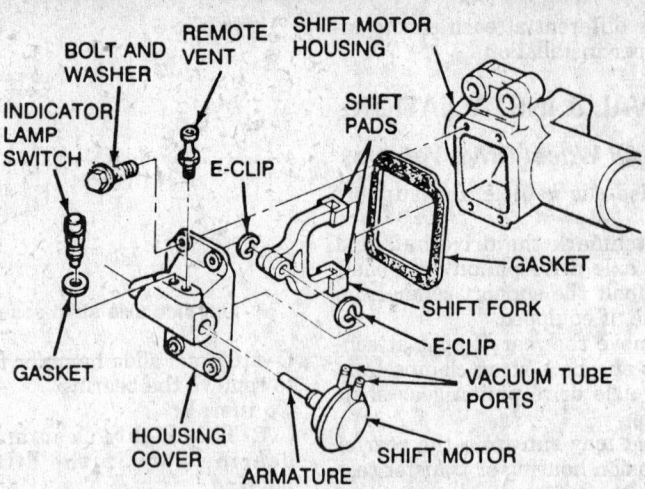

Shift motor and related parts—Dakota with 4WD

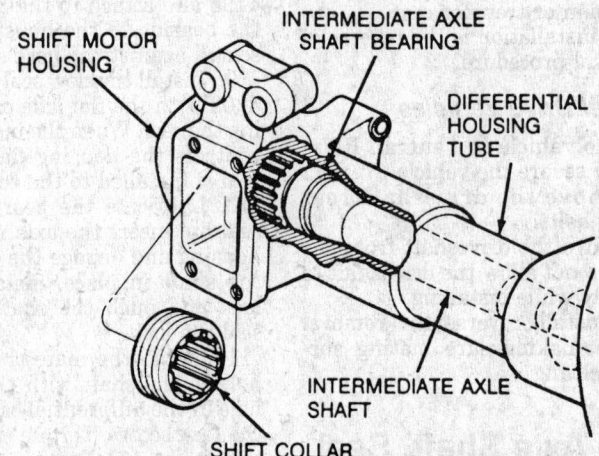

Shift collar and related parts—Dakota with 4WD

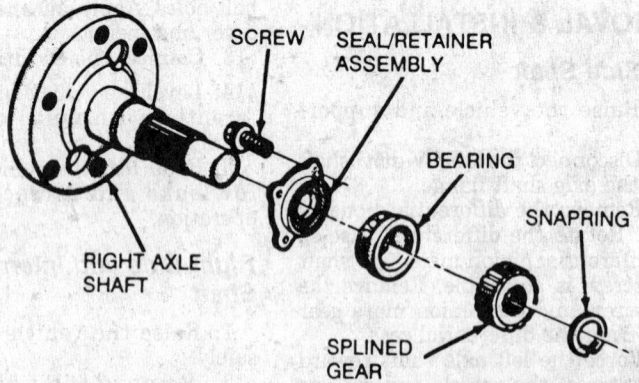

Right side axle shaft assembly—Dakota with 4WD

ward the center of the vehicle and remove the shaft C-clip lock from the recessed groove in the axle shaft. Remove the intermediate shaft from the differential housing and tube.

To install:

14. Insert the axle shaft into the tube and housing and engage the splines. With the shaft in place, install the C-clip lock and push the shaft outward to seat the lock.

15. Install the mate shaft, align the hole in the shaft with the lock screw hole in the differential case and install the lock screw. Torque the lock screw to 8 ft. lbs. (11 Nm).

16. Install the shift collar to the intermediate shaft.

17. Install a new seal to the seal retainer and install to the outer shaft. Press on the new bearing, install the splined gear and the snapring.

18. Insert the outer axle shaft into the shift motor housing. Torque the attaching screws to 17 ft. lbs. (23 Nm).

19. Install the shift motor housing cover and gasket. Ensure that the shift fork is correctly engaged in the collar groove. Install the cover bolts. Connect the vacuum tubes to the shift motor and the connector to the 4WD light switch.

20. Thoroughly clean and dry the case cover, mating surface, bolts and bolt holes. Apply silicone sealer to the cover and install.

21. Lubricate the contact surface area of the CV-driveshaft wear sleeve with grease and install the CV-driveshaft. Torque the hub nut to 190 ft. lbs. (258 Nm). Install the spring washer, nut lock and a new cotter pin.

22. Level the vehicle and fill the differential with multi-purpose gear lubricant.

23. Road test the vehicle and check for leaks and correct front axle operation.

Rear Axle Shaft, Bearing and Seal

REMOVAL & INSTALLATION

1. Raise the vehicle and support safely.

2. Remove the wheel and brake drum.

3. Remove the differential housing cover.

4. Rotate the differential case so the differential pinion mate gear shaft lock screw is accessible. Remove the lock screw and the pinion mate gear shaft from the differential case.

5. Force the axle shaft toward the center of the vehicle and remove the axle shaft C-clip lock from the recessed groove in the axle shaft.

the shift motor. Remove the shift motor, housing cover and gasket from the shift motor housing.

8. Remove the bearing seal/retainer attaching screws accessible via the holes in the axle shaft flange.

9. Remove the outer axle shaft from the differential housing.

10. Remove the snapring and the splined gear from the outer axle shaft.

Separate the bearing and seal from the outer axle shaft.

11. Remove the shift collar from the shift motor housing.

12. Rotate the differential case so the differential pinion mate gear shaft lock screw is accessible. Remove the lock screw and the pinion mate gear shaft from the differential case.

13. Force the intermediate shaft to-

6. Remove the axle shaft from the axle housing.

7. Pry the axle shaft seal from the end of the axle tube using a pry bar.

8. To remove the bearing, use tool C–4167 attached to a slide hammer.
To install:

9. Clean the bearing bore in the axle tube.

10. Insert the new axle shaft bearing onto the pilot of tool C–4198. The bearing is fully installed when it is seated firmly against the shoulder in the axle tube.

11. Install the new seal to the axle tube.

12. Lubricate the bearing bore and seal lip with grease and insert the axle shaft into the axle tube engaging its splines with the differential side gear splines.

13. Install the C–clip lock in the groove at the end of the shaft. Force the shaft outward to seat the C–clip.

14. Insert the differential pinion gear mate shaft into the case and through the thrust washers and pinion gears. Align the hole in the shaft with the lock screw hole in the differential case and install the lock screw. Torque the screw to 8 ft. lbs. (11 Nm).

15. Thoroughly clean and dry the case cover, mating surface, bolts and bolt holes. Apply silicone sealer to the cover and install.

16. Install the drum and wheel.

17. Fill the differential with the proper lubricant.

18. Road test the vehicle and check for leaks.

Front Wheel Hub, Knuckle and Bearings

REMOVAL & INSTALLATION

Caravan, Voyager and Town & Country

1. Raise the vehicle and support safely.

2. Remove the tire and wheel assembly. Remove the brake caliper from the adaptor and remove the adaptor. Remove the brake disc.

3. Remove the halfshaft.

4. Disconnect the tie rod from the knuckle.

5. Remove the 2 strut clamp bolts and remove the knuckle from the vehicle.

6. Attach the hub removal tool C–4811 or equivalent, and the triangular adapter, to the 3 rear threaded holes of the steering knuckle housing with the thrust button inside the hub bore.

7. Tighten the bolt in the center of the tool, to press the hub from the steering knuckle. Remove the tools.

8. Remove the bolts and bearing retainer from the outside of the steering knuckle.

9. Carefully pry the bearing seal from the machined recess of the steering knuckle and clean the recess.

10. Insert the tool C–4811 or equivalent, through the hub bearing and install bearing removal adapter to the outside of the steering knuckle. Tighten the tool to press the hub bearing

from the steering knuckle. Discard the bearing and the seal.

To install:

11. Use tool C–4811 or equivalent, and the bearing installation adapter to press in the hub bearing into the steering knuckle.

12. Install a new seal, the bearing retainer and the bolts to the steering knuckle.

13. Use the tool C–4811 or equivalent, and the hub installation adapter, to press the hub into the hub bearing.

14. Using the bearing installation tool C–4698 or equivalent, drive the new dust seal into the rear of the steering the hub and bearing from the knuckle as required.

15. Install the knuckle on the vehicle and insert lower ball joint nut into knuckle.

16. Install the tie rod through the knuckle and install attaching nut. Tighten nut to 35 ft. lbs. (47 Nm). Install new cotter pin into tie rod end.

17. Install the stub axle into the hub assembly. Install the 2 strut clamp bolts in place.

18. Install the rotor, adaptor and brake caliper.

19. Install the washer and hub nut onto the stub shaft and tighten nut to 180 ft. lbs. (244 Nm). Install spring washer, nut lock and new cotter pin.

20. Install the front wheel and tire assembly. Lower vehicle and align. Before moving vehicle, make sure to pump brake pedal to seat caliper piston against rotor.

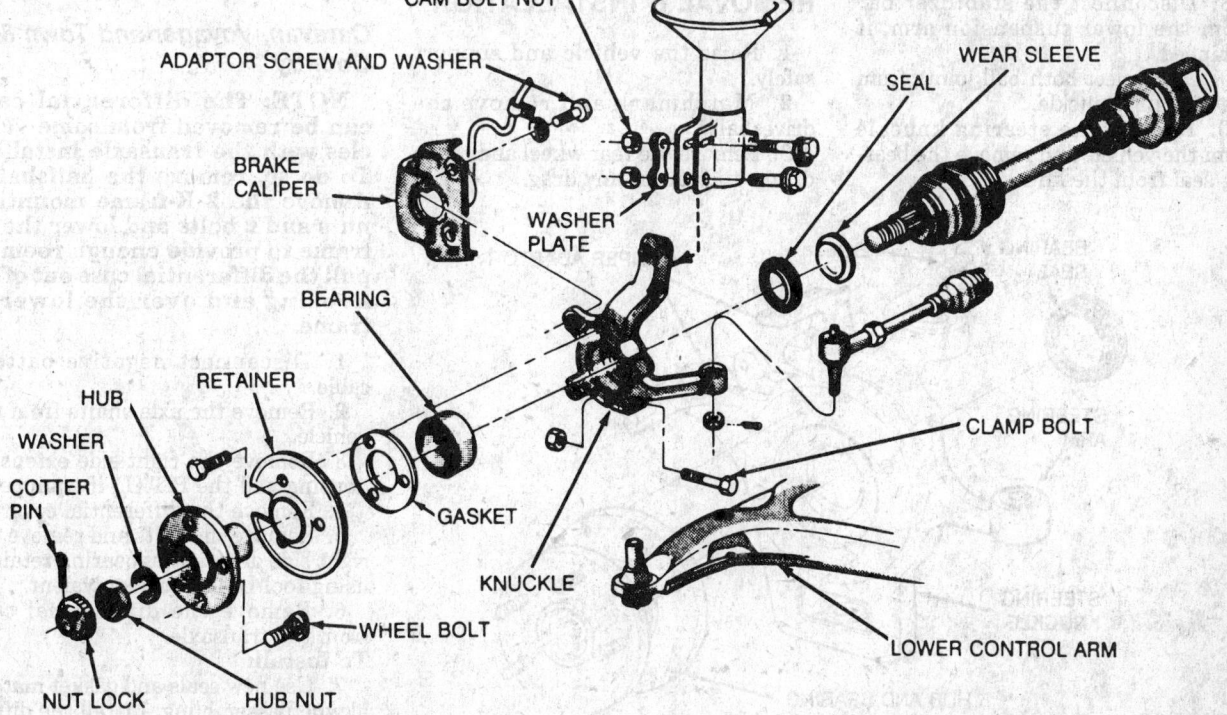

Front suspension components—Caravan, Voyager and Town & Country

Dakota

2WD

1. Raise the vehicle and support safely.
2. Remove the tire and wheel assembly.
3. Remove the brake caliper and rotor from the steering knuckle.
4. Disconnect the tie rod end from the steering knuckle.
5. Disconnect both ball joints from the steering knuckle.
6. Remove the steering knuckle from the vehicle. Remove the splash shield and the steering arm from the knuckle.
7. The installation is the reverse of the removal procedure. Torque the steering arm to knuckle nuts to 217 ft. lbs. (294 Nm).

4WD

1. Raise the vehicle and support safely. Remove the tire and wheel assembly.
2. Remove the cotter pin, nut lock and spring washer from the end of the CV-driveshaft. Remove the hub nut and washer.
3. Remove the brake caliper and the rotor from the hub.
4. Remove the bolts that attach the wheel hub to the steering knuckle.
5. Remove the wheel hub from the steering knuckle and the stub shaft.
6. Disconnect the tie rod end from the knuckle.
7. Remove the tension from the torsion bar, if equipped. Remove the lower shock absorber attaching bolt.
8. Disconnect the stabilizer bar from the lower suspension arm, if equipped.
9. Disconnect both ball joints from the steering knuckle.
10. Remove the steering knuckle from the vehicle and remove the bearing seal from the knuckle.

To install:

11. Install the wheel hub on the steering knuckle and torque the retaining bolts to 110 ft. lbs. (149 Nm).
12. Fill the cavity between the seal lip and the knuckle with multi-purpose lubricant. Position the new seal in the bore at the inner side of the knuckle and seat it with tool C–4698.
13. Lubricate the contact surface area of the CV-driveshaft wear sleeve with grease and install the steering knuckle, engaging the splines of the CV-driveshaft with those in the wheel hub. Torque the upper ball stud nut to 105 ft. lbs. (142 Nm) and the lower nut to 120 ft. lbs. (163 Nm).
14. Install the torsion bar.
15. Install the stabilizer bar and shock bolt.
16. Install the tie rod to the steering knuckle and torque the nut to 40 ft. lbs. (54 Nm).
17. Torque the hub nut to 190 ft. lbs. (258 Nm). Install the spring washer, nut lock and a new cotter pin.
18. Install the tire and wheel assembly.
19. Set the front suspension height so the difference between the distance from the surface that the tires are on to the lower suspension arm inner pivot and the distance that the tires are on to the outer end of the arm is 1–1½ in.
20. Align the front end.
21. Road test the vehicle.

Pinion Seal

REMOVAL & INSTALLATION

1. Raise the vehicle and support safely.
2. Matchmark and remove the driveshaft.
3. Remove the rear wheel and brake drums to prevent any drag.

4. Using an inch lb. torque wrench, measure the pinion bearing preload. Read the torque while the handle of the wrench is moving through several complete revolutions.
5. Using the proper tools, hold the companion flange and remove the drive pinion nut and washer.
6. Remove the companion flange. Lower the rear of the vehicle to prevent fluid loss.
7. Using a seal remover tool, remove the seal from the carrier and clean the seal seat.

To install:

8. The outside diameter of the seal is precoated with a special sealer so no sealing compound is required for installing. The seal is properly installed when the flange contacts the housing flange face.
9. Install the companion flange and the washer with the convex side out.
10. Tighten the pinion nut to 210 ft. lbs. (285 Nm) and check the pinion bearing preload. If the preload is less than the original preload measured, continue tightening the nut in very small increments until the proper preload is reached.
11. Install the driveshaft, drums and rear wheels.
12. Refill the differential with the proper lubricant.
13. Road test the vehicle and check for leaks.

Differential Carrier

REMOVAL & INSTALLATION

Caravan, Voyager and Town & Country

NOTE: The differential case can be removed from some vehicles with the transaxle installed. To do so, remove the halfshafts, remove the 2 K-frame mounting nuts and 2 bolts and lower the K-frame to provide enough room to pull the differential case out of its housing and over the lowered frame.

1. Disconnect negative battery cable.
2. Remove the axle shafts from the vehicle.
3. Remove the right side extension housing and the P.T.U., if equipped.
4. Remove the differential cover.
5. Remove the bolts and remove the right side differential bearing retainer using tool L–4435 or equivalent.
6. Remove the differential case from the transaxle.

To install:

7. Use new seals and gasket material when assembling. Install the differential case into the transaxle and in-

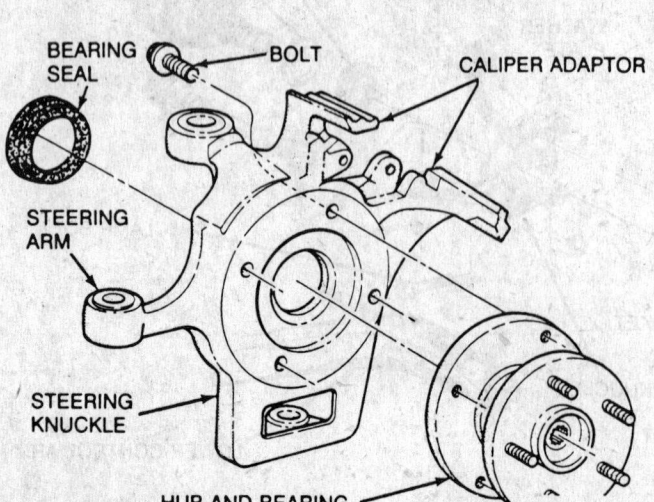

BEARING SEAL — BOLT — CALIPER ADAPTOR — STEERING ARM — STEERING KNUCKLE — HUB AND BEARING

Front wheel hub removal and installation—Dakota with 4WD

stall the right side differential bearing retainer. Torque bearing retainer bolts to 21 ft. lbs. (28 Nm).

8. Install the differential cover and fill with the proper lubricant.

9. Install the extension housing and P.T.U., if removed.

10. Install the axle shafts and connect negative battery cable.

Dakota

1. Raise the vehicle and support safely.

2. Remove the wheels. If removing the rear axle, remove the brake drums.

3. Remove the housing cover and drain the lubricant.

4. Remove the rear wheel anti-lock brake sensor, if equipped.

5. Remove both axle shafts.

6. Matchmark the bearing caps to the differential housing.

7. Remove the differential bearing threaded adjuster lock from each cap.

8. Loosen but do not remove the bearing caps.

9. Loosen the side adjusters using tool C–4164.

10. Remove the bearing caps, the threaded adjusters and the differential case.

To install:

11. Position the assembled differential case in the housing.

12. Install the bearing caps in their original positions according to the matchmarks made during the disassembly.

13. Torque the upper bolts to 10 ft. lbs. (14 Nm) and finger-tighten the bottom bolts.

14. Tighten the side adjusters until the proper side play specifications are reached.

15. Torque the bearing caps bolts to 45 ft. lbs. (61 Nm) for 7¼ in. rear or 100 ft. lbs. (136 Nm) for 8¼ in. rears.

16. Install both axle shafts.

17. Install the rear wheel anti-lock brake sensor, if equipped.

18. Install the housing cover and fill with the proper lubricant.

19. Install the drums if removed. Install the wheels.

20. Road test the vehicle.

Axle Housing

REMOVAL & INSTALLATION

Front Axle Housing

1. Disconnect the negative battery cable.

2. Raise vehicle and support safely. Remove skid plate, if equipped.

3. Remove the CV-driveshafts from the axle shaft flanges.

4. Remove the front propeller shaft from the transfer case.

5. Disconnect the vacuum hoses and wire connections from the shift motor.

6. Support the differential housing and remove the attaching bolts.

7. Remove the axle housing and the propeller shaft from the vehicle.

To install:

8. Install the axle housing and the propeller shaft into the vehicle. Install the attaching bolts.

9. Connect the vacuum hoses and wire connections to the shift motor.

10. Install the front propeller shaft from the transfer case.

11. Install the CV-driveshafts from the axle shaft flanges.

12. Install the skid plate and connect the negative battery cable.

Rear Axle Housing

1. Disconnect the negative battery cable.

2. Raise vehicle and support safely. Remove skid plate, if equipped.

3. Remove the rear wheel anti-lock brake sensor, if equipped.

4. Remove the rear wheels.

5. Disconnect the rear brake hose at the T-fitting.

6. Remove the driveshafts from the vehicle.

7. Disconnect the parking brake cables and reposition.

8. Support the weight of the assembly with the proper equipment. Disconnect the shock absorbers and remove the leaf spring nuts and U-bolts.

9. Remove the assembly from vehicle.

10. Install the axle housing into the vehicle. Install the shock absorbers and the leaf spring nuts and U-bolts.

11. Connect the parking brake cables. Install the drive shafts.

12. Connect the rear brake hose at the T-fitting.

13. Install the rear wheel anti-lock brake sensor, if equipped. Install the rear wheels.

14. Install the skid plate, if equipped.

15. Connect the negative battery cable.

MANUAL TRANSMISSION

For further information on transmissions/transaxles, please refer to "Chilton's Guide to Transmission Repair".

Transmission Assembly

REMOVAL & INSTALLATION

2WD Vehicles

1. Disconnect the negative battery cable.

2. Shift the transmission into Neutral and remove the upper shift lever. On 1992 models, do not remove the upper shift lever at this point.

3. Raise the vehicle and support safely. Remove the skid plate(s), if equipped. Drain the transmission lubricant.

4. Mark the location of the engine timing sensor, if equipped and remove sensor.

5. Matchmark and remove the driveshaft(s).

6. Disconnect the wires from the distance sensor, if equipped. Then loosen the sensor coupling and remove the sensor from the speedometer adaptor.

7. Disconnect the reverse light switch and reposition harness.

8. Install engine support fixture C–3487–A or equivalent, to support the engine. Raise the engine slightly with the support fixture.

9. Disconnect the insulator from the extension housing.

10. Using the proper equipment, support the transmission and remove the crossmember.

11. Disconnect and remove the Y-pipe from the exhaust system.

12. On 1992 models remove the upper shift lever as follows:

　a. Lower transmission 3–4 in.

　b. Reach up and around transmission case unseat shift lever dust boot from the transmission shift tower. Move boot upward on on lever.

　c. Press shift lever retainer downward with your finger.

　d. Turn retainer counterclockwise to release retainer. Lift lever and retainer out of shift tower.

13. Remove the slave cylinder from the clutch housing. Remove the transmission to clutch housing bolts.

14. Slide the transmission rearward until the input shaft clears the clutch disc.

15. Pull the transmission completely away from the clutch housing and remove the transmission from the vehicle.

To install:

16. Lubricate the pilot bushing and input shaft splines very lightly with high temperature lubricant.

17. Mount the transmission on transmission jack and lift it in place until the input shaft is centered in the

clutch housing opening. Roll the transmission forward until the input shaft splines fully engage with the clutch disc.

18. Install the transmission to clutch housing bolts. Torque the bolts to 28 ft. lbs. (38 Nm).

19. On 1992 models, lower transmission for access to the shift tower. Reach up and around the transmission and install the shift lever and retainer into the shift tower.

20. Press the retainer downward and turn clockwise to lock into position. Install the lever dust boot in place.

21. Install the transmission crossmember. Remove the transmission and engine support fixtures.

22. Connect the reverse light switch connector and clip all wiring to the transmission case.

23. Connect speedometer cable and distance sensor, if equipped.

24. Install the engine timing sensor in the position prior to removal.

25. Align U-joints and install the driveshaft(s).

26. Fill the transmission with the proper amount of 75W-90 or equivalent, API grade GL5 gear lubricant.

27. Apply silicone sealer to the perimeter of the shifter base and install the shifter assembly, if applicable.

28. Install the skid plate(s), if equipped.

29. Connect the negative battery cable and check the transmission for proper operation.

4WD Vehicles

1. Disconnect the negative battery cable.

2. Shift the transmission into Neutral and remove the upper shift lever. On 1992 models, do not remove the upper shift lever at this point.

3. Raise the vehicle and support safely. Remove the skid plate(s), if equipped. Drain the transmission lubricant.

4. Mark the location of the engine timing sensor, if equipped and remove sensor.

5. Matchmark and remove the driveshaft(s).

6. Disconnect the wires from the distance sensor, if equipped. Then loosen the sensor coupling and remove the sensor from the speedometer adaptor.

7. Disconnect the vacuum switch hoses at transfer case. Disconnect the transfer case shift linkage at transfer case range lever. Remove the shift linkage bracket at the transfer case and position aside.

8. Remove shift motor vacuum tank and bracket from rear crossmember. Remove transfer case from transmission adapter housing.

9. Disconnect the reverse light switch and reposition harness.

10. Install engine support fixture C-3487-A or equivalent, to support the engine. Raise the engine slightly with the support fixture.

11. Using the proper equipment, support the transmission and remove the crossmember.

12. Disconnect and remove the Y-pipe from the exhaust system.

13. On 1992 models remove the upper shift lever as follows:

  a. Lower transmission 3-4 in.

  b. Reach up and around transmission case unseat shift lever dust boot from the transmission shift tower. Move boot upward on on lever.

  c. Press shift lever retainer downward with your finger.

  d. Turn retainer counterclockwise to release retainer. Lift lever and retainer out of shift tower.

14. Remove the slave cylinder from the clutch housing. Remove the transmission to clutch housing bolts.

15. On some models, it may be necessary to remove the oil filter and the front axle struts to gain removal clearance. Remove only if necessary. Slide the transmission rearward until the input shaft clears the clutch disc.

16. Pull the transmission completely away from the clutch housing and remove the transmission from the vehicle.

To install:

17. Lubricate the pilot bushing and input shaft splines very lightly with high temperature lubricant.

18. Mount the transmission on transmission jack and lift it in place until the input shaft is centered in the clutch housing opening. Roll the transmission forward until the input shaft splines fully engage with the clutch disc.

19. Install the transmission to clutch housing bolts. Torque the bolts to 28 ft. lbs. (38 Nm).

20. On 1992 models, lower transmission for access to the shift tower. Reach up and around the transmission and install the shift lever and retainer into the shift tower.

21. Press the retainer downward and turn clockwise to lock into position. Install the lever dust boot in place.

22. Install the transmission crossmember. Remove the transmission and engine support fixtures.

23. Connect the reverse light switch connector and clip all wiring to the transmission case.

24. Install slave cylinder in clutch housing. Install strut and fasteners, if removed.

25. Install transfer case and tighten bolts to 35 ft. lbs. (47 Nm). Install and connect the transfer case shift linkage.

26. Connect vacuum harness connector to switch at transfer case. Install vacuum tank and bracket on rear crossmember.

27. Connect speedometer cable and distance sensor, if equipped.

28. Install the engine timing sensor in the position prior to removal.

29. Align U-joints and install the driveshaft(s).

30. Fill the transmission with the proper amount of 75W-90 or equivalent, API grade GL5 gear lubricant.

31. Apply silicone sealer to the pe-

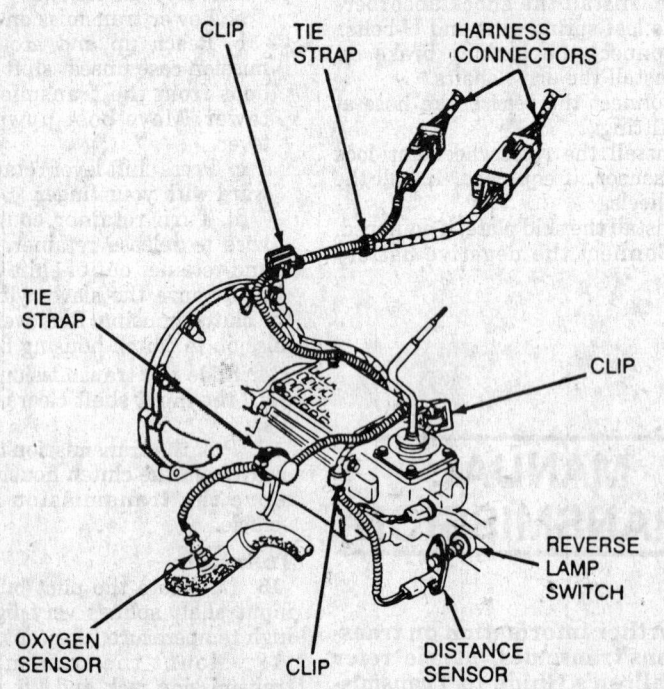

Manual transmission wiring harness routing — NP-2500 transfer case

rimeter of the shifter base and install the shifter assembly, if applicable.

32. Install the skid plate(s), if equipped.

33. Connect the negative battery cable and check the transmission for proper operation.

MANUAL TRANSAXLE

Transaxle Assembly

REMOVAL & INSTALLATION

NOTE: If vehicle is going to be rolled while the transaxle is out of the vehicle, obtain 2 outer CV-joints to install to the hubs. If vehicle is rolled without the proper torque applied to the front wheel bearings, the bearings will be destroyed.

1. Disconnect the negative battery cable.

2. Remove the air cleaner assembly if it is preventing access to the upper bell housing bolts. Remove the upper bell housing bolts. Disconnect the reverse light switch and the ground wire.

3. Remove the starter attaching nut and bolt at the top of the bell housing.

4. Raise the vehicle and support safely. Remove the tire and wheel assemblies. Remove the axle end cotter pins, nut locks, spring washers and axle nuts.

5. Remove the ball joint retaining bolts and pry the control arm from the steering knuckle. Position a drain pan under the transaxle where the axles enter the differential or extension housing. Remove the axles from the transaxle or center bearing. Unbolt the center bearing and remove the intermediate axle from the transaxle, if equipped.

6. Remove the anti-rotation link from the crossmember. Disconnect the shifter cables from the transaxle and unbolt the cable bracket.

7. Remove the speedometer cable adaptor bolt and remove the adaptor from the transaxle.

8. Remove the rear mount from the starter, unbolt the starter and position it to the side.

9. Using the proper equipment, support the weight of the engine.

10. Remove the front motor mount and bracket.

11. Position a transaxle jack under the transaxle.

12. Remove the lower bell housing bolts.

13. Remove the left side splash shield. Remove the transaxle mount bolts.

14. Carefully separate the transaxle from the engine.

15. Slide the transaxle rearward until the input shaft clears the clutch disc.

16. Pull the transaxle completely away from the clutch housing and remove from the vehicle.

17. To prepare the vehicle for rolling, support the engine or reinstall the front motor mount to the engine. Then reinstall the ball joints to the steering knuckle and install the retaining bolt. Install the obtained outer CV-joints to the hubs, install the washers and torque the axle nuts to 180 ft. lbs. (244 Nm). The vehicle may now be safely rolled.

To install:

18. Lubricate the pilot bushing and input shaft splines very lightly with high temperature lubricant.

19. Mount the transaxle securely on jack and and install the left side mount bolts. Lift it in place until the input shaft is centered in the clutch housing opening. Roll the transaxle forward until the input shaft splines fully engage with the clutch disc.

20. Install the transaxle to clutch housing bolts.

21. Raise the transaxle up and install the front motor mount and bracket.

22. Remove the engine and transaxle support fixtures.

23. Install the starter to the transaxle and install the lower bolt finger-tight.

24. Install a new O-ring to the speedometer cable adaptor and install to the extension housing; make sure it snaps in place. Install the retaining bolt.

25. Install the shift cable bracket and snap the cable ends in place. Install the anti-rotation link.

26. Install the axles and center bearing, if equipped. Install the ball joints to the steering knuckles and install new cotter pins. Fill the transaxle with SAE 5W–30 engine oil. Install the splash shield and install the wheels. Lower the vehicle.

27. Install the upper bell housing bolts.

28. Install the starter attaching nut and bolt at the top of the bell housing. Raise the vehicle again and tighten the starter bolt from under the vehicle. Lower the vehicle.

29. Connect the reverse light switch and the ground wire.

30. Install the air cleaner assembly, if removed.

31. Connect the negative battery cable and check the transaxle for proper operation.

CABLE ADJUSTMENT

1. Working over the left front fender, remove the lock pin from the transaxle selector shaft housing.

2. Reverse the lock pin (long end down) and insert it into the same threaded hole while pushing the selector shaft into the selector housing. A hole in the selector shaft will align with the lock pin, allowing the lock pin to be screwed into the housing. This operation locks the selector shaft in the neutral position between either 1st and 2nd gears or 3rd and 4th gears.

3. Remove the gearshift knob, the retaining nut and the pull-up ring from the gearshift lever.

4. If necessary, remove the shift lever boot and console to expose the gearshift linkage.

5. Fabricate 2 cable adjusting pins: $3/16$ in. diameter × 5 in. long with a $1/2$ in. 90 degrees bend at one end.

6. Place a pin in the hole provided at the right side and the other in the hole provided at the rear side of the shifting mechanism (make sure the alignment holes match). Torque the selector (right side) and the crossover (left side) adjusting bolts to 4–5 ft. lbs.

7. Remove the lock pin from the selector shaft housing and reinstall the lock pin (with the long end up) in the selector shaft housing. Torque the lock pin to 70 inch lbs. (8 Nm).

8. Check the first/reverse shifting and blockout into reverse.

9. Reinstall the console, boot, pull-up ring, retaining nut and knob.

CLUTCH

Clutch Assembly

REMOVAL & INSTALLATION

Caravan and Voyager

1. Disconnect the negative battery cable. Remove the transaxle.

2. Matchmark the clutch/pressure plate cover and flywheel. Insert a clutch plate alignment tool into the clutch disc hub.

3. Loosen the flywheel to pressure plate bolts gradually and evenly to avoid warpage.

4. Remove the pressure plate/clutch assembly from the flywheel.

5. Sand the flywheel or replace it, if scored, cracked or heat damaged.

To install:

6. Sparingly apply anti-sieze compound to the input shaft and clutch disc splines. Install a new release bearing.

7. Mount clutch assembly on flywheel taking care to align marks made before removal.

8. Install clutch alignment tool and apply pressure while tightening bolts to 21 ft. lbs. Make sure to step tighten bolts in a cross pattern to prevent any distortion of the clutch cover.

9. Install the transaxle and related parts that were removed.

10. Reconnect negative battery cable and check operation of the clutch.

Dakota

1. Disconnect the negative battery cable.

2. Raise the vehicle and support safely.

3. Remove the transmission and transfer case, if equipped.

4. Remove the clutch housing from the engine.

5. Remove the clutch cover bolts gradually in a cross pattern so the cover is not distorted.

6. Remove the clutch cover and disc.

To install:

7. Sparingly apply anti-sieze compound to the input shaft and clutch disc splines. Install a new release bearing.

8. Raise the clutch cover and disc into place and use a clutch aligning tool or spare input shaft to center the disc. Tighten all of the bolts finger-tight.

9. The cover bolts must be tightened gradually, evenly and to the proper torque to avoid distorting the cover. Torque $\frac{5}{16}$ in. diameter bolts to 17 ft. lbs. (23 Nm). Torque $\frac{3}{8}$ in. diameter bolts to 30 ft. lbs. (42 Nm).

10. Install the clutch housing, transmission and transfer case, if equipped.

11. Install the inspection cover, if removed.

12. Connect the negative battery cable and check the clutch for proper operation.

PEDAL FREE-PLAY ADJUSTMENT

NOTE: The Caravan and Voyager are equipped with a self-adjusting cable operated mechanism and no adjustment is provided. The Dakota has been equipped with a hydraulic clutch release system. There is no adjustment for free-play on this system.

Clutch Cable

REMOVAL & INSTALLATION

1. Disconnect the negative battery cable.

2. Remove the clip from the cable mounting bracket on the shock tower and remove the cable from the bracket.

3. Remove the retainer from the clutch release lever on the transaxle.

4. Pry out the ball end of the cable from the position adjuster inside the pedal.

5. The installation is the reverse of the removal procedure. After installing, push the clutch pedal 2 or 3 times to allow the self-adjuster mechanism to function.

Clutch Master Cylinder and Slave Cylinder

The clutch master cylinder, remote reservoir, slave cylinder and connecting lines are all serviced as a complete assembly. The cylinders and connecting lines are sealed units. They are prefilled with fluid from the factory and cannot be disassembled or serviced separately.

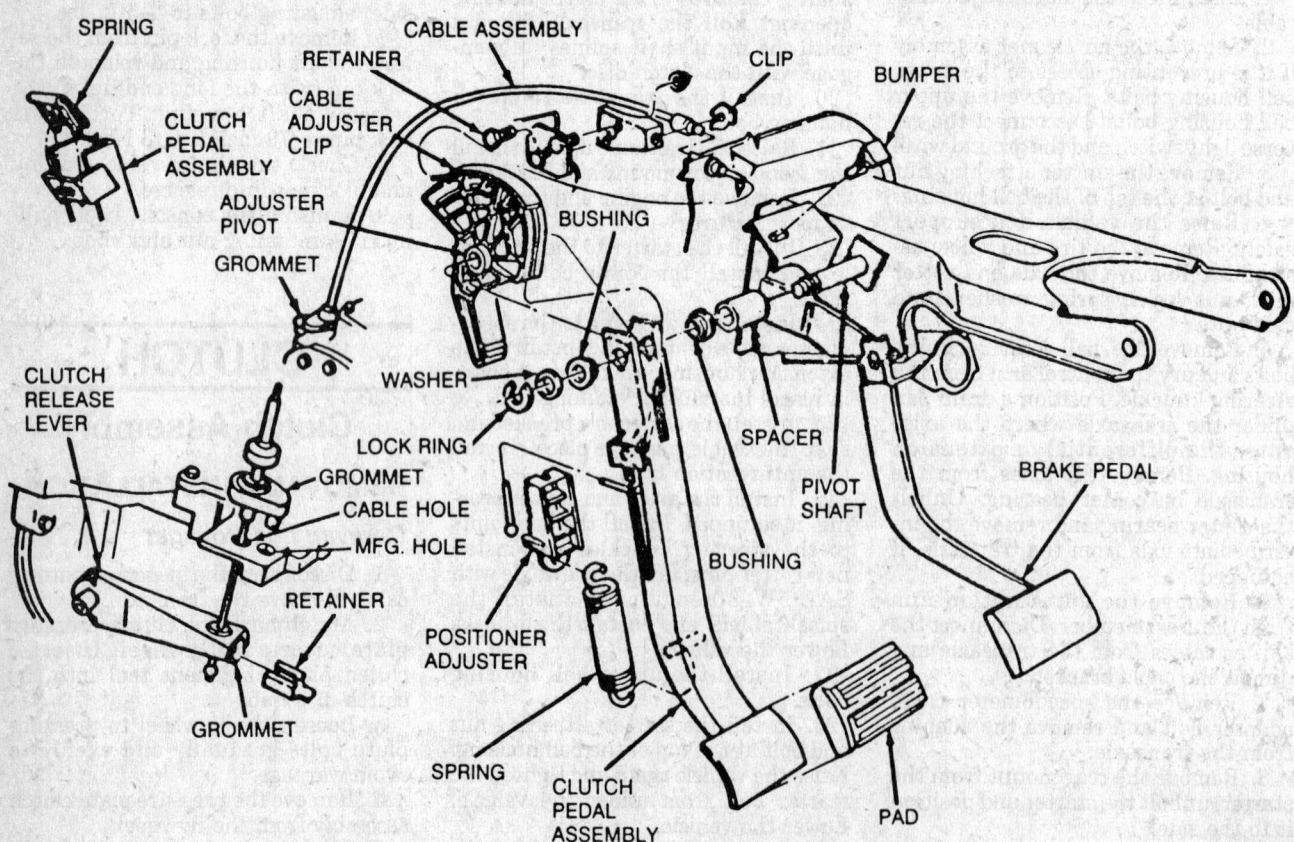

Clutch pedal, cable and related parts—Caravan and Voyager

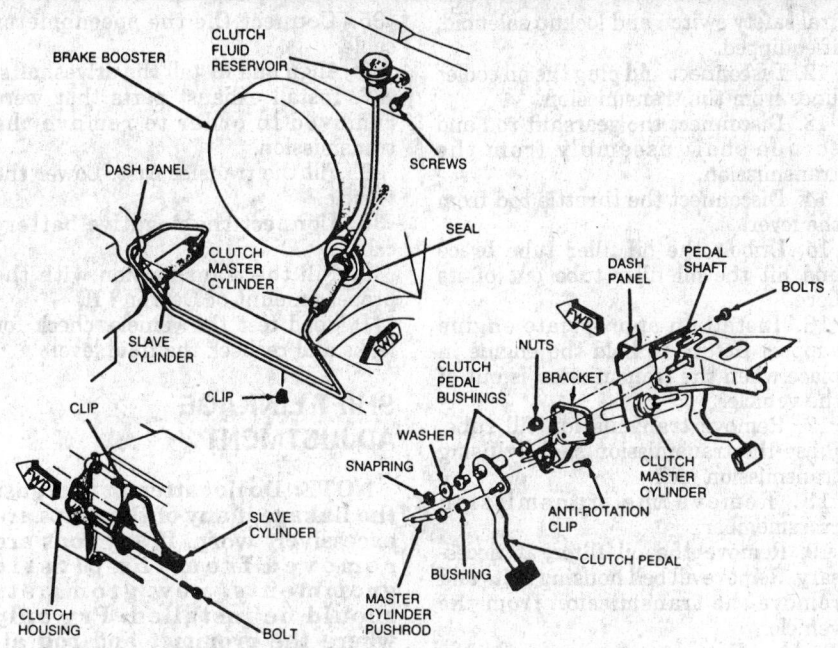

Clutch hydraulic linkage components—Dakota

REMOVAL & INSTALLATION

1. Disconnect the negative battery cable.
2. Raise the vehicle and support safely.
3. Remove the nuts attaching the slave cylinder to the bell housing.
4. Remove the slave cylinder and clip from the housing.
5. Lower the vehicle.
6. Remove the locating clip from the clutch master cylinder mounting bracket.
7. Remove the retaining ring, flat washer and wave washer that attach the clutch master cylinder pushrod to the clutch pedal. Slide the pushrod off of the pedal pin. Inspect the bushing on the pedal pin and replace, if excessively worn.
8. Verify that the cap on the clutch master cylinder reservoir is tight so fluid will not spill during removal.
9. Remove the screws attaching the reservoir and bracket to the dash panel and remove the reservoir.
10. Pull the clutch master cylinder rubber seal from the dash panel.
11. Rotate the clutch master cylinder 45 degrees to unlock it. Remove the cylinder from the dash panel.
12. Remove the clutch master cylinder, remote reservoir, slave cylinder and connecting lines from the vehicle.
To install:
13. Verify that the cap on the fluid reservoir is tight so fluid will not spill during installation.
14. Position the components in the replacement kit in their places on the vehicle.
15. Insert the master cylinder in the

dash. Rotate it 45 degrees to lock it in place.
16. Lubricate the rubber seal with lubricant to ease installation. Seat the seal around the cylinder in the dash.
17. Install the fluid reservoir and bracket, if equipped, to the dash panel.
18. Install the master cylinder pushrod to the clutch pedal pin. Secure the rod with the wave washer, flat washer and retaining ring. Install the locating clip. Do not remove the plastic shipping stop from the pushrod until the slave cylinder has been installed.
19. Raise the vehicle and support safely.
20. Insert the slave cylinder pushrod through the opening and make sure the cap on the end of the pushrod is securely engaged in the release lever before tightening the attaching nuts. Torque the nuts to 17 ft. lbs. (23 Nm).
21. Lower the vehicle. Remove the plastic shipping stop from the master cylinder pushrod.
22. Operate the clutch pedal a few times to verify proper operation of the system.
23. Connect the negative battery cable and road test the vehicle.

AUTOMATIC TRANSMISSION

For further information on transmissions/transaxles, please refer to "Chilton's Guide to Transmission Repair".

Transmission Assembly

REMOVAL & INSTALLATION

2WD Vehicles

1. Disconnect the negative battery cable.
2. Raise the vehicle and support safely. Drain the transmission and transfer case, if equipped.
3. Remove the exhaust crossover pipe from the exhaust system and reposition remaining exhaust pipes.
4. Remove the skid plates, if equipped. Remove the engine timing sensor marking original position for installation.
5. Matchmark and remove the driveshaft(s).
6. Disconnect the distance sensor, if equipped, and the speedometer cable.
7. Remove the engine to transmission struts.
8. Remove the starter and the fluid cooler lines bracket.
9. Remove the torque converter inspection cover.
10. Matchmark the converter to the flexplate. Remove the torque converter bolts.
11. Disconnect the wires to the neutral safety switch and lockup solenoid, if equipped.
12. Disconnect and plug the oil cooler lines from the transmission.
13. Disconnect the gearshift rod and torque shaft assembly from the transmission.
14. Disconnect the throttle rod from the lever.
15. Unbolt the oil filler tube brace and lift the oil filler tube out of its bore.
16. Install an appropriate engine support fixture to hold the engine in place when the transmission is out of the vehicle.
17. Raise the transmission slightly using transmission jack.
18. Remove the transmission crossmember.
19. Remove the oil filter, if necessary. Remove all bell housing bolts and remove the transmission from the vehicle.
To install:
20. Install the transmission securely on the transmission jack. Rotate the converter so it will align with the positioning of the flexplate.
21. Apply a coating of high temperature grease to the torque converter pilot hub.
22. Raise the transmission into place and push it forward until the dowels engage and the bell housing is flush with the block.
23. Install the oil filler tube. Install

the bell housing bolts and torque to 30 ft. lbs. (41 Nm). Install the oil filter, if removed.

24. Install the transmission crossmember. Remove the engine support fixture and the transmission jack.

25. Install the torque converter bolts. If new bolts are being installed, it is essential that the correct length bolt be used.

26. Connect the oil cooler lines.

27. Connect the throttle rod to the lever and adjust, if necessary.

28. Connect the gearshift rod and torque shaft assembly to the transmission and adjust if necessary.

29. Connect the wires to the neutral safety switch and lockup solenoid, if equipped. Make sure all wires are routed correctly and clipped in place.

30. Install the torque converter inspection cover, starter and transmission struts.

31. Connect the distance sensor, if equipped.

32. Connect the the speedometer cable.

33. Install the driveshaft(s).

34. Install exhaust parts that were removed in order to remove the transmission.

35. Lower the vehicle. Connect the negative battery cable.

36. Fill the transmission with the proper amount of Dexron® II.

37. Road test the vehicle, check for leaks and recheck the fluid level.

4WD Vehicles

1. Disconnect the negative battery cable.

2. Raise the vehicle and support safely. Drain the transmission and transfer case.

3. Remove the exhaust crossover pipe from the exhaust system and reposition remaining exhaust pipes.

4. Remove the skid plates, if equipped. Mark the location of the engine timing sensor and remove sensor.

5. Matchmark and remove the driveshafts.

6. Disconnect the vacuum switch hoses at transfer case. Disconnect the transfer case shift linkage at transfer case range lever. Remove the shift linkage bracket at the transfer case and position aside.

7. Remove shift motor vacuum tank and bracket from rear crossmember. Remove transfer case from transmission adapter housing.

8. Remove the starter and the fluid cooler lines bracket. Remove the engine to transmission struts.

9. Remove the torque converter inspection cover.

10. Matchmark the converter to the flexplate. Remove the torque converter bolts.

11. Disconnect the wires to the neutral safety switch and lockup solenoid, if equipped.

12. Disconnect and plug the oil cooler lines from the transmission.

13. Disconnect the gearshift rod and torque shaft assembly from the transmission.

14. Disconnect the throttle rod from the lever.

15. Unbolt the oil filler tube brace and lift the oil filler tube out of its bore.

16. Install an appropriate engine support fixture to hold the engine in place when the transmission is out of the vehicle.

17. Remove transmission fill tube. Raise the transmission slightly using transmission jack.

18. Remove the transmission crossmember.

19. Remove the oil filter, if necessary. Remove all bell housing bolts and remove the transmission from the vehicle.

To install:

20. Install the transmission securely on the transmission jack. Rotate the converter so it will align with the positioning of the flexplate.

21. Apply a coating of high temperature grease to the torque converter pilot hub.

22. Raise the transmission into place and push it forward until the dowels engage and the bell housing is flush with the block.

23. Install the oil filler tube. Install the bell housing bolts and torque to 30 ft. lbs. (41 Nm). Install the oil filter, if removed.

24. Install the transmission crossmember. Remove the engine support fixture and the transmission jack.

25. Install the torque converter bolts. If new bolts are being used, it is essential that the correct length be used.

26. Connect the oil cooler lines.

27. Install engine timing sensor to original position.

28. Connect the throttle rod to the lever and adjust, if necessary.

29. Connect the gearshift rod and torque shaft assembly to the transmission. Adjust if necessary.

30. Connect the wires to the neutral safety switch and lockup solenoid, if equipped. Make sure all wires are routed correctly and clipped in place.

31. Install the torque converter inspection cover, starter and transmission struts.

32. Install transfer case and tighten bolts to 35 ft. lbs. (47 Nm). Install and connect the transfer case shift linkage.

33. Connect vacuum harness connector to switch at transfer case. Install vacuum tank and bracket on rear crossmember.

34. Connect the distance sensor, if equipped.

35. Connect the the speedometer cable.

36. Align and install the driveshafts.

37. Install exhaust parts that were removed in order to remove the transmission.

38. Fill the transfer case. Lower the vehicle.

39. Connect the negative battery cable.

40. Fill the transmission with the proper amount of Dexron® II.

41. Road test the vehicle, check for leaks and recheck the fluid level.

SHIFT LINKAGE ADJUSTMENT

NOTE: Do not attempt to adjust the linkage if any of the parts are excessively worn. If any rods are removed from the plastic grommets, new grommets should be installed. Pry only where the grommet and rod attach, not on the rod itself. Use pliers to snap the rod into the new grommet.

1. Shift the transmission into **P**.

2. Raise the vehicle and support safely.

3. Loosen the shift rod adjusting swivel lock screw. Make sure the swivel turns freely on the rod.

4. Make sure the valve body is in the **P** position by moving it all the way rearward.

5. Adjust the swivel position on the shift rod to obtain a free pin fit in the toque shaft lever. Tighten the lock screw.

6. If the vehicle starts in any gear other than **P** or **N** or does not start in both **P** and **N**, then either the adjustment is wrong or another problem exists.

THROTTLE LINKAGE ADJUSTMENT

1. If carbureted, perform the adjustment with the engine at normal operating temperature and off of the fast idle cam.

2. If fuel injected, retract the ISC actuator by doing one of the following:

a. If the DRB II is available, connect its connector to the diagnostic connector. Start the engine and place the DRB II in the "Throttle Body Minimum Air Flow Test" mode. Disconnect the electrical connector on the ISC actuator. Shut off the engine and disconnect the DRB II; the actuator is now fully retracted.

b. If the DRB II is not available, 2 jumper wires may be used. With the engine off, disconnect the connector to the ISC actuator. Connect a pair of jumper wires to the battery. Connect the negative jumper to the top pin of the ISC actuator and the positive jumper to the other pin. Do not leave the jumpers connected for more than 5 seconds. Disconnect the jumpers and the ISC actuator is fully retracted.

3. Raise the vehicle and support safely.

4. Loosen the adjustable swivel lock screw on the throttle rod enough so the rod travels freely in the swivel.

5. Hold the throttle lever firmly forward against its internal stop and tighten the lock screw. Lower the vehicle.

6. Reconnect the ISC actuator, if equipped.

7. If equipped with fuel injection, turn the ignition key to the **RUN** position for at least 5 seconds, but do not start the engine. Turn the key to the **OFF** position.

8. Start the engine and road test the vehicle.

AUTOMATIC TRANSAXLE

For further information on transmissions/transaxles, please refer to "Chilton's Guide to Transmission Repair".

Transaxle Assembly

REMOVAL & INSTALLATION

NOTE: If vehicle is going to be rolled while the transaxle is out of the vehicle, obtain 2 outer CV-joints to install to the hubs. If vehicle is rolled without the proper torque applied to the front wheel bearings, the bearings will be destroyed. If equipped with All Wheel Drive, the Power Transfer Unit (PTU) must be removed before removing the transaxle.

1. Disconnect the negative battery cable. If equipped with 3.0L or 3.3L engine, drain the coolant. Remove the dipstick.

2. Remove the air cleaner assembly if it is preventing access to the upper bell housing bolts. Remove the upper bell housing bolts and water tube, where applicable. Unplug all electrical connectors from the transaxle.

3. Install engine support fixture.

4. If equipped with a 2.2L or 2.5L engine, remove the starter attaching nut and bolt at the top of the bell housing.

5. Raise the vehicle and support safely. Remove the tire and wheel assemblies.

6. Remove the P.T.U. as follows:

a. Position a drainpan under the transaxle where the axles enter the differential, P.T.U. or extension housing. Support the P.T.U.

b. Remove the ball joint retaining bolts and pry the control arm from the steering knuckle. Unbolt the center bearing and remove the intermediate axle from the transaxle, if equipped. Remove the axles.

c. Remove the power steering hose bracket at front crossmember.

d. Remove front Bobble strut bolt at front crossmember.

e. Remove crossmember bridge bolts.

f. Remove the mounting bolts and remove unit from vehicle.

7. Drain the transaxle. Disconnect and plug the fluid cooler hoses. Disconnect the shifter and kickdown linkage from the transaxle, if equipped.

8. Remove the speedometer cable adaptor bolt and remove the adaptor from the transaxle.

9. Remove the starter. Remove the torque converter inspection cover, matchmark the torque converter to the flexplate and remove the torque converter bolts.

10. Disconnect the electrical connector at the PRNDL switch and the neutral safety switch.

11. Remove the front motor mount and bracket.

12. If equipped with Distributorless Ignition System (DIS), remove the crankshaft position sensor.

13. Position a transaxle jack under the transaxle.

14. Remove the lower bell housing bolts.

15. Remove the left side splash shield. Remove the transaxle mount bolts.

16. Carefully separate the transaxle from the engine.

17. Slide the transaxle rearward until the dowels disengage from the mating holes in the transaxle case.

18. Pull the transaxle away from the engine and remove from the vehicle.

19. To prepare the vehicle for rolling, support the engine with a support or reinstall the front motor mount to the engine. Then reinstall the ball joints to the steering knuckle and install the retaining bolt. Install the obtained outer CV-joints to the hubs, install the washers and torque the axle nuts to 180 ft. lbs. (244 Nm). The vehicle may now be safely rolled.

To install:

20. Install the transmission securely on the transmission jack. Rotate the converter so it will align with the positioning of the flexplate.

21. Apply a coating of high temperature grease to the torque converter pilot hub.

22. Raise the transaxle into place and push it forward until the dowels engage and the bell housing is flush with the block.

23. Install the transaxle to bell housing bolts.

24. Jack the transaxle up and install the left side mount bolts. Install the torque converter bolts and torque to 55 ft. lbs. (74 Nm).

25. Install the front motor mount and bracket. Remove the engine and transaxle support fixtures.

26. Install the starter to the transaxle. Install the bolt finger-tight if equipped with a 2.2L or 2.5L engine.

27. Install the crankshaft position sensor to the bell housing.

28. Install a new O-ring to the speedometer cable adaptor and install to the extension housing; make sure it snaps in place. Install the retaining bolt.

29. Install the P.T.U. to the transaxle and secure the mounting bolts. Install the crossmember supports.

30. Connect the shifter and kickdown linkage to the transaxle, if equipped.

31. Install the axles and center bearing, if equipped. Install the ball joints to the steering knuckles. Torque the axle nuts to 180 ft. lbs. (244 Nm) and install new cotter pins.

32. Install the splash shield and install the wheels. Lower the vehicle.

33. Install the dipstick. Install the upper bell housing bolts.

34. Install the water pipe, if removed.

35. If equipped with 2.2L or 2.5L engine, install the starter attaching nut and bolt at the top of the bell housing. Raise the vehicle again and tighten the starter bolt from under the vehicle.

36. Lower the vehicle. Connect all electrical wiring to the transaxle.

37. Install the air cleaner assembly, if it was removed. Fill the transaxle with the proper amount of Dexron® II.

38. Connect the negative battery cable and check the transaxle for proper operation.

UPSHIFT AND KICKDOWN LEARNING PROCEDURE

A-604 Ultradrive Transaxle

In 1989, the A-604 4 speed, electronic transaxle was introduced; it is the first to use fully adaptive controls. The controls perform their functions based on real time feedback sensor information. Although, the transaxle is conventional in design, its functions are controlled by the ECM.

Since the A-604 is equipped with a learning function, each time the battery cable is disconnected, the ECM memory is lost. In operation, the transaxle must be shifted many times for the learned memory to be reinputed in the ECM; during this period, the vehicle will experience rough operation. The transaxle must be at normal operating temperature when learning occurs.

1. Maintain constant throttle opening during shifts. Do not move the accelerator pedal during upshifts.

2. Accelerate the vehicle with the throttle ⅛–½ open.

3. Make 15–20 1/2, 2/3 and 3/4 upshifts. Accelerating from a full stop to 50 mph each time at the aforementioned throttle opening is sufficient.

4. With the vehicle speed below 25 mph, make 5–8 wide open throttle kickdowns to 1st gear from either 2nd or 3rd gear. Allow at least 5 seconds of operation in 2nd or 3rd gear prior to each kickdown.

5. With the vehicle speed greater than 25 mph, make 5 part throttle to wide open throttle kickdowns to either 3rd or 2nd gear from 4th gear. Allow at least 5 seconds of operation in 4th gear, preferably at road load throttle prior to performing the kickdown.

SHIFT LINKAGE ADJUSTMENT

1. Place the shifter in the **P** detent.
2. Loosen the clamp bolt on the gearshift cable bracket.
3. Pull the shift lever all the way to the front detent position and tighten the lock screw.
4. Check for proper neutral safety switch operation.

THROTTLE PRESSURE CABLE ADJUSTMENT

1. Run the engine until it reaches normal operating temperature.
2. Loosen the cable mounting bracket lock screw.
3. Position the bracket so both alignment tabs are touching the transaxle case surface and tighten the lock screws.
4. Release the cross lock on the cable assembly by pulling the cross lock up.
5. To ensure proper adjustment, the cable must be free to slide all the way toward the engine against its stop after the cross lock is released.
6. Move the transaxle throttle control lever fully clockwise and press the cross lock down until it snaps into position.
7. Road test the vehicle and check the shift points.

THROTTLE PRESSURE ROD ADJUSTMENT

1. Run the engine until it reaches normal operating temperature.
2. Loosen the adjustment swivel lock screw.
3. To ensure proper adjustment, the swivel must be free to slide along the flat end of the throttle rod. Disassembly, clean and lubricate as required.
4. Hold the transaxle throttle control lever firmly toward the engine and tighten the swivel screw.
5. Road test the vehicle and check the shift points.

TRANSFER CASE

Transfer Case Assembly

REMOVAL & INSTALLATION

Except All Wheel Drive

1. Disconnect the negative battery cable.
2. Raise the vehicle and support safely.
3. Remove the skid plates, if equipped. Drain the transfer case fluid.
4. Disconnect the distance sensor, if equipped and disconnect the speedometer cable from the transfer case.
5. Matchmark and remove the driveshafts.
6. Disconnect the PTO, if equipped.
7. Disconnect the linkage, electrical connectors and vacuum lines from the transfer case.
8. Remove the vacuum tank and bracket on rear crossmember.
9. Using a jack, support the transfer case and remove the crossmember.
10. Unbolt the transfer case from the transmission and slide it backwards to remove it from the vehicle.

To install:

11. Install transfer case and retaining bolts. Install and connect the transfer case shift linkage.
12. Connect vacuum harness connector to switch at transfer case. Install vacuum tank and bracket on rear crossmember.
13. Connect the distance sensor, if equipped.
14. Connect the the speedometer cable.
15. Align and install the driveshafts.
16. Fill transfer case with Dexron® II to the correct level.

All Wheel Drive Power Transfer Unit (P.T.U.)

1. Raise the vehicle and support safely. Remove the tire and wheel assemblies.
2. Position a drainpan under the transaxle where the axles enter the P.T.U. Support the P.T.U.
3.. Remove the ball joint retaining bolts and pry the control arm from the steering knuckle. Unbolt the center bearing and remove the intermediate axle from the transaxle, if equipped. Remove the axles.
4. Remove front Bobble strut bolt at front crossmember.
5. Remove the mounting bolts and remove unit from vehicle.
6. The installation is the reverse of the removal procedure.

LINKAGE ADJUSTMENT

1. Move the transfer case shift lever boot aside for access to the shift lever and gate.
2. Move the shift lever into the **4H** position. Make sure the lever is against the appropriate gate. Insert a ⅛ inch spacer between the shift lever and gate and secure the lever with tape.
3. Raise the vehicle and support safely.
4. Loosen the shift rod clamp screw until the shift rod is free to slide in the swivel.
5. Verify that the lever is in the **4H** position and move it, if it has moved out of position.
6. Tighten the clamp screw.
7. Check the smoothness of operation of the transfer case.

FRONT SUSPENSION

Shock Absorbers
REMOVAL & INSTALLATION

1. Remove the upper shock nut, washer and bushing.
2. Raise and safely support the vehicle.
3. Remove the lower mounting bolts and remove the shock from the vehicle.
4. The installation is the reverse of the removal procedure.

MacPherson Strut
REMOVAL & INSTALLATION

1. Remove the 3 mounting nuts from the shock tower under the hood.

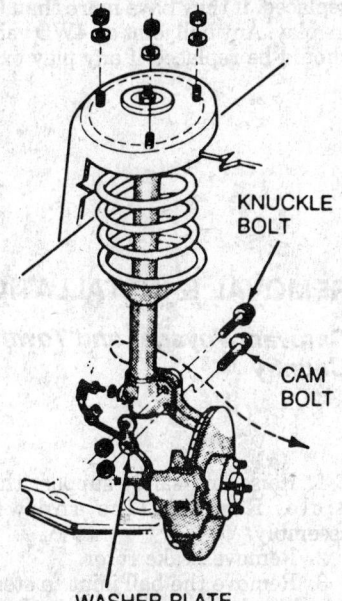

MacPherson strut assembly

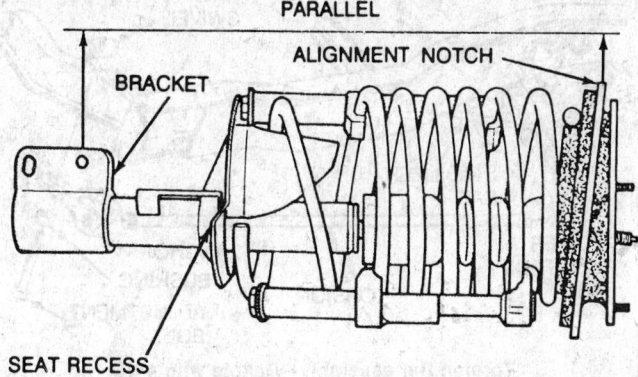

Installing and aligning the damper assembly— Caravan, Voyager and Town & Country

2. Raise the vehicle and support safely.

3. Remove the brake hose bracket screw from the strut.

4. Remove the strut to knuckle bolts, nuts and nut washer.

5. The installation is the reverse of the removal procedure. Torque the upper mounting nuts to 20 ft. lbs. (27 Nm).

6. Perform a front end alignment. Torque the strut to knuckle nuts to 75 ft. lbs. (100 Nm) plus ¼ turn.

Coil Spring

REMOVAL & INSTALLATION

Caravan, Voyager and Town & Country

1. Raise the vehicle and support safely. Remove the MacPherson strut from the vehicle.

2. Compress the coil spring using tool C-4838 or MacPherson spring compressor.

3. Remove the upper assembly retaining nut.

4. Remove the damper assembly and plastic strut tube, if equipped.

To install:

5. Install the compressed spring to the strut, making sure the spring end is seated in the recess.

6. Install the damper assembly to the strut rod.

7. Align the alignment notch or tab with the lower bracket.

8. Tighten the upper assembly retaining nut to 60 ft. lbs. (81 Nm). Make sure the spring and damper assembly are still aligned before releasing the compressor tool.

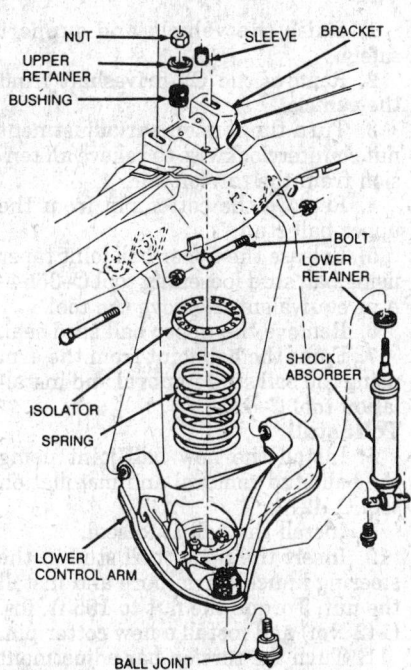

Front Suspension—Dakota with 2WD

9. Release the compressor tool.

10. Install the strut to the vehicle.

11. Perform a front end alignment.

Dakota With 2WD

1. Raise the vehicle and support safely.

2. Remove the shock absorber.

3. Disconnect the sway bar from the lower control arm, if equipped.

4. Install spring compressor tool DD-1278 or equivalent, to the coil spring and tighten the nut finger-tight, then back off half a turn.

5. Remove the cotter pin and lower ball joint nut.

6. Release the lower ball joint taper using ball stud loosening tool C-3564-A, or equivalent.

7. Remove the tool and remove the ball stud from the control arm. Release the compressor tool from the coil spring.

8. Pull the arm down and remove the spring with the rubber isolation pad from the vehicle.

To install:

9. Install the spring with the rubber isolator. Install the compressor tool and compress it enough so the lower ball joint can be inserted through the knuckle.

10. Torque the lower ball joint nut to 135 ft. lbs. (183 Nm). Install a new cotter pin. Remove the spring compressor.

11. Connect the sway bar to the lower control arm, if equipped.

12. Install the shock absorber.

Torsion Bar

REMOVAL & INSTALLATION

Dakota With 4WD

NOTE: The left and right side torsion bars are not interchangeable. The bars are identified by the letter R or L stamped into one end of the bar. The bars do not have a front or rear, though and can be installed with either end facing foward.

1. Remove the upper control arm jounce bumper.

2. Raise the vehicle with the front suspension hanging free and support safely.

3. Release the load from the torsion bar by turning the adjustment bolt counterclockwise.

4. Remove the adjustment bolt from the swivel and then remove the torsion bar and the anchor together from the vehicle. Remove the torsion bar from the anchor.

5. Remove any dirt, rust or pebbles from the hex shaped socket in the anchor and lower control arm.

6. Inspect the anchor and bolt and replace them, if any damage or corrosion exists.

To install:

7. Insert the torsion bar ends into the sockets.

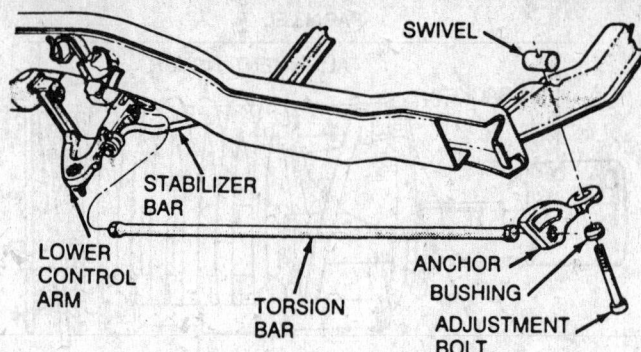

Torsion Bar assembly—Dakota with 4WD

8. Position the anchor and the bushing in the crossmember, insert the adjustment bolt and thread it into the swivel.

9. Turn the adjustment bolt clockwise to apply a load on the bar.

10. Lower the vehicle.

11. Set the front suspension height so the difference between the distance from the road surface to the lower control arm inner pivot and the distance from the road surface to the outer end of the arm is 1–1½ in.

12. Install the upper control arm jounce bumper.

Upper Ball Joint

INSPECTION

To inspect the ball joints, unload the suspension. Upper ball joints on 2WD vehicles or any ball joint on 4WD vehicles should be replaced if any play exists at all.

REMOVAL & INSTALLATION

Dakota

2WD

1. Raise the vehicle and support safely.

2. Position a support at the outer end of the lower control arm and lower the vehicle so the support compresses the coil spring.

3. Remove the tire and wheel assembly.

4. Release the upper ball joint taper using ball stud loosening tool C–3564–A or equivalent.

5. Unthread the ball joint from the control arm with tool C–3561 or equivalent.

To install:

6. Thread new ball joint into control arm with tool C–3561. Tighten ball stud with 125 ft. lbs. (170 Nm) torque. Install new seal over stud.

7. Install steering knuckle on upper ball stud. Tighten upper nut to 105 ft. lbs. (142 Nm) torque. Install new cotter pin.

8. Install wheel and tire assembly.

4WD

1. Raise the vehicle and support safely.

2. Remove the CV-driveshaft from the vehicle.

3. Turn the torsion bar adjustment nut counterclockwise to relieve all tension from the torsion bar.

4. Remove the cotter pin from the upper ball stud.

5. Release the upper ball joint taper using ball stud loosening tool C–3564–A or equivalent. Remove the tool.

6. Remove the upper ball stud seal.

7. Force the ball joint from the arm using the ball stud removal and installation tool C–4212.

To install:

8. Install the new ball joint using the ball stud removal and installation tool C–4212.

9. Install the ball stud seal.

10. Insert the upper ball stud in the steering knuckle arm bore and install the nut. Torque the nut to 105 ft. lbs. (142 Nm) and install a new cotter pin.

11. Turn the torsion bar adjustment nut clockwise to apply a load on the bar.

12. Install the CV-driveshaft and all related parts.

13. Lower the vehicle.

14. Set the front suspension height so the difference between the distance from the surface that the tires are on to the lower control arm inner pivot and the distance that the tires are on to the outer end of the arm is 1–1½ in.

Lower Ball Joint

INSPECTION

To inspect the ball joints on Caravan, Voyager and Town & Country, grasp the grease fitting by hand with the vehicle on the ground. If the grease fitting can be moved at all by hand, the ball joint should be replaced.

To inspect the lower ball joints on Dakota, unload the suspension. Lower ball joints on 2WD vehicles should be

replaced, if they have more than 0.020 in. play. Any ball joint on 4WD vehicles should be replaced if any play exists.

REMOVAL & INSTALLATION

Caravan, Voyager and Town & Country

1. Raise and safely support the vehicle. Remove tire and wheel assembly.

2. Remove brake rotor.

3. Remove the ball joint to steering knuckle clamp nut and bolt. Remove the lower ball joint stud from the steering knuckle.

4. Remove the seal from the joint. Install remover tool C–4699–2 over ball joint stud and against the joint upper housing.

5. Press the ball joint from the control arm.

To install:

6. Install ball joint onto the control arm and press into place using tool C–4699–2. Install seal on joint stud. Insert stud into steering knuckle and install knuckle clamp bolt and nut.

7. Install brake rotor, tire and wheel assembly.

Dakota

2WD

1. Raise the vehicle and support safely.

2. Remove the shock absorber.

3. Disconnect the sway bar from the lower control arm, if equipped.

4. Install spring compressor tool DD–1278 or equivalent, to the coil spring and tighten the nut finger-tight, then back off half a turn.

5. Remove the cotter pin and lower ball joint nut.

6. Release the lower ball joint taper using ball stud loosening tool C–3564–A or equivalent.

7. Remove the tool and remove the ball stud from the control arm. Release the compressor tool from the coil spring.

8. Pull the arm down and remove the spring with the rubber isolation pad from the vehicle. Remove the ball joint boot. Use tool C–4212, or an appropriate ball joint press to remove the ball joint from the arm.

To install:

9. Use the remover tool to press the

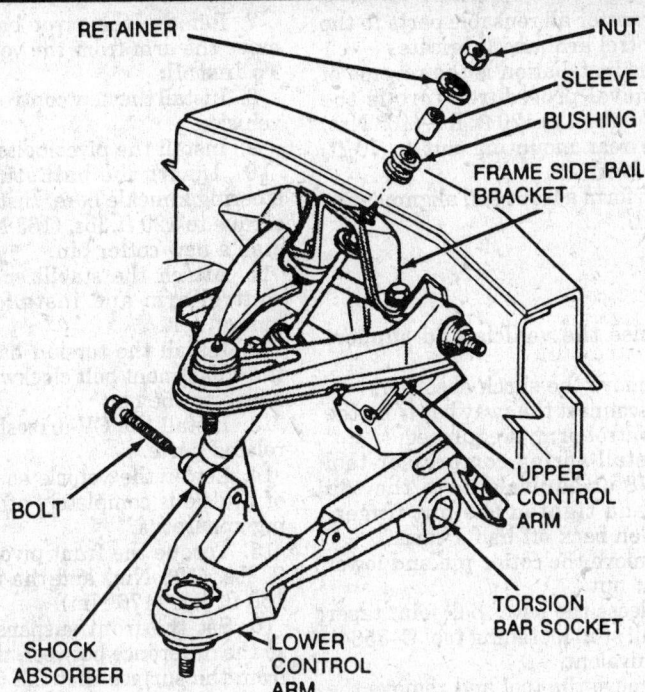

Front Suspension—Dakota with 4WD

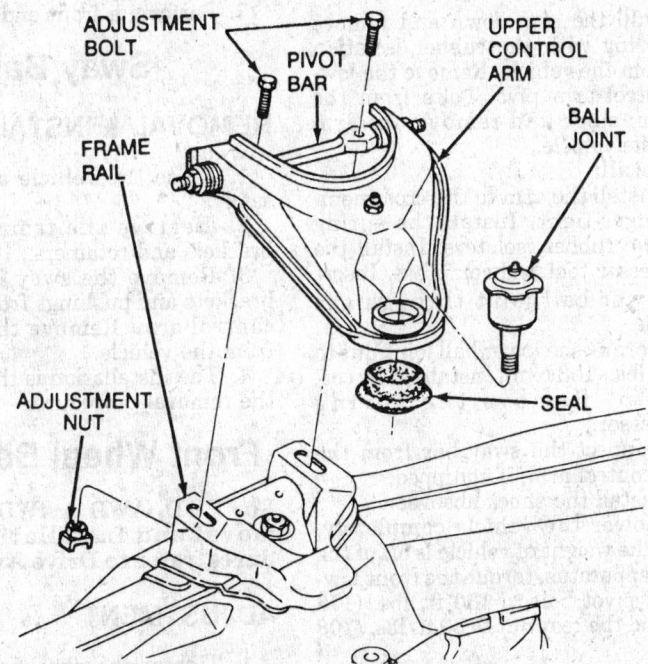

Upper control arm—Dakota with 2WD

the lower ball stud from the steering knuckle.

7. Pry the peened ball joint retainer sections upward from the lower control arm and remove the ball joint from the arm.

To install:

8. Install the new ball joint in the control arm. Peen the ball joint housing retainer over to secure the ball joint.

9. Install the grease seal.

10. Insert the ball stud into the steering knuckle bore. Torque the nut to 120 ft. lbs. (163 Nm) and install a new cotter pin.

11. Attach the stabilizer bar to the control arm and install the shock mount bolt.

12. Install the torsion bar and turn the adjustment bolt clockwise to apply a load to the bar.

13. Install the CV-driveshaft and all related parts.

14. Lower the vehicle.

15. Set the front suspension height so the difference between the distance from the surface that the tires are on to the lower control arm inner pivot and the distance that the tires are on to the outer end of the arm is 1–1½ in.

16. Perform a front end alignment.

Upper Control Arm

REMOVAL & INSTALLATION

Dakota

2WD

1. Raise the vehicle and support safely.

2. Remove the shock absorber.

3. Disconnect the sway bar from the lower control arm, if equipped.

4. Install spring compressor tool DD–1278 or equivalent, to the coil spring and tighten the nut finger-tight, then back off half a turn.

5. Remove the cotter pin and upper ball joint nut. Suspend the rotor assembly with a wire so there is not excessive pull on the brake hose.

6. Release the upper ball joint taper using ball stud loosening tool C–3564–A or equivalent.

7. Remove the tool and remove the ball stud from the control arm.

8. Remove the pivot bar retaining bolts and remove the arm from the vehicle.

To install:

9. Install the arm to the frame rail bracket and install the retaining bolts.

10. Torque the ball joint nut to 135 ft. lbs. (183 Nm). Install a new cotter pin. Remove the spring compressor.

11. Connect the sway bar from the lower control arm, if equipped.

12. Install the shock absorber.

13. Align the front end. When all set-

ball joint into the arm. Install a new rubber boot. Install the spring with the rubber isolators. Install the compressor tool and compress it enough so the lower ball joint can be inserted through the knuckle.

10. Torque the lower ball joint nut to 130 ft. lbs. (176 Nm). Install a new cotter pin. Remove the spring compressor.

11. Connect the sway bar to the lower control arm, if equipped.

12. Install the shock absorber.

4WD

1. Raise the vehicle and support safely.

2. Remove the CV-driveshaft. Remove the brake rotor, if necessary.

3. Remove the torsion bar.

4. Remove the shock absorber lower attaching bolt.

5. Disconnect the stabilizer bar from the lower control arm.

6. Remove the cotter pin and the nut from the lower ball stud. Separate

tings are at specifications, torque the pivot bar retaining bolts to 155 ft. lbs. (210 Nm).

4WD

1. Raise the vehicle and support safely.
2. Remove the CV-driveshaft from the vehicle.
3. Turn the torsion bar adjustment nut counterclockwise to relieve all tension from the torsion bar. Remove the screws that attach the brake hose bracket to the upper arm. Remove the shock absorber.
4. Remove the cotter pin from the upper ball stud.
5. Release the upper ball joint taper using ball stud loosening tool C–3564–A or equivalent. Remove the tool. Remove the ball stud from the steering knuckle.
6. Remove the pivot bar retaining bolts.
7. Remove the arm from the vehicle.

To install:

8. Position the arm at the frame rail bracket.
9. Install the pivot arm nuts and bolts.
10. Insert the upper ball stud in the steering knuckle arm bore and install the nut. Torque the nut to 125 ft. lbs. (170 Nm) and install a new cotter pin. Install the shock absorber and attach the brake hose bracket.
11. Turn the torsion bar adjustment nut clockwise to apply a load on the bar.
12. Install the CV-driveshaft and all related parts.
13. Lower the vehicle.
14. Set the front suspension height so the difference between the distance from the surface that the tires are on to the lower control arm inner pivot and the distance that the tires are on to the outer end of the arm is 1–1½ in.

Lower Control Arm

REMOVAL & INSTALLATION

Caravan, Voyager and Town & Country

1. Raise the vehicle and support safely. Remove the tire and wheel assembly.
2. Remove the sway bar.
3. Remove the ball joint stud retaining bolt and nut.
4. Pry the lower control arm from the steering knuckle.
5. Remove the control arm to crossmember bolts, nuts bushings and retainers.
6. Remove the control arm from the vehicle.

7. Transfer all reusable parts to the new control arm and lubricate.
8. The installation is the reverse of the removal procedure. Torque the front pivot bolt to 120 ft. lbs. (163 Nm) and the rear mounting nut to 70 ft. lbs. (95 Nm).
9. Perform a front end alignment as required.

Dakota

2WD

1. Raise the vehicle and support safely.
2. Remove the shock absorber.
3. Disconnect the sway bar from the lower control arm, if equipped.
4. Install spring compressor tool DD–1278 or equivalent, to the coil spring and tighten the nut finger-tight, then back off half a turn.
5. Remove the cotter pin and lower ball joint nut.
6. Release the lower ball joint taper using ball stud loosening tool C–3564–A or equivalent.
7. Remove the tool and remove the ball stud from the control arm. Release the compressor tool from the coil spring.
8. Pull the arm down and remove the spring with the rubber isolation pad from the vehicle. Remove the lower control arm pivot bolts from the crossmember and remove the arm from the vehicle.

To install:

9. Install the arm to the crossmember finger-tight. Install the spring with the rubber isolators. Install the compressor tool and compress. Insert the lower ball joint through the knuckle.
10. Torque the lower ball joint nut to 135 ft. lbs. (183 Nm). Install a new cotter pin. Remove the spring compressor.
11. Connect the sway bar from the lower control arm, if equipped.
12. Install the shock absorber.
13. Lower the vehicle completely. When the weight of vehicle is off of the lifting apparatus, torque the front lower arm pivot bolt to 130 ft. lbs. (176 Nm) and the rear nut to 80 ft. lbs. (108 Nm).
14. Align the front end as required.

4WD

1. Raise the vehicle and support safely.
2. Remove the CV-driveshaft.
3. Remove the torsion bar.
4. Remove the shock absorber lower attaching bolt.
5. Disconnect the stabilizer bar from the lower control arm.
6. Remove the cotter pin and the nut from the lower ball stud. Separate the lower ball stud from the steering knucle.

7. Remove the pivot bolts and remove the arm from the vehicle.

To install:

8. Install the new control arm to the vehicle.
9. Install the pivot bolts.
10. Insert the ball stud into the steering knuckle bore. Install the nut, torque to 120 ft. lbs. (163 Nm) and install a new cotter pin.
11. Attach the stabilizer bar to the control arm and install the shock mount bolt.
12. Install the torsion bar and turn the adjustment bolt clockwise to apply a load to the bar.
13. Install the CV-driveshaft and all related parts.
14. Lower the vehicle so the weight of vehicle is completely off of the lifting apparatus.
15. Torque the front pivot nut to 80 ft. lbs. (108 Nm) and the rear nut to 130 ft. lbs. (176 Nm).
16. Set the front suspension height so the difference between the distance from the surface that the tires are on to the lower control arm inner pivot and the distance that the tires are on to the outer end of the arm is 1–1½ in.
17. Perform a front end alignment.

Sway Bar

REMOVAL & INSTALLATION

1. Raise the vehicle and support safely.
2. Remove the front sway bar brackets and retainers.
3. Remove the sway bar support brackets and bushings from the lower control arm. Remove the sway bar from the vehicle.
4. The installation is the reverse of the removal procedure.

Front Wheel Bearings

For FWD, AWD & 4WD model removal and installation procedures, refer to Drive Axle section.

ADJUSTMENT

1. Tighten the wheel bearing nut to 20–25 ft. lbs. (27–34 Nm) while turning the rotor.
2. Loosen the wheel bearing adjusting nut completely.
3. Tighten the nut finger-tight.
4. Check the wheel bearing endplay. The specification is 0.0001–0.003 in.
5. Install the nut lock and cotter pin.

REMOVAL & INSTALLATION

1. Raise the vehicle and support safely.

2. Remove the tire and wheel assembly.

3. Remove the caliper and disc brake pads.

4. Remove the dust cap.

5. Remove the cotter pin, casteled nut lock, wheel bearing nut and washer from the spindle.

6. Remove the outer wheel bearing.

7. Remove the rotor with the inner wheel bearing from the spindle. Remove the grease seal.

To install:

8. Lubricate and install the inner wheel bearing. Install a new grease seal.

9. Install the rotor to the spindle.

10. Lubricate and install the outer wheel bearing, washer and nut. When the bearing preload is properly set, install the nut lock and a new cotter pin.

11. Install the grease cap.

12. Install the brake pads and caliper.

13. Install the wheel.

REAR SUSPENSION

Shock Absorber

REMOVAL & INSTALLATION

1. Raise the vehicle and support safely.

2. Remove the bolts that attach the shock to the frame or bracket.

3. Remove the shock from the vehicle.

4. The installation is the reverse of the removal procedure.

Leaf Springs

REMOVAL & INSTALLATION

Caravan, Voyager and Town & Country

1. Raise the vehicle and support safely.

2. Disconnect the actuator valve for the height sensing proportional valve.

3. Using floor stands under the axle assembly, raise the rear axle assembly to relieve weight on the rear springs.

4. Disconnect the shock absorbers from the axle brackets.

5. Remove the U-bolt nuts and washers and remove the U-bolts.

6. Lower the rear axle assembly allowing the spring to hang free.

7. Remove the bolts from the front hanger.

8. Remove the rear spring shackle

nuts and plate and remove the shackle from the spring.

9. Remove the front pivot bolt from the front spring hanger and remove the springs from the vehicle.

To install:

10. Assembly the shackle bushings and plate on the rear of the spring and ear spring hanger and start the shackle bolts.

11. Assembly the front spring hanger to the front of the spring eye and install the pivot bolt and nut. The pivot bolt must face inboard to prevent structural damage during installation.

12. Raise the front of the spring to the vehicle and install the 4 hanger bolts and torque to 45 ft. lbs. (61 Nm). Connect the actuator assembly for the height sensing proportioning valve.

13. Raise the axle assembly to the correct position with the axle centered under the spring center bolt.

14. Install the U-bolts, washers and nuts and torque the nuts to 60 ft. lbs. (81 Nm).

15. Install the shock absorbers.

16. Lower the vehicle so the full weight of vehicle is off of the lifting apparatus.

17. Torque the front pivot bolts to 105 ft. lbs. (142 Nm).

18. Torque the shackle nuts to 45 ft. lbs. (61 Nm).

19. Adjust the height sensing proportioning valve.

Dakota

1. Raise the vehicle and support safely.

2. Using the proper equipment, support the weight of the axle.

3. Remove the nuts, washers and U-bolts attaching the springs to the axle housing. Remove the spacer.

4. Remove the spring shackle bolts, shackle and spring front bolt.

5. Remove the springs and auxiliary spring, if equipped, from the vehicle.

6. The installation is the reverse of the removal procedure.

Rear Wheel Bearings

For rear wheel bearing removal and installation procedures on RWD, refer to Drive Axle section.

REMOVAL & INSTALLATION

Caravan, Voyager and Town & Country

EXCEPT ALL WHEEL DRIVE VEHICLE

1. Raise the vehicle and support safely.

2. Remove the tire and wheel assembly.

3. Remove the dust cap.

4. Remove the cotter pin, nut lock and nut.

5. Remove the thrust washer and the outer wheel bearing.

6. Remove the drum with the inner wheel bearing and the grease seal.

7. Remove the grease seal and remove the inner bearing.

To install:

8. Lubricate the inner bearing and install to the drum.

9. Install a new grease seal.

10. Install the drum to the vehicle.

11. Lubricate and install the outer wheel bearing to the spindle.

12. Install the thrust washer.

13. Install and tighten the wheel bearng nut to 20–25 ft. lbs. (27–34 Nm) while rotating the drum.

14. Back off the adjusting nut ¼ turn then tighten it finger-tight.

15. Install the nut lock and a new cotter pin.

ALL WHEEL DRIVE VEHICLE

The rear wheel bearings and hub are serviced as an assembly.

1. Raise the vehicle and support safely.

2. Remove the tire and wheel assembly.

3. Remove the halfshaft flange retaining bolts. Compress inner halfshaft joint and pull downward to clear rear carrier assembly. Remove the halfshaft.

NOTE: The halfshaft acts as a bolt and secures the hub/bearing assembly. If vehicle is to be supported or moved on its wheels, install and torque a bolt through the hub. This will assure no damage is done to the bearing/hub assembly.

4. Remove the wheel bearing retainer bolts and remove hub/bearing assembly from the vehicle.

To install:

5. Install new hub/bearing assembly with new gasket onto the vehicle.

6. Install the retaining screws and tighten to 96 ft. lbs. (130 Nm).

7. Install halfshaft, washer and hub nut after cleaning foreign material from threads. Torque nut to 180 ft. lbs. (244 Nm).

8. Install spring washer, nut lock and new cotter pin. Install the tire and wheel assembly.

Rear Axle Assembly

For rear axle assembly removal and installation procedures on RWD and AWD, refer to Drive Axle section.

REMOVAL & INSTALLATION

Caravan, Voyager and Town & Country

1. Raise the vehicle and support safely.
2. Disconnect the brake hose connection from the axle tube.
3. Unclip the brake tubes from the axle housing.
4. Support the axle tube with jack stands.
5. Unbolt the shock absorbers and the sway bar bushing retainers from the axle tube.
6. Unbolt the axle tube from the leaf springs.
7. Remove the axle from the vehicle.
8. The installation is the reverse of the removal procedure.

STEERING

Note: On 1992 models equipped with an Air bag, failure to disarm the system may result in deployment of the air bag and possible personal injury. To disarm the air bag, disconnect the negative battery. Allow system capacitor to discharge for 2 minutes before working on system.

Steering Wheel

REMOVAL & INSTALLATION

1. Disconnect the negative battery cable. Disarm Air bag, if equipped.
2. Straighten the steering wheel so the front tires are pointing straight forward. Remove the horn pad and related parts.
3. Remove the steering wheel holddown nut. Matchmark the steering wheel to the shaft.
4. Using a steering wheel puller, pull the steering wheel off of the shaft.
5. The installation is the reverse of the removal procedure.

Steering Column

REMOVAL & INSTALLATION

1. Disconnect the negative battery cable.
2. Remove trim bezel, steering column cover and lower reinforcement, as required.
3. Disconnect all wiring connectors from below the instrument panel that lead up into the steering column.
4. Disconnect gear indicator cable and lock bar from column as required.

5. Remove the nuts that attach the steering column assembly to the instrument panel support.
6. Firmly grasp the steering wheel and pull the steering column out, separating the stub shaft from the steering gear coupling.
7. The installation is the reverse of the removal procedure.
8. Connect the negative battery cable and check the steering column and all related components for proper operation.

Manual Steering Gear

REMOVAL & INSTALLATION

Dakota With 4WD

1. Disconnect the negative battery cable.
2. Raise the vehicle and support safely. Center the steering.
3. Disconnect the pitman arm from the center link.
4. Disconnect the steering gear to shaft coupling.
5. Remove the steering gear mount bolts and remove the gear from the vehicle.
6. Matchmark and remove the pitman arm from the sector shaft.
7. The installation is the reverse of the removal procedure. Torque the pitman arm nut to 175 ft. lbs. (237 Nm) and the mount bolts to 100 ft. lbs. (136 Nm).

ADJUSTMENT

1. If the vehicle wanders of the steering has too much play, the sector shaft can be adjusted.
2. Loosen the adjusting screw locknut and turn the screw all the down.
3. Back the screw off ¼–½ turn.
4. Tighten the locknut.
5. Road test the vehicle. If the steering wheel does not return easily after a turn, back the screw off until the wheel returns easily.

Manual Rack and Pinion

REMOVAL & INSTALLATION

Caravan, Voyager and Town & Country

1. Disconnect the negative battery cable.
2. Raise the vehicle and support safely. Remove the front wheel assemblies.
3. Remove the tie rod ends from the steering knuckles.
4. Remove the steering column from vehicle.

5. On All Wheel Drive models, remove the 2 bolts and nuts attaching the bridge to the front crossmember.
6. If equipped, remove the the air diverter valve bracket, from the left side of the crossmember.
7. Remove the front suspension crossmember attaching bolts and nuts.
8. Lower the crossmember.
9. Remove the lower steering column coupler from the steering gear. Remove the steering coupler roll pin and separate coupler from steering gear shaft.
10. Remove the axle boot shield, if equipped.
11. Remove the steering gear bolts from the front suspension crossmember.
12. Remove the steering gear from the left side of the vehicle.
To install:
13. Install the steering gear from the left side of the vehicle. Install the steering gear bolts in the front suspension crossmember.
14. Install the lower steering column coupler to the steering gear. Install the steering coupler roll pin.
15. Install the lower crossmember and the attaching bolts. The right rear crossmember bolt is a pilot bolt that correctly locates the crossmember, tighten it first. Torque the crossmember bolts to 90 ft. lbs.
16. On All Wheel Drive models, install the 2 bolts and nuts attaching the bridge to the front crossmember.
17. Install the steering column. Install the tie rod ends from the steering knuckles.
18. Install the tire and wheel assemblies on the vehicle.

Dakota With 2WD

1. Disconnect the negative battery cable.
2. Remove the tie rod ends from the steering knuckle.
3. Remove the coupling pin.
4. Remove the bolts attaching the rack to the crossmember.
5. Remove the gear from the vehicle.
6. The installation is the reverse of the removal procedure. Torque the mounting bolts to 150 ft. lbs. (203 Nm).

Power Steering Gear

REMOVAL & INSTALLATION

Dakota With 4WD

1. Disconnect the negative battery cable.
2. Raise the vehicle and support safely. Center the steering.

3. Disconnect the pitman arm from the center link.

4. Disconnect and plug the pressure and return hoses.

5. Disconnect the steering gear to shaft coupling.

6. Remove the steering gear mount bolts and remove the gear from the vehicle.

7. Matchmark and remove the pitman arm from the sector shaft.

8. The installation is the reverse of the removal procedure. Torque the pitman arm nut to 175 ft. lbs. (237 Nm) and the mount bolts to 100 ft. lbs. (136 Nm).

9. Refill the power steering pump.

ADJUSTMENT

1. If the vehicle wanders of the steering has too much play, the sector shaft can be adjusted.

2. Loosen the adjusting screw locknut and turn the screw all the down.

3. Back the screw off ¼–½ turn.

4. Tighten the locknut.

5. Road test the vehicle. If the steering wheel does not return easily after a turn, back the screw off until the wheel returns easily.

Power Rack and Pinion

REMOVAL & INSTALLATION

Caravan, Voyager and Town & Country

1. Disconnect the negative battery cable.

2. Raise the vehicle and support safely. Remove the front wheel assemblies.

3. Remove the tie rod ends from the steering knuckles.

4. Remove the steering column from vehicle.

5. On All Wheel Drive models, remove the 2 bolts and nuts attaching the bridge to the front crossmember.

6. If equipped, remove the the air diverter valve bracket, from the left side of the crossmember.

7. Disconnect and plug the oil lines from the rack.

8. Remove the front suspension crossmember attaching bolts and nuts. Lower the crossmember.

9. Remove the lower steering column coupler from the steering gear. Remove the steering coupler roll pin and separate coupler from steering gear shaft.

10. Remove the axle boot shield, if equipped.

11. Remove the steering gear bolts from the front suspension crossmember.

12. Remove the steering gear from the left side of the vehicle.

To install:

13. Install the steering gear from the left side of the vehicle. Install the steering gear bolts in the front suspension crossmember.

14. Install the lower steering column coupler to the steering gear. Install the steering coupler roll pin.

15. Install the lower crossmember and the attaching bolts. The right rear crossmember bolt is a pilot bolt that correctly locates the crossmember, tighten it first. Torque the crossmember bolts to 90 ft. lbs.

16. On All Wheel Drive models, install the 2 bolts and nuts attaching the bridge to the front crossmember.

17. Install the steering fluid lines. Refill the power steering pump.

18. Reinstall the steering column. Insert the tie rod ends into the steering knuckles.

19. Install the tire and wheel assemblies on the vehicle.

Dakota With 2WD

1. Disconnect the negative battery cable.

2. Remove the tie rod ends from the steering knuckle.

3. Disconnect and plug the power steering fluid lines.

4. Remove the coupling pin.

5. Remove the bolts attaching the rack to the crossmember.

6. Remove the gear from the vehicle.

7. The installation is the reverse of the removal procedure. Torque the mounting bolts to 150 ft. lbs. (203 Nm).

Power Steering Pump

REMOVAL & INSTALLATION

1. Disconnect the negative battery cable.

2. Position a drainpan under the power steering pump.

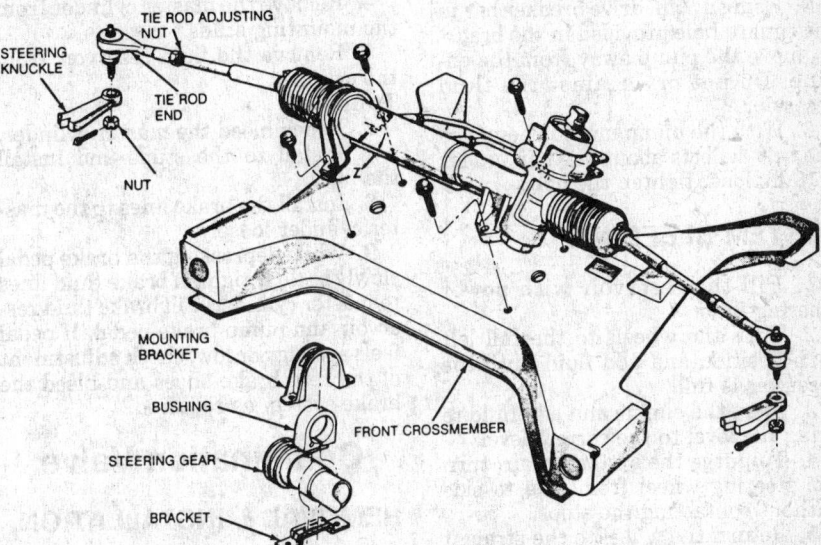

Rack and pinion steering gear mounting—Caravan, Voyager and Town & Country

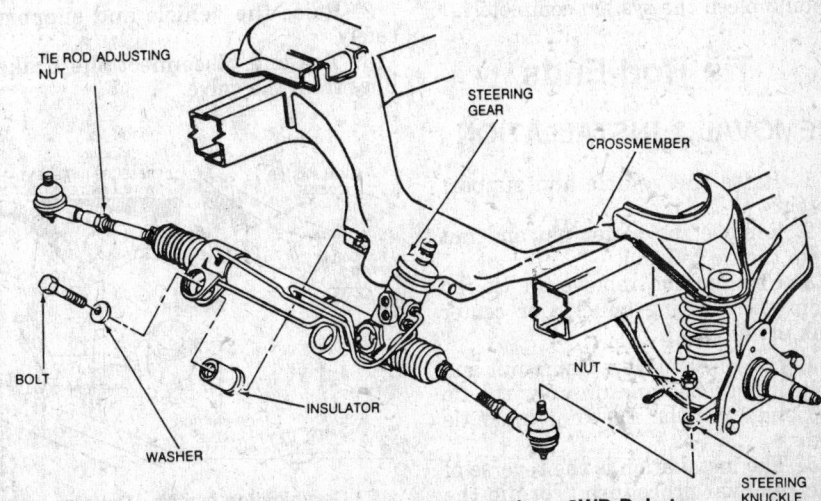

Rack and pinion steering gear mounting—2WD Dakota

3. Disconnect the fluid hoses from the pump and plug them.

4. Remove the front bracket attaching bolts and remove the belt from the pulley.

5. Remove the rear pump to bracket nut and remove the pump.

6. Remove the bracket from the pump.

7. Remove the pulley from the pump with the proper puller. Install the pulley on the new pump using the installation tools.

8. The installation is the reverse of the removal procedure.

BELT ADJUSTMENT

NOTE: The belt tension is automatically adjusted by a dynamic tensioner on the 3.0L and 3.3L engines and late 1991 3.9L and 5.2L engines. Adjustment is not possible.

1. Loosen the bracket mounting bolts.

2. Using a ½in. drive breaker bar in the square hole provided in the bracket, move the pump away from the engine. Do not pry against the fluid reservoir.

3. With the pump moved enough so the belt deflects about ¼–½ in. under a 10 lb. load, tighten the bolts.

SYSTEM BLEEDING

1. Fill the reservoir with power steering fluid.

2. Turn the wheels to the full left turn position and add fluid until the reservoir is full.

3. Start the engine and add fluid to bring the level to the correct level.

4. To purge the system of air, turn the steering wheel from side to side without contacting the stops.

5. Return the wheel to the straight ahead position and operate the engine for 2 minutes before road testing. This should bleed the system completely.

Tie Rod Ends

REMOVAL & INSTALLATION

1. Raise the vehicle and support safely.

2. Remove the cotter pin and nut from the tie rod end.

3. Using puller, remove the tie rod from the steering knuckle or center link, if equipped.

4. Loosen the sleeve clamp nut and bolt, if equipped and unscrew the tie rod end from the sleeve or inner tie rod.

5. The installation is the reverse of the removal procedure. Torque the

stud nuts to 45 ft. lbs. (61 Nm) and install a new cotter pin.

6. Perform a front end alignment as required.

BRAKES

For all brake system repair and service procedure not detained below, please refer to "Brakes" in the Unit Repair section.

Master Cylinder

REMOVAL & INSTALLATION

1. Disconnect the negative battery cable.

2. Disconnect and plug the brake lines from the master cylinder.

3. Remove the nuts attaching the master cylinder to the power booster.

4. Remove the master cylinder from the mounting studs.

5. Remove the fluid reservoir from the cylinder.

To install:

6. Bench bleed the master cylinder.

7. Install to the studs and install the nuts.

8. Install the brake lines to the master cylinder loosely.

9. While depressing the brake pedal slowly to floor tighten brake fluid lines to master cylinder. Fill brake fluid reservoir and pump brake pedal. If pedal feels spoongy or low, check adjustment of the rear brake shoes and bleed the brake system as required.

Combination Valve

REMOVAL & INSTALLATION

1. Disconnect the negative battery cable.

2. Raise the vehicle and support safely.

3. Tag and disconnect the brake lines from the valve.

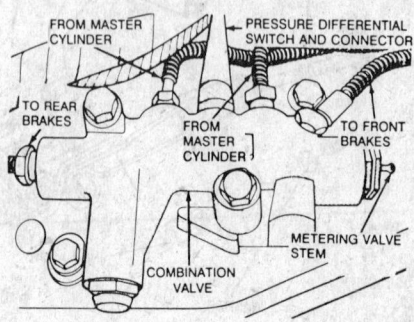

Combination switch—Dakota

4. Disconnect the wires to the pressure switch.

5. Remove the combination valve from the frame bracket.

6. The installation is the reverse of the removal procedure.

7. Bleed the brakes in the following order:

    a. Rear wheel anti-lock valve, if equipped

    b. Right rear wheel

    c. Left rear wheel

    d. Right front wheel

    e. Left front wheel

Height Sensing Proportioning Valve Assembly

REMOVAL & INSTALLATION

Caravan, Voyager and Town & Country

1. Disconnect the negative battery. Remove the rear stone shield.

2. Remove the actuator assembly.

3. Tag and disconnect the brake lines from the valve.

4. Remove the valve from the frame bracket.

5. The installation is the reverse of the removal procedure.

ADJUSTMENT

1. With the vehicle lifted so the rear suspension is hanging free, disconnect the shock absorbers.

2. Remove the rear tire and wheel assemblies and loosen both front spring hanger pivot bolts.

3. Make sure the actuator hook is properly seated on the valve lever. Loosen the adjustment nut on the actuator assembly.

4. Pull the actuator assembly toward the spring hanger until the valve lever bottoms on the valve body and hold the position.

5. Tighten the adjustment nut to 25 inch lbs. This will complete the adjustment.

6. Tighten the spring bolts, install the shocks and wheels and road test the vehicle.

Power Brake Booster

REMOVAL & INSTALLATION

1. Disconnect the negative battery cable. Disconnect the vacuum hose(s) from the booster.

2. Remove the nuts attaching the master cylinder to the booster and move the master cylinder to the side.

3. From inside of the vehicle, re-

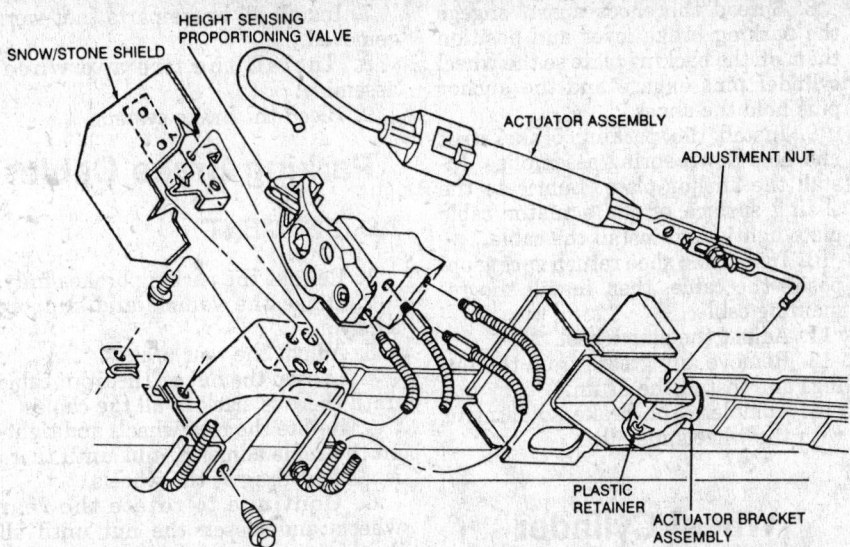

Height sensing proportioning valve—Caravan, Voyager and Town & Country

move the clip that secures the booster pushrod to the brake pedal.

4. Remove the nuts that attach the booster to the dash panel and remove it from the vehicle.

5. Transfer the check valve to the new booster.

6. The installation is the reverse of the removal procedure.

Brake Caliper

REMOVAL & INSTALLATION

1. Raise the vehicle and support safely.

2. Remove the tire and wheel assembly. Remove the hold-down spring on Dakota.

3. Remove the caliper mounting pin(s).

4. Lift the caliper off of the rotor. Remove the outer pad from the caliper.

5. Remove the brake hose retaining bolt from the caliper.

To install:

6. Install the brake hose to the caliper using new copper washers.

7. Position the caliper over the rotor so the caliper engages the adaptor correctly. Install the mounting pin(s). Install the hold-down spring, if equipped.

8. Fill the master cylinder and bleed the brakes.

Disc Brake Pads

REMOVAL & INSTALLATION

1. Remove ½ of the brake fluid from the master cylinder resrevoir.

2. Raise the vehicle and support safely. Remove the tire and wheel assemblies.

3. Remove the hold-down spring on Dakota. Remove the caliper and remove the outer pad from the caliper.

4. Remove the inner pad from the adaptor.

To install:

5. Use a large C-clamp to compress the piston back into the caliper bore.

6. Install the inner pad to the adaptor.

7. Position the caliper over the rotor so the caliper engages the adaptor correctly and install the retainer pin(s).

8. Install the hold-down spring, if equipped.

9. Refill the master cylinder.

Brake Rotor

REMOVAL & INSTALLATION

Caravan, Voyager and Town & Country

1. Raise the vehicle and support safely. Remove the tire and wheel assembly.

2. Remove the caliper and brake pads.

3. Remove the factory installed clips, if equipped. It is not necessary to reinstall these clips.

4. Remove the rotor from the hub.

5. The installation is the reverse of the removal procedure.

Dakota

1. Raise the vehicle and support safely.

2. Remove the wheel and tire assembly.

3. Remove the caliper and disc brake pads.

4. Remove the dust cap.

5. Remove the cotter pin, nut lock,

wheel bearing nut and washer from the spindle.

6. Remove the outer wheel bearing.

7. Remove the rotor and inner wheel bearing from the spindle. Remove the grease seal and inner wheel bearing from the back of the rotor.

To install:

8. Lubricate and install the inner wheel bearing into race in the rotor. Install a new grease seal.

9. Lubricate the spindle and install the rotor onto the spindle.

10. Lubricate and install the outer wheel bearing, washer and nut. When the bearing preload is properly set, install the nut lock and a new cotter pin.

11. Install the grease cap.

12. Install the brake pads and caliper.

13. Install the wheel and tire assembly.

Brake Drums

REMOVAL & INSTALLATION

Caravan, Voyager and Town & Country

1. Raise the vehicle and support safely.

2. Remove the wheel and tire assembly.

3. Remove the dust cap.

4. Remove the cotter pin and nut lock.

5. Remove the wheel bearing nut and washer from the spindle.

6. Remove the outer wheel bearing.

7. Remove the drum with the inner wheel bearing from the spindle. If the drum is difficult to remove, remove the plug from the rear of the backing plate and push the self adjuster lever away from the star wheel. Rotate the star wheel to retract the shoes. Remove the drum. Remove the grease seal from the drum.

To install:

8. Lubricate and install the inner wheel bearing in drum. Install a new grease seal.

9. Install the drum to the spindle.

10. Lubricate and install the outer wheel bearing, washer and nut. When the bearing preload is properly set, install the nut lock and a new cotter pin.

11. Install the grease cap.

12. Install the wheel and tire assembly. Adjust the rear brakes as required.

Dakota

1. Raise the vehicle and support safely.

2. Remove the wheel.

3. Remove the factory clips from the wheel studs, if equipped. It is not necessary to reinstall these clips.

4. Remove the drum. If the drum is

difficult to remove, remove the plug from the rear of the backing plate and push the self adjuster lever away from the star wheel. Rotate the star wheel to retract the shoes and remove the drum.

5. The installation is the reverse of the removal procedure.

6. Adjust the rear brakes as required.

Brake Shoes

REMOVAL & INSTALLATION

1. Raise the vehicle and support safely. Remove the wheel and tire assemblies and the brake drums.

2. Remove the primary and secondary shoe return springs from the anchor pin.

3. Lift the adjuster lever and disconnect the actuator cable.

4. Remove the shoe retainers and springs.

5. Remove the shoes (held together by the lower spring) while separating the parking brake actuating lever from the shoe with a twisting motion.

To install:

6. Thoroughly clean and dry the backing plate. To prepare the backing plate, lubricate the bosses, anchor pin and parking brake actuating lever pivot surface lightly with lithium based grease.

7. Remove, clean and dry all parts still on the old shoes. Lubricate the star wheel shaft threads with anti-sieze lubricant and transfer all parts to their proper locations on the new shoes.

8. Spread the shoes apart, engage the parking brake lever and position them on the backing plate so the wheel cylinder pins engage and the anchor pins hold the shoes.

9. Install the parking brake strut and hold-down spring assemblies. Install the anchor plate. Lubricate the sliding surface of the actuator cable plate lightly and install the cable.

10. Install the shoe return spring opposite the cable, then install the remaining cable.

11. Adjust the star wheel.

12. Remove any grease from the linings and install the drum.

13. Complete the brake adjustment with the wheels installed.

Wheel Cylinder

REMOVAL & INSTALLATION

1. Raise the vehicle and support safely.

2. Remove the wheel, drum and brake shoes.

3. Remove the brake line from the wheel cylinder.

4. Remove the wheel cylinder bolts and remove the cylinder from the backing plate.

To install:

5. Apply a very thin coating of silicone sealer to the cylinder mounting surface, install the cylinder to the backing plate and install the retaining bolts.

6. Connect the brake line to the wheel cylinder.

7. Install all brake parts that were removed.

8. Install the tire and wheel assembly.

9. Bleed the brake system.

Parking Brake Cable

ADJUSTMENT

1. Release the parking brakes fully.

2. Raise the vehicle and support safely.

3. Adjust the rear brakes.

4. Loosen the nut on the front cable until there is slack in all the cables.

5. Rotate the rear wheels and tighten the cable adjusting nut until there is a slight drag at the wheels.

6. Continue to rotate the rear wheels and loosen the nut until all drag is eliminated.

7. Back off the nut an additional 2 turns.

8. Apply and release the parking brake several times. Upon the least release, verify that there is no drag at the wheels.

9. To check the operation, make sure the parking brake holds on an incline.

REMOVAL & INSTALLATION

Front Parking Brake Cable

1. Raise the vehicle and support safely. Remove the front cable adjusting nut.

2. Remove the clip securing the cable to the anchor bracket, if equipped and slide the cable out of the bracket.

3. Remove the retainer attaching the cable to the pedal assembly frame. Disengage the cable from the pedal clevis.

4. Remove the cable grommet from the floor pan and remove the cable.

5. The installation is the reverse of the removal procedure.

Intermediate Brake Cable

1. Raise the vehicle and support safely.

2. Loosen the parking brake adjuster until cable is slack.

3. Disengage the adjuster hook from the frame rail.

4. Disengage the intermediate cable from the connectors and the right side cable guide.

5. Remove the cable.

6. The installation is the reverse of the removal procedure.

Rear Parking Brake Cable

1. Release the parking brake.

2. Raise the vehicle and support safely.

3. Remove the adjusting nut from the front cable.

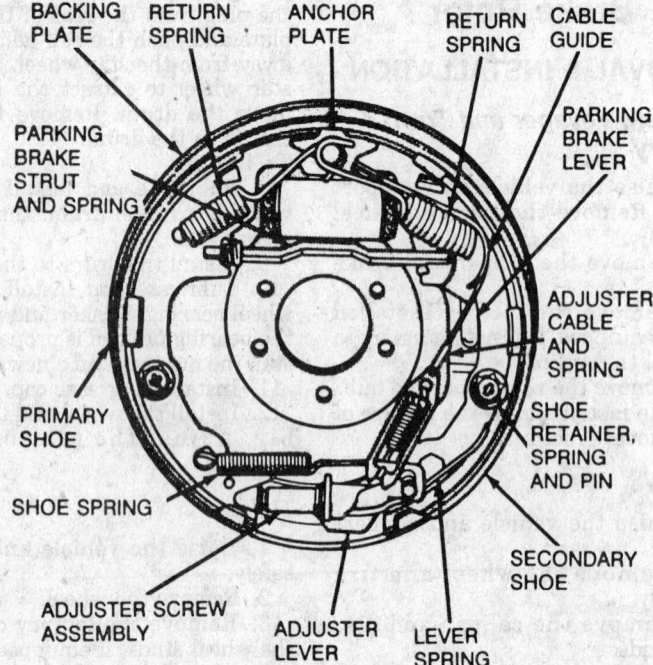

PARKING BRAKE STRUT AND SPRING

BACKING PLATE

RETURN SPRING

ANCHOR PLATE

RETURN SPRING

CABLE GUIDE

PARKING BRAKE LEVER

ADJUSTER CABLE AND SPRING

SHOE RETAINER, SPRING AND PIN

SECONDARY SHOE

PRIMARY SHOE

SHOE SPRING

ADJUSTER SCREW ASSEMBLY

ADJUSTER LEVER

LEVER SPRING

Typical rear brake shoe assembly

4. Remove the brake drums and the shoes, if necessary. Disconnect the cable from the lever and compress the cable retainer tabs. Remove the cable from the backing plate.

5. Remove the cable from the guides connectors or equalizer and remove the retaining clips from the frame bracket.

6. The installation is the reverse of the removal procedure.

Brake System Bleeding

Except Anti-Lock Brakes

NOTE: If using a pressure bleeder, follow the instructions furnished with the unit and choose the correct adaptor for the application. Do not substitute an adapter that "almost fits" as it will not work and could be dangerous.

MASTER CYLINDER

If the master cylinder is off the vehicle it can be bench bled.

1. Connect 2 short pieces of brake line to the outlet fittings, bend them until the free end is below the fluid level in the master cylinder reservoirs.

2. Fill the reservoir with fresh brake fluid. Pump the piston slowly until no more air bubbles appear in the reservoirs.

3. Disconnect the 2 short lines, refill the master cylinder and securely install the cylinder caps.

4. If the master cylinder is on the vehicle, it can still be bled, using a flare nut wrench.

5. Open the brake lines slightly with the flare nut wrench while pressure is applied to the brake pedal by a helper inside the vehicle.

6. Be sure to tighten the line before the brake pedal is released.

7. Repeat the process with both lines until no air bubbles come out.

CALIPERS AND WHEEL CYLINDERS

1. Fill the master cylinder with fresh brake fluid. Check the level often during the procedure.

2. Starting with the right rear wheel, remove the protective cap from the bleeder, and place where it will not be lost. Clean the bleed screw.

—— **CAUTION** ——

When bleeding the brakes, keep face away from the brake area. Spewing fluid may cause facial and/or visual damage. Do not allow brake fluid to spill on the car's finish; it will remove the paint.

3. If the system is empty, the most effecient way to get fluid down to the wheel is to loosen the bleeder about ½–¾ turn, place a finger firmly over the bleeder and have a helper pump the brakes slowly until fluid comes out the bleeder. Once fluid is at the bleeder, close it before the pedal is released inside the vehicle.

NOTE: If the pedal is pumped rapidly, the fluid will churn and create small air bubbles, which are almost impossible to remove from the system. These air bubbles will eventually congregate and a spongy pedal will result.

4. Once fluid has been pumped to the caliper or wheel cylinder, open the bleed screw again, have the helper press the brake pedal to the floor, lock the bleeder and have the helper slowly release the pedal. Wait 15 seconds and repeat the procedure (including the 15 second wait) until no more air comes out of the bleeder upon application of the brake pedal. Remember to close the bleeder before the pedal is released inside the vehicle each time the bleeder is opened. If not, air will be induced into the system.

5. If a helper is not available, connect a small hose to the bleeder, place the end in a container of brake fluid and proceed to pump the pedal from inside the vehicle until no more air comes out the bleeder. The hose will prevent air from entering the system.

6. Repeat the procedure on remaining wheel cylinders in order:
   a. Left rear
   b. Right front
   c. Left front

7. Hydraulic brake systems must be totally flushed if the fluid becomes contaminated with water, dirt or other corrosive chemicals. To flush, bleed the entire system until all fluid has been replaced with new fluid.

8. Install the bleeder cap(s) on the bleeder to keep dirt out. Always road test the vehicle after brake work of any kind is done.

Anti-Lock Brakes
PRESSURE BLEEDING

The brake lines may be pressure bled, using a standard diaphragm type pressure bleeder. Only diaphragm type pressure bleeding equipment should be used to bleed the system.

1. The ignition should be turned OFF and remain OFF throughout this procedure.

2. Depressurize the hydraulic accumulator.

—— **CAUTION** ——

Failure to depressurize the hydraulic accumulator, prior to performing this operation may result in personal injury and/or damage to the painted surfaces.

3. Remove the electrical connector from fluid level sensor on the reservoir cap(s) and remove the reservoir cap(s).

4. Install the pressure bleeder adapter.

5. Attach the bleeding equipment to the bleeder adapter. Charge the pressure bleeder to approximately 20 psi (138 kPa).

6. Connect a transparent hose to the bleed screw. Submerge the free end of the hose in a clear glass container, which is partially filled with clean, fresh brake fluid.

7. With the pressure turned ON, open the bleed screw ½–¾ turn and allow fluid to flow into the container. Leave the bleed screw open until clear, bubble-free fluid slows from the hose. If the reservoir has been drained or the hydraulic assembly removed from the vehicle prior to the bleeding operation, slowly pump the brake pedal 1–2 times while the bleed screw is open and fluid is flowing. This will help purge air from the hydraulic assembly. Tighten the bleeder screw to 7.5 ft. lbs. (10 Nm).

8. Repeat Step 7 at all wheels. Wheels should be bled in the following order:
   a. Left rear
   b. Right rear
   c. Left front
   d. Right front

9. After bleeding all 4 wheels, remove the pressure bleeding equipment and bleeder adapter by closing the pressure bleeder valve and slowly unscrewing the bleeder adapter from the hydraulic assembly reservoir. Failure to release pressure in the reservoir will cause spillage of brake fluid and could result in injury or damage to painted surfaces.

10. Using a syringe or equivalent method, remove excess fluid from the reservoir to bring the fluid level to full level.

11. Install the reservoir cap and connect the fluid level sensor connector. Turn the ignition ON and allow the pump to charge the accumulator.

MANUAL BLEEDING

1. Depressurize the hydraulic accumulator.

—— **CAUTION** ——

Failure to depressurize the hydraulic accumulator, prior to performing this operation may result in personal injury and/or damage to the painted surfaces.

2. Connect a transparent hose to the caliper bleed screw. Submerge the free end of the hose in a clear glass container, which is partially filled with clean, fresh brake fluid.

3. Slowly pump the brake pedal several times, using full strokes of the pedal and allowing approximately 5

seconds between pedal strokes. After 2 or 3 strokes, continue to hold pressure on the pedal, keeping it at the bottom of its travel.

4. With pressure on the pedal, open the bleed screw ½–¾ turn. Leave the bleed screw open until fluid no longer flows from the hose. Tighten the bleed screw and release the pedal.

5. Repeat this procedure until clear, bubble-free fluid flows from the hose.

6. Repeat all steps at each of the calipers. Calipers should be bled in the following order:
   a. Left rear
   b. Right rear
   c. Left front
   d. Right front

Anti-Lock Brake System Service

PRECAUTIONS

Failure to observe the following precautions may result in system damage.

• Before performing electric arc welding on the vehicle, disconnect the Electronic Brake Control Module (EBCM) and the hydraulic modulator connectors.

• When performing painting work on the vehicle, do not expose the Electronic Brake Control Module (EBCM) to temperatures in excess of 185°F (85°C) for longer than 2 hrs. The system may be exposed to temperatures up to 200°F (95°C) for less than 15 min.

• Never disconnect or connect the Electronic Brake Control Module (EBCM) or hydraulic modulator connectors with the ignition switch ON.

• Never disassemble any component of the Anti-Lock Brake System (ABS) which is designated non-servicable; the component must be replaced as an assembly.

• When filling the master cylinder, always use brake fluid which meets DOT-3 specifications; petroleum-based fluid will destroy the rubber parts.

DEPRESSURIZING THE HYDRAULIC ACCUMULATOR

1. With the ignition **OFF**, pump the brake pedal a minimum of 40 times, using approximately 50 lbs. (222 N) pedal force. A noticeable change in pedal feel will occur when the accumulator is discharged.

2. When a definite increase in pedal effort is felt, stroke the pedal a few additional times. This should remove all hydraulic pressure from the system.

Hydraulic Assembly

REMOVAL & INSTALLATION

1. Disconnect the negative battery cable. Depressurize the hydraulic accumulator.

— CAUTION —

Failure to depressurize the hydraulic accumulator, prior to performing this operation may result in personal injury and/or damage to the painted surfaces.

2. Remove the fresh air intake ducts. Remove the washer fluid tank from the vehicle.

3. Disconnect all electrical connectors from the hydraulic unit and pump/motor.

4. Remove as much of the fluid as possible from the reservoir on the hydraulic assembly.

5. Remove the pressure hose fitting (banjo bolt) from the hydraulic assembly. Use care not to drop the 2 washers used to seal the pressure hose fitting to the hydraulic assembly inlet.

6. Disconnect the return hose from the reservoir nipple. Cap the spigot on the reservoir.

7. Disconnect all brake tubes from the hydraulic assembly.

8. Remove the driver's side sound insulation panel.

9. Disconnect the pushrod from the brake pedal by using a small, flat tool to release the retainer clip on the brake pedal pin. The center tang on the clip must be moved back enough to allow the lock tab to clear the pin. Disconnect the pushrod from the pedal pin.

10. Remove the 4 underdash hydraulic assembly mounting nuts.

11. Remove the hydraulic assembly.
To install:

12. Position the hydraulic assembly on the vehicle.

13. Install and torque the mounting nuts to 21 ft. lbs. (28 Nm).

14. Using Lubriplate® or equivalent, coat the bearing surface of the pedal pin.

15. Connect the pushrod to the pedal and install a new retainer clip.

16. Install the brake tubes. If the proportioning valves were removed from the hydraulic assembly, reinstall valves and tighten to 20 ft. lbs. (27 Nm).

17. Install the return hose to the nipple on the reservoir.

18. Install the pressure hose to the hydraulic assembly; be sure the 2 washers are in there proper position. Tighten the bango bolt to 13 ft. lbs. (18 Nm).

19. Fill the reservoir to the top of the screen.

20. Connect all electrical connectors to the hydraulic assembly.

21. Bleed the entire brake system.

22. Install the cross brace, if disturbed. Install the fresh air intake duct and the washer tank.

23. Connect the negative battery cable and check the assembly for proper operation.

Wheel Speed Sensors

REMOVAL & INSTALLATION

Front Sensor

1. Raise the vehicle and support safely. Remove the wheel and tire assembly.

2. Remove the screw from the clip that holds the sensor to the fender shield.

3. Carefully pull the sensor assembly grommet from the fender shield.

4. Unplug the connector from the harness. Remove the retaininer clip from the strut damper bracket.

5. Remove the sensor mounting screw.

6. Carefully remove the sensor.
To install:

7. Coat the sensor with high temperature multi-purpose anti-corrosion compound before installing into the steering knuckle. Install the screw and tighten to 60 inch lbs. (7 Nm).

8. Connect the sensor connector to the harness and install the sensor connector lock.

9. Install the sensor assembly grommet and attach the clip to the fender shield.

NOTE: Proper installation of the wheel speed sensor cables is critical to continued system operation. Be sure the cables are installed in retainers. Failure to install the cables in the retainers may result in contact with moving parts and/or over-extension of the cables, resulting in an open circuit.

10. Install the wheel.

Rear Sensor

1. Raise the vehicle and support safely. Remove the wheel and tire assembly.

2. Carefully pull the sensor assembly grommet from the underbody and pull the harness through the hole.

3. Unplug the connector from the harness. Remove the retaininer clip from the strut damper bracket.

4. Remove the sensor spool grommet clip retaining screw from the body hose bracket, located in front of the inside of the trailing arm.

5. Remove the outboard sensor as-

sembly retaining nut and sensor mounting screw.

6. Carefully remove the sensor.

To install:

7. Coat the sensor with high temperature multi-purpose anti-corrosion compound before installing into the steering knuckle. Install the screw and tighten to 60 inch lbs. (7 Nm). Install the retaining nut.

8. Install the sensor spool grommet clip retaining screw.

9. Feed the sensor connector wire through the grommet and connect to the harness.

10. Install the sensor assembly grommet.

11. Install the wheel.

Hydraulic Bladder Accumulator

REMOVAL & INSTALLATION

1. Release pressure in the hydraulic accumulator. Disconnect the negative battery cable.

――――― **CAUTION** ―――――

Failure to depressurize the hydraulic accumulator, prior to performing this operation may result in personal injury and/or damage to the painted surfaces.

2. Using oil filter band wrench C–4065, loosen bladder accumulator.

3. Remove the bladder accumulator and brake fluid shield from the hydraulic assembly.

To install:

4. Install the brake fluid shield onto the hydraulic accumulator and install the bladder accumulator onto the hydraulic assembly.

5. Torque bladder to 35 ft. lbs. (48 Nm) using band wrench. Make sure the O-ring on the bladder is fulley seated on the hydraulic assembly.

6. Turn the ignition switch to the run position to energize the pump/motor assembly and pressurize the system. Check for leaks in the system.

7. Depressurize the accumulator again and check the brake fluid level in the reservoir. Add as required to correct the level.

Controller Anti-Lock Brake Module

REMOVAL & INSTALLATION

1. Turn the vehicle ignition to the OFF position.

2. Remove the speed control servo from vehicle.

3. Disconnect the wiring harness connector from the Controller Anti-Lock Brake (CAB) module. Remove

the 3 mounting bolts on the fender and remove the CAB from the vehicle.

To install:

4. Install the CAB module to the fender and tighten the attaching bolts to 105 inch lbs. (12 Nm).

5. Install the electrical connector and tignten the retaining bolt to 40 inch lbs. (5 Nm).

CHASSIS ELECTRICAL

Air Bag

DISARMING

To disarm the air bag, disconnect the negative battery. Allow system capacitor to discharge for 2 minutes before working on system. Failure to disarm the system may result in deployment of the air bag and possible personal injury.

Heater Blower Motor

REMOVAL & INSTALLATION

Without Air Conditioning

CARAVAN, VOYAGER AND TOWN & COUNTRY

1. Disconnect the negative battery cable.

2. Lower right side instrument panel trim cover and right cowl trim panel, as required. Disconnect the blower lead wire connector.

3. Remove the 2 screws at the top of the blower housing that secure it to the unit cover.

4. Remove the 5 screws from around the blower housing and separate the blower housing from the unit.

5. Remove the 3 screws that secure the blower assembly to the heater housing and remove the assembly from the unit. Remove the fan from the blower motor.

6. The installation is the reverse of the removal procedure.

7. Connect the negative battery cable and check the blower motor for proper operation.

EXCEPT― CARAVAN, VOYAGER AND TOWN & COUNTRY

1. Disconnect negative battery cable.

2. Remove the lower steering column cover, ash receiver and the intermittent wiper control module.

3. Remove the lower instrument panel module retaining screw to the

right of the steering column at cluster bezel.

4. Remove the center distribution duct retaining screws and panel support screw at the bottom of the module.

5. Remove the courtesy light at the lower right corner of the module and the screw near the ash receiver. Disconnect the blower motor wire harness.

6. Open the glovebox and remove the screws along the top edge.

7. Move the module rearward and down far enough to unclip the wiring harness and antenna cable and disconnect the speaker wire (if equipped with monaural radio) and glovebox light wire. Remove the module from the vehicle.

8. Remove the 2 screws at the top of the blower housing that secure it to the unit cover.

9. Remove the 5 screws from around the blower housing and separate the blower housing from the unit.

10. Remove the 3 screws that secure the blower assembly to the heater housing and remove the assembly from the unit. Remove the fan from the blower motor.

To install:

11. Install blower motor assembly to the heater housing and install retainer screws.

12. Install 2 screws at top of blower housing that secure it to cover. Connect the blower motor wire harness.

13. Connect radio wiring harness, antenna cable, the speaker wire (if equipped with monaural radio) and glovebox light wire.

14. Install module in vehicle. Open the glovebox and install the retaining screws along the top edge.

15. Install the courtesy light at the lower right corner of the module and the screw near the ash receiver.

16. Install the center distribution duct retaining screws and panel support screw at the bottom of the module.

17. Install the lower instrument panel module retaining screw to the right of the steering column at cluster bezel.

18. Install the lower steering column cover, ash receiver and the intermittent wiper control module.

19. Reconnect negative battery cable.

With Air Conditioning

CARAVAN, VOYAGER AND TOWN & COUNTRY

1. Disconnect the negative battery cable.

2. Lower right side instrument panel trim cover and right cowl trim panel, as required. Disconnect the blower lead wire connector.

3. Disconnect the 2 vacuum lines from the recirculating door actuator.

4. Remove the 2 screws at the top of the blower housing that secure it to the unit cover.

5. Remove the 5 screws from around the blower housing and separate the blower housing from the unit.

6. Remove the 3 screws that secure the blower assembly to the heater housing and remove the assembly from the unit. Remove the fan from the blower motor.

7. The installation is the reverse of the removal procedure.

8. Connect the negative battery cable and check the blower motor for proper operation.

EXCEPT— CARAVAN, VOYAGER AND TOWN & COUNTRY

1. Disconnect negative battery cable.

2. Remove the lower steering column cover, ash receiver and the intermittent wiper control module.

3. Remove the lower instrument panel module retaining screw to the right of the steering column at cluster bezel.

4. Remove the center distribution duct retaining screws and panel support screw at the bottom of the module.

5. Remove the courtesy light at the lower right corner of the module and the screw near the ash receiver.

6. Open the glovebox and remove the screws along the top edge.

7. Move the module rearward and down far enough to unclip the wiring harness and antenna cable and disconnect the speaker wire, if equipped with monaural radio, and glovebox light wire. Remove the module from the vehicle.

8. Disconnect 2 vacuum lines from the recirculating air door actuator and the blower motor wire harness. Remove the 2 screws at the top of the blower housing that secure it to the unit cover.

9. Remove the 5 screws from around the blower housing and separate the blower housing from the unit.

10. Remove the 3 screws that secure the blower assembly to the heater housing and remove the assembly from the unit. Remove the fan from the blower motor.

To install:

11. Install blower motor assembly to the heater housing and install retainer screws.

12. Install 2 screws at top of blower housing that secure it to cover.

13. Connect radio wiring harness, antenna cable, the speaker wire, if equipped with monaural radio, and glovebox light wire.

14. Reconnect the blower motor elec-

trical harness and the vacuum lines to the recirculating air door actuator. Install module in vehicle. Open the glovebox and install the retaining screws along the top edge.

15. Install the courtesy light at the lower right corner of the module and the screw near the ash receiver.

16. Install the center distribution duct retaining screws and panel support screw at the bottom of the module.

17. Install the lower instrument panel module retaining screw to the right of the steering column at cluster bezel.

18. Install the lower steering column cover, ash receiver and the intermittent wiper control module.

19. Reconnect negative battery cable.

Windshield Wiper Motor

REMOVAL & INSTALLATION

Caravan, Voyager and Town & Country

1. Disconnect the negative battery cable.

2. Remove the wiper arms.

3. Open the hood and remove the cowl top plenum grille and the plastic screen.

4. Remove the wiper pivot retaining screws and push the pivots down into the plenum chamber.

5. Remove the nut from the wiper motor output shaft and remove the linkage assembly from the motor.

6. Disconnect the wiper motor harness, remove the mounting nuts and remove the motor from the vehicle.

To install:

7. Install wiper motor and mounting screws and nuts tightening to 60–70 inch lbs. (8 Nm) torque. Connect wiper harness connector.

8. Install the linkage assembly to the motor. Install the nut onto the

wiper motor output shaft and tighten to 95 inch lbs. (10 Nm).

9. Install wiper pivot retaining screws and torque to 65 inch lbs. (8 Nm).

10. Install the plastic screen, the cowl top plenum grille and the wiper arms.

11. Reconnect the negative battery cable and check wiper operation.

Dakota

1. Disconnect the negative battery cable.

2. Disconnect the connector from the motor.

3. Remove 3 motor mounting nuts. Remove the wiper arms.

4. Remove the cowl panel mounting screws making sure to remove the screw in the center of the grille to the right of the right side pivot assembly.

5. Disengage the rear of the cowl grille from the windshield weather strip by pulling forward and remove cowel panel.

6. Pry up on clips in screen cover and remove.

7. Hold the drive crank with a wrench while remove the crank nut. Remove the drive crank.

8. Remove the motor from the vehicle.

To install:

9. Position motor on 3 studs on dash panel making certain rubber gasket and spacer between motor and dash panel are properly positioned.

10. Install 3 mounting nuts and torque to 65 inch lbs. (8 Nm).

11. Connect wiring harness. Align flats on the drive crank arm with flats on motor shaft and install tightening nut to 95 inch lbs. (11 Nm).

12. Connect negative battery cable and check wiper operation.

13. Install screen and hose in place and secure. Install cowl grille and wiper arms.

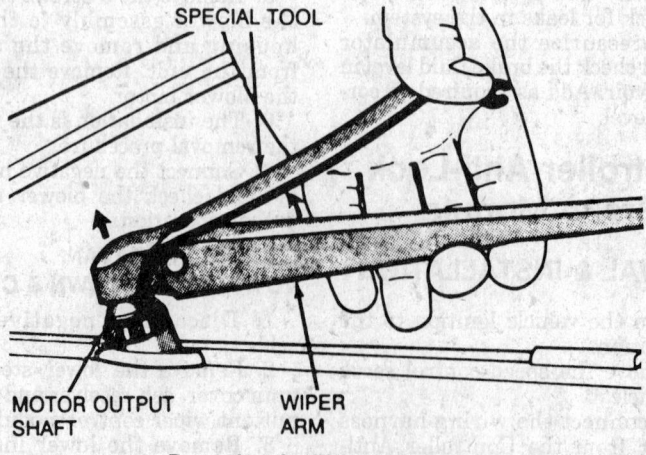

SPECIAL TOOL

MOTOR OUTPUT SHAFT

WIPER ARM

Removing the liftgate wiper arm

Liftgate Wiper Motor

REMOVAL & INSTALLATION

Caravan, Voyager and Town & Country

1. Disconnect the negative battery cable.
2. Remove the liftgate wiper using the special removal tool.

NOTE: Prying the arm off of the shaft with a prying device could damage the arm and possibly cause it to pop off of the shaft while in operation. This would cause poor rear vision for the driver and a dangerous situation for the driver behind. Do not bend or push the spring clip at the base of the arm to release the arm; it is self-releasing. Use only the indicated tool.

3. Open the liftgate and remove the inside trim panel.
4. Remove the motor mounting screws, disconnect the wiring harness and remove the motor from the liftgate.
5. The installation is the reverse of the removal procedure.
6. Connect the negative battery cable and check the rear wiper for proper operation.

Windshield Wiper Switch

REMOVAL & INSTALLATION

NOTE: On 1991–92 vehicles, the windshield wiper switch is part of the combination switch.

Except Tilt Wheel

1. Disconnect the negative battery cable.
2. Remove the lower steering column cover.
3. Remove the horn pad mounting screws from behind the steering wheel and remove the horn pad.
4. Remove the steering wheel nut, matchmark the steering wheel to the shaft and remove steering wheel.
5. Remove the plastic wiring channel from under the steering column.
6. Disconnect the wiper switch connector, intermittent wipe module connector and cruise control connector, if equipped.
7. Remove the side lock housing cover.
8. Remove the slotted hex-head screw that attaches the wiper switch to the turn signal switch and remove the switch.
9. Remove the control knob from

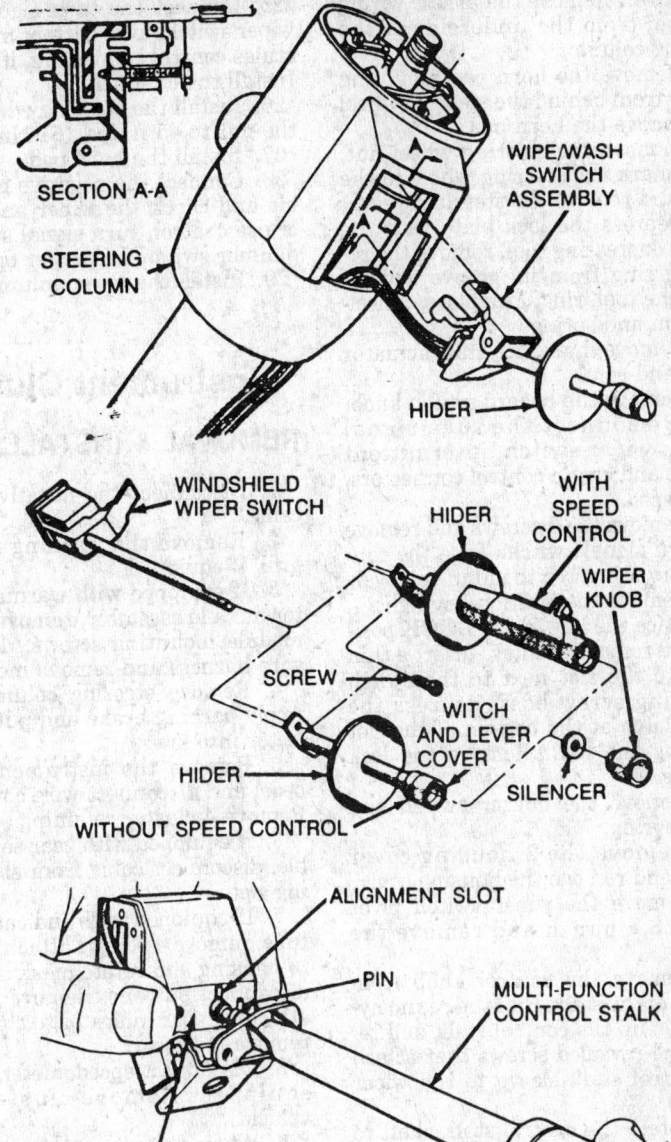

the end of the stalk. Pull the round nylon hider up the control stalk and remove the screws that attach the control stalk sleeve to the wiper switch.
10. Rotate the control stalk shaft to the full clockwise position and remove the shaft from the wiper switch by pulling it straight out.

To install:

11. Install the control shaft to the wiper switch, install the screws, the hider and the control knob.
12. Run the wiring through the opening and down the steering column, position the switch and install the hex-head screw. Make sure the dimmer switch rod is properly engaged.

13. Install the side lock housin cover.
14. Connect the wires and install t wiring channel.
15. Install the steering wheel torq the nut to 45 ft. lbs. (61 Nm).
16. Install the horn pad.
17. Connect the negative battery c ble and check the wiper and washe cruise control, turn signal switch ar dimmer switch for proper operation
18. Install the lower column cover.

Tilt Wheel

1. Disconnect the negative batter cable.
2. Remove the lower steering co

Figure labels (from illustration):

SECTION A-A
STEERING COLUMN
WIPE/WASH SWITCH ASSEMBLY
HIDER
WINDSHIELD WIPER SWITCH
HIDER
WITH SPEED CONTROL
WIPER KNOB
SCREW
WITCH AND LEVER COVER
SILENCER
HIDER
WITHOUT SPEED CONTROL
ALIGNMENT SLOT
PIN
MULTI-FUNCTION CONTROL STALK
WASH/WIPE SWITCH

Removing the wiper switch—except tilt wheel

umn cover. Remove the plastic wiring channel from the underside of the steering column.

3. Remove the horn pad mounting screws from behind the steering wheel and remove the horn pad.

4. Remove the steering wheel nut, matchmark the steering wheel to the shaft and remove the steering wheel.

5. Depress the lock plate with the proper depressing tool, remove the retaining ring from its groove and remove the tool, ring, lock plate, cancelling cam and spring.

6. Remove the switch stalk actuator screw and arm.

7. Remove the hazard switch knob.

8. Disconnect the turn signal switch, wiper switch, intermittent module and cruise control connectors, if equipped.

9. Remove the 3 screws and remove the turn signal switch. Tape the connector to the wires to aid in removal.

10. Remove the ignition key light.

11. Place the key in the **LOCK** position and remove the key. Insert a thin tool into the slot next to the switch mounting screw boss, depress the spring latch at the bottom of the slot releasing the lock. Remove the lock cylinder.

12. Remove the buzzer switch and wedge spring.

13. Remove the 3 housing cover screws and remove the housing cover.

14. Remove the wiper switch pivot pin with a punch and remove the switch.

15. Remove the control knob from the end of the stalk. Pull the round nylon hider up the control stalk and remove the revealed screws that attach the control stalk sleeve to the wiper switch.

16. Rotate the control stalk shaft to the full clockwise position and remove the shaft from the wiper switch by pulling it straight out.

To install:

17. Install the control shaft to the wiper switch, install the screws, the hider and the control knob.

18. Run the wiring through the opening and down the steering column, position the switch and install the wiper switch pivot pin.

19. Install the housing cover.

20. Install the buzzer switch and wedge spring.

21. Install the lock cylinder.

22. Install the ignition key light.

23. Install the turn signal switch, switch stalk actuator arm and hazard switch knob.

24. Install the spring, canceling cam, lock plate and ring on the steering shaft. Depress the plate with the depressing tool and install the ring securely in the groove. Remove the tool slowly.

25. Connect the turn signal switch, wiper switch, intermittent module and cruise control connectors, if equipped. Install the channel.

26. Install the steering wheel torque the nut to 45 ft. lbs. (61 Nm).

27. Install the horn pad.

28. Connect the negative battery cable and check the wiper and washer, cruise control, turn signal switch and dimmer switch for proper operation.

29. Install the lower column cover.

Instrument Cluster

REMOVAL & INSTALLATION

1. Disconnect the negative battery cable.

2. Remove the warning indicator grill, if equipped.

3. If equipped with warning indicator module assembly, remove warning module mounting screws, disconnect wire harness and remove module.

4. Remove steering column cover. Apply parking brake and put gear selector into low.

5. Remove the instrument cluster bezel and disconnect wire connectors. Remove 4 cluster retaining screws.

6. If equipped with gear selector cable, disconnect cable from shift housing slot.

7. If equipped with indicator guide tube, remove 4 screws attaching mask to housing and rotate mask down and disconnect buttons. Remove 2 screws attaching gear indicator to cluster and remove indicator.

8. Disconnect speedometer cable, if equipped. Remove cluster from vehicle.

9. When only removing gauge(s) or the speedometer, remove the trip odometer reset knob, if equipped, remove the mask and lens assembly and remove the desired gauge from the cluster. Disconnect the speedometer cable, if removing the speedometer.

To install:

10. Postion the cluster and feed the gear indicator cable through slot in shift housing. If equipped with an indicator guide tube, install onto cluster.

11. Connect all wiring and install the speedometer cable to the speedometer, if equipped. Make sure it is securely clicked in place.

12. Install the cluster retaining screws. Install warning indicator module and grille, if equipped.

13. Install the cluster bezel and steering column cover.

14. Connect the negative battery cable. Check all gauges and the speedometer for proper operation. Make sure the gearshift indicator is properly aligned.

Radio

REMOVAL & INSTALLATION

Caravan, Voyager and Town & Country

1. Disconnect the negative battery cable.

2. If equipped with a lower console, remove the ash tray and remove the 2 screws behind it. Remove the lower mounting screws on both sides of the console and pull the console forward. If not equipped with a lower console, open the ash tray and remove the tray.

3. Remove the 2 screws at the lower edge of the bezel. Remove the 3 screws from the top edge of the bezel and remove the bezel.

4. Remove the screws that attach the radio to the dash.

5. Pull the radio out, disconnect the connectors, ground cable and antenna and remove the radio.

6. The installation is the reverse of the removal procedure.

Dakota

1. Disconnect the negative battery cable.

2. Remove the steering column cover and remove the instrument panel bezel. Two screws are hidden behind the steering column cover.

3. Remove the screws that attach the radio to the instrument panel.

4. Pull the radio out, disconnect the connectors, ground cable and antenna and remove the radio.

5. The installation is the reverse of the removal procedure.

Headlight Switch

REMOVAL & INSTALLATION

Caravan, Voyager and Town & Country

1. Disconnect the negative battery cable.

2. Remove the headlight and accessory switch trim bezel.

3. Remove the switch plate from the

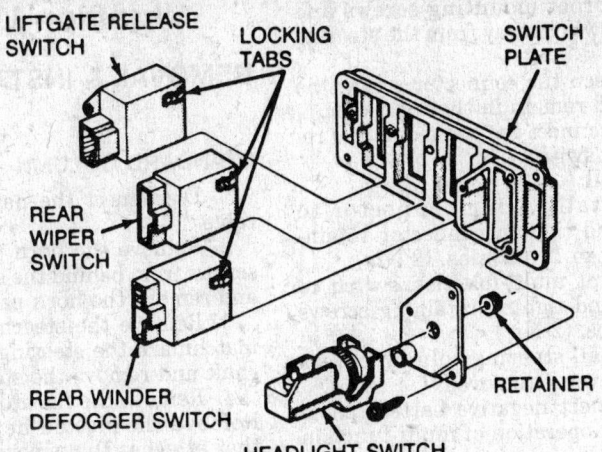

Headlight and accessory switch removal and Installation—Caravan, Voyager and Town & Country

lower panel, pull the assembly out and disconnect the wiring.

4. Depress the spring button and remove the headlight switch knob and stem.

5. Remove the headlight switch retainer and remove the switch.

To install:

6. Attach headlight switch to headlight switch plate with retainer and secure.

7. Install headlight switch knob and stem. Reconnect wiring connectors to switch.

8. Install 4 screws attaching switch to lower panel. Install trim bezel.

9. Connect negative battery cable and check operation of switch.

Dakota

1. Disconnect the negative battery cable.

2. Remove the steering column cover and remove the instrument panel bezel. Two screws are hidden behind the steering column cover.

3. Remove the screws from the

headlight switch bezel, pull the assembly out and disconnect the wiring.

4. Remove the nut retaining the bezel to the bracket. Depress the spring button on the right side of the switch and remove the headlight switch knob and stem.

5. Remove the spanner nut and remove the switch.

6. The installation is the reverse of the removal procedure.

Dimmer Switch

REMOVAL & INSTALLATION

NOTE: On 1991–92 vehicles, the dimmer switch is part of the combination switch.

1. Disconnect the negative battery cable.

2. Remove the lower steering column cover.

3. Unplug the switch, located on the lower portion of the steering column.

4. Holding the actuating rod against its upper seat, remove the bolts that attach the switch to the column and remove the switch.

5. The installation is the reverse of the removal procedure. Adjust the switch as required.

6. Connect the negative battery cable and check the switch for proper operation.

Turn Signal Switch

NOTE: On 1991–92 vehicles, the turn signal switch is part of the combination switch.

REMOVAL & INSTALLATION

STANDARD COLUMN

1. Disconnect the negative battery cable.

2. Remove the lower steering column cover.

3. Remove the horn pad mounting screws from behind the steering wheel and remove the horn pad.

4. Remove the steering wheel nut, matchmark the steering wheel to the shaft and remove the steering wheel.

5. Remove the plastic wiring channel from the underside of the steering column and disconnect the turn signal switch connector.

6. Remove the hazard switch knob. Remove the slotted hex-head screw that attaches the wiper switch to the turn signal switch.

7. Remove the 3 screws and pull the turn signal switch out of the column.

To install:

8. Run the wiring through the opening and down the steering column, position the switch and install the hex-head screw. Make sure the dimmer switch rod is properly engaged.

9. Install the 3 screws and the hazard switch knob.

10. Connect the wires and install the wiring channel.

11. Install the steering wheel. Torque the nut to 45 ft. lbs. (61 Nm).

12. Install the horn pad.

13. Connect the negative battery cable and check the turn signal switch and dimmer switch for proper operation.

14. Install the lower column cover.

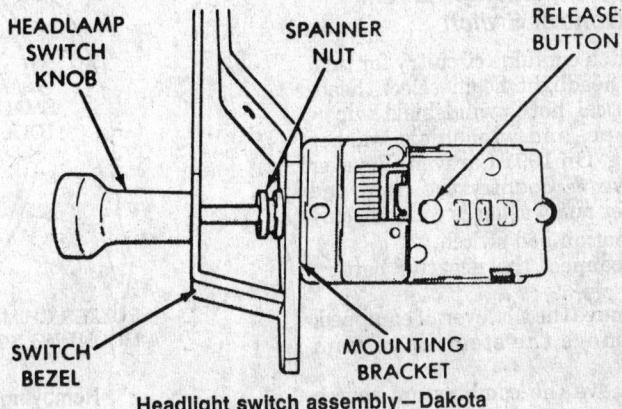

Headlight switch assembly—Dakota

TILT COLUMN

1. Disconnect the negative battery cable.

2. Remove the lower steering column cover. Remove the plastic wiring channel from the underside of the steering column.

3. Remove the horn pad mounting screws from behind the steering wheel and remove the horn pad.

4. Remove the steering wheel nut, matchmark the steering wheel to the shaft and remove the steering wheel.

5. Depress the lock plate with the proper depressing tool, remove the retaining ring from its groove and remove the tool, ring, lock plate, canceling cam and spring.

6. Remove the stalk actuator screw and arm.

7. Remove the hazard switch knob.

8. Disconnect the turn signal switch connector.

9. Remove the 3 screws and remove the turn signal switch. Tape the connector to the wires to aid in removal.

To install:

10. Run the wiring through the opening and down the steering column, install the turn signal switch, switch stalk actuator arm and hazard switch knob.

11. Install the spring, canceling cam, lock plate and ring on the steering shaft. Depress the plate with the depressing tool and install the ring securely in the groove. Remove the tool slowly.

12. Connect the turn signal switch connector and install the channel.

13. Install the steering wheel. Torque the nut to 45 ft. lbs. (61 Nm).

14. Install the horn pad.

15. Connect the negative battery cable and check the turn signal switch and dimmer switch for proper operation.

16. Install the lower column cover.

Combination Switch

REMOVAL & INSTALLATION

Turn Signal/Headlight Beam/ Wiper Control Switch

This switch contains circuitry for turn signals, headlight beam select, headlight optical horn, windshield wiper, pulse wipe, and windshield washer switching. On 1991 Caravan, Voyager and Town & Country the front and rear wiper and washer is controlled by a combination pod switch.

1. Disconnect the negative battery cable.

2. Remove the tilt lever, if equipped.

3. Remove the steering column covers.

4. Remove the combination switch tamper-proof mounting screws and pull the switch away from the steering column.

5. Loosen the connector screw; the screw will remain in the connector.

6. Disconnect the connector and remove the switch.

To install:

7. Install wiring connector to switch and tighten connector retaining screw to 17 inch lbs. (2 Nm).

8. Mount multi-function switch to column and tighten retaining screws 17 inch lbs. (2 Nm).

9. Install steering column covers and tilt lever if removed.

10. Connect negative battery cable and check operation of multi-function switch.

Front and Rear Wiper/Washer Pod Switch

1. Disconnect negative battery cable.

2. Remove 8 screws mounting cluster bezel to the instrument panel.

3. Gently pull cluster bezel out to gain access to the switch mounting fingers. Release fingers and push switch through mounting hole for pod switch.

4. Remove electrical connectors for switch.

To install:

5. Position cluster bezel for assembly and pull wiring connectors through mounting hole for pod switch assembly.

6. Install wiring connector to switch and verify proper function of the front and rear wipe/wash system.

7. Install switch into bezel making sure fingers snap into position. Assemble cluster bezel into instrument panel and tighten retaining screws.

Ignition Lock

REMOVAL & INSTALLATION

STANDARD COLUMN

1. Disconnect the negative battery cable.

2. Remove the horn pad mounting screws from behind the steering wheel and remove the horn pad.

3. Remove the steering wheel nut, matchmark the steering wheel to the shaft and remove the steering wheel.

4. Remove the hazard switch knob. Remove the slotted hex-head screw that attaches the wiper switch to the turn signal switch.

5. Remove the 3 screws and pull the turn signal switch out of the column as far as it will go. Unplug it below if necessary.

6. Remove the ignition switch key light.

7. Place the key in the **LOCK** position and remove the key.

8. Insert 2 small diameter tools into the release holes and push inward to release the spring loaded lock retainers while simultaneously pulling the key lock cylinder out of its bore.

To install:

9. Install the key cylinder.

10. Install the ignition switch key light.

11. Install the turn signal switch and hazard switch knob. Connect the wires if disconnected.

12. Install the steering wheel and torque the nut to 45 ft. lbs. (61 Nm).

13. Install the horn pad.

14. Connect the negative battery cable and check the lock cylinder for proper operation.

15. Install the lower column cover.

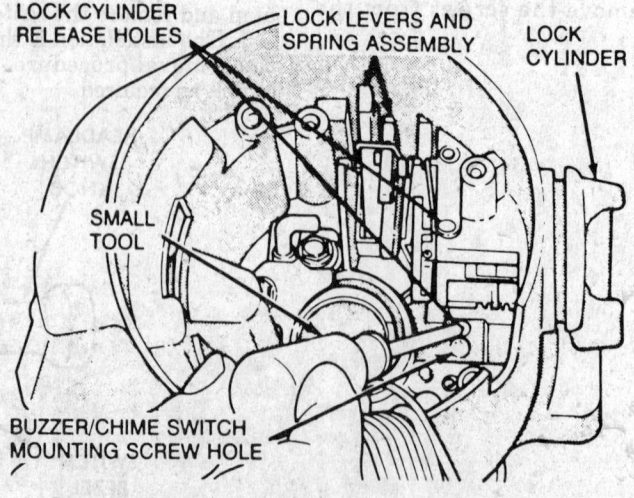

Removing the key lock cylinder—except tilt wheel

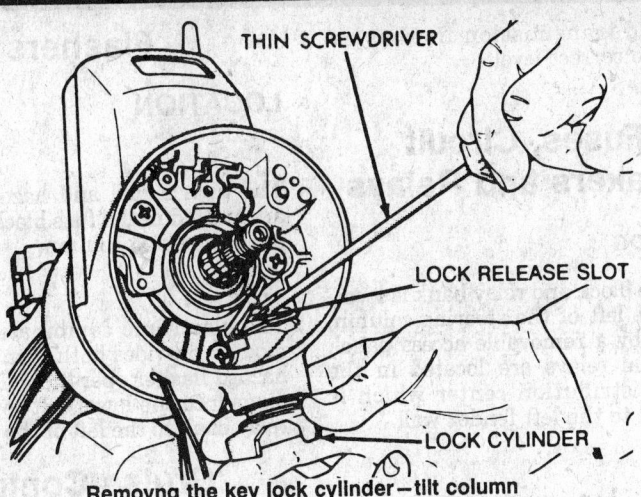

THIN SCREWDRIVER

LOCK RELEASE SLOT

LOCK CYLINDER

Removing the key lock cylinder—tilt column

TILT COLUMN

1. Disconnect the negative battery cable.

2. Remove the horn pad mounting screws from behind the steering wheel and remove the horn pad.

3. Remove the steering wheel nut, matchmark the steering wheel to the shaft and remove the steering wheel.

4. Depress the lock plate with the proper depressing tool, remove the retaining ring from its groove and remove the tool, ring, lock plate, canceling cam and spring.

5. Remove the stalk actuator screw and arm.

6. Remove the hazard switch knob.

7. Remove the 3 screws and pull the turn signal switch out of the column as far as it will go. Unplug it below if necessary.

8. Remove the ignition key light.

9. Place the key in the **LOCK** position and remove the key. Insert a thin tool into the slot next to the switch mounting screw boss, depress the spring latch at the bottom of the slot releasing the lock and remove the lock cylinder.

To install:

10. Install the lock cylinder.

11. Install the ignition key light.

12. Install the turn signal switch, switch stalk actuator arm and hazard switch knob.

13. Install the spring, canceling cam, lock plate and ring on the steering shaft. Depress the plate with the depressing tool and install the ring securely in the groove. Remove the tool slowly.

14. Connect the wires if disconnected.

15. Install the steering wheel and torque the nut to 45 ft. lbs. (61 Nm).

16. Install the horn pad.

17. Connect the negative battery cable and check the turn signal switch for proper operation.

18. Install the lower column cover, if removed.

Ignition Switch

REMOVAL & INSTALLATION

1. Disconnect the negative battery cable.

2. Remove the lower steering column cover.

3. Remove the steering column retaining nuts and allow the steering wheel to rest on the driver's seat.

4. Remove the 2 screws that attach the ignition switch to the column.

5. Rotate the switch 90 degrees and pull up to disengage it from the ignition switch rod.

To install:

6. Engage the switch with the rod, rotate the switch 90 degrees and push down until fully engaged.

7. Install the mounting screws finger-tight.

8. Place the key in the **LOCK** position and remove the key. Adjust switch by pushing up gently on the switch to take up all slack in the rod.

9. Tighten the mounting screws. Connect negative battery cable and check the switch for proper operation in all positions.

10. Install the steering column and cover.

Ignition Lock/Switch

REMOVAL & INSTALLATION

1. Disconnect the negative battery cable.

NOTE: If equipped with an airbag, it is imperative that the steering wheel removal and installation procedure under Steering is followed.

2. Remove the tilt lever, if equipped.

3. Remove the upper and lower column covers.

4. Remove the 3 ignition switch tamper-proof Torx screws; APEX 440–TX20H or equivalent is required.

5. Pull the switch away from the column. Release the connector locks on the 2 wiring connectors and disconnect them from the switch.

6. Remove the key lock cylinder from the ignition switch by performing the following:

a. Insert the key and turn the switch in the **LOCK** position. Using a small tool, depress the key cylinder retaining pin flush with the key cylinder surface.

b. Rotate the key clockwise to the **OFF** position to unseat the key cylinder from the ignition switch assembly. The cylinder bezel should be about ⅛in. above the ignition switch halo light ring. Do not attempt to remove the key cylinder at this point.

c. With the key cylinder in the unseated position, rotate the key counterclockwise to the **LOCK** position and remove the key.

d. Remove the key cylinder from the ignition switch.

To install:

7. Connect the wiring connectors.

8. Mount ignition switch to the column by performing the following:

a. Position the shifter in **PARK** position. The park lock dowel pin on the ignition switch assembly must engage with the column park lock slider linkage.

b. Verify that the ignition switch is in the **LOCK** position. The flag should be parallel to the ignition switch terminals. Apply a small amount of grease to the flag and pin.

c. Position the park lock link to mid-travel.

d. Align the locating pin hole and its pin and position the ignition switch against the lock housing face. Make sure the pin is inserted into the park lock link contour slot. Torque the retaining screws to 17 inch lbs.

9. With the key cylinder and ignition switch in the **LOCK** position, key not in cylinder, gently insert the key cylinder into the ignition switch until it bottoms.

10. Insert the key. Simultaneously push in on the cylinder and rotate the key to the **RUN** position. This action should fully seat the cylinder in the ignition switch.

11. Install the column covers and the tilt lever, if equipped.

12. Connect the negative battery cable and check the push-to-lock and park lock functions, halo lighting and all ignition switch positions for proper operation.

Stoplight Switch

ADJUSTMENT

1. Push the stoplight switch forward in the mounting bracket as far as it will go; the brake pedal should move forward slightly.

2. Pull back on the brake pedal bringing the striker toward the switch until the pedal will not go back any farther. This will cause the switch to ratchet backward into position and automatically adjust.

REMOVAL & INSTALLATION

1. Disconnect the negative battery cable.

2. Remove the switch mounting bracket assembly from the brake pedal bracket.

3. Remove the switch from its bracket.

4. The installation is the reverse of the removal procedure.

5. Connect the negative battery cable and check the switch for proper operation.

Neutral Safety Switch

REMOVAL & INSTALLATION

1. Disconnect negative battery cable.

2. Raise and safely support the vehicle. Position drain pan under switch.

3. Disconnect switch wiring harness.

4. Remove switch from case.

5. Verify that switch operating lever fingers are centered in the switch opening in the case when in Park and Neutral.

6. Install new seal on switch and install switch into case. Torque to 25 ft. lbs.

7. Test continuity of new switch. Connect wire harness and negative battery cable.

8. Add transmission fluid as required to correct level.

Fuses, Circuit Breakers and Relays

Location

The fuse block and relay bank is located to the left of the steering column, covered by a removable access panel. Additional relays are located in the power distribution center which is mounted to the left fender wall.

Computers

LOCATION

Rear Wheel Anti-Lock (RWAL) brake control module— located on the passenger side cowl panel under the instrument panel.

The Bendix Anti-Lock—10 Brake Controller (CAB)—located under the battery tray.
The Air Bag Diagnostic Module (ASDM)—located under the center console of the instrument panel.
Body Controller—located inside the passenger compartment, behind the right side kick panel.

Flashers

LOCATION

The turn signal and hazard flashers are located in the fuse block to the left of the steering column.

1992

The electronic combination flasher module provides both turn signal and hazard flasher operation. The module plugs into the lower left corner of the relay bank to the left of the fuse block.

Cruise Control

ADJUSTMENT

2.5L Engine

1. The clearance between the throttle stud and cable clevis should be $1/16$ in.

2. To adjust the cable, remove the retaining clip or loosen the retaining clamp nut at the throttle bracket.

3. Pull all slack out of the cable using a $1/16$ in. diameter tool to account for proper clearance. Make sure the curb idle position of the throttle blade is not affected.

4. Reinstall the retaining clip or nut.

3.0L Engine

1. Grip the cable core and lightly push toward the servo.

2. While holding the position, mark the core wire next to the protective sleeve.

3. Pull the core wire away from the servo. There should be a 0.24 in. (6mm) gap between the mark on the core wire and the protective sleeve.

4. If the gap is not correct, remove the adjustment clip from the throttle bracket and move the sleeve to bring the gap into specification.

5. Reinstall the clip.

Ford Full Size 3

BRONCO • PICK-UP • VAN

SPECIFICATIONS

VEHICLE IDENTIFICATION CHART

Code	Liters	Cu. In. (cc)	Cyl.	Fuel Sys.	Eng. Mfg.
		Engine Code			
A	2.3	140 (2300)	4	MFI	Ford
T	2.9	177 (2901)	6	MFI	Ford
W	3.0	181 (2967)	6	MFI	Nissan
U	3.0	183 (2983)	6	MFI	Ford
X	4.0	241 (3950)	6	MFI	Ford
Y	4.9	300 (4917)	6	MFI	Ford
N	5.0	302 (4950)	8	MFI	Ford
H	5.8	351 (5733)	8	MFI	Ford
R	5.8	351 (5733)	8	MFI	Ford
K	7.3	445 (7294)	8	DSL	Navistar
M	7.3	445 (7294)	8	DSL	Navistar
G	7.5	460 (7539)	8	MFI	Ford

MFI - Multiport fuel injection
DSL - Diesel

Code	Year
	Model Year
M	1991
N	1992
P	1993
R	1994
S	1995

ENGINE IDENTIFICATION

Year	Model	Engine Displacement Liters (cc)	Engine Series (ID/VIN)	Fuel System	No. of Cylinders	Engine Type
1991	Bronco	4.9 (4916)	Y	MFI	6	OHV
	Bronco	5.0 (4949)	N	MFI	8	OHV
	Bronco	5.8 (5752)	H	MFI	8	OHV
	E-150	4.9 (4916)	Y	MFI	6	OHV
	E-150	5.0 (4949)	N	MFI	8	OHV
	E-150	5.8 (5752)	H	MFI	8	OHV
	E-250	4.9 (4916)	Y	MFI	6	OHV
	E-250	5.0 (4949)	N	MFI	8	OHV
	E-250	5.8 (5752)	H	MFI	8	OHV
	E-250	7.3 (7292)	M	DDI	8	OHV
	E-250	7.5 (7538)	G	MFI	8	OHV
	E-350	4.9 (4916)	Y	MFI	6	OHV
	E-350	5.8 (5752)	H	MFI	8	OHV
	E-350	7.3 (7292)	M	DDI	8	OHV
	E-350	7.5 (7538)	G	MFI	8	OHV
	F-150	4.9 (4916)	Y	MFI	6	OHV
	F-150	5.0 (4949)	N	MFI	8	OHV
	F-150	5.8 (5752)	H	MFI	8	OHV
	F-250	4.9 (4916)	Y	MFI	6	OHV
	F-250	5.0 (4949)	N	MFI	8	OHV
	F-250	5.8 (5752)	H	MFI	8	OHV
	F-250	7.3 (7292)	M	DDI	8	OHV

ENGINE IDENTIFICATION

Year	Model	Engine Displacement Liters (cc)	Engine Series (ID/VIN)	Fuel System	No. of Cylinders	Engine Type
1991	F-250	7.5 (7538)	G	MFI	8	OHV
	F-350	4.9 (4916)	Y	MFI	6	OHV
	F-350	5.8 (5752)	H	MFI	8	OHV
	F-350	7.3 (7292)	M	DDI	8	OHV
	F-350	7.5 (7538)	G	MFI	8	OHV
	F-Super Duty	7.3 (7292)	M	DDI	8	OHV
	F-Super Duty	7.5 (7538)	G	MFI	8	OHV
1992	Bronco	4.9 (4916)	Y	MFI	6	OHV
	Bronco	5.0 (4949)	N	MFI	8	OHV
	Bronco	5.8 (5752)	H	MFI	8	OHV
	E-150	4.9 (4916)	Y	MFI	6	OHV
	E-150	5.0 (4949)	N	MFI	8	OHV
	E-150	5.8 (5752)	H	MFI	8	OHV
	E-250	4.9 (4916)	Y	MFI	6	OHV
	E-250	5.0 (4949)	N	MFI	8	OHV
	E-250	5.8 (5752)	H	MFI	8	OHV
	E-250	7.3 (7292)	M	DDI	8	OHV
	E-250	7.5 (7538)	G	MFI	8	OHV
	E-350	4.9 (4916)	Y	MFI	6	OHV
	E-350	5.8 (5752)	H	MFI	8	OHV
	E-350	7.3 (7292)	M	DDI	8	OHV
	E-350	7.5 (7538)	G	MFI	8	OHV
	F-150	4.9 (4916)	Y	MFI	6	OHV
	F-150	5.0 (4949)	N	MFI	8	OHV
	F-150	5.8 (5752)	H	MFI	8	OHV
	F-250	4.9 (4916)	Y	MFI	6	OHV
	F-250	5.0 (4949)	N	MFI	8	OHV
	F-250	5.8 (5752)	H	MFI	8	OHV
	F-250	7.3 (7292)	M	DDI	8	OHV
	F-250	7.5 (7538)	G	MFI	8	OHV
	F-350	4.9 (4916)	Y	MFI	6	OHV
	F-350	5.8 (5752)	H	MFI	8	OHV
	F-350	7.3 (7292)	M	DDI	8	OHV
	F-350	7.5 (7538)	G	MFI	8	OHV
	F-Super Duty	7.3 (7292)	M	DDI	8	OHV
	F-Super Duty	7.5 (7538)	G	MFI	8	OHV
1993	Bronco	4.9 (4916)	Y	MFI	6	OHV
	Bronco	5.0 (4949)	N	MFI	8	OHV
	Bronco	5.8 (5752)	H	MFI	8	OHV
	E-150	4.9 (4916)	Y	MFI	6	OHV
	E-150	5.0 (4949)	N	MFI	8	OHV
	E-150	5.8 (5752)	H	MFI	8	OHV
	E-250	4.9 (4916)	Y	MFI	6	OHV
	E-250	5.0 (4949)	N	MFI	8	OHV
	E-250	5.8 (5752)	H	MFI	8	OHV
	E-250	7.3 (7292)	M	IDI	8	OHV
	E-250	7.5 (7538)	G	MFI	8	OHV
	E-350	4.9 (4916)	Y	MFI	6	OHV
	E-350	5.8 (5752)	H	MFI	8	OHV
	E-350	7.3 (7292)	M	IDI	8	OHV

ENGINE IDENTIFICATION

Year	Model	Engine Displacement Liters (cc)	Engine Series (ID/VIN)	Fuel System	No. of Cylinders	Engine Type
1993	E-350	7.5 (7538)	G	MFI	8	OHV
	F-150	4.9 (4916)	Y	MFI	6	OHV
	F-150	5.0 (4949)	N	MFI	8	OHV
	F-150	5.8 (5752)	H	MFI	8	OHV
	F-250	4.9 (4916)	Y	MFI	6	OHV
	F-250	5.0 (4949)	N	MFI	8	OHV
	F-250	5.8 (5752)	H	EFI	8	OHV
	F-250	7.3 (7292)	M	IDI	8	OHV
	F-250	7.5 (7538)	G	EFI	8	OHV
	F-350	4.9 (4916)	Y	EFI	6	OHV
	F-350	5.8 (5752)	H	EFI	8	OHV
	F-350	7.3 (7292)	M	DDI	8	OHV
	F-350	7.5 (7538)	G	EFI	8	OHV
	Lightning Pick-up	5.8 (5752)	R	EFI	8	OHV
	F-Super Duty	7.3 (7292)	M	DDI	8	OHV
	F-Super Duty	7.5 (7538)	G	EFI	8	OHV
1994-95	Bronco	4.9 (4916)	Y	MFI	6	OHV
	Bronco	5.0 (4949)	N	MFI	8	OHV
	Bronco	5.8 (5752)	H	MFI	8	OHV
	E-150	4.9 (4916)	Y	MFI	6	OHV
	E-150	5.0 (4949)	N	MFI	8	OHV
	E-150	5.8 (5752)	H	MFI	8	OHV
	E-250	4.9 (4916)	Y	MFI	6	OHV
	E-250	5.0 (4949)	N	MFI	8	OHV
	E-250	5.8 (5752)	H	MFI	8	OHV
	E-250	7.3 (7292)	M	IDI	8	OHV
	E-250	7.5 (7538)	G	MFI	8	OHV
	E-350	4.9 (4916)	Y	MFI	6	OHV
	E-350	5.8 (5752)	H	MFI	8	OHV
	E-350	7.3 (7292)	M	IDI	8	OHV
	E-350	7.5 (7538)	G	MFI	8	OHV
	F-150	4.9 (4916)	Y	MFI	6	OHV
	F-150	5.0 (4949)	N	MFI	8	OHV
	F-150	5.8 (5752)	H	MFI	8	OHV
	F-250	4.9 (4916)	Y	MFI	6	OHV
	F-250	5.0 (4949)	N	MFI	8	OHV
	F-250	5.8 (5752)	H	EFI	8	OHV
	F-250	7.3 (7292)	M	IDI	8	OHV
	F-250	7.5 (7538)	G	EFI	8	OHV
	F-350	4.9 (4916)	Y	EFI	6	OHV
	F-350	5.8 (5752)	H	EFI	8	OHV
	F-350	7.3 (7292)	M	DDI	8	OHV
	F-350	7.5 (7538)	G	EFI	8	OHV
	Lightning Pick-up	5.8 (5752)	R	EFI	8	OHV
	F-Super Duty	7.3 (7292)	M	DDI	8	OHV
	F-Super Duty	7.5 (7538)	G	EFI	8	OHV

MFI - Multiport fuel injection
DDI - Direct diesel injection
OHV - Overhead valve
SOHC - Single overhead camshaft
MFI - Multiport fuel injection

GENERAL ENGINE SPECIFICATIONS

Year	Engine ID/VIN	Engine Displacement Liters (cc)	Fuel System Type	Net Horsepower @ rpm	Net Torque @ rpm (ft. lbs.)	Bore x Stroke (in.)	Compression Ratio	Oil Pressure @ rpm
1991	Y	4.9 (4916)	MFI	1	2	4.00x3.98	8.8:1	40-60@2000
	N	5.0 (4949)	MFI	185@3800	270@2400	4.00x3.00	9.0:1	40-60@2000
	H	5.8 (5752)	MFI	210@3800	315@2800 [3]	4.00x3.50	8.8:1	40-60@2000
	M	7.3 (7292)	DDI	180@3300 [4]	45@1400 [5]	11.00x4.18	21.5:1	40-70@2000
	G	7.5 (7538)	MFI	230@3600	390@2200	4.36x3.85	8.5:1	40-65@2000
1992	Y	4.9 (4916)	MFI	1	2	4.00x3.98	8.8:1	40-60@2000
	N	5.0 (4949)	MFI	185@3800	270@2400	4.00x3.00	9.0:1	40-60@2000
	H	5.8 (5752)	MFI	210@3800	315@2800 [3]	4.00x3.50	8.8:1	40-60@2000
	M	7.3 (7292)	DDI	180@3300 [4]	45@1400 [5]	11.00x4.18	21.5:1	40-70@2000
	G	7.5 (7538)	MFI	230@3600	390@2200	4.36x3.85	8.5:1	40-65@2000
1993	Y	4.9 (4916)	EFI	6	7	4.00x3.98	8.8:1	40-60@2000
	N	5.0 (4949)	EFI	185@3800	270@2400	4.00x3.98	9.0:1	40-60@2000
	H	5.8 (5752)	EFI	200@3800	310@2800	4.00x3.50	8.8:1	40-65@2000
	R	5.8 (5752)	EFI	240@4200	340@3200	4.00x3.50	8.8:1	40-65@2000
	C	7.3 (7292)	IDI	190@3000	395@1400	4.11x4.18	21.5:1	40-70@2000
	M	7.3 (7292)	IDI	185@3000 [5]	360@400 [6]	4.11x4.18	21.5:1	40-70@2000
	G	7.5 (7538)	EFI	245@4000	400@2200	4.36x3.85	8.5:1	40-65@2000
1994-95	Y	4.9 (4916)	EFI	6	7	4.00x3.98	8.8:1	40-60@2000
	N	5.0 (4949)	EFI	185@3800	270@2400	4.00x3.98	9.0:1	40-60@2000
	H	5.8 (5752)	EFI	200@3800	310@2800	4.00x3.50	8.8:1	40-65@2000
	R	5.8 (5752)	EFI	240@4200	340@3200	4.00x3.50	8.8:1	40-65@2000
	C	7.2 (7292)	IDI	190@3000	395@1400	4.11x4.18	21.5:1	40-70@2000
	M	7.3 (7292)	IDI	185@3000 [5]	360@1400 [6]	4.11x4.18	21.5:1	40-70@2000
	G	7.5 (7538)	EFI	245@4000	400@2200	4.36x3.85	8.5:1	40-65@2000

MFI - Multiport fuel injection
DDI - Direct diesel injection
1 Explorer and Aerostar: 155@4200
  Ranger: 160@4200
2 Aerostar: 215@2400
  Explorer: 220@2400
  Ranger: 225@2400
3 8500 lbs. GVWR-up: 310@2800
4 High altitude: 160@3300
5 High altitude: 305@1400
6 E-Series 3 speed automatic: 150@3400
  E-Series 4 speed OD automatic: 145@3400
  F-Series 5 speed manual or 4 speed OD automatic and 2.73 rear axle ratio: 145@3400
  F-Series 5 speed HD or 3 speed automatic: 150@3400
7 3 speed automatic: 260@2000
  4 speed OD automatic: 265@2000
  5 speed manual OD or 4 speed automatic OD: 265@2000
  5 speed manual HD or 3 speed automatic: 260@2000

GASOLINE ENGINE TUNE-UP SPECIFICATIONS

Year	Engine ID/VIN	Engine Displacement Liters (cc)	Spark Plugs Gap (in.)	Ignition Timing (deg.) MT	Ignition Timing (deg.) AT	Fuel Pump (psi)	Idle Speed (rpm) MT	Idle Speed (rpm) AT	Valve Clearance In.	Valve Clearance Ex.
1991	Y	4.9 (4916)	0.044	10B	10B	50-60	700	575	HYD	HYD
	N	5.0 (4949)	0.044	10B	10B	35-45	775	675	HYD	HYD
	H	5.8 (5752)	0.044	10B	10B	35-45	775	675	HYD	HYD
	G	7.5 (7538)	0.044	10B	10B	35-45	775	675	HYD	HYD
1992	Y	4.9 (4916)	0.044	10B	10B	50-60	700	575	HYD	HYD
	N	5.0 (4949)	0.044	10B	10B	35-45	775	675	HYD	HYD
	H	5.8 (5752)	0.044	10B	10B	35-45	775	675	HYD	HYD
	G	7.5 (7538)	0.044	10B	10B	35-45	775	675	HYD	HYD
1993	Y	4.9 (4916)	0.044	10B	10B	50-60	700	575	HYD	HYD
	N	5.0 (4949)	0.044	10B	10B	35-45	775	675	HYD	HYD
	H	5.8 (5752)	0.044	10B	10B	35-45	775	675	HYD	HYD
	R	5.8 (5752)	0.044	10B	10B	35-45	775	675	HYD	HYD
	G	7.5 (7538)	0.044	10B	10B	35-45	775	675	HYD	HYD
1994-95	Y	4.9 (4916)	0.044	10B	10B	50-60	700	575	HYD	HYD
	N	5.0 (4949)	0.044	10B	10B	35-45	775	675	HYD	HYD
	H	5.8 (5752)	0.044	10B	10B	35-45	775	675	HYD	HYD
	R	5.8 (5752)	0.044	10B	10B	35-45	775	675	HYD	HYD
	G	7.5 (7538)	0.044	10B	10B	35-45	775	675	HYD	HYD

NOTE: The Vehicle Emission Control Information label often reflects specification changes made during production. The label figures must be used if they differ from above

B - Before top dead center

HYD - Hydraulic

DIESEL ENGINE TUNE-UP SPECIFICATIONS

Year	Engine ID/VIN	Engine Displacement cu. in. (cc)	Valve Clearance Intake (in.)	Valve Clearance Exhaust (in.)	Intake Valve Opens (deg.)	Injection Pump Setting (deg.)	Injection Nozzle Pressure (psi) New	Injection Nozzle Pressure (psi) Used	Idle Speed (rpm)	Cranking Compression Pressure (psi)
1991	M	7.3 (7292)	HYD	HYD	-	8.5B [1]	1875	1425	[2]	190-440 [3]
1992	M	7.3 (7292)	HYD	HYD	-	8.5B [1]	1875	1425	[2]	190-440 [3]
1993	M	7.3 (7292)	HYD	HYD	-	8.5B [1]	1875	1425	[2]	190-440 [3]
1994-95	M	7.3 (7292)	HYD	HYD	-	8.5B [1]	1875	1425	[2]	[3]
	F	7.3 (7292)	HYD	HYD	-	[2]	NA	NA	[2]	[3]
	K	7.3 (7292)	HYD	HYD	-	[2]	NA	NA	[2]	[3]

NOTE: The Vehicle Emission Control Information label often reflects changes made during production. These figures must be used if they disagree from the above data

HYD - Hydraulic

B - Before top dead center

1 At 2000 rpm

2 See underhood emission label

3 Compression pressure in the lowest cylinder must be at least 75% of the highest cylinder

Minimum pressure: 195 psi; Maximum pressure: 440 psi

FIRING ORDERS

NOTE: To avoid confusion, always replace spark plug wires one at a time.

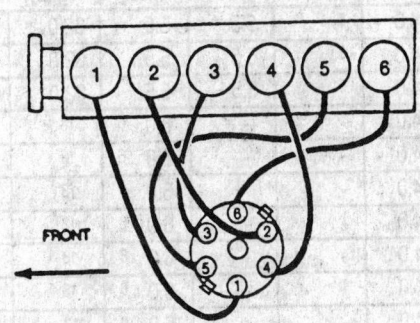

4.9L Engine
Engine Firing Order: 1–5–3–6–2–4
Distributor Rotation: Clockwise

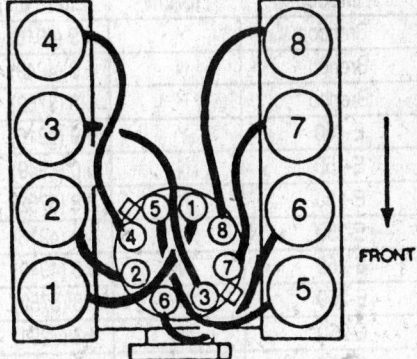

5.0L and 7.5L Engines
Engine Firing Order: 1–5–4–2–6–3–7–8
Distributor Rotation: Counterclockwise

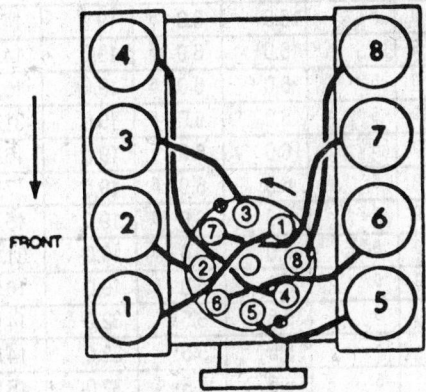

5.8L Engine
Engine Firing Order: 1–3–7–2–6–5–4–8
Distributor Rotation: Counterclockwise

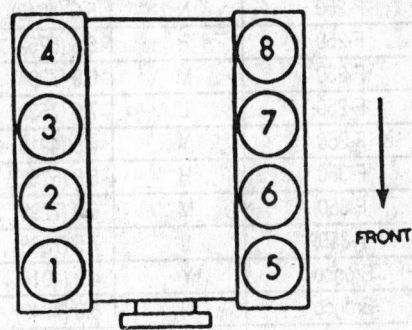

7.3L Diesel Engine
Engine Firing Order: 1–2–7–3–4–5–6–8

CAPACITIES

Year	Model	Engine ID/VIN	Engine Displacement Liters (cc)	Engine Crankcase with Filter	Transmission (pts.)			Transfer Case (pts.)	Drive Axle		Fuel Tank (gal.)	Cooling System (qts.)
					4-Spd	5-Spd	Auto.		Front (pts.)	Rear (pts.)		
1991	Bronco	Y	4.9 (4916)	6.0	7.0	7.0	24.0	5	5.5	5.5	32.0	14.0
	Bronco	N	5.0 (4949)	6.0	7.0 [4]	7.0	24.0	5	5.5	5.5	32.0	14.0
	Bronco	H	5.8 (5752)	6.0	7.0 [4]	7.0	24.0	5	5.5	5.5	32.0	15.0
	E-150	Y	4.9 (4916)	6.0	7.0 [4]	7.0	24.0	-	-	6.0 [6]	13	17.5
	E-150	N	5.0 (4949)	6.0	7.0 [4]	7.0	24.0	-	-	6.0 [6]	13	7
	E-150	H	5.8 (5752)	6.0	7.0 [4]	7.0	24.0	-	-	6.0 [6]	13	8
	E-250	Y	4.9 (4916)	6.0	7.0 [4]	7.0	24.0	-	-	6.0 [6]	13	17.5
	E-250	N	5.0 (4949)	6.0	7.0 [4]	7.0	24.0	-	-	6.0 [6]	13	7
	E-250	H	5.8 (5752)	6.0	7.0 [4]	7.0	24.0	-	-	6.0 [6]	13	8
	E-250	M	7.3 (7294)	10.0	7.0 [4]	7.0	24.0	-	-	6.0 [6]	13	31.0
	E-250	L	7.5 (7538)	6.0	7.0 [4]	7.0	24.0	-	-	6.0 [6]	13	28.0
	E-350	Y	4.9 (4916)	6.0	7.0 [4]	7.0	24.0	-	-	6.0 [6]	13	17.5
	E-350	H	5.8 (5752)	6.0	7.0 [4]	7.0	24.0	-	-	6.0 [6]	13	8
	E-350	M	7.3 (7294)	10.0	7.0 [4]	7.0	24.0	-	-	6.0 [6]	13	31.0
	E-350	L	7.5 (7538)	6.0	7.0 [4]	7.0	24.0	-	-	6.0 [6]	13	28.0
	F-150	Y	4.9 (4916)	6.0	7.0 [4]	7.0	24.0 [9]	5	6.0	6.0 [6]	19.0	14.0
	F-150	N	5.0 (4949)	6.0	7.0 [4]	7.0	24.0 [9]	5	6.0	6.0 [6]	19.0	14.0
	F-150	H	5.8 (5752)	6.0	7.0 [4]	7.0	24.0 [9]	5	6.0	6.0 [6]	19.0	15.0
	F-250	Y	4.9 (4916)	6.0	7.0 [4]	7.0	24.0 [9]	5	6.0	6.0 [6]	19.0	14.0
	F-250	N	5.0 (4949)	6.0	7.0 [4]	7.0	24.0 [9]	5	6.0	6.0 [6]	19.0	14.0
	F-250	H	5.8 (5752)	6.0	7.0 [4]	7.0	24.0 [9]	5	6.0	6.0 [6]	19.0	15.0
	F-250	M	6.9 (7294)	10.0	7.0 [4]	7.0	24.0 [9]	5	6.0	6.0 [6]	19.0	31.0
	F-250	L	7.5 (7538)	6.0	7.0 [4]	7.0	24.0 [9]	5	6.0	6.0 [6]	19.0	16.0
	F-350	Y	4.9 (4916)	6.0	7.0 [4]	7.0	24.0 [9]	5	6.0	6.0 [6]	19.0	17.5
	F-350	H	5.8 (5752)	6.0	7.0 [4]	7.0	24.0 [9]	5	6.0	6.0 [6]	19.0	15.0
	F-350	M	6.9 (7294)	10.0	7.0 [4]	7.0	24.0 [9]	5	6.0	6.0 [6]	19.0	31.0
	F-350	L	7.5 (7538)	6.0	7.0 [4]	7.0	24.0 [9]	5	6.0	6.0 [6]	19.0	16.0
1992	Bronco	Y	4.9 (4916)	6.0	7.0	7.0	24.0	5	5.5	5.5	32.0	14.0
	Bronco	N	5.0 (4949)	6.0	7.0 [4]	7.0	24.0	5	5.5	5.5	32.0	14.0
	Bronco	H	5.8 (5752)	6.0	7.0 [4]	7.0	24.0	5	5.5	5.5	32.0	15.0
	E-150	Y	4.9 (4916)	6.0	7.0	7.0	24.0	-	-	6.0 [6]	13	17.5
	E-150	N	5.0 (4949)	6.0	7.0 [4]	7.0	24.0	-	-	6.0 [6]	13	7
	E-150	H	5.8 (5752)	6.0	7.0 [4]	7.0	24.0	-	-	6.0 [6]	13	8
	E-250	Y	4.9 (4916)	6.0	7.0	7.0	24.0	-	-	6.0 [6]	13	17.5
	E-250	N	5.0 (4949)	6.0	7.0 [4]	7.0	24.0	-	-	6.0 [6]	13	7
	E-250	H	5.8 (5752)	6.0	7.0 [4]	7.0	24.0	-	-	6.0 [6]	13	8
	E-250	M	7.3 (7294)	10.0	7.0 [4]	7.0	24.0	-	-	6.0 [6]	13	31.0
	E-250	L	7.5 (7538)	6.0	7.0 [4]	7.0	24.0	-	-	6.0 [6]	13	28.0
	E-350	Y	4.9 (4916)	6.0	7.0	7.0	24.0	-	-	6.0 [6]	13	17.5
	E-350	H	5.8 (5752)	6.0	7.0 [4]	7.0	24.0	-	-	6.0 [6]	13	8
	E-350	M	7.3 (7294)	10.0	7.0 [4]	7.0	24.0	-	-	6.0 [6]	13	31.0
	E-350	L	7.5 (7538)	6.0	7.0 [4]	7.0	24.0	-	-	6.0 [6]	13	28.0
	F-150	Y	4.9 (4916)	6.0	7.0	7.0	24.0 [9]	5	6.0	6.0 [6]	19.0	14.0
	F-150	N	5.0 (4949)	6.0	7.0 [4]	7.0	24.0 [9]	5	6.0	6.0 [6]	19.0	14.0
	F-150	H	5.8 (5752)	6.0	7.0 [4]	7.0	24.0 [9]	5	6.0	6.0 [6]	19.0	15.0
	F-250	Y	4.9 (4916)	6.0	7.0	7.0	24.0 [9]	5	6.0 [6]	6.0 [6]	19.0	14.0

CAPACITIES

Year	Model	Engine ID/VIN	Engine Displacement Liters (cc)	Engine Crankcase with Filter	Transmission (pts.)			Transfer Case (pts.)	Drive Axle		Fuel Tank (gal.)	Cooling System (qts.)
					4-Spd	5-Spd	Auto.		Front (pts.)	Rear (pts.)		
	F-250	N	5.0 (4949)	6.0	7.0 [4]	7.0	24.0 [9]	5	6.0 [6]	6.0 [6]	19.0	14.0
	F-250	H	5.8 (5752)	6.0	7.0 [4]	7.0	24.0 [9]	5	6.0 [6]	6.0 [6]	19.0	15.0
	F-250	M	6.9 (7294)	10.0	7.0 [4]	7.0	24.0 [9]	5	6.0 [6]	6.0 [6]	19.0	31.0
	F-250	L	7.5 (7538)	6.0	7.0 [4]	7.0	24.0 [9]	5	6.0 [6]	6.0 [6]	19.0	16.0
	F-350	Y	4.9 (4916)	6.0	7.0	7.0	24.0 [9]	5	6.0 [6]	6.0 [6]	19.0	17.5
	F-350	H	5.8 (5752)	6.0	7.0	7.0	24.0 [9]	5	6.0 [6]	6.0 [6]	19.0	15.0
	F-350	M	6.9 (7294)	10.0	7.0 [4]	7.0	24.0 [9]	5	6.0 [6]	6.0 [6]	19.0	31.0
	F-350	L	7.5 (7538)	6.0	7.0 [4]	7.0	24.0 [9]	5	6.0 [6]	6.0 [6]	19.0	16.0
1993	Bronco	Y	4.9 (4916)	6.0	7.0	7.0	24.0	5	5.5	5.5	32.0	14.0
	Bronco	N	5.0 (4949)	6.0	7.0 [4]	7.0	24.0	5	5.5	5.5	32.0	14.0
	Bronco	H	5.8 (5752)	6.0	7.0 [4]	7.0	24.0	5	5.5	5.5	32.0	15.0
	E-150	Y	4.9 (4916)	6.0	7.0	7.0	24.0	-	-	6.0 [6]	[13]	17.5
	E-150	N	5.0 (4949)	6.0	7.0 [4]	7.0	24.0	-	-	6.0 [6]	[13]	7
	E-150	H	5.8 (5752)	6.0	7.0 [4]	7.0	24.0	-	-	6.0 [6]	[13]	8
	E-250	Y	4.9 (4916)	6.0	7.0 [4]	7.0	24.0	-	-	6.0 [6]	[13]	17.5
	E-250	N	5.0 (4949)	6.0	7.0 [4]	7.0	24.0	-	-	6.0 [6]	[13]	7
	E-250	H	5.8 (5752)	6.0	7.0 [4]	7.0	24.0	-	-	6.0 [6]	[13]	8
	E-250	M	7.3 (7294)	10.0	7.0 [4]	7.0	24.0	-	-	6.0 [6]	[13]	31.0
	E-250	L	7.5 (7538)	6.0	7.0 [4]	7.0	24.0	-	-	6.0 [6]	[13]	28.0
	E-350	Y	4.9 (4916)	6.0	7.0 [4]	7.0	24.0	-	-	6.0 [6]	[13]	17.5
	E-350	H	5.8 (5752)	6.0	7.0 [4]	7.0	24.0	-	-	6.0 [6]	[13]	8
	E-350	M	7.3 (7294)	10.0	7.0 [4]	7.0	24.0	-	-	6.0 [6]	[13]	31.0
	E-350	L	7.5 (7538)	6.0	7.0 [4]	7.0	24.0	-	-	6.0 [6]	[13]	28.0
	F-150	Y	4.9 (4916)	6.0	7.0 [4]	7.0	24.0 [9]	5	6.0	6.0 [6]	19.0	14.0
	F-150	N	5.0 (4949)	6.0	7.0 [4]	7.0	24.0 [9]	5	6.0	6.0 [6]	19.0	14.0
	F-150	H	5.8 (5752)	6.0	7.0 [4]	7.0	24.0 [9]	5	6.0	6.0 [6]	19.0	15.0
	F-250	Y	4.9 (4916)	6.0	7.0 [4]	7.0	24.0 [9]	5	6.0	6.0 [6]	19.0	14.0
	F-250	N	5.0 (4949)	6.0	7.0 [4]	7.0	24.0 [9]	5	6.0	6.0 [6]	19.0	14.0
	F-250	H	5.8 (5752)	6.0	7.0 [4]	7.0	24.0 [9]	5	6.0	6.0 [6]	19.0	15.0
	F-250	M	6.9 (7294)	10.0	7.0 [4]	7.0	24.0 [9]	5	6.0	6.0 [6]	19.0	31.0
	F-250	L	7.5 (7538)	6.0	7.0 [4]	7.0	24.0 [9]	5	6.0	6.0 [6]	19.0	16.0
	F-350	Y	4.9 (4916)	6.0	7.0 [4]	7.0	24.0 [9]	5	6.0	6.0 [6]	19.0	17.5
	F-350	H	5.8 (5752)	6.0	7.0 [4]	7.0	24.0 [9]	5	6.0	6.0 [6]	19.0	15.0
	F-350	M	7.3 (7294)	10.0	7.0 [4]	7.0	24.0 [9]	5	6.0	6.0 [6]	19.0	31.0
	F-350	L	7.5 (7538)	6.0	7.0 [4]	7.0	24.0 [9]	5	6.0	6.0 [6]	19.0	16.0
1994-95	Bronco	Y	4.9 (4916)	6.0	7.0	7.0	24.0	5	5.5	5.5	32.0	14.0
	Bronco	N	5.0 (4949)	6.0	7.0 [4]	7.0	24.0	5	5.5	5.5	32.0	14.0
	Bronco	H	5.8 (5752)	6.0	7.0 [4]	7.0	24.0	5	5.5	5.5	32.0	15.0
	E-150	Y	4.9 (4916)	6.0	7.0	7.0	24.0	-	-	6.0 [6]	[13]	17.5
	E-150	H	5.0 (4949)	6.0	7.0 [4]	7.0	24.0	-	-	6.0 [6]	[13]	7
	E-150	N	5.8 (5752)	6.0	7.0 [4]	7.0	24.0	-	-	6.0 [6]	[13]	8
	E-250	Y	4.9 (4916)	6.0	7.0 [4]	7.0	24.0	-	-	6.0 [6]	[13]	17.5
	E-250	N	5.0 (4949)	6.0	7.0 [4]	7.0	24.0	-	-	6.0 [6]	[13]	7
	E-250	H	5.8 (5752)	6.0	7.0 [4]	7.0	24.0	-	-	6.0 [6]	[13]	8
	E-250	M	7.3 (7294)	10.0	7.0 [4]	7.0	24.0	-	-	6.0 [6]	[13]	31.0
	E-250	L	7.5 (7538)	6.0	7.0 [4]	7.0	24.0	-	-	6.0 [6]	[13]	28.0

CAPACITIES

Year	Model	Engine ID/VIN	Engine Displacement Liters (cc)	Engine Crankcase with Filter	Transmission (pts.) 4-Spd	5-Spd	Auto.	Transfer Case (pts.)	Drive Axle Front (pts.)	Rear (pts.)	Fuel Tank (gal.)	Cooling System (qts.)
	E-350	Y	4.9 (4916)	6.0	7.0 4	7.0	24.0	-	-	6.0 6	13	17.5
	E-350	H	5.8 (5752)	6.0	7.0 4	7.0	24.0	-	-	6.0 6	13	8
	E-350	M	7.3 (7294)	10.0	7.0 4	7.0	24.0	-	-	6.0 6	13	31.0
	E-350	L	7.5 (7538)	6.0	7.0 4	7.0	24.0	-	-	6.0 6	13	28.0
	F-150	Y	4.9 (4916)	6.0	7.0 4	7.0	24.0 9	5	6.0	6.0 6	19.0	14.0
	F-150	N	5.0 (4949)	6.0	7.0 4	7.0	24.0 9	5	6.0	6.0 6	19.0	14.0
	F-150	H	5.8 (5752)	6.0	7.0 4	7.0	24.0 9	5	6.0	6.0 6	19.0	15.0
	F-250	Y	4.9 (4916)	6.0	7.0 4	7.0	24.0 9	5	6.0	6.0 6	19.0	14.0
	F-250	N	5.0 (4949)	6.0	7.0 4	7.0	24.0 9	5	6.0	6.0 6	19.0	14.0
	F-250	H	5.8 (5752)	6.0	7.0 4	7.0	24.0 9	5	6.0	6.0 6	19.0	15.0
	F-250	M	6.9 (7294)	10.0	7.0 4	7.0	24.0 9	5	6.0	6.0 6	19.0	31.0
	F-250	L	7.5 (7538)	6.0	7.0 4	7.0	24.0 9	5	6.0	6.0 6	19.0	16.0
	F-350	Y	4.9 (4916)	6.0	7.0 4	7.0	24.0 9	5	6.0	6.0 6	19.0	17.5
	F-350	H	5.8 (5752)	6.0	7.0 4	7.0	24.0 9	5	6.0	6.0 6	19.0	15.0
	F-350	M	7.3 (7294)	10.0	7.0 4	7.0	24.0 9	5	6.0	6.0 6	19.0	31.0
	F-350	L	7.5 (7538)	6.0	7.0 4	7.0	24.0 9	5	6.0	6.0 6	19.0	16.0

1 BW 13-50 manual shift: 3 pts. Dextron II
   BW 13-50 electric shift: 6.5 pts. Dextron II
   BW 13-54 mechanical shift: 3 pts. Dextron II
   BW 13-59 contains no lubricant and none should be added
2 6.75" ring gear: 3 pts.
   7.50" ring gear: 5 pts.
3 Mazda trans.: 3 pts.
   Mitsubishi trans.: 4.8 pts.
4 With OD: 4.5 pts.
5 New Process: 9 pts. Dextron II
   BW 1345: 6.5 pts. Dextron II
   BW 1356: 4 pts. Mercon
6 Heavy duty: 7.5 pts.
7 Manual trans.: 17.5 pts.
8 Manual trans.: 15 pts.
   Automatic trans.: 21 pts.
9 4WD: 27 pts.

10 Front axle Dana 28: 1.1 pts.
    Front axle Dana 35: 3.5 pts.
    Rear axle: 5.5 pts.
11 Short wheelbase: 17 gals.
    Long wheelbase: 17 or 21 gals.
    Ranger Supercab: 17 pr 21 gals.
12 2WD: 19.4 pts.
    4WD: 20.0 pts.
13 124" Wheelbase: 18 gals.
    138", 158" and 176" Wheelbase and front-mounted tank: 22 gals.
    138", 158" and 176" Wheelbase and rear-mounted tank: 16 gals.
14 Without cargo pkg, without rear heater: 11 3/8 qts.
    With cargo pkg, without rear heater: 12 1/4 qts.
    Without cargo pkg, with rear heater: 12 3/4 qts.
    With cargo pkg. and rear heater: 13 3/4 qts.

CAMSHAFT SPECIFICATIONS
All measurements given in inches.

Year	Engine ID/VIN	Engine Displacement Liters (cc)	Journal Diameter					Elevation		Bearing Clearance	Camshaft End Play
			1	2	3	4	5	In.	Ex.		
1991	Y	4.9 (4916)	2.017-2.018	2.017-2.018	2.017-2.018	2.017-2.018	NA	0.249-0.247	0.249-0.247	0.001-0.003	0.001-0.007
	N	5.0 (4949)	2.080	2.065	2.050	2.035	2.020	0.237	0.247	0.001-0.003	0.001-0.007
	H	5.8 (5752)	2.081	2.066	2.051	2.036	2.021	0.260	0.260	0.001-0.003	0.001-0.007
	M	7.3 (7294)	2.099-2.100	2.099-2.100	2.099-2.100	2.099-2.100	2.099-2.100	NA	NA	0.001-0.003	0.002-0.009
	G	7.5 (7538)	2.124-2.125	2.124-2.125	2.124-2.125	2.124-2.125	2.124-2.125	0.252	0.278	0.001-0.003	0.001-0.006
1992	Y	4.9 (4916)	2.017-2.018	2.017-2.018	2.017-2.018	2.017-2.018	NA	0.249-0.247	0.249-0.247	0.001-0.003	0.001-0.007
	N	5.0 (4949)	2.080	2.065	2.050	2.035	2.020	0.237	0.247	0.001-0.003	0.001-0.007
	H	5.8 (5752)	2.081	2.066	2.051	2.036	2.021	0.260	0.260	0.001-0.003	0.001-0.007
	M	7.3 (7294)	2.099-2.100	2.099-2.100	2.099-2.100	2.099-2.100	2.099-2.100	NA	NA	0.001-0.003	0.002-0.009
	G	7.5 (7538)	2.124-2.125	2.124-2.125	2.124-2.125	2.124-2.125	2.124-2.125	0.252	0.278	0.001-0.003	0.001-0.006
1993	Y	4.9 (4916)	2.017-2.018	2.017-2.018	2.017-2.018	2.017-2.018	NA	0.249-0.247	0.249-0.247	0.001-0.003	0.001-0.007
	N	5.0 (4949)	2.080	2.065	2.050	2.035	2.020	0.237	0.247	0.001-0.003	0.001-0.007
	H	5.8 (5752)	2.081	2.066	2.051	2.036	2.021	0.278	0.283	0.001-0.003	0.001-0.007
	R	5.8 (5752)	2.081	2.066	2.051	2.036	2.021	0.260	0.278	0.001-0.003	0.001-0.007
	M	7.3 (7294)	2.099-2.100	2.099-2.100	2.099-2.100	2.099-2.100	2.099-2.100	NA	NA	0.001-0.003	0.002-0.009
	G	7.5 (7538)	2.124-2.125	2.124-2.125	2.124-2.125	2.124-2.125	2.124-2.125	0.252	0.278	0.001-0.003	0.001-0.006
1994-95	Y	4.9 (4916)	2.017-2.018	2.017-2.018	2.017-2.018	2.017-2.018	NA	0.249-0.247	0.249-0.247	0.001-0.003	0.001-0.007
	N	5.0 (4949)	2.080	2.065	2.050	2.035	2.020	0.237	0.247	0.001-0.003	0.001-0.007
	H	5.8 (5752)	2.081	2.066	2.051	2.036	2.021	2.078	0.283	0.001-0.003	0.001-0.007
	R	5.8 (5752)	2.081	2.066	2.051	2.036	2.021	0.260	0.278	0.001-0.003	0.001-0.007
	M	7.3 (7294)	2.099-2.100	2.099-2.100	2.099-2.100	2.099-2.100	2.099-2.100	NA	NA	0.001-0.003	0.002-0.009
	G	7.5 (7538)	2.124-2.125	2.124-2.125	2.124-2.125	2.124-2.125	2.124-2.125	0.252	0.278	0.001-0.003	0.001-0.006

NA - Not Available

CRANKSHAFT AND CONNECTING ROD SPECIFICATIONS

All measurements are given in inches.

Year	Engine ID/VIN	Engine Displacement Liters (cc)	Crankshaft				Connecting Rod		
			Main Brg. Journal Dia.	Main Brg. Oil Clearance	Shaft End-play	Thrust on No.	Journal Diameter	Oil Clearance	Side Clearance
1991	Y	4.9 (4916)	2.3982-2.3990	0.0009-0.0028	0.004-0.008	5	2.1228-2.1236	0.0009-0.0027	0.006-0.013
	N	5.0 (4949)	2.2482-2.2490	0.0008-0.0015	0.004-0.008	3	2.1228-2.1236	0.0008-0.0015	0.010-0.020
	H	5.8 (5752)	2.9994-3.0002	0.0008-0.0015	0.004-0.008	3	2.3103-2.3111	0.0008-0.0026	0.010-0.020
	M	7.3 (7292)	3.1228-3.1236	0.0018-0.0046	0.025-0.085	3	2.4980-2.4990	0.0011-0.0036	0.012-0.024
	G	7.5 (7538)	2.9994-3.0002	0.0008-0.0026	0.004-0.008	3	2.4992-2.5000	0.0008-0.0025	0.010-0.020
1992	Y	4.9 (4916)	2.3982-2.3990	0.0009-0.0028	0.004-0.008	5	2.1228-2.1236	0.0009-0.0027	0.006-0.013
	N	5.0 (4949)	2.2482-2.2490	0.0008-0.0015	0.004-0.008	3	2.1228-2.1236	0.0008-0.0015	0.010-0.020
	H	5.8 (5752)	2.9994-3.0002	0.0008-0.0015	0.004-0.008	3	2.3103-2.3111	0.0008-0.0026	0.010-0.020
	M	7.3 (7292)	3.1228-3.1236	0.0018-0.0046	0.025-0.085	3	2.4980-2.4990	0.0011-0.0036	0.012-0.024
	G	7.5 (7538)	2.9994-3.0002	0.0008-0 0026	0.004-0.008	3	2.4992-2.5000	0.0008-0.0025	0.010-0.020
1993	Y	4.9 (4916)	2.3982-2.3990	0.0009-0.0028	0.004-0.008	5	2.1228-2.1236	0.0009-0.0027	0.006-0.013
	N	5.0 (4949)	2.2482-2.2490	0.0008-0.0015	0.004-0.008	3	2.1228-2.1236	0.0008-0.0015	0.010-0.020
	H	5.8 (5752)	2.9994-3.0002	0.0008-0.0015	0.004-0.008	3	2.3103-2.3111	0.0008-0.0026	0.010-0.020
	M	7.3 (7292)	3.1228-3.1236	0.0018-0.0046	0.025-0.085	3	2.4980-2.4990	0.0011-0.0036	0.012-0.024
	G	7.5 (7538)	2.9994-3.0002	0.0008-0.0026	0.004-0.008	3	2.4992-2.5000	0.0008-0.0025	0.010-0.020
1994-95	Y	4.9 (4916)	2.3982-2.3990	0.0009-0.0028	0.004-0.008	5	2.1228-2.1236	0.0009-0.0027	0.006-0.013
	N	5.0 (4949)	2.2482-2.2490	0.0008-0.0015	0.004-0.008	3	2.1228-2.1236	0.0008-0.0015	0.010-0.020
	H	5.8 (5752)	2.9994-3.0002	0.0008-0.0015	0.004-0.008	3	2.3103-2.3111	0.0008-0.0026	0.010-0.020
	M	7.3 (7292)	3.1228-3.1236	0.0018-0.0046	0.025-0.085	3	2.4980-2.4990	0.0011-0.0036	0.012-0.024
	G	7.5 (7538)	2.9994-3.0002	0.0008-0.0026	0.004-0.008	3	2.4992-2.5000	0.0008-0.0025	0.010-0.020

VALVE SPECIFICATIONS

Year	Engine ID/VIN	Engine Displacement Liters (cc)	Seat Angle (deg.)	Face Angle (deg.)	Spring Test Pressure (lbs. @ in.)	Spring Installed Height (in.)	Stem-to-Guide Clearance (in.) Intake	Exhaust	Stem Diameter (in.) Intake	Exhaust
1991	Y	4.9 (4916)	45	44	192@1.18	1	0.0010-0.0027	0.0010-0.0027	0.3415-0.3420	0.3415-0.3420
	N	5.0 (4949)	45	44	200@1.35	2	0.0010-0.0027	0.0015-0.0032	0.3415-0.3420	0.3415-0.3420
	H	5.8 (5752)	45	44	200@1.20	3	0.0010-0.0027	0.0015-0.0032	0.3415-0.3420	0.3415-0.3420
	M	7.3 (7292)	4	4	80@1.83	5	0.0055	0.0055	0.3716-0.3723	0.3716-0.3723
	G	7.5 (7538)	45	44	220@1.33	1.83	0.0010-0.0027	0.0010-0.0027	0.3415-0.3420	0.3415-0.3420
1992	Y	4.9 (4916)	45	44	192@1.18	1	0.0010-0.0027	0.0010-0.0027	0.3415-0.3420	0.3415-0.3420
	N	5.0 (4949)	45	44	200@1.35	2	0.0010-0.0027	0.0015-0.0032	0.3415-0.3420	0.3415-0.3420
	H	5.8 (5752)	45	44	200@1.20	3	0.0010-0.0027	0.0015-0.0032	0.3415-0.3420	0.3415-0.3420
	M	7.3 (7292)	4	4	80@1.83	5	0.0055	0.0055	0.3716-0.3723	0.3716-0.3723
	G	7.5 (7538)	45	44	220@1.33	1.83	0.0010-0.0027	0.0015-0.0032	0.3415-0.3420	0.3415-0.3420
1993	Y	4.9 (4916)	45	44	192@1.18	1	0.0010-0.0027	0.0010-0.0027	0.3415-0.3420	0.3415-0.3420
	N	5.0 (4949)	45	44	200@1.35	2	0.0010-0.0027	0.0015-0.0032	0.3415-0.3420	0.3415-0.3420
	H	5.8 (5752)	45	44	200@1.20	3	0.0010-0.0027	0.0015-0.0032	0.3415-0.3420	0.3415-0.3420
	R	5.8 (5752)	45	44	200@1.20	3	0.0010-0.0027	0.0015-0.0032	0.3415-0.3420	0.3415-0.3420
	M	7.3 (7292)	4	4	80@1.83	5	0.0055	0.0055	0.3716-0.3723	0.3716-0.3723
	G	7.5 (7538)	45	44	220@1.33	1.83	0.0010-0.0027	0.0010-0.0027	0.3415-0.3420	0.3415-0.3420
1994-95	Y	4.9 (4916)	45	44	192@1.18	1	0.0010-0.0027	0.0010-0.0027	0.3415-0.3420	0.3415-0.3420
	N	5.0 (4949)	45	44	200@1.35	2	0.0010-0.0027	0.0015-0.0032	0.3415-0.3420	0.3415-0.3420
	H	5.8 (5752)	45	44	200@1.20	3	0.0010-0.0027	0.0010-0.0027	0.3415-0.3420	0.3415-0.3420
	R	5.8 (5752)	45	44	200@1.20	3	0.0010-0.0027	0.0010-0.0027	0.3415-0.3420	0.3415-0.3420
	M	7.3 (7292)	4	4	80@1.83	5	0.0055	0.0055	0.3716-0.3723	0.3716-0.3723
	G	7.5 (7538)	45	44	220@1.33	1.83	0.0010-0.0027	0.0010-0.0027	0.3415-0.3420	0.3415-0.3420

1 Intake: 1.64 in.
  Exhaust: 1.47 in.
2 Intake: 1.68 in.
  Exhaust: 1.59 in.
3 Intake: 1.78 in.
  Exhaust: 1.59 in.
4 Intake: 3.00
  Exhaust: 3.75
5 Intake: 1.767 in.

PISTON AND RING SPECIFICATIONS

All measurements are given in inches.

Year	Engine ID/VIN	Engine Displacement Liters (cc)	Piston Clearance	Ring Gap			Ring Side Clearance		
				Top Compression	Bottom Compression	Oil Control	Top Compression	Bottom Compression	Oil Control
1991	Y	4.9 (4916)	0.0014-0.0022	0.0100-0.0200	0.0100-0.0200	0.0100-0.0350	0.0019-0.0036	0.0020-0.0040	SNUG
	N	5.0 (4949)	0.0018-0.0026	0.0100-0.0200	0.0100-0.0200	0.0100-0.0350	0.0020-0.0040	0.0020-0.0040	SNUG
	H	5.8 (5752)	0.0022-0.0030	0.0100-0.0200	0.0100-0.0200	0.0100-0.0350	0.0019-0.0036	0.0020-0.0040	SNUG
	M	7.3 (7292)	0.0055-0.0085	0.0130-0.0450	0.0600-0.0800	0.0100-0.0240	0.0020-0.0040	0.0020-0.0040	SNUG
	G	7.5 (7538)	0.0022-0.0300	0.0100-0.0200	0.0100-0.0200	0.0100-0.0350	0.0025-0.0045	0.0025-0.0045	SNUG
1992	Y	4.9 (4916)	0.0014-0.0022	0.0100-0.0200	0.0100-0.0200	0.0100-0.0350	0.0019-0.0036	0.0020-0.0040	SNUG
	N	5.0 (4949)	0.0018-0.0026	0.0100-0.0200	0.0100-0.0200	0.0100-0.0350	0.0020-0.0040	0.0020-0.0040	SNUG
	H	5.8 (5752)	0.0022-0.0030	0.0100-0.0200	0.0100-0.0200	0.0150-0.0350	0.0019-0.0036	0.0020-0.0040	SNUG
	M	7.3 (7292)	0.0055-0.0085	0.0130-0.0450	0.0600-0.0800	0.0100-0.0240	0.0020-0.0040	0.0020-0.0040	0.0010-0.0030
	G	7.5 (7538)	0.0022-0.0300	0.0100-0.0200	0.0100-0.0200	0.0100-0.0350	0.0025-0.0045	0.0025-0.0045	SNUG
1993	Y	4.9 (4916)	0.0014-0.0022	0.0100-0.0200	0.0100-0.0200	0.0100-0.0350	0.0019-0.0036	0.0020-0.0040	SNUG
	N	5.0 (4949)	0.0014-0.0022	0.0100-0.0200	0.0180-0.0280	0.0100-0.0400	0.0020-0.0040	0.0020-0.0040	SNUG
	H	5.8 (5752)	0.0018-0.0026	0.0100-0.0200	0.0100-0.0200	0.0150-0.0350	0.0019-0.0036	0.0020-0.0040	SNUG
	R	5.8 (5752)	0.0015-0.0023	0.0100-0.0200	0.0180-0.0280	0.0100-0.0400	0.0020-0.0040	0.0013-0.0033	SNUG
	M	7.3 (7292)	0.0055-0.0085	0.0130-0.0450	0.0600-0.0800	0.0100-0.0240	0.0020-0.0040	0.0020-0.0040	0.0010-0.0030
	G	7.5 (7538)	0.0022-0.0300	0.0100-0.0200	0.0100-0.0200	0.0100-0.0350	0.0025-0.0045	0.0025-0.0045	SNUG
1994-95	Y	4.9 (4916)	0.0014-0.0022	0.0100-0.0200	0.0100-0.0200	0.0100-0.0350	0.0019-0.0036	0.0020-0.0040	SNUG
	N	5.0 (4949)	0.0014-0.0022	0.0100-0.0200	0.0180-0.0280	0.0100-0.0400	0.0020-0.0040	0.0020-0.0040	SNUG
	H	5.8 (5752)	0.0018-0.0026	0.0100-0.0200	0.0100-0.0200	0.0150-0.0350	0.0019-0.0036	0.0020-0.0040	SNUG
	R	5.8 (5752)	0.0015-0.0023	0.0100-0.0200	0.0180-0.0280	0.0100-0.0400	0.0020-0.0040	0.0013-0.0033	SNUG
	M	7.3 (7292)	0.0055-0.0085	0.0130-0.0450	0.0600-0.0800	0.0100-0.0240	0.0020-0.0040	0.0020-0.0040	0.0010-0.0030
	G	7.5 (7538)	0.0022-0.0300	0.0100-0.0200	0.0100-0.0200	0.0100-0.0350	0.0025-0.0045	0.0025-0.0045	SNUG

TORQUE SPECIFICATIONS
All readings in ft. lbs.

Year	Engine ID/VIN	Engine Displacement Liters (cc)	Cylinder Head Bolts	Main Bearing Bolts	Rod Bearing Bolts	Crankshaft Damper Bolts	Flywheel Bolts	Manifold Intake	Manifold Exhaust	Spark Plugs	Lug Nut
1991	Y	4.9 (4916)	1	60-70	40-45	130-150	75-85	22-32	22-32	10-15	2
	N	5.0 (4949)	3	95-105	19-24	70-90	75-85	23-25	18-24	10-15	2
	H	5.8 (5752)	3	60-70	19-24	70-90	75-90	23-25	20-24	10-15	2
	M	7.3 (7292)	4	5	6	90	47	23-25	20-24	-	2
	G	7.5 (7538)	7	95-105	45-50	70-90	75-85	22-32	28-33	5-10	2
1992	Y	4.9 (4916)	1	60-70	40-45	130-150	75-85	22-32	22-32	10-15	2
	N	5.0 (4949)	2	95-105	19-24	70-90	75-85	23-25	18-24	10-15	2
	H	5.8 (5752)	3	60-70	19-24	70-90	75-90	23-25	20-24	10-15	2
	M	7.3 (7292)	4	5	6	90	47	23-25	20-24	-	2
	G	7.5 (7538)	7	95-105	45-50	70-90	75-85	22-32	28-33	5-10	2
1993	Y	4.9 (4916)	1	60-70	40-45	130-150	75-85	22-32	22-32	10-15	2
	N	5.0 (4949)	3	95-105	19-24	70-90	75-85	23-25	18-24	10-15	2
	H	5.8 (5752)	3	60-70	19-24	70-90	75-90	23-25	20-24	10-15	2
	R	5.8 (5752)	3	60-70	19-24	70-90	75-90	23-25	20-24	10-15	2
	M	7.3 (7292)	4	5	6	90	47	23-25	20-24	-	2
	G	7.5 (7538)	7	95-105	95-105	70-90	75-85	22-32	28-33	5-10	2
1994-95	Y	4.9 (4916)	1	60-70	40-45	130-150	75-85	22-32	22-32	10-15	2
	N	5.0 (4949)	3	95-105	19-24	70-90	75-85	23-25	18-24	10-15	2
	H	5.8 (5752)	3	60-70	19-24	70-90	75-90	23-25	20-24	10-15	2
	R	5.8 (5752)	3	60-70	19-24	70-90	75-90	23-25	20-24	10-15	2
	M	7.3 (7292)	4	5	6	90	47	23-25	20-24	-	2
	G	7.5 (7538)	7	95-105	45-50	70-90	75-85	22-32	28-33	5-10	2

1 Step 1: 55 ft. lbs.
  Step 2: 65 ft. lbs.
  Step 3: 85 ft. lbs.
2 E-F100, E-F150, E-F250: 90 ft. lbs.
  E-F350 with single rear wheels: 135 ft. lbs.
  F350 with dual rear wheels: 210 ft. lbs.
3 Step 1: 55-65 ft. lbs.
  Step 2: 65-72 ft. lbs.
4 Step 1: 65 ft. lbs.
  Step 2: 90 ft. lbs.
  Step 3: 100 ft. lbs.
5 Step 1: 75 ft. lbs.
  Step 2: 95 ft. lbs.
6 Step 1: 38 ft. lbs.
  Step 2: 50-55 ft. lbs.
7 Step 1: 80 ft. lbs.
  Step 2: 110 ft. lbs.
  Step 3: 130-140 ft. lbs.

BRAKE SPECIFICATIONS

All measurements in inches unless noted

Year	Model		Master Cylinder Bore	Brake Disc			Brake Drum Diameter			Minimum Lining Thickness	
				Original Thickness	Minimum Thickness	Maximum Runout	Original Inside Diameter	Max. Wear Limit	Maximum Machine Diameter	Front	Rear
1991	Bronco		NA	1.160	1.120	0.003	11.03	11.09	11.06	0.030	0.030
	E-150		NA	1.160	1.120	0.003	11.03	11.09	11.06	0.030	0.030
	E-250		NA	1.220	1.180	0.003	12.00	12.09	12.06	0.030	0.030
	E-350		NA	1.220	1.180	0.003	12.00	12.09	12.06	0.030	0.030
	F-150		NA	1.160	1.120	0.003	11.03	11.09	11.06	0.030	0.030
	F-250		NA	1.220	1.180	0.003	12.00	12.09	12.06	0.030	0.030
	F-350		NA	1.220	1.180	0.003	12.00	12.09	12.06	0.030	0.030
	Super Duty		NA	1.220	1.180	0.008	12.00	12.09	12.06	0.030	0.030
1992	Bronco		NA	1.160	1.120	0.003	11.03	11.09	11.06	0.030	0.030
	E-150		NA	1.160	1.120	0.003	11.03	11.09	11.06	0.030	0.030
	E-250		NA	1.220	1.180	0.003	12.00	12.09	12.06	0.030	0.030
	E-350		NA	1.220	1.180	0.003	12.00	12.09	12.06	0.030	0.030
	F-150		NA	1.160	1.120	0.003	11.03	11.09	11.06	0.030	0.030
	F-250		NA	1.220	1.180	0.003	12.00	12.09	12.06	0.030	0.030
	F-350		NA	1.220	1.180	0.003	12.00	12.09	12.06	0.030	0.030
	Super Duty		NA	1.220	1.180	0.008	12.00	12.09	12.06	0.030	0.030
1993	Bronco		NA	1.160	1.120	0.003	11.03	11.09	11.06	0.030	0.030
	E-150		NA	1.160	1.120	0.003	11.03	11.09	11.06	0.030	0.030
	E-250		NA	1.220	1.180	0.003	12.00	12.09	12.06	0.030	0.030
	E-350		NA	1.220	1.180	0.003	12.00	12.09	12.06	0.030	0.030
	F-150		NA	1.160	1.120	0.003	11.03	11.09	11.06	0.030	0.030
	F-250		NA	1.220	1.180	0.003	12.00	12.09	12.06	0.030	0.030
	F-350		NA	1.220	1.180	0.003	12.00	12.09	12.06	0.030	0.030
	Super Duty		NA	1.220	1.180	0.008	12.00	12.09	12.06	0.030	0.030
1994-95	Bronco		NA	1.160	1.120	0.003	11.03	11.09	11.06	0.030	0.030
	E-150		NA	1.160	1.120	0.003	11.03	11.09	11.06	0.030	0.030
	E-250		NA	1.220	1.180	0.003	12.00	12.09	12.06	0.030	0.030
	E-350		NA	1.220	1.180	0.003	12.00	12.09	12.06	0.030	0.030
	F-150		NA	1.160	1.120	0.003	11.03	11.09	11.06	0.030	0.030
	F-250		NA	1.220	1.180	0.003	12.00	12.09	12.06	0.030	0.030
	F-350		NA	1.220	1.180	0.003	12.00	12.09	12.06	0.030	0.030
	Super Duty		NA	1.220	1.180	0.008	12.00	12.09	12.06	0.030	0.030

NA - Not Available

WHEEL ALIGNMENT

Year	Model	Ride Height (in.)	Caster Range (deg.)	Caster Preferred Setting (deg.)	Camber Range (deg.)	Camber Preferred Setting (deg.)	Toe-in (in.)
1991	Bronco	3 1/4-3 1/2	6P-8P	3	1 3/4N-1/4N	3	1/32
		3 1/2-3 3/4	5P-7P	3	3/4N-3/4P	3	1/32
		4-4 1/4	4P-6P	3	1/4P-1 3/4P	3	1/32
		4 1/4-4 1/2	3P-5P	3	1 1/4P-2 3/4P	3	1/32
	E-150	4-4 1/2	7 1/2P-9 1/2P	3	1 1/4N-1/4P	3	1/32
		4 1/2-5	6 1/4P-8 1/4P	3	1/8N-1 1/4P	3	1/32
		5-5 1/2	5P-7P	3	7/8P-2 1/4P	3	1/32
		5 1/2-5 3/4	3 1/4P-5 1/4P	3	1 3/4P-3 1/4P	3	1/32
	E-250	3 3/4-4	7 5/8P-9 5/8P	3	3/4N-1/2P	3	1/32
		4 1/4-4 1/2	6 1/4P-8 1/4P	3	1/4P-1 1/2P	3	1/32
		4 3/4-5	5P-7P	3	1 1/4P-2 1/2P	3	1/32
		5 1/4-5 1/2	3 3/4P-5 3/4P	3	2 1/4P-3 1/2P	3	1/32
	E-350	3 3/4-4	7 5/8P-9 5/8P	3	3/4N-1/2P	3	1/32
		4 1/4-4 1/2	6 1/4P-8 1/4P	3	1/4P-1 1/2P	3	1/32
		4 3/4-5	5P-7P	3	1 1/4P-2 1/2P	3	1/32
		5 1/4-5 1/2	3 3/4P-5 3/4P	3	2 1/4P-3 1/2P	3	1/32
	F-150 4x2	3 1/4-3 1/2	5P-7P	3	3/4N-3/4P	0	1/32
		3 1/2-4	4P-6P	3	1/4N-1 1/4P	1P	1/32
		4-4 1/4	3 1/4P-5 1/4P	3	1/2P-2P	1P	1/32
		4 1/4-4 3/4	2 1/2P-4 1/2P	3	2P-3 1/2P	2 1/2P	1/32
		4 3/4-5	1 1/2P-3 1/2P	3	3P-4 1/2P	3 1/2P	1/32
	F-150 4x4	3 1/4-3 1/2	6P-8P	4	1 3/4N-1/4N	1N	1/32
		3 1/2-3 3/4	5P-7P	4	3/4N-3/4P	0	1/32
		4-4 1/4	4P-6P	4	1/4P-1 3/4P	1 1/4P	1/32
		4 1/4-4 1/2	3P-5P	4	1 1/4P-2 3/4P	1 3/4P	1/32
	F-250 4x2	2 1/2-2 3/4	5 1/4P-7 1/4P	4	1N-1/2P	0	3/32
		2 3/4-3	5P-7P	4	1/2N-3/4P	0	3/32
		3 1/4-3 1/2	4 1/2P-6 1/2P	4	1/8N-1 1/4P	3/4P	3/32
		3 3/4-4	4P-7P	4	1/2P-1 3/4P	1P	3/32
		4-4 1/4	3 1/2P-5 1/2P	4	3/4P-2 1/4P	1 1/2P	3/32
	F-250 4x4	5-5 1/4	3P-5P	5	1 3/4N-1/4N	1N	1/32
		5 1/2-5 3/4	3 1/8P-5 1/8P	5	3/4N-3/4P	0	1/32
		6-6 1/4	3 1/4P-5 1/4P	5	1/2P-2P	1P	1/32

WHEEL ALIGNMENT

Year	Model	Ride Height (in.)	Caster Range (deg.)	Caster Preferred Setting (deg.)	Camber Range (deg.)	Camber Preferred Setting (deg.)	Toe-in (in.)
		6 1/4-6 1/2	3 3/8P-5 3/8P	5	1 1/2P-3P	2P	1/32
		6 3/4-7	3 1/2P-5 1/2P	5	2 1/2P-4P	3P	1/32
	F-350 4x2	2 1/2-2 3/4	5 1/4P-7 1/4P	4	1N-1/2P	0	3/32
		2 3/4-3	5P-7P	4	1/2N-3/4P	0	3/32
		3 1/4-3 1/2	4 1/2P-6 1/2P	4	1/8N-1 1/4P	3/4P	3/32
		3 3/4-4	4P-7P	4	1/2P-1 3/4P	1P	3/32
		4-4 1/4	3 1/2P-5 1/2P	4	3/4P-2 1/4P	1 1/2P	3/32
	F-350 4x4	5-5 1/4	3P-5P	5	1 3/4N-1/4N	1N	1/32
		5 1/2-5 3/4	3 1/8P-5 1/8P	5	3/4N-3/4P	0	1/32
		6-6 1/4	3 1/4P-5 1/4P	5	1/2P-2P	1P	1/32
		6 1/4-6 1/2	3 3/8P-5 3/8P	5	1 1/2P-3P	2P	1/32
		6 3/4-7	3 1/2P-5 1/2P	5	2 1/2P-4P	3P	1/32
1992	Bronco	3 1/4-3 1/2	6P-8P	3	1 3/4N-1/4N	1N	1/32
		3 1/2-3 3/4	5P-7P	3	3/4N-3/4P	0	1/32
		4-4 1/4	4P-6P	3	1/4P-1 3/4P	1 1/4P	1/32
		4 1/4-4 1/2	3P-5P.	3	1 1/4P-2 3/4P	1 3/4P	1/32
	E-150	4-4 1/2	7 1/2P-9 1/2P	3	1 1/4N-1/4P	3	1/32
		4 1/2-5	6 1/4P-8 1/4P	3	1/8N-1 1/4P	3	1/32
		5-5 1/2	5P-7P	3	7/8P-2 1/4P	3	1/32
		5 1/2-5 3/4	3 1/4P-5 1/4P	3	1 3/4P-3 1/4P	3	1/32
	E-250	3 3/4-4	7 5/8P-9 5/8P	3	3/4N-1/2P	3	1/32
		4 1/4-4 1/2	6 1/4P-8 1/4P	3	1/4P-1 1/2P	3	1/32
		4 3/4-5	5P-7P	3	1 1/4P-2 1/2P	3	1/32
		5 1/4-5 1/2	3 3/4P-5 3/4P	3	2 1/4P-3 1/2P	3	1/32
	E-350	3 3/4-4	7 5/8P-9 5/8P	3	3/4N-1/2P	3	1/32
		4 1/4-4 1/2	6 1/4P-8 1/4P	3	1/4P-1 1/2P	3	1/32
		4 3/4-5	5P-7P	3	1 1/4P-2 1/2P	3	1/32
		5 1/4-5 1/2	3 3/4P-5 3/4P	3	2 1/4P-3 1/2P	3	1/32
	F-150 4x2	3 1/4-3 1/2	5P-7P	3	3/4N-3/4P	0	1/32
		3 1/2-4	4P-6P	3	1/4N-1 1/4P	1P	1/32

WHEEL ALIGNMENT

Year	Model	Ride Height (in.)	Caster Range (deg.)	Caster Preferred Setting (deg.)	Camber Range (deg.)	Camber Preferred Setting (deg.)	Toe-in (in.)
1992		4-4 1/4	3 1/4P-5 1/4P	3	1/2P-2P	1P	1/32
		4 1/4-4 3/4	2 1/2P-4 1/2P	3	2P-3 1/2P	2 1/2P	1/32
		4 3/4-5	1 1/2P-3 1/2P	3	3P-4 1/2P	3 1/2P	1/32
	F-150 4x4	3 1/4-3 1/2	6P-8P	4	1 3/4N-1/4N	1N	1/32
		3 1/2-3 3/4	5P-7P	4	3/4N-3/4P	0	1/32
		4-4 1/4	4P-6P	4	1/4P-1 3/4P	1 1/4P	1/32
		4 1/4-4 1/2	3P-5P	4	1 1/4P-2 3/4P	1 3/4P	1/32
	F-250 4x2	2 1/2-2 3/4	5 1/4P-7 1/4P	4	1N-1/2P	0	3/32
		2 3/4-3	5P-7P	4	1/2N-3/4P	0	3/32
		3 1/4-3 1/2	4 1/2P-6 1/2P	4	1/8N-1 1/4P	3/4P	3/32
		3 3/4-4	4P-7P	4	1/2P-1 3/4P	1P	3/32
		4-4 1/4	3 1/12P-5 1/2P	4	3/4P-2 1/4P	1 1/2P	3/32
	F-250 4x4	5-5 1/5	3P-5P	5	1 3/4N-1/4N	1N	1/32
		5 1/2-5 3/4	3 1/8P-5 1/8P	5	3/4N-3/4P	0	1/32
		6-6 1/4	3 1/4P-5 1/4P	5	1/2P-2P	1P	1/32
		6 1/4-6 1/2	3 3/8P-5 3/8P	5	1 1/2P-3P	2P	1/32
		6 3/4-7	3 1/2P-5 1/2P	5	2 1/2P-4P	3P	1/32
	F-350 4x2	2 1/2-2 3/4	5 1/4P-7 1/4P	4	1N-1/2P	0	3/32
		2 3/4-3	5P-7P	4	1/2N-3/4P	0	3/32
		3 1/4-3 1/2	4 1/2P-6 1/2P	4	1/8N-1 1/4P	3/4P	3/32
		3 3/4-4	4P-7P	4	1/2P-1 3/4P	1P	3/32
		4-4 1/4	3 1/12P-5 1/2P	4	3/4P-2 1/4P	1 1/2P	3/32
	F-350 4x4	5-5 1/5	3P-5P	5	1 3/4N-1/4N	1N	1/32
		5 1/2-5 3/4	3 1/8P-5 1/8P	5	3/4N-3/4P	0	1/32
		6-6 1/4	3 1/4P-5 1/4P	5	1/2P-2P	1P	1/32
		6 1/4-6 1/2	3 3/8P-5 3/8P	5	1 1/2P-3P	2P	1/32
		6 3/4-7	3 1/2P-5 1/2P	5	2 1/2P-4P	3P	1/32
1993	Bronco	8	2P-6P	-	-	1/4P	1/32
	E-150	8	2P-7 1/2P	-	1/4P-1/2P	1/4P	1/32

WHEEL ALIGNMENT

Year	Model	Ride Height (in.)	Caster Range (deg.)	Caster Preferred Setting (deg.)	Camber Range (deg.)	Camber Preferred Setting (deg.)	Toe-in (in.)
	E-250	8	2P-7 1/2P	-	0-1/2P	1/4P	1/32
	E-350	8	2P-7 1/2P	-	0-1/2P	1/4P	1/32
	F-150 4x2	8	2P-6P	-	-	1/4P	1/32
	F-150 4x4	8	2P-6P	-	1/4P-1/2P	1/4P	1/32
	F-250 4x2	8	2P-6P	-	-	1/4P	1/32
	F-250 4x4	8	2P-6P	-	-	1/4P	1/32
	F-350 4x2 ⑫	8	2P-6P	-	-	1/2P	1/32
	F-350 4x2 ⑬	8	2P-4 1/2P	-	-	1/2P	1/32
	F-350 4x4	8	2P-4 3/4P	-	1/4N-1/4P	0	1/32
	Super Duty 4x2 ⑭	8	2P-5P	3P	-	0	1/32
	Super Duty 4x2 ⑮	8	2P-5 1/2P	3P	-	5/8P	1/32
	F-150 4x4	8	2P-6P	4P	-	1/4P	1/32
	F-250 4x4	8	2P-5P	3P	-	1/4P	1/32
	F-350 4x4	8	2P-4 3/4P	3P	-	0	1/32
1994-95	Bronco	8	2P-6P	-	-	1/4P	1/32
	E-150	8	2P-7 1/2P	-	1/4P-1/2P	1/4P	1/32
	E-250	8	2P-7 1/2P	-	0-1/2P	1/4P	1/32
	E-350	8	2P-7 1/2P	-	0-1/2P	1/4P	1/32
	F-150 4x2	8	2P-6P	-	-	1/4P	1/32
	F-150 4x4	8	2P-6P	-	1/4P-1/2P	1/4P	1/32
	F-250 4x2	8	2P-6P	-	-	1/4P	1/32
	F-250 4x4	8	2P-5P	-	-	1/4P	1/32
	F-350 4x2 ⑫	8	2P-6P	-	-	1/2P	1/32
	F-350 4x2 ⑬	8	2P-4 1/2P	-	-	1/2P	1/32
	F-350 4x4 ⑭	8	2P-4 3/4P	-	1/4N-1/4P	0	1/32
	Super Duty 4x2 ⑮	8	2P-5P	3P	-	0	1/32
	Super Duty 4x2 ⑮	8	2P-5 1/2P	3P	-	5/8P	1/32
	F-150 4x4	8	2P-6P	4P	-	1/4P	1/32
	F-250 4x4	8	2P-5P	3P	-	1/4P	1/32
	F-350 4x4	8	2P-4 3/4P	3P	-	0	

WHEEL ALIGNMENT

Year	Model	Ride Height (in.)	Caster		Camber		Toe-in (in.)
			Range (deg.)	Preferred Setting (deg.)	Range (deg.)	Preferred Setting (deg.)	

1 With forged axle
2 With stamped axle
3 Not adjustable
4 Adjusted by placing shims between leaf springs and front axle
5 Adjusted by placing shims between leaf springs and front axle, except with Monobeam front suspension
6 Normal riding attitude with no more side to side clearance than 5/8 inch in front and 3/4 inch in rear
7 Caster and camber adjustment is possible with installation of service adjusters
8 Normal riding attitude with no more than side to side clearance than 5/8 inch in front and 3/4 inch in rear
9 Caster and camber adjustment is possible with installation of service adjusters
10 Front: 1/32; Rear: 0
11 Factory preset, not adjustable
12 SRW
13 DRW
14 Without stripped chassis
15 With stripped chassis

GASOLINE ENGINE MECHANICAL

Engine Assembly

REMOVAL & INSTALLATION

4.9L Engine

BRONCO AND F SERIES

1. Disconnect the negative battery cable and relieve the fuel system pressure. Drain the cooling system and the crankcase.

2. Mark the position of the hood on the hinges and remove the hood. Remove the throttle body inlet tubes.

3. Disconnect the heater hose from the water pump and coolant outlet housing. Disconnect the flexible fuel line from the fuel supply manifold.

4. Remove the radiator and shroud. Remove the cooling fan, water pump pulley and fan drive belt.

NOTE: The fan clutch/water pump hub has a right-hand thread.

5. If equipped with air conditioning, discharge the refrigerant from the system and remove the compressor and the condenser. Cap all openings.

6. Disconnect the accelerator cable at the throttle body. Remove the throttle return spring.

7. Disconnect the power brake booster vacuum hose at the intake manifold. If equipped with automatic transmission, disconnect the transmission kickdown cable at the throttle body.

8. Disconnect the exhaust pipe from the exhaust manifold. Disconnect the body ground strap and the battery ground cable from the engine.

9. Tag and disconnect the Electronic Engine Control (EEC) harness from all sensors.

10. Tag and disconnect the engine wiring harness at the ignition coil, the coolant temperature sending unit and the oil pressure sending unit. Position the wiring harness aside.

11. Remove the alternator mounting bolts and position the alternator aside, leaving the wires attached.

12. Remove the power steering pump from the mounting brackets and move aside, leaving the lines attached. If equipped with an air compressor, bleed the air system and disconnect the 2 air pressure lines at the compressor.

13. Raise and safely support the vehicle.

14. Remove the starter and automatic transmission filler tube bracket, if equipped. Remove the rear engine plate upper right bolt.

15. If equipped with a manual trans-

mission, remove the flywheel housing lower attaching bolts and disconnect the clutch return spring or clutch slave cylinder.

16. If equipped with an automatic transmission, remove the converter housing access cover assembly and remove the flywheel-to-converter attaching nuts. Secure the converter in the housing. Remove the transmission oil cooler lines from the retaining clip at the engine. Remove the lower converter housing-to-engine attaching bolts.

17. Remove the nut from each of the 2 front engine mounts.

18. Lower the vehicle and position a jack under the transmission to support it. Remove the remaining bell housing-to-engine attaching bolts.

19. Attach suitable engine lifting equipment. Raise the engine slightly and carefully pull it from the transmission. Lift the engine out of the vehicle.

To install:

20. Carefully lower the engine into the vehicle. Make sure the block dowels engage the holes in the bell housing.

21. If equipped with an automatic transmission, start the converter pilot into the crankshaft. Remove the retainer securing the converter in the housing.

22. If equipped with a manual transmission, start the transmission input shaft into the clutch disc. It may be

necessary to adjust the position of the transmission in relation to the engine if the transmission input shaft will not enter the clutch disc. If the engine hangs up after the shaft enters, turn the crankshaft slowly, with the transmission in gear, until the shaft splines mesh with the clutch disc splines.

23. Install the bell housing upper attaching bolts and remove the jack supporting the transmission. Lower the engine until it rests on the engine mounts and remove the lifting equipment.

24. Install the engine mount attaching nuts and washers and tighten to 54–74 ft. lbs. (73–100 Nm). Install the automatic transmission oil cooler lines bracket.

25. Install the remaining bell housing bolts and connect the clutch return spring.

26. Install the starter and connect the starter cable. Attach the automatic transmission fluid filler tube bracket, if equipped.

27. If equipped, install the transmission oil cooler lines in the bracket at the cylinder block.

28. Install the exhaust pipe to the exhaust manifold with the lockwashers and nuts.

29. Connect the engine ground strap and the battery ground cable. Connect the EEC harness to all sensors.

30. Connect the accelerator linkage to the throttle body and install the retracting spring. If equipped, connect the kickdown cable to the throttle body.

31. Reconnect the power brake vacuum hose to the intake manifold.

32. Connect the coil primary wire, oil pressure and coolant temperature sending unit wires, flexible fuel line, heater hoses and the positive battery cable.

33. Install the alternator and power steering pump.

34. Install the water pump pulley, cooling fan and drive belt. Tighten the fan bolts to 12–18 ft. lbs. (16–24 Nm).

NOTE: The fan clutch/water pump hub has a right-hand thread.

35. Install the radiator and shroud. Connect the lower radiator hose to the water pump and the upper radiator hose to the coolant outlet housing. Connect the air compressor lines. If removed, install the air conditioning compressor and condenser.

36. If equipped with an automatic transmission, connect the oil cooler lines.

37. Install the hood, aligning the marks that were made during removal.

38. Fill the crankcase with the proper type and quantity of engine oil. Fill and bleed the cooling system.

39. Connect the negative battery cable, start the engine and operate at fast idle. Check for leaks.

40. If equipped with a manual transmission, check for correct clutch operation. If equipped with an automatic transmission, adjust the transmission control linkage and check the fluid level.

41. Install the throttle body intake tubes. If equipped, evacuate and charge the air conditioning system.

E SERIES

1. Disconnect the negative battery cable and relieve the fuel system pressure. Remove the engine cover, drain the coolant and the crankcase and remove the air cleaner.

2. Remove the front bumper, then remove the grille and lower gravel deflector as an assembly.

3. Disconnect the upper and lower radiator hoses at the radiator. Disconnect the transmission oil cooler lines at the radiator, if equipped. Remove the radiator and shroud.

4. Disconnect the heater hoses at the engine. Disconnect the alternator and move aside.

5. Remove the power steering pump and support from the engine and move aside. If equipped with air conditioning, discharge the system and remove the air conditioning compressor.

6. Disconnect and plug the fuel line at the fuel rail. Tag and disconnect the distributor and sender unit wires from the engine.

7. Tag and disconnect the Electronic Engine Control (EEC) harness from all sensors.

8. Disconnect the brake booster hose at the engine and disconnect the accelerator cable and remove the bracket.

9. Disconnect the automatic transmission kickdown cable at the throttle body, if equipped.

10. Remove the exhaust manifold heat deflector and unbolt the exhaust pipe from the manifold.

11. Disconnect both ends of the automatic transmission vacuum line from the intake manifold and from the junction.

12. Remove the upper engine-to-transmission bolts. If equipped, remove the automatic transmission dipstick tube support bolt at the intake manifold.

13. Raise and safely support the vehicle.

14. Remove the starter. Remove the flywheel inspection cover. Remove the automatic transmission torque converter retaining nuts, then remove the front engine support nuts. Remove the oil filter.

15. Remove the rest of the transmission-to-engine fasteners, then lower the vehicle.

16. Install suitable engine lifting equipment and remove the engine from the vehicle.

To install:

17. Position the engine in the vehicle and start the mounting bolts.

18. Remove the engine lifting equipment and connect the exhaust pipe to the exhaust manifold.

19. Connect the transmission dipstick tube, if equipped, to the intake manifold. Install the manifold heat shield.

20. Connect the automatic transmission kickdown cable, then install the upper transmission-to-engine bolts.

21. Connect the transmission vacuum line at the junction. Install the accelerator cable and bracket assembly.

22. Connect the distributor and sender unit wires. Connect the brake booster hose.

23. Unplug and connect the fuel line to the fuel rail. Connect the transmission vacuum line to the manifold.

24. Install the alternator and the power steering pump and support bracket. Install the air conditioning compressor, if equipped.

25. Install the accessory drive belt. Connect the heater hoses and install the radiator and shroud assembly.

26. Position the grille and lower gravel deflector. Connect the upper radiator hose, then install the grille and deflector bolts and screws.

27. Install the bumper, then raise and safely support the vehicle.

28. Install the converter nuts, then install the flywheel inspection cover bolts.

29. Install the starter and the oil filter. Install the front engine support nuts and the lower engine-to-transmission bolts. Tighten to 60–80 ft. lbs. (82–108 Nm).

30. Connect the lower radiator hose and the transmission cooler lines. Install the alternator splash shield, if equipped, and lower the vehicle.

31. Fill the crankcase with the proper type and quantity of engine oil. Fill and bleed the cooling system.

32. Connect the negative battery cable, start the engine and check for leaks. If equipped, evacuate and charge the air conditioning system.

33. Install the engine cover.

5.0L and 5.8L Engines
BRONCO AND F SERIES

1. Disconnect the negative battery cable and relieve the fuel system pressure. Drain the cooling system and the crankcase.

2. Mark the position of the hood on the hinges and remove the hood.

3. Remove the air intake hoses, crankcase ventilation hose and carbon canister hose.

4. Disconnect the upper and lower radiator hoses and, if equipped, the automatic transmission oil cooler lines.

5. Disconnect and cap the power steering hoses. Disconnect the thermactor air pump hoses.

6. If equipped, discharge the air conditioning system and remove the condenser. Disconnect the air conditioning lines and the clutch wire at the compressor.

7. Remove the fan shroud and lay it over the fan. Remove the radiator and fan, shroud, pulley and belt.

8. Tag and disconnect the alternator wires. Remove the brackets for the air pump, alternator, power steering and compressor.

9. Disconnect the oil pressure sending unit lead from the sending unit. Disconnect and plug the flexible fuel line at the fuel tank line. Disconnect the evaporative emission hoses at the evaporative canister. Disconnect the chassis fuel line quick disconnects at the fuel rails.

10. Disconnect the accelerator cable at the throttle body. Disconnect the cruise control linkages, if equipped. Disconnect the automatic transmission kickdown rod and remove the return spring, if equipped.

11. Disconnect the power brake booster vacuum hose.

12. After disconnecting the accelerator and the TV cable from the throttle body, disconnect the throttle bracket from the upper intake manifold and swing out of the way with the cables still attached to the bracket.

13. Disconnect the heater hoses from the water pump and intake manifold or at the tee, if equipped. Disconnect the water temperature sending unit wire from the sending unit.

14. Remove the upper bell housing-to-engine attaching bolts.

15. Tag and disconnect the engine wire loom and position out of the way. Disconnect the ground strap from the cylinder block.

16. Raise and safely support the vehicle. Disconnect the starter cable from the starter. Remove the starter.

17. Disconnect the exhaust pipes and exhaust heat control valve, if equipped, from the exhaust manifolds.

18. Disconnect the engine mounts from the brackets on the frame.

19. If equipped with automatic transmission, remove the converter inspection plate and remove the torque converter-to-flywheel attaching bolts.

20. Remove the remaining bell housing-to-engine attaching bolts.

21. Lower the vehicle and support the transmission with a jack. Install suitable engine lifting equipment.

22. Raise the engine slightly and carefully pull it out from the transmission. Carefully lift the engine out of the engine compartment so the rear cover plate is not bent or other components damaged. Install the engine on a workstand.

To install:

23. Install suitable engine lifting equipment and remove the engine from the workstand. Lower the engine carefully into the engine compartment. Make sure the block dowels are through the rear cover plate, then engage the holes in the flywheel housing.

24. If equipped with a manual transmission, start the transmission input shaft into the clutch disc. It may be necessary to adjust the position of the transmission in relation to the engine if the transmission input shaft will not enter the clutch disc. If the engine hangs up after the shaft enters, turn the crankshaft slowly, with the transmission in gear, until the shaft splines mesh with the clutch disc splines.

25. Install the bell housing-to-engine upper bolts and the engine mount washers and attaching nuts. Remove the engine lifting equipment.

26. Raise and safely support the vehicle. Connect the exhaust pipes and the heat control valve, if equipped, to the manifolds.

27. Install the starter and the starter cable. Install the remaining bell housing-to-engine bolts.

28. If equipped with automatic transmission, install the converter-to-flywheel attaching bolts and install the converter inspection plate.

29. Remove the support from the transmission and lower the vehicle. Install the air conditioner and power steering bracket and components. Install the alternator/air pump bracket.

30. If equipped, connect the air conditioning compressor clutch lead wire. Connect the engine wire loom and the water temperature sending unit wire.

31. Connect the bellcrank/linkage bracket to the intake manifold. Connect the transmission shift rod or cable, if equipped and install the retracting spring. Connect the accelerator rod and cruise control linkage, if equipped.

32. Remove the plug from the fuel tank line and connect the fuel line and the oil pressure sending unit wire. Reconnect the evaporative emission hoses at the canister. Reconnect the chassis fuel lines to the fuel rail.

33. Install the pulley, clutch and fan and position the shroud over the fan. Install the accessory drive belt.

34. Position the alternator and install the mounting bolts. If equipped, connect the air conditioning lines to the compressor.

35. Install the radiator and connect the radiator hoses and transmission oil cooler lines, if equipped. Install the fan shroud.

36. If equipped, install the air conditioning condenser to the radiator. Connect the heater hoses at the water pump and intake manifold.

37. Fill the crankcase with the proper type and quantity of engine oil. Fill and bleed the cooling system. Connect the power brake booster vacuum hose.

38. Connect the negative battery cable, start the engine and bring to normal operating temperature. Check for leaks. Purge the power steering system of any air.

39. Install the air cleaner and intake duct assembly including the crankcase ventilation hose and carbon canister hose. If equipped, evacuate and charge the air conditioning system.

40. Install the hood, aligning the marks that were made during removal.

E SERIES

1. Disconnect the negative battery cable and relieve the fuel system pressure. Drain the cooling system and the crankcase. Remove the engine cover.

2. Remove the grille assembly including the gravel deflector. Remove the air cleaner, air ducts and closure hose.

3. Remove the upper grille support bracket, hood lock support and condenser upper mounting brackets, if equipped.

4. If equipped, discharge the air conditioning system and remove the condenser. Disconnect the air conditioning lines at the compressor and remove the accelerator cable bracket. Disconnect the cruise control linkage, if equipped.

5. Disconnect the radiator hoses at the radiator. Disconnect the heater hoses from the engine and at the heater core and water valve. If equipped with automatic transmission, disconnect the oil cooler lines at the radiator.

6. Remove the fan shroud and fan assembly. Remove the radiator and the water pump pulley and drive belt.

7. Disconnect the alternator lead wires and the air conditioner compressor clutch wire. Disconnect the power steering, thermactor air pump, alternator and air conditioning compressor. Plug the hoses as necessary.

8. Remove the accessory brackets complete with the accessories. Remove the air cleaner assembly. Disconnect the fuel line from the fuel rail.

9. Disconnect the throttle linkage at the throttle body and remove the accelerator cable bracket from the en-

gine. Disconnect the transmission shift rod, if equipped.

10. Raise and safely support the vehicle. Remove the oil filter.

11. Disconnect the exhaust pipes and exhaust heat control valve, if equipped, at the exhaust manifolds. Remove the 2 bolts retaining the transmission filler tube bracket to the right cylinder head.

12. Remove the engine mount attaching bolts and nuts. Remove the starter. If equipped with manual transmission, remove the housing-to-engine bolts.

13. If equipped with automatic transmission, remove the converter inspection cover bolts and remove the 4 nuts attaching the converter to the flexplate. Remove the 3 bolts retaining the adapter plate to the converter housing and remove the 4 converter housing-to-cylinder block lower bolts.

14. Remove the bolt retaining the ground cable to the cylinder block. Lower the vehicle and support the transmission with a floor jack.

15. Tag and disconnect the engine wire loom and position out of the way. Attach suitable engine lifting equipment and remove the remaining bell housing-to-engine bolts.

16. Carefully move the engine forward in the engine compartment so no components are damaged. Remove the engine from the vehicle and position on a workstand.

To install:

17. Install suitable engine lifting equipment and remove the engine from the workstand. Position the engine in the vehicle.

18. If equipped with automatic transmission, align the converter with the flexplate and the engine dowels with the converter housing.

19. If equipped with manual transmission, start the input shaft into the clutch disc. It may be necessary to adjust the position of the transmission in relation to the engine if the input shaft will not enter the clutch disc. If the engine hangs up after the shaft enters, turn the crankshaft slowly, with the transmission in gear, until the shaft splines mesh with the clutch disc splines. Align the housing on the engine and install the housing-to-engine bolts. Tighten to 40–50 ft. lbs. (55–67 Nm).

20. If equipped with automatic transmission, install 2 converter housing-to-cylinder block bolts.

21. Remove the engine lifting equipment and remove the floor jack from under the transmission.

22. Install the brackets and accessories. If equipped, connect the air conditioning compressor clutch wire.

23. Position the accelerator cable mounting bracket and install 3 attaching bolts.

24. Position the engine wire harness in the retainers and connect at respective locations. Connect the vacuum lines at the center of the upper intake manifold and connect the power brake booster vacuum hose.

25. Connect the heater hoses and connect the air conditioning compressor, thermactor air pump, alternator and power steering pump.

26. Install the water pump pulley, clutch and fan to the water pump and position the shroud over the fan assembly.

27. Raise and safely support the vehicle. Install the bolt attaching the ground cable to the cylinder block. If equipped with automatic transmission, install the remaining converter housing-to-engine bolts.

28. If equipped with automatic transmission, install the 3 adapter plate-to-converter housing bolts. Install the 4 nuts retaining the converter to the flexplate and install the converter inspection cover.

29. Install the starter and connect the cable. Install the engine mount attaching bolts and nuts and tighten to 50–70 ft. lbs. (68–95 Nm).

30. Install the 2 bolts that retain the transmission filler tube bracket to the right cylinder head.

31. Connect the exhaust pipes and, if equipped, the heat control valve at the exhaust manifolds.

32. Connect the throttle and transmission linkage and cruise control linkage, if equipped, at the throttle body. Connect the alternator wires to the alternator and the fuel line to the fuel supply manifold.

33. Install the accessory drive belt. If equipped, connect the air conditioning lines to the compressor. Install the accelerator cable bracket to the dash.

34. Position the radiator to the radiator support and install the attaching bolts. Install the shroud.

35. Connect the transmission oil cooler lines, if equipped, and connect the radiator hoses at the radiator.

36. Install the air conditioning condenser to the radiator support, if equipped. Install the grille upper support bracket, hood lock support, condenser upper mounting brackets and the grille.

37. Fill the crankcase with the proper type and quantity of engine oil. Fill and bleed the cooling system. Install the air cleaner, air ducts and closure hose.

38. Connect the negative battery cable, start the engine and bring to normal operating temperature. Check for leaks.

39. Evacuate and charge the air conditioning system, if equipped. Install the engine cover.

7.5L Engine
F SERIES

1. Disconnect the battery cables and relieve the fuel system pressure. Drain the cooling system and the crankcase.

2. Mark the position of the hood on the hinges and remove the hood.

3. Disconnect the thermactor air pump inlet hose at the front of the air cleaner housing. Disconnect the air outlet tube assembly at the throttle body and upper intake manifold. Remove the air cleaner assembly and air outlet tubes after disconnecting the air inlet tube at the air cleaner.

4. Remove the thermactor bypass valve hose connected to the air inlet tube and remove the air inlet tube.

5. Remove the air cleaner and intake duct assembly, including the crankcase ventilation hose and carbon canister hose.

6. Disconnect the radiator hoses at the radiator. If equipped with automatic transmission, disconnect the transmission oil cooler lines. If equipped with an engine oil cooler, disconnect the cooler lines at the oil filter adapter.

NOTE: Do not attempt to disconnect the engine oil cooler lines at the quick connect fittings behind or at the cooler, or damage to the fittings will occur.

7. If equipped, discharge the air conditioning system and remove the condenser. Disconnect the air conditioning lines at the compressor.

8. Remove the fan shroud and position it over the fan. Remove the radiator. Remove the fan shroud, fan clutch, belts and pulley.

9. Remove the alternator bolts and allow the alternator to swing down and out of the way.

10. Disconnect the throttle and transmission linkage at the throttle body and remove the accelerator cable bracket from the upper intake manifold. Remove the cruise control hardware, if equipped.

11. Disconnect the oil pressure sending unit wire from the sending unit. Disconnect the evaporative emission hoses at the evaporative canister.

12. Disconnect the fuel lines at the qick disconnect couplings. Tag and disconnect the vacuum lines at the intake manifold.

13. Disconnect the EGR tube at the left exhaust manifold and upper intake manifold. Plug the manifold opening.

14. Disconnect the power brake booster vacuum hose and disconnect the exhaust air supply bypass valve hose at the thermactor check valve. Disconnect the heater hoses from the water pump and intake manifold.

15. Remove the flywheel housing-to-engine upper bolts and disconnect the ground strap from the cylinder block. Remove the oil fill tube, dipstick and tube.

16. Raise and safely support the vehicle. Disconnect the starter cable and remove the starter.

17. Disconnect the exhaust pipes from the exhaust manifolds. Disconnect the engine mounts from the frame.

18. If equipped with automatic transmission, remove the converter inspection plate and remove the converter-to-flywheel attaching bolts. If equipped with manual transmission, remove the rear cover plate from the flywheel housing and remove the remaining flywheel housing-to-engine bolts.

19. If equipped, disconnect the air conditioner compressor clutch wire.

20. Lower the vehicle and support the transmission with a jack. Install suitable engine lifting equipment.

21. Raise the engine slightly and carefully pull it from the transmission. Carefully lift the engine out of the engine compartment so the rear cover plate is not bent or other components damaged. Install the engine on a workstand.

To install:

22. Attach the engine lifting equipment and remove the engine from the workstand. Lower the engine carefully into the engine compartment. Make sure the block dowels are through the rear cover plate, then engage the dowels with the holes in the flywheel housing.

23. If equipped with manual transmission, start the transmission input shaft into the clutch disc. It may be necessary to adjust the position of the transmission in relation to the engine if the input shaft will not enter the clutch disc. If the engine hangs up after the shaft enters, turn the crankshaft slowly, transmission in gear, until the shaft splines mesh with the clutch disc splines.

24. Install the flywheel housing upper bolts. Install the engine mount-to-frame nuts and washers and tighten to 60–80 ft. lbs. (81–108 Nm). Remove the engine lifting equipment.

25. Raise and safely support the vehicle. Connect the exhaust pipes to the exhaust manifolds. Install the starter and the starter cable and install the remaining flywheel housing-to-engine bolts.

26. If equipped with automatic transmission, install the converter-to-flywheel attaching bolts and the inspection plate. On manual transmissions, install the rear cover plate. Remove the support from the transmission and lower the vehicle.

27. If equipped, connect the air conditioning compressor clutch wire. Connect the water temperature sending unit wire.

28. Connect the acclerator cable bracket to the intake manifold. Connect the transmission and throttle linkage to the throttle body and cruise control linkage, if equipped.

29. Connect the fuel lines and connect the evaporative emission hoses at the evaporative canister. Install the pulley, belt, spacer and fan. Position the shroud over the fan.

30. Position the alternator and install the bolts. Install the drive belt.

31. If equipped, connect the 2 air conditioning lines to the compressor.

32. Install the radiator and connect the radiator hoses. If equipped, connect the transmission oil cooler lines and engine oil cooler. Install the fan shroud.

33. If equipped, install the air conditioning condenser.

34. Connect the heater hose at the water pump. Fill and bleed the cooling system.

35. Fill the crankcase with the proper type and quantity of engine oil. Connect the power brake booster vacuum hose.

36. Connect the negative battery cable, start the engine and bring to normal operating temperature. Check for leaks.

37. Install the air outlet tubes to the throttle body and upper intake manifold including the crankcase ventilation hose and carbon canister hose.

38. Evacuate and charge the air conditioning system, if equipped. Install the hood, aligning the marks that were made during removal.

E SERIES

1. Remove the engine cover. Disconnect the negative battery cable and relieve the fuel system pressure. Drain the cooling system and the crankcase.

2. Remove the grille assembly including the gravel deflector.

3. Remove the 4 bolts that secure the air cleaner and bracket assembly at the body cowl and radiator support. Disconnect the thermactor air pump inlet hose at the front of the air cleaner housing.

4. Disconnect the air outlet tube assembly at the throttle body and upper intake manifold. Remove the air cleaner assembly and air outlet tubes, after disconnecting the air inlet tube at the air cleaner.

5. Remove the thermactor bypass valve hose connected to the air inlet tube and remove the air inlet tube. Remove the upper grille support bracket, hood lock support and condenser upper mounting brackets.

6. If equipped, discharge the air conditioning system and remove the condenser. Disconnect the air conditioning lines at the compressor.

7. Remove the accelerator cable bracket from the dash and disconnect the 2 heater hoses from the engine.

9. If equipped, disconnect the engine oil cooler lines at the oil filter adapter. Do not attempt to disconnect the lines at the quick disconnect fittings at the cooler, as the fittings or the cooler may become damaged.

10. Disconnect the radiator hoses at the radiator. Remove the fan shroud, fan and radiator. Pivot the alternator inward and disconnect the alternator wires.

11. Disconnect the throttle and transmission linkage at the throttle body and remove the accelerator cable bracket from the upper intake manifold. Remove the cruise control hardware, if equipped.

12. Disconnect the fuel lines at the quick disconnect couplings and vacuum lines to the intake manifold. Disconnect the EGR tube at the left exhaust manifold and upper intake manifold. Plug the manifold opening.

13. Raise and safely support the vehicle. Remove the oil filter.

14. Disconnect the exhaust pipes from the exhaust manifolds. Remove the 2 bolts retaining the transmission filler tube bracket to the right cylinder head.

15. Disconnect the exhaust air supply bypass valve hose at the check valve.

16. Remove the engine mount attaching bolts and nuts. Remove the starter and disconnect the cable.

17. Remove the converter inspection cover bolts and remove the 4 nuts attaching the converter to the flexplate. Remove the 3 bolts retaining the adapter plate to the converter housing.

18. Remove the 4 converter housing-to-cylinder block lower bolts. Remove the bolt attaching the ground cable to the cylinder block. Lower the vehicle.

19. Loosen the belt tensioner and remove the drive belt. Disconnect the air conditioning lines and compressor clutch wire, if equipped. Disconnect and plug the power steering pump lines. Remove the pump and bracket from the front of the engine.

20. Tag and disconnect any vacuum lines at the rear of the intake manifold. Disconnect the engine wire loom and position aside.

21. Place a jack under the transmission and attach suitable engine lifting equipment.

22. Remove the remaining bell housing-to-cylinder block bolts and remove the engine from the vehicle.

To install:

23. Position the engine in the vehicle, aligning the converter to the

flexplate and the engine dowels to the transmission.

24. Install 2 bell housing-to-cylinder block upper bolts and remove the engine lifting equipment. Remove the floor jack from under the transmission.

25. If equipped, connect the air conditioning compressor clutch wire.

26. Position the accelerator cable mounting bracket and the cruise control servo to the intake manifold and install the attaching bolts.

27. Position the engine wire harness in the retainers and connect at the respective locations. Connect the vacuum lines at the rear of the intake manifold.

28. Connect the 2 heater hoses at the water outlet connector and at the water pump connector. Position the power steering bracket to the water pump and the left cylinder head and install the attaching bolts. Install the power steering pump to the bracket.

29. Position the power steering pump drive belt and install the automatic tensioner. If equipped, connect the 2 air conditioning lines to the compressor. Install the accelerator cable bracket to the dash.

30. Raise and safely support the vehicle. Install a new oil filter and install the bolt attaching the battery ground cable to the cylinder block.

31. Install the remaining bell housing-to-cylinder block bolts and install the adapter plate-to-converter housing bolts. Install the 4 nuts attaching the converter to the flexplate and install the converter inspection cover.

32. Install the starter motor and connect the cable. Install the engine mount attaching bolts and nuts and tighten to 50–70 ft. lbs. (68–94 Nm).

33. Install the 2 bolts attaching the transmission filler tube bracket to the right cylinder head. Connect the exhaust pipes at the exhaust manifolds.

34. Lower the vehicle. Connect the fuel lines and the vacuum lines. Connect the throttle and transmission linkage at the throttle body and connect the wires at the alternator.

35. Install the fan assembly to the water pump and position the shroud over the fan. Install the radiator, then install the shroud. Install the radiator hoses at the radiator.

36. Install the air conditioning condenser, if equipped. Connect the cruise control servo vacuum hose and secure the cable to the throttle body.

37. Install the grille upper support bracket, hood lock support and condenser upper mounting brackets. Install the grille assembly including gravel deflector.

38. Fill the crankcase with the proper type and quantity of engine oil. Fill and bleed the cooling system.

39. Connect the negative battery cable, start the engine and bring to normal operating temperature. Check for leaks.

40. Evacuate and charge the air conditioning system. Install the air cleaner assembly and the engine cover.

Engine Mounts

REMOVAL & INSTALLATION

4.9L Engine

BRONCO AND F SERIES—FRONT MOUNTS

1. Remove the upper and lower nut and washer from the mount. If only 1 mount is being removed, loosen the nut and washer on the opposite mount.

2. Using a jack and a wood block placed under the oil pan, raise the engine just enough to allow clearance for removal of the mount.

3. Remove the mount and shield, if equipped. Remove the upper and lower brackets.

To install:

4. Install the lower bracket and tighten the bolts and nuts to 48–62 ft. lbs. (65–85 Nm). Install the upper bracket and tighten the bolts to 59–81 ft. lbs. (80–110 Nm).

5. Align the locator pin with the hole and position the mount on the lower bracket. Install the lower nut and washer finger-tight.

6. Position the shield on the right mount.

7. Lower the engine carefully to make sure the mount stud and pin engage the upper bracket mounting hole.

8. Install the nut and washer on the upper mount stud. Tighten the upper and lower stud nuts to 71–93 ft. lbs. (96–127 Nm).

9. If only 1 mount was removed, tighten the other mount at the brackets.

BRONCO AND F SERIES—REAR MOUNT

1. Remove the attaching bolts, nut and washers.

2. Raise the transmission slightly to provide clearance and remove the mount and mount retainer.

3. Installation is the reverse of the removal procedure. Tighten the mount-to-transmission bolts to 64–71 ft. lbs. (87–96 Nm) and the mount-to-crossmember nuts to 60–80 ft. lbs. (81–109 Nm).

E SERIES—FRONT MOUNTS

1. Raise and safely support the vehicle.

2. Remove the front mount-to-support bracket nuts and washers from both mounts.

3. Place a jack and a wood block under the engine oil pan. Raise the engine just enough to take the weight off the support brackets.

4. The support brackets are mounted to the frame crossmember with 4 bolts and locknuts. If these are to be removed, loosen the 4 locknuts on each support bracket and remove from the crossmember.

5. Remove the mount-to-cylinder block bolts and remove the mounts.

6. Installation is the reverse of the removal procedure. Tighten the mount-to-cylinder block bolts and the support bracket locknuts to 60–80 ft. lbs. (81–108 Nm). Tighten the mount-to-support bracket nuts to 57–74 ft. lbs. (67–100 Nm).

E SERIES—REAR MOUNT

1. Raise and safely support the vehicle. Place a transmission jack under the transmission and raise it slightly to take the weight off of the rear mount.

2. Remove the mount-to-crossmember and transmission extension housing locknuts and bolts. Remove the rear mount.

3. Installation is the reverse of the removal procedure. Tighten the mount locknuts to 50–70 ft. lbs. (68–94 Nm).

5.0L and 5.8L Engines

BRONCO AND F SERIES—FRONT MOUNTS

1. Remove the bolts attaching the fan shroud to the radiator and position the shroud over the fan.

2. Remove the nut and washer attaching the mount to the chassis bracket.

3. Raise the engine. Remove the bolts and lockwashers attaching the mount to the cylinder block and remove the mount and the heat shield, if equipped.

4. Installation is the reverse of the removal procedure. Tighten the mount-to-engine bolts to 50–70 ft. lbs. (68–94 Nm) and the mount-to-chassis nut to 65–85 ft. lbs. (88–115 Nm).

BRONCO AND F SERIES—REAR MOUNT

1. Remove the mount-to-crossmember attaching nuts.

2. Raise the transmission with a floor jack to provide clearance and remove the 2 bolts, insulator and spacers.

3. Installation is the reverse of the removal procedure. Tighten the mount-to-transmission bolts to 60–80 ft. lbs. (81–108 Nm) and the mount-to-crossmember nuts to 50–70 ft. lbs. (68–94 Nm).

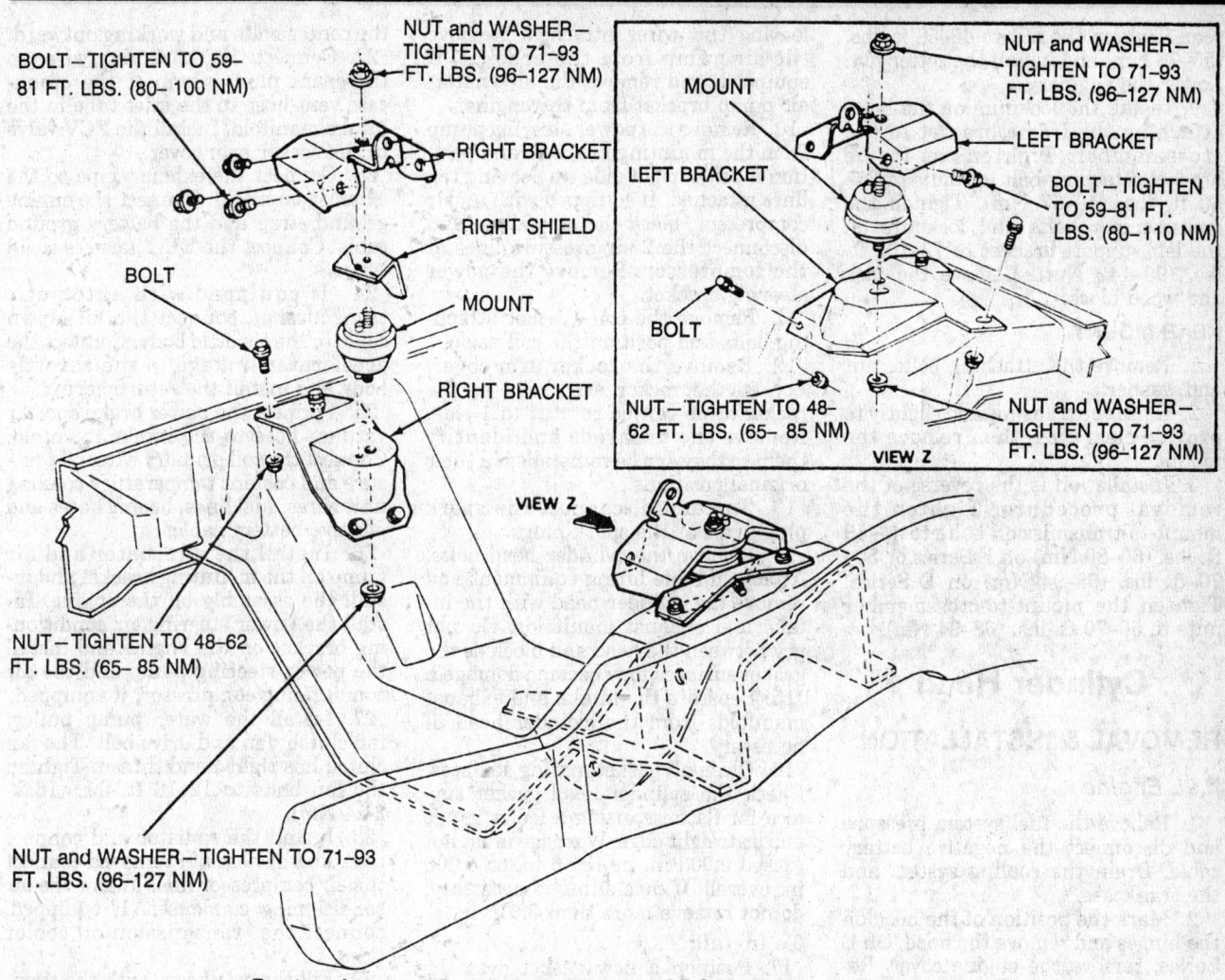

BOLT—TIGHTEN TO 59–81 FT. LBS. (80– 100 NM)

NUT and WASHER— TIGHTEN TO 71–93 FT. LBS. (96–127 NM)

RIGHT BRACKET

RIGHT SHIELD

BOLT

MOUNT

RIGHT BRACKET

NUT—TIGHTEN TO 48–62 FT. LBS. (65– 85 NM)

NUT and WASHER—TIGHTEN TO 71–93 FT. LBS. (96–127 NM)

VIEW Z

NUT and WASHER— TIGHTEN TO 71–93 FT. LBS. (96–127 NM)

MOUNT

LEFT BRACKET

LEFT BRACKET

BOLT—TIGHTEN TO 59–81 FT. LBS. (80–110 NM)

BOLT

NUT—TIGHTEN TO 48– 62 FT. LBS. (65– 85 NM)

NUT and WASHER— TIGHTEN TO 71–93 FT. LBS. (96–127 NM)

VIEW Z

Front engine mount Installation—Bronco and F Series with 4.9L engine

E SERIES—FRONT MOUNTS

1. Disconnect the negative battery cable.

2. Remove the fan shroud bolts and the mount-to-frame nuts.

3. Position a jack and a wood block under the engine. Lift the engine and then remove the starter.

4. Remove the mount-to-engine bolts and alternator splash shield. Remove the mounts.

To install:

5. Clear the slots in the engine block of debris using a $7/16$-14 bottoming tap.

6. Install the mounts using new bolts and tighten to 50–70 ft. lbs. (68–96 Nm).

7. Install the starter, then lower the engine. Move the jack out of the way.

8. Install the mount-to-frame nuts and tighten to 50–70 ft. lbs. (68–95 Nm). Install the fan shroud and bolts.

9. Connect the negative battery cable.

E SERIES—REAR MOUNT

NOTE: On 138 in. wheelbase ve- hicles, first remove the fuel reservoir to eliminate any part damage due to removal of rear transmission support crossmember.

1. Raise and safely support the vehicle.

2. Remove the nuts securing the mount to the crossmember.

3. Raise the transmission with a jack so the mount clears the crossmember.

4. Remove the bolts securing the mount to the transmission and remove the mount.

5. Installation is the reverse of the removal procedure. Tighten the mount-to-transmission bolts and mount-to-crossmember nuts to 50–70 ft. lbs. (68–94 Nm).

7.5L Engine

FRONT MOUNT

1. Support the engine with a jack and a wood block placed under the oil pan.

2. Remove the locknuts from the bolts attaching the support bracket to the frame crossmember and frame side rail.

3. Remove the through bolt attaching the engine support bracket to the mount.

4. Raise the engine with the jack until the mount is clear of the cup-shaped engine bracket.

5. Remove the mount and frame bracket as an assembly.

6. Remove the nuts attaching the mount to the frame bracket.

To install:

7. Assemble the mount to the frame bracket and install the attaching nuts. The mount should be installed so the word **TOP** is visible on either side of the engine. Tighten the attaching nuts to 50–70 ft. lbs. (68–94 Nm).

8. Position the mount and frame bracket assembly to the engine mount bracket and the frame crossmember. Install the through bolt attaching the mount to the engine bracket. The through bolt for the right mount must be installed from the front of the engine and the through bolt for the left mount must be installed from the

rear. Tighten the nut to 40–58 ft. lbs. (55–78 Nm) and install the cotter pin. Lower the engine.

9. Install the locknuts on the bolts attaching the frame bracket to the crossmember. Tighten the frame bracket attaching bolt locknuts to 35–50 ft. lbs. (48–67 Nm). Tighten the crossmember bolts and locknuts on the left support bracket to 72–105 ft. lbs. (98–142 Nm). Remove the jack and wood block.

REAR MOUNT

1. Remove the attaching bolts, nut and washer.

2. Raise the transmission slightly to provide clearance, then remove the mount.

3. Installation is the reverse of the removal procedure. Tighten the mount-to-transmission bolts to 44–59 ft. lbs. (60–80 Nm) on F Series or 50–70 ft. lbs. (68–94 Nm) on E Series. Tighten the mount-to-crossmember nuts to 50–70 ft. lbs. (68–94 Nm).

Cylinder Head

REMOVAL & INSTALLATION

4.9L Engine

1. Relieve the fuel system pressure and disconnect the negative battery cable. Drain the cooling system and the crankcase.

2. Mark the position of the hood on the hinges and remove the hood. On E Series, remove the engine cover. Remove the throttle body inlet tubes.

3. If equipped, properly discharge the air conditioning system and remove the compressor and the condenser.

4. Disconnect the heater hose from the water pump and coolant outlet housing. Disconnect the fuel lines.

5. Remove the radiator, cooling fan, water pump pulley and fan drive belt.

6. Disconnect the accelerator cable at the throttle body and remove the cable retracting spring. If equipped with automatic transmission, disconnect the transmission kickdown cable at the throttle body. Disconnect the power brake booster vacuum line at the manifold.

7. Disconnect the exhaust pipe from the exhaust manifold. Disconnect the body ground strap and the battery ground cable at the engine.

8. Tag and disconnect the Electronic Engine Control (EEC) harness from all sensors. Tag and disconnect the engine wiring harness at the ignition coil, coolant temperature sending unit and oil pressure sending unit. Position the harness aside.

·9. Remove the alternator mounting bolts and position the alternator aside,

leaving the wires attached. Remove the air pump from the bracket, if equipped, and remove the alternator/air pump bracket from the engine.

10. Remove the power steering pump from the mounting brackets and position it aside, right side up, leaving the lines attached. If equipped with an air compressor, bleed the air system and disconnect the 2 air pressure lines at the compressor. Remove the power steering bracket.

11. Remove the coil bracket attaching bolts and position the coil aside.

12. Remove the rocker arm cover. Loosen the rocker arm bolts so the rocker arms can be rotated to 1 side. Remove the pushrods and identify them so they can be reinstalled in their original positions.

13. Tag and disconnect the spark plug wires at the spark plugs.

14. Remove the cylinder head bolts. Attach suitable lifting equipment and remove the cylinder head with the intake and exhaust manifolds. Do not pry between the head and block as the gasket surfaces may become damaged.

15. Separate the intake and exhaust manifolds from the cylinder head, if necessary.

16. Clean all gasket mating surfaces. Check the cylinder head gasket surface for flatness, using a feeler gauge and a straight edge. Warpage must not exceed 0.003 in. in any 6 in. or 0.006 in. overall. If machining is necessary, do not remove more than 0.010 in.

To install:

17. Position a new gasket over the dowel pins on the cylinder block. Attach suitable lifting equipment to the cylinder head and lower it into position on the cylinder block and dowel pins. Remove the lifting equipment.

18. Coat the threads of the cylinder head bolts with clean engine oil and install. Tighten the bolts, in sequence, in 3 steps, first to 50–55 ft. lbs. (67–75 Nm), then to 60–65 ft. lbs. (82–88 Nm) and finally to 70–85 ft. lbs. (94–115 Nm).

19. Apply multi-purpose grease to both ends of the pushrods and install them in their original positions. Apply multi-purpose grease to the valve stem tips.

20. Apply multi-purpose grease to the rocker arm fulcrum seat and the fulcrum seat socket in the rocker arm. Position the rocker arms and tighten the rocker arm bolt just enough to hold the pushrod in position. Check the valve clearance, then tighten the rocker arm bolts to 17–23 ft. lbs. (24–31 Nm).

21. Clean the rocker arm cover and position a new gasket on the cylinder head. Install the rocker arm cover retaining bolts and tighten to 70–105 inch lbs. (7.9–11.9 Nm), starting with

the center bolts and working outward.

22. Connect the spark plug wires to the spark plugs. Connect the crankcase vent hose to the inlet tube in the intake manifold. Install the PCV valve in the rocker arm cover.

23. Connect the exhaust pipe to the exhaust manifold. Connect the engine ground strap and the battery ground cable. Connect the EEC harness to all sensors.

24. If equipped with automatic transmission, connect the kickdown cable to the throttle body. Connect the accelerator linkage to the throttle body and install the return spring.

25. Connect the power brake booster vacuum hose to the intake manifold. Connect the coil primary wire, oil pressure and coolant temperature sending unit wires, fuel lines, heater hoses and positive battery cable.

26. Install the alternator and air pump on the mounting bracket and install the assembly on the engine. Install the power steering/air conditioning bracket on the engine and install the power steering pump and the air conditioning compressor, if equipped.

27. Install the water pump pulley, fan clutch, fan and drive belt. The fan clutch has right-hand thread. Tighten the fan bolts to 12–18 ft. lbs. (16.2–24.4 Nm).

28. Install the radiator and connect the radiator hoses. If removed connect the air compressor lines and/or the air conditioning condenser. If equipped, connect the transmission oil cooler lines.

29. Fill the crankcase with the proper type and quantity of engine oil. Fill and bleed the cooling system.

30. If removed, install the hood, aligning the marks that were made during removal.

31. Connect the negative battery cable, start the engine and bring to normal operating temperature. Check for leaks.

32. Install the throttle body intake tubes. On E Series, install the engine cover. If equipped, evacuate and charge the air conditioning system.

5.0L and 5.8L Engines

1. Disconnect the negative battery cable and relieve the fuel system pressure. Drain the cooling system.

2. Remove the intake manifold and the rocker arm cover(s).

3. To remove the right cylinder

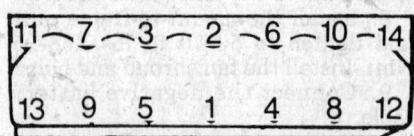

Cylinder head bolt torque sequence—4.9L engine

head, remove the drive belt and then remove the alternator and air pump mounting bracket complete with accessories. Swing the alternator down and out of the way. If necessary, remove the ignition coil and air cleaner inlet duct from the left cylinder head.

4. To remove the left cylinder head, remove the air conditioning compressor/power steering bracket at the front of the cylinder and position the accessories aside. Remove the oil dipstick and tube and cruise control bracket, if equipped.

5. Raise and safely support the vehicle. Disconnect the exhaust pipe(s) from the manifold(s).

6. Loosen the rocker arm fulcrum bolts so the rocker arms can be rotated to the side. Remove the pushrods and identify them so they can be reinstalled in their original positions.

7. On E Series, remove the bolts holding the thermactor air supply manifold to the rear of the cylinder head and disconnect the hose at the air pump. Remove the hose, pump valve and air supply manifold as an assembly.

8. On Bronco and F Series, disconnect the thermactor air supply hoses at the check valves and plug the check valve.

9. Remove the cylinder head bolts and remove the cylinder head(s). Remove the exhaust manifold(s) from the cylinder head(s), if necessary.

10. Clean all gasket mating surfaces. Check the cylinder head gasket surface for flatness, using a feeler gauge and a straight edge. Warpage must not exceed 0.003 in. in any 6 in. or 0.006 in. overall. If machining is necessary, do not remove more than 0.010 in.

To install:

11. Position a new cylinder head gasket over the dowels on the cylinder block. Install the cylinder head on the block. Coat the threads of the cylinder head bolts with clean engine oil and install, finger-tight.

12. On the 5.0L engine, tighten the cylinder head bolts, in sequence, in 2 steps, first to 55–65 ft. lbs. (75–88 Nm) and then to 65–72 ft. lbs. (88–97 Nm).

13. On the 5.8L engine, tighten the cylinder head bolts, in sequence, in 3 steps, first to 85 ft. lbs. (115 Nm), then to 95 ft. lbs. (129 Nm) and finally to 105–112 ft. lbs. (143–151 Nm).

14. Apply multi-purpose grease to both ends of the pushrods and install them in their original positions. Apply multi-purpose grease to the valve stem tips.

15. Apply multi-purpose grease to the rocker arm fulcrum seat and the fulcrum seat socket in the rocker arm. Position the rocker arms and tighten the rocker arm bolt just enough to hold the pushrod in position. Check

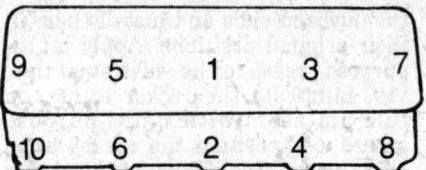

Cylinder head bolt torque sequence — 5.0L, 5.8L and 7.5L engines

the valve clearance, then tighten the rocker arm bolts to 18–25 ft. lbs. (25–33 Nm).

16. Position new gasket(s) on the exhaust pipe(s), if required. Connect the exhaust pipe(s) to the exhaust manifold(s) and tighten the nuts to 25–38 ft. lbs. (34–52 Nm).

17. If the right cylinder head was removed, install the air pump/alternator bracket and accessories. Install the ignition coil and air cleaner inlet duct, if necessary. Install the drive belt.

18. If the left cylinder head was removed, install the power steering/air conditioning compressor bracket with accessories at the front of the cylinder head. Install the oil dipstick and tube and cruise control bracket, if equipped.

19. Clean the rocker arm cover(s) and install new gasket(s). Install the rocker arm cover(s) on the cylinder head(s) and install the retaining bolts. Tighten to 3–5 ft. lbs. (4–6 Nm).

20. Install the intake manifold.

21. Install the thermactor air supply manifold at the rear of the cylinder heads. Reconnect the air supply hose to the air pump, if removed. Unplug the check valve and connect the thermactor air supply hose.

22. Fill and bleed the cooling system. Connect the negative battery cable. Start the engine and bring to normal operating temperature. Check for leaks.

7.5L Engine

1. Disconnect the negative battery cable and relieve the fuel system pressure. Drain the cooling system.

2. Remove the intake manifold.

3. Raise and safely support the vehicle as necessary. Disconnect the exhaust pipe from the exhaust manifold.

4. Remove the accessory drive belts.

5. Remove the thermactor air pump and alternator. Remove the air pump bracket from the right cylinder head.

6. If equipped with air conditioning, shut off the compressor at the service valves and remove the valves and hoses from the compressor. Remove the nuts attaching the compressor support bracket to the water pump. Remove the bolts attaching the compressor to the upper mounting bracket and position the compressor aside. Remove the compressor/power steering pump mounting bracket.

7. If not equipped with air conditioning, remove the bolts attaching the power steering pump bracket to the left cylinder head and position the pump and bracket out of the way.

8. Disconnect the oil filler tube on E Series.

9. Remove the rocker arm covers. Remove the rocker arm fulcrum bolt, rocker arm, oil deflector, fulcrum and pushrod assemblies. Identify components so they can be reinstalled in their original positions.

10. Remove the cylinder head bolts and remove the cylinder heads and exhaust manifolds as assemblies. Remove the exhaust manifolds from the cylinder heads, if necessary.

NOTE: If necessary to loosen the cylinder head gasket seal, pry at the forward corners of the cylinder heads against the casting bosses provided on the cylinder block. Do not damage the machined surfaces of the head or block.

11. Clean all gasket mating surfaces. Check the cylinder head gasket surface for flatness, using a feeler gauge and a straight edge. Warpage must not exceed 0.003 in. in any 6 in. or 0.006 in. overall. If machining is necessary, do not remove more than 0.010 in.

To install:

12. If the exhaust manifold was removed, coat the cylinder head and manifold port areas with a film of graphite grease and install the manifold to the cylinder head.

13. Place 2 long head bolts in the 2 rear lower bolt holes of the left cylinder head. Place a long head bolt in the rear lower bolt hole of the right cylinder head. Use rubber bands to retain the bolts in position, above the head-to-block mating surface, until the cylinder heads are installed.

14. Position new head gaskets on the block over the dowels. Do not apply sealer to the head gasket surfaces. Place the cylinder heads on the block, guiding the exhaust manifold studs into the exhaust pipe connections.

15. Install the remaining head bolts, with the longer bolts in the lower row of bolt holes. Tighten the bolts, in sequence, in 3 steps, first to 80–90 ft. lbs. (108–122 Nm), then to 100–110 ft. lbs. (136–149 Nm) and finally to 130–140 ft. lbs. (177–189 Nm).

16. Apply multi-purpose grease to the pushrod ends and install them in their original positions. Apply multi-purpose grease to the valve stem tips. Lubricate and install the rocker arms, tightening the fulcrum bolts to 18–25 ft. lbs. (25–33 Nm). Perform a valve clearance check.

17. Tighten the exhaust pipe stud nuts to 25–38 ft. lbs. (34–52 Nm).

18. Install the intake manifold.

19. If equipped with air conditioning, install the compressor mounting bracket, with the power steering pump in place, to the left cylinder head and water pump. Attach the compressor to the mounting bracket and connect the service valves and hoses to the compressor.

20. If not equipped with air conditioning, attach the power steering pump and bracket to the left cylinder head and water pump.

21. On E Series, connect the oil filler tube.

22. Install the thermactor supply manifold to the exhaust manifold. Install the bolt attaching the alternator and air pump bracket to the right cylinder head. Install the air pump, alternator and the drive belt.

23. Install the air conditioning and power steering pump drive belt.

24. Fill and bleed the cooling system. Fill and bleed the power steering reservoir, as necessary.

25. Connect the negative battery cable, start the engine and bring to normal operating temperature. Check for leaks.

26. If equipped, evacuate and charge the air conditioning system.

Valve Lifters

REMOVAL & INSTALLATION

Except 4.9L Engine

1. Disconnect the negative battery cable and relieve the fuel system pressure. Drain the cooling system.

2. Remove the intake manifold.

3. If necessary, disconnect the thermactor air supply hose at the air pump and place it aside.

4. Remove the rocker arm cover, then loosen the rocker arm bolts and rotate the rocker arms to the side.

5. Remove the pushrods and identify them so they can be reinstalled in their original positions.

6. Using lifter puller tool T70L–6500–A or equivalent, remove the lifters. If the lifters are to be reused, place them in a rack so they can be reinstalled in their original bores.

To install:

7. Clean the external surfaces and lubricate the lifters and lifter bores with clean engine oil. Install the lifters in their original positions. If new lifters are being installed, check each for a free fit in its bore.

8. Clean the pushrods in a suitable solvent and blow out the oil passages with compressed air. Check the pushrod ends for wear and/or damage. Check the pushrods for straightness by rolling them along a flat surface.

9. Apply multi-purpose grease to the pushrod ends and install them in their original positions. Apply multipurpose grease to the valve stem tips.

10. Lubricate the rocker arms and fulcrum seats with multi-purpose grease and position the rocker arms over the pushrods. Tighten the rocker arm bolts to 18–25 ft. lbs. (25–33 Nm). Perform a valve clearance check.

11. Clean the rocker arm cover and cylinder head gasket surfaces. Install new gaskets and the rocker arm covers. Tighten the rocker arm cover bolts to 3–5 ft. lbs. (4–6 Nm) on 5.0L and 5.8L engines or 6–9 ft. lbs. (8–12 Nm) on 7.5L engine.

12. Install the intake manifold.

13. Fill and bleed the cooling system. Connect the negative battery cable, start the engine and bring to normal operating temperature. Check for leaks.

4.9L Engine

1. Disconnect the negative battery cable and relieve the fuel system pressure.

2. Disconnect the inlet air hose at the crankcase filter cap. Remove the throttle body inlet tubes.

3. Disconnect the accelerator cable at the throttle body and remove the return spring. Remove the accelerator cable bracket from the upper intake manifold and position the cable and bracket assembly aside.

4. Remove the fuel line from the fuel rail.

5. Remove the upper intake manifold and throttle body as follows:

   a. Tag and disconnect the electrical connectors and vacuum lines at the upper intake manifold and throttle body.

   b. Disconnect the PCV hose from the fitting on the underside of the upper intake manifold.

   c. Disconnect the EGR tube from the EGR valve and the rear exhaust manifold. Remove the tube from the engine.

   d. Remove the thermactor tube from the lower intake manifold by removing the 2 nuts retaining the tube. Remove the nut attaching the thermactor bypass valve bracket to the lower intake manifold.

   e. On E Series, remove the nut attaching the transmission fill tube. Then remove the tube bracket off the intake manifold stud.

   f. Remove the 7 studs that retain the upper intake manifold. Remove the screw and washer attaching the upper intake manifold support bracket to the upper intake manifold.

   g. Remove the upper intake manifold and throttle body.

6. Remove the coil bracket attaching bolt and "E" core assembly attaching nuts and position the coil aside.

7. Remove the rocker arm cover. Tag and disconnect the spark plug wires at the spark plugs. Disconnect the coil wire and remove the distributor cap and spark plug wire assembly.

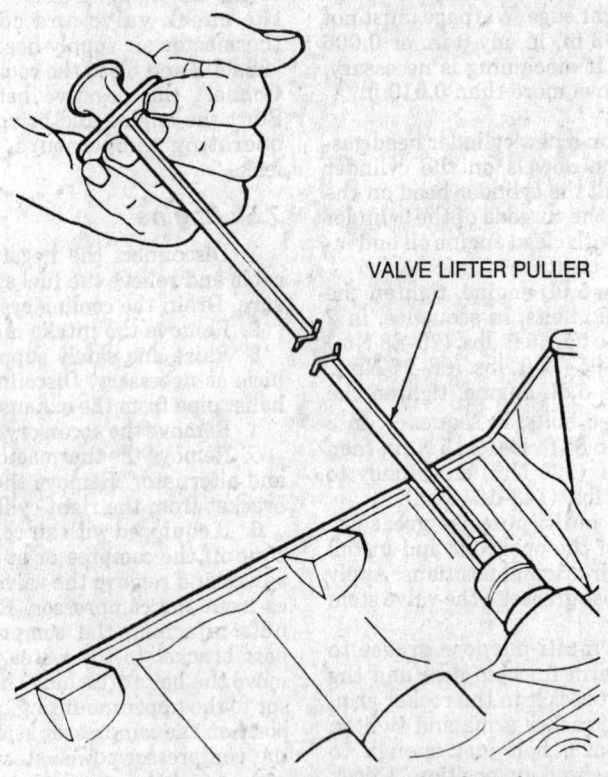

VALVE LIFTER PULLER

Valve lifter removal—5.0L, 5.8L and 7.5L engines

8. Remove the pushrod cover. Loosen the rocker arm bolts and rotate the rocker arms to the side. Remove the pushrods and identify them so they can be reinstalled in their original positions.

9. Remove the lifters. If the lifters are to be reused, place them in a rack so they can be reinstalled in their original bores.

To install:

10. Clean all gasket mating surfaces. Clean the external surfaces of the lifters.

11. Apply multi-purpose grease to the camshaft lobes. Coat the lifters and the lifter bores with clean engine oil. Install the lifters in their original positions. If new lifters are being installed, check each for a free fit in its bore.

12. Clean the pushrods in a suitable solvent and blow out the oil passages with compressed air. Check the pushrod ends for wear and/or damage. Check the pushrods for straightness by rolling them along a flat surface.

13. Apply multi-purpose grease to the pushrod ends and install them in their original positions. Apply multi-purpose grease to the valve stem tips.

14. Lubricate the rocker arms and fulcrum seats with multi-purpose grease and position the rocker arms over the pushrods. Tighten the rocker arm bolts to 17–23 ft. lbs. (24–31 Nm) and check the valve clearance.

15. Apply sealer to 1 side of a new pushrod cover gasket and install on the pushrod cover. Install the pushrod cover and tighten the screws to 25–35 inch lbs. (2.8–4.0 Nm).

16. Place a new gasket on the rocker arm cover and install on the cylinder head. Tighten the bolts to 70–105 inch lbs. (7.9–11.9 Nm). Install the PCV valve in the rocker arm cover.

17. Position the coil assembly on the cylinder head and install the attaching bolt or nuts. Install the distributor cap and spark plug wire assembly. Connect the spark plug wires at the spark plugs and the coil secondary high tension wire.

18. Install the upper intake manifold and throttle body assembly in the reverse order of removal. Tighten the attaching studs to 12–18 ft. lbs. (16.2–24.4 Nm) and the support screw to 22–32 ft. lbs. (29.8–43.4 Nm).

19. Install the accelerator cable bracket on the upper intake manifold and connect the cable to the throttle body.

20. Install the fuel line at the fuel rail. Install the throttle body inlet tubes. Connect the negative battery cable.

Valve Lash

ADJUSTMENT

Valve stem-to-valve rocker arm clearance should be within specification with the hydraulic lifter completely collapsed. Repeated machining operations, such as valve and/or valve seat refacing and cylinder head resurfacing or component replacement, will decrease the clearance to the point where if it is not compensated for, the hydraulic valve lifter will no longer function and the valve will be left open. For this reason, a 0.060 in. shorter or 0.060 in. longer pushrod is available to provide a means of compensating for dimensional changes in the valve train.

The positive stop rocker arm bolt eliminates the necessity to adjust the valve clearance. However, to obtain the proper valve clearance, all valve components must be in good condition and installed and tightened properly. Use the following service procedure to determine whether a shorter or longer pushrod is necessary.

4.9L Engine

1. Connect an auxiliary starter switch. Crank the engine with the ignition switch in the **OFF** position.

2. Make 2 marks on the crankshaft damper. Space the marks 120 degress apart so, with the timing mark, the damper is divided into 3 equal parts.

3. With No. 1 piston on TDC at the end of the compression stroke, tighten the rocker arm bolts of the No. 1 intake and exhaust valves to 17–23 ft. lbs. (24–31 Nm).

4. Using bleed down wrench T70P-6513-A or equivalent, slowly apply pressure to bleed down the lifter until the plunger is completely bottomed. Hold the lifter in this position and check the available clearance between the rocker arm and the valve stem tip with a feeler gauge. The clearance should be 0.125–0.175 in.

5. If the clearance is less than specified, install a shorter pushrod. If the clearance is greater than specified, install a longer pushrod.

6. Repeat this procedure for the rest of the valves, turning the crankshaft $\frac{1}{3}$ turn at a time in the direction of rotation. Adjust the valves in the firing order sequence 1-5-3-6-2-4.

5.0L and 5.8L Engines

1. Install an auxiliary starter switch. Crank the engine with the ignition switch **OFF** until No. 1 piston is on TDC after the compression stroke.

2. With the crankshaft in the positions designated in Steps 5a, 5b and 5c or Steps 6a, 6b and 6c, position the hydraulic lifter compressor tool T71P-6513-B or equivalent, on the rocker arm.

3. Slowly apply pressure to bleed down the hydraulic lifter until the plunger is completely bottomed. Hold

STEP 1 – SET NO. 1 PISTON ON TDC AT END OF COMPRESSION STROKE, ADJUST NO. 1 INTAKE and EXHAUST

STEP 4 – CHECK NO. 6 INTAKE and EXHAUST

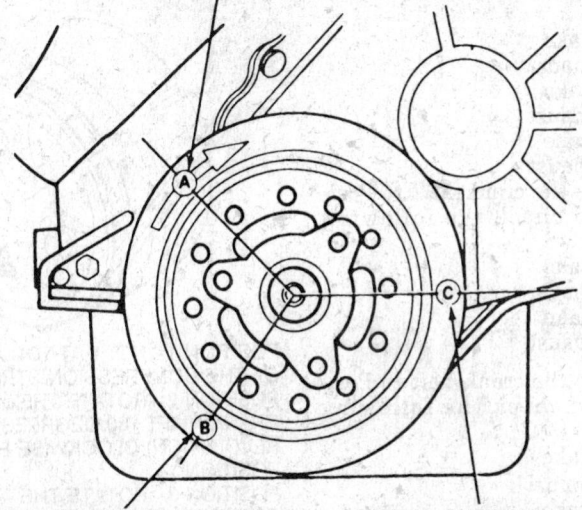

STEP 2 – CHECK NO. 5 INTAKE and EXHAUST

STEP 3 – CHECK NO. 3 INTAKE and EXHAUST

STEP 5 – CHECK NO. 2 INTAKE and EXHAUST

STEP 6 – CHECK NO. 4 INTAKE and EXHAUST

Valve clearance adjustment positions—4.9L engine

LIFTER BLEED DOWN WRENCH

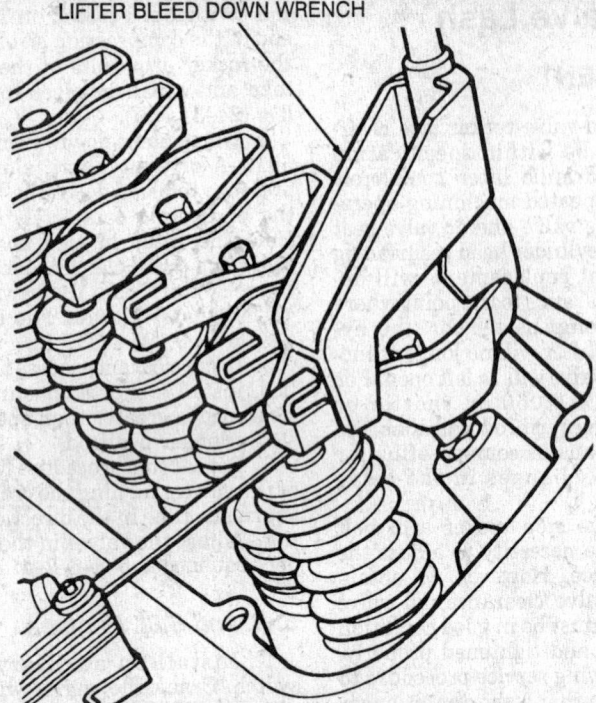

Checking valve clearance

the lifter in this position and check the available clearance between the rocker arm and the valve stem tip with a feeler gauge. The clearance should be 0.096–0.165 in. on 5.0L engine or 0.123–0.173 in. on 5.8L engine.

4. If the clearance is less than specified, install a shorter pushrod. If clearance is greater than specified, install a longer pushrod.

5. On 5.0L engine, clearance checking positions are as follows:

a. With the No. 1 piston on TDC at the end of the compression stroke, Position 1, check the following valves:
No. 1 intake
No. 1 exhaust
No. 7 intake
No. 5 exhaust
No. 8 intake
No. 4 exhaust

b. Rotate the crankshaft to Position 2 and check the following valves:
No. 5 intake
No. 2 exhaust
No. 4 intake
No. 6 exhaust

c. Rotate the crankshaft to Position 3 and check the following valves:
No. 2 intake
No. 7 exhaust
No. 3 intake
No. 3 exhaust
No. 6 intake
No. 8 exhaust

6. On 5.8L engine, clearance checking positions are as follows:

a. With the No. 1 piston on TDC at the end of the compression stroke, Position 1, check the following valves:
No. 1 intake
No. 1 exhaust
No. 4 intake
No. 3 exhaust

WITH NO. 1 AT TDC AT THE END OF THE COMPRESSION STROKE MAKE A CHALK MARK AT POINTS 2 and 3 APPROXIMATELY 90 DEGREES APART
TIMING POINTER

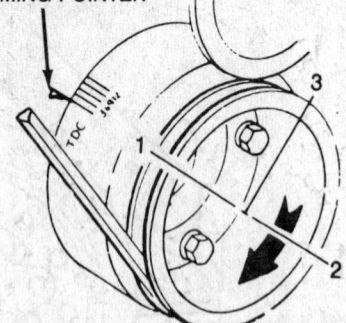

POSITION 1—NO. 1 AT TDC AT THE END OF THE COMPRESSION STROKE
POSITION 2—ROTATE THE CRANKSHAFT 180 DEGREES (½ REVOLUTION) CLOCKWISE FROM POSITION 1
POSITION 3—ROTATE THE CRANKSHAFT 270 DEGREES (¾ REVOLUTION) CLOCKWISE FROM POSITION 2

Valve clearance adjustment positions— 5.0L and 5.8L engines

No. 8 intake
No. 7 exhaust

b. Rotate the crankshaft to Position 2 and check the following valves:
No. 3 intake
No. 2 exhaust
No. 7 intake
No. 6 exhaust

c. Rotate the crankshaft to Position 3 and check the following valves:
No. 2 intake
No. 4 exhaust
No. 5 intake
No. 5 exhaust
No. 6 intake
No. 8 exhaust

7.5L Engine

1. Install an auxiliary starter switch. Crank the engine with the ignition switch **OFF** until No. 1 piston is on TDC after the compression stroke.

2. With the crankshaft in the positions designated in Steps 5 and 6, position hydraulic lifter compressor tool T71P–6513–B or equivalent, on the rocker arm.

3. Slowly apply pressure to bleed down the hydraulic lifter until the plunger is completely bottomed. Hold the lifter in this position and check the available clearance between the rocker arm and the valve stem tip with a feeler gauge. The clearance should be 0.100–0.150 in.

4. If the clearance is less than specified, install a shorter pushrod. If clearance is greater than specified, install a longer pushrod.

5. With the No. 1 piston on TDC at the end of the compression stroke, Position 1, check the following valves:
No. 1 intake

TIMING POINTER

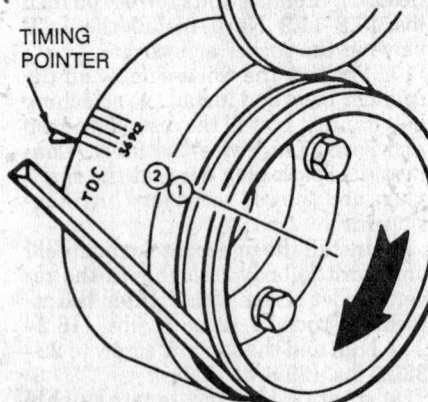

POSITION 1—NO. 1 AT TDC AT END OF COMPRESSION STROKE
POSITION 2—ROTATE THE CRANKSHAFT 360 DEGREES (1 REVOLUTION) CLOCKWISE FROM POSITION 1

Valve clearance adjustment positions— 7.5L engine

No. 1 exhaust
No. 3 intake
No. 4 exhaust
No. 7 intake
No. 5 exhaust
No. 8 intake
No. 8 exhaust

6. Rotate the crankshaft 360 degrees to Position 2 and check the following valves:

No. 2 intake
No. 2 exhaust
No. 4 intake
No. 3 exhaust
No. 5 intake
No. 6 exhaust
No. 6 intake
No. 7 exhaust

Rocker Arms

REMOVAL & INSTALLATION

4.9L Engine

1. Disconnect the negative battery cable and relieve the fuel system pressure. Disconnect the inlet air hose at the crankcase filter cap and remove the throttle body inlet tubes.

2. Disconnect the accelerator cable at the throttle body. Remove the cable retracting spring. Remove the accelerator cable bracket from the upper intake manifold and position the cable and bracket assembly aside.

3. Remove the fuel line from the fuel rail. Remove the upper intake and throttle body asembly as follows:

a. Tag and disconnect the electrical connectors and vacuum lines at the upper intake manifold and the throttle body.

b. Disconnect the PCV hose from the fitting on the underside of the upper intake manifold.

c. Disconnect the EGR tube from the EGR valve and the rear exhaust manifold. Remove the tube from the engine.

d. Remove the thermactor tube from the lower intake manifold by removing the 2 nuts retaining the tube. Remove the nut attaching the thermactor bypass valve bracket to the lower intake manifold.

e. On E Series, remove the nut attaching the transmission fill tube. Then remove the tube bracket off the intake manifold stud.

f. Remove the 7 studs that retain the upper intake manifold. Remove the screw and washer attaching the upper intake manifold support bracket to the upper intake manifold.

g. Remove the upper intake manifold and throttle body.

4. Remove the PCV valve from the rocker arm cover. Remove the crank-

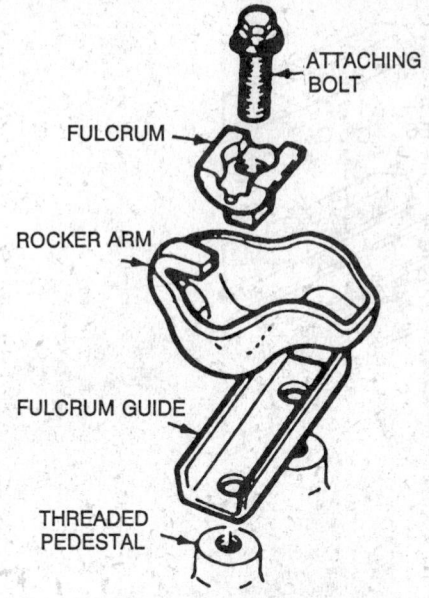

ATTACHING BOLT
FULCRUM
ROCKER ARM
FULCRUM GUIDE
THREADED PEDESTAL

Rocker arm assembly

case filter. Remove the cover bolts and remove the rocker arm cover.

5. Remove the rocker arm bolt, fulcrum seat and rocker arm.

To install:

6. Clean all gasket mating surfaces. Check the rocker arm cover gasket sealing surface for damage and/or distortion. Repair or replace the cover, as necessary.

7. Apply multi-purpose grease to the top of the valve stems, the rocker arm fulcrum seat and the fulcrum seat socket in the rocker arm.

8. Install the rocker arm, fulcrum seat and rocker arm bolt. Tighten the rocker arm bolt to 17–23 ft. lbs. (24–31 Nm). Check the valve clearance.

9. Place a new gasket on the cylinder head, making sure the tabs of the gasket face down towards the head. Install the rocker arm cover, making sure the gasket seats evenly all around the head. Partially tighten the cover bolts, starting at the middle and working outwards, then tighten the bolts to 4–7 ft. lbs. (5–9 Nm) in the same way.

10. Install the PCV valve. Install the upper intake manifold and throttle body assembly in the reverse order of removal. Tighten the attaching studs to 12–18 ft. lbs. (16.2–24.4 Nm) and the support screw to 22–32 ft. lbs. (29.8–43.4 Nm).

11. Install the accelerator cable bracket on the upper intake manifold and connect the cable to the throttle body. Connect the fuel line.

12. Connect the inlet air hose to the crankcase filter cap and install the throttle body inlet tubes. Connect the negative battery cable.

5.0L and 5.8L Engines

1. Disconnect the negative battery

cable. Remove the air cleaner and intake duct assembly, including the closed crankcase ventilation hoses and tubes.

2. Remove the coil and the solenoid brackets.

3. For removal of the right side rocker arm cover, remove the lifting eye and the thermactor tube.

4. Remove the oil filler pipe hose from the left side rocker arm cover, if equipped.

5. Tag and disconnect the spark plug wires. Remove the wires and bracket assemblies from the rocker arm cover attaching stud and position aside.

6. Remove the vacuum harness and the electrical connectors to the vacuum solenoids mounted on the rocker arm covers, if equipped, and position aside.

7. Disconnect the evaporative system hoses from the canister or the chassis and position aside. Remove the thermactor air supply hose, if equipped.

8. Remove the rocker arm cover bolts and remove the rocker arm covers.

9. Remove the rocker arm bolt, fulcrum seat and rocker arm.

To install:

10. Clean all gasket mating surfaces. Check the rocker arm cover gasket sealing surface for damage and/or distortion. Repair or replace the cover, as necessary.

11. Apply multi-purpose grease to the top of the valve stems, the rocker arm fulcrum seat and the fulcrum seat socket in the rocker arm.

12. Install the fulcrum guide, rocker arm, fulcrum seat and bolt. Tighten to 18–25 ft. lbs. (25–32 Nm). Check the valve clearance.

13. Place new gaskets in the rocker arm covers, making sure the tabs of the gasket engage the notches in the cover. Position the covers on the cylinder heads, install the bolts and tighten, in sequence, to 3–5 ft. lbs. (4–6 Nm). Wait 2 minutes, then tighten the bolts again, in sequence, to the same specifications.

14. Retighten all intake manifold bolts to 23–25 ft. lbs. (32–33 Nm) in the proper sequence. Install the vacuum harness bracket and attaching nut and tighten to 12–18 ft. lbs. (17–24 Nm).

15. Reconnect the vacuum harnesses and electrical connectors to the vacuum solenoids on the rocker arm covers, if equipped.

16. Connect the spark plug wires and bracket assembly to the attaching stud on the rocker arm cover and connect the wires at the spark plugs.

17. Connect the thermactor air supply hose, if equipped. Connect the oil

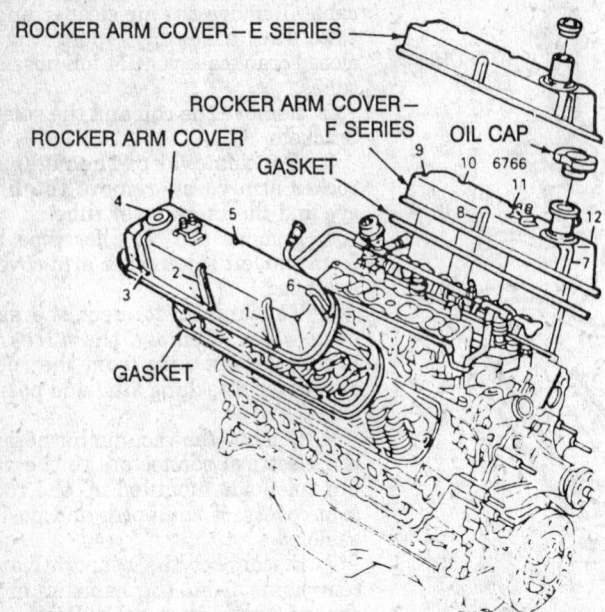

ROCKER ARM COVER—E SERIES

ROCKER ARM COVER—F SERIES

ROCKER ARM COVER

OIL CAP

GASKET

GASKET

Rocker arm cover bolt torque sequence—5.0L and 5.8L engines

filler pipe hose to the left rocker arm, if equipped.

18. Install the closed crankcase ventilation hoses or tubes. Connect the negative battery cable.

7.5L Engine

1. Disconnect the negative battery cable(s). Remove the air cleaner and intake duct assembly.

2. Remove the thermactor exhaust air supply control valve and bracket assembly and the ignition coil mounting bracket.

3. Disconnect the MTA hose at the thermactor valve. Disconnect the thermactor air control valve to air pump hose and tube inlet assembly.

4. Tag and disconnect the spark plug wires from the spark plugs and position the wires aside.

5. For the right side rocker arm cover, remove the PCV valve and position the wiring and vacuum harnesses to gain access in order to remove the cover.

6. Remove the rocker arm covers. Remove the fulcrum bolt, oil deflector, fulcrum seat and rocker arm.

To install:

7. Clean the rocker arm covers and the cylinder head sealing surfaces.

8. Apply multi-purpose grease to the top of the valve stems and the rocker arm and fulcrum seats.

9. Position the No. 1 piston at TDC at the end of the compression stroke. Install the rocker arm, fulcrum seat, oil deflector and fulcrum bolt on the following valves: No. 1 intake, No. 1 exhaust, No. 3 intake, No. 8 exhaust, No. 7 intake, No. 5 exhaust, No. 8 intake and No. 4 exhaust.

10. Rotate the crankshaft 1 revolu-

tion clockwise and install the rocker arm, fulcrum seat, oil deflector and fulcrum bolt on the following valves: No. 2 intake, No. 2 exhaust, No. 4 intake, No. 3 exhaust, No. 5 intake, No. 6 exhaust, No. 6 intake and No. 7 exhaust.

11. Make sure the fulcrum seat base is inserted into its slot on the cylinder head before tightening the fulcrum bolts. Tighten the fulcrum bolts to 18–25 ft. lbs. (25–33 Nm). Check the valve clearance.

12. Position a new seal in the cover seal groove, making sure the seal tang is aligned with the notch in the cover.

13. Position the covers on the cylinder heads. Tighten the 4 rocker arm cover bolts to 9 ft. lbs. (12 Nm) working from right-to-left. Install the PCV valve in the right side cover.

14. Install the ignition coil mounting bracket and connect the spark plug wires to the spark plugs.

15. Connect the thermactor air control valve and bracket assembly. Install the thermactor air supply tube and MTA hose at the thermactor valve.

16. Install the air cleaner and intake duct assembly. Connect the negative battery cable(s), start the engine and check for leaks.

Intake Manifold

REMOVAL & INSTALLATION

5.0L and 5.8L Engines

1. Disconnect the negative battery cable and relieve the fuel system pressure. Drain the cooling system. Remove the air intake duct assembly.

2. Tag and disconnect the electrical connectors at the air bypass valve, throttle position sensor and EGR position sensor.

3. Disconnect the throttle linkage at the throttle ball and the transmission linkage from the throttle body. Remove the bolts that secure the bracket to the intake and position the bracket and cables out of the way.

4. Tag and disconnect the upper intake manifold vacuum fitting connections by removing all the vacuum lines at the vacuum tree, EGR valve and fuel pressure regulator.

5. Disconnect the PCV system by disconnecting the hose from the fitting at the rear of the upper manifold.

6. Remove the canister purge lines from the fittings at the throttle body. Disconnect the water heater lines from the throttle body.

7. Disconnect the EGR tube from the EGR valve by removing the flange nut.

8. Remove the bolt from the upper intake support bracket to the upper manifold. Remove the upper manifold retaining bolts and remove the upper intake manifold and throttle body as an assembly.

9. Tag and disconnect the spark plug wires from the spark plugs. Disconnect the coil wire from the distributor cap and remove the cap and wires assembly.

10. Mark the position of the rotor and the distributor housing in relation to the engine, remove the distributor hold-down bolt and remove the distributor.

11. Tag and disconnect the electrical connections at the coolant temperature sensor, temperature sending unit, air charge temperature sensor, knock sensor, electrical vacuum regulator and thermactor solenoids.

12. Disconnect the injector wiring harness from the main harness assembly. Remove the oxygen sensor ground wire from the intake manifold stud.

NOTE: The plated stud and ground wire must be reinstalled in the same position.

13. Disconnect the fuel supply and return lines from the fuel supply manifold.

14. Remove the upper radiator hose from the thermostat housing. Remove the water bypass hose and remove the heater outlet hose at the intake manifold.

15. Remove the nut securing the coil bracket and move the bracket aside.

16. Remove the intake manifold mounting bolts and studs. Note the location of the bolts and studs for reinstallation. Remove the lower intake manifold assembly.

To install:

17. Clean all gasket mating surfaces.

18. Apply a $^1/_{16}$ in. bead of silicone sealer to the end seals–gaskets mating points. Install the end seals on the cylinder block and new gaskets on the cylinder heads. The gaskets must be interlocked with the seal tabs.

19. Install 2 locator pins into opposite corners and carefully lower the intake manifold assembly into position.

20. Install the intake manifold bolts and studs. Tighten all bolts and studs, in sequence, to 22–32 ft. lbs. (30–43 Nm). Wait 10 minutes and then tighten all bolts and studs again, in sequence, to specification.

21. Position the coil bracket and hold in place while tightening the nut. Install the coil and solenoid bracket to the intake manifold studs and exhaust manifold stud with the retaining nuts.

22. Connect the upper radiator hose and water bypass hose to the thermostat housing. Connect the heater outlet hose to the intake manifold.

23. Connect the fuel supply and return lines to the fuel supply manifold. Connect the fuel line retaining clips.

24. Connect the electrical connectors to the coolant temperature sensor, engine temperature sending unit, air charge temperature sensor, knock sensor, electrical vacuum regulator and thermactor solenoids.

25. Install the distributor, aligning the rotor and distributor housing with the marks that were made during removal. Install the hold-down bolt and the cap and wires assembly. Connect the coil and spark plug wires.

26. Position a new gasket on the lower intake manifold mounting face. Using alignment studs may be helpful.

27. Install the upper intake and throttle body assembly on the lower manifold, making sure the gasket remains in place. Install the 6 upper intake manifold retaining bolts and tighten to 12–18 ft. lbs. (16–24 Nm).

28. Install the upper intake support bracket to the upper manifold attaching bolt. Install the EGR tube.

29. Connect the canister purge lines and the water heater lines to the throttle body. Connect the PCV hose to the rear of the upper manifold.

30. Connect the vacuum lines to the vacuum tree, EGR valve and fuel pressure regulator.

31. Install the throttle linkage bracket with the 2 retaining bolts. Connect the throttle cable and transmission cable to the throttle body.

32. Connect the electrical connectors at the air bypass valve, throttle position sensor and EGR position sensor.

33. Fill and bleed the cooling system. Connect the negative battery cable, start the engine and bring to normal operating temperature. Check for leaks.

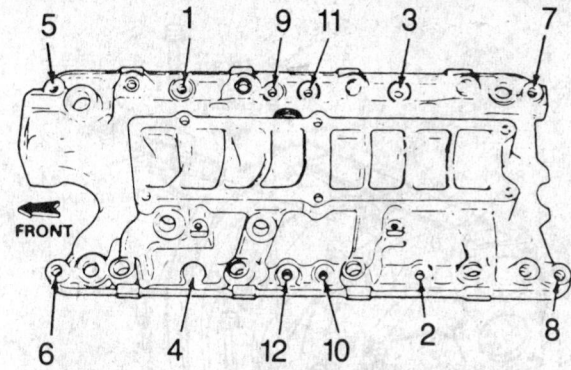

Lower intake manifold bolt torque sequence—5.0L and 5.8L engines

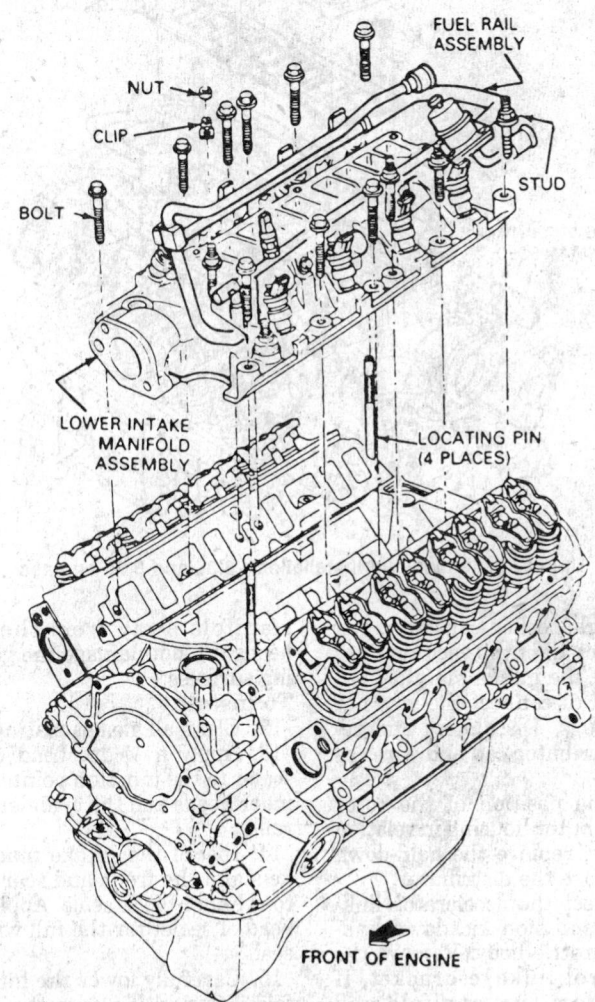

Lower intake manifold installation—5.0L and 5.8L engines

34. Check the ignition timing and adjust, if necessary.

7.5L Engine

1. Disconnect the negative battery cable(s) and relieve the fuel system pressure. Drain the cooling system and remove the air cleaner and intake duct assembly.

2. Remove the thermactor system from the right side of the engine. Re-move the ignition coil mounting bracket and remove the coil wire.

3. Disconnect and remove the external EGR tube.

4. Disconnect the upper radiator hose at the engine. Disconnect the heater hoses at the intake manifold and the water pump. Loosen the water pump bypass hose at the intake manifold.

5. Disconnect the PCV valve and

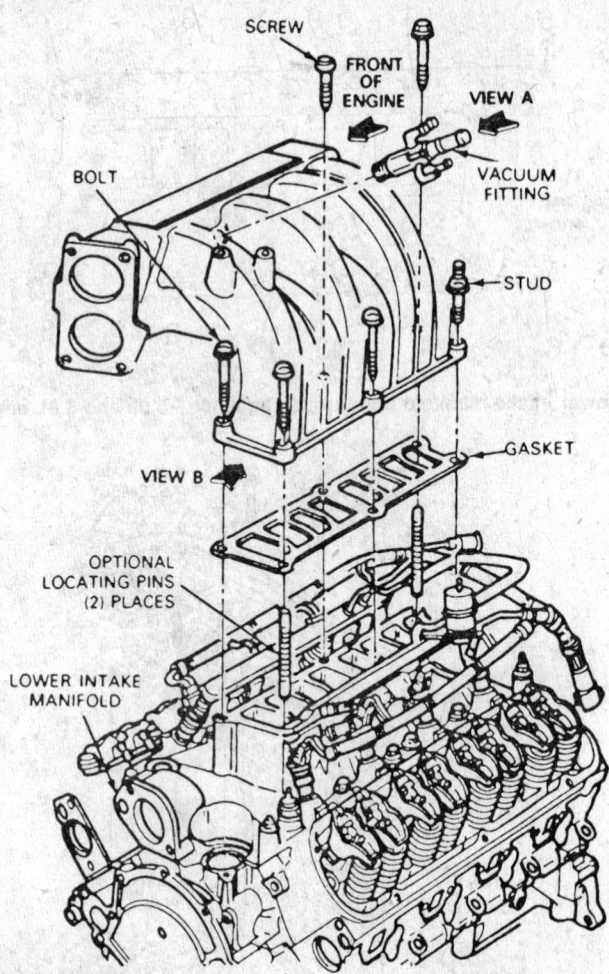

Upper intake manifold installation—5.0L and 5.8L engines

hose at the right rocker arm cover. Tag and disconnect all of the vacuum lines at the rear of the intake manifold.

6. Tag and disconnect the wires at the spark plugs. Disconnect and remove the distributor cap and wires as an assembly.

7. Mark the position of the rotor and the distributor housing in relation to the engine, remove the hold-down bolt and remove the distributor.

8. Disconnect the accelerator linkage and transmission kickdown linkage at the throttle body. Remove the cruise control linkage bracket, if equipped, from the manifold and throttle body. Remove the bolts holding the accelerator linkage cable and position the linkage aside.

9. Disconnect the fuel line at the fuel supply manifold.

10. Disconnect the wiring harness from the main wiring harness. Remove lower intake manifold and wiring as an assembly.

11. Remove the intake manifold attaching bolts and lift the manifold and throttle body from the engine as an assembly. It may be necessary to pry the

manifold away from the cylinder heads. Do not damage the gasket sealing surfaces.

To install:

12. Clean all gasket mating surfaces.

13. Apply a ⅛ in. bead of silicone sealer to the junction points of the cylinder heads and the cylinder block end rails.

14. Install the intake manifold gaskets and the front and rear manifold-to-cylinder block seals. Apply a ¹/₁₆ in. bead of sealer for the full width of the seal.

15. Carefully lower the intake manifold into position over the 4 studs in the ends of the cylinder heads. When the intake manifold is in place, run a finger around the seal area to make sure the seals are in place. If the seals have shifted, remove the manifold and re-position the seals.

16. Install the attaching bolts, stud bolts and nuts and tighten as follows:

    a. Tighten bolts 1–12, in sequence, to 8–12 ft. lbs. (11–16 Nm). Tighten nuts 13–16, in sequence, to 8–12 ft. lbs. (11–16 Nm).

    b. Tighten bolts and nuts 1–16, in

sequence, to 12–22 ft. lbs. (16–30 Nm).

    c. Tighten manifold bolts and nuts 1–16, in sequence, to 22–35 ft. lbs. (30–47 Nm).

17. Install the external EGR tube and thermactor exhaust air supply tubes.

18. Install the water pump bypass hose to the intake manifold fitting and heater hot water tube. Connect the radiator upper hose to the coolant outlet housing.

19. Install the distributor, aligning the rotor and the distributor housing with the marks that were made during removal. Install the hold-down bolt.

20. Connect the heater hoses at the intake manifold and water pump. Connect the PCV valve and hose to the right side rocker arm cover.

21. Connect the fuel lines to the fuel supply manifold and install the ignition coil mounting bracket. Connect all electrical connectors.

22. Position the accelerator linkage on the manifold and attach the accelerator linkage cable. Attach the cruise control linkage bracket to the intake manifold, if equipped. Connect the accelerator and kickdown linkage to the throttle body.

23. Connect the vacuum lines. Install the distributor cap and wires assembly. Connect the coil wire and the spark plug wires.

24. Fill and bleed the cooling system. Connect the negative battery cable(s), start the engine and bring to normal operating temperature. Check for leaks.

25. Check the ignition timing and adjust, if necessary.

26. While the engine is hot, tighten the manifold attaching nuts and bolts,

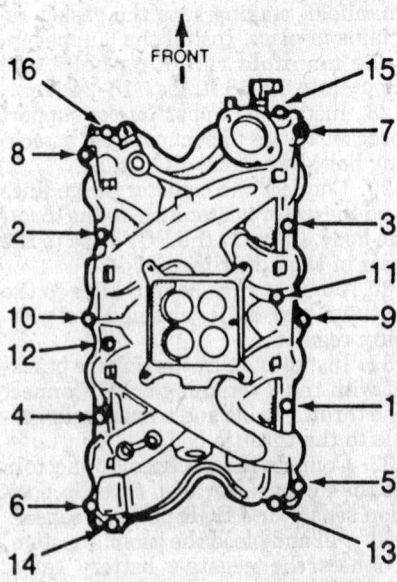

Intake manifold bolt torque sequence—7.5L engine

in sequence, to 22–35 ft. lbs. (30–47 Nm).

27. Install the air cleaner and intake duct assembly.

Exhaust Manifold

REMOVAL & INSTALLATION

5.0L and 5.8L Engines

1. Disconnect the negative battery cable.

2. Remove the air cleaner and intake duct assembly, including the crankcase ventilation hose.

3. Remove the bolts attaching the air cleaner inlet duct, if equipped.

4. Raise and safely support the vehicle. Disconnect the muffler inlet pipes.

5. Lower the vehicle.

6. Remove the exhaust manifold heat shields, if equipped, attaching bolts and flat washers. On the left exhaust manifold, remove the oil dipstick tube assembly, cruise control bracket and exhaust heat control valve, if equipped. Then remove the exhaust manifold.

To install:

7. Clean the exhaust manifold and cylinder head mating surfaces. Clean the mounting flange spherical seat of the exhaust manifolds and muffler inlet pipes.

8. Position the exhaust manifolds on the cylinder heads and install the attaching bolts and flat washers. On the left exhaust manifold, install the oil dipstick tube and the cruise control bracket, if equipped. Working from the center to the ends, tighten the bolts to 18–24 ft. lbs. (25–33 Nm).

9. Raise and safely support the vehicle. Position new gaskets, if equipped, on the muffler inlet pipes. Position the muffler inlet pipes and exhaust heat control valve, if equipped, into the manifolds. Install and tighten the attaching nuts to 25–38 ft. lbs. (34–52 Nm).

10. Lower the vehicle. Install the air cleaner and intake duct assembly, including the crankcase ventilation hose.

7.5L Engine

1. Disconnect the negative battery cable.

2. If removing the right exhaust manifold, remove the spark plug heat shield.

3. Tag and disconnect the spark plug wires from the spark plugs.

4. Disconnect the thermactor air supply manifold, if equipped.

5. Disconnect the external EGR tube on the left exhaust manifold.

6. Raise and safely support the ve-

hicle. Disconnect the muffler inlet pipes at the exhaust manifolds.

7. Lower the vehicle.

8. Remove the power steering pump support bracket. On E Series, remove the brace from the back of the power steering pump.

9. Remove the oil dipstick tube from the left exhaust manifold.

10. Remove the attaching bolts and remove the exhaust manifolds.

To install:

11. Clean the mating surfaces of the exhaust manifolds and cylinder heads. Clean the mounting flange of the manifolds and inlet pipes. Apply a light film of graphite grease to the exhaust manifolds.

12. Position the exhaust manifolds on the cylinder heads. Install the attaching bolts and washers, starting at the 4th bolt hole from the front of each manifold. Tighten the bolts to 22–30 ft. lbs. (30–41 Nm) working from the center of the manifold outwards.

13. Raise and safely support the vehicle. Position the inlet pipes to the manifolds and tighten the attaching nuts to 25–36 ft. lbs. (34–49 Nm).

14. Lower the vehicle. Install the power steering pump support bracket. Install the brace, if removed.

15. Connect the external EGR tube on the left exhaust manifold. Install the heat shield and spark plug wires.

16. Connect the negative battery cable. Start the engine and check for exhaust leaks.

Combination Manifold

REMOVAL & INSTALLATION

4.9L Engine

1. Disconnect the negative battery cable and relieve the fuel system pressure. Disconnect the inlet air hose at the crankcase filter cap. Remove the throttle body inlet hoses.

2. Disconnect the accelerator cable at the throttle body. Remove the cable retracting spring. Remove the accelerator cable bracket and position the cable and bracket aside.

3. Disconnect the fuel line from the fuel supply manifold. Remove the upper intake manifold and throttle body assembly as follows:

   a. Tag and disconnect the electrical connectors and vacuum lines at the upper intake manifold and the throttle body.

   b. Disconnect the PCV hose from the fitting on the underside of the upper intake manifold.

   c. Disconnect the EGR tube from the EGR valve and the rear exhaust manifold. Remove the tube from the engine.

   d. Remove the thermactor tube

from the lower intake manifold by removing the 2 nuts retaining the tube. Remove the nut attaching the thermactor bypass valve bracket to the lower intake manifold.

   e. On E Series, remove the nut attaching the transmission fill tube. Then remove the tube bracket off the intake manifold stud.

   f. Remove the 7 studs that retain the upper intake manifold. Remove the screw and washer attaching the upper intake manifold support bracket to the upper intake manifold.

   g. Remove the upper intake manifold and throttle body.

4. Tag and disconnect the vacuum lines. Raise and safely support the vehicle.

5. Disconnect the inlet pipe from the exhaust manifolds. Lower the vehicle.

6. Disconnect the power brake booster vacuum hose.

7. Remove the bolts attaching the manifolds to the cylinder head and lift the manifolds from the engine.

To install:

8. Clean the mating surfaces of the cylinder head and the manifolds.

9. If 1 of the manifolds is to be replaced, remove the tube fittings from the discarded manifolds and install them in the new manifolds, as required. Also install new studs in the new manifold.

10. Install a new intake manifold gasket.

NOTE: Combination intake/exhaust gasket is not to be used on a new exhaust manifold. Gasketing of the exhaust manifold is only recommended when the original exhaust manifold is reinstalled, in order to prevent leakage.

11. Coat the mating surfaces lightly with graphite grease. Place the manifold assemblies in position against the cylinder head, making sure the gasket has not become dislodged. Install the washers, bolts and nuts and tighten, in sequence, to 22–32 ft. lbs. (30–43 Nm).

12. Raise and safely support the vehicle. Connect the inlet pipe to the exhaust manifold and tighten the nuts to 25–36 ft. lbs. (34–49 Nm).

13. Lower the vehicle. Connect the crankcase vent hose to the intake manifold and position the hose clamp.

14. Install the upper intake manifold and throttle body assembly in the reverse order of removal. Tighten the attaching studs to 12–18 ft. lbs. (16.2–24.4 Nm) and the support screw to 22–32 ft. lbs. (29.8–43.4 Nm).

15. Connect the power brake booster vacuum hose and connect the vacuum lines. Connect the fuel line to the fuel supply manifold.

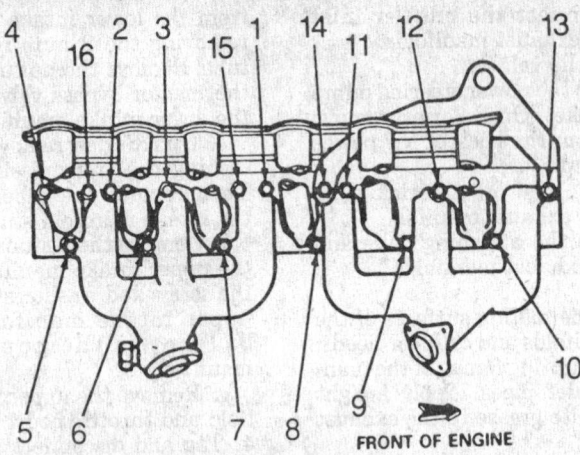

Combination manifold bolt torque sequence

16. Install the accelerator cable bracket on the upper intake and connect the accelerator cable.

17. Install the throttle body inlet hoses and connect the inlet air hose at the crankcase filter cap.

Cylinder Block Front Cover

REMOVAL & INSTALLATION

4.9L Engine

1. Disconnect the negative battery cable and drain the cooling system. Remove the shroud and radiator.

3. Remove the drive belt, power steering bracket bolts and swing aside. Remove the fan and pulleys.

4. Remove the bolt and washer from the end of the crankshaft and remove the damper using a suitable puller.

5. Remove the front oil pan and front cover attaching bolts. Loosen the first 6 bolts on each side of the oil pan. Lightly push the pan down so it does not exert any upeard force on the front cover, which may affect front seal alignment.

NOTE: Keep foreign material from entering the crankcase during this procedure or the engine oil will have to be changed.

6. Remove the front cover and discard the gasket. Remove the front cover oil seal using seal remover tool T70P-6B070-B or equivalent.
To install:

7. Clean the front cover and block gasket surfaces. Clean any oil from the oil pan gasket as it will be used over. Install a new front cover oil seal using a suitable seal installer.

8. Coat the front cover gasket surfaces of the block and cover with oil-re-

sistant sealer. Position a new front cover gasket on the block.

9. Apply silicone sealer to the block/pan junction and a small bead on the oil pan gasket sealing surface of the front cover. This provides an additional seal between the front cover and the used oil pan gasket.

NOTE: When applying silicone sealer, always use the specified bead size and join components within 15 minutes of application, before the sealer "sets-up".

10. Position the front cover over the end of the crankshaft and against the cylinder block. Start the cover and pan attaching screws. Slide front cover aligner tool T61P-6019-B or equivalent over the crank stub and into the seal bore of the cover. Tighten the oil pan bolts to 10–15 ft. lbs. (14–20 Nm) and the front cover bolts to 12–18 ft. lbs. (17–24 Nm).

NOTE: Tighten the front cover bolts first to obtain proper cover alignment.

11. Lubricate the crank stub, damper hub inside diameter and the seal rubbing surface with clean engine oil. Apply a ¼ in. bead of silicone to the inside of the keyway in the damper hub. Align the damper keyway with the crankshaft key and install the damper using damper replacer tool T52L-6306-AEE or equivalent.

12. Install the washer and damper retaining bolt and tighten to 130–150 ft. lbs. (177–203 Nm).

13. Install the pulleys, drive belt, fan, shroud, radiator and hoses.

14. Fill and bleed the cooling system. Change the engine oil, if necessary.

15. Connect the negative battery cable. Start the engine and bring to normal operating temperature. Check for leaks.

5.0L, 5.8L and 7.5L Engines

1. Disconnect the negative battery cable and drain the cooling system. Discharge the air conditioning system on 7.5L engine vehicles, if equipped.

2. Remove the radiator on all except Bronco and F Series with 5.0L and 5.8L engines.

3. Remove the fan, accessory drive belt, pulleys and shroud. Disconnect the hoses from the water pump.

4. On 5.0L and 5.8L engines, remove the air conditioning compressor/power steering pump bracket and accessories.

5. On 7.5L engine, remove the alternator, air pump and air conditioning compressor, if equipped, and their respective brackets.

6. Remove the damper attaching bolt and washer. Remove the damper using a suitable puller.

7. Remove the oil pan-to-front cover attaching bolts. Use a thin-bladed knife to cut the oil pan gasket flush with the cylinder block face prior to separating the cover from the block. Remove the front cover and water pump as an assembly.

8. Remove the front cover oil seal using seal remover tool T70P-6B070-B or equivalent.

To install:

9. Clean the cylinder block, front cover and oil pan gasket surfaces. Clean the oil pan gasket surface where the oil pan and front cover fasten.

10. Install a new front cover oil seal, using a suitable seal installer.

11. Coat the gasket surface of the oil pan with sealer, then cut and position the required new gasket on the oil pan and apply sealer at the corners. Coat the block and cover gasket surfaces with sealer and position a new gasket on the block.

12. Position the front cover on the block and install front cover-to-seal alignment tool T61P-6019-B or equivalent. It may be necessary to force the cover downward to slightly compress the pan gasket. This can be done using a suitable tool at the front cover bolt hole locations.

13. Coat the threads of the attaching bolts with sealer and install. While pushing on the alignment tool, tighten the oil pan-to-cover bolts on 5.0L and 5.8L engines to 12–18 ft. lbs. (17–24 Nm) or to 7–9 ft. lbs. (10–12 Nm) on 7.5L engine. Tighten the cover-to-cylinder block bolts to 12–18 ft. lbs. (17–24 Nm).

14. Apply multi-purpose grease to the oil seal rubbing surface of the damper inner hub. Line up the damper keyway with the crankshaft key and install the damper using damper replacer tool T79T-6316-A or equivalent. Install the damper bolt and wash-

er and tighten to 70–90 ft. lbs. (95–122 Nm).

15. Install the air conditioning compressor/power steering pump bracket and accessories on 5.0L and 5.8L engines. Install the alternator, air pump and air conditioning compressor and brackets on 7.5L engine.

16. Connect the hoses to the water pump. Install the pulleys, accessory drive belt, fan and shroud.

17. Install the radiator, if removed.

18. Fill and bleed the cooling system. Change the engine oil, if necessary.

19. Connect the negative battery cable, start the engine and bring to normal operating temperature. Check for leaks.

Front Cover Oil Seal

REMOVAL & INSTALLATION

1. Disconnect the negative battery cable(s).

2. Remove the fan, shroud and accessory drive belt(s).

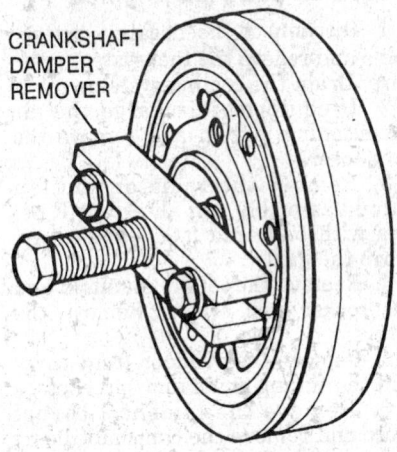

Crankshaft damper removal

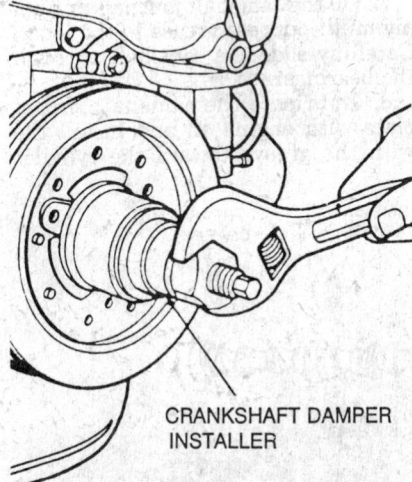

Crankshaft damper installation

3. Remove the crankshaft pulley.

4. Remove the crankshaft damper retaining bolt and washer. Remove the damper using a puller.

5. Remove the front cover oil seal using seal replacer tool T70P–6B070–B or equivalent.

To install:

6. Coat a new front cover oil seal with clean engine oil and install, using seal replacer tool T70P–6B070–A or equivalent.

7. Apply clean engine oil to the oil seal rubbing surface of the damper inner hub. Line up the damper keyway with the crankshaft key and install the damper, using damper replacer tool T52L–6306–AEE or T79T–6316–A or equivalent.

8. Install the damper retaining bolt and washer. Tighten to 130–150 ft. lbs. (177–203 Nm) on 4.9L engine or 70–90 ft. lbs. (95–122 Nm) on 5.0L, 5.8L and 7.5L engines.

9. Install the crankshaft pulley, accessory drive belt(s), fan and shroud. Connect the negative battery cable(s).

Timing Chain and/or Timing Sprockets

REMOVAL & INSTALLATION

4.9L Engine

1. Disconnect the negative battery cable and drain the cooling system and crankcase.

2. Remove the cylinder front cover assembly.

3. Crank the engine until the timing marks are aligned.

4. Install camshaft sprocket puller T82T–6256–A or equivalent, and remove the camshaft sprocket.

5. Install crankshaft damper remover T58P–6316–D or equivalent,

and remove the crankshaft sprocket. Remove the key from the crankshaft.

To install:

6. Make sure the key spacer and camshaft thrust plate are properly installed. Align the camshaft sprocket keyway with the camshaft key and install the sprocket using camshaft sprocket replacing adapter tool T65L–6306–A or equivalent.

7. Install the key in the crankshaft keyway. Install the crankshaft sprocket using crankshaft damper replacer tool T52L–6306–AEE or equivalent. Make sure the timing marks are properly aligned. Install the oil slinger.

8. Install the cylinder front cover. Fill the crankcase with the proper type and quantity of engine oil. Fill and bleed the cooling system.

9. Connect the negative battery cable. Start the engine and bring to normal operating temperature. Check for leaks. Check the ignition timing and curb idle speed and adjust, if necessary.

5.0L, 5.8L and 7.5L Engines

1. Disconnect the negative battery cable and drain the cooling system.

2. Remove the cylinder front cover assembly.

3. Rotate the crankshaft until the timing marks on the timing sprockets are aligned.

4. Remove the camshaft sprocket bolt, washer and eccentric, if equipped. Slide both sprockets and the timing chain forward and remove them as an assembly.

To install:

5. Install the sprockets and timing chain on the crankshaft and camshaft. Make sure the timing marks on the sprockets are positioned properly.

6. Install the eccentric, if equipped, the washer and camshaft retaining

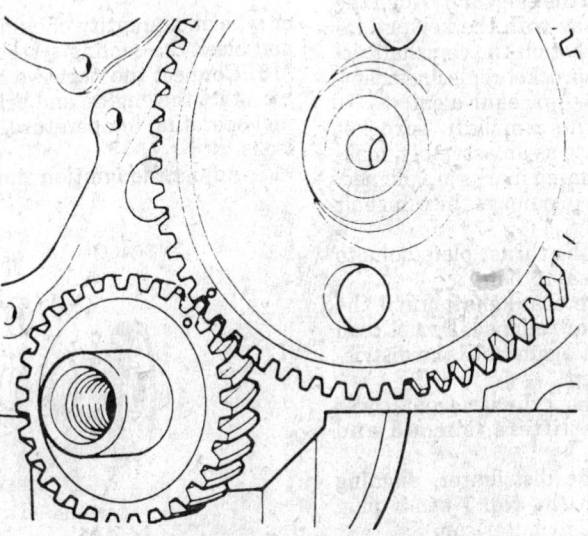

Timing sprocket alignment—4.9L engine

bolt. Tighten to 40–45 ft. lbs. (55–61 Nm).

7. Lubricate the timing chain and sprockets with heavy engine oil.

8. Install the cylinder front cover assembly.

9. Fill and bleed the cooling system. Connect the negative battery cable, start the engine and bring to normal operating temperature. Check for leaks.

10. Check the ignition timing and idle speed. Adjust, if necessary.

Camshaft

REMOVAL & INSTALLATION

4.9L Engine

1. Disconnect the negative battery cable and drain the cooling system and crankcase.

2. Remove the shroud, radiator, lifters and cylinder front cover.

3. Disconnect the primary wire at the coil and remove the distributor.

4. Check the camshaft endplay using a dial indicator. The endplay should not exceed 0.009 in. Replace the thrust plate if the endplay is excessive.

5. Turn the crankshaft to align the timing marks on the timing sprockets.

6. Remove the camshaft thrust plate bolts. Remove the camshaft sprocket using sprocket puller T82T-6256-A or equivalent. Remove the key, thrust plate and spacer.

7. Remove the camshaft, being careful to avoid damaging the camshaft bearings.

To install:

8. Oil the camshaft bearing journals and apply multi-purpose grease to the lobes.

9. Assemble the key, spacer and thrust plate to the camshaft. Align the sprocket keyway with the key and install the sprocket on the camshaft using camshaft sprocket replacing adapter T65L-6306-A or equivalent.

10. Install the camshaft, sprocket and thrust plate as an assembly, making sure the timing marks are aligned. Be careful not to damage the camshaft bearings.

11. Tighten the thrust plate bolts to 9–12 ft. lbs. (13–16 Nm).

12. Turn the crankshaft until the timing marks are aligned. Do not turn the crankshaft again until the distributor is installed.

13. Install the cylinder front cover and damper, lifters, shroud and radiator.

14. Install the distributor, aligning the rotor with the No. 1 spark plug tower in the distributor cap.

15. Fill the crankcase with the prop-

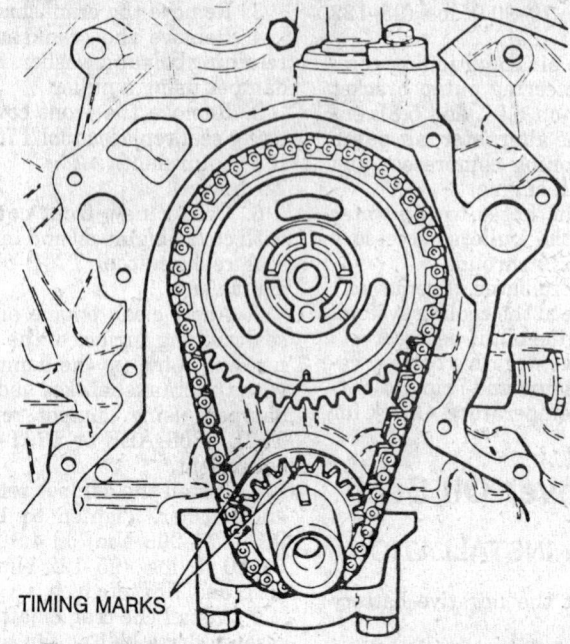

TIMING MARKS

Timing sprocket alignment—5.0L and 5.8L engines

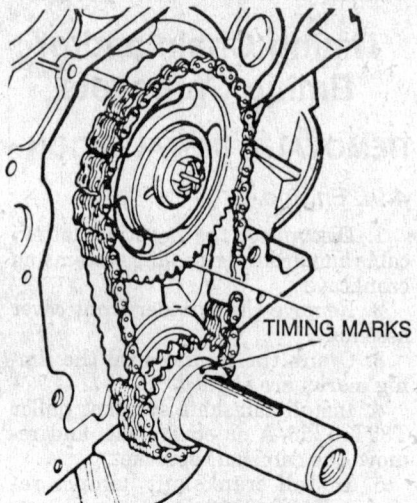

TIMING MARKS

Timing sprocket alignment—7.5L engine

er type and quantity of engine oil. Fill and bleed the cooling system.

16. Connect the negative battery cable, start the engine and bring to normal operating temperature. Check for leaks.

17. Adjust the ignition timing.

5.0L, 5.8L and 7.5L Engines

1. Disconnect the negative battery cable and relieve the fuel system pressure. Drain the cooling system.

2. If equipped, discharge the air conditioning system and remove the condenser.

3. Remove the radiator and fan shroud assembly. On all except F Series with 5.0L and 5.8L engines, remove the grille.

4. Remove the intake manifold and the rocker arm covers. Remove the lifters.

5. Remove the cylinder front cover and the timing chain and sprockets.

6. Remove the camshaft thrust plate and remove the camshaft, being careful not to damage the camshaft bearings.

To install:

7. Oil the camshaft journals and apply multi-purpose grease to the lobes. Carefully slide the camshaft through the bearings.

8. Lubricate the camshaft thrust plate with engine oil and install the with the groove toward the cylinder

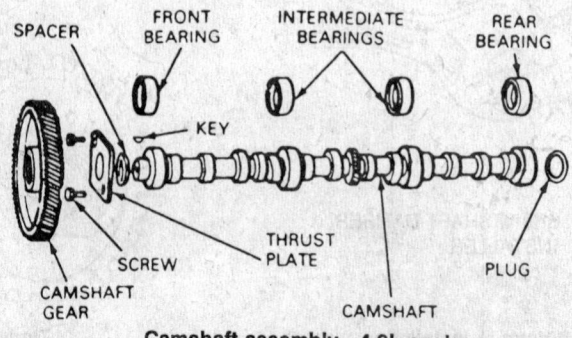

SPACER   FRONT BEARING   INTERMEDIATE BEARINGS   REAR BEARING

KEY

SCREW   THRUST PLATE   PLUG

CAMSHAFT GEAR   CAMSHAFT

Camshaft assembly—4.9L engine

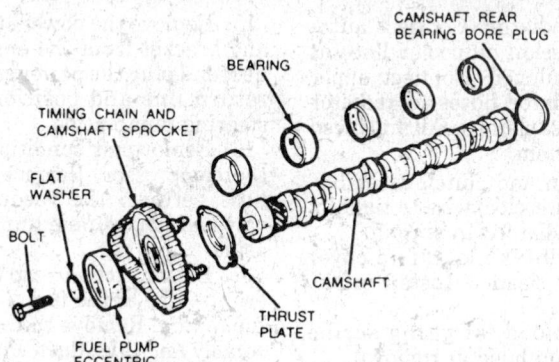

Typical camshaft assembly—5.0L, 5.8L and 7.5L engines

block. Tighten the thrust plate bolts to 9–12 ft. lbs. (13–16 Nm) on the 5.0L and 5.8L engines or 6–9 ft. lbs. (8–12 Nm) on 7.5L engine.

9. Check the camshaft endplay using a dial indicator. Endplay should not exceed 0.009 in. If endplay is excessive, replace the thrust plate.

10. Install the remaining components in the reverse order of their removal.

11. Fill and bleed the cooling system. Connect the negative battery cable, start the engine and bring to normal operating temperature. Check for leaks.

12. Check the ignition timing and idle speed; adjust if necessary. If equipped, evacuate and charge the air conditioning system.

Piston and Connecting Rod

POSITIONING

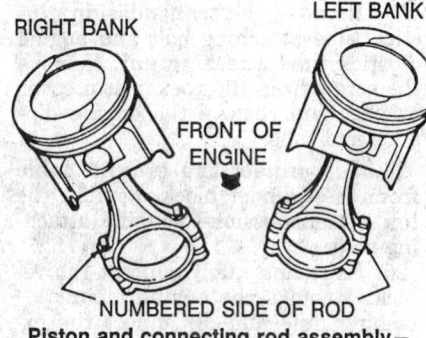

Piston and connecting rod assembly— 5.0L and 5.8L engines

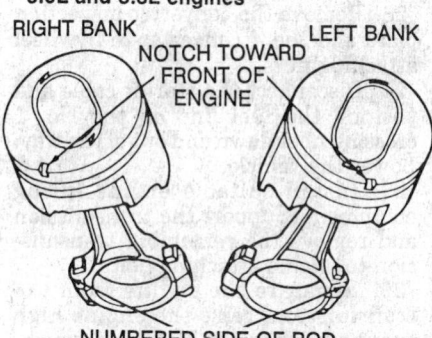

NUMBERED SIDE OF ROD 7.5L engine

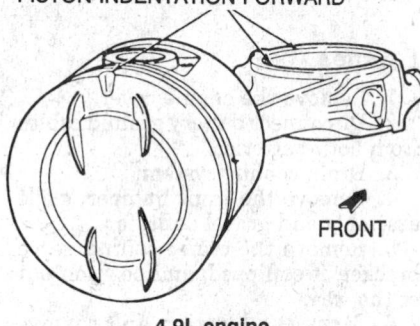

POSITION BEARING TANG SIDE OF ROD TO LEFT (TOWARDS CAMSHAFT) WITH PISTON INDENTATION FORWARD

4.9L engine

DIESEL ENGINE MECHANICAL

Engine Assembly

REMOVAL & INSTALLATION

F Series

1. Disconnect battery ground cables from both batteries.

2. Scribe alignment marks at hood hinges and remove hood.

3. Drain cooling system.

4. Remove air cleaner and intake duct assembly. Position a suitable cover over the air intake opening.

5. Remove radiator fan shroud.

6. Remove fan and clutch assembly. The fan clutch nut has left-hand thread and must be removed by turning clockwise.

7. Disconnect radiator upper and lower hoses from radiator.

8. Disconnect and plug automatic transmission oil cooler lines at radiator, if equipped.

9. Remove radiator.

10. Loosen the air conditioning compressor, if equipped and remove drive belt.

11. Remove the air conditioning compressor from its mounting, if equipped and position it on radiator upper support.

12. Loosen power steering pump and remove drive belt. Remove power steering pump and position out of the way on left side of engine compartment.

13. Disconnect fuel supply line heater, if equipped, and alternator wires at alternator. Disconnect oil pressure sending unit wire at sending unit located at the back of the engine.

14. Disconnect accelerator cable from injection pump. Disconnect cruise control cable from injection pump, if equipped. Remove accelerator cable bracket with cables attached, from intake manifold and position out of the way.

15. Disconnect transmission kickdown rod from injection pump, if equipped. Disconnect main wiring harness connector from right side of engine. Disconnect engine ground strap from rear of engine. Disconnect fuel return hose from left rear of engine. Remove the vacuum supply hose from the vacuum pump, if equipped.

16. Remove upper transmission-to-engine attaching bolts.

17. Disconnect heater hoses from water pump and right cylinder head. Disconnect water temperature sender wire from sender on left front of engine block. Disconnect water temperature overheat light switch wire from switch on top front of left cylinder head. Position wires out of the way.

18. Raise and safely support the vehicle.

19. Disconnect both battery ground cables from lower front of engine.

20. Disconnect and cap fuel inlet line at fuel supply pump.

21. Disconnect starter cables at starter motor.

22. Disconnect muffler inlet pipe at exhaust manifolds.

23. Disconnect engine insulators from No. 1 crossmember. Remove flywheel inspection plate. Remove converter-to-flywheel attaching nuts, if equipped. Lower vehicle.

24. Support transmission with a floor jack. Remove lower transmission to engine attaching bolts.

25. Attach suitable engine lifting equipment. Raise engine high enough to clear number 1 crossmember and pull forward.

26. Rotate the front of the engine approximately 45 degrees to the left and lift it out of the engine compartment. Be careful not to damage the windshield wiper motor when lifting the engine from the vehicle.

To install:

NOTE: If the engine to be in-

stalled has been overhauled or has been in storage, prime the entire engine lubricating system to fill the oil cooler, oil filter and cylinder block galleries with the proper type of oil. This will prevent piston and bearing scuffing.

27. Lower the engine into the vehicle, being careful not to damage the windshield wiper motor.

28. Start transmission main shaft into clutch disc. It may be necessary to adjust position of transmission in relation to engine if main shaft binds or will not enter clutch disc. If engine hangs up after main shaft enters clutch disc, rotate crankshaft slowly with the transmission in gear, until mainshaft splines mesh with clutch disc splines. Align convertor to flywheel studs, if equipped.

29. Lower engine onto engine insulator brackets on the number 1 crossmember.

30. Install lower transmission to engine attaching bolts and tighten. Remove engine lifting equipment. Raise and safely support the vehicle.

31. Install converter to flywheel attaching nuts, if equipped. Install flywheel inspection plate.

32. Install engine insulator support to crossmember bracket attaching nuts and washers. Connect muffler inlet pipes to exhaust manifolds. Connect both battery ground cables to the lower front of the engine. Connect starter cables to starter. Install fuel pump inlet line onto fuel pump. Lower vehicle.

33. Connect water temperature sender wire to sender on left front of engine block. Connect wire to water temperature overheat light switch on top of left cylinder head. Install heater hoses onto right cylinder head and water pump and tighten clamps.

34. Connect engine ground strap at rear of engine. Connect fuel return hose at left rear of engine and vacuum supply hose to vacuum pump. Connect transmission kickdown rod, if equipped.

35. Install accelerator cable bracket on intake manifold. Connect accelerator cable to injection pump. Connect cruise control cable, if equipped, to injection pump.

36. Connect oil pressure gauge sender wire to oil pressure sender.

37. Connect fuel supply line heater, if equipped, and alternator wires to alternator.

38. Install power steering pump and drive belt. Do not adjust belt at this time.

39. Install air conditioning compressor and drive belt. Adjust air conditioning compressor and power steering pump drive belts.

40. Install radiator. Connect automatic transmission oil cooler lines at radiator, if equipped. Connect upper and lower radiator hoses to radiator and tighten hose clamps. Fill and bleed the cooling system.

41. Install fan and clutch assembly. Turn nut counterclockwise to tighten.

42. Install radiator fan shroud.

43. Remove intake manifold cover and install air cleaner. Install intake duct assembly.

44. Install hood, aligning scribe marks drawn on hood at removal.

45. Connect battery ground cables at both batteries. Check the engine oil level and fill as required. Run engine and check for fuel, oil and coolant leaks.

E Series

1. Remove the engine cover.

2. Disconnect battery ground cables from both batteries.

3. Drain cooling system.

4. Remove the front bumper, grille assembly and gravel deflector.

5. Remove the cruise control servo bracket, if equipped, and position out of the way.

6. Mark the location and remove the hood latch and cable assembly from the grille upper support bracket. Remove the upper grille support.

7. Discharge the air conditioning system, if equipped. Disconnect the air conditioning lines from the condensor and remove the condensor.

8. Disconnect automatic transmission oil cooler lines at radiator, if equipped. Remove the transmission oil cooler and brackets.

9. Remove the radiator hoses at the engine and remove the shroud. Remove the radiator cooling fan. The fan clutch nut has left-hand thread and must be removed by turning clockwise.

10. Remove radiator attaching bolts and remove radiator.

11. Loosen and remove the vacuum pump and drive belt. Disconnect vacuum hose from the transmission shift modulator tube.

12. Loosen and remove the alternator adjusting arm, adjusting arm bracket and drive belt. Pivot the alternator inward toward the engine and disconnect the alternator wiring harness.

13. Disconnect the water temperature sender wire from the sender on the left front of the engine block and the water temperature overheat light switch wire from the switch on the top front of the left cylinder head. Position the wires aside.

14. Remove the 2 engine ground cables from the bottom front of the engine.

15. Remove the power steering pump and bracket from the engine. Disconnect and plug the power steering pump return line and position the power steering pump aside.

16. Remove air conditioning lines at the compressor. Remove the vacuum hose between the vacuum regulator valve and injection pump and position aside.

17. Disconnect and cap the fuel heater inlet line at the fuel filter filter and fuel pump. Remove the air cleaner assembly and inlet duct. Position a suitable cover over the air intake opening.

18. Disconnect and cap the fuel filter outlet line at the fuel filter ans injection pump. Cap the injection pump and fuel filter fittings. Remove the fuel filter return hose and remove the filter and bracket as an assembly.

19. Loosen the air conditioning compressor and rotate toward the engine. Remove and plug the fuel inlet hose at the fuel pump.

20. Remove the accelerator and cruise control cables from the injection pump and intake bracket and position aside. Remove the bracket. Remove the transmission kickdown rod.

21. Tag and disconnect the engine wiring harnesses from inside the vehicle and position aside. Disconnect the heater hose from the water pump and right cylinder head.

22. Remove the auxiliary heater and air conditioner hoses from the bracket at the left rear of the engine, if equipped. At the rear of the engine, disconnect the oil pressure sender wire from the sender and the fuel return line.

23. Remove the transmission dipstick tube attaching bolt and engine dipstick tube attaching nut. Remove the screw from the rocker arm cover bracket and remove the engine dipstick and tube.

24. Disconnect the ground cable from the cylinder block. Remove the top 4 transmission-to-engine attaching bolts.

25. Raise and safely support the vehicle. Remove the engine mount attaching nuts and disconnect the exhaust pipe at the manifolds.

26. Remove the converter inspection plate and the 4 converter-to-flywheel attaching nuts.

27. Disconnect the starter cable and position the fuel line on the No. 1 crossmember down and out of the way. Lower the vehicle.

28. Install suitable engine lifting equipment. Support the transmission and remove the remaining transmission-to-engine attaching bolts.

29. Separate the engine from the transmission, raise the engine high enough to clear the No. 1 crossmem-

ber, then pull the engine forward and out of the vehicle.

To install:

NOTE: If the engine to be installed has been overhauled or has been in storage, prime the entire engine lubricating system to fill the oil cooler, oil filter and cylinder block galleries with the proper type of oil. This will prevent piston and bearing scuffing.

30. Lower the engine into the vehicle and align the torque converter with the flywheel studs.

31. Position the transmission and install 2 lower transmission-to-engine attaching bolts.

32. Remove the engine lifting equipment and raise and safely support the vehicle.

33. Attach the starter cable. Install the 4 converter-to-flywheel nuts and the converter inspection plate.

34. Connect the exhaust pipe to the manifolds and install the engine mount attaching nuts.

35. Position the fuel line on the No. 1 crossmember and lower the vehicle. Install the top 4 transmission-to-engine attaching bolts.

36. Install the engine and transmission dipstick and tube assemblies. Connect the battery ground cable to the engine block.

37. At the left rear of the engine, connect the fuel return line and the auxiliary heater and air conditioner hoses, if equipped.

38. Install the transmission kickdown rod. Install the heater hoses on the right cylinder head and water pump.

39. Connect the engine wiring harness to the inside of the passenger compartment and the oil pressure sender wire to the sender at the rear of the engine.

40. Install the accelerator cable bracket to the intake manifold. Install the accelerator and cruise control cables to the bracket and the injection pump.

41. Remove the plug from the fuel inlet hose and install the hose to the fuel pump. Install the fuel filter and bracket.

42. Remove the caps from the fuel filter, outlet line and injection pump. Install the line between the fuel filter outlet and the injection pump. Connect the fuel filter return hose.

43. Remove the intake manifold cover and install the air cleaner and inlet duct assembly. Connect the fuel line between the fuel filter inlet and fuel pump.

44. Install the vacuum hose between the vacuum regulator valve and injection pump.

45. Connect the water temperature

sender wire to the sender on the left front of the engine block and the water temperature overheat light switch wire to the switch on top of the left cylinder head.

46. If equipped, install the air conditioner drive belt and refrigerant lines to the compressor.

47. Remove the plug from the power steering pump return line and install the line. Install the power steering pump drive belt and bracket.

48. Connect the 2 engine ground cables to the bottom front of the engine. Connect the alternator wiring harness to the alternator and fuel line heater.

49. Loosely attach the alternator adjusting arm and adjusting bracket to the alternator and engine. Install the drive belt.

50. Loosely install the vacuum pump to the engine. Install the vacuum pump drive belt and connect the vacuum hose transmission modulator tube.

51. Adjust all accessory drive belts.

52. Install the radiator, cooling fan and shroud. The fan clutch nut has left-hand thread; tighten by turning counterclockwise.

53. Connect the radiator hoses and the transmission oil cooler lines.

54. If equipped, install the air conditioning condenser and connect the lines.

55. Install the upper grille support and install the hood latch and cable assembly to the grille upper support bracket. Install the cruise control servo bracket, if equipped.

56. Install the front bumper, grille assembly and gravel deflector.

57. Fill and bleed the cooling system. Check the engine oil level and fill as needed with the proper type of oil.

58. Connect the ground cables to both batteries. Start the engine and bring to normal operating temperature. Check for leaks.

59. Evacuate and charge the air conditioning system, if equipped. Install the engine cover.

Engine Mounts

REMOVAL & INSTALLATION

Front Mounts

F SERIES

1. Disconnect both negative battery cables. Remove the fan shroud halves and raise and safely support the vehicle.

2. Remove the nuts attaching the mounts to the crossmember.

3. Disconnect the exhaust pipes at the manifolds.

4. Remove the bolts attaching the mounts to the engine block and lower the vehicle.

5. Attach suitable engine lifting equipment and raise the engine high enough for the mounts to clear the crossmember.

6. Remove the mount and bracket assemblies and remove the mount from the bracket.

To install:

7. Install the mount onto the mount bracket. Tighten the bolts to 60–80 ft. lbs. (81–108 Nm).

8. Install the mount and bracket assembly onto the engine block. Tighten the bolts to 75–95 ft. lbs. (102–129 Nm).

9. Lower the engine onto the crossmember and remove the engine lifting equipment. Raise and safely support the vehicle.

10. Install the mount-to-crossmember attaching nuts and washers. Tighten to 71–94 ft. lbs. (96–127 Nm).

11. Lower the vehicle and install the fan shroud. Connect both negative battery cables.

E SERIES

1. Disconnect both negative battery cables and remove the radiator fan shroud.

2. Loosen the vacuum pump and the alternator and remove the drive belts.

3. Disconnect the alternator wiring harness and remove the alternator adjusting bracket. Remove the alternator.

4. Remove the fuel filter/fuel heater/water separator inlet line from the fuel pump and fuel filter.

5. If equipped, discharge the air conditioning system, disconnect the refrigerant lines and remove the compressor and bracket.

6. Remove the engine cover. Remove the air cleaner and install a suitable cover over the air intake opening.

7. Remove and cap the fuel filter/fuel heater/water separator to injection pump fuel line. Cap the injection pump and fuel filter fittings.

8. Remove the fuel filter/fuel heater/water separator return line hose. Remove the fuel filter/fuel heater/water separator bracket bolts and remove the filter and bracket assembly.

9. Remove the kickdown rod from the injection pump. Raise and safely support the vehicle.

10. Disconnect the ground cables from the lower front of the engine and remove the nuts attaching the mounts to the No. 1 crossmember.

11. Disconnect and remove the transmission kickdown rod from the transmission. Lower the vehicle.

12. Attach suitable engine lifting equipment. Raise the engine until it contacts the body and remove the mount and bracket assemblies.

To install:

13. Install the mount to the bracket and tighten the bolts to 65–85 ft. lbs. (88–115 Nm).

14. Install the mount and bracket assembly onto the engine and tighten the bolts to 65–85 ft. lbs. (88–115 Nm).

15. Lower the engine and remove the lifting equipment. Raise and safely support the vehicle.

16. Install the mount to the No. 1 crossmember. Tighten the nuts to 54–74 ft. lbs. (73–100 Nm).

17. Install the transmission kickdown rod and connect to the transmission. Install the engine ground cables and lower the vehicle.

18. Install the fuel filter/fuel heater/water separator and bracket assembly. Uncap and install the fuel filter/fuel heater/water separator to injection pump fuel line. Install the fuel filter/fuel heater/water separator return hose.

19. Remove the intake opening cover and install the air cleaner.

20. If equipped, install the air conditioning compressor and bracket, connect the refrigerant lines and install the drive belt.

21. Install the fuel filter/fuel heater/water separator line.

22. Install the alternator, adjusting bracket and drive belt. Connect the alternator wiring.

23. Install the vacuum pump drive belt. Adjust all accessory drive belt tension.

24. Install the fan shroud. Connect both negative battery cables.

25. If equipped, evacuate and charge the air conditioning system. Install the engine cover.

Rear Mount

1. Remove the mount-to-support assembly bolt and locknut.

2. Remove the insulator-to-transmission housing bolts and lockwashers.

3. Raise the transmission with a jack and remove the mount and retainer.

4. Installation is the reverse of the removal procedure. Tighten the mount-to-transmission bolts and the mount-to-support bolts and nuts to 50–70 ft. lbs. (68–95 Nm).

Cylinder Head

REMOVAL & INSTALLATION

1. Disconnect battery ground cables from both batteries.

2. Drain cooling system.

3. Remove the radiator fan shroud.

4. Remove the radiator cooling fan and clutch assembly. The fan clutch nut has left-hand thread and must be removed by turning clockwise.

5. Disconnect wiring from alternator and fuel filter/fuel heater/water separator.

6. Remove the alternator belt and alternator.

7. Remove the vacuum pump drive belt and vacuum pump.

8. Disconnect and cap all fuel lines. Remove alternator and vacuum pump mounting bracket. On F series, remove fuel filter with bracket as an assembly, for right side only.

9. Remove the heater hose from the cylinder head.

10. Remove the injection pump.

11. Remove the intake manifold and valley cover.

12. Raise and safely support the vehicle.

13. Disconnect the muffler inlet pipe from the exhaust manifolds.

14. For right side cylinder head, remove bolt holding the transmission dipstick tube to cylinder head.

15. Lower the vehicle. For right cylinder head, remove engine oil dipstick tube fasteners.

16. Remove the exhaust manifolds from the engine. For right cylinder head, remove the engine oil dipstick, dipstick tube and O-ring

17. Remove the rocker arm cover, rocker arm and pushrods.

18. Remove the injection nozzles and glow pugs.

19. Remove the cylinder head bolts. Attach suitable lifting equipment and lift the cylinder head from the engine.

NOTE: The prechambers may fall out of the cylinder head upon removal.

20. Clean and inspect the cylinder head gasket surface, prechambers and ports for cracks. Cracks in the prechambers are acceptable as long as they do not extend past the fire ring mark.

21. Check the cylinder head gasket surface for flatness using a straight edge and a feeler gauge. Remove the prechambers using a brass drift and hammer prior to checking, to avoid false readings. Warpage must not exceed 0.006 in. If warpage is excessive, the cylinder head must be replaced.

To install:

22. Apply a light coating of lubricant to the mounting edge of the prechambers and install into the cylinder head. Lightly tap with a plastic-tipped hammer, if necessary.

23. Clean the engine block head gasket surface. Position a new gasket on the engine block over the dowel pins. Install gasket with the words **THIS SIDE UP** facing installer.

24. Install the cylinder head, being careful to prevent the prechambers from falling into the cylinder bores. Do not slide the head across the gasket.

25. Lubricate the head bolt threads, bolt heads and washers with clean engine oil. Do not use anti-seize or grease. Install the head bolts and washers and tighten, as follows:

   a. Tighten bolts, in numbered sequence, to 65 ft. lbs. (88 Nm).

   b. Tighten bolts, in numbered sequence, to 85 ft. lbs. (115 Nm).

   c. Tighten bolts, in line sequence, to 100 ft. lbs. (136 Nm).

   d. Tighten bolts, in line sequence, to 100 ft. lbs. (136 Nm).

26. Install the pushrods, with the copper colored ends toward the rocker arms, making sure the pushrods are fully seated into the lifters.

27. Install the rocker arms, rocker arm covers, valley pan, intake manifold and injection pump.

28. Connect the heater hose to the

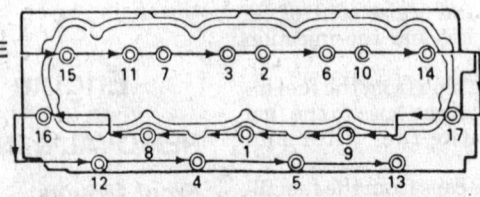

STEP 1: TIGHTEN BOLTS TO 65 FT. LBS. (88 NM) IN NUMBERED SEQUENCE SHOWN ABOVE
STEP 2: TIGHTEN BOLTS TO 85 FT. LBS. (115 NM) IN NUMBERED SEQUENCE SHOWN ABOVE
STEP 3: TIGHTEN BOLTS TO 100 FT. LBS. (136 NM) IN LINE SEQUENCE SHOWN ABOVE
STEP 4: REPEAT STEP NO. 3

Cylinder head bolt torque sequence—7.3L diesel engine

cylinder head. Install the fuel filter/fuel heater/water separator, if removed.

29. Install the alternator and vacuum pump bracket. Remove the protective caps and install the fuel lines.

30. Loosely install the engine oil dipstick tube and O-ring, if removed. Raise and safely support the vehicle.

31. Install the exhaust manifold(s). Install the fasteners holding the engine oil dipstick tube in position.

32. If removed, install the bolt attaching the transmission oil dipstick to the cylinder head. Connect the exhaust pipe to the manifolds and lower the vehicle.

33. Install the vacuum pump, alternator and drive belts. Adjust the belts.

34. Connect the alternator and the fuel filter/fuel heater/water expansion wiring harnesses.

35. Remove the intake manifold cover and install the air cleaner. Install the fan and clutch assembly. The fan clutch nut has left-hand thread; tighten by turning counterclockwise.

36. Install the fan shroud. Fill and bleed the cooling system.

37. Connect both negative battery cables. Start the engine, bring to normal operating temperature and check for leaks.

38. If necessary, purge high-pressure fuel lines of air by loosening connector ½–1 turn and cranking engine until bubble-free fuel flows from the connection.

— **CAUTION** —
Keep eyes and hands away from nozzle spray. Fuel spraying from the nozzle under high-pressure can penetrate the skin and cause infection. Medical attention should be provided immediately in the event of skin penetration.

Valve Lifters

REMOVAL & INSTALLATION

1. Disconnect both negative battery cables.

2. Remove the intake manifold, rocker arm covers, rocker arms and pushrods. Keep the rocker arms and pushrods in order so they may be reinstalled in their original positions.

3. Remove the Crankshaft Depression Regulator (CDR) tube and grommet from the valley pan.

4. Remove the bolts attaching the valley pan strap to the front of the engine block and remove the strap.

5. Remove the valley pan drain plug and remove the valley pan. Remove the lifter guide retainer.

6. Remove the lifter guides and the lifters. Keep them in order so they may

be reinstalled in their original positions.

To install:

7. Clean all gasket mating surfaces.

8. Lubricate the lifters and bores with clean engine oil and install the lifters. If the old lifters are being reused, make sure they are installed in their original positions.

9. Install the lifter guides and the lifter guide retainer.

10. Install the pushrods in their original positions, with the copper colored ends toward the rocker arms. Install the rocker arms and rocker arm covers.

11. Install the valley pan and drain plug and install the CDR tube with a new grommet into the valley pan. Install a new O-ring and new back-up ring on the CDR valve.

12. Install the valley pan strap onto the front of the valley pan. Install the intake manifold.

13. Connect both negative battery cables. Run the engine and check for oil and fuel leaks.

14. If necessary, purge high-pressure fuel lines of air by loosening connector ½–1 turn and cranking engine until bubble-free fuel flows from the connection.

— **CAUTION** —
Keep eyes and hands away from nozzle spray. Fuel spraying from the nozzle under high-pressure can penetrate the skin and cause infection. Medical attention should be provided immediately in the event of skin penetration.

Valve Lash

ADJUSTMENT

1. Remove the rocker arm covers.

2. Rotate the crankshaft until the lifter to be checked is on the base circle of the camshaft.

3. Position bleed-down wrench T83T–6500–A or equivalent, over the

rocker arm and push down, until the lifter is fully collapsed.

4. Check the gap between the valve tip and the rocker arm using a feeler gauge. The gap should not exceed 0.185 in.

5. Excessive clearance may be caused by loose rocker arm fulcrum bolts, or wear of the lifter roller, pushrod, rocker arm, rocker arm fulcrum or valve tip.

6. After checking is completed, reinstall the rocker arm covers.

Rocker Arms

REMOVAL & INSTALLATION

1. Disconnect both negative battery cables.

2. On E series, remove the fan shroud and engine cover.

3. If removing the right side rocker arm cover on E Series, proceed as follows:

   a. Remove engine oil dipstick tube fasteners and remove dipstick, tube assembly and rocker arm cover bracket.

   b. Remove transmission filler tube fasteners and remove filler tube and dipstick.

   c. Raise and safely support the vehicle.

   d. Remove nuts attaching right engine mount to frame. Slightly raise right side of the engine until fuel filter header touches vehicle sheet metal. Install suitable wood block between insulator and frame. Lower engine on block.

   e. Lower the vehicle.

4. Remove the rocker arm cover retaining bolts and remove the covers. Remove the rocker arm post mounting bolts.

5. Remove the rocker arms, posts and pushrods in order and identify so they may be installed in their original positions.

To install:

6. Clean all gasket mating surfaces.

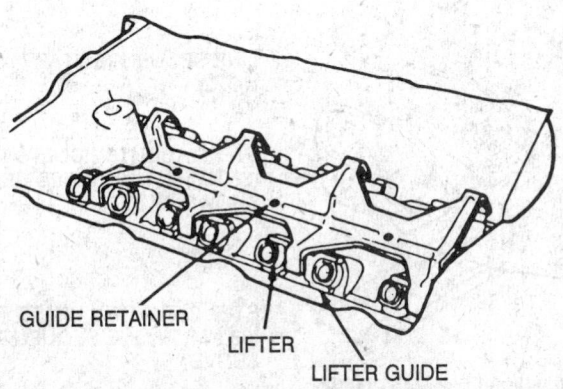

GUIDE RETAINER
LIFTER
LIFTER GUIDE

Valve lifter removal—7.3L diesel engine

7. Install the pushrods in their original positions with the copper colored end of the pushrod toward the rocker arm, making sure they are fully seated in the lifter pushrod seats.

8. Install the rocker arms and posts in their original positions. Apply multi-purpose grease to the valve stem tips.

9. Turn the engine over by hand until the timing mark is at the 11 o'clock position as viewed from the front of the engine. Install all rocker arm post attaching bolts and tighten to 20 ft. lbs. (27 Nm).

10. Install new gaskets and install the rocker arm covers. Tighten the attaching bolts to 6 ft. lbs. (8 Nm).

11. If the right side rocker arm cover on E Series was removed, proceed as follows:

  a. Raise and safely support the vehicle. Raise engine, remove wood block and lower engine onto crossmember.

  b. Install engine mount attaching washers and nuts.

  c. Lower vehicle. Install transmission filler tube and transmission oil dipstick.

  d. Install engine oil dipstick tube and rocker arm cover bracket. Install engine oil dipstick.

12. Install radiator fan shroud halves, if removed. Connect both negative battery cables.

13. Run the engine and inspect for oil leaks.

Intake Manifold

REMOVAL & INSTALLATION

1. Disconnect both negative battery cables.

2. On E series, remove the engine cover.

3. Remove the air cleaner and install a suitable cover over the air intake opening.

4. On E series, disconnect the fuel inlet line and fuel return line from the fuel filter. Remove the fuel filter bracket attaching bolts and remove the fuel filter and bracket as an assembly.

5. Remove the injection pump.

6. On F series, remove the fuel return hoses from No. 7 and No. 8 (rear) nozzles and remove the return hose to the fuel tank.

7. Remove the glow plug harness and controller. Remove the engine wiring harness from the engine. Remove the engine harness ground cable from the back of the left cylinder head.

8. Remove the intake manifold retaining bolts and remove the intake manifold.

To install:

9. Clean the cylinder block and intake manifold gasket surfaces of any silicone sealer or oil. Apply a 1/8 in. bead of silicone sealer to each end of the cylinder block.

10. Install the intake manifold and tighten to specifications using the 2-step procedure.

11. Install the engine wiring harness on the engine. Connect the engine wiring harness ground wire to the rear of the left cylinder head.

12. Install the glow plug controller and harness.

13. On E series, install the fuel filter and bracket as as assembly and tighten the bolts to 24–39 ft. lbs. (33–52 Nm). Install the fuel filter return and inlet fuel lines.

14. Install the injection pump.

15. Connect the fuel tank return hose and No. 7 and No. 8 nozzle fuel return hoses.

16. Remove the intake manifold cover and install the air cleaner. Install the engine cover on E series.

17. Connect both negative battery cables. Run the engine and check for oil and fuel leaks.

18. If necessary, purge high-pressure fuel lines of air by loosening connector 1/2–1 turn and cranking engine until bubble-free fuel flows from the connection.

> ## CAUTION
> *Keep eyes and hands away from nozzle spray. Fuel spraying from the nozzle under high-pressure can penetrate the skin and cause infection. Medical attention should be provided immediately in the event of skin penetration.*

Exhaust Manifold

REMOVAL & INSTALLATION

F Series

1. Disconnect both negative battery cables.

2. Raise and safely support the vehicle.

3. Disconnect the muffler inlet pipe from the exhaust manifolds.

4. If the right exhaust manifold is being removed, lower the vehicle. If the left exhaust manifold is being removed, leave the vehicle in the raised position.

5. Remove the exhaust manifold attaching bolts and the manifold.

To install:

6. Clean the mounting surfaces. Apply anti-seize compound on the exhaust manifold bolt threads and install the left manifold with a new gasket.

7. Tighten the bolts to specification, using the 2-step procedure.

8. Raise and safely support the vehicle. If the right manifold is being installed, repeat Steps 6 and 7.

9. Connect the muffler inlet pipe to the manifolds and tighten the nuts to 25–38 ft. lbs. (34–51 Nm). Lower the vehicle.

10. Connect the negative battery cables. Run the engine and check for exhaust leaks.

E Series

1. Remove the engine cover. Disconnect both negative battery cables.

2. For right side manifold removal only, remove the radiator fan shroud

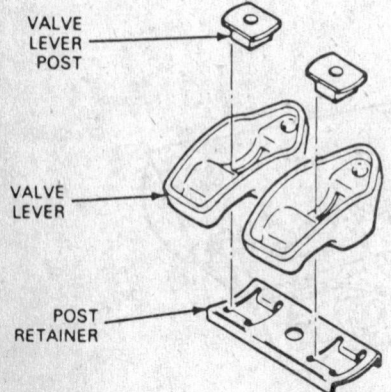

VALVE LEVER POST

VALVE LEVER

POST RETAINER

Rocker arm assembly—7.3L diesel engine

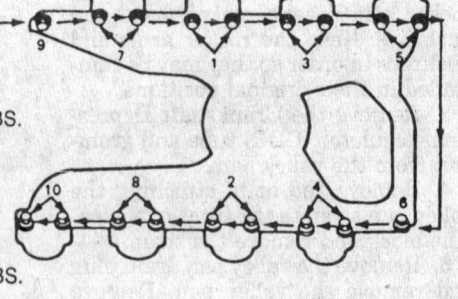

LINE SEQUENCE START HERE (STEP NO. 2)

STEP 1. TIGHTEN BOLTS TO 24 FT. LBS. (33 NM) IN NUMBERED SEQUENCE SHOWN ABOVE

STEP 2. TIGHTEN BOLTS TO 24 FT. LBS. (33 NM) IN LINE SEQUENCE SHOWN ABOVE

Intake manifold bolt torque sequence—7.3L diesel engine

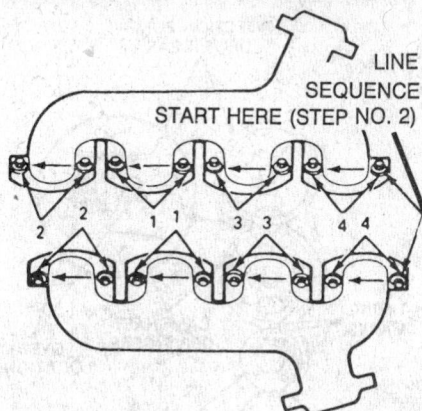

LINE SEQUENCE START HERE (STEP NO. 2)

2 2 1 1 3 3 4 4

STEP 1. TIGHTEN BOLTS TO 35 FT. LBS. (47 NM) IN NUMBERED SEQUENCE SHOWN ABOVE

STEP 2. TIGHTEN BOLTS TO 35 FT. LBS. (47 NM) IN LINE SEQUENCE SHOWN ABOVE

Exhaust manifold bolt torque sequence— 7.3L diesel engine

halves, engine oil dipstick and tube and transmission filler tube and dipstick.

3. Raise and safely support the vehicle.

4. For right side manifold removal only, remove the nuts attaching the engine mount to the frame. Slightly raise the right side of the engine until the fuel filter header touches vehicle sheet metal. Install a wood block between the mount and frame. Lower the engine onto the block.

5. Remove the muffler inlet pipe from the exhaust manifolds. Lower the vehicle.

6. Remove the exhaust manifold retaining bolts and exhaust manifold.

To install:

7. Clean the mounting surfaces. Apply anti-seize compound to the manifold retaining bolts and install the manifold with a new gasket.

8. Tighten the bolts to specification using the 2-step procedure.

9. Raise and safely support the vehicle.

10. If the right manifold was installed, raise the engine, remove the wood block and lower the engine onto the crossmember. Install the engine mount attaching washers and nuts.

11. Install the muffler inlet pipe to the exhaust manifolds and tighten the nuts to 25–38 ft. lbs. (34–51 Nm). Lower the vehicle.

12. If the right manifold was installed, install the transmission filler tube and dipstick, the engine oil dipstick tube and dipstick and the radiator fan shroud halves.

13. Connect the negative battery cables. Run the engine and check for exhaust leaks. Install the engine cover.

Cylinder Block Front Cover

REMOVAL & INSTALLATION

1. Disconnect both negative battery cables. Drain the cooling system.

2. Remove the air cleaner and install a suitable cover over the air intake opening.

3. Remove the fan shroud and fan and clutch assembly. Left-hand thread is used on the fan clutch nut; remove by turning the nut clockwise.

4. Remove the injection pump and injection pump adapter.

5. Remove the water pump.

6. Raise and safely support the vehicle.

7. Remove the crankshaft pulley and the damper attaching bolt. Remove the damper using a suitable puller.

8. Remove the ground cables at the front of the engine. Remove the bolts attaching the front cover to the engine block and oil pan.

9. Lower the vehicle. Remove the bolts attaching the engine front cover to the engine block and remove the front cover.

10. If the oil seal is to be replaced, support the front cover in a press and using drive handle tool T80T–4000–W or equivalent, and a 3¼ in. diameter spacer, drive the seal out of the cover.

To install:

11. Remove the old gasket material and clean all sealing surfaces.

12. Coat a new crankshaft oil seal with multi-purpose grease. Suuport the front cover and install the seal using seal replacer tool T83T–6700–A or equivalent, a suitable spacer and a press. When the tool is bottomed on the front cover surface, the seal is automatically installed at the proper depth.

13. Clean the outside surface of the front cover to remove any grease. Apply a ⅛ in. bead of silicone sealer around the outside diameter of the front seal and the edge of the front cover.

14. Install fabricated alignment dowels on the engine block and oil pan to align the front cover and gaskets.

15. Apply silicone sealer on the engine block and front cover sealing surfaces and install the gaskets on the engine block. Apply a ⅛ in. bead of silicone sealer to the rear corners of the oil pan and a ¼ in. bead of silicone sealer on the oil pan.

16. Install the front cover in position, on the oil pan dowels first, and install the attaching bolts. Remove the alignment dowels if required. Install and hand tighten the remaining front cover bolts.

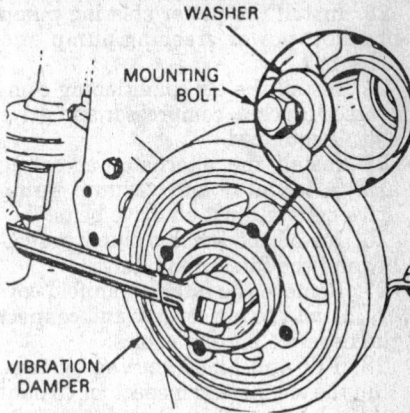

WASHER

MOUNTING BOLT

VIBRATION DAMPER

Vibration damper removal—7.3L diesel engine

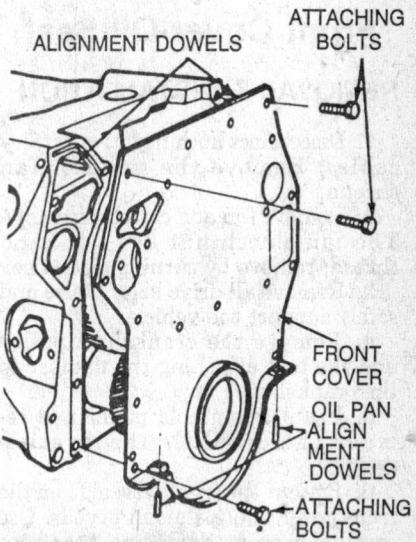

ALIGNMENT DOWELS

ATTACHING BOLTS

FRONT COVER

OIL PAN ALIGNMENT DOWELS

ATTACHING BOLTS

Cylinder block front cover installation— 7.3L diesel engine

17. Install the water pump gasket and water pump. Apply silicone sealer to the bolt threads and install the water pump attaching bolts. Remove the remaining alignment dowels.

18. Tighten the water pump bolts to 14 ft. lbs. (19 Nm). Tighten the front cover bolts. Install the injection pump adapter and injection pump.

19. Connect the heater hose to the water pump. Raise and safely support the vehicle.

20. Lubricate the damper seal nose with clean engine oil and install the crankshaft vibration damper, using damper replacer tool T83T–6316–B or equivalent. Add silicone sealer to the engine side of the retaining bolt washer to prevent oil leakage past the keyway. Install the vibration damper attaching bolt and tighten to 90 ft. lbs. (122 Nm).

21. Install the crankshaft pulley. Install both battery ground cables on the front of the engine and lower the vehicle.

22. Install the alternator adjusting arm bracket and water pump pulley.

23. Install the power steering pump bracket, power steering pump and drive belt.

24. Install the air conditioning compressor bracket, compressor and drive belt, if equipped.

25. Install the alternator adjusting arm, alternator and vacuum pump drive belts. Adjust all drive belts.

26. Install the fan and clutch assembly and the fan shroud halves.

27. Remove the intake manifold cover, install the air cleaner and connect the negative battery cables.

28. Fill and bleed cooling system. Run the engine and inspect for coolant and oil leaks. If equipped, evacuate and charge the air conditioning system.

Front Cover Oil Seal

REMOVAL & INSTALLATION

1. Disconnect both negative battery cables. Remove the radiator fan shroud.

2. Remove fan and clutch assembly. The fan clutch nut has left-hand thread; remove by turning clockwise.

3. Remove all drive belts. Raise and safely support the vehicle.

4. Remove the crankshaft pulley and the bolt attaching the damper to the crankshaft.

5. Install a suitable puller and remove the crankshaft vibration damper.

6. Pry out the front oil seal from the front cover using a small prybar. Use care to prevent damaging the front cover or crankshaft or breaking the oil pan seal by bending the front cover.

To install:

7. Coat a new seal with multi-purpose grease.

NOTE: It may be necessary to rotate the crankshaft to align the damper key with the seal installing tool.

8. On engines without 3 weldnuts on the front cover, place the seal into installation tool T83T-6700-A or equivalent, and install over the end of the crankshaft. Install damper replacer tool T83T-6316-B or equivalent, and tighten the nut against the washer and installation tool to force the seal into the front cover plate.

NOTE: Use care to prevent bending the front cover during oil seal installation and breaking the oil pan seal.

9. On engines with 3 weldnuts on the front cover, place the seal into installation tool T83T-6700-A or equivalent, install over the end of the crankshaft and attach the bridge to the weldnuts. Draw the seal into the front cover by rotating the center screw clockwise. The seal is automatically installed at the proper depth when the tool bottoms on the cover.

10. Clean the grease from the outside surfaces and apply a ⅛ in. bead of silicone sealer around the outside diameter of the front seal and the edge of the front cover.

11. Lubricate the damper seal nose with clean engine oil and install the crankshaft vibration damper using damper replacer tool T83T-6316-B or equivalent.

12. Apply silicone sealer to the engine side of the damper attaching bolt washer in the keyway area, to prevent oil leakage past the keyway. Install the bolt and tighten to 90 ft. lbs. (122 Nm).

13. Install the crankshaft pulley and lower the vehicle.

14. Install the drive belts and adjust the belt tension. Install the fan and clutch assembly. The fan clutch nut has left-hand thread; turn counterclockwise to tighten.

15. Install the fan shroud and connect the negative battery cables.

Crankshaft Drive Sprocket

REMOVAL & INSTALLATION

1. Disconnect both negative battery cables. Remove the air filter and install a suitable cover on the intake manifold opening.

2. Drain the cooling system and remove the radiator fan shroud.

3. Remove the fan and clutch assembly. The fan clutch nut has left-hand thread; turn clockwise to remove.

4. Remove the alternator and vacuum pump belts.

5. Remove the air conditioning compressor and position out of the way. Remove the mounting bracket.

6. Remove the power steering pump and position out of the way. Remove the power steering pump bracket.

7. Remove the water pump pulley and water pump.

8. Remove the cylinder block front cover.

9. Install a suitable puller and remove the crankshaft sprocket.

To install:

10. Install the crankshaft sprocket using crankshaft sprocket replacer tool T83T-6316-B or equivalent, aligning the crankshaft drive sprocket timing mark with the camshaft drive sprocket timing mark.

NOTE: If necessary, the sprocket may be heated in an oven to 300-350°F (149-177°C) for ease of

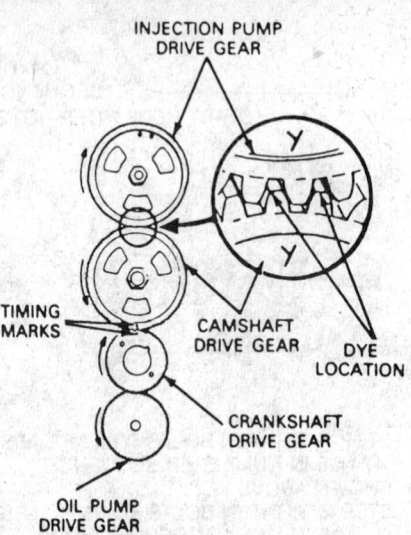

Timing sprocket positioning—7.3L diesel engine

installation. Do not use a torch to heat the sprocket.

11. Clean all sealing surfaces. Install the cylinder block front cover.

12. Install the crankshaft vibration damper using damper replacer tool T83T-6316-B or equivalent. Apply silicone sealer to the engine side of the washer, in area of keyway only, to prevent oil leakage past keyway. Install crankshaft vibration damper attaching bolt and tighten to 90 ft. lbs. (122 Nm).

13. Install the crankshaft pulley and water pump.

14. Install the injection pump adapter and injection pump.

15. Connect the heater hose to the water pump.

16. Install the power steering pump bracket, power steering pump and drive belt.

17. Install the air conditioning compressor bracket, compressor and drive belt, if equipped.

18. Install the alternator adjusting arm and the alternator and vacuum pump drive belts.

19. Adjust all drive belts. Check and set the injection pump timing.

20. Install the fan and clutch assembly and fan shroud halves. Fill and bleed the cooling system.

21. Remove the intake manifold cover. Install the air cleaner assembly and connect the negative cables to both batteries. Run the engine and inspect for oil and coolant leaks.

Injection Pump Drive Sprocket

REMOVAL & INSTALLATION

1. Disconnect both negative battery

cables. Remove the engine cover on E series.

2. Remove the air cleaner and install a suitable cover over the intake opening.

3. Remove the injection pump.

4. Remove the bolts attaching the injection pump drive sprocket cover to the engine block and remove the cover. Do not remove the drive sprocket yet.

5. Turn the engine over by hand to TDC on the compression stroke of No. 1 piston. Remove the glow plugs, to facilitate turning the engine over by hand.

6. To determine that No. 1 piston is at TDC on the compression stroke, position the injection pump drive sprocket dowel at the 4 o'clock position. The scribe line in the vibration damper should be at TDC.

NOTE: To aid aligning the timing marks, the pump drive sprocket and the camshaft sprocket are marked with "Y" timing marks. The crankshaft and camshaft sprockets are marked with "o" (dot) alignment marks. With the engine at TDC compression for No. 1 cylinder, the "Y" marks should be aligned.

7. Slide the injection pump sprocket back, but do not remove, to expose the top of the camshaft sprocket when looking down into the front cover. In addition to the "Y", the sprocket teeth adjacent to the "Y" on the camshaft sprocket are permanently dyed.

8. Remove the injection pump drive sprocket.

To install:

9. Clean all gasket and sealing surfaces.

NOTE: To determine that No. 1 piston is at TDC of compression stroke, position the injection pump drive sprocket dowel at the 4 o'clock position. The scribe line in the vibration damper should be at TDC.

10. With the drawn line on the drive sprocket at the 6 o'clock position, install the sprocket and align all drive sprocket timing marks. Use extreme care to avoid disturbing the injection pump drive gear, once it is in position.

11. Apply a ⅛ in. bead of silicone sealer along the bottom surface of the injection pump drive sprocket cover.

12. Install the injection pump drive sprocket cover and tighten the retaining bolts to 14 ft. lbs. (19 Nm). Apply sealing compound to the bolt threads before assembly. With the injection pump drive sprocket cover installed, the injection pump drive sprocket cannot jump timing.

13. Remove the intake manifold cov-

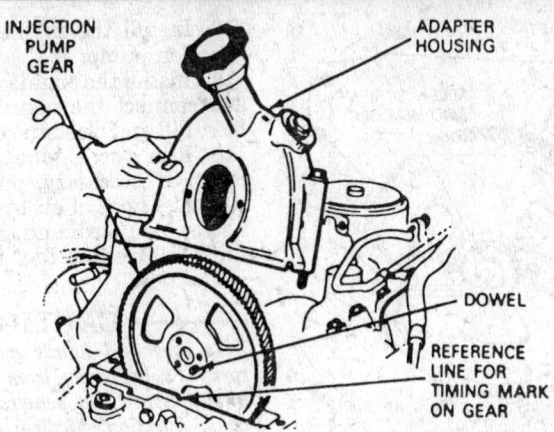

Injection pump drive sprocket cover—7.3L diesel engine

er. Install the air cleaner and connect both negative battery cables.

14. Run the engine and check for oil, fuel and coolant leaks.

15. If necessary, purge high-pressure fuel lines of air by loosening connector ½–1 turn and cranking engine until bubble-free fuel flows from the connection.

——— CAUTION ———
Keep eyes and hands away from nozzle spray. Fuel spraying from the nozzle under high-pressure can penetrate the skin and cause infection. Medical attention should be provided immediately in the event of skin penetration.

Camshaft Drive Sprocket

REMOVAL & INSTALLATION

1. Disconnect both negative battery cables. Remove the air cleaner and install a suitable cover over the air intake opening.

2. Drain the cooling system.

3. Remove the alternator and vacuum pump drive belts.

4. Remove the air conditioning compressor mounting bolts and position the compressor out of the way. Remove the compressor mounting bracket.

5. Remove the power steering pump and position out of the way. Remove the power steering pump bracket.

6. Remove the water pump pulley and water pump.

7. Remove the cylinder block front cover. Remove the camshaft Allen screw and washer.

8. Install a suitable sprocket puller and remove the sprocket.

NOTE: At this point of the service procedure, the fuel pump cam and thrust flange spacer can be removed by first removing the the fuel pump, then by using a

suitable puller to remove it from the camshaft.

To install:

9. Install a new thrust plate, if removed. Replace the fuel pump cam if damaged. If the fuel pump cam is replaced, replace the fuel pump.

10. Install the spacer and fuel pump cam against the camshaft thrust flange, using replacer tool T83T-6316–B or equivalent, if removed.

11. Install the camshaft drive sprocket against the fuel pump cam, aligning the timing mark with the mark on the crankshaft drive sprocket.

12. Install the camshaft Allen screw and tighten to 15 ft. lbs. (20 Nm).

13. Install the cylinder block front cover and water pump.

14. Install the injection pump sprocket and adapter. Install the injection pump.

15. Install the water pump pulley and power steering pump bracket. Install the power steering pump and drive belt.

16. Install the air conditioning compressor mounting bracket, compressor and drive belt. Install the alternator and vacuum pump drive belts. Adjust all drive belts.

17. Connect both negative battery cables. Remove the intake manifold cover and install the air cleaner.

18. Fill and bleed the cooling system. Run the engine and check for fuel, oil and coolant leaks.

19. If necessary, purge high-pressure fuel lines of air by loosening connector ½–1 turn and cranking engine until bubble-free fuel flows from the connection.

——— CAUTION ———
Keep eyes and hands away from nozzle spray. Fuel spraying from the nozzle under high-pressure can penetrate the skin and cause infection. Medical attention should be provided immediately in the event of skin penetration.

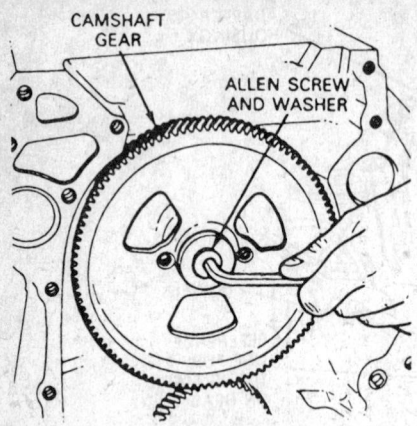

Camshaft drive sprocket removal—7.3L diesel engine

Camshaft

REMOVAL & INSTALLATION

1. Remove the engine from the vehicle and position on an engine stand.
2. Remove the injection pump and adapter, intake manifold, lifters, cylinder block front cover and fuel supply pump.
3. Remove the camshaft drive sprocket, fuel supply pump cam, spacer and thrust plate from the camshaft.
4. Remove the camshaft. Use care to avoid damaging the camshaft bearings.

To install:

5. Coat the camshaft lobes with multi-purpose grease and lubricate the journals with engine oil. Carefully slide the camshaft through the bearings. Install the camshaft thrust plate.
6. Install the spacer and fuel pump cam against the camshaft thrust flange using installation tool T83T-6316-B or equivalent.
7. Install the camshaft drive sprocket against the fuel pump cam, aligning the timing mark with the timing mark on the crankshaft drive sprocket, using installation tool T83T-6316-B, or equivalent.
8. Install the camshaft allen screw and tighten to 15 ft. lbs. (20 Nm).
9. Install the fuel supply pump. Install a new crankshaft oil seal in the front cover and install the front cover.
10. Install the water pump and injection pump adapter.
11. Lubricate the valve lifters and their bores with clean engine oil and install the lifters in their original positions. Install the lifter guides and lifter guide retainer.
12. Install the pushrods with the copper colored ends toward the rocker arms, making sure they are seated fully in the pushrod seats. Install the rocker arms and rocker arm covers.

13. Install the intake manifold and injection pump.
14. Install the engine into the vehicle and connect the negative battery cables. Fill and bleed the cooling system.
15. Run the engine and check for leaks. If necessary, purge high-pressure fuel lines of air by loosening connector ½–1 turn and cranking engine until bubble-free fuel flows from the connection.

— **CAUTION** —

Keep eyes and hands away from nozzle spray. Fuel spraying from the nozzle under high-pressure can penetrate the skin and cause infection. Medical attention should be provided immediately in the event of skin penetration.

16. Check the injection pump timing and road test the vehicle for proper operation.

Piston and Connecting Rod

POSITIONING

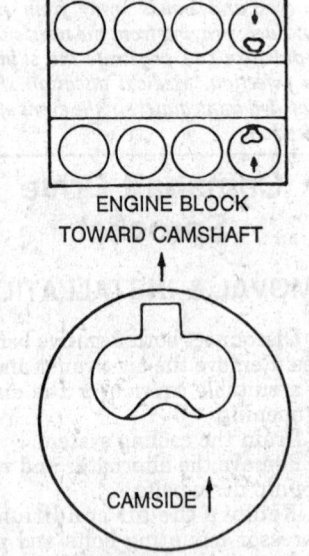

Piston orientation—7.3L diesel engine

ENGINE LUBRICATION

Oil Pan

REMOVAL & INSTALLATION

4.9L Engine

BRONCO AND F SERIES

1. Disconnect the negative battery cable. Drain the crankcase and the cooling system.
2. Remove the radiator. Remove the upper intake and throttle body assembly.
3. Raise and safely support the vehicle. Disconnect the starter cable and remove the starter.
4. Remove the engine front mount to support bracket nuts and washers. Use a transmission jack or equivalent, to raise the front of the engine, then install 1 in. thick wood blocks between the front mounts and support brackets. Lower the engine onto the blocks and remove the jack.
5. Remove the oil pan attaching bolts and lower the pan to the crossmember. Remove the 2 oil pump bolts and the 1 oil pump inlet tube nut, and drop the assembly in the oil pan. Remove the oil pan.

To install:

6. Clean all gasket mating surfaces and the oil pan.
7. Apply a pressure sensitive silicone based adhesive to the block rails and the front cover oil pan.
8. Apply a bead of silicone sealer to the parting lines between the front cover and block and to the rear bearing cap seal groove at the cylinder block junction. Immediately install the gasket to the block to facilitate pan installation.
9. Clean the inlet tube and screen assembly and place it in the oil pan.
10. Position the oil pan under the engine. Install the inlet tube and screen on the oil pump with a new gasket. Tighten the 2 screws to 10–15 ft. lbs. (14–20 Nm) and 1 nut to 22–32 ft. lbs. (30–43 Nm).
11. Raise the oil pan to the cylinder block and install the stiffener plates, if equipped, and attaching bolts. If not equipped with stiffener plates, tighten the bolts to 120–144 inch lbs. (14–16 Nm). If equipped with stiffener plates, tighten the bolts to 144–216 inch lbs. (16–24 Nm).

NOTE: Tighten the bolts in 2 steps. First, tighten all bolts except the 3 oil pan-to-front cover bolts installed through the curvature of the pan, then tighten those remaining 3 bolts.

12. Raise the engine with the jack and remove the wood blocks. Lower the engine until the mounts are positioned on the support brackets. Install the washers and nuts and tighten to 54–74 ft. lbs. (73–100 Nm).
13. Install the starter and connect the starter cable. Lower the vehicle.
14. Install the radiator. Install the upper intake and throttle body assembly.
15. Fill the crankcase with the prop-

er type and quantity of oil. Fill and bleed the cooling system.

16. Connect the negative battery cable. Start the engine and check for leaks.

E SERIES

1. Disconnect the negative battery cable. Remove the engine cover. Remove the air cleaner and disconnect the air inlet tubes from the throttle body.

2. Drain the cooling system and crankcase. If equipped with air conditioning, discharge the system and remove the compressor.

3. Remove the EGR valve. If equipped, disconnect the thermactor check valve inlet hose and remove the check valve.

4. Remove the upper intake and throttle body assembly.

5. Remove the upper radiator hose. Unbolt the fan shroud and position it over the fan. If equipped with automatic transmission, disconnect the oil filler tube.

6. Disconnect the exhaust inlet pipe at the manifold. Raise and support the vehicle safely. Remove the starter.

7. Remove the front engine mount nuts. Remove the power steering return line clip which is located in front of the crossmember.

8. Disconnect the lower radiator hose and, if equipped, the transmission oil cooler lines.

9. Raise the engine and place 3 in. wood blocks under the engine mounts. Remove the oil pan dipstick tube.

10. Remove the oil pan bolts and lower the oil pan. Remove the pickup tube and screen from the oil pump and lay them in the pan. Remove the oil pan.
To install:
11. Clean all gasket mating surfaces and clean the oil pan.

12. Apply a pressure sensitive silicone based adhesive to the block rails and the front cover oil pan.

13. Apply a bead of silicone sealer to the parting lines between the front cover and block and to the rear bearing cap seal groove at the cylinder block junction. Immediately install the gasket to the block to facilitate pan installation.

14. Clean the inlet tube and screen assembly and place it in the oil pan.

15. Position the oil pan under the engine. Install the inlet tube and screen on the oil pump with a new gasket. Tighten the 2 screws to 10–15 ft. lbs. (14–20 Nm) and 1 nut to 22–32 ft. lbs. (30–43 Nm).

16. Raise the oil pan to the cylinder block and install the stiffener plates, if equipped, and attaching bolts. If not equipped with stiffener plates, tighten the bolts to 120–144 inch lbs. (14–16 Nm). If equipped with stiffener plates,

tighten the bolts to 144–216 inch lbs. (16–24 Nm).

NOTE: Tighten the bolts in 2 steps. First, tighten all bolts except the 3 oil pan-to-front cover bolts installed through the curvature of the pan, then tighten those remaining 3 bolts.

17. Install the dipstick tube and lower the engine. Install the engine mount nuts.

18. Install the starter. Install the lower radiator hose and connect the transmission cooler lines, if equipped.

19. Install the power steering return line clip and position the line. Lower the vehicle.

20. Install the upper intake and throttle body. Install the EGR valve and connect the exhaust pipe. If equipped, install the thermactor check valve and connect the inlet hose.

21. Install the fan shroud and upper radiator hose. If equipped, install the air conditioning compressor.

22. Replace the oil filter and fill the crankcase with the proper type and quantity of oil. Fill and bleed the cooling system.

23. Connect the negative battery cable. Start the engine and check for leaks. If equipped, evacuate and charge the air conditioning system.

24. Install the air cleaner and air inlet tubes. Install the engine cover.

5.0L and 5.8L Engines
BRONCO AND F SERIES

1. Disconnect the negative battery cable. Remove the oil dipstick, on pan entry vehicles only. Remove the bolts attaching the fan shroud to the radiator and position the shroud over the fan.

2. Remove the upper intake manifold.

3. Remove the nuts and lockwashers attaching the engine mounts to the chassis bracket.

4. Disconnect the oil cooler lines, if equipped with automatic transmission.

5. Raise and safely support the vehicle. Remove the exhaust system.

6. Raise the engine and place wood blocks under the engine mounts. Drain the crankcase.

7. Support the transmission with a jack stand and remove the transmission crossmember.

8. Remove the oil pan bolts and lower the oil pan onto the crossmember.

9. Remove the oil pump pick-up tube and screen and lower into the oil pan. Remove the oil pan.
To install:
10. Clean all gasket mating surfaces, the oil pan and the inlet tube and screen. Inspect the oil pan gasket seal-

ing surfaces for damage and distortion due to overtightening. Repair as necessary.

11. Position a new oil pan gasket and seals on the cylinder block. Place the pickup tube and screen in the oil pan and position the pan on the crossmember.

12. Raise the pickup tube and screen into position. Attach the pickup tube to the pump with a new gasket and tighten the bolts to 12–18 ft. lbs. (16–24 Nm). Tighten the pickup tube nut to 22–32 ft. lbs. (30–43 Nm).

13. Raise the oil pan into position, install the attaching bolts and tighten to 9–11 ft. lbs. (13–14 Nm).

14. Install the transmission crossmember and remove the jackstand. Raise the engine and remove the wood blocks. Lower the engine, install the engine mount nuts and washers and tighten to 54–75 ft. lbs. (73–100 Nm).

15. Install the exhaust system and lower the vehicle.

16. If equipped, connect the transmission oil cooler lines. Install the fan shroud.

17. Install the upper intake manifold.

18. Fill the crankcase with the proper type and quantity of oil. Install the oil dipstick.

19. Connect the negative battery cable, start the engine and check for leaks.

E SERIES

1. Disconnect the battery and remove the engine cover. Remove the air cleaner and drain the cooling system.

2. If equipped with power steering, remove the pump and position it out of the way. If equipped, remove the air conditioning compressor retainers and clutch wire and position the compressor out of the way.

3. Disconnect the upper radiator hose. Remove the fan shroud bolts and oil filler tube. Remove the oil dipstick-to-exhaust manifold bolt. Raise and safely support the vehicle.

4. Remove the alternator splash shield. If equipped, disconnect the automatic transmission cooler lines at the radiator. Disconnect the lower radiator hose.

5. Disconnect and plug the fuel line at the fuel rail.

6. Remove the engine mount retaining nuts. Drain the crankcase and remove the dipstick tube.

7. Disconnect the muffler inlet pipe from the exhaust manifolds.

8. If equipped, remove the automatic transmission dipstick and tube. Disconnect the manual linkage at the transmission. Remove the center driveshaft support and remove the driveshaft from the transmission.

9. Place a transmission jack or

equivalent, under the oil pan and insert a wood block between the pan and jack.

NOTE: The engine and transmission assembly will pivot around the rear engine mount. The engine assembly must be raised 4 inches, measured from the front motor mounts. The engine must remain centered in the engine compartment to obtain this much lift.

10. Raise the engine and transmission assembly. Insert wood blocks to support the engine in the uppermost position.

11. Remove the oil pan bolts and lower the oil pan. Remove the oil pump and the oil pickup tube and lay them in the oil pan on 5.0L engine. Remove the oil pan from the vehicle.

To install:

12. Clean all gasket mating surfaces, the oil pan and the inlet tube and screen. Inspect the oil pan gasket sealing surfaces for damage and distortion due to overtightening. Repair as necessary.

13. Position a new oil pan gasket and seals to the cylinder block. Position the pan with the pump to the engine and install the pump. Attach the pickup tube to the No. 3 main bearing cap stud. Install the oil pan and tighten the attaching bolts to 9–11 ft. lbs. (13–14 Nm). Install the dipstick tube.

14. Position the transmission jack, raise the engine and remove the wood blocks. Lower the engine and remove the jack. Install the engine mount attaching nuts and tighten to 50–70 ft. lbs. (68–95 Nm).

15. Install the driveshaft and center driveshaft support. Connect the manual linkage at the transmission and install the transmission filler tube.

16. Install the muffler inlet pipe and tighten the nuts to 25–38 ft. lbs. (34–52 Nm). Connect the lower radiator hose.

17. If equipped, connect the transmission cooler lines to the radiator. Connect the fuel line to the fuel rail and install the alternator splash shield. Lower the vehicle.

18. Install the fan shroud and connect the upper radiator hose. If equipped, install the power steering pump and bracket. Install the bolt attaching the dipstick tube to the exhaust manifold.

19. If equipped, install the air conditioning compressor. Install the accessory drive belt(s) and adjust the tension, if necessary.

20. Install the oil filler tube and the air cleaner. Fill the crankcase with the proper type and quantity of oil. Fill and bleed the cooling system.

21. Connect the negative battery cable, start the engine and check for leaks. Install the engine cover.

7.5L Engine

1. Remove the engine cover on E Series. Disconnect the battery and drain the cooling system.

2. Disconnect the air inlet tube and remove the air cleaner assembly. Disconnect the throttle and transmission linkage at the throttle body. Disconnect the power brake vacuum lines.

3. Disconnect the fuel lines at the fuel rail. Disconnect the air tubes at the throttle body.

4. Disconnect the radiator hoses. If equipped, disconnect the oil cooler lines. Remove the fan, shroud and radiator. If equipped, remove the power steering pump and position it aside.

5. Remove the front engine mount attaching bolts. Remove the engine oil dipstick tube from the exhaust manifold. Remove the oil filler tube and bracket.

6. If equipped, rotate the air conditioning lines, at the rear of the compressor, down to clear the dash. If necessary, discharge the system and remove the lines.

7. Remove the upper intake manifold and throttle body as an assembly.

8. Raise and support the vehicle safely. Drain the crankcase and remove the oil filter.

9. Remove the muffler inlet pipe assembly. Disconnect the manual and kickdown linkage from the transmission. Remove the driveshaft and coupling shaft assembly. Remove the transmission dipstick tube.

10. Remove the dipstick and tube from the oil pan. Place a transmission jack or equivalent, under the engine oil pan. Insert a wood block between the jack surface and the oil pan. Jack the engine upward, pivoting about the rear mount until the transmission contacts the floor. Block the engine in position at the engine mounts. The engine must remain centralized to obtain the maximum height. The engine must be raised 4 inches at the mounts to remove the oil pan.

11. Remove the oil pan bolts and lower the oil pan. Remove the oil pump and pick up tube attachments and drop them into the oil pan. Remove the oil pan rearward from the vehicle.

To install:

12. Clean all gasket mating surfaces, the oil pan and the oil pickup tube and screen. Prime the oil pump by filling the inlet opening with oil and rotate the pump shaft until oil emerges from the outlet opening.

13. Apply gasket sealer to the cylinder block surface. Position the gaskets and end seals or 1-piece gasket on the block and press lightly until the gasket(s) sticks to the surface.

NOTE: Vehicles built prior to March 30, 1989 use a 4-piece oil pan gasket set. Vehicles built after this date use a 1-piece oil pan gasket.

14. Position the oil pan with the oil pump and pickup tube assembly to the chassis. Install the oil pump assembly to the cylinder block and the pickup tube to the main cap. Tighten to 22–32 ft. lbs. (30–43 Nm).

15. Raise the oil pan into place on the cylinder block and install the attaching bolts. Tighten the ¼ in. bolts to 7–9 ft. lbs. (10–12 Nm) and the 5/16 in. bolts to 8–11 ft. lbs. (11–15 Nm).

16. Place the jack under the engine and raise it enough to remove the blocks. Lower the engine and remove the jack.

17. Position the engine oil dipstick tube to the oil pan and exhaust manifold and install the attaching nut. Reposition the air conditioning lines, if equipped.

18. Install the engine mount bolts and connect the manual kickdown linkage at the tranmsission. Install the driveshaft and coupling shaft.

19. Install the muffler inlet pipe and install a new oil filter. Lower the vehicle and install the engine oil filler tube and bracket. Attach the dipstick tube to the exhaust manifold.

20. Install the radiator. Connect the transmission oil cooler lines, if equipped.

21. Install the fan and shroud, then connect the radiator hoses.

22. Install the upper intake manifold and throttle body. Connect the throttle and transmission linkage at the throttle body. Connect the fuel lines at the fuel rail and connect the power brake vacuum line.

23. Install the power steering pump and drive belt. Adjust the belt tension, if necessary. Install the transmission filler tube to the right cylinder head.

24. Install the air cleaner assembly and fresh air tube. Fill the crankcase with the proper type and quantity of oil. Fill and bleed the cooling system.

25. Connect the negative battery cable, start the engine and check for leaks. If necessary, evacuate and charge the air conditioning system. On E Series, install the engine cover.

7.3L Diesel Engine

1. Disconnect both negative battery cables.

2. Remove the engine oil dipstick and transmission dipstick, if equipped.

3. Remove the air cleaner and install a suitable cover over the intake opening.

4. Remove the fan and clutch asembly. The fan clutch nut has left-

hand thread; turn clockwise to remove.

5. Drain the cooling system and remove the radiator hoses.

6. Disconnect the power steering return hose from the power steering pump. Plug the hose and the pump.

7. Disconnect the alternator and air conditioning compressor wiring harness and fuel line heater connector from the alternator. Position the harness away from the engine.

8. Raise and safely support the vehicle. Disconnect and plug the transmission oil cooler lines from the radiator, if equipped.

9. Disconnect and plug the fuel pump inlet fuel hose. Drain the crankcase and remove the oil filter.

10. Remove the transmission filler tube assembly. Remove the front pipe from the exhaust manifold and muffler. Remove the upper inlet pipe mounting stud from the right exhaust manifold.

11. Remove the nuts and washers attaching the engine mounts to the crossmember. If equipped with automatic transmission, remove the 2 bolts securing the shift linkage bell crank to the transmission. Lower the vehicle.

12. Using suitable engine lifting equipment, raise the engine until the transmission housing contacts the body.

13. Install wood blocks 2–2¾ in. thick between the engine mounts and crossmember. Lower the engine onto the blocks.

14. Raise and safely support the vehicle. Remove the flywheel inspection plate and position the fuel pump inlet line at the rear of the crossmember. Remove the transmission oil cooler lines, if equipped, and position out of the way.

15. Remove the oil pan attaching bolts. Remove the transmission mount retaining nuts and raise the transmission approximately 1 in. using a transmission jack.

16. On F Series, remove the oil pump and pickup tube from the engine and lay it in the oil pan. Remove the oil pan. The crankshaft may have to be turned to reposition counterweights to aid in removal of the oil pan.

17. On E Series, the oil pump and pickup tube can be removed at this time.

To install:

18. Clean the oil pan, engine block and front and rear covers of old gasket material. If removed, clean the mating surface of the oil pickup tube and install on the oil pump with a new gasket.

19. If removed, prime the oil pump with clean engine oil, rotating the pump drive gear to distribute oil within the pump body. The oil pump and

pickup tube can be installed at this time, on E Series.

20. On F Series, place the oil pump and pickup tube in the oil pan and place the oil pan in position on the crossmember. Install the oil pump and pickup tube.

21. Apply a ⅛ in. bead of silicone sealer on the side rails of the engine block oil pan mating surface, and a ¼ in. bead of silicone sealer on the ends of the oil pan mating surface on the front and rear covers and in the mating corners.

22. Position the oil pan to the engine and install the attaching bolts.

23. Lower the transmission and install the transmission mount retaining nuts. Tighten to 70–94 ft. lbs. (96–128 Nm).

24. Install the flywheel inspection plate and lower the vehicle.

25. Raise the engine and remove the wood blocks. Lower the engine onto the crossmember and remove the engine lifting equipment. Raise and safely support the vehicle.

26. Install the automatic transmission filler tube with a new O-ring and install the retaining bolts. Install the nuts and washers attaching the engine mounts to the crossmember.

27. If equipped with automatic transmission, position the shift linkage bell crank to the transmission and tighten the bolts to 20–30 ft. lbs. (27–40 Nm).

28. Install the upper muffler inlet pipe mounting stud on the right exhaust manifold. Install the muffler inlet pipe.

29. Install the oil pan drain plug and new oil filter. Connect the fuel pump inlet hose to the fuel pump. Make sure the fuel line clip is installed in the crossmember.

30. Connect the transmission oil cooler lines, if equipped. Lower the vehicle.

31. Connect the alternator and air conditioning compressor wiring harness to the alternator and compressor. Connect the wiring harness to the top of the fuel filter/fuel heater/water separator.

32. Connect the power steering return hose to the power steering pump and connect the radiator hoses.

33. Install the fan and clutch assembly. The fan clutch nut has left-hand thread; tighten by turning counterclockwise.

34. Remove the intake manifold cover and install the air cleaner and intake tube. Install the engine oil and transmission oil dipsticks.

35. Fill the crankcase with the proper type and quantity of oil. Fill and bleed the cooling system.

36. Connect the negative battery cables. Run the engine and check for leaks. Check the power steering fluid and add, if necessary.

Oil Pump

REMOVAL & INSTALLATION

1. Disconnect the negative battery cable(s). Raise and safely support the vehicle.

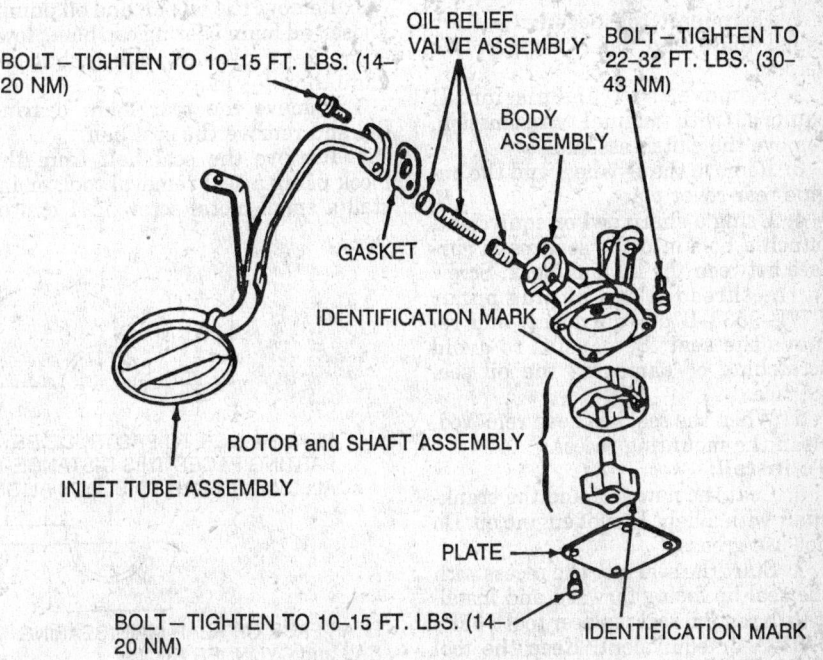

BOLT—TIGHTEN TO 10–15 FT. LBS. (14–20 NM)

OIL RELIEF VALVE ASSEMBLY

BOLT—TIGHTEN TO 22–32 FT. LBS. (30–43 NM)

BODY ASSEMBLY

GASKET

IDENTIFICATION MARK

INLET TUBE ASSEMBLY

ROTOR and SHAFT ASSEMBLY

PLATE

BOLT—TIGHTEN TO 10–15 FT. LBS. (14–20 NM)

IDENTIFICATION MARK

Oil pump assembly—5.0L and 5.8L engines

2. Remove the oil pan.

3. Remove the oil pump mounting bolts and pickup tube mounting nut and remove the oil pump and pickup tube.

To install:

4. Clean the pickup tube and screen assembly and install on the oil pump, using a new gasket.

5. Prime the pump by filling the inlet port with new engine oil. Rotate the pump shaft to distribute the oil within the pump body. If equipped, position the intermediate shaft into the oil pump.

6. Install the oil pump and pickup tube assembly. If equipped with an intermediate shaft, install the pump and shaft together.

NOTE: If equipped with an intermediate driveshaft, do not force the pump into position if it will not seat readily. The driveshaft hex may be misaligned with the distributor shaft. To align, rotate the intermediate driveshaft into a new position.

7. Install the oil pan and lower the vehicle.

8. Fill the crankcase with the proper type and quantity of engine oil. Connect the negative battery cable(s), start the engine and check for leaks.

Rear Main Bearing Oil Seal

REMOVAL & INSTALLATION

4.9L, 5.0L and 5.8L Engines

1. Disconnect the negative battery cable. Raise and safely support the vehicle.

2. Remove the transmission. If equipped with manual transmission, remove the clutch assembly.

3. Remove the flywheel and the engine rear cover plate.

4. Using a sharp awl or equivalent, punch a hole into the seal metal surface between the lip and block. Screw in the threaded end of plug puller T77L–9533–B or equivalent, and remove the seal. Be careful to avoid scratching or damaging the oil seal surface.

5. When the seal has been removed, clean the mounting recess.

To install:

6. Coat the new seal and the crankshaft with a light film of engine oil. Do not use grease.

7. Start the seal into the recess with the seal lip facing forward and install it with rear oil seal replacer tool T65P–6701–A or equivalent. Keep the tool straight with the centerline of the crankshaft and install the seal until

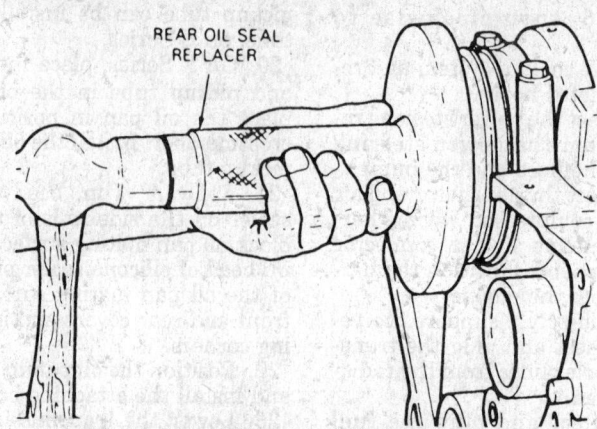

Checking oil pump inner-to-outer rotor tip clearance

the tool contacts the cylinder block surface.

8. After removing the tool, inspect the seal to make sure it was not damaged during installation.

9. Install the engine rear cover plate. Position the flywheel on the crankshaft flange. Coat the threads of the flywheel bolts with an oil resistant sealing compound and install. Tighten to 75–85 ft. lbs. (102–115 Nm).

10. If equipped with manual transmission, install the clutch assembly. Install the transmission and starter and lower the vehicle.

11. Connect the negative battery cable, start the engine and check for leaks.

7.5L Engine

1. Disconnect the negative battery cable. Raise and safely support the vehicle.

2. Remove the oil pan and oil pump. Loosen all main bearing cap bolts, lowering the crankshaft slightly, but not more than $\frac{1}{32}$ in.

3. Remove the rear main bearing cap and remove the seal half.

4. Remove the seal half from the block using a seal removal tool, or install a small metal screw in 1 end of

the seal and pull on the screw to remove the seal. Be careful to prevent damaging the crankshaft seal surfaces.

To install:

5. Carefully clean the seal grooves in the block and cap with a brush and solvent. Dry the area thoroughly.

6. Dip the seal halves in clean engine oil.

7. Carefully install the upper seal half into the cylinder block, with the lip facing toward the front of the engine, until approximately $\frac{3}{8}$ in. protrudes below the parting surface. Make sure no rubber has been shaved from the outside diameter of the seal by the bottom edge of the groove.

8. Tighten all but the rear main bearing cap bolts to 95–105 ft. lbs. (129–142 Nm).

9. Install the lower seal half in the rear main bearing cap with the lip facing toward the front of the engine. Allow the seal to protrude approximately $\frac{3}{8}$ in. above the parting surface to mate with the upper seal when the cap is installed.

10. Apply a $\frac{1}{16}$ in. bead of silicone sealer to the cylinder block, in the corner of the bearing cap recess next to the seal groove, and to the rear bearing

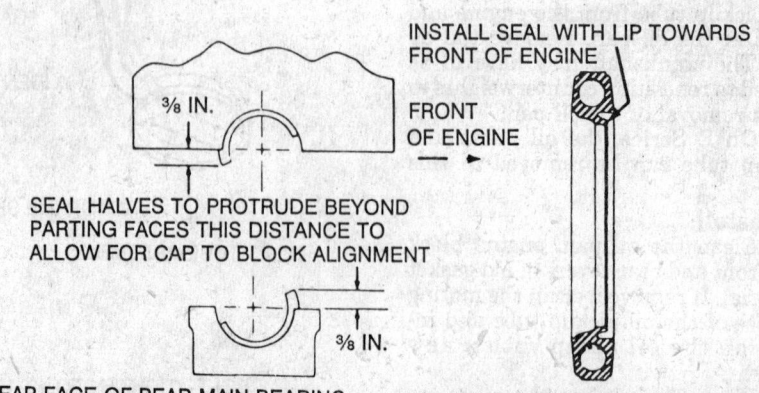

Checking oil pump rotor endplay

cap, on the mating surface in line with the seal groove, leaving a ⅛ in. gap between the sealer and the groove.

11. Install the rear main bearing cap and tighten the bolts to 95–105 ft. lbs. (129–142 Nm).

12. Install the oil pump and oil pan. Lower the vehicle.

13. Fill the crankcase with the proper type and quantity of oil. Connect the negative battery cable, start the engine and check for leaks.

7.3L Diesel Engine

1. Disconnect the negative battery cables. Raise and safely support the vehicle.

2. Remove the transmission. If equipped with manual transmission, remove the clutch housing and clutch assembly.

3. Remove the flywheel mounting bolts and spacer and remove the flywheel. Remove the rear engine cover.

4. Support the rear cover in a suitable press. Using a 4⅛ in. diameter spacer, remove the rear oil seal.

To install:

5. Clean the rear cover and engine block gasket surfaces. Remove old silicone sealer from the oil pan to rear cover sealing surface on the oil pan. Clean the sealing surfaces with suitable solvent and dry thoroughly.

6. Coat the rear engine cover seal bore inside diameter with gasket sealer. Support the rear engine cover in a suitable press and install the new seal, using seal replacer tool T83T–6701–A or equivalent.

NOTE: The seal must be installed from the engine block side of the rear cover flush with the seal bore inner surface.

7. Apply a ⅛ in. bead of silicone sealer around the outside diameter of the rear seal and the edge of the rear cover.

8. Install rear crankshaft seal pilot tool T83T–6701–B or equivalent, onto the crankshaft.

9. Apply gasket sealer to the engine block and rear cover gasket surfaces. Install the rear cover gasket to the engine block.

10. Apply a ¼ in. bead of silicone sealer at the corners of the oil pan and on the oil pan sealing surface.

11. Push the rear cover into position on the engine block and install the attaching bolts. Remove the seal pilot.

12. Position the flywheel on the crankshaft flange. Coat the threads of the flywheel bolts with sealing compound and install the spacer, bolts and washers. Tighten to 47 ft. lbs. (64 Nm) in a criss-cross pattern.

13. Install the clutch assembly, if equipped. Install the transmission and lower the vehicle.

14. Connect the negative battery cables, start the engine and check for leaks.

ENGINE COOLING

Radiator

REMOVAL & INSTALLATION

1. Disconnect the negative battery cable. Drain the cooling system by removing the cap and opening the drain cock located at the lower rear corner of the radiator.

2. Remove the overflow tube from the coolant recovery bottle and detach it from the radiator and shroud, if necessary.

3. Remove the radiator shroud retaining bolts, lift the shroud and drape it on the fan.

4. Disconnect the upper and lower radiator hoses from the radiator. If equipped, disconnect the heated water bypass hose from the radiator.

5. Disconnect the automatic transmission oil cooling lines from the radiator, if equipped.

6. Remove the radiator retaining bolts, tilt the radiator back to clear the radiator support and remove the radiator from the vehicle.

To install:

7. Position the radiator to the radiator support and install the attaching bolts. On F Series with diesel engine and E Series, tighten the bolts to 10–15 ft. lbs. (14–20 Nm). On Bronco and F Series with gasoline engine, tighten the bolts to 8–11 ft. lbs. (11–14 Nm).

8. Connect the automatic transmission oil cooling lines, if equipped.

9. Connect the upper and lower radiator hoses and tighten the clamps. Attach the heated water bypass hose, if equipped.

10. Position the shroud to the radiator and install the retaining bolts. Tighten to 4–6 ft. lbs. (5.4–8.0 Nm).

11. Attach the overflow tube from the coolant recovery bottle to the radiator.

12. Make sure the radiator drain cock is closed. Fill the cooling system with a 50/50 mixture of water and anti-freeze.

13. Connect the negative battery cable, start the engine and operate for 15 minutes. Check for coolant leaks.

14. Check the coolant level and bring it up to within 1½ in. of the radiator filler neck. Install the radiator cap.

Heater Core

REMOVAL & INSTALLATION

1. Disconnect the negative battery cable.

2. Drain the cooling system into a suitable container.

3. Disconnect the heater hoses from the heater core in the engine compartment.

4. On E Series, remove the 2 screws retaining the modesty panel to the underside of the instrument panel. Remove the modesty panel.

5. On Bronco and F Series, remove the glove compartment.

6. From inside the passenger compartment, remove the screws from the heater core cover and remove the cover. On Bronco and F Series, if necessary, disconnect the vacuum source but leave the vacuum harness attached to the cover.

7. On E Series, remove the screw and retaining bracket at the bottom of the heater core.

8. Remove the heater core and seal.

To install:

9. Position the heater core and seal into the heater case or plenum. On E Series, install the core retaining bracket and screw.

10. Position the heater core cover and install the attaching screws.

11. On E Series, install the modesty panel and the retaining screws. On Bronco and F Series, connect the vacuum harness to it's source connection and install the glove compartment.

12. Connect the heater hoses to the heater core.

13. Fill the cooling system to the proper level.

14. Connect the negative battery cable and check the system for proper operation and coolant leaks.

Water Pump

REMOVAL & INSTALLATION

4.9L Engine

1. Disconnect the negative battery cable. Drain the cooling system.

2. Remove the alternator drive belt. If equipped with an air compressor, remove the compressor belt.

3. Remove the fan and the pulley.

4. Disconnect the heater hose, radiator lower hose and radiator supply line at the water pump.

5. Remove the water pump attaching bolts and the water pump.

To install:

6. Clean the gasket mating surfaces of the water pump and engine block.

7. If a new water pump is being installed, remove the fittings from the

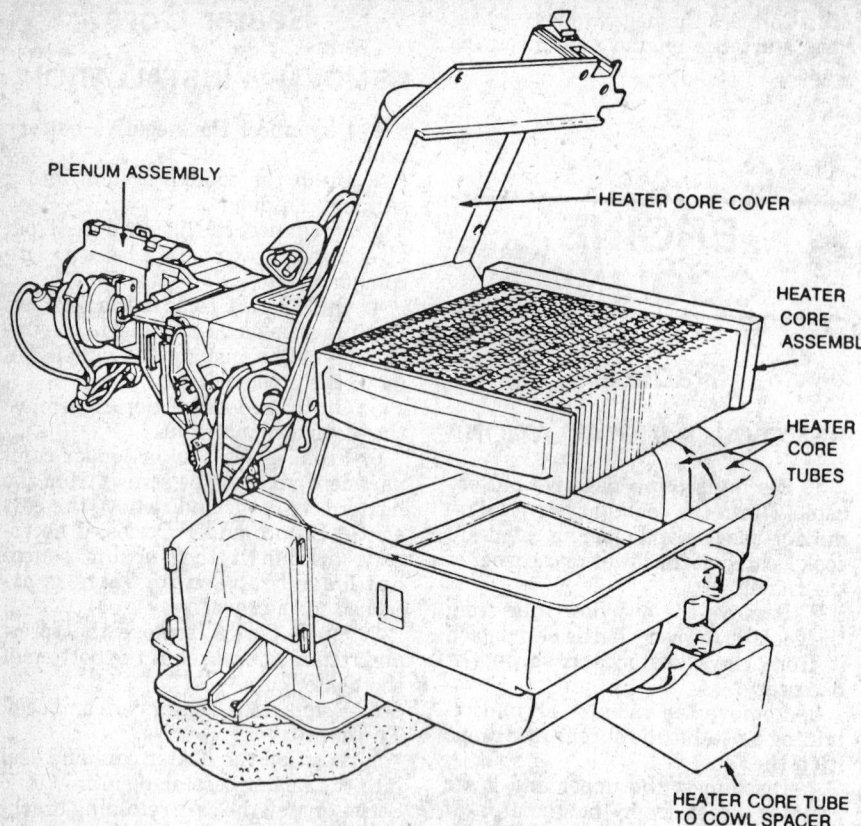

PLENUM ASSEMBLY

HEATER CORE COVER

HEATER CORE ASSEMBLY

HEATER CORE TUBES

HEATER CORE TUBE TO COWL SPACER

Heater core removal—Bronco and F Series

old pump and install them on the new pump.

8. Coat a new gasket on both sides with water-resistant sealer and install on the water pump. Position the water pump on the engine block.

9. Coat the threads of the mounting bolts with water-resistant sealer and install. Tighten the mounting bolts to 12–18 ft. lbs. (17–24 Nm).

10. Connect the hoses and install the pulley and fan.

11. Install the alternator belt. If equipped, install the compressor belt.

12. Connect the negative battery cable. Fill and bleed the cooling system. Operate the engine and check for leaks.

5.0L and 5.8L Engines

1. Disconnect the negative battery cable and drain the cooling system. On E series, remove the air cleaner and intake duct assembly.

2. Remove the bolts securing the fan shroud to the radiator, if equipped and position the shroud over the fan. On E series, remove the radiator.

3. Disconnect the lower radiator hose, heater hose and by-pass hose at the water pump. Remove the drive belts, fan, fan clutch and pulley. Remove the fan shroud, if equipped.

4. Remove the air conditioning compressor/power steering pump bracket and accessories to clear the stud bolt on the water pump housing.

5. Remove the bolts securing the water pump to the timing chain cover and remove the water pump.

To install:

6. Clean the gasket mating surfaces of the water pump and timing chain front cover.

7. Coat a new gasket on both sides with water-resistant sealer and position on the timing chain front cover. Install the pump with the attaching bolts and tighten to 12–18 ft. lbs. (17–24 Nm).

8. Connect the air conditioning compressor/power steering pump bracket and accessories to the cylinder head and water pump stud bolt.

9. Position the fan shroud over the water pump, if equipped. Install the pulley, fan clutch, fan and drive belts.

10. Install the radiator hose, heater hose and bypass hose. On E Series, install the radiator.

11. Install the fan shroud with the attaching bolts. Connect the negative battery cable.

12. Fill and bleed the cooling system. Operate the engine until normal operating temperature is reached. Check for leaks.

13. On E Series, install the air cleaner and intake duct assembly.

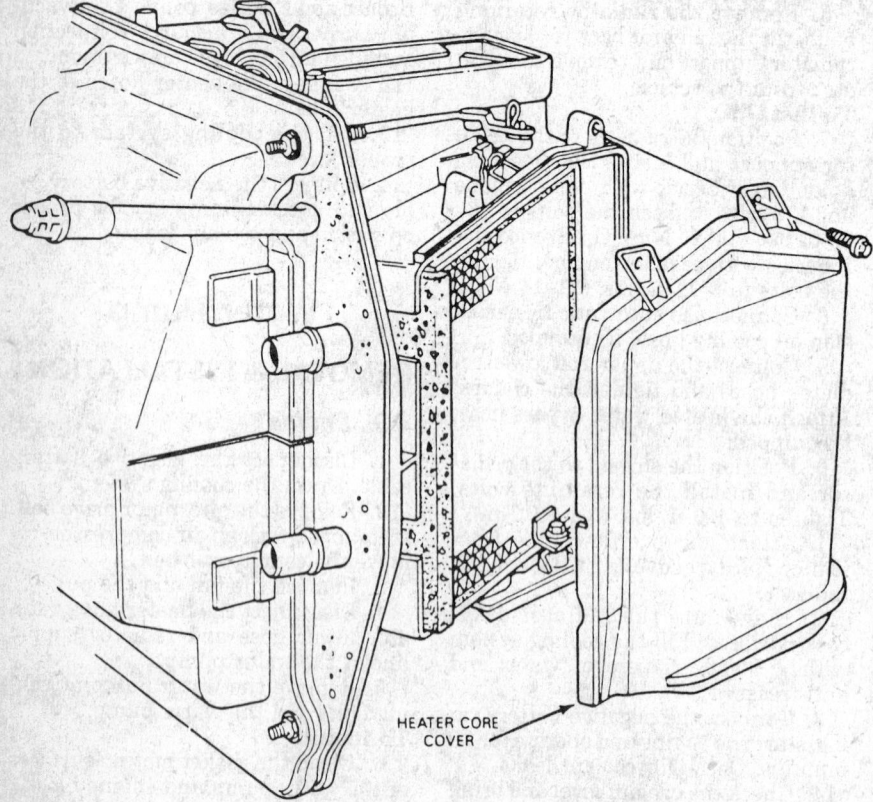

HEATER CORE COVER

Heater core removal—E Series

7.3L Engine

1. Disconnect the negative battery cable(s). Drain the cooling system and remove the fan shroud halves.

2. Remove the fan and clutch assembly using fan clutch pulley holder T83T–6312–A and fan clutch nut wrench T83T–6312–B or equivalent.

NOTE: The fan clutch nut has left hand thread. Remove by turning the nut clockwise.

3. Remove all drive belts and the water pump pulley.

4. Disconnect heater hose and remove the heater hose fitting from the water pump.

5. Remove the alternator adjusting arm and arm bracket.

6. Remove the air conditioning compressor and position the compressor out of the way. Remove the air conditioning compressor brackets.

7. Remove the power steering pump and bracket and position out of the way.

8. Remove water pump retaining bolts and remove the water pump from the engine.

To install:

9. Clean the gasket mating surfaces on the water pump and front cover. Clean the bolt threads.

10. Coat a new gasket on both sides with water-resistant sealer and install on the water pump. Position the pump on the front cover.

11. Coat the threads of the 2 top and 2 bottom water pump retaining bolts with water-resistant sealer and install along with the rest of the bolts. Tighten all bolts to 14 ft. lbs. (19 Nm).

12. Install the alternator adjusting arm bracket and the water pump pulley.

13. Coat the heater hose fitting with pipe sealant and install in the water pump. Connect the heater hose.

14. Install the power steering pump bracket, pump and drive belt. Install the air conditioning compressor bracket, compressor and drive belt.

15. Install the alternator adjusting arm and alternator drive belt.

16. Install the vacuum pump drive belt and adjust the tension on all accessory drive belts.

17. Install the fan and clutch assembly using the fan clutch pulley holder and the fan clutch nut wrench.

NOTE: The fan clutch nut has left hand thread. Tighten by turning the nut counterclockwise.

18. Install the radiator fan shroud halves. Fill and bleed the cooling system.

19. Connect the negative battery cables. Start the engine and bring to normal operating temperature. Check for leaks.

7.5L Engine

1. Disconnect the negative battery cable. Drain the cooling system and remove the fan shroud attaching bolts.

2. Remove the fan assembly attaching bolts and remove the shroud, fan and pulley.

3. Remove the power steering and air conditioning compressor, if equipped, drive belts.

4. If equipped with air conditioning, loosen the compressor attaching bolts and position the compressor aside. Remove the air conditioner and power steering pump bracket.

5. Loosen the alternator bolts and remove the alternator and air pump drive belt. Remove the air pump and bracket.

6. Disconnect the lower radiator hose, heater hose and bypass hose.

7. Remove the remaining water pump attaching bolts and remove the water pump from the cylinder front cover. Remove the separator plate from the water pump.

To install:

8. Clean the gasket mating surfaces on the water pump, cylinder front cover and separator plate.

9. Coat both sides of the new gaskets with water-resistant sealer. Install the gaskets, separator plate and water pump on the cylinder front cover with the attaching bolts. Tighten to 12–18 ft. lbs. (16–24 Nm).

10. Connect the bypass hose, heater hose and lower radiator hose.

11. Rotate the alternator and air pump bracket into position and install the attaching bolts. Install the air pump and alternator.

12. Install the air conditioning compressor bracket with the power steering pump in place. Install the compressor.

13. Install the pulley on the water pump and position the drive belts on their respective pulleys.

14. Lower the fan and shroud into position. Install the fan attaching bolts and tighten to 12–18 ft. lbs. (16–24 Nm).

15. Adjust the accessory drive belt tension. Fill and bleed the cooling system.

FAN CLUTCH NUT WRENCH

FAN CLUTCH PULLEY HOLDER

Fan clutch assembly removal and installation—7.3L diesel engine

THESE BOLTS ARE 2¾ IN. LONG. ALL OTHERS ARE 1½ IN. LONG.

WATER PUMP

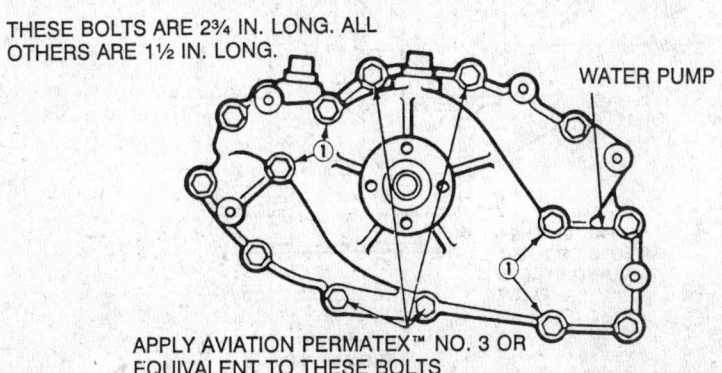

APPLY AVIATION PERMATEX™ NO. 3 OR EQUIVALENT TO THESE BOLTS

Water pump—7.3L diesel engine

16. Connect the negative battery cable. Start the engine and bring to normal operating temperature. Check for leaks.

Thermostat

REMOVAL & INSTALLATION

Gasoline Engine

1. Disconnect the negative battery cable. Drain the radiator so the coolant level is below the thermostat.

2. If equipped with a V8 engine, disconnect the bypass hoses at the water pump and the intake manifold. Remove the bypass tube.

3. Remove the water outlet housing attaching bolts and pull the housing away from the engine to gain access to the thermostat. Remove the thermostat.

To install:

4. Clean the gasket mating surfaces on the water outlet and intake manifold or cylinder head.

5. Coat both sides of a new water outlet housing gasket with water-resistant sealer and position on the cylinder head or intake manifold.

6. The water outlet housing on the 6 cylinder engine contains a locking recess into which the thermostat is turned and locked. Install the thermostat with the bridge section in the water outlet. Turn the thermostat clockwise to lock it in position on the flats cast into the water outlet.

7. If equipped with a V8 engine, install the thermostat in the intake manifold opening with the copper pellet or element toward the engine and the thermostat flange positioned in the recess.

8. Position the water outlet housing on the cylinder head or intake manifold. Install the attaching bolts and tighten to 12–15 ft. lbs. (17–20 Nm) on all except 7.5L engine. On 7.5L engine, tighten to 23–28 ft. lbs. (32–37 Nm).

9. If equipped with a V8 engine, install the bypass hose and tighten the clamps.

10. Fill and bleed the cooling system. Connect the negative battery cable, start the engine and bring to normal operating temperature. Check for leaks.

Diesel Engine

1. Disconnect ground cables from both batteries.

2. Drain the radiator so the coolant level is below the thermostat.

3. Remove the alternator and vacuum pump drive belts.

4. Remove the alternator and the vacuum pump and bracket. Position out of the way.

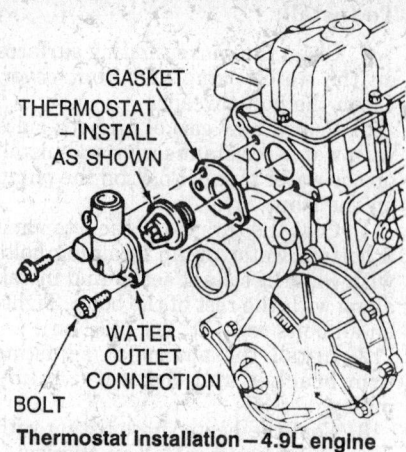

GASKET
THERMOSTAT
—INSTALL
AS SHOWN

WATER
OUTLET
CONNECTION

BOLT

Thermostat Installation—4.9L engine

5. Remove all but the lowest alternator/vacuum pump mounting casting bolt. Loosen the lowest bolt and pivot the alternator/vacuum pump casting outboard.

6. Remove the water outlet housing attaching bolts and pull the housing away from the engine to gain access to the thermostat. Remove the thermostat.

To install:

7. Clean the gasket mating surfaces on the water outlet housing and intake manifold.

8. Coat a new gasket with water-resistant sealer and position on the intake manifold opening.

9. Install the thermostat in the crankcase opening with the copper pellet or element toward the engine and the thermostat flange positioned in the recess.

10. Position the water outlet housing against the crankcase. Install the attaching bolts and tighten to 20 ft. lbs. (27 Nm).

11. Reposition the alternator/vacuum pump casting and install the attaching bolts.

12. Install the vacuum pump and bracket and the alternator. Install the drive belts.

13. Fill and bleed the cooling system. Connect the ground cables to both batteries.

14. Operate the engine until normal operating termperature is reached. Check the coolant level and check for leaks.

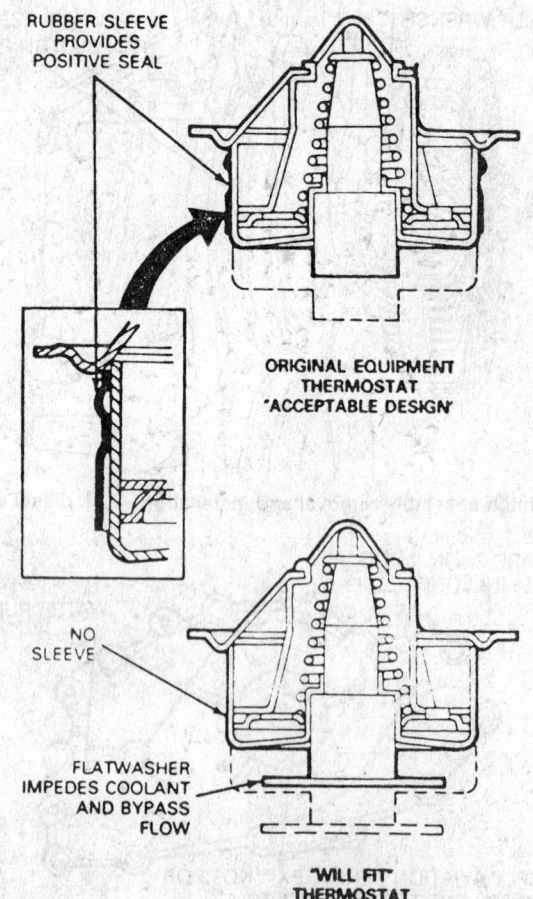

RUBBER SLEEVE
PROVIDES
POSITIVE SEAL

ORIGINAL EQUIPMENT
THERMOSTAT
"ACCEPTABLE DESIGN"

NO
SLEEVE

FLATWASHER
IMPEDES COOLANT
AND BYPASS
FLOW

"WILL FIT"
THERMOSTAT
"UNACCEPTABLE DESIGN"

Thermostat—7.3L diesel engine

COOLING SYSTEM BLEEDING

When the entire cooling system is drained, the following procedure should be used to remove air from the cooling system and ensure a complete fill.

1. Close the radiator draincock and install the cylinder block drain plug, if removed.

2. Fill the cooling system with a 50/50 mixture of anti-freeze and water. Allow several minutes for trapped air to escape. When filling a cross flow radiator, allow time for the coolant to flow through the radiator tubes to the other end of the tank to ensure the radiator is full.

3. Disconnect the heater outlet hose at the water pump to bleed or release trapped air in the system. When the coolant begins to escape, connect the heater outlet hose.

4. Install the radiator cap to the pressure relief position by installing the cap to the fully installed position and then backing off to the 1st stop. This will allow any air to escape and will minimize spillage.

5. Slide the heater temperature and mode selection levers to the maximum heat position.

6. Start the engine and allow to operate at fast idle for approximately 3–4 minutes.

7. With the engine shut off, wrap the radiator cap with a thick cloth, carefully remove the cap and add cool-ant to bring the coolant level up to 1½ in. below the cap seal.

8. Install the cap to the fully installed position. Then, back off to the 1st stop and operate the engine at fast idle until the thermostat opens and the upper radiator hose is warm. To check the coolant level, shut the engine off, wrap the cap with a thick cloth and cautiously remove the cap. Add additional coolant, if necessary. Install the cap to the fully installed position.

— **CAUTION** —

To avoid personal injury from scalding hot coolant or steam blowing out of the radiator, use extreme care when removing the cap from a hot radiator.

9. Fill the coolant recovery reservoir to the proper level with a 50/50 mix of anti-freeze and water.

ENGINE ELECTRICAL

NOTE: Disconnecting the negative battery cable on some vehicles may interfere with the functions of the on board computer systems and may require the computer to undergo a relearning process, once the negative battery cable is reconnected.

Distributor

REMOVAL

1. Disconnect the negative battery cable.

2. Mark the position of the No. 1 cylinder wire tower on the distributor base. Remove the distributor cap and position the cap and ignition wires to the side.

3. Disconnect the wiring harness plug from the distributor connector. Disconnect and plug the vacuum hoses from the vacuum diaphragm assembly, if equipped.

4. Rotate the engine, in normal direction of rotation, until No. 1 piston is on Top Dead Center (TDC) of the compression stroke. The TDC mark on the crankshaft pulley and the pointer on the timing cover should align. The rotor tip should be pointing at the No. 1 spark plug wire position on the distributor cap.

5. Scribe a mark on the distributor body and the engine to indicate the position of the rotor tip and the position of the distributor in the engine.

6. Remove the hold-down bolt and clamp located at the base of the distributor. Remove the distributor from the engine. Pay attention to the direction the rotor tip points if it moves from the No. 1 position when the drive gear disengages. For reinstallation purposes, the rotor should be at this point to insure proper gear mesh and timing.

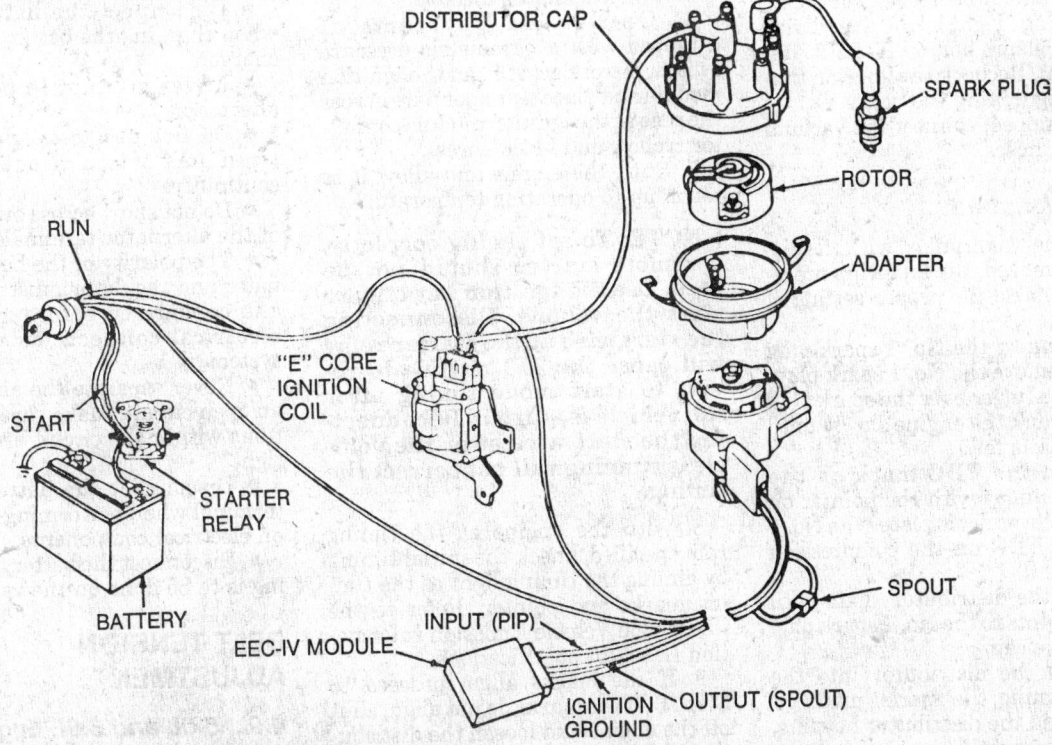

Thick Film Integrated (TFI) ignition system with universal distributor

7. Avoid turning the engine, if possible, while the distributor is removed. If the engine is turned from TDC position, TDC timing marks will have to be reset before the distributor is installed.

INSTALLATION

NOTE: Before installing, visually inspect the distributor. The drive gear should be free of nicks, cracks and excessive wear. The distributor drive shaft should move freely, without binding. If equipped with an O-ring, it should fit tightly and be free of cuts.

Timing Not Disturbed

1. Position the distributor in the engine with the rotor aligned to the marks made on the distributor or at the position the rotor pointed when the distributor was removed. Engage the oil pump intermediate shaft and insert the distributor until fully seated on the engine. If the distributor does not fully seat, turn the engine slightly to fully engage the intermediate shaft.

2. After the distributor has been fully seated into the block, recheck the timing mark and rotor alignment. Install the hold-down clamp and bolt. Snug the mounting bolt so the distributor can be turned for ignition timing purposes.

3. Install the distributor cap and connect the distributor to the wiring harness.

4. Connect the negative battery cable. Check and, if necessary, set the ignition timing. Tighten the distributor hold-down clamp bolt to 17–25 ft. lbs. (23–34 Nm). Recheck the ignition timing after tightening the bolt.

5. If equipped, connect the vacuum diaphragm hoses.

Timing Disturbed

If the engine was cranked with the distributor removed, the following procedure will enable the proper setting of the timing.

1. Disconnect the No. 1 spark plug wire and remove the No. 1 spark plug.

2. Place a finger over the spark plug hole and crank the engine slowly until compression is felt.

3. Align the TDC mark on the crankshaft pulley with the pointer on the timing cover. This places the No. 1 cylinder at TDC on the compression stroke.

4. Turn the distributor shaft until the rotor points to the No. 1 spark plug tower on the cap.

5. Install the distributor into the engine, aligning the marks made on the block and the distributor housing. Install the distributor hold-down clamp and bolt. Snug the bolt so the distributor housing can be moved for timing purposes.

6. Install the No. 1 spark plug and connect the spark plug wire. Install the distributor cap and connect the distributor to the wiring harness.

7. Connect the negative battery cable and set the ignition timing. Tighten the distributor hold-down clamp bolt to 17–25 ft. lbs. (23–34 Nm). Recheck the ignition timing after tightening the bolt.

8. If equipped, connect the vacuum diaphragm hoses.

Ignition Timing

ADJUSTMENT

1. Locate the timing marks and pointer on the crankshaft pulley and the timing cover. Clean the marks so they will be visible with a timing light. Apply chalk or bright-colored paint, if necessary.

2. Place the transaxle in **P** or **N**. The air conditioning and heater controls should be in the **OFF** position.

3. On Non-EEC systems, disconnect the vacuum hoses from the distributor vacuum advance connection at the distributor and plug the hoses.

4. Connect a suitable inductive timing light and a tachometer according to the manufacturer's instructions.

5. On EEC-IV systems, disconnect the single wire in-line spout connector or remove the shorting bar from the double wire spout connector.

6. On Non-EEC systems, if equipped with a barometric pressure switch, disconnect it from the ignition module and place a jumper wire across the pins at the ignition module connector (yellow and black wires).

7. Start the engine and allow it to warm up to operating temperature.

NOTE: To set timing correctly, a remote starter should not be used. Use the ignition key only to start the vehicle. Disconnecting the start wire at the starter relay will cause the TFI module to revert to start mode timing after the vehicle is started. Reconnecting the start wire after the vehicle is running will not correct the timing.

8. With the engine at the timing rpm specified, check the initial timing by aiming the timing light at the timing marks and pointer. Refer to the underhood Vehicle Emission Information Label for specifications.

9. If the marks align, proceed to Step 10. If the marks do not align, shut off the engine and loosen the distributor hold-down clamp bolt. Start the en-

gine, aim the timing light and turn the distributor until the timing marks align. Shut off the engine and tighten the distributor hold-down clamp bolt.

10. On EEC-IV systems, reconnect the single wire in-line spout connector or reinstall the shorting bar on the double wire spout connector. Check the timing advance to verify the distributor is advancing beyond the initial setting.

11. Remove the timing light and the tachometer.

12. On Non-EEC systems, unplug and reconnect the vacuum hoses. Remove the jumper wire from the ignition connector and reconnect, if applicable.

Alternator

PRECAUTIONS

Several precautions must be observed with alternator equipped vehicles to avoid damage to the unit.

● If the battery is removed for any reason, make sure it is reconnected with the correct polarity. Reversing the battery connections may result in damage to the one-way rectifiers.

● When utilizing a booster battery as a starting aid, always connect the positive to positive terminals and the negative terminal from the booster battery to a good engine ground on the vehicle being started.

● Never use a fast charger as a booster to start vehicles.

● Disconnect the battery cables when charging the battery with a fast charger.

● Never attempt to polarize the alternator.

● Do not use test lights of more than 12V when checking diode continuity.

● Do not short across or ground any of the alternator terminals.

● The polarity of the battery, alternator and regulator must be matched and considered before making any electrical connections within the system.

● Never separate the alternator on an open circuit. Make sure all connections within the circuit are clean and tight.

● Disconnect the battery ground terminal when performing any service on electrical components.

● Disconnect the battery if arc welding is to be done on the vehicle.

BELT TENSION ADJUSTMENT

4.9L, 5.0L and 5.8L Engines

Alternator belt tension is maintained

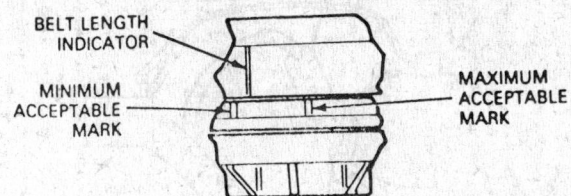

INDICATOR SHOULD BE BETWEEN MARKS

BELT LENGTH INDICATOR

MINIMUM ACCEPTABLE MARK

MAXIMUM ACCEPTABLE MARK

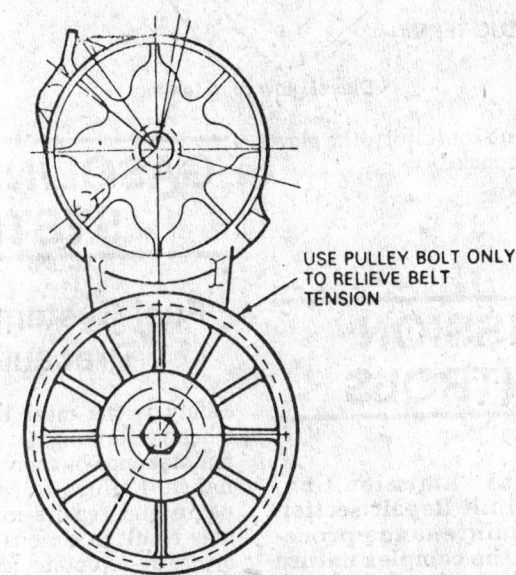

USE PULLEY BOLT ONLY TO RELIEVE BELT TENSION

Automatic belt tensioner—4.9L engine

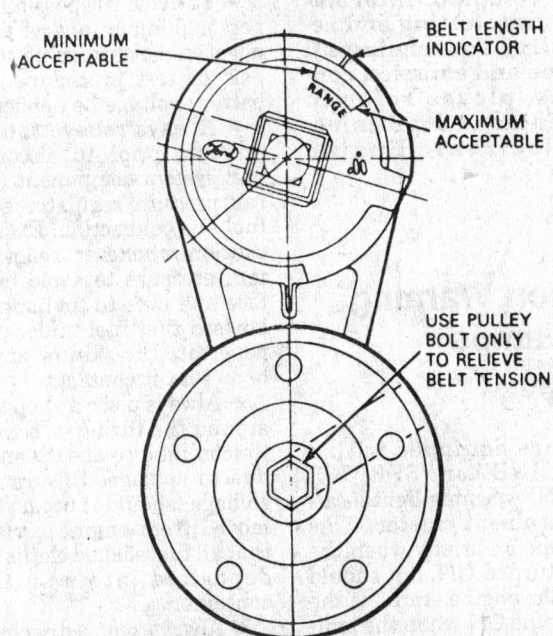

INDICATOR SHOULD BE BETWEEN MARKS

MINIMUM ACCEPTABLE

BELT LENGTH INDICATOR

RANGE

MAXIMUM ACCEPTABLE

USE PULLEY BOLT ONLY TO RELIEVE BELT TENSION

Automatic belt tensioner—5.0L and 5.8L engines

by an automatic tensioner on these engines. Correct belt tension will be maintained if the correct length belt is on the engine. The tensioner is working properly if the belt length indicator mark on the tensioner is between the maximum and minimum marks.

7.3L and 7.5L Engines

1. Position a suitable belt tension gauge midway on the longest accessible belt span.

2. Loosen the alternator adjustment and pivot bolts just enough to move the alternator.

3. Place a "C" clamp over the end of the adjusting arm and the adjusting bolt boss and tighten the belt to the correct tension. The correct tension is as follows:

7.3L engine with new belt—140–180 lbs.

7.3L engine with used belt (over 5 minutes engine operation)—95–115 lbs.

7.5L engine with new belt—160–200 lbs.

7.5L engine with used belt (over 5 minutes engine operation)—110–130 lbs.

4. Tighten the adjustment bolt to 30–40 ft. lbs. (40–55 Nm) and remove the "C" clamp.

5. Tighten the pivot bolt to 40–50 ft. lbs. (55–70 Nm). Check and readjust the belt tension, if necessary.

6. Start the engine and let it idle for 5 minutes.

7. Stop the engine and recheck the belt tension. If the tension is less than the used belt specification, then reset the belt tension to within the used belt tension limits.

8. If the belt will not hold tension, it must be replaced.

REMOVAL & INSTALLATION

1. Disconnect the negative battery cable.

2. Disconnect the electrical connectors from the alternator.

3. Remove the alternator belt.

4. Remove the alternator mounting bolts and the alternator. Remove the alternator fan shield, if equipped.

5. Installation is the reverse of the removal procedure. Adjust the belt tension, if necessary.

Voltage Regulator

REMOVAL & INSTALLATION

NOTE: Always disconnect the

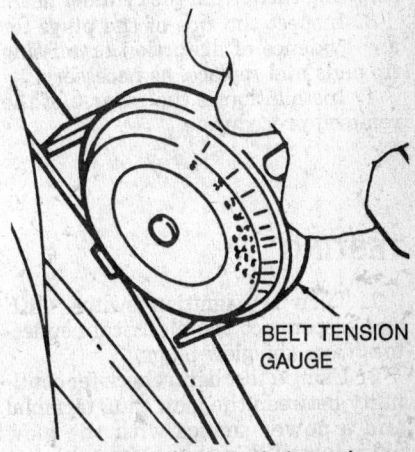

BELT TENSION GAUGE

Checking alternator belt tension

connector plug from the regulator before removing the regulator mounting screws. Removing the connector from an ungrounded regulator with the ignition switch in RUN will destroy the regulator.

1. Disconnect the negative battery cable.
2. Disconnect the regulator from the wiring harness.
3. Remove the regulator mounting screws.
4. Installation is the reverse of the removal procedure.

Starter

REMOVAL & INSTALLATION

1. Disconnect the negative battery cable.
2. Raise the vehicle and support it safely.
3. Disconnect the starter cable at the starter terminal.
4. Remove the starter mounting bolts and remove the starter.
To install:
5. Position the starter assembly to the flywheel housing and start the mounting bolts.
6. Snug all the bolts while holding the starter squarely against its mounting surface and fully inserted into the pilot hole. Tighten the bolts to 15–20 ft. lbs. (21–27 Nm).
7. Reconnect the starter cable at the starter terminal. Lower the vehicle. Connect the negative battery cable.

Diesel Glow Plugs

REMOVAL & INSTALLATION

1. Disconnect the negative battery cable. Disconnect the glow plug electrical leads.
2. Remove the glow plugs by unscrewing them from the cylinder head.
3. Inspect the tips of the plugs for any evidence of distortion or missing tip ends and replace, as necessary.
4. Installation is the reverse of the removal procedure.

TESTING

1. Turn the ignition switch **OFF** and disconnect the electrical connectors from the glow plugs.
2. Using a test light, check for continuity between the glow plug terminal and a power source with the glow plugs installed in the engine.

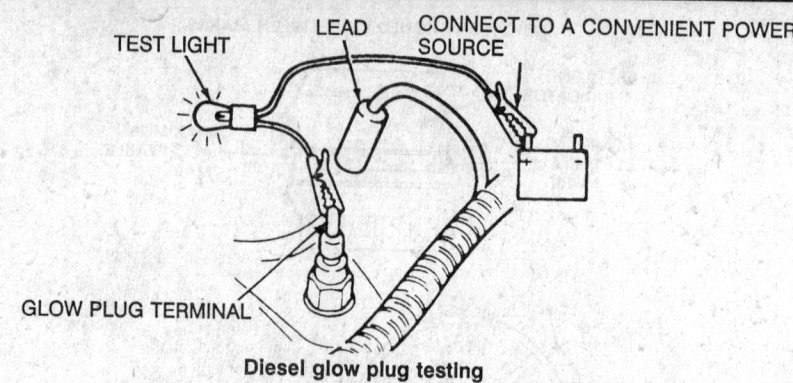

TEST LIGHT   LEAD   CONNECT TO A CONVENIENT POWER SOURCE

GLOW PLUG TERMINAL

Diesel glow plug testing

3. If there is no continuity, the glow plug must be replaced.

EMISSION CONTROLS

Please refer to "Emission Controls" in the Unit Repair section for system maintenance procedures. Due to the complex nature of modern electronic engine control systems, comprehensive diagnosis and testing procedures fall outside the confines of this repair manual. For complete information on diagnosis, testing and repair procedures concerning all modern engine and emission control systems, please refer to "Chilton's Guide to Fuel Injection and Electronic Engine Controls".

Emission Warning Lamps

RESETTING

All vehicles are equipped with a "CHECK ENGINE" or "SERVICE ENGINE SOON" warning light located on the instrument cluster. This light should come on briefly when the ignition key is turned **ON**, but should turn off when the engine starts. If the light does not come ON when the ignition key is turned **ON** or if it comes ON and stays ON when the engine is running, there is a malfunction in the electronic engine control system. After the malfunction has been remedied, using the proper procedures, the "CHECK ENGINE" or "SERVICE ENGINE SOON" light will go out.

GASOLINE FUEL SYSTEM

Fuel System Service Precautions

Safety is the most important factor when performing not only fuel system maintenance but any type of maintenance. Failure to conduct maintenance and repairs in a safe manner may result in serious personal injury or death. Maintenance and testing of the vehicle's fuel system components can be accomplished safely and effectively by adhering to the following rules and guidelines.

• To avoid the possibility of fire and personal injury, always disconnect the negative battery cable unless the repair or test procedure requires that battery voltage be applied.
• Always relieve the fuel system pressure prior to disconnecting any fuel system component (injector, fuel rail, pressure regulator, etc.), fitting or fuel line connection. Exercise extreme caution whenever relieving fuel system pressure to avoid exposing skin, face and eyes to fuel spray. Please be advised that fuel under pressure may penetrate the skin or any part of the body that it contacts.
• Always place a shop towel or cloth around the fitting or connection prior to loosening to absorb any excess fuel due to spillage. Ensure that all fuel spillage (should it occur) is quickly removed from engine surfaces. Ensure that all fuel soaked cloths or towels are deposited into a suitable waste container.
• Always keep a dry chemical (Class B) fire extinguisher near the work area.
• Do not allow fuel spray or fuel vapors to come into contact with a spark or open flame.
• Always use a backup wrench when loosening and tightening fuel line connection fittings. This will pre-

vent unnecessary stress and torsion to fuel line piping. Always follow the proper torque specifications.

• Always replace worn fuel fitting O-rings with new. Do not substitute fuel hose or equivalent where fuel pipe is installed.

RELIEVING FUEL SYSTEM PRESSURE

1. Locate and disconnect the electrical connection to either the fuel pump relay, the inertia switch or the in-line high pressure fuel pump.

2. Crank the engine for approximately 10 seconds. The engine may start and run for a short time. If so, crank the engine an additional 5 seconds after the engine stalls.

3. Connect the electrical connector that was disconnected in Step 1.

Fuel Line Couplings

REMOVAL & INSTALLATION

There are 3 methods in use to connect the fuel lines and fuel system compo-nents, the hairpin clip push connect fitting, the duck bill clip push connect fitting and the spring lock coupling. Each requires a different procedure to disconnect and connect.

Hairpin Clip Push Connect Fitting

1. Inspect the visible internal portion of the fitting for dirt accumulation. If more than a light coating of dust is present, clean the fitting before disassembly.

2. Some adhesion between the seals in the fitting and the tubing will occur with time. To separate, twist the fitting on the tube, then push and pull the fitting until it moves freely on the tube.

3. Remove the hairpin clip from the fitting by first bending and breaking the shipping tab. Next, spread the 2 clip legs by hand about ⅛ in. each to disengage the body and push the legs into the fitting. Lightly pull the triangular end of the clip and work it clear of the tube and fitting.

NOTE: Do not use hand tools to complete this operation.

4. Grasp the fitting and pull in an axial direction to remove the fitting from the tube. Be careful on 90 degree elbow connectors, as excessive side loading could break the connector body.

5. After disassembly, inspect and clean the tube end sealing surfaces. The tube end should be free of scratches and corrosion that could provide leak paths. Inspect the inside of the fitting for any internal parts such as O-rings and spacers that may have been dislodged from the fitting. Replace any damaged connector.

To install:

6. Install a new connector if damage was found. Insert a new clip into any 2 adjacent openings with the triangular portion pointing away from the fitting opening. Install the clip until the legs of the clip are locked on the outside of the body. Piloting with an index finger is necessary.

7. Before installing the fitting on the tube, wipe the tube end with a clean cloth. Inspect the inside of the fitting to make sure it is free of dirt and/or obstructions.

8. Apply a light coating of engine oil

Typical push connect fittings

to the tube end. Align the fitting and tube axially and push the fitting onto the tube end. When the fitting is engaged, a definite click will be heard. Pull on the fitting to make sure it is fully engaged.

Duck Bill Clip Push Connect Fitting

1. Inspect the visible internal portion of the fitting for dirt accumulation. If more than a light coating of dust is present, clean the fitting before disassembly.

2. Some adhesion between the seals in the fitting and the tubing will occur with time. To separate, twist the fitting on the tube, then push and pull the fitting until it moves freely on the tube.

3. Align the slot on push connect disassembly tool T82L–9500–AH or equivalent, with either tab on the clip, 90 degrees from the slots on the side of the fitting and insert the tool. This disengages the duck bill retainer from the tube.

4. Holding the tool and the tube with 1 hand, pull the fitting away from the tube.

NOTE: Use hands only. Only moderate effort is required if the tube has been properly disengaged.

5. After disassembly, inspect and clean the tube end sealing surfaces. The tube end should be free of scratches and corrosion that could provide leak paths. Inspect the inside of the fitting for any internal parts such as O-rings and spacers that may have been dislodged from the fitting. Replace any damaged connector.

6. Some fuel tubes have a secondary bead which aligns with the outer surface of the clip. These beads can make tool insertion difficult. If there is extreme difficulty, use the following disassembly method:

   a. Using pliers with a jaw width of 0.2 in. (5mm) or less, align the jaws with the openings in the side of the fitting case and compress the portion of the retaining clip that engages the fitting case. This disengages the retaining clip from the case. Often 1 side of the clip will disengage before the other. The clip must be disengaged from both openings.

   b. Pull the fitting off the tube by hand only. Only moderate effort is required if the retaining clip has been properly disengaged.

   c. After disassembly, inspect and clean the tube end sealing surfaces. The tube end should be free of scratches and corrosion that could provide leak paths. Inspect the inside of the fitting for any internal

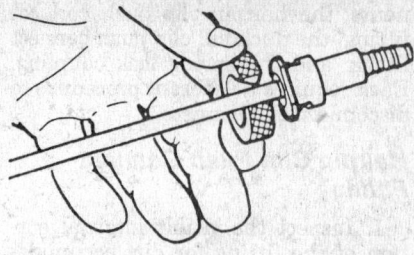

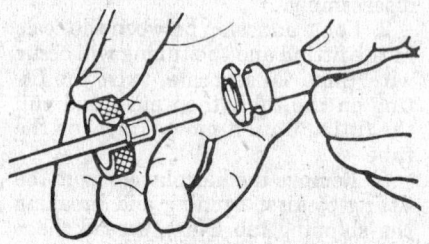

Duck bill push connect fitting removal

parts such as O-rings and spacers that may have been dislodged from the fitting. Replace any damaged connector.

   d. The retaining clip will remain on the tube. Disengage the clip from the tube bead and remove.

To install:

7. Install a new connector if damage was found. Install the new replacement clip into the body by inserting 1 of the retaining clip serrated edges on the duck bill portion into 1 side of the window openings. Push on the other side until the clip snaps into place.

8. Before installing the fitting on the tube, wipe the tube end with a clean cloth. Inspect the inside of the fitting to make sure it is free of dirt and/or obstructions.

9. Apply a light coating of engine oil to the tube end. Align the fitting and tube axially and push the fitting onto the tube end. When the fitting is engaged, a definite click will be heard. Pull on the fitting to make sure it is fully engaged.

Spring Lock Coupling

The spring lock coupling is a fuel line coupling held together by a garter spring inside a circular cage. When the coupling is connected together, the flared end of the female fitting slips behind the garter spring inside the cage of the male fitting. The garter spring and cage then prevent the flared end of the female fitting from pulling out of the cage. As an additional locking feature, most vehicles have a horseshoe shaped retaining clip that improves the retaining reliability of the spring lock coupling.

Fuel Tank

REMOVAL & INSTALLATION

Bronco

1. Relieve the fuel system pressure and disconnect the negative battery cable.

2. Drain the fuel into a suitable container at the fuel hose between the fuel pump and the fuel tube.

3. Raise and safely support the vehicle.

4. Loosen the clamp on the fuel filler pipe and disconnect the hose from the pipe by pulling along the internal fuel tube from the tank filler neck.

5. Disconnect the fuel lines at the fuel gauge sending unit.

6. Support the tank and remove the lower support bracket bolts or skid plate bolts. Remove the support assembly or skid plate attaching nut at each tank mounting strap. Lower the support assemblies and lower the tank enough to gain access to the tank vent hose.

7. Disconnect the fuel sending unit electrical connector, fuel tank vent hose and fuel tank-to-vapor separator lines at the fuel tank.

8. Remove the fuel tank.

To install:

9. Position the forward edge of the tank and skid plate, if equipped, to the frame crossmember, support the tank and connect the vent hose and fuel sender electrical connector.

10. Position the tank, skid plate and straps and install the attaching parts. Apply threadlocking compound and tighten the bolts and nuts to 27–37 ft. lbs. (37–50 Nm).

11. Connect the fuel lines to the fuel sender.

12. Insert the internal fuel tube into the tank filler neck. Connect the filler pipe-to-tank hose and vent tube at the filler pipe and install the hose clamp. Tighten to 25–35 inch lbs. (3–4 Nm).

13. Lower the vehicle. Fill the tank and check all connections for leaks.

14. Connect the negative battery cable.

F Series

MIDSHIP FUEL TANK—PLASTIC

1. Relieve the fuel system pressure and disconnect the negative battery cable(s).

2. Drain the fuel into a suitable container by siphoning through the fuel hose at the fuel pump-to-fuel tube connection.

3. Remove the skid plate and heat shields.

4. Disconnect the fuel gauge sending unit wire at the fuel tank.

5. Loosen the fuel filler hose clamp

SPRING LOCK COUPLING CHART

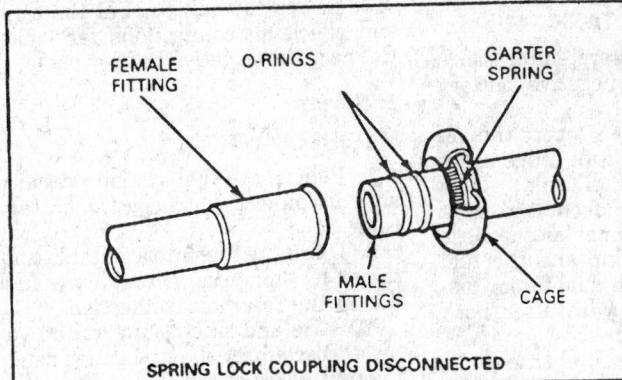

SPRING LOCK COUPLING DISCONNECTED

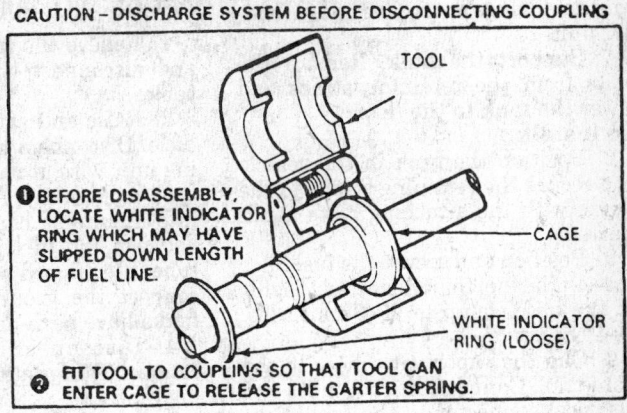

❶ BEFORE DISASSEMBLY, LOCATE WHITE INDICATOR RING WHICH MAY HAVE SLIPPED DOWN LENGTH OF FUEL LINE.

❷ FIT TOOL TO COUPLING SO THAT TOOL CAN ENTER CAGE TO RELEASE THE GARTER SPRING.

TO CONNECT COUPLING

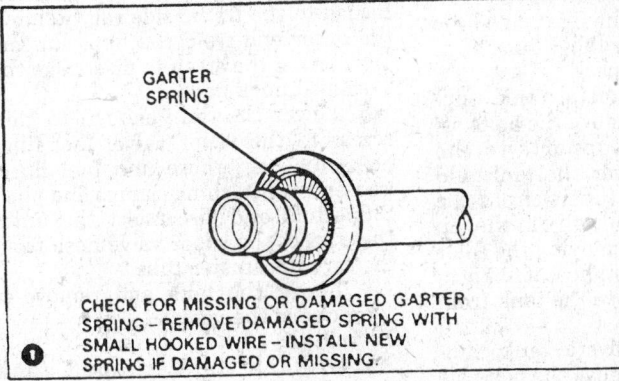

❶ CHECK FOR MISSING OR DAMAGED GARTER SPRING – REMOVE DAMAGED SPRING WITH SMALL HOOKED WIRE – INSTALL NEW SPRING IF DAMAGED OR MISSING.

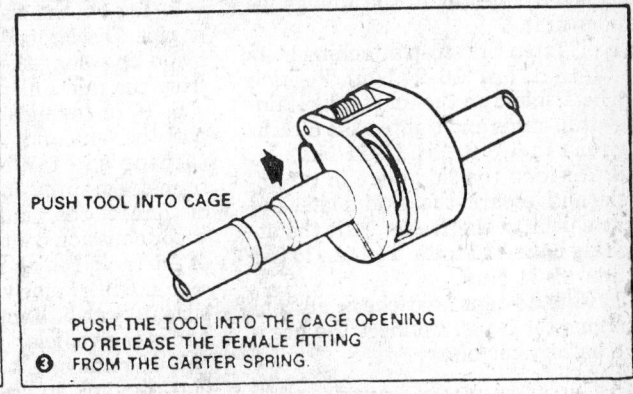

❸ PUSH THE TOOL INTO THE CAGE OPENING TO RELEASE THE FEMALE FITTING FROM THE GARTER SPRING.

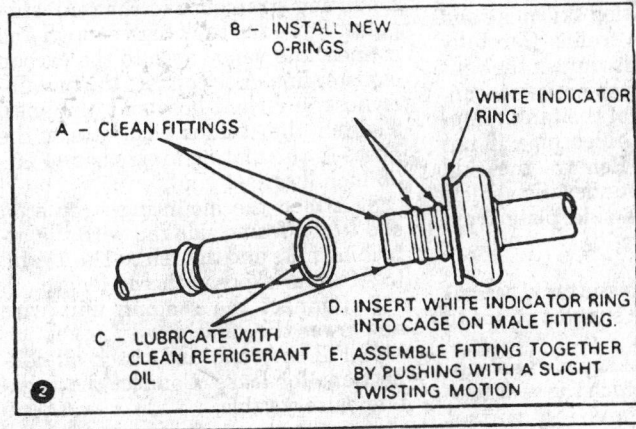

A – CLEAN FITTINGS
B – INSTALL NEW O-RINGS
C – LUBRICATE WITH CLEAN REFRIGERANT OIL
D. INSERT WHITE INDICATOR RING INTO CAGE ON MALE FITTING.
E. ASSEMBLE FITTING TOGETHER BY PUSHING WITH A SLIGHT TWISTING MOTION

❷

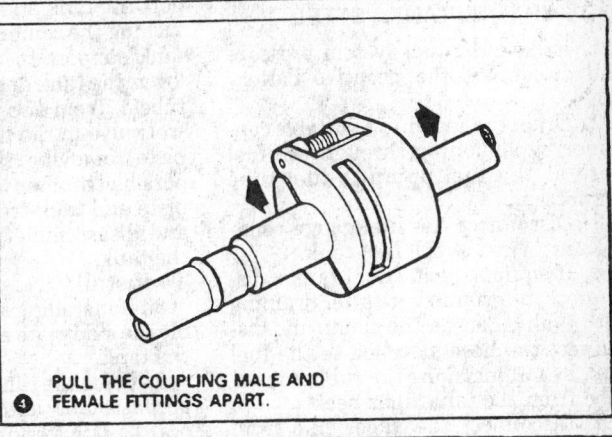

❹ PULL THE COUPLING MALE AND FEMALE FITTINGS APART.

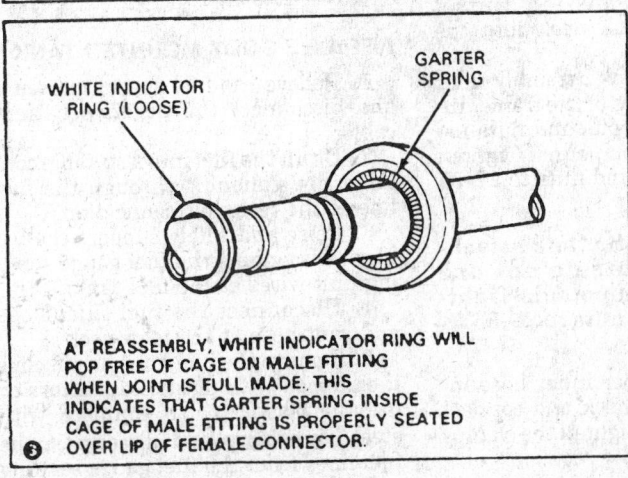

❸ AT REASSEMBLY, WHITE INDICATOR RING WILL POP FREE OF CAGE ON MALE FITTING WHEN JOINT IS FULL MADE. THIS INDICATES THAT GARTER SPRING INSIDE CAGE OF MALE FITTING IS PROPERLY SEATED OVER LIP OF FEMALE CONNECTOR.

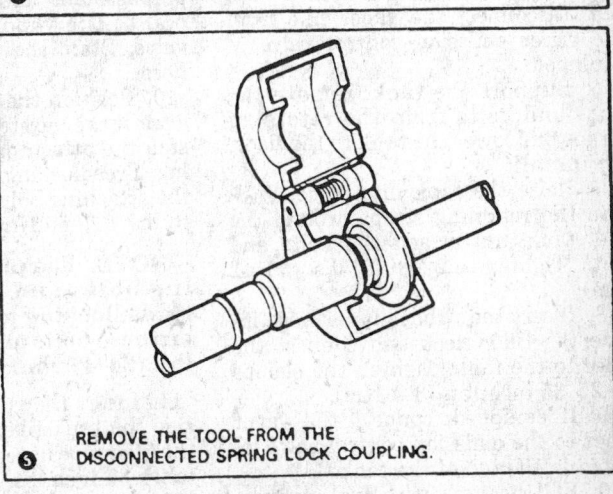

❺ REMOVE THE TOOL FROM THE DISCONNECTED SPRING LOCK COUPLING.

at the tank and disconnect the fuel filler hose.

6. Disconnect the fuel tube push connect fittings at the fuel gauge sending unit.

7. Support the tank. Remove the bolts from the retaining straps and lower the tank to the floor.

To install:

8. Position the tank in the vehicle and secure the retaining straps to the frame with the attaching bolts handstarted.

9. Position and secure the fuel filler hose on the fuel tank filler neck. Tighten the hose clamp to 27–35 inch lbs. (3–4 Nm).

10. Clip the vapor valve hose to the frame rail. Connect the fuel tubes and electrical connector to the fuel gauge sender unit.

11. Tighten the strap attaching bolts to 12–18 ft. lbs. (16–24 Nm). Position the heat shield on the fuel tank retaining strap studs and tighten the attaching nuts to 6–8 ft. lbs. (8–11 Nm).

12. Position the skid plate in the vehicle and secure the skid plate and heat shield to the frame with the attaching bolts and nuts. Tighten to 6–8 ft. lbs. (8–11 Nm).

13. Fill the tank and check all connections for leaks. Connect the negative battery cable(s).

MIDSHIP FUEL TANK—STEEL

1. Relieve the fuel system pressure and disconnect the negative battery cable.

2. Drain the fuel into a suitable container by siphoning through the fuel hose at the fuel pump-to-fuel tube connection.

3. Disconnect the fuel gauge sending unit wires at the fuel tank.

4. If equipped with dual tanks, disconnect the ground wire after draining both tanks. Loosen the clamp and disconnect the hose attached to the fuel tank by pulling along the rubber inner tube from the tank filler neck.

5. Disconnect the vapor line from the vapor emission control valve, if equipped.

6. Support the tank, remove the nuts and bolts from the retaining straps and lower the tank to the floor.

To install:

7. Raise the tank into place. Position the retaining straps around the fuel tank and attach the bolts and nuts. Tighten to 22–30 ft. lbs. (30–41 Nm).

8. Insert the rubber inner hose inside the filler neck and connect the hose to the tank. Tighten the clamps to 25–35 inch lbs. (3–4 Nm).

9. If equipped, connect the vapor lines to the emission control valve located in the top of the tank.

10. Fill the tank with fuel, connect the negative battery cable and start the engine. Check for leaks.

AFT-OF-AXLE FUEL TANK

1. Relieve the fuel system pressure and disconnect the negative battery cable.

2. Raise and safely support the vehicle. Disconnect the fuel gauge sending unit wire at the fuel tank.

3. Siphon the fuel from the tank into a suitable container at the hose between the fuel pump and the fuel tube. If equipped with dual tanks, disconnect the ground wire after both tanks have been drained.

4. Disconnect the fuel line push connect fittings at the fuel gauge sending unit.

5. Loosen the clamp on the fuel filler pipe. Disconnect the filler pipe hose by pulling along the rubber inner tube from the tank filler neck.

6. If removing a metal tank, support the tank and remove the bolts attaching the tank supports to the frame. Carefully lower the tank and disconnect the vent tube(s) from the vapor emission control valve in the top of the tank. Finish removing the filler pipe and filler pipe vent hose if not possible in Step 5. Remove the tank from under the vehicle.

7. If removing a plastic tank, support the tank and remove the bolts attaching the combination skid plate and tank support to the frame. Carefully lower the tank and disconnect the vent tube(s) from the vapor emission control valve in the top of the tank. Complete removing the filler pipe if not possible in Step 5. Remove the skid plate and tank from under the vehicle and disassemble the skid plate from the tank.

To install:

8. If installing a plastic tank, assemble the skid plate and support straps to the tank.

9. Raise the tank skid plate and support assembly and attach the vent hose(s) to the vapor emission control valve. Start the tank neck into the hose.

10. Position the tank assembly filler against the top straps of the frame. Install the attaching bolts and nuts using threadlocking compound. Tighten the attaching bolts and nuts to 27–37 ft. lbs. (37–50 Nm).

NOTE: The plastic tank attaching bolts and nuts do not use threadlocking compound. Tighten these bolts and nuts to 25–35 ft. lbs. (34–47 Nm).

11. Insert the rubber inner hose inside the tank filler neck and connect the filler pipe hose. Tighten the clamps to 25–35 inch lbs. (3–4 Nm).

12. Connect the sending unit electrical connector and the fuel lines. Install the drain plug, if equipped.

13. Lower the vehicle. Fill the tank and check all connections for leaks. Connect the negative battery cable.

E Series

MIDSHIP TANK

1. Relieve the fuel system pressure and disconnect the negative battery cable.

2. Drain the fuel into a suitable container by siphoning through the fuel line at the reservoir connection.

3. Raise and safely support the vehicle. Disconnect the fuel gauge sending unit wires at the fuel tank.

4. Support the tank in position. Disengage the mounting strap ends attached to the frame side rail. Remove the other end from the tank support by rotating the strap to disengage the T-shaped hook end.

5. Lower the tank enough to gain access to the vapor valve, fuel filler hose, fuel vent hose and fuel lines. Loosen the attaching clamps and push connectors and disconnect the lines. Disconnect the vapor valve hose from the carbon canister tube.

6. Lower the tank and remove it from under the vehicle.

To install:

7. Attach the T-shaped hook mounting strap ends in the tank supports. Raise the tank high enough and connect the vapor hose to the carbon canister line. Also connect the fuel filler hose, fuel vent hose and fuel lines. Tighten the clamps that attach the hoses to the mating tank parts to 25–35 inch lbs. (3–4 Nm).

8. Attach the mounting strap stud end to the frame side rail with the attaching nuts and tighten to 1¼–1½ in. (32–38mm) exposed thread length.

9. Connect the sending unit wire and lower the vehicle.

10. Fill the tank and check all connections for leaks. Connect the negative battery cable.

AFT AXLE BODY-MOUNTED TANK

1. Relieve the fuel system pressure and disconnect the negative battery cable.

2. Drain the fuel into a suitable container by siphoning through the fuel line at the reservoir connection.

3. Raise and safely support the vehicle. Disconnect the fuel gauge sending unit wires at the fuel tank.

4. Disconnect the fuel filler hose and vent hose at the tank neck.

5. Remove the fuel line hairpin clips if equipped with plastic connectors or use the plastic tool if equipped with steel connectors and disconnect the fuel lines from the fuel gauge sending unit.

6. Support the tank in position. Remove the nuts that attach the mounting straps to the T-bolts.

NOTE: The T-bolts are attached to the body brackets located at the rear of the tank. Disengage the straps from the T-bolts and the front body brackets. Lower the tank enough to gain access to the vapor valve.

7. Disconnect the carbon canister hose from the vapor control valve. Lower the fuel tank and remove it from under the vehicle.

To install:

8. Attach the front ends of the mounting straps to the front body brackets.

9. Raise the tank high enough to connect the carbon canister hose to the vapor valve and connect the hose.

10. Secure the strap ends to the T-bolts with the attaching nuts. Tighten the nuts so 1.65–1.85 in. (42–46mm) thread length is exposed.

11. Connect the fuel lnes to the sending unit. Connect the fuel filler hose and vent hose to the filler neck and vent neck at the tank. Tighten the hose clamps to 25–35 inch lbs. (3–4 Nm).

12. Connect the fuel gauge sending unit wires. Lower the vehicle.

13. Fill the tank and check all connections for leaks. Connect the negative battery cable.

AFT AXLE FRAME-MOUNTED TANK—EXCEPT 38 GALLON

1. Relieve the fuel system pressure and disconnect the negative battery cable.

2. Drain the fuel into a suitable container by siphoning through the fuel line at the reservoir connection.

3. Raise and safely support the vehicle. Disconnect the fuel gauge sending unit wires at the fuel tank.

4. Disconnect the fuel filler hose and vent hose from the tank. Remove the fuel line hairpin clips and disconnect the fuel lines from the fuel gauge sending unit.

5. Support the tank in position. Remove the nuts and bolts that attach the tank supports to the frame. Disengage the straps from the front tank support and the rear crossmember. Lower the tank enough to gain access to the vapor valve.

6. Disconnect the vapor hose from the vapor control valve. Lower the fuel tank and remove it from under the vehicle.

To install:

7. Attach the mounting straps to the rear crossmember. Raise the tank high enough to connect the vapor hose to the vapor valve and connect the hose.

8. Position the tank against the top

straps and install the support and attaching bolts and nuts. Tighten the nuts to 30–42 ft. lbs. (41–56 Nm).

9. Connect the fuel hose(s) to the fuel gauge sending unit. Connect the fuel filler hose to the filler neck and vent hose to the vent neck at the tank. Tighten the hose clamps to 25–35 inch lbs. (3–4 Nm).

10. Connect the fuel gauge sending unit wires and lower the vehicle.

11. Fill the tank and check all connections for leaks. Connect the negative battery cable.

AFT AXLE FRAME-MOUNTED TANK—38 GALLON

1. Relieve the fuel system pressure and disconnect the negative battery cable.

2. Drain the fuel into a suitable container by siphoning through the fuel line at the reservoir connection.

3. Raise and safely support the vehicle.

4. Disconnect the fuel filler hose and vent hose at the fuel filler pipe assembly.

5. Disconnect the fuel supply and return lines at the front fuel tank support; not at the fuel sender. Disconnect the vapor hose at the vapor tube.

6. Support the fuel tank in position and loosen the strap attachment nut at the rear that attaches the tank strap to the upper support. Disconnect the bolt that attaches the tank strap to the front crossmember and pivot the strap to the rear of the vehicle.

7. Lower the tank enough to gain access to the fuel gauge. Disconnect the fuel gauge sending unit wires at the fuel tank. If required, remove the front and rear attaching bolt and front heat shield.

8. Lower the fuel tank and remove it from under the vehicle.

To install:

9. Raise the fuel tank high enough to connect the vapor hose to the vapor tube and the fuel gauge sending unit wire.

10. Support the tank in position and tighten the strap attachment bolts at the front to 75–105 ft. lbs. (102–142 Nm). Tighten the rear nuts to 20–25 ft. lbs. (27–34 Nm). If required, attach the front and rear mounting bolts and reinstall the front heat shield.

11. Attach the fuel filler hose and vent hose at the pipe assembly and tighten the clamps to 20–25 ft. lbs. (34–47 Nm). Lower the vehicle.

12. Fill the tank and check all connections for leaks. Connect the negative battery cable.

Fuel Filter

REMOVAL & INSTALLATION

NOTE: If the fuel filter is being

serviced with the rear of the vehicle higher than the front, or if the tank is pressurized, fuel leakage or siphoning from the tank fuel lines could occur. To prevent this, keep the vehicle front end at or above the level of the rear of the vehicle. Also, relieve the tank pressure by loosening the fuel fill cap. The cap should be retightened after the pressure is relieved. If the vehicle is warm, change the fuel filter before the pressure rebuilds.

1. Relieve the fuel system pressure and disconnect the negative battery cable.

2. Raise and safely support the vehicle.

3. Loosen the screw clamp so the filter slides rearward easily.

4. Remove the push connect fittings at both ends of the filter. Install new retainer clips in each push connect fitting.

5. On E Series, remove the fuel filter from the metal bracket by rotating the fuel line and sliding the filter rearward. Be careful not to kink the fuel line.

6. On Bronco and F Series, gently pull the filter from the bracket, being careful not to kink the fuel line.

NOTE: Note the direction of the flow arrow prior to removal so the replacement filter may be installed in the same position.

To install:

7. On E Series, place the filter into the bracket with the flow arrow pointing toward the tab of the bracket. Slide the filter forward until it rests against the tab of the bracket.

8. On Bronco and F Series, position the filter and push to snap into the bracket.

9. On E Series, tighten the retaining screw of the screw clamp to 1.3–2.1 ft. lbs. (1.8–2.8 Nm).

10. Install the push connect fittings onto the filter ends.

11. Connect the negative battery cable. Turn the ignition switch from **OFF** to **RUN** position several times, without starting the engine. Check for fuel leaks.

12. Lower the vehicle.

Electric Fuel Pump

PRESSURE TESTING

High Pressure Pump

1. Make sure there is an adequate fuel supply.

2. Relieve the fuel system pressure.

3. Turn the ignition key **OFF**.

4. Connect a suitable fuel pressure gauge to the schrader valve on the fuel rail.

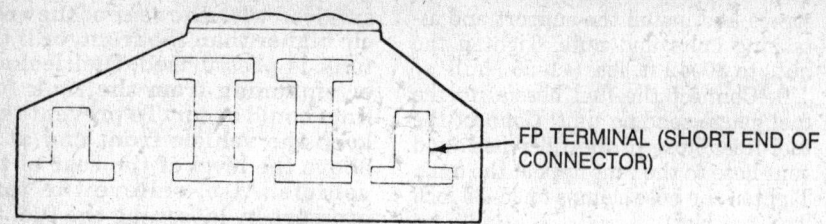

VIP test connector

FP TERMINAL (SHORT END OF CONNECTOR)

5. Install a test lead to the **FP** terminal on the VIP test connector.

6. Turn the ignition key to the **RUN** position, then ground the test lead to run the fuel pump.

7. Observe the fuel pressure reading on the pressure gauge. The fuel pressure should be 35–45 psi on all except the 4.9L engine. On the 4.9L engine, the fuel pressure should be 50–60 psi.

8. Relieve the fuel system pressure and turn the ignition key **OFF**. Remove the fuel pressure gauge and the test lead.

Low Pressure Pump

It has been found that if the low pressure pump is running, it is almost certain that it is operating correctly. Flow testing of the pump is only necessary if there is good reason to suspect pump damage that would cause the pump to run but not pump.

1. Attach a test lead to the **FP** terminal on the VIP test connector. Make sure the lead is long enough to reach the work area under the vehicle.

2. Turn the ignition key to the **RUN** position.

3. Raise and safely support the vehicle. Bring the test lead to a convenient point for grounding.

4. Ground the test lead and listen at the fuel tank for low pressure pump operation. Use a stethescope or other device to help hear the pump. The electrical connection to the high pressure pump may be disconnected to aid in hearing the low pressure pump run.

5. If the pump does not run, check the inertia switch, test connections, fuel pump relay and any other wiring problems that could prevent the pumps from running. Both pumps run from the same electrical circuit. If the high pressure pump runs, check the connections at the top of the fuel tank to pump for continuity and for voltage when the circuit is energized.

6. If the pump runs, remove the ground from the test lead and disconnect the pressure line from the pump at the reservoir inlet fitting.

7. Place the fuel line removed from the reservoir into a calibrated container of at least 1 qt. capacity. Ground the test lead and run the pump for 5 sec-

onds. The fuel level should be at least 6 oz. (180ml).

8. If the fuel level is not at least 6 oz. (180ml), momentarily restrict the line to provide back pressure to prime the pump. If there is still no flow, check for blocked lines and replace the pump if no problem is found.

REMOVAL & INSTALLATION

All E Series and 1988–89 Bronco and F Series are equipped with a low pressure fuel pump located inside the fuel tank and a high pressure fuel pump mounted on the frame. Beginning in 1990, Bronco and F Series are only equipped with a high pressure fuel pump located in the fuel tank.

Tank Mounted Pump

1. Relieve the fuel system pressure and disconnect the negative battery cable.

2. Raise and safely support the vehicle.

3. Remove the fuel tank.

4. Remove any dirt from the area around the sending unit/fuel pump assembly so it will not enter the tank.

5. Turn the locking ring counterclockwise and remove the locking ring, sending unit/fuel pump assembly and sealing gasket.

To install:

6. Clean the mounting surface on the fuel tank.

7. Place a new gasket in the groove of the fuel tank. Install the sending unit/fuel pump assembly into the tank

so the tabs are positioned into the tank slots. The gasket must remain in place during and after sending unit/fuel pump assembly installation.

8. Holding the sending unit/fuel pump assembly in place, install and rotate the locking ring clockwise. Metal locking rings should be rotated until the stop is against the retainer ring tab. Plastic locking rings should be tightened to 13–20 ft. lbs. (18–27 Nm) on 1988–89 vehicles or 40–55 ft. lbs. (54–75 Nm) on 1990–92 vehicles.

9. Install the fuel tank and lower the vehicle.

10. Fill the tank and check for leaks. Connect the negative battery cable.

Frame Mounted Pump

1. Relieve the fuel system pressure and disconnect the negative battery cable.

2. Raise and safely support the vehicle.

3. Disconnect and plug the inlet and outlet fuel lines. Disconnect the electrical connector at the pump.

4. Remove the pump from the mounting bracket.

6. Installation is the reverse of the removal procedure.

Fuel Injection

IDLE SPEED ADJUSTMENT

1988

1. Place the transmission in **P** or **N** and apply the parking brake.

2. Start the engine and bring to normal operating temperature. Make sure the air conditioner/heater and all accessories are **OFF**.

3. Check the ignition timing and adjust, if necessary.

4. Connect a tachometer according to the manufacturers instructions.

5. On all except 7.5L engine, check the curb idle speed. If it is not to the specification listed on the emission calibration decal, shut the engine **OFF**.

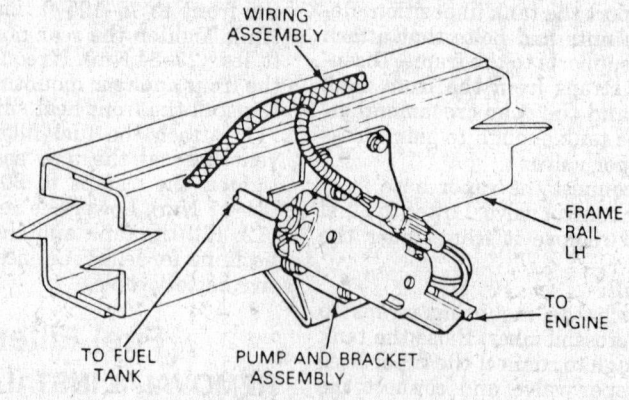

WIRING ASSEMBLY

FRAME RAIL LH

TO ENGINE

TO FUEL TANK

PUMP AND BRACKET ASSEMBLY

Frame mounted fuel pump

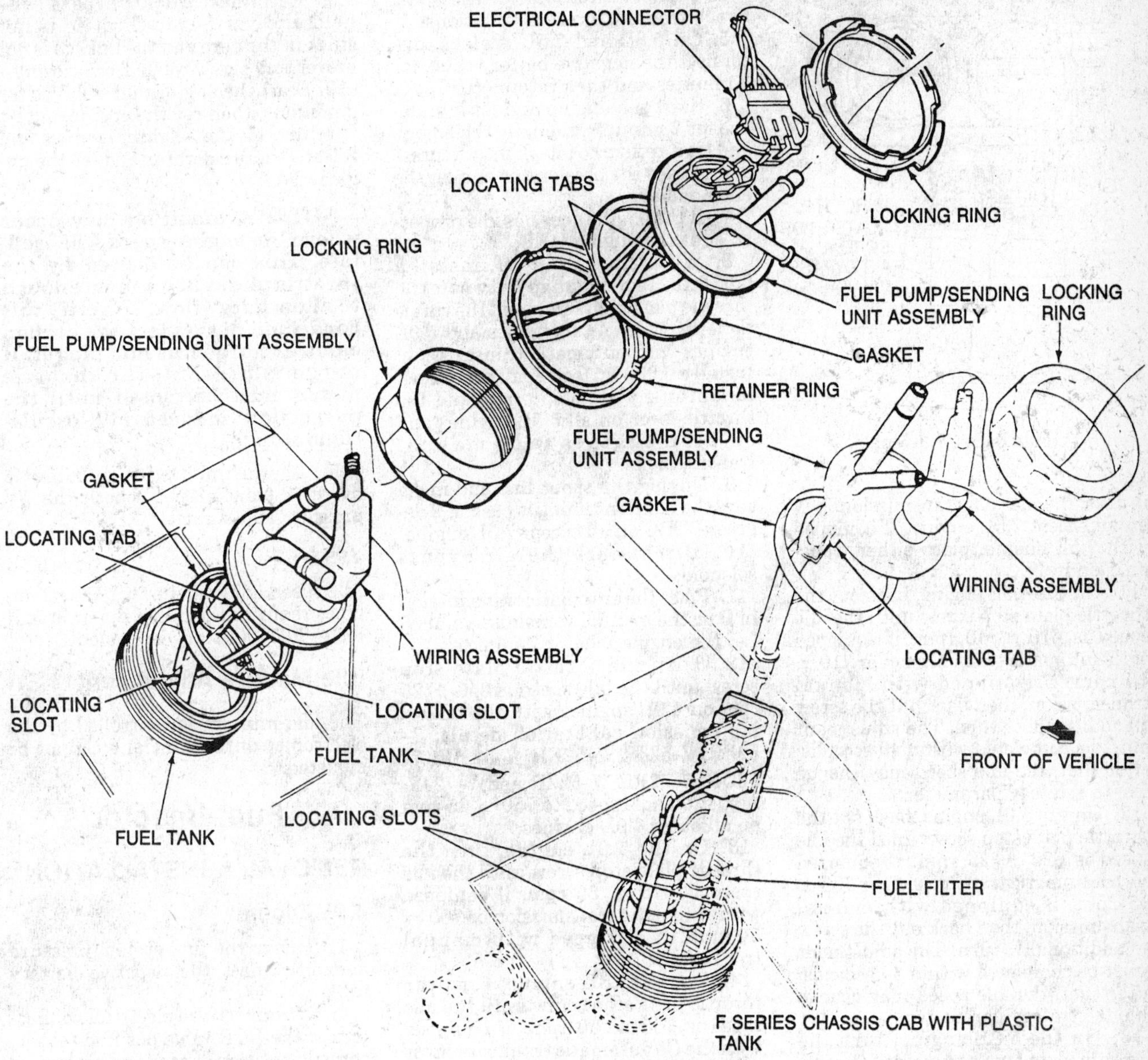

ELECTRICAL CONNECTOR

LOCATING TABS

LOCKING RING

LOCKING RING

FUEL PUMP/SENDING UNIT ASSEMBLY

FUEL PUMP/SENDING UNIT ASSEMBLY

GASKET

LOCKING RING

GASKET

LOCATING TAB

RETAINER RING

FUEL PUMP/SENDING UNIT ASSEMBLY

GASKET

WIRING ASSEMBLY

LOCATING SLOT

WIRING ASSEMBLY

LOCATING TAB

LOCATING SLOT

FUEL TANK

FRONT OF VEHICLE

FUEL TANK

LOCATING SLOTS

FUEL FILTER

F SERIES CHASSIS CAB WITH PLASTIC TANK

Tank mounted fuel pump installation

Disconnect the negative battery terminal for 3 minutes and then reconnect it. Start the engine and let it idle for 5 minutes with the automatic transmission in **D**.

6. Check the curb idle speed. If it is not to specification, shut the engine **OFF**.

7. If equipped with a 7.5L engine, disconnect the air bypass solenoid.

8. Place the transmission in **N** or **P**. Run the 4.9L and 5.0L engines at 1800 rpm for 30 seconds, the 5.8L engine at 1800 rpm for 60 seconds or the 7.5L engine at 2500 rpm for 30 seconds.

9. If equipped with 5.8L engine, let the engine idle. If it stalls, turn the throttle stop screw 1 turn clockwise and repeat Steps 8 and 9.

10. Place the automatic transmission

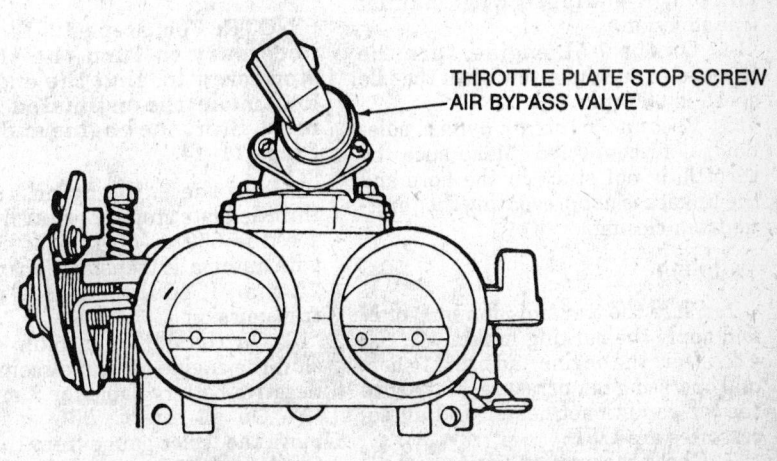

THROTTLE PLATE STOP SCREW
AIR BYPASS VALVE

Throttle body—4.9L, 5.0L and 5.8L engines

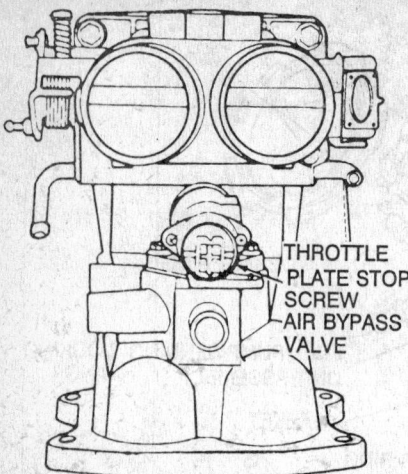

THROTTLE
PLATE STOP
SCREW
AIR BYPASS
VALVE

Throttle body—7.5L engine

in **D** or the manual transmission in **N** on all except 7.5L engine. If equipped with 7.5L engine, place either transmission in **N**.

11. On the 4.9L engine, back out the throttle plate stop screw until the idle speed is 610 ± 40 rpm, if equipped with automatic transmission or 710 ± 40 rpm, if equipped with manual transmission, then back out the screw an additional ½ turn. The adjustment must be completed within 40 seconds; after that, the idle speed may change due to strategy parameter.

12. On the 5.0L engine, back out the throttle plate stop screw until the idle speed is 625 ± 25 rpm, if equipped with automatic transmission or 700 ± 25 rpm, if equipped with manual transmission, then back out the screw an additional ½ turn. The adjustment must be completed within 40 seconds; after that, the idle speed may change due to strategy parameter.

13. On the 5.8L engine, turn the throttle plate stop screw until the idle speed is 625 ± 25 rpm, if equipped with automatic transmission or 650 ± 25 rpm, if equipped with manual transmission.

14. On the 7.5L engine, turn the throttle plate stop screw until the idle speed is 650 rpm.

15. Reconnect the air bypass solenoid, if disconnected. Make sure the throttle is not stuck in the bore and the linkage is not preventing the throttle from closing.

1989–90

1. Place the transmission in **P** or **N** and apply the parking brake.

2. Start the engine and bring to normal operating temperature. Make sure the air conditioner/heater and all accessories are **OFF**.

3. Check the ignition timing and adjust, if necessary.

4. Connect a tachometer according to the manufacturers instructions.

5. On 5.8L and 7.5L engines, disconnect the negative battery cable for 3 minutes and then reconnect it.

6. Start the engine and let it stabilize for 2 minutes, then rev the engine and let it return to idle, lightly depress and release the accelerator and let the engine idle.

7. If the engine does not idle properly, shut the engine **OFF**.

8. With the engine off, install a 0.050 in. feeler gauge between the throttle plate stop screw and the throttle lever on all 4.9L engines and 5.0L engines with automatic transmission. Install a 0.030 in. feeler gauge between the throttle plate stop screw and the throttle lever on all 5.8L engines and on 5.0L engines with manual transmission.

9. Unplug the spout line and make sure the ignition timing is base ± 2 degrees BTDC on all except 7.5L engine.

10. Disconnect the air bypass solenoid.

11. Place the automatic transmission in **P** or the manual transmission in **N**. Let the engine idle for 2 minutes.

12. Turn the throttle plate stop screw until the idle speed is 650 ± 25 rpm on 4.9L engines with the following emission calibration decals: 7-52ER, 7-52JR, 7-52KR, 7-52MR, 7-52QR, 7-52RR, 7-52ZR and 7-72JR. Adjust the idle speed to 750 ± 25 rpm on all other 4.9L engines.

13. On the 5.0L engine, turn the throttle plate stop screw until the idle speed is 675 ± 50 rpm, if equipped with automatic transmission or 700 ± 50 rpm, if equipped with manual transmission.

14. On the 5.8L engine, turn the throttle plate stop screw until the idle speed is 780 ± 50 rpm, if equipped with the C6 automatic tranmsission or 730 ± 50 rpm, if equipped with electronic overdrive or manual transmissions.

NOTE: For Steps 12–14, if it is necessary to turn the throttle stop screw in, shut the engine off and make the estimated adjustment. Start the engine and repeat Steps 11–14.

15. On the 7.5L engine, turn the throttle plate stop screw until the idle speed is 650 ± 50 rpm, if equipped with automatic transmission or 650 ± 25 rpm, if equipped with manual transmission.

16. On the 4.9L and 5.0L engines, shut the engine off and disconnect the negative battery cable for 3 minutes.

17. On all except 7.5L engine, remove the feeler gauge from the throttle plate stop screw and throttle lever pad. Reconnect the spout line.

18. Reconnect the air bypass solenoid and verify the throttle is not stuck in the bore and the linkage is not preventing the throttle from closing.

19. Start the engine and stabilize for 2 minutes, then rev the engine and let it return to idle. Lightly depress and release the accelerator and let the engine idle.

NOTE: A condition may occur where the engine speed will oscillate. This can be caused by the throttle plates being open enough to allow purge flow. To verify this condition, disconnect the carbon canister purge line and plug it. If purge is present, the throttle plates must be closed until the purge flow induced idle oscillations stop.

20. If equipped with automatic overdive transmission, check the TV pressure adjustment.

1991–92

Idle speed adjustment on 1991–92 engines requires the use of Super Star II tester 007–00028 or equivalent.

Idle Mixture Adjustment

The idle mixture is controlled by the electronic control unit and cannot be adjusted.

Fuel Injector

REMOVAL & INSTALLATION

4.9L Engine

1. Relieve the fuel system pressure and disconnect the negative battery cable.

2. Disconnect the electrical connectors at the EGR valve position sensor, throttle position sensor and air bypass valve.

3. Disconnect the vacuum lines at the EGR valve, the evaporative lines to the throttle body and electronic purge, if equipped and the vacuum lines at the upper intake manifold vacuum tree. Tag the lines to aid in reinstallation.

4. Disconnect the PCV hose from the fitting located on the underside of the upper intake manifold.

5. Remove the throttle linkage shield and disconnect the throttle linkage and cruise control cables. Unbolt the accelerator cable from the bracket and position the cable away from the engine.

6. Disconnect the air inlet hoses from the throttle body.

7. Disconnect the EGR tube from the EGR valve and the rear exhaust manifold. Remove the tube from the engine.

8. Remove the thermactor tube assembly from the lower intake manifold by removing the 2 nuts retaining the tube. Remove the nut attaching the thermactor bypass valve bracket to the lower intake manifold.

9. On E Series, remove the nut attaching the transmission fill tube, then remove the tube bracket off the intake manifold stud.

10. Remove the studs and screws that retain the upper intake manifold.

11. Remove the screw and washer assembly attaching the upper intake manifold support bracket to the upper intake manifold.

12. Remove the upper intake manifold and throttle body assembly from the lower intake manifold.

13. Disconnect the electrical connectors from the fuel injectors.

14. Disconnect the fuel lines from the fuel supply manifold.

15. Remove the injector cooling manifold, if equipped. Disconnect the electrical connector from the fuel manifold temperature switch, if equipped.

16. Disconnect the vacuum line to the pressure regulator.

17. Remove the strap surrounding the fuel manifold, injector electrical harness and the main vacuum harness.

18. Remove the 3 fuel supply manifold retaining studs.

19. Carefully disengage the fuel supply manifold from the fuel injectors and remove the manifold.

20. Grasping the injector body, pull up while gently rocking the injector from side-to side. Remove and discard the injector O-rings.
To install:

21. Lubricate new O-rings with clean light grade oil and install 2 on each injector.

NOTE: Never use silicone grease as it will clog the injectors.

22. Install the injectors, using a light, twisting, pushing motion.

23. Install the fuel supply manifold, pushing down to make sure all the fuel injector O-rings are fully seated in the fuel supply manifold cups and intake manifold. Install the 3 retaining bolts or studs.

24. Install the injector cooling manifold, if equipped. Connect the electrical connector to the fuel manifold temperature switch.

25. Connect the fuel lines to the fuel supply manifold.

26. Before connecting the electrical connectors to the fuel injectors, connect the negative battery cable and turn the ignition switch **ON**. This will cause the fuel pump to run for 2–3 seconds and pressurize the system.

27. Check for fuel leaks where the fuel injector is installed into the fuel supply manifold.

28. Turn the ignition switch **OFF** and disconnect the negative battery cable.

29. Attach the vacuum line to the fuel pressure regulator.

30. Secure the main vacuum harness and the injector electrical harness to the fuel supply manifold with a strap positioned between the No. 5 and No. 6 intake manifold runners.

31. Connect the electrical connector to the electronic purge valve, if equipped.

32. Position a new upper intake gasket on the lower manifold, using the lower manifold dowels to position the gasket.

33. Position the upper intake manifold onto the lower intake manifold, using the dowels of the lower intake to locate the manifold holes. Install the studs and screws and hand-tighten.

34. Tighten the studs to 12–18 ft. lbs. (16.2–24.4 Nm).

35. Position the upper intake manifold support onto the boss of the upper intake, located under the throttle body. Install the retaining screw and tighten to 22–32 ft. lbs. (29.8–43.4 Nm).

36. Install the EGR tube between the EGR valve and the rear exhaust manifold. The tube is routed between the lower intake runners No. 4 and No. 5. Tighten both fittings to 25–35 ft. lbs. (33.8–47.5 Nm).

37. Connect the PCV hose to the fitting, located on the underside of the upper intake manifold.

38. Position the thermactor tube assembly onto the studs of the lower intake manifold and tighten the attachment nuts to 8–12 ft. lbs. (10.8–16.3 Nm).

39. On E Series, position the transmission fill tube onto the stud of the lower intake manifold. Tighten the attachment nut to 8–12 ft. lbs. (10.8–16.3 Nm).

40. Install the accelerator cable and throttle linkage shield onto the accelerator bracket of the throttle body.

41. Connect the air inlet hoses to the throttle body.

42. Connect the vacuum line to the EGR valve, the evaporative lines to the throttle body and the vacuum lines to the vacuum tree.

43. Connect the electrical connectors to the throttle position sensor, air bypass valve, EGR valve position sensor and, if equipped, the electronic purge valve.

44. Connect the negative battery cable, start the engine and let it idle for 2 minutes.

45. Turn the engine **OFF** and check for fuel leaks.

5.0L and 5.8L Engines

1. Relieve the fuel system pressure and disconnect the negative battery cable.

2. Disconnect the electrical connectors at the air bypass valve, throttle position sensor and EGR position sensor.

3. Disconnect the throttle linkage at the throttle ball. Disconnect the automatic overdrive transmission linkage from the throttle body, if equipped. Remove the 2 bolts securing the throttle linkage bracket to the intake and position the bracket with the cables aside.

4. Disconnect the vacuum lines at the vacuum tree, EGR valve and fuel pressure regulator.

5. Disconnect the PCV hose from the fitting on the rear of the upper manifold.

6. Disconnect the canister purge line(s) from the fitting(s) on the throttle body.

7. Partially drain the cooling system and disconnect the water heater lines from the throttle body.

8. Disconnect the EGR tube from the EGR valve by removing the flange nut.

9. Remove the bolt from the upper intake support bracket to the upper manifold.

10. Remove the 6 upper intake manifold retaining bolts and remove the upper intake and throttle body assembly.

11. Disconnect the electrical connectors from the fuel injectors.

12. Disconnect the fuel lines from the fuel supply manifold.

13. Remove the 4 fuel supply manifold retaining bolts, carefully disengage the manifold from the fuel injectors and remove the manifold.

14. Grasping the injector body, pull up while gently rocking the injector from side-to-side. Remove and discard the injector O-rings.
To install:

15. Lubricate new O-rings with clean light grade oil and install 2 on each injector.

NOTE: Never use silicone grease at it will clog the injectors.

16. Install the injectors, using a light, twisting, pushing motion.

17. Install the fuel supply manifold, pushing down to make sure all the fuel injector O-rings are fully seated in the fuel supply manifold cups and intake manifold.

18. Install the 4 fuel supply manifold retaining bolts and tighten to 15–22 ft. lbs. (20–30 Nm).

19. Connect the fuel lines to the fuel supply manifold.

20. Before connecting the electrical connectors to the fuel injectors, con-

nect the negative battery cable and turn the ignition switch **ON**. This will cause the fuel pump to run for 2–3 seconds and pressurize the system.

21. Check for fuel leaks where the fuel injector is installed into the fuel supply manifold.

22. Turn the ignition switch **OFF** and disconnect the negative battery cable.

23. Clean and inspect the mounting faces of the upper and lower intake manifolds.

24. Position a new gasket on the lower intake mounting face. The use of alignment studs may be helpful.

25. Install the upper intake manifold and throttle body assembly to the lower manifold making sure the gasket remains in place.

26. Install the 6 upper intake manifold retaining bolts and tighten to 12–18 ft. lbs. (16–24 Nm).

27. Install the upper intake support bracket to the upper manifold attaching bolt.

28. Install the EGR tube.

29. Connect the canister purge lines to the fittings on the throttle body.

30. Connect the water heater lines to the throttle body.

31. Connect the PCV hose to the rear of the upper manifold.

32. Connect the vacuum lines to the vacuum tree, EGR valve and fuel pressure regulator.

33. Position the throttle linkage bracket with the cables to the upper intake manifold. Install the 2 retaining bolts and tighten to 8–10 ft. lbs. (11–13 Nm). Connect the throttle cable and transmission cable to the throttle body.

34. Connect the electrical connectors to the air bypass valve, throttle position sensor and EGR position sensor.

35. Fill and bleed the cooling system.

36. Connect the negative battery cable, start the engine and let it idle for 2 minutes.

37. Turn the engine **OFF** and check for fuel and coolant leaks.

7.5L Engine

1. Relieve the fuel system pressure and disconnect the negative battery cable.

2. Disconnect the throttle linkage at the throttle ball and the automatic transmission linkage, if equipped, from the throttle body. Remove the bolts securing the bracket to the intake and position the bracket with the cable out of the way.

3. Remove the 2 canister purge lines from the fittings on the throttle body, if equipped.

4. Disconnect the electrical connectors at the air bypass valve, throttle position sensor and EGR position sensor.

5. Disconnect the vacuum lines to the MAP sensor and EGR valve.

6. Disconnect the EGR valve flange nut and the PCV hose at the rear of the upper manifold.

7. Disconnect and plug the water heater lines at the throttle body, if equipped.

8. Disconnect the bypass valve clean air supply hose, connected to the bypass port, if equipped.

9. Remove the 4 upper intake manifold retaining bolts and remove the upper intake manifold and throttle body assembly.

10. Disconnect the vacuum hose at the fuel pressure regulator.

11. Disconnect the electrical connectors at the injectors.

12. Disconnect the fuel lines from the fuel supply manifold.

13. Remove the 4 fuel supply manifold retaining bolts, carefully disengage the manifold from the fuel injectors and remove the manifold.

14. Grasping the injector body, pull up while gently rocking the injector from side-to-side. Remove and discard the injector O-rings.

15. Inspect the injector plastic pintle protection cap and washer for signs of deterioration. Replace the complete injector as required. If the plastic pintle protection cap is missing, look for it in the intake manifold.

NOTE: The plastic pintle protection cap is not available as a separate part.

To install:

16. Lubricate new O-rings with clean light grade oil and install 2 on each injector.

NOTE: Never use silicone grease at it will clog the injectors.

17. Install the injectors, using a light, twisting, pushing motion.

18. Install the fuel supply manifold, pushing down to make sure all the fuel injector O-rings are fully seated in the fuel supply manifold cups and intake manifold.

19. Install the 4 fuel supply manifold retaining bolts and tighten to 15–22 ft. lbs. (20–30 Nm).

20. Connect the fuel lines to the fuel supply manifold.

21. Before connecting the electrical connectors to the fuel injectors, connect the negative battery cable and turn the ignition switch **ON**. This will cause the fuel pump to run for 2–3 seconds and pressurize the system.

22. Check for fuel leaks where the fuel injector is installed into the fuel supply manifold.

23. Turn the ignition switch **OFF** and disconnect the negative battery cable.

24. Connect the electrical connectors to the fuel injectors.

25. Connect the vacuum hose to the fuel pressure regulator.

26. Clean and inspect the mounting faces of the upper and lower intake manifolds.

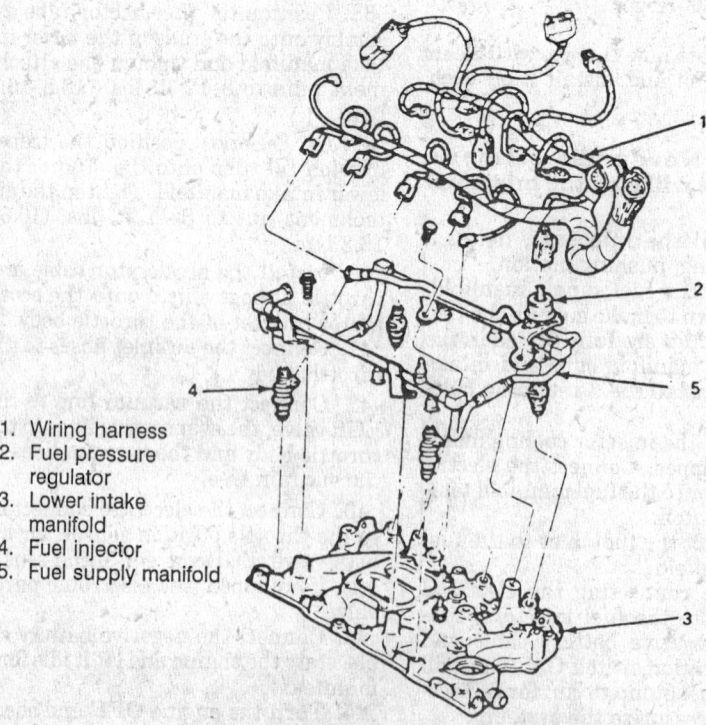

1. Wiring harness
2. Fuel pressure regulator
3. Lower intake manifold
4. Fuel injector
5. Fuel supply manifold

Fuel Injector removal–7.5L engine

27. Position a new gasket on the lower intake mounting face. The use of alignment studs may be helpful.

28. Install the upper intake manifold and throttle body assembly to the lower manifold making sure the gasket remains in place.

29. Install the 4 upper intake manifold retaining bolts and tighten to 12–18 ft. lbs. (16–24 Nm).

30. Install the EGR tube and connect the PCV hose to the rear of the upper manifold.

31. Connect the water heater lines to the throttle body, if equipped.

32. Connect the vacuum lines to the MAP sensor and EGR valve.

33. Position the throttle linkage bracket with the cable to the upper intake manifold. Install the 2 retaining bolts and tighten to 8–10 ft. lbs. (11–13 Nm). Connect the throttle cable and automatic transmission kickdown cable to the throttle body.

34. Connect the electrical connectors at the air bypass valve, throttle position sensor and EGR position sensor.

35. Connect the bypass valve clean air supply hose, if equipped.

36. Connect the negative battery cable, start the engine and let it idle for 2 minutes.

37. Turn the engine **OFF** and check for fuel leaks.

DIESEL FUEL SYSTEM

Fuel System Service Precautions

Safety is the most important factor when performing not only fuel system maintenance but any type of maintenance. Failure to conduct maintenance and repairs in a safe manner may result in serious personal injury or death. Maintenance and testing of the vehicle's fuel system components can be accomplished safely and effectively by adhering to the following rules and guidelines.

• To avoid the possibility of fire and personal injury, always disconnect the negative battery cable unless the repair or test procedure requires that battery voltage be applied.

• Always relieve the fuel system pressure prior to disconnecting any fuel system component (injector, fuel rail, pressure regulator, etc.), fitting or fuel line connection. Exercise extreme caution whenever relieving fuel system pressure to avoid exposing skin, face and eyes to fuel spray. Please be advised that fuel under pressure may penetrate the skin or any part of the body that it contacts.

• Always place a shop towel or cloth around the fitting or connection prior to loosening to absorb any excess fuel due to spillage. Ensure that all fuel spillage (should it occur) is quickly removed from engine surfaces. Ensure that all fuel soaked cloths or towels are deposited into a suitable waste container.

• Always keep a dry chemical (Class B) fire extinguisher near the work area.

• Do not allow fuel spray or fuel vapors to come into contact with a spark or open flame.

• Always use a backup wrench when loosening and tightening fuel line connection fittings. This will prevent unnecessary stress and torsion to fuel line piping. Always follow the proper torque specifications.

• Always replace worn fuel fitting O-rings with new. Do not substitute fuel hose or equivalent where fuel pipe is installed.

Fuel Filter

REPLACEMENT

1. Disconnect the battery ground cables from both batteries.

2. Place a container under the vehicle and drain the fuel from the fuel filter.

3. Remove the water drain tube from the bottom of the filter assembly.

4. Unscrew the water separator drain bowl and remove. Unscrew the filter element and discard.

To install:

5. Clean the gasket surfaces of the fuel filter adapter tp prevent contamination.

6. Lightly coat the filter sealing gaskets with clean diesel fuel.

NOTE: To avoid fuel contamination, do not add fuel directly to the new filter. Allow the engine to draw fuel through the filter.

7. Screw the new filter element

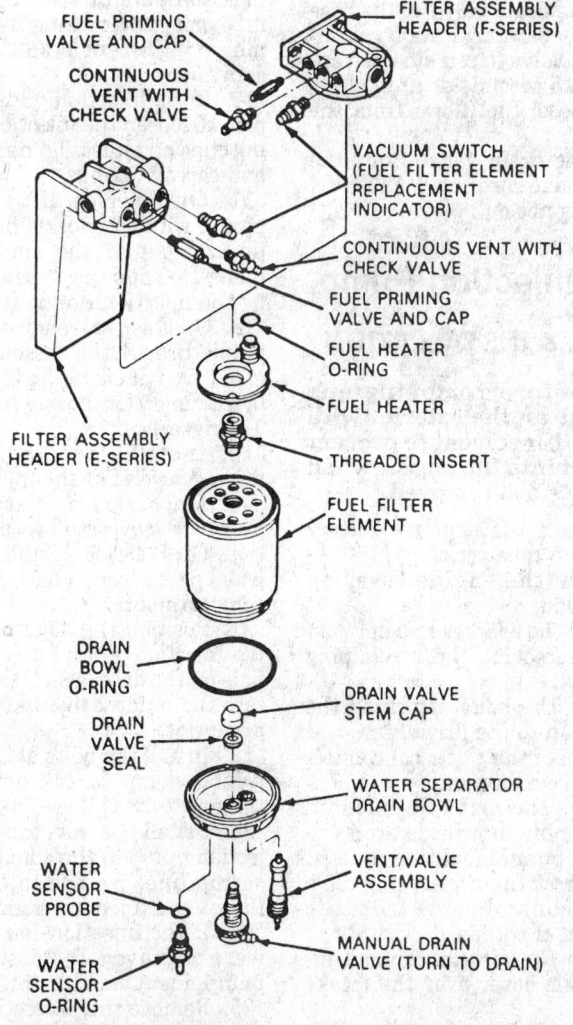

Fuel filter/water separator assembly–7.3L diesel engine

onto the filter base until the seal contacts the flange. Tighten another 180–300 degree turn.

8. Screw on the water separator drain bowl. Tighten another 180–300 degree turn.

9. Install the water drain tube and clean any spilled fuel from the top of the engine.

10. Connect the negative battery cables. Run the engine and check for leaks.

DRAINING WATER FROM THE SYSTEM

NOTE: Drain water from the water separator manual drain valve whenever the warning light comes on or every 5000 miles. The WATER IN FUEL light will glow when approximately 3.5 oz. of water accumulates in the separator.

1. Stop the vehicle and shut OFF the engine.

2. Place a container under the fuel filter/water separator drain tube to collect drain fluid.

3. Open the drain valve at the base of the water separator drain bowl. Allow the drain valve to remain open approximately 15 seconds or until clear, water-free diesel fuel flows from the drain tube.

4. Close the drain valve. Start the engine and make sure the WATER IN FUEL light is not on.

Diesel Injection Pump

REMOVAL & INSTALLATION

NOTE: Before removing any fuel lines, clean the exterior with clean fuel oil or solvent to prevent entry of dirt into the engine when the fuel lines are removed.

1. Disconnect the negative battery cables from both batteries.

2. Remove the engine cover on E250 and E350.

3. Remove the adapter housing cover plate by removing the 2 retaining bolts.

4. Remove the bolts attaching the injection pump to the drive gear.

5. Disconnect the electrical connectors to the injection pump.

6. Remove the fast idle solenoid bracket assembly to provide access to the injection pump mounting nuts.

7. Disconnect the accelerator cable and cruise control cable from the throttle lever, if equipped.

8. Remove the air cleaner and install a suitable cover over the intake opening.

9. Remove the accelerator cable

bracket, with the cables attached from the intake manifold and position aside.

10. On E series, disconnect the fuel inlet and return lines from the fuel filter and cap all lines.

11. On E series, remove the fuel filter bracket attaching bolts and remove the fuel filter and bracket as an assembly.

12. Remove the fuel return hose and clip from the 90 degree elbow at the governor cover. Cap the opening at the governor cover elbow.

13. Remove the fuel filter-to-injection pump fuel line and cap the fittings.

NOTE: It is not necessary to remove the injection lines from the injection pump to remove the injection pump. If the lines are to be removed, loosen the injection line fittings at the injection pump before removing it from engine.

14. Remove the fuel injection lines from the nozzles and cap the lines and nozzles to prevent entry of dirt into the system.

15. Remove the 3 nuts attaching the injection pump to the injection pump drive gear cover using injection pump mounting wrench T86T–9000–C or equivalent.

16. If the injection pump is to be replaced, loosen the injection line retaining clips and injection nozzle fuel lines and cap all fittings.

17. On F series, lift the injection pump, with the nozzle lines attached, up and out of the engine compartment. Do not carry the injection pump by the injection nozzle fuel lines.

18. On E series, remove the injection pump through the passenger compartment. Do not carry the injection pump by the injection nozzle fuel lines.

To install:

19. Install a new O-ring onto the drive gear end of the injection pump.

20. On F series, install injection pump by moving down and into position. On E series, install the injection pump from the passenger compartment.

21. Position the alignment dowel on the injection pump into the alignment hole on the drive gear. If necessary, rotate the pump driveshaft to align the drive slot.

22. Install the bolts attaching the injection pump to the drive gear and tighten to 25 ft. lbs. (34 Nm).

23. Install the nuts attaching the injection pump to the adapter. Align the scribe lines on the injection pump flange and injection pump adapter.

24. If the injection nozzle fuel lines were removed from the injection pump, install at this time.

25. Remove the protective caps from the nozzles and fuel lines. Install the

fuel line nuts onto the nozzles and tighten to 22 ft. lbs. (30 Nm).

26. Connect the fuel inlet line from the filter and fuel return line to the injection pump.

27. Install the injection pump fitting adapter with a new O-ring.

28. Clean the old sealant from the injection pump elbow threads using clean solvent and dry thoroughly. Start the elbow into the injection pump adapter and then apply a light coating of pipe sealant.

29. Tighten the elbow in the injection pump adapter to a minimum of 6 ft. lbs. (8 Nm). Then tighten further, if necessary, to align the elbow with the injection pump fuel inlet line, but do not exceed 360 degrees of rotation or 10 ft. lbs. (13 Nm).

30. Remove the caps and connect the fuel filter-to-injection pump fuel line.

31. On E Series, install the fuel filter and bracket as an assembly. Install the fuel filter inlet and return lines.

32. Install the accelerator cable bracket to the intake manifold. Remove the intake manifold cover and install the air cleaner.

33. Connect the accelerator and speed control cable, if equipped, to the throttle lever.

34. Install the fast idle solenoid bracket assembly. Reconnect the electrical connectors on the injection pump.

35. Clean the adapter housing cover plate sealing surfaces. Apply a ⅛ in. bead of sealant in the adapter housing grooves and install the adapter cover. Apply threadlocking compound to the retaining bolts and tighten to 14 ft. lbs. (19 Nm).

36. Connect the ground cables to both batteries. Start the engine and check for fuel leaks. If necessary, bleed the high pressure fuel lines of air by loosening the connector ½–1 turn and cranking the engine until bubble-free fuel flows from the connection.

─────── **CAUTION** ───────

Keep eyes and hands away from nozzle spray. Fuel spraying from the nozzle under high pressure can penetrate the skin and cause infection. Medical attention should be provided immediately in the event of skin penetration.

───────────────────────

37. Check and adjust the injection pump timing.

IDLE SPEED ADJUSTMENT

Curb Idle Speed

1. Place the transmission in N or P.

2. Start the engine and bring to normal operating temperature.

3. Idle speed is measured with manual tranmission in N or automatic transmission in D.

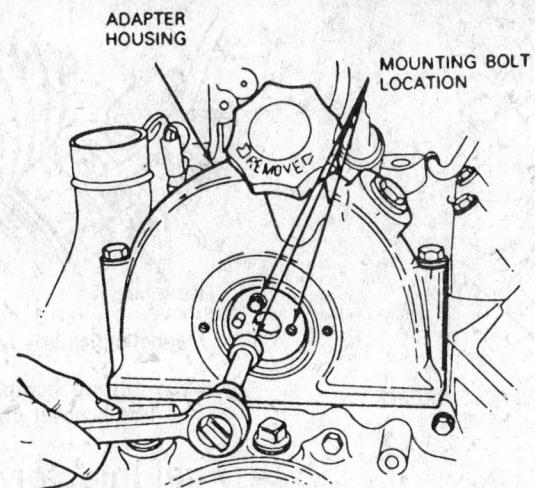

Injection pump drive gear attaching bolt removal—7.3L diesel engine

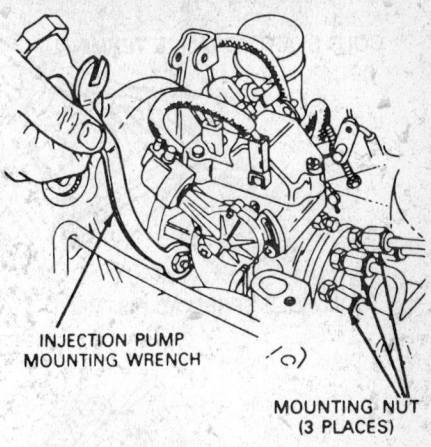

Injection pump removal and installation —7.3L diesel engine

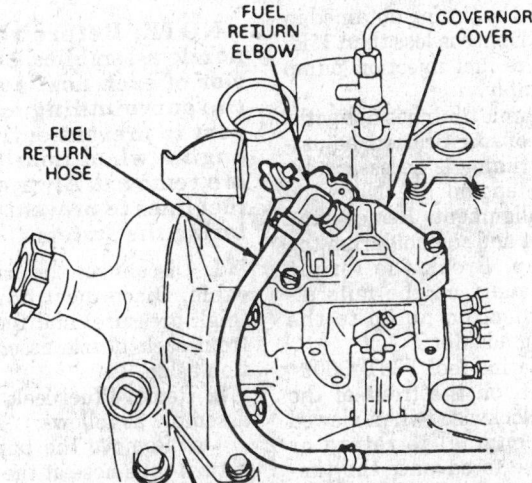

Fuel return line location—7.3L diesel engine

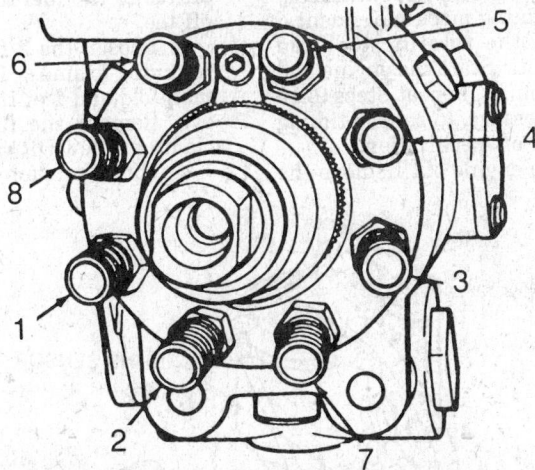

Injection pump cylinder numbering sequence—7.3L diesel engine

4. Make sure the curb idle adjusting screw is against the stop. If not, correct the vehicle linkage.

5. Connect tachometer tool 055–00108 or equivalent, according to the manufacturers instructions.

6. Set the curb idle speed to the specification listed on the vehicle emission control information decal, using the idle speed adjusting screw.

7. Place the transmission in **N** or **P**. Rev the engine momentarily. Place the

transmission in the specified gear and check the curb idle speed. Adjust again, if necessary.

Fast Idle Speed

1. Place the transmission in **N** or **P**.
2. Start the engine and bring to normal operating temperature.
3. Disconnect the fast idle solenoid from the wiring harness.
4. Apply battery voltage to the solenoid to activate it.
5. Rev the engine momentarily to set the solenoid to activate it.
6. Connect tachometer tool 055–00108 or equivalent, according to the manufacturers instructions.
7. Check the fast idle speed setting. The fast idle speed should be 825 ± 25 rpm. Adjust to specification by turning the solenoid plunger in or out.
8. Rev the engine momentarily and check the fast idle speed. Adjust as necessary.
9. Disconnect the battery voltage from the solenoid connector and connect to the wiring harness.

Diesel Injection Timing

ADJUSTMENT

Static Timing

1. Remove the fast idle bracket and solenoid from the injection pump.
2. Break the torque, keeping the nuts snug, on 3 nuts attaching the injection pump to pump mounting adapter using tool T83T–9000–B or equivalent.
3. Install rotating tool T83T–9000–C or equivalent, on the front of the pump and rotate the injection pump to align the timing mark on the injection pump mounting flange with the timing mark on the pump mounting adapter, to with ± 0.030 in.
4. Remove the rotating tool and tighten the nuts.

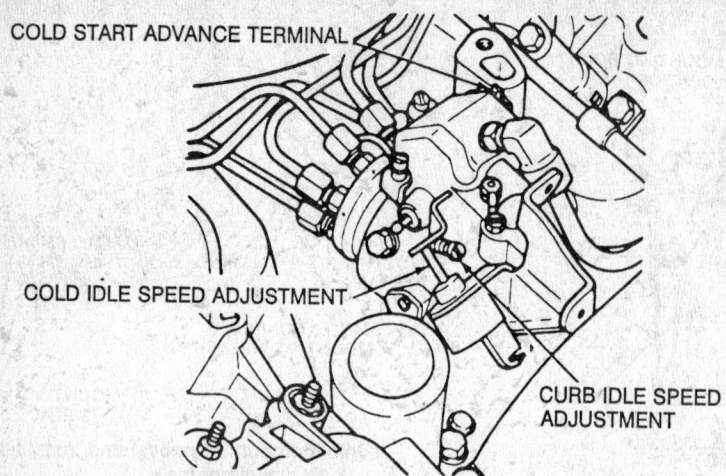

Idle speed adjusting screw location — 7.3L diesel engine

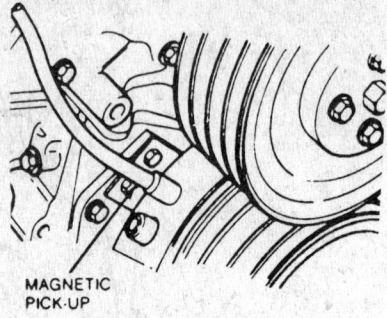

Magnetic pickup — dynamic timing

5. Visually check the timing to verify the timing marks are aligned. Install the fast idle bracket and solenoid.

Dynamic Timing

1. Bring the engine up to normal operating temperature. When checking or setting dynamic injection timing it is mandatory that the engine be stabilized at a normal operating temperature of 192–212°F (89–100°C). This temperature is needed to ensure proper fuel ignition in the precombustion chambers.

2. Stop the engine and install dynamic timing meter tool 078–00200 or equivalent, by placing the magnetic pickup in the timing pointer probe hole. Insert the pickup until it almost touches the vibration damper.

3. Attach the clamp from timing meter adapter tool 078–00201 or equivalent, to the line pressure sensor on the No. 1 injector nozzle (F series) or No. 4 injector nozzle (E series) and connect to the timing meter.

4. Connect the dynamic timing meter to the battery and dial in minus 20 degrees offset on the meter. Disconnect the cold start advance solenoid connector from the solenoid terminal.

5. With the transmission in N and the rear wheels raised off the ground, start the engine. Using throttle control tool D83T-9000-E or equivalent, set the engine speed to 2000 rpm with no accessory load. Observe the injection timing on the dynamic timing meter. The injection timing should be 8.5 degrees BTDC at 2000 rpm.

6. Apply battery voltage to the cold start advance solenoid terminal to activate it. Activating the cold start advance solenoid can result in engine speed increase. Adjust the throttle control to attain 2000 rpm, if necessary.

7. Check the timing at 2000 rpm. The timing should be advanced at least

1 degree before the timing obtained in Step 5. If the advance is less than 1 degree, replace the fuel injection pump top cover assembly.

8. If the dynamic timing is not within ± 2 degrees of specification, adjustment of pump timing is necessary.

9. Turn the engine off. Note the timing mark alignment. Remove the fast idle bracket and solenoid from the injection pump. Break the torque (keeping nuts snug) on the nuts attaching the injection pump to the pump mounting adapter.

10. Install rotating tool T83T-9000-C or equivalent, on the front of the pump. Rotate clockwise (when viewed from front of engine) to retard or counterclockwise to advance the timing, by lightly tapping the tool with a rubber mallet. A 2 degree movement of dynamic timing is approximately 0.030 in. of timing mark movement.

11. Remove the rotating tool and tighten the nuts. Start the engine and recheck the timing. Repeat Steps 9, 10 and 11 as necessary, to set the timing to ± 1 degree of specification.

12. Turn the engine off. Remove the

dynamic timing components. Install the fast idle bracket and solenoid.

Fuel Injector Nozzles

REMOVAL & INSTALLATION

NOTE: Before removing the nozzle assemblies, clean the exterior of each nozzle assembly and the surrounding area with solvent to prevent entry of dirt into engine when nozzle assemblies are removed. Always cap all open fuel lines to prevent dirt from entering the system.

1. Disconnect the negative battery cable. Disconnect nozzle fuel inlet (high pressure) and fuel leak off tees from each nozzle assembly and position aside.

2. Remove fuel leak off lines as an assembly as follows:

   a. Remove the pump to fuel return tube hose at the fuel return elbow. Cap the elbow at the pump. Disconnect the hose from the leak-off tee to the fuel filter at the leak-off tee.

   b. Loosen the 2 fuel return tube retaining clamps, 1 at the intake manifold and 1 at the engine lifting eye. Remove the fuel return hose clamp at the CDR valve bracket.

   c. With the clamps removed, re-

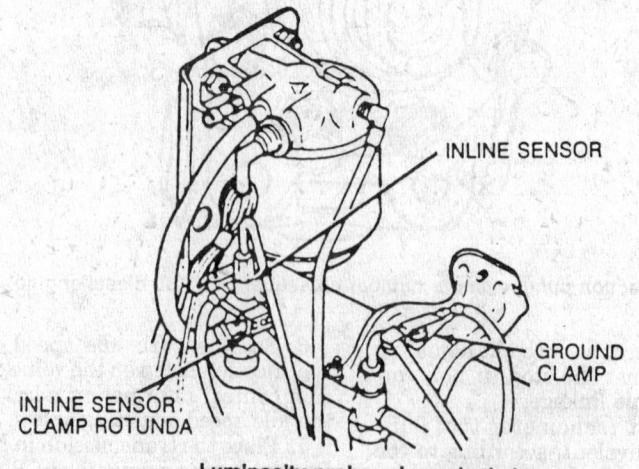

Luminosity probe — dynamic timing

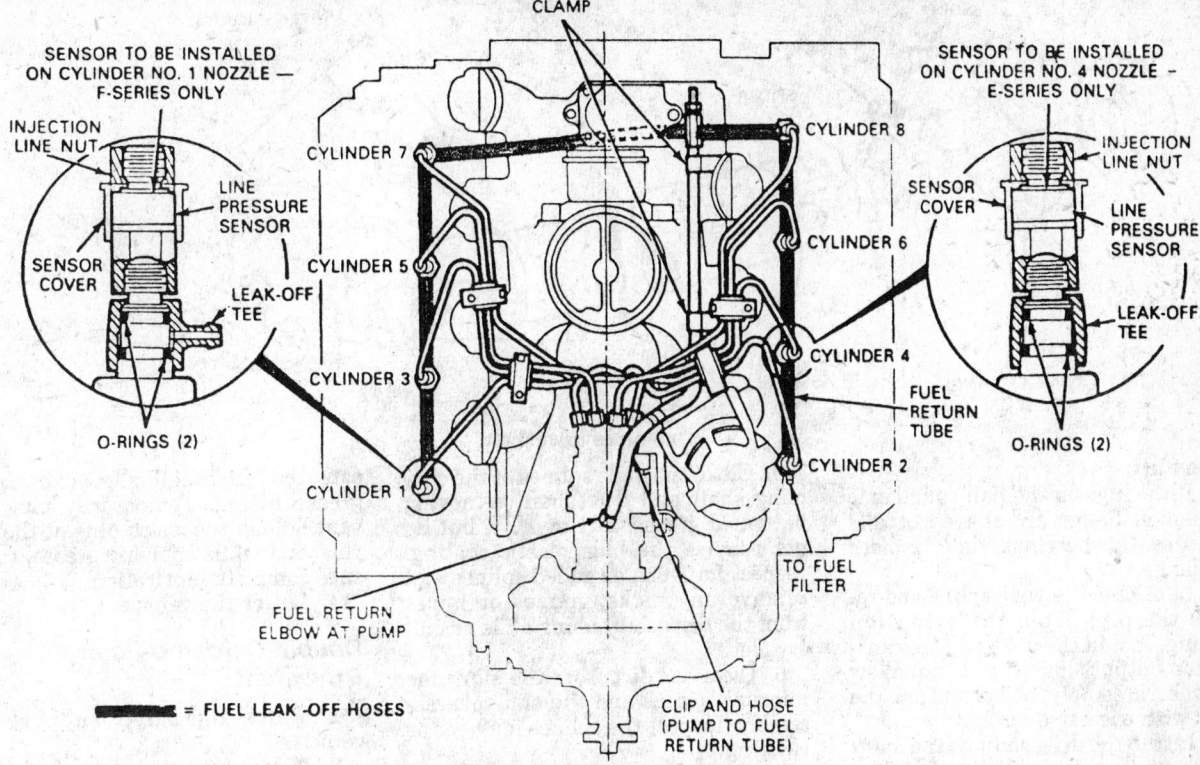

Fuel line routing and installation

move the return lines and tees as an assembly by lifting the tees off the nozzles.

3. Remove the injection nozzles by turning counterclockwise. Pull the nozzle assembly with the copper washer from the engine. Be careful not to strike the nozzle tip against any hard surface during removal. Note the position of the nozzles so they may be reinstalled in their original positions.

To install:

4. Thoroughly clean the nozzle bore in the cylinder head before reinserting the nozzle assembly. Pay particular attention to the seating surface, in order that no small particles of metal or carbon will cause the assembly to be cocked or permit blowby of combustion gases. Blow out the particles with compressed air.

5. Remove the protective cap and install a new copper gasket nozzle assembly with a small dab of multi-purpose grease. Apply anti-seize compound on the nozzle threads to aid installation and future removal.

6. Install the nozzle assembly into the cylinder head nozzle bore. Be careful that the nozzle tip does not strike against the recess wall. Tighten the nozzle assembly to 35 ft. lbs. (47 Nm).

7. Remove the protective caps from the nozzle assemblies and fuel lines. Install the leak-off tees and lines as an assembly by lowering onto the nozzles. Connect the clip and hose to the fuel return elbow at the pump. Install the line to retaining clamps.

NOTE: Install 2 new O-ring seals for each fuel return tee.

8. Connect the high pressure fuel lines. Start the engine. If necessary, bleed the fuel lines of air by loosening the connector ½–1 turn and cranking the engine until bubble-free fuel flows from the connection.

— **CAUTION** —

Keep eyes and hands away from nozzle spray. Fuel spraying from the nozzle under high pressure can penetrate the skin and cause infection. Medical attention should be provided immediately in the event of skin penetration.

9. Check for fuel leakage at the high-pressure connections.

DRIVE AXLE

Driveshaft and U-Joints

REMOVAL & INSTALLATION

One-Piece Driveshaft

1. Raise and safely support the vehicle.
2. To maintain driveline balance, if the yellow alignment marks are not visible, mark the relationship of the rear driveshaft yoke and the drive pinion flange of the axle, so they may be reinstalled in their original positions.

3. If equipped with a circular axle companion flange, proceed as follows:

　a. Remove the bolts retaining the axle flange yoke to the companion flange and disconnect the driveshaft from the axle.

　b. Lower the driveshaft and slide it rearward off the transmission output shaft.

　c. Install a plug in the transmission extension housing to prevent fluid loss.

4. If equipped with a half round axle companion flange, proceed as follows:

　a. Remove the nuts retaining the U-bolts to the axle companion flange.

　b. Remove the U-bolts and disconnect the U-joint from the axle companion flange being careful not to drop the U-joint bearing cups.

　c. Wrap tape around the U-joints to retain the bearing cups. Slide the driveshaft rearward off the transmission output shaft.

　d. Install a plug in the transmission extension housing to prevent fluid loss.

NOTE: On 4WD vehicles equipped with a slip between the center driveshafts, disconnect the driveshaft at the transfer case during removal.

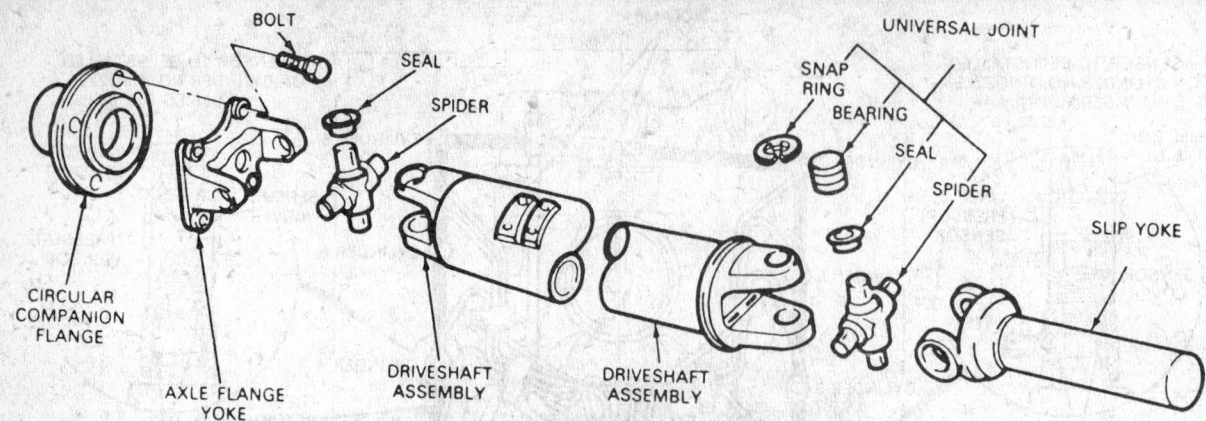

One-piece driveshaft

To install:

5. If the lugs on the half round axle companion flange are shaved or distorted so the bearings slide, replace the flange.

6. Lubricate the yoke spline and remove the plug from the extension housing. Install the yoke on the transmission output shaft, but do not allow the yoke assembly to bottom on the shaft with excessive force.

7. Install the driveshaft so the index marks or the yellow marks (if visible) on the yoke (light side) is in line with the yellow mark on the companion flange. This prevents vibration. If a vibration exists, the driveshaft should be disconnected from the axle, rotated 180 degrees, and reinstalled.

8. On half round axle companion flange, install the U-bolts and nuts attaching the U-joint to the companion flange. Tighten the U-bolt nuts to 8–15 ft. lbs. (11–20 Nm).

9. On circular axle companion flange, install the bolts retaining the axle flange yoke to the circular companion flange. Tighten to 70–95 ft. lbs. (95–129 Nm).

10. Lower the vehicle.

Driveshaft/Coupling Shaft

1. Raise and safely support the vehicle.

2. Mark the relationship of the rear driveshaft yoke and the rear axle companion flange before removal, so they can be reinstalled in their original positions.

3. Disconnect the driveshaft from the rear axle companion flange and disconnect the driveshaft slip yoke from the coupling shaft yoke. Wrap type around the loose bearing caps to prevent the bearings from falling off the universal joint spiders.

4. Remove the 2 center bearing support-to-crossmember attaching bolts. Remove the coupling shaft and wrap tape around the loose bearing caps. Install a plug in the transmission housing to prevent leakage.

5. Clean the male splines of the coupling shaft and driveshaft, removing hardened grease, dirt or rust, but do not remove the blue plastic coating. Inspect for worn or galled splines and remove any nicks, gouges or burrs from the driveshaft using a file or emery cloth.

6. Clean all dirt from the slip yoke internal splines and the slip yoke assembly. Inspect the splines for wear or twisting.

7. Clean all parts except the sealed center bearing and rubber insulator in solvent. Wipe the bearing and insulator clean with a cloth.

8. Inspect the slip yoke seal and replace, if necessary. Check the center support bearing for wear or rough action. Check the rubber insulator for hardening, cracking or deterioration. Replace if necessary.

To install:

9. Lubricate the coupling shaft slip yoke splines. Remove the plug from the transmission housing and install the front yoke of the coupling shaft assembly on the transmission ouput shaft. Do not allow the slip yoke assembly to bottom on the output shaft with excessive force.

10. Install the center bearing support bracket with the attaching bolts and spacers, if equipped.

11. Lubricate the splined stub shaft end of the coupling shaft assembly and the female splines of the slip yoke. Assemble the driveshaft slip yoke to the coupling shaft.

NOTE: If installing a new service driveshaft, align the factory made yellow paint mark at the rear of the driveshaft tube, with the factory made yellow paint mark on the outside diameter of the axle companion flange.

12. Connect the rear U-joint of the driveshaft to the rear axle companion flange and tighten the U-bolt nuts or strap bolts.

13. Using a hand grease gun, lubri-cate the driveshaft slip yoke at the grease fitting. Temporarily plug the vent hole in the welch plug at the slip yoke end while applying grease, to assure complete lubrication.

14. Lower the vehicle.

Double Cardan U-Joint Driveshaft

1. Raise and safely support the vehicle.

2. If removing the rear driveshaft on Bronco, mark the driveshaft in relation to the transfer case and rear axle companion flange. Disconnect the double cardan U-joint from the transfer case flange and the single U-joint from the rear axle flange. Tape the loose bearing caps and remove the driveshaft.

3. If removing the front driveshaft on F 350, mark the driveshaft in relation to the transfer case and front axle companion flange. Disconnect the double cardan U-joint from the transfer case flange and the single U-joint from the front axle flange. Remove the driveshaft.

To install:

4. Align the index marks and position the single U-joint end of the driveshaft to the axle and install the U-bolts and nuts. Position the U-joint to the transfer case with the index marks aligned and install the bolts and lockwashers.

5. Tighten the bolts at the transfer case to 20–25 ft. lbs. (28–33 Nm) and the nuts at the axle to 8–15 ft. lbs. (11–20 Nm).

6. Lower the vehicle.

Front Driveshaft

1. Raise and safely support the vehicle.

2. Mark the rear slip yoke in relation to the transfer case yoke for correct positioning during installation.

3. Remove the nuts and U-bolts, or bolts, that connect the rear slip yoke to the transfer case and the front yoke to the front axle.

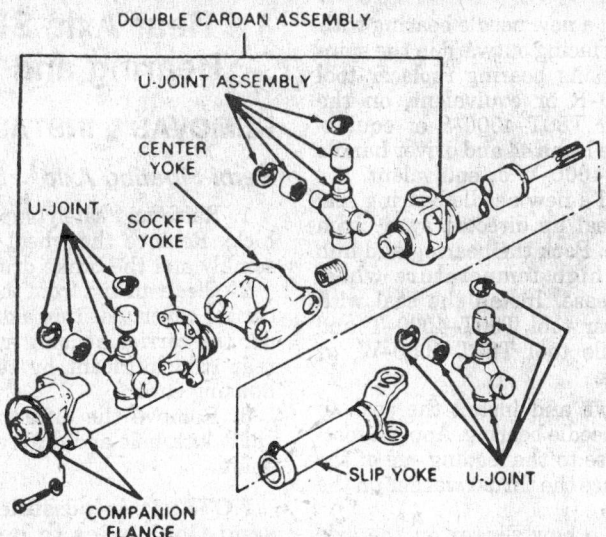

Double Cardan U-joint assembly

4. Remove the driveshaft. Wrap tape around the loose bearing caps to prevent the bearings from falling out out of the U-joint spiders.

To install:

5. Install the driveshaft, aligning the marks on the slip yoke and transfer case.

6. Install the nuts and U-bolts that retain the slip yoke to the transfer case yoke. Tighten the nuts to 8–15 ft. lbs. (11–20 Nm). On F 350, tighten the bolts to 20–28 ft. lbs. (28–33 Nm).

7. Install the nuts and U-bolts that retain the front driveshaft yoke to the front drive axle yoke. Tighten the nuts to 8–15 ft. lbs. (11–20 Nm).

8. Lower the vehicle.

Front Axle Shaft, Bearing and Seal

REMOVAL & INSTALLATION

Dana Model 60 Monobeam

1. Raise the vehicle and support it safely. Remove the front wheel and tire assembly.

2. Remove the disc brake caliper and wire it to the frame. Do not let the caliper hang by the brake hose.

3. Remove the 6 Allen head screws from the hub body and remove the cap. Remove the snapring that retains the axle shaft in the hub body.

4. Remove the lock ring seated in the groove of the hub and remove the body from the hub. If the body is difficult to remove, install 2 capscrews and pull the body assembly out of the hub.

5. Remove the outer locknut, lockwasher and the inner locknut from the spindle, using locknut wrench D85T–1197–A or equivalent. Remove the hub and rotor.

6. Remove the nuts retaining the spindle to the knuckle. Lightly tap the spindle with a rawhide hammer to remove it from the knuckle. Remove the splash shield and the caliper support from the knuckle.

7. Pull the axle shaft out of the steering knuckle. If required, remove the spacer from the axle shaft and the caged needle bearing from the spindle and remove the seal. Pull the bearing out of the spindle using collet D80L–100–T, actuator pin D80L–100–H and bridge assembly D80L–100–W or equivalents.

To install:

8. If removed, place a new caged needle bearing in the spindle bore. The writing on the bearing must face the rear of the spindle, toward the driving tool. Drive the bearing into the spindle with bearing replacer tool T80T–4000–R and driver handle tool T80T–4000–W or equivalents.

9. Pack the bearing with high temperature wheel bearing grease. Install the seal in the bore against the bearing. Pack the thrust face of the seal in the spindle bore and the V-seal on the axle shaft with high temperature wheel bearing grease.

10. Guide the axle shaft carefully through the knuckle and into the axle housing, making sure the axle shaft splines are engaged in the differential side gear splines.

11. Install the spacer on the axle shaft with the chamferred side inboard against the axle shaft. Install the splash shield and caliper support bracket on the steering knuckle.

12. Install the spindle on the steering knuckle and tighten the nuts to 50–60 ft. lbs. (68–81 Nm). Install the hub and rotor assembly on the spindle. Make sure the wheel bearings are adequately lubricated with high temperature wheel bearing grease.

13. Install the inner locknut on the spindle and seat the bearings by tightening the locknut to 50 ft. lbs. (68 Nm), with locknut wrench D85T–1197–A or equivalent.

14. Back off the inner locknut and retighten to 31–39 ft. lbs. (41–54 Nm) while rotating the hub and rotor. Back off the locknut 90 degrees.

15. Install the lockwasher so the key is positioned in the spindle groove. Tighten the inner locknut so the pin is

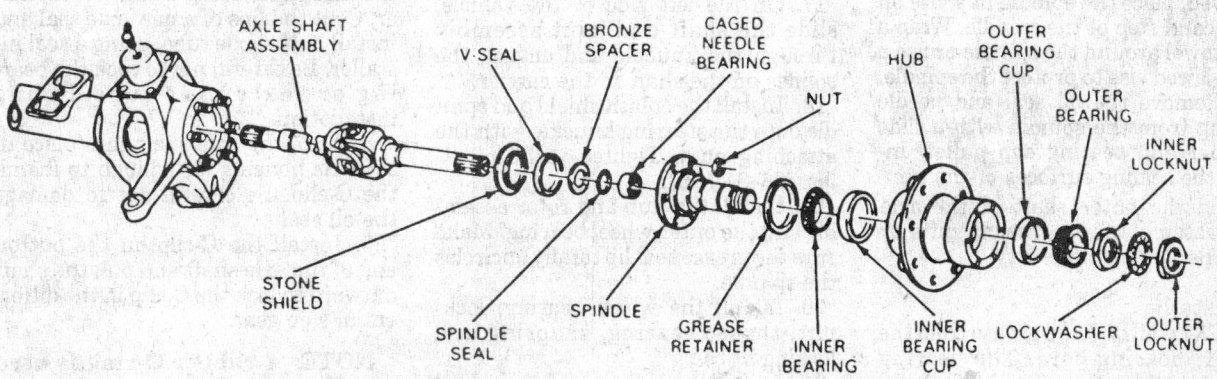

Dana model 60 Monobeam front drive axle assembly

aligned into the nearest hole in the lockwasher.

16. Install the outer locknut and tighten to 160–205 ft. lbs. (217–278 Nm), using locknut wrench D85T–1197–A or equivalent.

NOTE: The final wheel bearing end play should be 0–0.004 in. (0–0.15mm). The maximum allowable torque to rotate the hub is 20 inch lbs. (2.3 Nm).

17. Install the manual locking hub body assembly in the hub and rotor. Install the lock ring in the hub groove and the snapring on the end of the axle shaft.

18. Position the cap assembly with a new seal on the body. Install the 6 Allen-head screws and tighten to 35–45 inch lbs. (4–6 Nm).

19. Install the caliper and the front wheel and tire assembly. Lower the vehicle.

DanA Models 44 and 50

1. Raise the vehicle and support it safely. Remove the front wheel and tire assembly.

2. Remove the disc brake caliper and wire it to the frame. Do not let the caliper hang by the brake hose.

3. Remove the hub locks, wheel bearings and locknuts.

4. Remove the hub and rotor assembly and the outer bearing from the spindle.

5. Remove the nuts retaining the spindle to the steering knuckle. Tap the spindle with a nylon or rawhide hammer to jar the spindle from the knuckle. Remove the spash shield.

6. On the left side of the vehicle, remove the shaft and joint assembly by pulling the assembly out of the carrier.

7. On the right side of the vehicle, remove and discard the keystone clamp from the shaft and joint assembly and the stub shaft. Slide the rubber boot onto the stub shaft and pull the shaft and joint assembly from the splines of the stub shaft.

8. If the needle bearings are to be removed, place the spindle in a vise on the second step of the spindle. Wrap a shop towel around the spindle or use a brass-jawed vise to protect the spindle.

9. Remove the oil seal and needle bearing from the spindle with a slide hammer and bearing cup puller. Inspect the sealing surfaces of the spindle and the outer shaft of the axle shaft assembly for corrosion, pitts or wear. Replace as necessary.

To install:

10. Clean all dirt and grease from the spindle bearing bore. The bearing bores must be free from nicks and burrs.

11. Install a new needle bearing with the writing facing outward in the spindle bore, using bearing replacer tool T80T–4000–R or equivalent, on the Dana 50 or T80T–4000–S or equivalent, for the Dana 44 and driver handle tool T80T–4000–W or equivalent.

12. Install a new needle bearing seal with the seal lip directed away from the spindle. Pack the bearing and hub seal with high temperature wheel bearing grease. Install the seal with seal replacer tool T80T–4000–T and drive handle tool T80T–4000–W, or equivalents.

13. Remove and install the seal on top of the needle bearing. Apply a coating of grease to the leading ege of the seal lip. Place the thrust washer on the axle shaft.

14. Press a new slinger on the axle shaft using a suitable tool. Install the rubber V-seal on the slinger and axle shaft. The lip of the seal should face towards the spindle.

15. Install the plastic spacer on the axle shaft with the chamferred side inboard against the axle shaft. Pack the thrust face of the seal in the spindle bore and the V-seal on the axle shaft with grease.

16. On the right side of the vehicle, proceed as follows:

    a. Install the rubber boot and the new keystone clamps on the stub shaft slip yoke.

    b. Since the splines on the shaft are phased, there is only 1 way to assemble the right shaft and joint assembly through the knuckle and into the slip yoke.

    c. Align the missing spline in the slip yoke barrel with the gapless male spline on the shaft and joint assembly.

    d. Slide the right shaft and joint assembly through the knuckle and into the slip yoke making sure the splines are fully engaged.

    e. Slide the boot over the assembly and crimp the keystone clamp using keystone clamp pliers T63P–9171–A or equivalent.

17. On the left side of the vehicle, slide the shaft and joint assembly through the knuckle and engage the splines on the shaft in the carrier.

18. Install the splash shield and spindle onto the steering knuckle with the attaching nuts. Tighten to 50–60 ft. lbs. (68–81 Nm).

19. Install the hub and rotor assembly and the outer wheel bearing. Make sure the grease seal lip totally encircles the spindle.

20. Install the wheel bearing, locknut, thrust bearing, snapring and locking hubs.

21. Install the caliper and the wheel and tire assembly. Lower the vehicle.

Rear Axle Shaft, Bearing and Seal

REMOVAL & INSTALLATION

Semi-Floating Axle

1. Raise and safely support the vehicle. Remove the wheel and tire assembly and the brake drum.

2. Clean all dirt from the area of the carrier cover. Position a drain pan under the carrier housing and drain the rear axle lubricant by removing the housing cover.

3. Remove the differential pinion shaft lock bolt and differential pinion shaft.

NOTE: It is possible for Dana semi-float axles to be equipped with a lock bolt coated with either a locking compound treated thread or torque prevailing threads. The locking compound treated lock bolt has a $5/32$ in. hex socket head. The torque prevailing lock bolt has a 12-point drive head. The locking compound treated lock bolt must not be reused under any circumstances. The torque prevailing lock bolt may be reused up to 4 times, however, if there is any doubt concerning how many times it has been removed, replace the lock bolt.

4. Push the flanged end of the axle shaft toward the center of the vehicle and remove the C-clip from the button end of the shaft. If equipped, be careful not to damage the rubber O-ring in the axle shaft groove.

5. Remove the axle shaft from the housing, being careful not to damage the oil seal if it is to be reused.

6. Pry out the axle seal from the axle tube and discard. Use a suitable puller or slide hammer to remove the axle bearing from the tube.

To install:

7. Lubricate a new axle bearing with rear axle lubricant and install in the axle tube, using a suitable installer. Coat the lips of a new axle seal and install in the axle tube, using a seal installer. Be careful not to cock the bearing or seal in the tube during installation.

8. Slide the axle shaft into place in the axle housing far enough to install the C-clip. Be careful not to damage the oil seal.

9. Install the C-clip on the button end of the axle shaft and pull the shaft outward to lock the C-clip in the differential side gear.

NOTE: A rubber O-ring is used on all except Dana semi-float axles to hold the C-clip in position.

SPINDLE and LEFT SHAFT and JOINT
INSTALLATION—TYPICAL

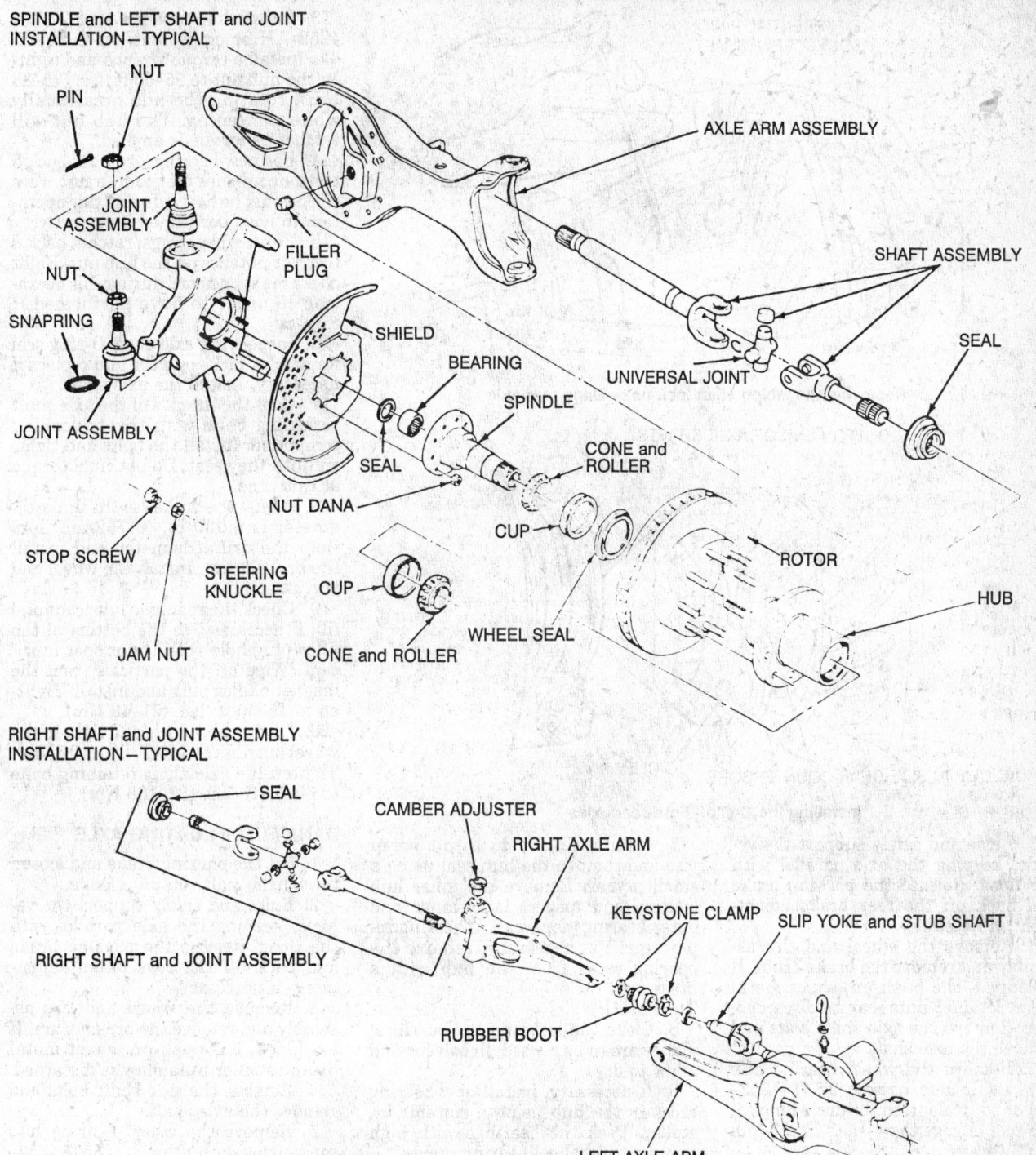

PIN

NUT

AXLE ARM ASSEMBLY

JOINT
ASSEMBLY

FILLER
PLUG

SHAFT ASSEMBLY

NUT

SNAPRING

SHIELD

BEARING

SPINDLE

UNIVERSAL JOINT

SEAL

JOINT ASSEMBLY

SEAL

NUT DANA

CONE and
ROLLER

STOP SCREW

CUP

ROTOR

STEERING
KNUCKLE

HUB

JAM NUT

CUP

CONE and ROLLER

WHEEL SEAL

RIGHT SHAFT and JOINT ASSEMBLY
INSTALLATION—TYPICAL

SEAL

CAMBER ADJUSTER

RIGHT AXLE ARM

KEYSTONE CLAMP

SLIP YOKE and STUB SHAFT

RIGHT SHAFT and JOINT ASSEMBLY

RUBBER BOOT

LEFT AXLE ARM

Dana models 44 and 50 axle shaft and joint assemblies

Be sure the O-ring is in the groove at the button end of the axle shaft before installing the C-clip.

10. Install the differential pinion shaft through the case and pinions, aligning the hole in the shaft with the lock bolt hole. On Ford rear axle assemblies, apply thread-locking compound and install the lock bolt. Tighten to 15–30 ft. lbs. (20–40 Nm). On Dana rear axle assemblies, install the

lock bolt and tighten to 20–25 ft. lbs. (27–34 Nm).

11. Clean the mating surface of the carrier and cover plate so they are free of any oil film or foreign material. Apply a $^1/_8$–$^3/_{16}$ in. bead of silicone sealer to the cover plate and install on the carrier with the attaching bolts. Tighten alternately and evenly to 30 ft. lbs. (41 Nm).

NOTE: Allow an hour cure time

before filling the carrier with lubricant.

12. Fill the axle housing with the proper type and quantity of lubricant. Install the brake drum and the wheel and tire assembly.

13. Lower the vehicle.

Full-Floating Axle

FORD FULL-FLOATING AXLE

1. Set the parking brake and loosen the 8 axle shaft retaining bolts.

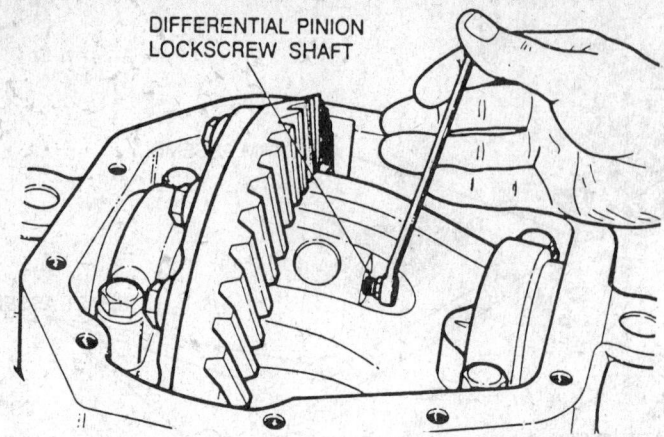

Removing the pinion shaft lock bolt—Dana rear axle

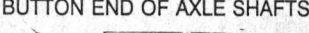

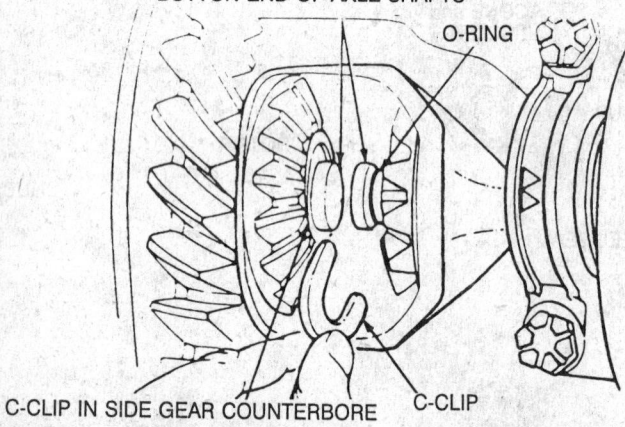

Installing the C-clip—Ford rear axles

2. Raise and safely support the vehicle, keeping the axle parallel with the floor. Release the parking brake and back off the rear brake adjustment, if necessary.

3. Remove the wheel and tire assembly and remove the brake drum. If equipped, the push-on, sheet metal drum retainer nuts may be discarded.

4. Remove the axle shaft bolts and remove the axle shaft.

5. Remove the wheel bearing hub nuts using hub wrench T85T-4252-AH or equivalent, so the drive tangs of the tool engage the 4 slots in the hub nut.

NOTE: The left hub nut has left hand thread and must be removed by turning clockwise. Each hub nut is stamped RH for right hand thread or LH for left hand thread. Do not use power impact tools on hub nuts.

6. Install step plate adapter tool D80L-630-7 and puller tool D80L-1002-L or equivalents, and loosen the hub to the point of removal. Remove the puller tool and step plate and remove the hub assembly. Be careful not to drop the outer bearing.

7. Install the hub in a soft jawed vise and remove the hub seal using a small prybar. Remove the inner hub bearing and inspect both inner and outer bearings for wear. If bearing replacement is necessary, remove the bearing races from the hub with a brass drift.

To install:

8. Clean the hub and the bearings, if they are to be reused, in solvent and allow to dry.

9. If necessary, install new bearing races in the hub using a suitable installer. Pack the bearings with high temperature wheel bearing grease.

10. Place the inner bearing in the hub and install a new oil seal, using a seal installer.

11. Clean the spindle thoroughly and then coat with axle lubricant. Push the hub and the outer bearing onto the spindle. Installing the hub this way causes the outer bearing to act as a pilot, making installation easier.

12. Install the hub nut on the spindle. Make sure the hub nut tab is located in the keyway prior to thread engagement. Turn the hub nut clockwise for right hand thread or counterclockwise for left hand thread.

13. Install hub wrench tool T85T-4252-AH or equivalent, on the spindle. Install a torque wrench and tighten the hub nut to 55–65 ft. lbs. (75–88 Nm), rotating the hub occasionally while tightening. The hub nut will ratchet as torque is applied.

14. For new bearings, ratchet back 5 teeth or notches on the hub nut. Five clicks must be heard during this operation to have performed it correctly.

15. For used bearings, ratchet back 8 teeth or notches on the hub nut. Eight clicks must be heard during this operation in order to have performed it correctly.

16. Inspect the axle shaft O-ring seal for cracks, nicks or wear and replace if necessary. Install the axle shaft.

17. Coat the threads of the axle shaft retaining bolts with thread-locking compound. Install the bolts and tighten until they seat. Do not final torque at this time.

18. Adjust the brakes so the brake diameter is 0.030 in. (0.762mm) less than the drum diameter and install the brake drum. Install the wheel and tire assembly.

19. Check the rear axle lubricant and fill, if necessary, to the bottom of the filler plug hole with the proper lubricant. Wipe off the particles from the magnetic filler plug and install. Tighten to 15–30 ft. lbs. (21–40 Nm).

20. Lower the vehicle and tighten the wheel lug nuts to 140 ft. lbs. (190 Nm). Tighten the axle shaft retaining bolts to 60–80 ft. lbs. (82–108 Nm).

DANA FULL-FLOATING AXLE

1. Set the parking brake and loosen the 8 axle shaft retaining bolts.

2. Raise and safely support the vehicle, keeping the axle parallel with the floor. Release the parking brake and back off the rear brake adjustment, if necessary.

3. Remove the wheel and tire assembly and remove the brake drum. If equipped, the push-on, sheet metal drum retainer nuts may be discarded.

4. Remove the axle shaft bolts and remove the axle shaft.

5. Remove the wheel bearing hub nuts using hub wrench T85T-4252-AH for E Series or T88T-4252-A for F Super Duty or equivalents, so the drive tangs of the tool engage the 4 slots in the hub nut.

NOTE: Do not use power impact tools on the hub nuts.

6. Remove the hub. Remove the hub seal using a small prybar. Remove the inner hub bearing and inspect both inner and outer bearings for wear. If bearing replacement is necessary, remove the bearing races from the hub with a brass drift.

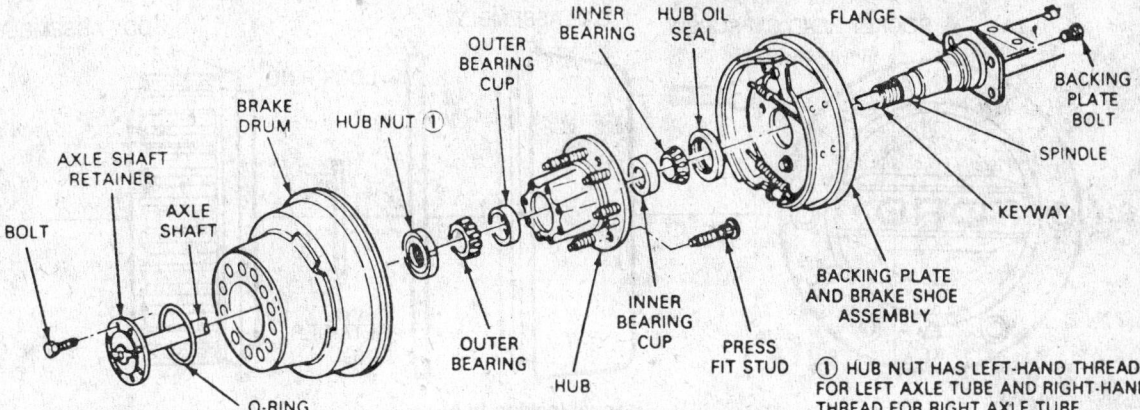

Rear axle shaft, bearing and hub assembly—Ford full-floating axle

Labels on figure:
- BOLT
- AXLE SHAFT RETAINER
- AXLE SHAFT
- O-RING
- BRAKE DRUM
- HUB NUT ①
- OUTER BEARING CUP
- OUTER BEARING
- HUB
- INNER BEARING
- INNER BEARING CUP
- PRESS FIT STUD
- HUB OIL SEAL
- BACKING PLATE AND BRAKE SHOE ASSEMBLY
- FLANGE
- BACKING PLATE BOLT
- SPINDLE
- KEYWAY
- ① HUB NUT HAS LEFT-HAND THREAD FOR LEFT AXLE TUBE AND RIGHT-HAND THREAD FOR RIGHT AXLE TUBE

To install:

7. Clean the hub and the bearings, if they are to be reused, in solvent and allow to dry.

8. If necessary, install new bearing races in the hub using a suitable installer. Check for proper seating by trying to insert a 0.0015 in. (0.038mm) feeler gauge between the races and the wheel hub in several places. The gauge should not enter under the race.

9. Pack the bearings with high temperature wheel bearing grease. Place the inner bearing in the hub and install a new seal with a seal installer.

10. Thoroughly clean the spindle of the axle housing. Wrap the threads of the spindle with electrical tape and carefully slide the hub onto the spindle. Remove the tape.

11. Install the outer wheel bearing and start the hub nut, making sure the hub tab is properly located in the keyway prior to thread engagement.

12. Install locknut wrench tool T85T–4252 for E Series or T88T–4252–A for F Super Duty or equivalents, so the drive tangs of the tool engage the 4 slots of the nut.

13. Install a torque wrench to the tool and tighten to 65–75 ft. lbs. (88–102 Nm) while rotating the hub. If using tool T88T–4252–A or equivalent, apply inward pressure while tightening to separate the ratcheting components of the nut.

14. After torquing, back off 90 degrees, then retighten to 15–20 ft. lbs. (20–27 Nm).

15. Install the axle shaft with a new flange gasket, lock washers and axle shaft retaining bolts. Coat the threads of the bolts with thread adhesive and tighten the bolts until they seat. Do not torque at this time.

16. Install the brake drum and the wheel and tire assembly. Lower the vehicle and adjust the brakes.

17. Tighten the wheel lug nuts. Tighten the axle shaft retaining bolts to 41–55 ft. lbs. (55–75 Nm) on E Series or 70–85 ft. lbs. (95–115 Nm) on F Super Duty.

Front Wheel Knuckle

REMOVAL & INSTALLATION

Dana Model 60 Monobeam

1. Raise and safely support the vehicle. Remove the axle shaft.

2. Alternately and evenly loosen the 4 bolts retaining the spindle cap to the knuckle; thereby relieving the spring compression. Remove the bolts.

3. Remove the spindle cap, compression spring and retainer. Remove and discard the gasket.

4. Remove the 4 bolts retaining the lower kingpin and retainer to the knuckle. Remove the lower kingpin and retainer from the knuckle.

5. Remove the tapered bushing from the top of the upper kingpin in the knuckle. Remove the knuckle from the axle yoke.

To install:

6. Install the knuckle on the axle yoke. Place the tapered bushing over the upper kingpin in the knuckle bore.

7. Place the lower kingpin and retainer assembly in the knuckle and axle yoke. Install the 4 bolts and alternately and evenly tighten to 70–90 ft. lbs. (95–122 Nm).

8. Place the retainer and compression spring on the tapered bushing. Install a new gasket and the spindle cap on the knuckle. Install the 4 bolts and alternately and evenly tighten to 70–90 ft. lbs. (95–122 Nm).

9. Install the axle shaft. Lubricate the upper kingpin through the grease fitting and the lower through the flush type fitting.

10. Lower the vehicle.

Dnan Models 44 and 50

1. Raise and safely support the vehicle.

2. Remove the axle shafts and spindles. Disconnect the steering linkage.

3. Remove the cotter pin from the top ball joint stud. Loosen the nut on the top stud and the bottom nut inside the knuckle. Remove the top nut.

4. Sharply hit the top stud with a plastic or rawhide hammer, to free the knuckle from the axle arm. Remove and discard the bottom nut.

5. Mark the position of the camber adjuster and remove it by hand. If difficult to remove, use pitman arm puller T64P–3590–F or equivalent.

6. Place the knuckle in a vise and remove the snapring from the bottom ball joint socket, if equipped.

7. Using a suitable ball joint remover, press the lower ball joint from the knuckle. Always remove the lower ball joint first, then remove the upper ball joint.

To install:

8. Clean the steering knuckle bore and insert the lower ball joint as straight as possible. Press the ball joint into position using a suitable ball joint press. Repeat the procedure for the upper ball joint.

9. Install the knuckle to the axle arm arm. Install the camber adjuster, aligning the marks that were made during removal.

10. Install a new nut on the bottom socket finger tight. Install and tighten the nut on the top socket finger tight, then tighten the bottom nut to 80 ft. lbs. (109 Nm).

11. Tighten the top nut to 100 ft. lbs. (136 Nm), then advance the nut until the castellation aligns with the cotter pin hole. Install the cotter pin. Do not loosen the top nut to install the cotter pin.

12. Retighten the bottom nut to 90–110 ft. lbs. (123–150 Nm). Install the axle shaft and spindle assembly and lower the vehicle.

Manual Locking Hubs

REMOVAL & INSTALLATION

1. Raise and support the vehicle safely.

2. Seperate the cap asembly from the body assembly by removing the 6 Allen head capscrews and remove the cap from the body.

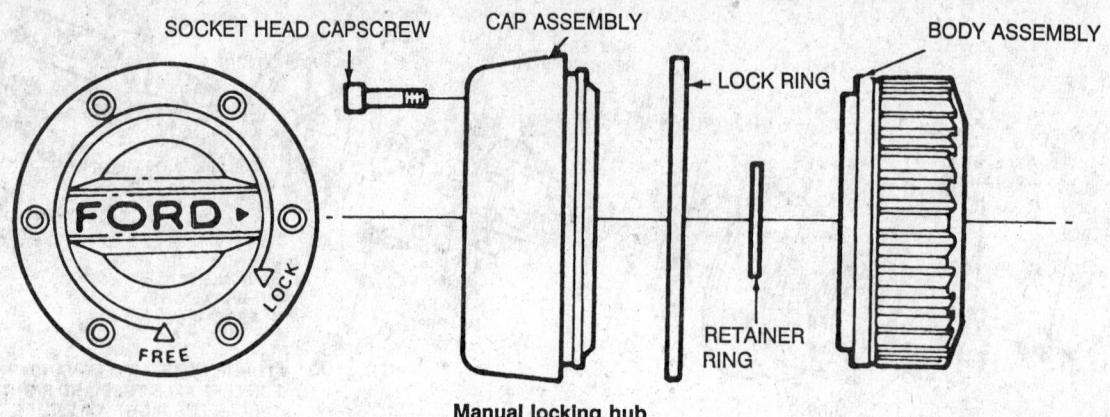

Manual locking hub

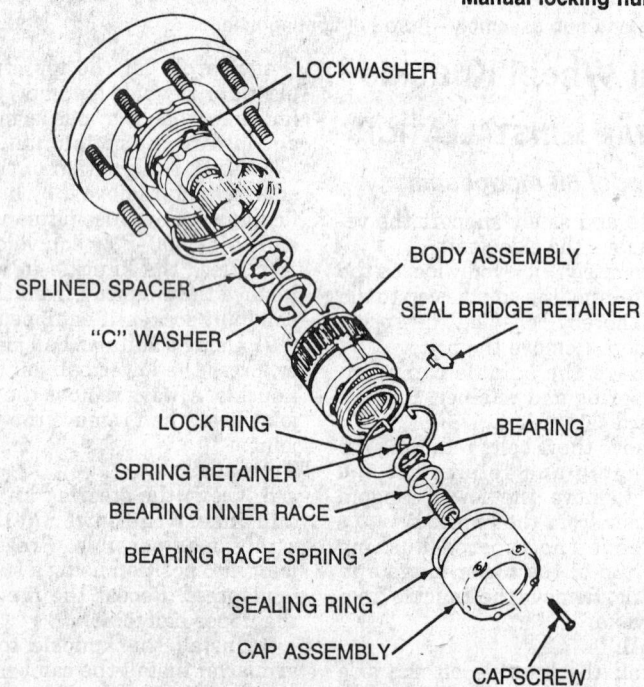

Automatic locking hub

3. Remove the snapring that retains the axle shaft in the hub body assembly.

4. Remove the lock ring seated in the groove of the wheel hub. Remove the body assembly from the hub. If the body is difficult to remove, install 2 capscrews and pull the body assembly out of the hub.

5. Installation is the reverse of the removal procedure. Tighten the Allen head retaining screws to 35–53 inch lbs. (4–6 Nm). Do not pack the cap assembly with grease, as excessive grease can cause excessive dialing effort.

Automatic Locking Hubs

REMOVAL & INSTALLATION

1. Raise and support the vehicle safely.

2. Remove the capscrews from the body assembly. Remove the cover, being careful not to drop the spring, ball bearing, bearing race or retainer.

3. Remove the rubber seal.

4. Remove the seal bridge retainer from the retainer ring spacer.

5. Remove the retaining ring by closing the ends with needle nose pliers while pulling the hub lock assembly from the hub.

6. Installation is the reverse of the removal procedure. Tighten the retaining screws to 40–50 inch lbs. (4.5–5.6 Nm).

Pinion Seal

REMOVAL & INSTALLATION

1. Raise and safely support the vehicle.

2. Mark the position of the driveshaft in relation to the pinion flange and remove the driveshaft.

3. On Ford 8.8 in. and 10.25 in. rear axles, proceed as follows:

a. Remove the wheel and tire assemblies.

b. Using and inch pound torque wrench on the pinion nut, record the torque required to rotate the pinion through several revolutions.

4. While holding the pinion flange with a suitable tool, remove the pinion nut. Remove the pinion flange. It may be necessary to use a puller to remove the flange.

5. Use a small prybar to remove the pinion seal.

To install:

6. Clean the pinion seal seat area. Coat the lips of a new seal with rear axle lubricant and install, using a seal installer.

7. Coat the inside of the pinion flange with a small amount of rear axle lubricant and install on the pinion. Install the pinion nut.

8. On Ford 8.8 in. and 10.25 in. rear axles, proceed as follows:

a. Gradually tighten the pinion nut while holding the pinion flange with a suitable tool.

b. Periodically check the torque required to rotate the pinion through several revolutions with the inch pound torque wrench.

c. Tighten the pinion nut only to the point where the torque required to rotate the pinion through several revolutions is the same as was recorded during the removal procedure.

9. On all other axle assemblies, tighten the pinion nut to the following torque values:

Dana 60 and 61 rear axles—250–270 ft. lbs. (339–366 Nm).

Dana 80 rear axle—440–500 ft. lbs. (596–677 Nm).

Dana 60 Monobeam front axle—220–280 ft. lbs. (298–379 Nm).

Dana 44 and 50 front axles—200–220 ft. lbs. (271–298 Nm).

10. Install the driveshaft, aligning the marks that were made during removal.

11. Install the wheel and tire assem-

blies, if removed, and lower the vehicle.

Axle Housing

REMOVAL & INSTALLATION

Dana Model 60 Monobeam

Front Axle

1. Raise and safely support the vehicle. Remove the front wheel and tire assemblies.

2. Remove the brake calipers and support with mechanics wire. Do not let the calipers hang by the brake hoses.

3. Disconnect the links from the stabilizer bar. If necessary, remove the stabilizer bar.

4. Remove the cotter pins and nuts securing the spindle connecting rod to the steering knuckles. Separate the connecting rods from the knuckles using pitman arm puller T64P-3590-F or equivalent, and wire the steering linkage to the springs.

5. Mark the position of the driveshaft on the pinion flange and remove the driveshaft. Disconnect the vent tube and plug the vent fitting at the axle housing.

6. On the right side of the vehicle, remove the nut and bolt and disconnect the tracking bar from the right spring cap.

7. Support the axle under the differential carrier with a suitable jack. Remove the retaining U-bolts and nuts and lower the axle from the vehicle.

To install:

8. Raise the axle into position with the jack. Make sure the retaining bolt head protruding from the leaf spring plate seats in the recessed portion of the axle spring mounting plate. Install the nuts, U-bolts and spring caps.

9. Connect the driveshaft to the pinion flange, aligning the marks that were made during removal.

10. On the right side of the vehicle, connect the tracking bar to the spring cap and tighten the nut and bolt to 163–203 ft. lbs. (221–275 Nm). Connect the vent tube.

11. Connect the spindle connecting rods to the steering knuckles. Install the nuts and tighten to 70–100 ft. lbs. (95–136 Nm). Install new cotter pins, advancing the nuts, if necessary, to align the castellations with the stud holes. Do not back off the nuts to line up the castellations and stud holes.

12. If removed, install the stabilizer bar. Connect the links to the stabilizer bar.

13. Install the brake calipers and the wheel and tire assemblies. Lower the vehicle.

Dana Models 44 and 50 Front Axles

WITH COIL SPRINGS

1. Raise and safely support the vehicle. Remove the front wheel and tire assemblies.

2. Remove the brake calipers and support with mechanics wire. Do not let the calipers hang by the brake hoses.

3. Disconnect the steering linkage from the steering knuckles.

4. Position a jack under the axle arm assembly and remove the upper coil spring retainers. Lower the jack and remove the coil spring, spring cushion and lower spring seat.

NOTE: The axle arm must be supported with the jack throughout spring removal and installation and must not be permitted to hang by the brake hose. If necessary, remove the caliper and suspend it with mechanics wire. Do not let the caliper hang by the brake hose.

5. Disconnect the shock absorber at the radius arm and upper mounting bracket. Remove the stud and spring seat at radius arm and axle arm. Remove the bolt securing the upper attachment to axle arm radius arm to lower attachment axle arm.

6. Disconnect the vent tube at the differential housing and discard the hose clamps. Remove the vent fitting and install a ⅛ in. pipe plug.

NOTE: The vent tube may have been temporarily plugged to stop vent lube blowout while the axle vent was relocated during assembly. Check for lubricant leakage through the pinion seal, axle seals or support arm to housing to make sure the plastic plug has been removed. If the plug has not been removed, remove it from the end of the vent tube.

7. Remove the pivot bolt securing the right axle arm to the crossmember. Remove and discard the keystone clamps and remove the boot from the shaft. Remove the right drive axle and pull the axle shaft from the slip shaft.

8. Position a jack under the differential housing. Remove the bolt securing the left axle assembly to the crossmember and remove the left drive axle assembly.

To install:

9. Position the left drive axle at the radius arm. Secure the drive axle to the crossmember with the pivot bolt. Tighten the bolt to 120–150 ft. lbs. (163–203 Nm).

10. Position the right axle assembly at the crossmember and radius arm. Align the axle shaft and install in the slip shaft. Install the boot on the shaft so the boot seats in the gooves. Position new keystone clamps over the grooves on the boot and crimp the clamp with keystone clamp pliers T63P-9171-A or equivalent. Secure the axle assembly to the crossmember with the pivot bolt and tighten to 120–150 ft. lbs. (163–203 Nm).

11. Install the vent fitting in the differential housing. Connect the vent tube to the vent fitting usinf new hose clamps.

12. Position the spring seat and install a new stud at the axle arm and upper radius arm. Install a new bolt at the axle assembly and lower radius arm. Tighten the bolts to 180–240 ft. lbs. (245–325 Nm).

13. Position the coil spring insulator and coil spring on the lower spring seat. Install the nut and tighten to 30–70 ft. lbs. (41–94 Nm). Position a jack under the axle assembly and raise the coil spring into position. Install the upper spring retainer and screw. Tighten to 13–18 ft. lbs. (18–24 Nm).

14. Lower the jack. Connect the shock absorbers to the upper and lower frame brackets.

15. Install the calipers and the wheel and tire assemblies. Connect the steering linkage.

16. Lower the vehicle and check the front end alignment.

WITH LEAF SPRINGS

1. Raise and safely support the vehicle. Remove the wheel and tire assemblies.

2. Remove the brake calipers. Suspend the calipers with mechanics wire. Do not let the calipers hang by the brake hose.

3. Disconnect the steering linkage from the steering knuckles.

4. Position a jack under the right axle assembly. Remove the U-bolts securing the shock absorber mounting plate and leaf springs to the tube and yoke assembly.

5. Disconnect the vent tube at the differential housing. Remove the vent fitting and install a ⅛ in. pipe plug.

6. Remove the pivot bolt that secures the right axle assembly to the crossmember and remove the right axle assembly. Remove and discard the keystone clamps and remove the boot from the shaft. Pull the axle out of the slip shaft.

7. Position the jack under the left axle assembly. Remove the U-bolts securing the shock absorber mounting plate and leaf spring to the tube and yoke assembly.

8. Position a jack under the differential housing. Remove the pivot bolt securing the left axle assembly to the crossmember and remove the axle assembly.

To install:

9. Position the left axle assembly and install the pivot bolt that secures the axle to the crossmember. Secure the shock absorber mounting plate to the leaf spring and axle assembly with the 2 U-bolts. Tighten the bolts to 85–120 ft. lbs. (116–162 Nm).

10. Position the right axle assembly at the crossmember. Install the boot on the shaft so the boot seats in the grooves. Position new keystone clamps over the grooves on the boot and crimp the clamp with keystone clamp pliers T63P-9171-A or equivalent.

11. Align the axle shaft and install in the slip shaft. Install the pivot bolt that secures the axle assembly to the crossmember and tighten to 120–150 ft. lbs. (123–203 Nm).

12. Install the shock absorber mounting plate, leaf spring and axle assembly using 2 U-bolts. Tighten to 85–120 ft. lbs. (116–162 Nm).

13. Install the vent fitting to the differential housing and connect the vent tube to the vent using a new hose clamp.

14. Connect the steering linkage. Install the brake calipers and the wheel and tire assemblies.

15. Lower the vehicle. Check the front end alignment.

Rear Axle

1. Raise and safely support the vehicle. Remove the wheel and tire assemblies and remove the brake drums.

2. Mark the position of the driveshaft on the pinion flange and remove the driveshaft.

3. Disconnect the parking brake cables from the brake shoes and backing plates. If equipped, disconnect the anti-lock brake sensor wiring.

4. Disconnect the flexible brake hose at the frame. Plug the hose and the line. Disconnect the axle vent hose.

5. Support the axle assembly with a suitable jack.

6. Disconnect the shock absorbers from the axle. Disconnect the stabilizer bar, if equipped.

7. Remove the nuts, U-bolts and spring seat caps and remove the axle assembly.

To install:

8. Raise the axle assembly into position and install the nuts, U-bolts and spring seat clamps. Connect the stabilizer bar, if equipped, and the shock absorbers. Remove the jack.

9. Connect the flexible brake hose and connect the parking brake cables at the backing plates and brake shoes. Connect the anti-lock brake sensor wiring, if equipped.

10. Install the driveshaft, aligning the marks that were made during removal. Connect the axle vent hose.

11. Install the brake drums and the wheel and tire assemblies. Check the rear axle lubricant level.

12. Lower the vehicle.

MANUAL TRANSMISSION

For further information on transmissions/transaxles, please refer to "Chilton's Guide to Transmission Repair".

Transmission Assembly

REMOVAL & INSTALLATION

Borg-Warner T-18 Four Speed

BRONCO AND F SERIES 4WD

1. Remove the 4 screws holding the floor mat. Remove the 11 screws holding the access cover to the floor pan. Place the shift lever in **R** and remove the cover.

2. Remove the insulator and the dust cover. Remove the transfer case shift lever, shift ball and boot as an assembly.

3. Remove the transmission shift lever, shift ball and boot as an assembly.

4. Raise and support the vehicle safely. Drain the transmission.

5. Disconnect the front and rear driveshafts from the transfer case and wire them out of the way.

6. Remove the the shift link and speedometer cable from the transfer case.

7. Position a transmission jack under the transfer case. Remove the 6 bolts holding the transfer case to the transmission and lower the transfer case from the vehicle.

8. Remove the 8 bolts that hold the rear support bracket to the transmission.

9. Position a transmission jack under the transmission and remove the rear support bracket and brace. Remove the 4 bolts that hold the transmission to the bell housing and remove the transmission.

To install:

10. Place the transmission on a transmission jack and install it in the vehicle. Install 2 guide pins in the bell housing top holes, to guide the transmission in place.

11. Install the 2 lower bolts, remove the guide pins and install the upper 2 bolts. Tighten to 35–50 ft. lbs. (47–67 Nm).

12. Place the rear support bracket in position and install the 8 retaining bolts. Install the 2 bolts at the rear mount bracket and remove the transmission jack.

13. Place the transfer case on the transmission jack and install the 6 retaining bolts and gasket. Position the transfer case on the transmission and tighten the bolts.

14. Install the transfer case shift link and the speedometer cable. Connect the front and rear driveshafts.

15. Fill the transmission and transfer case with the proper type and quantity of lubricant. Lower the vehicle.

16. Install the gasket and shift cover. Install the transfer case shift lever, shift ball and boot assembly and the transmission shift lever, shift ball and boot assembly.

17. Install the dust cover and insulator, the access cover-to-floor pan screws, and the floor mat screws.

F SERIES 2WD

1. Remove the floor mat, the body floor pan cover, the gearshift lever shift ball and boot as an assembly. and remove the transmission shift lever. Remove the isolator pad.

2. Raise and support the vehicle safely. Position a transmission jack under the transmission and disconnect the speedometer cable.

3. Disconnect the back-up light switch located at the rear of the gearshift housing cover.

4. Disconnect the driveshaft or coupling shaft and clutch actuator from the transmission and wire it to 1 side.

5. Disconnect the transmission mount and remove the transmission crossmember.

6. Remove the transmission attaching bolts.

7. Move the transmission to the rear until the input shaft clears the clutch housing and lower the transmission.

To install:

8. Place the transmission on a transmission jack, install guide studs in the clutch housing top holes, and raise the transmission until the input shaft splines are aligned with the clutch disc splines. The clutch release bearing and hub must be properly positioned in the release lever fork.

9. Slide the transmission forward on the guide studs until it is in position on the clutch housing. Install the top attaching bolts and tighten to 35–50 ft. lbs. (48–67 Nm). Remove the guide studs, install the 2 lower attaching bolts and tighten to 35–50 ft. lbs. (48–67 Nm).

10. Install the crossmember and transmission mount. Connect the speedometer cable and clutch actuator.

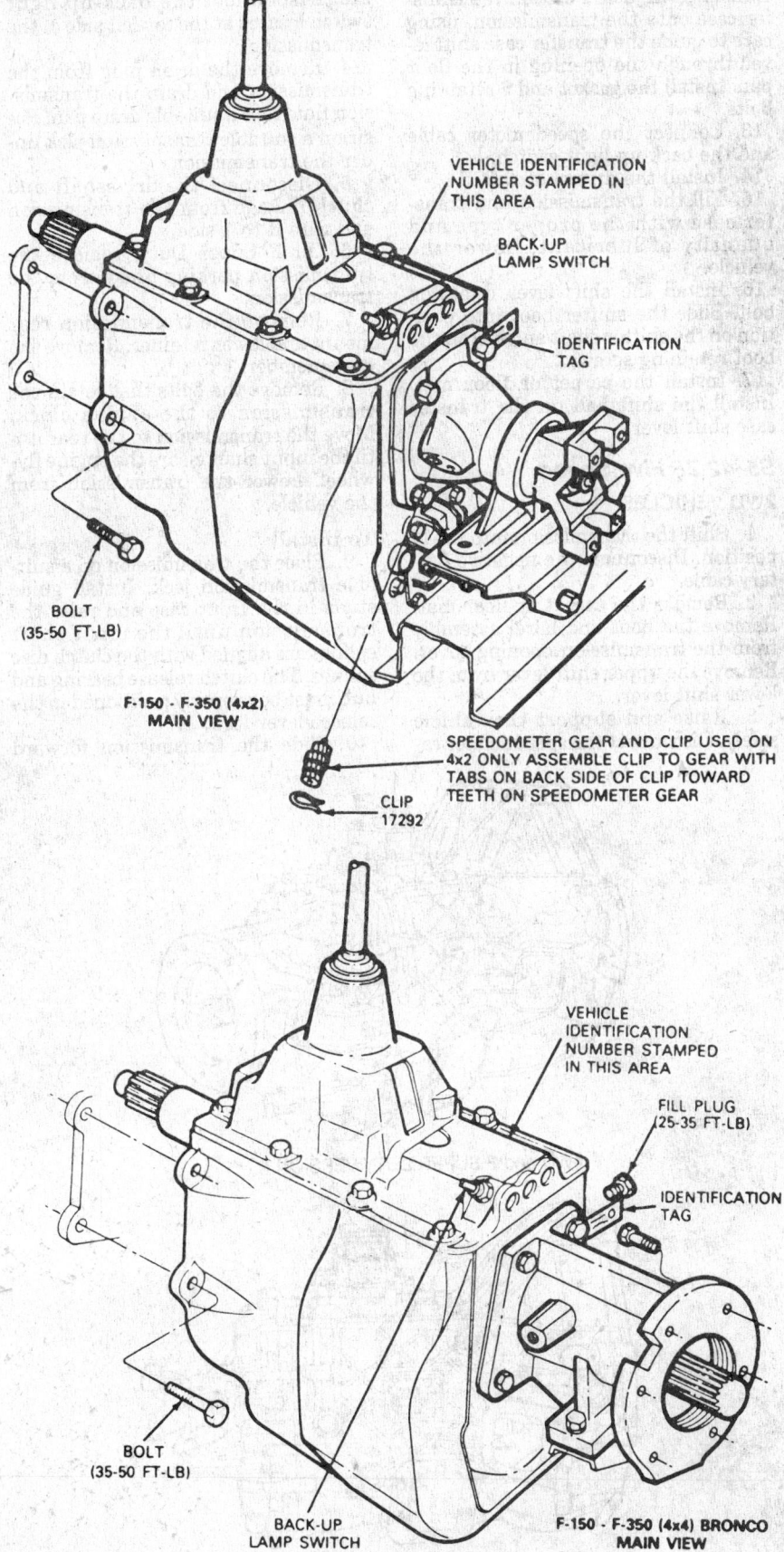

F-150 - F-350 (4x2) MAIN VIEW

VEHICLE IDENTIFICATION NUMBER STAMPED IN THIS AREA

BACK-UP LAMP SWITCH

IDENTIFICATION TAG

BOLT (35-50 FT-LB)

SPEEDOMETER GEAR AND CLIP USED ON 4x2 ONLY ASSEMBLE CLIP TO GEAR WITH TABS ON BACK SIDE OF CLIP TOWARD TEETH ON SPEEDOMETER GEAR

CLIP 17292

VEHICLE IDENTIFICATION NUMBER STAMPED IN THIS AREA

FILL PLUG (25-35 FT-LB)

IDENTIFICATION TAG

BOLT (35-50 FT-LB)

BACK-UP LAMP SWITCH

F-150 - F-350 (4x4) BRONCO MAIN VIEW

Borg-Warner T-18 manual transmission

11. Install the bolts attaching the front U-joint of the coupling shaft to the transmission output shaft. Connect the back-up light switch and lower the vehicle.

12. Install the shift lever, boot and shift ball as an assembly and lubricate the spherical ball seat. Install the isolator pad, floor pan cover and floor mat.

13. Road test the vehicle for proper operation.

Mazda M50D Five Speed

2WD VEHICLES

1. Shift the transmission into the **N** position. Disconnect the negative battery cable.

2. Remove the carpet or floor mat. Remove the shifter boot retainer screws and slide the boot up the shift lever shaft. Remove the shift lever retaining bolt and remove the shift lever.

3. Raise and support the vehicle safely. Disconnect the speedometer cable. Disconnect the back-up light switch located at the top left side of the transmission.

4. Remove the drain plug from the transmission and drain the transmission fluid into a suitable drain pan. Position a suitable transmission jack under the transmission.

5. Remove the driveshaft from the transmission. Disconnect the clutch slave cylinder hydraulic line.

6. Remove the transmission rear insulator and lower retainer. Remove the crossmember.

7. Remove the bolts that retain the transmission to the engine block. Move the transmission to the rear until the input shaft clears the clutch. Lower the transmission from the vehicle.

To install:

8. Place the transmission on a suitable transmission jack. Install guide studs into the engine block and raise the transmission up until the input shaft splines are aligned with the clutch disc splines.

9. Slide the transmission forward on the guide studs until the transmission is in position. Install the transmission retaining bolts and torque to 40–50 ft. lbs. (54–67 Nm). Remove the guide studs and install the remaining bolts. Tighten the 2 bolts for the lower plate to 9–12 ft. lbs. (12–16 Nm).

10. Install the crossmember. Position the mount and retainer between the transmission and crossmember. Install the bolts and tighten to 60–80 ft. lbs. (81–108 Nm). Install the nut retaining the mount and retainer to the crossmember. Tighten to 60–80 ft. lbs. (81–108 Nm) and remove the transmission jack.

11. Connect the speedometer cable and the clutch hydraulic line.

12. Connect the back-up light switch and install the driveshaft. Lower the vehicle.

13. Install the shift lever retaining bolt. Slide the shifter boot into position on the shifter shaft and install the boot retaining screws.

14. Install the carpet or floor mat. Connect the negative battery cable and road test for proper operation.

4WD VEHICLES

1. Shift the transmission into the **N** position. Disconnect the negative battery cable and remove the shift ball from the transfer case shift lever.

2. Remove the carpet or floor mat. Remove the shifter boot retainer screws and slide the boot up the shift lever shaft. Remove the shift lever retaining bolt and remove the shift lever.

3. Raise and support the vehicle safely. Remove the drain plugs and drain the transmission and transfer case.

4. Remove the front and rear driveshafts from the transfer case.

5. Disconnect the speedometer cable and the back-up light switch. If equipped, remove the skid pad from under the transfer case.

6. Support the transfer case, using a transmission jack. Remove the 6 bolts holding the transfer case to the transmission and carefully lower the transfer case from the vehicle, using care to ensure that the transfer case shift lever clears the opening in the floor pan.

7. Support the transmission with a transmission jack. Remove the transmission rear mount and lower retainer. Remove the crossmember.

8. Remove the bolts that retain the transmission to the engine block. Move the transmission to the rear until the input shaft clears the clutch. Lower the transmission from the vehicle.

To install:

9. Place the transmission on a transmission jack and install 2 guide studs in the transmission front case top holes. Raise the transmission up until the input shaft splines are aligned with the clutch disc splines.

10. Slide the transmission forward onto the guide studs until the transmission is in position. Install the 2 lower transmission retaining bolts and torque to 40–50 ft. lbs. (54–67 Nm). Remove the guide studs, install the upper bolts and tighten to 40–50 ft. lbs. (54–67 Nm).

11. Place the rear support bracket in position and install the retaining bolts. Install the crossmember. Position the mount between the transmission and crossmember. Remove the transmission jack.

12. Position the transfer case on the

transmission jack. Position the transfer case onto the transmission, using care to guide the transfer case shift lever through the opening in the floor pan. Install the gasket and 6 retaining bolts.

13. Connect the speedometer cable and the back-up light switch.

14. Install the driveshafts.

15. Fill the transmission and transfer case with the proper type and quantity of lubricant. Lower the vehicle.

16. Install the shift lever retaining bolt. Slide the shifter boot into position on the shifter shaft and install the boot retaining screws.

17. Install the carpet or floor mat. Install the shift ball on the transfer case shift lever.

S5–42 ZF Five Speed

2WD VEHICLES

1. Shift the transmission into the **N** position. Disconnect the negative battery cable.

2. Remove the carpet or floor mat. Remove the boot and bezel assembly from the transmission opening cover. Remove the upper shift lever from the lower shift lever.

3. Raise and support the vehicle safely. Disconnect the speedometer ca-

ble. Disconnect the back-up light switch located at the top left side of the transmission.

4. Remove the drain plug from the transmission and drain the transmission fluid into a suitable drain pan. Position a suitable transmission jack under the transmission.

5. Disconnect the driveshaft and clutch linkage from the transmission and wire it to 1 side.

6. On F Super Duty, remove the transmission parking brake from the transmission.

7. Remove the transmission rear mount and lower retainer. Remove the crossmember.

8. Remove the bolts that retain the transmission to the engine block. Move the transmission to the rear until the input shaft clears the engine flywheel. Lower the transmission from the vehicle.

To install:

9. Place the transmission on a suitable transmission jack. Install guide studs in the front case and raise the transmission until the input shaft splines are aligned with the clutch disc splines. The clutch release bearing and hub must be properly positioned in the release lever fork.

10. Slide the transmission forward

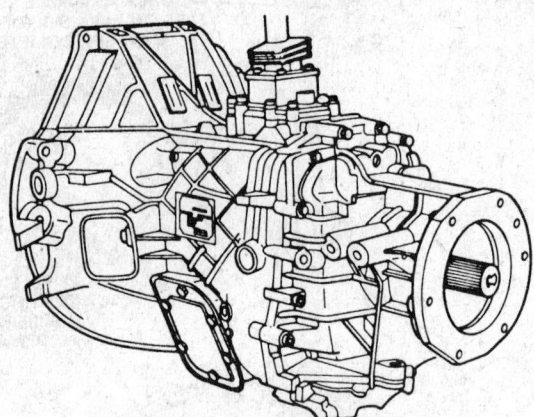

4WD and F SUPER DUTY VERSION

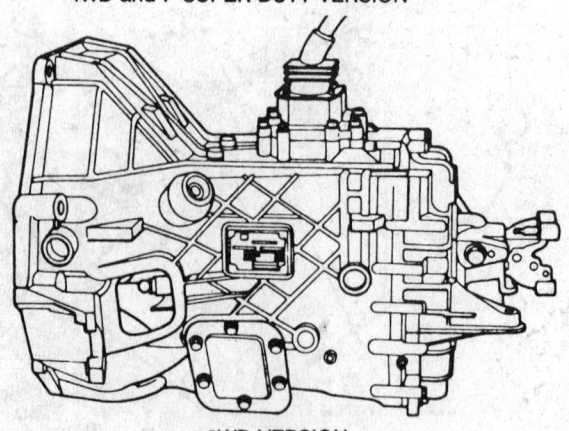

2WD VERSION

S5–42 ZF manual transmission

on the guide studs until the transmission is in position. Install the transmission retaining bolts and torque to 40–50 ft. lbs. (54–67 Nm). Remove the guide studs and install the 2 remaining bolts.

11. Install the crossmember. Position the mount and retainer between the transmission and crossmember. Install the bolts and tighten to 60–80 ft. lbs. (81–108 Nm). Install the nut retaining the mount and retainer to the crossmember. Tighten to 60–80 ft. lbs. (81–108 Nm) and remove the transmission jack.

12. Connect the speedometer cable, clutch linkage and back-up light switch.

13. On F Super Duty vehicles, install the transmission parking brake.

14. Connect the driveshaft. Fill the transmission with the proper type and quantity of lubricant and lower the vehicle.

15. Install the upper shift lever to the lower shift lever. Tighten the 2 retaining screws to 16–24 ft. lbs. (22–33 Nm). Install the boot and bezel assembly to the transmission opening cover.

16. Install the carpet or floor mat.

4WD VEHICLES

1. Shift the transmission into the **N** position. Disconnect the negative battery cable.

2. Remove the carpet or floor mat. Remove the boot and bezel from the transmission opening cover. Remove the upper shift lever from the lower shift lever.

3. Raise and support the vehicle safely. Remove the drain plugs and drain the transmission and transfer case.

4. Disconnect the front and rear driveshafts from the transfer case and wire them out of the way.

5. Disconnect the speedometer cable and back-up light switch. If equipped, remove the skid pan from under the transfer case.

6. Support the transfer case using a transmission jack. Remove the 6 bolts holding the transfer case to the transmission and carefully lower the transfer case from the vehicle, using care to ensure that the transfer case shift lever clears the opening in the floor pan.

7. Support the transmission with the transmission jack. Remove the transmission rear mount and lower retainer. Remove the crossmember.

8. Remove the bolts that retain the transmission to the engine block. Move the transmission to the rear until the input shaft clears the engine flywheel housing. Lower the transmission from the vehicle.

To install:

9. Place the transmission on a transmission jack and install 2 guide studs in the transmission front case top holes. Raise the transmission up until the input shaft splines are aligned with the clutch disc splines.

10. Slide the transmission forward on the guide studs until the transmission is in position. Install the transmission retaining bolts and torque to 40–50 ft. lbs. (54–67 Nm). Remove the guide studs and install the remaining bolts.

11. Place the rear support bracket in position and install the retaining bolts. Install the crossmember. Position the mount between the transmission and crossmember. Remove the transmission jack.

12. Position the transfer case on the transmission jack. Position the transfer case onto the transmission, using care to guide the transfer case shift lever through the opening in the floor pan. Install the gasket and 6 retaining bolts.

13. Connect the speedometer cable and the back-up light switch.

14. Install the front and rear driveshafts. Fill the transmission and transfer case with the proper type and quantity of lubricant. Lower the vehicle.

15. Install the upper shift lever to the lower shift lever and tighten the retaining bolts to 16–24 ft. lbs. (22–33 Nm). Install the boot and bezel assembly.

16. Install the carpet or floor mat. Install the shift ball on the upper shift lever and on the transfer case shift lever, if removed.

17. Road test the vehicle for proper shift operation.

CLUTCH

Clutch Assembly

REMOVAL & INSTALLATION

1. Disconnect the negative battery cable(s). Raise and safely support the vehicle.

2. Remove the clutch slave cylinder or the hydraulic line, as necessary.

3. Remove the dust cover and clutch release lever, if equipped.

4. Remove the transmission from the vehicle.

5. Mark the assembled position of the pressure plate to the flywheel, so they can be reinstalled in the same position.

6. Loosen the pressure plate attaching bolts evenly and remove the pressure plate and clutch disc.

7. Inspect the flywheel for wear, scoring and cracks. Machine or replace, as necessary. Inspect the clutch pilot bearing for wear and free movement. If replacement is necessary, remove using puller tool T58L–101–B or equivalent.

8. Inspect the clutch release bearing for wear and free movement; replace as necessary. If equipped with a concentric slave cylinder, remove the release bearing by twisting it until resistance is felt. Turning further will allow the preload spring to push the bearing assembly off the slave cylinder.

To install:

9. If the pilot bearing was removed, a new 1 must be installed. Install using replacer tool T74P–7137–A and clutch aligner tool T71P–7137–H or equivalent. Install the pilot bearing with the seal facing the transmission so the adapter is not cocked.

10. If the flywheel was removed, make sure the mating surfaces of the crank flange and flywheel are clean, and install the flywheel. Tighten the flywheel bolts to 75–85 ft. lbs. (102–115 Nm) on all except 7.3L diesel engine. On 7.3L diesel engine, apply locking compound to the bolts and tighten to 47 ft. lbs. (64 Nm).

11. Position the clutch disc on the flywheel so alignment tool D79T–7550–A or equivalent, can enter the pilot bearing and align the disc.

12. Install the pressure plate. If the original pressure plate is being reused, align the marks that were made during the removal procedure.

13. Install the pressure plate bolts and tighten in a gradual criss-cross pattern. The final torque should be 20–29 ft. lbs. (27–39 Nm) on all except 4.9L, 5.0L and 5.8L engines with 10 in. clutch and all 7.3L diesel and 7.5L engines, where the final torque should be 15–20 ft. lbs. (20–27 Nm).

14. Clean and lubricate the transmission bearing retainer. Clean and lubricate the bearing hub bore and install on the retainer. Clean and lubricate the release lever pivot stud, if equipped.

15. Install the transmission. If equipped, position the release lever in the bearing hub and align with the pivot stud. Push inward on the lever until it snaps into position. Install the dust boot, if equipped.

16. Install the slave cylinder, or attach the hydraulic line.

17. Connect the negative battery cable and bleed the hydraulic system, if necessary. Road test the vehicle for proper operation.

PEDAL HEIGHT/FREE PLAY ADJUSTMENT

The hydraulic clutch system provides

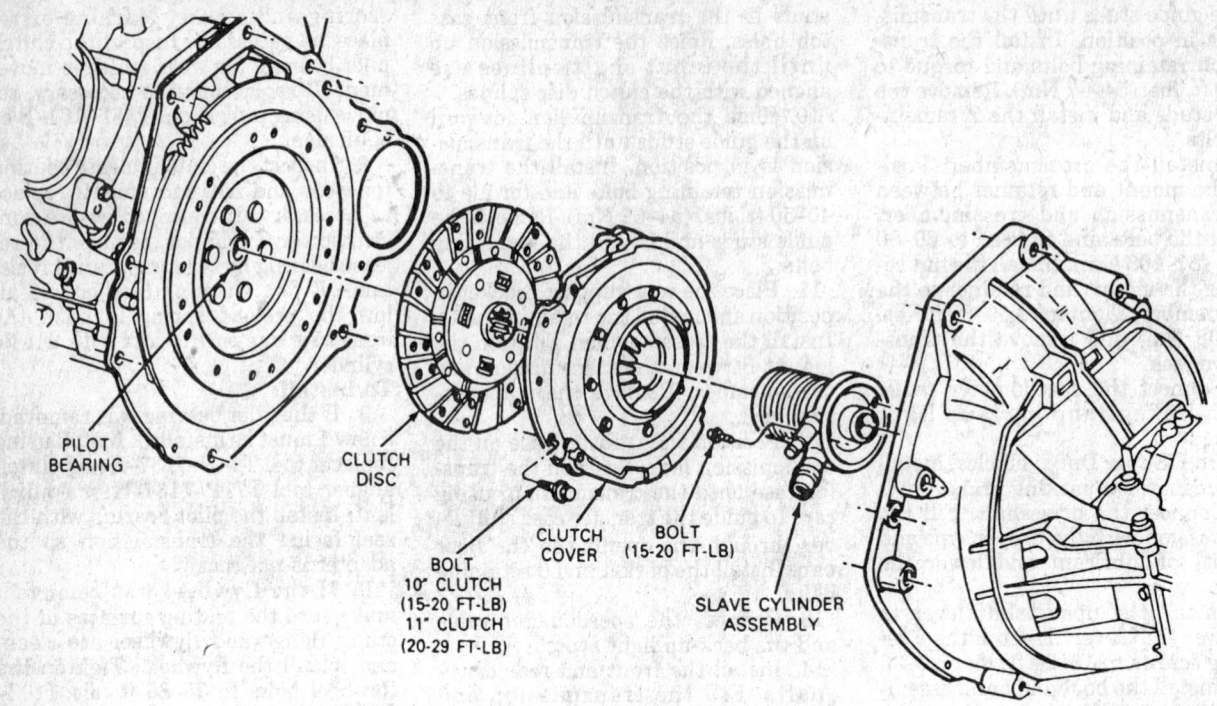

Clutch Installation—Bronco, F Series and E Series with 4.9L, 5.0L and 5.8L engines

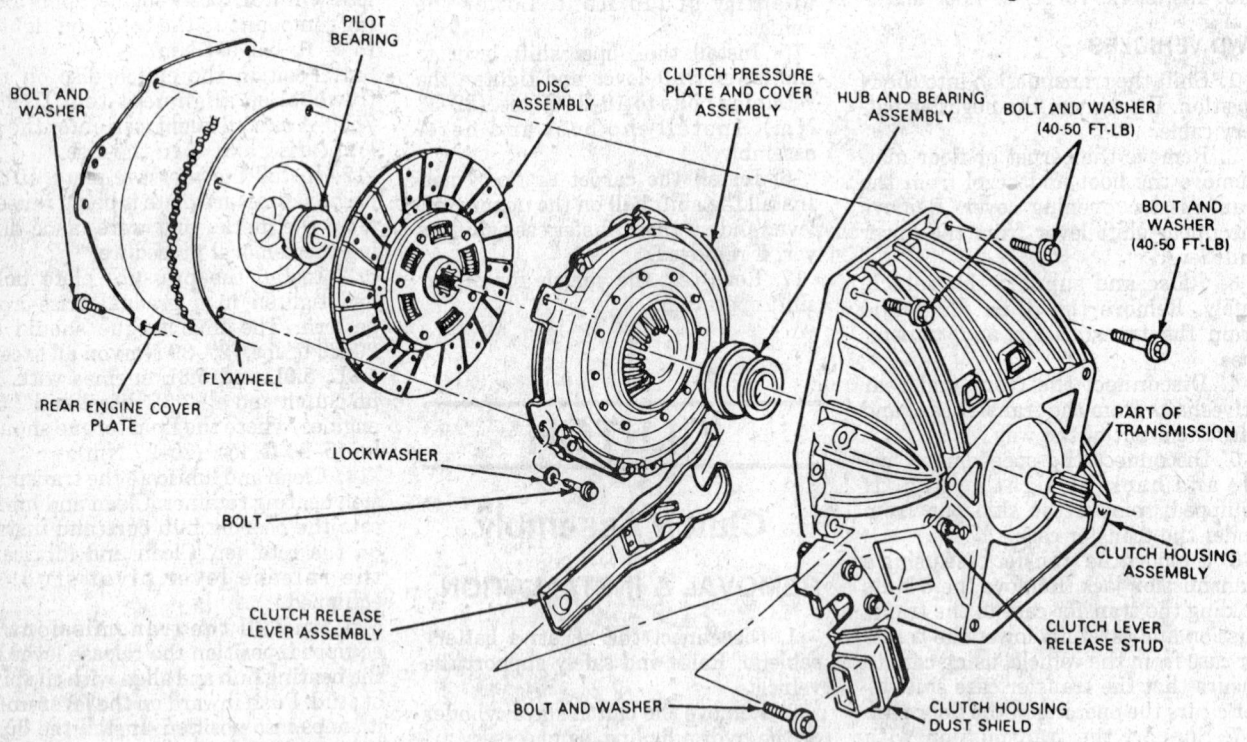

Clutch Installation—Bronco and F Series with 4.9L and 5.0L engines and Warner transmission

automatic adjustment. No adjustment of clutch linkage or pedal position is required.

Clutch Master Cylinder

REMOVAL & INSTALLATION

NOTE: The master cylinder is removed with the hydraulic line attached.

1. Disconnect the negative battery cable.

2. From inside the vehicle, carefully pry the pushrod and retainer bushing from the cross-shaft lever pin. Disconnect the interlock switch connector plug.

3. If equipped with an external slave cylinder, remove the slave cylinder and use a $^3/_{32}$ in. diameter punch to drive out the pin that holds the tube in the slave cylinder. Disconnect and plug the tube.

4. If equipped with a concentric slave cylinder, disconnect the coupling at the transmission with coupling tool

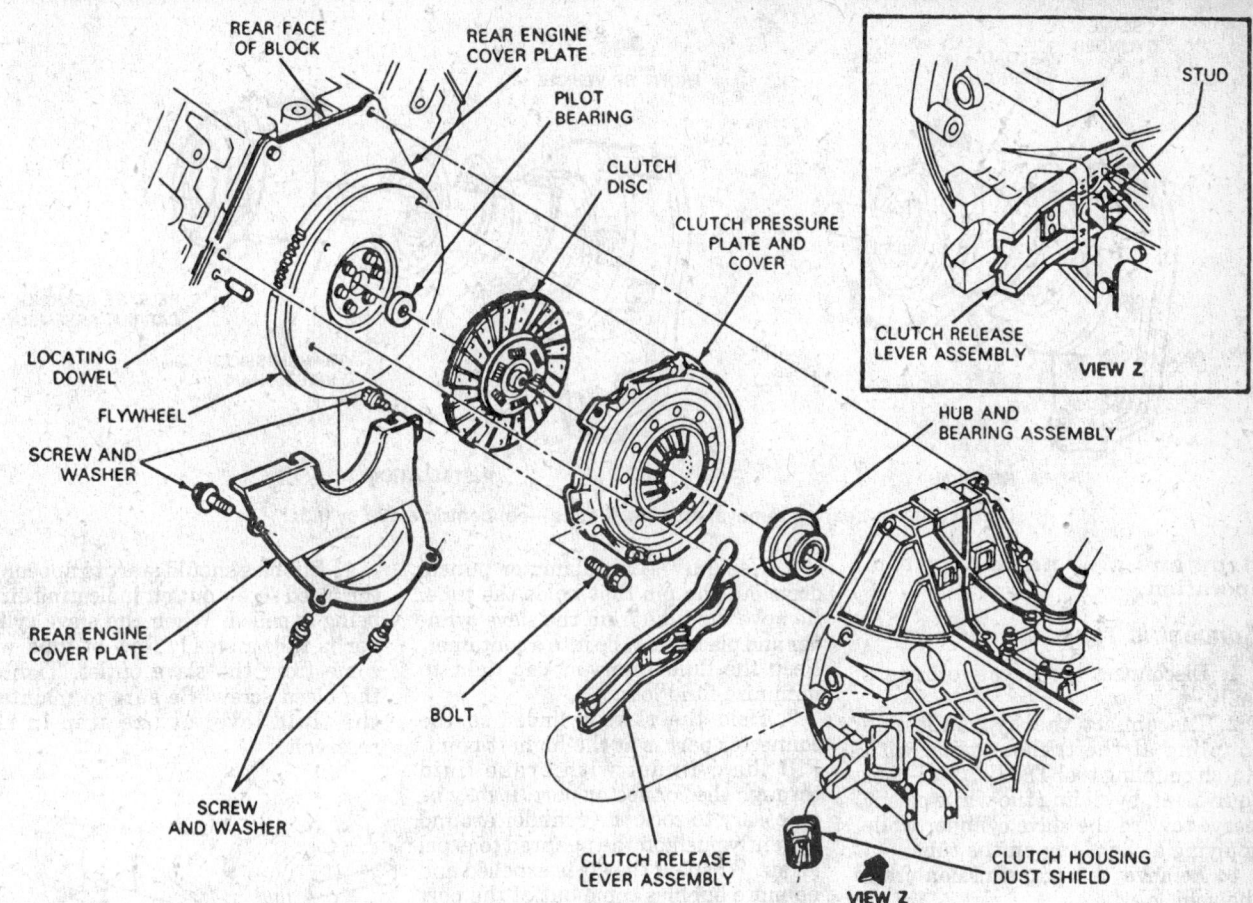

REAR FACE OF BLOCK

REAR ENGINE COVER PLATE

PILOT BEARING

CLUTCH DISC

CLUTCH PRESSURE PLATE AND COVER

STUD

CLUTCH RELEASE LEVER ASSEMBLY

VIEW Z

LOCATING DOWEL

FLYWHEEL

SCREW AND WASHER

REAR ENGINE COVER PLATE

SCREW AND WASHER

BOLT

HUB AND BEARING ASSEMBLY

CLUTCH RELEASE LEVER ASSEMBLY

VIEW Z

CLUTCH HOUSING DUST SHIELD

Clutch Installation—F Series with 7.3L diesel and 7.5L engines

T88T–70522–A or equivalent, by sliding the white plastic sleeve toward the slave cylinder while applying a slight tug on the tube. Plug the tube.

5. Note the clutch tube routing to the slave cylinder, then remove the attaching hardware for the hydraulic tube retaining clips.

6. Remove the 2 nuts and support bracket retaining the clutch reservoir and master cylinder assembly to the firewall and remove the assembly. On Bronco and F Series, when the master cylinder studs are free of the dash panel, rotate the cylinder 105 degrees counterclockwise to permit the interlock switch to pass through the dash panel.

7. Installation is the reverse of the removal procedure. Press the pushrod with the retainer bushing onto the cross-shaft lever pin until the bushing tabs snap into position in the groove. The flanged side of the bushing must be towards the cross-shaft lever. Bleed the system.

CLUTCH MASTER CYLINDER PUSHROD LENGTH ADJUSTMENT

To determine if the pushrod is adjust-

ed properly, disconnect it and see if it will reassemble to the pin. When the pushrod is disconnected from the pin, the master cylinder piston is fully retracted and the clutch pedal blade is contacting the rubber bumper stop. If the pushrod and lever pin are not aligned, proceed as follows, only in the sequence given.

1. Tighten the clutch pedal attaching nut on the left side.

2. Reinstall the master cylinder pushrod to the cross-shaft lever pin.

3. Stroke the clutch pedal several times to reset the position of the shaft to the pedal slot. This Step is mandatory.

4. Remove the pushrod from the lever pin and check the alignment. If the pushrod and lever pin still do not align, replace the cross-shaft lever.

Clutch Slave Cylinder

REMOVAL & INSTALLATION

External Type

NOTE: If the slave cylinder is disconnected with the hydraulic line attached, the master cylinder pushrod must be disconnected from the clutch pedal. If not dis-

connected, permanent damage to the slave cylinder will occur if the clutch pedal is depressed while the slave cylinder is disconnected.

1. Lift the 2 retaining tabs of the slave cylinder retaining bracket, using a small prybar. Disengage the tabs from the bell housing lugs and slide outward to remove.

2. Use a $^3/_{32}$ in. diameter punch to drive out the pin that holds the hydraulic tube in the slave cylinder. Disconnect and plug the tube.

To install:

3. Bleed the hydraulic system and connect the hydraulic tube. Install the retainer pin.

4. Push the slave cylinder pushrod into the cylinder. Engage the pushrod into the release lever and slide the slave cylinder into the bell housing lugs.

NOTE: A new slave cylinder contains a shipping strap that pre-positions the pushrod for installation and also provides a bearing insert. Following installation of the new slave cylinder, the first actuation of the clutch pedal will break the shipping

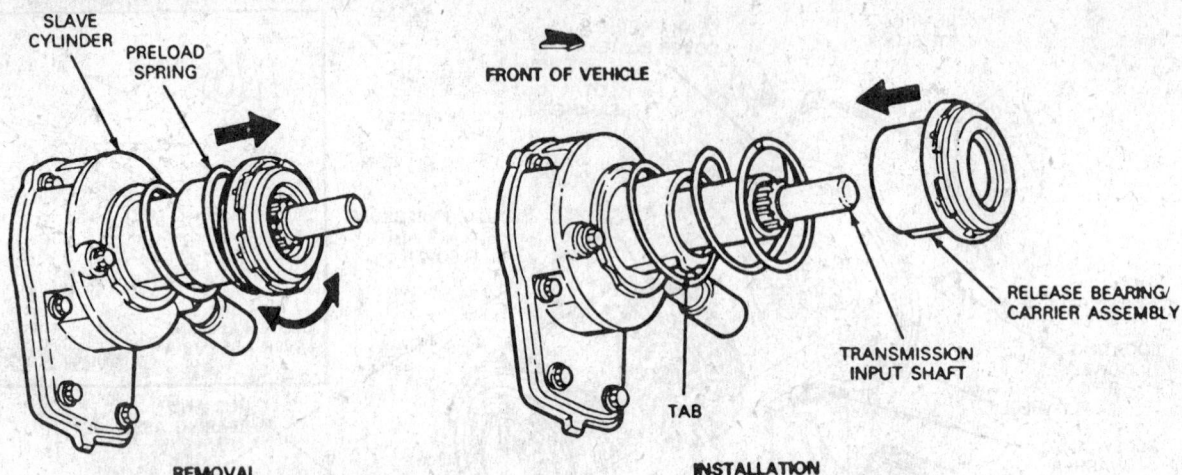

Clutch release bearing removal and installation—concentric slave cylinder

strap and give normal system operation.

Concentric Type

1. Disconnect the negative battery cable.

2. Disconnect the hydraulic line coupling at the transmission with clutch coupling tool T88T-70522–A or equivalent, by sliding the white plastic sleeve toward the slave cylinder while applying a slight tug on the tube.

3. Remove the transmission from the vehicle.

4. To remove the clutch release bearing from the slave cylinder, twist the bearing until resistance is felt. Turning further will allow the preload spring to push the bearing assembly off the slave cylinder.

5. Remove the bolts retaining the slave cylinder to the transmission and remove the slave cylinder from the transmission input shaft.

To install:

6. Position the slave cylinder over the transmission input shaft with the bleed screw and coupling facing the left side of the transmission.

7. Install the slave cylinder attaching bolts and tighten to 15–20 ft. lbs. (20–27 Nm).

8. If removed, install the release bearing by pushing into place.

9. Install the transmission.

10. Insert the male coupling into the female coupling on the slave cylinder and check that the connection is secure.

Hydraulic Clutch System Bleeding

With External Slave Cylinder

1. Clean the reservoir cap and slave cylinder in the area of the tube connection. Remove the slave cylinder from the transmission bell housing.

2. Using a $^3/_{32}$ in. diameter punch, drive out the pin that holds the tube. Remove the tube from the slave cylinder and place the tube into a container. Keep the fluid reservoir cap tight to minimize fluid loss.

3. Hold the slave cylinder so the connector port is at the highest point. Fill the cylinder with brake fluid through the connector port. It may be necessary to rock the cylinder around or gently push on the pushrod to expel all air. When all the air is expelled and no more bubbles come out of the port hole, reinstall the slave cylinder.

NOTE: Some fluid will be expelled from the connector port as the pushrod is compressed, attaching it to the transmission and lever.

4. Gravity fill the clutch master cylinder and tube as follows: Remove the reservoir cap and diaphragm. Fluid should flow out the open end of the tube into the container. Be sure to keep the reservoir full. When fluid is flowing out in a steady, uninterrupted flow and the fluid is level with the step in the reservoir, reinstall the cap and diaphragm. Install the end of the tube into the slave cylinder and replace the pin holding the tube in place.

With Concentric Slave Cylinder

1. Clean dirt and grease from the reservoir cap. Remove the cap and diaphragm and fill reservoir to the top with brake fluid.

2. Attach 1 end of a rubber tube to the slave cylinder bleed screw, located next to the inlet connection, and place the other end into a container.

3. Loosen the bleed screw, allowing fluid to move from the master cylinder down the tube to the slave cylinder. Be sure to keep the reservoir full to make sure no additional air enters the system.

4. Bubbles should start to appear at the bleed screw outlet, indicating air is being expelled. When the slave cylinder is full, a steady flow of fluid will come from the slave outlet. Tighten the bleed screw. Be sure to maintain the fluid level at the step in the reservoir.

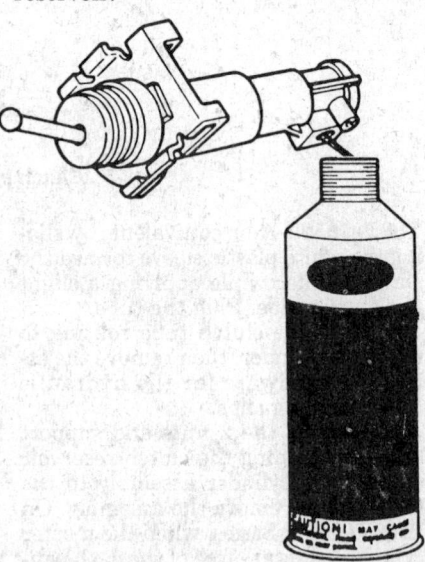

External slave cylinder bleeding

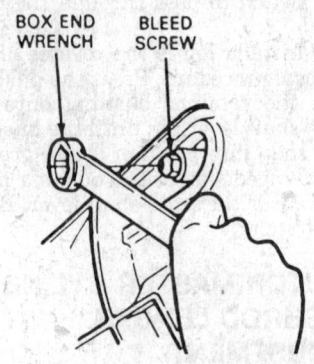

Concentric slave cylinder bleeding

5. Depress the clutch pedal to the floor and hold for 2 seconds, then release the pedal as rapidly as possible. The pedal must be released completely, then pause for 2 seconds. Repeat 10 times.

6. Check the fluid level in the reservoir. The fluid should be level with the step when the diaphragm is removed.

7. Repeat Steps 5 and 6 five times. Replace the reservoir diaphragm and cap.

8. Have an assistant hold the clutch pedal to the floor. Crack open the bleed screw to allow any additional air to escape. Close the screw and release the pedal. Check the fluid level.

AUTOMATIC TRANSMISSION

For further information on transmissions/transaxles, please refer to "Chilton's Guide to Transmission Repair".

Transmission Assembly

REMOVAL & INSTALLATION

Ford C6

BRONCO AND F SERIES

1. Disconnect the negative battery cable and disconnect the neutral switch wire at the plug connector. Raise and support the vehicle safely.

2. Place a drain pan under the transmission fluid pan. Starting at the rear of the pan and working toward the front, loosen the attaching bolts and allow the fluid to drain. Remove all the pan bolts except 2 at the front, to drain all of the fluid. After draining is completed, install 2 bolts at the rear of the pan to hold it in place.

3. Remove the converter access cover. Using a wrench on the crankshaft pulley, turn the engine over to gain access to and remove the converter-to-flywheel nuts. Place the drain pan under the converter and turn the crankshaft pulley to gain access to the converter drain plug. Remove the plug, drain the fluid and reinstall the plug.

4. On 2WD vehicles, remove the driveshaft.

5. Disconnect the speedometer cable from the extension housing.

6. Disconnect the throttle and manual linkage rods or cable controls from the levers at the transmission.

7. Disconnect the oil cooler lines from the transmission.

8. Remove the vacuum hose from the vacuum diaphragm unit. Remove the vacuum line from the retaining clip.

9. Disconnect the cable from the terminal on the starter motor. Remove the attaching bolts and remove the starter.

10. On 4WD vehicles, remove the transfer case.

11. Remove the 2 engine rear support and mount assembly attaching bolts.

12. Remove the 2 engine rear support and mount assembly-to-extension housing attaching bolts.

13. Remove the 6 bolts securing the No. 2 crossmember to the frame side rails.

14. Raise the transmission with a transmission jack and remove the crossmember.

15. Secure the transmission to the jack with the safety chain.

16. Remove the converter housing-to-engine attaching bolts.

17. Move the transmission away from the engine. Lower the transmission and remove it from under the vehicle.

To install:

18. Tighten the converter drain plug to 18–28 ft. lbs. (11–37 Nm). Position the converter on the transmission, making sure the converter drive flats are fully engaged in the pump gear.

19. With the converter properly installed, place the transmission on the jack. Secure the unit to the jack with the chain.

20. Rotate the converter so the studs and drain plug are in alignment with their holes in the flywheel.

21. Move the transmission forward into position, being careful not to damage the flywheel and the converter pilot.

NOTE: The converter must rest squarely against the flywheel. This indicates that the converter pilot is not binding in the engine crankshaft. Do not allow the converter drive flats to disengage from the pump gear.

22. Install the converter housing-to-engine bolts. On gasoline engines, tighten the retaining bolts to 40–50 ft. lbs. (55–67 Nm). On diesel engines, tighten the retaining bolts to 50–65 ft. lbs. (67–87 Nm).

23. Remove the transmission jack safety chain from around the transmission.

24. Position the No. 2 crossmember to the frame side rails. Install the attaching bolts.

25. On 4WD vehicles, install the transfer case.

26. Position the engine rear support and mount assembly above the crossmember. Install the rear support and mount assembly-to-extension housing mounting bolts.

27. Lower the transmission and remove the jack.

28. Secure the engine rear support and mount assembly to the crossmember with the attaching bolts.

29. Connect the vacuum line to the vacuum diaphragm, making sure the metal line is secured in the retaining clip.

30. Connect the oil cooler lines to the transmission.

31. Connect the throttle and manual linkage rods or cable controls to their respective levers on the transmission.

32. Connect the speedometer cable to the extension housing.

33. Secure the starter in place with the attaching bolts. Connect the cable to the terminal on the starter.

34. Install a new O-ring on the lower end of the transmission filler tube and insert the tube in the case.

35. Secure the converter-to-flywheel attaching nuts and tighten to 20–30 ft. lbs. (28–40 Nm).

36. Install the converter housing access cover and secure it with the attaching bolts.

37. Connect the driveshaft.

38. Clean the gasket mating surfaces on the transmission pan and transmission. Install a new pan gasket and install the pan. Tighten the retaining bolts to 8–12 ft. lbs. (11–16 Nm).

39. Lower the vehicle. Fill the transmission to the correct level with the proper type of fluid. Adjust the shift linkage.

40. Connect the neutral switch wire to the plug connector and connect the negative battery cable.

41. Start the engine and check for leaks. Road test the vehicle and check for proper operation.

E SERIES

1. Disconnect the negative battery cable. Remove the engine cover from inside the vehicle.

2. Disconnect the neutral start switch wires at the plug connector.

3. If equipped with 8 cylinder engine, remove the flex hose from the air cleaner heat tube.

4. Remove the upper converter housing to engine attaching bolts. Remove the bolt securing the filler tube to the engine.

5. Raise the vehicle and support it safely.

6. Place a drain pan under the transmission fluid pan. Starting at the rear of the pan and working toward the front, loosen the attaching bolts and allow the fluid to drain. Remove all the pan bolts except 2 at the front,

to drain all of the fluid. After draining is completed, install 2 bolts at the rear of the pan to hold it in place.

7. Remove the converter access cover. Using a wrench on the crankshaft pulley, turn the engine over to gain access to and remove the converter-to-flywheel nuts. Place the drain pan under the converter and turn the crankshaft pulley to gain access to the converter drain plug. Remove the plug, drain the fluid and reinstall the plug.

8. Disconnect the driveshaft and remove the fluid filler tube.

9. Disconnect the starter cable at the starter and remove the starter from the converter housing.

10. Position a suitable engine support bar to the frame and engine oil pan flanges.

11. Disconnect the cooler lines from the transmission. Disconnect the vacuum line from the diaphragm unit and remove the line from the retaining clip at the transmission.

12. Remove the speedometer driven gear from the extension housing.

13. Disconnect the manual and downshift linkage rods or cable controls from the control levers.

14. Position a transmission jack under the transmission. Install the safety chain to hold the transmission.

15. Remove the nuts securing the rear support and mount assembly to the crossmember. Remove the 6 bolts retaining the crossmember to the side rails and remove the 2 support gussets. Raise the transmission with the jack and remove the crossmember.

16. Remove the remaining converter housing-to-engine attaching bolts. Lower the jack and remove the converter and transmission assembly from under the vehicle.

To install:

17. Tighten the converter drain plug to 18–28 ft. lbs. (11–37 Nm). Position the converter on the transmission, making sure the converter drive flats are fully engaged in the pump gear.

18. With the converter properly installed, place the transmission on the jack. Secure the unit to the jack with the chain.

19. Rotate the converter so the studs and drain plug are in alignment with their holes in the flywheel.

20. Move the transmission forward into position, being careful not to damage the flywheel and the converter pilot.

NOTE: The converter must rest squarely against the flywheel. This indicates that the converter pilot is not binding in the engine crankshaft. Do not allow the converter drive flats to disengage from the pump gear.

21. Install the converter housing-to-engine bolts. On gasoline engines, tighten the retaining bolts to 40–50 ft. lbs. (55–67 Nm). On diesel engines, tighten the retaining bolts to 50–65 ft. lbs. (67–87 Nm).

22. Install the converter-to-flywheel attaching nuts and tighten to 20–30 ft. lbs. (28–40 Nm).

23. Install the crossmember. Install the rear support and mount assembly-to-crossmember attaching bolts and nuts.

24. Remove the safety chain and remove the jack from under the vehicle. Remove the engine support bar.

25. Install a new O-ring on the lower end of the transmission filler tube and insert the tube and dipstick in the case.

26. Connect the vacuum line to the vacuum diaphragm, making sure the line is secured in the retaining clip. Connect the oil cooler lines. Install the speedometer driven gear into the extension housing.

27. Connect the transmission linkage rods to the control levers. Be sure to use a new retaining ring and grommet. Attach the shift rod to the steering column shift lever. Align the flats of the adjusting stud with the flats of the rod slot and insert the stud through the rod. Assemble the adjusting stud nut and washer to a loose fit. Adjust the linkage.

28. Install the converter housing access cover and tighten the attaching bolts to 12–16 ft. lbs. (17–21 Nm).

29. Install the starter with the attaching bolts. Connect the starter cable.

30. Clean the gasket mating surfaces on the transmission pan and transmission. Install a new pan gasket and install the pan. Tighten the retaining bolts to 8–12 ft. lbs. (11–16 Nm).

31. Install the driveshaft and lower the vehicle.

32. If equipped with a V8 engine, install the flex hose to the air cleaner heat tube. Install the bolt that retains the filler tube to the cylinder block.

33. Connect the neutral start switch wires at the plug connector. Fill the transmission to the correct level with the proper type of fluid.

34. Connect the negative battery cable, start the engine and check for leaks. Adjust the throttle and manual linkage.

35. Install the engine compartment cover.

Ford AOD

1. Disconnect the negative battery cable. On E series, remove the engine cover. Raise the vehicle and support it safely.

2. Place the drain pan under the transmission fluid pan. Starting at the rear of the pan and working toward the front, loosen the attaching bolts and allow the fluid to drain. Finally remove all of the pan attaching bolts except 2 at the front, to allow the fluid to further drain. With the fluid drained, install 2 bolts on the rear side of the pan to temporarily hold it in place.

3. Remove the converter drain plug access cover from the lower end of the converter housing.

4. Remove the converter-to-flywheel attaching nuts. Place a wrench on the crankshaft pulley attaching bolt to turn the converter to gain access to the nuts.

5. Place a drain pan under the converter to catch the fluid. With the wrench on the crankshaft pulley bolt, turn the converter to gain access to the converter drain plug and remove the plug. After the fluid has been drained, reinstall the plug.

6. Mark the rear driveshaft yoke and axle companion flange so they can be reinstalled in their original positions. Disconnect the driveshaft from the rear axle and slide the shaft rearward from the transmission. Install a seal installation tool or equivalent, in the extension housing to prevent fluid leakage.

7. Disconnect the cable from the terminal on the starter motor. Remove the attaching bolts and remove the starter motor. Disconnect the neutral start switch wires at the plug connector.

8. Remove the rear mount-to-crossmember attaching bolts and the 2 crossmember-to-frame attaching bolts.

9. Remove the 2 engine rear support-to-extension housing attaching bolts.

10. Disconnect the TV cable from the transmission TV lever. Disconnect the manual rod from the transmission manual lever at the transmission.

11. Remove the bolt securing the bellcrank bracket to the converter housing.

12. On 4WD vehicles, remove the transfer case from the vehicle.

13. Raise the transmission with a transmission jack, to provide clearance to remove the crossmember. Remove the rear mount from the crossmember and remove the crossmember from the side supports.

NOTE: 4WD vehicles have a deep well oil pan. Use a transmission jack that will allow clearance for the oil pan depth and still provide support for the transmission at the oil pan rail.

14. Lower the transmission to gain access to the oil cooler lines.

15. Disconnect each oil line from the fittings on the transmission.

16. Disconnect the speedometer cable from the extension housing.

17. Remove the bolt that secures the transmission fluid filler tube to the cylinder block. Lift the filler tube and the dipstick from the transmission.

18. Secure the transmission to the jack with the chain.

19. Remove the converter housing-to-cylinder block attaching bolts.

20. Carefully move the transmission and converter assembly away from the engine and, at the same time, lower the jack to clear the underside of the vehicle.

21. Remove the converter and mount the transmission in a holding fixture.

To install:

22. Tighten the converter drain plug to 8–28 ft. lbs. (11–38 Nm). Position the converter on the transmission, making sure the converter drive flats are fully engaged in the pump gear by rotating the converter.

23. Lube the pilot with chassis grease. With the converter properly installed, place the transmission on the jack. Secure the transmission to the jack with a chain.

24. Rotate the converter until the studs and drain plug are in alignment with the holes in the flywheel.

25. Move the converter and transmission assembly forward into position, using care not to damage the flywheel and the converter pilot. The converter must rest squarely against the flywheel. This indicates that the converter pilot is not binding in the engine crankshaft.

26. Install and tighten the converter housing-to-engine attaching bolts to 40–50 ft. lbs. (55–68 Nm).

NOTE: Before installing the torque converter-to-flywheel nuts, a check should be made to ensure the converter is properly seated. The converter should move freely with respect to the flywheel. Graps the stud; movement back and forth should result in a metallic clank noise if the converter is properly seated. If the converter will not move, the transmission must be removed and the converter repositioned so the impeller hub is properly engaged in the pump gear.

27. Remove the safety chain from around the transmission. Install a new O-ring on the transmission filler tube, lube the O-ring with transmission fluid, and install the tube in the transmission case. Install the tube retaining bolt.

28. Connect the speedometer cable to the extension housing and the oil cooler lines to the transmission case.

29. Secure the engine rear support to the extension housing and tighten the bolts to 60–80 ft. lbs. (82–108 Nm).

30. Position the crossmember on the side supports. Position the rear mount on the crossmember and install the attaching bolt and nut.

31. On 4WD vehicles, install the transfer case.

32. Lower the transmission and remove the jack. Secure the crossmember to the side supports.

33. Position the bellcrank to the converter housing and install the attaching bolt. Connect the TV cable to the transmission TV lever. Connect the manual linkage rod to the manual lever at the transmission.

34. Install the converter-to-flywheel nuts and tighten to 20–34 ft. lbs. (27–46 Nm). Install the converter housing access cover.

35. Install the starter and conect the starter cable. Connect the neutral start switch wires.

36. Lubricate the driveshaft yoke splines and install the driveshaft. Align the marks on the companion flange and rear yoke made during removal.

37. Clean the gasket mating surfaces on the transmission pan and transmission. Install a new pan gasket and install the pan. Tighten the retaining bolts to 6–10 ft. lbs. (8–14 Nm).

38. Adjust the shift linkage and throttle linkage. Lower the vehicle.

39. Fill the transmission to the correct level with the proper type of fluid. Connect the negative battery cable, start the engine and check for leaks. Road test and check for proper operation. Check the fluid level.

Ford E4OD

1. Disconnect the negative battery cable. Remove the transmission dipstick. On E series, remove the engine cover.

2. Place the transmission selector in **N** position. Raise and safely support the vehicle.

3. On 4WD vehicles, remove the front driveshaft. Remove the rear driveshaft. On F Super Duty vehicles, remove the transmission mounted parking brake.

4. Disconnect the shift linkage. On 4WD vehicles, remove the shift linkage from transfer case shift lever.

5. Remove the manual lever position sensor connector by squeezing the connector tabs and pulling on connector.

NOTE: Do not attempt to pry the tab with a prybar. Remove the heat shield from the transmission before attempting to remove the connector.

6. Remove the solenoid body connector heat shield.

7. Remove the solenoid body connector by pushing on the center tab and pulling on the wire harness.

8. On 4WD vehicles, remove the 4WD switch connector from the transfer case. Use care not to overextend the tabs.

9. Pry the wire harness locator from the extension housing wire bracket. On 4WD vehicles, remove the wire harness locators from the left side of the crossmember.

10. Remove the speedometer cable and the lower converter cover bolts.

11. Remove the rear engine cover plate bolts. Remove the starter.

12. Using a $^{15}/_{16}$ in. socket, rotate the crankshaft bolt to gain access to converter nuts. Remove the 4 converter mounting nuts and discard.

13. Place transmission stand fixture tool 014–00763 or equivalent, on a universal transmission jack and position under the transmission. Use a safety strap to secure the transmission to the transmission stand fixture.

14. Loosen the 2 rear transmission mounting pad nuts. Remove the retaining bolts and remove the crossmember from the transmission.

15. Remove the transmission cooler lines from the transmission case. Cap cooling lines and plug fittings at transmission.

16. Remove the 6 bell housing bolts. Back out the converter pilot from the flywheel and gently lower the transmission while checking for obstructions.

17. Install torque converter handles, T81P-8902-C or equivalent, on the converter with handles in the 6 and 12 o'clock positions.

18. Remove the transmission filler tube from the stub tube. On 4WD vehicles, remove the transfer case vent hose from the detent bracket and the transfer case from the transmission. On F Super Duty models, remove the transmission mounted parking brake.

To install:

19. Place the transmission onto transmission stand fixture tool 014–00763 or equivalent.

20. On 4WD vehicles, install the transfer case to the transmission. On F Super Duty vehicles, install the transmission mounted parking brake.

21. Install the torque converter using torque converter handles T81P-7902-C or the equivalent. Carry the converter with the handles in the 6 and 12 o'clock positions. Push and rotate the converter onto the pump until it bottoms out.

NOTE: Check the seating of the converter by placing a straightedge across the bell housing. There must be a gap between the

converter pilot face and the straight edge.

22. Remove the converter handles. Check the condition of filler tube O-ring and replace if damaged or worn. Install the filler tube.

23. Rotate the converter studs to align with the flywheel mounting holes. Raise transmission into position while checking for any obstructions. Do not allow converter drive flats to disengage from pump gear. Use rubber converter drain plug cover to aid in the alignment of the converter studs.

NOTE: Use care not to damage the flywheel and converter pilot. The converter must rest squarely against the flywheel. This indicates that the converter pilot is not binding in the engine crankshaft.

24. Alternately snug up the bell housing bolts and final torque to 40–50 ft. lbs. (55–67 Nm).

25. Install the rubber converter drain plug cover and transmission cooling lines.

26. Install the crossmember and the transmission retaining bolts. Remove the safety strap and the transmission jack.

27. Rotate the crankshaft to gain access to converter studs. Install new stud nuts and torque to 20–30 ft. lbs. (28–40 Nm).

28. Install the starter, rear engine plate cover and lower dust cover.

29. Install the speedometer cable.

30. Completely seat the solenoid body connector into the solenoid valve body receptacle. An audible click sound indicates proper installation.

31. Install the solenoid body connector heat shield with offset bending inward.

32. On 4WD vehicles, install wire harness locators into crossmember.

33. Install the wire harness locator into extension housing wire bracket.

34. On 4WD vehicles, install the 4WD switch connector and connect the transfer case shift linkage.

35. Install the manual lever position sensor connector. An audible click sound indicates proper installation.

36. Install the shift linkage. On 4WD vehicles, install the shift rod to transfer case shift lever.

37. Install the rear driveshaft and on 4WD vehicles, install the front driveshaft.

38. Lower the vehicle and connect the negative battery cable. Refill the transmission with the proper type and quantity of fluid. Start the engine and check the fluid level.

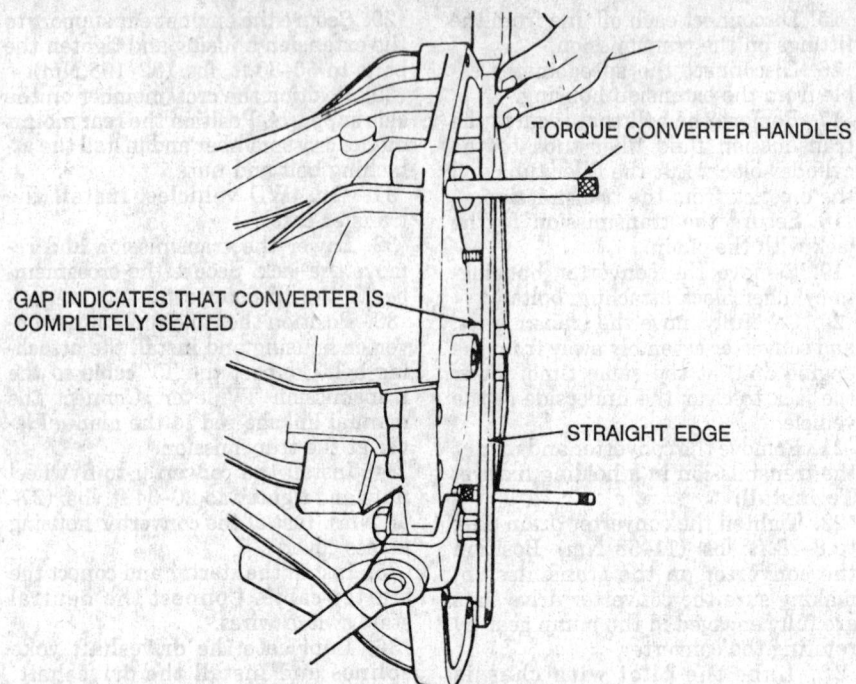

Torque converter installation—E4OD automatic transmission

39. Road test vehicle for proper operation.

SHIFT LINKAGE ADJUSTMENT

1. With the engine stopped and the parking brake applied, place the selector lever at the steering column in the **D** position for C6 transmissions and in the **OD** position for AOD and E40D transmissions. Hold it against the stop by applying an 8 pound weight to the selector lever knob.

2. Loosen the shift rod adjusting nut at point "A".

3. Shift the manual lever at the transmission into the **D** or **OD** position, by moving the lever all the way rearward, then forward 2 detents.

4. With the selector lever and the manual lever in position, tighten the nut at point "A" to 12–18 ft. lbs. (17–24 Nm). Use care to prevent motion between the stud and rod.

5. Remove the 8 pound weight from the steering column selector lever knob.

6. Operate the shift lever in all positions to make sure the manual lever at the transmission is in full detent in all gear ranges. Re-adjust the linkage, if required.

7. On Bronco and F Series, check for correct operation of the automatic transmission selector indicator.

THROTTLE VALVE CONTROL CABLE ADJUSTMENT

Automatic Overdrive Transmission

1988–89 4.9L ENGINE AND ALL 5.0L ENGINES

1. Set the parking brake and put the selector in the **N** position. Remove the protective cover over the the cable linkage, if equipped.

2. Verify that the throttle lever is at the idle stop. If it is not, check for binding or interference in the throttle system. Do not attempt to adjust the idle stop.

3. Verify that the cable routing is free of sharp bends or pressure points and that the cable operates freely. Lubricate the TV lever ball stud. Check for damage to cable or rubber boot.

4. Unlock the locking tab at the throttle body end by prying up with a small prybar to free the cable.

5. A retention spring must be installed on the TV control lever at the transmission, to hold it in the idle position, as far to the rear as the lever will travel, with about 10 pounds of force. If a suitable single spring is not available, two V8 TV return springs may be used. Attach retention spring(s) to the transmission TV lever and hook the rear end of the spring to the transmission case.

6. With the TV cable locking tab un-

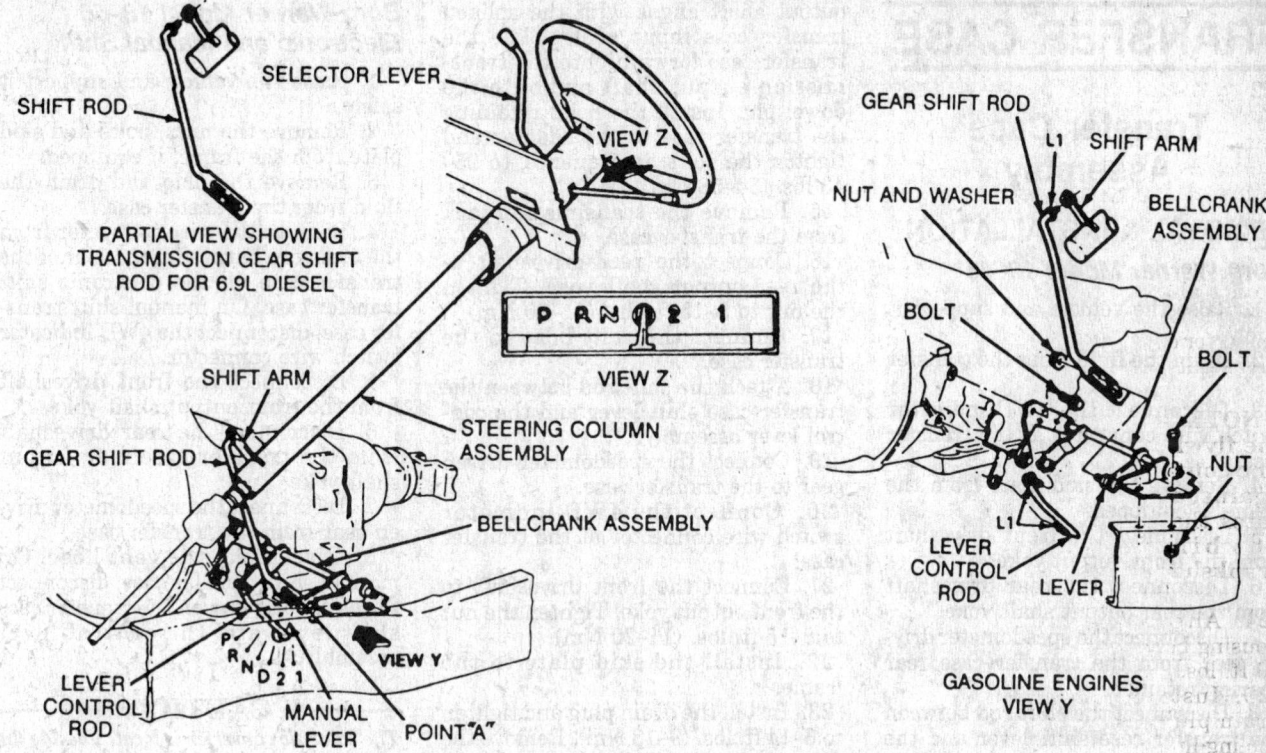

Shift linkage adjustment—C6 transmission

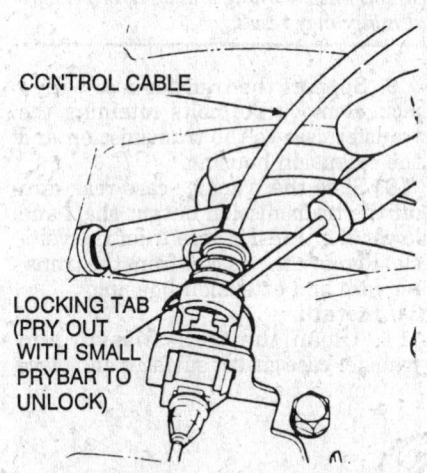

Unlocking TV cable locking tab— 5.0L engine

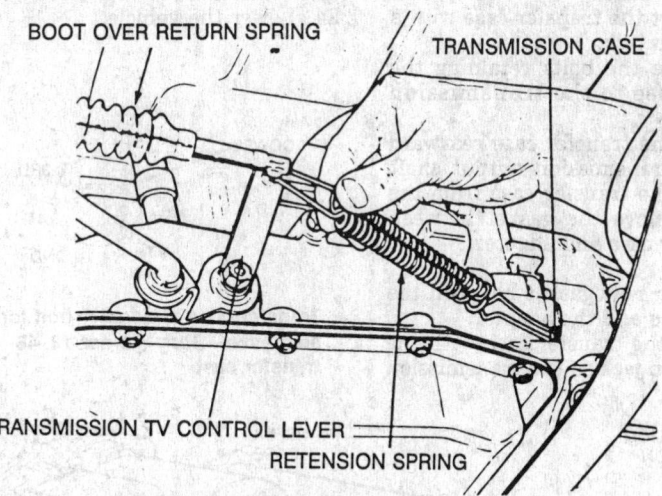

Installing retension springs—TV cable adjustment

locked and the retention spring in place, rotate the transmission outer TV lever 10–30 degrees and return slowly.

7. Push down on the locking tab until flush.

8. Remove the retension springs from the transmission TV lever.

TRANSMISSION SELECTOR INDICATOR ADJUSTMENT

1. Remove the shroud.
2. With the engine stopped and the parking brake applied, place the transmission selector lever at the steering column in the **OD** position, or **D** position if equipped with C6 transmission. Hold the lever against the stop by applying an 8 lb. weight to the selector lever.

NOTE: This adjustment must be made in the OD or on C6 transmission, the D position.

3. Place the cable loop over the retainer pin on the shift socket casting. Place the cable bracket in the "T" slot in the column outer tube.

4. Adjust the cable bracket in the slot until the indicator flag totally fills the rectangular adjustment under the **OD** or on C6 transmission, the **D**.

5. Tighten the bracket screw to 20 inch lbs. (2.25 Nm) without disturbing the flag's position. Release the lever.

6. The indicator flag should totally fill the shift position letters or numerals in each shift position.

NOTE: After the shift lever indicator flag has been adjusted, the manual linkage must be checked.

TRANSFER CASE

Transfer Case Assembly

REMOVAL & INSTALLATION

Borg Warner Model 13–45

1. Raise the vehicle and support it safely.
2. Drain the fluid from the transfer case.
3. Disconnect the 4WD indicator switch wire connector at the transfer case.
4. Remove the skid plate from the frame, if equipped.
5. Disconnect the front driveshaft from the front output yoke.
6. Disconnect the rear driveshaft from the rear output shaft yoke.
7. Disconnect the speedometer driven gear from the transfer case rear bearing retainer.
8. Disconnect the shift rod between the transfer case shift lever and the control lever assembly.
9. Disconnect the vent hose from the transfer case.
10. Support the transfer case with a transmission jack.
11. Remove the bolts retaining the transfer case to the transmission adapter.
12. Slide the transfer case rearward off of the transmission output shaft and lower the transfer case from the vehicle. Remove the gasket between the transfer case and adapter.
To install:
13. Place a new gasket between the transfer case and the adapter.
14. Raise the transfer case with the transmission jack so the transmission

output shaft aligns with the splined transfer case input shaft. Slide the transfer case forward onto the transmission output shaft and onto the dowel pin. Install the bolts retaining the transfer case to the adapter and tighten the bolts, in sequence, to 25–43 lbs. (34–58 Nm).
15. Remove the transmission jack from the transfer case.
16. Connect the rear driveshaft to the rear output shaft yoke. Tighten the nut to 8–15 ft. lbs. (11–20 Nm).
17. Connect the vent hose to the transfer case.
18. Attach the shift rod between the transfer case shift lever and the control lever assembly.
19. Connect the speedometer driven gear to the transfer case.
20. Connect the 4WD indicator switch wire connector at the transfer case.
21. Connect the front driveshaft to the front output yoke. Tighten the nut to 8–15 ft. lbs. (11–20 Nm).
22. Install the skid plate to the frame.
23. Install the drain plug and tighten to 6–14 ft. lbs. (9–18 Nm). Remove the filler plug and install 6.5 pints of the proper fluid. Install the filler plug and tighten to 15–25 ft. lbs. (21–33 Nm).
24. Lower the vehicle.

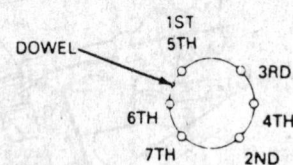

Transfer case-to-adapter bolt torque sequence—Borg Warner 13–45 transfer case

Borg Warner Model 13–56 Electronic and Manual Shift

1. Raise the vehicle and support it safely.
2. Remove the nuts, bolts and skid plate from the frame, if equipped.
3. Remove the plug and drain the fluid from the transfer case.
4. Remove the wire connector from the feed wire harness at the rear of the transfer case on the electronic shift transfer case. On manual shift transfer case, disconnect the 4WD indicator switch wire connector.
5. Disconnect the front driveshaft from the front output shaft yoke.
6. Disconnect the rear driveshaft from the transfer case rear output shaft yoke.
7. Disconnect the speedometer driven gear from the transfer case.
8. Disconnect the vent hose. On manual shift transfer case, disconnect the shift rod between the transfer case shift lever and the control lever assembly.

——— CAUTION ———

The catalytic converter is located beside the transfer case. Due to the extreme high temperatures generated by the converter, be careful when working around it, or personal injury may result.

9. Support the transfer case with a jack. Remove the bolts retaining the transfer case to the transmission and the extension housing.
10. Slide the transfer case rearward off the transmission output shaft and lower the transfer case from the vehicle. Remove the gasket from the transfer case and extension housing.
To install:
11. Clean the transmission and transfer case gasket surfaces and place

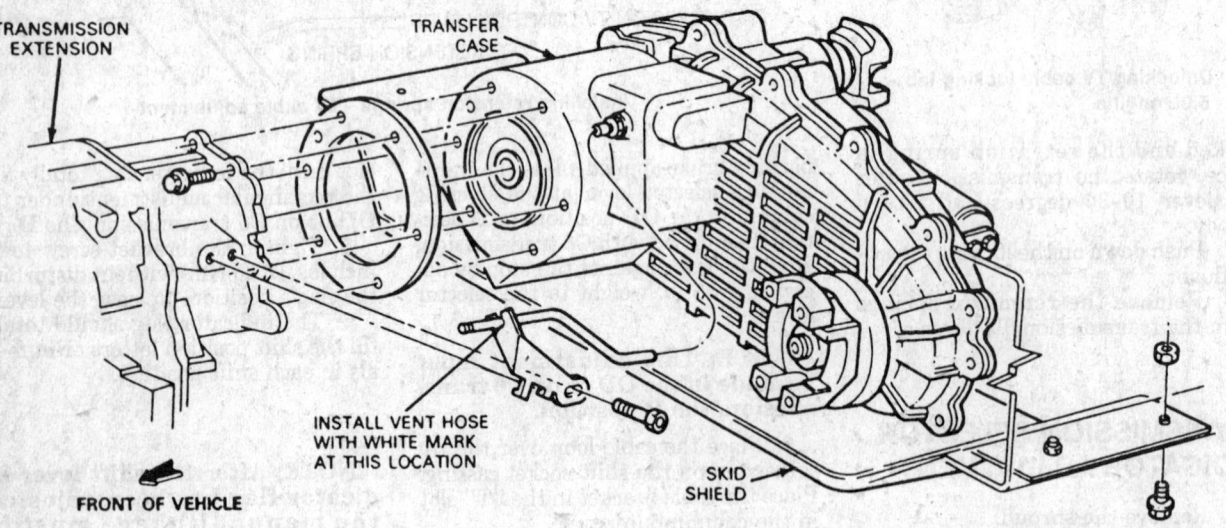

Transfer case installation—typical

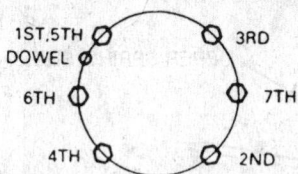

Transfer case-to-adapter bolt torque sequence—Borg Warner 13–56 electronic and manual shift transfer cases

a new gasket between the transfer case and extension housing.

12. Raise the transfer case with the transmission jack so the transmission output shaft aligns with the splined transfer case input shaft. Slide the transfer case forward onto the transmission output shaft and onto the dowel pin. Install the retaining bolts and tighten, in sequence, to 25–43 ft. lbs. (34–58 Nm).

13. Remove the transmission jack and connect the vent hose. Connect the speedometer cable.

14. Connect the rear driveshaft to the output shaft yoke. Tighten the bolts on the electronic shift transfer case and Bronco with manual shift transfer case, to 20–25 ft. lbs. (28–33 Nm). On F Series with manual shift transfer case, tighten the nut to 8–15 ft. lbs. (11–20 Nm).

15. If equipped with manual shift transfer case, attach the shift rod between the transfer case shift lever and the control lever assembly.

16. Connect the front driveshaft to the front output yoke. Tighten the nut to 8–15 ft. lbs. (11–20 Nm).

17. On electronic shift transfer case, connect the wire connectors on the rear of the transfer case, making sure the retaining tabs lock. On manual shift transfer case, connect the 4WD indicator switch wire connector at the transfer case.

18. Install the skid plate, if equipped.

19. Install the drain plug and tighten to 7–14 ft. lbs. (9–18 Nm). Remove the fill plug and install the proper type of fluid to the bottom of the fill hole plug. Install the fill plug and tighten to 14 ft. lbs. (18 Nm).

20. Lower the vehicle.

FRONT SUSPENSION

Shock Absorbers

REMOVAL & INSTALLATION

2WD Vehicles

1. Raise and safely support the vehicle.

2. Insert a wrench from the rear side of the spring upper seat to hold the shock absorber upper retaining nut. Loosen the stud by turning the hex provided on the exposed, lower, part of the stud and remove the nut and washer.

3. Disconnect the lower end of the shock absorber from the lower bracket by removing the nut and bolt.

4. Remove the shock absorber from the vehicle. On F Series, cut out the insulator from the upper spring seat.

5. Installation is the reverse of the removal procedure. Tighten the stud hex to 28 ft. lbs. (37 Nm) while holding the nut and tighten the bolt to 60 ft. lbs. (81 Nm).

4WD Vehicles

BRONCO AND F 150 WITH QUAD SHOCK ABSORBERS

1. Raise and safely support the vehicle.

2. Insert a wrench to hold either shock absorber upper retaining nut. Loosen the stud by turning the hex provided on the exposed, lower, part of the stud and remove the nut and washer.

3. Disconnect the lower end of the shock absorber from the lower bracket by removing the nut and bolt on the rear shock absorber, and nut and washer on the front shock absorber.

4. Compress the shock absorber and remove it from the vehicle. Cut out the insulators from the upper spring seat.

5. Installation is the reverse of the removal procedure. Tighten the stud hex to 25–35 ft. lbs. (34–47 Nm) while holding the nut. Tighten the bolt and nut to 52–74 ft. lbs. (70–100 Nm) on the rear shock absorber and tighten the nut to 70–100 ft. lbs. (95–136 Nm) on the front shock absorber.

F 250 AND F 350

1. Raise and safely support the vehicle.

2. Remove the nut and bolt that retains the shock absorber to the upper shock bracket.

3. Disconnect the lower end of the shock absorber from the U-bolt plate.

4. Compress the shock absorber and remove.

5. Reverse the removal procedure for installation. Tighten both nut and bolt assemblies to 52–74 ft. lbs. (70–100 Nm)

Coil Springs

REMOVAL & INSTALLATION

2WD Vehicles

1. Raise and safely support the ve-

hicle. Position a jack under the front axle.

NOTE: The axle must be supported on the jack throughout spring removal and installation and must not be permitted to hang by the brake hose. If necessary, remove the brake caliper and suspend with mechanics wire. Do not let the caliper hang by the brake hose.

2. Disconnect the shock absorber from the lower bracket.

3. Remove the spring upper retainer attaching bolts or screws from the top of the spring upper seat and remove the retainer.

4. Remove the nut attaching the spring lower retainer to the lower seat and lower insulator and axle and remove the retainer. Slowly lower the axle and remove the spring.

To install:

5. Position the spring and slowly raise the front axle. Install the lower retainer over the stud, lower insulator, if equipped, and lower seat and install the attaching nut.

6. Install the upper retainer over the spring and against the spring upper seat and install the attaching screw.

7. Tighten the upper retainer attaching bolt to 13–18 ft. lbs. (18–24 Nm) on F Series or 20–30 ft. lbs. (28–40 Nm) on E Series. Tighten the lower retainer attaching nut to 70–100 ft. lbs. (95–135 Nm).

8. Connect the shock absorber and remove the jack from the axle. Lower the vehicle.

4WD Vehicles

1. Raise and safely support the vehicle.

2. Remove the shock absorber-to-lower bracket attaching bolt and nut.

3. Remove the spring lower retainer attaching nut from the inside of the coil spring. Remove the spring upper retainer attaching screw and remove the upper retainer.

4. Position a jack under the axle and lower it enough to relieve the tension from the spring.

NOTE: The axle must be supported on the jack throughout spring removal and installation and must not be permitted to hang by the brake hose. If necessary, remove the brake caliper and suspend with mechanics wire. Do not let the caliper hang by the brake hose.

5. Remove the spring lower retainer and remove the spring.

To install:

6. Position the spring and slowly

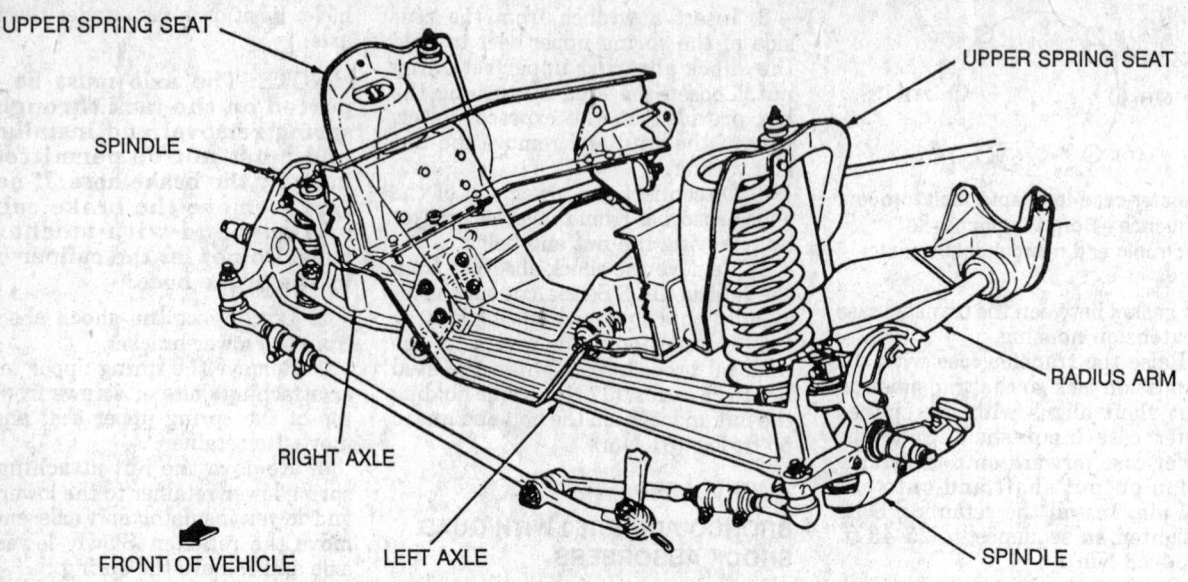

UPPER SPRING SEAT

SPINDLE

UPPER SPRING SEAT

RADIUS ARM

RIGHT AXLE

LEFT AXLE

SPINDLE

FRONT OF VEHICLE

Front suspension assembly—2WD F Series

raise the front axle. Make sure the spring is positioned correctly in the upper spring seats.

7. Install the lower retainer over the stud and lower seat and install the attaching nut. Tighten to 70–100 ft. lbs. (94–134 Nm).

8. Install the upper retainer over the spring and install the attaching screws. Tighten to 13–18 ft. lbs. (18–24 Nm).

9. Connect the shock absorber to the lower bracket and remove the jack from the axle. Lower the vehicle.

Leaf Springs

REMOVAL & INSTALLATION

1. Raise the vehicle frame until the weight is off the front spring with the wheels still touching the floor. Support the axle.

2. Disconnect the lower end of the shock absorber from the U-bolt spacer. Remove the U-bolts, cap and the spacer.

3. If equipped with Dana Model 60 Monobeam axle, remove the 2 bolts that retain the tracking bar to the right spring cap and tracking bar mounting bracket.

4. Remove the nut from the hanger bolt retaining the spring at the rear. Drive out the hanger bolt.

5. Remove the nut connecting the front shackle and the spring eye. Drive out the shackle bolt and remove the spring.

To install:

6. Position the new spring on the spring seat. Install the shackle bolt through the shackle and spring and tighten to 120–150 ft. lbs. (163–203 Nm).

7. Position the rear of the spring to allow the rear hanger bolt to be installed. Install the nut and tighten to 120–150 ft. lbs. (163–203 Nm).

8. Position the U-bolt spacer and place the U-bolts in position through the holes in the spring seat cap. Install, but do not tighten the U-bolt nuts. Make sure the spring center bolt is aligned with the indentation in the axle housing.

9. If equipped with a tracking bar, connect it to the crossmember mounting bracket and right spring cap.

10. Connect the lower end of the shock absorber to the U-bolt spacer. Lower the vehicle and tighten the U-bolt nuts to 85–120 ft. lbs. (115–163 Nm).

Upper Ball Joints

INSPECTION

NOTE: Always adjust the wheel

bearings before any ball joint inspection procedure.

1. Raise the vehicle and place safety stands under the I-Beam axle beneath the spring.

2. Grasp the upper edge of the tire and move the wheel in and out. Observe the upper spindle arm and the upper part of the axle jaw.

3. A $\frac{1}{32}$ in. (0.794mm) or greater movement between the upper part of the axle jaw and the upper spindle arm indicates that the upper ball joint must be replaced.

REMOVAL & INSTALLATION

2WD Vehicles

1. Raise and safely support the vehicle. Remove the front wheel and tire assembly.

2. Remove the brake caliper and suspend with mechanics wire. Do not let the caliper hang by the brake hose.

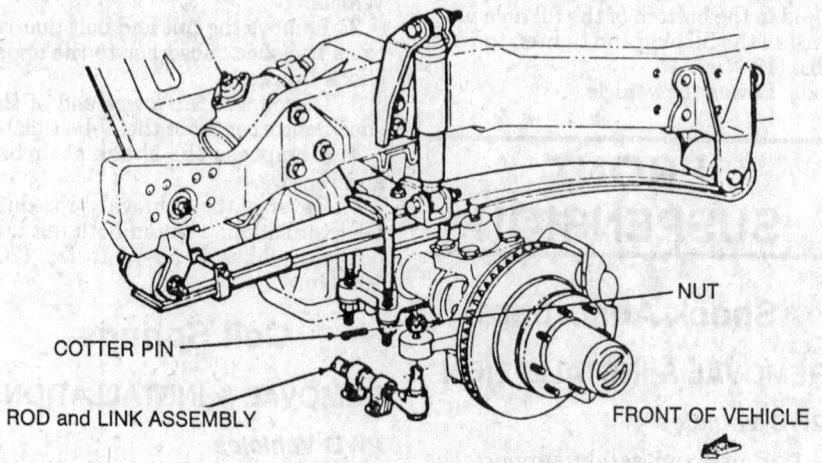

COTTER PIN

NUT

ROD and LINK ASSEMBLY

FRONT OF VEHICLE

Front suspension assembly—F 350

3. Remove the dust cap, cotter pin, nut retainer, nut, washer and outer bearing and remove the rotor from the spindle. Remove the brake dust shield.

4. Remove the cotter pin and nut and disconnect the tie rod end from the spindle.

5. Remove the cotter pin and nut from the lower ball joint stud. Remove the nut from the axle clamp bolt and remove the bolt from the axle.

6. Remove the camber adjuster from the upper ball joint stud and axle beam. Strike the inside area of the axle with a hammer to pop the lower ball joint loose from the axle beam.

NOTE: Do not use a pickle fork to separate the ball joint from the axle as this will damage the seal and the ball joint socket.

7. Remove the spindle and ball joint assembly from the axle.

8. Install the spindle assembly in a vise. Remove the snapring from the lower ball joint and remove the ball joint from the spindle using C-frame tool T74P-3044-A1 and receiver cup tool D81T-3010-A or equivalents. Turn the forcing screw clockwise until the ball joint is removed from the spindle.

9. Assemble the C-frame and receiver cup tools to the upper ball joint and turn the forcing screw clockwise until the ball joint is removed from the spindle.

NOTE: Always remove the lower ball joint first. Do not heat the ball joint or the spindle to aid in removal.

To install:

10. Position the lower ball joint and assemble the C-frame and receiver cup tools with installation cup tool D81T-3010-A4 or equivalents on the spindle. Turn the forcing screw clockwise until the ball joint is seated.

NOTE: Always install the lower ball joint first. Do not heat the ball joint or the spindle to aid in installation.

11. Install the snapring onto the lower ball joint. To install the upper ball joint, repeat Step 10.

12. Remove the spindle assembly from the vise and install on the axle.

13. Install the nut on the lower ball joint stud and tighten to 95-110 ft. lbs. (129-149 Nm). Continue tightening until the castellated nut lines up with the hole in the stud. Install the cotter pin.

14. Install the camber adjuster in the upper spindle over the ball joint stud. Make sure the adjuster is aligned properly. Install the clamp bolt and nut into the axle boss and tighten the nut to 48-65 ft. lbs. (65-88 Nm).

15. Install the dust shield and the rotor. Install the outer wheel bearing, washer and nut. Adjust the wheel bearings and install the nut retainer, cotter pin and dust cap.

16. Install the brake caliper. Connect the tie rod end to the spindle and tighten the nut to 70-100 ft. lbs. (94-135 Nm). Install a new cotter pin.

17. Install the wheel and tire assembly and lower the vehicle. Check the front end alignment.

4WD Vehicles

1. Raise and safely support the vehicle.

2. Remove the axle shafts and spindles. Disconnect the steering linkage.

3. Remove the cotter pin from the top ball joint stud. Loosen the nut on the top stud and the bottom nut inside the knuckle. Remove the top nut.

4. Sharply hit the top stud with a plastic or rawhide hammer, to free the knuckle from the axle arm. Remove and discard the bottom nut.

5. Mark the position of the camber adjuster and remove it by hand. If difficult to remove, use pitman arm puller T64P-3590-F or equivalent.

6. Place the knuckle in a vise and remove the snapring from the bottom ball joint socket, if equipped.

7. Using C-frame tool T74P-4635-C, plug tool T80T-3010-A4 and receiving cup tool T80T-3010-A2 or equivalents, turn the forcing screw clockwise until the bottom ball joint is removed from the steering knuckle.

NOTE: Always remove the lower ball joint first.

8. Repeat Step 7 to remove the upper ball joint.

To install:

9. Clean the steering knuckle bore and insert the lower ball joint as straight as possible. Assemble the C-frame tool with receiving cup tools T80T-3010-A3 and D81T-3010-A or equivalents to install the lower ball joint. Turn the forcing screw clockwise until the ball joint is firmly seated.

10. To install the upper ball joint, repeat Step 9, but use the C-frame tool with receiving cup tool T80T-3010-A3 and replacer tool T80T-3010-A1 or equivalents.

11. Install the knuckle to the axle arm arm. Install the camber adjuster, aligning the marks that were made during removal.

12. Install a new nut on the bottom socket finger tight. Install and tighten the nut on the top socket finger tight, then tighten the bottom nut to 80 ft. lbs. (109 Nm).

13. Tighten the top nut to 100 ft. lbs. (136 Nm), then advance the nut until the castellation aligns with the cotter pin hole. Install the cotter pin. Do not

loosen the top nut to install the cotter pin.

14. Retighten the bottom nut to 90-110 ft. lbs. (123-150 Nm). Install the axle shaft and spindle assembly and lower the vehicle.

Lower Ball Joints

INSPECTION

NOTE: Always adjust the wheel bearings before any ball joint inspection procedure.

1. Raise the vehicle and place safety stands under the I-Beam axle beneath the spring.

2. Have an assistant grasp the lower edge of the tire and move the wheel in and out. While the wheel is being moved, observe the lower spindle arm and the lower part of the axle jaw.

3. A $\frac{1}{32}$ in. (0.794) or greater movement between the lower part of the axle jaw and the lower spindle arm indicates that the lower ball joint must be replaced.

REMOVAL & INSTALLATION

2WD Vehicles

1. Raise and safely support the vehicle. Remove the front wheel and tire assembly.

2. Remove the brake caliper and suspend with mechanics wire. Do not let the caliper hang by the brake hose.

3. Remove the dust cap, cotter pin, nut retainer, nut, washer and outer bearing and remove the rotor from the spindle. Remove the brake dust shield.

4. Remove the cotter pin and nut and disconnect the tie rod end from the spindle.

5. Remove the cotter pin and nut from the lower ball joint stud. Remove the nut from the axle clamp bolt and remove the bolt from the axle.

6. Remove the camber adjuster from the upper ball joint stud and axle beam. Strike the inside area of the axle with a hammer to pop the lower ball joint loose from the axle beam.

NOTE: Do not use a pickle fork to separate the ball joint from the axle as this will damage the seal and the ball joint socket.

7. Remove the spindle and ball joint assembly from the axle.

8. Install the spindle assembly in a vise. Remove the snapring from the lower ball joint and remove the ball joint from the spindle using C-frame tool T74P-3044-A1 and receiver cup tool D81T-3010-A or equivalents. Turn the forcing screw clockwise until the ball joint is removed from the spindle.

NOTE: Do not heat the ball joint or the spindle to aid in removal.

To install:

9. Position the lower ball joint and ssssemble the C-frame and receiver cup tools with installation cup tool D81T-3010-A4 or equivalents on the spindle. Turn the forcing screw clockwise until the ball joint is seated.

NOTE: Do not heat the ball joint or the spindle to aid in installation.

10. Install the snapring onto the lower ball joint.
11. Remove the spindle assembly from the vise and install on the axle.
12. Install the nut on the lower ball joint stud and tighten to 95–110 ft. lbs. (129–149 Nm). Continue tightening until the castellated nut lines up with the hole in the stud. Install the cotter pin.
13. Install the camber adjuster in the upper spindle over the ball joint stud. Make sure the adjuster is aligned properly. Install the clamp bolt and nut into the axle boss and tighten the nut to 48–65 ft. lbs. (65–88 Nm).
14. Install the dust shield and the rotor. Install the outer wheel bearing, washer and nut. Adjust the wheel bearings and install the nut retainer, cotter pin and dust cap.
15. Install the brake caliper. Connect the tie rod end to the spindle and tighten the nut to 70–100 ft. lbs. (94–135 Nm). Install a new cotter pin.
16. Install the wheel and tire assembly and lower the vehicle. Check the front end alignment.

4WD Vehicles

1. Raise and safely support the vehicle.
2. Remove the axle shafts and spindles. Disconnect the steering linkage.
3. Remove the cotter pin from the top ball joint stud. Loosen the nut on the top stud and the bottom nut inside the knuckle. Remove the top nut.
4. Sharply hit the top stud with a plastic or rawhide hammer, to free the knuckle from the axle arm. Remove and discard the bottom nut.
5. Mark the position of the camber adjuster and remove it by hand. If difficult to remove, use pitman arm puller T64P-3590-F or equivalent.
6. Place the knuckle in a vise and remove the snapring from the bottom ball joint socket, if equipped.
7. Using C-frame tool T74P-4635-C, plug tool T80T-3010-A4 and receiving cup tool T80T-3010-A2 or equivalents, turn the forcing screw clockwise until the bottom ball joint is removed from the steering knuckle.

To install:

8. Clean the steering knuckle bore and insert the lower ball joint as straight as possible. Assemble the C-frame tool with receiving cup tools T80T-3010-A3 and D81T-3010-A or equivalents, to install the lower ball joint. Turn the forcing screw clockwise until the ball joint is firmly seated.
9. Install the knuckle to the axle arm arm. Install the camber adjuster, aligning the marks that were made during removal.
10. Install a new nut on the bottom socket finger tight. Install and tighten the nut on the top socket finger tight, then tighten the bottom nut to 80 ft. lbs. (109 Nm).
11. Tighten the top nut to 100 ft. lbs. (136 Nm), then advance the nut until the castellation aligns with the cotter pin hole. Install the cotter pin. Do not loosen the top nut to install the cotter pin.
12. Retighten the bottom nut to 90–110 ft. lbs. (123–150 Nm). Install the axle shaft and spindle assembly and lower the vehicle.

Stabilizer Bar

REMOVAL & INSTALLATION

2WD Vehicles

F SERIES

1. Disconnect the left and right ends of the bar from the link assembly attached to the spring seat.
2. Disconnect the retainer bolts and remove the stabilizer bar.
3. Disconnect the link assemblies by looosening the right and left locknuts from their I-beam brackets.

To install:

4. Loosely assemble the entire stabilizer bar system with both link assemblies outboard of the stabilizer bar. Force the stabilizer bar rearward to connect the bar ends to the link assemblies.
5. Tighten the nuts and bolts retaining the link assemblies to the stabilizer bar and spring seat to 52–74 ft. lbs. (71–100 Nm).
6. Make sure the insulators are seated in the retainers and the stabilizer bar is centered in the assembly. Tighten the stabilizer bar-to-frame retainer nuts and bolts to 27–37 ft. lbs. (37–50 Nm).

E SERIES

1. Disconnect the left and right ends of the bar from the link assembly attached to the I-beam bracket.
2. Disconnect the retainer bolts and remove the stabilizer bar.
3. Disconnect the link assemblies by looosening the right and left locknuts from their I-beam brackets.

To install:

4. Connect the right and left stabilizer links to the I-beam brackets by sliding the bolt with the washer through the link, I-beam bracket hole and toward the inside of the side rail. Tighten the washer and locknut to 40–60 ft. lbs. (55–81 Nm).

NOTE: The link must be installed with the bend facing forward.

5. Connect the stabilizer bar to the frame side rails by placing the 2 retainers on the bottom of the side rails and tightening the 4 through bolts to 15–25 ft. lbs. (21–33 Nm).
6. Connect the stabilizer bar ends to the link assemblies. Tighten the retaining nuts to 18–28 ft. lbs. (25–37 Nm).

4WD Vehicles

BRONCO AND F 150

1. Disconnect the stabilizer bar from the connecting links. Remove the nuts and bolts of the stabilizer bar retainer.
2. Remove the stabilizer bar retainer.
3. Remove the stabilizer bar and insulator. The stud does not have to be removed.
4. Installation is the reverse of the removal procedure. Tighten the retainer nuts to 27–37 ft. lbs. (37–50 Nm) and the link-to-stabilizer bolt and nut to 52–74 ft. lbs. (71–100 Nm).

F 250

1. Remove the bolts, washers and the nuts securing the links to the spring seat caps. If equipped with a Dana Model 60 Monobeam front axle, the links are secured to the mounting brackets.
2. Disconnect the links from the stabilizer bar and remove them.
3. Remove the retainers from the mounting bracket. Remove the stabilizer bar.
4. Installation is the reverse of the removal procedure. Tighten the nuts connecting the links to the spring seat caps to 52–74 ft. lbs. (70–100 Nm). Tighten the nuts connecting the links to the stabilizer bar to 15–25 ft. lbs. (21–33 Nm). Tighten the nuts and bolts connecting the retainers to the mounting bracket to 27–37 ft. lbs. (35–50 Nm).

F 350

1. Disconnect the left and right stabilizer bar ends from the link assembly.
2. Disconnect the retainer bolts and U-bolt and remove the stabilizer bar.
3. Disconnect the stabilizer links from the frame side rail mounting brackets.

To install:

4. Loosely assemble the entire stabilizer bar system with both link assemblies loosely attached to the frame mounting brackets and the stabilizer bar in position on the axle.

5. Make sure the stabilizer bar insulators are seated in the retainers and the stabilizer bar is centered between the leaf springs. Attach the stabilizer bar to the axle by assembling the retainers to the axle mounting brackets. Tighten the retaining bolts and U-bolt to 35–50 ft. lbs. (48–68 Nm).

6. Install the link assemblies to the frame mounting brackets and tighten the locknuts to 52–74 ft. lbs. (70–100 Nm).

7. Install the link assembly to the stabilizer bar and tighten the locknuts to 15–25 ft. lbs. (21–32 Nm).

King Pin and Bushings

REMOVAL & INSTALLATION

2WD Vehicles

1. Raise and support the vehicle safely. Remove the wheel and tire assembly.

2. Remove the brake caliper and suspend it with mechanics wire. Do not let the caliper hang from the brake hose.

3. Remove the rotor/hub and bearing assembly.

4. Disconnect the steering linkage from the spindle arm.

5. Remove the nut and lockwasher from the lock pin and remove the lock pin.

6. Remove the upper and lower king pin plugs, then drive the king pin out from the top of the axle. Remove the spindle and thrust bearing. Remove the king pin seal and thrust bearing.

7. Install the spindle in a vise. Remove and discard the seal from the bottom of the upper bushing bore of the upper spindle yoke.

8. Using suitable tools, drive the old bushings from the spindle bores.

To install:

9. Using a suitable driver, install a new top bushing. The bushing should be seated to a depth of 0.080 in. (2.03mm) minimum from the bottom of the upper spindle boss.

10. Using a suitable driver, install a new bottom spindle bushing. The bottom bushing should be seated to a depth of 0.130 in. (3.30mm) minimum from the top of the lower spindle boss.

11. Ream the new bushings to 0.001–0.003 in. (0.025–0.076mm) larger than the diameter of the new king pin. Clean all metal shavings from the bushings after reaming and coat the bushings with lubricant.

12. Make sure the king pin hole in the axle is free of nicks, burrs and dirt. Clean up the bore as necessary and lightly coat the surface with lithium grease.

13. Install a new king pin seal with the metal backing facing up towards the bushing into the spindle. Gently press the seal into position, being careful not to distort the casing.

14. Install a new thrust bearing with the lip flange facing down towards the lower bushing. Press until the bearing is firmly seated against the surface of the spindle.

15. Place the spindle in position on the axle with the thrust bearing.

16. Install the king pin with the **T** stamped on 1 end towards the top with the lock pin notch in the pin aligned with the lock pin hole in the axle. Insert the king pin through the bushings and axle from the top until the king pin notch is aligned with the lock pin hole.

17. Install a new lock pin with the threads pointing forward and the wedge groove facing the king pin notch. Firmly drive the lock pin into position and install the lock pin lockwasher and nut. Tighten the nut to 37–52 ft. lbs. (50–70 Nm). Install the king pin plugs at the top and bottom of the spindle and tighten to 35–50 ft. lbs. (48–67 Nm).

18. Lubricate the king pin and bushings with grease through both fittings until grease is visible seeping past the upper seal at the top and from the thrust bearing slip joint at the bottom. If grease does not appear, recheck the installation procedure.

19. Install the dust shield and the rotor/hub and bearing assembly. Adjust the wheel bearings.

20. Install the cailper assembly and connect the steering linkage to spindle. Install the wheel and tire assembly.

21. Lower the vehicle and check the front end alignment.

4WD Vehicles

1. Raise and safely support the vehicle. Remove the axle shaft.

2. Alternately and evenly loosen the 4 bolts retaining the spindle cap to the knuckle; thereby relieving the spring compression. Remove the bolts.

3. Remove the spindle cap, compression spring and retainer. Remove and discard the gasket.

4. Remove the 4 bolts retaining the lower kingpin and retainer to the knuckle. Remove the lower kingpin and retainer from the knuckle.

5. Remove the tapered bushing from the top of the upper kingpin in the knuckle. Remove the knuckle from the axle yoke.

6. Remove the upper king pin from the axle yoke with a piece of ⅞ in. hex-shaped case hardened metal stock or with a suitable ⅞ in. ex socket. Discard the upper king pin and seal.

NOTE: The upper king pin is tightened to 500–600 ft. lbs. (678–813 Nm).

7. Press the lower king pin grease retainer, bearing cup, bearing and seal from the axle yoke lower bore with a 2-jaw puller and suitable press plate. Discard the grease seal and retainer and lower bearing cup.

To install:

8. Coat the mating surface of a new lower king pin grease retainer with silicone sealer and install in the axle yoke bore so the concave portion of the retainer faces the upper king pin.

9. Using a suitable driver, drive a new bearing cup in the lower king pin bore until it bottoms against the grease retainer.

10. Pack the lower king pin bearing and the yoke bore with high-temperature wheel bearing grease. With a suitable driver, drive a new seal in the lower king pin bore.

11. Install a new seal, using installation tool T86T-3110–AH or equivalent. Install the upper king pin assembly in the yoke and tighten to 500–600 ft. lbs. (678–813 Nm).

12. Install the knuckle on the axle yoke. Place the tapered bushing over the upper kingpin in the knuckle bore.

13. Place the lower kingpin and retainer assembly in the knuckle and axle yoke. Install the 4 bolts and alternately and evenly tighten to 70–90 ft. lbs. (95–122 Nm).

14. Place the retainer and compression spring on the tapered bushing. Install a new gasket and the spindle cap on the knuckle. Install the 4 bolts and alternately and evenly tighten to 70–90 ft. lbs. (95–122 Nm).

15. Install the axle shaft. Lubricate the upper kingpin through the grease zerk and the lower through the flush type fitting.

16. Lower the vehicle.

Twin I-Beam Axle

REMOVAL & INSTALLATION

1. Raise and safely support the vehicle. Remove the front wheel spindle, the front spring and the stabilizer bar, if equipped.

2. Remove the spring lower seat from the radius arm and then remove the bolt and nut that attaches the radius arm to the front axle.

3. Remove the axle to frame pivot bracket bolt and nut. Remove the axle assembly.

To install:

4. Position the axle to the frame pivot bracket and install the bolt and nut finger-tight.

5. Position the opposite end of the of the axle to the radius arm. Install the attaching bolt from underneath through the bracket, the radius arm and the axle.

6. Install the spring lower seat on the radius arm so the hole in the seat indexes over the arm-to-axle bolt and the pin on the spring seat engages the slot in the radius arm.

7. Install the front spring, wheel spindle and, if equipped, the stabilizer bar.

8. Lower the vehicle. With the weight on the suspension, tighten the axle-to-frame pivot bolt to 120–150 ft. lbs. (163–203 Nm).

Front Wheel Bearings

ADJUSTMENT

2WD Vehicles

1. Raise the vehicle and support it safely.

2. Remove the wheel cover and the grease cap from the hub. Remove the cotter pin and the locknut.

3. Loosen the adjusting nut 3 turns. Try to obtain running clearance between the rotor and brake pads by rocking the wheel in and out or by tapping on the caliper housing.

NOTE: Do not pry on the caliper piston. If clearance cannot be obtained, remove the caliper and suspend it with mechanics wire. Do not let the caliper hang from the brake hose.

4. While rotating the wheel, tighten the adjusting nut to 17–25 ft. lbs. (23–34 Nm) to seat the bearings.

5. Back off the adjusting nut 120–180 degrees and install the retainer and new cotter pin without additional movement of the adjusting nut. Bearing endplay should be 0.00025–0.005 in. (0.006–0.127mm).

6. Check front wheel rotation. If the wheel rotates properly, reinstall the grease cap and wheel cover. If rotation is noisy or rough, remove and inspect the bearings.

7. Before driving the vehicle, pump the brake pedal several times to reseat the brake pads.

4WD Vehicles

BRONCO, F 150 AND F 250 WITH MANUAL LOCKING HUBS

1. Raise the vehicle and support it safely.

2. Remove the hub lock assembly.

3. Using a torque wrench and span-ner locknut wrench T86T–1197–A or equivalent, apply inward pressure to unlock the adjusting nut locking splines and turn the nut clockwise to tighten to 70 ft. lbs. (95 Nm) while rotating the wheel back and forth to seat the bearings.

4. Apply inward pressure to the locknut wrench to disengage the adjusting nut locking splines and back off the adjusting nut approximately 90 degress.

5. Retighten the adjusting nut to 15–20 ft. lbs. (20–27 Nm) and remove the tool and torque wrench. Check the final endplay of the hub and rotor on the spindle; it should be 0.00 in.

6. The torque required to rotate the hub and rotor assembly should not exceed 20 inch lbs. (2.3 Nm).

7. Install the hub lock assemblt and lower the vehicle.

BRONCO AND F 150 WITH AUTOMATIC LOCKING HUBS F 250 AND F 350 WITH MANUAL LOCKING HUBS

1. Raise and safely support the vehicle.

2. Remove the hub lock assembly.

3. Remove the outer locknut and lockwasher using spanner locknut wrench D85T–1197–A or equivalent.

4. Using the locknut wrench, tighten the inner locknut to 50 ft. lbs. (68 Nm) while rotating the hub back and forth, to seat the bearing.

5. Back off the inner locknut and retighten to 30–40 ft. lbs. (41–54 Nm).

6. Back off the locknut 90 degrees. Install the lockwasher so the key is positioned in the spindle groove. Tighten the inner locknut so the pin is aligned into the nearest lockwasher hole.

7. Install the outer locknut and tighten to 160–205 ft. lbs. (217–278 Nm). Check the final endplay of the spindle; it should be 0.000–0.004 in. (0.00–0.11mm).

8. The torque required to rotate the hub and rotor assembly should not exceed 20 inch lbs. (2.3 Nm).

9. Install the hub locks and lower the vehicle.

REMOVAL & INSTALLATION

2WD Vehicles

1. Raise and safely support the vehicle. Remove the front wheel and tire assembly.

2. Remove the brake caliper and suspend it with mechanics wire. Do not let the caliper hang from the brake hose.

3. Remove the grease cap, cotter pin, retainer adjusting nut and washer. Remove the outer bearing.

4. Pull the hub and rotor off the spindle. Remove and discard the grease seal.

5. Remove the inner bearing from the hub. Remove all traces of old lubricant from the bearings, hub and spindle with solvent and dry thoroughly.

6. Inspect the bearings and bearing races for scratches, pits, cracks or other wear. If the bearings or races are worn or damaged, remove the races using a brass drift.

To install:

7. If the inner or outer bearing races were removed, replace them in the hub with the proper tool. The races will be properly seated when they are fully bottomed.

8. Pack the inside of the hub with high temperature wheel bearing grease. Fill the hub until the grease is flush with the inside diameters of both bearing races.

9. Pack the bearings with wheel bearing grease, working as much lubricant as possible between the rollers and the cages.

10. Place the inner bearing in the inner race and install a new seal, using a seal installer.

11. Install the hub and rotor assembly on the spindle, being careful not to damage the seal. Install the outer bearing, washer and adjusting nut and adjust the wheel bearings.

12. Install the retainer, new cotter pin and grease cap. Install the caliper and the wheel and tire assembly.

13. Lower the vehicle. Before driving the vehicle, pump the brake pedal to reseat the brake pads.

REAR SUSPENSION

Shock Absorber

REMOVAL & INSTALLATION

1. Raise the vehicle and support it safely.

2. Remove the shock absorber lower attaching nut and bolt and swing the lower end free of the mounting bracket on the axle housing.

3. Remove the nut and bolt or nut from the upper shock absorber mount.
To install:

4. Position the shock absorber to the upper mount and install the nut, bolt and washers.

5. Swing the lower end of the shock absorber into the mounting bracket on the axle housing. Install the mounting bolt, nut and washers. Tighten the nut and bolt to 52–74 ft. lbs. (70–100 Nm).

6. Install the upper nut or nut and bolt and tighten to 40–60 ft. lbs. (54–81 Nm) on Bronco and F Series or 25–35 ft. lbs. (33–47 Nm) on E Series.

7. Lower the vehicle.

Leaf Spring

REMOVAL & INSTALLATION

Bronco and F Series

1. Raise and safely support the vehicle, so the weight is off the rear spring and the tires still touch the floor.

2. Remove the nuts from the spring U-bolts and drive the U-bolts from the U-bolt plate. If equipped, remove the auxiliary spring and spacer.

3. Remove the spring-to-bracket nut and bolt at the front of the spring.

4. Remove the shackle upper and lower nuts and bolts at the rear of the spring. Remove the spring and shackle assembly from the rear shackle bracket.

To install:

5. Position the spring in the shackle and install the upper shackle-to-spring bolt and nut with the bolt head facing outward.

6. Position the front end of the spring in the bracket and install the bolt and nut. Position the shackle in the rear bracket and install the bolt and nut.

7. Place the spring on top of the axle with the spring tie bolt centered in the hole provided in the seat. If equipped, install the auxiliary spring and spacer.

8. Install the spring U-bolts, U-bolt plate and nuts.

9. Lower the vehicle to the floor. Tighten the spring U-bolt nuts to 75–115 ft. lbs. (100–155 Nm) on Bronco, F 150 and F250 under 8500 lbs. or 150–210 ft. lbs. (200–280 Nm) on F 250 and F 350 over 8500 lbs. GVW.

10. Tighten the leaf spring-to-front bracket nut and bolt to 75–115 ft. lbs. (100–155 Nm) on 2WD F 150 or 150–177 ft. lbs. (200–240 Nm) on all other vehicles.

11. Tighten the leaf spring-to-rear shackle nut and bolt to 75–115 ft. lbs. (100–150 Nm) on all except 2WD F 250 and F 350, where the torque should be 150–210 ft. lbs. (200–280 Nm).

E 150

1. Raise and safely support the vehicle. Support the rear axle with a floor jack.

2. Disconnect the lower end of the shock absorber from the axle housing bracket. Remove the 2 U-bolts and plate.

3. Lower the axle and remove the

upper and lower rear shackle bolts. Pull the rear shackle assembly from the bracket and spring.

4. Remove the nut and mounting bolt that secures the front end of the spring. Remove the spring assembly from the front shackle bracket.

To install:

5. Assemble the front eye of the spring to the front bracket with the front mounting bolt and nut but do not tighten at this time.

6. Mount the rear end of the spring with the upper bolt of the rear shackle assembly passing through the eye of the spring. Insert the lower bolt through the rear spring and shackle bracket.

7. Assemble the spring center bolt in the pilot hole in the axle and install the plate. Install the U-bolts through the plate but do not tighten the attaching nuts at this time.

8. Raise the axle with the jack until the shock absorbers can be connected to the axle housing brackets.

9. Tighten the leaf spring-to-front bracket nut and bolt to 150–204 ft. lbs. (204–276 Nm). Tighten the leaf spring-to-axle U-bolt nut to 74–107 ft. lbs. (101–145 Nm). Tighten the leaf spring-to-rear shackle nut and bolt to 74–107 ft. lbs. (101–145 Nm).

10. Lower the vehicle.

E 250 and E 350

1. Raise and safely support the vehicle. Support the rear axle with a floor jack.

2. Disconnect the lower end of the shock absorber from the axle housing bracket. Remove the 2 U-bolts and plate.

3. Lower the axle and remove the spring front bolt from the hanger. Remove the 2 attaching bolts from the rear of the spring and remove the spring and shackle.

To install:

4. Assemble the upper end of the shackle to the spring with the attaching bolt. Connect the front of the spring to the front bracket with the attaching bolt.

5. Assemble the spring and shackle to the rear bracket with the attaching bolt. Place the U-bolt plate over the nut of the center bolt.

6. Raise the axle with a jack. Install the center bolt head through the pilot hole in the pad on the axle housing.

7. Install the spring U-bolts and attaching nuts and snug the nuts. Connect the shock absorber to the lower bracket.

8. Tighten the leaf spring-to-front bracket nut and bolt to 150–204 ft. lbs. (204–276 Nm). Tighten the leaf spring-to-axle U-bolt nut to 74–107 ft. lbs. (101–145 Nm) on light duty E 250 or 150–180 ft. lbs. (204–244 Nm) on heavy duty E 250 and E 350. Tighten the leaf spring-to-rear shackle nut and bolt to 74–107 ft. lbs. (101–145 Nm).

9. Lower the vehicle.

STEERING

Steering Wheel

REMOVAL & INSTALLATION

1. Disconnect the negative battery cable.

2. Remove 1 screw from the underside of each steering wheel spoke and lift the steering wheel pad from the steering wheel.

3. Disconnect the horn switch wires by pulling the spade terminal from the blade connectors. If equipped with cruise control, squeeze or pinch the "J" clip ground wire terminal firmly and pull it out of the hole in the steering wheel. Do not pull the ground terminal out of the threaded hole without squeezing the terminal clip to relieve the spring retention of the terminal in the threaded hole.

4. Remove the horn switch and the steering wheel retaining nut. If the steering wheel and steering shaft do not have mating flats, mark the posi-

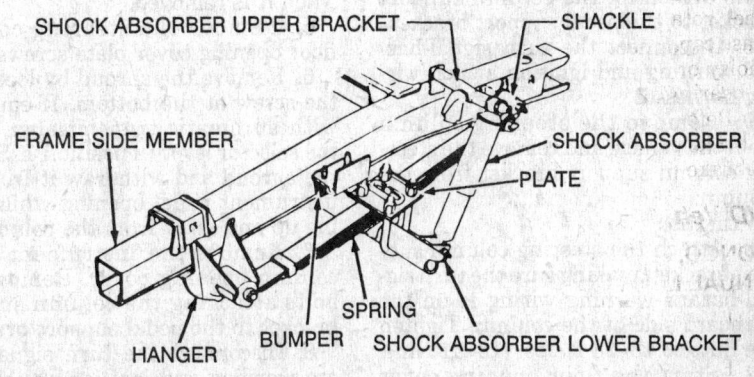

SHOCK ABSORBER UPPER BRACKET — SHACKLE

FRAME SIDE MEMBER — SHOCK ABSORBER

PLATE

HANGER   BUMPER   SPRING   SHOCK ABSORBER LOWER BRACKET

Rear suspension—typical

tion of the wheel on the shaft so it can be reinstalled in the same position.

5. Using a suitable puller, remove the steering wheel from the steering shaft. Do not hammer on the steering wheel or use a knock-off type puller as either will damage the steering column.

To install:

6. Position the steering wheel on the column shaft, aligning the mating flats or the marks that were made during removal.

7. Install the retaining nut and tighten to 30–42 ft. lbs. (41–56 Nm).

8. Connect the horn and cruise control wires, if equipped.

9. Install the steering wheel pad. Tighten the screws to 7–11 inch lbs. (0.8–1.2 Nm).

10. Connect the negative battery cable and test the steering column for proper operation.

Steering Column

REMOVAL & INSTALLATION

Bronco and F Series

1. Disconnect the negative battery cable and set the parking brake.

2. Remove the bolt and nut attaching the intermediate shaft to the steering column.

3. If equipped with automatic transmission, disconnect the shift linkage rod(s) from the column.

4. Remove the steering wheel. If equipped with tilt column, the steering wheel must be in the full up position when it is removed.

5. Remove the steering column floor opening cover plate screws.

6. Remove the shroud by loosening the screw at the bottom. If equipped with automatic transmission, place the selector lever in position 1. Spread the shroud and withdraw it from the instrument panel opening while pulling up and away from the column.

7. Remove the instrument panel column opening cover. Remove the bolts attaching the column support bracket to the pedal support bracket.

8. Disconnect the turn signal-hazard warning and ignition switch wiring harnesses.

9. Remove the steering column from the vehicle and remove the steering column support bracket from the column.

To install:

10. Attach the steering column support bracket, making sure the turn signal-hazard warning wiring is on the outboard side of the column. Tighten the nuts to 13–38 ft. lbs. (18–51 Nm).

11. Start the floor opening cover clamp bolt and press the plate until the clamp flats touch the stops on the column outer tube.

12. Insert the column into the engine compartment through the floor opening. Connect the turn signal-warning hazard and ignition switch wiring harnesses.

13. Raise the column up to the pedal support bracket and hand start the 2 bolts. Fasten the floor opening cover plate to the floor and tighten to 6–10 ft. lbs. (8–14 Nm).

14. Tighten the 2 support bracket bolts to 19–27 ft. lbs. (26–37 Nm) and tighten the cover plate clamp bolt to 8–18 ft. lbs. (11–24 Nm).

15. Install and adjust the transmission indicator actuation cable, if equipped with automatic transmission. Install the instrument panel steering column opening cover.

16. Mount the shroud by selecting position 1, if equipped with automatic transmission, and spreading the shroud around the steering column and through the opening in the instrument panel. Post on interior will index shroud when properly positioned.

17. Tighten the screw at the bottom of the shroud to 10–15 inch lbs. (1.1–1.7 Nm).

18. Attach the shift linkage rod to the column, if equipped with automatic transmission. Adjust the linkage, if required.

19. Fasten the intermediate shaft to the steering column and tighten to 35–50 ft. lbs. (46–68 Nm). Install steering wheel and connect the negative battery cable.

E Series

1. Disconnect the negative battery cable and set the parking brake.

2. Remove the 2 nuts attaching the flexible coupling to the steering shaft flange.

3. If equipped with automatic transmission, disconnect the shift linkage rod(s) from the column.

4. Remove the steering wheel. If equipped with tilt column, the steering wheel must be in the full up position when it is removed.

5. Remove the steering column floor opening cover plate screws.

6. Remove the shroud by loosening the screw at the bottom. If equipped with automatic transmission, place the selector lever in position 1. Spread the shroud and withdraw it from the instrument panel opening while pulling up and away from the column.

7. Remove the instrument panel column opening cover. Remove the bolts attaching the column support bracket to the pedal support bracket.

8. Disconnect the turn signal-hazard warning and ignition switch wiring harnesses.

9. Remove the steering column from the vehicle and remove the steering column support bracket from the column.

To install:

10. Attach the steering column support bracket and tighten the nuts to 13–38 ft. lbs. (18–51 Nm). Place the column in the vehicle.

11. Connect the turn signal-hazard warning and ignition switch wiring harnesses.

12. Insert the steering shaft flange through the floor opening so the flange engages the flexible coupling. Raise the column up to the pedal support bracket and loosely install the support bolts.

13. When the column has been removed, the flange will be fully telescoped into the column. Gently tap the flange down to engage the flexible coupling. Loosely install the flexible coupling flange nuts and floor plate fasteners.

14. Align the steering column and flexible coupling as follows:

  a. Make sure the coupling pin to flange cut-out clearance is 0.160 in. (4.06mm).

  b. Tighten the flange coupling nuts to 14–21 ft. lbs. (19–28 Nm).

  c. Tighten the steering column-to-support bracket bolts to 19–27 ft. lbs. (26–36 Nm).

  d. Tighten the steering column floor opening cover plate-to-dash panel bolts to 12–18 ft. lbs. (16–24 Nm).

  e. Tighten the steering column opening cover clamp bolt to 12–18 ft. lbs. (16–24 Nm).

15. Install the steering wheel. If equipped, attach the shift linkage rod and adjust the shift linkage, if necessary. Connect the negative battery cable.

Power Steering Gear

ADJUSTMENT

1. Make sure the wheels are in the straight ahead position. Remove the steering wheel pad. Disconnect the pitman arm from the sector shaft using a pitman arm puller.

2. Disconnect the fluid return line at the reservoir and cap the reservoir return line tube. Place the end of the return line in a clean container and turn the steering wheel back and forth several times to empty the steering gear. Discard the fluid.

3. Turn the steering wheel to 45 degrees from the right stop.

4. Attach an inch pound torque wrench to the steering wheel nut and record the torque required to rotate the shaft slowly approximately 1/8 turn

toward center from the 45 degree position.

5. Turn the steering gear back to center and record the torque required to rotate the shaft back and forth across the center position (± 90 degrees). If the vehicle has less than 5000 miles, the set torque measured rocking across center should be 14–18 inch lbs. (1.6–2.0 Nm) greater than that measured 45 degrees from the right stop. If the vehicle has more than 5000 miles, the set torque measured rocking across center should be 10–14 inch lbs. 1.13–1.6 Nm) greater than that measured 45 degrees from the right stop.

6. If reset is required, loosen the adjuster locknut and turn the sector shaft adjusting screw until the reading is the specified value greater than the torque at 45 degress from the stop. Hold the sector shaft screw in place and tighten the locknut.

7. Recheck the torque readings and install the pitman arm and steering wheel pad.

8. Connect the fluid return line to the reservoir and refill the system with fluid. Bleed the system. Adjust the belt tension, if necessary.

REMOVAL & INSTALLATION

1. Make sure the wheels are in the straight ahead position. Disconnect the negative battery cable.

2. Disconnect the pressure and return lines from the steering gear. Plug the lines and the ports in the gear to prevent entry of dirt.

3. If equipped, remove the splash shield from the flex coupling. Disconnect the flex coupling at the steering gear.

4. Raise the vehicle and support it safely. Remove the pitman arm nut and washer. Remove the pitman arm using a pitman arm puller. Be careful not to damage the seals.

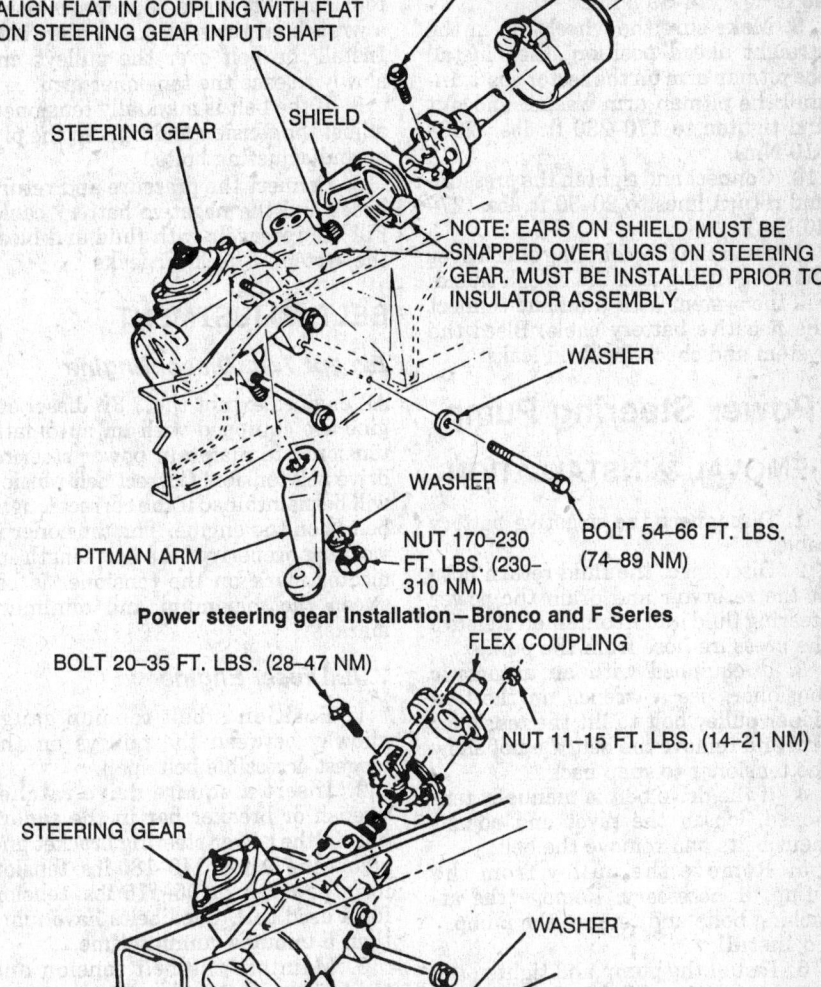

Power steering gear Installation—Bronco and F Series

Power steering gear Installation—E Series

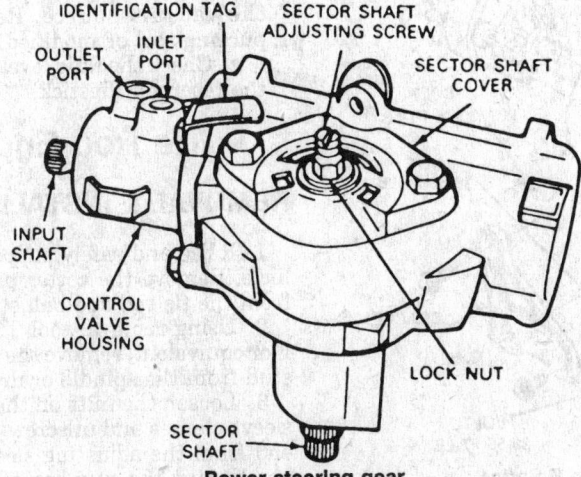

Power steering gear

5. Support the steering gear and remove the attaching bolts. Work the steering gear free from the flex coupling and remove the gear from the vehicle.

To install:

6. If equipped, install the splash shield onto the steering gear lugs.

7. Slide the flex coupling into place on the steering shaft assembly. Turn the steering wheel so the spokes are in the horizontal position.

8. Slide the steering gear input shaft into the flex coupling and into place on the frame side rail. Install the flex coupling attaching bolt and tighten to 25–34 ft. lbs. (34–46 Nm) on Bronco and F Series or 20–35 ft. lbs. (28–47 Nm) on E Series. Install the

gear to frame bolts and tighten to 54–66 ft. lbs. (74–89 Nm).

9. Make sure the wheels are in the straight ahead position, then install the pitman arm on the sector shaft. Install the pitman arm washer and nut and tighten to 170–230 ft. lbs. (230–310 Nm).

10. Connect and tighten the pressure and return lines to 20–30 ft. lbs. (27–40 Nm).

11. Snap the flex coupling shield over the hose fitting and the splash shield. Fill the system with fluid and connect the negative battery cable. Bleed the system and check for fluid leaks.

Power Steering Pump

REMOVAL & INSTALLATION

1. Disconnect the negative battery cable.

2. Disconnect the fluid return hose at the reservoir and drain the power steering fluid into a container. Remove the pressure hose from the pump.

3. If equipped with an automatic tensioner, use a wrench on the tensioner pulley bolt to lift the tensioner arm and remove the belt. Do not allow the tensioner to snap back.

4. If the drive belt is manually tensioned, loosen the pivot and adjustment bolts and remove the belt.

5. Remove the pulley from the pump, if necessary. Remove the attaching bolts and remove the pump.

To install:

6. Install the pump and tighten the attaching bolts. If the drive belt is manually tensioned, leave the pivot and adjusting bolts slightly snug.

7. Install the power steering pulley, if removed.

8. If equipped with an automatic tensioner, lift the tensioner arm using a wrench on the tensioner pulley bolt. Install the belt over the pulleys and slowly release the tensioner arm.

9. If the belt is manually tensioned, adjust the tension and tighten the pivot and adjusting bolts.

10. Connect the pressure and return hoses and the negative battery cable. Fill the reservoir with fluid and bleed the system. Check for leaks.

BELT ADJUSTMENT

Except 7.3L Diesel Engine

All engines except the 7.3L diesel engine are equipped with an automatic tensioner to maintain power steering drive belt tension. Correct belt tension will be maintained if the correct length belt is on the engine. The tensioner is working properly if the belt length indicator mark on the tensioner is between the maximum and minimum marks.

7.3L Diesel Engine

1. Position a belt tension gauge midway between the pulleys on the longest accessible belt span.

2. Insert a square drive ratchet wrench or breaker bar in the square hole in the power steering bracket and adjust the belt to 140–180 lbs. tension for a new belt or 95–115 lbs. tension for a used belt. Used belts have more than 5 minutes running time.

3. Maintain the belt tension and tighten the power steering pivot and adjustment bolts. Recheck the belt tension after tightening.

4. Remove the tension gauge and the ratchet or breaker bar.

SYSTEM BLEEDING

1. Fill the power steering fluid reservoir.

2. On all except 7.3L diesel engine, proceed as follows:

   a. Disconnect the coil wire.

   b. Crank the engine with the starter and continue adding fluid until the level remains constant. Do not prolong cranking as the battery may be drained and the starter damaged.

   c. Rotate the steering wheel approximately 30 degrees each side of center while continuing to crank the engine.

   d. Recheck the fluid level and fill, as required.

   e. Reconnect the coil wire.

3. On all vehicles, start the engine and allow it to run for several minutes.

4. Rotate the steering wheel from stop to stop.

5. Shut off the engine and recheck the fluid level. Add fluid, as required.

6. If air is still trapped in the system, proceed as follows:

   a. On Bronco and F Series, fabricate a purging tool. On E Series, it will be necessary to modify a pump reservoir cap.

   b. Make sure the reservoir fluid level is correct.

   c. On Bronco and F Series, insert the rubber stopper end of the fabricated purging tool. On E Series install the modified pump reservoir cap.

   d. Connect a suitable length of hose to the purging tool or modified cap. Connect the other end of the hose to an air conditioner vacuum pump or distributor machine. Do not use engine vacuum.

   e. Start the engine and let it idle for approximately 15 minutes. Turn the steering wheel 1 full cycle every 5 minutes but do not hit the stops. This will assist in removing trapped air.

   f. Stop the engine and disconnect the vacuum source. Remove the purging tool or modified cap.

   g. Check the fluid level and install the reservoir dipstick.

Tie Rod Ends

REMOVAL & INSTALLATION

1. Raise and safely support the vehicle. Remove the cotter pin and nut from the tie rod end ball stud.

2. Using removal tool T64P-3590-F or equivalent, remove the tie rod ball stud from the spindle or drag link.

3. Loosen the nuts on the adjusting sleeve clamps and unscrew the tie rod end from the adjusting sleeve. Count and record the number of turns re-

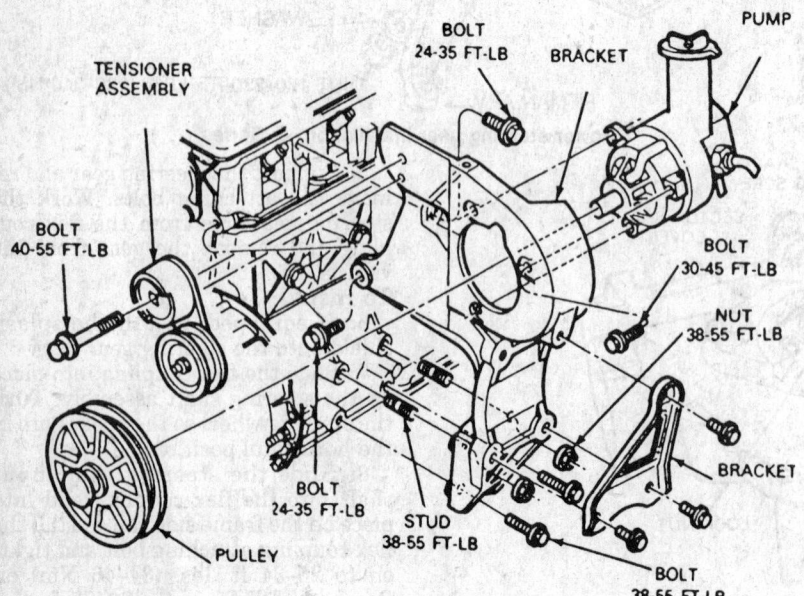

TENSIONER ASSEMBLY

BOLT 24-35 FT-LB

BRACKET

PUMP

BOLT 40-55 FT-LB

BOLT 30-45 FT-LB

NUT 38-55 FT-LB

BRACKET

BOLT 24-35 FT-LB

STUD 38-55 FT-LB

PULLEY

BOLT 38-55 FT-LB

Power steering pump installation—Bronco and F Series

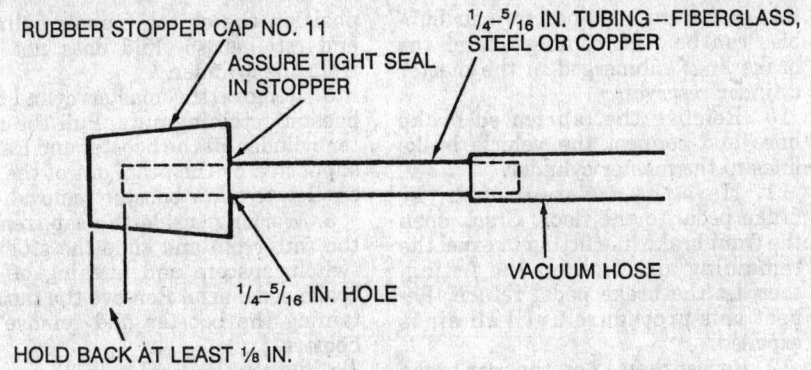

RUBBER STOPPER CAP NO. 11

ASSURE TIGHT SEAL IN STOPPER

¹/₄–⁵/₁₆ IN. TUBING—FIBERGLASS, STEEL OR COPPER

VACUUM HOSE

¹/₄–⁵/₁₆ IN. HOLE

HOLD BACK AT LEAST ¹/₈ IN.

Fabricated purging tool dimensions—Bronco and F Series

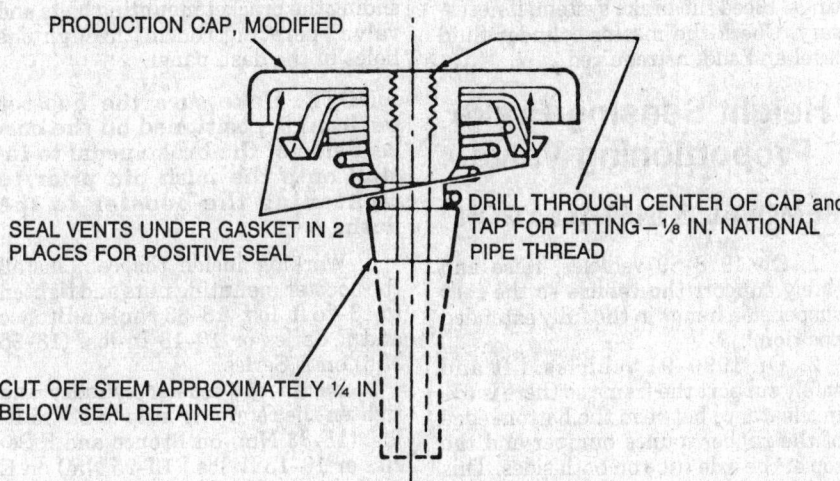

PRODUCTION CAP, MODIFIED

DRILL THROUGH CENTER OF CAP and TAP FOR FITTING—¹/₈ IN. NATIONAL PIPE THREAD

SEAL VENTS UNDER GASKET IN 2 PLACES FOR POSITIVE SEAL

CUT OFF STEM APPROXIMATELY ¹/₄ IN BELOW SEAL RETAINER

Modified pump reservoir cap—E Series

quired to remove the tie rod end from the sleeve.

To install:

4. Install the tie rod end in the adjusting sleeve, screwing in the same number of turns as was recorded during removal.

5. Tighten the adjusting sleeve clamp nuts to 30–42 ft. lbs. (40–57 Nm).

6. Insert the tie rod end ball stud into the spindle or drag link tapered hole. Install the nut and tighten to 50–75 ft. lbs. (68–101 Nm).

7. Install a new cotter pin.

NOTE: It may be necessary to further tighten the tie rod end ball stud nut to align the castellation in the nut with the hole in the ball stud, in order to install the cotter pin. Never loosen the nut to align the castellation and hole.

8. Lower the vehicle and check the toe setting.

BRAKES

For all brake system repair and service procedure not detailed below, please refer to "Brakes" in the Unit repair section.

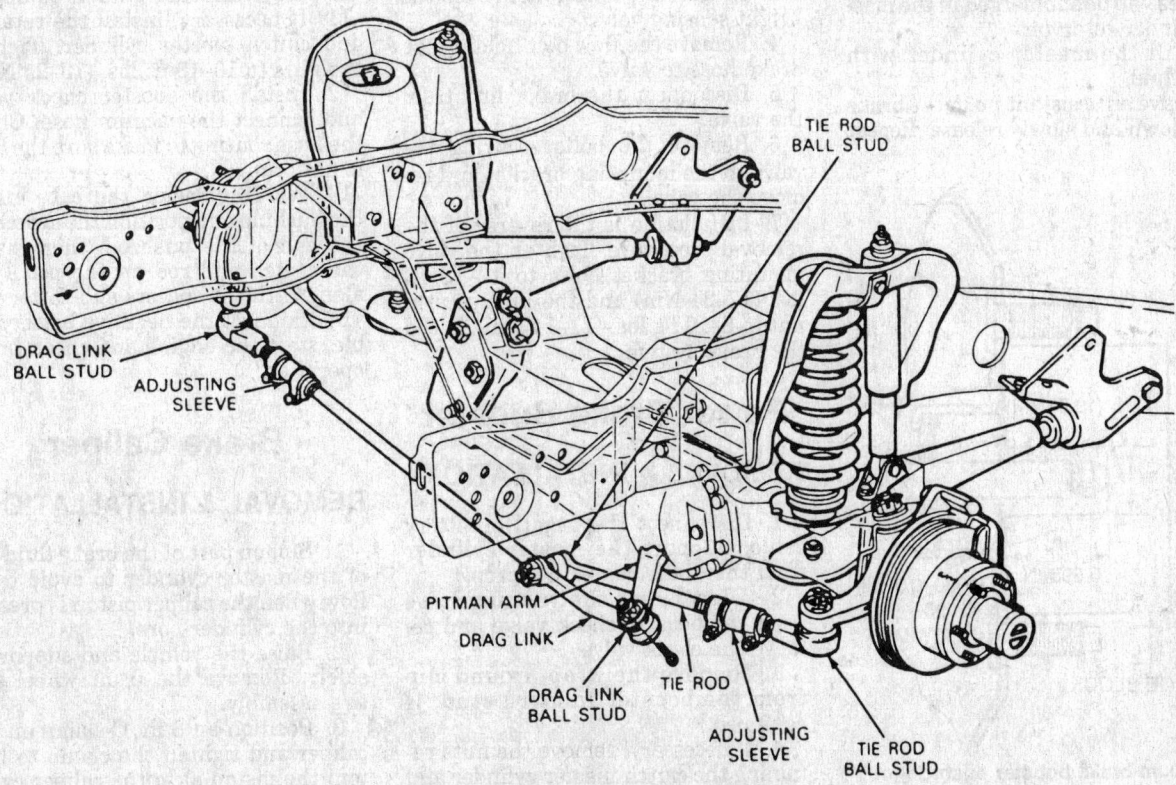

TIE ROD BALL STUD

DRAG LINK BALL STUD

ADJUSTING SLEEVE

PITMAN ARM

DRAG LINK

DRAG LINK BALL STUD

TIE ROD

ADJUSTING SLEEVE

TIE ROD BALL STUD

Steering linkage—4WD Bronco and F 150

Master Cylinder

REMOVAL & INSTALLATION

1. Disconnect the negative battery cable. Push the brake pedal down to expel vacuum from the brake booster system.

2. Disconnect the fluid level indicator switch connector from the master cylinder.

3. Disconnect and plug the hydraulic lines at the master cylinder. Plug the master cylinder ports.

4. Remove the brake booster to master cylinder retaining nuts and remove the master cylinder from the brake booster.

To install:

5. Before installing the master cylinder, check the distance from the outer end of the vacuum booster assembly pushrod to the front face of the vacuum brake booster assembly. The distance should be 0.980–0.995 in. (24.89–25.27mm). Turn the pushrod adjusting screw in or out, as required, to obtain the proper length.

6. Install the master cylinder with the attaching nuts. Tighten to 18–25 ft. lbs. (24–34 Nm). Connect the fluid level indicator switch.

7. Obtain 2 short lengths of brake line the same diameter as that used on the vehicle to connect to the master cylinder. Bend the lines, using a tubing bender, so when installed on the master cylinder ports, the other end of the lines will be submerged in the master cylinder reservoir.

8. Fill the master cylinder with brake fluid.

9. Have an assistant push the brake pedal down and slowly release. Repeat this procedure until no more air bubbles can be seen coming out of the brake lines submerged in the master cylinder reservoir.

10. Remove the fabricated brake lines and connect the vehicle brake lines to the master cylinder.

11. Have the assistant push the brake pedal to the floor. Crack open the front brake line fitting to expel the remaining air. Tighten the fitting, then let the brake pedal return. Repeat this procedure until all air is expelled.

12. Repeat Step 12 on the rear brake line fitting.

13. Final tighten the brake line fittings. Bleed the brake system, if necessary. Check the master cylinder fluid level and add, as required.

Height Sensing Brake Proportioning Valve

REMOVAL & INSTALLATION

1. On 1988–89 vehicles, raise and safely support the vehicle so the rear suspension hangs in the fully extended position.

2. On 1990–91 vehicles, lift and safely support the frame so there is 6⅝ in. clearance between the bottom edge of the rubber jounce bumper and the top of the axle tube on both sides. This is the correct position for installing the pre-indexed height sensing valve.

3. Remove the linkage arm from the height sensing valve.

4. Remove the flow bolt holding the brake hose to valve.

5. Disconnect the brake line from the valve.

6. Remove the bolts securing the valve to the mounting bracket and remove the valve.

7. Installation is the reverse of the removal procedure. Tighten the valve mounting bracket bolts to 12–18 ft. lbs. (17–24 Nm) and the linkage arm nut to 8–10 ft. lbs. (11–14 Nm). Bleed the brake system.

Power Brake Booster

REMOVAL & INSTALLATION

1. Disconnect the negative battery cable. Support the master cylinder from the underside with a prop.

2. Disconnect the vacuum hose from the booster check valve and remove the check valve.

3. Remove the wrap-around clip from the booster inboard stud, if equipped.

4. If necessary, remove the nuts retaining the clutch master cylinder and retainer to the booster. Remove and position the clutch master cylinder and retainer so fluid does not spill from the cylinder.

5. Remove the master cylinder-to-booster retaining nuts. Pull the master cylinder off the booster and leave it supported by the prop, out of the way enough to allow booster removal.

6. Working inside the cab, remove the cotter pin and slide the stoplight switch, spacers and bushing off the brake pedal arm. Remove the nuts retaining the booster and remove the booster.

To install:

7. Mount the booster assembly on the engine side of the dash panel by sliding the bracket mounting bolts and valve operating rod in through the holes in the dash panel.

NOTE: Make sure the booster pushrod is positioned on the correct side of the brake pedal to install onto the push pin prior to tightening the booster to the dash.

8. Working inside the cab, install the booster mounting nuts and tighten to 13–25 ft. lbs. (18–33 Nm) on Bronco and F Series or 10–18 ft. lbs. (13–25 Nm) on E Series.

9. Install the master cylinder and tighten the retaining nuts to 13–25 ft. lbs. (18–33 Nm) on Bronco and F Series or 10–18 ft. lbs. (13–25 Nm) on E Series.

10. Install the wrap-around clip on the booster inboard stud, if equipped.

11. If necessary, install the retainer and clutch master cylinder. Tighten the nuts to 10–18 ft. lbs. (13–25 Nm).

12. Install the booster check valve and connect the vacuum hose. Check the hose routing to make sure the hose is not crimped.

13. Working inside the cab, install the bushing and position the switch on the end of the pushrod, then install the switch and rod on the pedal pin along with the spacers and cotter pin.

14. Connect the negative battery cable, start the engine and check brake operation.

Brake Caliper

REMOVAL & INSTALLATION

1. Siphon part of the brake fluid out of the master cylinder to avoid overflow when the caliper piston is pressed into the cylinder bore.

2. Raise the vehicle and support it safely. Remove the front wheel and tire assembly.

3. Position an 8 in. C-clamp on the caliper and tighten the clamp to bottom the piston(s) in the caliper cylinder bore. Remove the clamp.

¾ IN. (19.05MM)

2¹⁵⁄₁₆ IN. (74.61MM)

0.995 IN.

0.980–0.995 IN. (24.98– 25.27MM)

GAUGE BLOCK

Vacuum brake booster pushrod-to-booster face dimension

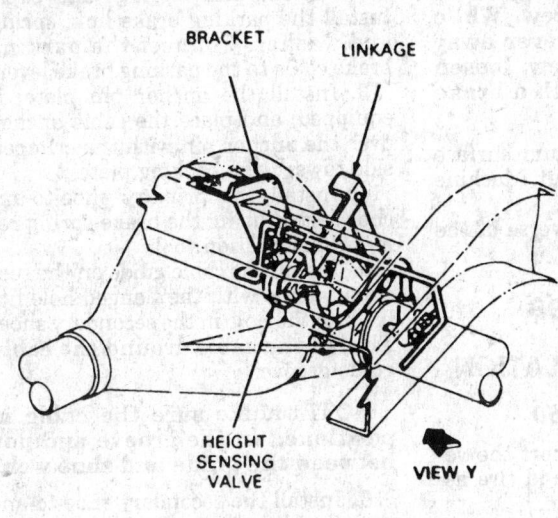

VIEW OF HEIGHT SENSING VALVE

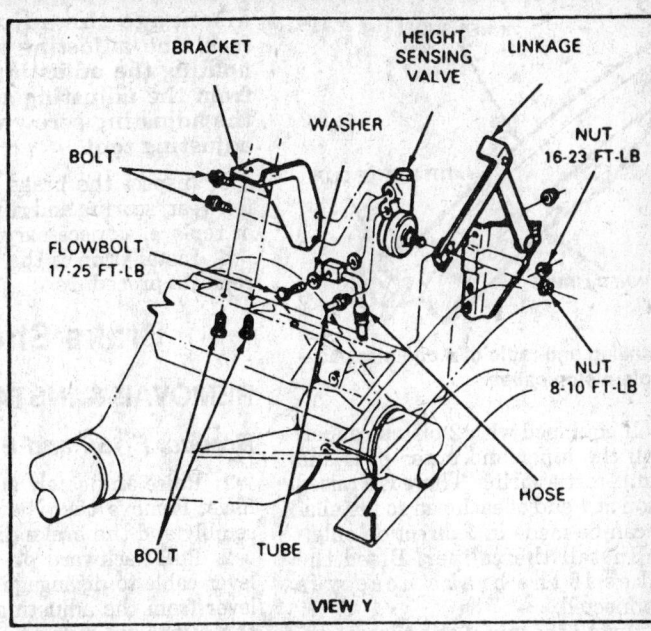

Height sensing brake proportioning valve—1988–89 E Series

NOTE: Do not pry the piston(s) away from the rotor.

4. Clean excess dirt from the area around the pin tabs. Tap the upper caliper pin towards the inboard side until the pin tabs touch the spindle face.

5. Insert a small prybar into the slot provided behind the pin tabs on the inboard side of the pin.

6. Use needle nose pliers to compress the outboard end of the pin while prying at the same time with the prybar, until the tabs slip into the spindle groove.

7. Place 1 end of a $^7/_{16}$ in. diameter punch against the end of the caliper pin and drive the pin out of the caliper slide groove.

8. Repeat the procedure for the lower pin.

9. Disconnect the brake hose and remove the caliper from the vehicle. Plug the brake hose.

To install:

10. Install the caliper on the spindle, making sure the mounting surfaces are free of dirt and lubricate the caliper grooves with disc brake caliper slide grease.

11. Install the pin with the pin retention tabs positioned next to the spindle groove. Tap the pin on the outboard end with a hammer and continue tapping inward until the retention tabs on the sides of the pin contact the spindle face. Repeat the procedure for the lower pin.

NOTE: Do not allow the tabs of the caliper pin to be tapped too far into the spindle groove or it will be necessary to tap the other end of the caliper pin until the

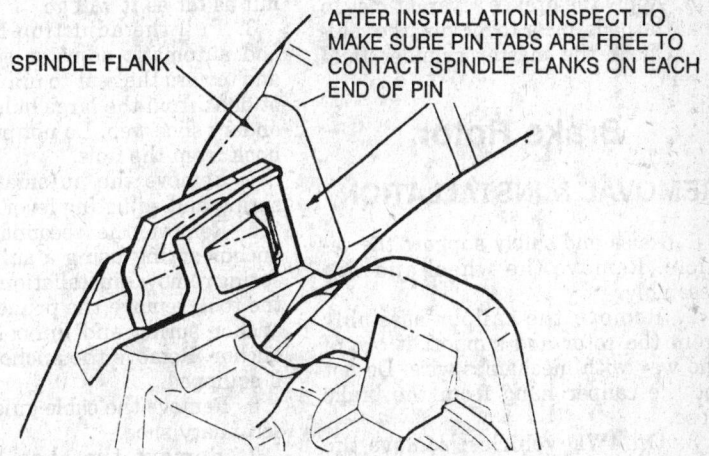

Correct caliper pin Installation

tabs snap into place. The tabs on each end of the caliper pin must be free to catch on the spindle flanks.

12. Connect the brake hose to the caliper and bleed the brake system.

13. Install the wheel and tire assembly and lower the vehicle.

Disc Brake Pads

REMOVAL & INSTALLATION

1. Raise and safely support the vehicle. Remove the caliper assembly.

NOTE: If the caliper is not being serviced, it is not necessary to disconnect the brake hose from the caliper. Suspend the caliper aside with mechanics wire; do not let it hang from the brake hose.

2. If equipped with single piston cal-

iper, remove the outer pad, then remove the anti-rattle clips and the inner pad.

3. If equipped with 2 piston caliper, remove the inner and outer pads and the anti-rattle spring.

To install:

4. If equipped with single piston caliper, proceed as follows:

a. Place a new anti-rattle clip on the lower end of the inner pad. Make sure the tabs on the clip are positioned properly and the clip is fully seated.

b. Position the inner pad and anti-rattle clip in the pad abutment with the anti-rattle clip tab against the pad abutment and the loop-type spring away from the rotor. Compress the anti-rattle clip and slide the upper end of the pad in position.

c. Install the outer pad. Crimp or bend the outer pad tabs to prevent the pads from rattling in the caliper.

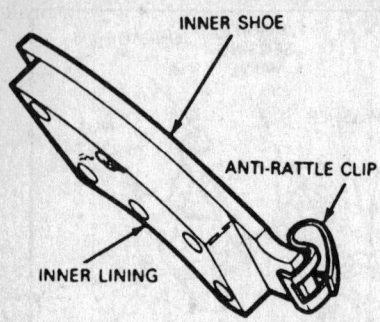

INNER SHOE

ANTI-RATTLE CLIP

INNER LINING

Installing anti-rattle clip on inner pad—single piston caliper

5. If equipped with 2 piston caliper, install the inner and outer pads and the anti-rattle spring. There is a raised section at 1 end of each pad so installation can be made in 1 direction only.

6. Install the caliper. Bleed the brakes if the brake hose was disconnected.

7. Install the wheel and tire assembly and lower the vehicle.

8. Apply the brakes several times to seat the pads, prior to moving the vehicle. Check the master cylinder fluid level.

Brake Rotor

REMOVAL & INSTALLATION

1. Raise and safely support the vehicle. Remove the wheel and tire assembly.

2. Remove the caliper assembly from the rotor and support it out of the way with mechanics wire. Do not let the caliper hang from the brake hose.

3. On 4WD vehicles, remove the locking hub assembly.

4. Remove the dust cap, cotter pin, locknut, washer and outer bearing and remove the rotor from the spindle.

5. Inspect the rotor for scoring, wear and runout; machine or replace as necessary.

6. Install in the reverse order of removal. Adjust the wheel bearings.

Brake Drum

REMOVAL & INSTALLATION

1. Raise and safely support the vehicle. Remove the wheel and tire assembly.

2. Remove the spring retaining nuts, if equipped, and remove the brake drum.

NOTE: If the brake drum will not come off, insert a narrow prybar through the brake adjusting hole in the backing plate and

disengage the adjusting lever from the adjusting screw. While holding the adjusting lever away from the adjusting screw, loosen the adjusting screw with a brake adjusting tool.

3. Inspect the brake drum surface for wear, scoring and runout. Machine or replace, as necessary.

4. Installation is the reverse of the removal procedure.

Brake Shoes

REMOVAL & INSTALLATION

Bronco, F 150 and E 150

1. Raise and safely support the vehicle. Remove the wheel and tire assembly and the brake drum.

2. Pull backward on the adjusting lever cable to disengage the adjusting lever from the adjusting screw. Move the outboard side of the adjusting screw upward and back off the pivot nut as far as it will go.

3. Pull the adjusting lever, cable and automatic adjuster spring down and toward the rear to unhook the pivot hook from the large hole in the secondary shoe web. Do not pry the pivot hook from the hole.

4. Remove the automatic adjuster spring and adjusting lever.

5. Remove the secondary shoe-to-anchor spring using a suitable brake spring removal/installation tool. Using the tool, remove the primary shoe-to-anchor spring and unhook the cable anchor. Remove the anchor pin plate, if equipped.

6. Remove the cable guide from the secondary shoe.

7. Remove the shoe hold-down springs, shoes, adjusting screw, pivot nut and socket. Note the color and position of each hold-down spring so they can be reassembled in the same position.

8. Remove the parking brake link and spring. Disconnect the parking brake cable from the parking brake lever.

9. Remove the secondary brake shoe and disassemble the parking brake lever from the shoe by removing the retaining clip and spring washer.
To install:
10. Clean the backing plate ledge pads and sand lightly. Apply a light coating of high temperature lithium grease to the points where the brake shoes touch the backing plate. Lubricate the adjusting cable eye and the anchor pin area.
11. Install the parking brake lever on the secondary shoe and secure with the spring washer and retaining clip.
12. Position the brake shoes on the backing plate and install the hold-

down spring pins, springs and cups. Install the parking brake link, spring and washer. Connect the parking brake cable to the parking brake lever.

13. Install the anchor pin plate, if equipped, and place the cable anchor over the anchor pin with the crimped side toward the backing plate.

14. Install the primary shoe-to-anchor spring using the brake spring removal/installation tool.

15. Install the cable guide on the secondary shoe with the flanged hole fitted into the hole in the secondary shoe. Thread the cable around the cable guide groove.

NOTE: Make sure the cable is positioned in the groove and not between the guide and shoe web.

16. Install the secondary shoe-to-anchor (long) spring.

NOTE: Make sure the cable end is not cocked or binding on the anchor pin when installed. All parts should be flat on the anchor pin.

17. Apply high temperature lithium grease to the threads and the socket end of the adjusting screw. Turn the adjusting screw into the adjusting pivot nut to the end of the threads and then loosen, ½ turn.

18. Place the adjusting socket on the screw and install the assembly between the shoe ends with the adjusting screw nearest the secondary shoe.

NOTE: Be sure to install the adjusting screw on the same side of the vehicle from which it came. To prevent incorrect installation, the socket end of each adjusting screw is stamped with R or L, to indicate installation on the right or left side of the vehicle. The adjusting pivot nuts have lines machined around the body of the nut, 2 lines indicating the right side nut and 1 line indicating the left side nut.

19. Hook the cable hook into the hole in the adjusting lever from the outboard plate side. The adjusting levers are also stamped with an R or L to indicate right or left side installation.

20. Place the hooked end of the adjuster spring in the large hole in the primary shoe web and connect the loop end of the spring to the adjuster lever hole.

21. Pull the adjuster lever, cable and automatic adjuster spring down toward the rear to engage the pivot hook in the large hole in the secondary shoe web.

22. After installation, check the action of the adjuster by pulling the section of the cable between the cable guide and the adjusting lever toward

the secondary shoe web far enough to lift the lever past a tooth on the adjusting screw wheel. The lever should snap into position behind the next tooth and releasing the cable should cause the adjuster spring to return the lever to its original position. This return action will turn the adjusting screw 1 tooth.

23. If pulling the cable does not produce the action described in Step 22 or if lever action is sluggish instead of positive and sharp, check the position of the lever on the adjusting screw toothed wheel. With the brake in a vertical position, anchor at the top, the lever should contact the adjusting wheel 1 tooth above the center line of the adjusting screw. If the contact point is below the center line, the lever will not lock on the adjusting screw wheel teeth and the screw will not turn as the lever is actuated by the cable.

24. To find the cause of the condition described in Step 23, proceed as follows:

a. Check the cable end fittings. The cable should completely fill or extend slightly beyond the crimped section of the fittings. If this does not happen, the cable assembly may be damaged and should be replaced.

b. Check the cable guide for damage. The cable groove should be parallel to the shoe web and the body of the guide should lie flat against the web. Replace the guide if it shows damage.

c. Check the pivot hook on the lever. The hook surfaces should be square with the body on the lever for proper pivoting. Repair the hook or replace the lever if the hook shows damage.

d. Be sure the adjusting screw socket is properly seated in the notch in the shoe web.

25. Adjust the brake shoes using either a brake adjustment gauge or manually with the drums installed.

26. If using a brake adjustment gauge, proceed as follows:

a. Measure the inside diameter of the brake drum with the gauge.

b. Reverse the tool and adjust the brake shoes until they touch the gauge. The gauge contact points on the shoes must be parallel to the vehicle with the center line through the center of the axle.

c. Install the drum and wheel and tire assembly. Lower the vehicle.

d. Apply the brakes sharply several times while driving the vehicle in reverse. Check brake operation by maling several stops while driving forward.

27. If manually adjusting the brakes, proceed as follows:

a. Install the brake drum and wheel and tire assembly.

b. Remove the cover from the adjusting hole at the bottom of the backing plate and turn the adjusting screw, using a suitable brake adjusting tool, to expand the brake shoes until they drag against the brake drum.

c. When the shoes are against the drum, insert a narrow prybar through the brake adjusting hole and disengage the adjusting lever from the adjusting screw. While holding the adjusting lever away from the adjusting screw, loosen the adjusting screw with the brake adjusting tool, until the drum rotates freely without drag.

d. Install the adjusting hole cover and lower the vehicle.

e. Apply the brakes. If the pedal travels more than halfway to the floor, there is too much clearance between the brake shoes and drums. Repeat the adjustment procedure.

F 250, F 350, E 250 and E 350

1. Raise and safely support the vehicle. Remove the wheel and tire assembly and the brake drum.
2. Remove the parking brake lever assembly retaining nut from behind the backing plate and remove the parking brake lever assembly.
3. Remove the adjusting cable assembly from the anchor pin, cable guide and adjusting lever.
4. Remove the brake shoe retracting springs. Remove the hold-down spring from each shoe.
5. Remove the brake shoes and the adjusting screw assembly. Disassemble the adjusting screw assembly.
To install:
6. Clean the backing plate ledge pads and sand lightly. Apply a light coating of high temperature lithium

grease to the points where the brake shoes touch the backing plate.
7. Apply high temperature lithium grease to the threads and socket end of the adjusting screw.
8. Install the upper retracting spring on the primary and secondary shoes and position the shoe assembly on the backing plate with the wheel cylinder pushrods in the shoe slots.
9. Install the brake shoe hold-down springs.
10. Install the brake shoe adjustment screw assembly with the slot in the head of the adjusting screw toward the primary shoe.
11. Install the lower retracting spring, adjusting lever spring, adjusting lever assembly and connect the adjusting cable to the adjusting lever. Position the cable in the cable guide and install the cable anchor fitting on the anchor pin.

NOTE: Be sure to install the adjusting screw on the same side of the vehicle from which it came. To prevent incorrect installation, the socket end of each adjusting screw is stamped with R or L, to indicate installation on the right or left side of the vehicle. The adjusting pivot nuts have lines machined around the body of the nut, 2 lines indicating the right side nut and 1 line indicating the left side nut.

12. Install the parking brake assembly in the anchor pin and washer and secure with the retaining nut behind the backing plate.
13. Adjust the brake shoes using either a brake adjustment gauge or manually with the drums installed.

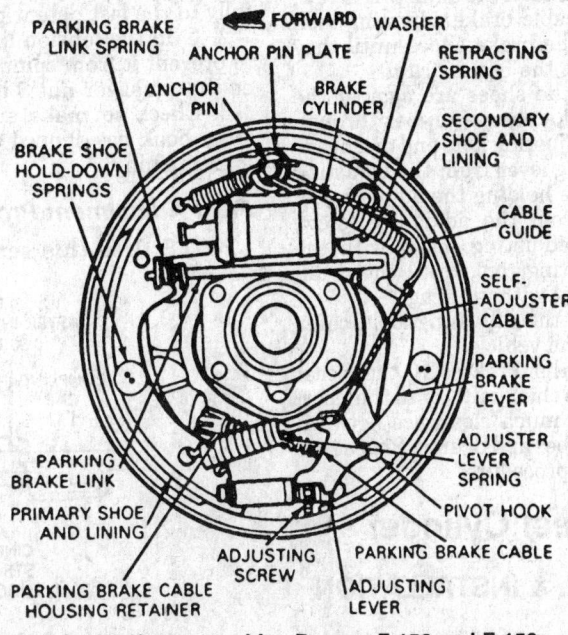

Rear brake shoe assembly—Bronco, F 150 and E 150

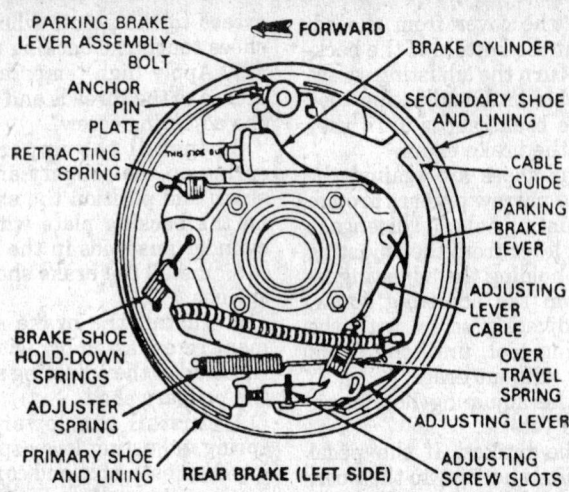

Rear brake shoe assembly—F 250, F 350, E 250 and E 350

14. If using a brake adjustment gauge, proceed as follows:

a. Measure the inside diameter of the brake drum with the gauge.

b. Reverse the tool and adjust the brake shoes until they touch the gauge. The gauge contact points on the shoes must be parallel to the vehicle with the center line through the center of the axle.

c. Install the drum and wheel and tire assembly. Lower the vehicle.

d. Apply the brakes sharply several times while driving the vehicle in reverse. Check brake operation by maling several stops while driving forward.

15. If manually adjusting the brakes, proceed as follows:

a. Install the brake drum and wheel and tire assembly.

b. Remove the cover from the adjusting hole at the bottom of the backing plate and turn the adjusting screw, using a suitable brake adjusting tool, to expand the brake shoes until they drag against the brake drum.

c. When the shoes are against the drum, insert a narrow prybar through the brake adjusting hole and disengage the adjusting lever from the adjusting screw. While holding the adjusting lever away from the adjusting screw, loosen the adjusting screw with the brake adjusting tool, until the drum rotates freely without drag.

d. Install the adjusting hole cover and lower the vehicle.

e. Apply the brakes. If the pedal travels more than halfway to the floor, there is too much clearance between the brake shoes and drums. Repeat the adjustment procedure.

Wheel Cylinder

REMOVAL & INSTALLATION

1. Raise and safely support the vehicle. Remove the wheel and tire assembly, brake drum and brake shoes.

2. Remove the cylinder to shoe connecting pins.

3. Disconnect the brake line from the wheel cylinder.

4. Remove the wheel cylinder retaining bolts and remove the cylinder from the brake backing plate.

5. Installation is the reverse of the removal procedure. Adjust the brakes and bleed the system.

Parking Brake Cable

ADJUSTMENT

Initial Adjustment Procedure

NOTE: Use this service adjustment procedure when a new tension limiter is installed.

1. Depress the parking brake pedal fully to the last detent position.

2. Grip the tension limiter housing to prevent it from spinning and tighten the equalizer nut 3 in. up the rod.

3. Check to make sure the cinch strap hook has slipped (less than 1½ in. remaining).

Field Adjustment Procedure

NOTE: Use this service adjust- ment procedure to correct a slack system, if a new tension limiter is not installed.

1. Make sure the brake drums are cold for correct adjustment. Depress the parking brake pedal fully, all the way to the floor.

2. Grip the tension limiter housing to prevent it from spinning and tighten the equalizer nut 6 full turns past its original position on the threaded rod until the cinch strap hooks begin to slip.

3. Using cable tension gauge 021–00018 or equivalent, behind the equalizer either toward the right or left drum, measure the cable tension. It should be at least 350 lbs. with the parking brake pedal fully in the last detent position. If the tension is low, repeat Step 2.

4. Release the parking brake and check for rear wheel drag. There should be no brake drag.

NOTE: The tension limiter will reset the parking brake tension any time the system is disconnected, as long as the distance between the bracket and the cinch strap hook is reduced during adjustment. When the cinch strap contacts the bracket, the system tension will greatly increase and over tension may result. If all available adjustment travel has been used, the tension limiter must be replaced.

REMOVAL & INSTALLATION

Equalizer To Control Assembly

1. Raise the vehicle and support it safely.

2. Back off the equalizer nut and remove the slug of front cable from the tension limiter.

3. Remove the parking brake cable from the cab mount on Bronco and F Series or crossmember on E Series and all retaining clips. Lower the vehicle.

4. Remove the forward ball end of the parking brake cable from the control assembly clevis.

5. Remove the cable from the con-

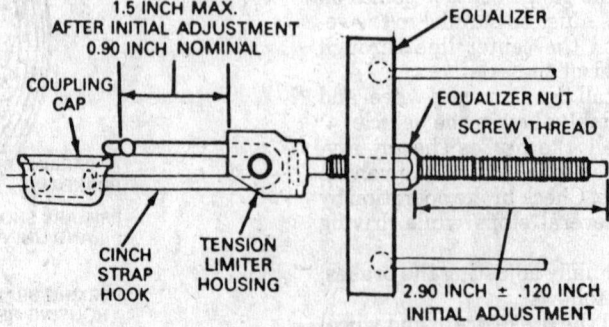

Parking brake cable tension limiter assembly

trol assembly by compressing the conduit end fitting prongs.

6. Using a cord attached to the control lever end of the cable, remove the cable from the vehicle.

To install:

7. Transfer the cord to the new cable. Position the cable in the vehicle, routing the cable through the dash panel. Remove the cord and secure the cable to the control.

8. Connect the forward ball end of the brake cable to the clevis of the control assembly. Raise and safely support the vehicle.

9. Route the cable through the cab mount on Bronco and F Series or through the crossmembers on E Series and secure in place with the retaining clips.

10. Connect the slug of the cable to the tension limiter connector. Adjust the parking brake cable at the equalizer using the initial adjustment or field adjustment procedure, as necessary.

11. Rotate both rear wheels to make sure the parking brake is not dragging.

Equalizer To Rear Wheel

1. Raise the vehicle and support it safely.

2. Remove the wheel and tire assembly, tension limiter and brake drum.

3. Remove the locknut on the threaded rod and disconnect the cable from the equalizer.

4. Compress the prongs that retain the cable housing to the frame bracket on Bronco and F Series or the crossmember on E Series, and pull the cable and housing out of the bracket or crossmember.

5. Working on the wheel side, compress the prongs on the cable retainer so they can pass through the hole in the brake backing plate. Draw the cable retainer out of the hole.

6. With the spring tension off of the parking brake lever, lift the cable out of the slot in the lever and remove the cable through the brake backing plate hole.

To install:

7. Pull the cable through the backing plate until the end is inserted over the parking brake lever slot. Pull the excess slack from the cable and insert the cable housing into the brake backing plate access hole until the retainer prongs expand.

8. Insert the front of the cable housing through the frame crossmember bracket until the prong expands. Insert the ball end of the cable into the key hole slots on the equalizer, rotate the equalizer 90 degrees and recouple the tension limiter threaded rod to the equalizer.

NOTE: On F 250, F 350, E 250

and E 350, check the clearance between the parking brake operating lever and the cam plate. The clearance should be 0.015 in. (0.38mm), when the brakes are fully released.

9. Install the brake drum and wheel and tire assembly. Adjust the brake shoes.

10. Adjust the parking brake tension using the initial adjustment or field adjustment procedure, as necessary.

11. Rotate both rear wheels to make sure the parking brake is not dragging.

Brake System Bleeding

1. Clean all dirt from the master cylinder filler cap.

2. If the master cylinder is known or suspected to have air in the bore, it must be bled before any of the wheel cylinders or calipers. To bleed the master cylinder, loosen the upper secondary left front outlet fitting approximately ¾ turn. Have an assistant depress the brake pedal slowly through it's full travel. Close the outlet fitting and let the pedal return slowly to the fully released position. Wait 5 seconds and then repeat the operation until all air bubbles disappear.

3. Repeat Step 2 with the right-hand front outlet fitting.

4. Continue to bleed the brake system by removing the rubber dust cap from the wheel cylinder bleeder fitting at the right-hand rear of the vehicle. Place a suitable box wrench on the bleeder fitting and attach a rubber drain tube to the fitting. The end of the tube should fit snugly around the bleeder fitting. Submerge the other end of the tube in a container partially filled with clean brake fluid and loosen the fitting ¾ turn.

5. Have an assistant push the brake pedal down slowly through it's full travel. Close the bleeder fitting and allow the pedal to slowly return to it's full release position. Wait 5 seconds and repeat the procedure until no bubbles appear at the submerged end of the bleeder tube. Secure the bleeder fitting and remove the bleeder tube. Install the rubber dust cap on the bleeder fitting.

6. Repeat the procedure in Steps 4 and 5 and bleed the rest of the system in the following sequence: left rear, right front and left front.

NOTE: If equipped with rear anti-lock brakes, bleed the Rear Anti-lock Brake System (RABS) valve in between the left rear wheel and the right front wheel in the sequence.

7. Refill the master cylinder reser-

voir after each wheel cylinder or caliper has been bled and install the master cylinder cover and gasket. When brake bleeding is completed, the fluid level should be filled to the maximum level indicated on the reservoir.

8. Always make sure the disc brake pistons are returned to their normal positions by depressing the brake pedal several times until normal pedal travel is established. If the pedal feels spongy, repeat the bleeding procedure.

Anti-lock Brake System Service

PRECAUTIONS

• Use caution when disassembling any hydraulic components as the system will contain residual pressure. Cover the area around the component to be removed with a shop cloth to catch any brake fluid spray. Do not allow brake fluid to come in contact with painted surfaces.

Rear Anti-lock Brake System (RABS) Module

REMOVAL & INSTALLATION

Bronco and F Series

1. Disconnect the negative battery cable.

2. Disconnect the wiring harness from the RABS module by depressing the plastic tab on the connector and pulling the connector off.

3. Remove the 2 screws that retain the module to the dash panel and remove the module.

4. Installation is the reverse of the removal procedure. Check the system for proper operation.

E Series

1. Disconnect the negative battery cable.

2. Remove the parking brake actuator assembly.

3. Remove the 2 screws that hold the module to the cowl panel. Remove the module.

4. Disconnect the wiring harness from the RABS module by depressing the plastic tab on the connector and pulling the connector off.

5. Installation is the reverse of the removal procedure. Check the RABS and parking brake systems for proper operation.

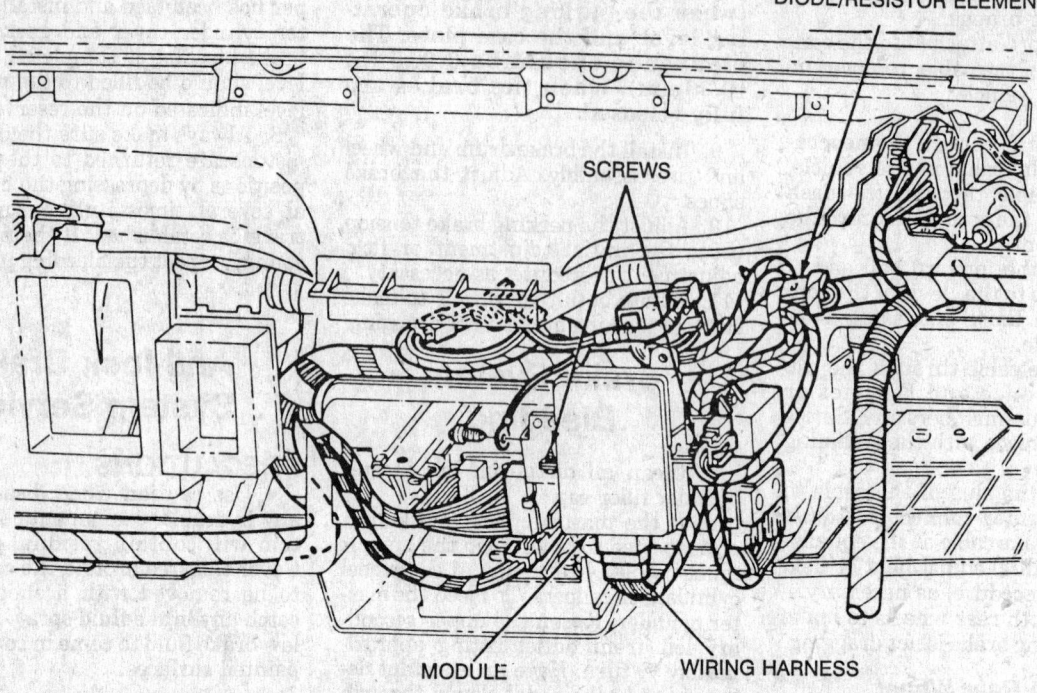

RABS module location—Bronco and F Series

RABS Valve

REMOVAL & INSTALLATION

1. Disconnect the negative battery cable.

2. Disconnect and plug the 2 brake lines connected to the RABS valve.

3. Disconnect the wiring harness from the valve harness.

4. Remove the screws or nuts retaining the valve and remove the valve.

To install:

5. Position the RABS valve and install the retaining screws or nuts. Tighten the retaining nuts on Bronco and F Series to 12–17 ft. lbs. (17–23 Nm). Tighten the retaining screws on E Series to 19–24 ft. lbs. (26–32 Nm).

6. Connect the brake valve wiring harness connector.

7. Connect the brake lines to the valve and tighten as follows:

  a. ½–20 threaded fittings—10–17 ft. lbs. (14–23 Nm).

  b. $^7/_{16}$–24 threaded fittings—10–15 ft. lbs. (14–20 Nm).

NOTE: Do not overtighten the fittings.

8. Bleed the brake system. It is not necessary to energize the valve electrically to bleed the rear brakes.

9. Connect the negative battery cable.

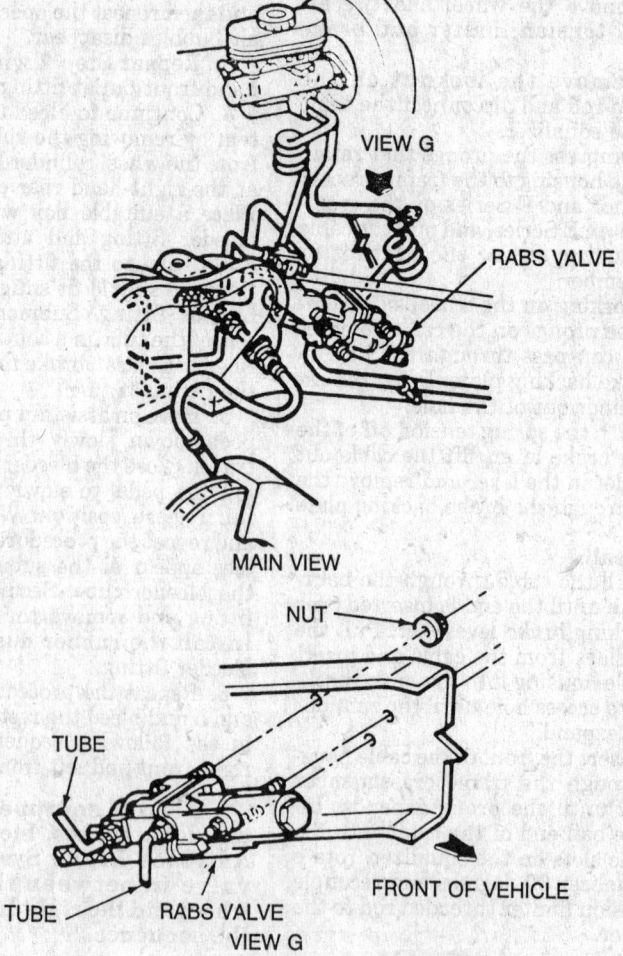

RABS valve installation—Bronco and F Series

RABS Sensor

REMOVAL & INSTALLATION

1. Disconnect the negative battery cable.

2. Pull the wiring harness connector off.

3. Remove the sensor hold down bolt and remove the sensor from the axle housing.

To install:

4. Clean the axle mounting surface. Use care to prevent dirt from entering the axle housing.

5. Inspect and clean the magnetized sensor pole piece to ensure that it is free from loose metal particles which could cause erratic system operation. Inspect the sensor O-ring for damage and replace, if necessary.

6. Lightly lubricate the sensor O-ring with motor oil, align the sensor bolt hole and install. Do not apply force to the plastic sensor connector. The sensor flange should slide to the mounting surface. This will insure the air gap setting is between 0.005–0.045 in. (0.127–1.14mm).

7. Install the hold down bolt and tighten to 25–30 ft. lbs. (34–40 Nm).

8. Inspect the blue sensor connector seal and replace if missing or damaged. Push the connector on the sensor.

9. Connect the negative battery cable.

Excitor Ring

INSPECTION

1. Remove the RABS sensor.

2. View the excitor ring teeth through the sensor hole. Rotate the rear axle and check the excitor ring teeth for damage or breakage. Dented or broken teeth could cause the RABS system to function when not required.

REMOVAL & INSTALLATION

To service the excitor ring, the differential case must be removed from the axle housing and the excitor ring pressed off the case.

NOTE: Upon removal, the excitor ring is to be discarded. It is not to be reused.

CHASSIS ELECTRICAL

Heater Blower Motor

REMOVAL & INSTALLATION

1. Disconnect the negative battery cable.

2. If equipped, remove the emission module forward of the blower motor.

3. Disconnect the blower motor wiring connector.

4. Disconnect the blower motor cooling tube at the blower motor.

5. Remove the screws attaching the blower motor and wheel to the blower housing.

6. Holding the cooling tube aside, pull the blower motor and wheel from the heater blower assembly and remove it from the vehicle.

7. Remove the hub clamp and remove the blower wheel from the blower motor shaft.

To install:

8. Install the blower wheel onto the blower motor shaft. Install the hub clamp.

9. Holding the cooling tube aside, position the blower motor and wheel in the blower housing and install the attaching screws.

10. Connect the blower motor cooling tube and the wire harness connector.

11. If equipped, install the emission module forward of the blower motor.

12. Connect the negative battery cable and check the blower motor for proper operation.

Windshield Wiper Motor

REMOVAL & INSTALLATION

1. Disconnect the negative battery cable.

2. On E series, remove the fuse panel and bracket assembly.

3. Remove both wiper arm and blade assemblies.

4. Disconnect the washer nozzle hose, if necessary and remove the cowl grille assembly.

5. Remove the wiper linkage clip from the motor output arm.

6. Disconnect the motor wiring connector.

7. Remove the 3 attaching screws and remove the motor.

8. Installation is the reverse of removal procedure. Always ensure the wiper motor is in the park position before installing. Torque the wiper motor retaining screws to 60–85 inch lbs. (6.7–9.5 Nm).

Windshield Wiper Switch

REMOVAL & INSTALLATION

Bronco and F Series

1. Disconnect the negative battery cable.

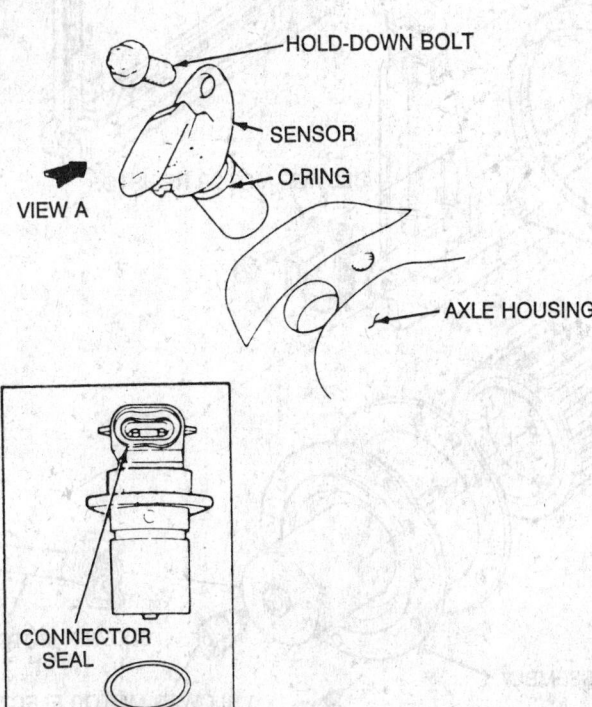

RABS sensor installation

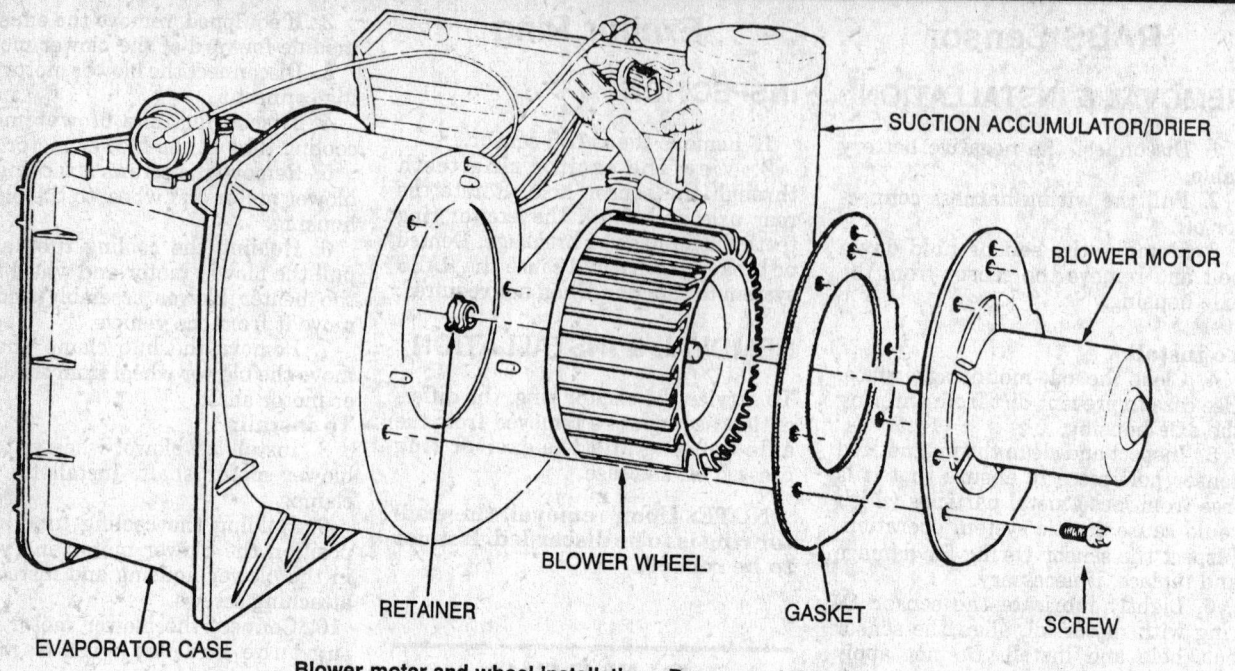

Blower motor and wheel Installation—Bronco and F Series

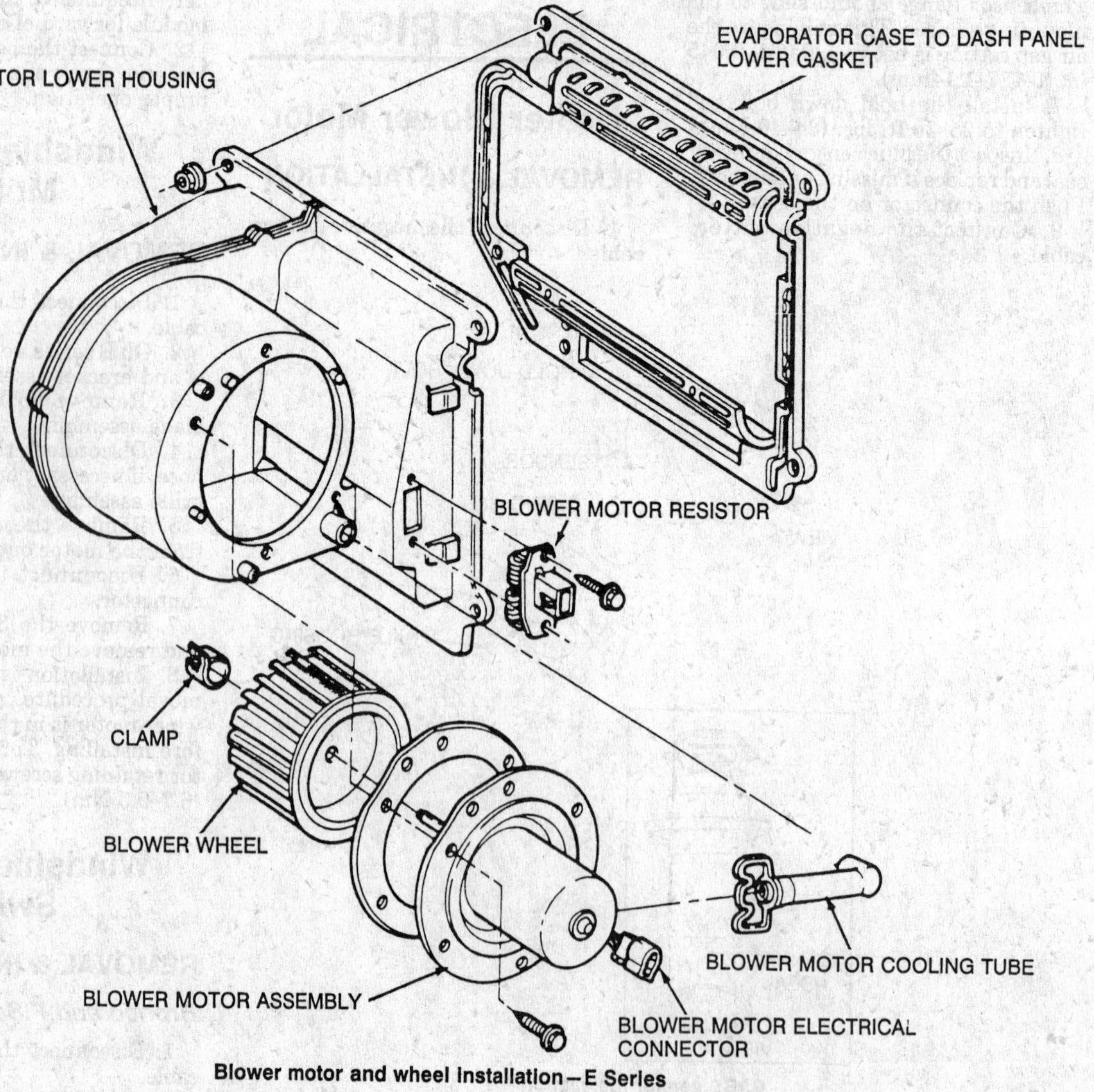

SUCTION ACCUMULATOR/DRIER

BLOWER MOTOR

BLOWER WHEEL

GASKET

SCREW

RETAINER

EVAPORATOR CASE

EVAPORATOR CASE TO DASH PANEL LOWER GASKET

EVAPORATOR LOWER HOUSING

BLOWER MOTOR RESISTOR

CLAMP

BLOWER WHEEL

BLOWER MOTOR COOLING TUBE

BLOWER MOTOR ASSEMBLY

BLOWER MOTOR ELECTRICAL CONNECTOR

Blower motor and wheel installation—E Series

2. Remove the wiper switch knob, bezel nut and bezel.

3. Pull out the switch from under the instrument panel.

4. Disconnect the electrical connector from the switch and remove the switch.

5. Installation is the reverse of the removal procedure.

E Series

1. Disconnect the negative battery cable.

2. Remove the windshield wiper switch knob.

3. Remove the ignition switch bezel.

4. Remove the headlight switch knob and shaft by pulling the switch to the **ON** position. Then, depress the button on the top of the switch and pull the knob and shaft out.

5. Remove the 2 screws at the bottom of the finish panel, then pry the 2 upper retainers away from the instrument panel.

6. Disconnect the connector from the switch.

7. Remove the attaching screws and remove the switch.

8. Installation is the reverse of the removal procedure.

Instrument Cluster

REMOVAL & INSTALLATION

Bronco and F Series

1. Disconnect the negative battery cable.

2. Remove the wiper/washer knob, using a hook tool to release each knob lock tab.

3. Remove the knob from the headlight switch. Remove the fog light switch knob, if equipped.

4. Remove the steering column shroud.

NOTE: If equipped with an automatic transmission, use care so as not to damage the transmission control selector indicator cable.

5. If equipped with an automatic transmission, remove the loop on the indicator cable assembly from the retainer pin. Remove the screw from the cable bracket and slide the bracket out of the slot in the tube.

6. Remove the cluster finish panel assembly.

7. Remove the 4 cluster attaching screws and disconnect the speedometer cable. Disconnect the wire connectors from the printed circuit and remove the cluster.

To install:

8. Position the cluster to the opening and connect the 2 connectors. Connect the speedometer cable and install the 4 cluster retaining screws.

9. If equipped with and automatic transmission, place the loop on the transmission indicator cable assembly over the retainer on the column.

10. Position the tab on the transmission indicator cable bracket into the slot on the column. Align the transmission indicator pointer and attach the screw. Adjust the transmission indicator as follows:

   a. With the engine off and the parking brake applied, place the transmission selector lever at the steering column in **D** or **OD** for automatic overdrive transmissions.

   b. Hold the lever against the **D** or **OD** stop using an approximate 8 lb. weight attached to the selector lever knob.

   c. Adjust the transmission indicator bracket to position the indicator in the rectangular adjustment band and attach the screw, taking care not to move the indicator.

   d. Change the transmission lever to **P** position and check the transmission indicator pointer. Shift the transmission lever to all shift positions and check the transmission indicator pointer position on each transmission shift position, verifying adjustment.

NOTE: The transmission indicator should only be adjusted using the adjustment window in the primary drive position and not adjusted in any other position.

11. Install the trim finish panel assembly and the column shroud.

12. Install the headlight switch knob. If equipped, install the fog light switch.

13. Install the wiper/washer control knobs.

14. Connect the negative battery cable. Check the operation of all gauges, lights, signals and transmission indicator pointer.

E Series

1. Disconnect the negative battery cable.

2. Remove the steering column shroud, if necessary.

3. If equipped with a tilt steering column, loosen the bolts which attach the column to the band support to provide sufficient clearance for cluster removal.

4. Remove the 7 cluster to panel retaining screws.

5. Position the cluster slightly away from the panel in order to disconnect the speedometer cable.

NOTE: If there is not sufficient access to disengage the speedometer cable from the speedometer, it may be necessary to remove the speedometer cable at the transmission and pull the cable through the cowl, to allow room to reach the speedometer quick disconnect.

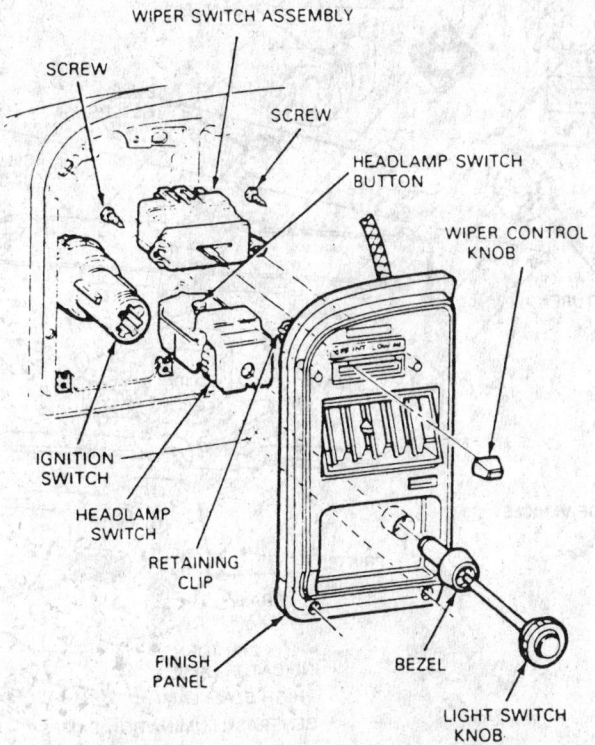

Windshield wiper switch installation – E Series

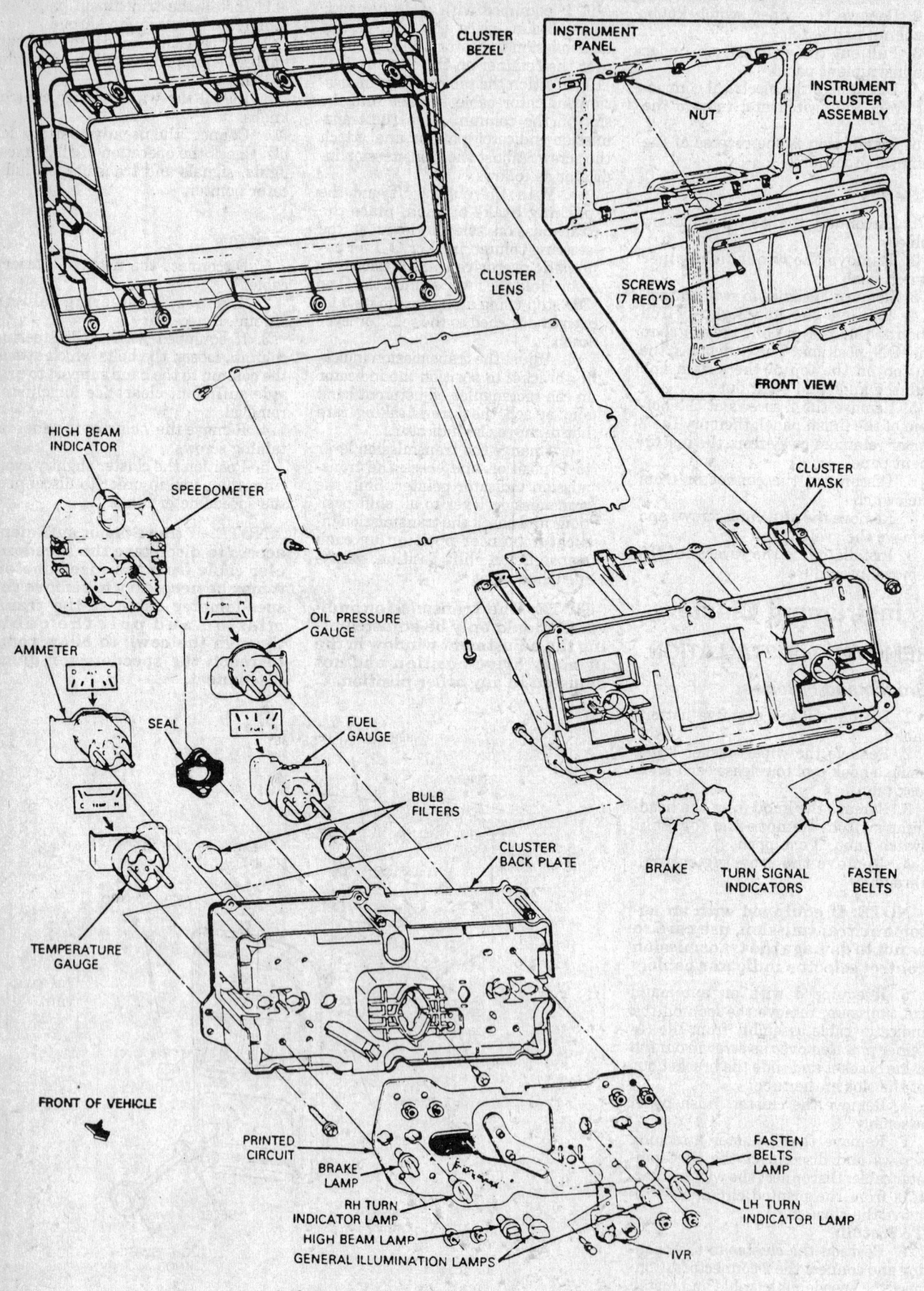

CLUSTER BEZEL

INSTRUMENT PANEL

NUT

INSTRUMENT CLUSTER ASSEMBLY

CLUSTER LENS

SCREWS (7 REQ'D)

FRONT VIEW

HIGH BEAM INDICATOR

SPEEDOMETER

CLUSTER MASK

OIL PRESSURE GAUGE

AMMETER

SEAL

FUEL GAUGE

BULB FILTERS

CLUSTER BACK PLATE

TEMPERATURE GAUGE

BRAKE

TURN SIGNAL INDICATORS

FASTEN BELTS

FRONT OF VEHICLE

PRINTED CIRCUIT

BRAKE LAMP

RH TURN INDICATOR LAMP

HIGH BEAM LAMP

GENERAL ILLUMINATION LAMPS

FASTEN BELTS LAMP

LH TURN INDICATOR LAMP

IVR

Instrument cluster exploded view—E Series

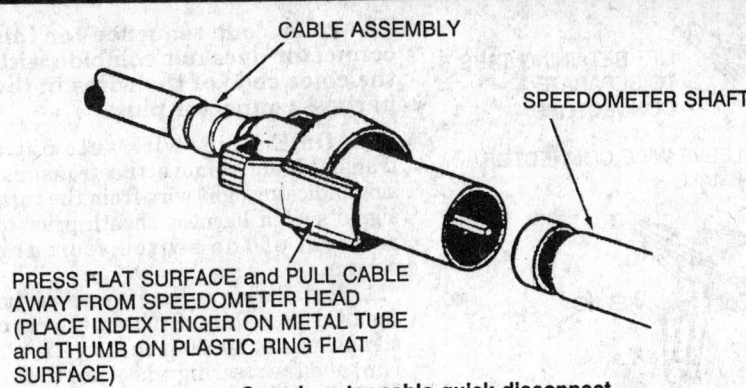

CABLE ASSEMBLY

SPEEDOMETER SHAFT

PRESS FLAT SURFACE and PULL CABLE AWAY FROM SPEEDOMETER HEAD (PLACE INDEX FINGER ON METAL TUBE and THUMB ON PLASTIC RING FLAT SURFACE)

Speedometer cable quick disconnect

6. Disconnect the harness connector plug from the printed circuit and remove the cluster assembly.

7. Installation is the reverse of the removal procedure. Before installation, apply an approximately $3/16$ in. diameter ball of silicone dielectric compound in the drive hole of the speedometer head.

Speedometer

REMOVAL & INSTALLATION

1. Disconnect the negative battery cable. Remove the instrument cluster.

2. Remove the lens and mask from the cluster.

3. Remove the 2 speedometer attaching screws and remove the speedometer.

4. Installation is the reverse of the removal procedure.

5. If a new speedometer is being installed, examine the square drive hole for sufficient lubrication. If lubrication is needed, apply a $3/16$ in. diameter dab of silicone grease in the drive hole.

Radio

REMOVAL & INSTALLATION

Bronco and F Series

1. Disconnect the negative battery cable.

2. Remove the bezel.

3. Remove the screws securing the radio mounting bracket to the instrument panel and pull out the radio.

4. Disconnect the antenna cable, speaker wires and power wire.

5. Installation is the reverse of the removal procedure.

E Series

1. Disconnect the negative battery cable.

2. Remove the heater and air conditioning control knobs. Remove the cigar lighter, if equipped.

3. If equipped with a cigar lighter,

snap out the name plate on the right side of the panel to gain access to 1 panel attaching screw. Remove the screw.

4. Remove the 5 remaining finish panel attaching screws. Being careful not to scratch the instrument panel, insert a small prybar and pop out the cluster panel.

5. Remove the 4 front radio retaining screws and remove the radio.

6. Disconnect the antenna cable, speaker wires and power wire.

7. Installation is the reverse of the removal procedure.

Headlight Switch

REMOVAL & INSTALLATION

Bronco and F Series

1. Disconnect the negative battery cable.

2. Remove the wiper/washer and headlight switch knobs using a hook tool to release each knob lock tab.

3. Remove the fog light switch

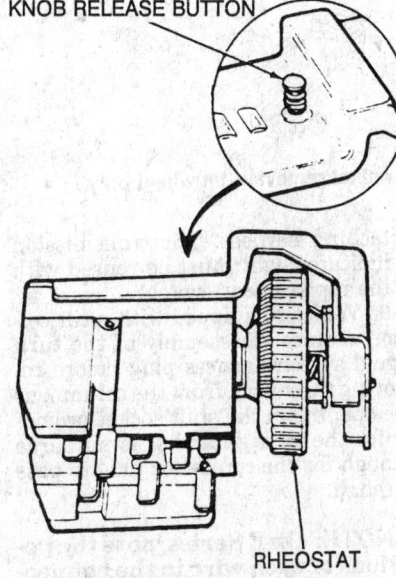

KNOB RELEASE BUTTON

RHEOSTAT

Headlight switch–E Series

knob, if equipped.

4. Remove the steering column shroud, if necessary and remove the cluster finish panel.

5. Unscrew the mounting nut and remove the switch from the instrument panel, then disconnect the wiring connector from the switch.

6. Installation is the reverse of the removal procedure.

E Series

1. Disconnect the negative battery cable.

2. Pull the headlight switch knob to the full **ON** position. Depress the shaft release button on the switch housing and pull the knob and shaft assembly out of the switch.

3. Unscrew the mounting nut. Remove the switch, then disconnect the wiring connector from the switch.

4. Installation is the reverse of the removal procedure. When installing the knob and shaft assembly, insert it all the way into the switch until a distinct click is heard. It may be necessary to rotate the shaft slightly until it engages the switch-contact carrier.

Dimmer Switch

REMOVAL & INSTALLATION

1. Disconnect the negative battery cable.

2. Pull back the floor mat or carpet in the area of the switch. Remove the dimmer switch retaining screws.

3. Disconnect the electrical connection from the switch.

4. Installation is the reverse of the removal procedure.

Turn Signal Switch

REMOVAL & INSTALLATION

1. Disconnect the negative battery cable.

2. Remove the horn switch.

3. Remove the steering wheel.

4. Unscrew the turn signal switch lever from the steering column.

5. Remove the steering column shroud. On E Series, remove the steering column opening cover.

6. Disconnect the switch wiring connector plug by lifting up on the tabs and separating. Remove the screws that secure the switch to the column.

7. On Bronco and F Series equipped with a fixed column, remove the switch by lifting it out of the column and guiding the connector plug through the opening in the shift socket.

8. On E series equipped with a fixed column, proceed as follows:

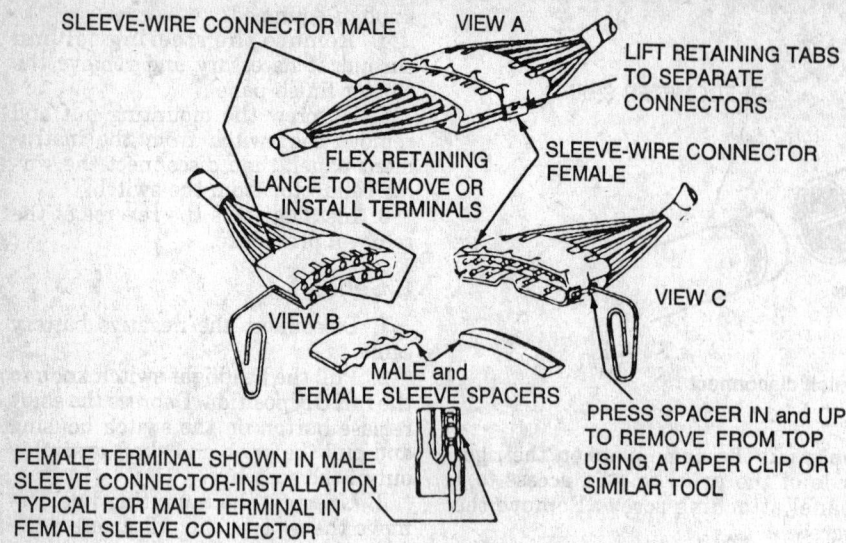

SLEEVE-WIRE CONNECTOR MALE

VIEW A

LIFT RETAINING TABS TO SEPARATE CONNECTORS

FLEX RETAINING LANCE TO REMOVE OR INSTALL TERMINALS

SLEEVE-WIRE CONNECTOR FEMALE

VIEW C

VIEW B

MALE and FEMALE SLEEVE SPACERS

PRESS SPACER IN and UP TO REMOVE FROM TOP USING A PAPER CLIP OR SIMILAR TOOL

FEMALE TERMINAL SHOWN IN MALE SLEEVE CONNECTOR-INSTALLATION TYPICAL FOR MALE TERMINAL IN FEMALE SLEEVE CONNECTOR

Turn signal switch wire connector removal

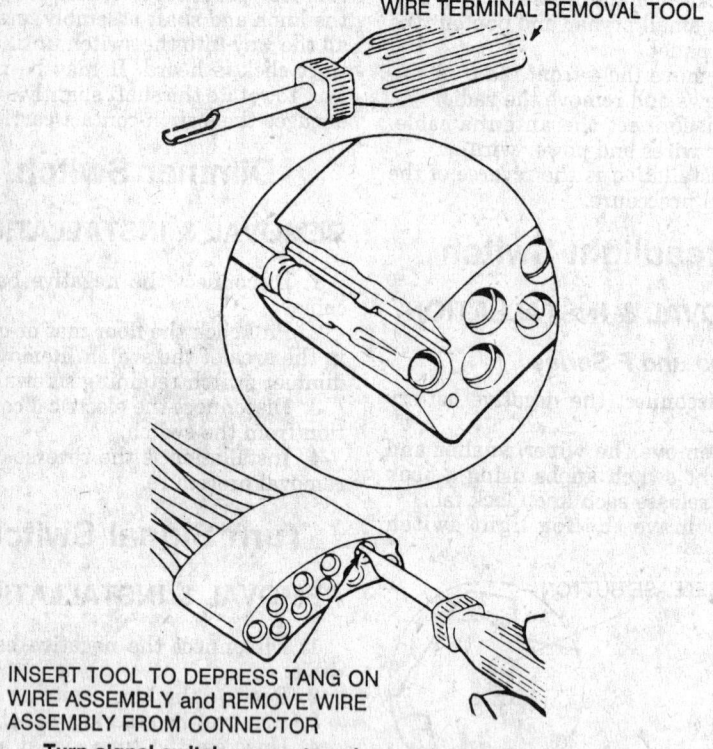

WIRE TERMINAL REMOVAL TOOL

INSERT TOOL TO DEPRESS TANG ON WIRE ASSEMBLY and REMOVE WIRE ASSEMBLY FROM CONNECTOR

Turn signal switch connector wire terminal removal—tilt wheel only

a. If equipped with an automatic transmission, remove the transmission indicator light assembly from the shift socket.

b. Lift the switch and, if equipped, the transmission indicator light harness out of the column and guide the connector plug through the openings in the brake and clutch pedal support bracket and the shift socket.

c. If the turn signal switch is to be replaced, the connector plug must be disassembled to remove the transmission indicator light and at-taching harness. The transmission indicator light must be reused with the replacement switch.

9. Vehicles equipped with a tilt column require disassembly of the turn signal switch harness plug before re-moving the switch from the column, as the opening in the shift socket provid-ed for the wiring harness is not large enough for the connector plug to pass through.

NOTE: On E Series, note the po-sitions of each wire in the connec-tor for reinstallation purposes.

The color code sequence for this connector does not coincide with the color code of the wires in the harness connector plug.

10. On E series with automatic transmission, remove the transmis-sion indicator light wire from the turn signal switch harness sheath prior to removal of the switch from the column.

11. Installation is the reverse of the removal procedure. Tighten the turn signal lever to 10–20 inch lbs. (1.2–2.2 Nm) and the steering wheel nut to 30–40 ft. lbs. (41–54 Nm).

Ignition Lock
REMOVAL & INSTALLATION
With Key

1. Disconnect the negative battery cable.

2. If equipped with a fixed column, remove the steering wheel.

3. If equipped with an automatic transmission, place the shift lever in P.

4. Turn the lock cylinder with the ignition key to the ON position.

5. Place a 1/8 in. (3.17mm) diameter wire pin or small drift punch and de-press the retaining pin while pulling out on the lock cylinder to remove it from the column housing. The pin is located inside the column near the base of the lock cylinder on fixed col-umn vehicles. On tilt column vehicles, the pin is located near the hazard warning button.

To install:

6. Lubricate the lock cylinder with grease.

7. Turn the lock cylinder to the ON position and depress the retaining pin, then insert the lock cylinder into its housing in the flange casting.

8. Make sure the cylinder is fully seated and aligned into the interlock-ing washer before turning the key to the OFF position. This will permit the cylinder retaining pin to extend into the cylinder casting housing hole.

9. Using the ignition key, rotate the lock cylinder to insure correct me-chanical operation in all positions.

10. Install the steering wheel, if re-moved. Connect the negative battery cable.

11. Check for proper start in P or N. Make sure the vehicle cannot be start-ed in D or R.

Without Key

The following procedure applies to ve-hicles where the ignition lock is inop-erative and the lock cylinder cannot be rotated due to a lost or broken ignition key and the key number not known or the lock cylinder cap is damaged and/

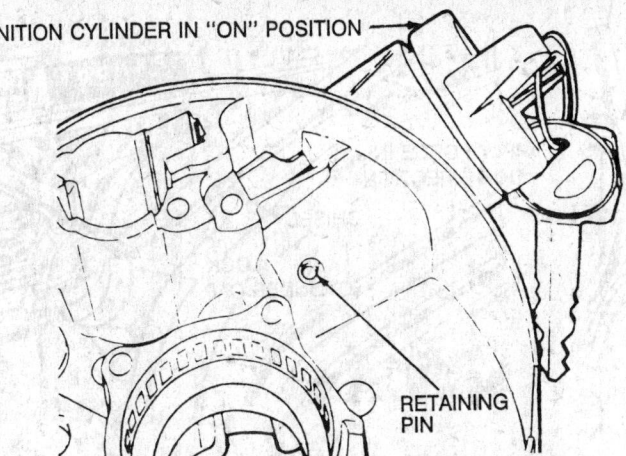

IGNITION CYLINDER IN "ON" POSITION

RETAINING PIN

Lock retaining pin access slot—fixed column

or broken to the extent that the lock cylinder cannot be rotated.

FIXED COLUMN

1. Disconnect the negative battery cable.

2. Remove the steering wheel and unscrew the turn signal lever.

3. Remove the steering column trim shrouds. Detach and lower the steering column assembly from the brake pedal support bracket.

4. Remove the ignition switch and key warning buzzer terminal and pin it in the **LOCK** position.

5. Remove the turn signal switch.

6. Remove the upper bearing snapring and the 2 T-bolt retaining nuts that secure the flange casting to the outer tube. Remove the entire flange casting assembly, the upper shaft bearing, the lock cylinder assembly, the ignition switch actuator and the actuator rod by pulling the assembly over the end of the steering column shaft.

7. If equipped with an automatic transmission, remove the lock actuator insert, the T-bolts and the transmission indicator insert. If equipped with a manual transmission, remove the key release lever assembly.

8. Replace the removed components with a new assembly consisting of the flange, lock cylinder assembly, steering column lock gear, steering column lock bearing, steering column upper bearing retainer and steering column lock actuator assembly.

9. If equipped with a manual transaxle, install the key release lever assembly. If equipped with an automatic transmission, install the transmission indicator insert. Install the T-bolts and lock actuator insert.

NOTE: Retain the ignition switch actuating rod from the removed casting assembly and use it with the new flange casting assembly.

To install:

10. Reassemble the parts listed in Step 8, install a new upper shaft bearing and set the actuator to drive gear— the last gear tooth aligns with the last tooth on the actuator when the actuator is fully rearward.

11. Install the turn signal switch and key warning buzzer.

12. Install the ignition switch and check and/or adjust for proper function.

13. Install the steering column assembly to the brake pedal support bracket.

14. Install the steering column trim shrouds, steering wheel and turn signal lever.

15. Using the ignition key, rotate the lock cylinder to insure correct mechanical operation in all positions.

16. Connect the negative battery cable. Check for proper start in **P** or **N**. Make sure the vehicle cannot be started in **D** or **R**.

TILT COLUMN

1. Disconnect the negative battery cable. Remove the steering column trim shrouds.

2. Tape the gap between the steering wheel hub and the cover casting. Cover the entire circumference of the casting.

3. Cover the adjacent seat and floor area with a suitable covering to protect the surrounding interior upholstery.

4. Pull out the hazard flasher switch and tape it down toward the floor to provide clearance for drilling out the lock cylinder retainer pin.

5. The tilt column lock cylinder retaining pin is located on the outside of the steering column cover casting near the hazard flasher button.

6. Tilt the steering column to the full up position and prepunch the lock cylinder retaining pin with a prick punch. Using a 1/8 in. (3.17mm) diameter drill with a right angle drive, drill

out the retaining pin, going no deeper than 1/2 in. (12.7mm).

NOTE: When drilling out the retaining pin, be careful not to damage the cover cast housing or the hazard flasher switch.

7. Tilt the steering column to the full down position. Place a chisel at the base of the ignition lock cylinder cap and, using a hammer, strike the chisel with sharp blows to break the cap away from the lock cylinder.

8. Using a 3/8 in. (9.8mm) diameter drill, drill down the middle of the ignition lock key slot approximately 1 3/4 in. (45mm) until the lock cylinder breaks loose from the steering column cover casting. Remove the lock cylinder and the drill shavings from the base of the cover cast housing.

9. Remove the steering wheel.

10. Remove the turn signal lever, then remove the 2 attaching screws from the turn signal switch and 1 attaching screw from the key warning buzzer terminal. Lift the turn signal switch up and over the end of the steering shaft but do not disconnect it from the wiring harness.

11. Remove the 4 attaching screws from the cover casting and lift the casting over the end of the steering shaft allowing the turn signal switch to pass through the cover casting. Removal of the cover casting will expose the upper actuator. Remove the upper actuator.

12. Remove the drive gear, snapring and washer from the cover casting along with the upper actuator. Thoroughly clean the components and inspect them for any damage resulting from the drilling operation. If any components show signs of damage, they must be replaced.

13. Clean the removed cover casting with compressed air to remove any drill shavings or foreign particles and carefully inspect it for damage. If the cover casting is damaged, it must be replaced.

To install:

14. Lubricate the tang of the lock cylinder with grease.

15. Attach the upper actuator to the lower actuator and lubricate the upper actuator. Reassemble the cover casting, upper actuator, turn signal switch and lever, lock drive gear, lock cylinder, steering wheel and the steering column trim shrouds.

16. Using the ignition key, rotate the lock cylinder to insure correct mechanical operation in all positions.

17. Connect the negative battery cable. Check for proper start in **P** or **N**. Make sure the vehicle cannot be started in **D** or **R**.

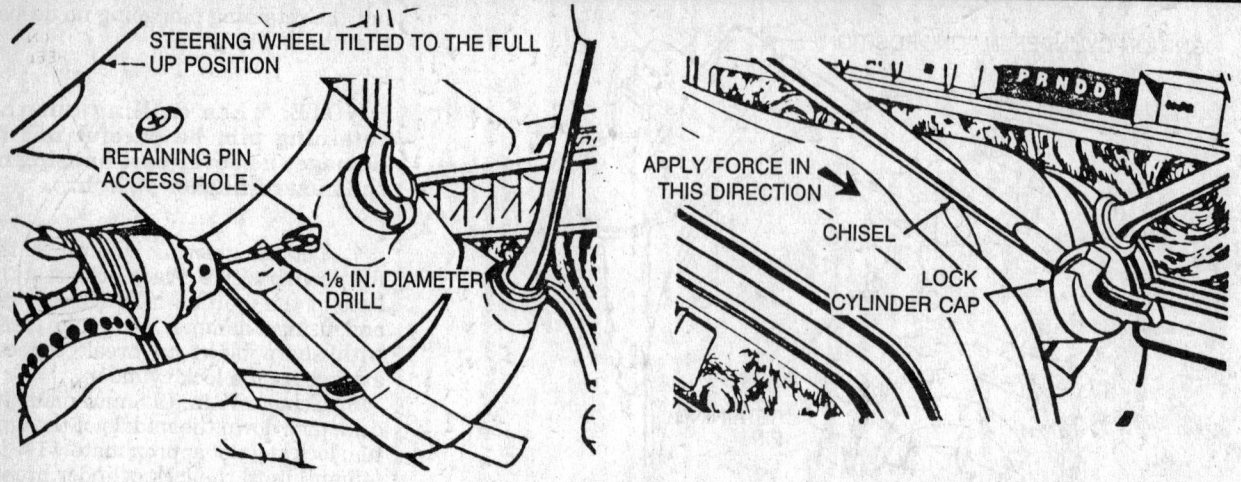

NOTE: DRILL DOWN THE MIDDLE OF THE LOCK CYLINDER REMAINING ON THE CENTER LINE FOR APPROXIMATELY 1¾ IN. (45MM) DEEP. REMOVE THE LOCK and DRILL SHAVINGS.

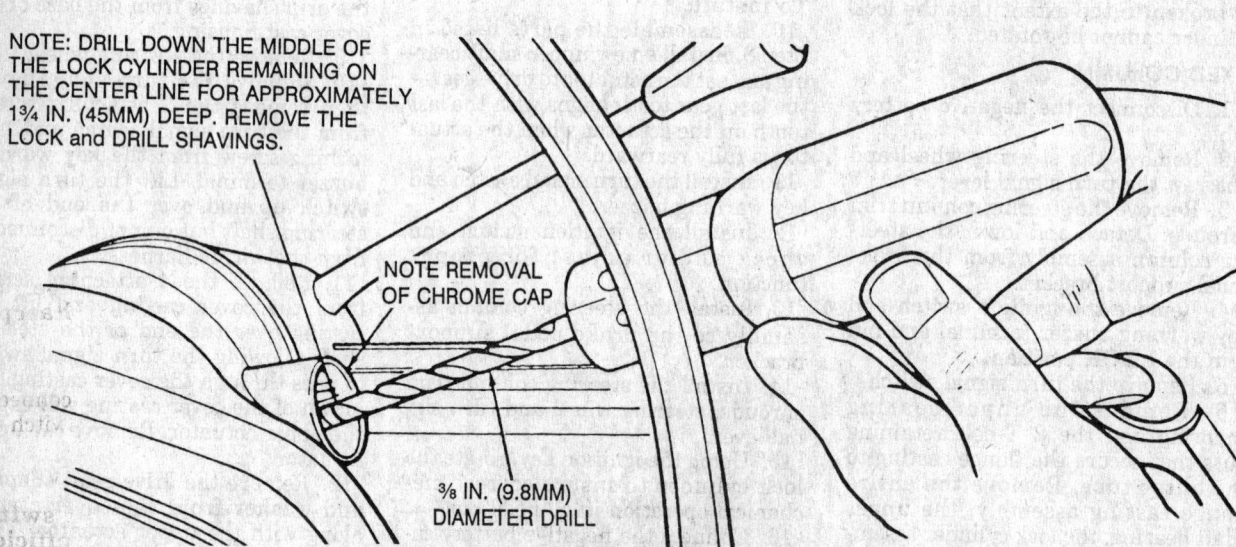

NOTE REMOVAL OF CHROME CAP

⅜ IN. (9.8MM) DIAMETER DRILL

Ignition lock removal—tilt column without key

Ignition Switch

REMOVAL & INSTALLATION

1. Disconnect the negative battery cable.

2. Remove the steering column shroud and lower the steering column.

3. Disconnect the switch wiring at the multiple plug.

4. Remove the 2 nuts that hold the switch to the steering column.

5. Lift the switch upward to disengage the actuator rod and remove the switch.

To install:

6. When installing the switch, both the locking mechanism at the top of the column and the switch itself must be in the **LOCK** position for correct adjustment.

7. To hold the mechanical parts of the column in the **LOCK** position, move the shift lever into **P**, if equipped with automatic transmission or **R**, if

equipped with manual transmission, turn the key to the **LOCK** position and remove the key. New replacement switches are already pinned in the **LOCK** position by a metal shipping pin inserted in a locking hole on the side of the switch.

8. Engage the actuator rod in the switch; it must be inserted in the slot of the sliding carrier.

9. Position the switch on the column and install the retaining nuts, but do not tighten them.

10. Move the switch up and down along the column to locate the mid-position of rod lash. Tighten the 2 ignition switch retaining nuts, top nut first to minimize rod binding, to 40–65 inch lbs. (4.5–7.3 Nm).

11. Remove the lockpin. Connect the negative battery cable and check for proper start in **P** or **N**. Make sure the starter cannot be actuated in **D** or **R** and that the engine will shut off in either **D**, **R** or **N**. If the engine will not shut off in these positions, the switch

is not adjusted properly and will have to be readjusted as follows:

a. Rotate the ignition key back and forth to either side of the lock, until a 5/64 in. (1.98mm) drill bit can be inserted through the lockpin hole as far as possible—minimum ⅜ in. (9.5mm). The lockpin hole is located on the right of the switch next to the steering column tube.

b. Loosen the 2 ignition switch mounting nuts.

c. Turn the ignition key to **LOCK**, feeling for the detent, and remove the ignition key.

d. Move the switch up and down along the column to locate the mid-position of rod lash. Tighten the 2 ignition switch mounting nuts, top nut first to minimize rod binding, to 40–65 inch lbs. (4.5–7.3 Nm).

e. Remove the drill bit from the ignition switch lockpin hole.

f. Plug in the electrical connector and operate the lock cylinder to

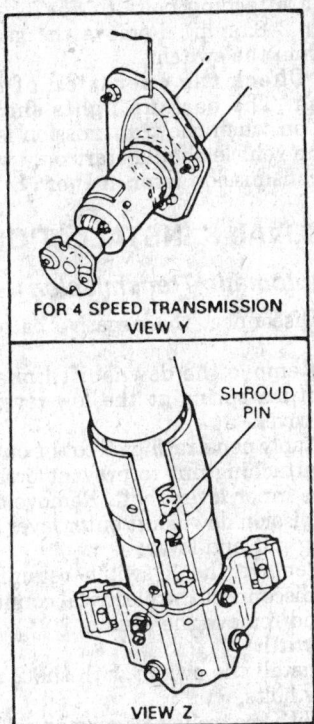

FOR 4 SPEED TRANSMISSION
VIEW Y

SHROUD
PIN

VIEW Z

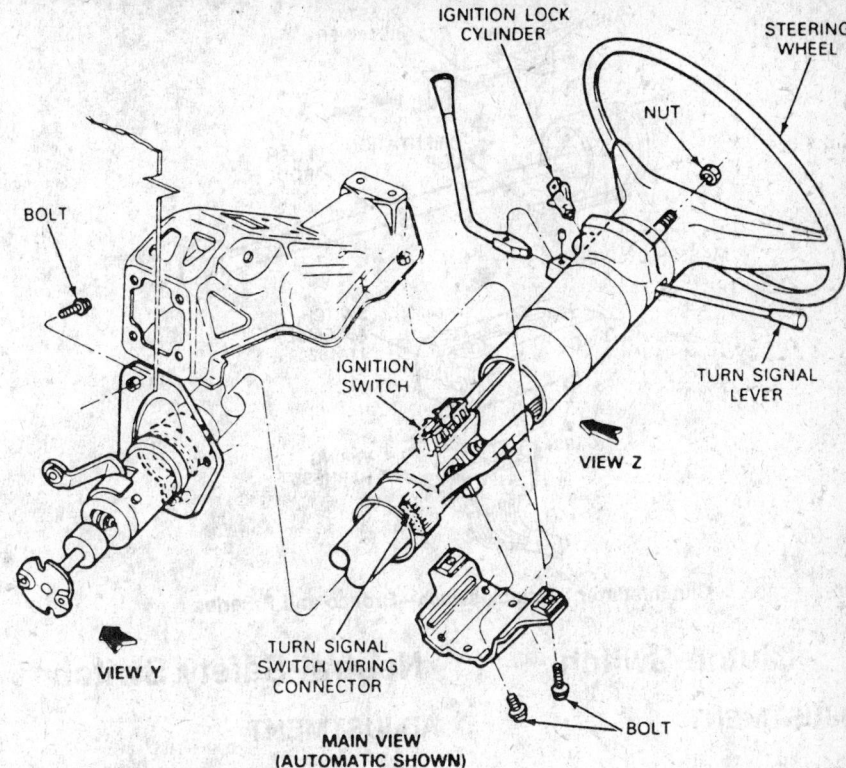

IGNITION LOCK
CYLINDER

STEERING
WHEEL

NUT

BOLT

IGNITION
SWITCH

TURN SIGNAL
LEVER

VIEW Z

VIEW Y

TURN SIGNAL
SWITCH WIRING
CONNECTOR

BOLT

MAIN VIEW
(AUTOMATIC SHOWN)
Ignition switch location

make sure the switch is positioned properly.

g. Check that all accessories are deactivated with the ignition switch in the **OFF** position and that all accessories are operable in **RUN** position.

12. Raise the steering column into position and install the steering column shroud.

13. Check that all accessories are deactivated with the ignition switch in the **OFF** position and that all accessories are operable in **RUN** position.

Stoplight Switch

REMOVAL & INSTALLATION

1. Disconnect the negative battery cable. Disconnect electrical connection from the switch. Locking tab must be lifted before connector can be removed.

2. Remove the hairpin retainer. Slide the switch, pushrod, nylon washer and bushing away from the pedal. Remove the washer and then the switch by sliding the switch up or down.

NOTE: Since the switch side plate nearest the brake pedal is slotted, it is not necessary to remove the master cylinder pushrod and bushing from the brake pedal pin. If equipped with cruise control, the spacer washer is re-

placed by the dump valve adapter washer assembly.

To install:

3. Position switch so the U-shaped side is nearest the pedal and directly over/under the pin. Then slide the switch up/down installing the master cylinder pushrod and bushing between the switch side plates.

4. Push the switch and pushrod assembly toward the brake pedal arm. Install the outside plastic washer to the pin. Install the hairpin retainer to hold the entire assembly.

NOTE: Do not substitute other

types of pin retainers. Use only the factory-supplied hairpin retainer.

5. Connect the electrical connector to the switch. Check the switch for proper operation.

NOTE: The stoplight switch wire harness must have sufficient length to travel with the switch during the full stroke of the pedal. If the wire length is too short, reroute or repair the harness, as necessary.

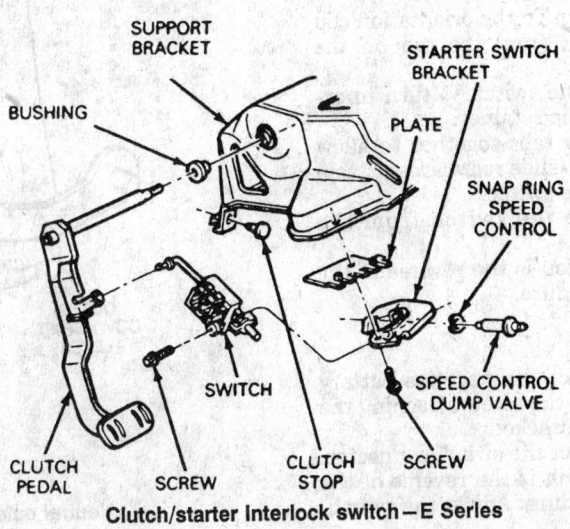

SUPPORT
BRACKET

STARTER SWITCH
BRACKET

BUSHING

PLATE

SNAP RING
SPEED
CONTROL

SPEED CONTROL
DUMP VALVE

CLUTCH
PEDAL

SCREW

SWITCH

CLUTCH
STOP

SCREW

Clutch/starter interlock switch—E Series

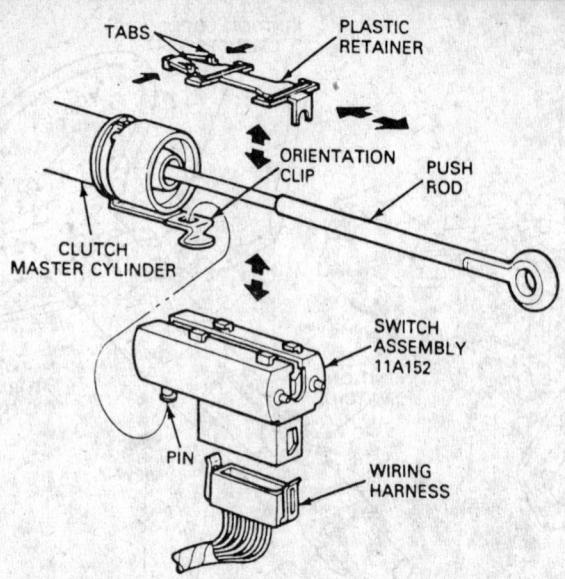

Clutch/starter interlock switch—Bronco and F Series

Clutch Switch

ADJUSTMENT

E Series

NOTE: The clutch/starter interlock switch on Bronco and F Series is not adjustable.

1. If the adjusting clip is out of position on the rod, remove both halves of the clip.
2. Position both halves of the clip closer to the switch and snap the clips together on the rod.
3. Depress the clutch pedal to the floor to adjust the switch.

REMOVAL & INSTALLATION

Bronco and F Series

1. Disconnect the negative battery cable. Disconnect the wiring harness from the switch.
2. Pull down on the orientation clip to separate it from the tab on the switch.
3. Rotate the switch ½ turn to expose the plastic retainer.
4. Push the tabs together to allow the retainer to slide rearward and separate from the switch.
5. Remove the switch from the pushrod.
6. Installation is the reverse of removal procedure.

E Series

1. Disconnect the negative battery cable. Remove the screw attaching the switch to the bracket.
2. Disconnect the switch connector.
3. Installation is the reverse of removal procedure. Adjust the switch.

Neutral Safety Switch

ADJUSTMENT

NOTE: The neutral safety switch is not adjustable on automatic overdrive transmissions.

C6 Automatic Transmission

1. Apply the parking brake.
2. With the automatic transmission linkage properly adjusted, loosen the 2 switch attaching bolts.
3. Place the transmission selector lever in **N**. Rotate the switch and insert the shank end of a No. 43 drill bit into the gauge pin holes of the switch. The gauge pin has to be inserted a full $^{31}/_{64}$ in. (12.303mm) into the 3 holes of the switch.
4. Tighten the 2 neutral start

switch attaching bolts to 55–75 inch lbs. (6.2–8.5mm). Remove the gauge pin from the switch.

5. Check the operation of the switch. The back-up lights should come on when the transmission is in **R**. The vehicle should start only with the transmission lever in **P** or **N**.

REMOVAL & INSTALLATION

C6 Automatic Transmission

1. Disconnect the negative battery cable.
2. Remove the downshift linkage rod return spring at the low-reverse servo cover.
3. Apply penetrating oil to the outer lever attaching nut to prevent breaking the inner lever shaft. Remove the transmission downshift outer lever attaching nut and lever.
4. Remove the 2 switch retaining bolts, disconnect the electrical connectors and remove the switch.
To install:
5. Install the switch with the 2 retaining bolts.
6. With the transmission manual lever in **N**, check the location of the switch with the gauge pin. Using a No. 43 drill bit as a gauge pin, insert the gauge pin into the 3 gauge pin holes.
7. Tighten the switch attaching bolts to 55–75 inch lbs. (6.3–8.4 Nm) and remove the gauge pin.
8. Install the outer downshift lever and retaining nut and tighten the nut. Install the downshift linkage rod return spring between the lever and retaining clip on the low-reverse servo cover.
9. Connect the electrical connectors. The red connectors must be connected together and the blue connectors must be connected together.
10. Check the operation of the

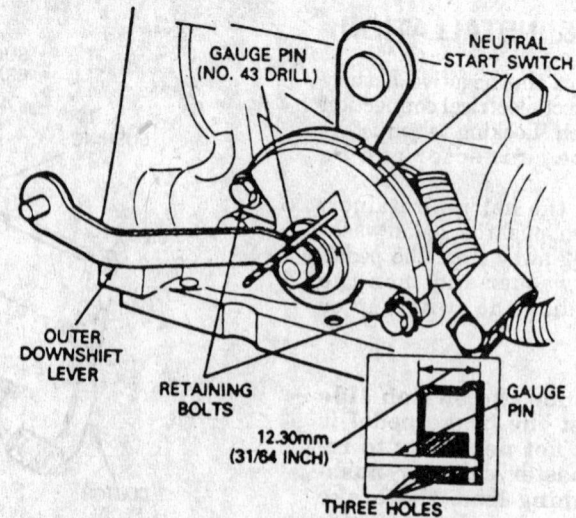

Neutral safety switch adjustment—C6 transmission

switch. The back-up lights should come on when the transmission is in **R**. The vehicle should start only with the transmission lever in **P** or **N**.

Automatic Overdrive Transmission

1. Disconnect the negative battery cable. Raise and safely support the vehicle.

2. Disconnect the switch electrical connector by lifting straight up off the switch.

3. Using neutral switch socket tool T74P-77247-A, remove the switch and the O-ring.

To install:

4. Install a new O-ring seal on a new neutral safety switch. Lube the seal with transmission fluid.

5. Install the new switch and seal into the case, using the neutral switch socket tool. Tighten to 8–11 ft. lbs. (11–15 Nm).

6. Connect the electrical connector to the switch.

7. Lower the vehicle and connect the negative battery cable.

Fuses, Circuit Breakers and Relays

LOCATION

Fuses

Fuses are located on the fuse panel. The fuse panel for the Bronco and F series vehicles is located on the firewall under the instrument panel left of the steering column. The fuse panel for the E series vehicles is located on the mounting bracket under the instrument panel left of the steering column.

NOTE: There is a 5 amp fuse, in-line, to protect the electric mirror circuit on Bronco and F Series, located on the harness near the fuse panel.

Fuse Links

A fuse link is a short length of special, Hypalon high temperature insulated wire, integral with the wiring harness. It is several wire gauges smaller than the circuit it protects. When heavy current flows, such as when a booster battery is connected incorrectly or when a short to ground occurs in the wiring harness, the fuse link burns out and protects the circuit.

The higher melting temperature and additional thickness of the Hypalon insulation will usually let the undersized internal fuse wire melt within the Hypalon casing with little damage to the high temperature insulation other than discoloration and/or

bubbling of the insulation surface. In extreme cases of excessive circuit current, the insulation may separate after the fuse wire has disintegrated. However, the bare wire will rarely be exposed. If it is difficult to determine if the fuse link is burned open, perform a continuity test.

BRONCO AND F SERIES

12 Gauge Fuse Link — used to protect the alternator circuit on vehicles with 100 amp alternators; located at the starter relay.

14 Gauge Fuse Link — used to protect the auxiliary battery circuit; located near the right fender apron and dash panel.

14 Gauge Fuse Link — used to protect the alternator circuit on vehicles with 70 amp alternator; located at the starter relay.

14 Gauge Fuse Link — used to protect the ignition switch and fuse panel feed circuit on all Bronco and on F Series with ammeter; located near right fender apron and dash panel.

16 Gauge Fuse Link — used to protect the ignition switch and fuse panel feed circuit on F Series without ammeter; located near right fender apron and dash panel.

16 Gauge Fuse Link — used to protect the alternator circuit on vehicles with 40 or 60 amp alternators; located at the starter relay.

16 Gauge Fuse Link — (2) used to protect the trailer circuit; located at starter relay.

16 Gauge Fuse Link — used to protect the heated rear window circuit on Bronco; located on left fender apron.

16 Gauge Fuse Link — used to protect the headlight switch and fuse panel feed circuit; located near the right fender apron and dash panel.

18 Gauge Fuse Link — used to protect the trailer or auxiliary lights circuit; located on the left fender apron.

18 Gauge Fuse Link — used to protect the electronic engine controls circuit; located at the starter relay.

18 Gauge Fuse Link — used to protect the headlight switch and fuse panel feed circuit on F Super Duty vehicles; located at the starter relay.

20 Gauge Fuse Link — used to protect the ignition switch to engine feed circuit on F Super Duty vehicles; located at the starter relay.

E SERIES

14 Gauge Fuse Link — used to protect the glow plug right bank circuit; located at the starter relay.

14 Gauge Fuse Link — used to protect the glow plug left bank circuit; located at the starter relay.

16 Gauge Fuse Link — used to protect the ignition switch and fuse panel feed circuit; located near the air conditioning case.

18 Gauge Fuse Link — used to protect the headlight switch and fuse panel feed circuit; located near the air conditioning case.

18 Gauge Fuse Link — used to protect the auxiliary air conditioner/heater circuit; located on the left fender at the dash.

20 Gauge Fuse Link — used to protect the oxygen sensor circuit; located near the master cylinder.

20 Gauge Fuse Link — used to protect the ignition circuit; located near the master cylinder.

20 Gauge Fuse Link — used to protect the blower motor circuit; located on the right fender apron.

20 Gauge Fuse Link — used to protect the diesel engine fuel heater; located on the right fender at the dash.

Circuit Breakers

BRONCO AND F SERIES

22 Amp Circuit Breaker — used to protect the headlight circuit; integral with light switch.

30 Amp Circuit Breaker — used to protect the power door locks, electric mirror and power tailgate circuits on 1988 vehicles; located on the fuse panel.

30 Amp Circuit Breaker — used to protect the power door locks and shift-on-the-fly circuits on 1989–92 vehicles; located on the fuse panel.

30 Amp Circuit Breaker — used to protect the power window circuit; located on the fuse panel.

E SERIES

7.5 Amp Circuit Breaker — used to protect the windshield wiper/washer circuit; located on the fuse panel.

20 Amp Circuit Breaker — used to protect the power door lock circuit; located on the fuse panel.

20 Amp Circuit Breaker — used to protect the power window circuit; located on the fuse panel.

Relays

BRONCO AND F SERIES

Charge Indicator Light Relay — located on the right fender apron, near the starter relay.

EEC Power Relay — located on the left side of engine compartment, on the relay assembly.

Fuel Pump Relay — located on the left side of engine compartment.

Horn Relay — located behind the right side of the instrument panel.

Starter Relay — located on the right fender apron.

Transmission Electronic Control Assembly Power Relay — located on the left side of the engine compartment.

E SERIES

Auxiliary Battery Relay—located on the left fender apron.

Auxiliary Blower Relay—located in the left rear corner of the engine compartment.

Blower Motor Relay—located in the right front corner of the engine compartment.

Door Lock/Unlock Control Relay—located behind the left side of the instrument panel, above the fuse panel.

EEC Power Relay—located on the right fender apron, near the battery.

Fuel Pump Relay—located on the right fender apron, near the battery.

Horn Relay—located behind the left side of the instrument panel, attached to the bottom of the steering column.

Starter Relay—located on the right front fender.

Transmission Electronic Control Assembly Power Relay—located on the right side of the firewall.

Trailer Exterior Lights Relay—located on the left rear of the vehicle, below the left tail light.

Computers

LOCATION

The Electronic Engine Control (EEC) module is located on the left side of the firewall on Bronco and F Series and on the right side fender apron, under the blower motor on E Series.

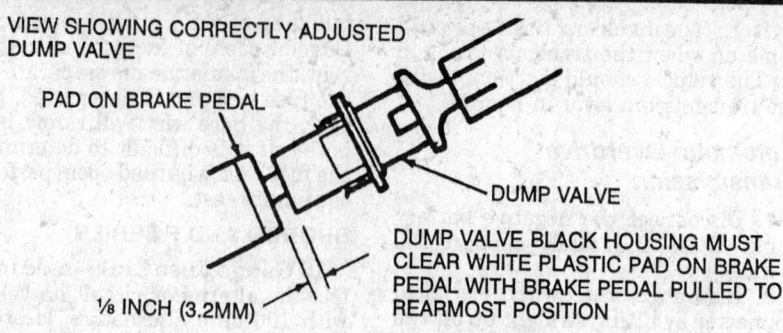

VIEW SHOWING CORRECTLY ADJUSTED DUMP VALVE

PAD ON BRAKE PEDAL

DUMP VALVE

⅛ INCH (3.2MM)

DUMP VALVE BLACK HOUSING MUST CLEAR WHITE PLASTIC PAD ON BRAKE PEDAL WITH BRAKE PEDAL PULLED TO REARMOST POSITION

Vacuum dump valve adjustment

Flashers

LOCATION

The turn signal flasher is located on the front of the fuse panel. The hazard flasher is located on the rear of the fuse panel behind the turn signal flasher.

Cruise Control

ADJUSTMENT

Actuator Cable

1. Snap the molded cruise control actuator cable retainer over the accelerator cable end fitting attached to the throttle ball stud.

2. Remove the adjuster retainer clip, if installed, from the adjuster mounting tab.

3. Insert the cruise control actuator cable adjuster mounting tab in the slot provided in the accelerator cable support bracket.

4. Pull the cable through the adjuster until a slight tension is felt without opening the throttle plate or increasing idle rpm.

5. Insert the adjuster retainer clip slowly until engagement is felt and then push downward until it locks in position.

Vacuum Dump Valve

The vacuum dump valve releases the vacuum in the servo assembly whenever the brake pedal is depressed. It should be checked whenever brake application does not disconnect the cruise control.

1. Move the valve forward in the retaining clip with the valve plunger contacting the brake pedal adapter and the pedal in the released position until ⅛ in. (3.2mm) or less of the plunger shows.

2. Make sure the brake pedal is against the stop in the release position after adjustment.

NOTE: If vacuum still does not release, replace the vacuum dump valve.

Ford Mid Size

AEROSTAR • EXPLORER • RANGER

SPECIFICATIONS

VEHICLE IDENTIFICATION CHART

		Engine Code					Model Year	
Code	Liters	Cu. In. (cc)	Cyl.	Fuel Sys.	Eng. Mfg.		Code	Year
A	2.3	140 (2300)	4	MFI	Ford		M	1991
T	2.9	177 (2901)	6	MFI	Ford		N	1992
W	3.0	181 (2967)	6	MFI	Nissan		P	1993
U	3.0	183 (2983)	6	MFI	Ford		R	1994
X	4.0	241 (3950)	6	MFI	Ford		S	1995

MFI - Multiport fuel injection
DSL - Diesel

ENGINE IDENTIFICATION

Year	Model	Engine Displacement Liters (cc)	Engine Series (ID/VIN)	Fuel System	No. of Cylinders	Engine Type
1991	Aerostar	3.0 (2999)	U	MFI	6	OHV
	Aerostar	4.0 (3998)	X	MFI	6	OHV
	Explorer	4.0 (3998)	X	MFI	6	OHV
	Ranger	2.3 (2294)	A	MFI	4	SOHC
	Ranger	2.9 (2900)	T	MFI	6	OHV
	Ranger	3.0 (2999)	U	MFI	6	OHV
	Ranger	4.0 (3998)	X	MFI	6	OHV
1992	Aerostar	3.0 (2999)	U	MFI	6	OHV
	Aerostar	4.0 (3998)	X	MFI	6	OHV
	Explorer	4.0 (3998)	X	MFI	6	OHV
	Ranger	2.3 (2294)	A	MFI	4	SOHC
	Ranger	2.9 (2900)	T	MFI	6	OHV
	Ranger	3.0 (2999)	U	MFI	6	OHV
	Ranger	4.0 (3998)	X	MFI	6	OHV
1993	Aerostar	3.0 (2999)	U	MFI	6	OHV
	Aerostar	4.0 (3998)	X	MFI	6	OHV
	Explorer	4.0 (3998)	X	MFI	6	OHV
	Ranger	2.3 (2294)	A	MFI	4	SOHC
	Ranger	2.9 (2900)	T	MFI	6	OHV
	Ranger	3.0 (2999)	U	MFI	6	OHV
	Ranger	4.0 (3998)	X	MFI	6	OHV
	Villager	3.0 (2967)	W	MFI	6	OHV
1994-95	Aerostar	3.0 (2999)	U	MFI	6	OHV
	Aerostar	4.0 (3998)	X	MFI	6	OHV
	Explorer	4.0 (3998)	X	MFI	6	OHV
	Ranger	2.3 (2294)	A	MFI	4	SOHC
	Ranger	2.9 (2900)	T	MFI	6	OHV
	Ranger	3.0 (2999)	U	MFI	6	OHV
	Ranger	4.0 (3998)	X	MFI	6	OHV
	Villager	3.0 (2967)	W	MFI	6	OHV

MFI - Multiport fuel injection
OHV - Overhead valve
SOHC - Single overhead camshaft

GENERAL ENGINE SPECIFICATIONS

Year	Engine ID/VIN	Engine Displacement Liters (cc)	Fuel System Type	Net Horsepower @ rpm	Net Torque @ rpm (ft. lbs.)	Bore x Stroke (in.)	Compression Ratio	Oil Pressure @ rpm
1991	A	2.3 (2294)	MFI	100@4600	133@2600	3.78x3.13	9.2:1	40-60@2000
	T	2.9 (2900)	MFI	140@4600	170@2600	3.66x2.83	9.0:1	40-60@2000
	U	3.0 (2999)	MFI	145@4800	165@3600	3.50x3.14	9.3:1	40-60@2500
	X	4.0 (3998)	MFI	1	2	3.95x3.32	9.0:1	40-60@2000
1992	A	2.3 (2294)	MFI	100@4600	133@2600	3.78x3.13	9.2:1	40-60@2000
	T	2.9 (2900)	MFI	140@4600	170@2600	3.66x2.83	9.0:1	40-60@2000
	U	3.0 (2999)	MFI	145@4800	165@3600	3.50x3.14	9.3:1	40-60@2500
	X	4.0 (3998)	MFI	1	2	3.95x3.32	9.0:1	40-60@2000
1993	A	2.3 (2294)	EFI	100@4600	133@2600	3.78x3.13	9.2:1	40-60@2000
	U	3.0 (2999)	EFI	145@4800	165@3600	3.50x3.14	9.3:1	40-60@2500
	W	3.0 (2967)	MFI	151@4800	174@4400	3.43x3.27	9.0:1	40-60@2000
	X	4.0 (3998)	EFI	1	2	3.95x3.32	9.0:1	40-60@2000
1994-95	A	2.3 (2294)	EFI	100@4600	133@2600	3.78x3.13	9.2:1	40-60@2000
	U	3.0 (2999)	MFI	145@4800	165@3600	3.50x3.14	9.3:1	40-60@2500
	W	3.0 (2967)	EFI	151@4800	174@4400	3.43x3.27	9.0:1	40-60@2000
	X	4.0 (3998)	EFI	1	2	3.95x3.32	9.0:1	40-60@2000

MFI - Multiport fuel injection
DDI - Direct diesel injection
1 Explorer and Aerostar: 155@4200
   Ranger: 160@4200
2 Aerostar: 215@2400
   Explorer: 220@2400
   Ranger: 225@2400

GASOLINE ENGINE TUNE-UP SPECIFICATIONS

Year	Engine ID/VIN	Engine Displacement Liters (cc)	Spark Plugs Gap (in.)	Ignition Timing (deg.) MT	Ignition Timing (deg.) AT	Fuel Pump (psi)	Idle Speed (rpm) MT	Idle Speed (rpm) AT	Valve Clearance In.	Valve Clearance Ex.
1991	A	2.3 (2294)	0.044	10B	10B	35-45	725	675	HYD	HYD
	T	2.9 (2900)	0.044	10B	10B	35-45	850	800	HYD	HYD
	U	3.0 (2999)	0.044	10B	10B	35-45	1	1	HYD	HYD
	X	4.0 (3998)	0.054	10B	10B	35-45	1	1	HYD	HYD
1992	A	2.3 (2294)	0.044	10B	10B	35-45	725	675	HYD	HYD
	T	2.9 (2900)	0.044	10B	10B	35-45	850	800	HYD	HYD
	U	3.0 (2999)	0.044	10B	10B	35-45	1	1	HYD	HYD
	X	4.0 (3998)	0.054	10B	10B	35-45	1	1	HYD	HYD
1993	A	2.3 (2294)	0.044	10B	10B	35-45	725	675	HYD	HYD
	U	3.0 (2999)	0.044	10B	10B	35-45	1	1	HYD	HYD
	W	3.0 92967)	0.034	-	15B	36-38	1	750	HYD	HYD
	X	4.0 (3998)	0.054	10B	10B	35-45	1	1	HYD	HYD
1994-95	A	2.3 (2294)	0.044	10B	10B	35-45	725	675	HYD	HYD
	U	3.0 (2999)	0.044	10B	10B	35-45	1	1	HYD	HYD
	W	3.0 (2967)	0.034	-	15B	36-38	-	750	HYD	HYD
	X	4.0 (3998)	0.054	10B	10B	35-45	1	1	HYD	HYD

NOTE: The Vehicle Emission Control Information label often reflects specification changes made during production. The label figures must be used if they differ from above

B - Before top dead center
HYD - Hydraulic
1 Idle speed is electronically controlled and cannot be adjusted

FIRING ORDERS

NOTE: To avoid confusion, always replace spark plug wires one at a time.

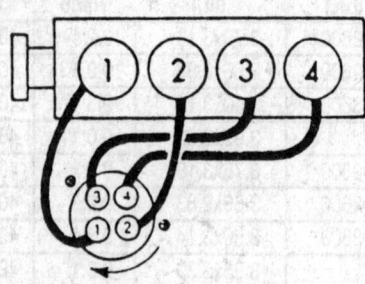

1988 2.0L and 2.3L Engines
Engine Firing Order: 1–3–4–2
Distributor Rotation: Clockwise

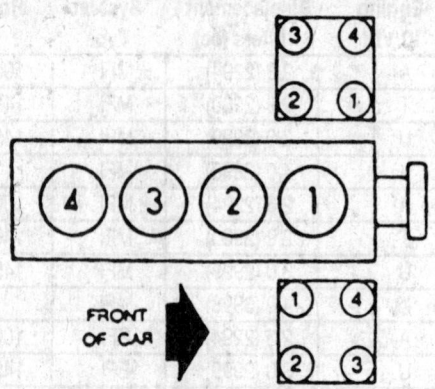

FRONT OF CAR

1989–92 2.3L Engine
Engine Firing Order: 1–3–4–2
Distributorless Ignition System

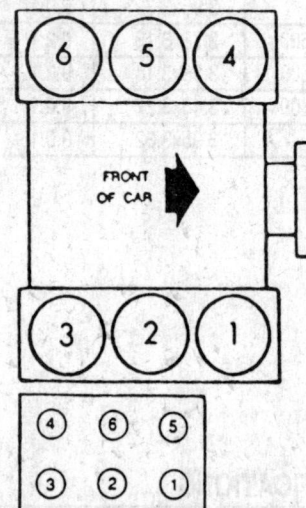

FRONT OF CAR

4.0L Engine
Engine Firing Order: 1–4–2–5–3–6
Distributorless Ignition System

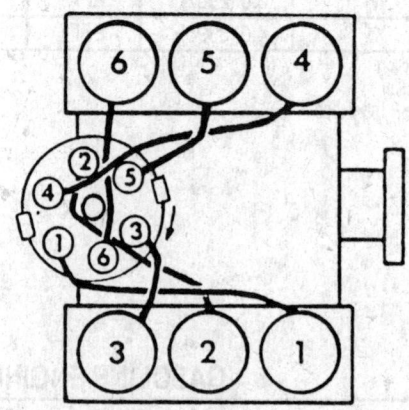

2.9L and 3.0L Engines
Engine Firing Order: 1–4–2–5–3–6
Distributor Rotation: Clockwise

CAPACITIES

Year	Model	Engine ID/VIN	Engine Displacement Liters (cc)	Engine Crankcase with Filter	Transmission (pts.)			Transfer Case (pts.)	Drive Axle		Fuel Tank (gal.)	Cooling System (qts.)
					4-Spd	5-Spd	Auto.		Front (pts.)	Rear (pts.)		
1991	Aerostar	U	3.0 (2999)	5.0	-	5.6	19.0	3.0	3.0	4	21.0	11.8
	Aerostar	X	4.0 (3998)	5.0	-	5.6	19.0	3.0	3.0	4	21.0	8.5
	Explorer	X	4.0 (3998)	5.0	-	5.6	19.0	3.0	4	4	21.0	8.5
	Ranger	A	2.3 (2294)	5.0	-	3	19.4	3.0	4	5.5	5	7.2
	Ranger	T	2.9 (2900)	5.0	-	3.0	6	1	2	2	5	7.5
	Ranger	U	3.0 (2999)	5.0	-	3.0	6	1	2	2	5	11.8
	Ranger	X	4.0 (3998)	5.0	-	3.0	6	1	2	2	5	8.5
1992	Aerostar	U	3.0 (2999)	5.0	-	5.6	19.0	3.0	3.0	4	21.0	11.8
	Aerostar	X	4.0 (3998)	5.0	-	5.6	19.0	3.0	3.0	4	21.0	8.5
	Explorer	X	4.0 (3998)	5.0	-	5.6	19.0	3.0	4	4	21.0	8.5
	Ranger	A	2.3 (2294)	5.0	-	3	19.4	3.0	4	5.5	5	7.2
	Ranger	T	2.9 (2900)	5.0	-	3.0	6	1	2	2	5	7.5
	Ranger	U	3.0 (2999)	5.0	-	3.0	6	1	2	2	5	11.8
	Ranger	X	4.0 (3998)	5.0	-	3.0	6	1	2	2	5	8.5

CAPACITIES

Year	Model	Engine ID/VIN	Engine Displacement Liters (cc)	Engine Crankcase with Filter	Transmission (pts.)			Transfer Case (pts.)	Drive Axle		Fuel Tank (gal.)	Cooling System (qts.)
					4-Spd	5-Spd	Auto.		Front (pts.)	Rear (pts.)		
1993	Aerostar	U	3.0 (2999)	5.0	-	5.6	19.0	3.0	3.0	4	21.0	11.8
	Aerostar	X	4.0 (3998)	5.0	-	5.6	19.0	3.0	3.0	4	21.0	8.5
	Explorer	X	4.0 (3998)	5.0	-	5.6	19.0	3.0	4	4	21.0	8.5
	Ranger	A	2.3 (2294)	5.0	-	3	19.4	3.0	4	5.5	5	7.2
	Ranger	U	3.0 (2999)	5.0	-	3.0	6	1	2	2	5	11.8
	Ranger	X	4.0 (3998)	5.0	-	3.0	6	1	2	2	5	8.5
	Villager	X	4.0 (3998)	5.0	-	-	17.1	-	-	-	20.0	7
1994-95	Aerostar	U	3.0 (2999)	5.0	-	5.6	19.0	3.0	3.0	4	21.0	11.8
	Aerostar	X	4.0 (3998)	5.0	-	5.6	19.0	3.0	3.0	4	21.0	8.5
	Explorer	X	4.0 (3998)	5.0	-	5.6	19.0	3.0	4	4	21.0	8.5
	Ranger	A	2.3 (2294)	5.0	-	3	19.4	3.0	4	5.5	5	7.2
	Ranger	U	3.0 (2999)	5.0	-	3.0	6	1	2	2	5	11.8
	Ranger	X	4.0 (3998)	5.0	-	3.0	6	1	2	2	5	8.5
	Splash	X	4.0 (3998)	5.0	-	3.0	6	1	2	2	5	8.5
	Villager	X	4.0 (3998)	5.0	-	-	17.1	-	-	-	20.0	7

1 BW 13-50 manual shift: 3 pts. Dextron II
  BW 13-50 electric shift: 6.5 pts. Dextron II
  BW 13-54 mechanical shift: 3 pts. Dextron II
  BW 13-59 contains no lubricant and none should be added
2 6.75" ring gear: 3 pts.
  7.50" ring gear: 5 pts.
3 Mazda trans.: 3 pts.
  Mitsubishi trans.: 4.8 pts.
4 Front axle Dana 28: 1.1 pts.
  Front axle Dana 35: 3.5 pts.
  Rear axle: 5.5 pts.

5 Short wheelbase: 17 gals.
  Long wheelbase: 17 or 21 gals.
  Ranger Supercab: 17 pr 21 gals.
6 2WD; 19.4 pts.
  4WD: 20.0 pts.
7 Without cargo pkg, without rear heater: 11 3/8 qts.
  With cargo pkg, without rear heater: 12 1/4 qts.
  Without cargo pkg, with rear heater: 12 3/4 qts.
  With cargo pkg. and rear heater: 13 3/4 qts.

CAMSHAFT SPECIFICATIONS
All measurements given in inches.

Year	Engine ID/VIN	Engine Displacement Liters (cc)	Journal Diameter					Elevation		Bearing Clearance	Camshaft End Play
			1	2	3	4	5	In.	Ex.		
1991	A	2.3 (2300)	1.7713-1.7720	1.7713-1.7720	1.7713-1.7720	1.7713-1.7720	NA	0.238	0.238	0.001-0.003	0.001-0.007
	T	2.9 (2900)	1.7285-1.7293	1.7135-1.7143	1.6985-1.6992	1.6835-1.6842	NA	0.359	0.370	0.001-0.003	0.001-0.004
	U	3.0 (2999)	2.0074-2.0084	2.0074-2.0084	2.0074-2.0084	2.0074-2.0084	NA	0.260	0.260	0.001-0.003	0.007
	X	4.0 (3998)	1.9510-1.9520	1.9370-1.9380	1.9220-1.9230	1.9070-1.9080	NA	0.402	0.402	0.001-0.006	0.009
1992	A	2.3 (2300)	1.7713-1.7720	1.7713-1.7720	1.7713-1.7720	1.7713-1.7720	NA	0.238	0.238	0.001-0.003	0.001-0.007
	T	2.9 (2900)	1.7285-1.7293	1.7135-1.7143	1.6985-1.6992	1.6835-1.6842	NA	0.359	0.370	0.001-0.003	0.001-0.004
	U	3.0 (2999)	2.0074-2.0084	2.0074-2.0084	2.0074-2.0084	2.0074-2.0084	NA	0.260	0.260	0.001-0.003	0.007
	X	4.0 (3998)	1.9510-1.9520	1.9370-1.9380	1.9220-1.9230	1.9070-1.9080	NA	0.402	0.402	0.001-0.006	0.009

CAMSHAFT SPECIFICATIONS

All measurements given in inches.

Year	Engine ID/VIN	Engine Displacement Liters (cc)	Journal Diameter					Elevation		Bearing Clearance	Camshaft End Play
			1	2	3	4	5	In.	Ex.		
1993	A	2.3 (2300)	1.7713-1.7720	1.7713-1.7720	1.7713-1.7720	1.7713-1.7720	NA	0.238	0.238	0.001-0.003	0.001-0.007
	U	3.0 (2999)	2.0074-2.0084	2.0074-2.0084	2.0074-2.0084	2.0074-2.0084	NA	0.260	0.260	0.001-0.003	0.007
	X	4.0 (3998)	1.9510-1.9520	1.9370-1.9380	1.9220-1.9230	1.9070-1.9080	NA	0.402	0.402	0.001-0.006	0.009
	W	3.0 (2967)	1.8504-1.8514	1.6732-1.6742	1.8898-1.8907	NA	NA	NA	NA	0.001-0.004	0.002
1994-95	A	2.3 (2300)	1.7713-1.7720	1.7713-1.7720	1.7713-1.7720	1.7713-1.7720	NA	0.238	0.238	0.001-0.003	0.001-0.007
	U	3.0 (2999)	2.0074-2.0084	2.0074-2.0084	2.0074-2.0084	2.0074-2.0084	NA	0.260	0.260	0.001-0.003	0.007
	W	3.0 (2967)	1.8504-1.8514	1.6732-1.6742	1.8898-1.8907	NA	NA	NA	NA	0.001-0.004	0.002
	X	4.0 (3998)	1.9510-1.9520	1.9370-1.9380	1.9220-1.9230	1.9070-1.9080	NA	0.402	0.402	0.001-0.006	0.009

NA - Not Available

CRANKSHAFT AND CONNECTING ROD SPECIFICATIONS

All measurements are given in inches.

Year	Engine ID/VIN	Engine Displacement Liters (cc)	Crankshaft				Connecting Rod		
			Main Brg. Journal Dia.	Main Brg. Oil Clearance	Shaft End-play	Thrust on No.	Journal Diameter	Oil Clearance	Side Clearance
1991	A	2.3 (2300)	2.2051-2.2059	0.0008-0.0026	0.004-0.008	3	2.0462-2.0472	0.0008-0.0026	0.004-0.010
	T	2.9 (2900)	2.2433-2.2441	0.0008-0.0015	0.0040-0.0080	3	2.1252-2.1260	0.0006-0.0016	0.004-0.011
	U	3.0 (2999)	2.5190-2.5198	0.0010-0.0014	0.0040-0.0080	3	2.1253-2.1261	0.0010-0.0014	0.006-0.014
	X	4.0 (3998)	2.2433-2.2441	0.0005-0.0019	0.012	3	2.1252-2.1260	0.0005-0.0022	0.0002-0.0025
1992	A	2.3 (2300)	2.2051-2.2059	0.0008-0.0026	0.004-0.008	3	2.0462-2.0472	0.0008-0.0026	0.004-0.010
	T	2.9 (2900)	2.2433-2.2441	0.0008-0.0015	0.0040-0.0080	3	2.1252-2.1260	0.0006-0.0016	0.004-0.011
	U	3.0 (2999)	2.5190-2.5198	0.0010-0.0014	0.0040-0.0080	3	2.1253-2.1261	0.0010-0.0014	0.006-0.014
	X	4.0 (3998)	2.2433-2.2441	0.0005-0.0019	0.012	3	2.1252-2.1260	0.0005-0.0022	0.0002-0.0025
1993	A	2.3 (2300)	2.2051-2.2059	0.0008-0.0026	0.004-0.008	3	2.0462-2.0472	0.0008-0.0026	0.004-0.010
	U	3.0 (2999)	2.5190-2.5198	0.0010-0.0014	0.0040-0.0080	3	2.1253-2.1261	0.0010-0.0014	0.006-0.014
	W	3.0 (2967)	2.4790-2.4793	0.0011-0.0022	0.0118	3	1.9667-1.9675	0.0011-0.0022	0.002-0.009
	X	4.0 (3998)	2.2433-2.2441	0.0005-0.0019	0.012	3	2.1252-2.1260	0.0005-0.0022	0.0002-0.0025

CRANKSHAFT AND CONNECTING ROD SPECIFICATIONS
All measurements are given in inches.

Year	Engine ID/VIN	Engine Displacement Liters (cc)	Crankshaft				Connecting Rod		
			Main Brg. Journal Dia.	Main Brg. Oil Clearance	Shaft End-play	Thrust on No.	Journal Diameter	Oil Clearance	Side Clearance
1994-95	A	2.3 (2300)	2.2051-2.2059	0.0008-0.0026	0.004-0.008	3	2.0462-2.0472	0.0008-0.0026	0.004-0.010
	U	3.0 (2999)	2.5190-2.5198	0.0010-0.0014	0.0040-0.0080	3	2.1253-2.1261	0.0010-0.0014	0.006-0.014
	W	3.0 (2967)	2.4790-2.4793	0.0011-0.0022	0.0118	3	1.9667-1.9675	0.0011-0.0022	0.002-0.009
	X	4.0 (3998)	2.2433-2.2441	0.0005-0.0019	0.012	3	2.1252-2.1260	0.0005-0.0022	0.0002-0.0025

VALVE SPECIFICATIONS

Year	Engine ID/VIN	Engine Displacement Liters (cc)	Seat Angle (deg.)	Face Angle (deg.)	Spring Test Pressure (lbs. @ in.)	Spring Installed Height (in.)	Stem-to-Guide Clearance (in.)		Stem Diameter (in.)	
							Intake	Exhaust	Intake	Exhaust
1991	A	2.3 (2300)	45	44	149@1.12	1.53-1.59	0.0010-0.0027	0.0015-0.0032	0.3416-0.3423	0.3411-0.3418
	T	2.9 (2900)	45	44	143@1.22	1.58-1.61	0.0008-0.0025	0.0018-0.0035	0.3159-0.3167	0.3149-0.3156
	U	3.0 (2999)	45	44	185@1.16	1.58-1.61	0.0010-0.0027	0.0015-0.0032	0.3126-0.3134	0.3121-0.3129
	X	4.0 (3998)	45	44	138@1.22	1.58-1.61	0.0008-0.0025	0.0018-0.0035	0.3159-0.3167	0.3149-0.3156
1992	A	2.3 (2300)	45	44	149@1.12	1.53-1.59	0.0010-0.0027	0.0015-0.0032	0.3416-0.3423	0.3411-0.3418
	T	2.9 (2900)	45	44	143@1.22	1.58-1.61	0.0008-0.0025	0.0018-0.0035	0.3159-0.3167	0.3149-0.3156
	U	3.0 (2999)	45	44	185@1.16	1.58-1.61	0.0010-0.0027	0.0015-0.0032	0.3126-0.3134	0.3121-0.3129
	X	4.0 (3998)	45	44	138@1.22	1.58-1.61	0.0008-0.0025	0.0018-0.0035	0.3159-0.3167	0.3149-0.3156
1993	A	2.3 (2300)	45	44	149@1.12	1.53-1.59	0.0010-0.0027	0.0015-0.0032	0.3416-0.3423	0.3411-0.3418
	U	3.0 (2999)	45	44	185@1.16	1.58-1.61	0.0010-0.0027	0.0015-0.0032	0.3126-0.3134	0.3121-0.3129
	W	3.0 (2967)	45	44	117@1.18	-	0.0008-0.0021	0.0016-0.0029	0.2742-0.2748	0.3136-0.3138
	X	4.0 (3998)	45	44	138@1.22	1.58-1.61	0.0008-0.0025	0.0018-0.0035	0.3159-0.3167	0.3149-0.3156
1994-95	A	2.3 (2300)	45	44	149@1.12	1.53-1.59	0.0010-0.0027	0.0015-0.0032	0.3416-0.3423	0.3411-0.3418
	W	3.0 (2967)	45	44	117@1.18	-	0.0008-0.0021	0.0016-0.0029	0.2742-0.2748	0.3136-0.3138
	U	3.0 (2999)	45	44	185@1.16	1.58-1.61	0.0010-0.0027	0.0015-0.0032	0.3126-0.3134	0.3121-0.3129
	X	4.0 (3998)	45	44	138@1.22	1.58-1.61	0.0008-0.0025	0.0018-0.0035	0.3159-0.3167	0.3149-0.3156

PISTON AND RING SPECIFICATIONS

All measurements are given in inches.

Year	Engine ID/VIN	Engine Displacement Liters (cc)	Piston Clearance	Ring Gap Top Compression	Ring Gap Bottom Compression	Ring Gap Oil Control	Ring Side Clearance Top Compression	Ring Side Clearance Bottom Compression	Ring Side Clearance Oil Control
1991	A	2.3 (2300)	0.0014-0.0022	0.0100-0.0200	0.0100-0.0200	0.015-0.055	0.0020-0.0040	0.0020-0.0040	SNUG
	T	2.9 (2900)	0.0011-0.0019	0.0150-0.0230	0.0150-0.0230	0.0150-0.0230	0.0020-0.0033	0.0020-0.0033	SNUG
	U	3.0 (2999)	0.0012-0.0023	0.0100-0.0200	0.0100-0.0200	0.0100-0.0490	0.0016-0.0037	0.0016-0.0037	SNUG
	X	4.0 (3998)	0.0008-0.0019	0.0150-0.0230	0.0150-0.0550	0.0150-0.0550	0.0020-0.0033	0.0020-0.0033	SNUG
1992	A	2.3 (2300)	0.0014-0.0022	0.0100-0.0200	0.0100-0.0200	0.015-0.055	0.0020-0.0040	0.0020-0.0040	SNUG
	T	2.9 (2900)	0.0011-0.0019	0.0150-0.0230	0.0150-0.0230	0.0150-0.0230	0.0020-0.0033	0.0020-0.0033	SNUG
	U	3.0 (2999)	0.0012-0.0023	0.0100-0.0200	0.0100-0.0200	0.0100-0.0490	0.0016-0.0037	0.0016-0.0037	SNUG
	X	4.0 (3998)	0.0008-0.0019	0.0150-0.0230	0.0150-0.0550	0.0150-0.0550	0.0020-0.0033	0.0020-0.0033	SNUG
1993	A	2.3 (2300)	0.0014-0.0022	0.0100-0.0200	0.0100-0.0200	0.015-0.055	0.0020-0.0040	0.0020-0.0040	SNUG
	T	2.9 (2900)	0.0011-0.0019	0.0150-0.0230	0.0150-0.0230	0.0150-0.0230	0.0020-0.0033	0.0020-0.0033	SNUG
	U	3.0 (2999)	0.0012-0.0023	0.0100-0.0200	0.0100-0.0200	0.0100-0.0490	0.0016-0.0037	0.0016-0.0037	SNUG
	W	3.0 (2967)	0.0006-0.0014	0.0083-0.0173	0.0071-0.0173	0.0079-0.0299	0.0016-0.0029	0.0012-0.0025	0.0390
	X	4.0 (3998)	0.0008-0.0019	0.0150-0.0230	0.0150-0.0550	0.0150-0.0550	0.0020-0.0033	0.0020-0.0033	SNUG
1994-95	A	2.3 (2300)	0.0014-0.0022	0.0100-0.0200	0.0100-0.0200	0.015-0.055	0.0020-0.0040	0.0020-0.0040	SNUG
	T	2.9 (2900)	0.0011-0.0019	0.0150-0.0230	0.0150-0.0230	0.0150-0.0230	0.0020-0.0033	0.0020-0.0033	SNUG
	U	3.0 (2999)	0.0012-0.0023	0.0100-0.0200	0.0100-0.0200	0.0100-0.0490	0.0016-0.0037	0.0016-0.0037	SNUG
	W	3.0 (2967)	0.0006-0.0014	0.0083-0.0173	0.0071-0.0173	0.0079-0.0299	0.0016-0.0029	0.0012-0.0025	0.0390
	X	4.0 (3998)	0.0008-0.0019	0.0150-0.0230	0.0150-0.0550	0.0150-0.0550	0.0020-0.0033	0.0020-0.0033	SNUG

TORQUE SPECIFICATIONS
All readings in ft. lbs.

Year	Engine ID/VIN	Engine Displacement Liters (cc)	Cylinder Head Bolts	Main Bearing Bolts	Rod Bearing Bolts	Crankshaft Damper Bolts	Flywheel Bolts	Manifold Intake	Manifold Exhaust	Spark Plugs	Lug Nut
1991	A	2.3 (2300)	1	1	2	100-120	56-64	14-21	14-21	5-10	3
	T	2.9 (2900)	6	65-75	19-24	85-96	47-52	18	25	15-18	3
	U	3.0 (2999)	4	65-81	5	141-169	54-64	18	25	8-10	3
	X	4.0 (3998)	7	66-77	19-24	8	59	9	19	10-15	3
1992	A	2.3 (2300)	1	1	2	100-120	56-64	14-21	14-21	5-10	3
	T	2.9 (2900)	6	65-75	19-24	85-96	47-52	18	25	15-18	3
	U	3.0 (2999)	4	65-81	5	141-169	54-64	18	25	8-10	3
	X	4.0 (3998)	7	66-77	19-24	8	59	9	19	10-15	3
1993	A	2.3 (2300)	1	1	2	100-120	56-64	14-21	14-21	5-10	3
	T	2.9 (2900)	6	65-75	19-24	85-96	47-52	18	25	15-18	3
	U	3.0 (2999)	4	65-81	5	141-169	54-64	18	25	8-10	3
	W	3.0 (2967)	10	67-74	18	90-98	33-43	129	13-16	10-15	13
	X	4.0 (3998)	7	66-77	19-24	8	59	16	19	10-15	3
1994-95	A	2.3 (2300)	1	1	2	100-120	56-64	14-21	14-21	5-10	3
	T	2.9 (2900)	6	65-75	19-24	85-96	47-52	18	25	15-18	3
	U	3.0 (2999)	4	65-81	5	141-169	54-64	18	25	8-10	3
	W	3.0 (2967)	10	67-74	11	90-98	33-43	12	13-16	10-15	13
	X	4.0 (3998)	7	66-77	19-24	8	59	9	19	10-15	3

1 Step 1: 50-60 ft. lbs.
Step 2: 80-90 ft. lbs.
2 Step 1: 25-30 ft. lbs.
Step 2: 30-36 ft. lbs.
3 Aerostar, Explorer and Ranger: 100 ft. lbs.
4 Step 1: 48-54 ft. lbs.
Step 2: 63-80 ft. lbs.
5 Step 1: 20-28 ft. lbs.
Step 2: Back off a minimum of two turns
Step 3: 20-25 ft. lbs.
6 Step 1: 22 ft. lbs.
Step 2: 51-55 ft. lbs.
Step 3: Turn 90 degrees
7 Tighten cylinder head bolts to 44 ft. lbs.
Tighten intake manifold bolts to 3-6 ft. lbs.
Tighten cylinder head bolts to 59 ft. lbs.
Tighten intake manifold bolts to 6-11 ft. lbs.
Tighten cylinder head bolts to 11-15 ft. lbs.
Tighten cylinder head bolts 85 degrees
Tighten intake manifold bolts to 11-15 ft. lbs.
Tighten intake manifold bolts to 15-18 ft. lbs.

8 Step 1: 30-37 ft. lbs.
Step 2: Turn 90 degrees
9 Step 1: 3-6 ft. lbs.
Step 2: 6-11 ft. lbs.
Step 3: 11-15 ft. lbs.
Step 4: 15-18 ft. lbs.
10 Step 1: 22 ft. lbs.
Step 2: 43 ft. lbs.
Step 3: Loosen all bolts completely
Step 4: 22 ft. lbs.
Step 5: 47 ft. lbs.
11 Step 1: 10-12 ft. lbs.
Step 2: 28-33 ft. lbs.
12 Step 1: Nuts and bolts: 3 ft. lbs.
Step 2: Bolts - 12-14 ft. lbs.; Nuts - 17-20 ft. lbs.
Step 3: Bolts: 12-14 ft. lbs.; Nuts - 17-20 ft. lbs.
13 72-87 ft. lbs.

BRAKE SPECIFICATIONS

All measurements in inches unless noted

Year	Model			Master Cylinder Bore	Brake Disc			Brake Drum Diameter			Minimum Lining Thickness	
					Original Thickness	Minimum Thickness	Maximum Runout	Original Inside Diameter	Max. Wear Limit	Maximum Machine Diameter	Front	Rear
1991	Aerostar	1		NA	0.850	0.810	0.003	9.00	9.09	9.06	0.030	0.030
	Aerostar	2		NA	0.850	0.810	0.003	10.00	10.09	10.06	0.030	0.030
	Bronco			NA	1.160	1.120	0.003	11.03	11.09	11.06	0.030	0.030
	Explorer	1		0.938	0.850	0.810	0.003	9.00	9.09	9.06	0.030	0.030
	Explorer	3		0.975	0.850	0.810	0.003	10.00	10.09	10.06	0.030	0.030
	Explorer	4		0.975	0.850	0.810	0.003	10.00	10.09	10.06	0.030	0.030
	Ranger	1		0.938	0.850	0.810	0.003	9.00	9.09	9.06	0.030	0.030
	Ranger	3		0.975	0.850	0.810	0.003	10.00	10.09	10.06	0.030	0.030
	Ranger	4		0.975	0.850	0.810	0.003	10.00	10.09	10.06	0.030	0.030
1992	Aerostar	1		NA	0.850	0.810	0.003	9.00	9.09	9.06	0.030	0.030
	Aerostar	2		NA	0.850	0.810	0.003	10.00	10.09	10.06	0.030	0.030
	Bronco			NA	1.160	1.120	0.003	11.03	11.09	11.06	0.030	0.030
	Explorer	1		0.938	0.850	0.810	0.003	9.00	9.09	9.06	0.030	0.030
	Explorer	3		0.938	0.850	0.810	0.003	10.00	10.09	10.06	0.030	0.030
	Explorer	4		0.975	0.850	0.810	0.003	10.00	10.09	10.06	0.030	0.030
	Ranger	1		0.938	0.850	0.810	0.003	9.00	9.09	9.06	0.030	0.030
	Ranger	3		0.938	0.850	0.810	0.003	10.00	10.09	10.06	0.030	0.030
	Ranger	4		0.975	0.850	0.810	0.003	10.00	10.09	10.06	0.030	0.030
1993	Aerostar	1		NA	0.850	0.810	0.003	9.00	9.09	9.06	0.030	0.030
	Aerostar	2		NA	0.850	0.810	0.003	10.00	10.09	10.06	0.030	0.030
	Bronco			NA	1.160	1.120	0.003	11.03	11.09	11.06	0.030	0.030
	Explorer	1		0.938	0.850	0.810	0.003	9.00	9.09	9.06	0.030	0.030
	Explorer	3		0.938	0.850	0.810	0.003	10.00	10.09	10.06	0.030	0.030
	Explorer	4		0.975	0.850	0.810	0.003	10.00	10.09	10.06	0.030	0.030
	Ranger	1		0.938	0.850	0.810	0.003	9.00	9.09	9.06	0.030	0.030
	Ranger	3		0.938	0.850	0.810	0.003	10.00	10.09	10.06	0.030	0.030
	Ranger	4		0.975	0.850	0.810	0.003	10.00	10.09	10.06	0.030	0.030
1994-95	Aerostar	1		NA	0.850	0.810	0.003	9.00	9.09	9.06	0.030	0.030
	Aerostar	2		NA	0.850	0.810	0.003	10.00	10.09	10.06	0.030	0.030
	Bronco			NA	1.160	1.120	0.003	11.03	11.09	11.06	0.030	0.030
	Explorer	1		0.938	0.850	0.810	0.003	9.00	9.09	9.06	0.030	0.030
	Explorer	3		0.938	0.850	0.810	0.003	10.00	10.09	10.06	0.030	0.030
	Explorer	4		0.975	0.850	0.810	0.003	10.00	10.09	10.06	0.030	0.030
	Ranger	1		0.938	0.850	0.810	0.003	9.00	9.09	9.06	0.030	0.030
	Ranger	3		0.938	0.850	0.810	0.003	10.00	10.09	10.06	0.030	0.030
	Ranger	4		0.975	0.850	0.810	0.003	10.00	10.09	10.06	0.030	0.030
	Villager			NA	0.974	0.945	0.003	9.84	9.90	-	0.030	0.030

NA - Not Available
1   With 9 inch brakes
2   With 10 inch brakes
3   4x2 with 10 inch brakes
4   4x4 with 10 inch brakes

WHEEL ALIGNMENT

Year	Model	Ride Height (in.)	Caster Range (deg.)	Caster Preferred Setting (deg.)	Camber Range (deg.)	Camber Preferred Setting (deg.)	Toe-in (in.)
1991	Aerostar	1	2 1/2P-4 1/2P	3 1/2P	5/16N-11/16P	3/16P	1/32
	Explorer 4x2	1 1/2-2	12P-14P	2	5 1/8N-3 1/8N	2	0
		2-2 1/2	12P-12P	2	3 1/2N-1 1/2N	2	0
		2 1/2-3	8P-10P	2	2 1/2N-1/2N	2	0
		3-3 1/2	6 1/4P-8 1/4P	2	1 1/2N-1/2P	2	0
		3 1/2-4	5 1/4P-7 1/4P	2	1/2N-1/2P	2	0
		4-4 1/2	3 3/4P-5 3/4P	2	1/2N-2 1/2P	2	0
		4 1/2-5	2 1/4P-4 1/4P	2	1 1/2P-3 1/2P	2	0
	Explorer 4x4	1 1/2-2	10P-12P	2	5 3/4N-3 3/4N	2	0
		2-2 1/2	8P-10P	2	4 1/2N-2 1/2N	2	0
		2 1/2-3	6P-8P	2	3N-1N	2	0
		3-3 1/2	4 1/2P-6 1/2P	2	1N-1P	2	0
		3 1/2-4	3 1/4P-5 1/4P	2	1/2P-2 1/2P	2	0
		4-4 1/2	2 3/4P-4 3/4P	2	1 3/4P-3 3/4P	2	0
		4 1/2-5	1/2P-2 1/2P	2	3 1/2P-5 1/2P	2	0
	Ranger 4x2	3 1/4-3 1/2	6 1/8P-8 3/4P	2	1 1/2N-1P	2	0
		3 1/2-3 3/4	5 1/2P-8 1/8P	2	1N-1 3/4P	2	0
		3 3/4-4	4 5/8P-7 1/2P	2	1/4N-2 3/8P	2	0
		4-4 1/4	3 3/4P-6 5/8P	2	3/8P-3P	2	0
		4 1/4-4 1/2	3 1/4P-5 3/4P	2	1P-4P	2	0
		4 1/2-4 3/4	2 1/2P-5 1/4P	2	2P-4 5/8P	2	0
	Ranger 4x4	3 1/4-3 1/2	4P-6 1/2P	2	7/8N-1 3/4P	2	0
		3 1/2-3 3/4	3 1/4P-6P	2	1/4N-2 3/8P	2	0
		4-4 1/4	1 7/8P-4 5/8P	2	7/8P-3 1/2P	2	0
1992	Aerostar	1	2 1/2P-4 1/2P	3 1/2P	5/16N-11/16P	3/16P	1/32
	Explorer 4x2	1 1/2-2	12P-14P	2	5 1/8N-3 1/8N	2	0
		2-2 1/2	10P-12P	2	3 1/2N-1 1/2N	2	0
		2 1/2-3	8P-10P	2	2 1/2N-1/2N	2	0
		3-3 1/2	6 1/4P-8 1/4P	2	1 1/2N-1/2P	2	0
		3 1/2-4	5 1/4P-7 1/4P	2	1/2N-1 1/2P	2	0
		4-4 1/2	3 3/4P-5 3/4P	2	1/2N-2 1/2P	2	0
		4 1/2-5	2 1/4P-4 1/4P	2	1 1/2P-3 1/2P	2	0
	Explorer 4x4	1 1/2-2	10P-12P	2	5 3/4N-3 3/4N	2	0
		2-2 1/2	8P-10P	2	4 1/2N-2 1/2N	2	0
		2 1/2-3	6P-8P	2	3N-1N	2	0
		3-3 1/2	4 1/2P-6 1/2P	2	1N-1P	2	0
		3 1/2-4	3 1/4P-5 1/4P	2	1/2P-2 1/2P	2	0
		4-4 1/2	2 3/4P-4 3/4P	2	1 3/4P-3 3/4P	2	0
		4 1/2-5	1/2P-2 1/2P	2	3 1/2P-5 1/2P	2	0
	Ranger 4x2	3 1/4-3 1/2	6 1/8P-8 3/4P	2	1 1/2N-1P	2	0
		3 1/2-3 3/4	5 1/2P-8 1/8P	2	1N-1 3/4P	2	0
		3 3/4-4	4 5/8P-7 1/2P	2	1/4N-2 3/8P	2	0
		4-4 1/4	3 3/4P-6 5/8P	2	3/8P-3P	2	0
		4 1/4-4 1/2	3 1/4P-5 3/4P	2	1P-4P	2	0
		4 1/2-4 3/4	2 1/2P-5 1/4P	2	2P-4 5/8P	2	0
	Ranger 4x4	3 1/4-3 1/2	4P-6 1/2P	2	7/8N-1 3/4P	2	0

WHEEL ALIGNMENT

Year	Model	Ride Height (in.)	Caster Range (deg.)	Caster Preferred Setting (deg.)	Camber Range (deg.)	Camber Preferred Setting (deg.)	Toe-in (in.)
		3 1/2-3 3/4	3 1/4P-6P	2	1/4N-2 3/8P	2	0
		4-4 1/4	1 7/8P-4 5/8P	2	7/8P-3 1/2P	2	0
1993	Aerostar	3	2 1/2P-4 1/2P	3 1/2P	5/16N-11/16P	3/16P	1/32
	Explorer 4x2	3	3P-8P	4	1/4N-3/4P	1/4P	1/32
	Explorer 4x4	3	2P-7P	4	1/4N-3/4P	1/4P	1/32
	Ranger 4x2	3	3P-8P	4	1/4N-3/4P	1/4P	1/32
	Ranger 4x4	3	2P-7P	4	1/4N-3/4P	1/4P	1/32
	Villager	3	-	6	-	6	5
1994-95	Aerostar	3	2 1/2P-4 1/2P	3 1/2P	5/16N-11/16P	3/16P	1/32
	Explorer 4x2	3	3P-8P	4	1/4N-3/4P	1/4P	1/32
	Explorer 4x4	3	2P-7P	4	1/4N-3/4P	1/4P	1/32
	Ranger 4x2	3	2P-6P	4	1/4N-3/4P	1/4P	1/32
	Ranger 4x4	3	2P-7P	4	1/4N-3/4P	1/4P	1/32
	Villager	3	-	6	-	6	5

1 Normal riding attitude with no more side to side clearance than 5/8 inch in front and 3/4 inch in rear
2 Caster and camber adjustment is possible with installation of service adjusters
3 Normal riding attitude with no more than side to side clearance than 5/8 inch in front and 3/4 inch in rear
4 Caster and camber adjustment is possible with installation of service adjusters
5 Front: 1/32; Rear: 0
6 Factory preset, not adjustable

ENGINE MECHANICAL

NOTE: Disconnecting the negative battery cable on some vehicles may interfere with the functions of the on board computer systems and may require the computer to undergo a relearning process, once the negative battery cable is reconnected.

Engine Assembly

REMOVAL & INSTALLATION

2.0L and 2.3L Engines

1. Disconnect the negative battery cable. On 2.3L engine, relieve the fuel system pressure.

2. Drain the cooling system. On 2.0L engine, remove the air cleaner and duct assembly. On 2.3L engine, disconnect the air cleaner tube at the throttle body. Disconnect the idle speed control hose and heat riser tube, if necessary.

3. Mark the location of the hinges on the hood and remove the hood.

4. Disconnect the radiator hoses and, if equipped, disconnect the transmission cooler lines. Remove the fan, fan shroud and radiator.

5. Remove the oil fill cap. Disconnect the engine wiring harness from the body wiring harness.

6. Disconnect the alternator wiring from the alternator, the starter cable from the starter and the accelerator cable from the carburetor or throttle body. If equipped, disconnect the transmission kickdown cable.

7. If equipped, remove the air conditioner compressor from the mounting bracket and position aside, leaving the refrigerant lines attached.

8. Disconnect the power brake vacuum hose. On 2.0L engine, disconnect the fuel line from the fuel pump. On 2.3L engine, disconnect the fuel lines from the fuel supply manifold.

9. Disconnect the heater hoses from the engine.

10. Remove the engine mount nuts. Raise and safely support the vehicle.

11. Drain the engine oil from the crankcase and remove the starter.

12. Disconnect the exhaust pipe at the exhaust manifold. If equipped with manual transmission, remove the dust cover. If equipped with automatic transmission, remove the converter

inspection plate, then remove the converter-to-flywheel bolts.

13. Remove the lower flywheel housing or converter housing attaching bolts and lower the vehicle.

14. Support the transmission and flywheel housing or converter housing with a jack.

15. Remove the flywheel housing or converter housing upper attaching bolts.

16. Attach suitable engine lifting equipment. Carefully lift the engine out of the vehicle and install on a workstand.

To install:

17. Remove the engine from the workstand and carefully lower it into the engine compartment. If equipped with automatic transmission, start the converter pilot into the crankshaft.

18. If equipped with manual transmission, start the transmission input shaft into the clutch disc. It may be necessary to adjust the position of the transmission in relation to the engine if the input shaft will not enter the clutch disc. If the engine hangs up after the shaft enters, turn the crankshaft in the clockwise direction slowly, transmission in gear, until the shaft splines mesh with the clutch disc splines.

19. Install the flywheel or converter housing attaching bolts and remove the engine lifting equipment.

20. Remove the jack from under the vehicle and raise and safely support the vehicle.

21. If equipped with automatic transmission, install the converter-to-flywheel attaching bolts. Install the lower flywheel housing or converter housing attaching bolts and install the dust plate or converter inspection cover.

22. Connect the exhaust pipe to the exhaust manifold. Install the starter and connect the starter cable.

23. Lower the vehicle and install the engine mount nuts. Tighten to 65–85 ft. lbs. (88–115 Nm).

24. Connect the heater hoses to the engine and the fuel lines to the fuel supply manifold or fuel pump. Connect the power brake vacuum hose.

25. Connect the wiring to the alternator and the accelerator cable to the carburetor or throttle body. If equipped, connect the transmission kickdown rod.

26. If equipped, install the air conditioning compressor in its mounting brackets.

27. Connect the engine wiring harness to the body wiring harness.

28. Install the fan, fan shroud and radiator. Connect the radiator hoses and, if equipped, the transmission cooler lines.

29. Install the hood, aligning the hinges with the marks that were made during removal.

30. On 2.0L engine, install the air cleaner assembly. On 2.3L engine, connect the air cleaner outlet tube at the throttle body. Connect the idle speed control hose and heat riser tube, if necessary.

31. Fill the crankcase with the proper type and quantity of engine oil. Install the oil cap.

32. Connect the negative battery cable. Fill and bleed the cooling system. Run the engine and check for leaks.

2.9L Engine

1. Disconnect the negative battery cable and relieve the fuel system pressure. Drain the cooling system.

2. Mark the position of the hood on the hinges and remove the hood. Remove the air cleaner intake hose.

3. Disconnect the radiator hoses and, if equipped with automatic transmission, disconnect the transmission cooler lines. Remove the fan shroud and radiator.

4. Remove the alternator and bracket and position the alternator aside. Disconnect the alternator ground wire from the engine block.

5. If equipped, remove the air conditioning compressor and power steering pump and position aside.

6. Disconnect the heater hoses at the intake manifold and water pump. Remove the ground wires from the cylinder block.

7. Disconnect the fuel lines from the fuel supply manifold. Disconnect the throttle cable shield and linkage at the throttle body and intake manifold.

8. Tag and disconnect all necessary vacuum lines and electrical connectors.

9. Raise and safely support the vehicle.

10. Disconnect the exhaust pipes from the exhaust manifolds. Disconnect the starter cable and remove the starter.

11. If equipped with manual transmission, remove the flywheel housing attaching bolts and remove the hydraulic clutch hose.

12. Remove the engine front mount-to-crossmember attaching nuts or through bolts.

13. If equipped with automatic transmission, remove the converter inspection cover and disconnect the converter from the flywheel.

14. Remove the cable. Remove the converter housing-to-engine bolts and the adapter plate-to-converter housing bolt.

15. Lower the vehicle and position a jack under the transmission. Install suitable engine lifting equipment.

16. Raise the engine slightly and carefully pull it from the transmission.

Carefully lift the engine out of the engine compartment so the rear cover plate is not bent or components damaged. Install the engine on a workstand.

To install:

17. Remove the engine from the workstand and lower it carefully into the engine compartment. Make sure the exhaust manifolds are properly aligned with the exhaust pipes.

18. If equipped with manual transmission, start the transmission input shaft into the clutch disc. It may be necessary to adjust the position of the transmission in relation to the engine if the input shaft will not enter the clutch disc. If the engine hangs up after the shaft enters, turn the crankshaft in the clockwise direction slowly, transmission in gear, until the shaft splines mesh with the clutch disc splines.

19. Install the flywheel housing or converter housing upper bolts, making sure the dowels in the engine block engage the flywheel housing or converter housing. Tighten the bolts to 33–45 ft. lbs. (45–61 Nm).

20. Install the clutch hose and remove the jack from under the transmission. Remove the engine lifting equipment.

21. If equipped with automatic transmission, position the kickdown rod on the transmission and engine. Raise and safely support the vehicle.

22. If equipped with automatic transmission, position the transmission linkage bracket and install the remaining converter housing bolts. Install the adapter plate-to-converter housing bolt. Install the converter-to-flywheel nuts and the inspection cover. Connect the kickdown rod on the transmission.

23. If equipped with manual transmission, install the flywheel housing attaching bolts.

24. Install the starter and connect the cable. Connect the exhaust pipes at the exhaust manifolds.

25. Install the engine front mount-to-crossmember attaching nuts or through bolts. Lower the vehicle.

26. Connect all vacuum hoses and electrical connectors to the locations marked during removal.

27. Install the throttle linkage and connect the fuel lines. Connect the heater hoses at the water pump and engine block.

28. Install the alternator and bracket and connect the ground wire to the engine block. Install the drive belt and adjust the tension.

29. Install the air conditioning compressor and power steering pump, if equipped.

30. Install the radiator and fan shroud. Connect the radiator hoses

and, if equipped, transmission cooler lines.

31. Connect the negative battery cable. Fill and bleed the cooling system. Run the engine and check for leaks.

32. If equipped, evacuate and charge the air conditioning system.

33. Install the intake hose. Install the hood, aligning the hinges with the marks that were made during removal.

3.0L Engine

AEROSTAR

1. Disconnect the negative battery cable and relieve the fuel system pressure. Drain the cooling system.

2. Disconnect the upper and lower radiator hoses.

3. Remove the air cleaner hose assembly. Removal should take place at the clamp retaining the hose to the air cleaner housing assembly.

4. Remove the engine fan retaining nut or bolts and remove the fan. Remove the fan shroud retaining screws and remove the shroud.

5. Disconnect the Barometric Manifold Absolute Pressure (BMAP) sensor electrical connector and vacuum line located on the dash panel.

6. Remove the shroud covering the throttle linkage and disconnect the linkage at the throttle body.

7. Loosen the idler arm and alternator jack screw retaining bolts and remove the accessory drive belts.

8. Disconnect the injector harness connector from the main harness. Disconnect the engine coolant temperature sensor, located near the thermostat housing and the engine coolant temperature sender.

9. Disconnect the canister purge solenoid hoses from both sides of the solenoid. If equipped with power steering, disconnect the pump pressure switch.

10. Mark the inlet and outlet heater hoses with chalk prior to removing them from the engine side of the ballast tube.

11. Remove the breather tubes from the air cleaner and rocker arm cover.

12. If equipped with automatic transmission, disconnect the transmission cooler lines from the radiator. Remove the radiator.

13. If equipped with air conditioning, remove the compressor retaining bolts and retain it to a sidemember with mechanics wire.

14. Disconnect the oil fill tube from the alternator bracket. If equipped with automatic transmission, disconnect the transmission fill tube from the manifold and gently pull out from the top of the vehicle.

15. Disconnect the electrical connectors from the alternator and the brake

booster vacuum line from the booster. Remove the bolt retaining the steering gear at the top of the shaft.

16. From inside the vehicle, remove the engine cover. Tag and disconnect the electrical connectors for the radio frequency interference suppressor, distributor TFI module and oil pressure sender.

17. Disconnect the fuel lines from the fuel supply manifold.

18. If equipped with manual transmission, place the shift lever in **N** and remove the screws retaining the shift lever boot to the floor. Slide the boot up the shift lever. Remove the bolt retaining the shift lever assembly to the transmission and remove the lever assembly.

19. Raise and safely support the vehicle. Disconnect the oil level sensor connector from the oil pan.

20. Mark the driveshaft to flange position. From the rear of the vehicle, remove the 4 bolts retaining the driveshaft. Slide the driveshaft forward, pull down, then pull out to release. Set the driveshaft aside.

21. Remove the bolt retaining the speedometer cable bracket to the transmission. Pull the speedometer out of the rear of the transmission.

22. Remove the starter ground strap and remove the battery connection to the starter. Remove the starter bolts and remove the starter.

23. If equipped with manual transmission, disconnect the clutch hydraulic hose as follows:

    a. On 1988 vehicles, remove the lockpin retaining the hose to the slave cylinder in the clutch housing. Remove and plug the hose.

    b. On 1989–92 vehicles, disconnect the coupling at the transmission with tool T88T–70522–A or equivalent, by sliding the white plastic sleeve toward the slave cylinder while applying a slight tug on the tube.

24. If equipped with manual transmission, disconnect the backup lamp switch and neutral sensing switch wires from the transmission.

25. If equipped with automatic transmission, proceed as follows:

    a. Disconnect the electrical connectors for the neutral safety switch and 3-4 shift solenoid connectors. Disconnect the selector and kickdown cable from the transmission lever.

    b. Disconnect the vacuum hose from the transmission vacuum modulator.

    c. Remove the converter access cover and adapter plate bolts from the lower end of the converter housing.

    d. Remove the flywheel-to-converter attaching nuts. Place a 22mm

socket and breaker bar on the crankshaft pulley bolt in order to turn the crankshaft and gain access to the flywheel-to-converter nuts.

    e. Disconnect the transmission cooler lines from the transmission.

26. Disconnect the oxygen sensor.

27. Position a jack under the transmission and a safety chain around the transmission. Slightly raise the transmission.

28. Remove the mount-to-crossmember attaching nuts. Remove the nuts and bolts attaching the crossmember to the 2 mounting brackets and remove the crossmember. If required, remove the bolts attaching the mount to the transmission and remove the mount.

29. Remove the converter housing-to-engine fasteners. Move the transmission to the rear so it disengages from the dowel pins and the converter disengages from the flywheel. Lower the transmission from the vehicle.

30. Disconnect and remove the exhaust pipe and catalytic converter. Remove both front wheel and tire assemblies.

31. Remove the engine block ground straps, 1 on the cylinder head just behind the power steering pump and the other just above where the exhaust manifold and exhaust pipe connect.

32. Remove the bar nuts and disconnect the stabilizer bar from the lower control arms. Discard the bar nuts.

33. Behind the spindles, disconnect and plug the brake lines at the bracket on the frame.

34. Position a jack under the lower control arm and raise the arm until tension is applied to the coil spring. Remove the bolt and nut retaining the spindle to the upper control arm ball joint. Slowly lower the jack to disconnect the spindle from the ball joint. Install safety chains around the lower control arm and spring seat.

35. Position drive train removal lift 109–00002 or equivalent, under the crossmember and engine assembly.

36. Slowly lower the vehicle until the crossmember rests on the removal lift. Place wood blocks under the front crossmember and rear of the engine block to keep the engine and crossmember assembly level. Install safety chains around the crossmember and lift.

37. With the engine and crossmember securely supported on the lift, remove the 3 nuts from the bolts that retain the engine crossmember assembly to the frame on each side of the vehicle.

38. Slowly lower the engine assembly out of the vehicle, making sure the air conditioning compressor and wiring harnesses do not interfere. When the

assembly is clear, roll the lift out from under the vehicle.

39. Separate the engine from the crossmember and position on a workstand.

To install:

40. Remove the engine from the workstand. With the front crossmember securely positioned on drive train removal lift 109-00002 or equivalent, slowly lower the engine until the motor mount studs enter the crossmember holes. Install the retaining nuts and tighten to 71-94 ft. lbs. (96-127 Nm).

NOTE: Install wood blocks under the oil pan and crossmember.

41. Roll the removal lift under the vehicle. Align the lift, engine and crossmember assembly so the 3 mounting bolts on each side of frame are in alignment with the holes in the crossmember.

42. Slowly lower the vehicle so the bolts are piloted in the crossmember holes. Raise the lift or lower the vehicle so the crossmember is against the frame. Install the nuts retaining the crossmember to the frame and tighten to 145-195 ft. lbs. (196-264 Nm). Raise the vehicle and remove the lift.

43. If equipped with automatic transmission, position the converter to the transmission making sure the converter hub is fully engaged in the pump gear. When the torque converter is fully installed, the distance between the converter pilot and the front of the converter housing should be $7/16-9/16$ in. Make sure the converter rotates freely and is not binding.

44. If equipped with manual transmission, install the transmission on a jack and lift into position. Make sure the transmission input shaft engages the pilot bearing in the flywheel and the flywheel housing holes are aligned with the engine block dowel pins. Install the flywheel housing-to-engine block bolts and tighten to 28-38 ft. lbs. (38-51 Nm).

45. If equipped with automatic transmission, place the transmission on a jack and secure with safety chains. Rotate the converter so the drive studs are aligned with the flywheel holes. Lift the transmission into position and connect the transmission cooler lines to the case. Move the transmission forward into position, making sure the converter housing holes align with the engine block dowel pins. Install the converter housing-to-engine block bolts and tighten to 28-38 ft. lbs. (38-51 Nm).

46. Position the crossmember in the 2 mounting brackets and install the mounting nuts and bolts. Slowly lower the transmission so the mount studs are installed in the proper slots in the

crossmember. Install the nuts and tighten to 71-94 ft. lbs. (97-127 Nm). Remove the safety chain and the jack.

47. Install the starter and tighten the mounting bolts to 15-20 ft. lbs. (21-27 Nm). Connect the starter cable.

48. If equipped with automatic transmission, connect the modulator vacuum hose. Position the selector cable in the case bracket and press the end of the cable on the ball stud on the lower portion of the selector lever. Install the retainer in the bracket.

49. If equipped with manual transmission, install the speedometer cable or connect the electronic speedometer wire. Connect the backup lamp and neutral sensor wires.

50. Remove the cap from the hydraulic clutch line. On 1988 vehicles, install the line and fitting in the slave cylinder and install the retaining clip. On 1989-92 vehicles, insert the male coupling into the female coupling on the slave cylinder and make sure the connection is secure.

51. Install the manual transmission shift lever to the shifter. Position the rubber shift boot on the floor and install the screws.

52. If equipped with automatic transmission, install the kickdown and selector cable. Adjust the cable. Connect the neutral safety switch, converter clutch solenoid and 3-4 shift solenoid connectors. Insert the speedometer driven gear into the transmission and retain with a clamp. Tighten the retaining screw to 20-25 inch lbs. (2.25-2.82).

53. Position a 22mm socket and breaker bar on the crankshaft pulley bolt. Rotate the pulley clockwise, as viewed from the front, to gain access to each torque converter studs. Install the nuts on the studs and tighten to 20-34 ft. lbs. (27-46 Nm).

54. Position the converter access cover and adapter plate on the converter housing. Install the bolts and tighten to 12-16 ft. lbs. (16-22 Nm).

55. Remove the safety chains from around the lower control arms and spring seat. Install a jack under the lower control arms. Slowly raise the control arm until the coil spring is under tension. Continue to raise the arm until the spindle is connected to the upper arm ball joint. Install a new nut and bolt and tighten to 27-37 ft. lbs. (37-50 Nm).

56. Connect the stabilizer bar to the lower control arms. Install new bar nuts and tighten to 12-18 ft. lbs. (16-24 Nm).

57. Connect the front brake lines to the caliper hoses at the frame brackets. Install the wheel and tire assemblies.

58. Install the driveshaft, aligning

the marks that were made during removal.

59. Install new gaskets on the exhaust manifold and catalytic converter. Install the exhaust pipe and catalytic converter. Tighten the converter-to-muffler nuts and bolts to 18-26 ft. lbs. (25-35 Nm). Tighten the exhaust pipe-to-exhaust manifold nuts to 25-34 ft. lbs. (34-46 Nm).

60. Connect the oxygen sensor connector and install the engine ground straps. Connect the fuel lines and the low-oil level sensor.

61. If equipped with automatic transmission, connect the transmission oil cooler lines. Install the transmission oil fill tube and lower the vehicle.

62. From inside the vehicle, connect the radio frequency interference suppressor, TFI module connector and oil pressure sender. Install the engine cover.

63. Connect the alternator electrical connectors and the brake booster vacuum hose. Connect the breather tube to the oil filler tube and attach the bolt retaining the steering gear to the top of the shaft.

64. If equipped, connect the power steering pressure switch. Connect the heater hoses to the ballast tube by matching the chalk lines to the specific inlet and outlet hoses.

65. Install the radiator. If equipped, connect the transmission oil cooler lines.

66. Connect the canister purge solenoid hoses from both sides of the solenoid. Connect the engine coolant temperature sensor, located near the thermostat housing and connect the engine coolant temperature sender.

67. If equipped with air conditioning, untie the compressor from the sidemember and position on the mounting bracket. Install the attaching bolts.

68. Install the alternator belt and tighten the idler arm. Place the injector harness behind the belt tension idler arm and tighten the idler arm.

69. Connect the throttle linkage to the ball stud located on the throttle body. Connect the shroud covering the throttle body.

70. Connect the BMAP sensor electrical connector and vacuum line located on the dash panel.

71. Install the fan and fan shroud. Connect the radiator hoses.

72. Connect the negative battery cable and install the air cleaner and duct assembly.

73. Bleed the brakes and the hydraulic clutch. Fill and bleed the cooling system. Check all fluid levels.

74. Run the engine and check for leaks and proper operation. Check the front alignment.

RANGER

1. Disconnect the negative battery cable and relieve the fuel system pressure. Drain the cooling system.

2. Mark the position of the hood on the hinges and remove the hood. Remove the air cleaner intake hose.

3. Disconnect the radiator hoses at the radiator. Remove the fan shroud attaching bolts and position the shroud over the fan. Remove the radiator, then remove the shroud.

4. Remove the alternator and bracket and position the alternator aside. Disconnect the alternator ground wire from the cylinder block.

5. Remove the air conditioner compressor and power steering pump and position aside, if equipped.

6. Disconnect the heater hoses at the intake manifold and water pump. Remove the ground wires from the cylinder block.

7. Disconnect the fuel lines at the chassis to engine connections. Disconnect the throttle cable shield and linkage at the throttle body and intake manifold.

8. Tag and disconnect the vacuum connections at the rear fitting in the upper intake manifold.

9. Tag and disconnect the wires at the ignition coil. Disconnect the 3 body wiring connectors on top of the right rocker arm cover. Disconnect the oil pressure and engine coolant temperature sender connectors.

10. Disconnect the injector harness, air charge temperature sensor and throttle position sensor. Disconnect the oxygen sensor connector at the rear of the engine and disconnect the brake booster vacuum hose.

11. Remove the engine front mount-to-crossmember attaching nuts.

12. Raise and safely support the vehicle.

13. Remove 2 lower air conditioner compressor bracket-to-engine bolts. Disconnect the wiring from the low oil level sensor and oil pressure sending unit.

14. Remove the retaining bracket holding the transmission cooling lines to the right side of the engine block.

15. Disconnect the exhaust pipes at the manifolds. Disconnect the starter cable and remove the starter.

16. If equipped with manual transmission, disconnect the hydraulic clutch line and remove the flywheel housing-to-engine block bolts.

17. If equipped with automatic transmission, remove the converter inspection cover and disconnect the converter from the flywheel.

18. Disconnect the kickdown and shift cables at the transmission. Remove the converter housing-to-engine block bolts and adapter plate-to-converter housing bolt.

19. Lower the vehicle. Remove the 2 bolts from the air conditioner compressor and position aside.

20. Attach suitable engine lifting equipment and position a jack under the transmission.

21. Raise the engine slightly and carefully pull it from the transmission. Carefully lift the engine out of the engine compartment so the rear cover plate is not bent or components damaged. Install the engine on a workstand.

To install:

22. Remove the engine from the workstand and carefully lower it into the engine compartment. Make sure the exhaust manifolds are aligned with the exhaust pipes.

23. If equipped with manual transmission, start the transmission input shaft into the clutch disc. It may be necessary to adjust the position of the transmission in relation to the engine, if the input shaft will not enter the clutch disc. If the engine hangs up after the shaft enters, turn the crankshaft in the clockwise direction slowly, transmission in gear, until the shaft splines mesh with the clutch disc splines.

24. If equipped with automatic transmission, start the converter pilot into the crankshaft. Make sure the converter rotates freely and is not binding. When the converter is fully installed in the transmission, the distance between the converter pilot and the edge of the converter housing should be $^7/_{16}$–$^9/_{16}$ in.

25. Install the flywheel housing or converter housing upper bolts, making sure the engine block dowels engage the housing. If equipped, install the clutch hydraulic line.

26. Remove the jack from under the transmission and remove the engine lifting equipment. If equipped with automatic transmission, position the kickdown cable on the transmission and engine.

27. Raise and safely support the vehicle. If equipped with automatic transmission, position the transmission linkage bracket and install the remaining converter housing bolts.

28. Install the adapter plate-to-converter housing bolt. Install the converter-to-flywheel nuts and install the inspection cover. Connect the kickdown cable at the transmission.

29. If equipped with manual transmission, install the flywheel housing attaching bolts.

30. Install the starter and connect the cables. Connect the exhaust pipes at the manifolds.

31. Install the engine front mount nuts and washers or through bolts. Lower the vehicle.

32. Install the ground wires to the engine block. Connect the coil wires, the 3 body wiring connectors and the oxygen wiring connector. Connect the coolant temperature sending unit and oil pressure sending unit. Connect the brake booster vacuum hose.

33. Install the throttle linkage and connect the fuel lines. Connect the heater hoses at the water pump and cylinder block.

34. Install the alternator and bracket. Connect the alternator ground wire to the engine block. Install the drive belt and adjust the tension.

35. Install the air conditioning compressor and power steering pump, if equipped.

36. Position the fan shroud over the fan. Install the radiator and connect the radiator hoses. Install the fan shroud.

37. Connect the negative battery cable. Fill and bleed the cooling system.

38. Bleed the hydraulic clutch, if necessary. Evacuate and charge the air conditioning system, if necessary.

39. Run the engine and check for leaks and proper operation. Install the intake hose. Install the hood, aligning the marks that were made during removal.

4.0L Engine

AEROSTAR

1. Disconnect the negative battery cable and relieve the fuel system pressure.

2. Remove the front grille. Remove the air cleaner tube and the air cleaner assembly.

3. Discharge the refrigerant from the air conditioning system, if equipped. Disconnect and remove the air conditioner compressor.

4. Drain the engine oil. Drain disconnect and remove the power steering oil cooler.

5. Remove the front bumper cover. Drain, disconnect and remove the transmission oil cooler.

6. Drain the radiator and disconnect the radiator hoses. Disconnect the transmission oil cooler lines from the radiator.

7. Disconnect the fan shroud and position it over the fan. Remove the radiator and the shroud.

8. Remove the accessory drive belt and the right front air diverter flap. Remove the center hood latch support and remove the alternator.

9. If equipped, discharge the air conditioning system and remove the compressor.

10. Remove the engine oil fill tube. From inside the vehicle, remove the engine cover.

11. Remove the ice/snow shield. Remove the power steering hoses and re-

move the power steering pump and bracket.

12. Remove the transmission oil and engine oil dipsticks and tubes. Disconnect the exhaust system from the exhaust manifolds.

13. Disconnect and remove the starter. Working through the starter opening in the converter housing, remove the converter to flexplate bolts.

14. Remove the transmission oil cooler line bracket retaining bolt and remove the engine mount to frame bolts.

15. Remove the converter housing-to-engine bolts, except the upper bolts.

16. Remove the left motor mount from the engine, then remove the upper converter housing bolts.

17. Disconnect the transmission and transfer case electrical connectors. Disconnect the fuel lines at the fuel supply manifold.

18. Disconnect the throttle linkage and bracket and position aside. Disconnect the heater hoses.

19. Tag and disconnect the vacuum lines from the vapor canister, lower intake manifold and upper intake manifold vacuum tee.

20. Remove the spark plug wires from the coil assembly. Disconnect and remove the coil assembly with mounting bracket.

21. Disconnect the throttle position sensor electrical connector. Remove the throttle body from the upper intake manifold.

22. Tag and disconnect the engine wiring harness main connectors. Install suitable engine lifting equipment and remove the engine from the vehicle. Position the engine on a workstand.

To install:

23. Remove the engine from the workstand and position it in the vehicle. Remove the engine lifting equipment.

24. Connect the engine wiring harness main connectors. Install the throttle body on the upper intake manifold.

25. Install the ignition coil assembly and connect the spark plug wires and coil assembly wiring connector. Connect the electrical connector for the throttle position sensor.

26. Connect the heater hoses. Connect the vacuum lines to the vapor canister, lower intake manifold and upper intake manifold vacuum tee.

27. Connect the throttle linkage and bracket. Connect the fuel lines to the fuel supply manifold.

28. Install the left motor mount. Install the converter housing-to-engine bolts and connect the transmission and transfer case electrical connectors.

29. Install the engine mount-to-

frame bolts and the transmission oil cooler line bracket retaining bolt.

30. Working through the starter opening in the converter housing, install the converter to flexplate bolts.

31. Install the starter and connect the starter wires. Connect the exhaust system to the exhaust manifolds.

32. Install the transmission oil and engine oil dipstick tubes and dipsticks.

33. Install the power steering pump and mounting bracket. Connect the power steering hoses.

34. Install the ice/snow shield and the engine oil fill tube. Install the engine cover.

35. Install and connect the air conditioner compressor, if equipped. Install and connect the alternator.

36. Install the center hood latch support and the right front air diverter flap. Install the accessory drive belt.

37. Install the radiator and fan shroud. Connect the transmission oil cooler lines and the radiator hoses.

38. Install the transmission oil cooler and the front bumper cover. Install and connect the power steering oil cooler.

39. Fill the engine with the proper type and quantity of engine oil.

40. Install and connect the air conditioner condenser, if equipped. Install the air cleaner assembly and air tube.

41. Install the front grille and connect the negative battery cable. Fill and bleed the cooling system.

42. If equipped, evacuate and charge the air conditioning system. Check all fluid levels.

43. Run the engine and check for leaks and for proper operation.

EXPLORER AND RANGER

1. Disconnect the negative battery cable and relieve the fuel system pressure. Drain the cooling system.

2. Mark the position of the hood on the hinges and remove the hood. Remove the air cleaner intake hose.

3. Disconnect the radiator hoses at the radiator. Disconnect the fan shroud and position it over the fan. Remove the radiator, then the shroud.

4. Remove the alternator and bracket and position the alternator aside. Disconnect the alternator ground wire from the cylinder block.

5. Remove the air conditioning compressor and power steering pump and position aside, if equipped.

6. Disconnect the heater hoses at the intake manifold and water pump. Remove the ground wires from the cylinder block.

7. Disconnect the fuel lines from the fuel supply manifold. Disconnect the throttle cable shield and linkage at the throttle body and intake manifold.

8. Tag and disconnect the vacuum

connections at the rear vacuum fitting in the upper intake manifold.

9. Disconnect the wiring from the ignition coil and oil pressure and engine coolant temperature senders. Disconnect the injector harness, air charge temperature sensor and throttle position sensor. Disconnect the brake booster vacuum hose.

10. Raise and safely support the vehicle. Disconnect the exhaust pipes at the manifolds. Disconnect the starter cable and remove the starter.

11. Remove the engine front mount-to-crossmember attaching nuts or through bolts.

12. Remove the converter inspection cover and disconnect the converter from the flywheel. Remove the cable.

13. Remove the converter housing-to-engine block bolts and the adapter plate-to-converter housing bolt. Lower the vehicle.

14. Position a jack under the transmission and install suitable engine lifting equipment.

15. Raise the engine slightly and carefully pull it from the transmission. Carefully lift the engine out of the engine compartment so the rear cover plate is not bent or components damaged. Install the engine on a workstand.

To install:

16. Remove the engine from the workstand and carefully lower it into the engine compartment. Make sure the exhaust manifolds are aligned with the exhaust pipes.

17. At the transmission, start the converter pilot into the crankshaft. Install the converter housing upper bolts, making sure the dowels in the cylinder block engage the flywheel housing. Tighten the bolts to 33–45 ft. lbs. (45–61 Nm).

18. Remove the jack from under the transmission and the engine lifting equipment.

19. Position the kickdown rod on the transmission and engine. Raise and safely support the vehicle.

20. Position the transmission linkage bracket and install the remaining converter housing bolts. Install the adapter plate-to-converter housing bolt. Install the converter-to-flywheel nuts and install the inspection cover. Connect the kickdown rod on the transmission.

21. Install the starter and connect the cable. Connect the exhaust pipes at the manifolds.

22. Install the engine front mount nuts and washers or through bolts. Lower the vehicle.

23. Install the ground wires to the engine block. Connect the ignition coil wiring, then connect the coolant temperature sending unit and oil pressure

sending unit. Connect the brake booster vacuum hose.

24. Install the throttle linkage and connect the fuel lines at the fuel supply manifold.

25. Connect the ground cable at the engine block. Connect the heater hoses to the water pump and cylinder block.

26. Install the alternator and bracket. Connect the alternator ground wire to the engine block. Install the accessory drive belt.

27. Install the air conditioner compressor and power steering pump, if equipped.

28. Position the shroud over the fan. Install the radiator and connect the radiator upper and lower hoses. Install the fan shroud attaching bolts.

29. Connect the negative battery cable. Fill and bleed the cooling system.

30. Run the engine and check for leaks and proper operation. If equipped, evacuate and charge the air conditioning system.

31. Install the intake hose. Install the hood, aligning the marks that were made during removal.

Engine Mounts

REMOVAL & INSTALLATION

Front Mounts

1. Remove the fan shroud attaching screws.

2. Support the engine using a wood block and a jack placed under the oil pan.

3. Remove the nuts and washers attaching the mounts to the engine bracket. Lift the engine enough to disengage the mount upper stud from the crossmember engine bracket.

4. Remove the bolt attaching the fuel pump shield to the left engine bracket, if necessary.

5. Remove the mount-to-crossmember attaching nut and washer assembly. Remove the engine mount.

To install:

6. Install the engine mount to the crossmember.

7. Install the bolt attaching the fuel pump shield to the left bracket, if necessary.

8. Lower the engine until the mount stud engages in the slot/hole of the engine bracket. Install the attaching nuts.

9. Remove the jack and wood block from the engine oil pan. Install the fan shroud attaching screws.

Rear Mount

1. Place a block of wood and a jack under the transmission.

2. Remove the 2 nuts attaching the

mount to the crossmember. Raise the transmission enough to lift the mount from the crossmember.

3. Remove the bolts and nuts attaching the crossmember to the frame side rails and remove the crossmember.

4. If equipped, remove the fasteners attaching the exhaust hanger to the rear engine mount.

5. Remove the 2 bolts attaching the mount to the transmission and remove the mount and retainer assembly.

To install:

6. Position the mount and retainer assembly to the transmission and install the 2 attaching bolts.

7. If equipped, install the fasteners attaching the exhaust hanger to the mount.

8. Install the crossmember to the frame side rails with the attaching nuts and bolts.

9. Lower the transmission and install the mount crossmember attaching nuts. Remove the jack and wood block.

Cylinder Head

REMOVAL & INSTALLATION

2.0L and 2.3L Engines

1. Disconnect the negative battery cable. Drain the cooling system.

2. Remove the air cleaner assembly. Remove the heater hose retaining screw(s) to the rocker arm cover.

3. If equipped, disconnect the distributor cap and spark plug wires and remove the assembly.

4. Remove the spark plugs. If equipped with distributorless ignition, remove the spark plug wire harnesses.

5. Remove the engine and alternator wiring harnesses. Disconnect the oxygen sensor at the exhaust manifold.

6. Tag and disconnect the required vacuum hoses. Remove the dipstick tube and bracket.

7. Remove the rocker arm cover attaching bolts and remove the cover. Remove the intake manifold attaching bolts.

8. Loosen the alternator retaining bolts and remove the belt from the pulley. Remove the mounting bracket-to-head retaining bolts.

9. Remove the upper radiator hose. Remove the timing belt cover bolt(s) and remove the cover. If equipped with power steering, move the power steering pump bracket.

10. Loosen the timing belt idler retaining bolts. Position the idler in the unloaded position and tighten the retaining bolts. Remove the timing belt from the camshaft pulley and auxiliary pulley.

11. Remove the 4 nuts and/or stud bolts retaining the heat stove to the exhaust manifold. Remove the 8 exhaust manifold retaining bolts.

12. Remove the timing belt idler and 2 bracket bolts. Remove the timing belt idler spring stop from the cylinder head.

13. Remove the cylinder head retaining bolts and remove the cylinder head.

14. Clean all gasket mating surfaces. Check the cylinder head for flatness using a straight edge and a feeler gauge. The cylinder head must not be warped more than 0.003 in. in any 6 in. or more than 0.006 in. overall.

To install:

15. Position a new head gasket on the block. Properly position the camshaft in the cylinder head and install the cylinder head on the block.

16. Install the cylinder head bolts and tighten, in sequence, in 2 steps, first to 50–60 ft. lbs. (68–81 Nm) and then to 80–90 ft. lbs. (108–122 Nm).

17. Install a new intake manifold gasket and position the intake manifold to the cylinder head. Install the retaining bolts.

18. Install the timing belt idler spring stop to the cylinder head. Position the timing belt idler to the cylinder head and install the retaining bolts.

19. Install the 8 exhaust manifold retaining bolts and the 4 nuts and/or stud bolts retaining the heat stove to the exhaust manifold.

20. If equipped, align the distributor rotor with the No. 1 spark plug location in the distributor cap.

21. Align the cam gear with the pointer and the crank pulley with the pointer on the timing belt cover.

22. Position the timing belt to the pulleys. Loosen the idler retaining, rotate the engine and check the timing alignment.

23. Adjust the belt tensioner and tighten the retaining bolts. Install the timing belt cover and the retaining bolt(s).

24. Install the upper radiator hose. Position the alternator bracket to the cylinder head and install the retainers. Install the drive belt and adjust the belt tension.

25. Install a new rocker arm cover gasket on the rocker arm cover. Install the rocker arm cover on the cylinder head and install the retaining bolts.

26. Install the spark plugs. Install the spark plug wires and the distributor cap, if equipped.

27. Install the dipstick tube and bracket. Connect the vacuum hoses. Install the retaining heater hose screw(s) to the rocker arm cover.

28. Connect the negative battery cable. Fill and bleed the cooling system.

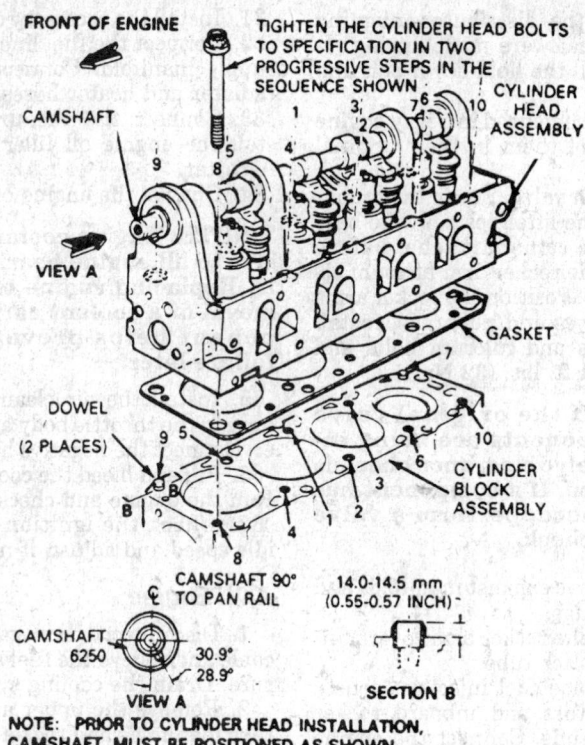

NOTE: PRIOR TO CYLINDER HEAD INSTALLATION, CAMSHAFT MUST BE POSITIONED AS SHOWN TO PROTECT PROTRUDING VALVES

Cylinder head installation—2.0L and 2.3L engines

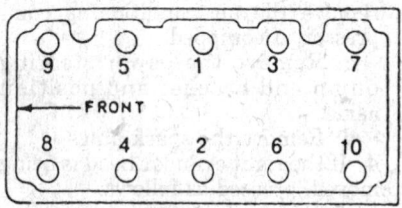

Cylinder head bolt torque sequence—2.0L and 2.3L engines

29. Start the engine and check for leaks. Install the air cleaner hose to the throttle body.

2.9L Engine

1. Disconnect the negative battery cable and relieve the fuel system pressure.

2. Drain the cooling system and remove the upper radiator hose.

3. Remove the intake tube from the throttle body and disconnect the throttle linkage and cover.

4. Tag the position of the spark plug wires and remove the distributor cap and wires as an assembly. Disconnect the distributor wiring harness.

5. Mark the position of the rotor in relation to the distributor housing and the position of the distributor housing in relation to the intake manifold. Remove the distributor hold-down bolt and clamp and remove the distributor.

6. Remove the rocker arm covers and rocker arm shafts.

7. Disconnect the fuel line from the fuel supply manifold. Remove the intake manifold.

8. Remove the pushrods, marking them so they can be reinstalled in their original positions.

9. Remove the exhaust manifolds.

10. Remove the cylinder head attaching bolts and remove the cylinder heads.

11. Clean all gasket mating surfaces. Check the cylinder head for flatness using a straight edge and a feeler gauge. The cylinder head must not be warped more than 0.003 in. in any 6 in. or more than 0.006 in. overall.

To install:

12. Position new cylinder head gaskets on the cylinder block.

NOTE: Gaskets are marked with the words "front" and "top" for correct positioning. Left and right cylinder head gaskets are not interchangeable.

13. Install fabricated alignment dowels in the cylinder block and install the cylinder heads.

14. Install new cylinder head bolts and remove the fabricated dowels. Tighten the bolts as follows:

Step 1: Tighten in sequence to 22 ft. lbs. (30 Nm).

Step 2: Tighten in sequence to 51–55 ft. lbs. (70–75 Nm).

Step 3: Wait 5 minutes.

Step 4: In sequence, turn all bolts 90 degrees.

15. Install the intake manifold.

16. Install the exhaust manifolds.

17. Apply engine oil to both ends of the pushrods and install them in their original positions.

18. Install the oil baffles and rocker arms.

19. Install the distributor, aligning the marks that were made during removal. Install the distributor hold-down bolt and clamp and connect the distributor wiring harness.

20. Adjust the valves and install the rocker arm covers.

21. Connect the fuel line to the fuel supply manifold.

22. Install the distributor cap and connect the spark plug wires.

23. Connect the negative battery cable. Fill and bleed the cooling system.

24. Run the engine and check for leaks. Check the ignition timing and idle speed and adjust, if necessary.

3.0L Engine

1. Disconnect the negative battery cable and relieve the fuel system pressure. Drain the cooling system.

2. Remove the air cleaner fresh air hose from the throttle body and air cleaner. If equipped, remove the engine oil filler adapter.

3. Disconnect the fuel lines. Tag and disconnect the necessary vacuum lines.

4. Disconnect the upper radiator hose and heater hose and position aside. On Ranger, disconnect the ignition coil electrical connector and remove the coil.

5. Remove the throttle body.

6. Remove the distributor cap. Mark the position of the rotor in relation to the distributor housing and the position of the distributor housing in relation to the intake manifold. Remove the distributor hold-down bolt and clamp and remove the distributor. Tag and disconnect the spark plug wires from the spark plugs and remove the distributor cap and wires assembly.

7. If removing the left cylinder head, proceed as follows:

  a. Remove the necessary accessory drive belt(s).

  b. On 1988 vehicles, remove the alternator adjusting arm.

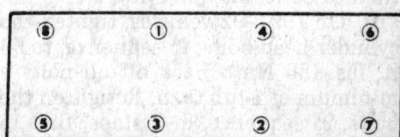

Cylinder head bolt torque sequence—2.9L engine

c. Remove the power steering pump and bracket, leaving the lines connected. Place the assembly aside in a position to prevent fluid leakage.

d. On Aerostar, remove the ignition coil and bracket.

e. Remove the engine oil dipstick tube.

f. Remove the fuel line retaining bracket bolt from the front of the cylinder head, if equipped.

8. If removing the right cylinder head, proceed as follows:

a. Remove the necessary accessory drive belt(s).

b. Remove the accessory drive belt idler or tensioner.

c. Remove the grounding strap throttle cable support bracket, if necessary.

d. On 1989–92 vehicles, disconnect the alternator electrical harnesses and remove the alternator and bracket assembly.

9. Remove the spark plugs.

10. Disconnect the exhaust pipes and remove the exhaust manifolds.

11. Remove the rocker arm covers. Loosen the rocker arm fulcrum retaining bolts enough to allow the rocker arm to be lifted off the pushrod and rotated to 1 side.

NOTE: Regardless of which head is to be removed, the No. 3 cylinder intake valve pushrod must be removed to allow removal of the intake manifold.

12. Remove the pushrods, marking them so they can be reinstalled in their original positions.

13. Remove the intake manifold.

14. Remove the cylinder head attaching bolts and remove the cylinder heads.

15. Clean all gasket mating surfaces. Check the cylinder head for flatness using a straight edge and a feeler gauge. The cylinder head must not be warped more than 0.003 in. in any 6 in. or more than 0.006 in. overall.

To install:

16. Position new head gasket(s) on the cylinder block, using the dowels for alignment.

17. Install the cylinder head(s) on the block. Oil the threads of new cylinder head bolts and hand tighten.

18. On 1988–90 vehicles, tighten the cylinder head bolts, in sequence, in 2 steps, first to 37 ft. lbs. (50 Nm) and then to 68 ft. lbs. (92 Nm).

19. On 1991–92 vehicles, tighten the cylinder head bolts, in sequence, to 59 ft. lbs. (80 Nm). Back off all bolts a minimum of 1 full turn. Retighten the bolts, in sequence, in 2 steps, first to 37 ft. lbs. (50 Nm) and then to 68 ft. lbs. (92 Nm).

20. Install the intake manifold.

21. Install the distributor, aligning the marks that were made during removal. Install the hold-down bolt and clamp.

22. Dip each pushrod in heavy engine oil and install them in their original positions.

23. For each valve, rotate the crankshaft until the lifter rests on the base circle of the camshaft lobe, before tightening the rocker arm fulcrum attaching bolts. Position the rocker arms over the valves and pushrods, install the fulcrums and fulcrum bolts and tighten to 24 ft. lbs. (32 Nm).

NOTE: If the original valve train components are being installed, a valve clearance check is not required. If a component has been replaced, perform a valve clearance check.

24. Install the exhaust manifolds and the spark plugs.

25. Install the rocker arm covers. Install the dipstick tube.

26. Install the fuel injector harness to the injectors and inboard rocker arm cover studs. Connect the engine harness to the main harness.

27. Install the distributor cap and connect the spark plug wires to the spark plugs.

28. Install the throttle body. Install the ignition coil and bracket, if necessary and connect the electrical connector.

29. Install the fuel line retaining bracket to the front of the cylinder head, if equipped. Tighten the retaining bolts to 26 ft. lbs. (35 Nm).

30. Install the power steering pump and bracket, if removed. Install the alternator and bracket assembly, if removed and connect the electrical harness.

31. Install the accessory drive belt(s).

32. Connect the fuel lines to the fuel supply manifold. Connect the upper radiator and heater hoses.

33. Connect the vacuum lines. Install the engine oil filler adapter on Aerostar.

34. Change the engine oil and filter.

NOTE: Engine coolant is corrosive to all engine bearing material. Replacing engine oil after removal of a coolant carrying component helps prevent engine failure later.

35. Install the air cleaner fresh air hose to the throttle body and air cleaner. Connect the negative battery cable.

36. Fill and bleed the cooling system. Run the engine and check for leaks.

37. Check the ignition timing and idle speed and adjust, if necessary.

4.0L Engine

1. Disconnect the negative battery cable and relieve the fuel system pressure. Drain the cooling system.

2. Remove the upper and lower intake manifolds and rocker arm covers.

3. If the left cylinder head is being removed, proceed as follows:

a. Remove the accessory drive belt.

b. Discharge the refrigerant and remove the air conditioning compressor, if equipped.

c. Remove the power steering pump and bracket and position aside.

d. Remove the spark plugs.

4. If the right cylinder head is being removed, proceed as follows:

a. Remove the accessory drive belt.

b. Remove the alternator and alternator bracket.

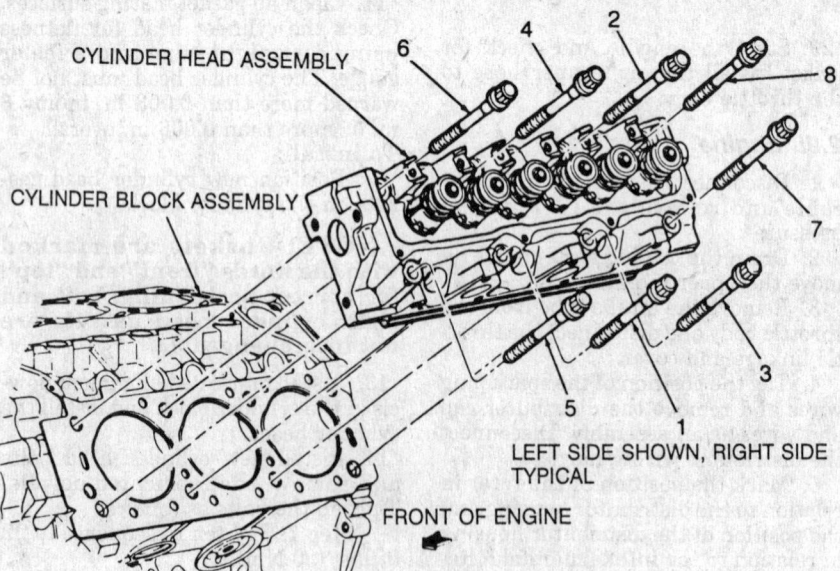

Cylinder head bolt torque sequence—3.0L engine

c. Remove the ignition coil and bracket assembly.

d. Remove the spark plugs.

5. Disconnect the exhaust pipe and remove the exhaust manifold(s).

6. Remove the rocker arm shaft assembly. Remove the pushrods, marking them so they can be reinstalled in the same positions.

7. Remove and discard the cylinder head attaching bolts and remove the cylinder heads.

8. Clean all gasket mating surfaces. Check the cylinder head for flatness using a straight edge and a feeler gauge. The cylinder head must not be warped more than 0.003 in. in any 6 in. or more than 0.006 in. overall.

To install:

9. Position new cylinder head gasket(s) on the cylinder block. Install cylinder head locating dowels.

NOTE: The cylinder head(s) and intake manifold are torqued alternately and in sequence to insure correct fit and gasket crunch.

10. Install new cylinder head bolts and tighten, in sequence, to 44 ft. lbs. (60 Nm).

11. Apply silicone sealer to the block and cylinder head mating surfaces at the 4 corners of the lifter valley opening. Install the intake manifold gasket and again apply sealer in the same locations.

12. Position the lower intake manifold on the 2 guide studs and install the nuts and bolts hand tight. Tighten the lower intake manifold bolts, in sequence, to 3–6 ft. lbs. (4–8 Nm).

13. Tighten the cylinder head bolts, in sequence, to 59 ft. lbs. (80 Nm).

14. Tighten the intake manifold, in sequence, to 6–11 ft. lbs. (8–15 Nm).

15. Turn the cylinder head bolts 80–85 degrees tighter, in sequence.

16. Tighten the intake manifold, in sequence, to 11–15 ft. lbs. (15–21 Nm) and then to 15–18 ft. lbs. (21–25 Nm), in sequence.

17. Dip both ends of each pushrod in clean engine oil and install in their original locations. Install the rocker arm and shaft assemblies and tighten the rocker arm shaft support bolts evenly to 46–52 ft. lbs. (62–70 Nm).

18. Apply silicone sealer to the 4 locations at the joint where the intake manifold and cylinder head meet. Install a new rocker arm cover gasket in each cover and install the rocker arm covers. Tighten the rocker arm cover bolts to 3–5 ft. lbs. (4–7 Nm), wait 2 minutes and then retighten to the same specification.

19. Install the upper intake manifold and tighten the nuts to 15–18 ft. lbs. (20–25 Nm).

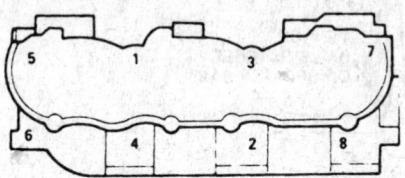

Cylinder head bolt torque sequence— 4.0L engine

20. Install the exhaust manifold(s) and connect the exhaust pipe.

21. Install the spark plugs and the ignition coil and bracket assembly.

22. Install the alternator and the accessory drive belt.

23. Install the power steering pump. Install the air conditioning compressor, if equipped.

24. Connect the negative battery cable. Fill and bleed the cooling system. Run the engine and check for leaks.

Valve Lifters

REMOVAL & INSTALLATION

Except 2.0L and 2.3L Engines

1. Disconnect the negative battery cable and relieve the fuel system pressure. Drain the cooling system.

2. Remove the intake manifold and rocker arm covers.

3. On 2.9L and 4.0L engines, loosen the rocker arm shaft support bolts 2 turns at a time until the rocker arm and shaft assembly can be removed.

4. On 3.0L engine, loosen the rocker arm fulcrum bolt enough so the rocker arm can be lifted from the pushrod and turned to 1 side.

5. Remove the pushrods, marking them so they can be reinstalled in their original positions.

6. On 2.9L engine, remove the cylinder heads.

7 Remove the lifters. Note the location of each lifter so it can be reinstalled in the same bore. If a lifter is stuck in the bore, use a suitable tool to rotate the lifter back and forth to loosen it from the gum and varnish that may have formed on the lifter.

NOTE: The 4.0L engine is equipped with roller lifters. Roller lifters have an alignment tab which fits into a locating groove in the lifter bore. Do not attempt to rotate a roller lifter in the bore.

To install:

8. Lubricate the lifters and bores with clean engine oil. Install each lifter in the same bore from which it was removed. On 4.0L engine, install the lifter with the alignment tab in the locating groove of the bore. If a new lifter is being installed, check for free fit in the bore.

9. On 2.9L engine, install the cylinder heads.

10. Check each pushrod for straightness and for damage, replace as necessary. Dip each pushrod end in clean engine oil and install in its original position.

11. On 2.9L and 4.0L engines, lubricate the rocker arm and shaft assembly and install. Draw the shaft support bolts down evenly, 2 turns at a time, until the shafts are fully down. Tighten the bolts to 43–50 ft. lbs. (59–67 Nm) on 2.9L engine or 46–52 ft. lbs. (62–70 Nm) on 4.0L engine.

12. On 3.0L engine, for each valve, rotate the crankshaft until the lifter rests on the base circle of the camshaft lobe, before tightening the rocker arm fulcrum attaching bolts. Position the rocker arms over the valves and pushrods, install the fulcrums and fulcrum bolts and tighten to 24 ft. lbs. (32 Nm).

13. Install the intake manifold and rocker arm covers.

14. Connect the negative battery cable. Fill and bleed the cooling system.

15. Run the engine and check for leaks.

2.0L and 2.3L Engines

1. Disconnect the negative battery cable and remove the air cleaner or air intake duct. On 2.3L engine, remove the throttle body and EGR supply tube.

2. Remove the rocker arm cover.

3. Rotate the crankshaft so the base circle of the cam is facing the applicable cam follower.

4. Using valve spring compressor lever tool T88T–6565–BH or equivalent, collapse the valve spring and slide the cam follower over the valve lifter and out.

5. Lift out the hydraulic valve lifter.

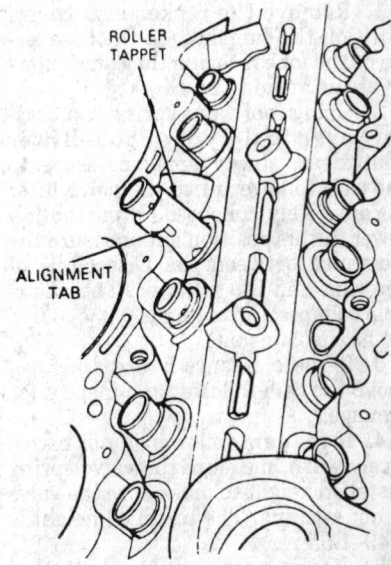

Valve lifter Installation—4.0L engine

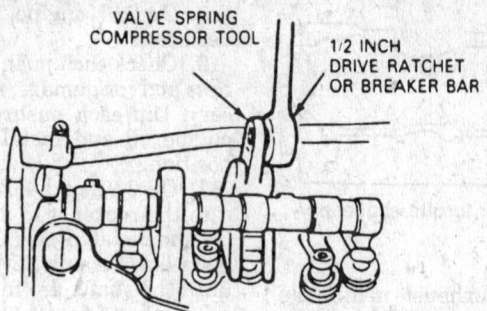

VALVE SPRING
COMPRESSOR TOOL

1/2 INCH
DRIVE RATCHET
OR BREAKER BAR

Cam follower removal—2.0L and 2.3L engines

To install:

6. Rotate the crankshaft so the base circle of the cam is facing the applicable cam follower.

7. Coat the hydraulic lifter with clean engine oil and install it in the bore.

8. Collapse the valve spring using valve spring compressor lever T88T-6565–BH or equivalent. Position the cam follower over the valve lifter and the valve stem.

9. Clean the gasket surfaces of the rocker arm cover and cylinder head.

10. Coat the rocker arm cover and a new gasket with gasket adhesive and install the gasket to the cover.

11. Install the cover and tighten the retaining bolts to 5–8 ft. lbs. (7–11 Nm).

12. Install the throttle body and EGR supply tube, if necessary. Install the air cleaner or air intake duct.

13. Connect the negative battery cable.

Valve Lash

ADJUSTMENT

2.0L and 2.3L Engines

1. Remove the rocker arm cover. Position the camshaft so the base circle of the lobe is facing the cam follower of the valve to be checked.

2. Using tool valve spring compressor lever tool T88T-6565–BH or equivalent, slowly apply pressure to the cam follower until the valve lifter is completely collapsed. Hold the follower in this position and measure the clearance between the base circle of the cam and the follower. The allowable collapsed lifter gap is 0.035–0.055 in. at the camshaft.

3. If the clearance is excessive, remove the cam follower and inspect for damage.

4. If the cam follower is not excessively worn, measure the valve spring installed height to make sure the valve is not sticking. The installed height is 1.49–1.55 in.

5. If the valve spring installed height is correct, check the camshaft

lobe lift. The lobe lift dimension is 0.2381 in.

6. If the cam follower, valve spring height and camshaft lobe lift are correct and the base circle-to-follower clearance is excessive, replace the valve lifter.

2.9L Engine

1. Remove the rocker arm cover assembly. On the cylinder to be adjusted, position the camshaft lobe so the lifters are on the base circle.

2. Loosen the adjusting screws until a distinct lash between the rocker arm pad and the valve tip end can be noticed. The plunger of the hydraulic lifter should now be fully extended under load of the internal spring.

3. Screw in the adjustment screws until the rocker arms slightly touch the valve stem.

4. To achieve the nominal working position of the plunger, turn in the adjusting screw 1½ turns, equivalent to 0.070 in.

3.0L Engine

1. Remove the rocker arm cover.

2. Rotate the crankshaft until the lifter is on the base circle of the cam on the valve to be checked.

3. Using a suitable tool, collapse the lifter fully and measure the clearance between the valve stem tip and rocker arm. The clearance should be 0.085–0.185 in. (2.15–4.69mm).

Rocker Arms/Shafts

REMOVAL & INSTALLATION

2.0L and 2.3L Engines

1. Disconnect the negative battery cable and remove the air cleaner or air intake duct. On 2.3L engine, remove the throttle body and EGR supply tube.

2. Remove the rocker arm cover.

3. Rotate the crankshaft so the base circle of the cam is facing the applicable cam follower.

4. Using valve spring compressor lever tool T88T-6565–BH or equivalent,

collapse the valve spring and slide the cam follower over the valve lifter and out.

To install:

5. Rotate the crankshaft so the base circle of the cam is facing the applicable cam follower.

6. Collapse the valve spring using valve spring compressor lever T88T-6565–BH or equivalent. Position the cam follower over the valve lifter and the valve stem.

7. Clean the gasket surfaces of the rocker arm cover and cylinder head.

8. Coat the rocker arm cover and a new gasket with gasket adhesive and install the gasket to the cover.

9. Install the cover and tighten the retaining bolts to 5–8 ft. lbs. (7–11 Nm).

10. Install the throttle body and EGR supply tube, if necessary. Install the air cleaner or air intake duct.

11. Connect the negative battery cable.

2.9L Engine

1. Disconnect the negative battery cable and relieve the fuel system pressure.

2. Tag and disconnect the spark plug wires. If equipped, disconnect the transmission kickdown linkage.

3. Disconnect the fuel lines. If equipped with air conditioning, remove the dipstick tube and bracket and remove the left lifting eye.

4. Remove the PCV valve hose and breather. Remove the rocker arm cover attaching screws and load distribution washers. Note the position of the washers so they can be reinstalled in their original positions.

5. If equipped, remove the electrical connections from the right rocker arm cover.

6. Using a light plastic hammer, tap the rocker arm covers to break the seal. Remove the rocker arm covers.

7. Remove the rocker arm shaft stand attaching bolts by loosening them 2 turns at a time, in sequence. Lift off the rocker arm and shaft assembly and oil baffle.

To install:

8. Loosen the valve lash adjusting screws a few turns. Apply engine oil to the assembly to provide initial lubrication.

9. Install the oil baffle and rocker arm shaft assembly to the cylinder head. Guide the adjusting screws onto the pushrods.

10. Install the rocker arm stand attaching bolts, running them down 2 turns at a time, in sequence, until the shaft assembly is seated. Tighten the rocker arm stand attaching bolts to 43–50 ft. lbs. (59–67 Nm).

11. Adjust the valve lash.

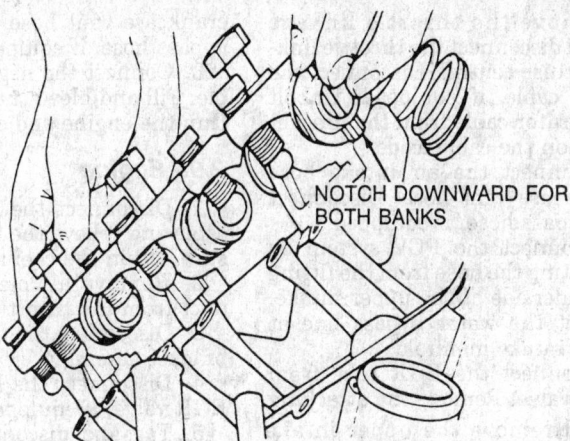

NOTCH DOWNWARD FOR BOTH BANKS

Rocker arm shaft assembly installation – 2.9L engine

12. Clean the gasket mating surfaces of the rocker arm covers and cylinder heads.

13. Install the rocker arm covers, using new gaskets. Install the attaching screws and load distribution washers, making sure the washers are installed in their original positions. Tighten to 3–5 ft. lbs. (4–7 Nm).

14. Connect the transmission kickdown linkage, if necessary. Connect the spark plug wires.

15. Connect the fuel lines and the negative battery cable. Start the engine and check for leaks.

3.0L Engine

1. Disconnect the negative battery cable. Remove the air cleaner fresh air hose, if necessary.

2. Tag and disconnect the spark plug wires from the spark plugs. Remove the spark plug wire/separator assembly from the rocker arm cover attaching bolt studs and position aside.

3. If the left rocker arm cover is being removed, proceed as follows:

a. On 1988 vehicles, remove the oil fill cap and disconnect the closure system hose.

b. On 1989–92 vehicles, remove the throttle body assembly and the PCV valve.

c. On 1991–92 vehicles, remove the fuel injector harness stand-offs from the inboard rocker arm cover studs. Move the harness aside.

4. If the right rocker arm cover is being removed, proceed as follows:

a. On 1988 vehicles, remove the PCV valve and disconnect the heater hoses.

b. On 1989–92 Aerostar, remove the oil filler tube assembly and disconnect the closure hose from the oil fill adapter.

c. On Ranger, disconnect the engine harness connectors and remove the air cleaner closure hose from the oil fill adapter.

d. On 1991–92 vehicles, remove the fuel injector harness stand-offs from the inboard rocker arm cover studs. Move the harness aside.

5. Remove the rocker arm cover attaching bolts and studs, noting their locations. Remove the rocker arm cover.

6. Remove the rocker arm fulcrum bolt and remove the rocker arm and fulcrum.

To install:

7. Lubricate the valve stem tip, pushrod end, fulcrum and rocker arm fulcrum seat with clean engine oil.

8. For each valve, rotate the crankshaft until the lifter rests on the base circle of the camshaft lobe, before tightening the rocker arm fulcrum attaching bolts. Position the rocker arms over the valves and pushrods, install the fulcrums and fulcrum bolts and tighten to 24 ft. lbs. (32 Nm).

9. Clean the rocker arm cover and cylinder head gasket mating surfaces of all gasket material and/or old silicone sealer.

10. Apply a bead of silicone sealer at the cylinder head to intake manifold rail step and position the rocker arm cover on the cylinder head.

11. Install the bolts/studs in their original locations and tighten to 9 ft. lbs. (12 Nm).

12. Install the remaining components in the reverse order of their removal. Start the engine and check for leaks.

4.0L Engine

1. Disconnect the negative battery cable and relieve the fuel system pressure.

2. Tag and disconnect the spark plug wires. Disconnect the fuel lines.

3. If the left rocker arm cover is being removed, upper intake manifold removal may be required.

4. If the right rocker arm cover is being removed, proceed as follows:

a. Remove the ignition coil and bracket assembly and remove the PCV valve hose and breather.

b. On 1991–92 vehicles, remove the air inlet duct and the hose attached to the oil fill tube. Remove the accessory drive belt and remove the alternator.

c. On 1991–92 vehicles, drain the cooling system and remove the upper radiator hose. Remove the low pressure air conditioner hose bracket from the upper intake, if still installed and remove the vacuum hose from the air cleaner.

5. Remove the rocker arm cover attaching screws and load distribution washers. Note the position of the washers so they can be reinstalled in their original positions.

6. Using a light plastic hammer, tap the rocker arm covers to break the seal. Remove the covers.

7. Remove the rocker arm shaft stand attaching bolts by loosening the bolts 2 turns at a time, in sequence. Lift off the rocker arm and shaft assembly.

To install:

8. Apply engine oil to the valve train assembly to provide initial lubrication.

9. Install the rocker arm shaft assembly to the cylinder head, guiding the rocker arms onto the pushrods.

10. Install the rocker arm stand attaching bolts, running them down 2 turns at a time, in sequence, until the shaft assembly is seated. Tighten the rocker arm stand attaching bolts to 46–52 ft. lbs. (62–70 Nm).

11. Clean all gasket material from the rocker arm cover and cylinder head.

12. Apply silicone sealer to the parting lines where the cylinder head and intake manifold seal. Install the rocker arm cover, using a new gasket.

13. Install the rocker arm cover attaching screws and load distribution washers, making sure the washers are in their original positions. Tighten to 3–5 ft. lbs. (4–7 Nm).

14. Install the remaining components in the reverse order of their removal. Start the engine and check for leaks.

Intake Manifold

REMOVAL & INSTALLATION

2.0L Engine

1. Disconnect the negative battery cable and drain the cooling system. Remove the air cleaner.

2. Disconnect the accelerator cable. Tag and disconnect the necessary vacuum hoses. Remove the hot water hose from the manifold cover nipple fitting.

3. Remove the engine oil dipstick and disconnect the heat tube at the EGR valve. Disconnect the fuel line at the carburetor fuel filter.

4. Remove the dipstick retaining nut from the EGR valve.

5. Disconnect and remove the PCV at the engine and intake manifold.

6. Remove the distributor cap and position the cap and wires aside. Remove the plastic spark plug connector from the rocker arm cover.

7. Remove the intake manifold retaining bolts and remove the manifold from the vehicle.

To install:

8. Clean all gasket mating surfaces.

9. Position a new gasket and install the intake manifold. Install the attaching bolts and tighten, in sequence, in 2 steps, first to 5–7 ft. lbs. (7–9 Nm) and then to 15–22 ft. lbs. (19–29 Nm).

10. Install the distributor cap and the plastic spark plug connector.

11. Install the dipstick tube retaining nut and connect the fuel line.

12. Connect the PCV and connect the heat tube at the EGR valve.

13. Install the dipstick and connect the vacuum hoses. Connect the accelerator cable.

14. Install the hot water hose to the manifold nipple fitting and install the air cleaner.

15. Connect the negative battery cable. Fill and bleed the cooling system. Run the engine and check for leaks.

2.3L Engine

1. Disconnect the negative battery cable and relieve the fuel system pressure. Drain the cooling system.

2. Tag and disconnect the electrical connectors at the throttle position sensor, air charge temperature sensor, engine coolant temperature sensor and air bypass valve, if equipped. Disconnect the knock sensor connector, if equipped.

3. Disconnect the injector wiring harness at the main engine harness and at the water temperature indicator sensor. Disconnect the ignition control assembly connector, if equipped.

4. Tag and disconnect the vacuum lines at the upper intake manifold vacuum tree, EGR valve, fuel pressure regulator and canister purge line.

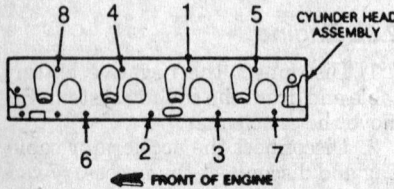

Intake manifold bolt torque sequence— 2.0L and 2.3L engines

5. Remove the throttle linkage shield and disconnect the throttle linkage and cruise control. Disconnect the kickdown cable, if equipped. Unbolt the accelerator cable from the bracket and position the cable aside.

6. Disconnect the air intake hose and crankcase vent hose. Disconnect the air bypass hose, if equipped.

7. Disconnect the PCV system by disconnecting the hose from the fitting on the underside of the upper intake. Disconnect the water bypass line at the lower intake manifold.

8. Disconnect the EGR tube from the EGR valve. Remove the attaching bolts and remove the upper intake manifold and throttle body assembly.

9. Remove the engine oil dipstick tube bracket attaching bolt. Disconnect the fuel lines from the fuel supply manifold.

10. Disconnect the electrical connectors from the fuel injectors and position aside. Remove the fuel supply manifold attaching bolts and remove the fuel supply manifold.

11. Remove the attaching bolts and remove the lower intake manifold.

To install:

12. Clean all gasket mating surfaces. Clean and oil the manifold bolt threads.

13. Position a new gasket and install the lower intake manifold. Install the attaching bolts and tighten, in sequence, in 2 steps, first to 5–7 ft. lbs. (7–9 Nm) and then to 15–22 ft. lbs. (20–30 Nm).

14. Install the fuel supply manifold and injectors with the 2 attaching bolts. Tighten to 15–22 ft. lbs. (20–30 Nm). Connect the electrical connectors to the injectors.

15. Position a new gasket on the lower intake manifold and install the upper intake manifold. Install the attaching bolts and tighten, in sequence, to 15–22 ft. lbs. (20–30 Nm).

16. Install the engine oil dipstick tube and retaining bolt. Connect the fuel lines to the fuel supply manifold.

17. Connect the EGR tube to the EGR valve. Tighten to 18–28 ft. lbs. (25–30 Nm).

18. Connect the water bypass line and connect the PCV hose. Connect the vacuum lines to the locations marked during removal.

19. Hold the accelerator cable bracket in position on the upper manifold and install the attaching bolts. Tighten to 10–15 ft. lbs. (13.5–20.5 Nm).

20. Install the accelerator cable to the bracket. Connect the accelerator cable and cruise control. Install the throttle linkage shield.

21. Connect the electrical connectors to the locations marked during removal.

22. Connect the air intake hose and

crankcase vent hose. Connect the air bypass hose, if equipped.

23. Connect the negative battery cable. Fill and bleed the cooling system. Run the engine and check for leaks.

2.9L Engine

1. Disconnect the negative battery cable and relieve the fuel system pressure. Drain the cooling system.

2. Remove air cleaner air intake duct from the throttle body.

3. Disconnect the throttle cable and bracket assembly.

4. Disconnect the EGR tube at the EGR valve, if equipped.

5. Tag and disconnect all vacuum hoses from the fittings on the upper intake manifold.

6. Disconnect the electrical connections at the throttle body, intake manifold upper and lower, distributor and EGR pressure sensor, if equipped. Also disconnect the fuel injector sub harness from the main EEC harness.

7. Remove the upper intake manifold assembly.

8. Disconnect and remove the hose from the water outlet to radiator and heater supply. Disconnect the fuel lines from the fuel supply manifold.

9. Tag and disconnect the spark plug wires. Remove the distributor cap and wires as an assembly.

10. Mark the position of the distributor rotor in relation to the distributor housing and the housing in relation to the engine. Remove the distributor hold-down screw and clamp and lift out the distributor.

11. Remove the rocker arm covers.

12. Remove the intake manifold at-

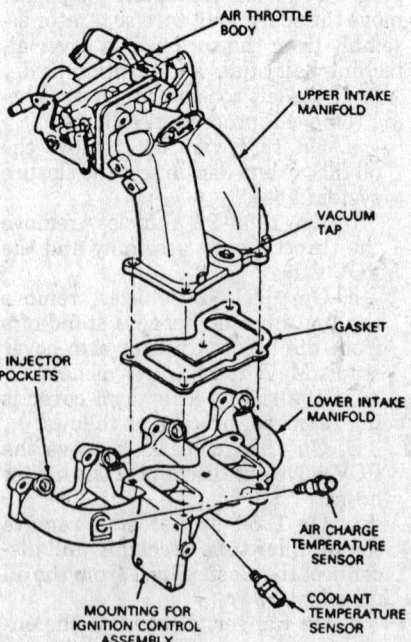

Intake manifold assembly—1989–92 2.3L engine

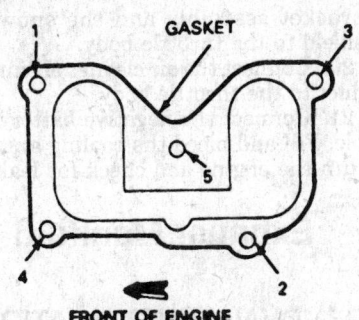

Upper intake manifold retaining bolt torque sequence—1989–92 2.3L engine

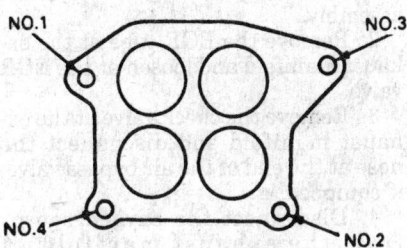

Upper intake manifold retaining bolt torque sequence—1988 2.3L engine

taching bolts and nuts. Note the length of the manifold attaching bolts during removal so they can be installed in their original positions. Tap the manifold lightly with a plastic mallet to break the gasket seal. Remove the manifold.

13. Remove all old gasket material and sealing compound.

To install:

14. Apply sealing compound to the joining surfaces. Place the intake manifold gasket in position, making sure the tab on the right bank cylinder head gasket fits into the cutout of the manifold gasket.

15. Apply sealing compound to the attaching bolt bosses on the intake manifold and position the intake manifold. Install the attaching bolts/nuts and tighten, in sequence, in 4 steps, first to 3–6 ft. lbs. (4–8 Nm), 2nd to 6–11 ft. lbs. (8–15 Nm), 3rd to 11–15 ft. lbs. (15–21 Nm) and 4th to 15–18 ft. lbs. (21–25 Nm).

16. Install the distributor, aligning the marks that were made during removal. Install the distributor clamp and attaching bolt.

17. Replace the rocker arm cover gaskets and install the rocker arm covers.

18. Install the distributor cap and spark plug wires. Connect the distributor wiring harness.

19. Apply sealing compound at the joining surfaces of the upper and lower intake manifold. Install new upper intake manifold gaskets.

20. Install the upper intake manifold

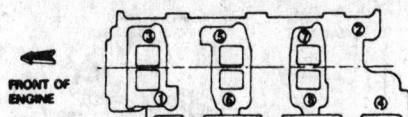

Intake manifold bolt torque sequence—2.9L engine

assembly and tighten the retaining bolts in sequence, center to end, in 2 steps, first to 7 ft. lbs. (10 Nm) and then to 15–18 ft. lbs. (21–25 Nm).

21. Connect all vacuum hoses to the fittings on the upper intake manifold.

22. Connect the electrical connectors at the throttle body, intake manifolds sub harness to EEC main harness and EGR pressure sensor, if equipped.

23. Install and adjust the throttle linkage bracket assembly and cover as required.

24. Connect the hoses from the water outlet to the radiator and the bypass hose from the thermostat housing rear cover to intake manifold.

25. Connect the fuel lines.

26. Connect the negative battery cable. Fill and bleed the cooling system.

27. Check the ignition timing and reset the engine idle speed to specification. Start the engine and check for coolant and oil leaks.

3.0L Engine

1. Disconnect the negative battery cable and relieve the fuel system pressure. Drain the cooling system.

2. Remove the air cleaner hoses to the throttle body and rocker arm cover. Disconnect the fuel lines from the fuel supply manifold.

3. Tag and disconnect the necessary vacuum lines.

4. Tag and disconnect the electrical connectors at the air charge temperature sensor, engine coolant temperature sensor, throttle position sensor, air bypass solenoid and coolant temperature sender.

5. Remove the snow shield from the power steering pump bracket and accelerator cable bracket.

6. Disconnect and remove the accelerator and cruise control cables from the accelerator mounting bracket and throttle lever.

7. Remove the alternator support brace.

8. Remove the throttle body-to-lower intake manifold retaining bolts and stud bolts and remove the throttle body assembly.

9. Disconnect the fuel injector harness stand-offs from the inboard rocker arm cover studs and each injector and remove from the engine.

10. Disconnect the upper radiator hose from the thermostat housing and disconnect the heater hoses.

11. Tag and disconnect the spark

plug wires. Remove the distributor cap and wires as an assembly.

12. Mark the position of the distributor rotor in relation to the distributor housing and the housing in relation to the engine. Remove the distributor hold-down screw and clamp and lift out the distributor.

13. Remove the ignition coil from the rear of the left cylinder head, if required.

14. Remove the rocker arm covers. Loosen the No. 3 cylinder intake valve rocker arm fulcrum bolt and rotate the rocker arm away from the valve. Remove the pushrod.

15. Remove the intake manifold bolts. Break the gasket seal by wedging a large prybar between the manifold an the block using the lug on the water pump as a leverage point. Be careful to prevent damage to machines surfaces.

16. Remove the intake manifold.

To install:

17. Clean all gasket mating surfaces.

18. Apply silicone sealer to the intersection of the cylinder block and cylinder head at the 4 corners of the lifter valley opening.

19. Install the front and rear intake manifold seals and secure with the retaining features. Position the intake manifold gaskets on the cylinder heads and insert the locking tabs on the cylinder head gaskets.

20. Carefully lower the intake manifold into position being careful not to disturb the silicone sealer. Install the intake manifold bolts and tighten, in sequence, in 2 steps, first to 11 ft. lbs. (15 Nm) and then to 19 ft. lbs. (26 Nm).

21. Install the No. 3 cylinder intake valve pushrod. Apply oil to the pushrod and rocker arm fulcrum and position the rocker arm over the valve and pushrod. Rotate the crankshaft to place the lifter on the base circle of the cam, then tighten the fulcrum bolt to 24 ft. lbs. (32 Nm).

22. Install the rocker arm covers and the fuel injector electrical harness.

23. Install the throttle body using a new gasket. Tighten the throttle body attaching bolts, in sequence, to 19 ft. lbs. (25 Nm).

24. Install the alternator brace. Tighten the nuts to 12 ft. lbs. (16 Nm).

25. Connect the PCV valve hose. Connect the engine coolant temperature sensor, air charge temperature sensor, throttle position sensor, air bypass solenoid and coolant temperature sender connectors.

26. Install the distributor, aligning the marks that were made during removal. Install the distributor cap and connect the spark plug wires. Connect the distributor electrical connector.

27. Install the ignition coil, if removed.

28. Connect the heater hoses and the upper radiator hose. Connect the vacuum lines to the locations marked during removal.

29. Connect the fuel lines to the fuel supply manifold. Change the engine oil and filter.

NOTE: Engine coolant is corrosive to all engine bearing material. Changing the oil after removal of a coolant carrying component helps prevent engine failure.

30. Connect the negative battery cable. Fill and bleed the cooling system. Install the air cleaner hose.

31. Run the engine and check for leaks. Check the ignition timing, idle speed, throttle linkage and cruise control and adjust, if necessary.

4.0L Engine

1. Disconnect the negative battery cable and relieve the fuel system pressure.

2. Remove the air cleaner air intake duct from the throttle body.

3. Remove the snow/ice shield and disconnect the throttle cable and bracket assembly.

4. Tag and disconnect the vacuum hoses from the fittings on the upper intake manifold.

5. Tag and disconnect the electrical connectors at the throttle body, upper intake manifold, lower intake manifold and injectors.

6. Disconnect the fuel lines from the fuel supply manifold.

7. Remove the ignition coil and bracket assembly.

8. Remove the mounting nuts and remove the upper intake manifold.

9. Remove the rocker arm covers.

10. Remove the intake manifold attaching bolts and nuts. Tap the manifold lightly with a plastic mallet the break the gasket seal and remove the manifold.

To install:

11. Clean all gasket mating surfaces.

12. Apply silicone sealer to the block and cylinder head mating surfaces at the 4 corners of the lifter valley opening. Install the intake manifold gaskets and again apply sealer to the same locations.

13. Position the intake manifold on the 2 guide studs and install the nuts and bolts hand tight. Tighten the bolts, in sequence, in 4 steps, first to 3–6 ft. lbs. (4–8 Nm), then to 6–11 ft. lbs. (8–15 Nm), then to 11–15 ft. lbs. (15–21 Nm) and finally to 15–18 ft. lbs. (21–25 Nm).

14. Apply silicone sealer to the 4 locations where the intake manifold and the cylinder heads meet. Install the

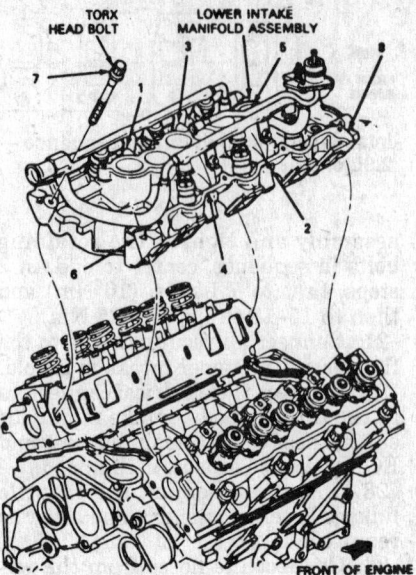

Intake manifold bolt torque sequence— 3.0L engine

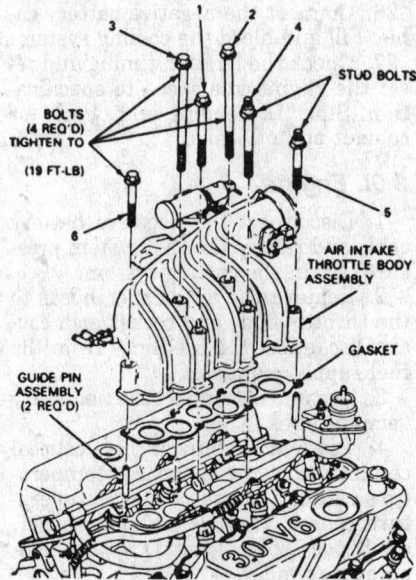

Air Intake throttle body bolt torque sequence—3.0L engine

rocker arm covers with new gaskets and tighten evenly to 3–5 ft. lbs. (4–7 Nm). Wait 2 minutes and tighten the bolts again to the same specification.

15. Install the upper intake manifold and tighten the nuts to 15–18 ft. lbs. (20–25 Nm).

16. Install the ignition coil and bracket assembly. Connect the fuel lines to the fuel supply manifold.

17. Connect the electrical connectors at the throttle body, upper intake manifold, lower intake manifold and injectors.

18. Connect the vacuum hoses to the fittings on the upper intake manifold.

19. Install the throttle cable and

bracket assembly and the snow/ice shield to the throttle body.

20. Connect the air cleaner air intake duct to the throttle body.

21. Connect the negative battery cable. Fill and bleed the cooling system. Run the engine and check for leaks.

Exhaust Manifold

REMOVAL & INSTALLATION

2.0L and 2.3L Engines

1. Disconnect the negative battery cable. Remove the air cleaner and duct assembly.

2. Remove the EGR tube at the exhaust manifold and loosen at the EGR valve.

3. Remove the check valve at the exhaust manifold and disconnect the hose at the end of the air bypass valve, if equipped.

4. Disconnect the oxygen sensor from the exhaust manifold, if equipped. Remove the sensor, if necessary.

5. Remove the screw attaching the heater hoses to the rocker arm cover. Disconnect the exhaust pipe from the exhaust manifold.

6. Remove the exhaust manifold mounting bolts and remove the manifold.

7. Installation is the reverse of the removal procedure. Tighten the exhaust manifold mounting bolts, in sequence, in 2 steps, first to 15–17 ft. lbs. (20–23 Nm) and then to 20–30 ft. lbs. (27–41 Nm).

2.9L Engine

1. Disconnect the negative battery cable.

2. Raise and safely support the vehicle, as necessary.

3. Disconnect the exhaust pipe from the exhaust manifold.

4. Disconnect the EGR tube at the manifold, if equipped.

5. Remove the manifold attaching bolts and remove the manifold.

6. Installation is the reverse of the removal procedure. Tighten the exhaust manifold bolts to 20–30 ft. lbs. (27–40 Nm).

3.0L Engine

1. Disconnect the negative battery cable.

2. Raise and safely support the vehicle, as necessary.

3. If removing the left exhaust manifold, remove the engine oil dipstick tube support bracket or retaining nut, as required. Rotate the tube out of the way or remove.

4. If removing the left exhaust man-

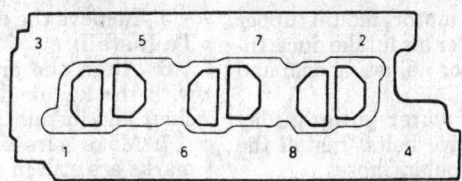

Intake manifold bolt torque sequence—4.0L engine

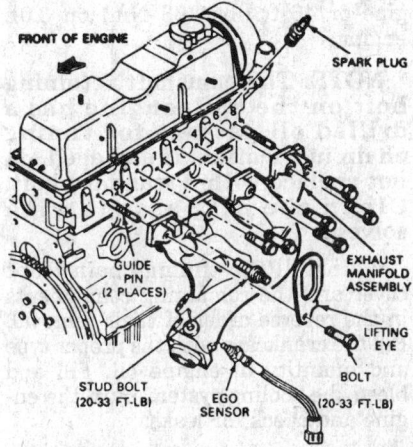

Exhaust manifold bolt torque sequence—2.0L and 2.3L engines

ifold on 1988–90 vehicles, remove the power steering pump pressure and return hoses.

5. If removing the right exhaust manifold on 1988–89 vehicles, remove the heater hose support bracket and disconnect the heater hoses.

6. Remove the spark plugs. Disconnect the exhaust pipe from the manifold.

7. Remove the exhaust manifold attaching nuts and remove the manifold.

8. Installation is the reverse of the removal procedure. Tighten the manifold attaching bolts to 18 ft. lbs. (25 Nm).

4.0L Engine

1. Disconnect the negative battery cable.

2. Raise and safely support the vehicle, as necessary.

3. If removing the left manifold, remove the engine oil dipstick tube support bracket. Remove the power steering pump pressure and return hoses, if necessary.

4. If removing the right manifold, remove the heater hose support bracket and disconnect the heater hoses.

5. Disconnect the exhaust pipe from the manifold.

6. Remove the manifold attaching bolts and remove the manifold.

7. Installation is the reverse of the removal procedure. Tighten the mounting bolts to 19 ft. lbs. (25 Nm).

Timing Chain Front Cover

REMOVAL & INSTALLATION

2.9L and 4.0L Engines

1. Disconnect the negative battery cable and drain the cooling system and crankcase.

2. Remove the oil pan and the radiator.

3. Remove the air conditioning compressor and power steering bracket, if equipped.

4. Remove the alternator and drive belt(s). Remove the fan.

5. Remove the water pump and heater and radiator hoses.

6. Remove the crankshaft pulley/damper assembly. On 4.0L engine, remove the crankshaft timing sensor.

7. Remove the front cover retaining bolts, noting their positions. If necessary, tap the cover lightly with a plastic hammer to break the gasket seal. Remove the front cover.

To install:

8. Clean all gasket mating surfaces. Apply sealer to the gasket surfaces on the cylinder block and the back side of the front cover plate. Install the guide sleeves.

9. Apply sealer to the front cover gasket surface and position a new gasket on the front cover.

10. Install the front cover with the retaining screws. Note the different bolt lengths on 4.0L engine. Tighten the bolts to 13–15 ft. lbs. (17–21 Nm).

11. On 4.0L engine, install the crankshaft timing sensor.

12. Install the crankshaft pulley/damper assembly. On 2.9L engine, tighten the attaching bolt to 85–96 ft. lbs. (115–130 Nm). On 4.0L engine, tighten the attaching bolt to 30–37 ft. lbs. (40–50 Nm), then tighten an additional 80–90 degrees.

13. Install the remaining components in the reverse order of their removal. Fill and bleed the cooling system. Run the engine and check for leaks.

3.0L Engine
AEROSTAR

1. Disconnect the negative battery

cable. Remove the air cleaner fresh air hose.

2. Drain the cooling system and the crankcase. Remove the cooling fan.

3. Loosen the water pump hub bolts and remove the accessory drive belts. Remove the water pump pulley.

4. Remove the alternator adjusting arm and brace. Move the alternator aside.

5. Remove the air conditioning compressor mounting bolts, if equipped. Tie the compressor aside with mechanics wire and remove the bracket.

6. Remove the crankshaft pulley and damper. Remove the water pump, if required.

NOTE: The timing cover can be removed with the water pump installed by not removing the 6mm water pump attaching bolts.

7. Disconnect the lower radiator hose and heater hose.

8. Remove the oil pan assembly. Disconnect the oil level sensor, if equipped, before removal.

9. Remove the front cover attaching bolts and remove the front cover.

To install:

10. Clean all gasket mating surfaces. Use a seal removal tool to remove the front cover oil seal.

11. Install a new front cover oil seal, using a seal installer. Position a new front cover gasket on the engine block dowel pins.

12. Install the front cover with the attaching bolts. Apply sealer to the 3 attaching bolts on the passenger side of the cover, prior to installation. Tighten the 8mm bolts to 19 ft. lbs. (25 Nm) and the 6mm bolts to 7 ft. lbs. (10 Nm).

13. Install the oil pan. Install the water pump, if removed.

14. Install the crankshaft pulley and damper. Tighten the damper attaching bolt to 141–169 ft. lbs. (190–230 Nm) on 1988 vehicles or 107 ft. lbs. (145 Nm) on 1989–92 vehicles.

15. Install the lower radiator hose and heater hose. Install the air conditioning bracket, compressor and brace, if equipped.

16. Install the alternator assembly, bracket, oil fill tube support and throttle body brace.

17. Install the water pump pulley and accessory drive belts. Install the cooling fan and shroud.

18. Fill the crankcase with the proper type and quantity of engine oil. Connect the negative battery cable.

19. Fill and bleed the cooling system. Run the engine and check for leaks. Install the air cleaner fresh air hose.

RANGER

1. Disconnect the negative battery

cable. Drain the cooling system and crankcase.

2. Remove the cooling fan and water pump pulley bolts. Remove the accessory drive belts and the water pump pulley.

3. Remove the alternator adjusting arm and the throttle body brace. Remove the heater air intake duct.

4. Remove the motor mount upper nuts. If equipped with automatic transmission and air conditioning, remove the air conditioning compressor upper bolts, then remove the front cover front nuts.

5. Remove the distributor assembly.

NOTE: Failure to remove the distributor assembly will result in a broken distributor.

6. Raise and safely support the vehicle. Remove the lower air conditioning compressor bolts and wire the compressor aside. Remove the compressor bracket.

7. Remove the crankshaft pulley and damper. Remove the oil pan. Disconnect the oil level sensor before pan removal.

8. Lower the vehicle and remove the lower radiator hose. Remove the water pump, if required.

NOTE: The timing cover can be removed with the water pump installed by not removing the 6mm water pump attaching bolts.

9. Remove the front cover attaching bolts and remove the front cover.
To install:

10. Clean all gasket mating surfaces. Use a seal removal tool to remove the front cover oil seal.

11. Install a new front cover oil seal, using a seal installer. Position a new front cover gasket on the engine block dowel pins.

12. Install the front cover with the attaching bolts. Apply sealer to the 3 attaching bolts on the passenger side of the cover, prior to installation. Tighten the 8mm bolts to 19 ft. lbs. (25 Nm) and the 6mm bolts to 7 ft. lbs. (10 Nm).

13. Raise and safely support the vehicle. Install the oil pan and connect the oil level sensor.

14. Install the water pump, if removed.

15. Install the crankshaft pulley and damper. Tighten the damper attaching bolt to 107 ft. lbs. (145 Nm).

16. Install the air conditioning compressor bracket, if equipped, position the compressor and install the lower bolts. Lower the vehicle.

17. Install the distributor. Install the front cover front nuts and the air conditioning compressor upper bolts, if equipped.

18. Install the motor mount upper nuts and the heater air intake duct. Install the alternator adjusting arm and brace.

19. Install the water pump pulley and accessory drive belts. Install the cooling fan and coolant hoses.

20. Fill the crankcase with the proper type and quantity of engine oil. Connect the negative battery cable.

21. Fill and bleed the cooling system. Run the engine and check for leaks. Check the ignition timing and adjust, if necessary.

Front Cover Oil Seal

REPLACEMENT

1. Disconnect the negative battery cable.

2. Drain the cooling system and remove the radiator, if necessary to provide access.

3. Remove the accessory drive belts and remove the crankshaft pulley and damper assembly.

4. Remove the seal from the front cover using a seal removal tool. Be careful not to damage the seal housing or crankshaft surfaces.
To install:

5. Coat a new oil seal with clean engine oil and install in the front cover, using a seal installer.

6. Install the crankshaft pulley and damper assembly. Tighten the damper attaching bolt to 85–96 ft. lbs. (115–130 Nm) on 2.9L engine, 141–169 ft. lbs. (190–230 Nm) on 1988 3.0L engine or 107 ft. lbs. (145 Nm) on 1989–92 3.0L engine. On 4.0L engine, tighten the attaching bolt to 30–37 ft. lbs. (40–50 Nm), then tighten an additional 80–90 degrees.

7. Install the accessory drive belts. Install the radiator, if removed.

8. Connect the negative battery cable. Fill and bleed the cooling system, if necessary.

Timing Chain and Sprockets

REMOVAL & INSTALLATION

2.9L and 3.0L Engines

1. Disconnect the negative battery cable and drain the cooling system.

2. Remove the timing chain front cover.

3. Rotate the crankshaft until No. 1 cylinder is at TDC and the crankshaft and camshaft sprocket timing marks are aligned.

4. Remove the camshaft sprocket retaining bolt and remove the sprocket and timing chain.

5. Remove the crankshaft sprocket.
To install:

6. Align the crankshaft sprocket with the key or dowel on the crankshaft and install the sprocket.

7. Make sure the sprocket timing marks are still in alignment.

8. Install the camshaft sprocket and timing chain. Install the camshaft sprocket retaining bolt and tighten to 19–28 ft. lbs. (26–38 Nm) on 2.9L engine or 46 ft. lbs. (63 Nm) on 3.0L engine.

NOTE: The camshaft retaining bolt on the 3.0L engine has a drilled oil passage for timing chain lubrication. If damaged, do not replace with a standard bolt. Clean the oil passage with solvent.

9. Install the timing chain front cover and the remaining components in the reverse order of their removal. Fill the crankcase with the proper type and quantity of engine oil. Fill and bleed the cooling system. Run the engine and check for leaks.

4.0L Engine

1. Disconnect the negative battery cable and drain the cooling system and crankcase.

2. Remove the oil pan and radiator. Remove the accessory drive belt and crankshaft damper.

3. Remove the water pump and timing chain front cover.

4. Remove the camshaft sprocket retaining bolt and the crankshaft sprocket key. Remove the sprockets with the timing chain.
To install:

5. Install the timing chain guide to the cylinder block with the pin of the guide inserted into the oil hole in the block. Install the 2 retaining bolts and tighten to 7–9 ft. lbs. (10–12 Nm).

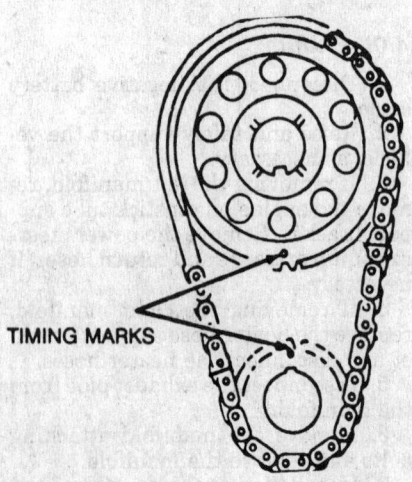

TIMING MARKS

Timing chain and sprocket alignment—3.0L engine

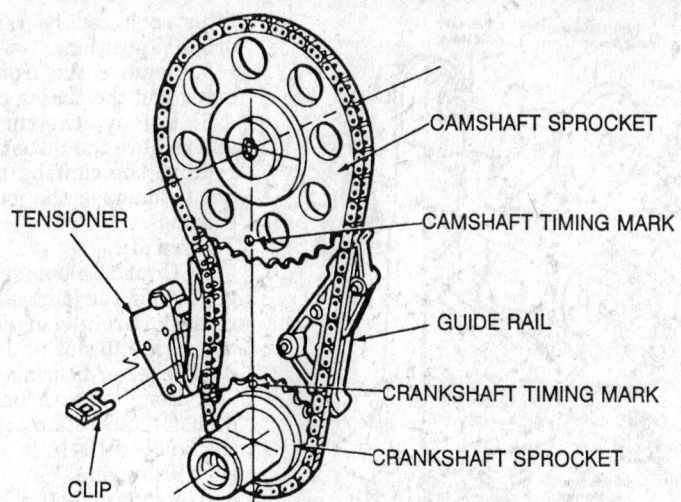

TENSIONER

CLIP

CAMSHAFT SPROCKET

CAMSHAFT TIMING MARK

GUIDE RAIL

CRANKSHAFT TIMING MARK

CRANKSHAFT SPROCKET

Timing chain and sprocket alignment—4.0L engine

6. Position the camshaft and crankshaft so the sprocket timing marks will align.

7. Install the sprockets and timing chain together. Install the timing chain tensioner with the clip in place to lock the tensioner in the retracted position.

8. Install the crankshaft key and check the timing marks on the sprockets for correct alignment. Make sure the tensioner side of the timing chain is held inward and the guide side of the chain is straight and tight.

9. Install the camshaft sprocket retaining bolt and tighten to 44–50 ft. lbs. (60–68 Nm). Remove the clip from the tensioner assembly.

10. Install the timing chain front cover and the remaining components in the reverse order of their removal. Fill the crankcase with the proper type and quantity of engine oil. Fill and bleed the cooling system. Run the engine and check for leaks.

Timing Belt Front Cover

REMOVAL & INSTALLATION

1. Disconnect the negative battery cable and drain the cooling system.

2. Loosen the thermactor pump bolts and remove the drive belt, if equipped.

3. Remove the fan blade and 4 water pump pulley bolts.

4. Loosen the alternator retaining bolts and remove the drive belt from the pulleys. Remove the upper radiator hose.

5. Remove the crankshaft pulley bolt and pulley. Remove the thermostat housing.

6. Loosen the power steering pump mounting bracket and position aside.

7. Remove the timing belt front cover retaining bolt(s). Release the cover interlocking tabs, if equipped. Remove the cover.
To install:

8. Install the front cover. If equipped, secure by snapping the interlocking tabs into place. Install the retaining bolt(s).

9. Install the power steering pump mounting bracket.

10. Install the thermostat housing and connect the upper radiator hose.

11. Install the crankshaft pulley and retaining bolt. Tighten to 103–133 ft. lbs. (140–180 Nm).

12. Position the alternator drive belt and adjust the belt tension. Install the water pump pulley and fan.

13. Position the thermactor pump drive belt, if equipped, and adjust the tension.

14. Connect the negative battery cable. Fill and bleed the cooling system. Run the engine and check for leaks.

OIL SEAL REPLACEMENT

1. Disconnect the negative battery cable.

2. Remove the timing belt front cover, timing belt and sprockets.

3. Use seal removal tool T74P-6700–B or equivalent, to remove the crankshaft, camshaft or auxiliary shaft seals. Make sure the jaws of the tool are gripping the thin edge of the seal very tightly before operating the jack-screw portion of the tool.
To install:

4. Coat the new seal with engine oil and install, using seal installation tool T74P-6150–A or equivalent.

5. Install the timing sprockets, timing belt and timing belt front cover. Connect the negative battery cable.

Timing Belt and Tensioner

REMOVAL & INSTALLATION

1. Disconnect the negative battery cable.

2. Remove the timing belt front cover.

3. Loosen the belt tensioner adjustment screw. Position belt tension adjusting tool T74P-6254–A or equivalent, on the tension spring rollpin and retract the belt tensioner. Tighten the adjustment screw to hold the tensioner in the retracted position.

4. On 1989–92 vehicles, remove the bolts holding the timing sensor in place and pull the sensor free of the dowel pin.

5. Remove the crankshaft pulley, hub and belt guide. Remove the timing belt.

6. If the timing belt tensioner is to be removed, remove the adjustment screw and the spring bolt and remove the tensioner.
To install:

7. If removed, install the timing belt tensioner. Install the spring bolt but do not tighten at this time. Position the tensioner in the fully retracted position and tighten the adjustment bolt.

8. Position the crankshaft sprocket to align with the TDC mark and the camshaft sprocket to align with the timing pointer. On 1988 vehicles, remove the distributor cap and set the distributor rotor to the No. 1 firing position by turning the auxiliary shaft.

9. Install the timing belt over the crankshaft sprocket and then counterclockwise over the auxiliary and camshaft sprockets. Align the belt fore and aft on the sprockets.

10. Loosen the tensioner adjustment bolt to allow the tensioner to move against the belt. If the spring does not have enough tension to move the roller against the belt and the belt hangs loose, it may be necessary to manually push the roller against the belt and tighten the bolt.

NOTE: The spring cannot be used to set belt tension. A wrench must be used on the tensioner assembly.

11. Remove a spark plug from each cylinder to make sure the engine does not jump time during Step 12.

12. Rotate the crankshaft 2 complete turns in the direction of normal rotation to remove the slack from the belt. Tighten the spring bolt to 28–40 ft. lbs. (38–54 Nm) and the adjustment bolt to 14–21 ft. lbs. (19–29 Nm).

13. Install the crankshaft belt guide.

14. On 1989–92 vehicles, proceed as follows:

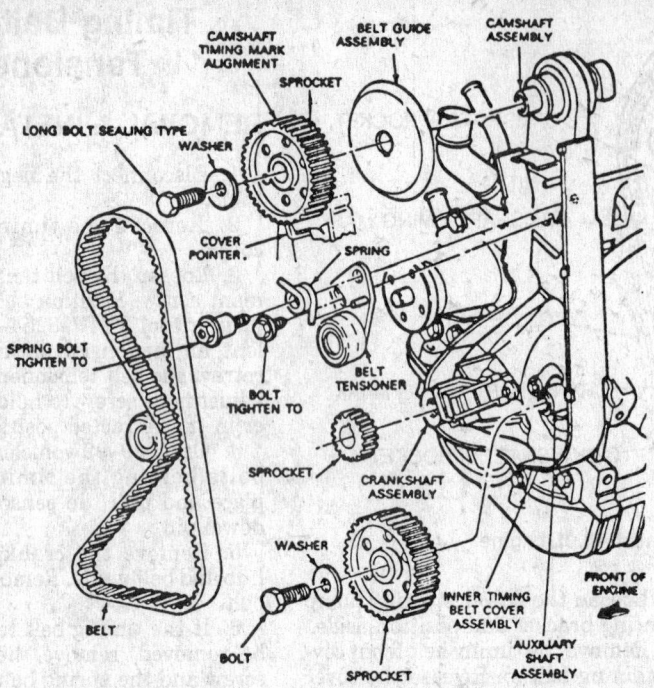

Timing belt, tensioner and sprockets installation—2.0L and 2.3L engines

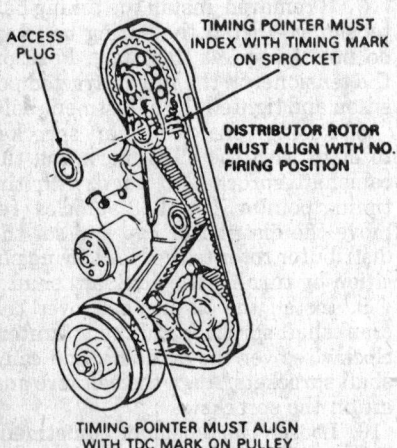

Timing belt and sprockets alignment—2.0L and 2.3L engines

a. Install the timing sensor onto the dowel pin and tighten the 2 longer bolts.

b. Rotate the crankshaft 45 degrees counterclockwise and install the crankshaft pulley and hub. Tighten the pulley bolt to 103–133 ft. lbs. (140–180 Nm).

c. Rotate the crankshaft 90 degrees clockwise so the vane of the crankshaft pulley engages with timing sensor positioner tool T89P-6316-A or equivalent. Tighten the 2 shorter sensor bolts.

d. Rotate the crankshaft 90 degrees counterclockwise and remove the sensor positioner tool.

e. Rotate the crankshaft 90 degrees clockwise and measure the outer vane-to-sensor air gap. The air

gap must be 0.018–0.039 in. (0.458–0.996mm).

15. Install the timing belt front cover and the remaining components in the reverse order of their removal.

Timing Sprockets

REMOVAL & INSTALLATION

1. Disconnect the negative battery cable.

2. Remove timing belt front cover and timing belt.

3. Remove timing sprockets retaining bolt(s). Remove the timing sprocket with a suitable puller.

4. Installation is the reverse of the remove procedure. Tighten the camshaft sprocket bolt to 52–70 ft. lbs. (70–95 Nm). Tighten the auxiliary shaft sprocket bolt to 30–40 ft. lbs. (40–54 Nm).

Camshaft

REMOVAL & INSTALLATION

Except 2.0L and 2.3L Engines

1. Disconnect the negative battery cable and relieve the fuel system pressure. Drain the crankcase and the cooling system.

2. Remove the rocker arm covers, rocker arms or rocker arm shaft assemblies and pushrods. Note the position of each component so it can be reinstalled in the same place.

3. Remove the intake manifold.

4. Remove the lifters. Identify each

lifter so it can be reinstalled in the original position.

5. Remove the front timing chain cover and the timing chain and gears.

6. Remove the thrust plate bolts and remove the thrust plate. Carefully remove the camshaft, being careful not to damage the journals, lobes or bearings.

To install:

7. Coat the camshaft lobes with grease and the journals with heavy engine oil. Carefully install the camshaft, being careful not to damage the journals, lobes or bearings.

8. Install the thrust plate and the thrust plate retaining bolts. Tighten the bolts to 13–16 ft. lbs. (17–21 Nm) on 2.9L engine, 7 ft. lbs. (10 Nm) on 3.0L engine or 7–10 ft. lbs. (10–13 Nm) on 4.0L engine.

9. Check the camshaft endplay using a dial indicator. The endplay should be 0.0008–0.004 in. on 2.9L and 4.0L engines or should not exceed 0.007 in. on 3.0L engine.

10. Install the remaining components in the reverse order of their removal. Fill the crankcase with the proper type and quantity of engine oil. Fill and bleed the cooling system. Run the engine and check for leaks.

2.0L and 2.3L Engines

1. Disconnect the negative battery cable and drain the cooling system. Remove the air cleaner assembly.

2. Tag and disconnect the spark plug wires at the plugs and rocker arm cover and position aside. Tag and disconnect the necessary vacuum lines.

3. Remove the rocker arm cover. Remove the alternator mounting bracket-to-cylinder head mounting bolts and position aside.

4. Disconnect and remove the upper radiator hose. Remove the radiator shroud.

5. Remove the timing belt front cover. If equipped with power steering, re-

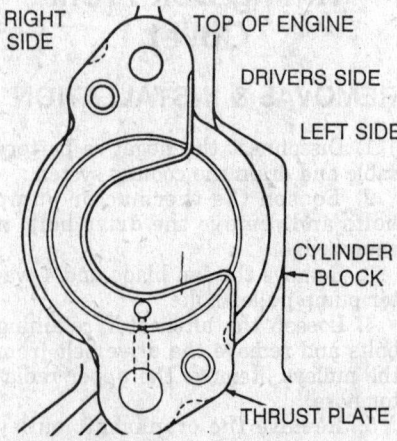

Camshaft thrust plate positioning— 2.9L engine

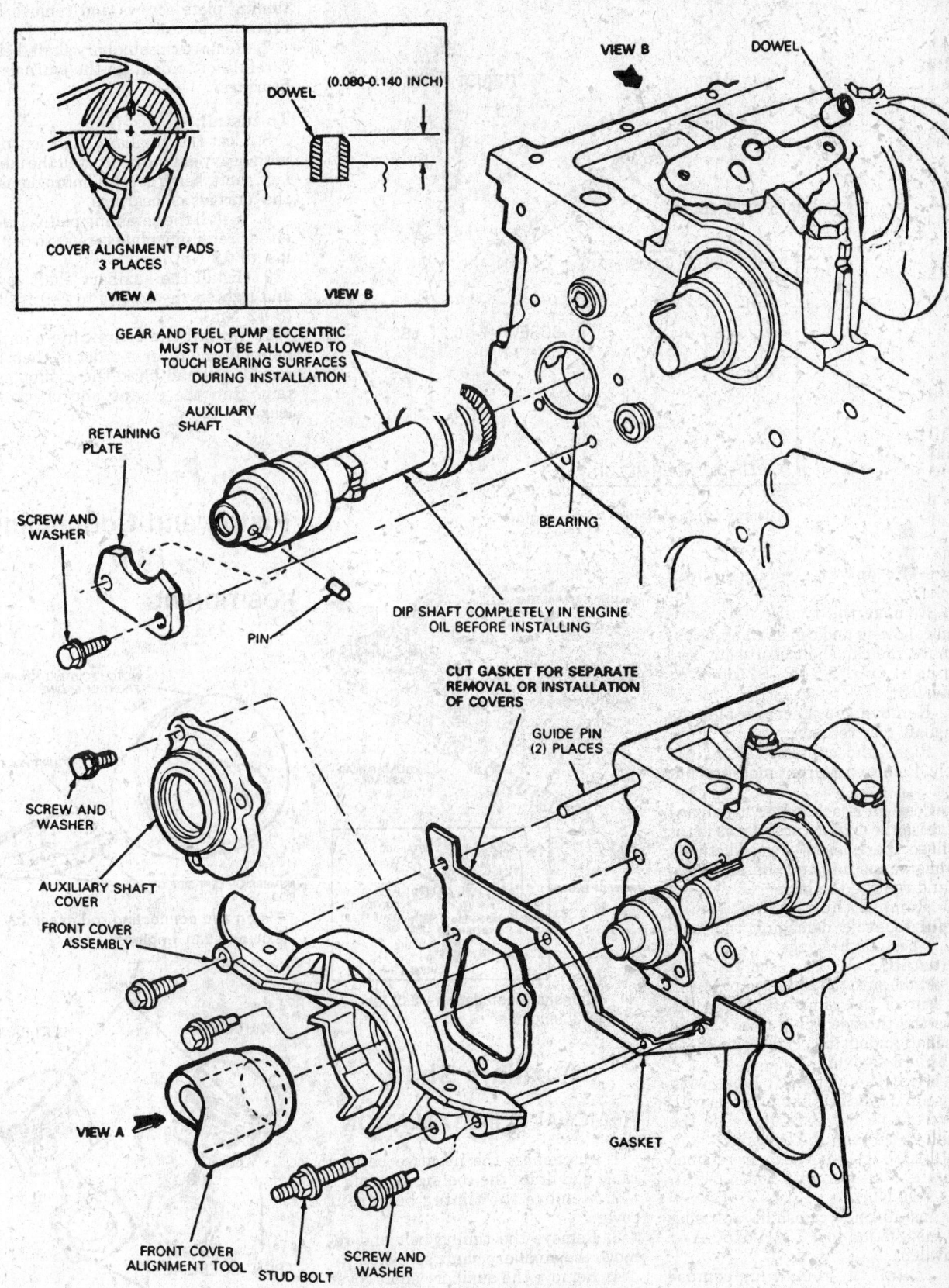

COVER ALIGNMENT PADS
3 PLACES

VIEW A

DOWEL (0.080-0.140 INCH)

VIEW B

VIEW B

DOWEL

GEAR AND FUEL PUMP ECCENTRIC
MUST NOT BE ALLOWED TO
TOUCH BEARING SURFACES
DURING INSTALLATION

RETAINING
PLATE

AUXILIARY
SHAFT

SCREW AND
WASHER

PIN

BEARING

DIP SHAFT COMPLETELY IN ENGINE
OIL BEFORE INSTALLING

CUT GASKET FOR SEPARATE
REMOVAL OR INSTALLATION
OF COVERS

GUIDE PIN
(2) PLACES

SCREW AND
WASHER

AUXILIARY SHAFT
COVER

FRONT COVER
ASSEMBLY

VIEW A

GASKET

FRONT COVER
ALIGNMENT TOOL

STUD BOLT

SCREW AND
WASHER

Auxiliary shaft installation—2.0L and 2.3L engines

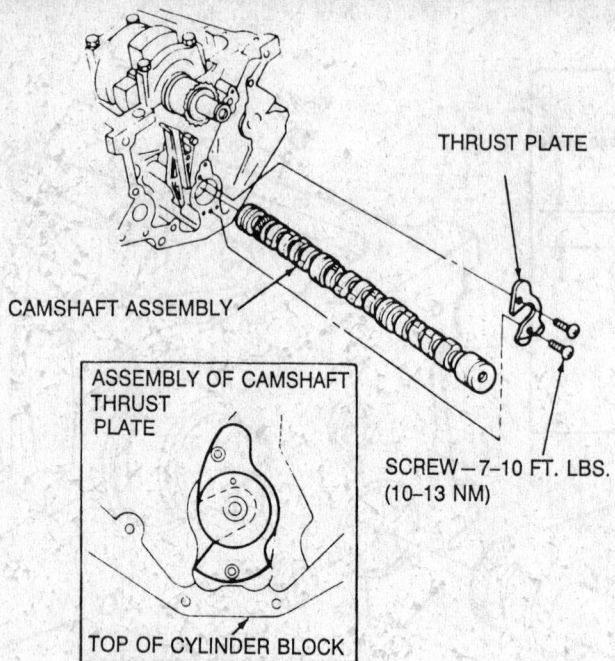

Camshaft installation—4.0L engine

6. Remove the auxiliary shaft retaining plate screws and remove the retaining plate.

7. Remove the auxiliary shaft, being careful not to damage the journals or bearings.

To install:

8. Coat the auxiliary shaft journals with heavy engine oil. Install the auxiliary shaft, being careful not to damage the journals or bearings.

9. Install the retaining plate. Tighten the retaining plate screws to 6–9 ft. lbs. (8–12 Nm).

10. Install the auxiliary shaft cover and tighten the screws to 6–9 ft. lbs. (8–12 Nm).

11. Install the remaining components in the reverse order of their removal. Fill and bleed the cooling system. Run the engine and check for leaks.

move the power steering pump bracket.

6. Remove the timing belt, camshaft followers and camshaft sprocket. Remove the camshaft seal using seal removal tool T74P-6700-B or equivalent.

7. Remove the 2 screws and the camshaft rear retainer.

8. Raise and support the vehicle safely. Remove the front motor mount bolts.

9. Position a jack under the engine and raise the engine carefully as far as it will go. Place blocks of wood between the engine mounts and chassis brackets and remove the jack.

10. Remove the camshaft, being careful to avoid damaging the journals, lobes and bearings.

To install:

11. Make sure the threaded plug is in the rear of the camshaft. If not, remove the threaded plug from the old camshaft and install. Tighten to 12–18 ft. lbs. (16–24 Nm).

12. Coat the camshaft lobes with grease and lubricate the journals with heavy engine oil. Carefully slide the camshaft through the bearings.

13. Install the camshaft rear retainer with the 2 screws. Tighten to 6–9 ft. lbs. (8–12 Nm).

14. Install a new camshaft seal using seal installation tool T74P-6150-A or equivalent.

15. Install the remaining components in the reverse order of their removal. Fill and bleed the cooling system. Run the engine and check for leaks.

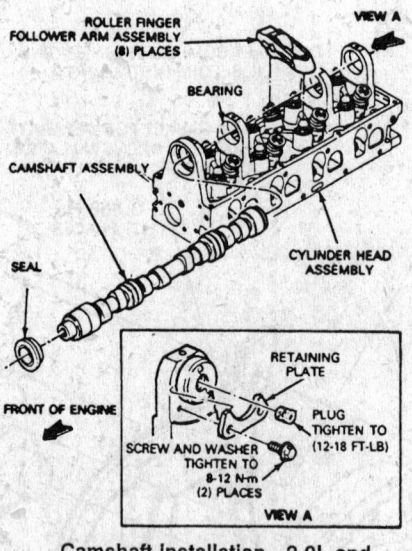

Camshaft installation—2.0L and 2.3L engines

Auxiliary Shaft

REMOVAL & INSTALLATION

1. Disconnect the negative battery cable and drain the cooling system.

2. Remove the timing belt front cover.

3. Remove the timing belt and remove the auxiliary shaft sprocket.

4. Remove the auxiliary shaft cover bolts and the cover.

5. On 1988 vehicles, remove the distributor assembly.

Piston and Connecting Rod

POSITIONING

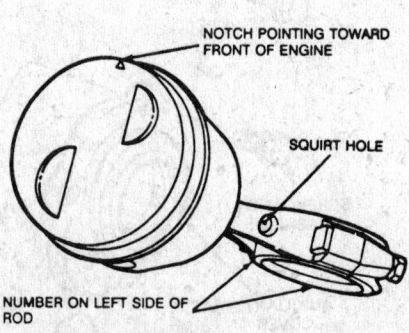

Piston and connecting rod assembly— 2.0L and 2.3L engines

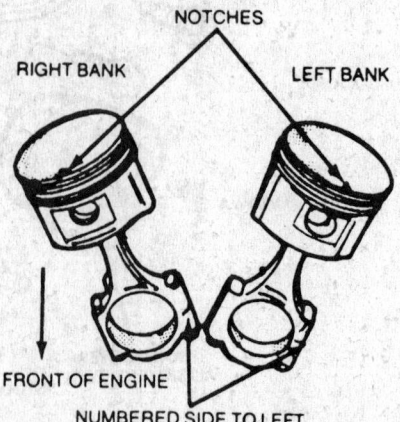

Piston and connecting rod assembly— 2.9L engine

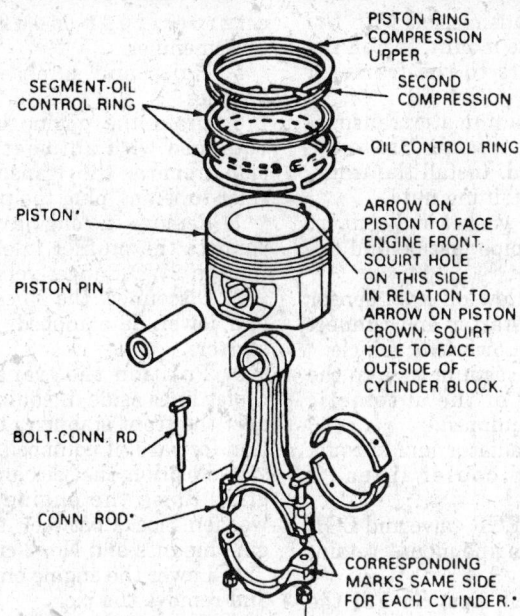

*RIGHT CYLINDER BANK PISTON AND ROD
ASSEMBLY SHOWN. LEFT BANK ASSEMBLY
IS SYMMETRICALLY OPPOSITE

Piston and connecting rod assembly—4.0L engine

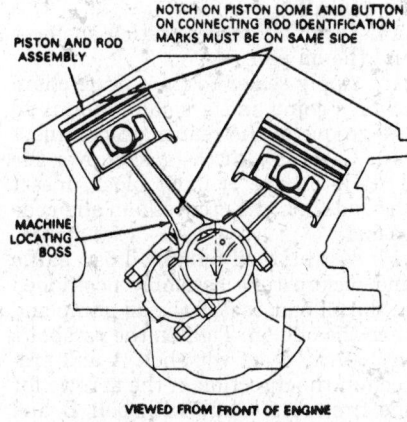

**Piston and connecting rod assembly—
3.0L engine**

ENGINE
LUBRICATION

Oil Pan

REMOVAL & INSTALLATION

2.0L Engine

1. Disconnect the negative battery cable and remove the air cleaner assembly.

2. Remove the engine oil dipstick and remove the engine mount retaining nuts.

3. Remove the fan shroud. Raise and safely support the vehicle.

4. Disconnect the exhaust manifold tube to the inlet pipe bracket below the thermactor check valve. Drain the engine oil.

5. Disconnect the starter cable and remove the starter. Disconnect the right front lower shock absorber mount.

6. Position a jack under the engine. Raise the engine and position suitable wood blocks between the engine mounts and frame brackets. Remove the jack.

7. Remove the oil pan retaining bolts and lower the pan to the crossmember. Remove the oil pump drive and pickup tube assembly and remove the oil pan out the rear of the vehicle.

To install:

8. Clean all gasket mating surfaces and the oil pan.

9. Apply sealer to the oil pan surfaces and install the left and right gaskets to the oil pan. Apply sealer to the front and rear bearing caps and install the front and rear seals. Press the seal tabs firmly into the block.

10. Install 2 guide pins and position the oil pan under the block so it rests on the crossmember. Install the oil pump drive and pickup tube assembly and install the oil pan to the block.

11. Position the jack under the engine. Raise the engine, remove the wood blocks and lower the engine into position.

12. Install the right front lower shock absorber mount. Install the starter and connect the starter cable.

13. Install a new oil filter. Install the exhaust manifold tube to the inlet pipe bracket located below the thermactor check valve.

14. Remove the jack and lower the vehicle.

15. Install the fan shroud and the engine mount retainer nuts.

16. Install the engine oil dipstick and the air cleaner assembly.

17. Fill the crankcase with the proper type and quantity of engine oil. Connect the negative battery cable.

18. Run the engine and check for leaks.

2.3L Engine

1. Disconnect the negative battery cable and remove the air cleaner outlet tube at the throttle body.

2. Remove the engine oil dipstick and remove the engine mount retaining nuts.

3. Disconnect the oil cooler lines at the radiator, if equipped. Remove the fan shroud.

4. If equipped with automatic transmission, remove the radiator retaining bolts and position the radiator upward and wire to the hood.

5. Raise and safely support the vehicle. Drain the engine oil.

6. Disconnect the starter cable and remove the starter. Disconnect the exhaust manifold tube to the inlet pipe bracket at the thermactor check valve.

7. Disconnect the catalytic converter at the inlet pipe.

8. Remove the insulator and retainer assembly at the transmission. Remove the transmission mount retaining nuts to the crossmember.

9. If equipped with automatic transmission, remove the oil cooler lines from the retainer at the block and remove the front crossmember.

10. If equipped with manual transmission, disconnect the right front lower shock absorber mount.

11. Position a jack under the engine. Raise the engine and position suitable wood blocks between the engine mounts and frame brackets. Remove the jack.

12. If equipped with automatic transmission, position a jack under the transmission and raise slightly.

13. Remove the oil pan retaining bolts and lower the pan to the chassis. Remove the low oil level sensor assembly and the oil pump drive and pickup tube assembly.

14. If equipped with automatic transmission, remove the oil pan out the front of the vehicle. If equipped

with manual transmission, remove the oil pan out from the rear.

To install:

15. Clean all gasket mating surfaces, the oil pan, oil pump exterior and pick-up tube screen.

16. Install the low oil level sensor assembly and tighten to 20–30 ft. lbs. (27–41 Nm).

17. Press a new gasket into the oil pan groove. Retain the gasket in the oil pan by press fit only.

18. Position the oil pan on the crossmember. Install the oil pump drive and pickup tube assembly.

19. Apply sealer in 6 places on the engine and install the oil pan. Install the oil pan flange bolts tight enough to compress the gasket to the point that the 2 transmission holes are aligned with the 2 tapped holes in the oil pan, but loose enough to allow movement of the pan relative to the block.

20. Install the 2 oil pan-to-transmission bolts and tighten to 30–39 ft. lbs. (40–50 Nm) to align the oil pan with the transmission, then loosen the bolts ½ turn.

21. Tighten all oil pan flange bolts to 90–120 inch lbs. (10–13.5 Nm), then retighten the 2 oil pan-to-transmission bolts to 30–39 ft. lbs. (40–50 Nm).

22. Install a new oil filter. Position a jack under the engine and raise it enough to remove the wood blocks. Shift the engine/transmission backward to its original position.

23. Install the mount/bracket assembly to the crossmember and lower the engine. Install the front crossmember, if removed.

24. Raise the transmission with the jack and install the mount. Install the stabilizer brackets to the frame, if removed.

25. Connect the automatic transmission oil cooler line retainer clip to the engine, if equipped. Install the transmission mount retaining nuts.

26. Install a new gasket and connect the rear exhaust pipe just behind the catalytic converter.

27. Connect the low oil level sensor wire. Install the starter and connect the starter cable. Lower the vehicle.

28. Connect the vacuum tube to the clip at the front of the automatic transmission, if equipped.

29. Install the radiator and shroud. Connect the oil cooler lines, if equipped.

30. Connect the EGR valve and EGR tube. Install the engine mount retaining nuts.

31. Install the oil dipstick. Fill the crankcase with the proper type and quantity of engine oil.

32. Connect the negative battery cable, start the engine and check for leaks.

2.9L Engine

1. Disconnect negative battery cable and remove the air intake tube.

2. Remove the fan shroud and position over the fan.

3. Remove the distributor cap and position forward of the dash panel. Remove the distributor and cover the bore opening.

4. Remove the nuts attaching the engine front mounts to the crossmember.

5. Raise and safely support the vehicle.

6. Drain the engine crankcase. If equipped with automatic transmission, remove the transmission fluid filler tube and plug the pan hole.

7. Remove the engine oil filter. Disconnect the muffler inlet pipe(s), except on 2WD Ranger vehicles.

8. Disconnect the oil cooler bracket and lower, if equipped. Remove the starter.

9. Position the transmission oil cooler lines aside, if equipped. Disconnect the front stabilizer bar and position forward, if equipped.

10. Position the jack under the engine. Raise the engine and install wooden blocks between the front engine mounts and No. 2 crossmember.

11. Lower the engine onto the blocks and remove the jack.

12. Remove the oil pan attaching bolts and lower the oil pan.

13. Remove the oil pump and pickup tube assembly, attached to the bearing cap. Lower into the oil pan, except on 2WD Ranger. Remove the oil pan.

To install:

14. Clean all gasket mating surfaces and the oil pan.

15. Apply sealer to the timing chain cover T-joint and a small amount to the groove in the rear main bearing.

16. Carefully lift the gasket over the 6 studs on the cylinder block, insert the ends into the grooves and align the gasket.

17. With the oil pump, oil pan baffle and pickup tube assembly positioned in the oil pan, install the oil pump and then the oil pan. Tighten the pan bolts in 2 steps. Start with bolt **A** and proceed in the direction of the arrows for the first step. Start with bolt **B** and proceed in the direction of the arrows for the second step. Final torque should be 4–6 ft. lbs. (5–8 Nm).

18. Position a jack under the engine and raise enough to remove the wooden blocks. Lower the engine and remove the jack.

19. Install the starter. Connect the muffler inlet pipes, if removed.

20. Connect the front stabilizer bar, if equipped. Position the transmission oil cooler lines and connect the cooler bracket, if equipped.

21. Install a new oil filter. If equipped with automatic transmission, unplug the oil pan and install the filler tube.

22. Lower the vehicle. Install the nuts attaching the engine front mounts to the crossmember.

23. Install the distributor assembly and cap. Install the fan shroud.

24. Fill the engine with the proper

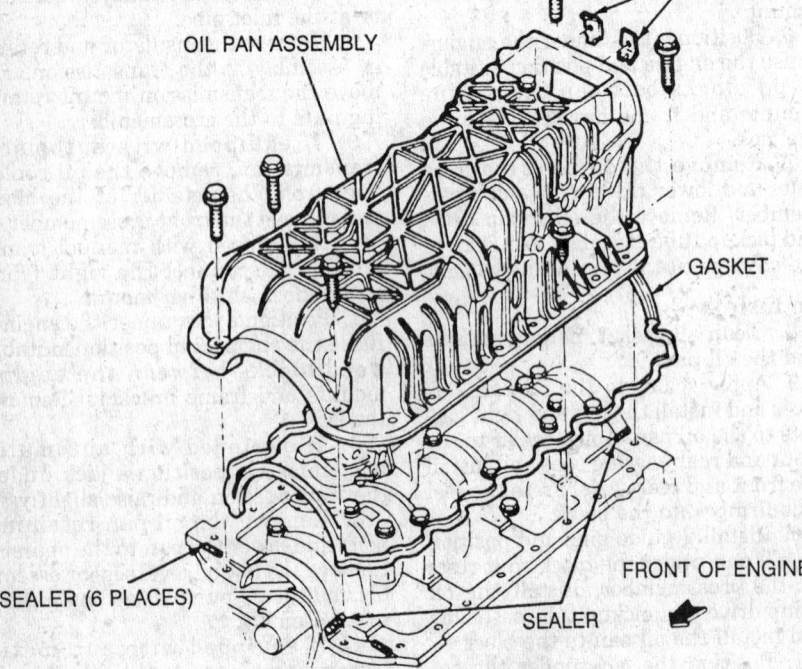

OIL PAN ASSEMBLY

SPACERS

GASKET

FRONT OF ENGINE

SEALER (6 PLACES)

SEALER

Oil pan installation—2.3L engine

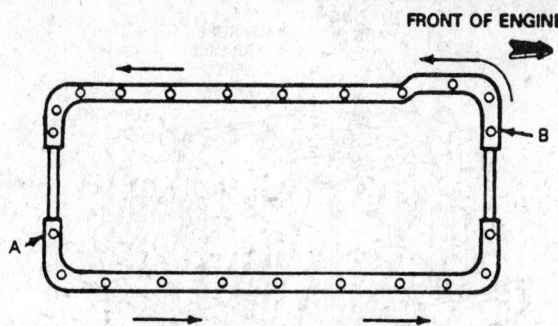

FRONT OF ENGINE

B

A

Oil pan bolt tightening procedure—2.9L engine

type and quantity of engine oil. Connect the negative battery cable.

25. Run the engine and check for leaks. Check the ignition timing and adjust, if necessary.

26. Install the air intake tube. Check the transmission fluid level.

3.0L Engine

AEROSTAR

1. Disconnect the negative battery cable.

2. Remove the engine oil level dipstick. Raise the vehicle and support it safely.

3. If equipped with an oil level sensor, remove the retaining clip at the sensor. Disconnect the electrical connector from the sensor.

4. Drain the crankcase. Remove the starter and transmission inspection cover.

5. On 1988 vehicles, loosen the transmission bolts and slide the transmission ¼ in. rearward.

6. Remove the oil pan attaching bolts and remove the pan.

To install:

7. Clean all gasket mating surfaces and the oil pan.

8. Apply a $^1/_5$ in. bead of silicone sealer to the junction of the rear main bearing cap and cylinder block and the junction of the front cover assembly and cylinder block.

9. Position the oil pan gasket to the oil pan and secure with sealer. Install the oil pan on the engine block with the attaching bolts and tighten to 9 ft. lbs. (12 Nm).

10. On 1988 vehicles, tighten the transmission bolts.

11. Install the starter and transmission inspection cover. Attach the low oil level sensor connector and install the retainer clip.

12. Lower the vehicle and install the engine oil level dipstick. Fill the crankcase with the proper type and quantity of engine oil.

13. Connect the negative battery cable, start the engine and check for leaks.

RANGER

1. Disconnect the negative battery

cable. Remove the engine oil level dipstick.

2. Disconnect the fan shroud and drape it over the fan. Remove the motor mount nuts from the frame.

3. Mark the position of the distributor rotor in relation to the distributor housing and the position of the housing in relation to the engine. Remove the distributor.

4. Raise and safely support the vehicle. Remove the low oil level sensor retainer clip at the sensor. Disconnect the electrical connector from the sensor.

5. Drain the crankcase and remove the starter. Remove the transmission inspection cover.

6. Remove the right axle beam on 2WD vehicles.

NOTE: The brake caliper must be removed and wired out of the way.

7. Remove the oil pan bolts. Position a jack under the engine and raise it approximately 2 in. Remove the oil pan.

NOTE: The oil pan fits tightly between the transmission spacer plate and oil pump pickup tube. Use care when removing to avoid damaging the pickup tube.

To install:

8. Clean all gasket mating surfaces and the oil pan.

9. Apply a $^1/_5$ in. bead of silicone sealer to the junction of the rear main bearing cap and cylinder block and the junction of the front cover assembly and cylinder block.

10. Position the oil pan gasket to the oil pan and secure with sealer. Install the oil pan on the engine block with the attaching bolts and tighten to 9 ft. lbs. (12 Nm).

11. Install the low oil level sensor connector and the retainer clip. Lower the engine assembly.

12. Install the right axle beam, if removed.

13. Install the transmission inspection cover and the starter. Lower the vehicle.

14. Install the fan shroud and the motor mount nuts. Install the distributor, aligning the marks that were made during removal.

15. Install the engine oil level dipstick. Fill the crankcase with the proper type and quantity of engine oil.

16. Connect the negative battery cable, start the engine and check for leaks.

4.0L Engine

AEROSTAR

1. Disconnect the negative battery cable.

2. Raise and safely support the vehicle. Remove the starter.

3. On 4WD vehicles, proceed as follows:

   a. Remove the front wheel and tire assemblies.

   b. Remove the pivot bolts and nuts from both lower control arms, to allow the control arms to hang.

   c. Remove the control arm rear pivot crossmember.

   d. Remove the lower nuts from both motor mounts.

   e. Remove the front drive axle assembly.

4. Drain the crankcase and remove the oil filter. Disconnect the low oil level sensor from the engine oil pan.

5. Remove the 2 transmission-to-engine oil pan bolts. Remove the oil pan retaining bolts and nuts.

6. On 4WD vehicles, raise the engine approximately 1 in.

7. Remove the oil pan.

To install:

8. Clean all gasket mating surfaces and the oil pan.

9. Place a small amount of silicone sealer on the block at the corner where the oil pan, rear seal and block mate.

10. Install a new crankshaft rear main bearing cap wedge seal. The seal should fit snugly into the sides of the rear main bearing cap.

11. Position a new oil pan gasket into the groove in the oil pan and position the 2 oil pan spacers on the oil pan locating pads.

NOTE: If the same oil pan is being reused, the existing spacers may be used. If a new pan is being installed, the pan-to-transmission gap must be measured to find the needed spacer thickness. Failure to use the correct spacer can result in improper clearance between the oil pan and transmission, resulting in oil pan damage and/or an oil leak.

12. If a new oil pan is being installed, find the correct spacer thickness as follows:

   a. Position the oil pan on the engine without the spacers and install the retaining nuts on the 4 locating studs.

b. Using a feeler gauge, measure the gap between the locating pads on the pan and the transmission converter housing.

c. If the measured gap is 0.011–0.020 in. (0.27–0.51mm), a 0.010 in. (0.254mm) spacer is required. If the measured gap is 0.021–0.029 in. (0.52–0.76mm), a 0.020 in. (0.508mm) spacer is required. If the measured gap is 0.030–0.039 in. (0.77–1.00mm), a 0.030 in. (0.762mm) spacer is required.

d. Remove the oil pan and position the correct spacers.

13. Install the oil pan with a new gasket on the engine and tighten the retaining bolts and nuts enough to compress the gasket so the transmission bolts align with the holes in the oil pan, but loose enough to allow the pan to move when the transmission bolts are installed.

14. Install the 2 transmission-to-oil pan bolts and tighten to 28–38 ft. lbs. (38–51 Nm) to align the oil pan with the transmission, then loosen the bolts ½ turn.

15. Tighten all the oil pan bolts and nuts evenly to 5–7 ft. lbs. (7–10 Nm), then retighten the 2 transmission-to-oil pan bolts to 28–38 ft. lbs. (38–51 Nm).

16. Connect the low oil level sensor. Install the drain plug and a new oil filter.

17. On 4WD vehicles, proceed as follows:

a. Lower the engine and install the lower motor mount nuts.

b. Install the front drive axle assembly.

c. Install the control arm rear pivot crossmember.

d. Reposition the lower control arms and install the pivot bolts and nuts.

e. Install the front wheel and tire assemblies.

18. Install the starter. Lower the vehicle and fill the crankcase with the proper type and quantity of engine oil.

19. Connect the negative battery cable, start the engine and check for leaks.

EXPLORER AND RANGER

1. Remove the engine assembly and install on a workstand with the oil pan facing up.

2. Remove the oil pan retaining bolts and remove the pan.

To install:

3. Clean all gasket mating surfaces and the oil pan.

4. Install a new crankshaft rear main bearing cap wedge seal. The seal should fit snugly into the sides of the rear main bearing cap.

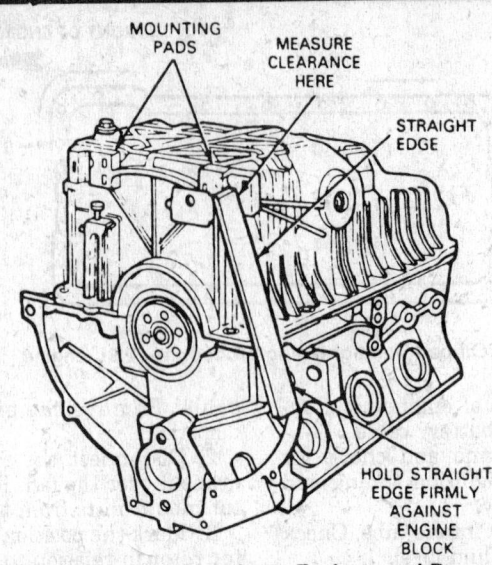

Spacer thickness measurement procedure—Explorer and Ranger with 4.0L engine

5. Position a new oil pan gasket to the engine block and place the oil pan in position on the 4 locating studs. Tighten the retaining nuts and bolts evenly to 5–7 ft. lbs. (7–10 Nm).

6. Measure the gap between the surface of the rear face of the oil pan, at the spacer locations, and the rear face of the engine block as follows:

a. With the oil pan installed on the engine, position a straight edge flat on the rear of the engine block so it extends over 1 of the oil pan/transmission bolt mounting pads.

b. Using a feeler gauge, measure the gap between the mounting pad and the straight edge. Repeat the procedure for the other mounting pad.

c. If the measured gap is 0.011–0.020 in. (0.27–0.51mm), a 0.010 in. (0.254mm) spacer is required. If the measured gap is 0.021–0.029 in. (0.52–0.76mm), a 0.020 in. (0.508mm) spacer is required. If the measured gap is 0.030–0.039 in. (0.77–1.00mm), a 0.030 in. (0.762mm) spacer is required.

d. Install the selected spacers to the mounting pads on the rear of the oil pan before bolting the engine and transmission together.

NOTE: Failure to use the correct spacer can result in improper clearance between the oil pan and transmission, resulting in oil pan damage and/or an oil leak.

7. Remove the engine from the workstand and install in the vehicle.

Oil Pump

REMOVAL & INSTALLATION

1. Disconnect the negative battery cable.

2. Remove the oil pan.

3. Remove the oil pump attaching bolts and, if equipped, remove the oil pump pickup tube retaining nut from the main bearing cap.

4. Remove the oil pump and oil pump driveshaft. Remove and clean the oil pump pickup tube and screen, as necessary.

To install:

5. Install the oil pump pickup tube and screen assembly, if removed.

6. Prime the oil pump by filling either the inlet or outlet port with clean engine oil. Rotate the pump shaft to distribute the oil within the pump body.

7. Insert the oil pump driveshaft into the opening in the block or main bearing cap. On 3.0L engine, assemble the shaft to the oil pump until the retainer clicks into place.

8. Install the oil pump, with a new gasket if equipped, and install the attaching bolts. Tighten the bolts to 14–21 ft. lbs. (19–29 Nm) on 2.0L and 2.3L engines, 6–10 ft. lbs. (9–13 Nm) on 2.9L engine, 35 ft. lbs. (48 Nm) on 3.0L engine or 13–15 ft. lbs. (17–21 Nm) on 4.0L engine.

9. On 2.0L and 2.3L engine, install the pickup tube retaining nut and tighten to 30–41 ft. lbs. (40–55 Nm).

10. Install the oil pan.

11. Fill the crankcase with the proper type and quantity of engine oil. Connect the negative battery cable, start the engine and check for leaks.

CHECKING

1. Remove the pump and disassemble. Thoroughly clean all parts in solvent and dry with compressed air.

2. Check the inside of the pump housing and the inner and outer gears for damage or excessive wear. Check the mating surfaces of the pump cover

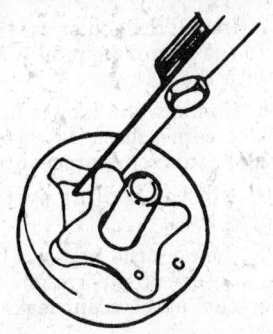

Measuring the oil pump inner-to-outer rotor tip clearance

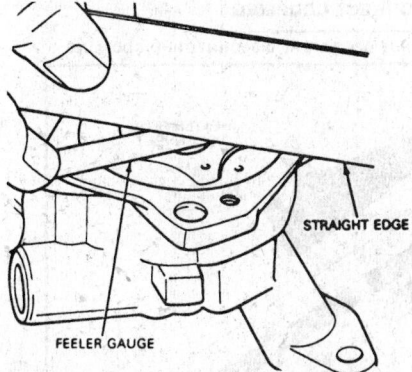

Meaasuring oil pump rotor endplay

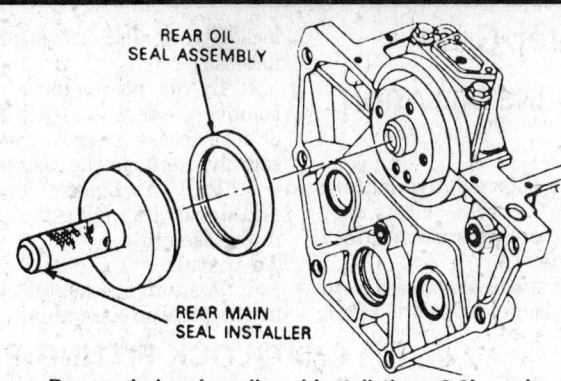

Rear main bearing oil seal installation—3.0L engine

for wear. Minor scuff marks are normal, but if the cover, gears or housing surfaces are excessively worn, scored or grooved, replace the entire pump.

3. Measure the inner to outer rotor tip clearance. With the rotor assembly removed from the pump and resting on a flat surface, the inner and outer rotor tip clearance must not exceed 0.012 in. (0.30mm) with a feeler gauge inserted ½ in. (13mm) minimum.

4. With the rotor assembly installed in the housing, place a straight edge over the rotor assembly and the housing. Measure the vertical clearance, the rotor endplay, between the straight edge and the inner rotor and outer race. Maximum clearance must not exceed 0.005 in. (0.13mm).

5. Inspect the relief valve spring for collapsed or worn condition. Check the spring tension. The tension should be 12.6–14.5 lbs. at 1.20 in. on 2.0L and 2.3L engines, 13.6–14.7 lbs. at 1.39 in. on 2.9L engine and 4.0L engines or 9.1–10.1 lbs. at 1.11 in. on 3.0L engine.

6. If any part of the oil pump requires replacement, replace the complete pump assembly.

Rear Main Bearing Oil Seal

REMOVAL & INSTALLATION

1. Disconnect the negative battery cable. Raise and safely support the vehicle.

2. Remove the transmission. If equipped with manual transmission, remove the clutch assembly.

3. Remove the flywheel and the engine rear cover plate.

4. Using a sharp awl or equivalent, punch a hole into the seal metal surface between the lip and block. Screw in the threaded end of plug puller T77L–9533–B or equivalent, and remove the seal. Be careful to avoid scratching or damaging the oil seal surface.

5. When the seal has been removed, clean the mounting recess.

To install:

6. Coat the new seal and the crankshaft with a light film of engine oil. Do not use grease.

7. Start the seal into the recess with the seal lip facing forward and install it with a suitable rear oil seal replacer tool. Keep the tool straight with the centerline of the crankshaft and install the seal until it is fully seated.

8. After removing the tool, inspect the seal to make sure it was not damaged during installation.

9. Install the engine rear cover plate. Position the flywheel on the crankshaft flange. Install the flywheel attaching bolts and tighten to 59 ft. lbs. (80 Nm) on all except 2.9L engine. Tighten the bolts on 2.9L engine to 47–52 ft. lbs. (64–70 Nm).

10. If equipped with manual transmission, install the clutch assembly. Install the transmission and starter and lower the vehicle.

11. Connect the negative battery cable, start the engine and check for leaks.

ENGINE COOLING

Radiator

REMOVAL & INSTALLATION

1. Disconnect the negative battery cable and remove the radiator cap.

— CAUTION —

Never remove the radiator cap while the engine is running or personal injury from scalding hot coolant or steam may result. If possible, wait until the engine has cooled to remove the radiator cap. If this is not possible, wrap a thick cloth around the radiator cap and turn it slowly to the first stop, to release the pressure in the cooling system. Step back while the pressure is released. After all pressure is released, remove the radiator cap completely.

2. Position a drain pan under the radiator and open the draincock to drain the radiator.

3. Disconnect the overflow hose from the radiator and the fan shroud, if necessary.

4. Remove the shroud or finger guard upper attaching screws. Lift the shroud out of the lower retaining clips and drape it on the fan.

5. Disconnect the radiator hoses from the radiator.

6. Disconnect and plug the automatic transmission oil cooling lines, if equipped.

7. Remove the radiator upper attaching screws, tilt the radiator back and lift directly upward, clear of the radiator support and cooling fan.

To install:

8. Make sure the radiator lower support rubber insulators are in place on the lower support.

9. Install the radiator, being careful to clear the fan. Make sure the mounting pins on the bottom of the radiator tanks are inserted into the holes in the lower support rubber insulators and the radiator is firmly seated on the insulators.

10. Install the radiator upper attaching screws. If equipped with automatic transmission, connect the transmission cooling lines.

11. Connect the radiator hoses to the radiator. Position the shroud in the retainer clips and install the attaching screws.

12. Connect the overflow hose and close the draincock. Connect the negative battery cable. Fill and bleed the cooling system.

Heater Core

REMOVAL & INSTALLATION

1. Disconnect the negative battery cable.
2. Drain the cooling system into a suitable container.
3. Disconnect the heater hoses from the heater core tubes. Use the snap-lock fitting disconnect procedure, if necessary.
4. In the passenger compartment, remove the screws attaching the heater core access cover to the plenum assembly. Remove the access cover.
5. Pull the heater core rearward and down, removing it from the plenum assembly.

To install:

6. Position the heater core and seal in the plenum assembly.
7. Install the heater core access cover to the plenum assembly and secure it with the screws.
8. Connect the heater hoses to the heater core tubes. Use the snap-lock fitting connection procedure.
9. Fill the cooling system to the proper level.
10. Connect the negative battery cable and check the system for proper operation and coolant leaks.

SNAP-LOCK FITTING PROCEDURES

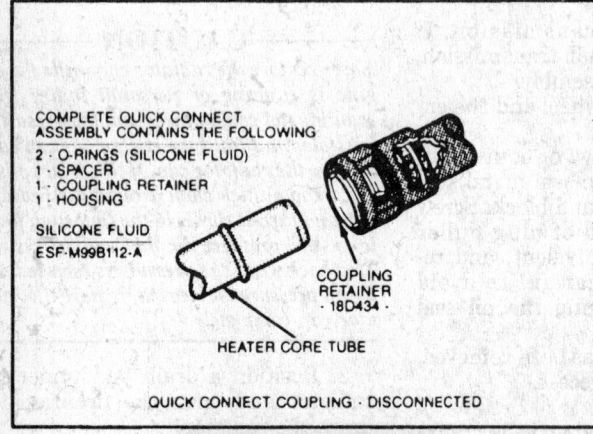

COMPLETE QUICK CONNECT ASSEMBLY CONTAINS THE FOLLOWING
2 - O-RINGS (SILICONE FLUID)
1 - SPACER
1 - COUPLING RETAINER
1 - HOUSING

SILICONE FLUID
ESF-M99B112-A

COUPLING RETAINER - 18D434

HEATER CORE TUBE

QUICK CONNECT COUPLING · DISCONNECTED

TO CONNECT COUPLING

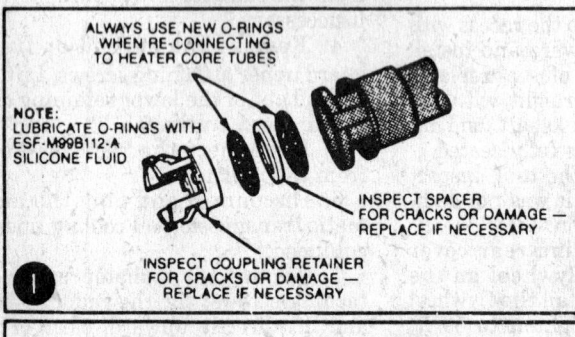

ALWAYS USE NEW O-RINGS WHEN RE-CONNECTING TO HEATER CORE TUBES

NOTE: LUBRICATE O-RINGS WITH ESF-M99B112-A SILICONE FLUID

INSPECT SPACER FOR CRACKS OR DAMAGE — REPLACE IF NECESSARY

INSPECT COUPLING RETAINER FOR CRACKS OR DAMAGE — REPLACE IF NECESSARY

①

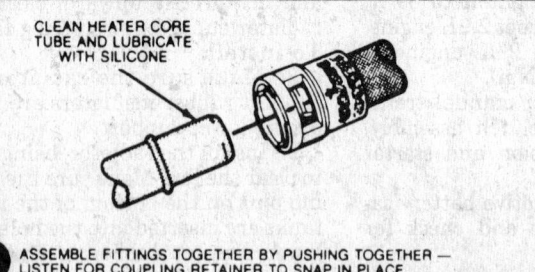

CLEAN HEATER CORE TUBE AND LUBRICATE WITH SILICONE

② ASSEMBLE FITTINGS TOGETHER BY PUSHING TOGETHER — LISTEN FOR COUPLING RETAINER TO SNAP IN PLACE

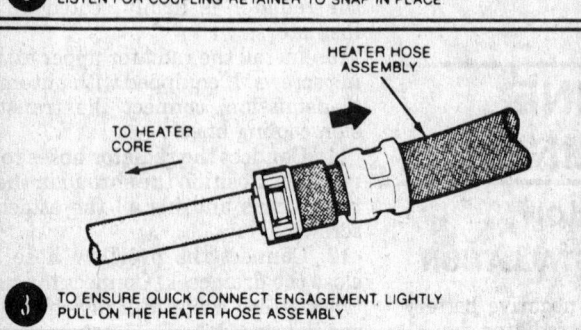

HEATER HOSE ASSEMBLY

TO HEATER CORE

③ TO ENSURE QUICK CONNECT ENGAGEMENT, LIGHTLY PULL ON THE HEATER HOSE ASSEMBLY

TO DISCONNECT COUPLING

CAUTION — ENGINE SHOULD BE OFF BEFORE DISCONNECTING COUPLING

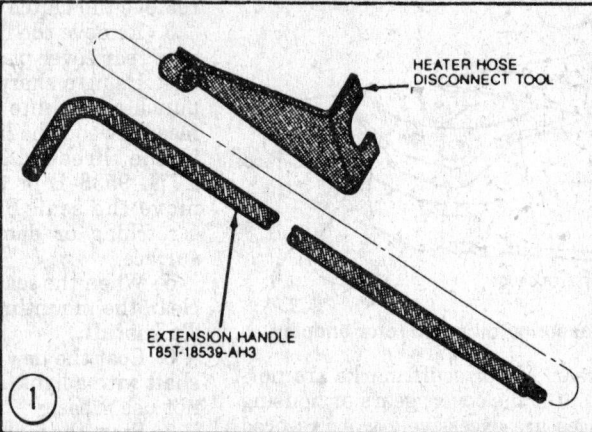

HEATER HOSE DISCONNECT TOOL

EXTENSION HANDLE T85T-18539-AH3

①

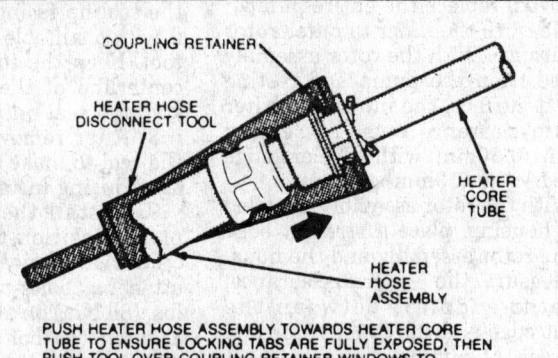

COUPLING RETAINER

HEATER HOSE DISCONNECT TOOL

HEATER CORE TUBE

HEATER HOSE ASSEMBLY

PUSH HEATER HOSE ASSEMBLY TOWARDS HEATER CORE TUBE TO ENSURE LOCKING TABS ARE FULLY EXPOSED, THEN PUSH TOOL OVER COUPLING RETAINER WINDOWS TO COMPRESS RETAINER LOCKING TABS — THEN PULL HOSE ASSEMBLY AWAY FROM HEATER CORE TUBE.
REMOVE TOOL THEN CONTINUE PULLING HOSE ASSEMBLY AWAY FROM HEATER CORE TUBE.

② NOTE: WHEN COMPRESSING WHITE COUPLING RETAINER, THE TOOL MUST BE PERPENDICULAR AND ON THE HIGHEST POINT OF THE COUPLING RETAINER AS SHOWN ABOVE.

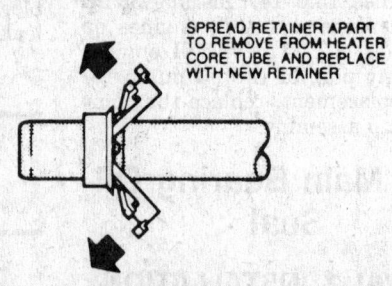

SPREAD RETAINER APART TO REMOVE FROM HEATER CORE TUBE, AND REPLACE WITH NEW RETAINER

③ WHEN THE QUICK CONNECT COUPLING IS DISCONNECTED — THE WHITE COUPLING RETAINER WILL REMAIN ON THE HEATER CORE TUBE INSTALL NEW COUPLING RETAINER, SPACER & NEW LUBRICATED O-RINGS INTO QUICK CONNECT ASSEMBLY HOUSING BEFORE RE-INSTALLING HEATER HOSE ASSEMBLY TO HEATER CORE TUBES.

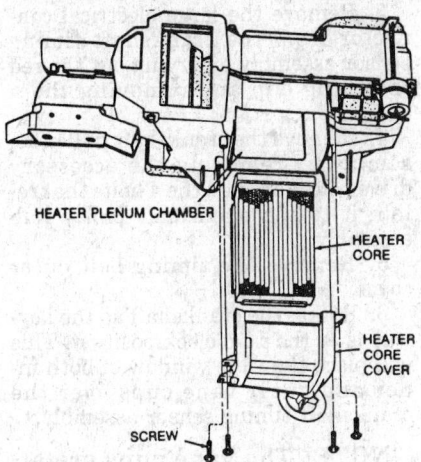

Heater core Installation—Explorer, 1990–92 Ranger and 1990 Bronco II

Water Pump

REMOVAL & INSTALLATION

2.0L and 2.3L Engines

1. Disconnect the negative battery cable and drain the cooling system.
2. Remove the bolts that retain the fan shroud and position the shroud back over the fan.
3. Remove the 4 bolts that retain the cooling fan, then remove the fan and shroud.
4. Remove the accessory drive belts.
5. Remove the water pump pulley and the vent hose to the canister.
6. Remove the heater hose at the water pump.
7. Remove the timing belt cover. Remove the lower radiator hose from the water pump.
8. Remove the water pump mounting bolts and the water pump. Clean all gasket mating surfaces.
9. Installation is the reverse of the removal procedure. Coat the threads of the mounting bolts with sealer before installation and tighten to 14–21 ft. lbs. (20–30 Nm). Fill and bleed cooling system.

2.9L and 4.0L Engines

1. Disconnect the negative battery cable and drain the cooling system.
2. Remove the lower radiator hose and the heater return hose from the pump.
3. Remove the fan and clutch assembly using fan clutch pulley holder T83T–6312–A and fan clutch nut wrench T83T–6312–B or equivalents. The fan clutch nut has left hand thread; remove by turning clockwise.
4. Loosen the alternator mounting bolts and remove the belt. If equipped with air conditioning, remove the alternator and bracket.
5. Remove the water pump pulley.

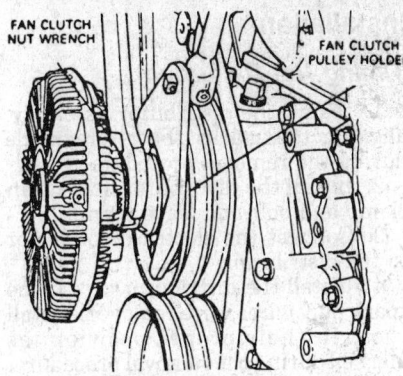

Fan and clutch assembly removal— 2.9L and 4.0L engines

6. Remove the water pump attaching bolts and remove the water pump. Note the length of the bolts when removing, so they can be reinstalled in the same positions.
7. Clean all gasket mating surfaces and install in the reverse order of removal. Tighten the water pump retaining bolts to 7–9 ft. lbs. (9–12 Nm) and the fan clutch nut to 30–100 ft. lbs. (40–135 Nm). Fill and bleed the cooling system.

3.0L Engine

1. Disconnect the negative battery cable and drain the cooling system.
2. Loosen the nut that attaches the fan clutch to the water pump shaft using a 22mm wrench. The nut has left-hand thread and must be turned clockwise to remove. Remove the fan and clutch assembly.
3. Loosen the 4 water pump pulley bolts, then remove the accessory drive belts. Remove the water pump pulley.
4. Remove the alternator adjusting arm and brace. Disconnect the heater hose and lower radiator hose from the pump.
5. Rotate the belt adjuster out of the way.
6. Remove the water pump attaching bolts, noting their positions for reinstallation and remove the water pump. Clean all gasket mating surfaces.
7. Installation is the reverse of the removal procedure. Apply sealer to the 8mm mounting bolt at the extreme passenger side, prior to installation. Tighten the 6mm mounting bolts to 7 ft. lbs. (10 Nm) and the 8mm mounting bolts to 19 ft. lbs. (25 Nm). Tighten the fan clutch nut to 30–100 ft. lbs. (40–135 Nm). Fill and bleed the cooling system.

Thermostat

REMOVAL & INSTALLATION

1. Disconnect the negative battery cable and drain the cooling system.

2. Disconnect the upper radiator hose and, if equipped, the heater hose from the water outlet.
3. Remove the water outlet retaining bolts and remove the water outlet. Remove the thermostat from the water outlet.
To install:
4. Clean all gasket mating surfaces.
5. Apply gasket sealer to a new water outlet gasket and install on the cylinder head or intake manifold.
6. Install the thermostat in the water outlet with the bridge section toward the radiator hose. Turn the thermostat clockwise to lock it in position on the flats cast into the water outlet.

NOTE: If the water outlet is equipped with a heater outlet tube opening, check that the full width of the opening is visible within the thermostat port in the assembly. The correct port alignment is required to provide maximum coolant flow to the heater.

7. Install the water outlet with the mounting bolts. Tighten the bolts to 14–21 ft. lbs. (20–30 Nm) on 2.0L and 2.3L engines, 7–10 ft. lbs. (9–13 Nm) on 2.9L and 4.0L engines or 18 ft. lbs. (25 Nm) on the 3.0L engine.
8. Connect the radiator hose and, if equipped, the heater hose to the water outlet.
9. Connect the negative battery cable. Fill and bleed the cooling system.

Cooling System Bleeding

When the entire cooling system is drained, the following procedure should be used to remove air from the cooling system and ensure a complete fill.

1. Close the radiator draincock and install the cylinder block drain plug, if removed.
2. Fill the cooling system with a 50/50 mixture of anti-freeze and water. Allow several minutes for trapped air to escape. When filling a cross flow radiator, allow time for the coolant to flow through the radiator tubes to the other end of the tank to ensure the radiator is full.
3. Install the radiator cap to the pressure relief position by installing the cap to the fully installed position and then backing off to the 1st stop. This will allow any air to escape and will minimize spillage.
4. Slide the heater temperature and mode selection levers to the maximum heat position.
5. Start the engine and allow to operate at fast idle for approximately 3–4 minutes.
6. With the engine shut off, wrap

the radiator cap with a thick cloth, carefully remove the cap and add coolant to bring the coolant level up to the filler neck seat.

7. Install the cap to the fully installed position. Then, back off to the 1st stop and operate the engine at fast idle until the thermostat opens and the upper radiator hose is warm. To check the coolant level, shut the engine off, wrap the cap with a thick cloth and cautiously remove the cap. Add additional coolant, if necessary. Install the cap to the fully installed position.

—— CAUTION ——

To avoid personal injury from scalding hot coolant or steam blowing out of the radiator, use extreme care when removing the cap from a hot radiator.

8. Fill the coolant recovery reservoir to the proper level with a 50/50 mix of anti-freeze and water.

ENGINE ELECTRICAL

NOTE: Disconnecting the negative battery cable on some vehicles may interfere with the functions of the on board computer systems and may require the computer to undergo a relearning process, once the negative battery cable is reconnected.

Distributor

REMOVAL

1. Disconnect the negative battery cable. If equipped with the 2.0L engine, remove 1 alternator mounting bolt and the drive belt; swing the alternator to 1 side.
2. Remove the distributor cap and position aside. If it is necessary to remove the spark plug wires, tag each wire and mark it's position on the distributor cap.
3. Disconnect the electrical connector from the distributor. If equipped, disconnect and plug the vacuum advance hose.
4. Mark the position of the rotor in relation to the distributor housing and mark the position of the distributor housing in relation to the intake manifold or cylinder block.
5. Remove the distributor hold-down clamp and bolt and remove the distributor.

Installation

Timing Not Disturbed

1. Install the distributor assembly, aligning the marks that were made during the removal procedure.
2. Install the distributor hold-down clamp and bolt and leave it snug.
3. Connect the electrical connector to the distributor.
4. Install the distributor cap. If the spark plug wires were removed, install them in their proper position, as marked during the removal procedure.
5. Connect the negative battery cable. Check the initial timing according to the proper procedure.
6. Adjust the timing, as necessary and tighten the distributor hold-down bolt. Connect the vacuum advance hose, if equipped.

Timing Disturbed

1. Disconnect the No. 1 spark plug wire and remove the No. 1 spark plug.
2. Place a finger over the spark plug hole and crank the engine slowly until compression is felt.
3. Align the TDC mark on the crankshaft pulley with the pointer on the timing cover. This places the No. 1 cylinder at TDC on the compression stroke.
4. Turn the distributor shaft until the rotor points to the No. 1 spark plug tower on the cap.
5. Install the distributor into the engine, aligning the marks made on the block or intake manifold and the distributor housing. Install the distributor hold-down clamp and bolt. Snug the bolt so the distributor housing can be moved for timing purposes.
6. Install the No. 1 spark plug and connect the spark plug wire. Install the distributor cap and connect the distributor electrical connector. If the spark plug wires were removed, install them in their proper position, as marked during the removal procedure.
7. Connect the negative battery cable and set the ignition timing. Tighten the distributor hold-down clamp bolt and recheck the ignition timing after tightening the bolt.
8. If equipped, connect the vacuum advance hose.

Distributorless Ignition

REMOVAL & INSTALLATION

Crankshaft Sensor

2.3L ENGINE

1. Disconnect the negative battery cable.
2. Disconnect the crankshaft timing sensor assembly electrical connectors from the engine harness.

3. Remove the large electrical connector from the crankshaft timing sensor assembly by prying out the red retaining clip and removing the 4 wires.
4. Remove the crankshaft pulley assembly by removing the accessory drive belts and then the 4 bolts that retain it to the crankshaft pulley hub assembly.
5. Remove the timing belt outer cover.
6. Rotate the crankshaft so the keyway is at the 10 o'clock position. This will place the vane window of both inner and outer vane cups over the crankshaft timing sensor assembly.

NOTE: The vane cups are attached to the crankshaft pulley hub assembly.

7. Remove the 2 crankshaft timing sensor assembly retaining bolts and the plastic wire harness retainer which secures the crankshaft timing sensor harness to it's mounting bracket. Then remove the crankshaft timing sensor assembly, sliding the electrical wires out from behind the inner timing belt cover.

To install:

8. Remove the large electrical connector from the new crankshaft timing sensor assembly.
9. Position the crankshaft timing sensor assembly. First slide the electrical wires behind the inner timing belt cover. Then, hold the sensor assembly loosely in place with the retaining bolts, but do not tighten at this time.
10. Install the large electrical connector onto the crankshaft timing sensor assembly.

NOTE: Make sure the 4 wires to the large electrical connector are installed in the proper locations. The sensor will not function properly if the wires are installed in the wrong locations.

11. Reconnect both of the crankshaft timing sensor electrical connectors to the engine harness.
12. Rotate the crankshaft so the outer vane on the crankshaft pulley hub assembly engages both sides of the crankshaft Hall effect sensor positioner tool T89P–6316–A or equivalent, and tighten the sensor assembly retaining bolts.
13. Rotate the crankshaft so the vane on the crankshaft pulley hub assembly is no longer engaged in the positioning tool. Remove the tool.
14. Install the new plastic wire harness retainer to secure the crankshaft timing sensor harness to it's mounting bracket. Trim off the excess.
15. Install the timing belt outer cover.

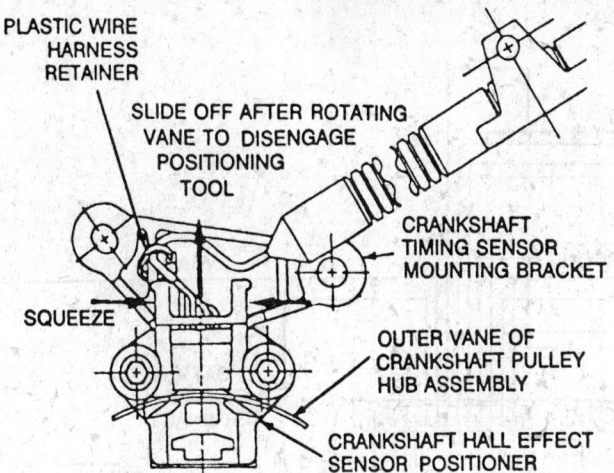

Positioning the crankshaft sensor—2.3L engine

PLASTIC WIRE HARNESS RETAINER

SLIDE OFF AFTER ROTATING VANE TO DISENGAGE POSITIONING TOOL

SQUEEZE

CRANKSHAFT TIMING SENSOR MOUNTING BRACKET

OUTER VANE OF CRANKSHAFT PULLEY HUB ASSEMBLY

CRANKSHAFT HALL EFFECT SENSOR POSITIONER

16. Install the crankshaft pulley assembly and tighten the 4 attaching bolts to 15–22 ft. lbs. (20–30 Nm). Install the accessory drive belts.

4.0L ENGINE

1. Disconnect the negative battery cable.
2. Disconnect the sensor electrical connector from the wiring harness.
3. Remove the crankshaft sensor mounting screws and remove the sensor.
4. Installation is the reverse of the removal procedure. Tighten the screws to 75–106 inch lbs. (8.5–12 Nm).

Ignition Module

2.3L ENGINE

1. Disconnect the negative battery cable.
2. Disconnect the electrical connectors at the module.
3. Remove the module retaining screws and remove the module from the lower intake manifold.

To install:

4. Apply an even coating of silicone dielectric compound WA-10, D7AZ-19A331–A or equivalent, to the mounting surface of the module.
5. Install the module and the retaining screws. Tighten the screws to 22–31 inch lbs. (2.5–3.5 Nm).
6. Connect the electrical connectors to the module and connect the negative battery cable.

4.0L ENGINE

1. On Aerostar, disconnect the negative battery cable.
2. On Explorer and Ranger, disconnect the battery cables and remove the battery.
3. Disconnect the electrical connector at the module.

4. Remove the module retaining bolt and remove the module. On Ranger, slide the assembly up; on Aerostar slide toward the front of the vehicle, to release the module from the tear-drop hole in the sheet metal.
5. Installation is the reverse of the removal procedure. Tighten the mounting bolt to 22–31 inch lbs. (2.5–3.5 Nm).

IGNITION COIL PACK

2.3L ENGINE

1. Disconnect the negative battery cable.
2. Squeeze the locking tabs of the coil wire retainer by hand and remove the spark plug wires with a twisting and pulling motion. Do not pull on the wire.
3. Disconnect the engine harness electrical connector from the ignition coil assembly.
4. Remove the ignition coil assembly by removing the 4 retaining screws.

NOTE: If equipped with power steering, it may be necessary to remove the intake coil and bracket as an assembly.

5. Installation is the reverse of the removal procedure.

4.0L ENGINE

1. Disconnect the negative battery cable.
2. Disconnect the electrical harness connector from the coil pack.
3. Remove the spark plug wires by squeezing the locking tabs to release the coil boot retainers.
4. Remove the coil pack retaining screws and remove the coil pack.
5. Installation is the reverse of the removal procedure. Tighten the screws to 40–62 inch lbs. (4.5–7.0 Nm).

Ignition Timing

ADJUSTMENT

NOTE: Always refer to the Vehicle Emission Information Label to verify the timing adjustment procedure.

Distributorless Ignition System

Base timing for distributorless engines is set from the factory at 10 degrees BTDC and is not adjustable.

Distributor Ignition System

1. Place automatic transmission in **P** or manual transmission in neutral. The air conditioning and heater controls should be in the **OFF** position.
2. On 2.0L engine, disconnect the distributor vacuum advance hose from the distributor and plug the hose.
3. Connect a suitable inductive timing light and a tachometer according to the manufacturer's instructions.
4. On all except 2.0L engine, disconnect the single wire in-line spout connector or remove the shorting bar from the double wire spout connector.
5. On 2.0L engine, if equipped with a barometric pressure switch, disconnect it from the ignition module and place a jumper wire across the pins at the ignition module connector (yellow and black wires).
6. Start the engine and bring to normal operating temperature.

NOTE: To set timing correctly, a remote starter should not be used. Use the ignition key only to start the vehicle. Disconnecting the start wire at the starter relay will cause the TFI module to revert to start mode timing after the vehicle is started. Reconnecting the start wire after the vehicle is running will not correct the timing.

7. With the engine at the timing rpm specified, check the initial timing by aiming the timing light at the timing marks and pointer. Refer to the underhood Vehicle Emission Information Label for specifications.
8. If the marks align, proceed to Step 9. If the marks do not align, shut off the engine and loosen the distributor hold-down clamp bolt. Start the engine, aim the timing light and turn the distributor until the timing marks align. Shut off the engine and tighten the distributor hold-down clamp bolt.
9. On all except 2.0L engine, reconnect the single wire in-line spout connector or reinstall the shorting bar on the double wire spout connector. Check the timing advance to verify the

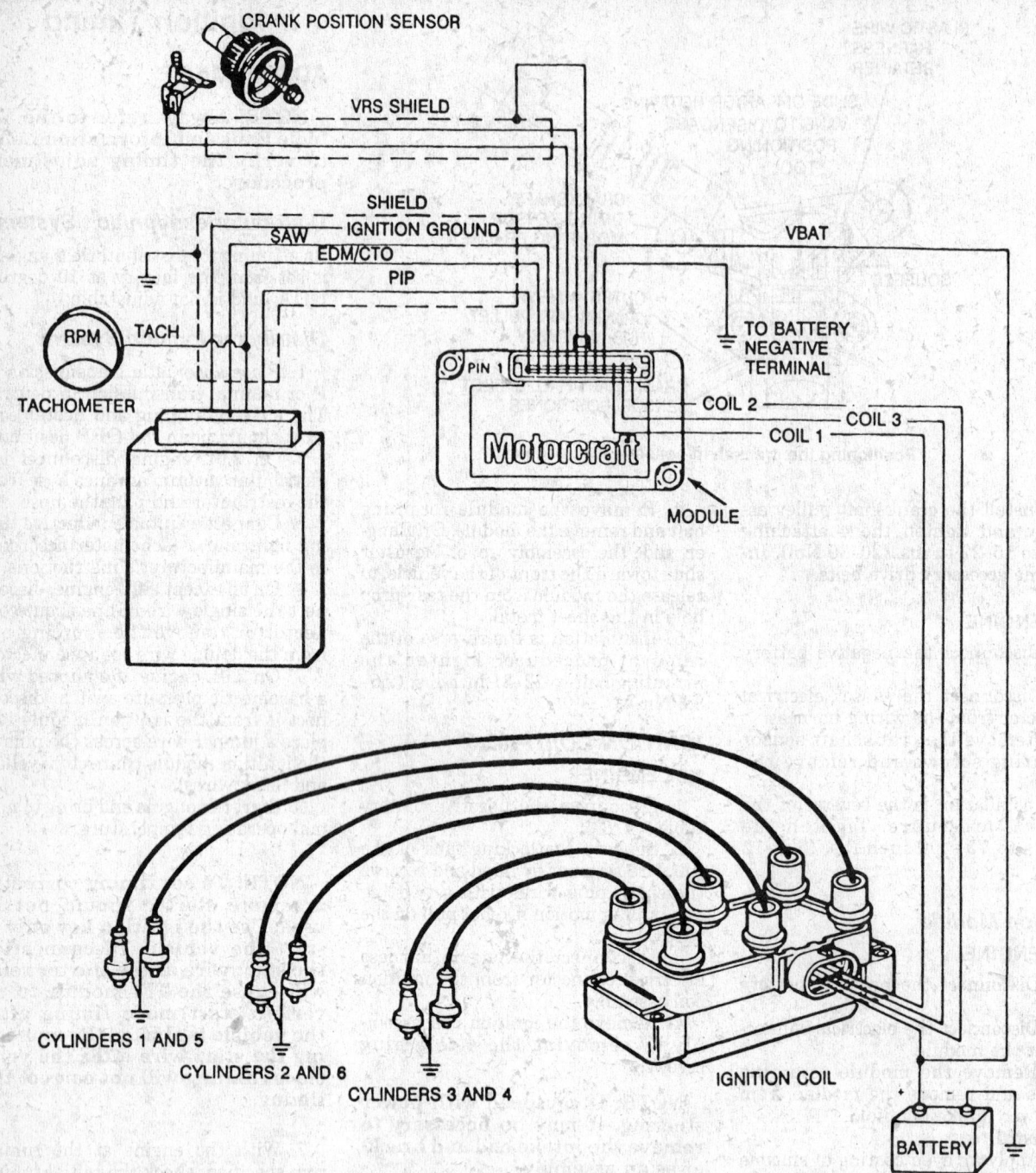

CRANK POSITION SENSOR

VRS SHIELD

SHIELD

SAW   IGNITION GROUND

EDM/CTO

PIP

VBAT

RPM

TACH

TO BATTERY
NEGATIVE
TERMINAL

TACHOMETER

PIN 1

COIL 2

COIL 3

Motorcraft

COIL 1

MODULE

CYLINDERS 1 AND 5

CYLINDERS 2 AND 6

CYLINDERS 3 AND 4

IGNITION COIL

BATTERY

Distributorless ignition system—4.0L engine

distributor is advancing beyond the initial setting.

10. Remove the timing light and tachometer.

11. On 2.0L engine, reconnect the vacuum advance hose at the distributor and, if equipped, remove the jumper wire and reconnect the barometric pressure switch.

Alternator

PRECAUTIONS

Several precautions must be observed

with alternator equipped vehicles to avoid damage to the unit.

• If the battery is removed for any reason, make sure it is reconnected with the correct polarity. Reversing the battery connections may result in damage to the one-way rectifiers.

• When utilizing a booster battery as a starting aid, always connect the positive to positive terminals and the negative terminal from the booster battery to a good engine ground on the vehicle being started.

• Never use a fast charger as a booster to start vehicles.

• Disconnect the battery cables when charging the battery with a fast charger.

• Never attempt to polarize the alternator.

• Do not use test lights of more than 12V when checking diode continuity.

• Do not short across or ground any of the alternator terminals.

• The polarity of the battery, alternator and regulator must be matched and considered before making any electrical connections within the system.

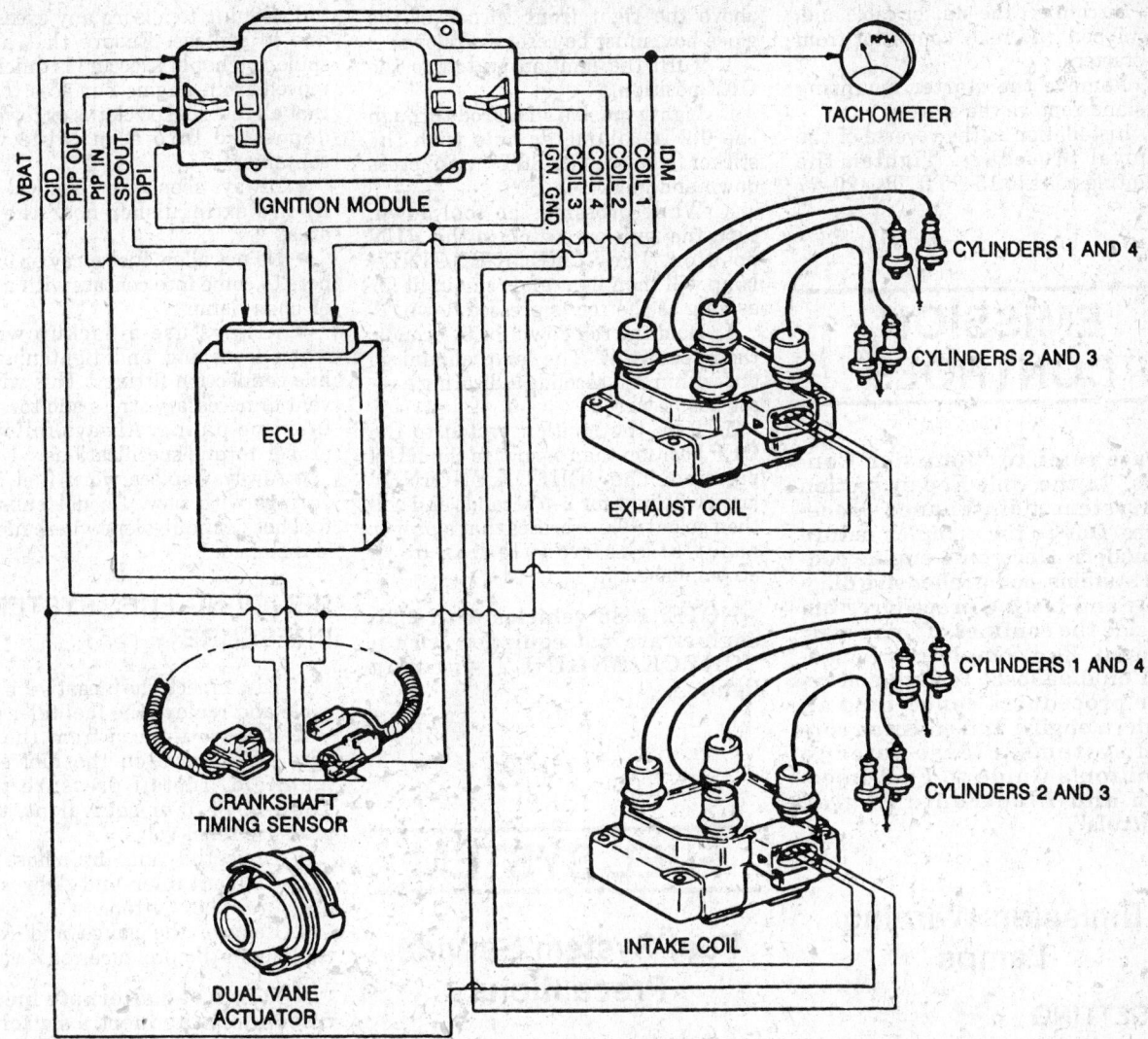

Distributorless ignition system—2.3L engine

• Never separate the alternator on an open circuit. Make sure all connections within the circuit are clean and tight.

• Disconnect the battery ground terminal when performing any service on electrical components.

• Disconnect the battery if arc welding is to be done on the vehicle.

BELT TENSION ADJUSTMENT

NOTE: On 4.0L engine, belt tension is maintained by an automatic tensioner. No adjustment is necessary.

1. Disconnect the negative battery cable. Loosen the alternator adjustment and pivot bolts.

2. Position a suitable belt tension gauge on an accessible belt span midway between the 2 pulleys.

3. Position the alternator housing using a suitable tool to attain correct belt tension. Be careful not to damage the alternator housing.

4. The correct belt tension for a new belt should be 150–190 lbs. on all except 2.9L engine. New belt tension on 2.9L engine should be 120–160 lbs. The correct belt tension for a used belt should be 140–160 lbs., on all except 2.9L engine. Used belt tension on 2.9L engine should be 110–130 lbs.

NOTE: A used belt has more than 10 minutes of operation.

5. Tighten the adjustment bolt and release the pressure. Tighten the pivot bolt.

6. Check the belt tension and reset, if necessary.

REMOVAL & INSTALLATION

1. Disconnect the negative battery cable.

2. Disconnect the electrical connectors from the alternator.

3. On all except 4.0L engine, loosen the alternator pivot bolt and remove the adjustment arm bolt. Remove the drive belt from the alternator pulley.

4. On 4.0L engine, loosen the drive belt tensioner and remove the drive belt from the alternator pulley.

5. Remove the pivot bolt or the mounting bolts and remove the alternator. Remove the alternator fan shield, if equipped.

6. Installation is the reverse of the removal procedure. Adjust the belt tension, if required.

Starter

REMOVAL & INSTALLATION

1. Disconnect the negative battery cable.

2. Raise the vehicle and support it safely.

3. Disconnect the starter cable and, if equipped, the relay connector from the starter.

4. Remove the starter mounting bolts and remove the starter.

5. Installation is the reverse of the removal procedure. Tighten the mounting bolts to 15–20 ft. lbs. (20–27 Nm).

EMISSION CONTROLS

Please refer to "Emission Controls" in the Unit Repair section for system maintenance procedures. Due to the complex nature of modern electronic engine control systems, comprehensive diagnosis and testing procedures fall outside the confines of this repair manual. For complete information on diagnosis, testing and repair procedures concerning all modern engine and emission control systems, please refer to "Chilton's Guide to Fuel Injection and Electronic Engine Controls".

Emission Warning Lamps

RESETTING

Except 1988 2.0L and 2.3L Engines

All vehicles are equipped with a "CHECK ENGINE" or "SERVICE ENGINE SOON" warning lamp located on the instrument cluster. This lamp should come on briefly when the ignition key is turned ON, but should turn off when the engine starts. If the lamp does not come ON when the ignition key is turned ON or if it comes ON and stays ON when the engine is running, there is a malfunction in the electronic engine control system. After the malfunction has been remedied, using the proper procedures, the "CHECK ENGINE" or "SERVICE ENGINE SOON" lamp will go out.

1988 2.0L Engine

The "CHECK ENGINE" lamp is used to indicate that maintenance of emission control devices is required. When the lamp stays on continuously, the required service should be performed. After the service is performed, the emission maintenance sensor, located

above the right front corner of the glove box, must be reset as follows:

1. Turn the ignition switch to the OFF position.

2. Lightly push a small rod through the 0.2 in. diameter hole with the sticker labeled "RESET" and press down and hold.

3. While pressing the tool down, turn the ignition switch to the RUN position. The "CHECK ENGINE" lamp will then light and remain lit for as long as the rod is pressed down.

4. Hold the rod down for 5 seconds, then remove it. The lamp should go out within 2–5 seconds indicating a reset has occurred.

5. Turn the ignition switch to the OFF position then turn it to the RUN position. The "CHECK ENGINE" lamp will light for 2–5 seconds and will then go out. This verifies that a proper reset of the module has been accomplished.

NOTE: 1988 vehicles with 2.3L engines are not equipped with a "CHECK ENGINE" warning light.

FUEL SYSTEM

Fuel System Service Precautions

Safety is the most important factor when performing not only fuel system maintenance but any type of maintenance. Failure to conduct maintenance and repairs in a safe manner may result in serious personal injury or death. Maintenance and testing of the vehicle's fuel system components can be accomplished safely and effectively by adhering to the following rules and guidelines.

• To avoid the possibility of fire and personal injury, always disconnect the negative battery cable unless the repair or test procedure requires that battery voltage be applied.

• Always relieve the fuel system pressure prior to disconnecting any fuel system component (injector, fuel rail, pressure regulator, etc.), fitting or fuel line connection. Exercise extreme caution whenever relieving fuel system pressure to avoid exposing skin, face and eyes to fuel spray. Please be advised that fuel under pressure may penetrate the skin or any part of the body that it contacts.

• Always place a shop towel or cloth around the fitting or connection prior

to loosening to absorb any excess fuel due to spillage. Ensure that all fuel spillage (should it occur) is quickly removed from engine surfaces. Ensure that all fuel soaked cloths or towels are deposited into a suitable waste container.

• Always keep a dry chemical (Class B) fire extinguisher near the work area.

• Do not allow fuel spray or fuel vapors to come into contact with a spark or open flame.

• Always use a backup wrench when loosening and tightening fuel line connection fittings. This will prevent unnecessary stress and torsion to fuel line piping. Always follow the proper torque specifications.

• Always replace worn fuel fitting O-rings with new. Do not substitute fuel hose or equivalent where fuel pipe is installed.

RELIEVING FUEL SYSTEM PRESSURE

1. Disconnect the negative battery cable and remove the fuel filler cap.

2. Remove the cap from the pressure relief valve on the fuel supply manifold. Install pressure gauge T80L–9974–B or equivalent, to the pressure relief valve.

3. Direct the gauge drain hose into a suitable container and depress the pressure relief button.

4. Remove the gauge and replace the cap on the pressure relief valve.

NOTE: As an alternate method, disconnect the inertia switch and crank the engine for 15–20 seconds until the pressure is relieved.

Fuel Line Couplings

REMOVAL & INSTALLATION

There are 3 methods in use to connect the fuel lines and fuel system components, the hairpin clip push connect fitting, the duck bill clip push connect fitting and the spring lock coupling.

FUEL PUMP SWITCH ASSEMBLY IN VEHICLE MUST HAVE RESET BUTTON ON TOP

INERTIA SWITCH
SCREW
HEATER
FLOOR (RH)

Typical inertia switch location

Each requires a different procedure to disconnect and connect.

Hairpin Clip Push Connect Fitting

1. Inspect the visible internal portion of the fitting for dirt accumulation. If more than a light coating of dust is present, clean the fitting before disassembly.

2. Some adhesion between the seals in the fitting and the tubing will occur with time. To separate, twist the fitting on the tube, then push and pull the fitting until it moves freely on the tube.

3. Remove the hairpin clip from the fitting by first bending and breaking the shipping tab. Next, spread the 2 clip legs by hand about ⅛ in. each to disengage the body and push the legs into the fitting. Lightly pull the triangular end of the clip and work it clear of the tube and fitting.

NOTE: Do not use hand tools to complete this operation.

4. Grasp the fitting and pull in an axial direction to remove the fitting from the tube. Be careful on 90 degree elbow connectors, as excessive side loading could break the connector body.

5. After disassembly, inspect and clean the tube end sealing surfaces. The tube end should be free of scratches and corrosion that could provide leak paths. Inspect the inside of the fitting for any internal parts such as O-rings and spacers that may have been dislodged from the fitting. Replace any damaged connector.

To install:

6. Install a new connector if damage was found. Insert a new clip into any 2 adjacent openings with the triangular portion pointing away from the fitting opening. Install the clip until the legs of the clip are locked on the outside of the body. Piloting with an index finger is necessary.

7. Before installing the fitting on the tube, wipe the tube end with a clean cloth. Inspect the inside of the fitting to make sure it is free of dirt and/or obstructions.

8. Apply a light coating of engine oil to the tube end. Align the fitting and tube axially and push the fitting onto the tube end. When the fitting is engaged, a definite click will be heard. Pull on the fitting to make sure it is fully engaged.

Duck Bill Clip Push Connect Fitting

1. Inspect the visible internal portion of the fitting for dirt accumulation. If more than a light coating of dust is present, clean the fitting before disassembly.

2. Some adhesion between the seals in the fitting and the tubing will occur with time. To separate, twist the fitting on the tube, then push and pull the fitting until it moves freely on the tube.

3. Align the slot on push connect disassembly tool T82L–9500–AH or equivalent, with either tab on the clip, 90 degrees from the slots on the side of the fitting and insert the tool. This disengages the duck bill retainer from the tube.

4. Holding the tool and the tube with 1 hand, pull the fitting away from the tube.

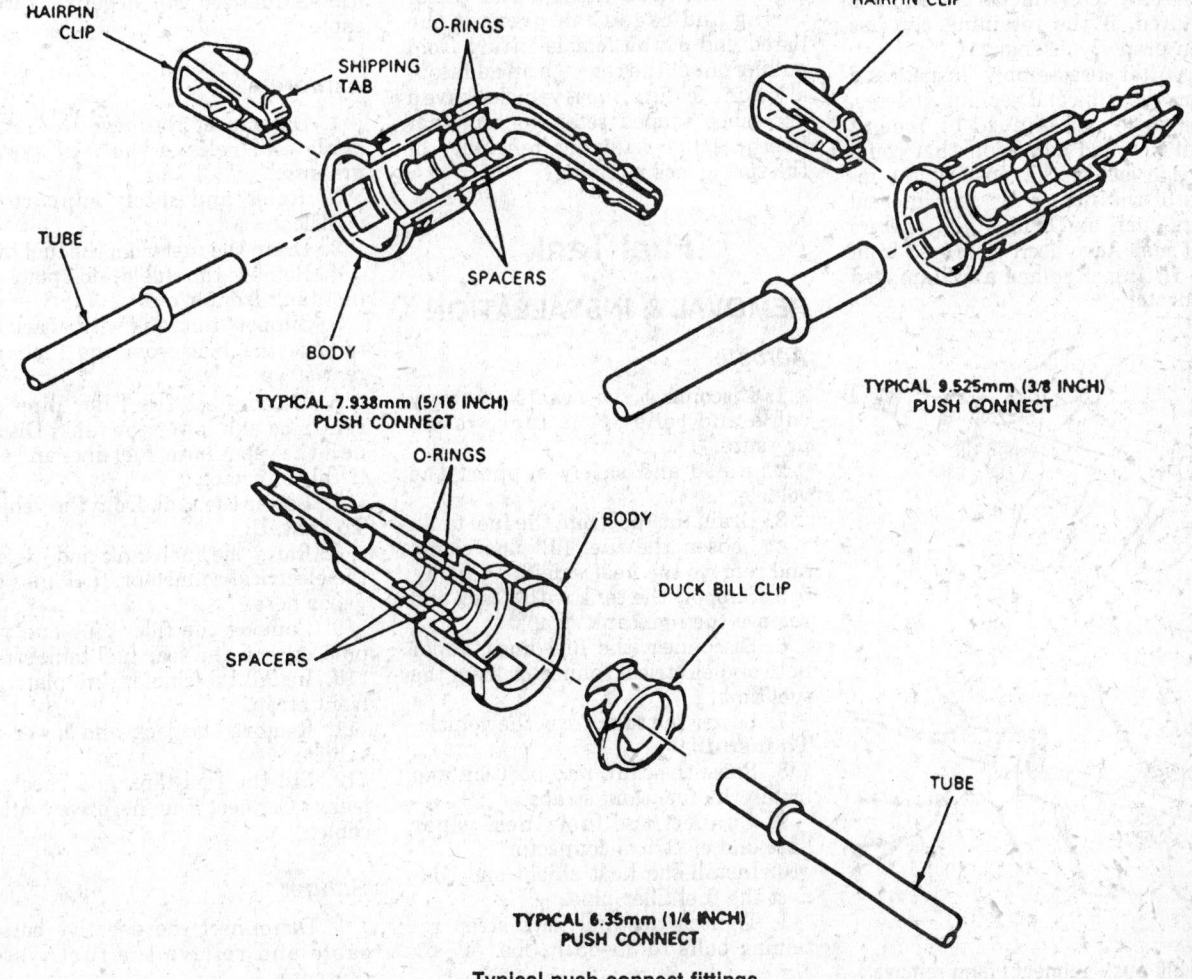

TYPICAL 7.938mm (5/16 INCH) PUSH CONNECT

TYPICAL 9.525mm (3/8 INCH) PUSH CONNECT

TYPICAL 6.35mm (1/4 INCH) PUSH CONNECT

Typical push connect fittings

NOTE: Use hands only. Only moderate effort is required, if the tube has been properly disengaged.

5. After disassembly, inspect and clean the tube end sealing surfaces. The tube end should be free of scratches and corrosion that could provide leak paths. Inspect the inside of the fitting for any internal parts such as O-rings and spacers that may have been dislodged from the fitting. Replace any damaged connector.

6. Some fuel tubes have a secondary bead which aligns with the outer surface of the clip. These beads can make tool insertion difficult. If there is extreme difficulty, use the following disassembly method:

a. Using pliers with a jaw width of 0.2 in. (5mm) or less, align the jaws with the openings in the side of the fitting case and compress the portion of the retaining clip that engages the fitting case. This disengages the retaining clip from the case. Often 1 side of the clip will disengage before the other. The clip must be disengaged from both openings.

b. Pull the fitting off the tube by hand only. Only moderate effort is required, if the retaining clip has been properly disengaged.

c. After disassembly, inspect and clean the tube end sealing surfaces. The tube end should be free of scratches and corrosion that could provide leak paths. Inspect the inside of the fitting for any internal parts such as O-rings and spacers that may have been dislodged from the fitting. Replace any damaged connector.

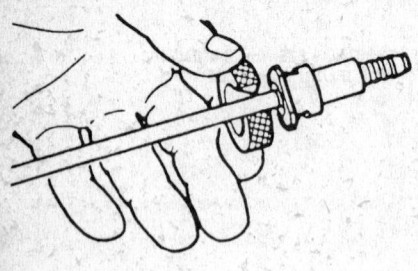

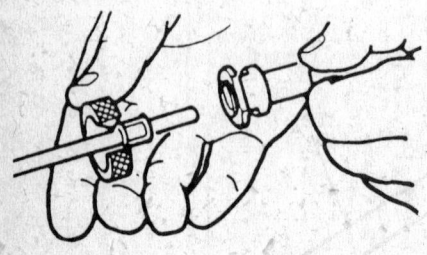

Duck bill push connect fitting removal

d. The retaining clip will remain on the tube. Disengage the clip from the tube bead and remove.

To install:

7. Install a new connector if damage was found. Install the new replacement clip into the body by inserting 1 of the retaining clip serrated edges on the duck bill portion into 1 side of the window openings. Push on the other side until the clip snaps into place.

8. Before installing the fitting on the tube, wipe the tube end with a clean cloth. Inspect the inside of the fitting to make sure it is free of dirt and/or obstructions.

9. Apply a light coating of engine oil to the tube end. Align the fitting and tube axially and push the fitting onto the tube end. When the fitting is engaged, a definite click will be heard. Pull on the fitting to make sure it is fully engaged.

Spring Lock Coupling

The spring lock coupling is a fuel line coupling held together by a garter spring inside a circular cage. When the coupling is connected together, the flared end of the female fitting slips behind the garter spring inside the cage of the male fitting. The garter spring and cage then prevent the flared end of the female fitting from pulling out of the cage. As an additional locking feature, most vehicles have a horseshoe shaped retaining clip that improves the retaining reliability of the spring lock coupling.

Fuel Tank

REMOVAL & INSTALLATION

Aerostar

1. Disconnect the negative battery cable and relieve the fuel system pressure.

2. Raise and safely support the vehicle.

3. Drain the fuel from the fuel tank.

4. Loosen the fuel fill pipe clamp and remove the heat shield.

5. Support the tank with a jack and remove the fuel tank straps.

6. Disconnect the fuel lines, vapor hose and electrical connector from the fuel tank.

7. Lower the tank from the vehicle.

To install:

8. Raise the tank into position and install the fuel tank straps.

9. Connect the fuel lines, vapor hose and electrical connector.

10. Install the heat shield and connect the fuel filler pipe.

11. Tighten the fuel tank strap retaining bolts to 35–45 ft. lbs. (47–61 Nm).

12. Remove the jack and lower the vehicle.

13. Fill the fuel tank and check for leaks. Connect the negative battery cable.

Bronco II

1. Disconnect the negative battery cable and relieve the fuel system pressure.

2. Raise and safely support the vehicle.

3. Drain the fuel from the fuel tank. Remove the skid plate, if equipped.

4. Support the fuel tank with a jack and remove the fuel tank straps.

5. Disconnect the fill pipe and lower the tank enough to disconnect the fuel lines, vapor hoses and electrical connector.

6. Lower the tank from the vehicle.

To install:

7. Raise the fuel tank and connect the electrical connector vapor hoses and fuel lines.

8. Connect the fuel filler pipe and install the fuel tank straps. Install the skid plate, if equipped.

9. Remove the jack and lower the vehicle.

10. Fill the fuel tank and check for leaks. Connect the negative battery cable.

Explorer

1. Disconnect the negative battery cable and relieve the fuel system pressure.

2. Raise and safely support the vehicle.

3. Drain the fuel from the fuel tank.

4. Remove the shield, skid plate and fuel tank front strap.

5. Support the tank with a jack and remove the bolt from the fuel tank rear strap.

6. Disconnect the filler pipe and vent pipe and lower the tank. Disconnect the vapor hose, fuel lines and electrical connector.

7. Lower the tank from the vehicle.

To install:

8. Raise the fuel tank and connect the electrical connector, fuel lines and vapor hose.

9. Connect the filler pipe and vent pipe. Attach the rear fuel tank strap.

10. Install the shield, skid plate and front strap.

11. Remove the jack and lower the vehicle.

12. Fill the fuel tank and check for leaks. Connect the negative battery cable.

Ranger

1. Disconnect the negative battery cable and relieve the fuel system pressure.

SPRING LOCK COUPLING CONNECT and DISCONNECT PROCEDURE

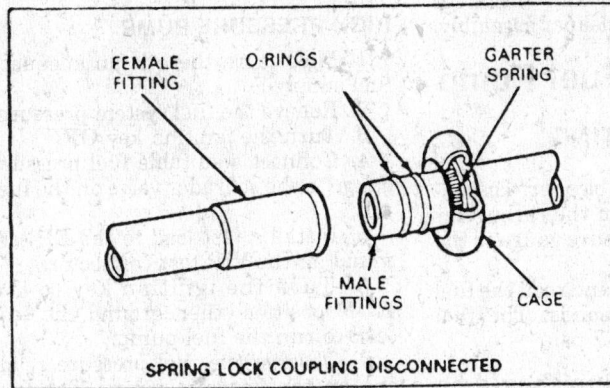

SPRING LOCK COUPLING DISCONNECTED

TO CONNECT COUPLING

TO DISCONNECT COUPLING
CAUTION - DISCHARGE SYSTEM BEFORE DISCONNECTING COUPLING

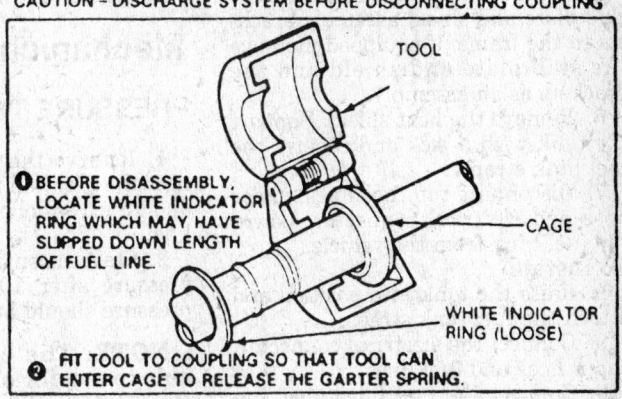

❶ BEFORE DISASSEMBLY, LOCATE WHITE INDICATOR RING WHICH MAY HAVE SLIPPED DOWN LENGTH OF FUEL LINE.

❷ FIT TOOL TO COUPLING SO THAT TOOL CAN ENTER CAGE TO RELEASE THE GARTER SPRING.

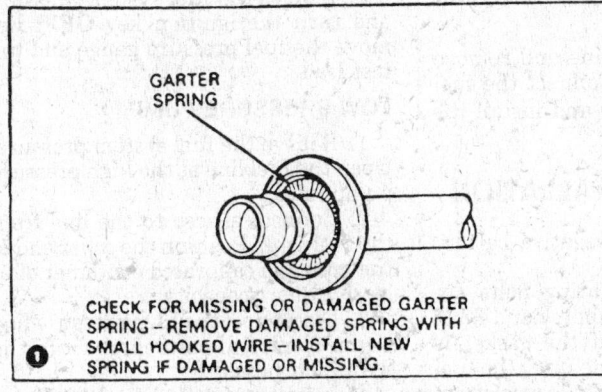

❶ CHECK FOR MISSING OR DAMAGED GARTER SPRING - REMOVE DAMAGED SPRING WITH SMALL HOOKED WIRE - INSTALL NEW SPRING IF DAMAGED OR MISSING.

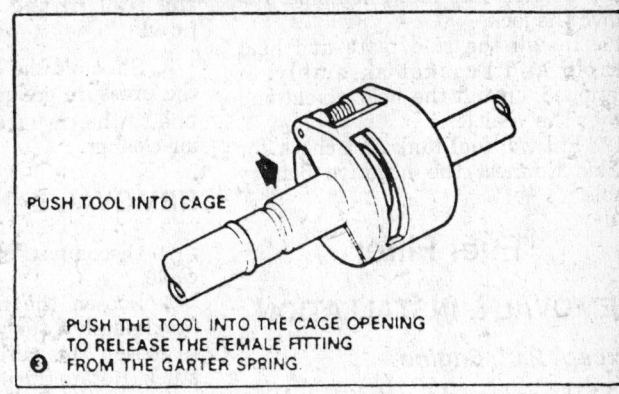

PUSH TOOL INTO CAGE

❸ PUSH THE TOOL INTO THE CAGE OPENING TO RELEASE THE FEMALE FITTING FROM THE GARTER SPRING.

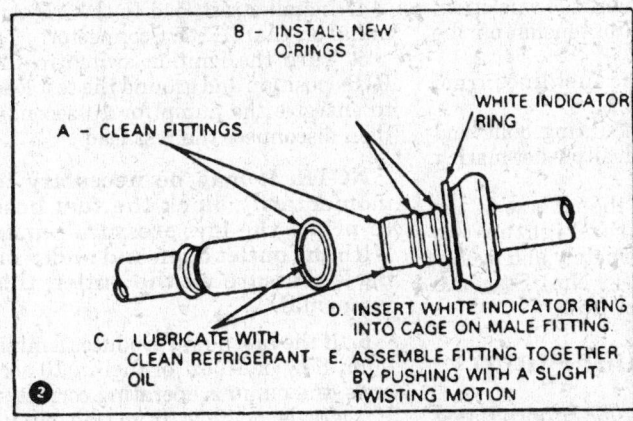

❷ A - CLEAN FITTINGS
B - INSTALL NEW O-RINGS
C - LUBRICATE WITH CLEAN REFRIGERANT OIL
D. INSERT WHITE INDICATOR RING INTO CAGE ON MALE FITTING.
E. ASSEMBLE FITTING TOGETHER BY PUSHING WITH A SLIGHT TWISTING MOTION

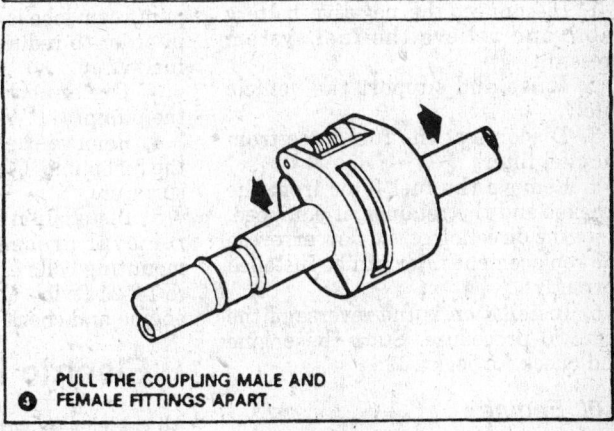

❹ PULL THE COUPLING MALE AND FEMALE FITTINGS APART.

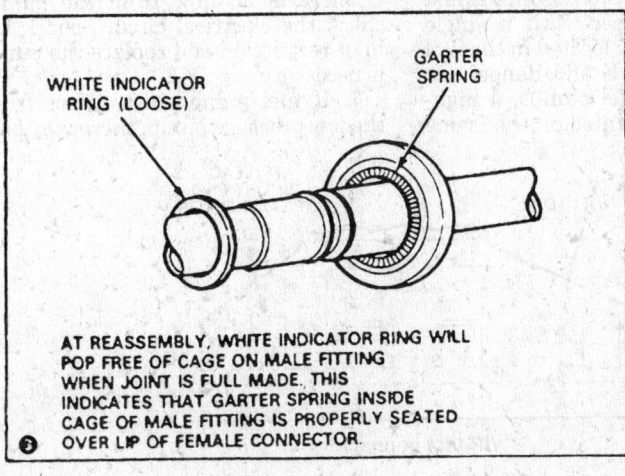

❸ AT REASSEMBLY, WHITE INDICATOR RING WILL POP FREE OF CAGE ON MALE FITTING WHEN JOINT IS FULL MADE. THIS INDICATES THAT GARTER SPRING INSIDE CAGE OF MALE FITTING IS PROPERLY SEATED OVER LIP OF FEMALE CONNECTOR.

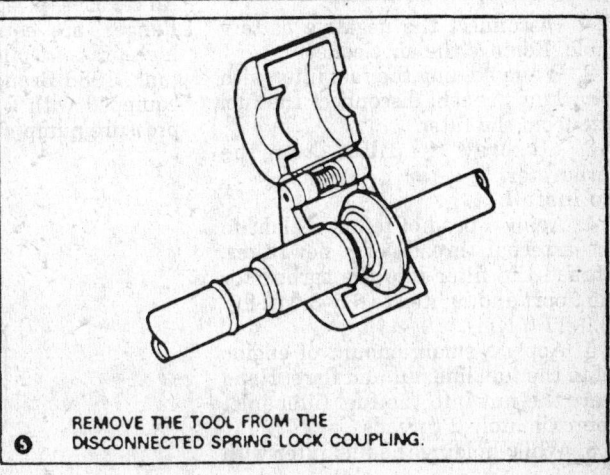

❺ REMOVE THE TOOL FROM THE DISCONNECTED SPRING LOCK COUPLING.

2. Raise and safely support the vehicle.

3. Drain the fuel from the fuel tank.

4. Loosen the filler pipe clamp.

5. Remove the bolts securing the skid plate and shield assembly brackets to the frame, if equipped. Remove the skid plate and shield and the brackets as an assembly.

6. Remove the heat shield. Support the tank with a jack and remove the fuel tank straps.

7. Disconnect the fuel lines, vapor hose and electrical connector. Lower the fuel tank from the vehicle.

To install:

8. Raise the tank into position and attach the fuel tank straps.

9. Connect the electrical connector, vapor hose and fuel lines.

10. Connect the fuel filler pipe. Remove the jack.

11. Install the skid plate and heat shield and bracket assembly, if equipped. Install the heat shield and lower the vehicle.

12. Fill the fuel tank and check for leaks. Connect the negative battery cable.

Fuel Filter

REMOVAL & INSTALLATION

Except 2.0L Engine

1. Disconnect the negative battery cable and relieve the fuel system pressure.

2. Raise and support the vehicle safely.

3. Disconnect the fuel lines from the fuel filter.

4. Remove the fuel filter from the bracket and the retainer, if equipped. Note the direction of the flow arrow so the replacement filter can be installed correctly.

5. Installation is the reverse of the removal procedure. Start the engine and check for leaks.

2.0L Engine

1. Disconnect the negative battery cable. Remove the air cleaner.

2. While holding the fuel filter with a backup wrench, disconnect the fuel line from the filter.

3. Unscrew the filter from the carburetor.

To install:

4. Apply a bead of teflon sealant to the external threads of a new filter. Thread the filter into the carburetor inlet port and tighten to 6.5–8.0 ft. lbs. (9.0–11.0 Nm).

5. Apply a small amount of engine oil to the fuel line nut and flare. Hand start the nut into the fuel filter inlet approximately 2 threads.

6. While holding the fuel filter with

a backup wrench, tighten the fuel line nut to 15–18 ft. lbs. (20–24 Nm).

7. Connect the negative battery cable, start the engine and check for leaks. Install the air cleaner assembly.

Mechanical Fuel Pump

PRESSURE TESTING

1. Remove the air cleaner. Disconnect the fuel line from the carburetor and attach a fuel pressure gauge to the line.

2. Start the engine and read the fuel pressure after 10 seconds. The fuel pressure should be 5–7 psi.

NOTE: The engine should be able to run for over 30 seconds on the fuel in the carburetor float bowl.

3. Shut off the engine and remove the pressure gauge. Connect the fuel line to the carburetor and install the air cleaner.

REMOVAL & INSTALLATION

1. Disconnect the negative battery cable.

2. Loosen the mounting bolts approximately 2 turns. Apply hand force to loosen the pump if the gasket is stuck. Rotate the engine until the fuel pump cam lobe is near its low position, in order to reduce the tension on the fuel pump.

3. Disconnect the fuel lines from the pump.

4. Remove the mounting bolts and the fuel pump. Clean all gasket mating surfaces.

5. Installation is the reverse of the removal procedure. Tighten the mounting bolts alternately and evenly to 14–21 ft. lbs. (19–29 Nm). Start the engine and check for leaks.

Electric Fuel Pump

All vehicles, except 1988 Bronco II and Ranger, are equipped with a single high-pressure pump located in the fuel tank. 1988 Bronco II and Ranger are equipped with 2 fuel pumps, a high-pressure pump mounted on the frame

rail and a low-pressure pump located in the fuel tank.

Pressure Testing

HIGH PRESSURE PUMP

1. Make sure there is an adequate fuel supply.

2. Relieve the fuel system pressure.

3. Turn the ignition key **OFF**.

4. Connect a suitable fuel pressure gauge to the schrader valve on the fuel rail.

5. Install a test lead to the **FP** terminal on the VIP test connector.

6. Turn the ignition key to the **RUN** position, then ground the test lead to run the fuel pump.

7. Observe the fuel pressure reading on the pressure gauge. The fuel pressure should be 35–45 psi.

8. Relieve the fuel system pressure and turn the ignition key **OFF**. Remove the fuel pressure gauge and the test lead.

LOW PRESSURE PUMP

1. Relieve the fuel system pressure. Open the fuel line at the high pressure pump inlet.

2. Connect a hose to the line from the fuel tank. Position the other end of the hose in a calibrated container of at least 1 quart capacity.

3. Disconnect the high pressure fuel pump electrical connector from the wiring harness.

4. Install a test lead to the **FP** terminal on the VIP test connector.

5. Turn the ignition switch to the **RUN** position and ground the test lead to energize the pump for 10 seconds, then disconnect the test lead.

NOTE: It may be necessary to momentarily block the fuel hose to prime the low pressure pump. With the outlet open and under no back pressure on the outlet, this is normal.

6. If the fuel pump produces a minimum flow of 16 oz. of fuel in 10 seconds, the pump is operating correctly. If there is no flow from the pump, check the electrical circuit, check for inlet restriction and replace the pump if necessary.

7. If fuel pump noise comes from the low pressure pump, there may be a

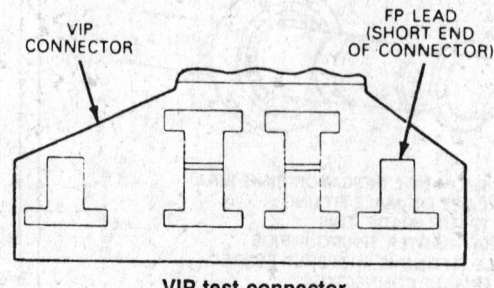

VIP test connector

malfunction in the high pressure pump. To check this, compare the vibration of the inlet and outlet fuel lines at the high pressure pump. If there is a large difference between the lines in vibration level, replace the high pressure pump and recheck for noise.

REMOVAL & INSTALLATION

Tank Mounted Pump

NOTE: On Ranger, the fuel pump may be accessed by removing the pickup box from the chassis, instead of removing the fuel tank.

1. Disconnect the negative battery cable and relieve the fuel system pressure.
2. Raise and safely support the vehicle.
3. Remove the fuel tank.
4. Remove any dirt that has accumulated around the fuel pump attaching flange so it will not enter the fuel tank during removal and installation.
5. Turn the fuel pump locking ring counterclockwise using a suitable tool. Remove the locking ring.
6. Remove the fuel pump and discard the seal ring. Separate the fuel pump from the sending unit, if required.

To install:

7. Clean the fuel pump mounting flange and tank mounting surface and seal ring groove.
8. Apply a light coating of grease on a new seal ring and install it in the groove.
9. Install the fuel pump to the sending unit, if removed. Install the fuel pump assembly in the tank, making sure the locating keys are in the keyways and the seal ring is in place.
10. Hold the fuel pump assembly and the seal ring in place and install the locking ring. Rotate the ring clockwise using a suitable tool. Find the fuel tank part number on the front bottom of the tank and proceed as follows:

   a. If equipped with part number E59A-9002-CAE tank, tighten the ring to 60–85 ft. lbs. (81–115 Nm), wait 5 minutes and retighten to the same specification. Use only the ring that was removed from the tank, do not replace with a ring from another tank.

   b. If equipped with plastic retaining ring E99A-9A307-D, tighten to 40–55 ft. lbs. (54–74 Nm). Use the same ring that was removed from the tank. If a new tank is installed, use a new ring.

   c. If equipped with part number E69A-9002-PA tank, tighten the locking ring once to 80–113 ft. lbs. (109–153 Nm).

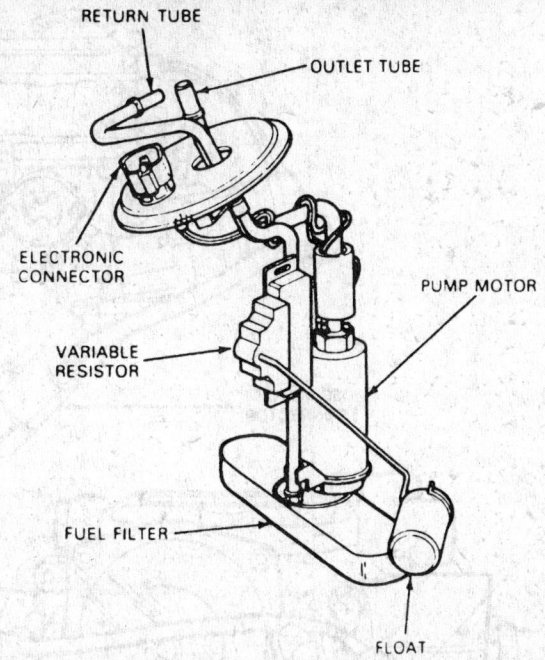

RETURN TUBE
OUTLET TUBE
ELECTRONIC CONNECTOR
PUMP MOTOR
VARIABLE RESISTOR
FUEL FILTER
FLOAT

Tank mounted electric fuel pump assembly

   d. On Ranger and Aerostar, tighten the polyethylene locking ring to 40–45 ft. lbs. (54–61 Nm).
11. Install the fuel tank in the vehicle.
12. Lower the vehicle and fill the fuel tank with at least 10 gallons of fuel. Connect the negative battery cable. Turn the ignition key to **RUN** for 3 seconds repeatedly, 5–10 times, to pressurize the system. Check for leaks.
13. Start the engine and check for leaks.

Frame Mounted Pump

1. Disconnect the negative battery cable and relieve the fuel system pressure.
2. Raise and support the vehicle safely.
3. Disconnect the fuel lines and electrical connector from the pump.
4. Remove the pump from the mounting bracket.
5. Installation is the reverse of the removal procedure. Start the engine and check for leaks.

Carburetor

REMOVAL & INSTALLATION

1. Disconnect the negative battery cable. Remove the air cleaner, disconnecting the emission control hoses and air intake hose, as required.
2. Disconnect the fuel line and the accelerator linkage. Tag and disconnect the vacuum hoses and electrical connections.

3. Remove the retaining nuts and the carburetor. Cover the intake manifold opening with a cloth.
4. Installation is the reverse of the removal procedure. Tighten the carburetor retaining nuts to 13–14 ft. lbs. (17.7–19.0 Nm).

IDLE SPEED ADJUSTMENT

1. Apply the parking brake and place the transmission in **N**.
2. Bring the engine to normal operating temperature. Turn off all accessories.
3. Check the engine idle speed as specified on the vehicle emission information label. If the idle speed is correct, no adjustment is required. If not, proceed to Step 4.
4. Turn the ignition key to **OFF**.
5. Disconnect the Idle Speed Control (ISC) motor connector. Disconnect the vacuum hose at the fast idle cam breaker and plug. Connect insulated jumper clip leads from the battery terminals to the top 2 ISC motor terminals to retract the motor.

NOTE: If the motor extends instead of retracts, reverse the polarity across the motor terminals. Battery voltage must not be applied to the bottom terminals of the motor, the Idle Tracking Switch (ITS) terminals. Isolate the ITS terminals with electrical tape during this procedure.

6. Restart the engine and check the closed throttle rpm; it should be 700 rpm or less. If adjustment is required,

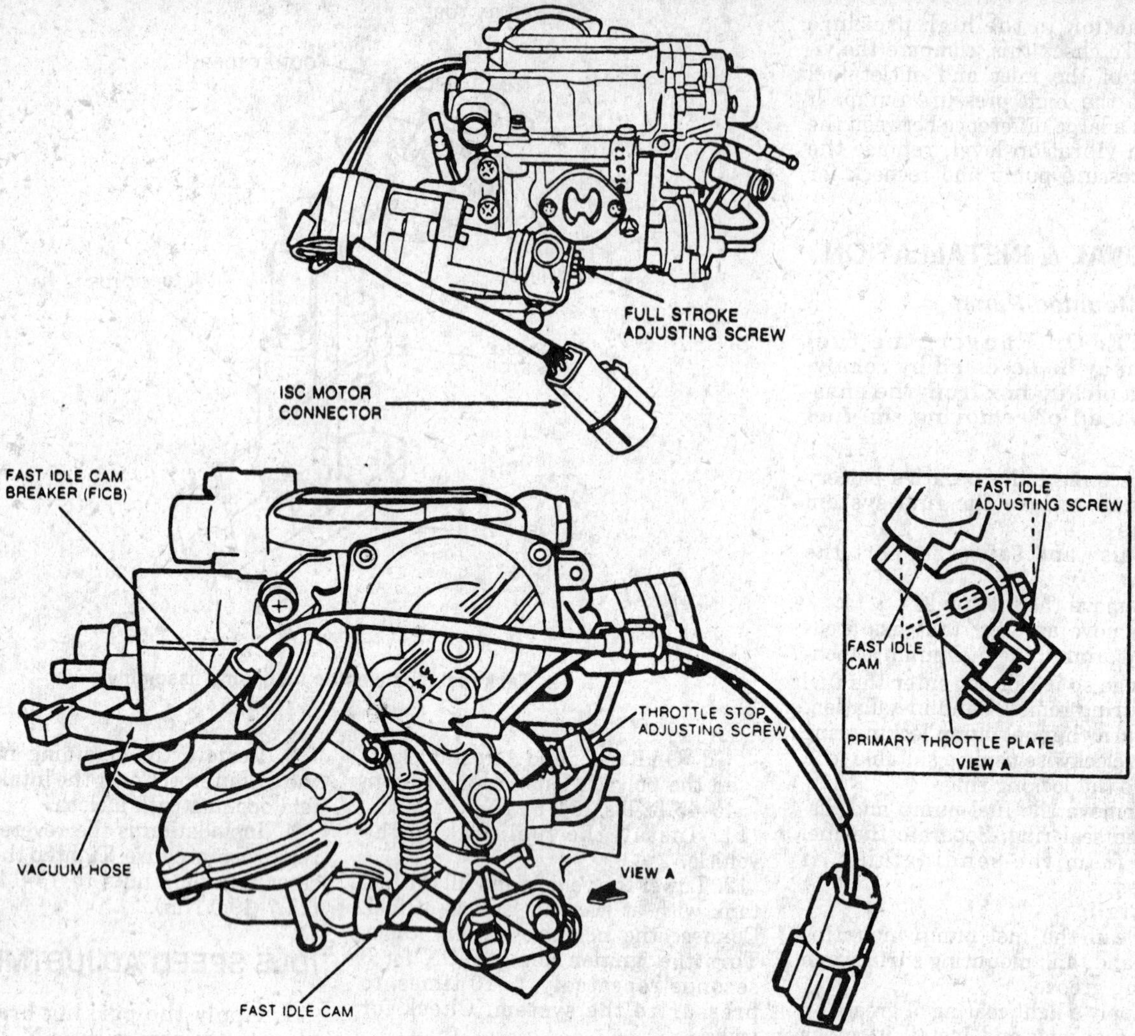

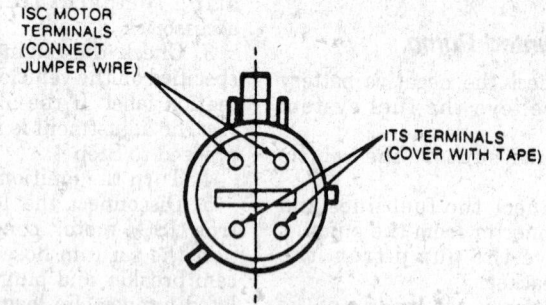

FULL STROKE
ADJUSTING SCREW

ISC MOTOR
CONNECTOR

FAST IDLE CAM
BREAKER (FICB)

VACUUM HOSE

THROTTLE STOP
ADJUSTING SCREW

VIEW A

FAST IDLE CAM

Carburetor component locations—2.0L engine

FAST IDLE
ADJUSTING SCREW

FAST IDLE
CAM

PRIMARY THROTTLE PLATE

VIEW A

ISC MOTOR
TERMINALS
(CONNECT
JUMPER WIRE)

ITS TERMINALS
(COVER WITH TAPE)

ISC motor connectors

remove the tamper proof cap and adjust the throttle stop adjusting screw.

7. Place the fast idle lever on the first step on the fast idle cam. Without touching the accelerator pedal, check/adjust the fast idle speed to specification with the fast idle adjusting screw.

8. Rev the engine momentarily, allowing it to return to closed throttle speed.

9. Reconnect the jumper leads the ISC motor terminals. If the motor retracts instead of extending, reverse polarity across the motor terminals. Dis-

connect the jumper leads after the motor has fully extended.

10. Check/adjust the ISC full stroke speed to 2000 ± 200 rpm using the full stroke speed adjusting screw.

11. Remove the electrical tape from the ITS terminals of the motor connector and reconnect the ISC motor to the engine wiring harness. Remove the plug from the fast idle cam breaker vacuum hose and reconnect.

12. Check the ISC function as follows:

   a. Turn off all accessories. The

idle speed should adjust to that specified on the vehicle emission information label.

   b. Turn the ignition switch **OFF**.

   c. Disconnect the engine coolant temperature switch.

   d. Restart the engine. The idle speed should adjust to 1200 ± 75 rpm.

   e. Turn the ignition switch **OFF**, then reconnect the engine coolant temperature switch.

13. After adjustments are completed, reconnect all parts.

IDLE MIXTURE ADJUSTMENT

Idle mixture adjustment requires propane enrichment tool T75L–9600–A or equivalent.

SERVICE ADJUSTMENTS

For all carburetor sevice adjustments procedures and specifications, please refer to "Carburetor

Service'' in the Unit Repair section.

Fuel Injection

IDLE SPEED ADJUSTMENT

1988–90

1. Apply the parking brake and place the transmission in **P** or **N**.

2. Start the engine and bring to normal operating temperature. Turn off the heater and all accessories.

3. Make sure the throttle lever is resting on the throttle plate stop screw and the ignition timing is set to specification.

4. Connect a tachometer according to the manufacturers instructions.

5. Shut the engine off and disconnect the negative battery cable for 5 minutes minimum, then reconnect it.

6. Start the engine and stabilize for 2 minutes, then rev the engine and let it return to idle. Lightly depress and release the accelerator and let the engine idle. If the engine does not idle properly, proceed to Step 7.

7. Disconnect the idle speed control-air bypass solenoid.

8. With the transmission in **N** or **P**, run the engine at 2500 rpm for 30 seconds.

9. On 1988 vehicles with 2.9L and 3.0L engines and 1989–90 vehicles with 3.0L engine, place the automatic transmission in **D**. On all other vehicles with automatic transmission, place the transmission in **P**.

10. Turn the throttle plate stop screw to adjust the idle speed to the following specifications:

　1988 2.3L engine—625 ± 25 rpm.
　1989–90 2.3L engines—575 ± 25 rpm.
　1988 2.9L engine—700 rpm.
　1989–90 2.9L engines—725 rpm.
　1988–90 3.0L engine with automatic transmission—625 ± 25 rpm.
　1988–90 3.0L engine with manual transmission—725 ± 25 rpm.
　1990 4.0L engine—675 rpm.

11. On all except 2.3L engine, shut the engine off and repeat Steps 8, 9 and 10.

12. On all except 2.3L engine, shut the engine off and disconnect the battery for 3 minutes minimum.

13. With the engine off, reconnect the idle speed control-air bypass solenoid. Make sure the throttle is not stuck in the bore and the linkage not preventing the throttle from closing.

14. Start the engine and let it stabilize for 2 minutes, then rev the engine and let it return to idle. Lightly depress and release the accelerator and let the engine idle.

15. Vehicles with 3.0L engine and

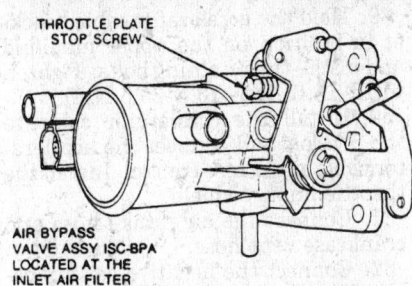

Throttle plate stop screw location— 2.3L engine

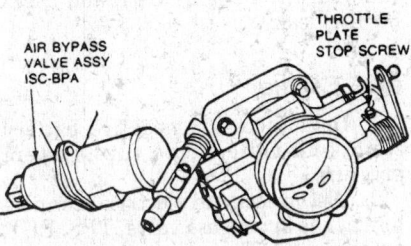

Throttle plate stop screw location— 2.9L engine

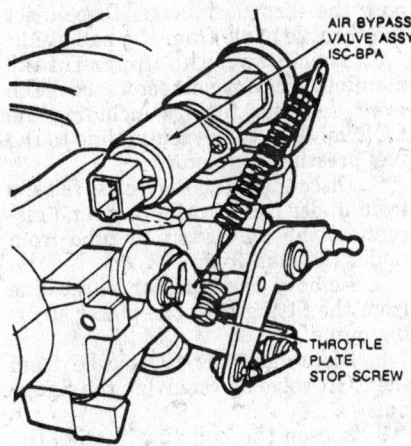

Throttle plate stop screw location— 3.0L engine

automatic transmission should now be idling at 700 ± 50 rpm. Vehicles with 3.0L engine and manual transmission should now be idling at 800 ± 50 rpm.

16. Remove all test equipment.

1991–92

Idle speed adjustment on 1991–92 engines requires the use of Super Star II tester 007–00028 or equivalent.

IDLE MIXTURE ADJUSTMENT

The idle mixture is controlled by the electronic control unit and cannot be adjusted.

Fuel Injector

REMOVAL & INSTALLATION

2.3L Engine

1. Disconnect the negative battery cable and relieve the fuel system pressure.

2. Disconnect the electrical connectors at the TPS, air bypass valve, injector wiring harness at the main engine harness and at the water temperature indicator sensor, ACT and ECT sensors and the ignition control assembly. Tag all lines prior to removal to aid reinstallation.

3. Remove the throttle linkage shield and disconnect the throttle linkage and cruise control. Unbolt the accelerator cable from the bracket and position out of the way.

4. Disconnect the vacuum lines at the upper intake manifold vacuum tree, EGR valve, fuel pressure regulator and canister purge line. Tag all lines prior to removal to aid reinstallation.

5. Disconnect the air intake hose and crankcase vent hose.

6. Disconnect the hose for the PCV system from the fitting on the underside of the upper intake manifold.

7. Disconnect the EGR tube from the EGR valve by removing the flange nut.

8. Remove the upper intake manifold retaining bolts and remove the upper intake manifold and throttle body assembly.

9. Disconnect the injector electrical connectors.

10. Disconnect the fuel lines from the fuel supply manifold.

11. Remove the 2 fuel supply manifold retaining bolts.

12. Carefully disengage the manifold and fuel injectors from the engine and remove the manifold and injectors.

13. Remove the injectors from the manifold by grasping the injector body and pulling while gently rocking the injector from side-to-side. Remove and discard the injector O-rings.
To install:

14. Lubricate new O-rings with clean light grade oil and install 2 on each injector.

NOTE: Never use silicone grease at it will clog the injectors.

15. Install the injectors, using a light, twisting, pushing motion.

16. Install the fuel supply manifold, pushing it down to make sure all the O-rings are seated in the fuel rail cups and intake manifold.

17. Install the manifold retaining bolts and tighten to 15–22 ft. lbs. (20–30 Nm) while holding the fuel manifold down.

18. Connect the fuel lines to the supply manifold.

19. After the fuel supply manifold has been installed and before the fuel injector wire connectors have been connected, connect the negative battery cable and turn the ignition **ON**. This will cause the fuel pump to run for 2–3 seconds and pressurize the fuel system.

20. Check for leaks where the fuel injector is installed into the fuel supply manifold.

21. Disconnect the negative battery cable.

22. Connect the fuel injector electrical connectors.

23. Make sure the gasket surfaces of the upper and lower intake manifolds are clean.

24. Place a new gasket on the lower intake manifold assembly and place the upper intake manifold and throttle body assembly in position.

25. Install the retaining bolts and tighten in sequence to 15–22 ft. lbs. (20–30 Nm).

NOTE: On 1989–92 vehicles, the 3 bolts with stud heads go in hole positions 2, 3 and 4.

26. Connect the EGR tube to the EGR valve and tighten to 18–28 ft. lbs. (25–30 Nm).

27. Connect the PCV system hose to the fitting on the underside of the upper intake manifold.

28. Connect all vacuum lines and electrical connectors according to the locations that were marked during the removal procedure.

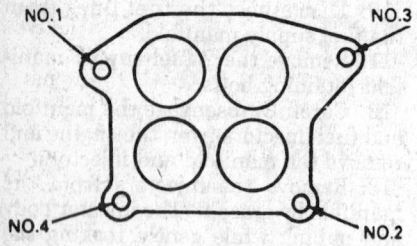

Upper Intake manifold bolt torque sequence—1988 2.3L engine

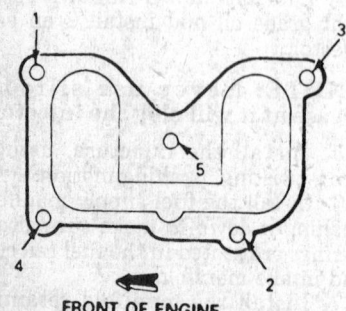

FRONT OF ENGINE

Upper Intake manifold bolt torque sequence—1989–92 2.3L engine

29. Hold the accelerator cable bracket in position on the upper manifold and install the retaining bolts. Tighten to 10–15 ft. lbs. (13.5–20.5 Nm).

30. Install the accelerator cable to the bracket and connect the accelerator cable and cruise control. Install the throttle linkage shield.

31. Connect the air intake hose and crankcase vent hose.

32. Connect the negative battery cable. Start the engine and let it idle for 2 minutes.

33. Turn the engine **OFF** and check for fuel leaks.

2.9L Engine

1. Disconnect the negative battery cable and relieve the fuel system pressure.

2. Disconnect the electrical connectors at the air bypass valve, TPS, EGR sensor and ACT.

3. Remove the air inlet tube from the air cleaner to throttle body.

4. Remove the snow/ice shield to expose the throttle linkage. Disconnect the throttle cable from the ball stud.

5. Disconnect the upper intake manifold vacuum connectors; both the front and rear fittings including the EGR valve and the vacuum line to the fuel pressure regulator.

6. Disconnect the PCV closure tube from under the throttle body and disconnect the PCV vacuum tube from under the manifold.

7. Remove the canister purge line from the fitting near the power steering pump.

8. Disconnect the EGR tube from the EGR valve by removing the flange nut.

9. Loosen the bolt that retains the air conditioning line at the upper rear of the upper manifold and disengage the retainer.

10. Remove the 6 upper intake manifold retaining bolts and remove the upper intake manifold and throttle body assembly.

11. Disconnect the fuel supply and return lines from the fuel supply manifold.

12. On 1990–92 vehicles, disconnect the fuel return line from the fuel pressure regulator as follows:

   a. Disengage the locking tabs on the connector retainer and separate the retainer halves.

   b. Inspect the visible internal portion of the fitting for dirt accumulation. Clean the fitting before disassembly.

   c. To disengage the fitting from the regulator, push the fitting toward the regulator, insert the fingers on fuel line coupling key T90P-9550-A or equivalent, into the slots in the coupling.

   d. Using the tool, pull the fitting from the regulator.

NOTE: If the fitting has been properly disengaged, the fitting should slide off the regulator with minimum effort.

13. Disconnect the electrical connectors from the fuel injectors.

14. Remove the 4 fuel supply manifold retaining bolts.

15. Carefully disengage the fuel supply manifold from the lower intake manifold. The fuel injectors are retained in the fuel supply manifold with clips.

16. Remove the retainer clips and inspect for corrosion and damage.

17. Remove the injector from the fuel supply manifold by grasping the injector body and pulling while gently rocking from side-to-side. Remove and discard the injector O-rings.

18. Inspect the injector plastic pintle protection cap and washer for signs of deterioration. Replace the complete injector as required. If the plastic pintle protection cap is missing, look for it in the intake manifold.

NOTE: The plastic pintle protection cap is not available as a separate part.

To install:

19. Lubricate new O-rings with clean light grade oil and install 2 on each injector.

NOTE: Never use silicone grease at it will clog the injectors.

20. Install the injectors, using a light, twisting, pushing motion.

21. Install the fuel supply manifold, pushing down to make sure all the fuel injector O-rings are fully seated in the fuel supply manifold cups and intake manifold.

22. Install the 4 retaining bolts and tighten to 71–97 inch lbs. (8–11 Nm) while holding the fuel rail assembly down.

23. Connect the fuel supply and return lines to the fuel supply manifold. On 1990–92 vehicles, tighten the line nut to 15–18 ft. lbs. (20–24 Nm).

24. On 1990–92 vehicles, install the fuel return line to the fuel pressure regulator by pushing it onto the fuel pressure regulator line up to the shoulder on the regulator line.

NOTE: The connector should grip the regulator line securely.

25. Install the connector retainer and snap the 2 halves of the retainer together.

26. Install the fuel injector retaining clips to the injectors.

27. After the fuel supply manifold has been installed and before the fuel

injector wire connectors have been connected, connect the negative battery cable and turn the ignition **ON**. This will cause the fuel pump to run for 2–3 seconds and pressurize the fuel system.

28. Check for leaks where the fuel injector is installed into the fuel supply manifold.

29. Disconnect the negative battery cable.

30. Connect the fuel injector electrical connectors.

31. Make sure the gasket surfaces of the upper and lower intake manifolds are clean.

32. Place a new gasket on the lower intake manifold assembly and place the upper intake manifold and throttle body assembly in position. The use of alignment studs may be helpful. Align the EGR tube in the valve.

33. Install the 6 upper intake manifold retaining bolts and tighten to 11–15 ft. lbs. (16–20 Nm).

34. Engage the air conditioner line retainer cup and tighten the bolt.

35. Tighten the EGR tube and flare fitting. Tighten the lower retainer nut at the exhaust manifold.

36. Connect the canister purge line, PCV vacuum hose and the PCV closure hose.

37. Connect the vacuum lines to the vacuum tree, EGR valve and fuel pressure regulator.

38. Connect the throttle cable to the throttle body and install the snow/ice shield.

39. Connect the electrical connectors at the air bypass valve, TPS and ACT sensor.

40. Install the air inlet tube from the throttle body to the air cleaner.

41. Connect the negative battery cable, start the engine and let it idle for 2 minutes.

42. Turn the engine **OFF** and check for fuel leaks.

3.0L Engine

1. Disconnect the negative battery cable and relieve the fuel system pressure.

2. Remove the air intake throttle body assembly as follows:

   a. Remove the engine air cleaner outlet tube between the air cleaner and throttle body.

   b. Remove the snow shield from the power steering pump bracket and accelerator cable bracket.

   c. Tag and disconnect the vacuum lines at the vacuum fittings on the intake manifold and PCV hose.

   d. Disconnect and remove the accelerator and cruise control cables from the accelerator mounting bracket and throttle lever.

   e. Remove the alternator support brace.

   f. Remove the throttle body-to-lower intake manifold retaining bolts and stud bolts. Remove the throttle body assembly.

3. Disconnect the fuel lines from the fuel supply manifold.

4. Disconnect the wiring harness from the injectors.

5. Disconnect the vacuum line from the fuel pressure regulator.

6. Remove the 4 fuel supply manifold retaining bolts.

7. Carefully disengage the fuel supply manifold from the fuel injectors by lifting and gently rocking the manifold.

8. Remove the injectors by lifting while gently rocking from side-to-side. Remove and discard the O-rings.

To install:

9. Lubricate new O-rings with clean light grade oil and install 2 on each injector.

NOTE: Never use silicone grease at it will clog the injectors.

10. Install the injectors in the fuel supply manifold, using a light, twisting, pushing motion.

11. Carefully install the fuel supply manifold and injectors into the lower intake manifold, 1 side at a time. Push the fuel supply manifold down to ensure that all injector O-rings are fully seated in the fuel supply manifold cups and intake manifold.

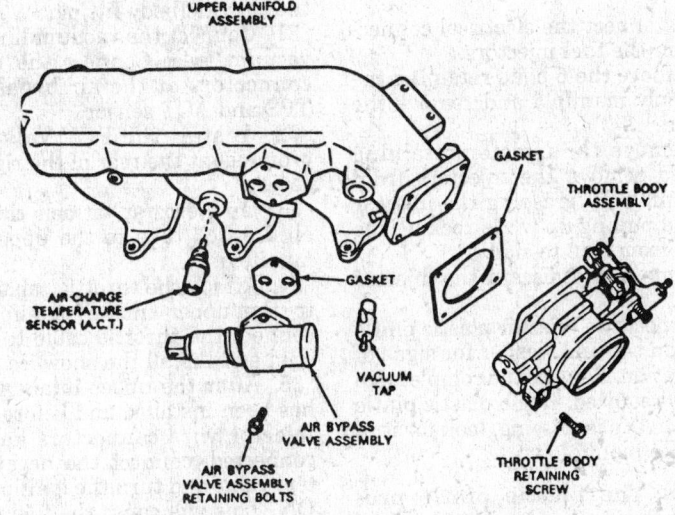

Upper intake manifold assembly—2.9L engine

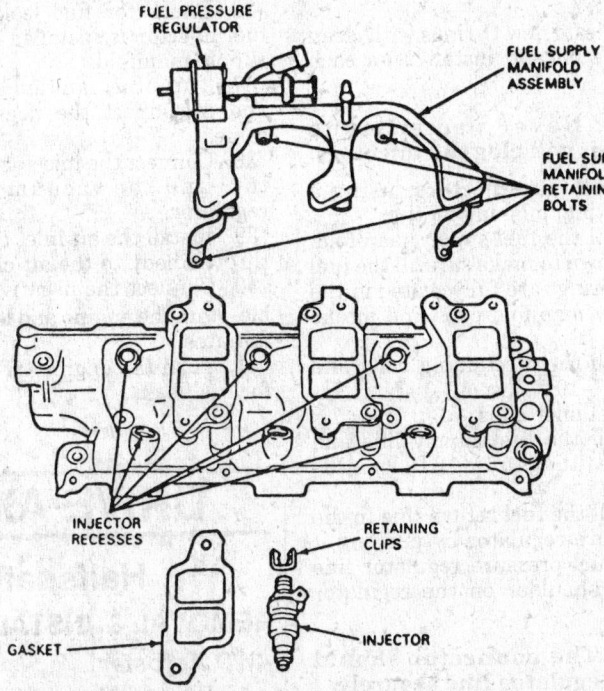

Fuel supply manifold assembly—2.9L engine

12. While holding the fuel supply manifold in place, install the 2 retaining bolts hand-tight and then tighten to 6–8 ft. lbs. (8–12 Nm).

13. Repeat Steps 11 and 12 for the other side of the fuel supply manifold.

14. Connect the fuel supply and return lines.

15. After the fuel supply manifold has been installed and before the fuel injector wire connectors have been connected, connect the negative battery cable and turn the ignition ON. This will cause the fuel pump to run for 2–3 seconds and pressurize the fuel system.

16. Check for leaks where the fuel injector is installed into the fuel supply manifold.

17. Disconnect the negative battery cable.

18. Connect the fuel injector electrical connectors.

19. Install the air intake throttle body in the reverse order of removal. Tighten the bolts, in sequence, to 19 ft. lbs. (25 Nm).

20. Connect the vacuum line to the fuel pressure regulator.

21. Connect the negative battery cable, start the engine and let it idle for 2 minutes.

22. Turn the engine OFF and check for fuel leaks.

4.0L Engine

1. Disconnect the negative battery cable and relieve the fuel system pressure.

2. Disconnect the electrical connectors at the air bypass valve, TPS and ACT sensor.

3. Remove the snow/ice shield to expose the throttle linkage. Remove the throttle cable bracket and disconnect the cable from the ball stud on the throttle body.

4. Remove the air inlet tube from the air cleaner to the throttle body.

5. Disconnect the PCV valve from the valve cover.

6. Disconnect the spark plug wires from the comb at the rear of the manifold.

7. Remove the canister purge line from the fitting in the throttle housing.

8. On Aerostar, remove the bolt retaining the engine oil dipstick tube.

9. Remove the bolt that retains the air conditioner line at the upper rear of the upper manifold.

10. Remove the 6 upper intake manifold retaining nuts and remove the upper intake and throttle body assembly.

11. Disconnect the fuel supply line fitting at the fuel manifold.

12. Disconnect the fuel return line from the fuel pressure regulator as follows:

a. Disengage the locking tabs on the connector retainer and separate the retainer halves.

b. Inspect the visible internal portion of the fitting for dirt accumulation. Clean the fitting before disassembly.

c. To disengage the fitting from the regulator, push the fitting toward the regulator, insert the fingers on fuel line coupling key T90P-9550–A or equivalent, into the slots in the coupling.

d. Using the tool, pull the fitting from the regulator.

NOTE: If the fitting has been properly disengaged, the fitting should slide off the regulator with minimum effort.

13. Disconnect the electrical connectors from the fuel injectors.

14. Remove the 6 bolts retaining the fuel supply manifold and remove the manifold.

15. Remove the injector retaining clips and remove the injectors from the manifold by grasping the injector body and pulling up while rocking the injector from side-to-side.

16. Remove and discard the injector O-rings.

17. Inspect the injector plastic pintle protection cap and washer for signs of deterioration. Replace the complete injector as required. If the plastic pintle protection cap is missing, look for it in the intake manifold.

NOTE: The plastic pintle protection cap is not available as a separate part.

To install:

18. Lubricate new O-rings with clean light grade oil and install 2 on each injector.

NOTE: Never use silicone grease at it will clog the injectors.

19. Install the injectors, using a light, twisting, pushing motion.

20. Install the fuel supply manifold, pushing down to make sure all the fuel injector O-rings are fully seated in the fuel supply manifold cups and intake manifold.

21. Install the 6 retaining bolts and tighten to 7–10 ft. lbs. (10–14 Nm). Install the retainer clips.

22. Install the fuel supply line and tighten the fitting to 15–18 ft. lbs. (20–24 Nm).

23. Install the fuel return line to the fuel pressure regulator by pushing it onto the fuel pressure regulator line up to the shoulder on the regulator line.

NOTE: The connector should grip the regulator line securely.

24. Install the connector retainer and snap the 2 halves of the retainer together.

25. Clean and inspect the mounting faces of the fuel manifold and upper intake manifold.

26. Position a new gasket on the mounting studs and install the upper intake manifold on the studs.

27. Install the 6 upper intake manifold retaining nuts and tighten to 15–18 ft. lbs. (20–25 Nm).

28. Connect the spark plug wires to the retainer comb at the rear of the intake manifold.

29. Attach the air conditioner line retainer and automatic transmission vacuum line retainer at the upper intake manifold.

30. Install the canister purge line on the throttle body fitting.

31. Connect the vacuum lines to the vacuum tree. Connect the electrical connectors at the air bypass valve, TPS and ACT sensor.

32. Install the PCV valve in the grommet at the rear of the right valve cover.

33. On Aerostar, attach the engine oil dipstick tube to the upper intake manifold.

34. Attach the throttle cable bracket to the upper intake manifold, then connect the throttle cable to the ball stud and install the snow/ice shield.

35. After the upper intake manifold has been installed and before the fuel injector wire connectors have been connected, connect the negative battery cable and turn the ignition switch ON. This will cause the fuel pump to run for 2–3 seconds and pressurize the system.

36. Check for fuel leaks where the fuel injector is installed into the fuel supply manifold.

37. Turn the ignition switch OFF and disconnect the negative battery cable.

38. Connect the injector wire connectors and the vacuum line to the regulator.

39. Install the air inlet tube from the throttle body to the air cleaner.

40. Connect the negative battery cable, start the engine and let it idle for 2 minutes.

41. Turn the engine OFF and check for fuel leaks.

DRIVE AXLE

Halfshaft

REMOVAL & INSTALLATION

4WD Aerostar

1. Raise and safely support the vehicle.

2. Remove the wheel and tire assembly and the hub retainer nut and washer. Discard the nut.

3. Mark the differential shaft flange in relation to the halfshaft so they can be reinstalled in their original position.

4. Remove the bolts and disconnect the halfshaft inboard flanges from the differential axle shaft flanges.

5. Support the end of the shaft by suspending it from an underbody component with mechanics wire. Do not allow the shaft to hang unsupported as damage to the outboard CV-joint may result.

6. Loosen the shock absorber on the lower control arm and move it to the side. Remove the rubber jounce bumper.

7. Separate the outboard CV-joint from the hub using front hub tools T81P–1104–C with T81P–1104–A and adapters T83P–1104–BH1 and T86P–1104–A1 or equivalents, and free the hub, bearing and knuckle assembly from the halfshaft by pushing in the CV-joint outer shaft until it is loose in the assembly.

NOTE: Never use a hammer to separate the outboard CV-joint stub shaft from the hub. Damage to the CV-joint threads and internal components may result.

8. Remove the halfshaft assembly from the vehicle. Once the halfshaft(s) have been removed, the vehicle must not be driven or rolled with the vehicle weight supported by the hub bearing.
To install:

9. Carefully align the splines of the outboard CV-joint stub shaft with the splines in the hub and push the shaft into the hub as far as possible.

10. Temporarily fasten the rotor to the hub with washers and 2 lug nuts. Insert a suitable steel rod into the rotor and rotate clockwise to contact the knuckle to prevent the rotor from turning during CV-joint installation.

11. Install the hub nut washer and a new hub retainer nut. Manually thread the nut onto the halfshaft as

far as possible, then tighten to 170–210 ft. lbs. (230–283 Nm).

12. Install the inboard flange of the halfshaft to the differential output flange, aligning the marks that were made during removal. Install the flange bolts and tighten to 22–29 ft. lbs. (30–40 Nm).

13. Install the wheel and tire assembly and lower the vehicle.

CV-Boot

REMOVAL & INSTALLATION

4WD Aerostar

1. Remove the halfshaft assembly from the vehicle.

2. Clamp the halfshaft in a vise equipped with jaw caps to prevent damage to machined surfaces. Do not allow the vise jaws to contact the boot or its clamp.

3. Remove the slip yoke boot clamps and separate the outboard shaft and joint assembly from the inboard slip yoke assembly. Remove and discard the slip yoke boot.

4. Cut the large boot clamp using suitable cutters and peel away from the boot. After removing the clamp, roll the boot back over the shaft and remove and discard the boot.

5. Clean all parts in suitable parts cleaning solvent.

NOTE: Do not submerge the slip yoke assembly U-joint in cleaning solvent.

6. Inspect the CV-joint and slip yoke assembly for excessive wear, pitting, rust and broken parts. If the CV-joint is no longer usable, the complete outer shaft assembly must be replaced.
To install:

7. Fill the CV-joint area around the balls with 2.8 oz. of suitable CV-joint grease. Then spread 1.4 oz. of grease evenly inside the large boot for a total combined fill of 4.2 oz.

8. Assemble the large outboard boot onto the outboard shaft and joint as-

sembly, making sure the boot is seated in the boot grooves. Tighten the clamps using suitable crimping pliers, but do not overtighten, as it may damage the clamp and/or boot.

9. Assemble the small boot and clamps onto the outboard shaft and joint assembly, but do not crimp the clamps at this time.

10. Coat the spline end of the outboard shaft and joint assembly with lubricant and assemble into the inner slip yoke assembly.

11. Slip the boot into place, making sure the boot is seated in the boot grooves. Tighten the clamps using a suitable tool, but do not overtighten, as it may damage the clamp and/or boot.

12. Install the halfshaft in the vehicle.

Driveshaft and U-Joints

REMOVAL & INSTALLATION

Rear Driveshaft

1. Raise and safely support the vehicle.

2. Mark the driveshaft in relation to the rear axle flange. If necessary, mark the relation of the driveshaft to the transfer case flange.

3. If equipped with a center bearing assembly, remove the retaining bolts and the spacers under the center bearing bracket, if installed.

4. Remove the attaching bolts and disconnect the driveshaft from the rear axle flange.

5. On 2WD vehicles, slide the driveshaft rearward until the slip yoke clears the transmission extension housing and remove the driveshaft. Plug the extension housing to prevent fluid leakage.

6. On 4WD vehicles, remove the bolts attaching the driveshaft to the transfer case flange and remove the driveshaft.

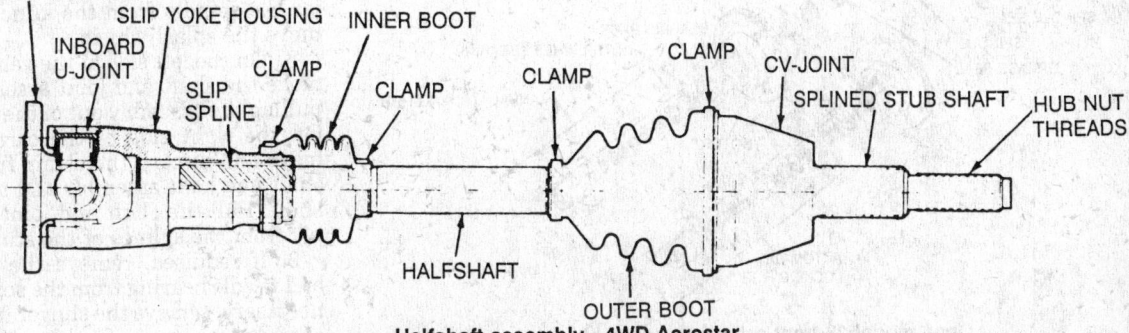

Halfshaft assembly—4WD Aerostar

CIRCULAR COMPANION FLANGE · SLIP YOKE HOUSING · INNER BOOT · INBOARD U-JOINT · CLAMP · SLIP SPLINE · CLAMP · CLAMP · CLAMP · CV-JOINT · SPLINED STUB SHAFT · HUB NUT THREADS · HALFSHAFT · OUTER BOOT

To install:

7. On 2WD vehicles, lubricate the splines of the slip yoke. Remove the plug from the extension housing and install the driveshaft assembly. Do not allow the slip yoke to bottom on the output shaft with excessive force.

8. On 4WD vehicles, install the driveshaft to the transfer case flange, aligning the marks that were made during removal. Install the attaching bolts and tighten to 61–87 ft. lbs. (83–118 Nm) if equipped with constant velocity U-joints or 12–16 ft. lbs. (17–22 Nm) if equipped with double cardan U-joints.

9. Connect the driveshaft to the rear axle flange, aligning the marks that were made during removal. In-stall the retaining bolts and tighten to 61–87 ft. lbs. (83–118 Nm) on 1988–89 vehicles or 70–95 ft. lbs. (95–129 Nm) on 1990–92 vehicles.

10. If equipped, install the center bearing attachment bolts and tighten to 27–37 ft. lbs. (37–50 Nm). Make sure the center bearing bracket is in-stalled "square" to the vehicle. If spac-ers were installed under the center bearing bracket, make sure they are reinstalled.

11. Lower the vehicle.

Front Driveshaft

1. Raise and safely support the vehicle.

2. Remove the bolts and straps or the flange bolts retaining the drive-shaft to the transfer case. If necessary, remove the boot from the transfer case to gain access to the slip yoke.

3. Remove the bolts and straps re-taining the front U-joint to the front axle and remove the front driveshaft.

To install:

4. If equipped, lubricate the slip yoke splines and the edge of the inner diameter of the rubber boot. Slide the driveshaft into the transfer case, mak-ing sure the wide-tooth splines are properly indexed. Reposition the boot and install the clamp.

5. If equipped, install the driveshaft to the transfer case flange and install the retaining bolts. Tighten to 12–16 ft. lbs. (17–22 Nm).

6. Install the driveshaft to the front axle flange with the straps and bolts. Tighten to 10–15 ft. lbs. (14–20 Nm).

7. Lower the vehicle.

Front Axle Shaft, Bearing and Seal

REMOVAL & INSTALLATION

Except Aerostar

1. Raise and safely support the ve-hicle. Remove the front wheel and tire assemblies.

2. Remove the disc brake caliper and wire it to the frame. Do not let the caliper hang by the brake hose.

3. Remove the hub locks, wheel bearings and locknuts.

4. Remove the hub, rotor and outer wheel bearing.

5. Remove the grease seal from the rotor with a seal removal tool. Remove the inner wheel bearing.

6. If the wheel bearings are to be re-placed, remove the inner and outer bearing races with a suitable puller or a hammer and brass drift.

7. Remove the nuts retaining the spindle to the steering knuckle. Tap the spindle with a plastic hammer to jar the spindle from the knuckle. Re-move the splash shield.

8. On the left side of the vehicle, re-move the shaft and joint assembly by pulling the assembly out of the carrier. On the right side of the carrier, re-move and discard the clamp from the shaft and joint assembly and the stub shaft. Pull the shaft and joint assem-bly from the splines of the stub shaft.

9. If required, remove the oil seal and needle bearing from the spindle. If necessary, remove the slinger from the shaft by driving it off with a hammer.

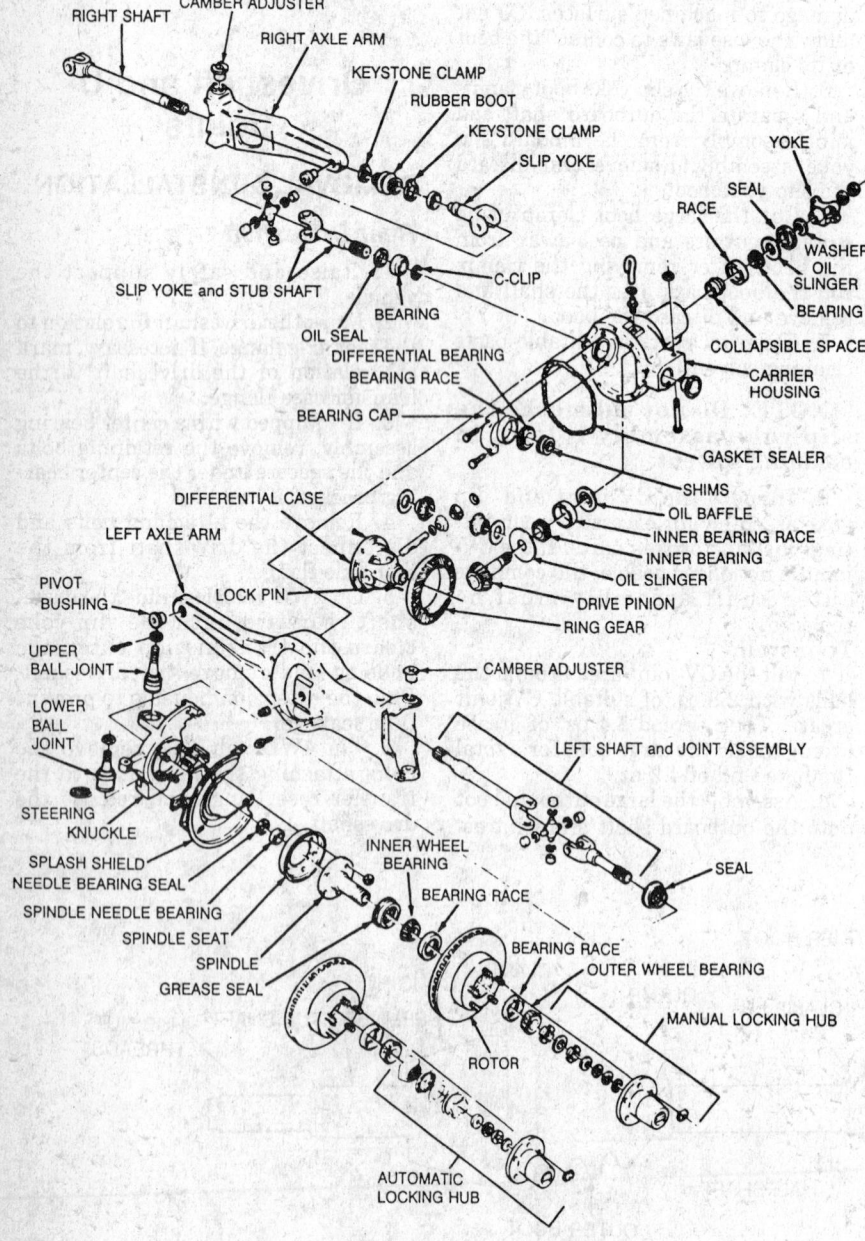

Dana model 28 front drive axle assembly

To install:

10. If removed, install a new bearing and seal in the spindle and/or press on a new shaft slinger.

11. On the right side of the carrier, install the rubber boot and new keystone clamps on the stub shaft slip yoke. Slide the right shaft and joint assembly into the slip yoke making sure the splines are fully engaged. Slide the boot over the assembly and crimp the keystone clamp using suitable pliers.

NOTE: The Dana model 28 axle has phased splines; there is only 1 way to assemble the right shaft and joint assembly into the slip yoke. The Dana model 35 axle does not have a blind spline, therefore pay special attention to make sure the yoke ears are in phase (in line) during assembly.

12. On the left side of the carrier, slide the shaft and joint assembly through the knuckle and engage the splines on the shaft in the carrier.

13. Install the splash shield and spindle onto the steering knuckle. Install and tighten the spindle nuts to 45 ft. lbs. (61 Nm).

14. If removed, drive the bearing races into the rotor using a suitable driver. Pack the inner and outer wheel bearings and the lip of a new seal with high-temperature wheel bearing grease.

15. Position the inner wheel bearing in the race and install the seal using a seal installer. Install the rotor on the spindle and install the outer wheel bearing in the race.

16. Install the wheel bearing, locknut, thrust bearing, snapring and locking hubs.

17. Install the caliper and the wheel and tire assemblies. Lower the vehicle.

Aerostar

1. Remove the axle assembly from the sub-frame and the vehicle.

2. Remove the cover plate and drain the lubricant. Mount the axle assembly in a suitable fixture.

3. Rotate the shafts so the open side of the snapring is exposed. Remove the snaprings.

4. Remove the axle shafts. Remove and discard the oil seal and needle bearings.

To install:

5. Make sure the bearing bore is free from nicks and burrs. Install a new caged needle bearing on installation tool T83T–1244–A or equivalent, with the manufacturer name and part

number facing outward towards the tool. Drive the needle bearing in until it is seated in the bore. The tool controls the installation depth of the seal and bearing.

6. Coat a new seal with lubricant and install using a seal installer.

7. Install the axle shafts into the carrier so the groove in the shaft is visible in the differential case.

8. Install the snapring in the axle shaft groove. Remove the axle assembly from the holding fixture.

9. Clean all old sealer from the carrier and cover surfaces. Apply a bead of RTV sealer ⅛–¼ in. wide. The bead should be continuous and should not pass through or outside the holes.

10. Install the cover and tighten the bolts to 20–25 ft. lbs. (27–34 Nm). Fill the carrier with the proper type and quantity of fluid.

11. Install the axle assembly in the vehicle.

Rear Axle Shaft, Bearing and Seal

REMOVAL & INSTALLATION

Ford 7.5 in. and 8.8 in. Rear Axles

1. Raise and safely support the vehicle.

2. Remove the rear wheel and tire assemblies and the brake drums.

3. Clean all dirt from the carrier cover area. Position a drain pan under the carrier, remove the cover and drain the rear axle.

NOTE: Whenever a plastic rear axle cover is removed, it must be replaced with a new cover and bolts. Steel rear axle covers may be reused.

4. For all 8.8 in. axles and all 7.5 in. axles except 3.73: 1 and 4.10: 1 ratio axles, proceed as follows:

   a. Remove the differential pinion shaft lock bolt and pinion shaft.

   b. Push the flanged end of the axle shafts toward the center of the vehicle and remove the C-lock from the button end of the axle shaft.

   c. Remove the axle shaft from the housing.

   d. Reinstall the pinion shaft and lock bolt to ensure the pinion gears remain in place.

5. On 7.5 in. axles equipped with 3.73: 1 and 4.10: 1 axle ratios, proceed as follows:

   a. Remove the pinion shaft lock bolt.

   b. Push out the pinion shaft until the step on the shaft contacts the ring gear.

   c. Remove the C-lock from the axle shaft.

   d. Remove the axle shaft from the housing.

   e. Reinstall the pinion shaft and lock bolt to ensure the pinion gears remain in place.

6. Using bearing remover T85L–1225–AH or equivalent and a suitable slide hammer, remove the axle bearing and seal as a unit.

To install:

7. Lubricate a new bearing with rear axle lubricant and install in the housing bore using a suitable driver.

8. Apply grease to the lips of a new axle seal and install, using a seal installer.

9. Remove the pinion shaft lock bolt and pinion shaft. On 7.5 in. axles equipped with 3.73:1 and 4.10:1 axle ratios, push out the pinion shaft until the step contacts the ring gear.

10. Slide the axle shaft into the axle housing, being careful not to damage the seal or axle bearing. Start the splines into the side gear and push firmly until the button end of the axle shaft can be seen in the differential case.

11. Install the C-lock on the button end of the axle shaft, then pull the shaft outboard until the shaft splines engage and the C-lock seats in the counterbore of the differential side gear.

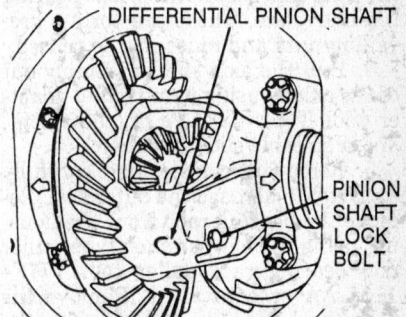

Differential pinion shaft and pinion shaft lock bolt location—Ford 7.5 in. and 8.8 in. rear axles.

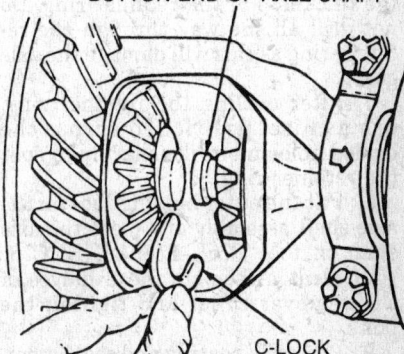

Axle shaft C-lock installation—Ford 7.5 in. and 8.8 in. rear axles.

NOTE: On 8.8 in. axles, a rubber O-ring is used to hold the C-lock in position on the axle shaft. Make sure the O-ring is in the groove at the button end of the axle shaft before installing the C-lock.

12. Slide the pinion shaft through the case and pinion gears, aligning the hole in the shaft with the lock bolt hole. Install the lock bolt and tighten to 15–30 ft. lbs. (20–40 Nm).

13. Clean all old sealer from the carrier and cover surfaces. Apply a bead of RTV sealer ⅛–¼ in. wide. The bead should be continuous and should not pass through or outside the holes.

14. Install the cover and tighten the bolts to 15–20 ft. lbs. (21–27 Nm) if equipped with a plastic cover or 25–35 ft. lbs. (34–37 Nm) if equipped with a steel cover. Fill the carrier with the proper type and quantity of fluid.

15. Install the brake drums and the wheel and tire assemblies. Lower the vehicle.

Dana 30 and 35–1A Rear Axles

1. Raise and safely support the vehicle.

2. Remove the rear wheel and tire assemblies and the brake drums.

3. Working through the hole provided in the axle shaft flange, remove the nuts that secure the wheel bearing retainer plate. These are torque prevailing nuts and must not be reused.

4. Pull the axle shaft assembly out of the axle housing using puller adapter tool T66L-4234-A or equivalent, and a suitable slide hammer.

5. Remove the aaxle shaft carefully so as not to damage the outer seal. Remove the bearing race from the housing using a slide hammer type puller and bearing race puller tool T77F-1102-AA or equivalent. Remove the brake backing plate and wire it to the chassis.

6. Mount the axle shaft in a suitable fixture. Drill a ¼–½ in. hole in the outside diameter of the inner retainer ring to a depth approximately ⅜ in. the thickness of the retainer ring. Do not drill all the way through the retainer ring as this will damage the axle shaft.

7. After drilling the retainer ring, use a chisel positioned across the drilled hole and strike sharply to split the retainer ring.

8. Put the outer bearing race on the axle shaft assembly and place the axle shaft in tool T75L-1165-A, B, C or equivalent. Assemble the 2 halves of the remover collet and tighten the bolts.

9. Press the bearing and seal assembly off of the shaft. Never use heat as this would damage the axle shaft. In-

spect the retainer plate for possible distortion and replace if damaged.

To install:

10. Install the outer retainer plate, if removed. Make sure it is not installed backwards.

11. Place a new lubricated seal and bearing on the axle shaft, making sure the race rib ring is facing the axle flange.

12. Press the tapered-bearing and seal assembly onto the axle shaft using tools T75L-1165-B, service plate and adapter tool T75L-1165-DA or equivalents. Apply enough pressure to seat the bearing against the axle shaft shoulder. Do not attempt to press on the bearing retainer at the same time.

13. Position a new bearing retainer on the shaft, then press it into position firmly against the bearing.

14. Apply lubricant to the outer diameter of the race and seal. Install the brake backing plate and attaching bolts.

15. Before sliding the shaft assembly into the axle housing, make sure the outer seal is fully mounted on the bearing.

16. Carefully slide the axle shaft into the housing, start the axle splines into the side gear and push the shaft in until the bearing bottoms in the housing.

17. Install the bearing retainer plate and nuts. Tighten the nuts to 25–35 ft. lbs. (34–47 Nm).

18. Install the brake drum and the wheel and tire assembly.

19. Add lubricant through the filler hole until the level reaches 3.8 ± ¼ in. below the bottom of the filler hole. The axle must be in the running position and the vehicle level.

20. Lower the vehicle.

Front Wheel Hub, Knuckle/Spindle and Bearings

REMOVAL & INSTALLATION

Except Aerostar

1. Raise and safely support the vehicle. Remove the front wheel and tire assemblies.

2. Remove the disc brake caliper and wire it to the frame. Do not let the caliper hang by the brake hose.

3. Remove the hub locks, wheel bearings and locknuts.

4. Remove the hub, rotor and outer wheel bearing.

5. Remove the grease seal from the rotor with a seal removal tool. Remove the inner wheel bearing.

6. If the wheel bearings are to be replaced, remove the inner and outer bearing races with a suitable puller or a hammer and brass drift.

7. Remove the nuts retaining the spindle to the steering knuckle. Tap the spindle with a plastic hammer to jar the spindle from the knuckle. Remove the splash shield.

8. On the left side of the vehicle, remove the shaft and joint assembly by pulling the assembly out of the carrier. On the right side of the carrier, remove and discard the clamp from the shaft and joint assembly and the stub shaft. Pull the shaft and joint assembly from the splines of the stub shaft.

9. Place the spindle in a vise on the second step of the spindle. Wrap a shop towel around the spindle or use a brass-jawed vise to protect the spindle.

10. Remove the oil seal and needle bearing from the spindle with a slide hammer and seal remover TOOL-1175-AC or equivalent. If necessary, remove the slinger from the shaft by driving off with a hammer.

11. Remove the cotter pin from the tie rod nut and then remove the nut. Tap on the tie rod stud to free it from the steering arm.

12. Remove the upper ball joint snapring and remove the upper ball joint pinch bolt. Loosen the lower ball joint nut to the end of the stud.

13. Strike the inside of the knuckle near the upper and lower ball joints to break the knuckle loose from the ball joint studs.

14. Remove the camber adjuster sleeve. Note the position of the slot in the camber adjuster so it can be reinstalled in the same position during assembly.

15. Remove the lower ball joint nut. Place the knuckle in a vise and remove the snapring from the bottom ball joint socket, if equipped.

16. Assemble C-frame T74P-4635-C and ball joint remover T83T-3050-A or equivalents on the lower ball joint. Turn the forcing screw clockwise until the lower ball joint is removed from the steering knuckle.

17. Assemble the C-frame and ball joint remover on the upper ball joint and remove in the same manner.

NOTE: Always remove the lower ball joint first.

To install:

18. Clean the steering knuckle bore and insert the lower ball joint in the knuckle as straight as possible.

19. Assemble C-frame T74P-4635-C, ball joint installer T83T-3050-A and receiver cup T80T-3010-A3 or equivalents to install the lower ball joint. Turn the forcing screw clockwise until the lower ball joint is firmly seated. Install the snapring on the lower ball joint.

NOTE: The lower ball joint must always be installed first.

20. Assemble the C-frame, ball joint installer and receiver cup to install the upper ball joint. Turn the forcing screw clockwise until the ball joint is firmly seated.

21. Install the camber adjuster into the support arm, making sure the slot is in the original position.

NOTE: The torque sequence in Steps 22 and 23 must be followed exactly when securing the knuckle. Excessive knuckle turning effort may result in reduced steering returnability if this procedure is not followed.

22. Install a new nut on the bottom ball joint stud. Tighten the nut to 90 ft. lbs. (122 Nm) minimum, then tighten to align the next slot in the nut with the hole in the stud. Install a new cotter pin.

23. Install the snapring on the upper ball joint stud. Install the upper ball joint pinch bolt and tighten to 48–65 ft. lbs. (65–88 Nm).

NOTE: The camber adjuster will seat itself into the knuckle at a predetermined position during the tightening sequence. Do not attempt to adjust this position.

24. Clean all dirt and grease from the spindle bearing bore. The bearing bores must be free from nicks and burrs.

25. Place the bearing in the bore with the manufacturers identification facing outward. Drive the bearing into the bore using spindle bearing replacer T80T–4000–S and driver handle T80T–4000–W or equivalents.

26. Install the grease seal in the bearing bore with the lip side of the seal facing towards the tool. Drive the seal in the bore using the same tools as in Step 25. Coat the bearing seal lip with high-temperature lubricant.

27. If removed, press on a new shaft slinger.

28. On the right side of the carrier, install the rubber boot and new keystone clamps on the stub shaft slip yoke. Slide the right shaft and joint assembly into the slip yoke making sure the splines are fully engaged. Slide the boot over the assembly and crimp the keystone clamp using suitable pliers.

NOTE: The Dana model 28 axle has phased splines; there is only 1 way to assemble the right shaft and joint assembly into the slip yoke. The Dana model 35 axle does not have a blind spline, therefore pay special attention to make sure the yoke ears are in phase (in line) during assembly.

29. On the left side of the carrier, slide the shaft and joint assembly through the knuckle and engage the splines on the shaft in the carrier.

30. Install the splash shield and spindle onto the steering knuckle. Install and tighten the spindle nuts to 45 ft. lbs. (61 Nm).

31. If removed, drive the bearing races into the rotor using a suitable driver. Pack the inner and outer wheel bearings and the lip of a new seal with high-temperature wheel bearing grease.

32. Position the inner wheel bearing in the race and install the seal using a seal installer. Install the rotor on the spindle and install the outer wheel bearing in the race.

33. Install the wheel bearing, locknut, thrust bearing, snapring and locking hubs.

34. Install the caliper and the wheel and tire assemblies. Lower the vehicle.

Aerostar

1. Place the front wheels in the straight ahead position. Raise and safely support the vehicle.

2. Raise and safely support the vehicle. Remove the wheel and tire assembly.

3. Remove the caliper and support it aside with mechanics wire. Do not let the caliper hang by the brake hose. Remove the brake rotor.

4. Remove the cotter pin and nut from the tie rod end stud. Disconnect the tie rod end from the steering knuckle.

5. Support the lower control arm. Remove the cotter pin and loosen the nut retaining the knuckle to the lower control arm ball joint. Disconnect the lower ball joint using tool T64P–3590–F or equivalent. Remove the tool and ball joint retaining nut.

6. With the vehicle body securely supported, pull down on the knuckle until the lower ball joint is disengaged from the steering knuckle.

7. Install hub remover T81P–1104–A with T81P–1104–C and hub knuckle adapters T83P–1104–BH1 and T88P–1104–A1 or equivalents, and free the hub, bearing and knuckle assembly from the halfshaft by pushing in the CV-joint outer shaft until it is loose in the assembly.

8. Remove the bolt and nut retaining the spindle to the upper control arm ball joint. While supporting the halfshaft, remove the knuckle by pulling the knuckle down and out of the upper ball joint and outward away from the halfshaft. Wire the halfshaft to the body to maintain it in a level position.

9. Remove the 3 bolts retaining the hub assembly to the steering knuckle and remove the hub assembly.

NOTE: The hub and bearing assembly is not serviceable. If the bearing requires replacement, it must be replaced as an assembly.

To install:

10. Install the hub assembly into the steering knuckle and tighten the 3 bolts.

11. Position the steering knuckle on the CV-joint outer shaft and into the lower ball joint. Make sure the splines of the CV-joint shaft are in proper mesh in the hub. Install the lower ball joint nut and tighten to 80–120 ft. lbs. (108–163 Nm). If required, continue turning the nut to the next castellation and install the cotter pin.

12. Temporarily install the rotor on the hub with washers and 2 lug nuts. Insert a steel rod into the rotor diameter and rotate clockwise to contact the knuckle.

13. Install the hub nut washer and a new hub nut. Rotate the nut clockwise to seat the CV-joint, but do not final torque at this time.

14. Position the upper control arm ball joint stud to the steering knuckle.

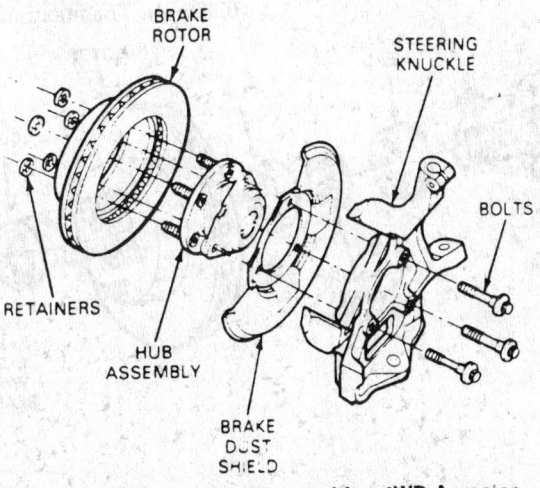

Steering knuckle and hub assembly—4WD Aerostar

Install the nut and bolt and tighten to 27–37 ft. lbs. (37–50 Nm).

15. Connect the tie rod end to the steering knuckle. Firmly seat the tie rod end stud into the tapered hole to prevent rotation while tightening. Install the nut and tighten to 52–74 ft. lbs. (70–100 Nm). If required, continue turning the nut to the next castellation and install the cotter pin.

16. Using a steel rod inserted into the rotor as in Step 12, tighten the hub nut to 170–210 ft. lbs. (230–285 Nm). Remove the wheel nuts and washers.

17. Install the dust shield, caliper, rotor and wheel and tire assembly. Lower the vehicle.

Manual Locking Hubs

REMOVAL & INSTALLATION

1. Raise and support the vehicle safely.

2. Remove the lug nuts and remove the wheel and tire assembly.

3. Remove the retainer washers from the lug nut studs and remove the manual locking hub assembly. To remove the internal hub lock assembly from the outer body assembly, remove the outer lock ring seated in the hub body groove. The internal assembly, spring and clutch gear will now slide out of the hub body. Do not remove the screw from the plastic dial.

4. Rebuild the hub assembly in the reverse order of disassembly.

5. Adjust the wheel bearing if necessary. Install the manual locking hub assembly over the spindle and place the retainer washers on the lug nut studs.

6. Install the wheel and tire assembly and lower the vehicle.

ADJUSTMENT

1. Raise and safely support the vehicle. Remove the wheel and tire assembly.

2. Remove the retainer washers from the lug nut studs and remove the manual locking hub assembly from the spindle.

3. Remove the snapring from the end of the spindle shaft.

4. On Dana model 28 axles, remove the axle shaft spacer, needle thrust bearing and bearing spacer. On Dana model 35 axles, remove the axle shaft spacer.

5. Remove the outer wheel bearing locknut from the spindle using locknut wrench T86T–1197–A or equivalent. Make sure the tabs on the tool engage the slots in the locknut.

6. Remove the locknut washer from the spindle.

7. Loosen the inner wheel bearing locknut using locknut wrench T86T–1197–A or equivalent. Make sure the tabs on the tool engage the slots in the locknut and the slot in the tool is centered over the locknut pin.

8. Tighten the inner locknut to 35 ft. lbs. (47 Nm) to seat the bearings.

9. Spin the rotor and back off the inner locknut ¼ turn. Retighten the inner locknut to 16 inch lbs. (1.8 Nm). Install the lockwasher on the spindle. It may be necessary to tighten the inner locknut slightly so the pin on the locknut aligns with the closest hole in the lockwasher.

10. Install the outer wheel bearing locknut using locknut wrench T86T–1197–A or equivalent. Tighten the locknut to 150 ft. lbs. (203 Nm).

11. On Dana model 28 axles, install the bearing thrust spacer, needle thrust bearing and axle shaft spacer. On Dana model 35 axles, install the axle shaft spacer.

12. Clip the snapring onto the end of the spindle. Install the manual hub assembly over the spindle and install the retainer washers.

13. Install the wheel and tire assembly. Check the endplay of the wheel and tire assembly on the spindle. Final endplay should be 0–0.003 in. (0–0.08mm). The maximum torque to ro-tate the hub should be 25 inch lbs. (2.8 Nm).

14. Lower the vehicle.

Automatic Locking Hubs

REMOVAL & INSTALLATION

1. Raise and support the vehicle safely. Remove the wheel lug nuts and remove the wheel and tire assembly.

2. Remove the retainer washers from the lug nut studs and remove the automatic locking hub assembly from the spindle.

3. Remove the snapring from the end of the spindle shaft.

4. On Dana model 28 axles, remove the axle shaft spacer, needle thrust bearing and the bearing spacer. On Dana model 35 axles, remove the axle shaft spacer.

5. Being careful not to damage the plastic moving cam, pull the cam assembly off the wheel bearing adjusting nut. On Dana model 28 axles, remove the thrust washer and needle thrust bearing from the adjusting nut. On Dana model 35 axles, remove the 2 plastic thrust spacers from the adjusting nut.

6. Using a magnet, remove the locking key. It may be necessary to rotate the adjusting nut slightly to relieve the pressure against the locking key, before the key can be removed.

NOTE: To prevent damage to the spindle threads, look into the spindle keyway under the adjusting nut and remove the separate locking key before removing the adjusting nut.

7. Loosen the wheel bearing adjusting nut from the spindle using a 2 ⅜ in. hex socket tool.

8. While rotating the hub and rotor assembly, tighten the wheel bearing adjusting nut to 35 ft. lbs. (47 Nm) to

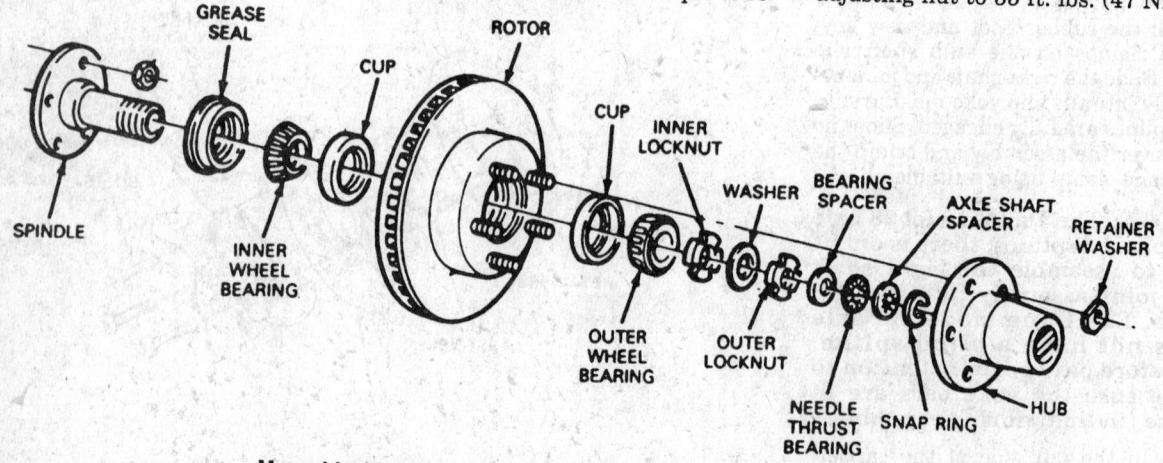

Manual locking hub assembly—Dana model 28 axle, model 35 similar

seat the bearings. Spin the rotor and back off the nut ¼ turn.

9. Retighten the adjusting nut to 16 inch lbs. (1.8 Nm) using a torque wrench. Align the closest hole in the wheel bearing adjusting nut with the center of the spindle keyway slot. Advance the nut to the next lug if required. Install the separate locking key in the spindle keyway under the adjusting nut.

NOTE: Extreme care must be taken when aligning the spindle nut adjustment lug with the center of the spindle keyway slot to prevent damage to the separate locking key.

10. On Dana model 28 axles, install the locknut needle bearing and thrust washer in the reverse order of removal. On Dana model 35 axles, install the 2 thrust spacers. Push or press the cam assembly onto the locknut by lining up the key in the fixed cam with the spindle keyway.

NOTE: Extreme care must be taken when aligning the fixed cam key with the spindle keyway to prevent damage to the fixed cam.

11. On Dana model 28 axles, install the bearing thrust washer, needle thrust bearing and axle shaft spacer. On Dana model 35 axles, install the axle shaft spacer.

12. Clip the snapring onto the end of the spindle.

13. Install the automatic locking hub assembly over the spindle by lining up the 3 legs in the hub assembly with the 3 pockets in the cam assembly. Install the retainer washers.

14. Install the wheel and tire assembly. Check the endplay of the wheel and tire assembly on the spindle. Final

endplay should be 0–0.003 in. (0–0.08mm). The maximum torque to rotate the hub should be 25 inch lbs. (2.8 Nm).

15. Lower the vehicle.

Pinion Seal
REMOVAL & INSTALLATION

NOTE: This service procedure disturbs the pinion bearing preload and this preload must be carefully reset when assembling.

1. Raise the vehicle and support it safely.

2. Remove the wheels and the brake drums.

3. Mark the driveshaft and the axle companion flange so the driveshaft can be reinstalled in the same position. Remove the driveshaft.

4. Using an inch pound torque wrench on the pinion nut, record the torque required to maintain rotation of the pinion through several revolutions.

5. While holding the companion flange with a suitable tool, remove the pinion nut. Mark the companion flange in relation to the pinion shaft so the flange can be reinstalled in the same position.

6. Using a suitable puller, remove the rear axle companion flange. Use a small prybar to remove the seal from the carrier.

To install:

7. Make sure the splines of the pinion shaft are free of burrs.

8. Apply grease to the lips of the pinion seal and install, using a seal installer.

9. Check the seal surface of the companion flange for scratches, nicks or a groove. Replace the companion

flange, as necessary. Apply a small amount of lubricant to the splines. Align the mark on the flange with the mark on the pinion shaft and install the companion flange.

NOTE: The companion flange must never be hammered on or installed with power tools.

10. Install a new nut on the pinion shaft. Hold the companion flange with a suitable tool while tightening the nut.

11. Tighten the pinion nut, rotating the pinion occasionally to ensure proper bearing seating. Take frequent pinion bearing torque preload readings until the original recorded preload reading is obtained or the following specification:

  Ford 7.5 in. and 8.8 in. rear axles: 8–14 inch lbs.
  Dana model 30 axle rear axles: 20–40 inch lbs.
  Dana model 35-1A rear axles: 15–35 inch lbs.
  Dana models 28 and Model 35 front axles: 15–25 inch lbs.
  Dana model 28-2 front axles (4WD Aerostar): 15–35 inch lbs.

NOTE: Under no circumstances should the pinion nut be backed off to reduce preload. If reduced preload is required, a new collapsible pinion spacer and pinion nut must be installed.

12. Install the driveshaft and check the fluid level in the carrier. Lower the vehicle.

Axle Housing
REMOVAL & INSTALLATION

Front Axle
AEROSTAR

1. Raise and safely support the vehicle.

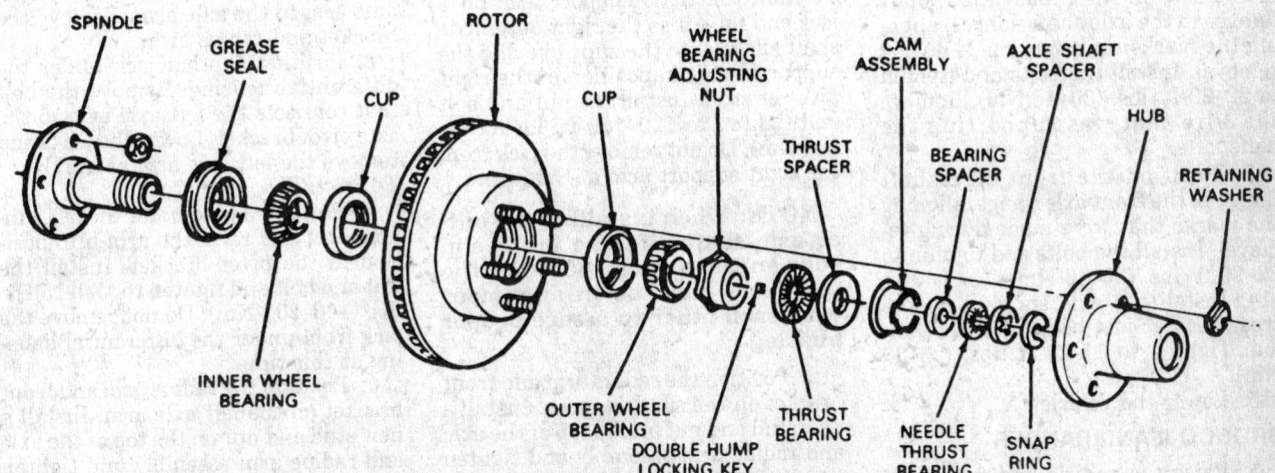

Automatic locking hub assembly—Dana model 28 axle, model 35 similar

2. Mark the driveshaft and axle flanges so they can be reassembled in their original positions. Disconnect the driveshaft from the companion flange.

3. Remove the bolts that connect the inboard halfshaft flanges to the axle shafts. Wire the halfshafts to the body in a level position.

4. Remove the vent hose from the vent fitting on the axle tube and cap the fitting to prevent any fluid leakage.

5. Remove the snubber from the crossmember rear lateral support, located below the axle pinion.

6. Support the front axle with a jack. Remove the locknuts that connect the axle mounting brackets to the crossmember.

7. Lower the jack and remove the front axle assembly.

8. If necessary, remove the screws holding the mounting bracket to the axle housing and remove the bracket.

To install:

9. If removed, attach the axle mounting bracket to the axle housing, using thread locking compound on the 2 bolts. Tighten to 70–80 ft. lbs. (95–108 Nm).

10. Raise the axle into position. Insert the axle mounting bushing studs through the axle mounting bracket. Install the locknuts using thread locking compound and tighten to 65–85 ft. lbs. (88–115 Nm). Tighten in the following order: left front, right front, pinion, rear.

NOTE: When installing the axle assembly with new mounting bushings, each nut must be tightened to full torque in 1 step to avoid coring of the locking patch on the stud, before the nut is fully torqued.

11. Attach the vent hose to the vent fitting on the axle tube. Make sure the hose is not kinked.

12. Position the inboard halfshaft flanges to the axle shaft flanges, aligning the marks that were made during removal. Install the bolts and tighten to 22–29 ft. lbs. (30–40 Nm). Remove the wire that was supporting the halfshafts.

13. Position the front driveshaft flange to the front axle flange, aligning the marks that were made during removal. Install the bolts and tighten to 22–29 ft. lbs. (30–40 Nm).

14. Install the axle snubber on the crossmember just below the axle pinion. Tighten to 17–24 ft. lbs. (23–33 Nm).

15. Lower the vehicle.

BRONCO II AND RANGER

1. Raise the vehicle and support it safely under the radius arm brackets.

2. Mark the front axle yoke and the driveshaft so they can be reassembled in the same position. Disconnect the driveshaft from the front axle yoke.

3. Remove the front wheel and tire assemblies. Remove the disc brake calipers and support them on a frame rail. Do not let the calipers hang by the brake hoses.

4. Remove the cotter pin and nut retaining the steering linkage to the knuckle. Disconnect the linkage from the knuckle.

5. Position a jack under the axle arm and slightly compress the coil spring. Remove the nut that retains the lower part of the spring to the axle arm. Slowly lower the jack and remove the coil spring, spacer, seat and stud.

6. Remove the nut and disconnect the shock absorber from the radius arm bracket. Remove the stud and bolts that connect the radius arm bracket and radius arm to the axle arm. Remove the bracket and radius arm.

7. Remove the pivot bolt that secures the right axle arm assembly to the crossmember.

8. Remove the clamps securing the axle shaft boot from the axle shaft slip yoke and axle shaft. Disconnect the outer axle shaft from the slip yoke assembly. Lower the jack and remove the right axle arm assembly.

9. Position another jack under the differential housing. Remove the bolt that connects the left axle arm to the crossmember. Lower the jacks and remove the left axle arm assembly.

To install:

10. Position a jack under the left support arm and raise the arm into position in the pivot bracket. Install the nut and bolt and tighten to 120–150 ft. lbs. (163–203 Nm). Do not remove the jack from under the differential housing at this time.

11. Place new clamps for the axle shaft boot on the axle shaft assembly. Position the right support arm on a jack and raise it so the right outer axle shaft slides onto the slip yoke and the support arm is in position in the right pivot bracket. Install the nut and bolt and tighten to 120–150 ft. lbs. (163–203 Nm). Do not remove the jack from the right support arm at this time.

NOTE: When installing the outer axle shaft into the inner slip yoke on Dana model 35 axles, the yoke ears must be in alignment with each other to assure proper phasing.

12. Position the radius arm and front bracket on the support arms. Install a new stud and nut on the top of the axle and radius arm assembly and tighten to 190–230 ft. lbs. (258–311 Nm). Install the bolts in the front of the brack-

et and tighten to 27–37 ft. lbs. (37–50 Nm).

13. Install the seat, spacer retainer and coil spring on the stud and nut. Raise the jack to compress the coil spring. Install the nut and tighten to 70–100 ft. lbs. (95–135 Nm).

14. Connect the shock absorber to the support arm assembly. Install the nut and tighten to 42–72 ft. lbs. (57–97 Nm).

15. Connect the tie rod ball joint to the knuckle. Install the nut and tighten to 50–75 ft. lbs. (68–101 Nm). Lower the jacks from the support arms.

16. Install the brake calipers and the wheel and tire assemblies.

17. Connect the driveshaft to the front axle yoke, aligning the marks that were made during removal. Install the U-bolts and tighten the nuts to 8–15 ft. lbs. (11–20 Nm).

18. Lower the vehicle.

EXPLORER

1. Raise and safely support the vehicle. Remove the front axle shaft and spindle assemblies.

2. Mark the front axle yoke and the driveshaft so they can be reassembled in the same position. Disconnect the driveshaft from the front axle yoke.

3. Remove the cotter pin and nut retaining the steering linkage to the knuckle. Disconnect the linkage from the knuckle.

4. Remove the left stabilizer bar link lower bolt and remove the link from the radius arm bracket.

5. Position a jack under the left axle arm and slightly compress the coil spring. Remove the shock absorber lower nut and disconnect the shock absorber from the radius arm bracket.

6. Remove the nut that retains the lower part of the spring to the axle arm. Slowly lower the jack and remove the coil spring, spacer, seat and stud.

7. Remove the stud and bolts that connect the radius arm bracket and radius arm to the axle arm. Remove the bracket and radius arm.

8. Position another jack under the differential housing. Remove the bolt that connects the left axle arm to the axle pivot bracket. Lower the jacks and remove the left axle arm assembly.

To install:

9. Position a jack under the left support arm and raise the arm into position in the pivot bracket. Install the nut and bolt and tighten to 120–150 ft. lbs. (163–203 Nm). Do not remove the jack from under the differential housing at this time.

10. Position the radius arm and front bracket on the left axle arm. Install a new stud and nut on the top of the axle and radius arm assembly and tighten to 190–230 ft. lbs. (258–311 Nm). Install the bolts in the front of the brack-

et and tighten to 27–37 ft. lbs. (37–50 Nm).

11. Install the seat, spacer retainer and coil spring on the stud and nut. Raise the jack to compress the coil spring. Install the nut and tighten to 70–100 ft. lbs. (95–135 Nm).

12. Connect the shock absorber to the radius arm. Install the nut and tighten to 42–72 ft. lbs. (57–97 Nm).

13. Connect the tie rod ball joint to the knuckle. Install the nut and tighten to 50–75 ft. lbs. (68–101 Nm). Install the stabilizer bar mounting bracket and tighten to 203–240 ft. lbs. (275–325 Nm).

14. Connect the front driveshaft to the front axle yoke, aligning the marks that were made during removal. Install the U-bolts and tighten the nuts to 8–15 ft. lbs. (11–20 Nm).

15. Install the spindle and axle shaft assemblies. Lower the vehicle.

Rear Axle

AEROSTAR

1. Raise and safely support the vehicle.

2. Release the parking brake cable tension by pulling rearward on the front cable approximately 2 in. Clamp the cable behind the crossmember to release the tension on the rear parking brake cables.

3. Remove the parking brake cables from the equalizer. Compress the tabs on the retainers and pull the cables through the rear crossmember.

4. Mark the driveshaft to the rear axle so they can be reassembled in the same position. Remove the driveshaft. Remove the wheel and tire assemblies.

5. Disconnect the brake hose from the chassis brake line. Plug the line to prevent fluid loss.

6. Disconnect the shock absorbers from the lower control arm. Discon-

nect the axle vent tube from the clip on the frame.

7. Lower the axle until the springs are no longer under compression. Remove the spring retainers and the coil springs.

8. Raise the axle to the normal load position and disconnect the control arms at the axle.

9. Remove the bolt and nut retaining the upper control arm to the rear axle. Remove the upper control arm from the axle. Mark the position of the cam adjuster in the axle bushing.

10. Lower the axle from the vehicle.

To install:

11. Raise the axle into position.

12. Position the upper control arm over the cam adjuster and bushing. Make sure the marks scribed on the bushing and adjuster are still in alignment. Install the bolt, nut and retainer and tighten until snug. Do not final torque at this time.

13. Lower the axle to the spring unloaded position. Place the lower insulator on the control arm and the upper insulator on top of the spring. The white colored tapered coil must face upward. Install the spring in position on the control arm and axle.

14. Install the lower retainer and nut and tighten to 41–64 ft. lbs. (55–88 Nm). Install the upper retainer and bolt and tighten to 30–40 ft. lbs. (40–55 Nm).

15. Raise the axle to the normal load position and tighten the bolt and nut retaining the lower control arm to the axle to 100–129 ft. lbs. (133–176 Nm).

16. Connect the shock absorbers to the lower control arm. Install the shock bolt nut on the inside of the lower control arm bracket. Install the nut and tighten to 40–60 ft. lbs. (54–82 Nm).

17. Attach the brake hose to the

frame and chassis brake line. Connect the axle vent hose, if equipped.

18. Install the wheel and tire assemblies. Install the driveshaft, aligning the marks that were made during removal.

19. Pull the parking brake cables and retainers through the clips on the vehicle underbody side rails and through the rear crossmember. Connect the brake cables to the equalizer. Unclamp the front parking brake cable to restore cable tension.

20. Bleed the brake system and lower the vehicle.

BRONCO II AND RANGER

1. Raise the vehicle and support it safely.

2. Remove the cover and drain the lubricant from the axle.

3. Remove the rear wheel and tire assemblies and remove the axle shafts.

4. Remove the 4 retaining nuts from each backing plate. Wire the backing plates to the underbody.

5. Disconnect the vent hose from the axle housing.

6. Remove the brake line from the clips that retain the line to the axle housing.

7. Remove the hydraulic brake T-fitting from the axle housing. Do not open the hydraulic brake system lines.

8. Mark the driveshaft and the axle companion flange so they can be reassembled in the same position. Remove the driveshaft.

9. Support the rear axle housing on a jack, then remove the spring clip U-bolt nuts. Remove the U-bolts and plates.

10. Disconnect the lower shock absorber studs or bolts from the mounting brackets on the axle housing. Lower the axle housing and remove it from under the vehicle.

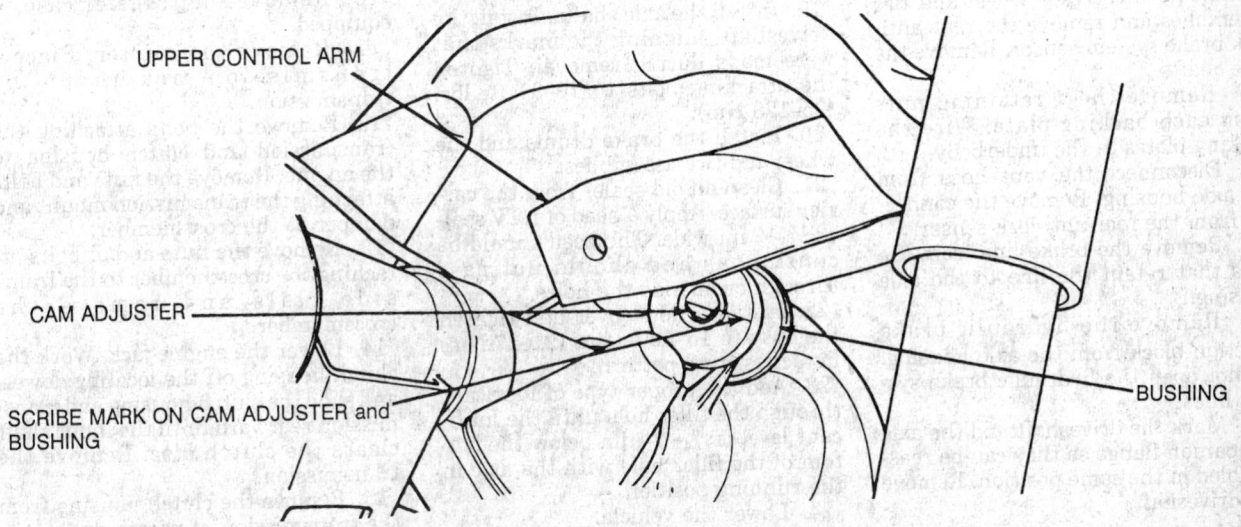

UPPER CONTROL ARM

CAM ADJUSTER

SCRIBE MARK ON CAM ADJUSTER and BUSHING

BUSHING

Rear axle housing removal—Aerostar

To install:

11. Raise the axle housing into position so the spring clip U-bolt plates can be installed. Tighten the spring clip U-bolt nuts to 55–75 ft. lbs. (70–101 Nm).

12. Connect the lower shock absorber studs or bolts to the axle housing mounting bracket. Tighten the nut/bolt to 40–60 ft. lbs. (54–82 Nm).

13. Position the brake lines to the axle housing and secure with the retaining clips. Install the T-fitting.

14. Install the brake backing plates on the axle housing flanges. Tighten the nuts to 20–40 ft. lbs. (28–54 Nm).

15. Apply stud and bearing mount compound to the threads on the vent. Install the vent and vent tube.

16. Install the axle shafts, brake drums and wheel and tire assemblies.

17. Install the driveshaft, aligning the marks that were made during removal.

18. Clean all old sealer from the carrier and cover surfaces. Apply a bead of RTV sealer 1/8–1/4 in. wide. The bead should be continuous and should not pass through or outside the holes.

19. Install the cover and tighten the bolts to 15–20 ft. lbs. (21–27 Nm) if equipped with a plastic cover or 25–35 ft. lbs. (34–37 Nm) if equipped with aa steel cover.

20. Add the proper type of rear axle lubricant until the level is 1/4–1/9 in. below the bottom of the filler hole with the axle in running position.

21. Lower the vehicle.

EXPLORER

1. Raise the vehicle and support it safely.

2. Remove the cover and drain the lubricant from the axle.

3. Remove the rear wheel and tire assemblies and remove the rear anti-lock brake system sensor. Remove the axle shafts.

4. Remove the 4 retaining nuts from each backing plate. Wire the backing plates to the underbody.

5. Disconnect the vent hose from the axle housing. Remove the connector from the rear anti-lock sensor.

6. Remove the brake line from the clips that retain the line to the axle housing.

7. Remove the hydraulic brake junction block from the axle housing. Do not open the hydraulic brake system lines.

8. Mark the driveshaft and the axle companion flange so they can be reassembled in the same position. Remove the driveshaft.

9. Support the rear axle housing on a jack, then remove the axle housing U-bolt nuts. Remove the U-bolts and shock absorber plates. Leave the shock absorbers attached to the plates.

10. Remove the stabilizer bar attaaching bracket bolts from the axle housing and position the stabilizer bar assembly away from the axle housing.

11. Raise the axle housing off the springs with the jack and move to the right side of the vehicle. Lower the left side of the axle housing below the left spring enough to clear the spring.

12. Remove the axle housing from the vehicle by lowering the axle and moving to the left until the right axle tube clears the right spring.

To install:

13. Install the axle housing on the transmission jack. Guide the right side of the axle housing over the right spring. Lift the left side of the axle housing over the left spring and position the axle housing on the spring center bolts.

14. Install the stabilizer bar to the axle housing. Tighten the stabilizer bar bracket bolts to 30–42 ft. lbs. (40–57 Nm).

15. Install the axle housing U-bolts over the axle tube. Position the shock absorber plates under the springs and install the U-bolts through the holes. Install the nuts and tighten to 88–108 ft. lbs. (119–146 Nm).

16. Install the axle vent tube to the axle vent fitting and secure with a clamp. Connect the rear anti-lock sensor connector.

17. Install the brake backing plates on the axle housing flanges. Tighten the nuts to 20–40 ft. lbs. (28–54 Nm).

18. Position the brake lines to the axle housing and secure with the retaining clips. Position the brake junction block to the axle housing and install the retaining screw.

19. Install the axle shafts. Install the driveshaft, aligning the marks that were made during removal. Tighten the attaching bolts to 70–95 ft. lbs. (95–128 Nm).

20. Install the brake drums and the wheel and tire assemblies.

21. Clean all old sealer from the carrier surface. Apply a bead of RTV sealer 1/8–1/4 in. wide. The bead should be continuous and should not pass through or outside the holes.

22. Install a new cover and tighten the bolts to 15–20 ft. lbs. (21–27 Nm) in a criss-cross pattern.

23. Add the proper type of lubricant through the filler hole until the lubricant level is 1/4–9/16 in. below the bottom of the filler hole with the axle in the running position.

24. Lower the vehicle.

For further information on transmissions/transaxles, please refer to "Chilton's Guide to Transmission Repair".

Transmission Assembly

REMOVAL & INSTALLATION

1. Disconnect the negative battery cable.

2. Place the gearshift lever in the **N** position.

3. Remove the shifter boot retainer screws and slide the boot up the shift lever shaft. Remove the shift lever attaching bolt(s) and remove the shift lever. Cover the opening in the transmission to prevent dirt from entering.

4. Raise and safely support the vehicle.

5. Mark the position of the driveshaft(s) on the flange(s) and remove the driveshaft(s). Plug the transmission or transfer case opening to prevent fluid leakage.

6. Disconnect the clutch hydraulic fluid line. Plug the line to prevent fluid leakage.

7. Disconnect the speedometer cable from the transmission or transfer case.

8. Disconnect the starter cable, backup lamp switch wire and neutral position switch wire.

9. Place a jack under the engine, with a wood block to protect the oil pan.

10. Remove the transfer case, if equipped.

11. Remove the starter. Place a transmission jack under the transmission.

12. Remove the bolts attaching the transmission and clutch housing to the engine. Remove the nuts and bolts attaching the transmission mount and damper to the crossmember.

13. Remove the nuts and/or bolts attaching the crossmember to the frame side rails and remove the crossmember.

14. Lower the engine jack. Work the clutch housing off the locating dowels and slide the clutch housing and transmission rearward until the input shaft clears the clutch disc. Remove the transmission.

15. Remove the clutch housing from the transmission, if necessary.

To install:

16. Make sure the machined mating surfaces and the locating dowels on the engine rear plate and the mating face of the clutch housing and locating dowel holes are free of burrs, dirt or paint. Install the clutch housing on the transmission, if removed.

17. Mount the transmission on a transmission jack and raise into position. Start the input shaft into the clutch disc, aligning the splines. Move the transmission forward until the clutch housing seats on the locating dowels.

18. Install the clutch housing-to-engine attaching bolts and tighten to 28–38 ft. lbs. (38–51 Nm). Remove the transmission jack.

19. Install the starter.

20. Raise the engine and install the crossmember, insulator and damper with the attaching nuts and bolts. Install the nuts and bolts attaching the transmission mount to the crossmember.

21. Install the transfer case, if equipped.

22. Remove the plug(s) and install the driveshaft(s), aligning the marks on the flange(s) that were made during the removal procedure.

23. Connect the starter cable, backup lamp switch wire, shift indicator wire and neutral position switch wire, if equipped.

24. Connect the hydraulic clutch line and bleed the system.

25. Connect the speedometer cable.

26. Remove the fill plug and check the fluid level. Add if necessary.

27. Lower the vehicle and remove the cover from the transmission opening.

28. Install the gearshift lever with the attaching bolt(s). Install the shifter boot.

29. Connect the negative battery cable. Check the transmission for proper operation.

CLUTCH

Clutch Assembly

REMOVAL & INSTALLATION

1. Disconnect the negative battery cable.

2. Disconnect the hydraulic clutch master cylinder from the clutch pedal.

3. Raise and safely support the vehicle. Remove the starter.

4. Use coupling disconnect tool T88T–70522–A or equivalent, to slide the white plastic sleeve toward the slave cylinder, then apply a slight tug on the tube to disconnect the hydraulic coupling. Plug the hose.

5. Remove the transmission.

6. Mark the position of the pressure plate on the flywheel so if the pressure plate is reused, it can be reinstalled in the same position.

7. Loosen the pressure plate attaching bolts evenly until the diaphragm spring is expanded. Remove the bolts, pressure plate and clutch disc.

8. Inspect the flywheel for wear, scoring and cracks. Machine or re-

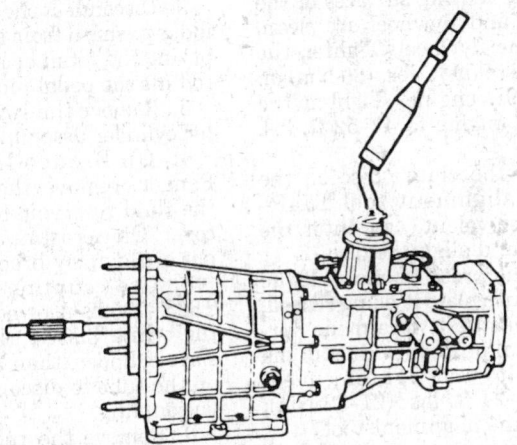

FM146 manual transmission

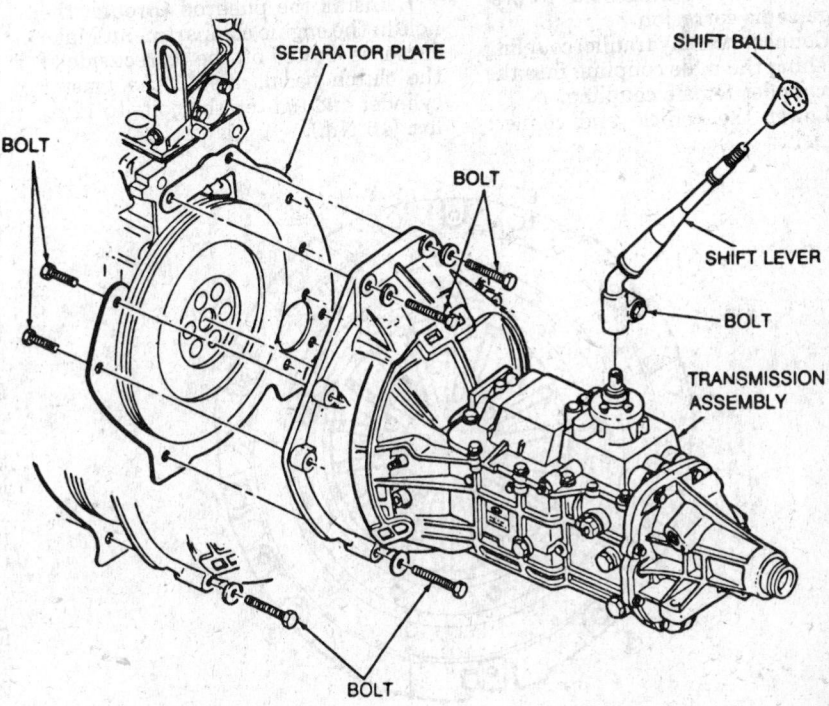

M50D manual transmission

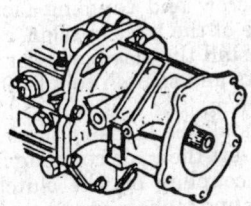

4WD SAME AS 2WD EXCEPT AS SHOWN

place, as necessary. Inspect the clutch pilot bearing for wear and free movement. If replacement is necessary, remove using puller tool T58L–101–B or equivalent.

9. Inspect the clutch release bearing for wear and free movement; replace as necessary. Remove the release bearing by twisting it until resistance is felt. Turning further will allow the preload spring to push the bearing assembly off the slave cylinder.

To install:

10. If the pilot bearing was removed, a new 1 must be installed. Install using replacer tool T71P–7137–C and clutch driver tool T71P–7137–H or equivalent. Install the pilot bearing with the seal facing the transmission so the adapter is not cocked.

11. If the flywheel was removed, make sure the mating surfaces of the crank flange and flywheel are clean, and install the flywheel. Tighten the flywheel bolts to 59 ft. lbs. (80 Nm) on all except 2.9L engine. Tighten the bolts on 2.9L engine to 47–52 ft. lbs. (64–70 Nm).

12. Position the clutch disc on the flywheel so alignment tool T74P–7137–K or equivalent, can enter the pilot bearing and align the disc.

13. Install the pressure plate. If the original pressure plate is being reused, align the marks that were made during the removal procedure. Install the attaching bolts and tighten, in sequence, to 15–24 ft. lbs. (21–32 Nm), then remove the alignment tool.

14. Install the transmission. If equipped, reuse the aluminum washers under the attaching bolts to prevent galvanic corrosion.

15. Connect the hydraulic coupling by pushing the male coupling into the slave cylinder female coupling.

16. Lower the vehicle and connect the clutch master cylinder to the brake pedal. Bleed the clutch system, if necessary.

17. Connect the negative battery cable.

PEDAL HEIGHT/FREE PLAY ADJUSTMENT

The hydraulic clutch system provides automatic adjustment. No adjustment of clutch linkage or pedal position is required.

Clutch Master Cylinder

REMOVAL & INSTALLATION

1. Disconnect the negative battery cable.

2. Disconnect the clutch master cylinder pushrod from the clutch pedal by prying the retainer bushing and pushrod off the pedal pin.

3. Remove the switch from the master cylinder assembly.

4. On Bronco II, Explorer and Ranger, remove the screw retaining the fluid reservoir to the cowl access cover. On Aerostar, slide the reservoir out of the relay bracket.

5. Use coupling disconnect tool T88T–70522–A or equivalent, to slide the white plastic sleeve toward the slave cylinder, then apply a slight tug on the tube to disconnect the hydraulic coupling.

6. Remove the retaining bolts and the clutch master cylinder.

To install:

7. Install the pushrod through the hole in the engine compartment. Make sure it is located on the correct side of the clutch pedal. Install the master cylinder and tighten the bolts to 12 ft. lbs. (16 Nm).

8. Insert the coupling end into the slave cylinder and install the tube into the clips.

9. On Bronco II, Explorer and Ranger, install the fluid reservoir on the cowl access cover with the retaining screw. On Aerostar, slide the reservoir into the relay bracket.

10. Replace the retainer bushing in the clutch master cylinder pushrod if worn or damaged. Install the retainer and pushrod on the clutch pedal pin. Make sure the flange of the bushing is against the pedal blade. Install the switch.

11. Connect the negative battery cable and bleed the clutch hydraulic system, if necessary.

Clutch Slave Cylinder

REMOVAL & INSTALLATION

NOTE: Before any vehicle service that requires slave cylinder removal, the clutch master cylinder pushrod must be disconnected from the clutch pedal. If not disconnected, permanent damage to the master cylinder will occur if the clutch pedal is depressed while the slave cylinder is disconnected.

1. Disconnect the negative battery cable.

2. Disconnect the coupling at the transmission using tool T88T–70522–A or equivalent, by sliding the white plastic sleeve toward the slave cylinder while applying a slight tug on the tube.

3. Raise and safely support the vehicle. Remove the transmission and clutch housing.

4. Remove the bolts retaining the slave cylinder to the transmission. Remove the slave cylinder from the transmission input shaft.

5. If necessary, remove the release bearing from the slave cylinder by twisting until resistance is felt, then turning further to allow the preload spring to push the bearing assembly off.

To install:

6. Push the release bearing into place, if removed.

7. Position the slave cylinder over the transmission input shaft with the bleed screw and coupling facing the left side of the transmission.

8. Install the slave cylinder attaching bolts and tighten to 13–19 ft. lbs. (18–26 Nm).

9. Install the transmission.

10. Insert the male coupling into the female coupling on the clutch slave cylinder and make sure the connection is secure.

11. Bleed the clutch hydraulic system, if necessary. Lower the vehicle

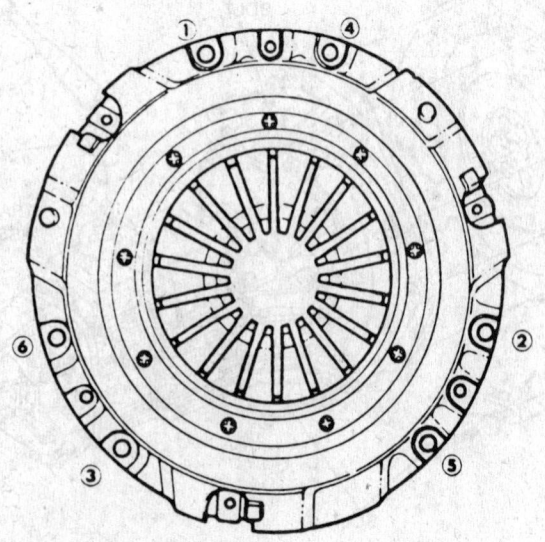

Clutch pressure plate bolt torque sequence

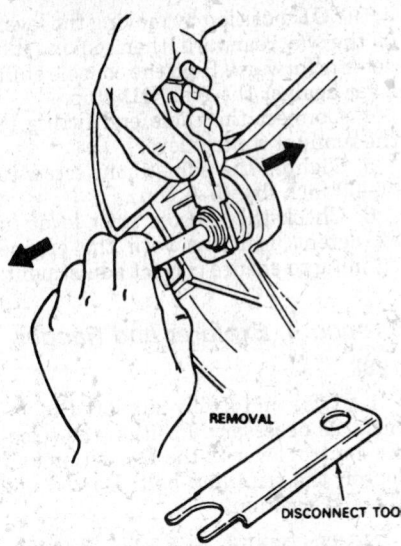

Disconnecting the hydraulic tube from the slave cylinder

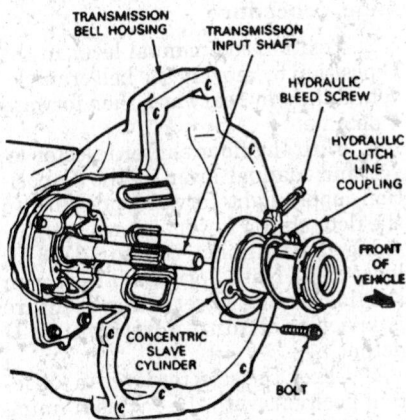

Slave cylinder installation

and connect the negative battery cable.

Hydraulic Clutch System Bleeding

NOTE: Under normal conditions, disconnecting the clutch coupling will not let air into the system. However, if there appears to be air in the system, indicated by a spongy pedal or insufficient bearing travel, the system must be bled.

1. Clean all dirt and grease from around the reservoir cap.
2. Remove the cap and fill the reservoir with heavy duty brake fluid.
3. Raise and safely support the vehicle, as necessary. Loosen the bleed screw, located in the slave cylinder body, next to the inlet connection.
4. Fluid will now begin to flow from the master cylinder, down the tube and into the slave cylinder.

NOTE: Keep the reservoir full at all times to make sure no additional air is drawn into the system.

5. Bubbles should begin to appear at the bleed screw outlet, indicating air is being expelled. When the slave cylinder is full, a steady stream of fluid will come from the slave cylinder outlet. Tighten the bleed screw.
6. Slowly depress the clutch pedal to the floor and hold. Loosen the bleed screw to allow air and excess fluid to be expelled. Retighten the bleed screw when fluid flow stops.
7. Depress and release the clutch pedal slowly, waiting 2 seconds between each cycle. Repeat 5 times.
8. Check the fluid level in the reservoir and add, if necessary. If evidence of air still exists, repeat Steps 6 and 7.

AUTOMATIC TRANSMISSION

For further information on transmissions/transaxles, please refer to "Chilton's Guide to Transmission Repair".

Transmission Assembly

REMOVAL & INSTALLATION

1. Disconnect the negative battery cable. Raise and safely support the vehicle.
2. Position a drain pan under the transmission fluid pan. On Explorer, pry the lower clips of the transmission heat shield back slightly to allow access to the pan bolts.
3. Starting at the rear of the transmission pan and working toward the front, loosen the attaching bolts and allow the fluid to drain. Remove all the bolts except the 2 at the front to allow the fluid to further drain. After all fluid has drained, reinstall 2 bolts at the rear of the pan to temporarily hold it in place.
4. Remove the converter access cover from the converter housing. On some applications, it may only be necessary to remove 1 bolt and swing the cover open.
5. Disconnect the starter cable and remove the starter.
6. Place a 22mm socket and breaker bar on the crankshaft pulley attaching bolt. Rotate the pulley clockwise, as viewed from the front, to gain access to each converter attaching nut. Remove the nuts.

NOTE: On 2.3L engine, the converter attaching nuts are accessed through the cover on the engine oil pan. On 2.9L and 4.0L engines, the converter attaching nuts are accessed through the starter mounting hole.

7. Mark the position of the driveshaft on the axle flange and remove the driveshaft. Plug the transmission to prevent fluid leakage.
8. Disconnect the speedometer cable from the transmission.
9. Disconnect the shift rod at the transmission manual lever. Remove the kickdown cable from the ball stud lever. Depress the tab on the cable downshift retainer and remove the cable from the bracket.
10. Disconnect the neutral safety switch wires and converter clutch solenoid connector. Disconnect the vacuum line from the vacuum modulator.
11. Position a transmission jack under the transmission and raise it slightly. Remove the engine rear support-to-crossmember bolts.
12. Remove the crossmember-to-frame side support attaching bolts and remove the crossmember insulator and support and damper.
13. Lower the jack under the transmission and allow the transmission to hang. On Bronco II, Explorer and

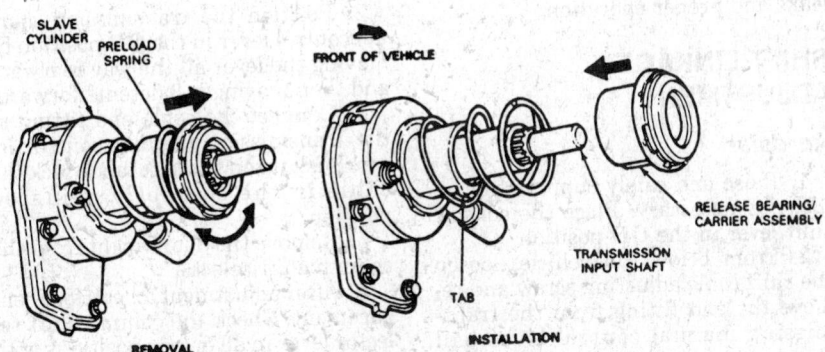

Clutch release bearing removal and installation

Ranger, position a jack to the front of the engine and raise it to gain access to the 2 upper converter housing-to-engine attaching bolts.

14. Disconnect the oil cooler lines at the transmission. Plug the lines and transmission to prevent the entrance of dirt.

15. Remove the lower converter housing-to-engine attaching bolts and remove the transmission filler tube.

16. Secure the transmission to the jack with a safety chain. Remove the 2 upper converter housing-to-engine attaching bolts. Move the transmission to the rear so it disengages from the dowel pins and the converter is disengaged from the flywheel. Lower the transmission from the vehicle.

NOTE: If the transmission is to be removed for an extended period, support the engine with a safety stand and wood block.

To install:

17. Position the converter to the transmission making sure the converter hub is fully engaged in the pump gear. To make sure the converter is fully engaged, push and rotate the converter until 2 "bumps" are felt. Keep pushing and rotating until the distance between the converter pilot and the edge of the converter housing is $7/16$–$9/16$ in.

18. Place the transmission on a transmission jack and secure with a safety chain. Rotate the converter so the drive studs are in alignment with the holes in the flywheel.

19. Raise the transmission and move it forward into position, being careful not to damage the flywheel and converter pilot.

NOTE: When moving the transmission, do not let the front of the transmission tilt downward. This will cause the converter to move forward and disengage from the pump gear. The converter must rest squarely against the flywheel. This indicates that the converter pilot is not binding in the engine crankshaft.

20. Install 2 converter housing-to-engine attaching bolts at the engine dowel locations and tighten to 28–38 ft. lbs. (38–51 Nm). Install the remaining attaching bolts and tighten to the same specification.

21. Remove the safety chain from the transmission.

22. Insert the filler tube in the stub tube and secure it to the cylinder block with the attaching bolt. Tighten the bolt to 28–38 ft. lbs. (38–51 Nm). If the stub tube is loosened or dislodged, it should be replaced.

23. Install the oil cooler lines in the retaining clip at the cylinder block.

Connect the lines to the transmission.

24. Remove the jack supporting the front of the engine.

25. Raise the transmission and position the crossmember, insulator and support and damper to the frame side supports. Install the attaching bolts and tighten to 20–30 ft. lbs. (27–41 Nm).

26. Lower the transmission and install the rear engine support-to-crossmember nut. Tighten the bolt to 60–80 ft. lbs. (82–108 Nm). Remove the transmission jack.

27. Install the vacuum hose on the vacuum modulator and attach the line to the clip. Connect the neutral safety switch plug and the converter clutch solenoid connector.

28. Install the flywheel-to-converter nuts and tighten to 20–34 ft. lbs. (27–46 Nm).

29. Install the converter access cover and adapter plate bolts and tighten to 12–16 ft. lbs. (16–22 Nm). On 2.3L engine, tighten the oil pan access cover bolts to 22–32 inch lbs. (2.5–3.6 Nm).

30. Install the starter and tighten the attaching bolts to 15–20 ft. lbs. (20–27 Nm). Connect the starter cable.

31. Connect the exhaust pipe to the exhaust manifold, if disconnected for removal.

32. Connect the shift rod to the manual lever and the downshift cable to the downshift lever. Connect the speedometer cable.

33. Install the driveshaft, aligning the marks on the axle flange. Adjust the manual and downshift linkage, as required.

34. Remove the bolts temporarily holding the transmission fluid pan and remove the pan. Discard the gasket and clean all old gasket material and dirt from the gasket mating surfaces.

35. Install the pan using a new gasket. Tighten the attaching bolts to 8–10 ft. lbs. (11–13.5 Nm).

36. Lower the vehicle and connect the negative battery cable. Fill the transmission with the proper type and quantity of fluid.

37. Run the vehicle and check for leaks and proper operation.

SHIFT LINKAGE ADJUSTMENT

Aerostar

1. Raise and safely support the vehicle, as necessary. Place the console shift lever in the **OD** position.

2. From below the vehicle, loosen the shift cable adjusting screw and remove the end fitting from the transmission manual control lever ball stud.

3. Position the manual control lever

in the **OD** position by moving the lever all the way rearward, then moving it 3 detents forward. Hold the console shift lever against the rear **OD** stop.

4. Connect the cable end fitting to the manual control lever.

5. Tighten the adjustment screw to 45–60 inch lbs. (5–6 Nm).

6. Check the console shift lever in all detent positions with the engine running to ensure correct adjustment.

Bronco II, Explorer and Ranger
1988

1. Raise and safely support the vehicle, as necessary. Position the selector control lever in the **D** position and loosen the trunnion bolt. Do not use the **OD** position.

NOTE: Make sure the shift lever detent pawl is held against the rearward Drive (D) detent stop during the linkage adjustment procedure.

2. Position the manual lever in the **D** position by moving the bell crank lever all the way rearward, then forward 4 detents.

3. With the floor shifter selector lever and manual lever in the **D** position, apply light forward pressure to the floor shifter lower arm while tightening the trunnion bolt to 13–23 ft. lbs. (18–31 Nm). Forward pressure on the floor shifter lower arm will ensure correct positioning within the **D** detent.

4. Check the selector lever in all detent positions with the engine running to ensure correct adjustment.

1989–92

1. Raise and safely support the vehicle, as necessary. From inside the vehicle, place the column shift selector lever in the **OD** position. Hang an 8 lb. weight on the selector lever.

2. From below the vehicle, pull down the lock tab on the shift cable and remove the fitting from the transmission manual control lever ball stud.

3. Position the transmission manual control lever in the **OD** position by moving the lever all the way rearward and then moving it 3 detents forward.

4. Connect the cable end fitting to the transmission manual control lever. Push up on the lock tab to lock the cable in the correctly adjusted position.

5. Remove the 8 lb. weight from the column shift selector.

6. After adjustment, check for **P** engagement. Check the column shift selector lever in all detent positions with the engine running to ensure correct adjustment.

SELECTOR INDICATOR ADJUSTMENT

1989–92 Bronco II, Explorer and Ranger

1. Remove the steering column shroud.

2. With the engine stopped and the parking brake applied, place the transmission selector lever at the steering column in the **OD** position.

3. Secure a 3 lb. weight to the end of the transmission selector lever.

4. Loosen the selector indicator screw on the column casting. Move the selector indicator adjustment until the orange pointer is completely within the letter "D" inside the **OD** graphic.

5. Tighten the selector indicator screw on the column to 10–15 inch lbs. (1.1–1.7 Nm).

6. Install the steering column shroud.

TRANSFER CASE

Transfer Case Assembly

REMOVAL & INSTALLATION

> ---------- **CAUTION** ----------
> *The catalytic converter is located beside the transfer case. Due to the extreme high temperatures generated by the converter, be careful when removing the transfer case or personal injury may result.*

Mechanical Shift Type

1. Disconnect the negative battery cable. Raise and safely support the vehicle.

2. If equipped, remove the skid plate from the frame. Remove the damper from the transfer case, if equipped.

3. Place a drain pan under the transfer case, remove the drain plug and drain the fluid. Disconnect the 4WD indicator switch wire connector at the transfer case.

4. If equipped with Borg Warner model 13–50 transfer case, disconnect the front driveshaft from the axle input yoke and pull the driveshaft and front boot assembly out of the transfer case front output shaft.

5. If equipped with Borg Warner model 13–54 transfer case, disconnect the front driveshaft from the transfer case output shaft yoke and wire the driveshaft out of the way.

6. Disconnect the rear driveshaft from the transfer case output shaft flange and wire the driveshaft out of the way.

7. Disconnect the speedometer driven gear from the transfer case rear cover. Disconnect the vent hose from the control lever.

8. Disconnect the nut from the shift lever and remove the shift lever, if necessary.

9. Remove the large and small bolts retaining the shifter to the extension housing. Remove the lever assembly and bushing.

10. Support the transfer case with a transmission jack. Remove the 5 bolts retaining the transfer case to the transmission and extension housing.

11. Slide the transfer case rearward off the transmission output shaft and lower the transfer case from the vehicle. Remove the gasket from between the transfer case and extension housing.

To install:

12. Install a new gasket on the front mounting face of the transfer case assembly.

13. Raise the transfer case with the transmission jack so the transmission output shaft aligns with the transfer case input shaft. Slide the transfer case forward onto the transmission output shaft and onto the dowel pin. Install the 5 retaining bolts and tighten, in sequence, to 25–35 ft. lbs. (34–48 Nm).

14. Remove the transmission jack.

15. Install and adjust the shifter. Always tighten the large bolt retaining the shifter to the extension housing before tightening the small bolt.

16. Install the vent assembly so the white marking on the hose is in position in the notch in the shifter. The upper end of the vent hose should be ¾ in. above the top of the shifter and positioned just below the floor pan.

17. Connect the speedometer driven gear to the transfer case rear cover. Tighten the screw to 20–25 inch lbs. (2.3–2.8 Nm).

18. Connect the rear driveshaft to the transfer case output shaft flange. Tighten the bolts to 61–87 ft. lbs. (83–118 Nm).

19. If equipped with Borg Warner model 13–50 transfer case, clean the transfer case front output shaft female splines. Apply suitable lubricant to the splines and insert the front driveshaft male spline. Connect the front driveshaft to the axle input yoke and tighten the bolts to 12–16 ft. lbs. (16–22 Nm). Push the driveshaft boot to engage the external groove on the transfer case front output shaft.

20. If equipped with Borg Warner model 13–54 transfer case, connect the front driveshaft to the transfer case

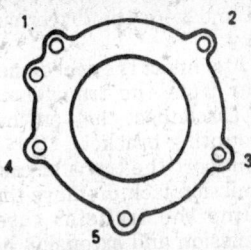

Transfer case-to-extension housing bolt torque sequence

output shaft yoke. Tighten the bolts to 12–16 ft. lbs. (16–22 Nm).

21. Connect the 4WD indicator switch wire connector at the transfer case.

22. Install the drain plug and tighten to 14–22 ft. lbs. (19–30 Nm). Remove the fill plug and fill the transfer case with the proper type of fluid to the bottom of the fill hole. Install the fill plug and tighten to 14–22 ft. lbs. (19–30 Nm).

23. Install the damper to the transfer case, if equipped. Using new damper bolts, tighten to 25–35 ft. lbs. (34–48 Nm).

24. Install the skid plate, if equipped. Tighten the nuts and bolts to 15–20 ft. lbs. (20–27 Nm).

25. Lower the vehicle and connect the negative battery cable.

Electronic Shift Type

1. Disconnect the negative battery cable. Raise and safely support the vehicle.

2. If equipped, remove the nuts, bolts and skid plate from the frame. Remove the damper from the transfer case, if equipped.

3. Place a drain pan under the transfer case, remove the drain plug and drain the fluid.

4. Remove the wire connector from the feed wire harness at the rear of the transfer case. First squeeze the locking tabs, then pull the connectors apart.

NOTE: Do not pull directly on the wires or pull outwardly on the locking tabs.

5. Remove the connector for the transfer case motor from the mounting bracket, if necessary.

6. If equipped with Borg Warner model 13–50 transfer case, disconnect the front driveshaft from the axle input yoke and pull the driveshaft and front boot assembly out of the transfer case front output shaft.

7. If equipped with Borg Warner model 13–54 transfer case, disconnect the front driveshaft from the transfer case ouput shaft yoke and wire the driveshaft out of the way.

8. Disconnect the rear driveshaft from the transfer case output shaft

flange and wire the driveshaft out of the way.

9. Disconnect the speedometer driven gear from the transfer case rear cover. Disconnect the vent hose from the mounting bracket.

10. Support the transfer case with a transmission jack. Remove the 5 bolts retaining the transfer case to the transmission and extension housing.

11. Slide the transfer case rearward off the transmission output shaft and lower the transfer case from the vehicle. Remove the gasket from between the transfer case and extension housing.

To install:

12. Install a new gasket on the front mounting face of the transfer case assembly.

13. Raise the transfer case with the transmission jack so the transmission output shaft aligns with the transfer case input shaft. Slide the transfer case forward onto the transmission output shaft and onto the dowel pin. Install the 5 retaining bolts and tighten, in sequence, to 25–35 ft. lbs. (34–48 Nm).

14. Remove the transmission jack.

15. Install the vent hose so the white marking on the hose aligns with the notch in the mounting bracket.

16. Connect the speedometer driven gear to the transfer case rear cover. Tighten the screw to 20–25 inch lbs. (2.3–2.8 Nm).

17. Connect the rear driveshaft to the transfer case output shaft flange. Tighten the bolts to 61–87 ft. lbs. (83–118 Nm).

18. If equipped with Borg Warner model 13-50 transfer case, clean the transfer case front output shaft female splines. Apply suitable lubricant to the splines and insert the front driveshaft male spline. Connect the front driveshaft to the axle input yoke and tight-en the bolts to 12–16 ft. lbs. (16–22 Nm). Push the driveshaft boot to engage the external groove on the transfer case front output shaft.

19. If equipped with Borg Warner model 13–54 transfer case, connect the front driveshaft to the transfer case output shaft yoke. Tighten the bolts to 12–16 ft. lbs. (16–22 Nm).

20. Attach the connector for the transfer case motor to the mounting bracket, if necessary.

21. Connect the wire connectors on the rear of the transfer case, making sure the retaining tabs lock.

22. Install the drain plug and tighten to 14–22 ft. lbs. (19–30 Nm). Remove the fill plug and fill the transfer case with the proper type of fluid to the bottom of the fill hole. Install the fill plug and tighten to 14–22 ft. lbs. (19–30 Nm).

23. Install the damper to the transfer case, if equipped. Using new damper

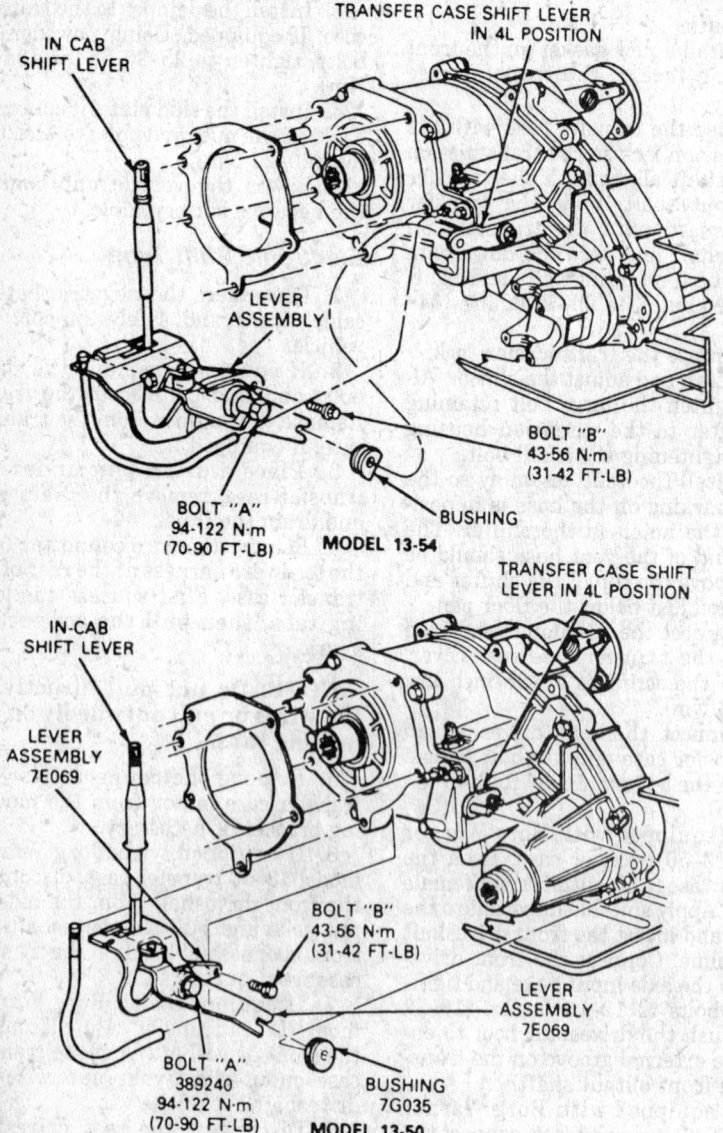

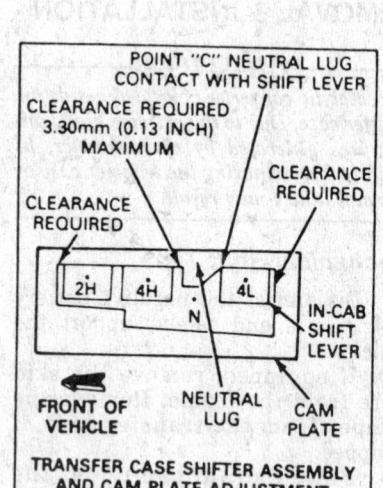

Transfer case linkage adjustment procedure

bolts, tighten to 25–35 ft. lbs. (34–48 Nm).

24. Install the skid plate, if equipped. Tighten the nuts and bolts to 15–20 ft. lbs. (20–27 Nm).

25. Lower the vehicle and connect the negative battery cable.

LINKAGE ADJUSTMENT

Mechanical Shift Type

1. Raise the shift boot to expose the top surface of the cam plate.

2. Loosen the bolts "A" and "B" on the control lever assembly approximately 2 turns. Move the transfer case shift lever to the **4L** position.

3. Move the cam plate rearward until the bottom chamfered corner of the neutral lug just contacts the forward right edge of the shift lever, point "C".

4. Hold the cam plate in this position and tighten bolt "A" first to 70–90 ft. lbs. (94–122 Nm), then tighten bolt "B" to 31–42 ft. lbs. (43–56 Nm).

5. Move the transfer case in-cab shift lever to all shift positions to check the positive engagement. There should be clearance, not exceeding 0.13 in. (3.30mm), between the shift lever and cam plate in **2H** front **4H** rear and **4L** shift positions.

6. Install the shift boot assembly.

FRONT SUSPENSION

Shock Absorbers

REMOVAL & INSTALLATION

1. Raise and safely support the vehicle. Remove the nut and the washer that attaches the shock absorber to the spring seat or coil spring upper bracket.

2. On Bronco II, Explorer and Ranger, remove the nut or nut and bolt that retains the shock absorber to the radius arm.

3. On Aerostar, remove the 2 bolts that retain the shock absorber to the bottom of the lower control arm.

3. Slightly compress the shock absorber, as necessary and remove it from the vehicle.

4. Installation is the reverse of the removal procedure. Tighten the upper retaining nut to 25–35 ft. lbs. (34–38 Nm). Tighten the lower retaining bolts or nuts on Aerostar to 16–24 ft. lbs. (22–33 Nm). Tighten the lower retaining nut or nut and bolt on Bronco II, Explorer and Ranger to 42–53 ft. lbs. (57–72 Nm).

Coil Springs

REMOVAL & INSTALLATION

Aerostar

1. Place the steering wheel and the steering system in the on center position.

2. Raise and safely support the vehicle. Remove the wheel and tire assembly.

3. Disconnect the stabilizer bar link bolt from the lower arm.

4. Remove the 2 nuts or bolts attaching the shock absorber to the lower arm. Remove the upper nut and washer and remove the shock absorber.

5. Using spring compressor tool D78P-5310-A or equivalent, install 1 plate with the pivot ball seat facing downward into the coils of the spring. Rotate the plate so it is flush with the upper surface of the lower arm.

6. Install the other plate with the pivot ball seat facing upward into the coils of the spring. Insert the upper ball nut through the coils of the spring, so the nut rests in the upper plate.

7. Insert the compression rod into the opening in the lower arm, through the upper and lower plate and upper ball nut. Insert the securing pin

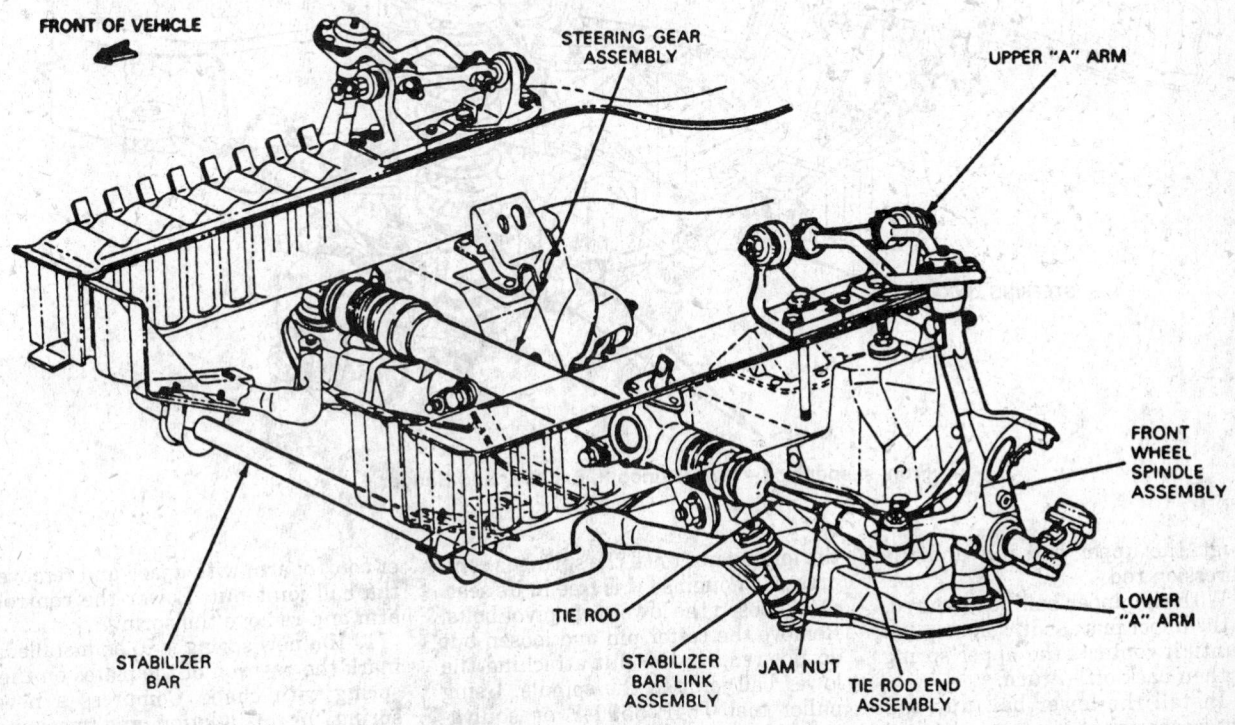

Front suspension—2WD Aerostar

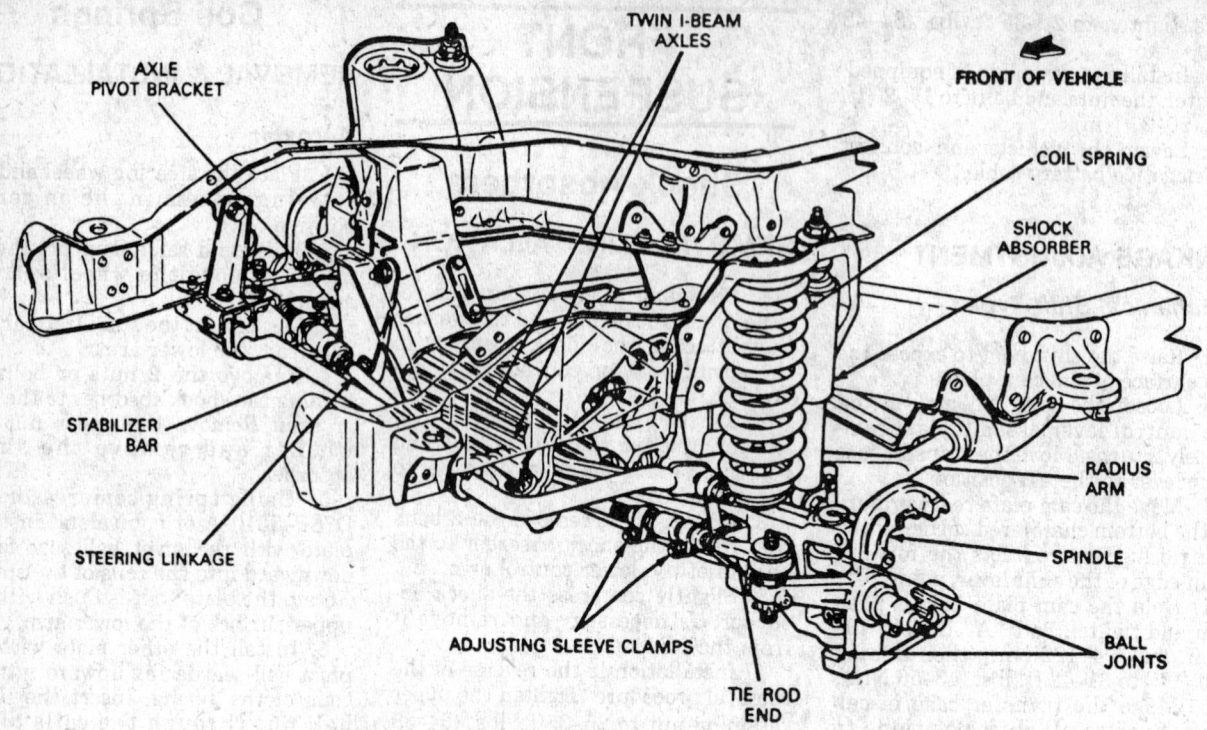

Front suspension—2WD Bronco II, Explorer and Ranger

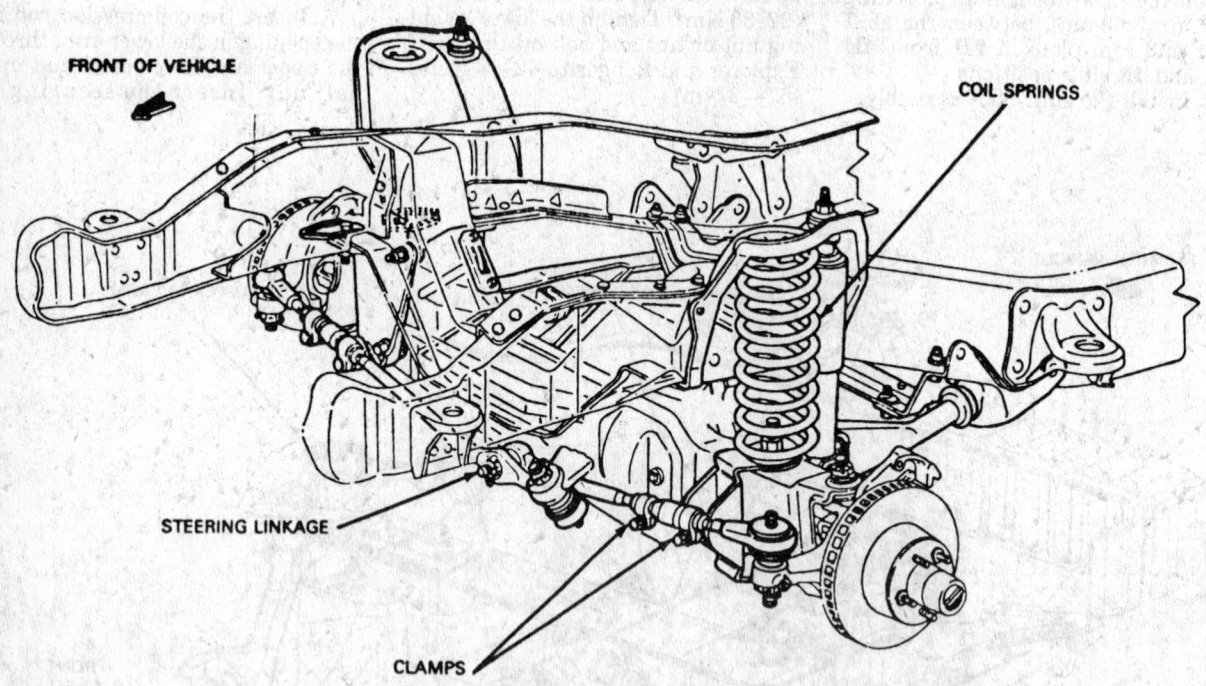

Front suspension—4WD Bronco II, Explorer and Ranger

through the upper ball nut and the compression rod.

8. With the upper ball nut secured, turn the upper plate so it walks up the coil until it contacts the upper spring seat, then back off ½ turn.

9. Install the lower ball nut and thrust washer on the compression rod and screw on the forcing nut. Tighten the forcing nut until the spring is compressed enough so it is free in its seat.

10. Loosen the lower arm pivot bolts. Remove the cotter pin and loosen but do not remove the nut attaching the lower ball joint to the spindle. Using puller tool T64P-3590-F or equivalent, loosen the lower ball joint. Remove the puller tool. Support the lower control arm with a jack and remove the ball joint nut. Lower the control arm and remove the spring.

11. If a new spring is to be installed, mark the position of the plates on the spring with chalk. Compress a new spring for installation and measure the compressed length of the old spring.

12. Loosen the forcing nut to relieve spring tension and remove the tools from the spring.

To install:

13. Assemble the spring compressor and locate in the same positions indicated in Step 11.

14. Before compressing the spring, make sure the upper ball nut securing pin is inserted properly.

15. Compress the coil spring until the spring height reaches the dimension obtained in Step 11.

16. Position the coil spring in the lower control arm. To complete installation, reverse the removal procedure. Tighten the lower ball joint nut to 80–120 ft. lbs. (108–163 Nm).

Bronco II, Explorer and Ranger

1. Raise and safely support the vehicle. Place a jack under the axle.

2. Remove the nut or nut and bolt attaching the shock absorber to the radius arm.

3. Remove the nut securing the spring to the axle and remove the retainer.

4. Slowly lower the axle to relieve the spring tension. Remove the spring by rotating the upper coil out of the tabs in the upper spring seat.

To install:

5. Install the top of the spring in the upper seat, rotating into position.

6. Raise the axle until the spring is seated in the lower spring seat. Install the lower retainer and tighten the nut to 70–100 ft. lbs. (95–136 Nm).

7. Connect the shock absorber to the radius arm and lower the vehicle.

Upper Ball Joints

INSPECTION

NOTE: Always check and adjust the wheel bearings before ball joint inspection.

1. Raise and safely support the vehicle.

2. On Aerostar, place a jack under the lower control arm and raise to slightly compress the spring. On Bronco II, Explorer and Ranger, place a jack under the axle beneath the coil spring.

3. Grasp the upper edge of the tire and move the wheel in and out. A $\frac{1}{32}$ in. or greater movement between the upper spindle arm and the upper control arm or upper part of the axle jaw indicates that the upper ball joint must be replaced.

REMOVAL & INSTALLATION

Aerostar

If upper ball joint replacement is nec-

essary, the entire upper control arm must be replaced.

2WD Bronco II, Explorer and Ranger

1988

1. Raise and safely support the vehicle. Remove the wheel and tire assembly.

2. Remove the brake caliper and support it aside with mechanics wire. Do not let the caliper hang by the brake hose.

3. Remove the dust cap, cotter pin, nut retainer, washer and outer bearing and remove the brake rotor from the spindle. Remove the brake dust shield.

4. Disconnect the steering linkage from the spindle and spindle arm by removing the cotter pin and nut. Remove the tie rod end from the spindle arm.

5. Remove the nut from the upper ball joint stud. Remove the cotter pin from the lower ball joint stud and loosen but do not remove the lower ball joint nut to the end of the stud.

6. Strike the inside of the spindle near the upper and lower ball joints to break the spindle loose from the ball joint studs. Remove the lower ball joint nut and remove the spindle.

7. Remove the snapring from the upper ball joint. Assemble C-frame tool T74P–4635–C and receiving cup tool D81T–3010–A or equivalent and press the upper ball joint from the axle.

NOTE: Do not heat the ball joint or axle to aid in removal.

To install:

8. Assemble the C-frame tool along with receiver and installation cups D81T–3010–A5, D81T–3010–A1 and D81T–3010–A4 or equivalents and press in the upper ball joint.

NOTE: Do not heat the ball joint or axle to aid in installation.

9. Install the snapring onto the upper ball joint.

10. If removed, install the camber adjuster in the spindle upper ball joint hole. Apply Loctite® 242 or equivalent, to the upper and lower ball joint stud threads.

11. Place the spindle over the ball joints. Install the lower ball joint stud nut and partially tighten to 35 ft. lbs. (47 Nm).

12. Install the upper ball joint stud nut and tighten to 85–110 ft. lbs. (116–149 Nm). Finish tightening the lower ball joint stud nut to 104–146 ft. lbs. (141–197 Nm). Advance the nut to the next castellation and install the cotter pin.

13. Install the remaining compo-

nents in the reverse order of their removal. Check and adjust the toe setting.

1989–92

1. Raise and safely support the vehicle. Remove the wheel and tire assembly.

2. Remove the brake caliper and support it aside with mechanics wire. Do not let the caliper hang by the brake hose.

3. Remove the dust cap, cotter pin, nut retainer, washer and outer bearing and remove the brake rotor from the spindle. Remove the brake dust shield.

4. Disconnect the steering linkage from the spindle and spindle arm by removing the cotter pin and nut. Remove the tie rod end from the spindle arm.

5. Remove the cotter pin and nut from the lower ball joint stud. Remove the axle clamp bolt from the axle.

6. Remove the camber adjuster from the upper ball joint stud and axle beam.

7. Strike the inside area of the axle to pop the lower ball joint loose from the axle beam. Remove the spindle and ball joint assembly from the axle.

NOTE: Do not use a pickle fork to separate the ball joint from the axle as this will damage the seal and ball joint socket.

To install:

8. Install the spindle assembly in a vise and remove the snapring from the lower ball joint. Remove the lower ball joint from the spindle using C-frame T74P–4635–C or equivalent and a suitable receiver cup to press the ball joint from the spindle.

NOTE: The lower ball joint must be removed first.

9. Repeat the procedure in Step 8 to remove the upper ball joint.

NOTE: Do not heat the ball joints or the spindle to aid in removal.

To install:

10. Assemble the C-frame and receiver cup and press in the upper ball joint.

11. Repeat the procedure in Step 10 to install the lower ball joint.

NOTE: Do not heat the ball joints or axle to aid in installation.

12. Install the snapring onto the ball joint.

13. Place the spindle and ball joints into the axle. Install the camber adjuster in the upper spindle over the ball joint stud making sure it is properly aligned.

14. Tighten the lower ball joint stud nut to 104–146 ft. lbs. (141–198 Nm). Continue tightening the castellated nut until it lines up with the hole in the stud, then install the cotter pin.

15. Install the clamp bolt into the axle boss and tighten to 48–65 ft. lbs. (65–88 Nm).

16. Install the remaining components in the reverse order of their removal.

4WD Bronco II, Explorer and Ranger

1. Raise and safely support the vehicle. Remove the front wheel and tire assemblies.

2. Remove the disc brake caliper and wire it to the frame. Do not let the caliper hang by the brake hose.

3. Remove the hub locks, wheel bearings and locknuts.

4. Remove the hub, rotor and outer wheel bearing.

5. Remove the nuts retaining the spindle to the steering knuckle. Tap the spindle with a plastic hammer to jar the spindle from the knuckle. Remove the splash shield.

6. On the left side of the vehicle, remove the shaft and joint assembly by pulling the assembly out of the carrier. On the right side of the carrier, remove and discard the clamp from the shaft and joint assembly and the stub shaft. Pull the shaft and joint assembly from the splines of the stub shaft.

7. Remove the cotter pin from the tie rod nut and then remove the nut. Tap on the tie rod stud to free it from the steering arm.

8. Remove the upper ball joint snapring and remove the upper ball joint pinch bolt. Loosen the lower ball joint nut to the end of the stud.

9. Strike the inside of the knuckle near the upper and lower ball joints to break the knuckle loose from the ball joint studs.

10. Remove the camber adjuster sleeve. Note the position of the slot in the camber adjuster so it can be reinstalled in the same position during assembly.

11. Remove the lower ball joint nut. Place the knuckle in a vise and remove the snapring from the bottom ball joint socket, if equipped.

12. Assemble C-frame T74P–4635–C and ball joint remover T83T–3050–A or equivalents on the lower ball joint. Turn the forcing screw clockwise until the lower ball joint is removed from the steering knuckle.

13. Assemble the C-frame and ball joint remover on the upper ball joint and remove in the same manner.

NOTE: Always remove the lower ball joint first.

To install:

14. Clean the steering knuckle bore and insert the lower ball joint in the knuckle as straight as possible.

15. Assemble C-frame T74P–4635–C, ball joint installer T83T–3050–A and receiver cup T80T–3010–A3 or equivalents to install the lower ball joint. Turn the forcing screw clockwise until the lower ball joint is firmly seated. Install the snapring on the lower ball joint.

NOTE: The lower ball joint must always be installed first.

16. Assemble the C-frame, ball joint installer and receiver cup to install the upper ball joint. Turn the forcing screw clockwise until the ball joint is firmly seated.

17. Install the camber adjuster into the support arm, making sure the slot is in the original position.

NOTE: The torque sequence in Steps 18 and 19 must be followed exactly when securing the knuckle. Excessive knuckle turning effort may result in reduced steering returnability if this procedure is not followed.

18. Install a new nut on the bottom ball joint stud. Tighten the nut to 90 ft. lbs. (122 Nm) minimum, then tighten to align the next slot in the nut with the hole in the stud. Install a new cotter pin.

19. Install the snapring on the upper ball joint stud. Install the upper ball joint pinch bolt and tighten to 48–65 ft. lbs. (65–88 Nm).

NOTE: The camber adjuster will seat itself into the knuckle at a predetermined position during the tightening sequence. Do not attempt to adjust this position.

20. On the right side of the carrier, install the rubber boot and new keystone clamps on the stub shaft slip yoke. Slide the right shaft and joint assembly into the slip yoke making sure the splines are fully engaged. Slide the boot over the assembly and crimp the keystone clamp using suitable pliers.

NOTE: The Dana model 28 axle has phased splines; there is only 1 way to assemble the right shaft and joint assembly into the slip yoke. The Dana model 35 axle does not have a blind spline, therefore pay special attention to make sure the yoke ears are in phase (in line) during assembly.

21. On the left side of the carrier, slide the shaft and joint assembly through the knuckle and engage the splines on the shaft in the carrier.

22. Install the splash shield and spindle onto the steering knuckle. Install and tighten the spindle nuts to 45 ft. lbs. (61 Nm).

23. Install the rotor on the spindle and install the outer wheel bearing in the race.

24. Install the wheel bearing, locknut, thrust bearing, snapring and locking hubs.

25. Install the caaliper and the wheel and tire assemblies. Lower the vehicle.

Lower Ball Joints

INSPECTION

NOTE: Always check and adjust the wheel bearings before ball joint inspection.

1. Raise and safely support the vehicle.

2. On Aerostar, place a jack under the lower control arm and raise to slightly compress the spring. On Bronco II, Explorer and Ranger, place a jack under the axle beneath the coil spring.

3. Grasp the lower edge of the tire and move the wheel in and out. A $1/32$ in. or greater movement between the lower control arm or lower axle and the spindle indicates that the lower ball joint must be replaced.

REMOVAL & INSTALLATION

Aerostar

If lower ball joint replacement is necessary, the entire lower control arm must be replaced.

2WD Bronco II, Explorer and Ranger

1988

1. Raise and safely support the vehicle. Remove the wheel and tire assembly.

2. Remove the brake caliper and support it aside with mechanics wire. Do not let the caliper hang by the brake hose.

3. Remove the dust cap, cotter pin, nut retainer, washer and outer bearing and remove the brake rotor from the spindle. Remove the brake dust shield.

4. Disconnect the steering linkage from the spindle and spindle arm by removing the cotter pin and nut. Remove the tie rod end from the spindle arm.

5. Remove the nut from the upper ball joint stud. Remove the cotter pin from the lower ball joint stud and loosen but do not remove the lower ball joint nut to the end of the stud.

6. Strike the inside of the spindle near the upper and lower ball joints to

break the spindle loose from the ball joint studs. Remove the lower ball joint nut and remove the spindle.

7. Remove the snapring from the upper ball joint. Assemble C-frame tool T74P–4635–C and receiving cup tool D81T–3010–A or equivalent and press the upper ball joint from the axle.

8. Repeat the procedure in Step 7 to remove the lower ball joint.

NOTE: Always remove the upper ball joint first. Do not heat the ball joint or axle to aid in removal.

To install:

9. Assemble the C-frame tool along with receiver and installation cups D81T–3010–A5, D81T–3010–A1 and D81T–3010–A4 or equivalents and press in the lower ball joint.

10. Repeat the procedure in Step 9 to install the upper ball joint.

NOTE: The lower ball joint must be installed first. Do not heat the ball joint or axle to aid in installation.

11. Install the snapring onto the ball joint.

12. If removed, install the camber adjuster in the spindle upper ball joint hole. Apply Loctite® 242 or equivalent, to the upper and lower ball joint stud threads.

13. Place the spindle over the ball joints. Install the lower ball joint stud nut and partially tighten to 35 ft. lbs. (47 Nm).

14. Install the upper ball joint stud nut and tighten to 85–110 ft. lbs. (116–149 Nm). Finish tightening the lower ball joint stud nut to 104–146 ft. lbs. (141–197 Nm). Advance the nut to the next castellation and install the cotter pin.

15. Install the remaining components in the reverse order of their removal. Check and adjust the toe setting.

1989–92

1. Raise and safely support the vehicle. Remove the wheel and tire assembly.

2. Remove the brake caliper and support it aside with mechanics wire. Do not let the caliper hang by the brake hose.

3. Remove the dust cap, cotter pin, nut retainer, washer and outer bearing and remove the brake rotor from the spindle. Remove the brake dust shield.

4. Disconnect the steering linkage from the spindle and spindle arm by removing the cotter pin and nut. Remove the tie rod end from the spindle arm.

5. Remove the cotter pin and nut from the lower ball joint stud. Remove the axle clamp bolt from the axle.

6. Remove the camber adjuster from the upper ball joint stud and axle beam.

7. Strike the inside area of the axle to pop the lower ball joint loose from the axle beam. Remove the spindle and ball joint assembly from the axle.

NOTE: Do not use a pickle fork to separate the ball joint from the axle as this will damage the seal and ball joint socket.

To install:

8. Install the spindle assembly in a vise and remove the snapring from the lower ball joint. Remove the lower ball joint from the spindle using C-frame T74P–4635–C or equivalent and a suitable receiver cup to press the ball joint from the spindle.

NOTE: Do not heat the ball joint or the spindle to aid in removal.

To install:

9. Assemble the C-frame and receiver cup and press in the lower ball joint.

NOTE: Do not heat the ball joint or axle to aid in installation.

10. Install the snapring onto the ball joint.

11. Place the spindle and ball joints into the axle. Install the camber adjuster in the upper spindle over the ball joint stud making sure it is properly aligned.

12. Tighten the lower ball joint stud nut to 104–146 ft. lbs. (141–198 Nm). Continue tightening the castellated nut until it lines up with the hole in the stud, then install the cotter pin.

13. Install the clamp bolt into the axle boss and tighten to 48–65 ft. lbs. (65–88 Nm).

14. Install the remaining components in the reverse order of their removal.

4WD Bronco II, Explorer and Ranger

1. Raise and safely support the vehicle. Remove the front wheel and tire assemblies.

2. Remove the disc brake caliper and wire it to the frame. Do not let the caliper hang by the brake hose.

3. Remove the hub locks, wheel bearings and locknuts.

4. Remove the hub, rotor and outer wheel bearing.

5. Remove the nuts retaining the spindle to the steering knuckle. Tap the spindle with a plastic hammer to jar the spindle from the knuckle. Remove the splash shield.

6. On the left side of the vehicle, re-move the shaft and joint assembly by pulling the assembly out of the carrier. On the right side of the carrier, remove and discard the clamp from the shaft and joint assembly and the stub shaft. Pull the shaft and joint assembly from the splines of the stub shaft.

7. Remove the cotter pin from the tie rod nut and then remove the nut. Tap on the tie rod stud to free it from the steering arm.

8. Remove the upper ball joint snapring and remove the upper ball joint pinch bolt. Loosen the lower ball joint nut to the end of the stud.

9. Strike the inside of the knuckle near the upper and lower ball joints to break the knuckle loose from the ball joint studs.

10. Remove the camber adjuster sleeve. Note the position of the slot in the camber adjuster so it can be reinstalled in the same position during assembly.

11. Remove the lower ball joint nut. Place the knuckle in a vise and remove the snapring from the bottom ball joint socket, if equipped.

12. Assemble C-frame T74P–4635–C and ball joint remover T83T–3050–A or equivalents on the lower ball joint. Turn the forcing screw clockwise until the lower ball joint is removed from the steering knuckle.

To install:

13. Clean the steering knuckle bore and insert the lower ball joint in the knuckle as straight as possible.

14. Assemble C-frame T74P–4635–C, ball joint installer T83T–3050–A and receiver cup T80T–3010–A3 or equivalents to install the lower ball joint. Turn the forcing screw clockwise until the lower ball joint is firmly seated. Install the snapring on the lower ball joint.

15. Install the camber adjuster into the support arm, making sure the slot is in the original position.

NOTE: The torque sequence in Steps 16 and 17 must be followed exactly when securing the knuckle. Excessive knuckle turning effort may result in reduced steering returnability if this procedure is not followed.

16. Install a new nut on the bottom ball joint stud. Tighten the nut to 90 ft. lbs. (122 Nm) minimum, then tighten to align the next slot in the nut with the hole in the stud. Install a new cotter pin.

17. Install the snapring on the upper ball joint stud. Install the upper ball joint pinch bolt and tighten to 48–65 ft. lbs. (65–88 Nm).

NOTE: The camber adjuster will seat itself into the knuckle at a predetermined position during

the tightening sequence. Do not attempt to adjust this position.

18. On the right side of the carrier, install the rubber boot and new keystone clamps on the stub shaft slip yoke. Slide the right shaft and joint assembly into the slip yoke making sure the splines are fully engaged. Slide the boot over the assembly and crimp the keystone clamp using suitable pliers.

NOTE: The Dana model 28 axle has phased splines; there is only 1 way to assemble the right shaft and joint assembly into the slip yoke. The Dana model 35 axle does not have a blind spline, therefore pay special attention to make sure the yoke ears are in phase (in line) during assembly.

19. On the left side of the carrier, slide the shaft and joint assembly through the knuckle and engage the splines on the shaft in the carrier.
20. Install the splash shield and spindle onto the steering knuckle. Install and tighten the spindle nuts to 45 ft. lbs. (61 Nm).
21. Install the rotor on the spindle and install the outer wheel bearing in the race.
22. Install the wheel bearing, locknut, thrust bearing, snapring and locking hubs.
23. Install the caaliper and the wheel and tire assemblies. Lower the vehicle.

Upper Control Arms

REMOVAL & INSTALLATION

Aerostar

1. Place the steering wheel and the steering system in the on center position.
2. Raise the vehicle and support it safely under the body rails.

NOTE: Only remove and install 1 upper control arm at a time. Never service both sides at the same time.

3. Remove the spindle or steering knuckle, as required.
4. Remove the bolt retaining the bolt retainer plate and remove the plate.
5. Mark the position of the control arm mounting brackets on the flat plate.
6. Remove the bolt and washer retaining the front mounting bracket to the flat plate.
7. From under the rail, remove the 3 nuts from the bolts retaining the 2 upper control arm mounting brackets to the body rail.
8. Remove the 3 long bolts retaining the mounting brackets to the body rail

by rotating the upper control arm out of position in order to remove the bolts.

9. Remove the upper control arm, upper ball joint and mounting bracket assembly and flat plate from the vehicle. If required, remove the damper assembly from the upper control arm.

To install:

10. If removed, install the damper assembly and tighten the retaining bolts to 22–29 ft. lbs. (30–39 Nm).
11. Place the flat plate for the mounting brackets in position on the body rail. Install the bolt and tighten to 10–14 ft. lbs. (14–18 Nm).
12. Place the mounting brackets and upper control arm assembly in position on the flat plate.
13. Install the 3 long bolts and washers retaining the mounting brackets to the body rail. Rotate or rock the upper control arm and mounting bracket assembly until the bolt heads rest against the mounting bracket and the studs extend through the body rail.
14. Move the mounting brackets into the position marked on the flat plate during removal.
15. Install and tighten the nuts retaining the mounting bracket bolts to the body rail to 145–195 ft. lbs. (196–264 Nm). Make sure the mounting brackets do not move from the marked position on the flat plate.

NOTE: The torque setting for the mounting bracket-to-body rail nuts and bolts is critical. They must be tightened to the specified torque.

16. Install and tighten the bolt retaining the front mounting bracket to the flat plate to 35–47 ft. lbs. (47–64 Nm).
17. Place the bolt retaining the plate in position on the mounting bracket and flat plate assembly. Install and tighten the bolt to 10–14 ft. lbs. (14–18 Nm).
18. Install the spindle or steering knuckle, as required.
19. Lower the vehicle and check the front end alignment.

Lower Control Arms

REMOVAL & INSTALLATION

Aerostar

1. Place the steering and the steering system in the on center position.
2. Raise the vehicle and support it safely under the frame.
3. Remove the coil spring.
4. Remove the bolts and nuts retaining the control arm to the No. 1 crossmember. Remove the lower control arm.

To install:

5. Position the control arm in the No. 1 crossmember and install the mounting bolts. Install the nuts and tighten until snug. Do not tighten to the specified torque at this time.
6. Install the coil spring.
7. With the vehicle in the normal ride position, tighten the lower control arm retaining nuts and bolts to 100–140 ft. lbs. (136–190 Nm).

Stabilizer Bar

REMOVAL & INSTALLATION

Aerostar

1. Raise and safely support the vehicle. Remove the nuts retaining the stabilizer bar to the lower control arm link.
2. Remove the insulators and disconnect the bar from the links. If required, remove the nuts retaining the links to the lower control arm. Remove the insulators and remove the links.
3. Remove the bolts retaining the bar mounting bracket to the frame and remove the stabilizer bar. If required, remove the insulators from the stabilizer bar.
4. Installation is the reverse of the removal procedure. Tighten the mounting bracket bolts to 16–24 ft. lbs. (22–33 Nm). Tighten the link nuts to 9–12 ft. lbs. (12–16 Nm).

2WD Bronco II and Explorer

1. Raise and safely support the vehicle.
2. Remove the nut and washer and disconnect the stabilizer link assembly from the front I-beam axle.
3. Remove the mounting bolts and remove the stabilizer bar retainers from the stabilizer bar assembly. Remove the stabilizer bar.
4. Installation is the reverse of the removal procedure. Tighten the retainer bolts to 35–50 ft. lbs. (47–68 Nm). Tighten the stabilizer bar link nuts to 30–44 ft. lbs. (40–60 Nm).

2WD Ranger

1. Raise and safely support the vehicle.
2. On 1988–89 vehicles, remove the nuts and U-bolts retaining the lower shock bracket/stabilizer bar bushing to the radius arm.
3. On 1990–92 vehicles, remove the nuts and bolts retaining the stabilizer bar to the end links.
4. Remove the retainers and remove the stabilizer bar and bushings.
5. Installation is the reverse of the removal procedure. Tighten the retainer bolts to 35–50 ft. lbs. (47–68 Nm). On 1988–89 vehicles, tighten the

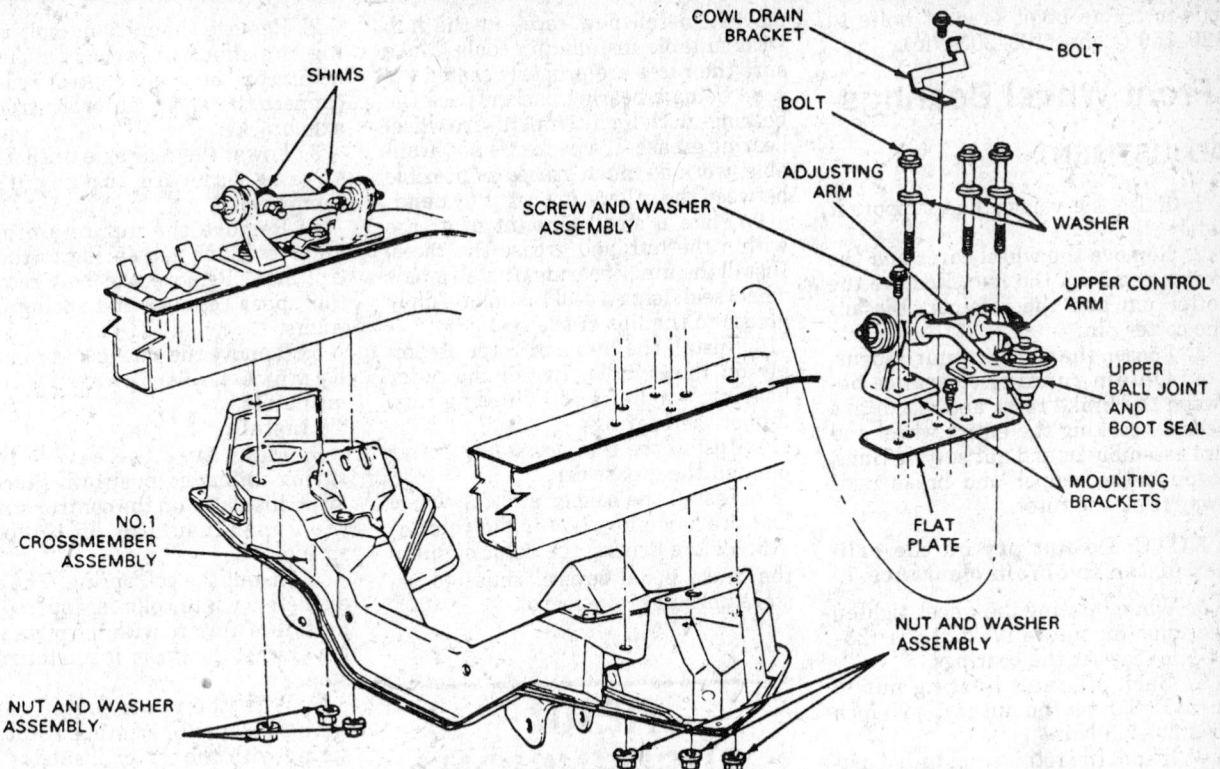

Upper control arm Installation—Aerostar

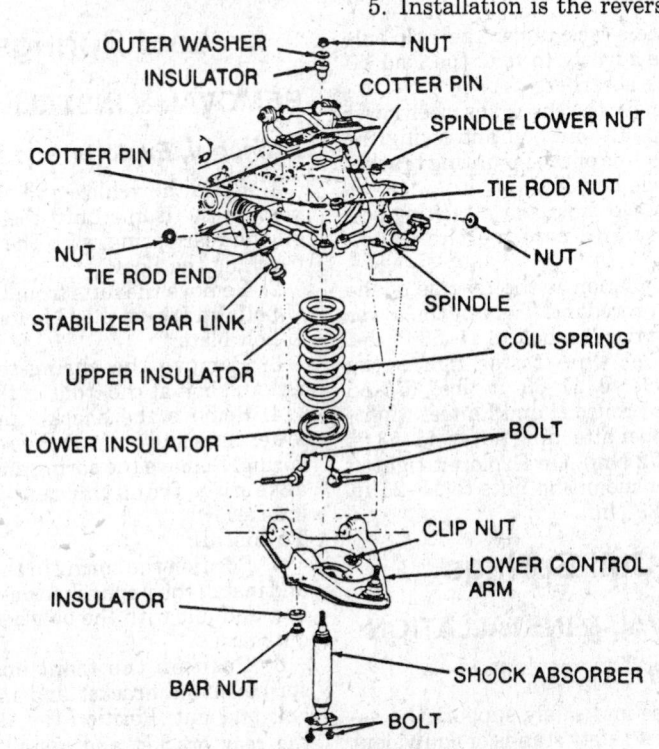

Lower control arm Installation—Aerostar

U-bolt nuts to 48–68 ft. lbs. (66–92 Nm). On 1990–92 vehicles, tighten the end link nuts to 30–44 ft. lbs. (40–60 Nm).

4WD Bronco II, Explorer and Ranger

1. Raise and safely support the vehicle.

2. Remove the bolts and retainers from the center and right end of the stabilizer bar.

3. Remove the nut, bolt and washer retaining the stabilizer bar to the stabilizer link.

4. Remove the stabilizer bar and bushings.

5. Installation is the reverse of the removal procedure. Tighten the retainer bolts to 35–50 ft. lbs. (48–68 Nm). Tighten the stabilizer bar link nut to 30–44 ft. lbs. (40–60 Nm).

I-Beam Axle

REMOVAL & INSTALLATION

1. Raise and safely support the vehicle. Remove the front wheel spindle, the front spring and the stabilizer bar, if equipped.

2. Remove the spring lower seat from the radius arm and then remove the bolt and nut that attaches the stabilizer bar bracket, if equipped, and radius arm to the front axle.

3. Remove the axle-to-frame pivot bracket bolt and nut.

To install:

4. Position the axle to the frame pivot bracket and install the bolt and nut finger tight.

5. Position the opposite end of the of the axle to the radius arm, install the attaching bolt from underneath through the bracket, the radius arm and the axle. Install the nut and tighten to 191–220 ft. lbs. (258–298 Nm).

6. Install the spring lower seat on the radius arm so the hole in the seat indexes over the arm-to-axle bolt. Install the front spring.

7. Install the front wheel spindle and stabilizer bar, if equipped.

8. Lower the vehicle and with the weight on the suspension, tighten the

axle-to-frame pivot bracket bolts to 120–150 ft. lbs. (163–203 Nm).

Front Wheel Bearings

ADJUSTMENT

1. Raise the vehicle and support it safely.
2. Remove the wheel cover and the grease cap from the hub. Remove the cotter pin and the retainer. Discard the cotter pin.
3. Loosen the adjusting nut 3 turns.
4. Obtain running clearance between the brake rotor and disc brake pads by rocking the entire wheel and tire assembly in and out several times to push the caliper and brake pads away from the rotor.

NOTE: Do not pry on the caliper piston to obtain clearance.

5. While rotating the wheel, tighten the adjusting nut to 17–25 ft. lbs. (23–34 Nm) to seat the bearings.
6. Back off the adjusting nut ½ turn. Retighten the nut to 18–20 inch lbs. (2.0–2.3 Nm).
7. Install the retainer on the adjusting nut so the castellations line up with the hole in the spindle without moving the nut. Install a new cotter pin.
8. Check the front wheel rotation. If the wheel rotates properly, reinstall the grease cap and the wheel cover. If rotation is noisy or rough, remove, inspect and lubricate the bearings and bearing races.
9. Before driving the vehicle, pump the brake pedal several times to restore normal brake travel.

REMOVAL & INSTALLATION

1. Raise and safely support the vehicle. Remove the wheel and tire assembly.
2. Remove the brake caliper and support it with mechanics wire. Do not let the caliper hang by the brake hose.
3. Remove the grease cap, cotter pin, retainer, adjusting nut and washer. Discard the cotter pin.
4. Remove the outer bearing and pull the hub and rotor off the spindle. Remove the grease seal using a seal removal tool. Discard the grease seal.
5. Remove the inner bearing from the hub. Remove all traces of old lubricant from the bearings, hub and spindle with solvent and dry thoroughly.
6. Inspect the bearings and bearing races for scratches, pits or cracks. If the bearings and/or races are worn or damaged, remove the races with a brass drift.
To install:
7. If the bearing races were re-

moved, Install new races in the hub with suitable installation tools. Make sure the races are properly seated.
8. Using a bearing packer, pack the bearings with high-temperature wheel bearing grease. If a packer is not available, work as much grease as possible between the rollers and cages by hand.
9. Place a small amount of grease within the hub and grease the races. Install the inner bearing. Install a new wheel seal using a seal installer. Apply grease to the lips of the seal.
10. Install the hub and rotor assembly on the spindle. Install the outer bearing, washer and adjusting nut. Adjust the bearings.
11. Install the retainer, a new cotter pin and the grease cap.
12. Install the caliper and the wheel and tire assembly. Lower the vehicle.
13. Before driving the vehicle, pump the brake pedal several times to restore normal brake travel.

REAR SUSPENSION

Shock Absorber

REMOVAL & INSTALLATION

Aerostar

1. Raise and safely support the vehicle.
2. Place a jack under the rear axle and raise slightly to take the load off the shock absorbers.
3. Remove the shock absorber lower attaching nut and bolt and swing the lower end free of the mounting bracket on the axle housing.
4. Remove the upper attaching bolt or nut(s) and remove the shock absorber.
5. Installation is the reverse of the removal procedure. Tighten the lower attaching nut and bolt to 41–53 ft. lbs. (55–72 Nm). On Aerostar, tighten the upper bolt to 41–63 ft. lbs. (55–85 Nm). On Bronco II and Ranger, tighten the upper attaching nut to 41–53 ft. lbs. (55–72 Nm). On Explorer, tighten the upper mounting nuts to 15–21 ft. lbs. (21–29 Nm).

Coil Springs

REMOVAL & INSTALLATION

Aerostar

1. Raise and safely support the vehicle. Place safety stands or equivalent on the frame rear lift points.

2. Remove the nut and bolt retaining the shock absorber to the axle mount on the lower control arm. Disconnect the shock absorber from the axle bracket.
3. Lower the rear axle until the coil springs are no longer under compression.
4. Remove the nut retaining the lower retainer and spring to the control arm. Remove the bolt retaining the upper retainer and spring to the frame.
5. Remove the spring and retainers. Remove the upper and lower insulators.
To install:
6. Make sure the axle is in the spring unloaded position. Place the lower insulator on the control arm and the upper insulator on top of the spring.
7. Install the coil spring. The small diameter, white colored, tapered coils must face upward with the pigtail resting against the upper insulator rubber stop.
8. With the upper pigtail resting against the rubber stop, rotate the spring with the upper insulator until the lower pigtail points in the 3 o'clock position. Install the upper retainer and bolt and tighten to 30–40 ft. lbs. (40–55 Nm).
9. Install the lower retainer and nut and tighten to 41–65 ft. lbs. (55–88 Nm).
10. Raise the axle to the normal ride position and install the shock absorber. Lower the vehicle.

Leaf Springs

REMOVAL & INSTALLATION

Bronco II, Explorer and Ranger

1. Raise the vehicle and safely support on the frame until the weight is off the rear spring, with the tires still touching the floor.
2. Remove the nuts from the spring U-bolts and drive the U-bolts from the U-bolt plate.
3. Remove the spring-to-bracket nut and bolt at the front of the spring.
4. Remove the shackle upper and lower nuts and bolts at the rear of the spring. Remove the spring and shackle assembly from the rear shackle bracket.
To install:
5. Position the spring in the shackle and install the upper shackle-to-spring bolt and nut with the bolt head facing outboard.
6. Position the front end of the spring in the bracket and install the bolt and nut. Position the shackle in the rear bracket and install the bolt and nut.

7. On Bronco II and Ranger, position the spring on top of the axle with the spring tie bolt centered in the hole provided in the seat. On Explorer, position the spring on the bottom of the axle with the spring tie bolt centered in the hole provided in the seat.

8. Install the spring U-bolts, U-bolt plate and nuts and lower the vehicle. On Bronco II and Ranger, tighten the spring U-bolt nuts to 65–75 ft. lbs. (88–102 Nm) and the front spring bolt and the rear shackle bolts and nuts to 75–115 ft. lbs. (100–155 Nm). On Explorer, tighten the spring U-bolt nuts to 88–108 ft. lbs. (119–146 Nm), the front spring bolt and nut to 64–91 ft. lbs. (87–123 Nm) and the rear shackle bolts and nuts to 75–115 ft. lbs. (100–155 Nm).

Rear Control Arms

REMOVAL & INSTALLATION

Aerostar

UPPER ARM

1. Raise the vehicle and support it safely. Place safety stands or equivalent on the frame rear lift points.

2. Remove the nut and bolt retaining the shock absorber to the lower axle bracket. Swing the lower end of the shock absorber free of the axle bracket.

3. Lower the rear axle assembly until the coil springs are no longer under compression.

4. Remove the bolt and nut retaining the upper control arm to the rear axle. Disconnect the upper control arm from the axle. Scribe a mark aligning the position of the cam adjuster in the axle bushing. The cam adjuster controls the rear axle pinion angle for driveline angularity.

5. Remove the bolt and nut retaining the upper control arm to the right frame bracket. Rotate the arm to disengage from the body bracket.

6. Remove the nut and washer retaining the upper control arm to the left frame bracket. Remove the outer insulator and spacer. Remove the control arm from the bracket. Remove the inner insulator from the control arm stud.

NOTE: If the left bracket attachments are loosened prior to disengaging the arm from the right bracket, the uncompressed left bushing will force the arm against the right bracket and make removal difficult.

To install:

7. Position the inner insulator on the control arm stud. Install the control arm so the stud extends through the left frame bracket. Install the spacer and outer insulator over the stud. Install nut and washer assembly and tighten until snug. Do not torque at this time.

8. Position the upper control arm in the right frame bracket. Install the bolt and nut and tighten until snug. Do not torque at this time.

9. Align the marks on the cam adjuster and axle bushing. Install the upper control arm to the axle housing. Install the nut and bolt and tighten until snug. Do not torque at this time.

10. Raise the rear axle to the normal ride position and install the shock absorber. With the axle in the normal ride position, tighten the control arm-to-left frame bracket nut to 60–100 ft. lbs. (81–135 Nm), control arm-to-right frame bracket nut and bolt to 100–133 ft. lbs. (135–170 Nm) and control arm-to-axle housing to 100–145 ft. lbs. (135–197 Nm) on 1988–89 vehicles or 155–210 ft. lbs. (210–284 Nm) on 1990–92 vehicles.

11. Remove the safety stands and lower the vehicle.

LOWER ARM

1. Raise and safely support the vehicle. Place safety stands under the frame rear lift points.

2. Remove the nut and bolt retaining the shock absorber to the lower axle bracket. Swing the lower end of the shock absorber free of the axle bracket.

3. Lower the rear axle until the coil springs are no longer under compression.

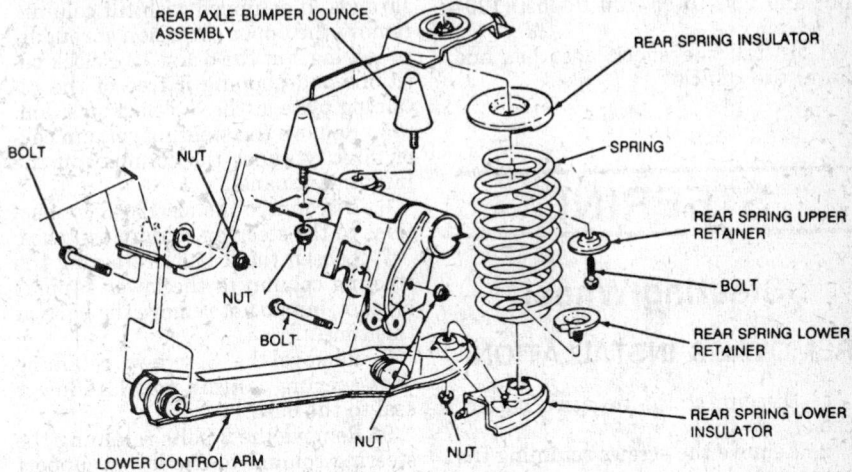

Rear lower control arm installation—Aerostar

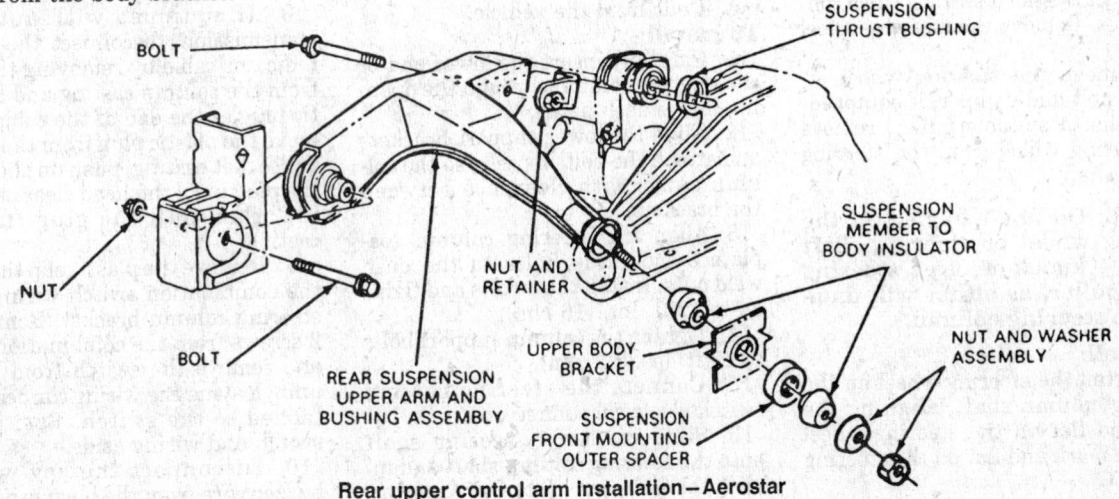

Rear upper control arm installation—Aerostar

4. Remove the nut attaching the lower retainer and coil spring to the lower control arm. Remove the insulator from the arm.

5. Remove the bolt and nut retaining the lower control arm to the axle housing.

6. Remove the nut and bolt retaining the lower control arm to the frame bracket and remove the lower control arm.

To install:

7. Position the lower control arm in the bracket on the axle housing. Install the bolt so the head is inboard on the axle bracket. Install the nut but do not tighten at this time.

8. Install the insulator on the lower control arm. With the axle in the spring unloaded position, install the coil spring and lower retainer on the lower control arm.

9. Install the nut attaching the retainer and spring to the lower control arm. Tighten the nut to 41–65 ft. lbs. (55–88 Nm).

10. Raise the axle to the normal ride position. Tighten the lower control arm-to-axle housing nut and bolt and lower control arm-to-frame bracket nut and bolt to 95–130 ft. lbs. (129–177 Nm).

11. Install the shock absorber and lower the vehicle.

STEERING

Steering Wheel

REMOVAL & INSTALLATION

1. Disconnect the negative battery cable.

2. Remove the screws retaining the steering wheel pad to the steering wheel spokes. Pull the pad back and disconnect the horn switch/cruise control wires. Remove the steering wheel pad.

3. Remove the steering wheel attaching bolt and damper, if equipped.

4. Using a suitable puller, remove the steering wheel from the steering column shaft.

NOTE: Do not hammer on the steering wheel or steering shaft or use a knock-off type steering wheel puller, as either will damage the steering column.

To install:

5. Install the steering wheel on the steering column shaft, aligning the mark and flat on the steering wheel with the mark and flat on the steering shaft.

6. If equipped, install the damper and align the locators with the hole in the wheel hub.

7. Install the steering wheel attaching bolt and tighten to 23–33 ft. lbs. (31–45 Nm).

8. Connect the horn switch/cruise control wires and install the steering wheel pad. Install the pad retaining screws and tighten to 8–11.5 inch lbs. (0.9–1.3 Nm).

9. Connect the negative battery and check the steering column for proper operation.

Steering Column

REMOVAL & INSTALLATION

Aerostar and 1988 Bronco II and Ranger

1. Disconnect the negative battery cable.

2. Remove the bolt attaching the steering column shaft to the lower shaft assembly and disengage the shaft.

3. Remove the steering wheel.

4. Remove the steering column trim shrouds. If equipped with tilt column, remove the upper extension shroud by squeezing it at the 6 and 12 o'clock positions and popping it free of the retaining plate at the 3 o'clock position.

5. Remove the steering column cover directly under the column, on the instrument panel.

6. Disconnect the electrical connections to the steering column switches.

7. Loosen the 2 bolts retaining the steering column to the lower support bracket, but do not remove the bolts at this time.

8. Remove the 3 screws retaining the steering column toeplate/lower seal to the dash.

9. Remove the 2 bolts retaining the steering column to the lower support bracket.

10. Lower the steering column and pull it out from the vehicle.

To install:

11. Carefully insert the lower end of the steering column through the opening in the dash panel.

12. Align the lower support brackets and attach the bolts loosely, so the column hangs with clearance between the brackets.

13. Align the steering column toeplate 3 mounting holes to the dash weld nuts. Install the 3 bolts and tighten to 12 ft. lbs. (16 Nm).

14. Tighten the column support bolts to 17 ft. lbs. (23 Nm).

15. Connect the steering column switch electrical connectors.

16. Slide the lower steering shaft onto the steering column shaft and install the bolt and nut. On Aerostar, tighten to 30–42 ft. lbs. (41–57 Nm). On 1988 Bronco II and Ranger, tighten to 19–28 ft. lbs. (26–38 Nm).

17. Attach the trim shrouds that cover the steering column upper end. If equipped with tilt column, snap the upper extension shroud in place.

18. Install the steering wheel and the steering column cover on the instrument panel.

19. Connect the negative battery cable and check the steering column for proper operation.

1989–92 Ranger, 1989–90 Bronco II and 1991–92 Explorer

1. Disconnect the negative battery cable and apply the parking brake. Place automatic transmission in **N**.

2. Remove the bolt that holds the intermediate shaft to the steering column shaft. Using a prybar, compress the intermediate shaft until it is clear of the steering column shaft.

3. If equipped with automatic transmission, remove the nuts from the studs and remove the shift cable bracket from the steering column bracket. Disconnect the shift cable from the column lever.

4. Remove the steering wheel. If equipped with tilt column, make sure the steering wheel is in the full up position before removal.

5. If equipped with tilt column, remove the tilt lever and remove the column collar by pressing on the collar from the top and bottom while removing the collar.

6. Remove the retaining screws and remove the panel trim cover.

7. Remove the 2 screws from the bottom of the column shroud. Remove the bottom half of the shroud by pulling the shroud down and toward the rear of the vehicle. If equipped with automatic transmission, move the shift lever as required to ease shroud removal. Lift the top half of the shroud from the column.

8. If equipped with automatic transmission, disconnect the selector indicator cable by removing the screw from the column casting and the plastic plug at the end of the cable. To remove the plastic plug from the shift lever socket casting, push on the nose of the plug until the head clears the casting, then pull the plug from the casting.

9. Remove the plastic clip that holds the combination switch wiring to the steering column bracket. Remove the 2 screws from the combination switch and remove the switch from the column, leaving the wiring connectors attached to the switch. Position the switch and wiring aside.

10. Disconnect the key warning buzzer wire from the horn brush wire.

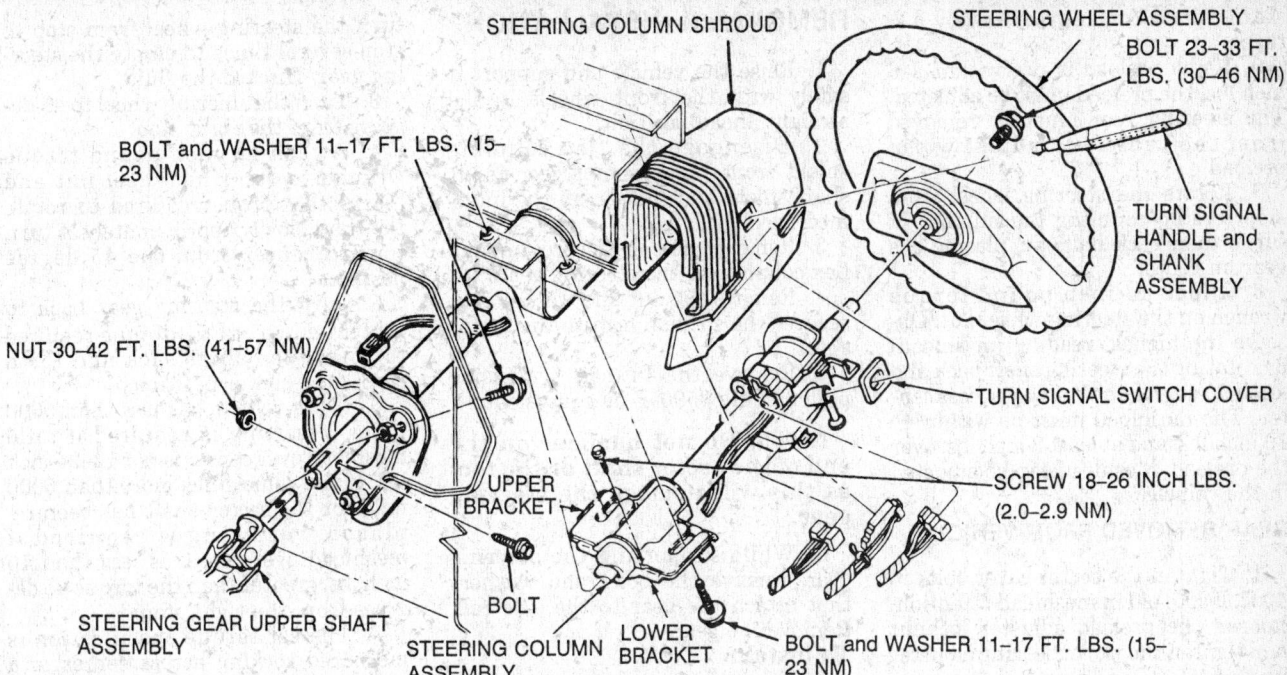

STEERING COLUMN SHROUD

STEERING WHEEL ASSEMBLY

BOLT 23–33 FT. LBS. (30–46 NM)

BOLT and WASHER 11–17 FT. LBS. (15–23 NM)

TURN SIGNAL HANDLE and SHANK ASSEMBLY

NUT 30–42 FT. LBS. (41–57 NM)

TURN SIGNAL SWITCH COVER

UPPER BRACKET

SCREW 18–26 INCH LBS. (2.0–2.9 NM)

STEERING GEAR UPPER SHAFT ASSEMBLY

BOLT

STEERING COLUMN ASSEMBLY

LOWER BRACKET

BOLT and WASHER 11–17 FT. LBS. (15–23 NM)

Steering column assembly—Aerostar

Remove the screw that holds the horn brush connector to the column and remove the connector.

11. Remove the 5 screws that hold the toe plate to the dash panel and loosen the toe plate clamp bolt.

12. Support the column and remove the bolts that hold the breakaway bracket to the pedal support bracket. Pry apart the locking tabs and disconnect the ignition switch wiring harness.

13. Carefully remove the column from the vehicle.

To install:

14. Carefully position the column in the hole in the vehicle floor. Connect the ignition switch wiring harness to the column connector.

15. Install the bolts that hold the breakaway bracket to the pedal support bracket, but do not tighten at this time.

16. Tighten the bolts that hold the toe plate to the floor to 8 ft. lbs. (11 Nm), then tighten the breakaway bracket-to-pedal support bracket bolts to 19–27 ft. lbs. (25–36 Nm). Tighten the toe plate clamp to 6–13 ft. lbs. (8–18 Nm).

17. Install the horn brush connector to the column and tighten the retaining screw to 21–29 inch lbs. (2.3–3.3 Nm). Attach the key warning buzzer wire connector to the horn brush wire. Route the wiring to prevent contact with moving parts.

18. Position the combinaation switch on the column with the attaching screws. Tighten to 18–26 inch lbs. (2–3 Nm). Install the plastic clip that holds the switch wiring to the steering column breakaway bracket.

19. If equipped with automatic transmission, connect the selector indicator cable by pushing the plastic plug at the end of the cable into the shift lever socket casting. When installed, the nose of the plug should be facing the steering wheel and the head of the plug away from the wheel. Install the cable retaining screw in the column and adjust the cable. If the shift lever was removed, install it at this time.

20. Position the top half of the shroud on the column so the screw moldings on the shroud seat in the mounting bores in the column. Place the automatic transmission shift lever in the lowest position to aid assembly.

21. Install the bottom half of the shroud by sliding the guides in the shroud bottom half into the tabs in the shroud top half. Install the shroud retaining screws and tighten to 6–10 inch lbs. (0.7–1.1 Nm).

22. If equipped with tilt column, install the column collar by pressing on the collar from the top and bottom while installing the collar on the column. Install the tilt lever and tighten to 2.2–3.6 ft. lbs. (3–5 Nm).

23. If equipped, place the automatic transmission selector lever in **N**. Install the steering wheel and the lower trim cover panel.

24. If equipped with automatic transmission, install the nuts on the studs and install the shift cable bracket on the steering column bracket. Connect the shift cable to the column lever.

25. Connect the column shaft to the intermediate shaft U-joint and tighten the pinch bolt to 25–35 ft. lbs. (34–47

Nm). The intermediate shaft must be in collapsed state to align, both shafts have a flat side, and then pulled up the column shaft until the bolt holes align. Make sure the intermediate shaft does not contact the plastic retainer at the base of the column. If it does, pull the lower shaft of the column slightly out of the column.

26. Connect the negative battery cable and check the adjustment of the selector indicator cable. Pull the shift lever toward the steering wheel until the **OD** detent in the transmission is felt. Release the shift lever, it should be against the detent wall in the column.

27. Release the parking brake lever and test drive the vehicle.

Manual Steering Gear

ADJUSTMENT

Preload and Meshload

GEAR IN VEHICLE

1. Make sure that the steering column is properly aligned and that the intermediate shaft flex coupling is not distorted.

2. Raise and safely support the vehicle. Disconnect the Pitman arm at the ball stud.

3. Lubricate the wormshaft seal with a drop of power steering fluid.

4. Remove the horn pad assembly from the steering wheel and turn the wheel slowly to 1 stop.

5. Using an inch pound torque wrench on the steering wheel nut, measure the torque (preload) required to rotate the steering wheel at a con-

stant speed for approximately 1½ turns.

6. If the preload is not within 2–6 inch lbs. the preload must be adjusted. The steering gear must be removed from the vehicle to adjust worm preload.

7. Rotate the steering wheel from stop to stop counting the number of turns then back halfway, placing the gear on center.

8. Place an inch pound torque wrench on the steering wheel nut. Observe the highest reading (meshload) by rotating the steering shaft back and forth 90 degrees either way across center. The meshload must be within 4–10 inch lbs. and at least 2 inch lbs. over the preload. Meshload can be adjusted in the vehicle.

GEAR REMOVED FROM VEHICLE

1. Tighten the sector cover bolts to 40 ft. lbs. to aid in meshload retention. Loosen the preload adjuster locknut and tighten the worm bearing adjuster nut until all endplay has been removed. Lubricate the wormshaft seal with a drop of power steering fluid.

2. Using an inch pound torque wrench and an $^{11}/_{16}$ in. 12-point socket, carefully turn the wormshaft all the way to the right. Measure the left turn torque (preload) required to rotate the wormshaft at a constant speed for approximately 1½ turns.

3. Tighten or loosen the adjuster nut as required until the correct preload of 5–6 inch lbs. is obtained. Tighten the adjuster locknut to 166–187 ft. lbs. (225–253 Nm).

4. Rotate the wormshaft from stop to stop, counting the total number of turns, then turn back halfway placing the gear on center.

5. Using the same torque wrench and socket, observe the highest reading (meshload) while the wormshaft is turned 90 degrees either way across center. If the reading is not within 9–11 inch lbs. and at least 4 inch lbs. over the preload, turn the sector shaft adjusting screw, as required.

6. Hold the sector shaft adjusting screw and tighten the locknut to 14–25 ft. lbs. (19–34 Nm).

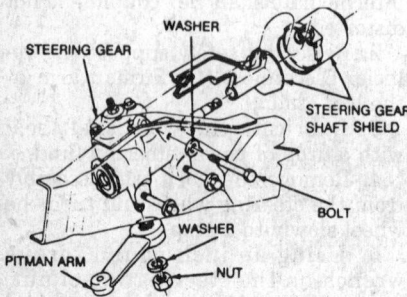

Manual steering gear installation

REMOVAL & INSTALLATION

1. Raise the vehicle and support it safely with the front wheels in the straight ahead position.

2. Disengage the flex coupling shield from the steering gear input shaft shield and slide it up the intermediate shaft.

3. Remove the bolt that retains the flex coupling to the steering gear.

4. Remove the nut and washer that secures the Pitman arm to the sector shaft.

5. Remove the Pitman arm using puller T64P–3590–F or equivalent.

NOTE: Do not hammer on the end of the sector shaft or the tool as this will damage the steering gear.

6. While supporting the steering gear, remove the bolts and washers that attach the gear to the side rail. Remove the gear.

To install:

7. Rotate the gear input shaft from stop-to-stop, counting the total number of turns, then turn back exactly half-way, placing the gear on center.

8. Slide the steering gear input shaft shield on the steering gear input shaft.

9. Position the flex coupling on the steering gear input shaft, making sure the flat on the gear input shaft is facing straight up and aligns with the flat on the flex coupling. Install the steering gear to the side rail with bolts and washers and tighten to 54–62 ft. lbs. (73–84 Nm).

10. Place the Pitman arm on the sector shaft and install the attaching washer and nut. Align the 2 blocked teeth on the Pitman arm with the 4 missing teeth on the steering gear sector shaft. Tighten the nut to 170–230 ft. lbs. (230–310 Nm).

11. Install the flex coupling to the steering gear input shaft with the attaching bolt and tighten to 25–35 ft. lbs. (34–47 Nm). Snap the flex coupling shield to the steering gear input shield.

12. Check the system to ensure equal turns from center to each lock position.

Power Steering Gear

ADJUSTMENT

1. Raise and safely support the vehicle. Disconnect the Pitman arm from the sector shaft using puller T64P–3590–F or equivalent.

2. Disconnect the fluid return line at the reservoir and cap the reservoir return line tube. Place the end of the return line in a clean container and

turn the steering wheel from stop to stop several times to empty the steering gear. Discard the fluid.

3. Turn the steering wheel to 45 degrees from the right stop.

4. Attach an inch pound torque wrench to steering wheel nut and record the torque required to rotate the shaft slowly approximately ⅛ turn toward center from the 45 degree position.

5. Turn the steering gear back to center and record the torque required to rotate the shaft back and forth across the center position.

6. If the vehicle has less than 5000 miles, resetting is required if total meshload over center is not 12–24 inch lbs. If the vehicle has more than 5000 miles or the sector shaft has been replaced, resetting is required if meshload over center is less than 10 inch lbs. greater than the torque 45 degrees from the right stop.

7. The set torque specification is measured rocking across center to a value of 9–13 inch lbs. greater than that measured 45 degress from the right stop.

8. If reset is required, loosen the locknut and turn the sector shaft adjusting screw until the reading is the specified value greater than the torque at 45 degrees from the stop.

9. Tighten the adjusting screw locknut and recheck. Install the Pitman arm and steering wheel cover.

10. Connect the fluid return line to the reservoir and refill the system with fluid. Bleed the system.

REMOVAL & INSTALLATION

1. Disconnect the pressure and return lines from the steering gear. Plug the lines and the ports in the gear to prevent entry of dirt.

2. Remove the upper and lower steering gear shaft U-joint shield from the flex coupling. Disconnect the flex coupling at the steering gear by removing the bolt.

3. Raise the vehicle and support it safely. Remove the Pitman arm attaching nut and washer. Remove the Pitman arm from the sector shaft using tool T64P–3590–F or equivalent. Be careful not to damage the seals.

4. Support the steering gear and remove the attaching bolts. Work the steering gear free from the flex coupling and remove the gear from the vehicle.

To install:

5. Install the lower U-joint shield onto the steering gear lugs. Slide the upper U-joint shield into place on the steering shaft assembly. Turn the steering wheel so the spokes are in the horizontal position.

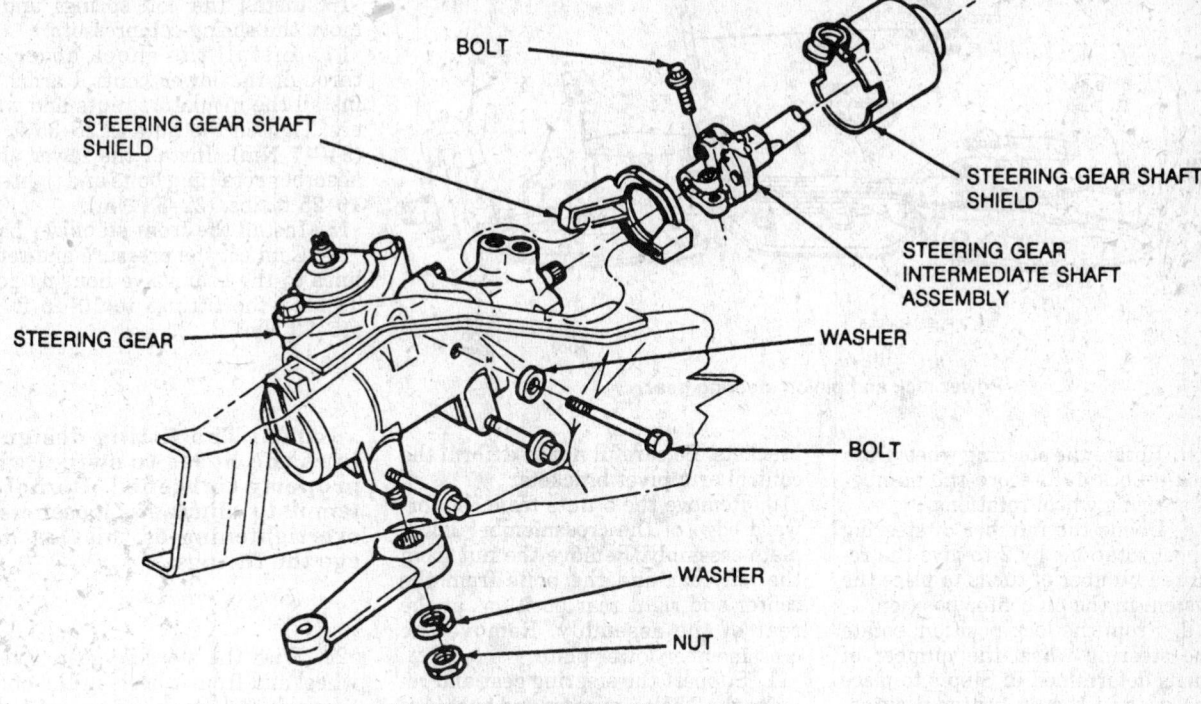

VIEW FOR INSTALLATION OF POWER STEERING

Power steering gear installation

6. Center the steering gear input shaft with the indexing flat facing down.

7. Slide the steering gear input shaft into the flex coupling and into place on the frame side rail. Install the attaching bolts and tighten to 50–62 ft. lbs. (68–84 Nm). Tighten the flex coupling bolt to 26–34 ft. lbs. (34–47 Nm).

8. Make sure the wheels are in the straight ahead position, then install the Pitman arm on the sector shaft. Install the attaching washer and nut and tighten to 170–228 ft. lbs. (230–310 Nm).

9. Connect and tighten the pressure and return lines to the steering gear to 20–30 ft. lbs. (27–40 Nm). Snap the upper and lower steering gear shaft U-joint shields together.

10. Fill and bleed the power steering system.

Power Rack and Pinion

ADJUSTMENT

Service adjustments are not required on the power rack and pinion gear.

REMOVAL & INSTALLATION

Except 4WD Aerostar

1. Place the steering system in the on center position as follows:

a. Start the engine.
b. Rotate the steering wheel from lock-to-lock and record the number of steering wheel rotations.
c. Divide the number of steering wheel rotations by 2 to give the required number of turns to place the system in the on center position.
d. From the lock position, rotate the steering wheel the number of turns determined in Step c to place the gear in the on center position. Make sure the wheels are in the straight ahead position.
e. Stop the engine.

2. Disconnect the negative battery cable and turn the ignition key to the **ON** position. Raise and safely support the vehicle.

3. Disconnect the pressure and return lines from the steering gear valve housing. Plug the lines and ports in the gear valve housing to prevent the entry of dirt.

4. Remove the bolt retaining the lower intermediate steering column shaft to the steering gear. Disconnect the shaft from the gear.

5. Remove the cotter pins and nuts from the tie rod ends. Separate the tie rod ends from the spindle arms.

6. Support the steering gear and remove the 2 nuts, bolts and washers retaining the gear to the crossmember. Remove the gear from the vehicle.

To install:

7. Position the steering gear on the crossmember. Install the retaining nuts, bolts and washers and tighten to 80–105 ft. lbs. (108–142 Nm).

8. Connect the pressure and return lines to the gear valve housing ports. Tighten the fittings to 10–15 ft. lbs. (15–20 Nm) on 1988–89 vehicles or 20–25 ft. lbs. (27–34 Nm) on 1990–92 vehicles.

NOTE: The fitting design allows the hoses to swivel when properly tightened. Do not attempt to eliminate looseness by overtightening as this can damage the fittings.

9. With the steering gear, steering wheel and front wheels in the on center position, attach the tie rod ends to the spindle arms. Install the nuts and tighten to 52–73 ft. lbs. (70–100 Nm). If required, advance the nuts to the next castellation and install new cotter pins.

10. Connect the steering column lower intermediate shaft to the gear. Install the bolt and tighten to 30–42 ft. lbs. (41–56 Nm).

11. Lower the vehicle and turn the ignition key to the **OFF** position. Connect the negative battery cable.

12. Fill and bleed the steering system. Check the toe setting and adjust, if necessary.

4WD Aerostar

1. Place the steering system in the on center position as follows:
a. Start the engine.

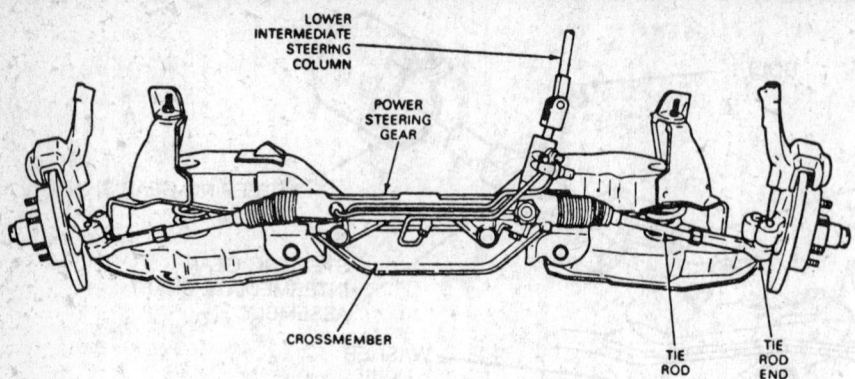

Power rack and pinion steering gear

b. Rotate the steering wheel from lock-to-lock and record the number of steering wheel rotations.

c. Divide the number of steering wheel rotations by 2 to give the required number of turns to place the system in the on center position.

d. From the lock position, rotate the steering wheel the number of turns determined in Step c to place the gear in the on center position. Make sure the wheels are in the straight ahead position.

e. Stop the engine.

2. Disconnect the negative battery cable and turn the ignition key to the **ON** position. Raise and safely support the vehicle and remove the wheel and tire assemblies.

3. Disconnect the pressure and return lines from the steering gear valve housing. Plug the lines and ports in the gear valve housing to prevent the entry of dirt.

4. Remove the bolt retaining the lower intermediate steering column shaft to the steering gear. Disconnect the shaft from the gear. Retain any dust seals which may be present.

5. Remove the cotter pins and nuts from the tie rod ends. Separate the tie rod ends from the spindle arms.

6. Remove the nut and washer with the insulator retaining the shock absorbers to the front crossmember brackets. Remove the bolts retaining the shock absorbers to the bottom of the lower control arms and remove the shock absorbers.

NOTE: Be careful during shock absorber removal as the shock absorber may extend somewhat due to gas pressure.

7. Remove the front stabilizer bar.

8. Install a suitable spring compressor through each coil spring to secure the spring and prevent extension when the control arm pivot bolts are removed.

9. Remove the lower control arm pivot bolts. Reposition and secure the lower control arms away from the

brackets. Be careful not to deform the control arm pivot brackets.

10. Remove the 5 nuts from the forward edge of the crossmember lower plate assembly. Remove the nut from the left rear and the bolts from the center and right rear positions at the rear of the assembly. Remove the crossmember lower plate.

11. Support the steering gear and remove the 2 bolts, spacers and bushings retaining the steering gear. Remove the gear from the vehicle.

To install:

12. Position the steering gear on the crossmember with the attaching bolts and tighten to 61–82 ft. lbs. (83–111 Nm). The large end of the spacer is installed facing the steering rack.

13. Install the crossmember lower plate with the nuts and bolt in the proper locations.

14. Position the lower control arms within the brackets and install the pivot bolts. Properly position the retaining flags.

15. Tighten all lower control arm pivot bolt nuts to 100–140 ft. lbs. (136–190 Nm).

16. Install the coil springs and remove the spring compressors.

17. Install the shock absorbers through the lower control arms and install the insulators, nuts and washers. Tighten the nuts to 25–35 ft. lbs. (34–47 Nm). Install the lower shock absorber retaining bolts and tighten to 16–25 ft. lbs. (22–33 Nm).

18. Install the front stabilizer bar.

19. Connect the pressure and return lines to the gear valve housing ports. Tighten the fittings to 10–15 ft. lbs. (15–20 Nm).

NOTE: The fitting design allows the hoses to swivel when properly tightened. Do not attempt to eliminate looseness by overtightening as this can damage the fittings.

20. With the steering gear, steering wheel and front wheels in the on center position, attach the tie rod ends to the spindle arms. Install the nuts and tighten to 52–73 ft. lbs. (70–100 Nm). If required, advance the nuts to the next castellation and install new cotter pins.

21. Position the intermediate shaft and any dust seals over the steering rack input shaft spline and make sure no rotation from the on center position has occurred. Install the bolt and tighten to 30–42 ft. lbs. (41–56 Nm).

22. Install the wheel and tire assemblies. Lower the vehicle and turn the ignition key to the **OFF** position. Connect the negative battery cable.

23. Fill and bleed the steering system. Check the toe setting and adjust, if necessary.

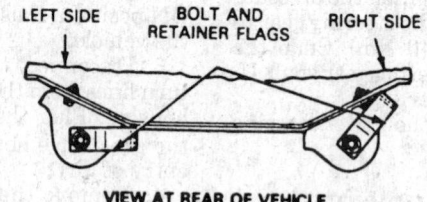

VIEW AT REAR OF VEHICLE

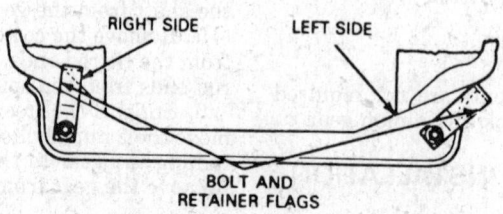

VIEW AT FRONT OF VEHICLE

Bolt and retainer flag positioning—4WD Aerostar

Power Steering Pump

REMOVAL & INSTALLATION

1. Remove the power steering fluid from the pump reservoir by disconnecting the fluid return hose at the reservoir and draining the fluid into a container.

2. Remove the pressure hose from the pump. If equipped, disconnect the power steering pump pressure switch.

3. If equipped with 2.0L, 2.3L and 3.0L engines, loosen the alternator or idler pulley assembly pivot and adjustment bolts to slacken belt tension. Remove the drive belt.

4. If equipped with 2.9L engine, loosen the adjustment nut and the slider bolts on the pump support to slacken belt tension. Remove the drive belt.

5. If equipped with 4.0L engine, slacken belt tension by lifting the tensioner pulley in a counterclockwise direction. Remove the drive belt from under the tensioner pulley and slowly lower the pulley to stop. Remove the drive belt.

6. Remove the engine oil dipstick tube, if necessary. Remove the power steering pump bracket support brace, if equipped.

7. Install steering pump pulley removal tool T69L–10300–B on the pulley. Hold the pump and rotate the tool nut counterclockwise to remove the pulley. Do not apply in and out pressure on the pump shaft as pressure will damage the internal thrust areas.

8. Remove the bolts attaching the pump to the bracket and remove the pump.

To install:

9. Install the pump on the bracket. Install and tighten the attaching bolts to 30–45 ft. lbs. (41–61 Nm). If equipped with 4.0L engine, tighten to 35–47 ft. lbs. (47–64 Nm), position the support on the bracket and install and tighten the mounting bolts to 35–47 ft. lbs. (47–64 Nm).

10. Install steering pump pulley replacement tool T65P–3A733–C and install the pulley. Remove the tool.

NOTE: Fore and aft location of the pulley on the pump shaft is critical for correct belt alignment. Make sure the pull-off groove on the pulley is facing front and flush with the end of the shaft ± 0.010 in. (0.254mm).

11. Install the drive belt.

12. If equipped with 4.0L engine, position and rotate the drive belt on the engine. While lifting the tensioner pulley in a counterclockwise direction, slide the belt under the tensioner pulley and lower the pulley to the belt.

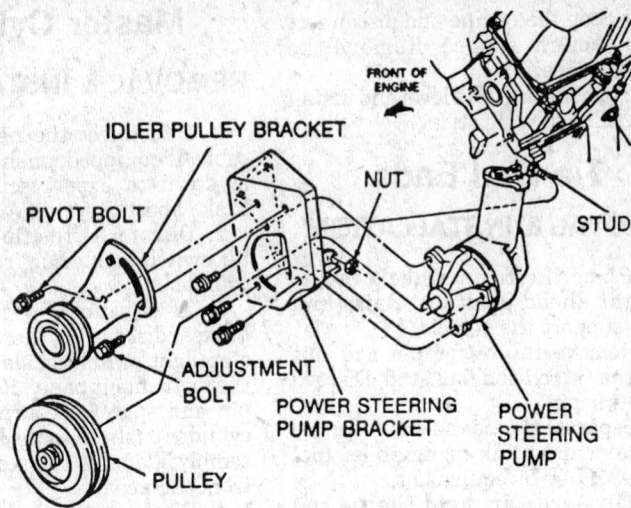

Power steering pump installation – 3.0L engine

13. If equipped with 2.0L, 2.3L or 3.0L engine, position the idler pulley or alternator to set the belt tension. Tighten the idler pulley or alternator pivot and adjustment bolts to 40 ft. lbs. (55 Nm) on the 2.0L and 2.3L engines or 35–47 ft. lbs. (47–64 Nm) on the 3.0L engine.

14. If equipped with the 2.9L engine, tighten the pump support adjustment nut to set the belt tension, then tighten the slider bolts to 35–47 ft. lbs. (47–64 Nm).

15. Install the power steering pump bracket support brace and/or engine dipstick tube, if removed.

16. Install the pressure hose to the pump fitting. Connect the return hose to the pump and tighten the clamp. If equipped, connect the power steering pump pressure switch.

17. Fill and bleed the power steering system. Check for leaks.

Belt Adjustment

Except 4.0L Engine

1. Loosen the accessory adjustment and pivot bolts.

2. Position a suitable belt tension gauge mid-way between the pulleys on the longest accessible belt span.

3. On 2.0L and 2.3L engines, move the alternator or idler pulley to apply tension to the belt. On 2.9L and 3.0L engines, move the power steering pump to apply tension to the belt.

4. On 2.0L, 2.3L and 3.0L engines, the tension should be 150–190 lbs. for a new belt or 140–160 lbs. for a used belt. On 2.9L engine, the tension should be 120–160 lbs. for a new belt or 110–130 lbs. for a used belt.

5. Tighten the adjustment bolt and release pressure, then tighten the pivot bolt.

6. Recheck the belt tension and adjust, if necessary.

4.0L ENGINE

Belt tension is maintained by an automatic tensioner. No adjustment is necessary.

SYSTEM BLEEDING

1. Fill the power steering fluid reservoir.

2. Disconnect the coil wire.

3. Crank the engine with the starter and continue adding fluid until the level remains constant. Do not prolong cranking as the battery may be drained and the starter damaged.

4. Rotate the steering wheel approximately 30 degrees each side of center while continuing to crank the engine.

5. Recheck the fluid level and fill, as required.

6. Reconnect the coil wire.

7. Start the engine and allow it to run for several minutes.

8. Rotate the steering wheel from stop to stop.

9. Shut off the engine and recheck the fluid level. Add fluid, as required.

10. If air is still trapped in the system, proceed as follows:

   a. Fabricate a purging tool.

   b. Make sure the reservoir fluid level is correct.

   c. Insert the rubber stopper end of the fabricated purging tool tightly into the filler tube.

   d. Connect a suitable length of hose to the purging tool. Connect the other end of the hose to an air conditioner vacuum pump or distributor machine. Do not use engine vacuum.

   e. Start the engine and let it idle for approximately 15 minutes. Turn the steering wheel 1 full cycle every 5 minutes but do not hit the stops. This will assist in removing trapped air.

f. Stop the engine and disconnect the vacuum source. Remove the purging tool.

g. Check the fluid level and install the filler tube dipstick.

Tie Rod Ends

REMOVAL & INSTALLATION

1. Place the front wheels in the straight ahead position. Raise and safely support the vehicle.

2. Remove the cotter pin and nut from the tie rod end ball stud. Discard the cotter pin.

3. Separate the tie rod end from the spindle or drag link using puller tool T64P-3590-F or equivalent.

4. On Aerostar, hold the tie rod with a wrench and loosen the tie rod jam nut. Mark the position of the tie rod end on the tie rod threads, grip the tie rod with suitable pliers and remove the tie rod end from the tie rod.

5. On Bronco II, Explorer and Ranger, loosen the bolts on the tie rod adjusting sleeve. Count the number of turns required to remove the tie rod from the tie rod adjusting sleeve and remove the tie rod.

To install:

6. On Aerostar, thread the replacement tie rod end onto the tie rod to the same location as the 1 that was removed. Hold the tie rod end with a wrench and tighten the jam nut to 35–50 ft. lbs. (48–68 Nm).

7. On Bronco II, Explorer and Ranger, install the tie rod into the adjusting sleeve the same number of turns required to remove it. With the adjusting sleeve clamps pointed down, tighten the adjusting sleeve nuts to 30–42 ft. lbs. (40–57 Nm).

8. Install the tie rod ball stud into the spindle or drag link. Install the nut and tighten to 50–75 ft. lbs. (70–100 Nm). Install a new cotter pin.

NOTE: If the cotter pin cannot be installed because the hole in the ball stud does not align with a castellation on the nut, continue to tighten the nut to align them. Never loosen the nut to align the hole and castellation.

9. Lower the vehicle. Check the toe-in setting and adjust, if necessary.

BRAKES

For all brake system repair and service procedure not detailed below, please refer to "Brakes" in the Unit repair section.

Master Cylinder

REMOVAL & INSTALLATION

1. Disconnect the negative battery cable. If equipped, push the brake pedal down to expel vacuum from the brake booster system.

2. Disconnect the fluid level indicator switch connector from the master cylinder.

3. If equipped with non-power brakes, disconnect the wires from the stop light switch inside the cab below the instrument panel. Remove the lock pin and spacers securing the master cylinder pushrod to the brake pedal assembly. Remove the stop light switch from the pedal.

4. Disconnect and plug the hydraulic lines at the master cylinder. Plug the master cylinder ports.

5. Remove the master cylinder-to-booster or master cylinder-to-dash panel retaining nuts and remove the master cylinder.

NOTE: On 1990–92 Aerostar, use care when removing the cartridge master cylinder so as not to scratch the exposed primary piston.

To install:

6. If equipped with power brakes, before installing the master cylinder on all except 1990–92 Aerostar, check the distance from the outer end of the vacuum booster assembly pushrod to the front face of the vacuum brake booster assembly. The distance should be 0.980–0.995 in. (24.89–25.27mm). Turn the pushrod adjusting screw in or out, as required, to obtain the proper length.

7. On 1990–92 Aerostar, make sure the interface seal is located in the an-nular groove in the cartridge master cylinder body where the cartridge master cylinder seals with the booster. Do not assemble the cartridge master cylinder to the booster without this square section interface seal.

8. If equipped with non-power brakes and the dash spacer was removed, coat the spacer with sealer and install on the dash panel.

9. Install the master cylinder with the attaching nuts. On 1990–92 Aerostar, be careful not to scratch the exposed primary piston and make sure the primary piston socket engages the booster pushrod before tightening. Tighten the nuts to 20 ft. lbs. (27 Nm). Connect the fluid level indicator switch.

10. If equipped with non-power brakes, secure the pushrod to the brake pedal assembly with the pin or shoulder bolt. Make sure the bushings and spacers are installed properly. In-

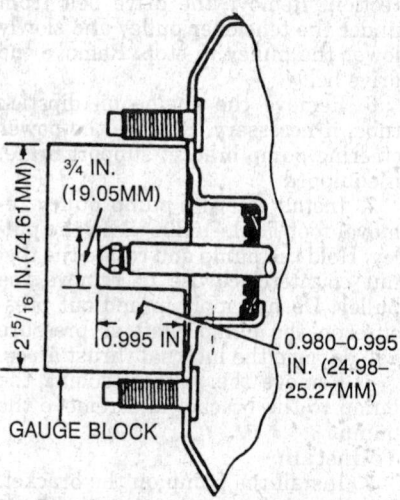

$2^{15}/_{16}$ IN.(74.61MM)
¾ IN. (19.05MM)
0.995 IN
0.980–0.995 IN. (24.98–25.27MM)
GAUGE BLOCK

Power brake booster dimensions

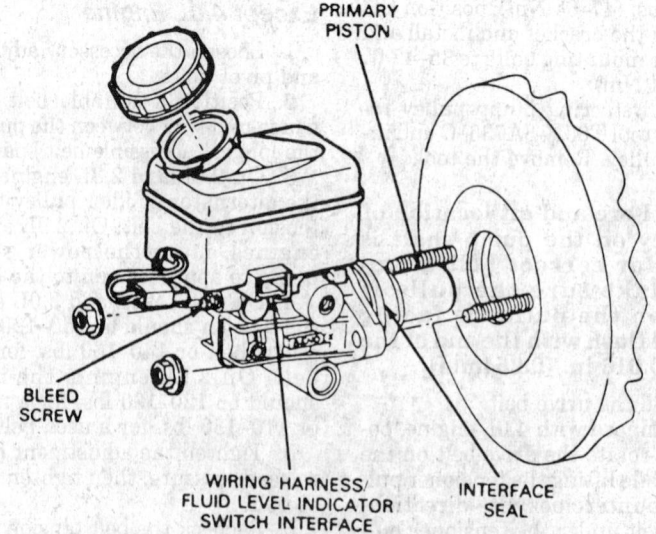

PRIMARY PISTON

BLEED SCREW

WIRING HARNESS/ FLUID LEVEL INDICATOR SWITCH INTERFACE

INTERFACE SEAL

Master cylinder installation – 1990–92 Aerostar

stall the lockpin and connect the wires to the stop light switch.

11. Connect the brake lines to the master cylinder and fill the master cylinder fluid reservoir. Wrap a shop cloth around the tubing below the fitting to be bled to absorb escaping brake fluid.

12. Have an assistant push the brake pedal to the floor. Crack open the brake line fitting to expel air trapped in the master cylinder. Tighten the fitting, then let the brake pedal return. Repeat this procedure until all air is expelled.

13. Repeat Step 12 on the remaining brake line fitting(s).

14. Final tighten the brake line fittings. Bleed the brake system, as required. Check the master cylinder fluid level.

Combination Valve

REMOVAL & INSTALLATION

1. Disconnect and plug the brake lines at the valve.

2. Remove the valve attaching bolt and remove the valve.

3. Installation is the reverse of the removal procedure. Bleed the brake system.

Power Brake Booster

REMOVAL & INSTALLATION

1. Disconnect the negative battery cable. Support the master cylinder from the underside with a prop.

2. Disconnect the vacuum hose from the booster check valve and remove the check valve.

3. Remove the master cylinder-to-booster retaining nuts. Pull the master cylinder off the booster and leave it supported by the prop, out of the way enough to allow booster removal.

4. Working inside the cab, remove the hairpin retainer and slide the stoplight switch, valve rod, spacers and bushing off the brake pedal arm. Remove the nuts retaining the booster and remove the booster.
To install:

5. Mount the booster assembly on the engine side of the dash panel by sliding the bracket mounting bolts and valve operating rod in through the holes in the dash panel.

6. Working inside the cab, install the booster mounting nuts and tighten to 13–25 ft. lbs. (18–33 Nm).

7. Before installing the master cylinder on all except 1990–92 Aerostar, check the distance from the outer end of the vacuum booster assembly pushrod to the front face of the vacuum brake booster assembly. The distance

should be 0.980–0.995 in. (24.89–25.27mm). Turn the pushrod adjusting screw in or out, as required, to obtain the proper length.

8. Install the master cylinder and tighten the retaining nuts to 20 ft. lbs. (27 Nm). Remove the prop from under the master cylinder.

9. Install the booster check valve and connect the vacuum hose. Check the hose routing to make sure the hose is not crimped.

10. Working inside the cab, install the bushing and position the switch on the end of the valve rod, then install the switch and rod on the pedal arm along with the spacers and hairpin retainer.

NOTE: Use only the factory supplied hairpin retainer. Do not substitute other types of retainers.

11. Connect the negative battery cable, start the engine and check brake operation.

Brake Caliper

REMOVAL & INSTALLATION

1. Siphon part of the brake fluid out of the master cylinder to avoid overflow when the caliper piston is pressed into the caliper bore.

2. Raise the vehicle and support it safely. Remove the wheel and tire assembly.

3. Position an 8 in. C-clamp on the caliper and tighten the clamp to move the caliper piston into the bore approximately ⅛ in. Avoid clamp contact with the outer shoe spring clip. Remove the clamp.

NOTE: Do not pry the piston away from the rotor.

4. Clean excess dirt from the pin tab area.

5. Using a ¼ in. drive socket, ⅜ in. deep and a light hammer, tap the upper caliper pin towards the outboard side until the pin tabs pass the spindle face.

6. Compress the inboard pin tab, if equipped, with pliers and, with a hammer, drive the pin out until the tab slips into the spindle groove.

7. Place 1 end of a ⁷⁄₁₆ in. diameter punch against the end of the caliper pin and tap the pin out of the caliper slide groove.

8. Repeat Steps 5, 6 and 7 to remove the lower pin.

9. Disconnect and plug the brake hose at the caliper. Remove the caliper from the rotor.
To install:

10. Make sure the caliper mounting surfaces are free of dirt. Lubricate the caliper grooves with disc brake caliper grease and install the caliper.

11. From the caliper outboard side, position the pin between the caliper and spindle grooves. The pin must be positioned so the tabs will be installed against the spindle outer face.

12. Tap the pin on the outboard end with a hammer until the retention tabs on the sides of the pin contact the spindle face.

13. Repeat Steps 11 and 12 for the lower pin.

NOTE: During installation, do not allow the tabs of the caliper pin to be tapped too far into the spindle groove. If this happens, it will be necessary to tap the other end of the caliper pin until the tabs snap in place. The tabs on each end of the pin must be free to catch on the spindle face.

14. Connect the brake hose to the caliper. Bleed the brake system.

15. Install the wheel and tire assembly and lower the vehicle. Check the brake fluid level and check the brakes for proper operation.

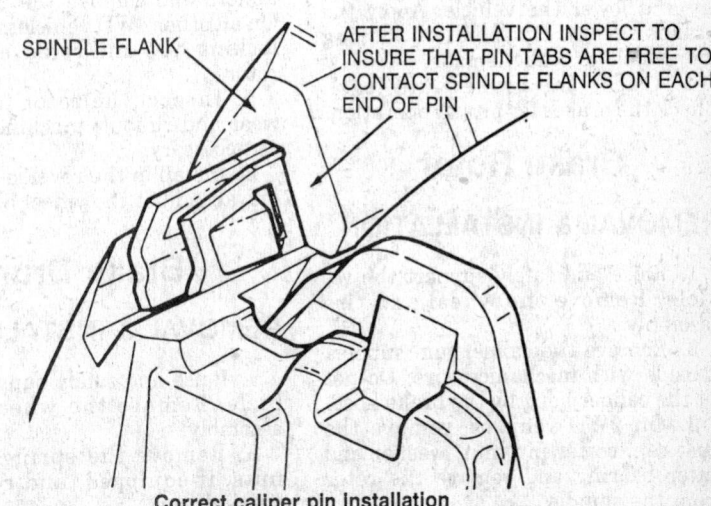

SPINDLE FLANK

AFTER INSTALLATION INSPECT TO INSURE THAT PIN TABS ARE FREE TO CONTACT SPINDLE FLANKS ON EACH END OF PIN

Correct caliper pin installation

Disc Brake Pads

REMOVAL & INSTALLATION

1. Siphon part of the brake fluid out of the master cylinder to avoid overflow when the caliper piston is pressed into the caliper bore.

2. Raise the vehicle and support it safely. Remove the wheel and tire assembly.

3. Remove the brake caliper, but do not disconnect the brake hose. Secure the caliper aside with mechanics wire.

NOTE: Do not let the caliper hang by the brake hose.

4. Compress the anti-rattle clip and remove the inner brake pad from the caliper.

5. Press each ear of the outer brake pad away from the caliper and slide the torque buttons out of the retention notches.

To install:

6. Bottom out the caliper piston in the caliper bore using an 8 in. C-clamp and a worn out inner brake pad or block of wood to push against the piston. Do not attempt to bottom out the piston with the outer brake pad installed.

7. Place a new anti-rattle clip on the lower end of the inner brake pad. Make sure the tabs on the clip are properly positioned and the clip is fully seated.

8. Position the inner brake pad and anti-rattle clip in the pad abutment with the ant-rattle clip tab against the pad abutment and the loop-type spring away from the rotor. Compress the anti-rattle clip and slide the upper end of the pad in position.

9. Install the outer pad, making sure the torque buttons on the pad are seated solidly in the matching holes in the caliper.

10. Install the caliper on the spindle.

11. Install the wheel and tire assembly and lower the vehicle. Apply the brakes several times before moving the vehicle to seat the pads.

12. Check the brake fluid level. Check the brakes for proper operation.

Brake Rotor

REMOVAL & INSTALLATION

1. Raise and safely support the vehicle. Remove the wheel and tire assembly.

2. Remove the caliper and support it aside with mechanics wire. Do not let the caliper hang by the brake hose.

3. On 2WD vehicles, remove the dust cap, cotter pin, nut, washer and outer bearing and remove the rotor from the spindle.

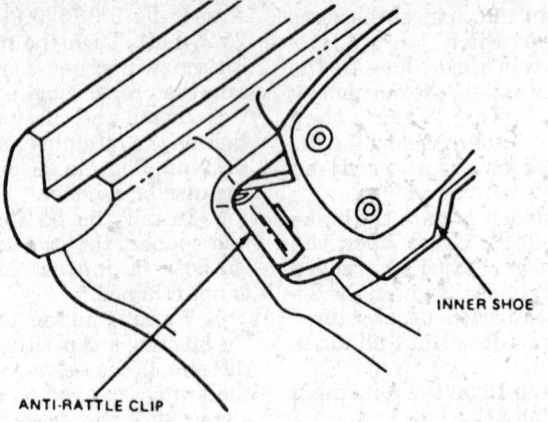

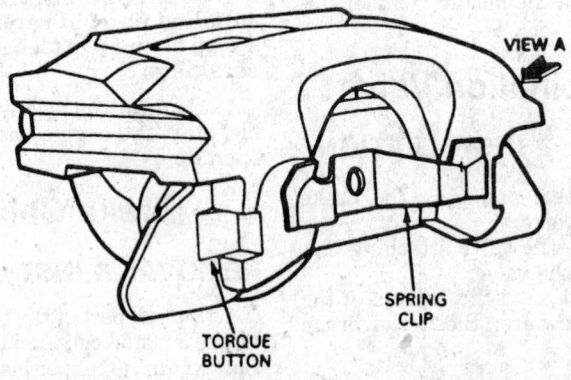

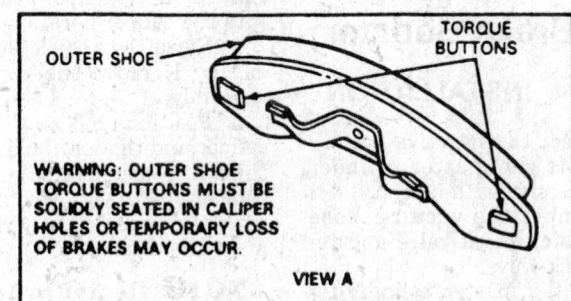

Correct brake pad installation

4. On 4WD Aerostar, remove the retainers and remove the brake rotor. On all other 4WD vehicles, remove the locking hub and remove the brake rotor.

5. Inspect the rotor for scoring, wear and runout; machine or replace as necessary.

6. Install in the reverse order of removal. Adjust the wheel bearings.

Brake Drum

REMOVAL & INSTALLATION

1. Raise and safely support the vehicle. Remove the wheel and tire assembly.

2. Remove the spring retaining nuts, if equipped, and remove the brake drum.

NOTE: If the brake drum will not come off, insert a narrow prybar through the brake adjusting hole in the backing plate and disengage the adjusting lever from the adjusting screw. While holding the adjusting lever away from the adjusting screw, loosen the adjusting screw with a brake adjusting tool.

3. Inspect the brake drum surface for wear, scoring and runout. Machine or replace, as necessary.

4. Installation is the reverse of the removal procedure.

Brake Shoes

REMOVAL & INSTALLATION

1. Raise and safely support the ve-

hicle. Remove the wheel and tire assembly and the brake drum.

2. Pull backward on the adjusting lever cable to disengage the adjusting lever from the adjusting screw. Move the outboard side of the adjusting screw upward and back off the pivot nut as far as it will go.

3. Pull the adjusting lever, cable and automatic adjuster spring down and toward the rear to unhook the pivot hook from the large hole in the secondary shoe web. Do not pry the pivot hook from the hole.

4. Remove the automatic adjuster spring and adjusting lever.

5. Remove the secondary shoe-to-anchor spring using a suitable brake spring removal/installation tool. Using the tool, remove the primary shoe-to-anchor spring and unhook the cable anchor. Remove the anchor pin plate, if equipped.

6. Remove the cable guide from the secondary shoe.

7. Remove the shoe hold-down springs, shoes, adjusting screw, pivot nut and socket. Note the color and position of each hold-down spring so they can be reassembled in the same position.

8. Remove the parking brake link and spring. Disconnect the parking brake cable from the parking brake lever.

9. Remove the secondary brake shoe. On 9 in. rear brakes, remove the parking brake lever from the shoe. On 10 in. rear brakes, remove the retainer clip and spring washer and remove the parking brake lever.

To install:

10. Clean the backing plate ledge pads and sand lightly. Apply a light coating of high temperature lithium grease to the points where the brake shoes touch the backing plate. Lubricate the adjusting cable eye and the anchor pin area.

11. Install the parking brake lever on the secondary shoe. On 10 in. brakes, secure with the spring washer and retaining clip.

12. Position the brake shoes on the backing plate and install the hold-down spring pins, springs and cups. Install the parking brake link, spring and washer. Connect the parking brake cable to the parking brake lever.

13. Install the anchor pin plate, if equipped, and place the cable anchor over the anchor pin with the crimped side toward the backing plate.

14. Install the primary shoe-to-anchor spring using the brake spring removal/installation tool.

15. Install the cable guide on the secondary shoe with the flanged hole fitted into the hole in the secondary shoe. Thread the cable around the cable guide groove.

NOTE: Make sure the cable is positioned in the groove and not between the guide and shoe web.

16. Install the secondary shoe-to-anchor (long) spring.

NOTE: Make sure the cable end is not cocked or binding on the anchor pin when installed. All parts should be flat on the anchor pin.

17. Apply high temperature lithium grease to the threads and the socket end of the adjusting screw. Turn the adjusting screw into the adjusting pivot nut to the end of the threads and then loosen, ½ turn.

18. Place the adjusting socket on the screw and install the assembly between the shoe ends with the adjusting screw nearest the secondary shoe.

NOTE: Be sure to install the adjusting screw on the same side of the vehicle from which it came. To prevent incorrect installation, the socket end of each adjusting screw is stamped with R or L, to indicate installation on the right or left side of the vehicle. The adjusting pivot nuts have lines machined around the body of the nut, 2 lines indicating the right side nut and 1 line indicating the left side nut.

19. Hook the cable hook into the hole in the adjusting lever from the outboard plate side. The adjusting levers are also stamped with an **R** or **L** to indicate right or left side installation.

20. Place the hooked end of the adjuster spring in the large hole in the primary shoe web and connect the loop end of the spring to the adjuster lever hole.

21. Pull the adjuster lever, cable and automatic adjuster spring down toward the rear to engage the pivot hook in the large hole in the secondary shoe web.

22. After installation, check the action of the adjuster by pulling the section of the cable between the cable guide and the adjusting lever toward the secondary shoe web far enough to lift the lever past a tooth on the adjusting screw wheel. The lever should snap into position behind the next tooth and releasing the cable should cause the adjuster spring to return the lever to its original position. This return action will turn the adjusting screw 1 tooth.

23. If pulling the cable does not produce the action described in Step 22 or if lever action is sluggish instead of positive and sharp, check the position of the lever on the adjusting screw toothed wheel. With the brake in a vertical position, anchor at the top, the lever should contact the adjusting wheel 1 tooth above the center line of the adjusting screw. If the contact point is below the center line, the lever will not lock on the adjusting screw wheel teeth and the screw will not turn as the lever is actuated by the cable.

24. To find the cause of the condition described in Step 23, proceed as follows:

a. Check the cable and fittings. The cable should completely fill or extend slightly beyond the crimped section of the fittings. If this does not happen, the cable assembly may be damaged and should be replaced.

b. Check the cable guide for damage. The cable groove should be parallel to the shoe web and the body of the guide should lie flat against the web. Replace the guide if it shows damage.

c. Check the pivot hook on the lever. The hook surfaces should be square with the body on the lever for proper pivoting. Repair the hook or replace the lever if the hook shows damage.

d. Be sure the adjusting screw socket is properly seated in the notch in the shoe web.

25. Adjust the brake shoes using either a brake adjustment gauge or manually with the drums installed.

26. If using a brake adjustment gauge, proceed as follows:

a. Measure the inside diameter of the brake drum with the gauge.

b. Reverse the tool and adjust the brake shoes until they touch the gauge. The gauge contact points on the shoes must be parallel to the vehicle with the center line through the center of the axle.

c. Install the drum and wheel and tire assembly. Lower the vehicle.

d. Apply the brakes sharply several times while driving the vehicle in reverse. Check brake operation by making several stops while driving forward.

27. If manually adjusting the brakes, proceed as follows:

a. Install the brake drum and wheel and tire assembly.

b. Remove the cover from the adjusting hole at the bottom of the backing plate and turn the adjusting screw, using a suitable brake adjusting tool, to expand the brake shoes until they drag against the brake drum.

c. When the shoes are against the drum, insert a narrow prybar through the brake adjusting hole and disengage the adjusting lever from the adjusting screw. While holding the adjusting lever away from the adjusting screw, loosen the adjusting screw with the brake adjusting tool, until the drum rotates freely without drag. .

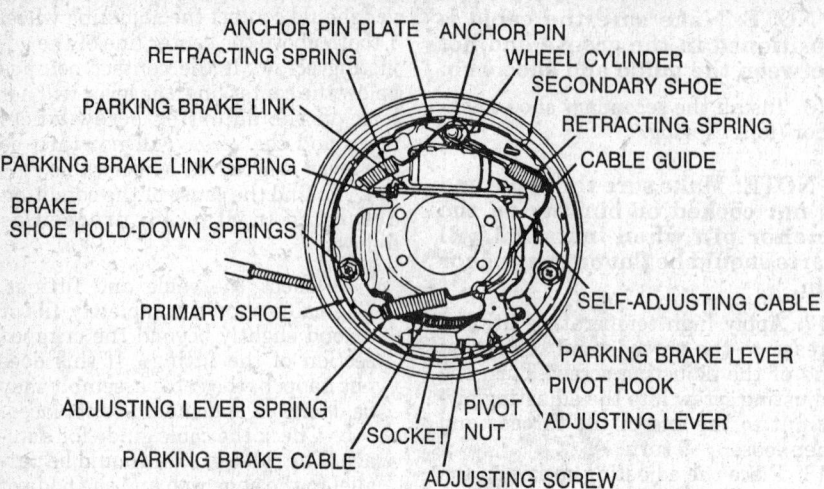

ANCHOR PIN PLATE — ANCHOR PIN
RETRACTING SPRING — WHEEL CYLINDER
— SECONDARY SHOE
PARKING BRAKE LINK — RETRACTING SPRING
PARKING BRAKE LINK SPRING — CABLE GUIDE
BRAKE
SHOE HOLD-DOWN SPRINGS
PRIMARY SHOE — SELF-ADJUSTING CABLE
— PARKING BRAKE LEVER
— PIVOT HOOK
ADJUSTING LEVER SPRING — ADJUSTING LEVER
SOCKET — PIVOT NUT
PARKING BRAKE CABLE
ADJUSTING SCREW

Rear brake shoe assembly

d. Install the adjusting hole cover and lower the vehicle.

e. Apply the brakes. If the pedal travels more than halfway to the floor, there is too much clearance between the brake shoes and drums. Repeat the adjustment procedure.

Wheel Cylinder

REMOVAL & INSTALLATION

1. Raise and safely support the vehicle. Remove the wheel and tire assembly, brake drum and brake shoes.

2. Remove the cylinder to shoe connecting pins.

3. Disconnect the brake line from the wheel cylinder.

4. Remove the wheel cylinder retaining bolts and remove the cylinder from the brake backing plate.

5. Installation is the reverse of the removal procedure. Adjust the brakes and bleed the system.

Parking Brake Cable

ADJUSTMENT

Aerostar

The parking brake system is self adjusting and requires no adjustment.

Bronco II, Explorer and Ranger

NOTE: Adjust the drum brakes before adjusting the parking brake. The brake drums must be cold for correct adjustment.

INITIAL ADJUSTMENT

Use this procedure when a new tension limiter is installed.

1. Apply the parking brake pedal to the fully engaged position.

2. Raise and safely support the vehicle, as necessary. Hold the threaded rod end of the right brake cable to keep it from spinning and thread the equalizer nut 2½ in. up the rod.

3. Check to make sure the cinch strap has slipped and there are less than 1⅜ in. remaining.

4. Release the parking brake and check for proper operation.

FIELD ADJUSTMENT

Use this procedure to correct a slack system if a new tension limiter is not installed.

1. Apply the parking brake pedal to the fully engaged position.

2. Raise and safely support the vehicle, as necessary. Grip the threaded rod to keep it from spinning and tighten the equalizer nut 6 full turns past its original position on the threaded rod.

3. Attach a suitable cable tension gauge in front of the equalizer assembly on the front cable and measure the cable tension. The cable tension should be 400–600 lbs. with the parking brake pedal in the last detent position. If tension is low, repeat Steps 2 and 3.

4. Release parking brake and check for rear wheel drag. There should be no brake drag.

REMOVAL & INSTALLATION

Front Cable

AEROSTAR

1. Place the parking brake control in the released position. Release the parking brake cable tension as follows:

a. Remove the boot cover from the parking brake control assembly.

b. Insert a steel pin through the pawl lockout pin hole. The pin must be inserted from the inboard side of the control (larger hole) at a slightly upward and forward angle then

moved downward and rearward to displace the self adjusting pawl to be inserted through the other hole. This locks out the self adjusting pawl.

c. Raise and safely support the vehicle with an assistant inside the vehicle. Pull rearward on the equalizer 1–2½ in. to rotate the self-adjuster reel backward.

d. Have the assistant insert a steel pin through the self-adjusting spring lock-out holes in the lever and control assembly. This locks the ratchet wheel in the cable released position.

NOTE: Do not remove the steel lock pin until the cables are connected to the equalizer. Pin removal releases the tension in the ratchet wheel causing the spring to unwind and release tension, requiring assembly removal to reset spring tension.

2. Raise and safely support the vehicle. Disconnect the rear parking brake cables from the equalizer. Remove the equalizer from the front cable.

3. Remove the bolts retaining the cover to the underbody reinforcement bracket and remove the cover. It may be necessary to loosen fuel tank straps and partially lower the fuel tank to gain access to the cover.

4. Remove the cable anchor pin from the pivot hole in the control assembly ratchet plate. Guide the front cable from the control assembly.

5. Insert a ½ in. box end 12-point distributor lock bolt wrench over the front fitting of the front cable. Push the wrench onto the cable retainer fitting in the crossmember. Compress the retainer fingers and push the retainer rearward through the hole.

6. Insert a ½ in. box end 12-point wrench over the rear fitting of the front cable. Push the wrench onto the cable retainer fitting in the crossmember. Compress the retainer fingers and push the retainer forward through the hole.

7. Pull the cable ends through the crossmembers and remove the cable. To install:

8. Feed the front cable through the holes in both crossmembers. Push the retainers through the holes so the fingers expand over each hole.

9. Route the front cable around the control assembly pulley and insert the cable anchor pin in the pivot hole in the ratchet plate.

10. Slide the return spring over the rear end of the front cable. Connect the equalizer to the front and rear cables.

11. Remove the lock pins from the control assembly to apply cable ten-

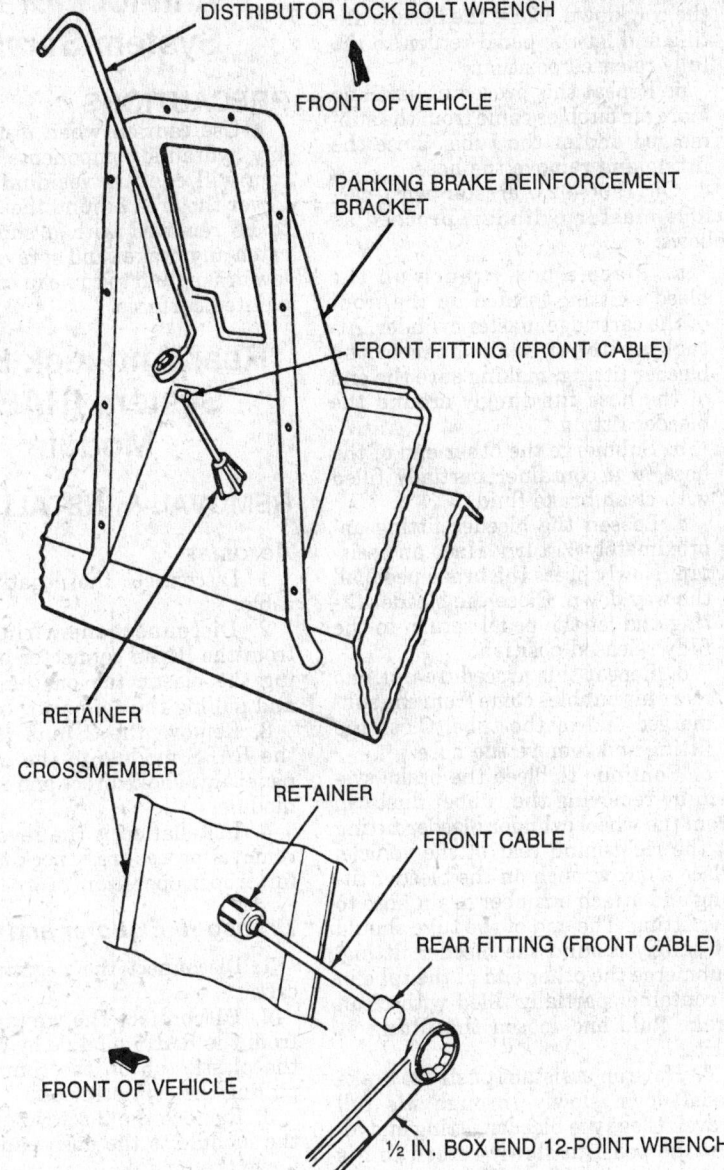

Front parking brake cable removal—Aerostar

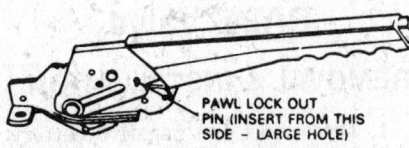

Releasing parking brake cable tension—Aerostar

sion. Position the cover on the reinforcement bracket. Install and tighten the bolts after visually checking to be sure the front cable is attached to the control.

12. Position the boot over the control. Install and tighten the screws.

13. Apply and release the control several times. Make sure the parking brakes are applied, and released and not dragging. Both pins must be removed for proper adjustment.

Bronco II, Explorer and Ranger

1. Raise and safely support the vehicle.

2. Back off the equalizer nut and remove slug of front cable on Bronco II and Ranger or intermediate cable on Explorer from the tension limiter.

3. On Bronco II and Ranger, remove the parking brake cable from the cable bracket. On Explorer, remove the intermediate cable from the bracket and disconnect the intermediate cable from the front cable.

4. Lower the vehicle. Remove the forward ball end of the parking brake cable from the control assembly clevis.

5. Remove the cable from the control assembly.

6. Using a cord attached to the control lever end of the cable, remove the cable from the vehicle pulling it up into the passenger compartment.

To install:

7. Transfer the cord to the new cable. Position the cable in the vehicle, routing the cable through the dash panel. Remove the cord and secure the cable to the control.

8. Connect the forward ball end of the brake cable to the clevis of the control assembly. Raise and safely support the vehicle.

9. Route the cable through the bracket. On Explorer, connect the front cable to the intermediate cable.

10. Connect the slug of the front or intermediate cable cable to the tension limiter connector. Adjust the parking brake cable at the equalizer using initial adjustment or field adjustment, as necessary.

11. Rotate both wheels to make sure the parking brakes are not dragging.

Rear Cable

1. Release parking brake control.

2. On Aerostar, to release tension on the rear cables, pull rearward on the equalizer assembly about 1–2 in. and place a clamp on the front cable behind the crossmember.

3. Raise and safely support the vehicle. Remove the wheel and tire assembly and the brake drum.

4. On Bronco II, Explorer and Ranger, remove the locknut on the threaded rod at the equalizer. Disconnect the rear parking brake cable from the equalizer.

5. Compress the prongs that retain the cable housing to the frame bracket or crossmember and pull out the cable and housing.

6. Working on the wheel side of the backing plate, compress the prongs on the cable retainer so they can pass through the hole in the brake backing plate.

7. Lift the cable out of the slot in the parking brake lever, attached to the secondary brake shoe, and remove the cable through the brake backing plate hole.

To install:

8. Route the cable through the hole in the backing plate. Insert the cable anchor behind the slot in the parking brake lever. Make sure the cable is securely engaged in the parking brake lever so the cable return spring is holding the cable in the parking brake lever.

9. Push the retainer through the hole in the backing plate so the retainer prongs engage the backing plate.

10. Properly route the cable and insert the front of the cable through the

frame bracket or crossmember until the prongs expand. Connect the rear cables to the equalizer.

11. On Bronco II, Explorer and Ranger, rotate the equalizer 90 degrees and recouple the threaded rod to the equalizer.

12. On Aerostar, remove the clamping device the reapply tension.

13. Install the brake drum and wheel and tire assembly. Adjust the rear brakes.

14. On Bronco II, Explorer and Ranger, adjust the parking brake tension using the initial adjustment or the field adjustment procedure, as necessary.

15. Apply and release the parking brake control several times. Rotate both wheels to make sure the parking brakes are applied and released and not dragging.

Brake System Bleeding

1. Clean all dirt from the master cylinder filler cap.

2. If the master cylinder is known or suspected to have air in the bore, it must be bled before any of the wheel cylinders or calipers. Proceed as follows:

a. Loosen the brake line fitting approximately ¾ turn. Wrap a shop cloth around the tubing below the fitting to absorb escaping brake fluid.

b. Have an assistant depress the brake pedal slowly through it's full travel to force air trapped in the master cylinder to escape at the fitting.

c. Tighten the fitting and let the pedal return slowly to the fully released position. Do not release the pedal until the fitting is tightened or air will re-enter the master cylinder.

d. Wait 5 seconds and then repeat the operation until all air bubbles disappear.

e. Repeat Steps a–d on the remaining master cylinder brake line fitting(s).

3. On Bronco II, Explorer and Ranger equipped with Rear Anti-lock Brakes (RABS), proceed as follows:

a. Place a box wrench on the bleeder fitting On the RABS valve. Attach a rubber drain hose to the bleeder fitting, making sure the end of the hose fits snugly around the bleeder fitting.

b. Submerge the other end of the hose in a container partially filled with clean brake fluid.

c. Loosen the bleeder fitting approximately ¾ turn. Have an assis-

tant slowly press the brake pedal all the way down. Close the bleeder fitting and let the pedal return to the fully released position.

d. Repeat this procedure until no more air bubbles come from the submerged end of the tube. Close the fitting and remove the hose.

4. On 1990–92 Aerostar with cartridge master cylinder, proceed as follows:

a. Place a box wrench on the bleeder fitting located on the front of the cartridge master cylinder. Attach a rubber drain hose to the bleeder fitting, making sure the end of the hose fits snugly around the bleeder fitting.

b. Submerge the other end of the hose in a container partially filled with clean brake fluid.

c. Loosen the bleeder fitting approximately ¾ turn. Have an assistant slowly press the brake pedal all the way down. Close the bleeder fitting and let the pedal return to the fully released position.

d. Repeat this procedure until no more air bubbles come from the submerged end of the tube. Close the fitting and remove the hose.

5. Continue to bleed the brake system by removing the rubber dust cap from the wheel cylinder bleeder fitting at the right-hand rear of the vehicle. Place a box wrench on the bleeder fitting and attach a rubber drain hose to the fitting. The end of the tube should fit snugly around the bleeder fitting. Submerge the other end of the tube in a container partially filled with clean brake fluid and loosen the fitting ¾ turn.

6. Have an assistant push the brake pedal down slowly through it's full travel. Close the bleeder fitting and allow the pedal to slowly return to it's full release position. Wait 5 seconds and repeat the procedure until no bubbles appear at the submerged end of the bleeder tube. Secure the bleeder fitting and remove the bleeder hose. Install the rubber dust cap on the bleeder fitting.

7. Repeat the procedure in Steps 5 and 6 and bleed the rest of the system in the following sequence: left rear, right front and left front.

8. Refill the master cylinder reservoir after each wheel cylinder or caliper has been bled and install the master cylinder cover and gasket. When brake bleeding is completed, the fluid level should be filled to the maximum level indicated on the reservoir.

9. Always make sure the disc brake pistons are returned to their normal positions by depressing the brake pedal several times until normal pedal travel is established. If the pedal feels spongy, repeat the bleeding procedure.

Anti-lock Brake System Service

PRECAUTIONS

• Use caution when disassembling any hydraulic components as the system will contain residual pressure. Cover the area around the component to be removed with a shop cloth to catch any brake fluid spray. Do not allow brake fluid to come in contact with painted surfaces.

Rear Anti-lock Brake System (RABS) Module

REMOVAL & INSTALLATION

Aerostar

1. Disconnect the negative battery cable.

2. Disconnect the wiring harness from the RABS connector by depressing the plastic tab on the connector and pulling the connector off.

3. Remove the 2 nuts that retain the RABS module to the instrument panel anti-shake brace and remove the module.

4. Installation is the reverse of the removal procedure. Check the system for proper operation.

Bronco II, Explorer and Ranger

1. Disconnect the negative battery cable.

2. Disconnect the wiring harness from the RABS module by depressing the plastic tab on the connector and pulling the connector off.

3. Remove the 2 screws that retain the module to the dash panel and remove the module.

4. Installation is the reverse of the removal procedure. Check the system for proper operation.

RABS Valve

REMOVAL & INSTALLATION

1. Disconnect the negative battery cable.

2. Disconnect and plug the 2 brake lines connected to the RABS valve.

3. Disconnect the wiring harness from the valve harness.

4. Remove the screw retaining the valve and remove the valve.

To install:

5. Position the RABS valve and install the retaining screw. Tighten the retaining screw on Bronco II, Explorer and Ranger to 11–14 ft. lbs. (15–20 Nm) or on Aerostar to 30–40 inch lbs. (3.3–4.5 Nm).

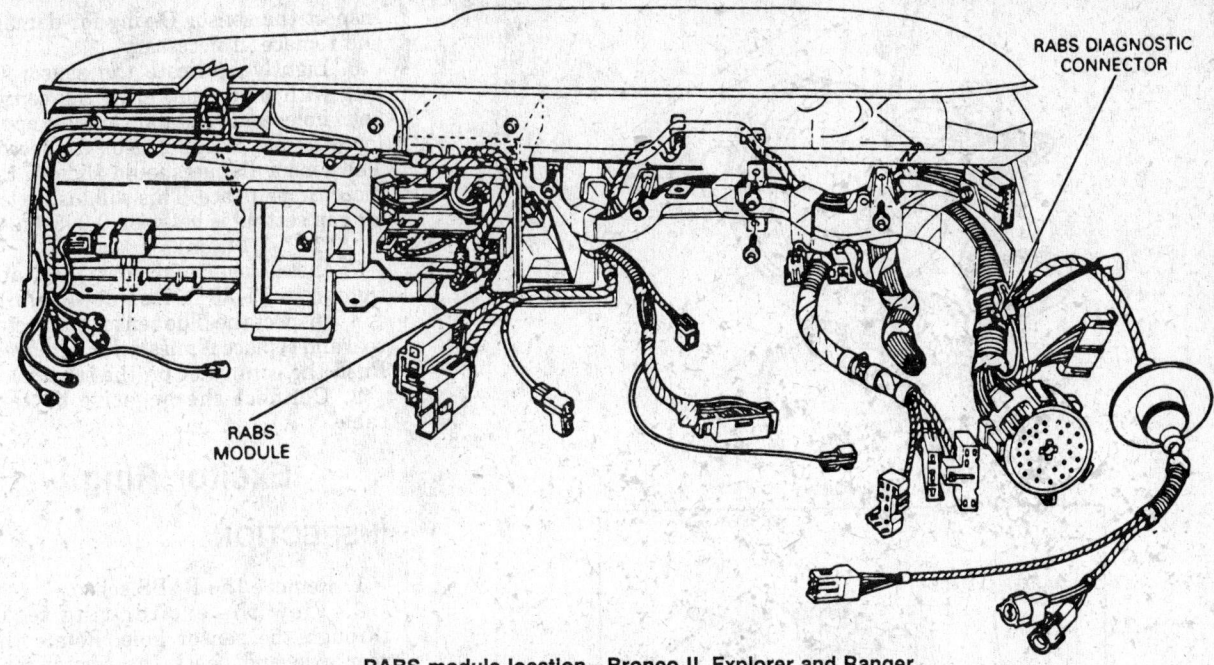

RABS module location—Bronco II, Explorer and Ranger

6. Connect the brake valve wiring harness connector.

7. Connect the brake lines to the valve and tighten as follows:

   a. ½–20 threaded fittings—10–17 ft. lbs. (14–23 Nm).

   b. ⁷/₁₆–24 threaded fittings—10–15 ft. lbs. (14–20 Nm).

   c. ⅜–24 threaded fittings—10–15 ft. lbs. (14–20 Nm).

NOTE: Do not overtighten the fittings.

8. Bleed the brake system. It is not necessary to energize the valve electrically to bleed the rear brakes.

9. Connect the negative battery cable.

RABS Sensor

REMOVAL & INSTALLATION

1. Disconnect the negative battery cable.

2. Pull the wiring harness connector off.

3. Remove the sensor hold-down bolt and remove the sensor from the axle housing.

To install:

4. Clean the axle mounting surface. Use care to prevent dirt from entering the axle housing.

5. Inspect and clean the magnetized sensor pole piece to ensure that it is free from loose metal particles which could cause erratic system operation.

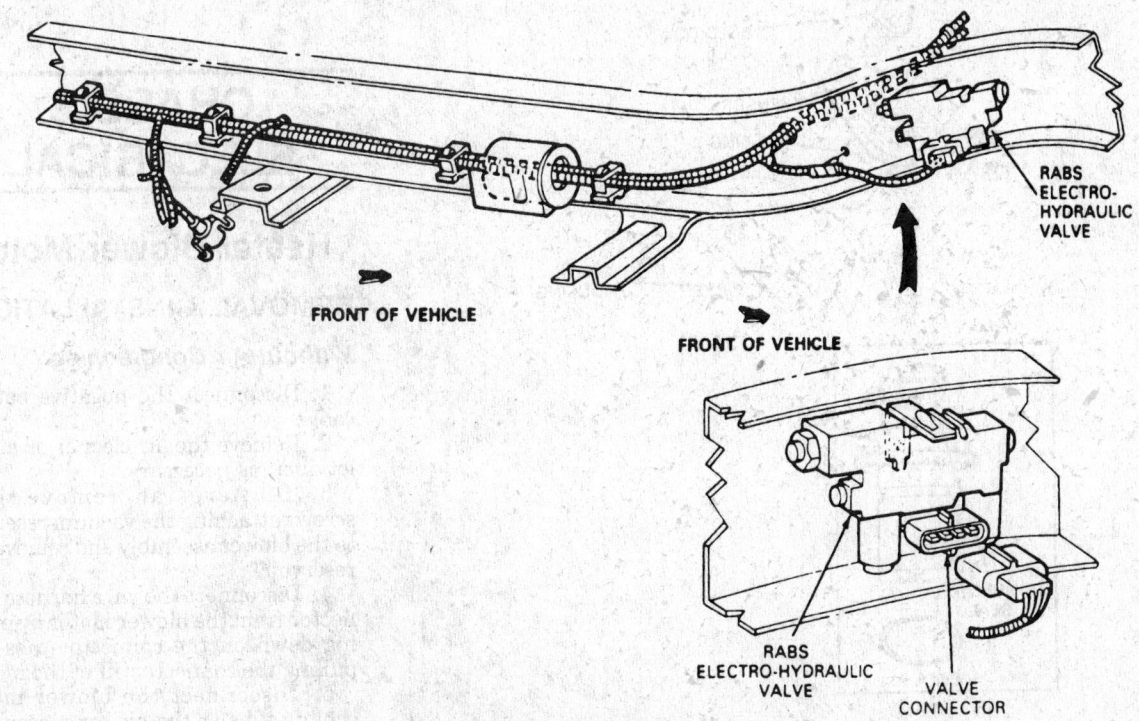

RABS valve location—Bronco II, Explorer and Ranger

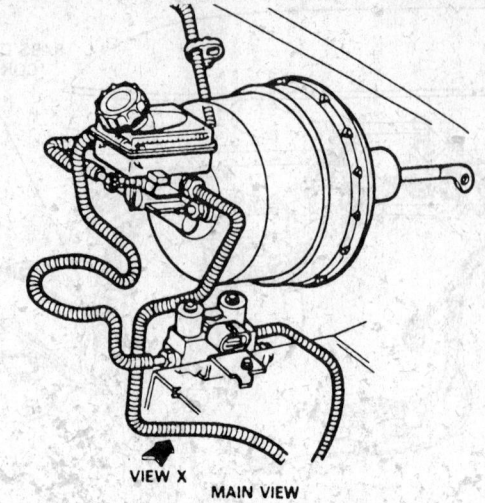

VIEW X
MAIN VIEW

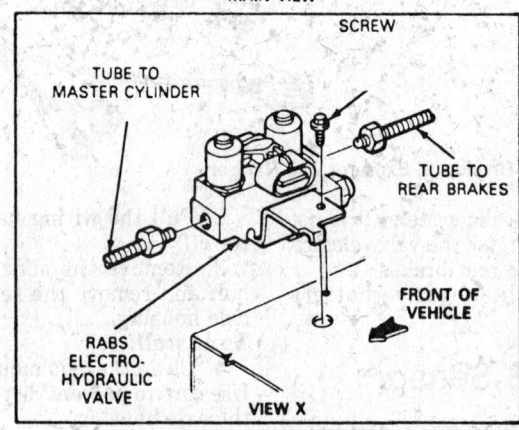

SCREW

TUBE TO
MASTER CYLINDER

TUBE TO
REAR BRAKES

FRONT OF
VEHICLE

RABS
ELECTRO-
HYDRAULIC
VALVE

VIEW X

RABS valve location—Aerostar

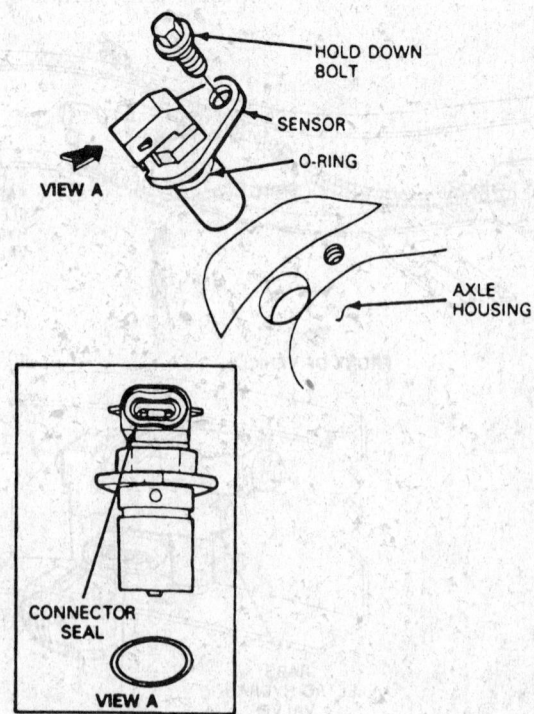

HOLD DOWN
BOLT

SENSOR

O-RING

VIEW A

AXLE
HOUSING

CONNECTOR
SEAL

VIEW A

RABS sensor Installation

Inspect the sensor O-ring for damage and replace, if necessary.

6. Lightly lubricate the sensor O-ring with motor oil, align the sensor bolt hole and install. Do not apply force to the plastic sensor connector. The sensor flange should slide to the mounting surface. This will insure the air gap setting is between 0.005–0.045 in. (0.127–1.14mm).

7. Install the hold down bolt and tighten to 25–30 ft. lbs. (34–40 Nm).

8. Inspect the blue sensor connector seal and replace if missing or damaged. Push the connector on the sensor.

9. Connect the negative battery cable.

Excitor Ring

INSPECTION

1. Remove the RABS sensor.
2. View the excitor ring teeth through the sensor hole. Rotate the rear axle and check the excitor ring teeth for damage or breakage. Dented or broken teeth could cause the RABS system to function when not required.

REMOVAL & INSTALLATION

To service the excitor ring, the differential case must be removed from the axle housing and the excitor ring pressed off the case.

NOTE: Upon removal, the excitor ring is to be discarded. It is not to be reused.

CHASSIS ELECTRICAL

Heater Blower Motor

REMOVAL & INSTALLATION

Without Air Conditioning

1. Disconnect the negative battery cable.
2. Remove the air cleaner or air inlet duct, as necessary.
3. On Aerostar, remove the 2 screws attaching the vacuum reservoir to the blower assembly and remove the reservoir.
4. Disconnect the wire harness connector from the blower motor by pushing down on the connector tabs and pulling the connector off of the motor.
5. Disconnect the blower motor cooling tube at the blower motor.
6. Remove the 3 screws attaching

the blower motor and wheel to the heater blower assembly.

7. Holding the cooling tube aside, pull the blower motor and wheel from the heater blower assembly and remove it from the vehicle.

8. Remove the blower wheel pushnut or clamp from the motor shaft and pull the blower wheel from the motor shaft.

To install:

9. Install the blower wheel on the blower motor shaft.

10. Install the hub clamp or pushnut.

11. Holding the cooling tube aside, position the blower motor and wheel on the heater blower assembly and install the 3 attaching screws.

12. Connect the blower motor cooling tube and the wire harness connector.

13. On Aerostar, install the vacuum reservoir on the hoses with the 2 screws.

14. Install the air cleaner or air inlet duct, as necessary.

15. Connect the negative battery cable and check the system for proper operation.

With Air Conditioning

1. Disconnect the negative battery cable.

2. In the engine compartment, disconnect the wire harness from the motor by pushing down on the tab while pulling the connection off at the motor.

3. Remove the air cleaner or air inlet duct, as necessary.

4. On Bronco II, Explorer and Ranger, remove the solenoid box cover retaining bolts and the solenoid box cover, if equipped.

5. Disconnect the blower motor cooling tube from the blower motor.

6. Remove the 3 blower motor mounting plate attaching screws and remove the motor and wheel assembly from the evaporator assembly blower motor housing.

7. Remove the blower motor hub clamp from the motor shaft and pull the blower wheel from the shaft.

To install:

8. Install the blower motor wheel on the blower motor shaft and install a new hub clamp.

9. Install a new motor mounting seal on the blower housing before installing the blower motor.

10. Position the blower motor and wheel assembly in the blower housing and install the 3 attaching screws.

11. Connect the blower motor cooling tube.

12. Connect the electrical wire harness hardshell connector to the blower motor by pushing into place.

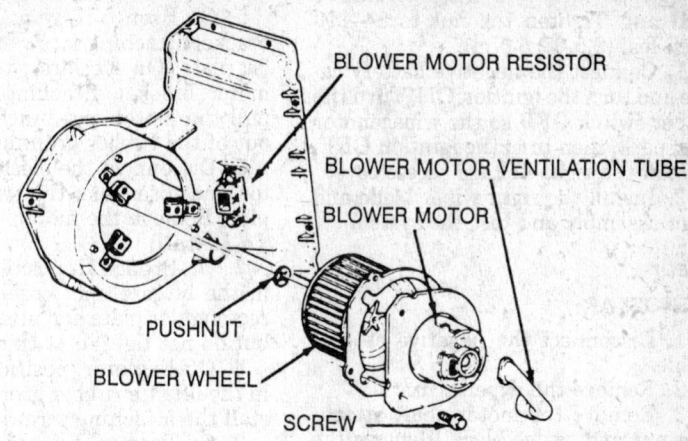

Blower motor installation–Bronco II, Explorer and Ranger

13. On Bronco II, Explorer and Ranger, position the solenoid box cover, if equipped, into place and install the 3 retaining screws.

14. Install the air cleaner or air inlet duct, as necessary.

15. Connect the negative battery cable and check the blower motor in all speeds for proper operation.

Window Wiper Motor

REMOVAL & INSTALLATION

Front

AEROSTAR

1. Turn the wiper switch **ON**. Turn the ignition switch to **RUN** until the wiper blades are in midpattern, then turn the ignition switch to **OFF** to keep the blades in this position.

2. Disconnect the negative battery cable, then disconnect the electrical connector from the wiper motor.

3. Remove both wiper arms and remove the cowl grille.

4. Remove the linkage retaining clip and disconnect the linkage from the motor crank arm.

5. Remove the motor retaining nuts while holding the motor to keep it from falling.

6. Installation is the reverse of the removal procedure.

BRONCO II, EXPLORER AND RANGER

1. Turn the wiper switch **ON**. Turn the ignition switch **ON** until the blades are straight up, then turn the ignition **OFF** to keep them there.

2. Disconnect the negative battery cable, then disconnect the electrical connector from the wiper motor.

3. Remove the right wiper arm and blade assembly. Remove the right pivot nut and allow the linkage to drop into the cowl.

4. Remove the linkage access cover, located on the right side of the dash panel, near the wiper motor.

5. Reach through the access cover opening and unsnap the wiper motor clip. Push the clip away from the linkage until it clears the nib on the crank pin, then push the clip off the linkage. Remove the linkage from the crank pin.

6. Remove the 3 attaching screws and remove the wiper motor.

To install:

7. Install the motor with the attaching screws. Tighten to 60–85 inch lbs. (6.8–9.6 Nm). Connect the motor electrical connector.

8. Install the clip completely onto the right linkage, making sure it is fully seated. Do not put the linkage on the motor crank pin and then try to install the clip.

9. Install the left and right linkage onto the wiper motor crank pin. Pull the linkage onto the crank pin until it snaps into place. The clip is properly installed if the nib is protruding through the center of the clip.

10. Install the right wiper pivot shaft

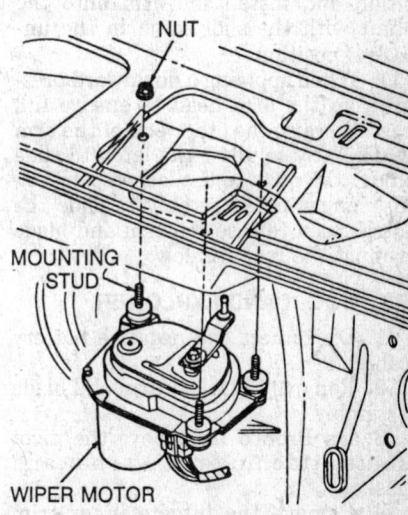

Front wiper motor installation–Aerostar

and nut. Tighten the nut to 84–110 inch lbs. (9.5–12.5 Nm).

11. Connect the negative battery cable and turn the ignition **ON**. Turn the wiper switch **OFF** so the wiper motor will park, then turn the ignition **OFF**. Install the right linkage access cover.

12. Install the right wiper blade and arm assembly and test the system.

Rear
AEROSTAR

1. Disconnect the negative battery cable.

2. Remove the wiper arm.

3. Remove the motor shaft attaching nut and wedge block. Remove the liftgate trim panel.

4. Disconnect the electrical connector and remove the motor wiring pins from the inner panel. Remove the motor.

To install:

5. Position the motor so the motor shaft protrudes through the hole in the outer panel. Attach the motor to the liftgate panel with the bracket attaching screws but do not tighten at this time.

6. Install the rubber seal and wedge block over the shaft and install the pivot attaching nut. Tighten to 5.0–5.8 ft. lbs. (6.8–7.9 Nm). Tighten the motor bracket attaching screw to 5.5–6.3 ft. lbs. (7.5–8.5 Nm).

7. Attach the motor wiring to the liftgate inner panel by installing the wiring pushpins in the holes.

8. Install the articulating arm onto the drive pivot shaft.

9. Connect the negative battery cable and turn the ignition switch **ON**. Operate the wiper switch to cycle and park the wiper system in order to ensure the system linkage is in the **PARK** position before the wiper arm is installed.

10. Locate the blade in the proper position and install the arm onto the shaft with the slide latch in the unlocked position.

11. While applying a downward pressure on the arm head to ensure full seating, raise the other end of the arm sufficiently to allow the latch to slide under the pivot to the locked position. Use finger pressure only to slide the latch, then release the arm and blade against the rear window.

BRONCO II AND EXPLORER

1. Disconnect the negative battery cable.

2. Remove the wiper arm and blade assembly.

3. On Bronco II, remove the pivot shaft attaching nut washer and gasket.

4. Remove the liftgate inner trim panel.

5. On Bronco II, remove the motor bracket attaching screw and rectangular plate. On Explorer, remove the 3 motor bracket attaching screws and pull the motor and bracket assembly out of the rubber grommet.

6. Disconnect the electrical connector and disengage the wiring locator pins. Remove the motor.

To install:

7. On Bronco II, position the motor in the liftgate and loosely install the rectangular plate and attaching screw, but do not tighten at this time.

8. On Explorer, position the motor in the liftgate rubber grommet and install the attaching screws.

9. On Bronco II, install the gasket, washer and nut. Tighten the nut to 60–69 inch lbs. (6.8–7.9 Nm). Tighten the motor bracket attaching screw to 5–6 ft. lbs. (7.5–8.5 Nm).

10. Connect the electrical connector and install the wiring locator pins in the holes provided.

11. Install the wiper arm and blade assembly. Connect the negative battery cable and check wiper operation.

12. Install the liftgate inner trim panel.

Window Wiper Switch

REMOVAL & INSTALLATION

Front
AEROSTAR

NOTE: The switch handle is an integral part of the switch and cannot be removed separately.

1. Disconnect the negative battery cable.

2. Remove the cluster finish panel retaining screws.

3. Remove the 3 left control pod retaining screws.

4. Remove the wiring connector from the switch.

5. Remove the 2 switch-to-control pod retaining screws and remove the switch.

6. Installation is the reverse of the removal procedure.

1988 BRONCO II AND RANGER

NOTE: The switch handle is an integral part of the switch and cannot be removed separately.

1. Disconnect the negative battery cable.

2. Remove the trim shrouds.

3. Disconnect the electrical connector.

4. Peel back the foam sight shield. Remove the 2 cross-recessed screws holding the switch and remove the switch.

5. Installation is the reverse of the removal procedure.

Rear
AEROSTAR

NOTE: The switch handle is an integral part of the switch and cannot be removed separately.

1. Disconnect the negative battery cable.

2. Remove the trim shrouds and the left switch pod.

3. Disconnect the electrical connector.

4. Remove the 2 cross-recessed screws holding the switch and remove the switch.

5. Installation is the reverse of the removal procedure.

1988 BRONCO II

1. Disconnect the negative battery cable.

2. Remove the headlight switch knob.

3. Pull the finish panel away from the instrument panel and disconnect the electrical connector.

4. Remove the screw attaching the wiper switch to the finish panel and remove the wiper switch.

5. Installation is the reverse of the removal procedure.

EXPLORER AND 1989–90 BRONCO II

1. Disconnect the negative battery cable.

2. Remove the 2 ash tray retaining screws and remove the ash tray.

3. Remove the cluster trim panel, which is held on by clips.

4. Remove the snap-in switch mounting bezel containing the switches and disconnect the electrical connector.

5. Remove the switch from the mounting bezel by pushing on the switch from the connector side until the snap-in mounting clips release.

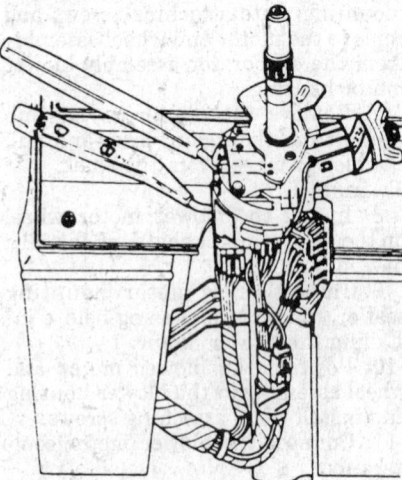

Front wiper switch location—1988 Bronco II and Ranger

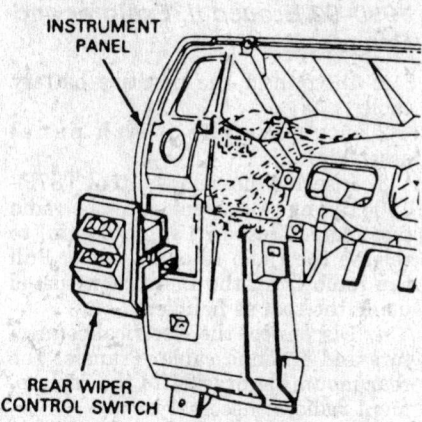

Rear wiper switch installation—
Explorer and 1988 Bronco II

6. Installation is the reverse of the removal procedure.

Instrument Cluster

REMOVAL & INSTALLATION

Aerostar

CONVENTIONAL CLUSTER

1. Disconnect the negative battery cable.

2. Remove the 7 cluster housing-to-panel retaining screws and remove the cluster housing.

3. Remove the 4 instrument cluster-to-panel retaining screws.

4. Disconnect the 2 wiring harness connectors from the backplate.

5. Disconnect the speedometer cable and remove the cluster assembly.

6. Installation is the reverse of the removal procedure. Apply an approximately $3/16$ in. diameter ball of silicone dialectric compound in the drive hole of the speedometer head prior to installation.

ELECTRONIC CLUSTER

1. Disconnect the negative battery cable.

2. Remove the cluster binnacle.

3. Remove the 4 cluster mounting screws.

4. Pull the top of the cluster toward the steering wheel, then reach behind the cluster and unplug the 3 connectors.

5. Swing the bottom of the cluster out and remove.

6. Installation is the reverse of the removal procedure.

Bronco II, Explorer and Ranger

1988

1. Disconnect the negative battery cable(s).

2. Remove the 2 steering column shroud-to-panel retaining screws and remove the shroud.

3. Remove the 2 lower instrument trim panels and rotate the headlight switch trim cover to gain access to the cluster trim cover screws.

NOTE: Removal of the headlight switch knob is not required to gain access to the cluster trim cover attaching screws.

4. Remove the 8 retaining screws and remove the cluster trim cover from the instrument panel.

5. Remove the 4 cluster-to-panel retaining screws and pull the cluster away from the panel enough to gain access to the speedometer cable. Disconnect the speedometer cable, electrical connectors and bulb and socket assemblies and remove the instrument cluster.

NOTE: If there is not enough room to disengage the cable from the speedometer, it may be necessary to disconnect the cable at the transmission and pull the cable through the cowl, to allow room to reach the speedometer quick disconnect.

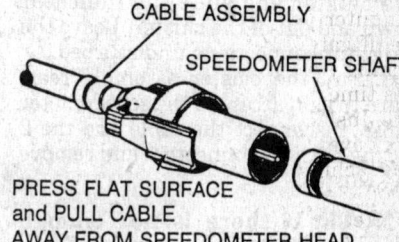

CABLE ASSEMBLY
SPEEDOMETER SHAFT
PRESS FLAT SURFACE
and PULL CABLE
AWAY FROM SPEEDOMETER HEAD

Speedometer cable quick disconnect

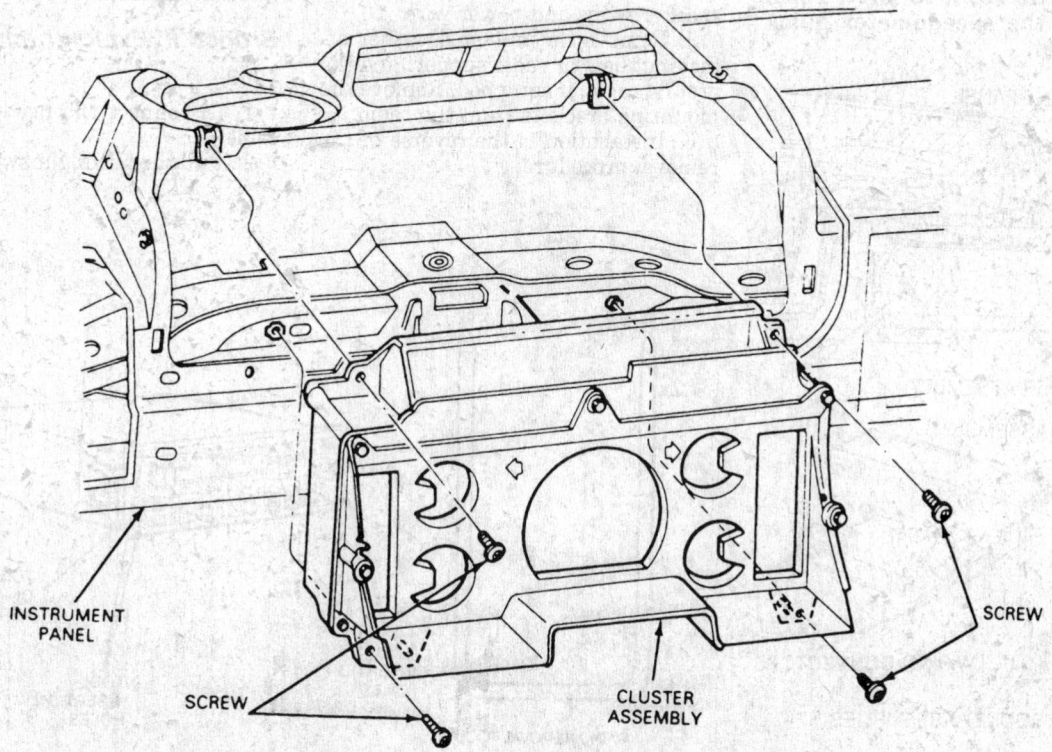

Instrument cluster installation—Aerostar with conventional cluster

6. Installation is the reverse of the removal procedure. Apply an approximately $^3/_{16}$ in. diameter ball of silicone dialectric compound in the drive hole of the speedometer head prior to installation.

1989–92

1. Disconnect the negative battery cable.

2. Open the ash tray and remove the 2 screws attaching the ash tray and instrument cluster trim panel. Remove the ash tray.

3. Unsnap the cluster trim panel by pulling rearward around the edge of the panel. Depress the hazard warning switch on the steering column and remove the cluster trim panel.

4. Remove the 4 screws securing the instrument cluster to the instrument panel.

5. If equipped with automatic transmission, remove the 2 screws attaching the shift position indicator to the cluster and slide the indicator down and out of the cluster. Leave the indicator connections undisturbed.

6. Pull the cluster assembly rearward to gain access to the speedometer cable. Disconnect the cable and the 2 wiring harness connectors and remove the cluster.

NOTE: If there is not enough room to disengage the cable from the speedometer, it may be necessary to disconnect the cable at the transmission and pull the cable through the cowl, to allow room to reach the speedometer quick disconnect.

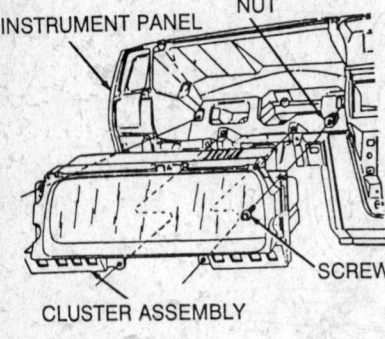

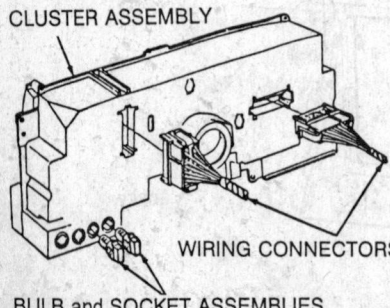

Instrument cluster installation—1989–92 Bronco II, Explorer and Ranger

7. Installation is the reverse of the removal procedure. Apply an approximately $^3/_{16}$ in. diameter ball of silicone dialectric compound in the drive hole of the speedometer head prior to installation.

Speedometer

REMOVAL & INSTALLATION

1. Disconnect the negative battery cable. Remove the instrument cluster.

2. Remove the lens and mask from the cluster.

3. Remove the 2 speedometer attaching screws and remove the speedometer.

4. Installation is the reverse of the removal procedure.

Radio

REMOVAL & INSTALLATION

Except 1990–92 Bronco II, Explorer and Ranger

1. Disconnect the negative battery cable.

2. Remove the finish panel assembly.

3. Remove the 4 screws securing the radio mounting brackets to the instrument panel. Remove the radio with the mounting brackets and rear bracket.

4. Disconnect the antenna cable, speaker wires and power wire.

5. Remove the nut and washer attaching the rear radio support and the ground cable, if equipped. Remove the mounting brackets from the radio.

6. Installation is the reverse of the removal procedure.

1990–92 Bronco II, Explorer and Ranger

1. Disconnect the negative battery cable.

2. Remove the finish panel assembly.

3. Insert radio removal tool T87P–19061–A or equivalent, into the radio face plate. Press in 1 in. (25.4mm) to release the radio retaining clips. Pull the radio from the instrument panel using the tool as handles.

4. Disconnect the electrical connectors and antenna cable. Transfer the rear mounting bracket to the replacement radio, if necessary.

5. Installation is the reverse of the removal procedure.

Headlight Switch

REMOVAL & INSTALLATION

Aerostar

1. Disconnect the negative battery cable.

2. Remove the 5 cluster finish panel assembly retaining screws and remove the cluster finish panel.

3. Remove the 3 left control pod assembly retaining screws.

4. Disconnect the electrical connector from the switch.

5. Remove the 2 switch-to-control pod retaining screws and remove the switch.

6. Installation is the reverse of the removal procedure.

Bronco II, Explorer and Ranger

1988

1. Disconnect the negative battery cable.

2. Pull the headlight switch knob to

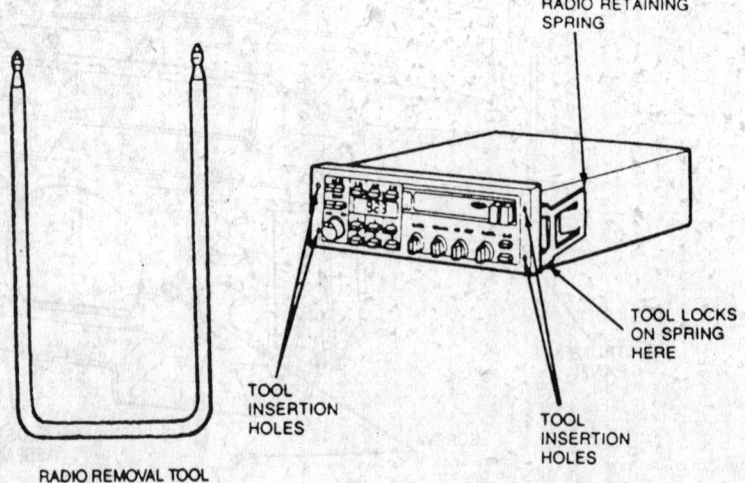

Radio removal—1990–92 Bronco II, Explorer and Ranger

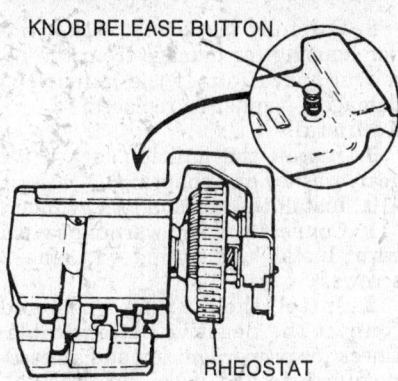

KNOB RELEASE BUTTON

RHEOSTAT

Headlight switch—Bronco II, Explorer and Ranger

the full **ON** position. Depress the shaft release button and remove the knob and shaft assembly.

3. Remove the instrument panel finish panel.

4. Unscrew the mounting nut. Remove the switch, then remove the wiring connector from the switch.

5. Installation is the reverse of the removal procedure.

1989–92

1. Disconnect the negative battery cable.

2. Remove the ash tray and the 2 finish panel retaining screws.

3. Remove the finish panel, which snaps off.

4. Remove the rear wiper switch and rear defrost switch on Bronco II and Explorer, or the storage bin on Ranger.

5. Pull the headlight switch knob to the full **ON** position. Reach through the opening below the headlight switch, depress the shaft release button on the switch and remove the knob and shaft assembly.

6. Remove the headlight switch retaining bezel nut. Pull the switch downward and through the opening to disconnect the connector.

7. Installation is the reverse of the removal procedure.

Combination Switch

The combination switch incorporates the turn signal and dimmer switch functions on Aerostar and 1988 Bronco II and Ranger. On 1989–90 Bronco II, 1991–92 Explorer and 1989–92 Ranger, the combination switch incorporates the turn signal, dimmer and windshield wiper/washer switch functions.

REMOVAL & INSTALLATION

Aerostar

1. Disconnect the negative battery cable.

2. Remove the steering wheel.

3. If equipped with tilt column, remove the upper extension shroud by squeezing it at the 6 and 12 o'clock positions and popping it free of the retaining plate at the 3 o'clock position.

4. Open the trim shroud by removing the 4 attaching screws. Swing the trim shroud down and remove the 2 screws attaching the shroud to the steering column assembly.

5. Remove the lock cylinder.

6. Remove the combination switch lever by grasping the lever and using a pulling and twisting motion, pull the lever straight out from the switch. Remove the shroud.

7. Peel back the foam sight shield from the switch. Remove the 2 self-tapping screws attaching the switch to the lock cylinder housing. Disengage the switch from the housing and disconnect the switch electrical connectors.

To install:

8. Connect the switch electrical connectors.

9. Install the switch with the self-tapping screws and tighten to 18–26 inch lbs. (2.0–2.9 Nm).

10. Stick the foam sight shield to the switch and install the trim shroud.

11. Install the lock cylinder. Install the tilt collar shroud, if equipped.

12. Install the lever into the switch by aligning the key on the lever with the keyway in the switch and pushing the lever into the switch to full engagement.

13. Install the steering wheel and connect the negative battery cable.

Bronco II, Explorer and Ranger

1988

1. Disconnect the negative battery cable.

2. Remove the 2 trim shroud halves by removing the 2 attaching screws.

3. If equipped with tilt column, remove the upper extension shroud by squeezing it at the 6 and 12 o'clock positions and popping it free of the retaining plate at the 3 o'clock position.

4. Remove the combination switch lever by grasping the lever and using a pulling and twisting motion, pull the lever straight out from the switch.

5. Peel back the foam sight shield from the switch. Remove the 2 self-tapping screws attaching the switch to the lock cylinder housing. Disengage the switch from the housing and disconnect the switch electrical connectors.

To install:

6. Connect the switch electrical connectors.

7. Install the switch with the self-tapping screws and tighten to 18–26 inch lbs. (2.0–2.9 Nm).

8. Stick the foam sight shield to the switch.

9. Install the lever into the switch by aligning the key on the lever with the keyway in the switch and pushing the lever into the switch to full engagement.

10. Install the steering column trim shrouds and connect the negative battery cable.

1989–92

1. Disconnect the negative battery cable.

2. Remove the steering column shroud.

3. Remove the 2 self-tapping screws that attach the combination switch to the steering column casting. Disengage the switch from the casting.

4. Disconnect the 3 electrical connectors, being careful not to damage

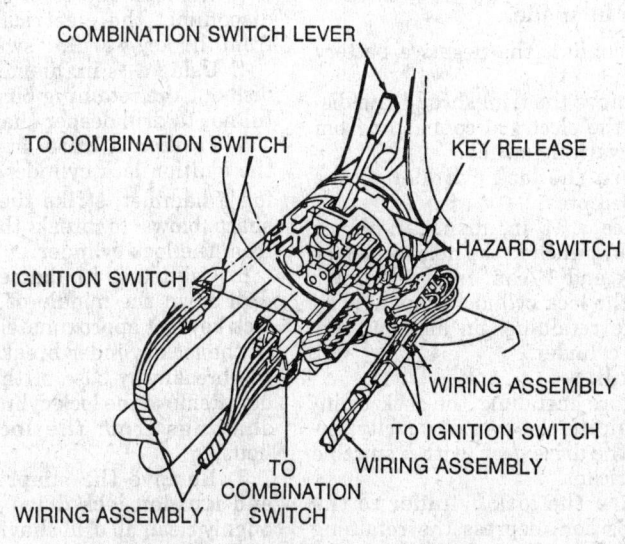

COMBINATION SWITCH LEVER

TO COMBINATION SWITCH

KEY RELEASE

IGNITION SWITCH

HAZARD SWITCH

WIRING ASSEMBLY

TO IGNITION SWITCH

WIRING ASSEMBLY

TO COMBINATION SWITCH

WIRING ASSEMBLY

Combination switch assembly—Aerostar

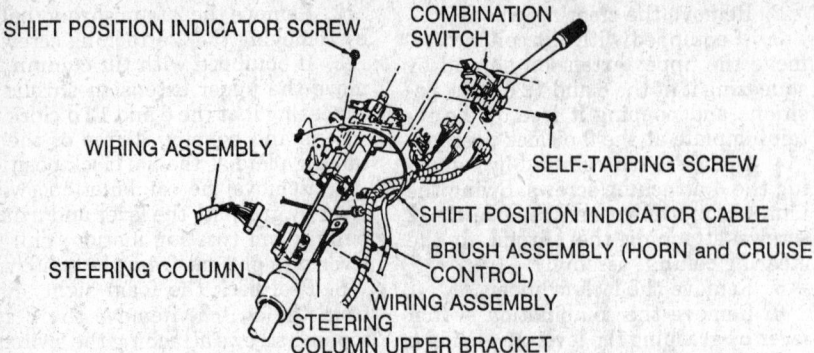

SHIFT POSITION INDICATOR SCREW

COMBINATION SWITCH

WIRING ASSEMBLY

SELF-TAPPING SCREW

SHIFT POSITION INDICATOR CABLE

STEERING COLUMN

BRUSH ASSEMBLY (HORN and CRUISE CONTROL)

WIRING ASSEMBLY

STEERING COLUMN UPPER BRACKET

Combination switch installation—1989–92 Bronco II, Explorer and Ranger

the locking tabs. Do not damage the shift position indicator cable.

To install:

5. Connect the 3 switch electrical connectors. The wiring for the switch is to be routed under the shift position indicator cable.

6. Install the switch with the self-tapping screws. Tighten the screws to 18–27 inch lbs. (2–3 Nm).

7. If equipped with automatic transmission, make sure the shift position indicator adjustment is correct.

8. Install the shroud and connect the negative battery cable. Check the steering column for proper operation.

Ignition Lock

REMOVAL & INSTALLATION

Aerostar and 1988 Bronco II and Ranger

FUNCTIONAL LOCK

NOTE: The following procedure should be used on vehicles with functional lock cylinders. Ignition keys are available for these vehicles or the ignition key numbers are known and the proper key can be made.

1. Disconnect the negative battery cable.

2. Remove the trim shroud and disconnect the electrical connector from the key warning switch.

3. Turn the lock cylinder to the **RUN** position.

4. Place a ⅛ in. diameter pin or small drift punch in the hole located at 4 o'clock and 1¼ in. from the outer edge of the lock cylinder housing. Depress the retaining pin and pull out the lock cylinder.

To install:

5. Before installing the lock cylinder, lubricate the cylinder cavity, including the drive gear, with a suitable lock lubricant.

6. Turn the lock cylinder to the **RUN** position, depress the retaining pin and insert it into the lock cylinder

housing. Make sure the cylinder is fully seated and aligned into the interlocking washer before turning the key to the **OFF** position. This will permit the cylinder retaining pin to extend into the hole in the lock cylinder housing.

7. Using the ignition key, rotate the lock cylinder to ensure correct mechanical operation in all positions. Connect the electrical connector to the key warning switch.

8. Connect the negative battery cable. Check for proper ignition functions and make sure the column is locked in the **LOCK** position.

9. Install the trim shrouds.

NON-FUNCTIONAL LOCK

NOTE: The following procedure should be used on vehicles where the ignition lock is inoperative and the lock cylinder cannot be rotated due to a lost or broken ignition key, the key number is not known or the lock cylinder cap is damaged and/or broken.

1. Disconnect the negative battery cable.

2. Remove the steering wheel.

3. Remove the trim shrouds and disconnect the electrical connector from the key warning switch.

4. Using a ⅛ in. diameter drill bit, drill out the retaining pin, being careful not to drill deeper than ½ in.

5. Position a chisel at the base of the ignition lock cylinder cap and using a hammer, strike the chisel with sharp blows to break the cap away from the lock cylinder.

6. Using a ⅜ in. diameter drill bit, drill down the middle of the ignition lock key slot approximately 1¾ in. until the lock cylinder breaks loose from the breakaway base of the lock cylinder. Remove the lock cylinder and drill shavings from the lock cylinder housing.

7. Remove the snapring, washer and ignition lock drive gear. Thoroughly clean all drill shavings and other foreign material from the casting.

8. Carefully inspect the lock cylinder housing for damage from the removal operation. If the housing is damaged, it must be replaced.

To install:

9. Install the ignition lock drive gear, washer and snapring.

10. Install the ignition lock cylinder.

11. Connect the key warning switch wire. Install the shroud with the 2 screws.

12. Install the steering wheel and connect the negative battery cable. Check for proper ignition and accessory functions and make sure the column locks in the **LOCK** position.

1989–92 Bronco II, Explorer and Ranger

FUNCTIONAL LOCK

NOTE: The following procedure should be used on vehicles with functional lock cylinders. Ignition keys are available for these vehicles or the ignition key numbers are known and the proper key can be made.

1. Disconnect the negative battery cable.

2. Remove the steering wheel and shroud.

3. Using the ignition key, turn the lock cylinder to the **ON** position. If equipped with an automatic transmission, the selector lever must first be placed in **P**.

4. Push down on the lock cylinder retaining pin with a ⅛ in. diameter wire pin or small punch. Pull the lock cylinder from the column housing.

5. Disconnect the lock cylinder wiring plug from the horn brush wiring connector.

To install:

6. Lubricate the lock cylinder with grease.

7. Turn the lock cylinder to the **ON** position and depress the retaining pin.

8. Insert the lock cylinder into its housing in the flange casting, making sure the tab at the end of the cylinder aligns with the slot in the ignition drive gear.

9. Turn the key to the **OFF** position. This will allow the cylinder retaining pin to extend into the cylinder casting housing hole.

10. Using the ignition key, rotate the lock cylinder to ensure correct mechanical operation in all positions.

11. Connect the key warning wire plug and install the steering column lower shroud.

12. Install the steering column opening trim panel and connect the negative battery cable.

13. Check for proper start in **P** or **N**. Make sure the vehicle cannot be started in **D** and **R**.

NON-FUNCTIONAL LOCK

NOTE: The following procedure should be used on vehicles where the ignition lock is inoperative and the lock cylinder cannot be rotated due to a lost or broken ignition key, the key number is not known or the lock cylinder cap is damaged and/or broken.

1. Disconnect the negative battery cable. If equipped with tilt wheel, tilt to the full up position.
2. Remove the steering wheel, tilt lever and steering column trim shrouds.
3. Punch the lock cylinder retaining pin with a prick punch, 1/8 in. maximum outside diameter. Using a 1/8 in. diameter drill, drill out the retaining pin going no deeper than 1/2 in. Be careful not to damage the cast housing.
4. Place a chisel at the base of the lock cylinder cap and, using a hammer, strike the chisel with sharp blows to break the cap away from the lock cylinder.
5. Using a 3/8 in. diameter drill bit, drill down the middle of the ignition lock key slot approximately 1¾ in. until the lock cylinder breaks loose from the steering column cover casting. Remove the lock cylinder and drill shavings from the base of the cover cast housing.
6. Remove the drive gear, bearing retainer and actuator from the casting. Toroughly clean and inspect all components. If any components or the casting are damaged, they must be replaced.

To install:
7. Lubricate the drive gear, bearing and retainer with grease and install. Lubricate the lock cylinder with grease.
8. Turn the lock cylinder to the **ON** position and depress the retaining pin. Insert the cylinder into its housing in the flange casting, making sure the tab at the end of the cylinder aligns with the slot in the ignition lock drive gear.
9. Turn the key to the **OFF** position. This will allow the cylinder retaining pin to extend into the cylinder casting housing hole.
10. Using the ignition key, rotate the lock cylinder to ensure correct mechanical operation in all positions.
11. Connect the key warning wire plug and install the steering column shrouds.
12. Install the steering wheel and connect the negative battery cable. Check for proper start in **P** or **N**. Make sure the vehicle cannot be started in **D** and **R**.

Ignition Switch

REMOVAL & INSTALLATION

1. Rotate the lock cylinder to the **LOCK** position. Disconnect the negative battery cable.
2. Remove the steering wheel.
3. If equipped with tilt wheel, remove the upper extension housing shroud by squeezing it at the 6 and 12 o'clock positions and popping it free of the retaining plate at the 3 o'clock position.
4. On all except Aerostar, remove the trim shroud halves.
5. On Aerostar, proceed as follows:
   a. Remove the panel to the right of the steering column.
   b. Remove the trim shroud by removing the 4 screws at the bottom of the shroud. Swing the bottom panel of the shroud open and remove the 2 screws attaching the shroud to the retaining plate.
   c. Remove the lock cylinder.
   d. Remove the shroud by first raising the left side so the window for the combination switch lever is up past the lever receptacle, then pull up on the right side of the shroud until the lock cylinder embossment clears the lock cylinder housing.
   e. Work the shroud past the instrument panel.
6. Disconnect the switch electrical connector.
7. If equipped, remove the retaining nuts and disengage the ignition switch from the actuator pin.
8. If equipped with break-off head bolts, remove them using an 1/8 in. drill bit, then remove the bolts using an "easy-out" tool or equivalent. Disengage the ignition switch from the actuator pin.

NOTE: An alternate method of removing the break-off head bolts is to use a hammer and chisel to turn the bolts 1 revolution counterclockwise. Then, using a pair of adjustable pliers, grab the bolt head and continue to rotate the bolt until removed.

To install:
9. Rotate the ignition key to the **RUN** position and align the holes in the switch casting base with the holes in the lock cylinder housing.
10. Install the nuts or if equipped, new break-off head bolts. Tighten the break-off head bolts until the heads break off.
11. Connect the electrical connector to the ignition switch and install the steering column trim shrouds.
12. On Aerostar, install the lock cylinder.

13. Install the steering wheel. Connect the negative battery cable and check the ignition switch for proper operation.

Stoplight Switch

REMOVAL & INSTALLATION

1. Disconnect the negative battery cable. Disconnect the electrical connector from the switch. The locking tab must be lifted before the connector can be removed.
2. Remove the hairpin clip and slide the switch, booster pushrod, nylon washer and bushing away from the pedal. Remove the washer, then the switch by sliding the switch up or down.

To install:
3. Position the switch so the U-shaped side is nearest the pedal and directly over/under the pin. Then slide the switch up/down installing the booster pushrod and bushing between the switch side plates.
4. Push the switch and pushrod assembly firmly toward the brake pedal arm. Install the outside plastic washer to the pin and install the hairpin clip. Do not substitute for this clip. Use only factory supplied hairpin clips.
5. Connect the electrical connector to the switch and connect the negative battery cable. Make sure the switch wire harness has sufficient length to travel with the switch during the full stroke of the brake pedal. Check the switch for proper operation.

Clutch Switch

ADJUSTMENT

1988 Aerostar

If the adjusting clip is out of position on the rod, remove both halves of the clip. Position the clip halves closer to the switch and snap the clips together on the rod. Depress the clutch pedal to the floor to adjust the switch.

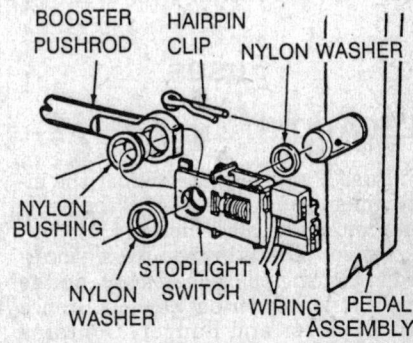

Stoplight switch assembly

REMOVAL & INSTALLATION

Except 1988 Aerostar

1. Disconnect the negative battery cable. Disconnect the wiring harness from the switch.
2. Pull the orientation clip away from the switch to separate it from the pin on the switch.
3. Rotate the switch to expose the plastic retainer.
4. Push the tabs together to allow the retainer to slide rearward and separate from the switch.
5. Remove the switch from the pushrod.
6. Installation is the reverse of the removal procedure.

1988 Aerostar

1. Disconnect the negative battery cable.
2. Disconnect the electrical connector and remove the switch.
3. Installation is the reverse of the removal procedure. Depress the clutch pedal to the floor to set the position of the self-adjusting clips.

Neutral Safety Switch

REMOVAL & INSTALLATION

1. Disconnect the negative battery cable.
2. Disconnect the electrical harness from the switch.
3. Remove the switch and O-ring using socket tool T74P-77247-A or equivalent.

NOTE: The use of other tools could crush or puncture the walls of the switch.

4. Installation is the reverse of the removal procedure. Tighten the switch to 7–10 ft. lbs. (9.5–13.6 Nm).
5. Check the operation of the switch with the parking brake applied. The engine should start only with the transmission selector lever in **N** or **P**. The back-up lights should illuminate only with the selector lever in **R**.

Fuses

LOCATION

Fuse Panel—located under the instrument panel, to the left of the steering column on all vehicles.

Power Distribution Box—located in the engine compartment, on the right inner fender on 1989–92 Bronco II, Explorer and Ranger. Contains only high current fuses.

Circuit Breakers

LOCATION

Aerostar

Front Wiper/Washer Circuit—6 amp circuit breaker located on the fuse panel.

Power Door Lock Circuit—30 amp circuit breaker located on the fuse panel on 1988–89 vehicles.

Power Window Circuit—20 amp circuit breaker located on the fuse panel.

Rear Cigar Lighter Circuit—30 amp circuit breaker located on the fuse panel on 1988–89 vehicles.

Bronco II, Explorer and Ranger

Combination Switch Circuit—20 amp circuit breaker located on the fuse panel on 1991–92 vehicles.

Cigar Lighter Circuit—20 amp circuit breaker located on the fuse panel on 1989–92 vehicles.

Flash-to-Pass Circuit—20 amp circuit breaker located on the fuse panel on 1989–92 vehicles.

Front Wiper/Washer Circuit—6 amp circuit breaker located on the fuse panel.

Headlight Circuit—22 amp circuit breaker, integral with the headlight switch.

Power Lumbar Support Circuit—20 amp circuit breaker located on the fuse panel on 1989–92 vehicles.

Power Window Circuit—20 amp circuit breaker located on the fuse panel on 1988–90 vehicles, 30 amp circuit breaker located on the fuse panel on 1991–92 vehicles.

Rear Wiper/Washer Circuit—4.5 amp circuit breaker located behind the instrument panel under the glove box on Bronco II and Explorer.

Relays

LOCATION

Aerostar

Auxiliary Air Conditioner-Heater Power Relay—located at the multi-relay bracket on the left fender apron.

Courtesy Lamp Relay—located at the left rear corner of the cargo area on 1990 vehicles and in front of the left "D" pillar, above the grommet on 1991–92 vehicles.

Day/Night Illumination Relay—located behind the left side of the instrument panel.

Door Ajar Relay—located at the bottom of the right "B" pillar.

Door Locks Control Relay—located at the right cowl panel.

EEC Power Relay—located on the multi-relay bracket on the left fender apron.

Fuel Pump Relay—located on the multi-relay bracket on the left fender apron.

Horn Relay—located behind the left side of the instrument panel.

Low Oil Level Relay—located behind the left side of the instrument panel.

Starter Relay—located on the left fender apron, next to the battery.

Trailer Tow Relay Module—located behind the left wheelhouse.

Wide Open Throttle Cut-Out Relay—located at the multi-relay bracket on the left fender apron.

Bronco II, Explorer and Ranger

Air Conditioning Relay—located on the right fender apron.

Choke Relay—located near the test connector take off.

EEC Power Relay—located on the lower cowl, near the Electronic Control Assembly, on 1988 vehicles or on the right fender apron on 1989–92 vehicles.

Fuel Metering Relay—located at the cigar lighter takeoff.

Fuel Pump Relay—located on the right fender apron.

Horn Relay—located at the top of the steering column on 1988 vehicles or behind the instrument panel, to the right of the steering column on 1989–92 vehicles.

Lock/Unlock Relays—located on the left fender apron.

Low Oil Level Relay—located behind the lower right side of the instrument panel.

Starter Relay—located on the right fender apron.

Stop/Turn Signal Relay—located on left rear quarter panel on Explorer.

Tail Light Relay—located on left rear quarter panel on Explorer.

Trailer Lamps Relay—located on top of the left radiator support.

Wide Open Throttle Cut-Out Relay—located at the right fender apron.

Computers

LOCATION

The engine Electronic Control Assembly (ECA) is located in the engine compartment, on the left side of the firewall on Aerostar or under the dash, at the right kick panel, on Bronco II, Explorer and Ranger.

Flashers

LOCATION

Both the turn signal and hazard flashers are attached to the fuse panel.

Cruise Control

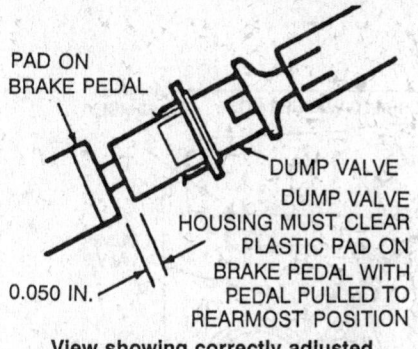

PAD ON
BRAKE PEDAL

DUMP VALVE

DUMP VALVE
HOUSING MUST CLEAR
PLASTIC PAD ON
BRAKE PEDAL WITH
PEDAL PULLED TO
REARMOST POSITION

0.050 IN.

**View showing correctly adjusted
dump valve**

ADJUSTMENT

Actuator Cable

1. Remove the cable retaining clip.
2. Disengage the throttle positioner.
3. Set the engine at hot idle.
4. Pull on the actuator cable to take up any slack. Maintain a light tension on the cable.
5. While holding the cable, insert the cable retaining clip and snap securely.

Vacuum Dump Valve

1. Firmly hold the brake pedal in the up, released, position.
2. Push in the dump valve until the valve bottoms against the pad on the brake pedal.
3. The dump valve housing must clear the plastic pad on the brake pedal by 0.050–0.10 in. (1.27–2.54mm) with the brake pedal pulled to the rearmost position.

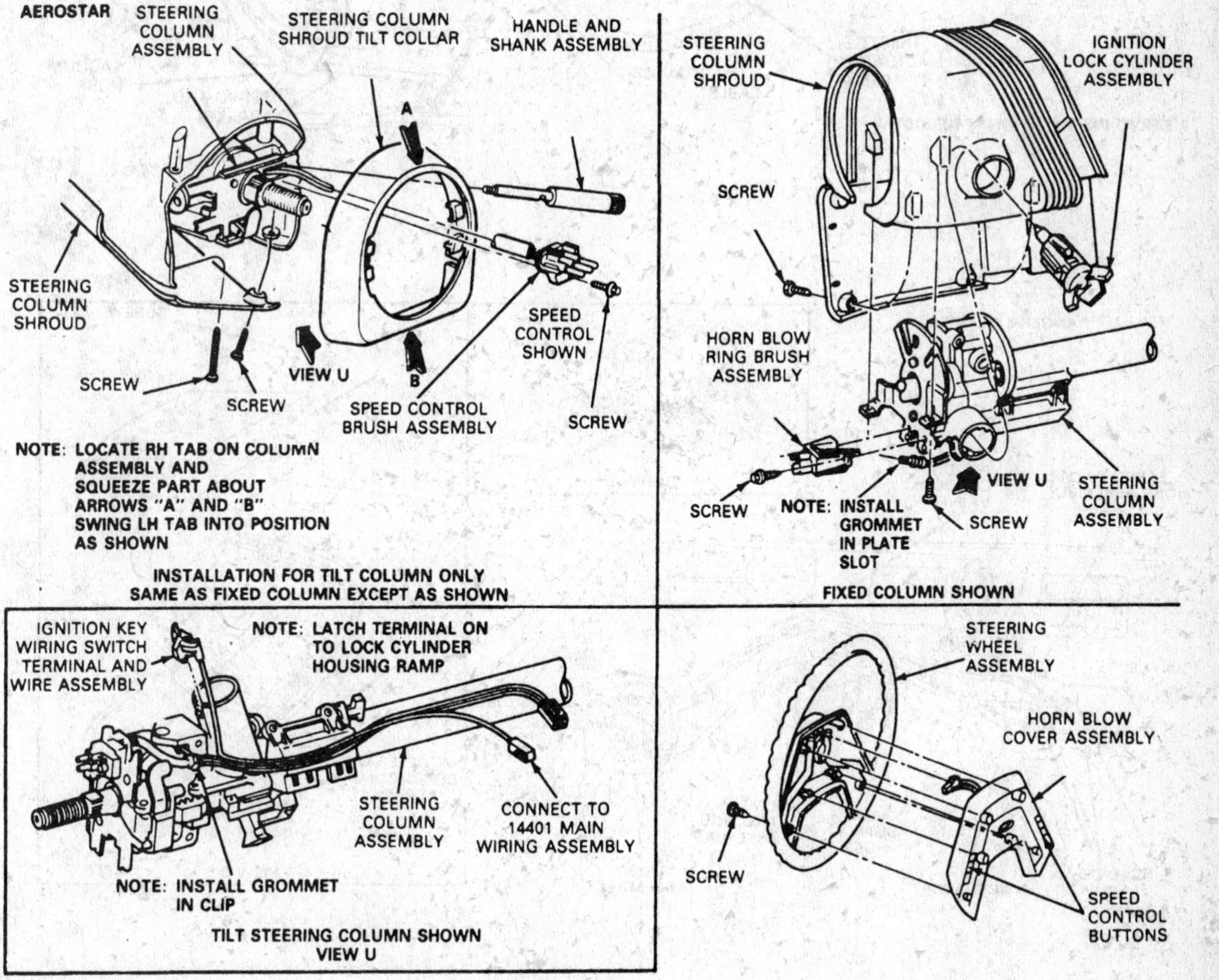

AEROSTAR STEERING COLUMN ASSEMBLY

STEERING COLUMN SHROUD TILT COLLAR

HANDLE AND SHANK ASSEMBLY

STEERING COLUMN SHROUD

IGNITION LOCK CYLINDER ASSEMBLY

A

STEERING COLUMN SHROUD

SCREW

SCREW

VIEW U

SPEED CONTROL BRUSH ASSEMBLY

SPEED CONTROL SHOWN

B

SCREW

SCREW

HORN BLOW RING BRUSH ASSEMBLY

SCREW

NOTE: INSTALL GROMMET IN PLATE SLOT

VIEW U

SCREW

STEERING COLUMN ASSEMBLY

NOTE: LOCATE RH TAB ON COLUMN ASSEMBLY AND SQUEEZE PART ABOUT ARROWS "A" AND "B" SWING LH TAB INTO POSITION AS SHOWN

**INSTALLATION FOR TILT COLUMN ONLY
SAME AS FIXED COLUMN EXCEPT AS SHOWN**

FIXED COLUMN SHOWN

IGNITION KEY WIRING SWITCH TERMINAL AND WIRE ASSEMBLY

NOTE: LATCH TERMINAL ON TO LOCK CYLINDER HOUSING RAMP

STEERING COLUMN ASSEMBLY

CONNECT TO 14401 MAIN WIRING ASSEMBLY

STEERING WHEEL ASSEMBLY

HORN BLOW COVER ASSEMBLY

SCREW

SPEED CONTROL BUTTONS

NOTE: INSTALL GROMMET IN CLIP

**TILT STEERING COLUMN SHOWN
VIEW U**

Cruise control switch assembly

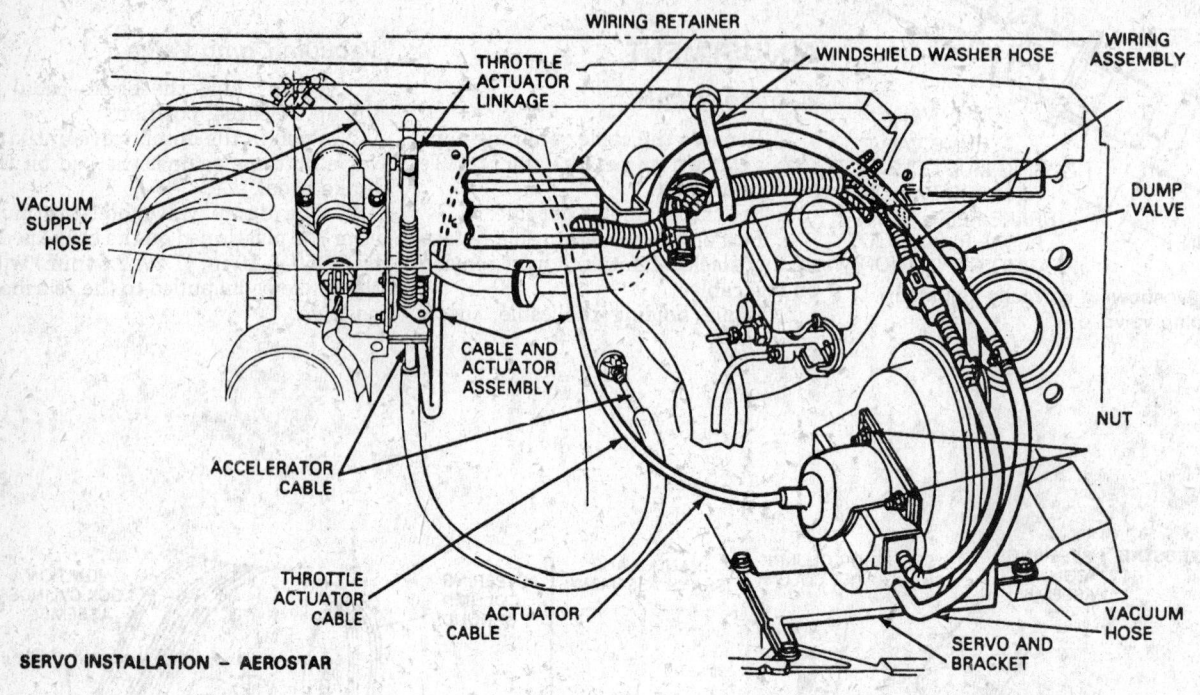

SERVO INSTALLATION – AEROSTAR

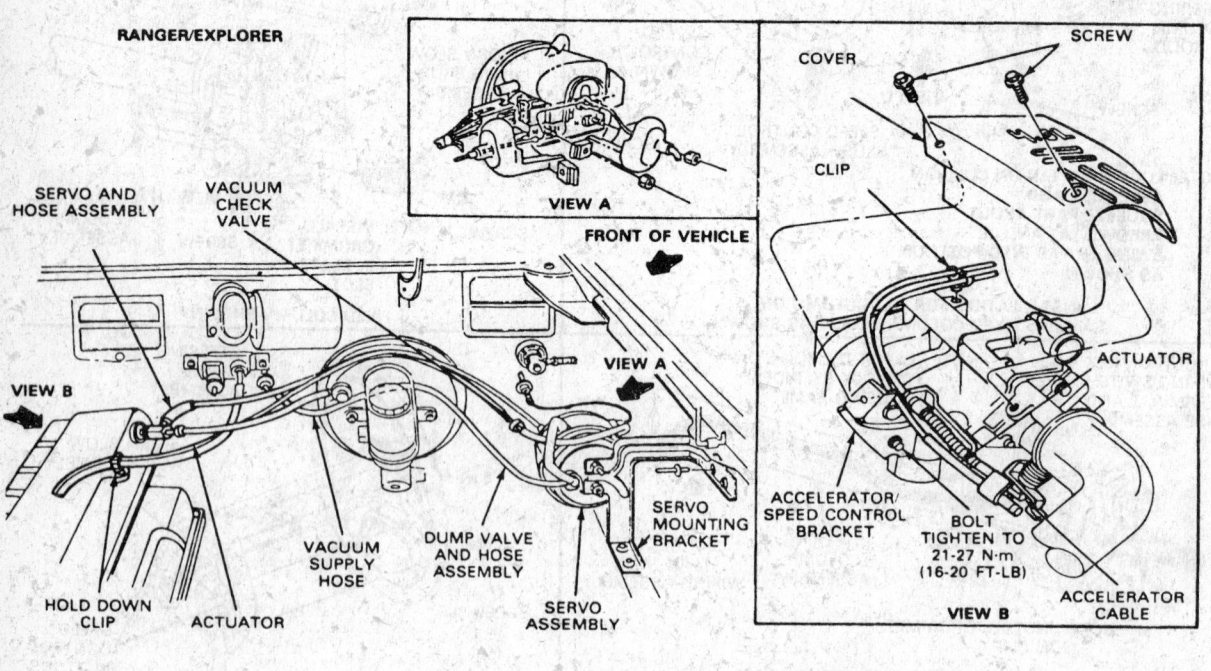

Cruise control component location

GENERAL MOTORS CORP. 4

GM Full-Size Trucks

Blazer • Jimmy • Pickup • Suburban • Van

VEHICLE IDENTIFICATION CHART

Engine Code							Model Year	
Code	Liters	Cu. In. (cc)	Cyl.	Fuel Sys.	Eng. Mfg.		Code	Year
4	2.2	134 (2189)	4	MFI	CPC		M	1991
A	2.5	151 (2474)	4	TFI	CPC		N	1992
E	2.5	151 (2474)	4	TFI	CPC-North		P	1993
R	2.8	173 (2835)	6	TFI	CPC		R	1994
D	3.1	191 (3130)	6	TFI	CPC		S	1995
L	3.8	231 (3785)	6	MFI	CPC			
B	4.3	263 (4293)	6	TFI	CPC			
W	4.3	263 (4293)	6	TFI	CPC			
Z	4.3	263 (4293)	6	TFI	CPC			
H	5.0	305 (4999)	8	TFI	CPC			
K	5.7	350 (5735)	8	TFI	CPC			
C	6.2	379 (6210)	8	DSL	DDA			
J	6.2	379 (6210)	8	DSL	DDA			
F	6.5	395 (6473)	8	DSL	CPC			
P	6.5	395 (6473)	8	DSL	CPC			
S	6.5	395 (6473)	8	DSL	CPC			
Y	6.5	395 (6473)	8	DSL	CPC			
N	7.4	454 (7440)	8	TFI	CPC			

TFI - Throttle body fuel injection
MFI - Multiport fuel injection
DSL - Diesel
CPC - Chevrolet/Pontiac/Canada
DDA - Detroit Diesel Allison

79113C01

ENGINE IDENTIFICATION

Year	Model	Engine Displacement Liters (cc)	Engine Series (ID/VIN)	Fuel System	No. of Cylinders	Engine Type
1991	Astro/Safari	4.3 (4293)	B	TFI	6	OHV
	Astro/Safari	4.3 (4293)	Z	TFI	6	OHV
	Lumina/Silhouette/Trans Sport	3.1 (3130)	D	TFI	6	OHV
	S10 Blazer/S15 Jimmy	2.8 (2835)	R	TFI	6	OHV
	S10 Blazer/S15 Jimmy	4.3 (4293)	Z	TFI	6	OHV
	Bravada	4.3 (4293)	Z	TFI	6	OHV
	S10 Pick-up/S15 Sonoma	2.5 (2474)	A	TFI	4	OHV
	S10 Pick-up/S15 Sonoma	2.5 (2474)	E	TFI	4	OHV
	S10 Pick-up/S15 Sonoma	2.8 (2835)	R	TFI	6	OHV
	S10 Pick-up/S15 Sonoma	4.3 (4293)	Z	TFI	6	OHV
	Syclone	4.3 (4293)	Z	TFI-Turbo	6	OHV
	Blazer/Jimmy	5.7 (5735)	K	TFI	8	OHV
	Blazer/Jimmy	6.2 (6210)	C	DSL	8	OHV
	Blazer/Jimmy	6.2 (6210)	J	DSL	8	OHV
	C1500	4.3 (4293)	Z	TFI	6	OHV
	C1500	5.7 (5735)	K	TFI	8	OHV
	C1500	6.2 (6210)	C	DSL	8	OHV
	C1500	6.2 (6210)	J	DSL	8	OHV
	C1500	7.4 (7440)	N	TFI	8	OHV
	C2500	4.3 (4293)	Z	TFI	6	OHV
	C2500	5.0 (4999)	H	TFI	8	OHV
	C2500	5.7 (5735)	K	TFI	8	OHV
	C2500	6.2 (6210)	C	DSL	8	OHV
	C2500	6.2 (6210)	J	DSL	8	OHV
	C2500	7.4 (7440)	N	TFI	8	OHV
	C3500	5.0 (4999)	H	TFI	8	OHV
	C3500	5.7 (5735)	K	TFI	8	OHV
	C3500	6.2 (6210)	C	DSL	8	OHV
	C3500	6.2 (6210)	J	DSL	8	OHV
	C3500	7.4 (7440)	N	TFI	8	OHV
	K1500	4.3 (4293)	Z	TFI	6	OHV
	K1500	5.0 (4999)	H	TFI	8	OHV
	K1500	5.7 (5735)	K	TFI	8	OHV
	K1500	6.2 (6210)	C	DSL	8	OHV
	K2500	4.3 (4293)	Z	TFI	6	OHV
	K2500	5.0 (4999)	H	TFI	8	OHV
	K2500	5.7 (5735)	K	TFI	8	OHV
	K2500	6.2 (6210)	C	DSL	8	OHV
	K2500	6.2 (6210)	J	DSL	8	OHV
	K3500	5.7 (5735)	K	TFI	8	OHV
	K3500	6.2 (6210)	J	DSL	8	OHV
	K3500	7.4 (7441)	N	TFI	8	OHV
	P30	4.3 (4293)	Z	TFI	6	OHV
	P30	5.7 (5735)	K	TFI	8	OHV
	P30	6.2 (6210)	J	DSL	8	OHV
	P30	7.4 (7440)	N	TFI	8	OHV
	Suburban	5.7 (5735)	K	TFI	8	OHV

79113Co2

ENGINE IDENTIFICATION

Year	Model	Engine Displacement Liters (cc)	Engine Series (ID/VIN)	Fuel System	No. of Cylinders	Engine Type
	Suburban	6.2 (6210)	C	DSL	8	OHV
	Suburban	6.2 (6210)	J	DSL	8	OHV
	G10	4.3 (4293)	Z	TFI	6	OHV
	G10	5.0 (4999)	H	TFI	8	OHV
	G10	5.7 (5735)	K	TFI	8	OHV
	G20	4.3 (4293)	Z	TFI	6	OHV
	G20	5.0 (4999)	H	TFI	8	OHV
	G20	5.7 (5735)	K	TFI	8	OHV
	G30	4.3 (4293)	Z	TFI	6	OHV
	G30	5.7 (5735)	K	TFI	8	OHV
	G30	6.2 (6210)	J	DSL	8	OHV
	G30	7.4 (7440)	N	TFI	8	OHV
1992	Astro/Safari	4.3 (4293)	W	TFI	6	OHV
	Astro/Safari	4.3 (4293)	Z	TFI	6	OHV
	Lumina/Silhouette/Trans Sport	3.1 (3130)	D	TFI	6	OHV
	Lumina/Silhouette/Trans Sport	3.8 (3785)	L	MFI	6	OHV
	S10 Blazer/S15 Jimmy	4.3 (4293)	W	TFI	6	OHV
	S10 Blazer/S15 Jimmy	4.3 (4293)	Z	TFI	6	OHV
	Bravada	4.3 (4293)	W	TFI	6	OHV
	Bravada	4.3 (4293)	Z	TFI	6	OHV
	S10 Pick-up/Sonoma	2.5 (2474)	A	TFI	4	OHV
	S10 Pick-up/Sonoma	2.8 (2835)	R	TFI	6	OHV
	S10 Pick-up/Sonoma	4.3 (4293)	Z	TFI	6	OHV
	Syclone	4.3 (4293)	Z	TFI-Turbo	6	OHV
	Typhoon	4.3 (4293)	Z	TFI-Turbo	6	OHV
	Blazer/Yukon	5.7 (5735)	K	TFI	8	OHV
	C1500	4.3 (4293)	Z	TFI	6	OHV
	C1500	5.0 (4999)	H	TFI	8	OHV
	C1500	5.7 (5735)	K	TFI	8	OHV
	C1500	6.2 (6210)	C	DSL	8	OHV
	C1500	6.2 (6210)	J	DSL	8	OHV
	C1500	7.4 (7440)	N	TFI	8	OHV
	C2500	4.3 (4293)	Z	TFI	6	OHV
	C2500	5.0 (4999)	H	TFI	8	OHV
	C2500	5.7 (5735)	K	TFI	8	OHV
	C2500	6.2 (6210)	C	DSL	8	OHV
	C2500	6.2 (6210)	J	DSL	8	OHV
	C2500	6.5 (6473)	F	DSL	8	OHV
	C2500	7.4 (7440)	N	TFI	8	OHV
	C3500	5.0 (4999)	H	TFI	8	OHV
	C3500	5.7 (5735)	K	TFI	8	OHV
	C3500	6.2 (6210)	C	DSL	8	OHV
	C3500	6.2 (6210)	J	DSL	8	OHV
	C3500	6.5 (6473)	F	DSL	8	OHV
	C3500	7.4 (7440)	N	TFI	8	OHV
	K1500	4.3 (4293)	Z	TFI	6	OHV
	K1500	5.0 (4999)	H	TFI	8	OHV

79113Ca2

ENGINE IDENTIFICATION

Year	Model	Engine Displacement Liters (cc)	Engine Series (ID/VIN)	Fuel System	No. of Cylinders	Engine Type
	K1500	5.7 (5735)	K	TFI	8	OHV
	K1500	6.2 (6210)	C	DSL	8	OHV
	K1500	7.4 (7440)	N	TFI	8	OHV
	K2500	4.3 (4293)	Z	TFI	6	OHV
	K2500	5.0 (4999)	H	TFI	8	OHV
	K2500	5.7 (5735)	K	TFI	8	OHV
	K2500	6.2 (6210)	C	DSL	8	OHV
	K2500	6.2 (6210)	J	DSL	8	OHV
	K2500	6.5 (6473)	F	DSL	8	OHV
	K2500	7.4 (7440)	N	TFI	8	OHV
	K3500	5.0 (4999)	H	TFI	8	OHV
	K3500	5.7 (5735)	K	TFI	8	OHV
	K3500	6.2 (6210)	C	DSL	8	OHV
	K3500	6.2 (6210)	J	DSL	8	OHV
	K3500	7.4 (7441)	N	TFI	8	OHV
	P30	4.3 (4293)	Z	TFI	6	OHV
	P30	5.7 (5735)	K	TFI	8	OHV
	P30	6.2 (6210)	J	DSL	8	OHV
	P30	7.4 (7440)	N	TFI	8	OHV
	Suburban	5.7 (5735)	K	TFI	8	OHV
	Suburban	7.4 (7440)	N	TFI	8	OHV
	G10	4.3 (4293)	Z	TFI	6	OHV
	G10	5.0 (4999)	H	TFI	8	OHV
	G10	5.7 (5735)	K	TFI	8	OHV
	G20	4.3 (4293)	Z	TFI	6	OHV
	G20	5.0 (4999)	H	TFI	8	OHV
	G20	5.7 (5735)	K	TFI	8	OHV
	G30	4.3 (4293)	Z	TFI	6	OHV
	G30	5.7 (5735)	K	TFI	8	OHV
	G30	6.2 (6210)	C	DSL	8	OHV
	G30	6.2 (6210)	J	DSL	8	OHV
	G30	7.4 (7440)	N	TFI	8	OHV
1993	Astro/Safari	4.3 (4293)	W	TFI	6	OHV
	Astro/Safari	4.3 (4293)	Z	TFI	6	OHV
	Lumina/Silhouette/Trans Sport	3.1 (3097)	D	TFI	6	OHV
	Lumina/Silhouette/Trans Sport	3.8 (3785)	L	MFI	6	OHV
	S10 Blazer/S15 Jimmy	4.3 (4293)	W	TFI	6	OHV
	S10 Blazer/S15 Jimmy	4.3 (4293)	Z	TFI	6	OHV
	Bravada	4.3 (4293)	W	TFI	6	OHV
	Bravada	4.3 (4293)	Z	TFI	6	OHV
	S10 Pick-up/Sonoma	2.5 (2474)	A	TFI	4	OHV
	S10 Pick-up/Sonoma	2.8 (2835)	R	TFI	6	OHV
	S10 Pick-up/Sonoma	4.3 (4293)	W	TFI	6	OHV
	Typhoon	4.3(4293)	Z	TFI-Turbo	6	OHV
	Blazer/Yukon	5.7 (5735)	K	TFI	8	OHV
	C1500	4.3 (4293)	Z	TFI	6	OHV
	C1500	5.0 (4999)	H	TFI	8	OHV

79113Cb2

ENGINE IDENTIFICATION

Year	Model	Engine Displacement Liters (cc)	Engine Series (ID/VIN)	Fuel System	No. of Cylinders	Engine Type
	C1500	5.7 (5735)	K	TFI	8	OHV
	C1500	6.2 (6210)	C	DSL	8	OHV
	C1500	6.2 (6210)	J	DSL	8	OHV
	C1500	7.4 (7440)	N	TFI	8	OHV
	C2500	4.3 (4293)	Z	TFI	6	OHV
	C2500	5.0 (4999)	H	TFI	8	OHV
	C2500	5.7 (5735)	K	TFI	8	OHV
	C2500	6.2 (6210)	C	DSL	8	OHV
	C2500	6.2 (6210)	J	DSL	8	OHV
	C2500	6.5 (6505)	F	DSL	8	OHV
	C2500	7.4 (7440)	N	TFI	8	OHV
	C3500	5.0 (4999)	H	TFI	8	OHV
	C3500	5.7 (5735)	K	TFI	8	OHV
	C3500	6.2 (6210)	C	DSL	8	OHV
	C3500	6.2 (6210)	J	DSL	8	OHV
	C3500	6.5 (6505)	F	DSL	8	OHV
	C3500	7.4 (7440)	N	TFI	8	OHV
	K1500	4.3 (4293)	Z	TFI	6	OHV
	K1500	5.0 (4999)	H	TFI	8	OHV
	K1500	5.7 (5735)	K	TFI	8	OHV
	K1500	6.2 (6210)	C	DSL	8	OHV
	K1500	6.2 (6210)	J	DSL	8	OHV
	K1500	7.4 (7440)	N	TFI	8	OHV
	K2500	4.3 (4293)	Z	TFI	6	OHV
	K2500	5.0 (4999)	H	TFI	8	OHV
	K2500	5.7 (5735)	K	TFI	8	OHV
	K2500	6.2 (6210)	C	DSL	8	OHV
	K2500	6.2 (6210)	J	DSL	8	OHV
	K2500	6.5 (6505)	F	DSL	8	OHV
	K2500	7.4 (7440)	N	TFI	8	OHV
	K3500	5.0 (4999)	H	TFI	8	OHV
	K3500	5.7 (5735)	K	TFI	8	OHV
	K3500	6.2 (6210)	C	DSL	8	OHV
	K3500	6.2 (6210)	J	DSL	8	OHV
	K3500	6.5 (6505)	F	DSL	8	OHV
	K3500	7.4 (7441)	N	TFI	8	OHV
	P30	4.3 (4293)	Z	TFI	6	OHV
	P30	5.7 (5735)	K	TFI	8	OHV
	P30	6.2 (6210)	J	DSL	8	OHV
	P30	7.4 (7440)	N	TFI	8	OHV
	Suburban	5.7 (5735)	K	TFI	8	OHV
	Suburban	7.4 (7440)	N	TFI	8	OHV
	G10	4.3 (4293)	Z	TFI	6	OHV
	G10	5.0 (4999)	H	TFI	8	OHV
	G10	5.7 (5735)	K	TFI	8	OHV
	G10	6.2 (6210)	C	DSL	8	OHV
	G10	6.2 (6210)	J	DSL	8	OHV

79113Cc2

ENGINE IDENTIFICATION

Year	Model	Engine Displacement Liters (cc)	Engine Series (ID/VIN)	Fuel System	No. of Cylinders	Engine Type
	G20	4.3 (4293)	Z	TFI	6	OHV
	G20	5.0 (4999)	H	TFI	8	OHV
	G20	5.7 (5735)	K	TFI	8	OHV
	G20	6.2 (6210)	C	DSL	8	OHV
	G20	6.2 (6210)	J	DSL	8	OHV
	G30	4.3 (4293)	Z	TFI	6	OHV
	G30	5.0 (4999)	H	TFI	8	OHV
	G30	5.7 (5735)	K	TFI	8	OHV
	G30	6.2 (6210)	C	DSL	8	OHV
	G30	6.2 (6210)	J	DSL	8	OHV
	G30	7.4 (7440)	N	TFI	8	OHV
1994-95	Astro/Safari	4.3 (4293)	z	TFI	6	OHV
	Astro/Safari	4.3 (4293)	Z	TFI	6	OHV
	Lumina/Silhouette/Trans Sport	3.1 (3097)	D	TFI	6	OHV
	Lumina/Silhouette/Trans Sport	3.8 (3785)	L	MFI	6	OHV
	S10 Blazer/S15 Jimmy	4.3 (4293)	W	TFI	6	OHV
	S10 Blazer/S15 Jimmy	4.3 (4293)	Z	TFI	6	OHV
	Bravada	4.3 (4293)	W	TFI	6	OHV
	Bravada	4.3 (4293)	Z	TFI	6	OHV
	S10 Pick-up/Sonoma	2.2 (2189)	4	TFI	4	OHV
	S10 Pick-up/Sonoma	4.3 (4293)	W	TFI	6	OHV
	S10 Pick-up/Sonoma	4.3 (4293)	Z	TFI	6	OHV
	Blazer/Yukon	5.7 (5735)	K	TFI	8	OHV
	Blazer/Yukon	6.5 (6505)	F	DSL	8	OHV
	C1500	4.3 (4293)	Z	TFI	6	OHV
	C1500	5.0 (4999)	H	TFI	8	OHV
	C1500	5.7 (5735)	K	TFI	8	OHV
	C1500	6.5 (6505)	F	DSL	8	OHV
	C1500	7.4 (7440)	N	TFI	8	OHV
	C2500	4.3 (4293)	Z	TFI	6	OHV
	C2500	5.0 (4999)	H	TFI	8	OHV
	C2500	5.7 (5735)	K	TFI	8	OHV
	C2500	6.5 (6505)	P	DSL	8	OHV
	C2500	6.5 (6505)	S	DSL	8	OHV
	C3500	5.7 (5735)	K	TFI	8	OHV
	C3500	6.5 (6505)	F	DSL	8	OHV
	C3500	6.5 (6505)	P	DSL	8	OHV
	C3500	6.5 (6505)	S	DSL	8	OHV
	C3500	7.4 (7440)	N	TFI	8	OHV
	K1500	4.3 (4293)	Z	TFI	6	OHV
	K1500	5.0 (4999)	H	TFI	8	OHV
	K1500	5.7 (5735)	K	TFI	8	OHV
	K1500	6.5 (6505)	F	DSL	8	OHV
	K1500	7.4 (7440)	N	TFI	8	OHV
	K2500	4.3 (4293)	Z	TFI	6	OHV
	K2500	5.0 (4999)	H	TFI	8	OHV
	K2500	5.7 (5735)	K	TFI	8	OHV

79113Cd2

ENGINE IDENTIFICATION

Year	Model	Engine Displacement Liters (cc)	Engine Series (ID/VIN)	Fuel System	No. of Cylinders	Engine Type
	K2500	6.5 (6505)	P	DSL	8	OHV
	K2500	6.5 (6505)	S	DSL	8	OHV
	K2500	6.5 (6505)	F	DSL	8	OHV
	K2500	7.4 (7440)	N	TFI	8	OHV
	K3500	5.7 (5735)	K	TFI	8	OHV
	K3500	6.5 (6505)	F	DSL	8	OHV
	K3500	6.5 (6505)	P	DSL	8	OHV
	K3500	6.5 (6505)	S	DSL	8	OHV
	K3500	7.4 (7441)	B	TFI	8	OHV
	P30	4.3 (4293)	Z	TFI	6	OHV
	P30	5.7 (5735)	K	TFI	8	OHV
	P30	6.5 (6505)	F	DSL	8	OHV
	P30	7.4 (7440)	N	TFI	8	OHV
	Suburban	5.7 (5735)	K	TFI	8	OHV
	Suburban	6.5 (6505)	P	DSL	8	OHV
	Suburban	6.5 (6505)	S	DSL	8	OHV
	Suburban	7.4 (7440)	N	TFI	8	OHV
	G10	4.3 (4293)	Z	TFI	6	OHV
	G10	5.0 (4999)	H	TFI	8	OHV
	G10	5.7 (5735)	K	TFI	8	OHV
	G10	6.5 (6505)	P	DSL	8	OHV
	G10	6.5 (6505)	Y	DSL	8	OHV
	G20	4.3 (4293)	Z	TFI	6	OHV
	G20	5.0 (4999)	H	TFI	8	OHV
	G20	5.7 (5735)	K	TFI	8	OHV
	G20	6.5 (6505)	P	DSL	8	OHV
	G20	6.5 (6505)	Y	DSL	8	OHV
	G30	4.3 (4293)	Z	TFI	6	OHV
	G30	5.0 (4999)	H	TFI	8	OHV
	G30	5.7 (5735)	K	TFI	8	OHV
	G30	6.5 (6505)	P	DSL	8	OHV
	G30	6.5 (6505)	Y	DSL	8	OHV
	G30	7.4 (7440)	N	TFI	8	OHV

TFI - Throttle body fuel injection
MFI - Multiport fuel injection
DSL - Diesel
OHV - Overhead valve

79113Ce2

GENERAL ENGINE SPECIFICATIONS

Year	Engine ID/VIN	Engine Displacement Liters (cc)	Fuel System Type	Net Horsepower @ rpm	Net Torque @ rpm (ft. lbs.)	Bore x Stroke (in.)	Compression Ratio	Oil Pressure @ rpm
1991	A	2.5 (2474)	TFI	105@4800	135@3200	4.00x3.00	8.3:1	41@2000
	E	2.5 (2474)	TFI	105@4800	135@3200	4.00x3.00	8.3:1	45@2000
	R	2.8 (2835)	TFI	125@4800	150@2200	3.56x3.04	8.5:1	50@2000
	D	3.1 (3130)	TFI	120@4400	175@2200	3.50x3.30	8.9:1	15@1100
	B	4.3 (4293)	TFI	170@4600	225@2200	4.00x3.48	9.3:1	18@2000
	Z	4.3 (4293)	TFI	145@4000	230@2400	4.00x3.48	9.3:1	18@2000
	H	5.0 (4999)	TFI	165@4400	240@2000	3.74x3.48	9.0:1	18@2000
	K	5.7 (5735)	TFI	3	4	4.00x3.48	8.5:1	18@2000
	C	6.2 (6210)	DSL	135@3600	240@2000	3.98x3.80	21.0:1	35@2000
	J	6.2 (6210)	DSL	135@3600	240@2000	3.98x3.80	21.0:1	35@2000
	N	7.4 (7440)	TFI	230@3600	385@1600	4.25x4.00	8.0:1	40@2000
1992	A	2.5 (2474)	TFI	105@4800	135@3200	4.00x3.00	8.3:1	41@2000
	R	2.8 (2835)	TFI	125@4800	150@2200	3.56x3.04	8.5:1	50@2000
	D	3.1 (3130)	TFI	120@4400	175@2200	3.50x3.30	8.9:1	15@1100
	L	3.8 (3785)	MFI	165@4300	220@3200	3.80x3.40	8.5:1	60@1850
	B	4.3 (4293)	TFI	170@4600	225@2200	4.00x3.48	9.3:1	18@2000
	W	4.3 (4293)	TFI	200@4500	260@3600	4.00x3.48	9.1:1	18@2000
	Z	4.3 (4293)	TFI	1	2	4.00x3.48	9.3:1	18@2000
	H	5.0 (4999)	TFI	165@4400	240@2000	3.74x3.48	9.0:1	18@2000
	K	5.7 (5735)	TFI	3	4	4.00x3.48	8.5:1	18@2000
	C	6.2 (6210)	DSL	130@3600	240@2000	3.98x3.80	21.3:1	35@2000
	J	6.2 (6210)	DSL	135@3600	240@2000	3.98x3.80	21.3:1	350@2000
	F	6.5 (6473)	DSL	5	6	4.05x3.80	21.0:1	40-45@2000
	N	7.4 (7440)	TFI	230@3600	385@1600	4.25x4.00	8.0:1	40@2000
1993	A	2.5 (2474)	TFI	105@4800	135@3200	4.00x3.00	8.3:1	41@2000
	R	2.8 (2835)	TFI	125@2800	150@2200	3.56x3.04	8.5:1	50@2000
	D	3.1 (3130)	TFI	120@4400	175@2200	3.50x3.30	8.9:1	15@1100
	L	3.8 (3785)	MFI	165@4300	220@3200	3.80x3.40	8.5:1	60@1850
	W	4.3 (4293)	TFI	200@4500	260@3600	4.00x3.48	9.1:1	18@2000
	Z	4.3 (4293)	TFI	1	2	4.00x3.48	9.3:1	18@2000
	H	5.0 (4999)	TFI	165@4400	240@2000	3.74x3.48	9.0:1	18@2000
	K	5.7 (5735)	TFI	3	4	4.00x3.48	8.5:1	18@2000
	C	6.2 (6210)	DSL	130@3600	240@2000	3.98x3.80	21.3:1	35@2000
	J	6.2 (6210)	DSL	135@3600	240@2000	3.98x3.80	21.3:1	35@2000
	F	6.5 (6473)	DSL	5	6	4.05x3.80	21.0:1	40-45@2000
	N	7.4 (7440)	TFI	230@3600	385@1600	4.25x4.00	8.0:1	40@2000

79113C03

GENERAL ENGINE SPECIFICATIONS

Year	Engine ID/VIN	Engine Displacement Liters (cc)	Fuel System Type	Net Horsepower @ rpm	Net Torque @ rpm (ft. lbs.)	Bore x Stroke (in.)	Compression Ratio	Oil Pressure @ rpm
1994-95	4	2.2 (2189)	MFI	118@5200	130@2800	3.50x3.46	9.0:1	56@3000
	D	3.1 (3130)	TFI	120@4400	175@2200	3.50x3.30	8.9:1	15@1100
	L	3.8 (3785)	MFI	165@4300	220@3200	3.80x3.40	8.5:1	60@1850
	W	4.3 (4293)	TFI	200@4500	260@3600	4.00x3.48	9.1:1	18@2000
	Z	4.3 (4293)	TFI	1	2	4.00x3.48	9.3:1	18@2000
	H	5.0 (4999)	TFI		240@2000	3.74x3.48	9.0:1	18@2000
	K	5.7 (5735)	TFI	3	4	4.00x3.48	8.5:1	18@2000
	F	6.5 (6473)	DSL	5	6	4.05x3.80	21.5:1	40-45@2000
	P	6.5 (6473)	DSL	190@3400	385@1700	4.06x3.82	21.5:1	40-45@2000
	S	6.5 (6473)	DSL	190@3400	275@1700	4.06x3.82	21.5:1	40-45@2000
	Y	6.5 (6473)	DSL	7	8	4.06x3.82	21.5:1	40-45@2000
	N	7.4 (7440)	TFI	230@3600	385@1600	4.25x4.00	8.0:1	40@2000

TFI - Throttle body fuel injection
MFI - Multiport fuel injection
DSL - Diesel
1 S10 and C/K Pick-ups: 160@4000
   C/K HD Pick-up: 155@4000
   G Van: 150@4000
1 S10: 230@2800
   C/K Pick-up: 235@2400
   C/K HD Pick-up and G-Van: 230@2400
3 Below 8500 GVWR: 210@4000
   Above 8500 GVWR: 190@4000

3 Below 8500 GVWR: 210@4000
   Above 8500 GVWR: 190@4000
4 Below 8500 GVWR: 300@2800
   Above 8500 GVWR: 300@2400
5 Below 15,000 GVWR: 180@3400
   Above 15,000 GVWR: 190@3400
6 Below 15,000 GVWR: 380@1700
   Above 15,000 GVWR: 360@1700

7 Below 8500 GVWR: 155@3600
   Above 8500 GVWR: 160@3600
8 Below 8500 GVWR: 275@1700
   Above 8500 GVWR: 290@1700

79113Ca3

GASOLINE ENGINE TUNE-UP SPECIFICATIONS

Year	Engine ID/VIN	Engine Displacement Liters (cc)	Spark Plugs Gap (in.)	Ignition Timing (deg.) MT	AT	Fuel Pump (psi)	Idle Speed (rpm) MT	AT	Valve Clearance In.	Ex.
1991	A	2.5 (2474)	0.060	1	1	9-13	1	1	HYD	HYD
	E	2.5 (2474)	0.060	1	1	9-13	1	1	HYD	HYD
	R	2.8 (2835)	0.040	1	1	9-13	1	1	HYD	HYD
	D	3.1 (3130)	0.045	1	1	9-13	1	1	HYD	HYD
	B	4.3 (4293)	0.035	1	1	9-13	1	1	HYD	HYD
	Z	4.3 (4293)	0.040	1	1	9-13	1	1	HYD	HYD
	H	5.0 (4999)	0.045	1	1	9-13	1	1	HYD	HYD
	K	5.7 (5735)	0.045	1	1	9-13	1	1	HYD	HYD
	N	7.4 (7440)	0.045	1	1	9-13	1	1	HYD	HYD
1992	A	2.5 (2474)	0.060	1	1	9-13	1	1	HYD	HYD
	R	2.8 (2835)	0.040	1	1	9-13	1	1	HYD	HYD
	D	3.1 (3130)	0.045	1	1	9-13	1	1	HYD	HYD
	L	3.8 (3785)	0.060	1	1	41-47 [2]	1	1	HYD	HYD
	B	4.3 (4293)	0.035	1	1	9-13	1	1	HYD	HYD
	W	4.3 (4293)	0.035	1	1	9-13	1	1	HYD	HYD
	Z	4.3 (4293)	0.035	1	1	9-13	1	1	HYD	HYD
	H	5.0 (4999)	0.045	1	1	9-13	1	1	HYD	HYD
	K	5.7 (5735)	0.045	1	1	9-13	1	1	HYD	HYD
	N	7.4 (7440)	0.045	1	1	9-13	1	1	HYD	HYD
1993	A	2.5 (2474)	0.060	1	1	9-13	1	1	HYD	HYD
	R	2.8 (2835)	0.040	1	1	9-13	1	1	HYD	HYD
	D	3.1 (3130)	0.045	1	1	9-13	1	1	HYD	HYD
	L	3.8 (3785)	0.060	1	1	41-47 [2]	1	1	HYD	HYD
	W	4.3 (4293)	0.035	1	1	9-13	1	1	HYD	HYD
	Z	4.3 (4293)	0.035	1	1	9-13	1	1	HYD	HYD
	H	5.0 (4999)	0.045	1	1	9-13	1	1	HYD	HYD
	K	5.7 (5735)	0.045	1	1	9-13	1	1	HYD	HYD
	N	7.4 (7440)	0.045	1	1	9-13	1	1	HYD	HYD
1994-95	4	2.2 (2189)	NA	1	1	9-13	1	1	HYD	HYD
	R	2.8 (2835)	0.040	1	1	9-13	1	1	HYD	HYD
	D	3.1 (3130)	0.045	1	1	9-13	1	1	HYD	HYD
	L	3.8 (3785)	0.060	1	1	41-47 [2]	1	1	HYD	HYD
	W	4.3 (4293)	0.035	1	1	9-13	1	1	HYD	HYD
	Z	4.3 (4293)	0.035	1	1	9-13	1	1	HYD	HYD
	H	5.0 (4999)	0.045	1	1	9-13	1	1	HYD	HYD
	K	5.7 (5735)	0.045	1	1	9-13	1	1	HYD	HYD
	N	7.4 (7440)	0.045	1	1	9-13	1	1	HYD	HYD

NOTE: The Vehicle Emission Control Information label often reflects specification changes made during production. The label figures must be used if they differ from those in this chart

1 Refer to underhood label for exact setting

2 With key on and engine off

79113C04

DIESEL ENGINE TUNE-UP SPECIFICATIONS

Year	Engine ID/VIN	Engine Displacement cu. in. (cc)	Valve Clearance Intake (in.)	Valve Clearance Exhaust (in.)	Intake Valve Opens (deg.)	Injection Pump Setting (deg.)	Injection Nozzle Pressure (psi) New	Injection Nozzle Pressure (psi) Used	Idle Speed (rpm)	Cranking Compression Pressure (psi)
1991	C	6.2 (6210)	HYD	HYD	2	1	1600	1500	2	NA
	J	6.2 (6210)	HYD	HYD	2	1	1600	1500	2	NA
1992	C	6.2 (6210)	HYD	HYD	2	1	1600	1500	2	NA
	J	6.2 (6210)	HYD	HYD	2	1	1600	1500	2	NA
	F	6.5 (6473)	HYD	HYD	2	2	1600	1500	2	NA
1993	C	6.2 (6210)	HYD	HYD	2	1	1600	1500	2	NA
	J	6.2 (6210)	HYD	HYD	2	1	1600	1500	2	NA
	F	6.5 (6473)	HYD	HYD	2	2	1600	1500	2	NA
1994-95	F	6.5 (6473)	HYD	HYD	3	3	1600	1500	3	NA
	P	6.5 (6473)	HYD	HYD	3	3	1800	1700	3	NA
	S	6.5 (6473)	HYD	HYD	3	3	1800	1700	3	NA
	Y	6.5 (6473)	HYD	HYD	3	3	1600	1500	3	NA

NOTE: The Vehicle Emission Control Information label often reflects specification changes made during production. The label figures must be used if they differ from those in this chart

HYD - Hydraulic

NA - Not Available

1 Set by aligning marks on top of engine front cover and injection pump flange

2 Refer to underhood label

3 Refer to underhood label

79113C05

FIRING ORDERS

NOTE: To avoid confusion, always replace spark plug wires one at a time.

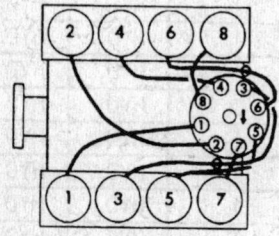

79114300

4.3L Engine
Engine Firing Order: 1–6–5–4–3–2
Distributor Rotation: Clockwise

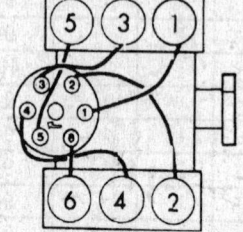

79114301

5.0L and 5.7L Engines
Engine Firing Order: 1–8–4–3–6–5–7–2
Distributor Rotation: Clockwise

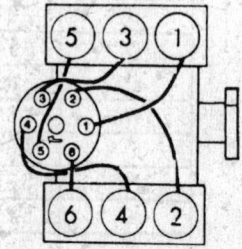

79114301

7.4L Engine
Engine Firing Order: 1–8–4–3–6–5–7–2
Distributor Rotation: Clockwise

CAPACITIES

Year	Model	Engine ID/VIN	Engine Displacement Liters (cc)	Engine Crankcase with Filter	Transmission (pts.)			Transfer Case (pts.)	Drive Axle		Fuel Tank (gal.)	Cooling System (qts.)
					4-Spd	5-Spd	Auto.		Front (pts.)	Rear (pts.)		
1991	Astro/Safari	B	4.3 (4293)	5.0	-	4.4	10.0	-	-	3.8	27.0	13.5 [8]
	Astro/Safari	Z	4.3 (4293)	5.0	-	4.4	10.0	-	-	3.8	27.0	13.5 [8]
	Lumina/Silhouette/Trans Sport	D	3.1 (3130)	4.5	-	-	8.0	-	-	-	18.0	13.4
	S10 Blazer/S15 Jimmy	R	2.8 (2835)	4.0	-	4.4	10.0	5	2.6	3.8	20.0	10.5
	S10 Blazer/S15 Jimmy	Z	4.3 (4293)	5.0	-	4.4	10.0	5	2.6	3.8	20.0	13.5
	Bravada	Z	4.3 (4293)	5.0	-	4.4	10.0	5	2.6	3.8	20.0	13.5
	S10 Pick-up/S15 Sonoma	A	2.5 (2474)	4.0	-	4.4	10.0	5	2.6	3.8	13.0 [12]	11.5
	S10 Pick-up/S15 Sonoma	E	2.5 (2474)	4.0	-	4.4	10.0	5	2.6	3.8	13.0 [12]	11.5
	S10 Pick-up/S15 Sonoma	R	2.8 (2835)	4.0	-	4.4	10.0	5	2.6	3.8	13.0 [12]	10.5
	S10 Pick-up/S15 Sonoma	Z	4.3 (4293)	5.0	-	4.4	10.0	5	2.6	3.8	20.0	13.5
	Syclone	Z	4.3 (4293)	5.0	-	-	10.0	-	2.6	3.8	20.0	13.5
	Blazer/Jimmy	K	5.7 (5735)	5.0	16	-	3	10	4.0	4	25.0 [5]	18.0
	Blazer/Jimmy	C	6.2 (6210)	7.0	16	-	3	10	4.0	4	27.0 [6]	25.0
	Blazer/Jimmy	J	6.2 (6210)	7.0	16	-	3	10	4.0	4	27.0 [6]	25.0
	C1500	Z	4.3 (4293)	5.0	16	3.6	3	-	-	4	9	10.9
	C1500	K	5.7 (5735)	5.0	16	3.6	3	-	-	4	9	18.0
	C1500	C	6.2 (6210)	7.0	16	3.6	3	-	-	4	9	25.0
	C1500	J	6.2 (6210)	7.0	16	3.6	3	-	-	4	9	25.0
	C1500	N	7.4 (7440)	6.0	16	3.6	3	-	-	4	9	25.0
	C2500	Z	4.3 (4293)	5.0	16	3.6	3	-	-	4	9	10.9
	C2500	H	5.0 (4999)	5.0	16	3.6	3	-	-	4	9	18.0
	C2500	K	5.7 (5735)	5.0	16	3.6	3	-	-	4	9	18.0
	C2500	C	6.2 (6210)	7.0	16	3.6	3	-	-	4	9	25.0
	C2500	J	6.2 (6210)	7.0	16	3.6	3	-	-	4	9	25.0
	C3500	H	5.0 (4999)	5.0	16	3.6	3	-	-	4	9	18.0
	C3500	K	5.7 (5735)	5.0	16	3.6	3	-	-	4	9	18.0
	C3500	C	6.2 (6210)	7.0	16	3.6	3	-	-	4	9	25.0
	C3500	J	6.2 (6210)	7.0	16	3.6	3	-	-	4	9	25.0
	C3500	N	7.4 (7440)	6.0	16	3.6	3	-	-	4	9	25.0
	K1500	Z	4.3 (4293)	5.0	16	-	3	-	-	4	9	10.9
	K1500	H	5.0 (4999)	5.0	16	-	3	-	-	4	9	18.0
	K1500	K	5.7 (5735)	5.0	16	-	3	-	-	4	9	18.0
	K1500	C	6.2 (6210)	7.0	16	-	3	-	-	4	9	25.0
	K2500	Z	4.3 (4293)	5.0	16	-	3	-	-	4	9	10.9
	K2500	H	5.0 (4999)	5.0	16	-	3	-	-	4	9	18.0
	K2500	K	5.7 (5735)	5.0	16	-	3	-	-	4	9	18.0
	K2500	C	6.2 (6210)	7.0	16	-	3	-	-	4	9	25.0
	K2500	J	6.2 (6210)	7.0	16	-	3	-	-	4	9	18.0
	K3500	K	5.7 (5735)	5.0	16	-	3	-	-	4	9	25.0
	K3500	N	5.7 (5735)	6.0	16	-	3	-	-	4	9	25.0
	K3500	J	6.2 (6210)	7.0	16	-	3	-	-	4	9	25.0
	R3500	K	5.7 (5735)	5.0	16	-	3	-	-	4	7	18.0
	R3500	J	6.2 (6210)	7.0	16	-	3	-	-	4	7	25.0
	R3500	N	7.4 (7440)	6.0	16	-	3	10	4.0	4	25.0 [10]	24.5
	P30	Z	4.3 (4293)	5.0	-	4.4	10.0	5	2.6	3.8	18.0	13.4
	P30	K	5.7 (5735)	5.0	16	3.6	3	-	-	4	7	18.0

79113C06

CAPACITIES

Year	Model	Engine ID/VIN	Engine Displacement Liters (cc)	Engine Crankcase with Filter	Transmission (pts.)			Transfer Case (pts.)	Drive Axle Front (pts.)	Rear (pts.)	Fuel Tank (gal.)	Cooling System (qts.)
					4-Spd	5-Spd	Auto.					
	P30	J	6.2 (6210)	7.0	16	3.6	3	-	-	13	14	25.0
	P30	N	7.4 (7440)	5.0	16	3.6	3	-	-	4	7	18.0
	Suburban	K	5.7 (5735)	5.0	16	-	3	10	4.0	4	25.0 [10]	18.0
	Suburban	C	6.2 (6210)	7.0	16	-	3	10	4.0	4	27.0 [11]	25.0
	Suburban	J	6.2 (6210)	7.0	16	-	3	10	4.0	4	27.0 [11]	25.0
	Suburban	N	7.4 (7440)	6.0	16	-	3	10	4.0	4	25.0 [10]	24.5
	G10	Z	4.3 (4293)	5.0	16	-	3	-	-	4	22.0 [11]	10.9
	G10	H	5.0 (4999)	5.0	16	-	3	-	-	4	22.0 [11]	17.0
	G10	K	5.7 (5735)	5.0	16	-	3	-	-	4	7	18.0
	G20	Z	4.3 (4293)	5.0	16	-	3	-	-	4	22.0 [11]	10.9
	G20	H	5.0 (4999)	5.0	16	-	3	-	-	4	22.0 [11]	17.0
	G20	K	5.7 (5735)	5.0	16	-	3	-	-	4	7	18.0
	G30	Z	4.3 (4293)	5.0	16	-	3	-	-	4	22.0 [11]	10.9
	G30	K	5.7 (5735)	5.0	16	-	3	-	-	4	7	18.0
	G30	J	6.2 (6210)	7.0	16	-	3	-	-	4	22.0 [11]	24.0
	G30	N	7.4 (7440)	6.0	16	-	3	-	-	13	14	24.5
1992	Astro/Safari	W	4.3 (4293)	5.0	-	4.4	10.0	-	-	3.8	27.0	13.5 [8]
	Astro/Safari	Z	4.3 (4293)	5.0	-	4.4	10.0	-	-	3.8	27.0	13.5 [8]
	Lumina/Silhouette/Trans Sport	D	3.1 (3130)	4.5	-	-	8.0	-	-	-	20.0	13.4
	Lumina/Silhouette/Trans Sport	L	3.8 (3785)	4.5	-	-	12.0	-	-	-	20.0	13.4
	S10 Blazer/S15 Jimmy	W	4.3 (4293)	4.5	-	-	10.0	-	3.5	3.5	20.0	12.0
	S10 Blazer/S15 Jimmy	Z	4.3 (4293)	5.0	-	4.4	10.0	5	3.5	3.5	20.0	12.1
	Bravada	W	4.3 (4293)	4.5	-	-	10.0	-	3.5	3.5	20.0	12.0
	Bravada	Z	4.3 (4293)	5.0	-	4.4	10.0	5	3.5	3.5	20.0	12.1
	S10 Pick-up/Sonoma	A	2.5 (2474)	4.0	-	4.4	10.0	5	3.5	3.5	13.0 [12]	11.5
	S10 Pick-up/Sonoma	Z	2.8 (2835)	4.0	-	4.4	10.0	5	3.5	3.5	13.0 [12]	10.5
	S10 Pick-up/Sonoma	R	4.3 (4293)	5.0	-	4.4	10.0	5	3.5	3.5	20.0	13.5
	Syclone	Z	4.3 (4293)	5.0	-	-	10.0	-	2.6	3.8	20.0	13.5
	Typhoon	Z	4.3 (4293)	5.0	-	-	10.0	-	2.6	3.8	20.0	13.5
	Blazer/Yukon	K	5.7 (5735)	5.0	-	-	3	10	4.0	4	25.0 [5]	18.0
	C1500	Z	4.3 (4293)	5.0	-	2	3	-	-	4	9	11.0
	C1500	H	5.0 (4999)	5.0	-	2	3	-	-	4	9	18.0
	C1500	K	5.7 (5735)	5.0	-	2	3	-	-	4	9	18.0
	C1500	C	6.2 (6210)	7.0	-	2	3	-	-	4	9	25.0
	C1500	J	6.2 (6210)	7.0	-	2	3	-	-	4	9	25.0
	C1500	N	7.4 (7440)	6.0	-	2	3	-	-	4	9	25.0
	C2500	Z	4.3 (4293)	5.0	-	2	3	-	-	4	9	11.0
	C2500	H	5.0 (4999)	5.0	-	2	3	-	-	4	9	18.0
	C2500	K	5.7 (5735)	5.0	-	2	3	-	-	4	9	18.0
	C2500	C	6.2 (6210)	7.0	-	2	3	-	-	4	9	25.0
	C2500	J	6.2 (6210)	7.0	-	2	3	-	-	4	9	25.0
	C2500	F	6.5 (6473)	7.0	-	2	3	-	-	4	9	26.5
	C3500	H	5.0 (4999)	5.0	-	2	3	-	-	4	9	18.0
	C3500	K	5.7 (5735)	5.0	-	2	3	-	-	4	9	18.0
	C3500	C	6.2 (6210)	7.0	-	2	3	-	-	4	9	25.0
	C3500	J	6.2 (6210)	7.0	-	2	3	-	-	4	9	25.0

79113Ca6

CAPACITIES

Year	Model	Engine ID/VIN	Engine Displacement Liters (cc)	Engine Crankcase with Filter	Transmission (pts.)			Transfer Case (pts.)	Drive Axle		Fuel Tank (gal.)	Cooling System (qts.)
					4-Spd	5-Spd	Auto.		Front (pts.)	Rear (pts.)		
	C3500	F	6.5 (6473)	7.0	-	2	3	-	-	4	9	26.5
	C3500	N	7.4 (7440)	6.0	-	2	3	-	-	4	9	25.0 [1]
	K1500	Z	4.3 (4293)	5.0	-	2	3	-	-	4	9	11.0
	K1500	H	5.0 (4999)	5.0	-	2	3	-	-	4	9	18.0
	K1500	K	5.7 (5735)	5.0	-	2	3	-	-	4	9	18.0
	K1500	C	6.2 (6210)	7.0	-	2	3	-	-	4	9	25.0
	K1500	N	7.4 (7440)	6.0	-	2	3	-	-	4	9	25.0 [1]
	K2500	Z	4.3 (4293)	5.0	-	2	3	-	-	4	9	11.0
	K2500	H	5.0 (4999)	5.0	-	2	3	-	-	4	9	18.0
	K2500	K	5.7 (5735)	5.0	-	2	3	-	-	4	9	18.0
	K2500	C	6.2 (6210)	7.0	-	2	3	-	-	4	9	25.0
	K2500	J	6.2 (6210)	7.0	-	2	3	-	-	4	9	25.0
	K2500	N	7.4 (7440)	6.0	-	2	3	-	-	4	9	25.0 [1]
	K3500	K	5.7 (5735)	5.0	-	2	3	-	-	4	9	18.0
	K3500	N	7.4 (7440)	6.0	-	2	3	-	-	4	9	25.0 [1]
	K3500	J	6.2 (6210)	7.0	-	2	3	-	-	4	9	25.0
	P30	Z	4.3 (4293)	5.0	-	4.4	10.0	5	2.6	3.8	18.0	13.4
	P30	K	5.7 (5735)	5.0	-	3.6	3	-	-	4	7	18.0
	P30	J	6.2 (6210)	7.0	-	3.6	3	-	-	13	14	25.0
	P30	N	7.4 (7440)	5.0	-	3.6	3	-	-	4	7	18.0
	Suburban	K	5.7 (5735)	5.0	-	-	3	10	4.0	4	25.0 [10]	18.0
	Suburban	N	7.4 (7440)	6.0	-	-	3	10	4.0	4	25.0 [10]	24.5
	G10	Z	4.3 (4293)	5.0	-	-	3	-	-	4	22.0 [11]	11.0
	G10	H	5.0 (4999)	5.0	-	-	3	-	-	4	22.0 [11]	17.0
	G10	K	5.7 (5735)	5.0	-	-	3	-	-	4	7	18.0
	G20	Z	4.3 (4293)	5.0	-	-	3	-	-	4	22.0 [11]	11.0
	G20	H	5.0 (4999)	5.0	-	-	3	-	-	4	22.0 [11]	17.0
	G20	K	5.7 (5735)	5.0	-	-	3	-	-	4	7	18.0
	G30	Z	4.3 (4293)	5.0	-	-	3	-	-	4	22.0 [11]	11.0
	G30	K	5.7 (5735)	5.0	-	-	3	-	-	4	7	18.0
	G30	C	6.2 (6210)	7.0	-	-	3	-	-	4	22.0 [11]	24.0
	G30	J	6.2 (6210)	7.0	-	-	3	-	-	4	22.0 [11]	24.0
	G30	N	7.4 (7440)	6.0	-	-	3	-	-	13	14	24.5
1993	Astro/Safari	W	4.3 (4293)	5.0	-	4.4	10.0	-	-	3.8	27.0	13.5 [8]
	Astro/Safari	Z	4.3 (4293)	5.0	-	4.4	10.0	-	-	3.8	27.0	13.5 [8]
	Lumina/Silhouette/Trans Sport	D	3.1 (3130)	4.5	-	-	8.0	-	-	-	20.0	13.4
	Lumina/Silhouette/Trans Sport	L	3.8 (3785)	4.5	-	-	12.0	-	-	-	20.0	13.4
	S10 Blazer/S15 Jimmy	W	4.3 (4293)	4.5	-	-	10.0	-	3.5	3.5	20.0	12.0
	S10 Blazer/S15 Jimmy	Z	4.3 (4293)	5.0	-	4.4	10.0	5	3.5	3.5	20.0	12.1
	Bravada	W	4.3 (4293)	4.5	-	-	10.0	-	3.5	3.5	20.0	12.0
	Bravada	Z	4.3 (4293)	5.0	-	4.4	10.0	5	3.5	3.5	20.0	12.1
	S10 Pick-up/Sonoma	A	2.5 (2474)	4.0	-	4.4	10.0	5	3.5	3.5	13.0 [12]	11.5
	S10 Pick-up/Sonoma	R	2.8 (2835)	4.0	-	4.4	10.0	5	3.5	3.5	13.0 [12]	10.5
	S10 Pick-up/Sonoma	W	4.3 (4293)	4.5	-	4.4	10.0	5	3.5	3.5	20.0	12.0
	Typhoon	Z	4.3 (4293)	5.0	-	-	10.0	-	2.6	3.8	20.0	13.5
	Blazer/Yukon	K	5.7 (5735)	5.0	-	-	3	10	4.0	4	25.0 [5]	18.0

79113Cb6

CAPACITIES

Year	Model	Engine ID/VIN	Engine Displacement Liters (cc)	Engine Crankcase with Filter	Transmission (pts.)			Transfer Case (pts.)	Drive Axle		Fuel Tank (gal.)	Cooling System (qts.)
					4-Spd	5-Spd	Auto.		Front (pts.)	Rear (pts.)		
	C1500	Z	4.3 (4293)	5.0	-	2	3	-	-	4	9	11.0
	C1500	H	5.0 (4999)	5.0	-	2	3	-	-	4	9	18.0
	C1500	K	5.7 (5735)	5.0	-	2	3	-	-	4	9	18.0
	C1500	C	6.2 (6210)	7.0	-	2	3	-	-	4	9	25.0
	C1500	J	6.2 (6210)	7.0	-	2	3	-	-	4	9	25.0
	C1500	N	7.4 (7440)	.6.0	-	2	3	-	-	4	9	25.0
	C2500	Z	4.3 (4293)	5.0	-	2	3	-	-	4	9	11.0
	C2500	H	5.0 (4999)	5.0	-	2	3	-	-	4	9	18.0
	C2500	K	5.7 (5735)	5.0	-	2	3	-	-	4	9	18.0
	C2500	C	6.2 (6210)	7.0	-	2	3	-	-	4	9	25.0
	C2500	J	6.2 (6210)	7.0	-	2	3	-	-	4	9	25.0
	C2500	F	6.5 (6473)	7.0	-	2	3	-	-	4	9	26.5
	C3500	H	5.0 (4999)	5.0	-	2	3	-	-	4	9	18.0
	C3500	K	5.7 (5735)	5.0	-	2	3	-	-	4	9	18.0
	C3500	C	6.2 (6210)	7.0	-	2	3	-	-	4	9	25.0
	C3500	J	6.2 (6210)	7.0	-	2	3	-	-	4	9	25.0
	C3500	F	6.5 (6473)	7.0	-	2	3	-	-	4	9	26.5
	C3500	N	7.4 (7440)	6.0	-	2	3	-	-	4	9	25.0 [1]
	K1500	Z	4.3 (4293)	5.0	-	2	3	-	-	4	9	11.0
	K1500	H	5.0 (4999)	5.0	-	2	3	-	-	4	9	18.0
	K1500	K	5.7 (5735)	5.0	-	2	3	-	-	4	9	18.0
	K1500	C	6.2 (6210)	7.0	-	2	3	-	-	4	9	25.0
	K1500	N	7.4 (7440)	6.0	-	2	3	-	-	4	9	25.0 [1]
	K2500	Z	4.3 (4293)	5.0	-	2	3	-	-	4	9	11.0
	K2500	H	5.0 (4999)	5.0	-	2	3	-	-	4	9	18.0
	K2500	K	5.7 (5735)	5.0	-	2	3	-	-	4	9	18.0
	K2500	C	6.2 (6210)	7.0	-	2	3	-	-	4	9	25.0
	K2500	J	6.2 (6210)	7.0	-	2	3	-	-	4	9	25.0
	K2500	F	6.5 (6473)	7.0	-	2	3	-	-	4	9	26.5
	K2500	N	7.4 (7440)	6.0	-	2	3	-	-	4	9	25.0 [1]
	K3500	H	5.0 (4999)	5.0	-	2	3	-	-	4	9	18.0
	K3500	K	5.7 (5735)	5.0	-	2	3	-	-	4	9	18.0
	K3500	C	6.2 (6210)	7.0	-	2	3	-	-	4	9	25.0
	K3500	J	6.2 (6210)	7.0	-	2	3	-	-	4	9	25.0
	K3500	F	6.5 (6473)	7.0	-	2	3	-	-	4	9	26.5
	K3500	N	7.4 (7440)	6.0	-	2	3	-	-	4	9	25.0 [1]
	P30	Z	4.3 (4293)	5.0	-	4.4	10.0	5	2.6	3.8	18.0	13.4
	P30	K	5.7 (5735)	5.0	-	3.6	3	-	-	4	7	18.0
	P30	J	6.2 (6210)	7.0	-	3.6	3	-	-	13	14	25.0
	P30	N	7.4 (7440)	5.0	-	3.6	3	-	-	4	7	18.0
	Suburban	K	5.7 (5735)	5.0	-	-	3	10	4.0	4	25.0 [10]	18.0
	Suburban	N	7.4 (7440)	6.0	-	-	3	10	4.0	4	25.0 [10]	24.5
	G10	Z	4.3 (4293)	5.0	-	-	3	-	-	4	22.0 [11]	11.0
	G10	H	5.0 (4999)	5.0	-	-	3	-	-	4	22.0 [11]	17.0
	G10	K	5.7 (5735)	5.0	-	-	3	-	-	4	7	18.0
	G10	C	6.2 (6210)	7.0	-	-	3	-	-	4	22.0 [11]	24.0

79113Cc6

CAPACITIES

Year	Model	Engine ID/VIN	Engine Displacement Liters (cc)	Engine Crankcase with Filter	Transmission (pts.)			Transfer Case (pts.)	Drive Axle		Fuel Tank (gal.)	Cooling System (qts.)
					4-Spd	5-Spd	Auto.		Front (pts.)	Rear (pts.)		
	G10	J	6.2 (6210)	7.0	-	-	3	-	-	4	22.0 11	24.0
	G20	Z	4.3 (4293)	5.0	-	-	3	-	-	4	22.0 11	11.0
	G20	H	5.0 (4999)	5.0	-	-	3	-	-	4	22.0 11	17.0
	G20	K	5.7 (5735)	5.0	-	-	3	-	-	4	7	18.0
	G20	C	6.2 (6210)	7.0	-	-	3	-	-	4	22.0 11	24.0
	G20	J	6.2 (6210)	7.0	-	-	3	-	-	4	22.0 11	24.0
	G30	Z	4.3 (4293)	5.0	-	-	3	-	-	4	22.0 11	11.0
	G30	K	5.7 (5735)	5.0	-	-	3	-	-	4	7	18.0
	G30	C	6.2 (6210)	7.0	-	-	3	-	-	4	22.0 11	24.0
	G30	J	6.2 (6210)	7.0	-	-	3	-	-	4	22.0 11	24.0
	G30	N	7.4 (7440)	6.0	-	-	3	-	-	13	14	24.5
1994-95	Astro/Safari	W	4.3 (4293)	5.0	-	4.4	10.0	-	-	3.8	27.0	13.5 8
	Astro/Safari	Z	4.3 (4293)	5.0	-	4.4	10.0	-	-	3.8	27.0	13.5 8
	Lumina/Silhouette/Trans Sport	D	3.1 (3130)	4.5	-	-	8.0	-	-	-	20.0	13.4
	Lumina/Silhouette/Trans Sport	L	3.8 (3785)	4.5	-	-	12.0	-	-	-	20.0	13.4
	S10 Blazer/S15 Jimmy	W	4.3 (4293)	4.5	-	-	10.0	-	3.5	3.5	20.0	12.0
	S10 Blazer/S15 Jimmy	Z	4.3 (4293)	5.0	-	4.4	10.0	5	3.5	3.5	20.0	12.1
	Bravada	W	4.3 (4293)	4.5	-	-	10.0	-	3.5	3.5	20.0	12.0
	Bravada	Z	4.3 (4293)	5.0	-	4.4	10.0	5	3.5	3.5	20.0	12.1
	S10 Pick-up/Sonoma	4	2.2 (2189)	4.0	-	4.4	10.0	5	3.5	3.5	13.0 12	11.5
	S10 Pick-up/Sonoma	W	4.3 (4293)	4.5	-	4.4	10.0	5	3.5	3.5	20.0	12.0
	S10 Pick-up/Sonoma	Z	4.3 (4293)	5.0	-	4.4	10.0	5	3.5	3.5	20.0	12.0
	Blazer/Yukon	K	5.7 (5735)	5.0	-	-	3	10	4.0	4	25.0 5	18.0
	Blazer/Yukon	F	6.5 (6473)	7.0	-	2	3	-	-	4	9	26.5
	C1500	Z	4.3 (4293)	5.0	-	2	3	-	-	4	9	11.0
	C1500	H	5.0 (4999)	5.0	-	2	3	-	-	4	9	18.0
	C1500	K	5.7 (5735)	5.0	-	2	3	-	-	4	9	18.0
	C1500	F	6.5 (6473)	7.0	-	2	3	-	-	4	9	26.5
	C1500	N	7.4 (7440)	6.0	-	2	3	-	-	4	9	25.0
	C2500	Z	4.3 (4293)	5.0	-	2	3	-	-	4	9	11.0
	C2500	H	5.0 (4999)	5.0	-	2	3	-	-	4	9	18.0
	C2500	K	5.7 (5735)	5.0	-	2	3	-	-	4	9	18.0
	C2500	P	6.5 (6473)	7.0	-	2	3	-	-	4	9	26.5
	C2500	S	6.5 (6473)	7.0	-	2	3	-	-	4	9	26.5
	C3500	H	5.0 (4999)	5.0	-	2	3	-	-	4	9	18.0
	C3500	K	5.7 (5735)	5.0	-	2	3	-	-	4	9	18.0
	C3500	F	6.5 (6473)	7.0	-	2	3	-	-	4	9	26.5
	C3500	P	6.5 (6473)	7.0	-	2	3	-	-	4	9	26.5
	C3500	S	6.5 (6473)	7.0	-	2	3	-	-	4	9	26.5
	C3500	N	7.4 (7440)	6.0	-	2	3	-	-	4	9	25.0 1
	K1500	Z	4.3 (4293)	5.0	-	2	3	-	-	4	9	11.0
	K1500	H	5.0 (4999)	5.0	-	2	3	-	-	4	9	18.0
	K1500	K	5.7 (5735)	5.0	-	2	3	-	-	4	9	18.0
	K1500	F	6.5 (6505)	7.0	-	2	3	-	-	4	9	25.0
	K1500	N	7.4 (7440)	6.0	-	2	3	-	-	4	9	25.0 1
	K2500	Z	4.3 (4293)	5.0	-	2	3	-	-	4	9	11.0

79113Cd6

CAPACITIES

Year	Model	Engine ID/VIN	Engine Displacement Liters (cc)	Engine Crankcase with Filter	Transmission (pts.) 4-Spd	5-Spd	Auto.	Transfer Case (pts.)	Drive Axle Front (pts.)	Rear (pts.)	Fuel Tank (gal.)	Cooling System (qts.)
	K2500	H	5.0 (4999)	5.0	-	2	3	-	-	4	9	18.0
	K2500	K	5.7 (5735)	5.0	-	2	3	-	-	4	9	18.0
	K2500	P	6.5 (6505)	7.0	-	2	3	-	-	4	9	25.0
	K2500	S	6.5 (6505)	7.0	-	2	3	-	-	4	9	25.0
	K2500	F	6.5 (6473)	7.0	-	2	3	-	-	4	9	26.5
	K2500	N	7.4 (7440)	6.0	-	2	3	-	-	4	9	25.0 [1]
	K2500	H	5.0 (4999)	5.0	-	2	3	-	-	4	9	18.0
	K3500	K	5.7 (5735)	5.0	-	2	3	-	-	4	9	18.0
	K3500	P	6.5 (6505)	7.0	-	2	3	-	-	4	9	25.0
	K3500	S	6.5 (6505)	7.0	-	2	3	-	-	4	9	25.0
	K3500	F	6.5 (6473)	7.0	-	2	3	-	-	4	9	26.5
	K3500	N	7.4 (7440)	6.0	-	2	3	-	-	4	9	25.0 [1]
	P30	Z	4.3 (4293)	5.0	-	4.4	10.0	5	2.6	3.8	18.0	13.4
	P30	K	5.7 (5735)	5.0	-	3.6	3	-	-	4	7	18.0
	P30	F	6.5 (6505)	7.0	-	3.6	3	-	-	13	14	25.0
	P30	N	7.4 (7440)	5.0	-	3.6	3	-	-	4	7	18.0
	Suburban	K	5.7 (5735)	5.0	-	-	3	10	4.0	4	25.0 [10]	18.0
	Suburban	P	6.5 (6505)	7.0	-	2	3	-	-	4	9	25.0
	Suburban	S	6.5 (6505)	7.0	-	2	3	-	-	4	9	25.0
	Suburban	N	7.4 (7440)	6.0	-	-	3	10	4.0	4	25.0 [10]	24.5
	G10	Z	4.3 (4293)	5.0	-	2	3	-	-	4	22.0 [11]	11.0
	G10	H	5.0 (4999)	5.0	-	2	3	-	-	4	22.0 [11]	17.0
	G10	K	5.7 (5735)	5.0	-	2	3	-	-	4	7	18.0
	G10	P	6.5 (6505)	7.0	-	2	3	-	-	4	22.0 [11]	24.0
	G10	S	6.5 (6505)	7.0	-	2	3	-	-	4	22.0 [11]	24.0
	G20	Z	4..3 (4293)	5.0	-	2	3	-	-	4	22.0 [11]	11.0
	G20	H	5.0 (4999)	5.0	-	2	3	-	-	4	22.0 [11]	17.0
	G20	K	5.7 (5735)	5.0	-	2	3	-	-	4	7	18.0
	G20	P	6.5 (6505)	7.0	-	2	3	-	-	4	22.0 [11]	24.0
	G20	Y	6.5 (6505)	7.0	-	2	3	-	-	4	22.0 [11]	24.0
	G30	Z	4.3 (4293)	5.0	-	2	3	-	-	4	22.0 [11]	11.0
	G30	K	5.7 (5735)	5.0	-	2	3	-	-	4	7	18.0
	G30	P	6.5 (6505)	7.0	-	2	3	-	-	4	22.0 [11]	24.0
	G30	Y	6.5 (6505)	7.0	-	2	3	-	-	4	22.0 [11]	24.0

1 3500HD: 28.5 qts. capacity
2 New Venture gear 4500: 8.0 pts.
  New Venture gear 5LM60: 4.4 pts.
3 350C trans.: 6.3 pts.
  THM400 and 4L80 trans.: 9.0 pts.
  THM700 R4 and 4L60 trans.: 10.0 pts.
4 8.5" ring gear: 4.2 pts.
  9.5" ring gear: 6.5 pts.
  9.75" ring gear: 6.0 pts.
  10.5" ring gear: 6.5 pts.

5 Available with optional 31 gallon tank
6 Available with optional 32 gallon tank
7 Short bed: 16 gals.; Long bed: 20 gals.
8 16.5 qts. with rear heater
9 Std. available with 25 and 34 gallon tanks
  Chassis cab available with 22, 30 and 34 gallon tanks
10 Available 31 and 40 gallon tanks
11 Available 32 and 41 gallon tanks
12 Available with 20 gallon tank

13 8.5" ring gear: 4.2 pts.
  9.5" ring gear: 6.5 pts.
  Chevrolet 10.5" ring gear: 6.5 pts.
  Dana 9.75" ring gear: 6.0 pts.
  Rockwell 12" ring gear: 12.5 pts.
14 Available with a variety of fuel tanks
15 13 qts. with rear heater
16 85mm: 3.6 pts.; 117mm: 8.4 pts.

79113Ce6

CAMSHAFT SPECIFICATIONS

All measurements given in inches.

Year	Engine ID/VIN	Engine Displacement Liters (cc)	Journal Diameter					Elevation		Bearing Clearance	Camshaft End Play
			1	2	3	4	5	In.	Ex.		
1991	A	2.5 (2474)	1.869	1.869	1.869	1.869	1.869	0.251	0.251	0.0007-0.0027	0.0015-0.0050
	E	2.5 (2474)	1.869	1.869	1.869	1.869	1.869	0.398	0.398	0.0007-0.0027	0.0015-0.0050
	R	2.8 (2835)	1.868-1.869	1.868-1.869	1.868-1.869	1.868-1.869	1.868-1.869	0.231	0.263	0.0010-0.0040	NA
	D	3.1 (3130)	1.868-1.882	1.868-1.882	1.868-1.882	1.868-1.882	NA	0.231	0.262	0.0010-0.0040	NA
	B	4.3 (4239)	1.8682-1.8692	1.8682-1.8692	1.8682-1.8692	1.8682-1.8692	NA	0.269	0.276	NA	0.004-0.012
	Z	4.3 (4293)	1.8682-1.8692	1.8682-1.8692	1.8682-1.8692	1.8682-1.8692	NA	0.357	0.390	0.0010-0.0030	0.004-0.012
	H	5.0 (4999)	1.8682-1.8692	1.8682-1.8692	1.8682-1.8692	1.8682-1.8692	NA	0.2484	0.2667	NA	0.004-0.012
	K	5.7 (5735)	1.8682-1.8692	1.8682-1.8692	1.8682-1.8692	1.8682-1.8692	NA	0.2600	0.2733	NA	0.004-0.012
	C	6.2 (6210)	2.1633-2.1642	2.1633-2.1642	2.1633-2.1642	2.1633-2.1642	2.0067-2.0089	2.808	2.808	1	NA
	J	6.2 (6210)	2.1633-2.1642	2.1633-2.1642	2.1633-2.1642	2.1633-2.1642	2.0067-2.0089	2.808	2.808	1	NA
	N	7.4 (7440)	1.9482-1.9492	1.9482-1.9492	1.9482-1.9492	1.9482-1.9492	1.9482-1.9492	0.2341-0.2345	0.2529-0.2531	NA	NA
1992	A	2.5 (2474)	1.869	1.869	1.869	1.869	1.869	0.251	0.251	0.0007-0.0027	0.0015-0.0050
	R	2.8 (2835)	1.868-1.870	1.868-1.870	1.868-1.870	1.868-1.870	1.868-1.870	0.262	0.273	0.0010-0.0040	NA
	D	3.1 (3130)	1.868-1.882	1.868-1.882	1.868-1.882	1.868-1.882	NA	0.231	0.262	0.0010-0.0040	NA
	L	3.8 (3785)	1.785-1.786	1.785-1.786	1.785-1.786	1.785-1.786	NA	0.250	0.255	0.0005-0.0035	NA
	B	4.3 (4293)	1.8682-1.8692	1.8682-1.8692	1.8682-1.8692	1.8682-1.8692	NA	0.269	0.276	NA	0.004-0.012
	W	4.3 (4293)	1.8682-1.8692	1.8682-1.8692	1.8682-1.8692	1.8682-1.8692	NA	0.288	0.294	NA	0.001-0.009
	Z	4.3 (4293)	1.8682-1.8692	1.8682-1.8692	1.8682-1.8692	1.8682-1.8692	NA	0.234	0.257	0.0010-0.0030	0.004-0.012
	H	5.0 (4999)	1.8682-1.8692	1.8682-1.8692	1.8682-1.8692	1.8682-1.8692	1.8682-1.8692	0.2336	0.2565	NA	0.004-0.012
	K	5.7 (5735)	1.8682-1.8692	1.8682-1.8692	1.8682-1.8692	1.8682-1.8692	1.8682-1.8692	0.2565	0.2690	NA	0.004-0.012
	C	6.2 (6210)	2.1633-2.1642	2.1633-2.1642	2.1633-2.1642	2.1633-2.1642	2.0067-2.0089	0.2808	0.2808	0.0010-0.0040	0.002-0.012
	J	6.2 (6210)	2.1633-2.1642	2.1633-2.1642	2.1633-2.1642	2.1633-2.1642	2.0067-2.0089	0.2808	0.2808	0.0010-0.0040	0.002-0.012
	F	6.5 (6473)	2.1642-2.1663	2.1642-2.1663	2.1642-2.1663	2.1642-2.1663	2.0067-2.0089	0.2808	0.2808	1	0.002-0.012

79113C07

CAMSHAFT SPECIFICATIONS
All measurements given in inches.

Year	Engine ID/VIN	Engine Displacement Liters (cc)	Journal Diameter					Elevation		Bearing Clearance	Camshaft End Play
			1	2	3	4	5	In.	Ex.		
	N	7.4 (7440)	1.9482-1.9492	1.9482-1.9492	1.9482-1.9492	1.9482-1.9492	1.9482-1.9492	0.2341-0.2345	0.2529-0.2531	NA	NA
1993	A	2.5 (2474)	1.869	1.869	1.869	1.869	1.869	0.251	0.251	0.0007-0.0027	0.0015-0.0050
	R	2.8 (2835)	1.868-1.870	1.868-1.870	1.868-1.870	1.868-1.870	1.868-1.870	0.262	0.273	0.0010-0.0040	NA
	D	3.1 (3130)	1.868-1.882	1.868-1.882	1.868-1.882	1.868-1.882	NA	0.231	0.262	0.0010-0.0040	NA
	L	3.8 (3785)	1.785-1.786	1.785-1.786	1.785-1.786	1.785-1.786	NA	0.250	0.255	0.0005-0.0035	NA
	W	4.3 (4293)	1.8682-1.8692	1.8682-1.8692	1.8682-1.8692	1.8682-1.8692	NA	0.288	0.294	NA	0.001-0.009
	Z	4.3 (4293)	1.8682-1.8692	1.8682-1.8692	1.8682-1.8692	1.8682-1.8692	NA	0.234	0.257	0.0010-0.0030	0.004-0.012
	H	5.0 (4999)	1.8682-1.8692	1.8682-1.8692	1.8682-1.8692	1.8682-1.8692	1.8682-1.8692	0.2336	0.2565	NA	0.004-0.012
	K	5.7 (5735)	1.8682-1.8692	1.8682-1.8692	1.8682-1.8692	1.8682-1.8692	1.8682-1.8692	0.2565	0.2690	NA	0.004-0.012
	C	6.2 (6210)	2.1633-2.1642	2.1633-2.1642	2.1633-2.1642	2.1633-2.1642	2.0067-2.0089	0.2808	0.2808	0.0010-0.0040	0.002-0.012
	J	6.2 (6210)	2.1633-2.1642	2.1633-2.1642	2.1633-2.1642	2.1633-2.1642	2.0067-2.0089	0.2808	0.2808	0.0010-0.0040	0.002-0.012
	F	6.5 (6473)	2.1642-2.1663	2.1642-2.1663	2.1642-2.1663	2.1642-2.1663	2.0067-2.0089	0.2808	0.2808	1	0.002-0.012
	N	7.4 (7440)	1.9482-1.9492	1.9482-1.9492	1.9482-1.9492	1.9482-1.9492	1.9482-1.9492	0.2341-0.2345	0.2529-0.2531	NA	NA
1994-95	4	2.2 (2189)	1.869	1.869	1.869	1.869	1.869	0.251	0.251	0.0007-0.0027	0.0015-0.0050
	D	3.1 (3130)	1.868-1.882	1.868-1.882	1.868-1.882	1.868-1.882	NA	0.231	0.262	0.0010-0.0040	NA
	L	3.8 (3785)	1.785-1.786	1.785-1.786	1.785-1.786	1.785-1.786	NA	0.250	0.255	0.0005-0.0035	NA
	W	4.3 (4293)	1.8682-1.8692	1.8682-1.8692	1.8682-1.8692	1.8682-1.8692	NA	0.288	0.294	NA	0.001-0.009
	Z	4.3 (4293)	1.8682-1.8692	1.8682-1.8692	1.8682-1.8692	1.8682-1.8692	NA	0.234	0.257	0.0010-0.0030	0.004-0.012
	H	5.0 (4999)	1.8682-1.8692	1.8682-1.8692	1.8682-1.8692	1.8682-1.8692	1.8682-1.8692	0.2336	0.2565	NA	0.004-0.012
	K	5.7 (5735)	1.8682-1.8692	1.8682-1.8692	1.8682-1.8692	1.8682-1.8692	1.8682-1.8692	0.2565	0.2690	NA	0.004-0.012
	F	6.5 (6473)	2.1642-2.1663	2.1642-2.1663	2.1642-2.1663	2.1642-2.1663	2.0067-2.0089	0.2808	0.2808	1	0.002-0.012
	P	6.5 (6473)	2.1642-2.1663	2.1642-2.1663	2.1642-2.1663	2.1642-2.1663	2.0067-2.0089	0.2808	0.2808	1	0.002-0.012

79113Ca7

CAMSHAFT SPECIFICATIONS

All measurements given in inches.

| Year | Engine ID/VIN | Engine Displacement Liters (cc) | Journal Diameter | | | | | Elevation | | Bearing Clearance | Camshaft End Play |
			1	2	3	4	5	In.	Ex.		
	S	6.5 (6473)	2.1642-2.1663	2.1642-2.1663	2.1642-2.1663	2.1642-2.1663	2.0067-2.0089	0.2808	0.2808	1	0.002-0.012
	Y	6.5 (6473)	2.1642-2.1663	2.1642-2.1663	2.1642-2.1663	2.1642-2.1663	2.0067-2.0089	0.2808	0.2808	1	0.002-0.012
	N	7.4 (7440)	1.9482-1.9492	1.9482-1.9492	1.9482-1.9492	1.9482-1.9492	1.9482-1.9492	0.2341-0.2345	0.2529-0.2531	NA	NA

1 Nos. 1-4: 0.00098-0.0046
No. 5: 0.00078-0.0044

79113Cb7

CRANKSHAFT AND CONNECTING ROD SPECIFICATIONS
All measurements are given in inches.

Year	Engine ID/VIN	Engine Displacement Liters (cc)	Crankshaft				Connecting Rod		
			Main Brg. Journal Dia.	Main Brg. Oil Clearance	Shaft End-play	Thrust on No.	Journal Diameter	Oil Clearance	Side Clearance
1991	A	2.5 (2474)	2.3000	0.0005-0.0022	0.0035-0.0085	5	2.000	0.0005-0.0026	0.0060-0.0020
	E	2.5 (2474)	2.3000	0.0005-0.0022	0.0035-0.0085	5	2.000	0.0005-0.0026	0.0060-0.0020
	R	2.8 (2835)	1	0.0016-0.0032	0.0024-0.0083	3	8	0.0014-0.0037	0.0063-0.0252
	D	3.1 (3130)	2.6473	0.0012-0.0027	0.0024-0.0083	3	1.9983-1.9994	0.0011-0.0032	0.014-0.027
	B	4.3 (4293)	2	3	0.0020-0.0060	3	2.2487-2.2497	0.0013-0.0035	0.0060-0.0140
	Z	4.3 (4293)	2	3	0.0020-0.0060	3	2.2487-2.2497	0.0013-0.0035	0.0060-0.0140
	H	5.0 (4999)	2	3	0.0020-0.0060	5	2.0988-2.0998	0.0013-0.0035	0.006-0.014
	K	5.7 (5735)	2	3	0.0020-0.0060	5	2.0988-2.0998	0.0013-0.0035	0.006-0.014
	C	6.2 (6210)	4	5	0.0020-0.0070	3	2.3980-2.3990	0.0017-0.0039	0.0070-0.0240
	J	6.2 (6210)	4	5	0.0020-0.0070	3	2.3980-2.3990	0.0017-0.0039	0.0070-0.0240
	N	7.4 (7440)	6	7	0.006-0.010	5	2.1990-2.2000	0.0009-0.0025	0.0130-0.0230
1992	A	2.5 (2474)	2.3000	0.0005-0.0022	0.0035-0.0085	5	2.000	0.0005-0.0026	0.0060-0.0020
	R	2.8 (2835)	1	0.0016-0.0032	0.0024-0.0083	3	8	0.0014-0.0037	0.0063-0.0252
	D	3.1 (3130)	2.6473-2.6483	0.0012-0.0027	0.0024-0.0083	3	1.9983-1.9994	0.0011-0.0032	0.014-0.027
	L	3.8 (3785)	2.4988-2.4998	0.0008-0.0022	0.003-0.011	3	2.2487-2.2499	0.0008-0.0022	0.003-0.015
	B	4.3 (4293)	2	3	0.0020-0.0060	3	2.2487-2.2497	0.0013-0.0035	0.0060-0.0140
	W	4.3 (4293)	9	3	0.0020-0.0060	3	2.2487-2.2497	0.0013-0.0035	0.0060-0.0140
	Z	4.3 (4293)	2	3	0.0020-0.0060	3	2.2487-2.2497	0.0013-0.0035	0.0060-0.0140
	H	5.0 (4999)	2	3	0.0020-0.0060	5	2.0988-2.0998	0.0013-0.0035	0.0060-0.0140
	K	5.7 (5735)	2	3	0.0020-0.0060	5	2.0988-2.0998	0.0013-0.0035	0.0060-0.0140
	C	6.2 (6210)	4	5	0.0020-0.0070	3	2.3980-2.3990	0.0017-0.0039	0.0070-0.0240
	J	6.2 (6210)	4	5	0.0020-0.0070	3	2.3980-2.3990	0.0017-0.0039	0.0070-0.0240
	F	6.5 (6473)	4	5	0.0020-0.0070	3	2.3980-2.3990	0.0017-0.0039	0.0070-0.0240

79113C08

CRANKSHAFT AND CONNECTING ROD SPECIFICATIONS

All measurements are given in inches.

Year	Engine ID/VIN	Engine Displacement Liters (cc)	Crankshaft				Connecting Rod		
			Main Brg. Journal Dia.	Main Brg. Oil Clearance	Shaft End-play	Thrust on No.	Journal Diameter	Oil Clearance	Side Clearance
	N	7.4 (7440)	6	7	0.006-0.010	5	2.1990-2.2000	0.0009-0.0025	0.0130-0.0230
1993	A	2.5 (2474)	2.3000	0.0005-0.0022	0.0035-0.0085	5	2.000	0.0005-0.0026	0.0060-0.0020
	R	2.8 (2835)	1	0.0016-0.0032	0.0024-0.0083	3	8	0.0014-0.0037	0.0063-0.0252
	D	3.1 (3130)	2.6473-2.6483	0.0012-0.0027	0.0024-0.0083	3	1.9983-1.9994	0.0011-0.0032	0.014-0.027
	L	3.8 (3785)	2.4988-2.4998	0.0008-0.0022	0.003-0.011	3	2.2487-2.2499	0.0008-0.0022	0.003-0.015
	W	4.3 (4293)	9	9	0.0020-0.0070	3	2.2487-2.2497	0.0013-0.0035	0.0060-0.0140
	Z	4.3 (4293)	2	2	0.0020-0.0060	3	2.2487-2.2497	0.0013-0.0035	0.0060-0.0140
	H	5.0 (4999)	2	2	0.0020-0.0060	3	2.2487-2.2497	0.0013-0.0035	0.0060-0.0140
	K	5.7 (5735)	2	2	0.0020-0.0060	3	2.2487-2.2497	0.0013-0.0035	0.0060-0.0140
	C	6.2 (6210)	4	4	0.0020-0.0070	3	2.3980-2.3990	0.0017-0.0039	0.0070-0.0240
	J	6.2 (6210)	4	4	0.0020-0.0070	3	2.3980-2.3990	0.0017-0.0039	0.0070-0.0240
	F	6.5 (6473)	4	4	0.0020-0.0070	3	2.3980-2.3990	0.0017-0.0039	0.0070-0.0240
	N	7.4 (7440)	6	6	0.006-0.010	5	2.1990-2.2000	0.0009-0.0025	0.0130-0.0230
1994-95	4	2.2 (2189)	2.4945-2.4954	0.0006-0.0019	0.0020-0.0070	5	1.9983-1.9994	0.00098-0.0031	0.0039-0.0149
	D	3.1 (3130)	2.6473-2.6483	0.0012-0.0027	0.0024-0.0083	3	1.9983-1.9994	0.0011-0.0032	0.014-0.027
	L	3.8 (3785)	2.4988-2.4998	0.0008-0.0022	0.003-0.011	3	2.2487-2.2499	0.0008-0.0022	0.003-0.015
	W	4.3 (4293)	8	3	0.0020-0.0070	3	2.2487-2.2497	0.0013-0.0035	0.0060-0.0140
	Z	4.3 (4293)	2	3	0.0020-0.0060	3	2.2487-2.2497	0.0013-0.0035	0.0060-0.0140
	H	5.0 (4999)	2	3	0.0020-0.0060	5	2.0988-2.0998	0.0013-0.0035	0.006-0.014
	K	5.7 (5735)	2	3	0.0020-0.0060	5	2.0988-2.0998	0.0013-0.0035	0.006-0.014
	F	6.5 (6473)	4	5	0.0040-0.0098	3	2.3980-2.3990	0.0017-0.0039	0.0070-0.0240
	P	6.5 (6473)	4	5	0.0040-0.0098	3	2.3980-2.3990	0.0017-0.0039	0.0070-0.0240
	S	6.5 (6473)	4	5	0.0040-0.0098	3	2.3980-2.3990	0.0017-0.0039	0.0070-0.0240

79113Ca8

CRANKSHAFT AND CONNECTING ROD SPECIFICATIONS

All measurements are given in inches.

Year	Engine ID/VIN	Engine Displacement Liters (cc)	Crankshaft				Connecting Rod		
			Main Brg. Journal Dia.	Main Brg. Oil Clearance	Shaft End-play	Thrust on No.	Journal Diameter	Oil Clearance	Side Clearance
	Y	6.5 (6473)	4	5	0.0040-0.0098	3	2.3980-2.3990	0.0017-0.0039	0.0070-0.0240
	N	7.4 (7440)	6	7	0.006-0.010	5	2.1990-2.2000	0.0009-0.0025	0.0130-0.0230

1 Three dots: 2.64728-2.64759
  Two dots: 2.64759-2.64790
  One dot: 2.64790-2.64822
2 No. 1: 2.4484-2.4493
  Nos. 2-3: 2.4481-2.4493
  No. 4: 2.4479-2.4488
3 No. 1: 0.0008-0.0020
  Nos. 2-3: 0.0011-0.0023
  No. 4: 0.0017-0.0032

4 Nos. 1-4: 2.9495-2.9504
  No. 5: 2.9493-2.9502
5 Nos. 1-4: 0.0083
  No. 5: 0.0055-0.0093
6 Nos. 1-4: 2.7481-2.7490
  No. 5: 2.7476-2.7486
7 Nos. 1-4: 0.0013-0.0025
  No. 5: 0.0024-0.0040

8 Two dots: 1.9983-1.9989
  One dot: 1.9989-1.9994
9 No. 1: 2.4488-2.4495
  Nos. 2-3: 2.4485-2.4494
  No. 4: 2.4480-2.4489

79113Cb8

VALVE SPECIFICATIONS

Year	Engine ID/VIN	Engine Displacement Liters (cc)	Seat Angle (deg.)	Face Angle (deg.)	Spring Test Pressure (lbs. @ in.)	Spring Installed Height (in.)	Stem-to-Guide Clearance (in.)		Stem Diameter (in.)	
							Intake	Exhaust	Intake	Exhaust
1991	A	2.5 (2474)	46	45	1.58-1.70@1.04	1.44	0.0010-0.0025	0.0013-0.0030	0.3133-0.3138	0.3128-0.3135
	E	2.5 (2474)	46	45	71-78@1.44	1.44	0.0010-0.0025	0.0013-0.0030	0.3133-0.3138	0.3128-0.3135
	R	2.8 (2835)	46	45	88@1.57	1.57	0.0010-0.0027	0.0010-0.0027	0.3410-0.3417	0.3410-0.3417
	D	3.1 (3130)	46	45	82@1.58	1.57	0.0010-0.0027	0.0010-0.0027	NA	NA
	B	4.3 (4293)	46	45	194-206@1.25	1.72	0.0010-0.0027	0.0010-0.0027	NA	NA
	Z	4.3 (4293)	46	45	194-206@1.25	1.72	0.0010-0.0027	0.0010-0.0027	0.3410-0.3417	0.3410-0.3417
	H	5.0 (4999)	46	45	76-84@1.70	1.72	0.0010-0.0027	0.0010-0.0027	0.3410-0.3417	0.3410-0.3417
	K	5.7 (5735)	46	45	76-84@1.70	1.72	0.0010-0.0027	0.0010-0.0027	0.3410-0.3417	0.3410-0.3417
	C	6.2 (6210)	46	45	80@1.81	1.81	0.0010-0.0027	0.0010-0.0027	0.3414	0.3414
	J	6.2 (6210)	46	45	80@1.81	1.81	0.0010-0.0027	0.0010-0.0027	0.3414	0.3414
	N	7.4 (7440)	46	45	74-86@1.80	1.80	0.0010-0.0027	0.0012-0.0029	0.3410-0.3417	0.3410-0.3417
	W	7.4 (7440)	46	45	74-86@1.80	1.80	0.0010-0.0027	0.0012-0.0029	0.3410-0.3417	0.3410-0.3417
1992	A	2.5 (2474)	46	45	1.58-1.70@1.04	1.44	0.0010-0.0025	0.0013-0.0030	0.3133-0.3138	0.3128-0.3135
	R	2.8 (2835)	46	45	88@1.57	1.57	0.0010-0.0027	0.0010-0.0027	0.3410-0.3417	0.3410-0.3417
	D	3.1 (3130)	46	45	82@1.58	1.57	0.0010-0.0027	0.0010-0.0027	NA	NA
	L	3.8 (3785)	46	45	210@1.315	1.69-1.71	0.0015-0.0035	0.0015-0.0032	NA	1
	B	4.3 (4293)	46	45	194-206@1.25	1.72	0.0010-0.0027	0.0010-0.0027	NA	NA
	W	4.3 (4293)	46	45	194-206@1.25	1.69-1.71	0.0011-0.0027	0.0011-0.0027	NA	NA
	Z	4.3 (4293)	46	45	194-206@1.25	1.72	0.0010-0.0027	0.0010-0.0027	NA	NA
	H	5.0 (4999)	46	45	76-84@1.70	1.72	0.0010-0.0027	0.0010-0.0027	NA	NA
	K	5.7 (5735)	46	45	76-84@1.70	1.72	0.0010-0.0027	0.0010-0.0027	NA	NA
	C	6.2 (6210)	46	45	230@1.39	1.81	0.0010-0.0027	0.0010-0.0027	NA	NA
	J	6.2 (6210)	46	45	230@1.39	1.81	0.0010-0.0027	0.0010-0.0027	NA	NA

79113C09

VALVE SPECIFICATIONS

Year	Engine ID/VIN	Engine Displacement Liters (cc)	Seat Angle (deg.)	Face Angle (deg.)	Spring Test Pressure (lbs. @ in.)	Spring Installed Height (in.)	Stem-to-Guide Clearance (in.) Intake	Stem-to-Guide Clearance (in.) Exhaust	Stem Diameter (in.) Intake	Stem Diameter (in.) Exhaust
	F	6.5 (6473)	46	45	230@1.39	1.81	0.0010-0.0027	0.0010-0.0027	NA	NA
	N	7.4 (7440)	46	45	74-86@1.80	1.80	0.0010-0.0027	0.0012-0.0029	NA	NA
1993	A	2.5 (2474)	46	45	1.58-1.70@1.04	1.44	0.0010-0.0025	0.0013-0.0030	0.3133-0.3138	0.3128-0.3135
	R	2.8 (2835)	46	45	88@1.57	1.57	0.0010-0.0027	0.0010-0.0027	0.3410-0.3417	0.3410-0.3417
	D	3.1 (3130)	46	45	82@1.58	1.57	0.0010-0.0027	0.0010-0.0027	NA	NA
	L	3.8 (3785)	46	45	210@1.315	1.69-1.72	0.0015-0.0035	0.0015-0.0032	NA	[1]
	W	4.3 (4293)	46	45	194-206@1.25	1.69-1.71	0.0011-0.0027	0.0011-0.0027	NA	NA
	Z	4.3 (4293)	46	45	194-206@1.25	1.72	0.0010-0.0027	0.0010-0.0027	NA	NA
	H	5.0 (4999)	46	45	76-84@1.70	1.72	0.0010-0.0027	0.0010-0.0027	NA	NA
	K	5.7 (5735)	46	45	76-84@1.70	1.72	0.0010-0.0027	0.0010-0.0027	NA	NA
	C	6.2 (6210)	46	45	230@1.39	1.81	0.0010-0.0027	0.0010-0.0027	NA	NA
	J	6.2 (6210)	46	45	230@1.39	1.81	0.0010-0.0027	0.0010-0.0027	NA	NA
	F	6.5 (6473)	46	45	230@1.39	1.81	0.0010-0.0027	0.0010-0.0027	NA	NA
	N	7.4 (7440)	46	45	74-86@1.80	1.80	0.0010-0.0027	0.0012-0.0029	NA	NA
1994-95	4	2.2 (2189)	46	45	228@1.278	1.71	0.0010-0.0020	0.0010-0.0030	NA	NA
	D	3.1 (3130)	46	45	82@1.58	1.57	0.0010-0.0027	0.0010-0.0027	NA	NA
	L	3.8 (3785)	46	45	210@1.315	1.69-1.72	0.0015-0.0035	0.0015-0.0032	NA	[1]
	W	4.3 (4293)	46	45	194-206@1.25	1.69-1.71	0.0011-0.0027	0.0011-0.0027	NA	NA
	Z	4.3 (4293)	46	45	194-206@1.25	1.72	0.0010-0.0027	0.0010-0.0027	NA	NA
	H	5.0 (4999)	46	45	76-84@1.70	1.72	0.0010-0.0027	0.0010-0.0027	NA	NA
	K	5.7 (5735)	46	45	76-84@1.70	1.72	0.0010-0.0027	0.0010-0.0027	NA	NA
	F	6.5 (6473)	46	45	230@1.39	1.81	0.0010-0.0027	0.0010-0.0027	NA	NA
	P	6.5 (6473)	46	45	230@1.39	1.81	0.0010-0.0027	0.0010-0.0027	NA	NA

79113Ca9

VALVE SPECIFICATIONS

Year	Engine ID/VIN	Engine Displacement Liters (cc)	Seat Angle (deg.)	Face Angle (deg.)	Spring Test Pressure (lbs. @ in.)	Spring Installed Height (in.)	Stem-to-Guide Clearance (in.)		Stem Diameter (in.)	
							Intake	Exhaust	Intake	Exhaust
	S	6.5 (6473)	46	45	230@1.39	1.81	0.0010-0.0027	0.0010-0.0027	NA	NA
	Y	6.5 (6473)	46	45	230@1.39	1.81	0.0010-0.0027	0.0010-0.0027	NA	NA
	N	7.4 (7440)	46	45	74-86@1.80	1.80	0.0010-0.0027	0.0012-0.0029	NA	NA

1 Upper: 0.3129-0.3137; Lower: 0.3118-0.3126

79113Cb9

PISTON AND RING SPECIFICATIONS
All measurements are given in inches.

Year	Engine ID/VIN	Engine Displacement Liters (cc)	Piston Clearance	Ring Gap			Ring Side Clearance		
				Top Compression	Bottom Compression	Oil Control	Top Compression	Bottom Compression	Oil Control
1991	A	2.5 (2474)	0.0014-0.0022	0.010-0.020	0.010-0.020	0.020-0.060	0.0020-0.0030	0.0010-0.0030	0.015-0.055
	E	2.5 (2474)	0.0014-0.0022	0.010-0.020	0.010-0.020	0.020-0.060	0.0020-0.0030	0.0010-0.0030	0.015-0.055
	R	2.8 (2835)	0.0007-0.0017	0.010-0.020	0.010-0.020	0.020-0.0550	0.0010-0.0030	0.0015-0.0037	0.008 MAX
	D	3.1 (3146)	0.0009-0.0022	0.010-0.020	0.020-0.028	0.010-0.030	0.0020-0.0035	0.0020-0.0035	0.008 MAX
	B	4.3 (4293)	0.0007-0.0017	0.010-0.020	0.010-0.025	0.015-0.055	0.0012-0.0032	0.0012-0.0032	0.002-0.007
	Z	4.3 (4293)	0.0007-0.0017	0.010-0.020	0.010-0.025	0.015-0.055	0.0012-0.0032	0.0012-0.0032	0.002-0.007
	H	5.0 (4999)	0.0007-0.0017	0.010-0.020	0.010-0.025	0.015-0.055	0.0012-0.0032	0.0012-0.0032	0.002-0.007
	K	5.7 (5735)	0.0007-0.0017	0.010-0.020	0.010-0.025	0.015-0.055	0.0012-0.0032	0.0012-0.0032	0.002-0.007
	C	6.2 (6210)	[1]	0.012-0.022	0.030-0.039	0.0098-0.020	0.0030-0.0070	0.030-0.040	0.002-0.004
	J	6.2 (6210)	[1]	0.012-0.022	0.030-0.039	0.0098-0.020	0.0030-0.0070	0.030-0.040	0.002-0.004
	N	7.4 (7440)	0.003-0.004	0.010-0.018	0.016-0.024	0.010-0.030	0.0012-0.0029	0.0012-0.0029	0.0050-0.0065
1992	A	2.5 (2474)	0.0015-0.0035	0.010-0.015	0.010-0.020	0.015-0.055	0.0015-0.0030	0.0015-0.0032	0.005-0.007
	R	2.8 (2835)	0.0007-0.0017	0.010-0.020	0.010-0.020	0.010-0.050	0.0010-0.0030	0.0015-0.0037	0.008 MAX
	D	3.1 (3146)	0.0009-0.0022	0.010-0.020	0.020-0.028	0.010-0.030	0.0020-0.0035	0.0020-0.0035	0.008 MAX
	L	3.8 (3785)	0.0004-0.0022 [2]	0.010-0.025	0.010-0.025	0.015-0.055	0.0013-0.0031	0.0013-0.0031	0.0011-0.0081
	B	4.3 (4293)	0.0007-0.0017	0.010-0.020	0.010-0.025	0.015-0.055	0.0012-0.0032	0.0012-0.0032	0.002-0.007
	W	4.3 (4293)	0.0007-0.0017	0.010-0.020	0.018-0.026	0.015-0.055	0.0014-0.0032	0.0014-0.0032	0.0014-0.0032
	Z	4.3 (4293)	0.0007-0.0017	0.010-0.020	0.010-0.025	0.015-0.055	0.0012-0.0032	0.0012-0.0032	0.002-0.007
	H	5.0 (4999)	0.0007-0.0017	0.010-0.020	0.018-0.026	0.010-0.030	0.0012-0.0032	0.0012-0.0032	0.002-0.007
	K	5.7 (5735)	0.0007-0.0017	0.010-0.020	0.018-0.026	0.010-0.030	0.0012-0.0032	0.0012-0.0032	0.002-0.007
	C	6.2 (6210)	[1]	0.012-0.022	0.030-0.039	0.0098-0.020	0.0030-0.0070	0.030-0.0400	0.002-0.004
	J	6.2 (6210)	[1]	0.012-0.022	0.030-0.039	0.0098-0.020	0.0030-0.0070	0.030-0.0400	0.002-0.004
	F	6.5 (6473)	[1]	0.010-0.020	0.030-0.039	0.0098-0.023	0.0030-0.0070	0.030-0.0400	0.002-0.004

79113C10

PISTON AND RING SPECIFICATIONS
All measurements are given in inches.

Year	Engine ID/VIN	Engine Displacement Liters (cc)	Piston Clearance	Ring Gap			Ring Side Clearance		
				Top Compression	Bottom Compression	Oil Control	Top Compression	Bottom Compression	Oil Control
	N	7.4 (7440)	0.003-0.0042	0.010-0.018	0.016-0.024	0.010-0.030	0.0012-0.0029	0.0012-0.0029	0.0050-0.0065
1993	A	2.5 (2474)	0.0015-0.0035	0.010-0.015	0.010-0.020	0.015-0.055	0.0015-0.0030	0.0015-0.0032	0.0005-0.007
	R	2.8 (2835)	0.0007-0.0017	0.010-0.020	0.010-0.020	0.010-0.050	0.0010-0.0030	0.0015-0.0037	0.008 MAX
	D	3.1 (3146)	0.0009-0.0023	0.0071-0.0161	0.0197-0.0280	0.098-0.0295	0.0020-0.0035	0.0020-0.0035	0.008 MAX
	L	3.8 (3785)	0.0004-0.0022 [3]	0.010-0.025	0.010-0.025	0.015-0.055	0.0013-0.0031	0.0013-0.0031	0.0011-0.0081
	W	4.3 (4293)	0.0007-0.0017	0.010-0.020	0.018-0.026	0.015-0.055	0.0014-0.0032	0.0014-0.0032	0.0014-0.0032
	Z	4.3 (4293)	0.0007-0.0017	0.010-0.020	0.010-0.025	0.015-0.055	0.0012-0.0032	0.0012-0.0032	0.002-0.007
	H	5.0 (4999)	0.0007-0.0021	0.010-0.020	0.018-0.026	0.010-0.030	0.0012-0.0032	0.0012-0.0032	0.002-0.007
	K	5.7 (5735)	0.0007-0.0021	0.010-0.020	0.018-0.026	0.010-0.030	0.0012-0.0032	0.0012-0.0032	0.002-0.007
	C	6.2 (6210)	[1]	0.010-0.020	0.030-0.039	0.0098-0.020	[4]	0.0015-0.0030	0.0015-0.0035
	J	6.2 (6210)	[1]	0.010-0.020	0.030-0.039	0.0098-0.020	[4]	0.0015-0.0030	0.0015-0.0035
	F	6.5 (6473)	[5]	0.010-0.020	0.030-0.039	0.0098-0.020	[4]	0.0015-0.0030	0.0015-0.0035
	N	7.4 (7440)	0.003-0.0042	0.010-0.018	0.016-0.024	0.010-0.030	0.0012-0.0029	0.0012-0.0029	0.0050-0.0065
1994-95	4	2.2 (2189)	0.0007-0.0017	0.010-0.020	0.010-0.020	0.010-0.050	0.0019-0.0027	0.0019-0.0027	0.0019-0.0082
	D	3.1 (3146)	0.0009-0.0023	0.0071-0.0161	0.0197-0.0280	0.098-0.0295	0.0020-0.0035	0.0020-0.0035	0.008 MAX
	L	3.8 (3785)	0.0004-0.0022 [2]	0.010-0.025	0.010-0.025	0.015-0.055	0.0013-0.0031	0.0013-0.0031	0.0011-0.0081
	W	4.3 (4293)	0.0007-0.0017	0.010-0.020	0.018-0.026	0.015-0.055	0.0014-0.0032	0.0014-0.0032	0.0014-0.0032
	Z	4.3 (4293)	0.0007-0.0017	0.010-0.020	0.010-0.025	0.015-0.055	0.0012-0.0032	0.0012-0.0032	0.002-0.007
	H	5.0 (4999)	0.0007-0.0021	0.010-0.020	0.018-0.026	0.010-0.030	0.0012-0.0032	0.0012-0.0032	0.002-0.007
	K	5.7 (5735)	0.0007-0.0021	0.010-0.020	0.018-0.026	0.010-0.030	0.0012-0.0032	0.0012-0.0032	0.002-0.007
	F	6.5 (6473)	[6]	0.010-0.020	0.030-0.039	0.0098-0.020	[4]	0.0015-0.0030	0.0015-0.0035
	P	6.5 (6473)	[6]	0.010-0.020	0.030-0.039	0.0098-0.020	[4]	0.0015-0.0030	0.0015-0.0035
	S	6.5 (6473)	[6]	0.010-0.020	0.030-0.039	0.0098-0.020	[4]	0.0015-0.0030	0.0015-0.0035

79113C1a

PISTON AND RING SPECIFICATIONS

All measurements are given in inches.

Year	Engine ID/VIN	Engine Displacement Liters (cc)	Piston Clearance	Ring Gap			Ring Side Clearance		
				Top Compression	Bottom Compression	Oil Control	Top Compression	Bottom Compression	Oil Control
	Y	6.5 (6473)	6	0.010-0.020	0.030-0.039	0.0098-0.020	4	0.0015-0.0030	0.0015-0.0035
	N	7.4 (7440)	0.0018-0.0030	0.010-0.018	0.016-0.024	0.010-0.030	0.0012-0.0029	0.0012-0.0029	0.0050-0.0065

1 Bohn piston Nos. 1-6: 0.0035-0.0045
  Bohn piston Nos. 7-8: 0.0040-0.0050
  Zollner piston Nos. 1-6: 0.0044-0.0054
  Zollner piston Nos. 7-8: 0.0049-0.0059
2 Measured 44mm from top of piston
3 Measured 44mm from top of piston

4 Keystone type ring
5 Bore Nos. 1-6: 0.0037-0.0047
  Bore Nos. 7-8: 0.0042-0.0052
6 Bore Nos. 1-6: 0.0037-0.0047
  Bore Nos. 7-8: 0.0042-0.0052

79113C1b

TORQUE SPECIFICATIONS
All readings in ft. lbs.

Year	Engine ID/VIN	Engine Displacement Liters (cc)	Cylinder Head Bolts	Main Bearing Bolts	Rod Bearing Bolts	Crankshaft Damper Bolts	Flywheel Bolts	Manifold Intake	Manifold Exhaust	Spark Plugs	Lug Nut
1991	A	2.5 (2474)	1	70	32	160	2	25	3	7-16	90
	E	2.5 (2474)	1	70	32	160	2	25	3	7-16	90
	R	2.8 (2835)	1	70	39	70	52	23	25	22	90
	D	3.1 (3130)	1	72	39	75	45	15	24	18	100
	B	4.3 (4293)	70	80	16	70	75	7	5	22	90
	Z	4.3 (4293)	70	80	16	70	75	7	5	22	90
	H	5.0 (4999)	68	11	45	70	74	35	17	22	18
	K	5.7 (5735)	68	11	45	70	74	35	17	22	18
	C	6.2 (6210)	9	10	48	200	65	31	26	-	18
	J	6.2 (6210)	9	10	48	200	65	31	26	-	18
	N	7.4 (7440)	81	100	48	85	65	30	40	22	18
1992	A	2.5 (2474)	1	70	32	160	2	25	3	7-15	90
	R	2.8 (2835)	4	70	39	70	52	15	25	22	90
	D	3.1 (3130)	4	72	39	75	45	15	24	18	100
	L	3.8 (3785)	20	21	22	8	13	7	38	12	100
	B	4.3 (4293)	65	80	16	70	75	19	6	22	90
	W	4.3 (4293)	65	75	16	70	75	35	6	22	90
	Z	4.3 (4293)	65	75	16	70	75	35	6	22	90
	H	5.0 (4999)	65	12	45	70	75	35	6	15	14
	K	5.7 (5735)	65	12	45	70	75	35	6	15	14
	C	6.2 (6210)	7	8	48	200	66	31	26	-	14
	J	6.2 (6210)	7	8	48	200	66	31	26	-	14
	F	6.5 (6473)	7	8	48	200	66	31	26	-	14
	N	7.4 (7440)	80	100	48	85	65	40	40	22	14
1993	A	2.5 (2474)	1	70	32	160	2	25	3	7-15	90
	R	2.8 (2835)	70	70	39	70	52	15	25	22	90
	D	3.1 (3130)	4	23	37	76	52	15	24	7	100
	L	3.8 (3785)	27	20	21	28	13	7	38	12	100
	W	4.3 (4293)	65	75	25	70	75	35	5	22	90
	Z	4.3 (4293)	65	75	25	70	75	35	5	22	90
	H	5.0 (4999)	65	12	45	70	75	35	5	15	14
	K	5.7 (5735)	65	12	45	70	75	35	5	15	14
	C	6.2 (6210)	26	10	48	200	66	31	26	-	14
	J	6.2 (6210)	26	10	48	200	66	31	26	-	14
	F	6.5 (6473)	26	10	48	200	66	31	26	-	14
	N	7.4 (7440)	80	100	48	85	65	40	40	22	14
1994-95	4	2.2 (2189)	29	70	38	77	55	24	10	3	100
	D	3.1 (3130)	31	23	37	76	52	15	25	3	100
	L	3.8 (3785)	27	20	21	30	13	7	38	11	100
	W	4.3 (4293)	65	75	25	70	75	35	5	22	90
	Z	4.3 (4293)	65	75	25	70	75	35	5	11	90
	H	5.0 (4999)	65	12	45	70	75	35	5	15	14
	K	5.7 (5735)	65	12	45	70	75	35	5	15	14
	F	6.5 (6473)	32	10	48	200	66	31	26	-	14
	P	6.5 (6473)	32	10	48	200	66	31	26	-	14
	S	6.5 (6473)	32	10	48	200	66	31	26	-	14

79113C11

TORQUE SPECIFICATIONS
All readings in ft. lbs.

Year	Engine ID/VIN	Engine Displacement Liters (cc)	Cylinder Head Bolts	Main Bearing Bolts	Rod Bearing Bolts	Crankshaft Damper Bolts	Flywheel Bolts	Manifold		Spark Plugs	Lug Nut
								Intake	Exhaust		
	Y	6.5 (6473)	32	10	48	200	66	31	26	-	14
	N	7.4 (7440)	80	100	48	85	65	35	40	22	14

1 Step 1: Tighten all head bolts to 18 ft. lbs.
  Step 2: Tighten all bolts to 26 ft. lbs. except No. 9
  Retorque No. 9 to 18 ft. lbs.
  Step 3: Tighten all an additional 90 degrees
2 Automatic trans.: 55 ft. lbs.
  Manual trans.: 65 ft. lbs.
3 Center bolts: 36 ft. lbs.; Outer bolts: 32 ft. lbs.
4 Coat threads with sealer
  Tighten all bolts to 40 ft. lbs.
  Tighten all an additional 90 degrees (1/4 turn)
5 Two center bolts: 26 ft. lbs.
  All others: 32 ft. lbs.
6 Two center bolts: 26 ft. lbs.
  All others: 20 ft. lbs.
7 Last bolt, left side & throttle bracket bolt: 41 ft. lbs.
  All others: 35 ft. lbs.
8 110 ft. lbs. plus 76 degrees
9 Coat threads with sealant
  Tighten all bolts to 20 ft. lbs.
  Retorque to 50 ft. lbs.
  Tighten all bolts an additional 90 degrees (1/4 turn)
10 Outer bolts: 100 ft. lbs.
  Inner bolts: 110 ft. lbs.
11 Outer bolts on caps 2-4: 68 ft. lbs.
  All others: 78 ft. lbs.
12 Outer bolts on caps 2-4: 70 ft. lbs.
  All others: 80ft. lbs.
13 11 ft. lbs. plus 50 degrees
14 All 5 & 6 stud single rear wheels: 110 ft. lbs.
  All 8 stud single rear wheels: 120 ft. lbs.
  All 8 stud dual rear wheels: 140 ft. lbs.
  All 10 stud dual wheels: 175 ft. lbs.
15 Tighten all bolts to 13 ft. lbs.
  Retorque to 19 ft. lbs.
16 20 ft. lbs plus 60 degrees
  With cast manifolds, Two center bolts: 26 ft. lbs.
  With cast manifolds, All others: 20 ft. lbs.
  With tubular manifolds: 26 ft. lbs/

18 All 5 & 6 stud single rear wheels: 110 ft. lbs.
  All 8 stud single rear wheels: 120 ft. lbs.
  All 8 stud dual rear wheels: 140 ft. lbs.
  All 10 stud dual wheels: 155 ft. lbs.
19 Last bolt, left side & throttle bracket bolt: 41 ft. lbs.
  All others: 35 ft. lbs.
20 Tighten bolts to 35 ft. lbs. then rotate 130 degrees
  Tighten four center bolts an additional 30 degrees
21 26 ft. lbs. plus 50 degrees
22 20 ft. lbs. plus 50 degrees
23 37 ft. lbs. plus 77 degrees
24 1st-time installation (new head): 22 ft. lbs.
  All other installations: 11 ft. lbs.
25 20 ft. lbs. plus 70 degrees
26 Coat threads with sealant
  Tighten all bolts to to 20 ft. lbs.
  Tighten all bolts an additional 90 degrees (1/4 turn)
27 Tighten all bolts to 35 ft. lbs.then rotate 130 degrees
  Tighten four center bolts an additional 30 degrees
28 110 ft. lbs. plus 76 degrees
29 Short bolts: 43 ft. lbs. plus 90 degrees
  Long bolts: 46 ft. lbs. plus 90 degrees
30 110 ft. lbs. plus 76 degrees
31 Coat threads with sealer
  Tighten all bolts to 33 ft. lbs.
  Tighten all an additional 90 degrees (1/4 turn)
32 Coat threads with sealant
  Tighten all bolts to 20 ft. lbs.
  Retorque to 50 ft. lbs.

79113Ca1

BRAKE SPECIFICATIONS
All measurements in inches unless noted

Year	Model	Master Cylinder Bore	Brake Disc			Brake Drum Diameter			Minimum Lining Thickness	
			Original Thickness	Minimum Thickness	Maximum Runout	Original Inside Diameter	Max. Wear Limit	Maximum Machine Diameter	Front	Rear
1991	Astro/Safari	NA	1	2	0.004	9.50	9.59	9.56	0.030	0.030
	Lumina/Silhouette/Trans Sport	0.944	1.043	0.972	0.004	8.863	8.909	8.877	0.030	0.030
	S10 Blazer/S15 Jimmy	NA	1.04	0.980	0.004	9.50	9.59	9.56	0.030	0.030
	Bravada	NA	1.04	0.980	0.004	9.50	9.59	9.56	0.030	0.030
	S10 Pick-up/S15 Sonoma	NA	1.04	0.980	0.004	9.50	9.59	9.56	0.030	0.030
	Syclone	NA	1.04	0.980	0.004	9.50	9.59	9.56	0.030	0.030
	Blazer/Jimmy	NA	1.50	1.48	0.004	5	6	7	0.030	0.030
	C1500	NA	1.25	1.23	0.004	5	6	7	0.030	0.030
	C2500	NA	1.50	1.48	0.004	5	6	7	0.030	0.030
	C3500	NA	1.50	1.48	0.004	5	6	7	0.030	0.030
	K1500	NA	1.50	1.48	0.004	5	6	7	0.030	0.030
	K2500	NA	15.0	1.48	0.004	5	6	7	0.030	0.030
	K3500	NA	1.50	1.48	0.004	5	6	7	0.030	0.030
	P30	NA	1.245	1.23	0.004	13.00	13.09	13.06	0.030	0.030
	R3500	NA	1.50	1.48	0.004	5	6	7	0.030	0.030
	Suburban	NA	1.50	1.48	0.004	5	6	7	0.030	0.030
	G10	NA	3	3	0.004	5	6	7	0.030	0.030
	G20	NA	3	3	0.004	5	6	7	0.030	0.030
	G30	NA	3	3	0.004	5	6	7	0.030	0.030
1992	Astro/Safari	NA	1	1	0.004	9.50	9.59	9.56	0.030	0.030
	Lumina/Silhouette/Trans Sport	0.944	1.043	0.972	0.004	8.863	8.909	8.877	0.030	0.030
	S10 Blazer/S15 Jimmy	NA	1.04	0.980	0.004	9.50	9.59	9.56	0.030	0.030
	Bravada	NA	1.04	0.980	0.004	9.50	9.59	9.56	0.030	0.030
	S10 Pick-up/Sonoma	NA	1.04	0.980	0.004	9.50	9.59	9.56	0.030	0.030
	Syclone	NA	1.04	0.980	0.004	9.50	9.59	9.56	0.030	0.030
	Typhoon	NA	1.04	0.980	0.004	9.50	9.59	9.56	0.030	0.030
	Blazer/Yukon	NA	1.50	1.48	0.004	5	6	7	0.030	0.030
	C1500	NA	1.25	1.23	0.004	5	6	7	0.030	0.030
	C2500	NA	1.50	1.48	0.004	5	6	7	0.030	0.030
	C3500	NA	1.50	1.48	0.004	5	6	7	0.030	0.030
	K1500	NA	1.50	1.48	0.004	5	6	7	0.030	0.030
	K2500	NA	1.50	1.48	0.004	5	6	7	0.030	0.030
	K3500	NA	1.50	1.48	0.004	5	6	7	0.030	0.030
	P30	NA	1.245	1.23	0.004	13.00	13.09	13.06	0.030	0.030
	Suburban	NA	1.50	1.48	0.004	5	6	7	0.030	0.030
	G10	NA	3	4	0.004	5	6	7	0.030	0.030
	G20	NA	3	4	0.004	5	6	7	0.030	0.030
	G30	NA	3	4	0.004	5	6	7	0.030	0.030
1993	Astro/Safari	NA	1	2	0.004	9.50	9.59	9.56	0.030	0.030
	Lumina/Silhouette/Trans Sport	0.944	1.043	0.972	0.004	8.863	8.909	8.877	0.030	0.030
	S10 Blazer/S15 Jimmy	NA	1.04	0.980	0.004	9.50	9.59	9.56	0.030	0.030
	Bravada	NA	1.04	0.980	0.004	9.50	9.59	9.56	0.030	0.030
	S10 Pick-up/Sonoma	NA	1.04	0.980	0.004	9.50	9.59	9.56	0.030	0.030
	Typhoon	NA	1.04	0.980	0.004	9.50	9.59	9.56	0.030	0.030
	Blazer/Yukon	NA	1.50	1.48	0.004	5	6	7	0.030	0.030

79113C12

BRAKE SPECIFICATIONS
All measurements in inches unless noted

Year	Model	Master Cylinder Bore	Brake Disc Original Thickness	Brake Disc Minimum Thickness	Brake Disc Maximum Runout	Brake Drum Diameter Original Inside Diameter	Brake Drum Diameter Max. Wear Limit	Brake Drum Diameter Maximum Machine Diameter	Minimum Lining Thickness Front	Minimum Lining Thickness Rear
	C1500	NA	1.25	1.23	0.004	5	6	7	0.030	0.030
	C2500	NA	1.50	1.48	0.004	5	6	7	0.030	0.030
	C3500	NA	1.50	1.48	0.004	5	6	7	0.030	0.030
	K1500	NA	1.50	1.48	0.004	5	6	7	0.030	0.030
	K2500	NA	1.50	1.48	0.004	5	6	7	0.030	0.030
	K3500	NA	1.50	1.48	0.004	5	6	7	0.030	0.030
	P30	NA	1.245	1.23	0.004	13.00	13.09	13.06	0.030	0.030
	Suburban	NA	1.50	1.48	0.004	5	6	7	0.030	0.030
	G10	NA	3	4	0.004	5	6	7	0.030	0.030
	G20	NA	3	4	0.004	5	6	7	0.030	0.030
	G30	NA	3	4	0.004	5	6	7	0.030	0.030
1994-95	Astro/Safari	NA	1	2	0.004	9.50	9.59	9.56	0.030	0.030
	Lumina/Silhouette/Trans Sport	0.944	1.043	0.972	0.004	8.863	8.909	8.877	0.030	0.030
	S10 Blazer/S15 Jimmy	NA	1.04	0.980	0.004	9.50	9.59	9.56	0.030	0.030
	Bravada	NA	1.04	0.980	0.004	9.50	9.59	9.56	0.030	0.030
	S10 Pick-up/Sonoma	NA	1.04	0.980	0.004	9.50	9.59	9.56	0.030	0.030
	Blazer/Yukon	NA	1.50	1.48	0.004	5	6	7	0.030	0.030
	C1500	NA	1.25	1.23	0.004	5	6	7	0.030	0.030
	C2500	NA	1.50	1.48	0.004	5	6	7	0.030	0.030
	C3500	NA	1.50	1.48	0.004	5	6	7	0.030	0.030
	K1500	NA	1.50	1.48	0.004	5	6	7	0.030	0.030
	K2500	NA	1.50	1.48	0.004	5	6	7	0.030	0.030
	K3500	NA	1.50	1.48	0.004	5	6	7	0.030	0.030
	P30	NA	1.245	1.23	0.004	13.00	13.09	13.06	0.030	0.030
	Suburban	NA	1.50	1.48	0.004	5	6	7	0.030	0.030
	G10	NA	3	4	0.004	5	6	7	0.030	0.030
	G20	NA	3	4	0.004	5	6	7	0.030	0.030
	G30	NA	3	4	0.004	5	6	7	0.030	0.030

1 Available with 1.04" and 1.25" rotors
2 1.04" rotors: 0.980
  1.25" rotors: 1.23
3 Available with 1.28" and 1.54" discs
4 1.28" disc: 1.23
  1.54" disc: 1.48
5 Available with 10", 11.15" and 13" drums

6 10" drum: 10.05
  11.15" drum: 11.240
  13" drum: 13.090
7 10" drum: 10.09
  11.15" drum: 11.210
  13" drum: 13.060

7911312a

WHEEL ALIGNMENT

Year	Model	Caster Range (deg.)	Caster Preferred Setting (deg.)	Camber Range (deg.)	Camber Preferred Setting (deg.)	Toe-in (in.)	Steering Axis Inclination (deg.)
1991	Astro/Safari [1]	1 11/16P-3 11/16P	2 11/16P	0-1 19/32P	13/16P	3/32P	NA
	Astro/Safari [2]	2 1/16P-4 1/16P	3 1/6P	0-1 3/4P	7/8P	3/32P	NA
	Lumina/Silhouette/Trans Sport	11/16P-2 11/16P	1 11/16P	1/2N-1/2P	0	0	NA
	S10 Blazer/S15 Jimmy	1P-3P	2P	0-1 5/8P	13/16P	1/8	NA
	Bravada	1P-3P	2P	0-1 5/8P	13/16P	1/8	NA
	S10 Pick-up/S15 Sonoma	1P-3P	2P	0-1 5/8P	13/16P	1/8	NA
	Blazer	-	3	3/4-2P	1 1/2P	0	NA
	C1500	2 3/4P-4 3/4P	3 3/4P	0-1P	1/2P	1/8P	NA
	C2500	2 3/4P-4 3/4P	3 3/4P	0-1P	1/2P	1/8P	NA
	C3500	2 3/4P-4 3/4P	3 3/4P	0-1P	1/2P	1/8P	NA
	K1500	3P-5P	4P	12/32P-1 13/32P	29/32P	1/16P	NA
	K2500	3P-5P	4P	12/32P-1 13/32P	29/32P	1/16P	NA
	K3500	3P-5P	4P	12/32P-1 13/32P	29/32P	1/16P	NA
	P30	-	7	1/2N-7/8P	3/16P	3/16P	NA
	R3500	-	4	1/2N-1P	3/4P	3/16P	NA
	Suburban	-	3	3/4P-2P	1 1/2P	0	NA
	G10	-	5	3/16N-1 3/16P	1/2P	3/16P	NA
	G20	-	5	3/16N-1 3/16P	1/2P	3/16P	NA
	G30	-	6	1/2N-7/8P	3/16P	3/16P	NA
1992	Astro/Safari [1]	1 11/16P-3 11/16P	2 11/16P	0-1 19/32P	13/16P	3/32P	NA
	Astro/Safari [2]	2 1/16P-4 1/16P	3 1/16P	0-1 3/4P	7/8P	3/32P	NA
	Lumina/Silhouette/Trans Sport	11/16P-2 11/16P	1 11/16P	1/2N-1/2P	0	0	NA
	S10 Blazer/S15 Jimmy	1P-3P	2P	0-1 5/8P	13/16P	1/8	NA
	Bravada	1P-3P	2P	0-1 5/8P	13/16P	1/8	NA
	S10 Pick-up/S15 Sonoma	1P-3P	2P	0-1 5/8P	13/16P	1/8	NA
	Syclone	1P-3P	2P	0-1 5/8P	13/16P	1/8	NA
	Typhoon	1P-3P	2P	0-1 5/8P	13/16P	1/8	NA
	Blazer/Yukon	-	3	3/4-2P	1 1/2P	1/8P	NA
	C1500	3 3/4P-5 3/4P	4 3/4P	0-1P	1/2P	1/8P	NA
	C2500	3 3/4P-5 3/4P	4 3/4P	0-1P	1/2P	1/8P	NA
	C3500	3 3/4P-5 3/4P	4 3/4P	0-1P	1/2P	1/8P	NA
	K1500	1P-5P	3P	0-1P	1/2P	1/4P	NA
	K2500	1P-5P	3P	0-1P	1/2P	1/4P	NA
	K3500	1P-5P	3P	0-1P	1/2P	1/4P	NA
	P30	-	6	1/2N-7/8P	3/16P	3/16P	NA
	Suburban	4	5	3/4P-2P	1 1/2P	0	NA
	G10	-	7	1/4N-1 1/4P	1/2P	3/16P	NA
	G20	-	7	1/4N-1 1/4P	1/2P	3/16P	NA
	G30	-	7	1/2N-1P	1/4P	3/16P	NA
1993	Astro/Safari [1]	1 11/16P-3 11/16P	2 11/16P	0-1 19/32P	13/16P	3/32P	NA
	Astro/Safari [2]	2 1/16P-4 1/16P	3 1/16P	0-1 3/4P	7/8P	3/32P	NA
	Lumina/Silhouette/Trans Sport	11/16P-2 11/16P	1 11/16P	1/2N-1/2P	0	0	NA
	S10 Blazer/S15 Jimmy	1P-3P	2P	0-1 5/8P	13/16P	1/8	NA
	Bravada	1P-3P	2P	0-1 5/8P	13/16P	1/8	NA
	S10 Pick-up/S15 Sonoma	1P-3P	2P	0-1 5/8P	13/16P	1/8	NA
	Typhoon	1P-3P	2P	0-1 5/8P	13/16P	1/8	NA

79113C13

WHEEL ALIGNMENT

Year	Model	Caster Range (deg.)	Caster Preferred Setting (deg.)	Camber Range (deg.)	Camber Preferred Setting (deg.)	Toe-in (in.)	Steering Axis Inclination (deg.)
	Blazer/Yukon	4	5	3/4-2P	1 1/2P	0	NA
	C1500	3 3/4P-5 3/4P	4 3/4P	0-1P	1/2P	1/8P	NA
	C2500	3 3/4P-5 3/4P	4 3/4P	0-1P	1/2P	1/8P	NA
	C3500	3 3/4P-5 3/4P	4 3/4P	0-1P	1/2P	1/8P	NA
	K1500	1P-5P	3P	0-1P	1/2P	1/4P	NA
	K2500	1P-5P	3P	0-1P	1/2P	1/4P	NA
	K3500	1P-5P	3P	0-1P	1/2P	1/4P	NA
	P30	-	6	1/2N-7/8P	3/16P	3/16P	NA
	Suburban	4	5	3/4P-2P	1 1/2P	0	NA
	G10	-	3	1/4N-1 1/4P	1/2P	3/16P	NA
	G20	-	3	1/4N-1 1/4P	1/2P	3/16P	NA
	G30	-	3	1/2N-1P	1/4P	3/16P	NA
1994-95	Astro/Safari [1]	1 11/16P-3 11/16P	2 11/16P	0-1 19/32P	13/16P	3/32P	NA
	Astro/Safari [2]	2 1/16P-4 1/16P	3 1/16P	0-1 3/4P	7/8P	3/32P	NA
	Lumina/Silhouette/Trans Sport	11/16P-2 11/16P	1 11/16P	1/2N-1/2P	0	0	NA
	S10 Blazer/S15 Jimmy	1P-3P	2P	0-1 5/8P	13/16P	1/8	NA
	Bravada	1P-3P	2P	0-1 5/8P	13/16P	1/8	NA
	S10 Pick-up/S15 Sonoma	1P-3P	2P	0-1 5/8P	13/16P	1/8	NA
	Blazer/Yukon	4	5	3/4-2P	1 1/2P	0	NA
	C1500	3 3/4P-5 3/4P	4 3/4P	0-1P	1/2P	1/8P	NA
	C2500	3 3/4P-5 3/4P	4 3/4P	0-1P	1/2P	1/8P	NA
	C3500	3 3/4P-5 3/4P	4 3/4P	0-1P	1/2P	1/8P	NA
	K1500	1P-5P	3P	0-1P	1/2P	1/4P	NA
	K2500	1P-5P	3P	0-1P	1/2P	1/4P	NA
	K3500	1P-5P	3P	0-1P	1/2P	1/4P	NA
	P30	-	6	1/2N-7/8P	3/16P	3/16P	NA
	Suburban	4	5	3/4P-2P	1 1/2P	0	NA
	G10	-	3	1/4N-1 1/4P	1/2P	3/16P	NA
	G20	-	3	1/4N-1 1/4P	1/2P	3/16P	NA
	G30	-	3	1/2N-1P	1/2P	3/16P	NA

1 2WD
2 4WD
3 With ride height 2 1/2-4, caster should be 8
4 Ride height and preferred camber settings
   2 1/2: 1 1/2 degrees
   3: 15/16 degrees
   3 1/2: 5/16 degrees
   3 3/4: 1/8 degrees
   4: 0 degrees
5 Ride height and preferred camber settings
   1 1/2: 3 1/2 degrees
   2: 3 1/8 degrees
   2 1/2: 2 11/16 degrees
   3: 2 3/8 degrees
   3 1/2: 2 1/8 degrees
   3 3/4: 1 15/16 degrees
   4: 1 7/8 degrees

6 Ride height and preferred camber settings
   1 1/2: 2 7/8 degrees
   2: 2 3/16 degrees
   2 1/2: 1 5/8 degrees
   3: 1 degree
   3 1/2: 1/2 degrees
   3 3/4: 3/16 degrees
   4: 0 degrees
7 Ride height and preferred camber settings
   2 1/2: 2 5/16 degrees
   3: 1 11/16 degrees
   3 1/2: 1 3/16 degrees
   3 3/4: 15/16 degrees
   4: 5/8 degrees

79113139

ENGINE ELECTRICAL

NOTE: Disconnecting the negative battery cable on some vehicles may interfere with the functions of the on board computer systems and may require the computer to undergo a relearning process, once the negative battery cable is reconnected.

Distributor

REMOVAL

1. Disconnect the negative battery cable.
2. Remove all necessary components in order to gain access to the distributor assembly.
3. Disconnect the distributor electrical connectors. If equipped, disconnect the vacuum line. Mark and remove the spark plug wires.
4. Remove the distributor cap. Position the engine at TDC. Matchmark the rotor and the distributor body. Matchmark the distributor assembly and the engine block.
5. Remove the distributor retaining bolt. Carefully remove the distributor from the vehicle.

NOTE: As the distributor is removed from the engine, the rotor will turn counterclockwise. Observe and mark the start and finish rotation of the rotor. When reinstalling, position the rotor at the last mark and set the distributor into the engine. As the distributor drops into place, the rotor should turn to its original position, providing the engine crankshaft has not been rotated with the distributor out.

INSTALLATION

Timing Not Disturbed

NOTE: To ensure correct ignition timing if the enginehas not been disturbed, the distributor must be installedwith the rotor in the same position as when removed.

1. Align the rotor to the last mark made and install the distributor in the engine.
2. The rotor should turn and end up at the first mark made. Ensure the distributor and oil pump rod are fully engaged.

3. Reconnect the distributor cap and wires.
4. Tighten the distributor hold-down bolt and set the timing.

Timing Disturbed

1. Remove the No. 1 spark plug. Place a finger over the sparkplug hole and rotate the engine in the normal direction of rotation slowly, until compression is felt.
2. Align the timing mark on the crankshaft pulley to the **0** onthe engine timing indicator by rotating the engine in the samedirection slowly.
3. Position the rotor between No. 1 and No. 8 spark plugtowers on the V8 engine or the No. 1 and No. 6 spark plugtowers on the V6 engine.
4. Install the distributor, distributor cap, spark plug, wiring and connectors.
5. Check the engine timing and adjust, as required.

Ignition Timing

Connect the timing light per manufacturers instructions to the engine. Ensure the timing light is connected to the No. 1 spark plug wire. Connect the tachometer to the engine per manufacturers instructions. If equipped with a diesel engine, a special timing light and a digital tachometer must be used.

ADJUSTMENT

1. Locate the timing marks on the crankshaft pulley and the front of the timing case cover.
2. Clean off the timing marks.
3. Use chalk or white paint to color the mark on the scale that will indicate the correct timing, when aligned with the mark on the pulley or the pointer.
4. Attach a tachometer to the engine. Attach a timing light to the engine.
5. Disconnect and plug the vacuum lines to the distributor, if equipped. Loosen the distributor lock bolt slightly to permit the distributor to be turned.
6. Adjust the idle to the correct specification.
7. With the timing light aimed at the pulley and the marks on the engine, turn the distributor in the direction of rotor rotation to retard the spark or in the opposite direction of rotor rotation to advance the spark. Align the marks on the pulley and the engine with the flashes of the timing light.

Alternator

PRECAUTIONS

Several precautions must be observed with alternator equippedvehicles to avoid damage to the unit.

- If the battery is removed for any reason, make sure it is reconnected with the correct polarity. Reversing the battery connections may result in damage to the 1-way rectifiers.
- When utilizing a booster battery as a starting aid, alwaysconnect the positive to positive terminals and the negative terminal from the booster battery to a good engine ground on thevehicle being started.
- Never use a fast charger as a booster to start vehicles.
- Disconnect the battery cables when charging the batterywith a fast charger.
- Never attempt to polarize the alternator.
- Do not use test lights of more than 12 volts when checkingdiode continuity.
- Do not short across or ground any of the alternator terminals.
- The polarity of the battery, alternator and regulator mustbe matched and considered before making any electrical connections within the system.
- Never separate the alternator on an open circuit. Make sureall connections within the circuit are clean and tight.
- Disconnect the battery ground terminal when performingany service on electrical components.
- Disconnect the battery if arc welding is to be done on thevehicle.

REMOVAL AND INSTALLATION

1. Disconnect the negative battery cable. Disconnect the electrical connectors at the alternator.
2. Remove the necessary components in order to gain access to the alternator assembly.
3. Remove the alternator belt. Remove the alternator retaining bolts. Remove the alternator from the vehicle.
4. Installation is the reverse of the removal procedure. Adjust the alternator belt to specification.

BELT TENSION ADJUSTMENT

V-Belt

1. Place a belt tension gauge at the center of the greatest span of a

warm not hot drive belt and measure the tension.

2. If the belt is below the specification, loosen the component mounting bracket and adjust to specification.

3. Run the engine at idle for 15 minutes to allow the belt to reseat itself in the pulleys.

4. Allow the drive belt to cool and re-measure the tension. Adjust as necessary to meet the following specifications:

    a. On the 4.3L, 5.0L, 5.7L, and 7.4L engines — old belt-90 lbs. (400 N) or new belt-135 lbs. (600 N).

    b. On the 6.2L diesel engine — old belt-67 lbs. (300 N) or new belt-146 lbs. (650 N).

Serpentine Belts

Serpentine belts use an automatic tensioner which is spring activated and can be turned to the left or the right to apply or release the pulley tension.

Starter

REMOVAL AND INSTALLATION1.

1. Disconnect the negative battery cable. As required, raise and support the vehicle safely.

2. Remove the flywheel cover. Remove the exhaust crossover pipe, as required.

3. Disconnect the electrical wiring harness and battery leads at the solenoid terminals.

4. Remove the starter mounting bolts and retaining nuts. Remove the starter assembly from the vehicle.

5. Installation is the reverse of the removal procedure. Install any shims that were removed with the starter and torque the bolts to specification.

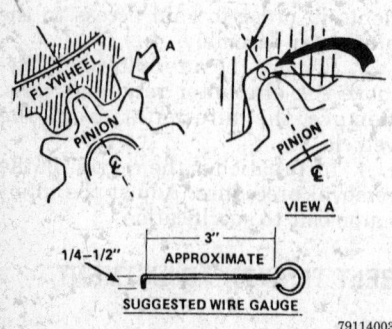

79114003

Measuring flywheel-to-starter pinion clearance during starter installation

Diesel Glow Plugs

REMOVAL AND INSTALLATION

1. Disconnect the negative battery cable.

2. Disconnect the electrical connection at the glow plug.

3. Using a suitable tool remove the glow plug.

4. Installation is the reverse of removal. Torque the glow plugs to 13 ft. lbs. (17 Nm).

TESTING

Inhibit Switch

1. Check the temperature controlled switch to make sure it is closed at low temperatures or open at temperatures above 125°F (52°C).

2. Remove the connector from the inhibit switch when the engine temperature is below 100°F (38°C).

3. Set the ohmmeter on a low range or use a self powered test light.

4. Test across the terminals. The switch should be closed (test light **ON** or a reading of less than 0.1 ohm on the meter).

5. Test terminals to ground with a test light or the ohmmeter on a high range. The light should be **OFF** or the meter show greater than 1.0 mega-ohm.

6. Replace the switch if it test open across the terminals or if either terminal is closed to the ground.

7. Disconnect the plug from the switch terminals when the engine is above 125°F.

8. Set the ohmmeter on the highest scale or use a self powered test light and test across the terminals. Test across each terminal to ground.

9. The switch should be open (test light **OFF** or high ohm reading of greater than 1 mega-ohm on the meter).

10. Replace the switch if it is closed. Use a socket wrench when installing the switch and torque to 17 ft. lbs. (21 Nm).

Controller

The glow plug controller provides glow plug operation after starting a cold engine.

1. With the engine cold 80°F (27°C), turn the engine control switch to the **RUN** position and let the glow plugs cycle.

2. After 2 minutes of letting the glow plugs cycle, crank the engine for 1 second; it is not important that the engine starts. Return the engine control switch to **RUN**. The glow plugs should cycle at least once after cranking.

3. If the plugs do not turn on, disconnect the controller connector and check terminal **B** with a grounded 12 volt test light. The light should be **OFF** with the engine control switch in **RUN**, and **ON** when the engine is cranked.

4. If the light does not operate as described, repair a short or open in the engine harness purple wire.

5. If the light works right but the afterstart glow plug feature does not, replace the controller.

CHASSIS ELECTRICAL

Heater Blower Motor

REMOVAL AND INSTALLATION

Without Air Conditioning

EXCEPT G-SERIES

1. Disconnect the negative battery terminal.

2. Mark the position of the blower motor in relation to its case.

3. Remove the electrical connection at the motor.

4. Remove the blower attaching screws and remove the assembly.

5. The blower wheel can be removed from the motor shaft by removing the retaining nut.

6. Installation is the reverse of the removal procedure. Apply a bead of sealer to the mounting flange before installation.

G-SERIES

1. Disconnect the negative battery cable.

2. Remove the coolant overflow bottle.

3. Unplug the motor wiring.

4. Remove the attaching screws and lift out the blower motor.

5. Installation is the reverse of the removal procedure.

With Air Conditioning

R/V-SERIES

1. Disconnect the negative battery terminal.

2. Remove the insulator attaching bolts and nuts. Remove the insulating shield from the case.

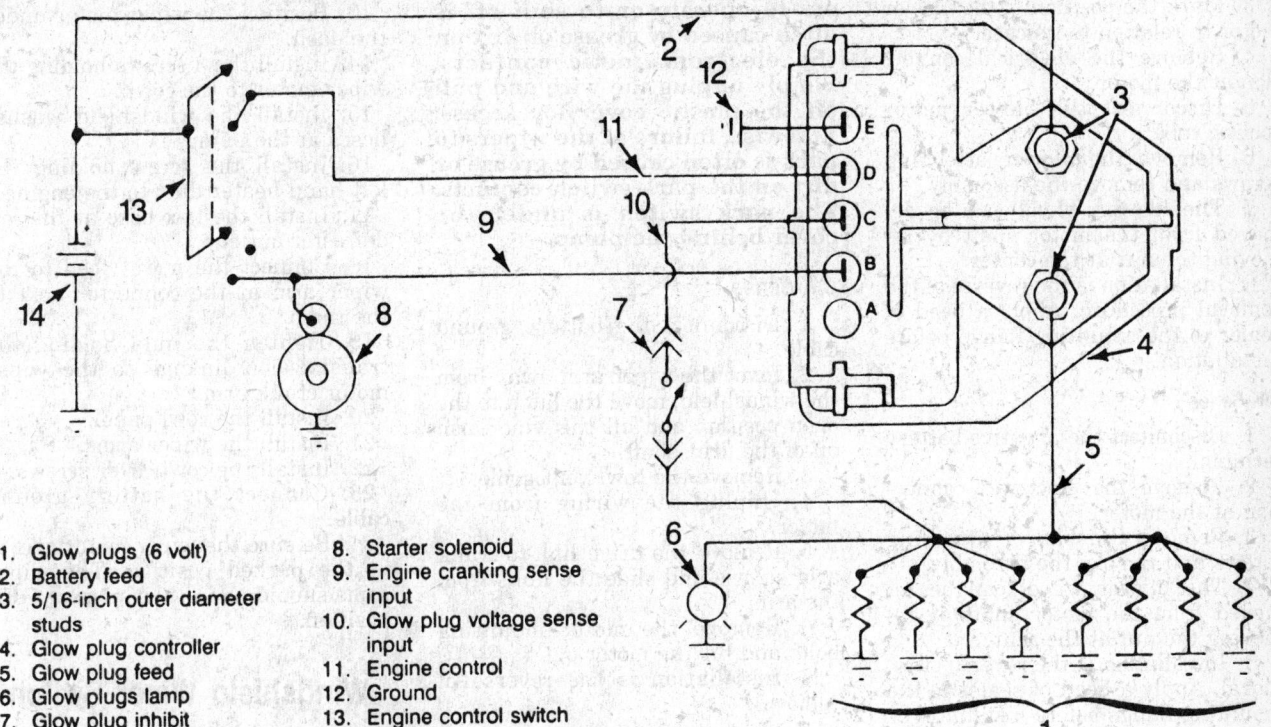

1. Glow plugs (6 volt)
2. Battery feed
3. 5/16-inch outer diameter studs
4. Glow plug controller
5. Glow plug feed
6. Glow plugs lamp
7. Glow plug inhibit temperature switch
8. Starter solenoid
9. Engine cranking sense input
10. Glow plug voltage sense input
11. Engine control
12. Ground
13. Engine control switch
14. Battery

79114004

Electronic glow plug system

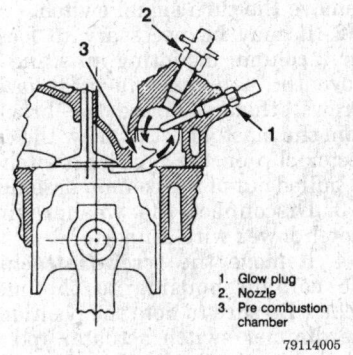

1. Glow plug
2. Nozzle
3. Pre combustion chamber

79114005

Diesel engine glow plug location

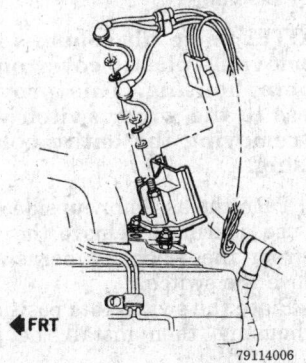

◄ FRT

79114006

Glow plug controller

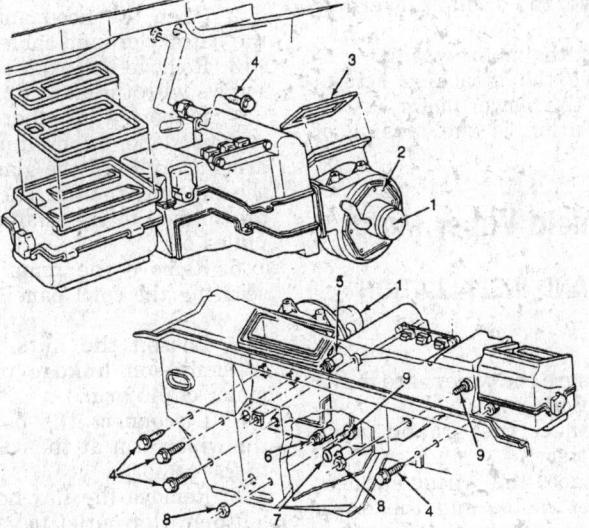

1. Blower motor
2. Screws
3. Gasket
4. Screw
5. Evaporator inlet line
6. Evaporator outlet line
7. Heater core tubes
8. Nut
9. Stud

79114007

Air conditioning blower motor installation — C/K series

3. Mark the position of the blower motor in relation to its case.

4. Remove the electrical connection at the motor.

5. Disconnect the blower motor cooling tube.

6. Remove the blower attaching screws and remove the assembly.

7. The blower wheel can be removed from the motor shaft by removing the nut at the center.

8. Installation is the reverse of the removal procedure. Apply a bead of sealer to the mounting flange before installation.

C/K-Series

1. Disconnect the negative battery terminal.

2. Remove the electrical connection at the motor.

3. Remove the blower attaching screws and remove the assembly.

4. The blower wheel can be removed from the motor shaft by removing the nut at the center.

5. Installation is the reverse of removal. Apply a bead of sealer to the mounting flange before installation.

G-Series

1. Disconnect the negative battery terminal.

2. Remove the coolant overflow bottle.

3. Unplug the motor wiring.

4. Remove the attaching screws and lift out the blower motor.

5. Installation is the reverse of removal.

Windshield Wiper Motor

REMOVAL AND INSTALLATION

R/V-Series

1. Make sure the wipers are in the parked position.

2. Disconnect the ground cable from the battery.

3. Disconnect the wiring harness at the wiper motor and the hoses from the washer pump.

4. Reach down through the access hole in the plenum and loosen the wiper drive rod attaching screws. Remove the drive rod from the wiper motor crank arm.

5. Remove the wiper motor attaching screws and the motor assembly and linkage.

6. To install, reverse the removal procedure.

NOTE: Lubricate the wiper motor crank arm pivot before re-installation. Failure of the wash-

ers to operate or to shut off is often caused by grease or dirt on the electromagnetic contacts. Simply unplug the wire and pull off the plastic cover for access. Likewise, failure of the wipers to park is often caused by grease or dirt on the park switch contacts. The park switch is under the cover behind the pump.

C/K-Series

1. Disconnect the battery ground cable.

2. Pivot the wiper arm away from the windshield, move the latch to the open position and lift the wiper arm off of the driveshaft.

3. Remove the cowl vent grille.

4. Unplug the wiring from the motor.

5. Remove the drive link-to-crank arm screws and slide the links from the arm.

6. Remove the motor mounting bolts and lift the motor out.

7. Installation is the reverse of removal.

G-Series

1. Make sure the wipers are parked. The wiper arms should be in their normal **OFF** position.

2. Open the hood and disconnect the battery ground cable.

3. Remove the exposed cowl cover screws with the hood up.

4. Remove the wiper arms. This can be done by pulling the wiper arms away from the glass to release the clip underneath. The wiper arms are splined to the shafts and can be pulled off.

5. Remove the remaining screws securing the cowl panel and remove it.

6. Loosen the nuts holding the transmission linkage to the wiper motor crank arm.

7. Disconnect the power feed to the wiper arm at the connector next to the radio.

8. Remove the flex hose from the left defroster outlet to gain access to the wiper motor screws.

9. Remove the screw holding the left hand heater duct to the engine shroud and move the heater duct down and out.

10. Remove the windshield washer hoses from the pump.

11. Remove the 3 screws holding the wiper motor to the cowl and lift the wiper motor out from under the dash.

To install:

12. Position the wiper motor in the park position.

13. Position the wiper motor under the dash.

14. Install the 3 screws holding the wiper motor to the cowl.

15. Install the windshield washer hoses at the pump.

16. Install the screw holding the left hand heater duct to the engine.

17. Install the flex hose at the left defroster outlet.

18. Connect the power feed to the wiper arm at the connector next to the radio.

19. Tighten the nuts holding the transmission linkage to the wiper motor crank arm.

20. Install the cowl panel.

21. Install the wiper arms.

22. Install the cowl cover screws.

23. Connect the battery ground cable.

24. Be sure the wiper motor arm is in the parked position. The wiper arms should be in their normal **OFF** position.

Windshield Wiper Switch

REMOVAL AND INSTALLATION

1. Disconnect the negative battery cable. Remove the steering wheel. Remove the turn signal switch.

2. It may be necessary to loosen the 2 column mounting nuts and remove the 4 bracket to mast jacket screws, then separate the bracket from the mast jacket to allow the connector clip on the ignition switch to be pulled out of the column assembly.

3. Disconnect the washer/wiper switch lower wire connector.

4. Remove the screws attaching the column housing to the mast jacket. Be sure to note the position of the dimmer switch actuator rod for reassembly in the same position. Remove the column housing and switch as an assembly.

NOTE: Some tilt columns have a removable plastic cover on the column housing. This provides access to the wiper switch without removing the entire column housing.

5. Turn the assembly upside down and use a drift to remove the pivot pin from the washer/wiper switch. Remove the switch.

6. Place the switch into position in the housing, then install the pivot pin.

7. Position the housing onto the mast jacket and attach by installing the screws. Install the dimmer switch actuator rod in the same position as

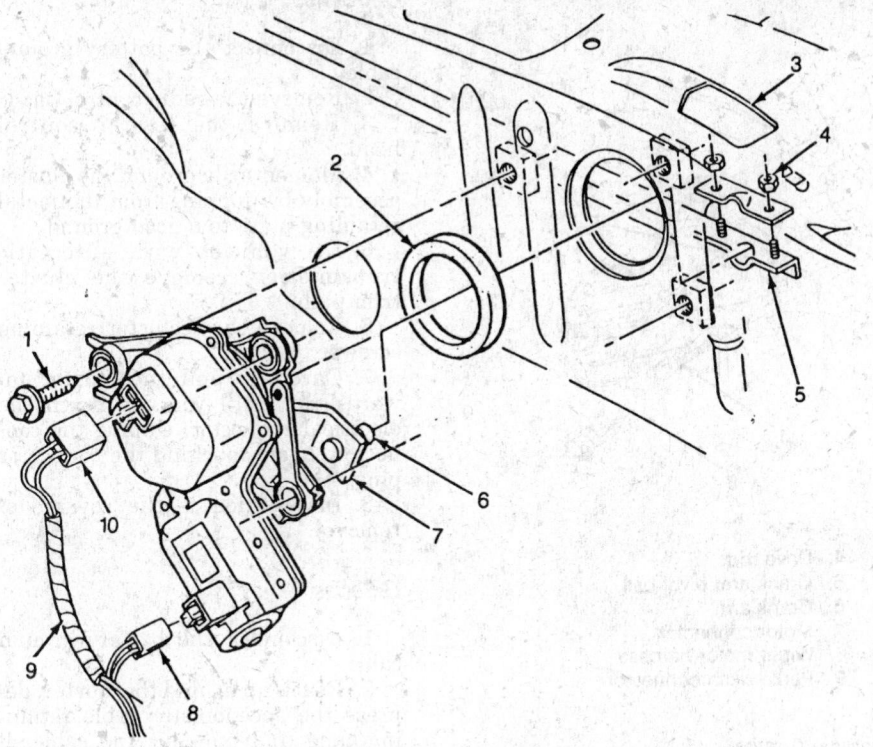

1. Screw
2. Seal
3. Access hole
4. Drive rod retaining cap nuts
5. Drive rod
6. Crank arm pivot ball
7. Crank arm
8. Motor connector
9. Wiper motor harness
10. Park switch connector

79114008

Wiper motor mounting — R/V series

1. Screw
2. Wiper motor
3. Nut
4. Bracket
5. Transmission
6. Bolt

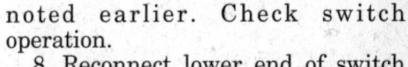

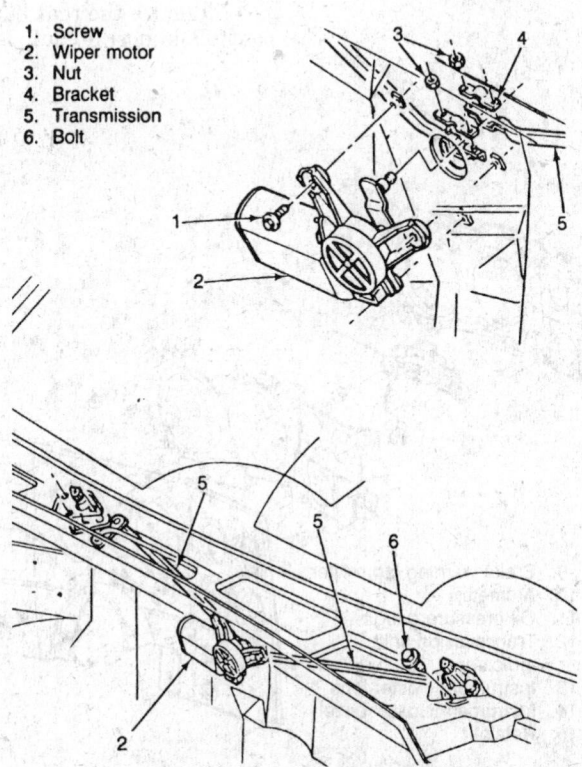

79114009

Wiper motor mounting — C/K series

noted earlier. Check switch operation.

8. Reconnect lower end of switch assembly.

9. Install remaining components in reverse order of removal. Be sure to attach column mounting bracket in original position.

Instrument Cluster

REMOVAL AND INSTALLATION

R/V-Series

1. Disconnect the battery ground cable.

2. Remove the headlight switch control knob.

3. Remove the radio control knobs, if required.

4. Remove the steering 4 column cover retaining screws.

5. Remove 8 screws and remove instrument bezel.

6. Reach under the dash, depress the speedometer cable tang, and remove the cable, if equipped.

7. Pull instrument cluster out just far enough to disconnect all lines and wires.

8. Remove the cluster.

9. Installation is the reverse of removal.

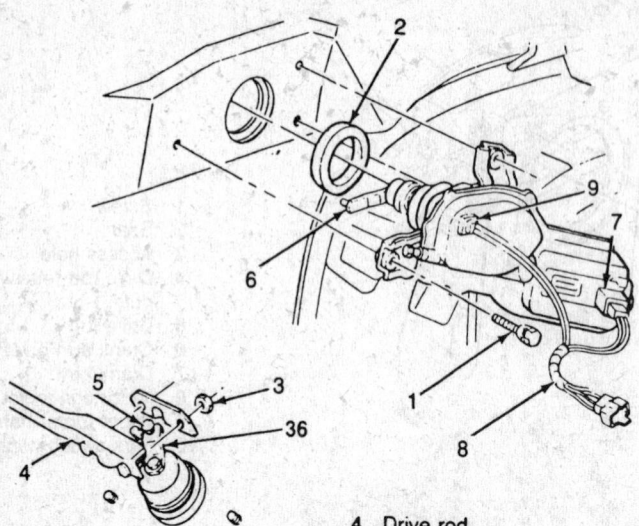

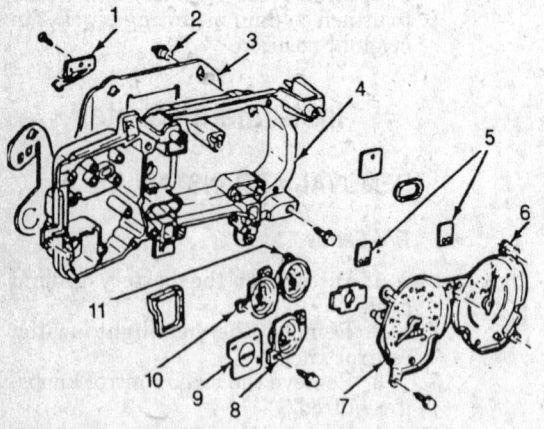

1. Screw
2. Seal
3. Drive rod retaining cup nuts
4. Drive rod
5. Crank arm pivot ball
6. Crank arm
7. Motor connector
8. Wiper motor harness
9. Park switch connector

79114010

Wiper motor mounting — G series

C/K-Series

1. Disconnect the battery ground cable.
2. Remove the radio control head.
3. Remove the heater control head.
4. Momentarily ground the cluster assembly by jumping from the metal retaining plate to a good ground.
5. On vehicles with automatic transmission, remove the cluster trim plate.
6. Remove the 4 cluster retaining screws.
7. Carefully pull the cluster towards you until you can reach the electrical connector. Unplug the connector. Avoid touching the connector pins.
8. Installation is the reverse of removal.

G-Series

1. Disconnect the battery ground cable.
2. Reach up behind the cluster, depress the speedometer cable retaining tang and pull out the cable, if equipped.
3. Remove the clock set stem knob, if equipped.
4. Remove the rear heater and fan controls knobs, if equipped.

1. Speedometer cable spring clip
2. Lamp bulb socket
3. Laminated circuit
4. Cluster case
5. Indicator lamp filter (turn signal)
6. Fuel gauge
7. Speedometer
8. Temperature gauge
9. Brake warning lamp filter
10. Ammeter
11. Oil pressure gauge
12. Transmission shift indicator
13. Instrument cluster lens
14. Instrument cluster bezel
15. Retainer

79114011

Exploded view of the cluster assembly — R/V series

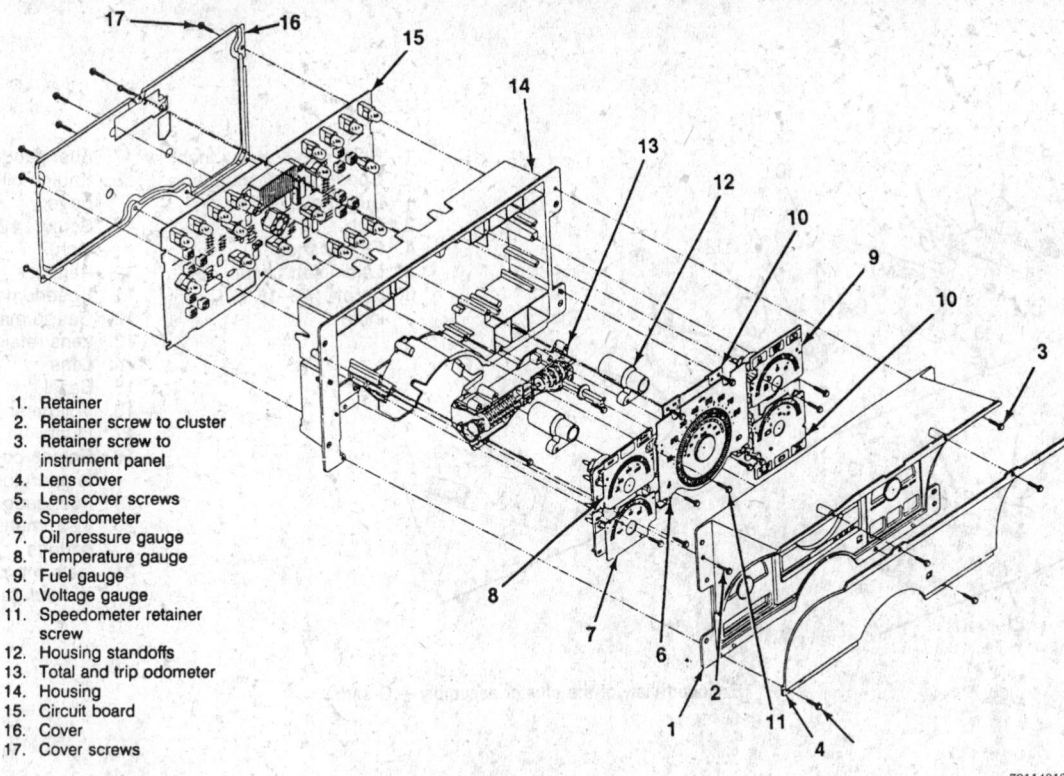

1. Retainer
2. Retainer screw to cluster
3. Retainer screw to instrument panel
4. Lens cover
5. Lens cover screws
6. Speedometer
7. Oil pressure gauge
8. Temperature gauge
9. Fuel gauge
10. Voltage gauge
11. Speedometer retainer screw
12. Housing standoffs
13. Total and trip odometer
14. Housing
15. Circuit board
16. Cover
17. Cover screws

79114012

Exploded view of the cluster assembly — C/K series

5. Remove the interior light and headlight switch bezel.

6. Remove the instrument panel bezel screws and remove the bezel.

7. Remove the cluster bezel retaining screws and remove the bezel.

8. Pull the top of the cluster away from the panel and lift out the bottom of the cluster. Pull the cluster out just far enough to unplug the wiring and remove the speedometer cable.

9. Installation is the reverse of removal.

Speedometer

REMOVAL AND INSTALLATION

R/V-Series

1. Disconnect the negative battery cable.

2. Remove the headlight switch knob assembly.

3. Remove the radio control knobs, if required.

4. Remove the clock adjuster stem.

5. Remove the instrument cluster bezel.

6. Remove the steering column cover.

7. Remove the instrument cluster lens.

8. Remove the transmission shift indicator.

9. Remove the cluster retainer.

10. Depress the spring clip and disconnect the speedometer cable.

11. Remove the cluster assembly and remove the speedometer assembly.

12. Installation is the reverse of removal.

C/K-Series

1. Disconnect the negative battery cable.

2. Remove the cluster.

3. Remove the speedometer mounting screws.

4. Carefully pull the speedometer from the circuit board. Avoid touching any of the circuit board pins.

5. Installation is the reverse of removal.

G-Series

1. Disconnect the negative battery cable.

2. Remove the cluster.

3. Remove the speedometer dial retaining screws.

4. Remove the 2 hex head screws and rubber grommets that hold the speedometer assembly to the cluster cover.

5. Disconnect the speedometer cable from the assembly and remove the speedometer.

6. Installation is the reverse of removal.

Radio

REMOVAL AND INSTALLATION

NOTE: Make certain the speaker is attached to the radio before the unit is turned ON. If not, the output transistors will be damaged.

R/V-Series

1. Disconnect the negative battery cable.

2. Remove the control knobs and the bezels from the radio control shafts.

3. Remove the nuts from the support shafts.

4. Remove the support bracket retaining screws.

5. Lifting the rear edge of the radio, push the radio forward until the control shafts clear the instrument panel then lower the radio far enough so the electrical connections can be disconnected.

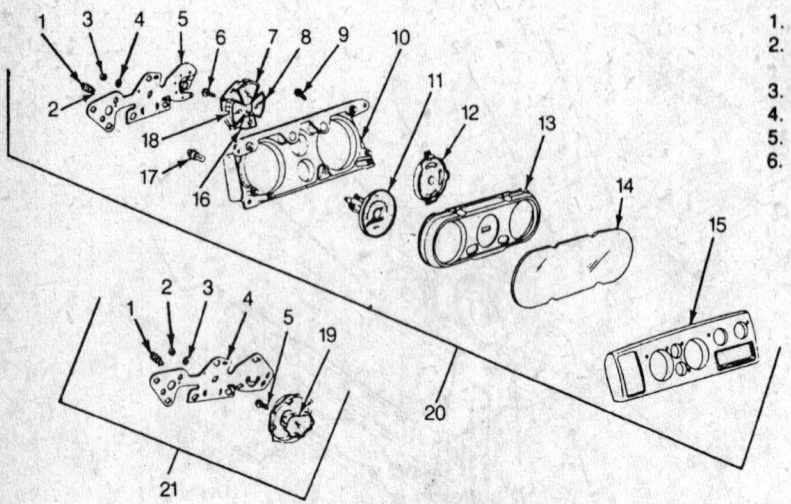

1. Bolt (#8–32 x $^7/_{16}$ inch)
2. Speedometer mounting bushing
3. Nut
4. Flat washer
5. Laminated circuit
6. Screw (#8–18 x $^7/_{16}$ inch)
7. Fuel gauge
8. Engine oil pressure gage
9. Screw (#8–16 x $^7/_{16}$ inch)
10. Retainer
11. Speedometer
12. Gauge mask
13. Lens retainer
14. Lens
15. Bezel
16. Voltmeter
17. Lamp
18. Engine coolant temperature gauge
19. Fuel gauge
20. Instrument cluster with gauges
21. Instrument cluster without gauges

79114013

Exploded view of the cluster assembly — G series

6. Remove the power lead, speaker, and antenna wires and then pull out the unit.

7. Installation is the reverse of removal.

C/K-Series

RECEIVER

1. Disconnect the negative battery cable.

2. Working behind the instrument panel, disconnect the wiring and antenna from the receiver.

3. Remove the receiver bracket screws.

4. Remove the receiver and bracket from the dash.

5. Remove the receiver-to-bracket nuts.

6. Installation is the reverse of removal. Torque the nuts to 14 ft. lbs.

TAPE PLAYER

1. Disconnect the negative battery cable.

2. Remove the control knobs and trim plate.

3. Remove the tape player-to-instrument panel screws.

4. Pull the unit towards you just far enough to disconnect the wiring. Then, remove the unit.

5. Installation is the reverse of removal.

RADIO/TAPE PLAYER CONTROL HEAD

1. Disconnect the negative battery cable.

2. Remove the instrument cluster trim and bezel.

3. Remove the mounting screws.

4. Pull the unit towards you just far enough to disconnect the wiring. Then, remove the unit.

5. Installation is the reverse of removal.

G-Series

1. Disconnect the negative battery cable.

2. Disconnect the ground cable from the battery.

3. Remove the engine cover.

4. Remove the air cleaner.

5. Cover the throttle body with a clean rag.

6. Remove the knobs, washers and nuts from the front of the radio.

7. Remove the rear bracket screw and bracket from the radio.

8. Remove the radio through the engine access area. Lower the radio far enough to detach the wiring.

9. Remove the radio.

10. Installation is the reverse of removal.

Headlight Switch

REMOVAL AND INSTALLATION

R/V-Series

1. Disconnect the negative battery cable.

2. Reaching up behind instrument cluster, depress shaft retaining button and remove switch knob and rod.

3. Remove instrument cluster bezel and trim panel screws.

4. Disconnect multiple wiring connectors at switch terminals.

5. Remove switch by rotating the switch locknut.

6. Installation is the reverse of removal.

C/K-Series

1. Disconnect the negative battery cable.

2. Remove the instrument cluster bezel.

3. Remove the headlight switch retaining screws.

4. Disconnect the wire connector and pull the switch away from the bezel.

5. Installation is the reverse of removal.

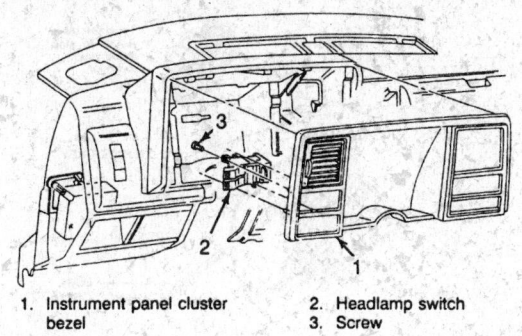

1. Instrument panel cluster bezel
2. Headlamp switch
3. Screw

79114014

Headlight switch installation — C/K series

G-Series

1. Disconnect the negative battery cable.
2. Press the switch knob retaining pin and remove the knob.
3. Remove the left instrument panel trim plate.
4. Remove the retaining nut securing the switch.
5. Disconnect the electrical connector from the back of the switch.
6. Remove the switch from the instrument panel.
7. Reverse the procedure for installation.

Dimmer Switch

REMOVAL AND INSTALLATION

1. Disconnect the negative battery cable.
2. Remove the steering wheel.
3. Disconnect the switch wire connector.
4. Remove the screws attaching the dimmer switch to the column housing. Be sure to note the position of the dimmer switch actuator rod for reassembly in the same position. Remove the switch by prying up and away from the housing.
5. Installation is the reverse of the removal procedure. Check and adjust the switch as necessary.

Turn Signal Switch

REMOVAL AND INSTALLATION

1. Disconnect the negative battery cable.
2. Remove the steering wheel.
3. Using a steering wheel lock plate compressor, remove the snapring from the groove on the shaft.
4. Remove the lockplate, turn signal cam, the upper bearing preload spring and the thrust washer from the upper steering shaft.
5. Remove the steering column lower cover.
6. Remove the turn signal lever from the column.
7. If equipped with cruise control, disconnect the cruise control wire from the harness near the bottom of the column. Remove the harness protector from the cruise control wire. Remove the turn signal lever. Do not remove the wire from the column.
8. Disconnect the turn signal wiring and remove the wires from the plastic protector.
9. Remove the turn signal switch mounting screws.
10. Slide the switch connector out of the bracket on the steering column.
11. If the switch is to be replaced, cut the wires near the top of the switch and discard the switch. Before cutting the wires, verify that the wire color codes are the same. Tape the connector of the new switch to the old wires, and pull the new harness down through the steering column while removing the old switch.
12. If the original switch is to be reused, tape a piece of wire or string around the connector and pull the harness up through the column, while pulling the string up through the column to help with reinstallation later.
13. After freeing the switch wiring protector from its mounting, pull the turn signal switch straight up and remove the switch, switch harness, and the connector from the column.
    To install:
14. Using the wire, string or switch wiring as previously described, pull the switch and wiring through the column and into position.
15. Position the switch connector onto the bracket on the column and reconnect the connector.
16. Install the turn signal switch mounting screws and turn signal lever.

17. Reconnect the cruise control wire connector, if equipped.
18. Install the steering column lower cover.
19. Install the lockplate, turn signal cam, the upper bearing preload spring and the thrust washer from the upper steering shaft.
20. Reinstall the lockplate and steering wheel.
21. Reconnect the negative battery cable.

Ignition Switch

REMOVAL AND INSTALLATION

1. Disconnect the negative battery cable. If equipped, remove the lower trim panel.
2. Remove the steering column retaining bolts. Lower the steering column assembly to gain access to the switch retaining screws. Extreme care is necessary to prevent damage to the collapsible column.
3. Make sure the switch is in the **LOCK** position. If the lock cylinder is out, pull the switch rod up to the stop, then go down 1 detent.
4. Remove the electrical connections from the switch. Remove the switch retaining screws. Remove the switch from the column.
5. Installation is the reverse of the removal procedure. Before installation, make sure the switch is in the **LOCK** position. Install the switch using the original screws. Use of screws that are too long could prevent the column from collapsing on impact.

Ignition Lock Cylinder

REMOVAL AND INSTALLATION

1. Disconnect the negative battery cable.
2. Remove the steering wheel.
3. Remove the turn signal switch. It is not necessary to completely remove the switch from the column. Pull the switch rearward far enough to slip it over the end of the shaft but do not pull the harness out of the column.
4. Turn the lock to **RUN**.
5. Remove the lock retaining screw and remove the lock cylinder.

NOTE: If the retaining screw is dropped on removal, it may fall into the column, requiring complete disassembly of the column to retrieve the screw.

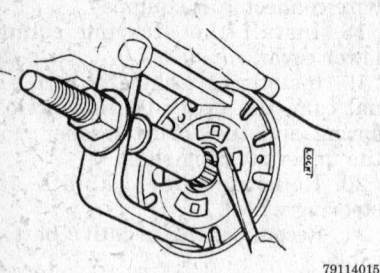

Removing the lock plate retaining ring

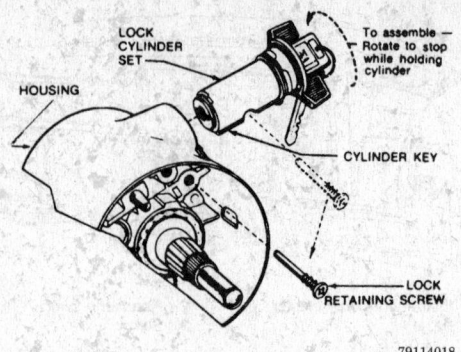

Ignition lock cylinder removal and installation

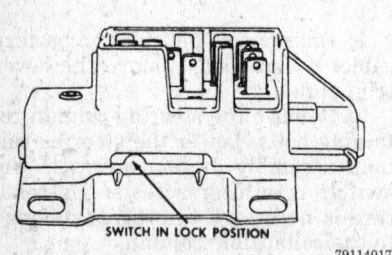

Ignition switch shown in the LOCK position for removal

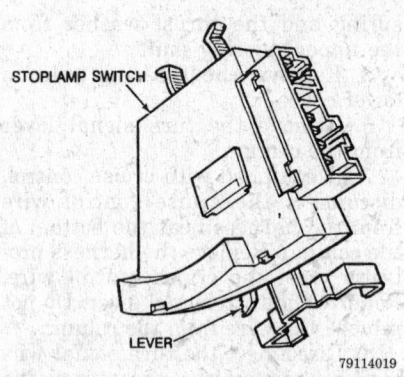

Stoplight switch adjustment lever — C/K series

To install:

6. Rotate the key to the stop while holding onto the cylinder.

7. Push the lock all the way in.

8. Install the screw. Tighten the screw to 40 inch lbs. for regular columns, 22 inch lbs. for adjustable columns.

9. Install the turn signal switch and the steering wheel.

Stoplight Switch

REMOVAL AND INSTALLATION

1. Disconnect the negative battery cable.

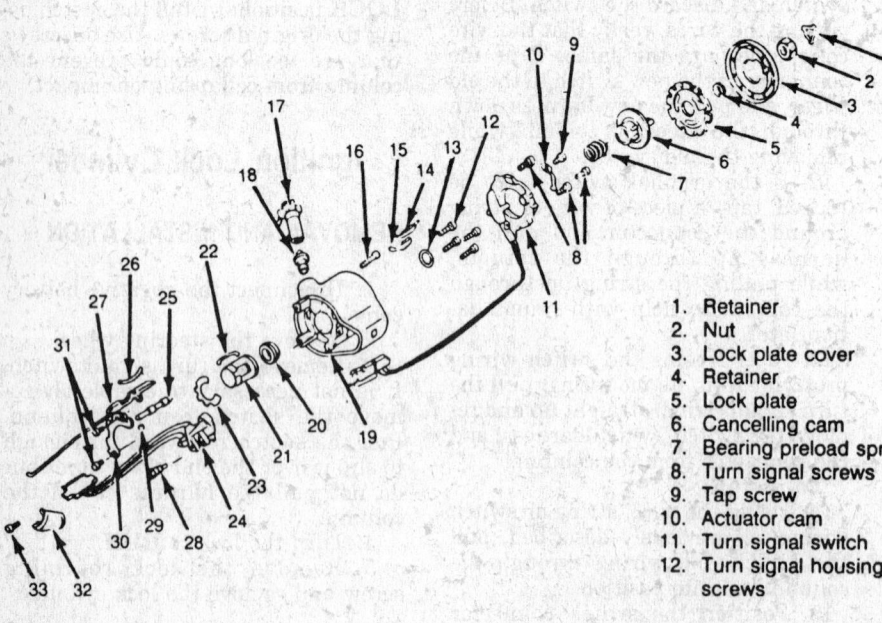

1. Retainer
2. Nut
3. Lock plate cover
4. Retainer
5. Lock plate
6. Cancelling cam
7. Bearing preload spring
8. Turn signal screws
9. Tap screw
10. Actuator cam
11. Turn signal switch
12. Turn signal housing screws
13. Washer
14. Tone alarm switch
15. Retainer clip
16. retainer screw
17. Ignition lock
18. Acuator sector
19. Housing assembly
20. Bearing
21. Bushing
22. Horn contact
23. Upper bearing retainer
24. Dimmer pivot and wiper switch
25. Shaft lock bolt
26. Switch rack preload spring
27. Actuator rack
28. Actuator pivot pin
29. Washer
30. Shift lever gate
31. Shift lever screw
32. Housing cover
33. Cover screw

Turn signal switch location

2. Disconnect the switch electrical connections. Remove the switch assembly from its mounting.

3. Installation is the reverse of the removal procedure. Adjust the switch, as required.

Adjustment

Except C/K-Series

1. Depress the brake pedal and press the switch in until it is firmly seated in its mounting.

2. Pull the brake pedal against the pedal stop until the switch does not make any noise.

3. Electrical contact should be made when the brake pedal is depressed from its fully released position.

C/K-Series

1. Depress the brake pedal fully.
2. Pull the lever on the brake switch back to its stop.
3. Pull the brake pedal back to its stop position.
4. Check for proper switch operation.

Fuses and Circuit Breakers

LOCATION

Fusible Links

In addition to circuit breakers and fuses, the wiring harness incorporates fusible links to protect the wiring. Links are used in place of a fuse in wiring circuits that are not normally fused, such as the ignition circuit. Fusible links are color coded red in the charging and load circuits to match the color of the circuits they protect. Each link is 4 gauges smaller than the cable it protects and the insulation is marked with the gauge size.The engine compartment wiring harness has several fusible links. The same size wire with a special hypalon insulation must be used when replacing a fusible link. The links are located in the following areas:

1. A molded splice at the starter solenoid BAT terminal, a 14 gauge red wire.

2. A 16 gauge red fusible link at the junction block to protect the unfused wiring of 12 gauge or larger wire. This link stops at the bulkhead connector.

3. The alternator warning light and field circuitry is protected by a 20 gauge fusible link. The link is in-

stalled as a molded splice in the circuit at the junction block.

4. The ammeter circuit is protected by two 20 gauge fusible links installed as molded splices in the circuit at the junction block and battery to starter circuit.

Circuit Breakers

A circuit breaker is an electrical switch which breaks the circuit in case of an overload. All models have a circuit breaker in the headlight switch to protect the headlight and parking light systems. An overload may cause the lights to flicker or flash ON and OFF, or in some cases, to remain OFF. Windshield wiper motors are protected by a circuit breaker at the motor.

Fuses

Fuses are located in the junction box below the instrument panel to the left of the steering column. Each fuse receptacle is marked as to the circuit it protects and the correct amperage of the fuse. Inline fuses are used on the underhood light and air conditioning.

Flashers

The turn signal flasher and the hazard/warning flasher plugs into the fuse block.

NOTE: A special heavy duty turn signal flasher is required to properly operate the turn signals when a trailer's lights are connected to the system.

Computers

LOCATION

Electronic Control Module (ECM): located in the passenger compartment by the right side kick panel or behind the glove box on the R/V and C/K-Series vehicles or on the drivers side under the seat on the G-Series vans.

Wiper Delay Module: R/V-Series: located on the steering column behind the lower cover. C/K-Series: located on the wiper motor. G-Series: located in the harness on the left side of the steering column.

Cruise Control Module: R/V and G-Series: located on the back of instrument panel next to the steering column. C/K-Series: located on the engine side of the cowl near the master cylinder.

Cruise Control

SERVO ADJUSTMENT

R/V and G-Series

1. Ensure ignition is **OFF** and the throttle valve fully closed before proceeding with adjustment.

2. If equipped with a gasoline engine, install the servo rod so the rod assembles over the stud per adjustment "B".

3. If equipped with a diesel engine, position the pin on the servo rod in the hole closest to the servo that allows for adjustment "C".

4. Install the servo rod retainer and verify that the system functions properly.

C/K-Series

1. Disconnect the engine end of the cruise control cable from the lever stud.

2. Pull lightly on the end of the cable.

3. If the cable does not pull out, the cable is adjusted properly. If the cable extends out, proceed as follows:

 a. Unlock the cable conduit lock mechanism.

 b. If equipped with a gasoline engine, move the cable until the throttle begins to open. Then move the cable in the opposite direction enough to close the throttle.

 c. If equipped with a diesel engine, move the cable until the injection pump lever moves off the idle stop screw. Then move the cable in the opposite direction far enough to return the lever to the idle stop screw.

 d. While holding the cable securely, push down on the cable conduit lock mechanism until it snaps into place.

 e. Verify the system functions properly.

ENGINE COOLING

Radiator

REMOVAL AND INSTALLATION

Except 6.2L Diesel Engine

1. Disconnect the negative battery cable.

2. Drain the cooling system into a suitable container.

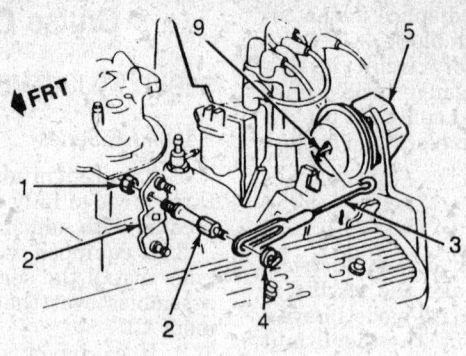

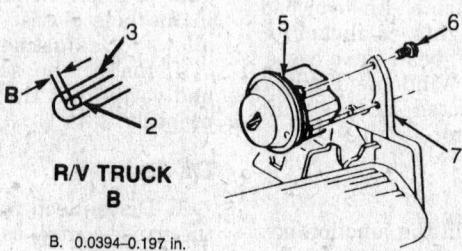

R/V TRUCK
B

B. 0.0394–0.197 in.
(1.0–5.0mm)
1. Nut
2. Lever
3. Stud
4. Rod
5. Retainer

6. Servo
7. Bolt
8. Bracket
9. Tab

79114021

Cruise control servo mounting and adjustment — R/V
and G series except diesel engine

3. Disconnect the radiator upper and lower hoses and, if applicable, the transmission or engine oil cooler lines. Remove the coolant recovery hose.

4. Remove the radiator upper panel if equipped.

5. Remove the fan shroud attaching screws and clutch fan as required.

6. Remove the radiator from the vehicle.

7. Installation is the reverse of the removal procedure.

6.2L Diesel Engine

1. Disconnect the negative battery cables.

2. Drain the cooling system.

3. Remove the air intake snorkel.

4. Remove the windshield washer bottle.

5. Remove the hood release cable.

6. Remove the upper fan shroud.

7. Disconnect the upper radiator hose.

8. Disconnect the transmission cooler lines.

9. Disconnect the low coolant sensor wire.

10. Disconnect the overflow hose.

11. Disconnect the engine oil cooler lines.

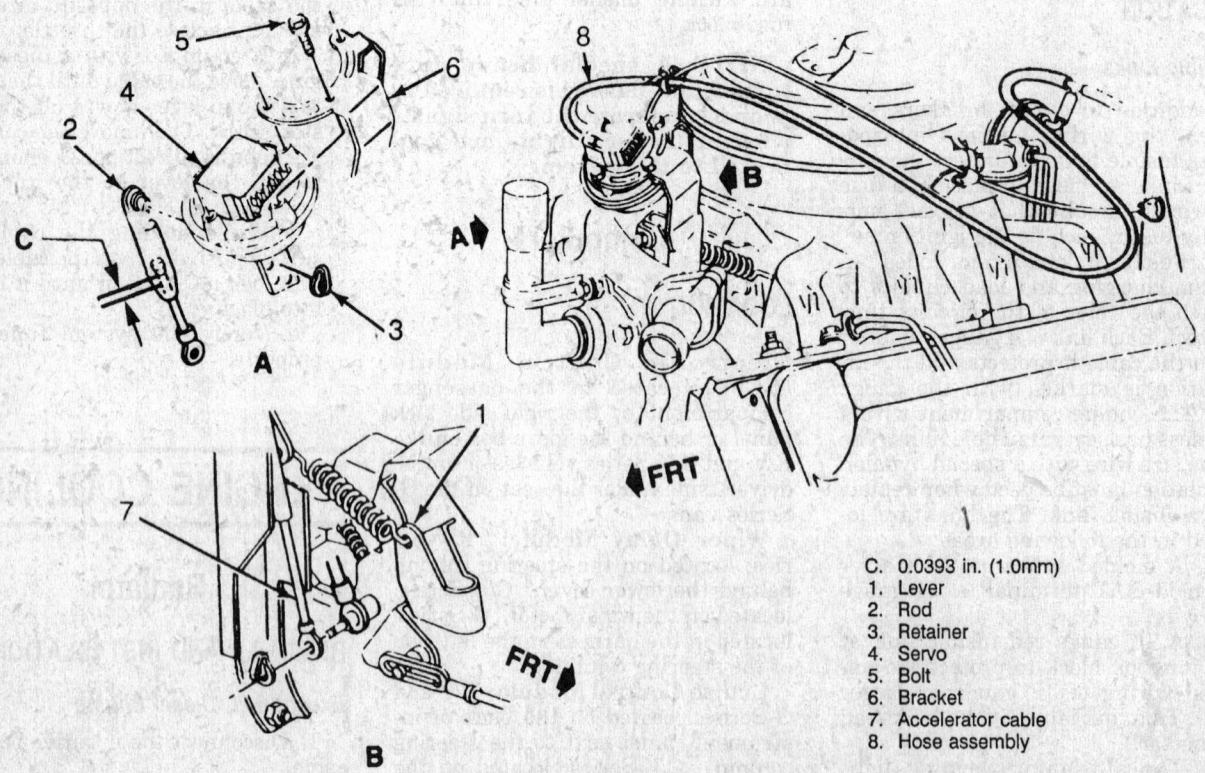

C. 0.0393 in. (1.0mm)
1. Lever
2. Rod
3. Retainer
4. Servo
5. Bolt
6. Bracket
7. Accelerator cable
8. Hose assembly

79114022

Cruise control servo mounting and adjustment — R/V and G series with diesel engine

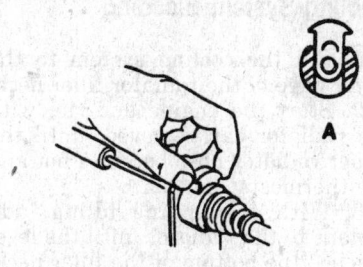

A. Unlocked Position

79114020

Unlocking cable for adjustment — C/K series

12. Disconnect the lower radiator hose.

13. On the G-Series, remove the brake master cylinder from the booster.

14. Unbolt and remove the radiator.

To install:

15. Install the radiator.

16. On G-Series Van, install the brake master cylinder on the booster.

17. Connect the lower radiator hose.

18. Connect the engine oil cooler lines.

19. Connect the overflow hose.

20. Connect the low coolant sensor wire.

21. Connect the transmission cooler lines.

22. Connect the upper radiator hose.

23. Install the upper fan shroud.

24. Install the hood release cable.

25. Install the windshield washer bottom.

26. Install the air intake snorkle.

27. Fill the radiator with the proper type and quantity of coolant and inspect the system for leaks.

Auxiliary Electric Cooling Fan

REMOVAL AND INSTALLATION

1. Disconnect the negative battery cable.

2. Remove the grille.

3. Unplug the fan harness connector.

4. Remove the fan-to-brace bolts and lift out the fan.

5. Installation is the reverse of removal. Torque the bolts to 53 inch lbs. (6 Nm).

Heater Core

REMOVAL AND INSTALLATION

R/V-Series

1. Disconnect the battery ground cable.

2. Disconnect the heater hoses at the core tubes and drain the engine coolant. Plug the core tubes to prevent spillage.

3. Remove the nuts securing the assembly from the engine firewall.

4. Remove the instrument panel.

5. Disconnect the air-defrost and temperature door cables.

6. Remove the floor outlet and remove the defroster duct-to-heater case screw.

7. Remove the heater case-to-instrument panel screws. Pull the assembly rearward to gain access to the wiring harness and disconnect the wires attached to the unit.

8. Remove the heater distributor from the vehicle.

9. Remove the heater core retaining straps and remove the core from the vehicle.

10. Installation is the reverse of removal. Be sure the core-to-case and case-to-dash panel sealer is intact. Fill the cooling system and check for leaks.

C/K-Series

1. Disconnect the negative battery cable.

2. Drain the cooling system into a suitable container.

3. Disconnect the heater hoses at the core tubes.

4. In the passenger compartment, remove the lower heater core cover-to-case screws.

5. Remove the bracket and screws securing the heater core to the case.

6. Remove the heater core.

7. Installation is the reverse of the removal procedure.

G-Series

1. Disconnect the negative battery cable.

2. Remove the coolant recovery bottle.

3. Place a pan under the vehicle and disconnect the heater intake and outlet hoses. Quickly remove and plug the hoses and support them in an upright position. Drain the coolant from the heater core into the pan.

4. Remove the heater distributor duct-to-case attaching screws and the duct-to-engine cover screw. Remove the duct.

5. Remove the engine housing cover.

6. Remove all the instrument panel attaching screws.

7. Carefully lower the steering column. Raise and support the right side of the instrument panel.

8. Remove the defroster duct-to-case attaching screws and the 2 screws attaching the distributor to the heater case.

9. Disconnect the temperature door cable. Carefully fold the cable back and out of the way.

10. Remove the 3 nuts from the engine compartment side of the heater case and the screw from the passenger compartment side.

11. Remove the heater case and core assembly.

12. Remove the core retaining straps and remove the core.

To install:

13. Install the core.

14. Install the core retaining straps.

15. Install the heater case and core assembly.

16. Install the 3 nuts on the engine compartment side of the distributor case and the screw on the passenger compartment side.

17. Connect the temperature door cable.

18. Install the defroster duct-to-case attaching screws and the 2 screws attaching the distributor to the heater case.

19. Install the steering column.

20. Install all the instrument panel attaching screws.

21. Install the engine cover.

22. Install the duct. Install the heater distributor duct-to-case attaching screws and the duct-to-engine cover screw.

23. Connect the heater intake and outlet hoses.

24. Fill the cooling system.

25. Install the coolant recovery bottom.

26. Connect the negative battery cable.

Water Pump

REMOVAL AND INSTALLATION

Except 6.2L Diesel Engine

1. Disconnect the negative battery cable.

2. Drain the engine coolant into a suitable container.

3. Loosen the alternator and other accessories at their adjusting points, and remove the fan belts from the fan pulley.

4. Remove the fan shroud, fan clutch and pulley.

5. Remove all accessory brackets that will interfere with water pump removal.

6. Disconnect the lower radiator, bypass and the heater hoses from the water pump.

7. Remove the bolts, pump assembly and old gasket from the engine.

To install:

8. Ensure the gasket surfaces on the pump and engine are clean.

9. Install the pump assembly with a new gasket. Tighten the bolts to 30 ft. lbs. (40 Nm).

10. Connect the hose between the water pump inlet and the nipple on the pump.

11. Install any accessory brackets.

12. Install the fan and pulley.

13. Install and adjust the alternator and other accessories.

14. Install the fan belts from the fan pulley.

15. Fill the cooling system.

16. Connect the battery.

6.2L Diesel Engine

1. Disconnect the negative battery cables.

2. Remove the fan and fan shroud.

3. Drain the engine coolant into a suitable container.

4. Remove the air conditioning hose bracket, as required.

5. Remove the oil filler tube, as required.

6. Remove the alternator pivot bolt and remove the alternator belt.

7. Remove the alternator upper and lower brackets.

8. Remove the power steering belt and pump assembly, secure it out of the way.

9. Remove the air conditioning belt, as required.

10. Disconnect the bypass hose and the lower radiator hose.

11. Remove the water pump bolts. Remove the water pump plate and gasket and water pump. If the pump gasket is to be replaced, remove the plate attaching bolts to the water pump and remove (and replace) the gasket.

To install:

12. When installing the pump, the flanges must be free of oil. Apply an anaerobic sealer GM part No. 1052357 or equivalent.

NOTE: The sealer must be wet to the touch when the bolts are torqued.

13. Attach the water pump and plate assembly. Torque the bolts to 30 ft. lbs. (41 Nm).

14. Connect the bypass hose and the lower radiator hose.

15. Install the air conditioning belt if equipped.

16. Install the power steering belt.

17. Install the alternator lower bracket.

18. Install the alternator pivot bolt.

19. Install the alternator belt.

20. Install the oil filler tube.

21. If equipped with air conditioning, install the air conditioning hose bracket nuts.

22. Fill the radiator with the proper type and quantity of antifreeze.

23. Install the fan and fan shroud.

24. Connect the batteries.

Thermostat

REMOVAL AND INSTALLATION

Except 6.2L Diesel Engine

1. Disconnect the negative battery cable.

2. Drain the radiator until the level is below the thermostat level (below the level of the intake manifold).

3. Remove the thermostat housing from the engine. Remove the thermostat from inside the housing.

4. Install new thermostat in the reverse order of removal, making sure the thermostat is inserted correctly. Clean the gasket surfaces on the thermostat housing and the intake manifold. Use a new gasket when installing the elbow to the manifold. Torque the housing bolts to 20 ft. lbs. (27 Nm).

5. Refill the cooling system with the proper type and quantity of antifreeze.

6.2L Diesel Engine

1. Disconnect the negative battery cables.

2. Remove the upper fan shroud.

3. Drain the cooling system to a point below the thermostat.

4. Remove the engine oil dipstick tube brace and the oil fill brace, as required.

5. Remove the upper radiator hose.

6. Remove the water outlet.

7. Remove the thermostat and gasket.

8. Installation is the reverse of removal. Use a new gasket coated with sealer. Torque the bolts to 35 ft. lbs. (47 Nm).

Cooling System Bleeding

1. Fill the cooling system to the lower edge of the radiator filler neck.

2. Start the engine and run with the radiator cap removed, until the upper radiator hose becomes hot and the thermostat is open.

3. With the engine idling, add coolant to the radiator until the level reaches the bottom of the filler neck.

4. Install the radiator cap and add coolant to the reservoir until the level reaches the **COLD** mark.

GASOLINE FUEL SYSTEM

Fuel System Service Precaution

When working with the fuel system certain precautions should be taken; always work in a well ventilated area, keep a dry chemical (Class B) fire extinguisher near the work area. Always disconnect the negative battery cable and do not make any repairs to the fuel system until all the necessary steps for repair have been reviewed.

RELIEVING FUEL SYSTEM PRESSURE

The 220 TBI unit used on the V6 and V8 engines has a constant bleed in the pressure regulator to relieve pressure any time the engine is turned off, however a small amount of fuel may be released when the fuel line is disconnected. As a precaution, cover the fuel line with a cloth and dispose of the cloth properly. Also, loosen the fuel filler cap to relieve tank vapor pressure.

Fuel Tank

REMOVAL AND INSTALLATION

1. Disconnect the negative battery cable.

2. Properly drain the fuel tank into a suitable container.

3. Raise and support vehicle safely.

4. Safely support the fuel tank and remove the retaining straps.

5. Disconnect the fuel tank sending unit wire, ground straps and hoses.

6. With the help of an assistant, lower the tank and remove the assembly from the vehicle.

7. Installation is the reverse of the removal procedure.

Fuel Filter

REMOVAL AND INSTALLATION

Fuel Injected Engine

1. Disconnect the negative battery cable.

2. Properly relieve the fuel system pressure.

3. Raise and support the vehicle safely.

4. Disconnect the fuel lines at the filter.

5. Remove the fuel filter from the retainer or mounting bolt.

6. To install, reverse the removal procedures ensuring the filter flow is in the correct direction. Start the engine and check for leaks.

NOTE: The filter has an arrow (fuel flow direction) on the side of the case, be sure to install it correctly in the system, the with arrow facing away from the fuel tank.

Electric Fuel Pump

TESTING

1. Ensure there is a sufficient quantity of fuel in the tank. Properly relieve the fuel system pressure.

2. Remove the air cleaner and plug the THERMAC vacuum port, as required.

3. Uncouple the fuel supply hose or line located in the engine compartment.

4. Install a fuel pressure gauge between the line. Tighten the line to ensure no leaks occur during testing.

5. Start the engine, check for leaks and observe the fuel pressure, it should be 9–13 psi.

6. Depressurize the fuel system, remove the testing tool, remove the plug from the THERMAC vacuum port, reconnect the fuel line, start the engine and check for fuel leaks.

REMOVAL AND INSTALLATION

1. With the engine turned off, properly relieve the fuel system pressure.

2. Disconnect the negative battery cable.

3. Raise and support the vehicle safely.

4. Drain the fuel tank and remove the assembly from the vehicle.

5. Using a suitable tool, turn the fuel lever sending unit and pump assembly locking ring (located on top of the fuel tank) counterclockwise, lift the assembly from the tank and remove the pump from the fuel lever sending device.

6. Pull the pump up into the attaching hose while pulling it outward away from the bottom support. Be careful not to damage the rubber insulator and strainer during removal. After the pump assembly is clear of the bottom support, pull it out of the rubber connector.

7. To install, reverse the removal procedures.

Fuel Injection

IDLE SPEED ADJUSTMENT

This procedure should be performed only if parts of the throttle body have been replaced. The engine should be at normal operating temperatu

1. Remove the air cleaner, adapter and gaskets. Discard the gaskets. Plug any vacuum line ports, as necessary.

2. Leave the Idle Air Control (IAC) valve connected and ground the diagnostic terminal (ALDL connector).

3. Turn the ignition switch to the **ON** position; do not start the engine. Wait for at least 10 seconds; this allows the IAC valve pintle to extend fully and seat in the throttle body.

4. With the ignition switch still in the **ON** position, disconnect IAC electrical connector.

5. Remove the ground from the diagnostic terminal and start the engine. Let the engine reach normal operating temperature.

6. Apply the parking brake and block the drive wheels. Remove the plug from the idle stop screw by piercing it first with a suitable tool, then applying leverage to lift the plug out.

7. Connect a suitable tachometer to the engine.

8. Ensure that the transmission is in the specified (N or D) position, with the ECM in "Open or Closed" loop as specified.

9. Adjust the idle stop screw to obtain the specified RPM reading.

10. Turn the ignition **OFF** and reconnect the IAC valve connector.

Unplug any plugged vacuum line ports and install the air cleaner, adapter and new gaskets.

11. Reset the IAC valve as follows:

NOTE: If installing a new IAC valve, measure and adjust the valve accordingly. If reinstalling a used IACvalve, do not push or pull on the pintle to adjust pintle length or damage to the IAC worm gear might occur.The valve is preset at the factory and will self-adjust when the following procedure is performed.

a. Set a new IAC valve by measuring the distance between the tip of the pintle and the valve mounting surface.

b. If greater than 1.10 in. (28mm), use light finger pressure to slowly retract the pintle. The force required to retract a new valve will not damage the valve.

c. Install the valve and connect the wire connector.

d. Reset a used IAC valve pintle position by depressing the accelerator pedal slightly, start the engine and run for 5 seconds, turn the key **OFF** for 10 seconds, then restart the vehicle and check for proper idle operation.

Fuel Injector

REMOVAL AND INSTALLATION

NOTE: When removing the injectors, be careful not to damage the electrical connector pins (on top of the injector), the injector fuel filter and the nozzle. The fuel injector is serviced as a complete assembly only. The injector is an electrical component and should not be immersed in any kind of cleaner.

1. Disconnect the negative battery cable.

2. Remove the air cleaner. Properly relieve the fuel system pressure.

3. Disconnect the injector electrical connector.

4. Remove the fuel meter cover and leave the cover gasket in place, if equipped with 2 injectors. Remove the retaining screw and injector retainer, if equipped with 1 injector.

5. Using a suitable tool or J–26868, carefully lift the injector until it is free from the fuel meter body.

6. Remove all O-rings and washers from the injector or housing.

7. Discard the fuel meter cover gasket.

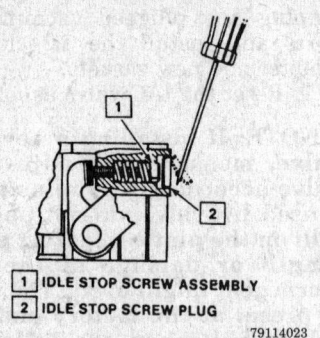

[1] IDLE STOP SCREW ASSEMBLY
[2] IDLE STOP SCREW PLUG

79114023

Removing idle stop screw plug for adjustment — TBI engines

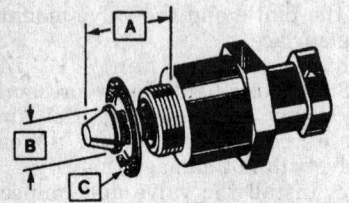

[A] DISTANCE OF PINTLE EXTENSION
[B] DIAMETER AND SHAPE OF PINTLE
[C] IAC VALVE GASKET

79114025

Typical IAC valve

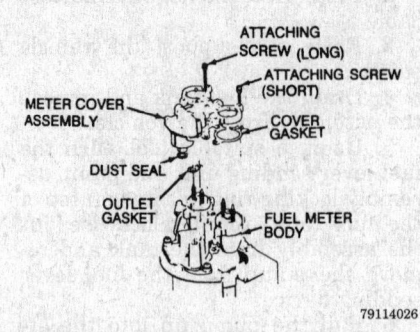

79114026

Replacing the fuel meter cover on the TBI 220

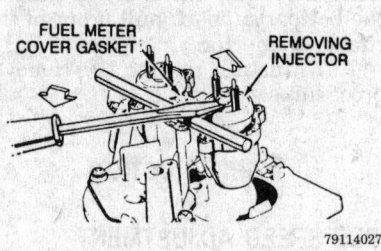

79114027

Removing the fuel injector on the TBI 220

8. To install, lubricate the O-rings with automatic transmission fluid and reverse the removal procedures.

DIESEL FUEL SYSTEM

Fuel Filter

REMOVAL AND INSTALLATION

1. Turn the ignition **OFF**. Remove the fuel tank cap to release any pressure or vacuum in the tank.
2. Drain the fuel from the fuel filter by opening both the air bleed and the water drain valve allowing the fuel to drain into a suitable container.
3. Unstrap both bail wires with a suitable tool and remove the filter.
   To install:
4. Before installing the new filter, ensure that both filter mounting plate fittings are clear of dirt.
5. Snap the filter into place with the bail wires.

MINIMUM IDLE SPEED

Engine	Transmission	Gear (D/N)	Engine Speed (RPM)**	Open/Closed Loop*
2.5L	Man.	N	600 ± 50	CL
	Auto.	N	500 ± 50	CL
2.8L	Man.	N	700 ± 50	OL
4.3L	Man.	N	400-525	CL
(under 8500	Auto.	D	400 ± 50	CL
GVW)	Auto.(1)	D	475 ± 50	CL
4.3L	Man.	N	400-525	CL
(Over 8500	Auto.	D	400 ± 50	CL
GVW)				
5.0L	Man.	N	500 ± 25	OL
	Auto.	D	425 ± 25	CL
5.7L	Man.	N	500 ± 25	OL
(under 8500 GVW)	Auto.	D	425 ± 25	CL
5.7L	Man.	N	550 ± 25	CL
(over 8500 GVW)	Auto.	D	450 ± 25	CL
7.4L	Man.	N	700 ± 25	OL
	Auto.	D	625 ± 25	OL

* Let engine idle until proper fuel control status (open/closed loop) is reached.

** If the engine has less than 500 miles or is checked at altitudes above 1500 feet, the idle rpm with a seated IAC valve should be lower than valves above.

(1) 4.3L High-Output ML Van Series

79114024

Minimum idle air rate — 1991 — 92 vehicles

6. Close the water drain valve and open the air bleed valve. Connect a ⅛ in. (3mm) I.D. hose to the air bleed port and place the other end into a suitable container.

7. Disconnect the fuel injection pump shutdown solenoid wire.

8. Crank the engine for 10–15 seconds, then wait 1 minute for the starter motor to cool. Repeat until clear fuel is observed coming from the air bleed.

NOTE: If the engine is to be cranked or started with the air cleaner removed, care must be taken to prevent dirt from being pulled into the air inlet manifold which could result in engine damage.

9. Close the air bleed valve, reconnect the injection pump solenoid wire and replace the fuel tank cap.

10. Start the engine, allow it to idle for 5 minutes and check the fuel filter for leaks.

DRAINING WATER FROM THE SYSTEM

Water is the worst enemy of the diesel fuel injection system. The injection pump, which is designed and constructed to extremely close toler-

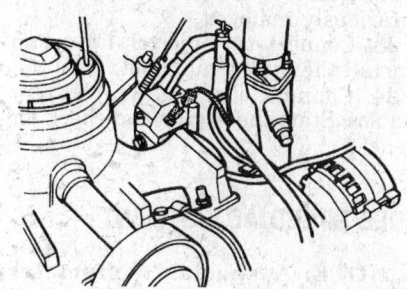

79114029

Diesel shutdown solenoid location

ances, and the injectors can be easily damaged if enough water is forced through them in the fuel. Engine performance will also be drastically affected and engine damage can occur. Diesel fuel is much more susceptible than gasoline to water contamination. Diesel engine vehicles are equipped with an indicator light system located in the instrument panel if water (1–2½ gallons) is detected in the fuel tank. The light will come ON for 2–5 seconds each time the ignition is turned ON, assuring the driver the light is working. If there is water in the fuel, the light will come back ON after a 15–20 second off delay, and then remain ON.

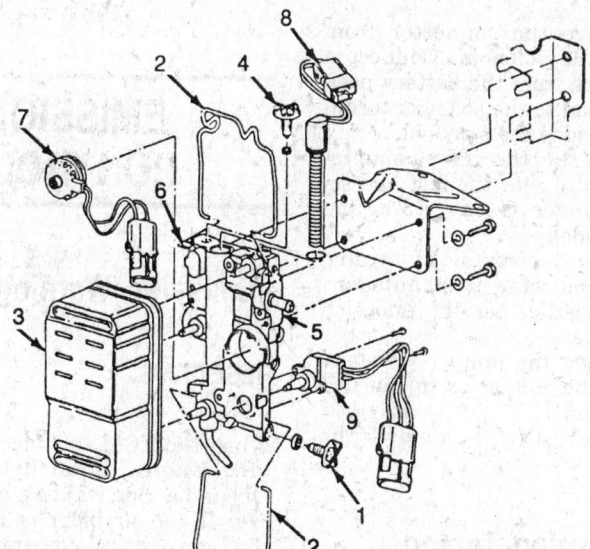

1. Water drain valve
2. Bail wires
3. Fuel meter element
4. Air bleed
5. Filter adapter
6. Air bleed port
7. Restriction switch
8. Fuel heater
9. Water sensor

79114028

Diesel engine fuel filter

Purging The Fuel Tank

The 6.2L diesel engines also use a water-in-fuel warning system. The fuel tank is equipped with a filter which screens out the water and lets it lay in the bottom of the tank below the fuel pickup. When the water level reaches a point where it could be drawn into the system, a warning light flashes in the cab. A built-in siphoning system starting at the fuel tank and going to the rear spring hanger on some models, and at the midway point of the right frame rail on other models permits you to attach a hose at the shut-off and siphon out the water. If it becomes necessary to drain water from the fuel tank, also check the primary fuel filter for water.

Draining The Fuel Tank

NOTE: Disconnect the negative battery cable before beginning the draining operation.

1. If the vehicle is not equipped with a drain plug, use the following procedure to remove the fuel:

2. Using a 10 ft. (305cm) piece of ⅜ in. (9.525mm) hose cut a flap slit 18 in. (457mm) from one end.

3. Install a pipe nipple, of slightly larger diameter than the hose, into the opposite end of the hose.

4. Install the nipple end of the hose into the fuel tank with the natural curve of the hose pointing downward. Keep feeding the hose in until the nipple hits the bottom of the tank.

5. Place the other end of the hose in a suitable container and insert a air hose pointing it in the downward direction of the slit and inject air into the line.

NOTE: If the vehicle is to be stored, always drain the fuel from the complete fuel system including, fuel pump supply, fuel injection pump, fuel lines and tank.

REMOVAL AND INSTALLATION

1. Disconnect the negative battery cable.

2. Drain the tank.

3. Raise and support the vehicle safely.

4. Remove the clamp on the filler neck and the vent tube hose.

5. Remove the gauge hose which is attached to the frame.

6. While supporting the tank securely, remove the support straps.

7. Lower the tank until the gauge wiring can be removed.

8. Remove the tank.

9. Install the unit by reversing the removal procedure. Make certain the anti-squeak material is replaced during installation.

Diesel Injection Pump

REMOVAL AND INSTALLATION

1. Disconnect both battery negative cables.

2. Remove the intake manifold.

3. Remove the fuel injection lines.

4. Disconnect the accelerator cable at the injection pump and the detent cable, if equipped.

5. Tag and disconnect the necessary wires and hoses at the injection pump.

6. Disconnect the fuel return line at the top of the injection pump.

7. Disconnect the fuel feed line at the injection pump.

8. Remove the air conditioning hose retainer bracket, if equipped.

9. Remove the oil fill tube, including the Crankcase Depression Regulator (CDR) valve vent hose assembly. Remove the grommet.

10. Scribe or paint a matchmark on the front cover and the injection pump flange.

11. Rotate the crankshaft and remove the injection pump drive gear bolts, accessing the bolts through the oil filler neck hole.

12. Remove the injection pump-to-front cover attaching nuts. Remove the pump and cap all open lines and nozzles.

To install:

13. Replace the pump gasket.

14. Align the locating pin on the pump hub with the slot in the injection pump driven gear. At the same time, align the timing marks.

15. Attach the injection pump to the front cover, aligning the timing marks before torquing the nuts to 30 ft. lbs. (40 Nm).

16. Install the drive gear to injection pump bolts, torquing the bolts to 20 ft. lbs. (25 Nm).

17. Install the grommet and oil fill tube.

18. Install the air conditioning bracket.

19. Connect the fuel feed line and torque to 20 ft. lbs. (25 Nm).

20. Connect the fuel return line to the pump.

21. Connect the accelerator and detent cables.

22. Connect all wires and hoses previously removed.

23. Connect the injector lines and install the intake manifold.

24. Connect the negative battery cables. Start the engine and check for leaks.

IDLE SPEED ADJUSTMENT

NOTE: A special tachometer suitable for diesel engines must be used. A gasoline engine type tachometer will not work with the diesel engine.

1. Set the parking brake and block the drive wheels.

2. Run the engine up to normal operating temperature. The air cleaner must be mounted and all accessories turned **OFF**.

3. Install the diesel tachometer as per the manufacturer's instructions.

4. Adjust the low idle speed screw on the fuel injection pump to manufacturer's specification per the emission control label.

NOTE: All idle speeds are to be set within 25 rpm of the specified values.

5. Adjust the fast idle speed as follows:

a. Remove the connector from the fast idle solenoid. Connect a jumper wire from the battery positive terminal to the solenoid terminal to energize the solenoid.

b. Open the throttle momentarily to ensure that the fast idle solenoid plunger is energized and fully extended.

c. Adjust the fast idle by turning the hex-head screw to manufacturers specification per the emission control label.

d. Remove the jumper wire and reinstall the connector to the fast idle solenoid.

6. Disconnect and remove the tachometer.

Injection Timing Adjustment

For the engine to be properly timed, the lines on the top of the injection pump adapter and the flange of the injection pump must be aligned.

The engine must be off for resetting the timing.

1. Loosen the 3 pump retaining nuts with tool J–26987, an injection pump intake manifold wrench or equivalent.

2. Align the mark on the injection pump with the marks on the front cover. Torque the nuts to 30 ft. lbs. (40 Nm). Use a ³/₄ in. open-end wrench on the boss at the front of the injection pump to aid in rotating the pump to align the marks.

Fuel Injector

REMOVAL AND INSTALLATION

1. Disconnect the negative cable on both batteries.

2. Disconnect the fuel line clip, and remove the fuel return hose.

3. Remove the fuel injection line.

4. Using GM special tool J–29873, remove the injector. Always remove the injector by turning the 30mm hex portion of the injector. Cap the injector and fuel lines when disconnected, to prevent contamination.

5. Install the injector with new gasket and torque to 50 ft. lbs. (70 Nm). Connect the injection line and torque the nut to 20 ft. lbs. (25 Nm). Install the fuel return hose, fuel line clips and connect the batteries.

EMISSION CONTROLS

Emission Warning Lamps

RESETTING

When the ECM sets a code, the **Service Engine Soon** light will come **ON** and a code will be stored in memory. If the problem is intermittent, the light will go out after 10 seconds when the fault goes away. However the code will stay in the memory for 50 starts or until the battery voltage to the ECM is removed. Removing battery voltage for 30 seconds will clear all stored codes.

NOTE: To prevent damage to the ECM, the key must be OFF when disconnecting or connecting power to the ECM.

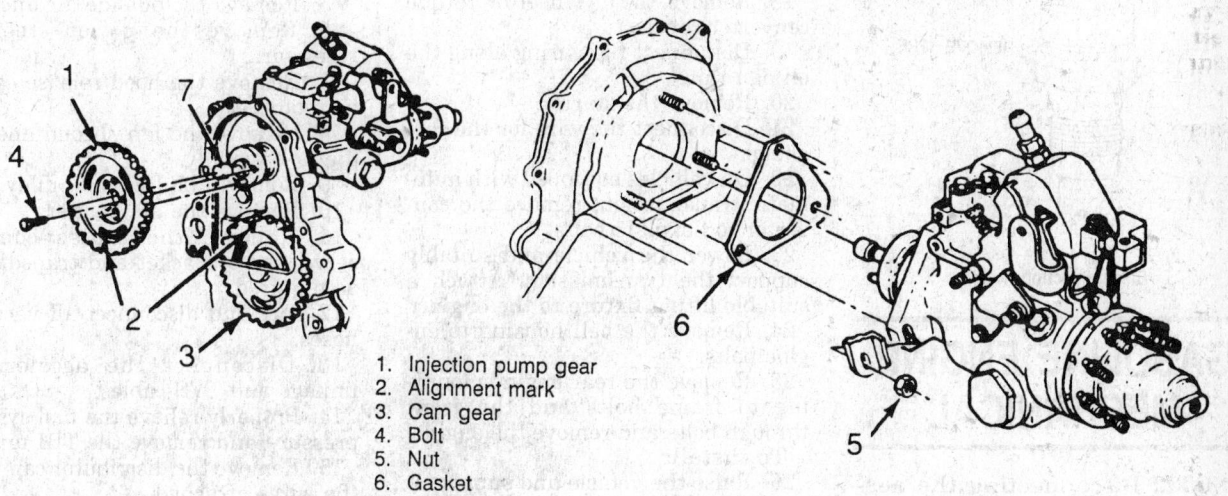

1. Injection pump gear
2. Alignment mark
3. Cam gear
4. Bolt
5. Nut
6. Gasket
7. Pump hub

79114030

Diesel injection pump installation

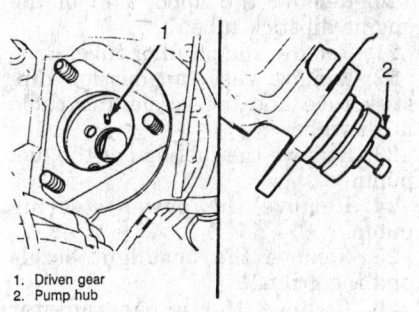

1. Driven gear
2. Pump hub

79114031

Diesel injection pump locating pin

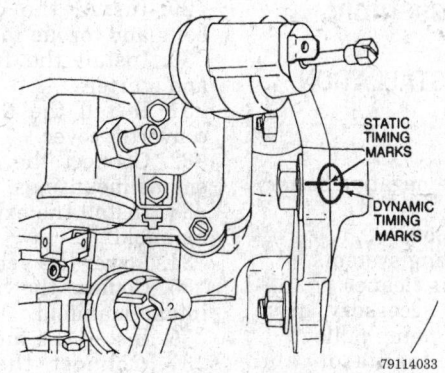

STATIC TIMING MARKS

DYNAMIC TIMING MARKS

79114033

Diesel injection timing alignment marks

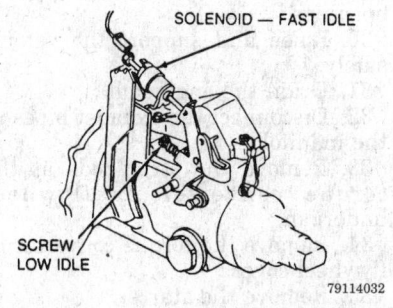

SOLENOID — FAST IDLE

SCREW LOW IDLE

79114032

Diesel injection pump idle adjustment locations

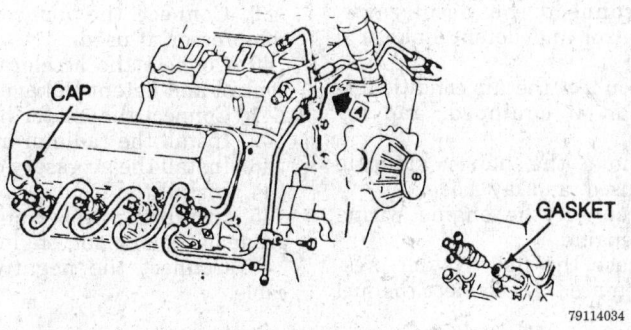

CAP

GASKET

79114034

Diesel injection nozzles

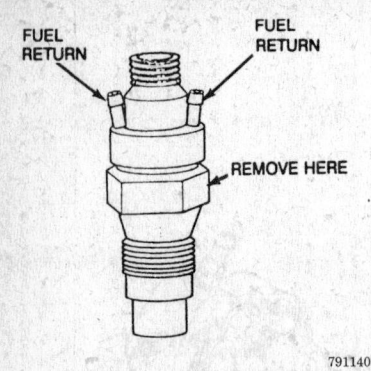

Diesel injection nozzle

GASOLINE ENGINE MECHANICAL

NOTE: Disconnecting the negative battery cable on some vehicles may interfere with the functions of the on board computer systems and may require the computer to undergo a relearning process, once the negative battery cable is reconnected.

Engine Assembly

REMOVAL AND INSTALLATION

Except G-Series

1. Disconnect the negative battery cable.
2. Remove the hood.
3. Drain the cooling system.
4. Remove the air cleaner.
5. Remove the accessory drive belt, fan and water pump pulley.
6. Remove the radiator and shroud.
7. Disconnect the heater hoses at the engine.
8. Disconnect the accelerator, cruise control and detent linkage if used.
9. Disconnect the air conditioning compressor, if equipped, and lay aside.
10. Remove the power steering pump, if used, and lay aside.
11. Disconnect the engine wiring from the engine.
12. Release the fuel system pressure as required. Disconnect the fuel line.
13. Disconnect the vacuum lines from the intake manifold.
14. Raise the vehicle and support it safely.
15. Drain the engine oil.

16. Disconnect the exhaust pipes from the manifold.
17. Disconnect the strut rods at the engine mountings, if used.
18. Remove the flywheel or torque converter cover.
19. Disconnect the wiring along the oil pan rail.
20. Remove the starter.
21. Disconnect the wire for the fuel gauge.
22. On vehicles equipped with automatic transmission, remove the converter to flexplate bolts.
23. Lower the vehicle and suitably support the transmission. Attach a suitable lifting fixture to the engine.
24. Remove the bell housing to engine bolts.
25. Remove the rear engine mounting to frame bolts and the front through bolts and remove the engine.

To install:

26. Raise the vehicle and support it safely.
27. Lower the engine and install the engine mounting bolts. Torque the rear engine mounting to frame bolts or nuts, the front through bolts and the front nuts to specifications.
28. Install the bell housing to engine bolts and torque to 35 ft. lbs. (44 Nm).
29. Install the converter to flex bolts and torque to specification.
30. Install the fuel gauge wiring and starter.
31. Install the flywheel or torque converter cover.
32. Connect the strut rods at the engine mountings, if used.
33. Install the exhaust pipes at the manifold.
34. Lower the vehicle.
35. Connect the vacuum lines to the intake manifold.
36. Install the fuel line.
37. Connect the engine wiring harness.
38. Install the power steering pump, if used.
39. Connect the air conditioning compressor, if used.
40. Connect the accelerator, cruise control and detent linkage.
41. Connect the heater hoses.
42. Install the radiator and shroud.
43. Install the accessory drive belts.
44. Install the hood.
45. Install the proper quantity and grade of coolant and engine oil.
46. Connect the negative battery cable.

G-Series

4.3L ENGINE

1. Disconnect the battery cables.

2. Remove the glove box, as required.
3. Drain the cooling system.
4. Remove the engine cover.
5. Remove the outside air duct.
6. Remove the power steering reservoir.
7. Remove the hood release cable, as required.
8. Remove the fan shroud and radiator assembly.
9. Remove the fan and pulley.
10. Remove the air cleaner.
11. Remove the cruise control servo, servo bracket and transducer, as required.
12. Tag and disconnect all vacuum hoses.
13. Disconnect the accelerator linkage and TVS cables.
14. Properly relieve the fuel system pressure and remove the TBI unit.
15. Remove the distributor cap with the wires attached.
16. Disconnect the heater hoses at the engine.
17. Remove the PCV valve, as required.
18. Properly discharge the air conditioning system and remove the air conditioning compressor and bracket.
19. Remove the alternator, as required.
20. Remove the upper half of the engine dipstick tube.
21. Remove the oil filler tube.
22. Remove the transmission dipstick tube and the accelerator cable at the tube.
23. Remove the fuel line at the fuel pump.
24. Remove the power steering pump.
25. Remove the headlight bezels and the grille.
26. Remove the upper radiator support.
27. Remove the lower fan shroud and filler panel.
28. Remove the hood latch support.
29. Remove the condenser. Cap all openings.
30. Raise and support the vehicle safely.
31. Drain the engine oil.
32. Disconnect the exhaust pipes at the manifolds.
33. Remove the strut rods at the torque converter or flywheel underpan.
34. Remove the torque converter or flywheel cover.
35. Remove the starter.
36. Remove the flexplate-to-torque converter bolts for automatic transmissions.
37. Remove the bell housing-to-engine bolts.

38. Remove the engine mounting through bolts.

39. Lower the vehicle and support the transmission using the proper equipment.

40. Attach an engine crane to the engine, pull the engine forward and upward and remove it from the vehicle.

To install:

41. Raise the engine into position.

42. Install the engine mount through bolts. Torque the bolts to specifications.

43. Install the bell housing-to-engine bolts. Torque the bolts to specifications.

44. Install the flexplate-to-torque converter bolts for automatic transmissions. Torque the bolts to specifications.

45. Install the starter.

46. Install the torque converter or flywheel cover.

47. Install the strut rods at the torque converter or flywheel underpan.

48. Connect the exhaust pipes at the manifolds.

49. Install the fuel line at the fuel pump.

50. Install the condenser.

51. Install the hood latch support.

52. Install the lower fan shroud and filler panel.

53. Install the transmission dipstick tube and the accelerator cable at the tube.

54. Install the PCV valve.

55. Install the distributor cap.

56. Install the cruise control servo, servo bracket and transducer.

57. Install the oil filler pipe and the engine dipstick tube.

58. Install the thermostat housing.

59. Connect the heater hoses at the engine.

60. Connect the engine wiring harness from the firewall connection.

61. Install the air conditioning compressor, mounting bracket and alternator assembly.

62. Install the radiator and the shroud.

63. Install the radiator support bracket.

64. Install the TBI unit.

65. Connect the accelerator linkage and TVS cable.

66. Install the windshield wiper jar and bracket.

67. Install the air conditioning compressor.

68. Install the air conditioning vacuum reservoir.

69. Charge the air conditioning system, using the proper equipment.

70. If equipped with an automatic transmission, install the cooler lines at the radiator.

71. Install the power steering reservoir.

72. Install the hood release cable.

73. Connect the radiator hoses at the radiator.

74. Install the headlight bezels and the grille.

75. Install the air cleaner.

76. Install the outside air duct.

77. Install the engine cover.

78. Fill the cooling system with the proper type and quantity of antifreeze.

79. Fill the crankcase.

80. Connect the battery cables.

81. Install the glove box.

5.0L AND 5.7L ENGINES

1. Disconnect the battery cables.

2. Drain the cooling system.

3. Remove the radiator coolant reservoir bottle.

4. Remove the upper radiator support.

5. Remove the grille and the lower grille valance.

6. Discharge the air conditioning system using the proper equipment.

7. Remove the air conditioning condenser from in front of the radiator.

8. If equipped with an automatic transmission, remove the fluid cooler lines from the radiator.

9. Disconnect the radiator hoses at the radiator.

10. Remove the radiator support bracket and remove the radiator and the shroud.

11. Remove the engine cover.

12. Remove the air cleaner.

13. Disconnect the accelerator linkage.

14. Disconnect all hoses and wires at the TBI unit.

15. Relieve the fuel system pressure and remove the TBI unit.

16. Disconnect the engine wiring harness from the firewall connection.

17. Tag and disconnect all vacuum lines.

18. Remove the power steering pump and lay the assembly aside.

19. Disconnect the heater hoses at the engine.

20. Remove the thermostat housing.

21. Remove the oil filler tube.

22. Remove the cruise control servo, servo bracket and transducer.

23. Raise and support the vehicle safely.

24. Drain the engine oil.

25. Disconnect the exhaust pipes at the manifolds.

26. Remove the driveshaft and plug the end of the transmission.

27. Disconnect the transmission shift linkage and the speedometer cable or speed sensor connector.

28. Disconnect the fuel and vapor return lines at the engine.

29. Remove the rear engine/transmission mount bolts.

30. Support the transmission and engine.

31. Remove the engine mount bracket-to-frame bolts.

32. Remove the engine mount through bolts.

33. Raise the engine slightly and remove the engine mounts.

34. Support the engine with a block of wood between the oil pan and the crossmember.

35. Lower the vehicle.

36. Install a suitable engine lifting device.

37. Remove the engine and transmission from the vehicle as an assembly.

38. Separate the engine and transmission as required.

To install:

39. Install the engine mount through bolts. Torque the bolts to specification.

40. Install the engine mount bracket-to-frame bolts. Torque the bolts to specification.

41. Remove the engine lifting device. Raise and support the vehicle safely.

42. Install the transmission mounting bolts. Torque the bolts to specification.

43. Connect the transmission shift linkage and the speedometer cable or speed sensor connector.

44. Install the driveshaft.

45. Install the condenser.

46. Install the hood latch support.

47. Install the lower fan shroud and filler panel.

48. Install the transmission dipstick tube and the accelerator cable at the tube.

49. Install the coolant hose at the intake manifold and the PCV valve.

50. Install the distributor cap and wires.

51. Install the cruise control servo, servo bracket and transducer.

52. Install the oil filler pipe and the engine dipstick tube.

53. Install the thermostat housing.

54. Connect the heater hoses at the engine.

55. Connect the engine wiring harness from the firewall connection.

56. Install the radiator and shroud.

57. Install the radiator support bracket.

58. Install the TBI unit.

59. Connect the accelerator linkage.

60. Install the windshield wiper jar and bracket.

61. Install the air conditioning compressor.

62. Connect the vacuum lines.

63. Charge the air conditioning system using the proper equipment.

64. If equipped with an automatic transmission, install the fluid cooler lines at the radiator.

65. Install the radiator coolant reservoir bottle.

66. Connect the radiator hoses at the radiator.

67. Install the upper radiator support the grille and the lower grille valance.

68. Install the air cleaner.

69. Install the air stove pipe.

70. Install the engine cover.

71. Fill the cooling system.

72. Connect the battery cables and inspect for leaks.

7.4L ENGINE

1. Disconnect the battery cables.

2. Drain the cooling system into a suitable container.

3. Remove the engine cover.

4. Remove the air cleaner and ducts.

5. Remove the cruise control servo, servo bracket and transducer, as required.

6. Remove the grille, lower grille valence, headlight bezels and all necessary sheetmetal to ease REMOVAL AND INSTALLATION of the engine and transmission assembly.

7. Remove the upper radiator support.

8. Discharge the air conditioning system using the proper equipment.

9. Remove the air conditioning condenser from in front of the radiator. Cap all lines.

10. Disconnect the radiator hoses at the radiator.

11. Disconnect the oil and transmission cooler lines.

12. Remove the radiator coolant reservoir bottle.

13. Remove the radiator support bracket and remove the radiator and the shroud.

14. Remove the power steering pump and position aside.

15. Remove the air conditioning compressor and brackets. Cap all open lines.

16. Disconnect the wiring, fuel lines and linkage at the TBI unit.

17. Properly relieve the fuel system pressure and remove the TBI unit.

18. Disconnect the ECS and MAP sensors with brackets. Disconnect the engine wiring harness from the firewall connection.

19. Disconnect the starter wires.

20. Disconnect the alternator wires.

21. Disconnect the temperature sensor wire.

22. Disconnect the oil pressure sender.

23. Disconnect the distributor and coil wiring.

24. Disconnect and plug the fuel supply and vapor lines.

25. Tag and disconnect all vacuum hoses.

26. Disconnect the heater hoses at the engine.

27. Remove the thermostat housing.

28. Remove the windshield wiper jar and bracket.

29. Remove the oil filler pipe and the engine dipstick tube.

30. Raise and support the vehicle safely.

31. Disconnect the exhaust pipes at the manifolds.

32. Remove the driveshaft and plug the end of the transmission.

33. Disconnect the transmission shift linkage and the speedometer cable.

34. Drain the engine oil.

35. Support the engine with a floor jack. Do not position the jack under the oil pan, crankshaft pulley or any sheetmetal!.

36. Attach a suitable lifting device to the engine to relieve the weight.

37. Remove the transmission mount bolts and remove the crossmember.

38. Raise the engine slightly and remove the engine mounts. Support the engine with wood between the oil pan and the crossmember.

39. Lower the vehicle.

40. Raise the engine as necessary and maneuver the engine/transmission assembly from the vehicle.

41. Separate the engine and transmission as follows:

   a. Support the engine properly.

   b. Remove the starter and converter housing underpan.

   c. Remove the flywheel-to-converter attaching bolts.

   d. Support the transmission on blocks.

   e. Disconnect the detent cable.

   f. Remove the transmission-to-engine mounting bolts.

   g. Remove the blocks from the engine only and guide the engine away from the transmission.

42. Mount the engine on a work stand.

To install:

43. Install the automatic transmission as follows:

   a. Position the transmission and engine together.

   b. Install the transmission-to-engine mounting bolts. Torque the bolts to specification.

   c. Connect the throttle linkage detent cable, if equipped.

   d. Install the flywheel-to-converter attaching bolts. Torque the bolts to specification.

   e. Install the starter and converter shield.

44. Raise the engine/transmission and guide the assembly into position in the vehicle.

45. Raise the engine slightly and install the engine mounts. Torque the mount-to-block bolts to specification.

46. Install the engine mount through bolts. Torque the bolts to specification.

47. Install the engine mount bracket-to-frame bolts. Torque the front and the rear mount bolts to specification.

48. Install the transmission mounting bolts. Torque the nuts to specification.

49. Connect the transmission shift linkage and the speedometer cable.

50. Install the driveshaft.

51. Connect the exhaust pipes at the manifolds.

52. Install the oil filler pipe, engine and transmission dipstick tube.

53. Install the windshield wiper jar and bracket.

54. Install the thermostat housing.

55. Connect the heater hoses at the engine.

56. Connect all vacuum hoses.

57. Connect the fuel supply and vapor lines.

58. Connect the distributor and coil wiring.

59. Connect the oil pressure sender.

60. Connect the temperature sensor wire.

61. Connect the alternator wires.

62. Connect the starter wires.

63. Connect the engine wiring harness at the firewall connection.

64. Install the TBI unit.

65. Connect the wiring, fuel lines and linkage at the TBI unit.

66. Install the air conditioning compressor.

67. Install the power steering pump.

68. Install the radiator and the shroud.

69. Install the radiator support bracket.

70. Install the radiator coolant reservoir bottle.

71. Install the fluid cooler lines at the radiator.

72. Connect the radiator hoses at the radiator.

73. Install the air conditioning condenser.

74. Install the air conditioning vacuum reservoir.

75. Charge the air conditioning system using the proper equipment.

76. Install the upper radiator support.

77. Connect the radiator hoses at the radiator.

78. Install the grille and the lower grille valance.

79. Install the cruise control servo, servo bracket and transducer.

80. Fill the crankcase the proper type and quantity of oil.

81. Install the air cleaner.

82. Install the engine cover.

83. Fill the cooling system.

84. Connect the battery cables.

Engine Mounts

REMOVAL AND INSTALLATION

1. Disconnect the negative battery cable.

2. Raise and safely support the vehicle.

3. Support the engine with a suitable jack.

NOTE: Do not position the jack under the oil pan, any sheetmetal or the crankshaft pulley, otherwise damage may occur.

4. Remove the engine mount through-bolt.

5. Raise the engine to gain sufficient clearance for the mount to be removed, being careful not to crush the distributor against the firewall.

6. Remove the engine mount attaching bolts, nuts and washers.

7. Remove the mount assembly.

8. Installation is the reverse of the removal procedure.

Cylinder Head

REMOVAL AND INSTALLATION

4.3L Engine

1. Disconnect the negative battery cable.

2. Remove the engine cover, as required. Drain the cooling system into a suitable container.

3. Relieve the fuel system pressure and remove the intake manifold.

4. Remove the exhaust manifold.

5. Remove the air pipe at the rear of the head, if equipped.

6. Remove the air pump bolt and spacer at the right cylinder head, if equipped.

7. Remove the power steering pump and brackets from the cylinder head, and lay them aside (left cylinder head).

8. Remove the air conditioner compressor, accessory brackets and alternator and position aside, as required.

9. Remove the spark plugs wires, ground straps and engine wiring harness connectors.

10. Remove the fuel lines and brackets.

11. Remove the rocker arm covers, the rocker arms and pushrods.

12. Remove the cylinder head bolts.

13. Remove the cylinder head.

To install:

14. Clean all gasket mating surfaces, install a new gasket and reinstall the cylinder head.

NOTE: Apply sealer to both sides of steel gasket. If a composition gasket is used, do not use sealer.

15. Clean the bolts, apply sealer to the threads, and install the bolts hand tight.

16. Tighten the bolts in the proper sequence and torque specifications.

17. Install the rocker arms and pushrods.

18. Install the wiring harness and accessory brackets.

19. Install the spark plugs.

20. Install the rocker arm cover.

21. Install the air conditioner compressor.

22. Install the power steering pump and brackets.

23. Install the alternator mounting bolt at the cylinder head.

24. Install the air pipe at the rear of the head.

25. Install the exhaust manifold.

26. Install the intake manifold.

27. Install the engine cover.

28. Connect the negative battery cable.

5.0L and 5.7L Engines

1. Disconnect the negative battery cable.

2. Properly relieve the fuel system pressure.

3. Drain the cooling system.

4. Remove the intake manifold.

5. Remove the exhaust manifolds and position out of the way.

6. Remove the air pump bolt and spacer.

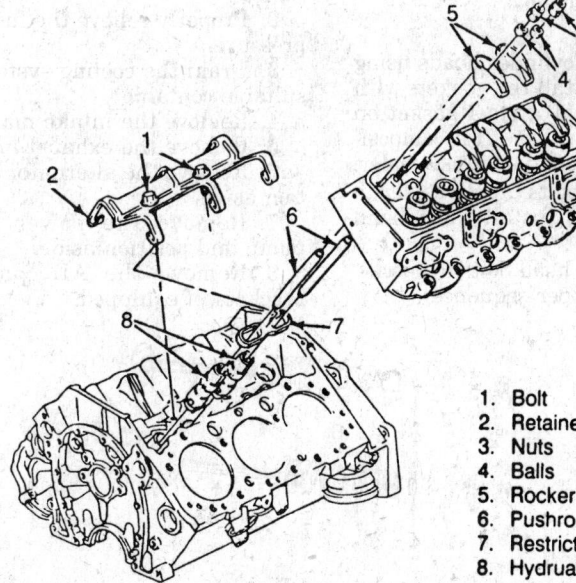

1. Bolt
2. Retainer
3. Nuts
4. Balls
5. Rocker arms
6. Pushrods
7. Restrictor
8. Hydrualic lifter

79114039

Cylinder head and components — 4.3L engine

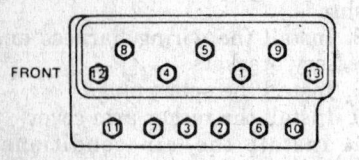

Cylinder head bolt torque sequence — 4.3L engine

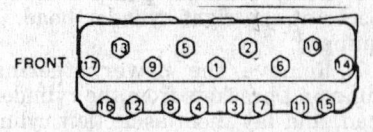

Cylinder head bolt torque sequence — 5.0L and 5.7L engines

7. If equipped with air conditioning, remove the compressor and the forward mounting bracket and lay the compressor aside. Do not disconnect any of the refrigerant lines.

8. Disconnect the wiring harness brackets, fuel lines and spark plug wires.

9. Remove the alternator, as required and position aside.

10. Remove the power steering pump, as required and position aside.

11. Remove the stud and nut securing the main accessory bracket to the cylinder head.

12. Remove the valve covers. Back off the rocker arm nuts and pivot the rocker arms out of the way and remove the pushrods. Mark the pushrods so they can be installed in their original positions.

13. Remove the cylinder head bolts and remove the heads.

To install:

14. Install the cylinder heads using new gaskets. Install the gaskets with the bead up. Coat a steel gasket on both sides with sealer. If a composition gasket is used, do not use sealer.

15. Clean the bolts, apply sealer to the threads, and install them hand tight.

16. Torque the head bolts to specification in the proper sequence .

17. Install the intake and exhaust manifolds.

18. Adjust the rocker arms and install the valve covers.

19. Install the exhaust manifolds.

20. Install the alternator, power steering pump and fuel lines.

21. Install the air pump and air conditioning compressor.

22. Connect all vacuum and electrical connectors. Install spark plugs and wires.

23. Fill the cooling system with the proper type and quantity of antifreeze.

24. Connect the negative battery cable.

7.4L Engine

1. Disconnect the negative battery cable.

2. Properly relieve the fuel system pressure.

3. Drain the cooling system into a suitable container.

4. Remove the intake manifold.

5. Remove the exhaust manifolds.

6. Remove the alternator and position aside.

7. Remove the power steering pump and position aside.

8. Remove the AIR pump and brackets, if equipped.

9. If equipped with air conditioning, remove the compressor and the forward mounting bracket and lay the compressor aside. Do not disconnect any of the refrigerant lines.

10. Remove the idler pulley and A/C compressor rear bracket on the vehicle, as required.

11. Remove the rocker arm covers.

12. Remove the spark plugs.

13. Remove the AIR pipes at the rear of the head, as required.

14. Disconnect the ground strap at the rear of the head.

15. Disconnect the sensor wire.

16. Back off the rocker arm nuts and pivot the rocker arms out of the way so the pushrods can be removed. Identify the pushrods so they can be installed in their original positions.

17. Remove the cylinder head bolts and remove the head.

To install:

18. Thoroughly clean the mating surfaces of the head and block. Clean the bolt holes thoroughly.

19. Install the cylinder heads using new gaskets. Install the gaskets with the bead up.

NOTE: Coat a steel gasket on both sides with sealer. If a composition gasket is used, do not use sealer.

20. Clean the bolts, apply sealer to the threads, and install the bolts hand tight.

21. Torque the head bolts in sequence to the proper torque specifications.

22. Install the intake and exhaust manifolds.

23. Install the pushrods.

24. Install the rocker arms and adjust properly.

25. Connect the sensor wire.

26. Connect the ground strap at the rear of the head.

27. Install the AIR pipes at the rear of the head.

28. Install the spark plugs.

29. Install the rocker arm cover.

30. Install the air conditioning compressor and the forward mounting bracket.

31. Install the AIR pump.

32. Install the alternator.

Valve Lifters

REMOVAL AND INSTALLATION

1. Disconnect the negative battery cable. Drain the cooling system into a suitable container. Properly relieve the fuel system pressure.

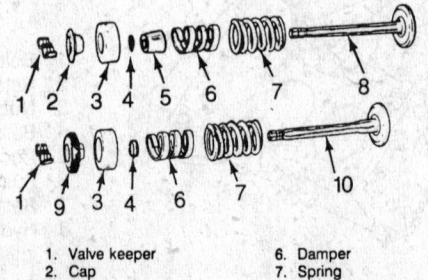

1. Valve keeper
2. Cap
3. Shield
4. O-Ring seal
5. Seal
6. Damper
7. Spring
8. Intake valve
9. Rotator
10. Exhaust valve

Valves and components — 4.3L, 5.0L and 5.7L engines

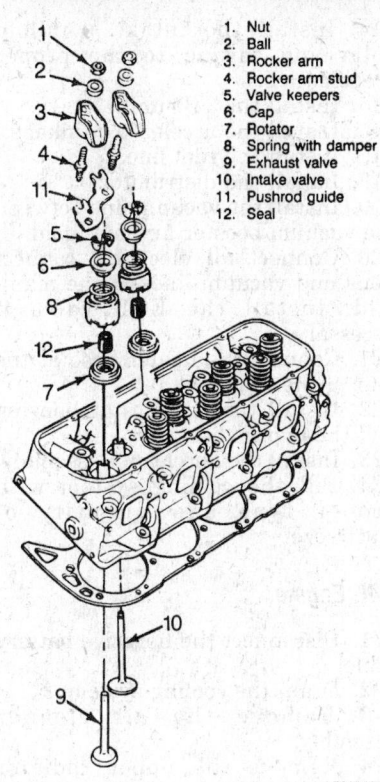

1. Nut
2. Ball
3. Rocker arm
4. Rocker arm stud
5. Valve keepers
6. Cap
7. Rotator
8. Spring with damper
9. Exhaust valve
10. Intake valve
11. Pushrod guide
12. Seal

79114040

Cylinder head and components — 7.4L engine

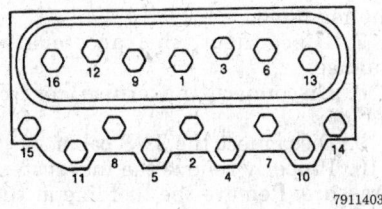

79114038

Cylinder head bolt torque sequence — 7.4L engine

2. Remove the air cleaner assembly.

3. Remove the distributor assembly.

4. Remove the intake manifold, valve covers and pushrods.

5. Remove the retainer bolts and the valve lifter retainer, as required.

6. Remove the valve lifter restrictor plate, as required.

7. Remove the lifters and place them in the exact order of removal to ensure lifter is installed in original bore upon reinstallation.

To install:

8. Lubricate the lifter contact and/or roller points with a suitable engine oil supplement or Molylube®.

9. Install the used lifters in its original bore or new lifters in any bore.

10. Install the valve lifter restrictor plate, if equipped.

11. Install the retainer bolts and the valve lifter retainer, if equipped.

12. Install the intake manifold and pushrods. Adjust the valve lifters to specification; install the valve covers.

13. Install the distributor assembly.

14. Install the air cleaner assembly.

15. Connect the negative battery cable. Fill the cooling system with the proper type and quantity of antifreeze.

Valve Lash

ADJUSTMENT

All engines use hydraulic lifters, which require no periodic adjustment. In the event of cylinder head removal or any operation that requires disturbing the rocker arms, the rocker arms will have to be adjusted on all vehicles, except the 7.4L engine, which use a positive stop bolt on the rocker arm instead of a stud and nut design.

1. Remove the valve covers and gaskets.

2. Crank the engine until the mark on the damper aligns with the **TDC** or **0** mark on the timing tab and the engine is in No. 1 cylinder firing position. This can be determined by placing a finger on the No. 1 cylinder valves as the marks align. If the valves do not move, it is in the No. 1 firing position. If the valves move, it is in No. 6 firing position (No. 4 on the V6 engine) and the crankshaft should be rotated 1 more revolution to the No. 1 firing position.

3. With the engine in No. 1 firing position, the following valves can be adjusted: **V6 Engine** Exhaust—1, 5, 6 Intake—1, 2, 3 **V8 Engine** Exhaust—1, 3, 4, 8 Intake—1, 2, 5, 7

4. To adjust the valves, proceed as follows:

   a. Back off the adjusting nut until lash is felt at the pushrod.

   b. Turn in the adjusting nut until all lash is removed and the pushrod is not capable of being rotated.

   c. Turn the adjusting nut in an additional ¾–1 turn to center the lifter plunger.

5. Crank the engine 1 full revolution until the marks are again in alignment. This is No. 6 firing position (No. 4 on the V6 engine). The following valves can now be adjusted: **V6 Engine** Exhaust—2, 3, 4 Intake—4, 5, 6 **V8 Engine** Exhaust—2, 5, 6, 7 Intake—3, 4, 6, 8

6. Repeat the adjustment procedure as directed in step 4.

7. Reinstall the valve covers using new gaskets.

Valve Arrangement

4.3L ENGINE

Left Side: E–I–E–I–E
  Right Side: E–I–I–E–I–E

5.0L AND 5.7L ENGINES

Left Side: E–I–I–E–E–I–I–E
  Right Side: E–I–I–E–E–I–I–E

7.4L ENGINE

Left Side: E–I–E–I–E–I–E–I
  Right Side: I–E–I–E–I–E–I

Rocker Arm

REMOVAL AND INSTALLATION

1. Disconnect the negative battery cable. On G-Series vehicle, remove the engine cover.

2. Remove all the necessary components in order to gain access to the engine valve covers. As required, properly relieve the fuel system pressure before disconnecting any fuel lines.

3. Remove the valve cover retaining bolts. Remove the valve cover from the engine.

4. Remove the rocker arm assemblies. Keep them in order for reinstallation.

5. Installation is the reverse of the removal procedure. Be sure to use new gaskets or RTV sealant, as necessary.

Intake Manifold

REMOVAL AND INSTALLATION

4.3L, 5.0L and 5.7L Engines

1. Disconnect the negative battery cable. Drain the cooling system.

2. Remove the air cleaner assembly.

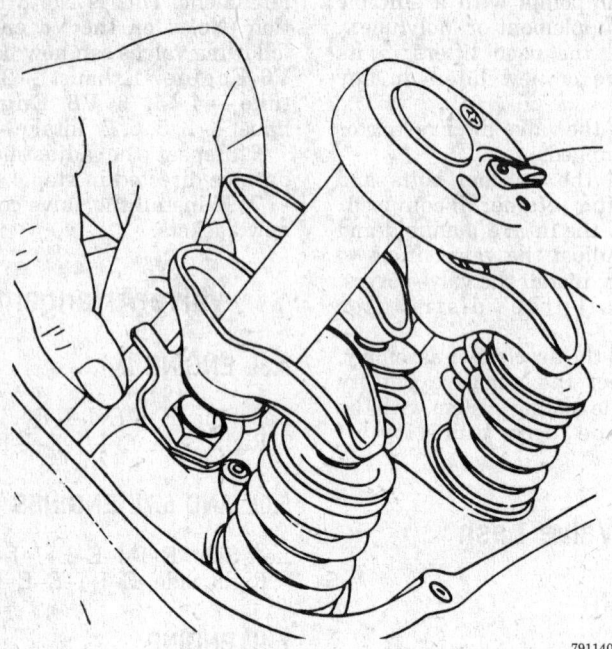

Adjusting hydraulic valves

79114042

3. Remove the thermostat housing and the bypass hose. It is not necessary to remove the top radiator hose from the thermostat housing.

4. Disconnect the heater hose at the rear of the manifold.

5. Disconnect all electrical connections and vacuum lines from the manifold. Remove the EGR valve if necessary.

6. Remove the power brake vacuum line from the vacuum booster to the manifold.

7. Remove the distributor.

8. Relieve the fuel system pressure and remove the fuel line at the TBI unit.

9. Remove the accelerator linkage.

10. Remove the TBI unit.

11. Remove the intake manifold bolts. Remove the manifold and the gaskets.

To install:

NOTE: Before installing the intake manifold, ensure the gasket surfaces are thoroughly clean.

12. Apply $^3/_{16}$ in. (5mm) bead of sealant to the front and rear sealing surfaces of the engine block manifold. Extend the bead $^1/_2$ in. (13mm) up each cylinder head to seal and retain the gaskets.

13. Install the manifold and the gaskets.

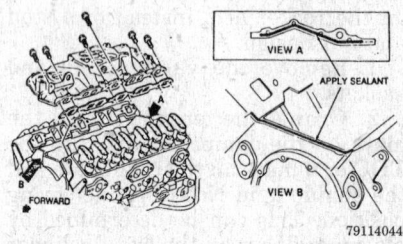

Intake manifold installation — 5.0L and 5.7L engines

79114044

14. Install the intake manifold bolts and torque to the proper specification.

15. Install the TBI unit.

16. Install the accelerator linkage.

17. Install the fuel line.

18. Install the distributor.

19. Install the vacuum line between the vacuum booster and manifold.

20. Connect all electrical connections and vacuum lines at the manifold. Install the EGR valve if necessary.

21. Connect the heater hose at the rear of the manifold.

22. Install the thermostat housing and the bypass hose.

23. Install the air cleaner assembly.

24. Fill the cooling system with proper type and quantity of antifreeze.

7.4L Engine

1. Disconnect the negative battery cable.

2. Drain the cooling system.

3. Remove the air cleaner assembly.

4. Remove the upper radiator hose, thermostat housing and the bypass hose.

5. Disconnect the heater hose and pipe at the TBI unit, if equipped.

6. Tag and disconnect all electrical connections and vacuum lines from the manifold.

7. Disconnect the accelerator linkage.

8. Disconnect the cruise control cable.

9. Disconnect the TVS cable.

10. Properly relieve the fuel system pressure. Remove the fuel line at the TBI unit.

11. Remove the TBI unit.

12. Remove the distributor assembly.

13. Remove the cruise control transducer.

14. Disconnect and remove the ignition coil.

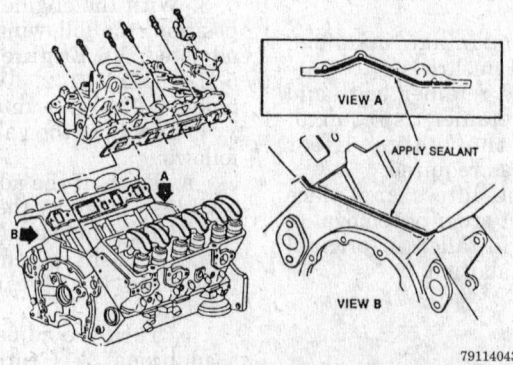

Intake manifold installation — 4.3L engine

79114043

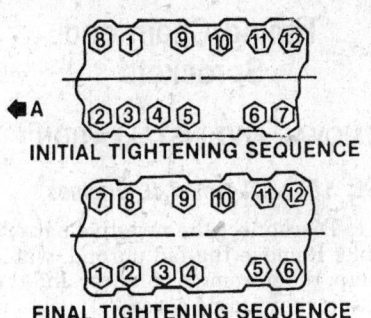

INITIAL TIGHTENING SEQUENCE

FINAL TIGHTENING SEQUENCE

A. Front Of Engine

79114045

Intake manifold torque sequence — 4.3L engine

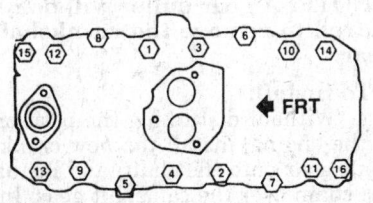

79114048

Intake manifold torque sequence — 1991-92 7.4L engine

15. Remove the EGR solenoid and bracket.
16. Remove the MAP sensor and bracket.
17. Remove the air conditioning compressor rear bracket.
18. Remove the front alternator/AIR pump bracket.
19. Remove the intake manifold bolts.
20. Remove the manifold, gaskets and seals.
 To install:
21. Install the manifold, gaskets and seals.

22. Install the intake manifold bolts. Torque the bolts to 30 ft. lbs. (40 Nm).
23. Install the front alternator/AIR pump bracket.
24. Install the air conditioning compressor rear bracket.
25. Install the MAP sensor and bracket.
26. Install the EGR solenoid and bracket.
27. Connect the ignition coil wires.
28. Install the cruise control transducer.
29. Install the distributor.
30. Install the TBI unit.
31. Install the fuel line at the TBI unit.
32. Connect the TVS cable.
33. Connect the cruise control cable.
34. Connect the accelerator linkage.
35. Connect all electrical connections and vacuum lines at the manifold.
36. Connect the heater hose and pipe.
37. Install the upper radiator hose, thermostat housing and the bypass hose.
38. Install the air cleaner assembly.
39. Fill the cooling system.
40. Connect the negative battery cable.

Exhaust Manifold

REMOVAL AND INSTALLATION

4.3L, 5.0L and 5.7L Engines

1. Disconnect the negative battery cable. Remove the air cleaner.
2. Remove the hot air shroud, if equipped.
3. Loosen the alternator and remove the lower bracket, as required.

4. Remove the air conditioner compressor rear bracket and the diverter valve and bracket.

NOTE: On models with air conditioning, it may be necessary to remove the compressor, and tie it out of the way. Do not disconnect the compressor lines.

5. Disconnect the oxygen sensor wire, if equipped.
6. Remove the power steering pump rear bracket and the AIR hose at the check valve, as required.
7. Raise and support the vehicle safely.
8. Disconnect the crossover pipe from both manifolds.
9. Remove the manifold bolts and remove the manifold(s). Some models have lock tabs on the front and rear manifold bolts which must be removed before removing the bolts.
10. Installation is the reverse of removal. Torque the bolts to the proper specification and sequence.

7.4L Engine

1. Disconnect the negative battery cable.
2. Remove the heat stove pipe.
3. Remove the dipstick tube.
4. Disconnect the AIR hose at the check valve and AIR lines at the manifold, as required.
5. Remove the spark plugs and wires.
6. Raise and support the vehicle safely.
7. Disconnect the exhaust pipe at the manifold. Remove the heat shield, if equipped. Disconnect the oxygen sensor wire, as required.
8. Lower the vehicle and remove the manifold bolts and spark plug heat shields.

NOTE: It may be necessary to raise the engine slightly to gain sufficient clearance for removal of the manifold on G-Series vehicles.

9. Remove the manifold assemblies.
 To install:
10. Clean the mating surfaces.
11. Clean the stud threads.
12. Install the manifold, spark plug heat shields and bolts. Tighten the bolts to 40 ft. lbs. (54 Nm). starting from the center bolts and working towards the outside.
13. Raise and safely support the vehicle. Connect the exhaust pipe, heat shields and oxygen sensor at the manifold.
14. Lower the vehicle. Install the spark plugs and wires.

FRONT

79114046

Intake manifold torque sequence — 5.0L and 5.7L engines

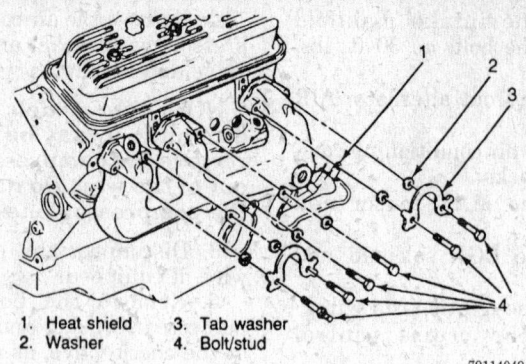

1. Heat shield   3. Tab washer
2. Washer   4. Bolt/stud

79114049

Exhaust manifold installation — typical

15. Connect the AIR hose and lines.
16. Install the dipstick tube.
17. Install the heat stove pipe.
18. Connect the battery negative cable.

Timing Chain Front Cover

REMOVAL AND INSTALLATION

1. Disconnect the negative battery cable.
2. Drain the cooling system. Remove the fan shroud assembly.
3. Remove the belts, pulleys and water pump assembly.
4. Remove the crankshaft pulley and damper.
5. Remove the oil pan-to-front cover bolts.
6. Remove the screws holding the timing chain cover to the block, pull the cover forward enough to cut the front oil pan seal. Cut the seal flush with the block on both sides.
7. Pull off the cover and gaskets.
8. Use a suitable tool to pry the old seal out of the front face of the cover.

To install:
9. Using seal driver J–22102 or equivalent, install the new seal so the open end is toward the inside of the cover.

NOTE: Coat the lip of the new seal with oil prior to installation.

10. Install a new front pan seal, cutting the tabs off.
11. Coat a new cover gasket with adhesive sealer and position it on the block.
12. Apply a 1/8 in. bead of RTV gasket material to the front cover. Install the cover carefully onto the locating dowels.
13. Tighten the attaching screws to specification.
14. Tighten the cover-to-pan bolts to specification.
15. Install the torsional damper.

16. Install the water pump assembly.
17. Connect the battery cables.
18. Fill the cooling system with the proper type and quantity of antifreeze.

Front Cover Oil Seal

REPLACEMENT

1. Disconnect the negative battery cable.
2. Remove the fan shroud, belts, fan and pulleys.
3. Remove the accessory drive pulley.
4. Remove the torsional damper bolt, then the damper using tool J–23523–E.
5. Pry the front crankshaft seal using a suitable tool, being careful not to damage the cover.

NOTE: Ensure the seal contact area on the crankshaft is not scored or grooved, repair or replace as necessary.

To install:
6. Install the seal using an appropriate seal installation tool. Ensure the seal open end is facing inside the engine. Coat the seal with engine oil.
7. Install the torsional damper using tool J–23523–E, then install the bolt and washer; torque to specification.
8. Install the accessory drive pulley, fan, shroud and pulleys.
9. Connect the negative battery cable.

Timing Chain and Sprockets

REMOVAL AND INSTALLATION

4.3L, 5.0L, 5.7L and 7.4L Engines

1. Disconnect the negative battery cable. Remove the fan shroud, water pump, the harmonic balancer and the timing chain front cover.
2. Crank the engine until the timing marks on both sprockets are nearest each other and in line between the shaft centers.
3. Remove the camshaft-to-gear attaching bolts. Remove the chain and the camshaft gear.

NOTE: A gear puller will be required to remove the crankshaft gear.

To install:
4. Without disturbing the position of the engine, install the new crankshaft gear onto the shaft, and install the chain over the camshaft gear. Install the camshaft gear with the timing marks aligned. Position the camshaft locating dowel in the dowel hole of the cam sprocket.
5. Install the camshaft securing bolts and torque to specification.
6. After the gears are in place, turn the engine 2 full revolutions to make certain the timing marks are in correct alignment between the shaft centers.

Camshaft

REMOVAL AND INSTALLATION

4.3L Engine

1. Disconnect the negative battery cable and properly relieve the fuel system pressure.
2. Remove the rocker arm covers.
3. Remove the intake manifold, pushrods and hydraulic lifters.
4. Remove the outside air duct.
5. Remove the power steering reservoir out of the way.
6. Remove the upper fan shroud bolts.
7. Drain and remove the radiator.
8. Disconnect the hood release cable at the latch.
9. Remove the upper fan shroud.
10. Remove the water pump.
11. Remove the torsional damper.
12. Remove the front cover.
13. Align the timing marks and remove the camshaft sprocket bolts.
14. Remove the camshaft sprocket and timing chain. Loosen the

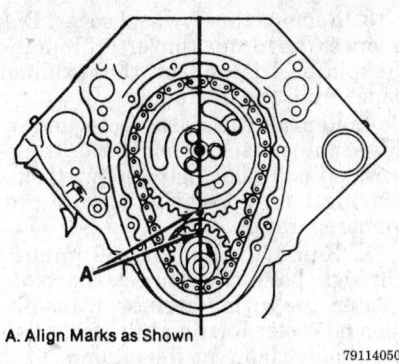

A. Align Marks as Shown

79114050

Timing alignment marks

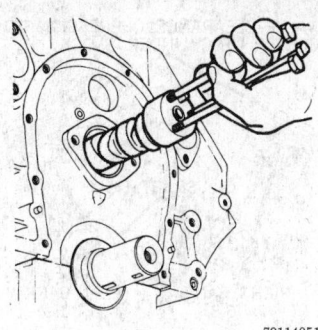

79114051

**Camshaft removal and installation —
4.3L, 5.0L, 5.7L and 7.4L engines**

sprocket by tapping on the lower edge. Remove the camshaft thrust plate.

15. Remove the crankshaft sprocket, as required.

16. Remove the front grill and condenser assembly, as required.

17. Install two or three 5/16–18 bolts 4–5 in. long into the camshaft threaded holes to handle the camshaft and carefully pull the camshaft from the block.

18. Inspect the shaft for signs of excessive wear or damage.

To install:

19. Liberally coat camshaft and bearings with heavy engine oil or engine assembly lubricant and insert the camshaft into the engine.

20. Install the condenser and grille assembly, if removed.

21. Align the timing marks on the camshaft and crankshaft gears.

22. Install the camshaft thrust-plate, sprocket and chain. Tighten the bolts to specification.

23. Install the front cover assembly.

24. Install the torsional damper.

25. Install the water pump.

26. Install the fan shroud and radiator.

27. Connect the hood release cable at the latch.

28. Install the outside air duct.

29. Install the hydraulic lifters and pushrods.

30. Adjust the valves.

31. Install the rocker arm covers.

32. Install the intake manifold.

33. Refill the cooling system.

34. Connect the battery cable.

5.0L and 5.7L Engines

1. Disconnect the negative battery cable, Drain the cooling system and properly relieve the fuel system pressure.

2. On the G-Series, remove the engine cover.

3. Remove the air cleaner.

4. On the G-Series, remove the grille and center support.

5. Remove the air conditioning condenser and swing the condenser forward from its mounting, if equipped.

6. Remove the fan, the shroud and the radiator.

7. Remove the valve covers.

8. Remove the water pump assembly.

9. Align the timing marks and remove the torsional damper.

10. Remove the timing chain cover.

11. Disconnect the electrical and vacuum connections at the intake manifold.

12. Mark the distributor rotor-to-housing location. Remove the distributor assembly.

13. Remove the intake manifold, pushrods and hydraulic lifters.

14. Remove the camshaft sprocket bolts, camshaft sprocket and timing chain. Tap the sprocket on its lower edge to loosen it.

15. Remove the crankshaft sprocket, as required.

16. As required, remove the front engine mount through bolts and raise the engine to gain sufficient clearance for camshaft removal.

17. Install two or three 5/16–18 bolts 4–5 in. long into the camshaft threaded holes; carefully pull the camshaft from the block.

18. Inspect the shaft for signs of excessive wear or damage.

To install:

19. Liberally coat camshaft and bearing with heavy engine oil or engine assembly lubricant and insert the cam into the engine.

20. Lower the engine and install the engine mount through bolts.

21. Align the timing marks on the camshaft and crankshaft gears.

22. Install the camshaft sprocket and chain and tighten the bolts to specification.

23. Install the hydraulic lifters and pushrods and adjust the valves.

24. Install the distributor assembly.

25. Install the timing chain cover.

26. Install the torsional damper.

27. Install the water pump.

28. Install the valve covers.

29. Install the fan, the shroud and radiator.

30. Install the air conditioning condenser, if equipped.

31. On the G-Series, install the grille and center support.

32. Install the air cleaner.

33. On the G-Series, install the engine cover.

34. Connect the battery cable and fill the cooling system.

7.4L Engine

1. Disconnect the negative battery cable. Properly relieve the fuel system pressure.

2. Remove the engine cover, G-Series. Remove the air cleaner assembly.

3. Remove the grille and center support section, as required.

4. Properly discharge the air conditioning system. Remove the air conditioning compressor, condenser and auxiliary fan, if equipped.

5. Drain the cooling system.

6. Remove the fan, the shroud and radiator.

7. Remove the alternator belt, remove the alternator assembly, as required.

8. Remove the valve covers.

9. Disconnect the hoses from the water pump.

10. Remove the water pump.

11. Align the timing marks at TDC. Remove the harmonic balancer and pulley.

12. Remove the engine front cover.

13. Mark the distributor rotor-to-housing location. Remove the distributor assembly.

14. Remove the intake manifold assembly.

15. Remove the lifters, pushrods, and rocker arms.

16. Rotate the camshaft so the timing marks align.

17. Remove the camshaft sprocket bolts. Remove the camshaft sprocket and timing.

18. Remove the engine mount through bolts. Raise and support the engine to aid in camshaft removal, as required.

19. Install two or three 5/16–18 bolts in the holes in the front of the camshaft and carefully pull the camshaft from the block.

To install:

20. Liberally coat camshaft and bearing with heavy engine oil or en-

gine assembly lubricant and insert the cam into the engine.

21. Align the timing marks on the camshaft sprocket and crankshaft gears.

22. Install the camshaft sprocket and chain and tighten the bolts to specification. Lower the engine and install the engine mount bolts.

23. Install the lifters and pushrods and adjust the valves.

24. Install the intake manifold.

25. Install the distributor using the locating marks made during removal.

26. Install the engine front cover.

27. Install the harmonic balancer and pulley.

28. Install the water pump.

29. Connect the hoses at the water pump.

30. Install the valve covers.

31. Install the alternator.

32. Install the fan shroud and radiator.

33. Fill the cooling system with the proper type and quantity of antifreeze.

34. Install the air conditioning condenser and compressor.

35. Install the grille and center support.

36. Install the air cleaner assembly.

37. Connect the battery.

Piston and Connecting Rod

POSITIONING

DIESEL ENGINE MECHANICAL

NOTE: Disconnecting the negative battery cable on some vehicles may interfere with the functions of the on board computer

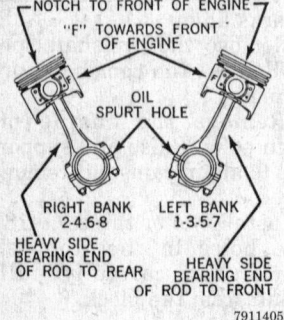

Correct relationship of the piston and rod — 5.0L and 5.7L engines

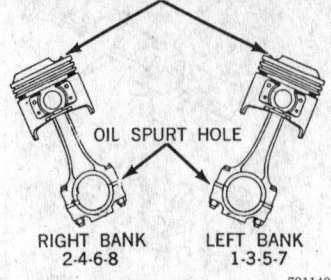

Correct relationship of the piston and rod — 7.4L engine

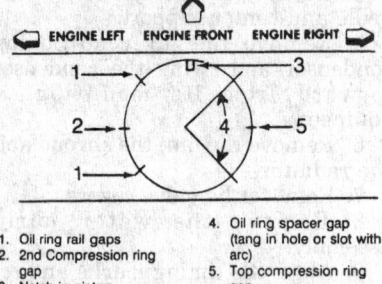

1. Oil ring rail gaps
2. 2nd Compression ring gap
3. Notch in piston
4. Oil ring spacer gap (tang in hole or slot with arc)
5. Top compression ring gap

Piston ring gap locations — 4.3L, 5.0L and 5.7L engines

systems and may require the computer to undergo a relearning process.

Engine

REMOVAL AND INSTALLATION

6.2L Engine

EXCEPT G-SERIES

1. Remove the hood. Disconnect and remove the batteries from the vehicle.

2. Raise the vehicle and support it safely. Drain the engine oil.

3. Remove the flywheel cover. Disconnect the torque converter from the flexplate. Disconnect the exhaust pipes from the manifolds.

4. Remove the starter bolts and remove the starter. Remove the transmission bell housing to engine bolts, leaving 1 or more bolts loose to prevent separation.

5. Remove the engine mount through bolts. Disconnect the block heater, the wiring harness, transmission oil cooler lines and the front battery cable clamp at the oil pan.

6. Lower the vehicle. Properly relieve the fuel pump pressure. Disconnect and plug the fuel lines and the oil cooler lines at the engine block. Remove the lower fan shroud bolts.

7. Drain the engine coolant. Remove the air cleaner assembly. Disconnect the ground cable from the alternator bracket. Disconnect the alternator wires and clips.

8. Disconnect the TPS, EGR–EPR and the fuel cut-off at the injection pump. Remove the harness from the clips at the rocker covers and disconnect the glow plugs.

9. Disconnect the EGR–EPR solenoids, glow plugs, controller, temperature sender and move the harness aside. Disconnect the ground strap on the left or right side.

10. Remove the fan assembly. Remove the upper radiator hoses at the engine. Remove the fan shroud.

11. Remove the power steering pump and belt. Remove the reservoir and lay the pump and reservoir aside. If equipped with air conditioning, remove the compressor and position aside.

12. Disconnect the vacuum lines at the cruise servo and accelerator cable at the injection pump. Disconnect the heater hoses at the engine.

13. Disconnect the lower radiator hose, the oil cooler lines, the heater hose and the overflow hose at the radiator.

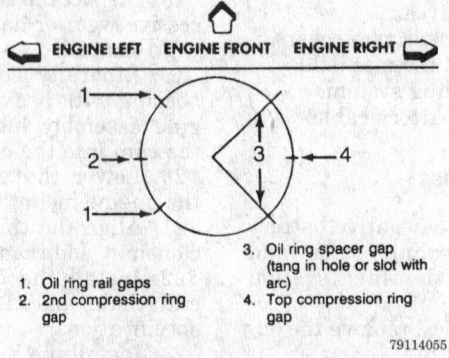

1. Oil ring rail gaps
2. 2nd compression ring gap
3. Oil ring spacer gap (tang in hole or slot with arc)
4. Top compression ring gap

Piston ring gap locations — 7.4L engine

14. Remove the radiator assembly. Remove the detent cable.

15. Install an engine lifting device, remove the loose bolts in the bell housing. Properly support the transmission. Carefully remove the engine from the vehicle.

To install:

16. Lower the engine and install the engine mounting bolts. Torque the bolts and nuts to specificati

17. Install the bell housing to engine bolts and torque to 30 ft. lbs. (40 Nm).

18. Remove the engine lifting fixture and transmission jack.

19. Raise the vehicle and support it safely.

20. Install the converter to flex bolts and torque to specification.

21. Install the starter assembly.

22. Install the flywheel or torque converter cover.

23. Install the starter.

24. Install the exhaust pipes at the manifold.

25. Connect the wiring harness, transmission cooler lines and a battery cable clamp at the oil pan.

26. Connect the fuel return lines at the engine.

27. Connect the oil cooler lines at the engine.

28. Lower the vehicle.

29. Install the radiator.

30. Install the heater hose to the engine.

31. Connect the accelerator, cruise control and detent cables at the injection pump.

32. Connect the power steering pump and reservoir.

33. Install the fan and the upper fan shroud.

34. Install the ground strap.

35. Connect the wiring to the rocker cover including the glow plug wires.

36. Connect the EGR-EPR solenoids, glow plug controller and temperature solenoid harness.

37. Connect the alternator wires and clips.

38. Connect the wiring at the injector pump.

39. Install the air cleaner and air conditioning compressor.

40. Install the hood.

41. Connect the negative battery cable.

42. Install the proper quantity and type of coolant and engine oil.

G-SERIES

1. Disconnect the negative battery cables. Properly relieve the fuel system pressure.

2. Remove the upper radiator support.

3. Remove the grille.

4. Remove the bumper.

5. Remove the lower grille valance.

6. Remove the hood latch and the upper tie bar.

7. Drain the cooling system.

8. Remove the radiator coolant reservoir bottle.

9. Remove the radiator support bracket, if equipped.

10. Remove the radiator and the fan shroud.

11. Remove the engine cover.

12. Remove the air cleaner, resonator and bracket.

13. Properly discharge the air conditioning system.

14. Remove the air conditioning condenser and cap all openings.

15. Disconnect the air cleaner bracket at the valve cover.

16. Remove the crankcase ventilator bracket and move aside.

17. Remove the intake manifold assembly. Remove the injector pump.

18. Raise and support the vehicle safely.

19. Remove the power steering pump and position aside.

20. Disconnect the exhaust pipes at the manifolds.

21. Remove the flywheel cover and remove the torque converter-to-flywheel attaching bolts22.

22. Remove the engine mount through bolts.

23. Disconnect the blocker heater wires.

24. Remove the starter assembly.

25. Disconnect the fuel line at the fuel pump (lift pump).

26. Lower the vehicle.

27. Remove the bell housing-to-engine bolts.

28. Remove the cruise control transducer, if equipped.

29. Remove the air conditioning compressor.

30. Remove the oil fill tube upper bracket.

31. Remove the wiring harness and connections from engine.

32. Unbolt the transmission dipstick tube and position aside.

33. Remove the heater hose at the engine.

34. Remove the alternator upper bracket.

35. Disconnect the glow plug temperature inhibit switch connector at the coolant crossover.

36. Remove the coolant crossover/thermostat assembly.

37. Connect a suitable engine lifting device to the center intake manifold bolt holes.

38. Support the transmission and remove the engine from the vehicle.

To install:

39. Install the engine into the vehicle. Raise and safely support the vehicle.

40. Install the engine mount through bolts and transmission bell housing bolts.

41. Install the flywheel-to-converter attaching bolts.

42. Install the starter and converter housing underpan.

43. Connect the block heater wires.

44. Connect the exhaust pipes to the manifold assembly.

45. Install the power steering pump. Lower the vehicle.

46. Install the coolant crossover/thermostat assembly.

47. Install the alternator upper bracket.

48. Install the heater hoses, transmission dipstick, air cleaner resonator and bracket assembly.

49. Install the wiring harness and connectors to the engine assembly.

50. Install the oil fill tube upper bracket.

51. Install the air conditioning compressor.

52. Install the injection pump and intake manifold.

53. Install the radiator, fan and lower shroud.

54. Install the air conditioning condenser.

55. Install the upper tie bar.

56. Install the fan lower shroud, coolant recovery bottle and battery cables.

57. Install the hood latch, lower valance, bumper, grille, and headlight bezels.

58. Fill the cooling system with the proper type and quantity of antifreeze. Check all fluid levels.

59. Evacuate and recharge the air conditioning system.

60. Install the engine cover. Inspect engine for leaks.

Engine Mounts

REMOVAL AND INSTALLATION

6.2L and 6.5L Engine

1. Disconnect the negative battery cable.

2. Raise and safely support the vehicle.

3. Properly support the engine.

NOTE: Do not position the jack under the oil pan, any sheetmetal or the crankshaft pulley, otherwise damage may occur.

4. Remove the engine mount through-bolt.

5. Raise the engine to gain sufficient clearance for the mount to be removed.

6. Remove the engine mount attaching bolts, nuts and washers.

7. Remove the mount assembly.

8. Installation is the reverse of the removal procedure.

Cylinder Head

REMOVAL AND INSTALLATION

6.2L and 6.5L Engine

RIGHT SIDE

1. Disconnect the negative battery cable, relieve the fuel system pressure and drain the coolant system. Properly discharge the air conditioning system, if equipped.

2. Remove the intake manifold. Remove the fan upper shroud. Remove the compressor assembly, if equipped.

3. Raise and support the vehicle safely.

4. Remove the exhaust manifold. Lower the vehicle.

5. Remove the valve cover, rocker arm assemblies and pushrods. Mark all components so they may be returned to their original location.

6. Remove the air cleaner resonator and bracket.

7. Remove the transmission and oil dipstick tube; remove the oil fill tube from the coolant crossover pipe.

8. Remove the heater, radiator and bypass hoses.

9. Remove the alternator upper bracket and alternator.

10. Remove the fuel bleeder valve at the coolant crossover pipe.

11. Remove the fuel return crossover line clamp bolts from both cylinder heads.

12. Disconnect the wire connector from the sensor in the coolant crossover pipe.

13. Remove the coolant crossover pipe/thermostat assembly.

14. Remove the head bolts and the cylinder head.

To install:

15. Clean the mating surfaces of the head and block thoroughly.

16. Install a new head gasket on the engine block. Do not coat the gaskets with any sealer on either engine. The gaskets have a special coating that eliminates the need for sealer. Install the cylinder head onto the block.

17. Clean the head bolts thoroughly. Coat the threads of the head bolts with sealing compound GM part 1052080 or equivalent, before installation. Tighten the head bolts to 20 ft. lbs. (25 Nm). in the proper sequence, next tighten all bolts to 50 ft. lbs. (65 Nm) in the proper sequence, and finally tighten all bolts an additional 90 degree (¼ turn).

18. Install the coolant crossover pipe and thermostat.

19. Install the fuel valve and alternator assembly.

20. Connect the bypass hose.

21. Connect the upper radiator hose.

22. Connect the heater hoses at the head.

23. Install the transmission and oil dipstick tube.

24. Install the air cleaner resonator and bracket.

25. Install the pushrods, hardened ends facing up.

26. Install the rocker arm assemblies.

27. Adjust the valves.

28. Install the valve cover. Install the alternator assembly.

29. Raise and support the vehicle safely.

30. Install the exhaust manifold. Lower the vehicle.

31. Install the upper fan shroud.

32. Install the intake manifold. Connect the negative battery cables.

33. Fill the cooling system with the proper type and quantity of antifreeze. Evacuate and recharge the air conditioning system.

LEFT SIDE

1. Disconnect the negative battery cables. Drain the cooling system into a suitable container. Properly discharge the air conditioning system.

2. Remove the intake manifold.

3. Remove the air conditioning compressor belt.

4. Remove the air conditioning compressor.

5. Remove the valve cover.

6. Remove the rocker arm assemblies. Mark the parts to ensure installation in their original location.

7. Remove the pushrods. Keep them in order.

8. Raise and support the vehicle safely.

9. Remove the exhaust manifold.

10. Remove the power steering pump and rear bracket, position the assembly aside. Lower the vehicle.

11. Remove the oil dipstick tube.

12. Disconnect the transmission detent cable.

13. Remove the glow plug controller and bracket.

14. Disconnect the wire connector from the sensor at the coolant crossover pipe.

15. Disconnect the oil fill tube, fuel bleeder valve and hoses from the coolant crossover pipe.

16. Disconnect the fuel crossover line from both cylinder heads.

17. Remove the air cleaner resonator and bracket.

18. Remove the alternator upper bracket.

19. Remove the coolant crossover pipe and thermostat.

20. Remove the head bolts.

21. Remove the cylinder head.

To install:

22. Clean the mating surfaces of the head and block thoroughly.

23. Install a new head gasket on the engine block. Do not coat the gaskets with any sealer on either engine. The gaskets have a special coating that eliminates the need for sealer. Install the cylinder head onto the block.

24. Clean the head bolts thoroughly. Coat the threads of the head bolts with sealing compound GM part 1052080 or equivalent, before installation. Install the rear cylinder head bolt first. Tighten the head bolts to 20

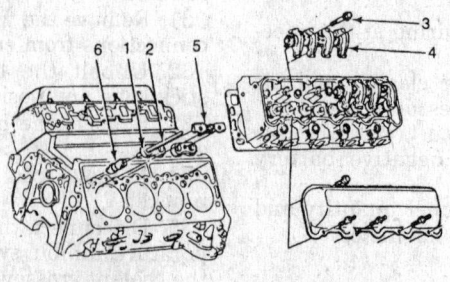

1. Hydraulic lifter
2. Pushrod
3. Bolt
4. Rocker arm assembly
5. Clamp
6. Guide plate

79114057

Cylinder head and components — 6.2L and 6.5L diesel engines

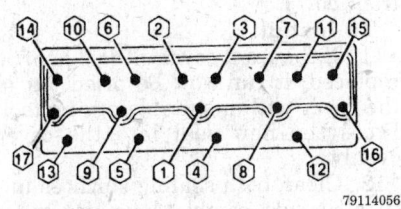

Cylinder head bolt torque sequence — 6.2L and 6.5L diesel engines

79114056

ft. lbs. (25 Nm). in the proper sequence, next tighten all bolts to 50 ft. lbs. (65 Nm) in the proper sequence, and finally tighten all bolts an additional 90 degree (¼ turn).

25. Install the coolant crossover pipe and thermostat.

26. Install the fuel valve and alternator assembly.

27. Connect the bypass hose.

28. Connect the upper radiator hose.

29. Connect the heater hoses at the head.

30. Install the transmission and oil dipstick tube.

31. Install the air cleaner resonator and bracket.

32. Install the pushrods, hardened ends facing up.

33. Install the rocker arm assemblies.

34. Adjust the valves.

35. Install the valve cover. Install the air conditioner compressor and bracket assembly.

36. Raise and support the vehicle safely.

37. Install the exhaust manifold. Lower the vehicle.

38. Install the upper fan shroud.

39. Install the intake manifold. Connect the negative battery cables.

40. Fill the cooling system with the proper type and quantity of antifreeze. Evacuate and recharge the air conditioning system.

Valve Lifters

REMOVAL AND INSTALLATION

6.2L and 6.5L Engine

1. Disconnect the negative battery cables.

2. Remove the valve covers, rocker arm shafts and pushrods. Position all components in the exact order which

they were removed, so they may be reinstalled in their original bore.

3. Remove the cylinder head, as required.

4. Remove the valve lifter clamps and guide plates.

5. Remove the hydraulic lifters. Position the lifters in the exact order which they were removed, so they may be reinstalled in their original bore.

To install:

6. Coat the roller tips with GM part 1052365 or equivalent.

NOTE: All new lifters must be primed by working the plunger while submerged in kerosene or diesel fuel. Some engines will have 2 sizes of lifters being used, standard and oversized. The oversized lifter will be etched "10" on the side and the block will be stamped O.S. on the cast pad adjacent to the lifter bore.

7. Install the lifters in their original bores.

8. Install the guide plates and clamps. Torque the clamps bolts to 18 ft. lbs. (26 Nm).

9. Turn the engine 2 full turns and verify the lifters move freely in the guide plates. If the engine does not turn freely, 1 or more lifters are binding in the guide plate.

10. Install the cylinder head assembly.

11. Install the rocker arm shafts and pushrods assembly. Position all components in the exact order they were removed.

12. Install the valve covers.

13. Connect the negative battery cables.

Valve Lash

ADJUSTMENT

All engines use hydraulic lifters, which require no periodic adjustment.

Rocker Arm and Shafts

REMOVAL AND INSTALLATION

1. Disconnect the negative battery cables. On the G-Series, remove the engine cover.

2. Remove all the necessary components in order to gain access to the engine valve covers. As required, properly relieve the fuel system pressure before disconnecting any fuel lines.

3. Remove the valve cover retaining bolts. Remove the valve cover from the engine.

4. Remove the rocker arm assemblies. Keep them in order for reinstallation.

5. Installation is the reverse of the removal procedure. Be sure to use new gaskets or RTV sealant, as necessary.

Intake Manifold

REMOVAL AND INSTALLATION

6.2L and 6.5L Engine

1. Disconnect both negative battery cables.

2. Drain the cooling system and properly relieve the fuel system pressure.

3. Remove the engine cover, G-Series.

4. Remove the air cleaner assembly.

5. Remove the EPR/EGR valve bracket from the intake manifold.

6. Remove the CDR valve.

7. Remove the crankcase ventilator hose and EGR.

8. Remove the air conditioning rear bracket, if equipped.

9. Remove the fuel line bracket and ground strap.

10. Remove the fuel filter bracket at the intake manifold.

11. Remove the intake manifold bolts. The injection line clips are retained by these bolts.

12. Remove the intake manifold.

NOTE: If the engine is to be further serviced with the manifold removed, install protective covers over the intake ports.

To install:

13. Clean the manifold gasket surfaces on the cylinder heads and install new gaskets before installing the manifold.

NOTE: The gaskets have an opening for the EGR valve on light duty installations. An insert covers this opening on heavy duty installations.

14. Install the intake manifold.

15. Install the intake manifold bolts and fuel injection line clips.

16. Install the fuel filter bracket at the intake manifold.

17. Install the fuel line bracket and ground strap.

18. Install the air conditioning rear bracket, if equipped.

19. Install the crankcase ventilator hose and EGR.

20. Install the CDR valve.
21. Install the EPR/EGR valve bracket from the intake manifold.
22. Install the air cleaner assembly.
23. Install the engine cover, G-Series.
24. Fill the cooling system with the proper type and quantity of antifreeze.
25. Connect both negative battery cables. Inspect for leaks.

Exhaust Manifold

REMOVAL AND INSTALLATION

6.2L and 6.5L Engines

1. Disconnect the batteries.
2. Raise and support the vehicle safely.
3. Disconnect the exhaust pipe from the manifold flange and lower the vehicle.
4. Remove the engine cover and disconnect the glow plug wires.
5. Remove the air cleaner duct bracket.
6. Remove the glow plugs. Remove the turbocharger assembly, as required.
7. Remove the air conditioner compressor rear bracket, as required.

8. Remove the manifold bolts and remove the manifold.
9. Installation is the reverse of the removal procedure. Torque the manifold bolts to 26 ft. lbs. (35 Nm).

Timing Chain Front Cover

REMOVAL AND INSTALLATION

6.2L and 6.5L Engines

1. Disconnect both negative battery cables. Drain the cooling system.
2. Remove the water pump and pulleys.
3. Rotate the crankshaft to align the marks on the torsional damper with the **0** mark on the timing tab.
4. Scribe a mark aligning the injection pump flange and the front cover, if not already marked.
5. Remove the crankshaft pulley and torsional damper.
6. Remove the front cover-to-oil pan bolts (4).
7. Remove the 2 fuel return line clips.
8. Remove the injection pump gear.
9. Remove the injection pump retaining nuts from the front cover.

10. Remove the baffle. Remove the remaining cover bolts and remove the front cover.

To install:

11. If the front cover oil seal is to be replaced, it can now be pried out of the cover with a suitable prying tool. Press the new seal into the cover evenly.
12. Clean both sealing surfaces until all traces of old sealer are gone. Apply a $\frac{3}{32}$ in. (2mm) bead of GM sealant 1052357 or equivalent to the sealing surface. Apply a $\frac{3}{16}$ in. (5mm) bead of RTV type sealer to the bottom portion of the front cover which attaches to the oil pan. Install the front cover.
13. Install the baffle.
14. Install the injection pump, making sure the scribe marks on the pump and front cover are aligned. Tighten the nuts to 31 ft. lbs. (42 Nm).
15. Install the injection pump driven gear, making sure the marks on the cam gear and pump are aligned. Torque the injection pump gear bolts to 17 ft. lbs. (23 Nm).

NOTE: Verify that there is a minimum clearance of 0.040 in. (1.0mm) between the injection pump gear and baffle or noise may be result.

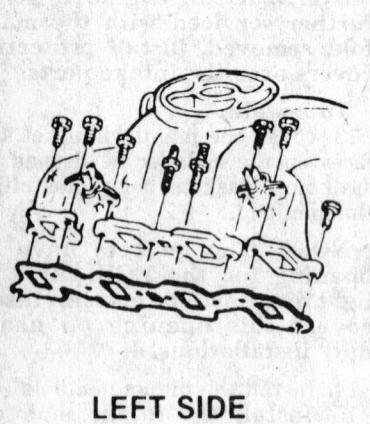

LEFT SIDE

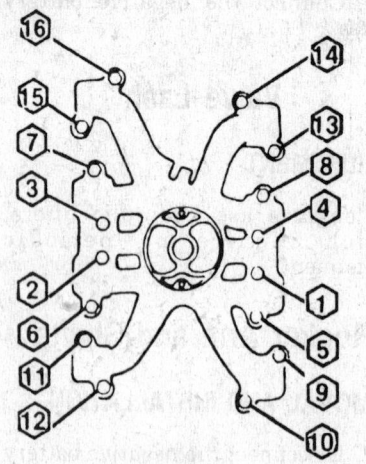

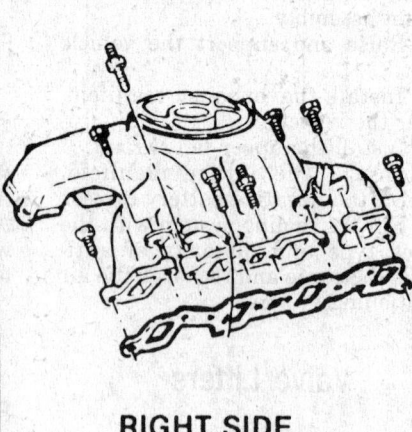

RIGHT SIDE

79114058

Intake manifold installation and torque sequence — 6.2L and 6.5L diesel engines

16. Install the fuel line clips, the front cover-to-oil bolts, and the torsional damper and crankshaft pulley. Torque the pan and the damper bolt to specification.

17. Install the water pump and pulley assembly.

18. Fill the cooling system with the proper type and quantity of antifreeze.

19. Connect the negative battery cables. Inspect engine for leaks.

Front Cover Oil Seal

REMOVAL AND INSTALLATION

NOTE: The oil seal can also be replaced with the front cover installed.

1. Disconnect both negative battery cables.

2. Remove the accessory drive belt.

3. Remove the crankshaft pulley and the torsional damper assembly.

4. Pry the old seal out of the cover using a suitable tool. Use care not to

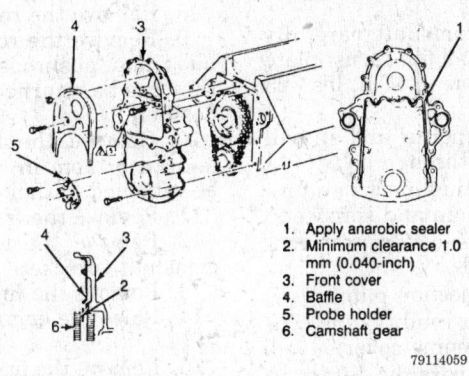

1. Apply anarobic sealer
2. Minimum clearance 1.0 mm (0.040-inch)
3. Front cover
4. Baffle
5. Probe holder
6. Camshaft gear

79114059

Front cover and components — 6.2L and 6.5L diesel engines

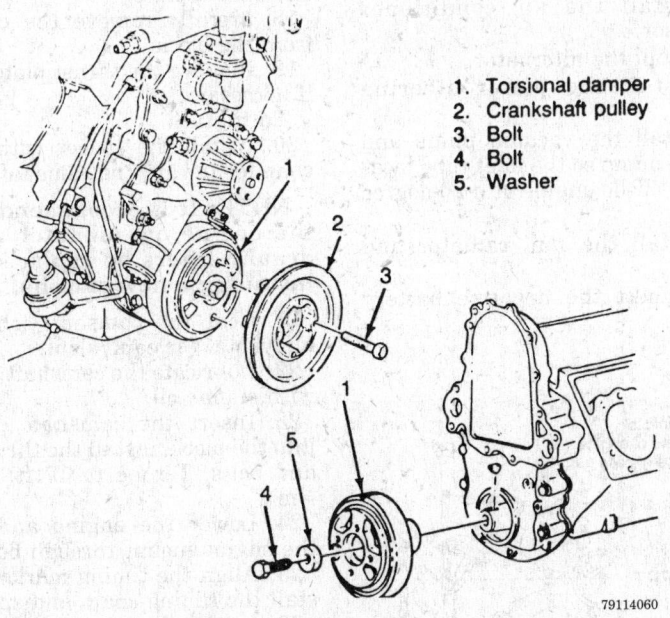

1. Torsional damper
2. Crankshaft pulley
3. Bolt
4. Bolt
5. Washer

79114060

Torsional damper and crankcase pulley installation — 6.2L and 6.5L diesel engines

damage the surface of the crankshaft. Install the new seal evenly into the cover.

5. The remaining installation is the reverse of the removal procedure.

Timing Chain

REMOVAL AND INSTALLATION

6.2L and 6.5L Engines

1. Disconnect the negative battery cables. Remove the front cover assembly.

2. Remove the injection pump gear.

3. Align the camshaft timing gear marks and remove the bolt and washer attaching the camshaft gear.

4. Remove the camshaft sprocket with the timing chain. Remove the crankshaft sprocket.

To install:

5. Install the cam sprocket, timing chain and crankshaft sprocket as a unit, aligning the timing marks on the sprockets.

6. Rotate the crankshaft to align the injection pump and camshaft gears. Install the injection pump gear.

7. Install the front cover. Connect the negative battery cables.

NOTE: The injection pump timing must be adjusted if a new chain, gears or sprockets were installed.

Camshaft

REMOVAL AND INSTALLATION

6.2L and 6.5L Engines

R/V-SERIES

1. Disconnect the battery cables.

2. Drain the cooling system.

3. Remove the radiator shrouds and fan.

4. Remove the vacuum pump.

NOTE: Do not run the engine while the vacuum pump is removed. The oil pump is powered by the vacuum pump drive gear and thus the engine will not have any oil pressure.

5. Remove the power steering pump.

6. Remove the alternator.

7. Remove the air conditioning compressor leaving the refrigerant lines connected and position the compressor aside.

8. Remove the rocker arm covers.

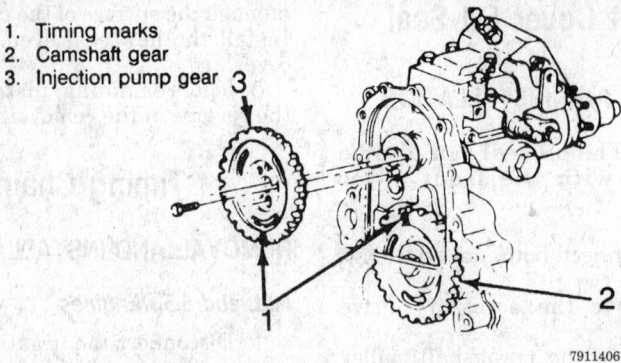

1. Timing marks
2. Camshaft gear
3. Injection pump gear

Injection pump gear and timing marks — 6.2L and 6.5L diesel engines

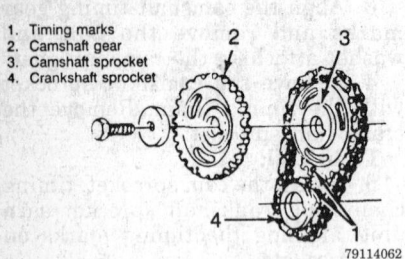

1. Timing marks
2. Camshaft gear
3. Camshaft sprocket
4. Crankshaft sprocket

Camshaft and sprockets — 6.2L and 6.5L diesel engines

9. Remove the rocker arm assemblies and pushrods. Mark the components so they can be reinstalled in their original position.

10. Remove the hydraulic lifters keeping them in order so they can be returned to their original bore.

11. Remove the front cover.

12. Remove the injection pump.

13. Disconnect the air conditioning condenser mounting bolts and lift the condenser out of the way.

14. Remove the timing chain and camshaft sprocket.

15. Remove the engine mounting through bolts.

16. Raise the engine and support it safely.

17. Remove the camshaft thrust plate bolts and the thrust plate.

18. Carefully remove the camshaft from the block.

19. Remove the thrust plate spacer, if necessary.

To install:

20. Install the spacer with the ID chamfer toward the camshaft.

NOTE: It is recommended that the engine oil, oil filter and hydraulic lifters be replaced when installing a new camshaft.

21. Coat the camshaft lobes with Molykote® or equivalent.

22. Lubricate the camshaft journals with engine oil.

23. Insert the camshaft carefully into the block, install the thrust plate and bolts and torque to 17 ft. lbs. (23 Nm).

24. Lower the engine and install the engine mount through bolts.

25. Align the timing marks and install the timing chain and sprockets.

26. Install the air conditioner condenser, if equipped.

27. Install the injection pump.

28. Install the air conditioning condenser and front timing cover.

29. Install the hydraulic lifters in the same bore as they were removed.

30. Install the rocker arm assemblies and pushrods in their original locations.

31. Install the rocker arm covers.

32. Install the air conditioner compressor.

33. Install the alternator.

34. Install the power steering pump.

35. Install the vacuum pump and align the pump so the inlet tube faces the front of the engine at a 20 degree angle.

36. Install the fan, radiator and shrouds.

37. Connect the negative battery cables.

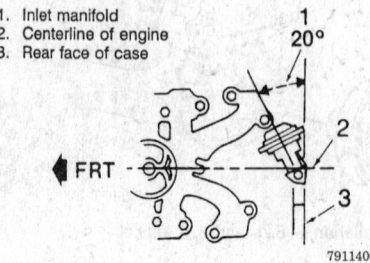

1. Inlet manifold
2. Centerline of engine
3. Rear face of case

Diesel engine vacuum pump positioning

38. Fill the cooling system with the proper type and quantity of antifreeze.

C/K-Series

1. Disconnect the battery cables and relieve the fuel system pressure.

2. Drain the cooling system.

3. Remove the radiator, the shroud and fan assembly.

4. Remove the grille and parking light assembly.

5. Remove the hood latch and brace assembly.

6. Remove the oil pump drive.

7. Remove the power steering pump, alternator and air conditioner compressor and position aside.

8. Remove the rocker arm covers.

9. Remove the rocker arm assemblies and pushrods. Mark them so they can be returned to their original position.

10. Remove the hydraulic lifters and keep them in order so they can be returned to their original bore.

11. Remove the front cover.

12. Remove the timing chain and camshaft sprocket.

13. Remove the injector pump.

14. Raise the engine and support it safely.

15. Remove the front engine mounting through bolts.

16. Remove the air conditioner condenser mounting bolts and lift the condenser out.

17. Remove the thrust plate bolts and thrust plate.

18. arefully remove the camshaft from the block.

19. Remove the thrust plate spacer, if necessary.

To install:

20. Install the spacer with the ID chamfer toward the camshaft.

NOTE: It is recommended that the engine oil, oil filter and hydraulic lifters be replaced when installing a new camshaft.

21. Coat the camshaft lobes with Molykote® or equivalent.

22. Lubricate the camshaft journals with engine oil.

23. Insert the camshaft carefully into the block, install the thrust plate and bolts. Torque to 17 ft. lbs. (23 Nm).

24. Lower the engine and install the engine mount through bolts.

25. Align the timing marks and install the timing chain and sprockets.

26. Install the air conditioner condenser, if equipped.

27. Install the injector pump.

28. Install the front cover.

29. Install the hydraulic lifters in the same bore as they were removed.

30. Install the rocker arm assemblies and pushrods in their original locations.
31. Install the rocker arm covers.
32. Install the power steering pump, alternator and air conditioner compressor.
33. Install the oil pump drive.
34. Install the hood latch and brace.
35. Install the grille and parking light assembly.
36. Install the radiator, the shroud and fan assembly.
37. Fill the cooling system with the proper type and quantity of antifreeze.
38. Connect the negative battery cables.

G-SERIES

1. Disconnect the battery cables and relieve the fuel system pressure.
2. Remove the headlight bezels.
3. Remove the grille, bumper and lower valence panel.
4. Remove the hood latch.
5. Remove the coolant recovery bottle.
6. Remove the upper tie bar.
7. Remove the air conditioner compressor.
8. Drain the cooling system and remove the radiator and fan.
9. Remove the oil pump drive.
10. Remove the cylinder heads to gain clearance for lifter removal.
11. Remove the alternator lower bracket.
12. Remove the water pump.
13. Remove the torsional damper.
14. Remove the front cover.
15. Remove the injection pump.
16. Remove the rocker arm covers.
17. Remove the rocker arm assemblies and pushrods. Mark them so they can be returned to their original position.
18. Remove the hydraulic lifters and keep them in order so they can be returned to their original bore.
19. Remove the timing chain and camshaft sprocket.
20. Remove the thrust plate bolts and thrust plate.
21. Carefully remove the camshaft from the block.
22. Remove the thrust plate spacer, if necessary.
To install:
23. Install the spacer with the ID chamfer toward the camshaft.

NOTE: It is recommended that the engine oil, oil filter and hydraulic lifters be replaced when installing a new camshaft.

24. Coat the camshaft lobes with Molykote or equivalent.
25. Lubricate the camshaft journals with engine oil.
26. Insert the camshaft carefully into the block, install the thrust plate and bolts and torque to 17 ft. lbs.
27. Align the timing marks and install the timing chain and sprockets.
28. Install the hydraulic lifters in the same bore as they were removed.
29. Install the rocker arm assemblies and pushrods in their original locations.
30. Install the rocker arm covers.
31. Install the fuel pump.
32. Install the front cover.
33. Install the torsional damper and water pump.
34. Install the alternator lower bracket.
35. Install the cylinder heads.
36. Install the oil pump drive.
37. Install the radiator and fan.
38. Install the air conditioner compressor.
39. Install the upper tie bar.
40. Install the coolant recovery bottle.
41. Install the hood latch.
42. Install the grille, bumper and lower valence panel.
43. Install the headlight bezels.
44. Install the battery cables.
45. Fill the cooling system.
46. Evacuate and charge the air conditioner system.

Piston and Connecting Rod

POSITIONING

ENGINE LUBRICATION

Oil Pan

REMOVAL AND INSTALLATION

4.3L Engine

1. Disconnect the negative battery cable. Raise and support the vehicle safely. Drain the engine oil into a suitable container.
2. Remove the exhaust crossover pipe.
3. Remove the torque converter cover, if equipped with automatic transmission.

4. Remove the strut rods at the flywheel cover, if equipped.
5. Remove the strut rod at the front engine mounts, if equipped.
6. Remove the starter assembly.
7. Remove the oil pan bolts, nuts and reinforcements.
8. Remove the oil pan and gaskets.
To install:
9. Thoroughly clean all gasket surfaces and install a new gasket, using only a small amount of sealer at the front and rear corners of the oil pan.
10. Install the oil pan and new gaskets.
11. Install the oil pan bolts, nuts and reinforcements. Torque the pan bolts to specification.
12. Install the starter assembly.
13. Install the strut rod brackets at the front engine mounts.
14. Install the strut rods at the flywheel cover.
15. Install the torque converter cover, if equipped with automatic transmission.
16. Install the exhaust crossover pipe. Lower the vehicle.
17. Connect the negative battery cable.
18. Fill the engine with the proper quantity and type of oil.

5.0L and 5.7L Engines

1. Disconnect the negative battery cable. Raise the vehicle, support it safely, and drain the engine oil.
2. Remove the exhaust crossover pipe.
3. Remove the flywheel or torque converter cover.
4. Remove the strut rods at the front engine mountings, if used.
5. Remove the oil pan bolts, nuts and reinforcements.
6. Remove the oil pan and gaskets.
To install:
7. Thoroughly clean all gasket surfaces and install a new gasket, using a small amount of sealer at the front and rear corners of the oil pan.
8. Install the oil pan and new gaskets.
9. Install the oil pan bolts, nuts and reinforcements. Torque the pan bolts to specification.
10. Install the strut rods at the front engine mountings.
11. Install the torque converter or flywheel cover.
12. Install the exhaust crossover pipe.
13. Connect the negative battery cable.
14. Fill the crankcase with the proper type and quantity of oil.

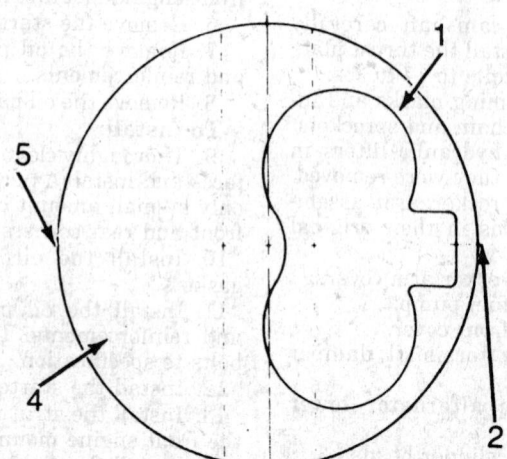

1. Oil control ring expander gap
2. Second compression ring gap
3. Centerline of piston pin
4. Oil control ring gap
5. Top compression ring gap

79114064

Piston ring gap locations — 6.2L and 6.5L diesel engines

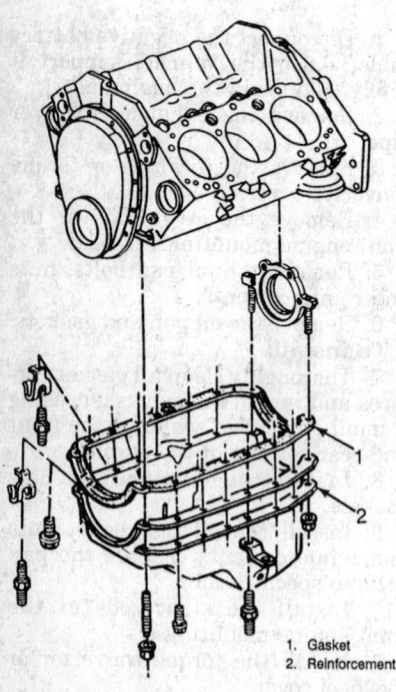

1. Gasket
2. Reinforcement

79114065

Oil pan assembly — 4.3L engine

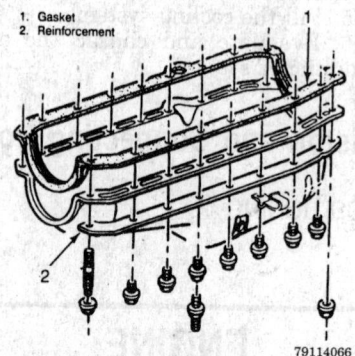

1. Gasket
2. Reinforcement

79114066

Oil pan assembly — 5.0L and 5.7L engines

6.2L and 6.5L Engines

R/V-SERIES

1. Disconnect the battery cables.
2. Raise and support the vehicle safely.
3. Drain the engine oil and remove the filter.
4. Remove the flywheel cover and oil pan dipstick.
5. Remove the left engine mount through bolts. Remove the exhaust pipes at the manifold, as required. Raise the engine.
6. Remove the strut rods, as required. Remove the battery cables, cooler lines and clamps from the oil pan.
7. Remove the oil pan bolts and remove the oil pan. Remove the oil pan gaskets and seals.

To install:

8. Clean the old RTV sealant from the oil pan and block.
9. Apply a 3/16 in. (5mm) bead of RTV sealant to the oil pan sealing surface, inboard of the bolt holes. The sealant must be wet to the touch when the oil pan is to be installed.
10. Install the oil pan rear seal.
11. Install the oil pan to the engine and install the retaining bolts. Torque the bolts to specification.
12. Lower the engine and install the oil pan dipstick.
13. Install the engine mount through bolt and nut.
14. Install the flywheel cover, strut rods.
15. Connect the battery cables, cooler lines and clamps to the oil pan and then lower the vehicle.
16. Refill with the proper grade and quantity of oil. Install the oil filter. Connect the negative battery cables.

C/K-SERIES

1. Disconnect the battery cables.
2. Raise the vehicle and support it safely.
3. Drain the engine oil. Remove the oil dipstick.
4. Remove the flywheel cover.
5. Disconnect the exhaust pipes from the manifolds.
6. Remove the front engine mount through bolts and raise the engine.
7. Remove the oil pan bolts and remove the oil pan.
8. Remove the oil pan rear seal.

To install:

9. Clean the old RTV sealant from the oil pan and block.
10. Apply a 3/16 in. (5mm) bead of RTV sealant to the oil pan sealing surface, inboard of the bolt holes. The sealant must be wet to the touch when the oil pan is to be installed.
11. Install the oil pan rear seal.
12. Install the oil pan to the engine and install the retaining bolts. Torque all except the rear 2 bolts to 84 inch lbs. Torque the rear 2 bolts to 17 ft. lbs.
13. Lower the engine.
14. Install the engine mounting through bolt and nut.
15. Install the oil dipstick. Install the exhaust pipes to the manifolds.
16. Install the flywheel cover and lower the vehicle.
17. Refill with the proper grade and quantity of oil.
18. Install the battery cables.

G-SERIES

1. Disconnect the battery cables.
2. Remove the engine cover.
3. Remove the engine oil dipstick.
4. Remove the engine oil dipstick tube at the rocker cover.
5. Raise the vehicle and support it safely.
6. Remove the transmission flywheel cover.
7. Drain the engine oil.
8. Disconnect the oil cooler lines at the block.
9. Remove the starter. Remove the battery cables, transmission cooler lines and attaching clamps from the oil pan.
10. Remove the oil pan bolts and remove the oil pan and oil pan rear seal.

To install:

11. Apply a 3/16 in. (5mm) bead of RTV sealant to the oil pan sealing surface, inboard of the bolt holes. The sealant must be wet to the touch when the oil pan is to be installed.
12. Install the oil pan rear seal.
13. Install the oil pan to the engine and install the retaining bolts. Torque all bolts to specifications.
14. Install the starter. Install the transmission, battery cables and attaching clamps to the oil pan.
15. Install the engine oil cooler lines.
16. Install the transmission flywheel cover.
17. Lower the vehicle.
18. Install the engine oil dipstick tube at the rocker cover.
19. Install the engine oil dipstick.
20. Install the engine cover.
21. Refill with the proper grade and quantity of oil.
22. Install the battery cables.

7.4L Engine

NOTE: Removal of the transmission may be necessary on the G-Series vehicles.

1. Disconnect the negative battery cable.
2. Remove the fan shroud.
3. Remove the air cleaner.
4. Remove the distributor cap.
5. Raise and support the vehicle safely.
6. Drain the engine oil. Remove the starter assembly, if equipped with manual transmission.
7. Remove the torque converter or clutch housing cover.
8. Remove the oil filter.
9. Remove the oil pressure line from the side of the block.
10. Support the engine with a floor jack.

11. Remove the engine mount through bolts.
12. Raise the engine just enough to remove the pan.
13. Remove the oil pan bolts, the oil pan and discard the gaskets.

To install:

14. Clean all mating surfaces thoroughly.
15. Apply RTV gasket material to the front and rear corners of the gaskets.
16. Coat the gaskets with adhesive sealer and position them on the block.
17. Install the rear pan seal in the pan with the seal ends mating with the gaskets.
18. Install the front seal on the bottom of the front cover, pressing the locating tabs into the holes in the cover.
19. Install the oil pan.
20. Install the pan bolts, clips and reinforcements. Torque the pan bolts to specification.
21. Lower the engine onto the mounts.
22. Install the engine mount through-bolts.
23. Install the oil pressure line.
24. Install the oil filter. Install the starter assembly, if removed.
25. Install the torque converter or clutch housing cover.
26. Install the distributor cap.
27. Install the air cleaner.
28. Install the fan shroud.
29. Connect the battery.
30. Fill the crankcase with the proper type and quantity of oil.

Oil Pump

REMOVAL AND INSTALLATION

1. Raise and support the vehicle safely and remove the oil pan.
2. Remove the bolt attaching the pump to the rear main bearing cap. Remove the pump and the extension shaft, which connects to the distributor/oil pump shaft.

To install:

3. If the pump has been disassembled and is being replaced or for any reason oil has been removed, it must be primed. It can either be filled with oil before installing the cover plate and oil kept within the pump during handling or the entire pump cavity can be filled with petroleum jelly.

NOTE: If the pump is not primed, the engine could be damaged before it receives adequate lubrication when the engine is started.

4. Engage the extension shaft with the oil pump shaft. Align the slot on the top of the extension shaft with the drive tang on the lower end of the distributor/oil pump shaft, and then position the pump at the rear main bearing cap so the mounting bolt can be installed. No gasket is used. Install the bolt and torque to 65 ft. lbs. (90 Nm)
5. Install the oil pan.

Rear Main Bearing oil Seal

REMOVAL AND INSTALLATION

4.3L, 5.0L and 5.7L Engines

1. Raise and support the vehicle safely.
2. Remove the transmission from the vehicle.
3. Remove the pressure plate/flywheel or flexplate assembly.
4. Insert a suitable tool into the notches provided in the seal retainer and pry the seal out.

NOTE: Care should be taken when removing the seal so as not to nick the crankshaft sealing surface.

5. Before installation lubricate the new seal with clean engine oil.
6. Install the seal on tool J–35621 or equivalent. Thread the tool into the rear of the crankshaft. Tighten the screws snugly. This is to insure that the seal will be installed squarely over the crankshaft. Tighten the tool wing nut until it bottoms.
7. Remove the tool from the crankshaft.
8. Install the flywheel or flexplate and transmission assembly.

6.2L and 6.5L Engines

2-PIECE SEAL

1. Engines are originally equipped with a 2-piece rope seal. This should be replaced with 2-piece lip seal, which is available as a service replacement.
2. Raise and support the vehicle safely. Drain the crankcase oil and remove the oil pan, pump and rear main bearing cap.
3. Using a suitable tool, remove the upper and lower rope seals.

To install:

4. Apply a light coat of oil to the seal lips where they contact the crankshaft. Roll 1 seal half into the block groove until a 1/2 in. (13mm) of one end is extending out of the block. Insert the other seal half into the op-

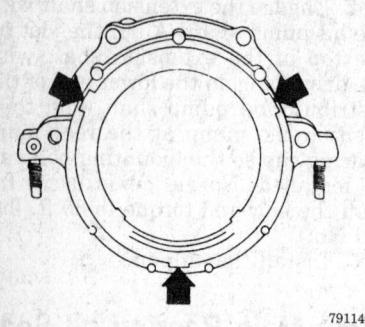

Rear main seal removal notches — 4.3L, 5.0L and 5.7L engines

79114067

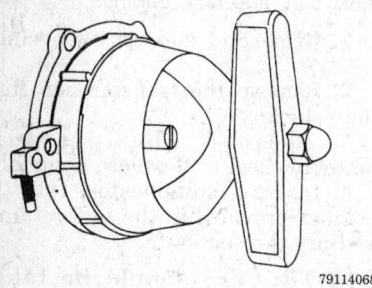

Rear main seal installation — 4.3L, 5.0L and 5.7L engines

79114068

posite side of the seal groove in the block.

5. The ends of the seal halves should now be positioned at the 8 and 2 o'clock or 4 and 10 o'clock positions.

6. Lightly coat the seal groove in the main bearing with adhesive Loctite 414® or equivalent.

7. Apply a thin film of anaerobic sealant or equivalent, to the main bearing cap as shown. Do not put sealant in the oil relief slot.

8. Tap the bearing cap into place using a brass or leather mallet.

9. Install the bearing cap bolts and torque to specification.

10. Install the oil pump and oil pan.

1-PIECE SEAL

1. The 1-piece rear main seal may be serviced without removing the oil pan and rear main bearing cap.

2. Raise and support the vehicle safely.

3. Remove the transmission from the vehicle.

4. Remove the pressure plate/flywheel or flexplate assembly.

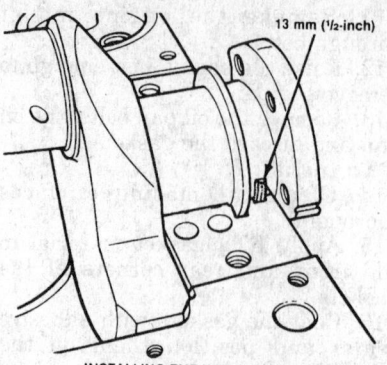

13 mm (½-inch)

INSTALLING THE UPPER SEAL HALF

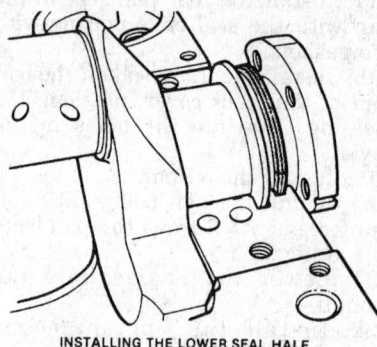

INSTALLING THE LOWER SEAL HALF

79114070

Rear main seal installation — 6.2L and 6.5L diesel engines

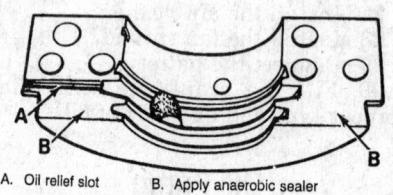

A. Oil relief slot   B. Apply anaerobic sealer

79114071

Applying sealer to the rear main bearing cap — 6.2L and 6.5L diesel engines

5. Insert a suitable tool into the notches provided in the seal retainer and pry the seal out.

NOTE: Care should be taken when removing the seal so as not to nick the crankshaft sealing surface.

To install:

6. Before installation lubricate the new seal with clean engine oil.

7. Install the seal with the spring cavity towards the crankshaft onto

tool J–39084 or equivalent. Thread the tool into the rear of the crankshaft. Tighten the screws snugly. This is to insure that the seal will be installed squarely over the crankshaft. Tighten the tool wing nut until it bottoms.

8. Remove the tool from the crankshaft.

9. Install the flywheel or flexplate and transmission assembly.

7.4L ENGINE

2-PIECE SEAL

1. The rear main bearing oil seal, both halves, can be removed without removal of the crankshaft. Always replace the upper and lower halves togeth

2. Raise and support the vehicle safely, drain the oil and remove the oil pan.

3. Remove the oil pump and the rear main bearing cap.

4. Using a suitable tool, pry the oil seal from the rear main bearing cap.

5. Using a small hammer and a brass pin punch, drive the top half of the oil seal from the rear main bearing.

6. Using a non-abrasive cleaner, clean the rear main bearing cap and the crankshaft.

7. Coat the new oil seal with engine oil; do not coat the ends of the seal.

8. Position the new upper half seal into the main bearing cap groove and carefully roll the seal around the cap groove.

NOTE: Make sure the seal lip is positioned toward the front of the engine.

9. Install the lower half onto the lower half of the rear main bearing cap.

10. Apply sealant to the cap-to-case mating surfaces and install the lower rear main bearing half to the engine; keep the sealant off of the seal's mating line.

11. Install the rear main bearing cap bolts and torque to 10 ft. lbs. (14 Nm). Using a lead hammer, tap the crankshaft forward and rearward, to line up the thrust surfaces. Torque the main bearing bolts to 110 ft. lbs. (150 Nm).

12. Install the oil pan and pump assembly. Refill the crankcase with the proper type and quantity of engine oil.

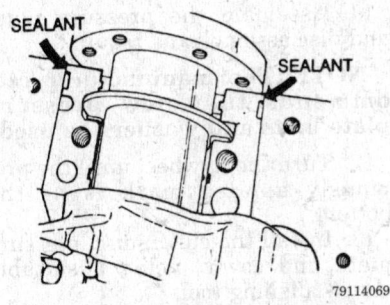

Applying sealer to the block — 7.4L engine

MANUAL TRANSMISSION

Transmission Assembly

REMOVAL AND INSTALLATION

R/V-Series

1. Raise and support the vehicle safely.
2. Drain the transmission.
3. Disconnect the speedometer cable/sensor and back-up light wire at transmission.
4. Remove the shifter boot. Remove the gearshift lever by pressing down firmly on the slotted collar plate with a pair of channel lock pliers and rotating counterclockwise. Plug the opening to keep out dirt.
5. Remove the transfer case, if equipped with 4 wheel drive.
6. Remove driveshaft after marking the position of the shaft to the flange.
7. Remove the exhaust and parking brake cables, as required. Position a transmission jack under the transmission to support it.
8. Remove the crossmember. Visually inspect to see if other equipment, brackets or lines, must be removed to permit removal of transmission.

NOTE: Mark position of crossmember when removing to prevent incorrect installation. The tapered surface should face the rear.

9. Remove the flywheel housing underpan.

10. Remove the top 2 transmission to housing bolts and insert 2 guide pins.

NOTE: The use of guide pins will not only support the transmission but will prevent damage to the clutch disc. Guide pins can be made by taking 2 bolts, the same as those just removed only longer, and cutting off the heads. Make a slot in the cut end of the bolt and use a suitable tool to install the special bolts. Be sure to support the clutch release bearing and support assembly during removal of the transmission. This will prevent the release from falling out of the flywheel housing.

11. Remove the 2 remaining bolts and slide transmission straight back from engine. Use care to keep the transmission drive gear straight in line with clutch disc hub.
12. Remove the transmission from the vehicle.

To install:

13. Place the transmission in high gear.
14. Coat the input shaft splines with high temperature grease.
15. Raise the transmission into position.
16. Install the guide pins in the top 2 bolt holes.
17. Roll the transmission forward and engage the clutch splines. Keep pushing the transmission forward until it mates with the engine.
18. Install the bolts, removing the guide pins. Torque the bolts to 75 ft. lbs.
19. Install the flywheel housing underpan.
20. Install the crossmember. Torque the crossmember-to-frame bolts to 55 ft. lbs.; the crossmember-to-transmission bolts to 40 ft. lbs.
21. Remove the transmission jack.
22. Install the driveshaft.
23. Install the gearshift lever.
24. Connect the speedometer cable/sensor and back-up light wire at transmission.
25. Fill the transmission with the proper quantity and type of oil.

NOTE: Do not force the transmission into the clutch disc hub. Do not let the transmission hang unsupported in the splined portion of the clutch disc.

C/K-Series

1. Disconnect the negative battery cable. Remove the shifter boot and lever.
2. Raise and support the vehicle safely. Drain the transmission.

3. Remove driveshaft after marking the position of the shaft to the axle flange.
4. Remove the exhaust pipes and parking brake cables.
5. Disconnect the wiring harness at the transmission.
6. Remove the transfer case, if equipped with 4 wheel drive.
7. Remove the clutch slave cylinder and support it out of the way.
8. On the 85mm (MG5) transmission, remove the flywheel housing inspection cover.
9. Position a transmission jack or equivalent, under the transmission for support.
10. Remove the crossmember. Visually inspect to see if other equipment, brackets or lines, must be removed to permit removal of transmission.

NOTE: Mark position of crossmember when removing to prevent incorrect installation. The tapered surface should face the rear.

11. Remove the top 2 transmission to housing bolts and insert two guide pins.

NOTE: The use of guide pins will not only support the transmission but will prevent damage to the clutch disc. Guide pins can be made by taking 2 bolts, the same as those just removed only longer, and cutting off the heads. Make an adjustment slot. Be sure to support the clutch release bearing and support assembly during removal of the transmission. This will prevent the release from falling out of the flywheel housing.

12. Remove the remaining bolts and slide transmission straight back from engine. Use care to keep the transmission drive gear straight in line with clutch disc hub.
13. Remove the transmission from the vehicle.

To install:

14. Place the transmission in high gear. Coat the input shaft splines with high temperature grease.
15. Raise the transmission into position.
16. Install the guide pins in the top 2 bolt holes.
17. Roll the transmission forward and engage the clutch splines. Keep pushing the transmission forward until it mates with the engine.
18. Install the bolts, removing the guide pins. Torque the bolts to specification.
19. Install the transfer case, if equipped.

20. Install the crossmember. Torque the crossmember bolts to specification. Remove the transmission jack.

21. Install the slave cylinder.

22. Install the driveshaft.

23. Install the exhaust system.

24. Install the inspection cover on the 85mm transmission.

25. Install the gearshift lever.

26. Connect the wiring harness at the transmission.

27. Fill the transmission with the proper type and quantity of oil.

NOTE: Do not force the transmission into the clutch disc hub. Do not let the transmission hang unsupported in the splined portion of the clutch disc.

G-Series

1. Raise and support the vehicle safely.

2. Drain the transmission.

3. Disconnect the speedometer cable, back-up light and TCS switch.

4. Remove the shift controls from the transmission.

5. Disconnect the driveshaft and remove it from the vehicle.

6. Support the transmission using the proper equipment.

7. Remove the transfer case, if equipped with 4 wheel drive.

8. Inspect the transmission to be sure all necessary components have been removed or disconnected.

9. Mark the front of the crossmember to be sure it is installed correctly.

10. Remove the flywheel housing under pan and transmission mounting bolts.

11. Move the transmission slowly away from the engine, keeping the mainshaft in alignment with the clutch disc hub. Support the clutch release bearing to prevent it from falling out of the flywheel housing when the transmission is removed.

12. Remove the transmission from the vehicle.

To install:

13. Lightly coat the mainshaft with high temperature grease. Do not use much grease, since, under normal operation, the grease will be thrown onto the clutch, causing it to fail.

14. Raise the transmission into position under the vehicle.

15. Roll the unit forward, engaging the spline of the mainshaft with the splines in the clutch hub. Continue pushing forward until the transmission mates with the bell housing.

16. Install and tighten the transmission-to-bell housing bolts. Install the transfer case, if equipped.

17. Install the flywheel housing under pan.

18. Install the crossmember. Torque the bolts to 50 ft. lbs.

19. Inspect the transmission to be sure all necessary components have been installed or connector.

20. Connect the driveshaft.

21. Install the shift controls.

22. Connect the speedometer cable.

23. Connect the back-up light switch.

24. Connect the TCS switch.

25. Fill the transmission with the proper quantity and type of oil.

26. Road test the vehicle.

CLUTCH

Clutch Assembly

REMOVAL AND INSTALLATION

1. Raise and support the vehicle safely. Remove the transmission and place a support under the engine.

2. Remove the slave cylinder or clutch adjusting rod.

3. Remove the bell housing cover.

4. Remove the bell housing from the engine.

5. Remove the throwout spring and fork.

6. Remove the ball stud from the bell housing.

7. Install a pilot tool to support the clutch while removing the assembly.

NOTE: Before removing the clutch from the flywheel, matchmark the flywheel, clutch cover and one pressure plate lug, so these parts may be assembled in their same relative positions and retain the factory balance.

8. Loosen the clutch attaching bolts one turn at a time to prevent distortion of the clutch cover until the tension is released.

9. Remove the clutch pilot tool and the clutch from the vehicle.

10. Check the clutch assembly, flywheel and pilot bearing for signs of wear, scoring, overheating, etc. If the clutch plate, flywheel or pressure plate is oil-soaked, inspect the engine rear main seal and the transmission input shaft seal and correct leakage as required. Replace any damaged parts.

To install:

11. Assemble the pressure plate and disc assembly, as required.

NOTE: The manufacturer recommends that new pressure plate bolts and washers be used.

12. Turn the flywheel until the previously applied mark is at the bottom.

13. Install the clutch disc, pressure plate and cover, using a suitable clutch aligning tool.

14. Turn the clutch until the matchmark on the clutch cover aligns with the mark on the flywheel.

15. Install the attaching bolts and tighten them a little at a time in a crossing pattern until the spring pressure is taken up.

16. Remove the aligning tool.

17. Coat the rounded end of the ball stud with high temperature wheel bearing grease.

18. Install the ball stud in the bell housing. Pack the ball stud from the lubrication fitting.

19. Pack the inside recess and the outside groove of the release bearing with high temperature wheel bearing grease and install the release bearing and fork.

20. Install the release bearing seat and spring.

21. Install the clutch housing. Torque the bolts to specification.

22. Install the cover.

23. Install the slave cylinder or clutch rod.

24. Install the transmission.

25. Bleed the hydraulic system or adjust the clutch as required.

Free Pedal Travel Adjustment

MECHANICAL CLUTCH

1. This adjustment is for the amount of clutch pedal free travel before the throwout bearing contacts the clutch release fingers. It is required periodically to compensate for clutch lining wear. Incorrect adjustment will cause gear grinding and clutch slippage or wear.

2. Raise and support the vehicle safely. Disconnect the clutch fork return spring at the fork on the clutch housing.

3. Loosen the outer locknut on the adjusting rod.

4. Move the clutch fork back until clutch spring pressure is felt.

5. Hold the clutch pedal against its bumper and turn the inner adjusting nut until it is 0.28 in. (7.1mm) from the cross lever.

6. Tighten the locknut against the cross lever.

1. Lubrication fitting (R/V and C/K models only)
2. Clutch fork
3. Spring washer
4. Screw
5. Flywheel housing
6. Screw
7. Ball stud
8. Boot
9. Retainer
10. Release bearing
11. Cover assembly
12. Driveplate
13. Cover
14. Screw
15. Pilot bearing
16. Screw
17. Strap
18. Pressure plate

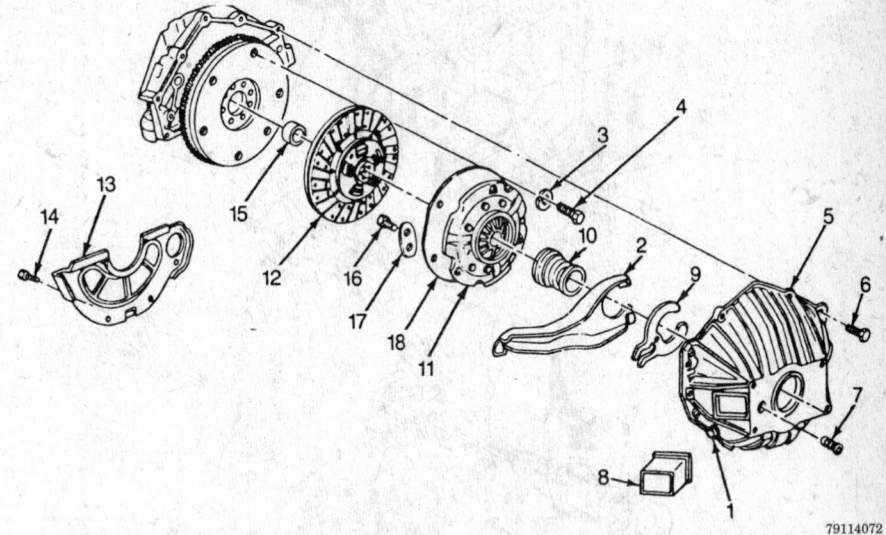

79114072

Exploded view of typical clutch assembly

7. Install the return spring.

8. Check the free travel at the pedal and readjust as necessary. It should be 1³⁄₈ in. (34mm).

Clutch Master Cylinder

REMOVAL AND INSTALLATION

1. Disconnect the negative battery cable.

2. Remove the lower steering column covers.

3. Remove the lower left side air conditioning duct, if equipped.

4. Disconnect the pushrod from the clutch pedal.

5. Disconnect the reservoir hose, if equipped.

6. Disconnect and plug the secondary cylinder hydraulic line to the master cylinder.

7. Remove the master cylinder retaining nuts and remove the master cylinder.

8. To install, use a new gasket, reverse the removal procedure and bleed the clutch system.

Clutch Slave Cylinder

REMOVAL AND INSTALLATION

1. Disconnect the negative battery cable.

2. Raise the vehicle and support it safely.

3. Disconnect the hydraulic line from the secondary cylinder.

4. Disconnect the hydraulic line from the master cylinder.

5. Remove the nut retaining the hydraulic line and the speedometer cable to the cowl, then install the nut to hold the speedometer cable in place, as required.

6. Cover all hydraulic lines to prevent dirt and moisture from entering the system.

7. Installation is the reverse of removal. Bleed the clutch hydraulic system.

Bleeding The Hydraulic Clutch System

1. Fill the clutch master cylinder with the proper grade and type fluid. Raise and support the vehicle safely.

2. Remove the slave cylinder retaining bolts. Hold the cylinder about 45 degrees with the bleeder at the highest point.

3. Fully depress the clutch pedal and open the bleeder screw. Repeat until all air is expelled from the system.

4. Be sure the fluid level remains full in the clutch master cylinder throughout the bleeding procedure.

5. Test the clutch for proper operation.

AUTOMATIC TRANSMISSION

Transmission Assembly

REMOVAL AND INSTALLATION

NOTE: It may be necessary to disconnect and remove the exhaust crossover pipe on 8 cylinder engines, and to disconnect the catalytic converter and remove its support bracket, if equipped.

1. Hydraulic line routing
2. Speedometer cable
3. Hydraulic line, secondary cylinder
4. Master cylinder
5. Nut
6. Nut
7. Bleeder Screw
8. Secondary cylinder

79114073

Typical secondary (slave) cylinder and hydraulic line

2 Wheel Drive

1. Disconnect the battery ground cable. Remove the air cleaner and disconnect the detent cable.
2. Raise and support the vehicle safely. Drain the transmission.
3. Matchmark the axle-to-driveshaft flanges and remove the driveshaft.
4. Disconnect the speedometer cable, downshift cable, vacuum modulator line, shift linkage, throttle linkage, fuel lines and fluid cooler lines at the transmission, as required. Remove the filler tube.
5. Support the transmission. Unbolt and remove the crossmember.
6. Remove the torque converter underpan, Matchmark the flywheel and converter, and remove the converter bolts.
7. Support the engine and lower the transmission slightly for access to the upper transmission to engine bolts.
8. Remove the transmission to engine bolts and pull the transmission back and out of the vehicle.

NOTE: Keep the front of the transmission up so the converter doesn't fall out.

9. Installation is the reverse of the removal procedure.

NOTE: Lubricate the internal yoke splines at the transmission end of the driveshaft with lithium base grease.

4 Wheel Drive

1. Disconnect the battery ground cable and remove the transmission dipstick. Detach the downshift cable. Remove the transfer case shift lever knob and boot.
2. Raise and support the vehicle safely.
3. Remove the skid plate, if any. Remove the flywheel cover.
4. Matchmark the flywheel and torque converter, remove the bolts and secure the converter so it doesn't fall out of the transmission.
5. Detach the shift linkage, speedometer cable, vacuum modulator line, downshift cable, throttle linkage and cooler times at the transmission. Remove the filler tube.
6. Remove the exhaust crossover pipe to manifold bolts.
7. Unbolt the transfer case adapter from the crossmember. Support the transmission and transfer case. Remove the crossmember.
8. Move the exhaust system aside. Detach the driveshafts after matchmarking their flanges. Disconnect the parking brake cable.
9. Unbolt the transfer case from the frame bracket. Support the engine. Unbolt the transmission from the engine, pull the assembly back, and remove.
10. Reverse the procedure for installation.

Shift Linkage Adjustment

1. Raise and support the vehicle safely.
2. Loosen the shift lever bolt or nut at the transmission lever so the lever is free to move on the rod.
3. Set the column shift lever to the neutral gate notch. Do not use the indicator pointer as a reference to position the shift lever, as this will not be accurate.
4. Set the transmission lever in the neutral position by moving it clockwise to the park detent, then counterclockwise 2 detents to neutral.
5. Hold the rod tightly in the swivel and tighten the nut or bolt to 17 ft. lbs. (23 Nm).
6. Move the column shifter to **P** and check that the engine starts. Check the adjustment by moving the selector to each gear position.

Neutral Start Switch Adjustment

1. This switch prevents the engine from being started unless the transmission is in **N** or **P**. It is located on the steering column. This switch also activates the back-up lights.

NOTE: The manual transmission back-up light switch is on the rear of the transmission.

2. Move the switch housing all the way toward the low gear position.
3. Move the gear selector to the **P** position. The main housing and housing back should ratchet, providing proper switch adjustment.

Throttle Valve Cable Adjustment

1. The adjustment is made at the engine end of the cable with the engine off, by rotating the throttle lever by hand. Do not use the accelerator pedal to rotate the throttle lever.
2. Remove the air cleaner.
3. Depress and hold down the metal adjusting tab at the end of the cable.
4. Move the slider until it stops against the fitting.
5. Release the adjusting tab.
6. Rotate the throttle lever to the full extent of it travel.

7. The slider must move towards the lever when the lever is at full travel. Make sure the cable moves freely.

NOTE: The cable may appear to function properly with the engine cold. Recheck it with the engine hot.

8. Road test the vehicle.

TRANSFER CASE

Transfer Case Assembly

REMOVAL AND INSTALLATION

1. Disconnect the negative battery cable. Raise and support the vehicle safely.

2. Drain the fluid from the transfer case. Disconnect the speedometer cable, if equipped. Remove the skid plate, the crossmember and supports, as required.

3. Matchmark the transfer case front output shaft yoke and driveshaft for reassembly. Disconnect the driveshaft from the transfer case.

4. Matchmark the rear axle yoke and the driveshaft for reassembly. Remove the driveshaft. Disconnect the shift linkage and electrical connections from the transfer case.

5. Remove the parking cables, as required.

6. Properly support the transfer case assembly. Remove the transfer case retaining bolts.

7. Remove the transfer case from the vehicle.

8. Installation is the reverse of the removal procedure.

Linkage

Adjustment

Model NP241

1. Position the transfer case lever in the 4H detent. Push the lower shift lever forward to the 4H stop.

2. Install the rod swivel in the shift lever hole. Hang a 0.200 in. (5.19mm) gauge rod behind the swivel.

3. Install the rear rod nut against the gauge with the shifter against the 4hi stop.

4. Remove the gauge tool. Push the rear of the swivel rearward against the nut.

5. Tighten the front rod nut against the swivel. Check for proper operation.

Model 205

1. Raise and support the vehicle safely. Remove the swivel from the transfer case shift lever.

2. Turn the swivel inward or outward to determine the correct shift detent.

3. Reinstall the swivel to the shift lever.

Model 1370

1. Position the transfer case lever in the 4H position.

2. Raise and support the vehicle safely.

3. Disconnect the linkage rod from the console shift lever.

4. Shift the transfer case into the 4H position (transfer shift lever in the forward detent).

5. Adjust the swivel to align with the hole in the console shift lever.

6. Lower the vehicle.

DRIVE AXLE

NOTE: The K-Series truck is the only vehicle using a halfshaft consisting of a CV-joint and boot; and is located only on the left front axle. The right side consists of an axle and tube assembly.

Halfshaft

REMOVAL AND INSTALLATION

K-Series

LEFT SIDE ONLY

1. Raise and support the vehicle safely.

2. Remove the wheel and tire assembly.

3. Remove the skid plate, as required.

4. Remove the drive axle hub nut and washer.

5. Remove the brake line support bracket from the upper control arm to allow extra travel of the control arm.

6. Remove the left outer tie rod attaching nut and cotter pin. Separate the tie rod from the steering knuckle using a suitable tie rod end splitter.

7. Position the tie rod aside and push steering linkage to the opposite side of the vehicle.

8. Remove the lower shock attaching nut and bolt; position the shock aside.

9. Remove the left stabilizer bar bracket and bushing at the frame. Remove the stabilizer bar bolt, spacer and bushings at the lower control arm.

10. Lower the vehicle, taking pressure off the upper control arm by placing a jack or jackstand below the lower control arm between the spring seat and the ball joint.

11. Remove the upper ball joint cotter pin and loosen (do not remove) the upper ball joint attaching nut. Using a suitable ball joint splitter, separate the ball joint stud from the steering knuckle. Remove the attaching nut.

NOTE: Cover the shock mounting bracket and lower ball joint stud with a towel to prevent the axle boot from tearing during REMOVAL AND INSTALLATION.

12. Separate the axle shaft from the hub and rotor using tool J–28733 or equivalent.

13. Remove the axle shaft inner flange bolts. Remove the shaft.

To install:

14. Lubricate the axle and hub splines with an approved high temperature wheel bearing grease. Position the shaft in the hub and install the inboard CV-joint-to-flange bolts.

15. Install the upper ball joint to steering knuckle and torque the stud nut to 61 ft. lbs. (83 Nm). Install a cotter pin through the upper ball joint stud and nut. Lubricate the ball joint as required.

16. Install the left stabilizer bar bracket and bushing at the frame. Install the stabilizer bar bolt, spacer and bushings at the lower control arm.

17. Position the lower shock in the mount bracket and install the attaching nut and bolt.

18. Connect the left tie rod end at the steering knuckle. Torque the nut to 35 ft. lbs. (47 Nm). Install a cotter pin through the tie rod stud and nut.

19. Connect the brake line bracket to the control arm, ensuring the line and/or hose is not twisted or kinked.

20. Install the skid plate, as required.

21. Install the axle hub washer and nut. Insert a drift through the rotor vanes to keep the axle from turning and torque the hub nut to 180 ft. lbs. (245 Nm) and the inboard CV-joint flange bolts to 60 ft. lbs. (80 Nm).

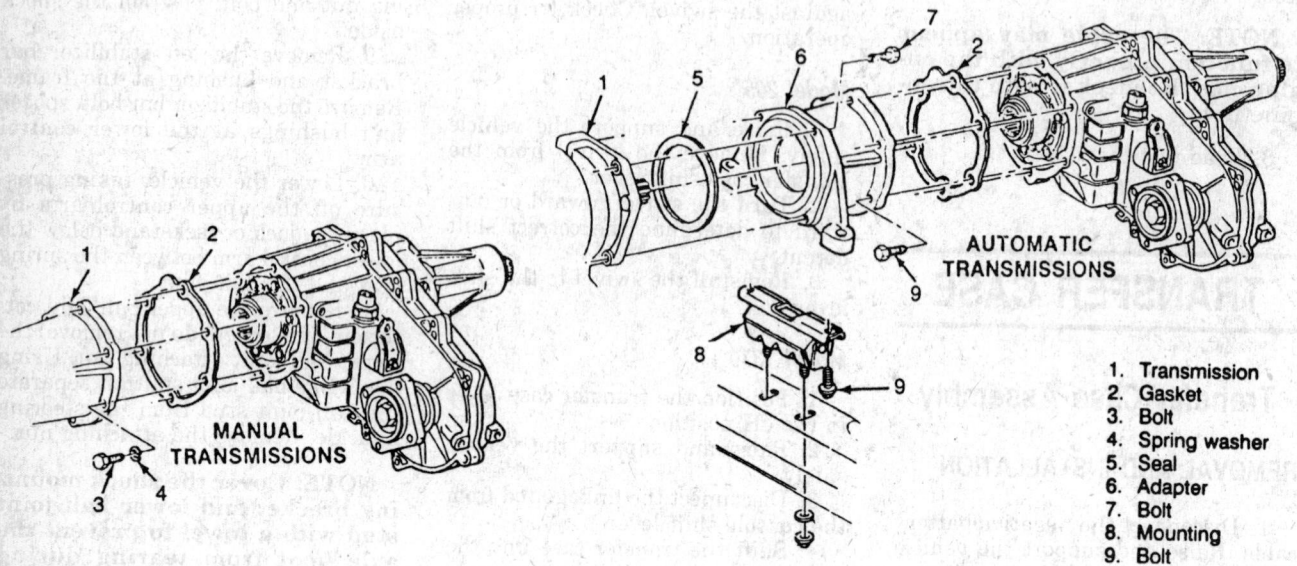

Typical transfer case installation

1. Transmission
2. Gasket
3. Bolt
4. Spring washer
5. Seal
6. Adapter
7. Bolt
8. Mounting
9. Bolt

79114074

22. Remove the drift, install the wheel and tire assembly. Remove stands and lower the vehicle.

CV-Boot

REMOVAL AND INSTALLATION

Outer

1. Raise and support the vehicle safely.
2. Remove the wheel and tire assembly.
3. Remove the skid plate, as required.
4. Remove the drive axle and place the assembly in a suitable holding fixture.
5. Remove the large (swage) ring using a chisel, being careful not to damage the housing surface.
6. Cut and remove the small clamp on the CV-boot.
7. Slide the boot back off of the joint housing. Clean excess grease from the joint area.
8. Spread the ears of the joint retaining clip and slide the joint off of the axle shaft.
9. Remove the boot from the axle shaft.

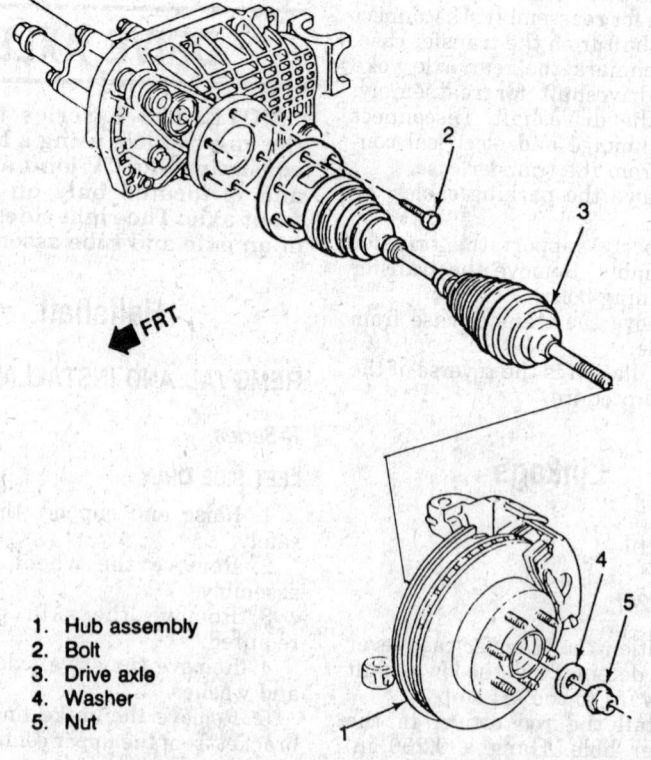

1. Hub assembly
2. Bolt
3. Drive axle
4. Washer
5. Nut

79114075

Drive axle removal and installation — K series

To install:

10. Place the small clamp on the small neck of the CV-boot. Do not crimp the clamp.

11. Slide the boot onto the axle shaft and into the groove on the axle shaft.

12. Crimp the small clamp using tool J–35910 or equivalent.

13. Disassemble the joint and thoroughly clean out old grease. Place a sufficient quantity of approved CV-joint grease inside the boot and in the joint.

14. Pinch the large CV-boot securing ring in an oval shape and slide the ring onto the boot large end.

15. Push the CV-joint onto the axle shaft until the retaining ring is seated in the groove on the axle shaft.

16. Slide the large diameter of the CV-boot with the ring in place over the outside of the CV-joint housing and locate the seal lip in the housing groove. Release any air trapped inside the boot using a thin flat blunt tool between the boot and housing.

17. Using a suitable ring installer tool, press the ring onto the housing until fully seated.

18. Install the halfshaft assembly, skid plate and the wheel and tire.

19. Lower the vehicle and road test.

Inner

1. Raise and support the vehicle safely.

2. Remove the wheel and tire assembly.

3. Remove the skid plate, as required.

4. Remove the drive axle and place the assembly in a suitable holding fixture.

5. Remove the large (swage) ring using a chisel, being careful not to damage the housing surface.

6. Cut and remove the small clamp on the CV-boot.

7. Slide the boot back off of the joint housing. Remove the joint and axle shaft from the joint housing.

8. Spread the ears of the joint retaining clip and slide the joint back on the axle shaft. Remove the forward retaining ring and slide the joint off of the shaft. Handle the joint with care to avoid disassembly.

9. Remove the rear retaining ring and remove the boot from the axle shaft.

To install:

10. Place the small clamp on the small neck of the CV-boot. Do not crimp the clamp.

11. Slide the boot onto the axle shaft and into the groove on the axle shaft.

12. Crimp the small clamp using tool J–35910 or equivalent. Install the forward joint retaining ring onto the axleshaft.

13. Thoroughly clean the old grease from the joint, boot and housing assembly. Place a sufficient quantity of approved CV-joint grease inside the boot.

14. Slide the joint assembly with the counterbore facing the axleshaft onto the shaft against the retaining ring.

15. Install the forward retaining ring in the groove on the axleshaft. Slide the joint towards the end of the shaft and move the inside retaining ring into the groove.

16. Pack the joint sufficient quantity of approved CV-joint grease. Pinch the large CV-boot securing ring in an oval shape and slide the ring onto the boot large end.

17. Push the CV-joint onto the axle shaft.

18. Slide the large diameter of the CV-boot with the ring in place over the outside of the CV-joint housing and locate the seal lip in the housing groove. Release any air trapped inside the boot using a thin flat blunt tool between the boot and housing.

19. Using a suitable ring installer tool, press the ring onto the housing until fully seated.

20. Install the halfshaft assembly, skid plate and the wheel and tire.

21. Lower the vehicle and road test.

Driveshaft and U-Joints

REMOVAL AND INSTALLATION

1. Raise and support the vehicle safely.

2. Mark the relationship of the driveshaft to the front axle and transfer case flange.

3. Remove the skid plate, if equipped.

4. Remove the bolts, nuts, washers and U-bolts/retainers from the front axle flange-to-driveshaft.

5. Remove the slip yoke from the front axle.

NOTE: Do not let the u-joint caps fall off the yoke, tape the bearing caps in place to avoid loss of the bearing rollers.

6. Remove the bolts at the transfer case flange, slide the driveshaft forward and disengage the assembly from the transfer case.

7. Place the driveshaft into a holding fixture.

8. Remove the retaining rings.

9. Using 2 sockets, one with a diameter just smaller than the bearing cap; the other with an opening large enough to accept a bearing cap, press out the bearing caps in a vice.

10. Remove the trunnion from the slip yoke.

To install:

11. Place the trunnion in between the yoke ears of the driveshaft and begin installing both bearing caps by hand, finish the installation using a vise.

12. Use a socket with a smaller diameter than the bearing cap and press the bearing cap in past the retainer ring groove. Install the retaining ring.

13. Turn the driveshaft over and repeat the procedure.

14. Install the slip yoke over the trunnion and repeat the procedure for the remaining bearing caps.

15. Install the driveshaft, U-bolt/retainers, nuts and bolts, aligning the shaft with the previous made marks on the transfer case and front axle.

16. Lower the vehicle and roadtest.

Front Axle Shaft, Bearing and Seal

REMOVAL AND INSTALLATION

V-Series

1. Raise and support the vehicle safely.

2. Remove the wheel and tire assembly.

3. Remove the brake caliper and position the assembly aside.

4. Remove the locking hub assembly. Remove the hub and bearing assembly as follows:

   a. Using either wheel bearing nut wrench (J–6893–D on ½ or ¾ ton vehicles or J–26878–A on 1 ton vehicles), remove the locknut, ring and adjusting nut.

   b. Remove the hub and rotor assembly, bearing careful not to let the bearing fall out of the hub.

5. Remove the splash shield and brake bracket.

6. Remove the spindle from the steering knuckle.

7. Remove the axle shaft as follows.

To install:

8. Lube the spindle and spindle bearing with the recommended lubricant.

9. Install the spindle seal and spacer onto the axle shaft with the

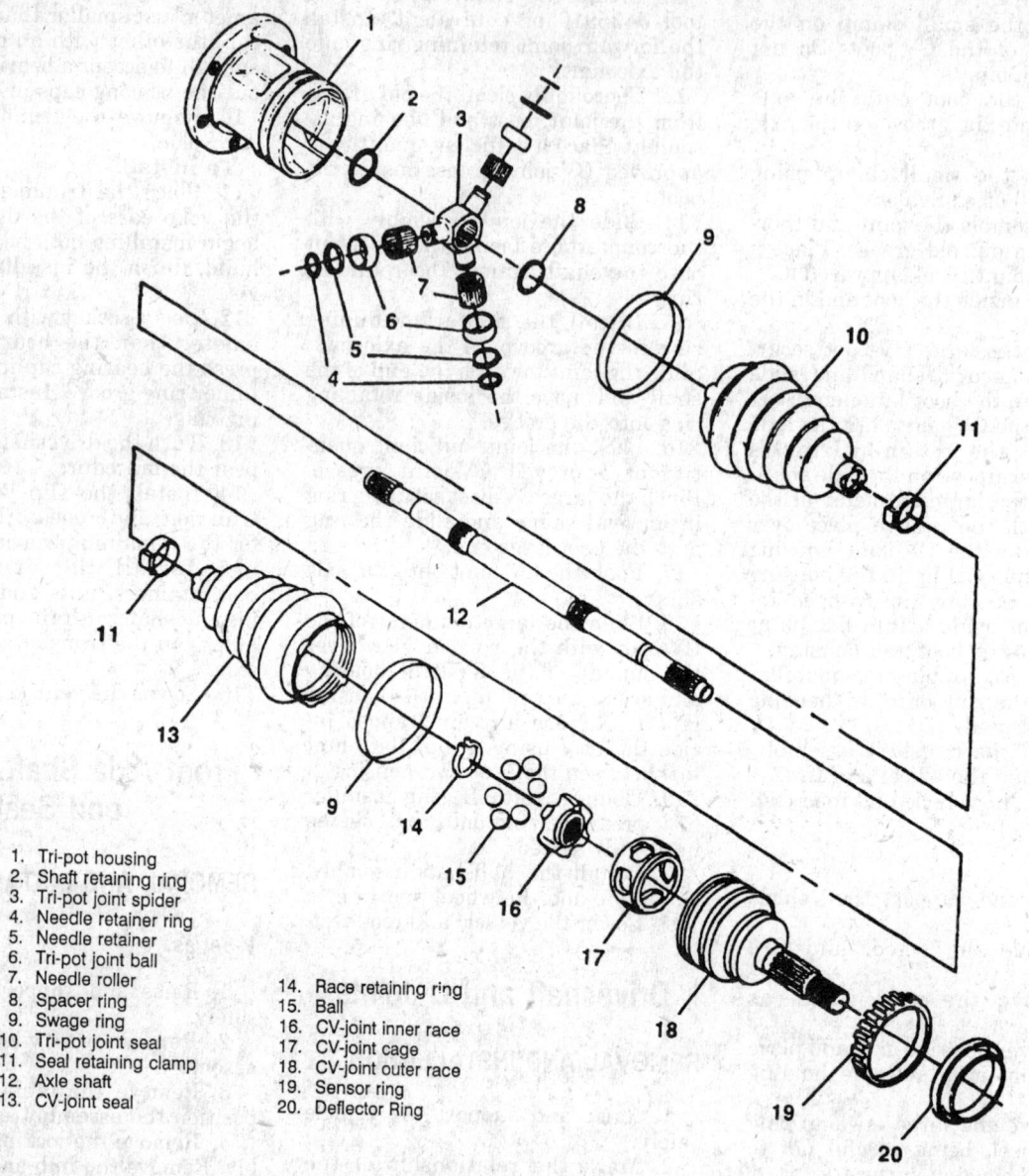

1. Tri-pot housing
2. Shaft retaining ring
3. Tri-pot joint spider
4. Needle retainer ring
5. Needle retainer
6. Tri-pot joint ball
7. Needle roller
8. Spacer ring
9. Swage ring
10. Tri-pot joint seal
11. Seal retaining clamp
12. Axle shaft
13. CV-joint seal
14. Race retaining ring
15. Ball
16. CV-joint inner race
17. CV-joint cage
18. CV-joint outer race
19. Sensor ring
20. Deflector Ring

79114076

Front drive axle components — K series

spacer's chamfer points facing toward the oil deflector.

10. Install the axle shaft into the housing.

11. Install the steering knuckle, brake caliper bracket and splash shield and secure with nuts and washers. Torque nuts to 65 ft. lbs. (88 Nm). Ensure the seal and oil deflector are in place on the steering knuckle.

12. Install the hub and rotor assembly as follows:

a. If replacing the bearing, drive the races out of the hub assembly.

b. Using a brass drift and hammer, drive the inner bearing and seal out of the hub.

c. Remove all grease from the bearings, hub and spindle assembly using a suitable solvent.

d. Inspect the bearings, races and spindle for heat stress, cracking or scores, repair or replace as necessary.

e. Using a bearing race installer, drive the new races into the hub assembly until fully seated.

f. Using an approved high-temperature wheel bearing grease pack the bearings, working the grease in between the rollers and apply a small quantity to the spindle and hub assembly.

g. Install the inner wheel bearing and a new seal into the hub.

h. Install the rotor/hub assembly onto the spindle.

i. Install the outer wheel bearing and push onto the spindle until fully seated against the race. Install the adjusting nut torque to 50 ft. lbs. (60 Nm) while spinning the rotor assembly.

j. Back off the adjusting nut and retorque the nut to 35 ft. lbs. (47 Nm) for automatic hubs or 50 ft. lbs. (60 Nm) for manual hubs.

k. Back off the adjusting nut a maximum of 3/8 of a turn for automatic hubs or just enough to free the bearing for manual hubs.

l. Install the ring and locknut, ensuring the tang on the inside di-

SWAGE CLAMP SIZE CHART

TOOL NO.	DESCRIPTION	APPLICATION
J 36652-1	Split Plate Swage Clamp	K 10/20
J 36652-2	Split Plate Swage Clamp	K 30 (Outboard)
J 36652-3	Split Plate Swage Clamp	K 30 (Inboard)

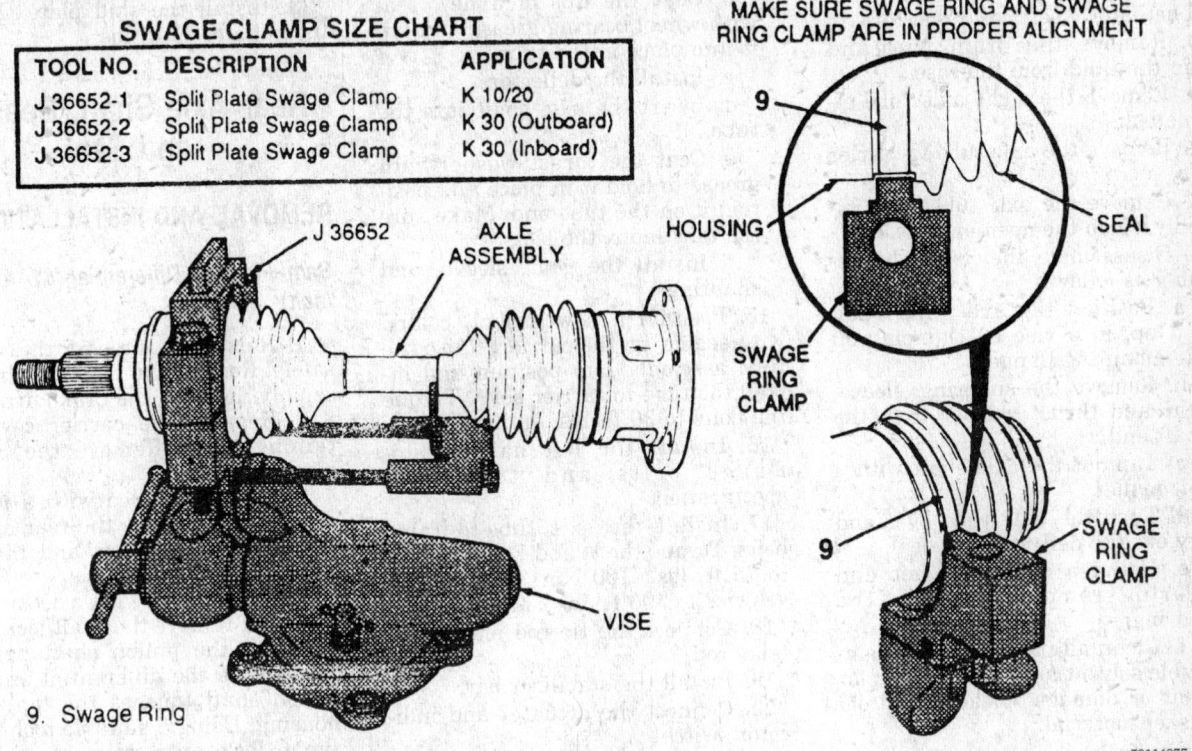

9. Swage Ring

Pressing axle boot in place — K series

MAKE SURE SWAGE RING AND SWAGE RING CLAMP ARE IN PROPER ALIGNMENT

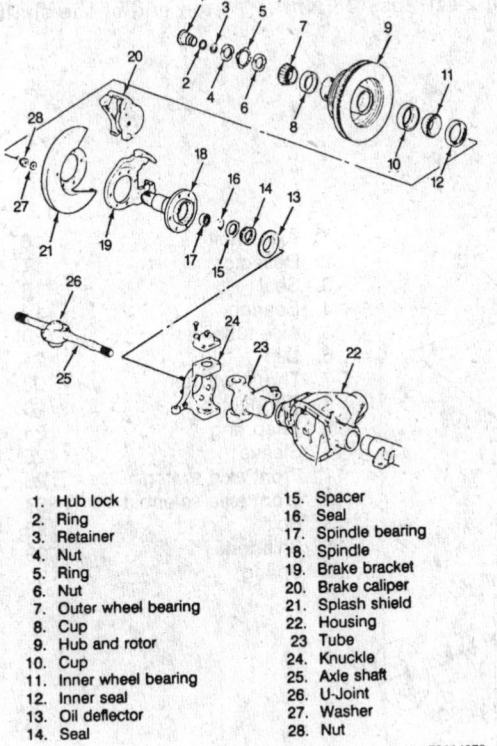

1. Hub lock	15. Spacer
2. Ring	16. Seal
3. Retainer	17. Spindle bearing
4. Nut	18. Spindle
5. Ring	19. Brake bracket
6. Nut	20. Brake caliper
7. Outer wheel bearing	21. Splash shield
8. Cup	22. Housing
9. Hub and rotor	23. Tube
10. Cup	24. Knuckle
11. Inner wheel bearing	25. Axle shaft
12. Inner seal	26. U-Joint
13. Oil deflector	27. Washer
14. Seal	28. Nut

79114079

Front driving axle components — except K series

ameter of the ring passes is aligned in the slot on the spindle.

m. The hole in the ring must align with the pin on the lock nut. Move the adjusting nut to align the pin.

n. Torque the locknut to 160 ft. lbs. (217 Nm).

13. Install the locking hub assembly.

14. Install the brake caliper assembly.

15. Install the wheel and tire assembly.

16. Lower the vehicle, check and add gear oil to the front axle as required. Road test the vehicle.

K-Series

1. Raise and support the vehicle safely.

2. Remove the right wheel.

3. Position a drain pan under the axle.

4. Remove the stabilizer bar.

5. Remove the skid plate.

6. Disconnect the right inner tie rod end at the relay rod.

7. Remove the left halfshaft-to-axle flange bolts, turn the wheel to loosen the axle from the axle tube, push the axle towards the front of the truck and tie the axle out of the way.

8. Unplug the indicator switch and actuator electrical connectors.

9. Remove the drain plug and drain the fluid from the case.

10. Remove the right axle tube-to-frame bolts.

11. Remove the axle tube-to-carrier bolts.

12. Remove the axle tube/shaft assembly. Keep the open end up.

13. Disassemble the axle tube assembly as follows:

a. Position the axle tube, open end up, in a vise by clamping on the mounting flange.

b. Remove the snapring, sleeve, gear and thrust washer from the shaft end.

c. Tap out the axle shaft with a soft mallet.

d. Turn the axle tube over and pry out the deflector and seal.

e. Using a slide hammer and bearing remover, remove the bearing.

f. Clean all parts in a non-flammable solvent and inspect them for wear or damage. Clean off all old gasket material.

To install:

14. To install, perform the following procedures:

a. Using a bearing driver, install the new outer bearing in the tube.

b. Coat the lips of a new seal with wheel bearing grease and tap it into place in the tube.

c. Install the deflector.

d. Insert the axle shaft into the tube.

e. Coat the thrust washer with grease to hold it in place and position it on the tube end. Make sure the tabs index the slots.

f. Install the gear, sleeve, and snapring.

15. Position a new gasket, coated with sealer, on the carrier. Raise the tube assembly into position and install the tube-to-carrier bolts. Torque the bolts to 30 ft. lbs. (40 Nm).

16. Install the left halfshaft-to-flange bolts and torque to specification.

17. Install the axle tube-to-frame bolts. Torque the 1/2 and 3/4 ton vehicle to 75 ft. lbs. (100 Nm) or the 1 ton vehicles to 106 ft. lbs. (145 Nm).

18. Connect the tie rod end to the relay rod.

19. Install the stabilizer bar.

20. Connect the actuator and indicator switch.

21. Install the drain plug. Tighten it to 24 ft. lbs. (33 Nm).

22. Fill the axle with the correct quantity and type of gear oil. Tighten the filler plug to 24 ft. lbs. (33 Nm).

23. Install the skid plate. Tighten the bolts to 25 ft. lbs.

Rear Axle Shaft, Bearing and Seal

REMOVAL AND INSTALLATION

Semi-Floating Differential (8½ AND 9½ Inch)

1. Raise and support the vehicle safely. Remove the tire and wheel assembly. Remove the brake drums.

2. Remove the carrier cover retaining bolts. Remove the carrier cover.

3. If not equipped with a locking differential, remove the rear axle pinion shaft lock screw and the rear axle pinion shaft.

4. If equipped with a locking differential, remove the shaft lock screw and pull the pinion shaft partially out. Rotate the differential until the pinion shaft touches the top of the housing. Using a suitable tool, rotate the C-lock until aligned with the thrust block.

5. Push the flanged end of the axle shaft toward the center of the vehicle. Remove the C-lock clip from the button end of the shaft.

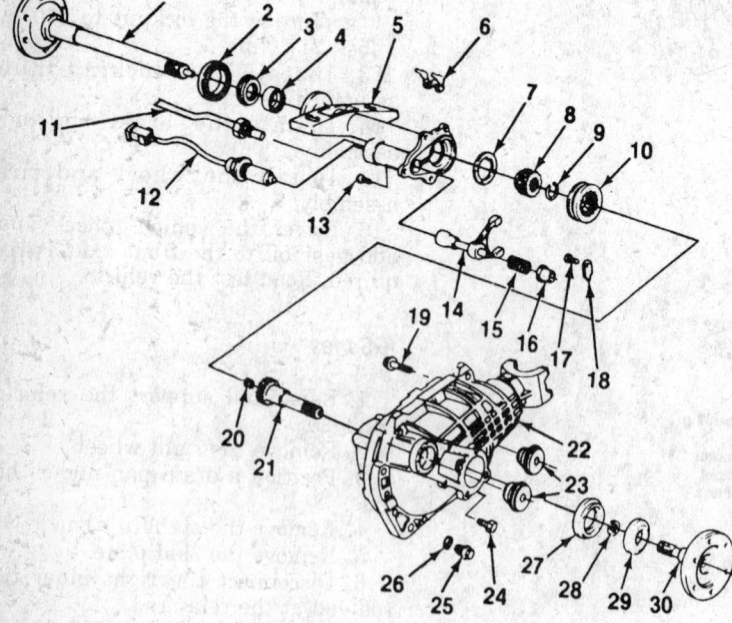

1. Axle shaft	16. Seal
2. Deflector	17. Bolt
3. Seal	18. Lock
4. Bearing	19. Bolt
5. Axle tube	20. Pilot bearing
6. Clip	21. Inner output shaft
7. Thrust washer	22. Carrier case
8. Connector	23. Carrier bushing
9. Snap ring	24. Drain plug
10. Sleeve	25. Fill plug
11. Front axle switch	26. Washer
12. Front axle solenoid	27. Seal
13. Bolt	28. Snap ring
14. Shift fork	29. Deflector
15. Spring	30. Output shaft

Front axle components — K series

79114078

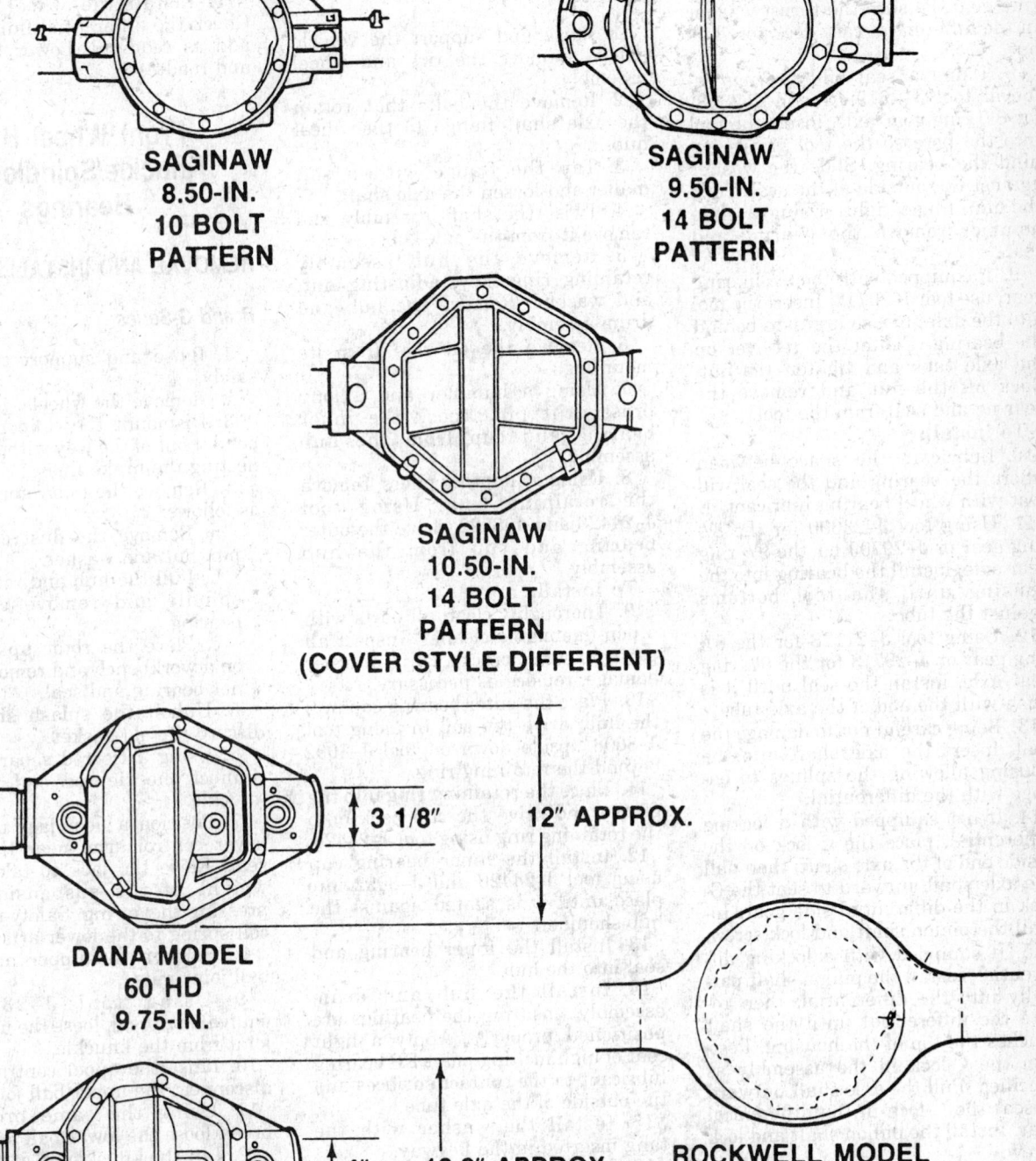

SAGINAW
8.50-IN.
10 BOLT
PATTERN

SAGINAW
9.50-IN.
14 BOLT
PATTERN

SAGINAW
10.50-IN.
14 BOLT
PATTERN
(COVER STYLE DIFFERENT)

DANA MODEL
60 HD
9.75-IN.

3 1/8"   12" APPROX.

DANA MODEL
70 OU AND HD
10.50-IN.

4"   12.8" APPROX.

ROCKWELL MODEL
12.0-IN.

Rear axle identification

6. Remove the axle shaft from the housing. Be careful not to damage the oil seal.

7. When removing the axle shaft on vehicles equipped with the 9½ in. ring gear, be sure the thrust washer in the differential case does not slide out.

8. Using a seal and bearing removal tool J–2619–01 and J–23689 for 8½ ring gear axle, insert the tool into the bore so the tool grasps behind the bearing. Slide the washer against the outside of the seal. Turn the nut finger-tight. Using a slide hammer remove the bearing and seal.

9. If equipped with the 9½ in. ring gear, use tool J–29712. Insert the tool into the axle tube so it grasps behind the bearing. Center the receiver on the axle tube and tighten the nut. Back off the nut and remove the bearing and seal from the tool.

To install:

10. Lubricate the space between where the bearing and the seal will seat with wheel bearing lubricant.

11. Using tool J–23690 for the 8½ ring gear or J–29709 for the 9½ ring gear axle, install the bearing into the housing until the tool bottoms against the tube.

12. Using tool J–21128 for the 8½ ring gear or J–29713 for the 9½ ring gear axle, install the seal until it is flush with the end of the axle tube.

13. Being careful not to damage the seal, insert the axle shaft into the housing allowing the splines to engage with the differential.

14. If not equipped with a locking differential, place the C-lock on the inside end of the axle shaft; then pull the axle shaft outward to seat the C-lock in the differential side gear. Install the pinion shaft and lock screw.

15. If equipped with a locking differential, install the pinion shaft partially into the differential; then rotate the differential until the shaft touches the top of the housing. Position the C-lock in the assembly as specified. Pull the axle shaft outward to seat the C-lock in the differential gear. Install the pinion shaft and lock screw.

16. Torque the lock screw to 25 ft. lbs. (34 Nm).

17. Install the carrier cover and gasket. Torque the bolts to 30 ft. lbs. (41 Nm)18.

18. Fill the axle assembly with the correct type of gear oil to the filler hole level.

19. Install the brake drum, wheel and tire assembly. Inspect the axle for leaks.

20. Lower the vehicle and road test.

Full Floating Differential (9¾ and 10½ Inch)

1. Raise and support the vehicle safely. Remove the tire and wheel assembly.

2. Remove the bolts that retain the axle shaft flange to the wheel hub.

3. Tap the flange with a soft mallet and loosen the axle shaft.

4. Twist the shaft assembly and remove it from the axle tube.

5. Remove the hub assembly retaining ring, key, adjusting nut, and washer. Remove the hub and drum assembly.

6. Remove the oil seal from its mounting.

7. Using a hammer and a long brass drift pin, knock the inner bearing and cup from the hub assembly.

8. Using a snapring pliers, remove the retaining ring. Using tool J–24426 and J–8092, drive the outer bearing and cup from the hub assembly.

To install:

9. Thoroughly clean all parts with a non-flammable cleaner. Inspect all parts for wear, cracks, chips or other damage, replace as necessary.

10. Place the outer bearing cup into the hub, drive the cup in using tool J–8608 upside down on tool J–8092 beyond the retaining ring.

11. Place the retaining ring into the groove and drive the cup back onto the retaining ring using tool J–24426.

12. Install the inner bearing cup using tool J–24426 and J–8092 into place until it is seated against the hub shoulder.

13. Install the inner bearing and seal into the hub.

14. Install the hub and drum assembly, ensuring the bearings are positioned properly. Apply a light coat of high melting point EP bearing lubricant to the contact surfaces and the outside of the axle tube.

15. Install the washer with the tang inserted in the keyway.

16. Install the nut and torque to 50 ft. lbs. (68 Nm), while rotating the hub and drum assembly. If the adjusting nut is not aligned with the keyway, back the nut off not more than 1 slot of the lock or the axle spindle using tool J–2222–C.

17. Place a new gasket or apply RTV sealant over the axle shaft flange. Position the shaft in the housing so the shaft splines enter the differential side gear.

18. Align the axle shaft flange holes with the hub and install the attaching bolts. Torque the bolts to 15 ft. lbs. (156 Nm).

19. Install the wheel assembly. Check the differential fluid level and add as required. Lower the vehicle and roadtest.

Front Wheel Hub, Knuckle/Spindle and Bearings

REMOVAL AND INSTALLATION

R and G-Series

1. Raise and support the vehicle safely.

2. Remove the wheels.

3. Dismount the caliper and suspend it out of the way without disconnecting the brake lines.

4. Remove the hub/rotor assembly as follows:

   a. Remove the dust cap, cotter pin, nut and washer.

   b. Pull the hub and rotor off the spindle and remove the outer bearing.

   c. Place the rotor upside down on a workbench and remove the inner bearing and seal.

5. Unbolt the splash shield and discard the old gasket.

6. Using a tie rod separator, disconnect the tie rod end from the knuckle.

7. Position a floor jack under the lower control arm, near the spring seat. Raise the jack to take up the weight of the suspension, compressing the spring. Safety-chain the coil spring to the lower arm.

8. Remove the upper and lower ball joint nuts.

9. Using tool J–23742 or equivalent, break loose the upper ball joint from the knuckle.

10. Raise the upper control arm to disconnect the upper ball joint.

11. Using the same procedure, break loose the lower ball joint.

12. Lift the knuckle off of the lower ball joint.

13. Inspect and clean the ball stud bores in the knuckle. Make sure there are no cracks or burrs. If the knuckle is damaged in any way, replace the assembly.

14. Check the spindle for wear, heat discoloration or damage. If at all damaged, replace the assembly.

To install:

15. Maneuver the knuckle onto both ball join

1. Drum
2. Bolt
3. Shaft
4. Lock
5. Seal
6. Bearing
7. Housing
8. Clip
9. Bolt
10. Carrier cover

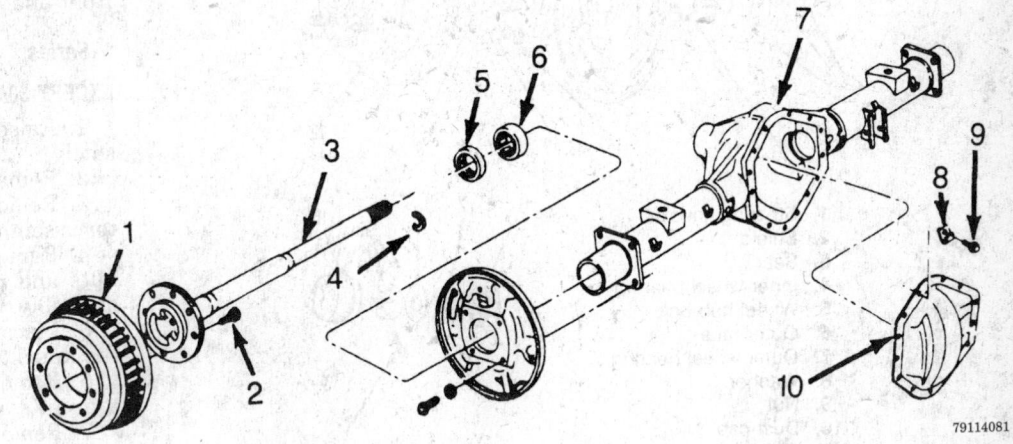

Semi-floating rear axle assembly

79114081

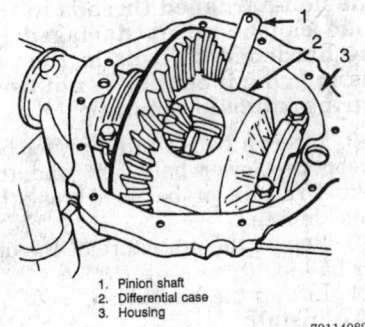

1. Pinion shaft
2. Differential case
3. Housing

79114082

Pinion shaft pin removal and installation —
semi-floating rear axle

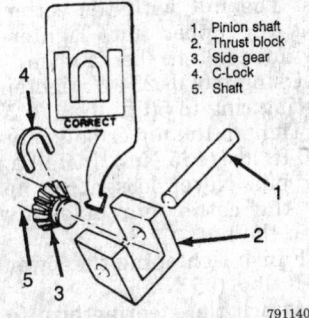

1. Pinion shaft
2. Thrust block
3. Side gear
4. C-Lock
5. Shaft

79114083

C-lock alignment for removal and
installation — semi-floating rear axle

A. Dana axles with disc
   brakes
B. Dana and saginaw axles
   with drum brakes

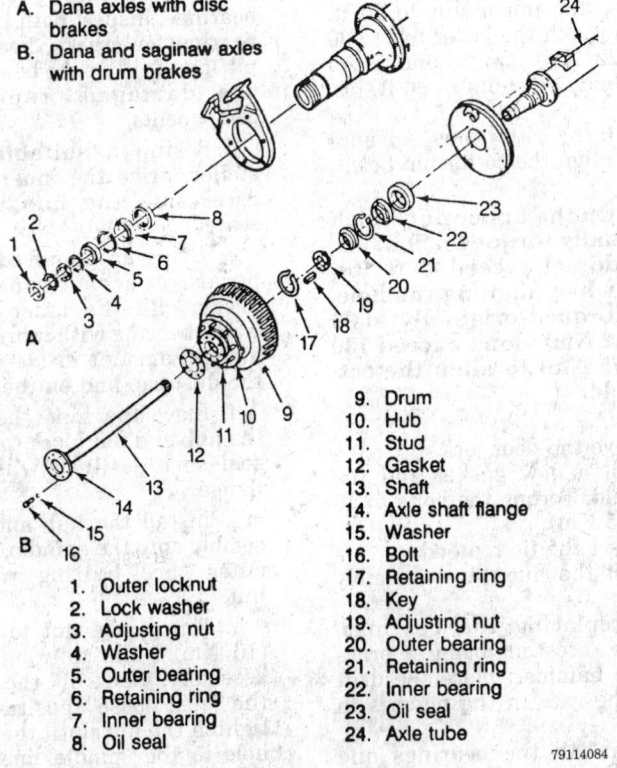

1. Outer locknut
2. Lock washer
3. Adjusting nut
4. Washer
5. Outer bearing
6. Retaining ring
7. Inner bearing
8. Oil seal

9. Drum
10. Hub
11. Stud
12. Gasket
13. Shaft
14. Axle shaft flange
15. Washer
16. Bolt
17. Retaining ring
18. Key
19. Adjusting nut
20. Outer bearing
21. Retaining ring
22. Inner bearing
23. Oil seal
24. Axle tube

79114084

Full floating axle and components

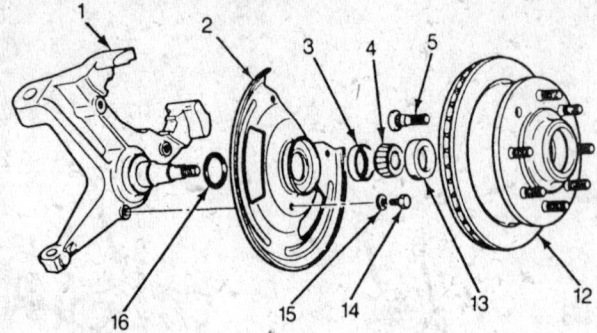

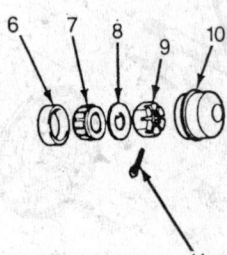

1. Steering knuckle
2. Shield
3. Seal
4. Inner wheel bearing
5. Wheel hub bolt
6. Outer race
7. Outer wheel bearing
8. Washer
9. Nut
10. Dust cap
11. Cotter pin
12. Wheel hub/rotor
13. Inner race
14. Bolt
15. Washer
16. Gasket

79114085

Hub and bearing assembly — R series

16. Install both nuts. On ½ ton series, torque the upper nut to 50 ft. lbs. (68 Nm) and the lower nut to 90 ft. lbs. (122 Nm). On ¾ and 1 ton series, torque both nuts to 90 ft. lbs. (122 Nm).

17. Install the cotter pins. Advance the nut to align the cotter pin hole.

NOTE: On the upper nut which was originally torque to 50 ft. lbs. (68 Nm), do not exceed 90 ft. lbs. (122 Nm) when aligning the hole. On nuts torqued originally to 90 ft. lbs. (122 Nm), don't exceed 130 ft. lbs. (175 Nm) to align the cotter pin hole.

18. Remove the floor jack.
19. Install a new gasket and the splash shield. Torque the bolts to 10 ft. lbs. (13.5 Nm).
20. Connect the tie rod end.
21. Install the hub/rotor assembly as follows:

   a. If replacing the bearings, drive the races out using a brass drift and hammer; place the drift behind the race in the notches in the hub.

   b. Clean all the bearings and hubs with a non-flammable solvent. Do not spin the bearings with compressed air to dry, otherwise damage may result.

   c. If reinstalling the original bearings, inspect both the race and bearing for cracks, heat stress or pitting; if either the bearing or race are damaged, replace both components.

   d. Using a suitable race installer, drive the inner and outer races into the hub until fully seated.

   e. Using an approved high temperature grease and bearing packer, fill the inner and outer wheel bearing with lubricant. Place a small amount inside the hub, in the dust cap and on the spindle.

   f. Place the inner bearing into the hub and flat block to install the seal so it is flush with the hub flange.

   g. Install the hub and rotor assembly onto the spindle, install the outer wheel bearing, washer and nut.

   h. Torque the nut to 12 ft. lbs. (16 Nm) while rotating the rotor assembly. Back off the nut until the "just loose" position. finger-tighten the nut until the cotter pin hole in the spindle lines up with the hole in the nut or cage. Do not back off the nut more than ½ of a flat.

   i. Install a new cotter pin.

   j. Rotor endplay should measure between 0.0012–0.0050 in. (0.03–0.13mm) when properly adjusted.

22. Install the caliper.
23. Install the wheels and check the front end alignment.

V-Series

EXCEPT 30/3500-SERIES

1. Raise and support the vehicle safely.
2. Remove the wheels.
3. Remove the caliper, hub and rotor assemblies.
4. Remove the spindle attaching nuts and plate. Remove the spindle assembly.
5. Disconnect the tie rod end from the knuckle.
6. Remove the knuckle-to-steering arm nuts and adapters.
7. Remove the steering arm from the knuckle.
8. Remove the cotter pins and nuts from the upper and lower ball joints.

NOTE: Do not remove the adjusting ring from the knuckle. If it is necessary to loosen the ring to remove the knuckle, don't loosen it more than 2 threads. The non-hardened threads in the yoke can be easily damaged by the hardened threads in the adjusting ring if caution is not used during knuckle removal.

9. Insert a ball joint separator between the lower ball joint and the yoke. Drive the prybar in to break the knuckle free.
10. Repeat the procedure at the upper ball joint.
11. Lift off the knuckle.
To install:
12. Position the knuckle on the yoke.
13. Start the ball joints into their sockets. Place the nuts onto the ball studs. The nut with the cotter pin slot is the upper nut. Tighten the lower nut to 30 ft. lbs. (40 Nm).
14. Using tool J–23447, tighten the adjusting ring to 50 ft. lbs. (70 Nm).
15. Tighten the upper ball joint nut to 100 ft. lbs. (135 Nm). Install a new cotter pin. Never loosen the nut to align the cotter pin hole, always tighten the nut.
16. Finish tightening the lower nut to 70 ft. lbs. (95 Nm).
17. Attach the steering arm to the knuckle using adapters and new nuts. Torque the nuts to 90 ft. lbs. (120 Nm).
18. Connect the tie rod end to the knuckle.

19. Install the spindle as follows:

  a. Mount the spindle in a holding fixture and remove the bearing seal and shaft bearing.

  b. Remove the spacer, seal and oil deflector from the shaft.

  c. Inspect all components for wear, heat stress and scoring. Replace as necessary.

  d. Lubricate the shaft bearing and spindle with a high melting point wheel bearing grease.

  e. Using tool J–8092 and J–23445–A, install the bearing and seal into the spindle.

  f. Install the oil seal onto the oil deflector with the deflector lip toward the spindle. Install the oil deflector and seal onto the axle shaft.

  g. Install the spacer onto the axle shaft with the chamfer pointing toward the oil deflector.

  h. Install the spindle onto the steering knuckle.

  i. Install the spindle plate. Install new nuts and washers. Torque the nuts to 65 ft. lbs. (88 Nm).

20. Install the hub/rotor assembly and wheel bearings. Adjust the bearings.

21. Install the locking hubs and caliper assembly.

22. Install the wheel and lower the vehicle.

30/3500-SERIES

1. Raise and support the vehicle safely.

2. Remove the wheel.

3. Remove the caliper assembly.

4. Remove the hub and rotor assembly.

5. Remove the spindle attaching nuts, washers plate and bracket. Remove the spindle assembly.

6. Remove the upper cap from the knuckle and/or steering arm from the knuckle, by loosening the bolts and/or nuts a little at a time in an alternating pattern. This will safely relieve spring pressure under the cap and/or

arm. Once spring pressure is relieved, remove the bolts and/or nuts, washers and cap and/or steering arm.

7. Remove the gasket and compression spring.

8. Remove the bolts and washers and remove the lower bearing cap and the lower kingpin.

9. Remove the upper kingpin bushing by pulling it out through the knuckle.

10. Remove the knuckle from the axle yoke.

11. Remove the retainer from the knuckle.

12. Using a large breaker bar and adapter J–26871, remove the upper kingpin from the axle yoke by applying 500–600 ft. lbs. (677–813 Nm) of torque to the kingpin to break it free.

13. Using a hammer and blunt drift, drive out the retainer, race bearing and seal from the axle yoke all at once.

To install:

14. Using tool J–7817, install a new retainer and race in the axle yoke.

15. Fill the recessed area in the retainer and race with an approved high temperature wheel bearing lubricant.

16. Completely pack the upper yoke roller bearing with wheel bearing grease. A cone-type bearing packer is preferable but the bearing may be packed by hand.

17. Install the bearing and a new seal in the upper axle yoke, using a bearing driver such as J–22301. Don't distort the seal. It should protrude slightly above the yoke when fully seated.

18. Using adapter tool J–28871, install the upper kingpin. The kingpin must be torque to 550 ft. lbs. (745 Nm).

19. Position the knuckle in the yoke. Working through the knuckle, install a new felt seal over the kingpin and position the knuckle on the kingpin.

20. Install the bushing over the kingpin.

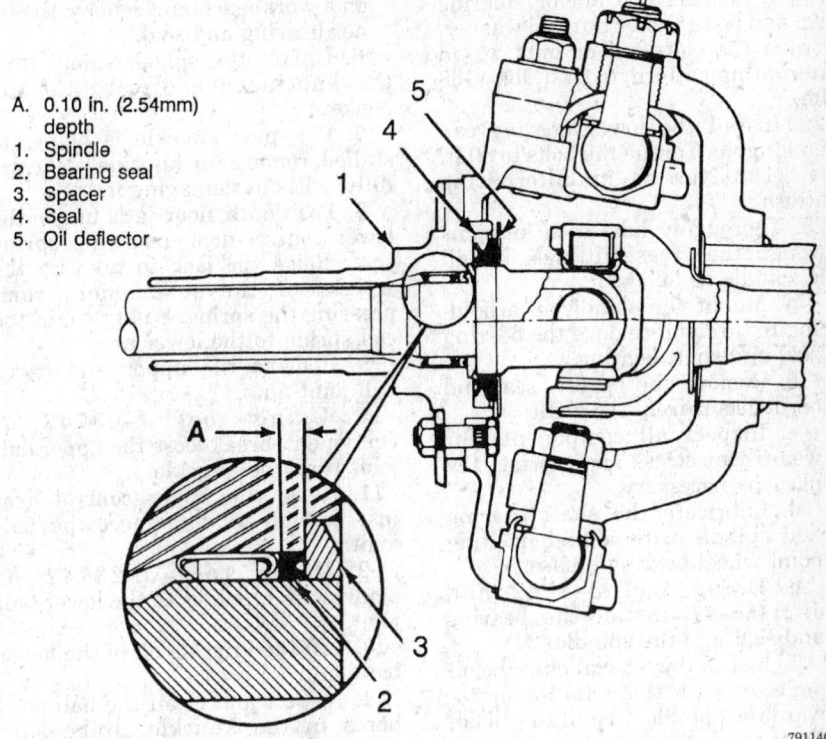

A. 0.10 in. (2.54mm) depth
1. Spindle
2. Bearing seal
3. Spacer
4. Seal
5. Oil deflector

Front spindle assembly — V series

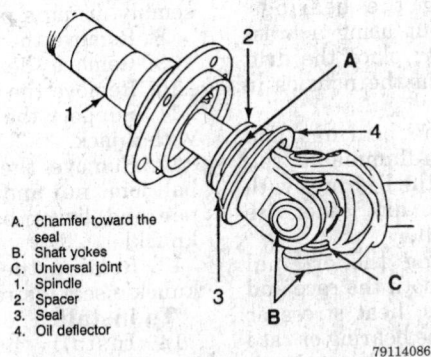

A. Chamfer toward the seal
B. Shaft yokes
C. Universal joint
1. Spindle
2. Spacer
3. Seal
4. Oil deflector

Front spindle assembly — V series

21. Install the compression spring, gasket, bearing cap and/or steering arm and bolts and/or nut and washer. Torque the bolts and/or nuts, in an alternating pattern, to 80 ft. lbs. (108 Nm).

22. Install the lower bearing cap and kingpin. Torque the bolts to 80 ft. lbs. (108 Nm) in an alternating pattern.

23. Thoroughly lube both kingpins through the grease fittings. Install the spindle as follows:

a. Mount the spindle in a holding fixture and remove the bearing seal and shaft bearing.

b. Remove the spacer, seal and oil deflector from the shaft.

c. Inspect all components for wear, heat stress and scoring. Replace as necessary.

d. Lubricate the shaft bearing and spindle with a high melting point wheel bearing grease.

e. Using tool J-8092 and J-21465-17, install the bearing and seal into the spindle.

f. Install the oil seal onto the oil deflector with the deflector lip toward the spindle. Install the oil deflector and seal onto the axle shaft.

g. Install the spacer onto the axle shaft with the chamfer pointing toward the oil deflector.

h. Install the spindle onto the steering knuckle.

i. Install the spindle bracket and plate. Install new nuts and washers. Torque the nuts to 65 ft. lbs. (88 Nm).

24. Install the hub and rotor assembly. Adjust the bearings.

25. Install the locking hubs and caliper assembly.

26. Install the wheel and tire assembly.

27. Lower the vehicle and check the front end alignment.

C-Series

1. Raise and support the vehicle safely. Let the control arms hang freely.

2. Remove the wheels.

3. Disconnect the tie rod end from the knuckle.

4. Dismount the caliper and suspend it out of the way without disconnecting the brake lines.

5. Remove the hub/rotor assembly as follows:

a. Remove the dust cap, cotter pin, nut and washer.

b. Pull the hub and rotor off the spindle and remove the outer bearing.

c. Place the rotor upside down on a workbench and remove the inner bearing and seal.

6. Unbolt the splash shield from the knuckle and discard the old gasket.

7. If a new knuckle is being installed, remove the knuckle seal carefully, without damaging it.

8. Position a floor jack under the lower control arm, near the spring seat. Raise the jack to take up the weight of the suspension, compressing the spring. Safety chain the coil spring to the lower arm.

9. Remove the upper and lower ball joint nuts.

10. Using tool J-23742 or equivalent, break loose the upper ball joint from the knuckle.

11. Raise the upper control arm just enough to disconnect the ball joint.

12. Using tool J-23742 or equivalent, break loose the lower ball joint.

13. Lift the knuckle off of the lower ball joint.

14. Inspect and clean the ball stud bores in the knuckle. Make sure there are no cracks or burrs. If the knuckle is damaged in any way, replace it.

15. Check the spindle for wear, heat discoloration or damage. If at all damaged, replace it.

To install:

16. Maneuver the knuckle onto both ball join

17. Install both nuts. Torque the nuts to 84 ft. lbs. (115 Nm).

18. Install new cotter pins. Always advance the nut to align the cotter pin hole.

19. Install the knuckle seal.

20. Remove the floor jack.

21. Install a new gasket and the splash shield. Torque the bolts to 19 ft. lbs. (26 Nm).

22. Connect the tie rod end.

23. Install the hub/rotor assembly as follows:

a. If replacing the bearings, drive the races out using a brass drift and hammer; place the drift behind the race in the notches in the hub.

b. Clean all the bearings and hubs with a non-flammable solvent. Do not spin the bearings with compressed air to dry, otherwise damage may result.

c. If reinstalling the original bearings, inspect both the race and bearing for cracks, heat stress or pitting; if either the bearing or race are damaged, replace both components.

d. Using a suitable race installer, drive the inner and outer races into the hub until fully seated.

e. Using an approved high temperature grease and bearing packer, fill the inner and outer wheel bearing with lubricant. Place a small amount inside the hub, in the dust cap and on the spindle.

f. Place the inner bearing into the hub and flat block to install the seal so it is flush with the hub flange.

g. Install the hub and rotor assembly onto the spindle, install the outer wheel bearing, washer and nut.

h. Torque the nut to 12 ft. lbs. (16 Nm) while rotating the rotor assembly. Back off the nut until the "just loose" position. Finger-tighten the nut until the cotter pin hole in the spindle lines up with the hole in the nut or cage. Do not back off the nut more than ½ of a flat.

i. Install a new cotter pin.

j. Rotor endplay should measure between 0.0012-0.0050 in. (0.03-0.13mm) when properly adjusted.

24. Install the caliper assembly.

25. Install the wheels.

26. Lower the vehicle and check the front end alignment.

K-Series

1. Disconnect the negative battery cable.

2. Remove ⅔ of the brake fluid from the reservoir.

3. Raise and support the vehicle safely.

4. Remove the wheel, caliper and brake disc.

5. Remove the drive axle nut and washer.

6. Remove the tie rod nut and remove the tie rod end from the end of the washer.

7. Remove the hub and bearing assembly, using a puller.

8. Remove the drive axle.

9. Remove the splash shield bolts.

10. Remove the splash shield.

11. Support the lower control arm with a jack.

12. Remove the upper and lower ball joint nut and disconnect the upper and lower ball joint from the knuckle.

13. Remove the knuckle and the knuckle seal, as required.

To install:

14. Install the seal into the knuckle, using tool J-36605 or equivalent.

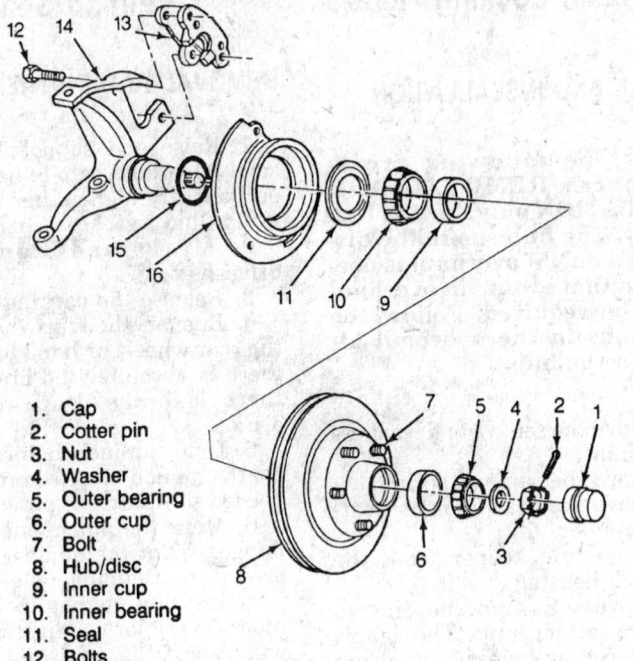

1. Cap
2. Cotter pin
3. Nut
4. Washer
5. Outer bearing
6. Outer cup
7. Bolt
8. Hub/disc
9. Inner cup
10. Inner bearing
11. Seal
12. Bolts
13. Caliper
14. Knuckle
15. Gasket
16. Shield

79114088

Hub and bearing assembly — C series

15. Install the knuckle to the upper and lower ball joints and tighten the nuts to 84 ft. lbs. (115 Nm). Tighten the nuts to align the new cotter pin, but do not tighten more than $1/6$ turn. Bend the pin ends against the nut flats.

16. Install the splash shield and tighten the bolts to 19 ft. lbs. (26 Nm)

17. Install the drive axle.

18. Install the hub and bearing assembly. Align the threaded holes and tighten the bolts to 133 ft. lbs. (180 Nm).

19. Install the tie rod end to the knuckle, and tighten the tie rod nut to 35 ft. lbs. (48 Nm).

20. Install the washer and axle nut and torque to 173 ft. lbs. (245 Nm).

21. Install the brake disc and caliper.

22. Install the axle joint cover.

23. Install the wheel and tire assembly.

24. Lower the vehicle and check the front end alignment.

Manual Locking Hubs

The engagement and disengagement of the hub is a manual operation which must be performed at each hub assembly. The hubs should be placed completely into the **LOCK** or **FREE**

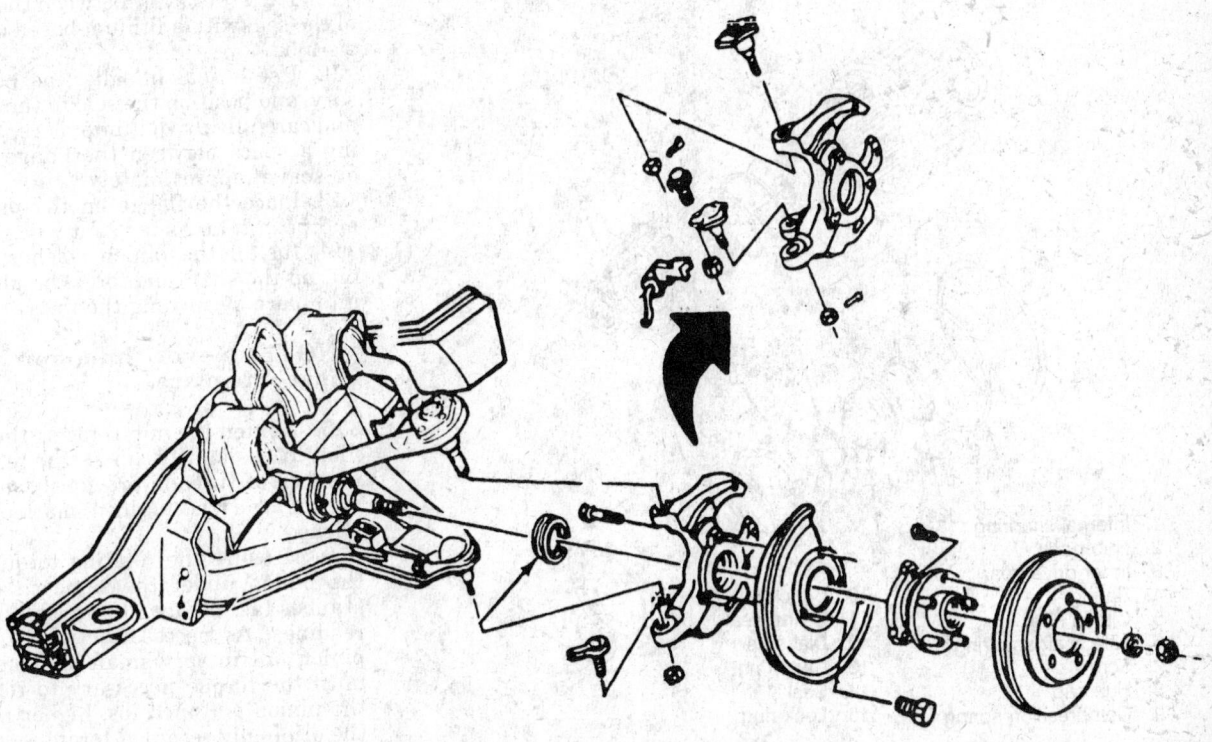

79114089

Hub, knuckle and ball joints — C/K series

position or damage will result. Locking hubs should be run in the Lock position periodically for a few miles to assure proper differential lubrication.

NOTE: Do not place the transfer case in either 4-wheel mode unless the hubs are in the Lock position!

REMOVAL AND INSTALLATION

1. Raise and support the vehicle safely.
2. Remove the wheel and tire assembly.
3. Remove the cap retaining bolts and take off the cap, O-ring, and knob assembly.
4. Remove the lock ring from the axle shaft.
5. Remove the internal snapring from the hub.
6. Remove the body assembly. Install 2 screws into the hub body and use as leverage to withdraw the assembly.
7. Installation is the reverse of the removal procedure.

Automatic Locking Hubs

REMOVAL AND INSTALLATION

NOTE: The following procedure covers REMOVAL AND INSTALLATION only, for the hub assembly. The hub should be disassembled only if overhaul is necessary. In that event, an overhaul kit will be required. Follow the instructions in the overhaul kit to rebuild the hub.

1. Remove the screws and washers from the hub.
2. Remove the cap and spring.
3. Remove the bearing race, bearing and retainer.
4. Remove the keeper from the outer clutch housing.
5. Remove the large snapring to release the locking unit. The snapring is removed by squeezing the ears of the snapring with needle-nose pliers.
6. Remove the locking unit from the hub by threading 2 hub cap screws into the outer clutch housing and pulling out the assembly.
7. Installation is the reverse of the removal procedure.

Pinion Seal

REMOVAL AND INSTALLATION

1. Raise and support the vehicle safely. It would help to have the front end slightly higher than the rear to avoid fluid loss.
2. Mark and remove the driveshaft.
3. Release the parking brake.
4. Remove the rear wheels. Rotate the rear wheels by hand to make sure there is absolutely no brake drag. If there is brake drag, remove the drums.
5. Using an inch lb. torque wrench on the pinion nut, record the force needed to rotate the pinion.
6. Mark the pinion shaft, nut and flange. Count the number of exposed threads on the pinion shaft.
7. Install a holding tool on the pinion. A very large adjustable wrench will do or if 1 is not available, put the drums back on and set the parking brake as tightly as possible.
8. Remove the pinion nut.
9. Slide the flange off of the pinion. A puller may be necessary.
10. Center punch the oil seal to distort it and pry the seal out, being careful to avoid scratching the bore.
To install:
11. Pack the cavity between the lips of the seal with a lithium-based chassis lube.
12. Use a seal installer, as necessary, and position the seal in the bore and carefully drive it into place, leaving a space between the flange and oil seal of approximately ⅛ in.
13. Place the flange on the pinion and push it on as far as it will go.
14. Install the pinion washer and nut on the shaft and force the pinion into place by turning the nut.

NOTE: Never hammer the flange into place.

15. Tighten the nut, rotating the pinion occasionally until the exact number of threads previously noted appear and the scribed marks are aligned.
16. Measure the rotating torque of the pinion under the same circumstances as before. Compare both readings. As necessary, tighten the pinion nut in very small increments until the torque necessary to rotate the pinion is 3 inch lbs. higher than the originally recorded torque.
17. Install the driveshaft and check rear fluid level.
18. Lower the vehicle and roadtest.

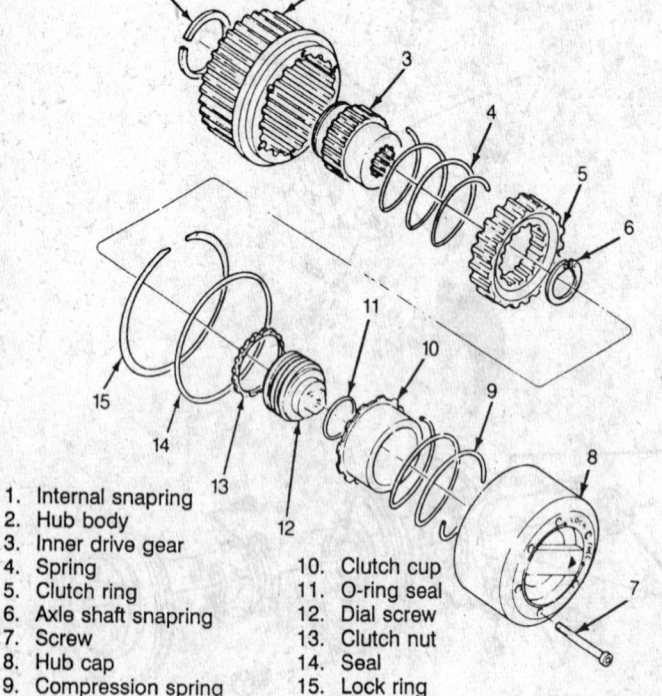

1. Internal snapring
2. Hub body
3. Inner drive gear
4. Spring
5. Clutch ring
6. Axle shaft snapring
7. Screw
8. Hub cap
9. Compression spring
10. Clutch cup
11. O-ring seal
12. Dial screw
13. Clutch nut
14. Seal
15. Lock ring

79114090

Manual hub components — R series

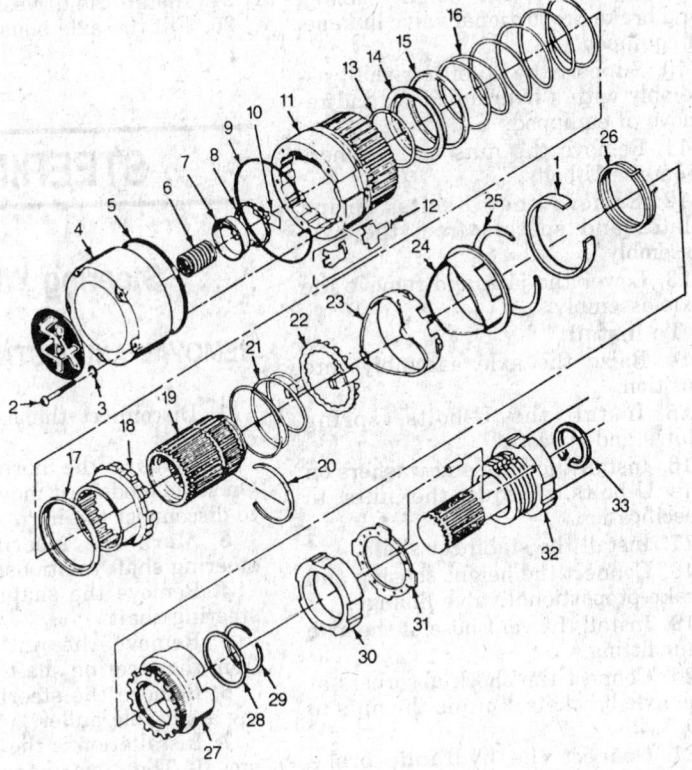

1. Brake band retainer
2. Screw
3. O-ring seal
4. Cover
5. Seal
6. Spring
7. Inner race
8. Bearing
9. Ring
10. Retainer
11. Outer clutch housing
12. Keeper
13. Ring
14. Retainer plate
15. Retainer
16. Return spring
17. Retainer
18. Clutch gear
19. Sleeve
20. Stop ring
21. Conical spring
22. Cam follower
23. Outer cage
24. Inner cage
25. Ring
26. Brake band
27. Drag sleeve
28. Spacer
29. Retaining ring
30. Adjusting nut
31. Lock ring
32. Lock ring
33. Nut with pin
34. Ring

79114091

Automatic hub components — R series

Front Axle Housing

REMOVAL AND INSTALLATION

V-Series

1. Raise and support the vehicle safely.
2. Matchmark and remove the driveshaft.
3. Disconnect the connecting rod from the steering arm.
4. Disconnect the brake caliper and position it out of the way, without disconnecting the brake line.
5. Disconnect the shock absorbers from the axle brackets.
6. Remove the front stabilizer bar.
7. Disconnect the axle vent tube clip at the differential housing.
8. Take up the weight of the axle assembly using a jack.
9. Remove the nuts, washers, U-bolts and plates from the axle and separate the axle from the springs. Remove the axle assembly from the vehicle.
   To install:
10. Position the axle under the vehices.
11. Install the plates, U-bolts, washers and nuts. Tighten the nuts to specification.
12. Remove the jack.

13. Connect the axle vent tube clip at the differential housing.
14. Install the front stabilizer bar.
15. Connect the shock absorbers at the axle brackets. Torque the bolts to specification.
16. Install the brake caliper assembly.
17. Connect the connecting rod at the steering arm.
18. Install the driveshaft and lower the vehicle.

K-Series

1. Raise and support the vehicle safely.
2. Remove the wheels.
3. Remove the skid plate.
4. Drain the carrier.
5. Matchmark and remove the front driveshaft.
6. Disconnect the right axle shaft at the tube flange.
7. Disconnect the left axle shaft at the carrier flange.
8. Wire both axle shafts out of the way.
9. Unplug the connectors at the indicator switch and actuator.
10. Disconnect the carrier vent hose.
11. Remove the axle tube-to-frame bolts, washers and nuts.

12. Remove the lower carrier mounting bolt.
13. Disconnect the right side inner tie rod end at the relay rod.
14. Depending on model, it may be necessary to remove the engine oil filter.
15. Support the carrier on a floor jack.
16. Remove the upper carrier mounting bolt.
17. Lower the carrier assembly from the truck.
   To install:
18. Raise the carrier into position.
19. Install the upper carrier mounting bolt, washers and nut. Then, install the lower carrier mounting bolt, washers and nut. Torque the bolts to 80 ft. lbs.
20. Remove the jack.
21. Install the oil filter.
22. Connect the tie rod end. Torque the nut to specification.
23. Install the axle tube-to-frame bolts, washers and nuts. Torque the nuts to specification.
24. Connect the vent hose.
25. Connect the wiring.
26. Connect the axle shafts at the flanges. Torque the bolts to specification.
27. Connect the driveshaft. Torque the bolts to specification.

28. Fill the carrier with SAE 85W-90 gear oil.

29. Install the wheels. Lower the vehicle.

30. Add any engine oil lost when the filter was removed.

Rear Axle Housing

REMOVAL AND INSTALLATION

All Series

1. Raise and support the vehicle safely.

2. For the 9¾ in. ring gear and the 10½ in. ring gear axles, place jackstands under the frame side rails for support.

3. Drain the lubricant from the axle housing and remove the driveshaft.

4. Remove the wheel, the brake drum or hub and the drum assembly.

5. Disconnect the parking brake cable from the lever and at the brake flange plate.

6. Disconnect the hydraulic brake lines from the connectors.

7. Disconnect the shock absorbers from the axle brackets.

8. Remove the vent hose from the axle vent fitting, if equipped.

9. Disconnect the height sensing and brake proportional valve linkage, if equipped.

10. Support the stabilizer shaft assembly with a hydraulic jack and remove, if equipped.

11. Remove the nuts and washers from the U-bolts.

12. Remove the U-bolts, spring plates and spacers from the axle assembly.

13. Lower the jack and remove the axle assembly.

To install:

14. Raise the axle assembly into position.

15. Install the U-bolts, spring plates and spacers.

16. Install the nuts and washers on the U-bolts. Torque the nuts to specification.

17. Install the stabilizer shaft.

18. Connect the height sensing and brake proportional valve linkage.

19. Install the vent hose at the axle vent fitting.

20. Connect the shock absorbers at the axle brackets. Torque the nuts to 80 ft. lbs.

21. Connect the hydraulic brake lines.

22. Connect the parking brake cable.

23. Install the wheels.

24. Install the driveshaft.

25. Fill the axle housing.

STEERING

Steering Wheel

REMOVAL AND INSTALLATION

1. Disconnect the battery ground cable.

2. Remove the horn button cap. On some models, it may be necessary to disconnect the horn wire.

3. Mark the steering wheel-to-steering shaft relationship.

4. Remove the snapring from the steering shaft.

5. Remove the nut and washer from the steering shaft.

6. Remove the steering wheel using a suitable puller.

7. Installation is the reverse of removal. The turn signal control assembly must be in the neutral position to prevent damaging the canceling cam and control assembly. Torque the nut to 30 ft. lbs. (40 Nm).

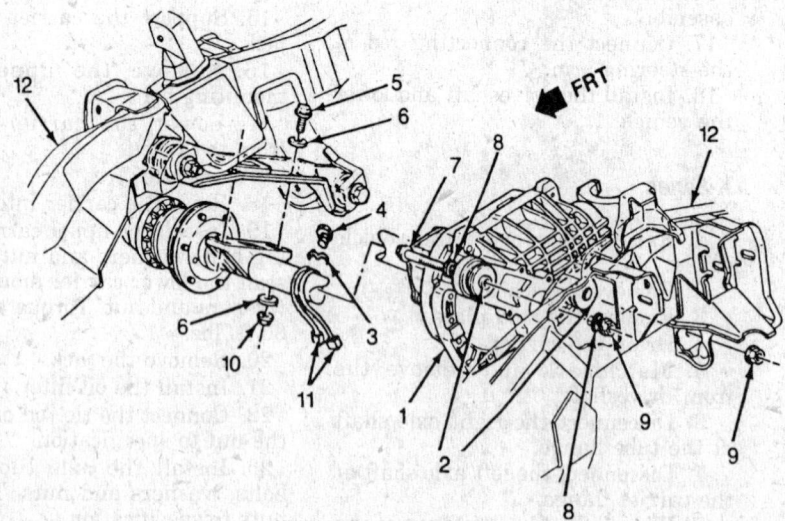

1. Differential carrier
2. Bushing
3. Clamp
4. Screw
5. Screw
6. Washer
7. Screw
8. Washer
9. Nut
10. Nut
11. Connectors
12. Frame

79114092

Front differential carrier mounting — K series

Manual Steering Gear

REMOVAL AND INSTALLATION

1. Set the front wheels in straight-ahead position by driving vehicle a short distance on a flat surface.

2. Raise and support the vehicle safely, as required.

3. Remove the steering shaft shield. Matchmark the relationship of the universal yoke to the wormshaft.

4. Remove the universal yoke pinch bolt.

5. Mark the relationship of the pitman arm to the pitman shaft.

6. Remove the pitman shaft nut and then remove the pitman arm from the pitman shaft, using puller J–6632 or J–29107.

7. Remove the steering gear to frame bolts and remove the gear assembly.

To install:

8. Ensure the steering gear is the centered position. Install the steering gear, guiding the steering gear shaft into the universal yoke.

9. Install the steering gear to frame bolts and torque to 100 ft. lbs. (135 Nm).

10. Install the yoke pinch bolt. Torque the pinch bolt to 22 ft. lbs. (30 Nm). Install the steering shaft shield.

11. Install the pitman arm onto the pitman shaft, lining up the marks made at removal. Install the pitman shaft nut torque to 185 ft. lbs. (250 Nm).

ADJUSTMENT

Before any steering gear adjustments are made, it is recommended that the front end of the vehicle be raised and supported safely and a thorough inspection be made for stiffness or lost motion in the steering gear, steering linkage and front suspension. Worn or damaged parts should be replaced, since a satisfactory adjustment of the steering gear cannot be obtained, if bent or badly worn parts exist.It is also very important that the steering gear be properly aligned in the vehicle. Misalignment of the gear places a stress on the steering worm shaft, therefore a proper adjustment is impossible.

To align the steering gear, loosen the steering gear-to-frame mounting bolts to permit the gear to align itself. Check the steering gear to frame mounting seat. If there is a gap at any of the mounting bolts, proper alignment may be obtained by plac-

ing shims where excessive gap appears. Tighten the steering gear-to-frame bolts. Alignment of the gear in the G-Series is very important and should be done carefully so a satisfactory, trouble-free gear adjustment may be obtained.

The steering gear is of the recirculating ball nut type. The ball nut, mounted on the worm gear, is driven by means of steel balls which circulate in helical grooves in both the worm and nut. Ball return guides attached to the nut serve to recirculate the 2 sets of balls in the grooves. As the steering wheel is turned to the right, the ball nut moves upward. When the wheel is turned to the left, the ball nut moves downward.

Before doing the adjustment procedures given below, ensure that the steering problem is not caused by faulty suspension components, bad front end alignment, etc. Then, proceed with the following adjustments.

BEARING DRAG

1. Mark the pitman arm-to-shaft relationship. Remove the pitman arm from the shaft.

2. Disconnect the battery ground cable.

3. Loosen the wormshaft nut and back the adjuster off a ¼ turn. Remove the horn cap.

4. Turn the steering wheel gently to the left stop, then back ½ turn.

5. Position an inch-pound torque wrench on the steering wheel nut and rotate it through a 90 degree arc. Note the torque required to turn the shaft. Proper torque is 5–8 inch lbs. (0.6–1 Nm).

NOTE: Do not use a torque wrench with a maximum torque reading of over 50 inch lbs. (6 Nm).

6. If the torque is incorrect, tighten the adjuster plug until the proper torque reading is achieved.

7. Hold the plug and tighten the adjuster locknut to 85 ft. lbs. (115 Nm).

NOTE: If the gear feels lumpy after adjustment this is probably due to a previous out of adjustment condition and damage has been done to the bearings. The gear will need to be replaced or rebuilt to correct the condition.

OVERCENTER PRELOAD

1. Turn the steering wheel lock-to-lock counting the total number of turns. Turn the wheel back ½ the to-

tal number of turns to the centered position.

2. Turn the lash (sector shaft) adjuster screw clockwise to remove all lash between the ball nut and sector teeth. Tighten the locknut to 22 ft. lbs. (30 Nm).

3. Using a torque wrench on the steering wheel nut, observe the highest reading while the gear is turned through the center position. It should be 16 inch lbs. (1.8 Nm) maximum torque.

4. If necessary repeat the adjustment procedure. Tighten the locknut to 22 ft. lbs. (30 Nm).

Power Steering Gear

REMOVAL AND INSTALLATION

R/V-Series

1. Set the front wheels in straight-ahead position by driving vehicle a short distance on a flat surface. Raise and support the vehicle safely as required.

2. Place a drain pan below the steering gear.

3. Disconnect the negative battery cable.

4. Disconnect the fluid lines. Cap the openings.

5. Mark the relationship of the pitman arm to the pitman shaft.

6. Remove the pitman shaft nut and then remove the pitman arm from the pitman shaft, using puller J–6632.

7. Remove the steering gear to frame bolts and remove the gear assembly.

To install:

8. Align the flat of the flexible coupling with the flat on the shaft. Push the coupling onto the shaft until the wormshaft bottoms against the end of the shaft.

9. Install the pinch bolt. Make sure the bolt passes through the shaft undercut. Tighten the pinch bolt to 75 ft. lbs. (102 Nm).

10. Place the steering gear into position, guiding the coupling bolts into the proper holes in the shaft flange.

11. Install the steering gear to frame bolts and torque to 66 ft. lbs. (90 Nm).

12. Install the coupling flange nuts and washers. Make sure the coupling alignment pins are centered in the flange slots. Tighten the nuts to 20 ft. lbs. (27 Nm). Maintain a coupling to flange dimension of 0.250–0.375 inches (6.4–9.5mm).

13. Install the pitman arm.

14. Install the hoses.

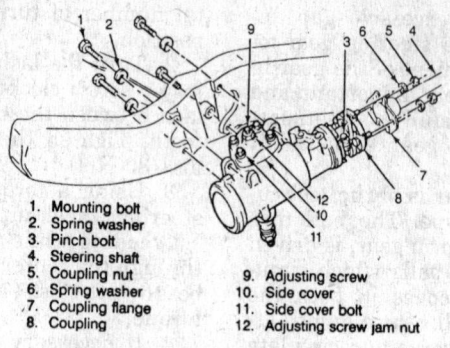

1. Mounting bolt
2. Spring washer
3. Pinch bolt
4. Steering shaft
5. Coupling nut
6. Spring washer
7. Coupling flange
8. Coupling
9. Adjusting screw
10. Side cover
11. Side cover bolt
12. Adjusting screw jam nut

79114093

Power steering gear installation — R series

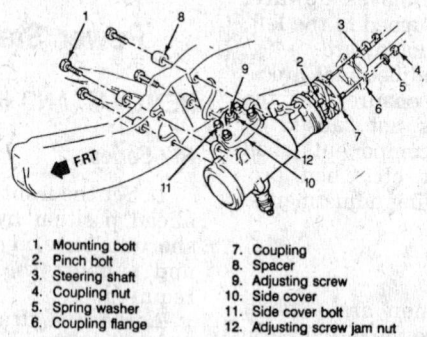

1. Mounting bolt
2. Pinch bolt
3. Steering shaft
4. Coupling nut
5. Spring washer
6. Coupling flange
7. Coupling
8. Spacer
9. Adjusting screw
10. Side cover
11. Side cover bolt
12. Adjusting screw jam nut

79114094

Power steering gear installation — V series

G-Series

1. Set the front wheels in straight-ahead position by driving vehicle a short distance on a flat surface. Raise and support the vehicle safely, as required.

2. Place a drain pan below the steering gear.

3. Disconnect the negative battery cable.

4. Disconnect the fluid lines. Cap the openings.

5. Matchmark the relationship of the universal yoke to the stubshaft.

6. Remove the universal yoke pinch bolt.

7. Mark the relationship of the pitman arm to the pitman shaft.

8. Remove the pitman shaft nut and then remove the pitman arm from the pitman shaft, using puller J–6632.

9. Remove the steering gear to frame bolts and remove the gear assembly.

To install:

10. Ensure the gear is in the centered position. Install the steering gear, guiding the steering gear shaft into the universal yoke.

11. Install the steering gear to frame bolts and torque to 66 ft. lbs. (90 Nm).

12. Install the yoke pinch bolt. Torque the pinch bolt to 46 ft. lbs. (62 Nm).

13. Install the pitman arm onto the pitman shaft, lining up the marks made at removal.

14. Connect the fluid lines and refill the reservoir. Bleed the system.

Adjustments

For proper adjustment, remove the gear from the vehicle, drain all the

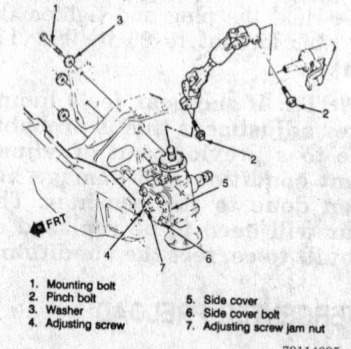

1. Mounting bolt
2. Pinch bolt
3. Washer
4. Adjusting screw
5. Side cover
6. Side cover bolt
7. Adjusting screw jam nut

79114095

Power steering gear installation — G series

fluid from the gear and place the gear in a holding fixture. It is important that the adjustments be made in the order given.

Worm Bearing Preload

1. Loosen the adjuster plug locknut.

2. Turn the adjuster plug in clockwise until firmly bottomed. Then, tighten it to 20 ft. lbs. (27 Nm).

3. Place an index mark on the gear housing in line with one of the holes in the adjuster plug.

4. Measure counterclockwise from the mark about ¼ inch and make another mark on the housing.

5. Rotate the adjuster plug counterclockwise until the hole is aligned with the second mark.

6. Install the locknut. Hold the plug and tighten the locknut to 81 ft. lbs. (110 Nm).

7. Place an inch pound torque wrench and 12-point deep socket on the stub shaft and measure the stub shaft rotating torque, starting with the torque wrench handle in a vertical position to a point ¼ turn to either side. Note your reading. The proper torque should be 4–10 inch lbs. (0.45–1.13 Nm). If the reading is incorrect, either your adjustment was done incorrectly or there is gear damage.

Overcenter Preload

1. Loosen the locknut and turn the pitman shaft adjuster screw counterclockwise until it is all the way out. Then, turn it in ½ turn.

2. Rotate the stub shaft from stop to stop, counting the total number of turns, then turn the shaft ½ that number to center the gear.

3. Place the torque wrench in a vertical position on the stub shaft and measure the torque necessary to rotate the shaft to a point 45 degrees to either side of center. Record the highest reading. On gears with less than 400 miles, the reading should be 6–10 inch lbs. higher than the worm bearing preload torque previously recorded, but not to exceed 18 inch lbs. (2 Nm). On gears with more than 400 miles, the reading should be 4–5 inch lbs. higher, but not to exceed 14 inch lbs. (1.5 Nm).

4. If necessary, adjust the torque reading by turning the adjuster screw.

5. When the adjustment is made, hold the screw and tighten the locknut to 35 ft. lbs. (47 Nm).

6. Install the gear and bleed the system.

Power Steering Pump

REMOVAL AND INSTALLATION

1. Place a drain pan under the pump.
2. Disconnect the negative battery cable.
3. Disconnect and cap the hoses at the pump.

NOTE: If equipped with a remote reservoir, disconnect and cap the reservoir hose at the pump.

4. Loosen the pump adjusting bolts and nuts and remove the pump belt.
5. Remove the pulley from the pump as required using a suitable pulley remover/installer tool J–29785–A or equivalent, as required.
6. Remove the adjusting bolts, nuts and brackets and remove the pump assembly.

To install:

7. Connect the brackets to the pump.
8. Place the pulley on the end of the pump shaft and install tool J–25033–B or equivalent.

NOTE: On models with a remote reservoir fill the pump housing with as much fluid as possible before mounting.

9. Install the pump assembly and attaching parts loosely to the engine.
10. Install the hoses to the pump and fill the reservoir. Bleed the pump by turning the pulley backwards (counterclockwise as viewed from the front) until the air bubbles cease to appear.
11. Tighten all retaining bolts and nuts.
12. Install the pump belt over the pulley and adjust.
13. Fill and bleed the system.

SYSTEM BLEEDING

1. Fill the reservoir to the proper level and let the fluid remain undisturbed for at least 2 minutes.
2. Start the engine and run it for only about 2 seconds.
3. Add fluid as necessary.
4. Repeat Steps 1–3 until the level remains constant.
5. Raise the front of the vehicle so the front wheels are off the ground. Set the parking brake and block both rear wheels front and rear. Manual transmissions should be in neutral;

automatic transmissions should be in **P.**

6. Start the engine and run it at approximately 1500 rpm.
7. Turn the wheels (off the ground) to the right and left, lightly contacting the stops.
8. Add fluid as necessary.
9. Lower the vehicle and turn the wheels right and left on the ground.
10. Check the level and refill as necessary.
11. If the fluid is extremely foamy, let the vehicle stand for a few minutes with the engine off and repeat the procedure. Check the belt tension and check for a bent or loose pulley. The pulley should not wobble with the engine running.
12. Check that no hoses are contacting any parts of the vehicle, particularly sheet metal.
13. Check the fluid level and refill as necessary.
14. Check for air in the fluid. Aerated fluid appears milky. If air is present, repeat the above operation. If it is obvious that the pump will not respond to bleeding after several attempts, a pressure test may be required.

Tie Rod Ends

REMOVAL AND INSTALLATION

C/K, R and G-Series

NOTE: Before servicing, note the position of the tie rod adjuster tube and the direction from which the bolts are installed. Do not attempt to disengage the tie rod ball stud using a wedge type tool, because seal damage could result.

1. Raise and support the vehicle safely. As required, remove the tire and wheel assembly.
2. Remove the cotter pins and nuts. Using the proper removal tool J–6627A separate the outer tie rod from the steering knuckle.
3. Disconnect the inner tie rod from the relay rod using tool J–6627A. Remove the tie rod ends from the adjuster tubes, counting the number of turns to aid in installation.

To install:

4. Grease the threads and turn the new tie rod end in as many turns as were needed to remove it. This will give approximately correct toe-in. Tighten the clamp bolts to 14 ft. lbs.
5. Secure the tie rod ends to the relay rod and steering knuckle with a new nut. Tighten the nuts to 40 ft.

lbs. (54 Nm) and install new cotter pins. Tighten the nut to align the cotter pin, do not loosen it.

6. Install the tire and wheel assembly.
7. Lower the vehicle and check the front end alignment.

V-Series

1. Raise and support the vehicle safely. As required, remove the tire and wheel assembly.
2. Remove the cotter pins and nuts from the rod assembly. Disconnect the shock absorber from the tie rod assembly.
3. Using the proper removal tool, J–6627A, separate the outer tie rods from the steering knuckle.

NOTE: Do not attempt to disengage the ball joint from the steering knuckle using a wedge type tool, because seal damage could result.

4. Disconnect the tie rod end bodies. Count the number of turns needed to remove the end bodies. Remove the tie rod ends from the adjuster tube.
5. Note the position of the adjuster tube and the direction from which the bolts are installed.

To install:

6. Install the tie rod ends the same number of turns as counted previously. Tighten the locknuts.
7. Install the tie rod assembly in the knuckles and tighten the castellated nuts to 40 ft. lbs. (55 Nm). Always advance the nut to align the cotter pin hole. Never back it off.
8. Tighten the tie rod end jam nuts to 175 ft. lbs. (237 Nm).
9. Install the tire and wheel assembly. Lower the vehicle and adjust the front alignment, as required.

BRAKES

For all brake system repair and service procedures not detailed below, please refer to "Brakes" in the Unit Repair section.

Master Cylinder

REMOVAL AND INSTALLATION

1. Disconnect the negative battery cable. Disconnect any electrical connections from the master cylinder, as

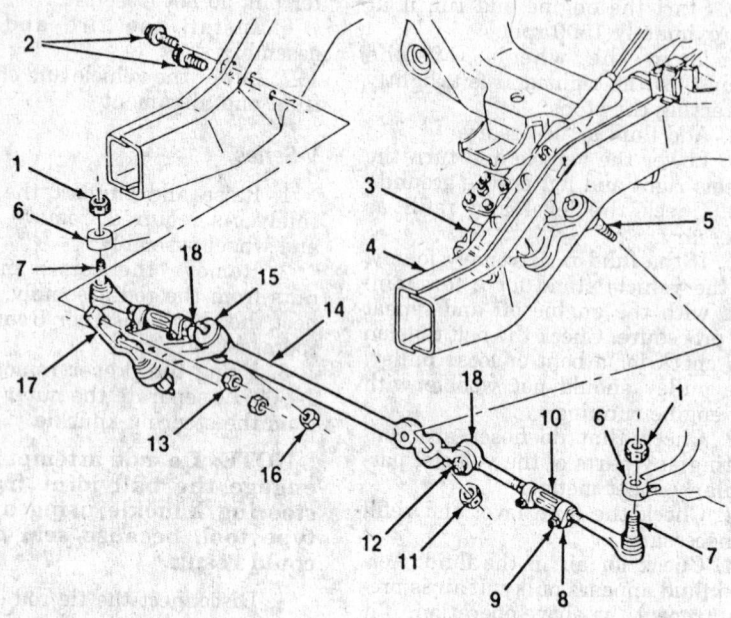

Steering linkage — C series

1. Tie rod outer ball joint nut
2. Idler arm frame bolts
3. Steering gear
4. Frame
5. Pitman arm ball stud
6. Knuckle
7. Tie rod ball stud
8. Clamp
9. Clamp nut
10. Adjuster tube
11. Pitman arm nut
12. Tie rod inner ball joint nut
13. Idler arm frame nut
14. Relay rod
15. Idler arm ball joint
16. Idler arm ball joint nut
17. Idler arm mounting bracket
18. Tie rod inner ball joint

79114099

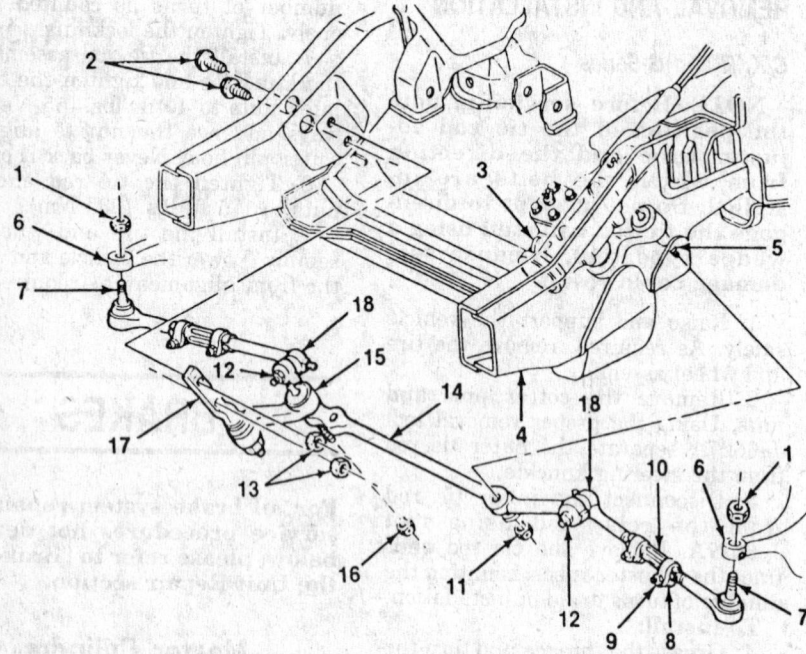

Steering linkage — K series

1. Tie rod outer ball joint nut
2. Idler arm frame bolts
3. Steering gear
4. Frame
5. Pitman arm ball stud
6. Knuckle
7. Tie rod ball stud
8. Clamp
9. Clamp nut
10. Adjuster tube
11. Pitman arm nut
12. Tie rod inner ball joint nut
13. Idler arm frame nut
14. Relay rod
15. Idler arm ball joint
16. Idler arm ball joint nut
17. Idler arm mounting bracket
18. Tie rod inner ball joint

79114100

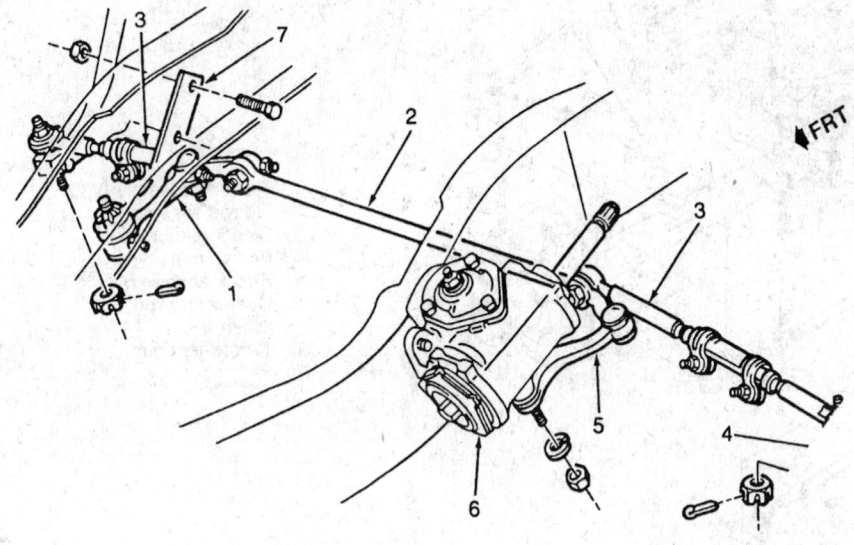

1. Idler arm
2. Relay rod
3. Tie rod assembly
4. Steering knuckle
5. Pitman arm
6. Steering gear
7. Idler arm frame support

Steering linkage — R series

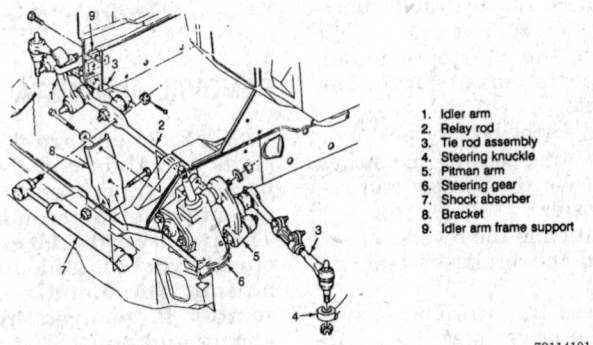

1. Idler arm
2. Relay rod
3. Tie rod assembly
4. Steering knuckle
5. Pitman arm
6. Steering gear
7. Shock absorber
8. Bracket
9. Idler arm frame support

Steering linkage — G series

required. Disconnect and plug the fluid lines.

2. If equipped with power brakes, remove the master cylinder to power booster retaining bolts. Remove the Rear Wheel Anti-Lock (RWAL) control module assembly, if equipped. Do not allow fluid to leak onto the module.

3. If equipped with manual brakes, remove the master cylinder pushrod from the brake pedal and remove the RWAL control module assembly.

4. Remove the master cylinder from the vehicle. If equipped with

power brakes remove the vacuum booster pushrod.

5. Installation is the reverse of the removal procedure. Bench bleed the master cylinder prior to installation. If equipped with power brakes be sure to install the vacuum booster pushrod. Bleed the system, as required.

Proportioning Valve

REMOVAL AND INSTALLATION

Except C/K-Series

1. Disconnect the negative battery cable. Disconnect the hydraulic lines and plug to prevent dirt from entering the system.

2. Disconnect the warning switch harness.

3. Remove the retaining bolts and remove the valve.

4. Installation is the reverse of removal procedure. Bleed the brake system.

C/K-Series

1. Disconnect the negative battery cable. Disconnect the hydraulic lines and plug to prevent dirt from entering the system.

2. Disconnect the warning switch harness.

3. Remove the RWAL control module assembly from the bracket.

4. Remove the bolts holding the Isolation/Dump Valve to the bracket.

5. Remove the nuts that hold the master cylinder and bracket to the brake booster.

6. Remove the bracket and combination valve assembly.

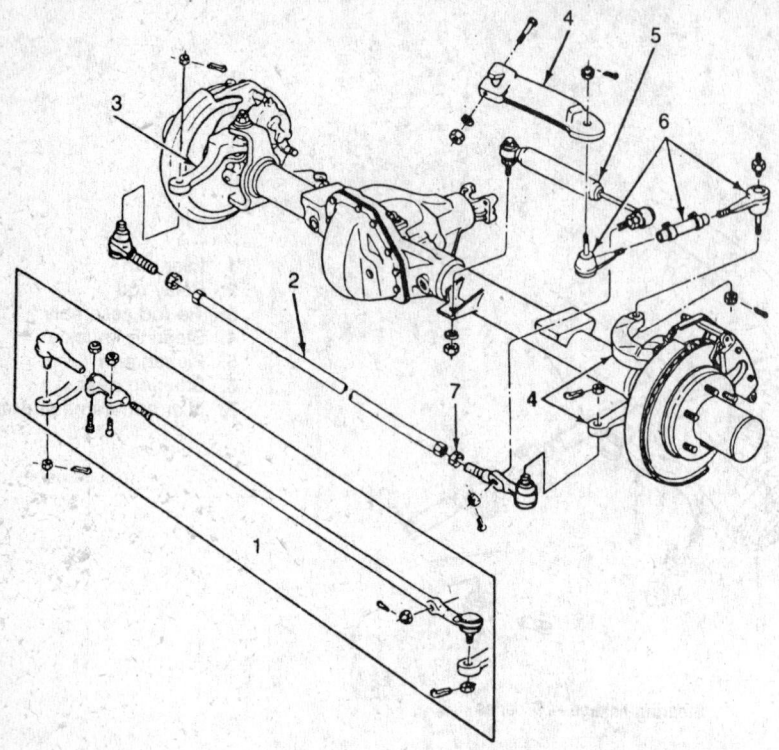

1. Tie rod assembly (V30)
2. Tie rod assembly
3. Steering knuckle
4. Pitman arm
5. Shock absorber
6. Connecting rod assembly
7. Tie rod jam nut

79114098

Steering linkage — V series

7. Installation is the reverse of removal. Tighten the nuts that hold the master cylinder and bracket to the brake booster to 20 ft. lbs. (27 Nm) and the bolts holding the Isolation/Dump Valve to the bracket to 17 ft. lbs. (24 Nm).

8. Bleed the brake system.

Power Brake Boosters

REMOVAL AND INSTALLATION

Vacuum Booster

1. Disconnect the negative battery cable. Apply the parking brakes.
2. Support the master cylinder and remove the master cylinder mounting nuts.
3. Disconnect the vacuum hose from the check valve.
4. Disconnect the booster pushrod.
5. Remove the booster mounting nuts from inside the vehicle and remove the booster.
6. Installation is the reverse of removal.

Hyfraulic Booster (Hydra-Boost)

1. Disconnect the negative battery cable. Apply the parking brakes.

2. Disconnect the hydraulic lines from the booster.
3. Support the master cylinder and remove the master cylinder mounting nuts.
4. Disconnect the booster pushrod.
5. Remove the booster mounting nuts from inside the vehicle and remove the booster.
6. Installation is the reverse of removal. Bleed the brake system, as follows:

a. To bleed the hydro-boost system, fill the power steering pump to the proper level. Allow the fluid to remain undisturbed for a few minutes.

b. Start the engine and add fluid until the level is constant with the engine running.

c. Raise and support the vehicle safely. Start the engine and turn the wheels from stop to stop, add fluid as required. Turn the engine **OFF** and lower the vehicle.

d. Start the engine and depress the brake pedal several times while rotating the steering wheel from stop to stop. Turn the engine off and pump the brake pedal 4 or 5 times.

e. If the power steering fluid is extremely foamy, allow the vehicle to sit for a short time and then perform the procedure again.

Brake Caliper

REMOVAL AND INSTALLATION

NOTE: There are 2 caliper designs and they can be identified by the method used to secure the assembly to the spindle bracket. The Delco 3000/3100 caliper is secured by a bolt and sleeve combination. The Bendix caliper assembly is secured by a slider, spring and bolt.

1. Remove the cover on the master cylinder and siphon enough fluid out of the reservoirs to bring the level to 1/3 full. This step prevents spilling fluid when the piston is pushed back.
2. Raise and support the vehicle safely. Remove the front wheels and tires.
3. Position a C-clamp around the outside pad and caliper; tighten the C-clamp until the caliper piston bottoms in its bore.
4. Remove the brake hose from the caliper by removing the inlet fitting.
5. Remove the bolt and sleeve or bolt and slider assemblies which hold the caliper and then lift the caliper off the rotor.
6. Remove the inboard and outboard shoe.

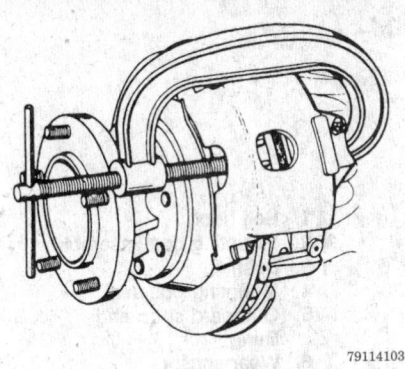

79114103

Compressing the caliper piston

To install:

7. Install the pads onto the caliper.

8. Position the caliper onto the knuckle/rotor assembly and secure the asseembly with the mounting bolts or sliders.

9. Reconnect the brake line to the caliper.

10. Pump the brake pedal and verify there is minimal brake pedal travel.

11. Check the brake fluid level. Install the tire and wheel assembly.

12. Lower the vehicle.

Disc Brake Pads

REMOVAL AND INSTALLATION

Delco Type

1. Remove the cover on the master cylinder and siphon out ⅔ of the fluid. This step prevents spilling fluid when the piston is pushed back into the caliper bore.

2. Raise and support the vehicle safely.

3. Remove the wheels.

4. Compress the brake piston back into its bore using a C-clamp.

5. Remove the 2 bolts which hold the caliper and then lift the caliper off the disc.

NOTE: Do not let the caliper assembly hang by the brake hose.

6. Remove the inboard and outboard shoe.

7. Remove the pad support spring from the piston, if equipped.

To install:

8. Thoroughly inspect, clean and lubricate all caliper slide points, bolts and hardware.

9. Position the retainer spring on the inner pad and insert the assembly into the center cavity of the piston.

10. Push down on the inner pad until it lays flat against the caliper. It is important to push the piston all the way into the caliper if new linings are installed or the caliper will not fit over the rotor.

11. Position the outboard pad with the ears of the pad over the caliper ears and the tab at the bottom engaged in the caliper cutout.

12. With the 2 pads in position, place the caliper over the brake disc and align the holes in the caliper with those of the mounting bracket.

NOTE: Make certain the brake hose is not twisted or kinked.

13. Install the mounting bracket bolts through the sleeves in the inboard caliper ears and through the mounting bracket, making sure the ends of the bolts pass under the retaining ears on the inboard pad.

14. Tighten the mounting bolts to 35 ft. lbs. (48 Nm). After both calipers are mounted pump the brake pedal to seat the pad against the rotor. Use a pair of channel lock pliers to bend over the upper ears of the outer pad so it isn't loose.

15. Install the wheels and lower the vehicle.

16. Add fluid to the master cylinder reservoirs so they are ¼ in. (6.35mm) from the top.

17. Test the brake pedal by pumping it to obtain a hard pedal. Check the fluid level again and add fluid as necessary. Do not move the vehicle until a pedal is obtained.

Bendix

1. Remove approximately ⅓ of the brake fluid from the master cylinder. Discard the used brake fluid.

2. Raise and support the vehicle safely and remove the wheel.

3. Push the piston back into its bore. This can be done by using a C-clamp.

4. Remove the bolt at the caliper slider. Use a brass drift pin to remove the slider and spring.

5. Rotate the caliper up and forward from the bottom and lift it off the caliper support.

6. Tie the caliper out of the way with a piece of wire. Be careful not to damage the brake line.

7. Remove the inner shoe from the caliper support. Discard the inner shoe clip.

8. Remove the outer shoe from the caliper.

To install:

9. Thoroughly clean, inspect and lubricate the caliper, slider and spring with silicone.

10. Install a new inboard shoe clip on the shoe.

11. Install the lower end of the inboard shoe into the groove provided in the support. Slide the upper end of the shoe into position. Be sure the clip remains in position.

12. Position the outboard shoe in the caliper with the ears at the top of the shoe over the caliper ears and the tab at the bottom of the shoe engaged in the caliper cutout. If assembly is difficult, a C-clamp may be used. Be careful not to damage the lining.

13. Position the caliper over the brake disc, top edge first. Rotate the caliper downward onto the support.

14. Place the spring over the caliper support key, install the assembly between the support and lower caliper groove. Tap into place until the key retaining screw can be installed.

15. Install the screw and torque to 15 ft. lbs. (20 Nm). The boss must fit fully into the circular cutout in the key.

16. Install the wheel and add brake fluid as necessary.

Brake Drums

REMOVAL AND INSTALLATION

Semi-Floating Axles

1. Raise and support the vehicle safely.

2. Mark the relationship of the wheel to the hub and remove the wheel.

3. Mark the relationship of the drum to the hub and pull the drum from the brake assembly. If the brake drums have been scored from worn linings, the brake adjuster must be backed off so the brake shoes will retract from the drum. The adjuster can be backed off by inserting a brake adjusting tool through the access hole provided. In some cases the access hole is provided in the brake drum. A metal cover plate is over the hole. This may be removed by using a hammer and chisel.

4. To install, reverse the removal procedure.

Full Floating Axles

To remove the drums from full floating rear axles, the axle shaft will have to be removed. Full floating rear axles can readily be identified by the bearing housing protruding through the center of the wheel.

1. Raise and support the vehicle safely.

2. Remove the wheel.

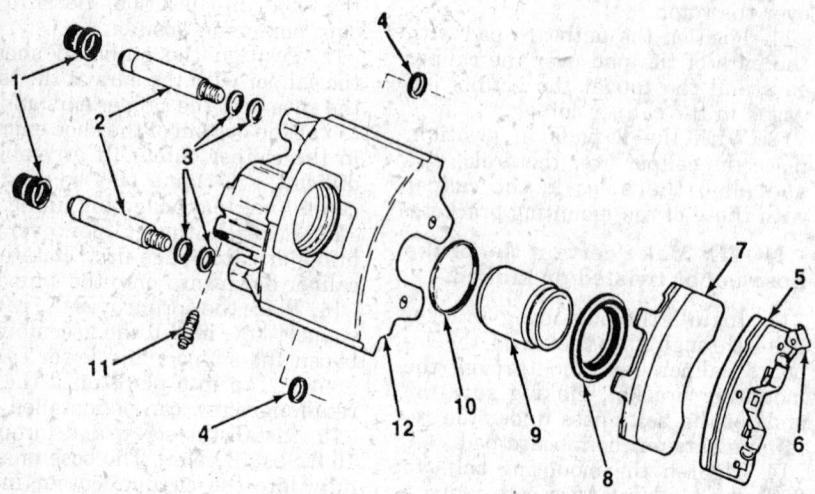

1. Bolt boot
2. Mounting bolt assembly
3. Bushing
4. Mounting bolt seal
5. Outboard shoe and lining
6. Wear sensor
7. Inboard shoe and lining
8. Boot
9. Piston
10. Piston seal
11. Bleeder valve
12. Caliper housing

Replacing the disc brake pads — Delco type

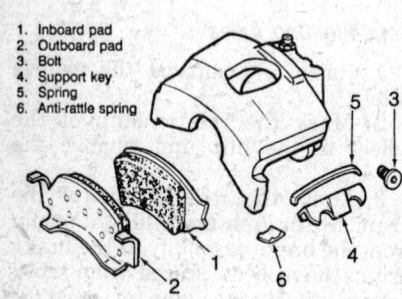

1. Inboard pad
2. Outboard pad
3. Bolt
4. Support key
5. Spring
6. Anti-rattle spring

Replacing the disc brake pads — Bendix type

3. Remove the axle shaft.
4. Remove the retaining ring, key and adjusting nut.
5. Remove the hub and drum.

To install:
6. Install the hub and drum to the tube.
7. Install the adjusting nut and torque to specification.
8. Install the key and retaining ring.
9. Install the axle shaft and wheel.

Brake Shoes

ADJUSTMENT

1. Raise and support the vehicle safely.
2. Remove the lanced area in the backing plate.
3. Turn the adjusting screw until the wheel can just be turned by hand.
4. Back the adjuster screw off 33 notches. There should be no brake drag felt after the adjuster is backed off 15 notches.
5. Install an adjuster hole plug in the backing plate.
6. Repeat the procedure for both wheels.

REMOVAL AND INSTALLATION

Leading/Trailing Brakes

NOTE: This brake system is used on the lower GVW rated vehicles.

1. Raise the vehicle and support it safely.
2. Remove the tire and wheel assembly.

3. Remove the brake drums.

NOTE: The brake pedal must not be depressed while the drums are removed.

4. Raise the lever arm of the actuator until the upper end is clear of the slot in the adjuster screw.
5. Slide the actuator off the adjuster pin. Disconnect the actuator spring from the shoe.
6. Remove the hold-down spring assemblies and pins.
7. Pull the bottom ends of the shoes apart and lift the lower return spring over the anchor plate. Allow the shoe ends to come together and remove the spring.
8. Remove the shoe assembly, along with the upper return spring and the adjusting screw assembly.
9. Remove the upper return spring and the adjusting screw assembly from the shoes.
10. Remove the retaining ring, pin, spring washer, and parking brake lever.

To install:
11. Clean adjuster wheel and the backing plates with a suitable cleaner. Lubricate the backing plate contact points, levers and adjuster with a suitable lubricant.
12. Assemble the parking lever, spring washer (concave side facing

the brake lever), pin, and retaining ring onto the rearward shoe.

13. Install the adjuster pin in the forward shoe with the pin projecting 0.276 in. (7mm) from the side of the shoe web where the adjuster actuator is installed.

14. With the brake shoes laying on a flat surface (the shoe with the parking lever to the rear of the vehicle), install the upper return spring.

15. Install the adjuster screw assembly with the spring clip facing the backing plate.

16. Place the shoes in position on the backing plate. Do not place the lower shoe webs under the anchor plate.

17. Install the lower return spring, spread the bottom of the shoes and position the shoe against the backing plate.

18. Install the hold-down pins and spring assemblies.

19. Install the adjuster actuator over the end of the adjuster pin so the top leg engages the notch in the adjuster screw.

20. Install the actuator spring, being careful not to over-stretch it more than 3.27 in. (83mm).

21. Install the parking brake cable to the lever.

22. Adjust the parking brake if the shoes will not totally retract.

23. Install the drum, tire and wheel assembly. Adjust the rear brakes and lower the vehicle.

Duo-Servo Brakes

1. Raise the vehicle and support it safely.
2. Remove the tire and wheel assembly.
3. Remove the brake drums.

NOTE: The brake pedal must not be depressed while the drums are removed.

4. Using a brake tool, remove the shoe return springs.
5. Remove the shoe guide.
6. Remove the hold-down springs and pins.
7. Remove the actuator lever and pivot.
8. Remove the lever return spring.
9. Remove the actuator link, parking brake strut, spring retaining ring.
10. Remove the parking brake lever and washer.
11. Remove the shoe assemblies.
12. Remove the adjuster screw and spring from the shoe assembly.
   To install:
13. Use a brake cleaning fluid to remove dirt from the brake drum.

Check the drums for scoring, cracks and for out-of-round; service the drums as necessary.

14. Check the wheel cylinders by carefully pulling the lower edges of the wheel cylinder boots away from the cylinders. If there is excessive leakage, the inside of the cylinder will drip fluid; repair or replace as necessary.

15. Check the flange plate, which is located around the axle, for leakage of differential lubricant.

16. Lightly lubricate the parking brake cable, parking brake lever where it enters the shoe and the backing plate-to-shoe contact points. Use high temperature, waterproof, grease or special brake lube.

17. Install the parking brake lever into the secondary shoe with the attaching bolt, spring washer, lockwasher, and nut. It is important that the lever move freely before the shoe is attached. Move the assembly and check for proper action.

18. Lubricate the adjusting screw and make sure it works freely.

19. Connect the adjuster screw and spring to the bottom portion of both shoes. Ensure the spring does not interfere with the adjuster rotation when installed. The primary (smaller shoe pad area) to the front and secondary shoe (larger shoe pad area) to the rear of the vehicle.

20. Install the shoe assembly. Ensuring the shoe webs are positioned correctly against the wheel cylinder.

21. Install the parking brake cable.

22. Secure the primary shoes with the holddown pin and spring.

23. Install the parking brake strut and the strut spring.

24. Install the actuator lever and pivot, securing the assembly with the holddown pin and spring. Install the actuator link and spring.

25. Install the return springs.

26. Check the operation of the self-adjusting mechanism by moving the actuating lever by hand.

27. Adjust the brakes and install the drum.

28. Adjust the parking brake.

29. Install the tire and wheel assembly.

30. Lower the vehicle.

Wheel Cylinders

REMOVAL AND INSTALLATION

1. Raise and support the vehicle safely.
2. Remove the wheel.

3. Back off the brake adjuster, if necessary, and remove the drum.
4. Disconnect and plug the brake line.
5. Remove the brake shoe upper return springs or remove the shoes, as necessary.
6. Remove the bolts securing the wheel cylinder to the backing plate.
7. Disengage the wheel cylinder pushrods from the brake shoes and remove the wheel cylinder.
8. Installation is the reverse of the removal procedure. Bleed the system.

Parking Brake Cable

REMOVAL AND INSTALLATION

Front Cable

1. Raise and support the vehicle safely.
2. Remove adjusting nut from equalizer.
3. Remove retainer clip from rear portion of front cable at frame and from lever arm.
4. Disconnect front brake cable from parking brake pedal or lever assemblies. Remove front brake cable. On some models, it may assist installation of the new cable if a heavy cord is tied to other end of cable in order to guide new cable through proper routing.
5. Installation is the reverse of the removal procedure.
6. Adjust the parking brake.

Center Cable

1. Raise and support the vehicle safely.
2. Remove adjusting nut from equalizer.
3. Unhook connector at each end and disengage hooks and guides.
4. Install new cable by reversing removal procedure.
5. Adjust parking brake.
6. Apply parking brake 3 times with heavy pressure and repeat adjustment.

Rear Cable

1. Raise and support the vehicle safely.
2. Remove rear wheel and brake drum.
3. Loosen adjusting nut at equalizer.
4. Disengage rear cable at connector.
5. Bend retainer fingers.
6. Disengage cable at brake shoe operating lever.

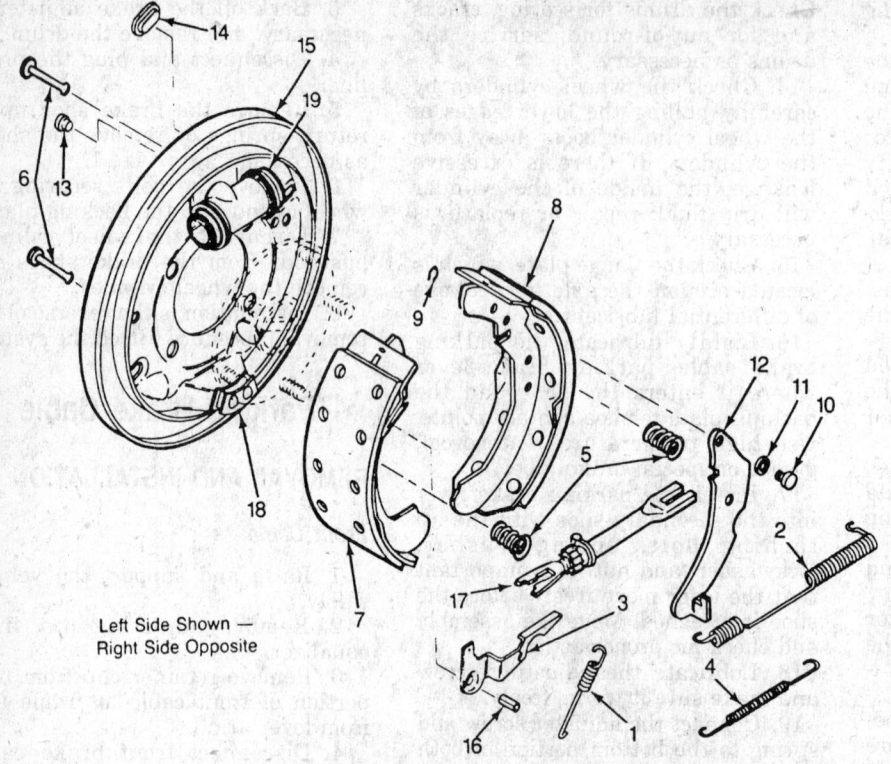

1. Actuator spring
2. Upper return spring
3. Adjuster actuator
4. Lower return spring
5. Hold-down spring assembly
6. Hold-down pin
7. Adjuster shoe and lining
8. Shoe and lining
9. Retaining ring
10. Pin
11. Spring washer
12. Park brake lever
13. Access hole plug
14. Inspection cover
15. Backing plate assembly
16. Adjuster pin
17. Adjusting screw assembly
18. Anchor plate
19. Wheel cylinder assembly

Left Side Shown —
Right Side Opposite

79114105

Exploded view of the rear drum brake assembly — Leading/Trailing type

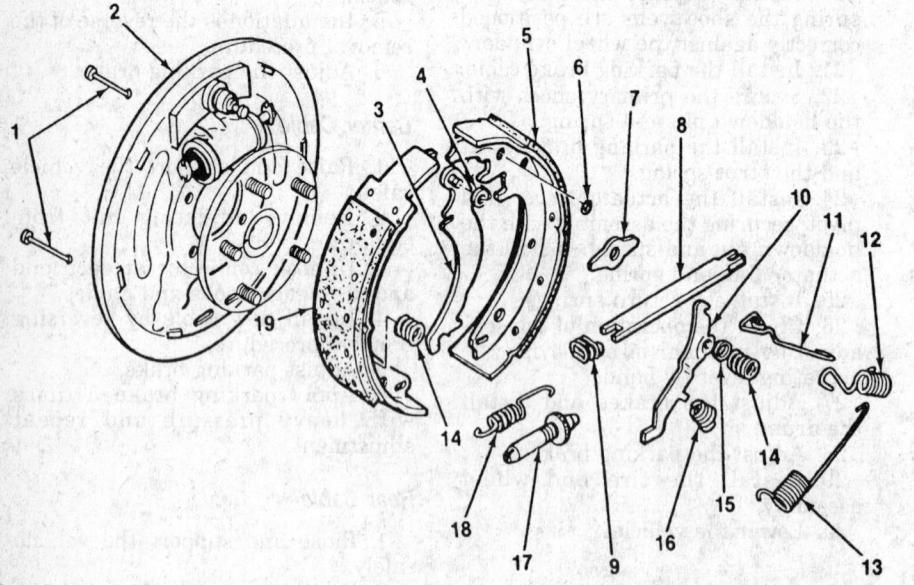

1. Hold-down pins
2. Backing plate
3. Parking brake lever
4. Washer
5. Secondary shoe
6. Retaining ring
7. Shoe guide
8. Parking brake strut
9. Strut spring
10. Actuator lever
11. Actuator link
12. Return spring
13. Return spring
14. Hold down spring
15. Lever pivot
16. Lever return spring
17. Adjusting screw assembly
18. Adjusting screw spring
19. Primary shoe

79114106

Exploded view of the rear drum brake assembly — Duo-Servo type

7. Install new cable by reversing removal procedure.

8. Adjust parking brake.

ADJUSTMENT

The rear brakes serve a dual purpose. They are used as service brakes and as parking brakes. To obtain proper adjustment of the parking brake, the service brakes must first be properly adjusted. Inspect the cables for binding or sticking.

R/V and G-Series

FOOT PEDAL TYPE

1. Apply the parking brake 4 notches from the fully released position.
2. Raise and support the vehicle.
3. Loosen the jam nut at the equalizer.
4. Tighten or loosen the adjusting nut until a light drag is felt when the rear wheels are rotated forward.
5. Tighten the check nut.
6. Release the parking brake and rotate the rear wheels. No drag should be felt. If even a light drag is felt, readjust the parking brake.
7. Lower the vehicle.

NOTE: If a new parking brake cable is being installed, pre-stretch it by applying the parking brake hard about three times before making adjustments.

LEVER TYPE

1. Raise and support the vehicle safely. Turn the adjusting knob on the parking brake lever counterclockwise until it stops.
2. Apply the parking brake. Loosen the equalizer nut.
3. Tighten the equalizer nut until light drag is felt while rotating the rear wheels in the forward motion.
4. Adjust the knob on the parking brake lever until a definite snap over center is felt.
5. Release the parking brake. Rotate the rear wheels in the forward motion. There should be no brake drag.

DRIVESHAFT TYPE

1. Raise and support the vehicle safely. Remove the clevis pin connecting the pull rod and the relay lever.
2. Rotate the brake drum to align the access hole with the adjusting screw. If equipped with manual transmission the access hole is located at the bottom of the backing plate. If equipped with automatic transmission the access hole is located at the top of the shoe.

3. For first time adjustment it will be necessary to remove the driveshaft and the drum in order to remove the lanced area from the drum and clean out the metal shavings.
4. Adjust the screw until the drum cannot be rotated by hand. Back off the adjusting screw 10 notches, the drum should rotate freely.
5. Position the parking brake lever in the fully released position. Take up the slack in the cable to overcome spring tension.
6. Adjust the clevis of the pull rod to align with the hole in the relay lever. Install the clevis pin. Install a new cover in the drum access hole.

C/K-Series

1. Raise and support the vehicle safely. Matchmark the wheel to the axle flange. Remove the tire and wheel assembly.
2. Matchmark the drum to the axle flange. Remove the brake drum.
3. Using tool J–21177A or equivalent, measure and record the brake drum inside diameter.
4. Turn the adjuster nut and adjust the shoe and lining to a diameter 0.010–0.020 in. less than the measured inside diameter of the brake drum.
5. Be sure the stops on the parking brake levers are against the edge of the brake shoe web. If not, loosen the parking brake cable adjustment.
6. Tighten the parking brake cable at the adjuster nut until the lever stops begin to move off of the shoe webs. Loosen the adjustment nut until the lever stops move back, barely touching the shoe webs. The final clearance between the stops and either web should be 0.5mm.
7. Install the drums and wheels. Align the assemblies with the matchmarks made during removal.
8. Apply and release the service brake pedal 30–35 times using normal pedal force. Pause about one second between each pedal application.
9. Depress the parking brake six clicks. Check the rear wheels they should not rotate.
10. Release the parking brake lever. Check for free wheel rotation.

BRAKE SYSTEM BLEEDING

1. The brake system must be bled when any brake line is disconnected or there is air in the system.

NOTE: Never bleed a wheel cylinder when a drum is removed.

2. Clean the master cylinder of excess dirt and remove the cylinder cover and the diaphragm.
3. Fill the master cylinder to the proper level. Check the fluid level periodically during the bleeding process and replenish it, as necessary. Do not allow the master cylinder fall below ½ full.
4. If the master cylinder is suspected or known to have air in the bore, bleed it before any wheel cylinder or caliper as follows:

   a. Disconnect the forward brake line connection at the master cylinder.

   b. Allow brake fluid to fill the master cylinder bore until it begins to flow from the forward line connector port.

   c. Connect the forward brake line to the master cylinder and tighten.

   d. Have an assistant depress the brake pedal slowly, 1 at a time and hold while loosening the forward brake line connection at the master cylinder to purge the air from the bore. Tighten the connection and then have an assistant release the brake pedal slowly. Wait 15 seconds and repeat the sequence. Repeat the sequence, including the 15 second wait until all air is removed from the bore.

   e. After all air is removed at the forward connection, repeat the above procedure for the rear connection at the master cylinder.

5. Bleed the individual wheel cylinders or calipers only after all air is removed from the master cylinder.
6. Attach the proper size box end wrench over the bleeder valve.
7. Attach a length of vinyl hose to the bleeder screw of the brake to be bled. Insert the other end of the hose into a clear jar half full of clean brake fluid, so the end of the hose is beneath the level of fluid. The correct sequence for bleeding is to work from the brake farthest from the master cylinder to the 1 closest; right rear, left rear, right front, left front.
8. Have an assistant depress and release the brake pedal 1 time and hold. Loosen the bleeder valve to purge the air from the cylinder. Tighten the bleeder screw and slowly release the pedal and wait 15 seconds. Repeat the sequence includ-

ing the 15 second wait until all air is removed.

NOTE: Make sure an assistant presses the brake pedal to the floor slowly. Rapid pumping of the brake pedal pushes the master cylinder secondary piston down the bore in a way that makes it difficult to bleed the rearside of the system.

9. Repeat this procedure at each of the wheels. Remember to check the master cylinder level occasionally. Use only fresh fluid to refill the master cylinder, not the fluid bled from the system.

10. When the bleeding process is complete, refill the master cylinder, install its cover and diaphragm and discard the fluid bled from the brake system.

Anti-Lock Brake System Service

There are 3 different systems used, Four Wheel Anti-Lock (4WAL) which is used on C/K-Series vehicles, Zero Pressure Rear Wheel Anti-Lock (ZPRWAL) which is used only on the 3500 HD-Series vehicle and Rear Wheel Anti-Lock (RWAL) which is used on all R/V and G-Series and C/K-Series except those equipped with 4WAL.

PRECAUTION

Failure to observe the following precautions may result in system damage.

- Before performing electric arc welding on the vehicle, disconnect the electronic brake control unit .
- When performing painting work on the vehicle, do not expose the electronic brake control unit to temperatures in excess of 185°F (85°C) for longer than 2 hours. The system may be exposed to temperatures up to 200°F (95°C) for less than 15 minutes.
- Never disconnect or connect the electronic brake control unit connector with the ignition switch ON.
- Never disassemble any component of the Anti-Lock Brake System (ABS) which is designated non-serviceable; the component must be replaced as an assembly.
- When filling the master cylinder, always use Delco Supreme 11 brake fluid or equivalent, which meets

DOT-3 specifications; petroleum base fluid will destroy the rubber parts. Do not allow fluid to be spilled on the Electronic Brake Control Unit.

Electronic Brake Control Module

REMOVAL AND INSTALLATION

Except 4WAL System

1. Disconnect the negative battery cable.
2. Disconnect the electrical connectors from the module.
3. Remove the module from the master cylinder/proportioning valve bracket.
4. Installation is the reverse of removal procedure.

4WAL System

1. Disconnect the negative battery cable.
2. Disconnect the electrical connectors from the module.
3. Disconnect and plug the brake lines at the module.
4. Remove the bolts attaching the module to the fenderwell.
5. Remove the module and bracket assembly from the vehicle.
6. Remove the bracket from the module.
7. Installation is the reverse of the removal procedure.

Isolation/Dump Valve

REMOVAL AND INSTALLATION

RWAL and 4WAL Systems

1. Disconnect the negative battery cable.
2. Disconnect and plug the brake line fittings at the isolation/dump valve located under the master cylinder.
3. Remove the master cylinder bracket bolts.
4. Disconnect the electronic brake control unit connector.
5. Remove the isolation/dump valve.
6. Installation is the reverse of the removal procedure.

Front Wheel Speed Sensor

REMOVAL AND INSTALLATION

4WAL System

2 WHEEL DRIVE

1. Disconnect the negative battery cable.
2. Raise and support the vehicle safely.
3. Remove the wheel and tire assembly.
4. Remove the brake caliper, hub and rotor assembly.
5. Remove the sensor wire from the clips on the upper control arm and disconnect the connector.
6. Remove the sensor and backing plate attaching bolts and remove the assembly.
7. Installation is the reverse of the removal procedure.

4 WHEEL DRIVE

1. Disconnect the negative battery cable.
2. Raise and support the vehicle safely.
3. Remove the wheel and tire assembly.
4. Disconnect the sensor wire connector.
5. Remove the bolts securing the sensor and sensor wire.
6. Remove the sensor from the spindle.
7. Installation is the reverse of the removal procedure.

FRONT SUSPENSION

Shock Absorbers

REMOVAL AND INSTALLATION

R-Series and G-Series

1. Raise and support the vehicle safely. Properly support the lower control arm assembly, as required. Remove the tire and wheel assembly.
2. Remove the upper shock absorber retaining bolt. Remove the lower shock absorber retaining bolt. Vehicles equipped with quad shocks have a spacer between them.
3. Remove the shock absorber from the vehicle.
4. Installation is the reverse of the removal procedure. On the R-Series vehicle, torque the bolts to specification.

V-Series

1. Raise and support the vehicle safely.

2. Remove the nuts and eye bolts securing the upper and lower shock absorber eyes. Quad shocks have a spacer between the lower end bushings.

3. Remove the shock absorber(s) and inspect the rubber eye bushings for wear or the shock for leaks, replace the shock absorber assembly as necessary.

4. Installation is the reverse of removal. Make sure the spacer is installed at the bottom end on quad shocks. Torque the upper end nut to 65 ft. lbs. (88 Nm). On dual shocks, torque the lower end to 65 ft. lbs. (88 Nm). On quad shocks, torque the lower end to 89 ft. lbs. (120 Nm).

C-Series

1. Remove the upper shock absorber retaining bolt. Raise and support the vehicle safely.

2. Properly support the lower control arm, as required. Remove the lower shock absorber retaining bolt.

3. Remove the shock absorber from the vehicle.

4. Installation is the reverse of removal. Tighten the upper nut to 100 inch lbs. (11 Nm); tighten the lower mounting bolts to 20 ft. lbs. (27 Nm).

K-Series

1. Raise and support the vehicle safely.

2. Remove the upper end bolt, nut and washer.

3. Remove the lower end bolt, nut and washer.

4. Remove the shock absorber and inspect the rubber bushings for wear and the shock for leaks, replace the shock absorber assembly necessary.

5. Installation is the reverse of removal. Torque the both nuts to 66 ft. lbs. (90 Nm). Make sure the bolts are inserted in the proper direction. The bolt head on the upper end should be forward; the bottom end bolt head is rearward.

Coil Springs

REMOVAL AND INSTALLATION

R-Series and G-Series

1. Raise and support the vehicle safely under the frame rails. The control arms should hang freely.

2. Remove the wheel.

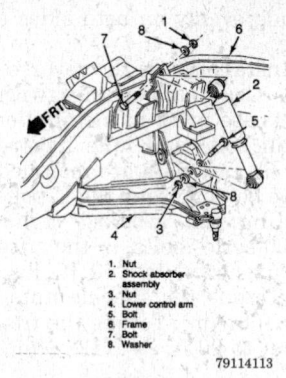

1. Nut
2. Shock absorber assembly
3. Nut
4. Lower control arm
5. Bolt
6. Frame
7. Bolt
8. Washer

79114113

Shock absorber installation — K series

3. Disconnect the shock absorber at the lower end and move it aside.

4. Disconnect the stabilizer bar from the lower control arm.

5. Support the lower control arm and install a spring compressor on the spring or chain the spring to the control arm as a safety precaution.

NOTE: On vehicles with an air cylinder inside the spring, remove the valve core from the cylinder and expel the air by compressing the cylinder with a prybar. With the cylinder compressed, replace the valve core so the cylinder will stay in the compressed position. Push the cylinder as far as possible towards the top of the spring.

6. Raise the jack to remove the tension from the lower control arm crossshaft and remove the 2 U-bolts securing the cross-shaft to the crossmember.

NOTE: The cross-shaft and lower control arm keeps the coil spring compressed. Use care when lowering the assembly.

7. Slowly release the jack and lower the control arm until the spring can be removed. Be sure all compression is relieved from the spring.

8. If the coil spring was chained, remove the chain and spring. If a compressor was used, remove the spring and slowly release the compressor.

9. Remove the air cylinder, if equipped.

To install:

10. Install the air cylinder so the protector plate is towards the upper control arm. The schrader valve should protrude through the hole in the lower control arm.

11. Install the chain and spring or compress the spring and install the assembly.

12. Slowly raise the jack and lower control arm. Line up the indexing hole in the shaft with the crossmember attaching studs.

13. Install the 2 U-bolts securing the cross-shaft to the crossmember. Torque the nuts to 85 ft. lbs. (115 Nm).

14. Remove the jack.

15. Connect the stabilizer bar to the lower control arm. Torque the nuts to 24 ft. lbs.

16. Connect the shock absorber at the lower end. Torque the bolt to specification.

17. If equipped with air cylinders, inflate the cylinder to 60 psi.

18. Install the wheel.

19. Lower the vehicle. Once the weight of the vehicle is on the wheels, reduce the air cylinder pressure to 50 psi.

20. Check the front end alignment.

C-Series

1. Raise and support the vehicle safely. Allow the control arms to hang free. Remove the tire and wheel assembly. Remove the shock absorber assembly, as required.

2. Install tool J–23028 under the lower control arm using a suitable jack. Install a safety chain around the spring and through the lower control arm.

3. Remove the stabilizer shaft from the lower control arm. Raise the jack and remove the tension on the lower control arm bolts.

4. Remove the lower control arm rear bolt, than remove the other retaining bolt.

5. Lower the jack and allow the lower control arm to hang free. Remove the spring assembly from the vehicle.

To install:

6. Install the chain and spring. If you used spring compressors, install the spring and compressors.

   a. Make sure the insulator is in place.

   b. Make sure the tape is at the lower end. New springs will have an identifying tape.

   c. Make sure the gripper notch on the top coil is in the frame bracket.

   d. Make sure 1 drain hole in the lower arm is covered by the bottom coil and the other is open.

7. Slowly raise the jack and lower control arm. Guide the control arm into place with a prybar.

8. Install the pivot shaft bolts, front one first. The bolts must be installed with the heads towards the

front of the vehicle. Remove the safety chain or spring compressors.

NOTE: Do not torque the bolts yet. The bolts must be torqued with the vehicle at its proper ride height.

9. Remove the jack.
10. Connect the stabilizer bar to the lower control arm. Torque the nuts to specification.
11. Install the shock absorber.
12. Install the wheel.
13. Lower the vehicle. Once the weight of the vehicle is on the wheels check the "Z" height as follows:

    a. Lift the front bumper about 1½ in. (38mm) and let it drop.

    b. Repeat this procedure 2 or 3 more times.

    c. Draw a line on the side of the lower control arm from the centerline of the control arm pivot shaft, dead level to the outer end of the control arm.

    d. Measure the distance between the lowest corner of the steering knuckle and the line on the control arm. Record the figure.

    e. Push down about 38mm on the front bumper and let it return. Repeat the procedure 2 or 3 more times.

    f. Re-measure the distance at the control arm.

    g. Determine the average of the 2 measurements.

    h. If the figure is correct, tighten the control arm pivot nuts to 121 ft. lbs. (165 Nm). Align the front end. If the figure is incorrect, replace the coil spring.

Leaf Spring

REMOVAL AND INSTALLATION

V-Series

1. Raise and support the vehicle and front axle safely so all tension is taken off of the front leaf springs.
2. Remove the shackle retaining bolts, nuts and spacers.
3. Remove the front spring-to-frame bracket bolt, washer and nut.
4. On the 10/1500 and 20/2500 both sides and the 30/3500 left side: remove the U-bolt nuts, washers, U-bolts, plate and spacers.
5. On the 30/3500 right side: remove the inboard spring plate bolts, U-bolt nuts, washers, U-bolt, plate and spacers.
6. To replace the bushing, place the spring in a press or vise and press out the bushing. Press in the new bushing. The new bushing should

protrude evenly on both sides of the spring.

7. Installation is the reverse of removal. Coat all bushings with silicone grease prior to installation. Install all bolts and nuts finger-tight. When all fasteners are installed, torque the bolts. Torque the U-bolt nuts, including the inboard right side 30/3500 series bolts, in the crisscross pattern shown, to 150 ft. lbs. (213 Nm). Torque the shackle nuts to 50 ft. lbs. (68 Nm). Torque the front eye bolt nut to 90 ft. lbs. (122 Nm).

Torsion Bar

REMOVAL AND INSTALLATION

K-Series

1. Raise and support the vehicle safely.
2. Remove the wheels.
3. Support the lower control arm with a floor jack.
4. Matchmark the both torsion bar adjustment bolt positions.
5. Using tool J-36202, increase the tension on the adjusting arm.
6. Remove the adjustment bolt and retaining plate.
7. Move the tool aside.
8. Slide the torsion bars forward.
9. Remove the adjusting arms.
10. Remove the nuts and bolts from the torsion bar support crossmember and slide the support crossmember rearward.
11. Matchmark the position of the torsion bars and note the markings on the front end of each bar. They are not interchangeable. Remove the torsion bars.
12. Remove the support crossmember.
13. Remove the retainer, spacer and bushing from the support crossmember.

To install:

14. Assemble the retainer, spacer and bushing on the support.
15. Position the support assembly on the frame, out of the way.
16. Align the matchmarks and install the torsion bars, sliding them forward until they are supported.
17. Install the adjuster arms on the torsion bars.
18. Bolt the support crossmember into position. Torque the center nut to 18 ft. lbs. (24 Nm); the edge nuts to 46 ft. lbs. (62 Nm).
19. Install the adjuster retaining plate and bolt on each torsion bar.
20. Using tool J-36202, increase tension on both torsion bars.

21. Install the adjustment retainer plate and bolt on both torsion bars.
22. Set the adjustment bolt to the marked position.
23. Release the tension on the torsion bar until the load is take up by the adjustment bolt.
24. Remove the tool.
25. Install the wheels.
26. Check the front end alignment and "Z" height.

Ball Joints

INSPECTION

1. Raise the vehicle and position floor stands under the left and right lower control arm as near as possible to each lower ball joint. There should be sufficient space between the upper control arm bumper and frame after the lower control arm has been supported.
2. Position a dial indicator against the wheel rim.
3. Grasp the front wheel; push in on bottom of the tire while pulling out at the top. Read the gauge, then reverse the push-pull procedure. Horizontal deflection on the dial indicator should not exceed 1.125 in. (3.18mm).
4. If the indicator exceeds 1.125 in. (3.18mm) or if the ball stud, when disconnected from the knuckle assembly, can be twisted in its socket by hand, replace the ball joint.

Upper Ball Joints

REMOVAL AND INSTALLATION

R-Series and G-Series

1. Raise and support the vehicle safely. Properly support the lower control arm.
2. Remove the wheel assembly. Remove the brake caliper and position it to the side.
3. Remove the cotter pin and the upper ball joint retaining bolt. Using the proper tool separate the upper joint from its mounting. Support the knuckle assembly so its weight will not damage the brake hose.
4. Remove the rivets from the ball joint assembly, using the proper tools. Remove the ball joint from the upper control arm.

To install:

5. Install the new ball joint into the control arm. Position the bleed vent in the rubber boot facing inward.

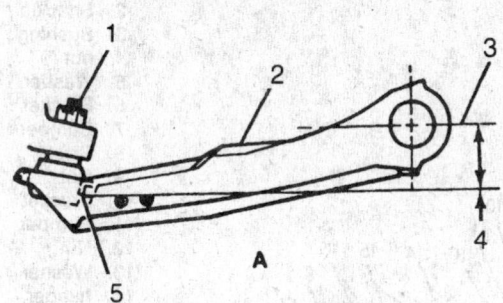

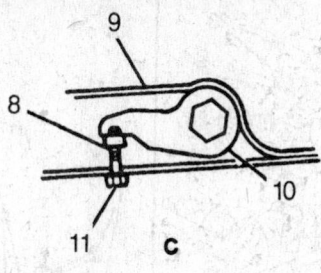

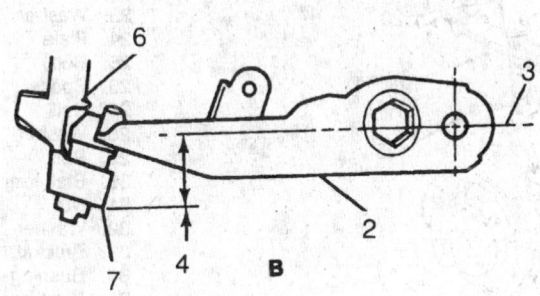

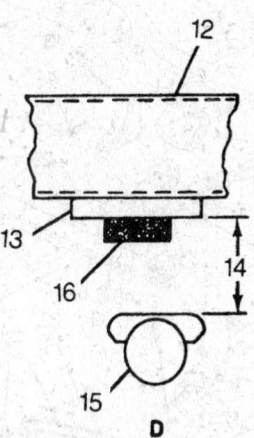

79114107

A. C-Series
B. K-Series
C. K-Series torsion bar
D. C/K-Series rear
   suspension
1. Lower ball joint
2. Lower control arm
3. Pivot bolt center line
4. "Z" Height
   C1, 2, 3—89–101mm
   K1, 2—151–163mm
   K3—139–151mm
5. Lower ball joint
   extrusion
6. Steering knuckle

7. Steering knuckle
   lower corner
8. Nut
9. Torsion bar support
   assembly
10. Torsion bar
    adjustment arm
11. Bolt—1 turn equals
    6mm height change
12. Frame
13. Bottom surface of
    jounce bracket
14. "D" Height
15. Rear axle
16. Jounce bumper

Trim height adjustement — C/K series

6. Secure the ball joint to the control arm using the new nuts and bolts.

7. Lower the upper arm ball joint stud into the steering knuckle.

8. Install the ball stud nut and torque to specification. Tighten the nut to align the cotter pin hole.

9. Install the brake caliper, if removed.

10. Install a new lube fitting and lubricate the new joint.

11. Install the tire and wheel.

12. Lower the vehicle.

V-Series

1. Raise and support the vehicle safely. Remove the tire and wheel assembly.

2. Remove the hub and rotor assembly. Remove the spindle.

3. Remove the steering knuckle assembly. If removing the left axle yoke ball joints, remove the steering arm. Position the steering knuckle assembly in a suitable vise.

4. The lower ball joint must be removed before service can be performed on the upper ball joint. first. Remove the snapring from the lower ball joint and press the ball joint from the knuckle assembly, using the proper tools.

5. Press the upper ball joint from the knuckle assembly, using the proper tools.

6. Installation is the reverse of the removal procedure. Check and adjust front alignment, as required.

C-Series

1. Raise and support the vehicle safely. Properly support the lower control arm, using the necessary equipment.

2. Remove the tire and wheel assembly. Remove the brake caliper and position it to the side.

3. Remove the cotter pin and the upper ball joint retaining bolt. Using the proper tool separate the upper

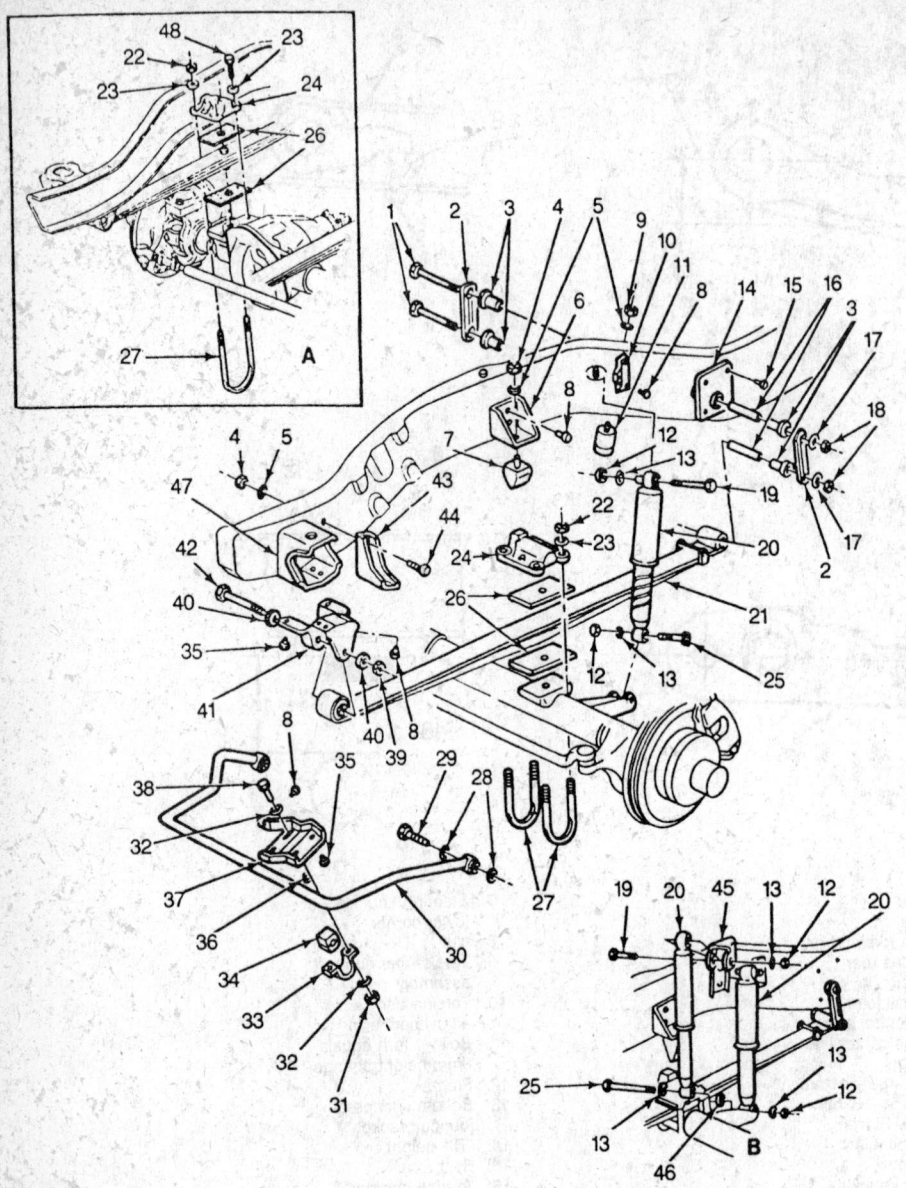

1. Bolt
2. Shackle
3. Bushing
4. nut
5. Washer
6. Bracket
7. Bumper
8. Rivet
9. Nut
10. Bracket
11. Bumper
12. Nut
13. Washer
14. hanger
15. Rivet
16. Spacer
17. Washer
18. Nut
19. Bolt
20. Shock absorber
21. Leaf spring
22. Nut
23. Washer
24. Plate
25. Bolt
26. Spacer
27. Bolt
28. Washer
29. Bolt
30. Stabilizer shaft
31. Nut
32. Washer
33. Bracket
34. Bushing
35. Rivet
36. Rivet
37. Bracket
38. Bolt
39. Nut
40. Washer
41. Hanger
42. Bolt
43. Reinforcement
44. Bracket
45. Spacer
46. Spacer
47. Bracket
48. Bolt
A. K3500 Series—right side
B. Quad shock

79114109

Front suspension assembly — V series

joint from its mounting. Support the knuckle assembly so its weight will not damage the brake hose.

4. Remove the rivets from the ball joint assembly, using the proper tools. Remove the ball joint from the upper control arm.

To install:

5. Install the replacement ball joint in the control arm, using the bolts and nuts supplied. Torque the nuts to 18 ft. lbs. (24 Nm).

6. Position the ball stud in the knuckle. Make sure it is squarely seated. Torque the ball stud nut to 84 ft. lbs. (115 Nm).

7. Install a new cotter pin.

8. Install a new lube fitting and lubricate the new joint.

9. If removed, install the brake caliper.

10. Install the wheel and lower the vehicle. Check and align the front end.

K-Series

1. Raise and support the vehicle safely.

2. Remove the wheel.

3. Unbolt the brake hose bracket from the control arm.

4. Using a 1/8 in. drill bit, drill a pilot hole through each ball joint rivet.

5. Drill out the rivets with a 1/2 in. drill bit. Punch out any remaining rivet material.

6. Remove the cotter pin and nut from the ball stud.

7. Support the lower control arm with a floor jack.

8. Using a ball joint separator, separate the stud from the knuckle.

To install:

9. Position the new ball joint on the control arm.

NOTE: Service replacement ball joints come with nuts and bolts to replace the rivets.

10. Install the bolts and nuts. Tighten the nuts to 17 ft. lbs. (23 Nm)

for 15 and 25-Series; 52 ft. lbs. (70 Nm) for 35-Series.

NOTE: The bolts are inserted from the bottom.

11. Start the ball stud into the knuckle. Make sure it is squarely seated. Install the ball stud nut and pull the ball stud into the knuckle with the nut. Torque the nut after the vehicle wheel are on the ground and the suspension is loaded.

12. Install the wheel.

13. Lower the vehicle. Once the weight of the vehicle is on the wheels tighten the nut to 84 ft. lbs. (115 Nm).

Lower Ball Joint

REMOVAL AND INSTALLATION

R-Series and G-Series

1. Raise and support the vehicle safely. Properly support the lower control arm, using the necessary equipment.

2. Remove the tire and wheel assembly. As required, remove the brake caliper and position it to the side.

3. Remove the cotter pin and the lower ball joint retaining bolt. Using the proper tool separate the ball joint from its mounting. Support the knuckle assembly so its weight will not damage the brake hose.

4. Press the ball joint out of the lower control arm, using the proper tool.

To install:

5. Start the new ball joint into the control arm. Position the bleed vent in the rubber boot facing inward.

6. Press the ball joint into the control arm until fully seated.

7. Lower the upper arm and insert the lower ball joint stud into the steering knuckle.

8. Install the brake caliper, if removed.

9. Install the ball stud nut and torque to 90 ft. lbs. (122 Nm) plus the additional torque necessary to align the cotter pin hole. Do not exceed 130 ft. lbs. (175 Nm) or back the nut off to align the holes with the pin.

10. Install a new lube fitting and lubricate the new joint.

11. Install the tire and wheel.

12. Lower the vehicle.

C-Series

1. Raise and support the vehicle safely. Properly support the lower

control arm, using the necessary equipment.

2. Remove the tire and wheel assembly. As required, remove the brake caliper and position it to the side.

3. Remove the cotter pin and the lower ball joint retaining nut. Using the proper tool separate the ball joint from its mounting. Support the knuckle assembly so its weight will not damage the brake hose.

4. Remove the rivets from the ball joint assembly, using the proper tools. Remove the ball joint from the control arm.

To install:

5. Secure the new ball joint to the control arm.

6. Position the ball joint into the knuckle. Install the nut and tighten it to 84 ft. lbs. (115 Nm).

7. Advance the nut to align the cotter pin hole and insert the new cotter pin.

8. Install the brake caliper, if removed.

9. Install a new lube fitting and lubricate the new joint.

10. Install the wheel.

11. Lower the vehicle.

12. Check the front end alignment.

K-Series

1. Raise and support the vehicle safely.

2. Remove the wheel.

3. Remove the splash shield from the knuckle.

4. Disconnect the inner tie rod end from the relay rod using a ball joint separator.

5. Remove the hub nut and washer. Insert a long drift or dowel through the vanes in the brake rotor to hold the rotor in place.

6. Remove the axle shaft inner flange bolts.

7. Using a puller, force the outer end of the axle shaft out of the hub. Remove the shaft.

8. Using a 1/8 in. drill bit, drill a pilot hole through each ball joint rivet.

9. Drill out the rivets with a 1/2 in. drill bit. Punch out any remaining rivet material.

10. Remove the cotter pin and nut from the ball stud.

11. Support the lower control arm with a floor jack.

12. Matchmark the both torsion bar adjustment bolt positions.

13. Using tool J-36202, increase the tension on the adjusting arm.

14. Remove the adjustment bolt and retaining plate.

15. Move the tool aside.

16. Slide the torsion bars forward.

17. Using a screw-type forcing tool, separate the ball joint from the knuckle.

To install:

18. Position the new ball joint on the control arm.

NOTE: Service replacement ball joints come with nuts and bolts to replace the rivets.

19. Install the bolts and nuts. Tighten the nuts to 45 ft. lbs.

NOTE: The bolts are inserted from the bottom.

20. Start the ball stud into the knuckle. Make sure it is squarely seated. Install the ball stud nut and pull the ball stud into the knuckle with the nut. Do not final-torque the nut yet.

21. Using tool J-36202, increase tension on both torsion bars.

22. Install the adjustment retainer plate and bolt on both torsion bars.

23. Set the adjustment bolt to the marked position.

24. Release the tension on the torsion bar until the load is take up by the adjustment bolt.

25. Remove the tool.

26. Position the shaft in the hub and install the washer and hub nut. Leave the drift in the rotor vanes and tighten the hub nut to 175 ft. lbs.

27. Install the flange bolts. Tighten them to 59 ft. lbs. Remove the drift.

28. Connect the inner tie rod end at the steering relay rod. Torque the nut to 35 ft. lbs.

29. Install the splash shield.

30. Install the wheel.

31. Lower the vehicle. Once the weight of the vehicle is on the wheels:

  a. Lift the front bumper about 1-1/2 in. (38mm) and let it drop.

  b. Repeat this procedure 2 or 3 more times.

  c. Draw a line on the side of the lower control arm from the centerline of the control arm pivot shaft, dead level to the outer end of the control arm.

  d. Measure the distance between the lowest corner of the steering knuckle and the line on the control arm. Record the figure.

  e. Push down about 1 1/2 in. (38mm) on the front bumper and let it return. Repeat the procedure 2 or 3 more times.

  f. Re-measure the distance at the control arm.

  g. Determine the average of the 2 measurements. The average distance should be as specified.

h. If the figure is correct, tighten the control arm pivot nuts to 94 ft. lbs. (128 Nm).

i. If the figure is not correct, tighten the pivot bolts to 94 ft. lbs. (128 Nm) and have the front end alignment corrected.

V-Series

1. Raise and support the vehicle safely.
2. Remove the wheels.
3. Remove the locking hubs.
4. Remove the spindle.
5. Disconnect the tie rod end from the knuckle.
6. Remove the knuckle-to-steering arm nuts and adapters.
7. Remove the steering arm from the knuckle.
8. Remove the cotter pins and nuts from the upper and lower ball joints.

NOTE: Do not remove the adjusting ring from the knuckle. If it is necessary to loosen the ring to remove the knuckle, don't loosen it more than 2 threads. The non-hardened threads in the yoke can be easily damaged by the hardened threads in the adjusting ring if caution is not used during knuckle removal

9. Insert the wedge-shaped end of the heavy prybar or wedge-type ball joint tool, between the lower ball joint and the yoke. Drive the prybar in to break the knuckle free.
10. Repeat the procedure at the upper ball joint.
11. Lift off the knuckle.
12. Secure the knuckle in a vise.
13. Remove the snapring from the lower ball joint. Using tools J–9519–30, J–23454–1 and J–23454–4 or their equivalent screw-type forcing tool, force the lower ball joint from the knuckle.
14. Using tools J–9519–30, J–23454–3 and J–23454–4 or their equivalent screw-type forcing tool, force the upper ball joint from the knuckle.
   To install:
15. Position the lower ball joint (the one without the cotter pin hole) squarely in the knuckle. Using tools J–9519–30, J–23454–2 and J–23454–3 or their equivalent screw-type forcing tool, force the lower ball joint into the knuckle until it is fully seated.
16. Install the snapring.
17. Position the upper ball joint (the one with the cotter pin hole) squarely in the knuckle. Using tools J–9519–30, J–23454–2 and

J–23454–3 or their equivalent screw-type forcing tool, force the upper ball joint into the knuckle until it is fully seated.
18. Position the knuckle on the yoke.
19. Start the ball joints into their sockets. Place the nuts onto the ball studs. The nut with the cotter pin slot is the upper nut. Tighten the lower nut to 30 ft. lbs. (40 Nm), for now.
20. Using tool J–23447, tighten the adjusting ring to 50 ft. lbs. (68 Nm).
21. Tighten the upper nut to specification. Install a new cotter pin. Never loosen the nut to align the cotter pin hole; always tighten it.
22. Tighten the lower nut to specification.
23. Attach the steering arm to the knuckle using adapters and new nuts.
24. Connect the tie rod end to the knuckle.
25. Install the spindle.
26. Install the hub/rotor assembly and wheel bearings. Adjust the bearings.
27. Install the locking hubs.
28. Install the wheel.
29. Check the front end alignment.

Upper Control Arm

REMOVAL AND INSTALLATION

R-Series and G-Series

1. Note and record the amount of shims. These shims must be installed in the same location as removed. Remove the nuts and the shims.
2. Raise and support the vehicle safely. Properly support the lower control arm, using the necessary equipment. The control arm must be supported so the spring and the control arm remain intact.
3. Remove the tire and wheel assembly. Remove the brake caliper assembly and position it to the side. Loosen the upper ball joint from the steering knuckle, using the proper tool. Support the hub assembly.
4. Remove the upper control arm retaining bolts. Remove the upper control arm from the vehicle.
   To install:
5. Place the control arm in position and install the nuts. Before tightening the nuts, insert the caster and camber shims in the same order as when installed.
6. Install the nuts securing the control arm shaft studs to the cross-member bracket. Tighten the nuts to 70 ft. lbs. (95 Nm) for 10/1500 series;

105 ft. lbs. (142 Nm) for all other series.
7. Install the ball stud nut. Torque the nut to 50 ft. lbs. (68 Nm) for 10/1500 series; 90 ft. lbs. (122 Nm) for all other series.
8. Install the cotter pin. Never back off the nut to install the cotter pin. Always advance it.
9. Install the brake caliper. Remove the spring compressor.
10. Install the wheel.
11. Check the front end alignment.

C/K-Series

1. Raise and support the vehicle safely.
2. Support the lower control arm with a floor jack.
3. Remove the wheel.
4. Unbolt the brake hose bracket from the control arm.
5. Remove the air cleaner extension.
6. Remove the cotter pin from the upper control arm ball stud and loosen the stud nut until the bottom surface of the nut is slightly below the end of the stud.
7. Install a spring compressor on the coil spring for safety.
8. Using a screw-type forcing tool, break loose the ball joint from the knuckle.
9. Remove the nuts and bolts securing the control arm to the frame brackets.
10. The 35-Series bushings are replaceable. The 15/25-Series bushings are welded in place.
   To install:
11. Place the control arm in position and install the shims, bolts and new nuts. Both bolt heads must be inboard of the control arm brackets. Tighten the nuts finger-tight for now.

NOTE: Do not torque the bolts yet. The bolts must be torque with the vehicle at its proper ride height.

12. Install the ball stud nut. Torque the nut to 84 ft. lbs. (115 Nm). Install the cotter pin. Never back off the nut to install the cotter pin.
13. Install the brake caliper.
14. Remove the spring compressor or safety chain.
15. Install the wheel.
16. Install the brake hose.
17. Install the air cleaner extension.
18. Install the battery ground cable.
19. Lower the vehicle. Once the weight of the vehicle is on the wheels:
   a. Lift the front bumper about 38mm and let it drop.

b. Repeat this procedure 2 or 3 more times.

c. Draw a line on the side of the lower control arm from the center-line of the control arm pivot shaft, dead level to the outer end of the control arm.

d. Measure the distance between the lowest corner of the steering knuckle and the line on the control arm. Record the figure.

e. Push down about 38mm on the front bumper and let it return. Repeat the procedure 2 or 3 more times.

f. Re-measure the distance at the control arm.

g. Determine the average of the 2 measurements. The average distance should be as specified.

h. If the figure is correct, tighten the control arm pivot nuts to 139 ft. lbs. (190 Nm).

i. If the figure is not correct, adjust or repair the ride height.

Lower Control Arm

REMOVAL AND INSTALLATION

R-Series and G-Series

1. Raise and support the vehicle safely. Properly support the lower control arm assembly. Remove the tire and wheel assembly. Remove the brake caliper and position it to the side.

2. Remove the control arm u-bolts and remove the coil spring. Remove the lower ball joint cotter pin and re-taining nut. Using the proper tool, separate the lower ball joint from the steering knuckle.

3. Remove the lower control arm from the vehicle.

To install:

4. Install the lower control arm and spring assembly. Torque the U-bolts to 85 ft. lbs. (115 Nm).

5. Install the ball stud nut. Torque the nut to 90 ft. lbs. (122 Nm). Install the cotter pin. Never back off the nut to install the cotter pin.

6. Install the brake caliper.

C-Series

1. Raise and support the vehicle safely. Remove the tire and wheel assembly. Properly support the lower control arm assembly. Remove the lower control arm bolts and coil spring.

2. Remove the lower ball joint cotter pin and retaining nut. Using proper tool, separate the lower ball joint from the steering knuckle.

3. Remove the lower control arm from the vehicle.

To install:

4. Slowly raise the jack and lower control arm with the coil spring. Guide the control arm into place with a prybar.

5. Install the pivot shaft bolts, front one first. The bolts must be in-stalled with the heads towards the front of the vehicle. Remove the safety chain or spring compressors.

NOTE: Do not torque the bolts yet. The bolts must be torque with the vehicle at its proper ride height.

6. Remove the jack.

7. Connect the stabilizer bar to the lower control arm. Torque the nuts to specification.

8. Install the shock absorber.

9. Install the wheel.

10. Lower the vehicle. Once the weight of the vehicle is on the wheels proceed as follows:

a. Lift the front bumper about 1½ in. (38mm) and let it drop.

b. Repeat this procedure 2 or 3 more times.

c. Draw a line on the side of the lower control arm from the center-line of the control arm pivot shaft, dead level to the outer end of the control arm.

d. Measure the distance between the lowest corner of the steering knuckle and the line on the control arm. Record the figure.

e. Push down about 1½ in. (38mm) on the front bumper and let it return. Repeat the procedure 2 or 3 more times.

f. Re-measure the distance at the control arm.

g. Determine the average of the 2 measurements. The average distance should be 73.6mm 6mm.

h. If the figure is correct, tighten the control arm pivot nuts to specification.

i. If the figure is not correct, tighten the pivot bolts and check the front end alignment corrected.

K-Series

1. Raise and support the vehicle safely.

2. Remove the wheel.

3. Remove the splash shield from the knuckle.

4. Disconnect the stabilizer bar from the control arm.

5. Remove the shock absorber.

6. Disconnect the tie rod end from the relay rod.

7. Remove the hub nut and washer. Insert a long drift or dowel

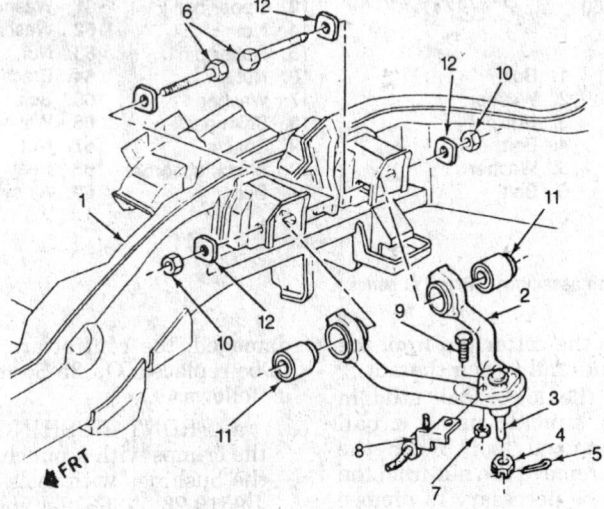

1.	Frame	7.	Nut
2.	Upper control arm	8.	Bracket
3.	Upper ball joint	9.	Screw
4.	Nut	10.	Nut
5.	Pin	11.	Bushing
6.	Bolt	12.	Washer

79114111

Upper control arm installation — K series

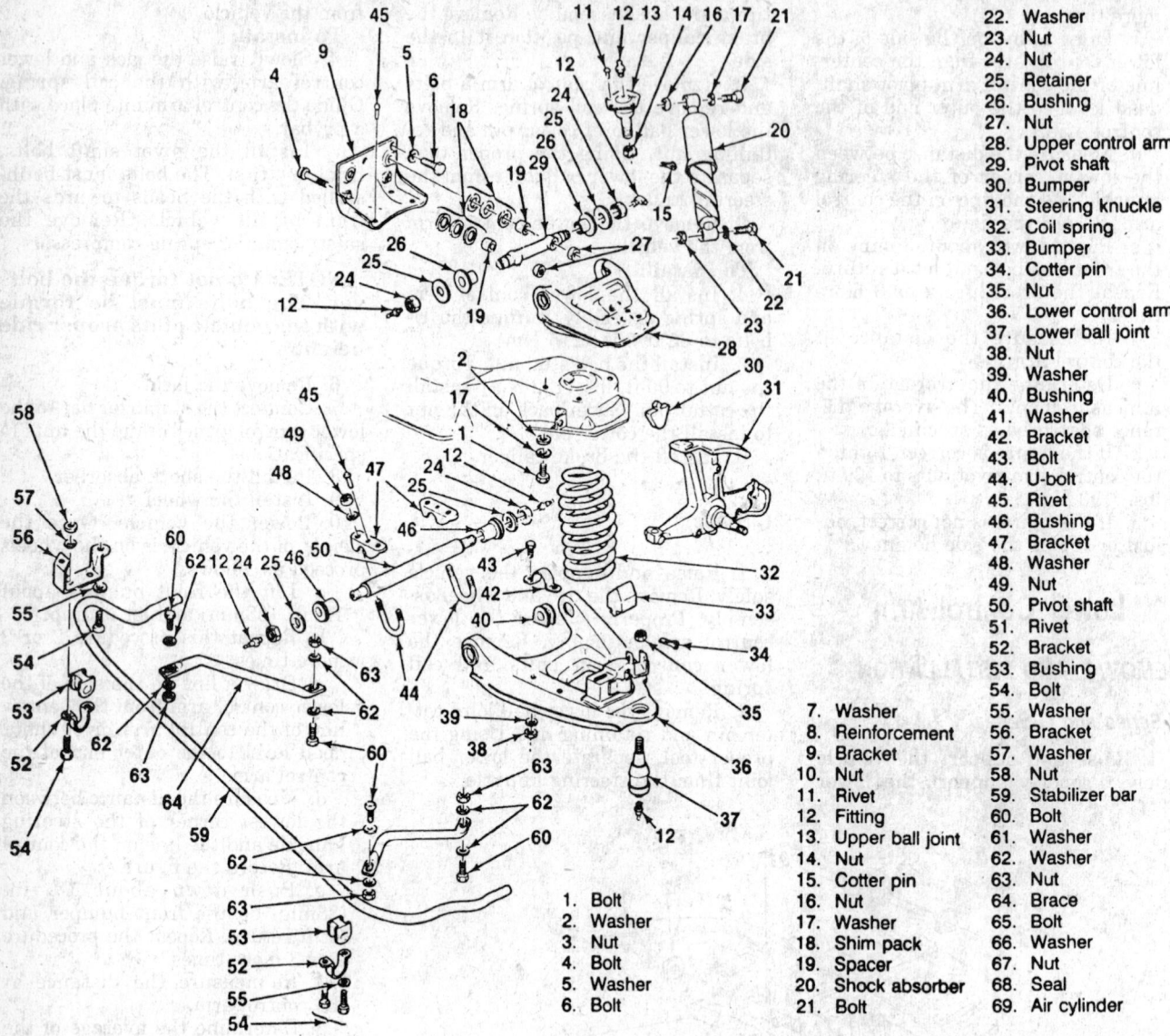

Front suspension assembly — R and G series

22. Washer	45. Rivet
23. Nut	46. Bushing
24. Nut	47. Bracket
25. Retainer	48. Washer
26. Bushing	49. Nut
27. Nut	50. Pivot shaft
28. Upper control arm	51. Rivet
29. Pivot shaft	52. Bracket
30. Bumper	53. Bushing
31. Steering knuckle	54. Bolt
32. Coil spring	55. Washer
33. Bumper	56. Bracket
34. Cotter pin	57. Washer
35. Nut	58. Nut
36. Lower control arm	59. Stabilizer bar
37. Lower ball joint	60. Bolt
38. Nut	61. Washer
39. Washer	62. Washer
40. Bushing	63. Nut
41. Washer	64. Brace
42. Bracket	65. Bolt
43. Bolt	66. Washer
44. U-bolt	67. Nut

7. Washer	68. Seal
8. Reinforcement	69. Air cylinder
9. Bracket	
10. Nut	
11. Rivet	
12. Fitting	
13. Upper ball joint	
14. Nut	
15. Cotter pin	
16. Nut	
17. Washer	
18. Shim pack	
19. Spacer	
20. Shock absorber	
21. Bolt	

1. Bolt	
2. Washer	
3. Nut	
4. Bolt	
5. Washer	
6. Bolt	

79114108

through the vanes in the brake rotor to hold the rotor in place.

8. Remove the axle shaft inner flange bolts.

9. Using a puller, force the outer end of the axle shaft out of the hub. Remove the shaft.

10. Support the lower control arm with a floor jack.

11. Matchmark the both torsion bar adjustment bolt positions.

12. Using tool J-36202, increase the tension on the adjusting arm.

13. Remove the adjustment bolt and retaining plate.

14. Move the tool aside.

15. Slide the torsion bars forward.

16. Remove the adjusting arm.

17. Remove the cotter pin from the lower ball stud and loosen the nut.

18. Loosen the lower ball stud in the steering knuckle using a ball joint stud removal tool. When the stud is loose, remove the nut from the stud. It may be necessary to remove the brake caliper and wire it to the frame to gain clearance.

19. Remove the control arm-to-frame bracket bolts, nuts and washers.

20. Remove the lower control arm and torsion bar as a unit.

21. Separate the control arm and torsion bar.

22. On 15 and 25-Series, the bushings are not replaceable. If they are

damaged, the control arm will have to be replaced. On 35-Series, proceed as follows:

a. FRONT BUSHING: Unbend the crimps with a punch. Force out the bushings with tools J-36618-2, J-9519-23, J-36618-4 and 36618-1.

b. REAR BUSHING: Force out the bushings with tools J-36618-5, J-9519-23, J-36618-3 and J-36618-2. There are no crimps.

To install:

23. On 35-Series, install a new front bushings, then a new rear bushing using the removal tools.

24. Assemble the control arm and torsion bar.

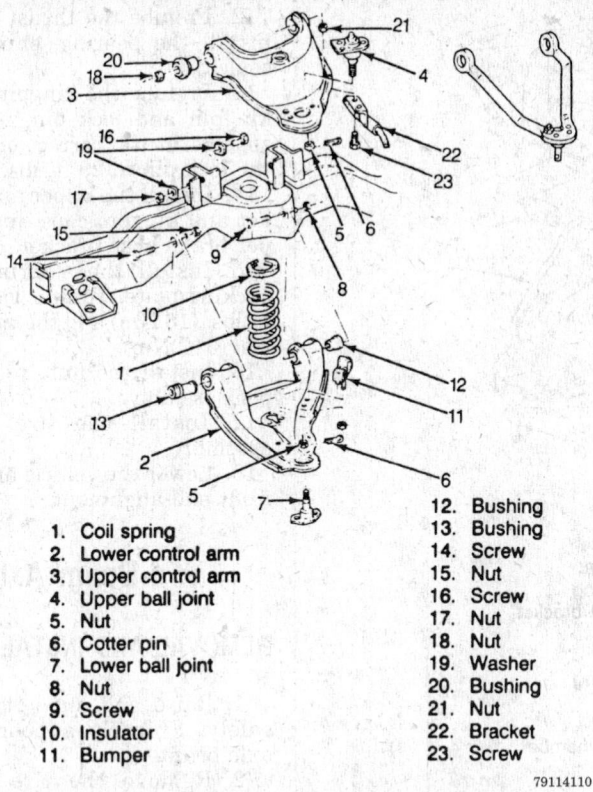

1. Coil spring
2. Lower control arm
3. Upper control arm
4. Upper ball joint
5. Nut
6. Cotter pin
7. Lower ball joint
8. Nut
9. Screw
10. Insulator
11. Bumper
12. Bushing
13. Bushing
14. Screw
15. Nut
16. Screw
17. Nut
18. Nut
19. Washer
20. Bushing
21. Nut
22. Bracket
23. Screw

79114110

Front suspension assembly — C series

25. Raise the control arm assembly into position. Insert the front leg of the control arm into the crossmember first, then the rear leg into the frame bracket.

26. Install the bolts, front one first. The bolts must be installed with the front bolt head heads towards the front of the vehicle and the rear bolt head towards the rear of the vehicle!

NOTE: Do not torque the bolts yet. The bolts must be torque with the vehicle at its proper ride height.

27. Start the ball joint into the knuckle. Make sure it is squarely seated. Tighten the nut to 96 ft. lbs. and install a new cotter pin. Always advance the nut to align the cotter pin hole. Never back it off.

28. Install the adjuster arm.

29. Using tool J-36202, increase tension on both torsion bars.

30. Install the adjustment retainer plate and bolt on both torsion bars.

31. Set the adjustment bolt to the marked position.

32. Release the tension on the torsion bar until the load is take up by the adjustment bolt.

33. Remove the tool.

34. Position the shaft in the hub and install the washer and hub nut.

Leave the drift in the rotor vanes and tighten the hub nut to specification.

35. Install the flange bolts. Tighten to specification. Remove the drift.

36. Connect the inner tie rod end at the steering relay rod. Torque the nut to specification.

37. Install the splash shield.

38. Connect the stabilizer bar to the lower control arm. Torque the nuts to specification.

39. Install the shock absorber.

40. Install the wheel.

41. Lower the vehicle. Once the weight of the vehicle is on the wheels proceed as follows:

   a. Lift the front bumper about 1½ in. (38mm) and let it drop.

   b. Repeat this procedure 2 or 3 more times.

   c. Draw a line on the side of the lower control arm from the center-line of the control arm pivot shaft, dead level to the outer end of the control arm.

   d. Measure the distance between the lowest corner of the steering knuckle and the line on the control arm. Record the figure.

   e. Push down about 1½ in. (38mm) on the front bumper and let it return. Repeat the procedure 2 or 3 more times.

   f. Re-measure the distance at the control arm.

   g. Determine the average of the 2 measurements. The average distance should be 73.6mm 6mm.

   h. If the figure is correct, tighten the control arm nuts to specification.

   i. If the figure is not correct, tighten the pivot bolts to specification and have the front end alignment corrected.

Sway Bar

REMOVAL AND INSTALLATION

R-Series and G-Series

1. Raise and support the vehicle safely. As required, remove the tire and wheel assemblies.

2. Properly support the stabilizer bar assembly. Remove the stabilizer shaft bushing retaining bolts. Remove the stabilizer link bushing nuts and bolts.

3. Remove the stabilizer bar from the vehicle.

4. Installation is the reverse of removal. Note, the split in the bushing faces forward. Coat the bushings with silicone grease prior to installation. Install all fasteners and torque to 24 ft. lbs. (33 Nm).

V-Series

1. Raise and support the vehicle safely.

2. Remove the wheels.

3. Remove the stabilizer bar-to-frame clamps.

4. Remove the stabilizer bar-to-spring plate bolts.

5. Remove the stabilizer bar and bushings.

6. Check the bushings for wear or splitting. Replace any damaged bushings.

7. Installation is the reverse of removal. The split in the bushing faces forward. Coat the bushings with silicone grease prior to installation. Torque the stabilizer bar-to-frame nuts to 52 ft. lbs. (70 Nm). Torque the stabilizer bar-to-spring plate bolts to 133 ft. lbs. (180 Nm).

C-Series

NOTE: The end link bushings, bolts and spacers are not interchangeable from left to right.

1. Raise and support the vehicle safely.

2. Remove the nuts from the end link bolts.

3. Remove the bolts, bushings and spacers.

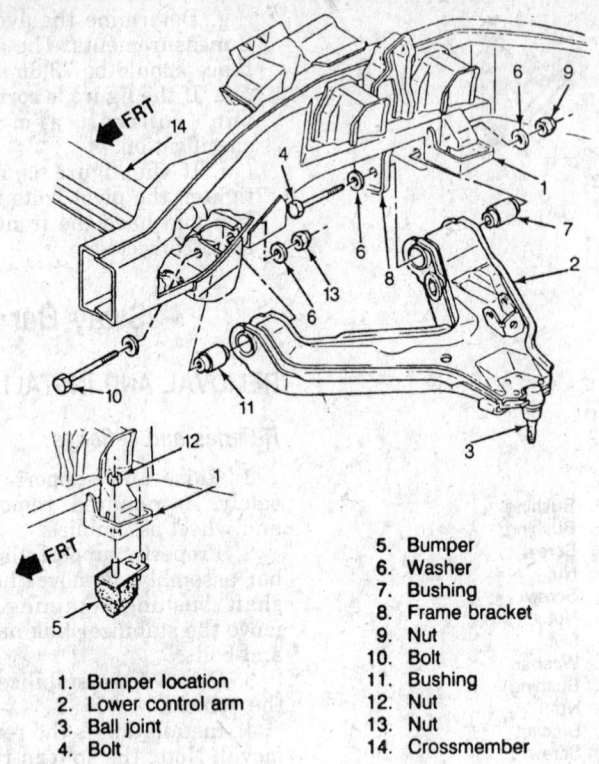

1. Bumper location
2. Lower control arm
3. Ball joint
4. Bolt
5. Bumper
6. Washer
7. Bushing
8. Frame bracket
9. Nut
10. Bolt
11. Bushing
12. Nut
13. Nut
14. Crossmember

79114114

Lower control arm installation — K series

4. Remove the bracket bolts and remove the stabilizer bar.

5. Inspect the bushings for wear or damage. Replace them as necessary.

6. Installation is the reverse of removal. Coat the bushings with silicone grease prior to assembly. The slit in the bushings faces the front of the vehicle. Torque the frame bracket bolts to 24 ft. lbs. (33 Nm); the end link nuts to 13 ft. lbs. (18 Nm).

K-Series

NOTE: The end link bushings, bolts and spacers are not interchangeable from left to right.

1. Raise and support the vehicle safely.

2. Remove the nuts, bolts, spacer and clamp from the stabilizer bar.

3. Remove the stabilizer bar and remove the insulator.

4. Inspect all parts for wear or damage and replace them as necessary.

5. Installation is the reverse of removal.

6. Unload the torsion bar tension using tool J-36202.

7. Coat the bushings with silicone grease prior to assembly. Torque the frame bracket bolts to 13 ft. lbs. (18 Nm); the end link nuts to 12 ft. lbs. (17 Nm).

Kingpins

REMOVAL AND INSTALLATION

1. Raise and support the vehicle safely.

2. Remove the tire and wheel assembly.

3. Remove the caliper, hub and rotor assembly.

4. Remove the backing plate and steering arm.

5. Remove the upper kingpin cap, gasket and brake hose bracket.

6. Remove the lower cap and gasket. Discard the gasket.

7. Remove the locking pin and nut from the axle.

8. Drive the kingpin out using a suitable drift punch. The spacer and bushings will also come out with the kingpin.

9. Remove the steering knuckle from the drive axle.

10. Remove the dust seal, shim and thrust bearing.

To install:

11. Install new bushings into the steering knuckle and ream the new bushings to 1.1804–1.1820 in. (29.982–30.022mm). Install the steering knuckle onto the axle.

12. Prelube the thrust bearing and install the bearing, shim, and dust seal.

13. Prelube the kingpin; install the kingpin and lock pin, inserting the spacers in the correct order. Torque the lock pin to 29 ft. lbs. (40 Nm).

14. Install the upper and lower gasket and kingpin caps. Install the upper brake hose bracket.

15. Install the steering arm and backing plate. Torque the bolts to 12 ft. lbs. (16 Nm) and the nuts to 230 ft. lbs. (312 Nm).

16. Install the hub, rotor and caliper assembly.

17. Install the tire and wheel assembly.

18. Lower the vehicle and check the front end alignment.

I-Beam Axle

REMOVAL AND INSTALLATION

1. Raise and support the vehicle safely. Properly support the front axle beam.

2. Remove the tire and wheel assembly.

3. Remove the steering arm, knuckle and spindle assembly.

4. Remove the shock absorber.

5. Remove the stabilizer link and bushings.

6. Remove the U-bolts from the axle.

7. Remove the U-bolt spacer and shock plate.

8. Remove the leaf spring, as required.

9. Remove the steering damper and remove the I-Beam axle.

10. Installation is the reverse of removal.

REAR SUSPENSION

Shock Absorbers

REMOVAL AND INSTALLATION

1. Raise and support the vehicle safely. Properly support the rear axle assembly. Remove the upper and lower shock absorber bolts.

2. Remove the shock absorber from the vehicle.

3. Installation is the reverse of the removal procedure.

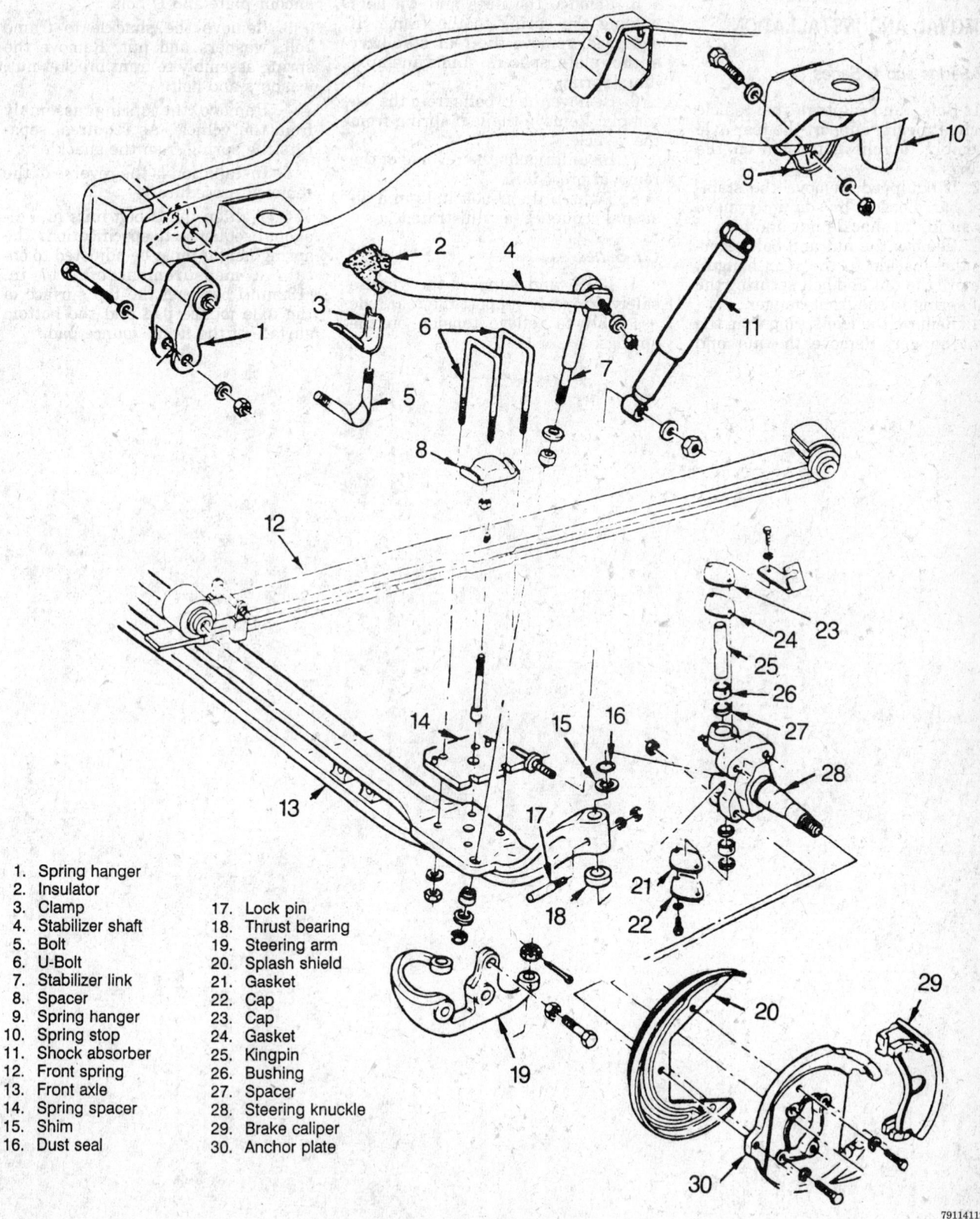

1. Spring hanger
2. Insulator
3. Clamp
4. Stabilizer shaft
5. Bolt
6. U-Bolt
7. Stabilizer link
8. Spacer
9. Spring hanger
10. Spring stop
11. Shock absorber
12. Front spring
13. Front axle
14. Spring spacer
15. Shim
16. Dust seal

17. Lock pin
18. Thrust bearing
19. Steering arm
20. Splash shield
21. Gasket
22. Cap
23. Cap
24. Gasket
25. Kingpin
26. Bushing
27. Spacer
28. Steering knuckle
29. Brake caliper
30. Anchor plate

79114112

Front I-Beam suspension — C3HD series

Leaf Spring

REMOVAL AND INSTALLATION

R/V-Series and G-Series

1. Raise and support the vehicle safely. Properly support the rear axle assembly to relieve tension on the springs.

2. If equipped, remove the stabilizer bar. Loosen, but do not remove the spring to shackle nut and bolt.

3. Remove the nut and bolt securing the shackle to the rear hanger. Remove the nut and bolt securing the leaf spring to the front hanger.

4. Remove the leaf spring from the front hanger. Remove the nut and bolt securing the shackle to the leaf spring. Remove the shackle.

5. Remove the nuts and washers holding the spring to the frame. If equipped, remove the rear stabilizer anchor plate, spacers, shims and auxiliary spring.

6. Remove the U-bolts from the assembly. Remove the leaf spring from the vehicle.

7. Installation is the reverse of the removal procedure.

8. Tighten the U-bolt nuts in a diagonal sequence, as illustrated.

C/K-Series

1. Raise and support the vehicle safely. Properly support the rear axle assembly to relieve tension on the springs.

2. Remove the shock absorber. Remove the U-bolt nuts, washers, anchor plate and U-bolt.

3. Remove the shackle to frame bolt, washers and nut. Remove the spring assembly to front bracket nut, washers and bolt.

4. Remove the spring assembly from the vehicle. As required, separate the spring from the shackle.

5. Installation is the reverse of the removal procedure.

6. Tighten the U-bolt nuts in a diagonal sequence to specification. The spring height must be adjusted to obtain a measurement of 7.17 in. (182mm) between the top surface of the axle jounce pad and the bottom surface of the frame jounce pad.

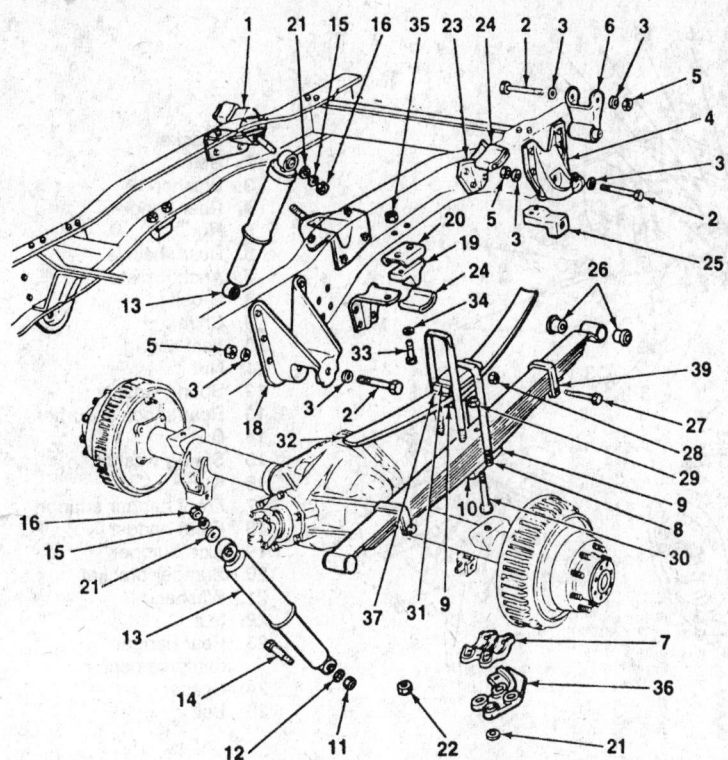

6. Rear shackle
7. Anchor plate
8. U-bolt
9. Shim
10. Leaf spring
11. Nut
12. Spring washer
13. Rear shock absorber
14. Bolt
15. Spring washer
16. Nut
17. Front hanger
18. Axle bumper
19. Bumper bracket
20. Washer
21. Nut
22. Bracket
23. Cushion
24. Rear hanger reinforcement
25. Leaf spring eye bushing
26. Bolt
27. Nut
28. Nut
29. Bolt
30. Spacer
31. Optional rear auxiliary spring
32. Bolt
33. Washer
34. Nut
35. Stabilizer bar anchor
36. Spacer
37. Spring clip

1. Bracket
2. Bolt
3. Washer
4. Rear hanger
5. Nut

79114116

Rear suspension assembly — R/V 30/3500 series

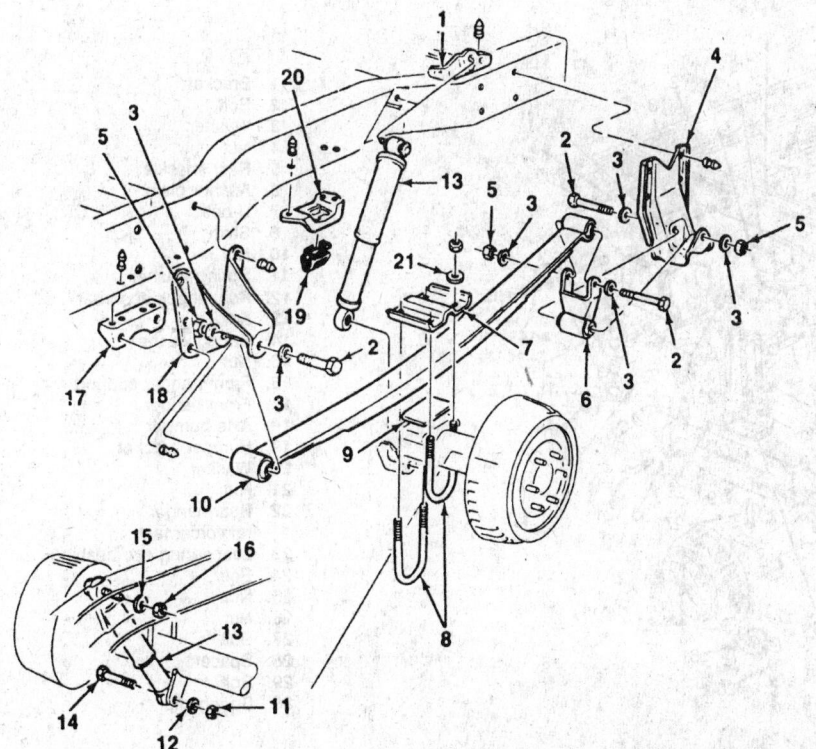

1. Bracket
2. Bolt
3. Washer
4. Rear hanger
5. Nut
6. Rear shackle
7. Anchor plate
8. U-bolt
9. Shim
10. Leaf spring
11. Nut
12. Spring washer
13. Rear shock absorber
14. Bolt
15. Spring washer
16. Nut
17. Front hanger support
18. Front hanger
19. Axle bumper
20. Bumper bracket
21. Washer
22. Nut

79114117

Rear suspension assembly — R/V 10/1500, 20/2500 series

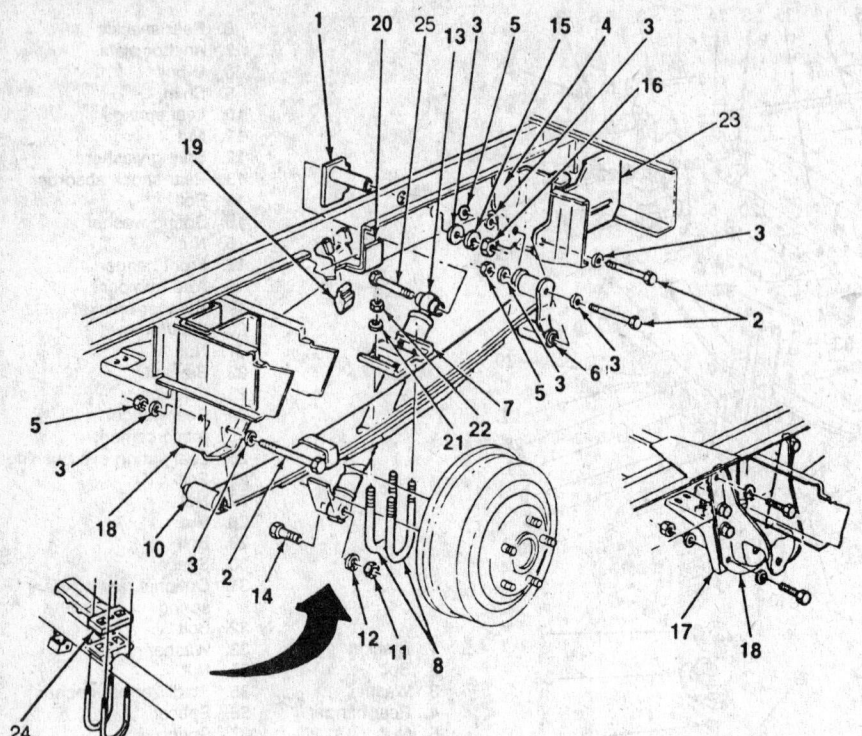

1. Bracket
2. Bolt
3. Washer
4. Rear hanger
5. Nut
6. Rear shackle
7. Anchor plate
8. U-bolt
9. Shim
10. Leaf spring
11. Nut
12. Spring washer
13. Rear shock absorber
14. Bolt
15. Spring washer
16. Nut
17. Front hanger support
18. Front hanger
19. Axle bumper
20. Bumper bracket
21. Washer
22. Nut
23. Rear hanger reinforcement
24. Spacer
25. Bolt

79114119

Rear suspension assembly — G10/1500 — G30/3500 series

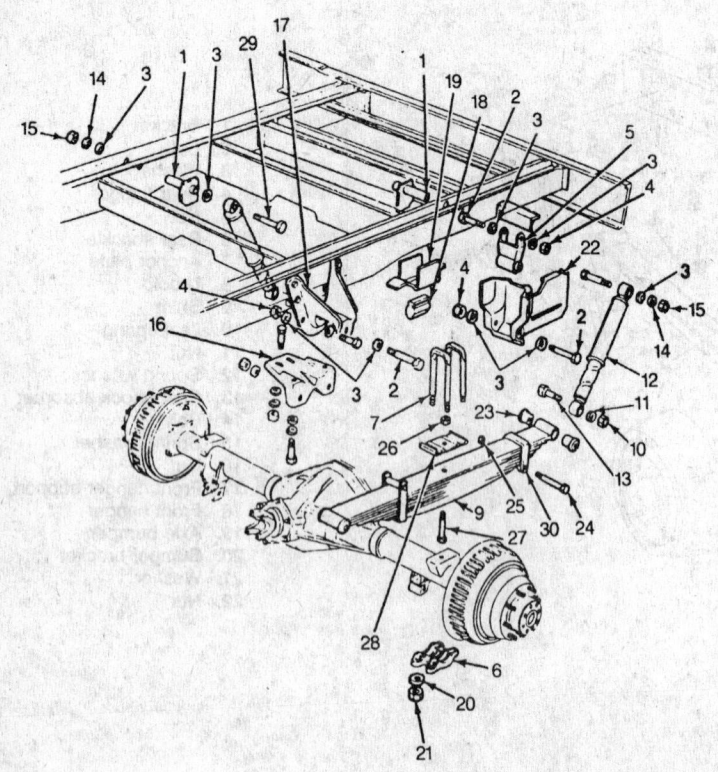

1. Bracket
2. Bolt
3. Washer
4. Rear hanger
5. Rear shackle
6. Anchor plate
7. U-bolt
8. Shim
10. Nut
11. Spring washer
12. Rear shock absorber
13. Bolt
14. Spring washer
15. Nut
16. Front hanger support
17. Front hanger
18. Axle bumper
19. Bumper bracket
20. Washer
21. Nut
22. Rear hanger reinforcement
23. Leaf spring eye bushing
24. Bolt
25. Nut
26. Nut
27. Bolt
28. Spacer
29. Bolt
30. Spring clip

79114120

Rear suspension assembly — G30/3500 series

GENERAL MOTORS CORP.

GM Mid-Size Trucks

5

CHEVY/GMC TRUCK—Astro/Safari • Blazer/Jimmy • Bravada • Cyclone • Lumina
APV • S10/S15 Pickup • Silhouette • Sonoma • TransSport • Typhoon

SPECIFICATIONS

VEHICLE IDENTIFICATION CHART

Code	Liters	Cu. In. (cc)	Cyl.	Fuel Sys.	Eng. Mfg.
4	2.2	134 (2189)	4	MFI	CPC
A	2.5	151 (2474)	4	TFI	CPC
E	2.5	151 (2474)	4	TFI	CPC-North
R	2.8	173 (2835)	6	TFI	CPC
D	3.1	191 (3130)	6	TFI	CPC
L	3.8	231 (3785)	6	MFI	CPC
B	4.3	263 (4293)	6	TFI	CPC
W	4.3	263 (4293)	6	TFI	CPC
Z	4.3	263 (4293)	6	TFI	CPC
H	5.0	305 (4999)	8	TFI	CPC
K	5.7	350 (5735)	8	TFI	CPC
C	6.2	379 (6210)	8	DSL	DDA
J	6.2	379 (6210)	8	DSL	DDA
F	6.5	395 (6473)	8	DSL	CPC
P	6.5	395 (6473)	8	DSL	CPC
S	6.5	395 (6473)	8	DSL	CPC
Y	6.5	395 (6473)	8	DSL	CPC
N	7.4	454 (7440)	8	TFI	CPC

Model Year	
Code	Year
M	1991
N	1992
P	1993
R	1994
S	1995

TFI - Throttle body fuel injection
MFI - Multiport fuel injection
DSL - Diesel
CPC - Chevrolet/Pontiac/Canada
DDA - Detroit Diesel Allison

79113C01

ENGINE IDENTIFICATION

Year	Model	Engine Displacement Liters (cc)	Engine Series (ID/VIN)	Fuel System	No. of Cylinders	Engine Type
1991	Astro/Safari	4.3 (4293)	B	TFI	6	OHV
	Astro/Safari	4.3 (4293)	Z	TFI	6	OHV
	Lumina/Silhouette/Trans Sport	3.1 (3130)	D	TFI	6	OHV
	S10 Blazer/S15 Jimmy	2.8 (2835)	R	TFI	6	OHV
	S10 Blazer/S15 Jimmy	4.3 (4293)	Z	TFI	6	OHV
	Bravada	4.3 (4293)	Z	TFI	6	OHV
	S10 Pick-up/S15 Sonoma	2.5 (2474)	A	TFI	4	OHV
	S10 Pick-up/S15 Sonoma	2.5 (2474)	E	TFI	4	OHV
	S10 Pick-up/S15 Sonoma	2.8 (2835)	R	TFI	6	OHV
	S10 Pick-up/S15 Sonoma	4.3 (4293)	Z	TFI	6	OHV
	Syclone	4.3 (4293)	Z	TFI-Turbo	6	OHV
	Blazer/Jimmy	5.7 (5735)	K	TFI	8	OHV
	Blazer/Jimmy	6.2 (6210)	C	DSL	8	OHV
	Blazer/Jimmy	6.2 (6210)	J	DSL	8	OHV
	C1500	4.3 (4293)	Z	TFI	6	OHV
	C1500	5.7 (5735)	K	TFI	8	OHV
	C1500	6.2 (6210)	C	DSL	8	OHV
	C1500	6.2 (6210)	J	DSL	8	OHV
	C1500	7.4 (7440)	N	TFI	8	OHV
	C2500	4.3 (4293)	Z	TFI	6	OHV
	C2500	5.0 (4999)	H	TFI	8	OHV
	C2500	5.7 (5735)	K	TFI	8	OHV
	C2500	6.2 (6210)	C	DSL	8	OHV
	C2500	6.2 (6210)	J	DSL	8	OHV
	C2500	7.4 (7440)	N	TFI	8	OHV
	C3500	5.0 (4999)	H	TFI	8	OHV
	C3500	5.7 (5735)	K	TFI	8	OHV
	C3500	6.2 (6210)	C	DSL	8	OHV
	C3500	6.2 (6210)	J	DSL	8	OHV
	C3500	7.4 (7440)	N	TFI	8	OHV
	K1500	4.3 (4293)	Z	TFI	6	OHV
	K1500	5.0 (4999)	H	TFI	8	OHV
	K1500	5.7 (5735)	K	TFI	8	OHV
	K1500	6.2 (6210)	C	DSL	8	OHV
	K2500	4.3 (4293)	Z	TFI	6	OHV
	K2500	5.0 (4999)	H	TFI	8	OHV
	K2500	5.7 (5735)	K	TFI	8	OHV
	K2500	6.2 (6210)	C	DSL	8	OHV
	K2500	6.2 (6210)	J	DSL	8	OHV
	K3500	5.7 (5735)	K	TFI	8	OHV
	K3500	6.2 (6210)	J	DSL	8	OHV
	K3500	7.4 (7441)	N	TFI	8	OHV
	P30	4.3 (4293)	Z	TFI	6	OHV
	P30	5.7 (5735)	K	TFI	8	OHV
	P30	6.2 (6210)	J	DSL	8	OHV
	P30	7.4 (7440)	N	TFI	8	OHV
	Suburban	5.7 (5735)	K	TFI	8	OHV

79113Co2

ENGINE IDENTIFICATION

Year	Model	Engine Displacement Liters (cc)	Engine Series (ID/VIN)	Fuel System	No. of Cylinders	Engine Type
	Suburban	6.2 (6210)	C	DSL	8	OHV
	Suburban	6.2 (6210)	J	DSL	8	OHV
	G10	4.3 (4293)	Z	TFI	6	OHV
	G10	5.0 (4999)	H	TFI	8	OHV
	G10	5.7 (5735)	K	TFI	8	OHV
	G20	4.3 (4293)	Z	TFI	6	OHV
	G20	5.0 (4999)	H	TFI	8	OHV
	G20	5.7 (5735)	K	TFI	8	OHV
	G30	4.3 (4293)	Z	TFI	6	OHV
	G30	5.7 (5735)	K	TFI	8	OHV
	G30	6.2 (6210)	J	DSL	8	OHV
	G30	7.4 (7440)	N	TFI	8	OHV
1992	Astro/Safari	4.3 (4293)	W	TFI	6	OHV
	Astro/Safari	4.3 (4293)	Z	TFI	6	OHV
	Lumina/Silhouette/Trans Sport	3.1 (3130)	D	TFI	6	OHV
	Lumina/Silhouette/Trans Sport	3.8 (3785)	L	MFI	6	OHV
	S10 Blazer/S15 Jimmy	4.3 (4293)	W	TFI	6	OHV
	S10 Blazer/S15 Jimmy	4.3 (4293)	Z	TFI	6	OHV
	Bravada	4.3 (4293)	W	TFI	6	OHV
	Bravada	4.3 (4293)	Z	TFI	6	OHV
	S10 Pick-up/Sonoma	2.5 (2474)	A	TFI	4	OHV
	S10 Pick-up/Sonoma	2.8 (2835)	R	TFI	6	OHV
	S10 Pick-up/Sonoma	4.3 (4293)	Z	TFI	6	OHV
	Syclone	4.3 (4293)	Z	TFI-Turbo	6	OHV
	Typhoon	4.3 (4293)	Z	TFI-Turbo	6	OHV
	Blazer/Yukon	5.7 (5735)	K	TFI	8	OHV
	C1500	4.3 (4293)	Z	TFI	6	OHV
	C1500	5.0 (4999)	H	TFI	8	OHV
	C1500	5.7 (5735)	K	TFI	8	OHV
	C1500	6.2 (6210)	C	DSL	8	OHV
	C1500	6.2 (6210)	J	DSL	8	OHV
	C1500	7.4 (7440)	N	TFI	8	OHV
	C2500	4.3 (4293)	Z	TFI	6	OHV
	C2500	5.0 (4999)	H	TFI	8	OHV
	C2500	5.7 (5735)	K	TFI	8	OHV
	C2500	6.2 (6210)	C	DSL	8	OHV
	C2500	6.2 (6210)	J	DSL	8	OHV
	C2500	6.5 (6473)	F	DSL	8	OHV
	C2500	7.4 (7440)	N	TFI	8	OHV
	C3500	5.0 (4999)	H	TFI	8	OHV
	C3500	5.7 (5735)	K	TFI	8	OHV
	C3500	6.2 (6210)	C	DSL	8	OHV
	C3500	6.2 (6210)	J	DSL	8	OHV
	C3500	6.5 (6473)	F	DSL	8	OHV
	C3500	7.4 (7440)	N	TFI	8	OHV
	K1500	4.3 (4293)	Z	TFI	6	OHV
	K1500	5.0 (4999)	H	TFI	8	OHV

79113Ca2

ENGINE IDENTIFICATION

Year	Model	Engine Displacement Liters (cc)	Engine Series (ID/VIN)	Fuel System	No. of Cylinders	Engine Type
	K1500	5.7 (5735)	K	TFI	8	OHV
	K1500	6.2 (6210)	C	DSL	8	OHV
	K1500	7.4 (7440)	N	TFI	8	OHV
	K2500	4.3 (4293)	Z	TFI	6	OHV
	K2500	5.0 (4999)	H	TFI	8	OHV
	K2500	5.7 (5735)	K	TFI	8	OHV
	K2500	6.2 (6210)	C	DSL	8	OHV
	K2500	6.2 (6210)	J	DSL	8	OHV
	K2500	6.5 (6473)	F	DSL	8	OHV
	K2500	7.4 (7440)	N	TFI	8	OHV
	K3500	5.0 (4999)	H	TFI	8	OHV
	K3500	5.7 (5735)	K	TFI	8	OHV
	K3500	6.2 (6210)	C	DSL	8	OHV
	K3500	6.2 (6210)	J	DSL	8	OHV
	K3500	7.4 (7441)	N	TFI	8	OHV
	P30	4.3 (4293)	Z	TFI	6	OHV
	P30	5.7 (5735)	K	TFI	8	OHV
	P30	6.2 (6210)	J	DSL	8	OHV
	P30	7.4 (7440)	N	TFI	8	OHV
	Suburban	5.7 (5735)	K	TFI	8	OHV
	Suburban	7.4 (7440)	N	TFI	8	OHV
	G10	4.3 (4293)	Z	TFI	6	OHV
	G10	5.0 (4999)	H	TFI	8	OHV
	G10	5.7 (5735)	K	TFI	8	OHV
	G20	4.3 (4293)	Z	TFI	6	OHV
	G20	5.0 (4999)	H	TFI	8	OHV
	G20	5.7 (5735)	K	TFI	8	OHV
	G30	4.3 (4293)	Z	TFI	6	OHV
	G30	5.7 (5735)	K	TFI	8	OHV
	G30	6.2 (6210)	C	DSL	8	OHV
	G30	6.2 (6210)	J	DSL	8	OHV
	G30	7.4 (7440)	N	TFI	8	OHV
1993	Astro/Safari	4.3 (4293)	W	TFI	6	OHV
	Astro/Safari	4.3 (4293)	Z	TFI	6	OHV
	Lumina/Silhouette/Trans Sport	3.1 (3097)	D	TFI	6	OHV
	Lumina/Silhouette/Trans Sport	3.8 (3785)	L	MFI	6	OHV
	S10 Blazer/S15 Jimmy	4.3 (4293)	W	TFI	6	OHV
	S10 Blazer/S15 Jimmy	4.3 (4293)	Z	TFI	6	OHV
	Bravada	4.3 (4293)	W	TFI	6	OHV
	Bravada	4.3 (4293)	Z	TFI	6	OHV
	S10 Pick-up/Sonoma	2.5 (2474)	A	TFI	4	OHV
	S10 Pick-up/Sonoma	2.8 (2835)	R	TFI	6	OHV
	S10 Pick-up/Sonoma	4.3 (4293)	W	TFI	6	OHV
	Typhoon	4.3 (4293)	Z	TFI-Turbo	6	OHV
	Blazer/Yukon	5.7 (5735)	K	TFI	8	OHV
	C1500	4.3 (4293)	Z	TFI	6	OHV
	C1500	5.0 (4999)	H	TFI	8	OHV

79113Cb2

ENGINE IDENTIFICATION

Year	Model	Engine Displacement Liters (cc)	Engine Series (ID/VIN)	Fuel System	No. of Cylinders	Engine Type
	C1500	5.7 (5735)	K	TFI	8	OHV
	C1500	6.2 (6210)	C	DSL	8	OHV
	C1500	6.2 (6210)	J	DSL	8	OHV
	C1500	7.4 (7440)	N	TFI	8	OHV
	C2500	4.3 (4293)	Z	TFI	6	OHV
	C2500	5.0 (4999)	H	TFI	8	OHV
	C2500	5.7 (5735)	K	TFI	8	OHV
	C2500	6.2 (6210)	C	DSL	8	OHV
	C2500	6.2 (6210)	J	DSL	8	OHV
	C2500	6.5 (6505)	F	DSL	8	OHV
	C2500	7.4 (7440)	N	TFI	8	OHV
	C3500	5.0 (4999)	H	TFI	8	OHV
	C3500	5.7 (5735)	K	TFI	8	OHV
	C3500	6.2 (6210)	C	DSL	8	OHV
	C3500	6.2 (6210)	J	DSL	8	OHV
	C3500	6.5 (6505)	F	DSL	8	OHV
	C3500	7.4 (7440)	N	TFI	8	OHV
	K1500	4.3 (4293)	Z	TFI	6	OHV
	K1500	5.0 (4999)	H	TFI	8	OHV
	K1500	5.7 (5735)	K	TFI	8	OHV
	K1500	6.2 (6210)	C	DSL	8	OHV
	K1500	6.2 (6210)	J	DSL	8	OHV
	K1500	7.4 (7440)	N	TFI	8	OHV
	K2500	4.3 (4293)	Z	TFI	6	OHV
	K2500	5.0 (4999)	H	TFI	8	OHV
	K2500	5.7 (5735)	K	TFI	8	OHV
	K2500	6.2 (6210)	C	DSL	8	OHV
	K2500	6.2 (6210)	J	DSL	8	OHV
	K2500	6.5 (6505)	F	DSL	8	OHV
	K2500	7.4 (7440)	N	TFI	8	OHV
	K3500	5.0 (4999)	H	TFI	8	OHV
	K3500	5.7 (5735)	K	TFI	8	OHV
	K3500	6.2 (6210)	C	DSL	8	OHV
	K3500	6.2 (6210)	J	DSL	8	OHV
	K3500	6.5 (6505)	F	DSL	8	OHV
	K3500	7.4 (7441)	N	TFI	8	OHV
	P30	4.3 (4293)	Z	TFI	6	OHV
	P30	5.7 (5735)	K	TFI	8	OHV
	P30	6.2 (6210)	J	DSL	8	OHV
	P30	7.4 (7440)	N	TFI	8	OHV
	Suburban	5.7 (5735)	K	TFI	8	OHV
	Suburban	7.4 (7440)	N	TFI	8	OHV
	G10	4.3 (4293)	Z	TFI	6	OHV
	G10	5.0 (4999)	H	TFI	8	OHV
	G10	5.7 (5735)	K	TFI	8	OHV
	G10	6.2 (6210)	C	DSL	8	OHV
	G10	6.2 (6210)	J	DSL	8	OHV

79113Cc2

ENGINE IDENTIFICATION

Year	Model	Engine Displacement Liters (cc)	Engine Series (ID/VIN)	Fuel System	No. of Cylinders	Engine Type
	G20	4.3 (4293)	Z	TFI	6	OHV
	G20	5.0 (4999)	H	TFI	8	OHV
	G20	5.7 (5735)	K	TFI	8	OHV
	G20	6.2 (6210)	C	DSL	8	OHV
	G20	6.2 (6210)	J	DSL	8	OHV
	G30	4.3 (4293)	Z	TFI	6	OHV
	G30	5.0 (4999)	H	TFI	8	OHV
	G30	5.7 (5735)	K	TFI	8	OHV
	G30	6.2 (6210)	C	DSL	8	OHV
	G30	6.2 (6210)	J	DSL	8	OHV
	G30	7.4 (7440)	N	TFI	8	OHV
1994-95	Astro/Safari	4.3 (4293)	z	TFI	6	OHV
	Astro/Safari	4.3 (4293)	Z	TFI	6	OHV
	Lumina/Silhouette/Trans Sport	3.1 (3097)	D	TFI	6	OHV
	Lumina/Silhouette/Trans Sport	3.8 (3785)	L	MFI	6	OHV
	S10 Blazer/S15 Jimmy	4.3 (4293)	W	TFI	6	OHV
	S10 Blazer/S15 Jimmy	4.3 (4293)	Z	TFI	6	OHV
	Bravada	4.3 (4293)	W	TFI	6	OHV
	Bravada	4.3 (4293)	Z	TFI	6	OHV
	S10 Pick-up/Sonoma	2.2 (2189)	4	TFI	4	OHV
	S10 Pick-up/Sonoma	4.3 (4293)	W	TFI	6	OHV
	S10 Pick-up/Sonoma	4.3 (4293)	Z	TFI	6	OHV
	Blazer/Yukon	5.7 (5735)	K	TFI	8	OHV
	Blazer/Yukon	6.5 (6505)	F	DSL	8	OHV
	C1500	4.3 (4293)	Z	TFI	6	OHV
	C1500	5.0 (4999)	H	TFI	8	OHV
	C1500	5.7 (5735)	K	TFI	8	OHV
	C1500	6.5 (6505)	F	DSL	8	OHV
	C1500	7.4 (7440)	N	TFI	8	OHV
	C2500	4.3 (4293)	Z	TFI	6	OHV
	C2500	5.0 (4999)	H	TFI	8	OHV
	C2500	5.7 (5735)	K	TFI	8	OHV
	C2500	6.5 (6505)	P	DSL	8	OHV
	C2500	6.5 (6505)	S	DSL	8	OHV
	C3500	5.7 (5735)	K	TFI	8	OHV
	C3500	6.5 (6505)	F	DSL	8	OHV
	C3500	6.5 (6505)	P	DSL	8	OHV
	C3500	6.5 (6505)	S	DSL	8	OHV
	C3500	7.4 (7440)	N	TFI	8	OHV
	K1500	4.3 (4293)	Z	TFI	6	OHV
	K1500	5.0 (4999)	H	TFI	8	OHV
	K1500	5.7 (5735)	K	TFI	8	OHV
	K1500	6.5 (6505)	F	DSL	8	OHV
	K1500	7.4 (7440)	N	TFI	8	OHV
	K2500	4.3 (4293)	Z	TFI	6	OHV
	K2500	5.0 (4999)	H	TFI	8	OHV
	K2500	5.7 (5735)	K	TFI	8	OHV

79113Cd2

ENGINE IDENTIFICATION

Year	Model	Engine Displacement Liters (cc)	Engine Series (ID/VIN)	Fuel System	No. of Cylinders	Engine Type
	K2500	6.5 (6505)	P	DSL	8	OHV
	K2500	6.5 (6505)	S	DSL	8	OHV
	K2500	6.5 (6505)	F	DSL	8	OHV
	K2500	7.4 (7440)	N	TFI	8	OHV
	K3500	5.7 (5735)	K	TFI	8	OHV
	K3500	6.5 (6505)	F	DSL	8	OHV
	K3500	6.5 (6505)	P	DSL	8	OHV
	K3500	6.5 (6505)	S	DSL	8	OHV
	K3500	7.4 (7441)	B	TFI	8	OHV
	P30	4.3 (4293)	Z	TFI	6	OHV
	P30	5.7 (5735)	K	TFI	8	OHV
	P30	6.5 (6505)	F	DSL	8	OHV
	P30	7.4 (7440)	N	TFI	8	OHV
	Suburban	5.7 (5735)	K	TFI	8	OHV
	Suburban	6.5 (6505)	P	DSL	8	OHV
	Suburban	6.5 (6505)	S	DSL	8	OHV
	Suburban	7.4 (7440)	N	TFI	8	OHV
	G10	4.3 (4293)	Z	TFI	6	OHV
	G10	5.0 (4999)	H	TFI	8	OHV
	G10	5.7 (5735)	K	TFI	8	OHV
	G10	6.5 (6505)	P	DSL	8	OHV
	G10	6.5 (6505)	Y	DSL	8	OHV
	G20	4.3 (4293)	Z	TFI	6	OHV
	G20	5.0 (4999)	H	TFI	8	OHV
	G20	5.7 (5735)	K	TFI	8	OHV
	G20	6.5 (6505)	P	DSL	8	OHV
	G20	6.5 (6505)	Y	DSL	8	OHV
	G30	4.3 (4293)	Z	TFI	6	OHV
	G30	5.0 (4999)	H	TFI	8	OHV
	G30	5.7 (5735)	K	TFI	8	OHV
	G30	6.5 (6505)	P	DSL	8	OHV
	G30	6.5 (6505)	Y	DSL	8	OHV
	G30	7.4 (7440)	N	TFI	8	OHV

TFI - Throttle body fuel injection
MFI - Multiport fuel injection
DSL - Diesel
OHV - Overhead valve

79113Ce2

GENERAL ENGINE SPECIFICATIONS

Year	Engine ID/VIN	Engine Displacement Liters (cc)	Fuel System Type	Net Horsepower @ rpm	Net Torque @ rpm (ft. lbs.)	Bore x Stroke (in.)	Compression Ratio	Oil Pressure @ rpm
1991	A	2.5 (2474)	TFI	105@4800	135@3200	4.00x3.00	8.3:1	41@2000
	E	2.5 (2474)	TFI	105@4800	135@3200	4.00x3.00	8.3:1	45@2000
	R	2.8 (2835)	TFI	125@4800	150@2200	3.56x3.04	8.5:1	50@2000
	D	3.1 (3130)	TFI	120@4400	175@2200	3.50x3.30	8.9:1	15@1100
	B	4.3 (4293)	TFI	170@4600	225@2200	4.00x3.48	9.3:1	18@2000
	Z	4.3 (4293)	TFI	145@4000	230@2400	4.00x3.48	9.3:1	18@2000
	H	5.0 (4999)	TFI	165@4400	240@2000	3.74x3.48	9.0:1	18@2000
	K	5.7 (5735)	TFI	3	4	4.00x3.48	8.5:1	18@2000
	C	6.2 (6210)	DSL	135@3600	240@2000	3.98x3.80	21.0:1	35@2000
	J	6.2 (6210)	DSL	135@3600	240@2000	3.98x3.80	21.0:1	35@2000
	N	7.4 (7440)	TFI	230@3600	385@1600	4.25x4.00	8.0:1	40@2000
1992	A	2.5 (2474)	TFI	105@4800	135@3200	4.00x3.00	8.3:1	41@2000
	R	2.8 (2835)	TFI	125@4800	150@2200	3.56x3.04	8.5:1	50@2000
	D	3.1 (3130)	TFI	120@4400	175@2200	3.50x3.30	8.9:1	15@1100
	L	3.8 (3785)	MFI	165@4300	220@3200	3.80x3.40	8.5:1	60@1850
	B	4.3 (4293)	TFI	170@4600	225@2200	4.00x3.48	9.3:1	18@2000
	W	4.3 (4293)	TFI	200@4500	260@3600	4.00x3.48	9.1:1	18@2000
	Z	4.3 (4293)	TFI	1	2	4.00x3.48	9.3:1	18@2000
	H	5.0 (4999)	TFI	165@4400	240@2000	3.74x3.48	9.0:1	18@2000
	K	5.7 (5735)	TFI	3	4	4.00x3.48	8.5:1	18@2000
	C	6.2 (6210)	DSL	130@3600	240@2000	3.98x3.80	21.3:1	35@2000
	J	6.2 (6210)	DSL	135@3600	240@2000	3.98x3.80	21.3:1	350@2000
	F	6.5 (6473)	DSL	5	6	4.05x3.80	21.0:1	40-45@2000
	N	7.4 (7440)	TFI	230@3600	385@1600	4.25x4.00	8.0:1	40@2000
1993	A	2.5 (2474)	TFI	105@4800	135@3200	4.00x3.00	8.3:1	41@2000
	R	2.8 (2835)	TFI	125@2800	150@2200	3.56x3.04	8.5:1	50@2000
	D	3.1 (3130)	TFI	120@4400	175@2200	3.50x3.30	8.9:1	15@1100
	L	3.8 (3785)	MFI	165@4300	220@3200	3.80x3.40	8.5:1	60@1850
	W	4.3 (4293)	TFI	200@4500	260@3600	4.00x3.48	9.1:1	18@2000
	Z	4.3 (4293)	TFI	1	2	4.00x3.48	9.3:1	18@2000
	H	5.0 (4999)	TFI	165@4400	240@2000	3.74x3.48	9.0:1	18@2000
	K	5.7 (5735)	TFI	3	4	4.00x3.48	8.5:1	18@2000
	C	6.2 (6210)	DSL	130@3600	240@2000	3.98x3.80	21.3:1	35@2000
	J	6.2 (6210)	DSL	135@3600	240@2000	3.98x3.80	21.3:1	35@2000
	F	6.5 (6473)	DSL	5	6	4.05x3.80	21.0:1	40-45@2000
	N	7.4 (7440)	TFI	230@3600	385@1600	4.25x4.00	8.0:1	40@2000

79113C03

GENERAL ENGINE SPECIFICATIONS

Year	Engine ID/VIN	Engine Displacement Liters (cc)	Fuel System Type	Net Horsepower @ rpm	Net Torque @ rpm (ft. lbs.)	Bore x Stroke (in.)	Compression Ratio	Oil Pressure @ rpm
1994-95	4	2.2 (2189)	MFI	118@5200	130@2800	3.50x3.46	9.0:1	56@3000
	D	3.1 (3130)	TFI	120@4400	175@2200	3.50x3.30	8.9:1	15@1100
	L	3.8 (3785)	MFI	165@4300	220@3200	3.80x3.40	8.5:1	60@1850
	W	4.3 (4293)	TFI	200@4500	260@3600	4.00x3.48	9.1:1	18@2000
	Z	4.3 (4293)	TFI	1	2	4.00x3.48	9.3:1	18@2000
	H	5.0 (4999)	TFI		240@2000	3.74x3.48	9.0:1	18@2000
	K	5.7 (5735)	TFI	3	4	4.00x3.48	8.5:1	18@2000
	F	6.5 (6473)	DSL	5	6	4.05x3.80	21.5:1	40-45@2000
	P	6.5 (6473)	DSL	190@3400	385@1700	4.06x3.82	21.5:1	40-45@2000
	S	6.5 (6473)	DSL	190@3400	275@1700	4.06x3.82	21.5:1	40-45@2000
	Y	6.5 (6473)	DSL	7	8	4.06x3.82	21.5:1	40-45@2000
	N	7.4 (7440)	TFI	230@3600	385@1600	4.25x4.00	8.0:1	40@2000

TFI - Throttle body fuel injection
MFI - Multiport fuel injection
DSL - Diesel
1 S10 and C/K Pick-ups: 160@4000
   C/K HD Pick-up: 155@4000
   G Van: 150@4000
1 S10: 230@2800
   C/K Pick-up: 235@2400
   C/K HD Pick-up and G-Van: 230@2400
3 Below 8500 GVWR: 210@4000
   Above 8500 GVWR: 190@4000

3 Below 8500 GVWR: 210@4000
   Above 8500 GVWR: 190@4000
4 Below 8500 GVWR: 300@2800
   Above 8500 GVWR: 300@2400
5 Below 15,000 GVWR: 180@3400
   Above 15,000 GVWR: 190@3400
6 Below 15,000 GVWR: 380@1700
   Above 15,000 GVWR: 360@1700

7 Below 8500 GVWR: 155@3600
   Above 8500 GVWR: 160@3600
8 Below 8500 GVWR: 275@1700
   Above 8500 GVWR: 290@1700

79113Ca3

GASOLINE ENGINE TUNE-UP SPECIFICATIONS

Year	Engine ID/VIN	Engine Displacement Liters (cc)	Spark Plugs Gap (in.)	Ignition Timing (deg.) MT	AT	Fuel Pump (psi)	Idle Speed (rpm) MT	AT	Valve Clearance In.	Ex.
1991	A	2.5 (2474)	0.060	1	1	9-13	1	1	HYD	HYD
	E	2.5 (2474)	0.060	1	1	9-13	1	1	HYD	HYD
	R	2.8 (2835)	0.040	1	1	9-13	1	1	HYD	HYD
	D	3.1 (3130)	0.045	1	1	9-13	1	1	HYD	HYD
	B	4.3 (4293)	0.035	1	1	9-13	1	1	HYD	HYD
	Z	4.3 (4293)	0.040	1	1	9-13	1	1	HYD	HYD
	H	5.0 (4999)	0.045	1	1	9-13	1	1	HYD	HYD
	K	5.7 (5735)	0.045	1	1	9-13	1	1	HYD	HYD
	N	7.4 (7440)	0.045	1	1	9-13	1	1	HYD	HYD
1992	A	2.5 (2474)	0.060	1	1	9-13	1	1	HYD	HYD
	R	2.8 (2835)	0.040	1	1	9-13	1	1	HYD	HYD
	D	3.1 (3130)	0.045	1	1	9-13	1	1	HYD	HYD
	L	3.8 (3785)	0.060	1	1	41-47 [2]	1	1	HYD	HYD
	B	4.3 (4293)	0.035	1	1	9-13	1	1	HYD	HYD
	W	4.3 (4293)	0.035	1	1	9-13	1	1	HYD	HYD
	Z	4.3 (4293)	0.035	1	1	9-13	1	1	HYD	HYD
	H	5.0 (4999)	0.045	1	1	9-13	1	1	HYD	HYD
	K	5.7 (5735)	0.045	1	1	9-13	1	1	HYD	HYD
	N	7.4 (7440)	0.045	1	1	9-13	1	1	HYD	HYD
1993	A	2.5 (2474)	0.060	1	1	9-13	1	1	HYD	HYD
	R	2.8 (2835)	0.040	1	1	9-13	1	1	HYD	HYD
	D	3.1 (3130)	0.045	1	1	9-13	1	1	HYD	HYD
	L	3.8 (3785)	0.060	1	1	41-47 [2]	1	1	HYD	HYD
	W	4.3 (4293)	0.035	1	1	9-13	1	1	HYD	HYD
	Z	4.3 (4293)	0.035	1	1	9-13	1	1	HYD	HYD
	H	5.0 (4999)	0.045	1	1	9-13	1	1	HYD	HYD
	K	5.7 (5735)	0.045	1	1	9-13	1	1	HYD	HYD
	N	7.4 (7440)	0.045	1	1	9-13	1	1	HYD	HYD
1994-95	4	2.2 (2189)	NA	1	1	9-13	1	1	HYD	HYD
	R	2.8 (2835)	0.040	1	1	9-13	1	1	HYD	HYD
	D	3.1 (3130)	0.045	1	1	9-13	1	1	HYD	HYD
	L	3.8 (3785)	0.060	1	1	41-47 [2]	1	1	HYD	HYD
	W	4.3 (4293)	0.035	1	1	9-13	1	1	HYD	HYD
	Z	4.3 (4293)	0.035	1	1	9-13	1	1	HYD	HYD
	H	5.0 (4999)	0.045	1	1	9-13	1	1	HYD	HYD
	K	5.7 (5735)	0.045	1	1	9-13	1	1	HYD	HYD
	N	7.4 (7440)	0.045	1	1	9-13	1	1	HYD	HYD

NOTE: The Vehicle Emission Control Information label often reflects specification changes made during production. The label figures must be used if they differ from those in this chart

1 Refer to underhood label for exact setting

2 With key on and engine off

79113C04

FIRING ORDERS

NOTE: To avoid confusion, always replace spark plug wires one at a time.

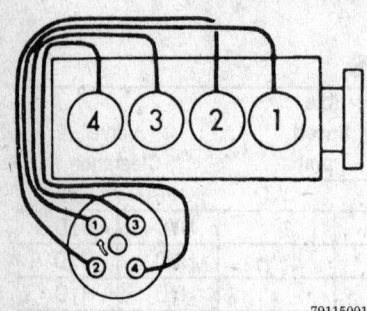

79115001

2.5L Engine
Engine Firing Order: 1 — 3 — 4 — 2
Distributor Rotation: Clockwise

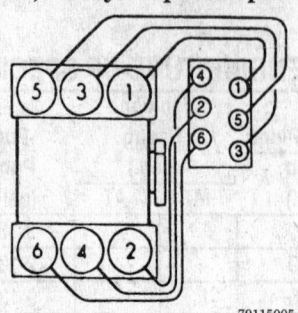

79115005

3.8L Engine
Engine Firing Order: 1 — 6 — 5 — 4 — 3 — 2
Distributorless Ignition System

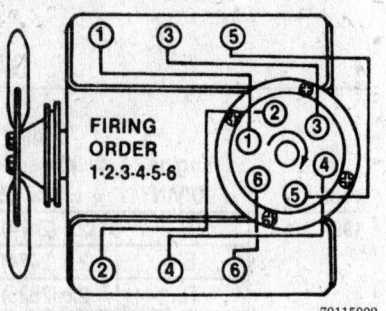

79115002

2.8L Engine
Engine Firing Order: 1 — 2 — 3 — 4 — 5 — 6
Distributor Rotation: Clockwise

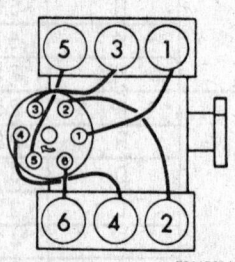

79115004

4.3L Engine
Engine Firing Order: 1 — 6 — 5 — 4 — 3 — 2
Distributor Rotation: Clockwise

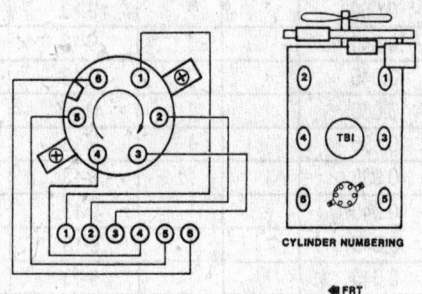

79115003

3.1L Engine
Engine Firing Order: 1 — 2 — 3 — 4 — 5 — 6
Distributor Rotation: Clockwise

CAPACITIES

Year	Model	Engine ID/VIN	Engine Displacement Liters (cc)	Engine Crankcase with Filter	Transmission (pts.)			Transfer Case (pts.)	Drive Axle		Fuel Tank (gal.)	Cooling System (qts.)
					4-Spd	5-Spd	Auto.		Front (pts.)	Rear (pts.)		
1991	Astro/Safari	B	4.3 (4293)	5.0	-	4.4	10.0	-	-	3.8	27.0	13.5 [8]
	Astro/Safari	Z	4.3 (4293)	5.0	-	4.4	10.0	-	-	3.8	27.0	13.5 [8]
	Lumina/Silhouette/Trans Sport	D	3.1 (3130)	4.5	-	-	8.0	-	-	-	18.0	13.4
	S10 Blazer/S15 Jimmy	R	2.8 (2835)	4.0	-	4.4	10.0	5	2.6	3.8	20.0	10.5
	S10 Blazer/S15 Jimmy	Z	4.3 (4293)	5.0	-	4.4	10.0	5	2.6	3.8	20.0	13.5
	Bravada	Z	4.3 (4293)	5.0	-	4.4	10.0	5	2.6	3.8	20.0	13.5
	S10 Pick-up/S15 Sonoma	A	2.5 (2474)	4.0	-	4.4	10.0	5	2.6	3.8	13.0 [12]	11.5
	S10 Pick-up/S15 Sonoma	E	2.5 (2474)	4.0	-	4.4	10.0	5	2.6	3.8	13.0 [12]	11.5
	S10 Pick-up/S15 Sonoma	R	2.8 (2835)	4.0	-	4.4	10.0	5	2.6	3.8	13.0 [12]	10.5
	S10 Pick-up/S15 Sonoma	Z	4.3 (4293)	5.0	-	4.4	10.0	5	2.6	3.8	20.0	13.5
	Syclone	Z	4.3 (4293)	5.0	-	-	10.0	-	2.6	3.8	20.0	13.5
	Blazer/Jimmy	K	5.7 (5735)	5.0	16	-	3	10	4.0	4	25.0 [5]	18.0
	Blazer/Jimmy	C	6.2 (6210)	7.0	16	-	3	10	4.0	4	27.0 [6]	25.0
	Blazer/Jimmy	J	6.2 (6210)	7.0	16	-	3	10	4.0	4	27.0 [6]	25.0
	C1500	Z	4.3 (4293)	5.0	16	3.6	3	-	-	4	9	10.9
	C1500	K	5.7 (5735)	5.0	16	3.6	3	-	-	4	9	18.0
	C1500	C	6.2 (6210)	7.0	16	3.6	3	-	-	4	9	25.0
	C1500	J	6.2 (6210)	7.0	16	3.6	3	-	-	4	9	25.0
	C1500	N	7.4 (7440)	6.0	16	3.6	3	-	-	4	9	25.0
	C2500	Z	4.3 (4293)	5.0	16	3.6	3	-	-	4	9	10.9
	C2500	H	5.0 (4999)	5.0	16	3.6	3	-	-	4	9	18.0
	C2500	K	5.7 (5735)	5.0	16	3.6	3	-	-	4	9	18.0
	C2500	C	6.2 (6210)	7.0	16	3.6	3	-	-	4	9	25.0
	C2500	J	6.2 (6210)	7.0	16	3.6	3	-	-	4	9	25.0
	C3500	H	5.0 (4999)	5.0	16	3.6	3	-	-	4	9	18.0
	C3500	K	5.7 (5735)	5.0	16	3.6	3	-	-	4	9	18.0
	C3500	C	6.2 (6210)	7.0	16	3.6	3	-	-	4	9	25.0
	C3500	J	6.2 (6210)	7.0	16	3.6	3	-	-	4	9	25.0
	C3500	N	7.4 (7440)	6.0	16	3.6	3	-	-	4	9	25.0
	K1500	Z	4.3 (4293)	5.0	16	-	3	-	-	4	9	10.9
	K1500	H	5.0 (4999)	5.0	16	-	3	-	-	4	9	18.0
	K1500	K	5.7 (5735)	5.0	16	-	3	-	-	4	9	18.0
	K1500	C	6.2 (6210)	7.0	16	-	3	-	-	4	9	25.0
	K2500	Z	4.3 (4293)	5.0	16	-	3	-	-	4	9	10.9
	K2500	H	5.0 (4999)	5.0	16	-	3	-	-	4	9	18.0
	K2500	K	5.7 (5735)	5.0	16	-	3	-	-	4	9	18.0
	K2500	C	6.2 (6210)	7.0	16	-	3	-	-	4	9	25.0
	K2500	J	6.2 (6210)	7.0	16	-	3	-	-	4	9	25.0
	K3500	K	5.7 (5735)	5.0	16	-	3	-	-	4	9	18.0
	K3500	N	5.7 (5735)	6.0	16	-	3	-	-	4	9	25.0
	K3500	J	6.2 (6210)	7.0	16	-	3	-	-	4	9	25.0
	R3500	K	5.7 (5735)	5.0	16	-	3	-	-	4	7	18.0
	R3500	J	6.2 (6210)	7.0	16	-	3	-	-	4	7	25.0
	R3500	N	7.4 (7440)	6.0	16	-	3	10	4.0	4	25.0 [10]	24.5
	P30	Z	4.3 (4293)	5.0	-	4.4	10.0	5	2.6	3.8	18.0	13.4
	P30	K	5.7 (5735)	5.0	16	3.6	3	-	-	4	7	18.0

79113C06

CAPACITIES

Year	Model	Engine ID/VIN	Engine Displacement Liters (cc)	Engine Crankcase with Filter	Transmission (pts.) 4-Spd	5-Spd	Auto.	Transfer Case (pts.)	Drive Axle Front (pts.)	Rear (pts.)	Fuel Tank (gal.)	Cooling System (qts.)
	P30	J	6.2 (6210)	7.0	16	3.6	3	-	-	13	14	25.0
	P30	N	7.4 (7440)	5.0	16	3.6	3	-	-	4	7	18.0
	Suburban	K	5.7 (5735)	5.0	16	-	3	10	4.0	4	25.0 10	18.0
	Suburban	C	6.2 (6210)	7.0	16	-	3	10	4.0	4	27.0 11	25.0
	Suburban	J	6.2 (6210)	7.0	16	-	3	10	4.0	4	27.0 11	25.0
	Suburban	N	7.4 (7440)	6.0	16	-	3	10	4.0	4	25.0 10	24.5
	G10	Z	4.3 (4293)	5.0	16	-	3	-	-	4	22.0 11	10.9
	G10	H	5.0 (4999)	5.0	16	-	3	-	-	4	22.0 11	17.0
	G10	K	5.7 (5735)	5.0	16	-	3	-	-	4	7	18.0
	G20	Z	4.3 (4293)	5.0	16	-	3	-	-	4	22.0 11	10.9
	G20	H	5.0 (4999)	5.0	16	-	3	-	-	4	22.0 11	17.0
	G20	K	5.7 (5735)	5.0	16	-	3	-	-	4	7	18.0
	G30	Z	4.3 (4293)	5.0	16	-	3	-	-	4	22.0 11	10.9
	G30	K	5.7 (5735)	5.0	16	-	3	-	-	4	7	18.0
	G30	J	6.2 (6210)	7.0	16	-	3	-	-	4	22.0 11	24.0
	G30	N	7.4 (7440)	6.0	16	-	3	-	-	13	14	24.5
1992	Astro/Safari	W	4.3 (4293)	5.0	-	4.4	10.0	-	-	3.8	27.0	13.5 8
	Astro/Safari	Z	4.3 (4293)	5.0	-	4.4	10.0	-	-	3.8	27.0	13.5 8
	Lumina/Silhouette/Trans Sport	D	3.1 (3130)	4.5	-	-	8.0	-	-	-	20.0	13.4
	Lumina/Silhouette/Trans Sport	L	3.8 (3785)	4.5	-	-	12.0	-	-	-	20.0	13.4
	S10 Blazer/S15 Jimmy	W	4.3 (4293)	4.5	-	-	10.0	-	3.5	3.5	20.0	12.0
	S10 Blazer/S15 Jimmy	Z	4.3 (4293)	5.0	-	4.4	10.0	5	3.5	3.5	20.0	12.1
	Bravada	W	4.3 (4293)	4.5	-	-	10.0	-	3.5	3.5	20.0	12.0
	Bravada	Z	4.3 (4293)	5.0	-	4.4	10.0	5	3.5	3.5	20.0	12.1
	S10 Pick-up/Sonoma	A	2.5 (2474)	4.0	-	4.4	10.0	5	3.5	3.5	13.0 12	11.5
	S10 Pick-up/Sonoma	Z	2.8 (2835)	4.0	-	4.4	10.0	5	3.5	3.5	13.0 12	10.5
	S10 Pick-up/Sonoma	R	4.3 (4293)	5.0	-	4.4	10.0	5	3.5	3.5	20.0	13.5
	Syclone	Z	4.3 (4293)	5.0	-	-	10.0	-	2.6	3.8	20.0	13.5
	Typhoon	Z	4.3 (4293)	5.0	-	-	10.0	-	2.6	3.8	20.0	13.5
	Blazer/Yukon	K	5.7 (5735)	5.0	-	-	3	10	4.0	4	25.0 5	18.0
	C1500	Z	4.3 (4293)	5.0	-	2	3	-	-	4	9	11.0
	C1500	H	5.0 (4999)	5.0	-	2	3	-	-	4	9	18.0
	C1500	K	5.7 (5735)	5.0	-	2	3	-	-	4	9	18.0
	C1500	C	6.2 (6210)	7.0	-	2	3	-	-	4	9	25.0
	C1500	J	6.2 (6210)	7.0	-	2	3	-	-	4	9	25.0
	C1500	N	7.4 (7440)	6.0	-	2	3	-	-	4	9	25.0
	C2500	Z	4.3 (4293)	5.0	-	2	3	-	-	4	9	11.0
	C2500	H	5.0 (4999)	5.0	-	2	3	-	-	4	9	18.0
	C2500	K	5.7 (5735)	5.0	-	2	3	-	-	4	9	18.0
	C2500	C	6.2 (6210)	7.0	-	2	3	-	-	4	9	25.0
	C2500	J	6.2 (6210)	7.0	-	2	3	-	-	4	9	25.0
	C2500	F	6.5 (6473)	7.0	-	2	3	-	-	4	9	26.5
	C3500	H	5.0 (4999)	5.0	-	2	3	-	-	4	9	18.0
	C3500	K	5.7 (5735)	5.0	-	2	3	-	-	4	9	18.0
	C3500	C	6.2 (6210)	7.0	-	2	3	-	-	4	9	25.0
	C3500	J	6.2 (6210)	7.0	-	2	3	-	-	4	9	25.0

79113Ca6

CAPACITIES

Year	Model	Engine ID/VIN	Engine Displacement Liters (cc)	Engine Crankcase with Filter	Transmission (pts.) 4-Spd	5-Spd	Auto.	Transfer Case (pts.)	Drive Axle Front (pts.)	Rear (pts.)	Fuel Tank (gal.)	Cooling System (qts.)
	C3500	F	6.5 (6473)	7.0	-	2	3	-	-	4	9	26.5
	C3500	N	7.4 (7440)	6.0	-	2	3	-	-	4	9	25.0 [1]
	K1500	Z	4.3 (4293)	5.0	-	2	3	-	-	4	9	11.0
	K1500	H	5.0 (4999)	5.0	-	2	3	-	-	4	9	18.0
	K1500	K	5.7 (5735)	5.0	-	2	3	-	-	4	9	18.0
	K1500	C	6.2 (6210)	7.0	-	2	3	-	-	4	9	25.0
	K1500	N	7.4 (7440)	6.0	-	2	3	-	-	4	9	25.0
	K2500	Z	4.3 (4293)	5.0	-	2	3	-	-	4	9	11.0
	K2500	H	5.0 (4999)	5.0	-	2	3	-	-	4	9	18.0
	K2500	K	5.7 (5735)	5.0	-	2	3	-	-	4	9	18.0
	K2500	C	6.2 (6210)	7.0	-	2	3	-	-	4	9	25.0
	K2500	J	6.2 (6210)	7.0	-	2	3	-	-	4	9	25.0
	K2500	N	7.4 (7440)	6.0	-	2	3	-	-	4	9	25.0 [1]
	K3500	K	5.7 (5735)	5.0	-	2	3	-	-	4	9	18.0
	K3500	N	7.4 (7440)	6.0	-	2	3	-	-	4	9	25.0 [1]
	K3500	J	6.2 (6210)	7.0	-	2	3	-	-	4	9	25.0
	P30	Z	4.3 (4293)	5.0	-	4.4	10.0	5	2.6	3.8	18.0	13.4
	P30	K	5.7 (5735)	5.0	-	3.6	3	-	-	4	7	18.0
	P30	J	6.2 (6210)	7.0	-	3.6	3	-	-	13	14	25.0
	P30	N	7.4 (7440)	5.0	-	3.6	3	-	-	4	7	18.0
	Suburban	K	5.7 (5735)	5.0	-	-	3	10	4.0	4	25.0 [10]	18.0
	Suburban	N	7.4 (7440)	6.0	-	-	3	10	4.0	4	25.0 [10]	24.5
	G10	Z	4.3 (4293)	5.0	-	-	3	-	-	4	22.0 [11]	11.0
	G10	H	5.0 (4999)	5.0	-	-	3	-	-	4	22.0 [11]	17.0
	G10	K	5.7 (5735)	5.0	-	-	3	-	-	4	7	18.0
	G20	Z	4.3 (4293)	5.0	-	-	3	-	-	4	22.0 [11]	11.0
	G20	H	5.0 (4999)	5.0	-	-	3	-	-	4	22.0 [11]	17.0
	G20	K	5.7 (5735)	5.0	-	-	3	-	-	4	7	18.0
	G30	Z	4.3 (4293)	5.0	-	-	3	-	-	4	22.0 [11]	11.0
	G30	K	5.7 (5735)	5.0	-	-	3	-	-	4	7	18.0
	G30	C	6.2 (6210)	7.0	-	-	3	-	-	4	22.0 [11]	24.0
	G30	J	6.2 (6210)	7.0	-	-	3	-	-	4	22.0 [11]	24.0
	G30	N	7.4 (7440)	6.0	-	-	3	-	-	13	14	24.5
1993	Astro/Safari	W	4.3 (4293)	5.0	-	4.4	10.0	-	-	3.8	27.0	13.5 [8]
	Astro/Safari	Z	4.3 (4293)	5.0	-	4.4	10.0	-	-	3.8	27.0	13.5 [8]
	Lumina/Silhouette/Trans Sport	D	3.1 (3130)	4.5	-	-	8.0	-	-	-	20.0	13.4
	Lumina/Silhouette/Trans Sport	L	3.8 (3785)	4.5	-	-	12.0	-	-	-	20.0	13.4
	S10 Blazer/S15 Jimmy	W	4.3 (4293)	4.5	-	-	10.0	-	3.5	3.5	20.0	12.0
	S10 Blazer/S15 Jimmy	Z	4.3 (4293)	5.0	-	4.4	10.0	5	3.5	3.5	20.0	12.1
	Bravada	W	4.3 (4293)	4.5	-	-	10.0	-	3.5	3.5	20.0	12.0
	Bravada	Z	4.3 (4293)	5.0	-	4.4	10.0	5	3.5	3.5	20.0	12.1
	S10 Pick-up/Sonoma	A	2.5 (2474)	4.0	-	4.4	10.0	5	3.5	3.5	13.0 [12]	11.5
	S10 Pick-up/Sonoma	R	2.8 (2835)	4.0	-	4.4	10.0	5	3.5	3.5	13.0 [12]	10.5
	S10 Pick-up/Sonoma	W	4.3 (4293)	4.5	-	4.4	10.0	5	3.5	3.5	20.0	12.0
	Typhoon	Z	4.3 (4293)	5.0	-	-	10.0	-	2.6	3.8	20.0	13.5
	Blazer/Yukon	K	5.7 (5735)	5.0	-	-	3	10	4.0	4	25.0 [5]	18.0

79113Cb6

CAPACITIES

Year	Model	Engine ID/VIN	Engine Displacement Liters (cc)	Engine Crankcase with Filter	Transmission (pts.)			Transfer Case (pts.)	Drive Axle		Fuel Tank (gal.)	Cooling System (qts.)
					4-Spd	5-Spd	Auto.		Front (pts.)	Rear (pts.)		
	C1500	Z	4.3 (4293)	5.0	-	2	3	-	-	4	9	11.0
	C1500	H	5.0 (4999)	5.0	-	2	3	-	-	4	9	18.0
	C1500	K	5.7 (5735)	5.0	-	2	3	-	-	4	9	18.0
	C1500	C	6.2 (6210)	7.0	-	2	3	-	-	4	9	25.0
	C1500	J	6.2 (6210)	7.0	-	2	3	-	-	4	9	25.0
	C1500	N	7.4 (7440)	.6.0	-	2	3	-	-	4	9	25.0
	C2500	Z	4.3 (4293)	5.0	-	2	3	-	-	4	9	11.0
	C2500	H	5.0 (4999)	5.0	-	2	3	-	-	4	9	18.0
	C2500	K	5.7 (5735)	5.0	-	2	3	-	-	4	9	18.0
	C2500	C	6.2 (6210)	7.0	-	2	3	-	-	4	9	25.0
	C2500	J	6.2 (6210)	7.0	-	2	3	-	-	4	9	25.0
	C2500	F	6.5 (6473)	7.0	-	2	3	-	-	4	9	26.5
	C3500	H	5.0 (4999)	5.0	-	2	3	-	-	4	9	18.0
	C3500	K	5.7 (5735)	5.0	-	2	3	-	-	4	9	18.0
	C3500	C	6.2 (6210)	7.0	-	2	3	-	-	4	9	25.0
	C3500	J	6.2 (6210)	7.0	-	2	3	-	-	4	9	25.0
	C3500	F	6.5 (6473)	7.0	-	2	3	-	-	4	9	26.5
	C3500	N	7.4 (7440)	6.0	-	2	3	-	-	4	9	25.0 [1]
	K1500	Z	4.3 (4293)	5.0	-	2	3	-	-	4	9	11.0
	K1500	H	5.0 (4999)	5.0	-	2	3	-	-	4	9	18.0
	K1500	K	5.7 (5735)	5.0	-	2	3	-	-	4	9	18.0
	K1500	C	6.2 (6210)	7.0	-	2	3	-	-	4	9	25.0
	K1500	N	7.4 (7440)	6.0	-	2	3	-	-	4	9	25.0 [1]
	K2500	Z	4.3 (4293)	5.0	-	2	3	-	-	4	9	11.0
	K2500	H	5.0 (4999)	5.0	-	2	3	-	-	4	9	18.0
	K2500	K	5.7 (5735)	5.0	-	2	3	-	-	4	9	18.0
	K2500	C	6.2 (6210)	7.0	-	2	3	-	-	4	9	25.0
	K2500	J	6.2 (6210)	7.0	-	2	3	-	-	4	9	25.0
	K2500	F	6.5 (6473)	7.0	-	2	3	-	-	4	9	26.5
	K2500	N	7.4 (7440)	6.0	-	2	3	-	-	4	9	25.0 [1]
	K2500	H	5.0 (4999)	5.0	-	2	3	-	-	4	9	18.0
	K3500	K	5.7 (5735)	5.0	-	2	3	-	-	4	9	18.0
	K3500	C	6.2 (6210)	7.0	-	2	3	-	-	4	9	25.0
	K3500	J	6.2 (6210)	7.0	-	2	3	-	-	4	9	25.0
	K3500	F	6.5 (6473)	7.0	-	2	3	-	-	4	9	26.5
	K3500	N	7.4 (7440)	6.0	-	2	3	-	-	4	9	25.0 [1]
	P30	Z	4.3 (4293)	5.0	-	4.4	10.0	5	2.6	3.8	18.0	13.4
	P30	K	5.7 (5735)	5.0	-	3.6	3	-	-	4	7	18.0
	P30	J	6.2 (6210)	7.0	-	3.6	3	-	-	13	14	25.0
	P30	N	7.4 (7440)	5.0	-	3.6	3	-	-	4	7	18.0
	Suburban	K	5.7 (5735)	5.0	-	-	3	10	4.0	4	25.0 [10]	18.0
	Suburban	N	7.4 (7440)	6.0	-	-	3	10	4.0	4	25.0 [10]	24.5
	G10	Z	4.3 (4293)	5.0	-	-	3	-	-	4	22.0 [11]	11.0
	G10	H	5.0 (4999)	5.0	-	-	3	-	-	4	22.0 [11]	17.0
	G10	K	5.7 (5735)	5.0	-	-	3	-	-	4	7	18.0
	G10	C	6.2 (6210)	7.0	-	-	3	-	-	4	22.0 [11]	24.0

79113Cc6

CAPACITIES

Year	Model	Engine ID/VIN	Engine Displacement Liters (cc)	Engine Crankcase with Filter	Transmission (pts.)			Transfer Case (pts.)	Drive Axle		Fuel Tank (gal.)	Cooling System (qts.)
					4-Spd	5-Spd	Auto.		Front (pts.)	Rear (pts.)		
	G10	J	6.2 (6210)	7.0	-	-	3	-	-	4	22.0 [11]	24.0
	G20	Z	4.3 (4293)	5.0	-	-	3	-	-	4	22.0 [11]	11.0
	G20	H	5.0 (4999)	5.0	-	-	3	-	-	4	22.0 [11]	17.0
	G20	K	5.7 (5735)	5.0	-	-	3	-	-	4	7	18.0
	G20	C	6.2 (6210)	7.0	-	-	3	-	-	4	22.0 [11]	24.0
	G20	J	6.2 (6210)	7.0	-	-	3	-	-	4	22.0 [11]	24.0
	G30	Z	4.3 (4293)	5.0	-	-	3	-	-	4	22.0 [11]	11.0
	G30	K	5.7 (5735)	5.0	-	-	3	-	-	4	7	18.0
	G30	C	6.2 (6210)	7.0	-	-	3	-	-	4	22.0 [11]	24.0
	G30	J	6.2 (6210)	7.0	-	-	3	-	-	4	22.0 [11]	24.0
	G30	N	7.4 (7440)	6.0	-	-	3	-	-	13	14	24.5
1994-95	Astro/Safari	W	4.3 (4293)	5.0	-	4.4	10.0	-	-	3.8	27.0	13.5 [8]
	Astro/Safari	Z	4.3 (4293)	5.0	-	4.4	10.0	-	-	3.8	27.0	13.5 [8]
	Lumina/Silhouette/Trans Sport	D	3.1 (3130)	4.5	-	-	8.0	-	-	-	20.0	13.4
	Lumina/Silhouette/Trans Sport	L	3.8 (3785)	4.5	-	-	12.0	-	-	-	20.0	13.4
	S10 Blazer/S15 Jimmy	W	4.3 (4293)	4.5	-	-	10.0	-	3.5	3.5	20.0	12.0
	S10 Blazer/S15 Jimmy	Z	4.3 (4293)	5.0	-	4.4	10.0	5	3.5	3.5	20.0	12.1
	Bravada	W	4.3 (4293)	4.5	-	-	10.0	-	3.5	3.5	20.0	12.0
	Bravada	Z	4.3 (4293)	5.0	-	4.4	10.0	5	3.5	3.5	20.0	12.1
	S10 Pick-up/Sonoma	4	2.2 (2189)	4.0	-	4.4	10.0	5	3.5	3.5	13.0 [12]	11.5
	S10 Pick-up/Sonoma	W	4.3 (4293)	4.5	-	4.4	10.0	5	3.5	3.5	20.0	12.0
	S10 Pick-up/Sonoma	Z	4.3 (4293)	5.0	-	4.4	10.0	5	3.5	3.5	20.0	12.0
	Blazer/Yukon	K	5.7 (5735)	5.0	-	-	3	10	4.0	4	25.0 [5]	18.0
	Blazer/Yukon	F	6.5 (6473)	7.0	-	2	3	-	-	4	9	26.5
	C1500	Z	4.3 (4293)	5.0	-	2	3	-	-	4	9	11.0
	C1500	H	5.0 (4999)	5.0	-	2	3	-	-	4	9	18.0
	C1500	K	5.7 (5735)	5.0	-	2	3	-	-	4	9	18.0
	C1500	F	6.5 (6473)	7.0	-	2	3	-	-	4	9	26.5
	C1500	N	7.4 (7440)	6.0	-	2	3	-	-	4	9	25.0
	C2500	Z	4.3 (4293)	5.0	-	2	3	-	-	4	9	11.0
	C2500	H	5.0 (4999)	5.0	-	2	3	-	-	4	9	18.0
	C2500	K	5.7 (5735)	5.0	-	2	3	-	-	4	9	18.0
	C2500	P	6.5 (6473)	7.0	-	2	3	-	-	4	9	26.5
	C2500	S	6.5 (6473)	7.0	-	2	3	-	-	4	9	26.5
	C3500	H	5.0 (4999)	5.0	-	2	3	-	-	4	9	18.0
	C3500	K	5.7 (5735)	5.0	-	2	3	-	-	4	9	18.0
	C3500	F	6.5 (6473)	7.0	-	2	3	-	-	4	9	26.5
	C3500	P	6.5 (6473)	7.0	-	2	3	-	-	4	9	26.5
	C3500	S	6.5 (6473)	7.0	-	2	3	-	-	4	9	26.5
	C3500	N	7.4 (7440)	6.0	-	2	3	-	-	4	9	25.0 [1]
	K1500	Z	4.3 (4293)	5.0	-	2	3	-	-	4	9	11.0
	K1500	H	5.0 (4999)	5.0	-	2	3	-	-	4	9	18.0
	K1500	K	5.7 (5735)	5.0	-	2	3	-	-	4	9	18.0
	K1500	F	6.5 (6505)	7.0	-	2	3	-	-	4	9	25.0
	K1500	N	7.4 (7440)	6.0	-	2	3	-	-	4	9	25.0 [1]
	K2500	Z	4.3 (4293)	5.0	-	2	3	-	-	4	9	11.0

79113Cd6

CAPACITIES

Year	Model	Engine ID/VIN	Engine Displacement Liters (cc)	Engine Crankcase with Filter	Transmission (pts.) 4-Spd	5-Spd	Auto.	Transfer Case (pts.)	Drive Axle Front (pts.)	Rear (pts.)	Fuel Tank (gal.)	Cooling System (qts.)
	K2500	H	5.0 (4999)	5.0	-	2	3	-	-	4	9	18.0
	K2500	K	5.7 (5735)	5.0	-	2	3	-	-	4	9	18.0
	K2500	P	6.5 (6505)	7.0	-	2	3	-	-	4	9	25.0
	K2500	S	6.5 (6505)	7.0	-	2	3	-	-	4	9	25.0
	K2500	F	6.5 (6473)	7.0	-	2	3	-	-	4	9	26.5
	K2500	N	7.4 (7440)	6.0	-	2	3	-	-	4	9	25.0 [1]
	K2500	H	5.0 (4999)	5.0	-	2	3	-	-	4	9	18.0
	K3500	K	5.7 (5735)	5.0	-	2	3	-	-	4	9	18.0
	K3500	P	6.5 (6505)	7.0	-	2	3	-	-	4	9	25.0
	K3500	S	6.5 (6505)	7.0	-	2	3	-	-	4	9	25.0
	K3500	F	6.5 (6473)	7.0	-	2	3	-	-	4	9	26.5
	K3500	N	7.4 (7440)	6.0	-	2	3	-	-	4	9	25.0 [1]
	P30	Z	4.3 (4293)	5.0	-	4.4	10.0	5	2.6	3.8	18.0	13.4
	P30	K	5.7 (5735)	5.0	-	3.6	3	-	-	4	7	18.0
	P30	F	6.5 (6505)	7.0	-	3.6	3	-	-	13	14	25.0
	P30	N	7.4 (7440)	5.0	-	3.6	3	-	-	4	7	18.0
	Suburban	K	5.7 (5735)	5.0	-	-	3	10	4.0	4	25.0 [10]	18.0
	Suburban	P	6.5 (6505)	7.0	-	2	3	-	-	4	9	25.0
	Suburban	S	6.5 (6505)	7.0	-	2	3	-	-	4	9	25.0
	Suburban	N	7.4 (7440)	6.0	-	-	3	10	4.0	4	25.0 [10]	24.5
	G10	Z	4.3 (4293)	5.0	-	2	3	-	-	4	22.0 [11]	11.0
	G10	H	5.0 (4999)	5.0	-	2	3	-	-	4	22.0 [11]	17.0
	G10	K	5.7 (5735)	5.0	-	2	3	-	-	4	7	18.0
	G10	P	6.5 (6505)	7.0	-	2	3	-	-	4	22.0 [11]	24.0
	G10	S	6.5 (6505)	7.0	-	2	3	-	-	4	22.0 [11]	24.0
	G20	Z	4..3 (4293)	5.0	-	2	3	-	-	4	22.0 [11]	11.0
	G20	H	5.0 (4999)	5.0	-	2	3	-	-	4	22.0 [11]	17.0
	G20	K	5.7 (5735)	5.0	-	2	3	-	-	4	7	18.0
	G20	P	6.5 (6505)	7.0	-	2	3	-	-	4	22.0 [11]	24.0
	G20	Y	6.5 (6505)	7.0	-	2	3	-	-	4	22.0 [11]	24.0
	G30	Z	4.3 (4293)	5.0	-	2	3	-	-	4	22.0 [11]	11.0
	G30	K	5.7 (5735)	5.0	-	2	3	-	-	4	7	18.0
	G30	P	6.5 (6505)	7.0	-	2	3	-	-	4	22.0 [11]	24.0
	G30	Y	6.5 (6505)	7.0	-	2	3	-	-	4	22.0 [11]	24.0

1 3500HD: 28.5 qts. capacity

2 New Venture gear 4500: 8.0 pts.
New Venture gear 5LM60: 4.4 pts.

3 350C trans.: 6.3 pts.
THM400 and 4L80 trans.: 9.0 pts.
THM700 R4 and 4L60 trans.: 10.0 pts.

4 8.5" ring gear: 4.2 pts.
9.5" ring gear: 6.5 pts.
9.75" ring gear: 6.0 pts.
10.5" ring gear: 6.5 pts.

5 Available with optional 31 gallon tank

6 Available with optional 32 gallon tank

7 Short bed: 16 gals.; Long bed: 20 gals.

8 16.5 qts. with rear heater

9 Std. available with 25 and 34 gallon tanks
Chassis cab available with 22, 30 and 34 gallon tanks

10 Available 31 and 40 gallon tanks

11 Available 32 and 41 gallon tanks

12 Available with 20 gallon tank

13 8.5" ring gear: 4.2 pts.
9.5" ring gear: 6.5 pts.
Chevrolet 10.5" ring gear: 6.5 pts.
Dana 9.75" ring gear: 6.0 pts.
Rockwell 12" ring gear: 12.5 pts.

14 Available with a variety of fuel tanks

15 13 qts. with rear heater

16 85mm: 3.6 pts.; 117mm: 8.4 pts.

79113Ce6

CAMSHAFT SPECIFICATIONS
All measurements given in inches.

Year	Engine ID/VIN	Engine Displacement Liters (cc)	Journal Diameter					Elevation		Bearing Clearance	Camshaft End Play
			1	2	3	4	5	In.	Ex.		
1991	A	2.5 (2474)	1.869	1.869	1.869	1.869	1.869	0.251	0.251	0.0007-0.0027	0.0015-0.0050
	E	2.5 (2474)	1.869	1.869	1.869	1.869	1.869	0.398	0.398	0.0007-0.0027	0.0015-0.0050
	R	2.8 (2835)	1.868-1.869	1.868-1.869	1.868-1.869	1.868-1.869	1.868-1.869	0.231	0.263	0.0010-0.0040	NA
	D	3.1 (3130)	1.868-1.882	1.868-1.882	1.868-1.882	1.868-1.882	NA	0.231	0.262	0.0010-0.0040	NA
	B	4.3 (4239)	1.8682-1.8692	1.8682-1.8692	1.8682-1.8692	1.8682-1.8692	NA	0.269	0.276	NA	0.004-0.012
	Z	4.3 (4293)	1.8682-1.8692	1.8682-1.8692	1.8682-1.8692	1.8682-1.8692	NA	0.357	0.390	0.0010-0.0030	0.004-0.012
	H	5.0 (4999)	1.8682-1.8692	1.8682-1.8692	1.8682-1.8692	1.8682-1.8692	NA	0.2484	0.2667	NA	0.004-0.012
	K	5.7 (5735)	1.8682-1.8692	1.8682-1.8692	1.8682-1.8692	1.8682-1.8692	NA	0.2600	0.2733	NA	0.004-0.012
	C	6.2 (6210)	2.1633-2.1642	2.1633-2.1642	2.1633-2.1642	2.1633-2.1642	2.0067-2.0089	2.808	2.808	1	NA
	J	6.2 (6210)	2.1633-2.1642	2.1633-2.1642	2.1633-2.1642	2.1633-2.1642	2.0067-2.0089	2.808	2.808	1	NA
	N	7.4 (7440)	1.9482-1.9492	1.9482-1.9492	1.9482-1.9492	1.9482-1.9492	1.9482-1.9492	0.2341-0.2345	0.2529-0.2531	NA	NA
1992	A	2.5 (2474)	1.869	1.869	1.869	1.869	1.869	0.251	0.251	0.0007-0.0027	0.0015-0.0050
	R	2.8 (2835)	1.868-1.870	1.868-1.870	1.868-1.870	1.868-1.870	1.868-1.870	0.262	0.273	0.0010-0.0040	NA
	D	3.1 (3130)	1.868-1.882	1.868-1.882	1.868-1.882	1.868-1.882	NA	0.231	0.262	0.0010-0.0040	NA
	L	3.8 (3785)	1.785-1.786	1.785-1.786	1.785-1.786	1.785-1.786	NA	0.250	0.255	0.0005-0.0035	NA
	B	4.3 (4293)	1.8682-1.8692	1.8682-1.8692	1.8682-1.8692	1.8682-1.8692	NA	0.269	0.276	NA	0.004-0.012
	W	4.3 (4293)	1.8682-1.8692	1.8682-1.8692	1.8682-1.8692	1.8682-1.8692	NA	0.288	0.294	NA	0.001-0.009
	Z	4.3 (4293)	1.8682-1.8692	1.8682-1.8692	1.8682-1.8692	1.8682-1.8692	NA	0.234	0.257	0.0010-0.0030	0.004-0.012
	H	5.0 (4999)	1.8682-1.8692	1.8682-1.8692	1.8682-1.8692	1.8682-1.8692	1.8682-1.8692	0.2336	0.2565	NA	0.004-0.012
	K	5.7 (5735)	1.8682-1.8692	1.8682-1.8692	1.8682-1.8692	1.8682-1.8692	1.8682-1.8692	0.2565	0.2690	NA	0.004-0.012
	C	6.2 (6210)	2.1633-2.1642	2.1633-2.1642	2.1633-2.1642	2.1633-2.1642	2.0067-2.0089	0.2808	0.2808	0.0010-0.0040	0.002-0.012
	J	6.2 (6210)	2.1633-2.1642	2.1633-2.1642	2.1633-2.1642	2.1633-2.1642	2.0067-2.0089	0.2808	0.2808	0.0010-0.0040	0.002-0.012
	F	6.5 (6473)	2.1642-2.1663	2.1642-2.1663	2.1642-2.1663	2.1642-2.1663	2.0067-2.0089	0.2808	0.2808	1	0.002-0.012

79113C07

CAMSHAFT SPECIFICATIONS
All measurements given in inches.

| Year | Engine ID/VIN | Engine Displacement Liters (cc) | Journal Diameter | | | | | Elevation | | Bearing Clearance | Camshaft End Play |
			1	2	3	4	5	In.	Ex.		
	N	7.4 (7440)	1.9482-1.9492	1.9482-1.9492	1.9482-1.9492	1.9482-1.9492	1.9482-1.9492	0.2341-0.2345	0.2529-0.2531	NA	NA
1993	A	2.5 (2474)	1.869	1.869	1.869	1.869	1.869	0.251	0.251	0.0007-0.0027	0.0015-0.0050
	R	2.8 (2835)	1.868-1.870	1.868-1.870	1.868-1.870	1.868-1.870	1.868-1.870	0.262	0.273	0.0010-0.0040	NA
	D	3.1 (3130)	1.868-1.882	1.868-1.882	1.868-1.882	1.868-1.882	NA	0.231	0.262	0.0010-0.0040	NA
	L	3.8 (3785)	1.785-1.786	1.785-1.786	1.785-1.786	1.785-1.786	NA	0.250	0.255	0.0005-0.0035	NA
	W	4.3 (4293)	1.8682-1.8692	1.8682-1.8692	1.8682-1.8692	1.8682-1.8692	NA	0.288	0.294	NA	0.001-0.009
	Z	4.3 (4293)	1.8682-1.8692	1.8682-1.8692	1.8682-1.8692	1.8682-1.8692	NA	0.234	0.257	0.0010-0.0030	0.004-0.012
	H	5.0 (4999)	1.8682-1.8692	1.8682-1.8692	1.8682-1.8692	1.8682-1.8692	1.8682-1.8692	0.2336	0.2565	NA	0.004-0.012
	K	5.7 (5735)	1.8682-1.8692	1.8682-1.8692	1.8682-1.8692	1.8682-1.8692	1.8682-1.8692	0.2565	0.2690	NA	0.004-0.012
	C	6.2 (6210)	2.1633-2.1642	2.1633-2.1642	2.1633-2.1642	2.1633-2.1642	2.0067-2.0089	0.2808	0.2808	0.0010-0.0040	0.002-0.012
	J	6.2 (6210)	2.1633-2.1642	2.1633-2.1642	2.1633-2.1642	2.1633-2.1642	2.0067-2.0089	0.2808	0.2808	0.0010-0.0040	0.002-0.012
	F	6.5 (6473)	2.1642-2.1663	2.1642-2.1663	2.1642-2.1663	2.1642-2.1663	2.0067-2.0089	0.2808	0.2808	1	0.002-0.012
	N	7.4 (7440)	1.9482-1.9492	1.9482-1.9492	1.9482-1.9492	1.9482-1.9492	1.9482-1.9492	0.2341-0.2345	0.2529-0.2531	NA	NA
1994-95	4	2.2 (2189)	1.869	1.869	1.869	1.869	1.869	0.251	0.251	0.0007-0.0027	0.0015-0.0050
	D	3.1 (3130)	1.868-1.882	1.868-1.882	1.868-1.882	1.868-1.882	NA	0.231	0.262	0.0010-0.0040	NA
	L	3.8 (3785)	1.785-1.786	1.785-1.786	1.785-1.786	1.785-1.786	NA	0.250	0.255	0.0005-0.0035	NA
	W	4.3 (4293)	1.8682-1.8692	1.8682-1.8692	1.8682-1.8692	1.8682-1.8692	NA	0.288	0.294	NA	0.001-0.009
	Z	4.3 (4293)	1.8682-1.8692	1.8682-1.8692	1.8682-1.8692	1.8682-1.8692	NA	0.234	0.257	0.0010-0.0030	0.004-0.012
	H	5.0 (4999)	1.8682-1.8692	1.8682-1.8692	1.8682-1.8692	1.8682-1.8692	1.8682-1.8692	0.2336	0.2565	NA	0.004-0.012
	K	5.7 (5735)	1.8682-1.8692	1.8682-1.8692	1.8682-1.8692	1.8682-1.8692	1.8682-1.8692	0.2565	0.2690	NA	0.004-0.012
	F	6.5 (6473)	2.1642-2.1663	2.1642-2.1663	2.1642-2.1663	2.1642-2.1663	2.0067-2.0089	0.2808	0.2808	1	0.002-0.012
	P	6.5 (6473)	2.1642-2.1663	2.1642-2.1663	2.1642-2.1663	2.1642-2.1663	2.0067-2.0089	0.2808	0.2808	1	0.002-0.012

79113Ca7

CAMSHAFT SPECIFICATIONS

All measurements given in inches.

Year	Engine ID/VIN	Engine Displacement Liters (cc)	Journal Diameter					Elevation		Bearing Clearance	Camshaft End Play
			1	2	3	4	5	In.	Ex.		
	S	6.5 (6473)	2.1642-2.1663	2.1642-2.1663	2.1642-2.1663	2.1642-2.1663	2.0067-2.0089	0.2808	0.2808	1	0.002-0.012
	Y	6.5 (6473)	2.1642-2.1663	2.1642-2.1663	2.1642-2.1663	2.1642-2.1663	2.0067-2.0089	0.2808	0.2808	1	0.002-0.012
	N	7.4 (7440)	1.9482-1.9492	1.9482-1.9492	1.9482-1.9492	1.9482-1.9492	1.9482-1.9492	0.2341-0.2345	0.2529-0.2531	NA	NA

1 Nos. 1-4: 0.00098-0.0046
No. 5: 0.00078-0.0044

79113Cb7

CRANKSHAFT AND CONNECTING ROD SPECIFICATIONS

All measurements are given in inches.

Year	Engine ID/VIN	Engine Displacement Liters (cc)	Crankshaft				Connecting Rod		
			Main Brg. Journal Dia.	Main Brg. Oil Clearance	Shaft End-play	Thrust on No.	Journal Diameter	Oil Clearance	Side Clearance
1991	A	2.5 (2474)	2.3000	0.0005-0.0022	0.0035-0.0085	5	2.000	0.0005-0.0026	0.0060-0.0020
	E	2.5 (2474)	2.3000	0.0005-0.0022	0.0035-0.0085	5	2.000	0.0005-0.0026	0.0060-0.0020
	R	2.8 (2835)	[1]	0.0016-0.0032	0.0024-0.0083	3	[8]	0.0014-0.0037	0.0063-0.0252
	D	3.1 (3130)	2.6473	0.0012-0.0027	0.0024-0.0083	3	1.9983-1.9994	0.0011-0.0032	0.014-0.027
	B	4.3 (4293)	[2]	[3]	0.0020-0.0060	3	2.2487-2.2497	0.0013-0.0035	0.0060-0.0140
	Z	4.3 (4293)	[2]	[3]	0.0020-0.0060	3	2.2487-2.2497	0.0013-0.0035	0.0060-0.0140
	H	5.0 (4999)	[2]	[3]	0.0020-0.0060	5	2.0988-2.0998	0.0013-0.0035	0.006-0.014
	K	5.7 (5735)	[2]	[3]	0.0020-0.0060	5	2.0988-2.0998	0.0013-0.0035	0.006-0.014
	C	6.2 (6210)	[4]	[5]	0.0020-0.0070	3	2.3980-2.3990	0.0017-0.0039	0.0070-0.0240
	J	6.2 (6210)	[4]	[5]	0.0020-0.0070	3	2.3980-2.3990	0.0017-0.0039	0.0070-0.0240
	N	7.4 (7440)	[6]	[7]	0.006-0.010	5	2.1990-2.2000	0.0009-0.0025	0.0130-0.0230
1992	A	2.5 (2474)	2.3000	0.0005-0.0022	0.0035-0.0085	5	2.000	0.0005-0.0026	0.0060-0.0020
	R	2.8 (2835)	[1]	0.0016-0.0032	0.0024-0.0083	3	[8]	0.0014-0.0037	0.0063-0.0252
	D	3.1 (3130)	2.6473-2.6483	0.0012-0.0027	0.0024-0.0083	3	1.9983-1.9994	0.0011-0.0032	0.014-0.027
	L	3.8 (3785)	2.4988-2.4998	0.0008-0.0022	0.003-0.011	3	2.2487-2.2499	0.0008-0.0022	0.003-0.015
	B	4.3 (4293)	[2]	[3]	0.0020-0.0060	3	2.2487-2.2497	0.0013-0.0035	0.0060-0.0140
	W	4.3 (4293)	[9]	[3]	0.0020-0.0060	3	2.2487-2.2497	0.0013-0.0035	0.0060-0.0140
	Z	4.3 (4293)	[2]	[3]	0.0020-0.0060	3	2.2487-2.2497	0.0013-0.0035	0.0060-0.0140
	H	5.0 (4999)	[2]	[3]	0.0020-0.0060	5	2.0988-2.0998	0.0013-0.0035	0.0060-0.0140
	K	5.7 (5735)	[2]	[3]	0.0020-0.0060	5	2.0988-2.0998	0.0013-0.0035	0.0060-0.0140
	C	6.2 (6210)	[4]	[5]	0.0020-0.0070	3	2.3980-2.3990	0.0017-0.0039	0.0070-0.0240
	J	6.2 (6210)	[4]	[5]	0.0020-0.0070	3	2.3980-2.3990	0.0017-0.0039	0.0070-0.0240
	F	6.5 (6473)	[4]	[5]	0.0020-0.0070	3	2.3980-2.3990	0.0017-0.0039	0.0070-0.0240

79113C08

CRANKSHAFT AND CONNECTING ROD SPECIFICATIONS

All measurements are given in inches.

| Year | Engine ID/VIN | Engine Displacement Liters (cc) | Crankshaft | | | | Connecting Rod | | |
			Main Brg. Journal Dia.	Main Brg. Oil Clearance	Shaft End-play	Thrust on No.	Journal Diameter	Oil Clearance	Side Clearance
	N	7.4 (7440)	6	7	0.006-0.010	5	2.1990-2.2000	0.0009-0.0025	0.0130-0.0230
1993	A	2.5 (2474)	2.3000	0.0005-0.0022	0.0035-0.0085	5	2.000	0.0005-0.0026	0.0060-0.0020
	R	2.8 (2835)	1	0.0016-0.0032	0.0024-0.0083	3	8	0.0014-0.0037	0.0063-0.0252
	D	3.1 (3130)	2.6473-2.6483	0.0012-0.0027	0.0024-0.0083	3	1.9983-1.9994	0.0011-0.0032	0.014-0.027
	L	3.8 (3785)	2.4988-2.4998	0.0008-0.0022	0.003-0.011	3	2.2487-2.2499	0.0008-0.0022	0.003-0.015
	W	4.3 (4293)	9	9	0.0020-0.0070	3	2.2487-2.2497	0.0013-0.0035	0.0060-0.0140
	Z	4.3 (4293)	2	2	0.0020-0.0060	3	2.2487-2.2497	0.0013-0.0035	0.0060-0.0140
	H	5.0 (4999)	2	2	0.0020-0.0060	3	2.2487-2.2497	0.0013-0.0035	0.0060-0.0140
	K	5.7 (5735)	2	2	0.0020-0.0060	3	2.2487-2.2497	0.0013-0.0035	0.0060-0.0140
	C	6.2 (6210)	4	4	0.0020-0.0070	3	2.3980-2.3990	0.0017-0.0039	0.0070-0.0240
	J	6.2 (6210)	4	4	0.0020-0.0070	3	2.3980-2.3990	0.0017-0.0039	0.0070-0.0240
	F	6.5 (6473)	4	4	0.0020-0.0070	3	2.3980-2.3990	0.0017-0.0039	0.0070-0.0240
	N	7.4 (7440)	6	6	0.006-0.010	5	2.1990-2.2000	0.0009-0.0025	0.0130-0.0230
1994-95	4	2.2 (2189)	2.4945-2.4954	0.0006-0.0019	0.0020-0.0070	5	1.9983-1.9994	0.00098-0.0031	0.0039-0.0149
	D	3.1 (3130)	2.6473-2.6483	0.0012-0.0027	0.0024-0.0083	3	1.9983-1.9994	0.0011-0.0032	0.014-0.027
	L	3.8 (3785)	2.4988-2.4998	0.0008-0.0022	0.003-0.011	3	2.2487-2.2499	0.0008-0.0022	0.003-0.015
	W	4.3 (4293)	8	3	0.0020-0.0070	3	2.2487-2.2497	0.0013-0.0035	0.0060-0.0140
	Z	4.3 (4293)	2	3	0.0020-0.0060	3	2.2487-2.2497	0.0013-0.0035	0.0060-0.0140
	H	5.0 (4999)	2	3	0.0020-0.0060	5	2.0988-2.0998	0.0013-0.0035	0.006-0.014
	K	5.7 (5735)	2	3	0.0020-0.0060	5	2.0988-2.0998	0.0013-0.0035	0.006-0.014
	F	6.5 (6473)	4	5	0.0040-0.0098	3	2.3980-2.3990	0.0017-0.0039	0.0070-0.0240
	P	6.5 (6473)	4	5	0.0040-0.0098	3	2.3980-2.3990	0.0017-0.0039	0.0070-0.0240
	S	6.5 (6473)	4	5	0.0040-0.0098	3	2.3980-2.3990	0.0017-0.0039	0.0070-0.0240

79113Ca8

CRANKSHAFT AND CONNECTING ROD SPECIFICATIONS
All measurements are given in inches.

| Year | Engine ID/VIN | Engine Displacement Liters (cc) | Crankshaft | | | | | Connecting Rod | | |
			Main Brg. Journal Dia.	Main Brg. Oil Clearance	Shaft End-play	Thrust on No.	Journal Diameter	Oil Clearance	Side Clearance
	Y	6.5 (6473)	④	⑤	0.0040-0.0098	3	2.3980-2.3990	0.0017-0.0039	0.0070-0.0240
	N	7.4 (7440)	⑥	⑦	0.006-0.010	5	2.1990-2.2000	0.0009-0.0025	0.0130-0.0230

1 Three dots: 2.64728-2.64759
  Two dots: 2.64759-2.64790
  One dot: 2.64790-2.64822
2 No. 1: 2.4484-2.4493
  Nos. 2-3: 2.4481-2.4493
  No. 4: 2.4479-2.4488
3 No. 1: 0.0008-0.0020
  Nos. 2-3: 0.0011-0.0023
  No. 4: 0.0017-0.0032

4 Nos. 1-4: 2.9495-2.9504
  No. 5: 2.9493-2.9502
5 Nos. 1-4: 0.0083
  No. 5: 0.0055-0.0093
6 Nos. 1-4: 2.7481-2.7490
  No. 5: 2.7476-2.7486
7 Nos. 1-4: 0.0013-0.0025
  No. 5: 0.0024-0.0040

8 Two dots: 1.9983-1.9989
  One dot: 1.9989-1.9994
9 No. 1: 2.4488-2.4495
  Nos. 2-3: 2.4485-2.4494
  No. 4: 2.4480-2.4489

79113Cb8

VALVE SPECIFICATIONS

Year	Engine ID/VIN	Engine Displacement Liters (cc)	Seat Angle (deg.)	Face Angle (deg.)	Spring Test Pressure (lbs. @ in.)	Spring Installed Height (in.)	Stem-to-Guide Clearance (in.)		Stem Diameter (in.)	
							Intake	Exhaust	Intake	Exhaust
1991	A	2.5 (2474)	46	45	1.58-1.70@1.04	1.44	0.0010-0.0025	0.0013-0.0030	0.3133-0.3138	0.3128-0.3135
	E	2.5 (2474)	46	45	71-78@1.44	1.44	0.0010-0.0025	0.0013-0.0030	0.3133-0.3138	0.3128-0.3135
	R	2.8 (2835)	46	45	88@1.57	1.57	0.0010-0.0027	0.0010-0.0027	0.3410-0.3417	0.3410-0.3417
	D	3.1 (3130)	46	45	82@1.58	1.57	0.0010-0.0027	0.0010-0.0027	NA	NA
	B	4.3 (4293)	46	45	194-206@1.25	1.72	0.0010-0.0027	0.0010-0.0027	NA	NA
	Z	4.3 (4293)	46	45	194-206@1.25	1.72	0.0010-0.0027	0.0010-0.0027	0.3410-0.3417	0.3410-0.3417
	H	5.0 (4999)	46	45	76-84@1.70	1.72	0.0010-0.0027	0.0010-0.0027	0.3410-0.3417	0.3410-0.3417
	K	5.7 (5735)	46	45	76-84@1.70	1.72	0.0010-0.0027	0.0010-0.0027	0.3410-0.3417	0.3410-0.3417
	C	6.2 (6210)	46	45	80@1.81	1.81	0.0010-0.0027	0.0010-0.0027	0.3414	0.3414
	J	6.2 (6210)	46	45	80@1.81	1.81	0.0010-0.0027	0.0010-0.0027	0.3414	0.3414
	N	7.4 (7440)	46	45	74-86@1.80	1.80	0.0010-0.0027	0.0012-0.0029	0.3410-0.3417	0.3410-0.3417
	W	7.4 (7440)	46	45	74-86@1.80	1.80	0.0010-0.0027	0.0012-0.0029	0.3410-0.3417	0.3410-0.3417
1992	A	2.5 (2474)	46	45	1.58-1.70@1.04	1.44	0.0010-0.0025	0.0013-0.0030	0.3133-0.3138	0.3128-0.3135
	R	2.8 (2835)	46	45	88@1.57	1.57	0.0010-0.0027	0.0010-0.0027	0.3410-0.3417	0.3410-0.3417
	D	3.1 (3130)	46	45	82@1.58	1.57	0.0010-0.0027	0.0010-0.0027	NA	NA
	L	3.8 (3785)	46	45	210@1.315	1.69-1.71	0.0015-0.0035	0.0015-0.0032	NA	[1]
	B	4.3 (4293)	46	45	194-206@1.25	1.72	0.0010-0.0027	0.0010-0.0027	NA	NA
	W	4.3 (4293)	46	45	194-206@1.25	1.69-1.71	0.0011-0.0027	0.0011-0.0027	NA	NA
	Z	4.3 (4293)	46	45	194-206@1.25	1.72	0.0010-0.0027	0.0010-0.0027	NA	NA
	H	5.0 (4999)	46	45	76-84@1.70	1.72	0.0010-0.0027	0.0010-0.0027	NA	NA
	K	5.7 (5735)	46	45	76-84@1.70	1.72	0.0010-0.0027	0.0010-0.0027	NA	NA
	C	6.2 (6210)	46	45	230@1.39	1.81	0.0010-0.0027	0.0010-0.0027	NA	NA
	J	6.2 (6210)	46	45	230@1.39	1.81	0.0010-0.0027	0.0010-0.0027	NA	NA

79113C09

VALVE SPECIFICATIONS

Year	Engine ID/VIN	Engine Displacement Liters (cc)	Seat Angle (deg.)	Face Angle (deg.)	Spring Test Pressure (lbs. @ in.)	Spring Installed Height (in.)	Stem-to-Guide Clearance (in.)		Stem Diameter (in.)	
							Intake	Exhaust	Intake	Exhaust
	F	6.5 (6473)	46	45	230@1.39	1.81	0.0010-0.0027	0.0010-0.0027	NA	NA
	N	7.4 (7440)	46	45	74-86@1.80	1.80	0.0010-0.0027	0.0012-0.0029	NA	NA
1993	A	2.5 (2474)	46	45	1.58-1.70@1.04	1.44	0.0010-0.0025	0.0013-0.0030	0.3133-0.3138	0.3128-0.3135
	R	2.8 (2835)	46	45	88@1.57	1.57	0.0010-0.0027	0.0010-0.0027	0.3410-0.3417	0.3410-0.3417
	D	3.1 (3130)	46	45	82@1.58	1.57	0.0010-0.0027	0.0010-0.0027	NA	NA
	L	3.8 (3785)	46	45	210@1.315	1.69-1.72	0.0015-0.0035	0.0015-0.0032	NA	[1]
	W	4.3 (4293)	46	45	194-206@1.25	1.69-1.71	0.0011-0.0027	0.0011-0.0027	NA	NA
	Z	4.3 (4293)	46	45	194-206@1.25	1.72	0.0010-0.0027	0.0010-0.0027	NA	NA
	H	5.0 (4999)	46	45	76-84@1.70	1.72	0.0010-0.0027	0.0010-0.0027	NA	NA
	K	5.7 (5735)	46	45	76-84@1.70	1.72	0.0010-0.0027	0.0010-0.0027	NA	NA
	C	6.2 (6210)	46	45	230@1.39	1.81	0.0010-0.0027	0.0010-0.0027	NA	NA
	J	6.2 (6210)	46	45	230@1.39	1.81	0.0010-0.0027	0.0010-0.0027	NA	NA
	F	6.5 (6473)	46	45	230@1.39	1.81	0.0010-0.0027	0.0010-0.0027	NA	NA
	N	7.4 (7440)	46	45	74-86@1.80	1.80	0.0010-0.0027	0.0012-0.0029	NA	NA
1994-95	4	2.2 (2189)	46	45	228@1.278	1.71	0.0010-0.0020	0.0010-0.0030	NA	NA
	D	3.1 (3130)	46	45	82@1.58	1.57	0.0010-0.0027	0.0010-0.0027	NA	NA
	L	3.8 (3785)	46	45	210@1.315	1.69-1.72	0.0015-0.0035	0.0015-0.0032	NA	[1]
	W	4.3 (4293)	46	45	194-206@1.25	1.69-1.71	0.0011-0.0027	0.0011-0.0027	NA	NA
	Z	4.3 (4293)	46	45	194-206@1.25	1.72	0.0010-0.0027	0.0010-0.0027	NA	NA
	H	5.0 (4999)	46	45	76-84@1.70	1.72	0.0010-0.0027	0.0010-0.0027	NA	NA
	K	5.7 (5735)	46	45	76-84@1.70	1.72	0.0010-0.0027	0.0010-0.0027	NA	NA
	F	6.5 (6473)	46	45	230@1.39	1.81	0.0010-0.0027	0.0010-0.0027	NA	NA
	P	6.5 (6473)	46	45	230@1.39	1.81	0.0010-0.0027	0.0010-0.0027	NA	NA

79113Ca9

VALVE SPECIFICATIONS

Year	Engine ID/VIN	Engine Displacement Liters (cc)	Seat Angle (deg.)	Face Angle (deg.)	Spring Test Pressure (lbs. @ in.)	Spring Installed Height (in.)	Stem-to-Guide Clearance (in.)		Stem Diameter (in.)	
							Intake	Exhaust	Intake	Exhaust
	S	6.5 (6473)	46	45	230@1.39	1.81	0.0010-0.0027	0.0010-0.0027	NA	NA
	Y	6.5 (6473)	46	45	230@1.39	1.81	0.0010-0.0027	0.0010-0.0027	NA	NA
	N	7.4 (7440)	46	45	74-86@1.80	1.80	0.0010-0.0027	0.0012-0.0029	NA	NA

1 Upper: 0.3129-0.3137; Lower: 0.3118-0.3126

79113Cb9

PISTON AND RING SPECIFICATIONS

All measurements are given in inches.

Year	Engine ID/VIN	Engine Displacement Liters (cc)	Piston Clearance	Ring Gap			Ring Side Clearance		
				Top Compression	Bottom Compression	Oil Control	Top Compression	Bottom Compression	Oil Control
1991	A	2.5 (2474)	0.0014-0.0022	0.010-0.020	0.010-0.020	0.020-0.060	0.0020-0.0030	0.0010-0.0030	0.015-0.055
	E	2.5 (2474)	0.0014-0.0022	0.010-0.020	0.010-0.020	0.020-0.060	0.0020-0.0030	0.0010-0.0030	0.015-0.055
	R	2.8 (2835)	0.0007-0.0017	0.010-0.020	0.010-0.020	0.020-0.0550	0.0010-0.0030	0.0015-0.0037	0.008 MAX
	D	3.1 (3146)	0.0009-0.0022	0.010-0.020	0.020-0.028	0.010-0.030	0.0020-0.0035	0.0020-0.0035	0.008 MAX
	B	4.3 (4293)	0.0007-0.0017	0.010-0.020	0.010-0.025	0.015-0.055	0.0012-0.0032	0.0012-0.0032	0.002-0.007
	Z	4.3 (4293)	0.0007-0.0017	0.010-0.020	0.010-0.025	0.015-0.055	0.0012-0.0032	0.0012-0.0032	0.002-0.007
	H	5.0 (4999)	0.0007-0.0017	0.010-0.020	0.010-0.025	0.015-0.055	0.0012-0.0032	0.0012-0.0032	0.002-0.007
	K	5.7 (5735)	0.0007-0.0017	0.010-0.020	0.010-0.025	0.015-0.055	0.0012-0.0032	0.0012-0.0032	0.002-0.007
	C	6.2 (6210)	[1]	0.012-0.022	0.030-0.039	0.0098-0.020	0.0030-0.0070	0.030-0.040	0.002-0.004
	J	6.2 (6210)	[1]	0.012-0.022	0.030-0.039	0.0098-0.020	0.0030-0.0070	0.030-0.040	0.002-0.004
	N	7.4 (7440)	0.003-0.004	0.010-0.018	0.016-0.024	0.010-0.030	0.0012-0.0029	0.0012-0.0029	0.0050-0.0065
1992	A	2.5 (2474)	0.0015-0.0035	0.010-0.015	0.010-0.020	0.015-0.055	0.0015-0.0030	0.0015-0.0032	0.005-0.007
	R	2.8 (2835)	0.0007-0.0017	0.010-0.020	0.010-0.020	0.010-0.050	0.0010-0.0030	0.0015-0.0037	0.008 MAX
	D	3.1 (3146)	0.0009-0.0022	0.010-0.020	0.020-0.028	0.010-0.030	0.0020-0.0035	0.0020-0.0035	0.008 MAX
	L	3.8 (3785)	0.0004-0.0022 [2]	0.010-0.025	0.010-0.025	0.015-0.055	0.0013-0.0031	0.0013-0.0031	0.0011-0.0081
	B	4.3 (4293)	0.0007-0.0017	0.010-0.020	0.010-0.025	0.015-0.055	0.0012-0.0032	0.0012-0.0032	0.002-0.007
	W	4.3 (4293)	0.0007-0.0017	0.010-0.020	0.018-0.026	0.015-0.055	0.0014-0.0032	0.0014-0.0032	0.0014-0.0032
	Z	4.3 (4293)	0.0007-0.0017	0.010-0.020	0.010-0.025	0.015-0.055	0.0012-0.0032	0.0012-0.0032	0.002-0.007
	H	5.0 (4999)	0.0007-0.0017	0.010-0.020	0.018-0.026	0.010-0.030	0.0012-0.0032	0.0012-0.0032	0.002-0.007
	K	5.7 (5735)	0.0007-0.0017	0.010-0.020	0.018-0.026	0.010-0.030	0.0012-0.0032	0.0012-0.0032	0.002-0.007
	C	6.2 (6210)	[1]	0.012-0.022	0.030-0.039	0.0098-0.020	0.0030-0.0070	0.030-0.0400	0.002-0.004
	J	6.2 (6210)	[1]	0.012-0.022	0.030-0.039	0.0098-0.020	0.0030-0.0070	0.030-0.0400	0.002-0.004
	F	6.5 (6473)	[1]	0.010-0.020	0.030-0.039	0.0098-0.023	0.0030-0.0070	0.030-0.0400	0.002-0.004

79113C10

PISTON AND RING SPECIFICATIONS
All measurements are given in inches.

| Year | Engine ID/VIN | Engine Displacement Liters (cc) | Piston Clearance | Ring Gap | | | Ring Side Clearance | | |
				Top Compression	Bottom Compression	Oil Control	Top Compression	Bottom Compression	Oil Control
	N	7.4 (7440)	0.003-0.0042	0.010-0.018	0.016-0.024	0.010-0.030	0.0012-0.0029	0.0012-0.0029	0.0050-0.0065
1993	A	2.5 (2474)	0.0015-0.0035	0.010-0.015	0.010-0.020	0.015-0.055	0.0015-0.0030	0.0015-0.0032	0.0005-0.007
	R	2.8 (2835)	0.0007-0.0017	0.010-0.020	0.010-0.020	0.010-0.050	0.0010-0.0030	0.0015-0.0037	0.008 MAX
	D	3.1 (3146)	0.0009-0.0023	0.0071-0.0161	0.0197-0.0280	0.098-0.0295	0.0020-0.0035	0.0020-0.0035	0.008 MAX
	L	3.8 (3785)	0.0004-[3] 0.0022	0.010-0.025	0.010-0.025	0.015-0.055	0.0013-0.0031	0.0013-0.0031	0.0011-0.0081
	W	4.3 (4293)	0.0007-0.0017	0.010-0.020	0.018-0.026	0.015-0.055	0.0014-0.0032	0.0014-0.0032	0.0014-0.0032
	Z	4.3 (4293)	0.0007-0.0017	0.010-0.020	0.010-0.025	0.015-0.055	0.0012-0.0032	0.0012-0.0032	0.002-0.007
	H	5.0 (4999)	0.0007-0.0021	0.010-0.020	0.018-0.026	0.010-0.030	0.0012-0.0032	0.0012-0.0032	0.002-0.007
	K	5.7 (5735)	0.0007-0.0021	0.010-0.020	0.018-0.026	0.010-0.030	0.0012-0.0032	0.0012-0.0032	0.002-0.007
	C	6.2 (6210)	[1]	0.010-0.020	0.030-0.039	0.0098-0.020	[4]	0.0015-0.0030	0.0015-0.0035
	J	6.2 (6210)	[1]	0.010-0.020	0.030-0.039	0.0098-0.020	[4]	0.0015-0.0030	0.0015-0.0035
	F	6.5 (6473)	[5]	0.010-0.020	0.030-0.039	0.0098-0.020	[4]	0.0015-0.0030	0.0015-0.0035
	N	7.4 (7440)	0.003-0.0042	0.010-0.018	0.016-0.024	0.010-0.030	0.0012-0.0029	0.0012-0.0029	0.0050-0.0065
1994-95	4	2.2 (2189)	0.0007-0.0017	0.010-0.020	0.010-0.020	0.010-0.050	0.0019-0.0027	0.0019-0.0027	0.0019-0.0082
	D	3.1 (3146)	0.0009-0.0023	0.0071-0.0161	0.0197-0.0280	0.098-0.0295	0.0020-0.0035	0.0020-0.0035	0.008 MAX
	L	3.8 (3785)	0.0004-[2] 0.0022	0.010-0.025	0.010-0.025	0.015-0.055	0.0013-0.0031	0.0013-0.0031	0.0011-0.0081
	W	4.3 (4293)	0.0007-0.0017	0.010-0.020	0.018-0.026	0.015-0.055	0.0014-0.0032	0.0014-0.0032	0.0014-0.0032
	Z	4.3 (4293)	0.0007-0.0017	0.010-0.020	0.010-0.025	0.015-0.055	0.0012-0.0032	0.0012-0.0032	0.002-0.007
	H	5.0 (4999)	0.0007-0.0021	0.010-0.020	0.018-0.026	0.010-0.030	0.0012-0.0032	0.0012-0.0032	0.002-0.007
	K	5.7 (5735)	0.0007-0.0021	0.010-0.020	0.018-0.026	0.010-0.030	0.0012-0.0032	0.0012-0.0032	0.002-0.007
	F	6.5 (6473)	[6]	0.010-0.020	0.030-0.039	0.0098-0.020	[4]	0.0015-0.0030	0.0015-0.0035
	P	6.5 (6473)	[6]	0.010-0.020	0.030-0.039	0.0098-0.020	[4]	0.0015-0.0030	0.0015-0.0035
	S	6.5 (6473)	[6]	0.010-0.020	0.030-0.039	0.0098-0.020	[4]	0.0015-0.0030	0.0015-0.0035

79113C1a

PISTON AND RING SPECIFICATIONS

All measurements are given in inches.

Year	Engine ID/VIN	Engine Displacement Liters (cc)	Piston Clearance	Ring Gap			Ring Side Clearance		
				Top Compression	Bottom Compression	Oil Control	Top Compression	Bottom Compression	Oil Control
	Y	6.5 (6473)	6	0.010-0.020	0.030-0.039	0.0098-0.020	4	0.0015-0.0030	0.0015-0.0035
	N	7.4 (7440)	0.0018-0.0030	0.010-0.018	0.016-0.024	0.010-0.030	0.0012-0.0029	0.0012-0.0029	0.0050-0.0065

1 Bohn piston Nos. 1-6: 0.0035-0.0045
 Bohn piston Nos. 7-8: 0.0040-0.0050
 Zollner piston Nos. 1-6: 0.0044-0.0054
 Zollner piston Nos. 7-8: 0.0049-0.0059
2 Measured 44mm from top of piston
3 Measured 44mm from top of piston

4 Keystone type ring
5 Bore Nos. 1-6: 0.0037-0.0047
 Bore Nos. 7-8: 0.0042-0.0052
6 Bore Nos. 1-6: 0.0037-0.0047
 Bore Nos. 7-8: 0.0042-0.0052

79113C1b

TORQUE SPECIFICATIONS
All readings in ft. lbs.

Year	Engine ID/VIN	Engine Displacement Liters (cc)	Cylinder Head Bolts	Main Bearing Bolts	Rod Bearing Bolts	Crankshaft Damper Bolts	Flywheel Bolts	Manifold Intake	Manifold Exhaust	Spark Plugs	Lug Nut
1991	A	2.5 (2474)	1	70	32	160	2	25	3	7-16	90
	E	2.5 (2474)	1	70	32	160	2	25	3	7-16	90
	R	2.8 (2835)	1	70	39	70	52	23	25	22	90
	D	3.1 (3130)	1	72	39	75	45	15	24	18	100
	B	4.3 (4293)	70	80	16	70	75	7	5	22	90
	Z	4.3 (4293)	70	80	16	70	75	7	5	22	90
	H	5.0 (4999)	68	11	45	70	74	35	17	22	18
	K	5.7 (5735)	68	11	45	70	74	35	17	22	18
	C	6.2 (6210)	9	10	48	200	65	31	26	-	18
	J	6.2 (6210)	9	10	48	200	65	31	26	-	18
	N	7.4 (7440)	81	100	48	85	65	30	40	22	18
1992	A	2.5 (2474)	1	70	32	160	2	25	3	7-15	90
	R	2.8 (2835)	4	70	39	70	52	15	25	22	90
	D	3.1 (3130)	4	72	39	75	45	15	24	18	100
	L	3.8 (3785)	20	21	22	8	13	7	38	12	100
	B	4.3 (4293)	65	80	16	70	75	19	6	22	90
	W	4.3 (4293)	65	75	16	70	75	35	6	22	90
	Z	4.3 (4293)	65	75	16	70	75	35	6	22	90
	H	5.0 (4999)	65	12	45	70	75	35	6	15	14
	K	5.7 (5735)	65	12	45	70	75	35	6	15	14
	C	6.2 (6210)	7	8	48	200	66	31	26	-	14
	J	6.2 (6210)	7	8	48	200	66	31	26	-	14
	F	6.5 (6473)	7	8	48	200	66	31	26	-	14
	N	7.4 (7440)	80	100	48	85	65	40	40	22	14
1993	A	2.5 (2474)	1	70	32	160	2	25	3	7-15	90
	R	2.8 (2835)	70	70	39	70	52	15	25	22	90
	D	3.1 (3130)	4	23	37	76	52	15	24	7	100
	L	3.8 (3785)	27	20	21	28	13	7	38	12	100
	W	4.3 (4293)	65	75	25	70	75	35	5	22	90
	Z	4.3 (4293)	65	75	25	70	75	35	5	22	90
	H	5.0 (4999)	65	12	45	70	75	35	5	15	14
	K	5.7 (5735)	65	12	45	70	75	35	5	15	14
	C	6.2 (6210)	26	10	48	200	66	31	26	-	14
	J	6.2 (6210)	26	10	48	200	66	31	26	-	14
	F	6.5 (6473)	26	10	48	200	66	31	26	-	14
	N	7.4 (7440)	80	100	48	85	65	40	40	22	14
1994-95	4	2.2 (2189)	29	70	38	77	55	24	10	3	100
	D	3.1 (3130)	31	23	37	76	52	15	25	3	100
	L	3.8 (3785)	27	20	21	30	13	7	38	11	100
	W	4.3 (4293)	65	75	25	70	75	35	5	22	90
	Z	4.3 (4293)	65	75	25	70	75	35	5	11	90
	H	5.0 (4999)	65	12	45	70	75	35	5	15	14
	K	5.7 (5735)	65	12	45	70	75	35	5	15	14
	F	6.5 (6473)	32	10	48	200	66	31	26	-	14
	P	6.5 (6473)	32	10	48	200	66	31	26	-	14
	S	6.5 (6473)	32	10	48	200	66	31	26	-	14

79113C11

TORQUE SPECIFICATIONS

All readings in ft. lbs.

Year	Engine ID/VIN	Engine Displacement Liters (cc)	Cylinder Head Bolts	Main Bearing Bolts	Rod Bearing Bolts	Crankshaft Damper Bolts	Flywheel Bolts	Manifold Intake	Manifold Exhaust	Spark Plugs	Lug Nut
Y	6.5 (6473)	32	10	48	200	66	31	26	-	14
N	7.4 (7440)	80	100	48	85	65	35	40	22	14

1 Step 1: Tighten all head bolts to 18 ft. lbs.
  Step 2: Tighten all bolts to 26 ft. lbs. except No. 9
  Retorque No. 9 to 18 ft. lbs.
  Step 3: Tighten all an additional 90 degrees

2 Automatic trans.: 55 ft. lbs.
  Manual trans.: 65 ft. lbs.

3 Center bolts: 36 ft. lbs.; Outer bolts: 32 ft. lbs.

4 Coat threads with sealer
  Tighten all bolts to 40 ft. lbs.
  Tighten all an additional 90 degrees (1/4 turn)

5 Two center bolts: 26 ft. lbs.
  All others: 32 ft. lbs.

6 Two center bolts: 26 ft. lbs.
  All others: 20 ft. lbs.

7 Last bolt, left side & throttle bracket bolt: 41 ft. lbs.
  All others: 35 ft. lbs.

8 110 ft. lbs. plus 76 degrees

9 Coat threads with sealant
  Tighten all bolts to 20 ft. lbs.
  Retorque to 50 ft. lbs.
  Tighten all bolts an additional 90 degrees (1/4 turn)

10 Outer bolts: 100 ft. lbs.
   Inner bolts: 110 ft. lbs.

11 Outer bolts on caps 2-4: 68 ft. lbs.
   All others: 78 ft. lbs.

12 Outer bolts on caps 2-4: 70 ft. lbs.
   All others: 80ft. lbs.

13 11 ft. lbs. plus 50 degrees

14 All 5 & 6 stud single rear wheels: 110 ft. lbs.
   All 8 stud single rear wheels: 120 ft. lbs.
   All 8 stud dual rear wheels: 140 ft. lbs.
   All 10 stud dual wheels: 175 ft. lbs.

15 Tighten all bolts to 13 ft. lbs.
   Retorque to 19 ft. lbs.

16 20 ft. lbs plus 60 degrees
   With cast manifolds, Two center bolts: 26 ft. lbs.
   With cast manifolds, All others: 20 ft. lbs.
   With tubular manifolds: 26 ft. lbs/

18 All 5 & 6 stud single rear wheels: 110 ft. lbs.
   All 8 stud single rear wheels: 120 ft. lbs.
   All 8 stud dual rear wheels: 140 ft. lbs.
   All 10 stud dual wheels: 155 ft. lbs.

19 Last bolt, left side & throttle bracket bolt: 41 ft. lbs.
   All others: 35 ft. lbs.

20 Tighten bolts to 35 ft. lbs. then rotate 130 degrees
   Tighten four center bolts an additional 30 degrees

21 26 ft. lbs. plus 50 degrees

22 20 ft. lbs. plus 50 degrees

23 37 ft. lbs. plus 77 degrees

24 1st-time installation (new head): 22 ft. lbs.
   All other installations: 11 ft. lbs.

25 20 ft. lbs. plus 70 degrees

26 Coat threads with sealant
   Tighten all bolts to to 20 ft. lbs.
   Tighten all bolts an additional 90 degrees (1/4 turn)

27 Tighten all bolts to 35 ft. lbs.then rotate 130 degrees
   Tighten four center bolts an additional 30 degrees

28 110 ft. lbs. plus 76 degrees

29 Short bolts: 43 ft. lbs. plus 90 degrees
   Long bolts: 46 ft. lbs. plus 90 degrees

30 110 ft. lbs. plus 76 degrees

31 Coat threads with sealer
   Tighten all bolts to 33 ft. lbs.
   Tighten all an additional 90 degrees (1/4 turn)

32 Coat threads with sealant
   Tighten all bolts to 20 ft. lbs.
   Retorque to 50 ft. lbs.

79113Ca1

BRAKE SPECIFICATIONS
All measurements in inches unless noted

Year	Model	Master Cylinder Bore	Brake Disc			Brake Drum Diameter			Minimum Lining Thickness	
			Original Thickness	Minimum Thickness	Maximum Runout	Original Inside Diameter	Max. Wear Limit	Maximum Machine Diameter	Front	Rear
1991	Astro/Safari	NA	1	2	0.004	9.50	9.59	9.56	0.030	0.030
	Lumina/Silhouette/Trans Sport	0.944	1.043	0.972	0.004	8.863	8.909	8.877	0.030	0.030
	S10 Blazer/S15 Jimmy	NA	1.04	0.980	0.004	9.50	9.59	9.56	0.030	0.030
	Bravada	NA	1.04	0.980	0.004	9.50	9.59	9.56	0.030	0.030
	S10 Pick-up/S15 Sonoma	NA	1.04	0.980	0.004	9.50	9.59	9.56	0.030	0.030
	Syclone	NA	1.04	0.980	0.004	9.50	9.59	9.56	0.030	0.030
	Blazer/Jimmy	NA	1.50	1.48	0.004	5	6	7	0.030	0.030
	C1500	NA	1.25	1.23	0.004	5	6	7	0.030	0.030
	C2500	NA	1.50	1.48	0.004	5	6	7	0.030	0.030
	C3500	NA	1.50	1.48	0.004	5	6	7	0.030	0.030
	K1500	NA	1.50	1.48	0.004	5	6	7	0.030	0.030
	K2500	NA	15.0	1.48	0.004	5	6	7	0.030	0.030
	K3500	NA	1.50	1.48	0.004	5	6	7	0.030	0.030
	P30	NA	1.245	1.23	0.004	13.00	13.09	13.06	0.030	0.030
	R3500	NA	1.50	1.48	0.004	5	6	7	0.030	0.030
	Suburban	NA	1.50	1.48	0.004	5	6	7	0.030	0.030
	G10	NA	3	3	0.004	5	6	7	0.030	0.030
	G20	NA	3	3	0.004	5	6	7	0.030	0.030
	G30	NA	3	3	0.004	5	6	7	0.030	0.030
1992	Astro/Safari	NA	1	1	0.004	9.50	9.59	9.56	0.030	0.030
	Lumina/Silhouette/Trans Sport	0.944	1.043	0.972	0.004	8.863	8.909	8.877	0.030	0.030
	S10 Blazer/S15 Jimmy	NA	1.04	0.980	0.004	9.50	9.59	9.56	0.030	0.030
	Bravada	NA	1.04	0.980	0.004	9.50	9.59	9.56	0.030	0.030
	S10 Pick-up/Sonoma	NA	1.04	0.980	0.004	9.50	9.59	9.56	0.030	0.030
	Syclone	NA	1.04	0.980	0.004	9.50	9.59	9.56	0.030	0.030
	Typhoon	NA	1.04	0.980	0.004	9.50	9.59	9.56	0.030	0.030
	Blazer/Yukon	NA	1.50	1.48	0.004	5	6	7	0.030	0.030
	C1500	NA	1.25	1.23	0.004	5	6	7	0.030	0.030
	C2500	NA	1.50	1.48	0.004	5	6	7	0.030	0.030
	C3500	NA	1.50	1.48	0.004	5	6	7	0.030	0.030
	K1500	NA	1.50	1.48	0.004	5	6	7	0.030	0.030
	K2500	NA	1.50	1.48	0.004	5	6	7	0.030	0.030
	K3500	NA	1.50	1.48	0.004	5	6	7	0.030	0.030
	P30	NA	1.245	1.23	0.004	13.00	13.09	13.06	0.030	0.030
	Suburban	NA	1.50	1.48	0.004	5	6	7	0.030	0.030
	G10	NA	3	4	0.004	5	6	7	0.030	0.030
	G20	NA	3	4	0.004	5	6	7	0.030	0.030
	G30	NA	3	4	0.004	5	6	7	0.030	0.030
1993	Astro/Safari	NA	1	2	0.004	9.50	9.59	9.56	0.030	0.030
	Lumina/Silhouette/Trans Sport	0.944	1.043	0.972	0.004	8.863	8.909	8.877	0.030	0.030
	S10 Blazer/S15 Jimmy	NA	1.04	0.980	0.004	9.50	9.59	9.56	0.030	0.030
	Bravada	NA	1.04	0.980	0.004	9.50	9.59	9.56	0.030	0.030
	S10 Pick-up/Sonoma	NA	1.04	0.980	0.004	9.50	9.59	9.56	0.030	0.030
	Typhoon	NA	1.04	0.980	0.004	9.50	9.59	9.56	0.030	0.030
	Blazer/Yukon	NA	1.50	1.48	0.004	5	6	7	0.030	0.030

79113C12

BRAKE SPECIFICATIONS
All measurements in inches unless noted

Year	Model	Master Cylinder Bore	Brake Disc Original Thickness	Brake Disc Minimum Thickness	Brake Disc Maximum Runout	Brake Drum Diameter Original Inside Diameter	Brake Drum Diameter Max. Wear Limit	Brake Drum Diameter Maximum Machine Diameter	Minimum Lining Thickness Front	Minimum Lining Thickness Rear
	C1500	NA	1.25	1.23	0.004	5	6	7	0.030	0.030
	C2500	NA	1.50	1.48	0.004	5	6	7	0.030	0.030
	C3500	NA	1.50	1.48	0.004	5	6	7	0.030	0.030
	K1500	NA	1.50	1.48	0.004	5	6	7	0.030	0.030
	K2500	NA	1.50	1.48	0.004	5	6	7	0.030	0.030
	K3500	NA	1.50	1.48	0.004	5	6	7	0.030	0.030
	P30	NA	1.245	1.23	0.004	13.00	13.09	13.06	0.030	0.030
	Suburban	NA	1.50	1.48	0.004	5	6	7	0.030	0.030
	G10	NA	3	4	0.004	5	6	7	0.030	0.030
	G20	NA	3	4	0.004	5	6	7	0.030	0.030
	G30	NA	3	4	0.004	5	6	7	0.030	0.030
1994-95	Astro/Safari	NA	1	2	0.004	9.50	9.59	9.56	0.030	0.030
	Lumina/Silhouette/Trans Sport	0.944	1.043	0.972	0.004	8.863	8.909	8.877	0.030	0.030
	S10 Blazer/S15 Jimmy	NA	1.04	0.980	0.004	9.50	9.59	9.56	0.030	0.030
	Bravada	NA	1.04	0.980	0.004	9.50	9.59	9.56	0.030	0.030
	S10 Pick-up/Sonoma	NA	1.04	0.980	0.004	9.50	9.59	9.56	0.030	0.030
	Blazer/Yukon	NA	1.50	1.48	0.004	5	6	7	0.030	0.030
	C1500	NA	1.25	1.23	0.004	5	6	7	0.030	0.030
	C2500	NA	1.50	1.48	0.004	5	6	7	0.030	0.030
	C3500	NA	1.50	1.48	0.004	5	6	7	0.030	0.030
	K1500	NA	1.50	1.48	0.004	5	6	7	0.030	0.030
	K2500	NA	1.50	1.48	0.004	5	6	7	0.030	0.030
	K3500	NA	1.50	1.48	0.004	5	6	7	0.030	0.030
	P30	NA	1.245	1.23	0.004	13.00	13.09	13.06	0.030	0.030
	Suburban	NA	1.50	1.48	0.004	5	6	7	0.030	0.030
	G10	NA	3	4	0.004	5	6	7	0.030	0.030
	G20	NA	3	4	0.004	5	6	7	0.030	0.030
	G30	NA	3	4	0.004	5	6	7	0.030	0.030

1 Available with 1.04" and 1.25" rotors
2 1.04" rotors: 0.980
  1.25" rotors: 1.23
3 Available with 1.28" and 1.54" discs
4 1.28" disc: 1.23
  1.54" disc: 1.48
5 Available with 10", 11.15" and 13" drums

6 10" drum: 10.05
  11.15" drum: 11.240
  13" drum: 13.090
7 10" drum: 10.09
  11.15" drum: 11.210
  13" drum: 13.060

7911312a

WHEEL ALIGNMENT

Year	Model	Caster		Camber		Toe-in (in.)	Steering Axis Inclination (deg.)
		Range (deg.)	Preferred Setting (deg.)	Range (deg.)	Preferred Setting (deg.)		
1991	Astro/Safari 1	1 11/16P-3 11/16P	2 11/16P	0-1 19/32P	13/16P	3/32P	NA
	Astro/Safari 2	2 1/16P-4 1/16P	3 1/6P	0-1 3/4P	7/8P	3/32P	NA
	Lumina/Silhouette/Trans Sport	11/16P-2 11/16P	1 11/16P	1/2N-1/2P	0	0	NA
	S10 Blazer/S15 Jimmy	1P-3P	2P	0-1 5/8P	13/16P	1/8	NA
	Bravada	1P-3P	2P	0-1 5/8P	13/16P	1/8	NA
	S10 Pick-up/S15 Sonoma	1P-3P	2P	0-1 5/8P	13/16P	1/8	NA
	Blazer	-	3	3/4-2P	1 1/2P	0	NA
	C1500	2 3/4P-4 3/4P	3 3/4P	0-1P	1/2P	1/8P	NA
	C2500	2 3/4P-4 3/4P	3 3/4P	0-1P	1/2P	1/8P	NA
	C3500	2 3/4P-4 3/4P	3 3/4P	0-1P	1/2P	1/8P	NA
	K1500	3P-5P	4P	12/32P-1 13/32P	29/32P	1/16P	NA
	K2500	3P-5P	4P	12/32P-1 13/32P	29/32P	1/16P	NA
	K3500	3P-5P	4P	12/32P-1 13/32P	29/32P	1/16P	NA
	P30	-	7	1/2N-7/8P	3/16P	3/16P	NA
	R3500	-	4	1/2N-1P	3/4P	3/16P	NA
	Suburban	-	3	3/4P-2P	1 1/2P	0	NA
	G10	-	5	3/16N-1 3/16P	1/2P	3/16P	NA
	G20	-	5	3/16N-1 3/16P	1/2P	3/16P	NA
	G30	-	6	1/2N-7/8P	3/16P	3/16P	NA
1992	Astro/Safari 1	1 11/16P-3 11/16P	2 11/16P	0-1 19/32P	13/16P	3/32P	NA
	Astro/Safari 2	2 1/16P-4 1/16P	3 1/16P	0-1 3/4P	7/8P	3/32P	NA
	Lumina/Silhouette/Trans Sport	11/16P-2 11/16P	1 11/16P	1/2N-1/2P	0	0	NA
	S10 Blazer/S15 Jimmy	1P-3P	2P	0-1 5/8P	13/16P	1/8	NA
	Bravada	1P-3P	2P	0-1 5/8P	13/16P	1/8	NA
	S10 Pick-up/S15 Sonoma	1P-3P	2P	0-1 5/8P	13/16P	1/8.	NA
	Syclone	1P-3P	2P	0-1 5/8P	13/16P	1/8	NA
	Typhoon	1P-3P	2P	0-1 5/8P	13/16P	1/8	NA
	Blazer/Yukon	-	3	3/4-2P	1 1/2P	1/8P	NA
	C1500	3 3/4P-5 3/4P	4 3/4P	0-1P	1/2P	1/8P	NA
	C2500	3 3/4P-5 3/4P	4 3/4P	0-1P	1/2P	1/8P	NA
	C3500	3 3/4P-5 3/4P	4 3/4P	0-1P	1/2P	1/8P	NA
	K1500	1P-5P	3P	0-1P	1/2P	1/4P	NA
	K2500	1P-5P	3P	0-1P	1/2P	1/4P	NA
	K3500	1P-5P	3P	0-1P	1/2P	1/4P	NA
	P30	-	6	1/2N-7/8P	3/16P	3/16P	NA
	Suburban	4	5	3/4P-2P	1 1/2P	0	NA
	G10	-	7	1/4N-1 1/4P	1/2P	3/16P	NA
	G20	-	7	1/4N-1 1/4P	1/2P	3/16P	NA
	G30	-	7	1/2N-1P	1/4P	3/16P	NA
1993	Astro/Safari 1	1 11/16P-3 11/16P	2 11/16P	0-1 19/32P	13/16P	3/32P	NA
	Astro/Safari 2	2 1/16P-4 1/16P	3 1/16P	0-1 3/4P	7/8P	3/32P	NA
	Lumina/Silhouette/Trans Sport	11/16P-2 11/16P	1 11/16P	1/2N-1/2P	0	0	NA
	S10 Blazer/S15 Jimmy	1P-3P	2P	0-1 5/8P	13/16P	1/8	NA
	Bravada	1P-3P	2P	0-1 5/8P	13/16P	1/8	NA
	S10 Pick-up/S15 Sonoma	1P-3P	2P	0-1 5/8P	13/16P	1/8	NA
	Typhoon	1P-3P	2P	0-1 5/8P	13/16P	1/8	NA

79113C13

WHEEL ALIGNMENT

Year	Model	Caster Range (deg.)	Caster Preferred Setting (deg.)	Camber Range (deg.)	Camber Preferred Setting (deg.)	Toe-in (in.)	Steering Axis Inclination (deg.)
	Blazer/Yukon	4	5	3/4-2P	1 1/2P	0	NA
	C1500	3 3/4P-5 3/4P	4 3/4P	0-1P	1/2P	1/8,	NA
	C2500	3 3/4P-5 3/4P	4 3/4P	0-1P	1/2P	1/8P	NA
	C3500	3 3/4P-5 3/4P	4 3/4P	0-1P	1/2P	1/8P	NA
	K1500	1P-5P	3P	0-1P	1/2P	1/4P	NA
	K2500	1P-5P	3P	0-1P	1/2P	1/4P	NA
	K3500	1P-5P	3P	0-1P	1/2P	1/4P	NA
	P30	-	6	1/2N-7/8P	3/16P	3/16P	NA
	Suburban	4	5	3/4P-2P	1 1/2P	0	NA
	G10	-	3	1/4N-1 1/4P	1/2P	3/16P	NA
	G20	-	3	1/4N-1 1/4P	1/2P	3/16P	NA
	G30	-	3	1/2N-1P	1/4P	3/16P	NA
1994-95	Astro/Safari [1]	1 11/16P-3 11/16P	2 11/16P	0-1 19/32P	13/16P	3/32P	NA
	Astro/Safari [2]	2 1/16P-4 1/16P	3 1/16P	0-1 3/4P	7/8P	3/32P	NA
	Lumina/Silhouette/Trans Sport	11/16P-2 11/16P	1 11/16P	1/2N-1/2P	0	0	NA
	S10 Blazer/S15 Jimmy	1P-3P	2P	0-1 5/8P	13/16P	1/8	NA
	Bravada	1P-3P	2P	0-1 5/8P	13/16P	1/8	NA
	S10 Pick-up/S15 Sonoma	1P-3P	2P	0-1 5/8P	13/16P	1/8	NA
	Blazer/Yukon	4	5	3/4-2P	1 1/2P	0	NA
	C1500	3 3/4P-5 3/4P	4 3/4P	0-1P	1/2P	1/8P	NA
	C2500	3 3/4P-5 3/4P	4 3/4P	0-1P	1/2P	1/8P	NA
	C3500	3 3/4P-5 3/4P	4 3/4P	0-1P	1/2P	1/8P	NA
	K1500	1P-5P	3P	0-1P	1/2P	1/4P	NA
	K2500	1P-5P	3P	0-1P	1/2P	1/4P	NA
	K3500	1P-5P	3P	0-1P	1/2P	1/4P	NA
	P30	-	6	1/2N-7/8P	3/16P	3/16P	NA
	Suburban	4	5	3/4P-2P	1 1/2P	0	NA
	G10	-	3	1/4N-1 1/4P	1/2P	3/16P	NA
	G20	-	3	1/4N-1 1/4P	1/2P	3/16P	NA
	G30	-	3	1/2N-1P	1/2P	3/16P	NA

1 2WD
2 4WD
3 With ride height 2 1/2-4, caster should be 8
4 Ride height and preferred camber settings
   2 1/2: 1 1/2 degrees
   3: 15/16 degrees
   3 1/2: 5/16 degrees
   3 3/4: 1/8 degrees
   4: 0 degrees
5 Ride height and preferred camber settings
   1 1/2: 3 1/2 degrees
   2: 3 1/8 degrees
   2 1/2: 2 11/16 degrees
   3: 2 3/8 degrees
   3 1/2: 2 1/8 degrees
   3 3/4: 1 15/16 degrees
   4: 1 7/8 degrees

6 Ride height and preferred camber settings
   1 1/2: 2 7/8 degrees
   2: 2 3/16 degrees
   2 1/2: 1 5/8 degrees
   3: 1 degree
   3 1/2: 1/2 degrees
   3 3/4: 3/16 degrees
   4: 0 degrees
7 Ride height and preferred camber settings
   2 1/2: 2 5/16 degrees
   3: 1 11/16 degrees
   3 1/2: 1 3/16 degrees
   3 3/4: 15/16 degrees
   4: 5/8 degrees

79113139

ENGINE ELECTRICAL

NOTE: Disconnecting the negative battery cable on some vehicles may interfere with the functions of the on board computer systems and may require the computer to undergo a relearning process.

Distributor

REMOVAL AND INSTALLATION

1. Disconnect the negative battery cable.
2. Remove all necessary components in order to gain access to the distributor assembly.
3. Disconnect the distributor electrical connectors. Mark and remove the spark plug wires.
4. Remove the distributor cap. Position the engine at TDC. Match Mark the rotor and the distributor body. Matchmark the distributor assembly and the engine block.
5. Remove the distributor retaining bolt. Carefully remove the distributor from the vehicle.

NOTE: As the distributor is removed from the engine, the rotor will turn counterclockwise. Observe and mark the start and finish rotation of the rotor. When reinstalling, position the rotor at the last mark and set the distributor into the engine. As the distributor drops into place, the rotor should turn to its original position, providing the engine crankshaft had not been rotated with the distributor out.

Installation

Timing Not Disyurbed

NOTE: To ensure correct ignition timing if the enginehas not been disturbed, the distributor must be installedwith the rotor in the same position as when removed.

1. Align the rotor to the last mark made and install the distributor in the engine.
2. The rotor should turn and end up at the first mark made. Ensure the distributor and oil pump rod are fully engaged.
3. Reconnect the distributor cap and wires.

4. Tighten the distributor hold-down bolt and set the timing.

Timing Disturbed

1. Remove the No. 1 spark plug. Place a finger over the sparkplug hole and rotate the engine in the normal direction of rotation slowly, until compression is felt.
2. Align the timing mark on the crankshaft pulley to the **0** on the engine timing indicator by rotating the engine in the samedirection slowly.
3. Turn the distributor shaft of the rotor until it points to the No. 1 spark plug tower on the cap.
4. Install the distributor, distributor cap, spark plug, wiringand connectors.
5. Check the engine timing and adjust, as required.

Distributorless Ignition System

REMOVAL AND INSTALLATION

Ignition Coils

NOTE: Any one of the 3 coils may be removed separately from the other 2 coils.

1. Disconnect the negative battery cable.
2. Remove the secondary ignition wires from the coil being removed.
3. Remove the coil mounting screws.
4. Remove the ignition coil assembly.
5. Installation is the reverse of the removal procedure.

Ignition Module

1. Disconnect the negative battery cable.
2. Disconnect the wire connector from the module.
3. Remove the screws securing the module and coils to the mounting plate.
4. Separate the ignition coils from the module by pulling straight up on the coils.
5. Position the coils and wiring so the assembly will not be hanging from the wires.
6. Installation is the reverse of the removal procedure.

Crankshaft Sensor

1. Disconnect the negative battery cable.
2. Remove the serpentine belt from the crankshaft pulley.

3. Raise and support the vehicle safely.
4. Remove the right front tire and wheel assembly.
5. Remove the right inner fender access cover.
6. Using a 28mm socket, remove the harmonic balancer retaining bolt.
7. Using a harmonic balancer pulley removal tool, remove the balancer assembly. Remove the sensor shield.
8. Disconnect the sensor electrical connector. Remove the sensor and pedestal from the engine block.
9. Remove the sensor from the pedestal.

To install:
10. Loosely install the crankshaft sensor on the pedestle.
11. Position the sensor with the pedestal attached onto tool J–37090.
12. Position the assembly onto the crankshaft.
13. Install the pedestal securing bolts and torque to 14–28 ft. lbs. (20–40 Nm).
14. Torque the pedestal pinch bolt to 36–40 inch lbs. (4–4.5 Nm).
15. Remove the tool. Install the crankshaft sensor shield.
16. Reconnect the wire connector.
17. Place tool J–37089 on the harmonic balancer; if any vane of the balancer touches the tool, replace the balancer assembly.
18. Position the balancer onto the crankshaft.
19. Apply sealer to the threads of the balancer bolt and torque the bolt to 110 ft. lbs. (150 Nm) plus turn the bolt an additional 76 degrees.
20. Install the fender shield, tire and wheel assembly.
21. Lower the vehicle.
22. Install the serpentine belt and connect the negative battery cable.

Ignition Timing

The 3.8L engine uses a Distributorless Ignition System (DIS) which is nonadjustable.

ADJUSTMENT

1. Locate and clean off the timing marks.
2. Use chalk or white paint to color the mark on the scale that will indicate the correct timing when aligned with the mark on the pulley or the pointer.
3. Attach a tachometer and timing light to the engine per manufacturers recommendation.
4. On some engines, it is necessary to disconnect the EST connector to

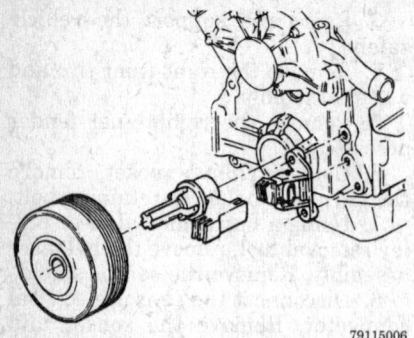

Install crankshaft sensor using tool J-37089 — 3.8L engine

79115006

set the timing. See the underhood emission control label.

5. Adjust the idle to the correct specification.

6. Loosen the distributor retaining bolt so the distributor can be turned with a little resistance.

7. With the timing light aimed at the timing marks, turn the distributor in the direction of rotor rotation to retard the timing, and in the opposite direction of rotor rotation to advance the timing. Set the timing marks to specification.

8. Turn the engine **OFF**, tighten the distributor holdown bolt and recheck the timing.

Alternator

PRECAUTIONS

Several precautions must be observed with alternator equipped vehicles to avoid damage to the unit.

• If the battery is removed for any reason, make sure it is reconnected with the correct polarity. Reversing the battery connections may result in damage to the 1-way rectifiers.

• When utilizing a booster battery as a starting aid, always connect the positive to positive terminals and the negative terminal from the booster battery to a good engine ground on the vehicle being started.

• Never use a fast charger as a booster to start vehicles.

• Disconnect the battery cables when charging the battery with a fast charger.

• Never attempt to polarize the alternator.

• Do not use test lamps of more than 12 volts when checking diode continuity.

• Do not short across or ground any of the alternator terminals.

• The polarity of the battery, alternator and regulator must be matched

and considered before making any electrical connections within the system.

• Never separate the alternator on an open circuit. Make sure all connections within the circuit are clean and tight.

• Disconnect the battery ground terminal when performing any service on electrical components.

• Disconnect the battery if arc welding is to be done on the vehicle.

REMOVAL AND INSTALLATION

1. Disconnect the negative battery cable. Disconnect the electrical connectors at the alternator.

2. Remove the necessary components in order to gain access to the alternator assembly. On Astro and Safari, remove the upper radiator fan shroud.

3. Remove the alternator belt. If equipped, relieve the tension on the serpentine drive belt and remove the serpentine drive belt.

4. Remove the alternator retaining bolts and remove the alternator from the vehicle.

5. Installation is the reverse of the removal procedure. Adjust the alternator belt, as required.

BELT TENSION ADJUSTMENT

Serpentine Drive Belt

The serpentine belt grooves must match the grooves in the pulleys. The tensioner is spring loaded. After removing the belt the tensioner will return the tension position.

1. Insert a ½ in. breaker bar into the tensioner pulley.

2. Rotate the tensioner to loosen the belt, then remove the belt.

3. Route the new belt over all the pulleys except the water pump.

4. Insert a ½ in. breaker bar into the tensioner pulley.

5. Rotate the tensioner pulley.

6. Install the belt over the water pump pulley.

7. Check the belt for correct V groove tracking.

Starter

REMOVAL AND INSTALLATION

Pickup, Blazer, Bravada, Jimmy, Sonoma, Cyclone and Typhoon

2WD VEHICLE

1. Disconnect the negative battery cable.

2. Raise and support the vehicle safely.

3. Disconnect the solenoid wiring.

4. On the 2.5L engine, disconnect the end mounting bracket and wiring.

5. Remove the 2 bolts and washers; then remove the starter and the shim.

6. Installation is the reverse of removal. Install the shim and torque the mounting bolts to specification.

4WD VEHICLE

1. Disconnect the negative battery cable.

2. Raise and support the vehicle safely.

3. Disconnect the solenoid wiring.

4. On the 2.5L engine, disconnect the end mounting bracket and wiring.

5. Remove the 4 bolts on the skid plate, if equipped and remove the skid plate.

6. Remove the bolts and 2 brackets holding the brake line to the crossmember.

7. Remove the 3 bolts on each side and remove the crossmember.

8. Remove the bracket holding the transmission cooler lines to the flywheel housing, brace rod to the flywheel housing, and the flywheel housing as necessary.

9. Remove the 2 bolts and washers and remove the starter and the shim.

10. Installation is the reverse of removal. Install the shim and torque the mounting bolts to specification.

Astro and Safari

1. Disconnect the negative battery cable.

2. Raise and support the vehicle safely.

3. Disconnect the solenoid wiring.

4. Remove the 2 bolts and washers and remove the starter and the shim.

5. Installation is the reverse of removal. Install the shim, if used, and torque the mounting bolts to specification.

Lumina APV, Silhouette and Trans Sport

3.1L ENGINE

1. Disconnect the negative battery cable.

2. Raise and support the vehicle safely.

3. Remove the air conditioning compressor brace.

4. Disconnect the oil pressure sensor electrical connector and remove the sensor.

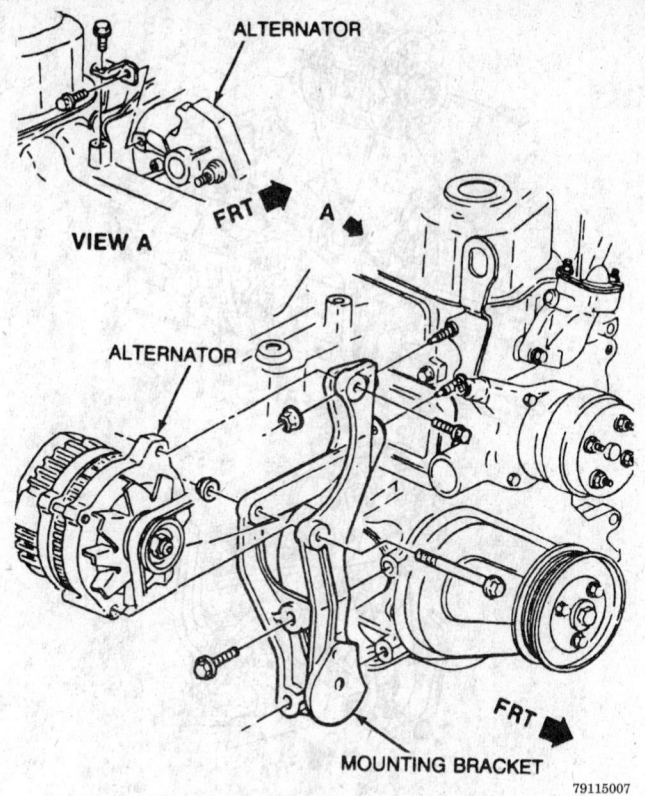

Alternator mounting — 2.5L engine

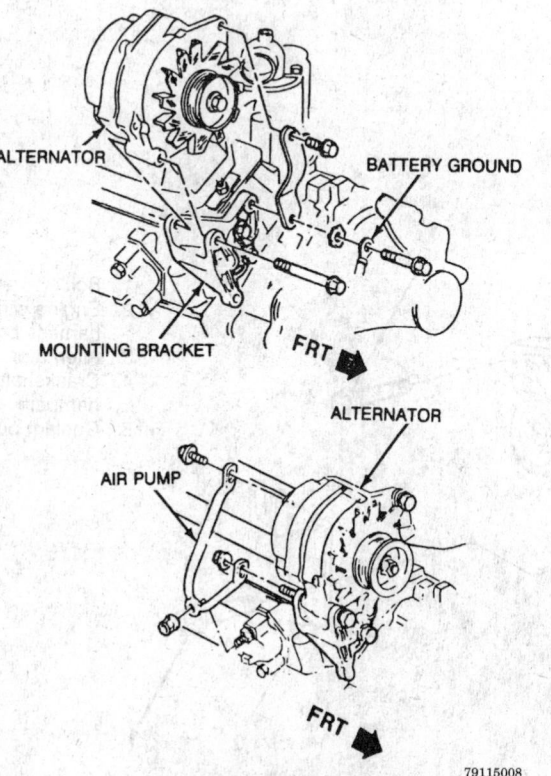

Alternator mounting — 2.8L engine

5. Remove the solenoid cover and disconnect the solenoid wiring.

6. Remove the 2 bolts and washers and remove the starter and the shim.

7. Installation is the reverse of removal. Install the shim, if used, and torque the mounting bolts to specification.

3.8L ENGINE

1. Disconnect the negative battery cable.

2. Raise and support the vehicle safely.

3. Disconnect the solenoid wiring and position aside.

4. Remove the flywheel cover. Remove the starter bolts and shim.

5. Remove the starter assembly.

6. Installation is the reverse of removal. Install the shim, if used, and torque the mounting bolts to specification.

CHASSIS ELECTRICAL

Heater Blower Motor

REMOVAL AND INSTALLATION

Pickup, Blazer, Bravada, Jimmy, Sonoma, Cyclone and Typhoon

1. Disconnect the battery ground cable.

2. Disconnect the blower motor electrical connections.

3. Remove the blower motor attaching screws.

4. Remove the blower motor from the vehicle.

5. Installation is the reverse of the removal procedure.

Astro and Safari

1. Disconnect the negative battery cable.

2. Remove the engine coolant bottle. Remove the 2 bolts from the windshield washer bottle and position the assembly aside.

3. Disconnect the electrical connections and cooling tube from the heater blower assembly. Remove the blower motor relay bracket, as required.

4. Remove the blower motor retaining screws and remove the motor from the vehicle.

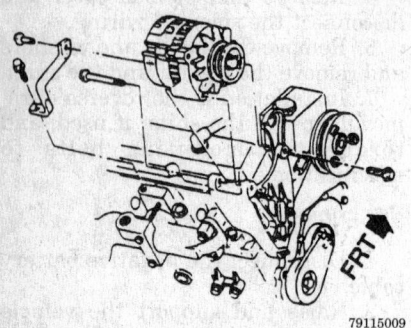

Alternator mounting — 3.1L engine

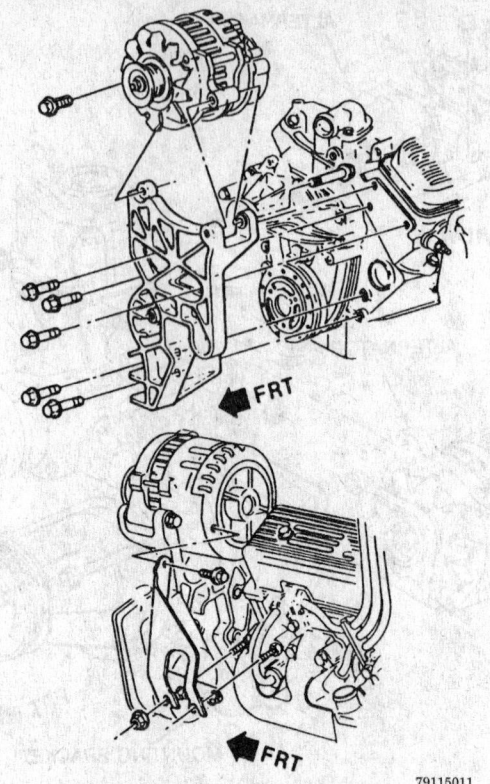

Alternator mounting — 4.3L engine

5. Installation is the reverse of the removal procedure. Transfer the blower motor cage, as required.

Lumina APV, Silhouette and Trans Sport

1. Disconnect the negative battery cable.
2. Remove the engine air cleaner.
3. Disconnect the left windshield wiper arm linkage.
4. Disconnect the blower motor electrical harness.
5. Remove the blower motor retaining screws and remove the blower motor assembly.

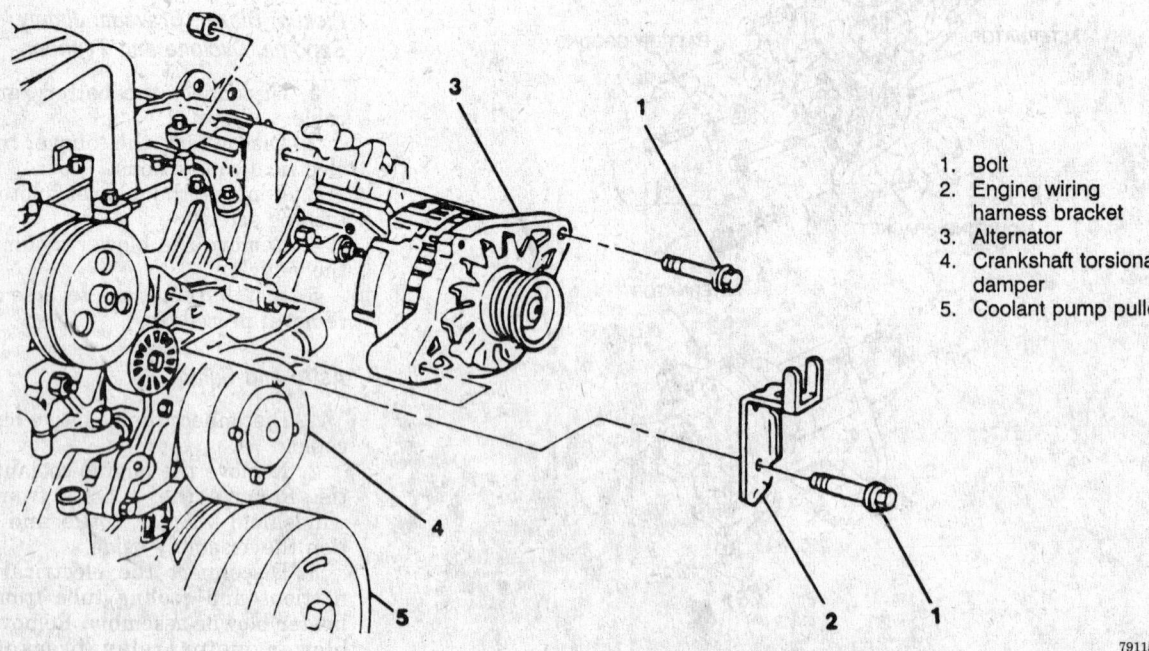

1. Bolt
2. Engine wiring harness bracket
3. Alternator
4. Crankshaft torsional damper
5. Coolant pump pulley

Alternator mounting — 3.8L engine

6. Installation is the reverse of removal.

Windshield Wiper Motor

REMOVAL AND INSTALLATION

Pickup, Blazer, Bravada, Jimmy, Sonoma, Cyclone and Typhoon

1. Disconnect the negative battery cable.
2. Remove the windshield wiper arms.
3. Remove the cowl vent grille and screen.
4. Mark the position of the wiper motor arm. Remove the wiper motor arm nut and remove the arm.
5. Disconnect the electrical wiring from the motor assembly.
6. Remove the motor retaining screws. Remove the windshield wiper motor while guiding it through the hole.
7. Installation is the reverse of the removal procedure. Install the arm in alignment with the previously made mark.

Astro and Safari

1. Disconnect the negative battery cable.
2. Remove the cowl vent grille.
3. Disconnect the transmission link from the crank arm on the motor by prying it toward the rear of the vehicle.
4. Remove the motor mounting bolts and remove the motor.
5. Installation is the reverse of removal.

Lumina APV, Silhouette and Trans Sport

1. Disconnect the negative battery cable.
2. Disconnect the wiper motor wiring harness connector.
3. Disconnect the transmission link from the crank arm on the motor by loosening the 2 crank arm screws.
4. Remove the motor mounting bolts and slide the motor out of its mounting.
5. Installation is the reverse of removal.

Windshield Wiper Switch

REMOVAL AND INSTALLATION

Except Trans Sport and Silhouette

The wiper switch is part of the multifunction switch on the column.
1. Disconnect the negative battery cable.
2. Remove the wiring protector cover under the column.
3. Disconnect the multifunction switch.
4. Disconnect the cruise control wire.
5. Remove the switch.
6. Installation is the reverse of removal.

Trans Sport and Silhouette

1. Disconnect the negative battery cable.
2. Grip the pod and carefully pull the wiper switch out to release the 2 spring retaining clips.
3. Disconnect the electrical connector and remove the switch.
4. Installation is the reverse of removal.

Instrument Cluster

REMOVAL AND INSTALLATION

Pickup, Blazer, Bravada, Jimmy, Sonoma, Cyclone and Typhoon

1. Disconnect the negative battery cable.
2. Remove the lamp switch trim plate screws and remove the trim plate.
3. Disconnect the lamp switch harness.
4. Remove the air conditioner and heater control assembly retaining screws and remove the control assembly.
5. Disconnect the air conditioner and heater control assembly harness.
6. Remove the filler panel screws and remove the filler panel.
7. Remove the instrument cluster housing nuts and remove the instrument cluster housing.
8. Remove the instrument cluster nuts. Disconnect the harness. Remove the instrument cluster.
9. Installation is the reverse of the removal procedure.

Astro and Safari

1. Disconnect the negative battery cable.

2. Remove the instrument panel cluster trim plate and screws.
3. Remove the instrument panel cluster and screws.
4. Disconnect the instrument panel cluster harness connectors and speedometer cable, as required.
5. Remove the instrument panel cluster.
6. Installation is the reverse of the removal procedure.

Lumina APV

1. Disconnect the negative battery cable.
2. Remove the 4 screws securing the instrument panel pad to the instrument panel lower trim pad.
3. On Canadian models, if equipped, disconnect the daytime running lights sensor attached to the pad under the front left hand speaker grille.
4. Lift the instrument panel trim pad up and pull the pad rearward to disengage the pad from the 4 slots in the lower trim pad and remove the pad assembly.
5. Grip the pod and carefully pull the wiper switch out to release the 2 spring retaining clips.
6. Disconnect the electrical connector and remove the switch.
7. Remove the 2 cluster housing screws attaching the trim to the instrument panel lower trim pad.
8. Feed the instrument panel harness through the switch pod in the housing.
9. Lift the housing up, pulling rearward to release the pad from the slots and remove the cluster.
10. Remove the cluster to the retainers.
11. Disconnect the panel and remove the cluster.

18. Install the instrument panel cluster.
19. Connect the negative battery cable.

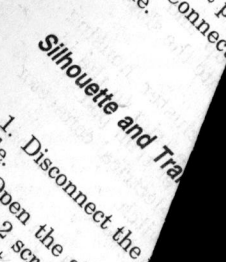

Silhouette and Trans Sport

1. Disconnect the negative battery cable.
2. Open the glove box to access the 2 screws securing the instrument panel trim plate to the instrument panel assembly.

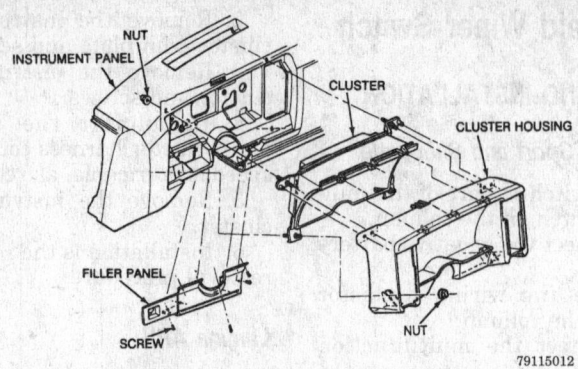

79115012

Instrument cluster assembly — Pick-Up, Blazer, Bravada and Jimmy

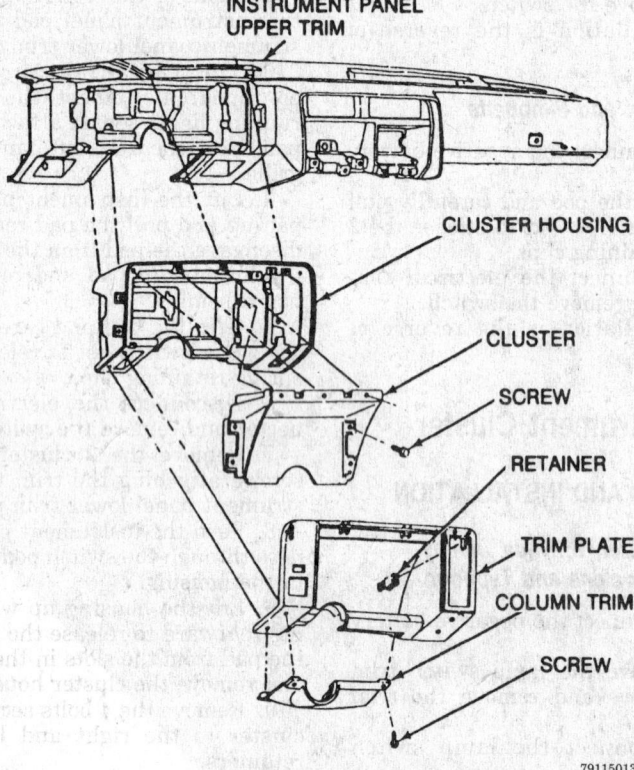

79115013

Instrument cluster assembly — Astro and Safari

ll the 4 screws securing
...ent panel pad to the in-
...nel lower trim pad.
... the negative battery

...the 2 tabs on the pad
...the housing.
...(Install the 2
...screws attaching t...
...strument panel...
...(Connect the...
...and install the s...

...negative battery

...ox door to ac-
...ing the lower
...d assembly
...anel pad
...grille

3. Remove the 2 screws securing
the instrument cluster trim panel to
the instrument panel pad.

4. Remove the screw retaining the
headlamp switch pod to the cluster
trim panel and remove the headlamp
switch pod. Disconnect the wiring
harness to the pod.

5. Remove the screw retaining the
windshield wiper switch pod to the
cluster trim panel and remove the
wiper switch pod. Disconnect the wiring harness to the pod.

6. Remove the 2 screws behind
each switch pod securing the cluster

trim panel to the left and right instrument cluster mounting brackets.

7. Remove the steering column
opening filler.

8. Disconnect the **PRNDL** cable
clip.

9. Disconnect the instrument
panel harness connector to the instrument cluster.

10. Remove the 2 screws on each
side from the left and right cluster
retainer brackets and remove the instruments cluster.

To install:

11. Connect the instrument panel
harness connector to the instrument
cluster.

12. Install the instrument cluster
and secure the cluster retainer brackets with 2 screws.

13. Connect the **PRNDL** cable clip.

14. Install the steering column
opening filler.

15. Install the 2 screws behind each
switch pod securing the cluster trim
panel to the left and right instrument
cluster mounting brackets.

16. Connect the wiring harness to
the pod. Install the wiper switch pod.

17. Connect the wiring harness to
the pod. Install the headlamp switch
pod.

18. Install the 2 screws securing
the instrument cluster trim panel to
the instrument panel pad.

19. Install the 2 screws securing
the lower instrument panel trim pad
assembly to the instrument panel
pad assembly, behind the glove box
door.

20. Connect the negative battery
cable.

Speedometer

The electronic controlled unit is non-
serviceable.

REMOVAL AND INSTALLATION

Pickup, Blazer and Jimmy

1. Remove the instrument cluster.
2. Remove the cluster case retaining screws.
3. Remove the speedometer to
cluster case retaining screws.
4. Remove the cluster panel from
the cluster case.
5. Remove the speedometer
mounting screws and remove the
speedometer.
6. Installation is the reverse of
removal.

Astro and Safari

1. Remove the instrument cluster.

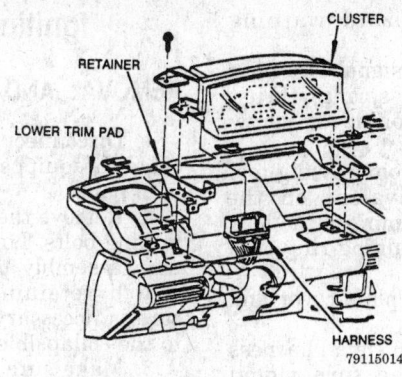

Instrument cluster assembly — Silhouette

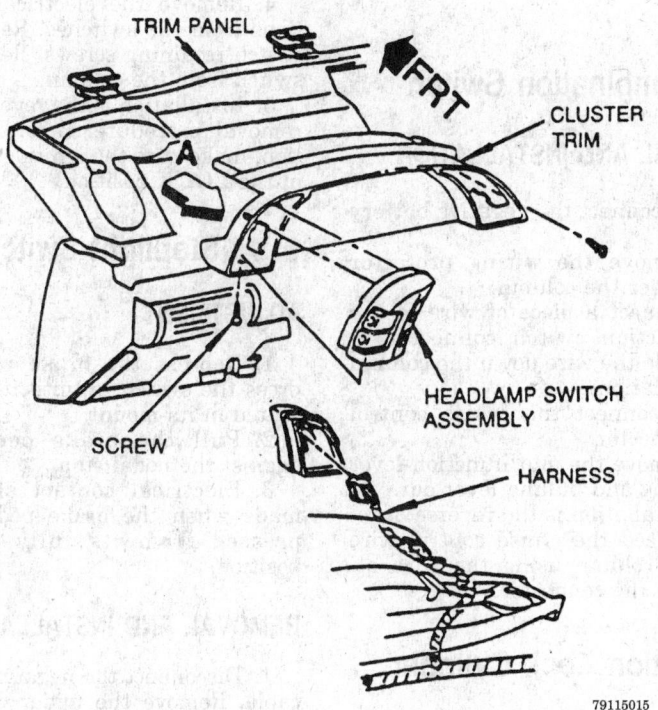

Instrument cluster assembly — Trans Sport

2. Remove the speedometer head retaining screws and remove the speedometer head.

3. Installation is the reverse of removal.

Radio

REMOVAL AND INSTALLATION

Pickup, Blazer, Bravada, Jimmy, Sonoma, Cyclone and Typhoon

1. Disconnect the negative battery cable.

2. Remove the ash tray and any wires needed for clearance.

3. Remove the trim plate.

4. Remove the nuts from the support clips.

5. Remove the support bracket retainer screws.

6. Pull the radio forward and disconnect the antenna cable, clock connector, speaker connectors and remove the radio.

NOTE: Do not let the antenna cable touch the clock connector.

7. Installation is the reverse of removal. Always connect the speaker wiring harness to the radio before applying power to the receiver to prevent damage to the receiver.

Astro and Safari

1. Disconnect the negative battery cable.

2. Remove the instrument panel compartment assembly.

3. Remove the accessory trim plate.

4. Disconnect the radio fasteners from the instrument panel.

5. Pull the radio forward and disconnect the antenna cable and speaker and electrical connectors and remove the radio.

6. Installation is the reverse of removal. Always connect the speaker wiring harness to the radio before applying power to the receiver to prevent damage to the receiver.

Lumina APV, Silhouette and Trans Sport

1. Disconnect the negative battery cable.

2. Remove the steering column opening filler panel.

3. Remove the right hand sound insulator panel.

4. Remove the accessory trim panel.

5. Loosen the 2 nuts at the bottom of the radio.

6. Remove the 2 screws attaching the radio bracket to the instrument panel lower trim pad assembly.

7. Slide the radio assembly out from the accessory housing.

8. Disconnect the antenna lead-in cable and the electrical connectors from the radio, then remove the radio.

9. Installation is the reverse of removal.

Dimmer Switch

REMOVAL AND INSTALLATION

1. Disconnect the negative battery cable. If equipped, remove the lower trim panel.

2. Remove the steering column retaining bolts. Lower the steering column assembly to gain access to the switch retaining screws. Extreme care is necessary to prevent damage to the collapsible column.

3. Make sure the ignition switch is in the **LOCK** position.

4. Remove the electrical connections from the switch. Remove the switch retaining screws. Remove the switch from the column.

To install:

5. Attach the switch to the column with the screws finger-tight. To adjust the dimmer switch, depress the switch slightly to allow insertion of a ⁵/₃₂ in. drill bit into the hole above the actuator rod. Force the switch upward to remove all lash, then tighten the screw.

6. Connect the switch wire harness.

7. Raise and secure the column the steering column. Install the lower trim plate, if removed.

8. Connect the negative battery cable.

Turn Signal Switch

REMOVAL AND INSTALLATION

1. Disconnect the negative battery cable.

2. Remove the steering wheel.

3. Remove the instrument panel trim cover. Remove the turn signal wire harness cover and disconnect the turn signal harness connector.

NOTE: To save time, attach a piece of wire or string to turn signal harness connector before pulling the harness up through the column. Then attach the string or wire to the new switch to help guide the harness back down through the column upon installation.

4. Insert a suitable tool in the slot in the lock plate cover and remove the cover.

5. Use lock plate compressing tool J–23653–A or equivalent, on the steering shaft and compress the lock plate. Pry the retaining ring off the shaft and remove the lock plate.

NOTE: If the column is being disassembled on a bench, the shaft could slide out of the end of the mast jacket when the snapring is removed.

6. Remove the turn signal canceling cam, upper bearing spring, thrust washer, multifunction lever screw and lever.

7. Press the hazard warning knob inward and then unscrew.

8. Remove the turn signal switch mounting screws. Pull the switch straight up, guiding the wiring harness and cover through the column housing.

To install:

9. Route the harness down through the column and secure the switch to the column.

10. Install the hazard warning knob and the screw.

11. Install the turn signal canceling cam, upper bearing spring, thrust washer, multifunction lever screw and lever.

12. Use lock plate compressing tool J–23653–A or equivalent on the steering shaft and compress the lock plate. Install the retaining ring onto the shaft.

13. Install the lock plate cover and steering wheel.

14. Connect the turn signal harness connector. Install the turn signal wire harness cover and the instrument panel trim cover.

15. Connect the negative battery cable.

Combination Switch

REMOVAL AND INSTALLATION

1. Disconnect the negative battery cable.

2. Remove the wiring protector cover under the column.

3. Connect a piece of wire to the multifunction switch connector to help guide the wire down the column on installation.

4. Disconnect the cruise control wire connector.

5. Remove the multifunction lever by turning and pulling lever out.

6. Installation is the reverse of removal. Feed the cruise control wire into the column using the wire attached to the connector.

Ignition Lock Cylinder

REMOVAL AND INSTALLATION

1. Disconnect the negative battery cable.

2. Place the lock cylinder in the **RUN** position.

3. Remove the steering wheel.

4. Remove the turn signal switch. Pull the switch rearward far enough to slip it over the shaft. Do not pull the harness out of the column.

5. Remove the retaining screw and remove the lock cylinder.

To install:

6. Align the cylinder and key with the keyway in the housing, rotate clockwise and push all the way in.

7. Install the retaining screw.

8. Install the turn signal switch and steering wheel.

9. Connect the negative battery cable.

Ignition Switch

REMOVAL AND INSTALLATION

1. Disconnect the negative battery cable. If equipped, remove the lower trim panel.

2. Remove the steering column retaining bolts. Lower the steering column assembly to gain access to the switch retaining screws. Extreme care is necessary to prevent damage to the collapsible column.

3. Make sure the switch is in the **LOCK** position. If the lock cylinder is out, pull the switch rod up to the stop, then go down one detent.

4. Remove the electrical connections from the switches. Remove the switch retaining screws. Remove the switch from the column.

5. Installation is the reverse of the removal procedure. Before installation, make sure the ignition switch is in the **LOCK** position.

Stoplight Switch

ADJUSTMENT

1. Depress the brake pedal and press the switch in until it is firmly seated in its mount.

2. Pull the brake pedal back against the pedal stop.

3. Electrical contact should be made when the brake pedal is depressed from its fully released position.

REMOVAL AND INSTALLATION

1. Disconnect the negative battery cable. Remove the under dash trim panel.

2. Disconnect the switch electrical connections. Remove the switch assembly from its mount.

3. Installation is the reverse of the removal procedure. Adjust the switch, as required.

Clutch Switch

ADJUSTMENT

The clutch switch is automatic adjusting upon installation and initial pedal depression.

REMOVAL AND INSTALLATION

1. Disconnect the negative battery cable. Remove the under dash trim panel.

2. Disconnect the switch electrical connections.

3. Remove the switch retaining screw, if equipped and remove the switch from the clutch pedal.

To install:

4. Move the slider to the rear of the shaft.

5. Push the clutch pedal to the floor.

6. Move the slider down the shaft.

7. Release the clutch pedal.

8. Install the electrical connector.

9. Install the trim panel. Reconnect the negative battery cable.

Neutral Safety Switch

ADJUSTMENT

1. Move the switch all the way toward the **L** gear position.

2. Move the selector to the **P** position.

3. The main housing and housing back should ratchet, providing proper switch adjustment.

REMOVAL AND INSTALLATION

1. Disconnect the negative battery cable.

2. Place the gear selector in **N** position.

3. Remove the steering column lower trim cover and disconnect the switch wire connector.

4. Spread the tangs on the housing and remove the switch.

To install:

5. Align the actuator on the switch with the hole in the shift tube.

6. Position the rear portion of the switch (wire connector side) to fit into the cutout in the lower jacket.

7. Push down on the front of the switch to engage the 2 tangs.

8. Move the gear selector to the **N** position and the switch is adjusted. Reconnect the switch wire connector.

9. Install the lower trim cover and connect the negative battery cable.

Fuses, Circuit Breakers and Relays

LOCATION

Astro, Blazer, Bravada, Jimmy, Pickup, Safari, Sonoma, Cyclone and Typhoon

Fuse Block — located at the far left side of the instrument panel.

Convenience Center — located at the far left side of the instrument panel.

Hazard Warning Relay — mounted on the convenience center.

Horn Relay — mounted on the convenience center.

Alarm Module or Buzzers — mounted on the convenience center.

Circuit Breakers — mounted on the convenience center.

A/C Low Fan Relay — mounted on the convenience center.

Lumina APV, Silhouette and Trans Sport

Fuse Panel — located inside the glove compartment.

Convenience Center — mounted to a bracket behind the glove compartment. It can be reached by removing the right sound insulator panel. It contains the hazard warning flasher, horn relay and circuit breakers, which can be removed by pulling straight out. It also contains the chime module and A/C low fan relay. To remove the chime module first release the locking tab.

Computers

LOCATION

Astro and Safari

Electronic Control Module (ECM) — located in the passenger compartment behind the right side cowl panel.

Electronic Spark Control (ESC) Module (4.3L engine) — located at the right side of the engine above the valve cover.

Cruise Control Module (Electronic Cluster) — located on the left, near the brake booster.

Cruise Control Module (Standard Cluster) — located on the left side of the instrument panel on the rear of the dash.

Blazer, Bravada, Jimmy, Pickup, Sonoma, Cyclone and Typhoon

Electronic Control Module (ECM) — located in the passenger compartment behind the right side of instrument panel.

Electronic Spark Control (ESC) Module — located at the center of the firewall.

Cruise Control Module — located behind the instrument panel on the left side of the steering column.

Lumina APV, Silhouette and Trans Sport

Electronic Control Module (ECM) — located below the right side of instrument panel, behind the convenience center.

Electronic Spark Control (ESC) Module — located behind the instrument panel, behind the convenience center.

Cruise Control Module — located behind the instrument panel on the left side.

Cruise Control

SERVO ADJUSTMENT

1. Remove the cruise control cable from the servo.

2. Ensure ignition and fast idle cam are **OFF** and throttle is fully closed before adjusting servo.

3. Connect the cable to the servo. The rod should have 0.039–0.150 in. (1–4mm) of clearance at the stud.

4. Install the cable retainer and check for proper operation.

ENGINE COOLING

Radiator

REMOVAL AND INSTALLATION

Astro, Blazer, Bravada, Jimmy, Pickup, Safari and Sonoma

1. Disconnect the negative battery cable.

2. Drain the coolant from the radiator.

3. Remove the overflow hose.

4. Remove the upper fan shroud.

5. Disconnect the hoses.

6. Disconnect and plug the transmission fluid cooler lines, if equipped.

7. Disconnect and plug the engine oil cooler lines, if equipped.

8. Remove the radiator retaining bolts and remove the radiator.

9. Installation is the reverse of the removal procedure.

Lumina APV, Silhouette and Trans Sport

1. Disconnect the negative battery cable.

2. Drain the coolant from the radiator.

3. Disconnect the engine forward strut bracket at the radiator, loosen the bolt at the other end and swing the strut rearward.

4. Disconnect the forward lamp harness from the fan frame and unplug the fan connector.

5. Remove the fan attaching bolts and remove the fan and frame assembly.

6. Scribe the latch location then remove the hood latch from the radiator support.

7. Disconnect the coolant hoses from the radiator and the coolant recovery tank hose from the radiator neck.

8. Disconnect and plug the transaxle oil cooler lines.

9. Remove the radiator to support attaching bolts and clamps and remove the radiator from the vehicle.

10. Installation is the reverse of removal.

Cyclone and Typhoon

1. Disconnect the battery cables; remove the battery and tray assembly.

2. Remove the turbocharger inlet elbow.

3. Drain the cooling system into a suitable container

4. Remove the heater and coolant reservoir hoses from the radiator.

5. Remove the upper fan shroud.

6. Remove the upper and lower radiator hoses from the radiator.

7. Disconnect and plug the transmission and oil cooler lines.

8. Remove the radiator assembly.

9. Installation is the reverse of removal procedure.

Electric Cooling Fan

REMOVAL AND INSTALLATION

Lumina APV, Silhouette and Trans Sport

1. Disconnect the negative battery cable.

2. Disconnect the engine forward strut bracket from the radiator frame and swing it rearward.

3. Disconnect the forward lamp harness from the fan frame.

4. Remove the fan attaching bolts.

5. Disconnect the fan wiring.

6. Remove the fan and frame assembly from the vehicle.

7. Installation is the reverse of removal procedure.

Heater Core

REMOVAL AND INSTALLATION

Blazer, Bravada, Jimmy, Pickup, Sonoma, Cyclone and Typhoon

1. Disconnect the negative battery cable.

2. Drain the cooling system.

3. Remove and plug the heater hoses at the heater core.

4. Remove the heater core cover attaching screws and remove the cover.

5. Remove the screw retainers at the end of the heater core.

6. Remove the core from under the dash assembly.

7. Installation is the reverse of the removal procedure.

Astro and Safari

1. Disconnect the negative battery cable. Drain the engine coolant.

2. Remove the engine coolant bottle. Remove the bolts from the windshield washer bottle and position it aside. Remove and plug the heater hoses at the heater core.

3. Remove the instrument panel lower right filler panel. Remove the air distributor duct. Remove the engine cover as needed.

4. Remove the air duct. Remove vacuum lines and control cables as required.

5. Remove the heater core assembly retaining screws.

6. Remove the heater core cover plate. Remove the heater core.

7. Installation is the reverse of the removal procedure.

Lumina APV, Silhouette and Trans Sport

1. Disconnect the negative battery cable.

2. Drain the cooling system.

3. Disconnect the heater hoses.

4. Remove the right side sound insulator from under the instrument panel.

5. Remove the glove box.

6. Disconnect the vacuum hoses and wiring from in front of the heater core cover.

7. Remove the heater core cover and remove the heater core.

8. Installation is the reverse of the removal procedure.

Water Pump

REMOVAL AND INSTALLATION

1. Disconnect the negative battery cable. Drain the coolant. Remove the fan shroud, as necessary. Remove the drive belts. Remove the fan shroud, fan and pulley assembly.

2. Remove the necessary components in order to gain access to the water pump retaining bolts.

3. Remove the water pump retaining bolts. Remove the water pump assembly from the engine.

4. Installation is the reverse of the removal procedure. Use a new gasket and torque all bolts to specification.

Thermostat

REMOVAL AND INSTALLATION

1. Disconnect the negative battery cable.

2. Drain the engine coolant.

3. Disconnect the upper radiator hose from the thermostat outlet.

4. Remove the thermostat housing bolts and remove the housing.

5. Remove the thermostat from the housing.

6. Installation is the reverse of the removal procedure. Torque the housing bolts to specification.

7. Bleed the cooling system.

COOLING SYSTEM BLEEDING

1. To bleed the system, start with the system cool, the radiator cap off and the radiator filled to about an inch below the filler neck.

2. Start the engine and run it at slightly above normal idle speed. If air bubbles appear and the coolant level drops, fill the system with an antifreeze/water mixture to bring the level back to the proper level.

3. Run the engine until the thermostat opens and coolant flow is visible.

4. At this point, air is often expelled and the level may drop. Keep refilling the system until the level is near the top of the radiator and remains constant.

5. Fill the radiator to the filler neck. Replace the radiator filler cap and ensure the coolant reservoir is filled to the correct level.

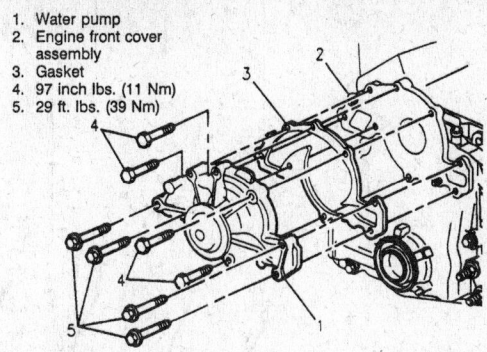

1. Water pump
2. Engine front cover assembly
3. Gasket
4. 97 inch lbs. (11 Nm)
5. 29 ft. lbs. (39 Nm)

79115018

Water pump assembly mounting — 3.8L engine

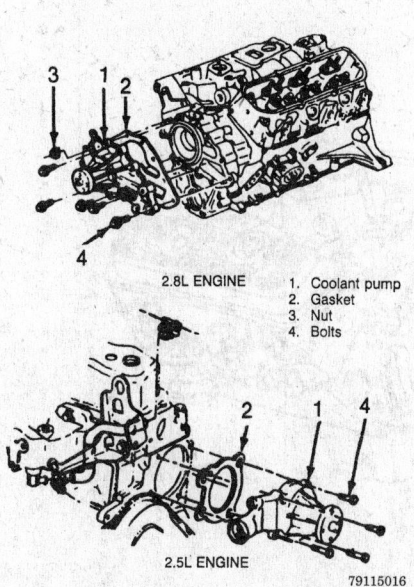

2.8L ENGINE

1. Coolant pump
2. Gasket
3. Nut
4. Bolts

2.5L ENGINE

79115016

Water pump assembly mounting — 2.5L and 2.8L engines

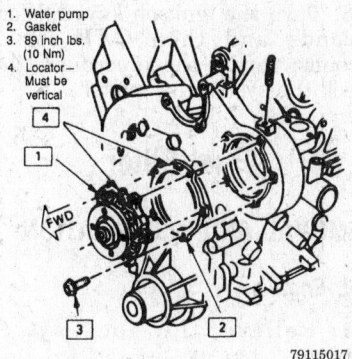

1. Water pump
2. Gasket
3. 89 inch lbs. (10 Nm)
4. Locator — Must be vertical

79115017

Water pump assembly mounting — 3.1L engine

FUEL SYSTEM

Fuel System Service Precaution

RELIEVING FUEL SYSTEM PRESSURE

Except 3.8L and 4.3L (MPI and CPI) Engines

The Throttle Body Injection (TBI) system used on the 2.5L, 2.8L, 3.1L and 4.3L (VIN Z) engines have a built in bleed feature which requires no pressure relief before servicing. The Multi-Port Injection (MPI) and Central Port Injection (CPI) systems used with the 3.8L and 4.3L (VIN W) engines do require pressure relief before servicing system components.

3.8L and 4.3L (MPI and CPI) Engines

1. Place the transmission selector in the **P** position for automatic transmission or neutral for a manual transmission.
2. Set the parking brake and block the drive wheels.
3. Disconnect the negative battery cable.
4. Connect a fuel pressure gauge to the fuel line. Install a bleed hose into an approved container and slowly open the valve to relieve the system pressure.
5. Service the system as required.

Fuel Tank

REMOVAL AND INSTALLATION

Except Lumina APV, Silhouette and Trans Sport

1. Disconnect the negative battery cable.
2. Properly relieve the fuel system pressure and, using an approved pump, drain the fuel from the tank into a suitable container.
3. Raise and support the vehicle safely.
4. Remove the fuel tank plastic shield.
5. Loosen the filler neck hose clamp at the tank and disconnect the filler neck.
6. Properly support the fuel tank and remove the tank brackets.
7. Lower the tank and remove the fuel/vapor hoses and electrical connections at the sender.
8. Remove the tank from the vehicle. Remove the fuel sender and seal ring using tool J–36608, as required.
   To install:
9. Install the fuel sender and seal ring using tool J–36608, if removed.
10. With the aid of an assistant, raise the tank into position.
11. With the tank in position, connect the fuel/vapor hoses and electrical connections at the sender.
12. Install the fuel tank and brackets.
13. Connect the filler neck and tighten the filler neck hose clamp at the tank.
14. Install the fuel tank plastic shield.
15. Connect the negative battery cable.
16. Lower the vehicle. Refill the tank and inspect for leaks.

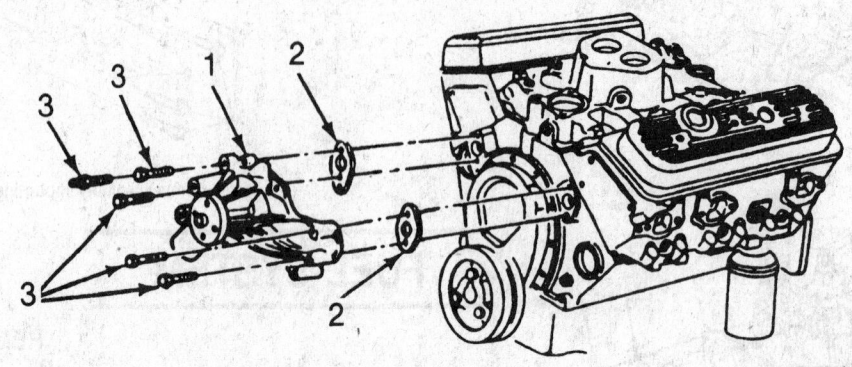

1. Water pump
2. Gasket
3. Bolt

79115019

Water pump assembly mounting — 4.3L engine

Lumina APV, Silhouette and Trans Sport

1. Disconnect the negative battery cable.

2. Properly relieve the fuel system pressure and, using an approved pump, drain the fuel from the tank into a suitable container.

3. Raise and support the vehicle safely.

4. Remove the tail pipe, muffler and converter hangers.

5. Remove the exhaust heat shield attaching bolts. Support the exhaust and move the heat shield to gain access to the right side fuel tank retaining strap attaching bolts.

6. Remove the in-line fuel filter body clips.

7. Grasp both ends of 1 fuel line at both the inlet side of the in-line fuel filter and the return line connections and twist a ¼ turn in each direction to loosen any dirt in the quick connect fitting.

8. Squeeze the plastic tabs of the male ends of the connectors and pull the connections apart.

9. Loosen the filler neck hose clamp at the tank and disconnect the filler neck.

10. Lower the tank and remove the fuel vapor/vent hoses and electrical connections at the sender.

11. Properly support the fuel tank and remove the strap bolts and the fuel tank straps.

12. Remove the tank from the vehicle. Remove the fuel sender and seal ring using tool J–35731, as required.

To install:

13. Install the fuel sender and seal ring using tool J–35731, if removed.

14. With the aid of an assistant, raise the tank into position.

15. With the tank in position, connect the fuel vapor/vent hoses and electrical connections at the sender.

16. Secure the fuel tank with the brackets.

17. Connect the filler neck and tighten the filler neck hose clamp at the tank.

18. Apply oil to the fuel feed/return line quick connect fitting and push connectors together until a snap is heard. Pull on both lines and verify they are secure.

19. Secure the in-line fuel filter with new body clips.

20. Install the exhaust heat shield, the tail pipe, muffler and converter hangers.

21. Lower the vehicle an fill the tank.

22. Connect the negative battery cable.

23. Turn the ignition key **ON** for 2 seconds and then **OFF** for 10 seconds, repeat the procedure; then check the system for leaks.

Fuel Filter

REMOVAL AND INSTALLATION

2.5L Engine

1. Relieve the fuel system pressure.

2. Disconnect the negative battery cable.

3. Remove the fuel line connections from the filter.

4. Remove the filter mounting clamp bolt and remove the filter.

5. Installation is the reverse of the removal procedure.

2.8L, 3.1L and 3.8L Engines

1. Disconnect the negative battery cable.

2. Remove the fuel line connections from the filter. If equipped with quick connect fittings, use tool J–37088A to separate the lines.

3. Remove the filter mounting clamp bolt and remove the filter.

4. Installation is the reverse of the removal procedure.

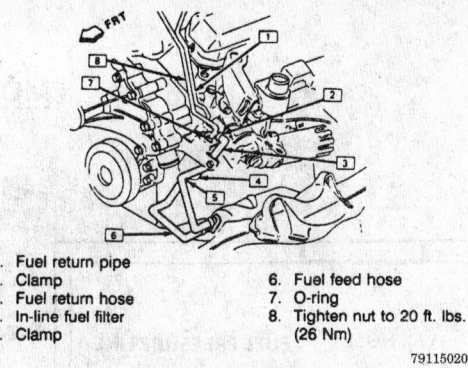

1. Fuel return pipe
2. Clamp
3. Fuel return hose
4. In-line fuel filter
5. Clamp
6. Fuel feed hose
7. O-ring
8. Tighten nut to 20 ft. lbs. (26 Nm)

79115020

Fuel filter removal and installation — 2.8L engine

4.3L Engine

1. The fuel filter is located along the frame rail of the vehicle.
2. Disconnect the negative battery cable.
3. Raise and safely support the vehicle.
4. Remove the fuel line connections from the filter.
5. Remove the filter mounting clamp bolt and remove the filter.
6. Installation is the reverse of the removal procedure.

Electric Fuel Pump

PRESSURE TESTING

THROTTLE BODY INJECTION (TBI)

1. Turn the engine **OFF** and relieve the fuel system pressure.
2. Disconnect the negative battery cable.
3. Disconnect the fuel supply line in the engine compartment. Install a suitable pressure gauge and T in the fuel line.
4. Connect the negative battery cable. Verify there is a sufficient quantity of fuel in the tank.

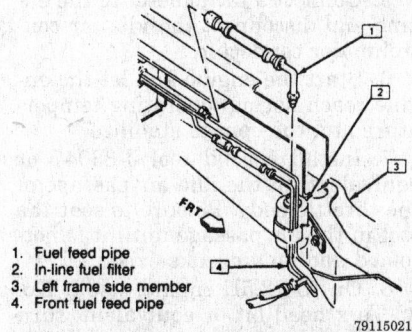

1. Fuel feed pipe
2. In-line fuel filter
3. Left frame side member
4. Front fuel feed pipe

79115021

Fuel filter removal and installation — 4.3L engine

5. Start the engine and observe the pressure reading. The fuel pressure should be 9–13 psi.
6. If the fuel pressure reading is not as specified, inspect the fuel pump for proper operation, the lines and filter for kinks or clogging.

Multi-Port Fuel Injection

3.8L Engine

1. Turn the engine **OFF** and relieve the fuel system pressure.
2. Disconnect the negative battery cable.
3. Connect a suitable pressure gauge to the pressure regulator service fitting.
4. Reconnect the negative battery terminal and verify there is a sufficient quantity of fuel in the tank.
5. Turn the ignition switch **ON**, the fuel pump should run for 2 seconds and turn **OFF**; the pressure gauge reading should be approximately 41–47 psi.
6. Turn the ignition switch **OFF** and observe the pressure gauge, the reading should not leak down.
7. Start the vehicle and observe the gauge, the pressure should be 3–10 psi lower because of vacuum applied to the regulator.
8. If not as specified, inspect the pump, regulator, filter and lines for proper operation, kinks or clogging.

4.3L (VIN Z) Turbocharged Engine

1. Turn the engine **OFF** and relieve the fuel system pressure.
2. Disconnect the negative battery cable.
3. Connect a suitable pressure gauge to the pressure regulator service fitting.
4. Reconnect the negative battery terminal and verify there is a sufficient quantity of fuel in the tank.
5. Turn the ignition switch **ON**, the fuel pump should run for 2

seconds and turn **OFF**; the pressure gauge reading should be approximately 35–38 psi.
6. Turn the ignition switch **OFF** and observe the pressure gauge, the reading should not leak down.
7. Start the vehicle and observe the gauge, the pressure should be 25–30 psi because of vacuum applied to the regulator.
8. If not as specified, inspect the pump, regulator, filter and lines for proper operation, kinks or clogging.

Central Port Injection (CPI)

1. Turn the engine **OFF** and relieve the fuel system pressure.
2. Disconnect the negative battery cable.
3. Connect a suitable pressure gauge to the pressure connection fitting.
4. Reconnect the negative battery terminal and verify there is a sufficient quantity of fuel in the tank.
5. Turn the ignition switch **ON**, the fuel pump should run for 2 seconds and turn **OFF**; the pressure gauge reading should be approximately 54–64 psi.
6. Turn the ignition switch **OFF** and observe the pressure gauge, the reading should not leak down.
7. Start the vehicle and observe the gauge, the pressure should be noticeably lower because of vacuum applied to the regulator.
8. If not as specified, inspect the pump, regulator, filter and lines for proper operation, kinks or clogging.

REMOVAL AND INSTALLATION

1. Properly relieve the fuel system pressure, as required.
2. Disconnect the negative battery cable.
3. Drain the fuel from the vehicle, into a suitable container.
4. Raise and safely support the vehicle.
5. Disconnect the fuel lines from the fuel tank. Disconnect the electrical leads from the fuel tank.
6. Disconnect the fuel filler neck from the tank.
7. Remove the tank mounting bolts and carefully lower the tank from the vehicle.
8. Remove the pump and Pickup assembly retaining ring and remove the pump from the tank.
9. Installation is the reverse of the removal procedure.

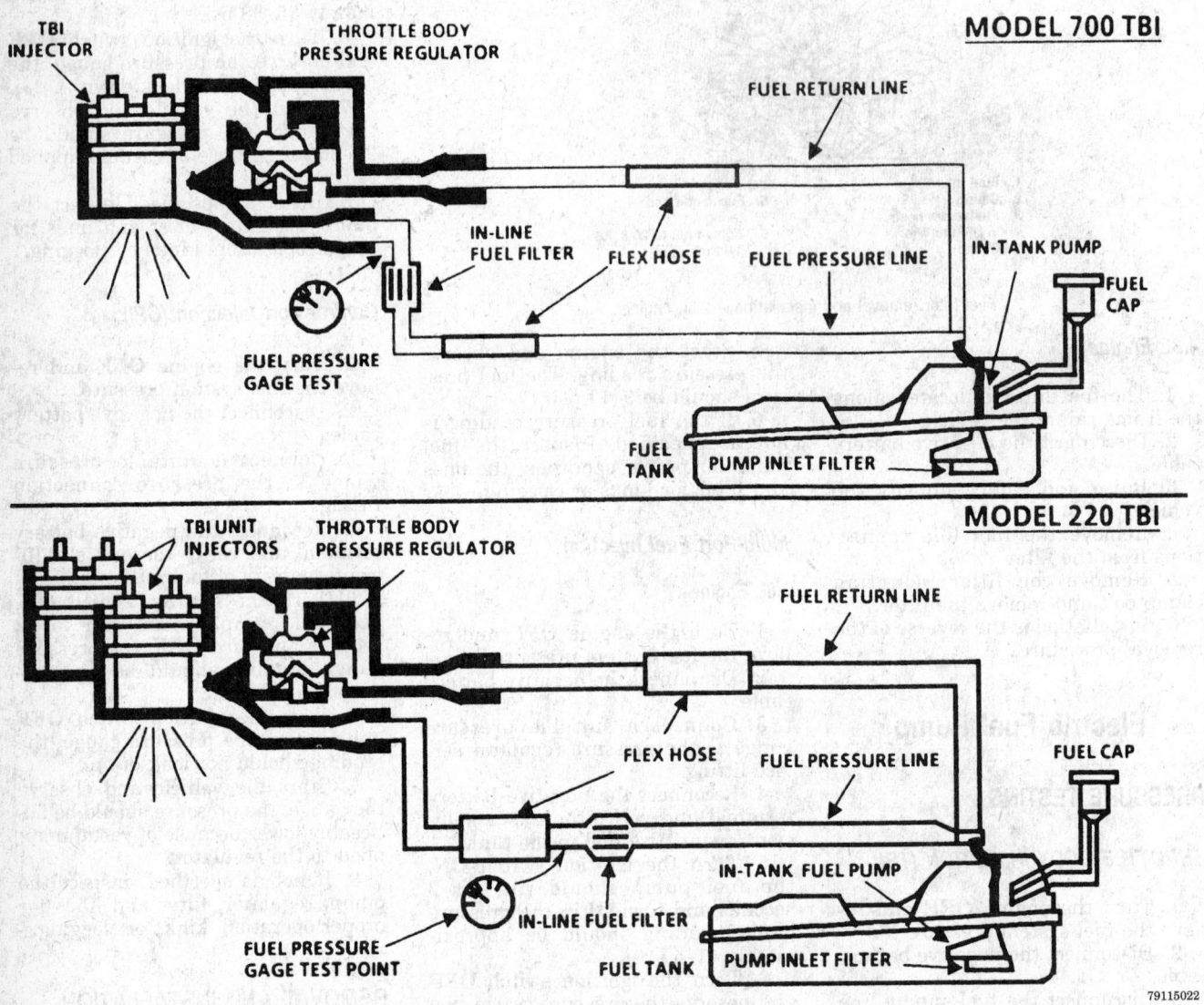

Fuel system pressure testing — 2.5L, 2.8L and 4.3L (VIN Z) TBI engines

Fuel Injection

IDLE SPEED ADJUSTMENT

2.5L Engine

The throttle stop screw that is used to adjust the idle speed of the vehicle, is preset at the factory. The throttle stop screw is then covered with a steel plug to prevent adjustment in the field. If it is necessary to gain access to the throttle stop screw, the following procedure will allow access to the throttle stop screw without removing the TBI unit from the manifo

1. Using a small punch or equivalent, mark the center line of the throttle stop screw. Drill a ⁵⁄₃₂ in. diameter hole through the casting of the hardened steel plug.

2. Using a ⁵⁄₁₆ in. diameter punch or equivalent, punch out the steel plug.

3. With the transmission in **P** for automatic transmission or neutral for manual transmission equipped vehicles, the parking brake applied and the drive wheels blocked, remove the air cleaner and plug the thermac vacuum port, as required.

4. If equipped with automatic transmission, remove the transmission detent cable from the throttle control bracket in order to gain access to the minimum air adjustment screw.

5. Connect a tachometer to the engine and disconnect the idle air control motor connector.

6. Start the engine and let the engine reach normal operating temperature and the rpm to stabilize.

7. Install special tool J-33047 or equivalent, in the idle air passage of the throttle body. Be sure to seat the tool in the air passage until it is bottomed and no air leaks exist.

8. On the 2.5L engine, use a No. 20 Torx head bit or equivalent, turn the throttle stop screws until the minimum idle rpm is within specification.

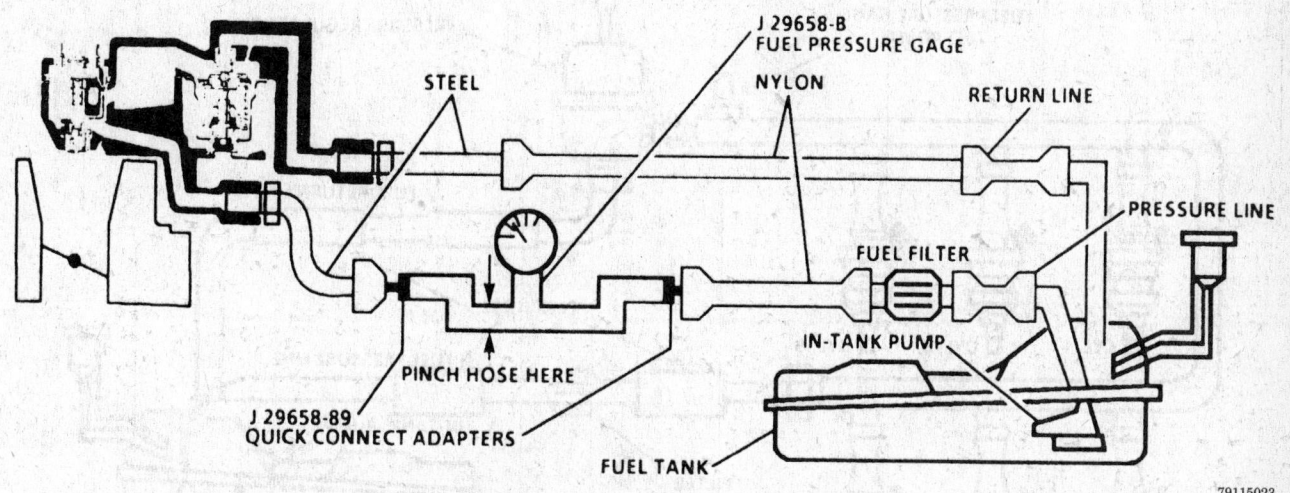

Fuel system pressure testing — 3.1L engine

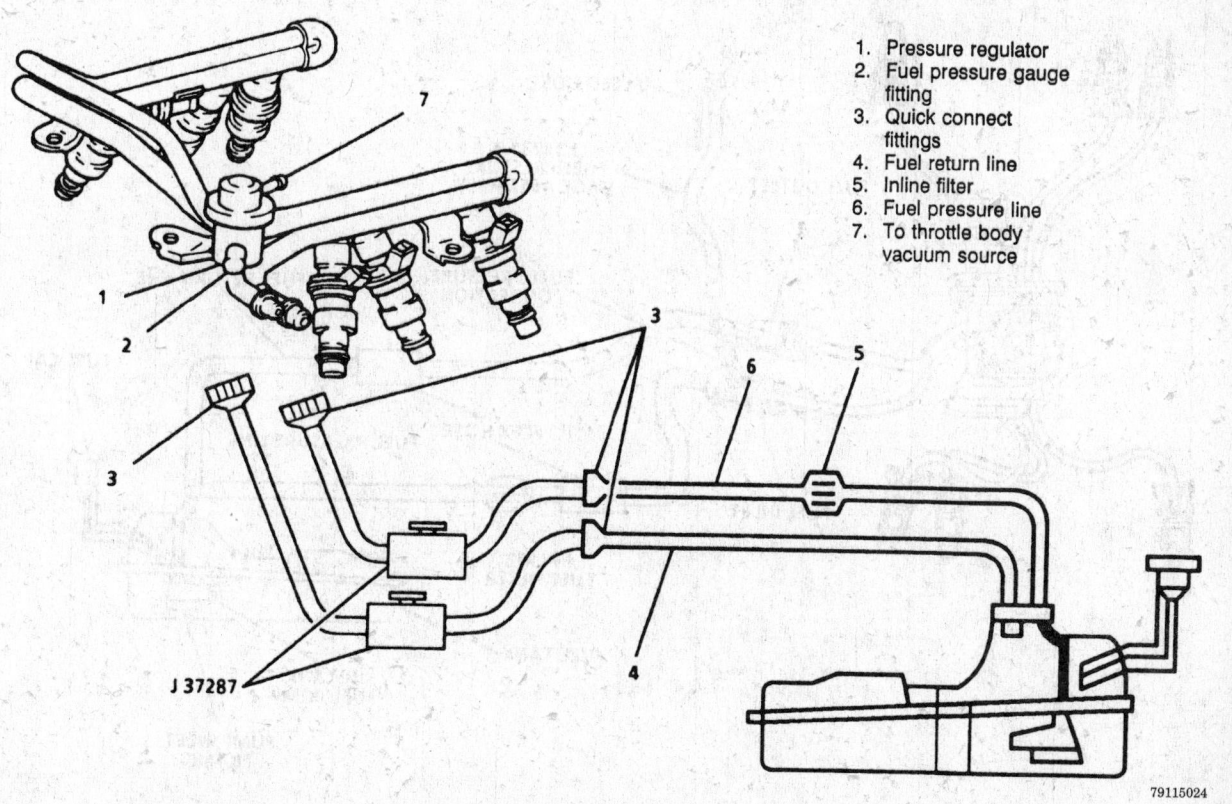

1. Pressure regulator
2. Fuel pressure gauge fitting
3. Quick connect fittings
4. Fuel return line
5. Inline filter
6. Fuel pressure line
7. To throttle body vacuum source

Fuel system pressure testing — 3.8L engine

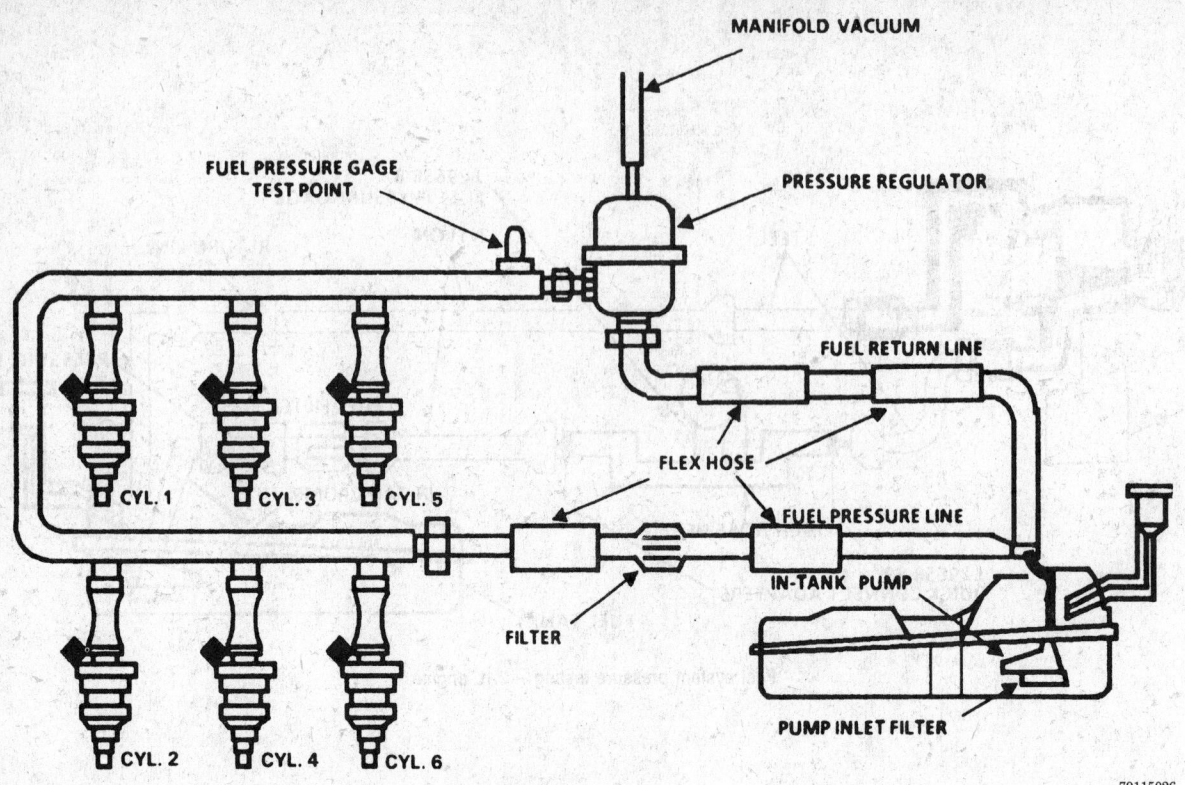

Fuel system pressure testing — 4.3L (VIN Z) Turbocharged engine

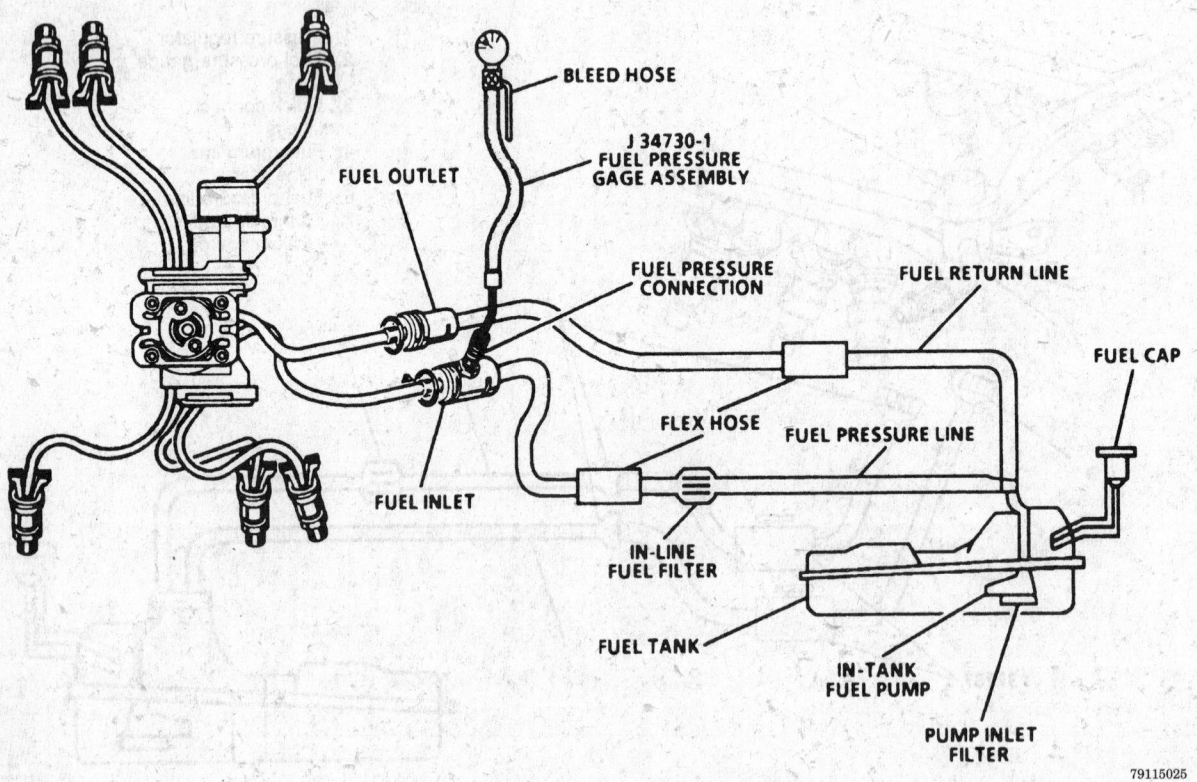

Fuel system pressure testing — 4.3L (VIN W) CPI engine

9. If removed install the isolator. Install the transmission detent cable, as required.

10. Shut down the engine and remove the special tool or equivalent from the throttle body.

11. Reconnect the idle air control motor connector and seal the hole drilled through the throttle body housing with silicone sealant or equivalent.

12. Check the throttle position sensor voltage as required. Install the air cleaner and thermac vacuum line.

2.8L, 3.1L and 4.3L (VIN Z) TBI Engines

1. Remove the air cleaner, adapter and gaskets. Discard the gaskets. Plug any vacuum line ports, as necessary.

2. Leave the Idle Air Control (IAC) valve connected and with the engine **OFF**, ground the diagnostic terminal (ALDL connector).

3. Turn the ignition switch to the **ON** position, do not start the engine. Wait for at least 30 seconds; this allows the IAC valve pintle to extend and seat in the throttle body.

4. With the ignition switch still in the **ON** position, disconnect IAC electrical connector.

5. Remove the ground from the diagnostic terminal and start the engine. Let the engine reach normal operating temperature.

6. Apply the parking brake and block the drive wheels. Remove the plug from the idle stop screw by piercing it with a suitable tool and then applying leverage to the tool to lift the plug out.

7. With the engine in the proper shift selector range, adjust the idle stop screw to set the minimum idle to specification.

8. Turn the ignition **OFF** and reconnect the IAC valve connector. Unplug any plugged vacuum line ports and install the air cleaner, adapter and new gaskets.

Fuel Injector

REMOVAL AND INSTALLATION

2.5L Engine

1. Relieve the fuel system pressure.

2. Disconnect the negative battery cable.

3. Remove the air cleaner assembly.

4. Remove the fuel injector wire. Remove the fuel injector retainer clip

screws. Remove the fuel injector retainer clip.

5. Using a suitable tool, gently pry the injector out.

6. Discard the upper and lower O-rings.

7. Installation is the reverse of the removal procedure. Lubricate both O-rings with light oil before installation.

2.8L, 3.1L and 4.3L (VIN Z) TBI Engines

1. Relieve the fuel system pressure.

2. Disconnect the negative battery cable.

3. Remove the air cleaner assembly.

4. Remove the fuel injector wire. Remove the fuel meter cover assembly.

5. Using a suitable tool, gently pry the injector out.

6. Discard the upper and lower O-rings.

7. Installation is the reverse of the removal procedure. Lubricate both O-rings with light oil before installation.

NOTE: Be sure the replacement injectors have an identical part number. The 4.3L (VIN Z) uses 2 different injectors with 2 different flow rates. Injectors with part no. 5235134 (color coded orange and green) are located on the throttle lever side, and those with part no. 5235342 (color coded pink and brown) are located on the TPS side.

3.8L and 4.3L (VIN Z) MPI Engines

1. Relieve the fuel system pressure.

2. Disconnect the negative battery cable.

3. Remove the upper intake plenum, if equipped with turbocharger.

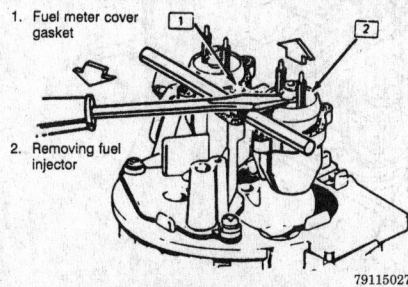

1. Fuel meter cover gasket

2. Removing fuel injector

79115027

Removing fuel injector — TBI

4. Remove the fuel supply and return lines from the fuel rail by squeezing the tabs and pulling the lines apart. Cap all open lines to prevent dirt and contaminants from entering.

5. Disconnect the vacuum line from the pressure regulator. Disconnect the fuel injector wire connectors.

6. Remove the fuel rail securing bolts and lift the rail up with equal force on both sides.

7. Remove the injector retaining clips and remove the injectors.

8. Installation is the reverse of the removal procedure.

4.3L (VIN W) CPI Engine

NOTE: The 4.3L (VIN W) engine is equipped with a non-repairable injector assembly which consists of the following: a fuel meter body, gasket seal, fuel pressure regulator, fuel injector and 6 poppet nozzles with fuel tubes. The assembly is housed in the lower manifold assembly.

1. Relieve the fuel system pressure.

2. Disconnect the negative battery cable.

3. Remove the upper intake plenum.

4. Disconnect the injector wire connector at the CPI unit.

5. Remove the fuel fitting clip and discard.

6. Remove the fuel inlet and return line.

7. Squeeze the poppet nozzle locking tabs together and lift the nozzles out of the casting.

8. Lift the CPI assembly out of the manifold.

To install:

9. Align the CPI assembly grommet with the casting grommet slot and push the assembly down until firmly seated in the bottom guide hole.

10. Push the poppet nozzles into the casting sockets. Ensure the nozzles are firmly seated and locked in their casting sockets. An unlocked nozzle may work loose and create a fuel leak in the manifold.

11. Install new o-rings on the fuel inlet and return lines and coat them lightly with engine oil. Connect the inlet and return fuel lines to the CPI unit and install a new fuel fitting clip.

12. Pressurize the fuel system and verify no fuel leakage.

13. Install the upper intake manifold assembly.

14. Connect the negative battery cable.

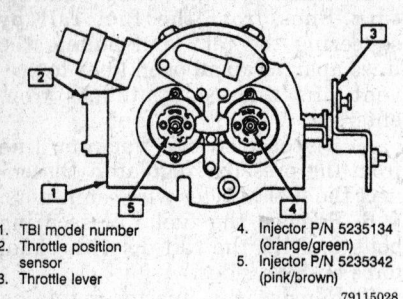

1. TBI model number
2. Throttle position sensor
3. Throttle lever
4. Injector P/N 5235134 (orange/green)
5. Injector P/N 5235342 (pink/brown)

79115028

Fuel injector location — 4.3L (VIN Z) engine

15. Inspect for leaks.

EMISSION CONTROLS

Emission Warning Lamps

These vehicles are equipped with an "Service Engine Soon" light. This will illuminate when the key is turned to the ON position and when a malfunction in the computer command control system occurs. The light cannot be reset until the malfunction is corrected and the ECM memory is cleared of the fault. A light that turns ON and OFF during engine operation indicates an intermittent problem and a light that remains ON indicates a permanent component failure.

RESETTING

1. Perform a diagnostic check of computer system codes.
2. Repair the failure.
3. Disconnect the negative battery cable for 15 seconds.
4. Reconnect the battery cable and check for failure codes.

ENGINE MECHANICAL

NOTE: Disconnecting the battery cable on some vehicles may interfere with the functions of the on board computer systems and may require the computer to undergo a relearning process.

Engine Assembly

REMOVAL AND INSTALLATION

Blazer, Bravada, Jimmy, Pickup and Sonoma

1. Disconnect the negative battery cable. Matchmark the hood hinges and remove the hood.
2. Drain the cooling system and disconnect the upper radiator hose at the radiator. Disconnect the coolant overflow hose.
3. Remove the upper fan shroud and disconnect the oil cooler lines.
4. Remove the radiator and the cooling fan. Disconnect the heater hoses.
5. Remove the air cleaner assembly and disconnect the vacuum hoses. Disconnect all necessary wires at the bulkhead and all of the main feed wires.
6. Disconnect the throttle cable and cruise control cable, if equipped. Remove the distributor cap.
7. Raise and safely support the vehicle. Disconnect the converter-to-exhaust pipe bolts. Remove the front drive shaft on 4WD vehicles.
8. Disconnect the exhaust pipes at the manifolds. Disconnect the strut

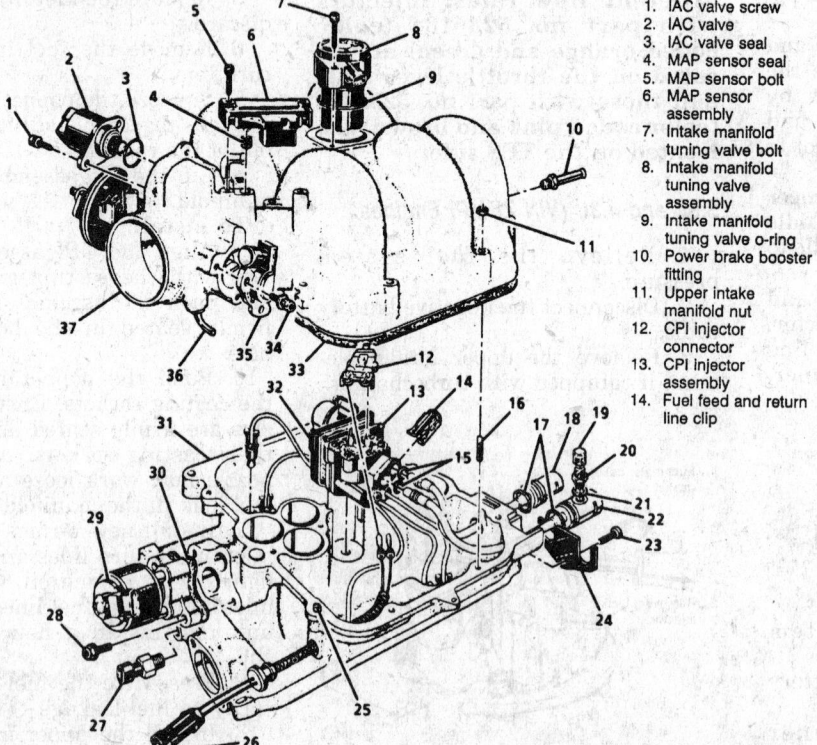

1. IAC valve screw
2. IAC valve
3. IAC valve seal
4. MAP sensor seal
5. MAP sensor bolt
6. MAP sensor assembly
7. Intake manifold tuning valve bolt
8. Intake manifold tuning valve assembly
9. Intake manifold tuning valve o-ring
10. Power brake booster fitting
11. Upper intake manifold nut
12. CPI injector connector
13. CPI injector assembly
14. Fuel feed and return line clip
15. Fuel feed and return line o-ring
16. Upper intake manifold stud
17. Lower intake manifold fuel feed and return line seal
18. Fuel feed line
19. Fuel pressure connection fitting cap
20. Fuel pressure connection fitting
21. Fuel pressure connection seal
22. Fuel return line
23. Fuel line retainer bolt
24. Fuel line returner
25. Upper intake manifold locating pin
26. CPI injection wiring
27. CTS assembly
28. EGR valve bolt
29. EGR valve
30. EGR valve gasket
31. Lower intake manifold assembly
32. CPI injector assembly seal
33. Upper intake manifold gasket
34. TPS assembly screw
35. TPS assembly
36. Fuel vapor canister purge tube
37. Upper intake manifold assembly with throttle body

79115029

Fuel injector location — 4.3L (VIN W) engine

rods at the bell housing. Remove the flywheel cover.

9. Remove the torque converter bolts on automatic transmission equipped vehicles. Remove the second crossmember on 4WD vehicles.

10. Remove the rear catalytic converter shield and disconnect the converter hanger at the exhaust pipe.

11. Remove the lower fan shroud and disconnect the fuel lines. Remove the 2 outer air dam bolts.

12. Remove the 2 left body mount bolts and, using a suitable jack, raise the body slightly. Remove the engine-to-transmission bolts.

13. Lower the body and remove the lower motor mount bolts. Lower the vehicle.

14. Remove the air conditioning compressor and power steering pump with their brackets from the engine. Do not disconnect the fluid or refrigerant lines.

15. Support the transmission with a suitable jack. Attach a suitable lifting device to the engine and remove the engine from the vehicle.

To install:

16. Carefully lower the engine into position in the vehicle.

17. Remove the transmission support. Raise and safely support the vehicle. Install the motor mount bolts, tighten to 52 ft. lbs. (70 Nm). Install the lower bell housing bolts. Tighten to 32 ft. lbs. (44 Nm).

18. Raise the body and install the upper bell housing bolts. Tighten to 32 ft. lbs. (44 Nm).

19. Lower the body and install the body mount bolts. Install the outer air dam bolts. Connect the fuel lines and install the lower fan shroud.

20. Connect the converter hanger at the exhaust pipe and install the shield at the converter. Install the torque converter bolts and install the flywheel cover.

21. Connect the strut rods at the bell housing. Connect the exhaust pipes to the manifolds. Connect the converter to the exhaust pipes.

22. Lower the vehicle. Install the air conditioning compressor and power steering pump. Install the distributor cap.

23. Connect the throttle cable and cruise control cable. Connect all electrical wires and vacuum hoses.

24. Install the air cleaner and connect the heater hoses. Install the fan and radiator. Connect the oil cooler lines. Install the upper fan shroud.

25. Connect the overflow hose and the radiator hose. Fill the cooling system.

26. Install the hood. Connect the negative battery cable and run the engine. Bleed the cooling system.

Astro and Safari

1. Disconnect the negative battery cable.

2. Drain the cooling system.

3. Raise and safely support the vehicle. Disconnect the exhaust pipes at the manifolds.

4. Disconnect the strut rods at the flywheel housing. Remove the flywheel cover. Remove the torque converter bolts. Drain the engine oil.

5. Remove the starter and oil filter. Disconnect the wires at the transmission. Disconnect the fuel lines.

6. Disconnect the oil cooler lines. Remove the lower fan shroud. Remove the motor mount bolts.

7. Lower the vehicle. Remove the headlight bezels and grille. Remove the radiator close out panel and the radiator support brace. Remove the lower tie bar.

8. Remove the hood latch mechanism. Remove the master cylinder and the upper fan shroud.

9. Remove the upper radiator core support and the radiator. If equipped, properly discharge the air conditioning system.

10. Remove the radiator filler panels. Remove the engine cover. Remove the air conditioning hose at the accumulator and remove the air conditioning compressor.

11. Remove the power steering pump. Disconnect the vacuum hoses at the intake manifold and the wiring at the bulkhead.

12. Remove the right kickpanel. Disconnect the harness at the ESC module and it through the bulkhead.

13. Remove the distributor cap. Remove the air conditioning accumulator, if equipped.

14. Disconnect the fuel lines. Remove the diverter valve. Remove the transmission dipstick tube.

15. Disconnect the heater hoses at the heater core. Remove the horn and remove the AIR system check valves. Remove the engine-to-transmission bolts.

16. Attach a suitable lifting device to the engine and remove the engine from the vehicle.

To install:

17. Install the engine into the vehicle. Install the bell housing bolts.

18. Install the AIR system check valves. Install the horn and connect the heater hoses, Install the transmission dipstick tube.

19. Install the diverter valve and connect the fuel lines. Install the accumulator and install the distributor cap.

20. Connect the harness to the ESC module. Install the right kick panel.

21. Connect the bulkhead wiring and the vacuum hoses. Install the power steering pump and the air conditioning compressor.

22. Connect the refrigerant line to the accumulator. Install the engine cover.

23. Install the radiator filler panel and the radiator. Install the radiator supports and the upper fan shroud.

24. Install the hood latch and the master cylinder. Install the lower tie bar. Install the lower radiator close-out panel.

25. Install the grille and headlight bezels. Raise and safely support the vehicle.

26. Install the engine mount bolts and tighten to 75 ft. lbs. (100 Nm). Connect the oil cooler lines. Connect the fuel hoses and the wires at the transmission.

27. Install the oil filter and the starter. Install the torque converter bolts. Install the flywheel cover.

28. Install the strut rods at the flywheel housing. Connect the exhaust pipes. Lower the vehicle.

29. Connect the negative battery cable. Fill the cooling system and the crankcase. Charge the air conditioning system. Fill and bleed the brake system.

Cyclone and Typhoon

1. Disconnect the negative battery cable. Matchmark the hood hinges and remove the hood.

2. Drain the coolant from the engine radiator and turbocharger air cooler. Remove the air cleaner and duct. Remove the turbocharger air inlet elbow.

3. Remove the upper fan shroud. Remove the serpentine belt and fan pulley.

4. Remove the battery tray and vacuum tank.

5. Raise and safely support the vehicle. Remove the front tire and wheel assemblies.

6. Remove the wheelhouse panels.

7. Disconnect the mufflers and tailpipe from the catalytic converter. Remove the catalytic converter support bolts.

8. Remove the turbocharger outlet pipe bracket and nuts. Move the outlet pipe and catalytic converter away from the turbocharger.

9. Disconnect the electrical connectors from the turbocharger air

cooler radiator temperature sensor and hoses.

10. Remove the turbocharger air cooler radiator.

11. Remove the exhaust crossover pipe and lower the vehicle.

12. Remove the oil cooler lines, overflow hoses and upper and lower hoses from the radiator. Remove the radiator assembly.

13. Remove the oil lines at the filter adapter.

14. Remove the power steering pump hoses from the steering gear.

15. Remove the engine coolant reservoir. Remove the air conditioning compressor and position aside. Do not discharge or disconnect the compressor lines.

16. Remove the turbocharger air cooler clamps, ducts and hoses. Remove the air cooler from the supports.

17. Remove the throttle body assembly from the upper intake plenum and position aside.

18. Remove the upper intake manifold. Disconnect the fuel rail from the fuel lines.

19. Disconnect all necessary vacuum and electrical connections from the engine.

20. Raise and safely support the vehicle and transmission. Remove the front and rear driveshaft.

21. Remove the transmission crossmember and mount. Remove the torque converter cover and remove the bolts securing the converter to the flywheel.

22. Disconnect the shift linkage from the transmission.

23. Disconnect the fuel hoses from the pipes near the transfer case. Remove the fuel line bracket from the transfer case.

24. Disconnect and remove the fuel line clips, electrical clips and connectors from the transmission and transfer case.

25. Remove the transfer case from the vehicle.

26. Disconnect the T.V. cable and cooler lines from the transmission.

27. Remove the transmission from the vehicle.

28. Remove the fuel line clips, oil line clips and wiring harness clips from the cylinder head.

29. Remove the starter motor.

30. Remove the engine mount through bolts and lower the vehicle.

31. Attach a suitable lifting device to the engine and remove the engine from the vehicle, disconnecting any necessary connections.

To install:

32. Position all wires and hoses out of the way during engine installation.

33. Carefully lower the engine into position in the vehicle.

34. Install the engine mount through bolts and lower the vehicle.

35. Install the starter motor.

36. Connect the fuel line clips, oil line clips and wiring harness clips to the cylinder head.

37. Install the transmission assembly.

38. Connect the T.V. cable and cooler lines to the transmission.

39. Install the transfer case assembly.

40. Connect the fuel line clips, electrical clips and connectors to the transmission and transfer case.

41. Connect the fuel hoses to the pipes near the transfer case. Connect the fuel line bracket to the transfer case.

42. Connect the shift linkage to the transmission.

43. Install the transmission crossmember and mount. Install the bolts securing the converter to the flywheel and install the torque converter cover.

44. Install the front and rear driveshafts

45. Connect all vacuum and electrical connections previously removed from the engine.

46. Connect the fuel rail to the fuel lines. Install the upper intake manifold.

47. Install the throttle body assembly onto the upper intake plenum.

48. Install the air cooler. Install the turbocharger air cooler clamps, ducts and hoses.

49. Install the engine coolant reservoir. Install the air conditioning compressor.

50. Connect the power steering pump hoses to the steering gear and the engine oil lines at the filter adapter.

51. Install the radiator assembly. Connect the oil cooler lines, overflow hoses and upper and lower hoses to the radiator.

52. Raise and support the vehicle safely. Install the exhaust crossover pipe and lower the vehicle.

53. Install the turbocharger air cooler radiator.

54. Install and connect the electrical connectors to the turbocharger air cooler radiator temperature sensor and hoses.

55. Raise and support the vehicle safely. Connect the turbocharger outlet pipe bracket and nuts.

56. Connect the mufflers and tailpipe to the catalytic converter. Install the catalytic converter support bolts.

57. Install the wheelhouse panels.

58. Install the front tire and wheel assemblies. Lower the vehicle.

59. Install the serpentine belt and fan pulley. Install the battery tray and vacuum tank. Install the upper fan shroud.

60. Fill the cooling system with the proper type and quantity of antifreeze. Install the air cleaner and duct. Install the turbocharger air inlet elbow.

61. Connect the negative battery cable. Install the hood, aligning the marks previously.

Lumina APV, Silhouette and Trans Sport

3.1L ENGINE

1. Disconnect the negative battery cable.

2. Drain the cooling system. Disconnect the air flow tube from the air cleaner.

3. Disconnect the electrical connector from the ECM and push it through to the engine compartment. Disconnect the harness from the clips on the body and lay it across the engine.

4. Disconnect the engine harness at the bulkhead connector. Disconnect the throttle and TV cables.

5. Disconnect the fuel lines. Disconnect the transaxle shift linkage.

6. Disconnect the cooler lines at the radiator. Disconnect the radiator and heater hoses.

7. Remove the air conditioning compressor from the bracket and support it aside. Remove the upper engine support strut.

8. Raise and safely support the vehicle. Remove the front wheel and tire assemblies.

9. Remove the stabilizer bar. Disconnect the tie rod ends and the lower control arm ball joints.

10. Disconnect the halfshafts and support them aside. Disconnect the steering shaft pinch bolt.

11. Remove the starter.

12. Disconnect the exhaust pipe at the manifold. Support the engine and sub-frame with a suitable jack.

13. Remove the sub-frame bolts and lower the engine/transaxle and subframe from the vehicle.

To install:

14. Raise the engine assembly into position and install the subframe bolts. Tighten to 35 ft. lbs.

15. Connect the exhaust pipe at the rear manifold. Install the starter.

16. Connect the steering shaft and install the pinch bolt. Connect the halfshafts to the transaxle.

17. Connect the lower control arm ball joints to the steering knuckles.

18. Install the stabilizer bar. Install the upper engine strut.

19. Install the wheel and tire assemblies. Lower the vehicle. Install the radiator and heater hoses.

20. Install the shift linkage. Connect the fuel lines and the throttle and TV cables.

21. Connect the harness to bulkhead connector. Connect the ECM harness to the ECM.

22. Connect the air cleaner hose and the radiator upper support.

23. Fill the cooling system. Install the air conditioning compressor.

24. Connect the negative battery cable.

3.8L ENGINE

1. Disconnect the negative battery cable. Properly relieve the fuel system pressure.

2. Remove the air cleaner and duct assembly. Remove the fuel lines from the fuel rail and the mounting bracket.

3. Disconnect the throttle cables from the throttle body and mounting bracket.

4. Drain the cooling system into a suitable container. Remove the radiator and heater hoses. Remove the engine cooling fan.

5. Remove the torque strut and brackets.

6. Disconnect the transaxle cooler lines and remove the radiator assembly.

7. Disconnect the relay center wiring.

8. Disconnect the fuel vapor canister hoses.

9. Disconnect the battery cables at the engine ground and starter assembly.

10. Disconnect the shift cables from the transaxle.

11. Disconnect and remove the ECM main harness and engine wiring.

12. Remove the serpentine belt. Remove the air conditioning compressor and position aside.

13. Disconnect the power steering line attachment at the lower right hand rail.

14. Raise and safely support the vehicle. Remove the front wheel and tire assemblies.

15. Remove the stabilizer shaft, tie rod ends and lower ball joints from

the steering knuckle. Remove the drive axles from the transaxle.

16. Remove the intermediate steering shaft pinch bolt.

17. Remove the starter assembly and flywheel cover. Remove the torque converter bolts.

18. Remove the exhaust pipe at the rear manifold.

19. Safely support the engine, frame and transaxle assembly. Remove the frame bolts.

20. Lower the engine, frame and transaxle assembly.

21. Remove the transaxle to engine mount bracket.

22. Remove the exhaust crossover, the left and right exhaust manifolds.

23. Remove the engine to frame front mount.

24. Remove the left and right engine mounts to frame.

25. Remove the transaxle to engine bolts and separate the assembly.

To install:

26. Position the assembly together and install the transaxle to engine bolts.

27. Install the left and right engine mounts to frame.

28. Install the engine to frame front mount.

29. Install the exhaust crossover, the left and right exhaust manifolds.

30. Install the transaxle to engine mount bracket.

31. Raise the engine, frame and transaxle assembly into position.

32. Safely support the engine, frame and transaxle assembly and install the frame bolts.

33. Install the exhaust pipe at the rear manifold.

34. Install the starter assembly and the torque converter bolts. Install the flywheel cover.

35. Install the intermediate steering shaft pinch bolt.

36. Install the drive axles into the transaxle. Install the stabilizer shaft, tie rod ends and lower ball joints from the steering knuckle.

37. Lower the vehicle. Install the front wheel and tire assemblies.

38. Connect the power steering line attachment at the lower right hand rack.

39. Install the air conditioning compressor and position aside. Install the serpentine belt.

40. Reconnect the ECM main harness and engine wiring.

41. Connect the shift cables to the transaxle.

42. Connect the battery cables at the engine ground and starter assembly.

43. Connect the fuel vapor canister hoses.

44. Connect the relay center wiring.

45. Connect the transaxle cooler lines and remove the radiator assembly.

46. Install the torque strut and brackets.

47. Install the engine cooling fan. Install the radiator and heater hoses. Fill the cooling system with the proper type and quantity of antifreeze.

48. Connect the throttle cables to the throttle body and mounting bracket.

49. Install the air cleaner and duct assembly. Connect the fuel lines to the fuel rail and the mounting bracket.

50. Connect the negative battery cable.

Engine Mounts

NOTE: When lifting or raising the engine for any reason, do not support the assembly under the oil pan, any sheet metal or the crankshaft pulley.

REMOVAL AND INSTALLATION

2.5L Engine

1. Disconnect the negative battery cable.

2. Raise and support the vehicle safely. Support the engine with a suitable lifting fixture.

3. Remove the engine mount through bolt and nut.

4. Raise the engine enough to permit removal of the engine mounting.

5. Remove the mount assembly attaching bolts, nuts and washers. Remove the mount assembly.

6. Installation is the reverse of the removal procedure.

2.8L Engine

1. Disconnect the negative battery cable.

2. Raise and support the vehicle safely. Support the engine with a suitable lifting fixture.

3. Remove the fan shroud and right front wheel.

4. Remove the engine mount through bolt and nut.

5. Raise the engine enough to permit removal of the engine mounting and block in position.

6. Remove the tie rod at the drag link. Remove the stabilizer link from the control arm.

7. Remove the lower shock absorber bolts.

8. Remove the lower control arm pivot bolts and position control aside.

9. Remove the mount assembly attaching bolts, nuts and washers. Remove the mount assembly.

10. Installation is the reverse of the removal procedure.

3.1L and 3.8L Engines

1. Disconnect the negative battery cable.

2. Raise and support the vehicle safely. Support the engine with a suitable lifting fixture.

3. Remove the engine mount nuts from below the engine frame mounting bracket.

4. Raise the engine enough to permit removal of the engine mounting and block in position.

5. Remove the mount assembly to engine bracket nuts and washers. Remove the mount assembly.

6. Installation is the reverse of the removal procedure.

4.3L Engine

EXCEPT ASTRO AND SAFARI

1. Disconnect the negative battery cable.

2. Raise and support the vehicle safely. Support the engine with a suitable lifting fixture.

3. Remove the cab or body mounting bolts. Raise the body and block in position.

4. Remove the engine mount through bolts.

5. Raise the engine enough to permit removal of the engine mounting and block in position.

6. Remove the mount assembly to frame bolts. Remove the mount assembly.

7. Installation is the reverse of the removal procedure.

ASTRO AND SAFARI

1. Disconnect the negative battery cable.

2. Raise and support the vehicle safely. Support the engine with a suitable lifting fixture.

3. Remove the engine mount through bolts.

4. Raise the engine enough to permit removal of the engine mounting and block in position.

5. Remove the mount assembly to frame bolts. Remove the mount assembly.

6. Installation is the reverse of the removal procedure.

Cylinder Head

REMOVAL AND INSTALLATION

2.5L Engine

1. Disconnect the negative battery cable. Drain the cooling system. On Astro and Safari vehicles, remove the engine cover. Remove the air cleaner assembly.

2. Remove the air conditioning compressor and position it aside. Remove the rocker arm cover. Remove the rocker arms and pushrods. Keep them in order for reinstallation.

3. Properly relieve the fuel system pressure. Disconnect the fuel line from the TBI unit. Disconnect all necessary electrical and vacuum lines.

4. Disconnect the accelerator cable, the cruise control cable and the TV cables. Remove the alternator and brackets.

5. Remove the water pump bypass hose. Disconnect the heater hoses at the intake manifold. Remove the upper radiator hose.

6. Disconnect the exhaust pipe from the exhaust manifold. Remove the fuel filter and filter brackets at the rear of the cylinder head assembly.

7. Remove the coil wire and spark plug wires. Disconnect the oxygen sensor electrical wire.

8. Remove the cylinder head retaining bolts. Remove the cylinder head along with the intake and exhaust manifold assembly.

To install:

9. Ensure the cylinder bolt threads in the block and threads on the bolts are cleaned, as dirt will affect bolt torque.

10. Coat the threads of the 2 cylinder head studs with sealing compound 1052080 or equivalent.

11. Position the new gasket over the dowel pins.

12. Install the cylinder head and retaining bolts. Torque the cylinder head bolts gradually to 18 ft. lbs. (25 Nm) in the proper sequence. Then tighten all bolts except stud (position No. 9) to 26 ft. lbs. (35 Nm); tighten No. 9 to 18 ft. lbs. (25 Nm). Finally, tighten all bolts and studs an additional 90 degree turn.

13. Connect the coil wire and spark plug wires.

14. Connect the oxygen sensor electrical wire.

15. Connect the exhaust pipe to the exhaust manifold. Connect the fuel filter and filter brackets to the rear of the cylinder head assembly.

16. Install the water pump bypass hose. Connect the heater hoses to the intake manifold. Install the upper radiator hose.

17. Connect the accelerator cable, the cruise control cable and the TV cables. Install the alternator and brackets.

18. Connect the fuel line from the TBI unit. Connect all electrical and vacuum lines.

19. Install the air conditioning compressor and position it aside. Install the rocker arm cover. Install the rocker arms and pushrods in the same order they were removed.

20. Connect the negative battery cable. Fill the cooling system with the proper type and quantity of antifreeze. Install the air cleaner assembly. On Astro and Safari vehicles, install the engine cover.

2.8L and 3.1L Engines

1. Disconnect the negative battery cable. Drain the radiator. Remove the intake manifold.

2. Remove the valve covers. Remove the rocker arms and pushrods. Keep them in order for reinstallation.

3. Raise and support the vehicle safely. Disconnect the exhaust manifolds from the exhaust pipes. On the left side disconnect the dipstick tube attachment. On the right side remove the alternator bracket. Lower the vehicle.

4. Remove the cylinder head retaining bolts. Remove the cylinder head from the engine along with the exhaust manifold.

To install:

5. Ensure that the cylinder bolt threads in the block and threads on the bolts are cleaned, as dirt will affect bolt torque.

6. Coat the threads of the cylinder head bolts with sealing compound 1052080 or equivalent.

7. Position the new gasket over the dowel pins with THIS SIDE UP showing.

8. Install the cylinder head onto the engine along with the exhaust manifold. Install the cylinder head retaining bolts. Torque the cylinder head bolts gradually in the proper sequence to 41 ft. lbs. (55 Nm). Then turn each bolt an additional 90 degrees. Cylinder head bolt torque sequence — engine

79115031

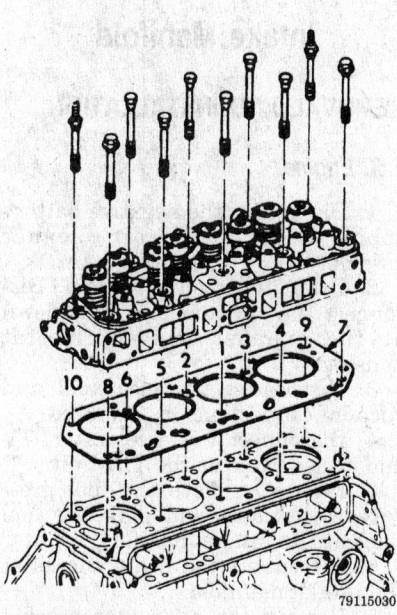

Cylinder head bolt torque sequence — engine

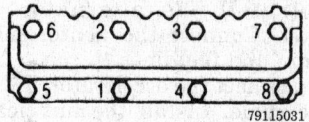

1. Raise and support the vehicle safely. Connect the exhaust manifolds to the exhaust pipes. On the left side, connect the dipstick tube. On the right side, install the alternator bracket. Lower the vehicle.

2. Install the valve covers. Install the rocker arms and pushrods keeping them in the same order as removed.

3. Install the intake manifold assembly.

4. Connect the negative battery cable. Fill the cooling system with the proper type and quantity of antifreeze.

3.8L Engine

1. Disconnect the negative battery cable. Remove the engine cover. Drain the coolant into a suitable container.

2. Remove the intake and exhaust manifolds.

3. Remove the alternator bracket and 1 air conditioner bracket bolt. Remove the power steering pump.

4. Remove the valve covers. Remove the rocker arms, pushrods and guide plates. Keep them in order for reinstallation. Remove the coil and spark plug wires.

5. Remove the belt tensioner assembly. Remove the fuel pump heat shield.

6. Remove the cylinder head retaining bolts. Remove the cylinder head from the engine.

To install:

7. Ensure that the cylinder bolt threads in the block and threads on the bolts are cleaned, as dirt will affect bolt torque.

8. Coat the underside of the bolt heads with sealing compound 1052080 or equivalent. Coat the threads of the bolts with thread lock.

9. Position the new gasket with the arrow pointing to the front of the engine.

10. Install the cylinder head onto the engine. Install the cylinder head retaining bolts and torque as follows:

   a. Torque the cylinder head bolts gradually in the proper sequence to 35 ft. lbs. (47 Nm).

   b. Then turn each bolt an additional 130 degree turn in sequence.

   c. Finally turn the 4 center bolts an additional 30 degrees.

11. Install the rocker arms, pushrods and guide plates in the same position from which they were removed. Torque the rocker arm bolts to 28 ft. lbs. (38 Nm).

12. Install the intake manifold and valve covers.

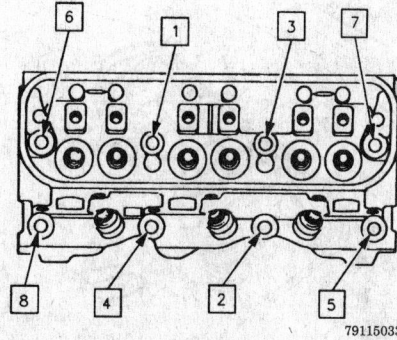

Cylinder head bolt torque sequence — engine

13. Raise and support the vehicle safely. Connect the exhaust manifolds to the exhaust pipes. Lower the vehicle.

14. Install the alternator and bracket. Install the air conditioner bracket bolt.

15. Install the ignition coil and spark plug wires.

16. Install the belt tensioner and power steering pump assembly.

17. Install the fuel pump heat shield.

18. Connect the negative battery cable. Fill the cooling system with the proper type and quantity of antifreeze.

4.3L Engine

1. Disconnect the negative battery cable. Remove the engine cover, if equipped. Drain the engine and turbocharger air cooler radiator, if equipped. Remove the intake manifold. Remove the required electrical and vacuum connections.

2. Remove the valve covers. Remove the rocker arms and pushrods. Keep them in order for reinstallation. Remove the spark plugs.

3. Raise and support the vehicle safely. Remove the exhaust manifolds. On the left side, disconnect the dipstick tube attachment. Lower the vehicle.

4. Remove the alternator, power steering and brackets. Remove the air conditioner compressor and belt tensioner bracket and position aside.

5. Remove the cylinder head retaining bolts. Remove the cylinder head from the engine.

To install:

6. Ensure that the cylinder bolt threads in the block and threads on the bolts are cleaned, as dirt will affect bolt torque.

7. If a steel gasket is used, be sure to coat both sides with sealer. Coat the threads of the cylinder head bolts with sealing compound 1052080 or equivalent.

8. Position the new gasket over the dowel pins with the bead up.

9. Install the cylinder head and retaining bolts. Torque the cylinder head bolts gradually in the proper sequence as follows:

   a. The first sequence to 25 ft. lbs. (34 Nm).

   b. The second sequence to 45 ft. lbs. (61 Nm).

   c. The final sequence to 65 ft. lbs. (90 Nm).

10. Install the alternator, power steering and brackets. Install the air conditioner compressor and belt tensioner bracket.

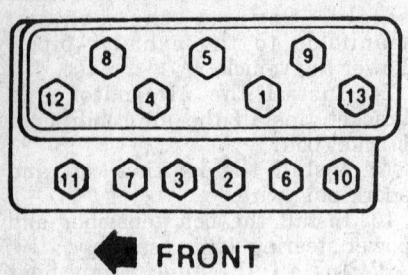

FRONT

79115032

Cylinder head bolt torque sequence — 4.3L engine

11. Raise and support the vehicle safely. Install the exhaust manifolds. On the left side, Connect the dipstick tube attachment. Lower the vehicle.

12. Install the rocker arms and pushrods in the same order as removed. Install the valve covers. Install the spark plugs and wires.

13. Install the intake manifold. Connect all electrical and vacuum connections.

14. Connect the negative battery cable. Install the engine cover. Fill the radiator with the proper type and quantity of antifreeze.

Valve Lash

Adjustment

2.8L and 4.3L (VIN Z) Engines

1. On Astro and Safari, remove the engine cover. Remove the rocker covers.

2. Crank the engine until the mark on the damper aligns with the **0** mark on the timing tab and the engine is in the No. 1 firing position. This can be achieved by placing a finger on No. 1 intake valve as the damper approaches TDC; if the valve is closed, the engine is on the firing stroke. If the valve moves when the damper is approaching TDC the engine is on No. 4 firing position and will need to be turned 1 additional turn to reach No. 1 firing position.

3. With the engine in the No. 1 firing position, the following valves may be adjusted on the 2.8L engine:
   a. Exhaust — 1, 2, 3
   b. Intake — 1, 5, 6

4. With the engine in the No. 1 firing position, the following valves may be adjusted on the 4.3L (VIN Z) engine:
   a. Exhaust — 1, 5, 6
   b. Intake — 1, 2, 3

5. Back out the adjusting nut until the pushrod can be turned with fin-

ger pressure and lash is felt. Then turn in the nut until all lash is removed and the pushrod cannot be turned. When all lash has been removed, turn the adjusting nut in 1½ turn on the 2.8L engine or 1 full turn on the 4.3L (VIN Z) engine.

6. Crank the engine until the mark on the damper aligns with the **0** mark on the timing tab. This is the No. 4 firing position.

7. With the engine in the No. 4 timing position, the following valves can be adjusted on the 2.8L engine:
   a. Exhaust — 4, 5, 6
   b. Intake — 2, 3, 4

8. With the engine in the No. 4 timing position, the following valves can be adjusted on the 4.3L (VIN Z) engine:
   a. Exhaust — 2, 3, 4
   b. Intake — 4, 5, 6

9. Back out the adjusting nut until lash is felt at the pushrod then turn in the nut until all lash is removed. When all lash has been removed, turn the adjusting nut in 1½ turn on the 2.8L engine or 1 full turn on the 4.3L (VIN Z) engine.

10. After all valves are adjusted, install the valve covers.

Rocker Arms

REMOVAL AND INSTALLATION

1. Disconnect the negative battery cable. On Astro and Safari, remove the engine cover.

2. Remove all the necessary components in order to gain access to the engine valve covers. Properly relieve the fuel system pressure before disconnecting any fuel lines.

3. Remove the valve cover retaining bolts. Remove the valve cover from the engine.

4. Remove the rocker arm assemblies. Keep them in order for reinstallation.

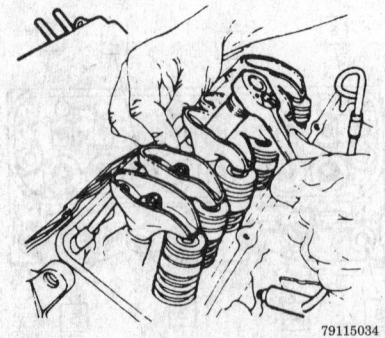

79115034

Valve adjustment — 2.8L and 4.3L (VIN Z) engines

5. Installation is the reverse of the removal procedure.

Intake Manifold

REMOVAL AND INSTALLATION

2.5L Engine

1. Disconnect the negative battery cable. Remove the air cleaner assembly.

2. Drain the cooling system. Disconnect the vacuum pipe hold-down at the exhaust and thermostat housing.

3. Disconnect the electrical and vacuum connections, as required.

4. Disconnect the accelerator, TV and cruise cable at the TBI unit.

5. Properly relieve the fuel pressure. Disconnect and plug the fuel lines at the intake manifold.

6. Disconnect the heater hoses at the intake manifold.

7. Remove the alternator bracket retaining bolts. Remove the alternator and position it aside. Remove the ignition coil wires.

8. Remove the intake manifold retaining bolts. Remove the intake manifold.

   To install:

9. Install the manifold and gasket into position. Install the retaining bolts and tighten in sequence to the correct torque.

10. Install the alternator and bracket. Connect the heater hoses, vacuum and fuel lines.

11. Connect the accelerator, TV and cruise cable. Install the air cleaner assembly. Install the coil wires.

12. Connect the negative battery cable. Fill the cooling system with the proper type and quantity of antifreeze. Run the engine, check the idle speed and check for leaks.

2.8L and 3.1L Engines

1. Disconnect the negative battery cable. Remove the air cleaner. Drain the cooling system. If equipped, remove the AIR pump and bracket.

2. Remove the distributor. Remove the heater and radiator hoses from the intake manifold. Remove the rocker arm covers.

3. Disconnect and label all vacuum hoses and electrical connections at the manifold. Remove the EFE pipe from the rear of the manifold. Remove the accelerator linkage. Disconnect and plug the fuel line.

4. As required, remove the TBI unit. Remove the intake manifold re-

taining bolts. Remove the intake manifold from the engine.

To install:

5. The gaskets are marked for right and left side installation. Do not interchange them. Clean the sealing surface of the engine block and apply a 5/16 in. bead of silicone sealer to each ridge.

6. Install the new gaskets onto the heads. The gaskets will have to be cut slightly to fit past the center pushrods. Do not cut any more material than necessary. Hold the gaskets in place by extending the ridge bead of sealer 1/4 in. onto the gasket ends. (When the intake manifold is installed, the area between the ridges and the manifold should be completely sealed.)

7. Install the intake manifold onto the engine. Install the intake manifold retaining bolts and torque in the proper sequence to specification.Intake manifold bolt torque sequence — 2.8L engine

79115035

1. As required, install the TBI unit.

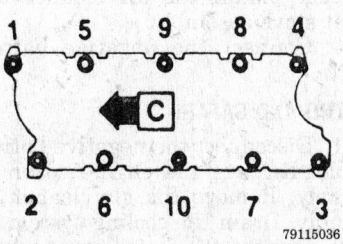

Intake manifold bolt torque sequence — 3.1L engine

2. Connect all vacuum hoses and electrical connections at the manifold. Install the EFE pipe from the rear of the manifold. Install the accelerator linkage. Connect the fuel line.

3. Install the distributor. Install the heater and radiator hoses from the intake manifold. Install the rocker arm covers. Install the AIR pump and bracket.

4. Fill the cooling system with the proper type and quantity of antifreeze. Install the air cleaner.

5. Connect the negative battery cable.

3.8L Engine

1. Disconnect the negative battery cable. Drain the cooling system into a suitable container.

2. Remove the spark plug wires on the right side of the engine.

3. Remove the air intake duct. Properly relieve the fuel system pressure and remove the fuel rail.

4. Remove the exhaust heat shield.

5. Remove the cable bracket to cylinder head mounting bolt.

6. Remove the upper intake manifold bolts and upper manifold.

7. Remove the power steering pump support bracket.

8. Loosen the alternator and move aside to obtain clearance.

9. Remove the heater pipes and bypass hose.

10. Remove the lower intake manifold bolts and remove the assembly.

To install:

11. Thoroughly clean all manifold mating surfaces, bolts and bolt holes. Apply sealant to the ends of the manifold seals and coat the bolt threads with Loctite®. Install the lower intake manifold, gasket and bolts. Torque the lower manifold bolts in sequence, twice, to 88 inch lbs. (10 Nm).

12. Install the heater pipes and bypass hose.

13. Install the alternator.

14. Install the power steering pump support bracket.

15. Install the upper intake manifold and torque the bolts to 22 ft. lbs. (30 Nm).

16. Install the cable bracket to cylinder head mounting bolt.

17. Install the exhaust heat shield.

18. Install the air intake duct. Install the fuel rail.

19. Install the spark plug wires on the right side of the engine.

20. Connect the negative battery cable. Fill the cooling system with the proper type and quantity of antifreeze.

4.3L (VIN W) Engine

1. Disconnect the negative battery cable. Drain the cooling system into a suitable container.

2. Remove the plastic cover on the top of the intake manifold.

3. Remove the wiring harness, which includes disconnecting the TPS, IAC motor, MAP sensor and the intake manifold tuning valve.

4. Remove the throttle and TV linkage from the upper intake manifold.

5. Remove the ignition coil and PCV hose from the manifold.

6. Remove the vacuum hoses from the manifold.

7. Mark the installation location of all studs. Remove the upper manifold bolts and studs.

8. Remove the distributor and wiring.

9. Properly relieve the fuel system pressure. Remove the fuel supply and return line at the rear of the manifold.

10. Remove the upper radiator and heater hoses.

11. Remove the air conditioner compressor bracket at the intake manifold.

12. Disconnect the wiring harness, including the fuel injectors, EGR valve and coolant temperature sensor.

13. Remove the lower intake manifold bolts and manifold assembly.

To install:

14. Thoroughly clean all gasket mating surfaces, including bolts. Apply GM sealer 1052080 or equivalent to the bolts. Apply a (5mm) bead of sealer to the front and rear sealing surfaces of the block, extending the bead 1/2 in. (13mm) up each cylinder head to seal and retain the gaskets.

15. Install the gaskets to the cylinder head with the port blocking plates facing the rear of the engine and the "THIS SIDE UP" stamping facing up.

16. Install the lower intake manifold and torque the bolts in sequence to 35 ft. lbs. (48 Nm).

17. Connect the wiring harness, including the fuel injectors, EGR valve and coolant temperature sensor.

18. Install the air conditioner compressor bracket at the intake manifold.

19. Install the upper radiator and heater hoses.

20. Connect the fuel supply and return line at the rear of the manifold.

21. Install the distributor and wiring.

22. Install the upper intake manifold and gaskets with the green seal-

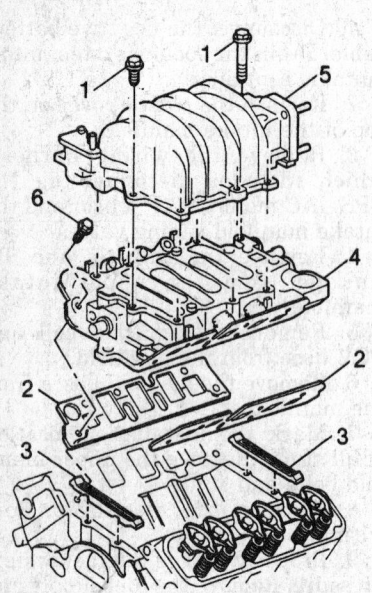

1. Intake manifold upper bolt
2. Intake manifold gasket
3. Intake manifold seal
4. Intake manifold lower
5. Intake manifold upper
6. Intake manifold lower bolt

79115037

Intake manifold assembly — 3.8L engine

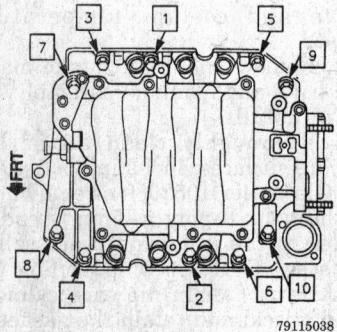

79115038

Intake manifold bolt torque sequence — 3.8L engine

ing lines facing up. Install the manifold bolts and studs in their previously marked positions. Torque the bolts and studs in sequence to 124 inch lbs. (14 Nm).

23. Install the vacuum hoses at the manifold.

24. Install the ignition coil and PCV hose at the manifold.

25. Install the throttle and TV linkage onto the upper intake manifold.

26. Install the wiring harness, which includes connecting the TPS, IAC motor, MAP sensor and the intake manifold tuning valve.

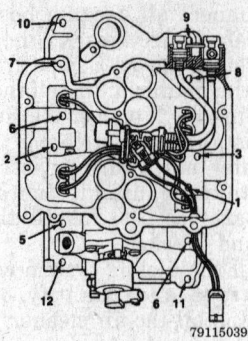

79115039

Lower intake manifold bolt torque sequence — 4.3L (VIN W) engine

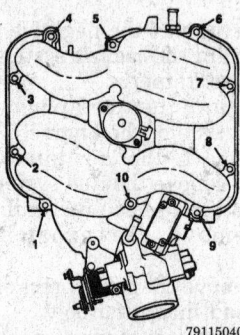

79115040

Upper intake manifold bolt torque sequence — 4.3L (VIN W) engine

27. Install the plastic cover on the top of the intake manifold.

28. Connect the negative battery cable. Fill the cooling system with the proper type and quantity of antifreeze.

4.3L (VIN Z) Except Turbocharged Engine

EXCEPT ASTRO AND SAFARI

1. Disconnect the negative battery cable. Remove the air cleaner and heat stove assembly. Drain the cooling system and remove the upper radiator hose.

2. Remove the 2 braces at the rear of the drive belt tensioner.

3. Remove the emission relays with brackets.

4. Remove the wiring harness from the clips and position aside.

5. Remove the ground cable from the manifold stud.

6. Remove the power brake vacuum pipe.

7. Remove the heater hose pipe at the manifold.

8. Properly relieve the fuel system pressure and disconnect the fuel lines at the manifold.

9. Remove the distributor cap, ignition wires and coil. Disconnect the ESC connector and remove the distributor.

10. Disconnect all necessary sensor wire connectors and vacuum lines at the manifold.

11. Disconnect the throttle, TV and cruise control cables at the throttle body.

12. Remove the intake manifold bolts and remove the assembly.

To install:

13. Thoroughly clean all gasket mating surfaces, including bolts. Apply GM sealer 1052080 or equivalent to the bolts.

14. Apply a (5mm) bead of sealer to the front and rear sealing surfaces of the block, extending the bead ½ in. (13mm) up each cylinder head to seal and retain the gaskets.

15. Install the gaskets to the cylinder head with the port blocking plates facing up and to the rear.

16. Install the intake manifold and torque the bolts in sequence to 35 ft. lbs. (48 Nm).

17. Connect the throttle, TV and cruise control cables at the throttle body.

18. Connect all necessary sensor wire connectors and vacuum lines at the manifold.

19. Install the distributor cap, ignition wires and coil. Connect the ESC connector and install the distributor.

20. Connect the fuel lines at the manifold.

21. Install the heater hose pipe at the manifold.

22. Install the power brake vacuum pipe.

23. Install the ground cable from the manifold stud.

24. Install the wiring harness onto the clips.

25. Install the emission relays with brackets onto the manifold.

26. Install the 2 braces at the rear of the drive belt tensioner.

27. Install the upper radiator hose and fill the cooling system with the proper type and quantity of antifreeze. Install the air cleaner and heat stove assembly.

28. Connect the negative battery cable.

ASTRO AND SAFARI

1. Disconnect the negative battery cable. Remove the engine cover assembly. Remove the air cleaner assembly. Drain the cooling system.

2. Remove the distributor cap and ignition wires. Disconnect the ESC connector and remove the distributor.

3. Remove the cruise control transducer, if equipped.

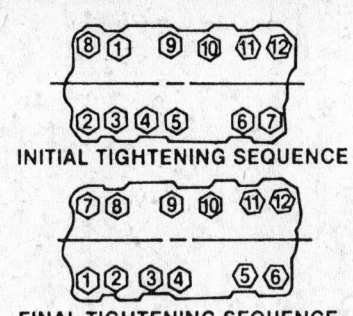

INITIAL TIGHTENING SEQUENCE

FINAL TIGHTENING SEQUENCE

79115042

Intake manifold bolt torque sequence — 4.3L (VIN Z) engine

4. Remove the detent, cruise and accelerator cables.

5. Remove the transmission and engine oil filler tubes at the alternator brace.

6. If equipped, remove the air conditioning compressor and idler pulley at the alternator brace. Remove the alternator brace.

7. Disconnect the fuel lines. Remove the necessary vacuum hoses and electrical wires.

8. Remove the AIR hoses and brackets, if equipped. Remove the upper radiator hose. Remove the heater hose at the manifold. As required, remove the TBI unit.

9. Remove the intake manifold retaining bolts. Remove the intake manifold from the engine.

To install:

10. The gaskets are marked for right and left side installation. Do not interchange them. Clean the sealing surface of the engine block and apply a $5/16$ in. bead of silicone sealer to each ridge.

11. Install the new gaskets onto the heads. The gaskets will have to be cut slightly to fit past the center pushrods. Do not cut any more material than necessary. Hold the gaskets in place by extending the ridge bead of sealer $1/4$ in. onto the gasket ends. (When the intake manifold is installed the area between the ridges and the manifold should be completely sealed.)

12. Install the intake manifold and torque the bolts in sequence to 35 ft. lbs. (47 Nm) except position No. 9 which is torqued to 41 ft. lbs. (56 Nm).

13. Install the AIR hoses and brackets, if equipped. Install the upper radiator hose. Install the heater hose at the manifold. Install the TBI unit.

14. Connect the fuel lines. Install the necessary vacuum hoses and electrical wires.

15. If equipped, install the air conditioning compressor and idler pulley at the alternator brace. Install the alternator brace.

16. Install the transmission and engine oil filler tubes at the alternator brace.

17. Install the detent, cruise and accelerator cables.

18. Install the cruise control transducer, if equipped.

19. Install the distributor, cap and ignition wires. Connect the ESC connector to the distributor.

20. Install the air cleaner assembly. Fill the cooling system with the proper type and quantity of antifreeze.

21. Connect the negative battery cable. Install the engine cover assembly.

4.3L (VIN Z) Turbocharged Engine

1. Disconnect the negative battery cable. Remove the air cleaner wing bolt and position the assembly aside. Drain the engine and turbocharger air cooler radiator antifreeze into a suitable container.

2. Remove the turbocharger air cooler, ducts, hoses and center support.

3. Disconnect all necessary vacuum and electrical connections from the upper and lower intake manifolds.

4. Remove the lower intake manifold left rear multi-use bracket.

5. Remove the throttle body, gasket and bracket from the upper intake manifold.

6. Remove the upper intake manifold bolts and remove the assembly.

7. Disconnect the heater hose and turbocharger coolant inlet pipe.

8. Disconnect the injector wire connectors and remove the fuel rail assembly.

9. Remove the air conditioner rear compressor bracket.

10. Disconnect the turbocharger coolant return line. Remove the upper radiator hose from the intake manifold.

11. Remove the distributor assembly.

12. Remove the lower intake manifold bolts and remove the assembly.

To install:

13. Thoroughly clean all gasket mating surfaces. Apply a $3/16$ in. (5mm) bead of GM sealant No. 1052366 or equivalent to the front and rear of the cylinder block; extend the bead up each cylinder head $1/2$ in. (13mm) to seal and retain the gaskets.

14. Lower the intake manifold onto the cylinder head and torque the bolts to 35 ft. lbs. (47 Nm).

15. Install the distributor assembly.

16. Connect the turbocharger coolant return line. Install the upper radiator hose at the intake manifold.

17. Install the air conditioner rear compressor bracket.

18. Install the fuel rail assembly and connect the injector wire connectors

19. Connect the heater hose and turbocharger coolant inlet pipe.

20. Install the upper intake manifold and bolts. Torque the bolts to 18 ft. lbs. (24 Nm) starting with the center bolts and working outward.

21. Install the throttle body, gasket and bracket at the upper intake manifold.

22. Install the lower intake manifold left rear multi-use bracket

23. Connect all vacuum and electrical connections previously removed from the upper and lower intake manifolds.

24. Install the turbocharger air cooler, ducts, hoses and center support.

25. Connect the negative battery cable. Install the air cleaner and duct assembly.

26. Fill the cooling system with the proper type and quantity of antifreeze.

Exhaust Manifold

REMOVAL AND INSTALLATION

2.5L Engine

1. Disconnect the negative battery cable.

2. Remove the exhaust stove pipe at the manifold. Disconnect the oxygen sensor wire.

3. Raise and safely support the vehicle. Disconnect the exhaust pipe at the exhaust manifold.

4. Remove the air conditioning compressor and brackets, if equipped.

5. Remove the exhaust manifold retaining bolts. Remove the manifold from the engine.

6. Installation is the reverse of the removal procedure. Torque the manifold bolts to specification.

2.8L Engine

1. Disconnect the negative battery cable. Raise and support the vehicle safely.

2. Disconnect the exhaust pipe from the manifold. Remove the 4 rear

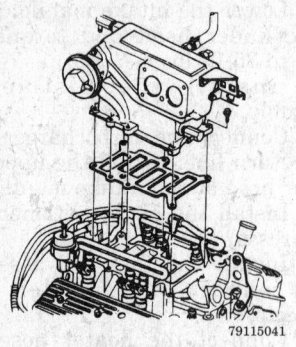

Intake manifold removal — 4.3L (VIN Z) turbocharged engine

manifold retaining bolts. Lower the vehicle.

3. Remove the air management hoses, pump and alternator brackets. Remove the heat stove tube.

4. If equipped, remove the power steering pump bracket.

5. Remove the remaining manifold attaching bolts. Remove the manifold from the engine.

6. Installation is the reverse of the removal procedure.

3.1L Engine

1. Disconnect the negative battery cable.

2. To remove the front exhaust manifold proceed as follows:

 a. Remove the serpentine belt and the air conditioning compressor.

 b. Remove the engine strut and bracket.

 c. Remove the crossover pipe.

 d. Remove the exhaust manifold attaching bolts and remove the manifold.

3. To remove the rear exhaust manifold proceed as follows:

 a. Disconnect the oxygen sensor wire.

 b. Remove the crossover pipe.

 c. Raise and support the vehicle safely.

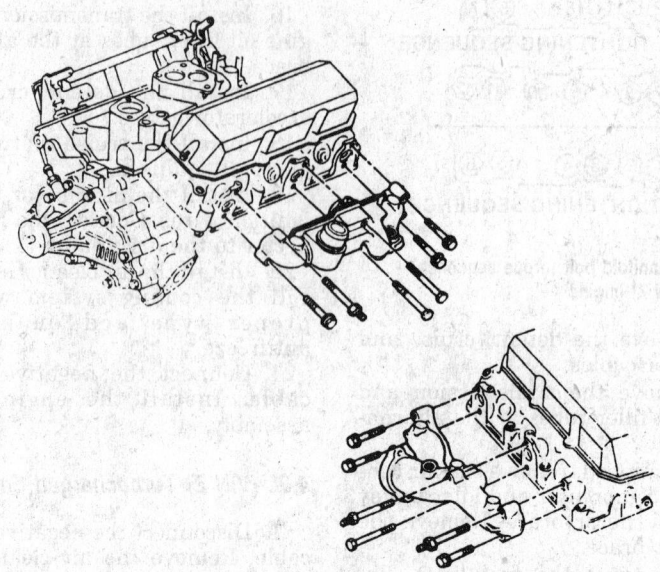

Exhaust manifold mounting — 2.8L engine

 d. Disconnect the exhaust pipe.

 e. Support the rear center of the frame.

 f. Remove the rear frame mount bolts.

 g. Lower the frame 8–10 in.

 h. Remove the exhaust manifold bolts and remove the assembly.

4. Installation is the reverse of the removal procedure.

3.8L Engine

1. Disconnect the negative battery cable.

2. To remove the front exhaust manifold proceed as follows:

 a. Remove the crossover pipe.

 b. Remove the spark plug wires.

 c. Remove the exhaust manifold bolts and dipstick tube.

 d. Remove the exhaust manifold assembly.

3. To remove the rear exhaust manifold proceed as follows:

 a. Disconnect the oxygen sensor wire.

 b. Remove the spark plug wires.

 c. Remove the throttle cable bracket.

 d. Remove the crossover pipe heat shield and remove the crossover pipe.

 e. Remove the transaxle dipstick and tube assembly.

 f. Disconnect the oxygen sensor lead.

 g. Remove the plastic vacuum tank mounted on the cowl.

 h. Raise and support the vehicle safely.

 i. Remove the catalytic converter heat shield and hanger.

 j. Remove the front exhaust pipe to the manifold attaching nuts.

 k. Remove the front exhaust pipe from the manifold. Lower the vehicle.

 l. Remove the engine lift bracket and remove the manifold attaching nuts.

 m. Remove the exhaust manifold assembly.

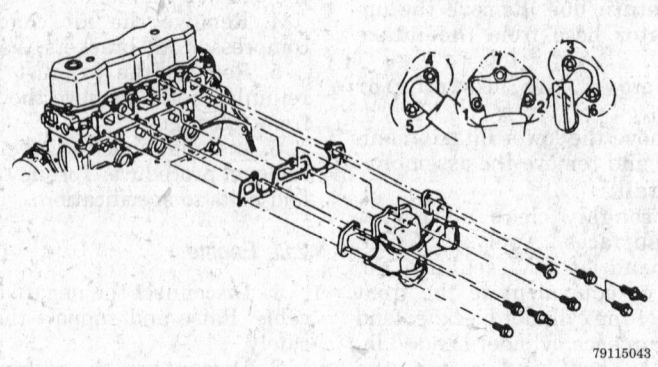

Exhaust manifold bolt torque sequence — 2.5L engine

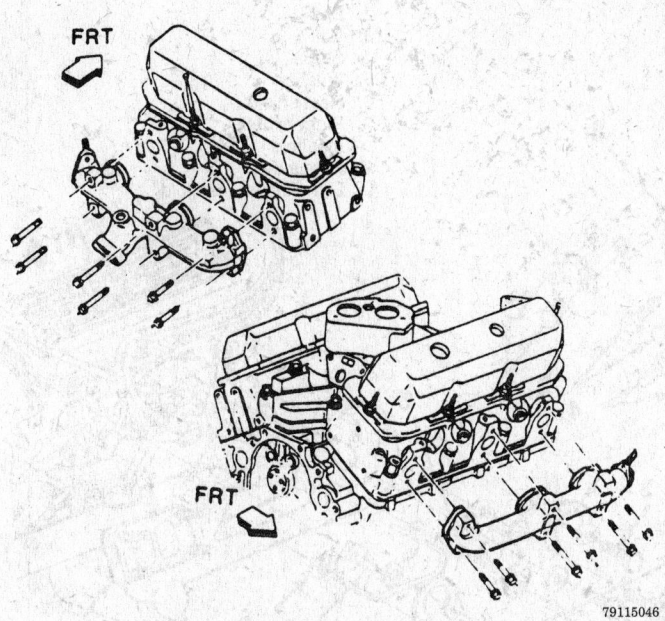

Exhaust manifold mounting — 3.1L engine

79115046

4. Installation is the reverse of the removal procedure.

4.3L Engine

NON-TURBOCHARGED ENGINE

1. Disconnect the negative battery cable. On Astro and Safari, remove the engine cover. Raise and support the vehicle safely.

2. Disconnect the exhaust pipes from the exhaust manifolds. Lower the vehicle.

3. To remove the right manifold, disconnect the heat stove pipe and the dipstick tube bracket.

4. To remove the left manifold, disconnect the oxygen sensor wire.

5. Disconnect the power steering bracket at the manifold, if equipped. Disconnect the alternator bracket at the manifold.

6. Remove the AIR hoses at the check valve, if equipped. Remove the exhaust manifold retaining bolts. Remove the manifold from the engine.

7. Installation is the reverse of the removal procedure.

TURBOCHARGED ENGINE

1. Disconnect the negative battery cable.

2. Raise and support the vehicle safely.

3. Remove the left and right front wheel assembly and wheelhouse.

4. Lower the vehicle and remove the left exhaust manifold as follows:

a. Remove the air cleaner and duct assembly.

b. Remove the turbocharger air inlet elbow.

c. Remove the upper fan shroud and loosen the fan nuts.

d. Remove the serpentine drive belt, fan and pulley assembly.

e. Remove the power steering pump pulley.

f. Remove the rear alternator brace.

g. Raise and support the vehicle safely.

h. Place a drain pan under the power steering pump and remove the power steering pump inlet and outlet hose.

i. Remove the oil filter lines bracket from the power steering pump.

j. Remove the intermediate shaft.

k. Remove the exhaust crossover pipe at the left manifold.

l. Remove the power steering pump and rear brace as an assembly from the front bracket.

m. Remove the spark plug wires and remove the spark plugs.

n. Remove the exhaust manifold bolts, studs, washers, lock tabs and heat shields.

o. Remove the exhaust manifold and lower the vehicle.

5. Remove the right exhaust manifold as follows:

a. Drain the engine coolant into a suitable container.

b. Remove the battery tray with the vacuum tank.

c. Disconnect the turbocharger oil feed hose and coolant return pipe.

d. Disconnect the oxygen sensor wire.

e. Raise and safely support the vehicle.

f. Remove the muffler and tailpipe from the catalytic converter.

g. Remove the catalytic converter support bolts and lower the vehicle.

h. Disconnect the turbocharger solenoid electrical connector.

i. Loosen the turbocharger coolant feed line.

j. Remove the turbocharger outlet pipe nuts.

k. Raise and safely support the vehicle.

l. Remove the turbocharger outlet pipe support bolt. Move the catalytic converter and outlet pipe away from the turbocharger.

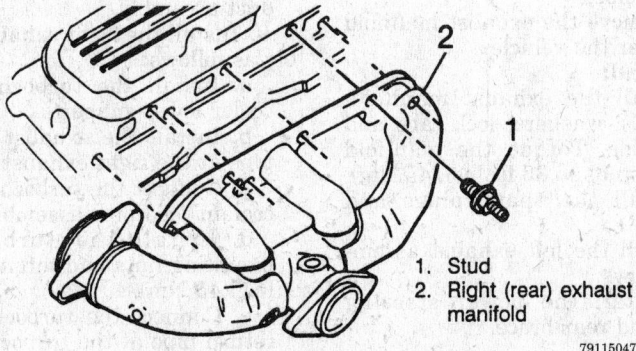

1. Stud
2. Right (rear) exhaust manifold

79115047

Exhaust manifold mounting — 3.8L engine

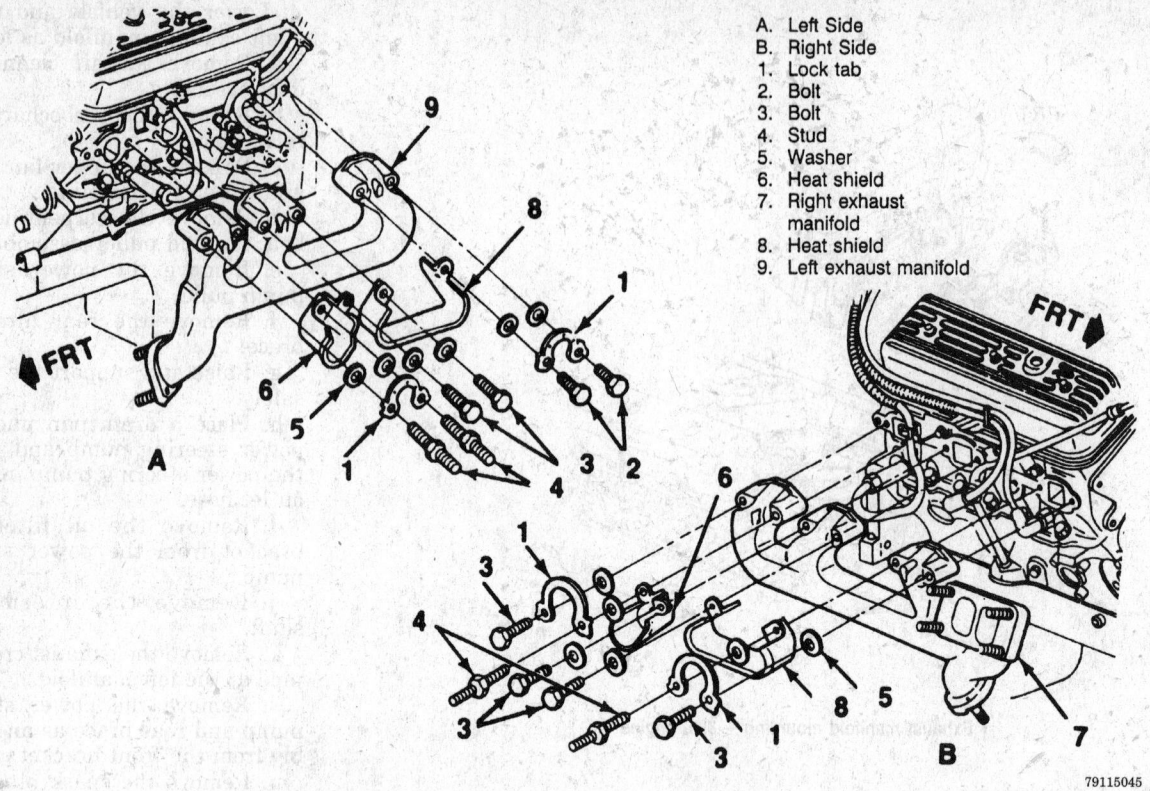

A. Left Side
B. Right Side
1. Lock tab
2. Bolt
3. Bolt
4. Stud
5. Washer
6. Heat shield
7. Right exhaust manifold
8. Heat shield
9. Left exhaust manifold

Exhaust manifold mounting — 4.3L engine

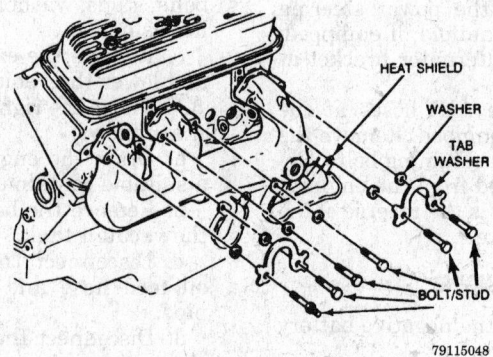

Exhaust manifold mounting — 4.3L engine

m. Disconnect the turbocharger oil return pipe from the turbocharger.

n. Remove the turbocharger mounting nuts.

o. Disconnect the turbocharger coolant feed pipe from the turbocharger and remove the assembly.

p. Remove the exhaust crossover pipe from the right exhaust manifold.

q. Remove the spark plug wires and remove the spark plugs.

r. Remove the turbocharger air cooler lower supports.

s. Remove the exhaust manifold bolts, studs, washers, lock tabs and heat shields.

t. Remove the exhaust manifold and lower the vehicle.

To install:

6. Install the exhaust manifold, bolts, studs, washers, lock tabs and heat shields. Torque the manifold bolts and studs to 33 ft. lbs. (45 Nm).

7. Install the spark plugs and wires.

8. Install the left exhaust assembly as follows:

a. Install the power steering pump and rear brace.

b. Install the exhaust crossover pipe at the left manifold and torque the bolts to 12 ft. lbs. (16 Nm).

c. Install the intermediate shaft.

d. Install the oil filter lines bracket to the power steering pump.

e. Install the power steering pump inlet and outlet hose.

f. Install the rear alternator brace.

g. Install the power steering pump pulley.

h. Install the serpentine drive belt, fan and pulley assembly.

i. Install the upper fan shroud.

j. Install the turbocharger air inlet elbow.

k. Install the air cleaner and duct assembly.

9. Install the right exhaust assembly as follows:

a. Install the turbocharger air cooler lower supports.

b. Install the exhaust crossover pipe at the right exhaust manifold.

c. Connect the turbocharger and coolant feed pipe assembly.

d. Install the turbocharger mounting nuts and torque to 33 ft. lbs. (45 Nm).

e. Connect the turbocharger oil return pipe at the turbocharger.

f. Install the turbocharger outlet pipe and nuts.

g. Install the turbocharger outlet pipe support bolt.

h. Connect the turbocharger solenoid electrical connector.

i. Install the catalytic converter support bolts.

j. Install the muffler and tailpipe at the catalytic converter and lower the vehicle.

k. Connect the oxygen sensor wire.

l. Connect the turbocharger oil feed hose and coolant return pipe.

m. Install the battery tray with the vacuum tank.

10. Raise and support the vehicle safely.

11. Install the left and right front wheel assembly and wheelhouse.

12. Connect the battery cables.

Turbocharger

REMOVAL AND INSTALLATION

1. Disconnect and remove the battery, tray and vacuum tank assembly.

2. Drain the antifreeze into a suitable container.

3. Remove the turbocharger air intake duct.

4. Remove the air intake duct.

5. Disconnect the solenoid wire connector at the turbocharger.

6. Raise and safely support the vehicle.

7. Remove the right wheel and tire assembly.

8. Remove the right wheelhouse panel.

9. Remove the turbocharger outlet pipe at the turbocharger. Lower the vehicle.

10. Remove the turbocharger coolant feed and return lines.

11. Remove the turbocharger oil feed and return lines.

12. Remove the turbocharger mounting bolts and remove the assembly.

To install:

13. Install the turbocharger, gasket and mounting bolts.

14. Install the turbocharger oil feed and return lines.

15. Install the turbocharger coolant feed and return lines.

16. Raise and safely support the vehicle. Install the turbocharger outlet pipe at the turbocharger.

17. Install the right wheelhouse panel.

18. Install the right wheel and tire assembly.

19. Lower the vehicle.

20. Connect the solenoid wire connector at the turbocharger.

21. Install the air intake duct.

22. Install the turbocharger air intake duct.

23. Fill the cooling system with the proper type and quantitiy of antifreeze.

24. Connect and Install the battery, tray and vacuum tank assembly.

25. Bleed the cooling system.

Timing Gear Front Cover

REMOVAL AND INSTALLATION

2.5L Engine

1. Disconnect the negative battery cable. Drain the cooling system.

2. Remove the power steering fluid reservoir from its mounting and support it aside.

3. Remove the upper fan shroud. Remove the accessory drive belt.

4. Disconnect the wiring from the alternator. Remove the alternator and its mounting brackets.

5. Using a suitable puller, remove the crankshaft pulley bolt and hub assembly. Disconnect the lower radiator hose at the water pump.

6. Remove the cover retaining bolts and remove the cover.

To install:

7. Lubricate the crankshaft seal with clean oil. Apply a 10mm bead of RTV sealer to oil pan lips. Apply a 6mm wide bead of RTV to the timing cover.

8. Install the cover in position on the engine. Tighten the retaining bolts to 90 inch lbs. (10 Nm).

9. Install the crankshaft pulley and hub. Connect the lower radiator hose to the water pump.

10. Install the alternator and its mounting brackets. Install the accessory drive belt.

11. Install the fan shroud and the power steering reservoir.

12. Refill the cooling system and connect the negative battery cable.

13. Run the engine and check for leaks.

Timing Chain Front Cover

REMOVAL AND INSTALLATION

2.8L and 4.3L Engines

1. Disconnect the negative battery cable.

2. Drain the cooling system. Remove the fan shroud assembly.

3. Remove the accessory drive belt and pulley.

4. Remove the water pump. Remove the power steering pump bracket.

5. Remove the crankshaft pulley and damper.

6. Disconnect the lower radiator hose. Remove the front cover retaining bolts.

7. Remove the front cover. Remove the old gasket material from the engine and the cover.

To install:

8. Install a new gasket in position. Install the front cover and tighten the retaining bolts to 18 ft. lbs. (24 Nm).

9. Install the water pump. Connect the lower radiator hose.

10. Install the crankshaft damper and pulley. Install the power steering mounting bracket.

11. Refill the cooling system to the correct level.

12. Connect the negative battery cable and bleed the cooling system.

3.1L Engine

1. Disconnect the negative battery cable.

2. Drain the cooling system.

3. Remove the accessory drive belt and tensioner.

4. Remove the power steering pump.

5. Raise and safely support the vehicle. Remove the inner splash shield.

6. Drain the engine oil. Remove the crankshaft pulley and damper. Remove the starter and support it aside.

7. Place a suitable jack under the engine-to-transaxle mount.

8. Remove the engine mount bolts and the engine mount. Raise the engine slightly.

9. Remove the lower front cover bolts and lower the oil pan. Remove the radiator hose at the water pump.

10. Remove the heater hose at the cooling system fill pipe. Remove the bypass and overflow hoses.

11. Remove the remaining front cover bolts and remove the front cover.

To install:

12. Clean all gasket mating surfaces. Install a new gasket in position on the engine block.

13. Apply sealer to the lower edges of the front cover. Install the front cover in position on the engine block. Tighten the upper bolts to 20 ft. lbs. (27 Nm).

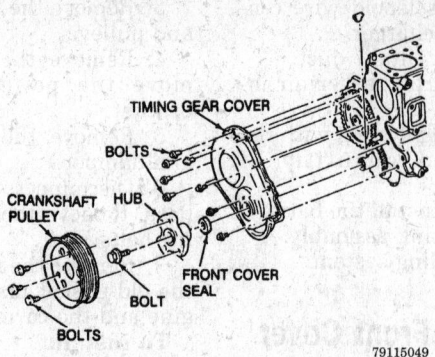

TIMING GEAR COVER
BOLTS
CRANKSHAFT PULLEY
HUB
FRONT COVER SEAL
BOLT
BOLTS

79115049

Front cover assembly — 2.5L engine

14. Install the oil pan in position. Install the lower cover bolts and tighten to 28 ft. lbs. (38 Nm).

15. Install the engine mount to the engine and lower the engine into position.

16. Install the crankshaft damper and pulley. Install the flywheel cover and inner splash shield.

17. Install the starter. Connect the heater bypass hose an the slower radiator hose to the water pump. Lower the vehicle.

18. Install the power steering pump bracket. Install the accessory drive belt and tensioner.

19. Refill the cooling system and the crankcase to the correct levels.

20. Connect the negative battery cable. Run the engine to normal operating temperature and check for leaks.

3.8L Engine

1. Disconnect the negative battery cable.

2. Drain the cooling system.

3. Remove the torque mount, accessory drive belt and tensioner.

4. Remove crankshaft damper using GM tool J–38197 or equivalent, while holding the flywheel in place using tool J–37096 or equivalent.

5. Remove the sensor shield.

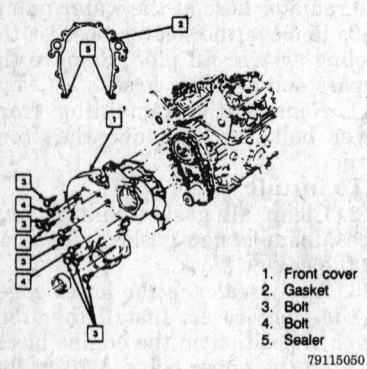

1. Front cover
2. Gasket
3. Bolt
4. Bolt
5. Sealer

79115050

Front cover assembly — 3.1L engine

6. Remove the oil pan to front cover bolts and the front cover attaching bolts.

7. Remove the front cover assembly.

8. Installation is the reverse of the removal procedure.

Front Cover Oil Seal

REPLACEMENT

The front cover seal can be replaced while the front cover is still on the vehicle. It is recommended by the manufacturer that the seal be replaced whenever the front cover is removed.

1. Disconnect the negative battery cable.

2. Remove the accessory drive belt.

3. Remove the crankshaft pulley and damper.

4. Using a suitable tool, pry the seal from the cover.

5. Lubricate the replacement seal with clean engine oil.

6. Using a suitable seal installer, position the seal on the crankshaft.

7. Install the crankshaft damper and pulley.

8. Install the accessory drive belt.

9. Connect the negative battery cable.

Timing Chain and Sprockets

REMOVAL AND INSTALLATION

2.8L Engine

1. Disconnect the negative battery cable. Drain the coolant.

2. Remove the front cover assembly.

3. Place the No. 1 piston at top dead center with the marks on the camshaft and crankshaft sprockets aligned (No. 4 cylinder firing).

4. Remove the bolts that hold the camshaft sprocket to the camshaft. This sprocket is a light press fit on the camshaft.

5. Remove the timing chain. Using a suitable puller, remove the crankshaft sprocket, as required.

To install:

6. Install the crankshaft sprocket using tool J–5590, ensuring the timing marks face outward.

7. Lubricate the thrust surface of the camshaft gear with Molykote® or equivalent. Install the chain onto camshaft sprocket.

8. Holding the sprocket vertically with the chain hanging down, align the marks on the camshaft and crankshaft sprockets and install the assembly onto the camshaft.

9. Install the camshaft to gear attaching bolts and torque to 17 ft. lbs. (23 Nm). After the sprockets are in place, turn the engine 2 full revolutions to make certain the timing marks are in correct alignment between the shaft centers.

10. Lubricate the chain with engine oil and install the front cover.

11. Connect the negative battery cable. Fill the cooling system with the proper type and quantity of antifreeze.

3.1L Engine

1. Disconnect the negative battery cable. Drain the coolant.

2. Remove the right front tire and wheel assembly.

3. Remove the front cover assembly. Ensure the marks on the crankshaft and camshaft gears are aligned using the marks on the damper stamping or cast alignment marks on cylinder and case.

4. Remove the bolts that hold the camshaft sprocket to the camshaft. This sprocket is a light press fit on the camshaft.

5. Remove the timing chain. Using a suitable puller, remove the crankshaft sprocket, as required.

To install:

6. Install the crankshaft sprocket.

7. Lubricate the camshaft thrust plate surface with Molykote® or equivalent. Install the chain onto camshaft sprocket.

8. Holding the sprocket vertically with the chain hanging down, align the marks on the camshaft and crankshaft sprockets and install the assembly onto the camshaft.

9. Install the camshaft to gear attaching bolts and torque to 18 ft. lbs. (24 Nm). After the sprockets are in

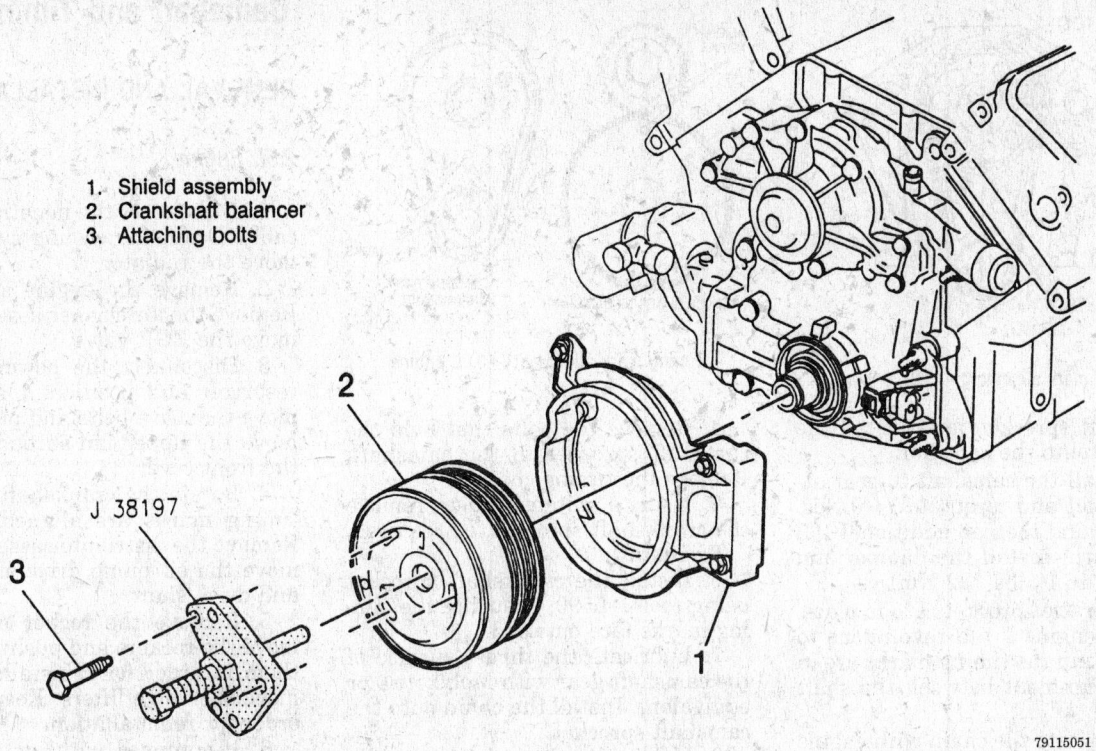

1. Shield assembly
2. Crankshaft balancer
3. Attaching bolts

J 38197

79115051

Removing crankshaft pulley hub assembly — 3.8L engine

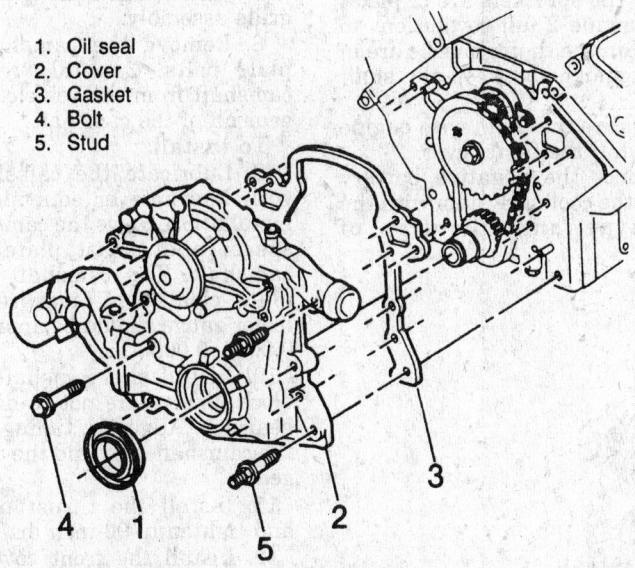

1. Oil seal
2. Cover
3. Gasket
4. Bolt
5. Stud

79115052

Front cover assembly — 3.8L engine

place, turn the engine 2 full revolutions to make certain the timing marks are in correct alignment between the shaft centers.

10. Lubricate the chain with engine oil and install the front cover.

11. Connect the negative battery cable. Fill the cooling system with the proper type and quantity of antifreeze.

3.8L Engine

1. Disconnect the negative battery cable. Drain the coolant.

2. Align the marks on the crankshaft damper and timing cover with the engine in the No. 1 firing position. Remove the front cover assembly.

3. Ensure the marks on the crankshaft and camshaft gears are aligned.

4. Remove the timing chain damper and camshaft sprocket.

5. Remove the timing chain. Using a suitable puller, remove the crankshaft sprocket, as required.

To install:

6. Install the crankshaft sprocket.

7. Install the chain onto camshaft sprocket.

8. Holding the sprocket vertically with the chain hanging down, align the marks on the camshaft and

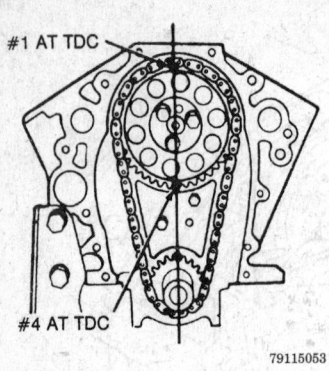

Timing mark alignment — 2.8L engine

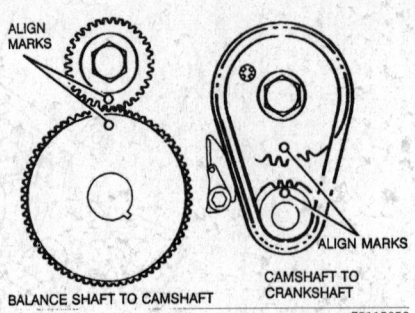

Timing mark alignment — 3.8L engine

Camshaft and Timing Gear

REMOVAL AND INSTALLATION

2.5L Engine

crankshaft sprockets and install the assembly onto the camshaft.

9. Install the camshaft to gear attaching bolt and torque to 74 ft. lbs. (100 Nm) and then an additional 105 degree turn. Install the damper and torque to 16 ft. lbs. (22 Nm).

10. After the sprockets are in place, turn the engine 2 full revolutions to make certain the timing marks are in correct alignment between the shaft centers.

11. Lubricate the chain with engine oil and install the front cover.

12. Connect the negative battery cable. Fill the cooling system with the proper type and quantity of antifreeze.

4.3L Engine

1. Disconnect the negative battery cable. Drain the coolant.

2. Align the marks on the crankshaft damper and timing cover with the engine in the No. 1 firing position. Remove the front cover assembly.

3. Ensure the marks on the crankshaft and camshaft gears are aligned.

4. Remove the bolts that hold the camshaft sprocket to the camshaft. Remove the timing chain.

5. Using a suitable puller, remove the crankshaft sprocket, as required.

To install:

6. Install the crankshaft sprocket using tool J–5590, ensuring the timing marks face outward.

7. Lubricate the thrust surface of the camshaft gear with Molykote® or equivalent. Install the chain onto the camshaft sprocket.

8. Holding the sprocket vertically with the chain hanging down, align the marks on the camshaft and crankshaft sprockets and install the assembly onto the camshaft.

9. Install the camshaft to gear attaching bolts and torque to specification. After the sprockets are in place, turn the engine 2 full revolutions to make certain the timing marks are in correct alignment between the shaft centers.

10. Lubricate the chain with engine oil and install the front cover.

11. Connect the negative battery cable. Fill the cooling system with the proper type and quantity of antifreeze.

1. Key
2. Damper assembly
3. Crankshaft sprocket
4. Bolt
5. Timing chain
6. Camshaft sprocket
7. Camshaft gear

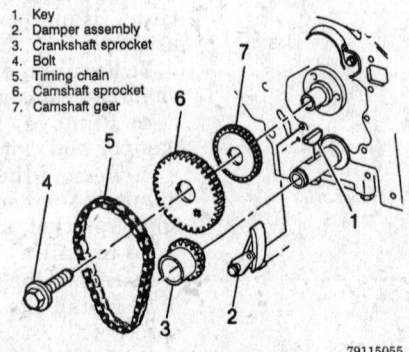

Exploded view of timing chain assembly — 3.8L engine

1. Disconnect the negative battery cable. Drain the cooling system. Remove the radiator.

2. Remove the engine side cover. Remove the air cleaner assembly. Remove the EGR valve.

3. Disconnect the power steering reservoir and position it aside. Remove the drive belts and pulleys. Remove the upper fan shroud. Remove the front cover.

4. Rotate the crankshaft until the timing marks are aligned at TDC. Remove the distributor assembly. Remove the oil pump drive shaft cover and drive shaft.

5. Remove the rocker cover. Remove the rockers and pushrods. Keep them in order for reinstallation. Remove the valve lifters. Keep them in order for reinstallation.

6. If equipped with air conditioning, it may be necessary to reposition the condenser in order to withdraw the camshaft from the engine. If the condenser must be removed, properly discharge the system before condenser removal.

7. Remove the headlight bezel and grille assembly.

8. Remove the camshaft thrust plate bolts. Carefully remove the camshaft from the vehicle. Press the gear off of the camshaft.

To install:

9. Lubricate the camshaft lobes with Molykote or equivalent before installation. Press the camshaft gear, spacer and thrust plate onto the camshaft. The camshaft to thrust plate clearance, measured with a feeler gauge should be approximately 0.0015–0.0050 in.

10. Install the camshaft into the block, using care not to damage the bearings. Align the timing marks on the camshaft gear and the crankshaft gear.

11. Install the thrust plate bolts and tighten to 90 inch lbs.

12. Install the front cover. Install the distributor and the oil pump shaft cover.

13. Install the lifters, pushrods, rocker arms and the engine side cover.

14. Install the remaining components in the reverse order of removal.

15. Refill the cooling system and connect the negative battery cable.

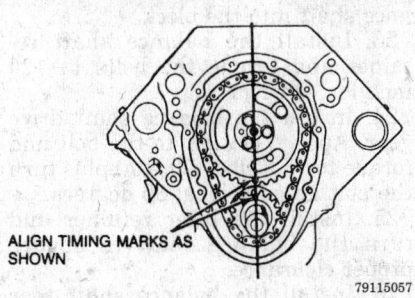

ALIGN TIMING MARKS AS
SHOWN

79115057

Timing mark alignment — 4.3L engine

Camshaft

REMOVAL AND INSTALLATION

2.8L Engine

1. Disconnect the negative battery cable.
2. Drain the cooling system.
3. Remove the upper fan shroud.
4. Remove the radiator.
5. Remove the valve cover and the valve train components. Remove the intake manifold assembly and remove the lifters.
6. Remove the front cover assembly.
7. Remove the timing chain and sprocket.
8. Remove the camshaft from the block. Insert 3 bolts approximately 3 in. long into the camshaft gear bolt holes to supply leverage while removing the camshaft. Use care not to damage the bearings.
   To install:
9. Lubricate the camshaft with Molykote or equivalent, before installation.
10. Install the camshaft into the cylinder block, use care not to damage the bearings.

11. Install the timing chain and sprocket. Make sure the timing marks align correctly.
12. Install the front cover assembly and hydraulic lifters.
13. Install the intake manifold assembly. Install the valve train components, adjust the valves and install the valve cover.
14. Install the radiator and the fan shroud.
15. Fill the cooling system to the correct level and connect the negative battery cable.

3.1L and 3.8L Engines

1. Disconnect the negative battery cable.
2. Drain the cooling system.
3. Remove the engine from the vehicle and support it in a suitable holding fixture.
4. Remove the intake manifold. Remove the valve cover and the valve train components.
5. Remove the front cover assembly.
6. Remove the timing chain and sprocket. Remove the thrust plate, if equipped.
7. Remove the camshaft from the block. Insert 3 bolts approximately 3 in. long into the camshaft gear bolt holes to supply leverage while removing the camshaft. Use care not to damage the bearings.
   To install:
8. Lubricate the camshaft with Molykote or equivalent, before installation.
9. Install the camshaft into the cylinder block, use care not to damage the bearings.
10. Install the thrust plate, if equipped. Install the timing chain and sprocket. Make sure the timing marks align correctly.
11. Install the front cover assembly.

12. Install the intake manifold assembly. Install the valve train components and the valve cover.
13. Install the engine into the vehicle.
14. Fill the cooling system to the correct level and connect the negative battery cable.

4.3L Engine

1. Disconnect the negative battery cable.
2. Drain the cooling system.
3. Remove the upper fan shroud, if required.
4. Remove the radiator.
5. Remove the valve cover and the valve train components. Remove the intake manifold assembly and remove the lifters.
6. Remove the front cover assembly.
7. Remove the timing chain and camshaft sprocket. Remove the balance shaft drive gear, if equipped. Remove the camshaft thrust plate.
8. Remove the camshaft from the block. Insert 3 bolts approximately 3 in. long into the camshaft gear bolt holes to supply leverage while removing the camshaft. Use care not to damage the bearings.
   To install:
9. Lubricate the camshaft with Molykote or equivalent, before installation.
10. Install the camshaft into the cylinder block, use care not to damage the bearings. Install the thrust plate, if equipped. Install the balance shaft drive gear, if equipped.
11. Install the timing chain and sprocket. Make sure the timing marks align correctly.
12. Install the front cover assembly and hydraulic lifters.
13. Install the intake manifold assembly. Install the valve train components, adjust the valves and install the valve cover.
14. Install the radiator and the fan shroud.
15. Fill the cooling system to the correct level and connect the negative battery cable.

Balance Shaft

REMOVAL AND INSTALLATION

3.8L Engine

1. Disconnect the negative battery cable.
2. Remove the engine from the vehicle.

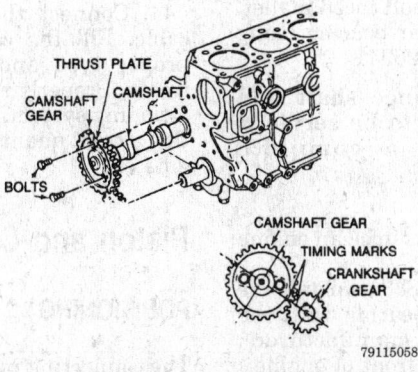

THRUST PLATE

CAMSHAFT
GEAR

CAMSHAFT

BOLTS

CAMSHAFT GEAR

TIMING MARKS

CRANKSHAFT
GEAR

79115058

**Timing gear assembly and timing mark
alignment — 2.5L engine**

3. Remove the flywheel and intake manifold.

4. Remove the lifter guide retainer.

5. Remove the timing chain cover.

6. Remove the balance shaft drive gear bolt, the camshaft sprocket and timing chain.

7. Remove the balance shaft retainer bolts, retainer and gear.

8. Using tool J–6125–B, remove the balance shaft.

9. Remove the balance shaft rear plug. Using tool J–36995–5, remove the balance shaft rear bearing.

NOTE: The balance shaft and bearings are only to be serviced or replaced as a complete assembly.

To install:

10. Dip the bearings in clean engine oil before installation.

11. Using tool J–36995–5, install the balance shaft rear bearing with the rolled edge facing into the engine and the manufacturer's markings facing the flywheel side.

12. Dip the front balance shaft bearing into engine oil. Using tool J–36996, install the balance shaft into the block.

13. Temporarily install the balance shaft retainer and bolts. Install the balance shaft drive gear. Apply Loctite® to the bolt and torque to 14 ft. lbs. (20 Nm) plus turn the bolt an additional 35 degrees.

14. Install the balance shaft rear plug.

15. Measure the balance shaft end play. End play should be 0–0.008 in. (0–0.203mm).

16. Measure the balance shaft radial play at both the front and rear. The front radial play should be 0–0.0011 in. (0–0.028mm). The rear radial play should be 0.0005–0.0047 in. (0.0127–0.119mm).

17. Temporarily install the camshaft gear and align the timing marks point straight down. Remove the camshaft gear and align the balance shaft marks point straight down. Install the camshaft gear and align the marks by turning the balance shaft.

18. Ensure the No. 1 piston is at TDC and install the camshaft sprocket and timing chain.

19. Measure the gear lash at 4 places, every ¼ turn. The lash should be 0.002–0.005 in. (0.050–0.127mm).

20. Tighten the balance shaft front bearing retainer to 22 ft. lbs. (30 Nm).

21. Install the timing chain cover.

22. Install the lifter guide retainer.

23. Install the flywheel and intake manifold.

24. Install the engine into the vehicle.

25. Connect the negative battery cable.

4.3L (VIN W) Engine

1. Disconnect the negative battery cable. Drain the cooling system into a suitable container. Properly relieve the fuel system pressure and discharge the air conditioning system.

2. Remove the air cleaner and air intake duct.

3. Remove the upper radiator shroud.

4. Remove the oil and transmission cooler lines at the radiator.

5. Remove the hoses from the radiator and remove the radiator.

6. Remove the air conditioner condenser and fan assembly.

7. Remove the serpentine belt.

8. Remove the brace at the coolant pump and remove the pump.

9. Remove the torsional damper.

10. Raise and support the vehicle safely.

11. Remove the flywheel cover. Drain the engine oil and loosen the oil pan bolts; remove the 2 nuts and bolts at the front of the oil pan.

12. Remove the front timing cover bolts and cover.

NOTE: Use care when removing the front cover, as not to damage the oil pan gasket.

13. Remove the crankshaft front cover oil seal.

14. Remove the camshaft sprocket, timing chain and balance shaft drive gear.

15. Remove the balance shaft retainer.

16. Remove the intake manifold assembly.

17. Remove the hydraulic lifter retainer.

18. Remove the balance shaft from the bearing using a soft faced mallet.

19. Remove the rear bearing using tool J–38834 and J–26941.

NOTE: The balance shaft and bearings are only to be serviced or replaced as a complete assembly.

To install:

20. Dip the bearings in clean engine oil before installation.

21. Using tool J–38834, install the balance shaft rear bearing with the flat edge and the manufacturer's markings facing the front of engine.

22. Dip the front balance shaft bearing into engine oil. Using tool J–36996 and J–8092, install the balance shaft into the block.

23. Install the balance shaft retainer and torque the bolts to 120 inch lbs. (14 Nm).

24. Install the balance shaft drive gear. Apply Loctite® to the bolt and torque to 15 ft. lbs. (20 Nm) plus turn the bolt an additional 35 degrees.

25. Install the lifter retainer and turn the balance shaft to ensure proper clearance.

26. Install the balance shaft rear plug.

27. Turn the camshaft so the balance shaft drive gear timing marks are aligned.

28. Install the balance shaft drive gear retaining stud and torque to 12 ft. lbs. (16 Nm).

29. Install the intake manifold assembly.

30. Install the timing chain and gears with the timing marks dot-to-dot; this is the No. 4 cylinder firing position, so set the distributor rotor to the No. cylinder firing position when installing. Torque the camshaft bolts and nut to 21 ft. lbs. (28 Nm).

31. Install the distributor assembly.

32. Install the timing cover oil seal and install the cover.

33. Raise and support the vehicle safely.

34. Install the flywheel cover.

35. Tighten the oil pan bolts; install the 2 nuts and bolts at the front of the oil pan.

36. Install the torsional damper.

37. Install the brace at the coolant pump and install the pump.

38. Install the serpentine belt.

39. Install the air conditioner condenser and fan assembly.

40. Install the radiator and hoses.

41. Install the oil and transmission cooler lines at the radiator.

42. Install the upper radiator shroud.

43. Install the air cleaner and air intake duct.

44. Connect the negative battery cable. Fill the cooling system with proper type and quantity of antifreeze. Properly recharge the air conditioning system. Fill the engine with the correct quantity and grade of engine oil.

Piston and Connecting Rod

POSITIONING

The connecting rod bearing tang slots must be on the side opposite of the camshaft, except for the 3.8L engine.

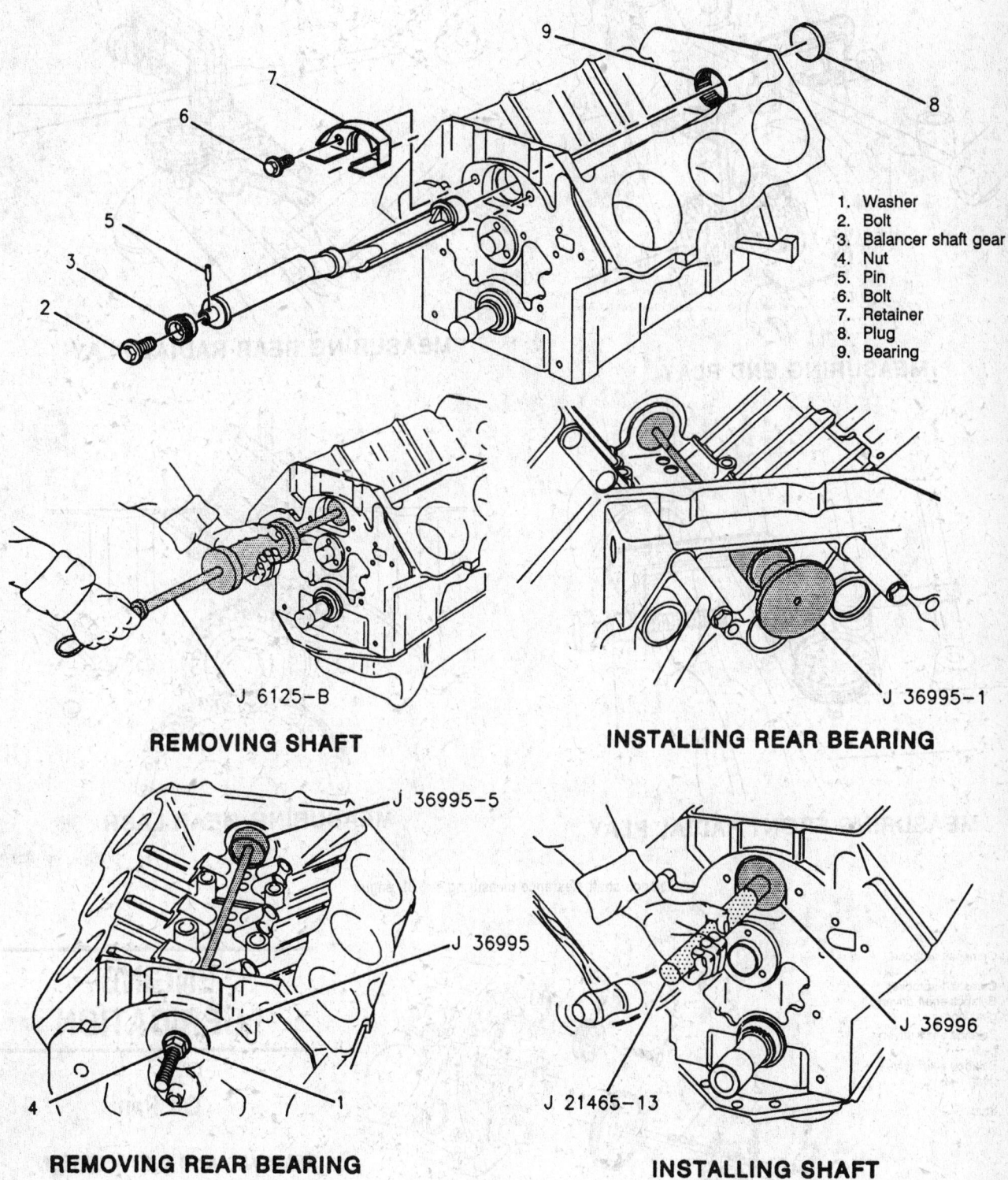

1. Washer
2. Bolt
3. Balancer shaft gear
4. Nut
5. Pin
6. Bolt
7. Retainer
8. Plug
9. Bearing

J 6125–B

REMOVING SHAFT

J 36995–1

INSTALLING REAR BEARING

J 36995–5

J 36995

J 21465–13

J 36996

REMOVING REAR BEARING

INSTALLING SHAFT

79115060

Balance shaft service — 3.8L engine

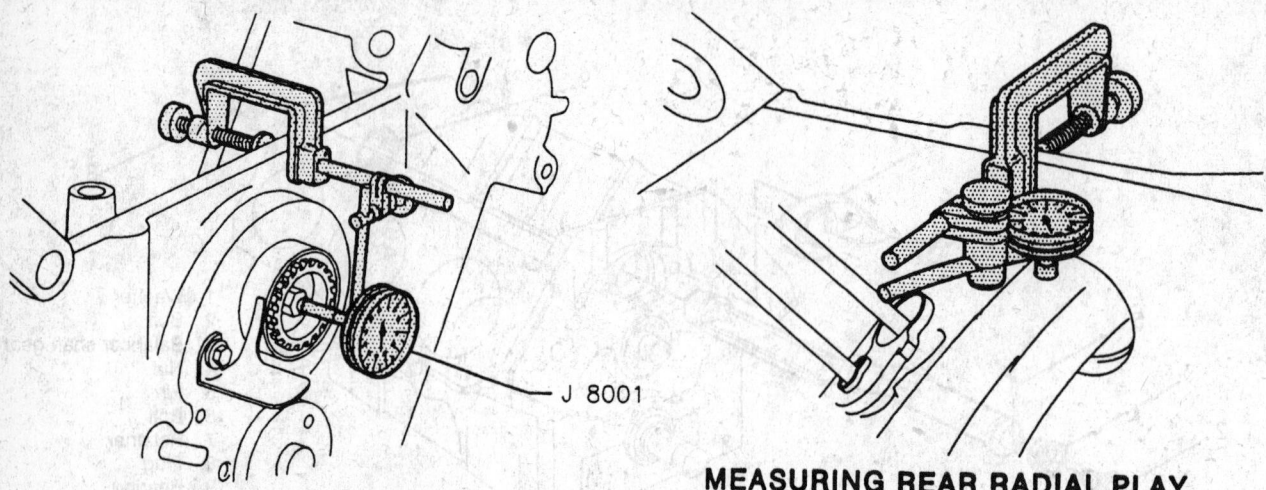

MEASURING END PLAY

J 8001

MEASURING REAR RADIAL PLAY

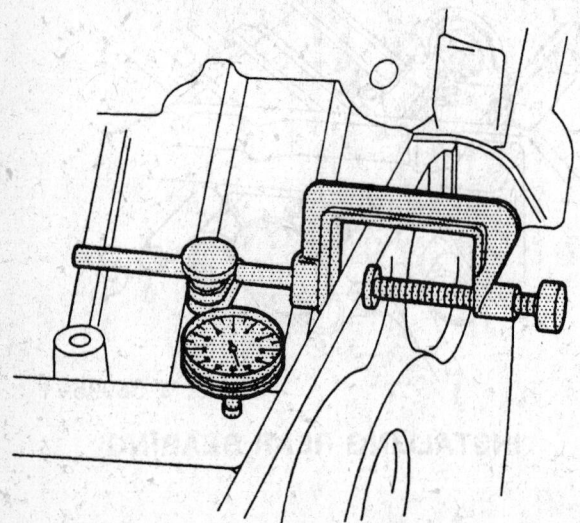

MEASURING FRONT RADIAL PLAY

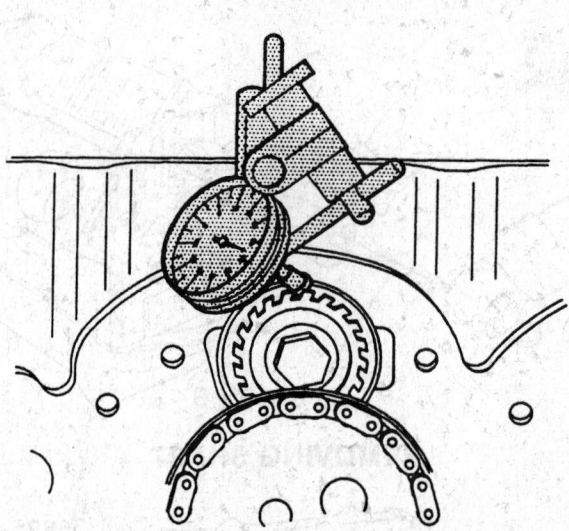

MEASURING GEAR LASH

79115059

Balance shaft clearance measuring — 3.8L engine

1. Camshaft sprocket bolt
2. Camshaft sprocket
3. Balance shaft driven gear bolt
4. Balance shaft driven gear
5. Balance shaft drive gear
6. Nut
7. Stud

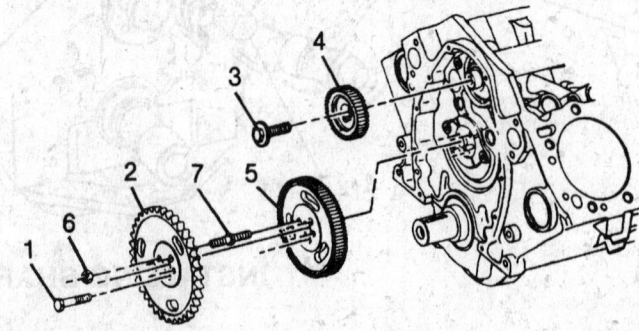

Balance shaft drive gears — 4.3L (VIN W) engine

79115061

ENGINE LUBRICATION

Oil Pan

REMOVAL AND INSTALLATION

2.5L Engine

2WD VEHICLE

1. Disconnect the negative battery cable.

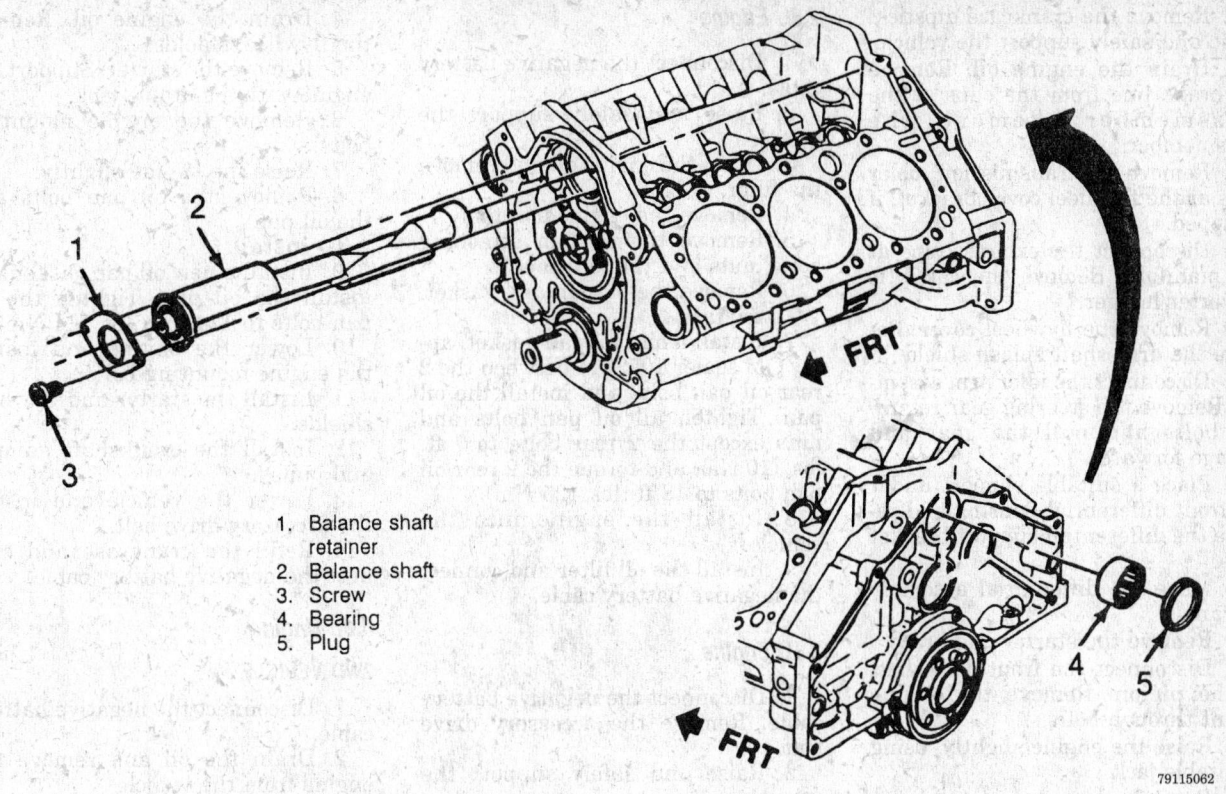

1. Balance shaft retainer
2. Balance shaft
3. Screw
4. Bearing
5. Plug

Balance shaft assembly — 4.3L (VIN W) engine

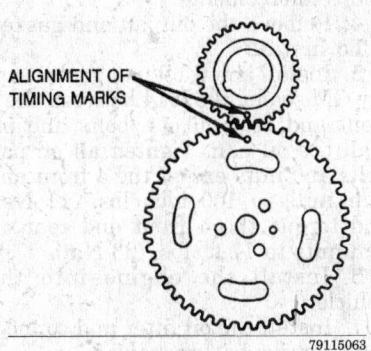

ALIGNMENT OF TIMING MARKS

Balance shaft alignment marks — 4.3L (VIN W) engine

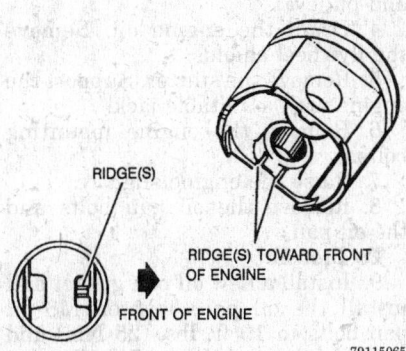

RIDGE(S)

RIDGE(S) TOWARD FRONT OF ENGINE

FRONT OF ENGINE

Piston installed position — 3.8L engine

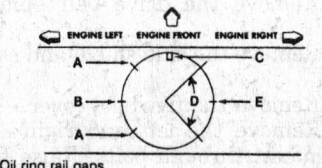

ENGINE LEFT   ENGINE FRONT   ENGINE RIGHT

a. Oil ring rail gaps
b. Second compression ring gap
c. Notch in piston
d. Oil ring spacer gap
e. Top compression ring gap

Piston and ring positioning

2. Remove the power steering reservoir and position aside.
3. Remove the radiator shroud.
4. Raise and safely support the vehicle.
5. Drain the engine oil.
6. Disconnect the strut rods. Disconnect the exhaust pipes at the manifolds.
7. Remove the catalytic converter and exhaust pipe.
8. Remove the flywheel cover. Remove the starter and brace.
9. Remove the brake lines from the crossmember. Remove the front engine mounting bolts.

10. Raise the engine slightly, with a suitable lifting device and insert wooden blocks to hold it.
11. Remove the oil pan retaining bolts and remove the oil pan.
   To install:
12. Apply RTV sealant to the oil pan flange and block. Install the oil pan into position.
13. Install the oil pan bolts and tighten to 90 inch lbs. (10 Nm).
14. Remove the blocks and lower the engine. Install the engine mount bolts.
15. Connect the brake line at the crossmember. Install the starter and brace.
16. Install the flywheel cover. Install the catalytic converter and the exhaust pipe.
17. Install the strut rods. Lower the vehicle. Install the radiator shroud.
18. Install the power steering reservoir. Refill the crankcase with the correct level of oil.
19. Connect the negative battery cable.

4WD VEHICLE

1. Disconnect the negative battery cable.
2. Remove the power steering pump from the fan shroud and remove the fan shroud.

3. Remove the crankcase dipstick. Raise and safely support the vehicle.

4. Drain the engine oil. Remove the brake line from the clips at the crossmember. Remove the crossmember.

5. Remove the transmission cooler lines at the flywheel cover bracket, if equipped.

6. Disconnect the exhaust pipe at the manifold. Remove the catalytic converter hanger.

7. Remove the flywheel cover. Remove the driveshaft splash shield.

8. Disconnect the idler arm assembly. Remove the steering gear retaining bolts and pull the gear and linkage forward.

9. Place a suitable support under the front differential housing and remove the differential support bracket bolts.

10. Move the differential assembly forward.

11. Remove the starter and brace.

12. Disconnect the front driveshaft at the pinion. Remove the engine mount through bolts.

13. Raise the engine slightly, using a suitable jack.

14. Remove the oil pan retaining bolts and remove the oil pan.

To install:

15. Clean all sealant from the block and the oil pan.

16. Apply a bead of RTV to the oil pan flange and to the sealing surfaces of the block.

17. Install the oil pan in position and install the retaining bolts. Tighten to 90 inch lbs.

18. Lower the engine and install the engine mount through bolts. Connect the front driveshaft at the drive pinion.

19. Install the starter and brace. Pull the differential back into position and install the retaining bolts.

20. Install the steering gear and idler arm assembly mounting bolts.

21. Install the driveshaft shield and the flywheel cover.

22. Install the catalytic converter hanger and the connect the exhaust pipe at the manifold.

23. Reconnect the transmission cooler lines to the clips at the flywheel housing.

24. Install the crossmember and reposition the brake lines. Lower the vehicle.

25. Install the dipstick. Refill the crankcase with oil. Install the upper fan shroud.

26. Install the power steering fluid reservoir. Connect the negative battery cable.

2.8L Engine

1. Disconnect the negative battery cable.

2. Raise and safely support the vehicle.

3. Drain the engine oil and remove the filter.

4. Remove the engine assembly.

5. Remove the oil pan attaching bolts, nuts and reinforcements.

6. Remove the oil pan and gasket.

To install:

7. Install a new oil pan gasket, apply GM sealer 1052914 between the 2 rear oil pan bolts and install the oil pan. Tighten all oil pan bolts and nuts except the 2 rear bolts to 7 ft. lbs. (10 Nm) and torque the 2 rear oil pan bolts to 18 ft. lbs. (25 Nm).

8. Install the engine into the vehicle.

9. Install the oil filter and connect the negative battery cable.

3.1L Engine

1. Disconnect the negative battery cable. Remove the accessory drive belt.

2. Raise and safely support the vehicle.

3. Remove the crankshaft damper and pulley.

4. Drain the engine oil. Remove the flywheel shields.

5. Remove the starter. Support the engine with a suitable jack.

6. Remove the engine mounting bolts.

7. Raise the engine slightly.

8. Remove the oil pan bolts and the oil pan.

To install:

9. Install a new oil pan gasket and install the oil pan. Tighten M8 oil pan bolts to 19 ft. lbs. (25 Nm) and the M6 oil pan bolts to 7 ft. lbs. (10 Nm).

10. Lower the engine and install the engine mounting bolts.

11. Install the starter and flywheel shields.

12. Install the crankshaft damper and pulley.

13. Lower the vehicle and install the accessory drive belt.

14. Refill the crankcase and connect the negative battery cable.

3.8L Engine

1. Disconnect the negative battery cable. Remove the accessory drive belt.

2. Raise and safely support the vehicle.

3. Remove the crankshaft damper and pulley.

4. Drain the engine oil. Remove the flywheel shields.

5. Remove the starter. Support the engine with a suitable jack.

6. Remove the engine mounting bolts.

7. Raise the engine slightly.

8. Remove the oil pan bolts and the oil pan.

To install:

9. Install a new oil pan gasket and install the oil pan. Tighten the oil pan bolts to 124 inch lbs. (14 Nm).

10. Lower the engine and install the engine mounting bolts.

11. Install the starter and flywheel shields.

12. Install the crankshaft damper and pulley.

13. Lower the vehicle and install the accessory drive belt.

14. Refill the crankcase and connect the negative battery cable.

4.3L Engine

2WD VEHICLE

1. Disconnect the negative battery cable.

2. Drain the oil and remove the engine from the vehicle.

3. Remove the oil pan bolts, nuts and reinforcements.

4. Remove the oil pan and gasket.

To install:

5. Install a new oil pan gasket, apply GM sealer 1052914 between the 2 front and rear oil pan bolts and install the oil pan. Tighten all oil pan bolts and nuts except the 4 front and rear nuts to 100 inch lbs. (11 Nm) and torque the 4 front and rear oil pan nuts to 17 ft. lbs. (23 Nm).

6. Install the engine into the vehicle.

7. Install the oil filter and connect the negative battery cable.

4WD Vehicle

1. Disconnect the negative battery cable.

2. Remove the crankcase dipstick. Raise and safely support the vehicle. Drain the engine oil.

3. Remove the drive belt splash shield.

4. Remove the axle shield and skid plate.

5. Remove the flywheel cover.

6. Remove the left and right engine mount through bolts. Raise the engine and block in position.

7. Remove the oil cooler lines and oil filter adapter.

8. Remove the pitman and idler arm.

9. Remove the front differential through bolts.

10. Remove the front propeller shaft and roll the differential forward.

11. Remove the starter and brace.

12. Remove the oil pan retaining bolts and remove the oil pan.

To install:

13. Clean all sealant from the block and the oil pan.

14. Apply a bead of RTV to the oil pan flange and to the sealing surfaces of the block. Install the oil pan gasket and seal.

15. Install the oil pan and the retaining bolts. Torque the bolts to 100 inch lbs. (11 Nm) and the nuts to 17 ft. lbs. (23 Nm).

16. Install the starter and brace.

17. Pull the differential back into position and install the retaining bolts.

18. Connect the front driveshaft at the drive pinion.

19. Lower the engine and install the engine mount through bolts.

20. Install the pitman and idler arm assembly.

21. Install the transfer case shield and the flywheel cover.

22. Install the front skid plate and axle shield.

23. Reconnect the oil cooler lines and oil filter adapter.

24. Install the drive belt splash shield. Lower the vehicle.

25. Install the dipstick. Refill the crankcase with oil.

26. Connect the negative battery cable.

Oil Pump

REMOVAL AND INSTALLATION

Except 3.8L Engine

1. Disconnect the negative battery cable.

2. Raise and safely support the vehicle.

3. Drain the engine oil. Remove the engine as required. Remove the oil pan.

4. Remove the oil pump retaining bolt and lower the pump from the block.

5. Remove the pump shaft.

6. Install the pump onto the block, ensure the pump shaft and driven shaft are properly aligned. No gasket is used.

7. Install the pump retaining bolt and torque to specification.

8. Install the oil pan. Install the engine, if removed.

9. Lower the vehicle. Refill the crankcase with oil.

10. Connect the negative battery cable. Disable the ignition system; crank engine for approximately 10 seconds to aid in priming the oil pump and reducing the risk of engine damage.

NOTE: If the oil pump does not build up oil pressure almost immediately, remove the pan and check for a loose oil pump to Pickup tube attachment. If necessary dismantle the pump and pack the pump cavity with petroleum jelly. Running the engine without measurable oil pressure will cause extensive damage.

3.8L Engine

1. Disconnect the negative battery cable.

2. Remove the front timing cover assembly.

3. Remove the oil pump cover attaching screws and remove the pump gears.

4. Remove the oil filter drip shield.

5. Remove the oil filter.

6. Remove the 4 bolts securing the adapter to the front cover.

7. Remove the adapter, gasket, the oil pressure valve and spring.

To install:

8. Clean all part in solvent and remove the old gaskets.

9. Check all parts for scoring, cracks or excessive wear. Check pressure regulator spring for loss of tension and replace as necessary.

10. Install the oil filter adapter, a new gasket, the oil pressure valve and spring.

11. Install the 4 adapter to front cover screws and torque to 22 ft. lbs. (30 Nm).

12. Install the oil filter.

13. Install the oil filter drip shield.

14. Install the pump gears into the cover assembly and pack with petroleum jelly. Install the oil pump cover attaching screws and torque to 97 inch lbs. (11 Nm).

15. Install the front timing cover assembly.

16. Connect the negative battery cable.

NOTE: If the oil pump Pickup screen is thought to be possibly clogged or dirty the oil pan should be removed and the screen removed and cleaned with solvent.

Checking

1. Disconnect the negative battery cable.

2. Remove the oil pressure sending unit.

3. Connect a suitable oil pressure gauge in place of the sending unit.

4. Reconnect the negative battery cable.

5. Ensure oil level is within specification.

6. Start the engine and set to specified test RPM.

7. Verify oil pressure reading meets specifications.

8. Remove the gauge and reinstall the sending unit.

Rear Main Oil Seal

REMOVAL AND INSTALLATION

1. Disconnect the negative battery cable.

2. Raise and safely support the vehicle.

3. Remove the engine and/or transmission from the vehicle, as required.

4. Remove the flywheel. Remove the clutch assembly, if equipped with manual transmission.

5. Remove the oil seal by prying it out with a suitable tool.

6. Coat the seal lips with oil and using a suitable tool, install the seal.

7. Install the flywheel and clutch assembly.

8. Install the transmission assembly.

9. Lower the vehicle. Connect the negative battery cable.

MANUAL TRANSMISSION

Transmission Assembly

REMOVAL AND INSTALLATION

1. Disconnect the negative battery cable.

2. Shift the transmission into neutral and remove the shift lever boot. Raise and support the vehicle safely. Drain the fluid.

3. Remove the shift lever assembly. Remove the parking brake cables for clearance.

4. Remove the driveshaft and skid plate. If equipped, drain and remove the transfer case.

5. Disconnect the speedometer cable at the transmission. Disconnect all electrical wires, as required.

6. Disconnect and remove exhaust components, as required. Remove the fuel lines, as required. Properly support the transmission assembly. Remove the clutch slave cylinder from its mounting.

7. Remove the transmission assembly retaining bolts. Remove the transmission crossmember retaining bolts. Remove the transmission mount retaining bolts. Remove the crossmember from the vehicle.

8. Properly support the clutch release bearing. Remove the engine-to-transmission bolts.

9. Properly support the engine assembly. Carefully remove the transmission assembly from the vehicle.

To install:

10. Put a thin layer of high temperature grease on the transmission main shaft. Shift the transmission into high gear (5th) before installing. Carefully install the transmission assembly onto the engine. Rotate the transmission clockwise onto the clutch hub splines.

11. Install the engine-to-transmission bolts and torque to specification.

12. Install the crossmember and retaining bolts.

13. Install the exhaust system. Install the fuel lines. Install the clutch slave cylinder.

14. Connect the speedometer cable at the transmission. Connect all electrical wires, as required.

15. Install the driveshaft and skid plate. Install the transfer case and fill with the proper type and quantity of oil. Fill the transmission with the proper type and quantity of oil.

16. Install the parking brake cables.

17. Lower the vehicle.

18. Install the shift lever assembly. Connect the negative battery cable. Check for proper clutch operation.

CLUTCH

Clutch Assembly

REMOVAL AND INSTALLATION

1. Disconnect the negative battery cable. Raise and support the vehicle safely. Remove the transmission as-

sembly. As required, remove the flywheel cover.

2. Remove the slave cylinder retaining bolts. Remove the slave cylinder from its mounting.

3. Remove the bell housing retaining bolts. Remove the bell housing from the vehicle, as required.

4. Slide the clutch fork from the ball stud. Inspect the ball stud and replace as required.

5. Install the clutch removal tool and support the clutch assembly. Matchmark the clutch and pressure plate for reassembly.

6. Loosen the clutch plate retaining bolts slowly and evenly one at a time until all pressure is released from the pressure plate assembly.

7. Remove the clutch, pressure plate and removal tool from the vehicle. Check the flywheel for damage, repair or replace, as required.

8. Inspect the pilot bearing and replace as necessary.

To install:

9. Lubricate the pilot bearing with a few drops machine oil. Install the flywheel, if removed.

10. Using a clutch alignment tool, support the clutch assembly. Align the clutch and pressure plate matchmarks, made previously.

11. Install the clutch pressure plate and disc assembly.

12. Tighten the clutch plate retaining bolts slowly and evenly one at a time to avoid warping the assembly. Remove the alignment tool.

13. Lubricate the ball stud with high temperature grease and slide the clutch fork onto the ball stud. The fork must be installed so the fingers and tabs fit into the groove of the release bearing.

14. Inspect the release bearing for roughness or heat stress, replace if necessary. Apply high temperature grease to the inside and outside groove of the release bearing.

15. Install the bell housing and retaining bolts, as required.

16. Install the slave cylinder and retaining bolts.

17. Install the transmission assembly. As required, install the flywheel cover.

18. Connect the negative battery cable. Lower the vehicle safely. Bleed the hydraulic clutch system, as required.

Clutch Master Cylinder

REMOVAL AND INSTALLATION

1. Disconnect the negative battery cable. Remove the under dash hush panel. Remove the lower steering column cover, as required. Remove the left side air conditioning duct work, as required.

2. Disconnect the pushrod from the clutch pedal. Disconnect the clutch master cylinder retaining nuts.

3. Disconnect and plug the reservoir hose at the clutch master cylinder assembly.

4. Disconnect and plug the fluid line to the slave cylinder at the clutch master cylinder assembly.

5. Remove the clutch master cylinder from the vehicle.

6. Installation is the reverse of the removal procedure. Bleed the hydraulic clutch system.

Clutch Slave Cylinder

REMOVAL AND INSTALLATION

1. Disconnect the negative battery cable.

2. Raise and support the vehicle safely. Disconnect and plug the fluid line at the slave cylinder.

3. Remove the slave cylinder retaining bolts. Remove the slave cylinder from the vehicle.

4. Installation is the reverse of the removal procedure. Bleed the system, as required.

Hydraulic Clutch System Bleeding

1. Fill the clutch master cylinder with the proper grade and type fluid. Raise and support the vehicle safely.

2. Remove the slave cylinder retaining bolts. Hold the cylinder at about a 45 degree angle with the bleeder at the highest point.

3. Fully depress the clutch pedal and open the bleeder screw. Close the bleeder screw after the pedal reaches the floor and repeat the procedure until all air is expelled from the system.

4. Ensure that the fluid level remains full in the clutch master cylinder throughout the bleeding procedure.

AUTOMATIC TRANSMISSION

Transmission Assembly

REMOVAL AND INSTALLATION

Except Turbocharged Engine

1. Disconnect the negative battery cable. Remove the air cleaner assembly. Disconnect the throttle valve cable and the throttle linkage.
2. Raise and support the vehicle safely.
3. Drain the transmission fluid. Disconnect the shift linkage. Properly relieve the fuel system pressure and remove the fuel lines.
4. Remove the driveshaft. If equipped with a transfer case, also remove the front driveshaft.
5. Disconnect and remove all required exhaust system components. Support the transmission using the proper equipment.
6. Remove the crossmember retaining bolts. Remove the transmission mount retaining bolts. Remove the crossmember from the vehicle and lower the transmission enough to gain access to other components.
7. Remove the transmission dipstick and tube. Disconnect the speedometer cable. If equipped, disconnect the vacuum modulator line.
8. Disconnect all electrical connections from the transmission assembly. Disconnect and plug the fluid cooler lines.
9. Remove the damper and support, as required. Remove the flywheel housing cover and torque converter to flywheel bolts. If equipped, remove the transfer case assembly.
10. Properly support the engine assembly. Remove the transmission to engine retaining bolts. Carefully remove the automatic transmission from the vehicle.

To install:

11. Install the transmission into position, making sure the converter is properly seated.
12. Install the engine-to-transmission and torque converter-to-flywheel bolts. Torque the bolts to specification.
13. Install the converter housing cover. Install the damper and support.
14. Connect all electrical leads to the transmission and connect the speedometer cable.

15. Install the dipstick tube. Install the crossmember and install the transmission mount retaining bolts. Reconnect the fuel lines.
16. Connect all exhaust system components. Install the front driveshaft on 4WD vehicles.
17. Connect the shift linkage and install the skid plate, if equipped. Lower the vehicle.
18. Connect the TV cable and throttle linkage.
19. Install the air cleaner assembly. Connect the negative battery cable.
20. Check the fluid level and operation of the transmission.

Turbocharged Engine

1. Disconnect the negative battery cable. Raise and safely support the vehicle.
2. Remove the transfer case assembly.
3. Remove the converter cover. Remove the torque converter-to-flywheel bolts.
4. Remove the shift control cable.
5. Remove the clip retaining the fuel line and speed sensor harness. Position the fuel line and harness aside.
6. Disconnect the T.V. cable.
7. Remove the oil cooler lines at the transmission.
8. Remove the transmission vent hose.
9. Remove the transmission dipstick, filler tube and bracket.
10. Remove the transmission retaining bolts and remove the assembly.

To install:

11. Install the transmission assembly and torque the bolts to 35 ft. lbs. (47 Nm).
12. Install the transmission dipstick, filler tube and bracket.
13. Install the transmission vent hose.
14. Install the oil cooler lines at the transmission.
15. Connect the T.V. cable.
16. Install the clip retaining the fuel line and speed sensor harness.
17. Install the shift control cable.
18. Tighten the torque converter-to-flywheel bolts to specifications. Install the converter cover.
19. Install the transfer case assembly. Fill the transfer case, as required.
20. Connect the negative battery cable. Lower the vehicle. Add transmission fluid, as necessary.

Shift Linkage Adjustment

1. As required, raise and support the vehicle safely.

2. Loosen the linkage rod retaining nut. Note the position of any washers, spacers and insulators. Position the transmission selector lever in the **N** position.
3. Do not use the indicator to determine the **N** position. Check the transmission shift lever bracket to determine that the transmission is in the **N** position.
4. Hold the selector rod and tighten the swivel to 17 ft. lbs. (23 Nm). Position the selector lever in the **P** detent.
5. Check the adjustment. The selector lever must go into all positions. The engine must only start in the **P** or **N** position. Align the selector indicator, as required.

TV Cable Adjustment

1. As required, remove the air cleaner assembly.
2. Depress and hold down the metal readjust tab at the engine end of the throttle valve cable.
3. Move the slider until it stops against the fitting. Release the readjustment tab.
4. Rotate the throttle lever to its full travel position. The slider must move toward the lever when the lever is rotated to its full travel position.
5. Check for proper operation. When the engine is cold, the cable may appear to be functioning properly, check the cable when the engine is hot.
6. Road test the vehicle.

AUTOMATIC TRANSAXLE

Transaxle Assembly

The automatic transaxle equipped in the Lumina APV, Silhouette and Trans Sport can only be removed as an assembly with the engine and transaxle/sub-frame.

REMOVAL AND INSTALLATION

1. Disconnect the negative battery cable.
2. Drain the cooling system. Disconnect the air flow tube from the air cleaner.
3. Disconnect the electrical connector from the ECM and push it through to the engine compartment.

Disconnect the harness from the clips on the body and lay it across the engine.

4. Disconnect the engine harness at the bulkhead connector. Disconnect the throttle and TV cables.

5. Disconnect the fuel lines. Disconnect the transaxle shift linkage.

6. Disconnect the cooler lines at the radiator. Disconnect the radiator and heater hoses.

7. Remove the air conditioning compressor from the bracket and support it aside. Remove the upper engine support strut.

8. Raise and safely support the vehicle. Remove the front wheel and tire assemblies.

9. Remove the stabilizer bar. Disconnect the tie rod ends and the lower control arm ball joints.

10. Disconnect the halfhsafts and support them aside. Disconnect the steering shaft pinch bolt.

11. Remove the starter. Remove the torque converter bolts.

12. Disconnect the exhaust pipe at the manifold. Support the engine and sub-frame with a suitable jack.

13. Remove the sub-frame bolts and lower the engine/transaxle and sub-frame from the vehicle. Remove the engine-to-transmission bolts.

To install:

14. Position the engine and transmission together in the subframe assembly and torque the engine-to-transmission bolts to 55 ft. lbs. (75 Nm). Raise the engine assembly into position and install the subframe bolts. Tighten to 38 ft. lbs. (52 Nm).

15. Connect the exhaust pipe at the rear manifold. Install the starter.

16. Connect the steering shaft and install the pinch bolt. Connect the halfshafts to the transaxle. Install the torque converter-to-flywheel bolts and torque to 35 ft. lbs. (47 Nm).

17. Connect the lower control arm ball joints to the steering knuckles.

18. Install the stabilizer bar. Install the upper engine strut.

19. Install the wheel and tire assemblies. Lower the vehicle. Install the radiator and heater hoses.

20. Install the shift linkage. Connect the fuel lines and the throttle and TV cables.

21. Connect the harness to bulkhead connector. Connect the ECM harness to the ECM.

22. Connect the air cleaner hose and the radiator upper support.

23. Fill the cooling system. Install the air conditioning compressor.

24. Connect the negative battery cable.

TV CABLE ADJUSTMENT

1. As required, remove the air cleaner assembly.

2. Depress and hold down the metal readjust tab at the engine end of the throttle valve cable.

3. Move the slider until it stops against the fitting. Release the readjustment tab.

4. Rotate the throttle lever to its full travel position. The slider must move toward the lever when the lever is rotated to its full travel position.

5. Check for proper operation. When the engine is cold, the cable may appear to be functioning properly, check the cable when the engine is hot.

6. Road test the vehicle.

NEUTRAL SAFETY SWITCH ADJUSTMENT

1. Place the transaxle in the **N** position.

2. Loosen the switch attaching screws.

3. Insert a 1/8 in. gauge pin into the service adjustment hole.

4. Rotate the switch on the shifter assembly to align the service hole in the switch with the hole in the carrier. When the gauge pin drops into the service adjustment hole, tighten the bolts and nut.

5. Remove the gauge pin and check the operation of the switch.

6. The vehicle should only start in the **N** or **P** position.

TRANSFER CASE

Transfer Case Assembly

REMOVAL AND INSTALLATION

1. Disconnect the negative battery cable. Shift the transfer case into the **4HI** position, as required.

2. Raise and support the vehicle safely. If equipped, remove the skid plate. Drain the fluid from the transfer case.

3. Matchmark the transfer case front output shaft yoke and driveshaft for reassembly. Disconnect the driveshaft from the transfer case.

4. Matchmark the rear axle yoke and the driveshaft for reassembly. Remove the driveshaft.

5. Disconnect the speedometer cable, as required. Disconnect the vacuum harness and electrical connections at the transfer case. Remove the catalytic converter hanger bolts at the converter assembly.

6. Support the transmission and transfer case assembly. Remove the transmission mount retaining bolts. Lower the transmission and transfer case assembly to gain additional clearance.

7. Remove the transfer case retaining bolts.

8. As required, remove the shift lever bracket mounting bolts from the transfer case adapter in order to remove the upper left transfer case retaining bolt.

9. Separate the transfer case from its mounting and remove it from the vehicle.

To install:

10. Install a new gasket to the transmission. Install the transfer case into the vehicle.

11. Tighten the transfer case mounting bolts to specification.

12. Install the shift lever bracket bolts. Support the transmission, raise the transmission slightly.

13. Install the converter mounting bracket. Install the rear transmission mounting bolts.

14. Connect the transfer case linkage. Connect the speedometer cable.

15. Install the driveshafts. Install the skid plate if equipped.

16. Fill the transfer case to the correct level.

17. Lower the vehicle. Connect the negative battery cable.

LINKAGE ADJUSTMENT

1. Remove the console assembly. Raise the shifter boot aside.

2. Loosen the selector lever retaining bolt. Loosen the shifter pivot bolt. Shift the transfer case into the **4HI** position.

3. Install a 5/16 in. (8mm) gauge pin through the shifter and into the bracket. Install a service bolt at the transfer case shift lever. This will lock the transfer case in the **4HI** position.

4. Tighten the selector lever retaining bolt 25–35 ft. lbs. Tighten the shifter pivot bolt 88–103 ft. lbs.

5. Remove the service bolt. Remove the gauge pin. Install the console and shifter boot.

6. Check for proper operation.

DRIVE AXLE

Halfshaft

REMOVAL AND INSTALLATION

Except Lumina APV, Silhouette and Trans Sport

1. Raise and support the vehicle safely. Remove the tire and wheel assemblies.
2. Remove the cotter pin, retainer, nut and washer. Remove the brake line support bracket.
3. Matchmark the axle tube-to-flange location. Loosen but do not remove the axle tube-to-flange bolts. Using the proper tools, remove the tie rods at the steering knuckles.
4. Remove the lower shock absorber retaining bolts and move the shock absorbers aside.
5. Separate the upper ball joint from the steering knuckle. Suspend the steering knuckle from the frame using a piece of wire.
6. As required, remove the skid plate. Remove the halfshaft to axle tube bolts.
7. Move the inner part of the halfshaft forward. Support it away from the frame. Using a suitable tool, remove the shaft from the hub and bearing assembly.
8. Remove the halfshaft from the vehicle.

To install:

9. Install the halfshaft into position. Push it into the hub.
10. Install the shaft retaining nut and washer. Tighten the halfshaft retaining nut to 160–200 ft. lbs. (220–270 Nm).
11. Install the cotter pin in the nut. Install the halfshaft-to-axle tube bolts and tighten to 60 ft. lbs. (80 Nm).
12. Install the lower shock absorber bolts. Connect the tie rods to the steering knuckle.
13. Install the skid plate, if equipped.
14. Install the wheel and tire assemblies. Lower the vehicle

Lumina APV, Silhouette and Trans Sport

1. Raise and safely support the vehicle.
2. Remove the tire and wheel assemblies.
3. Remove the halfshaft retaining nut and washer.

4. Remove the brake caliper from the rotor and support it aside.
5. Remove the brake rotor from the hub.
6. Disconnect the stabilizer shaft from the control arm and disconnect the ball joint from the steering knuckle.
7. Using a suitable axle seal protector to guard against possible boot damage. Remove the halfshaft from the transaxle using a suitable tool.
8. Remove the halfshaft from the hub and bearing assembly.

To install:

9. Install the halfshaft into the hub and bearing assembly.
10. Connect the lower ball joint to the steering knuckle.
11. Connect the stabilizer shaft to the control arm. Install the rotor.
12. Install the brake caliper. Install a new halfshaft nut and tighten the nut to 185 ft. lbs. (250 Nm).
13. Seat the halfshaft into the transaxle, by pushing it in firmly. Check that the shaft is seated by pulling on it.
14. Install the wheel and tire assemblies.
15. Lower the vehicle.

CV-Boot and Joint

REMOVAL AND INSTALLATION

Outer

1. Raise and safely support the vehicle.
2. Remove the tire and wheel assemblies.
3. Remove the halfshaft from the vehicle and place the assembly in a vice with soft jaws.
4. Cut the seal clamps and remove the boot. Slide the boot down away from the joint.
5. Clean the grease around the joint. Spread the snapring and remove the joint.
6. Discard the boot, if damaged.
7. Using a brass drift punch and hammer, tap on the joint cage until it moves enough to remove a ball, then remove the other balls in the same fashion.
8. Remove the cage and inner race by turning the assembly straight up 90 degrees to the center line of the outer housing. Maneuver the cage and inner race out of the outer race housing assembly. Then remove the inner race from the cage assembly.

To install:

9. Thoroughly clean all part in solvent. Apply grease to the ball grooves of the inner and outer race.

10. Install the inner race into the cage.
11. Install the cage assembly into the housing.
12. Tilt the cage up and install the ball into the assembly. Pack the assembly with grease.
13. Install the small boot clamp onto the axle shaft.
14. Install the axle boot onto the shaft and secure the small boot clamp using a suitable tool.
15. Install the joint assembly onto the axle shaft, the snapring will snap into place automatically. Pack the boot and outer joint with a sufficient quantity of grease.
16. Position the boot over the joint housing. Secure with the large clamp, using a suitable tool.
17. Reinstall the axle shaft assembly into the vehicle.
18. Install the tire and wheel assembly.
19. Lower the vehicle.

Inner

1. Raise and safely support the vehicle.
2. Remove the tire and wheel assemblies.
3. Remove the halfshaft from the vehicle and place the assembly in a vice with soft jaws.
4. Cut the seal clamps and remove the boot. Slide the joint housing off of the joint and shaft assembly.
5. Clean the grease around the joint. Spread the snapring and remove the joint.
6. Discard the boot, if damaged.

To install:

7. Thoroughly clean all part in solvent.
8. Install the small boot clamp onto the axle shaft.
9. Install the axle boot onto the shaft and secure the small boot clamp using a suitable tool.
10. Install the joint assembly with the snapring counterbore facing (out) the axle shaft housing, the snapring will snap into place automatically.
11. Install the spacer ring onto the axle shaft, ensuring the ring is fully seated in the groove.
12. Position the axle housing onto the joint and shaft assembly. Pack the boot and joint with a sufficient quantity of grease.
13. Position the boot over the joint housing. Secure with the large clamp, using a suitable tool.
14. Reinstall the axle shaft assembly into the vehicle.
15. Install the tire and wheel assembly.
16. Lower the vehicle.

Driveshaft and U-Joints

REMOVAL AND INSTALLATION

Front 1 Piece

EXCEPT ASTRO AND SAFARI

1. Raise and support the vehicle safely.
2. If equipped, remove the skid plate.
3. Matchmark the rear yoke of the driveshaft to the transfer case. Matchmark the front of the driveshaft to the differential housing.
4. Remove the bolts and retainers from the rear of the driveshaft. Remove the driveshaft from the vehicle.
5. Installation is the reverse of the removal procedure. Tighten the bolts to 15 ft. lbs. (20 Nm).

ASTRO AND SAFARI

1. Raise and support the vehicle safely.
2. Matchmark the rear of the driveshaft to the transfer case. Matchmark the front of the driveshaft to the differential housing.
3. Remove the bolts from the rear flange of the driveshaft. Remove the bolts from the front flange of the driveshaft. Lower the driveshaft and remove it from the vehicle.
4. Installation is the reverse of the removal procedure. Tighten the front flange bolts to 53 ft. lbs. (72 Nm) and the rear flange bolts to 92 ft. lbs. (125 Nm).

Rear 1 Piece

1. Raise and support the vehicle safely.
2. If equipped, remove the skid plate.
3. Matchmark the front yoke of the driveshaft to the transmission. Matchmark the rear of the driveshaft to the differential housing.
4. Remove the bolts and retainers from the rear of the driveshaft. Remove the driveshaft from the vehicle.
5. Installation is the reverse of the removal procedure. Tighten the bolts to 15 ft. lbs. (20 Nm).

Rear 2 Piece

1. Raise and support the vehicle safely.
2. As required, remove the skid plate.
3. Matchmark the front yoke of the driveshaft to the transmission. Matchmark the rear of the driveshaft to the differential housing.
4. Remove the center bearing bolts and washers. Remove the center bearing.
5. Remove the front driveshaft. Remove the rear driveshaft bolts and retainers. Remove the rear driveshaft.
6. Installation is the reverse of the removal procedure. When installing the center bearing, be sure to align it 90 degrees to the driveshaft center lines. Tighten the center bearing bolts to 25 ft. lbs. (34 Nm), and the driveshaft bolts to 15 ft. lbs. (20 Nm).

Rear Axle Shaft, Bearing and Seal

REMOVAL AND INSTALLATION

1. Raise and support the vehicle safely. Remove the tire, wheel and drum assemblies. Drain the lubricant.
2. Remove the carrier cover retaining bolts. Remove the carrier cover.
3. Remove the rear axle pinion shaft lock screw and the rear axle pinion shaft. Discard the lock screw.
4. Push the flanged end of the axle shaft toward the center of the vehicle. Remove the C-lock clip from the button end of the shaft.
5. Mark the axles left or right and remove the axle shaft from the housing. Be careful not to damage the oil seal.
6. Remove the rear wheel speed sensor, if equipped.
7. Remove the rear axle seal, using a suitable tool.
8. Remove the axle bearing from the housing using a suitable puller and slide hammer.
   To install:
9. Install the axle bearing into the housing until the bearing bottoms against the housing shoulder, using a suitable tool.
10. Using a suitable tool, install the seal until flush with the axle housing.
11. Lubricate the seal with gear oil. Install the rear wheel speed sensor, if equipped.
12. Slide the axle into position and install the C-lock clip, use care not to damage the oil seal. Pull the axle out to secure the C-lock in place.
13. Install the pinion shaft through the case and pinion.
14. Install a new pinion shaft lock bolt, tighten it to 25 ft. lbs. (34 Nm).
15. Install the carrier cover, using a new gasket. Tighten the cover bolts to 20 ft. lbs. (27 Nm).

16. Fill the differential to within $\frac{3}{8}$ in. of the filler opening.
17. Install the brake drum. Install the wheel and tire assembly.

Front Wheel Hub, Knuckle and Bearing

REMOVAL AND INSTALLATION

Except Lumina APV, Silhouette and Trans Sport

2WD VEHICLE

1. Raise and safely support the vehicle.
2. Remove the wheel and tire assemblies.
3. Remove the brake caliper and remove the brake rotor. Use care not to drop the bearings from the rotor.
4. Remove the splash shield attaching bolts and remove the splash shield from the knuckle.
5. Remove the tie rod end from the knuckle. Remove the ball studs from the knuckle.
6. Remove the steering knuckle from the lower ball stud and the knuckle from the vehicle.
   To install:
7. Install the upper and lower ball joint to the knuckle.
8. Attach the splash shield to the knuckle and tighten the bolts to 10 ft. lbs. (14 Nm).
9. Connect the tie rod end to the knuckle.
10. Install the rotor and bearings onto the spindle. Adjust the bearings.
11. Install the caliper. Install the wheel and tire assembly.
12. Lower the vehicle. Check the front end alignment.

4WD VEHICLE

1. Raise and safely support the vehicle.
2. Remove the wheel and tire assembly.
3. Remove the brake caliper and support it aside.
4. Remove the brake rotor. Remove the halfshaft retaining nut.
5. Disconnect the tie rod end from the knuckle.
6. Remove the hub and bearing assembly mounting bolts and remove the assembly from the steering knuckle.
7. Remove the splash shield from the knuckle. Remove the ball joints from the knuckle.
8. Remove the knuckle from the ball joints and remove the spacer from the knuckle.

To install:

9. Install the spacer on the knuckle. Install the knuckle in position on the ball joints.

10. Install the ball joint nuts and cotter pins. Install the splash shield to the knuckle.

11. Install the hub and bearing assembly to the knuckle. Tighten the hub mounting bolts to 86 ft. lbs. (116 Nm). Install the halfshaft retaining nut.

12. Connect the tie rod to the knuckle. Install the brake rotor and the brake caliper.

13. Install the wheel and tire assembly. Lower the vehicle.

14. Check the front end alignment.

Lumina APV, Silhouette and Trans Sport

1. Raise and safely support the vehicle.

2. Remove the wheel and tire assembly.

3. Remove the brake caliper and support it aside.

4. Remove the brake rotor. Remove the halfshaft retaining nut.

5. Disconnect the tie rod end from the knuckle.

6. Remove the hub and bearing assembly mounting bolts and remove the assembly from the steering knuckle.

7. Matchmark the strut-to-knuckle positioning. Remove the strut mounting bolts from the knuckle. Remove the splash shield from the knuckle. Remove the ball joint from the knuckle.

8. Remove the knuckle from the ball joint and remove the knuckle from the vehicle.

To install:

9. Install the knuckle in position on the ball joint. Slide the knuckle onto the halfshaft.

10. Install the ball joint nut and cotter pin. Install the splash shield to the knuckle.

11. Install the hub and bearing assembly to the knuckle. Tighten the hub mounting bolts to 86 ft. lbs. (116 Nm). Install the halfshaft retaining nut.

12. Connect the tie rod to the knuckle. Install the brake rotor and the brake caliper.

13. Install the wheel and tire assembly. Lower the vehicle.

14. Check the front end alignment.

Pinion Seal

REMOVAL AND INSTALLATION

1. Raise and support the vehicle safely. Remove the tire, wheel and drum assembly. Matchmark the driveshaft and the pinion flange.

2. Disconnect the driveshaft from the rear differential. Support the driveshaft aside.

3. Mark the position of the pinion flange, pinion shaft and nut. Using an inch lb. torque wrench, turn the pinion flange nut and record the amount of torque required to turn the pinion flange. Remove the pinion flange nut and washer.

4. Remove the pinion flange. Position a drain pan under the assembly to catch any excess lubricant.

5. Using the proper tool, remove the seal from its mounting.

6. Installation is the reverse of the removal procedure. Refill the differential to the correct level. Tighten the pinion flange nut until the torque required to turn the flange is 3–5 inch lbs. higher than the pre-disassembly reading.

Front Axle Housing

REMOVAL AND INSTALLATION

1. Unlock the steering column so the linkage is free to move. Disconnect the negative battery cable.

2. Raise and support the vehicle safely. Remove both front wheel and tire assemblies.

3. Insert a drift through the opening in the top of the brake caliper into the vanes of the brake rotor to keep the axle from turning. Remove the front driveshaft.

4. Remove the axle vent hose from the carrier fitting.

5. Disconnect the right and left halfshafts from the carrier by removing the retaining bolts.

6. Properly support the carrier assembly. Remove the carrier and axle tube to frame retaining bolts.

7. Remove the differential carrier assembly from the vehicle.

8. Installation is the reverse of the removal procedure.

Rear Axle Housing

REMOVAL AND INSTALLATION

Except Astro and Safari

1. Raise and support the vehicle safely. Drain the axle assembly.

2. Remove the tire, wheel and drum assemblies. Properly support the rear axle assembly.

3. Disconnect the shock absorbers. Matchmark the driveshaft and the pinion flange. Remove the driveshaft and position aside.

4. Remove the brake line at the axle housing and backing plate. Disconnect the rear wheel antilock speed sensor connector at the junction block, if equipped.

5. Remove the U bolts and anchor plates.

6. Remove the vent hose from the axle housing.

7. Move the axle housing from above the spring.

8. Installation is the reverse of the removal procedure. Fill and bleed the brake system.

Astro and Safari

1. Raise and support the vehicle safely. Remove the tire, wheel and drums assemblies. Properly support the rear axle assembly.

2. Disconnect the shock absorbers from the anchor plate. Matchmark the driveshaft and the pinion flange. Remove the driveshaft and position aside.

3. Disconnect the brake lines from the axle housing and backing plates. Disconnect the rear wheel anti-lock speed sensor connectors, if equipped.

4. Remove the stabilizer bar, if equipped. Remove the U bolts and anchor plates.

5. Lower the axle assembly and remove the lower spring shackle bolts.

6. Disconnect the vent hose from the axle housing. Remove the axle housing from the vehicle.

7. Installation is the reverse of the removal procedure. Fill and bleed the brake system.

STEERING

Steering Wheel

REMOVAL AND INSTALLATION

1. Disconnect the negative battery cable.
2. Remove the horn pad retaining screws. Remove the horn pad and disconnect the electrical lead.
3. Matchmark the steering wheel and the shaft.
4. Remove the steering wheel retaining clip and nut.
5. Remove the steering wheel, using a suitable puller.
6. Installation is the reverse of removal. Align the matchmarks made during removal and tighten the steering wheel retaining nut to 30 ft. lbs. (40 Nm).

Steering Column

REMOVAL AND INSTALLATION

EXCEPT Lumina APV, Silhouette and Trans Sport

1. Disconnect the negative battery cable.
2. Remove the transmission shift linkage from the column.
3. Remove the column upper clamp bolt; mark the relationship of the joint to the steering shaft.
4. Remove the steering column support bracket under the dash.
5. Remove the column to floor seal.
6. Remove the steering wheel. Disconnect the wire harness connector under the dash.
7. Rotate the column so the shift levers clear the floor opening and remove the assembly.
8. Installation is the reverse of the removal procedure.

Lumina APV, Silhouette and Trans Sport

1. Disconnect the negative battery cable.
2. Remove the left instrument panel sound insulator and trim pad. Remove the steering column trim collar.
3. Remove the steering wheel, if column is to be disassembled.

4. Remove the column upper clamp bolt; mark the relationship of the joint to the steering shaft.
5. Remove the steering column support bracket under the dash. Remove the shift indicator cable.
6. Disconnect the wire harness connector under the dash.
7. Remove the shift cable at the actuator and housing holder.
8. Remove the column assembly.
9. Installation is the reverse of the removal procedure.

Manual Steering Gear

REMOVAL AND INSTALLATION

1. Disconnect the negative battery cable. Raise and support the vehicle safely. Position the wheels in the straight ahead position.
2. If equipped, remove the steering gear coupling shield. Remove the steering gear lower coupling bolt. On some vehicles it may be necessary to separate the coupling at the upper and lower flange bolts.
3. Matchmark the pitman arm to the steering gear. Remove the pitman arm retaining nut and washer. Using the proper tool, separate the pitman arm from the steering gear assembly.
4. Remove the steering gear retaining bolts. Separate the steering gear from the intermediate shaft and remove the assembly from the vehicle.

To install:

5. Position the steering gear into in the vehicle.
6. Install the mounting bolts. Tighten the mounting bolts to 55 ft. lbs. (75 Nm).
7. Connect the steering gear and intermediate shaft, install the coupling flange bolts.
8. Connect the pitman arm to the steering gear in the previously marked position. Tighten the pitman arm retaining bolt to 185 ft. lbs. (250 Nm).
9. Install the coupling shield. Lower the vehicle.
10. Connect the negative battery cable.

ADJUSTMENT

1. Raise and safely support the vehicle.
2. Remove the coupling shield. Remove the pitman arm nut and washer.

3. Matchmark the pitman arm to the pitman shaft.
4. Remove the pitman arm, using a suitable puller.
5. Loosen the adjuster nut on the steering gear, then back out the adjuster a ¼ turn.
6. Lower the vehicle, keeping it just above the ground.
7. Remove the horn pad. Center the steering wheel by turning the wheel all the way in one direction until stopped by the gear, then turn the wheel back 1½ turns to the center position.
8. Install an inch lb. torque wrench on the steering wheel nut. Use a torque wrench with no more than a 50 inch lbs. reading capability. Check the thrust bearing preload as follows:
   a. Turn the torque wrench and steering wheel through a 90 degree arc.
   b. Tighten the adjuster plug until the proper preload is achieved, 5–8 inch lbs. (0.6–1.0 Nm).
   c. Tighten the adjuster nut to 85 ft. lbs.
   d. Turn the steering wheel to check the adjustment. The gear should turn smooth and not lumpy, from lock to lock.
9. To check the overcenter preload:
   a. Turn the steering wheel from lock to lock counting the total number of turns.
   b. Turn the wheel back, to exactly the ½ way point.
   c. Turn the over center adjuster screw clockwise, to take out all of the lash between the ball nut and the pitman shaft sector teeth.
   d. Tighten the jam nut to 22 ft. lbs. (30 Nm).
   e. Check the torque at the steering wheel, taking the highest reading as the wheel is turned.
   f. If necessary, loosen the adjuster nut and tighten the adjuster plug to obtain 4–10 inch lbs. (0.5–1.2 Nm).

To install:

10. Install the pitman arm onto the pitman shaft, aligning the matchmarks made during disassembly.
11. Install the pitman arm washer and nut. Install the coupling shield.
12. Lower the vehicle completely and install the horn pad.
13. Connect the negative battery cable.

Power Steering Gear

REMOVAL AND INSTALLATION

1. Disconnect the negative battery cable. Position the wheels in the straight ahead position.
2. Disconnect and cap the fluid lines.
3. If equipped, remove the steering gear coupling shield. Matchmark and remove the steering gear to lower coupling and bolt. On some vehicles, it may be necessary to separate the coupling at the upper and lower flange bolts.
4. Raise and support the vehicle safely. Matchmark the pitman arm to the steering gear. Remove the pitman arm retaining nut and washer. Using the proper tool, separate the pitman arm from the steering gear assembly.
5. Remove the steering gear retaining bolts. Remove the steering gear from the vehicle.

To install:

6. Install the steering gear in position in the vehicle.
7. Install the mounting bolts. Tighten the mounting bolts to 55 ft. lbs. (75 Nm).
8. Connect the steering gear coupling at the flange bolts.
9. Connect the pitman arm to the steering gear. Tighten the pitman arm retaining bolt to 185 ft. lbs. (250 Nm).
10. Connect the fluid lines. Install the coupling shield. Lower the vehicle.
11. Connect the negative battery cable. Fill and bleed the power steering system.

ADJUSTMENTS

For proper adjustment, remove the gear from the vehicle and drain all the fluid from the gear and place the gear in a holding fixture. It is important that the adjustments be made in the order given.

Worm Bearing Preload

1. Loosen the adjuster plug locknut.
2. Turn the adjuster plug in clockwise until firmly bottomed. Then, tighten it to 20 ft. lbs. (27 Nm).
3. Place an index mark on the gear housing in line with one of the holes in the adjuster plug.
4. Measure counterclockwise from the mark about ½ inch and place another mark on the housing.

5. Rotate the adjuster plug counterclockwise until the hole is aligned with the second mark.
6. Install the locknut. Hold the plug and tighten the locknut to 81 ft. lbs. (110 Nm).
7. Place an inch lb. torque wrench and 12-point deep socket on the steering gear stub shaft and measure the stub shaft rotating torque, starting with the torque wrench handle in a vertical position to a point ¼ turn to either side. Note your reading. The proper torque should be 4–10 inch lbs. (0.45–1.13 Nm). If the reading is incorrect, either the adjustment was done incorrectly or there is gear damage.

Overcenter Preload

1. Loosen the locknut and turn the pitman shaft adjuster screw counterclockwise until it is all the way out. Then, turn it in 1 turn.
2. Rotate the stub shaft from stop to stop, counting the total number of turns, then turn the shaft back ⅔ that number to center the gear. The stub shaft flat spot should face up.
3. Place the torque wrench in a vertical position on the stub shaft and measure the torque necessary to rotate the shaft to a point 45 degrees to either side of center. Record the highest reading.
4. Turn the adjuster in until the reading is 6–10 inch lbs. higher then the previous reading.
5. If necessary, adjust the torque reading by turning the adjuster screw.
6. When the adjustment is made, hold the screw and tighten the locknut to 20 ft. lbs. (27 Nm).
7. Install the gear and bleed the system.

Power Steering Rack and Pinion

ADJUSTMENT

Rack Bearing Preload

1. Raise and support the vehicle safely. Ensure the steering wheel is centered.
2. Loosen the adjuster plug locknut and turn the adjuster plug clockwise until it bottoms in the housing; then back off 50–70 degrees.
3. Check the returnability of the steering wheel after the adjustment.
4. Tighten the locknut to 50 ft. lbs. (70 Nm), while holding the adjuster plug stationary.
5. Lower the vehicle.

REMOVAL AND INSTALLATION

1. Disconnect the negative battery cable.
2. Remove the air cleaner assembly.
3. Remove the dust boot from the steering gear.
4. Remove the intermediate shaft lower pinch bolt and disconnect the intermediate shaft from the lower stub shaft.
5. Remove the fluid line retaining clips at the pump and disconnect the lines.
6. Raise and safely support the vehicle.
7. Remove the wheel and tire assemblies. Disconnect the tie rod ends at the steering knuckle.
8. Remove the remaining brackets and clips at the crossmember. Support the body safely with the appropriate equipment, to allow lowering of the subframe.
9. Remove the rear subframe mounting bolts and carefully lower the rear of the subframe approximately 5 in. (128mm).
10. Remove the rack and pinion mounting bolts and remove the rack through the left wheel opening.

To install:

11. Install the rack and pinion through the left wheel opening.
12. Install the rack and pinion mounting nuts, tighten to 70 ft. lbs. (95 Nm).
13. Raise the subframe assembly and install the rear mounting bolts.
14. Remove any supports and install the brackets and clips to the crossmember.
15. Install the wheel and tire assemblies. Lower the vehicle.
16. Connect the fluid lines at the pump and tighten to 18 ft. lbs. (25 Nm).
17. Install the line retaining clips. Connect the intermediate shaft to the stub shaft.
18. Install the dust boot over the steering gear.
19. Install the air cleaner assembly and connect the negative battery cable.
20. Fill and bleed the steering system.

Power Steering Pump

REMOVAL AND INSTALLATION

Except Turbocharged Engine

1. Disconnect the negative battery cable. Disconnect and cap the power

steering pump hoses. Remove the accessory drive belt.

2. As required, remove the power steering pump pulley using a suitable puller tool or equivalent.

3. Remove the pump mounting bolts. Remove the pump from the vehicle.

4. Installation is the reverse of the removal procedure. Bleed the power steering system.

Turbocharged Engine

1. Disconnect the negative battery cable.

2. Remove the air cleaner and duct assembly.

3. Remove the upper fan shroud.

4. Loosen the fan nuts and remove the serpentine belt.

5. Remove the fan and pulley assembly.

6. Remove the power steering pump pulley as follows:

   a. Install tool J–25034–B, ensure the pilot bolt bottoms in the pump shaft.

   b. Hold the pilot bolt with a suitable wrench.

   c. Turn the shaft locknut counterclockwise and remove the pulley.

7. Raise and safely support the vehicle.

8. Remove the left tire and wheel assembly.

9. Remove the left wheelhouse panel.

10. Remove the power steering hose bracket.

11. Place a drain pan below the pump and remove the power steering pressure and return lines, capping the lines to prevent dirt from entering.

12. Remove the bolts from the rear bracket at the alternator.

13. Lower the vehicle and remove the assembly.

14. Remove the bracket from the pump as necessary.

To install:

15. Install the bracket to the pump, if removed.

16. Install the pump assembly and torque the bolts to 37 ft. lbs. (50 Nm).

17. Install the bolts to the rear bracket at the alternator.

18. Install the power steering pressure and return lines.

19. Install the power steering hose bracket.

20. Raise and safely support the vehicle.

21. Install the left wheelhouse panel.

22. Install the left tire and wheel assembly. Lower the vehicle.

23. Install the power steering pump pulley as follows:

   a. Place the pulley on the shaft.

   b. Install tool J–25033–B, ensure the pilot bolt bottoms in the pump shaft.

   c. Hold the pilot bolt with a suitable wrench.

   d. Turn the shaft locknut clockwise and install the pulley.

24. Install the fan and pulley assembly.

25. Tighten the fan pulley nuts and install the serpentine belt.

26. Install the upper fan shroud.

27. Install the air cleaner and duct assembly.

28. Connect the negative battery cable. Fill and bleed the power steering system. Check system for leaks.

SYSTEM BLEEDING

1. Fill the fluid reservoir to the proper level.

2. Start the engine and let it run for at least 2 minutes. Turn the engine off.

3. Add fluid if necessary, then run the engine again. Repeat this until the fluid level remains constant.

4. Raise the front of the vehicle slightly, until the front wheels are just off the ground.

5. Start the engine and slowly turn the steering wheel from lock to lock, until the wheel contacts the stop.

6. Add fluid if necessary. Lower the vehicle to the ground.

7. Start the engine and again move the wheels side to side. Check the fluid level.

Tie Rod Ends

REMOVAL AND INSTALLATION

Except Lumina APV, Silhouette and Trans Sport

1. Raise and support the vehicle safely. Remove the tire and wheel assemblies.

2. Remove the cotter pins and nuts. Using the proper removal tool, separate the outer tie rod from the steering knuckle.

3. Disconnect the inner tie rod from the relay rod using the proper tool. Remove the tie rod ends from the adjuster tubes.

4. Installation is the reverse of the removal procedure. Tighten the inner tie rod ball stud nut to 35 ft. lbs. (47 Nm). Tighten the outer tie rod ball stud to the steering knuckle to 35 ft. lbs. (47 Nm). The number of threads on both the inner and outer tie rod ends must be equal within 3 threads.

5. Adjust the front end alignment, as required.

Lumina APV, Silhouette and Trans Sport

OUTER

1. Raise and support the vehicle safely. Remove the tire and wheel assemblies.

2. Remove the cotter pins and nuts. Using the proper removal tool, separate the outer tie rod from the steering knuckle.

3. Disconnect the inner tie rod from the relay rod using the proper tool. Remove the tie rod ends from the adjuster tubes.

4. Installation is the reverse of the removal procedure. Tighten the inner tie rod ball stud nut to 35 ft. lbs. (47 Nm). Tighten the outer tie rod ball stud to the steering knuckle to 35 ft. lbs. (47 Nm). The number of threads on both the inner and outer tie rod ends must be equal within 3 threads.

5. Adjust the front end alignment, as required.

INNER

1. Disconnect the negative battery cable.

2. Raise and support the vehicle safely.

3. Remove the rack and pinion assembly.

4. Place the assembly in a holding fixture.

5. Remove the outer tie rod assembly. Remove the inner tie rod jam nut.

6. Remove the tie rod end boot clamps. Remove the boot.

7. Remove the shock dampner from the inner tie rod assembly.

8. Remove the tie rod from the rack. Place a wrench on the flat of the rack assembly and another wrench on the flats of the inner tie rod housing.

9. Installation is the reverse of the removal procedure.

BRAKES

For all brake system repair and service procedures not detailed below, please refer to "Brakes" in the Unit Repair section.

Master Cylinder

REMOVAL AND INSTALLATION

1. Disconnect the negative battery cable.
2. Disconnect the electrical connections from the master cylinder, as required. Disconnect and plug the fluid lines.

NOTE: On Lumina APV, Silhouette and Trans Sport, the master cylinder reservoir can be removed to ease master cylinder removal.

3. Remove the master cylinder to power booster retaining bolts. Remove the RWAL control module assembly, if equipped with anti-lock brakes.
4. Remove the master cylinder from the vehicle. Remove the vacuum booster pushrod.
To install:
5. Install the master cylinder in position on the booster. Connect the booster pushrod.
6. Install the master cylinder retaining bolts and tighten to 20 ft. lbs. (27 Nm).
7. Connect the fluid lines to the master cylinder. Connect the RWAL control unit, if equipped with anti-lock brakes, to the bracket.
8. Connect the negative battery cable. Refill the master cylinder and bleed the brake system.

Combination Valve

REMOVAL AND INSTALLATION

Astro, Blazer, Jimmy, Pickup, Safari, Sonoma, Cyclone and Typhoon

The combination valve is mounted on the master cylinder bracket. On vehicles with anti-lock brakes, the combination valve is replaced with a dump/isolation valve. It is removed in the same manner.
1. Disconnect the negative battery cable.
2. Disconnect the brake lines from the combination valve.
3. Remove the mounting bolts and remove the valve from the bracket.
4. Installation is the reverse of the removal procedure. Bleed the brake system.

Proportioning Valve

REMOVAL AND INSTALLATION

Lumina APV, Silhouette and Trans Sport

1. Disconnect the negative battery cable.
2. Remove the electrical connector from the master cylinder.
3. Drain and remove the master cylinder reservoir.
4. Remove the proportioning valve caps from the master cylinder.
5. Remove the O-rings, springs and the valve pistons. Use care not to scratch the valves in any way.
6. Remove the valve seals from the valve pistons.
To install:
7. Install new seals on the valve pistons. Lubricate the seals and the pistons with silicon grease.
8. Install the valve pistons and O-rings into the master cylinder.
9. Install the valve cap assemblies and tighten to 20 ft. lbs. (27 Nm).
10. Install the reservoir assembly. Connect the electrical leads.
11. Connect the negative battery cable. Bleed the brake system.

Power Brake Booster

REMOVAL AND INSTALLATION

1. Disconnect the negative battery cable. Do not disconnect the master cylinder fluid lines, unless there is a clearance problem. Remove the master cylinder and position aside.
2. Remove the vacuum booster pushrod. Disconnect the vacuum hose from the booster assembly.
3. From inside the vehicle, remove the mounting studs which secure the vacuum booster to the fire wall.
4. Pull the booster away from the cowl and remove it from the vehicle.
5. Installation is the reverse of the removal procedure. Be sure to properly install the vacuum booster pushrod. Bleed the system.

Brake Caliper

REMOVAL AND INSTALLATION

1. Remove ⅔ of the brake fluid from the master cylinder reservoir.
2. Raise and support the vehicle safely. Remove the tire and wheel assembly.
3. Disconnect and plug the caliper fluid line. Remove the bolts retaining the caliper to the rotor. Remove the caliper from the rotor.

NOTE: If the caliper is being removed for brake pad replacement, the fluid line do not need to be disconnected.

4. Remove the disc brake pads from the caliper. Remove the disc brake pad retaining clips from inside the caliper.
To install:
5. Clean and lubricate the sleeves and bushings with silicon grease. Install the pads in the caliper.
6. Install the caliper in position over the rotor and install the mounting bolts. Tighten the mounting bolts to 38 ft. lbs. (51 Nm).
7. Connect the fluid lines to the caliper, if disconnected, and tighten to 33 ft. lbs. (45 Nm).
8. Install the wheel and tire assembly.
9. Lower the vehicle and refill the master cylinder to the correct level. Bleed the brake system if the fluid lines were disconnected from the caliper.

Disc Brake Pads

REMOVAL AND INSTALLATION

1. Remove ⅔ of the brake fluid from the master cylinder.
2. Raise and safely support the vehicle.
3. Place a C-clamp around the outer pad and caliper; tighten the C-clamp until the piston is fully compressed in the caliper. Remove the brake caliper.
4. Remove the inboard pad and retaining spring from the caliper.
5. Remove the outboard pad from the caliper.
6. Remove the sleeves and bushings.
To install:
7. Clean and lubricate the sleeves and bushing with silicon lubricant and install them in the caliper.
8. Clip the retaining spring onto the inboard pad and install the pad in the caliper.
9. Install the outboard pad into the caliper.
10. Install the caliper in position over the rotor and install the mounting bolts. Bend the tabs, on the outboard brake pad, over the caliper.
11. Install the wheel and tire assemblies.
12. Lower the vehicle, refill the master cylinder and pump pedal to

attain full brake pedal before road testing the vehicle.

Brake Rotor

REMOVAL AND INSTALLATION

Except Lumina APV, Silhouette and Trans Sport

2WD VEHICLE

1. Remove ⅔ of the brake fluid from the master cylinder.
2. Raise and support the vehicle safely. Remove the tire and wheel assembly.
3. Remove the bolts retaining the caliper to the rotor. Remove the caliper from the rotor and position it aside. Do not allow the caliper to hang unsupported.
4. Remove the dust cap, cotter pin, spindle nut, washer and outer wheel bearings.
5. Remove the rotor from the vehicle. On Astro and Safari with 4 wheel anti-lock brakes, remove the wheel speed sensor from behind the rotor.
6. Installation is the reverse of the removal procedure. Adjust the wheel bearings, as required.

4WD VEHICLE

1. Remove ⅔ of the brake fluid from the master cylinder.
2. Raise and safely support the vehicle.
3. Remove the wheel and tire assemblies.
4. Remove the brake caliper from the rotor and support it aside. Do not disconnect the brake lines.
5. Remove the rotor from the hub.
To install:
6. Install the rotor on the hub. Install the brake caliper.
7. Install the wheel and tire assemblies.
8. Lower the vehicle and refill the master cylinder.

Lumina APV, Silhouette and Trans Sport

1. Remove ⅔ of the brake fluid from the master cylinder. Raise and safely support the vehicle.
2. Remove the wheel and tire assemblies.
3. Remove the brake caliper from the rotor and support it aside. Do not disconnect the brake lines.
4. Remove the rotor from the hub.
To install:
5. Install the rotor on the hub. Install the brake caliper.
6. Install the wheel and tire assemblies.

7. Lower the vehicle and refill the master cylinder.

Brake Drums

REMOVAL AND INSTALLATION

1. Raise and safely support the vehicle.
2. Remove the wheel and tire assembly.
3. Remove the brake drum from the vehicle. If the drum will not pull of the axle, use a rubber mallet and tap it around the edge.
4. Install the drum on the axle and install the wheel and tire assembly.
5. Lower the vehicle.

Brake Shoes

REMOVAL AND INSTALLATION

Except Lumina APV, Silhouette and Trans Sport

1. Raise and safely support the vehicle.
2. Remove the wheel and tire assembly.
3. Remove the brake drum.
4. Remove the return springs from the brake shoes. Remove the shoe guide.
5. Remove the hold-down springs and pins. Remove the actuator lever and pivot.
6. Remove the lever return spring. Remove the actuator link.
7. Remove the parking brake strut and spring. Remove the parking brake lever.
8. Remove the brake shoes and the adjuster assembly.
To install:
9. Lubricate the contact points on the backing plate and the adjuster with lithium grease.
10. Install the parking brake lever, adjusting screw and spring assembly.
11. Install the shoe assembly onto the backing plate.
12. Install the parking brake lever, strut and strut spring.
13. Install the actuator lever and lever pivot. Install the actuator link.
14. Install the lever spring, the hold-down pins and springs.
15. Install the shoe guide. Install the return springs and install the brake drum in position.
16. Adjust the brakes as follows:
    a. Remove the knockout area in the backing plate, behind the adjuster assembly.

b. Ensure the parking brake system is adjusted properly with no tension on the cables or parking brake lever. The tops of the shoes should be firmly seated against the upper spring retaining anchor, if not as specified, loosen the parking brake cables.
    c. Install the drum and turn the brake adjuster until the wheels can just be turned by hand.
    d. Then, back the adjuster off 24 notches. No brake drag should be felt after 12 notches.
    e. Install an adjusting hole plug in the backing plate to prevent dirt and moisture from entering.
    f. Readjust the parking brake cable as necessary.
17. Install the wheel and tire assemblies.

Lumina APV, Silhouette and Trans Sport

1. Raise and safely support the vehicle.
2. Remove the wheel and tire assembly.
3. Remove the brake drum.
4. Remove the actuator spring from the brake shoes. Remove the retractor spring from the shoe web, being careful not to over stretch the spring.
5. Remove the adjuster shoe, adjuster actuator and adjusting screw assembly.
6. Do not remove the parking brake cable from the parking brake lever, unless the lever is being replaced. Remove the parking brake shoe.
7. Remove the retractor spring, as required.
To install:
8. Lubricate the contact points on the backing plate with lithium grease. Clean and lubricate the adjuster with lithium grease.
9. Install the retractor spring, if removed.
10. Install the parking brake shoe against the backing plate and snap the retractor spring into the slot on the brake shoe. Install the parking brake lever onto the parking brake shoe.
11. Install the adjuster shoe and adjusting screw assembly. Install the retractor spring into the slot on the adjuster shoe web.
12. Lubricate and install the adjuster actuator onto the adjuster shoe. Install the actuator spring.
13. Ensure the parking brake system is adjusted properly with no tension on the cables or parking brake lever. The tops of the shoes should be

firmly seated against the upper spring retaining anchor, if not as specified, loosen the parking brake cables.

14. Adjust the brakes using J–21177–A or equivalent. Turn the adjuster screw until the brake lining diameter is 0.050 in. (1.27mm) less than the inside diameter of the brake drum. Install the brake drum.

15. Install the wheel and tire assembly.

16. Lower the vehicle.

Wheel Cylinder

REMOVAL AND INSTALLATION

1. Raise and safely support the vehicle.

2. Remove the wheel and tire assemblies.

3. Remove the brake drum. Remove the brake shoes, as necessary.

4. Disconnect the brake fluid line from the wheel cylinder.

5. Remove the wheel cylinder retainer or bolt.

6. Remove the wheel cylinder from the backing plate.

To install:

7. Install the wheel cylinder in position on the backing plate.

8. Install the retainer or bolt. Connect the brake line to the wheel cylinder.

9. Install the brake linings and the brake drum.

10. Install the wheel and tire assembly. Lower the vehicle.

11. Bleed the brake system.

Parking Brake Cable

ADJUSTMENT

Except Lumina APV, Silhouette and Trans Sport

1. Raise and support the vehicle safely. Loosen the equalizer nut. Some vehicles may require the removal of the cable guide on the equalizer.

2. Set the parking brake pedal 2 clicks for 2WD vehicles or 3 clicks for 4WD vehicles.

3. Tighten the equalizer nut until the rear wheels will not rotate without excessive force in the forward motion.

4. Back off the equalizer nut until there is light drag when the wheels are rotated in the forward motion.

5. If removed, install the cable guide. Release the parking brake.

6. Rotate the rear wheels in the forward motion. There should be no brake drag.

Lumina APV, Silhouette and Trans Sport

1. Set the parking brake pedal 4 clicks.

2. Raise and support the vehicle safely. Remove the access plug in the backing plate.

3. Adjust the parking brake until an ⅛ in. drill can be inserted through the access hole into the space between the shoe web and park brake lever. Satisfactory adjustment will be obtained when an ⅛ in. drill bit will fit but a ¼ in. drill bit will not.

4. Release the parking brake and verify the rear wheels will rotate freely.

5. Replace the access plug and lower the vehicle. Check for proper operation of the parking brake.

REMOVAL AND INSTALLATION

Rear

1. Raise and support the vehicle safely. Remove the tire and wheel assembly. Remove the brake drum.

2. Loosen the equalizer and disconnect the cable at the center retainer.

3. Compress the plastic retainer fingers and remove the retainer from the frame bracket.

4. Remove the rear brake shoe assembly. Disconnect the parking brake cable. Remove the cable from the frame and from the brake backing plate.

5. Installation is the reverse of the removal procedure. Adjust the rear brakes, as required. Adjust the parking brake.

Front

1. Raise and support the vehicle safely. Loosen the adjuster nut and disconnect the front cable from the connector.

2. Compress the retainer fingers and loosen the assembly at the frame. Remove the supports.

3. Lower the vehicle. As required, remove dash trim panels to gain access to the parking brake pedal assembly.

4. Disconnect the cable from the parking brake pedal, compress the retainer fingers. Remove the cable from the vehicle.

5. Installation is the reverse of the removal procedure. Adjust the parking brake.

BRAKE SYSTEM BLEEDING

1. The brake system must be bled when any brake line is disconnected or there is air in the system.

NOTE: Never bleed a wheel cylinder when a drum is removed.

2. Clean the master cylinder of excess dirt and remove the cylinder cover and the diaphragm.

3. Fill the master cylinder to the proper level. Check the fluid level periodically during the bleeding process and replenish it, as necessary. Do not allow the master cylinder fall below ½ full.

4. If the master cylinder is suspected or known to have air in the bore, bleed it before any wheel cylinder or caliper as follows:

a. Disconnect the forward brake line connection at the master cylinder.

b. Allow brake fluid to fill the master cylinder bore until it begins to flow from the forward line connector port.

c. Connect the forward brake line to the master cylinder and tighten.

d. Have an assistant depress the brake pedal slowly, 1 depression at a time and hold while loosening the forward brake line connection at the master cylinder to purge the air from the bore. Tighten the connection and then have an assistant release the brake pedal slowly. Wait 15 seconds and repeat the sequence. Repeat the sequence, including the 15 second wait until all air is removed from the bore.

e. After all air is removed at the forward connection, repeat the above procedure for the rear connection at the master cylinder.

5. Bleed the individual wheel cylinders or calipers only after all air is removed from the master cylinder.

a. Attach the proper size box end wrench over the bleeder valve.

b. Attach a length of vinyl hose to the bleeder screw of the brake to be bled. Insert the other end of the hose into a clear jar half full of clean brake fluid, so the end of the hose is under the level of fluid. The correct sequence for bleeding is to work from the brake farthest from the master cylinder to the 1 closest; right rear, left rear, right front, left front.

6. Have an assistant depress and release the brake pedal 1 time and hold. Loosen the bleeder valve to purge the air from the cylinder. Tighten the bleeder screw and slowly

release the pedal and wait 15 seconds. Repeat the sequence including the 15 second wait until all air is removed.

NOTE: Make sure an assistant presses the brake pedal to the floor slowly. Rapid pumping of the brake pedal pushes the master cylinder secondary piston down the bore in a way that makes it difficult to bleed the rearside of the system.

7. Repeat this procedure at each of the wheels. Remember to check the master cylinder level occasionally. Use only fresh fluid to refill the master cylinder, not the fluid bled from the system.

8. When the bleeding process is complete, refill the master cylinder, install its cover and diaphragm and discard the fluid bled from the brake system.

Anti-Lock Brake System Service

There are 3 systems used, first, the Four Wheel Anti-Lock (4WAL) which is used on all vehicles, except the Lumina APV, Silhouette and Trans Sport. Second, the Rear Wheel Anti-Lock (RWAL) which is used only on all vehicles except the Bravada, Lumina APV, Silhouette and Trans Sport and third, the Anti-Lock Brake System (ABS) which is used only on the Lumina APV, Silhouette and Trans Sport.

PRECAUTION

Failure to observe the following precautions may result in system damage.
• Before performing electric arc welding on the vehicle, disconnect the Electronic Brake Control Unit.
• When performing painting work on the vehicle, do not expose the Electronic Brake Control Unit to temperatures in excess of 185°F (85°C) for longer than 2 hours. The system may be exposed to temperatures up to 200°F (95°C) for less than 15 minutes.
• Never disconnect or connect the Electronic Brake Control Unit connector with the ignition switch ON.
• Never disassemble any component of the Anti-Lock Brake System which is designated non-serviceable; the component must be replaced as an assembly.

• When filling the master cylinder, always use Delco Supreme 11 brake fluid or equivalent, which meets DOT-3 specifications; petroleum base fluid will destroy the rubber parts. Do not allow fluid to be spilled on the Electronic Brake Control Unit.

Control Module

REMOVAL AND INSTALLATION

RWAL System

1. Disconnect the negative battery cable.
2. Disconnect the electrical connectors from the module.
3. Remove the module from the master cylinder/proportioning valve bracket.
4. Installation is the reverse of the removal procedure.

4WAL System

1. Disconnect the negative battery cable.
2. Disconnect the electrical connectors from the module.
3. Disconnect and plug the brake lines at the module.
4. Remove the bolts attaching the module to the fenderwell.
5. Remove the module and bracket assembly from the vehicle.
6. Remove the bracket from the module.
7. Installation is the reverse of the removal procedure.

Electronic Brake Control Module (EBCU)

1. Disconnect the negative battery cable.
2. Remove the lower sound insulator under the steering column.
3. Remove the screws attaching the module to the dash panel.
4. Disconnect the electrical connectors from the module.
5. Remove the module from the vehicle.
6. Installation is the reverse of the removal procedure.

Isolation/Dump Valve

REMOVAL AND INSTALLATION

RWAL System

1. Disconnect the negative battery cable.
2. Disconnect and plug the brake line fittings at the isolation/dump

valve located under the master cylinder.
3. Remove the master cylinder bracket bolts.
4. Disconnect the Electronic Control Unit connector.
5. Remove the isolation/dump valve.
6. Installation is the reverse of the removal procedure.

Front Wheel Speed Sensor

REMOVAL AND INSTALLATION

4WAL System

2 WHEEL DRIVE

1. Disconnect the negative battery cable.
2. Raise and support the vehicle safely.
3. Remove the wheel and tire assembly.
4. Remove the brake caliper, hub and rotor assembly.
5. Remove the sensor wire from the clips on the upper control arm and disconnect the connector.
6. Remove the sensor and backing plate attaching bolts and remove the assembly.
7. Installation is the reverse of the removal procedure.

4 WHEEL DRIVE

1. Disconnect the negative battery cable.
2. Raise and support the vehicle safely.
3. Remove the wheel and tire assembly.
4. Remove the hub and rotor assembly.
5. Disconnect the sensor wire connector.
6. Remove the bolts securing the sensor and sensor wire.
7. Remove the sensor from the spindle.
8. Installation is the reverse of the removal procedure.

Anti-lock Brake System (ABS)

1. Disconnect the negative battery cable.
2. Raise and support the vehicle safely.
3. Remove the wheel and tire assembly.
4. Remove the hub and rotor assembly.
5. Disconnect the sensor wire connector.

6. Remove the sensor from the hub assembly.

NOTE: The speed sensor cannot be removed from the hub assembly without damaging the sensor, therefore the sensor must be replaced. The sensor has 2 parts, which must be replaced as an assembly.

7. Installation is the reverse of the removal procedure.

Rear Wheel Speed Sensor

REMOVAL AND INSTALLATION

4WAL System

1. Disconnect the negative battery cable.
2. Raise and support the vehicle safely.
3. Remove the wheel, tire and drum assembly.
4. Remove the primary (forward) brake shoe.
5. Disconnect the sensor wire connector and remove the sensor wire from the rear axle clips.
6. Remove the bolts securing the sensor and sensor wire.
7. Remove the sensor from the backing plate.
8. Installation is the reverse of the removal procedure.

Anti-lock Brake System (ABS)

1. Disconnect the negative battery cable.
2. Raise and support the vehicle safely.
3. Remove the wheel, tire and drum assembly.
4. Remove the bolts and nuts attaching the rear wheel bearing and speed sensor assembly.
5. Remove the wheel bearing and speed sensor assembly.
6. Disconnect the sensor wire connector.
7. Installation is the reverse of the removal procedure.

FRONT SUSPENSION

Shock Absorbers

REMOVAL AND INSTALLATION

1. Remove the top shock mounting nut and grommet.

2. Raise and safely support the vehicle.
3. Remove the wheel and tire assembly.
4. Remove the lower shock mounting bolts and remove the shock absorber.
5. Installation is the reverse of the removal procedure.

MacPherson Strut Assembly

All front wheel drive vehicles came equipped with a MacPherson strut front suspension.

REMOVAL AND INSTALLATION

Lumina APV, Silhouette and Trans Sport

NOTE: Do not remove the top center nut from the strut assembly. This nut should only be removed when the strut assembly is out of the vehicle, mounted in a holding fixture and the coil spring is in a compressed position using the proper strut coil spring compressor.

1. Remove the 3 nuts that retain the top of the strut assembly.
2. Raise and safely support the vehicle.
3. Remove the wheel and tire assembly. Remove the brake line bracket from the strut mount.
4. Remove the lower strut mounting bolts.
5. Remove the strut assembly from the vehicle and place the strut in a suitable holding fixture.
6. Disassemble the strut as follows:
   a. With the strut coil spring in a compressed position approximately ½ its normal length, remove the nut from the top of the strut.
   b. Place tool J–34013–27 or equivalent guide rod on top of the damper shaft. Use the rod to guide the damper shaft straight down through the bearing cap while decompressing the spring.
   c. Remove the coil spring and other components.
7. Installation is the reverse of the removal procedure. Ensure the spring seat flat should face 10 degrees forward of the centerline of the strut assembly spindle.
8. Tighten the strut lower bolts to 140 ft. lbs. (190 Nm) and the upper mounting nuts to 18 ft. lbs. (25 Nm).

Coil Springs

All 2 wheel rear drive vehicles came equipped with a coil spring front suspension.

REMOVAL AND INSTALLATION

Astro, Blazer, Bravada, Jimmy, Pickup, Safari and Sonoma

2WD VEHICLE

1. Raise and support the vehicle safely. Remove the wheel and tire assembly. Remove the shock absorber lower retaining bolts.
2. Push the shock absorber through the control arm and into the spring.
3. With the vehicle supported so the control arms hang free, install tool J–23028 or equivalent, onto a suitable jack and into the lower control arm bushings. Remove the stabilizer bar from the control arm.
4. Remove the stabilizer to lower control arm attachment. Raise the jack and remove the tension on the lower control arm bolts.
5. Install a safety chain around the spring and through the lower control arm. Remove the lower control arm rear pivot bolt, than remove the other pivot bolt.
6. Lower the jack and allow the lower control arm to hang free. Remove the spring assembly from the vehicle.
7. Installation is the reverse of the removal procedure. When positioning the spring in the lower control arm, be sure the spring insulator is in the proper position before lifting the control arm in place.

Torsion Bars and Support

All 4 wheel drive vehicles came equipped with a torsion bar front suspension instead of coil spring.

REMOVAL AND INSTALLATION

Astro, Blazer, Jimmy, Pickup, Safari, Sonoma, Cyclone and Typhoon

1. Raise and safely support the vehicle. Remove the wheel and tire assemblies.
2. Remove the torsion bar adjusting bolt using tool J–36202 or equivalent. Count the number of tool turns required to remove the bolt.
3. Remove the torsion bar support retainer plate and insulator.

4. Remove the torsion bar by sliding it forward into the control arm and lowering it from the vehicle.

5. Remove the torsion bar support from the vehicle. Remove the adjusting arm and the adjusting arm bolt.

To install:

6. Install the adjusting arm to the support and loosely install the adjusting bolt.

7. Install the support to the frame and the insulator to the frame end.

8. Install the retainer to the support. Install the retainer mounting bolts and tighten to 26 ft. lbs. (35 Nm).

9. Tighten the center retainer bolt to 25 ft. lbs. (34 Nm).

10. Install the torsion bar to the lower control arm and raise and slide the torsion bar into the adjusting arm. The torsion bar should have 6mm clearance at the support.

11. Attach tool J–36202 to the support and tighten it against the adjusting arm the recorded number of turns.

12. Install the adjusting bolt and turn it in until it contacts the adjusting arm. Remove the tool.

13. Install the wheel and tire assemblies. Lower the vehicle.

Upper Ball Joint

INSPECTION

1. Raise and safely support the lower control arm.

2. Wipe the ball joint clean and check the seal for cuts or tears.

3. Check the wheel bearings for proper adjustment.

4. Position a dial indicator against the lowest outside point of the tire. Rock the wheel in and out.

5. Check the reading on the dial indicator. The reading should be no more than 0.125 inch (3.18mm).

REMOVAL AND INSTALLATION

1. Raise and support the vehicle safely. Properly support the lower control arm.

NOTE: The control arm must be supported so the spring and the control arm remain intact.

2. Remove the tire and wheel assembly. As required, remove the brake caliper and position it aside.

3. Remove the cotter pin and the upper ball joint retaining bolt. Using the proper tool separate the upper joint from its mounting. Support the

knuckle assembly so its weight will not damage the brake hose.

4. Remove the rivets from the ball joint assembly, using a drill with an ⅛ in. and then ½ in. bit. Remove the ball joint from the upper control arm.

5. Installation is the reverse of the removal procedure. Tighten the replacement bolts to 17 ft. lbs. (23 Nm). Check and adjust the front end alignment, as required.

Lower Ball Joint

Inspection

EXCEPT Lumina APV, Silhouette and Trans Sport

1. Raise and safely support the vehicle.

2. The ball joint wear is indicated by the position of the grease fitting on the bottom of the joint.

3. The round portion of the grease nipple must protrude from the bottom of the joint. If the nipple is flush with or inside of the joint, it must be replaced.

Lumina APV, Silhouette and Trans Sport

1. Raise and safely support the lower control arm.

2. Wipe the ball joint clean and check the seal for cuts or tears.

3. Check the wheel bearings for proper adjustment.

4. Position a dial indicator against the lowest outside point of the tire. Rock the wheel in and out.

5. Check the reading on the dial indicator. The reading should be no more than 0.125 inch (3.18mm).

REMOVAL AND INSTALLATION

Except Lumina APV, Silhouette and Trans Sport

1. Raise and support the vehicle safely. Properly support the lower control arm.

NOTE: The control arm must be supported so the spring and the control arm remain intact.

2. Remove the tire and wheel assembly. As required, remove the brake caliper and position it aside.

3. Remove the cotter pin and the lower ball joint retaining bolt. Using the proper tool separate the ball joint from its mounting. Support the knuckle assembly so its weight will not damage the brake hose.

4. Remove the rivets from the ball joint assembly, using a drill with a ⅛ in. bit. Remove the ball joint from the control arm.

5. Installation is the reverse of the removal procedure. Be sure to use the nuts and bolts that are supplied with the replacement ball joint assembly. Tighten the replacement bolts to 17 ft. lbs. (23 Nm). Check and adjust the front end alignment, as required.

Lumina APV, Silhouette and Trans Sport

1. Raise and support the vehicle safely. Properly support the lower control arm.

NOTE: The control arm must be supported so the spring and the control arm remain intact.

2. Remove the tire and wheel assembly. As required, remove the brake caliper and position it aside.

3. Remove the pinch bolt from the lower ball joint. Using the proper tool separate the upper joint from the steering knuckle. Support the knuckle assembly so its weight will not damage the brake hose.

4. Remove the rivets from the ball joint assembly, using a drill with an ⅛ in. and then ½ in. bit.

5. Remove the stabilizer shaft bushing assembly nut.

6. Remove the ball joint from the lower control arm.

7. Installation is the reverse of the removal procedure. Tighten the ball joint pinch bolt to 33 ft. lbs. (45 Nm). Check and adjust the front end alignment, as required.

Upper Control Arms

REMOVAL AND INSTALLATION

1. Note and record the amount of shims used at the control arm retaining bolts. These shims must be installed in the same location as removed. Remove the nuts and the shims.

2. Raise and support the vehicle safely. Properly support the lower control arm. The control arm must be supported so the spring and the control arm remain intact.

3. Remove the wheel and tire assembly. Separate the upper ball joint from the steering knuckle, using the proper tool. Support the hub assembly.

4. Remove the upper control arm retaining bolts. Remove the upper control arm from the vehicle.

5. Installation is the reverse of the removal procedure. Tighten the upper control arm nuts to 65 ft. lbs. (88 Nm). Check and adjust the front end alignment, as required.

Lower Control Arms

REMOVAL AND INSTALLATION

Except Lumina APV, Silhouette and Trans Sport

1. Raise and support the vehicle safely.
2. Remove the wheel and tire assemblies. Properly support the lower control arm assembly.
3. Remove the coil spring.
4. Remove the lower ball joint cotter pin and retaining nut.
5. Using the proper tool, separate the lower ball joint from the steering knuckle.
6. Remove the lower control arm from the vehicle.
7. Installation is the reverse of the removal procedure. Check and adjust front alignment, as required.

Lumina APV, Silhouette and Trans Sport

1. Raise and safely support the vehicle so the suspension hangs freely.
2. Remove the wheel and tire assemblies.
3. Remove the stabilizer shaft-to-control arm mounting bolt. Remove the lower ball joint pinch bolt.
4. Separate the steering knuckle from the lower ball joint.
5. Remove the lower control arm mounting bolts and remove the control arm from the vehicle.
To install:
6. Install the control arm in position on the vehicle frame. Do not tighten the control arm bolts at this time.
7. Install the stabilizer shaft to the control arm, do not tighten the bolts at this time.
8. Connect the steering knuckle to the control arm using a new pinch bolt. Tighten to 33 ft. lbs. (45 Nm). Lower the vehicle so the weight is supported by the control arms.
9. Tighten the control arm bolts to 61 ft. lbs. (83 Nm). Tighten the stabilizer shaft bolts to 32 ft. lbs. (43 Nm).
10. Install the wheel and tire assemblies. Lower the vehicle completely.

Stabilizer Shaft

REMOVAL AND INSTALLATION

1. Raise and safely support the vehicle.
2. Remove the wheel and tire assembly.
3. Remove the left and right side stabilizer mounting bolts. Keep the sides separate for installation.
4. Remove the center stabilizer insulators and lower the stabilizer from the vehicle.
To install:
5. Install the stabilizer in position in the vehicle. Tighten the left and right mounting bolts to 24 ft. lbs. and the center bushing supports to 35 ft. lbs.
6. Install the wheel and tire assemblies and lower the vehicle.

Front Wheel Bearings

ADJUSTMENT

Astro, Blazer, Jimmy, Pickup, Safari and Sonoma

1. Raise and safely support the vehicle.
2. Remove the brake caliper and the dust cap from the wheel hub.
3. Remove the cotter pin from the castle nut.
4. Tighten the castle nut to 12 ft. lbs. (16 Nm) while turning the rotor forward.
5. Loosen the castle nut, then tighten it, by hand to a just loose position.
6. Back the nut off slightly, until the hole in the spindle aligns with a slot in the nut. Do not back the nut off more than ½ of a flat.
7. Install a new cotter pin. Measure the rotor end play at the castle nut, it should not exceed 0.0005 in. (0.13mm).
8. Install the bearing dust cover and install the wheel and tire assembly.
9. Lower the vehicle.

REMOVAL AND INSTALLATION

Astro, Blazer, Jimmy, Pickup, Safari and Sonoma

NOTE: All front and 4 wheel drive vehicles, refer to the drive axle section.

1. Raise and support the vehicle safely. Remove the tire and wheel assembly.

2. Remove the brake caliper and position it aside. Remove the dust cover, cotter pin, washer and spindle nut.
3. Remove the outer wheel bearing assembly. Remove the brake rotor. Remove the inner wheel bearing assembly.
4. Installation is the reverse of the removal procedure. When installing new bearings, be sure to install new bearing races inside the rotor.
5. Pack new bearings with the proper grade and type wheel bearing grease. Adjust the wheel bearings.

REAR SUSPENSION

Shock Absorbers

REMOVAL AND INSTALLATION

Except Lumina APV, Silhouette and Trans Sport

1. Raise and support the vehicle safely.
2. Properly support the rear axle assembly.
3. Remove the upper shock absorber retaining bolt.
4. Remove the lower shock absorber bolt.
5. Remove the shock absorber from the vehicle.
To install:
6. Install the shock in position and install the mounting bolts.
7. Tighten the top mounting bolts to 17 ft. lbs. (23 Nm) and the lower mounting bolts to 47 ft. lbs. (64 Nm).
8. Lower the vehicle.

Lumina APV, Silhouette and Trans Sport

1. Open the lift gate and open the trim cover.
2. Remove the upper shock mounting nut and grommet.
3. Raise and safely support the vehicle. Properly support the rear axle assembly.
4. If equipped with electronic level control suspension, remove the air line from the shock absorber. Allow the air to bleed off.
5. Remove the lower mounting bolt and remove the shock from the vehicle.
To install:
6. Install the shock in position and install the lower mounting bolt. Tighten to 44 ft. lbs. (59 Nm). Con-

nect the air line to the shock, if equipped.

7. Lower the vehicle and install the upper shock retaining nut, tighten it to 16 ft. lbs. (22 Nm). Install the trim cover.

Coil Springs

REMOVAL AND INSTALLATION

Lumina APV, Silhouette and Trans Sport

1. Raise and safely support the vehicle.
2. Safely support the rear axle assembly.
3. Remove the right and left brake line-to-axle attaching screws. Allow the brake lines to hang freely.
4. Disconnect the track bar-to-axle attaching bolt.
5. Disconnect the lower shock absorber mounting bolts.
6. Slowly lower the rear axle and remove the springs and insulators.
7. Installation is the reverse of the removal procedure.

Leaf Springs

REMOVAL AND INSTALLATION

Astro, Blazer, Jimmy, Pickup, Safari, Sonoma, Cyclone and Typhoon

1. Raise and support the vehicle safely. Properly support the rear axle assembly to relieve tension on the springs.
2. Remove the shock absorbers. Remove the U bolt nuts, washers, anchor plates and the U bolts.
3. Remove the shackle to frame bolt, washers and nut. Remove the spring assembly to front bracket nut, washers and bolt.
4. Remove the spring assembly from the vehicle. As required, separate the spring from the shackle.
5. Installation is the reverse of the removal procedure. Tighten the U bolts to 85 ft. lbs. (115 Nm) in 2 gradual steps. Tighten the front and rear shackle nuts to 92 ft. lbs. (125 Nm).

Rear Wheel Hub and Bearing Assembly

REMOVAL AND INSTALLATION

Lumina APV, Silhouette and Trans Sport

1. Raise and safely support the vehicle.
2. Remove the wheel and tire assemblies.
3. Remove the brake drum.
4. Remove the hub and bearing assembly mounting bolts and remove it from the vehicle. Support the brake assembly with a wire.
5. Installation is the reverse of the removal procedure. Tighten the hub assembly mounting bolts to 45 ft. lbs. (60 Nm).

Track Bar

REMOVAL AND INSTALLATION

Lumina APV, Silhouette and Trans Sport

1. Raise and safely support the vehicle.
2. Remove the track bar mounting bolts from the body and the axle.
3. Lower the track bar from the vehicle.
   To install:
4. Install the track bar at the axle, loosely install the bolt.
5. Connect the other end of the track bar at the frame.
6. Tighten the bolt at the axle to 44 ft. lbs. (60 Nm) and the track bar-to-frame bolt to 35 ft. lbs. (47 Nm).
7. Lower the vehicle.

Rear Axle Assembly

REMOVAL AND INSTALLATION

Lumina APV, Silhouette and Trans Sport

1. Raise and safely support the vehicle.

2. Properly support the rear axle assembly.
3. Remove the wheel and tire assemblies. Remove the brake drums.
4. Disconnect the parking brake cable from the brake system.
5. Disconnect the brake line brackets from the axle assembly.
6. Disconnect the shock absorbers from the axle.
7. Disconnect the track bar bolt at the rear axle.
8. Lower the rear axle slightly and remove the coil springs and insulator.
9. Remove the hub and bearing assemblies, support the backing plate and brake assemblies with wire from the vehicle.

NOTE: Do not allow the backing plates to hang by the brake lines.

10. Remove the control arm to frame bracket attaching bolts and lower the axle assembly from the vehicle.
    To install:
11. Raise the axle assembly to the vehicle and install the control arm bolts. Torque control arm bracket bolts to 84 ft. lbs. (115 Nm). Torque the control arm to bracket pivot bolts to 28 ft. lbs. (38 Nm).
12. Install the backing plate, hub and bearing assembly. Use car not to kink the brake lines.
13. Install the brake line brackets to the axle assembly. Connect the parking brake cable to the brake assembly.
14. Position the coil springs and insulators on the axle and raise the axle.
15. Connect the shock absorber lower bolts to the axle.
16. Connect the track bar to the rear axle.
17. Install the brake drums. Install the wheel and tire assemblies.
18. Lower the vehicle.

Jeep 6

Cherokee • Comanche • Grand Cherokee • Grand Wagoneer • Wrangler

SPECIFICATIONS

VEHICLE IDENTIFICATION CHART

Engine Code							Model Year	
Code	Liters	Cu. In. (cc)	Cyl.	Fuel Sys.	Eng. Mfg.		Code	Year
E	2.5	150 (2458)	4	TFI	Chrysler		L	1990
P	2.5	150 (2458)	4	MFI	Chrysler		M	1991
L	4.0	243 (3983)	6	MFI	Chrysler		N	1992
M	4.2	258 (4228)	6	2BBL	Chrysler		P	1993
S	4.0	242 (3966)	6	MFI	Chrysler		R	1994
Y	5.2	318 (5211)	8	MFI	Chrysler		S	1995
7	5.9	360 (5899)	8	2BBL	Chrysler			

TFI-Throttle Body Injection
MFI-Multiport Fuel Injection

79114C01

ENGINE IDENTIFICATION

Year	Model	Engine Displacement Liters (cc)	Engine Series (ID/VIN)	Fuel System	No. of Cylinders	Engine Type
1991	Comanche	2.5 (2458)	P	MFI	4	OHV
	Comanche	4.0 (3966)	S	MFI	6	OHV
	Wrangler	2.5 (2458)	P	MFI	4	OHV
	Wrangler	4.0 (3966)	S	MFI	6	OHV
	Cherokee	2.5 (2458)	P	MFI	4	OHV
	Cherokee	4.0 (3966)	S	MFI	6	OHV
	Grand Wagoneer	5.9 (5899)	7	2BBL	8	OHV
1992	Comanche	2.5 (2458)	P	MFI	4	OHV
	Comanche	4.0 (3966)	S	MFI	6	OHV
	Wrangler	2.5 (2458)	P	MFI	4	OHV
	Wrangler	4.0 (3966)	S	MFI	6	OHV
	Cherokee	2.5 (2458)	P	MFI	4	OHV
	Cherokee	4.0 (3966)	S	MFI	6	OHV
1993	Wrangler	2.5 (2458)	P	MFI	4	OHV
	Wrangler	4.0 (3966)	S	MFI	6	OHV
	Cherokee	2.5 (2458)	P	MFI	4	OHV
	Cherokee	4.0 (3966)	S	MFI	6	OHV
	Grand Cherokee	4.0 (3966)	S	MFI	6	OHV
	Grand Cherokee	5.2 (5211)	Y	MFI	8	OHV
	Grand Wagoneer	5.2 (5211)	Y	MFI	8	OHV
1994-95	Wrangler	2.5 (2458)	P	MFI	4	OHV
	Wrangler	4.0 (3966)	S	MFI	6	OHV
	Cherokee	2.5 (2458)	P	MFI	4	OHV
	Cherokee	4.0 (3966)	S	MFI	6	OHV
	Grand Cherokee	4.0 (3966)	S	MFI	6	OHV
	Grand Cherokee	5.2 (5211)	Y	MFI	8	OHV

TFI-Throttle Body Injection
MFI-Multiport Fuel Injection
OHV-Overhead Valve

79114C02

GENERAL ENGINE SPECIFICATIONS

Year	Engine ID/VIN	Engine Displacement Liters (cc)	Fuel System Type	Net Horsepower @ rpm	Net Torque @ rpm (ft. lbs.)	Bore x Stroke (in.)	Compression Ratio	Oil Pressure @ rpm
1991	P	2.5 (2458)	MFI	123@5250 [1]	139@3250 [2]	3.876x3.188	9.1:1	37@1600 [4]
	S	4.0 (3966)	MFI	180@4750	220@2500 [3]	3.880x3.440	8.8:1	37@1600 [4]
	7	5.9 (5899)	2BBL	144@3200	280@1500	4.080x3.440	8.25:1	37@1600 [4]
1992	P	2.5 (2458)	MFI	123@5250 [1]	139@3250 [2]	3.876x3.188	9.1:1	37@1600 [5]
	S	4.0 (3966)	MFI	180@4750	220@2500 [3]	3.880x3.440	8.8:1	37@1600 [5]
1993	P	2.5 (2458)	MFI	130@5250	139@3250 [1]	3.876x3.188	9.1:1	37@1600 [5]
	S	4.0 (3966)	MFI	180@4750	220@2500 [2]	3.880x3.440	8.8:1	37@1600 [5]
	Y	5.2 (5211)	MFI	220@4800	285@3600	3.910x3.310	9.1:1	30@3000 [5]
1994-95	P	2.5 (2458)	MFI	130@5250	139@3250 [1]	3.876x3.188	9.1:1	37@1600 [5]
	S	4.0 (3966)	MFI	180@4750	220@2500 [2]	3.880x3.440	8.8:1	37@1600 [5]
	Y	5.2 (5211)	MFI	220@4800	285@3600	3.19x3.310	9.1:1	30@3000 [5]

[1] Cherokee and Comanche 149@3250
[2] Cherokee, Comanche and Grand Wagoneer 225@4000
[3] Above 3000 rpm, pressure can vary to a maximum of 80 psi.
[4] Above 1600 rpm, pressure can vary to a maximum of 75 psi..
[5] Above 3000 rpm, pressure can vary to a maximum of 80 psi.

79114C03

GASOLINE ENGINE TUNE-UP SPECIFICATIONS

Year	Engine ID/VIN	Engine Displacement Liters (cc)	Spark Plugs Gap (in.)	Ignition Timing (deg.) MT	Ignition Timing (deg.) AT	Fuel Pump (psi)	Idle Speed (rpm) MT	Idle Speed (rpm) AT	Valve Clearance In.	Valve Clearance Ex.
1991	P	2.5 (2458)	0.035	[1]	[1]	39-41	[1]	[1]	HYD	HYD
	S	4.0 (3966)	0.035	[1]	[1]	39-41	[1]	[1]	HYD	HYD
	7	5.9 (5899)	0.035	8B	8B	5-6.5	600	600 [2]	HYD	HYD
1992	P	2.5 (2458)	0.035	[1]	[1]	39-41	[1]	[1]	HYD	HYD
	S	4.0 (3966)	0.035	[1]	[1]	39-41	[1]	[1]	HYD	HYD
1993	P	2.5 (2458)	0.035	[1]	[1]	39-41	[1]	[1]	HYD	HYD
	S	4.0 (3966)	0.035	[1]	[1]	39-41	[1]	[1]	HYD	HYD
	Y	5.2 (5211)	0.035	[1]	[1]	39-41	[1]	[1]	HYD	HYD
1994-95	P	2.5 (2458)	0.035	[1]	[1]	39-41	[1]	[1]	HYD	HYD
	S	4.0 (3966)	0.035	[1]	[1]	39-41	[1]	[1]	HYD	HYD
	Y	5.2 (5211)	0.035	[1]	[1]	39-41	[1]	[1]	HYD	HYD

NOTE: The Vehicle Emission Control Information label often reflects specification changes made during production. The label figures must be used if they differ from those in this chart

B - Before top dead center

HYD - Hydraulic

[1] Not Adjustable
[2] 500 with solenoid deenergized

79114C04

CAPACITIES

Year	Model	Engine ID/VIN	Engine Displacement Liters (cc)	Engine Crankcase with Filter	Transmission (pts.) 4-Spd	5-Spd	Auto.	Transfer Case (pts.)	Drive Axle Front (pts.)	Rear (pts.)	Fuel Tank (gal.)	Cooling System (qts.)
1991	Comanche	P	2.5 (2458)	4.0	7.0	7.4 [5]	17.0	2.2	2.5	2.5 [3]	18.5 [6]	10.0
	Comanche	S	4.0 (3966)	6.0	7.0	6.7	17.0	2.2	2.5	2.5 [3]	18.5 [6]	12.0
	Wrangler	P	2.5 (2458)	4.0	-	7.4 [5]	16.0	3.3	2.5	2.5 [3]	15.0 [4]	9.0
	Wrangler	S	4.0 (3966)	6.0	-	6.7	16.0	3.3	2.5	2.5 [3]	15.0 [4]	12.0
	Cherokee	P	2.5 (2458)	4.0	-	7.4 [5]	17.0	3.0 [2]	2.5	2.5 [3]	20.2	10.0
	Cherokee	S	4.0 (3966)	6.0	-	6.7	17.0	3.0 [2]	2.5	2.5 [3]	20.2	12.0
	Grand Wagoneer	7	5.9 (5899)	5.0	-	-	17.0	6.0	3.8	2.5 [3]	20.3	15.5
1992	Comanche	P	2.5 (2458)	4.0	7.0	7.4 [5]	17.0	2.2	2.5	2.5 [3]	18.5 [6]	10.0
	Comanche	S	4.0 (3966)	6.0	7.0	6.7	17.0	2.2	2.5	2.5 [3]	18.5 [6]	12.0
	Wrangler	P	2.5 (2458)	4.0	-	7.4 [5]	16.0	3.3	2.5	2.5 [3]	15.0 [4]	9.0
	Wrangler	S	4.0 (3966)	6.0	-	6.7	16.0	3.3	2.5	2.5 [3]	15.0 [4]	10.0
	Cherokee	P	2.5 (2458)	4.0	-	7.4 [5]	17.0	3.0 [2]	2.5	2.5 [3]	20.2	10.5
	Cherokee	S	4.0 (3966)	6.0	-	6.7	17.0	3.0 [2]	2.5	2.5 [3]	20.2	12.0
1993	Wrangler	P	2.5 (2458)	4.0	-	7.4 [5]	16.0	3.3	2.5	2.5 [3]	15.0 [4]	9.0
	Wrangler	S	4.0 (3966)	6.0	-	6.7	16.0	3.3	2.5	2.5 [3]	15.0 [4]	10.5
	Cherokee	P	2.5 (2458)	4.0	-	7.4 [5]	17.0	3.0 [2]	2.5	2.5 [3]	20.2	10.0
	Cherokee	S	4.0 (3966)	6.0	-	6.7	17.0	3.0 [2]	2.5	2.5 [3]	20.2	12.0
	Grand Cherokee	S	4.0 (3966)	6.0	-	6.5	17.0	3.2 [7]	3.1	3.4	23.0	9.3
	Grand Cherokee	Y	5.2 (5211)	5.0	-	6.5	17.0	3.2 [7]	3.1	3.4	23.0	14.9
	Grand Wagoneer	Y	5.2 (5211)	5.0	-	-	17.0	3.2 [7]	3.1	3.4	23.0	14.9
1994-95	Wrangler	P	2.5 (2458)	4.0	-	7.4 [5]	16.0	3.3	2.5	2.5 [3]	15.0 [4]	9.0
	Wrangler	S	4.0 (3966)	6.0	-	6.7	16.0	3.3	2.5	2.5 [3]	15.0 [4]	10.5
	Cherokee	P	2.5 (2458)	4.0	-	7.4 [5]	17.0	3.0 [2]	2.5	2.5 [3]	20.2	10.0
	Cherokee	S	4.0 (3966)	6.0	-	6.7	17.0	3.0 [2]	2.5	2.5 [3]	20.2	12.0
	Grand Cherokee	S	4.0 (3966)	6.0	-	6.5	17.0	3.2 [7]	3.1	3.4	23.0	9.3
	Grand Cherokee	Y	5.2 (5211)	5.0	-	6.5	17.0	3.2 [7]	3.1	3.4	23.0	14.9

1 Aisan Warner - 7.4 qts.
2 Command-Trac - 2.2 pts.
3 Heavy Duty - 3.0 pts.
4 Optional - 20 gal.
5 2WD - 7.4 pts.
6 Long Bed - 23.5 gal.
7 NP242 - 2.9; NP249 - 3.0

79114C06

CAMSHAFT SPECIFICATIONS

All measurements given in inches.

Year	Engine ID/VIN	Engine Displacement Liters (cc)	Journal Diameter					Elevation		Bearing Clearance	Camshaft End Play
			1	2	3	4	5	In.	Ex.		
1991	P	2.5 (2458)	2.0300-2.0290	2.0200-2.0190	2.1000-2.0090	2.0000-1.9990	-	0.2650	0.2650	0.0010-0.0030	0
	S	4.0 (3996)	2.0300-2.0290	2.0200-2.0190	2.1000-2.0090	2.0000-1.9990	-	0.2530	0.2530	0.0010-0.0030	0
	7	5.9 (5899)	2.1195-2.1205	2.0895-2.0905	2.0595-2.0605	2.0295-2.0305	1.9995-2.0005	0.2660	0.2660	0.0010-0.0030	0
1992	P	2.5 (2548)	2.0300-2.0290	2.0200-2.0190	2.1000-2.0090	2.0000-1.9990	-	0.2650	0.2650	0.0010-0.0030	0
	S	4.0 (3966)	2.0300-2.0290	2.0200-2.0190	2.1000-2.0090	2.0000-1.9990	-	0.2530	0.2530	0.0010-0.0030	0
1993	P	2.5 (2458)	2.0300-2.0290	2.0200-2.0190	2.1000-2.0090	2.0000-1.9990	-	0.2650	0.2650	0.0010-0.0030	0
	S	4.0 (3966)	2.0300-2.0290	2.0200-2.0190	2.1000-2.0090	2.0000-1.9990	-	0.2530	0.2530	0.0010-0.0030	0
	Y	5.2 (5211)	1.9990-1.9980	1.9830-1.9820	1.9680-1.9670	1.9520-1.9510	1.5615-1.5605	NA	NA	0.0010-0.0030	0.002-0.010
1994-95	P	2.5 (2458)	2.0300-2.0290	2.0200-2.0190	2.1000-2.0090	2.0000-1.9990	-	0.2650	0.2650	0.0010-0.0030	0
	S	4.0 (3966)	2.0300-2.0290	2.0200-2.0190	2.1000-2.0090	2.0000-1.9990	-	0.2530	0.2530	0.0010-0.0030	0
	Y	5.2 (5211)	1.9990-1.9980	1.9830-1.9820	1.9680-1.9670	1.9520-1.9510	1.5615-1.5605	NA	NA	0.0010-0.0030	0.002-0.010

79114C07

CRANKSHAFT AND CONNECTING ROD SPECIFICATIONS

All measurements are given in inches.

Year	Engine ID/VIN	Engine Displacement Liters (cc)	Crankshaft				Connecting Rod		
			Main Brg. Journal Dia.	Main Brg. Oil Clearance	Shaft End-play	Thrust on No.	Journal Diameter	Oil Clearance	Side Clearance
1991	P	2.5 (2458)	2.4996-2.5001	0.0010-0.0025	0.0015-0.0065	2	2.0934-2.0955	0.0010-0.0025	0.0100-0.0190
	S	4.0 (3966)	2.4996-2.5001	0.0010-0.0025	0.0015-0.0065	3	2.0934-2.0955	0.0010-0.0025	0.0100-0.0190
	7	5.9 (5899)	2.7474-2.7589 [1]	0.0010-0.0030 [2]	0.0030-0.0080	3	2.0934-2.0955	0.0010-0.0030	0.0060-0.0180
1992	P	2.5 (2458)	2.4996-2.5001	0.0010-0.0025	0.0015-0.0065	2	2.0934-2.0955	0.0010-0.0025	0.0100-0.0190
	S	4.0 (3966)	2.4996-2.5001	0.0010-0.0025	0.0015-0.0065	3	2.0934-2.0955	0.0010-0.0025	0.0100-0.0190
1993	P	2.5 (2458)	2.4996-2.5001	0.0010-0.0025	0.0015-0.0065	2	2.0934-2.0955	0.0010-0.0025	0.0100-0.0190
	S	4.0 (3966)	2.4996-2.5001	0.0010-0.0025	0.0015-0.0065	3	2.0934-2.0955	0.0010-0.0025	0.0100-0.0190
	Y	5.2 (5211)	2.4995-2.5005	0.0005-0.0025 [3]	0.0020-0.0100	3	2.1240-2.1250	0.0005-0.0022	0.0060-0.0140
1994-95	P	2.5 (2458)	2.4996-2.5001	0.0010-0.0025	0.0015-0.0065	2	2.0934-2.0955	0.0010-0.0025	0.0100-0.0190
	S	4.0 (3966)	2.4996-2.5001 [4]	0.0010-0.0025	0.0015-0.0065	3	2.0934-2.0955	0.0010-0.0025	0.0100-0.0190
	Y	5.2 (5211)	2.4995-2.5005	0.0005-0.0025 [5]	0.0020-0.0100	3	2.1240-2.1250	0.0005-0.0022	0.0060-0.0140

1 Rear main: 2.7464-2.7479
2 Rear main: 0.002-0.004
3 No. 1: 0.0005-0.0015
4 No. 7: 2.4980-2.4995
5 No. 1: 0.0005-0.0015

79114C08

VALVE SPECIFICATIONS

Year	Engine ID/VIN	Engine Displacement Liters (cc)	Seat Angle (deg.)	Face Angle (deg.)	Spring Test Pressure (lbs. @ in.)	Spring Installed Height (in.)	Stem-to-Guide Clearance (in.)		Stem Diameter (in.)	
							Intake	Exhaust	Intake	Exhaust
1991	P	2.5 (2458)	44.5	45	200@1.216	1.640	0.0010-0.0030	0.0010-0.0030	0.3110-0.3120	0.3110-0.3120
	S	4.0 (3966)	44.5	45	210@1.200	1.625	0.0010-0.0030	0.0010-0.0030	0.3110-0.3120	0.3110-0.3120
	7	5.9 (5899)	1	2	200@1.356	1.786	0.0010-0.0030	0.0010-0.0030	0.3715-0.3725	0.3715-0.3725
1992	P	2.5 (2458)	44.5	45	200@1.216	1.640	0.0010-0.0030	0.0010-0.0030	0.3110-0.3120	0.3110-0.3120
	S	4.0 (3966)	44.5	45	210@1.200	1.625	0.0010-0.0030	0.0010-0.0030	0.3110-0.3120	0.3110-0.3120
1993	P	2.5 (2458)	44.5	45	200@1.216	1.640	0.0010-0.0030	0.0010-0.0030	0.3110-0.3120	0.3110-0.3120
	S	4.0 (3966)	44.5	45	210@1.200	1.625	0.0010-0.0030	0.0010-0.0030	0.3110-0.3120	0.3110-0.3120
	Y	5.2 (5211)	44.25-44.75	43.25-43.75	200@1.212	1.640	0.0010-0.0030	0.0010-0.0030	0.3110-0.3120	0.3110-0.3120
1994-95	P	2.5 (2458)	44.5	45	200@1.216	1.640	0.0010-0.0030	0.0010-0.0030	0.3110-0.3120	0.3110-0.3120
	S	4.0 (3966)	44.5	45	210@1.200	1.625	0.0010-0.0030	0.0010-0.0030	0.3110-0.3120	0.3110-0.3120
	Y	5.2 (5211)	44.25-44.75	43.25-43.75	200@1.212	1.640	0.0010-0.0030	0.0010-0.0030	0.3110-0.3120	0.3110-0.3120

1 Intake: 30 degrees
Exhaust: 44.5 degrees
2 Intake: 29 degrees
Exhaust: 44 degrees

79114C09

PISTON AND RING SPECIFICATIONS

All measurements are given in inches.

Year	Engine ID/VIN	Engine Displacement Liters (cc)	Piston Clearance	Ring Gap			Ring Side Clearance		
				Top Compression	Bottom Compression	Oil Control	Top Compression	Bottom Compression	Oil Control
1991	P	2.5 (2458)	0.0013-0.0021	0.0100-0.0200	0.0100-0.0200	0.0150-0.0550	0.0010-0.0032	0.0010-0.0032	0.0010-0.0021
	S	4.0 (3966)	0.0013-0.0017	0.0100-0.0200	0.0100-0.0200	0.0100-0.0250	0.0017-0.0032	0.0017-0.0032	0.0010-0.0080
	7	5.9 (5899)	0.0012-0.0020	0.0100-0.0200	0.0100-0.0200	0.0150-0.0450	0.0015-0.0035	0.0015-0.0035	0.0010-0.007
1992	P	2.5 (2458)	0.0013-0.0021	0.0100-0.0200	0.0100-0.0200	0.0150-0.0550	0.0010-0.0032	0.0010-0.0032	0.0010-0.0021
	S	4.0 (3966)	0.0009-0.0017	0.0100-0.0200	0.0100-0.0200	0.0100-0.0250	0.0017-0.0032	0.0017-0.0032	0.0010-0.0080
1993	P	2.5 (2458)	0.0013-0.0021	0.0100-0.0200	0.0100-0.0200	0.0150-0.0550	0.0010-0.0032	0.0010-0.0032	0.0010-0.0021
	S	4.0 (3966)	0.0013-0.0017	0.0100-0.0200	0.0100-0.0200	0.0100-0.0250	0.0017-0.0032	0.0017-0.0032	0.0010-0.0080
	Y	5.2 (5211)	0.0005-0.0015	0.0100-0.0200	0.0100-0.0200	0.0100-0.0500	0.0015-0.0030	0.0015-0.0030	0.0020-0.0080
1994-95	P	2.5 (2458)	0.0013-0.0021	0.0100-0.0200	0.0100-0.0200	0.0150-0.0550	0.0010-0.0032	0.0010-0.0032	0.0010-0.0021
	S	4.0 (3966)	0.0013-0.0017	0.0100-0.0200	0.0100-0.0200	0.0100-0.0250	0.0017-0.0032	0.0017-0.0032	0.0010-0.0080
	Y	5.2 (5211)	0.0005-0.0015	0.0100-0.0200	0.0100-0.0200	0.0100-0.0500	0.0015-0.0030	0.0015-0.0030	0.0020-0.0080

79114C10

TORQUE SPECIFICATIONS
All readings in ft. lbs.

Year	Engine ID/VIN	Engine Displacement Liters (cc)	Cylinder Head Bolts	Main Bearing Bolts	Rod Bearing Bolts	Crankshaft Damper Bolts	Flywheel Bolts	Manifold Intake	Manifold Exhaust	Spark Plugs	Lug Nut
1991	P	2.5 (2458)	8	80	33	80	50	4	30	27	80-110
	S	4.0 (3966)	2	80	33	80	100-110	5	5	27	80-110
	7	5.9 (5899)	110	100	33	90	95	43	9	27	80-110
1992	P	2.5 (2458)	8	80	33	80	50	4	30	27	80-110
	S	4.0 (3966)	2	80	33	80	105	10	10	27	80-110
1993	P	2.5 (2458)	1	80	33	80	50	4	30	27	80-110
	S	4.0 (3966)	2	80	33	80	105	10	10	27	80-110
	Y	5.2 (5211)	11	85	45	135	105	12	20	30	80-110
1994-95	P	2.5 (2458)	8	80	33	80	50	4	30	27	80-110
	S	4.0 (3966)	2	80	33	80	105	10	10	27	80-110
	Y	5.2 (5211)	11	85	45	135	105	12	20	30	80-110

1 Step 1: Torque all bolts to 22 ft. lbs.
Step 2: Torque all bolts to 45 ft. lbs.
Step 3: Torque bolts 1-6 to 110 ft. lbs.
Step 4: Torque bolt 7 to 100 ft. lbs.
Step 5: Torque bolts 8-10 to 100 ft. lbs.
2 Step 1: Torque all bolts to 22 ft. lbs.
Step 2: Torque all bolts to 45 ft. lbs.
Step 3: Torque bolts 1-10 and 12-14 to 110 ft. lbs.
Step 4: Torque bolt 11 to 100 ft. lbs.
3 Plus 60 degrees
4 Bolts 1, 6 & 7: 30 ft. lbs.
Bolts 2-5: 23 ft. lbs.
5 Bolts 1-5 and 8-11: 24 ft. lbs.
Bolts 6, 7: 17 ft. lbs.
6 Middle bolts: 30 ft. lbs.
Outer bolts: 23 ft lbs.

7 Outer bolts: 15 ft. lbs.
8 Step 1: Torque all bolts to 22 ft. lbs.
Step 2: Torque all bolts to 45 ft. lbs.
Step 3: Torque bolts 1-6 to 110 ft. lbs.
Step 4: Torque bolt 7 to 100 ft. lbs.
Step 5: Torque bolts 8-10 to 110 ft. lbs.
9 Outer bolts: 15 ft. lbs.
Inner bolts: 25 ft. lbs.
10 Bolt 1: 30 ft. lbs.
Bolts 2-7: 23 ft. lbs.
11 Step 1: Torque all bolts to 50 ft. llbs.
Step 2: Torque all bolts to 105 ft. lbs.
12 Step 1: Torque bolts 1-4 to 72 in. lbs.
Step 2: Torque bolts 5-12 to 72 in. lbs.
Step 3: Torque all bolts to 12 ft. lbs.

79114C11

BRAKE SPECIFICATIONS

All measurements in inches unless noted

Year	Model	Master Cylinder Bore	Brake Disc Original Thickness	Brake Disc Minimum Thickness	Maximum Runout	Brake Drum Diameter Original Inside Diameter	Brake Drum Diameter Max. Wear Limit	Brake Drum Diameter Maximum Machine Diameter	Minimum Lining Thickness Front	Minimum Lining Thickness Rear
1991	Comanche	0.937	NA	0.890 [1]	0.005	9.0 [4]	[5]	NA	0.031	0.031
	Wrangler	0.937	NA	0.890	0.005	9.0	9.05	NA	0.031	0.031
	Cherokee	0.937	NA	0.890 [1]	0.005	9.0	9.05	NA	0.031	0.031
	Grand Wagoneer	1.125	NA	1.215	0.005	11.0	11.06	NA	0.031	0.031
1992	Comanche	0.937	NA	[1]	0.003	NA	[7]	NA	NA	NA
	Wrangler	0.937	NA	[1]	0.003	NA	[7]	NA	NA	NA
	Cherokee	0.937	NA	[1]	0.003	NA	[7]	NA	NA	NA
1993	Wrangler	NA	NA	0.890	0.005	NA	[7]	NA	NA	NA
	Cherokee	NA	NA	0.890 [1]	0.005	NA	[7]	NA	NA	NA
	Grand Cherokee	NA	NA	0.890 [8]	0.005	-	-	-	NA	NA
	Grand Wagoneer	NA	NA	0.890 [8]	0.005	-	-	-	NA	NA
1994-95	Wrangler	NA	NA	0.890	0.005	NA	[7]	NA	NA	NA
	Cherokee	NA	NA	0.890	0.005	NA	[7]	NA	NA	NA
	Grand Cherokee	NA	NA	0.890 [8]	0.005	-	-	-	NA	NA

NA - Not Available

1 2WD - 0.866 inch
2 1 ton - 10.5 inches
3 1 ton - 10.56 inches
4 1 ton - 10.0 inches
5 1 ton - 10.06 inches
6 Minumum useable thickness is either cast or stamped on rotor hub face
7 Maximum diameter is listed on outside of drum
8 Rear rotors have minimum allowable thickness listed on edge of parking brake drum

79114C12

WHEEL ALIGNMENT

Year	Model	Caster Range (deg.)	Caster Preferred Setting (deg.)	Camber Range (deg.)	Camber Preferred Setting (deg.)	Toe-in (in.)	Steering Axis Inclination (deg.)
1991	Comanche	5P-9P	6P	3/4N-1/2P	0	0	NA
	Wrangler	3	4	1/2N-1/2P	0	0	NA
	Cherokee	5P-9P	6P	3/4N-1/2P	0	0	NA
	Grand Wagoneer	4P-5P	4P	0-1/2P	0	0-3/16P	8 1/2
1992	Comanche	5P-9P	6P	3/4N-1/2P	0	0	NA
	Wrangler	3	4	1/2N-1/2P	0	0	NA
	Cherokee	5P-9P	6P	3/4N-1/2P	0	0	NA
1993	Wrangler	3	4	1/2N-1/2P	0	0	NA
	Cherokee	5P-9P	6P	3/4N-1/2P	0	0	NA
	Grand Cherokee	6 1/2P-7 1/2P	7P	3/4N-1/2P	1/4N	0-1/4P	NA
	Grand Wagoneer	6 1/2P-7 1/2P	7P	3/4N-1/2P	1/4N	0-1/4P	NA
1994-95	Wrangler	3	4	1/2N-1/2P	0	0	NA
	Cherokee	5P-9P	6P	3/4N-1/2P	0	0	NA
	Grand Cherokee	6 1/2P-7 1/2P	7P	3/4N-1/2P	1/4N	0-1/4P	NA

N - Negative

P - Positive

1 With manual transmission: 7 1/2P-8 1/2P
With automatic transmission: 6P-7P

2 With manual transmission: 8P
With automatic transmission: 6 1/2P

3 With manual transmission: 6 1/2P-9P
With automatic transmission: 5 1/4P-7 1/4P

4 With manual transmission: 6 1/2P
With automatic transmission: 8P

79114C13

FIRING ORDERS

NOTE: To avoid confusion, always replace spark plug wires one at a time.

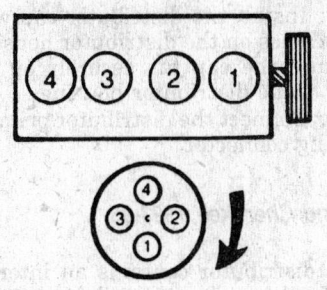

79114001

4 Cylinder Engine
Engine Firing Order: 1-3-4-2
Distributor Rotation: Clockwise

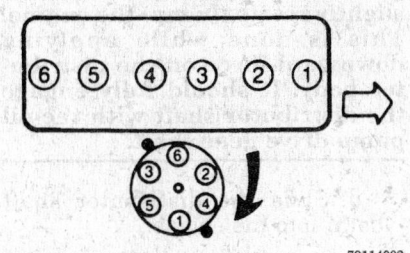

79114002

6 Cylinder Engines
Engine Firing Order: 1-5-3-6-2-4
Distributor Rotation: Clockwise

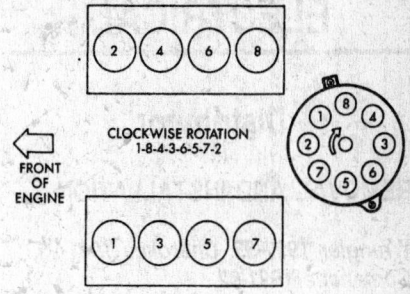

CLOCKWISE ROTATION
1-8-4-3-6-5-7-2

FRONT
OF
ENGINE

47293

V8 Engine
Engine Firing Order: 1-8-4-3-6-5-7-2
Distributor Rotation: Clockwise

ENGINE ELECTRICAL

Distributor

REMOVAL AND INSTALLATION

Wrangler 1991-95, Cherokee 1991-95, Comanche 1991-92

1. Disconnect the negative battery cable.
2. Unfasten the distributor cap retaining screws. Remove the distributor cap with the coil and spark plug wires attached and position them aside.
3. Disconnect the distributor primary wiring connector.
4. Scribe a mark on the distributor housing in line with the tip of the rotor.
5. Note the position of the rotor and distributor housing in relation to the surrounding engine components as reference points for installing the distributor.
6. Remove the distributor hold-down bolt and clamp.
7. Lift the distributor straight up and out of the engine.
   To install:
8. Clean the distributor mounting area of the cylinder block.
9. Install a new distributor mounting gasket.

NOTE: There is a fork on the distributor housing where the housing seats against the engine block. The slot in the fork aligns with the distributor hold-down bolt hole in the engine block. The distributor is correctly installed when the rotor is correctly positioned. This is the slot in the fork aligned with the hold-down bolt hole in the cylinder block. Because of the fork in the distributor housing initial ignition timing is not adjustable (the distributor cannot be rotated).

10. If the engine was not rotated while the distributor was removed, perform the following:
    a. Position the distributor shaft in the cylinder block.
    b. Align the rotor tip with the scribe mark on the distributor housing during removal.
    c. Turn the rotor approximately 1/8 turn counterclockwise past the scribe mark.

d. Slide the distributor shaft down into the engine.

NOTE: It may be necessary to move the rotor and shaft (slightly) to engage the distributor shaft with the slot in the oil pump shaft. the same may have to be done to engage the distributor gear with the camshaft gear. However, the rotor should align with the scribe mark when the distributor shaft is down in place.

11. If the engine was rotated while the distributor was removed, perform the following:
    a. Remove the No. 1 spark plug.
    b. Hold a finger over the spark plug hole and rotate the engine until compression pressure is felt. Slowly continue to rotate the engine until the timing index on the vibration damper pulley aligns the Top Dead Center (TDC) mark (0 degree) on the timing degree scale.

NOTE: Always rotate the engine in the direction of normal rotation. Do not turn the engine backward to align the timing marks.

c. Using a flat blade screwdriver, rotate the oil pump gear to position the slot in the oil pump shaft slightly before the 11 O'clock position.

NOTE: With the distributor cap removed, install the distributor with the rotor located just past the 2 o'clock position.

d. With the distributor fully engaged in its correct position, the rotor should be just past the 3 o'clock position.
e. Install the spark plug and cable.

— **WARNING** —
If the distributor cap is incorrectly positioned on the distributor housing, the cap or rotor may be damaged when the engine is started.

— **WARNING** —
Ensure the distributor shaft fully engages into the oil pump drive gear shaft. It may be necessary to slightly rotate (bump) the engine. This is done while applying downward force on the distributor body. It should fully engage the distributor shaft with the oil pump drive gear shaft.

12. Install the distributor clamp and hold-down bolts. Torque the bolt to 17 ft. lbs. (23 Nm).
13. Install the distributor cap with the cables on the distributor housing. Ensure the cap fits securely on the rim of the distributor housing.
14. Connect the distributor primary wiring connector.

Grand Cherokee 1993-95

The distributor contains an internal oil seal that prevents oil from entering the distributor housing. The seal is not serviceable.

1. Disconnect the negative battery cable.
2. If equipped with A/C, removing the cooling fan and shroud to gain access to the vibration damper bolt.
3. Label and remove the high tension wires from the distributor cap.
4. Remove the primary lead from the terminal post at the side of the distributor.
5. Turn the engine clockwise, using a socket on the end of the crankshaft damper bolt, until the rotor is pointing to the No. 1 spark plug wire post and the timing mark on the damper aligns with the 0 on the timing scale; No. 1 cylinder is at TDC on the compression stroke.

NOTE: The timing mark is on the edge of the vibration damper, closest to the front engine cover.

6. Remove the distributor cap. Note the position of the rotor and distributor. Scribe a mark on the base of the distributor and the engine as an installation reference.
7. Remove the bolt for the distributor hold-down clamp.
8. Remove the distributor from the engine.
   To install:
9. Using a flat bladed tool, turn the oil pump gear shaft, located in the distributor mounting hole, until the slot is slightly past the 11 o'clock position.
10. Install the rotor.
11. Without engaging the distributor gear into the cam gear, position the distributor into the hole in the engine block.
12. Visually line up the hold-down ear of the distributor housing with the hold-down clamp hole.
13. Turn the rotor to the 4 o'clock position.
14. Slide the distributor into the block until it seats keeping the hold-

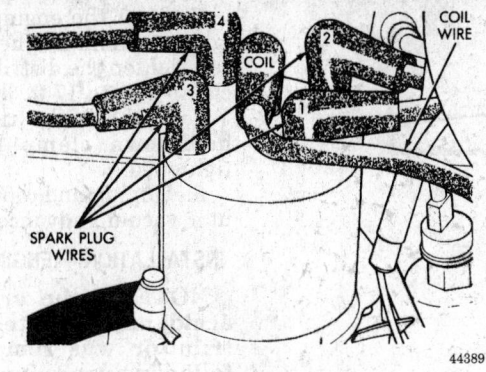

Spark plug cable positions

44389

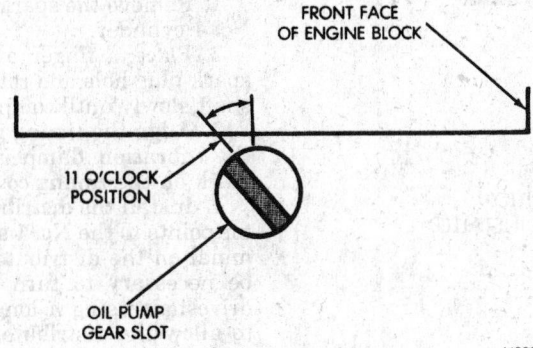

Oil pump gear slot positioning

44390

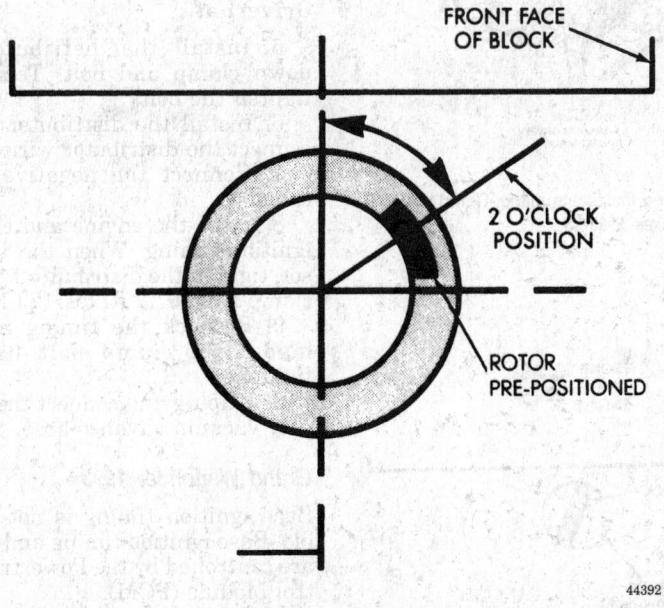

Distributor installation

44392

down ear aligned with the hole in the block.

NOTE: The rotor should be in the 5 o'clock position with the trailing edge of the rotor blade lined up with the No. 1 spark plug post position.

15. Install the hold-down clamp and torque the bolt to 17 ft. lbs. (23 Nm).
16. Install the distributor cap.
17. Connect the distributor electrical connector.
18. If removed, install the cooling fan and shroud.
19. Connect the battery cable.

Grand Wagoneer 1991

REMOVAL

1. Disconnect the negative battery cable.
2. Disconnect the distributor wiring connector.
3. Remove the distributor cap and move it aside, leaving the wires attached.
4. Disconnect and plug the distributor vacuum advance hose.
5. Mark the position of the rotor in relation to the distributor housing and the position of the distributor housing in relation to the cylinder block.
6. Remove the distributor hold-down bolt and the clamp.
7. Remove the distributor.

NOTE: Do not rotate the crankshaft while the distributor is removed from the engine.

INSTALLATION — ENGINE NOT DISTURBED

1. Clean the distributor mounting area on the cylinder block and install a new distributor mounting gasket.
2. Lower the distributor into the engine, aligning the rotor and distributor housing with the marks made during the removal procedure.

NOTE: Make sure the distributor shaft fully engages the oil pump driveshaft. It may be necessary to slightly rotate the engine while applying downward hand force on the distributor body to fully engage the distributor shaft with the oil pump driveshaft.

3. Install the distributor hold-down clamp and bolt. Temporarily tighten the bolt.
4. Install the distributor cap and connect the distributor wiring.
5. Connect the negative battery cable.

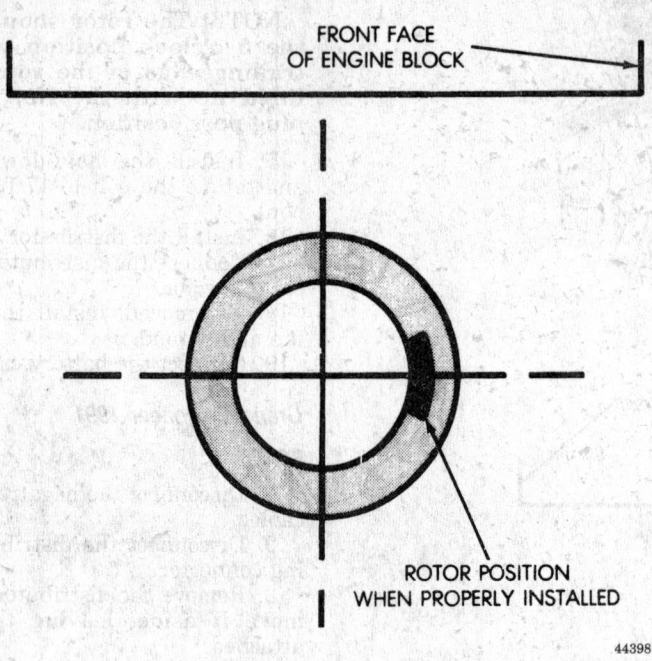

Distributor rotor position

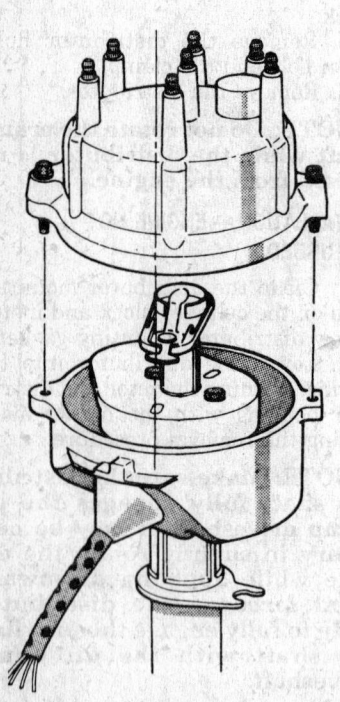

Distributor cap removal-Grand
Cherokee 1993-95

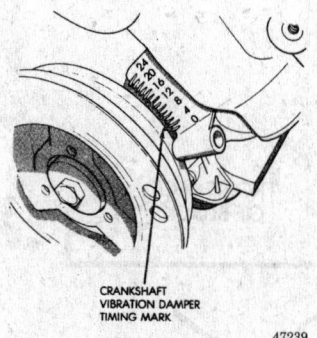

Aligning the timing marks-Grand
Cherokee 1993-95

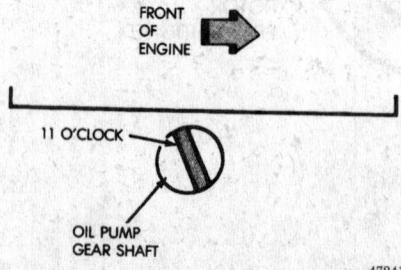

Oil pump gear shaft alignment-Grand Cherokee
1993-95

6. Start the engine and check the ignition timing. When the timing is set, tighten the distributor hold-down clamp bolt to 17 ft. lbs. (23 Nm).

7. Recheck the timing after the hold-down clamp bolt has been tightened.

8. Unplug and connect the distributor vacuum advance hose.

INSTALLATION — ENGINE DISTURBED

NOTE: If the crankshaft was accidentally rotated after the distributor was removed, use the following procedure to install the distributor.

1. Remove the spark plug from the No. 1 cylinder.

2. Place a finger over the No. 1 spark plug hole and rotate the crankshaft slowly until compression is felt.

3. Align the timing index mark on the vibration damper with the 0° mark on the timing cover scale.

4. Install the distributor so the rotor points to the No. 1 spark plug terminal on the distributor cap. It may be necessary to turn the oil pump driveshaft using a long screwdriver, to allow the distributor to drop into position.

NOTE: Make sure the distributor shaft fully engages the oil pump driveshaft. It may be necessary to slightly rotate the engine while applying downward hand force on the distributor body to fully engage the distributor shaft with the oil pump driveshaft.

5. Install the distributor hold-down clamp and bolt. Temporarily tighten the bolt.

6. Install the distributor cap and connect the distributor wiring.

7. Connect the negative battery cable.

8. Start the engine and check the ignition timing. When the timing is set, tighten the distributor hold-down clamp bolt to 17 ft. lbs. (23 Nm).

9. Recheck the timing after the hold-down clamp bolt has been tightened.

10. Unplug and connect the distributor vacuum advance hose.

Grand Wagoneer 1993

Base ignition timing is not adjustable. Base ignition timing and advance are controlled by the Powertrain Control Module (PCM).

1. Disconnect the negative battery cable.

2. Loosen the 2 screws and remove the distributor cap and position it aside.

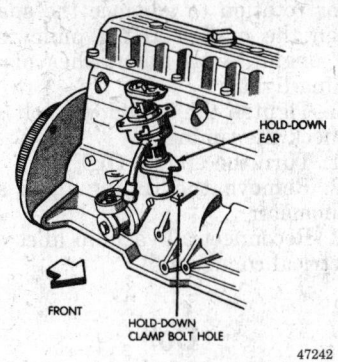

Distributor installation-Grand Cherokee 1993-95

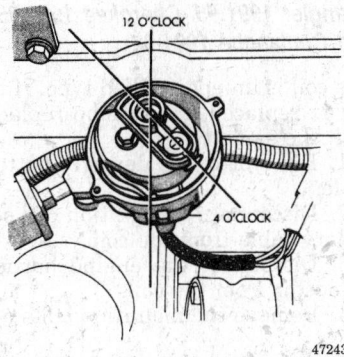

Rotor alignment-Grand Cherokee 1993-95

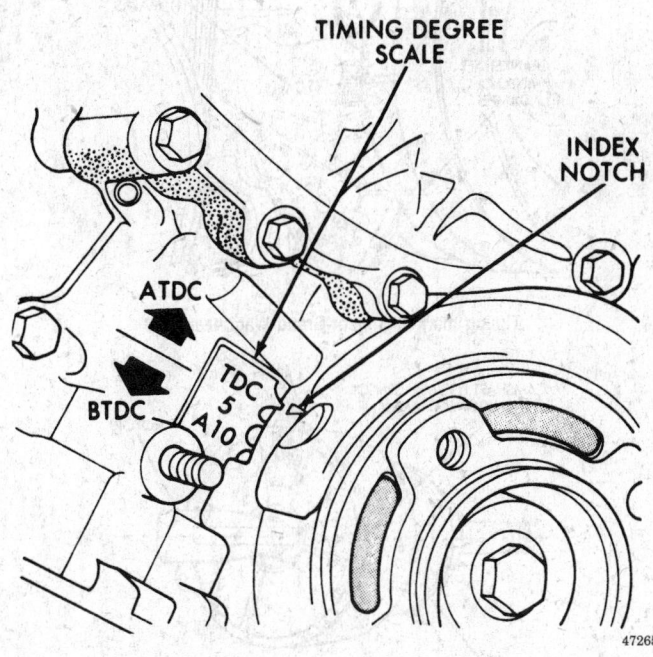

Timing mark location-Grand Wagoneer 1991

3. Matchmark the distributor housing in relation to the engine.

4. Attach a socket to the vibration damper and rotate the crankshaft clockwise until the indicating mark on the vibration damper aligns with the 0 degree mark (TDC) on the timing chain cover.

5. The distributor rotor should now be aligned to the **CYL. NO. 1** alignment mark stamped into the camshaft position sensor. If not, rotate the crankshaft through another complete 360 degree turn. Note the position of the number 1 cylinder spark plug cable (on the cap) in relation to the rotor. The rotor should now be aligned to this position.

6. Disconnect the camshaft position sensor wiring harness from the main engine wiring harness.

7. Remove the rotor.

8. Remove the distributor hold-down bolt and clamp.

9. Remove the distributor.

── WARNING ──

Do not crank the engine with the distributor removed.

To install:

10. If the engine has been cranked while the distributor is removed, establish the relationship between the distributor shaft and No. 1 piston position as follows:

    a. Remove the No. 1 spark plug.

    b. Rotate the crankshaft in a clockwise direction, as viewed from the front, until the No. 1 cylinder piston is at the top of its compression stroke. Continue to slowly rotate the engine until the indicating mark is aligned to the 0 degree (TDC) mark on the timing chain cover.

    c. Install the spark plug and connect the cable.

11. Clean the top of the cylinder block.

12. Lubricate the oil seal on the distributor with engine oil.

13. Install the rotor.

14. Position the distributor into the block and engage the tongue of the distributor shaft with the slot in the distributor oil pump drive gear. Position the rotor to the No. 1 spark plug terminal position.

15. Install the hold-down clamp and loosely install the bolt.

16. Rotate the distributor housing until the rotor is aligned to the **CYL. NO. 1** alignment mark on the camshaft position sensor.

17. Torque the hold-down clamp bolt to 200 inch lbs. (22 Nm).

18. Connect the camshaft position sensor wiring harness to the main engine harness.

19. Install the distributor cap and tighten the screws.

20. Ensure the spark plug wires are firmly connected to their terminals.

21. Connect the negative battery cable. Start the engine and check for proper operation.

REMOVAL AND INSTALLATION

Wrangler 1991-95, Cherokee 1991-95, Comanche 1991-92, Grand Cherokee 1993-95 and Grand Wagoneer 1993

Base ignition timing is not adjustable. The distributor does not have a built-in centrifugal or vacuum assisted advance. Base ignition timing and timing advance are controlled by the Engine Control Module (ECM) which monitors inputs from various sensors to determine and adjust correct ignition timing.

Grand Wagoneer 1991

1. Locate the timing marks on the crankshaft pulley and the front of the timing case cover.

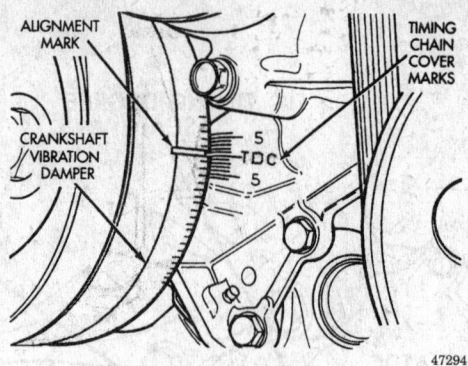

Timing mark alignment-Grand Wagoneer 1993

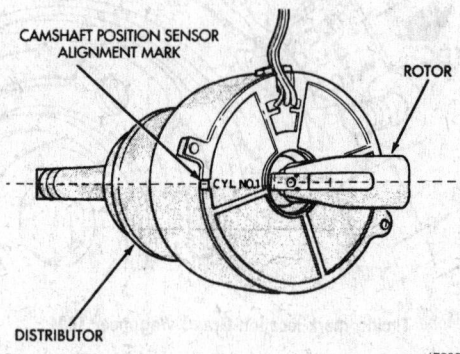

Rotor alignment mark-Grand Wagoneer 1993

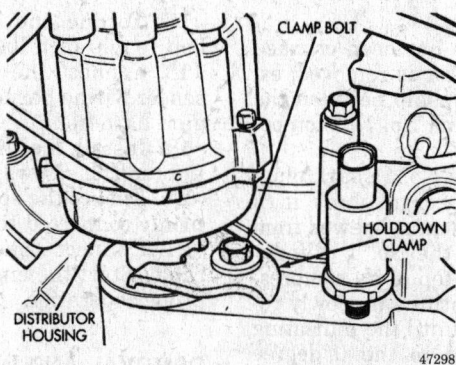

Distributor hold-down and clamp-Grand Wagoneer 1993

rotor rotation to advance the spark. Align the marks on the pulley and the engine with the flashes of the timing light.

10. Tighten the hold-down bolt and recheck the timing.

11. Turn the engine **OFF**.

12. Remove the timing light and tachometer.

13. Reconnect all vacuum lines and electrical connectors.

Coil

REMOVAL AND INSTALLATION

Wrangler 1991-95, Cherokee 1991-95 and Comanche 1991-92

The coil is an epoxy filled type. If the coil is replaced, it must be replaced with the same type.

1. Disconnect the negative battery cable.

2. Disconnect the ignition coil secondary cable from the coil.

3. Disconnect the engine harness connector from the coil.

4. Remove the mounting bolts and coil.

To install:

5. Install the ignition coil to the bracket on the cylinder block and tighten the mounting bolts.

6. Connect the engine harness and high tension lead to the coil.

7. Connect the negative battery cable.

Grand Cherokee 1993-95

The coil is an epoxy filled type. If the coil is replaced, it must be replaced with the same type.

1. Disconnect the negative battery cable.

2. Disconnect the ignition coil secondary cable from the coil.

3. Disconnect the engine harness connector from the coil.

4. Remove the mounting bolts and coil.

To install:

5. Install the ignition coil to the bracket on the cylinder block and tighten the mounting bolts.

6. Connect the engine harness and high tension lead to the coil.

7. Connect the negative battery cable.

Grand Wagoneer 1991

1. Disconnect the negative battery cable.

2. Disconnect the high tension lead from the coil.

2. Clean the marks so they are clearly visible.

3. Use chalk or white paint to color the mark on the scale that will indicate the correct timing, when aligned with the mark on the pulley or the pointer.

4. Attach a tachometer to the negative side of the coil and ground. Attach a timing light to the battery and connect the inductive probe to the No. 1 plug wire. Place the probe as close to the spark plug as possible.

5. Start the engine and allow it to reach operating temperature. Place

the transmission in **P** and turn OFF all accessories.

6. Disconnect and plug the vacuum lines to the distributor.

7. Loosen the distributor lock bolt just enough so the distributor can be turned with a little resistance.

8. Start the engine and check the idle; if necessary adjust it to the correct specification.

9. With the timing light aimed at the pulley and the marks on the engine, turn the distributor in the direction of rotor rotation to retard the spark, and in the opposite direction of

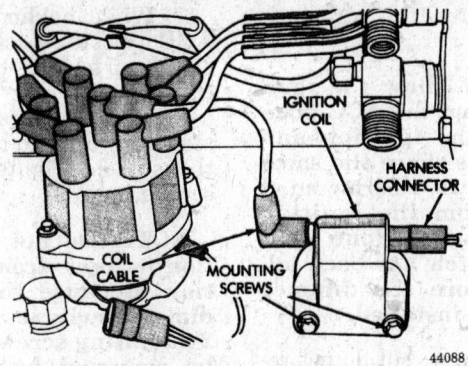

Ignition coil-6 cylinder shown (4 cylinder similar)-
Grand Cherokee 1993-95

3. Remove the coil electrical connector harness by pulling the assembly straight out.

4. Remove the hold-down bolt and coil.

To install:

5. Install the coil and tighten the hold-down bolt.

6. Connect the electrical connector and high tension lead to the coil.

7. Connect the negative battery cable.

Grand Wagoneer 1993

The ignition coil is mounted to a bracket near the front of the right cylinder head.

1. Disconnect the negative battery cable.

2. Disconnect the high tension lead from the coil.

3. Disconnect the electrical connectors from the coil.

—————— **CAUTION** ——————

Do not remove the coil mounting bracket-to-cylinder head mounting bolts. The coil mounting bracket is under accessory drive belt tension. If this bracket is removed all accessory belt tensoin will be relieved.

4. Remove the 2 bolts securing the coil to the bracket and remove the coil.

To install:

5. Install the coil and torque the mounting bolts to 100 inch lbs. (11Nm).

6. Connect the high tension lead and electrical connectors tp the coil.

7. Connect the negative battery cable.

Ignition Switch

REMOVAL AND INSTALLATION

Wrangler 1991, Cherokee 1991, Comanche 1991 and Grand Wagoneer 1991

1. Disconnect the negative battery cable.

2. Place the ignition lock in the **LOCK** position.

3. Remove the 2 switch mounting screws.

4. Disconnect the switch from the remote rod.

5. Disconnect the wiring and remove the switch.

To install:

6. With the actuator rod disconnected, position the ignition switch.

7. Move the slider to the extreme left **ACC** position.

8. Ignition switch installation for non-tilt columns:

   a. Position the actuator rod in the slider hole and install the switch to the steering column. Be careful not to move the slider out of detent.

   b. Secure the key in the **ACC** position and push the switch down the column slightly to remove the slack in the actuator rod.

9. Ignition switch installation for tilt columns:

   a. Position the actuator rod into the slider hole.

   b. Install the ignition switch to the column but do not tighten the attaching screws.

   c. Push the switch down the steering column to remove the lash in the actuator rod while holding the key in the **ACC** position. Be careful not to move the slider out of detent.

10. Tighten the attaching screws.

11. Connect the white connector and then the black connector to the switch.

12. Connect the negative battery cable.

Wrangler 1992-95

1. Disconnect the negative battery cable.

2. If equipped, remove the windshield wiper intermittent control module and its bracket.

3. Turn the ignition to the **ACC** position.

4. Remove the headlamp dimmer switch attaching nuts.

5. Lift the switch from the steering column while disengaging the actuator rod.

6. Tape the ignition and dimmer switch actuator rods to the steering column to prevent disengagement from the upper position of the steering column.

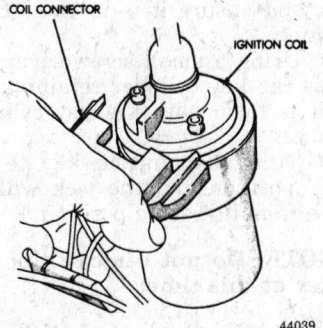

Removing the ignition coil connector-
Grand Wagoneer 1991

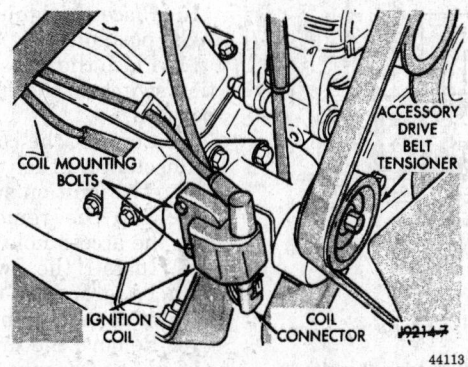

Ignition coil-GRand Wagoneer 1993

7. Remove the ignition switch-to-steering column retaining screws.

8. Disengage the ignition switch from the remote actuator rod by lifting straight up and remove the switch from the column.

9. Disconnect the black connector and then the other connector from the switch.

To install:

10. Place the ignition switch in the **ACC** position.

11. Place the slider bar in the ignition switch to the to the **ACC** detent position.

12. Connect the colored (non-black) connector and then the black connector to the ignition switch.

13. Slip the remote actuator rod into the access hole on the switch.

14. Install the switch to the column, be careful not to move the slider bar out of the detent position.

15. Remove the tape from the rods.

16. Loosely install the ignition switch-to-steering column screws.

17. While holding the key in the **ACC** position, slide the ignition switch up towards the steering wheel (non-tilt steering wheel) or down away from the steering wheel (tilt steering wheel) to remove slack from

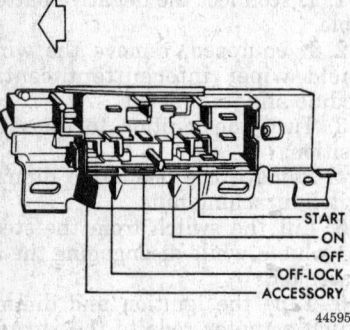

Ignition switch - non-tilt steering-Wrangler 1992-95

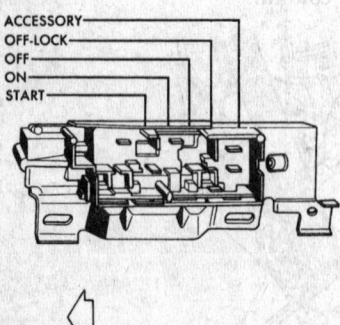

Ignition switch - tilt steering-Wrangler 1992-95

the switch. Tighten the attaching screws.

NOTE: Do not allow the ignition to move from the ACC position. Because the ignition and dimmer switches share the same 2 mounting screws, 1 screw must be removed from the ignition switch. This must be done after the ignition switch has been adjusted and before the dimmer switch has been installed.

18. Remove 1 screw, but do not remove the stud/nut.

19. Install and adjust the dimmer switch.

20. If equipped, install the intermittent wiper control module and bracket.

21. Connect the negative battery cable.

Cherokee 1992-95 and Comanche 1992

1. Disconnect the negative battery cable.

2. Remove the lower instrument panel.

3. Turn the ignition to the **ACC** position.

4. Remove the headlamp dimmer switch attaching nuts.

5. Lift the switch from the steering column while disengaging the actuator rod.

6. Tape the ignition and dimmer switch actuator rods to the steering column to prevent disengagement from the upper position of the steering column.

7. Remove the ignition switch-to-steering column retaining screws.

8. Disengage the ignition switch from the remote actuator rod by lifting straight up and remove the switch from the column.

9. Disconnect the black connector and then the other connector from the switch.

To install:

10. Place the ignition switch in the **ACC** position.

11. Place the slider bar in the ignition switch to the to the **ACC** detent position.

12. Connect the colored (non-black) connector and then the black connector to the ignition switch.

13. Slip the remote actuator rod into the access hole on the switch.

14. Install the switch to the column, be careful not to move the slider bar out of the detent position.

15. Remove the tape from the rods.

16. Loosely install the ignition switch-to-steering column screws.

17. While holding the key in the **ACC** position, slide the ignition switch up towards the steering wheel (non-tilt steering wheel) or down away from the steering wheel (tilt steering wheel) to remove slack from the switch. Tighten the attaching screws.

NOTE: Do not allow the ignition to move from the ACC position. Because the ignition and dimmer switches share the same 2 mounting screws, 1 screw must be removed from the ignition switch. This must be done after the ignition switch has been adjusted and before the dimmer switch has been installed.

18. Remove 1 screw, but do not remove the stud/nut.

19. Install and adjust the dimmer switch.

20. If equipped install the lower instrument panel.

21. Connect the negative battery cable.

Grand Cherokee 1993-95 and Grand Wagoneer 1993

1. Disconnect the negative battery cable.

2. If equipped, remove the tilt lever.

3. Remove the upper and lower steering column covers with a suitable Torx® driver.

4. Using Snap-On tamper-proof bit TTXR20BO or equivalent, remove the ignition switch screws.

5. Pull the ignition switch away from the column.

6. Release the 2 connector locks on the 7-terminal wiring connector and remove the connector from the ignition switch.

7. Release the connector lock on the key-in-switch and halo light 4-terminal connector and remove the connector from the ignition switch.

8. Insert the key into the ignition lock and ensure it is in the **LOCK** position.

9. Using a small screwdriver, depress the key cylinder retaining pin so it is flush with the key cylinder surface.

10. Turn the ignition key to the **OFF** position and the lock will release from its seated position.

NOTE: Do not remove the cylinder at this time.

11. Turn the key to the **LOCK** position and remove the key.

12. Remove the ignition lock.

To install:

13. Install the electrical connectors to the switch. Ensure the switch locking tabs are fully seated in the wiring connectors.

14. Mount the ignition switch to the column. The dowel pin on the ignition switch assembly must engage with the column park-lock slider linkage. Ensure the ignition switch is in the lock position (flag is parallel with the ignition switch terminals).

15. Apply a dab of grease to the flag and pin. Position the park-lock link and slider to mid-travel. Position the ignition lock against the lock housing face. Ensure the pin is inserted into the park-lock link contour slot and tighten the retaining screw.

16. With the ignition lock and switch in the **LOCK** position, insert the lock into the switch assembly until it bottoms.

17. Assemble the column covers.

18. If equipped, install the tilt wheel lever.

19. Connect the negative battery cable.

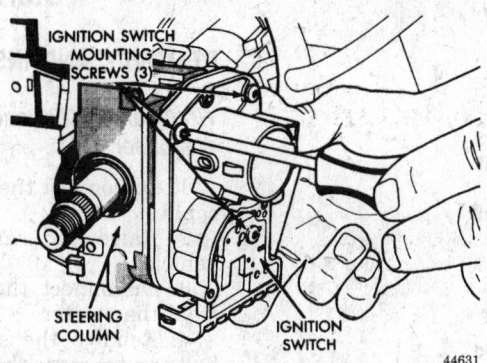

Ignition switch removal - Grand Cherokee 1993-95 and Grand Wagoneer 1993

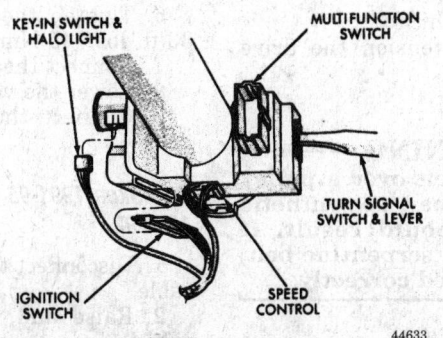

Key-in switch and halo lamp connector - Grand Cherokee 1993-95 and Grand Wagoneer 1993

Alternator

REMOVAL AND INSTALLATION

Wrangler 1991-95 and Grand Wagoneer 1991

1. Disconnect the negative battery cable.
2. Loosen and remove the drive belt.
3. Disconnect the electrical connectors from the alternator.
4. Remove the mounting bolts and alternator.

To install:

5. Attach the alternator to the mounting bracket and loosely install the bolts/nuts.
6. Connect the electrical connectors to the alternator.
7. Install and tension the drive belt.

------ **WARNING** ------

Never force a belt over a pulley rim using a prybar, as synthetic fiber damage could result. If equipped with a serpentine belt, ensure it is routed correctly.

8. Torque the alternator mounting bolts/nuts to 20 ft. lbs. (27 Nm).
9. Connect the negative battery cable.

Cherokee 1991-95 and Comanche 1991-92

1. Disconnect the negative battery cable.
2. Loosen and remove the drive belt.
3. Raise and safely support the vehicle.
4. Disconnect the electrical connectors from the alternator.
5. Remove the mounting bolts and alternator.

To install:

6. Attach the alternator to the mounting bracket and loosely install the bolts/nuts.
7. Connect the electrical connectors to the alternator.
8. Lower the vehicle.
9. Install and tension the drive belt.

------ **WARNING** ------

Never force a belt over a pulley rim using a prybar, as synthetic fiber damage could result. If equipped with a serpentine belt, ensure it is routed correctly.

10. Torque the alternator mounting bolts/nuts to 20 ft. lbs. (27 Nm).
11. Connect the negative battery cable.

1. Disconnect the negative battery cable.
2. Loosen and remove the drive belt.
3. Disconnect the electrical connectors from the alternator.
4. Remove the mounting bolts and alternator.

To install:

5. Attach the alternator to the mounting bracket and loosely install the bolts/nuts.
6. Connect the electrical connectors to the alternator.
7. Install and tension the drive belt.

------ **WARNING** ------

Never force a belt over a pulley rim using a prybar, as synthetic fiber damage could result. If equipped with a serpentine belt, ensure it is routed correctly.

8. Torque the alternator mounting bolts/nuts to 20 ft. lbs. (27 Nm).
9. Connect the negative battery cable.

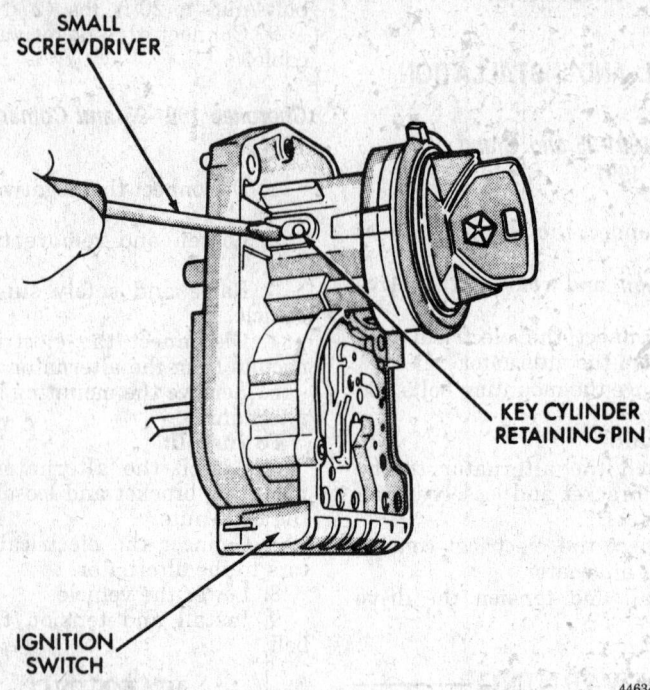

Key cylinder retaining ring - Grand Cherokee 1993-95 and Grand Wagoneer 1993

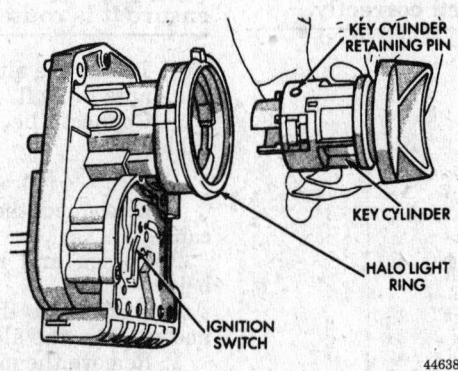

Key cylinder removal - Grand Cherokee 1993-95 and Grand Wagoneer 1993

Grand Cherokee 1993-95

1. Disconnect the negative battery cable.
2. Loosen and remove the drive belt.
3. Raise and safely support the vehicle.
4. Disconnect the electrical connectors from the alternator.
5. Remove the mounting bolts and alternator.
To install:
6. Attach the alternator to the mounting bracket and loosely install the bolts/nuts.

7. Connect the electrical connectors to the alternator.
8. Lower the vehicle.
9. Install and tension the drive belt.

— **WARNING** —
Never force a belt over a pulley rim using a prybar, as synthetic fiber damage could result. If equipped with a serpentine belt, ensure it is routed correctly.

10. Torque the alternator mounting bolts/nuts to 20 ft. lbs. (27 Nm).

11. Connect the negative battery cable.

Grand Wagoneer 1993

1. Disconnect the negative battery cable.
2. Loosen and remove the drive belt.
3. Disconnect the electrical connectors from the alternator.
4. Remove the mounting bolts and alternator.
To install:
5. Attach the alternator to the mounting bracket and loosely install the bolts/nuts.
6. Connect the electrical connectors to the alternator.
7. Install and tension the drive belt.

— **WARNING** —
Never force a belt over a pulley rim using a prybar, as synthetic fiber damage could result. If equipped with a serpentine belt, ensure it is routed correctly.

8. Torque the alternator mounting bolts/nuts to 30 ft. lbs. (41 Nm).
9. Connect the negative battery cable.

Starter

REMOVAL AND INSTALLATION

Wrangler 1991-95 and Grand Wagoneer 1991

1. Disconnect the negative battery cable.
2. Raise and safely support the vehicle.
3. Disconnect the electrical leads from the starter.
4. Remove the starter mounting bolts and remove the starter from the vehicle.
To install:
5. Install the starter in position.
6. Tighten the mounting bolts to 33 ft. lbs. (45 Nm).
7. Connect the electrical leads.
8. Lower the vehicle.
9. Connect the negative battery cable.

Cherokee 1991-95 and Comanche 1991-92

1. Disconnect the negative battery cable.
2. Raise and safely support the vehicle.
3. Disconnect the electrical leads from the starter.

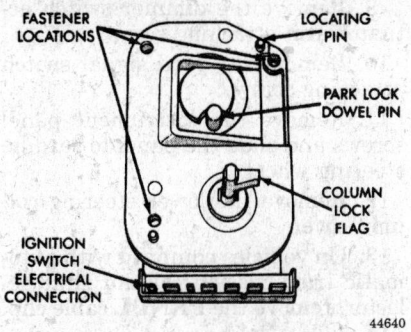

Ignition switch view from the column - Grand Cherokee 1993-95 and Grand Wagoneer 1993

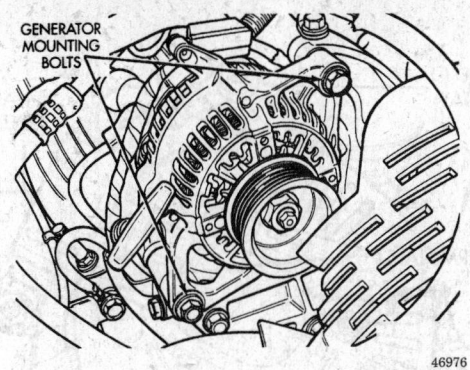

Alternator mounting bolts - Grand Wagoneer 1993

4. Remove the starter mounting bolts and remove the starter from the vehicle.

To install:

5. Install the starter in position.

6. Tighten the mounting bolts to 33 ft. lbs. (45 Nm).

7. Connect the electrical leads.

8. Lower the vehicle.

9. Connect the negative battery cable.

Grand Cherokee 1993-94

1. Disconnect the negative battery cable.

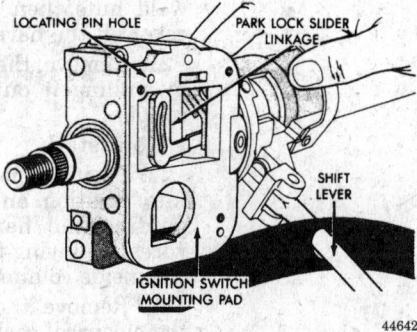

Ignition switch mounting pad - Grand Cherokee 1993-95 and Grand Wagoneer 1993

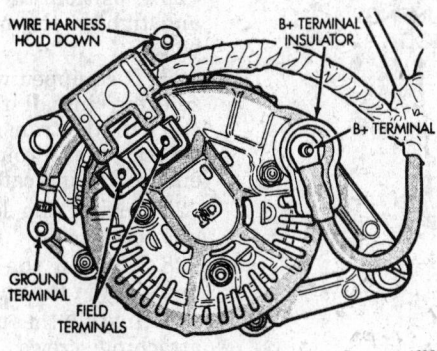

Rear view of the alternator- Cherokee 1991-95 and Comanche 1991-92

2. Raise and safely support the vehicle.

3. Disconnect the electrical leads from the starter.

4. Remove the starter mounting bolts and remove the starter from the vehicle.

To install:

5. Install the starter in position.

6. Tighten the mounting bolts to 33 ft. lbs. (45 Nm).

7. Connect the electrical leads.

8. Lower the vehicle.

9. Connect the negative battery cable.

Grand Wagoneer 1993

1. Disconnect the negative battery cable.

2. Raise and safely support the vehicle.

3. Disconnect the electrical leads from the starter.

4. Remove the starter mounting bolts and remove the starter from the vehicle.

To install:

5. Install the starter.

6. Tighten the mounting bolts to 50 ft. lbs. (68 Nm).

7. Connect the eletrical connectors.

8. Connect the electrical leads and lower the vehicle.

CHASSIS ELECTRICAL

Combination Switch-Wiper

REMOVAL AND INSTALLATION

Wrangler 1991-95

1. Disconnect the negative battery cable.

2. Remove the steering wheel.

3. Remove the lockplate cover by compressing with tool C–4156 or equivalent.

4. Release and discard the steering shaft retaining snapring.

5. Remove the compressor tool.

6. Remove the lockplate, canceling cam, upper bearing preload spring and thrust washer from the steering column.

7. Remove the horn button components from the canceling cam.

8. Remove the hazard warning switch knob.

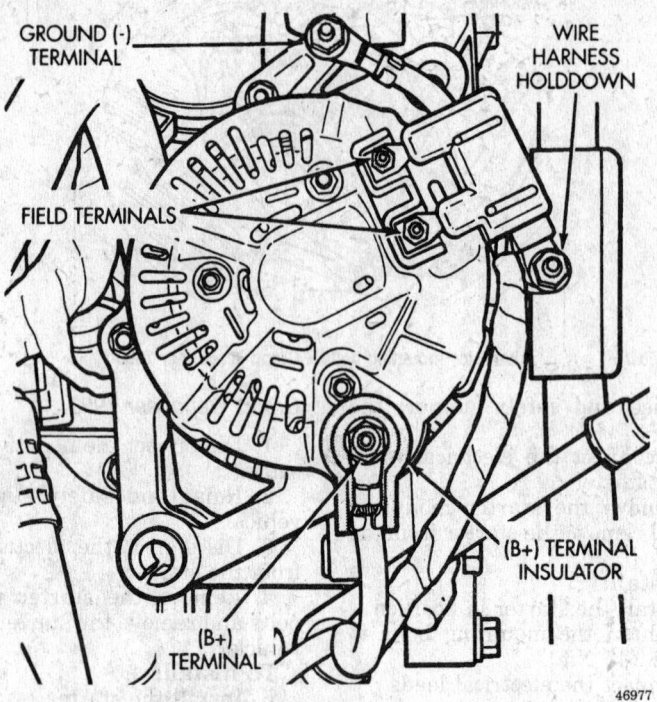

GROUND (-) TERMINAL

WIRE HARNESS HOLDDOWN

FIELD TERMINALS

(B+) TERMINAL INSULATOR

(B+) TERMINAL

46977

Alternator electrical connectors - Grand Wagoneer 1993

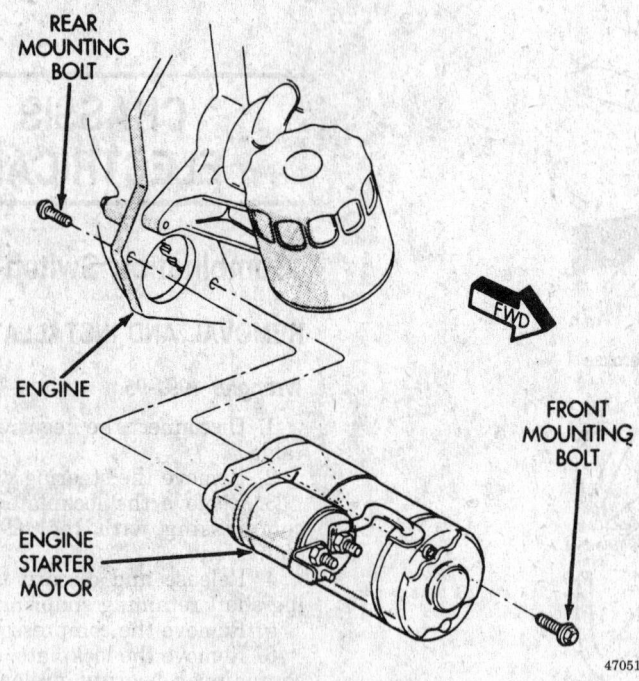

REAR MOUNTING BOLT

ENGINE

ENGINE STARTER MOTOR

FWD

FRONT MOUNTING BOLT

47051

Starter removal and installation

9. Remove the dimmer switch actuator arm attaching screw.

10. Remove the turn signal switch attaching screws.

11. Remove the instrument panel screws and slide the panel toward he steering wheel.

12. Remove the lower steering column cover.

13. On vehicles equipped with automatic transmission column shift selector, remove the **PRNDL** cable clip.

14. Remove the nuts securing steering column bracket to the brake sled.

15. Remove the bolts holding the steering column bracket to the column.

16. Loosen the column brace mounting nut at the drivers side kick panel and allow the column to drop.

17. Push the turn signal connector up and out of the steering column connector.

18. Pry up the locking tabs of the steering column connector and remove the connector from the column bracket.

19. Tape the connector flat against the wire harness to prevent it from hanging up during removal.

20. Remove the plastic harness cover by pulling it up and over the weld nuts then open and slide the cover off the harness.

21. Remove the combination lever by pulling it out straight from the column.

To install:

22. Position the combination switch into position on the column while guiding the harness. Ensure the wires are laying flat on the bottom of the inside column.

23. Remove the tape and connect the electrical connector.

24. Connect the turn signal connector.

25. Position the steering column and tighten the mounting nuts and bolts.

26. If equipped with a column shift selector, install the **PRNDL** cable clip with the transmission in **N**. Move the selector through all ranges and ensure the indicator is aligned.

27. Install the lower steering column cover.

28. Position the instrument panel and tighten the screws.

29. Install the turn signal switch attaching screws.

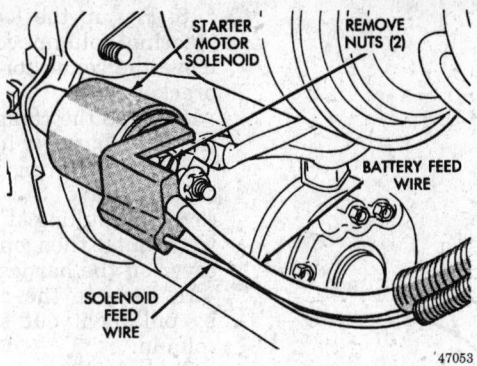

Solenoid harness removal

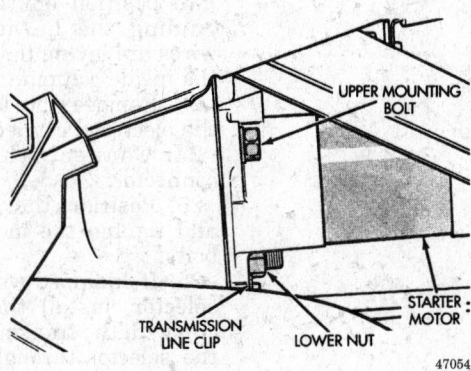

Starter mounting nut and bolt

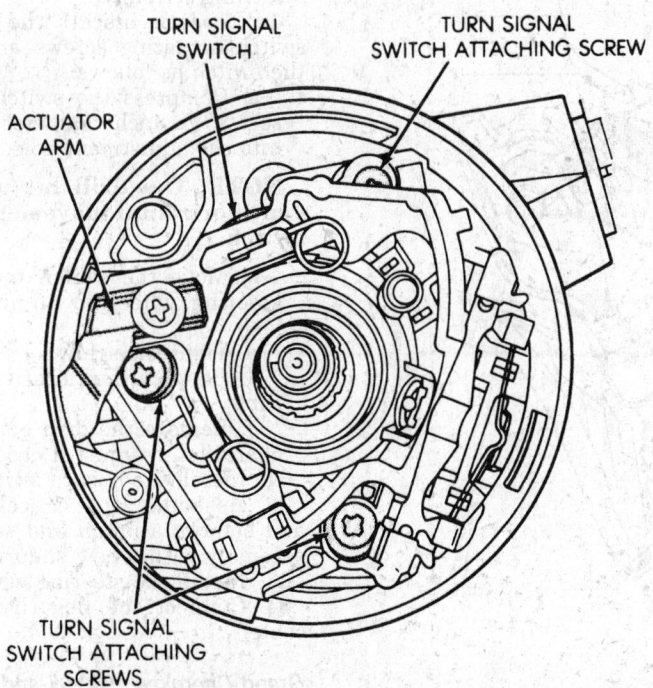

Combination switch actuating lever screws

30. Loosely install the dimmer switch actuating screws and adjust the switch as follows:

a. Compress the switch and insert a ³/₃₂ inch diameter drill bit into the adjustment hole.

NOTE: The drill bit will prevent horizontal movement of the switch.

b. Move the switch toward the steering wheel to eliminate rod lash.

c. Tighten the screw.

d. Connect the negative battery cable.

e. Remove the drill bit and test operation, readjust if necessary.

31. Install the hazard switch knob.

32. Assemble the canceling cam and steering column and secure the assembly with a new snapring.

33. Install the steering wheel.

34. Connect the negative battery cable.

Cherokee 1991-95, Comanche 1991-92 and Grand Wagoneer 1991

1. Disconnect the negative battery cable.

2. Remove the steering wheel.

3. Remove the lockplate cover by compressing with tool C-4156 or equivalent.

4. Release and discard the steering shaft retaining snapring.

5. Remove the compressor tool.

6. Remove the lockplate, canceling cam, upper bearing preload spring and thrust washer from the steering column.

7. Remove the horn button components from the canceling cam.

8. Remove the hazard warning switch knob.

9. Remove the dimmer switch actuator arm attaching screw.

10. Remove the turn signal switch attaching screws.

11. Remove the lower instrument panel cover trim panel.

12. Remove the lower steering column cover.

13. On vehicles equipped with automatic transmission column shift selector, remove the **PRNDL** cable clip.

14. Remove the nuts securing the steering column bracket to the brake sled.

15. Remove the bolts holding the steering column bracket to the column.

16. Loosen the column brace mounting nut at the drivers side kick panel and allow the column to drop.

17. Push the turn signal connector up and out of the steering column connector.

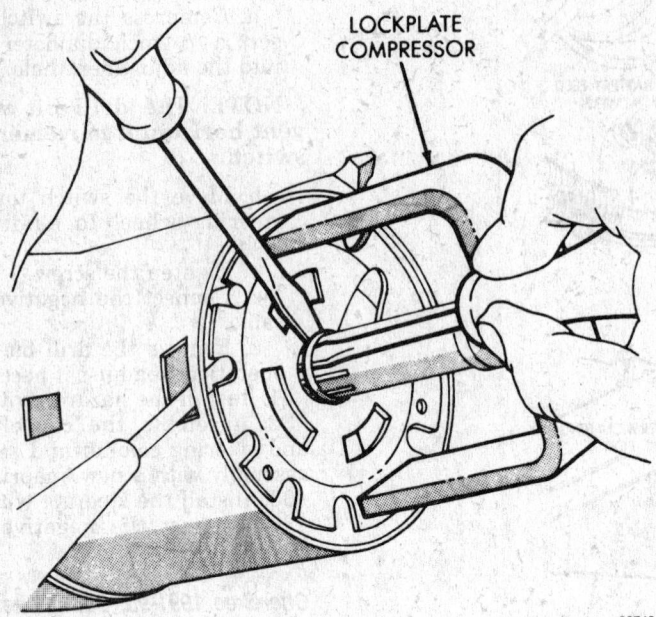

Lockplate removal

38742

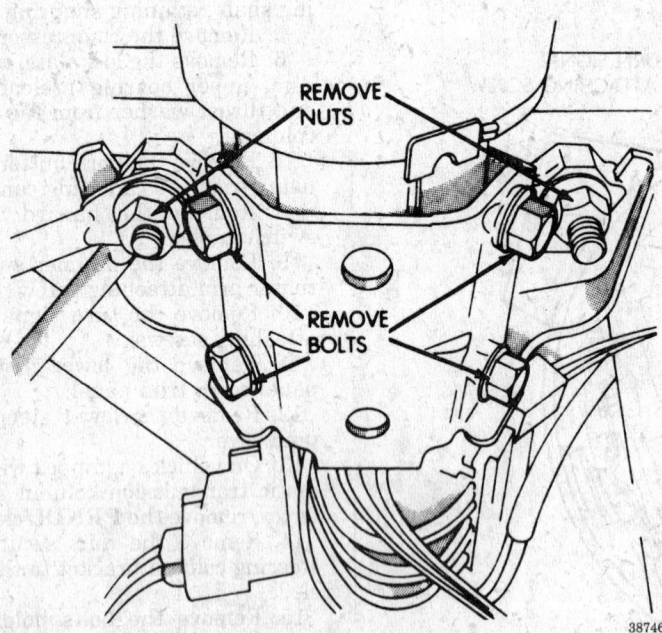

Lower steering column

38746

18. Pry up the locking tabs of the steering column connector and remove the connector from the column bracket.

19. Tape the connector flat against the wire harness to prevent it from hanging up during removal.

20. Remove the plastic harness cover by pulling it up and over the weld nuts, then open and slide the cover off the harness.

21. Remove the combination lever by pulling it out straight from the column.

To install:

22. Position the combination switch into position on the column while guiding the harness. Ensure the wires are laying flat on the bottom of the inside column.

23. Remove the tape and connect the electrical connector.

24. Connect the turn signal connector.

25. Position the steering column and tighten the mounting nuts and bolts.

26. If equipped with a column shift selector, install the **PRNDL** cable clip with the transmission in **N**. Move the selector through all ranges and ensure the indicator is aligned.

27. Install the lower steering column cover.

28. Install the lower instrument panel cover trim panel.

29. Install the turn signal switch attaching screws.

30. Loosely install the dimmer switch actuating screws and adjust the switch as follows:

    a. Compress the switch and insert a $3/32$ inch diameter drill bit into the adjustment hole.

NOTE: The drill bit will prevent horizontal movement of the switch.

    b. Move the switch toward the steering wheel to eliminate rod lash.

    c. Tight the screw.

    d. Connect the negative battery cable.

    e. Remove the drill bit and test operation, readjust if necessary.

31. Install the hazard switch knob.

32. Assemble the canceling cam and steering column and secure the assembly with a new snapring.

33. Install the steering wheel.

34. Connect the negative battery cable.

Grand Cherokee 1993-95 and Grand Wagoneer 1993

1. Disconnect the negative battery cable.

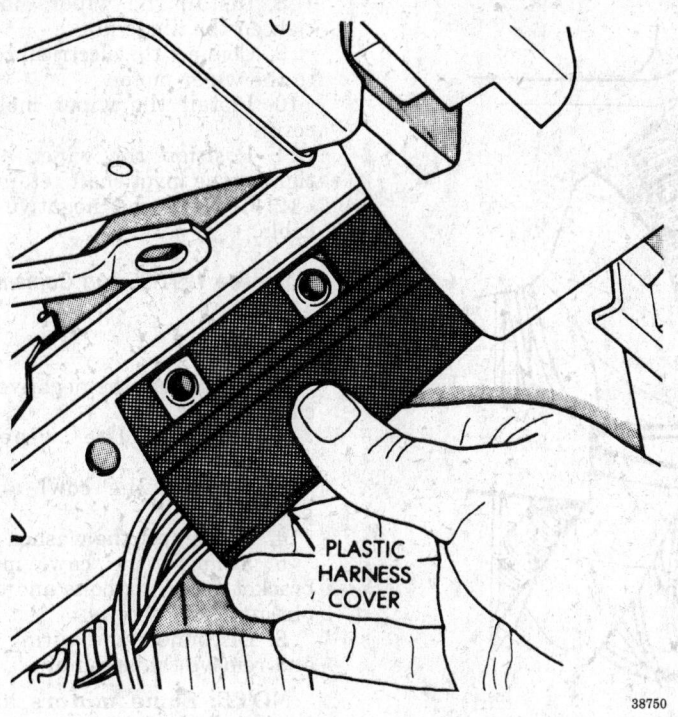

PLASTIC HARNESS COVER

38750

Plastic harness cover

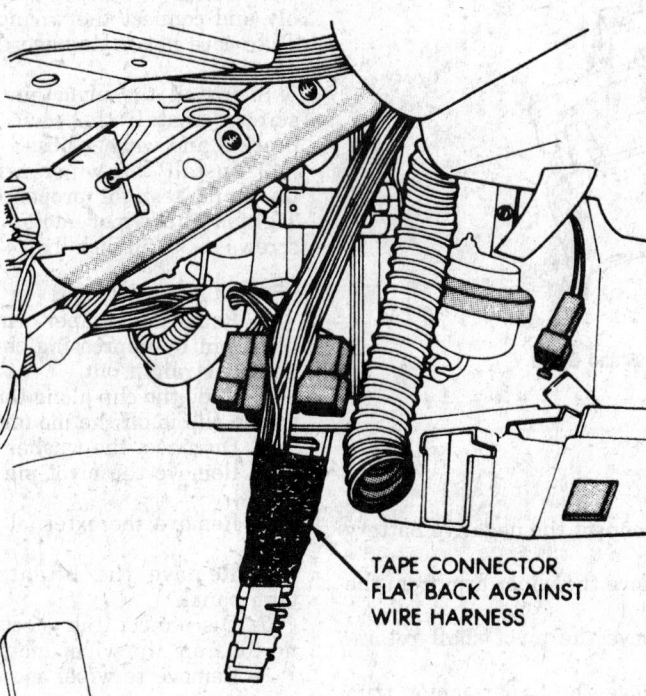

TAPE CONNECTOR
FLAT BACK AGAINST
WIRE HARNESS

38748

Taped switch connector

2. If equipped, remove the tilt lever.

3. Remove the upper and lower steering column covers with a suitable Torx® driver.

4. Remove the steering column trim panel.

5. Remove the knee blocker.

6. Remove the steering column retaining nut and lower the column.

7. Using Snap-On tamper-proof bit TTXR20B2 or equivalent, remove the multi-function switch screws.

8. Pull the switch away from the column, loosen the connector screw (which will remain in the connector) and unplug the electrical connector.

To install:

9. Connect the electrical connector to the multi-function switch and tighten the retaining screw.

10. Mount the multi-function switch to the steering column and tighten the screws.

11. Position the steering column and tighten the retaining nuts.

12. Install the knee blocker, lower trim panel and steering column covers.

13. If equipped, install the tilt steering lever.

14. Connect the negative battery cable.

Wiper Motor

REMOVAL AND INSTALLATION

Wrangler 1991-95

FRONT

1. Disconnect the negative battery cable.

2. Remove the necessary hard or soft top components from the windshield frame.

3. Remove the left and right windshield hold-down knobs and fold the windshield forward.

4. Remove the left access hole cover.

5. Disconnect the drive link from the left wiper pivot.

6. Disconnect the wiper motor harness from the switch.

7. Grasp the motor and pull the drive arm out of the access hole. Pry the arm off the motor pivot.

NOTE: Do not remove the pivot attaching nut.

8. Remove the screws holding the intermittent wiper module bracket to the bottom of the instrument panel. Remove the motor.

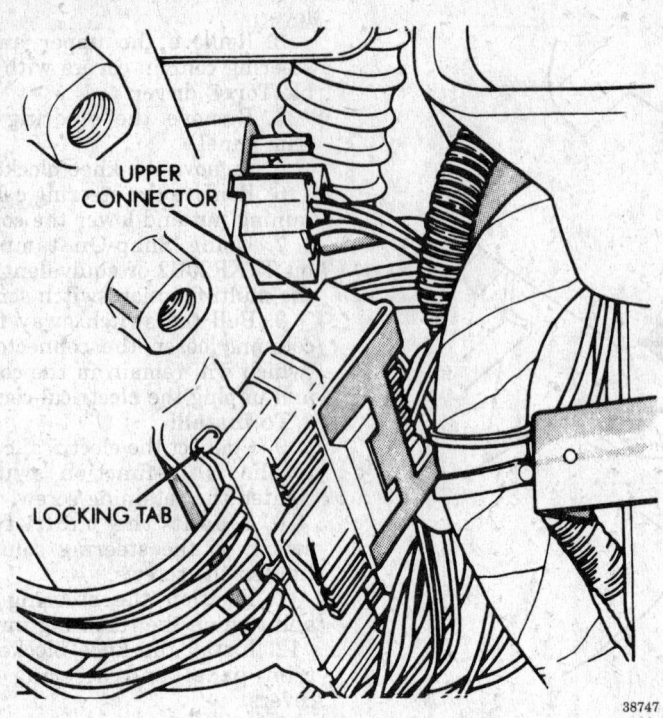

Turn signal connector

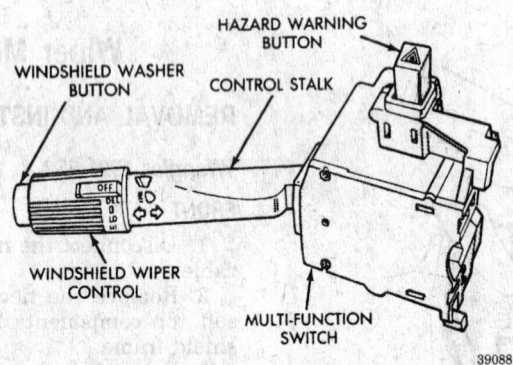

Combination switch - Grand Cherokee 1993-95 and Grand Wagoneer 1993

To install:

9. Install and connect the wiring harness.

10. Install the screws holding the intermittent wiper module.

11. Turn the motor **ON** and check for proper operation.

12. Install the drive arm on the motor pivot. Install the motor and tighten the attaching bolts to 96 inch lbs. (11 Nm).

13. Connect the drive link at the left wiper pivot and install the harness clips. Install the windshield and hard or soft top assembly.

REAR

1. Disconnect the negative battery cable.

2. Remove the wiper arm from the wiper motor.

3. Remove the pivot shaft retaining nut.

4. Remove the wiper motor trim cover.

5. Disconnect the electrical connector from the wiper motor.

6. Remove the top hinge nut securing the wiper motor.

7. Remove the wiper motor.

To install:

8. Install the wiper motor and tighten the hinge nut.

9. Connect the electrical connector to the wiper motor.

10. Install the wiper motor trim cover.

11. Position the wiper arm and tighten the pivot shaft retaining nut.

12. Connect the negative battery cable.

Cherokee 1991-95 and Comanche 1991-92

FRONT

1. Disconnect the negative battery cable.

2. Remove the wiper arm assemblies.

3. Remove the cowl and trim panel.

4. Disconnect the washer hose.

5. Remove the cowl mounting bracket attaching bolts and the pivot pin attaching screws.

6. Disconnect the wiring harness and remove the assembly.

NOTE: Some motors are protected by a rubber case, care should be used so as not to damage this protective coat.

To install:

7. Install the wiper motor assembly and connect the wiring harness. Take care not to damage the rubber case.

8. Install the pivot pin attaching screws. Install the cowl mounting bracket and washer hose.

9. Install the wiper arm assemblies and test for proper operation. Tighten the wiper motor attaching screws to 35–50 inch lbs. (47–67 Nm).

REAR

1. Remove the wiper arm from the pivot pin by depressing the tab and pulling straight out.

2. Slide the clip along the hose until the clip is off the mounting.

3. Disconect the washer hose.

4. Remove the pivot pin retaining nut.

5. Remove the external bezel and seal.

6. Remove the liftgate interior trim panel.

7. Disconnect the electrical connector from the wiper motor.

8. Remove te wiper motor mounting screws.

9. Remove the wiper motor.

To install:

10. Position the wiper motor into the liftgate cavity with the pivot pin protruding through the hole in the liftgate.

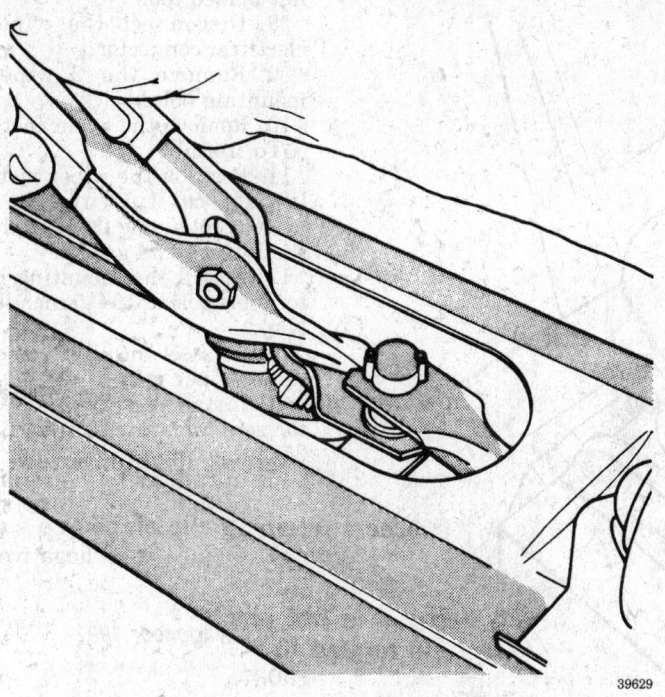

Connecting the front drive arm to the pivot shaft - Wrangler

39629

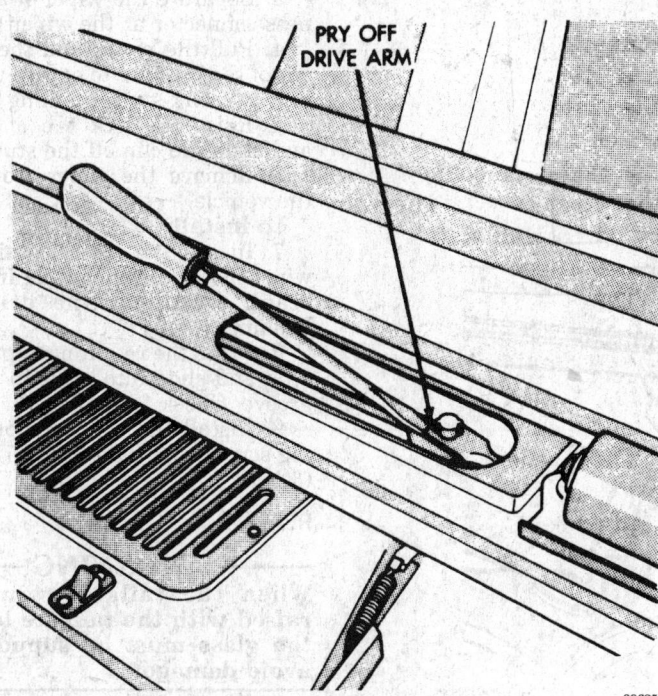

PRY OFF
DRIVE ARM

39625

Disconnecting the front drive arm - Wrangler

11. Install the mounting screws.
12. Connect the electrical connector to the wiper motor.
13. Install the pivot pin, seal bezel and attaching nut. Torque the nut to 32 inch lbs. (4 Nm).
14. Lubricate the male end of the bezel with water and connect the washer hose.
15. Install the liftgate panel trim.
16. Install the wiper arm assembly and connect the extenal washer hose to the bezel.
17. Slide the clip along the hose until it is over the hose mount.
18. Position the arm so the blade is parallel to the window and comes no closer than 5mm to the window seal when operating on a wet surface.
19. Connect the negative battery cable.

Grand Cherokee 1993-95 and Grand Wagoneer 1993

FRONT

1. Disconnect the negative battery cable.
2. Lift the wiper arms upward, slide the tab up and remove the wiper arms.
3. Remove the cowl grille screws, disconnect the washer hose and remove the grille.
4. Remove the bolts securing the wiper linkage.
5. Turn the linkage over and remove the nut securing the crank arm to the motor.
6. Remove the screws holding the linkage to the wiper motor and remove the motor.
   To install:
7. Install the wiper motor and tighten the screws and nut.
8. Install the wiper linkage.
9. Connect the washer hose to the cowl grille and install the grille.
10. Install the wiper arm assemblies.
11. Connect the negative battery cable.

REAR

1. Disconnect the negative battery cable.
2. Lift the wiper arm and insert a 1/8 inch pin into the arm hole.
3. Remove the wiper arm assembly from the pivot pin by depressing the tab and pulling the blade straight out of the arm.
4. Remove the wiper motor retaining nut.
5. Remove the external panel.
6. Remove the 5 screws holding the liftgate interior panel.

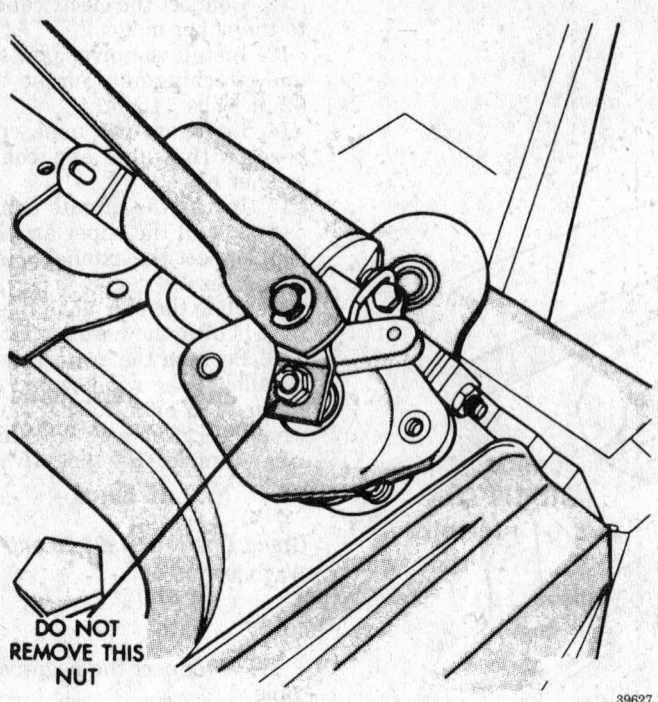

DO NOT REMOVE THIS NUT

39627

Front drive arm removal - Wrangler

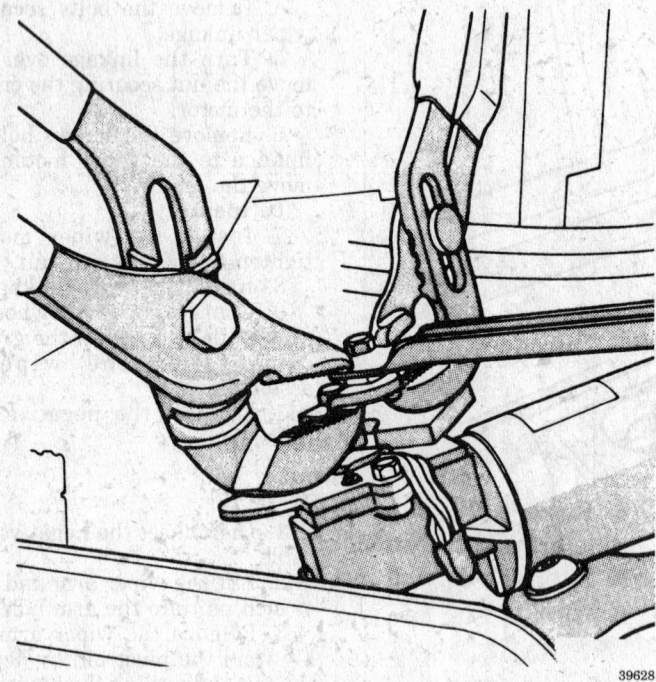

39628

Installing the front drive arm onto the motor - Wrangler

7. Remove the panel with a wide flat bladed tool.

8. Disconnect the wiper motor electrical connector.

9. Remove the 2 wiper motor mounting bolts.

10. Remove the wiper motor.

To install:

11. Position the wiper motor in the liftgate cavity with the knurled driver protruding through the hole in the liftgate and gasket.

12. Install the mounting bolts and torque them to 10–15 inch lbs. (1–1.7 Nm).

13. Connect the electrical connector to the wiper motor.

14. Install the bezel and wiper motor retaining nut. Torque the nut to 35–50 inch lbs. (4–5.6 Nm).

15. Install the liftgate trim panel.

16. Install and position the wiper arm.

17. Connect the negative battery cable.

Grand Wagoneer 1991

FRONT

1. Disconnect the negative battery cable.

2. Remove the screws attaching the motor adapter plate to the dash panel.

3. Separate the wiper wiring harness connector at the wiper motor.

4. Pull the motor and the linkage out of the opening to expose the drive link to crank stud retaining clip.

5. Raise the lock tab of the clip and slide the clip off the stud.

6. Remove the wiper motor from the vehicle.

To install:

7. Install the motor, position the wiper motor assembly and insert the crank stud into the drive link bushing.

8. Press the retaining clip onto the stud and slide into place in the stud groove. Check for proper fit.

9. Install the wiper motor attaching screws and tighten to 25 inch lbs. (3 Nm).

REAR

───── **WARNING** ─────

When the tailgate window is raised with the tailgate lowered, the glass must be supported to avoid damage.
─────────────────────

1. Disconnect the negative battery cable.

2. Lift the plastic cover on the wiper arm, remove the nut and arm.

3. Remove the nut holding the pivot assembly to the tailgate panel.

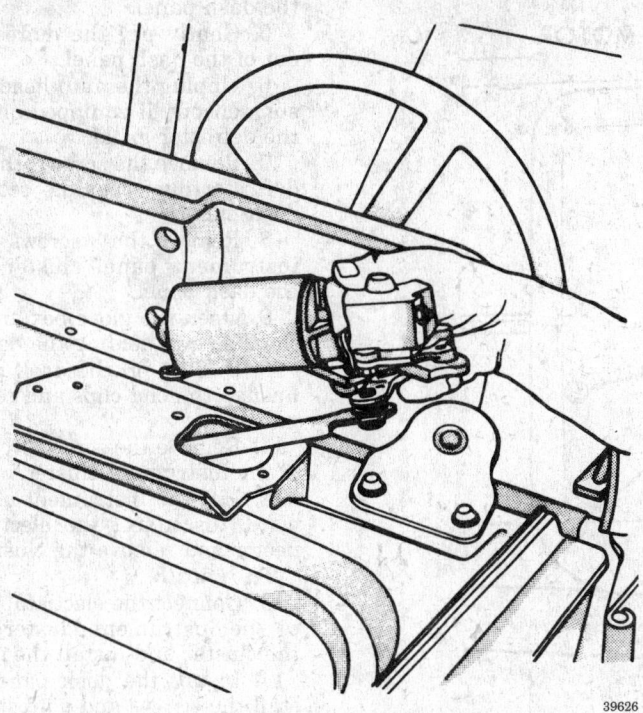

Removing the front wiper motor and arm drive - Wrangler

39626

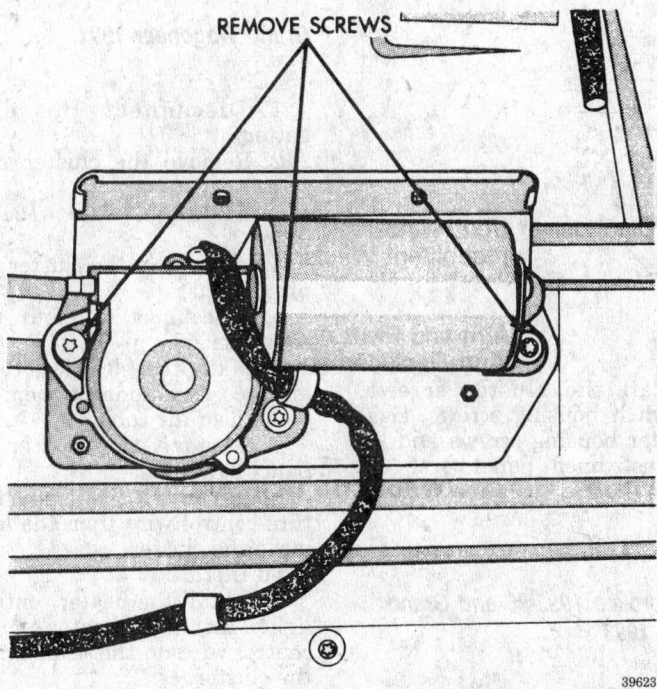

REMOVE SCREWS

39623

Front wiper motor mounting bolts - Wrangler

4. Remove the tailgate and access cover.

5. With the tailgate lowered, raise the window by operating the the window switch and depressing the safety switch on the left side of the tailgate.

6. Disconnect the electrical connector from the wiper motor.

7. Remove the bolt securing the wiper motor.

8. Remove the wiper motor.

To install:

9. Install the wiper motor and connect the electrical connector.

10. Install the nut to the tailgate panel and torque it to 8 ft. lbs. (11 Nm).

11. Install the wiper arm and adjust it to park on the rubber tailgate window, just off the glass. Torque the nut to 8 ft. lbs. (11 Nm) and install the plastic cover.

Instrument Cluster

REMOVAL AND INSTALLATION

Wrangler 1991-95

1. Disconnect the negative battery cable.

2. Remove the 6 shroud screws.

3. Slide the shroud toward the steering wheel.

4. Remove the 3 screws holding the right side switch panel.

5. Remove the 3 screws holding the left side switch panel.

6. Remove the 2 screws holding the instrument cluster in place.

7. Lift up the top of the cluster. Roll the cluster out between the between the steering column and instrument panel far enough to disconnect the connector located behind the tachometer.

8. Disconnect the cluster connectors and speedometer cable. Remove the cluster.

To install:

9. Install the cluster after connecting all electrical connectors and speedometer cable.

10. Install the instrument cluster retaining screws.

11. Install the left and right switch panels.

12. Install the shroud.

13. Connect the negative battery cable.

Cherokee 1991-95 and Comanche 1991-92

1. Disconnect the negative battery cable.

2. Remove the instrument panel bezel screws and unsnap the bezel.

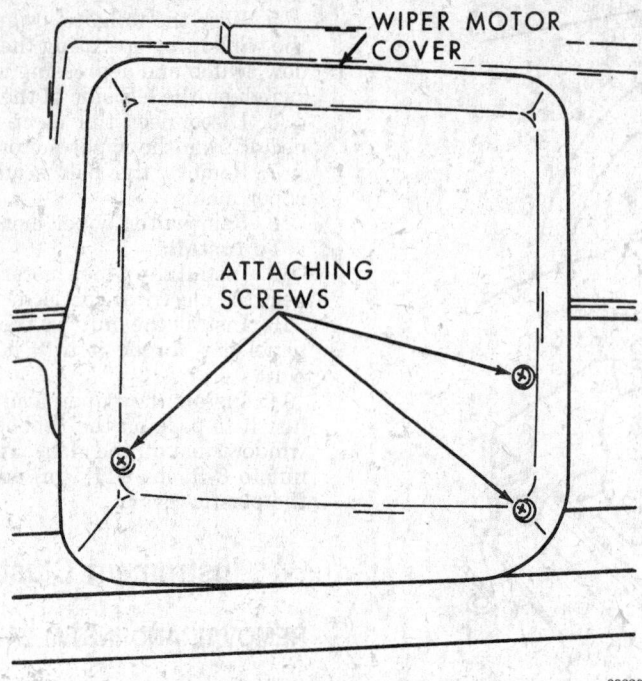

Rear wiper motor cover

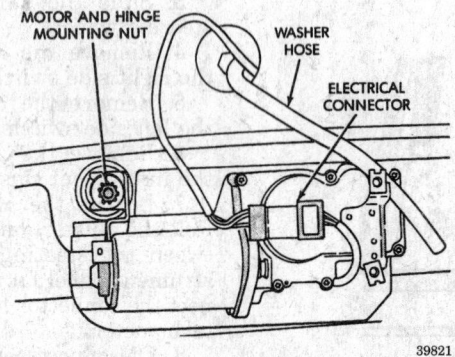

Rear wiper motor

3. Remove the cigarette lighter housing screws.

4. If equipped, remove the switch housing screws.

5. Remove the instrument panel cluster screws.

6. Disconnect the speedometer cable, pull the cluster out slowly and disconnect the electrical connectors at the cluster back. Remove the cluster.

To install:

7. Install the cluster after connecting all electrical connectors and speedometer cable.

8. Install the cluster screws, rocker switch housing screws, cigarette lighter housing screws and install the instrument panel bezel.

9. Connect the negative battery cable.

Grand Cherokee 1993-95 and Grand Wagoneer 1993

1. Disconnect the negative battery cable.

2. Remove the ash tray.

3. Remove the screws holding the center cluster bezel and remove the bezel.

4. Remove the 2 screws holding the dash panel.

5. Gently pry the defroster grille out of the dash panel.

6. Unplug the auto headlamp and sun sensors, if equipped and remove the defroster grille.

7. Remove the screws, through the defroster duct opening, securing the dash panel.

8. Remove the 3 screws above the instrument panel cluster securing the dash panel.

9. Open the glove box and remove the 2 screws holding the dash panel.

10. Pull up on the dash panel and unsnap the end clips and remove the panel.

11. Remove the screws from the top of the instrument cluster.

12. Lift the instrument cluster upward, disconnect the electrical connector and remove the cluster.

To install:

13. Connect the electrical connector to the instrument cluster, position the cluster and install the screws.

14. Install the dash panel and install the screws and defroster grille.

15. Install the center cluster bezel.

16. Install the ash tray.

17. Connect the negative battery cable.

Grand Wagoneer 1991

1. Disconnect the negative battery.

2. Remove the cluster retaining screws.

3. Disconnect the speedometer cable.

4. Disconnect the cluster terminal pin plug.

5. Disconnect the four terminal connector.

6. Mark the electrical connectors and hoses, disconnect them and the blend door air cable.

7. Remove the heater control panel lights.

8. Disconnect the heater temperature control wire from the lever. Remove the cluster.

To install:

9. Install the heater control panel lights. Install the heater temperature control wire on the lever and install the cluster.

10. Install the blend door hoses and all electrical connectors.

11. Install the speedometer cable.

12. Install the cluster retaining screws and connect the negative battery cable.

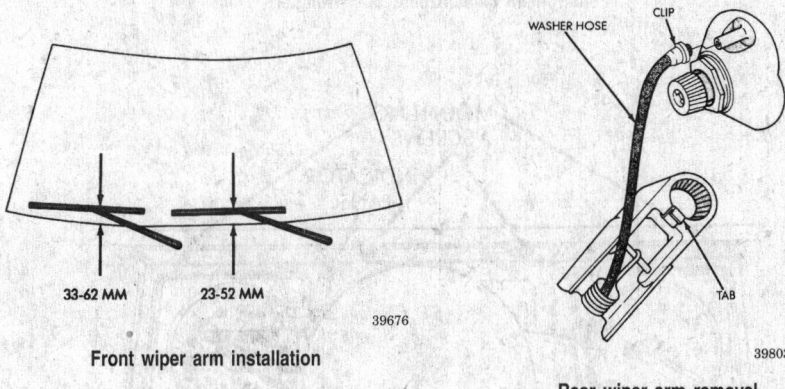

Front pivot assembly removal

PIVOT PIN SCREWS (4)

BRACKET NUTS

39672

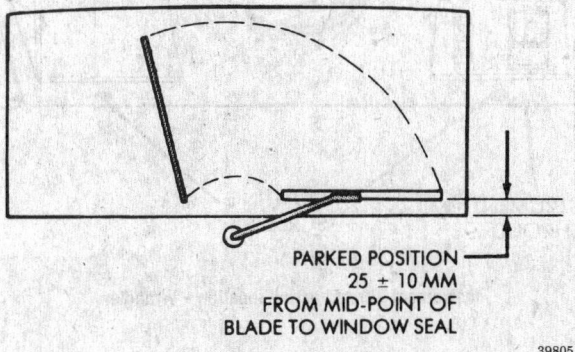

Front wiper arm installation

33-62 MM   23-52 MM

39676

WASHER HOSE   CLIP

TAB

Rear wiper arm removal

39803

PARKED POSITION
25 ± 10 MM
FROM MID-POINT OF
BLADE TO WINDOW SEAL

39805

Rear wiper arm positioning

Speedometer

REMOVAL AND INSTALLATION

Wrangler 1991-95, Cherokee 1991-95, Comanche 1991-92, Grand Cherokee 1993-95, Grand Wagoneer 1991 and 1993

1. Disconnect the negative battery cable.
2. Remove the instrument cluster.
3. Remove the lens attaching screws.

—— **WARNING** ——
Do not touch the face of a gauge or the back of the lens. It may leave a permanent mark.

4. Remove the screw(s) from the speedometer and pull it from the cluster.
5. Pull the speedometer out of the circuit board carefully.
   To install:
6. Install the speedometer and screws.
7. Install the instrument cluster lens.
8. Install the instrument cluster.
9. Connect the negative battery cable.

Combination Switch

REMOVAL AND INSTALLATION

Wrangler 1991-95

1. Disconnect the negative battery cable.
2. Remove the steering wheel.
3. Remove the lockplate cover by compressing with tool C–4156 or equivalent.
4. Release and discard the steering shaft retaining snapring.
5. Remove the compressor tool.
6. Remove the lockplate, canceling cam, upper bearing preload spring and thrust washer from the steering column.
7. Remove the horn button components from the canceling cam.
8. Remove the hazard warning switch knob.
9. Remove the dimmer switch actuator arm attaching screw.
10. Remove the turn signal switch attaching screws.
11. Remove the instrument panel screws and slide the panel toward he steering wheel.
12. Remove the lower steering column cover.

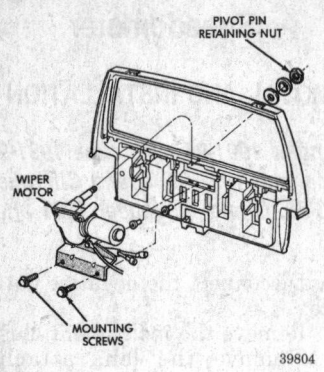

Rear wiper motor removal and installation

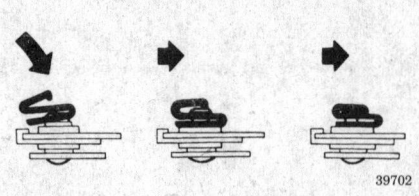

Retaining clip installation

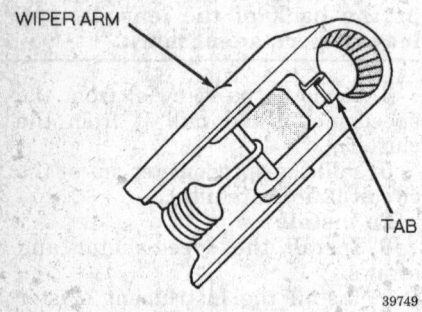

Rear wiper arm removal - Grand Cherokee and Grand Wagoneer

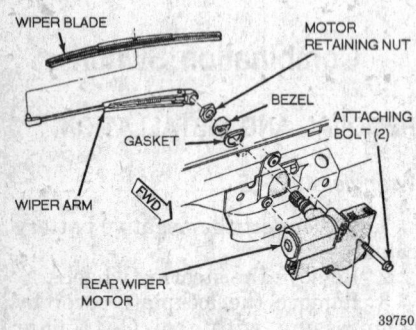

Rear wiper motor removal and installation - Grand Cherokee and Grand Wagoneer

13. On vehicles equipped with automatic transmission column shift selector, remove the **PRNDL** cable clip.

14. Remove the nuts securing steering column bracket to the brake sled.

15. Remove the bolts holding the steering column bracket to the column.

16. Loosen the column brace mounting nut at the drivers side kick panel and allow the column to drop.

17. Push the turn signal connector up and out of the steering column connector.

18. Pry up the locking tabs of the steering column connector and remove the connector from the column bracket.

19. Tape the connector flat against the wire harness to prevent it from hanging up during removal.

20. Remove the plastic harness cover by pulling it up and over the weld nuts then open and slide the cover off the harness.

21. Remove the combination lever by pulling it out straight from the column.

To install:

22. Position the combination switch into position on the column while

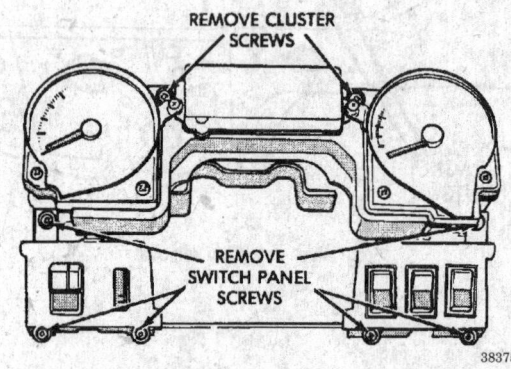

Instrument cluster removal - Wrangler

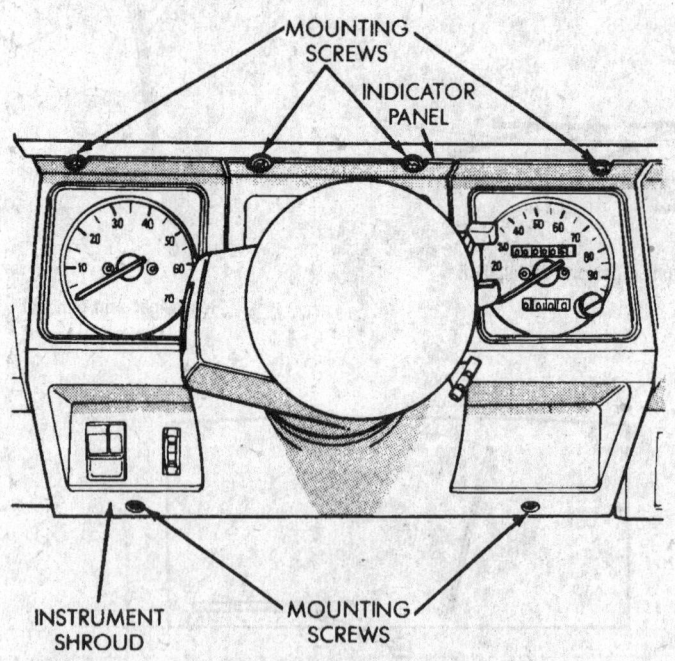

Instrument shroud screw location - Wrangler

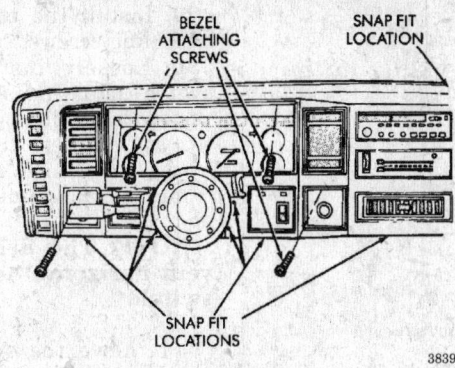

Instrument cluster bezel removal and installation - Cherokee

38392

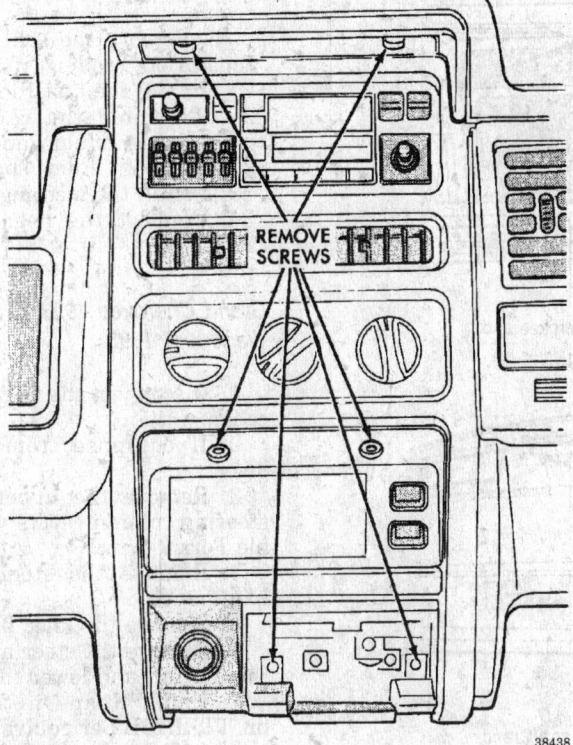

Center bezel retaining screw location - Grand Cherokee and Grand Wagoneer 1993-95

38438

guiding the harness. Ensure the wires are laying flat on the bottom of the inside column.

23. Remove the tape and connect the electrical connector.

24. Connect the turn signal connector.

25. Position the steering column and tighten the mounting nuts and bolts.

26. If equipped with a column shift selector, install the **PRNDL** cable clip with the transmission in **N**. Move the selector through all ranges and ensure the indicator is aligned.

27. Install the lower steering column cover.

28. Position the instrument panel and tighten the screws.

29. Install the turn signal switch attaching screws.

30. Loosely install the dimmer switch actuating screws and adjust the switch as follows:

a. Compress the switch and insert a 3/32 inch diameter drill bit into the adjustment hole.

NOTE: The drill bit will prevent horizontal movement of the switch.

b. Move the switch toward the steering wheel to eliminate rod lash.

c. Tighten the screw.

d. Connect the negative battery cable.

e. Remove the drill bit and test operation, readjust if necessary.

31. Install the hazard switch knob.

32. Assemble the canceling cam and steering column and secure the assembly with a new snapring.

33. Install the steering wheel.

34. Connect the negative battery cable.

Cherokee 1991-95, Comanche 1991-92 and Grand Wagoner 1991

1. Disconnect the negative battery cable.

2. Remove the steering wheel.

3. Remove the lockplate cover by compressing with tool C-4156 or equivalent.

4. Release and discard the steering shaft retaining snapring.

5. Remove the compressor tool.

6. Remove the lockplate, canceling cam, upper bearing preload spring and thrust washer from the steering column.

7. Remove the horn button components from the canceling cam.

8. Remove the hazard warning switch knob.

9. Remove the dimmer switch actuator arm attaching screw.

10. Remove the turn signal switch attaching screws.

11. Remove the lower instrument panel cover trim panel.

12. Remove the lower steering column cover.

13. On vehicles equipped with automatic transmission column shift selector, remove the **PRNDL** cable clip.

14. Remove the nuts securing the steering column bracket to the brake sled.

15. Remove the bolts holding the steering column bracket to the column.

16. Loosen the column brace mounting nut at the drivers side kick panel and allow the column to drop.

17. Push the turn signal connector up and out of the steering column connector.

18. Pry up the locking tabs of the steering column connector and remove the connector from the column bracket.

19. Tape the connector flat against the wire harness to prevent it from hanging up during removal.

20. Remove the plastic harness cover by pulling it up and over the

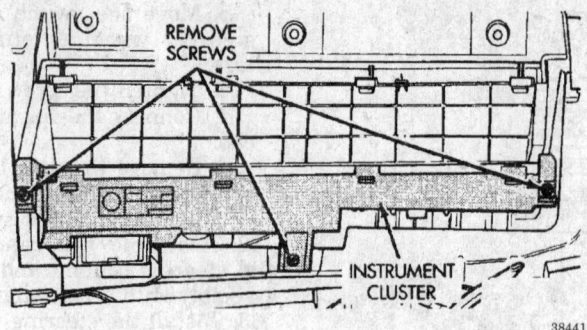

Instrument cluster attaching screws - Grand Cherokee and Grand Wagoneer 1993-95

Instrument cluster-to-dash pad screw location - Grand Cherokee and Grand Wagoneer 1993-95

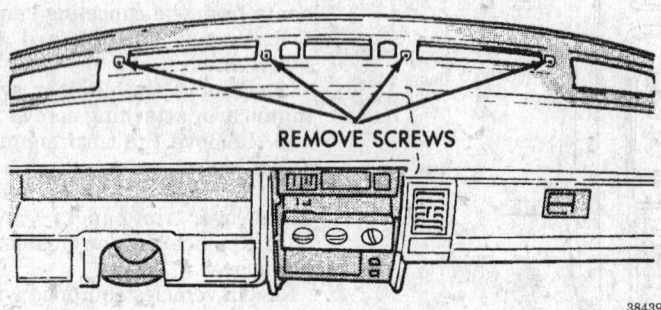

Upper dash pad attaching screw location - Grand Cherokee and Grand Wagoneer 1993-95

weld nuts, then open and slide the cover off the harness.

21. Remove the combination lever by pulling it out straight from the column.

To install:

22. Position the combination switch into position on the column while guiding the harness. Ensure the wires are laying flat on the bottom of the inside column.

23. Remove the tape and connect the electrical connector.

24. Connect the turn signal connector.

25. Position the steering column and tighten the mounting nuts and bolts.

26. If equipped with a column shift selector, install the **PRNDL** cable clip with the transmission in **N**. Move the selector through all ranges and ensure the indicator is aligned.

27. Install the lower steering column cover.

28. Install the lower instrument panel cover trim panel.

29. Install the turn signal switch attaching screws.

30. Loosely install the dimmer switch actuating screws and adjust the switch as follows:

a. Compress the switch and insert a ³/₃₂ inch diameter drill bit into the adjustment hole.

NOTE: The drill bit will prevent horizontal movement of the switch.

b. Move the switch toward the steering wheel to eliminate rod lash.

c. Tight the screw.

d. Connect the negative battery cable.

e. Remove the drill bit and test operation, readjust if necessary.

31. Install the hazard switch knob.

32. Assemble the canceling cam and steering column and secure the assembly with a new snapring.

33. Install the steering wheel.

34. Connect the negative battery cable.

Grand Cherokee 1993-95 and Grand Wagoneer 1993

1. Disconnect the negative battery cable.

2. If equipped, remove the tilt lever.

3. Remove the upper and lower steering column covers with a suitable Torx® driver.

4. Remove the steering column trim panel.

5. Remove the knee blocker.

6. Remove the steering column retaining nut and lower the column.

7. Using Snap-On tamper-proof bit TTXR20B2 or equivalent, remove the multi-function switch screws.

8. Pull the switch away from the column, loosen the connector screw (which will remain in the connector) and unplug the electrical connector.

To install:

9. Connect the electrical connector to the multi-function switch and tighten the retaining screw.

10. Mount the multi-function switch to the steering column and tighten the screws.

11. Position the steering column and tighten the retaining nuts.

12. Install the knee blocker, lower trim panel and steering column covers.

13. If equipped, install the tilt steering lever.

14. Connect the negative battery cable.

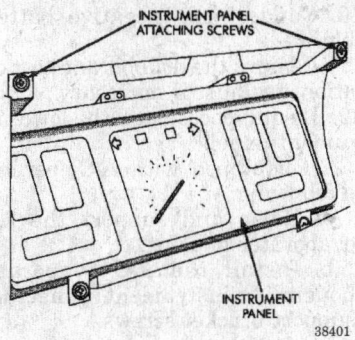

Instrument cluster removal and installation - Grand Wagoneer 1991

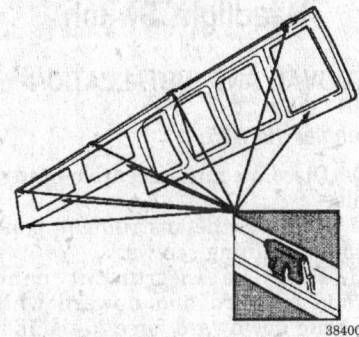

Instrument panel bezel retaining clips - Grand Wagoneer 1991

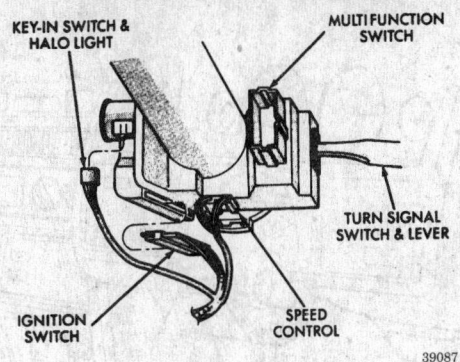

Steering column electrical connectors

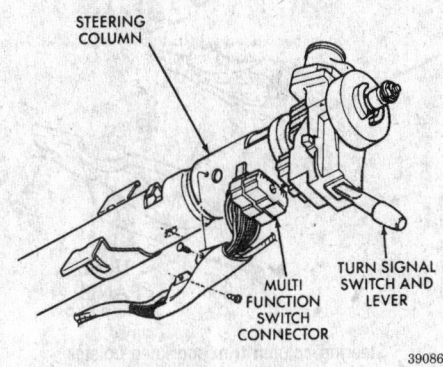

Combination switch connector

Dimmer Switch

REMOVAL AND INSTALLATION

Wrangler 1991-95

1. Disconnect the negative battery cable.
2. If equipped, remove the lower instrument panel cover.
3. Remove the instrument housing screws.
4. Slide the housing towards the steering wheel and apply upward pressure to the housing and downward pressure to the indicator to release the holding tabs.

5. Remove the instrument housing.
6. If equipped with A/C, perform the following:
   a. Support the A/C evaporator housing.
   b. Remove the A/C evaporator housing-to-instrument panel attaching screws.
   c. Remove the A/C evaporator housing support bracket.
   d. Lower the evaporator.
7. Disconnect the dimmer switch electrical connector.
8. Tape the actuator rod to the steering column.
9. Remove the dimmer switch.

To install:

10. Force the dimmer switch onto the actuator rod and loosely install the screws.
11. Compress the switch and insert a 3/32 inch diameter drill bit into the adjustment hole.

NOTE: The drill bit will prevent horizontal movement of the switch.

12. Move the switch toward the steering wheel to eliminate rod lash, then tighten the screw.
13. Connect the negative battery cable.
14. Remove the drill bit and test operation, readjust if necessary.
15. If equipped, install the lower instrument panel.
16. If equipped with A/C, perform the following:
   a. Raise and support the A/C evaporator housing.
   b. Install the A/C evaporator housing-to-instrument panel and support bracket screws.
17. Position the instrument panel shroud under the steering column and slide the holding tabs into the shroud notches.
18. Place the instrument panel shroud into position and install the screws.

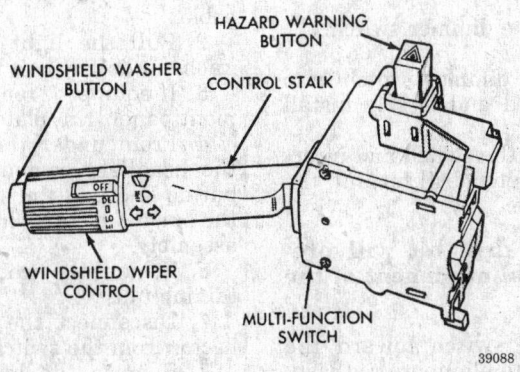

Combination switch

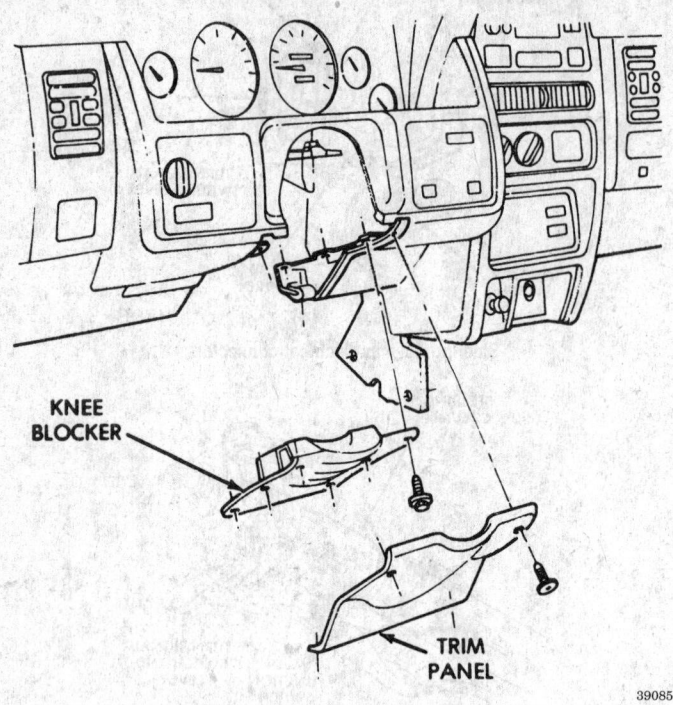

KNEE
BLOCKER

TRIM
PANEL

39085

Steering column trim and knee bolster

19. Connect the negative battery cable.

Cherokee 1991-95 and Comanche 1991-92

1. Disconnect the negative battery cable.
2. If equipped, remove the lower instrument panel cover.
3. Disconnect the dimmer switch electrical connector.
4. Tape the actuator rod to the steering column.
5. Remove the dimmer switch.
To install:
6. Force the dimmer switch onto the actuator rod and loosely install the screws.
7. Compress the switch and insert a $3/32$ inch diameter drill bit into the adjustment hole.

NOTE: The drill bit will prevent horizontal movement of the switch.

8. Move the switch toward the steering wheel to eliminate rod lash, then tight the screw.
9. Connect the negative battery cable.
10. Remove the drill bit and test operation, readjust if necessary.
11. If equipped, install the lower instrument panel.

Grand Wagoneer 1991

1. Disconnect the negative battery cable.
2. If equipped, remove the lower instrument panel cover.
3. If equipped with A/C, perform the following:
   a. Support the A/C evaporator housing.
   b. Remove the A/C evaporator housing-to-instrument panel attaching screws.
   c. Remove the A/C evaporator housing support bracket.
   d. Lower the evaporator.
4. Disconnect the dimmer switch electrical connector.
5. Tape the actuator rod to the steering column.
6. Remove the dimmer switch.
To install:
7. Force the dimmer switch onto the actuator rod and loosely install the screws.
8. Compress the switch and insert a $3/32$ inch diameter drill bit into the adjustment hole.

NOTE: The drill bit will prevent horizontal movement of the switch.

9. Move the switch toward the steering wheel to eliminate rod lash, then tighten the screw.

10. Connect the negative battery cable.
11. Remove the drill bit and test operation, readjust if necessary.
12. If equipped, install the lower instrument panel.
13. If equipped with A/C, perform the following:
   a. Raise and support the A/C evaporator housing.
   b. Install the A/C evaporator housing-to-instrument panel and support bracket screws.
14. Connect the negative battery cable.

Headlight Switch

REMOVAL AND INSTALLATION

Wrangler 1991-95

1. Disconnect the negative battery cable.
2. Remove the instrument panel shroud retaining screws.
3. Pull the instrument panel shroud outward and upward while applying downward force to the indicator panel and remove the shroud.
4. Remove the screws retaining the switch, pull the switch from the instrument panel cavity and disconnect the electrical connector.
To install:
5. Connect the electrical connectors to the headlight switch and install the retaining screws.
6. Position the instrument panel shroud under the steering column and slide the holding tabs into the shroud notches.
7. Place the instrument panel shroud into position and install the screws.
8. Connect the negative battery cable.

Cherokee 1991-95 and Comanche 1991-92

1. Disconnect the negative battery cable.
2. Pull the light switch control knob out as far as it will go.
3. If equipped, remove the instrument panel trim plate.
4. From under the dash depress the headlight switch shaft retainer button and pull the shaft along with the knob from the headlight switch assembly.
5. Remove the headlight switch retaining nut.
6. Disconnect the electrical connector from the switch.
7. Remove the headlight switch from the vehicle.

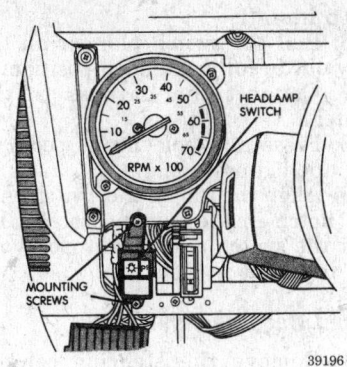

Headlight switch

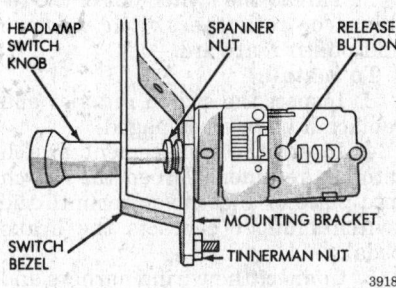

Headlight switch

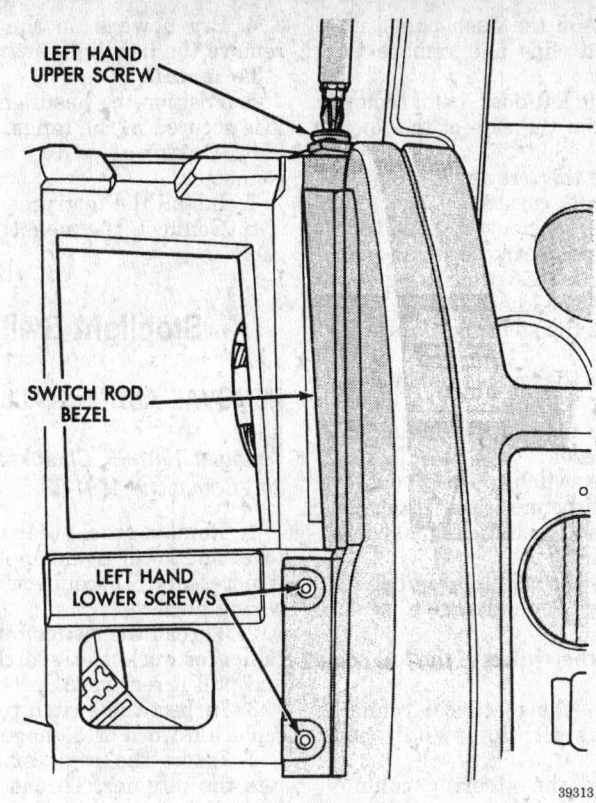

Left switch pod bezel screws

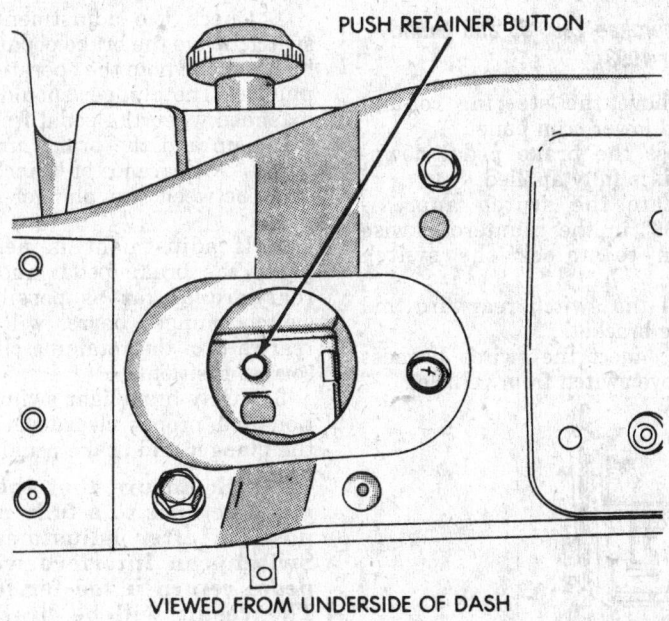

VIEWED FROM UNDERSIDE OF DASH

Headlight switch shaft removal

To install:

8. Connect the electrical connector, install the headlamp switch and tighten the nut.

9. Insert the shaft and knob into the switch.

10. Connect the negative battery cable.

Grand Cherokee 1993-95 and Grand Wagoneer 1993

1. Disconnect the negative battery cable.

2. Remove the ash tray.

3. Remove the screws holding the center cluster bezel and remove the bezel.

4. Remove the 2 screws holding the dash panel.

5. Gently pry the defroster grille out of the dash panel.

6. Unplug the auto headlamp and sun sensors, if equipped and remove the defroster grille.

7. Remove the screws, through the defroster duct opening, securing the dash panel.

8. Remove the 3 screws above the instrument panel cluster securing the dash panel.

9. Open the glove box and remove the 2 screws holding the dash panel.

10. Pull up on the dash panel, un-snap the end clips and remove the panel.

11. With the left door open, remove the screw from the side of the lower trim panel.

12. Remove the screws securing the steering column covers.

13. Remove the screw from the bottom of the lower trim panel and un-snap it from the instrument panel.

14. Remove the knee blocker.

15. Remove the steering column retaining nuts.

16. Remove the screws holding the bezels.

17. Remove the screws holding the switch pod bezel.

18. Pull the switch pod bezel out far enough to disconnect the electrical connectors and remove the switch pod assemblies.

19. Remove the headlight switch retaining screws and switch.

To install:

20. Install the switch to the left pod assembly.

21. Connect the electrical connectors and install the switch pod assemblies.

22. Position the steering column and install the retaining nuts.

23. Install the knee blocker.

24. Install the lower trim panel.

25. Install the steering column cover screws.

26. Install the screw securing the left side of the lower trim panel.

27. Position the dash panel and install the screws and defroster grille.

28. Install the center cluster bezel.

29. Install the ash tray.

30. Connect the negative battery cable.

Grand Wagoneer 1991

1. Disconnect the negative battery cable.

2. Remove the instrument cluster.

3. Disconnect the electrical connector from the headlight switch.

4. Pry upward on the tangs and remove the headlight switch.

To install:

5. Position the headlight switch so it is secured in the tangs.

6. Connect the electrical connector.

7. Install the instrument cluster.

8. Connect the negative battery cable.

Stoplight Switch

REMOVAL AND INSTALLATION

Wrangler 1991-95, Cherokee 1991-95 and Comanche 1991-92

1. Remove the steering column cover and lower trim panel for access, if necessary. Disconnect the switch wiring harness.

2. Thread the switch out of the retainer or rock the switch up/down and pull it rearward.

3. Inspect the switch retainer and replace if worn or damaged.

4. Insert the replacement switch into the retainer. Thread the switch into place or rock up/down until the switch plunger contacts the brake pedal.

5. Connect the wiring harness and adjust the switch.

Grand Cherokee 1993-95 and Grand Wagoneer 1993

1. Remove the steering column cover and lower trim panel.

2. Press the brake pedal downward so it is fully applied.

3. Rotate the switch approximately 30° in the counterclockwise direction to unlock the switch retainer.

4. Pull the switch rearward and out of the bracket.

5. Disconnect the switch harness and remove switch from vehicle.

To install:

6. Pull the switch plunger all the way out to fully extended position.

7. Connect the harness to the switch.

8. Press and hold brake pedal in the applied position.

9. Align the tab on the switch with the notch in the switch bracket. Insert the switch and turn it clockwise 30° to lock it in place.

Grand Wagoneer 1991

1. Remove the steering column cover and lower trim panel for access, if necessary. Disconnect the switch wiring harness.

2. Thread the switch out of the retainer or rock the switch up/down and pull it rearward.

To install:

3. Inspect the switch retainer and replace if worn or damaged.

4. Insert the replacement switch into the retainer. Thread the switch into place or rock up/down until the switch plunger contacts the brake pedal.

5. Connect the wiring harness and adjust the switch.

ADJUSTMENT

Wrangler 1991-95, Cherokee 1991-95 and Comanche 1991-92

1. Check the adjustment of the switch. Move the brake pedal forward by hand and note the operation of the plunger. The plunger should be fully extended when the pedal free-play is taken up and the brake application begins. A clearance of 1/8 inch should exist between the plunger and the pedal.

2. If adjustment is necessary, grasp the brake pedal and pull it rearward as far as possible. The switch plunger barrel will ratchet rearward in the retaining clip to the correct position.

3. Verify brakelight switch operation and proper clearance between the plunger and brake pedal.

NOTE: Ensure that the brake pedal returns to a fully released position after adjustment. The Switch can interfere with full pedal return if too far forward. The result will be brake drag caused by partial brake application.

Grand Cherokee 1993-95 and Grand Wagoneer 1993

1. Depress and then release the brake pedal.

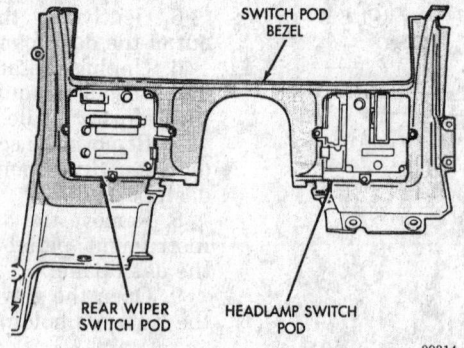

SWITCH POD BEZEL

REAR WIPER SWITCH POD

HEADLAMP SWITCH POD

39314

Rear view of the switch pod bezel

2. Pull the brake pedal fully rearward.

3. The brake pedal will rachet the plunger to the correct position as it is pushed into the switch body.

Grand Wagoneer 1991

1. Hold the brake pedal in the applied position.

2. Push the stoplight switch through the mounting bracket until it stops against the brake pedal bracket. Release the pedal to set the switch in the proper position.

3. Check the position of the switch. The switch plunger should be in the ON position and activate the brake lights after a brake pedal travel of $3/8$–$5/8$ inch (9.5–15.5mm).

Radio

REMOVAL AND INSTALLATION

Wrangler 1991-95, Cherokee 1991-95, Comanche 1991-92, Grand Cherokee 1993-95, Grand Wagoner 1991 and 1993

1. Disconnect the negative battery cable.

2. Remove the instrument panel attaching screws and remove the bezel.

3. Remove the radio attaching screws.

4. Slide the radio out of its housing and disconnect the electrical connector, ground wire and antenna lead.

To install:

5. Connect the electrical connectors and antenna lead to the back of the radio.

6. Slide the radio into its cavity, carefully guiding the wires and if equipped, install the clip to the underside of the dash.

7. Install the radio attaching screws.

8. Install the instrument panel bezel.

9. Connect the negative battery cable.

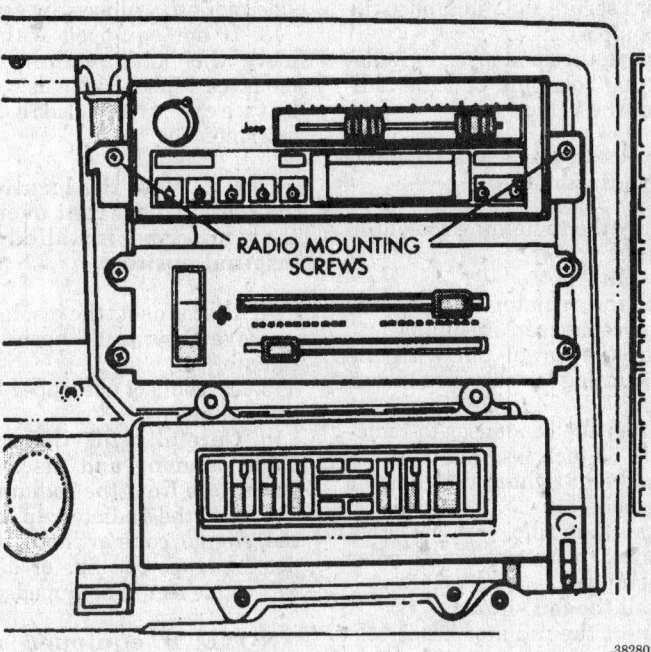

38280

Radio mounting screw location

ENGINE COOLING

Radiator

REMOVAL AND INSTALLATION

Wrangler 1991-95 and Grand Wagoneer 1991

1. Disconnect the negative battery cable.

2. Drain the cooling system.

3. Remove the radiator upper and lower hoses.

4. Disconnect the overflow tube from the radiator.

5. If equipped, remove the transmission cooler lines.

6. Remove the fan shroud mounting bolts and pull the fan shroud back to the engine.

7. Remove all attaching bolts and screws that secure the radiator to the radiator support.

8. Remove the condenser to radiator mounting bolts.

9. Pull the radiator out of the vehicle taking care not to damage the radiator fins.

To install:

10. Slide the radiator into position behind the condenser, if equipped and torque the mounting screws 6 ft. lbs. (8 Nm).

11. Close the radiator drain.

12. Install the fan shroud.

13. Connect the transmission cooler lines, if equipped.

14. Connect the hoses to the radiator.

15. Connect the negative battery cable.

16. Fill the cooling system.

17. Connect the reserve bottle hose and install the radiator cap.

Cherokee 1991-95 and Comanche 1991-92

1. Disconnect the negative battery cable.

2. Remove the front grille mounting screws and grille, as necessary.

3. Drain the cooling system.

4. Remove the radiator upper and lower hoses.

5. If equipped, remove the transmission cooler lines.

6. Remove the fan shroud mounting bolts and pull the fan shroud back to the engine.

7. Remove the alignment dowel E-clip from the lower radiator mounting bracket.

8. Disconnect the overflow tube from the radiator.

9. Remove all attaching bolts and screws that secure the radiator to the radiator support.

10. Remove the condenser to radiator mounting bolts and pull the radiator out of the vehicle.

NOTE: Take care not to damage the radiator fins.

11. Empty the remaining coolant in the radiator.

To install:

12. Slide the radiator into position behind the condenser, if equipped.

13. Align the dowel pin with the bottom mounting bracket and install the E-clip.

14. Tighten the condenser-to-radiator bolts to 55 inch lbs. (6.2 Nm).

15. Install and tighten the radiator mounting bolts.

16. Install the grille.

17. Connect the transmission cooler lines, if equipped.

18. Install the fan shroud.

19. Connect the radiator hoses.

20. Connect the negative battery cable.

21. Fill the cooling system to the correct level.

Grand Cherokee 1993-95

1. Disconnect the negative battery cable.

2. Open the radiator valve and drain the cooling system.

3. Remove the fan and shroud assembly.

4. If equipped, disconnect the automatic transmission cooling line quick-fit connections.

5. Matchmark the upper radiator crossmember and adjust the crossmemeber to the left or right.

6. Eight clips are used to retain a rubber seal to the body. Gently pry up the outboard clips (2 per side) until the rubber seal can be removed. Do not remove the seals entirely. Fold back the seal on both sides to access the grille opening reinforcement mounting bolts and remove the bolts.

7. Remove the grille.

8. Remove the upper brace bolt from each of the 2 radiator braces.

9. Remove the crossmember-to-radiator mounting nuts.

10. Working through the grille opening, remove the lower bracket bolt securing the lower part of the hood latch or hood latch cable from the crossmember.

11. Lift the crossmember straight up and position it aside.

12. If equipped with A/C, remove the 2 A/C condenser-to-radiator mounting bolts which also retain the side mounted rubber air seals.

13. If not equipped with A/C, remove the bolts retaining the side mounted rubber air seals compressed between the radiator and crossmember.

NOTE: Note the location of the air seals. To prevent overheating, they must be installed in their original positon.

14. Disconnect the coolant reservoir/overflow tank hose from the radiator.

15. Disconnect the upper hose from the radiator.

16. Carefully lift the radiator a slight amount and disconnect the lower hose from the radiator.

17. Lift the radiator up and out of the engine compartment, take care not to scrape the fins or disturb the A/C condenser if equipped.

NOTE: If equipped with an auxiliary automatic transmission oil cooler, use caution during radiator removal. The oil cooler lines are routed through a rubber air seal on the left side of the radiator. Do not cut or tear this seal.

To install:

18. Lower the radiator into the vehicle. Guide the alignment dowels into the hoses in the rubber air seals and then through the A/C support brackets, if equipped. Continue to guide the radiator through the rubber grommets located in the lower crossmember.

NOTE: On vehicles equipped with A/C, the L-shaped brackets, located on the bottom of the condenser, must be positioned between the bottom of the rubber air seals and top of rubber grommets.

19. Connect the lower radiator hose to the radiator.

20. Connect the upper radiator hose to the radiator.

21. If equipped with A/C, install the bolts condenser-to-radiator mounting bolts.

22. If not equipped with A/C, install the rubber air seal retaining bolts.

23. Connect the reservoir/overflow tank hose to the radiator.

24. If the radiator-to-upper crossmember rubber insulators were removed, install them.

25. Install the hood latch support bracket-to-lower frame crossmember bolt.

26. Install the bolts securing the upper radiator crossmember to the body.

27. Install the radiator-to-upper crossmember nuts.

28. Install a bolt to each upper radiator brace.

29. Install the grille.

30. Position the rubber seal and push down on the clips until seated.

31. If equipped, connect the transmission cooling lines.

32. Install the fan shroud with the fan.

33. Install the fan and shroud.

34. Rotate the fan blades and ensure they do not interfere with the shroud and at least 1.0 inch (25mm) of clearance is allowed. Correct as necessary.

35. Fill the cooling system.

Grand Wagoneer 1993

1. Disconnect the negative battery cable.

2. Open the radiator valve and drain the cooling system.

3. Remove the fan and shroud assembly.

4. If equipped, disconnect the automatic transmission cooling line quick-fit connections.

5. Matchmark the upper radiator crossmember and adjust the crossmemeber to the left or right.

6. Eight clips are used to retain a rubber seal to the body. Gently pry up the outboard clips (2 per side) until the rubber seal can be removed. Do not remove the seals entirely. Fold back the seal on both sides to access the grille opening reinforcement mounting bolts and remove the bolts.

7. Remove the grille.

8. Remove the upper brace bolt from each of the 2 radiator braces.

9. Remove the crossmember-to-radiator mounting nuts.

10. Working through the grille opening, remove the lower bracket bolt securing the lower part of the hood latch or hood latch cable from the crossmember.

11. Lift the crossmember straight up and position it aside.

12. If equipped with A/C, remove the 2 A/C condenser-to-radiator mounting bolts which also retain the side mounted rubber air seals.

13. If not equipped with A/C, remove the bolts retaining the side mounted rubber air seals compressed

between the radiator and crossmember.

NOTE: Note the location of the air seals. To prevent overheating, they must be installed in their original positon.

14. Disconnect the coolant reservoir/overflow tank hose from the radiator.

15. Disconnect the upper hose from the radiator.

16. Carefully lift the radiator a slight amount and disconnect the lower hose from the radiator.

17. Lift the radiator up and out of the engine compartment, take care not to scrape the fins or disturb the A/C condenser if equipped.

NOTE: If equipped with an auxiliary automatic transmission oil cooler, use caution during radiator removal. The oil cooler lines are routed through a rubber air seal on the left side of the radiator. Do not cut or tear this seal.

To install:

18. Lower the radiator into the vehicle. Guide the alignment dowels into the hoses in the rubber air seals and then through the A/C support brackets, if equipped. Continue to guide the radiator through the rubber grommets located in the lower crossmember.

NOTE: On vehicles equipped with A/C, the L-shaped brackets, located on the bottom of the condenser, must be positioned between the bottom of the rubber air seals and top of rubber grommets.

19. Connect the lower radiator hose to the radiator.

20. Connect the upper radiator hose to the radiator.

21. If equipped with A/C, install the bolts condenser-to-radiator mounting bolts.

22. If not equipped with A/C, install the rubber air seal retaining bolts.

23. Connect the reservoir/overflow tank hose to the radiator.

24. If the radiator-to-upper crossmember rubber insulators were removed, install them.

25. Install the hood latch support bracket-to-lower frame crossmember bolt.

26. Install the bolts securing the upper radiator crossmember to the body.

27. Install the radiator-to-upper crossmember nuts.

28. Install a bolt to each upper radiator brace.

29. Install the grille.

30. Position the rubber seal and push down on the clips until seated.

31. If equipped, connect the transmission cooling lines.

32. Install the fan shroud with the fan.

33. Install the fan and shroud.

34. Rotate the fan blades and ensure they do not interfere with the shroud and at least 1.0 inch (25mm) of clearance is allowed. Correct as necessary.

35. Fill the cooling system.

Cooling Fan

REMOVAL AND INSTALLATION

Wrangler 1991-95, Cherokee 1991-95, Comanche 1991-92 and Grand Wagoneer 1991

VISCOUS DRIVE FAN

1. Remove the upper fan shroud bolts and lift the shroud from its lower securing tabs.

2. Remove the accessory drive belts.

3. Remove the fan flange-to-pulley mounting nuts.

4. Remove the fan and viscous drive as an assembly.

5. Remove the fan blade-to-viscous drive bolts and separate the assembly.

To install:

6. Position the fan on the viscous drive. Install the bolts and torque them to 187 inch lbs. (24 Nm).

7. Position the mounting flange of the viscous drive assembly onto the pulley. Install the nuts and torque them to 18 ft. lbs. (24 Nm).

8. Install the accessory drive belts.

9. Insert the shroud into its retaining tabs and install the upper bolts.

AUXILIARY ELECTRIC FAN

1. Disconnect the negative battery cable.

2. Disconnect the electrical connector.

3. Remove the upper fan shroud bolts.

4. Lift the fan assembly up and out of the engine compartment.

To install:

5. Insert the shroud into its retaining tabs and install the upper bolts.

6. Connect the electrical connector.

7. Connect the negative battery cable.

Grand Cherokee 1993-95

VISCOUS DRIVE FAN

1. Remove the upper fan shroud bolts and lift the shroud from its lower securing tabs.

2. Remove the accessory drive belts.

3. Remove the fan flange-to-pulley mounting nuts.

4. Remove the fan and viscous drive as an assembly.

5. Remove the fan blade-to-viscous drive bolts and separate the assembly.

To install:

6. Position the fan on the viscous drive. Install the bolts and torque them to 187 inch lbs. (24 Nm).

7. Position the mounting flange of the viscous drive assembly onto the pulley. Install the nuts and torque them to 18 ft. lbs. (24 Nm).

8. Install the accessory drive belts.

9. Insert the shroud into its retaining tabs and install the upper bolts.

AUXILIARY ELECTRIC FAN

1. Disconnect the negative battery cable.

2. Disconnect the electrical connector.

3. Remove the upper fan shroud bolts.

4. Lift the fan assembly up and out of the engine compartment.

To install:

5. Insert the shroud into its retaining tabs and install the upper bolts.

6. Connect the electrical connector.

7. Connect the negative battery cable.

Grand Wagoneer 1993

1. Disconnect the negative battery cable.

2. The viscous fan drive and blade assembly is threaded into the water pump hub shaft. Remove the fan drive and blade assembly from the water pump by turning the mounting nut counterclockwise as viewed from the front while securing the water pump pulley. Do not remove or unbolt the fan drive and blade at this time.

NOTE: The threads on the viscous fan drive are right hand threaded.

3. Remove the 2 fan shroud-to-upper crossmember nuts.

4. Remove the fan drive, blade and shroud as an assembly.

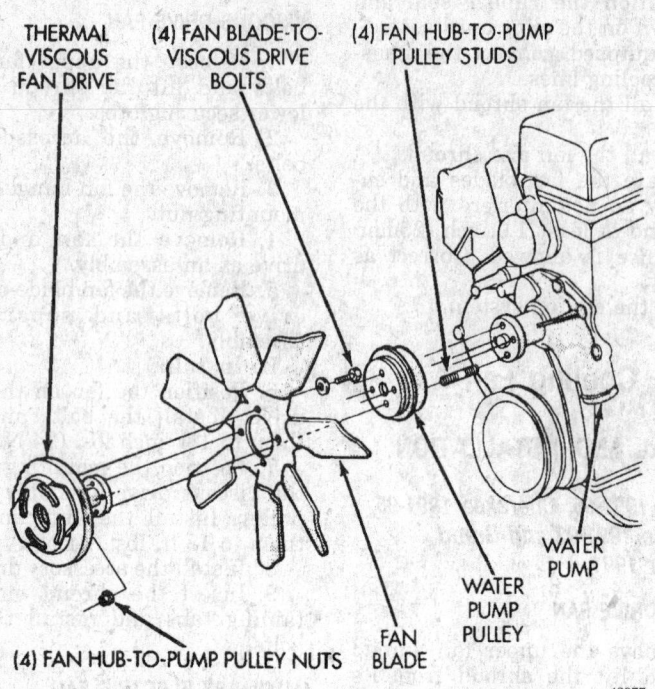

THERMAL VISCOUS FAN DRIVE

(4) FAN BLADE-TO-VISCOUS DRIVE BOLTS

(4) FAN HUB-TO-PUMP PULLEY STUDS

WATER PUMP

WATER PUMP PULLEY

FAN BLADE

(4) FAN HUB-TO-PUMP PULLEY NUTS

42977

Viscous fan removal and installation - Grand Cherokee

───── WARNING ─────

Do not place the viscous fan drive in a horizontal position. If stored horizontally, silicone fluid in the viscous fan drive could drain into its bearing assembly and the assembly would have to be replaced.

───── CAUTION ─────

Do not remove the water pump pulley-to-water pump bolts. The pulley is under spring tension.

5. Remove the 4 bolts securing the fan blade assembly to the viscous fan drive.

To install:

6. Install the fan blade on the viscous drive. Install the bolts and torque them to 17 ft. lbs. (23 Nm).

7. Position the fan shroud, viscous fan and blade into the engine compartment as an assembly.

8. Position the fan shroud to the radiator. Insert the lower slots of the shroud into the crossmember. Install the upper attaching nuts.

NOTE: Ensure the upper and lower portions of the fan shroud are firmly connected. All air must flow through the radiator.

9. Install the fan drive and blade assembly to the water pump shaft and tighten the nut.

NOTE: Ensure there is at least 1 inch (25mm) between the tips of the fan blades and shroud.

10. Connect the negative battery cable.

Water Pump

REMOVAL AND INSTALLATION

Wrangler 1991-95, Cherokee 1991-95 and Comanche 1991-92

NOTE: Some vehicles use a serpentine drive belt and have a reverse rotating water pump coupled with a viscous fan drive assembly. The components are identified by the words REVERSE stamped on the cover of the viscous drive and on the inner side of the fan. The word REV is also cast into the body of the water pump.

1. Disconnect the negative battery cable.
2. Drain the cooling system.
3. Disconnect the hoses at the pump.
4. Remove the drive belts.

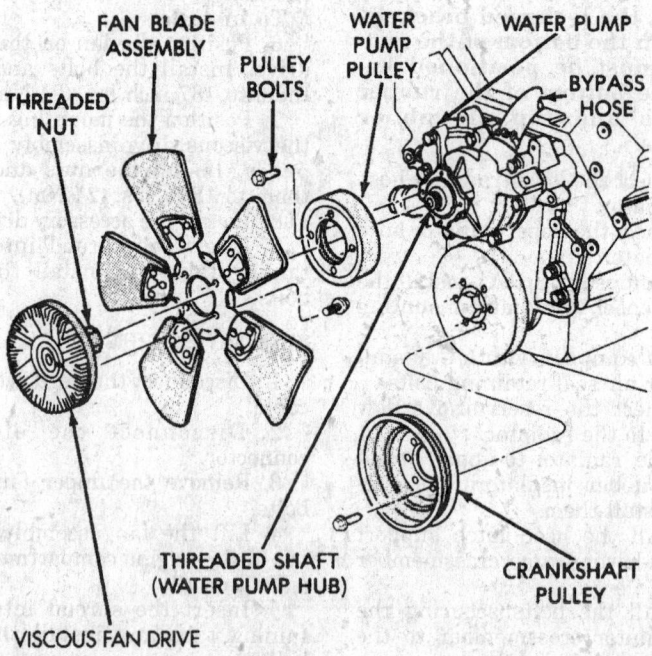

FAN BLADE ASSEMBLY

PULLEY BOLTS

WATER PUMP PULLEY

WATER PUMP

BYPASS HOSE

THREADED NUT

THREADED SHAFT (WATER PUMP HUB)

CRANKSHAFT PULLEY

VISCOUS FAN DRIVE

42987

Exploded view of the fan drive and blade - Grand Wagoneer

5. Remove the power steering pump bracket.

6. Remove the fan and shroud.

7. If equipped, remove the idler pulley to gain clearance for pump removal.

8. Unbolt and remove the pump.

To install:

9. Clean the mating surfaces thoroughly.

10. Using a new gasket, install the pump and torque the bolts to 13 ft. lbs. (18 Nm).

11. If removed, install the idler pulley.

12. Reconnect the hoses at the pump and install accessory drive belt.

13. Install the power steering pump bracket. Install the fan and shroud.

14. Adjust the belt tension and fill the cooling system to the correct level.

15. Operate the engine with the heater control valve in the **HEAT** position until the thermostat opens to purge air from the system. Check coolant level and fill as required.

Grand Cherokee 1993-95

NOTE: **Some vehicles use a serpentine drive belt and have a reverse rotating water pump coupled with a viscous fan drive assembly. The components are identified by the words REVERSE stamped on the cover of the viscous drive and on the inner side of the fan. The word REV is also cast into the body of the water pump.**

1. Disconnect the negative battery cable.

2. Drain the cooling system.

3. Disconnect the hoses at the pump.

4. Remove the drive belts.

5. Remove the power steering pump bracket.

6. Remove the fan and shroud.

7. If equipped, remove the idler pulley to gain clearance for pump removal.

8. Unbolt and remove the pump.

To install:

9. Clean the mating surfaces thoroughly.

10. Using a new gasket, install the pump and torque the bolts to 13 ft. lbs. (18 Nm).

11. If removed, install the idler pulley.

12. Reconnect the hoses at the pump and install accessory drive belt.

13. Install the power steering pump bracket. Install the fan and shroud.

14. Adjust the belt tension and fill the cooling system to the correct level.

15. Operate the engine with the heater control valve in the **HEAT** position until the thermostat opens to purge air from the system. Check coolant level and fill as required.

Grand Wagoneer 1991

1. Disconnect the negative battery cable.

2. Drain the cooling system.

3. Disconnect the upper radiator hoses at the radiator.

4. Loosen the accessory drive belts.

5. Separate the fan shroud from the radiator and remove the fan assembly from the water pump.

6. Remove the water pump pulley.

7. Remove the fan shroud.

8. Remove the alternator front mounting bracket.

9. If equipped with air conditioning, remove the compressor bolts and pivot the compressor aside. Do not disconnect the compressor lines.

10. If equipped with power steering remove the stud/nut retaining the bracket and pivot the power steering/air pump assembly forward.

11. Disconnect the heater hose and bypass hose at the water pump.

12. Disconnect the lower radiator hose at the water pump.

13. Remove the water pump mounting bolts and remove the pump from the engine.

To install:

14. Clean the gasket mating surfaces.

15. Install the water pump and gasket in position. Tighten the water pump-to-timing case bolts to 48 inch lbs. (5 Nm) and water pump-to-block bolts to 28 ft. lbs. (38 Nm).

16. Install the power steering pump, air pump and bracket assembly.

17. Connect the heater hose, bypass hose and lower radiator hose.

18. Pivot the air conditioning compressor into position.

19. Install the alternator front mounting brackets.

20. Position the shroud against the engine and install the fan assembly.

21. Install the shroud to the radiator and install the accessory drive belts.

22. Connect the upper radiator hose.

23. Connect the negative battery cable.

24. Fill the cooling system to the correct level and bleed the system.

Grand Wagoneer 1993

1. Disconnect the negative battery cable.

2. Open the radiator valve and drain the cooling system.

3. Remove the cooling fan and shroud as an assembly.

4. Remove the accessory drive belt.

5. Remove the water pump pulley from the hub.

6. Disconnect the hoses from the water pump.

7. Loosen the heater hose coolant return tube mounting bolt and nut and remove the tube. Discard the O-ring.

8. Remove the water pump mounting bolts.

9. Loosen the clamp at the water pump end of the bypass hose. Slip the bypass hose from the water pump while removing the pump from the engine. Discard the gasket.

To install:

10. Clean all gasket mating surfaces.

11. Guide the water pump and new gasket into position while connecting the bypass hose to the pump. Torque the water pump bolts to 30 ft. lbs. (40 Nm).

12. Install the bypass hose clamp.

13. Spin the water pump to ensure the pump impeller does not rub against the timing chain cover.

14. Coat a new O-ring with coolant and install it to the heater hose coolant return tube.

15. Install the coolant return tube to the engine. Ensure the slot in the tube bracket is bottomed to the mounting bolt. This will properly position the return tube.

16. Connect the radiator hose to the water pump.

17. Connect the heater hose and clamp to the return tube.

18. Install the water pump pulley and torque the bolts to 20 ft. lbs. (27 Nm).

19. Install the accessory drive belt.

20. Install the cooling fan and shroud.

21. Fill the cooling system.

22. Connect the negative battery cable.

23. Start the engine and check for leaks.

Thermostat

REMOVAL AND INSTALLATION

Wrangler 1991-95, Cherokee 1991-95, Comanche 1991-92, Grand Cherokee 1993-95 and Grand Wagoneer 1993

1. Disconnect the negative battery cable.
2. Remove the necessary hoses from the thermostat housing.
3. If necessary, disconnect the coolant temperature sensor electrical connector.
4. Remove the 2 attaching screws and lift the housing from the engine.
5. Remove the thermostat and gasket.
To install:
6. Clean all gasket surfaces thoroughly.
7. Place the thermostat in the housing with the spring inside the engine.
8. Install a new gasket with a small amount of sealing compound applied to both sides.
9. Install the water outlet and tighten the attaching bolts to 30 ft. lbs. (41 Nm).
10. Connect the hoses and if disconnected, the coolant temperature sensor connector to the housing.
11. Refill the cooling system.

Grand Wagoneer 1991

1. Disconnect the negative battery cable.
2. If equipped with A/C, perform the following:
   a. Remove the accessory drive belt.
   b. Remove the alternator without disconnecting the wiring and position it aside.
3. Disconnect the upper radiator hose from the thermostat housing.
4. Position the wiring harness behind the thermostat housing.
5. Remove the thermostat housing bolts, housing and thermostat. Discard the gasket.
To install:
6. Clean the gasket mating surfaces.
7. Install the thermostat with the spring side down.
8. Install the gasket over thermostat.
9. Install the thermostat housing with the word FRONT embossed on the housing placed toward the front of the vehicle. Torque the bolts to 200 inch lbs. (23 Nm).

CAUTION

The housing must be tightened evenly and the thermostat must be centered into the recessed groove in the intake manifold. Failure to do so could result in a cracked housing, damaged intake manifold threads or coolant leak.

---

10. Install the upper radiator hose to the thermostat housing.
11. If equipped with A/C, install the alternator and drive belt.
12. Fill the cooling system.
13. Connect the negative battery cable.
14. Start the engine and check for leaks.

Blower Motor

REMOVAL AND INSTALLATION

Wrangler 1991-95

NOTE: It is not necessary to discharge the refrigerant system.

1. Disconnect the negative battery cable.
2. Remove the hose clamps and dash grommet retaining screws.
3. Remove the evaporator housing-to-instrument panel screws and the housing mounting bracket screw.
4. Lower the evaporator housing to gain access to the blower motor attaching screws.
5. Remove the blower motor attaching screws and remove the blower motor. Disconnect the blower motor wiring and remove the blower motor from the vehicle.
To install:
6. Install the blower motor and connect the wiring. Install the blower motor attaching screws.
7. Position the evaporator housing and install the housing-to-instrument panel screws and housing mounting bracket screw.
8. Install the dash grommet retaining screws and hose clamps.
9. Connect the negative battery cable.

Cherokee 1991-95 and Comanche 1991-92

NOTE: The blower motor is removed from the engine compartment.

1. Disconnect the negative battery cable.
2. If equipped with 4.0L engine, proceed as follows:
   a. Remove the coolant bottle retaining strap and move the bottle

aside. Remove the coolant bottle bracket.
   b. If equipped with anti-lock brakes, remove the anti-lock brake pump and bracket and position it aside.
3. Unplug the blower motor wiring connector.
4. Remove the blower motor mounting screws and lift out the motor.
To install:
5. Install the blower motor into position and connect the electrical leads.
6. If equipped with 4.0L engine, proceed as follows:
   a. Install the anti-lock brake pump and bracket assembly, if equipped.
   b. Install the coolant bottle bracket and coolant bottle.
7. Connect the negative battery cable.

Grand Cherokee 1993-95 and Grand Wagoneer 1993

1. Disconnect the negative battery cable.
2. Disconnect the blower motor cooling tube.
3. Unplug the blower motor wiring connector.
4. Remove the blower motor mounting screws and lift out the motor.
To install:
5. Make sure the seal is installed on the blower motor housing.
6. Install the blower motor into position and install the mounting screws.
7. Connect the wiring connector.
8. Connect the blower motor cooling tube.
9. Connect the negative battery cable.

Grand Wagoneer 1991

1. Disconnect the negative battery cable.
2. Disconnect the blower motor wiring connector.
3. Remove the blower motor attaching screws and remove the blower motor.
4. Installation is the reverse of the removal procedure.

Heater Core

REMOVAL AND INSTALLATION

Wrangler 1991-95

1. Disconnect the negative battery cable.

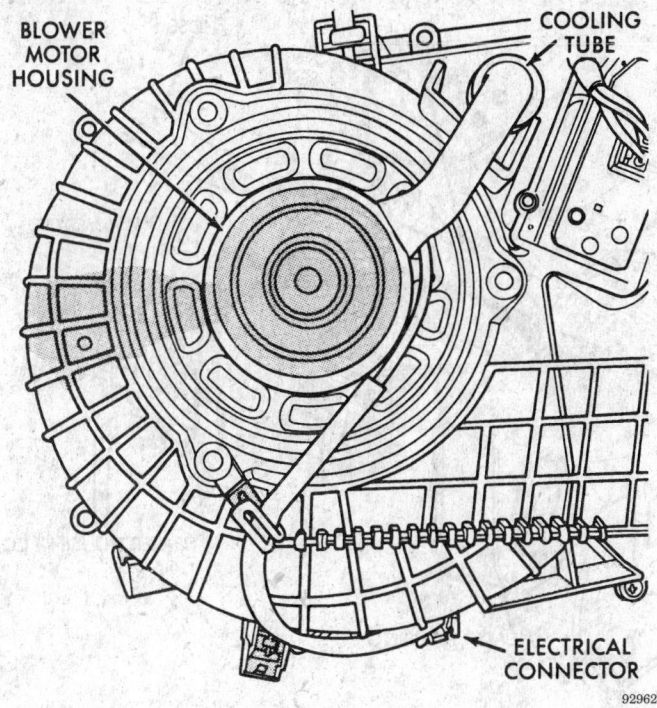

BLOWER
MOTOR
HOUSING

COOLING
TUBE

ELECTRICAL
CONNECTOR

92962

Blower motor

2. Drain the coolant.

3. Disconnect the heater hoses at the core tubes.

4. Disconnect the vent door cables and the blower motor electrical connectors.

5. Disconnect the defroster duct.

6. Remove the nuts attaching the heater housing studs to the engine compartment side of the dash.

7. Remove the heater housing assembly by tilting it downward, to disengage it from the defroster duct, and pulling it rearward and out from under the instrument panel.

8. Remove the heater housing cover and remove the heater core.

To install:

9. Install the heater core in the housing and install the cover.

10. Position the heater housing on the dash panel and install the seals on the heater core outlet and inlet tubes, and over the blower housing.

11. Install the attaching nuts and tighten alternately until 2 threads are visible beyond each nut.

NOTE: Overtightening the housing nuts can cause the housing to distort, resulting in air leaks.

12. Connect the defroster duct to the housing and the blower motor electrical connection.

13. Connect the vent door control cables and heater hoses.

14. Fill and bleed the cooling system.

15. Connect the negative battery cable and check the system for proper operation.

Cherokee 1991-95 and Comanche 1991-92

1. Disconnect the negative battery cable.

2. Drain the coolant.

3. Disconnect the heater hoses at the core tubes.

4. If equipped with air conditioning, properly discharge the refrigerant.

5. Disconnect the air conditioning hose from the expansion valve and cap all openings. Always use a backup wrench on the fitting.

6. Disconnect the blower motor wires and vent tube.

7. Remove the center console, if equipped.

8. Remove the lower instrument panel.

9. Disconnect the wiring at the air conditioner relay, blower motor resistors and air conditioner thermostat. Disconnect the vacuum hoses at the vacuum motor.

10. Cut the plastic retaining strap that retains the evaporator housing to the heater core housing.

11. Disconnect and remove the heater control cable.

12. Remove the 3 clips at the rear blower housing flange and remove the retaining screws.

13. Remove the housing attaching nuts from the studs on the engine compartment side of the firewall.

14. Remove the evaporator drain tube.

15. Remove the right kick panel and the instrument panel support bolt.

16. Gently pull out on the right side of the dash and rotate the housing down and toward the rear to disengage the mounting studs from the firewall. Remove the housing.

17. Unbolt and remove the core from the housing.

To install:

18. Install the core in the housing.

19. Position the housing on the mounting studs on the firewall.

20. Install the right kick panel and the instrument panel support bolt.

21. Install the evaporator drain tube.

22. Install the housing attaching nuts from the studs on the engine compartment side of the firewall.

23. Install the 3 clips at the rear blower housing flange and install the retaining screws.

24. Connect the heater control cable.

25. Install a new plastic retaining strap that retains the evaporator housing to the heater core housing.

26. Connect the wiring at the air conditioner relay, blower motor resistors and air conditioner thermostat.

27. Connect the vacuum hoses at the vacuum motor.

28. Install the lower half of the instrument panel.

29. Install the center console, if equipped.

30. Connect the blower motor wires and vent tube.

31. Connect the air conditioning hose at the expansion valve. Always use a backup wrench.

32. Connect the heater hoses at the core tubes.

33. Fill the cooling system.

34. Evaluate, charge and leak test the refrigerant system.

Grand Cherokee 1993-95 and Grand Wagoneer 1993

1. Disconnect the negative battery cable.

2. Properly discharge the A/C system.

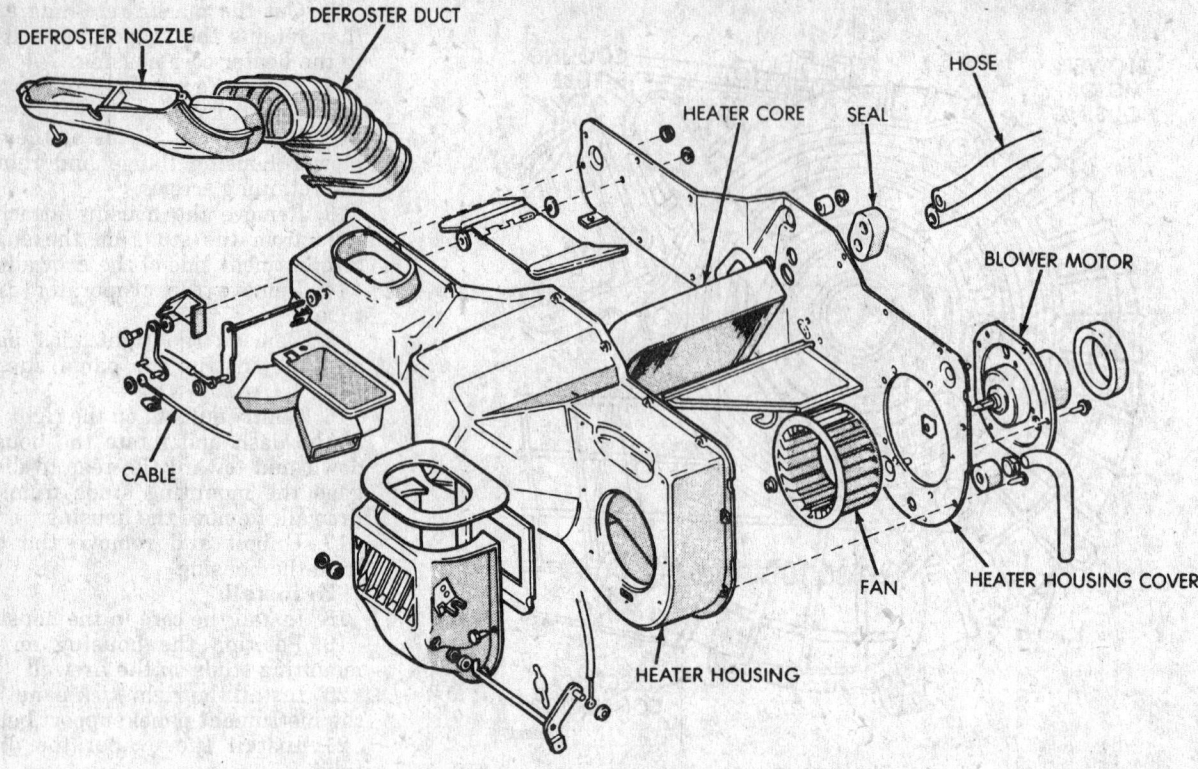

DEFROSTER NOZZLE

DEFROSTER DUCT

HOSE

HEATER CORE

SEAL

BLOWER MOTOR

CABLE

FAN

HEATER HOUSING COVER

HEATER HOUSING

47146

Heating system components - Wrangler

3. Disconnect the A/C hoses from the evaporator lines. Cap the openings to prevent the entrance of dirt or moisture.

4. Drain the cooling system.

5. Disconnect the heater hoses from the heater core tubes.

6. Remove the coolant reservoir/overflow tank.

7. Separate the Powertrain Control Module (PCM) from its mounting bracket and position it aside. Do not disconnect the wire harness.

8. Remove the heater-A/C unit attaching nuts from the studs on the engine compartment side of the dash panel.

9. Working inside the vehicle, remove the defroster duct bezel from the instrument panel.

10. Remove the speaker grilles.

11. Remove the upper instrument panel retaining nuts.

12. Remove the screws retaining the lower left side panel at the instrument panel. Remove the mounting bolt for the instrument panel at the left side cowl through the access hole provided.

13. Remove the ashtray.

14. Remove the instrument panel mounting screw located behind the ashtray.

15. Remove the instrument panel mounting bolt located on the right side cowl.

16. Fold down the carpet at the left side of the console and remove the 2 mounting screws.

17. Remove the lower column cover.

18. Remove the knee bolster.

19. Remove the tilt lever and both steering column covers.

20. Disconnect the steering column wiring.

21. Remove the nuts at the steering column mount and lower the column.

22. Remove the instrument panel mounting bolt above the steering column.

23. Disconnect the bulkhead connect at the left side of the dash panel.

24. Disconnect the wiring cluster at the lower left side of the dash panel.

25. Remove the right side kick panel access door and disconnect the wiring at the kick panel.

26. Pull back and lower the instrument panel. Disconnect the A/C vacuum line and antenna.

27. Remove the instrument panel.

28. Remove the defroster duct.

29. Disconnect the rear floor heat duct from the center adapter heat duct.

30. Disconnect the electrical connectors.

31. Remove the attaching nuts from the studs in the passenger compartment side of the dash panel.

32. Remove the heater-A/C unit from the vehicle.

33. Remove the heater core retaining screws.

34. Pull the heater core straight out from the housing.

To install

35. Install the heater core into the housing.

36. Position the clips over the heater core tubes. Install and tighten the screws.

37. Position the heater A/C unit into the dash panel. Be sure to position the drain tube into its hole.

38. Install the passenger compartment attaching nuts.

39. Install the attaching nuts on the engine compartment side of the dash panel.

40. Connect the heater hose to the heater core tubes.

41. Connect the A/C hoses to the evaporator lines.

42. Install the reservoir/overflow tank.

43. Install the PCM.

44. Install the defroster duct.

45. Connect the rear floor heat duct to the adaptor. Ensure the carpet is not interfering with heat duct outlets.

46. Connect the electrical connectors.

47. Connect the A/C vacuum lines and antenna to the instrument panel.

48. Position the instrument panel into place.

49. Connect the right side kick panel wiring harness.

50. Connect the wiring cluster at the lower left side of the instrument panel.

51. Connect the bulk head connector at the left side of the dash.

52. Install the instrument panel mounting screw above the steering column.

53. Position the steering column into place and install the mounting nuts.

54. Install the steering column covers and tilt wheel lever.

55. Install the knee bolster.

56. Install the lower column cover.

57. Install the screws located under the left side console carpet.

58. Install the instrument panel bolt located near the right side cowl.

59. Install the instrument panel bolt located behind the ashtray and install the ashtray.

60. Install the instrument panel bolt through the left side cowl.

61. Install the lower left side instrument panel mounting screws.

62. Install the upper instrument panel mounting bolts.

63. Install the speaker and defroster grilles.

64. Fill the cooling system.

65. Evacuate and recharge the A/C system.

66. Connect the negative battery cable.

67. Start the engine and check heater-A/C operations.

Grand Wagoneer 1991

1. Disconnect the negative battery cable.

2. Drain the cooling system.

3. Disconnect the temperature control cable from the blend-air door.

4. Disconnect the heater hoses from the heater core tubes.

5. Disconnect the blower motor resistor wires.

6. Remove the heater core housing-to-dash panel attaching screws.

7. Remove the heater core housing assembly.

8. Remove the attaching screws holding the halves together and separate the housing.

9. Remove the heater core-to-housing attaching screws and remove the heater core.

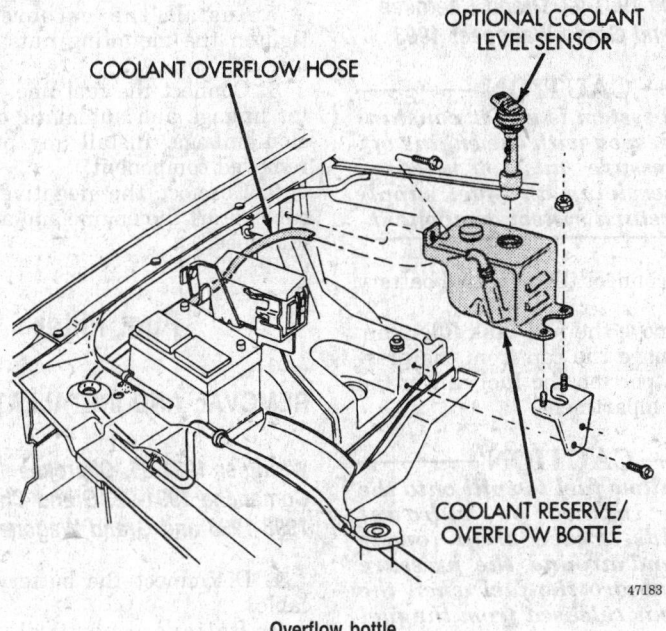

OPTIONAL COOLANT
LEVEL SENSOR

COOLANT OVERFLOW HOSE

COOLANT RESERVE/
OVERFLOW BOTTLE

47183

Overflow bottle

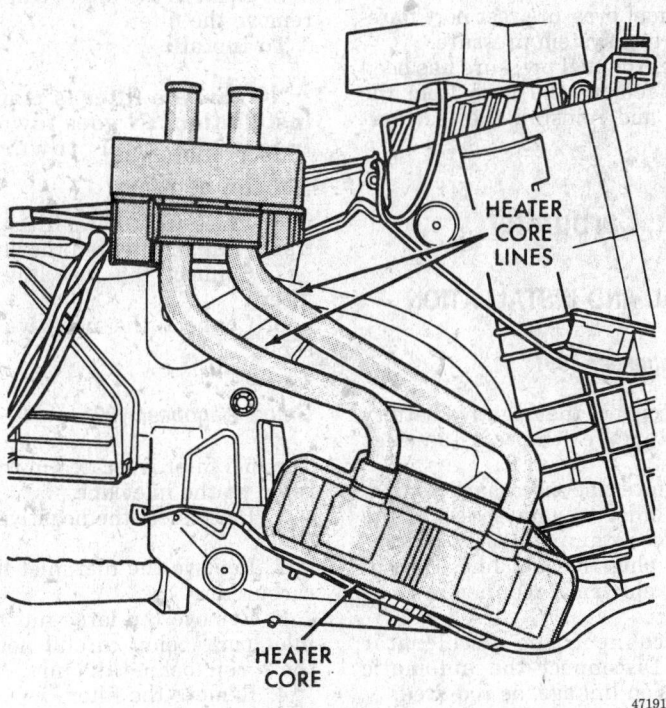

HEATER
CORE
LINES

HEATER
CORE

47191

Heater core

To install:

10. Install the heater core and tighten the attaching screws.

11. Assemble the housing together and tighten the screws.

12. Install the heater core housing assembly and tighten the mounting screws.

13. Connect the blower motor resistor wires.

14. Connect the heater hoses.

15. Connect the temperature control cable.

16. Fill the cooling system to the proper level.

17. Connect the negative battery cable.

FUEL SYSTEM

Idle Speed

ADJUSTMENT

Grand Wagoneer 1991

1. If equipped with automatic transmission, lock the parking brake, block the wheels and place the selector lever in **D** before adjusting the idle speed.

2. Connect a tachometer to the ignition coil negative terminal.

3. Start the engine and allow it to reach normal operating temperature.

4. Turn the hex head adjustment screw on the solenoid carriage to obtain the correct engine speed.

5. Disconnect the solenoid wire connector and adjust the curb idle speed screw to specification.

6. Reconnect the solenoid wire connector and stop the engine.

7. If equipped with a dashpot, position the throttle at the curb idle position and depress the dashpot stem.

8. Measure the clearance between the stem and the throttle lever. A clearance of 0.032 inch should exist.

9. Adjust the clearance as required by loosening the locknut and turning the dashpot until the correct clearance is obtained. Tighten the dashpot locknut.

REMOVAL AND INSTALLATION

Wrangler 1991-95, Cherokee 1991-95, Comanche 1991-92, Grand Cherokee 1993-95 and Grand Wagoneer 1993

——————— **CAUTION** ———————
The fuel system is under constant pressure, even with the engine off. Fuel pressure must be released before servicing any fuel supply or fuel return system component.

1. Disconnect the negative battery cable.

2. Remove the fuel tank filler cap.

3. Remove the cap from the pressure test port on the fuel rail in the engine compartment.

——————— **CAUTION** ———————
Do not allow fuel to spill onto the engine intake or exhaust manifolds. Place shop towels under and around the pressure port to absorb the fuel when the pressure is released from the fuel rail.

4. Place one end of the hose from a suitable fuel pressure gauge into an approved gasoline container.

5. Screw the other end of the hose onto the fuel pressure test port to relieve the fuel system pressure.

6. After the fuel pressure has been released, remove the hose from the test port and reinstall the test port cap.

Carburetor

REMOVAL AND INSTALLATION

Grand Wagoneer 1991

1. Disconnect the negative battery cable. Remove the air cleaner assembly.

2. Remove the necessary components in order to gain access to the carburetor retaining bolts. Disconnect and plug the fuel line. Disconnect all electrical connectors, as required.

3. Disconnect the accelerator linkage. Disconnect the automatic transmission linkage, as required.

4. Remove the carburetor retaining bolts. Remove the carburetor from the vehicle.

To install:

5. Clean the gasket mating surfaces thoroughly.

6. If equipped, position the spacer and gasket. Tighten the bolts alternately in a criss-cross pattern to 7 ft. lbs. (9 Nm).

7. Install the carburetor and tighten the mounting nuts to 16 ft. lbs. (22 Nm).

8. Connect the fuel line, accelerator linkage and automatic transmission linkage. Install any previously removed component.

9. Connect the negative battery cable, start the engine and adjust the idle speed.

Fuel Filter

REMOVAL AND INSTALLATION

Wrangler 1991-95, Cherokee 1991-95, Comanche 1991-92, Grand Cherokee 1993-1995 and Grand Wagoner 1993

1. Disconnect the battery ground cable.

2. Relieve the fuel system pressure.

3. Raise and support the rear of the vehicle safely.

4. Disconnect the fuel lines from the filter.

5. Remove the filter strap bolt and remove the filter.

To install:

NOTE: The filter is marked for installation. IN goes towards the fuel tank; OUT towards the engine.

6. Place the new filter on the frame rail and tighten the strap bolt.

7. Connect the fuel lines to the filter.

8. Connect the negative battery cable.

Grand Wagoneer 1991

The fuel filter is located in the carburetor, at the inlet line.

1. Disconnect the negative battery cable.

2. Remove the fuel inlet line from carburetor.

3. Remove the large nut from the inlet port, being careful not to lose the spring behind the nut.

4. Remove the filter.

To install:

5. Position the filter with the spring and install the nut.

6. Connect the fuel line.

7. Connect the negative battery cable.

Fuel Injector

REMOVAL AND INSTALLATION

Wrangler 1991-95, Cherokee 1991-95 and Comanche 1991-92

1. Disconnect the negative battery cable.
2. Relieve fuel system pressure.
3. Disconnect the fuel lines at the ends of the fuel rail assembly.
4. Mark and disconnect the injector wire harness connectors.
5. Remove the fuel rail retaining bolts.
6. Disconnect the vacuum line from the fuel pressure regulator.
7. Remove the fuel rail assembly from the engine.

NOTE: On models with automatic transmission, it may be necessary to remove the automatic transmission throttle pressure cable and bracket to remove the fuel rail assembly.

8. Remove the clips that retain the injectors to the fuel rail and remove the injectors.
   To install:
9. Install the injectors and clips.
10. Install the fuel rail and tighten the fuel rail mounting bolts to 20 ft. lbs. (27 Nm).
11. Connect the vacuum line to the fuel pressure regulator.
12. Connect the fuel injector electrical connectors.
13. Connect the fuel lines to the injectors.
14. Connect the negative battery cable.

Grand Cherokee 1993-95

1. Disconnect the negative battery cable.
2. Relieve the fuel system pressure.

3. Remove the air duct from the throttle body.
4. Disconnect the Manifold Absolute Pressure (MAP) sensor, Idle Air Control (IAC) motor and Throttle Position Sensor (TPS) electrical connectors from the throttle body.
5. Disconnect the vacuum line from the throttle body.
6. Disconnect (unsnap) the control cables from the throttle body (lever) arm.
7. Remove the throttle body from the intake manifold. Discard the gasket.
8. If equipped with A/C, disconnect the compressor-to-intake manifold support bracket.
9. Disconnect the electrical connectors from the fuel injectors.

NOTE: The fuel injector wiring harness is numerically tagged (INJ. 1, INJ. 2, etc.) for injector position identification.

10. Remove the EVAP canister purge solenoid/bracket assembly from the intake manifold.

NOTE: Do not attempt to disconnect the fuel line/tubes at the rear of the fuel rail. Fuel rail connections are made under the vehicle at the frame rail.

11. Raise and support the vehicle safely.
12. Disconnect the fuel rail quick-connect fittings at the fuel lines leading to the rear of the vehicle.
13. Lower the vehicle.
14. Remove the remaining fuel rail mounting bolts.

NOTE: Due not attempt to separate the fuel rail halves at the connecting hoses and due not attempt to install clamps to the hoses. When removing the fuel rail do not bend or kink these hoses.

15. Carefully rock the left fuel rail until the fuel injectors start to clear

the intake manifold. Repeat the procedure on the right side.
16. Remove the fuel rail, with the fuel injectors attached, from the engine.
17. Remove the clips retaining the injector to the fuel rail and remove the injectors.
    To install:
18. Coat each injector O-ring with engine oil.
19. Install the injectors and clips to the fuel rail.
20. Position each injector to the intake manifold.
21. Push the right side of the fuel rail down, taking care not to tear the O-ring, until the injector bottom. Repeat the procedure on the left side.
22. Install the fuel rail mounting bolts.
23. Install the EVAP canister purge solenoid to the intake manifold.
24. Connect the air temperature sensor electrical connector.
25. Connect the electrical connectors to the injectors.
26. If equipped with A/C, install the support bracket.
27. Clean the mating surfaces on the throttle body.
28. Install the throttle body with a new gasket and torque the bolts to 200 inch lbs. (23 Nm).
29. Connect the control cables. If equipped with automatic transmission, the throttle cable must be adjusted.
30. Connect the vacuum line and electrical connectors to the throttle body.
31. Install the air duct to the throttle body.
32. Connect the vacuum line to the fuel pressure regulator.
33. Raise and support the vehicle safely.
34. Connect the fuel rail lines.
35. Lower the vehicle.
36. Connect the negative battery cable.
37. Start the engine and check for leaks.

Grand Wagoneer 1993

1. Disconnect the negative battery cable.
2. Relieve the fuel system pressure.
3. Remove the air duct from the throttle body.
4. Disconnect the Manifold Absolute Pressure (MAP) sensor, Idle Air Control (IAC) motor and Throttle Position Sensor (TPS) electrical connectors from the throttle body.
5. Disconnect the vacuum line from the throttle body.

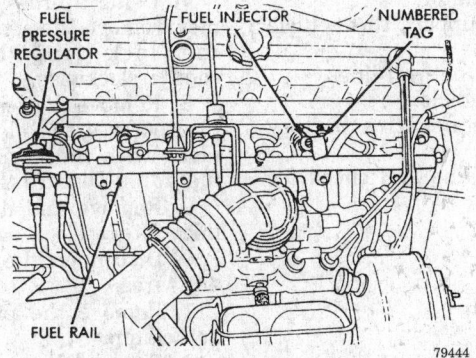

FUEL PRESSURE REGULATOR — FUEL INJECTOR — NUMBERED TAG

FUEL RAIL

79444

Fuel rail

6. Disconnect (unsnap) the control cables from the throttle body (lever) arm.

7. Remove the throttle body from the intake manifold. Discard the gasket.

8. If equipped with A/C, disconnect the compressor-to-intake manifold support bracket.

9. Disconnect the electrical connectors from the fuel injectors.

NOTE: The fuel injector wiring harness is numerically tagged (INJ. 1, INJ. 2, etc.) for injector position identification.

10. Remove the EVAP canister purge solenoid/bracket assembly from the intake manifold.

NOTE: Do not attempt to disconnect the fuel line/tubes at the rear of the fuel rail. Fuel rail connections are made under the vehicle at the frame rail.

11. Raise and support the vehicle safely.

12. Disconnect the fuel rail quick-connect fittings at the fuel lines leading to the rear of the vehicle.

13. Lower the vehicle.

14. Remove the remaining fuel rail mounting bolts.

NOTE: Do not attempt to separate the fuel rail halves at the connecting hoses and do not attempt to install clamps to the hoses. When removing the fuel rail do not bend or kink these hoses.

15. Carefully rock the left fuel rail until the fuel injectors start to clear the intake manifold. Repeat the procedure on the right side.

16. Remove the fuel rail, with the fuel injectors attached, from the engine.

17. Remove the clips retaining the injector to the fuel rail and remove the injectors.

To install:

18. Coat each injector O-ring with engine oil.

19. Install the injectors and clips to the fuel rail.

20. Position each injector to the intake manifold.

21. Push the right side of the fuel rail down, taking care not to tear the O-ring, until the injector bottom. Repeat the procedure on the left side.

22. Install the fuel rail mounting bolts.

23. Install the EVAP canister purge solenoid to the intake manifold.

24. Connect the air temperature sensor electrical connector.

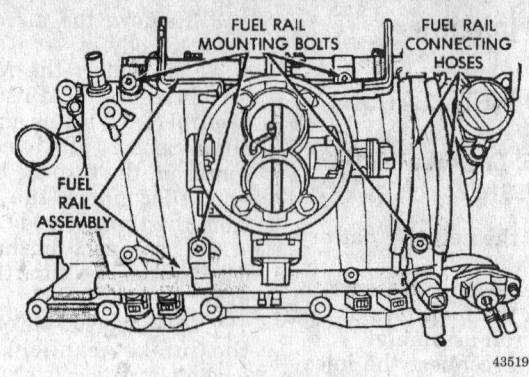

Fuel rail mounting bolts

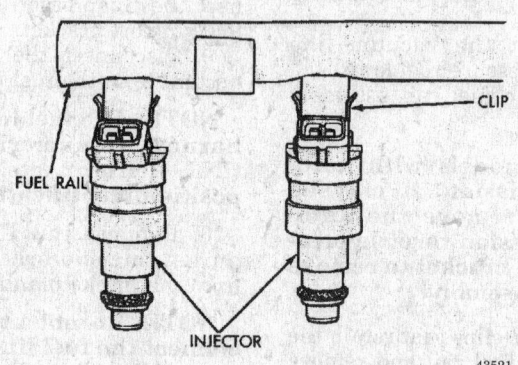

Fuel injector and rail assembly

25. Connect the electrical connectors to the injectors.

26. If equipped with A/C, install the support bracket.

27. Clean the mating surfaces on the throttle body.

28. Install the throttle body with a new gasket and torque the bolts to 200 inch lbs. (23 Nm).

29. Connect the control cables. If equipped with automatic transmission, the throttle cable must be adjusted.

30. Connect the vacuum line and electrical connectors to the throttle body.

31. Install the air duct to the throttle body.

32. Connect the vacuum line to the fuel pressure regulator.

33. Raise and support the vehicle safely.

34. Connect the fuel rail lines.

35. Lower the vehicle.

36. Connect the negative battery cable.

37. Start the engine and check for leaks.

Fuel Pump

REMOVAL AND INSTALLATION

Wrangler 1991-95, Cherokee 1991-95 and Comanche 1991-92

1. Disconnect the negative battery cable.

2. Remove the fuel tank filler cap.

3. Drain the fuel from the fuel tank.

4. Raise and safely support the rear of the vehicle.

5. Remove the fuel inlet and outlet hoses from the sending unit.

6. Remove the sending unit wires.

7. Using a brass punch and hammer, remove the sending unit retaining lock ring by tapping it counterclockwise.

8. Remove the sending unit, which incorporates the electric fuel pump, along with the O-ring seal from the fuel tank. Discard the O-ring.

9. Remove and discard the pump inlet filter.

10. Disconnect the fuel pump terminal wires.

11. Remove the pump outlet hose and clamp.

12. Remove the pump top mounting bracket nut and remove the pump.

To install:

13. Install a new inlet filter on the pump.

14. Assemble the pump and bracket.

15. Connect the hose and wiring.

16. Install the unit and new O-ring in the tank. The rubber stopper on the end of the fuel return tube must be inserted into the cup in the fuel tank reservoir.

17. Install the lock ring. Carefully tap it into place until it seats against the stop on the tank.

18. Connect the hoses.

19. Connect the wiring.

20. Lower the vehicle.

21. Refill the fuel tank.

22. Run the engine and check for leaks.

Grand Cherokee 1993-95 and Grand Wagoneer 1993

1. Disconnect the negative battery cable.

2. Remove the fuel tank filler cap.

3. Drain the fuel from the fuel tank.

4. Raise and safely support the rear of the vehicle.

5. Remove the fuel inlet and outlet hoses from the sending unit.

6. Remove the sending unit wires.

7. Using a brass punch and hammer, remove the sending unit retaining lock ring by tapping it counterclockwise.

8. Remove the sending unit, which incorporates the electric fuel pump, along with the O-ring seal from the fuel tank. Discard the O-ring.

NOTE: The fuel pump cannot be replaced separately. The sending unit and pump must be replaced as an assembly.

To install:

9. Install the unit and new O-ring in the tank. The rubber stopper on the end of the fuel return tube must be inserted into the cup in the fuel tank reservoir.

10. Install the lock ring. Carefully tap it into place until it seats against the stop on the tank.

11. Connect the hoses.

12. Connect the wiring.

13. Lower the vehicle.

14. Refill the fuel tank.

15. Run the engine and check for leaks.

Grand Wagoneer 1991

1. Disconnect the negative battery cable.

2. Disconnect the fuel lines from the pump.

3. Remove the attaching bolts holding the fuel pump to the engine and lift the fuel pump off of the engine.

To install:

4. Thoroughly clean all of the mating surfaces.

5. Cement a new gasket to the mating surface of the fuel pump.

6. Position the fuel pump on the cylinder block so the cam lever of the pump rests on the camshaft.

7. Secure the pump to the engine with the retaining bolts and lock washers. Tighten bolts to 13 ft. lbs. (18 Nm).

8. Connect the fuel lines to the fuel pump. Tighten the fuel lines to 18 ft. lbs. (25 Nm).

Fuel Tank

REMOVAL AND INSTALLATION

Wrangler 1991-95

1. Disconnect the negative battery cable.

2. Relieve the fuel system pressure.

3. Remove the fuel filler cap and drain the fuel tank.

4. Raise and support the vehicle safely.

5. Disconnect the fuel fill hose and vent hose from the filler neck.

6. Disconnect the fuel gauge electrical connectors.

7. Label and disconnect all other fuel and vent hoses attached to the fuel tank.

NOTE: The fuel tank and skid plate are removed as a unit. Do not loosen the fuel tank strap nuts. Remove the assembly by loosening the skid plate/fuel tank assembly mounting nuts.

8. Place a floorjack under the fuel tank.

9. Remove the strap nuts and pull the straps away from the tank.

10. Lower the tank from the vehicle.

To install:

11. Raise the fuel tank into position and connect all fuel filler and vent hoses.

12. Wrap the straps around the fuel tank and tighten the skid plate/fuel tank assembly mounting nuts to 65 inch lbs. (7 Nm).

13. Connect the fuel supply and vent hoses.

14. Connect the fuel gauge electrical connectors.

15. Lower the vehicle.

16. Fill the fuel tank.

17. Connect the negative battery cable.

18. Run the vehicle and check for leaks.

Cherokee 1991-95

1. Disconnect the negative battery cable.

2. Relieve the fuel system pressure.

3. Remove the fuel filler cap and drain the fuel tank.

4. Raise and support the vehicle safely.

5. Disconnect the fuel fill hose and vent hose from the filler neck.

6. Disconnect the fuel pump/gauge electrical connectors.

7. Label and disconnect all other fuel and vent hoses attached to the fuel tank.

8. Remove the skid plate.

9. Place a floorjack under the fuel tank.

10. Remove the strap nuts and pull the straps away from the tank.

11. Lower the tank from the vehicle.

To install:

12. Raise the fuel tank into position and connect all fuel filler and vent hoses.

13. Wrap the straps around the fuel tank and tighten the nuts to 100 inch lbs. (11 Nm).

14. Install the fuel tank skid plate.

15. Connect the fuel supply and vent hoses.

16. Connect the fuel pump/gauge electrical connectors.

17. Lower the vehicle.

18. Fill the fuel tank.

19. Connect the negative battery cable.

20. Run the vehicle and check for leaks.

Comanche 1991-92

1. Disconnect the negative battery cable.

2. Relieve the fuel system pressure.

3. Remove the fuel filler cap and drain the fuel tank.

4. Raise and support the vehicle safely.

5. Disconnect the fuel fill hose and vent hose from the filler neck.

6. Disconnect the fuel pump/gauge electrical connectors.

7. Label and disconnect all other fuel and vent hoses attached to the fuel tank.

8. If equipped, remove the skid plate.

9. Remove the fuel tank shield, if equipped.

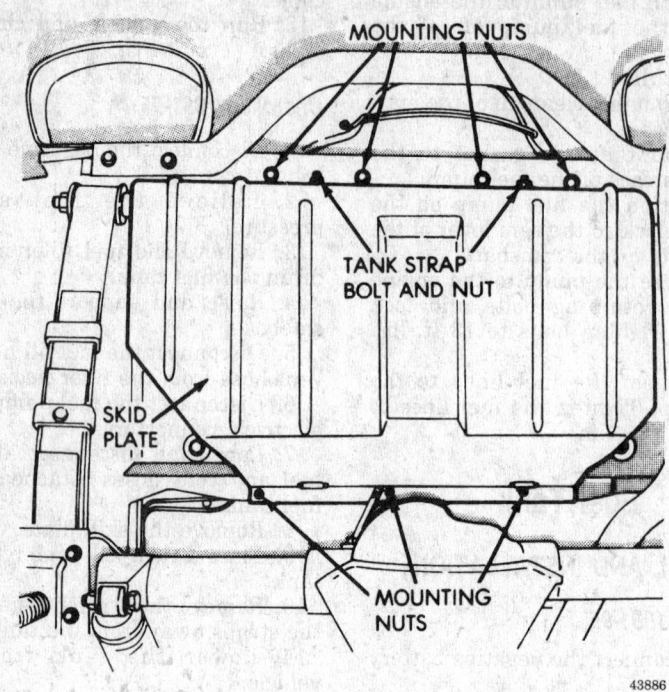

MOUNTING NUTS

TANK STRAP
BOLT AND NUT

SKID
PLATE

MOUNTING
NUTS

43886

Fuel tank mounting nut locations - Wrangler

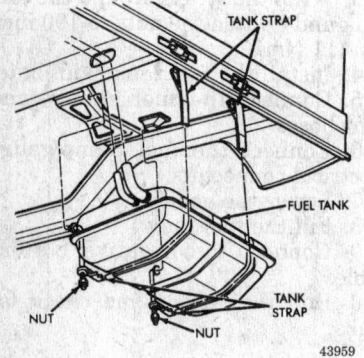

TANK STRAP

FUEL TANK

NUT   NUT

TANK
STRAP

43959

Fuel tank removal and installation - Cherokee

10. Place a floorjack under the fuel tank.

11. Remove the strap nuts and pull the straps away from the tank.

12. Lower the tank from the vehicle.

To install:

13. Raise the fuel tank into position and connect all fuel filler and vent hoses.

14. Wrap the straps around the fuel tank and tighten the nuts to 65 inch lbs. (7 Nm) for the 2 outer straps and 43 inch lbs. (5 Nm) for the center strap.

15. Install the tank shield and skid plate, if equipped.

16. Connect the fuel supply and vent hoses.

17. Connect the fuel pump/gauge electrical connectors.

18. Lower the vehicle.

19. Fill the fuel tank.

20. Connect the negative battery cable.

21. Run the vehicle and check for leaks.

Grand Cherokee 1993-95 and Grand Wagoneer 1993

1. Disconnect the negative battery cable.

2. Relieve the fuel system pressure.

3. Remove the fuel filler cap and drain the fuel tank.

4. Raise and support the vehicle safely.

5. Remove the 2 tow hooks.

6. Disconnect the fuel fill hose and vent hose from the filler neck.

7. Disconnect the fuel pump/gauge electrical connectors.

8. Label and disconnect all other fuel and vent hoses attached to the fuel tank.

9. If equipped, remove the skid plate.

10. Remove the trailer hitch, if equipped and exhaust tail pipe heat shield.

11. Remove the fuel tank shield.

12. Place a floorjack under the fuel tank.

13. Remove the strap nuts and pull the straps away from the tank.

14. Lower the tank from the vehicle.

NOTE: **The right side of the fuel tank must be lowered first to gain access to the 2 fuel filler hose clamps located on the left side of the tank.**

To install:

15. Raise the fuel tank into position and connect all fuel filler and vent hoses.

16. Tighten the 2 mounting nuts until 3.149 inches (80mm) is attained between the end of the mounting bolt and bottom of the strap.

17. Install the tank shield, skid plate, trailer hitch, exhaust shield and tow hooks, as required.

18. Connect the fuel supply and vent hoses.

19. Connect the fuel pump/gauge electrical connectors.

20. Lower the vehicle.

21. Fill the fuel tank.

22. Connect the negative battery cable.

23. Run the vehicle and check for leaks.

Grand Wagoneer 1991

1. Disconnect the negative battery cable.

2. Remove the fuel filler cap and drain the fuel tank.

3. Raise and support the vehicle safely.

4. Disconnect the fuel fill hose and vent hose from the filler neck.

5. Disconnect the fuel gauge electrical connectors.

6. Label and disconnect all other fuel and vent hoses attached to the fuel tank.

NOTE: **The fuel tank and skid plate are removed as a unit.**

7. Place a floorjack under the fuel tank.

8. Remove the fuel tank/skid plate assembly mounting bolts, then lower the assembly.

9. Lower the tank from the vehicle.

10. Remove the tank straps to remove the tank from the skid plate.

To install:

11. Place the tank onto the skid plate and wrap the straps around the tank.

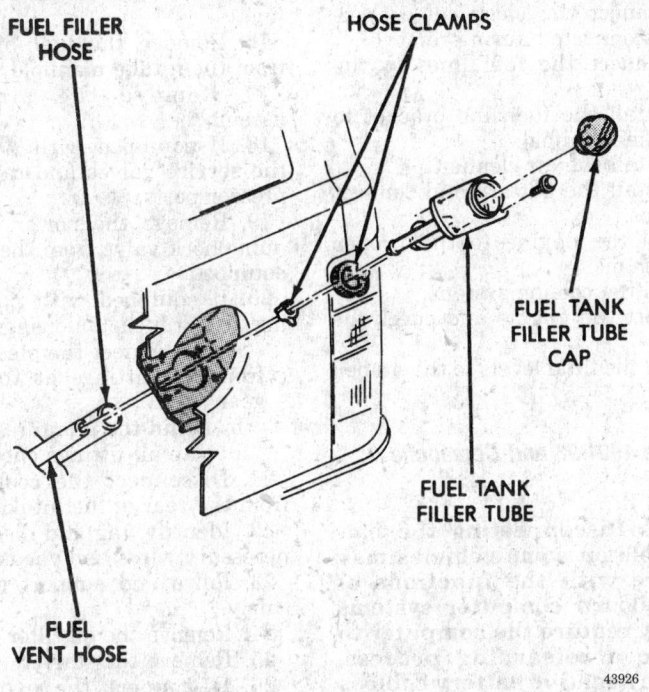

FUEL FILLER HOSE

HOSE CLAMPS

FUEL TANK FILLER TUBE CAP

FUEL TANK FILLER TUBE

FUEL VENT HOSE

43926

Fuel filler tube and hoses - Grand Cherokee and Grand Wagoneer

12. Raise the fuel tank into position and connect all fuel filler and vent hoses.

13. Install and tighten the fuel tank mounting bolts.

14. Connect the fuel supply and vent hoses.

15. Connect the fuel gauge electrical connectors.

16. Lower the vehicle.

17. Fill the fuel tank.

18. Connect the negative battery cable.

19. Run the vehicle and check for leaks.

EMISSION CONTROLS

Emission Warning Lamp

RESETTING

Wrangler 1991-93, Cherokee 1991-93 and Comanche 1991-92

1. Connect DRB-II or equivalent, to the diagnostic connector, located next to the electronic control unit in the engine compartment.

2. Follow the DRB-II Function Flow Diagram to reset the EMR light.

ENGINE MECHANICAL

Engine Assembly

REMOVAL AND INSTALLATION

Wrangler 1991-95

1. Place a protective cloth on the windshield frame. Raise the hood and rest it on the frame.

2. Disconnect the battery cables and remove the battery.

3. Properly relieve the fuel system pressure.

NOTE: Label all electrical connectors and vacuum lines prior to disconnecting them, so they can be reinstalled in their proper locations.

4. Drain the cooling system.

5. Disconnect the wires from the alternator.

6. Disconnect the ignition coil and distributor wire connections.

7. Disconnect the oil pressure sending unit connector.

8. Disconnect the wires from the starter.

9. Disconnect the fuel injection wires.

10. Disconnect the fuel lines from the fuel rails.

11. Remove the fuel line bracket from the intake manifold.

12. Disconnect the engine ground strap.

13. Remove the air cleaner assembly.

14. Disconnect the canister purge hose from the vapor canister "T" connector.

15. Disconnect the idle speed actuator wire connector.

16. Disconnect the throttle cable and remove it from the bracket.

17. Disconnect the throttle rod from the bellcrank.

18. If equipped, disconnect the cruise control cable.

19. Disconnect the oxygen sensor electrical connector.

20. Disconnect the upper and lower hoses from the radiator.

21. Disconnect the coolant hoses from the rear of the intake manifold and thermostat housing.

22. Disconnect the heater hoses.

23. Remove the fan shroud screws.

24. Remove the radiator and fan shroud.

25. Remove the engine cooling fan.

26. Remove the engine cooling fan and install a $5/16$ x $1/2$ inch capscrew through the fan pulley into the water pump flange. This will maintain the pulley and water pump in alignment when the crankshaft is rotated.

27. If equipped, disconnect the check valve from the power brake booster.

28. If equipped with power steering, perform the following:

   a. Disconnect the steering hoses from the fittings at the steering gear.

   b. Drain the pump reservoir.

   c. Cap all fittings once removed.

29. Raise and support the vehicle safely.

30. Remove the oil filter.

31. Remove the starter.

32. Remove the flywheel access cover.

33. Remove the engine support cushion-to-bracket through bolts.

34. Disconnect the exhaust pipe from the manifold.

35. Remove the upper flywheel housing bolts and loosen the bottom bolts.

36. Remove the engine shock damper bracket from the sill.

37. Lower the vehicle.

38. Attach a lifting device to the engine.

39. Place a support under the bellhousing.

40. Remove the remaining flywheel bolts.

41. Lift the engine from the vehicle.

42. Install the oil filter to keep foreign material out of the engine.

To install:

43. Remove the oil filter.

44. Lower the engine into the vehicle. To ease installation, remove the engine support cushions to aid in engine-to-transmission alignment.

45. Insert the transmission shaft into the clutch spline.

46. Align the flywheel housing with the engine.

47. Install and tighten the flywheel housing bolts finger-tight.

48. If removed, install the engine support cushions.

49. Lower the engine into place and remove the lifting device.

50. Raise and support the vehicle safely.

51. Attach the engine shock damper bracket to the sill.

52. Attach the exhaust pipe to the manifold and torque the nuts to 23 ft. lbs. (31 Nm).

53. Install the flywheel access cover.

54. Install the remaining flywheel bolts and torque them to 28 ft. lbs. (38 Nm).

55. Install the starter.

56. Install the oil filter.

57. Lower the vehicle.

58. Connect the coolant lines and tighten the clamps.

59. If equipped with power steering.

   a. Connect the hoses to the steering gear and torque the nut to 38 ft. lbs. (52 Nm).

   b. Fill the pump reservoir with fluid.

60. Remove the alignment capscrew and install the fan assembly.

61. Install the accessory drive belt.

62. Install the radiator and shroud.

63. Connect the radiator hoses.

64. Connect the oxygen sensor electrical connector.

65. Connect the throttle valve rod and retainer. Connect the throttle cable and install the rod and spring.

66. If equipped, connect the speed control cable.

67. Install the vacuum hose and check valve to the brake booster.

68. Connect the electrical connections disconnected during removal.

69. Connect the fuel lines to the fuel rail.

70. Install the fuel line bracket to the intake manifold.

71. Install the air cleaner.

72. Install the battery and connect the cables.

73. Fill the engine to the proper level with oil.

74. Fill the cooling system.

75. Start the engine and check for leaks.

76. Fill the fluid levels to the proper level.

Cherokee 1991-93 and Comanche 1991-92

NOTE: Disconnecting the battery cable on some vehicles may interfere with the functions of the on board computer systems and may require the computer to undergo a relearning process, once the negative battery cable is disconnected.

1. Disconnect the negative battery cable.

2. Properly relieve the fuel system pressure.

3. If equipped with A/C, properly discharge the system.

4. Matchmark the hood and hinges and remove the hood.

5. Drain the cooling system.

NOTE: Label all electrical connectors and vacuum lines prior to disconnecting them, so they can be reinstalled in their proper locations.

6. Remove the upper, lower and coolant recovery hoses.

7. Remove the fan shroud.

8. If equipped with an automatic transmission, disconnect the fluid cooler lines.

9. Remove the radiator and if equipped, A/C condenser.

10. Remove the engine cooling fan and install a $5/16$ x $1/2$ inch capscrew through the fan pulley into the water pump flange. This will maintain the pulley and water pump in alignment when the crankshaft is rotated.

11. Disconnect the heater hoses.

12. Disconnect the throttle linkages, speed control cable, if equipped and throttle valve rod.

13. Disconnect the oxygen sensor electrical connector.

14. Disconnect the fuel injection harness connectors.

15. Disconnect the quick-connection fuel lines at the fuel rail and return line.

16. Remove the fuel line bracket from the intake manifold.

17. Remove the air cleaner assembly.

18. If equipped with A/C, remove the service valves and cap the compressor ports.

19. Remove the power brake vacuum check valve from the booster, if equipped.

20. If equipped with power steering, perform the following:

   a. Disconnect the steering hoses from the fittings at the steering gear.

   b. Drain the pump reservoir.

   c. Cap all fittings once removed.

21. Disconnect the coolant hoses from the rear of the intake manifold.

22. Identify, tag and disconnect all necessary wires and vacuum lines.

23. Raise and support the vehicle safely.

24. Remove the oil filter.

25. Remove the starter.

26. Disconnect the exhaust pipe from the manifold.

27. Remove the flywheel/converter housing access cover.

28. If equipped with an automatic transmission, matchmark the converter to the drive plate and remove the bolts.

29. Remove the upper flywheel/converter housing bolts and loosen the bottoms bolts.

30. Remove the engine mount-to-engine compartment bracket bolts.

31. Remove the engine shock damper bracket from the sill.

32. Lower the vehicle.

33. Attach a lifting device to the engine.

34. Raise the engine slightly off the front supports.

35. Place a support stand under the transmission housing.

36. Remove the remaining flywheel bolts.

37. Lift the engine out of the vehicle.

38. Install the oil filter to keep foreign material out of the engine.

To install:

39. Remove the oil filter.

40. Lower the engine into the vehicle. To ease installation, remove the engine mounts to aid in engine-to-transmission alignment.

41. If equipped with a manual transmission, perform the following.

   a. Insert the transmission shaft into the clutch spline.

   b. Align the flywheel housing with the engine.

c. Install and tighten the fly-wheel housing bolts finger-tight.

42. If equipped with an automatic transmission, perform the following.

   a. Align the torque converter housing with the engine.

   b. Loosely install the converter housing lower bolts and install the next higher nut and bolt on each side.

   c. Tighten all 4 bolts finger-tight.

43. If removed, install the engine mounts.

44. Lower the engine into place and remove the lifting device.

45. Raise and support the vehicle safely.

46. If equipped with an automatic transmission, perform the following.

   a. Align the torque converter to the drive plate.

   b. Install the bolts and torque them to 40 ft. lbs. (54 Nm).

   c. Install the access cover.

   d. Install the exhaust pipe support.

47. Install the remaining converter/flywheel bolts finger-tight.

48. Install the starter.

49. Tighten the engine support cushion bolts/nuts.

50. Torque the loose converter/flywheel bolts to 28 ft. lbs. (38 Nm).

51. Install the oil filter.

52. Connect the exhaust pipe to the manifold.

53. Lower the vehicle.

54. Connect the coolant hoses and tighten the clamps.

55. If equipped with power steering, perform the following:

   a. Unplug the lines and connect them to the steering gear. Torque the fittings to 38 ft. lbs. (52 Nm).

   b. Fill the pump reservoir with fluid.

56. Remove the alignment cap screw and install the fan.

57. Install the radiator, condenser, if equipped and fan shroud.

58. Connect the radiator hoses.

59. If equipped with an automatic transmission, connect the cooling lines.

60. Connect the oxygen sensor electrical connector.

61. Connect the throttle valve rod and retainer. Connect the throttle cable and install the rod and spring.

62. If equipped, connect the cruise control cable.

63. Connect the fuel lines to the throttle body.

64. Connect all vacuum lines and electrical connectors disconnected during removal.

65. If equipped with A/C, connect the service valves to the compressor ports.

66. Install the air cleaner.

67. Install the hood.

68. Connect the battery cables.

69. Fill the cooling system.

70. Start the engine and check for leaks.

71. If equipped, recharge the A/C system.

72. Check and top off fluid levels.

Cherokee 1994-95

NOTE: Disconnecting the battery cable on some vehicles may interfere with the functions of the on board computer systems and may require the computer to undergo a relearning process, once the negative battery cable is disconnected.

1. Disconnect the negative battery cable.

2. Properly relieve the fuel system pressure.

3. If equipped with A/C, properly discharge the system.

4. Matchmark the hood and hinges and remove the hood.

5. Drain the cooling system.

NOTE: Label all electrical connectors and vacuum lines prior to disconnecting them, so they can be reinstalled in their proper locations.

6. Remove the upper, lower and coolant recovery hoses.

7. Remove the fan shroud.

8. If equipped with an automatic transmission, disconnect the fluid cooler lines.

9. Remove the radiator and if equipped, A/C condenser.

10. Remove the engine cooling fan and install a $5/16$ x $1/2$ inch capscrew through the fan pulley into the water pump flange. This will maintain the pulley and water pump in alignment when the crankshaft is rotated.

11. Disconnect the heater hoses.

12. Disconnect the throttle linkages, speed control cable, if equipped and throttle valve rod.

13. Disconnect the oxygen sensor electrical connector.

14. Disconnect the fuel injection harness connectors.

15. Disconnect the quick-connection fuel lines at the fuel rail and return line.

16. Remove the fuel line bracket from the intake manifold.

17. Remove the air cleaner assembly.

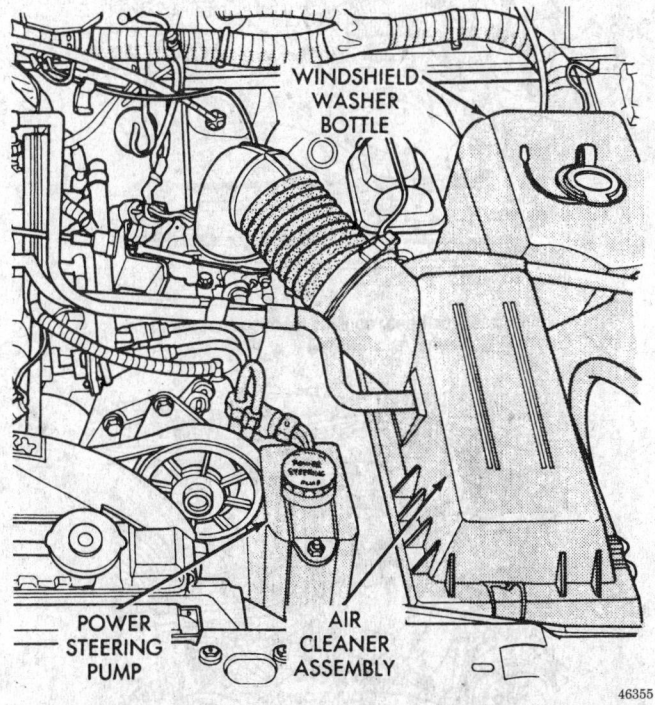

WINDSHIELD WASHER BOTTLE

POWER STEERING PUMP

AIR CLEANER ASSEMBLY

46355

Power steering pump - Cherokee and Comanche

18. If equipped with A/C, remove the service valves and cap the compressor ports.

19. Remove the power brake vacuum check valve from the booster, if equipped.

20. If equipped with power steering, perform the following:

  a. Disconnect the steering hoses from the fittings at the steering gear.

  b. Drain the pump reservoir.

  c. Cap all fittings once removed.

21. Disconnect the coolant hoses from the rear of the intake manifold.

22. Identify, tag and disconnect all necessary wires and vacuum lines.

23. Raise and support the vehicle safely.

24. Remove the oil filter.

25. Remove the starter.

26. Disconnect the exhaust pipe from the manifold.

27. Remove the flywheel/converter housing access cover.

28. If equipped with an automatic transmission, matchmark the converter to the drive plate and remove the bolts.

29. Remove the upper flywheel/converter housing bolts and loosen the bottoms bolts.

30. Remove the engine mount-to-engine compartment bracket bolts.

31. Remove the engine shock damper bracket from the sill.

32. Lower the vehicle.

33. Attach a lifting device to the engine.

34. Raise the engine slightly off the front supports.

35. Place a support stand under the transmission housing.

36. Remove the remaining flywheel bolts.

37. Lift the engine out of the vehicle.

38. Install the oil filter to keep foreign material out of the engine.

To install:

39. Remove the oil filter.

40. Lower the engine into the vehicle. To ease installation, remove the engine mounts to aid in engine-to-transmission alignment.

41. If equipped with a manual transmission, perform the following.

  a. Insert the transmission shaft into the clutch spline.

  b. Align the flywheel housing with the engine.

  c. Install and tighten the flywheel housing bolts finger-tight.

42. If equipped with an automatic transmission, perform the following.

  a. Align the torque converter housing with the engine.

  b. Loosely install the converter housing lower bolts and install the next higher nut and bolt on each side.

  c. Tighten all 4 bolts finger-tight.

43. If removed, install the engine mounts.

44. Lower the engine into place and remove the lifting device.

45. Raise and support the vehicle safely.

46. If equipped with an automatic transmission, perform the following.

  a. Align the torque converter to the drive plate.

  b. Install the bolts and torque them to 40 ft. lbs. (54 Nm).

  c. Install the access cover.

  d. Install the exhaust pipe support.

47. Install the remaining converter/flywheel bolts finger-tight.

48. Install the starter.

49. Tighten the engine support cushion bolts/nuts.

50. Torque the loose converter/flywheel bolts to 28 ft. lbs. (38 Nm).

51. Install the oil filter.

52. Connect the exhaust pipe to the manifold.

53. Lower the vehicle.

54. Connect the coolant hoses and tighten the clamps.

55. If equipped with power steering, perform the following:

  a. Unplug the lines and connect them to the steering gear. Torque the fittings to 38 ft. lbs. (52 Nm).

  b. Fill the pump reservoir with fluid.

56. Remove the alignment cap screw and install the fan.

57. Install the radiator, condenser, if equipped and fan shroud.

58. Connect the radiator hoses.

59. If equipped with an automatic transmission, connect the cooling lines.

60. Connect the oxygen sensor electrical connector.

61. Connect the throttle valve rod and retainer. Connect the throttle cable and install the rod and spring.

62. If equipped, connect the cruise control cable.

63. Connect the fuel lines to the throttle body.

64. Connect all vacuum lines and electrical connectors disconnected during removal.

65. If equipped with A/C, connect the service valves to the compressor ports.

66. Install the air cleaner.

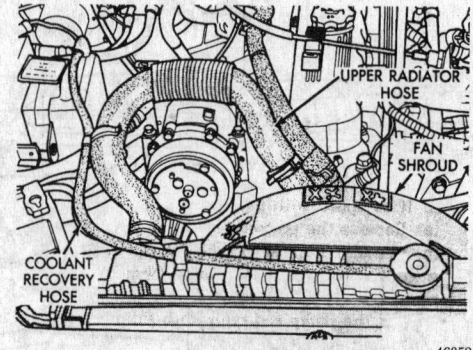

Upper radiator hose, coolant recovery hose and fan shroud

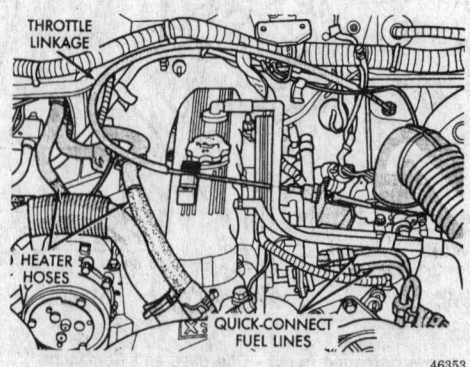

Heater hoses, throttle linkage and fuel lines

67. Install the hood.
68. Connect the battery cables.
69. Fill the cooling system.
70. Start the engine and check for leaks.
71. If equipped, recharge the A/C system.
72. Check and top off fluid levels.

Grand Cherokee 1993-95

1. Matchmark the hood to the hinges and remove the hood.
2. Remove the battery.
3. Properly relieve the fuel system pressure.
4. Drain the cooling system.
5. Remove the air cleaner and tube.
6. Remove the radiator.
7. Remove the heater hoses.
8. Label and disconnect the necessary vacuum lines.
9. Remove the distributor cap and wiring.
10. Disconnect the accelerator linkage.
11. Remove the air duct from the throttle body.
12. Label and disconnect the Manifold Absolute Pressure (MAP) sensor, Idle Air Control (IAC) motor and Throttle Position Sensor (TPS) electrical connectors from the throttle body.
13. Disconnect the vacuum line from the throttle body.
14. Disconnect (unsnap) the control cables from the throttle body (lever) arm.
15. Remove the throttle body from the intake manifold. Discard the gasket.
16. Disconnect the oil pressure electrical connector.
17. If equipped, properly discharge the A/C system.
18. Disconnect the A/C lines from the compressor.
19. If equipped with power steering, disconnect the lines from the pump.
20. Remove the starter.
21. Remove the alternator.
22. Raise and support the vehicle safely.
23. Disconnect the fuel line connections coming from the fuel rail.
24. Disconnect the exhaust pipe from the manifold.
25. Support the transmission with a stand.
26. Remove the bell housing bolts and inspection plate.
27. Attach a C-clamp to the bottom of the torque converter housing to prevent the torque converter from coming out.

28. Matchmark the torque converter to the drive plate and remove the bolts.
29. Disconnect the engine from the torque converter drive plate.
30. Install a suitable lifting device to the engine.

WARNING

Do not lift the engine by the intake manifold.

31. Remove the front engine mount thru-bolts.
32. Lower the vehicle.
33. Remove the engine from the vehicle and mount on a suitable workstand.
   To install:
34. Remove the engine from the workstand and position it in the engine compartment.
35. Raise and support the vehicle.
36. Position the torque converter and drive plate. Torque the bolts to 271 inch lbs. (31 Nm).
37. Install the front engine mount thru-bolts.
38. Install the bell housing bolts and torque them to 30 ft. lbs. (41 Nm).
39. Remove the C-clamp and install the inspection plate. Remove the stand from the transmission.
40. Connect the exhaust pipe to the manifold.
41. Connect the fuel rail lines.
42. Lower the vehicle.
43. Install the starter.
44. Install the alternator.
45. If equipped, install the power steering hoses.
46. If equipped, connect the A/C hoses.
47. Connect the accelerator linkage.
48. Connect the starter wires.
49. Connect the oil pressure electrical connector.
50. Install the distributor cap and wires.
51. Connect the vacuum lines.
52. Install the radiator, radiator hoses and heater hoses.
53. Install the fan shroud into position.
54. Install the air cleaner.
55. Install the battery.
56. Fill the cooling system.
57. Start the engine and check for leaks.
58. If equipped, recharge the A/C system.
59. Install the hood.
60. Road test the vehicle.

Grand Wagoneer 1991

1. Disconnect the negative battery cable.

2. Remove the air cleaner.
3. Drain the cooling system.
4. Disconnect the upper and lower radiator hoses.
5. If equipped with an automatic transmission, disconnect the cooler lines from the radiator.
6. Remove the radiator and the fan.
7. If equipped with air conditioning, evacuate the system, remove the condenser and cap all openings immediately.
8. If equipped, remove the power steering pump and the drive belt, place the unit aside. Do not remove the power steering hoses.
9. If equipped with air conditioning, remove the compressor.
10. Label and disconnect all wires, lines, linkage and hoses that are connected to the engine.
11. Raise and safely support the vehicle.
12. Remove both of the front engine mount-to-frame retaining nuts.
13. Disconnect the exhaust pipes at the support bracket and exhaust manifold.
14. Support the weight of the engine with a lifting device.
15. Remove the front mount and bracket assemblies from the engine.
16. Remove the transfer case shift lever boot and the transmission access cover.
17. If equipped with an automatic transmission, remove the upper bolts securing the transmission bell housing to the engine.
18. If equipped with a manual transmission, remove the upper bolts that secure the clutch housing to the engine.
19. Remove the starter motor.
20. If equipped with an automatic transmission:
   a. Remove the engine to transmission adapter plate inspection covers.
   b. Mark the assembled position of the converter and flexplate and remove the converter-to-flexplate retaining screws.
   c. Remove the remaining bolts securing the transmission bell housing to the engine.
21. If equipped with a manual transmission, remove the lower cover of the clutch housing and remaining bolts securing the clutch housing to the engine.
22. Lower the vehicle.
23. Support the transmission with a jack.
24. Attach a lifting device to the engine and lift the engine upward and

forward at the same time, removing it from the vehicle.

To install:

25. Lower the engine into the vehicle and slide it rearward to engage the transmission.

26. If equipped with an automatic transmission, install the upper bolts securing the transmission bell housing to the engine. Torque them to 27 ft. lbs. (37 Nm).

27. If equipped with a manual transmission, install the upper bolts that secure the clutch housing to the engine. Torque them to 27 ft. lbs. (37 Nm).

28. If equipped with a manual transmission, install the lower cover of the clutch housing and the remaining bolts that secure the clutch housing to the engine. Torque the clutch housing spacer-to-bolts to 12–15 ft. lbs. (16–20 Nm). Torque the clutch housing lower bolts to 43 ft. lbs. (59 Nm).

29. If equipped with an automatic transmission:

a. Install the remaining bolts securing the transmission bell housing to the engine. Torque the bell housing lower bolts to 43 ft. lbs. (59 Nm).

b. Mark the assembled position of the converter and flexplate and

install the converter-to-flexplate retaining screws. Torque the bolts to 20–25 ft. lbs. (27–34 Nm).

c. Install the engine to transmission adapter plate inspection covers.

30. Install the starter motor. Torque the mounting bolts to 18 ft. lbs. (25 Nm).

31. Install the transfer case shift lever boot and the transmission access cover.

32. Install the front mount and bracket assemblies to the engine.

33. Remove the lifting device.

34. Connect the exhaust pipes at the support bracket and exhaust manifold. Torque the nuts to 20 ft. lbs. (27 Nm).

35. Install both of the engine front mount-to-frame retaining nuts. Torque them to 35 ft. lbs. (48 Nm).

36. Connect all wires, lines, linkage, and hoses to the engine.

37. If equipped, install the power steering pump and the drive belt.

NOTE: If the vehicle has air conditioning, mount and connect the air conditioning compressor.

38. Install the radiator, fan and the condenser.

39. Evacuate, charge and leak test the refrigerant system.

40. If equipped with an automatic transmission, connect the cooler lines to the radiator.

41. Connect the upper and lower radiator hoses.

42. Fill the cooling system and connect the negative battery cable.

43. Install the air cleaner.

Grand Wagoneer 1993

1. Matchmark the hood to the hinges and remove the hood.

2. Remove the battery.

3. Properly relieve the fuel system pressure.

4. Drain the cooling system.

5. Remove the air cleaner and tube.

6. Remove the radiator.

7. Remove the heater hoses.

8. Label and disconnect the necessary vacuum lines.

9. Remove the distributor cap and wiring.

10. Disconnect the accelerator linkage.

11. Remove the air duct from the throttle body.

12. Label and disconnect the Manifold Absolute Pressure (MAP) sensor, Idle Air Control (IAC) motor and Throttle Position Sensor (TPS) electrical connectors from the throttle body.

13. Disconnect the vacuum line from the throttle body.

14. Disconnect (unsnap) the control cables from the throttle body (lever) arm.

15. Remove the throttle body from the intake manifold. Discard the gasket.

16. Disconnect the oil pressure electrical connector.

17. If equipped, properly discharge the A/C system.

18. Disconnect the A/C lines from the compressor.

19. If equipped with power steering, disconnect the lines from the pump.

20. Remove the starter.

21. Remove the alternator.

22. Raise and support the vehicle safely.

23. Disconnect the fuel line connections coming from the fuel rail.

24. Disconnect the exhaust pipe from the manifold.

25. Support the transmission with a stand.

26. Remove the bell housing bolts and inspection plate.

27. Attach a C-clamp to the bottom of the torque converter housing to prevent the torque converter from coming out.

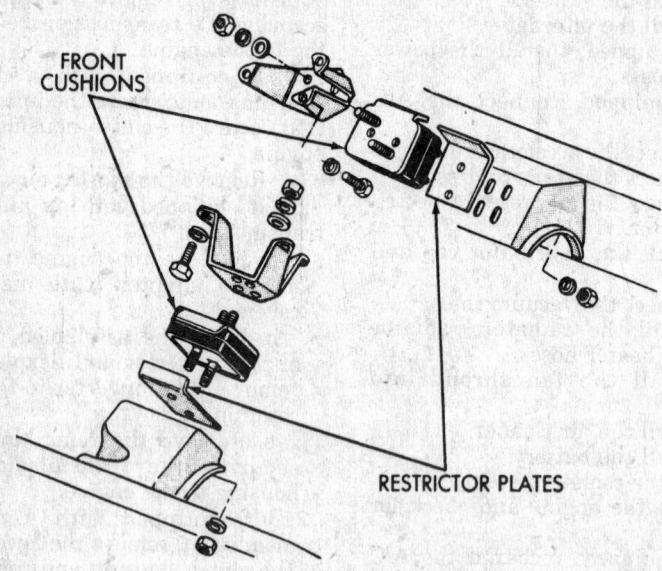

FRONT CUSHIONS

RESTRICTOR PLATES

46229

Front engine mounts - Grand Wagoneer 1991

28. Matchmark the torque converter to the drive plate and remove the bolts.

29. Disconnect the engine from the torque converter drive plate.

30. Install a suitable lifting device to the engine.

――――― WARNING ―――――

Do not lift the engine by the intake manifold.

31. Remove the front engine mount thru-bolts.

32. Lower the vehicle.

33. Remove the engine from the vehicle and mount on a suitable workstand.

To install:

34. Remove the engine from the workstand and position it in the engine compartment.

35. Raise and support the vehicle.

36. Position the torque converter and drive plate. Torque the bolts to 271 inch lbs. (31 Nm).

37. Install the front engine mount thru-bolts.

38. Install the bell housing bolts and torque them to 30 ft. lbs. (41 Nm).

39. Remove the C-clamp and install the inspection plate. Remove the stand from the transmission.

40. Connect the exhaust pipe to the manifold.

41. Connect the fuel rail lines.

42. Lower the vehicle.

43. Install the starter.

44. Install the alternator.

45. If equipped, install the power steering hoses.

46. If equipped, connect the A/C hoses.

47. Connect the accelerator linkage.

48. Connect the starter wires.

49. Connect the oil pressure electrical connector.

50. Install the distributor cap and wires.

51. Connect the vacuum lines.

52. Install the radiator, radiator hoses and heater hoses.

53. Install the fan shroud into position.

54. Install the air cleaner.

55. Install the battery.

56. Fill the cooling system.

57. Start the engine and check for leaks.

58. If equipped, recharge the A/C system.

59. Install the hood.

60. Road test the vehicle.

Engine Mount

REMOVAL AND INSTALLATION

Wrangler 1991-93

1. Disconnect the negative battery cable.

2. Raise and safely support the vehicle.

3. Position a floor jack under the oil pan with a block of wood between the pan and the jack. Support the engine.

4. Remove the nut from the mount through bolt, but do not remove the through bolt.

5. Remove the engine mount-to-frame bracket retaining bolt and nut.

6. Remove the through bolt and the engine mount.

To install:

7. If the engine support bracket was removed, position the support bracket and install the attaching bolts. Tighten to 45 ft. lbs. (61 Nm).

8. Install the engine mount into position and install the through bolt and nut.

9. Install the mount-to-frame bracket bolt and nut. On Cherokee and Comanche, tighten to 30 ft. lbs. (41 Nm). On Wrangler, tighten to 48 ft. lbs. (65 Nm).

10. Tighten the through bolt to 48 ft. lbs. (65 Nm).

11. Remove the engine support and lower the vehicle.

12. Connect the negative battery cable.

Wrangler 1994-95

1. Disconnect the negative battery cable.

2. Raise and safely support the vehicle.

3. Position a floor jack under the oil pan with a block of wood between the pan and the jack. Support the engine.

4. Remove the nut from the mount through bolt, but do not remove the through bolt.

5. Remove the engine mount-to-frame bracket retaining bolt and nut.

6. Remove the through bolt and the engine mount.

To install:

7. If the engine support bracket was removed, position the support bracket and install the attaching bolts. Tighten to 46 ft. lbs. (62 Nm).

8. Install the engine mount into position.

9. Install the mount-to-frame bracket bolt and nut. Tighten to 38 ft. lbs. (52 Nm).

10. Install the through bolt and nut and tighten to 51 ft. lbs. (69 Nm).

11. Remove the engine support and lower the vehicle.

12. Connect the negative battery cable.

Cherokee 1991-93 and Comanche 1991-92

1. Disconnect the negative battery cable.

2. Raise and safely support the vehicle.

3. Position a floor jack under the oil pan with a block of wood between the pan and the jack. Support the engine.

4. Remove the nut from the mount through bolt, but do not remove the through bolt.

5. Remove the engine mount-to-frame bracket retaining bolt and nut.

6. Remove the through bolt and the engine mount.

To install:

7. If the engine support bracket was removed, position the support bracket and install the attaching bolts. Tighten to 45 ft. lbs. (61 Nm).

8. Install the engine mount into position and install the through bolt and nut.

9. Install the mount-to-frame bracket bolt and nut. On Cherokee and Comanche, tighten to 30 ft. lbs. (41 Nm). On Wrangler, tighten to 48 ft. lbs. (65 Nm).

10. Tighten the through bolt to 48 ft. lbs. (65 Nm).

11. Remove the engine support and lower the vehicle.

12. Connect the negative battery cable.

Cherokee 1994-95

1. Disconnect the negative battery cable.

2. Raise and safely support the vehicle.

3. Position a floor jack under the oil pan with a block of wood between the pan and the jack. Support the engine.

4. Remove the nut from the mount through bolt, but do not remove the through bolt.

5. Remove the engine mount-to-frame bracket retaining bolt and nut.

6. Remove the through bolt and the engine mount.

To install:

7. If the engine support bracket was removed, position the left bracket and the right bracket with generator brace onto the cylinder block. Install the bolts and stud nuts.

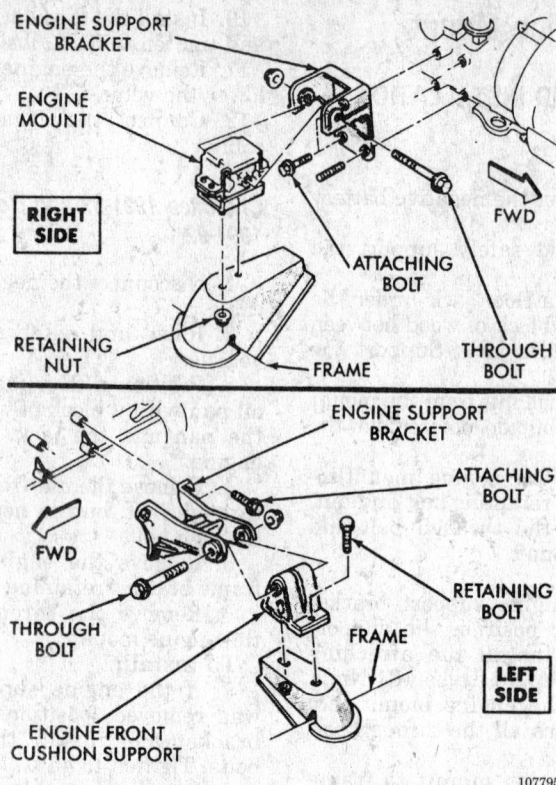

Front engine mounts - Wrangler

107795

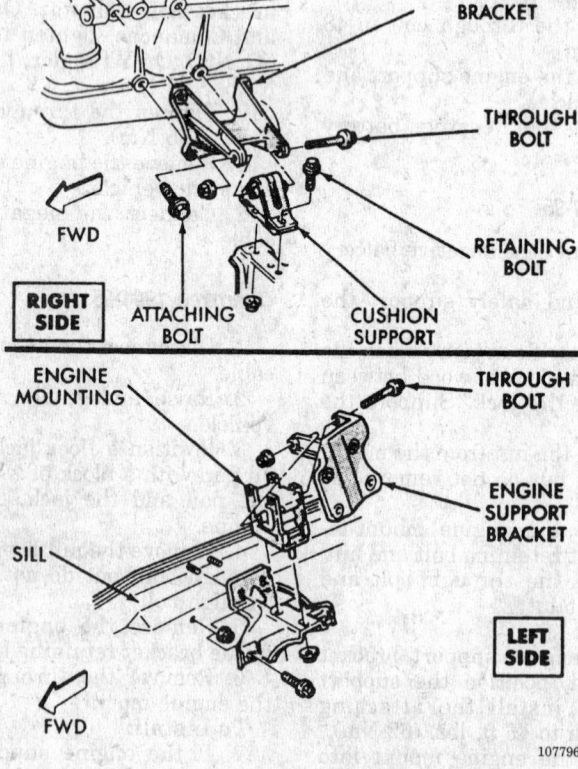

Front engine mounts - Cherokee and Comanche

107796

8. On the right side, tighten the bolts to 45 ft. lbs. (61 Nm) and the stud nuts to 34 ft. lbs. (46 Nm). On the left side, tighten the bolts to 45 ft. lbs. (61 Nm).

9. If the frame support bracket was removed, position the brackets onto the lower front sill and install the bolts and stud nuts. Tighten the bolts to 40 ft. lbs. (54 Nm) and the stud nuts to 30 ft. lbs. (41 Nm).

10. Install the engine mount into position. Tighten the right mount nuts to 48 ft. lbs. (65 Nm) and the left mount bolt/nut to 30 ft. lbs. (41 Nm).

11. Install the through bolt and nut and tighten to 48 ft. lbs. (65 Nm).

12. Remove the engine support and lower the vehicle.

13. Connect the negative battery cable.

Grand Cherokee 1993-95

1. Disconnect the negative battery cable.

2. Raise and support the vehicle safely.

3. Support the engine using a floor jack and a block of wood under the oil pan.

4. Remove the nut from the through-bolt but do not remove the through-bolt.

5. Remove the retaining bolts and nuts from the engine support cushions.

6. Remove the engine support cushions.

To install:

7. If the engine support bracket was removed, position the bracket and install the attaching bolts. Tighten the bolts to 45–48 ft. lbs. (61–65 Nm).

8. Install the engine support cushion and the through-bolt. Install the engine support cushion attaching bolts and tighten to 30–48 ft. lbs. (41–48 Nm). Tighten the through-bolt to 89 ft. lbs. (121 Nm).

9. Remove the engine support and lower the vehicle.

10. Connect the negative battery cable.

Grand Wagoneer 1991

1. Disconnect the negative battery cable.

2. Raise and safely support the vehicle.

3. Position a floor jack under the oil pan with a block of wood between the pan and the jack.

4. Remove the engine mount retaining nuts.

5. Raise the engine with the jack and remove the engine mounts.

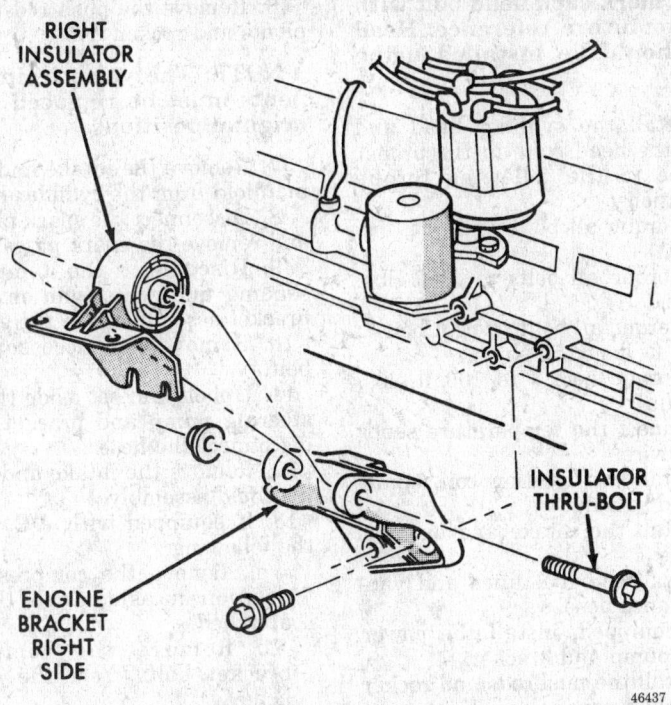

Right side engine mount - Grand Cherokee

To install:

6. Install the engine mounts into position and lower the engine.

7. Install the retaining nuts and washers and tighten to 33 ft. lbs. (45 Nm).

8. Remove the supports and lower the vehicle. connect the negative battery cable.

Grand Wagoneer 1993

1. Disconnect the negative battery cable.

2. Position the fan to assure clearance for the radiator top tank and hose.

3. Raise and support the vehicle safely.

4. Support the engine using a floor jack and a block of wood under the oil pan.

5. Remove the engine support insulator through-bolts and nuts.

6. Raise the engine slightly.

7. Remove the engine support insulator bolts and remove the insulator.

To install:

8. If removed, install the sill bracket assembly.

9. Install the right side bracket onto the sill. Torque the bolts to 48 ft. lbs. (65 Nm).

10. Install the left side bracket onto the sill. Torque the top bolts to 48 ft. lbs. (65 Nm), side bolts to 70 ft. lbs. (95 Nm) and bottom bolts to 89 ft. lbs. (121 Nm).

11. With the engine raised slightly, position the support insulator assembly onto the engine block. Torque the bolts to 65 ft. lbs. (88 Nm).

12. Lower the engine while aligning the engine support insulator into the sill bracket.

13. Install the through-bolt and nut. Torque the right side nut to 48 ft. lbs. (65 Nm) and left side to 89 ft. lbs. (121 Nm).

14. Remove the engine support and lower the vehicle.

15. Connect the negative battery cable.

Cylinder Head

REMOVAL AND INSTALLATION

Wrangler 1991-95, Cherokee 1991-95 and Comanche 1991-92

1. Disconnect the negative battery cable.

2. Properly relieve the fuel system pressure.

3. Drain the cooling system.

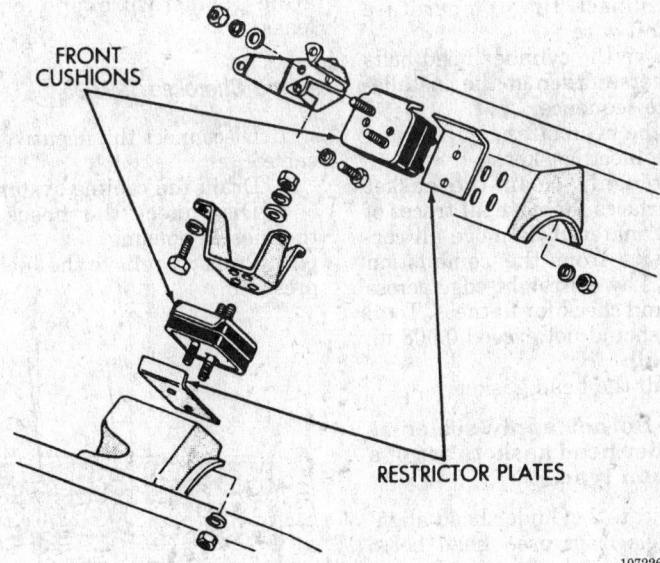

Front engine mounts - Grand Wagoneer 1991

4. Disconnect the hoses at the thermostat housing.

5. Remove the air cleaner.

6. Remove the rocker arm cover.

NOTE: The rocker arm cover has a cured gasket attached to it. This gasket should not be removed. If sections of the gasket are missing or are compressed, replace the cover. However, minor damage such as small cracks, cuts or chips can be repaired with liquid gasket material.

7. Remove the rocker arms and the pushrods. Keep them in their original order for installation.

8. Remove the power steering pump bracket. Suspend the pump aside.

9. Remove the intake and exhaust manifolds.

10. If equipped with A/C, perform the following:

   a. Remove the compressor and position it aside with the lines attached.

   b. Remove the compressor bracket bolts from the cylinder head.

   c. Loosen the through bolt at the bottom of the bracket.

11. Disconnect the fuel lines and vacuum advance hose.

12. Remove the intake and exhaust manifolds.

13. Remove the spark plugs and wires.

14. Remove the ignition coil and bracket assembly.

15. Disconnect the temperature sending unit wire.

16. Remove the cylinder head bolts in the reverse order of the installation torque sequence.

17. Lift the head off the engine and remove the head gasket.

18. Thoroughly clean the gasket mating surfaces. Remove all traces of old gasket material. Remove all carbon deposits from the combustion chambers. Lay a straight-edge across the head and check for flatness. Total deviation should not exceed 0.008 in.

   To install:

19. Install the head gasket.

NOTE: Do not apply sealer as the cylinder head gaskets are of a composition type.

20. Fabricate 2 cylinder head alignment dowels from used head bolts. Using the longest bolts, trim the hex head off and cut slots into the top.

NOTE: Cylinder head bolts should be reused only once. Replace the head bolts which were previously used or are marked

with paint. If head bolts are to be reused, mark each head bolt with paint for future reference. Head bolts should be installed using sealer.

21. Install the cylinder head and tighten the head bolts in the proper sequence to the following torque specifications:

   a. Torque all bolts to 22 ft. lbs. (30 Nm).

   b. Torque all bolts to 45 ft. lbs. 61 Nm).

   c. Torque all bolts, except bolt 7, to 110 ft. lbs. (150 Nm)

   d. Torque bolt 7 to 100 ft. lbs. (136 Nm).

22. Connect the temperature sending unit wire.

23. Install the ignition coil, spark plugs and wires.

24. Install the intake and exhaust manifolds.

25. Install the fuel lines and vacuum advance hose.

26. If equipped, install the power steering pump and bracket.

27. Install the pushrods and rocker arms in their original positions.

28. Install the rocker arm cover.

29. Install the A/C compressor.

30. Install the accessory drive belt.

31. Install the air cleaner.

32. Connect the hoses at the thermostat housing.

33. Fill the cooling system.

34. Connect the negative battery cable.

35. Run the engine to normal operating temperature and check for leaks.

Grand Cherokee 1993-95

1. Disconnect the negative battery cable.

2. Drain the cooling system.

3. Disconnect the hoses at the thermostat housing.

4. Properly relieve the fuel system pressure.

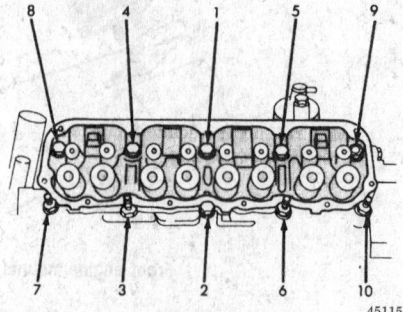

Cylinder head torque sequence - Wrangler, Cherokee and Comanche

5. Remove the cylinder head cover.

6. Remove the push rods, bridges, pivots and rocker arms.

NOTE: The valve train components must be replaced in their original positions.

7. Remove the intake and exhaust manifold from the cylinder head.

8. Disconnect the spark plug wires and remove the spark plugs.

9. Disconnect the temperature sending unit wire, ignition coil and bracket assembly from the engine.

10. Remove the accessory drive belt(s).

11. Unbolt and set aside the power steering pump and bracket. Do not disconnect the hoses.

12. Remove the intake and exhaust manifold assembly.

13. If equipped with A/C, perform the following:

   a. Remove the compressor and position it aside with the lines attached.

   b. Remove the compressor bracket bolts from the cylinder head.

   c. Loosen the through bolt at the bottom of the bracket.

14. Remove the alternator.

15. Remove the cylinder head bolts, the cylinder head and gasket from the block.

NOTE: Bolt No. 14 cannot be removed until the head is moved forward. Pull the bolt out as far as it will go and suspend in place by wrapping with tape.

16. Discard the gasket. Thoroughly clean the head and block mating surfaces. Check them for warpage with a straight-edge. Deviation should not exceed 0.002 in. in a 6 in. span.

   To install:

17. Coat a new head gasket with suitable sealing compound and place it on the block. Most replacement gaskets will have the word **TOP** stamped on them.

NOTE: Apply sealing compound only to the cylinder head gasket. Do not allow sealing compound to enter the cylinder bore.

18. Install the cylinder head and bolts. The threads of bolt No. 11 must be coated with Loctite® 592 sealant before installation. Torque the bolts in 3 steps, using the correct sequence:

   a. Torque all bolts to 22 ft. lbs. (30 Nm).

   b. Torque all bolts to 45 ft. lbs. (61 Nm).

   c. Retorque all bolts to 45 ft. lbs. (61 Nm).

d. Torque all bolts, except bolt 11, in sequence to 110 ft. lbs. (150 Nm).

e. Torque bolt 11 to 100 ft. lbs. (136 Nm)

NOTE: Cylinder head bolts should be reused only once. Replace the head bolts which were previously used or are marked with paint. If head bolts are to be reused, mark each head bolt with paint for future reference. Head bolts should be installed using sealer.

19. Install the ignition coil.
20. Install the air conditioning compressor.
21. Install the alternator.
22. Install the intake and exhaust manifold assembly.
23. Install the power steering pump and bracket.
24. Install the accessory drive belt(s).
25. Connect the temperature sending unit wire, ignition coil and bracket.
26. Install the spark plugs and wires.
27. Install the pushrods, rocker arm assembly, gasket and cylinder head cover.
28. Connect the hoses at the thermostat housing.
29. Fill the cooling system.
30. Connect the negative battery cable.
31. Run the engine to normal operating temperature and check for leaks.

Grand Wagoneer 1991

1. Disconnect the negative battery cable.
2. Drain the cooling system.
3. When removing the right cylinder head, it may be necessary to re-

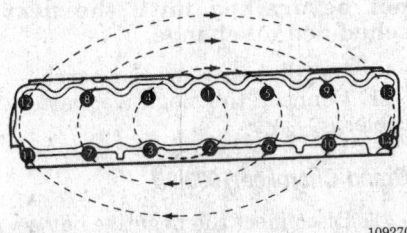

109270

Cylinder head torque sequence - Grand Cherokee

move the heater core housing from the firewall.
4. Remove the valve cover(s) and gasket(s).
5. Remove the rocker arm assemblies and pushrods.

NOTE: The valve train components must be replaced in their original positions.

6. Remove the spark plugs.
7. Remove the intake manifold with the carburetor still attached.
8. Remove the exhaust pipes at the flange of the exhaust manifold..
9. Loosen the accessory drive belts.
10. Disconnect the alternator bracket, air conditioner compressor bracket and remove the negative battery cable from the right cylinder head.
11. Disconnect the air pump and power steering bracket from the left cylinder head.
12. Disconnect the air pump and power steering pump brackets from the left cylinder head.
13. Remove the cylinder head bolts and lift the head(s) from the cylinder block.
14. Remove the cylinder head gasket from the head or the block.
15. Thoroughly clean the gasket mating surfaces. Remove all traces of old gasket material. Remove all carbon deposits from the combustion chambers. Lay a straight-edge across the head and check for flatness. Total deviation should not exceed 0.008 in.

To install:

16. Evenly coat both sides of the gasket with a non-hardening sealing compound. The gasket should be stamped **TOP** for installation. Place the gasket on the block.

NOTE: Do not apply sealant to the head or block.

17. Install the cylinder head on the block.
18. Install the bolts and tighten, in sequence, to 80 ft. lbs. Then re-tighten, in sequence, to 110 ft. lbs.
19. Connect the air pump and power steering pump brackets to the left cylinder head.
20. Connect the alternator bracket to the right cylinder head.
21. Adjust all the accessory drive belts.
22. Install the exhaust pipes at the flange of the exhaust manifold with new gaskets.
23. Install the intake manifold with the carburetor attached.

24. Install the spark plugs.
25. Install the rocker arm assemblies and the pushrods.

NOTE: The valve train components must be replaced in their original positions.

26. Install the valve cover(s) and gasket(s).
27. Install the heater core housing on the firewall, if removed.
28. Fill the cooling system.
29. Connect the negative battery cable.
30. Run the engine to normal operating temperature and check for leaks.

Grand Wagoneer 1993

1. Disconnect the negative battery cable.
2. Properly relieve the fuel system pressure.
3. Drain the cooling system.
4. Remove the alternator.
5. Disconnect the PCV valve.
6. Disconnect the EVAP fuel lines.
7. Remove the air cleaner and disconnect the fuel lines.
8. Disconnect the accelerator linkage, speed control cable, if equipped, and transmission kickdown cables.
9. Remove the return spring.
10. Remove the distributor cap and wires.
11. Disconnect the coil wires.
12. Disconnect the heat indicator sending unit wire.
13. Disconnect the heater and by-pass hoses.
14. Remove the cylinder head covers, discard the gaskets.
15. Remove the intake manifold and throttle body.
16. Remove the exhaust manifolds.
17. Remove the rocker arm assemblies and push rods.

NOTE: Identify the rocker arms and push rods for installation purposes.

18. Remove the spark plugs.
19. Remove the cylinder head bolts.
20. Remove the heads and discard the gaskets.
21. Thoroughly clean the gasket mating surfaces.

To install:

22. Apply Perfect Sealer No. 5 or equivalent, to the inner corners of the new head gaskets.
23. Position the head gaskets onto the block.
24. Position the cylinder heads onto the cylinder block.

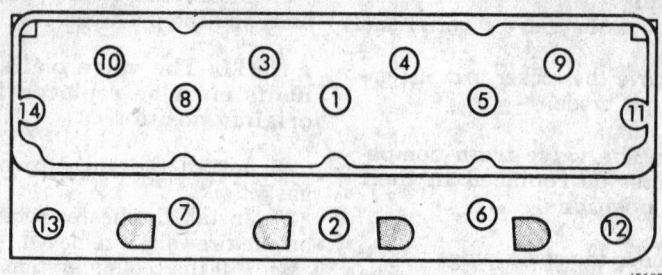

Cylinder head torque sequence - Grand Wagoneer 1991

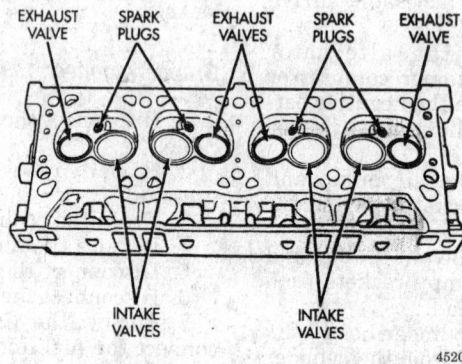

Cylinder head assembly - Grand Wagoneer

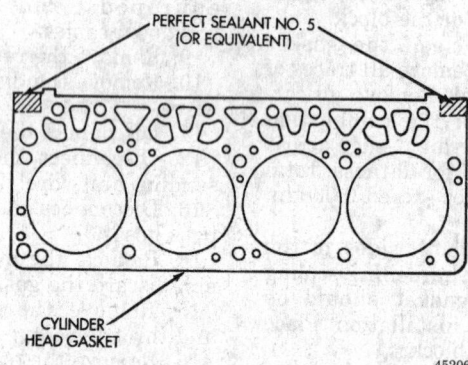

Cylinder head gasket sealant locations - Grand Wagoneer

25. Install the cylinder head bolts as follows:

 a. Torque all bolts, in sequence, to 50 ft. lbs. (68 Nm).

 b. Torque all bolts, in sequence, to 105 ft. lbs. (143 Nm).

 c. Repeat Step b to ensure the torque is correct.

26. Install the pushrods and rocker arms to their original positions.

27. Install the intake and exhaust manifolds.

28. Install the spark plugs.

29. Install the coil wires.

30. Connect the heat indicating sending unit wire.

31. Connect the heater and bypass hoses.

32. Install the distributor cap and wires.

33. Hook up the return spring.

34. Connect the accelerator linkage, speed control cable, if equipped and transmission kickdown cables.

35. Install the fuel lines.

36. Install the alternator.

37. Install the intake manifold-to-alternator bracket support rod.

38. Place new cylinder head cover gaskets into position and install the cylinder head covers. Torque the bolts to 95 inch lbs. (11 Nm).

39. Install the PCV valve.

40. Connect the EVAP lines.

41. Install the air cleaner.

42. Fill the cooling system.

43. Connect the negative battery cable. Start the engine and check for leaks.

Valve Lifter

REMOVAL AND INSTALLATION

Wrangler 1991-95, Cherokee 1991-95 and Comanche 1991-92

1. Disconnect the negative battery cable.

2. Remove the rocker arm cover.

3. Remove the rocker arm assembly by alternately loosening the bolts 1 turn at a time.

4. Remove the pushrods.

NOTE: Keep all components in the order they were removed so they can be reinstalled in their original positions.

5. Remove the lifter through the pushrod opening in the cylinder head using a hydraulic lifter removal tool.

To install:

6. Dip each lifter in MOPAR engine oil supplement or equivalent, and install into the lifter bore. If reusing the old lifters, make sure they are reinstalled in their original bores.

7. Install the pushrods in their original positions.

8. Install the rocker arm assembly components in their original positions. Gradually and alternately tighten the rocker arm bolts to 19 ft. lbs. (26 Nm).

9. Pour the remaining engine oil supplement into the engine.

NOTE: The engine oil supplement must remain in the engine for at least 1000 miles but need not be drained until the next scheduled oil change.

10. Install the rocker arm cover.

11. Connect the negative battery cable.

Grand Cherokee 1993-95

1. Disconnect the negative battery cable.

2. Remove the rocker arm cover.

3. Remove the rocker arm assembly by alternately loosening the bolts 1 turn at a time.

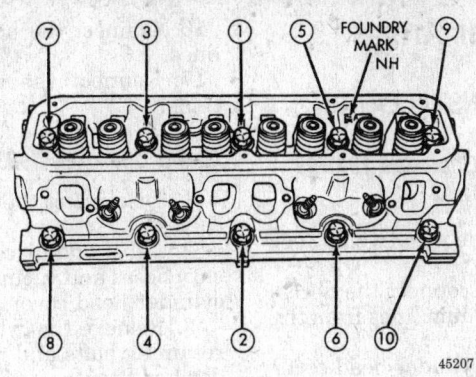

Cylinder head bolt torque sequence - Grand Wagoneer

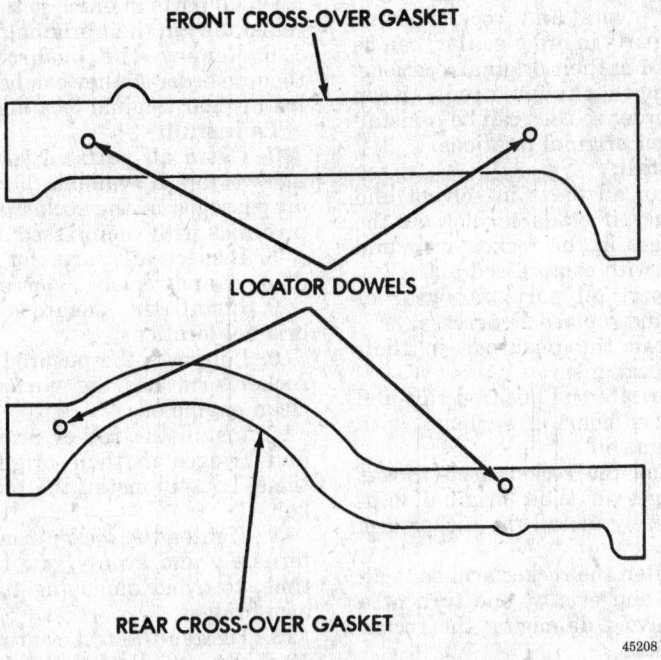

FRONT CROSS-OVER GASKET

LOCATOR DOWELS

REAR CROSS-OVER GASKET

Cross-over gaskets and locator dowels - Grand Wagoneer

4. Remove the pushrods.

NOTE: Keep all components in the order they were removed so they can be reinstalled in their original positions.

5. Remove the lifter through the pushrod opening in the cylinder head using a hydraulic lifter removal tool.
   To install:
6. Dip each lifter in MOPAR engine oil supplement or equivalent, and install into the lifter bore. If reusing the old lifters, make sure they are reinstalled in their original bores.

7. Install the pushrods in their original positions.
8. Install the rocker arm assembly components in their original positions. Gradually and alternately tighten the rocker arm bolts to 19 ft. lbs. (26 Nm).
9. Pour the remaining engine oil supplement into the engine.

NOTE: The engine oil supplement must remain in the engine for at least 1000 miles but need not be drained until the next scheduled oil change.

10. Install the rocker arm cover.

11. Connect the negative battery cable.

Grand Wagoneer 1991

1. Disconnect the negative battery cable.
2. Remove the rocker arm cover.
3. Remove the rocker arm assembly by alternately loosening the bolts 1 turn at a time.
4. Remove the pushrods.

NOTE: Keep all components in the order they were removed so they can be reinstalled in their original positions.

5. Remove the intake manifold.
6. Remove the lifters using a hydraulic lifter removal tool.
   To install:
7. Dip each lifter in MOPAR engine oil supplement or equivalent and install into the lifter bore. If reusing the old lifters, make sure they are reinstalled in their original bores.
8. Install the rocker arm assembly components in their original positions. Tighten the rocker arm bolts to 19 ft. lbs. (26 Nm).
9. Install the intake manifold.
10. Pour the remaining engine oil supplement into the engine.

NOTE: The engine oil supplement must remain in the engine for at least 1000 miles but need not be drained until the next scheduled oil change.

11. Install the rocker arm covers with new gaskets.
12. Connect the negative battery cable.

Grand Wagoneer 1993

1. Disconnect the negative battery cable.
2. Remove the air cleaner.
3. Remove the cylinder head covers.
4. Remove the rocker assembly and push rods.

NOTE: Keep all components in the order they were removed.

5. Remove the intake manifold.
6. Remove the yoke retainer and aligning yokes.
7. Install valve lifter tool C-4129-A or equivalent, through the opening in the cylinder head and seat the tool firmly onto the valve lifter.
8. Pull the valve lifter out of the cylinder block using a twisting motion. If all lifters are to be removed, identify each one for installation.
9. If the lifter bore in the cylinder block is scored, scuffed or shows signs of sticking, ream the bore to the next

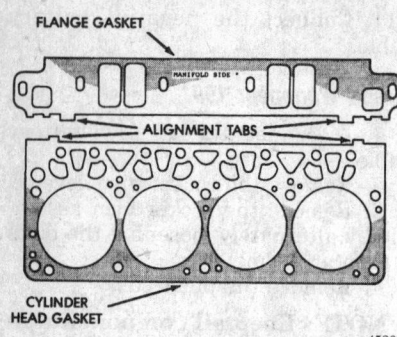

Intake manifold flange gasket alignment -
Grand Wagoneer

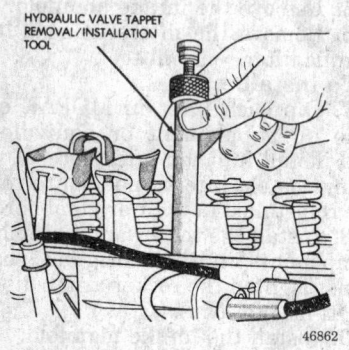

Removing the valve lifter through the
pushrod bore

oversize and replace the lifters with
oversized.

To install:

10. Lubricate the lifters with clean
engine oil or suitable assembly lube.

11. Install the valve lifters and
push rods into their original positions. Ensure the oil feed hole in the
side of the valve lifter body faces up
away from the crankshaft.

12. Install the aligning yokes with
the arrow pointing toward the
crankshaft.

13. Install the yoke retainer. Torque the bolts to 200 inch lbs. (23 Nm).

14. Install the rocker arms.

15. Install the intake manifold.

16. Install the cylinder head covers
with new gaskets.

17. Connect the negative battery
cable.

18. Start the engine and allow it to
idle until the valve lifters fill with oil
and become quiet. Check for leaks.

———— WARNING ————
**To prevent valve mechanism
damage, do not run the engine
above fast idle until all lifters fill
with oil and become quiet.**

Rocker Arm

REMOVAL AND INSTALLATION

Wrangler 1991-95, Cherokee 1991-95 and Comanche 1991-92

1. Disconnect the negative battery
cable.

2. Label and disconnect the necessary hoses and vacuum lines from the
cylinder head cover.

3. Remove the cylinder head cover
retaining bolts and remove the cylinder head cover.

4. Alternately loosen the rocker
arm bolts, one turn at a time, to avoid
damaging the rocker arm bridge.

5. Remove the rocker arm bolts,
bridges, pivots and rocker arms.
Keep all parts in order so they can be
reinstalled in their original locations.

6. Remove the pushrods. Keep
them in order so they can be reinstalled in their original locations.

To install:

7. Clean all parts in solvent and
allow to dry. If available, blow out the
oil passages in the rocker arms and
pushrods with compressed air.

8. Inspect all parts for wear or
damage and replace as necessary.

9. Install the pushrods in their
original locations.

10. Lubricate the pushrod tips and
rocker arm bearing surfaces with
clean engine oil.

11. Install the rocker arms, pivots
and bridges in their original locations. Loosely install the rocker arm
bolts.

12. Tighten the rocker arm bolts alternately and evenly, one turn at a
time, to avoid damaging the rocker
arm bridges.

13. Tighten the rocker arm bolts to
19 ft. lbs. (26 Nm) on 1985-91 vehicles or 21 ft. lbs. (28 Nm) on 1992-94
vehicles.

14. Clean the cylinder head cover
and cylinder head cover mating
surfaces.

**NOTE: Some vehicles are
equipped with a cylinder head
cover containing an integral gasket. This gasket should not be removed. If sections of this gasket
material are missing or damaged,
the cylinder head cover must be
replaced. Sections of this type
cover with minor damage may be
repaired with liquid gasket
material.**

15. Install the cylinder head cover
using a new gasket or sealant, as
required.

16. Connect the hoses and vacuum
lines.

17. Connect the negative battery
cable.

Grand Cherokee 1993-95

1. Disconnect the negative battery
cable.

2. Label and disconnect the necessary hoses and vacuum lines from the
cylinder head cover.

3. Remove the cylinder head cover
retaining bolts and remove the cylinder head cover.

4. Alternately loosen the rocker
arm bolts, one turn at a time, to avoid
damaging the rocker arm bridge.

5. Remove the rocker arm bolts,
bridges, pivots and rocker arms.
Keep all parts in order so they can be
reinstalled in their original locations.

6. Remove the pushrods. Keep
them in order so they can be reinstalled in their original locations.

To install:

7. Clean all parts in solvent and
allow to dry. If available, blow out the
oil passages in the rocker arms and
pushrods with compressed air.

8. Inspect all parts for wear or
damage and replace as necessary.

9. Install the pushrods in their
original locations.

10. Lubricate the pushrod tips and
rocker arm bearing surfaces with
clean engine oil.

11. Install the rocker arms, pivots
and bridges in their original locations. Loosely install the rocker arm
bolts.

12. Tighten the rocker arm bolts alternately and evenly, one turn at a
time, to avoid damaging the rocker
arm bridges.

13. Tighten the rocker arm bolts to
19 ft. lbs. (26 Nm) on 1985-91 vehicles or 21 ft. lbs. (28 Nm) on 1992-94
vehicles.

14. Clean the cylinder head cover
and cylinder head cover mating
surfaces.

**NOTE: Some vehicles are
equipped with a cylinder head
cover containing an integral gasket. This gasket should not be removed. If sections of this gasket
material are missing or damaged,
the cylinder head cover must be
replaced. Sections of this type
cover with minor damage may be
repaired with liquid gasket
material.**

15. Install the cylinder head cover
using a new gasket or sealant, as
required.

16. Connect the hoses and vacuum
lines.

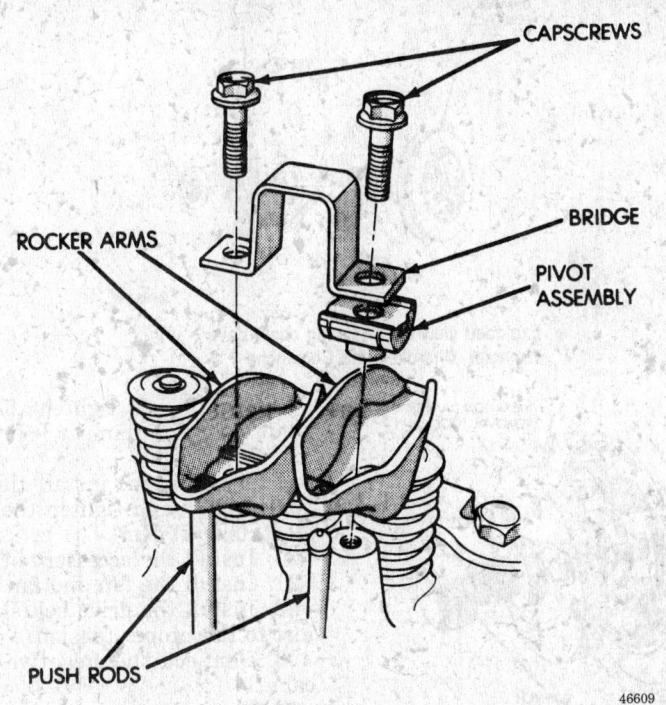

CAPSCREWS

ROCKER ARMS

BRIDGE

PIVOT ASSEMBLY

PUSH RODS

46609

Exploded view of the rocker arm assembly

17. Connect the negative battery cable.

Grand Wagoneer 1991

1. Disconnect the negative battery cable.
2. Remove the air cleaner.
3. Disconnect the air hose from the air injection manifold.
4. If removing the left side cylinder head cover, disconnect the power brake booster vacuum hose, if equipped.
5. If equipped, disconnect the throttle stop solenoid wire on the left side.
6. Remove the Thermostatically controlled Air Cleaner (TAC) hot air duct from the right side.
7. Label and disconnect the spark plug wires and remove the wire separator from the cylinder head cover bracket.
8. Remove the cylinder head cover retaining bolts and washers. Strike the cover with a rubber mallet to break it loose from the cylinder head. Remove the cylinder head cover.
9. Alternately loosen the rocker arm bolts, one turn at a time, to avoid damaging the rocker arm bridge.
10. Remove the rocker arm bolts, bridge, pivots and rocker arms. Keep

all parts in order so they can be reinstalled in their original locations.
11. Remove the pushrods. Keep them in order so they can be reinstalled in their original locations.
To install:
12. Clean all parts with solvent and allow to dry. If available, use compressed air to clean out the oil passages in the rocker arms and pushrods.
13. Inspect all parts for wear or damage and replace as necessary.
14. Install the pushrods in their original locations. Make sure they are seated in the valve lifters.
15. Lubricate the pushrod tips and rocker arm bearing surfaces with clean engine oil.
16. Install the rocker arms, pivots and bridges in their original locations. Loosely install the rocker arm bolts.
17. Alternately tighten the rocker arm bolts, one turn at a time, to avoid damaging the bridge. Final tighten to 19 ft. lbs. (26 Nm).
18. Thoroughly clean the cylinder head cover and cylinder head mating surfaces.
19. Apply a bead of RTV sealant to the cylinder head cover gasket surface area and install the cover. In-

stall the retaining bolts and tighten to 50 inch lbs. (6 Nm).
20. Install the remaning components in the reverse order of removal.

Grand Wagoneer 1993

1. Disconnect the negative battery cable.
2. Label and disconnect the necessary hoses from the cylinder head cover.
3. If removing the left cylinder head cover, remove the coolant tube bracket.
4. Remove the spark plug wires from the holders and disconnect them from the spark plugs.
5. Remove the cylinder head cover and gasket. The steel backed silicon gasket can be used again if not damaged.
6. Remove the rocker arm bolts and remove the rocker arm pivots and rocker arms. Keep the rocker arm assemblies in order so they can be reinstalled in their original locations.
7. Remove the pushrods, keeping them in order so they can be reinstalled in theor original locations.
To install:
8. Rotate the crankshaft until the **V8** mark lines up with the TDC mark on the timing chain cover (located 17.5° ATDC from the No. 1 firing mark).

——— WARNING ———
Do not rotate or crank the engine during or immediately after rocker arm installation. Allow about 5 minutes for the hydraulic lifters to bleed down.

9. Install the pushrods in their original locations. Make sure they are seated in the lifters.
10. Lubricate the pushrod tips, rocker arm bearing surfaces and rocker arm pivots with clean engine oil.
11. Install the rocker arm and pivot assemblies in their original locations. Tighten the bolts to 21 ft. lbs. (28 Nm).
12. Install the cylinder head cover and gasket. On the left cover, install the coolant tube bracket. Tighten the cylinder head cover retaining bolts to 95 inch lbs. (11 Nm).
13. Connect the spark plug wires to the spark plugs and install the wires in the holders.
14. Connect all hoses that were disconnected during the removal procedure.
15. Connect the negative battery cable.

Timing Chain Front Cover

REMOVAL AND INSTALLATION

Wrangler 1991-95, Cherokee 1991-95 and Comanche 1991-92

1. Disconnect the negative battery cable.

2. Remove the drive belt(s), fan and fan shroud. If equipped, remove the accessory drive belt pulley.

3. Remove the vibration damper retaining bolt and washer. Remove the vibration damper using a suitable puller.

4. Remove the accessory drive brackets attached to the timing cover.

5. Remove the A/C compressor, if equipped, and alternator bracket from the cylinder head and move to one side.

6. Remove the oil pan-to-timing case cover bolts and the cover-to-cylinder block bolts.

7. Remove the timing case cover front seal and gasket from the engine.

8. Cut off the oil pan side gasket end tabs and oil pan front seal tabs flush with the front face of the cylinder block. Remove the gasket tabs.

9. Clean the timing case cover, oil pan and cylinder block gasket surfaces.

10. Remove the crankshaft seal oil seal from the cover by prying it out with a suitable tool.

   To install:

11. Install a new seal in the timing cover using a suitable seal installation tool.

12. Apply sealer to both sides of the replacement cover gasket and position the gasket on the cylinder block. Cut the end tabs off the replacement oil pan gasket corresponding to those cut off the original gasket. Attach the end tabs to the oil pan with sealer.

13. Coat the front cover seal end tab recesses generously with sealer and position the seal on the timing cover.

14. Apply engine oil to the seal-oil pan contact surface, then position the cover on the cylinder block.

15. Insert timing case cover alignment tool J22248 or equivalent, in the crankshaft opening. Install the cover bolts and tighten the cover-to-cylinder block bolts to 62 inch lbs. (7 Nm); the oil pan-to-cover bolts to 11 ft. lbs. (13 Nm).

16. Remove the cover alignment tool and position a replacement oil seal on the tool with the seal lip fac-

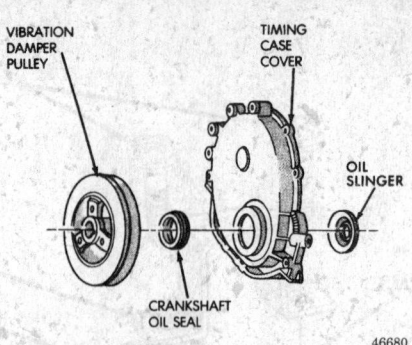

Exploded view of the timing chain cover - Wrangler, Cherokee and Comanche

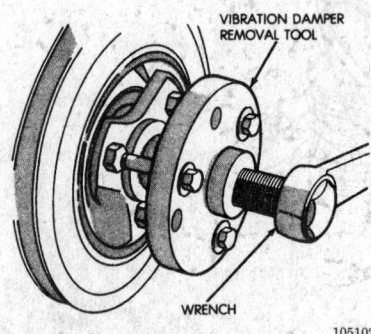

Vibration damper removal - Wrangler, Cherokee and Comanche

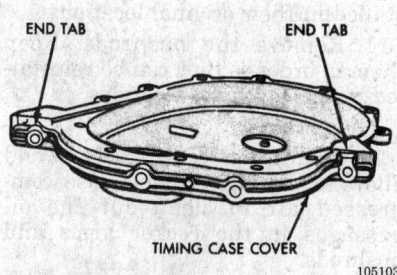

Timing case cover end tabs - Wrangler, Cherokee and Comanche

ing outward. Apply a light coat of sealer to the seal and a light coat of oil to the crankshaft. Install the seal on the timing cover.

17. Apply a light film of oil to the vibration damper hub seal contact surface. Install the vibration damper using a suitable installation tool.

   NOTE: Do not hammer the damper into place as damage may result to the damper or engine.

18. Install and tighten the crankshaft vibration damper bolt to 80 ft. lbs. (108 Nm).

19. If equipped, install the crankshaft pulley and tighten the bolts to 20 ft. lbs. (27 Nm).

20. Install the accessory brackets.

21. Install the fan and fan shroud.

22. Install the drive belt(s) and adjust to the proper tension.

23. Connect the negative battery cable.

24. Start the engine and check for leaks.

Grand Cherokee 1993-95

1. Disconnect the negative battery cable.

2. Remove the drive belt(s), fan and fan shroud. If equipped, remove the accessory drive belt pulley.

3. Remove the vibration damper retaining bolt and washer. Remove the vibration damper using a suitable puller.

4. Remove the accessory drive brackets attached to the timing cover.

5. Remove the A/C compressor, if equipped, and alternator bracket from the cylinder head and move to one side.

6. Remove the oil pan-to-timing case cover bolts and the cover-to-cylinder block bolts.

7. Remove the timing case cover front seal and gasket from the engine. Make sure the tension spring and thrust pin do not fall out of the camshaft sprocket retaining preload bolt.

8. Cut off the oil pan side gasket end tabs and oil pan front seal tabs flush with the front face of the cylinder block. Remove the gasket tabs.

9. Clean the timing case cover, oil pan and cylinder block gasket surfaces.

10. Remove the crankshaft seal oil seal from the cover by prying it out with a suitable tool.

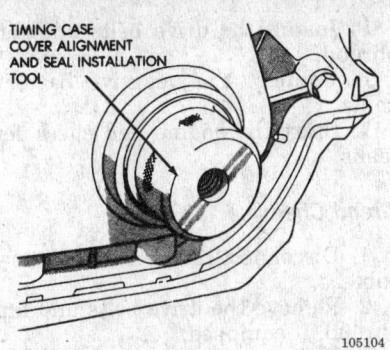

TIMING CASE COVER ALIGNMENT AND SEAL INSTALLATION TOOL

105104

Seal installation - Wrangler, Cherokee and Comanche

To install:

11. Install a new seal in the timing cover using a suitable seal installation tool.

12. Apply sealer to both sides of the replacement cover gasket and position the gasket on the cylinder block. Cut the end tabs off the replacement oil pan gasket corresponding to those cut off the original gasket. Attach the end tabs to the oil pan with sealer.

13. Coat the front cover seal end tab recesses generously with sealer and position the seal on the timing cover.

14. Apply engine oil to the seal-oil pan contact surface. Make sure the tension spring and thrust pin are in place in the preload bolt, then position the cover on the cylinder block.

15. Insert timing case cover alignment tool J22248 or equivalent, in the crankshaft opening. Install the cover bolts and tighten the cover-to-cylinder block bolts to 62 inch lbs. (7 Nm); the oil pan-to-cover bolts to 11 ft. lbs. (13 Nm).

16. Remove the cover alignment tool and position a replacement oil seal on the tool with the seal lip facing outward. Apply a light coat of sealer to the seal and a light coat of oil to the crankshaft. Install the seal on the timing cover.

17. Apply a light film of oil to the vibration damper hub seal contact surface. Install the vibration damper using a suitable installation tool.

NOTE: Do not hammer the damper into place as damage may result to the damper or engine.

18. Install and tighten the crankshaft vibration damper bolt to 80 ft. lbs. (108 Nm).

19. If equipped, install the crankshaft pulley and tighten the bolts to 20 ft. lbs. (27 Nm).

20. Install the accessory brackets.

21. Install the fan and fan shroud.

22. Install the drive belt(s) and adjust to the proper tension.

23. Connect the negative battery cable.

24. Start the engine and check for leaks.

Grand Wagoneer 1991

1. Disconnect the negative battery cable.

2. Drain the cooling system.

3. Disconnect the radiator hoses and bypass hose.

4. Remove all of the drive belts and fan and spacer assembly.

5. If equipped, remove the A/C compressor and bracket assembly from the engine and position aside. Do not disconnect the A/C hoses.

6. Remove the alternator, alternator mounting bracket and back idler pulley.

7. Disconnect the heater hose.

8. Remove the power steering pump and bracket assembly. Remove the air pump and bracket as an assembly. Do not disconnect the power steering hoses.

9. Remove the distributor cap. Mark the position of the rotor in relation to the distributor housing and mark the position of the distributor housing in relation to the timing cover.

10. Remove the distributor.

11. Remove the fuel pump.

12. Remove the vibration damper pulley and vibration damper.

13. Remove the 2 front oil pan bolts and the bolts which secure the timing chain cover to the engine block.

NOTE: The timing chain cover retaining bolts vary in length and must be installed in the same locations from which they were removed.

14. Remove the cover by pulling forward until it is free of the locating dowel pins.

15. Clean the gasket surface of the cover and the engine block.

16. Pry out the original seal from inside the timing chain cover and clean the seal bore.

To install:

17. Install a new seal using a suitable seal installer.

18. Apply a light film of motor oil to the lips of the new seal.

19. Before reinstalling the timing gear cover, remove the lower locating dowel pin from the engine block.

NOTE: The pin is required for correct alignment of the cover and must be installed after the cover is in position.

20. Cut both sides of the oil pan gasket flush with the engine block with a razor blade.

21. Trim a new gasket to correspond to the amount cut off at the oil pan.

22. Apply sealer to both sides of the new gasket and install the gasket on the timing case cover.

23. Install the new front oil pan seal to the bottom of the timing chain cover.

24. Align the tongues of the new oil pan gasket pieces with the oil pan seal and cement them into place on the cover.

25. Apply a bead of sealer to the cutoff edges of the original oil pan gaskets.

26. Place the timing case cover into position and install the front oil pan bolts. Tighten the bolts slowly and evenly until the cover aligns with the upper locating dowel.

27. Install the lower dowel through the cover and drive it into the corresponding hole in the engine block.

28. Install the cover retaining bolts in the same locations from which they were removed. Tighten to 25 ft. lbs. (34 Nm).

29. Apply a light coating of clean engine oil to the vibration damper hub seal contact surface. Install the damper and tighten the retaining bolt to 90 ft. lbs. (122 Nm).

30. Install the damper pulley and tighten the retaining bolts to 30 ft. lbs. (41 Nm).

31. Install the distributor, aligning the rotor and housing with the marks made during removal.

32. Assemble the remaining components in the reverse order of removal.

33. Start the engine and check for leaks. Adjust the ignition timing.

Grand Wagoneer 1993

1. Disconnect the negative battery cable.

2. Properly relieve the fuel system pressure.

3. Drain the cooling system.

4. Remove the serpentine belt.

5. Remove the cooling fan shroud and position it on the engine.

6. Remove the water pump.

7. Remove the power steering pump.

8. Remove the vibration damper using puller C-3688 or equivalent.

9. Disconnect the fuel lines.

10. Loosen the oil pan bolts and remove the front bolt at each side.

11. Remove the timing chain cover bolts. Remove the chain cover and gasket using extreme caution to avoid damaging the oil pan gasket.

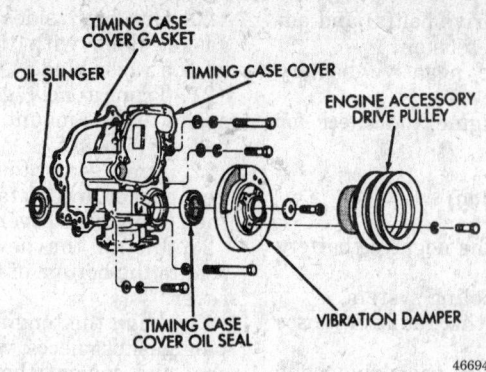

Exploded view of the timing case cover and components - Grand Wagoneer

To install:

12. Install a new timing chain cover gasket to the chain cover. Apply a small amount of Mopar silicone rubber adhesive sealant or equivalent, at the joint where the chain cover and oil pan gasket meet.

13. Install the timing chain cover taking care not to damage to oil pan. Torque the timing chain cover bolts to 30 ft. lbs. (41 Nm) and oil pan bolts to 215 inch lbs. (24 Nm).

14. Install the vibration damper. Torque the crankshaft bolt to 135 ft. lbs. (183 Nm) and pulley bolt to 200 inch lbs. (23 Nm).

15. Connect the fuel lines.

16. Install the water pump.

17. Install the power steering pump.

18. Install the serpentine belt.

19. Install the cooling fan shroud.

20. Fill the cooling system.

21. Connect the negative battery cable.

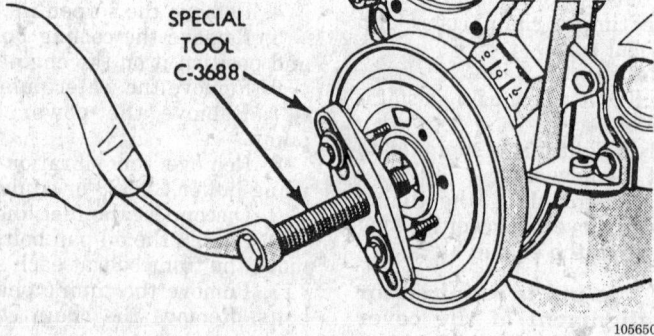

Vibration damper removal

Front Cover Seal

REMOVAL AND INSTALLATION

Wrangler 1991-95, Cherokee 1991-95 and Comanche 1991-92

1. Disconnect the negative battery cable.

2. Remove the drive belts and fan shroud, if equipped.

3. If equipped, remove the crankshaft pulley. Remove the vibration damper retaining bolt.

4. Remove the vibration damper using a suitable puller.

5. Using a suitable tool, pry the oil seal from the front cover. Take care not to damage the front cover or crankshaft.

To install:

6. Position the replacement oil seal on seal installation tool 6139 or equivalent. Install the seal in the cover.

7. Apply a light coat of oil to the seal lip and the vibration damper seal contact surface. Install the vibration damper and tighten the damper bolt to 80 ft. lbs. (108 Nm).

8. Install the crankshaft pulley and torque the bolts to 20 ft. lbs. (27 Nm).

9. Install the drive belts and fan shroud.

10. Connect the negative battery cable.

11. Start the engine and check for leaks.

Grand Cherokee 1993-95

1. Disconnect the negative battery cable.

2. Remove the drive belts and fan shroud, if equipped.

3. If equipped, remove the crankshaft pulley. Remove the vibration damper retaining bolt.

4. Remove the vibration damper using a suitable puller.

5. Using a suitable tool, pry the oil seal from the front cover. Take care not to damage the front cover or crankshaft.

To install:

6. Position the replacement oil seal on seal installation tool 6139 or equivalent. Install the seal in the cover.

7. Apply a light coat of oil to the seal lip and the vibration damper seal contact surface. Install the vibration damper and tighten the damper bolt to 80 ft. lbs. (108 Nm).

8. Install the crankshaft pulley and torque the bolts to 20 ft. lbs. (27 Nm).

9. Install the drive belts and fan shroud.

10. Connect the negative battery cable.

11. Start the engine and check for leaks.

Grand Wagoneer 1991

1. Disconnect the negative battery cable.

2. Remove the drive belts and vibration damper pulley.

3. Remove the vibration damper bolt, then remove the vibration damper using a suitable puller.

4. Using a suitable tool, pry the oil seal from the front cover. Take care not to damage the front cover.

To install:

5. Wipe the crankshaft sealing area clean. Apply sealer to the outer metal surface of the replacement seal.

6. Install the seal using seal installer tool 7778 or equivalent.

7. Apply a light coat of oil to the vibration damper seal contact surface and install the damper. Tighten the damper bolt to 90 ft. lbs. (122 Nm).

8. Install the crankshaft pulley and torque the bolts to 30 ft. lbs. (41 Nm).

9. Install the drive belts.

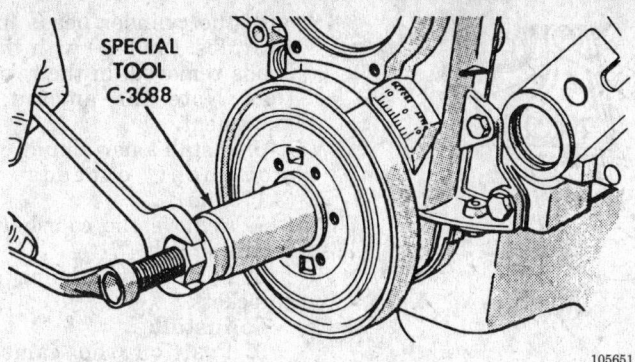

Vibration damper installation

10. Connect the negative battery cable.

11. Start the engine and check for leaks.

Grand Wagoneer 1993

1. Disconnect the negative battery cable.

2. Remove the fan shroud retaining bolts and set the shroud back over the engine.

3. Remove the cooling fan.

4. Remove the accessory drive belt and the vibration damper pulley.

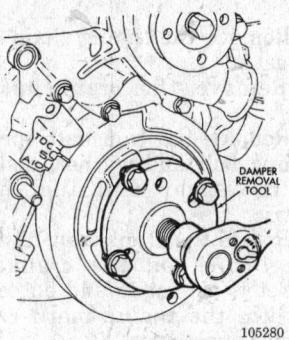

Vibration damper removal - Grand Wagoneer 1991

Oil seal installation - Grand Wagoneer 1991

5. Remove the vibration damper bolt and washer. Remove the vibration damper using puller tool C-3688 or equivalent.

6. Using a suitable tool, pry the oil seal from the front cover. Take care not to damage the front cover.

To install:

7. Position the replacement oil seal on seal installation tool 6635 or equivalent. Install the seal in the cover using the tool.

8. Apply a light coat of oil to the vibration damper hub seal contact surface and install the damper using a suitable installation tool. Tighten the damper bolt to 135 ft. lbs. (183 Nm).

9. Install the crankshaft pulley and torque the bolts to 200 inch lbs. (23 Nm).

10. Install the drive belt, fan and fan shroud.

11. Connect the negative battery cable.

12. Start the engine and check for leaks.

Timing Chain and Sprockets

REMOVAL AND INSTALLATION

Wrangler 1991-95, Cherokee 1991-95 and Comanche 1991-92

1. Disconnect the negative battery cable.

2. Remove the drive belt(s), fan and fan shroud.

3. If equipped, remove the crankshaft pulley. Remove the vibration damper retaining bolt and washer.

4. Using a puller, remove the vibration damper.

5. Remove the timing case cover.

6. Rotate the crankshaft until the **O** timing mark on the crankshaft sprocket is closest to and on a center line with the timing mark on the camshaft sprocket.

7. Remove the oil slinger from the crankshaft.

8. Remove the camshaft retaining bolt and remove the sprockets and chain as an assembly. If the timing chain tensioner is to be replaced, the oil pan must be removed.

To install:

9. Turn the timing chain tensioner lever to the unlock (down) position. Pull the tensioner block toward the tensioner lever to compress the spring. Hold the block and turn the tensioner lever to the lock (up) position.

10. Install the sprockets and timing chain. Ensure the timing marks on the sprockets are properly aligned.

11. Install the camshaft sprocket retaining bolt and washer and tighten to 80 ft. lbs. (108 Nm).

12. To verify correct alignment, turn the crankshaft to position the camshaft sprocket timing mark at the 1 o'clock position. This positions the crankshaft sprocket timing mark where the adjacent tooth meshes with the chain at the 3 o'clock position. Count the number of chain pins between the timing marks of both sprockets; there must be 20 pins.

13. Install the oil slinger, timing case cover and all other components removed in reverse order.

14. Connect the negative battery cable.

15. Start the engine, check the ignition timing and check for leaks.

Grand Cherokee 1993-95

1. Disconnect the negative battery cable.

2. Remove the timing chain cover.

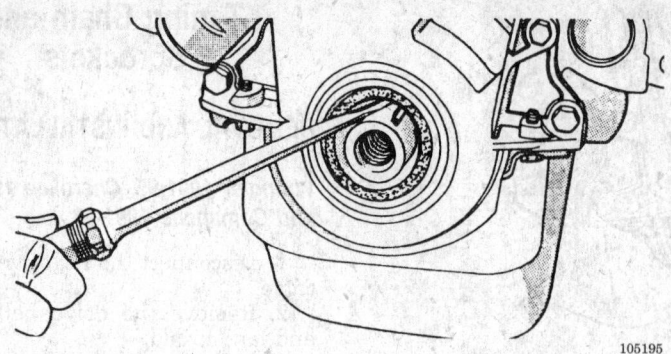

105195

Front cover oil seal removal - Grand Wagoneer 1992-95

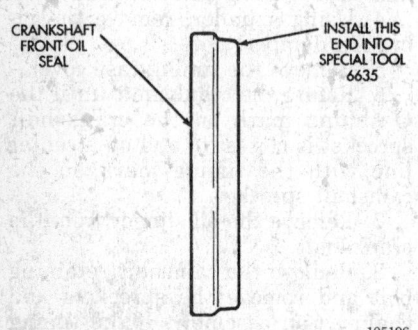

CRANKSHAFT FRONT OIL SEAL

INSTALL THIS END INTO SPECIAL TOOL 6635

105196

Position the seal on the installation tool - Grand Wagoneer 1992-95

3. Place a scale next to the timing chain so any movement of the chain can be measured.

4. Place a torque wrench and socket over the camshaft sprocket attaching bolt. Apply 30 ft. lbs. (41 Nm) with the cylinder heads installed or 15 ft. lbs. (20 Nm) with the cylinder heads removed.

NOTE: With the torque applied the crankshaft sprocket should not be permitted to move, but it may be necessary to block the crankshaft to prevent rotation.

5. Hold the scale with the dimension reading even with the edge of a chain link. Apply 30 ft. lbs. (41 Nm)

with the cylinder heads installed or 15 ft. lbs. (20 Nm) with the cylinder heads removed, in the reverse direction. Note the amount of chain movement.

6. Install a new timing chain if the movement exceeds $\frac{1}{8}$ inch (3.175mm).

7. Remove the camshaft sprocket retaining bolt.

8. Remove the timing chain and sprockets.

To install:

9. Position the camshaft and crankshaft sprockets on a bench with the timing marks facing each other.

10. Position the timing chain onto the sprockets.

11. Turn the crankshaft and camshaft to line up with the keyway location in the crankshaft and camshaft sprockets.

12. Keeping tension on the chain, slide the sprocket and chain assembly onto the engine.

13. Ensure the timing marks are still aligned by using a straight edge.

14. Install the camshaft bolt and torque it to 50 ft. lbs. (68 Nm).

15. Install the timing chain cover.

16. Connect the negative battery cable.

Grand Wagoneer 1991

1. Disconnect the negative battery cable.

2. Remove the timing chain cover and gasket.

3. Remove the crankshaft oil slinger.

4. Remove the camshaft sprocket retaining bolt and washer, distributor drive gear and fuel pump eccentric.

5. Rotate the crankshaft until the timing mark on the crankshaft sprocket is closest to and on centerline with, the timing mark on the camshaft sprocket.

6. Remove the crankshaft sprocket, camshaft sprocket and timing chain as an assembly.

7. Disassemble the chain and sprockets.

To install:

8. Assemble the timing chain, crankshaft sprocket and camshaft sprocket with the timing marks on both sprockets aligned.

9. Install the assembly to the crankshaft and the camshaft.

10. Install the fuel pump eccentric. The fuel pump eccentric must be installed with the stamped word **REAR** facing the camshaft sprocket.

11. Install the distributor drive gear, washer and retaining bolt. Tighten the bolt to 30 ft. lbs. (41 Nm).

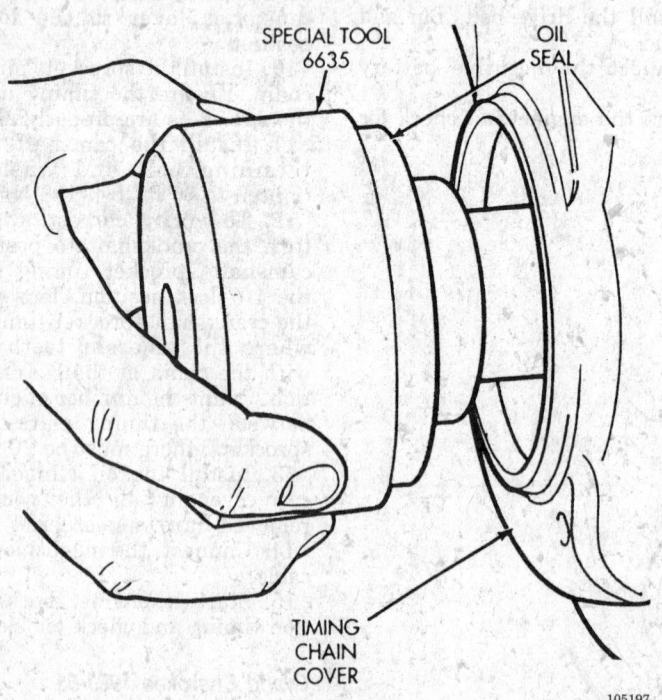

SPECIAL TOOL 6635

OIL SEAL

TIMING CHAIN COVER

105197

Install the tool and seal on the crankshaft - Grand Wagoneer 1992-95

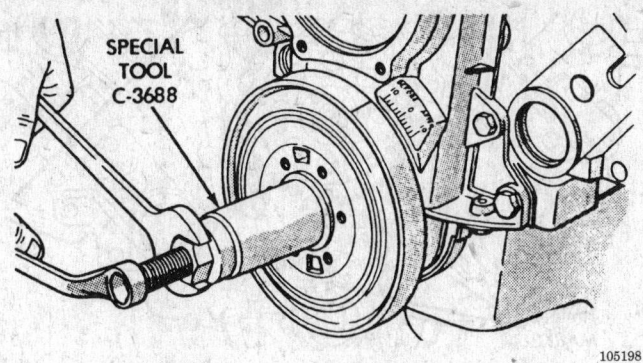

SPECIAL
TOOL
C-3688

105198

Vibration damper installation - Grand Wagoneer 1992-95

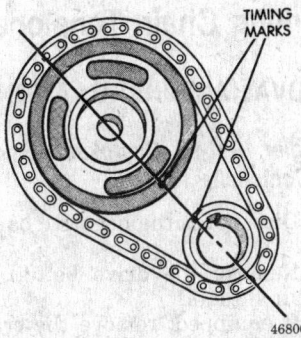

TIMING
MARKS

46800

Crankshaft-to-camshaft sprocket alignment - Wrangler, Cherokee and Comanche

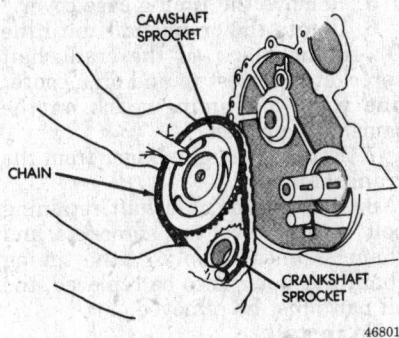

CAMSHAFT
SPROCKET

CHAIN

CRANKSHAFT
SPROCKET

46801

Removing the timing chain and sprockets - Wrangler, Cherokee and Comanche

12. Verify proper timing chain installation as follows:

  a. Rotate the crankshaft until the timing mark on the camshaft sprocket is on a horizontal line at the 3 o'clock position.

  b. Beginning with the pin directly adjacent to the camshaft sprocket timing mark, count the number of pins down to the timing mark on the crankshaft sprocket.

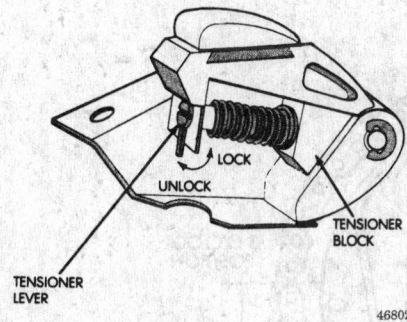

LOCK

UNLOCK

TENSIONER
LEVER

TENSIONER
BLOCK

46802

Loading the timing chain tensioner - Wrangler, Cherokee and Comanche

  c. There must be 20 pins between these 2 points; the crankshaft sprocket timing mark must be between pins 20 and 21.

13. Install the crankshaft oil slinger.

14. Install the timing chain cover using a new gasket and oil seal.

Grand Wagoneer 1993

1. Disconnect the negative battery cable.

2. Remove the timing chain cover.

3. Place a scale next to the timing chain so any movement of the chain can be measured.

4. Place a torque wrench and socket over the camshaft sprocket attaching bolt. Apply 30 ft. lbs. (41 Nm) with the cylinder heads installed or 15 ft. lbs. (20 Nm) with the cylinder heads removed.

NOTE: With the torque applied the crankshaft sprocket should not be permitted to move, but it may be necessary to block the crankshaft to prevent rotation.

5. Hold the scale with the dimension reading even with the edge of a chain link. Apply 30 ft. lbs. (41 Nm) with the cylinder heads installed or 15 ft. lbs. (20 Nm) with the cylinder heads removed, in the reverse direction. Note the amount of chain movement.

6. Install a new timing chain if the movement exceeds $1/8$ inch (3.175mm).

7. Remove the camshaft sprocket retaining bolt.

8. Remove the timing chain and sprockets.

  To install:

9. Position the camshaft and crankshaft sprockets on a bench with the timing marks facing each other.

10. Position the timing chain onto the sprockets.

11. Turn the crankshaft and camshaft to line up with the keyway location in the crankshaft and camshaft sprockets.

12. Keeping tension on the chain, slide the sprocket and chain assembly onto the engine.

13. Ensure the timing marks are still aligned by using a straight edge.

14. Install the camshaft bolt and torque it to 50 ft. lbs. (68 Nm).

15. Install the timing chain cover.

16. Connect the negative battery cable.

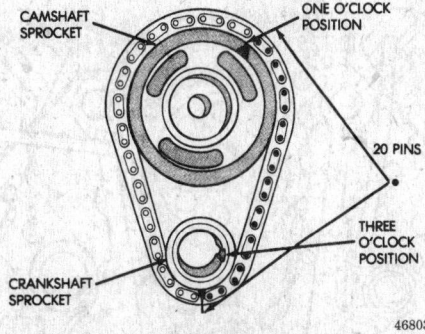

CAMSHAFT
SPROCKET

ONE O'CLOCK
POSITION

20 PINS

THREE
O'CLOCK
POSITION

CRANKSHAFT
SPROCKET

46803

Properly installed timing chain and sprockets - Wrangler, Cherokee and Comanche

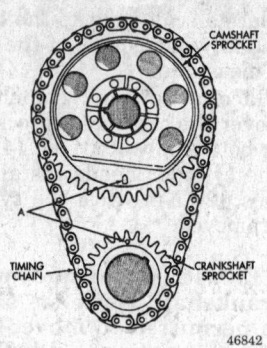

Aligning the sprocket timing marks - Grand Wagoneer

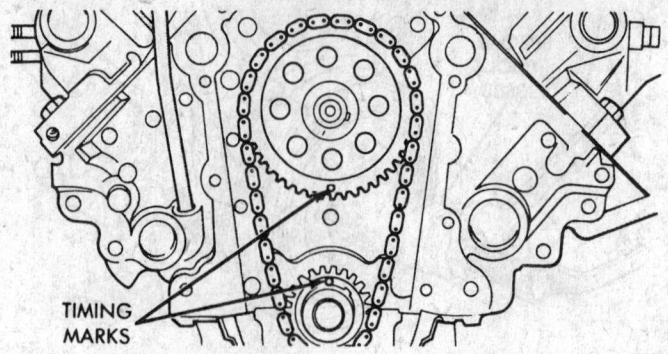

Aligning the sprocket timing marks - Grand Wagoneer

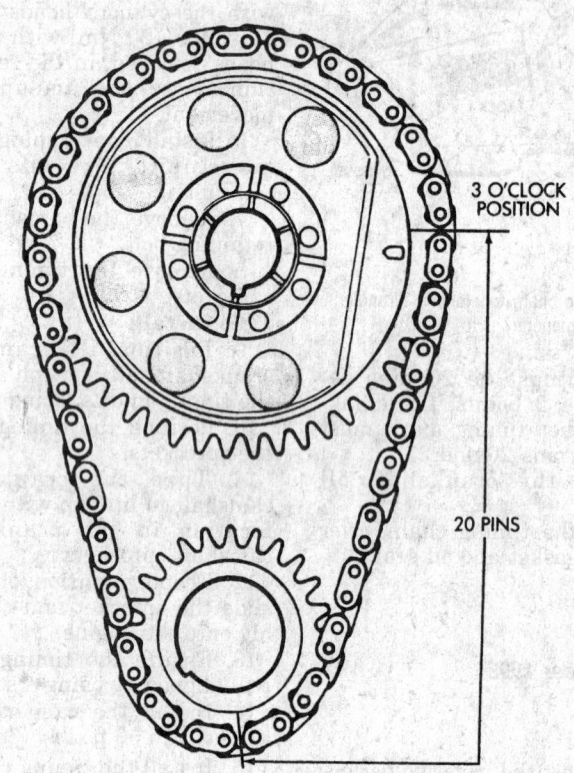

Properly installed timing chain and sprockets - Grand Wagoneer

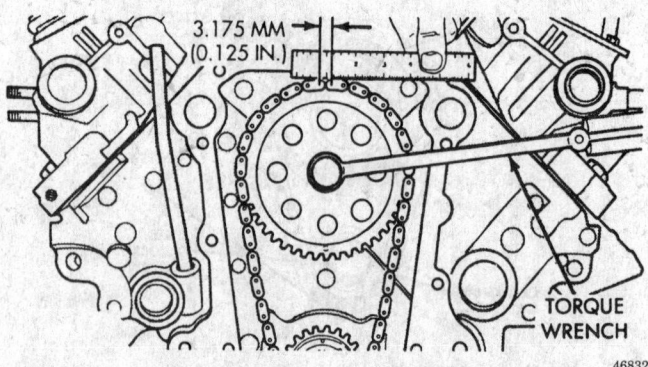

Measuring the timing chain wear - Grand Wagoneer

Timing Chain Tensioner

REMOVAL AND INSTALLATION

Wrangler 1991-95, Cherokee 1991-95 and Comanche 1991-92

1. Disconnect the negative battery cable.
2. Remove the drive belt(s), fan and fan shroud.
3. If equipped, remove the crankshaft pulley. Remove the vibration damper retaining bolt and washer.
4. Using a puller, remove the vibration damper.
5. Remove the timing case cover.
6. Rotate the crankshaft until the O timing mark on the crankshaft sprocket is closest to and on a center line with the timing mark on the camshaft sprocket.
7. Remove the oil slinger from the crankshaft.
8. Remove the camshaft retaining bolt and remove the sprockets and chain as an assembly. If the timing chain tensioner is to be replaced, the oil pan must be removed.

To install:

9. Turn the timing chain tensioner lever to the unlock (down) position. Pull the tensioner block toward the tensioner lever to compress the spring. Hold the block and turn the tensioner lever to the lock (up) position.
10. Install the sprockets and timing chain. Ensure the timing marks on the sprockets are properly aligned.
11. Install the camshaft sprocket retaining bolt and washer and tighten to 80 ft. lbs. (108 Nm).
12. To verify correct alignment, turn the crankshaft to position the camshaft sprocket timing mark at the 1 o'clock position. This positions the crankshaft sprocket timing mark where the adjacent tooth meshes with the chain at the 3 o'clock position. Count the number of chain pins

between the timing marks of both sprockets; there must be 20 pins.

13. Install the oil slinger, timing case cover and all other components removed in reverse order.

14. Connect the negative battery cable.

15. Start the engine, check the ignition timing and check for leaks.

Camshaft

REMOVAL AND INSTALLATION

Wrangler 1991

1. Disconnect the negative battery cable.

2. If equipped with air conditioning, properly discharge the system.

3. Drain the cooling system.

4. Remove the radiator and air conditioning condenser, if equipped.

5. Matchmark the distributor and engine for installation.

6. Matchmark the rotor position by marking it on the distributor body.

7. Remove the distributor and wires.

8. Remove the rocker arm cover.

9. Remove the rocker arm assemblies.

10. Remove the pushrods.

NOTE: Keep all valve train components in order for installation.

11. Using tool J–21884 or equivalent, remove the hydraulic lifters.

12. Remove the timing case cover.

NOTE: If the camshaft sprocket appears to have been rubbing against the cover, check the oil pressure relief holes in the rear cam journal for debris.

13. Remove the timing chain and sprockets.

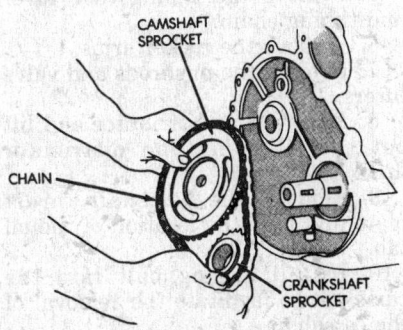

Removing the timing chain and sprockets - Wrangler, CHerokee and Comanche

14. Slide the camshaft from the engine.

To install:

15. Inspect the camshaft for wear and damage.

16. Lubricate all moving components with engine oil.

17. Install the camshaft, taking care not to damage the cam bearings.

18. Install the timing chain and sprockets. Ensure that the timing marks are correctly positioned. Torque the camshaft sprocket bolt to 50 ft. lbs.

19. Install the timing case cover, hydraulic lifters, pushrods, rocker arm assemblies, cylinder head cover, distributor and ignition wires.

20. Install the distributor. The distributor rotor should align with the position of the No. 1 spark plug terminal on the distributor cap when the distributor shaft is down in place.

NOTE: It may be necessary to rotate the oil pump shaft with a long flat-blade screwdriver to engage the oil pump drive tang.

21. Install all other components in reverse order of removal.

22. Start the engine and allow it to reach operating temperature. Check the ignition timing.

Wrangler 1992-95, Cherokee 1991-95 and Comanche 1991-92

1. Disconnect the negative battery cable.

2. If equipped with air conditioning, properly discharge the system.

3. Drain the cooling system.

4. Remove the radiator and air conditioning condenser, if equipped.

5. Matchmark the distributor and engine for installation.

6. Matchmark the rotor position by marking it on the distributor body.

7. Remove the distributor and wires.

8. Remove the rocker arm cover.

9. Remove the rocker arm assemblies.

10. Remove the pushrods.

NOTE: Keep all valve train components in order for installation.

11. Using tool J–21884 or equivalent, remove the hydraulic lifters.

12. Remove the timing case cover.

NOTE: If the camshaft sprocket appears to have been rubbing against the cover, check the oil pressure relief holes in the rear cam journal for debris.

13. Remove the timing chain and sprockets.

14. Slide the camshaft from the engine.

To install:

15. Inspect the camshaft for wear and damage.

16. Lubricate all moving components with engine oil.

17. Install the camshaft, taking care not to damage the cam bearings.

18. Install the timing chain and sprockets. Ensure that the timing marks are correctly positioned. Torque the camshaft sprocket bolt to 50 ft. lbs.

19. Install the timing case cover, hydraulic lifters, pushrods, rocker arm assemblies, cylinder head cover, distributor and ignition wires.

20. Install the distributor. The distributor rotor should align with the position of the No. 1 spark plug terminal on the distributor cap when the distributor shaft is down in place.

NOTE: It may be necessary to rotate the oil pump shaft with a long flat-blade screwdriver to engage the oil pump drive tang.

21. Install all other components in reverse order of removal.

22. Start the engine and allow it to reach operating temperature. Check the ignition timing.

Grand Cherokee 1993-95

1. Disconnect the negative battery cable.

2. Drain the cooling system and remove the radiator.

3. Properly relieve the fuel system pressure.

4. Remove the condenser and receiver/drier as a charged unit.

5. Remove the valve cover and gasket, rocker assemblies, pushrods and lifters.

NOTE: The pushrods must be replaced in their original locations.

6. Remove the drive belts, cooling fan, fan hub assembly, vibration damper and timing chain cover.

7. Remove the distributor assembly, including the spark plug wires.

8. Remove the cylinder head.

9. Remove the valve lifters. Keep them in order for installation.

10. Rotate the crankshaft until the timing mark of the crankshaft sprocket is adjacent to and on a center line with the timing mark of the camshaft sprocket.

11. Remove the crankshaft sprocket, camshaft sprocket and timing chain as an assembly.

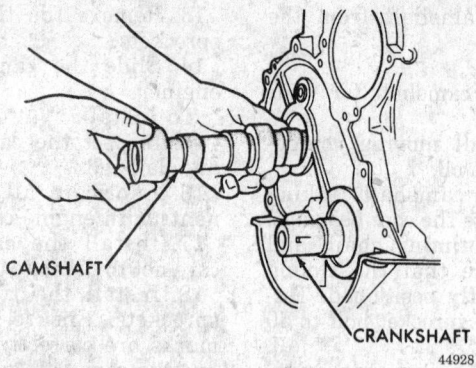

CAMSHAFT

CRANKSHAFT
44928

Camshaft removal - Wrangler, Cherokee and Comanche

12. Remove the front bumper or grille as required and carefully slide out the camshaft.

To install:

13. Lubricate the camshaft with an engine oil supplement.

14. Slide the camshaft into the block carefully to avoid damage to the bearings.

15. Install the crankshaft sprocket, camshaft sprocket and the timing chain as an assembly.

16. Make sure the timing mark of the crankshaft sprocket is adjacent to and on a centerline with, the timing mark of the camshaft sprocket. Torque the camshaft sprocket bolt to 80 ft. lbs. (109 Nm).

17. Lubricate the tension spring, thrust pin and pin bore in the preload bolt. Install the assembly on the preload bolt. Install the timing cover. Install the vibration damper.

18. Install the valve lifters, cylinder head, pushrods, rocker assemblies, valve cover and gasket.

19. Install the distributor assembly. The rotor should be aligned with the No. 1 cylinder spark plug terminal on the cap when the distributor is fully seated on the cylinder block.

20. Install the condenser and receiver/drier as a charged unit.

21. Install the radiator and fill the cooling system.

22. Connect the negative battery cable.

23. Start the engine and allow it to reach operating temperature.

24. Check the ignition timing.

Grand Wagoneer 1991

1. Disconnect the negative battery cable.

2. Drain the radiator and both banks of the block.

3. Remove the lower hose at the radiator, bypass hose at the pump, thermostat housing and radiator.

4. If equipped with air conditioning, remove the condenser and receiver assembly as a charged unit.

5. Remove the distributor, all wires and coil from the manifold.

6. Remove the intake manifold.

7. Remove the valve covers, rocker arms and pushrods.

8. Remove the lifters.

NOTE: The valve train components must be replaced in their original locations.

9. Remove the cooling fan and hub assembly.

10. Remove the fuel pump.

11. Remove the heater hose from the water pump.

12. Remove the alternator and bracket as an assembly. Move it aside, do not disconnect the wiring.

13. Remove the crankshaft pulley and damper.

14. Remove the lower radiator hose at the water pump.

15. Remove the timing chain cover.

NOTE: Reinstall the vibration damper as a means of rotating the engine during servicing.

16. Rotate the engine until the timing marks are properly aligned. Remove the distributor/oil pump drive gear, fuel pump eccentric, sprockets and the timing chain.

17. Remove the grille.

18. Remove the camshaft carefully by sliding it forward out of the engine.

To install:

19. Coat all parts with engine oil supplement.

20. Carefully slide the camshaft into the engine.

21. Install the grille.

22. Install the distributor/oil pump drive gear, fuel pump eccentric, sprockets and the timing chain. Tighten the camshaft bolt to 30 ft. lbs. (41 Nm). Install the timing chain cover.

23. Install the crankshaft pulley and the damper. Tighten the damper bolt to 90 ft. lbs. (122 Nm); the pulley bolts to 30 ft. lbs. (41 Nm).

24. Install the lower radiator hose at the water pump.

25. Install the alternator and bracket as an assembly.

26. Install the cooling fan and hub assembly.

27. Install the fuel pump.

28. Install the heater hose to the water pump.

29. Install the lifters.

NOTE: The valve train components must be replaced in their original locations.

30. Install the valve covers, rocker arms and pushrods.

31. Install the intake manifold as an assembly.

32. Install the distributor, all wires and the coil on the manifold.

33. Install the radiator.

34. If equipped with air conditioning, install the condenser and receiver assembly.

35. Install the thermostat and housing.

36. Install the bypass hose at the pump.

37. Install the lower hose at the radiator.

38. Fill the cooling system.

39. Connect the negative battery cable.

40. Start the engine and allow to reach operating temperature.

41. Check the ignition timing.

42. Check for leaks.

Grand Wagoneer 1993

1. Disconnect the negative battery cable.

2. Remove the engine from the vehicle and mount on a suitable workstand.

3. Remove the intake manifold.

4. Remove the valve covers.

5. Remove the timing case cover and timing chain.

6. Remove the rocker arms.

7. Remove the pushrods and valve lifters.

8. Remove the distributor and lift out the oil pump and distributor drive shaft.

9. Remove the camshaft thrust plate and note the location of the oil tab.

10. Install a long bolt into the camshaft to facilitate the removal of the camshaft.

11. Remove the camshaft, being careful not to damage the cam bearings with the cam lobes.

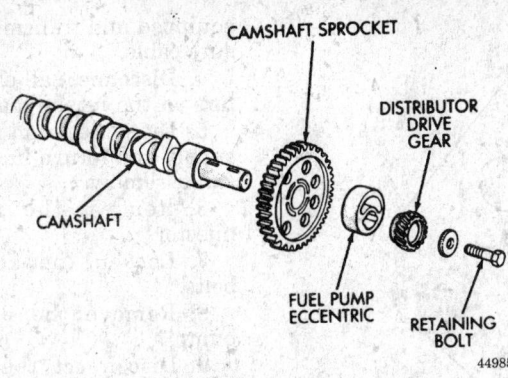

Exploded view of the camshaft and sprocket - Grand Wagoneer 1991

To install:

12. Lubricate the camshaft lobes and bearings with engine oil and insert the camshaft to within 2 inches (51mm) of its final position in the cylinder block.

13. Install camshaft gear installer tool C-3509 or equivalent, with the tool secured in the distributor drive gear position. Hold the tool in position with a distributor lockplate bolt.

NOTE: The tool will restrict the camshaft from being pushed in too far and prevent knocking out the plug in the rear of the cylinder block. The tool should remain installed until the camshaft and crankshaft sprockets with the chain are installed.

14. Install the camshaft thrust plate and chain oil tab. Ensure the tang enter the lower right hole in the thrust plate and torque the bolts to 210 inch lbs. (24 Nm). The top edge of the tab should be flat against the thrust plate in order to catch oil for chain lubrication.

15. Install the timing chain and sprockets.

16. Remove installer tool C-3509.

17. Ensure the camshaft end-play is 0.002–0.010 inch (0.051–0.76mm); if not, install a new thrust plate.

18. Install the drive shaft and distributor.

19. Install the lifters and pushrods.

20. Install the rocker arms.

21. Install the valve covers with new gaskets.

22. Install the intake manifold.

23. Install the engine.

24. Connect the negative battery cable.

Piston and Connecting Rod

POSITIONING

Intake Manifold

REMOVAL AND INSTALLATION

Wrangler 1991-95, Cherokee 1991-95 and Comanche 1991-92

NOTE: The intake and exhaust manifold are mounted externally on the left side of the engine and are attached to the cylinder head. They are removed as a unit.

1. Disconnect the negative battery cable.

2. Remove the air cleaner assembly.

Exploded view of the camshaft and sprocket - Grand Wagoneer

3. Disconnect the accelerator cable, cruise control cable, if equipped and transmission line pressure cable.

4. Disconnect all electrical connectors on the intake manifold.

5. Disconnect and remove the fuel supply and return lines from the fuel rail assembly.

6. Remove the fuel rail and injectors.

7. Loosen the accessory drive belts.

8. Remove the power steering pump.

9. Disconnect the exhaust pipe from the manifold and discard the seal.

10. Remove the intake and exhaust manifold.

To install:

11. Clean the gasket mating surfaces thoroughly. Install a new gasket over the alignment dowels and position the exhaust manifold to the cylinder head. Install bolt No. 3 finger-tight.

12. Install the intake manifold and the remaining bolts and washers.

13. Tighten bolts, in sequence, to the following torque specifications:
   a. Bolts 1–5 — 23 ft. lbs. (31 Nm)
   b. Bolts 6 and 7 — 17 ft. lbs. (23 Nm)
   c. Bolts 8–11 — 23 ft. lbs. (31 Nm)

14. Install the fuel rail and injectors.

15. Install the power steering pump and tension the accessory belt to specification.

16. Using new O-rings, install the fuel supply and return lines.

17. Connect all electrical connectors, vacuum connectors, throttle cable, cruise control cable and transmission lines pressure cable.

18. Install the air cleaner assembly.

19. Using a new seal, connect the exhaust pipe to the manifold and torque the bolts to 23 ft. lbs. (31 Nm).

20. Connect the negative battery cable.

21. Start the engine and check for leaks.

Grand Cherokee 1993-95

NOTE: The intake and exhaust manifold are mounted externally on the left side of the engine and are attached to the cylinder head. They are removed as a unit.

1. Disconnect the negative battery cable.

2. Remove the air cleaner assembly.

3. Disconnect the accelerator cable, cruise control cable, if

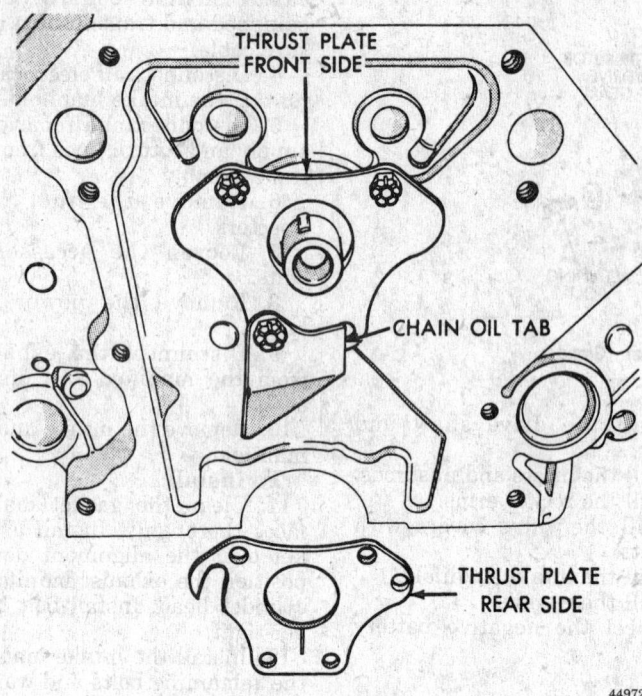

Timing chain oil tab installation - Grand Wagoneer

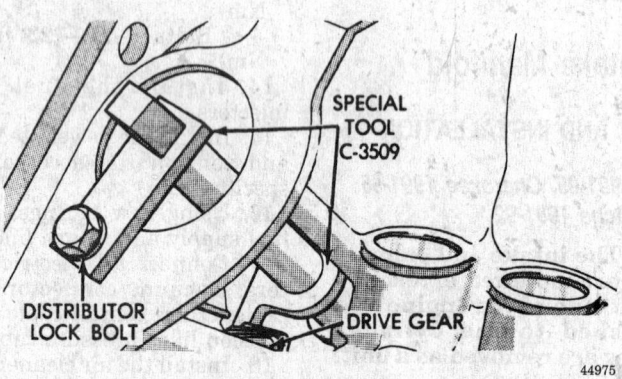

Camshaft holding tool

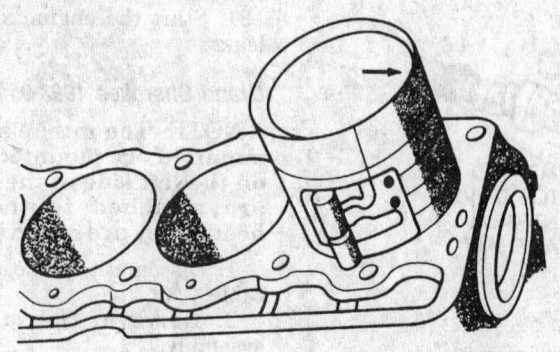

Connecting rod and piston installation

equipped and transmission line pressure cable.

4. Disconnect all electrical connectors on the intake manifold.

5. Disconnect and remove the fuel supply and return lines from the fuel rail assembly.

6. Remove the fuel rail and injectors.

7. Loosen the accessory drive belts.

8. Remove the power steering pump.

9. Disconnect the exhaust pipe from the manifold and discard the seal.

10. Remove the intake and exhaust manifold.

To install:

11. Clean the gasket mating surfaces thoroughly. Install a new gasket over the alignment dowels and position the exhaust manifold to the cylinder head. Install bolt No. 3 finger-tight.

12. Install the intake manifold and the remaining bolts and washers.

13. Tighten bolts, in sequence, to the following torque specifications:

    a. Bolts 1–5 — 23 ft. lbs. (31 Nm)

    b. Bolts 6 and 7 — 17 ft. lbs. (23 Nm)

    c. Bolts 8–11 — 23 ft. lbs. (31 Nm)

14. Install the fuel rail and injectors.

15. Install the power steering pump and tension the accessory belt to specification.

16. Using new O-rings, install the fuel supply and return lines.

17. Connect all electrical connectors, vacuum connectors, throttle cable, cruise control cable and transmission lines pressure cable.

18. Install the air cleaner assembly.

19. Using a new seal, connect the exhaust pipe to the manifold and torque the bolts to 23 ft. lbs. (31 Nm).

20. Connect the negative battery cable.

21. Start the engine and check for leaks.

Grand Wagoneer 1991

1. Disconnect the negative battery cable.

2. Drain the coolant from the radiator.

3. Remove the air cleaner assembly.

4. Disconnect the spark plug wires.

5. Remove the spark plug wire separators from the valve cover brackets.

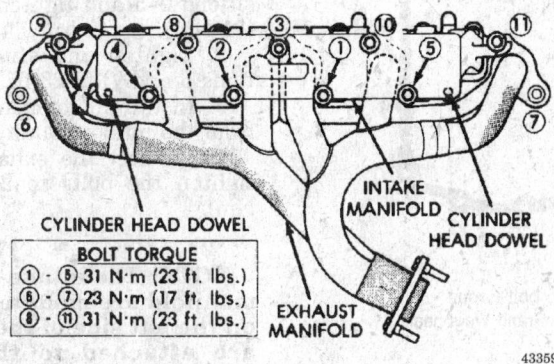

Correct piston and rod alignment

Intake and exhaust manifold mounting bolts - Grand Cherokee

Grand Wagoneer 1993

The aluminum intake manifold is a single plane design with equal length runners. The manifold is sealed by flange side gaskets with front and rear cross-over gaskets. The intake manifold has an internal EGR.

1. Disconnect the negative battery cable.
2. Properly relieve the fuel system pressure.
3. Drain the cooling system.
4. Remove the air cleaner.
5. Remove the alternator.
6. Remove the fuel lines and fuel rail.
7. Disconnect the accelerator linkage and, if equipped, the cruise control and transmission kickdown cables.
8. Remove the return spring.
9. Remove the distributor cap and wires.
10. Disconnect the coil wires.
11. Disconnect the heat indicator sending unit wire.
12. Disconnect the heater and by-pass hoses.
13. Disconnect the PCV and EVAP lines.
14. If equipped with A/C, remove the compressor and position it aside with the lines attached.
15. Remove the support bracket from the mounting bracket and intake manifold.
16. Remove the intake manifold and discard the gaskets.
17. Remove the throttle body and discard the gasket.
18. Turn the intake manifold upside down and support it. Remove the bolts and lift the plenum pan off the manifold. Discard the gasket.
19. Clean all gasket mating surfaces. Clean the intake manifold with solvent and blow dry with compressed air. The plenum pan rail must be clean, dry and free of all foreign material.

To install:

20. Place a new plenum pan gasket onto the seal rail of the intake manifold.
21. Position the pan over the gasket and align the holes. Hand-tighten the bolts.
22. Torque the bolts as follows:
   a. Torque all bolts, in sequence, to 24 inch lbs. (2.7 Nm).
   b. Torque all bolts, in sequence, to 48 inch lbs. (5.4 Nm).
   c. Torque all bolts, in sequence, to 84 inch lbs. (9.5 Nm).
   d. Repeat Step c to ensure proper torque.
23. Using a new gasket, install the throttle body onto the intake mani-

6. Disconnect the upper radiator hose and the bypass hose from the intake manifold.
7. Disconnect the wiring connector from the coolant temperature sending unit.
8. Disconnect the heater hose from the rear of the manifold.
9. Disconnect the ignition coil bracket and lay the coil aside.
10. Label and disconnect all lines, hoses, linkages and wires from the carburetor and intake manifold.
11. Disconnect the air delivery hoses at the air distribution manifolds.
12. Disconnect the air pump diverter valve and lay the valve and the bracket assembly, including the hoses, aside.
13. Remove the intake manifold retaining bolts and remove the intake manifold.
14. Remove the carburetor.
15. Remove the intake manifold, metal gasket and end seals.

To install:

16. Thoroughly clean the mating surfaces of the intake manifold, cylinder block and cylinder head.
17. Apply a non-hardening sealer to both sides of the replacement manifold gasket. Position the gasket by aligning the locators at the rear of

the cylinder head. While holding the rear of the gasket in place, align the front locators.
18. Install the 2 end seals. Apply Permatex® No. 2 or equivalent, to the seal ends.
19. Install the intake manifold and retaining bolts. Tighten the bolts starting at the center and working outwards, in steps, to 43 ft. lbs. (58 Nm).
20. Install the carburetor.
21. Connect the air pump diverter valve.
22. Connect the air delivery hoses at the air distribution manifolds.
23. Connect all lines, hoses, linkages and wires to the carburetor and intake manifold.
24. Install the ignition coil bracket.
25. Connect the upper radiator hose and the bypass hose to the intake manifold.
26. Connect the heater hose to the rear of the manifold.
27. Connect the spark plug wires.
28. Install the spark plug wire separators on the valve cover brackets.
29. Install the air cleaner assembly.
30. Fill the cooling system.
31. Connect the negative battery cable. Start the engine and check for leaks

fold. Tighten the bolts to 200 inch lbs. (23 Nm).

24. Place the 4 plastic locator dowels into the holes in the block.

25. Apply Mopar rubber adhesive sealant or equivalent, to the 4 corner joints.

NOTE: An excessive amount of sealant is not required to ensure a leak proof seal, however, and excessive amount of sealant may reduce the effectiveness of the flange gasket. The sealant should be slightly higher than the crossover gaskets (approximately 0.2 inch).

26. Install the front and rear crossover gaskets onto the dowels.

27. Install the flange gaskets. Ensure the vertical port alignment tab is resting on the deck face of the block. Also, the horizontal mating alignment tabs must be in position with the mating cylinder head gasket tabs. The words MANIFOLD SIDE should be visible on the center of each flange gasket.

28. Carefully lower the intake manifold into place. Use the alignment dowels in the crossover gaskets to position the manifold. Once in place ensure the gaskets are still in position.

29. Torque the manifold bolts in sequence to the following specifications:

a. Bolts 1–4 — 72 inch lbs. (8 Nm) in 12 inch lbs. (1.4 Nm) intervals

b. Bolts 5–12 — 72 inch lbs. (8 Nm)

c. Repeat Steps a-b to ensure proper torque.

d. All bolts — 12 ft. lbs. (16 Nm)

e. Repeat Step d to ensure proper torque.

30. Connect the PCV and EVAP lines.

31. Install the coil wires.

32. Connect the heat indicator sending unit wire.

33. Connect the heater and bypass hoses.

34. Install the distributor cap and wires.

35. Hook up the return spring.

36. Connect the accelerator linkage and, if equipped, cruise control and transmission kick down cables.

37. Install the fuel lines and fuel rail.

38. Install the support bracket.

39. Install the alternator and drive belt.

40. If equipped with A/C, install the compressor.

41. Install the air cleaner.

42. Fill the cooling system.

43. Connect the negative battery cable.

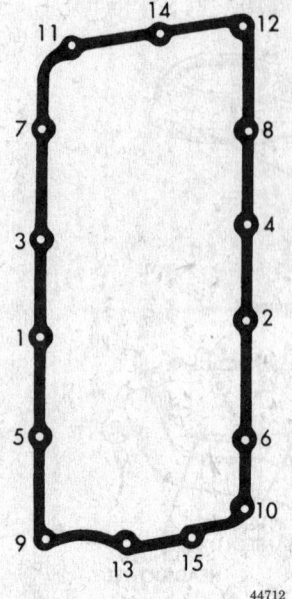

Plenum pan bolt torque sequence - Grand Wagoneer

44712

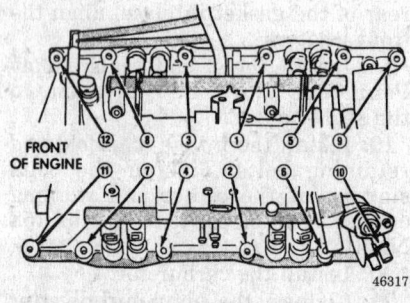

Intake manifold bolt torque sequence - Grand Wagoneer

46317

44. Start the engine and check for leaks.

Exhaust Manifold

REMOVAL AND INSTALLATION

Wrangler 1991-95, Cherokee 1991-95 and Comanche 1991-92

1. Disconnect the negative battery cable.

2. Remove the intake manifold.

3. Disconnect the exhaust pipe at the manifold.

4. Remove the fasteners and exhaust manifold.

To install:

5. Clean the intake manifold and cylinder head mating surfaces.

6. Using a new intake manifold gasket, position the intake and exhaust manifolds on the cylinder head and place spacers over the end studs to center the exhaust manifold. Install the end stud nuts and washer clamps but do not tighten.

7. Install washer clamp and bolt at position 1 and tighten to 30 ft. lbs. (41 Nm).

8. Install bolts and washers at positions 2–5 and tighten to 23 ft. lbs. (31 Nm).

9. Tighten end stud nuts (positions 6 and 7) to 23 ft. lbs. (31 Nm).

10. Install all components removed from the intake manifold.

11. Connect the exhaust pipe and tighten the bolts to 23 ft. lbs. (31 Nm).

NOTE: The intake and exhaust manifold are mounted externally on the left side of the engine and are attached to the cylinder head. They are removed as a unit.

1. Disconnect the negative battery cable.

2. Remove the air cleaner assembly.

3. Disconnect the accelerator cable, cruise control cable, if equipped and transmission line pressure cable.

4. Disconnect all electrical connectors on the intake manifold.

5. Disconnect and remove the fuel supply and return lines from the fuel rail assembly.

6. Remove the fuel rail and injectors.

7. Loosen the accessory drive belts.

8. Remove the power steering pump.

9. Disconnect the exhaust pipe from the manifold and discard the seal.

10. Remove the intake and exhaust manifold.

To install:

11. Clean the gasket mating surfaces thoroughly. Install a new gasket over the alignment dowels and position the exhaust manifold to the cylinder head. Install bolt No. 3 finger-tight.

12. Install the intake manifold and the remaining bolts and washers.

13. Tighten bolts, in sequence, to the following torque specifications:

a. Bolts 1–5 — 23 ft. lbs. (31 Nm)

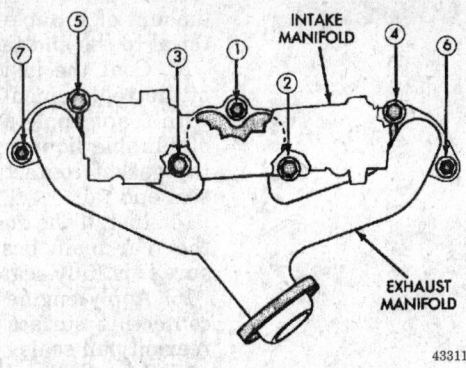

INTAKE MANIFOLD

EXHAUST MANIFOLD

43311

Exhaust manifold mounting bolts - Wrangler, Cherokee and Comanche

b. Bolts 6 and 7 — 17 ft. lbs. (23 Nm)

c. Bolts 8–11 — 23 ft. lbs. (31 Nm)

14. Install the fuel rail and injectors.

15. Install the power steering pump and tension the accessory belt to specification.

16. Using new O-rings, install the fuel supply and return lines.

17. Connect all electrical connectors, vacuum connectors, throttle cable, cruise control cable and transmission lines pressure cable.

18. Install the air cleaner assembly.

19. Using a new seal, connect the exhaust pipe to the manifold and torque the bolts to 23 ft. lbs. (31 Nm).

20. Connect the negative battery cable.

21. Start the engine and check for leaks.

Grand Cherokee 1993-95

1. Disconnect the negative battery cable.

2. Remove the exhaust manifold heat shields.

3. Remove the spark plug wire loom and cables from the mounting stud at the rear of the valve cover and position the cables at the top of the valve cover.

4. Label and disconnect the 2 hoses from the EGR valve.

5. Disconnect the electrical connector and hoses from the EGR transducer.

6. Remove the EGR valve and discard the gasket.

7. Disconnect the oil pressure sending unit electrical connector.

8. Using oil pressure sending unit remover C-4597 or equivalent, remove the sending unit.

9. Loosen the EGR mounting nut from the intake manifold.

10. Remove the mounting bolts and EGR tube. Discard the gasket.

11. Raise and safely support the vehicle.

12. Disconnect the exhaust pipes from the manifolds.

13. Lower the vehicle.

14. Remove the fasteners and exhaust manifold.

To install:

NOTE: If the manifold mounting studs came out with the fasteners, replace the studs.

15. Position the manifold and install the conical washers on the studs.

16. Install new bolt and washer assemblies into the remaining holes. Working from the center outward, torque the fasteners to 20 ft. lbs. (27 Nm).

17. Raise and support the vehicle safely.

18. Connect the exhaust pipes to the manifolds and torque the fasteners to 23 ft. lbs. (31 Nm).

19. Lower the vehicle.

20. Clean the EGR and tube gasket mating surfaces.

21. Install a new gasket onto the exhaust manifold ends of the EGR tube and install the tube. Tighten the tube nut to the intake manifold and torque the tube-to-exhaust manifold bolts to 204 inch lbs. (14 Nm).

22. Coat the threads of the oil pressure sending unit with sealer taking care not to apply sealant to the opening.Install the sending unit and torque it to 130 inch lbs. (14 Nm) and connect the electrical connector.

23. Install the EGR valve and new gasket to the intake manifold and torque the bolts to 200 inch lbs. (23 Nm).

24. Position the EGR transducer and connect the vacuum lines and electrical connector.

25. Position the spark plug cables and loom into place and connect the cables.

26. Install the exhaust heat shields and torque the bolts to 20 ft. lbs. (27 Nm).

27. Connect the negative battery cable.

Grand Wagoneer 1991

1. Disconnect the negative battery cable.

2. Disconnect the spark plug wires.

3. Disconnect the air delivery hose at the distribution manifold.

4. Remove the air distribution manifold and injection tubes.

5. Disconnect the exhaust pipe at the manifold.

6. Remove the exhaust manifold retaining bolts along with the spark plug shields.

7. Remove the exhaust manifold from the vehicle.

To install:

8. Using a new gasket, install the exhaust manifold and tighten the 2 center bolts to 25 ft. lbs. (34 Nm) and 4 outer bolts to 15 ft. lbs. (20 Nm).

9. Tighten the exhaust pipe nuts to 20 ft. lbs. (27 Nm).

10. Install the air injection manifold and injection tubes.

11. Connect the air delivery hose.

12. Connect the spark plug cables.

13. Connect the negative battery cable.

Grand Wagoneer 1993

1. Disconnect the negative battery cable.

2. Remove the exhaust manifold heat shields.

3. Remove the spark plug wire loom and cables from the mounting stud at the rear of the valve cover and position the cables at the top of the valve cover.

4. Label and disconnect the 2 hoses from the EGR valve.

5. Disconnect the electrical connector and hoses from the EGR transducer.

6. Remove the EGR valve and discard the gasket.

7. Disconnect the oil pressure sending unit electrical connector.

8. Using oil pressure sending unit remover C-4597 or equivalent, remove the sending unit.

9. Loosen the EGR mounting nut from the intake manifold.

10. Remove the mounting bolts and EGR tube. Discard the gasket.

11. Raise and safely support the vehicle.

12. Disconnect the exhaust pipes from the manifolds.

13. Lower the vehicle.

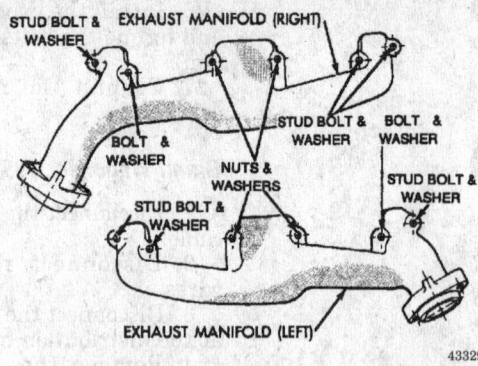

Right and left exhaust manifolds - Grand Cherokee

14. Remove the fasteners and exhaust manifold.

To install:

NOTE: If the manifold mounting studs came out with the fasteners, replace the studs.

15. Position the manifold and install the conical washers on the studs.

16. Install new bolt and washer assemblies into the remaining holes. Working from the center outward, torque the fasteners to 20 ft. lbs. (27 Nm).

17. Raise and support the vehicle safely.

18. Connect the exhaust pipes to the manifolds and torque the fasteners to 23 ft. lbs. (31 Nm).

19. Lower the vehicle.

20. Clean the EGR and tube gasket mating surfaces.

21. Install a new gasket onto the exhaust manifold ends of the EGR tube and install the tube. Tighten the tube nut to the intake manifold and torque the tube-to-exhaust manifold bolts to 204 inch lbs. (14 Nm).

22. Coat the threads of the oil pressure sending unit with sealer taking care not to apply sealant to the opening.Install the sending unit and torque it to 130 inch lbs. (14 Nm) and connect the electrical connector.

23. Install the EGR valve and new gasket to the intake manifold and torque the bolts to 200 inch lbs. (23 Nm).

24. Position the EGR transducer and connect the vacuum lines and electrical connector.

25. Position the spark plug cables and loom into place and connect the cables.

26. Install the exhaust heat shields and torque the bolts to 20 ft. lbs. (27 Nm).

27. Connect the negative battery cable.

ENGINE LUBRICATION

Oil Pan

REMOVAL AND INSTALLATION

Wrangler 1991, Cherokee 1991 and Comanche 1991

1. Disconnect the negative battery cable.

2. Raise and support the vehicle safely.

3. Drain the engine oil.

4. Disconnect the exhaust pipe at the manifold.

5. Disconnect the exhaust hanger at the catalytic converter and lower the pipe.

6. Remove the starter.

7. Remove the bellhousing access cover.

8. Position a jackstand directly under the vibration damper. Place a piece of wood between the jack and the vibration damper.

9. Remove the engine mount bolts/nuts. Using the jack, raise engine until there is enough room to remove the oil pan.

10. Remove the oil pan bolts and remove the oil pan by sliding it to the rear.

11. Clean all sealant and old gasket material from the oil pan and cylinder block mating surfaces. Thoroughly clean the oil pan.

To install:

12. Install a replacement front seal on the timing case cover and apply a generous amount of suitable liquid gasket material to the recesses in the end tabs.

13. Cement the replacement oil pan side gaskets into position on the cylinder block. Apply a generous amount of suitable liquid gasket material to the end tabs of the gaskets.

14. Coat the inside curved surface of the replacement oil pan rear seal with soap. Apply a generous amount of suitable liquid gasket material to the gasket contacting surface of the seal end tabs.

15. Install the seal in the recess of the rear main bearing cap, making sure it is fully seated.

16. Apply engine oil to the oil pan contacting surface of the front and rear oil pan seals.

17. Install the oil pan and bolts. Tighten the 1/4 in. bolts to 80 inch lbs. (9 Nm) and the 5/16 in. bolts to 11 ft. lbs. (15 Nm).

18. Lower the engine and install the engine mount bolts/nuts. Lower the jack and remove the piece of wood.

19. Install the bellhousing access cover.

20. Install the starter.

21. Connect the exhaust pipe to the hanger and the exhaust manifold.

22. Install the oil pan drain plug and tighten to 25 ft. lbs. (34 Nm).

23. Lower the vehicle.

24. Fill the crankcase to the proper level with the recommended oil.

25. Connect the negative battery cable. Start the engine and check for leaks.

Wrangler 1992-95, Cherokee 1992-95 and Comanche 1992

1. Disconnect the negative battery cable.

2. Raise and support the vehicle safely.

3. Drain the engine oil.

4. Disconnect the exhaust pipe at the manifold.

5. Disconnect the exhaust hanger at the catalytic converter and lower the pipe.

6. Remove the starter.

7. Remove the bellhousing access cover.

8. Position a jackstand directly under the vibration damper. Place a piece of wood between the jack and the vibration damper.

9. Remove the engine mount through bolts. Using the jack, raise the engine until there is enough room to remove the oil pan.

10. Remove the oil pan bolts and remove the oil pan.

11. Clean all sealant and old gasket material from the oil pan and cylinder block mating surfaces. Thoroughly clean the oil pan.

To install:

12. Fabricate 4 alignment dowels from 1 1/2 in. x 1/4 in. bolts. Cut the

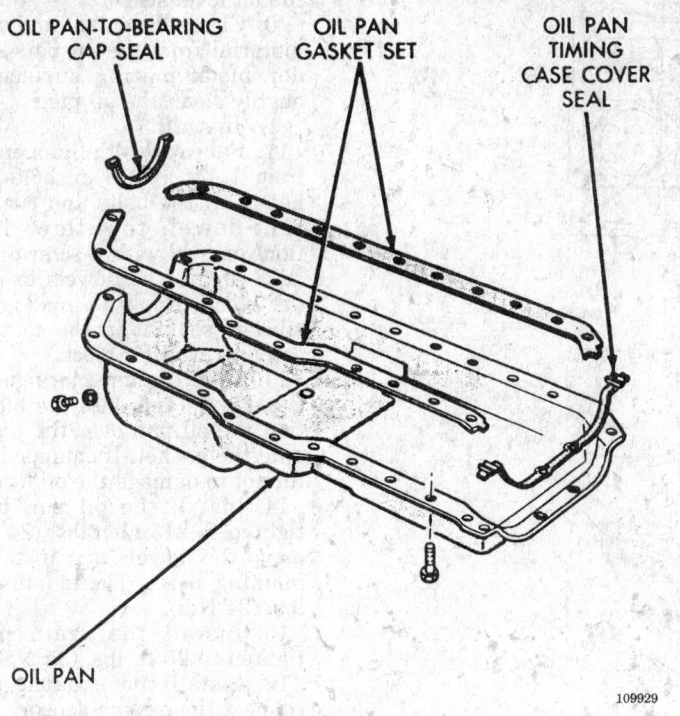

Oil pan, gaskets and seals

OIL PAN-TO-BEARING CAP SEAL

OIL PAN GASKET SET

OIL PAN TIMING CASE COVER SEAL

OIL PAN

109929

heads off the bolts and cut a slot into the top of the dowel to allow installation/removal with a screwdriver.

13. Install 2 dowels in the timing case cover and the other 2 in the cylinder block. Slide the one-piece gasket over the dowels and onto the block and timing case cover.

14. Position the oil pan over the dowels and onto the gasket. Install the 1/4 in. pan bolts and tighten to 120 inch lbs. (14 Nm). Install the 5/16 in. pan bolts and tighten to 156 inch lbs. (18 Nm).

15. Remove the dowels and install the remaining 1/4 in. pan bolts. Tighten to 120 inch lbs. (14 Nm).

16. Lower the engine and install the engine mount through bolts and nuts. Lower the jack and remove the piece of wood.

17. Install the bellhousing access cover.

18. Install the starter.

19. Connect the exhaust pipe to the hanger and the exhaust manifold.

20. Install the oil pan drain plug and tighten to 25 ft. lbs. (34 Nm).

21. Lower the vehicle.

22. Fill the crankcase to the proper level with the recommended oil.

23. Connect the negative battery cable. Start the engine and check for leaks.

Grand Cherokee 1993-95

1. Disconnect the negative battery cable.

2. Raise and safely support the vehicle.

3. Remove the oil pan drain plug and drain the engine oil.

4. Remove the oil filter.

5. Remove the starter.

6. If equipped, disconnect the oil level sensor.

7. Position the oil cooler lines out of the way.

8. Disconnect the oxygen sensor and remove the exhaust pipe.

9. Remove the oil pan bolts and carefully slide the oil pan to the rear.

If equipped, be careful not to damage the oil level sensor.

10. Clean all sealant and old gasket material from the oil pan and cylinder block mating surfaces. Thoroughly clean the oil pan.

To install:

11. Fabricate 4 alignment dowels from 1 1/2 x 5/16 in. bolts. Cut the heads off the bolts and cut a slot in the dowel to allow installation/removal with a screwdriver.

12. Install the dowels in the cylinder block. Apply a small amount of silicone sealant in the corner of the cap and cylinder block.

13. Slide the one-piece gasket over the dowels and onto the block. Position the oil pan over the dowels and onto the gasket. If equipped, be careful not to damage the oil level sensor.

14. Install the oil pan bolts and tighten to 215 inch lbs. (24 Nm). Remove the dowels and install the remaining bolts. Tighten to 215 inch lbs. (24 Nm).

15. Install the drain plug and tighten to 25 ft. lbs. (34 Nm).

16. Install the exhaust pipe and connect the oxygen sensor.

17. Install the oil filter. If equipped, connect the oil level sensor.

18. Install the starter. Move the oil cooler lines back into position.

19. Lower the vehicle and connect the negative battery cable.

20. Fill the engine with the proper type and quantity of oil. Start the engine and check for leaks.

Grand Wagoneer 1991

1. Disconnect the negative battery cable.

2. Raise and safely support the vehicle.

3. Remove the oil pan drain plug and drain the engine oil.

4. Remove the starter.

NOTE: It may be necessary to raise the engine in order to remove the oil pan.

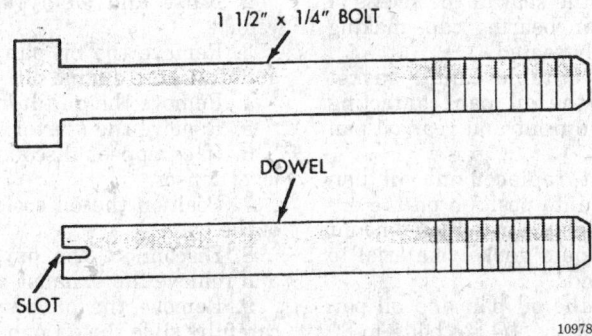

1 1/2" x 1/4" BOLT

DOWEL

SLOT

109788

Fabricate alignment dowels from 1/4 in. bolts - Wrangler, Cherokee and Comanche

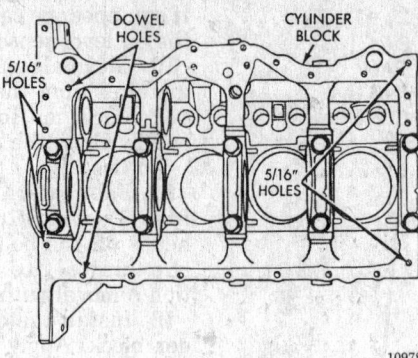

Position of dowels in cylinder block - Wrangler, Cherokee and Comanche

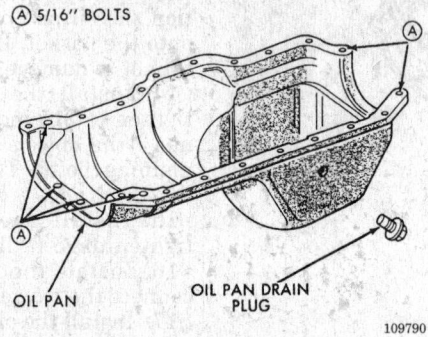

Install 5/16 in. oil pan bolts at position "A" - Wrangler, Cherokee and Comanche

5. Remove the oil pan bolts and remove the oil pan.

6. Clean all sealant and old gasket material from the oil pan and cylinder block mating surfaces. Thoroughly clean the oil pan.

To install:

7. Install a replacement oil pan front seal for the timing case cover. Apply a generous amount of suitable liquid gasket material to the end tabs.

8. Coat the inside curved surface of the replacement oil pan rear seal with soap. Apply a generous amount of suitable liquid gasket material to the gasket contacting surface of the seal end tabs.

9. Install the seal in the recess of the rear main bearing cap, making sure it is fully seated.

10. Apply suitable liquid gasket material to the oil pan contacting surface of the front and rear oil pan seals.

11. Cement replacement oil pan side gaskets into position on the engine block. Apply a generous amount of suitable liquid gasket material to the gasket ends.

12. Install the oil pan and oil pan bolts. Tighten the 1/4 in. bolts to 80 inch lbs. (9 Nm) and the 5/16 in. bolts to 11 ft. lbs. (15 Nm).

13. Lower and secure the engine if it was raised.

14. Install and tighten the oil pan drain plug.

15. Install the starter.

16. Lower the vehicle. Fill the engine with the proper type and quantity of oil.

17. Connect the negative battery cable. Start the engine and check for leaks.

Grand Wagoneer 1993

1. Disconnect the negative battery cable.

2. Raise and safely support the vehicle.

3. Remove the oil pan drain plug and drain the engine oil.

4. Remove the oil filter.

5. Remove the starter.

6. If equipped, disconnect the oil level sensor.

7. Position the oil cooler lines out of the way.

8. Disconnect the oxygen sensor and remove the exhaust pipe.

9. Remove the oil pan bolts and carefully slide the oil pan to the rear.

If equipped, be careful not to damage the oil level sensor.

10. Clean all sealant and old gasket material from the oil pan and cylinder block mating surfaces. Thoroughly clean the oil pan.

To install:

11. Fabricate 4 alignment dowels from 1 1/2 x 5/16 in. bolts. Cut the heads off the bolts and cut a slot in the dowel to allow installation/removal with a screwdriver.

12. Install the dowels in the cylinder block. Apply a small amount of silicone sealant in the corner of the cap and cylinder block.

13. Slide the one-piece gasket over the dowels and onto the block. Position the oil pan over the dowels and onto the gasket. If equipped, be careful not to damage the oil level sensor.

14. Install the oil pan bolts and tighten to 215 inch lbs. (24 Nm). Remove the dowels and install the remaining bolts. Tighten to 215 inch lbs. (24 Nm).

15. Install the drain plug and tighten to 25 ft. lbs. (34 Nm).

16. Install the exhaust pipe and connect the oxygen sensor.

17. Install the oil filter. If equipped, connect the oil level sensor.

18. Install the starter. Move the oil cooler lines back into position.

19. Lower the vehicle and connect the negative battery cable.

20. Fill the engine with the proper type and quantity of oil. Start the engine and check for leaks.

Oil Pump

REMOVAL AND INSTALLATION

Wrangler 1991-95, Cherokee 1991-95 and Comanche 1991-92

1. Disconnect the negative battery cable. Raise and safely support the vehicle.

2. Drain the engine oil and remove the oil pan.

3. Unbolt and remove the pump assembly from the block. Discard the gasket.

—————— **WARNING** ——————

If the oil pump is not to be serviced, do not disturb the position of the oil inlet tube and strainer assembly in the pump body. If the tube is moved within the pump body, a replacement tube and strainer assembly must be installed to assure an airtight seal.

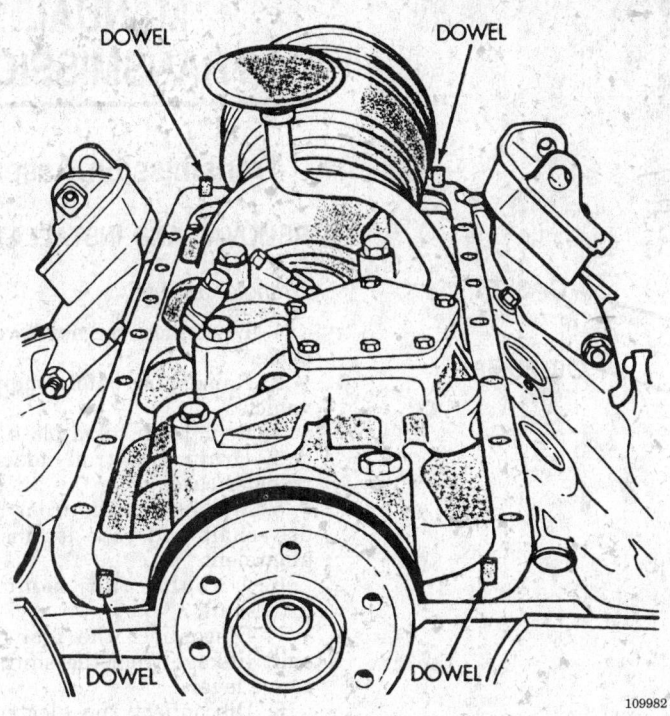

DOWEL   DOWEL

DOWEL   DOWEL

109982

Position of dowels in cylinder block - Grand Cherokee

2. Remove the retaining bolts and separate the oil pump cover, gasket and oil filter as an assembly from the timing case cover.

3. Remove the drive gear and idler gear by sliding them out of the body.

4. Remove the oil pressure relief valve from the pump cover for cleaning by removing the retaining cap and spring.

To install:

5. Install the oil pressure relief valve in the pump cover with the spring and retainer cap.

6. Install the idler shaft, idler gear and drive gear assembly.

NOTE: To ensure self priming of the oil pump, fill the pump with petroleum jelly prior to the installation of the oil pump cover. Do not use grease of any type.

7. Install the pump cover and oil filter assembly with a replacement gasket. Tighten the retaining bolts to 55 inch lbs. (6 Nm).

8. Connect the negative battery cable.

To install:

4. If a new pump is being installed, prime the pump by submerging the strainer in clean engine oil and turning the pump gears until oil emerges from the pump feed hole.

5. Using a new gasket, install the pump on the cylinder block. Torque the short bolt to 10 ft. lbs. and the long bolt to 17 ft. lbs.

6. Install the oil pan and lower the vehicle.

7. Fill the engine with the proper type and quantity of oil.

8. Connect the negative battery cable. Start the engine and check for proper oil pressure.

Grand Cherokee 1993-94

1. Disconnect the negative battery cable. Raise and safely support the vehicle.

2. Drain the engine oil and remove the oil pan.

3. Unbolt and remove the pump assembly from the block. Discard the gasket.

WARNING

If the oil pump is not to be serviced, do not disturb the position of the oil inlet tube and strainer assembly in the pump body. If the tube is moved within the pump body, a replacement tube and strainer assembly must be installed to assure an airtight seal.

To install:

4. If a new pump is being installed, prime the pump by submerging the strainer in clean engine oil and turning the pump gears until oil emerges from the pump feed hole.

5. Using a new gasket, install the pump on the cylinder block. Torque the short bolt to 10 ft. lbs. and the long bolt to 17 ft. lbs.

6. Install the oil pan and lower the vehicle.

7. Fill the engine with the proper type and quantity of oil.

8. Connect the negative battery cable. Start the engine and check for proper oil pressure.

Grand Wagoneer 1991

1. Disconnect the negative battery cable.

Grand Wagoneer 1993

1. Disconnect the negative battery cable.

2. Raise and safely support the vehicle.

3. Drain the engine oil and remove the oil pan.

4. Unbolt and remove the pump assembly from the rear main bearing cap.

To install:

5. If a new pump is being installed, prime the pump by submerging the pickup in clean engine oil and turning the pump gears until oil emerges from the pump feed hole.

6. Install the oil pump. During installation, slowly rotate the pump body to ensure driveshaft-to-pump rotor shaft engagement.

7. Hold the oil pump base flush against the mating surface of the rear main bearing cap and finger tighten the pump mounting bolts. Tighten the mounting bolts to 30 ft. lbs. (41 Nm).

8. Install the oil pan and lower the vehicle. Fill the engine with the proper type and quantity of oil.

9. Connect the negative battery cable. Start the engine and check for proper oil pressure.

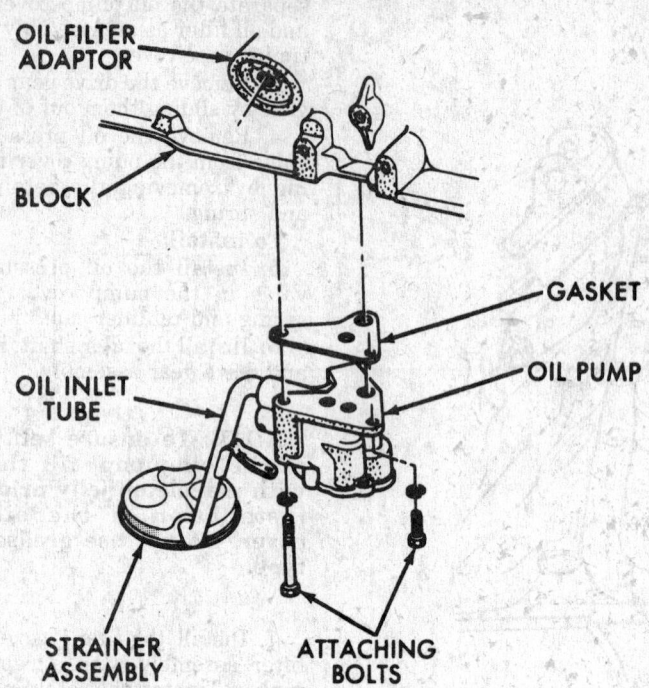

OIL FILTER ADAPTOR

BLOCK

GASKET

OIL INLET TUBE

OIL PUMP

STRAINER ASSEMBLY

ATTACHING BOLTS

44850

Oil pump installation - Grand Cherokee

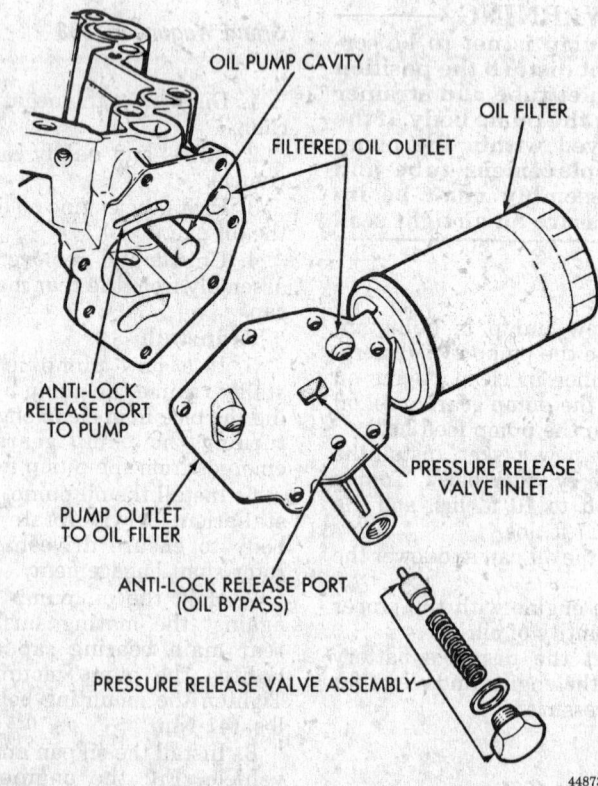

OIL PUMP CAVITY

FILTERED OIL OUTLET

OIL FILTER

ANTI-LOCK RELEASE PORT TO PUMP

PUMP OUTLET TO OIL FILTER

PRESSURE RELEASE VALVE INLET

ANTI-LOCK RELEASE PORT (OIL BYPASS)

PRESSURE RELEASE VALVE ASSEMBLY

44873

Oil pump cover - Grand Wagoneer 1991

MANUAL TRANSMISSION

Transmission Assembly

REMOVAL AND INSTALLATION

Wrangler 1991-95

1. Disconnect the negative battery cable.
2. Raise and safely support the vehicle.
3. Remove the skid plate.
4. Drain the transmission and transfer case.
5. Matchmark the front and rear driveshaft and yoke for installation alignment.
6. Unbolt and remove the driveshafts.
7. Disconnect the transfer case shift linkage from the shift lever or range lever.
8. Disconnect the electrical connectors and vent hose from the transmission and differential.
9. Position a jack under the transmission and take up the weight slightly.
10. Unbolt and remove the rear crossmember.
11. Remove the transfer case from the transmission.
12. Lower the transmission enough to provide access to the shift lever.
13. Reach up and around the transmission case and unseat the shift lever dust boot from the transmission shift tower. Reposition the boot to access the lever retainer.
14. Press the shift lever retainer downward and turn it clockwise to release it.
15. Lift the lever and retainer out of the shift tower.

NOTE: It is not necessary to remove the shift lever from the floor pan boot. Leave the lever in place for installation.

16. If equipped, remove the engine timing sensor.
17. If equipped, disconnect the speedometer cable from the transmission.
18. Disconnect the slave cylinder from the transmission and position it aside.
19. Secure the transmission to the jack.
20. Unbolt the transmission from the engine and lower the jack while pulling back.

To install:

21. Lubricate the pilot bearing and transmission input splines with high temperature grease.

22. Align the transmission input shaft and clutch disc splines and install the transmission. Torque the bolts to 45 ft. lbs. (61 Nm).

23. Reach up around the transmission and insert the shift lever in the shift tower. Press the lever retainer downward and turn it clockwise to lock it in place. Install the dust lever on the dust boot.

24. Install the transfer case and torque the bolts to 26 ft. lbs. (35 Nm).

25. Install the crossmember. Torque the bolts to 26–33 ft. lbs. (37–45 Nm).

26. Install the slave cylinder to the transmission housing.

27. If equipped, install the engine timing sensor.

28. If equipped, connect the speedometer cable.

29. Connect the electrical connectors and vent hose to the transmission and transfer case.

30. Connect the transfer case shift rod to the lever.

31. Install the driveshafts using new strap bolts. Torque the strap bolt nuts to 14 ft. lbs.; the flange-to-transfer case bolts to 35 ft. lbs.

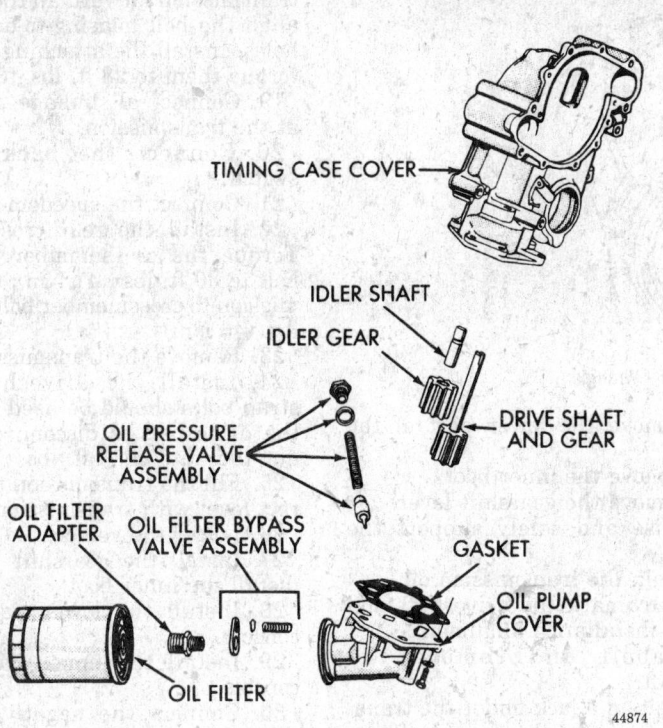

Exploded view of the oil pump and filter - Grand Wagoneer 1991

ITEM	TORQUE
A	41-68 N·m (30-50 ft. lbs.)
B	27-47 N·m (20-35 ft. lbs.)

Rear transmission mounting - Wrangler

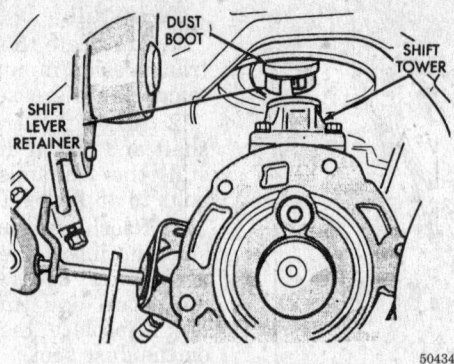

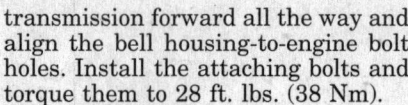

Removing and installing the shift lever - Wrangler

Matchmarking the propeller shafts - Wrangler

32. Fill the transmission and transfer case.
33. Lower the vehicle.
34. Connect the negative battery cable.

Cherokee 1991-95 and Comanche 1991-92

2WD Vehicles

1. Disconnect the negative battery cable.
2. Raise the outer gearshift lever boot and remove the upper part of the console.

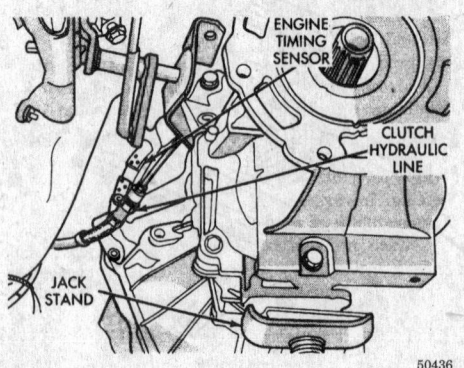

Hydraulic line and timing sensor location - Wrangler

3. Remove the lower part of the console.
4. Remove the inner boot.
5. Remove the gearshift lever.
6. Raise and safely support the vehicle.
7. Drain the transmission oil.
8. Matchmark the driveshaft and yoke for installation alignment.
9. Unbolt and remove the driveshaft.
10. Position a jack under the transmission and take the weight off the transmission slightly.
11. Unbolt and remove the rear crossmember.
12. Disconnect the speedometer cable.
13. Disconnect the backup light switch.
14. Disconnect all linkage and hoses from the transmission.
15. Unbolt the transmission from the engine and lower the transmission while pulling it back.

To install:
16. Lightly grease the input shaft splines.
17. Raise the transmission into position.
18. Roll the transmission forward and engage the input shaft and clutch disc spline. Wiggle the output shaft yoke to get the splines to mesh. Once the splines mesh, push the

transmission forward all the way and align the bell housing-to-engine bolt holes. Install the attaching bolts and torque them to 28 ft. lbs. (38 Nm).
19. Connect all linkage and hoses at the transmission.
20. Connect the backup light switch.
21. Connect the speedometer cable.
22. Install the rear crossmember. Torque the crossmember attaching bolt to 30 ft. lbs. (41 Nm); the transmission to crossmember bolts to 33 ft. lbs. (45 Nm)
23. Remove the transmission jack.
24. Install the driveshaft. New strap bolts should be used whenever the driveshaft is disconnected. Torque the nuts to 14 ft. lbs. (19 Nm).
25. Fill the transmission to the correct level with transmission oil.
26. Lower the vehicle.
27. Install the gearshift lever and install the inner boot.
28. Install the lower part of the console.
29. Install the upper part of the console.
30. Connect the negative battery cable.

4WD Vehicles

1. Disconnect the negative battery cable.
2. Raise and safely support the vehicle.
3. If equipped, remove the skid plate.
4. Drain the transmission and transfer case.
5. Matchmark the front and rear driveshaft and yoke for installation alignment.
6. Unbolt and remove the driveshafts.
7. Disconnect the transfer case shift linkage from the shift lever or range lever.
8. Disconnect the electrical connectors and vent hose from the transmission and differential.
9. Position a jack under the transmission and take up the weight slightly.
10. Unbolt and remove the rear crossmember.
11. Remove the transfer case from the transmission.
12. Lower the transmission enough to provide access to the shift lever.
13. Reach up and around the transmission case and unseat the shift lever dust boot from the transmission shift tower. Reposition the boot to access the lever retainer.
14. Press the shift lever retainer downward and turn it clockwise to release it.

15. Lift the lever and retainer out of the shift tower.

NOTE: It is not necessary to remove the shift lever from the floor pan boot. Leave the lever in place for installation.

16. If equipped, remove the engine timing sensor.

17. If equipped, disconnect the speedometer cable from the transmission.

18. Disconnect the slave cylinder from the transmission and position it aside.

19. Support the transmission to the jack.

20. Unbolt the transmission from the engine and lower the jack while pulling back.

To install:

21. Lubricate the pilot bearing and transmission input splines with high temperature grease.

22. Align the transmission input shaft and clutch disc splines and install the transmission. Torque the bolts to 45 ft. lbs. (61 Nm).

23. Reach up around the transmission and insert the shift lever in the shift tower. Press the lever retainer downward and turn it clockwise to lock it in place. Install the dust lever on the dust boot.

24. Install the transfer case and torque the bolts to 26 ft. lbs. (35 Nm).

25. Install the crossmember. Torque the bolts to 26–33 ft. lbs. (37–45 Nm).

26. Install the slave cylinder to the transmission housing.

27. If equipped, install the engine timing sensor.

28. If equipped, connect the speedometer cable.

29. Connect the electrical connectors and vent hose to the transmission and transfer case.

30. Connect the transfer case shift rod to the lever.

31. Install the driveshafts using new strap bolts. Torque the strap bolt nuts to 14 ft. lbs.; the flange-to-transfer case bolts to 35 ft. lbs.

32. Fill the transmission and transfer case.

33. Lower the vehicle.

34. Connect the negative battery cable.

Grand Cherokee 1993-95 and Grand Wagoneer 1993

2WD Vehicles

1. Disconnect the negative battery cable.

2. Raise the outer gearshift lever boot and remove the upper part of the console.

3. Remove the lower part of the console.

4. Remove the inner boot.

5. Remove the gearshift lever.

6. Raise and safely support the vehicle.

7. Drain the transmission oil.

8. Matchmark the driveshaft and yoke for installation alignment.

9. Unbolt and remove the driveshaft.

10. Position a jack under the transmission and take the weight off the transmission slightly.

11. Unbolt and remove the rear crossmember.

12. Disconnect the speedometer cable.

13. Disconnect the backup light switch.

14. Disconnect all linkage and hoses from the transmission.

15. Unbolt the transmission from the engine and lower the transmission while pulling it back.

To install:

16. Lightly grease the input shaft splines.

17. Raise the transmission into position.

18. Roll the transmission forward and engage the input shaft and clutch disc spline. Wiggle the output shaft yoke to get the splines to mesh. Once the splines mesh, push the transmission forward all the way and align the bell housing-to-engine bolt holes. Install the attaching bolts and torque them to 28 ft. lbs. (38 Nm).

19. Connect all linkage and hoses at the transmission.

20. Connect the backup light switch.

21. Connect the speedometer cable.

22. Install the rear crossmember. Torque the crossmember attaching bolt to 30 ft. lbs. (41 Nm); the transmission to crossmember bolts to 33 ft. lbs. (45 Nm).

23. Remove the transmission jack.

24. Install the driveshaft. New strap bolts should be used whenever the driveshaft is disconnected. Torque the nuts to 14 ft. lbs. (19 Nm).

25. Fill the transmission to the correct level with transmission oil.

26. Lower the vehicle.

27. Install the gearshift lever and install the inner boot.

28. Install the lower part of the console.

29. Install the upper part of the console.

30. Connect the negative battery cable.

4WD Vehicles

1. Disconnect the negative battery cable.

2. Raise and safely support the vehicle.

3. If equipped, remove the skid plate.

4. Drain the transmission and transfer case.

5. Matchmark the front and rear driveshaft and yoke for installation alignment.

6. Unbolt and remove the driveshafts.

7. Disconnect the transfer case shift linkage from the shift lever or range lever.

8. Disconnect the electrical connectors and vent hose from the transmission and differential.

9. Position a jack under the transmission and take up the weight slightly.

10. Unbolt and remove the rear crossmember.

11. Remove the transfer case from the transmission.

12. Lower the transmission enough to provide access to the shift lever.

13. Reach up and around the transmission case and unseat the shift lever dust boot from the transmission shift tower. Reposition the boot to access the lever retainer.

14. Press the shift lever retainer downward and turn it clockwise to release it.

15. Lift the lever and retainer out of the shift tower.

NOTE: It is not necessary to remove the shift lever from the floor pan boot. Leave the lever in place for installation.

16. If equipped, remove the engine timing sensor.

17. If equipped, disconnect the speedometer cable fro the transmission.

18. Disconnect the slave cylinder from the transmission and position it aside.

19. Support the transmission to the jack.

20. Unbolt the transmission from the engine and lower the jack while pulling back.

To install:

21. Lubricate the pilot bearing and transmission input splines with high temperature grease.

22. Align the transmission input shaft and clutch disc splines and install the transmission. Torque the bolts to 45 ft. lbs. (61 Nm).

23. Reach up around the transmission and insert the shift lever in the shift tower. Press the lever retainer downward and turn it clockwise to lock it in place. Install the dust lever on the dust boot.

ITEM	TORQUE
A	54-75 N•m (40-55 FT.LBS.)
B	33-49 N•m (24-36 FT.LBS.)
C	33-60 N•m (24-44 FT.LBS.)

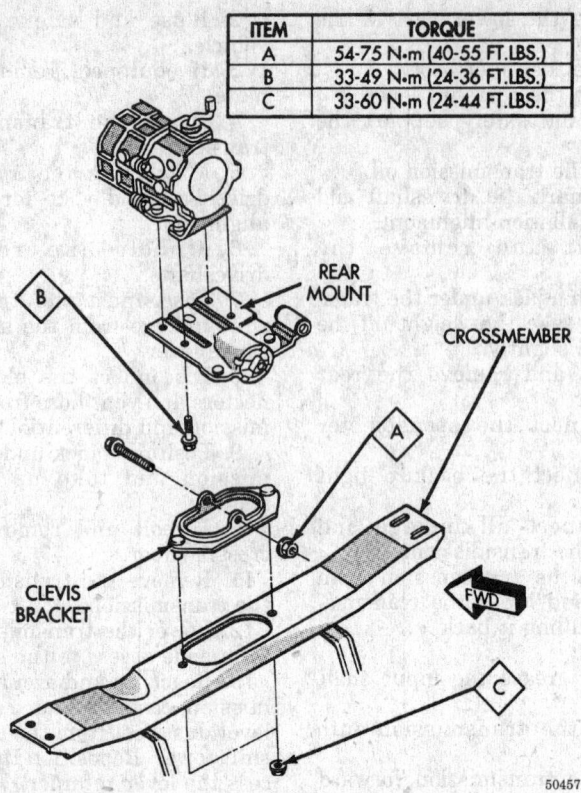

Rear transmission mounting - Grand Cherokee and Grand Wagoneer

50457

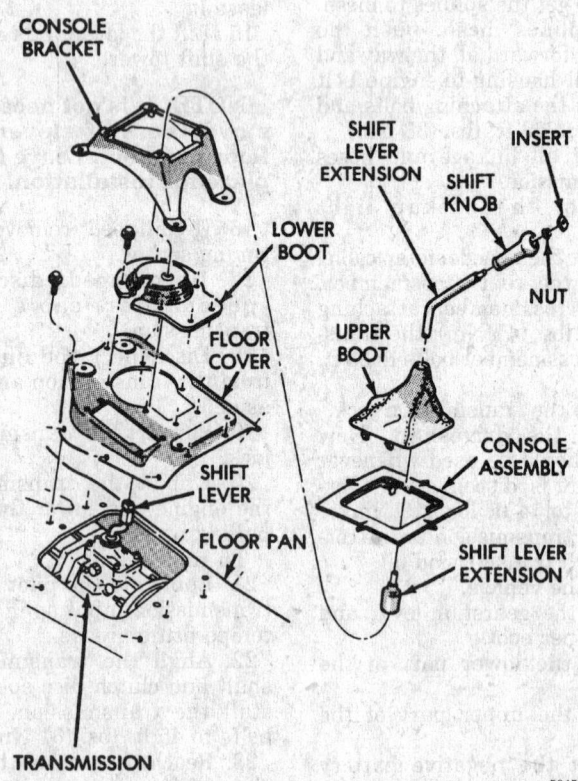

Shift lever attachment - Grand Cherokee and Grand Wagoneer

50458

24. Install the transfer case and torque the bolts to 26 ft. lbs. (35 Nm).

25. Install the crossmember. Torque the bolts to 26–33 ft. lbs. (37–45 Nm).

26. Install the slave cylinder to the transmission housing.

27. If equipped, install the engine timing sensor.

28. If equipped, connect the speedometer cable.

29. Connect the electrical connectors and vent hose to the transmission and transfer case.

30. Connect the transfer case shift rod to the lever.

31. Install the driveshafts using new strap bolts. Torque the strap bolt nuts to 14 ft. lbs.; the flange-to-transfer case bolts to 35 ft. lbs.

32. Fill the transmission and transfer case.

33. Lower the vehicle.

34. Connect the negative battery cable.

CLUTCH

Clutch Disc and Pressure Plate

REMOVAL AND INSTALLATION

Wrangler 1991-95, Cherokee 1991-95 and Comanche 1991-92

1. Raise and safely support the vehicle.

2. Remove the transmission or transmission/transfer case assembly.

3. Matchmark the pressure plate and flywheel. Loosen the pressure plate bolts, a little at a time, in rotation, to avoid warpage.

4. Remove the pressure plate and clutch disc.

5. Inspect the flywheel for scoring, cracks, warpage or other wear; resurface or replace as necessary.

6. Inspect the pilot bearing for excessive wear or damage and replace as necessary.

To install:

7. If removed, install the pilot bearing after lubricating with grease. Seat the bearing in the crankshaft with a clutch alignment tool.

8. Check the clutch disc runout by installing the disc on the transmission input shaft. Runout should not exceed 0.020 in. (0.5mm) when measured ¼ inch from the outer edge of the facing.

9. Install the clutch alignment tool in the pilot bearing.

10. Install the clutch disc on the tool.

11. Install the pressure plate and tighten the bolts finger-tight. The pressure plate bolts must be tightened a little at a time, in rotation, to avoid warpage. Torque the pressure plate bolts as follows:

- 2.5L engine — 23 ft. lbs. (31 Nm)
- 4.0L engine — 40 ft. lbs. (54 Nm)

12. Install the transmission or transmission/transfer case assembly and lower the vehicle.

Grand Cherokee 1993-95 and Grand Wagoneer 1993

1. Raise and safely support the vehicle.

2. Remove the transmission or transmission/transfer case assembly.

3. Matchmark the pressure plate and flywheel. Loosen the pressure plate bolts, a little at a time, in rotation, to avoid warpage.

4. Remove the pressure plate and clutch disc.

5. Inspect the flywheel for scoring, cracks, warpage or other wear; resurface or replace as necessary.

6. Inspect the pilot bearing for excessive wear or damage and replace as necessary.

To install:

7. If removed, install the pilot bearing after lubricating with grease. Seat the bearing in the crankshaft with a clutch alignment tool.

8. Check the clutch disc runout by installing the disc on the transmission input shaft. Runout should not exceed 0.020 in. (0.5mm) when measured 1/4 inch from the outer edge of the facing.

9. Install the clutch alignment tool in the pilot bearing.

10. Install the clutch disc on the tool.

11. Install the pressure plate and tighten the bolts finger-tight. The pressure plate bolts must be tightened a little at a time, in rotation, to avoid warpage. Torque the pressure plate bolts as follows:

- 4.0L engine — 40 ft. lbs. (54 Nm)
- 5.2L engine — 5/16 inch bolts: 17 ft. lbs. (23 Nm) and 3/8 inch bolts 30 ft. lbs. (41 Nm)

12. Install the transmission or transmission/transfer case assembly and lower the vehicle.

Clutch Cable

Clutch Master Cylinder

REMOVAL AND INSTALLATION

Wrangler 1991-93

1. Raise and support the vehicle safely.

2. Disconnect the quick-disconnect fitting by pushing the round disc inward to unsnap and separate the fittings.

3. Lower the vehicle.

4. Remove the cotter pin and washer securing the master cylinder push rod to the clutch pedal and slide the rod off the arm.

5. Unbolt the master cylinder from the firewall.

NOTE: The top bolt is installed from the engine compartment. The bottom bolt in installed from the passenger compartment.

To install:

6. Position the master cylinder and install the top bolt. Install the bottom bolt. Install the nuts and torque them to 19 ft. lbs. (26 Nm).

7. Connect the brake line. Ensure the fittings snap together securely.

8. Connect the push rod to the clutch pedal arm and install the washer and a new cotter pin.

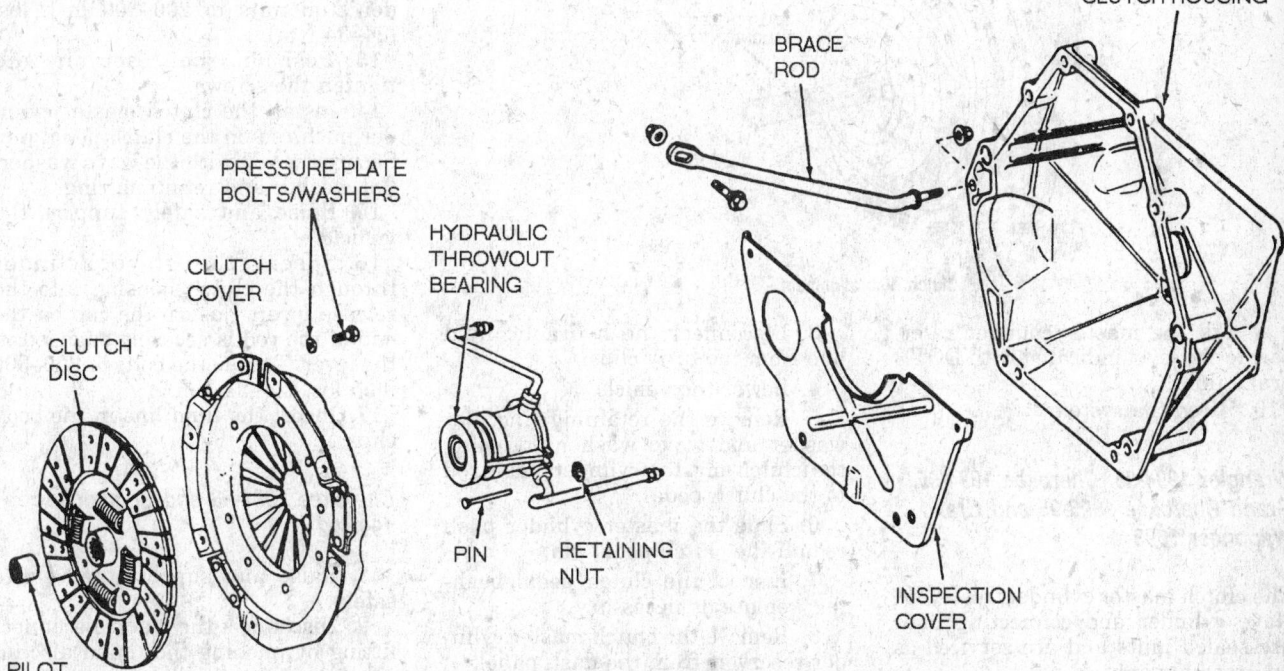

Clutch components - 2.5L engine

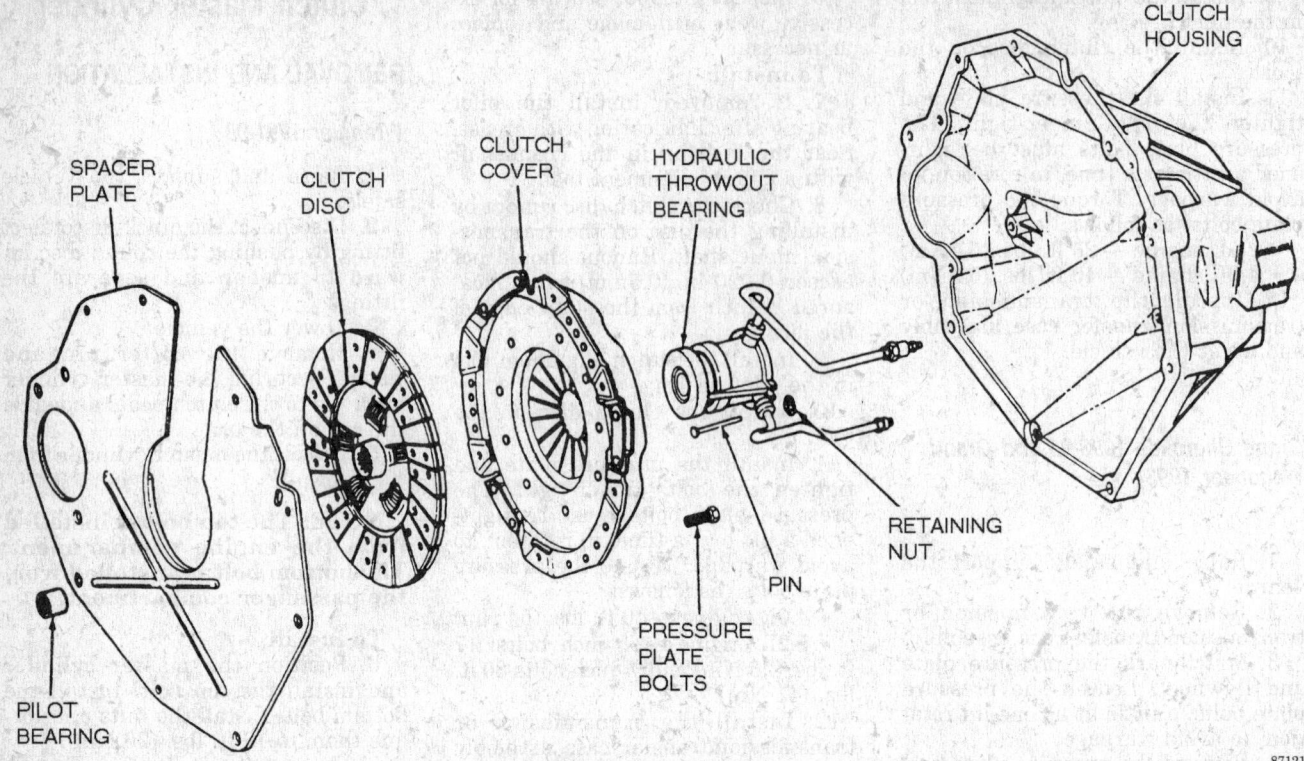

Clutch components - 4.0L engine

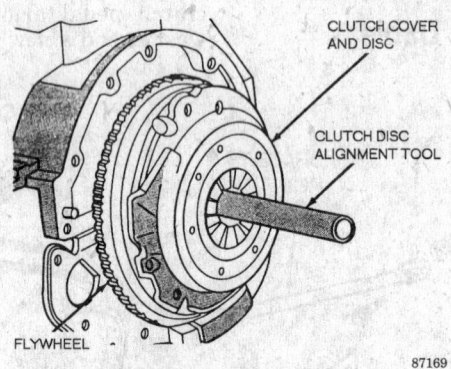

Clutch disc alignment

9. Fill the master cylinder reservoir to the level indicated with DOT 3 brake fluid.

10. Bleed the system.

Wrangler 1994-95, Cherokee 1994-95, Grand Cherokee 1993-95 and Grand Wagoneer 1993

The clutch master cylinder, reservoir, slave cylinder and connecting lines are sealed units and are serviced as an assembly only.

1. Raise and safely support the vehicle.

2. Remove the slave cylinder and clip from the clutch housing.

3. Disconnect the hydraulic fluid line from the body clips.

4. Lower the vehicle.

5. Remove the retaining ring, flat washer and wave washer attaching the clutch master cylinder push rod to the clutch pedal.

6. Slide the master cylinder push rod off the clutch pedal pin.

7. Inspect the clutch pedal bushing, replace as necessary.

8. Remove the clutch master cylinder reservoir from the dash panel.

9. Remove the clutch master cylinder stud nuts.

10. Remove the assembly from the vehicle.

To install:

11. Position the assembly into the vehicle.

12. Torque the clutch master cylinder stud nuts to 200–300 inch lbs. (24–34 Nm).

13. Position the reservoir and tighten the screws.

14. Install the clutch master cylinder push rod on the clutch pedal pin. Secure the rod with the wave washer, flat washer and retaining ring.

15. Raise and safely support the vehicle.

16. Insert the slave cylinder through the clutch housing into the release lever. Ensure the cap on the end of the rod is securely engaged in the lever. Torque the bolts to 200–300 inch lbs. (24–34 Nm).

17. Insert the fluid line in the body clips.

Cherokee 1991-93 and Comanche 1991-92

1. Raise and support the vehicle safely.

2. Disconnect the quick-disconnect fitting by pushing the round disc inward to unsnap and separate the fittings.

3. Lower the vehicle.

4. Remove the lower instrument panel trim cover.

5. Remove the cotter pin and washer securing the master cylinder push rod to the pivot arm and slide the rod off the arm.

6. Unbolt the master cylinder from the firewall.

NOTE: The top bolt is installed from the engine compartment. The bottom bolt in installed from the passenger compartment.

To install:

7. Position the master cylinder and install the top bolt. Install the bottom bolt. Install the nuts and torque them to 19 ft. lbs. (26 Nm).

8. Connect the brake line. Ensure the fittings snap together securely.

9. Connect the push rod to the pivot arm and install the washer and a new cotter pin.

10. Fill the master cylinder reservoir to the level indicated with DOT 3 brake fluid.

11. Bleed the system.

Clutch Slave Cylinder

REMOVAL AND INSTALLATION

Wrangler 1991-93, Cherokee 1991-93 and Comanche 1991-92

The hydraulic concentric bearing is serviced as an assembly only. The release bearing portion of the assembly is permanently attached to the piston. The hydraulic lines are also permanently attached.

1. Disconnect the negative battery cable.

2. Raise and support the vehicle safely.

3. Remove the transmission.

4. Disconnect the clutch master cylinder fluid line.

5. Remove the insulator plate bolts and slide the plate off the bleed line.

6. Remove the concentric bearing retaining nut.

7. Remove the concentric bearing from the transmission input shaft. If the bearing will be reused, secure the bearing and piston with rubber bands.

To install:

8. Inspect the bearing mounting pin and replace the front cover if the pin is damaged. Install the concentric bearing on the transmission input shaft.

9. Guide the bearing fluid and bleed lines through the openings in the clutch housing.

10. Position the bearing boss on the mounting pin and seat the bearing against the transmission. Install a

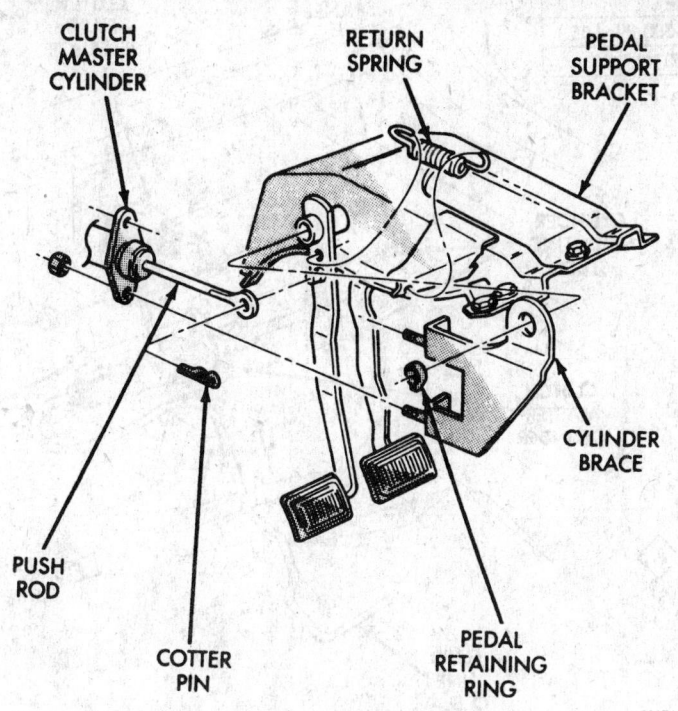

Push rod attachment - Wrangler

50174

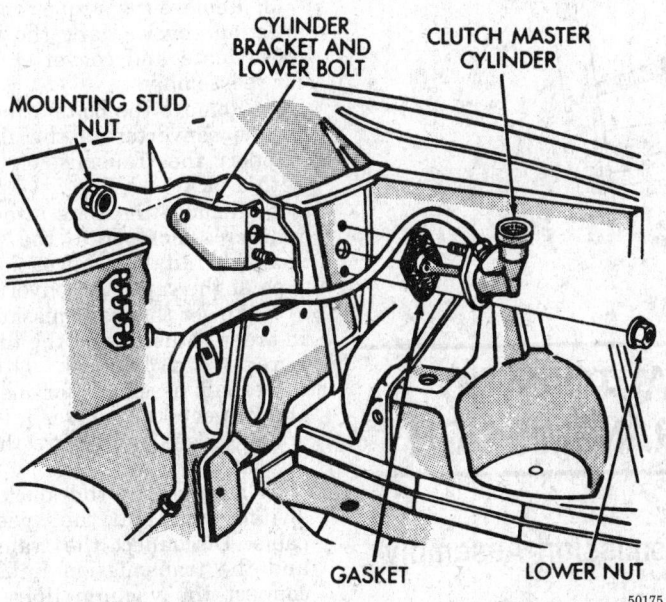

Exploded view of the master cylinder mounting components - Wrangler

50175

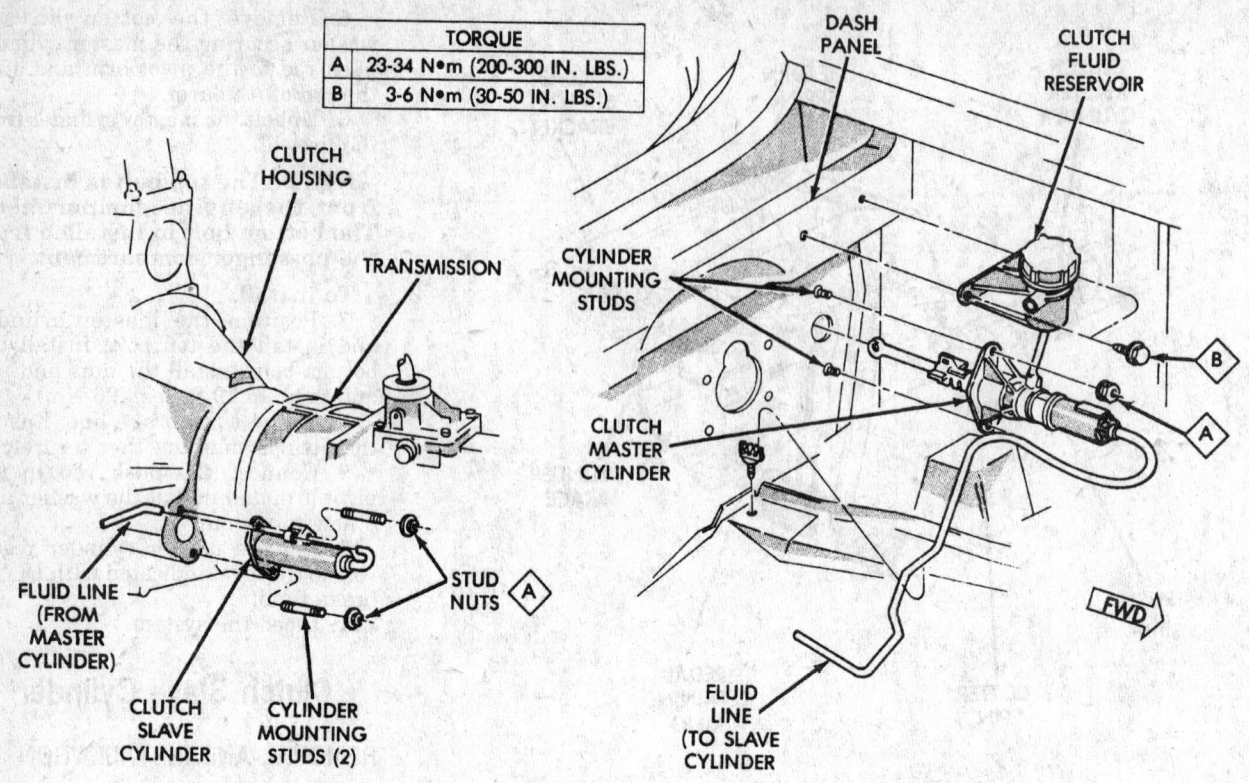

	TORQUE
A	23-34 N•m (200-300 IN. LBS.)
B	3-6 N•m (30-50 IN. LBS.)

Hydraulic clutch components

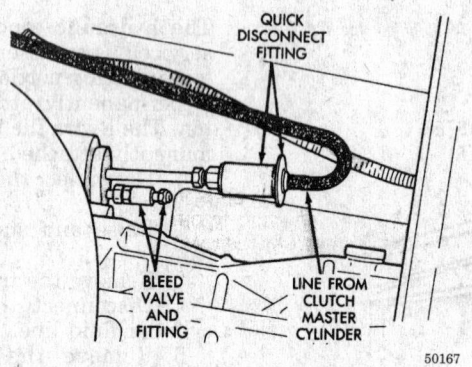

Hydraulic fluid line fitting

new retaining nut and unhook the T-handle straps retaining the bearing.

11. Install the insulator and plate.

12. Install the transmission and transfer case and connect the clutch master cylinder fluid line.

13. Fill and bleed the clutch hydraulic system.

AUTOMATIC TRANSMISSION

Transmission Assembly

REMOVAL AND INSTALLATION

Wrangler 1991

1. Disconnect the negative battery cable. Raise and support the vehicle safely.

2. Matchmark the rear driveshaft and yoke for reassembly. Disconnect and remove the rear driveshaft.

3. Remove the torque converter inspection cover. Mark the converter drive plate and converter assembly for reassembly.

4. Remove the bolts attaching the torque converter to the flex plate. Support the transmission assembly with a jack.

5. Remove the bolts attaching the rear crossmember to the transmission side rail. Disconnect the exhaust pipe at the catalytic converter.

6. Lower the transmission slightly in order to disconnect the fluid cooler lines. Matchmark the front driveshaft assembly for installation. Disconnect the driveshaft at the transfer case and secure the assembly out of the way.

7. Disconnect the backup light switch wire and the speedometer cable. Disconnect the transfer case and the transmission linkage. Disconnect the vacuum lines and the vent hose.

8. Remove the bolts attaching the transmission assembly to the engine. Move the transmission assembly and the torque converter rearward to clear the crankshaft.

9. Carefully lower the transmission assembly from the vehicle. Re-

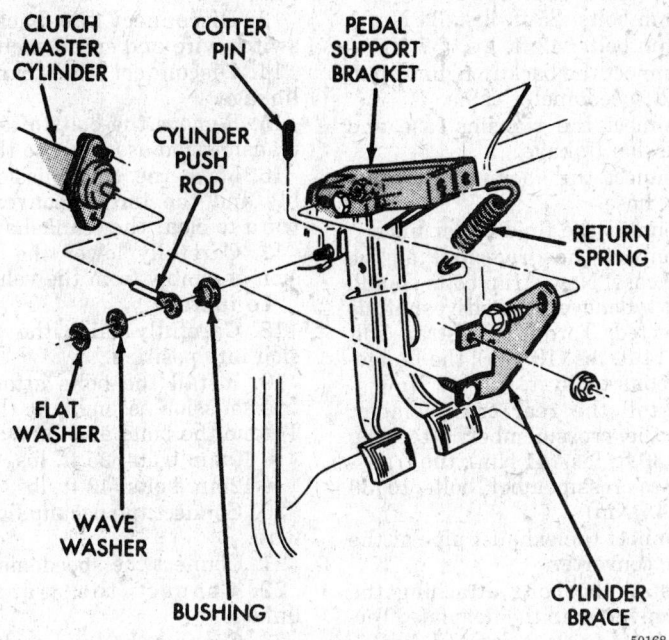

Push rod attachment

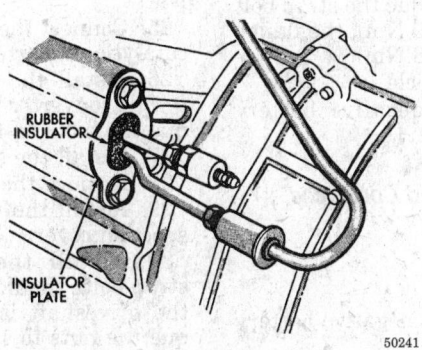

Insulator plate and bleed screw-6 cylinder
engine - Wrangler, Cherokee and Comanche

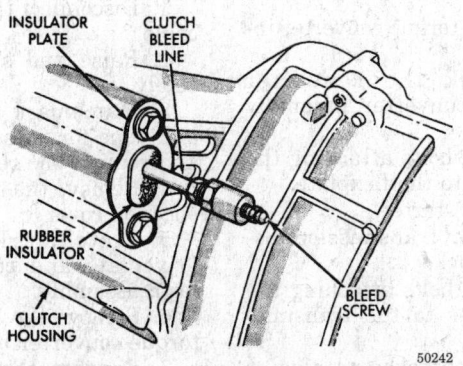

Insulator plate and bleed screw-4 cylinder engine

move the transfer case retaining bolts from the transmission assembly.

To install:

10. If the transmission and transfer case were separated, re-attach them and torque the bolts to 26 ft. lbs.

11. Carefully raise the transmission into position.

12. Install the bolts attaching the transmission assembly to the engine. Torque the bolts to 25 ft. lbs.

13. Connect the backup light switch wire and the speedometer cable.

14. Connect the transfer case and the transmission linkage.

15. Connect the vacuum lines and the vent hose.

16. Connect the fluid cooler lines.

17. Connect the driveshaft at the transfer case. New strap bolts should be used whenever the driveshaft is disconnected. Torque the strap bolt nuts to 14 ft. lbs; the flange-to-case bolts to 35 ft. lbs.

18. Install the rear crossmember. Torque the crossmember attaching bolts to 30 ft. lbs.; the transmission-to-crossmember bolts to 33 ft. lbs.

19. Connect the exhaust pipe at the catalytic converter.

20. Install the bolts attaching the torque converter to the flex plate. Torque the bolts to 40 ft. lbs.

21. Remove the floor jack.

22. Install the torque converter inspection cover.

23. Install the rear driveshaft. Use new strap bolts. Torque the strap bolt nuts to 14 ft. lbs.; the flange bolts to 35 ft. lbs.

24. Lower the truck.

25. Connect the negative battery cable.

Wrangler 1992-95

1. Disconnect the negative battery cable.

2. Raise and support the vehicle safely.

3. Matchmark the rear driveshaft and yoke for reassembly. Disconnect and remove the rear driveshaft.

4. Remove the torque converter inspection cover.

5. Matchmark the converter driveplate and converter assembly for reassembly.

6. Remove the bolts attaching the torque converter to the flexplate.

7. Support the transmission assembly with a jack.

8. Remove the bolts attaching the rear crossmember to the transmission side rail.

9. Disconnect the exhaust pipe at the catalytic converter.

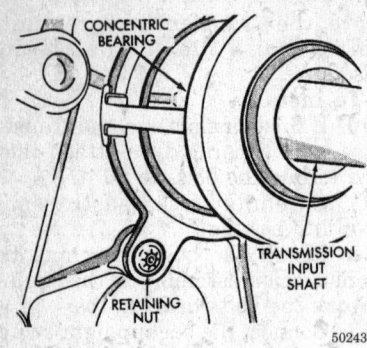

Retaining nut removal and installation - Wrangler, Cherokee and Comanche

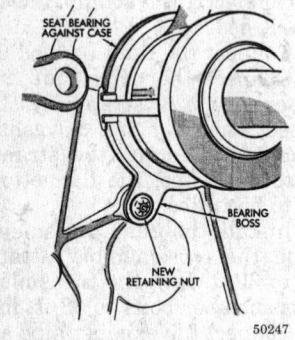

Concentric bearing installation - Wrangler, Cherokee and Comanche

10. Lower the transmission slightly in order to disconnect the fluid cooler lines.

11. Matchmark the front driveshaft assembly for installation.

12. Disconnect the driveshaft at the transfer case and secure the assembly aside.

13. Disconnect the backup light switch wire and speedometer cable.

14. Disconnect the transfer case and transmission linkage.

15. Disconnect the vacuum lines and vent hose.

16. Remove the bolts attaching the transmission assembly to the engine.

17. Move the transmission assembly and torque converter rearward to clear the crankshaft.

18. Carefully lower the transmission assembly from the vehicle.

19. Remove the transfer case retaining bolts from the transmission assembly.

To install:

20. If the transmission and transfer case were separated, re-attach them and torque the bolts to 26 ft. lbs.

21. Carefully raise the transmission into position.

22. Install the bolts attaching the transmission assembly to the engine. Torque the bolts as follows:
- 10mm bolts: 25 ft. lbs. (34 Nm)
- 12mm bolts: 42 ft. lbs. (57 Nm)

23. Connect the backup light switch wire and speedometer cable.

24. Connect the transfer case and transmission linkage.

25. Connect the vacuum lines and the vent hose.

26. Connect the fluid cooler lines.

27. Connect the driveshaft at the transfer case. New strap bolts should be used whenever the driveshaft is disconnected. Torque the strap bolt nuts to 14 ft. lbs (19 Nm); the flange-to-case bolts to 35 ft. lbs. (48 Nm).

28. Install the rear crossmember. Torque the crossmember attaching bolts to 30 ft. lbs. (41 Nm); the transmission-to-crossmember bolts to 33 ft. lbs. (45 Nm).

29. Connect the exhaust pipe at the catalytic converter.

30. Install the bolts attaching the torque converter to the flexplate. Torque the bolts to 40 ft. lbs. (61 Nm).

31. Remove the floor jack.

32. Install the torque converter inspection cover.

33. Install the rear driveshaft. Use new strap bolts. Torque the strap bolt nuts to 14 ft. lbs. (19 Nm); the flange bolts to 35 ft. lbs. 48 Nm).

34. Lower the vehicle.

35. Connect the negative battery cable.

Cherokee 1991-95 and Comanche 1991-92

2WD

1. Disconnect the negative battery cable.

2. Raise and support the vehicle safely.

3. Matchmark the rear driveshaft and yoke for reassembly.

4. Disconnect and remove the rear driveshaft.

5. Remove the torque converter inspection cover.

6. Matchmark the converter driveplate and converter assembly for reassembly.

7. Remove the bolts attaching the torque converter to the flexplate.

8. Remove the starter.

9. Support the transmission assembly using a jack.

10. Remove the bolts attaching the rear crossmember to the transmission side rail.

11. Disconnect the exhaust pipe at the catalytic converter.

12. Lower the transmission slightly in order to disconnect the fluid cooler lines.

13. Disconnect the backup light switch wire and speedometer cable.

14. Disconnect the transmission linkage.

15. Remove the bolts attaching the transmission assembly to the engine.

16. Move the transmission assembly and the torque converter rearward to clear the crankshaft.

17. Carefully lower the transmission assembly from the vehicle.

To install:

18. Carefully raise the transmission into position.

19. Install the bolts attaching the transmission assembly to the engine. Torque the bolts as follows:
- 10mm bolts: 25 ft. lbs. (34 Nm)
- 12mm bolts: 42 ft. lbs. (57 Nm)

20. Connect the backup light switch wire.

21. Connect the speedometer cable.

22. Connect the transmission linkage.

23. Connect the fluid cooler lines.

24. Install the rear crossmember. Torque the crossmember bolts to 30 ft. lbs. (41 Nm); the transmission-to-crossmember bolts to 33 ft. lbs. (45 Nm)

25. Connect the exhaust pipe at the catalytic converter.

26. Install the bolts attaching the torque converter to the flexplate. Torque the bolts to 40 ft. lbs. (54 Nm).

27. Install the starter.

28. Remove the transmission jack.

29. Install the torque converter inspection cover.

30. Install the driveshaft. New strap bolts should be used every time the driveshaft is disconnected. Torque the nuts to 14 ft. lbs. (19 Nm).

31. Lower the vehicle.

32. Connect the negative battery cable.

4WD

1. Disconnect the negative battery cable.

2. Raise and support the vehicle safely.

3. Matchmark the rear driveshaft and yoke for reassembly. Disconnect and remove the rear driveshaft.

4. Remove the torque converter inspection cover.

5. Matchmark the converter driveplate and converter assembly for reassembly.

6. Remove the bolts attaching the torque converter to the flexplate.

7. Support the transmission assembly with a jack.

8. Remove the bolts attaching the rear crossmember to the transmission side rail.

9. Disconnect the exhaust pipe at the catalytic converter.

10. Lower the transmission slightly in order to disconnect the fluid cooler lines.

11. Matchmark the front driveshaft assembly for installation.

12. Disconnect the driveshaft at the transfer case and secure the assembly aside.

13. Disconnect the backup light switch wire and speedometer cable.

14. Disconnect the transfer case and transmission linkage.

15. Disconnect the vacuum lines and vent hose.

16. Remove the bolts attaching the transmission assembly to the engine.

17. Move the transmission assembly and torque converter rearward to clear the crankshaft.

18. Carefully lower the transmission assembly from the vehicle.

19. Remove the transfer case retaining bolts from the transmission assembly.

To install:

20. If the transmission and transfer case were separated, re-attach them and torque the bolts to 26 ft. lbs.

21. Carefully raise the transmission into position.

22. Install the bolts attaching the transmission assembly to the engine. Torque the bolts as follows:
- 10mm bolts: 25 ft. lbs. (34 Nm)
- 12mm bolts: 42 ft. lbs. (57 Nm)

23. Connect the backup light switch wire and speedometer cable.

24. Connect the transfer case and transmission linkage.

25. Connect the vacuum lines and the vent hose.

26. Connect the fluid cooler lines.

27. Connect the driveshaft at the transfer case. New strap bolts should be used whenever the driveshaft is disconnected. Torque the strap bolt nuts to 14 ft. lbs (19 Nm); the flange-to-case bolts to 35 ft. lbs. (48 Nm).

28. Install the rear crossmember. Torque the crossmember attaching bolts to 30 ft. lbs. (41 Nm); the transmission-to-crossmember bolts to 33 ft. lbs. (45 Nm).

29. Connect the exhaust pipe at the catalytic converter.

30. Install the bolts attaching the torque converter to the flexplate. Torque the bolts to 40 ft. lbs. (61 Nm).

31. Remove the floor jack.

32. Install the torque converter inspection cover.

33. Install the rear driveshaft. Use new strap bolts. Torque the strap bolt nuts to 14 ft. lbs. (19 Nm); the flange bolts to 35 ft. lbs. 48 Nm).

34. Lower the vehicle.

35. Connect the negative battery cable.

Grand Cherokee 1993-95 and Grand Wagoneer 1993

2WD

1. Disconnect the negative battery cable.

2. Raise and support the vehicle safely.

3. If equipped, remove the skid plate.

4. If the transmission is being removed for repair, drain the fluid and reinstall the pan.

5. Matchmark the drive shaft yoke and remove the driveshaft.

6. Disconnect the vehicle speed wires, transmission solenoid wires and park-neutral position switch wires.

7. Disconnect the wires from the transmission speed sensor at the rear of the overdrive unit.

8. Remove the exhaust Y-pipe.

9. Unclip the wire harness from the transmission clips.

10. Disconnect the throttle valve and gearshift cables from the levers on the valve body manual shaft. Position the cables aside and secure them to the underbody.

11. Remove the dust cover from the transmission converter housing.

12. Remove the starter.

13. Remove the bolts attaching the converter to the driveplate.

14. Disconnect the cooler fluid lines from the transmission.

15. Support the transmission with a jack.

16. Remove the nuts and bolts securing the rear crossmember to the insulator and remove the crossmemeber.

17. Lower the jack to gain access to the upper portion of the transmission.

18. Remove the crankshaft position sensor.

19. Remove the transmission fill tube and discard the O-ring.

20. Remove the bolts attaching the transmission to the engine.

21. Slide the transmission back and secure a C-clamp to the converter.

22. Remove the transmission.

To install:

23. Ensure the torque converter hub and hub drive are free from sharp edges, scratches or nicks. Polish with 400 grit sandpaper if necessary.

24. Lubricate the converter hub and pump seal with high temperature grease.

25. Secure the C-clamp to the converter.

26. Ensure the dowel pins are seated in the engine block and protrude far enough to align the transmission.

27. Align the transmission with the engine dowels and converter with the driveplate. Install 2 transmission bolts to keep it in place.

28. Remove the C-clamp and install the torque converter bolts. Torque the bolts as follows:
- 3 lug converter — 40 ft. lbs. (54 Nm)
- 4 lug converter — 270 inch lbs. (31 Nm)

29. Install and tighten the remaining transmission-to-engine bolts.

30. Install the crankshaft position sensor.

31. Install the dust cover on the converter housing.

32. Install the starter.

33. Connect the transmission shift and throttle valve cables to the transmission.

34. Fasten the wire harness to the transmission.

35. Connect the harness conectors disconnected during removal.

36. Install the transmission filler tube with a new O-ring.

37. Install the rear crossmember.

38. Connect the fluid cooler lines to the transmission.

39. Align and install the drive shaft.

40. Install the exhaust system components.

41. Lower the vehicle.

42. Connect the negative battery cable.

43. Check the transmission control cables; adjust if necessary.

44. If the transmission fluid was drained, fill the transmission with fluid to the proper level.

4WD

1. Disconnect the negative battery cable.

2. Raise and support the vehicle safely.

3. If equipped, remove the skid plate.

4. If the transmission is being removed for repair, drain the fluid and reinstall the pan.

5. Matchmark the drive shaft yokes and remove both driveshafts.

6. Disconnect the vehicle speed wires, transmission solenoid wires and park-neutral position switch wires.

7. Unclip the wire harness from the transmission clips.

8. Disconnect the transfer case shift linkage from the lever. Remove the linkage and bracket from the transfer case. Position the linkage aside.

9. Remove the nuts attaching the transfer case to the overdrive unit gear case.

10. Place a jack under the transfer case and remove the case.

11. Support the transmission with the jack.

12. Remove the rear transmission crossmemeber.

13. Remove the exhaust Y-pipe.

14. Remove the crankshaft position sensor.

15. Disconnect the gearshift linkage from the lever on the transmission.

16. Remove the transmission shift linkage torque shaft assembly from the transmission and frame rail. Position it aside.

17. Remove the transmission-to-engine brackets.

18. Remove the dust cover from the transmission converter housing.

19. Remove the starter.

20. Remove the bolts attaching the converter to the driveplate.

21. Disconnect the cooler fluid lines from the transmission.

22. Disconnect the solenoid and park/neutral position wire switch wires.

23. Remove the transmission fill tube and discard the O-ring.

24. Lower the jack to gain access to the upper portion of the transmission.

25. Remove the bolts attaching the transmission to the engine.

26. Slide the transmission back and secure a C-clamp to the converter.

27. Move the transmission rearward until it clears the engine block dowels.

NOTE: On some models, part of the flange joining the vehicle cab and dash panel may interfere with transmission removal. If necessary peen this part of the flange over with a mallet.

28. Remove the transmission.

To install:

29. Ensure the torque converter hub and hub drive are free from sharp edges, scratches or nicks. Polish with 400 grit sandpaper if necessary.

30. Lubricate the converter hub and pump seal with high temperature grease.

31. Secure the C-clamp to the converter.

32. Ensure the dowel pins are seated in the engine block and protrude far enough to align the transmission.

33. Align the transmission with the engine dowels and converter with the driveplate. Install 2 transmission bolts to keep it in place.

34. Remove the C-clamp and install the torque converter bolts. Torque the bolts as follows:
- 3 lug converter — 40 ft. lbs. (54 Nm)
- 4 lug, except 10.75 inch converter — 55 ft. lbs. (74 Nm)
- 4 lug, 10.75 inch converter — 270 inch lbs. (31 Nm)

35. Install the starter.

36. Install the strut brackets securing the transmission to the engine and front axle.

37. Install and tighten the remaining transmission-to-engine bolts.

38. Install the crankshaft position sensor.

39. Install the transmission filler tube with a new O-ring.

40. Install the exhaust system components.

41. Install the shift linkage torque bracket.

42. Connect the shift linkage to the transmission.

43. Connect the harness connectors disconnected during removal.

44. Install the rear crossmember.

45. Install the transfer case. Torque the nuts as follows:
- 3/8 stud nuts — 35 ft. lbs. (47 Nm)
- 5/16 stud nuts — 26 ft. lbs. (35 Nm)

46. Install the damper on the transfer case rear retainer if removed. Torque the nuts to 40 ft. lbs. (54 Nm).

47. Connect the transfer case shift linkage.

48. Connect the fluid cooler lines to the transmission.

49. Align and install the drive shafts. Torque the U-joint clamp bolts to 170 inch lbs. (19 Nm).

50. Fill the transfer case to the proper level with fluid.

51. Lower the vehicle.

52. Connect the negative battery cable.

53. Fill the transmission to the proper level with Mopar ATF Plus or equivalent.

54. Check the transmission control cables, adjust if necessary.

55. Check and adjust the transmission and transfer case shift linkage.

Grand Wagoneer 1991

1. Disconnect the negative battery cable.

2. Raise and support the vehicle safely.

3. Matchmark the rear driveshaft and yoke for reassembly. Disconnect and remove the rear driveshaft.

4. Remove the torque converter inspection cover.

5. Matchmark the converter driveplate and converter assembly for reassembly.

6. Remove the bolts attaching the torque converter to the flexplate.

7. Support the transmission assembly with a jack.

8. Remove the bolts attaching the rear crossmember to the transmission side rail.

9. Disconnect the exhaust pipe at the catalytic converter.

10. Lower the transmission slightly in order to disconnect the fluid cooler lines.

11. Matchmark the front driveshaft assembly for installation.

12. Disconnect the driveshaft at the transfer case and secure the assembly aside.

13. Disconnect the backup light switch wire and speedometer cable.

14. Disconnect the transfer case and transmission linkage.

15. Disconnect the vacuum lines and vent hose.

16. Remove the bolts attaching the transmission assembly to the engine.

17. Move the transmission assembly and torque converter rearward to clear the crankshaft.

18. Carefully lower the transmission assembly from the vehicle.

19. Remove the transfer case retaining bolts from the transmission assembly.

To install:

20. If the transmission and transfer case were separated, re-attach them and torque the bolts to 26 ft. lbs.

21. Carefully raise the transmission into position.

22. Install the bolts attaching the transmission assembly to the engine. Torque the bolts to 30 ft. lbs. (41 Nm).

23. Connect the backup light switch wire and speedometer cable.

24. Connect the transfer case and transmission linkage.

25. Connect the vacuum lines and the vent hose.

26. Connect the fluid cooler lines.

27. Connect the driveshaft at the transfer case. New strap bolts should be used whenever the driveshaft is disconnected. Torque the strap bolt nuts to 14 ft. lbs (19 Nm); the flange-to-case bolts to 35 ft. lbs. (48 Nm).

28. Install the rear crossmember. Torque the crossmember attaching

bolts to 30 ft. lbs. (41 Nm); the transmission-to-crossmember bolts to 33 ft. lbs. (45 Nm).

29. Connect the exhaust pipe at the catalytic converter.

30. Install the bolts attaching the torque converter to the flexplate. Torque the bolts to 40 ft. lbs. (61 Nm).

31. Remove the floor jack.

32. Install the torque converter inspection cover.

33. Install the rear driveshaft. Use new strap bolts. Torque the strap bolt nuts to 14 ft. lbs. (19 Nm); the flange bolts to 35 ft. lbs. 48 Nm).

34. Lower the vehicle.

35. Connect the negative battery cable.

Shift Linkage

REMOVAL AND INSTALLATION

Wrangler 1991

1. Shift the transmission into **P** and lock the steering column.

2. Raise and support the vehicle safely.

3. Check the condition of the shift rods, bell crank, bell crank brackets, and linkage bushings. Repair as necessary.

4. Loosen the shift rod trunnion jamnuts.

5. Remove the lockpin that retains the shift rod trunnion to the bell crank. Disengage the trunnion and shift rod at the bell crank.

6. Move the transmission lever rearward into the **P** detent. Be sure the lever is as far rearward as it will go.

7. Check the engagement of the park detent by trying to rotate the driveshaft with the rear wheels off of the ground. The shaft will not rotate if the park detent is engaged.

8. Adjust the trunnion until it will fit in the bell crank arm freely. Prevent the shift rod from turning while tightening the the bolt or nut. Tighten the jamnuts.

9. Install the lock pin.

NOTE: Gearshift linkage lash must be eliminated to obtain proper adjustment. Eliminate lash by pulling down on the shift rod and pressing up on the bell crank.

10. Check engine starting in **P** and **N**; be sure it will not start in any other gear.

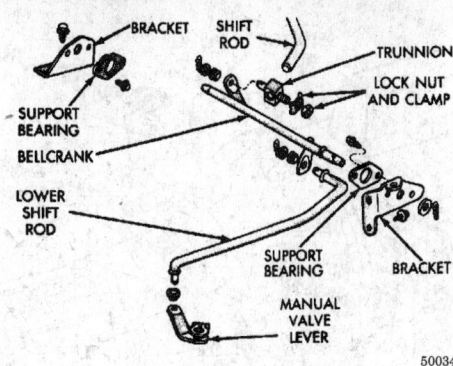

Exploded view of the shift linkage - Wrangler 1991

Wrangler 1992-95

1. Shift the transmission into **P** and lock the steering column.

2. Raise and support the vehicle safely.

3. Check the condition of the shift rods, bell crank, bell crank brackets, and linkage bushings. Repair as necessary.

4. Loosen the shift rod trunnion lock bolt or nut.

5. Remove the lockpin that retains the shift rod trunnion to the bell crank. Disengage the trunnion and shift rod at the bell crank.

6. Move the transmission lever rearward into the **P** detent. Be sure the lever is as far rearward as it will go.

7. Check the engagement of the park detent by trying to rotate the driveshaft with the rear wheels off of the ground. The shaft will not rotate if the park detent is engaged.

8. Adjust the trunnion until it will fit in the bell crank arm freely. Prevent the shift rod from turning while tightening the the bolt or nut. Tighten the bolt or nut.

9. Install the lock pin.

NOTE: Gearshift linkage lash must be eliminated to obtain proper adjustment. Eliminate lash by pulling down on the shift rod and pressing up on the bell crank.

10. Check engine starting in **P** and **N**; be sure it will not start in any other gear.

Cherokee 1991-95, Comanche 1991-92 and Grand Cherokee 1993

1. Place the gear shift lever in the **P** position.

2. Raise and safely support the vehicle.

3. Release the cable adjuster clamp to unlock the cable.

4. Unsnap the cable from the bracket.

5. Move the transmission lever rearward into the **P** detent. Be sure the lever is as far rearward as it will go.

6. Check the engagement of the park detent by trying to rotate the driveshaft with the rear wheels off of the ground. The shaft will not rotate if the park detent is engaged.

7. Snap the cable onto the bracket.

8. Lock the cable by pressing the adjuster clamp down until it snaps into place.

9. Check engine starting in **P** and **N**; be sure it will not start in any other gear.

Grand Cherokee 1994-95

Do not attempt linkage adjustment if any components are worn or damaged. If either linkage rod must be disconnected, the plastic grommet securing the rod in the lever must be replaced. Disconnect the rod with a pry tool. Pry only where the grommet and rod attach and not on the rod itself. Then cut away the old grommet. Use pliers to snap the new grommet into the lever and to snap the rod into the grommet.

1. Shift the transmission into **P**.

2. Raise and support the vehicle safely.

3. Check the condition of the shift rods, control lever, bushings, washers and torque shaft. Repair as necessary.

4. Loosen the lock bolt in the rear shift rod adjusting swivel.

5. Slide the swivel off the torque shaft arm. Ensure the swivel turns freely on the rear shift rod.

6. Move the transmission lever rearward into the **P** detent. Be sure the lever is as far rearward as it will go.

7. Adjust the swivel position on the rear shift rod to obtain free pin fit in the torque shaft lever. Torque the swivel lock bolt to 90 inch lbs. (10 Nm).

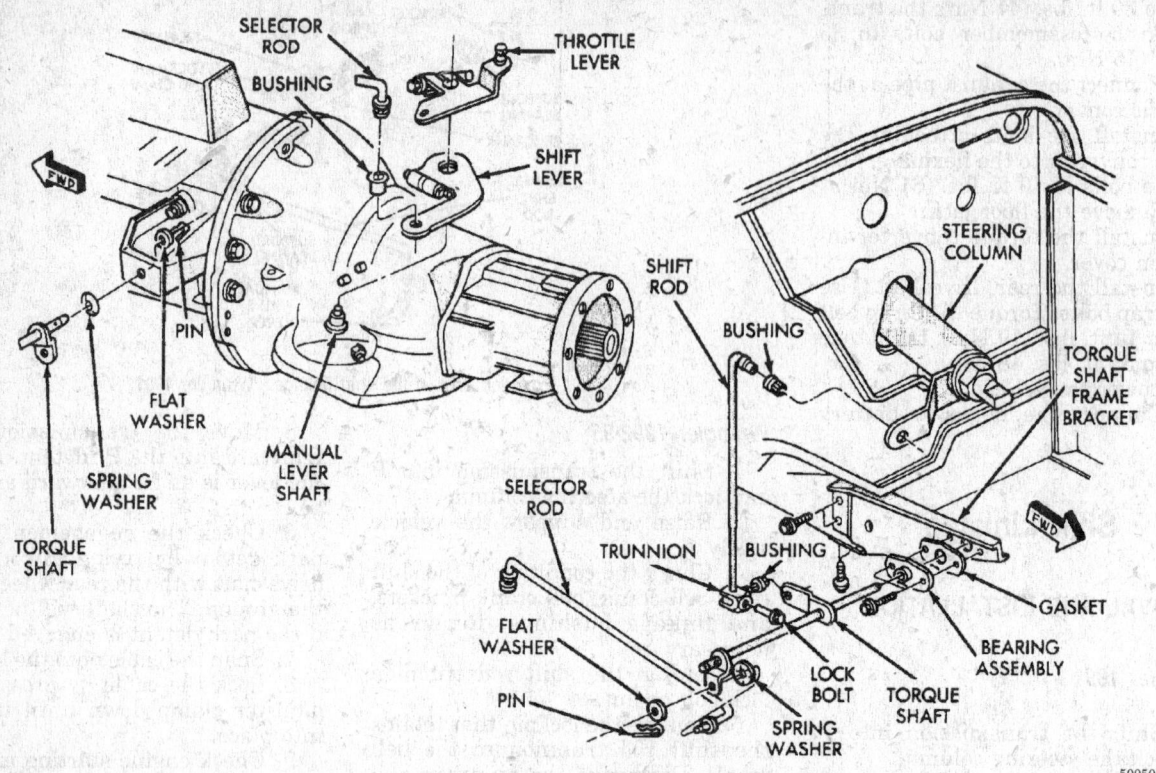

Exploded view of the shift linkage - Wrangler 1992-95

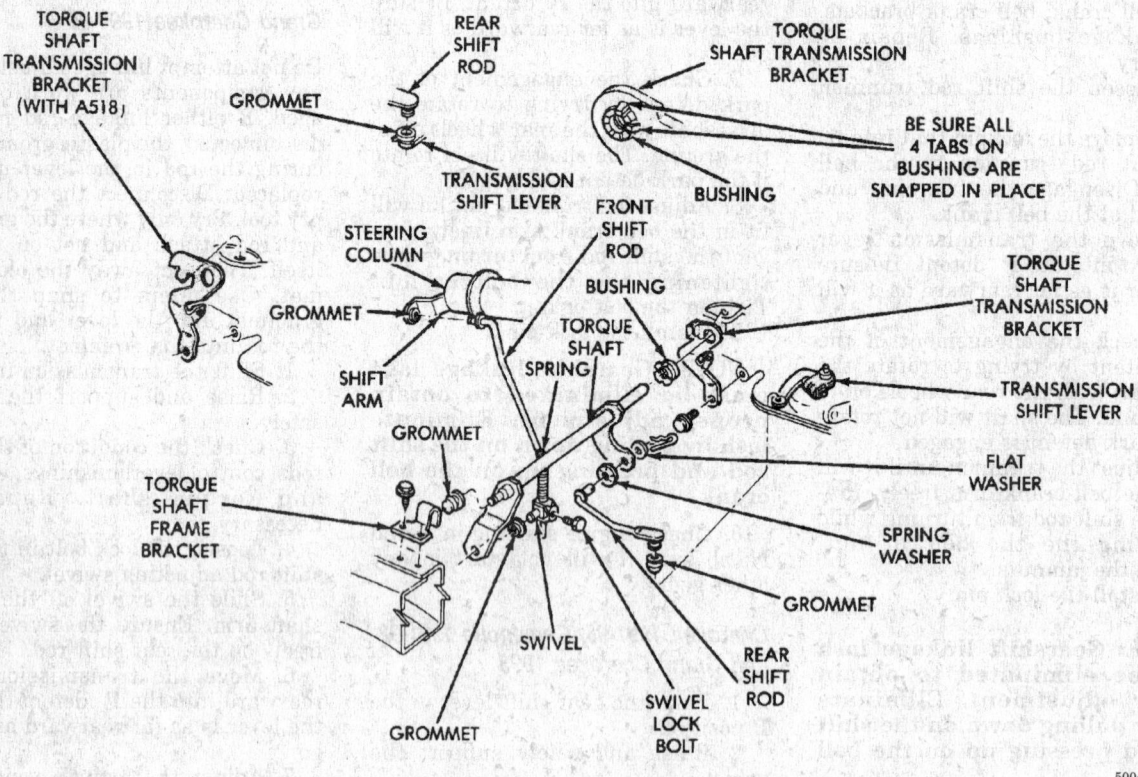

Exploded view of the shift linkage - Grand Cherokee

8. Check engine starting in **P** and **N**; be sure it will not start in any other gear.

Grand Wagoneer 1991

1. Shift the transmission into **P** and lock the steering column.
2. Raise and support the vehicle safely.
3. Check the condition of the shift rods, bell crank, bell crank brackets, and linkage bushings. Repair as necessary.
4. Loosen the shift rod trunnion jamnuts.
5. Remove the lockpin that retains the shift rod trunnion to the bell crank. Disengage the trunnion and shift rod at the bell crank.
6. Move the transmission lever rearward into the **P** detent. Be sure the lever is as far rearward as it will go.
7. Check the engagement of the park detent by trying to rotate the driveshaft with the rear wheels off of the ground. The shaft will not rotate if the park detent is engaged.
8. Adjust the trunnion until it will fit in the bell crank arm freely. Prevent the shift rod from turning while tightening the the bolt or nut. Tighten the jamnuts.
9. Install the lock pin.

NOTE: Gearshift linkage lash must be eliminated to obtain proper adjustment. Eliminate lash by pulling down on the shift rod and pressing up on the bell crank.

10. Check engine starting in **P** and **N**; be sure it will not start in any other gear.

Grand Wagoneer 1993

Do not attempt linkage adjustment if any components are worn or damaged. If either linkage rod must be disconnected, the plastic grommet securing the rod in the lever must be replaced. Disconnect the rod with a pry tool. Pry only where the grommet and rod attach and not on the rod itself. Then cut away the old grommet. Use pliers to snap the new grommet into the lever and to snap the rod into the grommet.

1. Shift the transmission into **P**.
2. Raise and support the vehicle safely.
3. Check the condition of the shift rods, control lever, bushings, washers and torque shaft. Repair as necessary.
4. Loosen the lock bolt in the rear shift rod adjusting swivel.

5. Slide the swivel off the torque shaft arm. Ensure the swivel turns freely on the rear shift rod.
6. Move the transmission lever rearward into the **P** detent. Be sure the lever is as far rearward as it will go.
7. Adjust the swivel position on the rear shift rod to obtain free pin fit in the torque shaft lever. Torque the swivel lock bolt to 90 inch lbs. (10 Nm).
8. Check engine starting in **P** and **N**; be sure it will not start in any other gear.

TRANSFER CASE

Transfer Case Assembly

REMOVAL AND INSTALLATION

Wrangler 1991-95, Cherokee 1991-95, Comanche 1991-92, Grand Cherokee 1993-95 and Grand Wagoneer 1993

NP 207

1. Shift the case into **4H**.
2. Raise and safely support the vehicle.
3. Drain the transfer case.
4. Matchmark the rear driveshaft and remove it.
5. Disconnect the speedometer cable, vacuum hoses and vent hose from the case.
6. Support the transmission with a transmission jack.
7. Remove the crossmember.
8. Matchmark the front driveshaft and remove it.
9. Disconnect the shift lever linkage rod at the case.
10. Remove the shift lever bracket bolts.
11. Support the transfer case with a jack and remove the attaching bolts.
12. Pull the case out of the vehicle.
To install:
13. Raise the transfer case into position. Torque the attaching bolts to 26 ft. lbs.
14. Connect the shift lever linkage rod at the case.
15. Install the shift lever bracket bolts.
16. Install the front driveshaft. New strap bolts should be used. Torque the nuts to 14 ft. lbs. Torque the flange bolts to 35 ft. lbs.
17. Install the crossmember. Torque the bolts to 30 ft. lbs.
18. Remove the support jack.

19. Connect the speedometer cable, vacuum hoses and vent hose at the case.
20. Install the rear driveshaft. Use new strap bolts, torqued to 14 ft. lbs. Torque the flange bolts to 35 ft. lbs.
21. Fill the transfer case to the correct level.
22. Lower the vehicle.

NP 228

1. Raise and safely support the vehicle.
2. Drain the transfer case.
3. Disconnect the speedometer cable, vacuum hoses and vent hose from the case.
4. Disconnect the shift lever linkage rod at the case.
5. Support the transmission with a jack.
6. Remove the rear crossmember.
7. Matchmark the rear driveshaft and remove it.
8. Matchmark the front driveshaft and remove it.
9. Remove the shift lever bracket bolts.
10. Support the transfer case with a jack and remove the attaching bolts.
11. Pull the case rearward out of the vehicle.
To install:
12. Position the transfer case in the vehicle.
13. Torque the attaching bolts to 40 ft. lbs.
14. Install the shift lever bracket bolts.
15. Install the front and rear driveshafts. New strap bolts should be used. Torque the strap bolt nuts to 14 ft. lbs.; the flange nuts to 35 ft. lbs.
16. Install the rear crossmember. Torque the bolts to 30 ft. lbs.
17. Connect the shift lever linkage rod at the case.
18. Connect the speedometer cable, vacuum hoses and vent hose from the case.
19. Fill the transfer case to the correct level.
20. Lower the vehicle.

NP 229

1. Raise and safely support the vehicle.
2. Drain the transfer case.
3. Disconnect the speedometer cable and vent hose.
4. Disconnect the shift lever link at the operating lever.
5. Support the transmission with a jack.
6. Remove the rear crossmember.
7. Matchmark the driveshafts and remove them.
8. Disconnect the shift motor vacuum hoses.

9. Disconnect the shift linkage at the case.

10. Support the transfer case with a floor jack or transmission jack and remove the attaching bolts.

11. Pull the case rearward and remove it.

12. Clean the gasket mating surfaces and use new gasket material for installation.

To install:

13. Raise the transfer case into position. Make certain the case and transmission are mated without binding, before torquing the attaching bolts. Torque the bolts to 26 ft. lbs. (35 Nm).

14. Connect the shift linkage at the case.

15. Connect the shift motor vacuum hoses.

16. Install the driveshafts. New strap bolts should be used. Torque the nuts to 14 ft. lbs. (19 Nm); torque the flange nuts to 35 ft. lbs. (48 Nm).

17. Install the rear crossmember. Torque the bolts to 30 ft. lbs. (41 Nm).

18. Remove the transmission support jack.

19. Connect the speedometer cable and vent hose.

20. Connect the shift lever link at the operating lever.

21. Fill the transfer case.

22. Lower the vehicle.

NP 231, 242 and 249

1. Shift the transfer case into **N**.

2. Raise and support the vehicle safely.

3. Drain the lubricant.

4. Matchmark and remove the front and rear driveshafts.

5. Support the transmission with a jack.

6. Remove the rear crossmember.

7. Disconnect the speedometer cable or speed sensor connector.

8. Disconnect the linkage.

9. Disconnect the vent and vacuum hoses and the indicator wire.

10. Support the transfer case with a transmission jack.

11. Remove the transfer case-to-transmission bolts.

12. Pull the case rearward to disengage it and lower it from the truck.

To install:

13. Raise the transfer case into position. Make certain the case and transmission are mated without binding, before torquing the attaching bolts. Torque the bolts to 26 ft. lbs. (35 Nm).

14. Connect the shift linkage at the case.

15. Connect the vacuum hoses.

16. Install the driveshafts. New strap bolts should be used. Torque the nuts to 14 ft. lbs. (19 Nm); torque the flange nuts to 35 ft. lbs. (48 Nm).

17. Install the rear crossmember. Torque the bolts to 30 ft. lbs. (41 Nm).

18. Remove the transmission floor jack.

19. Connect the speedometer cable or speed sensor connector and vent hose.

20. Connect the shift lever link at the operating lever.

21. Fill the transfer case to the correct lever.

22. Lower the vehicle.

Grand Wagoneer 1991

1. Raise and safely support the vehicle.

2. Drain the transfer case.

3. Disconnect the speedometer cable and vent hose.

4. Disconnect the shift lever link at the operating lever.

5. Support the transmission with a jack.

6. Remove the rear crossmember.

7. Matchmark the driveshafts and remove them.

8. Disconnect the shift motor vacuum hoses.

9. Disconnect the shift linkage at the case.

10. Support the transfer case with a floor jack or transmission jack and remove the attaching bolts.

11. Pull the case rearward and remove it.

12. Clean the gasket mating surfaces and use new gasket material for installation.

To install:

13. Raise the transfer case into position. Make certain the case and transmission are mated without binding, before torquing the attaching bolts. Torque the bolts to 26 ft. lbs. (35 Nm).

14. Connect the shift linkage at the case.

15. Connect the shift motor vacuum hoses.

16. Install the driveshafts. New strap bolts should be used. Torque the nuts to 14 ft. lbs. (19 Nm); torque the flange nuts to 35 ft. lbs. (48 Nm).

17. Install the rear crossmember. Torque the bolts to 30 ft. lbs. (41 Nm).

18. Remove the transmission support jack.

19. Connect the speedometer cable and vent hose.

20. Connect the shift lever link at the operating lever.

21. Fill the transfer case.

22. Lower the vehicle.

DRIVE AXLE

Halfshaft

REMOVAL AND INSTALLATION

Grand Cherokee 1993-95 and Grand Wagoneer 1993

Extreme care must be used to avoid puncturing or tearing the boots. Also avoid damage to the ABS tone ring pressed onto the CV-joint.

The most common failure of CV-joints is torn or ripped boots and subsequent loss or contamination of the lubricant. Look for lubricant around the exterior of the boot. Check for loose clamps and punctured or torn boots, replace as necessary. If the joint was operating satisfactorily and the grease does not appear contaminated, the boot may be reinstalled. When the CV drive shaft is removed from the vehicle for service, the boot should be properly cleaned.

The rubber material in the CV-joint boots is not compatible with oil, gasoline or petrolium-based cleaning solvents. Use only soap and water to clean the boots. After cleaning, the rubber boot must be thoroughly rinsed and dried.

1. Remove the front axle shaft.

2. Remove and discard the CV-joint boot retaining rings.

3. Slide the boot off the axle shaft.

To install:

4. Clean the boot and check for cracks, tears, and scuffed areas on the surface. If any of these conditions exist, replace the boot.

5. Apply CV-joint lubricant to the CV-joint, as necessary.

6. Install the boot onto the shaft with the large end of the boot facing the joint and fill the boot with axle shaft lubricant.

NOTE: Ensure the boot is not twisted.

7. Ensure the clamp sealing area is in the grooved section of the housing and install new clamps.

8. Install the axle shaft.

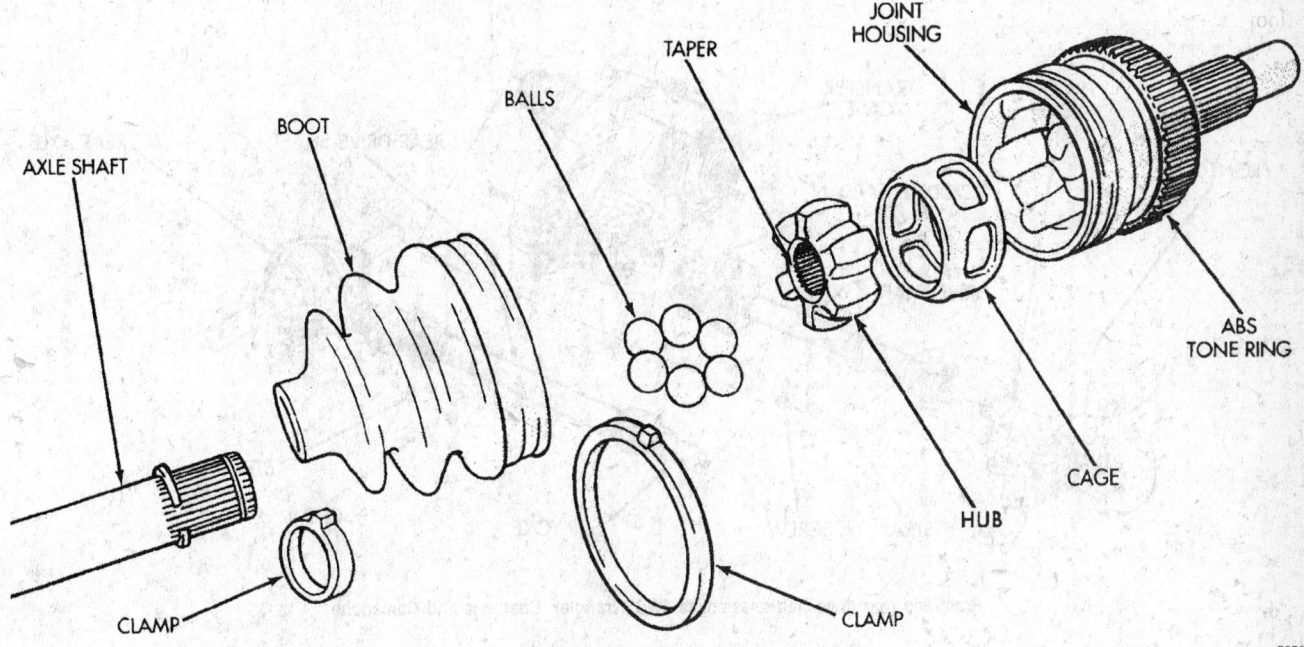

Exploded view of the CV-joint - Grand Cherokee and Grand Wagoneer

50500

Driveshaft

REMOVAL AND INSTALLATION

Wrangler 1991-95, Cherokee 1991-95, Comanche 1991-92, Grand Cherokee 1993-95, Grand Wagoneer 1991 and 1993

FRONT

NOTE: These vehicles may come equipped with one of 2 different type front driveshafts. The first type has a conventional universal joint at the axle but a double offset joint at the transfer case. The second type has a conventional universal joint at the axle and a double cardan joint at the transfer case.

1. Place the transmission and transfer case in **N**.
2. Raise and safely support the vehicle.
3. Matchmark the shaft ends, axle and transfer case.
4. Remove the U-joint strap bolts at the front axle yoke.
5. Remove the double offset joint flange nuts at the transfer case.

To install:
6. Position the driveshaft on the transfer case output shaft and axle yoke aligning the marks.
7. Torque the new U-joint strap-to-axle yoke bolts to 14 ft. lbs. (19 Nm).
8. Torque the new U-joint strap-to-transfer case bolts to 20 ft. lbs. (27 Nm).
9. Lower the vehicle.

REAR

NOTE: Two different driveshafts are used on these vehicles. With Command-Trac®, the driveshaft has welded yokes at each end. With Selec-Trac®, a welded yoke is used at the rear and a splined slip yoke is used at the front.

1. Place the transmission in **N**.
2. Raise and safely support the vehicle.
3. Matchmark the yokes and flanges.
4. On vehicles with Command-Trac®, the driveshaft may be removed by disconnecting it at the axle and sliding it from the front yoke, leaving the front yoke attached to the transfer case. If done this way, matchmark the driveshaft and front yoke before separation.

5. On vehicles with Selec-Trac®, disconnect the yokes from the axle and transfer case. Remove the driveshaft.
6. Installation is the reverse of removal. Torque the U-joint strap nuts to 19 ft. lbs. (26 Nm).

NOTE: New U-joint straps should be used.

Front Axle Shaft

REMOVAL AND INSTALLATION

Wrangler 1991, Cherokee 1991 and Comanche 1991

1. Raise and safely support the vehicle.
2. Remove the front wheels.
3. Remove the hub and bearing assemblies.
4. Remove the left side axle shaft from the axle shaft tube.
5. If equipped with Command-Trac®remove the shift motor and housing from the axle shaft tube.
6. Remove the right side axle from the shaft tube.

NOTE: If equipped with Command-Trac® ensure the shift collar remains on the intermediate shaft.

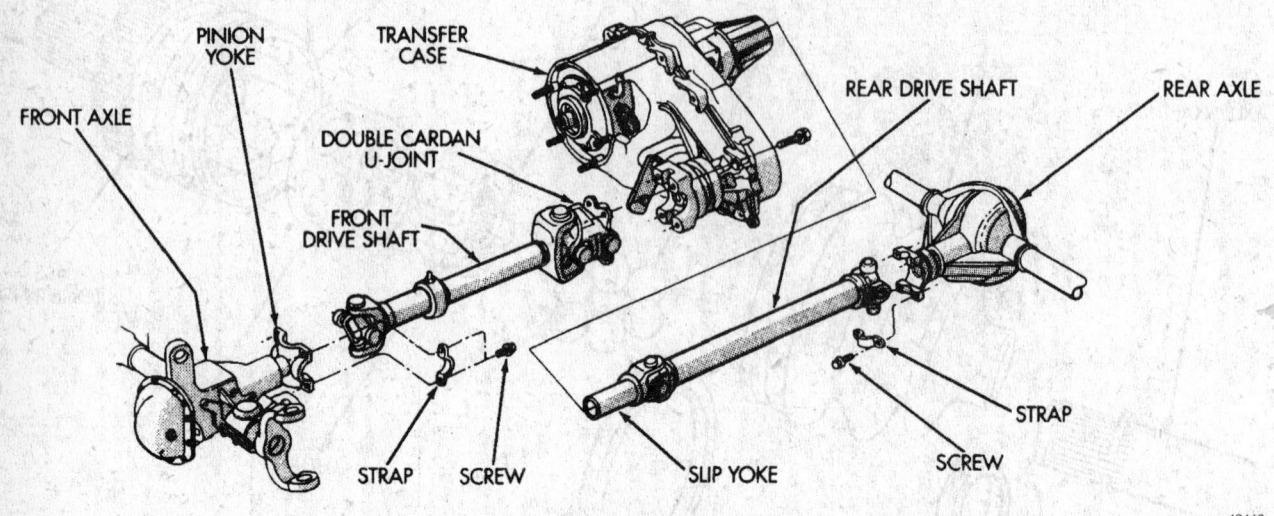

Front and rear driveshaft assemblies-4WD Wrangler, Cherokee and Comanche

To install:

7. Install the axle shafts into the housing.

8. If equipped with Comand-Trac®, ensure the shift collar is correctly positioned on the intermediate axle shaft. Install the shift motor and housing, ensure the fork engages with the shift collar and torque the bolts to 8 ft. lbs. (11 Nm).

9. Install the hub and bearing assemblies.

10. Install the wheels.

11. Lower the vehicle.

Wrangler 1992-95

1. Raise and support the vehicle safely.

2. Remove the wheel.

3. Remove the caliper and wheel.

4. Remove the hub and bearing assembly.

5. If equipped with a shift motor, perform the following:

    a. Disconnect the vacuum and wiring connector from the shift housing.

    b. Remove the indicator switch.

    c. Remove the shift motor housing cover, gasket and shield from the housing.

    d. Remove the E-clips from the shift motor housing and shaft. Re-move the shift motor and fork from the housing.

    e. Remove the O-ring seal from the shift motor shaft.

6. Remove the axle from the housing, being careful to avoid damaging the axle shaft oil seals in the differential.

To install:

7. Thoroughly clean the axle shaft and apply a thin film of wheel bearing grease to the shaft splines, seal contact surface and hub bore.

8. Carefully install the axle shaft, being careful to avoid damaging the differential seals.

9. If equipped with a shift motor, perform the following:

    a. Install a new O-ring seal on the shift motor.

    b. Insert the shift motor shaft through the hole in the housing and shift fork. The shift fork offset should be toward the differential.

    c. Install the E-clips on the shift motor shaft and housing.

    d. Install the shift motor housing gasket and cover. Ensure the shift fork is correctly guided into the shift collar groove.

    e. Install the shift motor shield and torque the bolts to 101 inch lbs. (11 Nm).

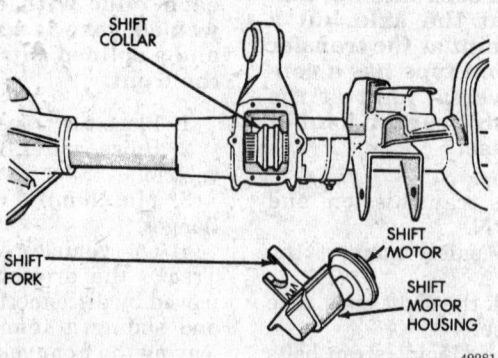

Shift motor housing and shift collar - Wrangler

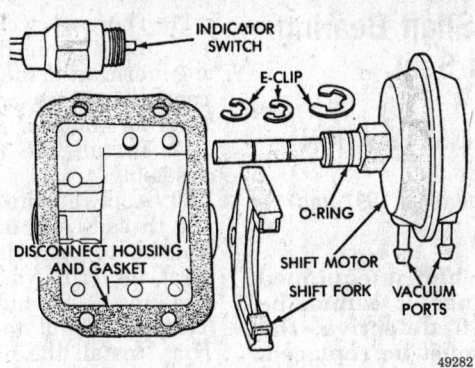

Shift motor components - Wrangler

f. Add 5 ounces of API grade 5 hydraulic gear lubricant or equivalent, into the shift motor housing through the indicator switch mounting hole.

g. Install the indicator switch.

h. Connect the electrical connector and vacuum hoses.

10. Install the wheel bearing and hub assembly.

11. Install the rotor and caliper.

12. Install the wheel.

Cherokee 1992-95, Comanche 1992, Grand Cherokee 1993-95 and Grand Wagoneer 1993

1. Raise and support the vehicle safely.

2. Remove the wheel.

3. Remove the caliper and disc brake rotor.

4. Remove the hub and bearing assembly.

5. Remove the disc brake rotor shield.

6. Remove the axle shaft from the housing, being careful to avoid damaging the axle shaft oil seals in the differential.

To install:

7. Thoroughly clean the axle shaft and apply a thin film of wheel bearing grease to the shaft splines, seal contact surface and hub bore.

8. Carefully install the axle shaft, being careful to avoid damaging the differential oil seals.

9. Install the brake dust shield.

10. Install the wheel bearing and hub assembly.

11. Install the rotor and caliper.

12. Install the wheel and lower the vehicle.

Grand Wagoneer 1991

1. Raise and support the vehicle safely.

2. Remove the wheel and disc brake caliper.

3. Remove the disc brake rotor/hub cap.

4. Remove the axle shaft retaining snapring, drive gear, pressure spring and spring retainer.

5. Remove the inner and outer locknut and washer using nut wrench tool J6893D or equivalent.

6. Remove the disc brake rotor. The spring retainer and outer bearing will be retained with the rotor.

7. Remove the nuts and bolts attaching the spindle and support shield, and remove them from the

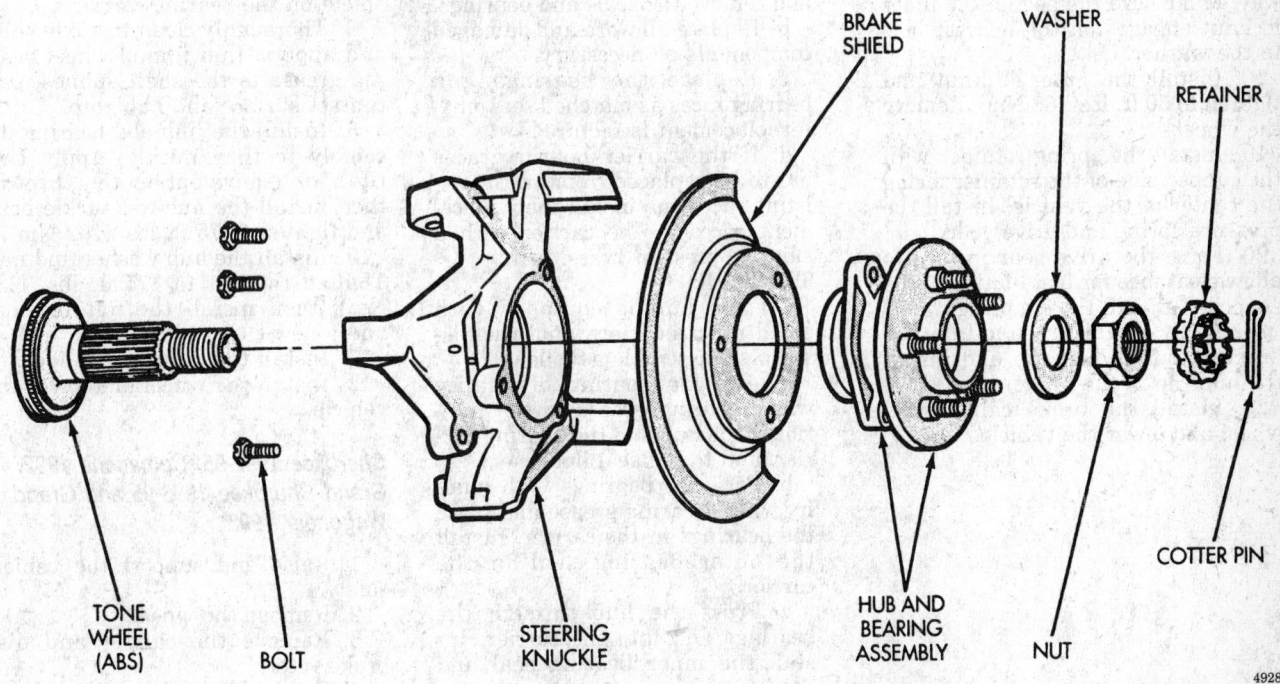

Exploded view of the hub, knuckle and shaft assembly - Wrangler

steering knuckle. If necessary, tap the spindle with a soft mallet to remove it from the knuckle.

8. Remove the axle shaft, seal and bronze thrust washer.

To install:

9. Clean all components thoroughly. If damaged replace the bronze thrust washer.

10. Install the bronze thrust washer with the chamfered side inward. Install the replacement seal using a seal driver.

11. Pack wheel bearing grease around the inner face of the axle shaft and the seal. Apply lubricant to the seal area adjacent to the thrust washer and spidle yoke.

12. Install the axle shaft into the tube and gently force the shaft into the differential side gear.

13. Position the spindle on the steering knuckle and install the retaining nuts. Tighten to 28 ft. lbs. (38 Nm).

14. Install the support shield and disc brake rotor.

15. Install the inner wheel bearing locknut. Tighten the locknut just enough to remove the endplay.

16. Install the wheel and tighten the lug nuts hand tight. Tighten the inner locknut to 50 ft. lbs. (68 Nm), then loosen 45–65 degrees. Rotate the wheel whiel tightening the locknut to seat the bearings evenly.

17. Install the washer so the inner tab is aligned with the spindle keyway. Make sure the peg on the inner locknut engages on the nearest hole in the washer.

18. Install the outer locknut and tighten to 50 ft. lbs. (68 Nm). Remove the wheel.

19. Install the spring retainer with the cupped side of the retainer facing the center of the vehicle. Install the pressure spring and drive gear.

20. Force the drive gear inward to allow clearance for installation of the snapring. Install the snapring.

21. Apply suitable sealant to the brake rotor/hub cap rim and install the hub cap in the brake rotor.

22. Install the brake caliper and wheel and lower the vehicle.

Front Axle Shaft Bearing and Seal

REMOVAL AND INSTALLATION

Wrangler 1991, Cherokee 1991 and Comanche 1991

NOTE: If the hub is equipped with ball bearings, it cannot be serviced and, if defective, the complete unit must be replaced. If the hub is equipped with tapered roller bearings, its internal components can be serviced and replaced as necessary.

1. Raise and support the vehicle safely.

2. Remove the wheel.

3. Remove the brake caliper, leaving the brake line connected and safely support the caliper aside.

4. Remove the brake rotor.

5. Remove the cotter pin, nut retainer and axle hub nut.

6. Remove the hub-to-steering knuckle attaching bolts. Remove the hub from the steering knuckle.

7. Remove the axle shaft dust slinger and rotor shield from the bearing carrier.

8. If the hub is equipped with tapered roller bearings and disassembly is necessary, procedd as follows:

   a. Press the hub from the carrier and remove the seals and bearings.

   b. Replace all worn and damaged components as necessary.

   c. Replace the bearings and bearing races as matched sets only, if replacement is required.

   d. If the carrier bearing races are to be replaced, remove the existing races and install the replacement races in the carrier with a suitable press or brass drift.

To install:

9. If the hub is equipped with tapered roller bearings and was disassembled, assemble as follows:

   a. Apply a coating of quality wheel bearing grease to the interior of the hub and the bearing carrier, and to the seal lips.

   b. Pack the bearings with quality wheel bearing grease and place the bearings in the carrier. Install the outer bearing seal in the carrier.

   c. Press the hub through the bearings and into the carrier. Install the inner bearing seal and carrier seal.

10. Position the disc brake rotor shield and axle dust slinger on the bearing carrier.

11. Thoroughly clean the axle shaft and apply a thin film of wheel bearing lubricant to the shaft splines, seal contact surfaces, and hub bore in the steering knuckle.

12. Install the hub and bearing assembly.

13. Apply Loctite® or equivalent, to the threads and install the steering knuckle-to-hub bolts. Torque them to 75 ft. lbs. (102 Nm).

14. Install the hub washer and nut. Torque the nut to 175 ft. lbs. (237 Nm). Install the nut retainer and a new cotter pin.

15. Clean the brake rotor contact surfaces and install the rotor.

16. Install the brake caliper.

17. Install the wheel.

18. Lower the vehicle.

Wrangler 1992-95

1. Raise and support the vehicle safely.

2. Remove the wheel.

3. Remove the caliper and disc brake rotor.

4. Remove the cotter pin, nut retainer and axle hub nut.

5. Remove the hub-to-knuckle bolts and remove the hub and bearing assembly from the steering knuckle and axle shaft.

6. Remove the disc brake rotor shield from the bearing carrier.

To install:

7. Install the disc brake rotor shield on the bearing carrier.

8. Thoroughly clean the axle shaft and apply a thin film of wheel bearing grease to the shaft splines, seal contact surface and hub bore.

9. Install the hub abd bearing assembly to the knuckle. Apply Loctite® or equivalent to the threads, then install the hub-to-knuckle bolts and tighten to 75 ft. lbs. (102 Nm).

10. Install the hub washer and nut. Tighten the nut to 175 ft. lbs. (237 Nm), then install the nut retainer and a new cotter pin.

11. Install the rotor and caliper.

12. Install the wheel and lower the vehicle.

Cherokee 1992-95, Comanche 1992, Grand Cherokee 1993-95 and Grand Wagoneer 1993

1. Raise and support the vehicle safely.

2. Remove the wheel.

3. Remove the caliper and disc brake rotor.

4. Remove the cotter pin, nut retainer and axle hub nut.

5. Remove the hub-to-knuckle bolts and remove the hub and bear-

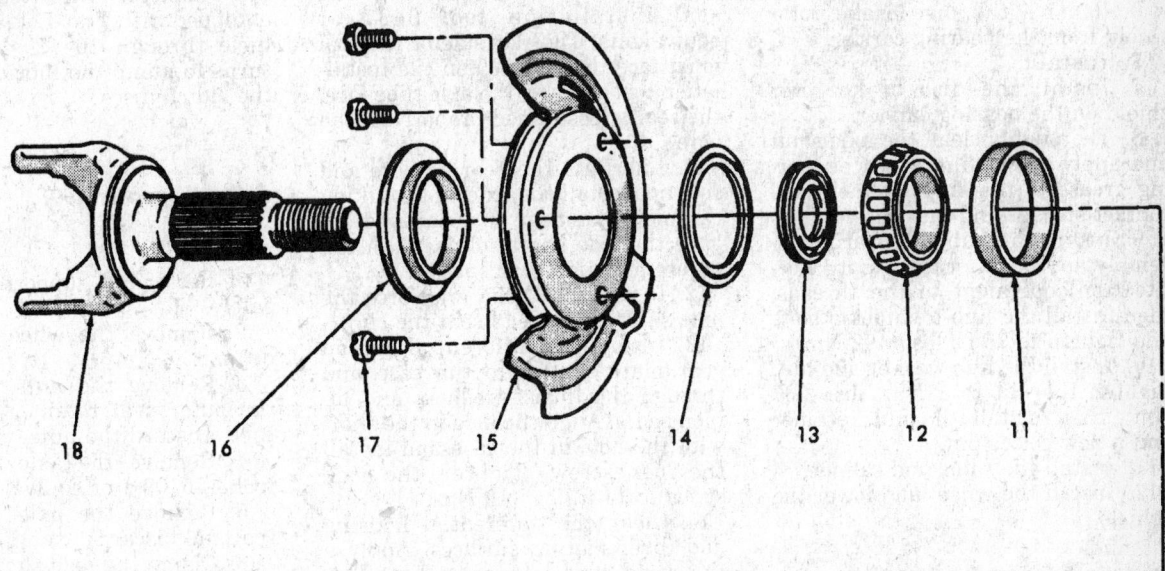

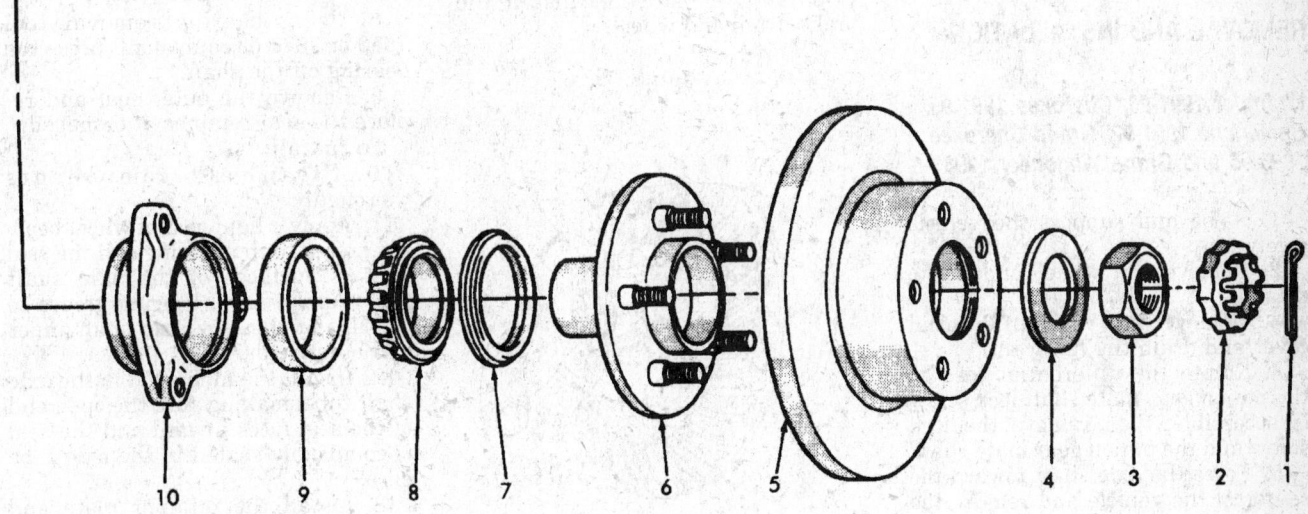

1. COTTER PIN
2. NUT RETAINER
3. NUT
4. WASHER
5. BRAKE ROTOR
6. HUB
7. OUTER BEARING SEAL
8. OUTER BEARING
9. OUTER BEARING RACE

10. BEARING CARRIER
11. INNER BEARING RACE
12. INNER BEARING
13. INNER BEARING SEAL
14. CARRIER SEAL
15. ROTOR SHIELD
16. AXLE SHAFT DUST SLINGER
17. BEARING CARRIER BOLTS
18. AXLE SHAFT

42357

Exploded view of the front hub and bearing assembly - Wrangler, Cherokee and Comanche

ing assembly from the steering knuckle and axle shaft.

6. Remove the disc brake rotor shield from the bearing carrier.

To install:

7. Install the disc brake rotor shield on the bearing carrier.

8. Thoroughly clean the axle shaft and apply a thin film of wheel bearing grease to the shaft splines, seal contact surface and hub bore.

9. Install the hub abd bearing assembly to the knuckle. Apply Loctite® or equivalent to the threads, then install the hub-to-knuckle bolts and tighten to 75 ft. lbs. (102 Nm).

10. Install the hub washer and nut. Tighten the nut to 175 ft. lbs. (237 Nm), then install the nut retainer and a new cotter pin.

11. Install the rotor and caliper.

12. Install the wheel and lower the vehicle.

Rear Axle Shaft, Bearing and Seal

REMOVAL AND INSTALLATION

Wrangler 1991-95, Cherokee 1991-95, Comanche 1991-92, Grand Cherokee 1993-95 and Grand Wagoneer 1993

1. Raise and support the vehicle safely.

2. Remove the wheel and brake drum.

3. Remove the rear differential cover and drain the lubricant.

4. Rotate the differential case so the pinion gear mate shaft lock screw is accessible, then remove the lock screw and the pinion gear mate shaft.

5. Force the axle shaft toward the center of the vehicle and remove the C-clip.

6. Remove the axle shaft. Inspect the axle shaft for evidence of wear or damage; replace the axle shaft if necessary.

7. Pry the axle shaft seal from the axle tube using a small prybar. Remove the axle bearing using bearing removal tool 6310 or equivalent.

To install:

8. Wipe the axle shaft bearing bore in the axle tube clean.

9. Install a new bearing using driver handle tool C-4171 and bearing installation tool 6437 or equivalents. Insert the bearing in the axle tube, making sure it is not cocked. Tap the driver handle tool until the bearing is firmly seated.

10. Install a new axle shaft seal using driver handle tool C-4171 and seal installation tool 6437 or equivalents. The flat side of the tool must face the seal. When the installation tool contacts the axle tube face, the seal is positioned properly in the bore.

11. Lubricate the bearing bore and seal lip. Install the axle shaft, taking care not to damage the seal, and engage the axle shaft splines with the differential side gear splines.

12. Install the C-clip and force the axle shaft outward to seat the clip.

13. Insert the differential pinion gear mate shaft into the case and through the thrust washers and pinion gears. Align the hole in the shaft with the hole in the case and install the lock screw. Tighten the lock screw to 14 ft. lbs. (19 Nm).

14. Clean the differential housing and cover mating surfaces. Apply a thin bead of sealant around the bolt circle on the housing and on the cover.

15. Install the differential housing cover and tighten the attaching bolts to 35 ft. lbs. (47 Nm). Install the brake drum and wheel.

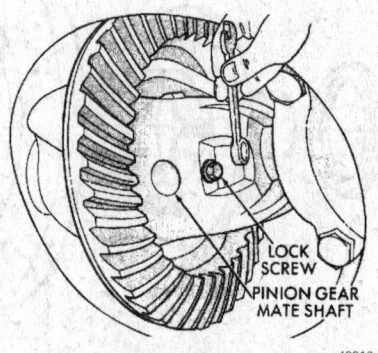

Mate shaft lock screw

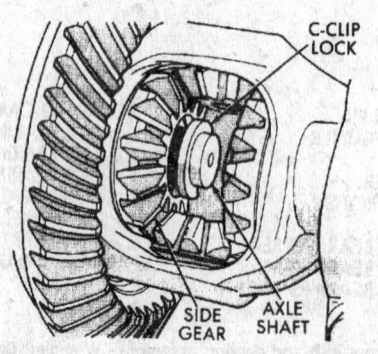

Axle shaft C-clip lock

16. Fill the differential with suitable lubricant and test drive. If equipped with Trac-Lok, drive the vehicle through 10–12 slow figure 8 turns to pump the lubricant through the clutch discs.

Grand Wagoneer 1991

1. Raise and support the vehicle safely.

2. Remove the wheel and brake drum.

3. Remove the nuts that attach the outer seal retainer to the axle tube. Discard the nuts.

4. Remove the axle shaft using tool J-25109-1 or equivalent.

5. Remove the axle shaft inner seal and discard.

6. Clamp the axle shaft in a vise.

7. Center punch then drill a ¼ inch hole in the bearing retaining ring. Drill approximately 3/4 through the retaining ring. Cut the retaining ring with a chisel and remove.

8. Using bearing removal tool 7650 or 7671 or equivalent, press the bearing off the shaft.

9. Remove the outer seal and replace the seal retainer, if damaged.

To install:

10. Clean all components thoroughly.

11. Apply a light coat of wheel bearing grease to the bearing and the seal contact surfaces of the axle shaft tube. Apply wheel bearing grease to the lips of the new axle shaft inner and outer seals.

12. Install the inner seal in the axle shaft tube, making sure the open end of the seal faces inward and the seal is completely seated in the axle tube bore.

13. Install the retainer plate and outer seal on the axle shaft.

14. Pack the axle shaft bearing with wheel bearing grease and position it on the axle shaft.

15. Press the replacement bearing on the shaft with a press and installer tool 7650 or 7671 or equivalent. When the bearing is completely seated on the axle shaft, press a new retaining ring onto the axle shaft and against the bearing.

16. Install the axle shaft into the axle tube. Position and align the seal retainer and brake support plate on the axle tube flange studs. Install the nuts and tighten to 32 ft. lbs. (43 Nm).

17. Install the brake drum and wheel. Lower the vehicle.

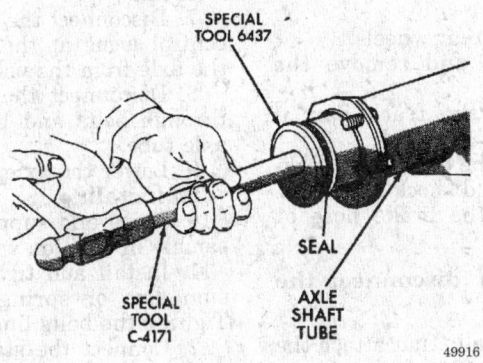

AXLE SHAFT
TUBE

NUT

GUIDE
PLATE

GUIDE

THREADED
ROD

FOOT   ADAPTER

49915

Axle shaft bearing removal tool

SPECIAL
TOOL 6437

SEAL

SPECIAL
TOOL
C-4171

AXLE
SHAFT
TUBE

49916

Axle shaft seal installation

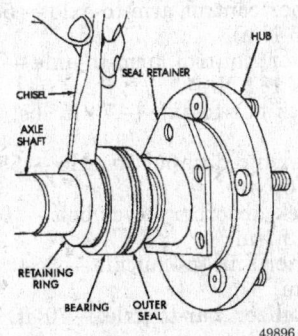

HUB

SEAL RETAINER

CHISEL

AXLE
SHAFT

RETAINING
RING

BEARING   OUTER
SEAL

49896

Bearing retaining ring removal -
Grand Wagoneer

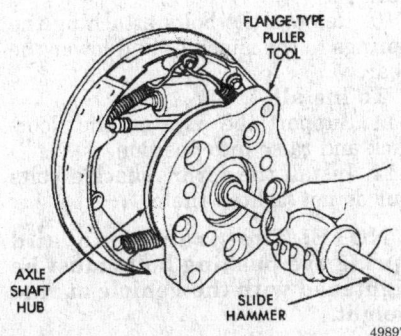

FLANGE-TYPE
PULLER
TOOL

AXLE
SHAFT
HUB

SLIDE
HAMMER

49897

Axle shaft removal - Grand Wagoneer

U-Joint

REMOVAL AND INSTALLATION

Wrangler 1991-95, Cherokee 1991-95, Comanche 1991-92, Grand Cherokee 1993-95, Grand Wagoneer 1991 and 1993

Most Jeep vehicles use a conventional universal joint at both ends of both driveshafts. All models have C-type retainer rings on the inside of the bearing caps. The constant velocity joint is also assembled with snaprings on the outside.

The U-joints are not serviceable and if defective, must replaced as a unit. If the socket yokes, balls, springs, bearings, seals, thrust washers, spiders or bearing caps are damaged or worn, replace the complete U-joint.

SINGLE CROSS CARDAN JOINT

1. Clamp the yoke, not the tube, in a vise.
2. Remove the bearing cap C-retainers. Tap on the bearing caps to relieve pressure as necessary.
3. Support the yoke on the vise jaws.
4. Tap one bearing cap in until the opposite one comes out.
5. Turn the yoke around and tap the exposed end of the spider to drive the remaining bearing cap out.
   To install:
6. Lubricate all needle bearings, bearing caps, and bearing surfaces with chassis grease.
7. Place the seals on the spider.
8. Install one cap and needle bearing assembly partway into the shaft yoke.
9. Install the spider and the opposite bearings and cap.
10. Support the yoke and seat both caps with a hammer.
11. Install the retainer C-clips. Tap the bearing caps as necessary.
12. Install the other two cap and bearing assemblies. Hold them in place with tape until the shaft is reinstalled.

CONSTANT VELOCITY (DOUBLE CARDAN) JOINT

1. Remove the bearing cap retainer snaprings.
2. Mark all components for reassembly.
3. Use a ⅝ in. socket as a bearing cap driver and a 1¹/₁₆ in. socket as a bearing cap receiver. Squeeze the assembly in a vise to force out the bearing caps.

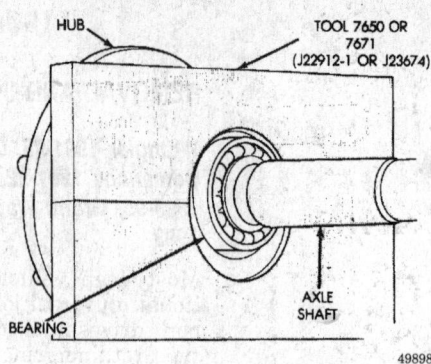

Axle shaft bearing removal - Grand Wagoneer

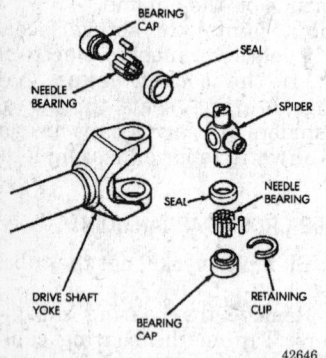

Exploded view of a single cardan U-joint

4. Repeat the operation of step 3 to remove the bearing caps at the other end of the joint.

5. Clean all parts in solvent and dry.

NOTE: Do not disassemble the socket yoke, centering ball, spring, needle bearings, retainer, and thrust washers. These parts are sold as an assembly only.

To install:

6. Lubricate all bearings and contact surfaces with chassis grease.

7. Install the bearing caps on the transfer case yoke ends of the rear spider. Tape them in place.

8. Assemble the socket yoke and the rear spider.

9. Place the rear spider in the link yoke and install the bearing caps. Press them into place with the ⅝ in. socket. Install the snaprings.

10. Install the front spider, bearing caps, and snaprings in the driveshaft yoke.

11. Install the thrust washer and socket spring in the ball socket bearing bore. Install the thrust washer on the ball socket bearing boss on the driveshaft yoke. Align the ball socket bearing boss with the ball socket

bearing bore and insert the boss into the bore.

12. Align the front spider with the link yoke and install the bearing caps and snaprings.

Drive Axle Housing

REMOVAL AND INSTALLATION

Wrangler 1991-95, Cherokee 1991-95 and Comanche 1991-92

REAR

1. Raise and support the vehicle safely.

2. Remove the rear wheels.

3. Matchmark and remove the driveshaft.

4. Disconnect the track bar, if equipped.

5. Disconnect the vent hose, parking brake cable and shock absorbers.

6. Disconnect the brake hose at the junction.

NOTE: Do not disconnect the wheel cylinders.

7. Loosen the bolts that attach the spring front eyes to the frame.

8. Remove the spring U-bolts and support the axle with a floorjack.

9. Raise the rear just enough to relieve the axle weight from the springs.

10. Remove the bolts attaching the springs to the shackles and lower the rear.

To install:

11. Support the axle on the floorjack and raise into position.

12. Install the spring shackle bolts but do not tighten them.

NOTE: The track bar and spring eye bushing bolts must be tightened with the vehicle at ride height.

13. Align the spring center bolts with the locating holes and lower the

axle onto the springs. Install the U-bolts and brackets.

14. Tighten the U-bolts to the 44–52 ft. lbs. (60–70 Nm).

15. Connect the axle vent tube, parking brake cables and brake hose.

16. Align and install the driveshaft. Tighten the bolts to 14 ft. lbs. (19 Nm).

17. Connect the track bar and torque the bolts to 74 ft. lbs. (100 Nm).

18. Install the shock absorbers.

19. Bleed the brakes.

20. Check the fluid level in the differential.

21. Install the wheels.

22. Lower the vehicle.

FRONT

1. Raise and support the vehicle safely.

2. Remove the wheels, calipers and rotors.

3. Disconnect all vacuum hoses at the axle.

4. Matchmark and remove the driveshaft.

5. Disconnect the stabilizer bar, tie rod and drag/center link, shock absorbers, steering damper, track bar and ABS brake sensors.

6. Support the front axle with a floorjack.

7. Disconnect the upper and lower control arms at the axle and lower the axle from the vehicle.

8. Disconnect the spring bushing through bolts and U-bolts from the axle tube.

9. Lower the axle.

To install:

10. Raise and support the axle assembly in position with a floor jack.

11. Install and tighten all suspension arm or spring shackle bolts. Tighten the bolts finger-tight.

12. Connect the stabilizer bar, rod and center link, shock absorbers, steering damper, track bar and ABS brake sensors. Lower the vehicle to normal ride height and tighten the suspension bolts as follows:

Upper control arm-to-axle — 55 ft. lbs. 75 Nm)

Lower control arm-to-axle — 133 ft. lbs. 181 Nm)

Track bar-to-axle — 74 ft. lbs. (101 Nm)

Steering damper-to-axle — 55 ft. lbs. (75 Nm)

Shock absorber lower bolt — 14 ft. lbs. (19 Nm)

Center link-to-knuckle — 35 ft. lbs. (48 Nm)

Stabilizer bar-to-axle — 70 ft. lbs. (95 Nm)

Stabilizer bar-to-connecting links — 55 ft. lbs. (75 Nm)

U-joint strap nuts — 14 ft. lbs. (19 Nm)

U-bolt retaining nuts — 100 ft. lbs. (136 Nm)

NOTE: Discard the U-joint straps. New replacement straps must be used whenever the straps are removed.

13. Align and install the driveshaft.
14. Connect all vacuum hoses at the axle.
15. Install the wheels, calipers and rotors.
16. Check the front wheel alignment.

Grand Cherokee 1993-95 and Grand Wagoneer 1993

REAR

1. Raise and support the vehicle safely.
2. Remove the rear wheels.
3. Matchmark and remove the driveshaft.
4. Disconnect the track bar, if equipped.
5. Disconnect the vent hose, parking brake cable and shock absorbers.
6. Disconnect the brake hose at the junction.

NOTE: Do not disconnect the wheel cylinders.

7. Loosen the bolts that attach the spring front eyes to the frame.
8. Remove the spring U-bolts and support the axle with a floorjack.
9. Raise the rear just enough to relieve the axle weight from the springs.
10. Remove the bolts attaching the springs to the shackles and lower the rear.

To install:

11. Support the axle on the floorjack and raise into position.
12. Install the spring shackle bolts but do not tighten them.

NOTE: The track bar and spring eye bushing bolts must be tightened with the vehicle at ride height.

13. Align the spring center bolts with the locating holes and lower the axle onto the springs. Install the U-bolts and brackets.
14. Tighten the U-bolts to the following specifications:
 • Lower suspension arm bolts: 130 ft. lbs. (177 Nm)
 • Upper suspension arm bolts: 55 ft. lbs. (75 Nm)
15. Connect the axle vent tube, parking brake cables and brake hose.

16. Align and install the driveshaft. Tighten the bolts to 14 ft. lbs. (19 Nm).
17. Connect the track bar and torque the bolts to 74 ft. lbs. (100 Nm).
18. Install the shock absorbers.
19. Bleed the brakes.
20. Check the fluid level in the differential.
21. Install the wheels.
22. Lower the vehicle.

FRONT

1. Raise and support the vehicle safely under the frame rails behind the lower suspension arm brackets.
2. Remove the front wheels.
3. Remove the brake components and if equipped, the ABS sensor.
4. Disconnect the axle vent hose and driveshaft.
5. Disconnect the stabilizer bar link at the axle bracket.
6. Disconnect the axle brackets from the axle bracket.
7. Disconnect the track bar from the axle bracket.
8. Disconnect the tie rod and drag link from the steering knuckles.
9. Disconnect the steering dampener from the axle bracket.
10. Support the axle with a hydraulic jack.
11. Disconnect the upper and lower suspension arms from the axle bracket.
12. Lower the axle with the jack.

NOTE: The coil springs will drop with the axle.

13. Remove the coil springs.
To install:

---WARNING---

All suspension components using rubber bushings should be tightened with the vehicle at ride height. It is important to have the springs supporting the vehicles weight when the fasteners are torqued. If the springs are not at their normal ride position, vehicle ride comfort could be affected along with premature bushing wear. The rubber bushings must never be lubricated.

14. Install the springs, retainer clip and bolts.
15. Position the axle into the vehicle and align it with the straight pads.
16. Position the upper and lower suspension arm at the axle bracket. Install the bolts finger-tight.
17. Connect the tack bar to the axle bracket and install the bolt. Do not tighten at this time.

18. Install the shock absorbers and torque the nuts to 20 ft. lbs. (27 Nm).
19. Install the stabilizer link to the axle bracket and torque the nut to 70 ft. lbs. (95 Nm).
20. Install the drag link and tie rod to the steering knuckles. Torque the nuts to 35 ft. lbs. (47 Nm).
21. Install the steering dampener to the steering axle bracket and torque the nut to 55 ft. lbs. (75 Nm).
22. Install the brake components and if equipped ABS sensor.
23. Connect the vent hose and driveshaft.
24. Check differential lubricant, add if necessary.
25. Install the wheels and tires.
26. Lower the vehicle.
27. Torque the upper suspension arm nuts to 55 ft. lbs. (75 Nm) and lower suspension arm nuts to 85 ft. lbs. (115 Nm).
28. Torque the track bar bolt at the axle bracket to 55 ft. lbs. (75 Nm).
29. Check the front wheel alignment.

Grand Wagoneer 1991

REAR

1. Raise and support the vehicle safely.
2. Remove the rear wheels.
3. Matchmark and remove the driveshaft.
4. Disconnect the track bar, if equipped.
5. Disconnect the vent hose, parking brake cable and shock absorbers.
6. Disconnect the brake hose at the junction.

NOTE: Do not disconnect the wheel cylinders.

7. Loosen the bolts that attach the spring front eyes to the frame.
8. Remove the spring U-bolts and support the axle with a floorjack.
9. Raise the rear just enough to relieve the axle weight from the springs.
10. Remove the bolts attaching the springs to the shackles and lower the rear.

To install:

11. Support the axle on the floorjack and raise into position.
12. Install the spring shackle bolts but do not tighten them.

NOTE: The track bar and spring eye bushing bolts must be tightened with the vehicle at ride height.

13. Align the spring center bolts with the locating boles and lower the axle onto the springs. Install the U-bolts and brackets.

14. Tighten the U-bolts to the 100 ft. lbs. (136 Nm).

15. Connect the axle vent tube, parking brake cables and brake hose.

16. Align and install the driveshaft. Tighten the bolts to 14 ft. lbs. (19 Nm).

17. Connect the track bar and torque the bolts to 74 ft. lbs. (100 Nm).

18. Install the shock absorbers.

19. Bleed the brakes.

20. Check the fluid level in the differential.

21. Install the wheels.

22. Lower the vehicle.

FRONT

1. Raise and support the vehicle safely.

2. Remove the wheels, calipers and rotors.

3. Disconnect all vacuum hoses at the axle.

4. Matchmark and remove the driveshaft.

5. Disconnect the stabilizer bar, tie rod and drag/center link, shock absorbers, steering damper, track bar and ABS brake sensors.

6. Support the front axle with a floorjack.

7. Disconnect the upper and lower control arms at the axle and lower the axle from the vehicle.

8. Disconnect the spring bushing through bolts and U-bolts from the axle tube.

9. Lower the axle.

To install:

10. Raise and support the axle assembly in position with a floor jack.

11. Install and tighten all suspension arm or spring shackle bolts. Tighten the bolts finger-tight.

12. Connect the stabilizer bar, rod and center link, shock absorbers, steering damper, track bar and ABS brake sensors. Lower the vehicle to normal ride height and tighten the suspension bolts as follows:

Upper control arm-to-axle — 55 ft. lbs. 75 Nm)

Lower control arm-to-axle — 133 ft. lbs. 181 Nm)

Track bar-to-axle — 74 ft. lbs. (101 Nm)

Steering damper-to-axle — 55 ft. lbs. (75 Nm)

Shock absorber lower bolt — 14 ft. lbs. (19 Nm)

Center link-to-knuckle — 35 ft. lbs. (48 Nm)

Stabilizer bar-to-axle — 70 ft. lbs. (95 Nm)

Stabilizer bar-to-connecting links — 55 ft. lbs. (75 Nm)

U-joint strap nuts — 14 ft. lbs. (19 Nm)

U-bolt retaining nuts — 100 ft. lbs. (136 Nm)

NOTE: Discard the U-joint straps. New replacement straps must be used whenever the straps are removed.

13. Align and install the driveshaft.

14. Connect all vacuum hoses at the axle.

15. Install the wheels, calipers and rotors.

16. Check the front wheel alignment.

Pinion Seal

REMOVAL AND INSTALLATION

Wrangler 1991-95, Cherokee 1991-95, Comanche 1991-92, Grand Cherokee 1993-95 and Grand Wagoneer 1993

1. Raise and support the vehicle safely.

2. Matchmark and remove the driveshaft.

3. Rotate the pinion gear 3–4 times and measure the amount of torque necessary to rotate the pinion.

4. Measure and note the amount of torque necessary to to rotate the pinion gear with a torque wrench.

NOTE: It is necessary to note the torque to properly adjust the pinion gear bearing preload torque after seal installation.

5. Remove the pinion yoke nut and washer.

6. Using a suitable puller, remove the pinion yoke.

7. Matchmark the pinion yoke and gear for installation alignment reference.

8. Carefully remove the pinion gear seal.

To install:

9. Lubricate and install the new pinion seal using a seal driver.

10. Align and install the yoke, washer and pinion nut. Tighten the pinion nut only enough to remove the shaft end-play.

—————— WARNING ——————
Do not over-tighten or loosen and re-tighten the nut.

11. Using a torque wrench, note the amount of torque necessary to rotate the pinion. The required pinion preload torque is equal to the measurement noted in the removal steps, plus 5 inch lbs.

12. Tighten the pinion nut in small increments and continue measuring preload torque until correct.

13. Align and install the driveshaft using new strap bolts. Torque the bolts to 14 ft. lbs. (19 Nm).

Grand Wagoneer 1991

1. Raise and support the vehicle safely.

2. Matchmark and remove the driveshaft.

3. Using a holding tool and socket wrench, remove the pinion yoke nut and washer. Discard the nut.

4. Using a puller, remove the yoke.

5. Punch the seal with a pin punch and pry it from the seal bore.

To install:

6. Drive the new seal into place using a seal driver.

7. Install the yoke, washer and a new nut. Torque the nut to 210 ft. lbs. (286 Nm).

8. Check the amount of torque required to rotate the pinion gear, it should be 10–20 inch lbs.

NOTE: New strap bolts should be used whenever the driveshaft is disconnected.

9. Install the driveshaft. Torque the strap bolt nuts to 14 ft. lbs.

STEERING

Steering Wheel

REMOVAL AND INSTALLATION

Wrangler 1991-95, Cherokee 1991-95, Comanche 1991-92 and Grand Wagoneer 1991

1. Disconnect the negative battery cable.

2. Set the front tires in a straight-ahead position.

3. Pull the horn button from the steering wheel.

4. If equipped with a standard wheel, remove the trim cover attaching screws and the trim cover. If equipped with a sport wheel, remove the horn contact and flex-plate.

5. Remove the steering wheel nut and, if equipped, vibration damper.

6. Scribe a line mark on the steering wheel and steering shaft if there is not one already.

7. Remove the steering wheel using a suitable puller.

To install:

8. To install, align the scribe marks on the steering shaft with the steering wheel.

9. If equipped, install the vibration damper on the hub. Install the steering wheel retaining nut and tighten to 25 ft. lbs. (34 Nm).

10. Install the internal components, horn flex-plate and the trim cover/horn button.

11. Connect the negative battery cable.

Grand Cherokee 1993-95 and Grand Wagoneer 1993

> ### CAUTION
> *Before removing the steering wheel, disconnect the negative battery cable, then wait at least 2 minutes for the system capacitor to discharge, in order to disable the air bag system. Failure to do so could result in accidental air bag deployment and possible injury.*

1. Disconnect the negative battery cable. Wait at least 2 minutes for the reserve capacitor to discharge.

2. Set the front tires in a straight-ahead position.

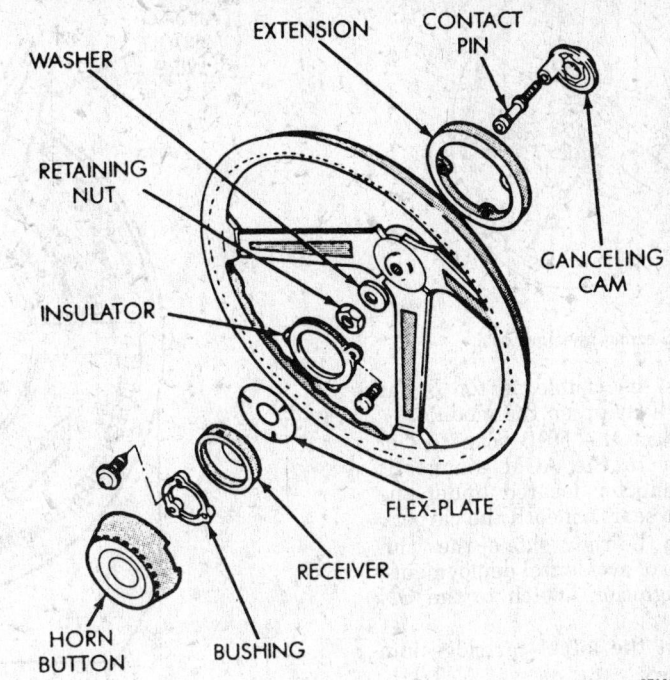

Sport wheel removal

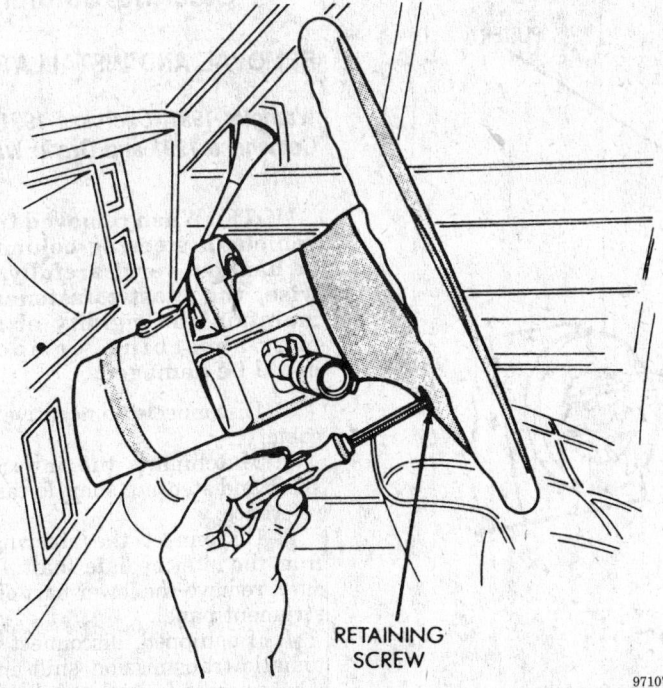

Trim cover removal

3. Remove the air bag module retaining nuts from behind the steering wheel.

4. Remove the air bag module.

5. If equipped, remove the cruise control switch.

6. Disconnect the horn wiring.

7. Remove the steering wheel retaining nut.

8. Scribe a line mark on the steering wheel and steering shaft if there is not one already.

9. Remove the steering wheel using a suitable puller.

To install:

10. To install, align the scribe marks on the steering shaft with the steering wheel.

11. Ensure the wheel compresses the 2 lock tabs on the clockspring.

12. Pull the air bag and if equipped, cruise control wires through the larger hole and horn wire through the smaller hole. Ensure they are not pinched.

13. Install the steering wheel nut and torque it to 45 ft. lbs. (61 Nm) while forcing the steering wheel don the shaft with the nut.

14. Connect the wire feed to the horn buttons.

15. Connect the feed wires and torque the air bag retaining nuts to 90 inch lbs. (10 Nm). Ensure the air bag is completely seated. the latching clip

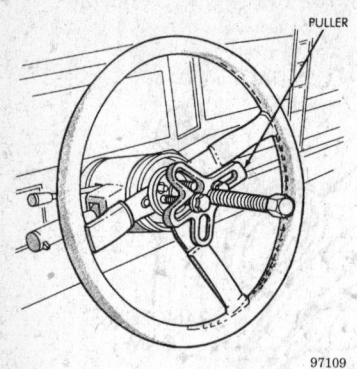

Steering wheel removal

97109

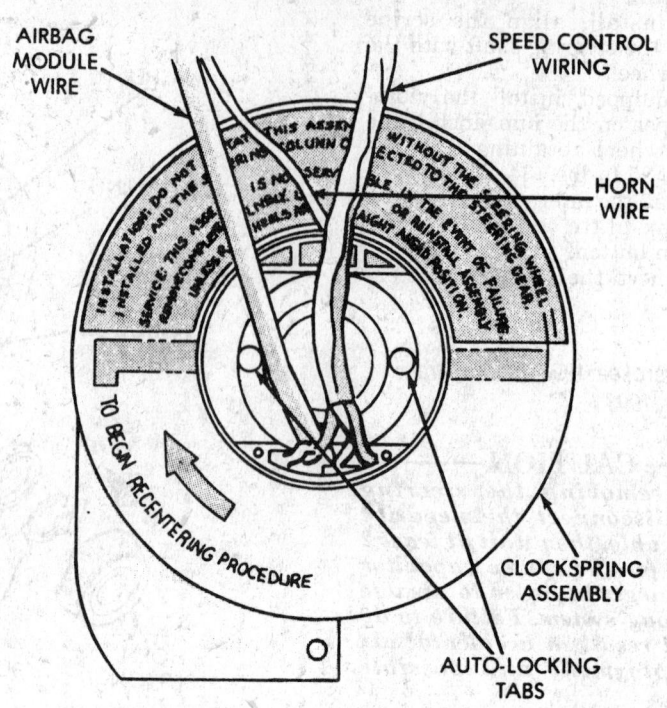

Clockspring (auto-locking) - Grand Wagoneer and Grand Cherokee

47546

arms must be visible on top of the connector housing on the module.

16. Connect the DRB scan tool or equivalent to the ACM diagnostic 6-way connector located under the right front seat, beneath the carpet.

17. From the right side of the vehicle (in case of accidental deployment), turn the ignition switch to the **ON** position.

18. Insert the latest cartridge into the DRB.

19. Connect the negative battery cable.

20. Using the DRB, read and record any active or stored Diagnostic Trouble Code (DTC), repair as necessary.

21. If no DTC is found, remove the DRB.

Steering Column

REMOVAL AND INSTALLATION

Wrangler 1991, Cherokee 1991, Comanche 1991 and Grand Wagoneer 1991

NOTE: When removed from the vehicle, a steering column must be handled very carefully. Otherwise, the plastic fasteners that maintain the rigidity of the energy-absorbing components could be damaged.

1. Disconnect the negative battery cable.

2. Matchmark the intermediate shaft and steering shaft for assembly reference.

3. Disconnect the steering shaft from the intermediate shaft. If necessary, remove the lower part of the instrument panel.

4. If equipped, disconnect the automatic transmission shift indicator pointer cable from the shift housing.

5. Disconnect the column support bracket.

6. Lower the steering column.

Steering wheel removal - Grand Wagoneer and Grand Cherokee

47545

7. Disconnect the following wiring harnesses, if equipped from the steering column:
- Ignition switch
- Dimmer switch
- Turn signal
- Windshield wiper
- Cruise control
- Park lock cable

8. Disconnect the steering column toe plate from the instrument panel and remove the column.

To install:

9. Position the steering column in vehicle paying close attention to the alignment marks.

10. Connect the bracket to the steering column.

11. Connect all wiring harnesses.

12. Connect the park lock cable.

13. Connect the toe plate to the instrument panel and shift indicator to the transmission.

14. Install the lower part of the instrument panel.

15. Check the operation of the shift lever and indicator and adjust as necessary.

16. Connect the negative battery cable.

Wrangler 1992-95

NOTE: Bumping, jolting and hammering on the steering column shaft and gear shift tube must be avoided during all service procedures.

1. Position the front wheels in the straight ahead position.

2. Disconnect the negative battery cable.

3. Remove the steering wheel.

4. If equipped with a column shift lever, disconnect the shift cable grommet by prying it from the shift lever.

5. Disconnect the column shaft-to-steering gear coupler upper bolt.

6. Remove the lower portion of the instrument panel.

7. Remove the nuts securing the column bracket to the brake sled.

8. Remove the bolts securing the steering column bracket to the column.

9. Disconnect the following items from the steering column connectors:
- Ignition switch
- Dimmer switch
- Turn signal switch
- Windshield wiper
- Cruise control, if equipped
- Park-lock cable, if equipped

10. Remove the bolts attaching the toe plate to the floor pan.

11. Remove the steering column from the vehicle.

To install:

12. Position the steering column into the vehicle and align the gear coupling. Torque the bolt to 25 ft. lbs. (34 Nm).

13. Connect the electrical harnesses disconnected during removal.

14. Install the support bracket on the column and torque the bolts to 180 inch lbs. (20 Nm).

15. Raise the column to the brake sled studs and connect the brace to the column. Torque the brake sled nuts or bolts to 22 ft. lbs. (30 Nm).

16. If equipped with a column shift lever, install the shift cable grommet on the shift lever.

17. Install the toe plate-to-floor pan bolts and torque them to 66 inch lbs. (8 Nm).

18. Install the lower portion of instrument panel.

19. Install the steering wheel.

20. Connect the negative battery cable.

Cherokee 1992-95 and Comanche 1992

NOTE: Bumping, jolting and hammering on the steering column shaft and gear shift tube must be avoided during all service procedures.

1. Position the front wheels in the straight ahead position.

2. Disconnect the negative battery cable.

3. Remove the steering wheel.

4. If equipped with a column shift lever, disconnect the shift cable grommet by prying it from the shift lever.

5. Disconnect the column shaft-to-steering gear coupler upper bolt.

6. Remove the lower portion of the instrument panel.

7. If equipped with a column shift lever, disconnect the shift indicator cable from the shift housing.

8. Remove the nuts securing the column bracket to the brake sled.

9. Remove the bolts securing the steering column bracket to the column.

10. Loosen the column brace mounting nut at the drivers side kick panel and allow the column to drop.

11. Disconnect the following items from the steering column connectors:
- Ignition switch
- Dimmer switch
- Turn signal switch
- Windshield wiper
- Cruise control, if equipped
- Park-lock cable, if equipped

12. Remove the bolts attaching the toe plate to the floor pan.

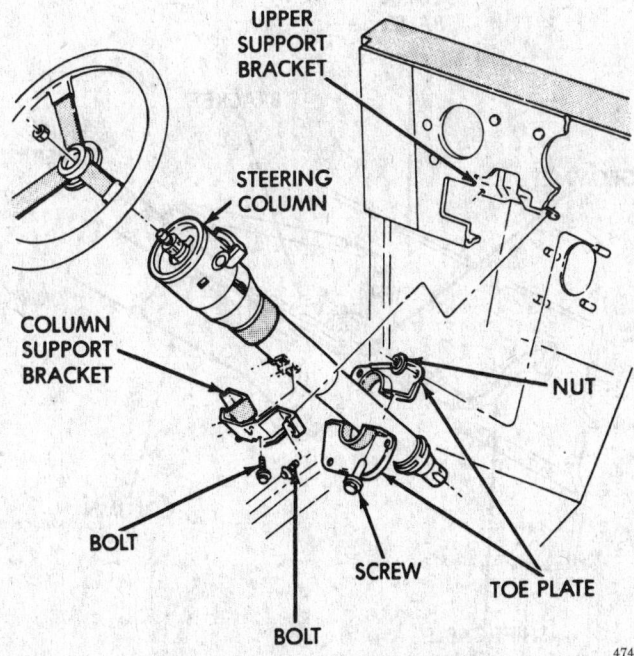

Steering column removal and installation

47491

13. Remove the steering column from the vehicle.

To install:

14. Position the steering column into the vehicle and align the gear coupling. Torque the bolt to 25 ft. lbs. (34 Nm).

15. Connect the electrical harnesses disconnected during removal.

16. Install the support bracket on the column and torque the bolts to 180 inch lbs. (20 Nm).

17. Raise the column to the brake sled studs and connect the brace to the column. Torque the brake sled nuts or bolts to 22 ft. lbs. (30 Nm).

18. If equipped with a column shift lever, install the shift cable grommet on the shift lever.

19. Install the toe plate-to-floor pan bolts and torque them to 66 inch lbs. (8 Nm).

20. If equipped with a column shift lever, install the shift indicator cable on the housing.

21. Install the lower portion of the instrument panel.

22. Install the steering wheel.

23. Connect the negative battery cable.

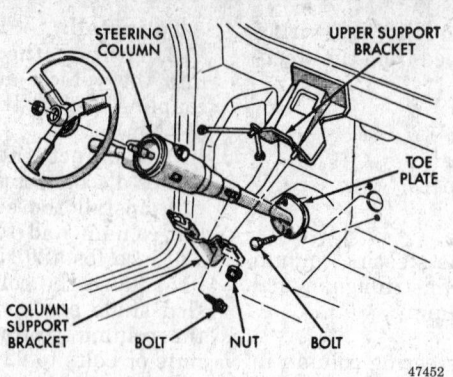

Steering column removal and installation - Cherokee and Comanche

Grand Cherokee 1993-95 and Grand Wagoneer 1993

— **CAUTION** —

Before removing the steering wheel or air bag module, disconnect the negative battery cable, then wait at least 2 minutes for the system capacitor to discharge, in order to disable the air bag system. Failure to do so could result in accidental air bag deployment and possible injury.

1. Position the front wheels in the straight ahead position.

2. Disconnect the negative battery cable.

3. Remove the steering wheel.

4. Remove the column coupler upper pinch bolt.

5. Remove the trim panel column cover and support plate.

6. If equipped, remove the tilt lever from the column.

7. Remove the upper and lower lock housing shrouds.

8. Remove the heater cross over tube from under the column.

9. Loosen the panel bracket nuts and studs and allow the column to drop.

10. Remove the wiring harness from the column.

11. Remove the interlock cable from the column.

12. Remove the toe plate-to-dash panel nuts.

13. Remove the panel bracket nuts and studs. Remove the steering column from the vehicle.

To install:

14. Position the steering column into the vehicle and align the column to the coupler.

NOTE: Do not apply force to the top of the column.

15. Ensure the ground clip is on the left spacer slot.

16. Install the interlock cable to the steering column.

17. Install the wiring harness connections to the steering column.

18. Install the shaft coupler pinch bolt finger-tight and position the column to the panel bracket.

19. Ensure both spacers are fully seated in the column support bracket and torque the nuts and studs to 105 inch lbs. (12 Nm). Ensure the nut is installed on the short threaded side of the stud.

20. Torque the toe plate attaching nuts to 105 inch lbs. (12 Nm).

21. Torque the coupler pinch bolt to 35 ft. lbs. (47 Nm).

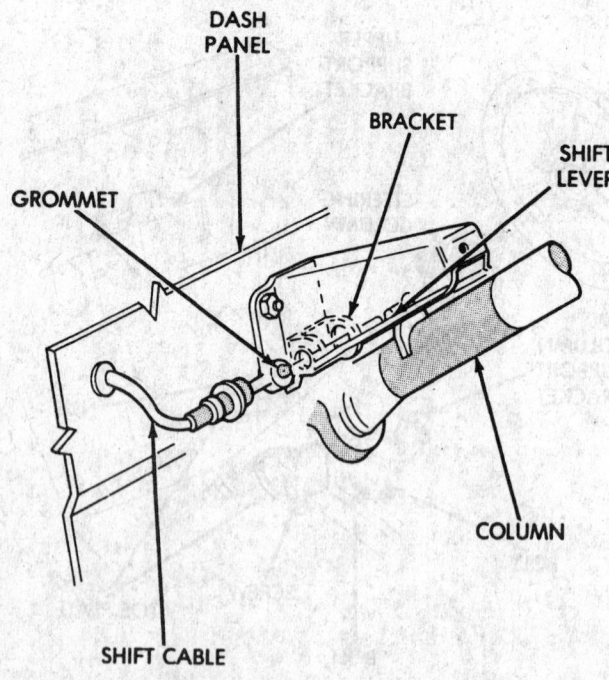

Shift cable grommet-automatic transmission - Cherokee and Comanche

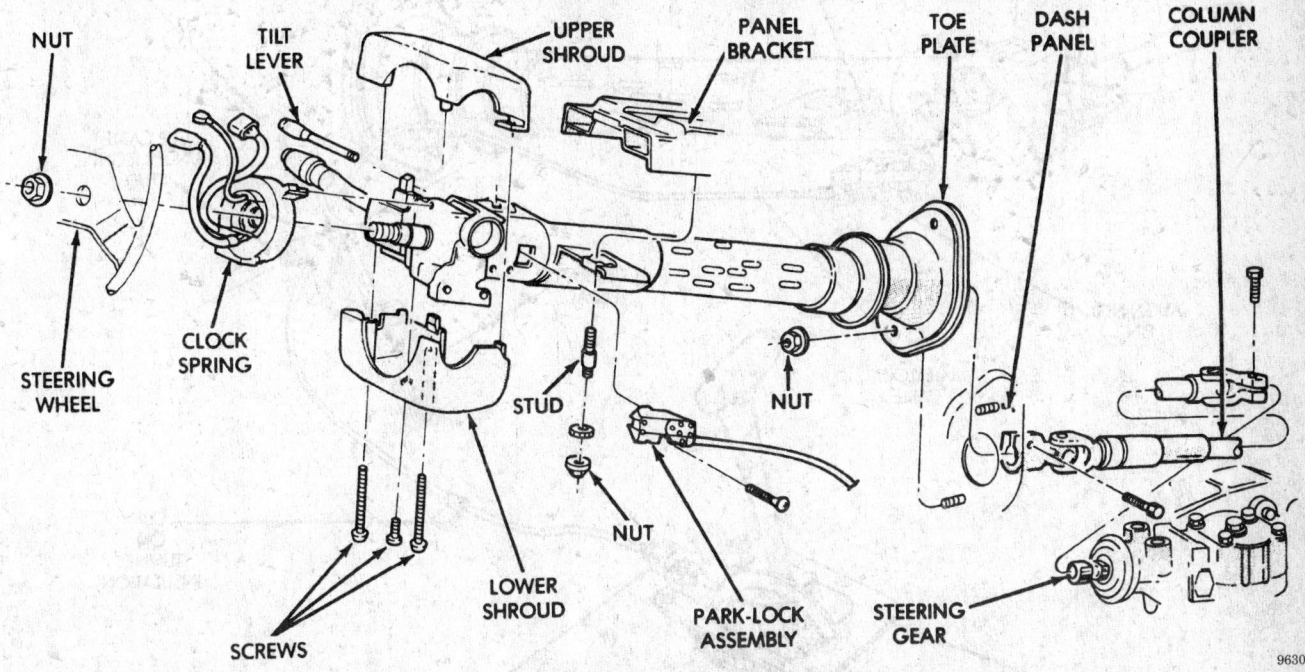

NUT   TILT LEVER   UPPER SHROUD   PANEL BRACKET   TOE PLATE   DASH PANEL   COLUMN COUPLER

STEERING WHEEL   CLOCK SPRING   STUD   NUT

SCREWS   LOWER SHROUD   NUT   PARK-LOCK ASSEMBLY   STEERING GEAR

96301

Exploded view of the steering column - Grand Cherokee and Grand Wagoneer

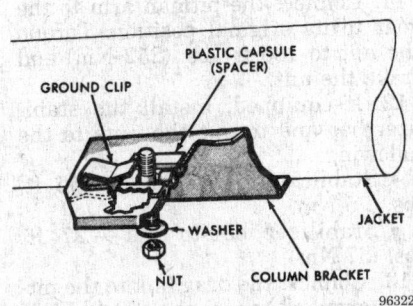

PLASTIC CAPSULE (SPACER)

GROUND CLIP

JACKET

WASHER

NUT   COLUMN BRACKET

96322

Ground clip and spacer installation - Grand Cherokee and Grand Wagoneer

22. Install the heater cross over tube.
23. Install the upper and lower shrouds.
24. If equipped, install the tilt lever.
25. Install the trim panel column cover and support plate.
26. Install the steering wheel.
27. If installing a new column, remove the shaft shipping lock pin.
28. Connect the negative battery cable.

Manual Steering Gear

REMOVAL AND INSTALLATION

Wrangler 1991-95, Cherokee 1991-95, Comanche 1991-92 and Grand Wagoneer 1991

1. Place the front wheels in the straight ahead position.
2. Disconnect the steering shaft from the gear.
3. Raise and safely support the vehicle.
4. Disconnect the center link from the pitman arm.
5. If necessary, remove the front stabilizer bar.
6. Remove the pitman arm nut, matchmark the arm and shaft and remove the arm with a puller.
7. Unbolt and remove the gear.
To install:
8. Install the steering gear. Torque the steering gear to frame bolts to 65–78 ft. lbs. (88–106 Nm).
9. Install the pitman arm to its original position. Torque the nut to 185 ft. lbs. (252 Nm) and stake the nut securely.

10. If equipped, install the stabilizer bar and bolts as follows:
• Stabilizer bar-to-frame — 55 ft. lbs. (75 Nm)
• Stabilizer bar-to-link — 27 ft. lbs. (37 Nm)
11. Connect the center link to the pitman arm and torque the nut to 55 ft. lbs. (75 Nm).
12. Lower the vehicle.
13. Connect the steering shaft to the gear and tighten the nuts.

ADJUSTMENT

Wrangler 1991-95, Cherokee 1991-95, Comanche 1991-92 and Grand Wagoneer 1991

1. Raise and safely support the vehicle.
2. Check the steering gear mounting bolt torque.
3. Matchmark the pitman arm and shaft and remove the pitman arm nut. Remove the arm with a puller.
4. Loosen the pitman adjusting screw locknut, then back off the adjusting screw 2–3 turns.
5. Remove the horn button and cover. Slowly turn the steering wheel in one direction as far as it will go, then back ½ turn.

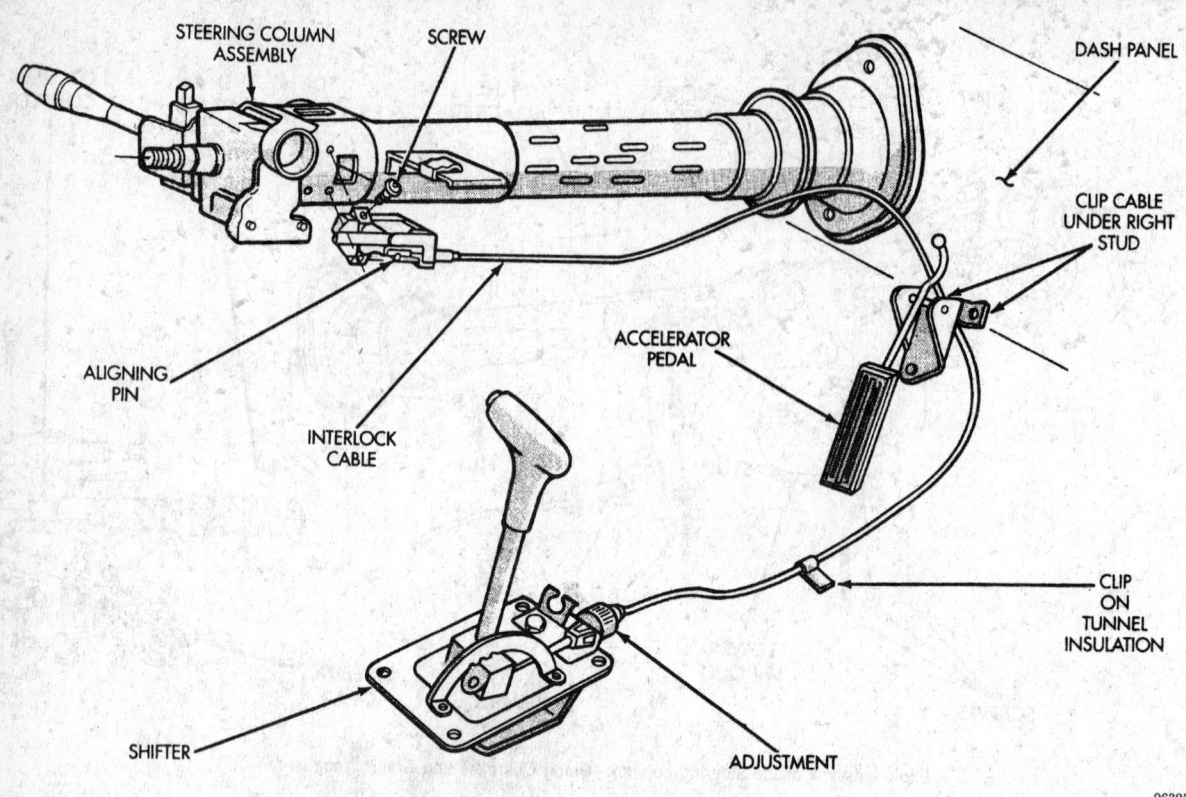

STEERING COLUMN ASSEMBLY

SCREW

DASH PANEL

CLIP CABLE UNDER RIGHT STUD

ACCELERATOR PEDAL

ALIGNING PIN

INTERLOCK CABLE

CLIP ON TUNNEL INSULATION

SHIFTER

ADJUSTMENT

96325

Ignition interlock cable routing - Grand Cherokee and Grand Wagoneer

6. Install a socket and torque wrench in the steering wheel nut. Measure the worm bearing preload by turning the wheel through a 90 degree arc (¼ turn) with the wrench. Preload should be 5–8 inch lbs.

7. If preload is not within specifications, turn the adjuster screw clockwise to increase or counterclockwise to decrease the preload.

8. When the desired preload is attained, tighten the adjuster locknut to 50 ft. lbs. and recheck the adjustment.

9. Rotate the steering wheel slowly from lock-to-lock, counting the number of turns. Turn the wheel back, ½ the number of turns to center the gear, then turn the wheel ½ turn off center.

10. Install the torque wrench and socket on the steering wheel nut. Measure the torque required to turn the gear through the center point of travel. The drag should equal the worm bearing preload torque plus 4–10 inch lbs. but not exceed a total of 18 inch lbs.

11. If adjustment is required, loosen the pitman shaft screw locknut and turn the adjusting screw to obtain the desired torque. Tighten the locknut to 25 ft. lbs. and recheck the overcenter drag.

12. Install all parts and check steering wheel alignment.

Power Steering Gear

REMOVAL AND INSTALLATION

Wrangler 1991, Cherokee 1991, Comanche 1991 and Grand Wagoneer 1991

1. Place the wheels in the straight ahead position.
2. Place a drain pan under the steering gear.
3. Disconnect the fluid lines from the steering gear, plug them and position them aside.
4. Disconnect the intermediate shaft from the stub shaft.
5. Raise and safely support the vehicle.
6. Disconnect the drag link from the pitman arm.
7. If necessary, remove the front stabilizer bar.
8. Remove the pitman arm nut, matchmark the arm and shaft and remove the arm with a puller.
9. Unbolt and remove the gear.
   To install:
10. Install the gear and torque the bolts to 65 ft. lbs. (88 Nm).

11. Connect the pitman arm to the gear to its original position. Torque the nut to 185 ft. lbs. (252 Nm) and stake the nut.
12. If equipped, install the stabilizer bar and torque the nuts to the following:
   • Stabilizer bar-to-frame — 55 ft. lbs. (75 Nm)
   • Stabilizer bar-to-link — 27 ft. lbs. (37 Nm)
13. Connect the drag link to the pitman arm and torque the nut to 55 ft. lbs. (75 Nm).
14. Lower the vehicle.
15. Connect the intermediate shaft to the stub shaft.
16. Connect the hydraulic lines and tighten the fittings.
17. Fill the power steering system to the correct level and bleed the system.

Wrangler 1992-95

1. Place the wheels in the straight ahead position.
2. Disconnect and cap the power steering lines from the steering gear.
3. Remove the column coupler shaft from the gear.
4. Matchmark the pitman arm to the gear and remove the pitman arm.
5. Remove the steering gear retaining bolts and remove the gear.

To install:

6. Align the column coupler shaft to the steering gear.

7. Position the steering gear and bracket on the frame rail and install the bolts. Torque the bolts to 78 ft. lbs. (105 Nm).

8. Align and install the pitman arm.

9. Connect the power steering lines.

10. Fill the power steering reservoir to the proper level with fluid and bleed the system.

Cherokee 1992-95 and Comanche 1992

1. Place the wheels in the straight ahead position.

2. Disconnect and cap the power steering lines from the steering gear.

3. Remove the column coupler shaft from the gear.

4. Matchmark the pitman arm to the gear and remove the pitman arm.

5. Remove the steering gear retaining bolts and remove the gear.

To install:

6. Align the column coupler shaft to the steering gear.

7. Position the steering gear and bracket on the frame rail and install

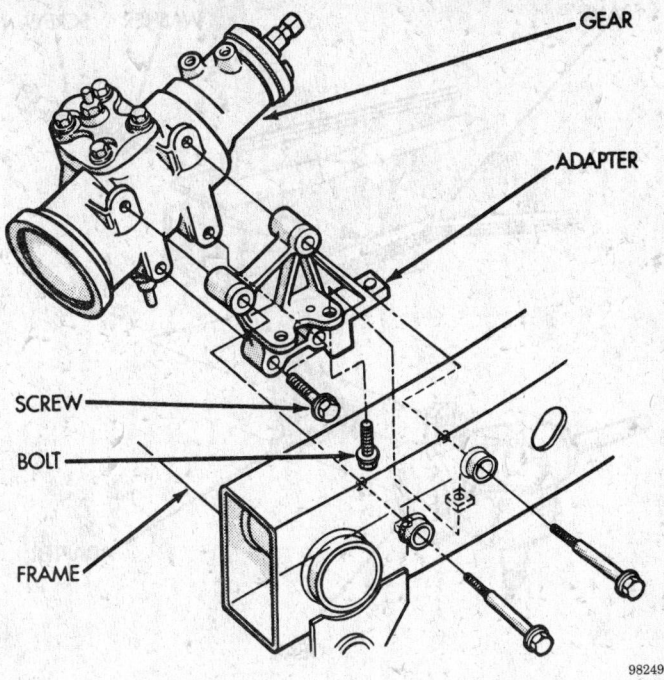

Steering gear installation - Wrangler 1992-95

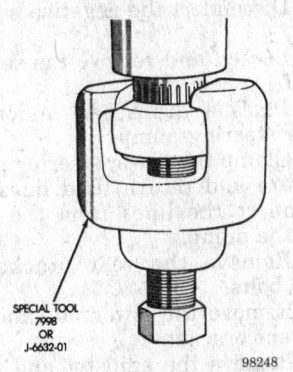

Pitman arm removal - Wrangler 1992-95

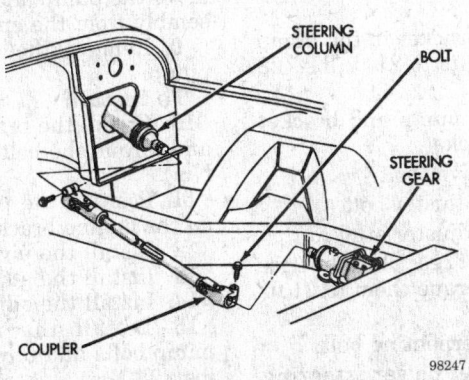

Column shaft removal - Wrangler 1992-95

the bolts. Torque the bolts to 70 ft. lbs. (95 Nm).

8. Align and install the pitman arm.

9. Connect the power steering lines.

10. Fill the power steering reservoir to the proper level with fluid and bleed the system.

Grand Cherokee 1993-95 and Grand Wagoneer 1993

1. Place the wheels in the straight ahead position.

2. Disconnect and cap the power steering lines from the steering gear.

3. Remove the column coupler shaft from the gear.

4. Matchmark the pitman arm to the gear and remove the pitman arm.

5. Remove the steering gear retaining bolts and remove the gear.

To install:

6. Align the column coupler shaft to the steering gear.

7. Position the steering gear and bracket on the frame rail and install the bolts. Torque the bolts to 65 ft. lbs. (88 Nm).

8. Align and install the pitman arm.

9. Connect the power steering lines.

10. Fill the power steering reservoir to the proper level with fluid and bleed the system.

Steering Pump

REMOVAL AND INSTALLATION

Wrangler 1991-95, Cherokee 1991-95, Comanche 1991-92 and Grand Wagoneer 1991

SERPENTINE DRIVE BELT

1. Disconnect the negative battery cable.

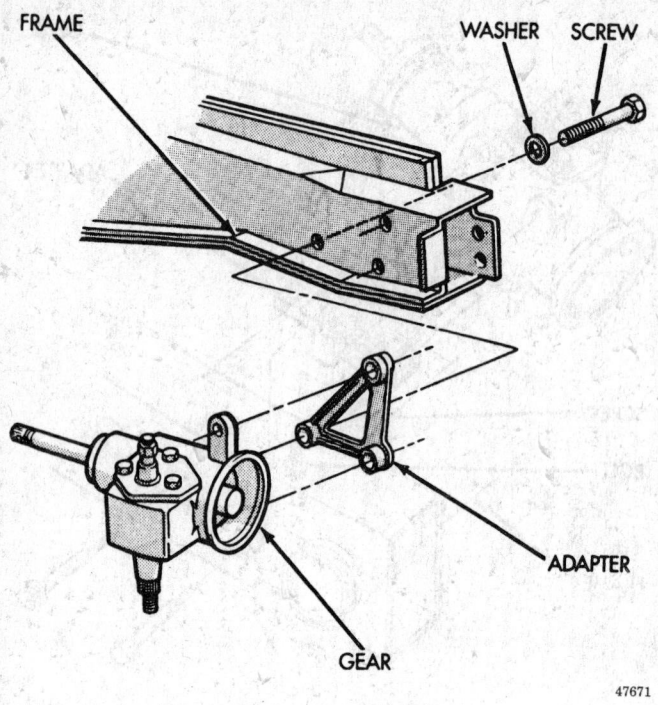

Steering gear installation

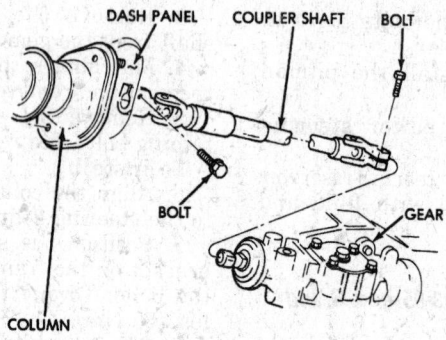

Coupling shaft removal

2. Loosen and remove the serpentine drive belt.

3. Place a drain pan under the power steering pump.

4. Clamp the power steering pump pressure and return fluid lines and disconnect the lines from the hose from the pump.

5. Remove the rear bracket-to-pump bolts.

6. Remove the lower nut and adjustment bracket.

7. Remove the adjuster and pivot bolts.

8. Tilt the pump forward and remove the pump and front bracket assembly from the engine bracket.

9. Remove the bracket from the pump.

To install:

10. Install the bracket to the pump and torque the bolts to 21 ft. lbs. (28 Nm).

11. Position the pump and bracket on the engine bracket.

12. Install the pivot bolt.

13. Install the adjuster bolt.

14. Install the adjuster stud nut.

15. Install the rear bracket-to-pump bolts and torque them to 21 ft. lbs. (28 Nm).

16. Install the serpentine belt.

17. Connect the power steering lines and remove the clamps.

18. Fill the pump reservoir to the proper level with fluid.

19. Connect the negative battery cable.

20. Bleed the system.

V-TYPE DRIVE BELT

1. Disconnect the negative battery cable.

2. Remove the drive belt.

3. Remove the air cleaner.

4. Disconnect the hoses at the pump and cap the hose ends.

5. Remove the front bracket-to-engine bolts.

6. Support the pump and remove the pump-to-rear bracket nuts.

7. Lift out the pump.

To install:

8. Install the pump into position and torque the pump-to-bracket nuts to 28 ft. lbs.; the bracket-to-engine bolts to 33 ft. lbs.

9. Connect the hoses.

10. Install the air cleaner.

11. Install the drive belt.

12. Connect the negative battery cable.

13. Bleed the system.

Grand Cherokee 1993-95 and Grand Wagoneer 1993

4.0L ENGINE

1. Disconnect the negative battery cable.

2. Loosen and remove the serpentine drive belt.

3. Place a drain pan under the power steering pump.

4. Clamp the power steering pump pressure and return fluid lines and disconnect the lines from the hose from the pump.

5. Remove the rear bracket-to-pump bolts.

6. Remove the lower nut and adjustment bracket.

7. Remove the adjuster and pivot bolts.

8. Tilt the pump forward and remove the pump and front bracket assembly from the engine bracket.

9. Remove the bracket from the pump.

To install:

10. Install the bracket to the pump and torque the bolts to 21 ft. lbs. (28 Nm).

11. Position the pump and bracket on the engine bracket.

12. Install the pivot bolt.

13. Install the adjuster bolt.

14. Install the adjuster stud nut.

15. Install the rear bracket-to-pump bolts and torque them to 21 ft. lbs. (28 Nm).

16. Install the serpentine belt.

17. Connect the power steering lines and remove the clamps.

18. Fill the pump reservoir to the proper level with fluid.

19. Connect the negative battery cable.

20. Bleed the system.

5.2L ENGINE

1. Disconnect the negative battery cable.

2. Loosen and remove the serpentine drive belt.

3. Place a drain pan under the power steering pump.

4. Disconnect the return and pressure lines from the pump.

5. Remove the bolts attaching the pump to the bracket on the engine block.

6. If necessary, remove the bracket-to-engine block bolts.

To install:

7. Install the bracket to the engine block and torque the bolts to 30 ft. lbs. (41 Nm).

8. Mount the pump on the bracket and torque the bolts 20 ft. lbs. (27 Nm).

9. Install the serpentine belt.

10. Connect the fluid lines to the pump and remove the clamps.

11. Fill the power steering reservoir to the proper level with fluid.

12. Connect the negative battery cable.

13. Bleed the system.

Tie Rod Ends-Power Steering

REMOVAL AND INSTALLATION

Wrangler 1991-95

1. Raise and safely support the vehicle.

2. Remove the cotter pins and nuts from the tie rod end ball studs, drag link end ball stud, and steering damper piston rod ball stud.

3. Use a suitable puller to loosen the ball studs, then remove the tie rod.

4. If necessary, losen the adjustment sleeve clamp bolts and remove the tie rod end from the tie rod. Count the number of turns required to remove the tie rod end so the replacement can be reinstalled in approximately the same position.

To install:

5. If removed, install the tie rod end to the tie rod. Thread the tie rod end on the same number of turns that was required to remove the old one. Position the adjustment sleeve bolts so the threaded ends of the bolts face rearward and are angled upward.

6. Attach the tie rod ends to the steering knuckles and the drag link ball stud to the tie rod. Install the nuts and tighten to 35 ft. lbs. (47 Nm).

7. Install new cotter pins in the ball studs. If the ball stud hole does not align with the nut castellation, tighten the nut further until the cotter pin can be installed.

8. Attach the steering damper to the tie rod and tighten the nut to 53 ft. lbs. (71 Nm). Install a new cotter pin. If the ball stud hole does not align with the nut castellation, tighten the nut further until the cotter pin can be installed.

9. Lower the vehicle and check the toe adjustment.

Cherokee 1991-95 and Comanche 1991-92

TIE ROD

1. Raise and safely support the vehicle.

2. Remove the cotter pins at the steering knuckle and drag link.

3. Loosen the ball studs using a suitable puller.

4. If necessary, loosen the end clamp bolts and remove the tie rod ends from the tube. Count the num-

FASTENER TORQUE			
LETTER	N•m	IN. LBS.	FT. LBS.
A	57	—	42
B	28	250	21
C	47	—	35

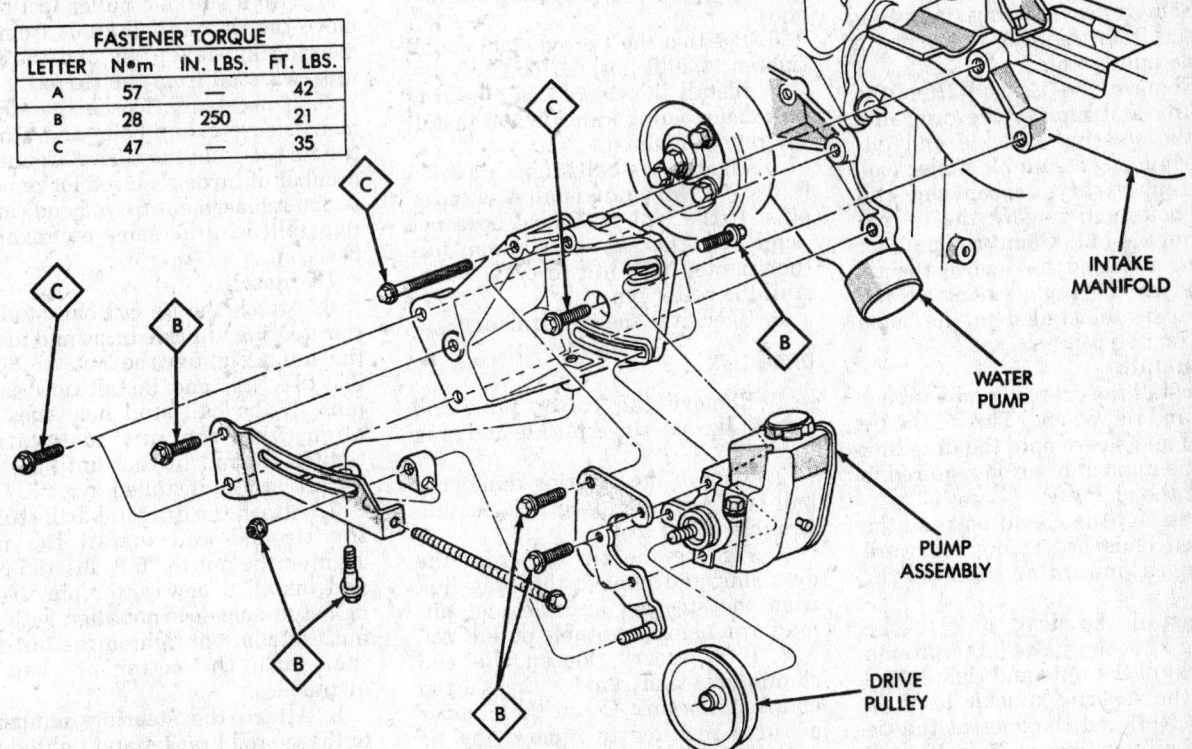

Steering pump installation-4.0L engine

INTAKE MANIFOLD

WATER PUMP

PUMP ASSEMBLY

DRIVE PULLEY

47753

ber of turns required to remove the tie rod ends so the replacement tie rod ends can be reinstalled in the same approximate position.

To install:

5. If removed, install the tie rod ends in the tube. Thread the tie rod ends into the tube the same number of turns required to remove the old ones.

6. Position the tie rod clamp bolts so the bolt heads face the rear of the vehicle and tighten to 22 ft. lbs. (30 Nm).

7. Install the tie rod on the drag link and steering knuckle and install the retaining nuts.

8. Tighten the ball stud-to-steering knuckle nut to 35 ft. lbs. (47 Nm) and the ball stud-to-drag link nut to 35 ft. lbs. (47 Nm) on 1985–90 vehicles or 55 ft. lbs. (75 Nm) on 1991–94 vehicles. Install new cotter pins. If the ball stud hole does not align with the nut castellation, further tighten the nut in order to install the cotter pin.

9. Lower the vehicle and check the toe adjustment.

DRAG LINK

1. Raise and safley support the vehicle.

2. Remove the cotter pins and nuts at the steering knuckle, tie rod and pitman arm.

3. Remove the steering dampener ball stud from the drag link using a suitable puller tool.

4. Remove the tie rod from the drag link and remove the drag link from the steering knuckle and pitman arm using a suitable puller tool.

5. If necessary, loosen the end clamp bolts and remove the tie rod end from the link. Count the number of turns required to remove the tie rod end so the replacement tie rod end can be reinstalled in the same approximate position.

To install:

6. Install the drag link adjustment sleeve and tie rod end. Thread the tie rod end and sleeve onto the drag link the same number of turns required to remove the old ones.

7. Position the clamp bolts so the threaded ends are facing rearward and angled upward and tighten the bolts.

8. Install the drag link to the steering knuckle, tie rod and pitman arm. Install the nuts and tighten the one at the steering knuckle to 35 ft. lbs. (47 Nm) and the ones at the tie rod and pitman arm to 35 ft. lbs. (47 Nm) on 1985–90 vewhicles or to 55 ft. lbs. (75 Nm) on 1991–94 vehicles. In-

stall new cotter pins. If the ball stud hole does not align with the nut castellation, further tighten the nut in order to install the cotter pin.

9. Install the steering dampener onto the drag link and tighten the nut to 35 ft. lbs. (47 Nm) on 1985–90 vehicles or 55 ft. lbs. (75 Nm) on 1991–94 vehicles. Install a new cotter pin. If the ball stud hole does not align with the nut castellation, further tighten the nut in order to install the cotter pin.

10. Lower the vehicle and check the toe adjustment.

Grand Cherokee 1993-95 and Grand Wagoneer 1993

TIE ROD

1. Remove the cotter pins at the steering knuckle and drag link.

2. Loosen the ball studs using a suitable puller.

3. If necessary, loosen the end clamp bolts and remove the tie rod ends from the tube. Count the number of turns required to remove the tie rod ends so the replacement tie rod ends can be reinstalled in the same approximate position.

To install:

4. If removed, install the tie rod ends in the tube. Thread the tie rod ends into the tube the same number of turns required to remove the old ones.

5. Position the tie rod clamps and tighten to 20 ft. lbs. (27 Nm).

6. Install the tie rod on the drag link and steering knuckle and install the retaining nuts.

7. Tighten the ball stud nuts to 55 ft. lbs. (75 Nm) and install new cotter pins. If the ball stud hole does not align with the nut castellation, further tighten the nut in order to install the cotter pin.

8. Check the toe adjustment.

DRAG LINK

1. Remove the cotter pins and nuts at the steering knuckle and drag link.

2. Remove the steering dampener ball stud from the drag link using a suitable puller tool.

3. Remove the tie rod from the drag link and remove the drag link from the steering knuckle and pitman arm using a suitable puller tool.

4. If necessary, loosen the end clamp bolts and remove the tie rod end from the link. Count the number of turns required to remove the tie rod end so the replacement tie rod end can be reinstalled in the same approximate position.

To install:

5. Install the drag link adjustment sleeve and tie rod end. Thread the tie rod end and sleeve onto the drag link the same number of turns required to remove the old ones.

6. Position the clamp bolts and tighten to 36 ft. lbs. (49 Nm).

7. Install the drag link to the steering knuckle, tie rod and pitman arm. Install the nuts and tighten to 55 ft. lbs. (75 Nm). Install new cotter pins. If the ball stud hole does not align with the nut castellation, further tighten the nut in order to install the cotter pin.

8. Install the steering dampener onto the drag link and tighten the nut to 55 ft. lbs. (75 Nm). Install a new cotter pin. If the ball stud hole does not align with the nut castellation, further tighten the nut in order to install the cotter pin.

9. Check the toe adjustment.

Grand Wagoneer 1991

1. Raise and safely support the vehicle.

2. Remove the cotter pins and nuts from the tie rod and drag link ball studs.

3. Remove the steering dampener piston rod-to-tie rod bracket and move the dampener aside.

4. Use a suitable puller tool to remove the tie rod ball studs from the steering knuckle arms and the drag link ball stud from the tie rod.

5. If necessary, loosen the adjustment sleeve clamp bolts and remove the tie rod from the sleeve. Count the number of turns required for removal so the replacement tie rod end can be reinstalled in the same approximate position.

To install:

6. Attach the tie rod ball studs to the steering knuckle arms and install the nuts. Tighten the nuts to 60 ft. lbs. (81 Nm) and install new cotter pins. If the ball stud hole does not align with the nut castellation, tighten the nut further until the cotter pin can be installed.

7. Attach the drag link ball stud to the tie rod and install the nut. Tighten the nut to 70 ft. lbs. (95 Nm) and install a new cotter pin. If the ball stud hole does not align with the nut castellation, tighten the nut further until the cotter pin can be installed.

8. Attach the steering dampener to the tie rod bracket and tighten the nut to 30 ft. lbs. (41 Nm).

9. Lower the vehicle and check the toe adjustment.

FASTENER TORQUE			
LETTER	N•m	IN. LBS.	FT. LBS.
◇A	251	—	185
◇B			
◇C	74	—	55
◇D			
◇E	49	—	36
◇F	27	—	20

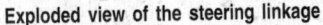

Exploded view of the steering linkage

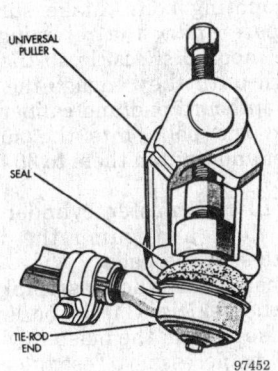

Ball stud removal

BRAKES

Brake Caliper

REMOVAL AND INSTALLATION

Wrangler 1991-95, Cherokee 1991-95, Comanche 1991-92, Grand Cherokee 1993-95, Grand Wagoneer 1991 and 1993

1. Drain ⅔ of the brake fluid from the front reservoir. Use the bleeder screw at the front outlet port to drain the fluid. If equipped with anti-lock brakes, relieve the system pressure.

2. Raise and safely support the vehicle.

3. Remove the wheels.

4. Place a C-clamp on the caliper so the solid end contacts the back of the caliper and screw end contacts the metal part of the outboard brake pad.

5. Tighten the clamp until the caliper moves far enough to force the piston to the bottom of the piston bore. This will back the brake pads off of the rotor surface to facilitate the removal and installation of the caliper assembly.

6. Remove the C-clamp.

NOTE: Do not push down on the brake pedal or the piston and brake pads will return to their original positions up against the rotor.

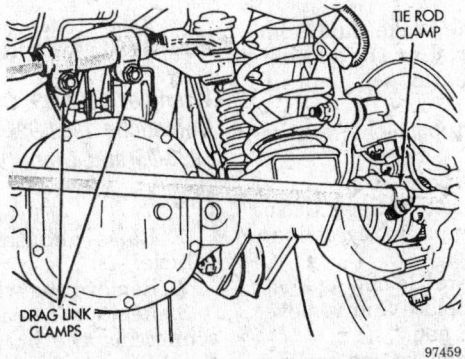

Tie rod and drag link clamp positioning

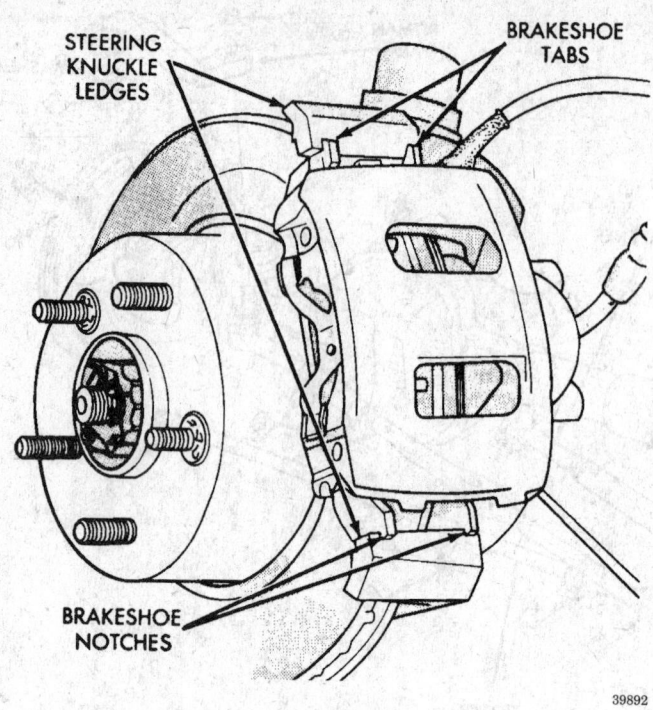

STEERING KNUCKLE LEDGES

BRAKESHOE TABS

BRAKESHOE NOTCHES

39892

Caliper installation

7. Remove both of the Allen head mounting bolts and lift the caliper off the rotor.

NOTE: If just the brake pads are being replaced, it is not necessary to remove the caliper assembly entirely from the vehicle. Do not remove the brake line. Rest the caliper on the front spring or other support. Do not allow the brake hose to support the weight of the caliper.

8. If the caliper is being removed, it is necessary to disconnect the brake fluid hose. Clean the brake fluid hose-to-caliper connection thoroughly. Remove the hose-to-caliper bolt. Cap or tape the open ends to keep dirt out. Discard the copper gaskets.

To install:

9. Connect the brake line to the caliper with new sealing washers and hand-tighten the fitting bolt.

10. Position the caliper into place over the rotor.

11. Coat the caliper mounting bolt with silicone grease and torque them to 7–15 ft. lbs. (10–20 Nm).

12. Position the brake line clear of all chassis components, untwisted and free of kinks. Torque the fitting bolt to 23 ft. lbs. (31 Nm).

13. Install the wheels.

14. Fill the master cylinder with fluid and bleed the brake system.

15. Before driving the vehicle, pump the brakes several times to seat the pads.

Disc Brake Pads

REMOVAL AND INSTALLATION

Wrangler 1991-95, Cherokee 1991-95, Comanche 1991-92, Grand Cherokee 1993-95, Grand Wagoneer 1991 and 1993

1. Raise and safely support the vehicle.

2. Drain ⅔ of the brake fluid from the front reservoir. Use the bleeder screw at the front outlet port to drain the fluid.

3. Raise and support the vehicle safely.

4. Remove the wheels.

5. Remove the brake caliper. Use a suitable tool to compress the caliper piston into the bore.

6. Hold the anti-rattle clip against the caliper anchor plate and remove the outboard brake pad.

7. Remove the inboard pad and its anti-rattle clip.

To install:

8. Clean all the mounting holes and bushing grooves in the caliper ears. Clean the mounting bolts. Replace the bolts if they are corroded or if the threads are damaged. Wipe the inside of the caliper clean, including the exterior of the dust boot. Inspect the dust boot for cuts or cracks and for proper seating in the piston bore. If evidence of fluid leakage is noted, the caliper should be rebuilt.

NOTE: Do not use abrasives on the bolts in order not to destroy their protective plating. Do not use compressed air to clean the inside of the caliper, as it may unseat the dust boot seal.

9. Install the inboard anti-rattle clip on the trailing end of the anchor plate. The split end of the clip must face away from the rotor.

10. Install the inboard pad in the caliper. The pad must lay flat against the piston.

11. Install the outboard pad in the caliper while holding the anti-rattle clip.

12. With the pads installed, position the caliper over the rotor. Line up the mounting holes in the caliper and the support bracket and insert the mounting bolts. Make sure the bolts pass under the retaining ears on the inboard shoes. Push the bolts through until they engage the holes of the outboard pad and caliper ears. Thread the bolts into the support bracket and tighten them to 30 ft. lbs. 41 Nm).

13. Fill the master cylinder with brake fluid and pump the brake pedal to seat the pads.

14. Install the wheel assembly and lower the vehicle. Check the level of the brake fluid in the master cylinder and fill as necessary. Test the operation of the brakes before taking the vehicle onto the road.

Brake Rotor

REMOVAL AND INSTALLATION

Wrangler 1991-95, Cherokee 1991-95, Comanche 1991-92, Grand Cherokee 1993-95 and Grand Wagoneer 1993

2WD

1. Raise and safely support the vehicle.

2. Remove the wheels.

3. Remove the caliper without disconnecting the brake line. Suspend the caliper aside; do not let it hang by the brake hose.

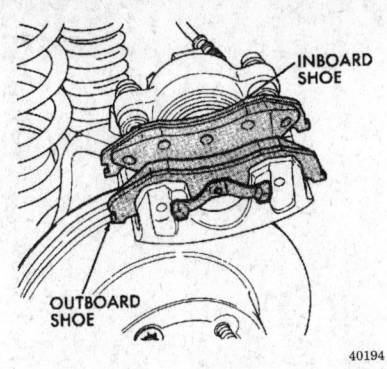

Disc brake pads installed in the caliper

4. Remove the grease cap, cotter pin, nut cap, nut and washer from the spindle.

5. Pull slowly on the hub and catch the outer bearing as it falls.

6. Remove the hub and rotor. The inner bearing and seal can be removed by prying out and discarding the inner seal.

To install:

7. Clean and repack the hub and bearings, install the inner bearing and a new seal.

8. Position the hub and rotor on the spindle and install the outer bearing.

9. Install the washer and nut.

10. While turning the rotor, torque the nut to 25 ft. lbs. (34 Nm) to seat the bearings.

11. Back off the nut ½ turn and, while turning the rotor, torque the nut to 19 inch lbs.

12. Install the nut cap and a new cotter pin. Install the grease cap.

13. Install the caliper.

14. Install the wheels and lower the vehicle. Before moving the vehicle, pump the brake pedal several times to seat the brake pads against the rotor.

4WD

1. Loosen the lug nuts on the front wheel.

2. Raise and safely support the vehicle.

3. Remove the front wheel.

4. Remove the caliper but leave the brake line connected. Suspend the caliper aside; do not let the caliper hang by the brake hose.

5. Remove the spring nuts (if equipped) and rotor.

To install:

6. Install the rotor with new spring nuts.

7. Install the caliper.

8. Install the front wheel.

9. Lower the vehicle.

10. Before moving the vehicle, pump the brake pedal several times to seat the disc brake pads against the rotor.

Grand Wagoneer 1991

1. Raise and safely support the vehicle.

2. Remove the wheel.

3. Remove the caliper without disconnecting the brake line. Suspend the caliper aside; do not let it hang by the brake hose.

4. Remove the hub cap and snapring.

5. Remove the drive gear, pressure spring and spring cup.

6. Remove the outer nut, nut washer and inner nut.

7. Remove the rotor, grease seal and rear bearing. Remove the rotor shield, if necessary.

To install:

8. Clean and repack the bearings with high temperature bearing grease. Also apply grease to the rotor hub cavity and spindle.

9. Install the rear bearing and new grease seal in the rotor. Install the rotor on the spindle.

10. Install the inner nut. Ensure that the locating peg on one side of the nut is facing outward.

11. Install the wheel but do not tighten the wheel nuts. Tighten the inner nut to 50 ft. lbs. (68 Nm). while

rotating the wheel to seat the bearings. Continue rotating the wheel and back off the nut 45–65 degrees.

12. Install the nut washer and outer nut. Ensure the locating peg on the inner nut is seated in one of the washer holes before installing the outer nut.

13. Tighten the outer nut to a minimum of 50 ft. lbs. (68 Nm). Install the spring cup and pressure spring.

14. Install the drive gear then push the gear inward and install the snapring.

15. Remove the wheel and install the caliper.

16. Reinstall the wheel and lower the vehicle. Before moving the vehicle, pump the brake pedal several times to seat the brake pads against the rotor.

Brake Drum

REMOVAL AND INSTALLATION

Wrangler 1991-95, Cherokee 1991-95, Comanche 1991-92, Grand Cherokee 1993-95, Grand Wagoneer 1991 and 1993

1. Raise and safely support the vehicle.

2. Remove the wheel.

3. Remove the spring nuts (if installed) from the lug bolts and remove the drum from the vehicle.

NOTE: It may be necessary to back off the brake adjusters to remove the drum.

To install:

4. Ensure the contacting surfaces are clean and flat. Install the drum on the hub.

5. Adjust the brake shoes, if necessary.

6. Install the spring nuts on the lug bolts.

7. Install the wheel.

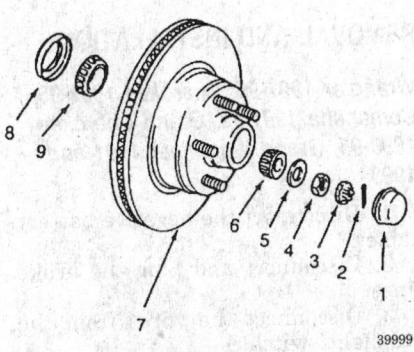

Disc brake rotor-2WD vehicles

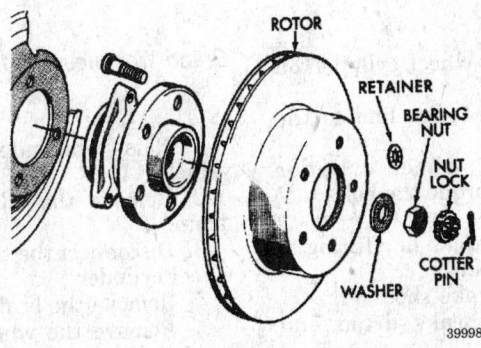

Disc brake rotor-4WD vehicles

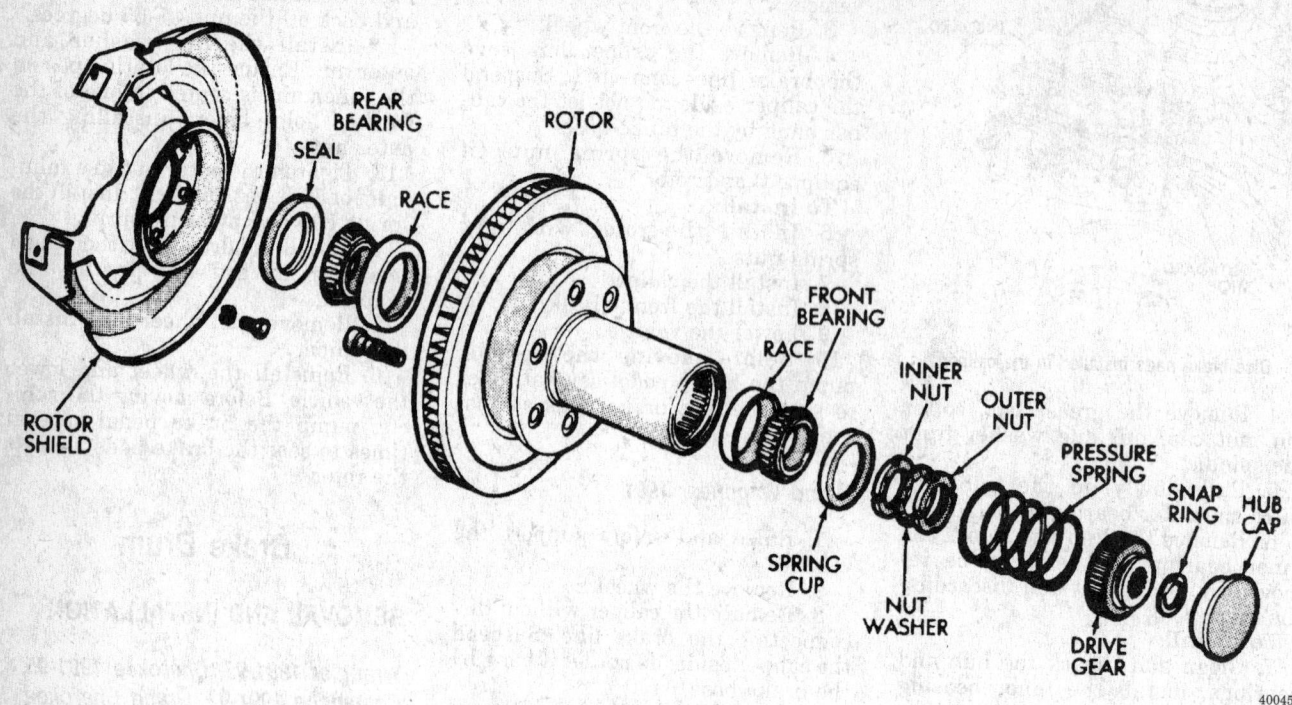

SEAL

REAR BEARING

ROTOR

RACE

FRONT BEARING

RACE

INNER NUT

OUTER NUT

PRESSURE SPRING

SNAP RING

HUB CAP

ROTOR SHIELD

SPRING CUP

NUT WASHER

DRIVE GEAR

40045

Brake rotor assembly - Grand Wagoneer 1991

Wheel Cylinder

REMOVAL AND INSTALLATION

Wrangler 1991-95, Cherokee 1991-95, Comanche 1991-92, Grand Cherokee 1993-95 and Grand Wagoneer 1993

1. Raise and safely support the vehicle. If equipped with anti-lock brakes, relieve the system pressure.
2. Remove the wheel and brake drum.
3. Disconnect the brake line at the wheel cylinder.
4. Remove the brake shoes.
5. Remove the wheel cylinder attaching bolts and remove the wheel cylinder.

To install:
6. Position the wheel cylinder on the backing plate.
7. Connect the brake line to the cylinder fitting.
8. Install the wheel cylinder mounting bolts. Torque the bolts to 7 ft. lbs. (10 Nm).
9. Tighten the brake line fitting to 140 inch lbs. (16 Nm).
10. Install the brake shoes.
11. Install the brake drum and wheel.
12. Lower the vehicle.

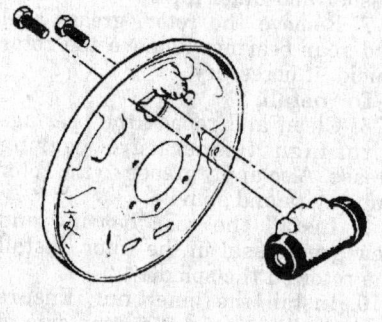

40457

Wheel cylinder installation

13. Fill the brake system to the correct level and bleed the brakes.

Grand Wagoneer 1991

1. Raise and safely support the vehicle.
2. Remove the wheel and brake drum.
3. Disconnect the brake line at the wheel cylinder.
4. Remove the brake shoes.
5. Remove the wheel cylinder attaching bolts and remove the wheel cylinder.

To install:
6. Position the wheel cylinder on the backing plate.
7. Connect the brake line to the cylinder fitting.
8. Install the wheel cylinder mounting bolts. Torque the bolts to 18 ft. lbs. (24 Nm).
9. Tighten the brake line fitting to 140 inch lbs. (16 Nm).
10. Install the brake shoes.
11. Install the brake drum and wheel.
12. Lower the vehicle.
13. Fill the brake system to the correct level and bleed the brakes.

Master Cylinder

REMOVAL AND INSTALLATION

Wrangler 1991-95, Cherokee 1991-95, Comanche 1991-92, Grand Cherokee 1993-95, Grand Wagoneer 1991 and 1993

1. Disconnect the negative battery cable.
2. Disconnect and plug the brake lines.
3. Disconnect the wires from the stoplight switch.
4. Remove the attaching nuts and lift the assembly from the vehicle.

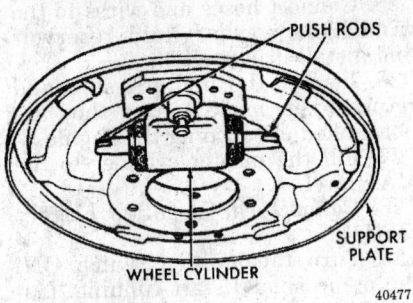

Wheel cylinder mounting - Grand Wagoneer 1991

To install:

5. Fill the master cylinder with brake fluid and operate the pushrod until fluid squirts from the ports.

6. Install the master cylinder. Torque the mounting nuts to 15–21 ft. lbs. (20–29 Nm).

7. Connect the stoplight switch.

8. Connect the brake lines.

9. Bleed the brake system.

Power Brake Booster

REMOVAL AND INSTALLATION

Wrangler 1991-95, Cherokee 1991-95, Comanche 1991-92 and Grand Wagoneer 1991

1. Loosen but do not remove the nuts attaching the master cylinder to booster.

2. Remove the instrument panel lower trim cover.

3. Remove the retaining clip attaching the booster pushrod to the brake pedal.

4. Disconnect the stoplight switch electrical connector.

5. Remove the bolts and nuts attaching the booster to the dash panel and remove the vacuum hose.

6. Remove the master cylinder from the booster and carefully position it aside, being careful not to damage the brake lines. Remove the booster.

To install:

7. Install the check valve and grommet in the booster. Position the booster on the vehicle and install the attaching nuts and bolts.

8. Attach the booster pushrod to the brake pedal. Tighten the booster mounting bolts and nuts to 22–30 ft. lbs. (30–41 Nm).

9. Connect the stoplight switch electrical connector and reinstall the trim panels.

10. Install the master cylinder on the booster and tighten the nuts to

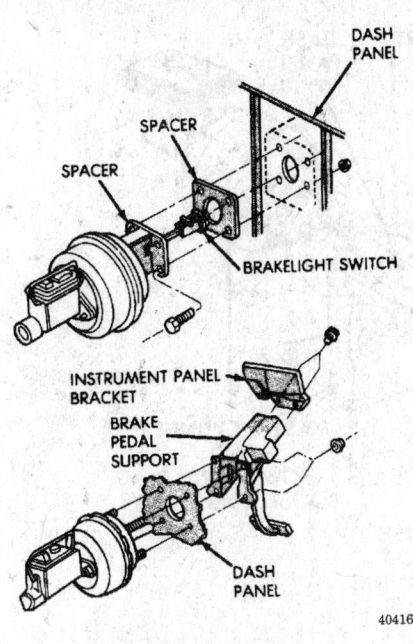

Power brake booster mounting

15–25 ft. lbs. (21–34 Nm). Top off the master cylinder, connect the vacuum hose and check for proper operation of the system.

Grand Cherokee 1993-95 and Grand Wagoneer 1993

1. Loosen but do not remove the nuts attaching the master cylinder to booster.

2. Remove the instrument panel lower trim cover.

3. Remove the air conditioner duct and lower trim panel. Remove the cotter pin, washer and bushing attaching the booster pushrod to the pedal.

4. Remove the bolts and nuts attaching the booster to the dash panel and remove the vacuum hose.

5. Remove the stoplight switch connector.

6. Remove the master cylinder from the booster and carefully position it aside, being careful not to damage the brake lines. Remove the booster.

To install:

7. Install the check valve and grommet in the booster. Position the booster on the vehicle and install the attaching nuts and bolts.

8. Attach the booster pushrod to the brake pedal. Tighten the booster

mounting bolts and nuts to 22–30 ft. lbs. (30–41 Nm).

9. Connect the stoplight electrical connector and reinstall the trim panels and air conditioner ducts.

10. Install the master cylinder on the booster and tighten the nuts to 15–25 ft. lbs. (21–34 Nm). Top off the master cylinder, connect the vacuum hose and check for proper operation of the system.

Proportioning Valve

REMOVAL AND INSTALLATION

Comanche 1991-92

1. Disconnect the linkage and spring.

2. Disconnect and cap the brake lines.

3. Unbolt and remove the valve.

To install:

4. Install the proportioning valve. Torque the valve bracket-to-frame bolts to 155 inch lbs.; the valve-to-bracket bolts to 118 inch lbs.

5. Connect the brake lines.

6. Connect the linkage and spring.

7. Bleed the brakes.

Master Cylinder-ABS

REMOVAL AND INSTALLATION

Wrangler 1991, Cherokee 1991, Comanche 1991 and Grand Wagoneer 1991

NOTE: The master cylinder, modulator and accumulator are serviced as an assembly only. Do not attempt to disassemble or repair these components.

1. Disconnect the negative battery cable.

2. Remove the screws holding the windshield washer fluid reservoir. Disconnect the hoses and wires and remove the reservoir.

3. Remove the air cleaner assembly.

4. Disconnect the ABS ECU wiring connectors at the pressure modulator.

5. Disconnect the wiring at the proportioning valve/differential switch.

6. Disconnect the brake line at the coupling on the underside of the proportioning valve.

7. Disconnect the high pressure line at the accumulator block and immediately cap the port to prevent entry of dirt.

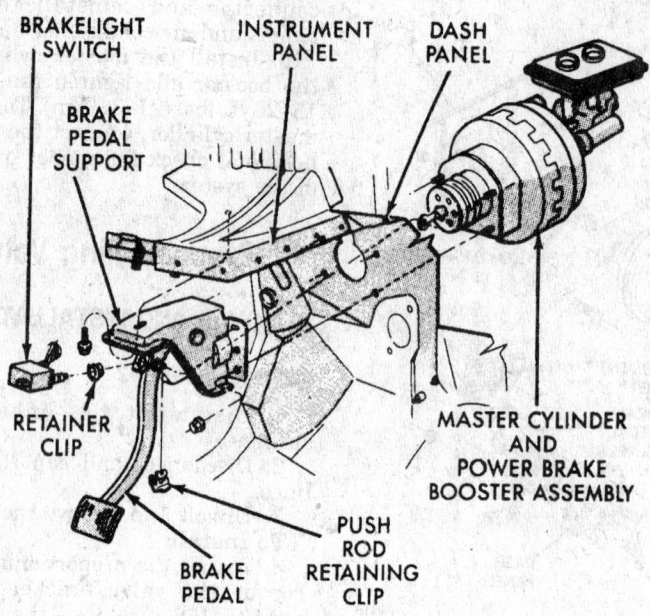

BRAKELIGHT SWITCH
INSTRUMENT PANEL
DASH PANEL
BRAKE PEDAL SUPPORT
RETAINER CLIP
BRAKE PEDAL
PUSH ROD RETAINING CLIP
MASTER CYLINDER AND POWER BRAKE BOOSTER ASSEMBLY

40427

Power brake booster installation - Grand Cherokee and Grand Wagoneer

8. Place a catch pan below the supply line at the reservoir. Disconnect the hose at the reservoir and discard the fluid drained from the line.

9. Remove the wires from the low pressure switch on the accumulator block, the modulator boost pressure switch and the fluid level switch.

10. Disconnect the front brake lines at the outboard side of the pressure modulator.

11. Inside the vehicle, disconnect the wiring to the brake pedal switch. It may be necessary to remove the instrument panel lower trim for access.

12. Remove the master cylinder pushrod bolt; disconnect the pushrod from the pedal. Immediately discard the nuts from the pushrod bolt; they are not reusable.

13. Remove the nuts from the brake mounting bracket studs.

14. Under the hood, carefully pull the hydraulic components and bracket forward until the studs clear the firewall.

15. Lift the assembly up and out of the engine compartment.

To install:

16. Install the pad onto the new bracket and place the bracket with the brake components in place against the firewall. Make certain

the studs are correctly seated in the holes.

17. Position the master cylinder/modulator/accumulator on the firewall and tighten the mounting nuts to 27 ft. lbs. (31 Nm).

18. Connect the wiring harness to the low pressure switch, the differential switch, the pressure modulator, the low fluid and boost pressure switches.

19. Inside the vehicle, install the nuts on the mounting studs and tighten them to 31 ft. lbs. (42 Nm).

20. Align the brake pedal, the brake light switch and the master cylinder pushrod. Install the pushrod bolt. The pushrod bolt must be correctly installed to avoid interference with the bracket. The bolt head must be on the left side of the brake pedal.

21. Use new nuts on the pushrod bolt. The inner locknut should be tightened to 25 ft. lbs. (34 Nm); the outer jam nut should be tightened to 75 inch lbs. (8.5 Nm).

22. Connect the wiring to the brake switch and replace the lower dashboard trim if removed. In the engine compartment, connect the brake lines to the proportioning valve.

23. Connect the pressure and return lines to the accumulator.

24. Install the air cleaner assembly.

25. Connect hoses and wires to the windshield washer fluid reservoir and install it in position.

26. Inspect the high-pressure and return lines; make certain they are not kinked or touching the engine.

27. Fill the master cylinder to the MAX level.

28. Connect the negative battery cable.

29. Turn the ignition switch **ON**; the pump should start running. Listen for a drop in pump rpm; this shows the pump is pressurizing the system. If there is no drop in pump rpm within 20 seconds, immediately shut the ignition **OFF** and check for hydraulic leaks.

NOTE: Severe pump damage will result if the pump runs without pressurizing the system for any length of time.

30. Add fluid to the reservoir as necessary. Do not overfill. Bleed the brake system.

Wrangler 1992-95, Cherokee 1992-95, Comanche 1992, Grand Cherokee 1993-95 and Grand Wagoneer 1993

NOTE: The master cylinder, modulator and accumulator are serviced as an assembly only. Do not attempt to disassemble or repair these components.

1. Disconnect the negative battery cable.

2. Remove the windshield washer reservoir.

3. Pump the brake pedal to exhaust all vacuum from the power brake booster.

—————— **WARNING** ——————
It is very important that all vacuum be exhausted from the booster. Failure to do so could result in damage to the master cylinder-to-booster seal when the master cylinder is removed.

4. Disconnect the anti-lock electrical harness connectors and position them aside.

5. Remove the clamps securing the reservoir hoses to the Hydraulic Control Unit (HCU) pipes.

6. Position a small drain pan under the master cylinder reservoir hoses.

7. Disconnect the reservoir hoses from the HCU pipes and allow the fluid to drain into the container. Discard the fluid.

8. Remove the combination valve.

9. Disconnect the brake lines from the master cylinder.

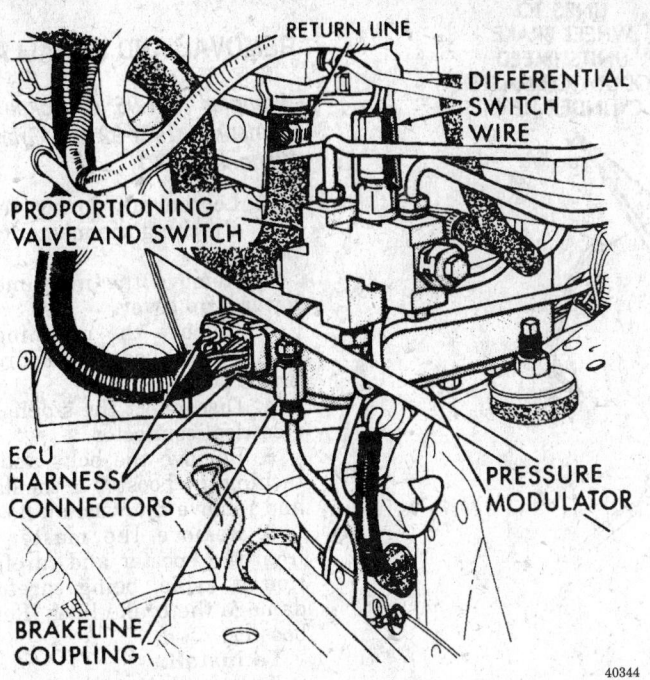

RETURN LINE

DIFFERENTIAL SWITCH WIRE

PROPORTIONING VALVE AND SWITCH

ECU HARNESS CONNECTORS

PRESSURE MODULATOR

BRAKELINE COUPLING

40344

ECU harness and fluid line connections

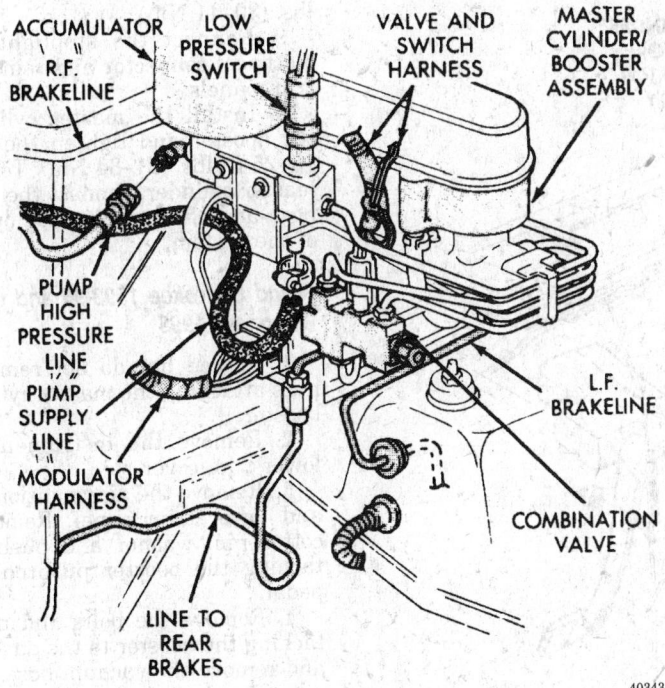

ACCUMULATOR R.F. BRAKELINE

LOW PRESSURE SWITCH

VALVE AND SWITCH HARNESS

MASTER CYLINDER/ BOOSTER ASSEMBLY

PUMP HIGH PRESSURE LINE

PUMP SUPPLY LINE

MODULATOR HARNESS

L.F. BRAKELINE

COMBINATION VALVE

LINE TO REAR BRAKES

40343

Master cylinder components

10. Remove the nuts attaching the master cylinder to the brake booster mounting studs.

11. Pull the master cylinder forward and out of the studs.

To install:

12. If installing a new master cylinder, fill the reservoir with fluid and depress the push rod until brake fluid comes out of the fluid line connections.

—————— **WARNING** ——————

The seal between the master cylinder and brake booster can be damaged if the master cylinder is improperly installed. To avoid damage, install the master cylinder only as described.

13. Have an assistant press the brake pedal until the brake booster push rod is visible in the opening at the front of the booster. Have the assistant hold the brake pedal depressed.

14. Guide the master cylinder onto the mounting studs and onto the push rod.

NOTE: Ensure the push rod is properly aligned and seated in the master cylinder.

15. Have the assistant slowly release the brake pedal while seating the master cylinder on the booster mounting studs. Keep the push rod centered in the master cylinder while seating the cylinder.

16. Install the stud nuts and torque them to 220 inch lbs. (25 Nm).

17. Connect the brake lines and torque them to 132 inch lbs. (25 Nm).

18. Connect the reservoir hoses to the HCU pipes and tighten the clamps.

19. Ensure the master cylinder and booster are properly seated before proceeding.

20. Install the combination valve.

21. Connect the electrical connectors.

22. Fill the master cylinder reservoir and bleed the brakes.

23. Install the washer reservoir.

24. Install the air cleaner.

25. Connect the negative battery cable.

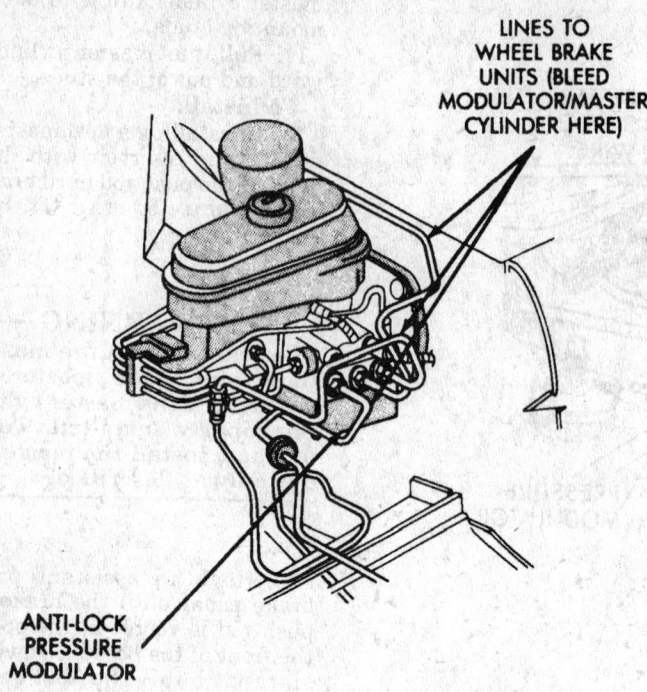

LINES TO
WHEEL BRAKE
UNITS (BLEED
MODULATOR/MASTER
CYLINDER HERE)

ANTI-LOCK
PRESSURE
MODULATOR

40345

Modulator brake line connections

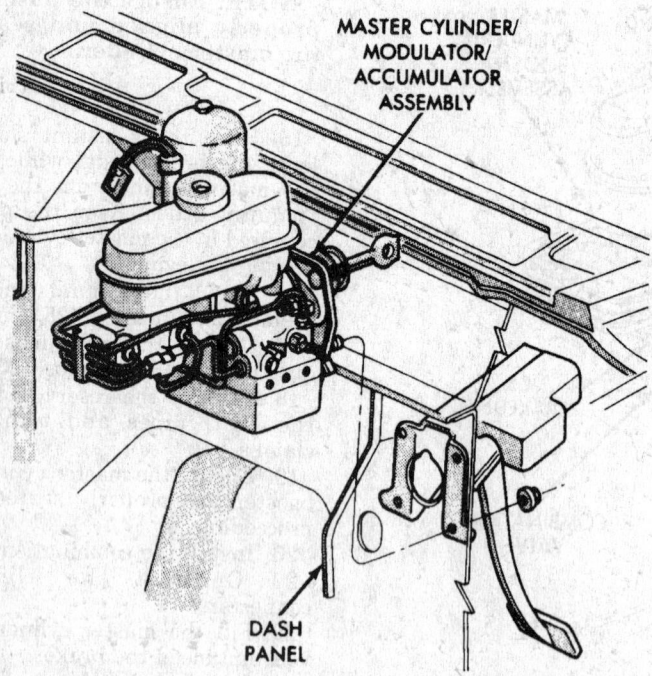

MASTER CYLINDER/
MODULATOR/
ACCUMULATOR
ASSEMBLY

DASH
PANEL

40346

Modulator, accumulator and master cylinder assembly

Power Brake Booster-ABS

REMOVAL AND INSTALLATION

Wrangler 1991-95, Cherokee 1991-95, Comanche 1991-92 and Grand Wagoneer 1991

1. Loosen but do not remove the nuts attaching the master cylinder to booster.
2. Remove the instrument panel lower trim cover.
3. Remove the retaining clip attaching the booster pushrod to the brake pedal.
4. Disconnect the stoplight switch electrical connector.
5. Remove the bolts and nuts attaching the booster to the dash panel and remove the vacuum hose.
6. Remove the master cylinder from the booster and carefully position it aside, being careful not to damage the brake lines. Remove the booster.

To install:

7. Install the check valve and grommet in the booster. Position the booster on the vehicle and install the attaching nuts and bolts.
8. Attach the booster pushrod to the brake pedal. Tighten the booster mounting bolts and nuts to 22–30 ft. lbs. (30–41 Nm).
9. Connect the stoplight switch electrical connector and reinstall the trim panels.
10. Install the master cylinder on the booster and tighten the nuts to 15–25 ft. lbs. (21–34 Nm). Top off the master cylinder, connect the vacuum hose and check for proper operation of the system.

Grand Cherokee 1993-95 and Grand Wagoneer 1993

1. Loosen but do not remove the nuts attaching the master cylinder to booster.
2. Remove the instrument panel lower trim cover.
3. Remove the air conditioner duct and lower trim panel. Remove the cotter pin, washer and bushing attaching the booster pushrod to the pedal.
4. Remove the bolts and nuts attaching the booster to the dash panel and remove the vacuum hose.
5. Remove the stoplight switch connector.
6. Remove the master cylinder from the booster and carefully position it aside, being careful not to damage the brake lines. Remove the booster.

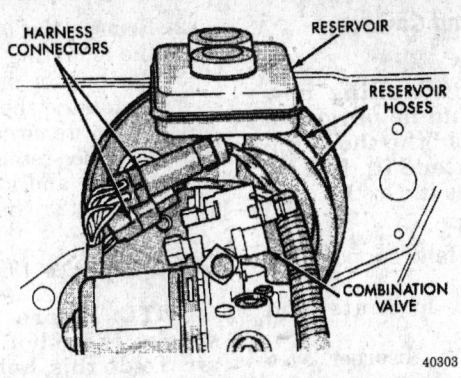

HARNESS CONNECTORS — RESERVOIR — RESERVOIR HOSES — COMBINATION VALVE

40303

Harness, reservoir hose and valve position

To install:

7. Install the check valve and grommet in the booster. Position the booster on the vehicle and install the attaching nuts and bolts.

8. Attach the booster pushrod to the brake pedal. Tighten the booster mounting bolts and nuts to 22–30 ft. lbs. (30–41 Nm).

9. Connect the stoplight electrical connector and reinstall the trim panels and air conditioner ducts.

10. Install the master cylinder on the booster and tighten the nuts to 15–25 ft. lbs. (21–34 Nm). Top off the master cylinder, connect the vacuum hose and check for proper operation of the system.

Brake System Bleeding

BLEEDING

Wrangler 1991-95, Cherokee 1991-95, Comanche 1991-92, Grand Cherokee 1993-95, Grand Wagoneer 1991 and 1993

Bleeding the anti-lock braking system is basically a 3 step process consisting of a conventional manual brake bleed, a 2nd brake bleed using scan tool DRB-II or equivalent, to run the pump and a repeat of the conventional manual brake bleed. The procedure is as follows:

1. If a new master cylinder is to be installed, bench bleed the cylinder before installation.

2. Clean the master cylinder reservoir caps and reservoir exterior.

3. Fill the master cylinder with DOT 3 standard brake fluid.

4. Bleed the brake system in the following sequence:

  a. Master cylinder

  b. Hydraulic Control Unit (HCU) valve body (at fluid lines)

  c. Right rear wheel

  d. Left rear wheel

  e. Right front wheel

  f. Left front wheel

5. Bleed the master cylinder and HCU at the brake fittings.

6. Attach a bleed hose to the bleed screw on the first wheel brake unit to be bled. Immerse the end of the bleed hose in a container partially filled with brake fluid.

7. Have an assistant apply and hold the brake pedal.

8. Open the bleed screw ½ turn and close the screw once the brake pedal reaches the floor.

NOTE: Do not pump the brake pedal at any time while bleeding. This compresses air into small bubbles which are distributed throughout the system.

9. Repeat the bleeding operation 5–7 more times at each wheel or until the fluid entering the container is free of air bubbles. Check the reservoir fluid lever frequently and add fluid if necessary.

NOTE: Do not allow the master cylinder reservoir to run dry while bleeding the system. Running the reservoir dry will allow air to reenter the system necessitating a second bleeding operation.

10. Connect the DRB-II scan tool or equivalent, to the diagnostic connector.

11. Perform the "Bleed Brake" operation as described in the scan tool manual.

12. Repeat Steps 4–9.

13. Verify proper brake operation before moving the vehicle.

BRAKELIGHT SWITCH — INSTRUMENT PANEL — DASH PANEL — BRAKE PEDAL SUPPORT — RETAINER CLIP — BRAKE PEDAL — PUSH ROD RETAINING CLIP — MASTER CYLINDER AND POWER BRAKE BOOSTER ASSEMBLY

40427

Power brake booster installation

FRONT SUSPENSION

Shock Absorber

REMOVAL AND INSTALLATION

Wrangler 1991-95

NOTE: Before installing new shocks, they should be purged of air. To do this, hold the shock upright and fully extend it, then invert and compress it. Do this several times.

1. Raise and safely support the front of the vehicle.
2. Remove the locknuts and washers.
3. Pull the shock absorber eyes and rubber bushings from the mounting pins.
4. Install the shock in the reverse order of the removal procedure. Torque the retaining nuts and bolts as follows:

- Lower — 45 ft. lbs.
- Upper — 13 ft. lbs.

Cherokee 1991-95 and Comanche 1991-92

NOTE: Before installing new shocks, they should be purged of air. To do this, hold the shock upright and fully extend it, then invert and compress it. Do this several times.

1. Raise and safely support the front of the vehicle.
2. Remove the locknuts and washers.
3. Pull the shock absorber eyes and rubber bushings from the mounting pins.
4. Install the shocks in the reverse order of the removal procedure. Torque the retaining nuts and bolts as follows:

- Lower — 14 ft. lbs.
- Upper — 8 ft. lbs.

Grand Cherokee 1993-95 and Grand Wagoneer 1993

NOTE: Before installing new shocks, they should be purged of air. To do this, hold the shock upright and fully extend it, then invert and compress it. Do this several times.

1. Remove the upper nut, retainer and grommet from the engine compartment,

2. Remove the lower bolt and nut from the mounting bracket.
3. Remove the shock absorber.
4. Install the shocks in the reverse order of the removal procedure. Torque the upper retaining nut to 14 ft. lbs. (19 Nm) and lower bolt and nut to 17 ft. lbs. (23 Nm).

Grand Wagoneer 1991

NOTE: Before installing new shocks, they should be purged of air. To do this, hold the shock upright and fully extend it, then invert and compress it. Do this several times.

1. Raise and safely support the front of the vehicle.
2. Using a suitable jack, relieve the axle weight from the springs.
3. Remove the upper locknut and washers that attach the shock absorber to frame rail bracket stud.
4. Remove the lower bolt and nut attaching the shock to the axle shaft tube bracket.
5. Remove the shock absorber and washer from the frame rail bracket stud.

To install:

6. Install the washer and upper shock absorber eye on the frame rail bracket stud, then position the lower

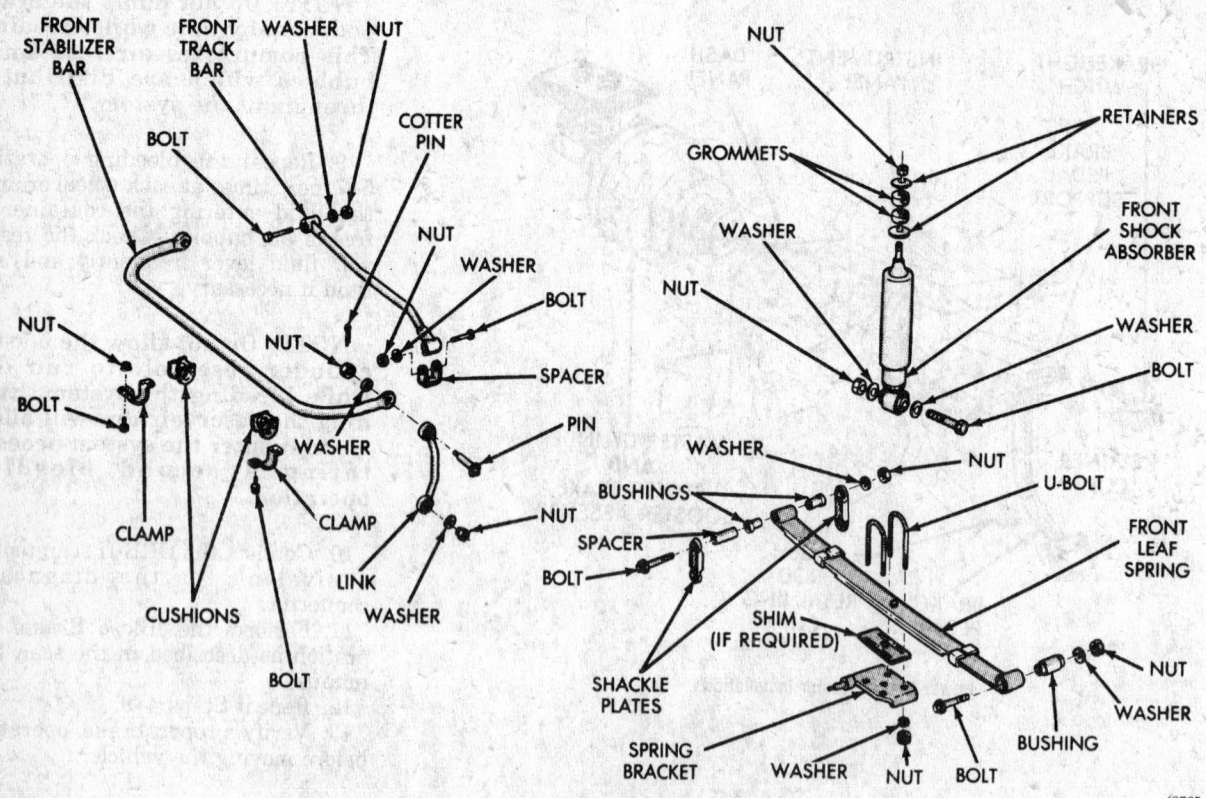

Exploded view of the front suspension - Wrangler

48765

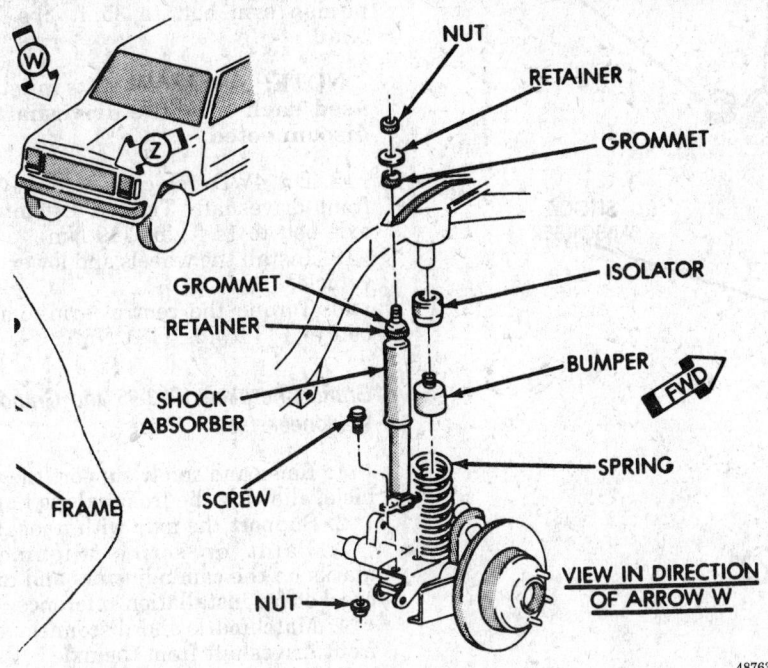

Shock absorber installation

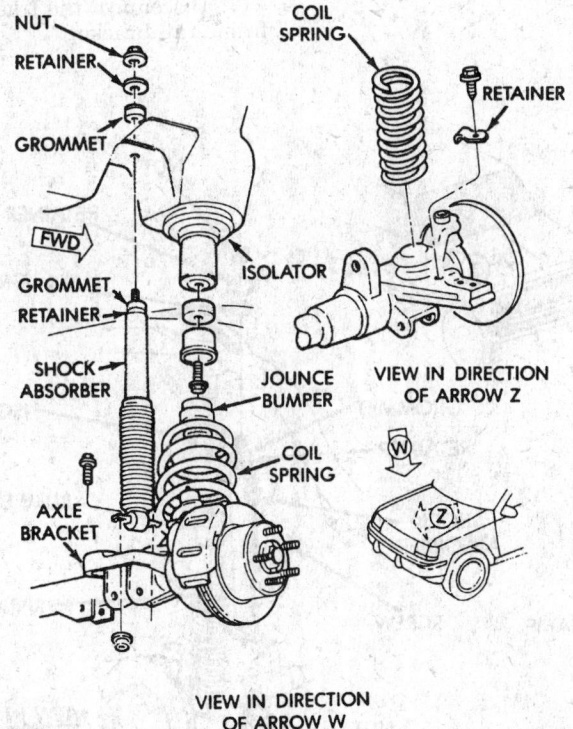

VIEW IN DIRECTION
OF ARROW W

Exploded view of the shock absorber and mounting components -
Grand Cerokee and Grand Wagoneer

shock absorber eye in the axle shaft tube bracket.

7. Install the upper washer and locknut on the frame rail bracket stud, then install the bolt and locknut attaching the shock absorber to the axle shaft tube bracket. Tighten the locknuts to 45 ft. lbs. (61 Nm).

8. Remove the supports and lower the vehicle.

Coil Spring

REMOVAL AND INSTALLATION

Cherokee 1991-95 and Comanche 1991-92

1. Raise and safely support the vehicle, allowing the front axle to hang.
2. Support the axle with a jack.
3. Remove the wheels.
4. On 4WD vehicles, matchmark and disconnect the front driveshaft from the axle.
5. Disconnect the lower control arm at the axle.
6. Disconnect the stabilizer bar links and the shock absorbers at the axle.
7. Disconnect the track bar at the sill bracket.
8. Disconnect the tie rod at the pitman arm.
9. Lower the axle until tension is removed from the spring, then loosen the spring retainer and remove the spring.
10. If necessary, remove the jounce bumper from the upper spring mount.

To install:

11. If removed, install the jounce bumper and torque the bolts to 31 ft. lbs. (42 Nm).
12. Position the replacement spring on the retainer, tighten the spring retainer bracket screw and lift the axle into position.
13. Connect the lower control arm to the axle. Do not tighten at this time.
14. Remove the support jack. It is important that the front springs are supporting the weight of the vehicle when the track bar attaching bolts are tightened. Vehicle ride comfort could be adversely affected.
15. Connect the stabilizer bar links and the shock absorbers at the axle. tighten the shock absorber-to-axle bolt to 14 ft. lbs. (19 Nm); the stabilizer bar-to-axle bolt to 70 ft. lbs. (95 Nm).
16. Connect the track bar at the sill bracket. Tighten the track bar-to-frame rail bolt to 35 ft. lbs. (48 Nm).

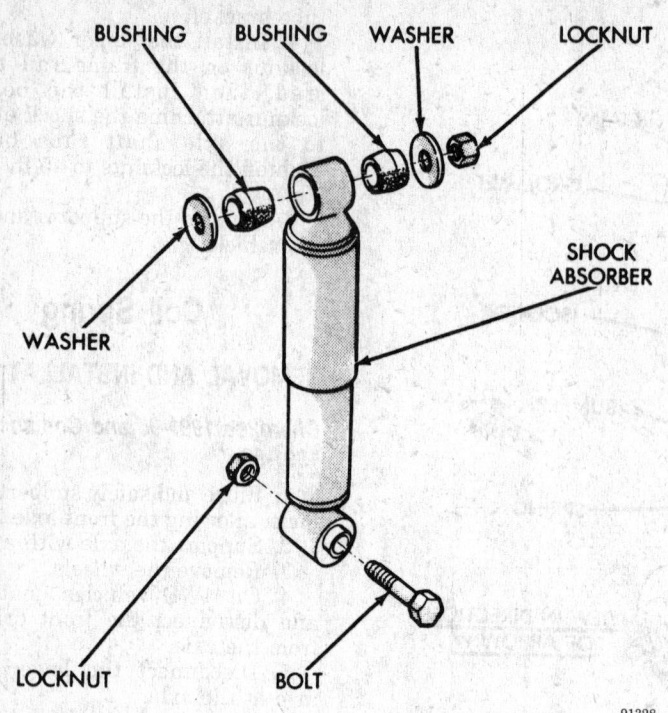

Exploded view of the shock absorber mounting components - Grand Wagoneer 1991

17. Connect the tie rod at the pitman arm. Tighten the center link-to-pitman arm bolt to 35 ft. lbs. (48 Nm).

NOTE: New strap bolts must be used each time the driveshaft is disconnected.

18. On 4WD vehicles, connect the front driveshaft. Tighten U-joint-to-axle bolt to 14 ft. lbs. (19 Nm).
19. Install the wheels and lower the vehicle.
20. Torque the control arm-to-axle bolt to 133 ft. lbs. (181 Nm).

Grand Cherokee 1993-95 and Grand Wagoneer 1993

1. Raise and safely support the vehicle, allowing the front axle to hang.
2. Support the axle with a jack.
3. Paint or scribe alignment marks on the cam adjusters and axle bracket for installation reference.
4. Matchmark and disconnect the front driveshaft from the axle.
5. Disconnect the lower suspension arm nut, cam and cam bolt from the axle.
6. Disconnect the stabilizer bar links and shock absorbers at the axle.
7. Disconnect the track bar at the frame rail bracket.

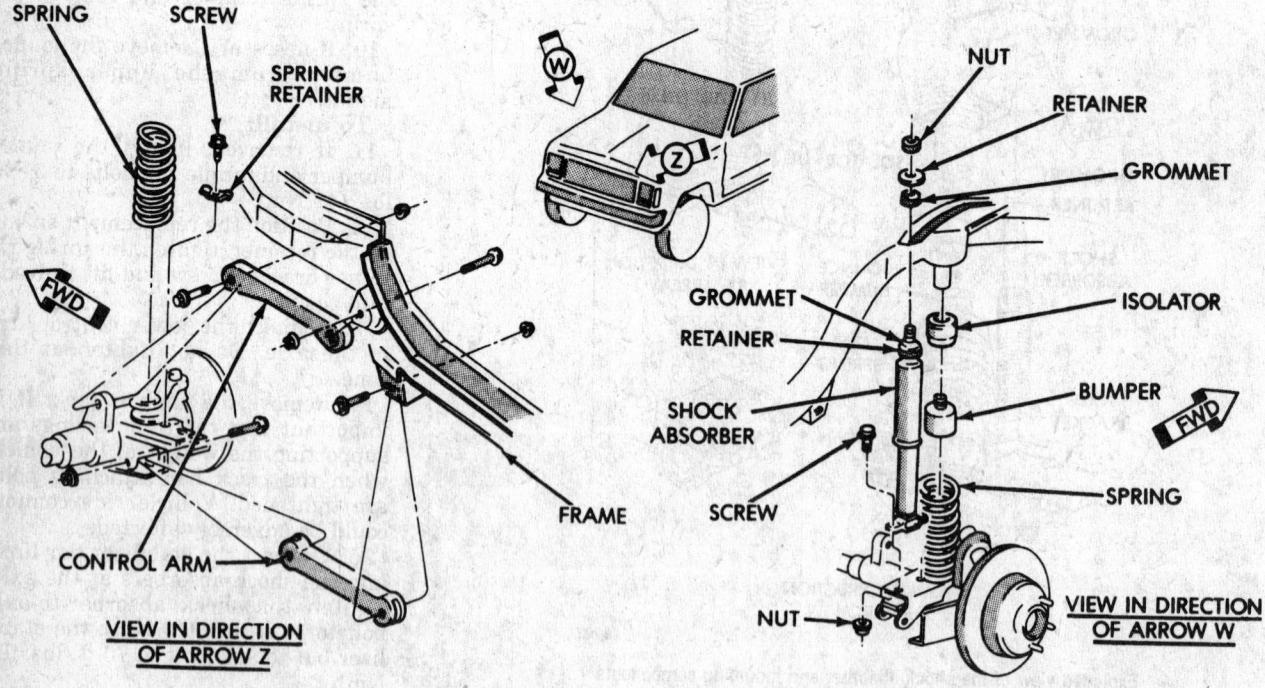

Front coil spring and shock absorber - Cherokee and Comanche

8. Disconnect the drag link at the pitman arm.

9. Lower the axle until spring is free from the upper mount, then remove the coil spring clip screw and remove the spring.

10. If necessary, remove the jounce bumper from the upper spring mount.

To install:

11. If removed, install the jounce bumper and torque the bolts to 31 ft. lbs. (42 Nm).

12. Position the replacement spring on axle pad. Install the spring clip and torque the screw to 16 ft. lbs. (21 Nm).

13. Raise the axle into position until the spring seats in the upper mount.

14. Connect the stabilizer bar links and shock absorbers to the axle bracket. Connect the track bar to the frame rail bracket.

15. Install the lower suspension arm to the axle.

16. Connect the driveshaft to the yoke.

17. Lower the vehicle.

Leaf Spring

REMOVAL AND INSTALLATION

Wrangler 1991-95

1. Raise and support the vehicle safely.

2. Support the front axle so the weight is relieved.

3. Remove the stabilizer bar link attaching nut.

4. Remove the nuts, U-bolts and bracket from the axle.

5. Remove the nut and bolt attaching the spring front eye to the shackle.

6. Remove the spring from the vehicle.

To install:

7. Position the spring front eye in the shackle.

8. Position the rear eye in the hanger bracket.

9. Install the spring bracket, U-bolts and nuts. Torque the nuts to 90 ft. lbs. (122 Nm).

10. Attach the stabilizer bar links.

11. Remove the axle support.

12. Lower the vehicle.

13. Torque the front shackle plate nut to 100 ft. lbs. (135 Nm).

14. Torque the rear eye bracket nut to 105 ft. lbs. (142 Nm).

Grand Wagoneer 1991

1. Raise and support the vehicle safely.

2. Support the front axle so the weight is relieved.

3. If equipped, remove the stabilizer bar links attaching nut.

4. Remove the nuts, U-bolts and bracket from the axle.

5. Remove the nut and bolt attaching the spring front eye to the shackle.

6. Remove the spring from the vehicle.

To install:

7. Position the spring front eye in the shackle.

8. Position the rear eye in the hanger bracket.

9. Install the spring bracket, U-bolts and nuts. Torque the nuts to 90 ft. lbs. (122 Nm).

10. If equipped, attach the stabilizer bar links.

11. Remove the axle support.

12. Lower the vehicle.

13. Torque the front shackle plate nut to 100 ft. lbs. (135 Nm).

14. Torque the rear eye bracket nut to 105 ft. lbs. (142 Nm).

Torsion Bar-Front

Upper Ball Joint

REMOVAL AND INSTALLATION

Wrangler 1991-95, Cherokee 1991-95, Comanche 1991-92, Grand Cherokee 1993-95 and Grand Wagoneer 1993

1. Remove the steering knuckle.

2. Position receiver tool J34503-1 or equivalent, on top of the upper ball joint.

3. Place adapter tool J34503-3 or equivalent, in a C-clamp and position the clamp and adapter under the upper ball joint.

4. Tighten the clamp screw to remove the ball joint.

To install:

5. Place adapter tool J34503-5 on top of the replacement upper ball joint.

6. Place angled receiver tool J34503-12 or equivalent, between the C-clamp and the yoke.

7. Tighten the clamp screw and seat the ball joint.

8. Install the steering knuckle.

Grand Wagoneer 1991

1. Remove the steering knuckle.

2. Remove both arms from the frame of a suitable puller tool.

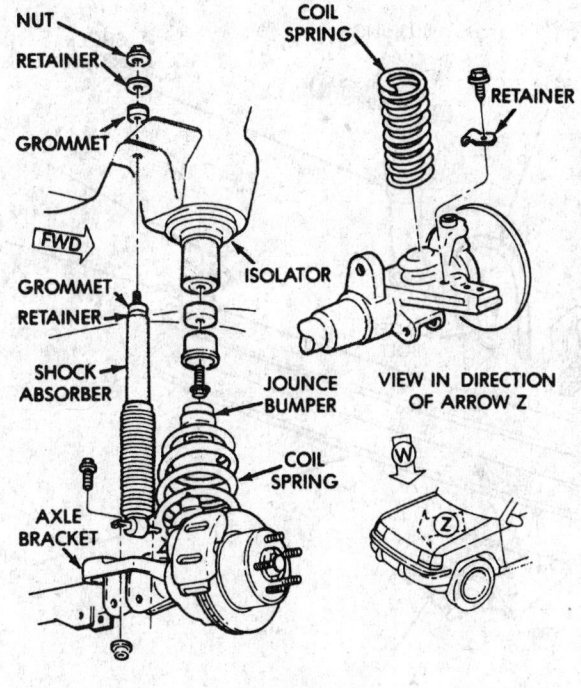

Coil spring and shock absorber - Grand Cherokee and Grand Wagoneer

3. Place button tool J25211-3 or equivalent, on the upper ball joint.

4. Thread the puller frame halfway onto the puller screw. Insert the nut end of the puller screw up through the lower ball joint hole in the steering knuckle. Position the puller frame up against the knuckle and the bottom of the puller screw against the button tool.

5. Tighten the puller screw to press the upper ball joint from the steering knuckle.

6. Remove the tools from the steering knuckle.

To install:

7. Install both the arms on the puller tool frame.

8. Position the replacement upper ball joint in the steering knuckle.

9. Install plate tool J25211-1 or equivalent, on the spindle mounting studs and installer cup tool J25211-2 or equivalent, on the upper ball stud.

10. Install the assembled puller tool on the steering knuckle. Hook one puller arm in the plate and hook the opposite arm in the knuckle. Make sure the puller screw is centered on the installer cup.

11. Tighten the puller screw to press the ball joint into the steering knuckle, then remove the installation tools.

12. Install the steering knuckle.

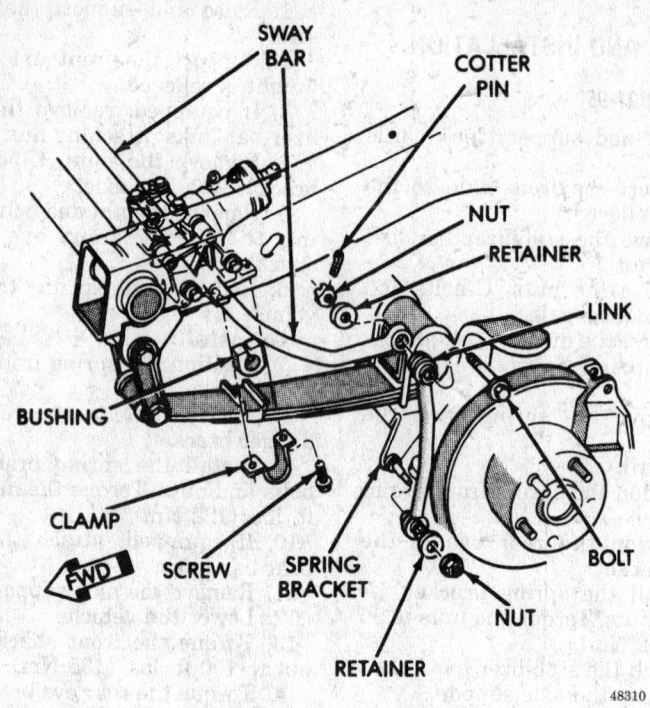

Front suspension - Wrangler

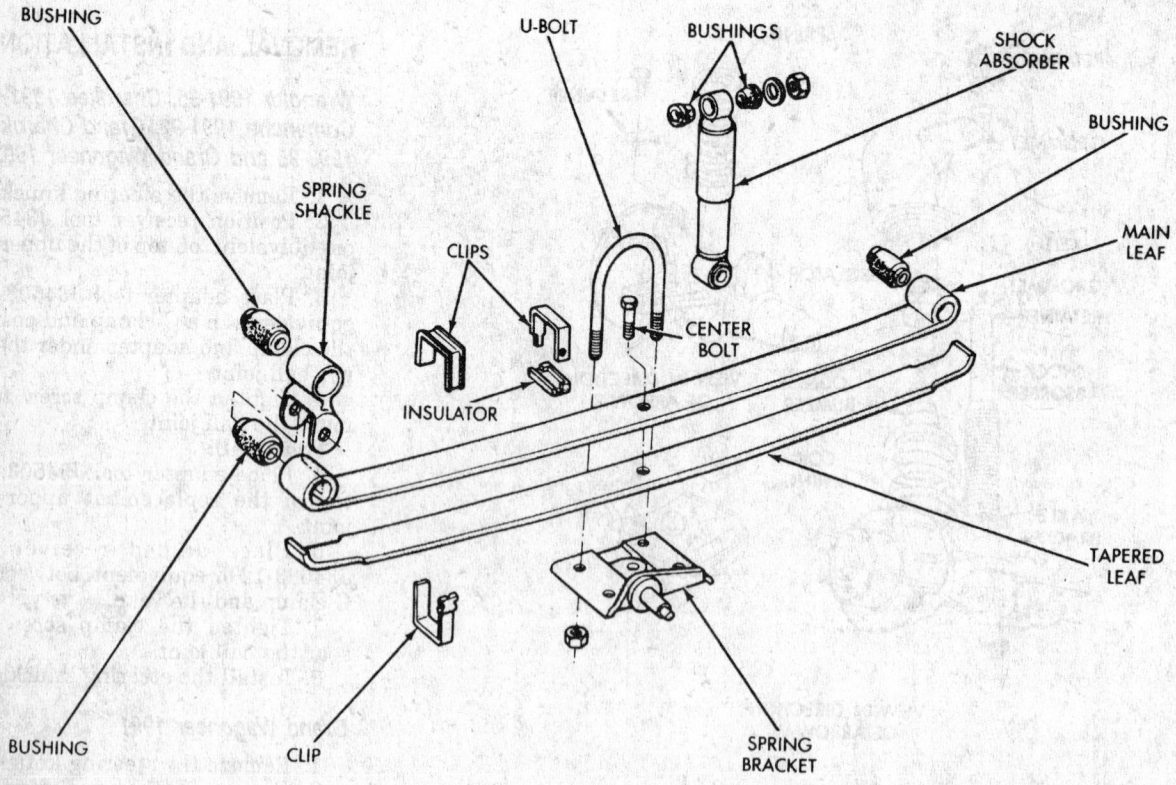

Front suspension - Grand Wagoneer 1991

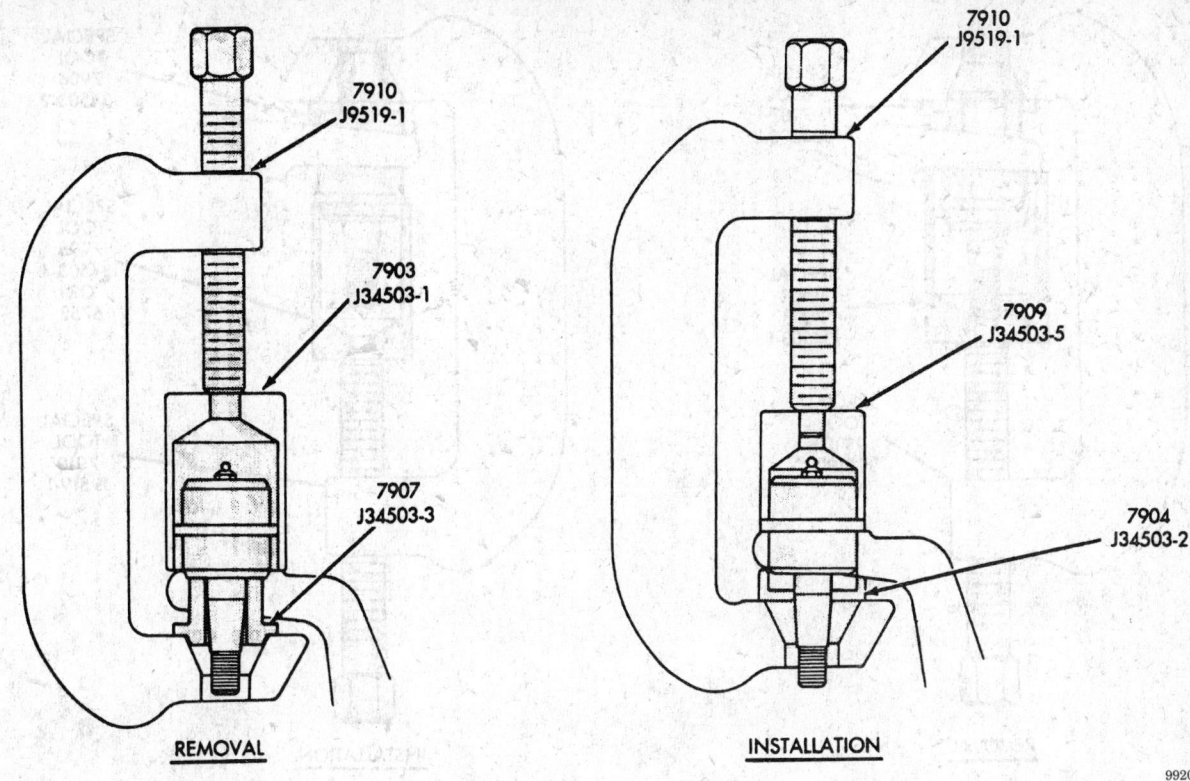

REMOVAL

INSTALLATION

99209

Upper ball joint removal and installation

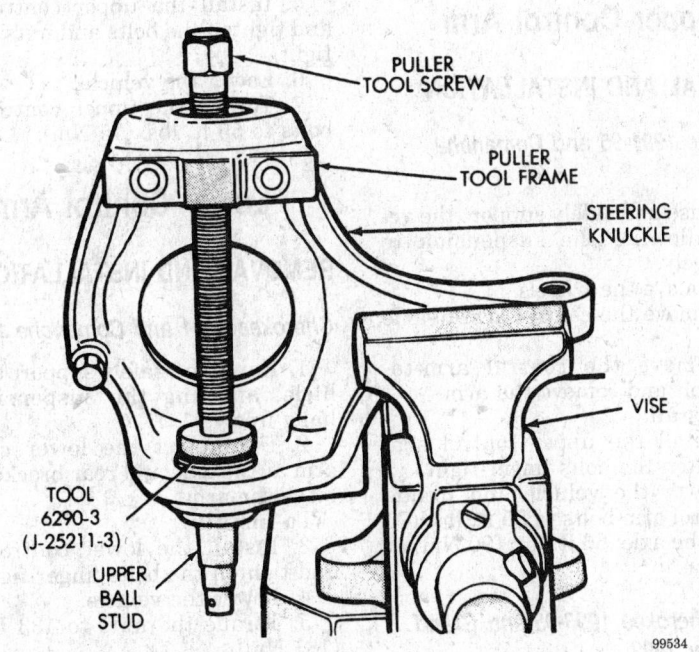

PULLER
TOOL SCREW

PULLER
TOOL FRAME

STEERING
KNUCKLE

VISE

TOOL
6290-3
(J-25211-3)

UPPER
BALL
STUD

99534

Upper ball joint removal - Grand Wagoneer 1991

Lower Ball Joint

REMOVAL AND INSTALLATION

Wrangler 1991-95, Cherokee 1991-95, Comanche 1991-92, Grand Cherokee 1993-95 and Grand Wagoneer 1993

1. Remove the steering knuckle.
2. Position receiver tool J34503-1 or equivalent, on a C-clamp screw and adapter tool J34503-3 or equivalent between the top of the yoke and the C-clamp.
3. Tighten the clamp screw to remove the ball joint.

To install:

4. Position installer tool J34503-4 or equivalent, on the C-clamp screw and receiver tool J34503-2 or equivalent, between the top of the yoke and the C-clamp.
5. Tighten the clamp screw until the ball joint is fully seated.
6. Install the steering knuckle.

Grand Wagoneer 1991

1. Remove the steering knuckle.
2. Remove the lower ball joint snapring. Clamp the steering knuckle securely in a vise with the upper ball joint facing downward.

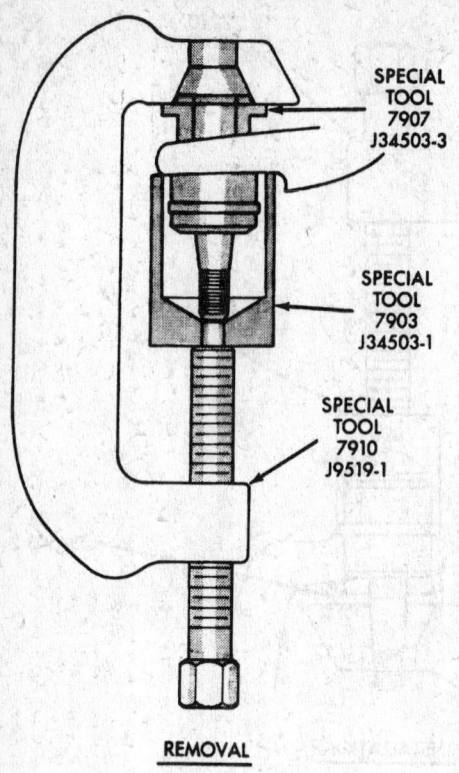

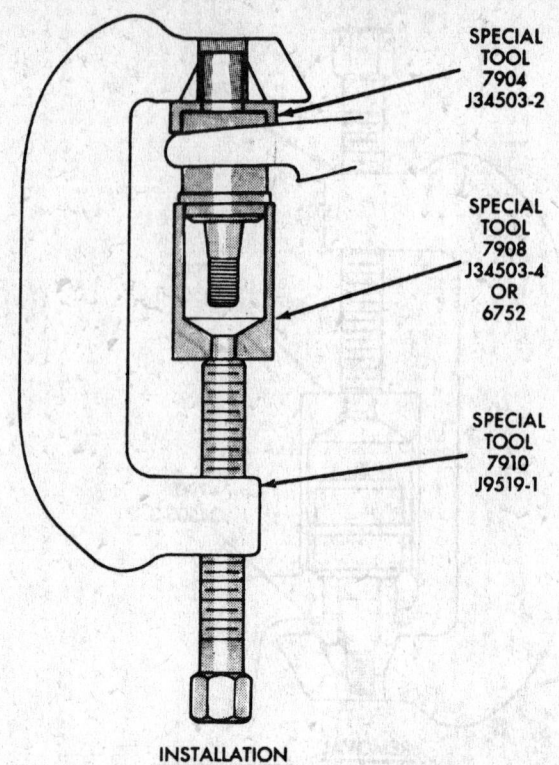

REMOVAL

INSTALLATION

48392

Lower ball joint removal and installation

3. Attach plate tool J25211-1 or equivalent, to the spindle mating surface on the steering knuckle.

4. Position button tool J2511-3 or equivalent, on the lower ball joint.

5. Assemble and install a puller tool on the steering knuckle. Hook one puller arm in the plate tool and hook the opposite arm in the steering knuckle.

6. Tighten the puller screw to press the lower ball joint out of the steering knuckle, then remove the tools from the knuckle.

To install:

7. Install the steering knuckle in the vise, then position the replacement ball joint in the knuckle.

8. Place adapter tool J25211-4 or equivalent, over the nut end of the puller screw and against the puller frame.

9. Insert the nut end of the puller screw through the upper ball joint hole in the knuckle and hold the adapter and frame against the knuckle.

10. Place installer cup tool J25211-2 or equivalent, on the ball joint.

11. Tighten the puller screw to press the lower ball stud into the steering knuckle.

12. Install the replacement snapring and remove the tools.

13. Install the steering knuckle.

Upper Control Arm

REMOVAL AND INSTALLATION

Cherokee 1991-95 and Comanche 1991-92

1. Raise and safely support the vehicle, allowing the suspension to hang freely.

2. Remove the wheels.

3. Remove the control arm-to-axle bolt.

4. Remove the control arm-to-frame bolt and remove the arm.

To install:

5. Install the upper control arm and tighten the bolts finger-tight.

6. Lower the vehicle and torque the control arm bolts to 55 ft. lbs. (75 Nm) at the axle; 66 ft. lbs. (90 Nm) at the frame.

Grand Cherokee 1993-95 and Grand Wagoneer 1993

1. Raise and support the vehicle safely.

2. Remove the upper control arm-to-axle bracket nut and bolt.

3. Remove the upper control arm-to-frame nut and bolt.

4. Remove the arm.

To install:

5. Install the upper control arm and tighten the bolts and nuts finger-tight.

6. Lower the vehicle.

7. Torque the upper control arm bolts to 55 ft. lbs. (75 Nm).

Lower Control Arm

REMOVAL AND INSTALLATION

Cherokee 1991 and Comanche 1991

1. Raise and safely support the vehicle, allowing the suspension to hang freely.

2. Disconnect the lower control arm at the axle and rear bracket. Remove the arm.

To install:

3. Install the lower control arm and tighten the bolts finger-tight.

4. Lower the vehicle.

5. Torque the nuts to 133 ft. lbs. (181 Nm).

Cherokee 1992-95 and Comanche 1992

1. Raise and safely support the vehicle, allowing the suspension to hang freely.

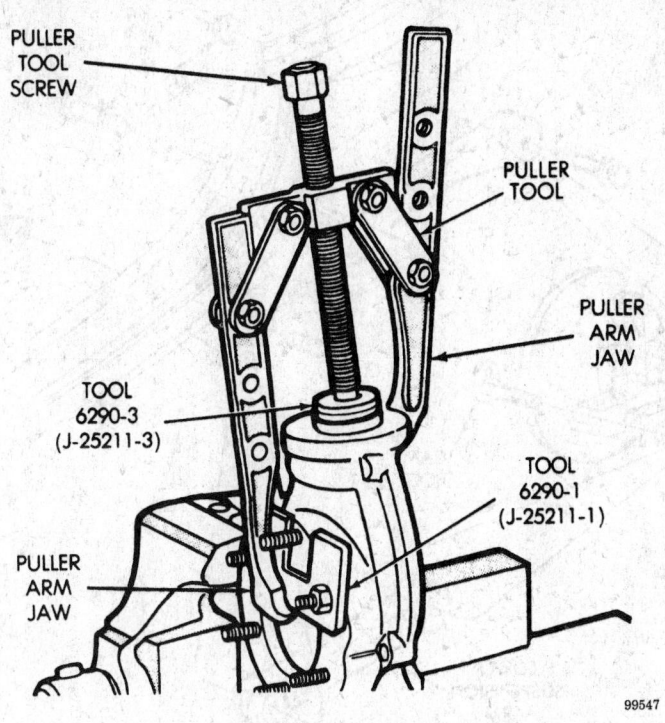

Lower ball joint removal - Grand Wagoneer 1991

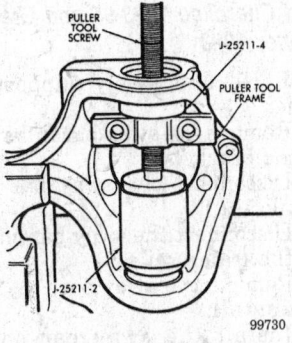

Lower ball joint installation - Grand Wagoneer 1991

2. Disconnect the lower control arm at the axle and rear bracket. Remove the arm.

To install:

3. Install the lower control arm and tighten the bolts finger-tight.

4. Lower the vehicle.

5. Torque the nuts to 85 ft. lbs. (115 Nm).

Grand Cherokee 1993-95 and Grand Wagoneer 1993

1. Raise and safely support the vehicle, allowing the suspension to hang freely.

2. Paint or scribe alignment marks on the cam adjusters and suspension arm for installation reference.

3. Remove the lower control arm nut, cam and cam bolt from the axle.

4. Remove the nut and bolt from the frame rail bracket.

5. Remove the arm.

To install:

6. Install the lower control arm.

7. Tighten the rear bolts finger-tight.

8. Install the cam bolt, cam and nut in the axle while realigning the reference marks.

9. Lower the vehicle.

10. Torque the nuts to 130 ft. lbs. (176 Nm).

Sway Bar

REMOVAL AND INSTALLATION

Wrangler 1991-95

1. Raise and safely support the vehicle.

2. Remove the retaining nut from the connecting link bolt.

3. Disconnect the bracket retaining nuts and brackets.

4. Remove the stabilizer bar.

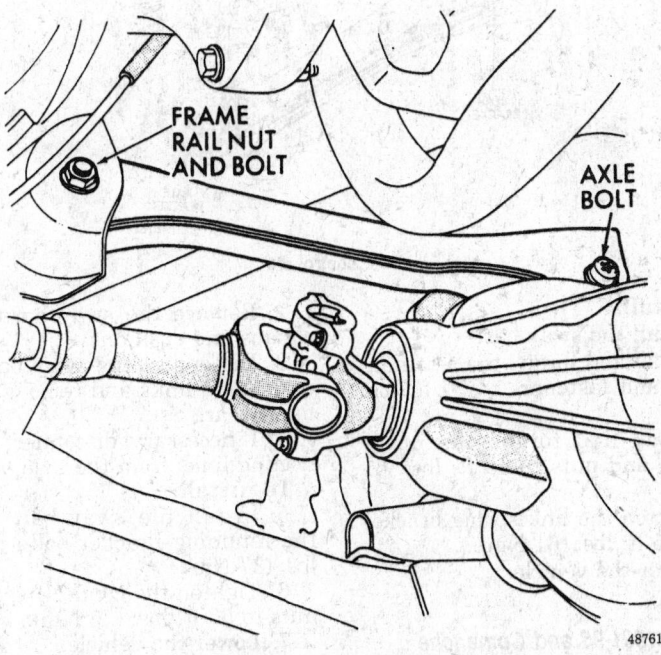

Upper control arm removal and installation - Cherokee and Comanche

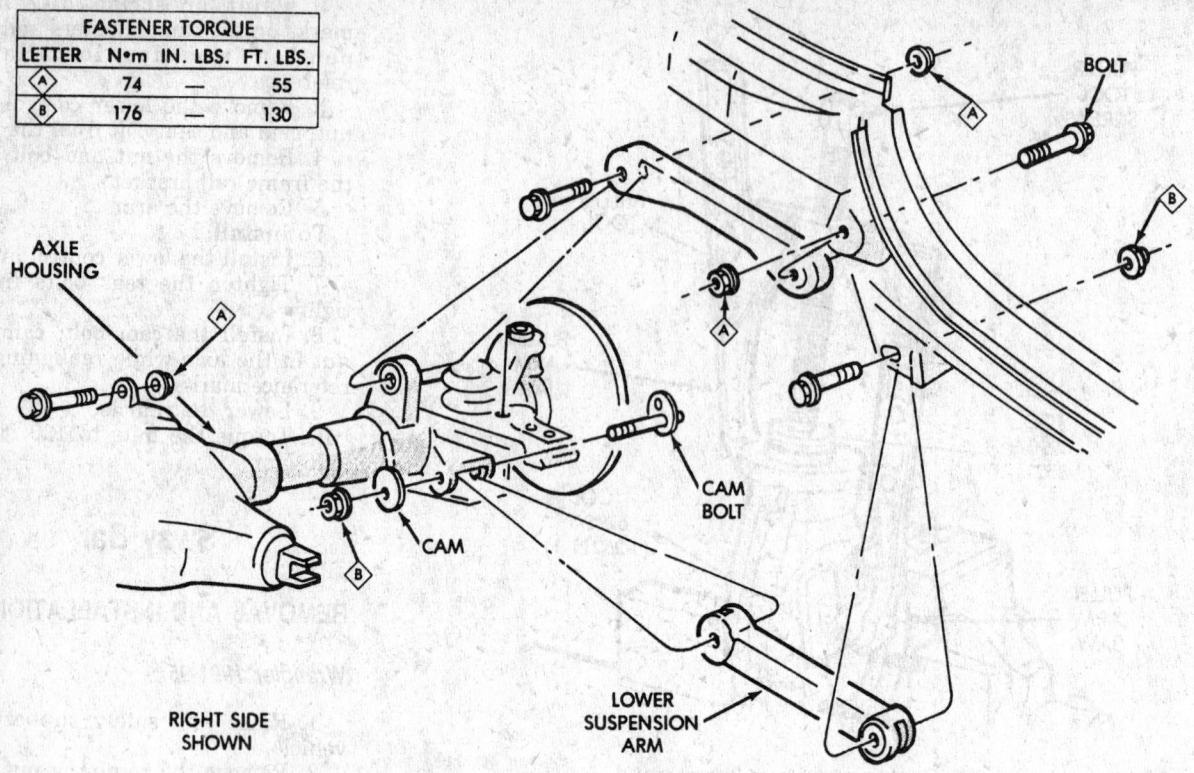

FASTENER TORQUE			
LETTER	N•m	IN. LBS.	FT. LBS.
A	74	—	55
B	176	—	130

Exploded view of the suspension arms - Grand Cherokee and Grand Wagoneer

100435

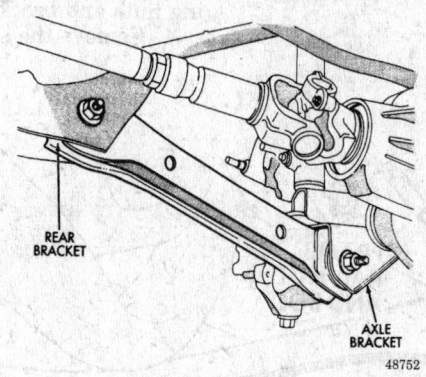

Lower control arm

48752

To install:

5. Install the sway bar.

6. Install and torque the retaining brackets and fasteners to 30 ft. lbs. (41 Nm).

7. Install and torque the upper link bolts and nuts to 45 ft. lbs. (61 Nm).

8. Tighten the link spring bracket nuts to 45 ft. lbs. (61 Nm).

9. Lower the vehicle.

Cherokee 1991-95 and Comanche 1991-92

1. Raise and safely support the vehicle.

2. Remove the sway bar-to-frame clamps and cushions.

3. Disconnect the sway bar at the connecting links and remove the stabilizer bar.

4. If necessary, disconnect the connecting links from the axle bracket.

To install:

5. Install the sway bar. Tighten the retaining bracket bolts to 27 ft. lbs. (37 Nm).

6. Tighten the connecting rod link nuts to 55 ft. lbs. (75 Nm).

7. Lower the vehicle.

Grand Cherokee 1993-95 and Grand Wagoner 1993

1. Raise and safely support the vehicle.

2. Remove the sway bar links from the axle brackets.

3. Disconnect the sway bar from the links.

4. Disconnect the sway bar clamps from the frame rails.

5. Remove the sway bar.

To install:

6. Install the sway bar on the frame rail and install the clamps and bolts. Ensure the bar is centered with equal spacing on both sides. Torque the retaining bracket bolts to 55 ft. lbs. (75 Nm).

7. Install the connecting rod link and grommets onto the stabilizer bar and brackets. Torque the nuts to 70 ft. lbs. (95 Nm).

8. Torque the sway bar-to-connecting link nut to 27 ft. lbs. (36 Nm).

9. Lower the vehicle.

Grand Wagoneer 1991

1. Raise and safely support the vehicle.

2. Remove the sway bar-to-frame clamps and cushions.

3. Disconnect the sway bar at the connecting links and remove the stabilizer bar.

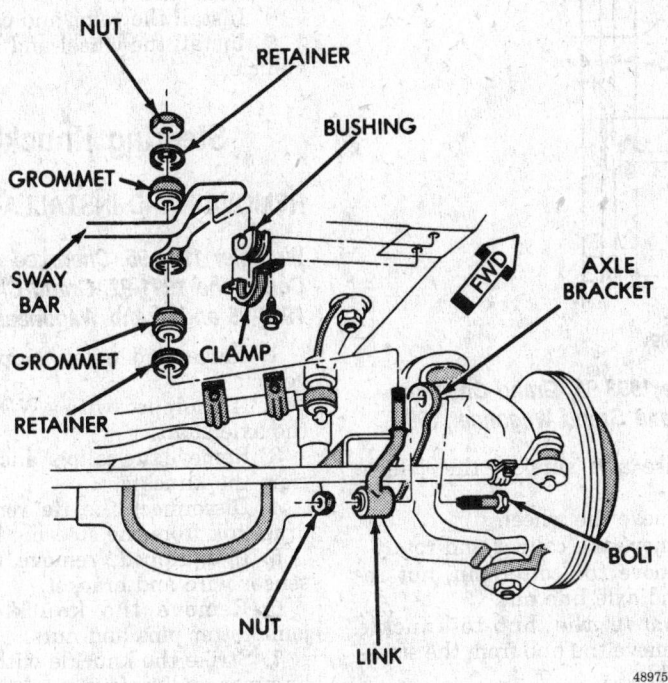

Exploded view of the sway bar and mounting components

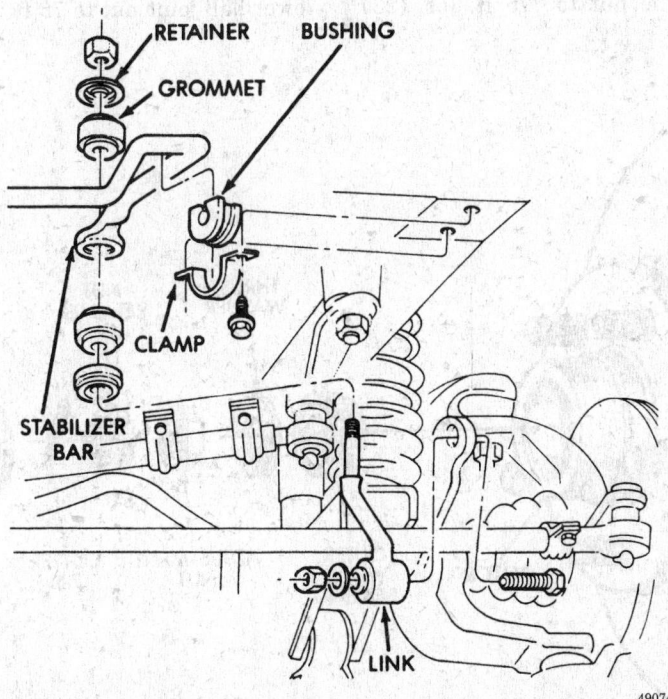

Exploded view of the sway bar and mounting components - Grand Cherokee and Grand Wagoneer

4. If necessary, disconnect the connecting links from the axle bracket.

To install:

5. Install the sway bar. Tighten the retaining bracket bolts to 35 ft. lbs. (47 Nm).

6. Tighten the connecting rod link nuts to 55 ft. lbs. (75 Nm).

7. Lower the vehicle.

Front Wheel Bearing and Hub

REMOVAL AND INSTALLATION

Cherokee 1991-92 and Comanche 1991-92

2WD VEHICLES

1. Loosen the lug nuts on the front wheels.

2. Raise and safely support the vehicle.

3. Remove the front wheels.

4. Remove the calipers but don't disconnect the brake lines. Suspend the calipers aside.

5. Remove the dust cap, cotter pin, nut retainer, nut and thrust washer from the spindle.

6. Remove the rotor. Be ready to catch the outer bearing.

7. Carefully drive out the inner bearing and seal from the hub, using a wood block.

8. Inspect the bearing races for excessive wear, pitting or grooves. If they are cracked or grooved or if pitting and excess wear is present, drive them out with a drift or punch.

To install:

9. Check the bearing for excess wear, pitting or cracks or excess looseness.

NOTE: If it is necessary to replace either the bearing or the race, replace both. Never replace just a bearing or a race. These parts wear in a mating pattern. If just one is replaced, premature failure of the new part will result.

10. If the old parts are retained, thoroughly clean them in solvent and allow them to dry on a clean towel. Never spin dry them with compressed air.

11. Thoroughly clean the spindle.

12. Thoroughly clean the inside of the hub.

13. Pack the inside of the hub with EP wheel bearing grease. Add grease to the hub until it is flush with the inside diameter of the bearing cup.

14. Pack the bearing with the same grease.

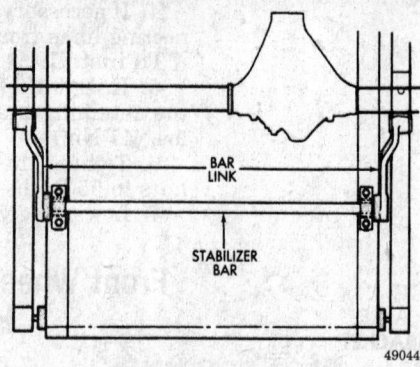

Sway bar - Grand Wagoneer 1991

15. If a new race is being installed, very carefully drive it into position until it bottoms all around, using a brass drift. Be careful to avoid scratching the surface.

16. Place the inner bearing in the race and install a new grease seal.

17. Position the hub and rotor on the spindle and install the outer bearing.

18. Install the washer and nut. Adjust the wheel bearing.

19. Install the nut cap and a new cotter pin. Install the grease cap.

20. Install the caliper and wheels. Lower the vehicle.

Cherokee 1993-95, Grand Cherokee 1993-95 and Grand Wagoneer 1993

1. Raise and support the vehicle safely.

2. Remove the wheel.

3. Remove the caliper and rotor.

4. Remove the cotter pin, nut retainer and axle hub nut.

5. Remove the hub-to-knuckle bolts. Remove the hub from the steering knuckle.

To install:

6. Install the hub to the steering knuckle. Torque the bolts to 75 ft. lbs. (102 Nm).

7. Install the hub washer and nut. Torque the nut to 175 ft. lbs. (237

Nm). Install the nut retainer and a new cotter pin.

8. Install the rotor and caliper.

9. Install the wheel and lower the vehicle.

Steering Knuckle

REMOVAL AND INSTALLATION

Wrangler 1991-95, Cherokee 1991-95, Comanche 1991-92, Grand Cherokee 1993-95 and Grand Wagoneer 1993

1. Raise and safely support the vehicle.

2. If equipped with 4WD, remove the axle shaft.

3. Remove the caliper anchor plate from the knuckle.

4. Disconnect the tie rod end or drag link from the steering knuckle.

5. If equipped, remove the ABS sensor wire and bracket.

6. Remove the knuckle-to-ball joint cotter pins and nuts.

7. Strike the knuckle with a brass hammer and remove it.

To install:

8. Position the knuckle over the ball joint studs and install the nuts.

9. Tighten the upper ball joint nut to 75 ft. lbs. (101 Nm). Tighten the lower ball joint nut to 75 ft. lbs. (101

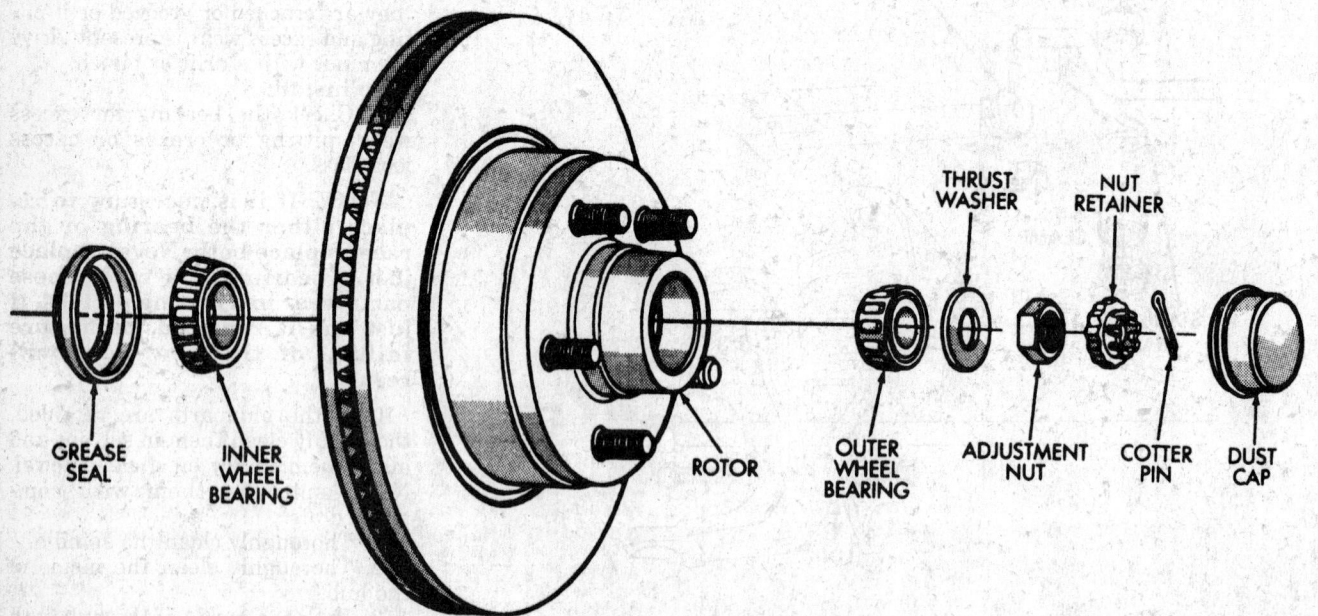

Exploded view of the hub and bearings-2WD vehicles

Nm) on 1990-91 vehicles or 80 ft. lbs. (109 Nm) on 1992-94 vehicles.

10. Install new cotter pins. If the ball stud hole does not align with the nut castellation, further tighten the nut until the cotter pin can be installed.

11. Connect the tie rod end or drag link.

12. If equipped, install the ABS sensor wire and bracket.

13. Install the caliper anchor plate on the knuckle. Torque the caliper anchor bolts to 77 ft. lbs. (105 Nm).

14. Install the axle shaft.

Grand Wagoneer 1991

1. Raise and support the vehicle safely.

2. Remove the axle shaft.

3. Disconnect the tie rod end at the steering knuckle arm.

4. Remove and discard the lower ball joint nut.

5. Remove the cotter pin from the upper ball joint. Loosen the upper ball joint nut until the top edge of the nut is flush with the top of the stud.

6. Unseat the upper and lower ball joints using a lead hammer.

7. Remove the upper ball joint stud nut and the steering knuckle.

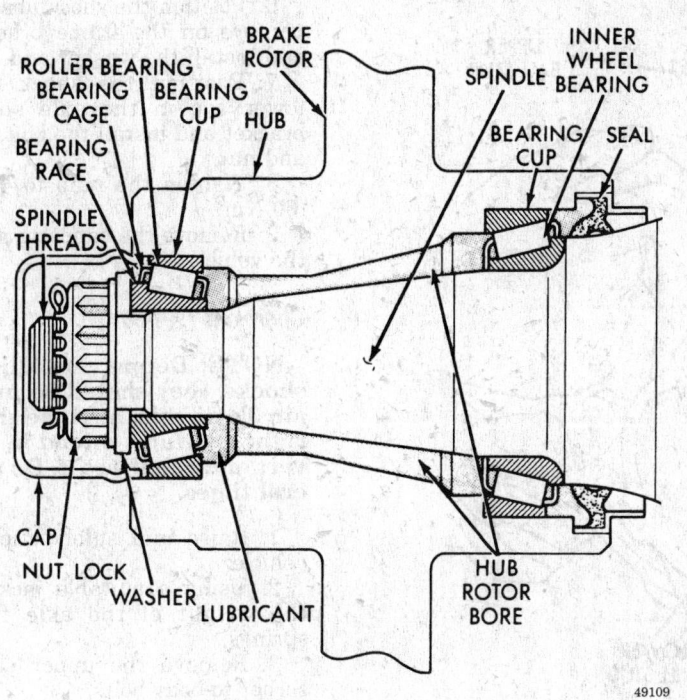

Cut-away view of the hub and bearings-2WD vehicles

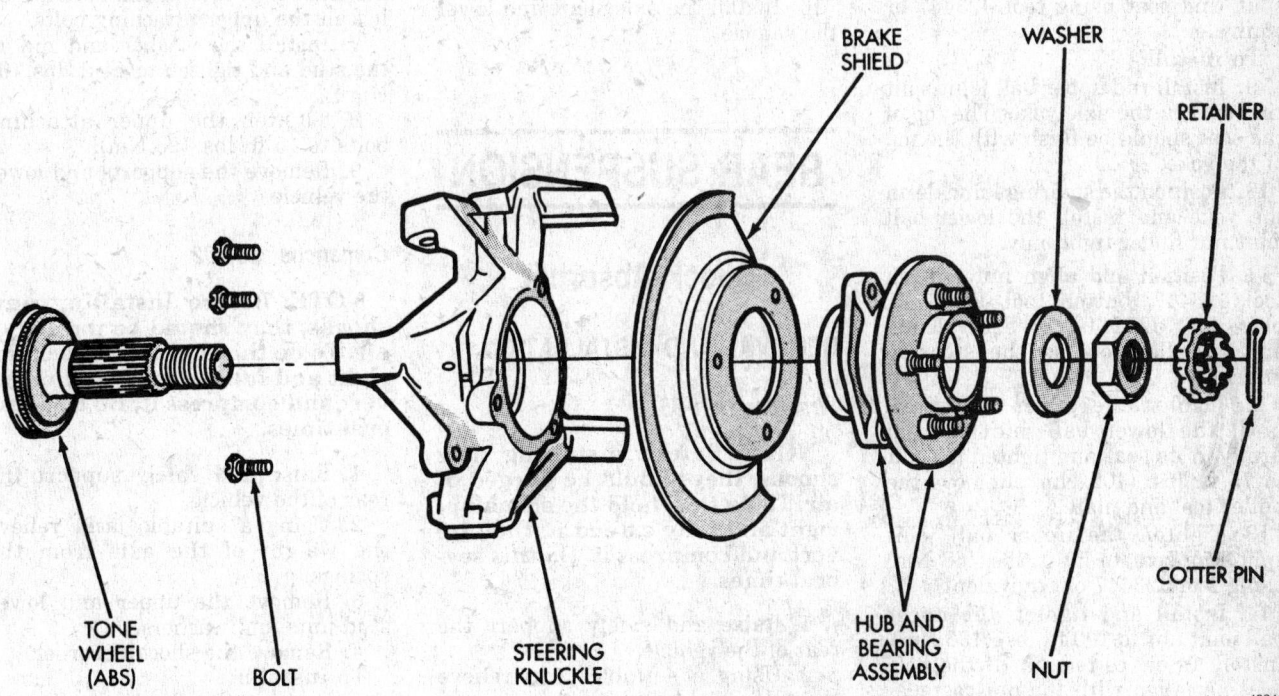

Exploded view of the hub, knuckle and axle shaft

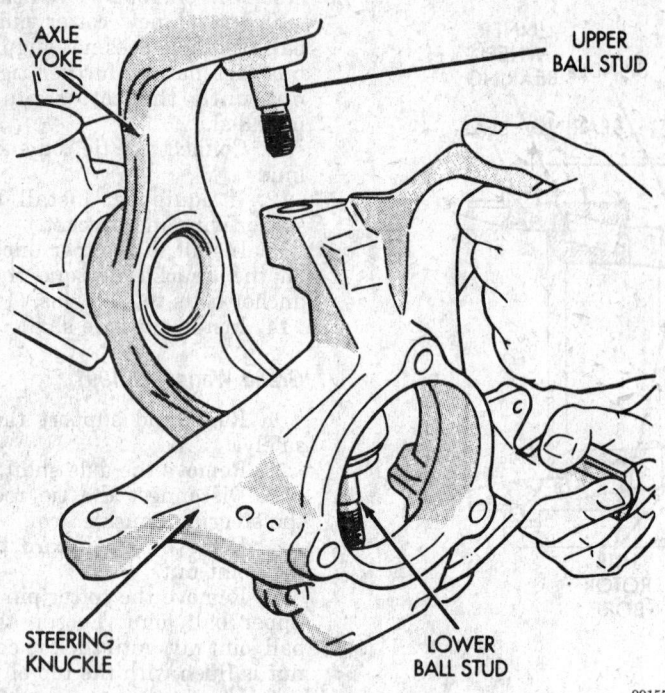

AXLE YOKE

UPPER BALL STUD

STEERING KNUCKLE

LOWER BALL STUD

99155

Steering knuckle removal and installation

8. Remove the upper ball joint split ring seat using tool J23447 or equivalent.

To install:

9. Install the upper ball joint split ring seat in the axle yoke. The top of the seat should be flush with the top of the yoke.

10. Position the steering knuckle on the yoke and install the lower ball joint nut finger-tight only.

11. Position and align nut wrench tool J23447, button tool J25211-3, plate tool J25211-1 or equivalents, and a puller tool on the steering knuckle.

12. Tighten the puller tool screw until the lower ball joint is held firmly in its seat and tighten the nut to 75 ft. lbs. (102 Nm). Remove the puller tool and plate.

13. Tighten the upper ball joint split ring seat to 50 ft. lbs. (68 Nm) using tool J23447 or equivalent.

14. Install and tighten the upper ball joint nut to 100 ft. lbs. (136 Nm). Install a new cotter pin. If the hole does not align with the nut castellation, further tighten the nut until the cotter pin can be installed.

NOTE: Never loosen the nut to align the holes.

15. Connect the tie rod end.
16. Install the axle shaft and lower the vehicle.

REAR SUSPENSION

Shock Absorber

REMOVAL AND INSTALLATION

Wrangler 1991-95

NOTE: Before installing new shocks, they should be purged of air. To do this, hold the shock upright and fully extend it, then invert and compress it. Do this several times.

1. Raise and safely support the rear of the vehicle.
2. Using a suitable jack, relieve the axle weight from the springs.
3. Remove the upper locknut and washer from the frame bracket stud.
4. Remove the lower bolt, nut and washers from the axle shaft tube bracket.
5. Remove the shock absorber.

To install:
6. Position the shock absorber upper eye on the frame bracket stud and install the washer and nut.
7. Position the shock absorber lower eye in the axle shaft tube bracket and install the bolt, washers and nut.
8. Tighten the nuts to 44 ft. lbs. (60 Nm).
9. Remove the supports and lower the vehicle.

Cherokee 1991-95

NOTE: Before installing new shocks, they should be purged of air. To do this, hold the shock upright and fully extend it, then invert and compress it. Do this several times.

1. Raise and safely support the vehicle.
2. Using a suitable jack, relieve the weight of the axle from the springs.
3. Remove the upper shock absorber-to-body bolts.
4. Remove the lower stud nut and washer.
5. Remove the shock absorber.

To install:
6. Position the shock absorber on the axle shaft tube bracket stud, then install the upper attaching bolts.
7. Install the washer and nut to the stud and tighten to 44 ft. lbs. (60 Nm).
8. Tighten the upper attaching bolts to 15 ft. lbs. (20 Nm).
9. Remove the supports and lower the vehicle.

Comanche 1991-92

NOTE: Before installing new shocks, they should be purged of air. To do this, hold the shock upright and fully extend it, then invert and compress it. Do this several times.

1. Raise and safely support the rear of the vehicle.
2. Using a suitable jack, relieve the weight of the axle from the springs.
3. Remove the upper and lower stud nuts and washers.
4. Remove the shock absorber.

To install:
5. Position the shock absorber on the studs and install the washers and nuts.
6. Tighten the nuts to 44 ft. lbs. (60 Nm).
7. Remove the supports and lower the vehicle.

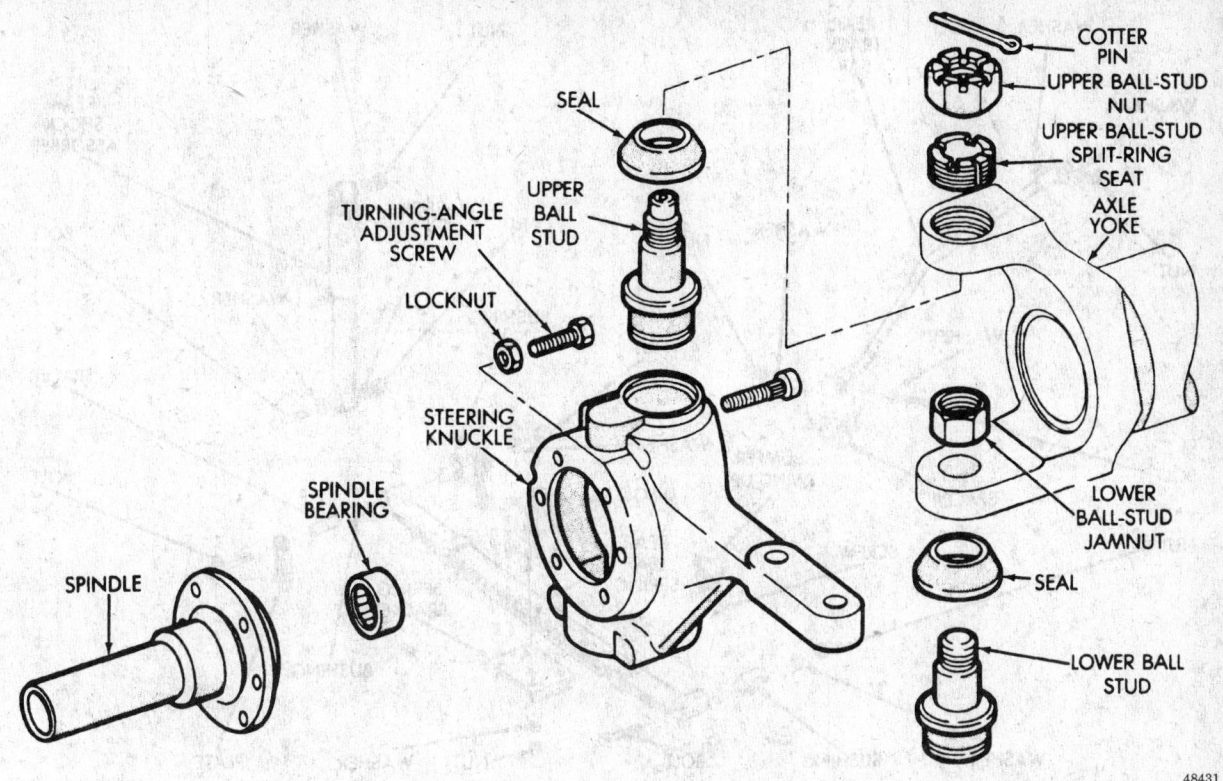

Exploded view of the steering knuckle

48431

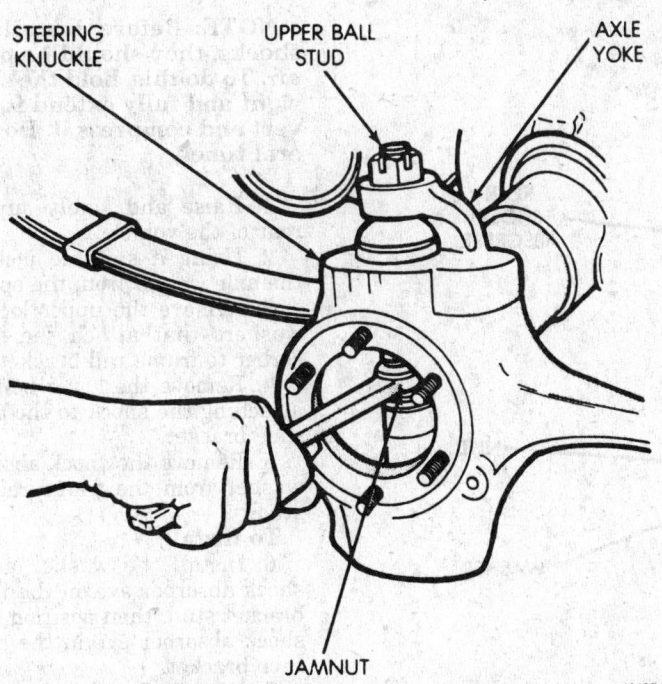

Removing the lower ball joint nut

48433

Grand Cherokee 1993-95 and Grand wagoneer 1993

NOTE: Before installing new shocks, they should be purged of air. To do this, hold the shock upright and fully extend it, then invert and compress it. Do this several times.

1. Raise and safely support the rear of the vehicle.
2. Using a suitable jack, relieve the weight of the axle from the springs.
3. Remove the upper stud nut and washer from the frame rail stud.
4. Remove the lower nut and bolt from the axle bracket.
5. Remove the shock absorber.

To install:

6. Position the shock absorber on the frame rail stud and axle bracket.
7. Install the nut and retainer on the stud and tighten to 52 ft. lbs. (70 Nm).
8. Install the bolt and nut to the axle bracket and tighten to 68 ft. lbs. (92 Nm).
9. Remove the supports and lower the vehicle.

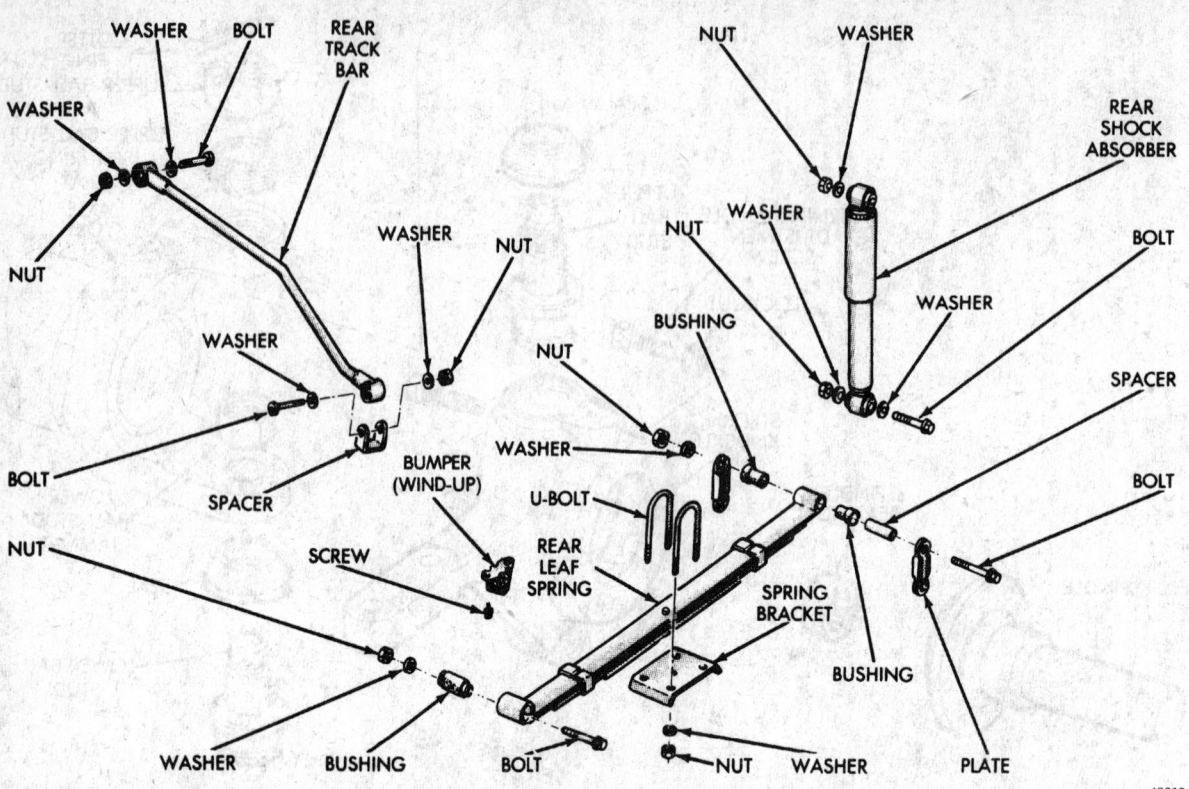

Exploded view of the rear suspension - Wrangler

48819

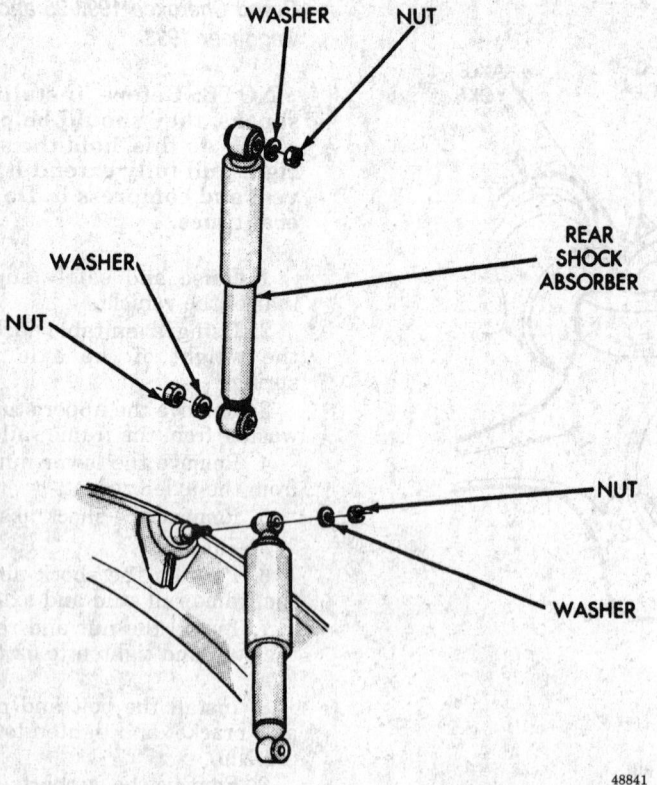

Exploded view of the shock absorber mounting components - Comanche

48841

Grand Wagoneer 1991

NOTE: Before installing new shocks, they should be purged of air. To do this, hold the shock upright and fully extend it, then invert and compress it. Do this several times.

1. Raise and safely support the rear of the vehicle.
2. Using a suitable jack, relieve the axle weight from the springs.
3. Remove the upper locknut and washers that attach the shock absorber to frame rail bracket stud.
4. Remove the lower bolt and nut attaching the shock to the axle shaft tube bracket.
5. Remove the shock absorber and washer from the frame rail bracket stud.

To install:

6. Install the washer and upper shock absorber eye on the frame rail bracket stud, then position the lower shock absorber eye in the axle shaft tube bracket.
7. Install the upper washer and locknut on the frame rail bracket stud, then install the bolt and locknut attaching the shock absorber to the axle shaft tube bracket. Tighten the locknuts to 45 ft. lbs. (61 Nm).

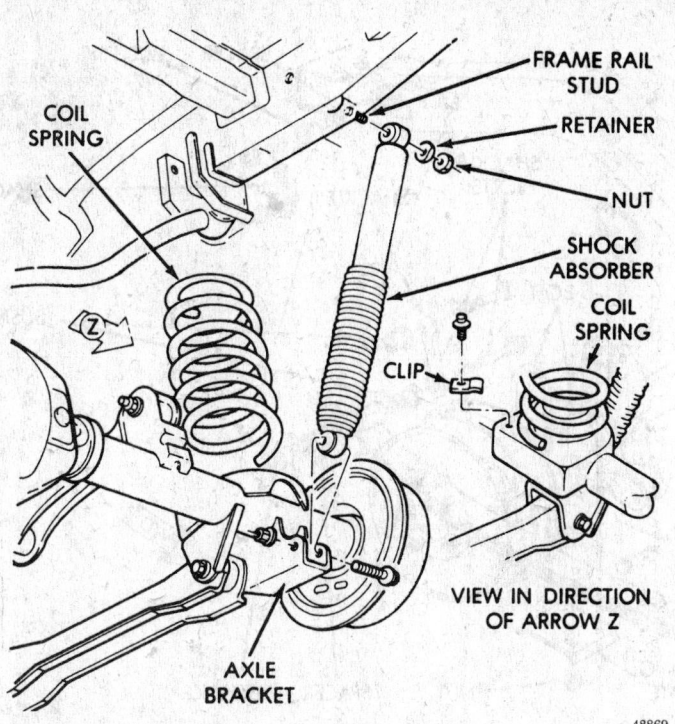

48869

Exploded view of the shock absorber and mounting components - Grand Cherokee and Grand Wagoneer

8. Remove the supports and lower the vehicle.

Coil Spring

REMOVAL AND INSTALLATION

Grand Cherokee 1993-95 and Grand Wagoneer 1993

1. Raise and safely support the vehicle.
2. Support the axle with a suitable jack.
3. Disconnect the sway bar links and shock absorbers from the axle bracket.
4. Disconnect the track bar from the frame rail bracket.
5. Lower the axle until the spring is free from the upper mount seat. Remove the coil spring clip screw and remove the spring.
   To install:
6. Position the coil spring on the axle pad, then install the spring clip and screw. Tighten the screw to 16 ft. lbs. (22 Nm).
7. Raise the axle into position until the spring seats in the upper mount.
8. Connect the sway bar links and shock absorbers to the axle bracket.

Connect the track bar to the frame rail bracket.
9. Remove the supports and lower the vehicle.

Leaf Spring

REMOVAL AND INSTALLATION

Wrangler

1. Raise the vehicle and support it safely.
2. Position a jack under the axle. Raise the axle to relieve the springs of the axle weight.
3. Remove the wheels.
4. Remove the spring U-bolts and tie plates.
5. Remove the bolt attaching the spring rear eye to the shackle.
6. Remove the bolt attaching the spring front eye to the shackle.
7. Remove the spring from its mounting.
   To install:
8. Position the replacement spring front eye in the shackle. Loosely install the attaching bolt. Perform the same procedure for the rear spring eye.
9. Position and align the spring with the axle tube.

10. Lower the rear axle until it is completely supported by the spring.
11. Install the spring bracket, U-bolts and nuts. Torque the nuts to 90 ft. lbs. (122 Nm).
12. Install the wheels.
13. Lower the vehicle and tighten the spring eye bolts to 100 ft. lbs. (136 Nm).

Cherokee 1991-95 and Comanche 1991-92

1. Raise the vehicle and support it safely.
2. Position a jack under the axle. Raise the axle to relieve the springs of the axle weight.
3. Remove the wheels.
4. If equipped, disconnect the sway bar links at the spring plates.
5. Remove the spring U-bolts and spring plates.
6. Remove the bolt attaching the spring rear eye to the shackle.
7. Remove the bolt attaching the spring front eye to the shackle.
8. Lower the axle and remove the spring from its mounting.
   To install:
9. Position the replacement spring front eye in the shackle. Loosely install the attaching bolt. Perform the same procedure for the rear spring eye.
10. Position and align the spring with the axle tube.
11. Install the spring plates, U-bolts and nuts. Torque the nuts to 52 ft. lbs. (70 Nm) on Cherokee and Wagoneer or 65 ft. lbs. (88 Nm) on Comanche.
12. If equipped, connect the sway bar link to the spring bracket and tighten the nut to 70 ft. lbs. (95 Nm).
13. Install the wheels.
14. Lower the vehicle.
15. Torque the spring eye bolts to 111 ft. lbs. on 1985-88 vehicles, 105 ft. lbs. on 1989 vehicles or 65 ft. lbs. on 1990-94 vehicles.

Grand Wagoneer 1991

1. Raise the vehicle and support it safely.
2. If the left side spring is to be replaced, remove the fuel tank skid plate.
3. Position a jack under the axle. Raise the axle to relieve the springs of the axle weight.
4. Disconnect the shock absorber from the axle.
5. Remove the wheels.
6. Remove the spring U-bolts and tie plates.
7. Remove the bolt attaching the spring rear eye to the shackle.

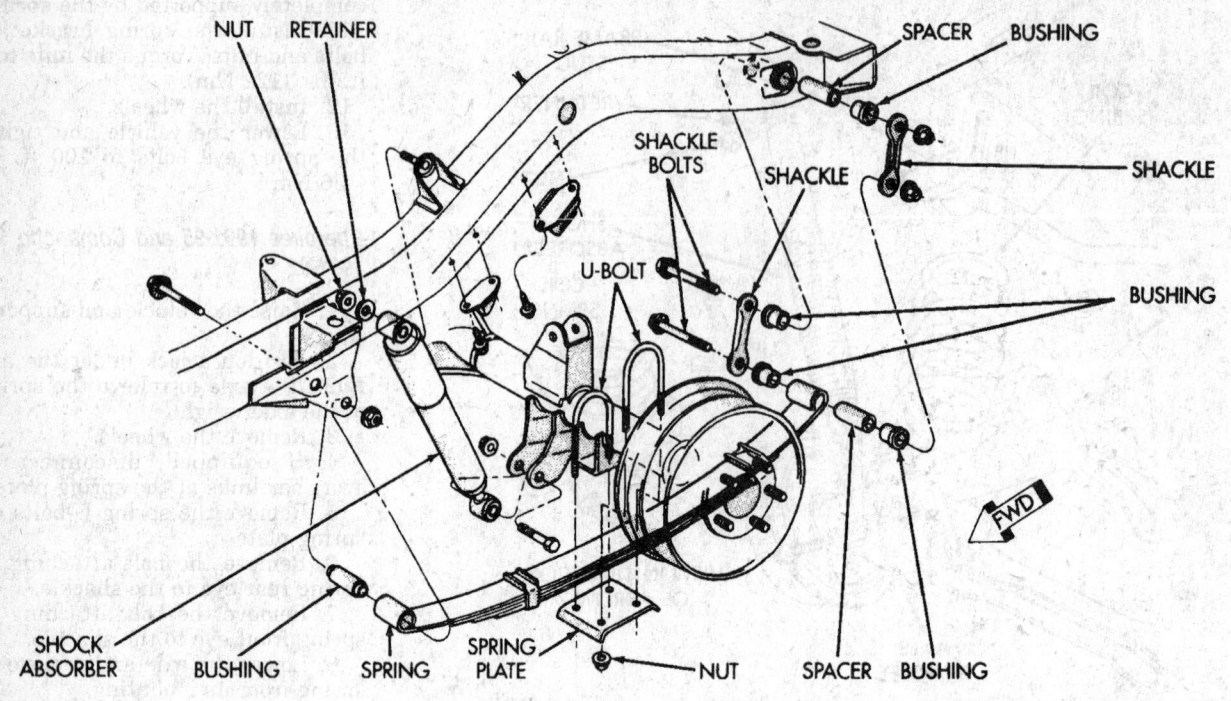

NUT · RETAINER · SPACER · BUSHING · SHACKLE BOLTS · SHACKLE · SHACKLE · U-BOLT · BUSHING · BUSHING · FWD · SHOCK ABSORBER · BUSHING · SPRING · SPRING PLATE · NUT · SPACER · BUSHING

49740

Exploded view of the rear suspension - Wrangler

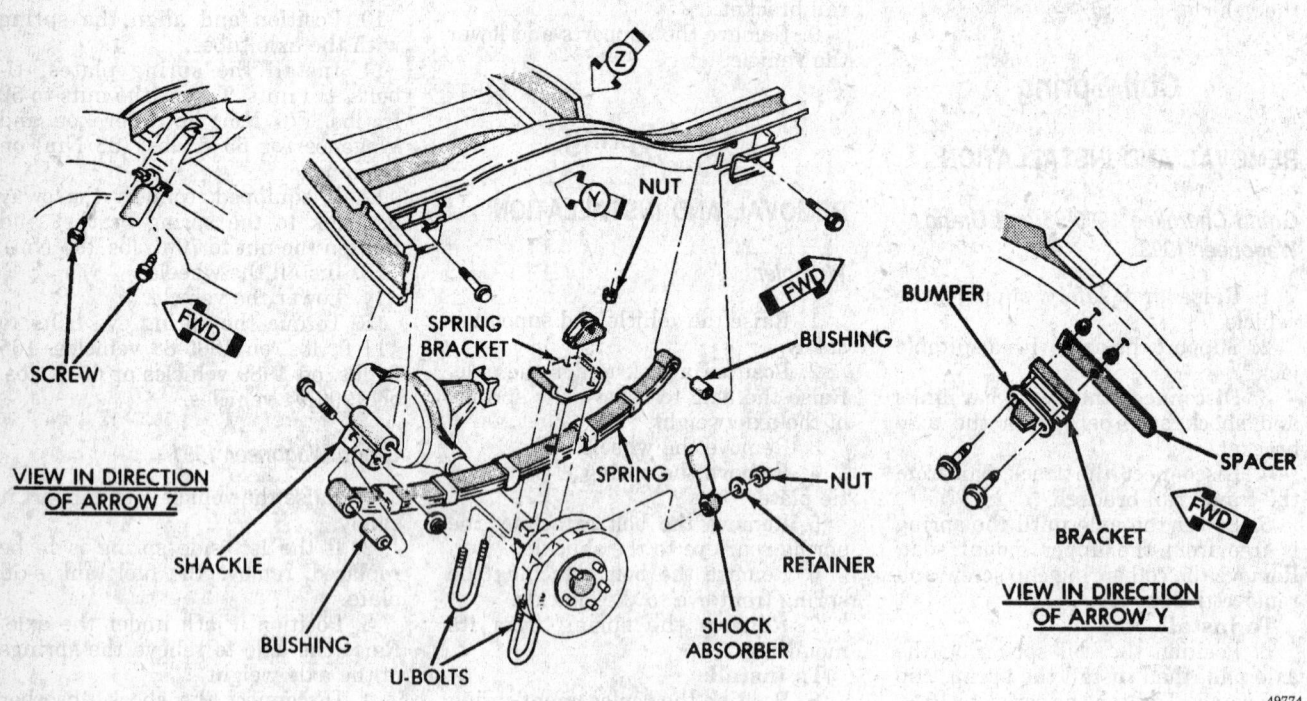

Z · Y · NUT · FWD · SCREW · VIEW IN DIRECTION OF ARROW Z · SHACKLE · BUSHING · U-BOLTS · SPRING BRACKET · SPRING · BUSHING · NUT · RETAINER · SHOCK ABSORBER · BUMPER · SPACER · BRACKET · FWD · VIEW IN DIRECTION OF ARROW Y

49774

Exploded view of the rear suspension - Cherokee and Comanche

8. Remove the bolt attaching the spring front eye to the shackle.

9. Remove the spring from its mounting.

To install:

10. Position the replacement spring front eye in the shackle. Loosely install the attaching bolt. Perform the same procedure for the rear spring eye.

11. Position and align the spring with the axle tube.

12. Install the spring bracket, U-bolts and nuts. Torque the nuts to 100 ft. lbs. (136 Nm).

13. Install the wheels.

14. Connect the shock absorber to the axle.

15. If removed, install the fuel tank skid plate.

16. Lower the vehicle and tighten the spring eye bolts to 100 ft. lbs. (136 Nm).

Lower Control Arm

REMOVAL AND INSTALLATION

Grand Cherokee 1993-95 and Grand Wagoneer 1993

1. Raise and support the vehicle safely.

2. Remove the lower control arm nut and bolt at the axle bracket.

3. Remove the lower control arm nut and bolt at the frame rail.

4. Remove the lower control arm.

To install:

5. Install the lower control arm and tighten the bolts finger-tight.

6. Lower the vehicle.

7. Torque the bolts and nuts to 130 ft. lbs. (177 Nm).

Sway Bar

REMOVAL AND INSTALLATION

Cherokee 1991-95 and Comanche 1991-92

1. Raise and support the vehicle safely.

2. Disconnect the sway bar links from the springs.

3. Disconnect the sway bar from the frame rails.

4. Remove the sway bar.

To install:

5. Install the sway bar and torque the bar link bolts to 55 ft. lbs. (74 Nm).

6. Connect the sway bar to the frame rail and torque the bolts to 40 ft. lbs. (54 Nm).

7. Lower the vehicle.

Grand Cherokee 1993-95 and Grand Wagoneer 1993

1. Raise and support the vehicle safely.

2. Remove 1 wheel.

3. Disconnect the stabilizer bar links from the axle brackets.

4. Lower the exhaust by disconnecting the muffler and tail pipe hangers.

5. Disconnect the stabilizer bar from the links.

6. Disconnect the stabilizer bar clamps from the frame rails. Remove the stabilizer bar.

To install:

7. Position the stabilizer bar on the frame rail and install the clamps and bolts.

8. Ensure the bar is centered with equal spacing on both sides. Torque the bolts to 40 ft. lbs. (54 Nm).

9. Install the links and grommets onto the stabilizer bar and axle brackets. Install the nuts and torque them to 27 ft. lbs. (36 Nm).

10. Connect the muffler and tail pipe to their hangers.

11. Install the wheel and lower the vehicle.

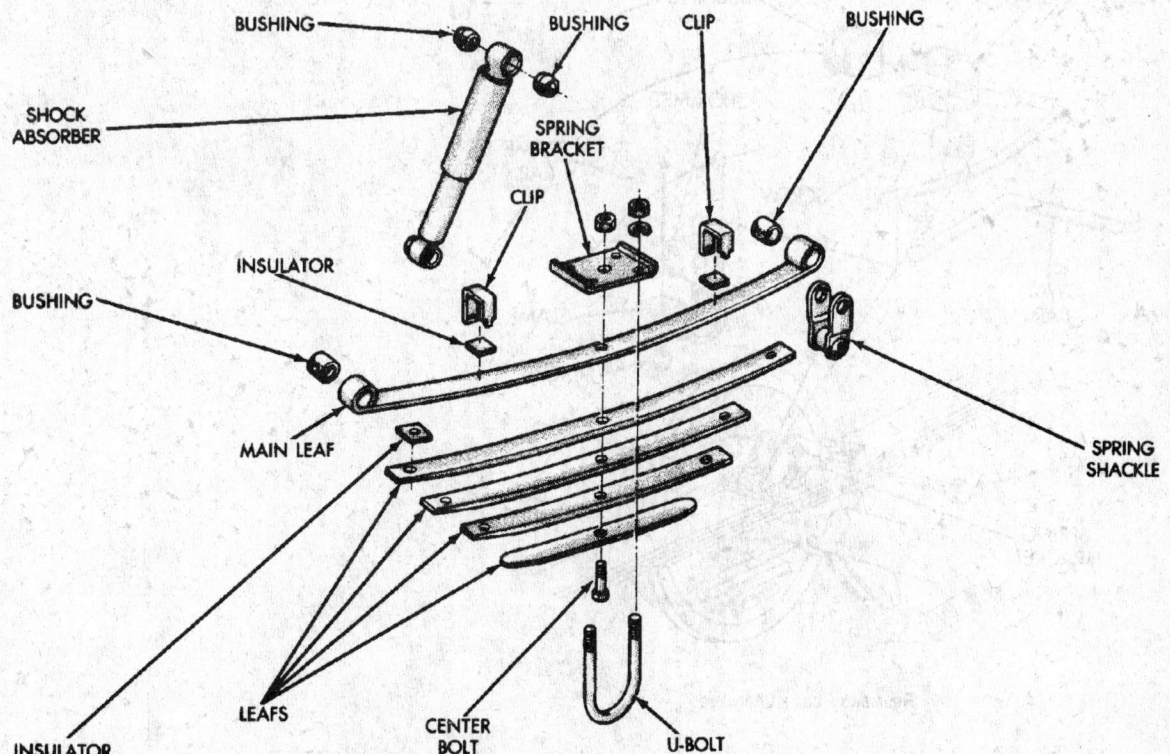

Exploded view of the rear suspension - Grand Wagoneer 1991

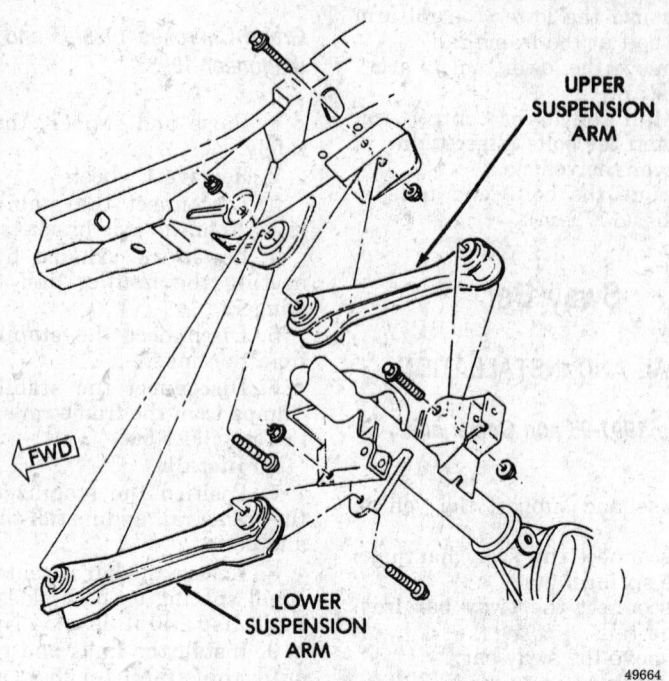

Upper and lower control arms - Grand Cherokee and Grand Wagoneer

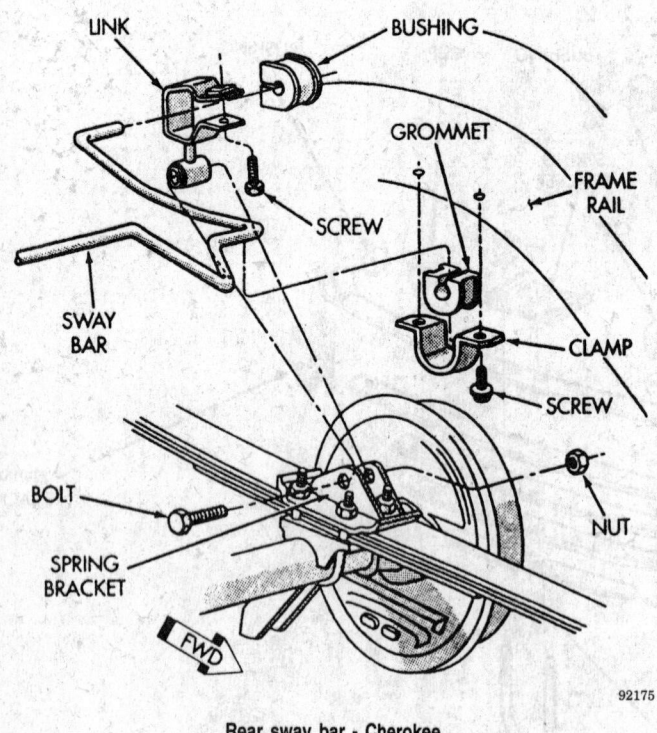

Rear sway bar - Cherokee

SPECIFICATIONS

ENGINE IDENTIFICATION

Year	Model	Engine Displacement Liters (cc)	Engine Series (ID/VIN)	Fuel System	No. of Cylinders	Engine Type
1991	Tracker	1.6 (1590)	U	EFI	4	SOHC
1992	Tracker	1.6 (1590)	U	EFI	4	SOHC
1993	Tracker	1.6 (1590)	U	EFI	4	SOHC
1993	Tracker	1.6 (1590)	U	EFI	4	SOHC
1994-95	Tracker	1.6 (1590)	U	EFI	4	SOHC
	Tracker	1.6 (1590)	U	MFI	4	SOHC

SOHC - Single overhead camshaft

EFI - Electronic fuel injection

MFI - Multiport fuel injection

7911aC02

GENERAL ENGINE SPECIFICATIONS

Year	Engine ID/VIN	Engine Displacement Liters (cc)	Fuel System Type	Net Horsepower @ rpm	Net Torque @ rpm (ft. lbs.)	Bore x Stroke (in.)	Compression Ratio	Oil Pressure @ rpm
1991	U	1.6 (1590)	EFI	80@5400	94@3000	2.95x3.54	8.9:1	51-62@3000
1992	U	1.6 (1590)	EFI	80@5400	94@3000	2.95x3.54	8.9:1	51-62@3000
1993	U	1.6 (1590)	EFI	80@5400	94@3000	2.95x3.54	8.9:1	51-62@3000
1994-95	U	1.6 (1590)	EFI	80@5400	94@3000	2.95x3.54	8.9:1	47-61@3000
	U	1.6 (1590)	1	95@5600	98@4000	2.95x3.54	8.9:1	47-61@3000

1 California and New York models

EFI - Electronic fuel injection

MFI - Multiport fuel injection

7911aC03

GASOLINE ENGINE TUNE-UP SPECIFICATIONS

Year	Engine ID/VIN	Engine Displacement Liters (cc)	Spark Plugs Gap (in.)	Ignition Timing (deg.)		Fuel Pump (psi)	Idle Speed (rpm)		Valve Clearance	
				MT	AT		MT	AT	In.	Ex.
1991	U	1.6 (1590)	0.030	8B	8B	34-40	800	800	0.009- [1] 0.011	0.0102- 0.0118
1992	U	1.6 (1590)	0.030	8B	8B	34-40	800	800	0.009- [1] 0.011	0.0102- 0.0118
1993	U	1.6 (1590)	0.030	8B	8B	34-40	800	800	0.009- [1] 0.011	0.0102- 0.0118
1994-95	U	1.6 (1590)	0.030	8B	8B	34-40	800	800	0.009- [1] 0.011	0.0102- 0.0118
	U	1.6 (1590) [2]	0.030	8B	8B	36-43	800	800	0.005- 0.007	0.005- 0.007

NOTE: The Vehicle Emission Control Information label often reflects specification changes made during production. The label figures must be used if they differ from above

B - Before top dead center

1 Specifications for hot engine.

2 California and New York models

7911aC04

FIRING ORDERS

NOTE: To avoid confusion, always replace spark plug wires one at a time.

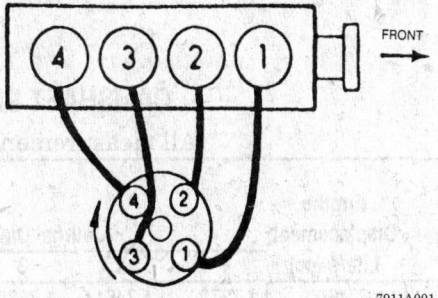

7911A001

1.6L Engine
Engine Firing Order: 1 — 3 — 4 — 2
Distributor Rotation: Clockwise

CAPACITIES

Year	Model	Engine ID/VIN	Engine Displacement Liters (cc)	Engine Crankcas with Filter	Transmission (pts.)			Transfe Case (pts.)	Drive Axle		Fuel Tank (gal.)	Cooling System (qts.)
					4-Spd	5-Spd	Auto.		Front (pts.)	Rear (pts.)		
1991	Tracker	U	1.6 (1590)	4.5	-	3.2	9.8 [1]	4	4.6	2.1	11.1	5.5 [2]
1992	Tracker	U	1.6 (1590)	4.5	-	3.2	9.8 [1]	4	4.6	2.1	11.1	5.5 [2]
1993	Tracker	U	1.6 (1590)	4.5	-	3.2	10.2 [1]	4	3.6	2.2	11.1	5.5 [2]
1994-95	Tracker	U	1.6 (1590)	4.5	-	3.2	10.2 [1]	4	3.6	2.2	11.1	5.5 [2]

1 Automatic transmission - Specification is after complete overhaul. Drain and fill will be less

2 Manual transmission: 5.6 qts.

7911aC06

CAMSHAFT SPECIFICATIONS

All measurements given in inches.

Year	Engine ID/VIN	Engine Displacement Liters (cc)	Journal Diameter					Elevation		Bearing Clearance	Camshaft End Play
			1	2	3	4	5	In.	Ex.		
1991	U	1.6 (1590)	1.7372-1.7381	1.7451-1.7460	1.7530-1.7539	1.7609-1.7618	1.7687-1.7697	1.7687-1.7697	1.4763-1.4724	0.002-0.0036	0.0039
1992	U	1.6 (1590)	1.7372-1.7381	1.7451-1.7460	1.7530-1.7539	1.7609-1.7618	1.7687-1.7697	1.4763-1.4724	1.4749-1.4724	0.0020-0.0036	0.0039
1993	U	1.6 (1590)	1.7372-1.7381	1.7451-1.7460	1.7530-1.7539	1.7609-1.7618	1.7687-1.7697	1.4763-1.4724	1.4749-1.4724	0.0020-0.0036	0.0039
1994-95	U	1.6 (1590)	1.7372-1.7381	1.7451-1.7460	1.7530-1.7539	1.7609-1.7618	1.7687-1.7697	1.4763-1.4724	1.4749-1.4724	0.0020-0.0036	0.0039
	U	1.6 (1590)	1.1000-1.1008	1.1000-1.1008	1.1000-1.1008	1.1000-1.1008	1.1000-1.1008	1.4241-1.4303	1.4314-1.4376	0.0016-0.0032	0.0039

7911aC07

CRANKSHAFT AND CONNECTING ROD SPECIFICATIONS

All measurements are given in inches.

Year	Engine ID/VIN	Engine Displacement Liters (cc)	Crankshaft				Connecting Rod		
			Main Brg. Journal Dia.	Main Brg. Oil Clearance	Shaft End-play	Thrust on No.	Journal Diameter	Oil Clearance	Side Clearance
1991	U	1.6 (1590)	2.0465-2.0472	0.0012-0.0023	0.0100-0.0149	3	1.7316-1.7323	0.0008-0.0031	0.0039-0.0078
1992	U	1.6 (1590)	2.0465-2.0472	0.0012-0.0023	0.0100-0.0149	3	1.7316-1.7323	0.0008-0.0031	0.0039-0.0078
1993	U	1.6 (1590)	2.0465-2.0472	0.0012-0.0023	0.0100-0.0149	3	1.7316-1.7323	0.0008-0.0031	0.0039-0.0078
1994-95	U	1.6 (1590)	2.0465-2.0472	0.0008-0.0016	0.0010-0.0049	3	1.7316-1.7323	0.0008-0.0019	0.0039-0.0078

7911aC08

VALVE SPECIFICATIONS

Year	Engine ID/VIN	Engine Displacement Liters (cc)	Seat Angle (deg.)	Face Angle (deg.)	Spring Test Pressure (lbs. @ in.)	Spring Installed Height (in.)	Stem-to-Guide Clearance (in.)		Stem Diameter (in.)	
							Intake	Exhaust	Intake	Exhaust
1991	U	1.6 (1590)	45	45	54.7-64.3@ 1.63	1.91	0.0008-0.0019	0.0014-0.0025	0.2742-0.2748	0.2737-0.2742
1992	U	1.6 (1590)	45	45	54.7-64.3@ 1.63	1.91	0.0008-0.0019	0.0014-0.0025	0.2742-0.2748	0.2737-0.2742
1993	U	1.6 (1590)	45	45	54.7-64.3@ 1.63	1.91	0.0008-0.0019	0.0014-0.0025	0.2742-0.2748	0.2737-0.2742
1994-95	U	1.6 (1590)	45	45	50.2-64.3@ 1.63	1.63	0.0008-0.0019	0.0014-0.0025	0.2742-0.2748	0.2737-0.2742
	U [1]	1.6 (1590)	45	45	23.6-27.5@ 1.24	1.24	0.0008-0.0018	0.0018-0.0028	0.2152-0.2157	0.2142-0.2148

[1] California and New York models

7911aC09

PISTON AND RING SPECIFICATIONS
All measurements are given in inches.

| Year | Engine ID/VIN | Engine Displacement Liters (cc) | Piston Clearance | Ring Gap | | | Ring Side Clearance | | |
				Top Compression	Bottom Compression	Oil Control	Top Compression	Bottom Compression	Oil Control
1991	U	1.6 (1590)	0.0008-0.0015	0.0079-0.0129	0.0079-0.0137	0.0079-0.0275	0.0012-0.0027	0.0008-0.0023	NA
1992	U	1.6 (1590)	0.0008-0.0015	0.0079-0.0137	0.0079-0.0137	0.0079-0.0275	0.0012-0.0027	0.0008-0.0023	NA
1993	U	1.6 (1590)	0.0008-0.0015	0.0079-0.0137	0.0079-0.0137	0.0079-0.0275	0.0012-0.0027	0.0008-0.0023	NA
1994-95	U	1.6 (1590)	0.0008-0.0015	0.0079-0.0129	0.0079-0.0137	0.0079-0.0275	0.0012-0.0027	0.0008-0.0023	NA

7911aC10

TORQUE SPECIFICATIONS
All readings in ft. lbs.

| Year | Engine ID/VIN | Engine Displacement Liters (cc) | Cylinder Head Bolts | Main Bearing Bolts | Rod Bearing Bolts | Crankshaft Damper Bolts | Flywheel Bolts | Manifold | | Spark Plugs | Lug Nut |
								Intake	Exhaust		
1991	U	1.6 (1590)	54	39	26	81 [1]	58	17	17	18	87
1992	U	1.6 (1590)	54	39	26	81 [1]	58	17	17	18	87
1993	U	1.6 (1590)	54	39	26	81 [1]	58	17	17	18	87
1994-95	U	1.6 (1590)	52 [2]	40	26	81 [1]	58	17	17	18	87

1 Crankshaft timing sprocket belt

2 Tighten in three steps

7911aC11

BRAKE SPECIFICATIONS
All measurements in inches unless noted

Year	Model		Master Cylinder Bore	Brake Disc			Brake Drum Diameter			Minimum Lining Thickness	
				Original Thickness	Minimum Thickness	Maximum Runout	Original Inside Diameter	Max. Wear Limit	Maximum Machine Diameter	Front	Rear
1991	Tracker	1	NA	0.394	0.315	0.006	8.66	8.74	8.74	0.236	0.210
1992	Tracker	1	NA	0.394	0.315	0.006	8.66	8.74	8.74	0.236	0.210
1993	Tracker	1	NA	0.394	0.315	0.006	8.66	8.74	8.74	0.236	0.210
1994-95	Tracker	1	NA	0.394	0.315	0.006	8.66	8.74	8.74	0.236	0.210

1 Minimum lining thickness includes pad/shoe backing

7911aC12

WHEEL ALIGNMENT

Year	Model		Caster		Camber		Toe-in (in.)	Steering Axis Inclination (deg.)
			Range (deg.)	Preferred Setting (deg.)	Range (deg.)	Preferred Setting (deg.)		
1991	Tracker		1/2P-2 1/2P	1 1/2P	1/2N-1 1/2P	1/2P	3/32-1/4	31
1992	Tracker		1 1/2P-2 1/2P	2P	1/2N-1 1/2P	1/2P	5/64-1/4	31
1993	Tracker		1 1/2P	2P	1/2N-1 1/2P	1/2P	5/64-1/4	31
1994-95	Tracker		1/2P-2 1/2P	1 1/2P	1/2N-1 1/2P	1/2P	5/64-1/4	31

7911aC13

ENGINE ELECTRICAL

NOTE: Disconnecting the battery cable on some vehicles may interfere with the functions of the on board computer systems and may require the computer to undergo a relearning process, once the negative battery cable is disconnected.

Distributor

REMOVAL

1. Disconnect the negative battery terminal from the battery.
2. Disconnect vacuum hose and electrical connections.

NOTE: Do not bend or twist the spark plug wires to avoid internal damage. Grip the the wire boot when removing or installing the wires.

3. Remove the distributor cap.
4. Mark the rotor position on the distributor housing and the distributor housing position on the engine.
5. Remove the distributor flange bolt and remove the distributor.

NOTE: Do not crank the engine with the distributor removed.

INSTALLATION

Timing Not Disturbed

1. Install the distributor and align the marks on the distributor housing and on the engine.
2. Tighten the flange bolt to 11 ft. lbs. and install the distributor cap.
3. Connect the electrical connections and the vacuum hoses.
4. Connect the negative battery cable and set the timing to specification.

Timing Disturbed

1. Rotate the crankshaft in a clockwise position until the specified timing mark on the flywheel aligns with the timing matchmark on the engine.

NOTE: After aligning the 2 marks, remove the cylinder head cover to visually check so the rocker arms are not riding on the camshaft cams at No. 1 cylinder. If the arms are found to be riding

on the cams, turn the crankshaft 360 degrees to realign the 2 marks.

2. Position the rotor to the No. 1 cylinder position and install the distributor tightening the flange bolt to 11 ft. lbs.
3. Connect the vacuum and electrical connections.
4. Connect the negative battery cable and set the timing to specification.

Ignition Timing

ADJUSTMENT

1. Start the engine and warm to normal operating temperature. Prior to any adjustment, be sure all the electrical accessories are OFF.
2. After warming, make sure the idle speed is 800 rpm.
3. Remove cap from monitor coupler next to battery and connect terminals C and D with a jumper wire. Connect the timing light to the No. 1 cylinder spark plug wire.
4. With the engine running at the specified idle speed, direct the timing light to the crankshaft pulley. If the specified timing mark on the timing tab is aligned with the timing notch on the crankshaft pulley the ignition is properly timed.
5. If the timing is out of adjustment, loosen the distributor flange bolt and turn the distributor housing to advance or retard the timing. Turn the distributor counterclockwise to advance the timing and clockwise to retard the timing.
6. After the adjustment, tighten the flange bolt and recheck the timing.

Alternator

BELT TENSION ADJUSTMENT

1. Disconnect the negative battery cable.
2. Loosen the alternator bolt and pivot bolts.
3. Adjust the belt tension to 0.24–0.32 in. of deflection using a belt tension guage.

REMOVAL AND INSTALLATION

1. Disconnect the negative battery cable.
2. Disconnect the wire coupler and white lead wire from the alternator.
3. Remove the charcoal cannister mounting bracket, if necessary.

4. Remove the alternator mounting bolt and alternator drive belt adjusting bolt.
5. Remove the alternator from the vehicle.

To install:

6. Install the alternator and tighten the mounting bolt and drive belt adjusting bolt. Adjust the drive belt tension.
7. Install the shroud under the radiator and connect the wire coupler to the alternator.
8. Connect the negative battery cable.

Starter

REMOVAL AND INSTALLATION

1. Disconnect the negative battery cable.
2. Raise and support the vehicle safely.
3. Disconnect the lead wire and battery cable from the starter motor.
4. Support the starter and remove the 2 mounting bolts.
5. Remove the starter.

To install:

6. Install the starter and tighten the 2 mounting bolts to 22 ft. lbs.
7. Connect the lead wire to the starter.
8. Lower the vehicle and connect the negative battery cable.

CHASSIS ELECTRICAL

Heater Blower Motor

REMOVAL AND INSTALLATION

1. Disconnect the negative battery cable.
2. Remove the glove box assembly.
3. Disconnect the blower motor and resistor wire connectors.
4. Disconnect the fresh air control cable from the blower motor case.
5. Loosen but do not remove the blower housing fastener bolts.
6. Remove the 3 blower motor mounting screws.
7. Remove the blower motor.

To install:

8. Install the blower motor and secure with the 3 screws.
9. Tighten the blower housing fastener bolts.

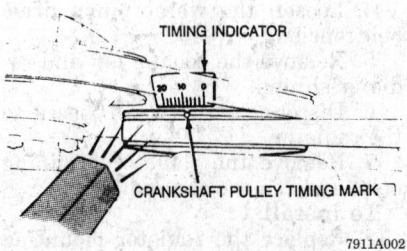

TIMING INDICATOR

CRANKSHAFT PULLEY TIMING MARK

7911A002

Timing mark location — Tracker

10. Connect the fresh air control cable to the blower motor case.
11. Connect the electrical connectors and install the glove box assemb
12. Connect the negative battery cable.

Windshield Wiper Motor

REMOVAL AND INSTALLATION

1. Disconnect the negative battery cable.
2. Remove the right and left cowl grilles.
3. Remove the wiper linkage to wiper mounting nut.
4. Disconnect the wire connector from the wiper motor.
5. Remove the 4 wiper motor mounting bolts and remove the wiper motor.
To install:
6. Install the wiper motor and install the 4 mounting bolts.
7. Connect the electrical connector and connect the wiper linkage.
8. Install the cowl grilles and connect the negative battery cable.

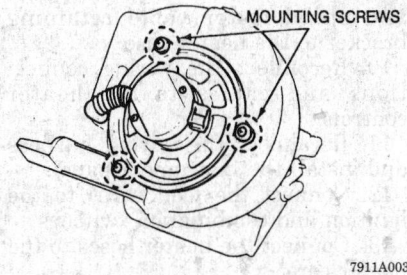

MOUNTING SCREWS

7911A003

Heater blower motor mounting screws

Instrument Cluster

REMOVAL AND INSTALLATION

1. Disconnect the negative battery cable.
2. Remove the 4 screws from the front shroud.
3. Remove the front shroud.
4. Remove the 4 cluster mounting screws and slide the cluster outwards.
5. Disconnect the lead wires and speedometer cable from the speedometer.
6. Remove the instrument cluster assembly.
To install:
7. Install the lead wires and speedometer cable.
8. Install the 4 cluster screws and tighten.
9. Install the front shroud and install the 2 upper and 2 lower screws.
10. Connect the negative battery cable.

Radio

REMOVAL AND INSTALLATION

1. Disconnect the negative battery cable.
2. Remove the ashtray and the ashtray guide assembly.
3. Remove the 1 bolt that secures radio to the bracket.
4. Disconnect the electrical connections and antenna lead from the radio.
5. Remove the radio from the vehicle.
To install:
6. Install the radio and connect the antenna wire and electrical connector.
7. Install the 1 bolt between the radio and the bracket assembly.
8. Install the ashtray and guide assembly and connect the negative battery cable.

Combination Switch

The combination switch incorporates the turn signal, windshield wiper, dimmer and headlight switches into one switch.

REMOVAL AND INSTALLATION

1. Disconnect the negative battery cable.

2. Remove the horn button and the steering wheel retaining nut and remove the steering wheel.
3. Remove the upper and lower column cover screws and remove the covers.
4. Disconnect the lead wires from the combination switch at the connector.
5. Remove the combination switch assembly screws.
6. Remove the combination switch from the steering column.
To install:
7. Install the combination switch assembly and install the screws.
8. Connect the electrical connector.
9. Install the upper and lower column covers.
10. Install the steering wheel and tighten the shaft nut to 24 ft. lbs.
11. Install the horn button and connect the negative battery cable.

Ignition Switch

REMOVAL AND INSTALLATION

1. Disconnect the negative battery cable.
2. Remove the steering wheel from the vehicle.
3. Disconnect the wire connector at the ignition switch.
4. With the ignition switch in the **OFF** position, remove the mounting bolts and remove the switch.
To install:
5. Install the ignition switch and replace the mounting bolts.
6. Connect the electrical connector to the ignition switch.
7. Install the steering wheel and tighten the shaft nut to 24 ft. lbs.
8. Install the horn button and connect the negative battery cable.

Ignition Lock

REMOVAL AND INSTALLATION

1. Disconnect the negative battery cable.
2. Remove the steering wheel.
3. Disconnect the lead wires to the ignition and combination switches.
4. Disconnect the steering joint by removing the joint bolt.
5. Remove the steering column fastening bolts.
6. Remove the steering column assembly.
7. Using a center punch, loosen and remove the steering lock mounting bolts.

8. Turn the key to **ACC** or **ON** position and remove the steering lock assembly from the steering column.

To install:

9. To install the switch, position the oblong hole of the steering shaft in the center of the hole in the steering column.

10. Turn the ignition key to the **ACC** or **ON** position and install the steering lock assembly onto the column.

11. Turn the ignition key to the **LOCK** position and remove it.

12. Align the hub on the lock with oblong hole of the steering shaft.

13. Rotate the shaft to assure that the steering shaft is locked.

14. Tighten the 2 new steering lock mounting bolts.

15. Turn the ignition key to the **ACC** or **ON** position and check to be sure the steering shaft rotates smoothly. Also check the lock mechanism.

16. Install the steering column assembly by attaching the steering joint and joint bolt.

17. Install the steering column fastening bolts.

18. Reconnect the lead wires to the ignition and combination switch

19. Replace the steering wheel and negative battery cable.

Stoplight Switch

REMOVAL AND INSTALLATION

1. Disconnect the negative battery cable. Remove the under dash trim panel, if equipped.

2. Push the brake pedal down and remove the stoplight switch locknut.

3. Disconnect the wire connector at the switch.

4. Remove the switch from the bracket.

To install:

5. Install the new switch on the bracket.

6. Adjust the switch so there is 0.02–0.04 in. clearance between the switch and the brake pedal.

7. Replace the locknut and tighten to 7.5–10.5 ft. lbs.

8. Replace the negative battery cable.

Clutch Switch

REMOVAL AND INSTALLATION

1. Disconnect the negative battery cable. Remove the under dash trim panel, if equipped.

2. Disconnect the clutch/start switch connector beside the clutch pedal bracket.

3. Remove the locknut and clutch/start switch.

To install:

4. Install the clutch/start switch and tighten the locknut.

5. Connect the electrical connector, connect the negative battery cable and install the under dash trim panel.

Fuses and Relays

Location

The fuse box and turn signal relay are located under the driver side of the dashboard.

Computer

Location

The Electronic Control Module (ECM) computer is located under the dash on the driver side of the vehicle.

Cruise Control

ADJUSTMENT

1. Remove the top cover from the actuator assembly.

2. Loosen the lock nuts on the cable.

3. Adjust the cable so the play range is between 0.04–0.08 in. when the actuator lever is fully closed.

4. Tighten the lock nut to 4 ft. lbs. and replace the top cover.

Flashers

Location

The flashers are located under the driver side of the dashboard.

ENGINE COOLING

Radiator

REMOVAL AND INSTALLATION

1. Disconnect the negative battery cable.

2. Drain the cooling system.

3. Disconnect and plug the transmission cooler lines, if equipped.

4. Loosen the water pump drive belt tension.

5. Remove the cooling fan and radiator shroud.

6. Disconnect the water hoses to the radiator.

7. Remove the radiator retaining bolts and remove the radiator.

To install:

8. Replace the radiator mounting bolts and install the radiator.

9. Reconnect the water hoses to the radiator.

10. Install the cooling fan and radiator shro

11. Tighten the water pump drive belt to the proper tension.\sc19\®MDFL\sc158\®MDNM\sc158\

12. Reconnect the transmission cooler lines, if necessary to the radiator.

13. Refill the cooling system.

14. Reconnect the negative battery cable.

Heater Core

REMOVAL AND INSTALLATION

1. Disconnect the negative battery cable. Drain the cooling system.

2. Disconnect the 2 heater hoses from the heater case.

3. Remove the steering wheel. Disconnect the lead wires from the ignition and combination switches.

4. Disconnect the wiring harness and remove the instrument panel.

5. Disconnect all the wiring connections and cables from the heater controls.

6. Disconnect the defroster duct and speedometer cable retaining bracket from the heater case.

7. Remove the heater assembly and remove the heater core.

To install:

8. Install the heater core and install the heater assembly.

9. Reconnect the defroster duct and speedometer cable retaining bracket to the heater case.

10. Reconnect the wiring connections and cables to the heater controls.

11. Reconnect the wiring harness and install the instrument panel.

12. Connect the lead wires to the ignition and combination switches.

13. Connect the heater hoses to the heater core.

14. Refill the cooling system with the proper coolant and replace the negative battery cable.

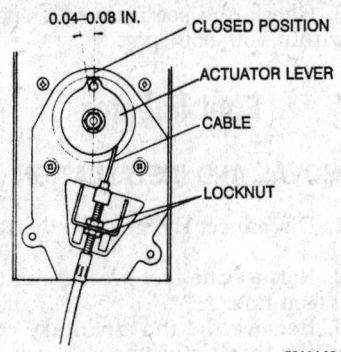

0.04–0.08 IN.

CLOSED POSITION

ACTUATOR LEVER

CABLE

LOCKNUT

7911A004

Cruise control adjustment

Water Pump

REMOVAL AND INSTALLATION

1. Disconnect the negative battery cable.
2. Drain the cooling system.
3. Loosen the drive belt tension and remove the drive belt. If equipped, loosen the air conditioning belt tension and remove the air conditioning belt.
4. Remove the radiator fan shroud mounting bolts and radiator fan mounting bolts. Remove the radiator shroud, fan and water pump pulley from the vehicle.
5. Remove the crankshaft pulley bolts and remove the crankshaft pulley.

NOTE: The crankshaft pulley bolt can be removed without removing the center crankshaft bolt.

6. Remove the timing belt cover mounting bolts and remove the cover.
7. Loosen the timing belt tensioner adjusting bolt and pivot nut. Hold the tensioner to loosen the timing belt and remove the belt from the camshaft pulley.
8. Remove the timing belt tensioner mounting bolts and remove the tensioner plate and spring.
9. Remove the water pump mounting bolts and remove the water pump assembly.
10. Remove the dipstick tube and the alternator bracket from the vehicle.
   To install:
11. Clean and inspect the surface of the engine before installation.
12. Using a new gasket, install the new water pump on the engine. Torque the mounting bolts to 7.5–9.0 ft. lbs.

13. Install the dipstick tube and the alternator bracket to the vehicle.
14. Install the rubber seals between the water pump to cylinder head and water pump to oil pump.
15. Install the timing belt tensioner plate, tensioner and spring.
16. Align the marks on the timing belt and the camshaft sprocket. Install the timing belt in the same position on the camshaft sprocket as when removed.
17. Adjust the timing belt to be free of any slack. Torque the tensioner bolts to 7.5–9.0 ft. lbs.
18. Install the crankshaft and water pump pulleys. Torque the crankshaft and water pulley bolts to 7.5–9.0 ft. lbs.
19. Install the timing belt cover, cooling fan/clutch, shroud and drive belt. Adjust the belt tension.
20. As required, adjust the intake and exhaust valve lash.
21. Refill the cooling system with the proper coolant and replace the negative battery cable.

Thermostat

REMOVAL AND INSTALLATION

1. Disconnect the negative battery cable.
2. Drain the cooling system.
3. Disconnect the thermostat cap from the intake manifold and remove the thermostat.
   To install:
4. Clean and inspect the surfaces of the housing and the engine.
5. Install the new thermostat with the spring facing towards the engine.
6. Install a new gasket and the thermostat cap to the intake manifold.
7. Refill the cooling system with the proper coolant and replace the negative battery cable.

Cooling System Bleeding

In the top portion of the thermostat, an air bleed valve is provided. This valve is for venting air, if any, that is accumulated in the system.

FUEL SYSTEM

Fuel System Service Precaution

When working with the fuel system, certain precautions should be taken; always work in a well ventilated area, keep a dry chemical (Class B) fire extinguisher near the work area. Always disconnect the negative battery cable and do not make any repairs to the fuel system until all the necessary steps for repair have been reviewed.

Relieving Fuel System Pressure

The fuel pressure must be relieved before performing any service on the fuel system.
1. Disconnect the negative battery cable.
2. Remove the fuel filler cap from the fuel filler neck to release the fuel vapor pressure in the fuel tank. Reinstall the fuel cap.
3. Raise and support the vehicle safely.
4. Place an appropriate container under the fuel filter.
5. Cover the plug bolt on the fuel filter inlet union bolt with a rag and loosen the plug bolt slowly to release the fuel pressure gradually.
6. When the pressure has been released, tighten the plug bolt to 7.5 ft. lbs. so the fuel does not leak.
7. Lower the vehicle.
8. Reconnect the negative battery cable.

Fuel Filter

REMOVAL AND INSTALLATION

1. Disconnect the negative battery cable.
2. Remove the fuel filler cap to release the fuel vapor pressure in the fuel tank. After releasing pressure, reinstall filler cap.
3. Raise and support the vehicle safely.
4. Release the fuel pressure.
5. Disconnect the inlet and outlet hoses from the fuel filter.
6. Remove the fuel filter from the chassis frame.

To install:

7. Install the new fuel filter to the frame and connect the inlet and outlet hoses.

8. Lower the vehicle and connect the negative battery terminal.

Electric Fuel Pump

Pressure Testing

1. Remove the fuel filler cap to release the fuel vapor pressure in the fuel tank. Reinstall the cap.

2. Raise and support the vehicle safely.

3. Release the fuel pressure in the fuel feed line.

4. Remove the plug bolt on the fuel filter union bolt and connect the proper fuel pressure gauge to the fuel filter inlet union bolt.

NOTE: If the pressure in the fuel tank is not released prior to system service, the fuel in the fuel tank may be forced out through the fuel hoses during disconnection.

5. Start the engine and warm to the proper operating temperature.

6. Measure the fuel pressure. At a specified idle speed or with the fuel pump operating and the engine stopped the fuel pressure should read: 34.1–39.8 psi. Within 1 minute after the fuel pump has stopped the fuel pressure should read: 21.3 psi.

7. Release the fuel pressure and remove the fuel pressure gauge.

8. Install the plug bolt to the fuel filter inlet bolt. Use a new gasket.

9. Start the engine and check for leaks.

10. Lower the vehicle.

REMOVAL AND INSTALLATION

NOTE: The fuel pump is located in the fuel tank. The fuel pump and the fuel gauge sending unit are located on the upper part of the fuel tank.

1. Disconnect the negative battery cable.

2. Remove the rear bumper assembly.

3. Disconnect the fuel gauge sending unit and fuel pump electrical connectors from the fuel tank.

4. Relieve the fuel system pressure.

5. Disconnect the fuel tank filler hose cover, filler hose and fuel tank inlet valve (breather hose).

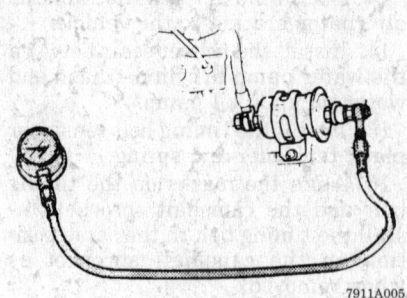

Checking fuel pump pressure

6. Disconnect the inlet pipe from the fuel filter and remove the fuel vapor and return hoses.

7. If necessary, drain the fuel from the tank using a hand operating pump.

8. Disconnect the fuel tank protector and remove the fuel tank and cover from the vehicle.

9. Remove the fuel pump from the fuel tank.

To install:

10. Install the fuel pump to the fuel tank using a new gasket.

11. Install the fuel tank and cover to the vehicle and connect the fuel tank protector.

12. Replace the fuel vapor and return hoses and connect the inlet pipe to the fuel filter.

13. Reconnect the fuel tank inlet valve (breather hose), filler hose and fuel tank filler hose cover.

14. Connect the fuel pump and fuel gauge sending sending unit electrical connectors to the fuel tank.

15. Replace the bumper assembly and reconnect the negative battery cable.

16. If necessary, refill the fuel tank with any fuel that had been drained.

17. Start the engine and check for leaks when finished.

Fuel Injection

IDLE SPEED ADJUSTMENT

NOTE: Before starting the engine, place the gear shift lever in neutral for manual transmission or P for automatic transmission. Set the parking brake and block the drive wheels.

1. Warm the engine to the normal operating temperature.

2. Race the engine until the engine speed exceeds 1500 rpm and than let it slow down to idle speed.

3. Check and see if the idle speed is within 750–850 rpm.

Fuel Injector

REMOVAL AND INSTALLATION

1. Disconnect the negative battery cable.

2. Release the fuel pressure in the fuel feed line.

3. Remove the the air intake case from the throttle body.

4. Remove the fuel feed pipe clamp from the intake manifold and disconnect the fuel feed pipe from the throttle body.

5. Remove the injector cover.

6. Disconnect the injector coupler, release its wire harness from the clamp and remove its grommet from the throttle body.

7. Place a cloth over the injector and a hand on top of it. Using an air gun, blow low pressure compressed air into the fuel inlet port of the throttle body, and the injector can be removed.

NOTE: Be precise about the pressure of the compressed air. Using excessively high pressure may force the injector to jump out and may cause damage, not only to the injector itself but also to other parts.

To install:

8. Apply thin coating of gasoline to O-rings and then install the fuel injector to the throttle body.

9. Connect fuel injector wire connector and install injector cover.

10. Connect fuel feed pipe to throttle body.

11. Connect the negative battery cable and turn the ignition **ON** for 3 seconds and then **OFF** until fuel pressure is felt at the return hose.

12. Install air intake case assembly.

13. Secure all wires and check for any leaks in the system.

EMISSION CONTROLS

Emission Warning Lamps

The CHECK ENGINE light automatically comes on at 50,000, 80,000 and 100,000 miles.

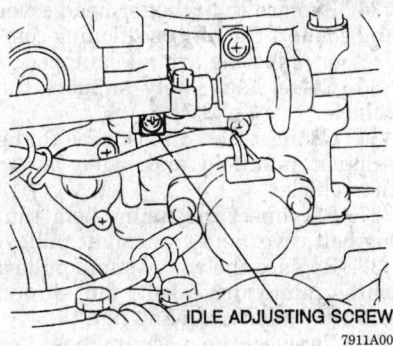

IDLE ADJUSTING SCREW

7911A006

Idle adjustment

RESETTING

The lamp reset switch is located on the left side of the dashboard, mounted on the steering column support. The lamp can be reset by moving the switch upwards and then downwards. If the switch remains on after being reset, check the system.

ENGINE MECHANICAL

NOTE: Disconnecting the battery cable on some vehicles may interfere with the functions of the on board computer systems and may require the computer to undergo a relearning process, once the negative battery cable is disconnected.

Engine

REMOVAL AND INSTALLATION

1. Disconnect the negative and positive cables at the battery.

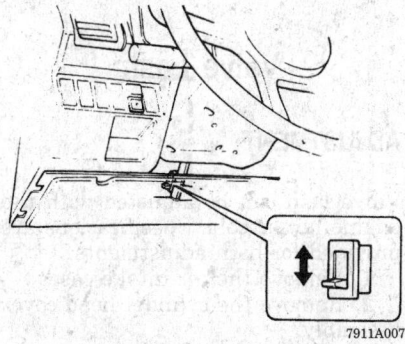

7911A007

Resetting the check engine light

2. Remove the hood from the vehicle and drain the cooling system.

3. Remove the radiator reservoir tank, fan shroud, cooling fan and radiator.

4. Properly discharge the air conditioning system and remove the air conditioning condenser, if equipped.

5. Remove the air cleaner outlet hose.

6. Disconnect the accelerator and the automatic transmission kickdown cable from the throttle body, if equipped.

7. Disconnect and tag the throttle opener VSV and EGR VSV wires at the coupler.

8. Disconnect and tag the water temperature, oil pressure, air temperature and ground cable wires at the intake manifold.

9. Disconnect and tag the injector, throttle position sensor and idle speed control solenoid valve wires at their couplers, if equipped.

10. Disconnect the PTC heater wires at the coupler for automatic transmission vehicle, if equipped.

11. Disconnect and tag the wires at the starter and alternator.

12. Disconnect and tag the oxygen sensor and distributor wires at their couplers. Remove the coil wire.

13. Remove the starter motor and disconnect the ground wires from the distributor assembly.

14. Remove the fuel tank filler cap to relieve the pressure. Reinstall the cap.

15. Relieve the fuel pressure in the fuel feed line. Disconnect and tag the fuel feed and return hoses.

16. Remove the gear shift lever mounting bolts and remove the shifter.

17. Disconnect the canister purge hose and remove the pressure sensor hose from the fuel filter, if equipped.

18. Disconnect the brake booster hose from the intake manifold.

19. Remove the vacuum hose for the automatic transmission from the intake manifold, if equipped.

20. Disconnect the heater hoses from the heater core outlet pipe and the intake manifold.

21. Raise and safely support the vehicle.

22. Drain the engine oil and remove the exhaust pipe from the exhaust manifold and muffler.

23. Disconnect the clutch cable, if equipped with manual transmission.

24. Drain the automatic transmission fluid, if equipped.

25. Disconnect the clutch (torque converter) housing lower plate.

26. Remove the lock drive plate, using the special tool, if equipped with automatic transmission.

27. Lower the vehicle.

28. Remove the nuts and bolts fastening the cylinder block and transmission.

29. Support the transmission, using a stand or jack.

30. Support the engine from the top using a chain type hoist or equivalent means.

31. Remove the engine mounts with the chassis side mounting brackets.

NOTE: Before lifting the engine, check to ensure all the hoses, wires and cables are disconnected from the engine.

32. Remove the engine assembly from the chassis and transmission by sliding towards the front side and carefully hoist the engine.

To install:

33. Install the engine into the engine compartment and connect to the transmission.

34. Replace the engine mounts with the chassis side mounting brackets. Tighten to 41 ft. lbs.

35. Remove the lifting device.

36. Install the nuts and bolts fastening the cylinder block and transmission.

37. Raise and safely support the vehicle.

38. Replace the lock driveplate, if equipped.

39. Connect the clutch (torque converter) housing lower plate.

40. Connect the automatic transmission lines, if equipped.

41. Reconnect the clutch cable to the bracket, if equipped with manual transmission.

42. Replace the exhaust pipe to the exhaust manifold and muffler.

43. Lower the vehicle.

44. Reconnect the heater hoses to the heater core outlet pipe and the intake manifold.

45. Install the vacuum hose for the automatic transmission to the intake manifold, if equipped.

46. Reconnect the brake booster hose to the intake manifold.

47. Replace the pressure sensor hose to the fuel filter and connect the canister purge hose, if equipped.

48. Replace the fuel return and feed hoses.

49. Install the starter motor and connect the ground wire to the distributor.

50. Replace the coil wire and connect the oxygen sensor and distributor wires to their couplers.

51. Reconnect the wires at the alternator and starter motor.

52. Replace the PTC heater wires at the coupler, if equipped with automatic transmission.

53. Install the idle speed control solenoid valve, throttle position sensor and injector wires at their couplers, if equipped.

54. Connect the ground cable, air temperature, oil pressure and water temperature wires to the intake manifold.

55. Reconnect the throttle opener VSR and EGR VSR wires to their couplers, if equipped.

56. Connect the accelerator cable and the automatic transmission kickdown cable, if equipped, to the throttle body.

57. Replace the air cleaner outlet hose. Refill the engine oil and the transmission oil to the proper lever.

58. Replace the air conditioning condenser and properly recharge the air conditioning system, if equipped.

59. Replace the radiator, cooling fan, fan shroud and radiator reservoir tank.

60. Replace the hood and refill the cooling system with the proper coolant.

61. Reconnect the positive and negative battery cables.

62. Adjust the accelerator, clutch and the automatic transmission kickdown cable, if equipped.

63. Before starting the engine, check to see if all the parts disassembled are back in place securely.

64. Start the engine and check the timing. Check for any oil leaks.

Cylinder Head

REMOVAL AND INSTALLATION

1. Disconnect the negative battery cable. Drain the cooling system.

2. Disconnect the air intake case.

3. Disconnect the brake booster hose from the intake manifold and the air valve water hose from the throttle body.

4. Disconnect the accelerator and the automatic transmission kickdown cable, if equipped, from the throttle body.

5. Disconnect and tag the throttle opener VSV and EGR VSV wires at the coupler.

6. Disconnect and tag the water temperature, oil pressure, air temperature sensor and ground cable wires at the intake manifold.

7. Disconnect and tag the injector, throttle position sensor and idle speed control solenoid valve wires at their couplers, if equipped.

8. Disconnect the PTC heater wires at the coupler, if equipped with automatic transmission.

9. Disconnect and tag the oxygen sensor and distributor wires at their couplers. Remove the coil wire.

10. Disconnect the ground wires from the distributor assembly.

11. Remove the fuel tank filler cap to relieve the pressure. Reinstall the cap.

12. Release the fuel pressure in the fuel feed line.

13. Disconnect the fuel feed and return hoses.

14. Disconnect the canister purge hose and remove the pressure sensor hose from the fuel filter, if equipped.

15. Remove the vacuum hose for the automatic transmission from the intake manifold, if equipped.

16. Disconnect the radiator cooling fan, fan shroud, water pump drive belt and water pump pulley.

17. Disconnect the crankshaft pulley, timing belt cover and the timing belt.

18. Raise and safely support the vehicle.

19. Disconnect the exhaust pipe from the exhaust manifold and lower the vehicle.

20. Disconnect the air conditioning compressor adjusting arm, if equipped.

21. Remove the cylinder head cover mounting bolts and remove the cylinder head cover.

22. Loosen all the valve adjusting screw locknuts, turn the adjusting screws back all the way to allow all the valves to close.

23. Remove the intake manifold from the cylinder head.

24. Remove the distributor and distributor housing from the cylinder head.

25. Remove the cylinder head mounting bolts and remove the cylinder head from the engine.

26. Remove any oil and water in the cylinder bores and on top of the pistons.

27. Clean and inspect the sealing surfaces of the cylinder head and the engine block. **To install:**

28. Install the cylinder head using a new gasket.

29. Replace the cylinder head mounting bolts, in sequence, and tighten to 46–54 ft. lbs.

30. Install the distributor and distributor housing to the cylinder head.

31. Replace the intake manifold on the cylinder head.

32. Tighten all the valve adjusting screw nuts.

33. Adjust the valve lash.

34. Replace the cylinder head cover and connect the air conditioning compressor adjusting arm, if equipped.

35. Raise and safely support the vehicle.

36. Connect the exhaust pipe to the exhaust manifold and safely lower the vehicle.

37. Reconnect the timing belt, timing belt cover and crankshaft pulley.

38. Replace the water pump pulley, water pump drive belt, fan shroud and the radiator cooling fan.

39. Replace the vacuum hose for the automatic transmission to the intake manifold, if equipped.

40. Replace the pressure sensor hose to the fuel filter and connect the canister purge hose, if equipped.

41. Replace the fuel return and feed hoses.

42. Connect the ground wires to the distributor assembly.

43. Replace the coil wire. Connect the oxygen sensor and distributor wires to their couplers.

44. Connect the PTC wires to the coupler.

45. Connect the idle speed control valve, throttle position sensor and injector wires to their couplers, if equipped.

46. Reconnect the ground cable, air temperature, oil pressure and water temperature sensor wires to the intake manifold.

47. Replace the throttle opener VSV and EGR VSV wires to their couplers.

48. Connect the accelerator and the automatic transmission kickdown cables, if equipped, to the throttle body.

49. Replace the air valve water hose to the throttle body and connect the brake booster hose to intake manifold.

50. Reconnect the air intake case.

51. Refill the cooling system with the proper coolant and connect the negative battery.

52. Check for any water and oil leaks when finished.

Valve Lash

ADJUSTMENT

Valve lash can be adjusted with the engine hot or cold. Specifications are provided for both adjustments.

1. Remove the air intake case.

2. Remove the cylinder head cover assembly.

3. Turn the crankshaft clockwise so the **V** mark on the crankshaft pul-

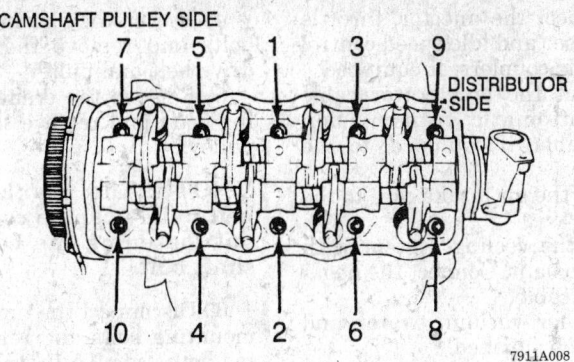

CAMSHAFT PULLEY SIDE

DISTRIBUTOR SIDE

7911A008

Cylinder head and bolt tightening sequence

the cylinder head by placing the rocker arms and springs on the shafts as they are installed. With the rocker arms, springs and shafts installed, torque the rocker shaft mounting screws to 7.0–8.5 ft. lbs.

14. Install the timing belt pulley and inside cover by locking the camshaft with the proper size rod.

15. With the camshaft locked, tighten the camshaft pulley bolt to 41–46 ft. lbs.

16. The remainder of the installation is the reverse of the removal procedure.

ley is aligned with the **0** mark on the timing belt cover.

4. Remove the distributor cap and confirm that the rotor is facing the No. 1 firing position. If the rotor is out of place, turn the crankshaft 360 degrees.

5. With the engine in this position, check the valve lash at valves 1, 2, 5 and 7. Clearance should be: Intake 0.0051–0.0067 in. (0.13–0.17mm) COLD or 0.009–0.011 in. (0.23–0.27mm) HOT. Exhaust 0.0063–0.0075 in. (0.15–0.19mm) COLD or 0.0102–0.0114 in. (0.25–0.29mm.) HOT.

NOTE: The valves are adjusted by loosening the locknut on the valve adjuster and turning the adjusting screw to obtain the proper clearance. Once the proper clearance is obtained, the locknut must be torqued to 11–13 ft. lbs. while holding the adjusting screw. Check the clearance after the locknut is torqued.

6. Rotate the crankshaft 360 degrees and check the valve lash at valves 3, 4, 6 and 8.

7. After adjusting and checking all the valves, install the cylinder head cover, distributor cover and intake hose.

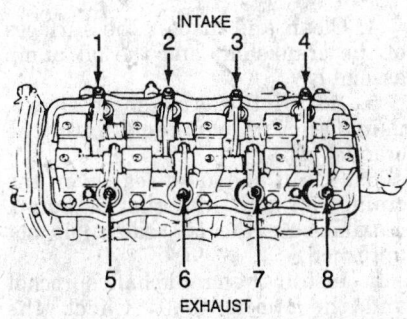

INTAKE

EXHAUST

7911A009

Valve lash adjustment

Rocker Arms/Shafts

REMOVAL AND INSTALLATION

NOTE: The rocker arm shafts are not identical and must be kept in the proper order for installation. If the shafts get mixed up before installation, the intake rocker shaft has a 14mm stepped end and the exhaust rocker shaft has a 13mm stepped end. The stepped end of the intake rocker shaft faces the front of the engine and the stepped end of the exhaust rocker shaft faces the rear of the engine.

1. Disconnect the negative battery cable. Drain the cooling system.

2. Remove the front grille.

3. Remove the radiator cooling fan and fan shroud. Properly discharge the air conditioning system and remove the air conditioning flexible suction hose, if equipped.

4. Remove the radiator.

5. Remove the water pump drive belt and the water pump pulley.

6. Remove the timing belt outside cover, timing belt and the tensioner.

7. Disconnect the air intake case and remove the cylinder head cover.

8. Remove the camshaft timing belt pulley and timing belt inside cover. Insert the proper size rod into the hole in the camshaft to lock the camshaft and loosen the pulley nut.

9. Loosen all the valve lash adjusting screw locknuts and turn the adjusting screws out all the way.

10. Remove the rocker arm shaft screws.

11. Remove the intake and exhaust rocker arm shafts, rocker arms and springs.

12. Keep all the valve train parts in the order that they were removed.

To install:

13. Apply engine oil to the rocker arms, springs and shafts. Install the shafts in the correct direction, into

Intake Manifold

REMOVAL AND INSTALLATION

1. Disconnect the negative battery cable. Drain the cooling system.

2. Remove the air intake case from the manifold.

3. Disconnect the accelerator cable and the automatic transmission kickdown cable, if equipped.

4. Disconnect the injector, throttle position sensor and idle speed control solenoid valve wires at their couplers, if equipped.

5. Disconnect the vacuum hoses from the throttle body and throttle opener. Remove the water hose from the air valve, if equipped.

6. Remove the fuel filler cap to release the fuel pressure in the fuel tank. Reinstall the cap.

7. Release the fuel pressure in the fuel line.

8. Disconnect the fuel line from the throttle body and intake manifold.

9. Remove the fuel return hose.

10. Disconnect the throttle body from the intake manifold and remove the PCV hose from the cylinder head cover.

11. Disconnect the pressure sensor hose from the fuel filter, if equipped, and remove the brake booster hose from the intake manifold.

12. Disconnect the vacuum hose for the automatic transmission from the intake manifold, if equipped.

13. Remove the VSV (for throttle opener) hose from the intake manifold.

14. Disconnect the water hose from the thermostat cap. Remove the heater inlet and water bypass hose from the intake manifold.

15. Disconnect the hoses at the EGR valve and remove the ground wire from the intake manifold.

16. Disconnect the wire couplers from the air temperature sensor,

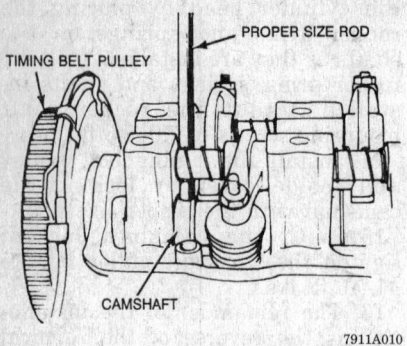

PROPER SIZE ROD
TIMING BELT PULLEY
CAMSHAFT

7911A010

Locking the camshaft in position

water temperature sensor, water temperature gauge and PTC heater, if equipped with automatic transmission.

17. Disconnect the wire harnesses from their clamps.

18. Remove the intake manifold mounting bolts and remove the intake manifold from the cylinder head.

19. Remove the PCV valve, EGR valve, fuel filter, thermostat from the intake manifold.

20. Clean and inspect the sealing surfaces of the intake manifold and the cylinder head.**To install:**

21. Install the thermostat, fuel filter, EGR valve and PCV valve to the intake manifold.

22. Using a new gasket, install the intake manifold and tighten the mounting bolts to 13–20 ft. lbs.

23. Reconnect the wiring harnesses to their clamps and fasten the wire couplers to the air temperature sensor, water temperature sensor, water temperature gauge and PTC heater, if equipped with automatic transmission.

24. Reconnect the ground wire to the intake manifold and replace the hoses at the EGR valve.

25. Install the water heater inlet and bypass hose to the intake manifold. Replace the water hose to the thermostat cap.

26. Replace the VSV (for throttle opener) hose to the intake manifold.

27. Connect the automatic transmission vacuum hose to the intake manifold, if equipped.

28. Replace the brake booster hose to the intake manifold and connect the pressure sensor to the fuel filter, if equipped.

29. Reconnect the fuel line to the throttle body to the intake manifold.

30. Replace the water hose to the air valve. Connect the vacuum hoses to the throttle body and throttle opener, if equipped.

31. Reconnect the injector, throttle position sensor and idle speed control wires to their couplers, if equipped.

32. Connect the accelerator cable and the automatic transmission kickdown cable, if equipped, to the throttle body.

33. Install the air intake case to the manifold.

34. Refill the cooling system with the proper coolant. Connect the negative battery cable.

35. Check for vacuum, water and oil leaks when finished.

Exhaust Manifold

REMOVAL AND INSTALLATION

1. Disconnect the negative cable.

2. Raise and safely support the vehicle.

3. Disconnect the exhaust pipe from the exhaust manifold and lower the vehicle.

4. Disconnect the oxygen sensor lead wire at the coupler and remove the air intake case bracket.

5. Disconnect the exhaust manifold upper and lower covers or heat shields from the exhaust manifold.

6. Remove the exhaust manifold mounting bolts and remove the exhaust manifold from the cylinder head.**To install:**

7. Clean and inspect the sealing surfaces of the exhaust manifold and the cylinder head.

8. Using new gaskets, install the exhaust manifold to the cylinder head and tighten the mounting bolts to 13–20 ft. lbs.

9. The remainder of the installation is the reverse of the removal procedure.

10. Check for exhaust leaks when finished.

Timing Belt Front Cover

REMOVAL AND INSTALLATION

1. Disconnect the negative battery cable.

2. Remove the radiator cooling fan and fan shroud.

3. If equipped, disconnect the air conditioning compressor drive belt, properly discharge the air conditioning system and remove the air conditioning compressor flexible suction hose.

4. Loosen the alternator mounting bolts and remove the water pump drive belt and pulley.

5. Remove the crankshaft mounting bolts and remove the crankshaft pulley.

NOTE: The crankshaft drive belt pulley can be removed without loosening the center crankshaft bolt.

6. Disconnect the timing belt cover mounting bolts and remove the timing belt cover.**To install:**

7. Clean and inspect all mounting surfaces.

8. Install the timing belt cover and tighten the bolts to 7.0–9.0 ft. lbs.

9. Install the crankshaft pulley. Replace the 5 mounting bolts and tighten to 7.0–9.0 ft. lbs.

10. Install the water pump drive belt pulley and replace the drive belt

11. Adjust the belt tension and tighten the alternator mounting bolts.

12. Install the air conditioning flexible suction hose and replace the air conditioning compressor drive belt. Adjust the belt tension and properly recharge the air conditioning system, if equipped.

13. Replace the radiator cooling fan and fan shroud.

14. Connect the negative battery cable.

Oil Seal Replacement

1. Disconnect the negative battery cable.

2. Remove the the timing belt and the crankshaft sprocket.

3. Insert a suitable tool between the crankshaft and the oil seal and pull the seal outwards to remove it.

NOTE: Use care when removing or installing the oil seal, not to damage the crankshaft or the oil pump sealing surfaces.

4. Clean and inspect the surfaces of the crankshaft and the oil pump assembly.

5. Install the crankshaft sleeve, using the proper tool, onto the crankshaft.

6. Install the new seal over the crankshaft and into the oil pump, making sure the oil seal lip is not upturned.

7. Install the crankshaft sprocket and the timing belt. Check the timing.

8. Connect the negative battery cable.

Timing Belt and Tensioner

ADJUSTMENT

1. Disconnect the negative battery cable.

2. Remove the radiator cooling fan and fan shroud.

3. If equipped, disconnect the air conditioning compressor drive belt, properly discharge the air conditioning system and remove the air conditioning compressor flexible suction hose.

4. Loosen the alternator mounting bolts and remove the water pump drive belt and pulley.

5. Remove the crankshaft mounting bolts and disconnect the crankshaft pulley.

NOTE: The crankshaft drive belt pulley can be removed without loosening the center crankshaft bolt.

6. Disconnect the timing belt cover mounting bolts and remove the timing belt cover. Loosen but do not remove the tensioner bolt.

7. Disconnect the air intake case from the intake manifold.

8. Remove the cylinder head cover and loosen all the valve adjusting screws to permit free rotation of the camshaft.

9. Turn the camshaft pulley clockwise and align the timing marks.

10. Turn the crankshaft clockwise, using a 17mm wrench to crank the timing belt pulley bolt.

11. Align the punch mark on the timing belt pulley with the arrow mark on the oil pump.

12. With the 4 marks aligned, remove any slack from the drive side of the belt. Tighten the tensioner bolt to 17.5–21.5 ft. lbs.

13. To allow the belt to be free of any slack, turn the crankshaft clockwise 2 full rotations. Confirm that the 4 marks are aligned.

14. Replace the timing cover and tighten the bolts to 7.0–8.5 ft. lbs.

15. Adjust the valve lash and install the cylinder head cover, using a new gasket.

16. Connect the air intake case to the throttle body.

17. Install the crankshaft pulley and replace the 5 mounting bolts. Tighten to 7.0–8.5 ft. lbs.

18. Replace the water pump pulley, water pump drive belt and tighten the alternator mounting bolts.

19. Install the air conditioning compressor flexible suction hose and the air compressor belt. Properly recharge the air conditioning system, if equipped.

20. Replace the radiator cooling fan and fan shroud.

21. Connect the negative battery cable.

22. Properly recharge the air conditioning system, if equipped. Run the engine and check for any leaks.

REMOVAL AND INSTALLATION

1. Disconnect the negative battery cable.

2. Remove the radiator cooling fan and fan shroud.

3. If equipped, disconnect the air conditioning compressor drive belt, properly discharge the air conditioning system and remove the air conditioning compressor flexible suction hose.

4. Loosen the alternator mounting bolts and remove the water pump drive belt and pulley.

5. Remove the crankshaft mounting bolts and disconnect the crankshaft pulley.

NOTE: The crankshaft drive belt pulley can be removed without loosening the center crankshaft bolt.

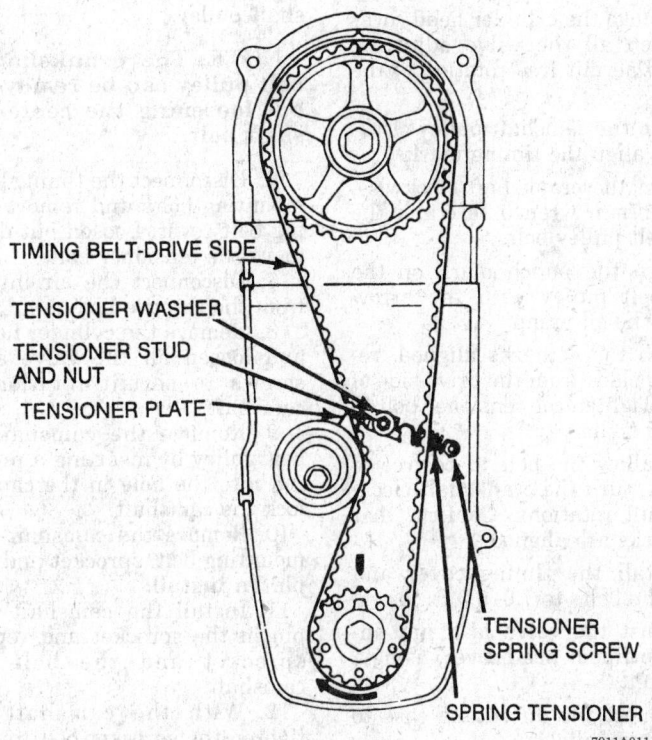

Exploded view of the timing belt and the tensioner assembly

7911A011

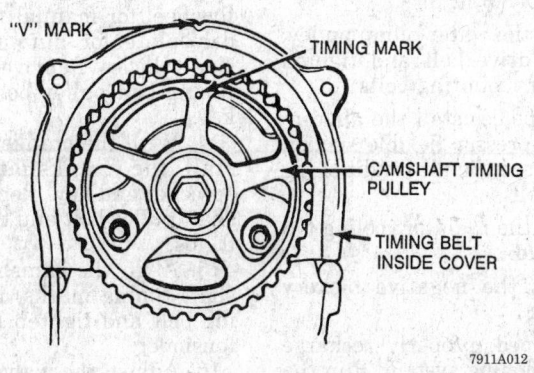

Timing marks on the camshaft pulley

7911A012

6. Disconnect the timing belt cover mounting bolts and remove the timing belt cover. Loosen but do not remove the tensioner bolt.

7. Disconnect the air intake case from the intake manifold.

8. Loosen the timing belt tensioner adjusting bolt and pivot nut. Hold pressure on the tensioner to loosen the timing belt and remove the timing belt from the camshaft and crankshaft pulleys.

9. Remove the timing belt tensioner, tensioner plate and tensioner spring. **To install:**

10. Install the timing belt tensioner, plate and spring. Hand tighten the tensioner bolt and stud only at this time.

11. Remove the cylinder head cover and loosen all the valve adjusting screws to permit free rotation of the camshaft.

12. Turn the camshaft pulley clockwise and align the timing marks.

13. Turn the crankshaft clockwise, using a 17mm wrench to crank the timing belt pulley bolt.

14. Align the punch mark on the timing belt pulley with the arrow mark on the oil pump

15. With the 4 marks aligned, remove any slack from the drive side of the belt. Tighten the tensioner bolt to 17.5–21.5 ft. lbs.

16. To allow the belt to be free of any slack, turn the crankshaft clockwise 2 full rotations. Confirm that the 4 marks are aligned.

17. Install the timing cover and tighten the bolts to 7.0–8.5 ft. lbs.

18. Adjust the valve lash and install the cylinder head cover, using a new gasket.

19. Connect the air intake case to the intake manifold.

20. Install the crankshaft pulley and replace the mounting bolts. Tighten to 7.0–8.5 ft. lbs.

21. Replace the water pump pulley, water pump drive belt and tighten the alternator mounting bolts.

22. If equipped, install the air conditioning compressor flexible suction hose and replace the air conditioning compressor belt.

23. Replace the radiator cooling fan and fan shroud.

24. Connect the negative battery cable.

25. If equipped, properly recharge the air conditioning system. Run the engine and check for any leaks.

Timing Sprockets

REMOVAL AND INSTALLATION

1. Disconnect the negative battery cable.

2. Remove the radiator cooling fan/clutch and fan shroud.

3. If equipped, disconnect the air conditioning compressor drive belt, properly discharge the air conditioning system and remove the air conditioning compressor flexible suction hose.

4. Loosen the alternator mounting bolts and remove the water pump drive belt and pulley.

5. Remove the crankshaft mounting bolts and disconnect the crankshaft pulley.

NOTE: The crankshaft drive belt pulley can be removed without loosening the center crankshaft bolt.

6. Disconnect the timing belt cover mounting bolts and remove the timing belt cover. Loosen but do not remove the tensioner bolt.

7. Disconnect the air intake case from the throttle body.

8. Remove the cylinder head cover and loosen all the valve adjusting screws to permit rotation of the camshaft.

9. Remove the camshaft timing belt pulley by inserting a proper size rod into the hole in the camshaft to lock the camshaft.

10. Remove the camshaft sprocket mounting bolt, sprocket and sprocket pin. **To install:**

11. Install the camshaft sprocket pin in the sprocket and replace the sprocket and the bolt on the camshaft.

12. With the camshaft locked, tighten the sprocket bolt to 41–46 ft. lbs.

13. Remove the crankshaft sprocket by using a gear stopper to hold the flywheel for manual transmission or driveplate for automatic transmission vehicles. Remove the crankshaft timing belt pulley bolt, sprocket and key.

14. With the crankshaft locked, install the crankshaft timing belt sprocket and key. Replace the crankshaft pulley bolt and tighten to 47–54 ft. lbs.

15. Align the camshaft and crankshaft timing marks, install the timing belt and tighten the timing belt tensioner.

16. Adjust the valve lash and replace the cylinder head cover.

17. Reconnect the air intake hose to the throttle body.

18. If equipped, install the air conditioning compressor flexible suction hose and replace the air conditioning compressor belt.

19. Install the water pump pulley and replace the water pump drive belt.

20. Replace the radiator cooling fan and fan shroud.

21. Install the water pump pulley and replace the water pump drive belt.

22. Connect the negative battery cable.

23. If equipped, properly recharge the air conditioning system. Run the engine and check for any leaks.

Camshaft

REMOVAL AND INSTALLATION

1. Disconnect the negative battery cable. Drain the cooling system.

2. Disconnect the air intake case.

3. Disconnect the brake booster hose from the intake manifold and the air valve water hose from the throttle body.

4. Disconnect the accelerator and the automatic transmission kickdown cable, if equipped, from the throttle body.

5. Disconnect and tag the throttle opener VSV and EGR VSV wires at the coupler.

6. Disconnect and tag the water temperature, oil pressure, air temperature sensor and ground cable wires at the intake manifold.

7. Disconnect and tag the injector, throttle position sensor and idle speed control solenoid valve wires at their couplers, if equipped.

8. Disconnect the PTC heater wires at the coupler, if equipped with automatic transmission.

9. Disconnect and tag the oxygen sensor and distributor wires at their couplers. Remove the coil wire.

10. Disconnect the ground wires from the distributor assembly.

11. Remove the fuel tank filler cap to relieve the pressure. Reinstall the cap.

12. Release the fuel pressure in the fuel feed line.

13. Disconnect the fuel injector connections and fuel hoses.

14. Disconnect the canister purge hose and remove the pressure sensor hose from the fuel filter, if equipped.

15. Remove the vacuum hose for the automatic transmission from the intake manifold, if equipped.

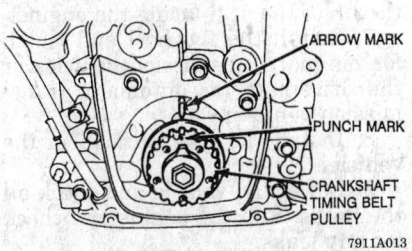

Timing marks on the crankshaft pulley

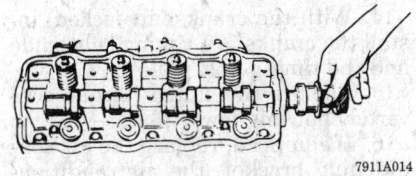

Camshaft removal

53. Connect the accelerator and the automatic transmission kickdown cables, if equipped, to the throttle body.

54. Install the air valve water hose to the throttle body and connect the brake booster hose to intake manifold.

55. Reconnect the air intake case.

56. Refill the cooling system with the proper coolant and connect the negative battery.

57. Check for any water and oil leaks when finished.

ENGINE LUBRICATION

Oil Pan

REMOVAL AND INSTALLATION

1. Raise and safely support the vehicle.

2. Drain and remove the front differential assembly from the chassis.

3. Drain the engine oil.

4. Remove the clutch housing lower plate or the torque converter housing lower plate.

5. Remove the oil pan mounting bolts and remove the pan.

To install:

6. Clean and inspect the sealing surfaces on the oil pan and the engine block.

7. Using new gaskets, install the oil pan and tighten the oil pan bolts to 7.0–8.5 ft. lbs.

NOTE: Tightening should begin at the center moving outward on both sides.

8. Install the oil drain plug and tighten to 22.0–28.5 ft. lbs.

9. Replace the clutch housing lower plate or the torque converter.

10. Install the front differential assembly to the chassis and fill with differential oil.

11. Lower the vehicle.

12. Refill the engine with engine oil. Run the engine and check for leaks.

Oil Pump

REMOVAL AND INSTALLATION

1. Disconnect the negative battery cable.

16. Disconnect the radiator cooling fan, fan shroud, water pump drive belt and water pump pulley.

17. Disconnect the crankshaft pulley, timing belt cover and the timing belt.

18. Raise and safely support the vehicle.

19. Disconnect the exhaust pipe from the exhaust manifold and lower the vehicle.

20. Disconnect the air conditioner compressor adjusting arm, if equipped.

21. Remove the cylinder head cover mounting bolts and remove the cylinder head cover.

22. Loosen all the valve adjusting screw locknuts, turn the adjusting screws back all the way to allow all the valves to close.

23. Remove the intake manifold from the cylinder head.

24. Remove the distributor and distributor housing from the cylinder head.

25. Remove the cylinder head mounting bolts and remove the cylinder head from the engine.

26. Remove any oil and water in the cylinder bores and on top of the pistons.

27. Clean and inspect the sealing surfaces of the cylinder head and the engine block.

28. Disconnect the timing belt gear from the camshaft and remove the rocker arms, springs and rocker arm shafts.

29. Keep all the valve train parts in the order that they were removed.

30. Remove the camshaft from the rear of the cylinder head.

To install:

31. To install, lubricate the lobes and journals of the camshaft and the oil seal on the cylinder head with engine oil.

32. Install the camshaft to the cylinder head from the transmission side.

33. Install the cylinder head using a new gasket.

34. Install the cylinder head mounting bolts and tighten to 46–54 ft. lbs.

35. Install the distributor and distributor housing to the cylinder head.

36. Install the intake manifold to the cylinder head.

37. Tighten all the valve adjusting screw nuts.

38. Adjust the valve lash.

39. Install the cylinder head cover and connect the air conditioning compressor adjusting arm, if equipped.

40. Raise and safely support the vehicle.

41. Connect the exhaust pipe to the exhaust manifold and safely lower the vehicle.

42. Reconnect the timing belt, timing belt cover and crankshaft pulley.

43. Install the water pump pulley, water pump drive belt, fan shroud and the radiator cooling fan.

44. Install the vacuum hose for the automatic transmission to the intake manifold, if equipped.

45. Install the pressure sensor hose to the fuel filter and connect the canister purge hose, if equipped.

46. Install the fuel return and feed hoses.

47. Connect the ground wires to the distributor assembly.

48. Install the coil wire. Connect the oxygen sensor and distributor wires to their couplers.

49. Connect the PTC wires to the coupler.

50. Connect the idle speed control valve, throttle position sensor and injector wires to their couplers, if equipped.

51. Reconnect the ground cable, air temperature, oil pressure and water temperature sensor wires to the intake manifold.

52. Install the throttle opener VSV and EGR VSV wires to their couplers.

2. Remove the radiator cooling fan, fan shroud, water pump pulley and water pump drive belt.

3. Remove the timing belt outside cover, timing belt and timing belt tensioner.

4. Disconnect the alternator and remove the bracket, if neccessary.

5. If equipped, properly discharge the air conditioning system and remove the air compressor and bracket.

6. Disconnect the crankshaft timing belt pulley and the timing belt guide.

NOTE: To lock the crankshaft, engage a special tool (gear stopper) with the flywheel ring gear for manual transmission or driveplate ring gear for automatic transmission vehicles. With the crankshaft locked, remove the crankshaft timing belt pulley bolt.

7. Raise and safely support the vehicle. Drain the engine oil.

8. Remove the clutch (torque converter) housing lower plate.

9. Remove the oil pan mounting bolts and remove the oil pan. Disconnect the oil pump strainer.

10. Remove the oil pump mounting bolts and remove the oil pump.

To install:

11. Clean and inspect the sealing surfaces of the oil pump and the engine block.

12. Using a new gasket, install the oil pump to the engine.

13. Install the No. 1 and No. 2 mounting bolts and tighten to 7.0–8.5 ft. lbs.

NOTE: To prevent the oil seal lip from being damaged when installing the oil pump to the crankshaft, use a proper seal guide tool when installing. After installing the oil pump, check to be sure the oil lip is not up-turned, then remove the special

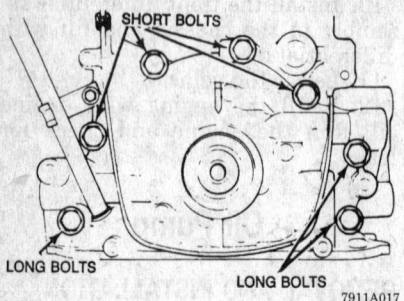

Oil pump installation — bolt location

tool. The edge of the oil pump gasket might bulge out. If it does, cut off bulge with knife.

14. With the crankshaft locked, install the crankshaft timing belt guide and the timing belt pulley.

15. Replace the clutch (torque converter) housing lower plate.

16. If equipped, replace the air conditioning bracket, the air conditioning compressor and properly rechargew the air conditioning system.

17. Replace the alternator and the alternator bracket, if necessary.

18. Reinstall the timing belt tensioner, timing belt and timing belt outside cover.

19. Replace the water pump drive belt, water pump pulley, radiator cooling fan and fan shroud.

20. Refill the engine with engine oil and connect the negative battery cable.

21. Run the engine and check for any leaks.

CHECKING

With the oil pump removed from the engine, certain clearances must be checked on the oil pump, if it is being reused.

1. The radial clearance is the clearance between the oil pump outer rotor and the oil pump case. The maximum clearance is 0.0122 in.

2. The side clearance is measured with a straight-edge across the mounting surface of the oil pump. The measurement is taken between the oil pump gears and the straight-edge. The maximum clearance is 0.0059 in.

Rear Main Bearing Oil Seal

REMOVAL AND INSTALLATION

1. Raise and safely support the vehicle.

2. Support the engine and remove the transmission from the vehicle.

3. Disconnect the clutch and flywheel for manual transmission vehicles or the driveplate for automatic transmission vehicles.

4. Using a suitable tool, remove the seal by pulling it outwards. Use care not to damage the sealing surface of the crankshaft.

To install:

5. Clean and inspect the sealing surfaces of the crankshaft and seal housing.

6. Using a seal driver, install the new seal into the seal housing with the lip of the seal facing the engine.

7. Install the flywheel and clutch for manual transmission vehicles or the driveplate for automatic transmission vehicles.

8. Install the transmission in the vehicle.

9. Lower the vehicle and check all the fluids. Run the engine and check for any leaks.

MANUAL TRANSMISSION

Transmission Assembly

REMOVAL AND INSTALLATION

1. Disconnect the negative battery cable.

2. Remove the console cover, remove the 4 gear shift boot mounting bolts and slide the boot upwards on the gear shifter.

3. Remove the boot clamp and remove the second boot from the shift lever case.

4. Push the gear shift lever control case down with fingers, turn it counterclockwise and remove the shift control lever.

5. Remove the transfer case shift control lever the same way.

6. Disconnect the breather hose and clamp at the rear of the cylinder head.

7. Remove the clamp at the rear of the intake manifold to free up the wiring harness. Disconnect the harness coupler.

8. Raise and safely support the vehicle. Drain the oil from the transmission and the transfer case.

9. Disconnect the starter lead wires and mounting bolts and remove the starter.

10. Remove the fuel line clamp on the transmission. Disconnect the bolts fastening the engine to the transmission.

11. Disconnect the flange bolts from the front driveshaft and remove and mark the shaft.

12. Disconnect the flange bolts from the rear driveshaft and remove and mark the shaft.

13. Disconnect the clutch cable and remove the clutch housing lower plate.

14. Disconnect the center exhaust pipe and remove the nuts from the joint with the engine.

15. Disconnect the speedometer cable from the transfer case.

16. Position a transmission jack and remove the engine rear mounting member from the vehicle. Move the transmission and transfer case rearwards and lower.

17. Disconnect the wiring harness and breather hose at the transmission.

18. Separate the gear shift lever case and transfer case from the transmission. Remove the transmission from the vehicle.

To install:

19. Install the gear shift lever case and transfer case to the transmission.

20. Install the transmission wiring harness and breather hose.

21. Raise the transmission and transfer case on the transmission jack and place under the vehicle.

22. Install the engine rear mounting member and tighten the bolts.

23. Connect the speedometer cable to the transfer case and joint with the engine.

24. Remove the transmission jack and replace the center exhaust pipe.

25. Replace the clutch housing lower plate and connect the clutch cable.

26. Install the front and rear driveshaft and replace the flange bolts and tighten to 37 ft. lbs.

27. Connect the bolts attaching the engine to the transmission and tighten to 37 ft. lbs. and replace the fuel line clamp on the transmission.

28. Install the starter and replace the mounting bolts and the lead wires to the starter.

29. Lower the vehicle. Refill the transmission and transfer case with the recommended gear oil.

30. Connect the wiring coupler and replace the clamp holding the wiring harness at the rear of the intake manifold.

31. Replace the breather hose and clamp at the rear of the cylinder head.

32. Install the transfer case and the gear shift control levers. Replace the gear shift lever boots and the console cover.

33. Connect the negative battery cable. Run the engine and check for any leaks.

CLUTCH

Clutch Assembly

REMOVAL AND INSTALLATION

1. Disconnect the negative battery cable.

2. Raise and safely support the vehicle.

3. Support the engine and remove the transmission from the vehicle.

4. Support the pressure plate and remove the 6 pressure plate to flywheel mounting bolts.

5. Remove the pressure plate and the clutch disc from the flywheel.

6. Inspect the condition of the flywheel, pressure plate and clutch disc and replace as necessary.

7. Remove the flywheel mounting bolts and remove the flywheel if necessary.

To install:

8. Install the new flywheel and replace the mounting bolts, tighten to 41–47 ft. lbs.

9. Inspect the condition of the clutch release bearing and the input shaft bearing and replace if necessary.

10. Align the clutch disc to the flywheel using a clutch disc alignment tool.

11. Install the pressure plate and evenly tighten the pressure plate bolts to 13–20 ft. lbs.

NOTE: Before assembling, make sure the clutch disc and the pressure plate are clean and dry.

12. With the engine supported, install the transmission in the vehicle.

13. Lower the vehicle. Connect the negative battery cable.

14. Check and adjust the clutch pedal height. Check the clutch operation when finished.

Pedal Height/Free-play
ADJUSTMENT

The only adjustment possible is the clutch pedal free-play. The free play is adjusted by moving the joint nut on the release bearing arm. Clutch linkage free-play should be between 0.6–1.1 in.

Clutch Cable

REMOVAL AND INSTALLATION

1. Disconnect the negative battery cable.

2. Raise and safely support the vehicle.

3. Disconnect the cable adjusting nuts on the clutch release and remove the adjusting nuts on the cable support bracket.

4. Disconnect the clutch cable from the 3 cable clamps.

5. Remove the clutch cable support bolts. Lower the vehicle.

6. Disconnect the cable hook at the clutch pedal shaft arm and remove the clutch cable.

To install:

7. Install the cable after first applying grease to the cable end hook and the joint pin.

8. Connect the cable to the 3 cable clamps and install the clutch cable support bolts.

9. Lower the vehicle and connect the negative battery cable.

ADJUSTMENT

Adjust the clutch pedal height by adjusting the bolt located on the pedal bracket so the clutch pedal exceeds the height of the brake pedal by 0.2 in. (5mm). Tighten the locknut.

AUTOMATIC TRANSMISSION

Transmission Assembly

REMOVAL AND INSTALLATION

1. Disconnect the negative battery cable.

2. Disconnect the transmission shift control lever and remove the transfer case shift control lever knob.

3. Disconnect the breather hose from the clamp at the rear of the cylinder head.

4. Disconnect the wiring harness clamp at the rear end of the intake manifold to free up the harness.

5. Disconnect the wiring harness coupler and remove the detent cable at the throttle body, if equipped.

6. Remove the vacuum modulator hose at the intake manifold.

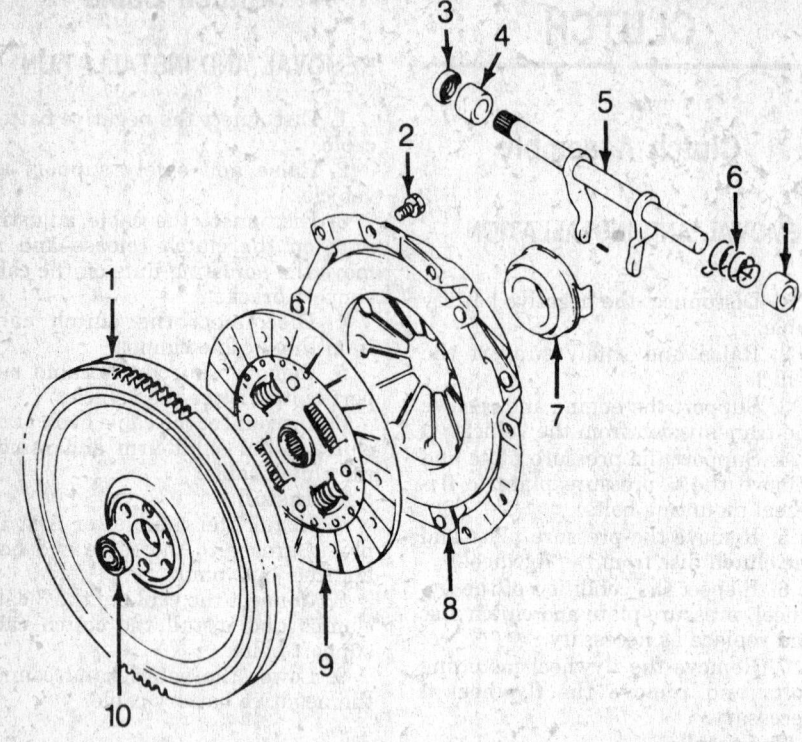

1. Flywheel
2. Cover bolt
3. Release shaft seal
4. Release shaft bushing
5. Clutch release shaft
6. Release return spring
7. Clutch release bearing
8. Clutch cover
9. Clutch disc
10. Input shaft bearing

7911A018

Exploded view of the clutch assembly

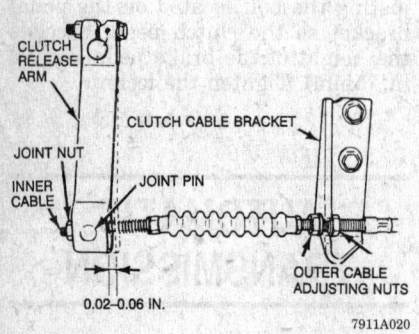

0.02–0.06 IN.

7911A020

Clutch cable free travel adjustment

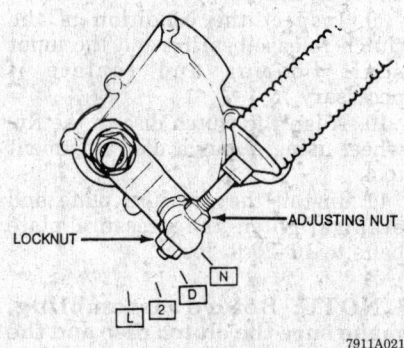

7911A021

Adjusting the shift cable assembly

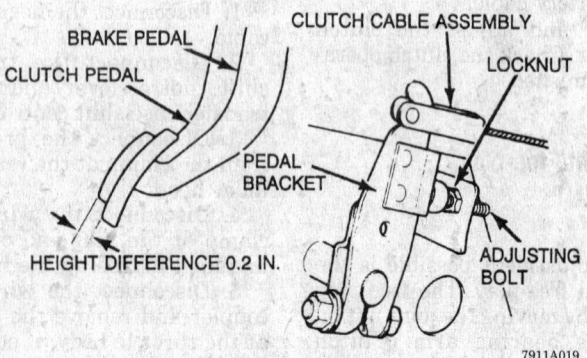

7911A019

Clutch pedal height adjustment

7. Raise and safely support the vehicle.

8. Disconnect the starter lead wires and mounting bolts, remove the starter motor.

9. Drain the oil from the transmission and transfer case.

10. Disconnect the flange bolts from the front driveshaft, remove and mark the shaft.

11. Disconnect the flange bolts from the rear driveshaft, remove and mark the shaft.

12. Disconnect the select cable from the transmission and the speedometer cable from the transfer case.

13. Remove the torque converter housing lower plate and disconnect and plug the oil cooler lines.

14. Disconnect the transfer case skid plate and remove the kickdown cable from the transmission.

15. Remove the exhaust bracket at the catalytic converter and at the transmission. Disconnect the center exhaust pipe, if necessary.

16. Remove the transmission to engine retaining bolts and nutsa.

17. Support the transmission using a transmission jack or equivalent. Disconnect the transmission crossmember mounting bolts and remove the crossmember

18. Disconnect the lead wires and breather hose at the transmission

and lower the transmission with the transfer case from the vehicle. Remove the torque converter to flywheel bolts.

19. Disconnect the transmission-to-transfer case bolts and separate the transmission from the transfer case.

To install:

20. Connect the transfer case to the transmission and tighten the bolts to 20 ft. lbs. Replace the torque converter to flywheel bolts.

21. Raise the transmission, connect the lead wires and breather to the transmission.

22. Replace the transmission to engine retaining bolts and tighten to 62 ft. lbs.

23. Install the transmission cross member and tighten the bolts to 62 ft. lbs.

24. Connect the exhaust bracket to the catalytic converter and the transmission.

25. Reconnect the center exhaust pipe, if necessary.

26. Connect the transfer case skid plate and replace the kickdown cable to the transmission.

27. Replace the torque converter housing lower plate and connect the transmission cooler lines.

28. Reconnect the select cable to the transmission and the speedometer cable to the transfer case.

29. Replace the front and rear driveshafts and tighten the flange bolts to 37 ft. lbs.

30. Install the starter and connect the lead wires to the starter.

31. Lower the vehicle and connect the vacuum modulator hose to the intake manifold.

32. Connect the wiring coupler at the rear of the intake manifold and replace the wiring harness clamp in the proper position.

33. Connect the breather hose to the clamp at the rear of the cylinder head.

34. Install the transmission shift control lever and replace the transfer case shift control lever knob.

35. Connect the negative battery cable and refill the transmission. Run the engine and check for leaks.

SHIFT LINKAGE ADJUSTMENT

1. Adjust the shift cable assembly by moving the shift selector to the **N** position and placing a pin in the selector to hold that position.

2. Put the manual select lever in **L** position and set the cable to the cable bracket with the E-clip.

3. Adjust the manual select lever back to the **N** position and tighten the cable end locknut to 5 ft. lbs.

NOTE: When the manual selector is moved to the N position, there should be a little clearance between the lever and the adjusting nut.

THROTTLE LINKAGE ADJUSTMENT

Adjust the cable by loosening the cable locknut and adjust the cable to specifications. The cable endplay should be 0.12–0.20 in. when the throttle valve is in the idle position.

DETENT CABLE ADJUSTMENT

1. Make sure accelerator play is within specifications.

2. Loosen the kickdown cable nut and adjusting nut.

3. With the accelerator pedal fully depressed and the cable pulled towards the firewall, adjust the locknut-to-bracket clearance to 0.039 in. by turning the locknut.

NOTE: When adjusting the clearance make sure the adjusting nut does not rub against the bracket.

4. Release the accelerator pedal and adjust the locknut-to-bracket clearance by tightening the adjusting nut. Tighten the locknut securely.

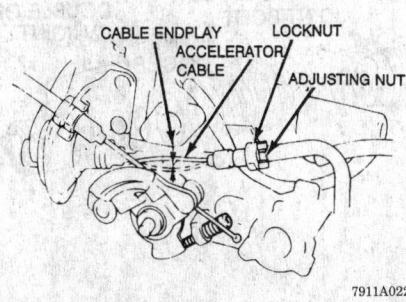

7911A022

Accelerator cable adjustment

TRANSFER CASE

Transfer Case Assembly

REMOVAL AND INSTALLATION

1. Disconnect the negative battery cable.

2. Raise and safely support the vehicle.

3. Support the transmission and transfer case using a transmission jack.

4. Disconnect the speedometer cable from the transfer case.

5. Remove the transmission and transfer case from the vehicle.

6. Remove the 10 transmission-to-transfer case mounting bolts and separate the transfer case from the transmission.

To install:

7. Install the 10 transmission-to-transfer case mounting bolts and tighten the bolts to 20 ft. lbs.

8. Install the transmission and transfer case into the vehicle and tighten the mounting bolts to 37 ft. lbs.

9. Connect the speedometer cable to the transfer case.

10. Lower the vehicle and connect the negative battery cable.

DRIVE AXLE

Halfshaft

REMOVAL AND INSTALLATION

1. Raise and safely support the vehicle. Drain the transmission.

2. Remove the front wheels and disconnect the locking hubs.

3. Disconnect the halfshaft snaprings and remove the stabilizer bar links.

4. Disconnect the tie rod ends and remove the brake caliper and brake disc. Support the brake caliper.

5. Disconnect the wheel hub and remove the steering knuckle.

6. Disconnect the differential side joint snapring and remove the right side halfshaft by prying the joint away from the differential assembly.

7. Disconnect the left side halfshaft bolts and remove the halfshaft from the left inner axle flange.

To install:

8. Install the left side halfshaft and the halfshaft bolts and tighten to 37 ft. lbs.

9. Install the right side halfshaft and install the snapring.

10. Install the steering knuckle and wheel hub assembly.

11. Install the brake caliper and connect the tie rod end.

12. Install the left side driveshaft snapring and connect the stabilizer bar link.

13. Install the front wheels and lower the vehicle.

14. Refill the transmission with the proper fluid.

Driveshaft and U-Joints

REMOVAL AND INSTALLATION

1. Raise and safely support the vehicle.

2. Matchmark the driveshafts to the yokes on the transfer case and the differential.

3. Drain the transfer case oil when servicing the front driveshaft shaft.

4. Support the driveshaft and remove the attaching bolts.

5. Remove the driveshaft from the vehicle.

To install:

6. Install the driveshaft, by aligning the matchmarks to the vehicle.

7. Tighten the mounting bolts to 37 ft. lbs.

8. Refill the transfer case and lower the vehicle.

Rear Axle Shaft, Bearing and Seal

REMOVAL AND INSTALLATION

1. Raise and safely support the vehicle.

2. Remove the rear wheels and remove the rear brake drums from the vehicle.

3. Drain the gear oil from the rear axle housing.

4. Remove the rear wheel bearing retainer nuts from the rear axle housing.

5. Using a suitable tool, remove the axle shaft from the housing.

NOTE: Do not remove the backing plate with the axle. This may cause damage to the inner seal.

6. If the axle, axle bearing or backing plate is being replaced, support

the axle in a vise with additional support under the shaft next to the bearing.

— CAUTION —

Eye protection must be worn during the 3 steps. Failure to do so could cause injury.

7. With the axle shaft supported properly, grind the top and bottom of the axle bearing retainer to remove it without damaging the axle shaft.

8. Using a chisel, finish removing the retainer from the axle shaft.

9. Using a press or suitable bearing puller, remove the axle shaft bearing from the axle shaft.

10. Using a suitable tool, remove the seal from the axle housing.

To install:

11. Using a seal driver, install the new seal with the lip facing the housing to the same depth as the old seal.

12. Install the new bearing and the retainer on the axle shaft using a suitable press.

13. Install the axle shaft into the rear axle housing and replace the rear wheel bearing retaining nuts, tighten to 17 ft. lbs.

14. Replace the rear brake drums and replace the rear tires on the vehicle.

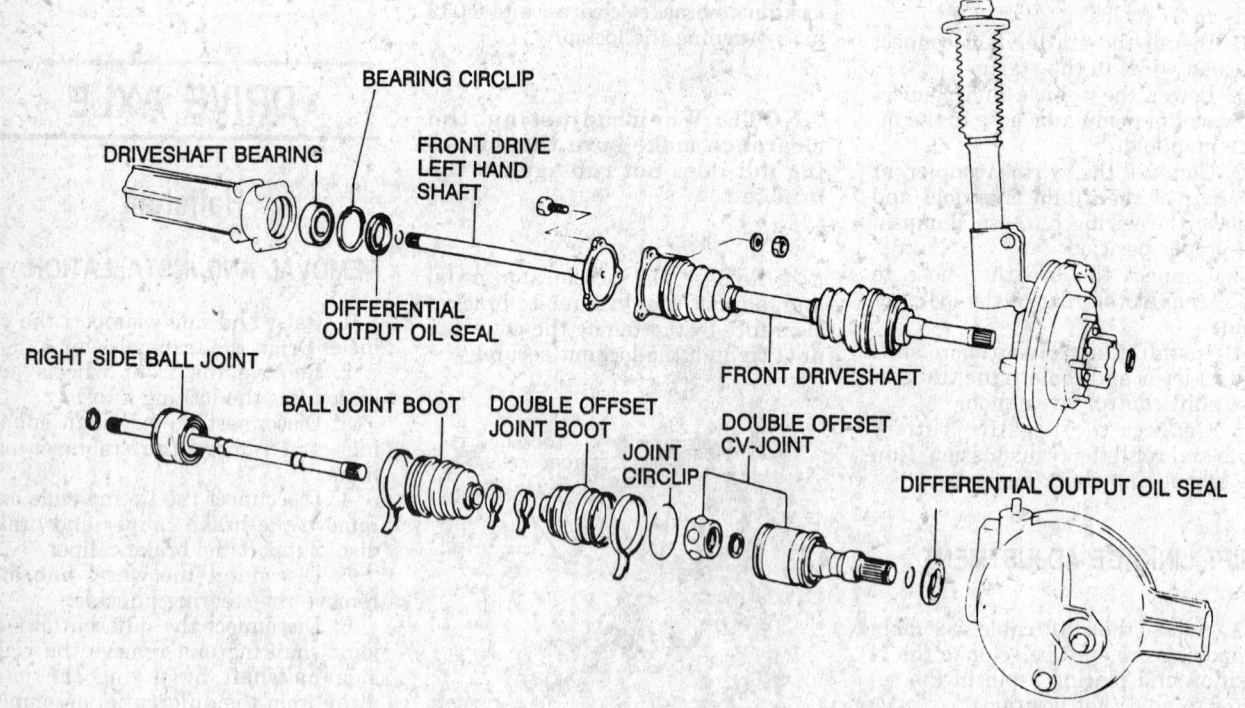

Exploded view of the halfshaft assembly

7911A023

15. Refill the rear axle housing with the proper gear oil and safely lower the vehicle.

Front Wheel Hub, Knuckle and Bearings

REMOVAL AND INSTALLATION

1. Disconnect the negative battery cable. Raise and safely support the vehicle.
2. Remove the wheels and disconnect the locking hub assembly.
3. Remove the caliper mounting bracket and position the caliper out of the way. Support the caliper.
4. Disconnect the front wheel bearing lock plate and washer and remove the wheel hub complete with bearings and seals.
5. Remove the oil seal and race from the wheel hub.
6. Clean and inspect the hub and bearing seats. Install the new bearing, race and grease seal in the same position.
7. Support the suspension with a jack and remove the dust cover.
8. Disconnect the spindle by tapping with a hammer.
9. Remove the strut bracket bolts from the steering knuckle.
10. Disconnect the tie rod end from the knuckle and the knuckle from the control arm.
11. Remove the dust seal and the spindle from the knuckle.
   To install:
12. Install the new seal and the spindle to the knuckle.
13. Connect the knuckle to the ball joint and tighten the nut to 40 ft. lbs.
14. Connect the knuckle to the strut damper and tighten the bolts to 66 ft. lbs.
15. Install the tie rod end to the knuckle and tighten the nut 30 ft. lbs.
16. Install the spindle and replace the dust cover and remove the jack from under the suspension.
17. Install the wheel hub assembly and connect the locking hub.
18. Install the wheels and lower the vehicle.
19. Connect the negative battery cable.

Manual Locking Hubs

REMOVAL AND INSTALLATION

1. Disconnect the negative battery cable and remove the locking hub cover.
2. Raise and support the vehicle safely.
3. Remove the locking hub body assembly.

NOTE: The manual locking hubs must not be packed with grease.

To install:
4. Install a new O-ring to the locking hub assembly.
5. Install the locking hub body assembly to the wheel hub flange and tighten the hub body bolts to 18 ft. lbs.
6. Install a new gasket in the manual locking hub cover.
7. Replace the locking hub cover and connect the negative battery cable.
8. Lower the vehicle.

NOTE: The O mark on the hub knob must be in the FREE position. Tighten the locking hub cover bolts to 106 inch lbs.

Automatic Locking Hubs

REMOVAL AND INSTALLATION

1. Disconnect the negative battery cable and remove the automatic locking hub cover.
2. Disconnect the automatic hub body assembly.
   To install:
3. Install a new O-ring to the automatic locking hub body assembly.
4. Connect the automatic locking hub body assembly to the wheel hub flange. Tighten the hub body assembly bolts to 18 ft. lbs.
5. Install a new gasket in the automatic locking hub cover and replace the hub cover.
6. Connect the negative battery cable.

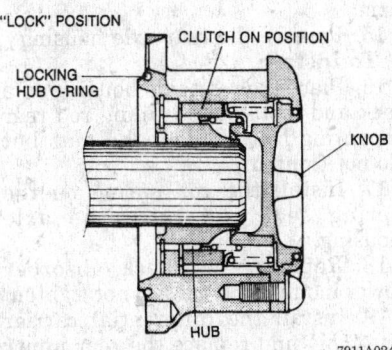

Wheel hub "LOCK" position

7911A024

Pinion Seal

REMOVAL AND INSTALLATION

1. Raise and safely support the vehicle.
2. Matchmark and remove the driveshaft.
3. Check the turning torque of the pinion before proceeding.
4. Using a pinion flange holding tool, remove the pinion nut and the washer.
5. Remove the pinion flange from the differential carrier and pry out the pinion seal.
   To install:
6. Clean and inspect the sealing surface of the carrier.
7. Using a seal driver, install the new seal into the carrier until the flange of the seal is flush with the carrier.

NOTE: Tightening the flange nut will preload the pinion bearings. Exceeding the preload specifications will compress the collapsible spacer to far and require the spacer to be replaced.

8. Install the pinion flange and using the pinion flange holding tool, replace the pinion nut and washer. Tighten the pinion nut to the same torque as before.
9. Align the matchmarks and install the driveshaft.
10. Check the level of the differential fluid when finished.

Differential Carrier

REMOVAL AND INSTALLATION

1. Raise and safely support the vehicle and disconnect the negative battery cable. Drain the oil from the rear differential.
2. Remove the left and right axle shafts.
3. Disconnect the driveshaft.
4. Remove the 4 mounting bolts to the upper rear suspension arm.
5. Support the differential carrier with a proper jack.
6. Remove the differential case nuts and lower the differential assembly from the rear housing.
   To install:
7. Clean and inspect the sealing surfaces of the carrier and the housing.
8. Using a liquid sealant on the carrier, install the carrier in the housing and tighten the nuts to 13–20 ft. lbs.

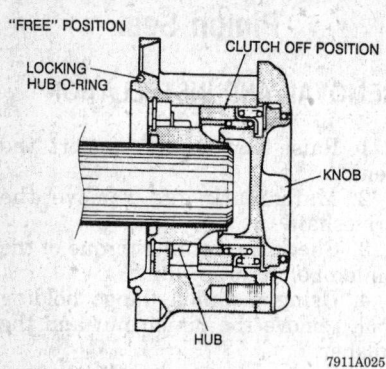

Wheel hub "FREE" position

7911A025

9. Replace the 4 mounting bolts to the upper rear suspension arm and tighten to 41 ft. lbs.

10. Connect the rear driveshaft and replace the left and rear axles.

11. Refill the rear differential to the proper level with SAE 75W–90W API GL5 Hypoid Gear Oil.

12. Lower the vehicle and connect the negative battery cable.

Front Axle Housing

REMOVAL AND INSTALLATION

1. Raise and safely support the vehicle. Remove the front wheels.

2. Drain the oil from the front differential.

3. Remove caliper assembly from brake disc and support caliper assembly.

4. Remove the tie rod end from the steering knuckle using a tie rod end remover.

5. Remove the 8 oil seal cover bolts and separate the felt pad, seal and retainer.

6. Remove the 4 kingpin bolts from the top and bottom and remove the kingpins from the knuckle.

7. Remove axle shafts from housing by pulling outward gradually.

8. Remove the bolts between the flange yoke and companion flange and disconnect the front drive shaft.

9. Remove the 8 U-joint nuts and remove the housing assembly.

To install:

10. Install the housing and torque the housing U-bolt nuts to 43 ft. lbs.

11. Connect the front drive shaft and install the bolts.

12. Install the front axle shafts and install the kingpins. Apply Loctite® to the kingpin bolts.

13. Install oil seal, felt pad and retainer and tighten the 8 cover bolts.

14. Install the tie rod end to the steering knuckle.

15. Install the front caliper to the disc and install the front wheels.

16. Refill the differential with the proper lubricant and bleed the brake system when finished.

Rear Axle Housing

REMOVAL AND INSTALLATION

1. Raise and safely support the vehicle. Drain the rear differential assembly.

2. Make sure the parking brake is in the released position.

3. Remove the rear wheels and remove the rear brake drums from the vehicle.

4. Disconnect the parking brake cables from the levers. Remove the parking brake lever stop plates.

5. Disconnect and plug the brake lines to the wheel cylinders.

6. Remove the rear wheel bearing retainer nuts and remove the axle shafts from the vehicle. Do not remove the axle shafts with the backing plates attached.

7. Remove the rear axle carrier assembly.

8. Disconnect the brake line from the flexible hose and remove the E-clip. Remove the brake lines from the rear housing.

9. Remove the breather hose from the axle housing and disconnect the rear driveshaft.

10. Support the rear axle housing with a jack.

11. Remove the ball joint bracket from the differential carrier and remove the carrier assembly.

12. Loosen the rear mount nut of the trailing rod. Do not remove it.

13. Disconnect the shock absorber lower mount bot.

14. Lower the jack to relieve the tension of the coil springs and remove the rear mount bolt of the trailing arm.

15. Remove the rear axle housing.

To install:

16. Place the rear axle housing on a jack and install the trailing rod rear mounting bolts. Mount the nuts but do not tighten.

17. Install the coil spring on the spring seat and raise the axle housing.

18. Replace the shock absorber lower mounting bolts. Do not tighten.

19. Install the differential carrier assembly and replace the rear upper ball joint bracket onto the carrier assembly and tighten to 37 ft. lbs.

20. Install the rear driveshaft and remove the jack from under the axle housing.

21. Replace the breather hose and brake lines to axle housing. Tighten them securely.

22. Connect the flexible brake hose to the bracket on the axle housing and secure with the E-clip.

23. Tighten the trailing rod nuts and the shock absorber nuts to 66 ft. lbs.

24. Install the brake line to the flexible hose and replace the axle shafts.

25. Install the brake lines to the wheel cylinders and replace the brake drums.

26. Install the wheels and refill the rear differential assembly.

27. Bleed the brake system and lower the vehicle.

STEERING

Steering Wheel

REMOVAL AND INSTALLATION

1. Disconnect the negative battery cable.

2. Disconnect the horn button and remove the steering wheel shaft nut.

3. Make matchmarks on the steering wheel and the shaft to use as a guide during reinstallation.

4. Remove the steering wheel, using a suitable steering wheel puller.

To install:

5. Install the steering wheel onto the shaft, aligning the matchmarks.

6. Install and tighten the shaft nut to 24 ft. lbs.

7. Install the horn button and connect the negative battery cable.

Manual Steering Gear

REMOVAL AND INSTALLATION

1. As required, raise and support the vehicle safely. Disconnect the steering lower shaft mounting bolts.

2. Disconnect the center link end from the pitman arm.

3. Remove the 3 steering gear box mounting bolts.

4. Disconnect the steering lower shaft joint and remove the steering gear.

To install:

5. Install the steering gear box by connecting to the lower shaft joint.

NOTE: Align the flat part of the steering gear worm shaft with the bolt hole of the lower shaft joint.

6. Replace the steering gear box mounting bolts and tighten to 50–72 ft. lbs.

7. Attach the center link to the pitman arm and tighten the nut to 22–50 ft. lbs.

8. Connect the lower shaft mounting bolts and tighten to 14–22 ft. lbs.

ADJUSTMENT

1. Check the worm shaft to make sure it is free from thrust play.

2. Place the pitman arm in a position that is nearly parallel with the worm shaft.

3. With the pitman arm in this position, the front wheels are in a straight forward position.

4. Measure the worm shaft starting torque from it's straight foward position. The torque should be 0.4–0.7 ft. lbs.

5. Adjust the worm shaft adjusting bolt to specifications.

Power Steering Gear

REMOVAL AND INSTALLATION

1. Disconnect the negative battery cable.

2. Remove the coolant reservoir tank from the radiator.

3. Disconnect the steering column lower shaft from the gear box.

4. Raise and safely support the vehicle.

5. Remove the center link nut and lock washer and disconnect the center link from the pitman arm, using a piman arm puller.

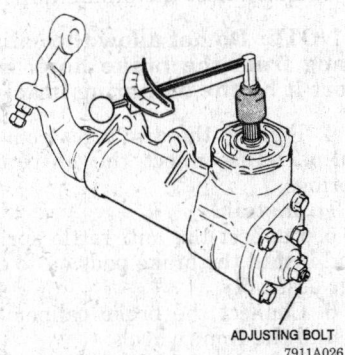

ADJUSTING BOLT
7911A026

Adjusting the steering gear assembly

6. Lower the vehicle. Remove the pressure hose from the power steering gear assembly and plug.

7. Disconnect the return hose and plug. Remove the 3 power steering gear mounting bolts.

8. Remove the power steering gear. Disconnect the pitman arm from the gear assembly, note the alignment marks.

To install:

9. Align the matchmarks on the pitman arm and the power steering gear sector shaft. Install the pitman arm to the gear assembly and tighten the nut to 95 ft. lbs.

10. Install the power steering gear assembly on the vehicle and tighten the mounting bolts to 72 ft. lbs.

11. Connect the power steering pressure and return hoses.

12. Raise and safely support the vehicle.

13. Install the center link to the pitman arm and tighten the nut to 40 ft. lbs. Lower the vehicle.

14. Connect the steering column lower shaft to the gear assembly and tighten the bolts to 29 ft. lbs.

15. Install the coolant reservoir tank to the radiator. Refill the power steering pump.

16. Connect the negative battery cable. Run the engine and operate the power steering. Recheck the fluid level and for any leaks.

ADJUSTMENT

1. Check the worm shaft to make sure it is free from thrust play.

2. Place the pitman arm in a position that is nearly parallel with the worm shaft.

3. With the pitman arm in this position, the front wheels are in a straight forward state.

4. Measure the worm shaft starting torque from it's straight forward state. The torque should be 0.4–0.7 ft. lbs.

5. Adjust the worm shaft adjusting bolt to specifications.

Power steering Pump

REMOVAL AND INSTALLATION

NOTE: Before disconnecting the power steering pressure and return line at the pump assembly, make sure any dirt or grease is removed.

1. Disconnect the negative battery cable. Remove the coolant reservoir tank from the radiator.

2. Loosen the air conditioning compressor adjusting and pivot bolts, if equipped.

3. Loosen the power steering pump adjusting and mounting bolts, if not equipped with air conditioning.

4. Remove the power steering belt.

5. Disconnect the power steering pressure and return hose and plug.

6. Disconnect the power steering pressure switch lead wire at the switch terminal.

7. Remove the engine oil filter.

8. Remove the power steering pump mounting and adjusting bolts.

9. Remove the power steering pump.

To install:

10. Install the power steering pump and replace the pump mounting bolts. Do not tighten.

11. Install the power steering pump pressure switch lead wire to the switch terminal.

12. Replace the power steering pressure and return hoses.

13. Install the power steering belt and tighten the power steering pump mounting bolts to 21 ft. lbs.

14. Tighten the air conditioning mounting bolts to 21 ft. lbs., if equipped.

15. Replace the coolant reservoir tank to the radiator and connect the negative battery cable. Refill the power steering pump.

16. Replace the oil filter and fill the crankcase to the proper level.

17. Run the engine and operate the power steering. Recheck the fluid level and for any leaks.

Belt ADJUSTMENT

1. To adjust the power steering belt tension, loosen the adjusting bolt of the air conditioning compressor, if equipped, or that of the power steering pump for vehicles without air conditioning.

2. Adjust the belt tension to 0.24–0.35 in. deflection, using the proper belt tension gauge.

3. Tighten the proper adjusting and mounting bolts to the specified torque.

System Bleeding

1. Raise and support the vehicle safely.

2. Fill the power steering reservoir to the specified level.

3. Run the engine for 3 to 5 minutes, stop it and add fluid if necessary to reach specified level.

4. With the engine stopped, turn the steering wheel to the left and to

the right as far as it turns. Repeat a few times and refill the reservoir.

5. With the engine running at idle speed, bleed air from the system by loosening the bleeder valve at the gear assembly.

6. Repeat the stop to stop turn of the steering wheel until all the foam is gone.

7. Tighten the bleed valve securely. Recheck the fluid level in the reservoir.

NOTE: When air bleeding is not complete, it is indicated by a foaming fluid on the level indicator or a humming noise from the power steering pump.

Tie Rod Ends

REMOVAL AND INSTALLATION

1. Raise and safely support the vehicle and remove the wheels.

2. Remove the tie rod end from the steering knuckle, using a suitable tie rod end remover tool.

3. Mark the tie rod end locknut position on the tie rod thread.

4. Loosen the locknut and remove the tie rod end from the tie rod.

To install:

5. Install the tie rod end locknut and the tie rod end to the tie rod. Align the locknut with the mark on the tie rod thread and tighten the locknut to 48 ft. lbs.

6. Connect the tie rod end to the steering knuckle. Tighten the castle nut until the holes of the split pin are aligned but only within the specified torque 33 ft. lbs.

BRAKES

Master Cylinder

REMOVAL AND INSTALLATION

1. Disconnect the negative battery cable. Remove the air cleaner case.

2. Disconnect the reservoir lead wire.

3. Clean the outside of the reservoir and remove the fluid from the reservoir.

4. Disconnect and plug the brake fluid lines at the master cylinder.

5. Remove the master cylinder to booster mounting bolts and remove the master cylinder.

To install:

6. Install the new master cylinder and tighten the mounting bolts to 12 ft. lbs.

7. Install the hydraulic brake lines to the master cylinder and tighten and tighten the flare nuts to 13 ft. lbs.

8. Replace the reservoir lead wire, if equipped.

9. Fill the reservoir with the specified brake fluid. Replace the air cleaner case.

10. Bleed the air from the brake hydraulic system and check the brake pedal play.

Proportioning Valve

REMOVAL AND INSTALLATION

1. Raise and safely support the vehicle.

2. Disconnect and plug the hydraulic brake lines from the proportioning valve assembly.

3. Remove the proportioning valve from the vehicle body.

NOTE: The proportioning valve should be removed with the spring attached.

To install:

4. Install the proportioning valve to the vehicle body and tighten the mounting bolts to 20 ft. lbs.

5. Connect the hydraulic brake lines to the proportioning valve and tighten the flare nuts to 13 ft. lbs.

6. Fill the brake reservoir to the proper level and bleed the brake hydraulic system. Lower the vehicle.

NOTE: Bleed the air from the proportioning valve bleeder valve.

Power Brake Booster

REMOVAL AND INSTALLATION

1. Disconnect the negative battery cable. Remove the air cleaner case.

2. Remove the fluid from the brake reservoir and remove the master cylinder.

3. Disconnect the vacuum hose from the booster. Remove the pushrod clevis pin and cotter pin from the brake pedal arm.

4. Disconnect the brake booster mounting nuts and remove the brake booster from the vehicle.

5. Disconnect the pedal attachment from the booster.

To install:

6. Connect the pedal attachment to the booster and install the booster to the vehicle. Tighten the mounting nuts to 7.5–11.5 ft. lbs.

7. Install the pushrod clevis pin and cotter pin to the brake pedal arm.

8. Connect the vacuum hose to the booster and install the master cylinder to the booster.

9. Replace the air cleaner case and connect the negative battery cable.

10. Fill the brake reservoir and bleed the brake hydraulic system.

Brake Caliper

REMOVAL AND INSTALLATION

1. Disconnect the negative battery cable. Raise and safely support the vehicle.

2. Remove the wheels. Disconnect and plug the brake line.

3. Remove the caliper mounting bolts and remove the caliper from the vehicle.

To install:

4. Install the caliper on the vehicle. Tighten the mounting bolts to 65 ft. lbs.

5. Connect the hydraulic brake line, using 2 new washers. Replace the front wheels.

6. Lower the vehicle. Connect negative battery cable.

7. Fill the brake reservoir and bleed the hydraulic brake system.

Disc Brake Pads

REMOVAL AND INSTALLATION

1. Disconnect the negative battery cable. Raise and safely support the vehicle.

2. Remove the wheels.

3. Disconnect the brake caliper.

NOTE: Do not allow the caliper hang from the brake hose. Support it by the mounting bracket.

4. Remove the disc pads from the caliper. Disconnect the anti-rattle springs.

To install:

5. Connect the anti rattle springs and install the brake pads on to caliper assembly.

6. Connect the brake caliper and install the front wheels.

7. Lower the vehicle. Connect the negative battery cable.

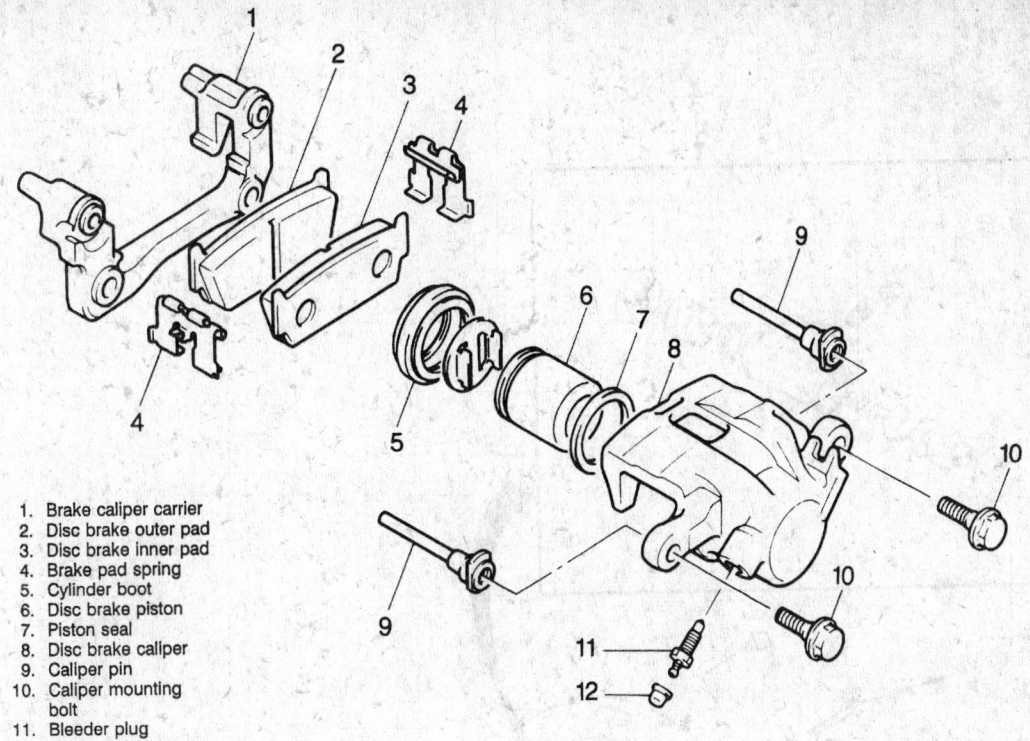

1. Brake caliper carrier
2. Disc brake outer pad
3. Disc brake inner pad
4. Brake pad spring
5. Cylinder boot
6. Disc brake piston
7. Piston seal
8. Disc brake caliper
9. Caliper pin
10. Caliper mounting bolt
11. Bleeder plug
12. Bleeder plug cap

7911A027

Exploded view of the front disc brake assembly

Brake Rotor

REMOVAL AND INSTALLATION

1. Disconnect the negative battery cable.

2. Raise and safely support the vehicle. Remove the wheel assembly.

3. Remove the brake caliper mounting bracket, with the caliper and the brake line attached, move it out of the way and support it.

NOTE: Do not allow the caliper hang from the brake hose. Support it by the mounting bracket.

4. Install 2 bolts into the threaded holes in the brake rotor and tighten them evenly. This will remove the rotor from the hub assembly.
To install:
5. Install the new brake rotor on the hub and install the caliper assembly.

6. Install the front wheels and tighten to 65 ft. lbs.

7. Lower the vehicle and connect the negative battery cable.

Brake Drums

REMOVAL AND INSTALLATION

1. Disconnect the negative battery cable. Raise and safely support the vehicle.

2. Make sure the parking brake is released.

3. Remove the wheels and the rear drum nuts from the vehicle.

4. Remove the brake drum, using a slide hammer puller.
To install:
5. Install the brake drum, tighten the brake drum nuts to 58 ft. lbs.

6. Install the rear wheels and lower the vehicle. Connect the negative battery cable.

Brake Shoes

REMOVAL AND INSTALLATION

1. Disconnect the negative battery cable. Raise and safely support the vehicle.

2. Remove the rear wheels from the vehicle. Remove the rear brake drums.

3. Remove the brake shoe hold-down springs.

4. Disconnect the parking brake cable from the parking brake shoe lever and remove the brake shoes.
To install:
5. Install the brake shoes and install the parking brake lever and cable to the assembly.

6. Install the brake shoe hold-down springs and install the rear brake drums.

7. Install the rear wheels, lower the vehicle and connect the negative battery terminal.

Wheel Cylinder

REMOVAL AND INSTALLATION

1. Disconnect the negative battery cable. Raise and safely support the vehicle.

2. Remove the rear wheels and brake drums from the vehicle.

3. Remove the rear brake shoes and disconnect the brake line from the rear of the wheel cylinder. Plug the brake line.

4. Remove the 2 rear wheel cylinder mounting bolts. Remove the rear wheel cylinder.
To install:
5. Install the wheel cylinder and tighten the mounting bolts to 6.0–8.5 ft. lbs.

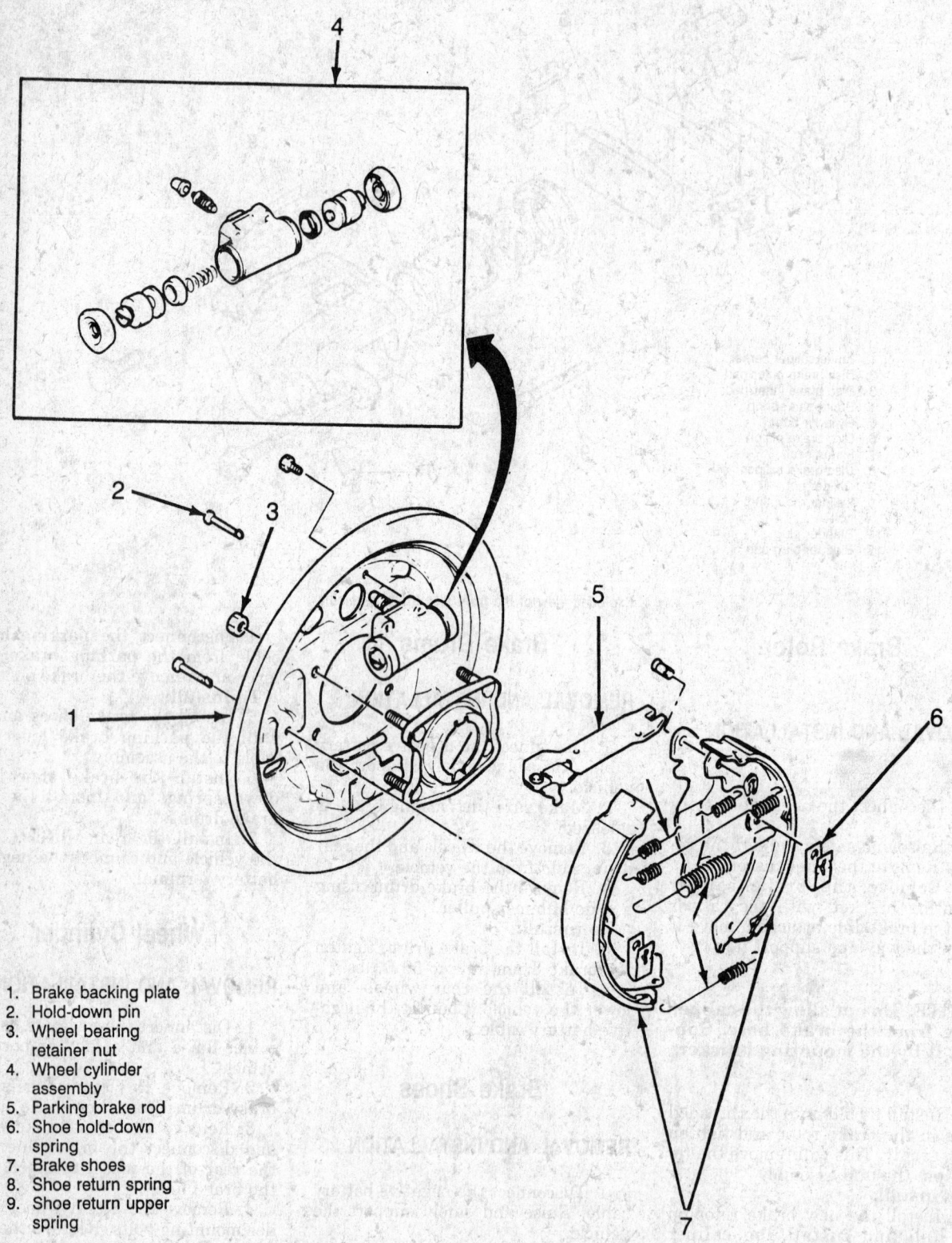

1. Brake backing plate
2. Hold-down pin
3. Wheel bearing
   retainer nut
4. Wheel cylinder
   assembly
5. Parking brake rod
6. Shoe hold-down
   spring
7. Brake shoes
8. Shoe return spring
9. Shoe return upper
   spring

7911A028

Exploded view of the rear brake drum assembly

6. Connect the brake line to the wheel cylinder and install the rear brake shoes.

7. Replace the rear brake drums and the rear wheels.

8. Lower the vehicle. Connect the negative battery cable.

9. Bleed the brake system and check for any leaks when finished.

Parking Brake Cable

REMOVAL AND INSTALLATION

1. Disconnect the parking brake cable from the parking lever.

2. Raise and safely support the vehicle. Remove the rear wheels and brake drums from the vehicle.

3. Remove the rear brake shoes and disconnect the parking brake cable from the parking brake shoe lever.

4. Remove the cable from the brake backing plate by squeezing the parking brake cable stop ring.

To install:

5. Install the cable to the backing plate and to the brake shoe lever.

6. Install the brake shoes and install the rear brake drums.

7. Install the rear wheels and connect the cable to the parking brake lever.

ADJUSTMENT

1. Adjust the parking brake lever by loosening or tightening the self locking nut at the park brake lever.

2. The proper adjustment is when the parking brake lever is within 7–9 notches, when the lever is pulled up at 44 lbs.

3. Ckeck the rear drum for dragging after adjustment.

FRONT SUSPENSION

MacPherson Strut

REMOVAL AND INSTALLATION

1. Disconnect the negative battery cable.

2. Raise and safely support the vehicle. Allow the front suspension to hang free.

3. Remove the front wheel. Disconnect the E-clip mounting the brake hose and remove the brake hose from the strut bracket.

4. Remove the strut bracket to steering knuckle bolts.

5. Lower the vehicle. Remove the strut support nuts, while holding the strut by hand. Remove the strut assembly.

To install:

6. Install the strut assembly and tighten the strut support nuts to 18 ft. lbs.

7. Raise and safely support the vehicle.

8. Install the strut bracket to steering knuckle bolts and tighten to 66 ft. lbs.

9. Connect the brake hose to the strut bracket using the E-clip.

10. Replace the front wheels and lower the vehicle.

11. Connect the negative battery cable.

Coil Springs

REMOVAL AND INSTALLATION

1. Disconnect the negative battery cable.

2. Raise and safely support the vehicle. Allow the front suspension to hang free.

3. Remove the front wheel and the locking hub assembly.

4. Remove the front axle circlip and washer.

5. Remove the brake caliper mounting bracket, with the caliper and the brake line attached, move it out of the way and support it.

6. Remove the brake disc and disconnect the stabilizer link from the control arm.

7. Disconnect the tie rod end and support the lower control arm using a jack.

8. Disconnect the strut bracket and remove the ball joint castle nut.

9. Remove the steering knuckle and the wheel hub assembly while lowering the jack.

10. Remove the coil spring from the vehicle.

To install:

11. Install the coil spring to the vehicle and install the wheel hub and steering knuckle while raising the jack.

12. Connect the strut bracket and install the ball joint castle nut.

13. Connect the tie rod end and connect the stabilizer link to the control arm.

14. Install the brake disc and connect the caliper assembly.

15. Install the front axle circlip and washer and install the locking hub assembly.

16. Install the front wheels, lower the vehicle and connect the negative battery cable.

Lower Ball Joints

INSPECTION

1. Inspect for the smoothness of the rotation.

2. Check the ball stud for damage.

3. Inspect the dust shield for damage.

REMOVAL AND INSTALLATION

1. Disconnect the negative battery cable. Remove the coil spring from the vehicle.

2. Remove the control arm mounting bolts.

3. Remove the control arm.

4. Disconnect the ball joint from the control arm and remove the ball joint.

To install:

5. Install the ball joint to the control arm and tighten the bolts to 63 ft. lbs.

6. Install the control arm to the chassis and tighten the bolts to 74 ft. lbs.

7. Install the coil spring and connect the negative battery cable.

Lower Control Arms

REMOVAL AND INSTALLATION

1. Disconnect the negative battery cable. Remove the coil spring from the vehicle.

2. Remove the control arm mounting bolts.

3. Remove the control arm.

To install:

4. Install the control arm to the chassis and tighten the bolts to 74 ft. lbs.

5. Install the coil spring and connect the negative battery cable.

Stabilizer Bar

REMOVAL AND INSTALLATION

1. Disconnect the negative battery cable. Raise and safely support the vehicle.

2. Disconnect the left and the right stabilizer ball joints from the front control arms.

3. Remove the stabilizer bar mount bushing bracket bolts and nuts.

4. Remove the stabilizer bar.

5. Remove the stabilizer links.

To install:

6. Install the new stabilizer bar, using new bushings.

7. Torque the stabilizer link nuts to 21 ft. lbs. and the stabilizer bar bracket bolts and nuts to 37 ft. lbs.

8. Lower the vehicle and connect the negative battery cable.

REAR SUSPENSION

Shock Absorbers

REMOVAL AND INSTALLATION

1. Raise and safely support the vehicle. Support the rear axle housing.

2. Remove the upper and lower shock absorber mounting bolts.

3. Remove the rear shock absorber.

To install:

4. Install the the rear shock absorber.

5. Replace the upper mounting bolts and tighten to 21 ft. lbs. Replace the lower mounting bolts and tighten to 63 ft. lbs.

Coil Springs

REMOVAL AND INSTALLATION

1. Raise and safely support the vehicle. Remove the rear wheels.

2. Support the rear axle housing, using a floor jack.

3. Remove the shock absorber lower mounting bolt.

4. Lower the rear axle housing so the coil spring can be removed.

5. Remove the coil spring from the vehicle.

To install:

6. Install the coil spring to the spring seat and raise the axle housing.

7. Install the lower shock absorber mounting bolt but do not tighten.

8. Connect the parking brake cable hangers and install the rear wheels.

9. Lower the vehicle and tighten the lower shock absorber nuts to 64 ft. lbs.

ISUZU 8

Amigo • Pickup • Rodeo • Trooper • Trooper II

SPECIFICATIONS

VEHICLE IDENTIFICATION CHART

Code	Liters	Cu. In. (cc)	Cyl.	Fuel Sys.	Eng. Mfg.
		Engine Code			
L	2.3	137 (2254)	4	2 BBL	Isuzu
E	2.6	156 (2559)	4	MFI	Isuzu
R	2.8	173 (2828)	6	TFI	GM
Z	3.1	189 (3098)	6	TFI	Isuzu
V	3.2	193 (3165)	6	MFI	Isuzu
W	3.2	193 (3165)	6	MFI	Isuzu

MFI - Multiport fuel injection

TFI - Throttle body fuel injection

Code	Year
	Model Year
M	1991
N	1992
P	1993
R	1994
S	1995

7911bC01

ENGINE IDENTIFICATION

Year	Model	Engine Displacement Liters (cc)	Engine Series (ID/VIN)	Fuel System	No. of Cylinders	Engine Type
1991	Amigo	2.3 (2254)	4ZD1	2 BBL	4	SOHC
	Amigo	2.6 (2559)	4ZE1	MFI	4	SOHC
	Pick-up	2.3 (2254)	4ZD1	2 BBL	4	SOHC
	Pick-up	2.6 (2559)	4ZE1	MFI	4	SOHC
	Pick-up	3.1 (3098)	CPC	TFI	6	OHV
	Rodeo	2.6 (2559)	4ZE1	MFI	4	SOHC
	Rodeo	3.1 (3098)	CPC	TFI	6	OHV
	Trooper	2.6 (2559)	4ZE1	MFI	4	SOHC
	Trooper	2.8 (2828)	CPC	TFI	6	OHV
1992	Amigo	2.3 (2254)	4ZD1	2 BBL	4	SOHC
	Amigo	2.6 (2559)	4ZE1	MFI	4	SOHC
	Pick-up	2.3 (2254)	4ZD1	2 BBL	4	SOHC
	Pick-up	2.6 (2559)	4ZE1	MFI	4	SOHC
	Pick-up	3.1 (3098)	CPC	TFI	6	OHV
	Rodeo	2.6 (2559)	4ZE1	MFI	4	SOHC
	Rodeo	3.2 (3165)	6VD1	MFI	6	SOHC
	Trooper	3.2 (3165)	6VD1	MFI	6	SOHC
	Trooper	3.2 (3165)	6VD1	MFI	6	DOHC
1993	Amigo	2.3 (2254)	4ZD1	2 BBL	4	SOHC
	Amigo	2.6 (2559)	4ZE1	MFI	4	SOHC
	Pick-up	2.3 (2254)	4ZD1	2 BBL	4	SOHC
	Pick-up	2.6 (2559)	4ZE1	MFI	4	SOHC
	Pick-up	3.1 (3098)	CPC	TFI	6	OHV
	Rodeo	2.6 (2559)	4ZE1	MFI	4	SOHC
	Rodeo	3.2 (3165)	6VD1	MFI	6	SOHC
	Trooper	3.2 (3165)	6VD1	MFI	6	SOHC
	Trooper	3.2 (3165)	6VD1	MFI	6	DOHC
1994-95	Amigo	2.6 (2559)	4ZE1	MFI	4	SOHC
	Pick-up	2.3 (2254)	4ZD1	2 BBL	4	SOHC
	Pick-up	2.3 (2254)	4ZD1	MFI	4	SOHC
	Pick-up	2.6 (2559)	4ZE1	MFI	4	SOHC
	Pick-up	3.1 (3098)	CPC	TFI	6	OHV
	Rodeo	2.6 (2559)	4ZE1	MFI	4	SOHC
	Rodeo	3.2 (3165)	6VD1	MFI	6	SOHC
	Trooper	3.2 (3165)	6VD1	MFI	6	SOHC
	Trooper	3.2 (3165)	6VD1	MFI	6	DOHC

MFI - Multiport fuel injection
TFI - Throttle body fuel injection
DOHC - Doubel overhead camshaft
SOHC - Single overhead camshaft
OHV - Overhead valve

7911bC02

GENERAL ENGINE SPECIFICATIONS

Year	Engine ID/VIN	Engine Displacement Liters (cc)	Fuel System Type	Net Horsepower @ rpm	Net Torque @ rpm (ft. lbs.)	Bore x Stroke (in.)	Compression Ratio	Oil Pressure @ rpm
1991	4ZD1	2.3 (2254)	2 BBL	96@4600	123@2600	3.52x3.54	8.3:1	57@3000
	4ZE1	2.6 (2559)	MFI	120@4600	150@2600	3.65x3.74	8.6:1	57-71@4000
	CPC	2.8 (2828)	TFI	125@4800	150@2400	3.50x2.99	8.9:1	30-55@2000
	CPC	3.1 (3098)	TFI	120@4400	165@2800	3.50x3.31	8.5:1	30-55@2000
1992	4ZD1	2.3 (2254)	2 BBL	96@4600	123@2600	3.52x3.54	8.3:1	57@3000
	4ZE1	2.6 (2559)	MFI	120@4600	150@2600	3.65x3.74	8.6:1	57-71@4000
	CPC	3.1 (3098)	TFI	120@4400	165@2800	3.50x3.31	8.5:1	30-55@2000
	6VD1 1	3.2 (3165)	MFI	175@5200	188@4000	3.67x3.03	9.3:1	57-80@3000
	6VD1 2	3.2 (3165)	MFI	190@5600	195@3800	3.67x3.03	9.3:1	57-80@3000
1993	4ZD1	2.3 (2243)	2 BBL	96@4600	123@2600	3.52x3.54	8.3:1	57@3000
	4ZE1	2.6 (2559)	MFI	120@4600	150@2600	3.65x3.74	8.6:1	57-71@4000
	CPC	3.1 (3098)	TFI	120@4400	165@2600	3.50x3..31	8.5:1	30-55@2000
	6VD1 1	3.2 (3165)	MFI	175@5200	188@4000	3.67x3.03	9.3:1	57-80@3000
	6VD1 2	3.2 (3165)	MFI	190@5600	195@3800	3.67x3.03	9.3:1	57-80@3000
1994-95	4ZD1	2.3 (2254)	2 BBL	96@4600	123@2600	3.52x3.54	8.3:1	57@3000
	4ZD1	2.3 (2254)	MFI	100@4600	125@2600	3.52x3.54	8.3:1	57@3000
	4ZE1	2.6 (2559)	MFI	120@4600	150@2600	3.65x3.74	8.6:1	57-71@4000
	CPC	3.1 (3098)	TFI	120@4400	165@2800	3.50x3.31	8.5:1	30-55@2000
	6VD1 1	3.2 (3165)	MFI	175@5200	188@4000	3.67x3.03	9.3:1	57-80@3000
	6VD1 2	3.2 (3165)	MFI	190@5600	195@3800	3.67x3.03	9.3:1	57-80@3000

MFI - Mutliport fuel injection
TFI - Throttle body fuel injection
MFI - Multiport fuel injection
1 Single overhead camshaft
2 Double overhead camshaft

7911bC03

GASOLINE ENGINE TUNE-UP SPECIFICATIONS

Year	Engine ID/VIN	Engine Displacement Liters (cc)	Spark Plugs Gap (in.)	Ignition Timing (deg.)		Fuel Pump (psi)	Idle Speed (rpm)		Valve Clearance	
				MT	AT		MT	AT	In.	Ex.
1991	4ZD1	2.3 (2254)	0.040	6B	6B	3.5	850	950	0.006	0.010
	4ZE1	2.6 (2559)	0.040	12B	12B	35	850	950	0.008	0.008
	CPC	2.8 (2828)	0.045	10B	10B	9-13	800	800	4	4
	CPC	3.1 (3098)	0.040	10B	10B	9-13	800	800	4	4
1992	4ZD1	2.3 (2254)	0.040	6B	6B	3.5	850	950	0.006	0.010
	4ZE1	2.6 (2559)	0.040	12B	12B	35	850	950	0.008	0.008
	CPC	3.1 (3098)	0.040	10B	10B	9-13	800	800	4	4
	6VD1	3.2 (3165)	0.040-0.043	5B	5B	41-46	750	750	NA	NA
1993	4ZD1	2.3 (2254)	0.040	6B	6B	3.5	850	950	0.006	0.010
	4ZE1	2.6 (2559)	0.040	12B	12B	35	850	950	0.008	0.008
	CPC	3.1 (3098)	0.040	10B	10B	9-13	800	800	4	4
	6VD1	3.2 (3165)	0.040-0.043	5B	5B	41-46	750	750	NA	NA
1994-95	4ZD1	2.3 (2254)	0.040	1	1	2	850	950	3	3
	4ZE1	2.6 (2559)	0.040	12B	12B	35	850	950	0.008	0.008
	CPC	3.1 (3098)	0.040	10B	10B	9-13	800	800	4	4
	6VD1	3.2 (3165)	0.040-0.043	5B	5B	41-46	750	750	NA	NA

NOTE: The Vehicle Emission Control Information label reflects changes made during production. The label s must be used if it differs from above.

B - Before top dead center

NA - Non-adjustable

1 Carbureted: 6B
  Fuel-injected: 12B

2 Carbureted: 3.5 psi
  Fuel-injected: 35 psi

3 Carbureted, Intake: 0.006
  Carbureted. Exhaust: 0.010
  Fuel-injected, Intake: 0.008
  Fuel injected, Exhaust: 0.008

4 Zero lash, plus 1 1/4 turns

7911bC04

FIRING ORDERS

NOTE: To avoid confusion, always replace spark plug wires one at a time.

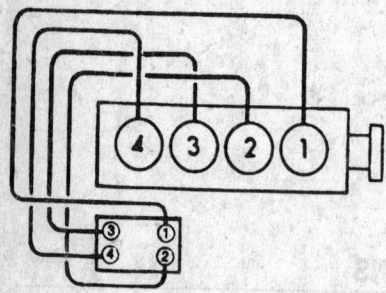

7911B001

4 Cylinder Engines
Engine Firing Order:1-3-4-2
Distributor Rotation: Counterclockwise

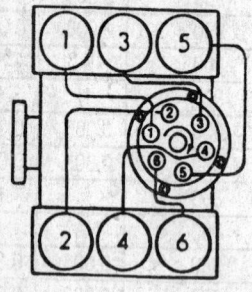

7911b002

V6 Engines (Except 3.2L)
Engine Firing Order: 1-2-3-4-5-6
Distributor
Rotation: Clockwise

CAPACITIES

Year	Model	Engine ID/VIN	Engine Displacement Liters (cc)	Engine Crankcase with Filter	Transmission (pts.) 4-Spd	5-Spd	Auto.	Transfer Case (pts.)	Drive Axle Front (pts.)	Rear (pts.)	Fuel Tank (gal.)	Cooling System (qts.)
1991	Amigo	4ZD1	2.3 (2254)	4.2	-	3.2	-	-	-	3.2	21.9	9.5
	Amigo	4ZE1	2.6 (2559)	5.2	-	6.2	13.8	3.0	3.2	3.8	21.9	9.5
	Amigo	CPC	3.1 (3098)	4.5	-	6.2	-	3.0	3.2	3.8	21.9	11.4
	Pick-up	4ZD1	2.3 (2254)	4.2	-	3.2	-	-	-	3.2	[2]	9.5
	Pick-up	4ZE1	2.6 (2559)	5.2	-	6.2	13.8	3.0	3.2	3.8	[2]	9.5
	Pick-up	CPC	3.1 (3098)	4.5	-	6.2	-	3.0	3.2	3.8	21.9	11.4
	Rodeo	4ZE1	2.6 (2559)	5.8	-	6.2	19.0	3.0	3.2	3.9	21.9	9.5
	Rodeo	CPC	3.1 (3098)	4.5	-	4.8	19.0	3.0	3.2	3.9	21.9	11.4
	Trooper	4ZE1	2.6 (2559)	5.2	-	6.2	24.0	[1]	3.2	3.8	21.9	8.5
	Trooper	CPC	2.8 (2828)	4.5	-	6.2	18.6	3.0	3.2	3.8	21.9	10.6
					-	3.2		-	-	3.2	21.9	9.5
1992	Amigo	4ZD1	2.3 (2254)	4.2	-	3.2	-	-	-	3.2	21.9	9.5
	Amigo	4ZE1	2.6 (2559)	5.2	-	6.2	13.8	3.0	3.2	3.8	21.9	9.5
	Pick-up	4ZD1	2.3 (2254)	4.2	-	3.2	-	-	-	3.2	[2]	9.5
	Pick-up	4ZE1	2.6 (2559)	5.2	-	6.2	13.8	3.0	3.2	3.8	[2]	9.5
	Pick-up	CPC	3.1 (3098)	4.5	-	6.2	-	3.0	3.2	3.8	21.9	11.4
	Rodeo	4ZE1	2.6 (2559)	5.8	-	[3]	19.0	3.0	3.2	3.9	21.9	[4]
	Rodeo	6VD1	3.2 (3165)	6.2	-	[3]	18.2	3.0	3.2	3.9	21.9	[5]
	Trooper	6VD1	3.2 (3165)	6.3	-	6.2	18.2	3.0	3.2	3.9	22.5	[5]
	Trooper	6VD1	3.2 (3165)	6.3	-	6.2	18.2	3.0	3.2	3.9	22.5	[5]
					-	3.2		-	-	3.2	21.9	9.5
1993	Amigo	4ZD1	2.3 (2254)	4.2	-	3.2	-	-	-	3.2	21.9	9.5
	Amigo	4ZE1	2.6 (2559)	5.2	-	6.2	13.8	3.0	3.2	3.8	21.9	9.5
	Pick-up	4ZD1	2.3 (2254)	4.2	-	3.2	13.8	-	-	3.2	[2]	9.5
	Pick-up	4ZE1	2.6 (2559)	5.2	-	6.2	13.8	3.0	3.2	3.8	[2]	9.5
	Pick-up	CPC	3.1 (3098)	4.5	-	6.2	-	3.0	3.2	3.8	21.9	11.4
	Rodeo	4ZE1	2.6 (2559)	5.8	-	[3]	19.0	3.0	3.2	3.9	21.9	[4]
	Rodeo	6VD1	3.2 (3165)	6.2	-	[3]	18.2	3.0	3.2	3.9	22.5	[5]
	Trooper	6VD1	3.2 (3165)	6.3	-	6.2	18.2	3.0	3.2	3.9	22.5	[5]
	Trooper	6VD1	3.2 (3165)	6.3	-	6.2	18.2	3.0	3.2	3.9	22.5	[5]
1994-95	Amigo	4ZE1	2.6 (2559)	5.2	-	6.2	13.8	3.0	3.2	3.8	21.9	9.5
	Pick-up	4ZD1	2.3 (2254)	4.2	-	3.2	-	-	-	3.2	[2]	9.5
	Pick-up	4ZE1	2.6 (2559)	5.2	-	6.2	13.8	3.0	3.2	3.2	[2]	9.5
	Pick-up	CPC	3.1 (3098)	4.5	-	6.2	-	3.0	3.2	3.2	21.9	11.4
	Rodeo	4ZE1	2.6 (2559)	5.8	-	[3]	19.0	3.0	3.2	3.9	21.9	[4]
	Rodeo	6VD1	3.2 (3165)	6.2	-	[3]	18.2	3.0	3.2	3.9	21.9	[4]
	Trooper	6VD1	3.2 (3165)	6.3	-	6.2	18.2	3.0	3.2	3.9	22.5	[5]
	Trooper	6VD1	3.2 (3165)	6.3	-	6.2	18.2	3.0	3.2	3.9	22.5	[5]

[1] Manual transaxle: 7.3
Automatic transaxle: 7.8
[2] Standard bed: 14.0
Spacecab and long bed: 19.8
[3] MUA transmission: 6.2
Borg-Warner transmission: 4.8
[4] Manual transmission: 9.7
Automatic transmission: 9.3
[5] Manual transmission: 9.3
Automatic transmission: 9.0

7911bC06

CAMSHAFT SPECIFICATIONS
All measurements given in inches.

Year	Engine ID/VIN	Engine Displacement Liters (cc)	Journal Diameter					Elevation		Bearing Clearance	Camshaft End Play
			1	2	3	4	5	In.	Ex.		
1991	4ZD1	2.3 (2254)	1.339	1.339	1.339	1.339	1.339	1.451	1.451	0.0033-0.0051	0.0002-0.0059
	4ZE1	2.6 (2559)	1.339	1.339	1.339	1.339	1.339	NA	NA	0.0026-0.0043	0.0080
	CPC	2.8 (2828)	1.867-1.881	1.867-1.881	1.867-1.881	1.867-1.881	NA	0.262-0.273	0.262-0.273	0.0010-0.0040	NA
	CPC	3.1 (3098)	1.867-1.881	1.867-1.881	1.867-1.881	1.867-1.881	NA	0.230-0.267	0.230-0.267	0.0010-0.0040	NA
1992	4XD1	2.3 (2254)	1.331-1.339	1.331-1.339	1.331-1.339	1.331-1.339	1.331-1.339	1.432-1.456	1.432-1.456	0.0033-0.0051	0.0002-0.0059
	4ZE1	2.6 (2559)	1.331-1.339	1.331-1.339	1.331-1.339	1.331-1.339	1.331-1.339	NA	NA	0.0026-0.0043	0.0080 [5]
	CPC	3.1 (3098)	1.867-1.881	1.867-1.881	1.867-1.881	1.867-1.881	NA	0.230-0.267	0.230-0.267	0.0010-0.0040	NA
	6VD1 [1]	3.2 (3165)	1.7634-1.7701	1.7634-1.7701	1.7634-1.7701	1.7634-1.7701	NA	3	3	0.0016-0.0197	0.0028-0.0098
	6VD1 [2]	3.2 (3165)	1.0555-1.0618	1.0555-1.0618	1.0555-1.0618	1.0555-1.0618	NA	4	4	0.0019-0.0059	0.0020-0.0079
1993	4ZD1	2.3 (2254)	1.331-1.339	1.331-1.339	1.331-1.339	1.331-1.339	1.331-1.339	1.432-1.456	1.432-1.456	0.0033-0.0051	0.0002-0.0059
	4ZE1	2.6 (2559)	1.331-1.339	1.331-1.339	1.331-1.339	1.331-1.339	1.331-1.339	NA	NA	0.0026-0.0043	0.0080 [5]
	CPC	3.1 (3098)	1.7634-1.7701	1.7634-1.7701	1.7634-1.7701	1.7634-1.7701	NA	3	3	0.0016-0.0197	0.0028-0.0098
	6VD1 [1]	3.2 (3165)	1.0555-1.0618	1.0555-1.0618	1.0555-1.0618	1.0555-1.0618	NA	4	4	0.0019-0.0059	0.0020-0.0079
	6VD1 [2]	3.2 (3165)	1.050-1.051	1.050-1.051	1.050-1.051	1.050-1.051	1.050-1.051	1.503	1.503	0.0011-0.0031	0.0020-0.0060
1994-95	4ZD1	2.3 (2254)	1.331-1.339	1.331-1.339	1.331-1.339	1.331-1.339	1.331-1.339	1.432-1.456	1.432-1.456	0.0033-0.0051	0.0002-0.0059
	4ZE1	2.6 (2559)	1.331-1.339	1.331-1.339	1.331-1.339	1.331-1.339	1.331-1.339	NA	NA	0.0026-0.0043	0.0080
	CPC	3.1 (3098)	1.867-1.881	1.867-1.881	1.867-1.881	1.867-1.881	NA	0.230-0.267	0.230-0.267	0.0010-0.0040	NA
	6VD1 [1]	3.2 (3165)	1.7634-1.7701	1.7634-1.7701	1.7634-1.7701	1.7634-1.7701	NA	3	3	0.0016-0.0197	0.0028-0.0098
	6VD1 [2]	3.2 (3165)	1.0555-1.0618	1.0555-1.0618	1.0555-1.0618	1.0555-1.0618	NA	4	4	0.0019-0.0059	0.0020-0.0079

NA - Not Available

1 Single overhead camshaft
2 Double overhead camshaft
3 Intake or exhaust: 1.6732-1.6870
4 Intake: 1.7441-1.7579
  Exhaust: 1.7429-1.7567
5 Limit

7911bC07

CRANKSHAFT AND CONNECTING ROD SPECIFICATIONS

All measurements are given in inches.

Year	Engine ID/VIN	Engine Displacement Liters (cc)	Crankshaft				Connecting Rod		
			Main Brg. Journal Dia.	Main Brg. Oil Clearance	Shaft End-play	Thrust on No.	Journal Diameter	Oil Clearance	Side Clearance
1991	4ZD1	2.3 (2254)	2.1819-2.2016	0.0009-0.0047	0.0024-0.0118	3	1.9065-1.9262	0.0012-0.0024	0.0078-0.0130
	4ZE1	2.6 (2559)	2.1819-2.2016	0.0009-0.0047	0.0024-0.0118	3	1.9065-1.9262	0.0008-0.0020	0.0078-0.0130
	CPC	2.8 (2828)	2.6473-2.6483	0.0016-0.0033	0.0020-0.0080	3	1.9983-1.9993	0.0013-0.0026	0.0060-0.0170
	CPC	3.1 (3098)	2.6473-2.6483	0.0012-0.0027	0.0024-0.0083	3	1.9983-1.9994	0.0011-.0032	0.0140-0.0267
1992	4ZD1	2.3 (2254)	2.1819-2.2016	0.0009-0.0047	0.0024-0.0118	3	1.9065-1.9262	0.0012-0.0470	0.0078-0.0130
	4ZE1	2.6 (2559)	2.1819-2.2016	0.0009-0.0047	0.0024-0.0118	3	1.9065-1.9262	0.0012-0.0470	0.0078-0.0130
	CPC	3.1 (3098)	2.6473-2.6483	0.0012-0.0027	0.0024-0.0083	3	1.9983-1.9994	0.0011-0.0032	0.0140-0.0267
	6VD1	3.2 (3165)	2.5165-2.5170	0.0010-0.0050	0.0020-0.0120	3	2.2434-2.2441	0.0010-0.0047	0.0060-0.0160
1993	4ZD1	2.3 (2254)	2.1819-2.2016	0.0009-0.0047	0.0024-0.0118	3	1.9065-1.9262	0.0012-0.0470	0.0078-0.0130
	4ZE1	2.6 (2559)	2.1819-2.2016	0.0009-0.0047	0.0024-0.0118	3	1.9065-1.9262	0.0012-0.0470	0.0078-0.0130
	CPC	3.1 (3098)	2.6473-2.6483	0.0012-0.0027	0.0024-0.0083	3	1.9983-1.9994	0.0011-0.0032	0.0140-0.0267
	6VD1	3.2 (3165)	2.5165-2.5170	0.0010-0.0050	0.0020-0.0120	3	2.2434-2.2441	0.0010-0.0047	0.0060-0.0160
1994-95	4ZD1	2.3 (2254)	2.1819-2.2016	0.0009-0.0047	0.0024-0.0118	3	1.9065-1.9262	0.0012-0.0470	0.0078-0.0130
	4ZE1	2.6 (2559)	2.1819-2.2016	0.0009-0.0047	0.0024-0.0118	3	1.9065-1.9262	0.0012-0.0470	0.0078-0.0130
	CPC	3.1 ((3098)	2.6473-2.6483	0.0012-0.0027	0.0024-0.0083	3	1.9983-1.9994	0.0011-0.0032	0.0140-0.0267
	6VD1	3.2 (3165)	2.5165-2.5170	0.0010-0.0050	0.0020-0.0120	3	1.2434-2.2441	0.0010-0.0047	0.0060-0.0160

7911bC08

VALVE SPECIFICATIONS

Year	Engine ID/VIN	Engine Displacement Liters (cc)	Seat Angle (deg.)	Face Angle (deg.)	Spring Test Pressure (lbs. @ in.)	Spring Installed Height (in.)	Stem-to-Guide Clearance (in.)		Stem Diameter (in.)	
							Intake	Exhaust	Intake	Exhaust
1991	4ZD1	2.3 (2245)	45	45	49-56@ 1.61	1.61	0.0009- 0.0080	0.0015- 0.0098	0.3102- 0.3134	0.3091- 0.3124
	4ZE1	2.6 (2559)	45	45	49-56@ 1.61	1.61	0.0009- 0.0080	0.0015- 0.0098	0.3102- 0.3134	0.3091- 0.3124
	CPC	2.8 (2828)	46	45	175@1.26	1.72	0.0010- 0.0027	0.0010- 0.0027	0.3410- 0.3420	0.3410- 0.3420
	CPC	3.1 (3098)	46	45	82@1.58	1.58	0.0010- 0.0027	0.0010- 0.0027	0.3410- 0.3420	0.3410- 0.3420
1992	4ZD1	2.3 (2254)	45	45	49-56@ 1.61	1.61	0.0009- 0.0080	0.0015- 0.0098	0.3102- 0.3134	0.3091- 0.3124
	4ZE1	2.6 (2559)	45	45	49-56@ 1.61	1.61	0.0009- 0.0080	0.0015- 0.0098	0.3102- 0.3134	0.3091- 0.3124
	CPC	3.1 (3098)	46	45	82@1.58	1.58	0.0010- 0.0027	0.0010- 0.0027	0.3410- 0.3420	0.3410- 0.3420
	6VD1	3.2 (3165)	45	45	45-55@ 1.54	1.54	0.0009- 0.0078	0.0012- 0.0078	0.2323- 0.2346	0.2323- 0.2350
1993	4ZD1	2.3 (2254)	45	45	49-56@ 1.61	1.61	0.0009- 0.0080	0.0015- 0.0098	0.3102- 0.3134	0.3091- 0.3124
	4ZE1	2.6 (2559)	45	45	49-56@ 1.61	1.61	0.0009- 0.0080	0.0015- 0.0098	0.3102- 0.3134	0.3191- 0.3124
	CPC	3.1 (3098)	46	45	82@1.58	1.58	0.0010- 0.0027	0.0010- 0.0027	0.3410- 0.3420	0.3410- 0.3420
	6VD1	3.2 (3165)	45	45	45-55@ 1.54	1.54	0.0009- 0.0078	0.0012- 0.0078	0.2323- 0.2346	0.2323- 0.2350
1994-95	4ZD1	2.3 (2254)	45	45	49-56@ 1.61	1.61	0.0009- 0.0080	0.0015- 0.0098	0.3102- 0.3134	0.3091- 0.3124
	4ZE1	2.6 (2559)	45	45	49-56@ 1.61	1.61	0.0009- 0.0080	0.0015- 0.0098	0.3102- 0.3134	0.3191- 0.3124
	CPC	3.1 (3098)	46	45	82@1.58	1.58	0.0010- 0.0027	0.0010- 0.0027	0.3410- 0.3420	0.3410- 0.3420
	6VD1	3.2 (3165)	45	45	45-55@ 1.54	1.54	0.0009- 0.0078	0.0012- 0.0078	0.2323- 0.2346	0.2323- 0.2350

7911bC09

PISTON AND RING SPECIFICATIONS

All measurements are given in inches.

Year	Engine ID/VIN	Engine Displacement Liters (cc)	Piston Clearance	Ring Gap Top Compression	Ring Gap Bottom Compression	Ring Gap Oil Control	Ring Side Clearance Top Compression	Ring Side Clearance Bottom Compression	Ring Side Clearance Oil Control
1991	4ZD1	2.3 (2254)	0.0008-0.0016	0.0120-0.0590	0.0240-0.0590	0.0080-0.0590	0.0010-0.0059	0.0008-0.0059	NA
	4ZE1	2.6 (2559)	0.0010-0.0018	0.0120-0.0590	0.0240-0.0590	0.0080-0.0590	0.0010-0.0059	0.0008-0.0059	NA
	CPC	2.8 (2828)	0.0007-0.0017	0.0098-0.0196	0.0098-0.0196	0.0020-0.0550	0.0011-0.0027	0.0015-0.0037	0.0080 MAX
	CPC	3.1 (3098)	0.0009-0.0022	0.0100-0.0200	0.0200-0.0280	0.0100-0.0300	0.0020-0.0035	0.0020-0.0035	0.0080 MAX
1992	4ZD1	2.3 (2254)	0.0008-0.0016	0.0120-0.0590	0.0240-0.0590	0.0080-0.0590	0.0010-0.0059	0.0008-0.0059	NA
	4ZE1	2.6 (2559)	0.0010-0.0018	0.0120-0.0590	0.0240-0.0590	0.0080-0.0590	0.0010-0.0059	0.0008-0.0059	NA
	CPC	3.1 (3098)	0.0009-0.0022	0.0100-0.0200	0.0200-0.0280	0.0100-0.0300	0.0020-0.0035	0.0020-0.0035	NA
	6VD1	3.2 (3165)	1	0.0138-0.0590	0.0177-0.0590	0.0059-0.0590	0.0006-0.0059	0.0006-0.0059	NA
1993	4ZD1	2.3 (2254)	0.0008-0.0016	0.0120-0.0590	0.0240-0.0590	0.0080-0.0590	0.0010-0.0059	0.0008-0.0059	NA
	4ZE1	2.6 (2559)	0.0010-0.0018	0.0120-0.0590	0.0240-0.0590	0.0080-0.0590	0.0010-0.0059	0.0008-0.0059	NA
	CPC	3.1 (3098)	0.0009-0.0022	0.0100-0.0200	0.0200-0.0280	0.0100-0.0300	0.0020-0.0035	0.0020-0.0035	NA
	6VD1	3.2 (3165)	1	0.0138-0.0590	0.0177-0.0590	0.0059-0.0590	0.0006-0.0059	0.0006-0.0059	NA
1994-95	4ZD1	2.3 (2254)	0.0008-0.0016	0.0120-0.0590	0.240-0.0590	0.0080-0.0590	0.0010-0.0059	0.0008-0.0059	NA
	4ZE1	2.6 (2559)	0.0010-0.0018	0.0120-0.0590	0.240-0.0590	0.0080-0.0590	0.0010-0.0059	0.0008-0.0059	NA
	CPC	3.1 (3098)	0.0009-0.0022	0.0100-0.0200	0.0200-0.0280	0.0100-0.0300	0.0020-0.0035	0.0020-0.0035	NA
	6VD1	3.2 (3165)	1	0.0138-0.0590	0.0177-0.0590	0.0059-0.0590	0.0006-0.0059	0.0006-0.0059	NA

NA - Not Available

1 Size mark A pistons:
    0.47 inch (12mm) from top of cylinder block: 0.0016-0.0024
    1.89 inch (48mm) from top cylinder block : 0.0019-0.0027
    3.74 inch (95mm) from top of cylinder block: 0.0024-0.0031
Size mark B pistons:
    0.47 inch (12mm) from top of cylinder block: 0.0012-0.0020
    1.89 inch (48mm) from top cylinder block : 0.0016-0.0023
    3.74 inch (95mm) from top of cylinder block: 0.0020-0.0027
Size mark C pistons:
    0.47 inch (12mm) from top of cylinder block: 0.0008-0.0016
    1.89 inch (48mm) from top cylinder block : 0.0012-0.0019
    3.74 inch (95mm) from top of cylinder block: 0.0016-0.0023

7911bC10

TORQUE SPECIFICATIONS
All readings in ft. lbs.

Year	Engine ID/VIN	Engine Displacement Liters (cc)	Cylinder Head Bolts	Main Bearing Bolts	Rod Bearing Bolts	Crankshaft Damper Bolts	Flywheel Bolts	Manifold		Spark Plugs	Lug Nut
								Intake	Exhaust		
1991	4ZD1	2.3 (2254)	2	65-80	42-45	76-98	40-47	14-18	14-18	10-17	4
	4ZE1	2.6 (2559)	2	65-80	42-45	79-102	40-47	14-18	14-18	10-17	4
	CPC	2.8 (2828)	3	70	39	70	52	23	25	22	4
	CPC	3.1 (3098)	3	72	39	70	52	19	25	14	4
1992	4ZD1	2.3 (2254)	2	72	43	87	40	16	16	14	4
	4ZE1	2.6 (2559)	2	72	43	87	40	16	16	14	4
	CPC	3.1 (3098)	3	72	39	70	52	19	25	14	4
	6VD1	3.2 (3165)	1	29	39	123	40	17	42	13	87
1993	4ZD1	2.3 (2254)	2	72	43	87	40	16	16	14	4
	4ZE1	2.6 (2559)	2	72	43	87	40	16	16	14	4
	CPC	3.1 (3098)	3	72	39	70	52	19	25	14	4
	6VD1	3.2 (3165)	1	29	49	123	40	17	42	13	87
1994-95	4ZD1	2.3 (2254)	2	72	43	87	40	16	16	14	4
	4ZE1	2.6 (2559)	2	72	43	87	40	16	16	14	4
	CPC	3.1 (3098)	3	72	39	70	52	19	25	14	4
	6VD1	3.2 (3165)	1	29	49	123	40	17	42	13	87

1 8x1.25 bolts: 15 ft. lbs.
  11x1.5 bolts: 47 ft. lbs.
2 Step 1: 58 ft. lbs.
  Step 2: 72 ft. lbs.

3 Step 1: 41 ft. lbs.
  Step 2: Turn an additional 90 degrees
4 Steel wheels: 58-72 ft. lbs.
  Aluminum wheels: 80-94 ft. lbs.

7911bC11

BRAKE SPECIFICATIONS
All measurements in inches unless noted

Year	Model	Master Cylinder Bore	Brake Disc Original Thickness	Brake Disc Minimum Thickness	Brake Disc Maximum Runout	Brake Drum Diameter Original Inside Diameter	Brake Drum Diameter Max. Wear Limit	Brake Drum Diameter Maximum Machine Diameter	Minimum Lining Thickness Front	Minimum Lining Thickness Rear
1991	Amigo	1	3	7	0.005	10.00	10.06	NA	0.039	0.039
	Pick-up	1	3	7	0.005	10.00	10.06	NA	0.039	0.039
	Rodeo	1.000	0.866	0.811	0.005	10.00	10.06	NA	0.039	0.039
	Trooper	1.000	3	7	0.005	-	-	-	0.039	0.039
1992	Amigo	1.000	5	9	0.005	8.27 [13]	8.32 [13]	NA	0.039	0.039 [15]
	Pick-up	0.938	3	8	12	10.01 [13]	10.06 [13]	NA	0.039	[14]
	Rodeo [2]	1.000	0.866	10	0.005	10.00	10.06	NA	0.039	0.039
	Rodeo [4]	1.000	6	11	0.005	8.27 [13]	8.32 [13]	NA	0.039	0.039 [15]
	Trooper	1.000	6	11	0.005	8.27 [13]	8.27 [13]	NA	0.039	0.039 [15]
1993	Amigo	1.000	5	9	0.005	8.27 [13]	8.32 [13]	NA	0.039	0.039 [15]
	Pick-up	0.938	3	8	12	10.01	10.06	NA	0.039	[14]
	Rodeo [2]	1.000	0.866	10	0.005	10.00	10.06	NA	0.039	0.039
	Rodeo [4]	1.000	6	11	0.005	8.27 [13]	8.32 [13]	NA	0.039	0.039 [15]
	Trooper	1.000	6	11	0.005	8.27 [13]	8.27 [13]	NA	0.039	0.039 [15]
1994-95	Amigo	1.000	5	7	0.005	8.27 [13]	8.32 [13]	NA	0.039	0.039 [13]
	Pick-up	0.938	4	6	12	10.01	10.06	NA	0.039	[14]
	Rodeo [2]	1.000	0.866	8	0.005	10.00	10.06	NA	0.039	0.039
	Rodeo [4]	1.000	6	9	0.005	8.27 [13]	8.32 [13]	NA	0.039	0.039 [15]
	Trooper	1.000	6	9	0.005	8.27 [13]	8.27 [13]	NA	0.039	0.039 [15]

NA - Not Available

1 2.3L engine: 0.938
   Except 2.3L engine: 1.000
2 2.6L engine
3 Front: 0.866
   Rear: 0.472
4 3.2L engine
5 Front: 1.026
   Rear: 0.709
6 Front: 1.020
   Rear: 0.710
7 Front: 0.811
   Rear: 0.417
8 Front: 0.811
   Rear: 0.417 (Minimum machine diameter: 0.417)
9 Front: 0.970
   Rear: 0.654 (Minimu machine diameter: 0.668)

10 0.811 (Minimum machine diameter: 0.826)
11 Front: 0.969 (Minimu machine diameter: 0.983)
   Rear: 0.654 (Minimum machine diameter: 0.668)
12 Front: 0.0050
   Rear: 0.0051
13 Emergency brake drum surface
14 Disc: 0.040
   Drum: 0.039
15 Specification includes disc pads and parking brake shoes

7911bC12

WHEEL ALIGNMENT

Year	Model		Caster Range (deg.)	Caster Preferred Setting (deg.)	Camber Range (deg.)	Camber Preferred Setting (deg.)	Toe-in (in.)	Steering Axis Inclination (deg.)
1991	Amigo		1 3/4P-3 1/4P	2 1/2P	1/2N-1 1/2P	1/2P	0-3/8P	10
	Pickup	1	7/8P-2 3/8P	1 5/8P	1/2N-1 1/2P	1/2P	0-3/8P	10
	Pickup	2	1 1/8P-2 5/8P	1 7/8P	1/2N-1 1/2P	1/2P	0-3/8P	10
	Pickup	3	1 3/16P-2 11/16P	1 15/16P	1/2N-1 1/2P	1/2P	0-3/8P	10
	Pick-up	4	1 7/16P-2 15/16P	2 3/16P	1/2N-1 1/2P	1/2P	0-3/8P	10
	Rodeo		1 9/16P-2 1/16P	2 1/3P	1/2N-1 1/2P	1/2P	0-3/8P	10
	Trooper		2P-3P	2 1/2P	0-1P	1/2P	0-5/32P	10
1992	Amigo		1 3/4P-3 1/4P	2 1/2P	1/2N-1 1/2P	1/2P	0-3/16P	10
	Pick-up	1	7/8P-2 3/8P	1 5/8P	1/2N-1 1/2P	1/2P	0-3/16P	10
	Pick-up	2	1 1/8P-2 5/8P	1 7/8P	1/2N-1 1/2P	1/2P	0-3/16P	10
	Pick-up	3	1 3/16P-2 11/16P	1 15/16P	1/2N-1 1/2P	1/2P	0-3/16P	10
	Pick-up	4	1 7/16P-2 15/16P	2 3/16P	1/2N-1 1/2P	1/2P	0-3/16P	10
	Rodeo		1 9/16P-2 1/16P	2 1/3P	1/2N-1 1/2P	1/2P	0-3/16P	10
	Trooper	5	1 1/4P-2 3/4P	2P	1/2N-1/2P	0	3/32N-3/16P	12
	Trooper	6	1 7/16P-2 15/16P	2 3/16P	1/2N-1/2P	0	3/32N-3/32P	12
1993	Amigo		1 3/4P-3 1/4P	2 1/2P	1/2N-1 1/2P	1/2P	0-3/16P	10
	Pick-up	1	7/8P-2 3/8P	1 5/8P	1/2N-1 1/2P	1/2P	0-3/16P	10
	Pick-up	2	1 1/8P-2 5/8P	1 7/8P	1/2N-1 1/2P	1/2P	0-3/16P	10
	Pick-up	3	1 3/16P-2 11/16P	1 15/16P	1/2N-1 1/2P	1/2P	0-3/16P	10
	Pick-up	4	1 7/16P-2 15/16P	2 3/16P	1/2N-1 1/2P	1/2P	0-3/16P	10
	Rodeo		1 9/16P-2 1/16P	2 1/3P	1/2N-1 1/2P	1/2P	0-3/16P	10
	Trooper	5	1 1/4P-2 3/4P	2P	1/2N-1/2P	0	3/32N-3/16P	12
	Trooper	6	1 7/16P-2 15/16P	2 3/16P	1/2N-1/2P	0	3/32N-3/32P	12
1994-95	Amigo		1 3/4P-3 1/4P	2 1/2P	1/2N-1 1/2P	1/2P	0-3/16P	10
	Pick-up	1	7/8P-2 3/8P	1 5/8P	1/2N-1 1/2P	1/2P	0-3/16P	10
	Pick-up	2	1 1/8P-2 5/8P	1 7/8P	1/2N-1 1/2P	1/2P	0-3/16P	10
	Pick-up	3	1 3/16P-2 11/16P	1 15/16P	1/2N-1 1/2P	1/2P	0-3/16P	10
	Pick-up	4	1 7/17P-2 15/16P	2 3/16P	1/2N-1 1/2P	1/2P	0-3/16P	10
	Rodeo		1 9/16P-2 1/16P	2 1/3P	1/2N-1 1/2P	1/2P	0-3/16P	10
	Trooper	5	1 1/4P-2 3/4P	2P	1/2N-1/2P	0	3/32P-3/16P	12
	Trooper	6	1 7/16P-2 15/16P	2 3/16P	1/2N-1/2P	0	3/32P-3/16P	12

NA - Not Available

1 2WD short wheelbase
2 2WD long wheelbase
3 4WD short wheelbase
4 4WD long wheelbase
5 2 door
6 4 door

7911bC13

ENGINE ELECTRICAL

NOTE: Disconnecting the negative battery cable on some vehicles may interfere with the functions of the on board computer systems and may require the computer to undergo a relearning process, once the negative battery cable is reconnected.

Distributor

REMOVAL

2.3L and 2.6L Engines

1. Disconnect the battery ground cable and the distributor wiring connector.
2. Turn the engine to TDC of cylinder No. 4. Remove the distributor cap and mark the position of the rotor to the distributor body and the distributor body to the engine block. This allows for easier installation.
3. Remove the distributor holddown bolt and remove the distributor.

2.8L and 3.1L Engines

1. Disconnect the negative battery cable.
2. Disconnect and label the spark plug wires at the distributor cap.
3. Remove the distributor cap from the distributor. Disconnect and label the distributor wiring.
4. Using a piece of chalk, matchmark the rotor to the distributor housing and the housing to the engine.
5. Remove the distributor holddown nut, the clamp and lift the distributor from the engine.

NOTE: When removing the distributor, it may be necessary to rotate the distributor shaft slightly to disengage it from the drive gear.

INSTALLATION

Timing Not Disturbed

1. Lower the distributor into the engine and align the rotor and distributor housing to the matchmarks.
2. Check and/or adjust the ignition timing when finished.

Timing Disturbed

2.8L AND 3.1L ENGINES

1. Remove the No. 1 spark plug.
2. Rotate the crankshaft in the normal direction of rotation until compression is felt at the spark plug hole.
3. Continue rotating the engine in the same direction while observing the timing marks at the indicator line up when No. 1 cylinder is at TDC.
4. Install the distributor and align the rotor with the No. 1 lug on the distributor cap.
5. Check and/or adjust the ignition timing when finished.

2.3L AND 2.6L ENGINES

1. Remove the valve cover. Turn the engine so cylinder No. 4 is on TDC of the compression stroke and the camshaft pulley setting mark is aligned with the setting mark on the front plate.
2. Lubricate the distributor O-ring with engine oil. Align the distributor setting mark with the distributor shaft setting mark.
3. Install the distributor on the cylinder head aligning the mark on the cylinder head with the mark on the distributor shaft.
4. Install the distributor holddown bolt and valve cover. Install the distributor cap and the distributor wiring. Connect the negative battery terminal and time the engine. Torque the hold-down bolt to 14 ft. lbs. (19 Nm).

Ignition Timing

ADJUSTMENT

NOTE: The following procedures are basic outlines for timing the ignition system. As emission controls change and are updated, the timing procedures will change. Check the underhood emissions sticker for updated timing instructions and follow them exactly.

2.3L and 2.6L Engines

NOTE: Set the air gap in the distributor before timing the engine. The timing marks are located near the front crankshaft pulley and consist of a pointer with graduations attached to the engine block and a mark on the crankshaft pulley.

1. Check and correct the air gap in the distributor.
2. Locate and clean the timing marks on the crankshaft pulley and the front of the engine.
3. Using an inductive pick-up timing light, connect it to the No. 1 spark plug wire. Attach a timing light to the ignition coil.
4. If the distributor is equipped with a vacuum advance, disconnect and plug the vacuum line.
5. Make sure all wires from the timing light and tachometer are clear of the fan and belts. Start the engine.
6. Adjust the idle to the correct rpm.
7. Aim the timing light at the timing marks. Adjust the distributor until the timing marks are aligned.
8. Tighten the distributor mounting bolt and check the timing again.
9. Turn the engine OFF and remove the timing light and tachometer. Connect the distributor vacuum line.

2.8L and 3.1L Engines

1. Set the parking brake and block the drive wheels.
2. Operate the engine to normal operating temperatures and turn the air conditioning OFF, if equipped.
3. Verify that the Check Engine light is not turned ON.
4. Place the Electronic Spark Timing (EST) into the bypass mode by disconnect the timing connector.

NOTE: The EST is a single wire connector, located under the center console in the passenger compartment. Do not disconnect the 4-wire connector from the distributor.

5. Using an inductive pick-up timing light, connect it to the No. 1 spark plug wire.
6. Check and/or adjust the engine speed to 800 rpm.
7. Loosen the distributor holddown bolt and turn the distributor until the timing mark, on the crankshaft pulley, is aligned with the 10 degree BTDC timing mark on the timing cover.
8. Tighten the distributor holddown bolt.
9. Reconnect the timing connector and clear the ECM trouble code(s).

Air Gap Setting

The air gap setting in the distributor should be checked and adjusted before the ignition timing is adjusted.

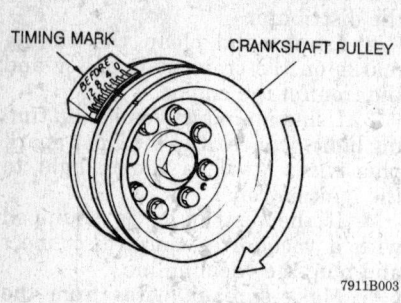

TIMING MARK   CRANKSHAFT PULLEY

7911B003

View of the ignition timing marks

ADJUSTMENT

1. Remove the distributor cap, O-ring and rotor.

2. Use a feeler gauge to measure the air gap at the pick-up coil projection. The gap should be 0.008–0.016 in. for 3.1L engine or 0.012–0.020 in. for 2.3L and 2.6L engines; adjust it, if necessary.

3. Loosen the screws and move the signal generator until the gap is correct. Tighten the screws and recheck the gap.

NOTE: The electrical parts in this system are not repairable. If found to be defective, they must be replaced.

Alternator

BELT TENSION ADJUSTMENT

2.3L and 2.6L Engines

1. Loosen the alternator pivot bolt.
2. Rotate the alternator to produce a belt deflection of 0.40 in. (10mm).

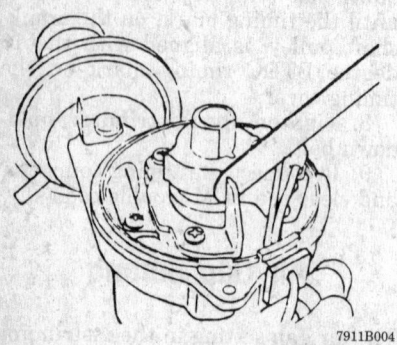

7911B004

Adjusting the air gap — 2.3L engine

3. Tighten the alternator pivot bolt.

2.8L and 3.1L Engines

The 2.8L and 3.1L engines uses a single serpentine belt to drive all engine accessories. The belt tension is maintained by a spring loaded tensioner. The belt tensioner has the ability to control the belt tension over a broad range of belt lengths as the belt ages and stretches.

REMOVAL AND INSTALLATION

2.3L and 2.6L Engines

1. Disconnect the negative battery cable. If equipped with an air pump, it may be necessary to remove it.

2. Disconnect and label the alternator wiring.

3. Remove the alternator pivot bolt on the lower part of the alternator. Remove the drive belt from the pulley.

4. Remove the alternator mounting bolt(s) and the alternator from the engine.

To install:

5. Install the alternator.

6. Adjust the belt tension and tighten the alternator mounting bolts.

7. Reconnect the alternator's wiring connectors. Connect the negative battery cable.

2.8L and 3.1L Engines

1. Disconnect the negative battery cable.

2. Remove the terminal plug and the battery lead from the rear of the alternator.

3. Remove the drive belt.

4. Remove the bracket bolt from the rear of the alternator.

5. Remove the mounting bolts from the front of the alternator and the alternator from the vehicle.

To install:

6. Install the alternator and the mounting bolts.

7. Install the drive belt.

8. Torque the lower mounting bolt to 26 ft. lbs. (35 Nm), the upper mounting bolt to 18 ft. lbs. (25 Nm) and the air pump bracket bolt to 18 ft. lbs. (25 Nm).

9. Connect the terminal connector and the battery lead to the rear of the alternator. Reconnect the negative battery cable.

Starter

REMOVAL AND INSTALLATION

1. Disconnect the negative battery cable. Remove the skid plate if necessary.

2. On 4 cylinder engines, it may be necessary to disconnect and remove the EGR pipe.

3. Disconnect and label the starter wiring at the starter.

4. Remove the starter-to-engine bolts and the shims. Remove the starter from the vehicle.

To install:

5. Install the starter and shims to the engine and replace the bolts. Torque the bolts to 30 ft. lbs. (40 Nm).

6. Connect the electrical leads to the starter.

7. If the EGR pipe was removed, install it. Replace the skid plate, if removed.

8. Connect the negative battery cable.

CHASSIS ELECTRICAL

Heater Blower Motor

The heater blower motor is located under the right side of the dash.

REMOVAL AND INSTALLATION

1. Disconnect the negative battery cable.

2. Disconnect and label the blower motor electrical leads.

3. Remove the blower-to-heater unit screws and lower the blower motor assembly.

To install:

4. Install the heater blower motor, into the heater unit and install the screws.

5. Connect the electrical connectors to the heater blower motor.

6. Connect the negative battery cable.

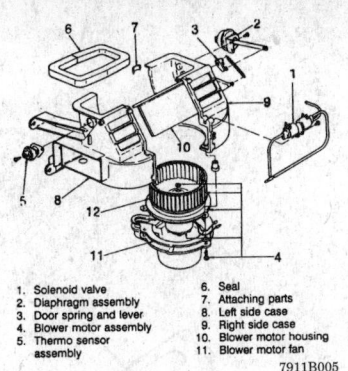

1. Solenoid valve
2. Diaphragm assembly
3. Door spring and lever
4. Blower motor assembly
5. Thermo sensor assembly
6. Seal
7. Attaching parts
8. Left side case
9. Right side case
10. Blower motor housing
11. Blower motor fan

7911B005

Exploded view of the heater blower motor

Windshield Wiper Motor

REMOVAL AND INSTALLATION

Front

1. Disconnect the negative battery cable.
2. Disconnect the electrical connector from the wiper motor.
3. Remove the wiper motor bracket-to-chassis bolts.
4. Disconnect the wiper motor from the wiper linkage at the ball joint.

To install:

5. Connect the wiper motor to the wiper linkage at the ball joint.
6. Install the wiper motor-to-chassis bolts.
7. Connect the electrical connector to the wiper motor.
8. Connect the negative battery cable.

Rear

1. Disconnect the negative battery cable.
2. Remove the back door panel to expose the wiper motor and disconnect the motor wiring.
3. Remove the cover from the wiper arm retaining nut and remove the nut. Pull off the arm and remove the motor shaft mounting hardware.
4. Remove the motor from the door cavity.
5. Installation is reverse of removal.

Windshield Wiper Switch

REMOVAL AND INSTALLATION

Except Pick-Up and Amigo

The windshield wiper switch is a part of the combination switch located on the steering column.

Pick-Up and Amigo

The windshield wiper switch is a part of the switch cluster located on the left side of the instrument cluster.

1. Disconnect the negative battery cable.
2. Remove the instrument cluster-to-dash screws.
3. Pull the instrument cluster forward and disconnect the electrical connector from the switch cluster.
4. From the rear of the instrument cluster, loosen the switch cluster-to-instrument cluster bolts.
5. Separate the windshield wiper switch from the switch cluster.

To install:

6. Install the windshield wiper switch to the switch cluster.
7. Tighten the switch cluster-to-instrument cluster bolts.
8. Connect the electrical connector to the switch cluster.
9. Install the instrument cluster.
10. Connect the negative battery cable.

Instrument Cluster

REMOVAL AND INSTALLATION

Trooper/Trooper II

1. Disconnect the negative battery cable.
2. Remove the steering wheel.
3. Remove the knob from the light switch.
4. Remove the cover-to-instrument panel screws and the cover.
5. Remove the instrument panel-to-dash screws and pull the panel forward.
6. Disconnect the electrical connectors and the speedometer cable from the rear of the instrument panel.
7. Remove the instrument panel from the vehicle.

To install:

8. Connect the speedometer cable and the electrical connectors to the rear of the instrument cluster.
9. Install the instrument cluster to the dash.
10. Install the cover to the instrument cluster.
11. Install the knob to the light switch.
12. Install the steering wheel.
13. Connect the negative battery cable.

Pick-Up, Amigo and Rodeo

TILT STEERING WHEEL

1. Disconnect the negative battery cable.
2. Move the steering wheel to the fully down position.
3. Remove the instrument cluster-to-dash screws and pull the instrument cluster forward.
4. Disconnect the electrical connectors and the speedometer cable from the instrument cluster.

To install:

5. Connect the speedometer cable and electrical connectors to the instrument cluster.
6. Install the instrument cluster to the dash.
7. Connect the negative battery cable.

EXCEPT TILT STEERING WHEEL

1. Disconnect the negative battery cable.
2. Remove the steering wheel and the steering wheel cowl.
3. Remove the instrument cluster-to-dash screws and pull the instrument cluster forward.
4. Disconnect the electrical connectors and the speedometer cable from the instrument cluster.

To install:

5. Connect the speedometer cable and electrical connectors to the instrument cluster.
6. Install the instrument cluster to the dash.
7. Install the steering column cowl and the steering wheel.
8. Connect the negative battery cable.

Auxiliary Gauge Panel

REMOVAL AND INSTALLATION

Trooper/Trooper II

1. Disconnect the negative battery cable.
2. Remove the 2 screws holding the auxiliary gauge panel face plate and remove the face plate.
3. Remove the screws holding the gauges and pull the gauges forward.
4. Disconnect the wiring to the gauges and remove the gauge.
5. Installation is the reverse of removal.

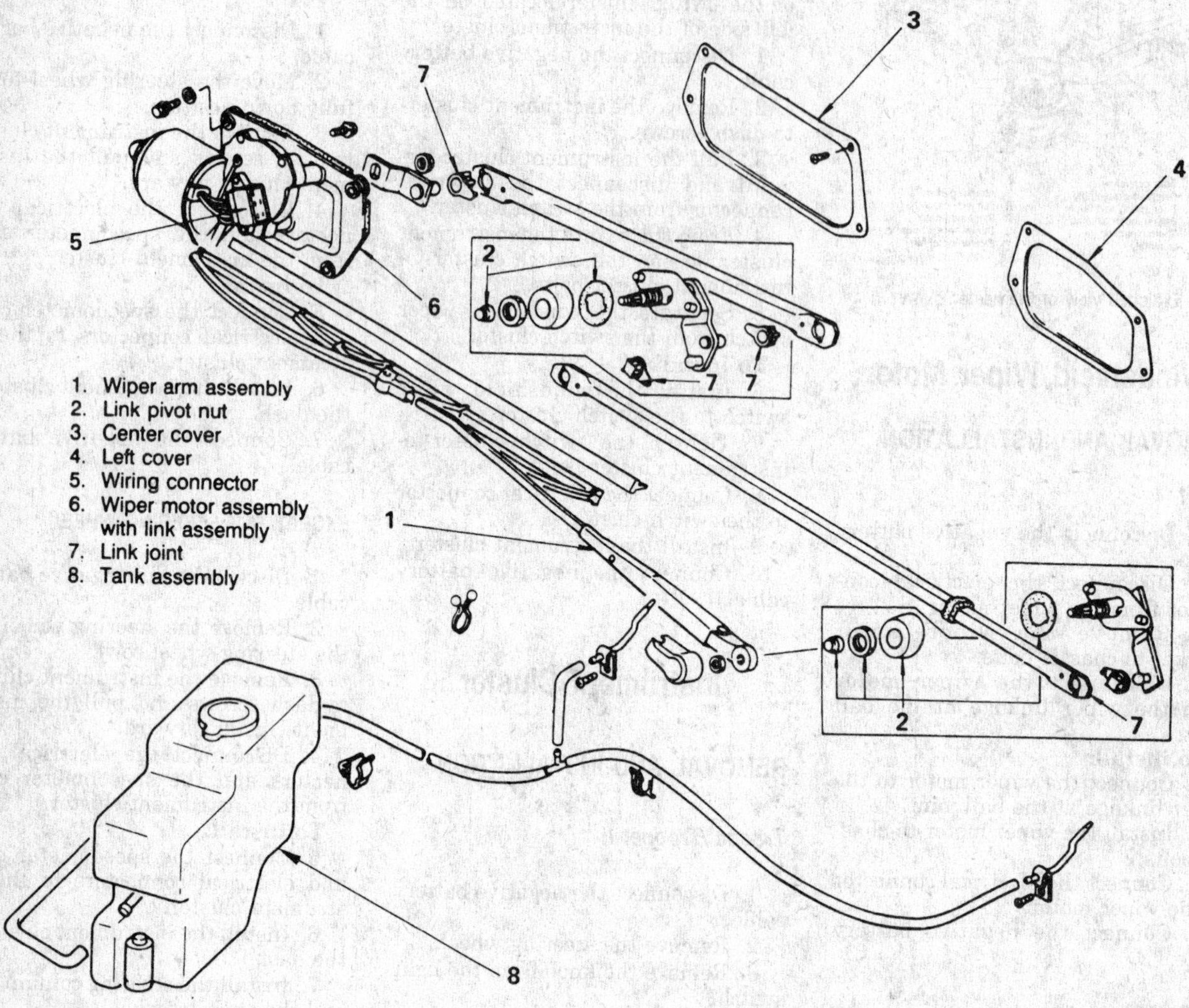

1. Wiper arm assembly
2. Link pivot nut
3. Center cover
4. Left cover
5. Wiring connector
6. Wiper motor assembly with link assembly
7. Link joint
8. Tank assembly

7911B006

Exploded view of the wiper motor and linkage assembly

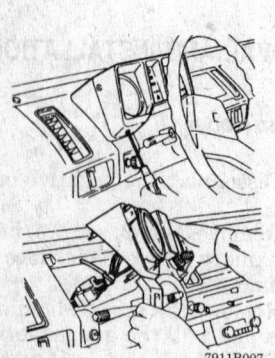

7911B007

Removing the instrument cluster — Trooper/Trooper II shown — Pick-Up similar

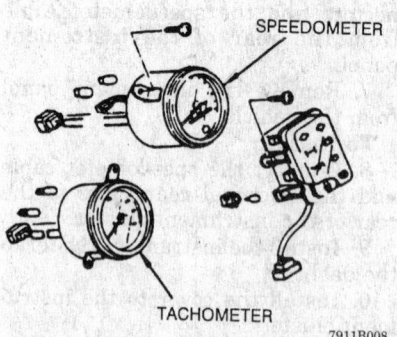

SPEEDOMETER

TACHOMETER

7911B008

Exploded view of the speedometer — Trooper/Trooper II

Speedometer

REMOVAL AND INSTALLATION

Except Trooper/Trooper II

1. Disconnect the negative battery terminal.
2. Remove the instrument cluster. Remove the instrument cluster glass.
3. Pull the indicator needles off the shafts of the speedometer and tachometer.
4. Remove the speedometer fastening screws and remove the speedometer.
5. Installation is the reverse of removal. Be sure to align the instru-

ment indicator needles when pressing them onto their shafts.

Trooper/Trooper II

1. Disconnect the negative battery cable.
2. Remove the instrument cluster.
3. Remove the speedometer holding screws from the instrument cluster and remove the speedometer.

To install:

4. Install the speedometer to the instrument cluster.
5. Install the instrument cluster.
6. Connect the negative battery cable.

Radio

REMOVAL AND INSTALLATION

Trooper/Trooper II

1. Disconnect the negative battery cable.
2. Remove the knobs, the nuts and washers.
3. Remove the face plate from the radio.
4. Remove the radio shroud screws and the shroud.
5. Disconnect the harness connector and the feeder cable.
6. Remove the radio-to-bracket screws and the radio.

To install:

7. Install the radio and the radio-to-bracket screws.
8. Connect the feeder cable and the harness connector.
9. Install the radio shroud and screws.
10. Install the radio faceplate.
11. Install the washers, nuts and knobs.
12. Connect the negative battery cable.

Pick-Up, Rodeo and Amigo

1. Disconnect the negative battery cable.
2. Remove the radio console-to-dash screws and the console.
3. Disconnect the harness connector and the feeder cable, if equipped.
4. Remove the radio bracket-to-radio screws.
5. Remove the radio.

To install:

6. Install the radio and the radio bracket-to-radio screws.
7. Connect the harness connector and the feeder cable, if equipped.
8. Install the console and the radio console-to-dash screws.
9. Connect the negative battery cable.

Headlight Switch

REMOVAL AND INSTALLATION

Trooper/Trooper II

The headlight switch is located on the lower left side of the dash.

1. Disconnect the negative battery cable.
2. Remove the headlight switch knob.
3. Disconnect and label the headlight switch wiring under the dashboard.
4. Remove the headlight switch locknut and the switch from the dash.

To install:

5. Install the switch to the dash and secure with the locknut.
6. Connect the electrical connector to the headlight switch.
7. Install the headlight switch knob.
8. Connect the negative battery cable.

Pick-Up, Rodeo and Amigo

The push button headlight switch, is a part of the switch cluster located on the left side of the dash.

1. Disconnect the negative battery cable.
2. Remove the instrument cluster-to-dash screws.
3. Pull the instrument cluster forward and disconnect the electrical connector from the switch cluster.
4. From the rear of the instrument cluster, loosen the switch cluster-to-instrument cluster bolts.
5. Separate the headlight switch from the switch cluster.

To install:

6. Install the headlight switch to the switch cluster.
7. Tighten the switch cluster-to-instrument cluster bolts.
8. Connect the electrical connector to the switch cluster.
9. Install the instrument cluster.
10. Connect the negative battery cable.

Combination Switch

REMOVAL AND INSTALLATION

1. Disconnect the negative battery cable.
2. From the rear of the steering wheel, remove the horn pad screw and lift the horn pad upward to remove it.
3. Disconnect the electrical connector from the horn pad.

4. Remove the steering wheel-to-steering column nut.
5. Matchmark the steering wheel to the steering shaft for reinstallation purposes.

NOTE: Never apply a blow to the steering wheel shaft with a hammer or other impact tool, to remove the steering wheel, for the steering shaft may become damaged.

6. Using a steering wheel puller, press the steering wheel from the steering column.
7. Remove the contact ring.
8. Remove the steering column covers.
9. Disconnect the electrical connector from the combination switch.
10. Remove the combination switch-to-steering column screws and the switch.

To install:

11. Install the combination switch to the steering column and secure it with screws.
12. Connect the combination switch electrical connector.
13. Install the steering column cover.
14. Install the contact ring.
15. Align the steering wheel-to-steering column matchmarks and torque the nut to 22–29 ft. lbs. (30–40 Nm).
16. Connect the electrical connector to the horn pad.
17. Install the horn pad and the horn pad screw.
18. Connect the negative battery cable.

Ignition Lock/Switch

REMOVAL AND INSTALLATION

1. Disconnect the negative battery cable.
2. From the rear of the steering wheel, remove the horn pad screw and lift the horn pad upward to remove it.
3. Remove the steering wheel-to-steering column nut.
4. Matchmark the steering wheel to the steering shaft for ease of installation purposes.

NOTE: Never apply a blow to the steering wheel shaft with a hammer or other impact tool, to remove the steering wheel, for the steering shaft may become damaged.

5. Using the proper steering wheel puller, remove the steering wheel from the steering column.

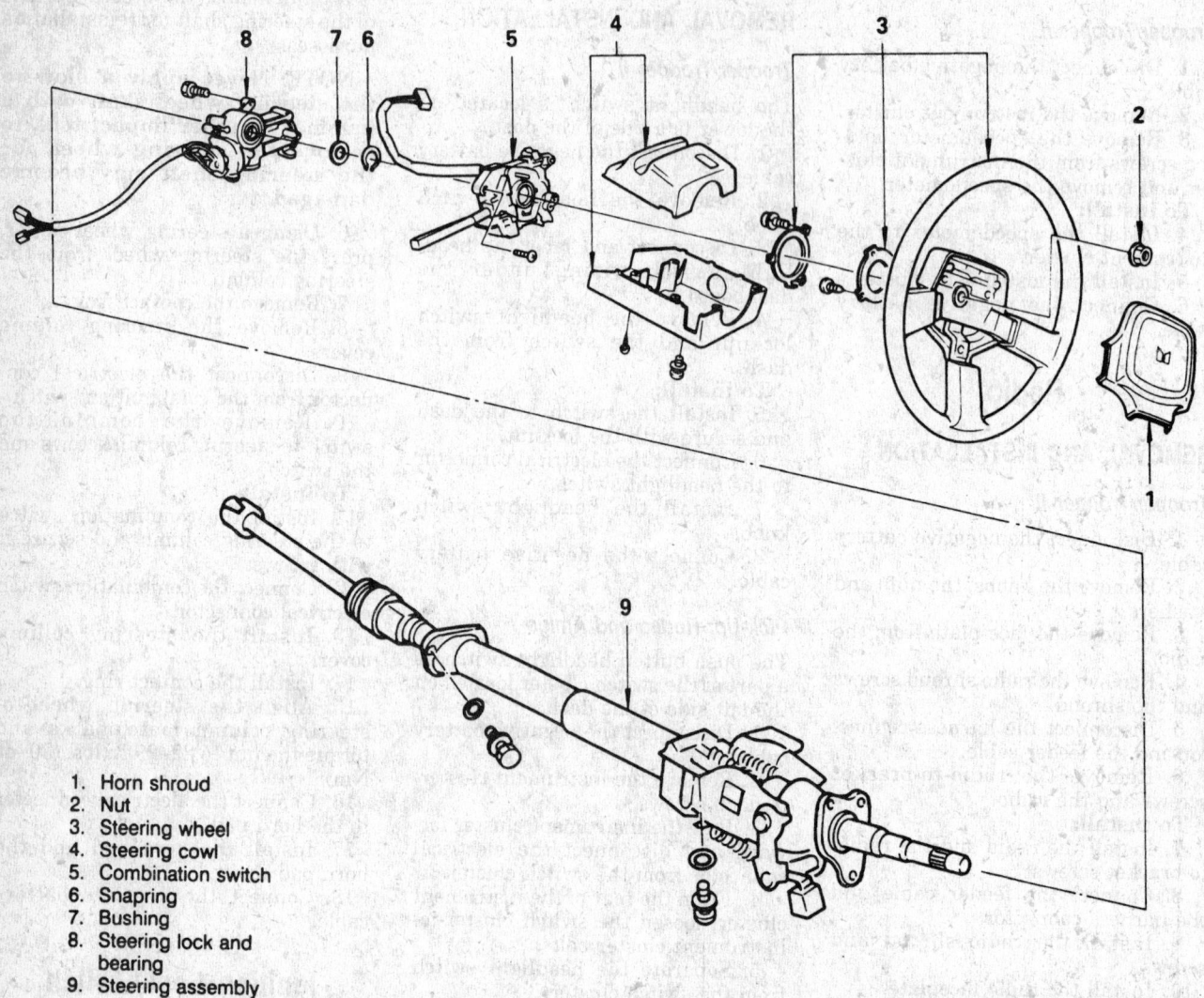

1. Horn shroud
2. Nut
3. Steering wheel
4. Steering cowl
5. Combination switch
6. Snapring
7. Bushing
8. Steering lock and bearing
9. Steering assembly

7911B009

Exploded view of the steering column — Amigo and Pick-Up 1988 — 90 — others are similar

6. Remove the horn contact ring.

7. Remove the steering column covers.

8. Disconnect the electrical connector from the combination switch.

9. Remove the combination switch-to-steering column screws and the switch.

10. On the Rodeo, Amigo and Pick-Up, remove the ignition lock/switch-to-steering column snapring and bushing.

11. Disconnect the electrical connector from the ignition lock/switch assembly.

12. On the Rodeo, Amigo and Pick-Up, remove the ignition lock/switch-to-steering column bolts and the lock/switch assembly. On the Trooper/Trooper II, remove the lock/switch pinch bolts and remove the lock/switch unit.

To install:

13. Install the ignition lock/switch-to-steering column bolts, the bushing and the snapring on the Rodeo, Amigo and Pick-Up. Install the lock/switch pinch bolts on the Trooper/Trooper II.

14. Install the combination switch to the steering column and tighten the screws.

15. Connect the combination switch electrical connector.

16. Install the steering column covers.

17. Install the contact ring.

18. Align the steering wheel-to-steering column matchmarks and torque the nut to 22–29 ft. lbs. (30–40 Nm) on the Trooper/Trooper II or 26 ft. lbs. (35 Nm) on the Rodeo, Pick-Up and Amigo.

19. Install the horn pad and the horn pad screw.

20. Connect the negative battery cable.

Stoplight Switch

ADJUSTMENT

1. Loosen the locknut on the stoplight switch.
2. Rotate the body off the switch to adjust the gap between the switch plunger and the brake pedal.
3. The gap should be 0.020–0.040 in. (0.5–1.0mm).
4. Once the gap has been set, hold the body of the stoplight switch to prevent it from rotating and tighten the locknut.

REMOVAL AND INSTALLATION

1. Disconnect the negative battery cable.
2. Locate the stoplight switch on the brake pedal support.
3. Disconnect the electrical connector from the stoplight switch.
4. Remove the locknut and the stoplight switch.
   To install:
5. Install the switch on the support and adjust the switch so there is 0.020–0.040 in. (0.5–1.0mm) clearance between the switch and the brake pedal.
6. Hold the body of the stoplight switch to prevent it from turning and tighten the locknut.
7. Connect the wiring to the switch and the negative battery cable.
8. Check the operation of the switch.

Ignition Enable Clutch Switch

ADJUSTMENT

No adjustment is possible on the ignition enable clutch switch.

REMOVAL AND INSTALLATION

1. Disconnect the negative battery cable. Remove the electrical leads at the clutch switch.
2. Remove the fastening screws on the switch body and remove the switch from the linkage.
3. Install the switch and tighten the mounting screws.
4. Connect the negative battery cable and check the operation of the switch.

Neutral Safety Switch

ADJUSTMENT

Adjustment of the neutral safety switch should be done when the engine will start in any transmission range other than **N** or **P**.

1. Loosen the hold-down bolt for the neutral safety switch on the transmission housing. Set the shifter to the **N** range.
2. Turn the switch so the groove and the neutral basic line are aligned.
3. Tighten the hold-down bolt 9 ft. lbs. (13 Nm).
4. While holding the brake pedal and with the wheels chocked, check that the engine will not start in any other range than **N** or **P**.

REMOVAL AND INSTALLATION

1. Disconnect the negative battery cable.
2. Disconnect the electrical connector to the neutral safety switch.
3. Disconnect the linkage and remove the hold-down bolt.
4. Remove the switch from the transmission body.
5. Installation is in reverse order. Adjust the switch for proper operation.

Fuses and Circuit Breakers

LOCATION

The main fuse box is located at the lower left side of the instrument panel on vehicles other than the Trooper/Trooper II. On the Trooper/Trooper II, the fuse box is located on the left side valance panel inside the engine compartment. On the Amigo, Rodeo and the Pick-Up

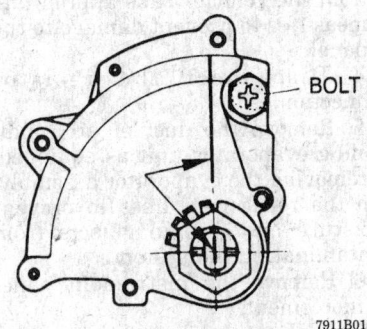

BOLT

7911B010

View of the neutral start switch

the fuse/relay box is located at the right side of the engine compartment incorporating the fusible links within it. Fusible links are in a box located beside the battery on the Trooper/Trooper II.

Computers

LOCATIONS

Amigo, Pick-Up and Rodeo

The ECM is located below the left side of the dash at the kick panel. The rear wheel anti-lock controller is located under the right, front seat on Pick-Up and Rodeo. On Amigo, it is located behind the right side of the dash above the kick panel.

Trooper/Trooper II

The ECM is located on the center of the floor below the console. The electronic spark control module is located on the right side of the engine compartment on the inner fender panel. The electronic transmission controller is located behind the left side of the instrument panel, near the kick panel.

Flashers

LOCATIONS

The flasher unit of the Amigo, Rodeo and Pick-Up is located behind the left side of the dash above the kick panel. The hazard relay on the Trooper/Trooper II is located below the left side of the instrument panel on the brake pedal support.

Cruise Control

ADJUSTMENT

ACTUATOR CABLE

1. The engine should be at normal idle speed when performing this adjustment.
2. Loosen the 2 nuts at the actuator end of the cruise control actuator cable.
3. Pull the outer cable towards the accelerator pedal and make sure there is no play in the inner cable.
4. Turn the nut on the accelerator side of the actuator so it contacts the bracket and tighten the nut on the other side of the bracket to lock the cable adjustment.

BRAKE AND CLUTCH SWITCHES

1. Loosen the locknut on the appropriate switch.

2. Rotate the body of the switch so the plunger is fully compressed and the body of the switch makes contact with the pedal arm.

3. On the brake switch back off the switch ½ turn. Back off the clutch switch 1 full turn for the Rodeo or ½ turn except for the Rodeo.

4. Tighten the locknut and check the operation of the switch.

ENGINE COOLING

Radiator

REMOVAL AND INSTALLATION

1. Disconnect the negative battery cable.

2. Drain the cooling system.

3. Remove the upper, lower and reservoir hoses from the radiator. Remove the air intake duct assembly, if necessary for Rodeo.

4. Remove the fan shroud-to-radiator bolts and the shroud. If equipped with a 2 piece shroud, unclip and remove the lower shroud section first, then unbolt and remove the main upper section.

5. Remove the automatic transmission cooling lines, if equipped.

6. Remove the radiator-to-chassis bolts and the radiator.

To install:

7. Install the radiator and the radiator-to-chassis bolts.

8. Install the automatic transmission cooling lines, if equipped.

9. Install the fan shroud and the shroud-to-radiator bolts.

10. Reconnect the radiator hoses. Replace the air intake duct.

11. Refill the cooling system.

12. Connect the negative battery cable.

Heater Core

REMOVAL AND INSTALLATION

Rodeo, Amigo and Pick-Up

1. Disconnect the negative battery cable. Drain the coolant into a clean container for reuse or dispose of the coolant in a safe and environmentaly sound way. If equipped with air conditioning, discharge the system using an appropriate refrigerant recycling unit.

2. Disconnect the heater hoses at the inlet of the heater unit. Plug the hoses and inlets to prevent excess coolant from dripping into the passenger compartment.

3. Remove the instrument panel:

   a. Remove the steering wheel and its cowling.

   b. Remove the vent by inserting an appropriate prybar and carefully prying on the unit.

   c. Remove the instrument panel nuts under the vents.

   d. If equipped with an automatic transmission, remove the driving pattern indicator panel.

   e. Remove the screws holding the gauge hood and remove.

   f. Pull out the gauge cluster and remove the clips from the gauge hood.

   g. Disconnect the wiring connector blocks from the back of the cluster and remove the speedometer connection. Remove the gauge cluster.

   h. Remove the hood release handle by removing the screws on the underside of the handle.

   i. Remove the steering column lower cover, the use box, the side trim, ECM box, the front console and the lower reinforcement.

   j. Remove the screws holding the speaker grill and remove the speaker.

   k. Remove the glove box hinge pins. Remove the heater control knobs and remove the bezel.

   l. Remove the control lever assembly after disconnecting the control cables from the heater and blower units.

   m. Remove the illumination controller from the instrument panel.

   n. Remove the instrument panel nuts and bolts. Disconnect the instrument harness at the ECM location.

   o. Remove the instrument panel from the vehicle. Take appropriate measures to prevent damage to the panel.

4. Disconnect the resistor connection.

5. Remove the duct or air conditioning evaporator unit as equipped. If removing the evaporator assembly, cap the refrigerant lines to prevent moisture and foreign objects from contaminating the system.

6. Remove the instrument panel reinforcement.

7. Remove the heater unit after unclipping the harness and removing the nuts.

8. Remove the case section with the control levers.

9. Split the heater core case and remove the heater core.

To install:

10. Install the heater core in the heater core case and reassemble the heater unit.

11. Install the heater unit in the vehicle and replace the harness holding clip.

12. Install the instrument panel reinforcement and the duct or air conditioning evaporator unit as equipped. Connect the resistor wiring connection.

13. Install the instrument panel. Installation is the reverse of removal.

14. Connect the heater hoses and refill the cooling system. Recharge the air conditioning system, if equipped.

15. Connect the negative battery terminal.

Trooper/Trooper II

1. Disconnect the negative battery cable. Drain the coolant into a clean container for reuse or dispose of the coolant in a safe and environmentally sound way. If equipped with air conditioning, discharge the system using an appropriate refrigerant recycling unit.

2. Disconnect the heater hoses at the inlet of the heater unit. Plug the hoses and inlets to prevent excess coolant from dripping into the passenger compartment.

3. Remove the instrument panel:

   a. Remove the gauge cluster.

   b. Remove the screws on the underside of the gauge hood and remove the gauge assembly.

   c. Remove the glove compartment and radio.

   d. Remove the side vents by turning the vents and prying out.

   e. Remove the 10 bolts holding the instrument panel. The bolt heads are covered by a clip–on cover. Pry the covers off to expose the bolt heads.

   f. Remove the instrument panel taking care not to damage it or the interior of the vehicle.

4. Disconnect the refrigerant lines to the evaporator unit, if equipped.

5. Remove the vacuum line and the evaporator relay connection.

6. Remove the evaporator unit nuts and remove the evaporator unit, if equipped. If not air conditioner equipped, remove the duct.

7. Disconnect the air ducts from the unit.

8. Disconnect the heater unit relay connector.

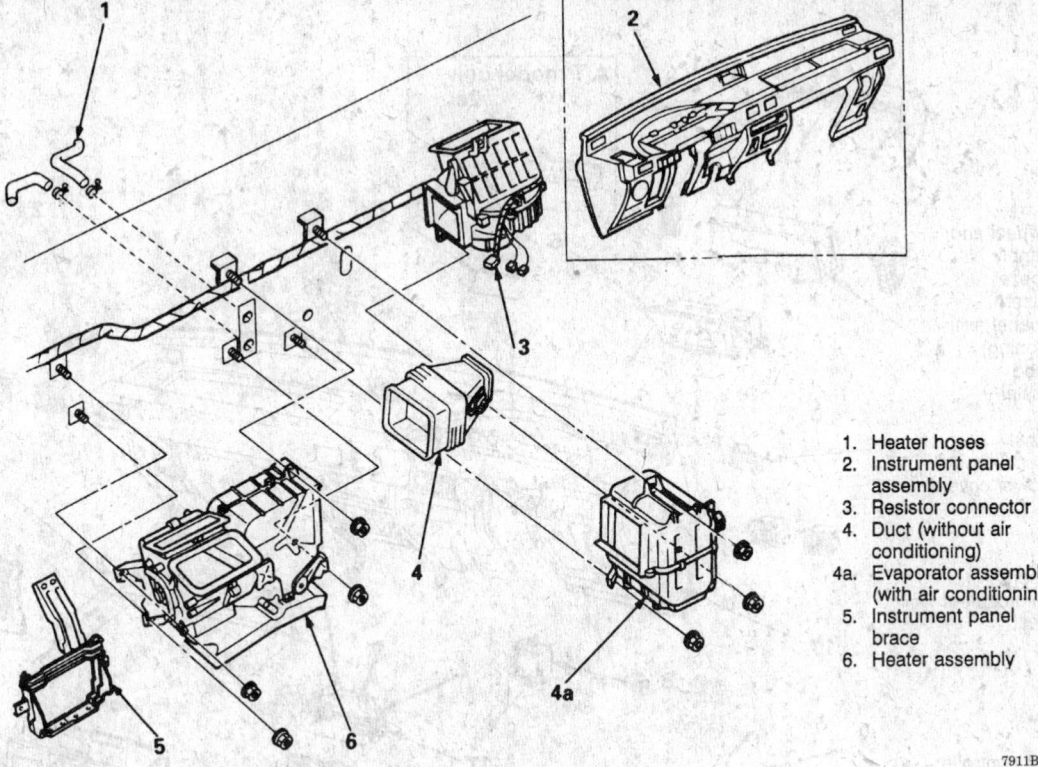

1. Heater hoses
2. Instrument panel assembly
3. Resistor connector
4. Duct (without air conditioning)
4a. Evaporator assembly (with air conditioning)
5. Instrument panel brace
6. Heater assembly

7911B011

Heater unit parts removal order — Amigo, Rodeo and Pick-Up

9. Remove the heater unit nuts and remove the heater unit from the vehicle.

10. Remove the defroster door spring, rod and lever.

11. Remove the battery relay and air conditioning resistor.

12. Remove the plate and seal and the water valve rod.

13. Remove the core assembly.

To install:

14. Install the core assembly in the heater unit.

15. Reassemble the heater unit and install the unit in the vehicle.

16. Connect the electrical connector to the unit and install the ducts.

17. Install the evaporator unit and connect the refrigerant lines, vacuum and electric connections.

18. Install the instrument panel in reverse order of removal.

19. Connect the heater hoses and refill the cooling system. Recharge the air conditioning system, if equipped.

20. Connect the negative battery terminal.

Water Pump

REMOVAL AND INSTALLATION

Except 2.8L engines

1. Disconnect the negative battery terminal.

2. Drain the coolant from the system. Collect the coolant in a clean pan for reuse or dispose of the old coolant in an environmentaly safe and legally compliant way.

3. Disconnect the radiator hoses and remove the air duct assembly.

4. Unclip the lower fan shroud from the upper portion of the fan shroud and remove.

5. Remove the 4 bolts holding the upper fan shroud and remove the shroud from the engine compartment.

6. Remove the 4 nuts holding the fan, fan clutch and pulley. Remove the assembly.

7. Remove the power steering drive belt, if equipped. Loosen the bolts at the pump bracket to remove tension from the belt.

8. Loosen the air pump adjustment and remove the drive belt. Loosen the idler for the air conditioning compressor drive belt and remove the belt.

9. Remove the 2 idler pulley bolts and remove the idler.

10. Loosen and remove the alternator drive belt.

11. Remove the 4 bolts holding the water pump pulley to the water pump hub. Remove the pulley.

12. Remove the 2 bolts holding the starter. Disconnect the starter electrical leads. Remove the starter.

13. Hold the crankshaft stationary with a flywheel lock or equivalent, and loosen the crank pulley bolt. Remove the crank pulley.

14. Remove the the upper and lower timing cover.

15. Remove the 4 bolts and 1 nut holding the water pump and remove the water pump.

To install:

16. Installation is the reverse of removal. Torque the water pump bolts to 14 ft. lbs. (19 Nm). Torque the water pump nut to 20 ft. lbs. (25 Nm).

17. Torque the lower and upper timing cover bolts to 4 ft. lbs. (6 Nm). Torque the crank pulley to 90 ft. lbs. (122 Nm).

18. Torque the power steering pump bracket bolts to 32 ft. lbs. (43 Nm). Torque the fan and fan clutch nuts to 20 ft. lbs. (25 Nm).

19. Refill the cooling system and run the engine. Check for leaks.

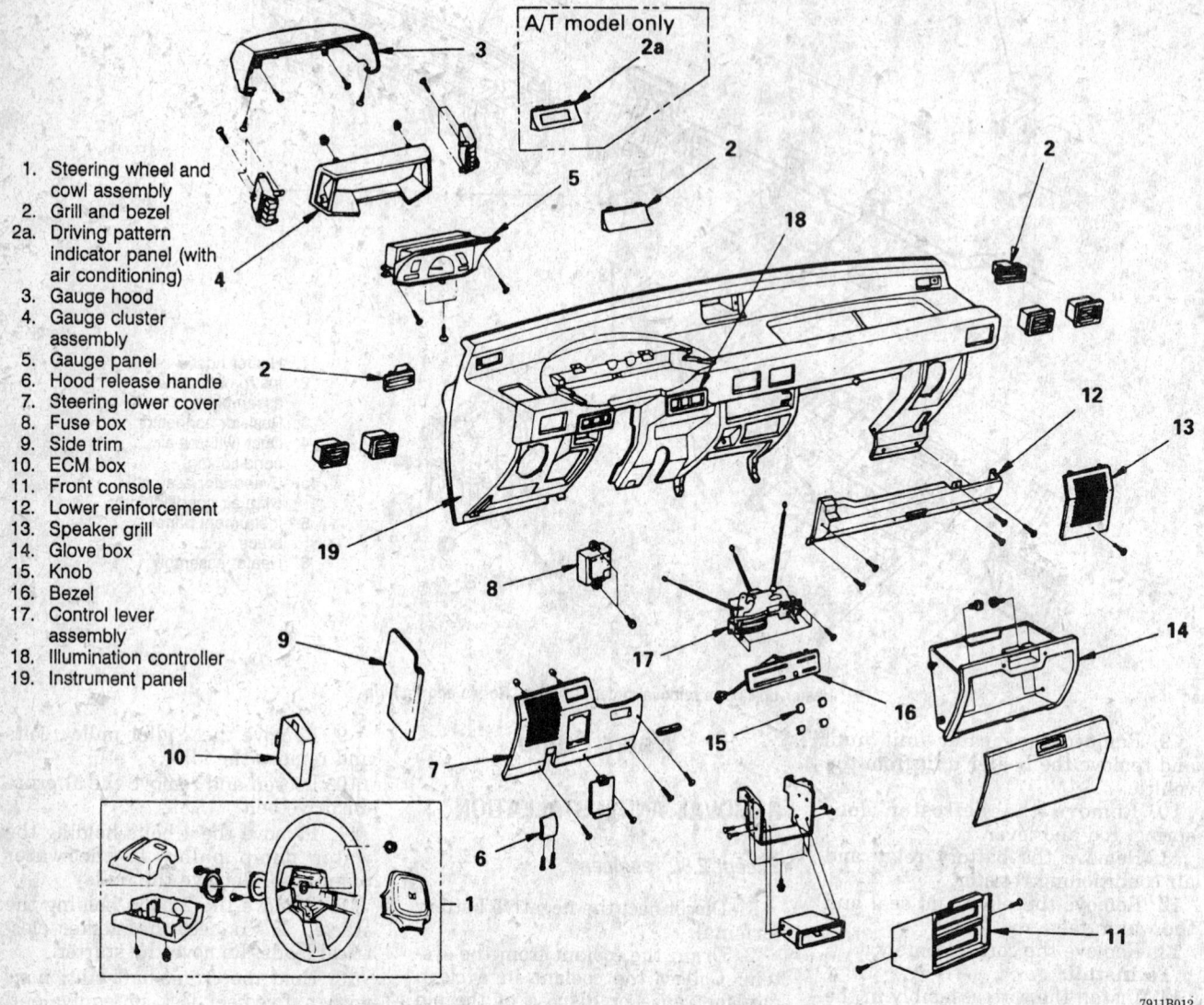

A/T model only
2a

1. Steering wheel and cowl assembly
2. Grill and bezel
2a. Driving pattern indicator panel (with air conditioning)
3. Gauge hood
4. Gauge cluster assembly
5. Gauge panel
6. Hood release handle
7. Steering lower cover
8. Fuse box
9. Side trim
10. ECM box
11. Front console
12. Lower reinforcement
13. Speaker grill
14. Glove box
15. Knob
16. Bezel
17. Control lever assembly
18. Illumination controller
19. Instrument panel

7911B012

Instrument panel parts removal order — Amigo, Rodeo and Pick-Up

2.8L and 3.1L Engines

1. Disconnect the negative battery terminal. Drain the coolant from the system. Collect the coolant in a clean pan for reuse or dispose of the old coolant in an environmentaly safe and legally compliant way.

2. Remove the serpentine drive belt.

3. Remove the upper radiator hose. Remove the air conditioner line bracket from the radiator bracket.

4. Unclip the lower fan shroud from the upper portion of the fan shroud and remove.

5. Remove the 4 bolts holding the upper fan shroud and remove the shroud from the engine compartment

6. Remove the 4 nuts holding the fan, fan clutch and pulley. Remove the assembly.

7. Remove the water pump pulley and the air conditioner compressor along with its bracket.

8. Remove the power steering pump and its bracket.

9. Remove the water pump bolts and remove the water pump.

To install:

10. Installation is the reverse of removal.

11. Coat the water pump bolts with sealant and torque the water pump bolts to 22 ft. lbs. (30 Nm).

12. Torque the power steering pump bolts to 37 ft. lbs. (50 Nm).

13. Torque the air conditioning compressor bolts to 78 ft. lbs. (58 Nm).

14. Torque the fan and fan clutch bolts to 17 ft. lbs. (23 Nm).

15. Refill the cooling system and check the system for leaks.

Thermostat

REMOVAL AND INSTALLATION

The thermostat is located, under the thermostat housing, on top of the in-

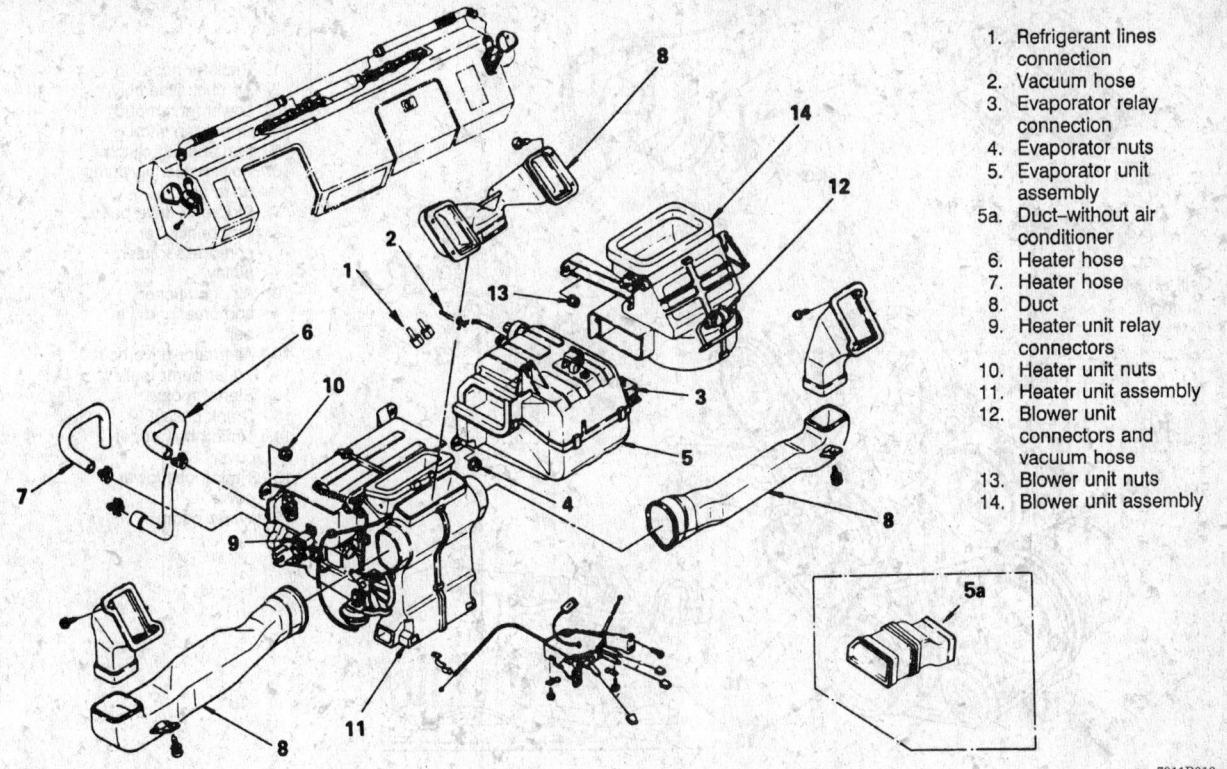

1. Refrigerant lines connection
2. Vacuum hose
3. Evaporator relay connection
4. Evaporator nuts
5. Evaporator unit assembly
5a. Duct–without air conditioner
6. Heater hose
7. Heater hose
8. Duct
9. Heater unit relay connectors
10. Heater unit nuts
11. Heater unit assembly
12. Blower unit connectors and vacuum hose
13. Blower unit nuts
14. Blower unit assembly

7911B013

Heater unit parts removal order — Trooper/Trooper II

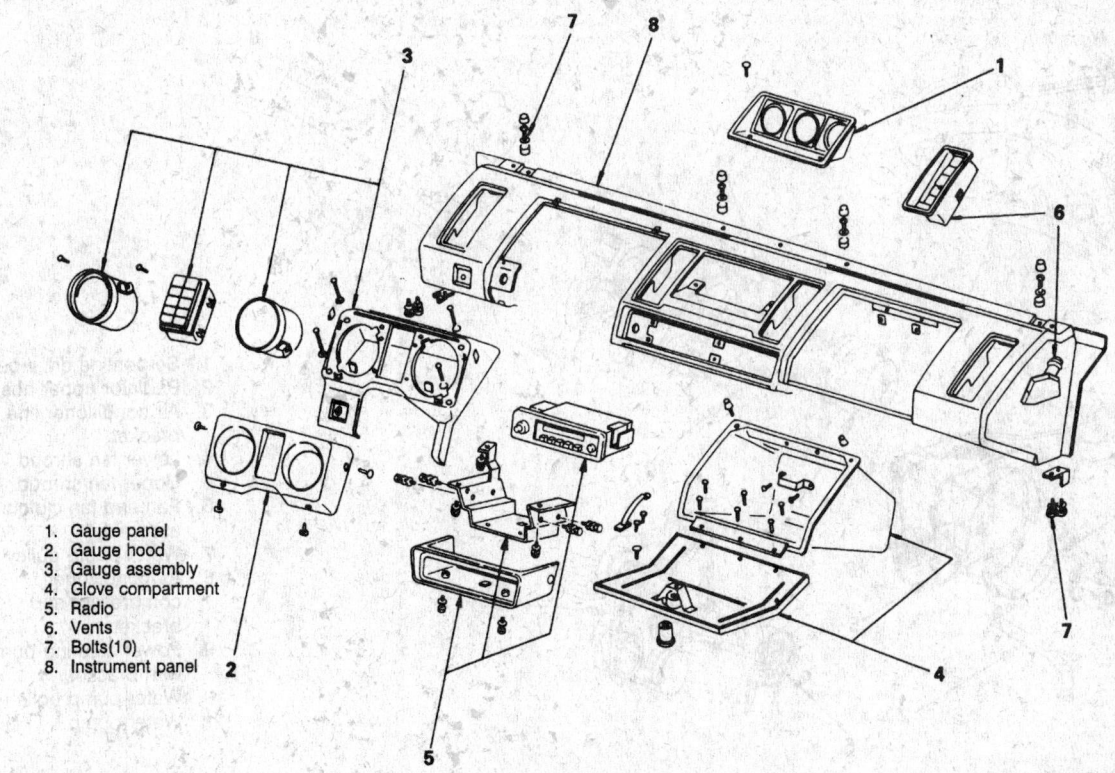

1. Gauge panel
2. Gauge hood
3. Gauge assembly
4. Glove compartment
5. Radio
6. Vents
7. Bolts(10)
8. Instrument panel

7911B014

Instrument panel parts removal order — Trooper/Trooper II

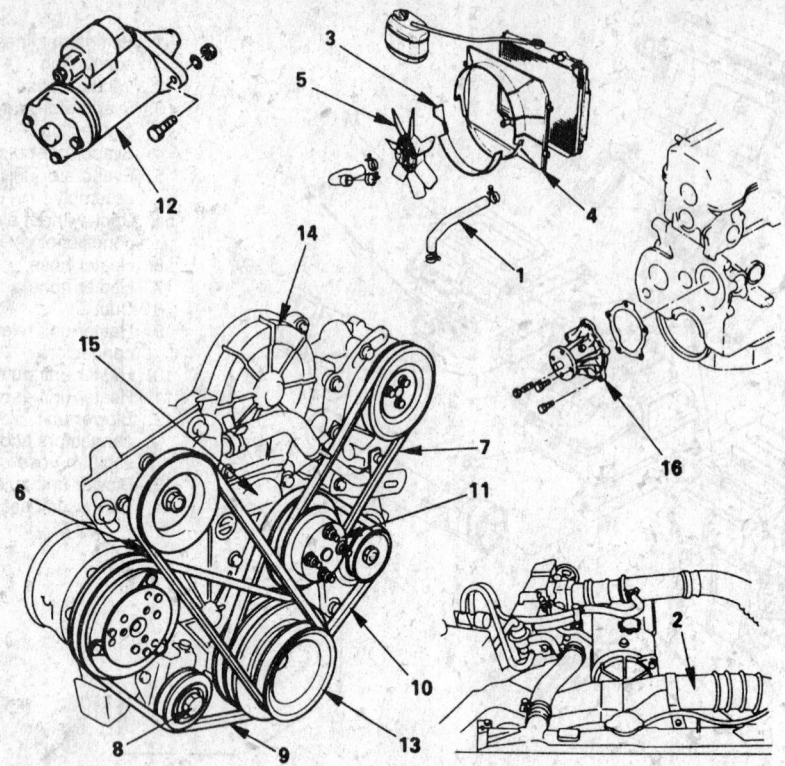

1. Radiator hose
2. Air duct assembly
3. Lower fan shroud
4. Upper fan shroud
5. Fan and fan clutch
6. Power steering pump drive belt
7. Air pump drive belt
8. Air conditioner compressor idler pulley
9. Air conditioner compressor drive belt
10. Alternator drive belt
11. Water pump pulley
12. Starter motor
13. Crank pulley
14. Timing belt upper cover
15. Timing belt lower cover
16. Water pump

7911B015

Water pump removal — 2.3L and 2.6L Engines

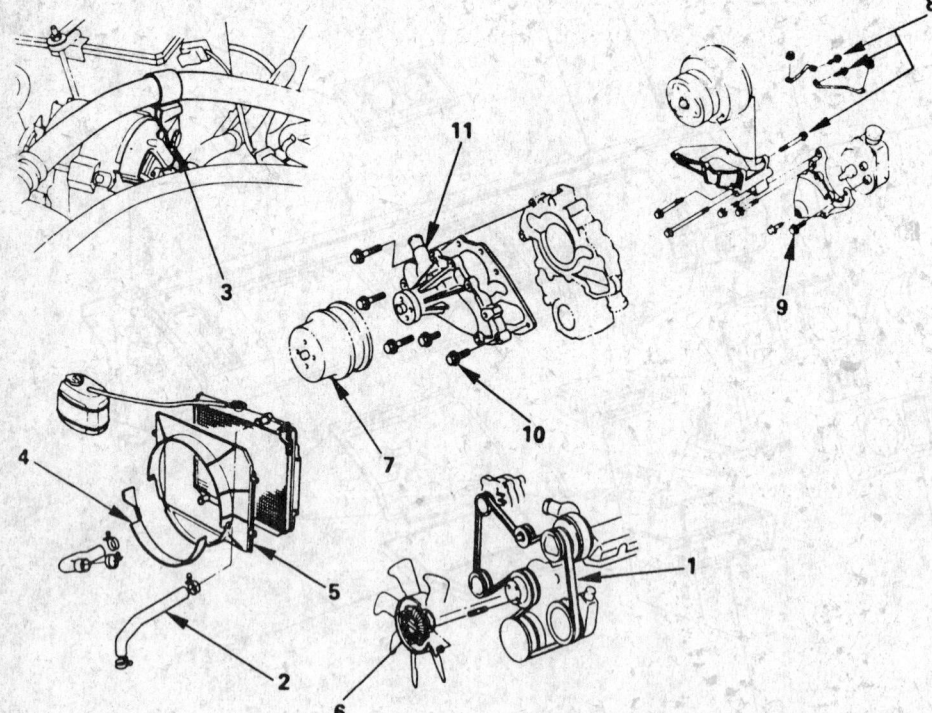

1. Serpentine drive belt
2. Radiator upper hose
3. Air conditioner line bracket
4. Lower fan shroud
5. Upper fan shroud
6. Fan and fan clutch assembly
7. Water pump pulley
8. Air conditioner compressor and bracket
9. Power steering pump with bracket
10. Water pump bolts
11. Water pump

7911B016

Water pump removal — 2.8L and 3.1L Engines

take manifold at the front of the engine.

1. Disconnect the negative battery cable.

2. Drain the cooling system. Disconnect the upper radiator hose from the thermostat housing.

3. Remove the air cleaner assembly.

4. Remove the thermostat housing from the intake manifold.

5. Remove the gasket and the thermostat.

To install:

6. Install the thermostat, with the spring facing the engine.

7. Using a new gasket, install the thermostat housing.

8. Connect the radiator hose to the thermostat housing and refill the cooling system.

9. Install the air cleaner.

10. Connect the negative battery cable.

11. Operate the engine until normal operating temperatures are reached and check the thermostat operation.

Cooling System Bleeding

2.6L Engine

1. Check that the jiggle valve is located at the top of the thermostat.

2. Remove the thermal valve from the thermostat housing. Remove the radiator cap.

3. Set the heater on the highest setting. Fill the cooling system until coolant begins to flow from the thermal valve opening.

4. Apply Loctite® 262 or equivalent, to the thermal valve threads. Install the valve on the housing until the it stops. Using a wrench on the metal portion of the valve, tighten 1 full turn.

5. Continue to turn the thermal valve until the ports on the valve are parallel with the centerline of the thermostat housing mounting bolts.

6. Run the engine until it reaches operating temperature. When bubbles stop forming at the neck of the radiator, install the radiator cap.

7. Shut off engine and check the level of coolant in the reservoir tank.

FUEL SYSTEM

Fuel System Service Precautions

Disconnect the negative battery cable. Keep a Class B dry chemical fire extinguisher available. Always relieve the fuel pressure before disconnecting a fuel line. Wrap a shop cloth around the fuel line when disconnecting a fuel line. Always use new O-rings. Do not replace the fuel pipes with fuel hoses. Always use a backup wrench when opening or closing a fuel line.

Relieving Fuel System Pressure

CARBURETED ENGINE

1. Release the fuel vapor pressure in the fuel tank by removing the fuel tank cap and reinstalling it.

2. Cover the fuel line with an absorbent shop cloth and loosen the connection slowly to release the fuel pressure gradually.

FUEL INJECTED ENGINES

1. Allow the engine to cool. Then, remove the fuel pump fuse from the fuse block.

2. Crank the engine, it will start and run until the fuel supply remaining in the fuel lines is exhausted. When the engine stops, engage the starter again for 3.0 seconds to assure dissipation of any remaining pressure.

3. With the ignition **OFF**, replace the fuel pump fuse.

Fuel Filter

On the Rodeo, Amigo and Pick-Up, the fuel filer is located directly in front of the fuel tank. On the Trooper/Trooper II, the fuel filter is located along the inner side of the right frame rail, near the rear of the vehicle.

REMOVAL AND INSTALLATION

1. Properly relieve the fuel system pressure.

2. Raise and safely support the vehicle.

3. Remove the fuel hose clamps and the fuel hoses from each side of the filter.

4. Cap the open fuel lines to prevent spillage and the entry of contaminants into the fuel system.

5. Remove the mounting bolt and the filter.

To install:

6. Using a new filter, install it to the vehicle; be sure its directional arrow matches the fuel flow.

7. Install the fuel hoses and fuel hose clamps to the filter.

8. Start the engine and check for leaks at the filter.

Mechanical Fuel Pump

REMOVAL AND INSTALLATION

2.3L Engine

The fuel pump is located at the right side of the engine, directly under the intake manifold.

1. Relieve the fuel pressure.

2. Disconnect the negative battery cable.

3. Remove the air cleaner assembly.

4. Remove the intake manifold assembly.

5. Disconnect and plug the fuel lines at the fuel pump.

6. Remove the fuel pump-to-engine bolts and the pump assembly.

To install:

7. Remove the cylinder head cover.

8. Rotate the engine to position the No. 4 cylinder at TDC.

9. Lift the fuel pump pushrod toward the camshaft and hold it in the raised position.

10. Using a new gasket, install the fuel pump on the engine; torque the bolts to 15–25 ft. lbs. (20–4 Nm).

11. Connect the fuel hoses to the fuel pump.

12. Using a new gasket, install the intake manifold.

13. Install the air cleaner assembly.

14. Connect the negative battery cable.

15. Start the engine and check for fuel leaks.

Electric Fuel Pump

Pressure Testing

2.6L, 2.8L and 3.1L Engines

1. Relieve the fuel pressure.

2. Disconnect the fuel line near the engine and install fuel pressure gauge T-connector in the line.

3. Connect the fuel pressure gauge to the T-connector.

4. Start the engine and check the fuel pressure; it should be 43 psi for 2.6L and 3.1L engines or 9–13 psi for 2.8L engine.

5. After checking, turn the engine **OFF**.

6. Relieve the fuel pressure.

7. Remove the pressure gauge from the fuel line and reconnect the fuel line.

8. Start the engine and check for leaks.

REMOVAL AND INSTALLATION

The electric fuel pump is located in the fuel tank.

2.6L, 2.8L and 3.1L Engines

1. Relieve the fuel pressure.

2. Disconnect the negative battery cable.

3. Raise and safely support the vehicle.

4. Drain the fuel from the fuel tank.

5. Disconnect the electrical connectors and the fuel lines from the fuel tank.

6. Remove the fuel tank-to-vehicle supports and lower the tank from the vehicle.

7. Remove the fuel sending unit cover-to-fuel tank screws and lift the assembly from the tank.

8. Remove the fuel pump from the fuel sending unit.

To install:

9. Install the fuel pump to the fuel sending unit.

10. Using a new gasket, install the fuel sending unit to the fuel tank and tighten the screws.

11. Raise the fuel tank into the vehicle and install the tank-to-chassis connectors.

12. Connect the electrical connectors and the fuel lines to the fuel sending unit.

13. Lower the vehicle.

14. Refill the fuel tank.

15. Connect the negative battery cable.

16. Start the engine and check for fuel leaks.

Carburetor

REMOVAL AND INSTALLATION

1. Disconnect the negative battery cable.

2. Remove the air cleaner wing nut and disconnect the rubber hoses from the clips on the air cleaner cover.

3. Remove the bracket bolts, if equipped, at the air cleaner and remove the air cleaner cover and filter element.

4. Disconnect the hot air hose (to the hot air duct), the air hose to the air pump at the air cleaner and the vacuum hose at the joint nipple side of the intake manifold.

5. Loosen the bolt clamping the air cleaner to the carburetor. Separate the air cleaner body from the carburetor but do not remove it completely.

6. Disconnect the PCV hose (to the camshaft cover), the rubber hoses to the check and relief valve. Remove the air cleaner body.

7. Disconnect the vacuum hoses from the EGR valve.

8. Disconnect the choke control wire.

9. Disconnect the lead from the throttle solenoid.

10. Disconnect the throttle linkage return spring.

11. Disconnect the accelerator linkage.

12. Disconnect the fuel line at the carburetor.

13. Remove the carburetor-to-manifold nuts and the carburetor.

To install:

14. Using a new gasket, install the carburetor to the intake manifold.

15. Connect the fuel line and the accelerator linkage to the carburetor.

16. Connect the throttle linkage return spring, the throttle solenoid lead, the choke control wire and the vacuum hoses to the EGR and the PCV.

17. Install the air cleaner and any necessary hoses.

18. Start the engine and check for fuel leaks.

Idle Speed Adjustment

1. Firmly set the parking brake and block the drive wheels.

2. Place the transmission in **N**.

3. Operate the engine until it reaches normal operating temperatures. Be sure the choke is fully open and the air cleaner is installed. If equipped, turn the air conditioning **OFF**.

4. Disconnect and plug the distributor vacuum, the canister purge and EGR vacuum lines. Shut off the vacuum to the idle compensator by bending the rubber hose.

5. Turn the throttle adjusting screw to the required idle speed.

6. If equipped with air conditioning, turn air conditioning control to **MAX COLD** and the blower on **HIGH**.

7. Open throttle to approximately ⅓ opening and allow it to close.

NOTE: The speed-up solenoid should reach full travel.

8. Adjust the speed-up solenoid screw to 850–950 rpm.

Idle Mixture Adjustment

1. Firmly set the parking brake and block the drive wheels.

2. Place the transmission in **N**.

3. Remove the carburetor assembly.

4. Using a drill, drill a hole through the sealing plug covering the idle mixture screw and pry the plug from the carburetor.

5. Reinstall the carburetor.

6. Operate the engine until it reaches normal operating temperatures. Be sure the choke is fully open and the air cleaner is installed. If equipped, turn the air conditioning **OFF**.

7. Disconnect and plug the distributor vacuum, the canister purge and EGR vacuum lines. Shut off the vacuum to the idle compensator by bending the rubber hose.

8. Connect a dwell meter (4 cyl. scale) or duty meter to the duty monitor lead.

9. Turn the idle mixture screw all the way in and back out 1½ turns.

10. Turn the throttle adjusting screw until the engine speed is 900 rpm.

11. Adjust the idle mixture screw to achieve an average dwell of 36 degrees or duty of 40 percent.

NOTE: The dwell or duty reading specified is the average of the most constant variation.

12. Reset the throttle adjusting screw until the engine speed is 850–950 rpm.

13. Reinstall a mixture adjustment plug.

Service Adjustments

For all carburetor service adjustment procedures and Specifications, please refer to Carburetor Service in the Unit Repair section.

Fuel Injection

IDLE SPEED ADJUSTMENT

2.6L Engine

1. Firmly set the parking brake and block the drive wheels.

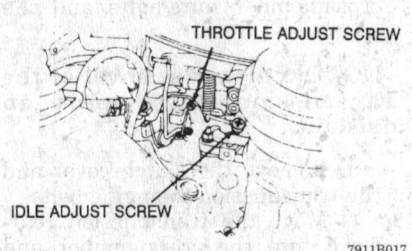

THROTTLE ADJUST SCREW

IDLE ADJUST SCREW

7911B017

View of the idle and throttle adjusting screws — carbureted models

2. Place the transmission in **N**.

3. Set the engine tachometer.

4. Make sure the throttle valve is fully closed.

5. If equipped with air conditioning, turn **OFF** the air conditioning.

6. Place the manual transmission in neutral or the automatic transmission in **P**.

7. Disconnect the electrical connector from the Vacuum Switching Valve (VSV) on the pressure regulator; the idle speed should 850–950 rpm.

8. If the idle speed is not correct, turn the adjusting screw **A** on the throttle body.

2.8L and 3.1L Engines

The idle speed is controlled by the ECM and no adjustment is necessary or possible.

Idle Mixture Adjustment

No idle mixture adjustment is necessary or possible.

7911B018

Location of the vacuum switching valve (VSV) electrical connector — 2.6L engine

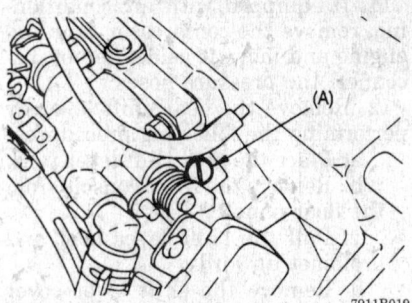

(A)

7911B019

View of the throttle body adjusting screw to alter the idle speed — 2.6L engine

Fuel Injector

REMOVAL AND INSTALLATION

2.6L Engine

The engine is equipped 4 fuel injectors, with 1 located at each cylinder.

1. Relieve the fuel pressure.

2. Disconnect the negative battery cable.

3. Label and disconnect the electrical connectors from the fuel injectors.

4. Disconnect the fuel rail from the fuel system.

5. Remove the fuel rail from the intake manifold; pull the fuel rail with the injectors connected from the intake manifold.

6. Separate the fuel injectors from the fuel rail.

To install:

7. Replace the fuel injector O-rings.

8. Install the fuel injectors to the fuel rail.

9. Lubricate the fuel injector O-rings with automatic transmission fluid and press them, with the fuel rail, into the intake manifold.

10. Install the fuel rail-to-intake manifold bolts.

11. Connect the fuel rail to the fuel system.

12. Connect the electrical connectors to the fuel injectors.

13. Connect the negative battery cable.

14. Turn the ignition switch **ON** and check for fuel leaks at the fuel rail.

2.8L and 3.1L Engines

The engine is equipped with 2 fuel injectors, both are located in the throttle body.

1. Relieve the fuel pressure.

2. Disconnect the negative battery cable.

3. Remove the air cleaner.

4. At the injector electrical connectors, squeeze the 2 tabs together and pull them straight upward.

5. Remove the fuel meter cover and leave the cover gasket in place.

6. Using a small prybar, carefully pry the injectors upward until they are free of the throttle body.

7. Remove the small O-ring from the nozzle end of the injector. Carefully rotate the injector's fuel filter back-and-forth to remove it from the base of the injector.

8. Discard the fuel meter cover gasket.

9. Remove the large O-ring and backup washer from the top of the counterbore of the fuel meter body injector cavity.

To install:

10. Lubricate the O-rings with automatic transmission fluid and push them into the fuel injector cavities.

11. Install the new fuel meter cover gasket and cover.

12. Install the electrical connectors to the injectors.

13. Install the air cleaner and connect the negative battery cable.

EMISSION CONTROLS

Emission Warning Lamps

RESETTING

Once the problem in the system is corrected, the trouble codes must be cleared from the ECM memory. The Check Engine light can be reset by disconnecting the negative battery cable or the ECM fuse at the fuse box for at least 10 seconds and then reconnecting the cable.

ENGINE MECHANICAL

NOTE: Disconnecting the negative battery cable on some vehicles may interfere with the functions of the on board computer systems and may require the computer to undergo a relearning process, once the negative battery cable is reconnected.

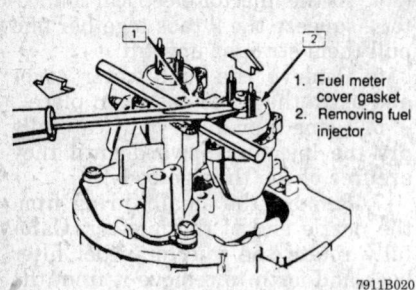

1. Fuel meter cover gasket
2. Removing fuel injector

7911B020

Removing the fuel injectors from the throttle body — 2.8L and 3.1L Engines

Engine

REMOVAL AND INSTALLATION

2WD Vehicles

1. Relieve the fuel pressure. Disconnect both battery cables, the negative cable first. Remove the battery.

2. Matchmark the hood-to-hinges and remove the hood.

3. Remove the undercover, if equipped. Open the drain plugs on the radiator and the cylinder and drain the cooling system.

4. Remove the air cleaner for 2.3L, 2.8L and 3.1L engines or air cleaner duct and hose for 2.6L engine. Using a clean shop cloth, cover the air cleaner port to prevent dirt from entering the engine.

5. Label and disconnect the necessary hoses, electrical connectors, control cables and control rods from the engine.

6. Label and disconnect the following items:
   a. Air switch valve hose.
   b. Oxygen sensor wire.
   c. Vacuum switch valve hose.
   d. Thermal vacuum switching valve hose.
   e. Pressure regulator vacuum hose.
   f. Canister hose.
   g. ECM harness.
   h. Fuel hose(s).

7. Remove the clutch return spring, if equipped, the clutch control cable, if equipped, the backup light switch connector and the speedometer cable from the transmission.

8. Remove the radiator grille from the deflector panel.

9. Disconnect the upper and lower radiator hoses and the reservoir tank hose.

10. Remove the fan shroud, fan blade assembly and the radiator.

11. If equipped with air conditioning, remove the compressor from the engine and move it aside; do not disconnect the pressure hoses.

12. Remove the gear shift lever by performing the following procedures:
   a. Place the gear shift lever in **N**.
   b. Remove the front console from the floor panel.
   c. Pull the shift lever boot and grommet upward.
   d. Remove the shift lever cover bolts and the shift lever.

13. Raise and safely support the vehicle. Remove the front wheels.

14. Drain the oil from the engine and the transmission fluid.

15. If equipped with an automatic transmission, perform the following procedures:
   a. Remove the oil level gauge and the tube.
   b. Disconnect the shift select control link rod from the select lever.
   c. Disconnect the downshift cable from the transmission.
   d. Disconnect and plug the fluid coolant lines from the transmission.

16. If equipped with a 1-piece driveshaft, remove the driveshaft flange-to-pinion nuts, lower the driveshaft and pull it from the transmission.

17. If equipped with a 2-piece driveshaft, perform the following procedures:
   a. Remove the rear driveshaft flange-to-pinion nuts.
   b. Remove the rear driveshaft flange-to-front driveshaft flange bolts and the rear driveshaft.
   c. Remove the center bearing-to-chassis bolts, move the front driveshaft rearward and from the transmission.

18. Remove the starter-to-engine bolts and the starter.

19. Remove the exhaust pipe-to-exhaust manifold nuts, the exhaust pipe bracket-to-transmission bolts, the front exhaust pipe-to-2nd exhaust pipe bolts and the front exhaust pipe from the vehicle.

20. Attach an engine hanger to the rear of the exhaust manifold.

21. Using an engine hoist, connect it to the engine hangers and support the engine.

22. If equipped with a manual transmission, perform the following procedures:
   a. Using a transmission jack, place it under the transmission; do not support it.
   b. Remove the rear mount-to-transmission nuts.

c. Remove the rear mount-to-crossmember nuts/bolts and the mount.

NOTE: Further removal of the transmission may require an assistant.

   d. Remove the clutch cover and the transmission-to-engine bolts.
   e. Move the transmission rearward into the crossmember and floor pan area; the transmission may rest on the crossmember.
   f. Lower the front of the transmission toward the jack.
   g. Firmly, grasp the transmission the rear cover while the assistant raises the jack toward the transmission.
   h. Carefully lower the transmission onto the jack and center it.
   i. Lower the jack and move the transmission rearward.

23. If equipped with an automatic transmission, perform the following procedures:

NOTE: Removal of the transmission will require an assistant.

   a. Remove the torque converter-to-flexplate bolts through the starter hole.
   b. Using a transmission jack, place it under the transmission; do not support it.
   c. Remove the rear mount-to-transmission nuts.
   d. Remove the rear mount-to-crossmember nuts/bolts and the mount.
   e. Remove the transmission-to-engine bolts.
   f. Move the transmission rearward into the crossmember and floor pan area; the transmission may rest on the crossmember.
   g. Lower the front of the transmission toward the jack.
   h. Firmly, grasp the transmission the rear cover while the assistant raises the jack toward the transmission.
   i. Carefully, lower the transmission onto the jack and center it.
   j. Lower the jack and move the transmission rearward.

24. Remove the engine-to-mount nuts/bolts.

25. Using the hoist, slowly, lift the engine; be sure to hold the front of the engine higher than the rear.

26. Place the engine on a workstand.

   To install:

27. Using the hoist, slowly, lower the engine into the vehicle; be sure to hold the front of the engine higher than the rear.

28. Install the engine-to-mount nuts/bolts.

29. If equipped with an automatic transmission, perform the following procedures:

NOTE: Installation of the transmission will require an assistant.

a. Raise the transmission into position.

b. Raise the rear of the transmission and move it into position on the crossmember.

c. Move the transmission forward and engage it with the engine.

d. Install the engine-to-transmission bolts.

e. Install the mount and the rear mount-to-crossmember nuts/bolts.

f. Install the rear mount-to-transmission nuts.

g. Install the torque converter-to-flexplate bolts through the starter hole.

30. If equipped with a manual transmission, perform the following procedures:

NOTE: Installation of the transmission may require an assistant.

a. Raise the transmission into position.

b. Raise the rear of the transmission and move it into position on the crossmember.

c. Move the transmission forward and engage it with the engine.

d. Install the engine-to-transmission bolts.

e. Install the mount and the rear mount-to-crossmember nuts/bolts.

f. Install the rear mount-to-transmission nuts.

31. Remove the engine hoist and the engine hanger from the rear of the exhaust manifold.

32. Install the front exhaust pipe, exhaust pipe-to-exhaust manifold nuts, the exhaust pipe bracket-to-transmission bolts, the front exhaust pipe-to-2nd exhaust pipe bolts.

33. Install the starter and the starter-to-engine bolts.

34. If equipped with a 2-piece driveshaft, perform the following procedures:

a. Install the front driveshaft into the transmission and the center bearing-to-chassis bolts.

b. Install the rear driveshaft and the rear driveshaft flange-to-front driveshaft flange bolts.

c. Install the rear driveshaft flange-to-pinion nuts.

35. If equipped with a 1-piece driveshaft, install the driveshaft into the transmission and the driveshaft flange-to-pinion nuts.

36. If equipped with an automatic transmission, perform the following procedures:

a. Connect the fluid coolant lines to the transmission.

b. Connect the downshift cable to the transmission.

c. Connect the shift select control link rod to the select lever.

d. Install the oil level gauge and the tube.

37. Install the front wheels and lower the vehicle.

38. Install the gear shift lever by performing the following procedures:

a. Install the shift lever and the shift lever cover bolts.

b. Push the grommet and shift lever boot downward.

c. Install the front console to the floor panel.

39. If equipped with air conditioning, install the compressor to the engine.

40. Install the radiator, the fan blade assembly and the fan shroud.

41. Connect the upper and lower radiator hoses and the reservoir tank hose.

42. Install the radiator grille to the deflector panel.

43. Install the clutch return spring, if equipped, the clutch control cable, if equipped, the backup light switch connector and the speedometer cable to the transmission.

44. Connect the following items:

a. Air switch valve hose.

b. Oxygen sensor wire.

c. Vacuum switch valve hose.

d. Thermal vacuum switching valve hose.

e. Pressure regulator vacuum hose.

f. Canister hose.

g. ECM harness.

h. Fuel hose(s).

45. Connect the necessary hoses, electrical connectors, control cables and control rods to the engine.

46. Install the air cleaner for 2.3L engine or air cleaner duct and hose for 2.6L engine.

47. Refill the engine, the transmission and the cooling system. Install the undercover, if equipped.

48. Install the hood.

49. Install the battery and connect both battery cables, the positive cable first.

50. Adjust the belt tension. Start the engine, check for leaks.

51. Check and/or adjust the idle speed and ignition timing.

4WD Vehicles

1. Relieve the fuel pressure. Disconnect both battery cables, the negative cable first. Remove the battery.

2. Matchmark the hood-to-hinges and remove the hood.

3. Remove the undercover, if equipped. Open the drain plugs on the radiator and the cylinder and drain the cooling system.

4. Remove the air cleaner for 2.3L, 2.8L and 3.1L engines or air cleaner duct and hose for 2.6L engine. Using a clean shop cloth, cover the air cleaner port to prevent dirt from entering the engine.

5. Label and disconnect the necessary hoses, electrical connectors, control cables and control rods from the engine.

6. Label and disconnect the following items:

a. Air switch valve hose.

b. Oxygen sensor wire.

c. Vacuum switch valve hose.

d. Thermal vacuum switching valve hose.

e. Pressure regulator vacuum hose.

f. Canister hose.

g. ECM harness.

h. Fuel hose(s).

7. Remove the clutch return spring, if equipped, the clutch control cable, if equipped, the backup light switch connector and the speedometer cable from the transmission.

8. Remove the radiator grille from the deflector panel.

9. Disconnect the upper and lower radiator hoses and the reservoir tank hose.

10. Remove the fan shroud, fan blade assembly and the radiator.

11. If equipped with air conditioning, remove the compressor from the engine and move it aside; do not disconnect the pressure hoses.

12. If equipped with a 2.8L or 3.1L engine, perform the following procedures:

a. Remove the power steering pump-to-engine brackets and move the pump aside.

b. Remove the spark plug wire from the No. 1 spark plug.

c. Remove the distributor cap with the No. 1 spark plug wire.

d. Remove the ignition coil.

13. Remove the gear shift lever by performing the following procedures:

a. Place the gear shift lever in **N**.

b. Remove the front console from the floor panel.

c. Pull the shift lever boot and grommet upward.

d. Remove the shift lever cover bolts and the shift lever.

14. Remove the transfer shift lever by performing the following procedures:

a. Place the transfer shift lever in **H** for 2.3L or 2.6L engine or **2H** for 2.8L or 3.1L engine.

b. Pull the shift lever boot and dust cover upward.

c. Remove the shift lever retaining bolts.

d. Pull the shift lever from the transfer case.

15. Raise and safely support the vehicle. Remove the front wheels. Drain the oil from the engine.

16. Drain the transmission and transfer case fluid.

17. If equipped with an automatic transmission, perform the following procedures:

a. Remove the oil level gauge and the tube.

b. Disconnect the shift select control link rod from the select lever.

c. Disconnect the downshift cable from the transmission.

d. Disconnect and plug the fluid coolant lines from the transmission.

18. If equipped with a 1-piece driveshaft, remove the driveshaft flange-to-pinion nuts, lower the driveshaft and pull it from the transmission.

19. If equipped with a 2-piece driveshaft, perform the following procedures:

a. Remove the rear driveshaft flange-to-pinion nuts.

b. Remove the rear driveshaft flange-to-front driveshaft flange bolts and the rear driveshaft.

c. Remove the center bearing-to-chassis bolts, move the front driveshaft rearward and from the transmission.

20. Remove the front driveshaft's splined yoke flange-to-transfer case bolts and separate the front driveshaft from the transfer case; do not allow the splined flange to fall away from the driveshaft.

21. Remove the starter-to-engine bolts and the starter.

22. If equipped with a clutch slave cylinder, remove it from the transmission and move it aside.

23. Remove the exhaust pipe-to-exhaust manifold nuts, the exhaust pipe bracket-to-transmission bolts, the front exhaust pipe-to-2nd exhaust pipe bolts and the front exhaust pipe from the vehicle.

24. Attach an engine hanger to the rear of the exhaust manifold.

25. Using an engine hoist, connect it to the engine hangers and support the engine.

26. If equipped with a 2.8L engine, remove the catalytic converter and the parking brake cable bracket.

27. Remove the transmission/transfer case assembly by performing the following procedures:

a. Using a transmission jack, place it under the transmission and support the assembly.

b. Remove the rear mount-to-transmission nuts.

c. Remove the rear mount-to-side mount member nuts/bolts and the mount.

d. Remove the transmission-to-engine bolts.

e. Move the transmission assembly rearward.

f. Carefully lower the transmission.

28. Remove the engine-to-mount nuts/bolts.

29. Using the hoist, slowly, lift the engine; be sure to hold the front of the engine higher than the rear.

30. Place the engine on a work stand.

To install:

31. Using the hoist, slowly lower the engine into the vehicle; be sure to hold the front of the engine higher than the rear.

32. Install the engine-to-mount nuts/bolts.

33. Install the transmission/transfer assembly by performing the following procedures:

a. Raise the transmission into position.

b. Move the transmission forward and engage it with the engine.

c. Install the engine-to-transmission bolts.

d. Install the rear mount and the rear mount-to-side mount member nuts/bolts.

e. Install the rear mount-to-transmission nuts.

f. Remove the transmission jack.

34. If equipped with a 2.8L or 3.1L engine, install the catalytic converter and the parking brake cable bracket.

35. Remove the engine hoist and the engine hanger from the rear of the exhaust manifold.

36. Install the front exhaust pipe, exhaust pipe-to-exhaust manifold nuts, the exhaust pipe bracket-to-transmission bolts, the front exhaust pipe-to-2nd exhaust pipe bolts.

37. If equipped with a clutch slave cylinder, install it onto the transmission.

38. Install the starter and the starter-to-engine bolts.

39. Install the front driveshaft's splined yoke flange-to-transfer case bolts.

40. If equipped with a 2-piece driveshaft, perform the following procedures:

a. Install the front driveshaft into the transmission and the center bearing-to-chassis bolts.

b. Install the rear driveshaft and the rear driveshaft flange-to-front driveshaft flange bolts.

c. Install the rear driveshaft flange-to-pinion nuts.

41. If equipped with a 1-piece driveshaft, install the driveshaft into the transmission and the driveshaft flange-to-pinion nuts.

42. If equipped with an automatic transmission, perform the following procedures:

a. Connect the fluid coolant lines to the transmission.

b. Connect the downshift cable to the transmission.

c. Connect the shift select control link rod to the select lever.

d. Install the oil level gauge and the tube.

43. Install the front wheels and lower the vehicle.

44. Install the transfer shift lever by performing the following procedures:

a. Position the shift lever into the transfer case.

b. Install the shift lever retaining bolts.

c. Push the dust cover and the shift lever boot downward.

45. Install the gear shift lever by performing the following procedures:

a. Install the shift lever and the shift lever cover bolts.

b. Push the grommet and shift lever boot downward.

c. Install the front console to the floor panel.

46. If equipped with a 2.8L or 3.1L engine, perform the following procedures:

a. Install the ignition coil.

b. Install the distributor cap with the No. 1 spark plug wire.

c. Install the spark plug wire from the No. 1 spark plug and reconnect the wires to the distributor cap.

d. Install the power steering pump-to-engine brackets.

47. If equipped with air conditioning, install the compressor to the engine.

48. Install the radiator, the fan blade assembly and the fan shroud.

49. Connect the upper and lower radiator hoses and the reservoir tank hose.

50. Install the radiator grille to the deflector panel.

51. Install the clutch return spring, if equipped, the clutch control cable, if equipped, the backup light switch connector and the speedometer cable to the transmission.

52. Connect the following items:
   a. Air switch valve hose.
   b. Oxygen sensor wire.
   c. Vacuum switch valve hose.
   d. Thermal vacuum switching valve hose.
   e. Pressure regulator vacuum hose.
   f. Canister hose.
   g. ECM harness.
   h. Fuel hose(s).

53. Connect the necessary hoses, electrical connectors, control cables and control rods to the engine.

54. Install the air cleaner for 2.3L, 2.8L and 3.1L engines or air cleaner duct and hose for 2.6L engine.

55. Refill the engine, the transmission, the transfer case and the cooling system. Install the undercover, if equipped.

56. Install the hood.

57. Install the battery and connect both battery cables, the positive cable first.

58. Adjust the belt tension. Start the engine, check for leaks.

59. Check and/or adjust the idle speed and ignition timing.

Engine Mounts

REMOVAL AND INSTALLATION

1. Disconnect the negative battery cable. Using a hoist, lift up on the engine hoisting tabs to remove weight from the engine mounts.

2. Loosen the engine mount hardware and remove the mounts from the engine and crossmember.

3. Install the mount and replace the hardware. Torque the bolts and nuts to 38 ft. lbs. (52 Nm).

4. Lower the engine and connect the negative battery cable.

Cylinder Head

REMOVAL AND INSTALLATION

2.3L and 2.6L Engines

1. Relieve the fuel pressure. Disconnect the negative battery cable. Drain the cooling system.

2. Remove the drive belts from the power steering pump, the air pump, the air conditioning compressor, if equipped, and the cooling fan.

3. Rotate the engine to position the No. 1 cylinder on TDC.

4. Remove the distributor cap, high tension cables and the distributor.

5. Remove the exhaust manifold-to-exhaust pipe bolts.

6. Label and disconnect the electrical connectors and vacuum hoses which may be in the way.

7. On the 2.3L engine, remove the carburetor.

8. Remove the coolant hoses. It is not necessary to remove the radiator and the cooling fan assembly.

9. Remove the crankshaft pulley bolt and the pulley.

10. Remove the upper and lower timing belt covers, the tension spring and the timing belt.

11. Remove the camshaft pulley bolt, the pulley and the camshaft boss.

12. Remove the timing belt guide plate and the cylinder head front plate.

13. Remove the rocker arm cover and gasket.

14. Remove the cylinder head-to-engine bolts, the cylinder head and gasket.

15. Clean the gasket mounting surfaces.

To install:

16. Using a new gasket, install the cylinder head and torque the bolts, in sequence to 58 ft. lbs. in the 1st step, and to 65–79 ft. lbs. in the final step.

17. Install the camshaft pulley.

18. Using a new gasket, install the rocker arm cover.

19. Align the camshaft pulley mark with the mark on the front plate. Make sure the keyway on the crankshaft if facing upward, aimed at the pointer on the engine block.

20. Install the timing belt in the following order: crankshaft pulley, the

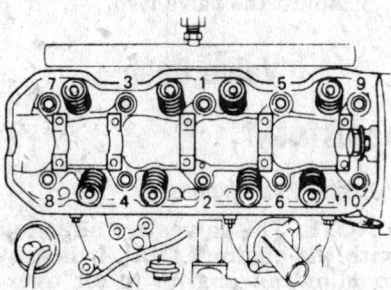

7911B021

View of the cylinder head bolt torquing sequence — 2.3L and 2.6L Engines

oil pump pulley, the camshaft and the tensioner.

21. Install the timing belt covers, using a new gasket.

22. Install the crankshaft pulley.

23. Install the cooling fan assembly, the radiator and the coolant hoses.

24. On the 2.3L engine, install the carburetor.

25. Connect the electrical connectors and vacuum hoses.

26. Install the exhaust manifold-to-exhaust pipe bolts.

27. Install the distributor, the distributor cap and the high tension cables.

28. Install the drive belts to the power steering pump, the air pump, the air conditioning compressor, if equipped, and the cooling fan.

29. Disconnect the negative battery cable. Refill the cooling system.

30. Start the engine and check for leaks.

2.8L and 3.1L Engines

Left Side

1. Relieve the fuel pressure. Disconnect the negative battery cable. Drain the cooling system.

2. Remove the intake manifold.

3. Raise and safely support the vehicle.

4. Disconnect the exhaust pipe from the exhaust manifold and remove the exhaust manifold-to-cylinder head bolts.

5. Remove the dipstick tube from the engine.

6. Lower the vehicle.

7. Loosen the rocker arm nuts, turn the rocker arms and remove the pushrods; keep the pushrods in the same order as removed.

8. Remove the cylinder head bolts in stages and in the reverse order of torquing.

9. Remove the cylinder head; do not pry on the head to loosen it.

10. Clean the gasket mounting surfaces.

To install:

11. Position a new cylinder head gasket over the dowel pins with the words **This Side Up** facing upwards. Carefully, guide the cylinder head into place.

12. Install the cylinder head bolts and torque, in sequence to 41 ft. lbs. in the 1st step, and then an additional ¼ turn in the final step.

13. Install the pushrods; make sure the lower ends are in the lifter heads. Torque the rocker arm nuts to 14–20 ft. lbs. (20–27 Nm).

14. Install the intake manifold.

15. Install the dipstick tube to the engine.

16. Install the exhaust manifold-to-cylinder head bolts and the exhaust pipe-to-exhaust manifold nuts.

17. Refill the cooling system. Start the engine and check for leaks.

Right Side

1. Relieve the fuel pressure. Disconnect the negative battery cable. Drain the cooling system.

2. Remove the intake manifold.

3. If equipped, remove the cruise control servo bracket, the air management valve and hose.

4. Raise and safely support the vehicle.

5. Disconnect the exhaust pipe from the exhaust manifold and remove the exhaust manifold-to-cylinder head bolts.

6. Remove the exhaust pipe at crossover, the crossover and the heat shield, if equipped.

7. Lower the vehicle.

8. Label and disconnect the electrical wiring and vacuum hoses that may interfere with the removal of the right cylinder head.

9. Loosen the rocker arm nuts, turn the rocker arms and remove the pushrods; keep the pushrods in the same order as removed.

10. Remove the cylinder head bolts in stages and in the reverse order of torquing.

11. Remove the cylinder head; do not pry on the head to loosen it.

12. Clean the gasket mounting surfaces.

To install:

13. Position a new cylinder head gasket over the dowel pins with the words **This Side Up** facing upwards. Carefully, guide the cylinder head into place.

14. Install the cylinder head bolts and torque, in sequence to 41 ft. lbs. in the 1st step, and then an additional ¼ turn in the final step.

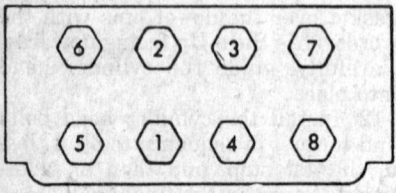

View of the cylinder head bolt torquing sequence — 2.8L and 3.1L Engines

7911B022

15. Install the pushrods; make sure the lower ends are in the lifter heads. Torque the rocker arm nuts to 14–20 ft. lbs. (20–27 Nm).

16. Install the intake manifold.

17. Install the exhaust pipe at crossover, the crossover and the heat shield, if equipped.

18. Install the exhaust manifold-to-cylinder head bolts and the exhaust pipe-to-exhaust manifold nuts.

19. Connect the electrical wiring and vacuum hoses to the right cylinder head.

20. If equipped, install the cruise control servo bracket, the air management valve and hose.

21. Refill the cooling system. Start the engine and check for leaks.

Valve Lifters

REMOVAL AND INSTALLATION

2.8L and 3.1L Engines

1. Disconnect the negative battery cable.

2. Remove the rocker arm covers.

3. Remove the rocker arm nut, the pivot balls, the rocker arm and the pushrods. Keep all components separated so they may be reinstalled in the same location.

4. Remove the lifter from their bores using a hydraulic valve lifter J–929901 or equivalent.

NOTE: The intake and exhaust pushrods are of different lengths.

To install:

5. Lubricate the lifters with Molykote or equivalent, and install in the bores. Install the pushrods in their original location; be sure the lower ends are seated in the lifter.

6. Coat the bearing surfaces of the rocker arms and pivot balls with Molykote or equivalent.

7. Install the rocker arm nuts and torque them to 14–20 ft. lbs. (20–27 Nm).

8. Adjust the valve lash.

Valve Lash

ADJUSTMENT

2.3L and 2.6L Engines

NOTE: The valves are adjusted with the engine COLD. It is best to allow an engine to sit overnight before beginning a valve adjustment. While all valve adjustments must be made as accurately as possible, it is better to

have the valve adjustment slightly loose rather than slightly tight. A burned valve may result from overly tight valve adjustments.

1. Remove the rocker arm cover and discard the gasket.

2. Make sure both the cylinder head and camshaft retaining bolts are tightened to the proper torque.

3. Rotate the crankshaft pulley until the No. 1 piston is at TDC of the compression stroke.

NOTE: To make sure the piston is on the correct stroke, remove the spark plug and place a finger over the hole. Feel for air being forced out of the spark plug hole. Both valves on No. 1 cylinder will be closed. Stop turning the crankshaft when the TDC timing mark on the crankshaft pulley is directly aligned with the timing mark pointer.

4. With the No. 1 piston at TDC of the compression stroke, adjust the clearances of the following valves: Intake 1 and 2; Exhaust 1 and 3

5. Adjust the clearance by loosening the locknut and turning the adjusting screw. Retightening the locknut when the proper thickness feeler gauge passes between the camshaft or valve stem and has a slight drag when the clearance is corrected.

6. Rotate the crankshaft 1 complete revolution (360 degrees) to position the No. 4 piston at TDC of its compression stroke and adjust the clearances of the following valves: Intake 3 and 4; Exhaust 2 and 4

7. After adjustment, use a new gasket, sealant and install the rocker arm cover.

2.8L and 3.1L Engines

1. Remove the valve covers.

2. Rotate the crankshaft until the No. 1 cylinder is on the TDC of its compression stroke.

NOTE: When the notch on the damper pulley is aligned with the 0 timing mark and the rocker arms of the No. 1 cylinder do not move, the engine is at the TDC of the compression stroke of the No. 1 cylinder.

3. With the engine at TDC of the No. 1 cylinder, adjust the following valves: Exhaust 1, 2 and 3; Intake: 1, 5 and 6

4. Back out the adjusting nut until lash is felt.

5. Tighten the adjusting nut until the lash is removed, then, turn the

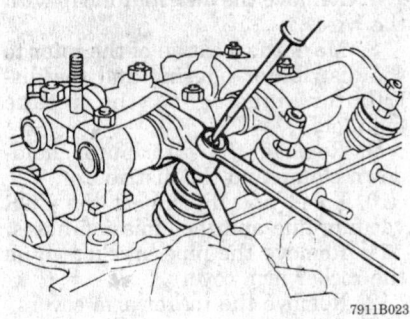

Adjusting the valves — 2.3L and 2.6L Engines

nut 1½ additional turns to center the lifter plunger.

6. Rotate the engine 1 complete revolution and reposition the notch on the damper pulley with the **0** mark on the timing tab; this is the No. 4 cylinder firing position.

7. With the engine at TDC of the No. 4 cylinder, adjust the following valves: Exhaust: 4, 5 and 6; Intake 2, 3 and 4

8. Back out the adjusting nut until lash is felt.

9. Tighten the adjusting nut until the lash is removed, plus, turn the nut 1½ additional turns to center the lifter plunger.

10. Using a new gasket and sealant, install the rocker arm covers.

Rocker Arms/Shafts

REMOVAL AND INSTALLATION

2.3L and 2.6L Engines

1. Disconnect the negative battery cable. Remove the rocker cover. Remove the accessory drive belts and the engine cooling fan and water pump pulley. Remove the starter and crankshaft pulley. Remove the timing covers and timing belt.

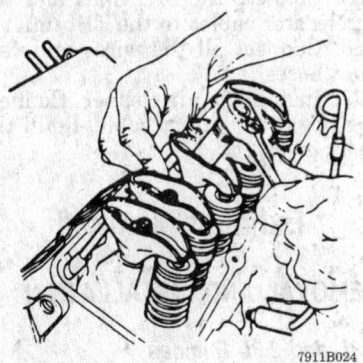

Adjusting the valves — 2.8L and 3.1L Engines

2. Loosen the rocker arm shaft bracket nuts a little at a time, in sequence, starting with the outer nuts.

3. Remove the nuts from the rocker arm shaft brackets. Remove shaft assembly.

4. To disassemble the rockers and shafts; remove the spring from the rocker arm shaft, the rocker brackets and arms. Keep parts in order for reassembly.

5. Before installing apply a generous amount of clean engine oil to the rocker arm shaft, rocker arms and valve stems.

To install:

6. Install the longer shaft on the exhaust valve side and the shorter shaft on the intake side so the aligning marks on the shafts are turned on the front side of the engine.

7. Assemble the rocker arm shaft brackets and rocker arms to the shafts so the cylinder number, on the upper face of the brackets, points toward the front of the engine.

8. Align the mark on the No. 1 rocker arm shaft bracket with the mark on the intake and exhaust valve side rocker arm shaft.

9. Make certain the amount of projection of the rocker arm shaft beyond the face of the No. 1 rocker arm

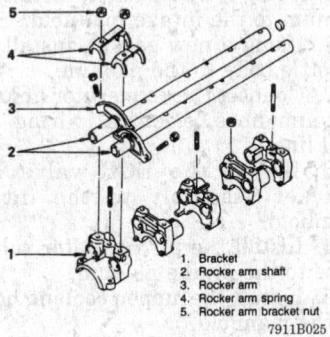

1. Bracket
2. Rocker arm shaft
3. Rocker arm
4. Rocker arm spring
5. Rocker arm bracket nut

Exploded view of the rocker arm/shaft assembly — 2.3L and 2.6L Engines

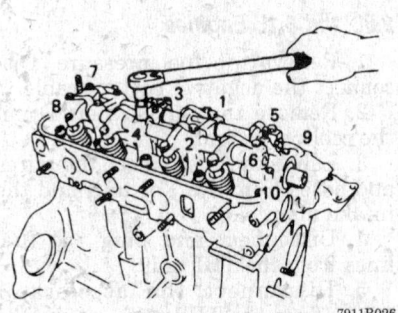

View of the rocker arm/shaft assembly torquing sequence — 2.3L and 2.6L Engines

shaft bracket, is longer on the exhaust side shaft than on the intake shaft when the rocker arm shaft stud holes are aligned with the rocker arm shaft bracket stud holes.

10. Place the rocker arm shaft springs in position between the shaft bracket and rocker arm.

11. Check that the punch mark on the rocker arm shaft is facing upward, then, install the rocker arm shaft bracket assembly onto the cylinder head studs. Align the mark on the camshaft with the mark on the No. 1 rocker arm shaft bracket.

12. Torque the rocker arm shaft brackets-to-cylinder head nuts to 16 ft. lbs. and bolts to 6 ft. lbs.

NOTE: Hold the rocker arm springs while torquing the nuts to prevent damage to the spring. Start with the center nut and work outward.

13. Adjust the valves and install the camshaft cover, with a new gasket and sealer. Install the timing covers, accessories and drive belts. Check the ignition timing.

2.8L and 3.1L Engines

1. Disconnect the negative battery cable.

2. Remove the rocker arm covers.

3. Remove the rocker arm nut, the pivot balls, the rocker arm and the pushrods. Keep all components separated so they may be reinstalled in the same location.

NOTE: The intake and exhaust pushrods are of different lengths.

To install:

4. Install the pushrods in their original location; be sure the lower ends are seated in the lifter.

5. Coat the bearing surfaces of the rocker arms and pivot balls with Molykote or equivalent.

6. Install the rocker arm nuts and torque them to 14–20 ft. lbs. (20–27 Nm).

7. Adjust the valve lash.

Intake Manifold

REMOVAL AND INSTALLATION

2.3L Engine

1. Relieve the fuel pressure. Disconnect the negative battery cable and remove the air cleaner assembly.

2. Remove the EGR pipe clamp bolt at the rear of the cylinder head.

3. Raise and support the vehicle safely. Remove the EGR pipe from the intake and exhaust manifolds.

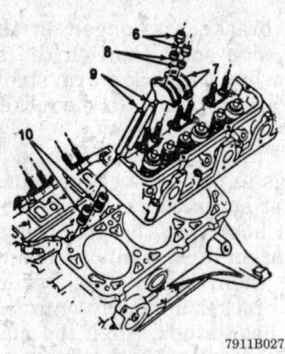

Exploded view of the cylinder head/rocker arm assembly — 2.8L and 3.1L Engines

4. Remove the EGR valve and bracket assembly from the intake manifold.

5. Lower the vehicle and drain the cooling system.

6. Remove the upper coolant hoses from the manifold.

7. Disconnect the accelerator linkage, vacuum lines, electrical wiring and fuel line from the intake manifold.

8. Remove the intake manifold mounting nuts and remove the manifold from the cylinder head.

9. Remove the lower heater hose while holding the manifold away from the engine. Remove the manifold from the vehicle.

To install:

10. Connect the lower heater hose to the manifold. Using a new gasket, install the intake manifold.

11. Connect the accelerator linkage, vacuum lines, electrical wiring and fuel line to the intake manifold.

12. Connect the upper coolant hose to the intake manifold.

13. Install the EGR valve and bracket assembly to the intake manifold.

14. Install the EGR pipe to the intake and exhaust manifolds. Lower the vehicle.

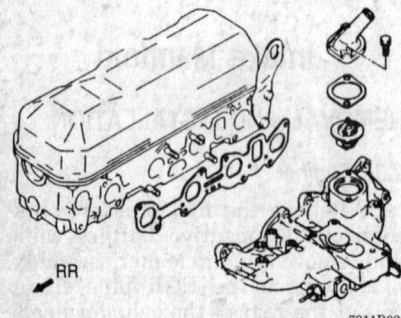

Exploded view of the intake manifold — 2.3L engine

15. Install the EGR pipe clamp bolt to the rear of the cylinder head.

16. Install the air cleaner and connect the negative battery cable.

2.6L Engine

1. Relieve the fuel pressure. Disconnect the negative battery cable and remove the air duct.

2. Drain the cooling system. Remove the upper coolant hoses from the manifold.

3. Remove the air regulator rubber hose from the intake plenum.

4. Remove the EGR valve and bracket assembly from the intake manifold.

5. Disconnect the accelerator linkage, vacuum lines, electrical wiring and fuel line from the throttle body.

6. Remove the throttle body-to-plenum nuts and the throttle body.

7. Remove the plenum-to-intake manifold bolts and the plenum. Remove the fuel injector rail with the fuel injectors.

8. Remove the intake manifold-to-cylinder head nuts and the manifold from the cylinder head.

To install:

9. Using a new gasket, install the intake manifold to the cylinder head. Install the fuel injector rail with the injectors.

10. Using a new gasket, install the plenum to the intake manifold.

11. Using a new gasket, install the throttle body to the plenum.

12. Connect the accelerator linkage, vacuum lines, electrical wiring and fuel line.

13. Install the EGR valve and bracket assembly to the intake manifold.

14. Install the air regulator rubber hose to the intake plenum.

15. Install the upper coolant hoses to the manifold.

16. Install the air duct. Connect the negative battery cable. Refill the cooling system.

2.8L and 3.1L Engines

1. Relieve the fuel pressure. Disconnect the negative battery cable.

2. Remove the air cleaner. Drain the cooling system.

3. Label and disconnect the wires and hoses from the TBI unit and the intake manifold.

4. Disconnect and plug the fuel lines from the TBI unit.

5. Disconnect the accelerator cables from the TBI unit.

6. Disconnect the ignition wires from the spark plugs and the wires from the coil.

7. Remove the distributor cap with the wires.

8. Mark the location of the rotor to the distributor housing and the distributor housing to the intake manifold.

9. Remove the distributor hold-down clamp and the distributor.

10. Label and disconnect the EGR vacuum line and the emission hoses.

11. Remove the pipe brackets from the rocker arm covers.

12. Remove the rocker arm covers.

13. Remove the upper radiator hose and the heater hose.

14. Disconnect the electrical connectors from the coolant sensors.

15. Remove the intake manifold nuts/bolts, the manifold and gaskets.

16. Clean the gasket mounting surfaces.

To install:

17. Using RTV sealant, apply an ⅛ in. bead to the front and rear of the block; make sure no water or oil is present.

18. Using new gaskets, marked right and left side, apply a ¼ in. bead of sealant to hold them in place and install them onto the cylinder heads; the gaskets may have to be cut to be installed around the pushrods.

19. Install the intake manifold and torque the nuts/bolts, in sequence, to 19 ft. lbs. (26 Nm) and retorque using the same sequence.

NOTE: Make sure the areas between the case ridges and the intake manifold are completely sealed.

20. Install the heater hose and the radiator to the manifold.

21. Using new gaskets, install the rocker arm covers.

22. Connect the electrical connectors to the coolant sensors.

23. Install the pipe brackets.

24. Align the matchmarks and install the distributor and the distributor cap.

25. Connect the fuel lines and the accelerator cables to the TBI unit.

26. Connect all the wires and vacuum hoses.

27. Install the air cleaner. Connect the negative battery cable. Refill the cooling system.

Exhaust Manifold

REMOVAL AND INSTALLATION

2.3L and 2.6L Engines

1. Disconnect the negative battery cable and remove the air duct.

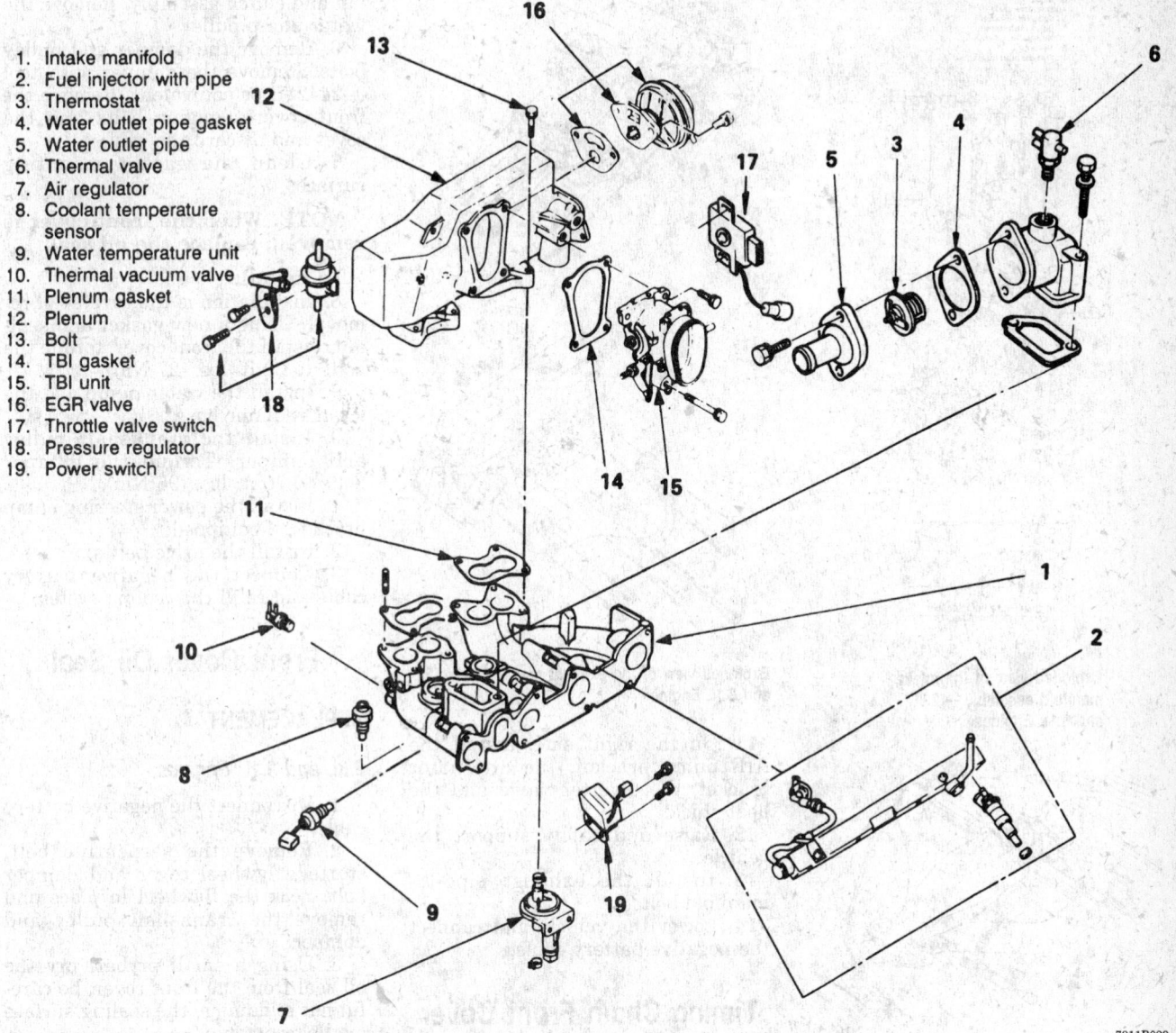

1. Intake manifold
2. Fuel injector with pipe
3. Thermostat
4. Water outlet pipe gasket
5. Water outlet pipe
6. Thermal valve
7. Air regulator
8. Coolant temperature sensor
9. Water temperature unit
10. Thermal vacuum valve
11. Plenum gasket
12. Plenum
13. Bolt
14. TBI gasket
15. TBI unit
16. EGR valve
17. Throttle valve switch
18. Pressure regulator
19. Power switch

Exploded view of the intake manifold assembly — 2.6L engine

7911B029

2. Remove the hoses from the air pump.

3. Remove the air pump bolts, remove the drive belt and the air pump.

4. Remove the EGR pipe clamp bolt at the rear of the cylinder head.

5. Raise and safely support the vehicle. Remove the EGR pipe from the intake and exhaust manifolds. If necessary, remove the dipstick and tube.

6. Disconnect the exhaust pipe from the exhaust manifold. Disconnect the electrical connector from the oxygen sensor.

7. Remove the manifold shield.

8. Remove the manifold-to-cylinder head nuts and the manifold from the engine.

To install:

9. Using a new gasket, install the exhaust manifold and torque the nuts to 33 ft. lbs. (44 Nm).

10. Install the heat shield. If the dipstick was removed, install the tube and the dipstick.

11. Connect the exhaust pipe to the exhaust manifold. Torque the nuts to 49 ft. lbs. (67 Nm). Connect the electrical connector to the oxygen sensor.

12. Install the EGR pipe to the intake and exhaust manifolds and lower the vehicle. Torque to 33 ft. lbs. (44 Nm).

13. Install the EGR pipe clamp bolt to the rear of the cylinder head.

14. Install the air duct. Connect the negative battery cable.

2.8L and 3.1L Engines

1. Disconnect the negative battery cable.

2. Raise and safely support the vehicle.

3. Remove the exhaust pipe from the manifold.

4. Lower the vehicle and remove the rear manifold bolts.

5. On the right side, remove the diverter valve, the heat shield, the AIR pump bracket and alternator bracket.

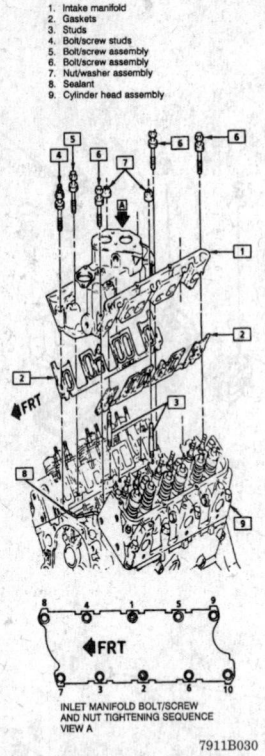

1. Intake manifold
2. Gaskets
3. Studs
4. Bolt/screw studs
5. Bolt/screw assembly
6. Bolt/screw assembly
7. Nut/washer assembly
8. Sealant
9. Cylinder head assembly

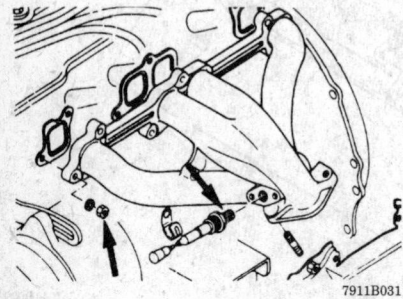

INLET MANIFOLD BOLT/SCREW
AND NUT TIGHTENING SEQUENCE
VIEW A

7911B030

Exploded view of the intake manifold assembly — 2.8L and 3.1L Engines

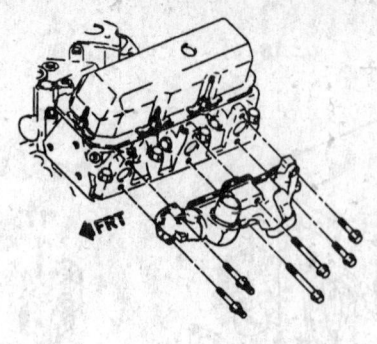

7911B032

Exploded view of the exhaust manifolds — 2.8L and 3.1L Engines

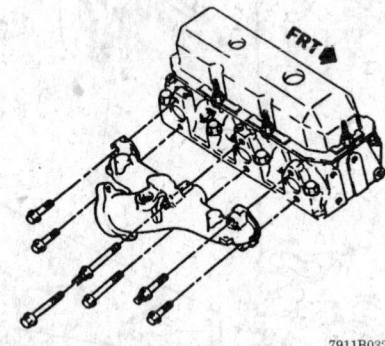

7911B031

Exploded view of the exhaust manifold — 2.6L engine

6. On the left side, remove the heat stove tube and the power steering bracket.

7. Remove the exhaust manifold-to-cylinder head bolts and the manifold.

8. Clean the gasket mounting surfaces.

To install:

9. Using a new gasket, install the exhaust manifold-to-cylinder head bolts and torque the bolts to 25 ft. lbs. (34 Nm).

10. On the left side, install the power steering bracket and heat stove tube.

11. On the right side, install the AIR pump bracket, the alternator bracket, the diverter valve and the heat shield.

12. Raise and safely support the vehicle.

13. Install the exhaust pipe-to-manifold bolts.

14. Lower the vehicle and connect the negative battery cable.

Timing Chain Front Cover

REMOVAL AND INSTALLATION

2.8L and 3.1L Engines

1. Disconnect the negative battery cable.

2. Drain the cooling system. Remove the lower radiator hose from the front cover. Raise and support the vehicle.

3. Remove the stone guard and radiator undercover from beneath the vehicle. Remove the serpentine belt and suspension crossmember.

4. Remove the starter and the flywheel cover. Loosen the oil pan nuts and bolts but do not remove them. Remove the power steering bracket, if equipped.

5. Remove the upper radiator hose and the air conditioner pipe bracket.

Remove the fan shrouds. Remove the fan and clutch assembly. Remove the water pump pulley.

6. Remove the damper and pulley bolts. Remove the damper with tool J–24420B or equivalent. Remove the front cover-to-engine bolts and the cover and discard the gasket.

7. Clean the gasket mounting surfaces.

NOTE: When the front cover is removed, replace the oil seal.

To install:

8. Installation is the reverse of removal. Using a new gasket and sealant, install the front cover. torque the bolts to 20 ft. lbs. (27 Nm).

9. Install the water pump and the lower radiator hose.

10. Install the crankshaft pulley and damper. Torque the damper bolts to 70 ft. lbs. (95 Nm).

11. Install the power steering pump bracket, if equipped.

12. Install the drive belt(s).

13. Connect the negative battery cable and refill the cooling system.

Front Cover Oil Seal

REPLACEMENT

2.8L and 3.1L Engines

1. Disconnect the negative battery cable.

2. Remove the serpentine belt, starter, flywheel cover and damper bolt. Lock the flywheel in place and remove the crankshaft pulley and damper.

3. Using a small prybar, pry the oil seal from the front cover; be careful not to damage the sealing surface or the crankshaft.

4. Using an oil seal installation tool, lubricate the new seal with engine oil and drive it into the front cover; be careful not to cut the seal lip.

5. Install the crankshaft pulley and damper. Install the starter and flywheel cover.

6. Connect the negative battery cable.

Timing Chain and Sprockets

REMOVAL AND INSTALLATION

2.8L and 3.1L Engines

1. Disconnect the negative battery cable.

2. Rotate the crankshaft to position the No. 1 cylinder at the TDC of its compression stroke.

3. Remove the front cover.

4. Inspect the sprocket for chipped teeth and wear.

5. Inspect the timing chain for wear; if the chain can be pulled out more than 0.374 in. (9.5mm) from the damper, replace the chain.

6. Remove camshaft sprocket-to-camshaft bolts, the sprocket and the timing chain; if necessary, use a mallet to tap the sprocket from the camshaft.

7. Using a puller tool, press the crankshaft sprocket from the crankshaft.

To install:

8. Using an installation tool and a hammer, drive the crankshaft sprocket onto the crankshaft; make sure the timing mark faces outward.

9. Using Molykote or equivalent, lubricate the camshaft sprocket thrust surface and install the timing chain onto the sprocket.

10. While holding the camshaft sprocket and chain vertically, align the marks on the camshaft and crankshaft sprockets.

11. Align the camshaft dowel with the camshaft sprocket hole. Install the camshaft sprocket and torque the bolts to 17 ft. lbs. (23 Nm).

12. Lubricate the timing chain with engine oil.

13. Install the front cover and crankshaft pulley.

14. Connect the negative battery cable.

15. Start the engine, then, check and/or adjust the timing.

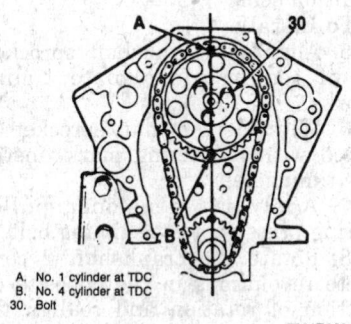

A. No. 1 cylinder at TDC
B. No. 4 cylinder at TDC
30. Bolt

7911B033

View of the timing chain alignment marks — 2.8L and 3.1L Engines

Timing Belt Front Cover

REMOVAL AND INSTALLATION

2.3L and 2.6L Engines

1. Disconnect the negative battery cable.

2. Remove all accessory drive belts and the cooling fan assembly.

3. Remove the crankshaft pulley bolt and pulley.

4. Remove the upper timing belt cover.

5. Remove the lower timing belt cover.

6. To install, reverse the removal procedures.

Oil Seal Replacement

2.3L and 2.6L Engines

1. Disconnect the negative battery cable. Remove the crankshaft pulley.

2. Remove the upper and lower timing belt covers.

3. Rotate the crankshaft to align the camshaft sprocket with the mark on the rear timing cover and the crankshaft sprocket keyway with the mark on the oil seal retainer cover.

NOTE: With the timing marks aligned, the engine is positioned on the TDC of the No. 4 cylinder's compression stroke.

4. Loosen the timing belt tensioner and relax the tension and remove the timing belt from the crankshaft sprocket.

5. Remove the crankshaft sprocket bolt, the sprocket, the key and deflector shield.

6. Using a small prybar, pry the oil seal from the oil seal retainer.

To install:

7. Using a new oil seal, lubricate it with engine oil and tap it into the retainer with an oil seal installation tool.

8. Install the deflector, the key, the crankshaft sprocket and bolt.

9. With the crankshaft sprocket aligned with the timing mark, install the timing belt.

10. Apply the tensioner pulley spring pressure to the timing belt.

11. Rotate the crankshaft 2 complete revolutions in the opposite direction of rotation and realign the timing marks.

12. Loosen the tensioner pulley bolt to allow the spring to adjust the correct tension. Torque the tensioner pulley bolt to 14 ft. lbs. (19 Nm).

13. Install the timing covers and the crankshaft pulley.

14. To complete the installation, reverse the removal procedures.

Timing Belt and Tensioner

ADJUSTMENT

2.3L and 2.6L Engines

1. Disconnect the negative battery cable. Remove the crankshaft pulley.

2. Remove the upper and lower timing belt covers.

3. Loosen the timing belt tensioner and relax the belt tension.

4. Apply the tensioner pulley spring pressure to the timing belt.

5. Rotate the crankshaft 2 complete revolutions in the opposite direction of rotation and realign the timing marks.

6. Loosen the tensioner pulley bolt to allow the spring to adjust the correct tension. Torque the tensioner pulley bolt to 14 ft. lbs. (19 Nm).

7. Install the timing covers and the crankshaft pulley.

8. To complete the installation, reverse the removal procedures.

REMOVAL AND INSTALLATION

2.3L and 2.6L Engines

1. Disconnect the negative battery cable. Remove the cranshaft pulley.

2. Remove the upper and lower timing belt covers.

3. Rotate the crankshaft to align the camshaft sprocket with the mark on the rear timing cover and the crankshaft sprocket keyway with the mark on the oil seal retainer cover.

NOTE: With the timing marks aligned, the engine is positioned on the TDC of the No. 4 cylinder's compression stroke.

4. Loosen the timing belt tensioner and relax the tension and remove the timing belt from the crankshaft sprocket.

To install:

5. With the crankshaft and the camshaft sprockets aligned with the timing marks, install the timing belt. Install the timing belt using the following sequence: the crankshaft sprocket, the oil pump sprocket and the camshaft sprocket.

6. Apply the tensioner pulley spring pressure to the timing belt.

7. Rotate the crankshaft 2 complete revolutions in the opposite direction of rotation and realign the timing marks.

8. Loosen the tensioner pulley bolt to allow the spring to adjust the cor-

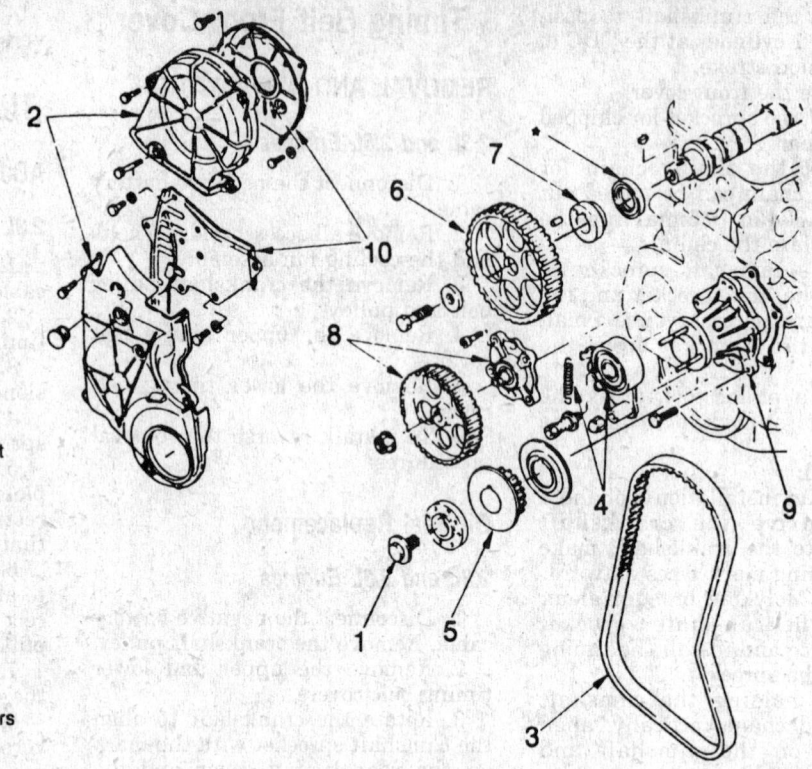

1. Crankshaft pulley bolt
2. Timing belt cover
3. Timing belt
4. Tensioner pulley and spring
5. Crankshaft timing sprocket
6. Camshaft timing sprocket
7. Camshaft boss
8. Oil pump and pulley
9. Water pump
10. Rear timing belt covers

Exploded view of the timing belt assembly — 2.3L and 2.6L Engines

rect tension. Torque the tensioner pulley bolt to 14 ft. lbs.

9. Install the timing covers and the crankshaft pulley.

10. To complete the installation, reverse the removal procedures.

Timing Sprockets

REMOVAL AND INSTALLATION

2.3L and 2.6L Engines

Camshaft Sprocket

1. Disconnect the negative battery cable.
2. Remove the timing belt.

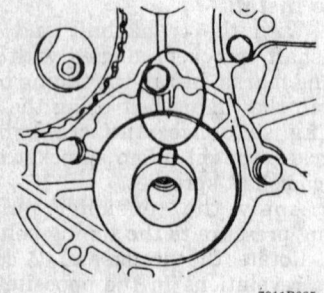

View of the crankshaft sprocket alignment mark — 2.3L and 2.6L Engines

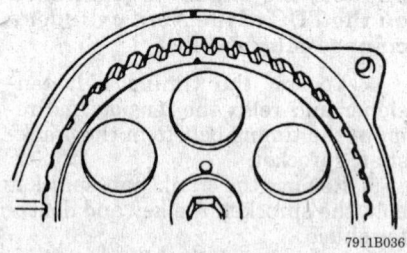

View of the camshaft sprocket alignment mark — 2.3L and 2.6L Engines

3. Remove the camshaft sprocket-to-camshaft bolt and the sprocket.

NOTE: It may be necessary to use a mallet to tap the sprocket from the camshaft.

4. Remove and replace the camshaft oil seal.
To install:
5. Align the camshaft sprocket-to-rear plate timing marks. With the crankshaft sprocket aligned with the timing mark, install the timing belt.
6. Apply the tensioner pulley spring pressure to the timing belt.
7. Rotate the crankshaft 2 complete revolutions in the opposite di-

rection of rotation and realign the timing marks.

8. Loosen the tensioner pulley bolt to allow the spring to adjust the correct tension. Torque the tensioner pulley bolt to 14 ft. lbs.

9. Install the timing covers and the crankshaft pulley.

10. To complete the installation, reverse the removal procedures.

Crankshaft Sprocket

1. Disconnect the negative battery cable.
2. Remove the timing belt.
3. Using a puller, press the sprocket from the crankshaft.
4. Remove and replace the crankshaft oil seal.
To install:
5. Align the crankshaft sprocket-to-oil seal retainer plate timing marks.
6. With the camshaft sprocket aligned with its timing mark, install the timing belt.
7. Apply the tensioner pulley spring pressure to the timing belt.
8. Rotate the crankshaft 2 complete revolutions in the opposite direction of rotation and realign the timing marks.
9. Loosen the tensioner pulley bolt to allow the spring to adjust the cor-

rect tension. Torque the tensioner pulley bolt to 14 ft. lbs.

10. Install the timing covers and the crankshaft pulley.

11. To complete the installation, reverse the removal procedures.

Camshaft

REMOVAL AND INSTALLATION

2.3L and 2.6L Engines

1. Disconnect the negative battery cable.

2. Rotate the crankshaft to position the No. 4 cylinder on the TDC of its compression stroke.

3. Remove the distributor cap and move it aside. Matchmark the rotor to the distributor housing and the distributor housing to the engine. Remove the distributor.

4. Remove the rocker arm cover, the timing belt cover and the timing belt.

5. Remove the rocker arm assembly-to-cylinder head bolts, the rocker arm assembly and the camshaft. If necessary, remove the camshaft sprocket-to-camshaft bolt and the sprocket.

To install:

6. Lubricate the camshaft with engine oil and position it onto the cylinder head.

7. Install the rocker arm assembly and torque the bolts to bolts to 6 ft. lbs. (8 Nm) and the nuts to 16 ft. lbs. (22 Nm).

8. Align the timing marks and install the timing belt.

9. Using a new gasket, install the rocker arm cover.

10. Install the timing belt cover.

11. Align the matchmarks and install the distributor to the cylinder head.

12. To complete the installation, reverse the removal procedures.

13. Rotate the engine manually 2 times to check that there is no piston-to-valve interference. With the timing marks aligned, start the engine, then, check and/or adjust the engine timing.

2.8L and 3.1L Engines

1. Relieve the fuel pressure. Disconnect the negative battery cable.

2. Drain the cooling system.

3. Remove the upper fan shroud and the radiator.

4. Disconnect the fuel line(s), the accelerator linkage, the vacuum hoses and electrical connectors from the throttle body unit.

5. Remove the rocker arm covers.

6. Loosen the valves, rotate them 90 degrees and remove the pushrods; be sure to keep them aligned so they may be installed in their original positions.

7. Remove the intake manifold. Using a hydraulic lifter removal tool, pull the valve lifters from the engine.

8. Remove the damper, front cover, timing chain and sprocket.

9. Using 3 long bolts, thread them into the camshaft holes. Grasp the bolts and carefully, pull the camshaft from the front of the engine.

NOTE: All the camshaft bearing journals are the same diameter; exercise care in removing the camshaft so the bearings do not become damaged.

10. Lubricate the camshaft with engine oil and install it into the engine.

11. Using a hydraulic lifter installation tool, install the hydraulic lifters into the engine.

12. Using new gaskets and sealant, install the intake manifold.

13. Install the pushrods and the rocker arms.

14. Install the camshaft sprocket, the timing chain and the front cover; be sure the timing marks are aligned.

15. Adjust the valves.

16. Using new gaskets, install the rocker arm covers.

17. To complete the installation, reverse the removal procedures. Refill the cooling system.

18. Start the engine and allow it to reach normal operating temperatures. Check and/or adjust the timing.

Piston and Connecting Rod

POSITIONING

ENGINE LUBRICATION

Oil Pan

REMOVAL AND INSTALLATION

2.3L and 2.6L Engines

2WD VEHICLES

1. Disconnect the negative battery cable.

2. Raise and safely support the vehicle.

3. Drain the engine oil. Remove the dipstick and the dipstick tube.

4. Remove the front splash shield, if equipped.

5. Remove the flywheel cover.

6. Disconnect the engine mount nuts and bolts. Raise the engine off the mounts to provide clearance for pan removal.

7. Remove the oil pan bolts and remove the oil pan.

8. Clean the gasket mounting surfaces.

9. Using a new gasket and sealant, install the oil pan. Torque the oil pan-to-engine bolts to 13 ft. lbs. (18 Nm).

10. To complete the installation, reverse the removal procedure. Torque the engine mount bolts to 41 ft. lbs. (55 Nm). Refill the crankcase.

4WD VEHICLES — UPPER OIL PAN

1. Disconnect the negative battery cable.

2. Raise and safely support the vehicle.

3. Drain the engine oil. Remove the dipstick and the dipstick tube.

4. Remove the front splash shield, if equipped.

5. Remove the flywheel cover.

6. Disconnect the engine mount nuts and bolts. Raise the engine off the mounts to provide clearance for pan removal.

7. Remove the oil pan bolts and remove the oil pan.

8. Clean the gasket mounting surfaces.

9. Using a new gasket and sealant, install the oil pan. Torque the oil pan-to-engine bolts to 13 ft. lbs. (18 Nm).

10. To complete the installation, reverse the removal procedure. Torque the engine mount bolts to 41 ft. lbs. (55 Nm). Refill the crankcase.

4WD VEHICLES — LOWER OIL PAN

1. Raise and safely support the vehicle.

2. Drain the crankcase.

3. Remove the lower oil pan-to-upper oil pan bolts and the lower pan.

4. Clean the gasket mounting surfaces.

5. Using a new gasket and sealant, install the lower oil pan and torque the bolts to 47–94 inch lbs.

6. Refill the crankcase.

2.8L and 3.1L Engines

1. Disconnect the negative battery cable.

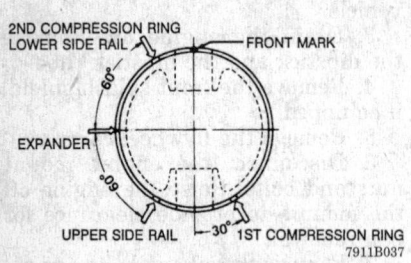

Positioning of the piston and compression rings — 2.3L and 2.6L Engines

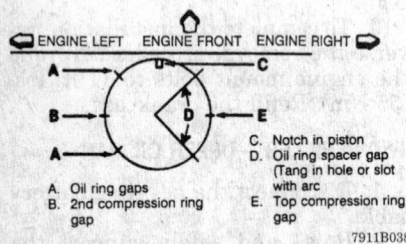

Positioning of the piston and compression rings — 2.8L and 3.1L Engines

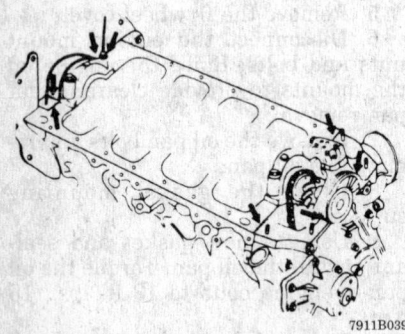

Use sealant at the indicated points when installing the pan gasket — 2.3L engine

2. Remove the dipstick. Raise and safely support the vehicle. Drain the crankcase.

3. Remove the front skid plate and the crossmember.

4. Remove the exhaust pipe-to-catalytic converter bolts, the exhaust pipe-to-manifold bolts and the Y-exhaust pipe.

5. Remove the front driveshaft from the front differential.

6. Remove the braces from the flywheel cover.

7. Disconnect the electrical connectors from the starter. Remove the

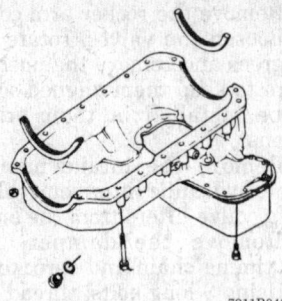

Exploded view of the oil pan used with 4WD — 2.3L and 2.6L engines — 1988 — 90

starter-to-engine bolts and the starter.

8. Remove the flywheel inspection cover.

9. Matchmark the pitman arm-to-pitman shaft for reassembly. Remove the pitman arm-to-pitman arm shaft nut and separate the pitman arm from the pitman shaft.

10. Remove the idler arm-to-shaft nut and separate the idler arm from the shaft.

11. Remove the rubber hose from the front axle vent and support the axle housing assembly.

12. Remove both bolts from the left axle housing isolator and the right axle housing isolator, then, lower the front axle housing assembly.

13. Remove the oil pan-to-engine bolts, the oil pan and discard the gasket.

14. Clean the gasket mounting surfaces.

To install:

15. Using a new gasket and sealant, install the oil pan. Torque both rear pan-to-engine bolts to 18 ft. lbs. (25 Nm) and the other bolts/nuts/studs to 7 ft. lbs. (10 Nm).

16. To complete the installation, reverse the removal procedures. Torque the following fasteners: Pitman arm-to-pitman shaft nut — 159 ft. lbs. (215 Nm); Idler arm-to-shaft nut — 86 ft. lbs. (117 Nm); Front drive axle shaft bolts — 46 ft. lbs. (62 Nm).

17. Refill the crankcase. Connect the negative battery cable.

18. Start the engine and check for leaks.

Oil Pump

REMOVAL AND INSTALLATION

2.3L and 2.6L Engines

The oil pump is attached to the front, lower right side of the engine and is driven by the timing be

1. Remove the upper and lower timing belt covers.

2. Remove the timing belt from the crankshaft and oil pump sprockets.

3. Remove the oil pump sprocket-to-oil pump nut and the sprocket from the oil pump.

4. Using a 6mm Allen wrench, remove the oil pump-to-engine bolts and the oil pump.

To install:

5. Using petroleum jelly, pack the oil pump.

6. Using a new O-ring, install the oil pump and torque the bolts to 10–17 ft. lbs. (13–23 Nm).

7. Install the sprocket to the oil pump and torque the nut to 48–62 ft. lbs.

8. Align the timing marks on the camshaft and crankshaft sprockets and install the timing belt.

9. To complete the installation, reverse the removal procedures.

2.8L and 3.1L Engines

The oil pump is attached to the cylinder block and is located in the oil pan.

1. Disconnect the negative battery cable. Raise and safely support the vehicle.

2. Drain the crankcase. Remove the oil pan.

3. Remove the oil pump-to-engine bolts and the oil pump.

To install:

4. Align the oil pump shaft with the hexagon socket and install the pump. Torque the oil pump-to-engine bolts to 30 ft. lbs. (41 Nm).

5. Install the oil pan.

6. To complete the installation, reverse the removal procedures.

7. Connect the negative battery cable. Start the engine and check for leaks.

CHECKING

2.3L and 2.6L Engines

1. Visually inspect the oil pump for wear, damage or other abnormal conditions.

2. Insert the oil pump vane into the cylinder block.

3. Place a straight-edge across the oil pump opening and a feeler gauge

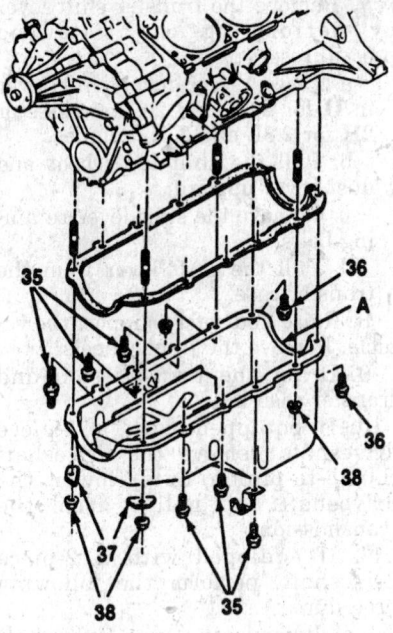

A.	Apply sealant here
35.	Bolts
36.	Bolts
37.	Reinforcements
38.	Nuts

7911B041

Exploded view of the oil pan assembly — 2.8L and 3.1L Engines

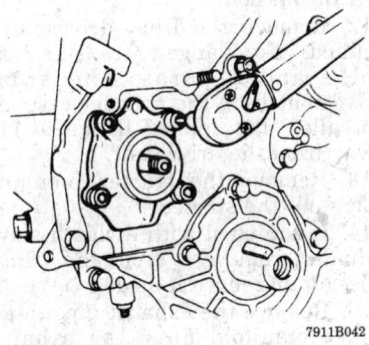

7911B042

View of the oil pump — 2.3L and 2.6L Engines

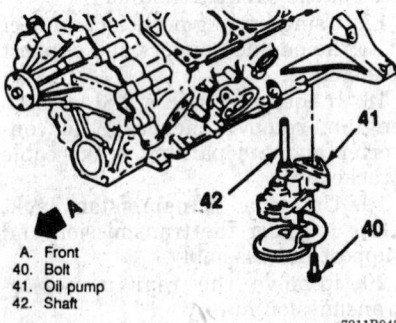

A.	Front
40.	Bolt
41.	Oil pump
42.	Shaft

7911B043

Exploded view of the oil pump assembly — 2.8L and 3.1L Engines

between the straight-edge and the vane; the clearance between the vane-to-cylinder block surface should be 0.002–0.004 in. (0.04–0.09mm), if not, replace the vane.

4. Using a feeler gauge, measure the side clearance between the cylinder block and the vane; it should be 0.009–0.0014 in. (0.24–0.36mm), if not, replace the vane.

5. Position the vane onto the rotor shaft.

6. Using a feeler gauge, measure the clearance between the rotor and the vane; it should be 0.005–0.006 in. (0.13–0.15mm), if not, replace the rotor and/or vane.

2.8L and 3.1L Engines

1. Visually inspect the oil pump for wear, damage or other abnormal conditions.

2. Check that the regulator valve moves freely. Soak in solvent to free the piston, if stuck.

3. Lay a straight-edge across the body of the pump and the faces of the gears and measure the side lash with a feeler gauge between the straight-edge and the pump. The side lash should be 0.002–0.005 in. (0.05–0.13mm).

4. Check the gear lash between the body and the gear with a feeler gauge. The lash should be 0.009–0.105 in. (0.23–0.38mm).

5. Check the gear lash between the gears with a feeler gauge. The lash should be 0.003–0.004 in. (0.08–0.10mm).

6. Replace the pump if any of the measurements are out of specification.

Rear Main Bearing Oil Seal

REMOVAL AND INSTALLATION

1. Disconnect the negative battery cable. Raise and safely support the vehicle.

2. If equipped with an automatic transmission, remove the transmission. If equipped with a manual transmission, remove the transmission and clutch assembly.

3. Remove the starter without disconnecting the wires and secure it aside.

4. Remove the flywheel-to-crankshaft bolts and the flywheel.

5. Using a small prybar, carefully, remove the oil seal work the tool around the diameter of the seal until the seal begins to lift out. Use care not to damage the seat and area around the seal.

6. Using a new oil seal, lubricate the seal lips with clean engine oil.

7. Using an oil seal installation tool, install the new oil seal.

8. To complete the installation, reverse the removal procedures.

MANUAL TRANSMISSION

Transmission

REMOVAL AND INSTALLATION

2WD Vehicles

1. Disconnect the negative battery cable.

2. Remove the undercover, if equipped.

3. Remove the air cleaner for 2.3L, 2.8L and 3.1L engines or air cleaner duct and hose for 2.6L engine. Using a clean shop cloth, cover the air cleaner port to prevent dirt from entering the engine.

4. Label and disconnect the necessary hoses and electrical connectors.

5. Remove the clutch return spring, the clutch control cable, the backup light switch connector and the speedometer cable from the transmission.

6. Remove the gear shift lever by performing the following procedures:

  a. Place the gear shift lever in **N**.

  b. Remove the front console from the floor panel.

  c. Pull the shift lever boot and grommet upward.

  d. Remove the shift lever cover bolts and the shift lever.

7. Raise and safely support the vehicle. Remove the front wheels.

8. Drain the transmission fluid.

9. If equipped with a 1-piece driveshaft, remove the driveshaft flange-to-pinion nuts, lower the driveshaft and pull it from the transmission.

10. If equipped with a 2-piece driveshaft, perform the following procedures:

  a. Remove the rear driveshaft flange-to-pinion nuts.

  b. Remove the rear driveshaft flange-to-front driveshaft flange bolts and the rear driveshaft.

  c. Remove the center bearing-to-chassis bolts, move the front driveshaft rearward and from the transmission.

11. Remove the starter-to-engine bolts and the starter.

12. Remove the exhaust pipe-to-exhaust manifold nuts, the exhaust pipe bracket-to-transmission bolts, the front exhaust pipe-to-2nd exhaust pipe bolts and the front exhaust pipe from the vehicle.

13. Attach an engine hanger to the rear of the exhaust manifold.

14. Using an engine hoist, connect it to the engine hangers and support the engine.

15. Using a transmission jack, place it under the transmission; do not support it.

16. Remove the rear mount-to-transmission nuts.

17. Remove the rear mount-to-crossmember nuts/bolts and the mount.

NOTE: Further removal of the transmission may require an assistant.

18. Remove the clutch cover and the transmission-to-engine bolts.

19. Move the transmission rearward into the crossmember and floor pan area; the transmission may rest on the crossmember.

20. Lower the front of the transmission toward the jack.

21. Firmly, grasp the transmission the rear cover while the assistant raises the jack toward the transmission.

22. Carefully, lower the transmission onto the jack and center it.

23. Lower the jack and move the transmission rearward.

To install:

NOTE: Installation of the transmission may require an assistant.

24. Raise the transmission into position.

25. Raise the rear of the transmission and move it into position on the crossmember.

26. Move the transmission forward and engage it with the engine.

27. Install the engine-to-transmission bolts.

28. Install the mount and the rear mount-to-crossmember nuts/bolts.

29. Install the rear mount-to-transmission nuts.

30. Remove the engine hoist and the engine hanger from the rear of the exhaust manifold.

31. Install the front exhaust pipe, exhaust pipe-to-exhaust manifold nuts, the exhaust pipe bracket-to-transmission bolts, the front exhaust pipe-to-2nd exhaust pipe bolts.

32. Install the starter and the starter-to-engine bolts.

33. If equipped with a 2-piece driveshaft, perform the following procedures:

   a. Install the front driveshaft into the transmission and the center bearing-to-chassis bolts.

   b. Install the rear driveshaft and the rear driveshaft flange-to-front driveshaft flange bolts.

   c. Install the rear driveshaft flange-to-pinion nuts.

34. If equipped with a 1-piece driveshaft, install the driveshaft into the transmission and the driveshaft flange-to-pinion nuts.

35. Install the front wheels and lower the vehicle.

36. Install the gear shift lever by performing the following procedures:

   a. Install the shift lever and the shift lever cover bolts.

   b. Push the grommet and shift lever boot downward.

   c. Install the front console to the floor panel.

37. Install the clutch return spring, the clutch control cable, the backup light switch connector and the speedometer cable to the transmission.

38. Connect the necessary hoses and electrical connectors.

39. Install the air cleaner for 2.3L, 2.8L and 3.1L engines or air cleaner duct and hose for 2.6L engine.

40. Refill the transmission. Install the undercover, if equipped.

41. Connect the negative battery cable.

4WD Vehicles

1. Disconnect the negative battery cable.

2. Remove the undercover, if equipped.

3. Remove the air cleaner for 2.3L, 2.8L and 3.1L engines or air cleaner duct and hose for 2.6L engine. Using a clean shop cloth, cover the air cleaner port to prevent dirt from entering the engine.

4. Label and disconnect the necessary hoses and electrical connectors.

5. Remove the clutch return spring, the clutch control cable, the backup light switch connector and the speedometer cable from the transmission.

6. Remove the gear shift lever by performing the following procedures:

   a. Place the gear shift lever in **N**.

   b. Remove the front console from the floor panel.

   c. Pull the shift lever boot and grommet upward.

   d. Remove the shift lever cover bolts and the shift lever.

7. Remove the transfer shift lever by performing the following procedures:

   a. Place the transfer shift lever in **H** for 2.3L and 2.6L engines or **2H** for 2.8L and 3.1L engines.

   b. Pull the shift lever boot and dust cover upward.

   c. Remove the shift lever retaining bolts.

   d. Pull the shift lever from the transfer case.

8. Raise and safely support the vehicle. Remove the front wheels.

9. Drain the transmission and transfer case fluid.

10. If equipped with a 1-piece driveshaft, remove the driveshaft flange-to-pinion nuts, lower the driveshaft and pull it from the transmission.

11. If equipped with a 2-piece driveshaft, perform the following procedures:

   a. Remove the rear driveshaft flange-to-pinion nuts.

   b. Remove the rear driveshaft flange-to-front driveshaft flange bolts and the rear driveshaft.

   c. Remove the center bearing-to-chassis bolts, move the front driveshaft rearward and from the transmission.

12. Remove the front driveshaft's splined yoke flange-to-transfer case bolts and separate the front driveshaft from the transfer case; do not allow the splined flange to fall away from the driveshaft.

13. Remove the starter-to-engine bolts and the starter.

14. If equipped with a clutch slave cylinder, remove it from the transmission and move it aside.

15. Remove the exhaust pipe-to-exhaust manifold nuts, the exhaust pipe bracket-to-transmission bolts, the front exhaust pipe-to-2nd exhaust pipe bolts and the front exhaust pipe from the vehicle.

16. Attach an engine hanger to the rear of the exhaust manifold.

17. Using an engine hoist, connect it to the engine hangers and support the engine.

18. If equipped with a 2.8L or 3.1L engine, remove the catalytic converter and the parking brake cable bracket.

19. Using a transmission jack, place it under the transmission and support the assembly.

20. Remove the rear mount-to-transmission nuts.

21. Remove the rear mount-to-side mount member nuts/bolts and the mount.

22. Remove the transmission-to-engine bolts.

23. Move the transmission assembly rearward.

24. Carefully lower the transmission.

To install:

25. Raise the transmission into position.

26. Move the transmission forward and engage it with the engine.

27. Install the engine-to-transmission bolts.

28. Install the rear mount and the rear mount-to-side mount member nuts/bolts.

29. Install the rear mount-to-transmission nuts.

30. Remove the transmission jack.

31. If equipped with a 2.8L or 3.1L engine, install the catalytic converter and the parking brake cable bracket.

32. Remove the engine hoist and the engine hanger from the rear of the exhaust manifold.

33. Install the front exhaust pipe, exhaust pipe-to-exhaust manifold nuts, the exhaust pipe bracket-to-transmission bolts, the front exhaust pipe-to-2nd exhaust pipe bolts.

34. If equipped with a clutch slave cylinder, install it onto the transmission.

35. Install the starter and the starter-to-engine bolts.

36. Install the front driveshaft's splined yoke flange-to-transfer case bolts.

37. If equipped with a 2-piece driveshaft, perform the following procedures:

    a. Install the front driveshaft into the transmission and the center bearing-to-chassis bolts.

    b. Install the rear driveshaft and the rear driveshaft flange-to-front driveshaft flange bolts.

    c. Install the rear driveshaft flange-to-pinion nuts.

38. If equipped with a 1-piece driveshaft, install the driveshaft into the transmission and the driveshaft flange-to-pinion nuts.

39. Install the front wheels and lower the vehicle.

40. Install the transfer shift lever by performing the following procedures:

    a. Position the shift lever into the transfer case.

    b. Install the shift lever retaining bolts.

    c. Push the dust cover and the shift lever boot downward.

41. Install the gear shift lever by performing the following procedures:

    a. Install the shift lever and the shift lever cover bolts.

    b. Push the grommet and shift lever boot downward.

    c. Install the front console to the floor panel.

42. Install the clutch return spring, the clutch control cable, the backup light switch connector and the speedometer cable to the transmission.

43. Connect the necessary hoses and electrical connectors.

44. Install the air cleaner for 2.3L, 2.8L and 3.1L engines or air cleaner duct and hose for 2.6L engine.

45. Refill the transmission and the transfer case. Install the undercover, if equipped.

46. Install the negative battery cable.

Linkage Adjustment

No adjustments are possible on the transmission or transfer case linkage.

CLUTCH

Clutch Assembly

REMOVAL AND INSTALLATION

1. Raise and support the vehicle safely.

2. On 2WD vehicles, remove the transmission. On 4WD models, remove the transmission and transfer case as an assembly.

3. Matchmark the clutch assembly to the flywheel so the clutch assembly can be reassembled in the same position. Lock the flywheel in place to prevent it from turning.

4. Loosen the pressure plate-to-flywheel bolts, 1 turn at a time in an alternating sequence, until the spring tension is relieved to avoid distorting or bending the pressure plate.

5. Using a clutch alignment tool, support the pressure plate and cover assembly and remove the bolts and clutch assembly.

To install:

6. Apply a thin coat of grease to the pressure plate fingers, diaphragm spring, clutch cover grooves and the drive bosses on the pressure plate.

7. Apply a thin coat of lubricant to the splines in the clutch disc.

8. Using a clutch alignment tool, assemble the clutch disc and pressure plate onto the flywheel.

9. Align the matchmarks and install the pressure plate-to-flywheel bolts and torque the bolts to 12–14 ft. lbs. (16–19 Nm) using a star pattern tightening sequence. Remove the aligning tool.

10. On 2WD vehicles, install the transmission. On 4WD models, install the transmission and transfer case as an assembly.

11. To complete the installation, reverse the removal procedures. Adjust the clutch cable linkage, if equipped.

Pedal Height Adjustment

The clutch pedal height is the distance from the center of the clutch pedal pad to the firewa

1. Locate the clutch switch (with cruise control) or clutch pedal stop bolt (without cruise control) at the top of the clutch pedal under the dash.

2. Loosen the clutch switch or stop bolt as equipped. Loosen the clutch master cylinder pushrod yoke nut.

3. Adjust the clutch master cylinder pushrod to obtain a clutch pedal height of 6.7–7.1 in. (171–181mm) for Rodeo with 2.6L engine, 7.6–8.0 in. (192–202mm) for Rodeo with 3.1L engine or 9.15–9.55 in. (232–242mm) for Trooper/Trooper II, Pick-Up and Amigo.

4. After adjusting the pedal height, tighten the pushrod yoke nut. Screw in the clutch switch until the plunger is fully depressed and then unscrew 1 full turn and tighten the locknut. Tighten the pedal stop bolt so it just touches the pedal and tighten the locknut.

Clutch Cable

REMOVAL AND INSTALLATION

1. Loosen the clutch cable lock and adjusting nuts. Remove the clutch cable clip in the engine compartment.

2. Raise and safely support the vehicle. Remove the spring from the shift fork end.

3. Disconnect the cable end from the shift fork and pull the cable assembly through the bracket.

4. Lower the vehicle enough to disengage the hooked part of the clutch pedal from the cable eye. Pull the cable assembly towards the engine compartment and remove the cable from the vehicle.

5. To install, reverse the removal procedures. Adjust the cable when finished.

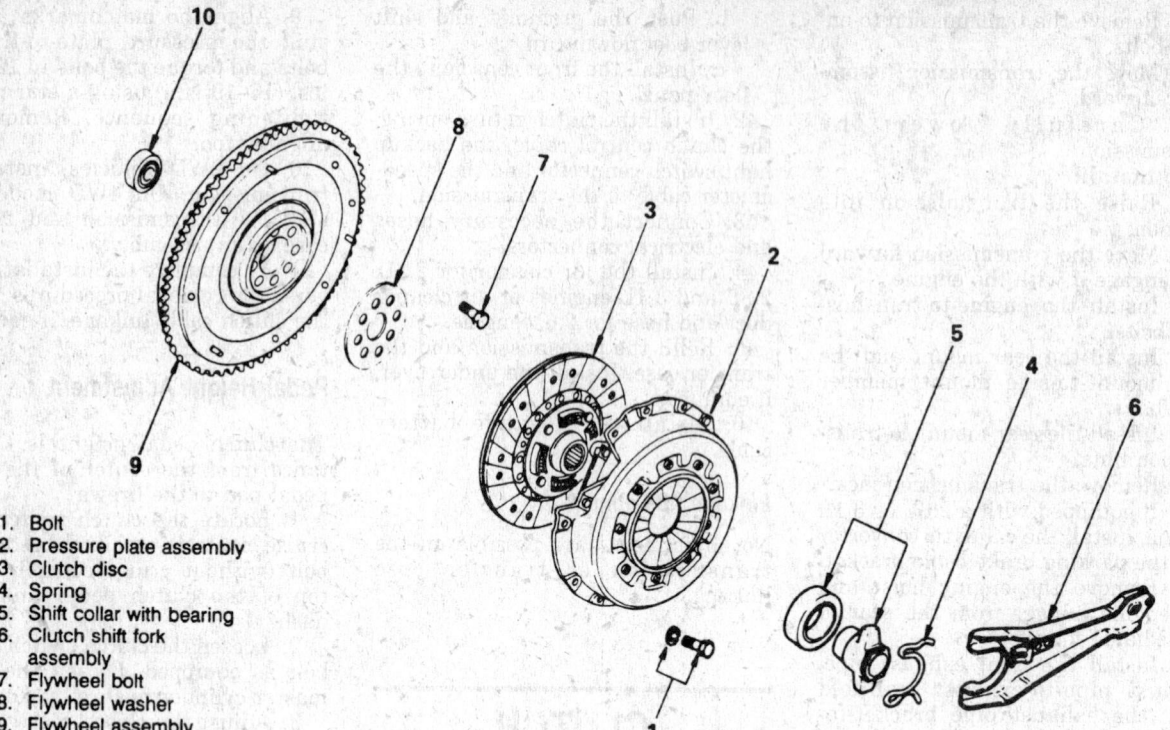

1. Bolt
2. Pressure plate assembly
3. Clutch disc
4. Spring
5. Shift collar with bearing
6. Clutch shift fork assembly
7. Flywheel bolt
8. Flywheel washer
9. Flywheel assembly
10. Crankshaft ball bearing

Exploded view of the clutch assembly

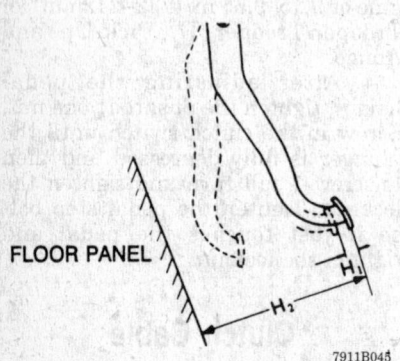

FLOOR PANEL

Measuring the clutch pedal height

ADJUSTMENT

1. Pull the outer cable, located under the hood, forward as far as possible and secure it.
2. Turn the adjusting nut inward it touches the damper rubber washer, located at the firewall.
3. Depress and release the clutch pedal 3 times.
4. Tighten the adjusting nut again.
5. Pull the outer cable forward again and fully tighten the adjusting nut.

6. Loosen the nut to provide a 1/8 in. clearance between the adjusting nut and the damper washer.
7. Release the outer cable and tighten the locknut to secure the adjusting nut.

Clutch Master Cylinder

The clutch master cylinder is located on the firewall inside the engine compartment.

REMOVAL AND INSTALLATION

1. Disconnect the negative battery cable. Remove the hydraulic line from the clutch master cylinder.
2. Disconnect the master cylinder pushrod from the clutch pedal.
3. Remove the master cylinder-to-firewall nuts and remove the master cylinder.
4. To install, reverse the removal procedures. Torque the master cylinder-to-firewall nuts to 8–15 ft. lbs. (12–20 Nm).
5. On master cylinders that use a banjo fitting, use new washers and torque the hydraulic line-to-master cylinder bolt to 22–29 ft. lbs. (30–39 Nm).

6. On master cylinders that use a flare fitting, torque the flare nut to 14.5 ft. lbs. (20 Nm).
7. Bleed the hydraulic clutch system.

Clutch Slave Cylinder

The clutch slave cylinder is attached to the bell housing.

REMOVAL AND INSTALLATION

1. Disconnect the negative battery cable. Remove the hydraulic line from the clutch slave cylinder.
2. Remove the slave cylinder-to-bell housing bolts and remove the slave cylinder.
3. To install, reverse the removal procedures.
4. On cylinders with flare fittings, torque the hydraulic line-to-slave cylinder fitting to 11–17 ft. lbs. (16–25 Nm).
5. On cylinders with banjo fittings, use new washers and torque the hydraulic line-to-slave cylinder fitting to 25 ft. lbs. (35 Nm).
6. Bleed the hydraulic clutch system.

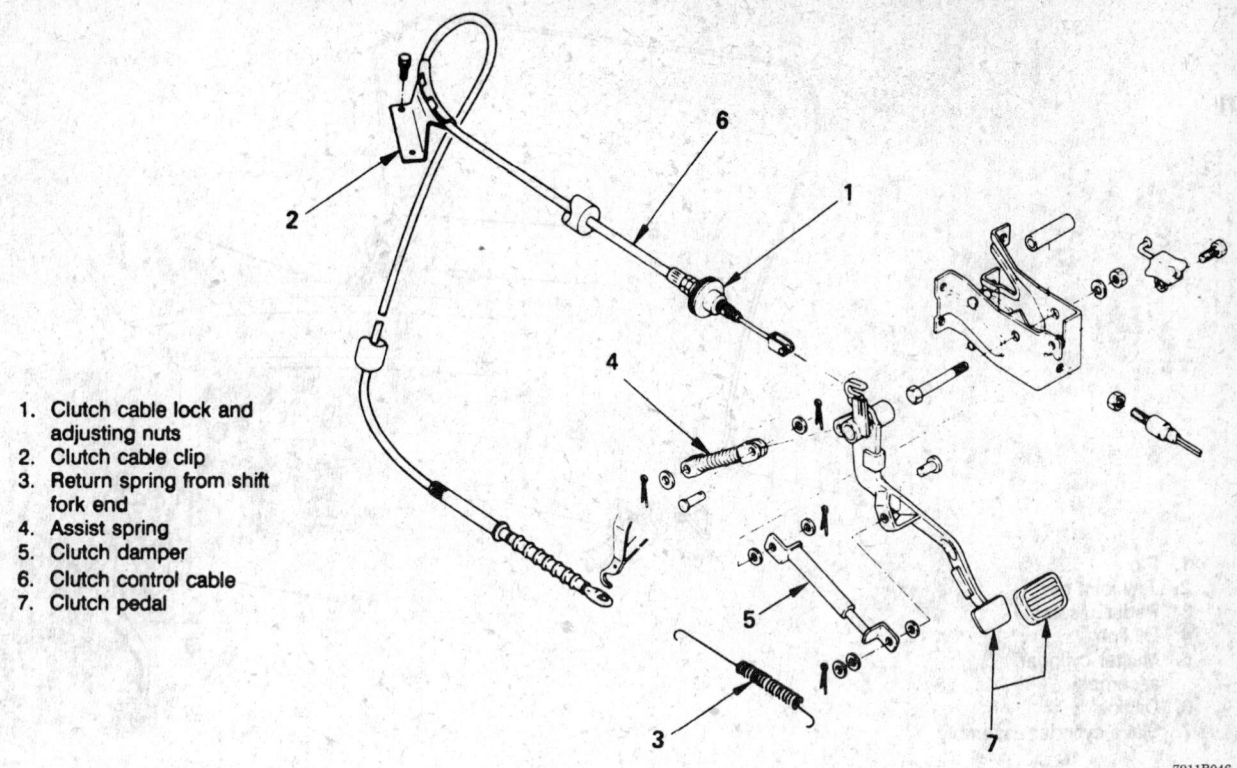

1. Clutch cable lock and adjusting nuts
2. Clutch cable clip
3. Return spring from shift fork end
4. Assist spring
5. Clutch damper
6. Clutch control cable
7. Clutch pedal

Exploded view of the clutch cable assembly

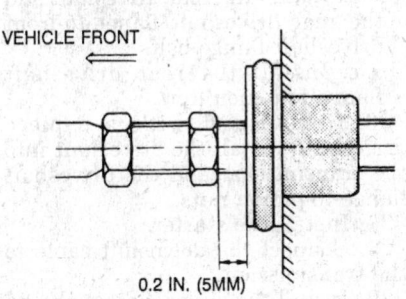

VEHICLE FRONT

0.2 IN. (5MM)

7911B047

View of the clutch cable adjustment

Clutch Damper Cylinder

REMOVAL AND INSTALLATION

6 Cylinder Vehicles

Locate the clutch hydraulic damper in the clutch hydraulic line to the slave cylinder.

1. Disconnect the hydraulic fitting from the damper body.
2. Unbolt the damper from the mounting position bracket.
3. Remove the damper and cap the hydraulic lines to prevent dirt and moisture from contaminating the system.

4. Installation is the reverse of removal. Bleed the hydraulic system.

Bleeding the Hydraulic Clutch System

1. Firmly, set the parking brake.
2. Check the reservoir fluid level and refill, if necessary.
3. Bleed the damper cylinder first, if equipped, then the slave cylinder. The bleeding procedure, as outlined in steps 4 through 9, is the same for both the slave cylinder and the damper cylinder.
4. Using a vinyl tube, connect it to the bleeder screw and submerge the other end in a transparent container of brake fluid.
5. Have an assistant pump the clutch pedal several times and hold it.
6. Loosen the bleeder screw and allow the air bubble fluid to flow into the container, then, tighten the bleeder screw.
7. Release the clutch pedal.
8. Repeat this operation until the fluid is clear of air bubbles.
9. Refill the reservoir. Remove the vinyl tube and replace the rubber cap on the bleeder screw.

AUTOMATIC TRANSMISSION

Transmission Assembly

REMOVAL AND INSTALLATION

2WD Vehicles

1. Disconnect the negative battery cable. Raise and safely support the vehicle.
2. Remove the undercover, if equipped.
3. Drain the transmission fluid from the oil pan.
4. Remove the throttle cable at the engine end. Remove the transmission dipstick.
5. Unbolt the starter and place it aside in a safe location. Support the starter so it does not strain the electrical connections.
6. If equipped with a 1-piece driveshaft, remove the driveshaft flange-to-pinion nuts, lower the driveshaft and pull it from the transmission.

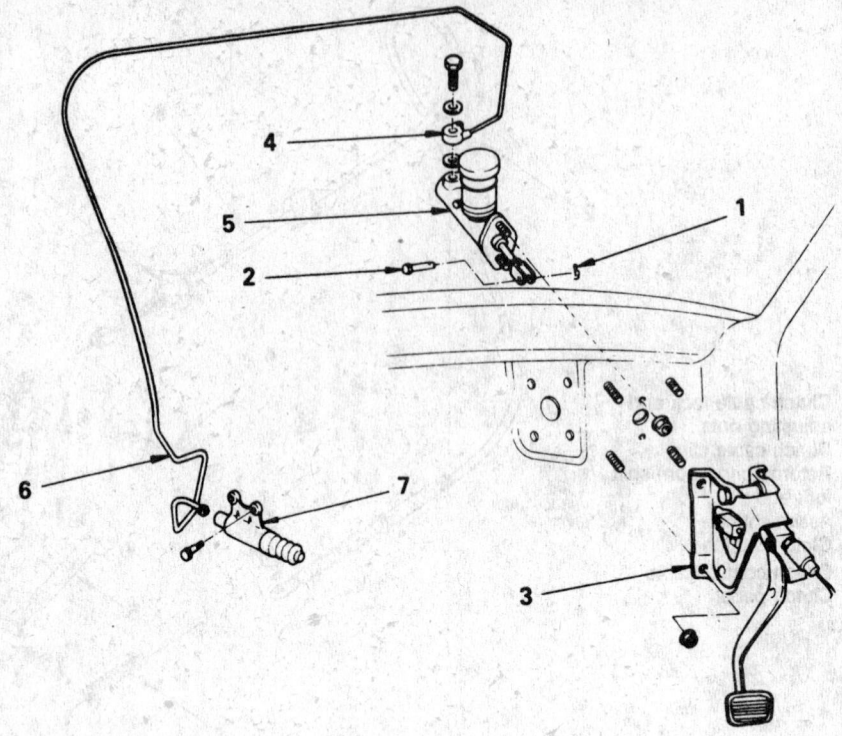

1. Pin
2. Jaw joint pin
3. Pedal assembly
4. Oil line
5. Master cylinder assembly
6. Oil line
7. Slave cylinder assembly

7911B048

Exploded view of the hydraulic system

7. If equipped with a 2-piece driveshaft, perform the following procedures:

a. Remove the rear driveshaft flange-to-pinion nuts.

b. Remove the rear driveshaft flange-to-front driveshaft flange bolts and the rear driveshaft.

c. Remove the center bearing-to-chassis bolts, move the front driveshaft rearward and from the transmission.

8. Disconnect the shift lever at the shifter end.

9. Disconnect the speedometer cable

10. Disconnect the oil cooler lines and place the cooler bypass line close to the transmission case to prevent damage during transmission removal.

11. Remove the torque converter-to-flexplate bolts through the starter hole.

12. Using a transmission jack, place it under the transmission and raise it slightly.

13. Remove the rear mount-to-transmission nuts.

14. Remove the rear mount-to-crossmember nuts/bolts and the mount.

15. Remove the transmission-to-engine bolts.

16. Move the transmission back and lower the transmission out of the vehicle.

To install:

NOTE: Installation of the transmission will require an assistant.

17. Raise the transmission into position.

18. Raise the rear of the transmission and move it into position on the crossmember.

19. Move the transmission forward and engage it with the engine.

20. Install the engine-to-transmission bolts and torque to 47 ft. lbs. (64 Nm).

21. Install the mount and the rear mount-to-crossmember nuts/bolts.

22. Install the rear mount-to-transmission nuts.

23. Install the torque converter-to-flexplate bolts through the starter hole and torque to 22 ft. lbs. (30 Nm).

24. Connect the oil cooler lines, speedometer cable and shift linkage.

25. If equipped with a 2-piece driveshaft, perform the following procedures:

a. Install the front driveshaft into the transmission and the center bearing-to-chassis bolts.

b. Install the rear driveshaft and the rear driveshaft flange-to-front driveshaft flange bolts.

c. Install the rear driveshaft flange-to-pinion nuts.

26. If equipped with a 1-piece driveshaft, install the driveshaft into the transmission and the driveshaft flange-to-pinion nuts.

27. Install the starter.

28. Connect the downshift cable to the transmission.

29. Install the oil level dipstick and the tube.

30. Refill the transmission. Install the undercover, if equipped.

31. Connect the negative battery cable.

32. Start the engine, check for leaks.

4WD Vehicles

TROOPER/TROOPER II

1. Disconnect the negative battery cable. Raise and safely support the vehicle.

2. Remove the undercover, if equipped.

3. Drain the transmission fluid from the oil pan.

4. Remove the throttle cable at the engine end. Remove the transmission dipstick.

5. Unbolt the starter and place it aside in a safe location. Support the starter so it does not strain the electrical connections.

6. Remove the front driveshaft.

7. To remove the 1-piece rear driveshaft, remove the driveshaft flange-to-pinion nuts and driveshaft-to-output flange bolts, lower the driveshaft.

8. Disconnect the shift lever at the shifter end.

9. Disconnect the speedometer cable.

10. Disconnect the oil cooler lines and place the cooler bypass line close to the transmission case to prevent damage during transmission removal.

11. Remove the torque converter-to-flexplate bolts through the starter hole.

12. Using a transmission jack, place it under the transmission and raise it slightly.

13. Remove the rear mount-to-transmission nuts.

14. Remove the rear mount-to-crossmember nuts/bolts and the mount.

15. Remove the transmission-to-engine bolts.

16. Move the transmission back and lower the transmission out of the vehicle.

To install:

NOTE: Installation of the transmission will require an assistant.

17. Raise the transmission into position.

18. Raise the rear of the transmission and move it into position on the crossmember.

19. Move the transmission forward and engage it with the engine.

20. Install the engine-to-transmission bolts and torque to 47 ft. lbs. (64 Nm).

21. Install the mount and the rear mount-to-crossmember nuts/bolts.

22. Install the rear mount-to-transmission nuts.

23. Install the torque converter-to-flexplate bolts through the starter hole and torque to 22 ft. lbs. (30 Nm).

24. Connect the oil cooler lines, speedometer cable and shift linkage.

25. Install the front driveshaft.

26. Install the rear driveshaft.

27. Install the starter.

28. Connect the downshift cable to the transmission.

29. Install the oil level dipstick and the tube.

30. Refill the transmission. Install the undercover, if equipped.

31. Connect the negative battery cable.

32. Start the engine, check for leaks.

RODEO

1. Disconnect both battery cables.

2. Remove the center console housing. Remove the transfer case control lever.

3. Disconnect the transmission shifter control linkage.

4. Remove the undercover to expose the transmission housing.

5. Remove the rear driveshaft and the front driveshaft.

6. Disconnect the parking brake cable from the relay link.

7. Disconnect the speedometer cable from the transmission housing.

8. Remove the left side exhaust pipe.

9. Remove the transmission oil cooler lines.

10. Remove the starter motor and place off to the side.

11. Remove the mode switch from the side of the transmission.

12. Remove the breather hose, dipstick, flywheel cover and the torque converter bolts.

13. Remove the rear crossmember and the exhaust Y-assembly.

14. Number the transmission mounting bolts as they are removed and note the position from which they came. The bolts are of different lengths and this will ease reassembly.

15. Using a transmission jack, remove the transmission assembly.

To install:

16. Install the transmission assembly and replace the mounting bolts into their original locations. Torque the bolts to 55 ft. lbs. (75 Nm).

17. Install the dipstick, exhaust pipe Y-assembly and crossmember.

18. Install the torque converter bolts and torque to 40 ft. lbs. (54 Nm).

19. Install the flywheel cover and torque the M6 bolts to 6 ft. lbs. (8 Nm) and the M10 bolts to 25 ft. lbs. (34 Nm).

20. Install the breather hose.

21. Install the mode selector switch. Place the lever in the **N** position and install the switch on the transmission body. Align the slots on the support and back. Insert a 0.09 in. (2.34mm) pin into the ports and torque the mounting bolts to 104 inch lbs. (12 Nm). Remove the pin before moving the lever.

22. Install the starter motor and torque the mounting bolts to 30 ft. lbs. (40 Nm).

23. Install the oil cooler lines and torque the fittings to 33 ft. lbs. (44 Nm).

24. Install the left front exhaust pipe, the speedometer cable, the parking brake cable and the driveshafts.

25. Install the undercover, the shift linkage, the transfer control linkage and the center console.

26. Connect the battery cables and refill the transmission. Check for proper operation.

Shift Linkage Adjustment

1. Loosen the shift linkage adjusting nut.

2. Push the shift lever fully rearward.

3. Return the shift lever 2 notches to the **N** position.

4. While holding the selector lever lightly toward the **R** range side, tighten the shift linkage nut.

Throttle Linkage Adjustment

1. Depress the accelerator pedal all the way and check that the throttle valve opens fully.

NOTE: If the valve does not open fully, adjust the accelerator link.

2. Fully depress the accelerator.

3. Loosen the adjustment nuts.

4. Adjust the cable housing so the distance between the end of the boot and stopper on the cable is the 0.03–0.06 in. (0.8–1.5mm).

5. Tighten the adjusting nuts.

6. Recheck the adjustment.

TRANSFER CASE

Transfer Case Assembly

REMOVAL AND INSTALLATION

The transfer case is an integral part of the transmission housing. Although the 2 cases can be separated, the transfer case should be removed with the transmission. The transfer case linkage is not adjustable.

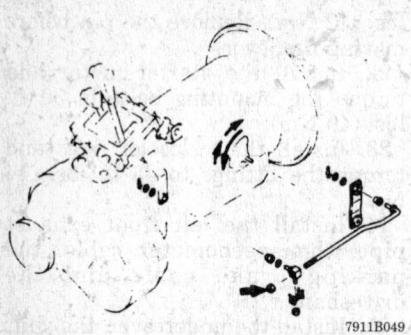

Adjusting the automatic transmission shift linkage

7911B049

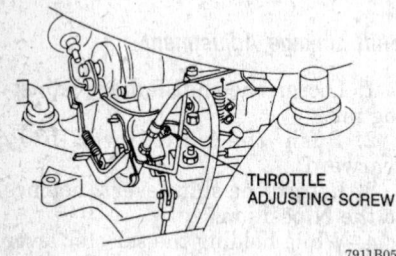

THROTTLE ADJUSTING SCREW

7911B050

Location of the throttle adjusting screw — 1988 engine

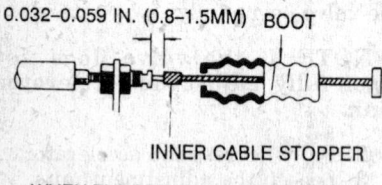

0.032–0.059 IN. (0.8–1.5MM) BOOT

INNER CABLE STOPPER

WHEN THROTTLE VALVE IS FULLY CLOSED

7911B051

View of the throttle cable adjustment — 1988 engine

DRIVELINE

Driveshaft

REMOVAL AND INSTALLATION

2WD Driveshaft

1. Raise and support the vehicle safely.

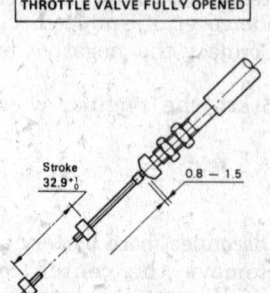

THROTTLE VALVE FULLY OPENED

Stroke 32.9⁺⁰

0.8 – 1.5

7911B052

View of the throttle cable adjustment — 1988 — 90 gasoline engine

2. Matchmark the driveshaft to the yokes.

3. Remove the driveshaft flange retaining bolts and remove the driveshaft.

4. To install, reverse the removal procedures. Torque the bolts 46 ft. lbs. (63 Nm).

4WD Driveshaft

REAR

1. Raise and safely support the vehicle.

2. Matchmark the driveshaft flange-to-differential pinion flange.

3. If equipped with a 1-piece driveshaft, remove the driveshaft flange-to-pinion nuts, lower the driveshaft and pull it from the transmission.

4. If equipped with a 2-piece driveshaft, perform the following procedures:

    a. Remove the rear driveshaft flange-to-pinion nuts.

    b. Remove the rear driveshaft flange-to-front driveshaft flange bolts and the rear driveshaft.

    c. Remove the center bearing-to-chassis bolts, move the front driveshaft rearward and from the transmission.

To install:

5. If equipped with a 2-piece driveshaft, perform the following procedures:

    a. Install the front driveshaft into the transmission and the center bearing-to-chassis bolts.

    b. Install the rear driveshaft and the rear driveshaft flange-to-front driveshaft flange bolts.

    c. Install the rear driveshaft flange-to-pinion nuts.

6. If equipped with a 1-piece driveshaft, install the driveshaft into the transmission and the driveshaft flange-to-pinion nuts.

7. To complete the installation, reverse the removal procedures. Torque the retaining bolts to 46 ft. lbs. (63 Nm).

FRONT

1. Raise and safely support the vehicle.

2. Matchmark the driveshaft flange-to-transfer case flange and the driveshaft-to-differential pinion flange.

3. Remove the front driveshaft's splined yoke flange-to-transfer case bolts and separate the front driveshaft from the transfer case; do not allow the splined flange to fall away from the driveshaft.

4. Remove the driveshaft flange-to-differential pinion flange bolts and separate the driveshaft from the front differential.

5. To install, align the matchmarks and reverse the removal procedures. Torque the retaining bolts to 46 ft. lbs. (63 Nm).

U-Joints

REMOVAL AND INSTALLATION

1. Raise and support the vehicle safely. Remove the driveshaft.

2. If the front yoke is to be disassembled, matchmark the driveshaft and sliding splined yoke so driveline balance is preserved upon reassembly. Remove the snaprings that retain the bearing caps.

3. Select 2 press components, with one small enough to pass through the yoke holes for the bearing caps and the other large enough to receive the bearing cap.

4. Use a vise or a press and position the small and large press components on either side of the U-joint. Press in on the smaller press component so it presses the opposite bearing cap out of the yoke and into the larger press component. If the cap does not come all the way out, grasp it with a pair of pliers and work it out.

5. Reverse the position of the press components so the smaller press component presses on the cross. Press the other bearing cap out of the yoke.

6. Repeat the procedure on the other bearings.

7. To install, grease the bearing caps and needles thoroughly if they are not pregreased. Start a new bear-

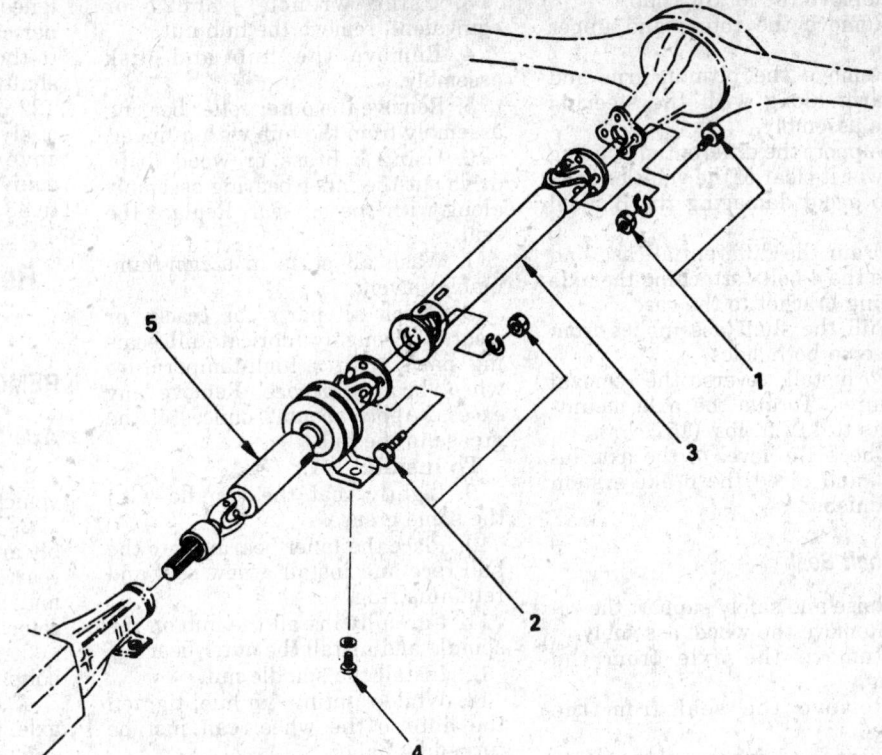

1. Differential side bolt
2. Flange bolt
3. 2nd propeller shaft assembly
4. Center bearing bracket bolt
5. 1st propeller shaft assembly

7911B053

Exploded view of the rear driveshaft assembly

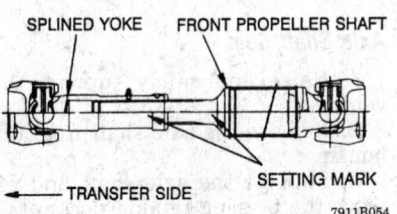

SPLINED YOKE   FRONT PROPELLER SHAFT

TRANSFER SIDE   SETTING MARK

7911B054

View of the front driveshaft — 4WD

ing cap into a side of the yoke. Position the cross in the yoke.

NOTE: Some U-joints have a grease fitting that must be installed in the joint before assembly. When installing the fitting, make sure once the driveshaft is installed in the vehicle that the fitting is accessible to be greased at a later date.

8. Select 2 press components small enough to pass through the yoke holes. Put the press components against the cross and the cap and press the bearing cap ¼ in. below the

surface of the yoke. If there is a sudden increase in the force needed to press the cap into place or if the cross starts to bind, the bearings are cocked. They must be removed and restarted in the yoke. Failure to do so will cause premature bearing failure.

9. Install a new snapring.
10. Start the new bearing into the opposite side. Place a press component on it and press in until the opposite bearing contacts the snapring.
11. Install a new snapring.
12. Install the other bearings in the same manner.
13. Check the joint for free movement. If binding exists, smack the yoke ears with a brass or plastic faced hammer to seat the bearing needles. If binding still exists, disassemble the joint and check to see if the needles are in place. Do not strike the bearings unless the shaft is supported firmly. Do not install the driveshaft until free movement exists at all joints.

Front Axle Shaft, Bearing and Seal

REMOVAL AND INSTALLATION

Axle Shaft

1. Raise and safely support the vehicle.
2. Disconnect the front driveshaft from the differential.
3. Remove the wheels and skid plate.
4. Loosen the torsion bar completely with the height control adjusting bolts.
5. Remove the strut bars.
6. Disconnect the stabilizer bars from the lower control arms.
7. Remove the caliper assemblies and wire them to the frame; do not disconnect the brake lines.
8. Remove the ball joints from the tie rods.
9. Disconnect the upper control arms from the frame; make sure to note the number and positions of the shims.
10. Remove the steering link ends from the lower control arms.
11. Disconnect the shock absorbers from the lower control arms.
12. Disconnect the lower control arms from the frame.

13. Remove the locking hub.

14. Remove the rotors and upper links.

15. Remove the pitman arm and idler arm along with the steering linkage assembly.

16. Support the differential housing and lower it clear of the vehicle. Take care to avoid damaging the Birfield joints.

17. Drain the differential case and remove the 4 bolts attaching the axle mounting bracket to the case.

18. Pull the shaft assemblies from the case on both sides.

19. To install, reverse the removal procedures. Torque the axle mounting nuts to 112 ft. lbs. (152 Nm).

20. Check the level of the axle lubricant and bleed the brake system when finished.

Axle Shaft Seal

1. Raise and safely support the vehicle. Remove the wheel assembly.

2. Remove the axle from the housing.

3. Remove the seal from the housing.

4. Clean and inspect the sealing surfaces of the housing and axle.

5. Using a seal installer tool, drive the new seal into the housing with the lip of the seal facing the housing.

6. Lightly coat the lip of the seal with oil and install the axle in the housing.

7. To complete the installation, reverse the removal procedures.

8. Check the level of the axle lubricant when finished.

Axle Shaft Bearings

1. Raise and safely support the vehicle.

2. Remove the axle shaft from the housing.

3. Support the axle shaft and remove the bearing retainer locknut and washer.

4. Remove the retainer, bearing and seal from the axle shaft.

5. To install, reverse the removal procedures.

6. Always replace the seal and lock washer when removing the axle shaft from the housing. Torque the bearing retainer nut to 188–195 ft. lbs.

Wheel Bearings

1. Place the transfer case in **2H**. Raise and safely support the vehicle.

2. Remove the free wheeling hub cover assembly. Remove the brake caliper and hang aside.

3. Using wrench J-36827 or equivalent, remove the hub nut.

4. Remove the hub and disk assembly.

5. Remove the outer roller bearing assembly from the hub with a finger.

6. Using a brass or wood drift, drive out the inner bearing assembly along with the oil seal. Replace the seal.

7. Wash all parts in a non-flammable solvent.

8. Check all parts for cracks or wear. Thoroughly lubricate all bearing parts with a high-temperature wheel bearing grease. Remove any excess. Apply about 2 ounces of the grease to the hub.

To install:

9. Lightly coat the spindle with the same grease.

10. Place the inner bearing into the hub race and install a new seal and retaining ring.

11. Carefully install the hub on the spindle and install the outer bearing.

12. Install the spindle nut.

13. While rotating the hub, tighten the hub so the wheel can just be turned by hand.

14. Turn the hub 2–3 turns and back off the nut just enough so it can be loosened with the fingers.

15. Finger-tighten the nut so all play is taken up at the bearing.

16. Attach a pull scale to a lug nut and check the amount of pull needed to start the wheel turning. Initial pull should be 2.6–4.0 lbs. When performing this test, make sure the brake pads are not touching the rotor. If the rotating torque is not correct, tighten the spindle nut until it is.

17. Install the lockwasher so the bolt holes align. If the bolt holes do not align, reverse the position of the lock washer. The bolt head should be able to sink below the surface of the washer when tightened.

18. Clean the hub flange surface and areas surrounding the spindle and with the transfer case lever in the **2H** position, install the inner cam with the gear teeth facing out.

19. Lower the vehicle and install tools J-36835-2 and J-36836 or equivalent, on the axle shaft until it comes in contact with the lock washer.

20. Measure the gap between the tool and the snapring tool and select shims to reduce the gap to 0.00–0.10mm. Remove the tool.

21. Apply grease to the hub splines and lubrication grooves. Install the drive clutch assembly. The cut portion of the drive clutch must be aligned with the concave part of the inner cam. Match the teeth of the cam to the inner cam by turning the axle shaft.

22. Install the snapring and previously selected shims, gasket and cover. Apply Loctite® 515 or equivalent, and torque the cover bolts to 43 ft. lbs. (59 Nm).

Rear Axle Shaft, Bearing and Seal

REMOVAL AND INSTALLATION

Axle Shaft

1. Raise and safely support the vehicle.

2. Remove the rear wheel assembly and brake drum. If equipped with rear disk brakes, remove the caliper and hang aside. Remove the brake rotor.

3. Remove the 4 axle retainer bolts.

4. Using a slide hammer on the axle, pull the axle out of the housing.

5. To install, reverse the removal procedures.

6. If equipped with drum brakes, torque the axle retention bolts to 55 ft. lbs. (75 Nm). If equipped with disk brakes, torque the axle retention bolts to 75 ft. lbs. (103 Nm).

Axle Shaft Seal

1. Raise and safely support the vehicle.

2. Remove the axle shaft from the housing.

3. Support the axle shaft and remove the bearing retainer locknut.

4. Remove the retainer, bearing and seal from the axle shaft.

5. Using a seal driver tool, remove the seal and install the new seal in the retainer.

6. To install, reverse the removal procedures.

7. Torque the bearing retainer nut to 188–195 ft. lbs. (250–260 Nm).

8. Check the level of the axle lubricant when finished.

Front Wheel Hub, Knuckle and Bearings

REMOVAL AND INSTALLATION

1. Raise and safely support the vehicle. Remove the wheel assembly.

2. Remove the brake caliper and support it on a wire. Remove the rotor and dust shield.

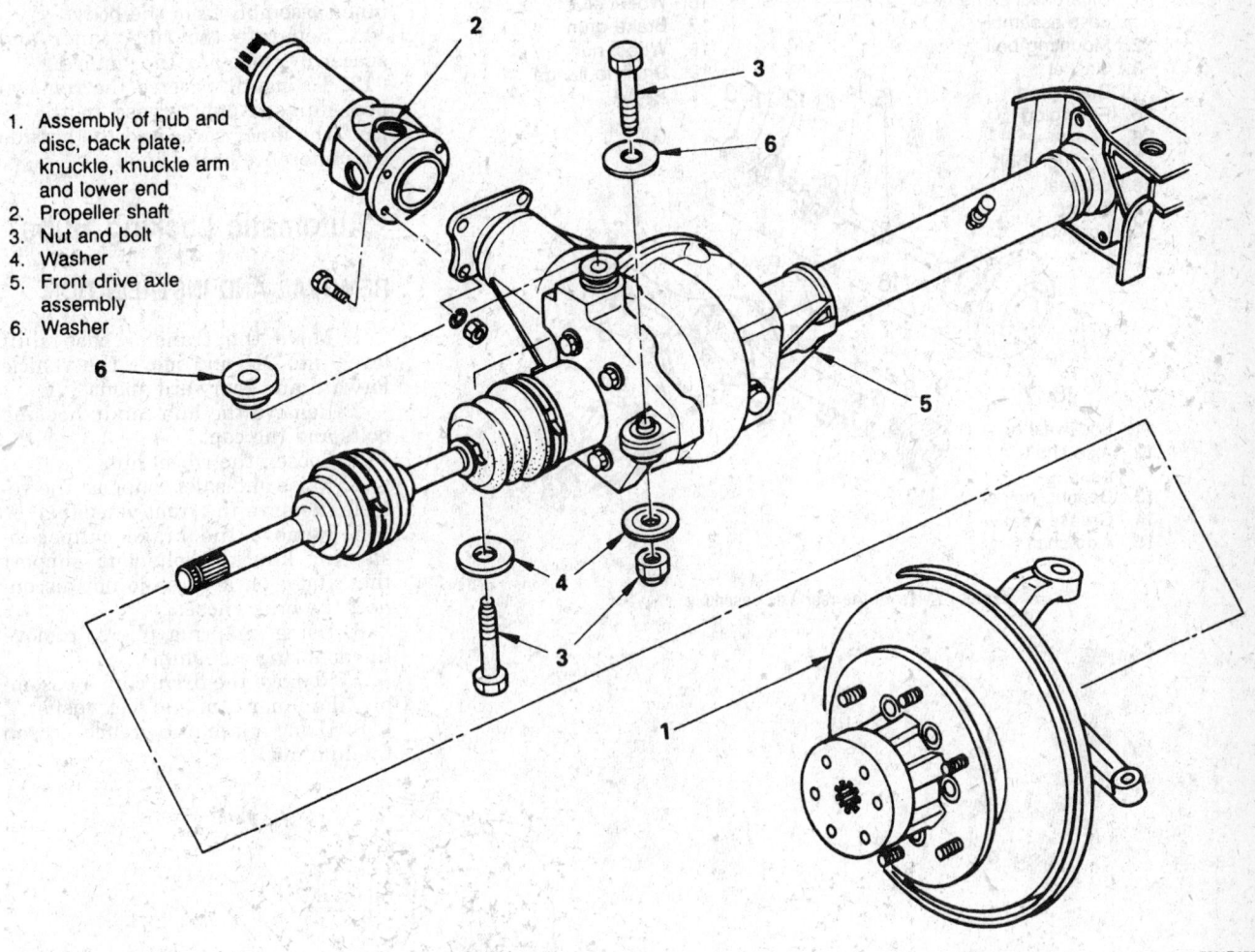

1. Assembly of hub and disc, back plate, knuckle, knuckle arm and lower end
2. Propeller shaft
3. Nut and bolt
4. Washer
5. Front drive axle assembly
6. Washer

Exploded view of the front axle assembly — 4WD

7911B055

3. If equipped with 4WD, remove the axle shaft from the hub.

4. Remove the tie rod end-to-steering knuckle nut and separate the tie rod from the steering knuckle.

5. Support the lower control arm and separate the steering knuckle from the lower ball joint.

6. Separate the steering knuckle from the upper ball joint.

7. Remove the steering knuckle from the vehicle.

8. To install, reverse the removal procedures.

9. Torque the upper ball joint nut to 75 ft. lbs. (100 Nm), the lower ball joint nut to 95 ft. lbs. (127 Nm) and

the steering linkage ball nut to 75 ft. lbs. (100 Nm).

Manual Locking Hubs

REMOVAL AND INSTALLATION

1. Place the transfer case in the **2H** position. Raise and safely the vehicle.

2. Set the hubs in the **FREE** position.

3. Remove the hub cover bolts and the hub cover.

4. While pushing the follower toward the knob, turn the clutch as-

sembly clockwise and then remove the clutch assembly from the knob.

5. Remove the snapring and the knob from the cover. Do not loose the detent ball.

6. Remove the ball and spring from the knob.

7. Remove the X-ring from the knob by pressing it off.

NOTE: Do not use a sharp instrument to remove this ring because it may scratch the ring.

8. Remove the compression spring, retaining spring and the follower from the clutch assembly.

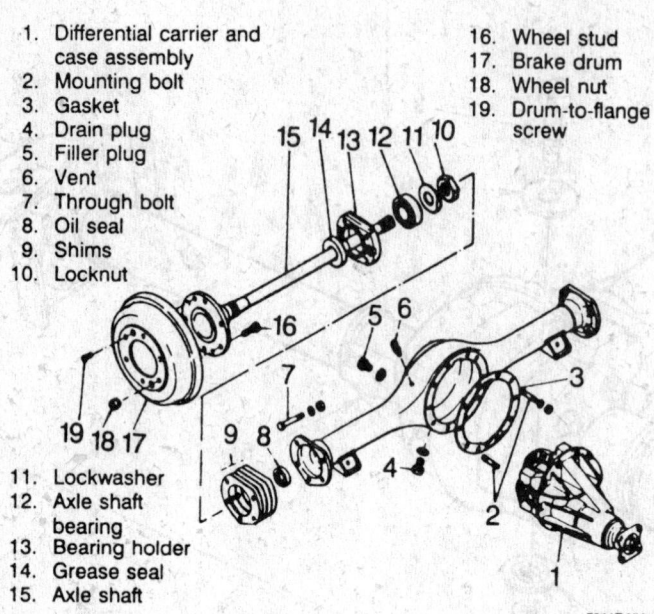

1. Differential carrier and case assembly
2. Mounting bolt
3. Gasket
4. Drain plug
5. Filler plug
6. Vent
7. Through bolt
8. Oil seal
9. Shims
10. Locknut
11. Lockwasher
12. Axle shaft bearing
13. Bearing holder
14. Grease seal
15. Axle shaft
16. Wheel stud
17. Brake drum
18. Wheel nut
19. Drum-to-flange screw

Exploded view of the rear axle assembly

9. Remove the retaining spring from the clutch assembly by turning it counterclockwise.

10. Remove the snapring and the inner assembly from the body.

11. Separate the ring, inner and spacer by removing the snapring.

12. To install, reverse the removal procedures. Apply grease to the X-ring, the inner cover and the outside circumference of the knob.

Automatic Locking Hubs

REMOVAL AND INSTALLATION

1. Move the transfer case shift lever into **2H** and move the vehicle forward and rearward about 3 ft.

2. Remove the hub cap-to-housing bolts and the cap.

3. Loosen the wheel nuts.

4. Raise and safely support the vehicle. Remove the front wheel(s).

5. Remove the brake caliper-to-steering knuckle bolts and support the caliper on a wire; do not disconnect the brake hose.

6. Using snapring pliers, remove the snapring and shims.

7. Remove the drive clutch assembly, the inner cam and lockwasher.

8. Using a hub nut wrench, loosen the hub nut.

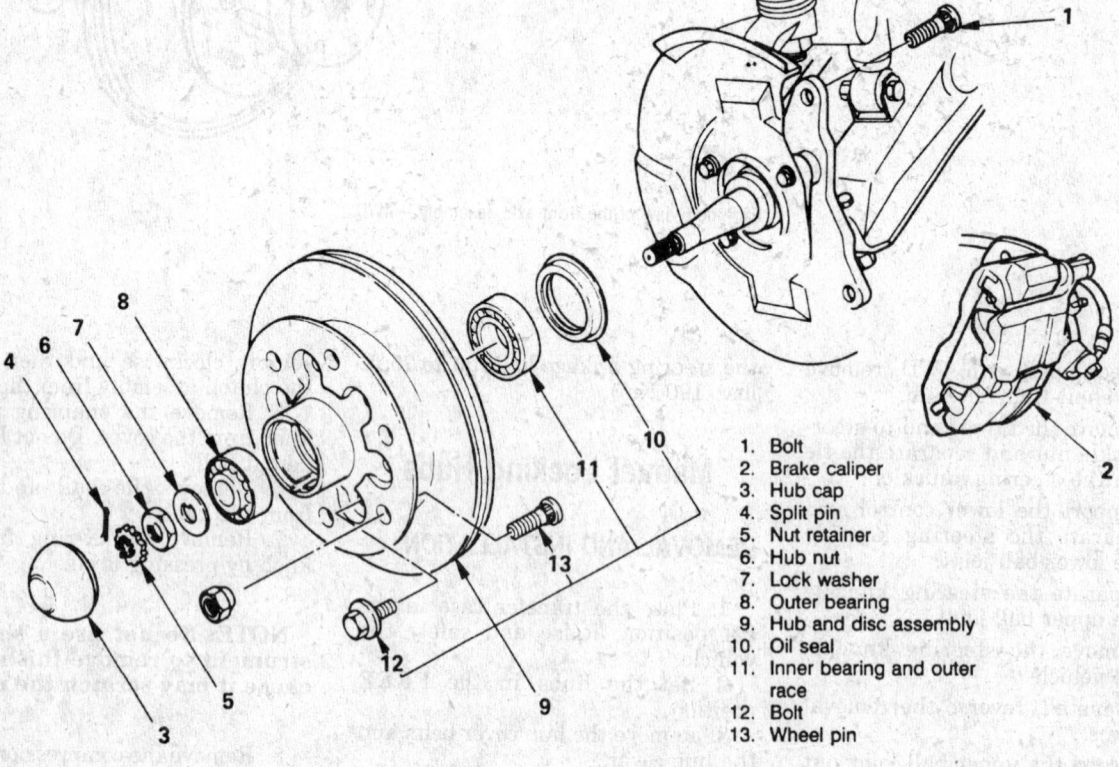

1. Bolt
2. Brake caliper
3. Hub cap
4. Split pin
5. Nut retainer
6. Hub nut
7. Lock washer
8. Outer bearing
9. Hub and disc assembly
10. Oil seal
11. Inner bearing and outer race
12. Bolt
13. Wheel pin

Exploded view of the front wheel assembly — 2WD Pick-Up

1. Bolt
2. Brake caliper
3. Bolt
4. Cover
5. Lock washer
6. Hub nut
7. Hub and disc assembly
8. Outer bearing
9. Oil seal
10. Inner bearing
11. Bolt
12. Wheel pin

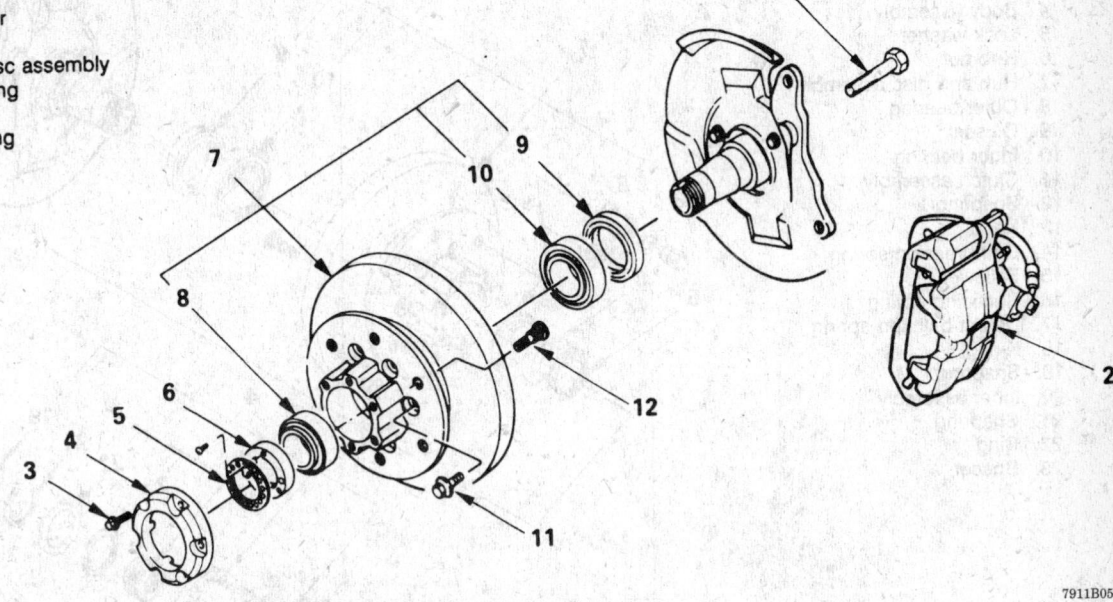

7911B058

Exploded view of the front wheel assembly — 2WD Amigo

9. Pull the hub from the spindle.

10. If necessary, use a brass drift and a hammer to drive the wheel bearings from the hub.

11. If removing the disc from the hub, scribe matchmarks, remove the disc-to-hub bolts and separate the disc from the hub.

To install:

12. To install, reverse the removal procedur

13. When installing the hub nut, perform the following procedures:

a. Torque the hub nut to 22 ft. lbs. and loosen the nut.

b. Using a spring gauge, connect it to the stud bolt at 90 degrees.

c. Retorque the hub nut until the spring gauge measures a bearing preload of 4.4–5.5 lbs. for new bearing and oil seal or 2.6–4.0 lbs. for used bearing and new oil seal.

d. Adjust the snapring clearance by performing the following procedures:

e. Install the special adjusting tool onto the hub until it comes in contact with the lock washer.

f. Using a feeler gauge, measure the clearance **t** between the hub and the snapring groove on the axle shaft.

g. If the clearance is larger than the snapring groove, install shims on the shaft so clearance **t** is 0–0.039 in. (0–0.1mm).

14. To complete the installation, reverse the removal procedures. Apply Loctite® to the hub cap bolts and torque the hub cap-to-hub assembly bolts to 43 ft. lbs. (60 Nm).

Pinion Seal

REMOVAL AND INSTALLATION

1. Raise and safely support the vehicle. If necessary, remove the skid plate.

2. Matchmark and remove the front driveshaft.

3. Check the turning torque of the pinion before proceeding. This is the torque that must be reached during installation of the pinion nut.

NOTE: The amount of turning torque required to move the pinion gear should be 20–30 ft. lbs. of torque.

4. Using a pinion flange holding tool, remove the pinion nut and washer.

5. Remove the pinion flange from the pinion gear.

6. Pry the pinion seal out of the differential carrier.

7. Clean and inspect the sealing surface of the carrier.

To install:

8. Using a seal driver tool, drive the new seal into the carrier until the flange on the seal is flush with the carrier.

9. With the seal installed, the pinion bearing preload must be set.

10. Tighten the pinion nut while holding the flange, until the turning torque is the same as before removal of the nut.

11. Align the matchmarks and install the driveshaft.

12. Check the level of the differential lubricant when finished.

Differential Carrier

REMOVAL AND INSTALLATION

Front Drive Axle

1. Raise and safely support the vehicle.

2. Drain the differential oil.

3. Matchmark and remove the front driveshaft.

4. Remove the axle shafts from the differential.

5. Remove the differential carrier mounting bolts and remove the carrier.

1. Bolt
2. Housing assembly
3. Snapring and shims
4. Body assembly
5. Lock washer
6. Hub nut
7. Hub and disc assembly
8. Outer bearing
9. Oil seal
10. Inner bearing
11. Clutch assembly
12. Snapring
13. Knob
14. Compression spring
15. Follower
16. Retaining spring
17. Detent ball and spring
18. X-ring
19. Snapring
20. Inner assembly
21. Snapring
22. Ring
23. Spacer

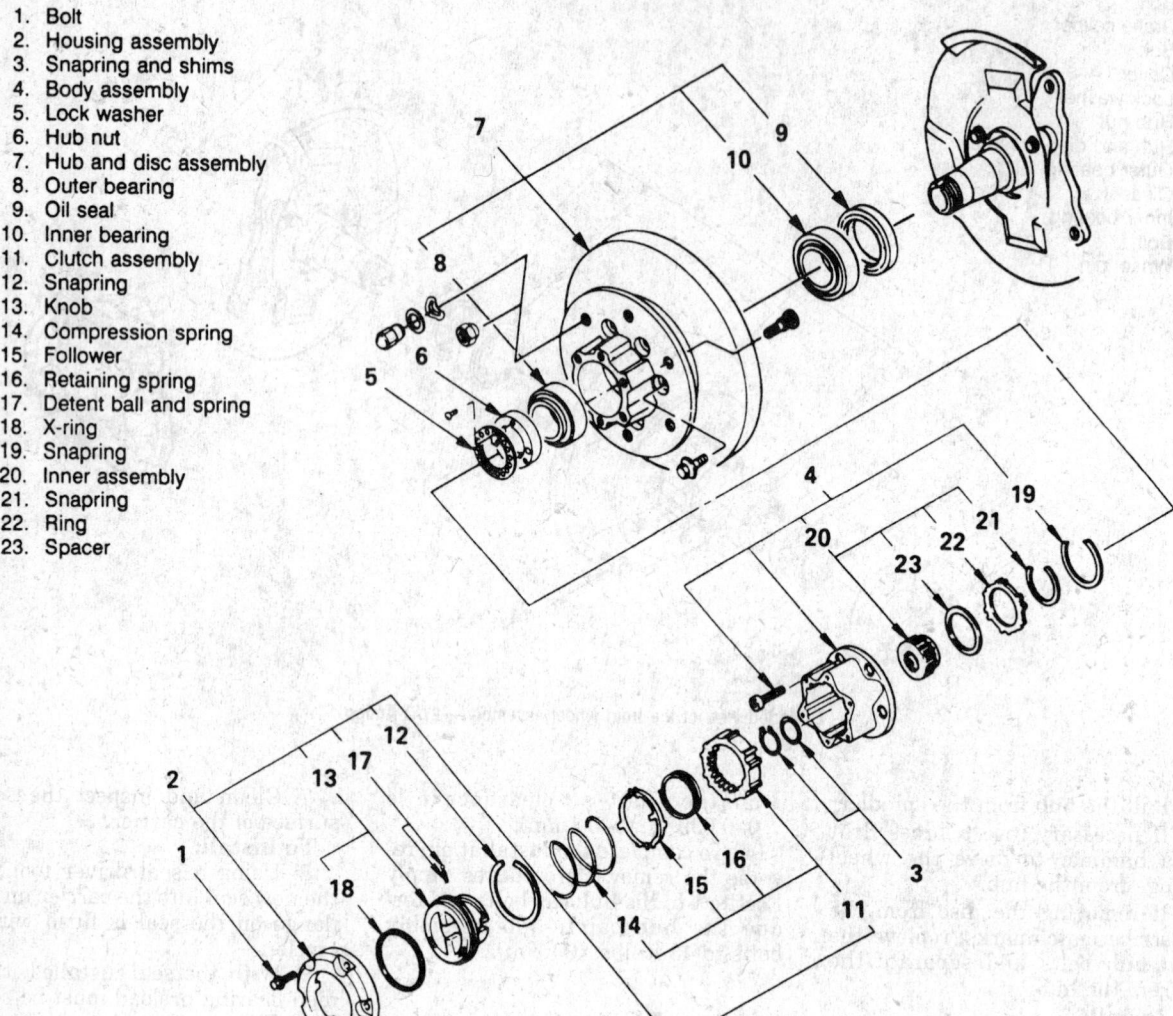

Exploded view of the manual locking hub assembly — 4WD

7911B059

To install:

6. To install, reverse the removal procedures. Use a new gasket when installing. Torque the bolts to 19 ft. lbs. (26 Nm).

7. Fill the differential to the correct level when finished.

Rear Drive Axle

1. Raise and safely support the vehicle.

2. Drain the differential oil.

3. Matchmark and remove the rear driveshaft. Remove the drum or caliper and rotor assemblies.

4. Remove the axle retainer bolts and remove the axle shafts from the axle housing.

5. Remove the differential carrier mounting bolts and the carrier.

To install:

6. To install, reverse the removal procedures. Use a new gasket when installing. Torque the differential mounting bolts to 47 ft. lbs. (64 Nm). Torque the bearing retainer bolts to 75 ft. lbs. (103 Nm) on disk brake models or 54 ft. lbs. (74 Nm) on drum brake models.

7. Fill the differential to the correct level when finished.

Axle Housing

REMOVAL AND INSTALLATION

Front Housing

1. Raise and safely support the vehicle.

2. Matchmark and disconnect the front driveshaft from the differential.

3. Remove the wheels and skid plate.

4. Loosen the torsion bar completely with the height control adjusting bolts.

5. Remove the strut bars.

6. Disconnect the stabilizer bars from the lower control arms.

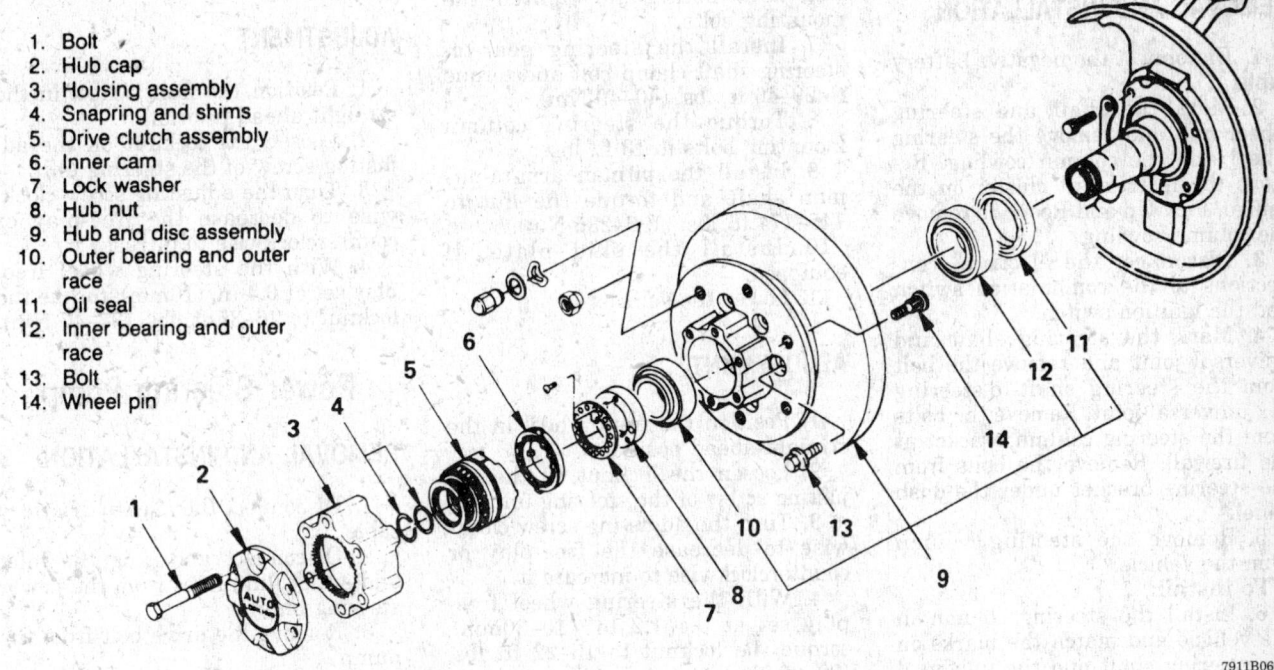

1. Bolt
2. Hub cap
3. Housing assembly
4. Snapring and shims
5. Drive clutch assembly
6. Inner cam
7. Lock washer
8. Hub nut
9. Hub and disc assembly
10. Outer bearing and outer race
11. Oil seal
12. Inner bearing and outer race
13. Bolt
14. Wheel pin

Exploded view of the automatic locking hub assembly — 4WD

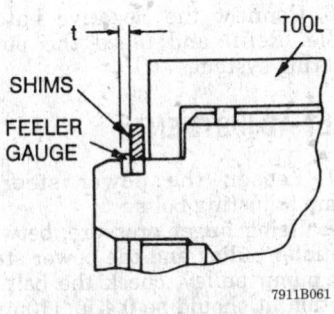

Using a feeler gauge to measure the shim clearance on the automatic locking hub assembly — 4WD

7. Remove the caliper assemblies and suspend them on a wire; do not disconnect the brake lines.
8. Remove the tie rod ends from the steering knuckles.
9. Disconnect the upper control arms from the frame; note the number and positions of the shims.
10. Remove the steering link ends from the lower control arms.
11. Disconnect the shock absorbers from the lower control arms.
12. Disconnect the lower control arms from the frame.
13. Remove the locking hub.

14. Remove the rotors and upper links.
15. Remove the pitman arm and idler arm along with the steering linkage assembly.
16. Support the differential housing and lower it clear of the vehicle. Take care to avoid damaging the Birfield joints.
17. To install, reverse the removal procedures. Check the level of the axle lubricant when finished.

Rear Housing

1. Raise and safely support the vehicle. Remove the rear wheels.
2. Disconnect the shock absorbers from the spring plates.
3. Disconnect and plug the brake lines on the rear axle housing.
4. Disconnect the parking brake cables from the rear axle housing.
5. Support the rear axle housing and remove the housing to leaf spring U-bolts.
6. Remove the rear axle housing from the vehicle.
To install:
7. To install, reverse the removal procedures.
8. Torque the housing U-bolts to 36–43 ft. lbs. and the shock absorber bolts to 27–30 ft. lbs.

9. Bleed the brake system and check the level of the axle lubricant when finished.

STEERING

Steering Wheel

REMOVAL AND INSTALLATION

1. Disconnect the negative battery cable.
2. From the rear of the steering wheel, remove the horn pad screw. Pry the horn pad upward from the steering wheel.
3. Remove the steering wheel-to-steering column nut.
4. Matchmark the steering wheel-to-steering shaft.
5. Using a steering wheel puller, press the steering wheel from the steering shaft.
6. To install, align the matchmarks and reverse the removal procedures. Torque the steering wheel nut to 22–29 ft. lbs. (30–39 Nm).

Steering Column

REMOVAL AND INSTALLATION

1. Disconnect the negative battery cable.

2. Mark the shaft and steering wheel hub and remove the steering wheel and the column cowling. Remove the instrument cluster on the Amigo, Pick-Up and Rodeo to remove the column cowling.

3. Disconnect the electrical connections to the combination switch and the ignition switch.

4. Mark the steering shaft and universal joint and remove the bolt from the steering shaft-to-steering box universal joint. Remove the bolts from the steering column bracket at the firewall. Remove the bolts from the steering bracket under the dash panel.

5. Remove the steering column from the vehicle.

To install:

6. Install the steering column in the vehicle and match the marks on the steering shaft and the universal joint. Connect the electrical connections to the combination switch and the ignition switch.

7. Insert the bolts into the steering bracket under the dash. Torque the bolts to 11–14 ft. lbs. (1.5–2.0 Nm).

8. Insert the bolts into the firewall bracket and torque to 10–17 ft. lbs. (1.4–2.4 Nm).

9. Install the universal joint bolt and torque to 14–22 ft. lbs. (2.0–3.0 Nm).

10. Install the steering wheel, matching the marks made during removal. Torque the nut to 22–29 ft. lbs. (3.0–4.0 Nm). Install the cowling and the instrument panel. Connect the battery.

Manual Steering Gear

REMOVAL AND INSTALLATION

1. Disconnect the negative battery cable. Raise and safely support the vehicle. Remove the skid plate, if equipped.

2. Remove pitman arm nut and washer. Matchmark the pitman arm-to-pitman shaft.

3. Using a puller tool, press the pitman arm from the pitman shaft.

4. Remove the steering gear-to-steering shaft clamp bolt.

5. Remove the steering gear-to-frame bolts and the steering gear from vehicle.

To install:

6. Place the steering gear in position and install and tighten the mounting bolts.

7. Install the steering gear-to-steering shaft clamp bolt and torque to 29–40 ft. lbs. (40–49 Nm).

8. Torque the steering column mounting bolts to 13 ft. lbs.

9. Install the pitman arm-to-pitman shaft and torque the nut to 145–174 ft. lbs. (196–236 Nm).

10. Install the skid plate, if equipped.

11. Lower the vehicle.

ADJUSTMENT

1. Position the front wheel in the straight ahead position.

2. Loosen the locknut on the adjusting screw of the steering unit.

3. Turn the adjusting screw clockwise to decrease the free-play or counterclockwise to increase it.

4. With the steering wheel free-play set at 0.4–1.2 in. (10–30mm), torque the locknut to 15–22 ft. lbs. (20–29 Nm).

Power Steering Gear

REMOVAL AND INSTALLATION

1. Raise and safely support the vehicle. Remove the skip plate, if equipped.

2. Remove pitman arm nut and washer. Matchmark the pitman arm-to-pitman shaft.

3. Using a puller tool, press the the pitman arm from the pitman shaft.

4. Disconnect and plug the power steering lines at the steering gear.

5. Remove the steering gear-to-steering shaft clamp bolt.

6. Remove the steering gear-to-frame bolts and the steering gear from vehicle.

To install:

7. Place the steering gear in position and install and tighten the mounting bolts. Connect the power steering lines to the steering gear.

8. Install steering gear-to-steering shaft bolts and torque to 29–40 ft. lbs. (40–49 Nm).

9. Torque steering column mounting bolts to 13 ft. lbs.

10. Install the pitman arm to the pitman shaft. Install washer and torque nut to 145–174 ft. lbs. (196–236 Nm).

11. Install the skid plate, if equipped.

12. Lower the vehicle. Refill and bleed the power steering system.

ADJUSTMENT

1. Position the front wheel in the straight ahead position.

2. Loosen the locknut on the adjusting screw of the steering unit.

3. Turn the adjusting screw clockwise to decrease the free-play or counterclockwise to increase it.

4. With the steering wheel free-play set at 0.4 in. (10mm), torque the locknut to 26–35 ft. lbs. (37–47 Nm).

Power Steering Pump

REMOVAL AND INSTALLATION

1. Disconnect the negative battery cable.

2. Disconnect and plug the inlet and outlet fluid lines from the power steering pump.

3. Remove the drive belt from the pump.

4. Remove the pump-to-bracket bolts and the pump from the brackets.

5. To install, reverse the removal procedures.

6. Connect the negative battery cable. Refill and bleed the power steering system.

BELT ADJUSTMENT

1. Loosen the power steering pump adjusting bolts.

2. Using finger pressure, between the idler pulley and the power steering pump pulley, check the belt deflection; it should be 0.4 in. (10mm).

3. With the power steering pump adjusted to the correct belt deflection, tighten the pump bolts.

SYSTEM BLEEDING

1. Fill the power steering reservoir to the proper level when cold.

2. Start and operate the engine until it reaches normal operating temperatures.

3. Turn the engine **OFF** and check the fluid level. If necessary, fill the reservoir to the proper level.

4. Run the engine and turn the steering wheel from lock-to-lock, in both directions, 3–4 times; do not hold the steering wheel at the lock position for more than 5 seconds or temperature rise will result.

5. Return the steering wheel to center, turn the engine **OFF** and al-

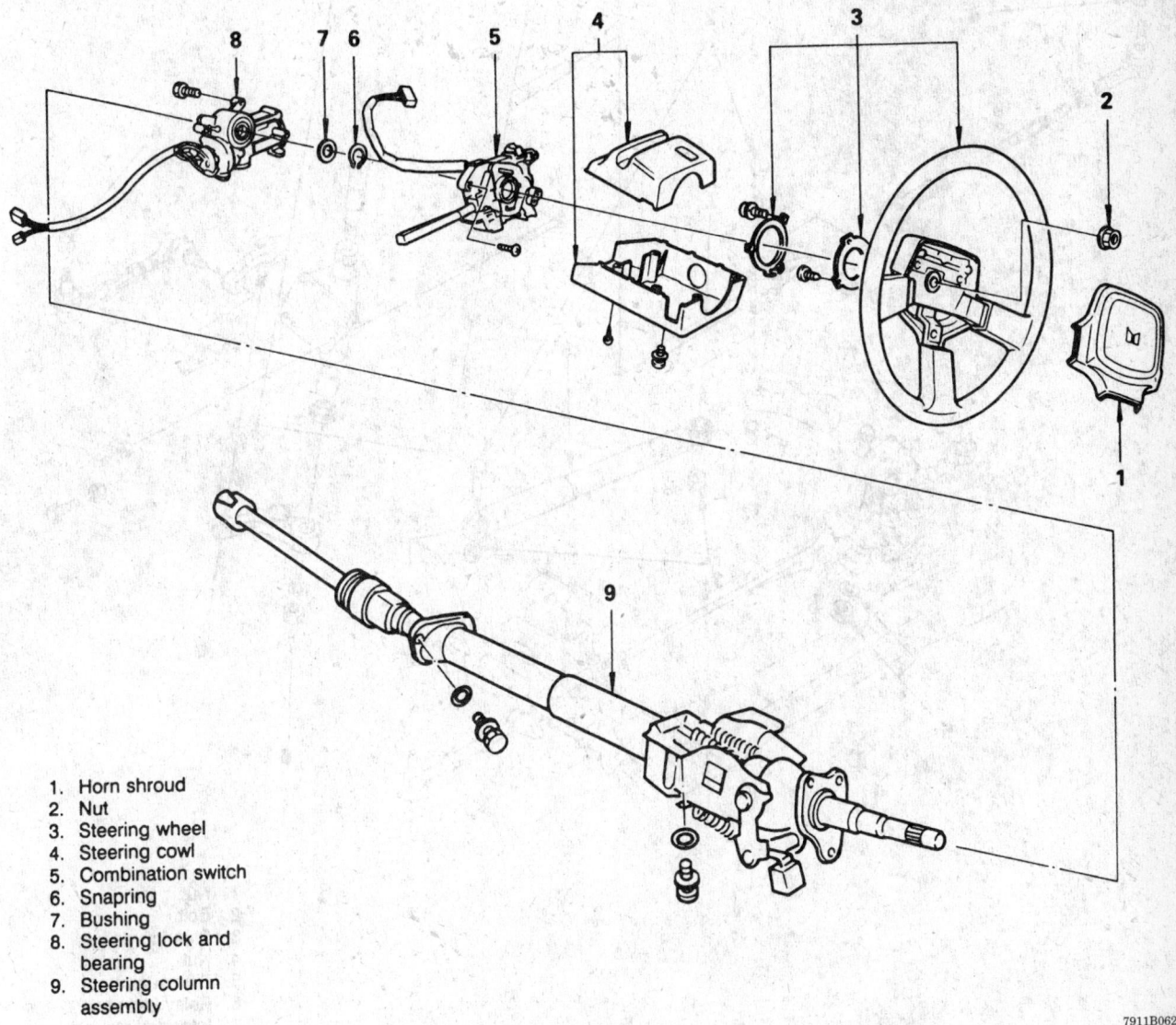

1. Horn shroud
2. Nut
3. Steering wheel
4. Steering cowl
5. Combination switch
6. Snapring
7. Bushing
8. Steering lock and bearing
9. Steering column assembly

7911B062

Exploded view of the steering column assembly — 1988 — 90

low the fluid to sit for 5 minutes before adding any more.

6. If necessary, repeat the bleeding procedure until the air bubbles are removed from the system.

7. Fill the system to the proper level when finished.

Tie Rod Ends

REMOVAL AND INSTALLATION

1. Raise and safely support the vehicle.

2. Matchmark the tie rod ends to the tie rod shaft for reinstallation purposes.

3. Remove the cotter pin and nut from the tie rod end and loosen the clamping bolts on the sleeve.

4. Using a tie rod end puller, separate the tie rod from the steering knuckle.

5. Unscrew the tie rod while counting the number of turns required to remove it.

6. Check the tie rod end for damage and replace it, if necessary.

To install:

7. Install the tie rod end in the sleeve the same number of turns as when removing it.

8. Install the tie rod end in the steering knuckle. Install the nut and new cotter pin.

9. Check the toe in when finished.

Intermediate Rod and Tie Rods

REMOVAL AND INSTALLATION

1. Raise and safely support the vehicle.

2. Remove cotter pin from the ball studs connecting tie rods-to-intermediate rod and the steering damper. Remove the castellated nuts. Using a ball joint separator tool, separate the parts.

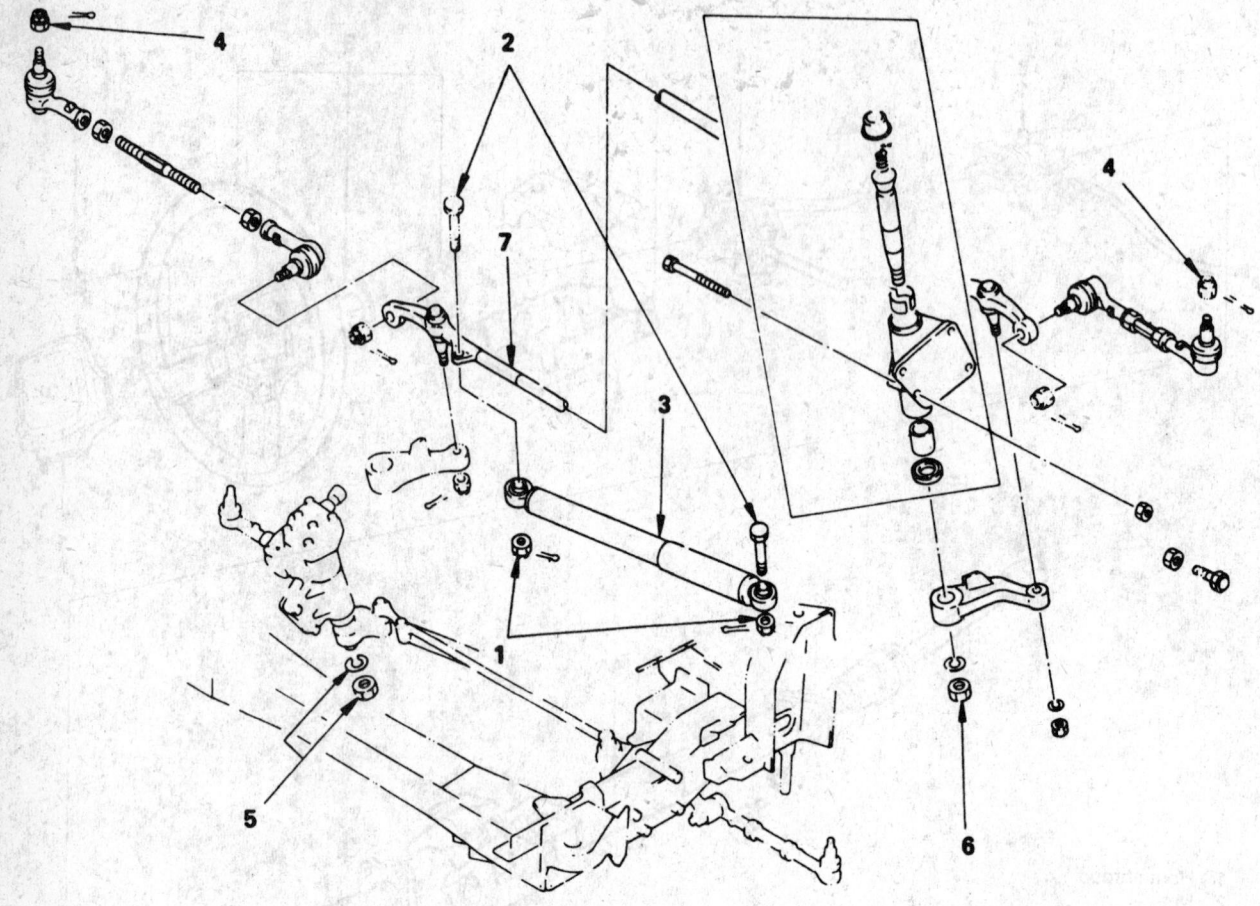

1. Nut
2. Bolt
3. Steering damper
4. Nut
5. Pitman arm nut
6. Relay lever nut
7. Intermediate rod

7911B063

Exploded view of the steering linkage assembly

3. Remove the nut and lockwasher on ball stud connecting the intermediate rod to idler arm. Using a ball joint separator tool, separate the intermediate rod from the idler arm.

4. Remove the intermediate rod with tie rods.

5. If the tie rod is replaced, disconnect the intermediate rod from tie rod.

To install:

6. Make sure the threads on the ball studs and nuts are clean and smooth.

7. Install the intermediate rod-to-idler arm and torque the nut to 50 ft. lbs. (67 Nm).

8. Raise the end of the rod and install it on the pitman arm. Torque the nut to 44 ft. lbs. (59 Nm). Tighten the nut just enough to insert cotter pin and install new cotter pin.

9. Install intermediate rod to steering damper end. Torque nut to 87 ft. lbs. (117 Nm), then, advance nut just enough to insert cotter pin and install new cotter pin.

10. Install the tie rods to adapter, torque nut to 44 ft. lbs. (59 Nm), then, advance nut just enough to insert cotter pin and install new cotter pin and lubricate tie rod ball studs.

BRAKES

For all brake system repair and service procedures not detailed below, please refer to Brakes in the Unit Repair section.

Master Cylinder

REMOVAL AND INSTALLATION

1. Disconnect the negative battery cable. Firmly, set the parking brake and block the wheels.

2. Draw off the brake fluid from the reservoir with a clean syringe. Disconnect and plug the brake lines from the master cylinder.

NOTE: Be careful not to spill any brake fluid on any painted surface. Brake fluid acts exactly like paint remover.

3. Remove the master cylinder-to-power brake unit nuts.

4. Remove the master cylinder from the booster.

5. Bleed the master cylinder before installing.

6. Install the master cylinder onto the booster.

7. Connect the fluid lines, refill the master cylinder with the proper brake fluid and bleed the brake system. Torque the flare fittings to 6.5–11 ft. lbs. (8.8–15 Nm). Torque the master cylinder-to-booster nuts to 7–1 ft. lbs. (10–16 Nm).

BLEEDING

1. Set the parking brake and perform the bleeding operation with the engine running to prevent damage to the pushrod seal. Route the exhaust outside if working in an enclosed area. Carefully clean all dirt from around the master cylinder filler cap.

2. If a bleeder tank is used, follow the manufacturer's instructions.

3. Remove the filler cap and fill the master cylinder to the lower edge of the filler neck.

4. Clean off the bleeder connections at all of the wheel cylinders or disc brake calipers. Attach the bleeder hose and fixture to the right rear wheel cylinder bleeder screw and place the end of the tube in a glass jar, submerged in brake fluid.

5. Have an assistant pump the pedal several times and hold it down. Open the bleeder valve 1/2–3/4 of a turn. Close the valve before the pressure is completely released. Repeat this procedure until bubbles cease to appear at the end of the bleeder hose.

6. Check the level of the brake fluid in the master cylinder and add fluid, if necessary. Repeat this procedure at the left rear wheel, then right front wheel and finishing at the left front. It is a good opportunity while the bleeding equipment is available to bleed the clutch hydraulic system using the same procedure.

7. After the bleeding operation at each caliper or wheel cylinder has been completed, refill the master cyl-inder reservoir, retract the filler cap diaphragm and replace the filler cap.

NOTE: Never reuse brake fluid which has been removed from the lines through the bleeding process because it contains moisture and dirt.

Proportioning Valve

The proportioning valve is located directly under the master cylinder.

REMOVAL AND INSTALLATION

1. Disconnect the negative battery cable. Firmly, set the parking brake and block the wheels.

2. Disconnect the electrical connector from the proportioning valve.

3. Disconnect and plug the fluid lines from the proportioning valve.

4. Remove the proportioning valve-to-chassis bolts and the valve.

5. To install, reverse the removal procedures.

6. Bleed the brake system.

Power Brake Booster

REMOVAL AND INSTALLATION

1. Disconnect the negative battery cable. Firmly, set the parking brake and block the wheels.

2. Disconnect the vacuum hose to the vacuum booster.

3. Disconnect and plug the brake fluid lines at the master cylinder. Place rags under the master cylinder to catch any leaking fluid.

NOTE: Be careful not to spill any brake fluid on any painted surface. Brake fluid will damage painted surfaces.

4. Inside the vehicle, remove the snapring from the clevis pin and separate the clevis pin from the brake pedal.

5. Remove the vacuum booster mounting nuts at the firewall and lift out the power unit and master cylinder/reservoir as an assembly.

To install:

6. Check the distance from the flange face of the vacuum booster to the end of the pushrod before installation of the master cylinder. The distance should be 0.709–0.717 in. (18.0–18.2mm). If the measurement deviates from the specified range, make an adjustment with the locknut at the end of the pushrod.

7. Apply sealer to the dashboard fitting face plate and mount the booster assembly and torque the mounting nuts to 16–23 ft. lbs. (22–32 Nm).

8. Connect the brake pedal clevis, brake lines and the vacuum hose. Torque the brake lines to 6.5–11 ft. lbs. (10–15 Nm).

9. Bleed the brake system.

Brake Caliper

REMOVAL AND INSTALLATION

1. Raise and safely support the vehicle. Remove the wheel assembly.

2. Disconnect and plug the brake fluid line from the caliper. On rear calipers disconnect parking brake cables.

3. Remove the brake caliper mounting bolts and the caliper from the mount.

4. Remove the brake pads and clips from the caliper. Inspect the brake pads for wear; replace them, if necessary.

To install:

5. Fill the brake caliper with brake fluid and connect the fluid line to the caliper using new washers. Torque the brake line banjo fitting to 22–29 ft. lbs. (30–40 Nm). Install the brake pads and clips onto the caliper.

6. Install the caliper on the mounting bracket. Torque the caliper-to-mounting bracket bolts to 20–27 ft. lbs. (28–38 Nm) for front calipers or 12–17 ft. lbs. (16–24 Nm) for rear calipers.

7. Bleed the brake system. Install the wheel assembly and lower the vehicle.

Disc Brake Pads

REMOVAL AND INSTALLATION

Front

1. Raise and safely support the vehicle. Remove the wheel assembly.

2. Remove the brake caliper mounting bolts and remove the caliper without disconnecting the brake fluid line. Support the caliper so it does not hang on the brake line.

3. Remove the brake pads and retaining clips from the caliper.

4. Use a C-clamp and press the brake caliper piston into the caliper until it bottoms out.

To install:

5. Install the new brake pads and clips in the caliper and install the caliper in the mounting bracket.

6. Install the wheel assembly. Check the brake fluid level.

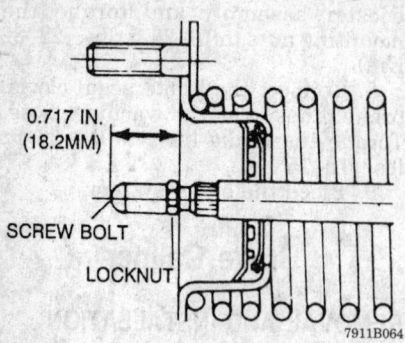

0.717 IN.
(18.2MM)

SCREW BOLT

LOCKNUT

7911B064

View of the power booster pushrod adjustment

7. Pump the brake pedal until pressure is felt before moving the vehicle.

Rear

1. Raise and safely support the vehicle. Remove the wheel assembly.

2. Remove the brake caliper mounting bolts and remove the caliper without disconnecting the brake fluid line. Support the caliper so it does not hang on the brake line.

3. Remove the brake pads and retaining clips from the caliper.

4. Using tool J-37617 or equivalent, rotate the piston clockwise while retracting the piston into the bore. Align the notches of the piston face so the centerline of the notches is perpendicular to the centerline of the mounting bosses.

To install:

5. Install the new brake pads and clips in the caliper and install the caliper in the mounting bracket.

6. Install the wheel assembly. Check the brake fluid level.

7. Pump the brake pedal until pressure is felt before moving the vehicle.

Brake Rotor

REMOVAL AND INSTALLATION

Front

2WD VEHICLES

1. Raise and safely support the vehicle. Remove the wheel assembly.

2. Remove the brake caliper without disconnecting the fluid line. Support the caliper aside.

3. Remove the brake caliper mounting bracket.

4. Remove the dust cover, cotter pin and locknut from the rotor.

5. Place a hand over the outer wheel bearing in the rotor and remove the rotor from the spindle.

6. Remove the rotor-to-hub bolts and pull off the rotor.

To install:

7. Install the rotor on the hub and torque the bolts to 75 ft. lbs. (105 Nm).

8. Torque the hub nut to 22 ft. lbs. (29 Nm) and loosen the nut. Using a spring gauge, connect it to the stud bolt at 90 degrees. Retorque the hub nut until the spring gauge measures a bearing preload of 1.8–2.2 lbs.

9. Install the caliper mounting bracket and torque the bolts to 103–126 ft. lbs. (142–174 Nm).

10. Install the caliper and torque the mounting bolts to 20–27 ft. lbs. (28–38 Nm).

11. Install the wheel and lower the vehicle.

4WD VEHICLE

1. Place the transfer case in **2H**. Raise and safely support the vehicle.

2. Remove the free wheeling hub cover assembly. Remove the brake caliper and hang aside.

3. Using wrench J-36827 or equivalent, remove the hub nut.

4. Remove the hub and disk assembly. Unbolt the rotor from the hub.

5. Remove the outer roller bearing assembly from the hub with a finger.

6. Using a brass or wood drift, drive out the inner bearing assembly along with the oil seal. Replace the seal.

7. Wash all parts in a non-flammable solvent.

8. Check all parts for cracks or wear. Thoroughly lubricate all bearing parts with a high-temperature wheel bearing grease. Remove any excess. Apply about 2 ounces of the grease to the hub.

To install:

9. Bolt the rotor to the hub and torque the bolts to 75 ft. lbs. (105 Nm).

10. Lightly coat the spindle with the same grease. Place the inner bearing into the hub race and install a new seal and retaining ring.

11. Carefully install the hub on the spindle and install the outer bearing.

12. Install the spindle nut.

13. While rotating the hub, tighten the hub so the wheel can just be turned by hand.

14. Turn the hub 2–3 turns and back off the nut just enough so it can be loosened with the fingers.

15. Finger-tighten the nut so all play is taken up at the bearing.

16. Attach a pull scale to a lug nut and check the amount of pull needed to start the wheel turning. Initial pull should be 2.6–4.0 lbs. When performing this test, make sure the brake pads are not touching the ro-

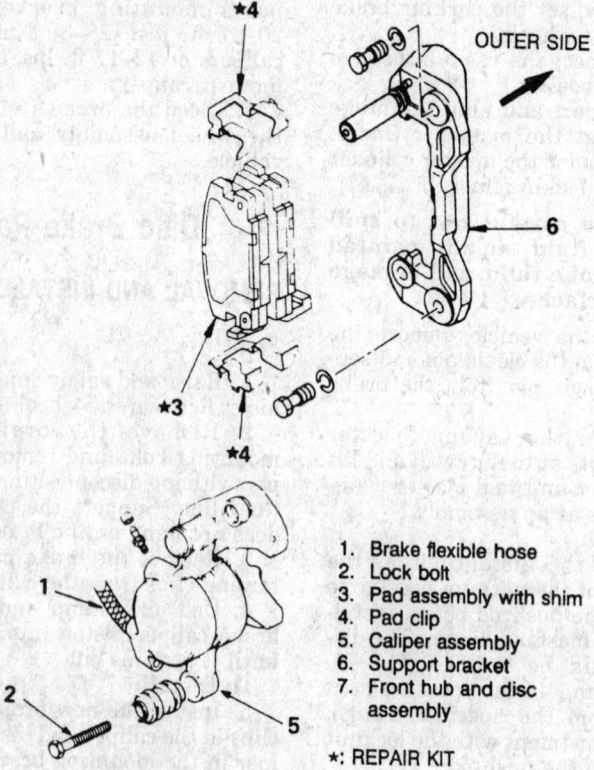

OUTER SIDE

1. Brake flexible hose
2. Lock bolt
3. Pad assembly with shim
4. Pad clip
5. Caliper assembly
6. Support bracket
7. Front hub and disc assembly

*: REPAIR KIT

7911B066

Exploded view of the front disc brake assembly

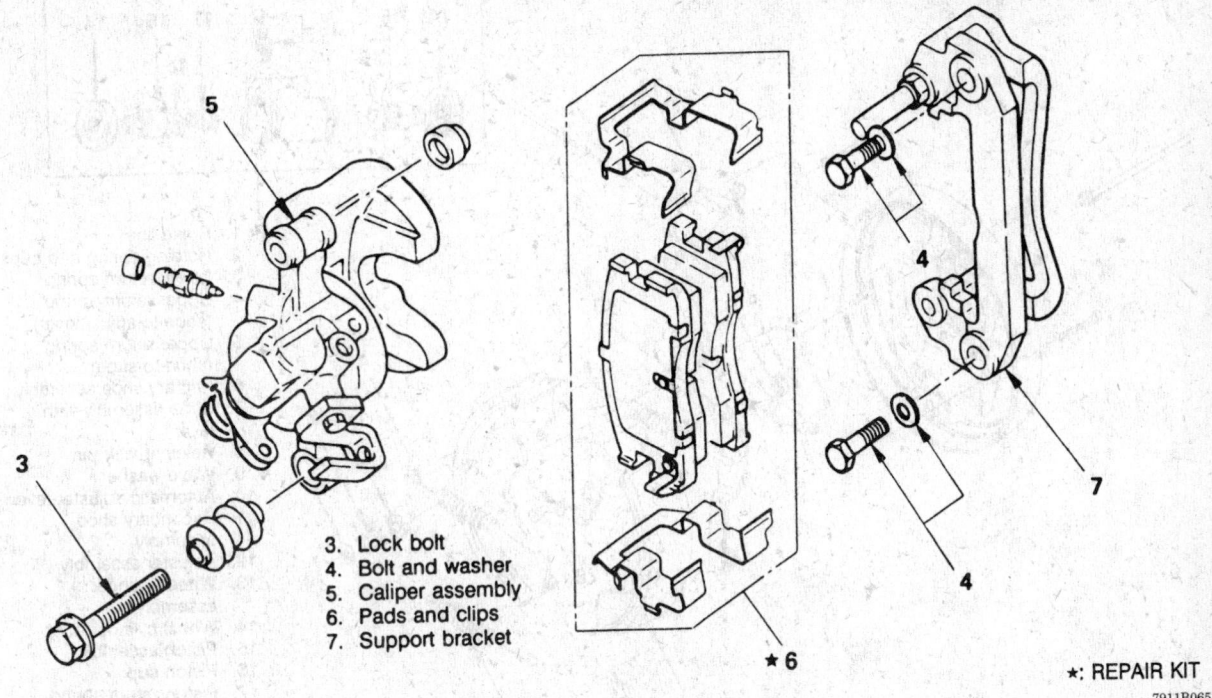

3. Lock bolt
4. Bolt and washer
5. Caliper assembly
6. Pads and clips
7. Support bracket

★: REPAIR KIT

7911B065

Exploded view of the rear disc brake assembly

tor. If the rotating torque is not correct, tighten the spindle nut until it is.

17. Install the lockwasher so the bolt holes lineup. If the bolt holes do not line up, reverse the position of the lock washer. The bolt head should be able to sink below the surface of the washer when tightened down.

18. Clean the hub flange surface and areas surrounding the spindle and with the transfer case lever in the **2H** position, install the inner cam with the gear teeth facing out.

19. Lower the vehicle and install tools J-36835-2 and J-36836 or equivalent, on the axle shaft until it comes in contact with the lock washer.

20. Measure the gap between the tool and the snapring tool and select shims to reduce the gap to 0.00–0.10mm. Remove the tool.

21. Apply grease to the hub splines and lubrication grooves. Install the drive clutch assembly. The cut portion of the drive clutch must be aligned with the concave part of the inner cam. Match the teeth of the cam to the inner cam by turning the axle shaft.

22. Install the snapring and previously selected shims, gasket and cover. Apply Loctite® 515 or equivalent, and torque the cover bolts to 43 ft. lbs. (59 Nm).

23. Install the caliper mounting bracket and torque the bolts to 103–126 ft. lbs. (142–174 Nm).

24. Install the caliper and torque the mounting bolts to 20–27 ft. lbs. (28–38 Nm).

25. Mount the wheel assembly and lower the vehicle. Pump the brakes before moving the vehicle.

Rear

1. Raise and safely support the vehicle. Remove the wheel assembly.

2. Remove the brake caliper without disconnecting the fluid line. Support the caliper aside.

3. Remove the brake caliper from the mounting bracket. Remove the brake caliper mounting bracket.

4. Remove the rotor from the axle shaft.

5. To install, reverse the removal procedures. Torque the mounting bracket bolts to 69–84 ft. lbs. (95–116 Nm). Torque the caliper mounting bolt to 12–17 ft. lbs. (16–24 Nm).

Brake Drums

REMOVAL AND INSTALLATION

1. Raise and safely support the vehicle.

2. Remove the wheel and the brake drum.

3. To install, reverse the removal procedures.

Brake Shoes

REMOVAL AND INSTALLATION

1. Raise and safely support the vehicle. Remove the tire and wheel assembly.

2. Remove the brake drum.

3. Remove the return springs, the hold-down springs and lift the brake shoe assembly from the backing plate.

4. Disconnect the parking brake cable from the adjuster.

5. To install, reverse the removal procedures.

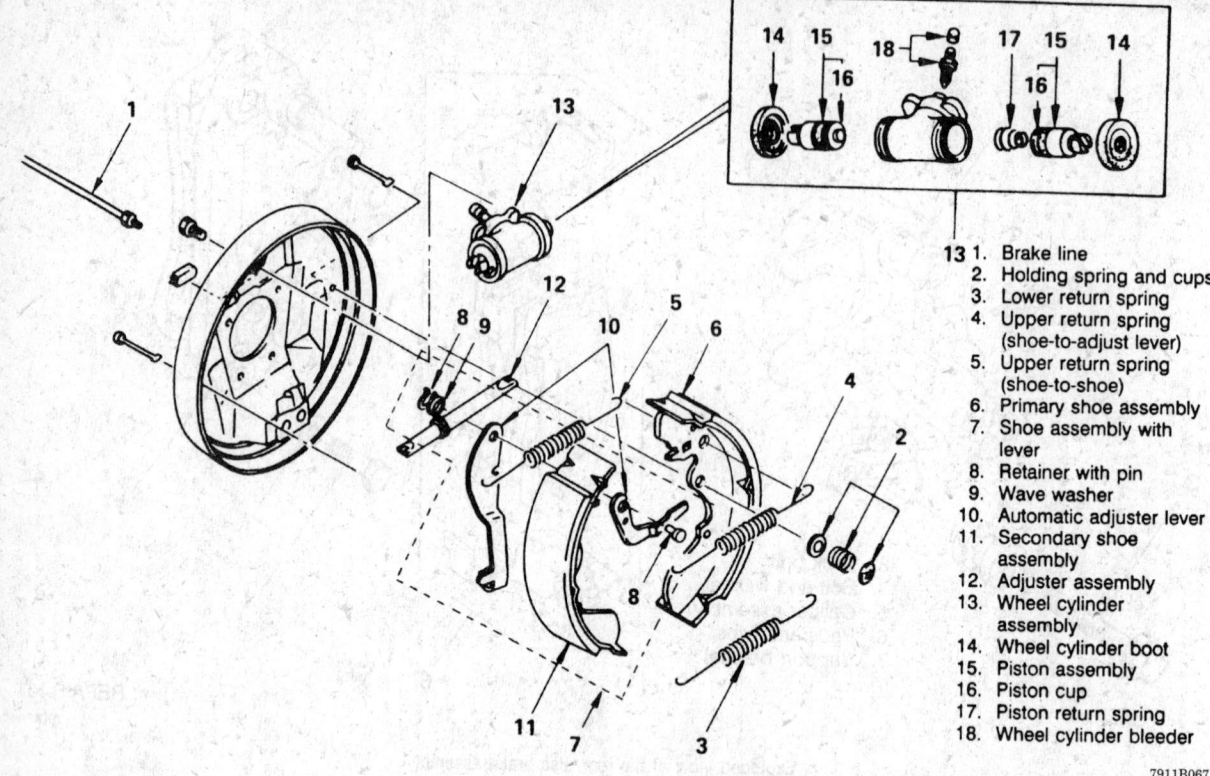

13 1. Brake line
2. Holding spring and cups
3. Lower return spring
4. Upper return spring (shoe-to-adjust lever)
5. Upper return spring (shoe-to-shoe)
6. Primary shoe assembly
7. Shoe assembly with lever
8. Retainer with pin
9. Wave washer
10. Automatic adjuster lever
11. Secondary shoe assembly
12. Adjuster assembly
13. Wheel cylinder assembly
14. Wheel cylinder boot
15. Piston assembly
16. Piston cup
17. Piston return spring
18. Wheel cylinder bleeder

7911B067

Exploded view of the rear brake drum/shoe assembly

Wheel Cylinder

REMOVAL AND INSTALLATION

1. Raise and safely support the vehicle. Remove the wheel assembly, brake drums and shoes.
2. Disconnect and plug the brake line at the wheel cylinder.
3. Remove the wheel cylinder-to-backing plate bolts and the cylinder.
4. Cap the openings of the brake line and the wheel cylinder.
5. To install, reverse the removal procedures. Bleed the brake system.

Parking Brake Cable

REMOVAL AND INSTALLATION

1. Raise and safely support the vehicle.
2. Loosen the cable adjusting nut and remove the lever return spring. Remove the adjusting nut.
3. Remove the cotter pin from the retaining pin on the 2nd lever assembly and remove the front cable.
4. Remove the 2 cotter pins from the retaining pins on the intermediate cable and remove the cable.

5. Remove the retaining clips from the rear fixing brackets and lower the rear brake cables.
6. Remove the rear wheel assemblies and brake drums. Remove the rear brake shoes and disconnect the rear brake cables from the lever in the rear brake shoes.
7. To install, reverse the removal procedures. Adjust the cables when finished.

ADJUSTMENT

NOTE: Adjustment of the parking brake is necessary every time the rear brake cables are disconnected or after overhauling the rear brake assembly.

1. Fully, release the parking brake lever and check the cable for free movement.
2. Firmly, grab the 2nd relay lever rod. Rotate the adjusting nut until all the slack is removed from the cable. Tighten the adjusting nut.
3. Apply the parking brake to the fully set position 3–4 times.
4. If the parking brake is properly adjusted, the traveling range should be between 12–14 notches. If the travel is incorrect, readjust to specifications.

Load Sensing Proportioning Valve

ADJUSTING

1. Connect a continuity meter between the mounting bracket and the linkage arm. The arm is insulated so be sure to cut through to the metal arm when attaching the lead.
2. Place weight into the vehicle bed to end up with a 1257 lbs. (570 kg) rear axle weight. Have an assistant sit in the drivers seat.
3. Loosen the locknut on the side of the valve body. Raise the valve body and allow to drop slowly. At the point where continuity is achieved, tighten the locknut. Torque the nut to 12 ft. lbs. (17 Nm).
4. If continuity is broken while tightening the nut, loosen the nut and lower the valve body slightly. Tighten the nut.
5. Check the adjustment by pressing down slightly on the linkage near to the valve. If continuity is broken then the adjustment is acceptable.

REMOVAL AND INSTALLATION

1. Disconnect the fluid lines to the valve body.

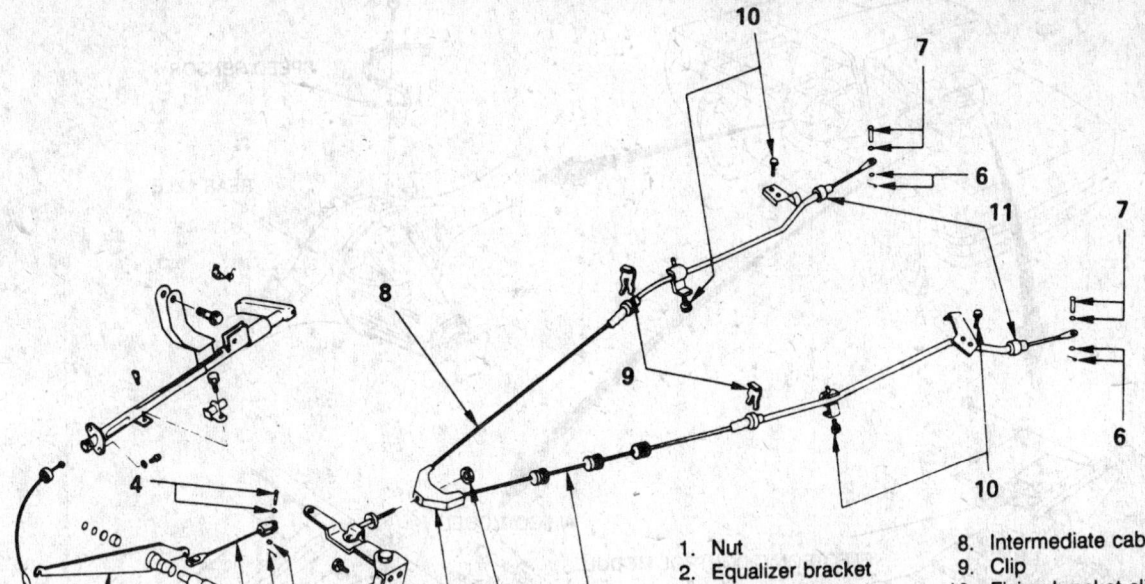

1. Nut
2. Equalizer bracket
3. Split pin with plain washer
4. Pin with curved washer
5. Front lower cable
6. Split pin with plain washer
7. Pin with curved washer
8. Intermediate cable
9. Clip
10. Fixing bracket clip
11. Rear cable assembly
12. 2nd relay lever assembly with return spring
13. 1st relay lever assembly

7911B068

Exploded view of the parking brake cable assembly

2. Disconnect the linkage at the axle housing.

3. Remove the mounting bolts at the bracket and remove the valve.

4. Mount the valve, attach the linkage and lines. Bleed the system.

Rear Wheel Anti-Lock Brake Valve

REMOVAL AND INSTALLATION

1. Disconnect the negative battery cable. Locate the valve body on the right side frame rail.

2. Disconnect the wiring harness connector from the valve body lead.

3. Disconnect the brake lines at the valve body. Cap the lines to prevent the ingress of dirt and moisture.

4. Remove the bolts holding the valve to the frame rail bracket and remove the valve.

5. Installation is the reverse of removal. Torque the brake line fittings to 11 ft. lbs. (15 Nm).

Speed Sensor

REMOVAL AND INSTALLATION

1. Disconnect the negative battery terminal. Locate the speed sensor on the top of the rear axle housing.

2. Disconnect the wiring harness lead from the sensor.

3. Remove the speed sensor bolt and remove the speed sensor.

4. Installation is the reverse of removal. Torque the bolt to 17 ft. lbs. (23 Nm).

Rear Wheel Anti-Lock Brake System ECM

REMOVAL AND INSTALLATION

1. Disconnect the negative battery terminal. Remove the 4 bolts holding the passenger seat assembly and remove the seat.

2. Disconnect the wiring harness from the ECM.

3. Unbolt the ECM from the floor and remove with the bracket.

4. Installation is the reverse of removal.

FRONT SUSPENSION

Shock Absorbers

REMOVAL AND INSTALLATION

1. Raise and support the vehicle safely.

2. Hold the upper stem of the shock absorber from turning and remove the upper stem retaining nut, retainer and rubber grommet. On some vehicles, it may be necessary to remove the bumpstops to gain access to the mounting bolts.

3. Remove the bolt retaining the lower shock absorber pivot to the lower control arm and remove the shock absorber from the vehicle.

To install:

4. Install the shock absorber by first installing the lower retainer and rubber grommet over the upper stem and then, installing the shock fully extended up through the upper control arm so the upper stem passes through the mounting hole in the frame bracket.

5. Install the upper rubber grommet, retainer and attaching nut over the shock absorber upper stem.

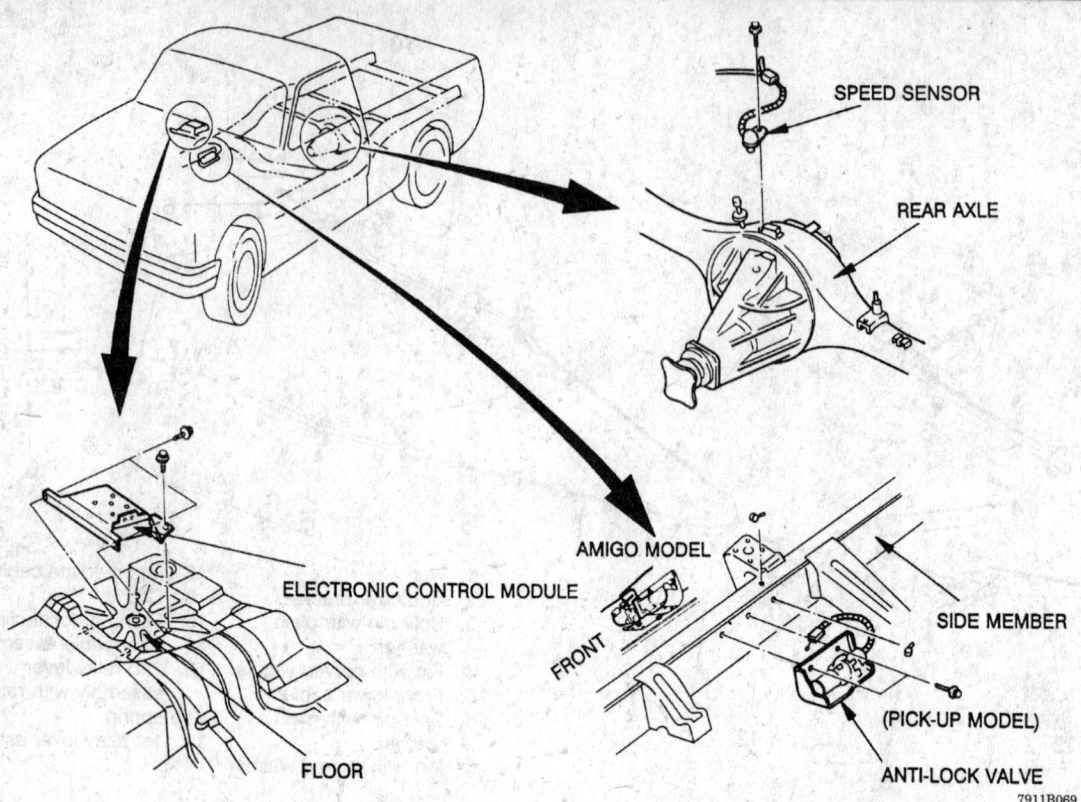

SPEED SENSOR

REAR AXLE

ELECTRONIC CONTROL MODULE

AMIGO MODEL

FRONT

SIDE MEMBER

(PICK-UP MODEL)

ANTI-LOCK VALVE

FLOOR

7911B069

Rear wheel anti-lock brake component locations — Amigo and Pick-Up — Rodeo similar

6. Hold the upper stem of the shock absorber from turning and tighten the retaining nut. Torque to 15 ft. lbs. (20 Nm).

7. Install the retainers attaching the shock absorber lower pivot to the lower control arm and tighten them. Torque the mounting bolt to 60 ft. lbs. (84 Nm).

8. Install the bumpstops if removed. Lower the vehicle.

Torsion Bars

REMOVAL AND INSTALLATION

1. Raise and safely support the vehicle.
2. Mark the location of the height adjustment bolt and remove from the height control arm.
3. Mark the location and remove the height control arm from the torsion bar and the third crossmember.
4. Mark the location and withdraw the torsion bar from the lower control arm.
   To install:
5. To install, apply a generous amount of grease to the serrated ends of the torsion bar.
6. Hold the rubber bumpers in contact with the lower control arm.

Raise the vehicle up under the lower control arm to accomplish this.

7. Insert the front end of the torsion bar into the control arm.
8. Install the height control arm in position so it's end is reaching the adjusting bolt. Be sure to lubricate the part of the height control arm that fits into the chassis with grease.
9. Install a new cotter pin in the control arm.
10. Turn the adjusting bolt to the location marked before removal.
11. Lower the vehicle and check the vehicle height.

Upper Ball Joints

INSPECTION

Grasp the top of the front wheel and pull it in and out several times to check for excessive movement of the ball joint; if no movement exist, the joint is in good shape.

REMOVAL AND INSTALLATION

1. Raise and safely support the vehicle. Remove the wheel and tire assembly.

2. Mark the position of the torsion bar adjuster and remove the tension from the torsion bar.
3. Remove the upper ball joint-to-steering knuckle nut.
4. Using a ball joint separator tool, separate the upper ball joint from the steering knuckle.
5. Remove the upper ball joint-to-upper control arm bolts and the ball joint.
6. To install, reverse the removal procedures. Torque the upper ball joint-to-upper control arm bolts to 21–25 ft. lbs. (29–35 Nm) and the upper ball joint-to-steering knuckle nut to 72–87 ft. lbs. (96–117 Nm) for 2WD or to 65–80 ft. lbs. (88–108 Nm) for 4WD.
7. Adjust the tension on the torsion bar to its original position and lower the vehicle.

Lower Ball Joints

INSPECTION

1. Raise and safely support the front of the vehicle.
2. Using a large prybar, place it under the front wheel and try to pry the wheel upwards.

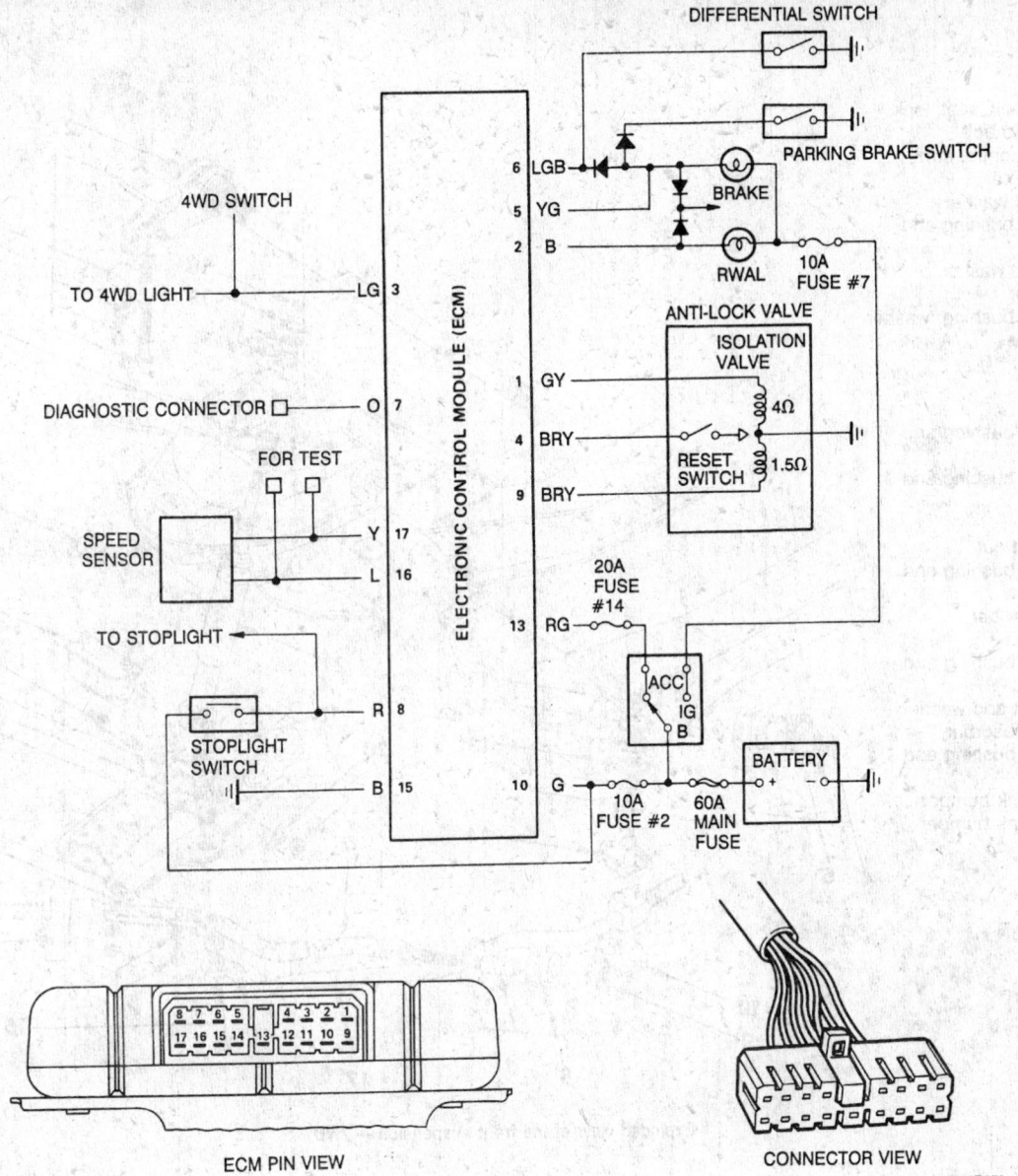

Rear wheel anti-lock brake system schematic

3. If excessive upward movement or clunking is noticed, the ball joint is damaged and requires replacement.

REMOVAL AND INSTALLATION

1. Raise and safely support the vehicle.
2. Remove the wheel and tire assembly.
3. Mark the position of the torsion bar adjuster and release the torsion bar tension.
4. Remove the cotter pin and castellated nut which retains the ball joint to the steering knuckle.
5. Remove the lower ball joint-to-lower control arm and strut rod.

6. Remove the ball joint.
To install:
7. Install the lower ball joint by mounting the joint to the lower control arm and torque the bolts to 45–56 ft. lbs. (61–76 Nm) for 2WD or to 68–83 ft. lbs. (93–113 Nm) for 4WD.
8. Install the ball joint stud into the steering knuckle and install the castellated nut and torque it to 101–116 ft. lbs. (137–157 Nm) for 2WD or to 87–111 ft. lbs. (117–137 Nm) for 4WD and just enough additional torque to align the cotter pin hole with a castellation on the nut. Install a new cotter pin.

9. Lubricate the lower ball joint through the grease fitting.
10. Adjust the torsion bar tension to its original position.
11. Install the wheel assembly and lower the vehicle.

Upper Control Arms

REMOVAL AND INSTALLATION

NOTE: The upper control arm and ball joint are replaced as an assembly.

1. Raise and safely support the vehicle.

1. Adjust bolt, seat, lock plate and bolt
2. Height control arm
3. Torsion bar
4. Nut and washer
5. Rubber bushing and washer
6. Bolt and washer
7. Strut bar
8. Rubber bushing, washer and tube
9. Bolt
10. Bracket
11. Nut
12. Rubber bushing and washer
13. Rubber bushing and washer
14. Bracket
15. Bolt and nut
16. Rubber bushing and washer
17. Stabilizer bar
18. Nut
19. Rubber bushing and washer
20. Bolt, nut and washer
21. Shock absorber
22. Rubber bushing and washer
23. Lower link bumper
24. Upper link bumper

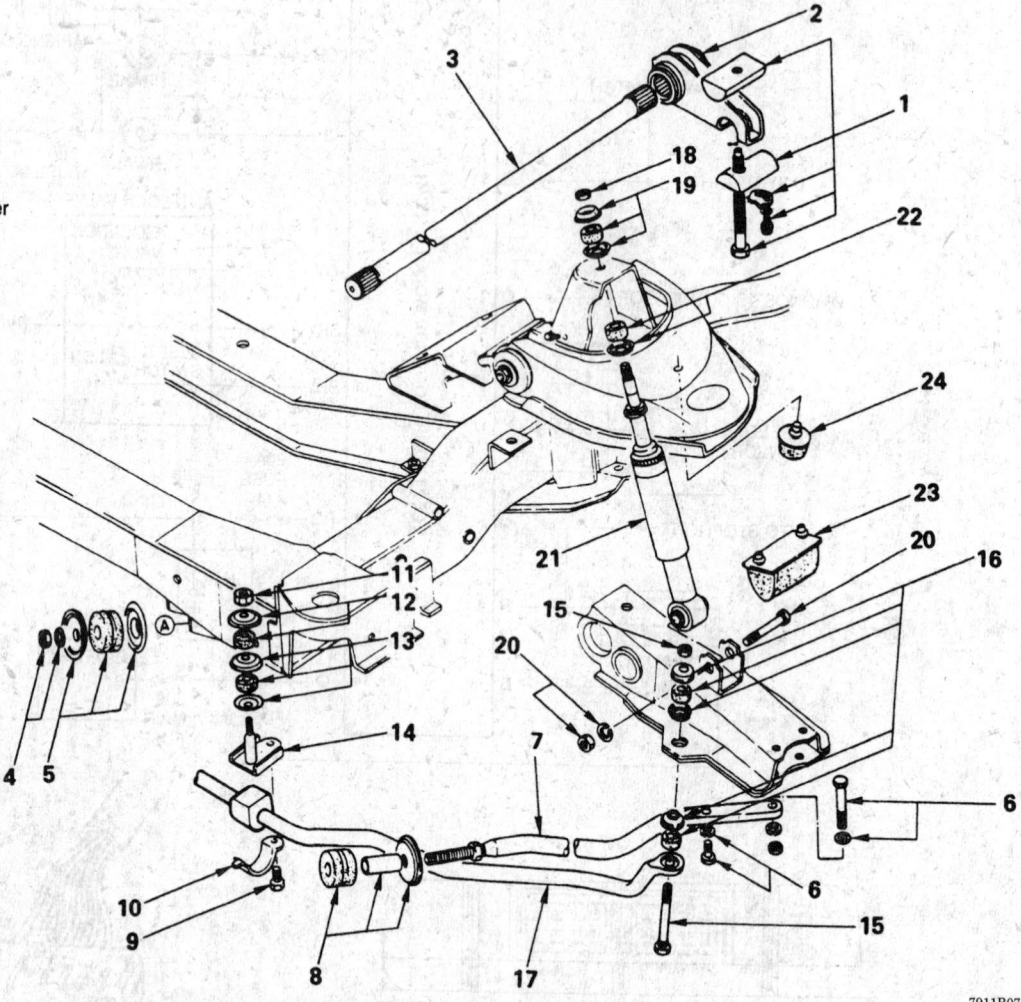

Exploded view of the front suspension — 2WD

7911B071

2. Remove the wheel and tire assembly. Mark the position of the torsion bar adjuster and release the torsion bar tension.

3. Remove the cotter pin nut fastening the upper control arm and upper ball joint assembly and disconnect the upper control arm from the steering knuckle.

NOTE: Do not allow the steering knuckle to hang by the flexible brake line. Wire the steering knuckle up to the frame temporarily.

4. Remove the bolts from the upper pivot shaft and remove the upper control arm from the bracket. Be sure to note the position and number of shims used for adjusting the camber and caster angles when removing the upper control arm. The shims must be replaced in their original position.

5. To remove the pivot shaft and bushings from the upper control arm assembly, remove the bushing nuts from the pivot shaft by loosening them alternately, then remove the pivot shaft.

To install:

6. To install the upper control arm and ball joint assembly, first install the pivot shaft boots to the pivot shaft.

7. Fill the internal part of the bushings with grease and screw the bushings into the pivot shaft. Be sure to screw the right side and the left side bushings alternately into the pivot shafts carefully avoiding getting grease on the outer face of the bushings. Tighten the nuts to 250 ft. lbs. (333 Nm).

NOTE: Be sure the control arm and bushings are centered properly and the control arm rotates with resistance but not binding on the pivot shaft when tightened to the proper torque.

8. Install the grease fittings and lubricate the parts with grease through the grease fittings.

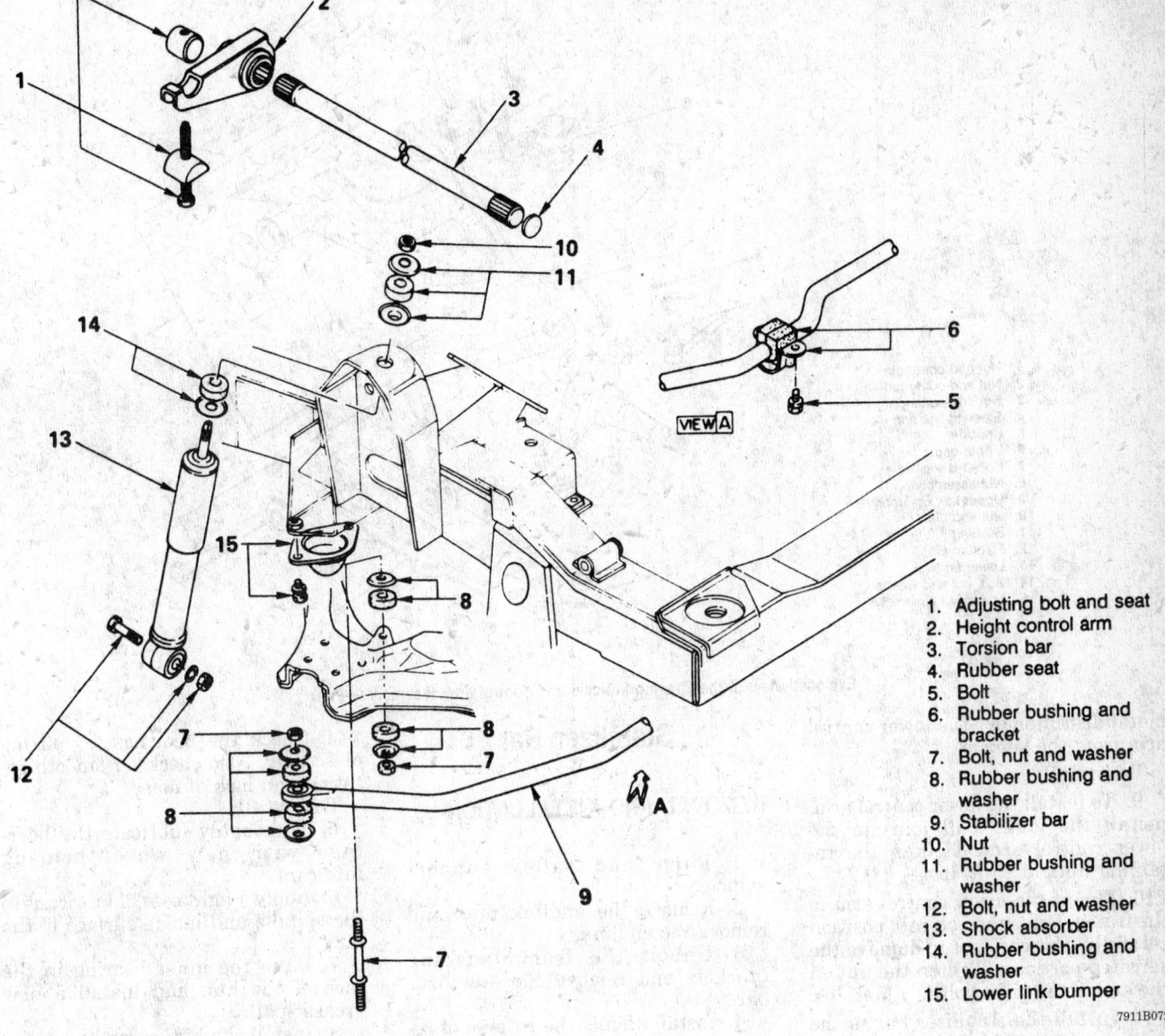

1. Adjusting bolt and seat
2. Height control arm
3. Torsion bar
4. Rubber seat
5. Bolt
6. Rubber bushing and bracket
7. Bolt, nut and washer
8. Rubber bushing and washer
9. Stabilizer bar
10. Nut
11. Rubber bushing and washer
12. Bolt, nut and washer
13. Shock absorber
14. Rubber bushing and washer
15. Lower link bumper

7911B072

VIEW A

Exploded view of the front suspension — 4WD

9. Install the ball joint stud through the steering knuckle. Install the castellated nut and tighten it to 75 ft. lbs. and just enough additional torque to install the cotter pin. Use a new cotter pin.

10. Mount the upper control arm to the chassis frame and install the shims in their original positions between the pivot shaft and bracket. Tighten the pivot shaft attaching nuts to 55 ft. lbs.

NOTE: Tighten the thinner shim pack's nut first for improved shaft-to-frame clamping force and torque retention.

11. Install the dust cover. Adjust the torsion bar to its original position.

12. Install the wheel assembly and lower the vehicle.

Lower Control Arms

REMOVAL AND INSTALLATION

1. Raise and safely support the vehicle.

2. Remove the wheel and tire assembly. Mark the position of the torsion bar adjuster and release the torsion bar tension.

3. Remove the strut bar by removing the frame side bracket and the double nuts, washer and the rubber bushing from the front side of the strut bar. Remove the strut bar-to-lower control arm bolts and remove the bar.

4. Disconnect the stabilizer bar from the lower control arm.

5. Remove the torsion bar.

6. Disconnect the shock absorber from the lower control arm.

7. Remove the lower ball joint from the lower control arm joint.

8. Remove the retaining nut and drive out the bolt holding the lower control arm to the chassis with a soft

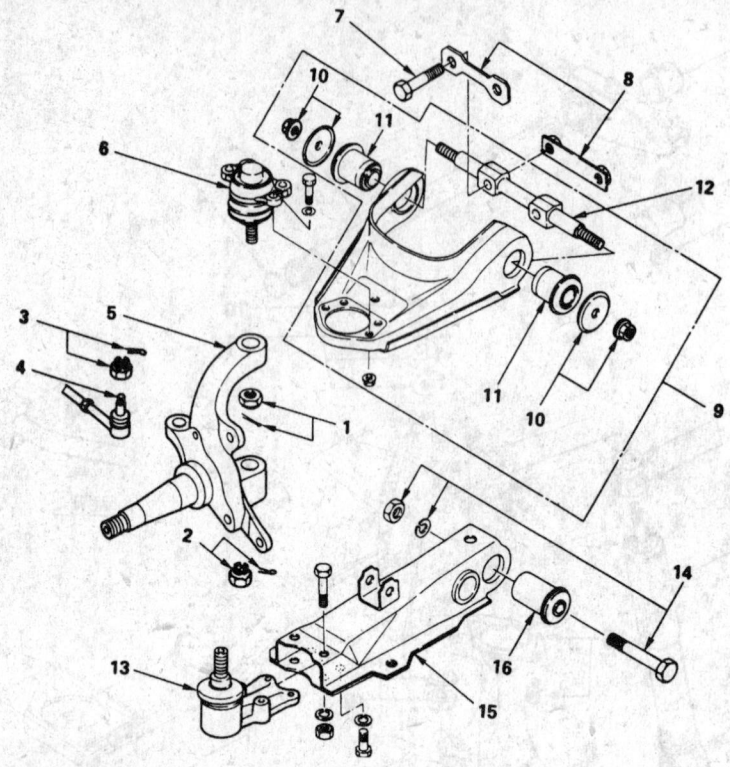

1. Nut and cotter pin
2. Nut and cotter pin
3. Nut and cotter pin
4. Steering link end
5. Knuckle
6. Upper end
7. Bolt and washer
8. Nut assembly
9. Upper link assembly
10. Nut and plate
11. Bushing
12. Fulcrum pin
13. Lower end
14. Bolt, nut and washer
15. Lower link assembly
16. Bushing

7911B073

Exploded view of the steering knuckle and control arm assembly — 2WD

metal drift. Remove the lower control arm from the vehicle.

To install:

9. To install the lower control arm, install the lower ball joint to the lower control arm. Tighten the retaining nuts to 45 ft. lbs.

10. Mount the lower control arm to the frame. Drive the bolt into position carefully. Use care not to damage the serrated portions. Tighten the nut on the end of the pivot bolt to 135 ft. lbs.

11. Install the stabilizer bar to the lower control arm.

12. Place the washers and bushings on the strut rod and install it through the frame bracket. Install the second set of washers and bushings on the strut rod together with the lockwashers and nut. Leave the nut loose temporarily.

13. Install the strut rod to the lower control arm and tighten the bolts to 45 ft. lbs.

14. Assemble the lower ball joint to the steering knuckle. Adjust the torsion bar to the original position.

15. Install the wheel assembly and lower the vehicle.

16. Tighten the 1st strut bar-to-chassis frame attaching nut to 175 ft. lbs. and the 2nd locknut to 55 ft. lbs. with the vehicle on the ground.

Stabilizer Bar

REMOVAL AND INSTALLATION

1. Raise and safely support vehicle.

2. Remove the endlink nuts and remove the endlinks

3. Unbolt the frame bushing brackets and remove the stabilizer bar.

4. Installation is the reverse of removal. Torque the bracket bolts to 20 ft. lbs. (28 Nm) and the endlink nuts to 10 ft. lbs. (14 Nm).

Front Wheel Bearings — 2WD

REMOVAL AND INSTALLATION

1. Raise and safely support the vehicle. Remove the wheel assembly. Remove the hub assembly.

2. Remove the outer roller bearing assembly from the hub. Pry out the inner bearing lip seal and remove the inner bearing assembly.

3. Wash all parts in a cleaning solvent and allow to air dry.

4. Check the bearings for pitting or scoring. Also check for smooth rotation and lack of noise.

To install:

5. Thoroughly lubricate the bearings with new wheel bearing lubricant.

6. Apply a light coat of lubricant to the spindle and inside surface of the hub.

7. Place the inner bearing in the race of the hub and install a new grease seal.

8. Install the hub assembly on the spindle.

9. Install the outer wheel bearing, washer and adjust nut.

10. Adjust the wheel bearings.

11. Install the dust cap on the hub.

12. Install the brake caliper and support assembly.

13. Install the wheel assembly.

ADJUSTMENT

1. With the wheel raised, remove the hub cap and dust cap and then remove the cotter pin and nut retainer from the end of the spindle.

2. While rotating the wheel, tighten the spindle nut to 22 ft. lbs. (29 Nm).

3. Turn the hub 2–3 turns and loosen the nut just enough so it can be turned by hand.

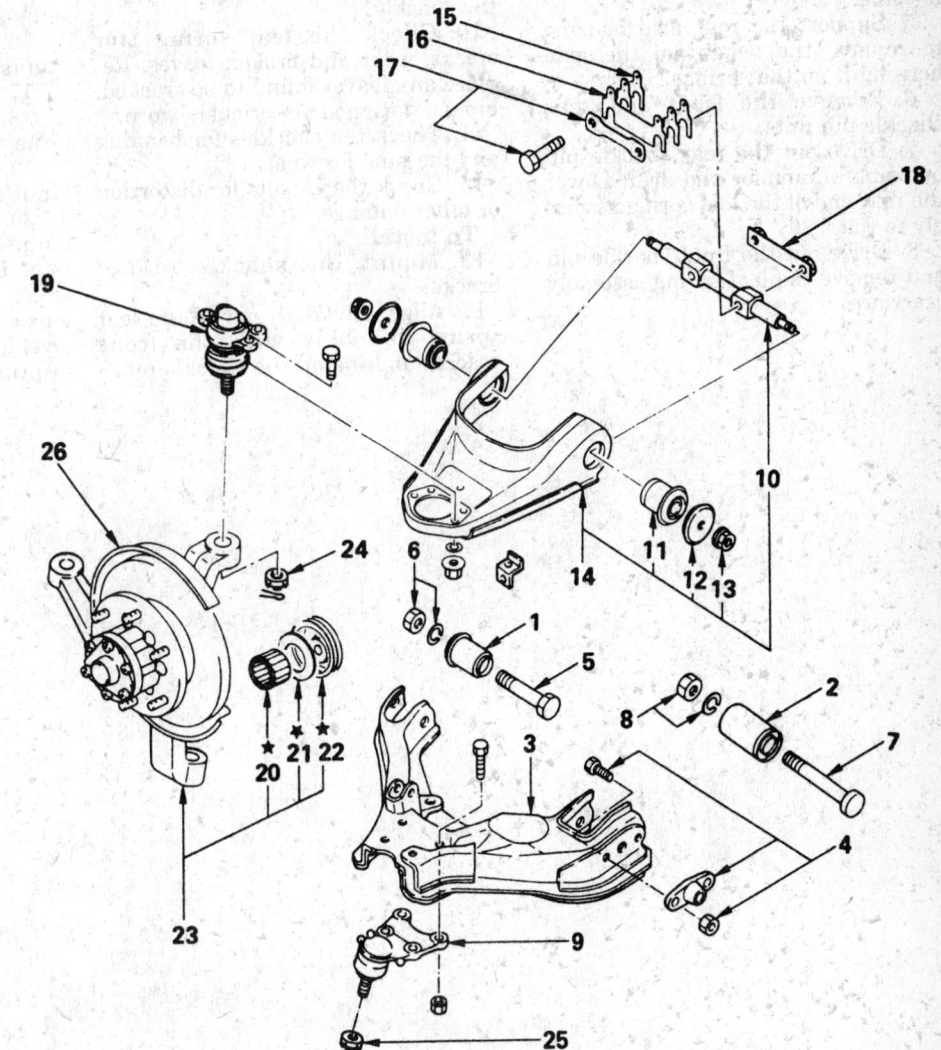

1. Front bushing
2. Rear bushing
3. Lower link assembly
4. Torsion bar arm
5. Bolt
6. Nut and washer
7. Bolt
8. Nut and washer
9. Lower end
10. Fulcrum pin
11. Bushing
12. Plate
13. Nut
14. Upper link assembly
15. Caster shims
16. Camber shims
17. Bolt and plate
18. Nut assembly
19. Upper end
20. Needle bearing
21. Washer
22. Oil seal
23. Knuckle assembly
24. Nut and cotter pin
25. Nut and cotter pin
26. Back plate

Exploded view of the steering knuckle and control arm assembly — 4WD

7911B074

4. Turn the nut all the way hand tight and check to be sure the hub has no free-play.

5. Measure the starting torque by pulling a wheel hub stud with a pull scale. Tighten the spindle nut so the pull scale reads 1.1–2.6 lbs. when the hub begins to rotate.

NOTE: Make sure the brake pads are not in contact with the drum when measuring rotating torque.

6. Install the nut retainer, new cotter pin, dust cap and hub cap.

REAR SUSPENSION

Shock Absorbers

REMOVAL AND INSTALLATION

1. Raise and safely support the vehicle.
2. Remove the shock absorber-to-lower mount nut, washers and bushings.
3. Remove the shock absorber-to-chassis nut, washers and bushings.

4. Remove the shock absorber.
5. To install, reverse the removal procedures. Torque the mounting nuts to 30 ft. lbs. (40 Nm.)

Leaf Springs

REMOVAL AND INSTALLATION

1. Raise and safely support the vehicle so the leaf springs are hanging freely. Remove the wheel assemblies.
2. Remove the rear shock absorbers.
3. Remove the parking brake cable clips.

4. Remove the nuts from the U-bolts holding the springs to the axle housing.

5. Support the rear axle housing to remove the weight of the axle housing from the springs.

6. Remove the front and rear shackle pin nuts.

7. Drive out the rear shackle pin by using a hammer and drift. Lower the rear end of the leaf spring assembly to the floor.

8. Drive out the front shackle pin and remove the leaf spring assembly rearward.

9. Remove the shackle pin from the rear spring bracket and remove the shackle.

10. Check the leaf springs for cracks, wear and broken leaves. Replace any leaves found to be cracked, broken, fatigued or seriously worn.

11. Check the shackles for bending and the pins for wear.

12. Check the U-bolts for distortion or other damage.

To install:

13. Mount the shackle to the bracket.

14. Align the front end of the leaf spring assembly with the front bracket and install the shackle pin.

15. Align the rear end of the leaf spring assembly with the shackle and install the shackle pin.

16. Loosely install the shackle pin nuts and install the U-bolts.

17. Install the shock absorbers.

18. Clip the parking brake cable to the bracket.

19. Tighten the front shackle pin nut to 112 ft. lbs. (152 Nm).

20. Tighten the rear shackle pin nuts to 72 ft. lbs. (98 Nm). Tighten the U-bolt nuts to 48 ft. lbs. (67 Nm).

21. Install the wheels. Remove the axle housing support and lower the vehicle so the weight is on the leaf springs.

MAZDA 9

B-Series • MPV • Navajo

9-1

SPECIFICATIONS

ENGINE IDENTIFICATION

Year	Model	Engine Displacement Liters (cc)	Engine Series (ID/VIN)	Fuel System	No. of Cylinders	Engine Type
1991	2200	2.2 (2184)	F2	EFI	4	SOHC
	2200	2.2 (2184)	F2	2 BBL	4	SOHC
	2600i	2.6 (2606)	G6	EFI	4	SOHC
	MPV	2.6 (2606)	G6	EFI	4	SOHC
	MPV	3.0 (2954)	JE	EFI	6	SOHC
	Navajo	4.0 (4016)	X	EFI	6	OHV
1992	B2200	2.2 (2184)	F2	EFI	4	SOHC
	B2200	2.2 (2184)	F2	2 BBL	4	SOHC
	B2600i	2.6 (2606)	G6	EFI	4	SOHC
	MPV	2.6 (2606)	G6	EFI	4	SOHC
	MPV	3.0 (2954)	JE	EFI	6	SOHC
	Navajo	4.0 (4016)	X	EFI	6	OHV
1993	B2200	2.2 (2184)	F2	2 BBL	4	SOHC
	B2200	2.2 (2184)	F2	EFI	4	SOHC
	B2600i	2.6 (2606)	G6	EFI	4	SOHC
	MPV	2.6 (2606)	G6	EFI	4	SOHC
	MPV	3.0 (2954)	JE	EFI	6	SOHC
	Navajo	4.0 (4016)	X	EFI	6	OHV
1994-95	B2300	2.3 (2298)	A	EFI	4	SOHC
	B3000	3.0 (2968)	U	EFI	6	OHV
	B4000	4.0 (4016)	X	EFI	6	OHV
	MPV	2.6 (2606)	G6	EFI	4	SOHC
	MPV	3.0 (2954)	JE	EFI	6	SOHC
	Navajo	4.0 (4016)	X	EFI	6	OHV

EFI - Electronic fuel injection

SOHC - Single overhead camshaft

OHV - Overhead valve

7911cC02

GENERAL ENGINE SPECIFICATIONS

Year	Engine ID/VIN	Engine Displacement Liters (cc)	Fuel System Type	Net Horsepower @ rpm	Net Torque @ rpm (ft. lbs.)	Bore x Stroke (in.)	Compression Ratio	Oil Pressure @ rpm
1991	F2	2.2 (2184)	EFI	110@4700	130@3000	3.39x3.70	8.6:1	43-57@3000
	F2	2.2 (2184)	2 BBL	85@4500	118@2500	3.39x3.70	8.6:1	43-57@3000
	G6	2.6 (2606)	EFI	121@4600	149@3500	3.62x3.86	8.4:1	44-58@3000
	JE	3.0 (2954)	EFI	190@5600	191@4500	3.54x3.05	8.4:1	46-71@3000
	X	4.0 (4016)	EFI	155@4200	220@2400	3.95x3.32	9.0:1	40-60@2000
1992	F2	2.2 (2184)	EFI	110@4700 [1]	130@3000 [2]	3.39x3.70	8.6:1	43-57@3000
	F2	2.2 (2184)	2 BBL	85@4500	118@2500	3.39x3.70	8.6:1	43-57@3000
	G6	2.6 (2606)	EFI	121@4600	149@3500	3.62x3.86	8.4:1	45-58@3000
	JE	3.0 (2954)	EFI	150@5000	165@4000	3.54x3.05	9.2:1	46-71@3000
	X	4.0 (4016)	EFI	160@4500	225@2500	3.95x3.32	9.1:1	40-60@2000
1993	F2	2.2 (2184)	EFI	145@4300	190@3500	3.39x3.70	7.8:1	43-57@3000
	F2	2.2 (2184)	2 BBL	110@4700	130@3000	3.39x3.70	8.6:1	43-56@3000
	G6	2.6 (2606)	EFI	121@4600	149@3500	3.62x3.86	8.4:1	45-58@3000
	JE	3.0 (2954)	EFI	150@5000	165@4000	3.54x3.05	8.5:1	53-75@3000
	X	4.0 (4016)	EFI	160@4500	[3]	3.94x3.31	9.0:1	40-60@2000
1994-95	A	2.3 (2298)	EFI	98@4600	130@2600	3.78x3.13	9.2:1	40-60@2000
	G6	2.6 (2606)	EFI	121@4600	149@3500	3.60x3.90	8.4:1	45-58@3000
	JE	3.0 (2954)	EFI	155@5000	169@4000	3.50x3.00	8.5:1	53-75@3000
	U	3.0 (2968)	EFI	140@4800	160@3000	3.50x3.14	9.3:1	40-60@2500
	X	4.0 (4016)	EFI	160@4500	225@2500 [4]	3.95x3.32	9.1:1	40-60@2000

EFI - Electronic fuel injection
2 BBL - 2 barrel carburetor
1 B2200 models: 91@4500
2 B2200 models: 118@2000
3 Manual transmission: 220@2500
   Automatic transmission: 220@2200
4 Automatic transmission: 220@2800

7911cC03

GASOLINE ENGINE TUNE-UP SPECIFICATIONS

Year	Engine ID/VIN	Engine Displacement Liters (cc)	Spark Plugs Gap (in.)	Ignition Timing (deg.) MT	Ignition Timing (deg.) AT	Fuel Pump (psi)	Idle Speed (rpm) MT	Idle Speed (rpm) AT	Valve Clearance In.	Valve Clearance Ex.
1991	F2 [1]	2.2 (2184)	0.041	6B [10]	6B [10]	64-85 [4]	750 [5]	750 [5]	HYD	HYD
	G6	2.6 (2606)	0.041	6B [2]	6B [2]	64-85	750 [3]	770 [3]	HYD	HYD
	JE [7]	3.0 (2954)	0.041	11B [3]	11B [3]	64-85	800 [3]	800 [3]	HYD	HYD
	X	4.0 (4016)	[11]	10B [8]	10B [8]	35-45	12	12	HYD	HYD
1992	F2	2.2 (2184)	0.041	6B [6]	6B [6]	64-85 [9]	750 [13]	750 [14]	HYD	HYD
	G6	2.6 (2606)	0.041	5B [2]	5B [2]	64-85	750 [15]	770 [15]	HYD	HYD
	JE	3.0 (2954)	0.041	11B [2]	11B [2]	64-85	800 [15]	800 [15]	HYD	HYD
	X	4.0 (4016)	[11]	10B [8]	10B [8]	35-45	16	16	HYD	HYD
1993	F2	2.2 (2184)	0.039-0.043	5-7B [6]	5-7B [6]	64-85 [18]	730-770 [6,19]	750-790 [6,19]	HYD	HYD
	G6	2.6 (2606)	0.039-0.043	4-6B [6]	4-6B [6]	64-85	730-770 [6]	750-790 [6]	HYD	HYD
	JE	3.0 (2954)	0.039-0.043	10-12B [6]	10-12B [6]	64-85	780-820	780-820	HYD	HYD
	X	4.0 (4016)	0.039-0.043	10B [17]	10B [17]	35-45	750-850	750-850	HYD	HYD
1994-95	A	2.3 (2298)	[11]	8-12B [20]	8-12B [20]	125 [21]	475-575	475-575	HYD	HYD
	G6	2.6 (2606)	0.039-0.043	4-6B [6]	4-6B [6]	64-85	750-800	750-800	HYD	HYD
	JE	3.0 (2954)	0.039-0.043	10-12B [6]	10-12B [6]	64-85	780-820	780-820	HYD	HYD
	U	3.0 (2968)	[11]	8-12B [20]	8-12B [20]	125 [21]	22	22	HYD	HYD
	X	4.0 (4016)	[11]	8-12B [8]	8-12B [8]	125 [21]	22	22	HYD	HYD

NOTE: The Vehicle Emission Control Information label often reflects changes made during production. Use the label figures if they disagree with the above data.

B - Before top dead center
HYD - Hydraulic

1 Single overhead camshaft
2 Plus or minus 1 degree
3 Test connector grounded
4 Canadian carbureted models:
   Manual transmission: 3.7-4.7
   Automatic transmission: 2.8-3.6
5 Canadian B2200 models: 800-850 rpm
6 Data link connector terminal 10 grounded
7 MPV
8 Base timing, not adjustable
9 Canadian carbureted models:
   Manaual Transmission 3.4-4.7
   Automatic Transmission 2.8-3.6
10 Test connector grounded with Electronic fuel injection

11 Refer to Vehicle Emissions label in engine compartment
12 Not adjustable
13 Canadian B2200 models: 800-850 rpm
14 B2200 models with automatic transmission: 770 rpm
15 Plus or minus 20 rpm
16 Not adjustable
17 Plus or minus 2 degrees
18 Carbureted models: 3.7 to 4.7 manual transmission; 2.8 to 3.6 automatic trans
19 Carbureted models: 800-850 rpm
20 With SPOUT shorting bar disconnected
21 Maximum
22 Automatically adjusted

7911cC04

FIRING ORDERS

NOTE: To avoid confusion, always replace spark plug wires one at a time.

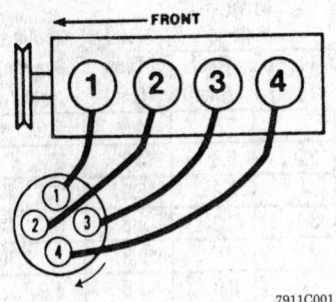

7911C001

2.2L Engine
Engine Firing Order: 1 — 3 — 4 — 2
Distributor Rotation: Clockwise

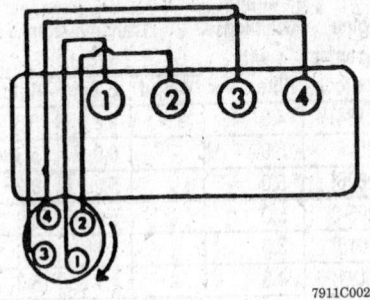

7911C002

2.6L (2606cc) Engine
Engine Firing Order: 1 — 3 — 4 — 2
Distributor Rotation: Clockwise

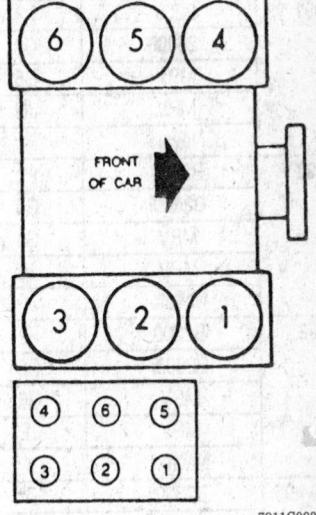

7911C003

4.0L Engine
Engine Firing Order: 1-4-2-5-3-6
Distributorless Ignition System

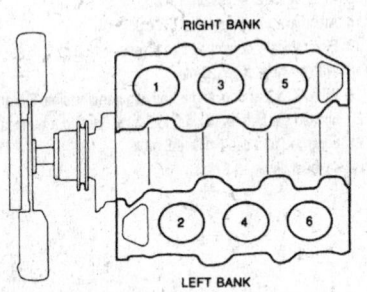

7911C004

3.0L Engine
Engine Firing Order: 1 — 2 — 3 — 4 — 5 — 6
Distributor Rotation: Clockwise

CAPACITIES

Year	Model	Engine ID/VIN	Engine Displacement Liters (cc)	Engine Crankcase with Filter	Transmission (pts.)			Transfer Case (pts.)	Drive Axle		Fuel Tank (gal.)	Cooling System (qts.)
					4-Spd	5-Spd	Auto.		Front (pts.)	Rear (pts.)		
1991	2200	F2	2.2 (2184)	4.3	3.6	4.2	15.8 [8]	-	-	2.6	14.8 [10]	7.9 [11]
	2600i	G6	2.6 (2606)	5.0	-	6.0 [7]	15.8 [8]	4	3.2	3.6 [9]	14.8	7.2
	MPV	G6	2.6 (2606)	5.0	-	5.2 [12]	18.2	-	-	3.2	15.9	7.6 [16]
	MPV	JE	3.0 (2954)	5.0	-	5.2 [12]	18.2	3	3.6	3.2 [13]	19.6 [14]	10.3 [15]
	Navajo	X	4.0 (4016)	5.0	-	5.6	20.0	3	17	17	19.3	8.1 [18]
1992	B2200	F2	2.2 (2184)	4.3	-	4.2	15.8 [12]	-	-	2.6	14.8 [10]	7.9
	B2600i	G6	2.6 (2606)	5.0	-	6.0	15.8 [12]	4	3.2	3.6	14.8 [10]	7.2
	MPV	G6	2.6 (2606)	5.0	-	6.0	15.8 [6]	-	-	3.2	15.9	7.2 [8]
	MPV	JE	3.0 (2954)	5.0	-	6.0	15.8 [6]	3	3.6	3.2	16.3 [7]	10.1 [9]
	Navajo	X	4.0 (4016)	5.0	-	5.6	20.0	3	3.5	5.3	19.3	7.8 [6]
1993	B2200	F2	2.2 (2184)	4.3		4.2	15.8 [9]	-	-	2.6	14.8 [10]	7.9
	B2600i	G6	2.6 (2606)	5.0	-	6.0	15.8 [9]	4	3.2	3.6	14.8 [10]	7.9
	MPV	G6	2.6 (2606)	5.0	-	-	15.8 [6]	-	-	3.2	15.9	7.6
	MPV	JE	3.0 (2954)	5.0	-	-	15.8 [6]	3	3.6	3.2	19.6 [7]	10.3
	Navajo	X	4.0 (4016)	5.0	-	5.6	20.0	3	3.5	5.3	19.3	7.8 [6]
1994-95	B2300	A	2.3 (2298)	5.0	-	3.6	19.4	-	-	5.0 [8]	16.3 [9]	7.2
	B3000	U	3.0 (2968)	4.5	-	3.6	19.4 [2]	3	5.0	5.0 [8]	16.3 [9]	11.8
	B4000	X	4.0 (4016)	5.0	-	3.6	19.4 [2]	3	5.0	5.0 [8]	16.3 [9]	8.1 [7]
	MPV	G6	2.6 (2606)	5.0	-	-	15.8 [6]	-	-	3.2	19.6	7.6
	MPV	JE	3.0 (2954)	5.0	-	-	15.8 [6]	3	3.6	3.2	19.6 [10]	10.3
	Navajo	X	4.0 (4016)	5.0	-	5.6	20.0	3	3.5	5.3	19.3	8.0

1 4 wheel drive: 6.8 pts.
2 Electronically controlled transmission: 18.2 pts.
3 4 wheel drive: 3.2 pts.
4 Long bed carbureted: 17.4 gals.
5 Without heater: 7.4 qts.
6 4 wheel drive: 5.6 pts.
7 4 wheel drive: 3.6 pts.
8 4 wheel drive: 19.8 gals.
9 Manual transmission: 10.1 qts.
10 Manual transmission: 7.2 qts.
11 Dana model 35 axles: 7.0 pts.
    Limited slip models: 5.3 pts. plus four ounces of friction modifier
    Models without limited slip: 5.0-5.5 pts.plus four ounces of friction modifier
12 With supercool AC and automatic transmission: 8.5 qts.

13 With AC: 8.6 qts.
14 not used
15 Automatic transmission: 7.6 qts.
16 Automatic transmission: 10.3 qts.
17 Electronically controlled transmission: 18.2 pts.
18 With 4 wheel drive: 20 pts.
19 With AC super cool and automatic transmission: 8.5 qts.
20 Limited slip differential: 5.0 to 5.3 pts. plus four ounces of friction modifier
21 Long bed and cab plus: 19.6 gals.
22 4 wheel drive: 19.8 gals.

7911cC06

CRANKSHAFT AND CONNECTING ROD SPECIFICATIONS
All measurements are given in inches.

Year	Engine ID/VIN	Engine Displacement Liters (cc)	Crankshaft				Connecting Rod		
			Main Brg. Journal Dia.	Main Brg. Oil Clearance	Shaft End-play	Thrust on No.	Journal Diameter	Oil Clearance	Side Clearance
1991	F2	2.2 (2184)	2.3597-2.3604	1	0.0031-0.012	3	2.0055-2.0061	0.0011-0.0039	0.004-0.012
	G6	2.6 (2606)	2.3597-2.3604	0.0010-0.0031	0.0031-0.0118	4	2.0055-2.0061	0.0011-0.0039	0.0043-0.012
	JE	3.0 (2954)	2.4385-2.4392	0.0010-0.0031	0.0031-0.0118	4	2.0842-2.0848	0.0009-0.004	0.007-0.016
	X	4.0 (4016)	2.2433-2.2441	0.0005-0.0019	0.016-0.0126	3	2.1252-2.1260	0.0003-0.0024	0.0002-0.0025
1992	F2	2.2 (2184)	2.3597-2.3604	1	0.0031-0.012	3	2.0055-2.0061	0.0011-0.0039	0.004-0.012
	G6	2.6 (2606)	2.3597-2.3604	0.0010-0.0031	0.0031-0.0118	4	2.0055-2.0061	0.0011-0.0039	0.0043-0.012
	JE	3.0 (2954)	2.4385-2.4392	0.0010-0.0031	0.0031-0.0118	4	2.0842-2.0848	0.0009-0.004	0.007-0.016
	X	4.0 (4016)	2.2433-2.2441	0.0005-0.0019	0.016-0.0126	3	2.1252-2.1260	0.0003-0.0024	0.0002-0.0025
1993	F2	2.2 (2184)	2.3597-2.3604	1	0.0032-0.0118	3	2.0056-2.0060	0.0011-0.0039	0.004-0.012
	G6	2.6 (2606)	2.3598-2.3604	0.0010-0.0031	0.0032-0.0118	4	2.0055-2.0060	0.0011-0.0039	0.0044-0.012
	JE	3.0 (2954)	2.4385-2.4392	0.0010-0.0031	0.0031-0.0118	4	2.0842-2.0848	0.0009-0.004	0.007-0.016
	X	4.0 (4016)	2.2433-2.2441	0.0005-0.0019	0.016-0.0126	3	2.1252-2.1260	0.0003-0.0024	0.0002-0.0025
1994-95	A	2.3 (2299)	2.2059-2.2051	0.0008-0.0026	0.004-0.012	3	2.0462-2.0472	0.0008-0.0026	0.0035-0.014
	G6	2.6 (2606)	2.3598-2.3604	0.0010-0.0031	0.0032-0.0118	3	2.0055-2.0060	0.0011-0.0039	0.0044-0.012
	JE	3.0 (2954)	2.4385-2.4392	0.0010-0.0031	0.0032-0.0118	4	2.0847-2.0848	0.0009-0.004	0.007-0.016
	U	3.0 (2954)	2.5190-2.5198	0.0005-0.0023	0.004-0.008	4	2.1253-2.1261	0.0007-0.0027	0.006-0.014
	X	4.0 (4016)	2.2433-2.2441	0.0005-0.0019	0.016-0.0126	3	2.1252-2.1260	0.0003-0.0024	0.000-0.002

1 Nos. 1, 2, 4 and 5: 0.0010-0.0031
Nos. 3: 0.0012-0.0031

7911cC08

VALVE SPECIFICATIONS

Year	Engine ID/VIN	Engine Displacement Liters (cc)	Seat Angle (deg.)	Face Angle (deg.)	Spring Test Pressure (lbs. @ in.)	Spring Installed Height (in.)	Stem-to-Guide Clearance (in.)		Stem Diameter (in.)	
							Intake	Exhaust	Intake	Exhaust
1991	F2	2.2 (2184)	45	45	2	1	0.0010-0.0024	0.0012-0.0016	7	8
	G6	2.6 (2606)	45	45	0.069	1.963-1.970	0.0010-0.0024	0.0012-0.0016	0.2744-0.2750	0.2742-0.2748
	JE	3.0 (2954)	45	45	2	3	0.0010-0.0024	0.0012-0.0016	0.2744-0.2750	0.3159-0.3165
	X	4.0 (4016)	45	44	0.078	1.91	0.0008-0.0025	0.0018-0.0035	0.3159-0.3167	0.3149-0.3156
1992	F2	2.2 (2184)	45	45	4	5	0.0010-0.0024	0.0012-0.0026	0.3161-0.3167	0.3159-0.3165
	G6	2.6 (2606)	45	45	0.069	1.970	0.0010-0.0024	0.0012-0.0026	0.2744-0.2750	0.2742-0.2748
	JE	3.0 (2954)	45	45	2	6	0.0010-0.0024	0.0012-0.0026	0.2744-0.2750	0.3159-0.3165
	X	4.0 (4016)	45	44	0.078	1.91	0.0008-0.0025	0.0018-0.0035	0.3159-0.3167	0.3149-0.3156
1993	F2	2.2 (2184)	45	45	4	5	0.0010-0.0024	0.0012-0.0026	0.3162-0.3167	0.3160-0.3165
	G6	2.6 (2606)	45	45	0.069	1.963	0.0010-0.0023	0.0012-0.0025	0.2744-0.2749	0.2743-0.2748
	JE	3.0 (2954)	45	45	2	6	0.0010-0.0023	0.0012-0.0025	0.2745-0.2750	0.3160-0.3165
	X	4.0 (4016)	45	44	0.078	1.91	0.0008-0.0025	0.0018-0.0035	0.3159-0.3167	0.3149-0.3156
1994-95	A	2.3 (2299)	45	44	0.078	1.877	0.0010-0.0055	0.0015-0.0055	0.3416-0.3423	0.3411-0.3418
	G6	2.6 (2606)	45	45	0.069	1.963	0.0010-0.0023	0.0012-0.0025	0.2744-0.2749	0.2743-0.2748
	JE	3.0 (2954)	45	45	2	6	0.0010-0.0023	0.0012-0.0025	0.2745-0.2750	0.3160-0.3165
	U	3.0 (2968)	44	44	-	1.84	0.0010-0.0027	0.0015-0.0032	0.3134-0.3126	0.3129-0.3121
	X	4.0 (4016)	45	44	0.078	1.91	0.0008-0.0025	0.0018-0.0035	0.3159-0.3167	0.3149-0.3156

1 Inner: 1.681-1.732
   Outer: 1.984-2.047
2 Intake:
   Inner: 0.063
   Outer: 0.071
   Exhaust:
   Inner: 0.073
   Outer: 0.080
3 Intake:
   Inner: 1.821-1.840
   Outer: 1.984-2.003
   Exhaust:
   Inner: 2.071-2.092
   Outer: 2.274-2.296

4 Inner: 0.06
   Outer: 0.07
5 Inner: 1.681
   Outer: 1.984
6 Intake:
   Inner: 1.56
   Outer: 1.73
   Exhaust:
   Inner: 1.59
   Outer: 1.77
7 B2200: 0.0010-0.0024
8 B2200: 0.0010-0.0024

7911cC09

PISTON AND RING SPECIFICATIONS

All measurements are given in inches.

Year	Engine ID/VIN	Engine Displacement Liters (cc)	Piston Clearance	Ring Gap			Ring Side Clearance		
				Top Compression	Bottom Compression	Oil Control	Top Compression	Bottom Compression	Oil Control
1991	F2	2.2 (2184)	0.0014-0.0030	0.008-0.014	0.006-0.012	0.008-0.028	0.0012-0.0028	0.0012-0.0028	NA
	G6	2.6 (2606)	0.0023-0.0029	0.008-0.014	0.010-0.016	0.008-0.028	0.0012-0.0028	0.0012-0.0028	NA
	JE	3.0 (2954)	0.0019-0.0026	0.008-0.014	0.006-0.012	0.008-0.028	0.0012-0.0028	0.0012-0.0028	NA
	X	4.0 (4016)	0.0008-0.0019	0.015-0.023	0.015-0.023	0.015-0.055	0.0020-0.0033	0.0020-0.0033	NA
1992	F2	2.2 (2184)	0.0017-0.0024	0.008-0.013	0.006-0.011	0.008-0.027	0.0012-0.0027	0.0012-0.0027	NA
	G6	2.6 (2606)	0.0023-0.0029	0.008-0.014	0.010-0.016	0.008-0.027	0.0012-0.0028	0.0012-0.0028	NA
	JE	3.0 (2954)	0.0009-0.0020	0.008-0.014	0.006-0.012	0.008-0.028	0.0012-0.0028	0.0012-0.0028	NA
	X	4.0 (4016)	0.0008-0.0019	0.015-0.023	0.015-0.023	0.015-0.055	0.0020-0.0033	0.0020-0.0033	NA
1993	F2	2.2 (2184)	0.0017-0.0024	0.008-0.013	0.006-0.011	0.008-0.027	0.0012-0.0027	0.0012-0.0027	NA
	G6	2.6 (2606)	0.0023-0.0029	0.008-0.013	0.010-0.015	0.008-0.027	0.0012-0.0027	0.0012-0.0027	NA
	JE	3.0 (2954)	0.0010-0.0020	0.008-0.013	0.006-0.012	0.008-0.027	0.0012-0.0027	0.0012-0.0027	NA
	X	4.0 (4016)	0.0008-0.0019	0.015-0.023	0.015-0.023	0.015-0.055	0.0020-0.0033	0.0020-0.0033	NA
1994-95	A	2.3 (2299)	0.0024-0.0034	0.010-0.020	0.010-0.020	0.015-0.049	0.0016-0.0033	0.0016-0.0033	NA
	G6	2.6 (2606)	0.0023-0.0029	0.008-0.013	0.010-0.015	0.008-0.027	0.0012-0.0027	0.0012-0.0027	NA
	JE	3.0 (2954)	0.0010-0.0020	0.008-0.013	0.006-0.012	0.008-0.027	0.0012-0.0027	0.0012-0.0027	NA
	U	3.0 (2968)	0.0012-0.0023	0.010-0.020	0.010-0.020	0.010-0.049	0.0016-0.0037	0.0016-0.0037	NA
	X	4.0 (4016)	0.0008-0.0019	0.015-0.023	0.015-0.023	0.015-0.055	0.0020-0.0033	0.0020-0.0033	NA

7911cC10

TORQUE SPECIFICATIONS
All readings in ft. lbs.

Year	Engine ID/VIN	Engine Displacement Liters (cc)	Cylinder Head Bolts	Main Bearing Bolts	Rod Bearing Bolts	Crankshaft Damper Bolts	Flywheel Bolts	Manifold		Spark Plugs	Lug Nut
								Intake	Exhaust		
1991	F2	2.2 (2184)	59-64	61-65	48-51	116-123	71-76	14-22	25-36	11-17	65-87 [2]
	G6	2.6 (2606)	[1]	48-51	61-65	130-145	67-72	14-19	16-21	11-17	65-87 [2]
	JE	3.0 (2954)	[3]	[4]	[5]	116-123	76-81	14-19	16-21	11-17	65-87
	X	4.0 (4016)	[6]	66-77	18-24	30-37	59	[6]	18	11-17	100
1992	F2	2.2 (2184)	59-64	61-65	48-51	116-123	71-76	14-22	25-36	11-17	65-87
	G6	2.6 (2606)	[1]	61-65	48-51	130-145	67-72	14-19	16-21	11-17	65-87
	JE	3.0 (2954)	[8]	[9]	[10]	116-123	76-81	14-19	16-21	10-13	65-87
	X	4.0 (4016)	[6]	66-77	18-24	30-37	59	[6]	18	11-17	100
1993	F2	2.2 (2184)	59-64	61-65	48-51	116-123	71-76	14-22	25-36	11-17	65-87
	G6	2.6 (2606)	[11]	61-65	48-51	130-145	67-72	14-19	16-21	11-17	65-87
	JE	3.0 (2954)	[8]	[9]	[10]	116-123	76-81	14-19	16-21	10-13	65-87
	X	4.0 (4016)	[6]	66-77	18-24	30-37	59	[6]	18	11-17	100
1994-95	A	2.3 (2298)	[12]	[14]	30-36	15-22	60	19-28	[13]	7-15	100
	G6	2.6 (2606)	[11]	61-65	48-51	130-145	67-72	14-19	16-21	11-17	65-87
	JE	3.0 (2954)	[8]	[9]	[10]	116-123	76-81	14-19	16-21	10-13	65-87
	U	3.0 (2968)	[15]	59	23-28	24	59	[16]	18	7-15	100
	X	4.0 (4016)	[6]	66-77	18-24	30-37	59	[6]	18	11-17	100

1 Step 1: 59-64 ft. lbs.
  Step 2: Tighten two bolts nearest gear to 12-17 ft. lbs.
2 Truck-styled wheels: 87-108 ft. lbs.
3 Step 1: 14 ft. lbs.
  Step 2: Turn each bolt 90 degrees
  Step 3: Repeat Step 2
4 Step 1: 14 ft. lbs.
  Step 2: Turn each bolt 90 degrees
  Step 3: Turn each bolt 45 degrees
5 Step 1: 22 ft. lbs.
  Step 2: Turn each nut 40 degrees
6 Step 1: Tighten cylinder head to 44 ft. lbs.
  Step 2: Tighten intake manifold to 3-6 ft. lbs.
  Step 3: Tighten cylinder head to 59 ft. lbs.
  Step 4: Tighten intake to 6-11 ft. lbs.
  Step 5: Tighten cylinder head 80 to 85 degrees
  Step 6: Tighten intake manifold 11-15 ft. lbs., then 15-18 ft. lbs.
7 not used
8 Step 1: 14 ft. lbs.
  Step 2: Turn each bolt 90 degrees
  Step 3: Repeat Step 2
9 Step 1: 14 ft. lbs.
  Step 2: Turn each bolt 90 degrees
  Step 3: Turn each bolt 45 degrees
10 Step 1: 22 ft. lbs.
   Step 2: Turn each nut 40 degrees

11 Step 1: 59-64 ft. lbs.
   Step 2: Tighten two bolts nearest gear 12-17 ft. lbs.
   Step 1: 51-59 ft. lbs.
12 Step 2: 80-89 ft. lbs.
   Step 1: 15-17 ft. lbs.
   Step 2: 6-8 ft. lbs.
14 Step 1: Tighten by hand until seated
   Step 2: 50-59 ft. lbs.
   Step 3: 74-84 ft. lbs.
15 Step 1: 59 ft. lbs.
   Step 2: Back off one full turn
   Step 3: 37 ft. lbs.
   Step 4: 68 ft. lbs.
16 Step 1: 11 ft. lbs.
   Step 2: 19 ft. lbs.

7911cC11

BRAKE SPECIFICATIONS
All measurements in inches unless noted

Year	Model	Master Cylinder Bore	Brake Disc			Brake Drum Diameter			Minimum Lining Thickness	
			Original Thickness	Minimum Thickness	Maximum Runout	Original Inside Diameter	Max. Wear Limit	Maximum Machine Diameter	Front	Rear
1991	B2200	0.875	0.790 [1]	0.710 [2]	0.006	10.24	10.30	NA	0.118	0.040
	B2600	0.875	0.790 [1]	0.710 [2]	0.006	10.24	10.30	NA	0.118	0.040
	MPV	0.940	0.940	0.870	0.004	10.24	10.30	NA	0.040	0.040
	Navajo	-	-	0.810	0.010	-	[3]	0.060	-	-
1992	B2200	0.875	0.790 [1]	0.710 [2]	0.006	10.24	10.30	NA	0.118	0.040
	B2600	0.875	0.790 [1]	0.710 [2]	0.006	10.24	10.30	NA	0.118	0.040
	MPV	0.940	1.180 [4]	1.100 [5]	0.004	10.24	10.30	NA	0.080	0.040
	Navajo	NA	NA	0.810	0.010	NA	[3]	0.06	NA	NA
1993	B2200	0.875	0.790 [1]	0.710 [2]	0.006	10.24	10.30	NA	0.118	0.040
	B2600	0.875	0.790 [1]	0.710 [2]	0.006	10.24	10.30	NA	0.118	0.040
	MPV	0.940	1.180 [4]	1.100 [5]	0.004	10.24	10.30	NA	0.080	0.040
	Navajo	NA	NA	0.810	0.010	NA	[3]	0.06	NA	NA
1994-95	B2300	NA	NA	0.81	0.003	NA	[3]	0.003	0.012	0.003
	B3000	NA	NA	0.81	0.003	NA	[3]	0.003	0.012	0.003
	B4000	NA	NA	0.81	0.003	NA	[3]	0.003	0.012	0.003
	MPV	0.940	1.18 [1]	1.10 [5]	0.004	10.24	10.30	NA	0.080	0.040
	Navajo	NA	NA	0.81	0.010	NA	[3]	0.060	NA	NA

NA - Not Available
1 4x4: 0.870
2 4x4: 0.790
3 Refer to the maximum diameter stamped on drum
4 4x4: 1.10
5 4x4: 1.02

7911cC12

WHEEL ALIGNMENT

Year	Model			Caster Range (deg.)		Caster Preferred Setting (deg.)	Camber Range (deg.)	Camber Preferred Setting (deg.)	Toe-in (in.)	Steering Axis Inclination (deg.)
1991	B2200			0-1 2/3P	5	1 5/6P	1/3P-1 1/3P	3/4P	0-1/4P	NA
	B2600	1		0-1 2/3P	5	1 5/6P	1/3P-1 1/3P	3/4P	0-1/4P	NA
	B2600	2		0-1 2/3P	6	1 5/6P	1/3P-1 1/2P	1P	0-1/4P	NA
	MPV	1		4 11/16P-6 3/16P		5 7/16P	3/8N-1 1/8P	3/8P	1/32P-9/32P	NA
	MPV	2		4 3/4P-6 1/4P		5 1/2P	5/16N-11/16P	3/16P	1/32P-9/32P	NA
	Navajo			2 1/2P-6P		-	3/4N-1 1/4P	-	1/8N-1/8P	NA
1992	B2200			0-1 2/3P	3	1 5/16P	1/3P-1 1/3P	3/4P	0-1/4P	NA
	B2600	1		0-1 2/3P	3	1 5/16P	1/3P-1 1/3P	3/4P	0-1/4P	NA
	B2600	2		0-1 2/3P	4	1 5/6P	1/3P-1 1/2P	1P	0-1/4P	NA
	MPV	1		4 1/16P-5 9/16P		4 13/16P	1/8N-7/8P	3/8P	5/16P	NA
	MPV	2		4 5/16P-5 13/16P		6 1/16P	5/16N-11/16P	3/16P	5/16P	NA
	Navajo			2 1/2P-6P		-	3/4N-1 1/4P	-	1/8N-1/8P	NA
1993	B2200			0-1 2/3P	3	1 5/16P	1/3P-1 1/3P	3/4P	0-1/4P	NA
	B2600	1		0-1 2/3P	3	1 5/16P	1/3P-1 1/3P	3/4P	0-1/4P	NA
	B2600	2		0-1 2/3P	4	1 5/16P	1/3P-1 1/2P	1P	0-1/4P	NA
	MPV	1		4 1/16P-5 9/16P		4 13/16P	7/8N-7/8P	3/8P	5/16P	NA
	MPV	2		4 5/16P-5 13/16P		5 1/16P	5/16N-11/16P	3/16P	5/16P	NA
	Navajo			2 1/2P-6P		-	3/4N-1 1/4P	-	1/8N-1/8P	NA
1994-95	B2300			2P-6P		4P	1/4N-3/4P	1/4P	1/32	-
	B3000			2P-6P		4P	1/4N-3/4P	1/4P	1/32	-
	B4000	1		2P-6P		4P	1/4N-3/4P	1/4P	1/32	-
	B4000	2		2P-7P		5P	1/4N-3/4P	1/4P	1/32	-
	MPV	1		4 1/6P-5 9/16P		4 13/16P	1/8N-7/8P	3/8P	5/16P	NA
	MPV	2		4 5/16P-5 13/16P		5 1/16P	5/16N-11/16P	3/16P	5/16P	NA
	Navajo			2 1/2P-6P		-	3/4N-1/14P	-	1/8N-1/8P	NA

NA - Not Available
F - Front
R - Rear
1 2WD
2 4WD
3 2WD: 1P-2 7/8P
  4WD: 1 5/8P-3 1/2P

4 2WD: 1 5/16P
  4WD: 2 9/16P
5 If equipped with Power Steering: 1 1/3P-2 1/3P
6 If equipped with Power Steering: 2 3/4P

7911cC13

ENGINE ELECTRICAL

NOTE: Disconnecting the negative battery cable on some vehicles may interfere with the functions of the on board computer systems and may require the computer to undergo a relearning process, once the negative battery cable is reconnected.

Distributor

REMOVAL

1. Disconnect the negative battery cable.
2. Remove the distributor cap from the distributor, leaving the spark plug wires attached. If spark plug wire removal is necessary to remove the distributor cap, tag the wires prior to removal so they can be reinstalled in the correct position.
3. Disconnect the electrical connectors and vacuum hose(s), if equipped, from the distributor.
4. Mark the position of the rotor in relation to the distributor housing and the position of the distributor housing on the cylinder head.
5. Remove the distributor hold-down bolt(s) and remove the distributor.
6. Check the distributor O-ring for cuts or other damage and replace, if necessary.

INSTALLATION

Timing Not Disturbed

1. Lubricate the distributor O-ring with clean engine oil.
2. Install the distributor with the hold-down bolt(s), aligning the marks that were made during removal. Tighten the hold-down bolt(s) to 14–19 ft. lbs. (19–25 Nm).
3. Connect the electrical connectors and vacuum hose(s), if equipped.
4. Install the distributor cap on the distributor. Connect the spark plug wires, if removed.
5. Connect the negative battery cable. Start the engine and check the ignition timing.

Timing Disturbed

1. Disconnect the spark plug wire from the No. 1 cylinder spark plug and remove the spark plug. Place a finger over the spark plug hole.

2. Turn the crankshaft in the normal direction of rotation until compression is felt at the spark plug hole.
3. Align the mark on the crankshaft pulley with the TDC mark on the timing belt cover.
4. Lubricate the distributor O-ring with clean engine oil.
5. Turn the distributor shaft until the rotor points to the No. 1 spark plug tower on the distributor cap and install the distributor. Install the distributor hold-down bolt(s) and align the distributor housing with the mark made on the cylinder head during removal. Snug the bolt(s).
6. Connect the electrical connectors and vacuum hose(s), if equipped.
7. Install the distributor cap on the distributor. Connect the spark plug wires, if removed.
8. Install the spark plug in the No. 1 cylinder and connect the spark plug wire.
9. Connect the negative battery cable. Start the engine and adjust the ignition timing. Tighten the distributor hold-down bolt(s) to 14–19 ft. lbs. (19–25 Nm) after the timing has been set.

Distributorless Ignition

REMOVAL AND INSTALLATION

Crankshaft Sensor

1. Disconnect the negative battery cable.
2. Disconnect the sensor electrical connector from the wiring harness.
3. Remove the crankshaft sensor mounting screws and remove the sensor.
4. Installation is the reverse of the removal procedure. Tighten the screws to 75–106 inch lbs. (8.5–12 Nm).

Ignition Module

1. Disconnect the battery cables and remove the battery.
2. Disconnect the electrical connector at the module.
3. Remove the module retaining bolt and remove the module.
4. Installation is the reverse of the removal procedure. Tighten the mounting bolt to 22–31 inch lbs. (2.5–3.5 Nm).

Ignition Coil Pack

1. Disconnect the negative battery cable.
2. Disconnect the electrical harness connector from the coil pack.

3. Remove the spark plug wires by squeezing the locking tabs to release the coil boot retainers.
4. Remove the coil pack retaining screws and remove the coil pack.
5. Installation is the reverse of the removal procedure. Tighten the screws to 40–62 inch lbs. (4.5–7.0 Nm).

Ignition Timing

ADJUSTMENT

Except 4.0L Engine

1. Before starting the engine, clean and mark the timing marks on the timing belt cover and crankshaft pulley.
2. Connect a timing light and tachometer according to the manufacturers instructions.
3. Start the engine and bring to normal operating temperature. Turn all electric loads **OFF**.
4. On fuel injected vehicles, connect a jumper wire between the green 1-pin test connector and ground.
5. Check the idle speed and adjust, if necessary, to the specification on the underhood emission information label.
6. Aim the timing light and verify that the timing marks on the crankshaft pulley and timing belt cover are aligned. If the marks are aligned, proceed to Step 8.
7. If the marks are not aligned, loosen the distributor hold-down bolts and turn the distributor housing to adjust the timing. Tighten the hold-down bolts to 14–19 ft. lbs. (19–25 Nm) and recheck the timing.
8. Shut off the engine. Remove the jumper wire and all test equipment.

4.0L Engine

Timing on the 4.0L engine is preset from the factory at 10 degrees BTDC and is not adjustable.

Alternator

PRECAUTIONS

Several precautions must be observed with alternator equipped vehicles to avoid damage to the unit.

If the battery is removed for any reason, make sure it is reconnected with the correct polarity. Reversing the battery connections may result in damage to the one-way rectifiers.

When utilizing a booster battery as a starting aid, always connect the

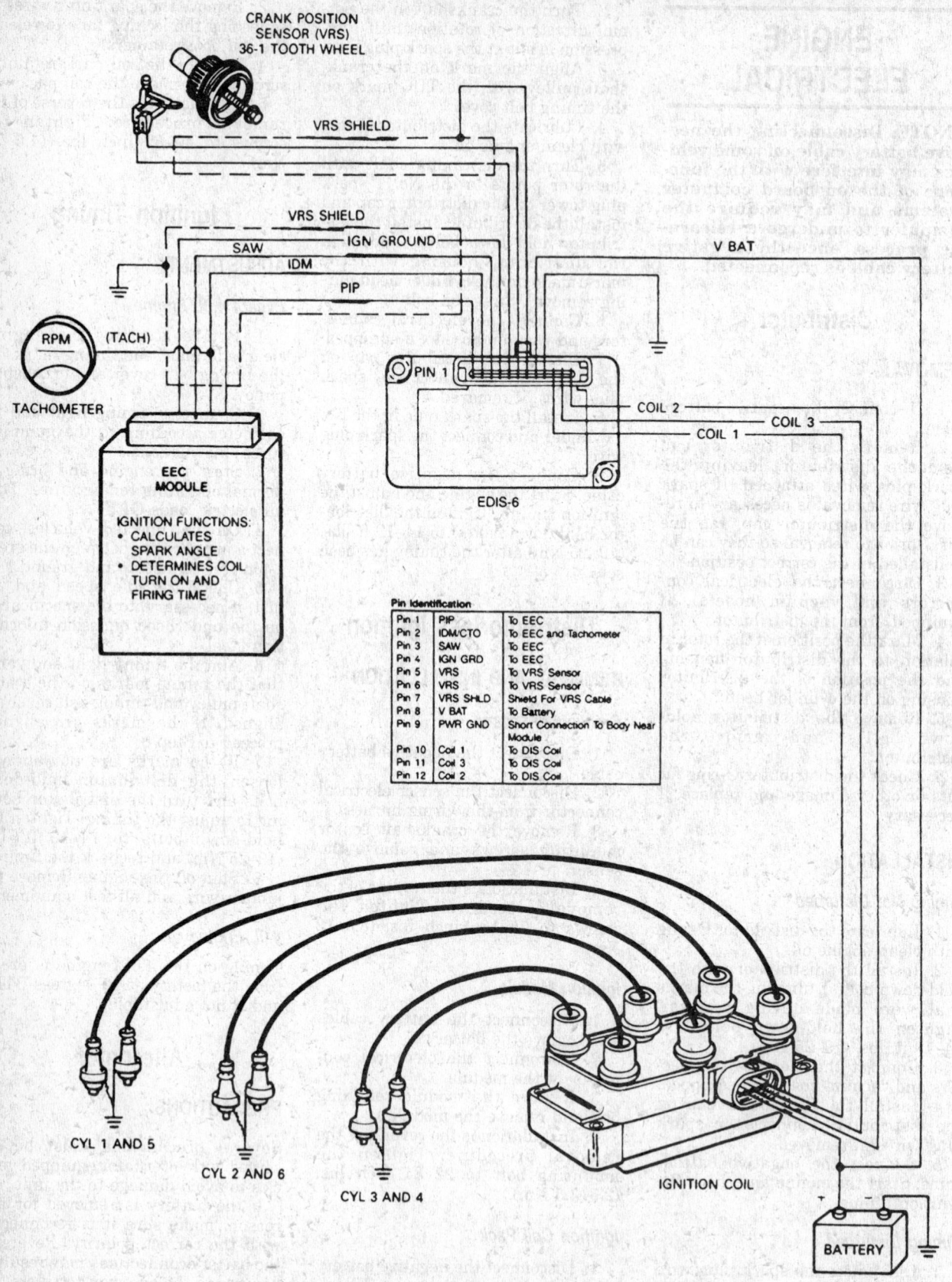

CRANK POSITION
SENSOR (VRS)
36-1 TOOTH WHEEL

VRS SHIELD

VRS SHIELD
SAW
IDM
PIP
IGN GROUND
V BAT

RPM (TACH)

TACHOMETER

PIN 1

COIL 2
COIL 3
COIL 1

EDIS-6

EEC
MODULE

IGNITION FUNCTIONS:
• CALCULATES
  SPARK ANGLE
• DETERMINES COIL
  TURN ON AND
  FIRING TIME

Pin Identification

Pin 1	PIP	To EEC
Pin 2	IDM/CTO	To EEC and Tachometer
Pin 3	SAW	To EEC
Pin 4	IGN GRD	To EEC
Pin 5	VRS	To VRS Sensor
Pin 6	VRS	To VRS Sensor
Pin 7	VRS SHLD	Shield For VRS Cable
Pin 8	V BAT	To Battery
Pin 9	PWR GND	Short Connection To Body Near Module
Pin 10	Coil 1	To DIS Coil
Pin 11	Coil 3	To DIS Coil
Pin 12	Coil 2	To DIS Coil

CYL 1 AND 5

CYL 2 AND 6

CYL 3 AND 4

IGNITION COIL

BATTERY

7911C005

Distributorless ignition system — 4.0L engine

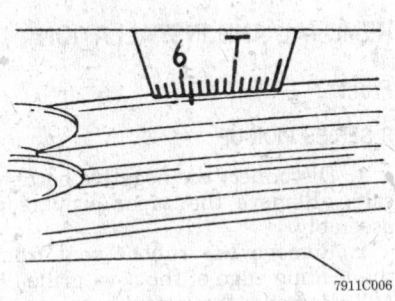

Timing mark location — 2.2L and 2.6L engines

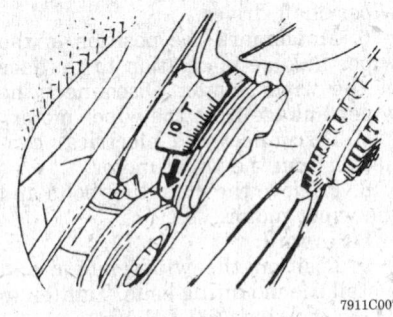

Timing mark location — 3.0L engine

positive to positive terminals and the negative terminal from the booster battery to a good engine ground on the vehicle being started.

Never use a fast charger as a booster to start vehicles.

Disconnect the battery cables when charging the battery with a fast charger.

Never attempt to polarize the alternator.

Do not use test lights of more than 12V when checking diode continuity.

Do not short across or ground any of the alternator terminals.

The polarity of the battery, alternator and regulator must be matched and considered before making any electrical connections within the system.

Never separate the alternator on an open circuit. Make sure all connections within the circuit are clean and tight.

Disconnect the battery ground terminal when performing any service on electrical components.

Disconnect the battery if arc welding is to be done on the vehicle.

BELT TENSION ADJUSTMENT

Except 4.0L Engine

1. Disconnect the negative battery cable.
2. Check the belt tension by applying approximately 22 lbs. pressure mid-way between the pulleys and measuring the belt deflection. Belt deflection specifications are as follows: 2.2L engines — New belt: 0.28–0.31 in., Used belt: 0.31–0.35 in.   2.6L engine — New belt: 0.39–0.47 in., Used belt: 0.43–0.51 in.   3.0L engine — New belt: 0.35–0.39 in., Used belt: 0.39–0.47 in.
3. If belt tension is incorrect, loosen the alternator pivot and adjusting bolts. Position the alternator housing using a suitable tool to attain correct belt tension. Be careful not to damage the alternator housing.
4. When correct belt tension is achieved, tighten the alternator adjusting bolt to 14–19 ft. lbs. (19–25 Nm) on all except 2.2L engine. On 2.2L engine, tighten the adjusting bolt to 23–34 ft. lbs. (31–46 Nm).
5. Tighten the alternator pivot bolt to 27–38 ft. lbs. (37–52 Nm).
6. Connect the negative battery cable.

4.0L Engine

Belt tension is maintained by an automatic tensioner; no adjustment is necessary.

REMOVAL AND INSTALLATION

1. Disconnect the negative battery cable.
2. Disconnect the electrical connectors at the alternator. Remove the wiring connector bracket on 4.0L engine.
3. On the 3.0L engine, loosen the power steering pump drive belt adjusting locknut and remove the drive belt. Remove the power steering pump pulley.
4. On all except 4.0L engine, loosen the alternator adjusting and pivot bolts and remove the drive belt. On 4.0L engine, loosen the drive belt tensioner and remove the drive belt.

5. Remove the alternator bolts and remove the alternator.
6. Installation is the reverse of the removal procedure. Tighten the alternator mounting bolts on 4.0L engine to 22–30 ft. lbs. (30–40 Nm). Adjust the belt tension on all except 4.0L engine.
7. Tighten the power steering pump pulley nut to 29–43 ft. lbs. (39–59 Nm).

Starter

REMOVAL AND INSTALLATION

Except 4WD MPV

1. Disconnect the negative battery cable. Raise and safely support the vehicle.
2. Disconnect the electrical connectors from the starter.
3. Remove the starter mounting bolts and remove the starter.
4. Installation is the reverse of the removal procedure. Tighten the starter mounting bolts to 27–38 ft. lbs. (37–52 Nm) except 4.0L engine or 15–20 ft. lbs. (21–27 Nm) on 4.0L engine.

4WD MPV

1. Disconnect the negative battery cable.
2. Remove the alternator.
3. Raise and safely support the vehicle. Remove the splash shields.
4. Remove the power steering pump mounting bolts and position the pump aside, without disconnecting the power steering hoses.
5. Remove the automatic transmission cooler line brackets.
6. Mark the position of the driveshaft on the axle flange and remove the front driveshaft.
7. Remove the wiring harness bracket and the automatic transmission cooler line bracket that is next to the starter.
8. Disconnect the electrical connectors from the starter.
9. Remove the fuel and brake line shield.
10. Remove the starter mounting bolts and remove the starter.
11. Installation is the reverse of the removal procedure. Tighten the starter mounting bolts to 27–38 ft. lbs. (37–52 Nm).

CHASSIS ELECTRICAL

Heater Blower Motor

REMOVAL AND INSTALLATION

Except Navajo

1. Disconnect the negative battery cable.
2. On B Series Pick-Up, remove the ECU. On MPV, remove the passenger side lower panel and undercover.
3. Disconnect the electrical connector from the blower motor.
4. Remove the attaching screws and remove the blower motor.
5. Installation is the reverse of the removal procedure.

Navajo

1. Disconnect the negative battery cable.
2. Remove the air cleaner.
3. Disconnect the wire harness connector from the blower motor by pushing down on the tab while pulling the connector off at the motor.
4. If equipped with air conditioning, remove the 3 solenoid box cover retaining bolts and the solenoid box cover.
5. Disconnect the blower motor cooling tube at the blower motor.
6. Remove the 3 attaching screws and remove the blower motor.
7. If necessary, remove the blower wheel hub push-nut and remove the blower wheel.
8. Installation is the reverse of the removal procedure.

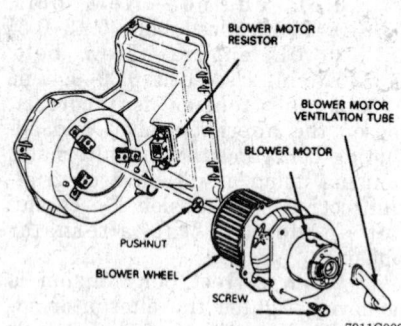

Blower motor assembly exploded view — Navajo

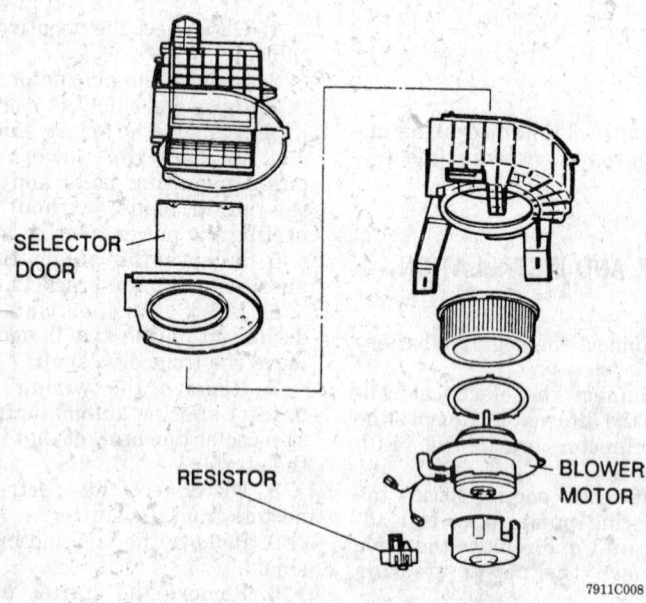

Blower motor assembly exploded view — MPV

Window Wiper Motor

REMOVAL AND INSTALLATION

Front

B SERIES PICK-UP

1. Disconnect the negative battery cable. Remove the wiper arm/blade assembly.
2. Remove the rubber seal from the leading edge of the cowl grille.
3. Remove the attaching screws and remove the cowl grille.
4. Remove the access hole covers.
5. Remove the bolts attaching the wiper shaft drives.
6. Matchmark the position of the wiper linkage in relation to the face of the wiper motor. Disconnect the wiper linkage from the wiper motor.
7. Disconnect the electrical connector from the wiper motor.
8. Remove the mounting bolts and the wiper motor.
   To install:
9. Position the wiper motor and install the mounting bolts. Tighten to 61–87 inch lbs. (6.9–9.8 Nm).
10. Connect the electrical connector.
11. Attach the wiper linkage to the motor, aligning the mark that was made during removal.
12. Install the bolts attaching the wiper shaft drives and tighten to 61–87 inch lbs. (6.9–9.8 Nm).
13. Install the access hole covers, cowl grille and seal.
14. Install the wiper arm/blade assemblies. Adjust the arm height to 0.8 in. (20mm) from the lower windshield moulding and tighten the arm retaining nuts to 7.2–10 ft. lbs. (9.8–14 Nm).
15. Connect the negative battery cable and check wiper operation.

MPV

1. Disconnect the negative battery cable.
2. Disconnect the electrical connector from the wiper motor.
3. Disconnect the wiper linkage from the motor crank arm. If necessary to remove the crank arm from the motor, matchmark the arm to the bracket prior to removal.
4. Remove the mounting bolts and remove the wiper motor.
   To install:
5. Position the wiper motor and install the mounting bolts. Tighten to 61–87 inch lbs. (6.9–9.8 Nm).
6. If removed, install the motor crank arm, aligning the marks that were made during removal.

7. Connect the wiper linkage to the motor crank arm. Connect the electrical connector.

8. Make sure when in the park position, the wiper blades are 0.98–1.38 in. (25–35mm) from the lower windshield moulding.

NAVAJO

1. Turn the wiper switch **ON**. Turn the ignition switch **ON** until the blades are straight up, then turn the ignition **OFF** to keep them there.

2. Disconnect the negative battery cable, then disconnect the electrical connector from the wiper motor.

3. Remove the right wiper arm and blade assembly. Remove the right pivot nut and allow the linkage to drop into the cowl.

4. Remove the linkage access cover, located on the right side of the dash panel, near the wiper motor.

5. Reach through the access cover opening and unsnap the wiper motor clip. Push the clip away from the linkage until it clears the nib on the crank pin, then push the clip off the linkage. Remove the linkage from the crank pin.

6. Remove the 3 attaching screws and remove the wiper motor.

To install:

7. Install the motor with the attaching screws. Tighten to 60–85 inch lbs. (6.8–9.6 Nm). Connect the motor electrical connector.

8. Install the clip completely onto the right linkage, making sure it is fully seated. Do not put the linkage on the motor crank pin and then try to install the clip.

9. Install the left and right linkage onto the wiper motor crank pin. Pull the linkage onto the crank pin until it snaps into place. The clip is properly installed if the nib is protruding through the center of the clip.

10. Install the right wiper pivot shaft and nut. Tighten the nut to 84–110 inch lbs. (9.5–12.5 Nm).

11. Connect the negative battery cable and turn the ignition **ON**. Turn the wiper switch **OFF** so the wiper motor will park, then turn the ignition **OFF**. Install the right linkage access cover.

12. Install the right wiper blade and arm assembly and test the system.

Rear

MPV

1. Disconnect the negative battery cable.

2. Remove the wiper arm cover, retaining nut and the wiper arm/blade assembly.

3. Remove the seal cap and outer bushing.

4. Remove the liftgate trim panel and screen.

5. Disconnect the electrical connector, remove the mounting bolts and remove the wiper motor.

To install:

6. Position the wiper motor and install the attaching bolts. Tighten to 61–87 inch lbs. (6.9–9.8 Nm).

7. Connect the electrical connector.

8. Install the screen and liftgate trim panel.

9. Install the seal cap and outer bushing.

10. Connect the negative battery cable. Set the motor shaft to the park position by turning the rear wiper switch from **ON** to **OFF**.

11. Install the wiper arm/blade assembly so the blade is 0.98–1.38 in. (25–35mm) from the lower window moulding.

12. Install the arm retaining nut and tighten to 52–87 inch lbs. (5.9–9.8 Nm). Install the wiper arm cover.

NAVAJO

1. Disconnect the negative battery cable.

2. Remove the wiper arm and blade assembly.

3. Remove the liftgate inner trim panel.

4. Remove the 3 motor bracket attaching screws and pull the motor and bracket assembly out of the rubber grommet.

5. Disconnect the electrical connector and disengage the wiring locator pins. Remove the motor.

To install:

6. Position the motor in the liftgate rubber grommet and install the attaching screws.

7. Connect the electrical connector and install the wiring locator pins in the holes provided.

8. Install the wiper arm and blade assembly. Connect the negative battery cable and check wiper operation.

9. Install the liftgate inner trim panel.

Window Wiper Switch

REMOVAL AND INSTALLATION

Front

The function of the front wipers is controlled by the combination switch.

Rear

MPV

1. Disconnect the negative battery cable.

2. Remove the front cluster panel.

3. Remove the attaching screws from the left cluster switch, disconnect the electrical connectors and remove the rear wiper switch.

4. Installation is the reverse of the removal procedure.

NAVAJO

1. Disconnect the negative battery cable.

2. Remove the 2 ash tray retaining screws and remove the ash tray.

3. Remove the cluster trim panel, which is held on by clips.

4. Remove the snap-in switch mounting bezel containing the switches and disconnect the electrical connector.

5. Remove the switch from the mounting bezel by pushing on the switch from the connector side until the snap-in mounting clips release.

6. Installation is the reverse of the removal procedure.

Instrument Cluster

REMOVAL AND INSTALLATION

B Series Pick-Up

1. Disconnect the negative battery cable.

2. Remove the screws and the instrument cluster hood.

3. Remove the instrument cluster attaching screws.

4. Pull the cluster rearward enough to gain access to the rear of the cluster. Disconnect the electrical connectors and speedometer cable.

5. Remove the instrument cluster.

6. Installation is the reverse of the removal procedure.

MPV

1. Disconnect the negative battery cable.

2. Remove the screws attaching the front cluster assembly. Disconnect the electrical connectors from the cluster assembly switches and remove the cluster assembly.

3. Remove the instrument cluster attaching screws.

4. Pull the cluster rearward enough to gain access to the rear of the cluster. Disconnect the electrical connectors and speedometer cable.

5. If equipped with automatic transmission, remove the lock pin

from the shift position indicator wire end.

6. Remove the instrument cluster.

7. Installation is the reverse of the removal procedure. If equipped with automatic transmission, adjust the shift position indicator as follows:

a. Using moderate force, pull the shift position indicator wire end out fully to set the indicator position. Only pull the wire until the indicator is felt to bottom. Do not use excessive force.

b. Using a small prybar, properly align the spring hook. Pull the indicator wire until it just stops. Be careful not to move the position indicator after aligning.

c. Install a suitable adjusting pin in the position indicator wire end.

d. Install the instrument cluster.

e. Make sure the transmission selector lever is in **P**.

f. Mount the shift position indicator wire housing to the outer bracket using the clip. Hook the position indicator wire to the selector.

g. Turn the ignition switch to **ACC**. Move the selector lever from **P** to **L** to **P**.

h. Remove the adjusting pin and install the lock pin.

Navajo

1. Disconnect the negative battery cable.

2. Open the ash tray and remove the 2 screws attaching the ash tray and instrument cluster trim panel. Remove the ash tray.

3. Unsnap the cluster trim panel by pulling rearward around the edge of the panel. Depress the hazard warning switch on the steering column and remove the cluster trim panel.

4. Remove the 4 screws securing the instrument cluster to the instrument panel.

5. If equipped with automatic transmission, remove the 2 screws attaching the shift position indicator to the cluster and slide the indicator down and out of the cluster. Leave the indicator connections undisturbed.

6. Pull the cluster assembly rearward to gain access to the speedometer cable. Disconnect the cable and the 2 wiring harness connectors and remove the cluster.

NOTE: If there is not enough room to disengage the cable from the speedometer, it may be necessary to disconnect the cable at the transmission and pull the

cable through the cowl, to allow room to reach the speedometer quick disconnect.

7. Installation is the reverse of the removal procedure. Apply an approximately 3/16 in. diameter ball of silicone dielectric compound in the drive hole of the speedometer head prior to installation.

Speedometer

REMOVAL AND INSTALLATION

1. Disconnect the negative battery cable. Remove the instrument cluster.

2. Remove the trip meter and/or clock adjusting knob(s), if necessary.

3. Separate the lens and mask assembly from the cluster.

4. Remove the mounting screws and remove the speedometer.

5. Installation is the reverse of the removal procedure.

Radio

REMOVAL AND INSTALLATION

B Series Pick-Up

1. Disconnect the negative battery cable.

2. Remove the shift lever knob(s). Remove the attaching screws and the front console.

3. Remove the ashtray.

4. Remove the attaching screws and the radio housing.

5. Disconnect the electrical connectors and antenna lead from the radio.

6. Remove the screws from the radio brackets and remove the radio.

7. Installation is the reverse of the removal procedure.

MPV

1. Disconnect the negative battery cable.

2. Remove the radio face plate.

3. Remove the attaching screws and pull the radio rearward enough to gain access to the electrical connectors and antenna cable.

4. Disconnect the electrical connectors and antenna cable and remove the radio.

5. Installation is the reverse of the removal procedure.

Navajo

1. Disconnect the negative battery cable.

2. Remove the finish panel assembly.

3. Insert radio removal tool 49 UN01 050 or equivalent, into the radio face plate. Press in 1 in. (25.4mm) to release the radio retaining clips. Pull the radio from the instrument panel using the tool as handles.

4. Disconnect the electrical connectors and antenna cable. Transfer the rear mounting bracket to the replacement radio, if necessary.

5. Installation is the reverse of the removal procedure.

Headlight Switch

REMOVAL AND INSTALLATION

Navajo

1. Disconnect the negative battery cable.

2. Remove the ash tray and the 2 finish panel retaining screws.

3. Remove the finish panel, which snaps off.

4. Remove the rear wiper switch and rear defrost switch.

5. Pull the headlight switch knob to the full **ON** position. Reach through the opening below the headlight switch, depress the shaft release button on the switch and remove the knob and shaft assembly.

6. Remove the headlight switch retaining bezel nut. Pull the switch downward and through the opening to disconnect the connector.

7. Installation is the reverse of the removal procedure.

Combination Switch

REMOVAL AND INSTALLATION

Except Navajo

1. Disconnect the negative battery cable.

2. Remove the steering wheel.

3. Remove the upper and lower column shroud halves.

4. Disconnect the electrical connectors to the switch.

5. Remove the attaching screw(s), release the lock, if required and remove the switch.

6. Installation is the reverse of the removal procedure.

Navajo

1. Disconnect the negative battery cable.

2. Remove the steering column shroud.

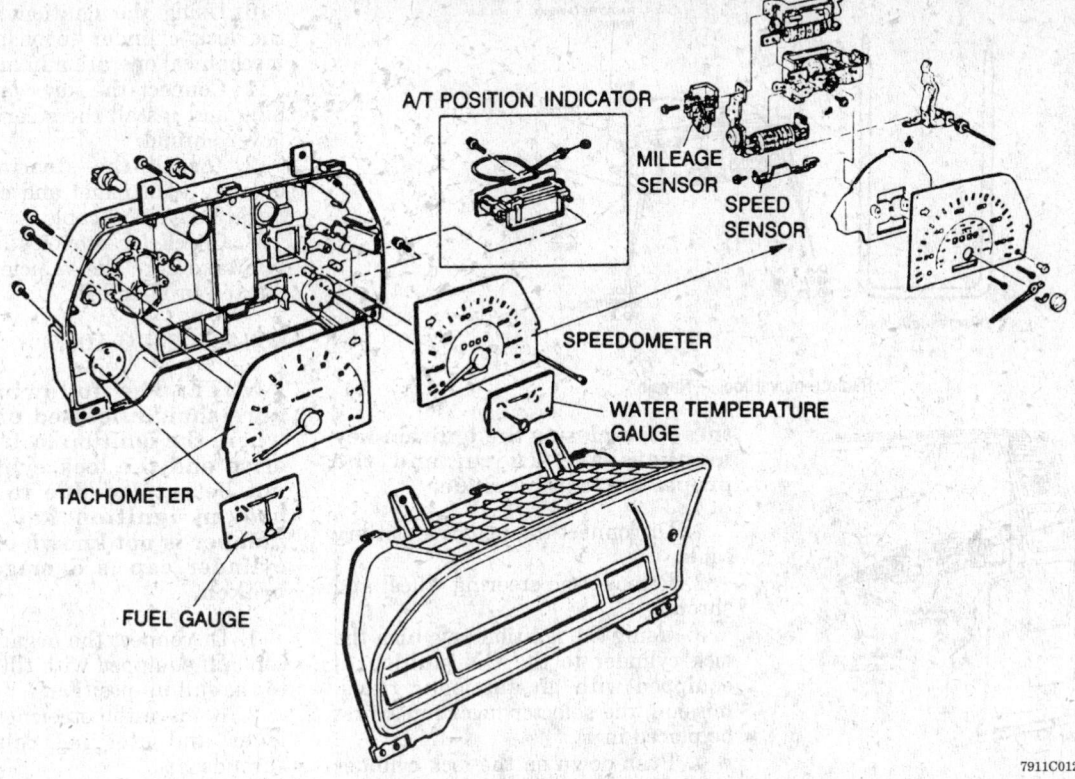

A/T POSITION INDICATOR

MILEAGE SENSOR

SPEED SENSOR

SPEEDOMETER

WATER TEMPERATURE GAUGE

TACHOMETER

FUEL GAUGE

7911C012

Instrument cluster exploded view — MPV

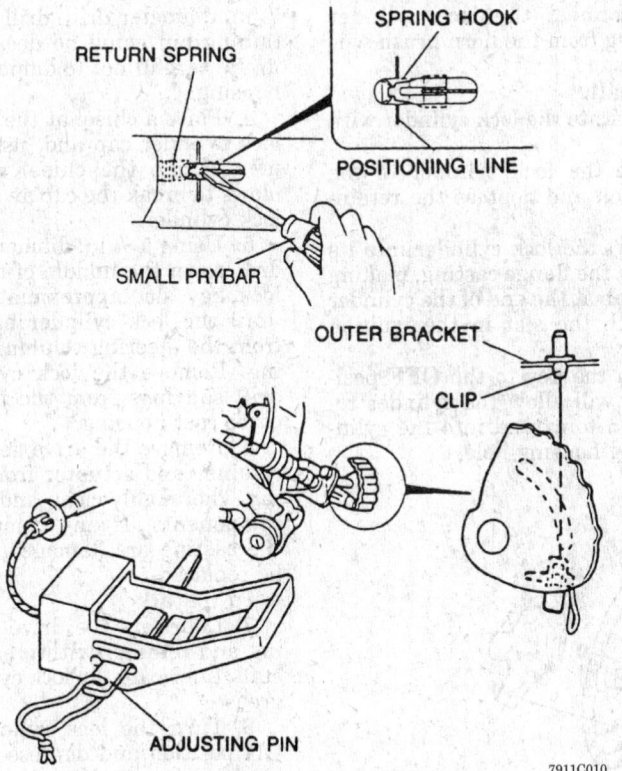

RETURN SPRING

SPRING HOOK

POSITIONING LINE

SMALL PRYBAR

OUTER BRACKET

CLIP

ADJUSTING PIN

7911C010

Shift position indicator adjustment — MPV

3. Remove the 2 self-tapping screws that attach the combination switch to the steering column casting. Disengage the switch from the casting.

4. Disconnect the 3 electrical connectors, being careful not to damage the locking tabs. Do not damage the shift position indicator cable.

To install:

5. Connect the 3 switch electrical connectors. The wiring for the switch is to be routed under the shift position indicator cable.

6. Install the switch with the self-tapping screws. Tighten the screws to 18–27 inch lbs. (2–3 Nm).

7. If equipped with automatic transmission, make sure the shift position indicator adjustment is correct.

8. Install the shroud and connect the negative battery cable. Check the steering column for proper operation.

Ignition Lock

REMOVAL AND INSTALLATION

Except Navajo

1. Disconnect the negative battery cable.

2. Remove the combination switch.

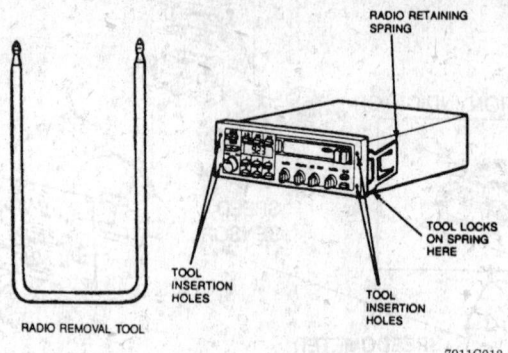

Radio removal tool — Navajo

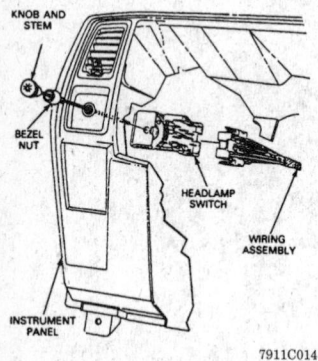

Headlight switch installation — Navajo

3. Use a chisel to make a groove in the head of each lock retaining bolt, then remove the bolts using a suitable tool.

4. Remove the lock assembly.

5. Installation is the reverse of the removal procedure. Install new lock retaining bolts and tighten them until the heads break off.

Navajo

FUNCTIONAL LOCK

NOTE: The following procedure should be used on vehicles with functional lock cylinders. Ignition keys are available for these vehicles or the ignition key numbers are known and the proper key can be made.

1. Disconnect the negative battery cable.

2. Remove the steering wheel and shroud.

3. Using the ignition key, turn the lock cylinder to the **ON** position. If equipped with an automatic transmission, the selector lever must first be placed in **P**.

4. Push down on the lock cylinder retaining pin with a 1/8 in. diameter wire pin or small punch. Pull the lock cylinder from the column housing.

5. Disconnect the lock cylinder wiring plug from the horn brush wiring connector.

To install:

6. Lubricate the lock cylinder with grease.

7. Turn the lock cylinder to the **ON** position and depress the retaining pin.

8. Insert the lock cylinder into its housing in the flange casting, making sure the tab at the end of the cylinder aligns with the slot in the ignition drive gear.

9. Turn the key to the **OFF** position. This will allow the cylinder retaining pin to extend into the cylinder casting housing hole.

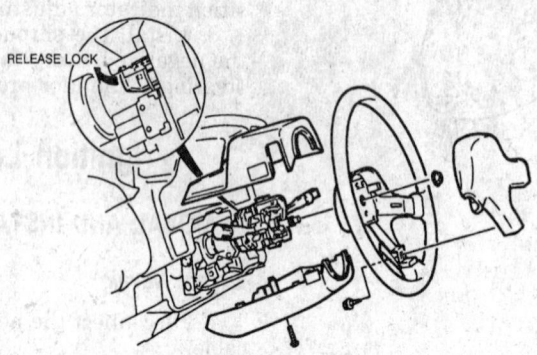

Combination switch removal — MPV; B Series Pick-Up similar

10. Using the ignition key, rotate the lock cylinder to ensure correct mechanical operation in all positions.

11. Connect the key warning wire plug and install the steering column lower shroud.

12. Install the steering column opening trim panel and connect the negative battery cable.

13. Check for proper start in **P** or **N**. Make sure the vehicle cannot be started in **D** and **R**.

NON-FUNCTIONAL LOCK

NOTE: The following procedure should be used on vehicles where the ignition lock is inoperative and the lock cylinder cannot be rotated due to a lost or broken ignition key, the key number is not known or the lock cylinder cap is damaged and/or broken.

1. Disconnect the negative battery cable. If equipped with tilt wheel, tilt to the full up position.

2. Remove the steering wheel, tilt lever and steering column trim shrouds.

3. Punch the lock cylinder retaining pin with a prick punch, 1/8 in. maximum outside diameter. Using a 1/8 in. diameter drill, drill out the retaining pin going no deeper than 1/2 in. Be careful not to damage the cast housing.

4. Place a chisel at the base of the lock cylinder cap and, using a hammer, strike the chisel with sharp blows to break the cap away from the lock cylinder.

5. Using a 3/8 in. diameter drill bit, drill down the middle of the ignition lock key slot approximately 1 3/4 in. until the lock cylinder breaks loose from the steering column cover casting. Remove the lock cylinder and drill shavings from the base of the cover cast housing.

6. Remove the drive gear, bearing retainer and actuator from the casting. Thoroughly clean and inspect all components. If any components or the casting are damaged, they must be replaced.

To install:

7. Lubricate the drive gear, bearing and retainer with grease and install. Lubricate the lock cylinder with grease.

8. Turn the lock cylinder to the **ON** position and depress the retaining pin. Insert the cylinder into its housing in the flange casting, making sure the tab at the end of the cylinder aligns with the slot in the ignition lock drive gear.

9. Turn the key to the **OFF** position. This will allow the cylinder retaining pin to extend into the cylinder casting housing hole.

10. Using the ignition key, rotate the lock cylinder to ensure correct mechanical operation in all positions.

11. Connect the key warning wire plug and install the steering column shrouds.

12. Install the steering wheel and connect the negative battery cable. Check for proper start in **P** or **N**. Make sure the vehicle cannot be started in **D** and **R**.

Ignition Switch

REMOVAL AND INSTALLATION

Except Navajo

1. Disconnect the negative battery cable.

2. Remove the steering column covers.

3. Disconnect the electrical connectors from the switch.

4. Remove the attaching screw and remove the switch.

5. Installation is the reverse of the removal procedure.

Navajo

1. Rotate the lock cylinder to the **LOCK** position. Disconnect the negative battery cable.

2. Remove the steering wheel.

3. If equipped with tilt wheel, remove the upper extension housing shroud by squeezing it at the 6 and 12 o'clock positions and popping it free of the retaining plate at the 3 o'clock position.

4. Remove the trim shroud halves.

5. Disconnect the switch electrical connector.

6. Remove the retaining nuts and disengage the ignition switch from the actuator rod.

7. Installation is the reverse of the removal procedure.

Stoplight Switch

ADJUSTMENT

Except Navajo

1. Disconnect the negative battery cable.

2. Disconnect the electrical connector from the switch.

3. Loosen the switch locknut and turn the switch until it does not contact the pedal.

4. Loosen the booster pushrod locknut and turn the rod to adjust the pedal height to 7.09–7.28 in. (180–185mm) on B Series Pick-Up or 7.52–7.91 in. (191–201mm) on MPV.

5. Depress the brake pedal a few times to eliminate the vacuum in the system. Gently depress the pedal and check the free-play. Turn the booster pushrod to adjust the free-play to 0.16–0.28 in. (4–7mm).

6. Tighten the booster pushrod locknut.

7. Turn the stoplight switch until it contacts the pedal, then turn an additional ½ turn. Tighten the switch locknut.

8. Connect the electrical connector and the negative battery cable.

Navajo

The stoplight switch is not adjustable.

REMOVAL AND INSTALLATION

Except Navajo

1. Disconnect the negative battery cable.

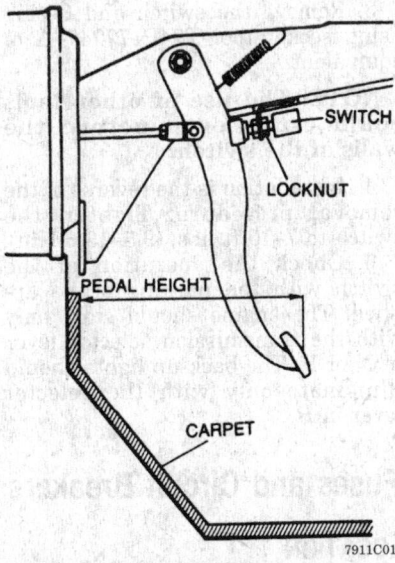

Pedal height measurement location for stoplight switch and clutch switch adjustment

2. Disconnect the electrical connector from the switch.

3. Loosen the switch locknut and remove the switch.

4. Installation is the reverse of the removal procedure. Adjust the switch.

Navajo

1. Disconnect the negative battery cable. Disconnect the electrical connector from the switch. The locking tab must be lifted before the connector can be removed.

2. Remove the hairpin clip and slide the switch, booster pushrod, nylon washer and bushing away from the pedal. Remove the washer, then the switch by sliding the switch up or down.

To install:

3. Position the switch so the U-shaped side is nearest the pedal and directly over/under the pin. Then slide the switch up/down installing the booster pushrod and bushing between the switch side plates.

4. Push the switch and pushrod assembly firmly toward the brake pedal arm. Install the outside plastic washer to the pin and install the hairpin clip. Do not substitute for this clip. Use only factory supplied hairpin clips.

5. Connect the electrical connector to the switch and connect the negative battery cable. Make sure the switch wire harness has sufficient length to travel with the switch during the full stroke of the brake pedal. Check the switch for proper operation.

Clutch Switch

ADJUSTMENT

Except Navajo

1. Disconnect the negative battery cable.

2. Disconnect the electrical connector from the switch.

3. Loosen the switch locknut and turn the switch until the clutch pedal height is 7.13–7.52 in. (181–191mm) on B2200, 7.52–7.91 in. (191–201mm) on B2600 or 8.19–8.58 in. (208–218mm) on MPV.

4. Tighten the locknut.

5. Connect the electrical connector and the negative battery cable.

Navajo

The clutch switch is not adjustable.

REMOVAL AND INSTALLATION

Except Navajo

1. Disconnect the negative battery cable.
2. Disconnect the electrical connector from the switch.
3. Loosen the switch locknut and remove the switch.
4. Installation is the reverse of the removal procedure. Adjust the switch.

Navajo

1. Disconnect the negative battery cable. Disconnect the wiring harness from the switch.
2. Pull the orientation clip away from the switch to separate it from the pin on the switch.
3. Rotate the switch to expose the plastic retainer.
4. Push the tabs together to allow the retainer to slide rearward and separate from the switch.
5. Remove the switch from the pushrod.
6. Installation is the reverse of the removal procedure.

Neutral Safety Switch

ADJUSTMENT

Except Navajo

HYDRAULICALLY-CONTROLLED TRANSMISSION

1. Raise and safely support the vehicle.
2. Move the manual shaft to the **N** position.
3. Loosen the switch mounting bolts.
4. Remove the screw on the switch body and move the switch so the screw hole is aligned with the small hole inside the switch. Check the alignment by inserting an approximately 0.079 in. (2mm) diameter pin through the holes.
5. Tighten the mounting bolts to 43–61 inch lbs. (4.9–6.9 Nm) and remove the alignment pin.
6. Install the screw in the switch body and check for proper operation. The starter should operate only when the transmission is in the **P** or **N** range.

ELECTRONICALLY-CONTROLLED TRANSMISSION

1. Raise and safely support the vehicle.
2. Move the manual shaft to the **N** position.

3. Loosen the switch mounting bolts.
4. Align the holes of the switch and the manual shaft lever and insert an approximately 0.157 in. (4mm) diameter pin through the holes.
5. Tighten the mounting bolts to 22–35 inch lbs. (2.5–3.9 Nm) and remove the pin.
6. Check the switch for proper operation. The starter should operate only when the transmission is in the **P** or **N** range.

Navajo

The neutral safety switch is not adjustable.

REMOVAL AND INSTALLATION

Except Navajo

1. Disconnect the negative battery cable. Raise and safely support the vehicle.
2. Disconnect the manual shaft lever from the transmission.
3. Remove the mounting bolts and the switch.
4. Installation is the reverse of the removal procedure. Adjust the switch.

Navajo

1. Disconnect the negative battery cable.
2. Disconnect the electrical harness from the switch.
3. Remove the switch and O-ring using socket tool T74P–77247–A or equivalent.

NOTE: The use of other tools could crush or puncture the walls of the switch.

4. Installation is the reverse of the removal procedure. Tighten the switch to 7–10 ft. lbs. (9.5–13.6 Nm).
5. Check the operation of the switch with the parking brake applied. The engine should start only with the transmission selector lever in **N** or **P**. The back-up lights should illuminate only with the selector lever in **R**.

Fuses and Circuit Breakers

LOCATION

B Series Pick-Up

The fuse box is located under the instrument panel, to the left of the steering column. There is also a small fuse box under the hood, on the right

inner fender, that contains the main vehicle fuses.

MPV

The main fuse block is located on the right side of the engine compartment. There is also a fuse box located above the left kick panel.

Navajo

The fuse panel is located under the instrument panel to the left of the steering column. There is also a power distribution box located in the engine compartment on the right inner fender, next to the starter relay, which contains mostly high-current fuses.

NOTE: All circuit breakers on Navajo are located in the fuse panel except for a 4.5 amp circuit breaker for the rear wiper/washer and a 22 amp circuit breaker for the high beams which is integral with the headlight switch.

Relays

LOCATION

B Series Pick-Up

Circuit Open Relay — located on the left kick panel.
Daytime Running Lights (DRL) Relay — located on the firewall in the engine compartment on Canadian vehicles.
EGI Main Relay — located at the right front of the engine compartment.
Fuel Cut Relay — located under the left side of the dash.
Horn Relay — located on the right inner fender in the engine compartment.

MPV

Air Conditioning Relay — located on the right inner fender in the engine compartment.
ALL Relay — located on the left inner fender in the engine compartment.
Blower Motor Relay — located on the left inner fender in the engine compartment.
Change Motor Relay No. 1 — located on the firewall in the engine compartment.
Change Motor Relay No. 2 — located on the firewall in the engine compartment.

Circuit Opening Relay — located in the front of the engine compartment.

Condenser Fan Relay — located on the right inner fender in the engine compartment.

Daytime Running Light (DRL) Relay — located on the left inner fender in the engine compartment.

EGI Main Relay — located on the left inner fender in the engine compartment.

Horn Relay — located on the left inner fender in the engine compartment.

Kickdown Relay — located on the left inner fender in the engine compartment.

Rear Blower Motor Relay — located on the rear blower unit.

Rear Cooler Relay No. 1 — located on the rear cooling unit.

Rear Cooler Relay No. 2 — located on the rear cooling unit.

Rear Cooler Relay No. 3 — located on the rear cooling unit.

Rear Wheel Anti-Lock Brake Relay — located on the left inner fender in the engine compartment.

Navajo

Air Conditioning Cutoff Relay — located under the power distribution box at the right fender apron.

Fuel Pump Relay — located under the power distribution box at the right fender apron.

Stop/Turn Signal Relays — located on the left rear quarter panel.

Low Oil Level Relay — located on right side of dash panel.

Starter Relay — located on the fender apron in the right front of the engine compartment.

Tail light Relay — located at the left rear quarter panel.

Computers

LOCATION

The Engine Control Unit (ECU) is located at the right kick panel on B Series Pick-Ups and Navajo or under the center of the dash on MPV.

Flashers

LOCATION

The turn signal/hazard flasher unit is located under the left side of the dash on B Series Pick-Up or behind the driver's side kick panel on MPV. On Navajo, both the turn signal and haz-ard flashers are attached to the fuse panel.

Cruise Control

ADJUSTMENT

Actuator Cable

EXCEPT NAVAJO

1. Remove the clamp.
2. Adjust the nut so the actuator cable free-play is 0.04–0.12 in. (1–3mm) when the cable is pressed lightly.
3. Reinstall the clamp.

NAVAJO

1. Remove the cable retaining clip.
2. Disengage the throttle positioner.
3. Set the engine at hot idle.
4. Pull on the actuator cable to take up any slack. Maintain a light tension on the cable.
5. While holding the cable, insert the cable retaining clip and snap securely.

Vacuum Dump Valve

NAVAJO

1. Firmly hold the brake pedal in the up, released, position.

(B2200)

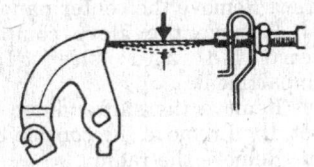

(B2600I)

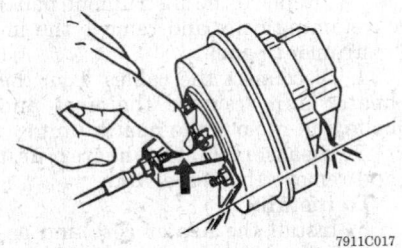

7911C017

Actuator cable free-play adjustment — B series Pick-Up

2. Push in the dump valve until the valve bottoms against the pad on the brake pedal.

Clutch and Stoplight Switches

EXCEPT NAVAJO

When replacing these switches, be sure to adjust the pedal height to the proper specification.

ENGINE COOLING

Radiator

REMOVAL AND INSTALLATION

Except Navajo

1. Disconnect the negative battery cable and remove the radiator cap.

—————— **CAUTION** ——————
Never remove the radiator cap while the engine is running or personal injury from scalding hot coolant or steam may result. If possible, wait until the engine has cooled to remove the radiator cap. If this is not possible, wrap a thick cloth around the radiator cap and turn it slowly to the first stop, to release the pressure in the cooling system. Step back while the pressure is released. After all pressure is released, remove the radiator cap completely.

2. Position a drain pan under the radiator and open the draincock to drain the radiator.
3. On MPV, remove the fresh air duct.
4. Disconnect the upper and lower radiator hoses and the coolant reservoir hose.
5. If equipped with automatic transmission, disconnect and plug the oil cooler lines at the radiator.
6. Remove the cooling fan retaining nuts and the fan.
7. Remove the fan shroud attaching bolts and remove the shroud. Disconnect the thermoswitch electrical connector, if necessary.
8. Remove the radiator attaching bolts and remove the radiator.
   To install:
9. Position the radiator and install the attaching bolts. Tighten to 16–22 ft. lbs. (22–30 Nm).
10. Install the fan shroud with the attaching bolts and tighten to 69–95 inch lbs. (7.8–11.0 Nm). Connect the

thermoswitch electrical connector, if necessary.

11. Install the cooling fan with the retaining nuts and tighten to 69–95 inch lbs. (7.8–11.0 Nm).

12. If equipped with automatic transmission, unplug and connect the oil cooler lines at the radiator.

13. Connect the coolant reservoir hose and the upper and lower radiator hoses.

14. On MPV, install the fresh air duct.

15. Close the draincock and connect the negative battery cable. Fill and bleed the cooling system.

16. Run the engine and check for leaks.

Navajo

1. Disconnect the negative battery cable and remove the radiator cap.

CAUTION

Never remove the radiator cap while the engine is running or personal injury from scalding hot coolant or steam may result. If possible, wait until the engine has cooled to remove the radiator cap. If this is not possible, wrap a thick cloth around the radiator cap and turn it slowly to the first stop, to release the pressure in the cooling system. Step back while the pressure is released. After all pressure is released, remove the radiator cap completely.

---

2. Position a drain pan under the radiator and open the draincock to drain the radiator.

3. Disconnect the overflow hose from the radiator and the fan shroud, if necessary.

4. Remove the shroud or fan guard upper attaching screws. Lift the shroud out of the lower retaining clips and drape it on the fan.

5. Disconnect the radiator hoses from the radiator.

6. Disconnect and plug the automatic transmission oil cooling lines, if equipped.

7. Remove the radiator upper attaching screws, tilt the radiator back and lift directly upward, clear of the radiator support and cooling fan.

To install:

8. Make sure the radiator lower support rubber insulators are in place on the lower support.

9. Install the radiator, being careful to clear the fan. Make sure the mounting pins on the bottom of the radiator tanks are inserted into the holes in the lower support rubber in-sulators and the radiator is firmly seated on the insulators.

10. Install the radiator upper attaching screws. If equipped with automatic transmission, connect the transmission cooling lines.

11. Connect the radiator hoses to the radiator. Position the shroud in the retainer clips and install the attaching screws.

12. Connect the overflow hose and close the draincock. Connect the negative battery cable. Fill and bleed the cooling system.

Heater Core

REMOVAL AND INSTALLATION

B Series Pick-Up

1. Disconnect the negative battery cable. Drain the cooling system.

2. Disconnect the heater hoses from the heater unit and remove the grommet.

3. Remove the instrument panel as follows:

   a. Remove the steering wheel.

   b. Remove the upper and lower column covers and the combination switch.

   c. Remove the instrument cluster hood and the instrument cluster.

   d. Remove the left side panel from the instrument panel.

   e. Remove the hole cover from the upper right corner of the instrument panel center panel to expose a center panel attaching screw. Remove the center panel.

   f. Remove the glove compartment lid and the glove compartment.

   g. Remove the shifter knob and boot, then remove the console box.

   h. Remove the radio.

   i. Remove the instrument panel retaining bolt covers from the top and sides of the instrument panel.

   j. Remove the instrument panel retaining bolts and remove the instrument panel.

4. Disconnect the cables from the heater unit, remove the nuts and bolts and remove the heater unit.

5. Disassemble the heater unit and remove the heater core.

To install:

6. Install the heater core and assemble the heater unit.

7. Install the heater unit and tighten the retaining nuts and bolts to 69–95 inch lbs. (7.8–11.0 Nm). Connect the control cables.

8. Install the instrument panel in the reverse order of removal.

9. Install the grommet and connect the heater hoses.

10. Connect the negative battery cable. Fill and bleed the cooling system.

11. Run the engine and check heater operation. Check for leaks.

MPV

1. Disconnect the negative battery cable. Drain the cooling system.

2. Remove the instrument panel as follows:

   a. Disconnect the hood release knob from the instrument panel.

   b. Remove the steering wheel.

   c. Remove the upper and lower column covers.

   d. Disconnect the electrical connectors from the combination switch. Loosen the screw, release the lock and remove the switch.

   e. Remove the front cluster assembly and the instrument cluster.

   f. Remove the covers from the left and right side of the instrument panel.

   g. If equipped, remove the undercover from the lower right side of the instrument panel.

   h. Remove the right and left lower panels.

   i. Remove the duct from the driver's side under the steering column.

   j. Remove the ashtray and radio trim panel.

   k. Remove the radio housing and the radio.

   l. Remove the knobs and the switch panel assembly.

   m. Remove the temperature and blower controls and the airflow mode control.

   n. Remove the bolt covers, 3 bolts and the upper garnish.

   o. Remove the retaining bolts and remove the instrument panel.

3. Disconnect the heater hoses from the heater core.

4. Remove the nuts and the instrument panel stay.

5. Remove the nuts and the heater unit case. Use caution to prevent spilling coolant from the heater core.

6. Disassemble the heater unit and remove the heater core.

To install:

7. Install the heater core and assemble the heater unit.

8. Install the heater unit case with the attaching nuts.

9. Install the instrument panel stay with the attaching nuts.

10. Connect the heater hoses to the heater core.

11. Install the instrument panel in the reverse order of removal. When

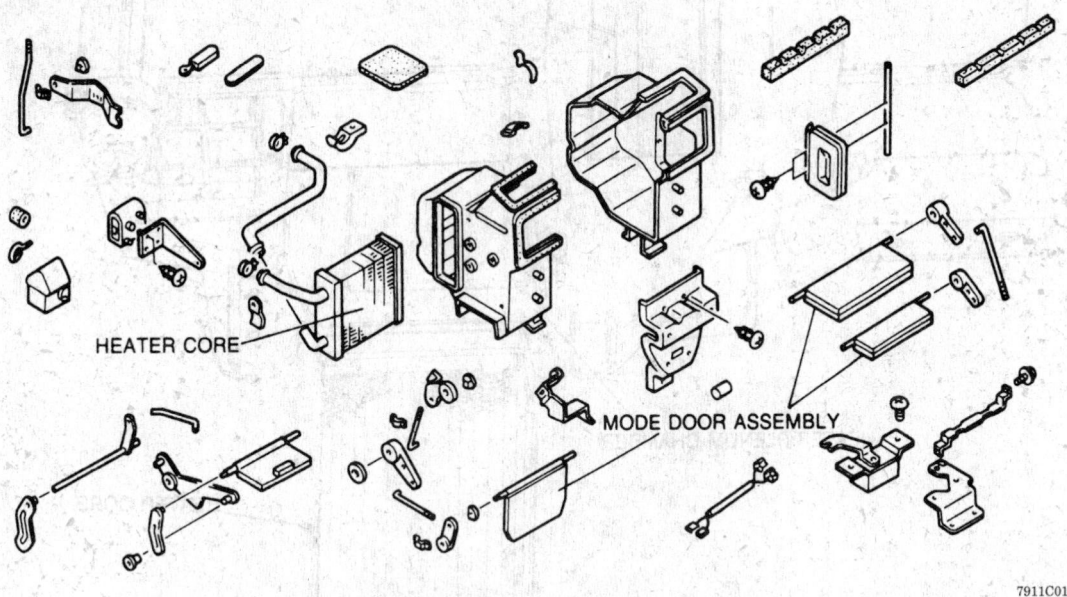

HEATER CORE

MODE DOOR ASSEMBLY

7911C018

Heater unit disassembled view — B Series Pick-Up

attaching the airflow mode control wire, proceed as follows:

   a. Set the control knob to the far left position.

   b. Set the mode control link to the extreme stop position, then install the wire loop on the link.

   c. Clamp the wire into position with the clip.

   d. Move the mode lever to be sure the wire is attached. Make sure it can move the full stroke between DEF and VENT.

12. When attaching the Rec-Fresh control wire, proceed as follows:

   a. Set the control knob to the far right position.

   b. Set the air-mix control link to the extreme stop position, then install the wire loop on the link.

   c. Clamp the wire into position with the clip.

   d. Move the Rec-Fresh lever to be sure the wire is attached. Make sure it can move the full stroke between REC and FRESH.

13. When attaching the temperature control wire, proceed as follows:

   a. Set the control knob to the full clockwise position.

   b. Set the temperature control link to the extreme stop position, then install the wire loop on the link.

   c. Clamp the wire into position with the clip.

   d. Move the temperature control lever to be sure the wire is attached. Make sure it can move the full stroke between HOT and COLD.

14. Connect the negative battery cable. Fill and bleed the cooling system.

15. Run the engine and check heater operation. Check for leaks.

Navajo

1. Disconnect the negative battery cable.

2. Drain the cooling system into a suitable container.

3. Disconnect the heater hoses from the heater core tubes. Use the snap-lock fitting disconnect procedure, if necessary.

4. In the passenger compartment, remove the screws attaching the heater core access cover to the plenum assembly. Remove the access cover.

5. Pull the heater core rearward and down, removing it from the plenum assembly.

   To install:

6. Position the heater core and seal in the plenum assembly.

7. Install the heater core access cover to the plenum assembly and secure it with the screws.

8. Connect the heater hoses to the heater core tubes. Use the snap-lock fitting connection procedure.

9. Fill the cooling system to the proper level.

10. Connect the negative battery cable and check the system for proper operation and coolant leaks.

Water Pump

REMOVAL AND INSTALLATION

2.2L Engine

1. Disconnect the negative battery cable. Drain the cooling system.

2. Remove the cooling fan and the fan shroud.

3. Remove the water pump drive belt and the water pump pulley.

4. Remove the timing belt front covers and the timing belt and tensioner. Remove the timing belt idler pulley.

5. Remove the coolant inlet pipe, if necessary.

6. Remove the mounting bolts and the water pump.

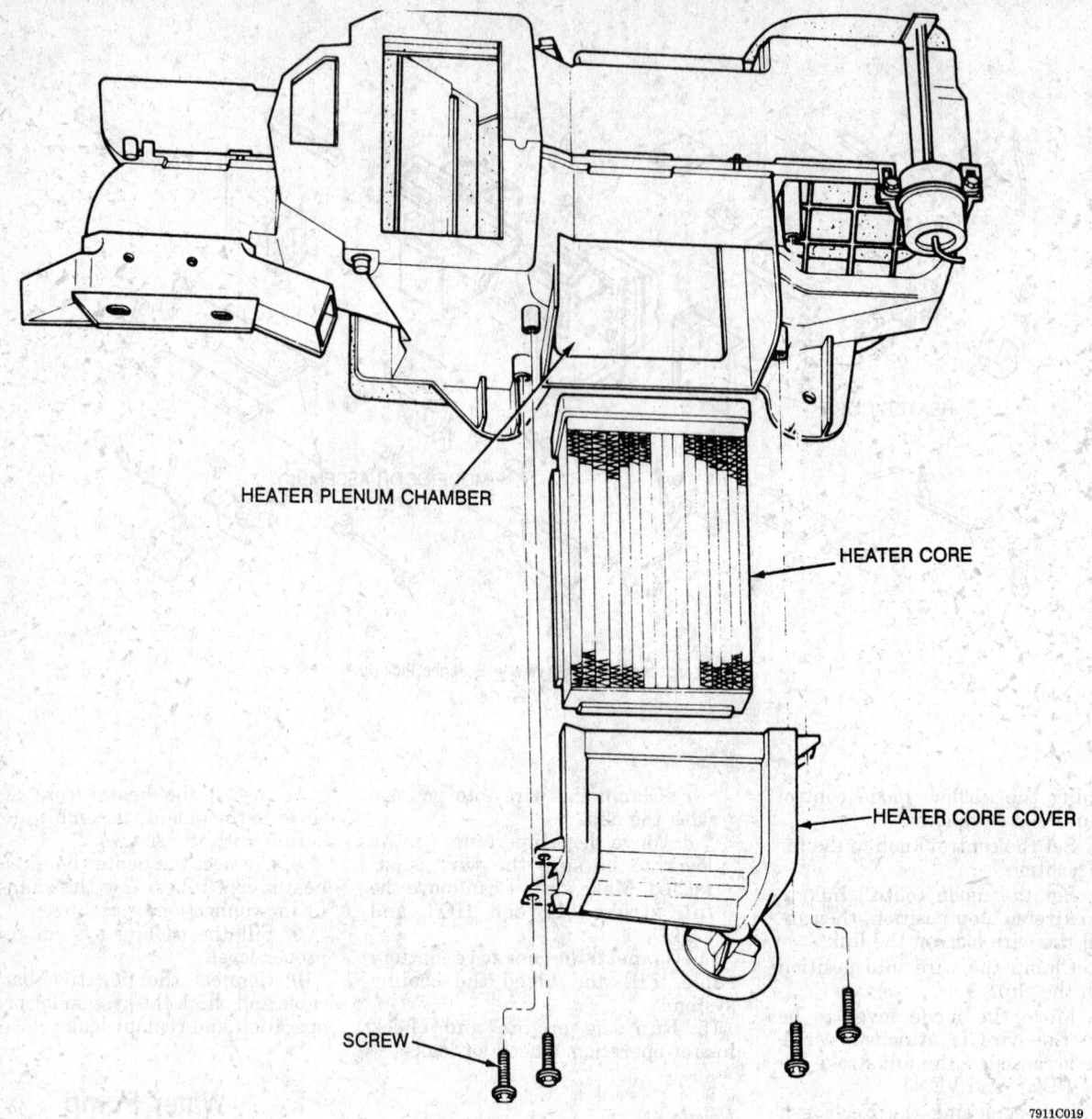

HEATER PLENUM CHAMBER

HEATER CORE

HEATER CORE COVER

SCREW

7911C019

Heater core installation — Navajo

To install:

7. Clean all gasket mating surfaces.

8. Install a new water pump gasket and the water pump. Tighten the water pump mounting bolts to 14–19 ft. lbs. (19–25 Nm).

9. Install a new gasket and the coolant inlet pipe, if removed.

10. Install the timing belt idler pulley, timing belt and timing belt front cover.

11. Install the water pump pulley and drive belt. Install the cooling fan and shroud.

12. Adjust the drive belt tension. Connect the negative battery cable.

13. Fill and bleed the cooling system. Run the engine and check for leaks.

2.6L Engine

1. Disconnect the negative battery cable. Drain the cooling system.

2. Remove the cooling fan and the fan shroud.

3. Remove the water pump drive belt and the water pump pulley.

4. Remove the mounting bolts/nuts and the water pump.

To install:

5. Clean all gasket mating surfaces.

6. Install a new gasket and the water pump. Tighten the mounting bolts/nuts to 14–19 ft. lbs. (19–25 Nm).

7. Connect the water bypass hose, if necessary.

8. Install the water pump pulley and the drive belt. Install the cooling fan and shroud.

9. Adjust the drive belt tension. Connect the negative battery cable.

10. Fill and bleed the cooling system. Run the engine and check for leaks.

3.0L Engine

1. Disconnect the negative battery cable. Drain the cooling system.

2. Remove the fresh air duct.

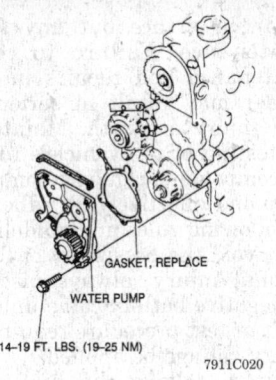

GASKET, REPLACE

WATER PUMP

14–19 FT. LBS. (19–25 NM)

7911C020

Water pump installation — 2.2L engine

3. Remove the cooling fan and the fan shroud. Loosen the drive belt and remove the water pump pulley.

4. Remove the timing belt front covers and the timing belt.

5. Remove the water pump mounting bolts/nuts and the water pump.

To install:

6. Clean all gasket mating surfaces.

7. Install a new gasket and the water pump. Tighten the mounting bolts/nuts to 14–19 ft. lbs. (19–25 Nm).

8. Install the timing belt and the timing belt front covers.

9. Install the water pump pulley and drive belt. Install the cooling fan and shroud.

10. Adjust the drive belt tension. Install the fresh air duct.

11. Connect the negative battery cable. Fill and bleed the cooling system.

12. Run the engine and check for leaks.

4.0L Engine

1. Disconnect the negative battery cable and drain the cooling system.

2. Remove the lower radiator hose and the heater return hose from the pump.

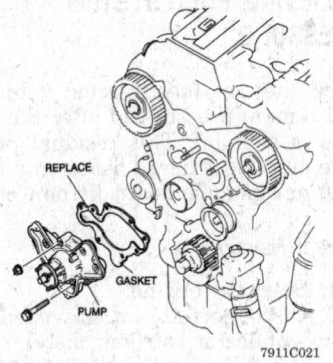

REPLACE

GASKET

PUMP

7911C021

Water pump installation — 3.0L engine

3. Remove the fan and clutch assembly using fan clutch pulley holder 49 UN01 007 and fan clutch nut wrench 49 UN01 008 or equivalents. The fan clutch nut has left hand thread; remove by turning clockwise.

4. Loosen the alternator mounting bolts and remove the belt. If equipped with air conditioning, remove the alternator and bracket.

5. Remove the water pump pulley.

6. Remove the water pump attaching bolts and remove the water pump. Note the length of the bolts when removing, so they can be reinstalled in the same positions.

7. Clean all gasket mating surfaces and install in the reverse order of removal. Tighten the water pump retaining bolts to 7–9 ft. lbs. (9–12 Nm) and the fan clutch nut to 30–100 ft. lbs. (40–135 Nm). Fill and bleed the cooling system.

Thermostat

REMOVAL AND INSTALLATION

Except 4.0L Engine

1. Disconnect the negative battery cable. Drain the cooling system.

2. On 2.2L and 2.6L engines, disconnect the upper radiator hose from the water outlet. On 3.0L engine, disconnect the lower radiator hose from the water outlet.

3. Remove the water outlet and the thermostat.

To install:

4. Clean all gasket mating surfaces. On 3.0L engine, make sure the thermostat O-ring is not damaged.

5. Install the thermostat. If equipped, the jiggle pin should be on the side facing the water outlet.

6. On 2.2L and 2.6L engines, install a new gasket with the seal print facing the cylinder head.

7. Install the water outlet. On 3.0L engine, face the mark on the

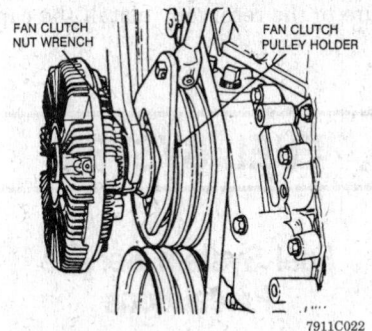

FAN CLUTCH NUT WRENCH   FAN CLUTCH PULLEY HOLDER

7911C022

Fan clutch-to-water pump installation — 4.0L engine

water outlet to the front of the engine.

8. Install the water outlet retaining bolts/nuts and tighten to 14–19 ft. lbs. (19–25 Nm).

9. Connect the radiator hose and the negative battery cable.

10. Fill and bleed the cooling system. Run the engine and check for leaks.

4.0L Engine

1. Disconnect the negative battery cable and drain the cooling system.

2. Disconnect the coolant hose from the water outlet.

3. Remove the water outlet retaining bolts and remove the water outlet. Remove the thermostat from the water outlet.

To install:

4. Clean all gasket mating surfaces.

5. Apply gasket sealer to a new water outlet gasket and install on the engine.

6. Install the thermostat in the water outlet with the bridge section toward the coolant hose. Turn the thermostat clockwise to lock it in position on the flats cast into the water outlet.

NOTE: Make sure the thermostat is aligned so the full width of the heater outlet tube opening is visible within the thermostat port in the assembly. The correct port alignment is required to provide maximum coolant flow to the heater.

7. Install the water outlet with the mounting bolts. Tighten the bolts to 7–10 ft. lbs. (9–13 Nm).

8. Connect the coolant hose.

9. Connect the negative battery cable. Fill and bleed the cooling system.

Cooling System Bleeding

EXCEPT NAVAJO

1. Use the following steps to remove air from the system and ensure a complete fit.

2. Close the radiator draincock and install the cylinder block drain plug, if removed.

3. Slowly pour a 50/50 mixture of water and antifreeze into the radiator up to the coolant filler port.

4. Fill the coolant reservoir with the same mixture up to the FULL level.

5. Install the radiator cap securely and start the engine.

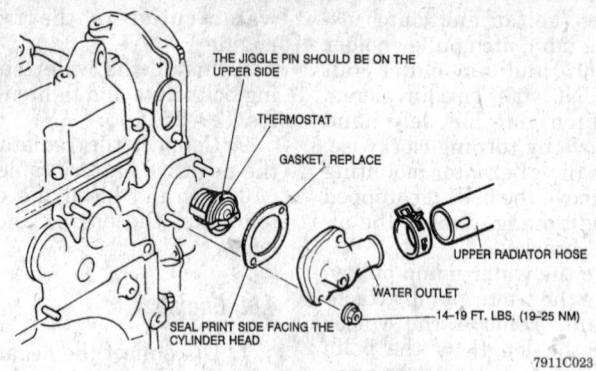

Thermostat installation — 2.2L engine

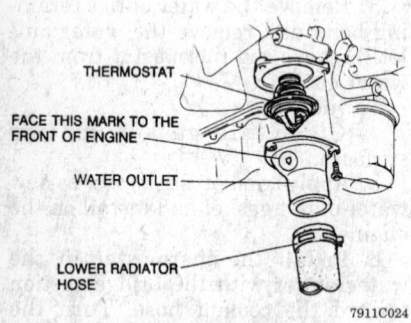

Thermostat installation — 3.0L engine

6. Run the engine at idle speed until it reaches normal operating temperature.

NOTE: If the temperature increases beyond normal, there is excessive air in the system. Stop the engine, allow it to cool and repeat Steps 2–4.

7. Run the engine up to 2200–2800 rpm for 5 seconds and return to idle. Repeat several times.

8. Stop the engine and wait until the system has cooled down. Remove the radiator cap and check the coolant level. If the coolant level has dropped, repeat the procedure from Step 2.

NAVAJO

1. Use the following steps to remove air from the system and ensure a complete fit.

2. Close the radiator draincock and install the cylinder block drain plug, if removed.

3. Fill the cooling system with a 50/50 mixture of water and antifreeze. Allow several minutes for trapped air to escape and for coolant mixture to flow through the radiator.

4. Install the radiator cap to the fully installed position, then back off to the first stop.

5. Slide the heater temperature and mode selection levers to maximum heat position.

6. Start the engine and operate at approximately 2000 rpm for 3–4 minutes.

7. Shut the engine off. Wrap the radiator cap with a thick cloth and carefully remove the cap.

8. Add coolant mixture to bring the coolant level up to the filler neck seat.

—— CAUTION ——
Use caution when adding coolant to the radiator to avoid hot coolant or steam blowing out from the radiator and possibly causing personal injury.

9. Install the radiator cap to the fully installed position, then back off to the first stop.

10. Run the engine at fast idle until the upper radiator hose is warm, indicating the thermostat is open.

11. Shut the engine off. Wrap the radiator cap with a thick cloth and carefully remove the cap. Add coolant, if necessary and reinstall the cap to the fully installed position.

12. Remove the coolant reservoir cap and add 1.1 qts. of coolant mixture to the reservoir. Install the cap.

FUEL SYSTEM

Fuel System Service Precautions

Safety is the most important factor when performing not only fuel system maintenance but any type of maintenance. Failure to conduct maintenance and repairs in a safe manner may result in serious personal injury or death. Maintenance and testing of the vehicle's fuel system components can be accomplished safely and effectively by adhering to the following rules and guidelines.

To avoid the possibility of fire and personal injury, always disconnect the negative battery cable unless the repair or test procedure requires that battery voltage be applied.

Always relieve the fuel system pressure prior to disconnecting any fuel system component (injector, fuel rail, pressure regulator, etc.), fitting or fuel line connection. Exercise extreme caution whenever relieving fuel system pressure to avoid exposing skin, face and eyes to fuel spray. Please be advised that fuel under pressure may penetrate the skin or any part of the body that it contacts.

Always place a shop towel or cloth around the fitting or connection prior to loosening to absorb any excess fuel due to spillage. Ensure that all fuel spillage (should it occur) is quickly removed from engine surfaces. Ensure that all fuel soaked cloths or towels are deposited into a suitable waste container.

Always keep a dry chemical (Class B) fire extinguisher near the work area.

Do not allow fuel spray or fuel vapors to come into contact with a spark or open flame.

Always use a backup wrench when loosening and tightening fuel line connection fittings. This will prevent unnecessary stress and torsion to fuel line piping. Always follow the proper torque specifications.

Always replace worn fuel fitting O-rings with new. Do not substitute fuel hose or equivalent where fuel pipe is installed.

RELIEVING FUEL SYSTEM PRESSURE

Fuel lines on fuel injected vehicles will remain pressurized after the engine is shut off. This residual pressure must be relieved before any fuel lines or components are disconnected.

Except Navajo

1. Start the engine.

2. Disconnect the circuit opening relay connector, airflow meter connector or fuel pump connector.

3. After the engine stalls, turn **OFF** the ignition switch.

4. Reconnect the electrical connector.

NOTE: After releasing fuel system pressure, the system must be primed before starting the engine to avoid excessive cranking. To prime the system, connect the terminals of the yellow 2-pin test connector with a jumper wire and turn the ignition switch ON for approximately 10 seconds. Check for fuel leaks, then turn the ignition switch OFF and remove the jumper wire.

Navajo

1. Disconnect the negative battery cable and remove the fuel filler cap.

2. Remove the cap from the pressure relief valve on the fuel supply manifold. Install pressure gauge 49 UN01 010 or equivalent, to the pressure relief valve.

3. Direct the gauge drain hose into a suitable container and depress the pressure relief button.

4. Remove the gauge and replace the cap on the pressure relief valve.

NOTE: As an alternate method, disconnect the inertia switch and crank the engine for 15–20 seconds until the pressure is relieved.

Fuel Line Couplings

REMOVAL AND INSTALLATION

Navajo

There are 2 methods in use to connect the fuel lines and fuel system components on Navajo, the hairpin clip push connect fitting and the spring lock coupling. Each requires a different procedure to disconnect and connect.
\

Hairpin Clip Push Connect Fitting

1. Inspect the visible internal portion of the fitting for dirt accumulation. If more than a light coating of dust is present, clean the fitting before disassembly.

2. Some adhesion between the seals in the fitting and the tubing will occur with time. To separate, twist the fitting on the tube, then push and pull the fitting until it moves freely on the tube.

3. Remove the hairpin clip from the fitting by first bending and breaking the shipping tab. Next, spread the 2 clip legs by hand about ⅛ in. each to disengage the body and

push the legs into the fitting. Lightly pull the triangular end of the clip and work it clear of the tube and fitting.

NOTE: Do not use hand tools to complete this operation.

4. Grasp the fitting and pull in an axial direction to remove the fitting from the tube. Be careful on 90 degree elbow connectors, as excessive side loading could break the connector body.

5. After disassembly, inspect and clean the tube end sealing surfaces. The tube end should be free of scratches and corrosion that could provide leak paths. Inspect the inside of the fitting for any internal parts such as O-rings and spacers that may have been dislodged from the fitting. Replace any damaged connector.

To install:

6. Install a new connector if damage was found. Insert a new clip into any 2 adjacent openings with the triangular portion pointing away from the fitting opening. Install the clip until the legs of the clip are locked on the outside of the body. Piloting with an index finger is necessary.

7. Before installing the fitting on the tube, wipe the tube end with a clean cloth. Inspect the inside of fitting to make sure it is free of dirt and/or obstructions.

8. Apply a light coating of engine oil to the tube end. Align the fitting and tube axially and push the fitting onto the tube end. When the fitting is engaged, a definite click will be heard. Pull on the fitting to make sure it is fully engaged.

Spring Lock Coupling

The spring lock coupling is a fuel line coupling held together by a garter spring inside a circular cage. When the coupling is connected together, the flared end of the female fitting slips behind the garter spring inside the cage of the male fitting. The gar-

ter spring and cage then prevent the flared end of the female fitting from pulling out of the cage. As an additional locking feature, most couplings have a horseshoe shaped retaining clip that improves the retaining reliability of the spring lock coupling.

Fuel Tank

REMOVAL AND INSTALLATION

B Series Pick-Up

1. Relieve the fuel system pressure and disconnect the negative battery cable.

2. Remove the fuel filler cap. Raise and safely support the vehicle.

3. Position a suitable container under the fuel tank. Remove the drain plug and drain the tank.

4. Disconnect the electrical connector from the sending unit or sending unit/fuel pump assembly.

5. Disconnect the fuel filler hose, evaporative hoses, breather hose and fuel lines.

6. Position a jack under the fuel tank and remove the tank attaching nuts. Lower the tank from the vehicle.

To install:

7. Raise the tank into position and install the attaching nuts. Remove the jack.

8. Connect the fuel lines and evaporative hoses, making sure they are pushed onto the fuel tank fittings at least 1 in. (25mm). Connect the breather hose.

9. Connect the fuel filler hose, making sure the hose is pushed onto the fuel tank pipe and filler pipe at least 1.4 in. (35mm).

10. Connect the electrical connector to the sending unit or sending unit/fuel pump assembly.

11. Install the drain plug and lower the vehicle.

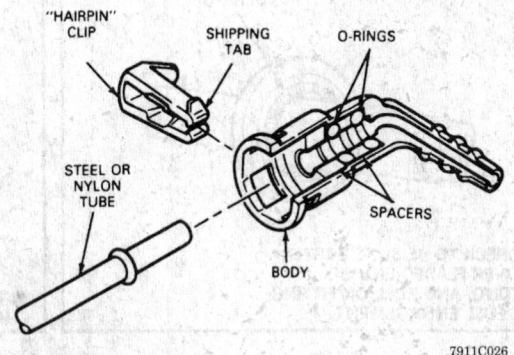

7911C026

Hairpin clip push connect fitting — Navajo

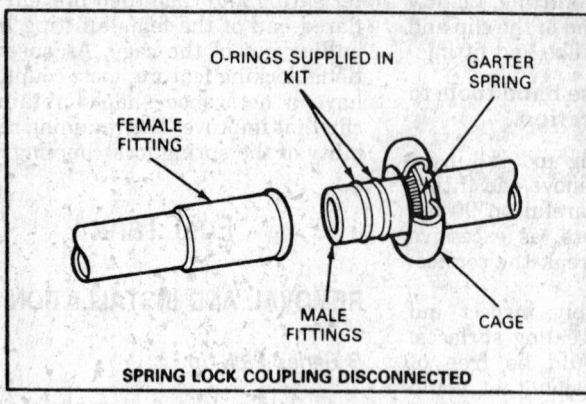

FEMALE FITTING

O-RINGS SUPPLIED IN KIT

GARTER SPRING

MALE FITTINGS

CAGE

SPRING LOCK COUPLING DISCONNECTED

TO DISCONNECT COUPLING
CAUTION — DISCHARGE SYSTEM BEFORE DISCONNECTING COUPLING

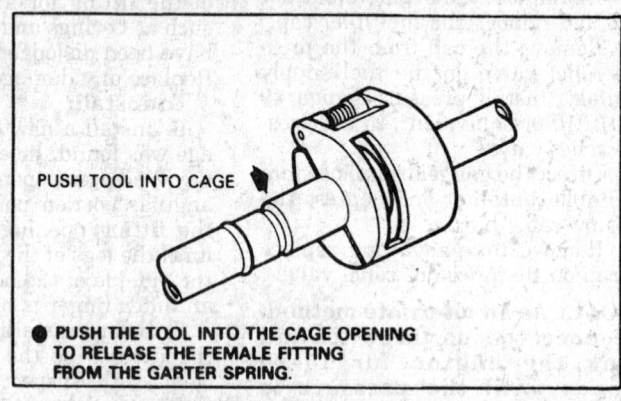

TOOL

CAGE

● FIT TOOL TO COUPLING SO THAT TOOL CAN ENTER CAGE TO RELEASE THE GARTER SPRING.

TO CONNECT COUPLING

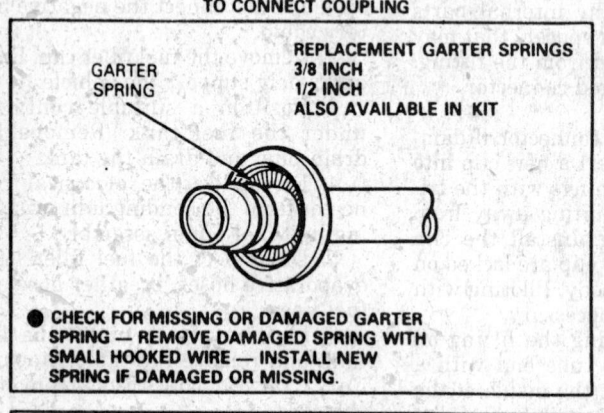

GARTER SPRING

REPLACEMENT GARTER SPRINGS
3/8 INCH
1/2 INCH
ALSO AVAILABLE IN KIT

● CHECK FOR MISSING OR DAMAGED GARTER SPRING — REMOVE DAMAGED SPRING WITH SMALL HOOKED WIRE — INSTALL NEW SPRING IF DAMAGED OR MISSING.

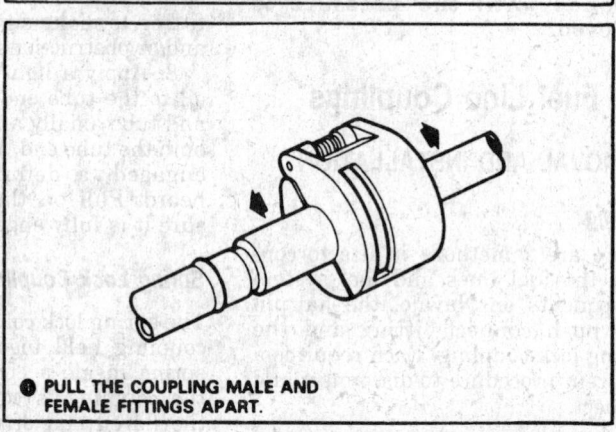

PUSH TOOL INTO CAGE

● PUSH THE TOOL INTO THE CAGE OPENING TO RELEASE THE FEMALE FITTING FROM THE GARTER SPRING.

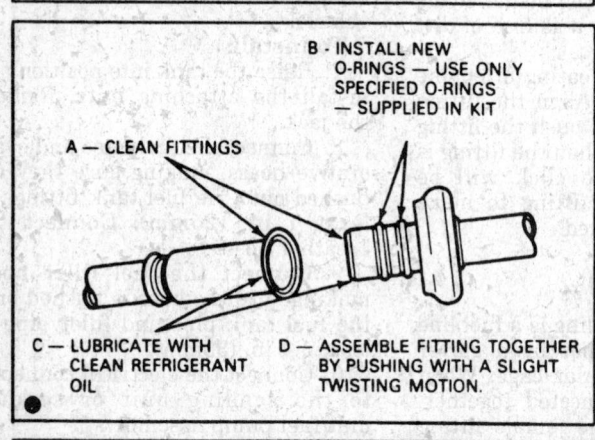

A — CLEAN FITTINGS

B - INSTALL NEW O-RINGS — USE ONLY SPECIFIED O-RINGS — SUPPLIED IN KIT

C — LUBRICATE WITH CLEAN REFRIGERANT OIL

D — ASSEMBLE FITTING TOGETHER BY PUSHING WITH A SLIGHT TWISTING MOTION.

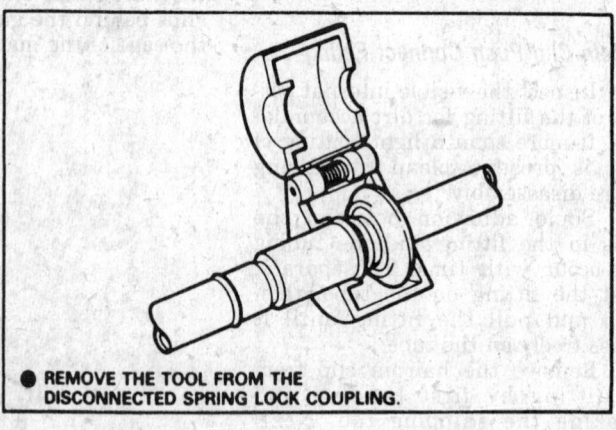

● PULL THE COUPLING MALE AND FEMALE FITTINGS APART.

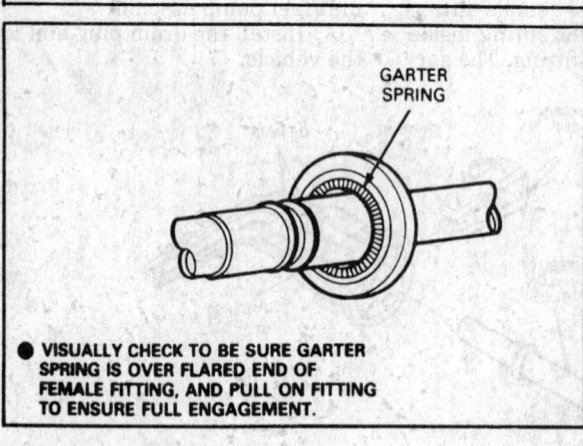

GARTER SPRING

● VISUALLY CHECK TO BE SURE GARTER SPRING IS OVER FLARED END OF FEMALE FITTING, AND PULL ON FITTING TO ENSURE FULL ENGAGEMENT.

● REMOVE THE TOOL FROM THE DISCONNECTED SPRING LOCK COUPLING.

7911C025

Spring lock coupling connect and disconnect procedure

12. Fill the fuel tank and install the filler cap. Check for leaks.

13. Start the engine and check for leaks.

MPV

1. Relieve the fuel system pressure and disconnect the negative battery cable.

2. Remove the fuel filler cap. Raise and safely support the vehicle.

3. Position a suitable container under the fuel tank. Remove the drain plug and drain the tank.

4. Disconnect the fuel pump electrical connector.

5. Disconnect the fuel lines, evaporative hoses, breather hose and fuel filler hose.

6. Support the tank with a jack. Remove the retaining bolts and the fuel tank straps.

7. Lower the fuel tank from the vehicle.

To install:

8. Raise the fuel tank into position and install the straps and retaining bolts. Tighten to 32–44 ft. lbs. (43–61 Nm). Remove the jack.

9. Connect the fuel lines and evaporative hoses, making sure they are pushed onto the fuel tank fittings at least 1 in. (25mm). Connect the breather hose.

10. Connect the fuel filler hose, making sure the hose is pushed onto the fuel tank pipe and filler pipe at least 1.4 in. (35mm).

11. Connect the fuel pump electrical connector.

12. Install the drain plug and lower the vehicle.

13. Fill the fuel tank and install the filler cap. Check for leaks.

14. Start the engine and check for leaks.

Navajo

1. Disconnect the negative battery cable and relieve the fuel system pressure.

2. Raise and safely support the vehicle.

3. Drain the fuel from the fuel tank.

4. Remove the shield, skid plate and fuel tank front strap.

5. Support the tank with a jack and remove the bolt from the fuel tank rear strap.

6. Disconnect the filler pipe and vent pipe and lower the tank. Disconnect the vapor hose, fuel lines and electrical connector.

7. Lower the tank from the vehicle.

To install:

8. Raise the fuel tank and connect the electrical connector, fuel lines and vapor hose.

9. Connect the filler pipe and vent pipe. Attach the rear fuel tank strap.

10. Install the shield, skid plate and front strap.

11. Remove the jack and lower the vehicle.

12. Fill the fuel tank and check for leaks. Connect the negative battery cable.

Fuel Filter

REMOVAL AND INSTALLATION

B Series Pick-Up

The fuel filter is located in the engine compartment.

1. Relieve the fuel system pressure. Disconnect the negative battery cable.

2. Raise and safely support the vehicle, if necessary.

3. Disconnect the fuel lines from the fuel filter.

4. Remove the fuel filter or, if equipped, remove the fuel filter and bracket assembly.

5. Installation is the reverse of the removal procedure. Make sure the flow arrow on the fuel filter is facing in the proper direction of fuel flow.

MPV

1. The fuel filter is located in the engine compartment, next to the pulsation damp

2. Relieve the fuel system pressure. Disconnect the negative battery cable.

3. Disconnect the fuel lines from the filter.

4. Remove the filter bracket bolts and remove the filter and bracket assembly.

5. Remove the fuel filter from the mounting bracket, if necessary.

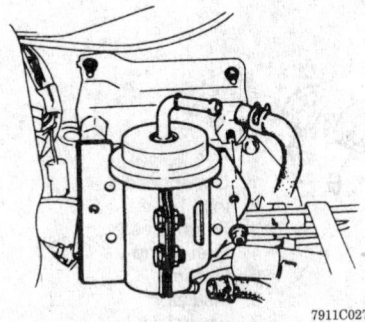

7911C027

Fuel filter location — B Series Pick-Up with fuel injection

6. Installation is the reverse of the removal procedure. Make sure the flow arrow on the fuel filter is facing in the proper direction of fuel flow.

Navajo

1. The fuel filter is located on the underside of the vehicle, attached to the frame rail.

2. Disconnect the negative battery cable and relieve the fuel system pressure.

3. Raise and support the vehicle safely.

4. Disconnect the fuel lines from the fuel filter.

5. Remove the fuel filter from the bracket. Note the direction of the flow arrow so the replacement filter can be installed correctly.

6. Installation is the reverse of the removal procedure. Start the engine and check for leaks.

Mechanical Fuel Pump

PRESSURE TESTING

1. Disconnect the negative battery cable.

2. Disconnect the fuel line from the carburetor.

3. Connect a fuel pressure gauge to the fuel line.

4. Disconnect the fuel return hose from the fuel pump and plug the fuel pump return outlet.

5. Connect the negative battery cable and start the engine.

6. Check the fuel pressure while the engine is idling. The fuel pressure should be 3.7–4.7 psi.

7. Replace the pump if the pressure is not as specified.

8. Shut off the engine and disconnect the negative battery cable. Remove the fuel pressure gauge and reconnect the fuel lines.

9. Connect the negative battery cable.

REMOVAL AND INSTALLATION

1. Disconnect the negative battery cable.

2. Disconnect and plug the inlet, outlet and return hoses at the fuel pump.

3. Remove the fuel pump mounting bolts and remove the fuel pump, insulator and gaskets.

To install:

4. Clean all gasket mating surfaces.

5. Install new gaskets and the insulator and fuel pump. Install the

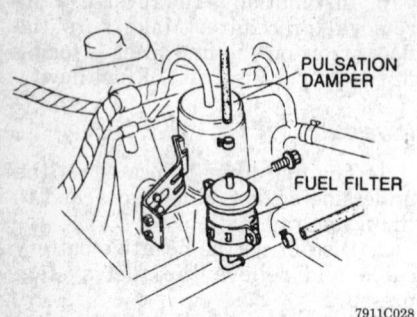

Fuel filter location — MPV

7911C028

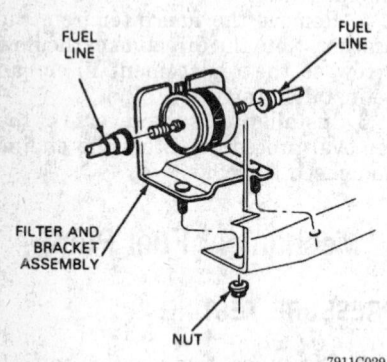

Fuel filter location — Navajo

7911C029

mounting bolts and tighten to 14–19 ft. lbs. (19–25 Nm).

6. Unplug and connect the fuel lines.

7. Connect the negative battery cable, start the engine and check for leaks.

Electric Fuel Pump

PRESSURE TESTING

Carbureted Engines

1. Turn the ignition switch **OFF** and disconnect the negative battery cab

2. Disconnect the main fuel hose and connect a pressure gauge to it.

3. Connect the negative battery cable. Connect a jumper wire between the **B** and **D** terminals of the fuel pump control unit.

4. Turn the ignition switch **ON** and check the fuel pressure. It should be 2.8–3.6 psi.

5. If the fuel pressure is not as specified, replace the fuel pump.

6. Turn the ignition switch **OFF** and disconnect the negative battery cable.

7. Remove the fuel pressure gauge and reconnect the main fuel hose. Remove the jumper wire from the fuel pump control unit.

8. Connect the negative battery cable.

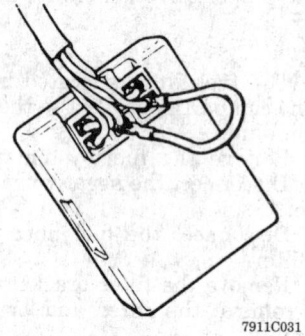

7911C031

Connecting the jumper wire to the fuel pump control unit

Fuel Injected Engines

EXCEPT NAVAJO

1. Relieve the fuel system pressure and disconnect the negative battery cable.

2. Disconnect the fuel line from the fuel filter outlet. Connect a fuel pressure gauge to the fuel filter outlet.

3. Connect the negative battery cable. Connect the terminals of the yellow 2-pin test connector with a jumper wire.

4. Turn the ignition switch **ON** to operate the fuel pump and check the fuel pressure. It should be 64–85 psi.

5. If the fuel pressure is not as specified, replace the fuel pump.

6. Turn the ignition switch **OFF** and disconnect the negative battery cable. Remove the jumper wire from the test connector.

7. Remove the fuel pressure gauge and reconnect the fuel line to the fuel filter outlet.

8. Connect the negative battery cable.

NAVAJO

1. Make sure there is an adequate fuel supply.

2. Relieve the fuel system pressure.

3. Turn the ignition key **OFF**.

4. Connect a suitable fuel pressure gauge to the schrader valve on the fuel rail.

5. Install a test lead to the **FP** terminal on the VIP test connector.

6. Turn the ignition key to the **RUN** position, then ground the test lead to run the fuel pump.

7. Observe the fuel pressure reading on the pressure gauge. The fuel pressure should be 35–45 psi.

8. Relieve the fuel system pressure and turn the ignition key **OFF**. Remove the fuel pressure gauge and the test lead.

REMOVAL AND INSTALLATION

NOTE: The fuel pump is located inside the fuel tank, attached to the tank sending unit assembly.

B Series Pick-Up

1. Relieve the fuel system pressure and disconnect the negative battery cable.

2. Remove the fuel tank.

3. Remove any dirt that has accumulated around the sending unit/fuel pump assembly so it will not enter the fuel tank during removal and installation.

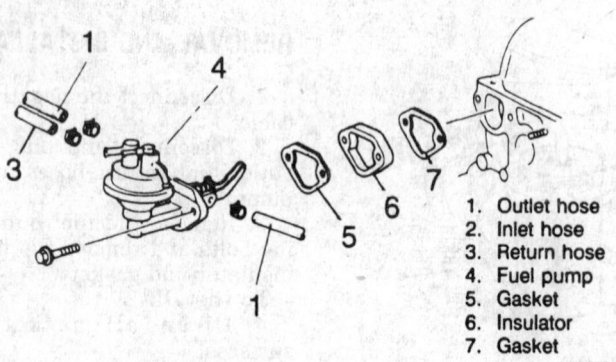

1. Outlet hose
2. Inlet hose
3. Return hose
4. Fuel pump
5. Gasket
6. Insulator
7. Gasket

7911C030

Mechanical fuel pump installation

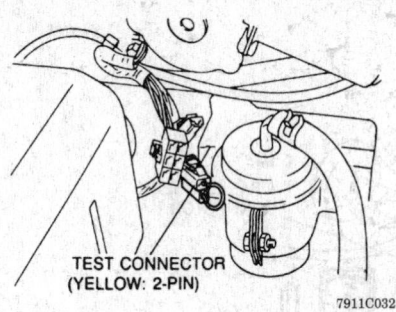

TEST CONNECTOR
(YELLOW: 2-PIN)

7911C032

Test connector location — B Series Pick-Up

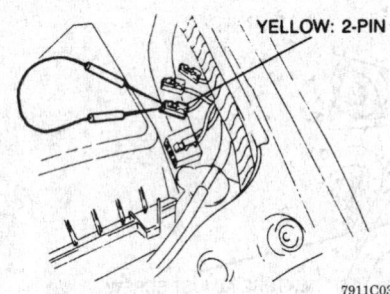

YELLOW: 2-PIN

7911C033

Test connector location — MPV

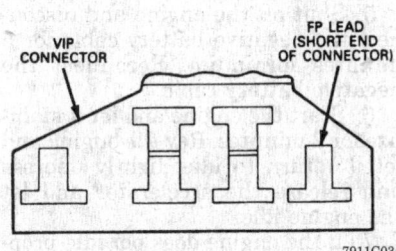

VIP
CONNECTOR

FP LEAD
(SHORT END
OF CONNECTOR)

7911C034

FP terminal location on VIP connector — Navajo

4. Remove the attaching screws and remove the sending unit/fuel pump assembly.

5. If necessary, disconnect the electrical connectors and the fuel hose and remove the pump from the sending unit assembly.

6. Installation is the reverse of the removal procedure. Be sure to install a new seal rubber gasket.

MPV

1. Relieve the fuel system pressure and disconnect the negative battery cable.

2. Remove the rear seat and lift up the rear floormat. Remove the fuel pump cover.

3. Disconnect the sending unit/fuel pump assembly electrical connector and the fuel lines.

4. Remove any dirt that has accumulated around the sending unit/fuel pump assembly so it will not enter the fuel tank during removal and installation.

5. Remove the attaching screws and remove the sending unit/fuel pump assembly.

6. If necessary, disconnect the electrical connectors and the fuel hose and remove the pump from the sending unit assembly.

7. Installation is the reverse of the removal procedure. Be sure to install a new seal rubber gasket.

Navajo

1. Disconnect the negative battery cable and relieve the fuel system pressure.

2. Raise and safely support the vehicle.

3. Remove the fuel tank.

4. Remove any dirt that has accumulated around the fuel pump attaching flange so it will not enter the fuel tank during removal and installation.

5. Turn the fuel pump locking ring counterclockwise using a suitable tool. Remove the locking ring.

6. Remove the fuel pump and discard the seal ring. Separate the fuel pump from the sending unit, if required.

To install:

7. Clean the fuel pump mounting flange and tank mounting surface and seal ring groove.

8. Apply a light coating of Molybdenum grease on a new seal ring and install it in the groove.

9. Install the fuel pump to the sending unit, if removed. Install the fuel pump assembly in the tank, making sure the locating keys are in the keyways and the seal ring is in place.

10. Hold the fuel pump assembly and the seal ring in place and install the locking ring. Rotate the ring clockwise using a suitable tool. Tighten the locking ring to 40–45 ft. lbs. (54–61 Nm).

11. Install the fuel tank in the vehicle.

12. Lower the vehicle and fill the fuel tank with at least 10 gallons of fuel. Connect the negative battery cable. Turn the ignition key to **RUN** for 3 seconds repeatedly, 5–10 times,

to pressurize the system. Check for leaks.

13. Start the engine and check for leaks.

Carburetor

REMOVAL AND INSTALLATION

1. Disconnect the negative battery cable.

2. Remove the air cleaner assembly.

3. Disconnect the accelerator cable and, if equipped, the cruise control cable.

4. Tag and disconnect the vacuum hoses and electrical connectors.

5. Disconnect the fuel lines.

6. Remove the attaching nuts and remove the carburetor.

7. Installation is the reverse of the removal procedure. Be sure to use a new gasket. Tighten the carburetor attaching nuts in a criss-cross pattern.

IDLE SPEED ADJUSTMENT

1. Make sure the ignition timing, spark plugs, and carburetor float level are all in normal operating condition. Turn off all lights and other unnecessary electrical loads.

2. Connect a tachometer to the engine.

3. Start the engine and allow it to reach normal operating temperature. Make sure the choke valve has fully opened.

4. Check the idle speed. If necessary, adjust it to specification by turning the throttle adjusting screw. The idle speed should be 800–850 rpm with manual transmission in neutral or automatic transmission in **P**.

IDLE MIXTURE ADJUSTMENT

2.2L Engine

1. Start the engine and allow it to reach normal operating temperature. Let the engine run at idle.

2. Connect a dwell meter (90 degrees, 4 cylinder) to the air/fuel check connector (Br/Y).

3. Check the idle mixture at the specified idle speed. The idle mixture should be 20–70 degrees. If the idle mixture is not as specified, adjust as follows:

a. Remove the carburetor and knock out the spring pin. Reinstall the carburetor.

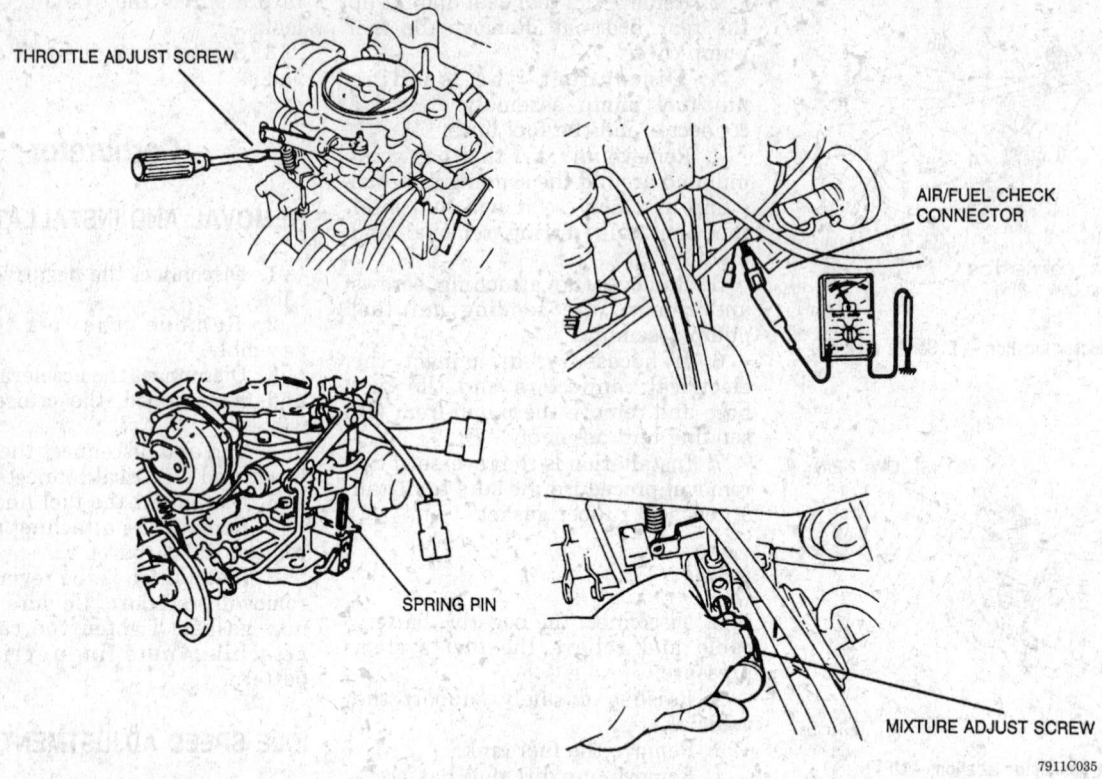

THROTTLE ADJUST SCREW

AIR/FUEL CHECK CONNECTOR

SPRING PIN

MIXTURE ADJUST SCREW

7911C035

Idle speed and idle mixture adjustment — 2.2L engine

b. Install the air cleaner and make sure the idle compensator is closed. Make sure all vacuum hoses are properly connected.

c. Connect a tachometer to the engine.

d. Warm the engine and run it at idle. Make sure the idle speed is correct.

e. Reconnect the dwell meter to the air/fuel check connector.

f. Adjust the idle mixture to 27–45 degrees by turning the mixture adjust screw.

g. Tap in the spring pin.

Service Adjustments

For all carburetor service adjustment procedures and specifications, please refer to "Carburetor Service" in the Unit Repair section.

Fuel Injection

IDLE SPEED ADJUSTMENT

Except Navajo

1. Place manual transmission in neutral or automatic transmission in **P**.

2. Make sure all accessories are **OFF**.

3. Connect a tachometer and timing light to the engine.

4. Warm the engine to normal operating temperature.

5. Check the ignition timing and adjust, if necessary.

6. Ground the green 1-pin test connector to the body with a jumper wire.

7. Check the idle speed. Specifications are as follows: 2.2L and 2.6L engines — Manual transmission: 730–770 rpm. Automatic transmission: 750–790 rpm. 3.0L engine — Manual and automatic transmission: 780–820 rpm.

8. If the idle speed is not within specification, adjust by turning the air adjusting screw.

9. After adjustment, disconnect the jumper wire from the test connector. Recheck the ignition timing.

Navajo

1. Place manual transmission in neutral or automatic transmission in **P**. Apply the parking brake.

2. Make sure the heater and accessories are **OFF**.

3. Start the engine and bring to normal operating temperature. Make sure the throttle lever is resting on the throttle plate stop screw.

4. Check the ignition timing and adjust, if necessary.

5. Shut off the engine and disconnect the negative battery cable for 5 minutes minimum. Reconnect the negative battery cable.

6. Start the engine and let it stabilize for 2 minutes. Rev the engine and let it return to idle, lightly depress and release the accelerator and let the engine idle.

7. If the engine does not idle properly, shut off the engine and disconnect the idle speed control-air bypass solenoid.

8. Run the engine at 2500 rpm for 30 seconds, then let it idle for 2 minutes.

9. Check/adjust the idle rpm to 675 rpm by turning the throttle plate stop screw.

NOTE: If the screw must be turned in, shut the engine off and make the estimated adjustment, then start the engine and repeat Steps 8 and 9.

10. Shut the engine off and repeat Steps 8 and 9.

11. Shut the engine off and disconnect the negative battery cable for 5 minutes minimum.

12. With the engine off, reconnect the idle speed control-air bypass solenoid. Make sure the throttle is not

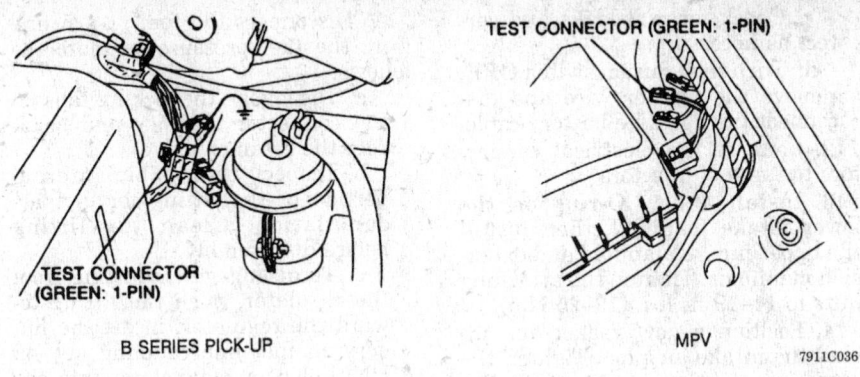

Test connector locations

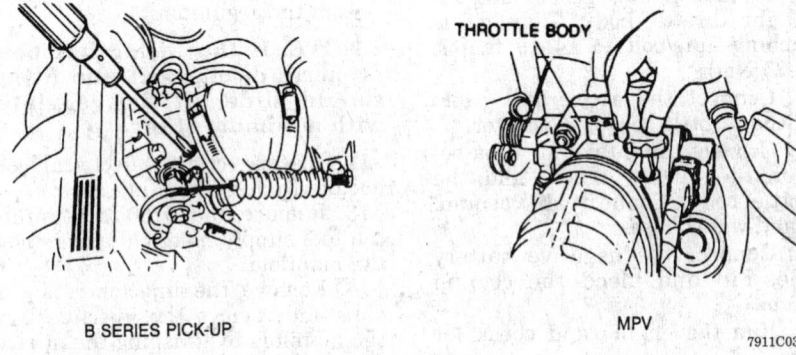

Air adjusting screw locations

stuck in the bore and the linkage is not preventing the throttle from closing.

13. Start the engine and let it stabilize for 2 minutes. Rev the engine and let it return to idle, lightly depress and release the accelerator and let the engine idle.

NOTE: A condition may occur where the engine rpm will oscillate. This can be caused by the throttle plates being open enough to allow purge flow. To make sure of this condition, disconnect the carbon canister purge line and plug it. If purge is present, the throttle plates must be closed until the purge flow induced idle oscillations stop.

Idle Mixture Adjustment

The idle mixture is controlled by the Electronic Control Unit (ECU) and is not adjustable.

Fuel Injector

REMOVAL AND INSTALLATION

2.2L and 2.6L Engines

1. Relieve the fuel system pressure and disconnect the negative battery cable. Drain the cooling system.
2. Disconnect the intake air hose and the ventilation hose.
3. Remove the air pipe and resonance chamber on 2.6L engine.
4. Disconnect the accelerator cable from the throttle lever.
5. Disconnect the water hoses.
6. Tag and disconnect the electrical connectors from the idle speed control valve, throttle sensor and idle switch.
7. Remove the attaching nuts and the throttle body.
8. Remove the upper intake manifold brackets.
9. Tag and disconnect the vacuum hoses and PCV hose. Tag and disconnect the intake air thermosensor connector and ground wire.
10. Remove the injector harness bracket. Remove the nuts and bolts and the upper intake manifold.

11. Disconnect the vacuum hose and fuel lines from the fuel supply manifold.
12. Remove the fuel supply manifold with the pressure regulator.
13. Disconnect the electrical connector(s) from the injector(s). Remove the grommet(s), injector(s) and insulator(s).

To install:

14. Clean all gasket mating surfaces.
15. Apply a small amount of clean engine oil to new O-ring(s) and install on the injector(s). Install the injector(s) and insulator(s) and connect the electrical connector(s).
16. Install the fuel supply manifold and tighten the attaching bolts to 14–19 ft. lbs. (19–25 Nm).
17. Connect the fuel lines and vacuum hose.
18. Position a new gasket on the lower intake manifold and install the upper intake manifold. Tighten the attaching nuts and bolts to 14–19 ft. lbs. (19–25 Nm).
19. Install the injector harness bracket.
20. Connect the ground wire and the intake air thermosensor connector. Connect the PCV hose and the vacuum hoses.
21. Install the upper intake manifold brackets and tighten the bolts to 14–19 ft. lbs. (19–25 Nm).
22. Position a new gasket and install the throttle body. Tighten the attaching nuts to 14–19 ft. lbs. (19–25 Nm).
23. Connect the electrical connectors to the idle speed control valve, throttle sensor and idle switch.
24. Connect the water hoses. Connect the accelerator cable to the throttle lever.
25. Install the air pipe and resonance chamber, if equipped.
26. Connect the ventilation hose and intake air hose.
27. Connect the negative battery cable. Fill and bleed the cooling system.
28. Run the engine and check for leaks.

3.0L Engine

1. Relieve the fuel system pressure and disconnect the negative battery cable. Drain the cooling system.
2. Tag and disconnect all necessary vacuum, air and water hoses. Remove the airflow meter-to-throttle body intake air tube.
3. Disconnect the accelerator cable and the throttle sensor connector.
4. Remove the throttle body attaching nuts and the throttle body.

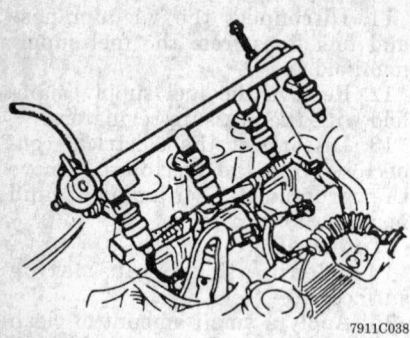

Fuel supply manifold installation — 2.2L and 2.6L engines

5. Remove the bypass air control valve and remove the throttle body-to-upper intake manifold intake air pipe.

6. Remove the extension manifolds. Remove the upper intake manifold along with the shutter valve actuator.

7. Disconnect the fuel lines and remove the fuel supply manifold. Disconnect the electrical connector(s) from the injector(s). Remove the injector(s) and insulator(s).

To install:

8. Clean all gasket mating surfaces.

9. Apply a small amount of clean engine oil to new O-ring(s) and install on the injector(s). Install the injector(s) and insulator(s).

10. Install the fuel supply manifold and tighten the attaching nuts to 14–19 ft. lbs. (19–25 Nm).

11. Connect the fuel lines to the fuel supply manifold. Before installing the remaining components, check for fuel leaks as follows:

a. Connect the negative battery cable.

b. Connect the terminals of the fuel pump test connector with a jumper wire and turn the ignition switch **ON** to run the fuel pump and pressurize the system.

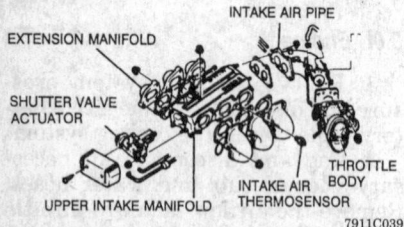

Upper intake manifold, throttle body and related components — 3.0L engine

c. Check for fuel leaks and correct as necessary.

d. Turn the ignition switch **OFF**, remove the jumper wire and disconnect the negative battery cable.

12. Connect the electrical connectors to the fuel injectors.

13. Install a new O-ring on the lower intake manifold, then install the upper intake manifold and extension manifolds. Tighten the attaching nuts to 14–19 ft. lbs. (19–25 Nm).

14. Position a new gasket and install the intake air pipe. Tighten the attaching nuts/bolt to 14–19 ft. lbs. (19–25 Nm).

15. Position a new gasket and install the throttle body. Tighten the attaching nuts/bolt to 14–19 ft. lbs. (19–25 Nm).

16. Connect the accelerator cable and the throttle sensor connector.

17. Connect the intake air tube between the airflow meter and the throttle body. Connect all vacuum, air and water hoses.

18. Connect the negative battery cable. Fill and bleed the cooling system.

19. Run the engine and check for leaks.

4.0L Engine

1. Disconnect the negative battery cable and relieve the fuel system pressure.

2. Disconnect the electrical connectors at the air bypass valve, TPS and ACT sensor.

3. Remove the snow/ice shield to expose the throttle linkage. Remove the throttle cable bracket and disconnect the cable from the ball stud on the throttle body.

4. Remove the air inlet tube from the air cleaner to the throttle body.

5. Disconnect the PCV valve from the valve cover.

6. Disconnect the spark plug wires from the comb at the rear of the manifold.

7. Remove the canister purge line from the fitting in the throttle housing.

8. Remove the bolt that retains the air conditioner line at the upper rear of the upper manifold.

9. Remove the 6 upper intake manifold retaining nuts and remove the upper intake and throttle body assembly.

10. Disconnect the fuel supply line fitting at the fuel manifold.

11. Disconnect the fuel return line from the fuel pressure regulator as follows:

a. Disengage the locking tabs on the connector retainer and separate the retainer halves.

b. Inspect the visible internal portion of the fitting for dirt accumulation. Clean the fitting before disassembly.

c. To disengage the fitting from the regulator, push the fitting toward the regulator, insert the fingers on fuel line coupling key 49 UN01 006 or equivalent, into the slots in the coupling.

d. Using the tool, pull the fitting from the regulator.

NOTE: If the fitting has been properly disengaged, the fitting should slide off the regulator with minimum effort.

12. Disconnect the electrical connectors from the fuel injectors.

13. Remove the 6 bolts retaining the fuel supply manifold and remove the manifold.

14. Remove the injector retaining clips and remove the injectors from the manifold by grasping the injector body and pulling up while rocking the injector from side-to-side.

15. Remove and discard the injector O-rings.

16. Inspect the injector plastic pintle protection cap and washer for signs of deterioration. Replace the complete injector as required. If the plastic pintle protection cap is missing, look for it in the intake manifold.

NOTE: The plastic pintle protection cap is not available as a separate part.

To install:

17. Lubricate new O-rings with clean light grade oil and install 2 on each injector.

NOTE: Never use silicone grease at it will clog the injectors.

18. Install the injectors, using a light, twisting, pushing motion.

19. Install the fuel supply manifold, pushing down to make sure all the fuel injector O-rings are fully seated in the fuel supply manifold cups and intake manifold.

20. Install the 6 retaining bolts and tighten to 7–10 ft. lbs. (10–14 Nm). Install the retainer clips.

21. Install the fuel supply line and tighten the fitting to 15–18 ft. lbs. (20–24 Nm).

22. Install the fuel return line to the fuel pressure regulator by pushing it onto the fuel pressure regulator

line up to the shoulder on the regulator line.

NOTE: The connector should grip the regulator line securely.

23. Install the connector retainer and snap the 2 halves of the retainer together.

24. Clean and inspect the mounting faces of the fuel manifold and upper intake manifold.

25. Position a new gasket on the mounting studs and install the upper intake manifold on the studs.

26. Install the 6 upper intake manifold retaining nuts and tighten to 15–18 ft. lbs. (20–25 Nm).

27. Connect the spark plug wires to the retainer comb at the rear of the intake manifold.

28. Attach the air conditioner line retainer and automatic transmission vacuum line retainer at the upper intake manifold.

29. Install the canister purge line on the throttle body fitting.

30. Connect the vacuum lines to the vacuum tree. Connect the electrical connectors at the air bypass valve, TPS and ACT sensor.

31. Install the PCV valve in the grommet at the rear of the right valve cover.

32. Attach the throttle cable bracket to the upper intake manifold, then connect the throttle cable to the ball stud and install the snow/ice shield.

33. After the upper intake manifold has been installed and before the fuel injector wire connectors have been connected, connect the negative battery cable and turn the ignition switch **ON**. This will cause the fuel pump to run for 2–3 seconds and pressurize the system.

34. Check for fuel leaks where the fuel injector is installed into the fuel supply manifold.

35. Turn the ignition switch **OFF** and disconnect the negative battery cable.

36. Connect the injector wire connectors and the vacuum line to the regulator.

37. Install the air inlet tube from the throttle body to the air cleaner.

38. Connect the negative battery cable, start the engine and let it idle for 2 minutes.

39. Turn the engine **OFF** and check for fuel leaks.

EMISSION CONTROLS

Emission Warning Lamps

RESETTING

B Series Pick-Up

On Federal and Canadian vehicles, the Malfunction Indicator Light (MIL) will come on at 60,000 and 80,000 mile intervals to indicate the need for scheduled maintenance of the emission control system. On California vehicles, the MIL will come on any time an engine management input device malfunctions, indicating that service is necessary. After the required service or maintenance has been performed, the MIL can be reset by changing the connector connections.

MPV

Federal and Canadian vehicles are equipped with a mileage sensor which is linked to the odometer. At every 80,000 miles, the mileage sensor will cause the Malfunction Indicator Light (MIL) to illuminate, indicating that the oxygen sensor must be replaced. After replacing the oxygen sensor, remove the instrument cluster and reset the MIL by reversing the position of the MIL set screw.

Navajo

All vehicles are equipped with a "CHECK ENGINE" or warning light located on the instrument cluster. This light should come on briefly when the ignition key is turned **ON**, but should turn off when the engine starts. If the light does not come **ON** when the ignition key is turned **ON**

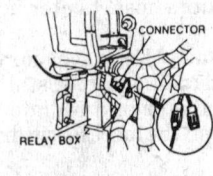

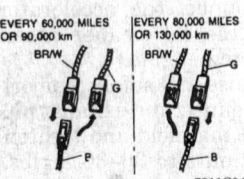

7911C040

Resetting the MIL — B Series Pick-Up

or if it comes ON and stays ON when the engine is running, there is a malfunction in the electronic engine control system. After the malfunction has been remedied, using the proper procedures, the "CHECK ENGINE" light will go out.

ENGINE MECHANICAL

NOTE: Disconnecting the negative battery cable on some vehicles may interfere with the functions of the on board computer systems and may require the computer to undergo a relearning process, once the negative battery cable is reconnected.

Engine Assembly

REMOVAL AND INSTALLATION

B Series Pick-Up

1. Relieve the fuel system pressure, disconnect the battery cables and remove the battery.

2. Raise and safely support the vehicle. Drain the engine oil and coolant. Remove the splash shields, as necessary.

3. Remove the starter and the transmission.

4. Disconnect the exhaust system from the exhaust manifold. Lower the vehicle.

5. Remove the air cleaner assembly, if carburetor equipped. Disconnect the accelerator cable.

6. Remove the cooling fan and the radiator shroud. Disconnect the radiator hoses and transmission oil cooler lines, if equipped and remove the radiator.

7. Disconnect the fuel lines, heater hoses and brake vacuum hose.

8. Tag and disconnect the necessary electrical connectors and vacuum hoses.

9. If carburetor equipped, disconnect the secondary air pipe assembly. On 2.6L engine, remove the resonance chamber.

10. Remove the accessory drive belt(s). If equipped, remove the power steering pump pulley and the power steering pump. Position the pump aside, leaving the hoses connected.

11. If equipped, remove the air conditioning compressor and position aside, leaving the hoses attached.

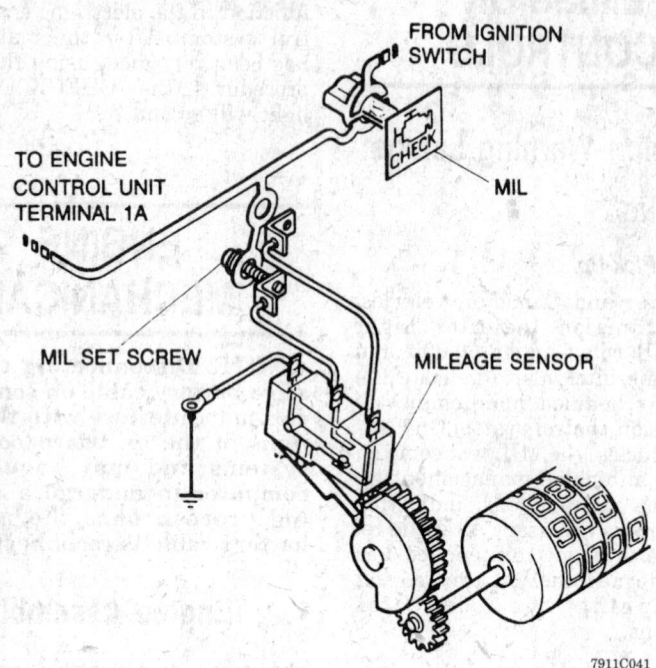

FROM IGNITION SWITCH

TO ENGINE CONTROL UNIT TERMINAL 1A

CHECK

MIL

MIL SET SCREW

MILEAGE SENSOR

7911C041

Mileage sensor assembly — MPV

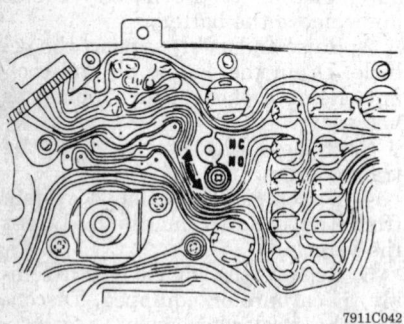

7911C042

MIL reset screw location at rear of instrument cluster — MPV

12. Remove the gusset plates, if equipped. Remove the transmission oil cooler line retainers, if equipped.

13. Attach suitable engine lifting equipment to the engine. Remove the engine mount nuts and remove the engine from the vehicle.

14. Install the engine on a workstand.

To install:

15. Remove the engine from the workstand and position in the vehicle. Install the engine mount nuts and tighten to 30–36 ft. lbs. (40–49 Nm).

16. Install the gusset plates, if equipped. Attach the transmission oil cooler line retainers, if equipped.

17. Install the air conditioning compressor, if equipped. Tighten the mounting bolts to 29–40 ft. lbs. (39–54 Nm).

18. Install the power steering pump, if equipped. Tighten the mounting bolts to 23–34 ft. lbs. (31–46 Nm). Install the power steering pump pulley and tighten the nut to 36–43 ft. lbs. (49–59 Nm).

19. Install the secondary air pipe, if equipped. Install the resonance chamber, if equipped.

20. Connect all vacuum lines and electrical connectors. Connect the brake vacuum hose, heater hoses and fuel lines.

21. Install the accessory drive belt(s) and cooling fan. Install the fan shroud, radiator and radiator hoses.

22. Adjust the accessory drive belt tension.

23. Connect the accelerator cable. Install the air cleaner assembly on carbureted engine.

24. Raise and safely support the vehicle. Connect the exhaust pipe to the exhaust manifold and tighten the attaching nuts to 30–36 ft. lbs. (40–49 Nm).

25. Install the starter and the transmission assembly. Install the splash shields and lower the vehicle.

26. Fill the crankcase with the proper type and quantity of engine oil. Install the battery and connect the battery cables.

27. Fill and bleed the cooling system. Run the engine and check for leaks and proper operation.

MPV

1. Relieve the fuel system pressure, disconnect the battery cables and remove the battery.

2. Raise and safely support the vehicle. Drain the engine oil and coolant. Remove the splash shield.

3. Remove the starter and the transmission.

4. Disconnect the exhaust pipes from the exhaust manifolds and lower the vehicle.

5. Remove the fresh air duct and the radiator hoses. Disconnect the transmission oil cooler lines from the radiator, if equipped.

6. Remove the radiator, fan shroud and cooling fan. Remove the accessory drive belts.

7. Tag and disconnect the necessary electrical connectors and vacuum lines. Disconnect the brake vacuum hose, heater hoses and fuel lines.

8. On 2.6L engine, disconnect the accelerator cable and remove the resonance chamber and air cleaner. On 3.0L engine, disconnect the accelerator cable and remove the air cleaner and airflow meter.

9. Remove the shroud upper panel and the air conditioning pipe bracket. On 3.0L engine, remove the protector cover from the front of the engine.

10. If equipped, remove the power steering pump and position aside, leaving the hoses attached. It is necessary to remove the power steering pulley prior to removing the pump.

11. If equipped, remove the air conditioning compressor and position aside, leaving the hoses attached.

12. Remove the lower grille and radiator grille. Remove the shroud upper plate and the additional condenser fan, if equipped.

13. Attach suitable engine lifting equipment to the engine. Remove the engine mount nuts and remove the engine from the vehicle.

14. Install the engine on a workstand.

To install:

15. Remove the engine from the workstand. Lower the engine into the

vehicle, being careful not to damage the piping.

NOTE: Lean the air conditioning condenser forward to ease engine installation.

16. Install the engine mount nuts and tighten to 25–36 ft. lbs. (34–49 Nm). Install the additional condenser fan, if equipped.

17. Apply a bead of sealer to each side of the front support, then install the shroud upper plate. Tighten the mounting bolts to 61–87 inch lbs. (6.9–9.8 Nm).

18. Install the radiator grille and lower grille.

19. Install the air conditioning compressor, if equipped. Tighten the mounting bolts to 13–20 ft. lbs. (18–26 Nm). Install the air conditioner pipe bracket and tighten the mounting nuts to 61–87 inch lbs. (6.9–9.8 Nm).

20. Install the power steering pump, if equipped. Tighten the mounting bolts to 23–34 ft. lbs. (31–46 Nm). Install the pump pulley and tighten the nut to 29–43 ft. lbs. (39–59 Nm).

21. Install the shroud upper panel and tighten the bolts to 69–95 inch lbs. (7.8–11.0 Nm).

22. Connect the accelerator cable. On 2.6L engine, install the resonance chamber and air filter. On 3.0L engine, install the air cleaner and airflow meter.

23. Connect all electrical connectors and vacuum hoses.

24. Connect the brake vacuum hose, heater hoses and fuel lines.

25. Install the accessory drive belts and the cooling fan. Install the fan shroud and the radiator. Adjust the drive belt tension.

26. Install the radiator hoses and fresh air duct. If equipped, connect the transmission oil cooler lines.

27. Raise and safely support the vehicle. Connect the exhaust pipes to the exhaust manifolds. Tighten the nuts to 25–36 ft. lbs. (34–49 Nm).

28. Install the starter and transmission assembly. Install the splash shield and lower the vehicle.

29. Install the battery and connect the negative battery cables. Fill the crankcase with the proper type and quantity of engine oil.

30. Fill and bleed the cooling system. Run the engine and check for leaks and proper operation.

Navajo

1. Disconnect the negative battery cable and relieve the fuel system pressure. Drain the cooling system.

2. Mark the position of the hood on the hinges and remove the hood. Remove the air cleaner intake hose.

3. Disconnect the radiator hoses at the radiator. Disconnect the fan shroud and position it over the fan. Remove the radiator, then the shroud.

4. Remove the alternator and bracket and position the alternator aside. Disconnect the alternator ground wire from the cylinder block.

5. Remove the air conditioning compressor and power steering pump and position aside, if equipped.

6. Disconnect the heater hoses at the intake manifold and water pump. Remove the ground wires from the cylinder block.

7. Disconnect the fuel lines from the fuel supply manifold. Disconnect the throttle cable shield and linkage at the throttle body and intake manifold.

8. Tag and disconnect the vacuum connections at the rear vacuum fitting in the upper intake manifold.

9. Disconnect the wiring from the ignition coil and oil pressure and engine coolant temperature senders. Disconnect the injector harness, air charge temperature sensor and throttle position sensor. Disconnect the brake booster vacuum hose.

10. Raise and safely support the vehicle. Disconnect the exhaust pipes at the manifolds. Disconnect the starter cable and remove the starter.

11. Remove the engine front mount-to-crossmember attaching nuts or through bolts.

12. Remove the converter inspection cover and disconnect the converter from the flywheel. Remove the cable.

13. Remove the converter housing-to-engine block bolts and the adapter plate-to-converter housing bolt. Lower the vehicle.

14. Position a jack under the transmission and install suitable engine lifting equipment.

15. Raise the engine slightly and carefully pull it from the transmission. Carefully lift the engine out of the engine compartment so the rear cover plate is not bent or components damaged. Install the engine on a workstand.

To install:

16. Remove the engine from the workstand and carefully lower it into the engine compartment. Make sure the exhaust manifolds are aligned with the exhaust pip

17. At the transmission, start the converter pilot into the crankshaft. Install the converter housing upper

bolts, making sure the dowels in the cylinder block engage the flywheel housing. Tighten the bolts to 33–45 ft. lbs. (45–61 Nm).

18. Remove the jack from under the transmission and the engine lifting equipment.

19. Position the kickdown rod on the transmission and engine. Raise and safely support the vehicle.

20. Position the transmission linkage bracket and install the remaining converter housing bolts. Install the adapter plate-to-converter housing bolt. Install the converter-to-flywheel nuts and install the inspection cover. Connect the kickdown rod on the transmission.

21. Install the starter and connect the cable. Connect the exhaust pipes at the manifolds.

22. Install the engine front mount nuts and washers or through bolts. Lower the vehicle.

23. Install the ground wires to the engine block. Connect the ignition coil wiring, then connect the coolant temperature sending unit and oil pressure sending unit. Connect the brake booster vacuum hose.

24. Install the throttle linkage and connect the fuel lines at the fuel supply manifold.

25. Connect the ground cable at the engine block. Connect the heater hoses to the water pump and cylinder block.

26. Install the alternator and bracket. Connect the alternator ground wire to the engine block. Install the accessory drive belt.

27. Install the air conditioner compressor and power steering pump, if equipped.

28. Position the shroud over the fan. Install the radiator and connect the radiator upper and lower hoses. Install the fan shroud attaching bolts.

29. Connect the negative battery cable. Fill and bleed the cooling system.

30. Run the engine and check for leaks and proper operation. If equipped, evacuate and charge the air conditioning system.

31. Install the intake hose. Install the hood, aligning the marks that were made during removal.

Engine Mounts

REMOVAL AND INSTALLATION

Front Mounts

1. Remove the fan shroud attaching bolts.

2. Support the engine using a wood block and a jack under the oil pan.

3. Remove the engine mount bracket nut(s). Raise the engine with the jack just enough to remove the mount.

4. Remove the attaching nuts/bolts and remove the mount.

5. Installation is the reverse of the removal procedure.

Rear Mounts

1. Raise and safely support the vehicle.

2. Support the transmission with a jack.

3. Remove the mount-to-crossmember nuts/bolts and the mount-to-transmission bolts.

4. Raise the transmission just enough to remove the mount.

5. Installation is the reverse of the removal procedure.

Cylinder Head

REMOVAL AND INSTALLATION

2.2L Engine

1. Relieve the fuel system pressure and disconnect the negative battery cable.

2. Remove the splash shield and drain the cooling system.

3. If carburetor equipped, remove the air cleaner assembly. If fuel injected, remove the air intake hose.

4. Disconnect the accelerator cable.

5. Remove the cooling fan and fan shroud.

6. Disconnect the fuel lines. If equipped with carburetor and manual transmission, remove the fuel pump.

7. Disconnect the heater hoses and brake vacuum hose. Tag and disconnect the necessary electrical connectors and vacuum hoses.

8. Tag and disconnect the spark plug wires and remove the spark plugs. Remove the distributor.

9. If carburetor equipped, disconnect the secondary air pipe assembly.

10. Disconnect the upper radiator and water bypass hoses.

11. Remove the intake and exhaust manifolds.

12. Remove the timing belt front cover and the timing belt tensioner and timing belt.

13. Remove the rocker arm cover.

14. Loosen the cylinder head bolts in 2–3 steps in the proper sequence.

Remove the bolts and remove the cylinder head.

15. Clean all gasket mating surfaces. Measure the cylinder head for distortion in 6 directions using a straight-edge. The maximum allowable distortion is 0.006 in. (0.15mm).

16. If distortion is excessive, resurface or replace the cylinder head. If resurfacing, do not remove more than 0.008 in. (0.20mm). The cylinder head height should be 3.620–3.624 in. (91.95–92.05mm).

NOTE: If resurfacing the cylinder, it will be necessary to remove the rocker arm/shaft assemblies.

To install:

17. Position a new head gasket on the cylinder block and install the cylinder head. Apply oil to the threads and seat faces of the cylinder head bolts and install.

18. Tighten the cylinder head bolts in 2–3 steps in the proper sequence. The final torque specification is 59–64 ft. lbs. (80–86 Nm).

19. Apply silicone sealer to each side of the front and rear camshaft bearing caps where the cap meets the cylinder head. Install the rocker arm cover and tighten the bolts to 26–35 inch lbs. (2.9–3.9 Nm).

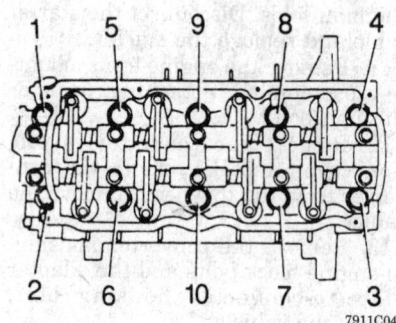

Cylinder head bolt removal sequence — 2.2L engine

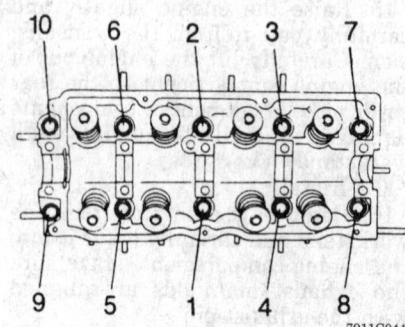

Cylinder head bolt torque sequence — 2.2L engine

20. Install the timing belt and tensioner. install the timing belt front cover.

21. Install the intake and exhaust manifolds.

22. Install the upper radiator and water bypass hoses. If carburetor equipped, install the secondary air pipe.

23. Install the distributor. Install the spark plugs and connect the spark plug wires.

24. Connect all vacuum hoses and electrical connectors. Connect the heater hoses and the brake vacuum hose.

25. If equipped, install the fuel pump. Connect the fuel lines.

26. Install the cooling fan and shroud.

27. Connect the accelerator cable. Install the air cleaner or air intake hose.

28. Install the splash shield. Connect the negative battery cable.

29. Fill and bleed the cooling system. Run the engine and check for leaks.

30. Check and adjust the ignition timing.

2.6L Engine

1. Relieve the fuel system pressure and disconnect the negative battery cable. Drain the cooling system.

2. Disconnect the accelerator cable and remove the air intake pipe and resonance chamber.

3. Remove the air conditioning drive belt and idler.

4. Tag and disconnect the spark plug wires. Remove the spark plugs.

5. Remove the upper radiator and oil cooler water hoses. Disconnect the brake vacuum hose, canister hose and fuel line.

6. Tag and disconnect the oxygen sensor connector and emission harness connectors. Remove the solenoid valves.

7. Remove the rocker arm cover. Turn the crankshaft pulley until the timing mark on the camshaft sprocket is 90 degrees to the right. The mark should be parallel to the top of the cylinder head surface. Make sure the yellow crankshaft pulley timing mark is aligned with the indicator pin.

8. Mark the position of the distributor rotor in relation to the distributor housing and the distributor housing in relation to the cylinder head. Remove the distributor.

NOTE: Do not rotate the engine after distributor removal.

9. Hold the crankshaft pulley with a suitable tool and remove the distributor drive gear/camshaft pulley retaining bolt and the drive gear.

10. Remove the service cover on the timing chain cover. Push the chain adjuster sleeve in toward the left and insert a 0.08 in. (2mm) diameter by 1.77 in. (45mm) long pin into the lever hole to hold it. Be careful that the pin does not fall.

11. Secure the camshaft sprocket and chain with a wire to hold the chain in place on the sprocket. Do not allow the sprocket and chain to fall down into the engine and cause the chain to become disengaged from the crankshaft sprocket.

12. Loosen the cylinder head bolts in 2–3 steps in the proper sequence. Remove the bolts and remove the cylinder head.

13. If necessary, remove the intake and exhaust manifolds from the cylinder head.

14. Clean all gasket mating surfaces. Measure the cylinder head for distortion in 6 directions using a straight-edge. The maximum allowable distortion is 0.006 in. (0.15mm).

15. If distortion is excessive, resurface or replace the cylinder head. If resurfacing, do not remove more than 0.008 in. (0.20mm). The cylinder head height should be 3.541–3.545 in. (89.95–90.05mm).

NOTE: It will be necessary to remove the rocker arm/shaft assembly if resurfacing is necessary.

To install:

16. If removed, install the intake and exhaust manifolds.

17. Apply silicone sealer to the cylinder block head gasket surface at the mating junction of the cylinder block and the timing chain cover.

18. Position a new head gasket and install the cylinder head. Apply clean engine oil to the head bolt threads and seat faces and install the bolts.

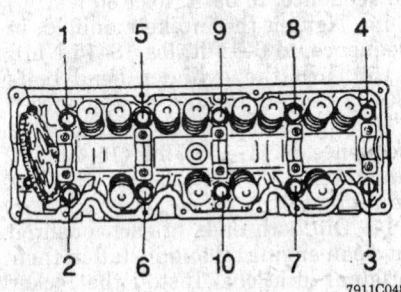

Cylinder head bolt removal sequence — 2.6L engine

19. Tighten the cylinder head bolts, in sequence, in 2–3 steps. The final torque specification is 59–64 ft. lbs. (80–86 Nm). Tighten the 2 cylinder head-to-timing chain cover bolts to 12–17 ft. lbs. (16–23 Nm).

20. Install the camshaft sprocket onto the camshaft dowel pin, then remove the wire securing the camshaft sprocket and chain.

21. Remove the retaining pin from the chain adjuster and install the service cover with a new gasket. Tighten the mounting bolts to 69–95 inch lbs. (7.8–11.0 Nm) and the mounting nuts to 61–87 inch lbs. (6.9–9.8 Nm).

22. Install the distributor drive gear with a new washer and lock bolt. Position a shop cloth to protect the cylinder head and place a small prybar between the cylinder head and the small tab on the camshaft between the lobes, to hold the camshaft in place. Tighten the lock bolt to 36–45 ft. lbs. (49–61 Nm).

23. Apply sealer to the cylinder head and install the half circle seals.

24. Make sure the timing mark on the camshaft sprocket is still positioned 90 degrees to the right and the yellow crankshaft pulley timing mark is aligned with the indicator pin.

25. Apply clean engine oil to a new O-ring and install on the distributor.

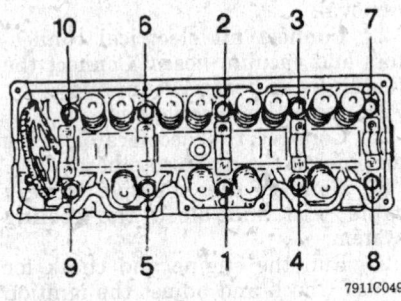

Cylinder head bolt torque sequence — 2.6L engine

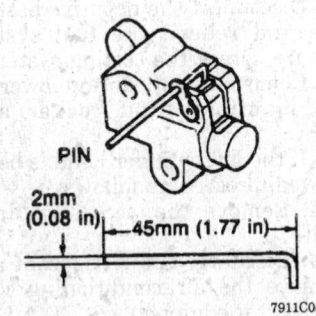

Retaining the chain adjuster — 2.6L engine

Apply clean engine oil to the distributor driven gear and install the distributor, aligning the marks that were made during removal.

26. Apply clean engine oil to the valves, rocker arms and timing chain.

27. Coat a new rocker arm cover gasket with sealer and install on the rocker arm cover.

28. Apply silicone sealer to the front and rear half circle seals and install the rocker arm cover. Tighten the attaching bolts to 52–78 inch lbs. (5.9–8.8 Nm).

29. Install the solenoid valves. Connect all electrical connectors and vacuum hoses.

30. Connect the fuel line, brake vacuum hose, upper radiator hose and oil cooler water hose.

31. Install the spark plugs and the spark plug wires.

32. Install the air conditioner drive belt and idler, if equipped. Install the air intake pipe and resonance chamber and connect the accelerator cable.

33. Connect the negative battery cable. Fill and bleed the cooling system.

34. Run the engine and check for leaks. Check and adjust the ignition timing.

3.0L Engine

1. Relieve the fuel system pressure and disconnect the negative battery cable. Drain the cooling system.

2. Remove the air intake hose and disconnect the accelerator cable.

3. Turn the crankshaft to align the timing marks on the timing belt sprockets. Mark the direction of rotation on the belt and remove the timing belt.

4. Disconnect the brake vacuum hose and the fuel line. Tag and disconnect all necessary electrical connectors and vacuum hoses.

5. Mark the position of the distributor rotor in relation to the distributor housing and the distributor housing in relation to the cylinder head. Remove the distributor.

6. Remove the extension manifolds. Loosen the intake manifold bolts, in sequence, in 2–3 steps and remove the bolts and intake manifold.

7. Remove the rocker arm cover. Remove the center exhaust pipe insulator and center exhaust pipe.

8. Remove the exhaust manifold insulator and exhaust manifold. Remove the seal plate.

9. Loosen the cylinder head bolts, in sequence, in 2–3 steps and remove the bolts and cylinder head.

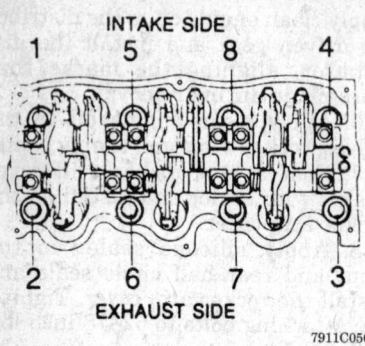

INTAKE SIDE

EXHAUST SIDE

7911C050

Cylinder head bolt removal sequence —
3.0L engine

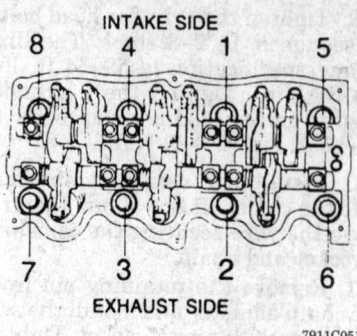

INTAKE SIDE

EXHAUST SIDE

7911C051

Cylinder head bolt torque sequence — 3.0L
engine

10. Clean all gasket mating surfaces. Measure the cylinder head for distortion in 6 directions using a straight-edge. The maximum allowable distortion is 0.004 in. (0.10mm).

11. If distortion is excessive, resurface or replace the cylinder head. If resurfacing, do not remove more than 0.006 in. (0.15mm). The cylinder head height should be 4.931–4.935 in. (125.25–125.35mm).

NOTE: It will be necessary to remove the rocker arm/shaft assembly if resurfacing is necessary.

To install:

12. Check the oil control plug projection from the cylinder block. It should be 0.209–0.224 in. (5.3–5.7mm).

13. Apply clean engine oil to a new O-ring and install on the oil control plug. Install a new cylinder head gasket, making sure the **L** mark, for left bank or **R** mark, for right bank, is facing upward.

14. Install the cylinder head, being careful not to damage the oil control plug O-ring.

15. Measure the length of the cylinder head bolts prior to installation and replace if not within specification. The intake side bolts should be 4.25 in. (108mm) and should not exceed 4.29 in. (109mm). The exhaust side bolts should be 5.43 in. (138mm) and should not exceed 5.47 in. (139mm).

16. Apply clean engine oil to the threads and the seat face of the cylinder head bolts and install. Tighten the bolts as follows: Step 1: Tighten each bolt, in sequence, to 14 ft. lbs. (20 Nm). Step 2: Tighten each bolt, in sequence, an additional 90 degree turn. Step 3: Repeat Step 2.

17. Install the seal plate, exhaust manifold and exhaust manifold insulator.

18. Install the center exhaust pipe and exhaust pipe insulator.

19. Coat a new rocker arm cover gasket with silicone sealant and install onto the rocker arm cover. Install the rocker arm cover with new seal washers and tighten the bolts to 30–39 inch lbs. (3.4–4.4 Nm).

20. Connect the injector harness and position a new intake manifold gasket. Install the intake manifold and tighten the bolts, in sequence in 2–3 steps. The final torque specification is 14–19 ft. lbs. (19–25 Nm).

21. Apply clean engine oil to new O-rings and install the extension manifolds with new gaskets. Tighten to 14–19 ft. lbs. (19–25 Nm).

22. Install the distributor, aligning the marks that were made during removal.

23. Connect all electrical connectors and vacuum hoses. Connect the fuel line and brake vacuum hose.

24. Install the timing belt.

25. Connect the accelerator cable and install the air intake hose.

26. Connect the negative battery cable. Fill and bleed the cooling system.

27. Run the engine and check for leaks. Check and adjust the ignition timing.

4.0L Engine

1. Disconnect the negative battery cable and relieve the fuel system pressure. Drain the cooling system.

2. Remove the upper and lower intake manifolds and rocker arm covers.

3. If the left cylinder head is being removed, proceed as follows:
   a. Remove the accessory drive belt.
   b. Discharge the refrigerant and remove the air conditioning compressor, if equipped.
   c. Remove the power steering pump and bracket and position aside.

d. Remove the spark plugs.

4. If the right cylinder head is being removed, proceed as follows:
   a. Remove the accessory drive belt.
   b. Remove the alternator and alternator bracket.
   c. Remove the ignition coil and bracket assembly.
   d. Remove the spark plugs.

5. Disconnect the exhaust pipe and remove the exhaust manifold(s).

6. Remove the rocker arm shaft assembly. Remove the pushrods, marking them so they can be reinstalled in the same positions.

7. Remove and discard the cylinder head attaching bolts and remove the cylinder heads.

8. Clean all gasket mating surfaces. Check the cylinder head for flatness using a straight-edge and a feeler gauge. The cylinder head must not be warped more than 0.003 in. in any 6 in. or more than 0.006 in. overall.

To install:

9. Position new cylinder head gasket(s) on the cylinder block. Install cylinder head locating dowels.

NOTE: The cylinder head(s) and intake manifold are torqued alternately and in sequence to insure correct fit and gasket crunch.

10. Install new cylinder head bolts and tighten, in sequence, to 44 ft. lbs. (60 Nm).

11. Apply silicone sealer to the block and cylinder head mating surfaces at the 4 corners of the lifter valley opening. Install the intake manifold gasket and again apply sealer in the same locations.

12. Position the lower intake manifold on the 2 guide studs and install the nuts and bolts hand tight. Tighten the lower intake manifold bolts, in sequence, to 3–6 ft. lbs. (4–8 Nm).

13. Tighten the cylinder head bolts, in sequence, to 59 ft. lbs. (80 Nm).

14. Tighten the intake manifold, in sequence, to 6–11 ft. lbs. (8–15 Nm).

15. Turn the cylinder head bolts 80–85 degrees tighter, in sequence.

16. Tighten the intake manifold, in sequence, to 11–15 ft. lbs. (15–21 Nm) and then to 15–18 ft. lbs. (21–25 Nm), in sequence.

17. Dip both ends of each pushrod in clean engine oil and install in their original locations. Install the rocker arm and shaft assemblies and tighten the rocker arm shaft support bolts evenly to 46–52 ft. lbs. (62–70 Nm).

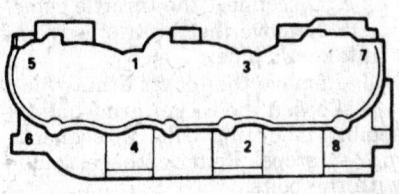

Cylinder head bolt torque sequence — 4.0L engine

18. Apply silicone sealer to the 4 locations at the joint where the intake manifold and cylinder head meet. Install a new rocker arm cover gasket in each cover and install the rocker arm covers. Tighten the rocker arm cover bolts to 3–5 ft. lbs. (4–7 Nm), wait 2 minutes and then retighten to the same specification.

19. Install the upper intake manifold and tighten the nuts to 15–18 ft. lbs. (20–25 Nm).

20. Install the exhaust manifold(s) and connect the exhaust pipe.

21. Install the spark plugs and the ignition coil and bracket assembly.

22. Install the alternator and the accessory drive belt.

23. Install the power steering pump. Install the air conditioning compressor, if equipped.

24. Connect the negative battery cable. Fill and bleed the cooling system. Run the engine and check for leaks.

Valve Lifters

REMOVAL AND INSTALLATION

Except 4.0L Engine

All engines except 4.0L are equipped with hydraulic lash adjusters located in the rocker arms. The hydraulic lash adjusters can be removed by hand or with suitable pliers after the rocker arm/shaft assembly is removed.

4.0L Engine

1. Disconnect the negative battery cable and relieve the fuel system pressure. Drain the cooling system.

2. Remove the intake manifold and rocker arm covers.

3. Loosen the rocker arm shaft support bolts 2 turns at a time until the rocker arm and shaft assembly can be removed.

4. Remove the pushrods, marking them so they can be reinstalled in their original positions.

5. Remove the lifters. Note the location of each lifter so it can be reinstalled in the same bore. The roller lifters have an alignment tab which fits into a locating groove in the lifter bore. Do not attempt to rotate a roller lifter in the bore in an effort to remove it.

To install:

6. Lubricate the lifters and bores with clean engine oil. Install each lifter in the same bore from which it was removed. Install the lifter with the alignment tab in the locating groove of the bore. If a new lifter is being installed, check for free fit in the bore.

7. Check each pushrod for straightness and for damage, replace as necessary. Dip each pushrod end in clean engine oil and install in its original position.

8. Lubricate the rocker arm and shaft assembly and install. Draw the shaft support bolts down evenly, 2 turns at a time, until the shafts are fully down. Tighten the bolts to 46–52 ft. lbs. (62–70 Nm).

9. Install the intake manifold and rocker arm covers.

10. Connect the negative battery cable. Fill and bleed the cooling system.

11. Run the engine and check for leaks.

Rocker Arms/Shafts

REMOVAL AND INSTALLATION

2.2L Engine

1. Disconnect the negative battery cable.

2. Remove the air cleaner assembly or air intake hose, as required.

3. Remove the rocker arm cover.

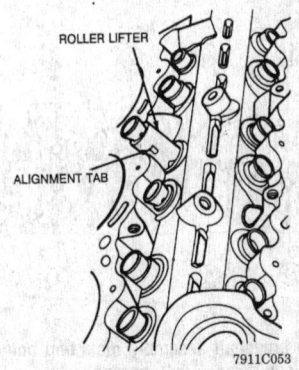

Valve lifters — 4.0L engine

4. Loosen the rocker arm/shaft assembly mounting bolts in 2–3 steps in the proper sequence. Remove the rocker arm/shaft assembly together with the bolts.

5. If necessary, disassemble the rocker arm/shaft assembly, noting the position of each component to ease reassembly.

6. Check for wear or damage to the contact surfaces of the shafts and rocker arms; replace as necessary.

7. Measure the rocker arm inner diameter, it should be 0.6300–0.6310 in. (16.000–16.027mm). Measure the rocker arm shaft diameter, it should be 0.6286–0.6293 in. (15.966–15.984mm).

8. Subtract the shaft diameter from the rocker arm diameter to get the oil clearance. The oil clearance should be 0.0006–0.0024 in. (0.016–0.061mm) and should not exceed 0.004 in. (0.10mm). Replace parts, as necessary, if the oil clearance is not within specification.

To install:

9. Apply clean engine oil to the rocker arm shafts and rocker arms and assemble the rocker arm/shaft assembly in the reverse order of disassembly. Make sure the rocker arm shaft oil holes in the center camshaft cap face each other.

NOTE: Use the mounting bolts for alignment.

10. Apply silicone sealant to the cylinder head on the front and rear camshaft cap mounting surface. Apply clean engine oil to the camshaft journals and valve stem tips.

11. Install the rocker arm/shaft assembly and tighten the bolts, in sequence, in 2–3 steps to a maximum torque of 13–20 ft. lbs. (18–26 Nm).

12. Apply silicone sealant to each side of the front and rear camshaft cap and the cylinder head in the area where the caps meet the cylinder head.

13. Install the rocker arm cover and tighten the mounting bolts to 26–35 inch lbs. (2.9–3.9 Nm).

14. Install the air cleaner assembly or air intake tube. Connect the negative battery cable, start the engine and check for leaks and proper operation.

2.6L Engine

1. Disconnect the negative battery cable.

2. Remove the air intake hose.

3. Remove the rocker arm cover.

4. Loosen the rocker arm/shaft assembly mounting bolts in 2–3 steps in the proper sequence. Remove the

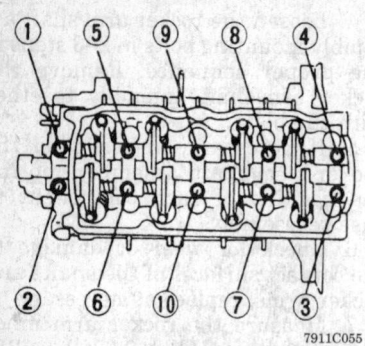

Rocker arm/shaft assembly mounting bolts removal sequence — 2.2L engine

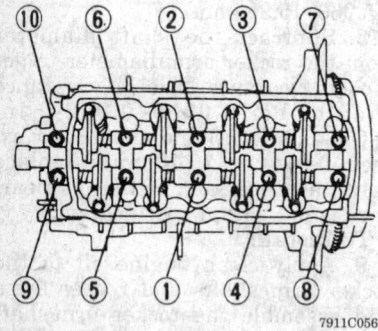

Rocker arm/shaft assembly mounting bolts torque sequence — 2.2L engine

rocker arm/shaft assembly together with the bolts.

5. If necessary, disassemble the rocker arm/shaft assembly, noting the position of each component to ease reassembly.

6. Check for wear or damage to the contact surfaces of the shafts and rocker arms; replace as necessary.

7. Measure the rocker arm inner diameter, it should be 0.8268–0.8281 in. (21.000–21.033mm). Measure the rocker arm shaft diameter, it should be 0.8252–0.8260 in. (20.959–20.980mm).

8. Subtract the shaft diameter from the rocker arm diameter to get the oil clearance. The oil clearance should be 0.0008–0.0029 in. (0.020–0.074mm) and should not exceed 0.004 in. (0.10mm). Replace parts, as necessary, if the oil clearance is not within specification.

To install:

9. Apply clean engine oil to the rocker arm shafts and rocker arms and assemble the rocker arm/shaft assembly in the reverse order of disassembly, noting the following:

   a. The intake side shaft has twice as many oil holes as the exhaust side shaft.

   b. The No. 4 camshaft cap has an oil hole from the cylinder head; make sure it is installed correctly.

10. Apply clean engine oil to the camshaft journals and valve stem tips.

11. Install the rocker arm/shaft assembly and tighten the mounting bolts, in sequence, in 2–3 steps to a maximum torque of 14–19 ft. lbs. (19–25 Nm).

12. Coat a new gasket with silicone sealer and install on the rocker arm cover. Apply sealer to the cylinder head in the area of the half circle seals and install the rocker arm cover. Install the mounting bolts and tighten to 52–78 inch lbs. (5.9–8.8 Nm).

13. Install the air intake hose. Connect the negative battery cable, start the engine and check for leaks and proper operation.

3.0L Engine

1. Disconnect the negative battery cable.

2. If removing the driver's side rocker arm/shaft assembly, proceed as follows:

   a. Remove the air inlet tube.

   b. Disconnect the necessary electrical connectors and vacuum hoses

from the throttle body and intake air pipe.

   c. Disconnect the throttle cable.

   d. Remove the throttle body and intake air pipe.

3. Remove the rocker arm cover.

4. Loosen the rocker arm/shaft assembly mounting bolts in sequence, in 2–3 steps. Remove the assembly with the bolts.

5. If necessary, disassemble the rocker arm/shaft assembly, noting the position of each component to ease reassembly.

6. Check for wear or damage to the contact surfaces of the shafts and rocker arms; replace as necessary.

7. Measure the rocker arm inner diameter, it should be 0.7480–0.7493 in. (19.000–19.033mm). Measure the rocker arm shaft diameter, it should be 0.7464–0.7472 in. (18.959–18.980mm).

8. Subtract the shaft diameter from the rocker arm diameter to get the oil clearance. The oil clearance should be 0.0008–0.0029 in. (0.020–0.074mm) and should not exceed 0.004 in. (0.10mm). Replace parts, as necessary, if the oil clearance is not within specification.

To install:

9. Apply clean engine oil to the rocker arm shafts and rocker arms and assemble the rocker arm/shaft assembly in the reverse order of disassembly, noting the following. The intake side shaft has twice as many oil holes as the exhaust side shaft.

10. Apply clean engine oil to the camshaft journals and valve stem tips.

11. Install the rocker arm/shaft assembly and tighten the mounting bolts, in sequence, in 2–3 steps to a maximum torque of 14–19 ft. lbs. (19–25 Nm).

NOTE: Be careful that the rocker arm shaft spring does not get caught between the shaft and mounting boss during installation.

12. Coat a new gasket with silicone sealant and install on the rocker arm cover. Install the rocker arm cover with new seal washers and tighten the bolts to 30–39 inch lbs. (3.4–4.4 Nm).

13. Install the intake air pipe, throttle body and air intake tube, if removed. Connect the throttle cable and the necessary electrical connectors and vacuum hoses.

14. Connect the negative battery cable, start the engine and check for leaks and proper operation.

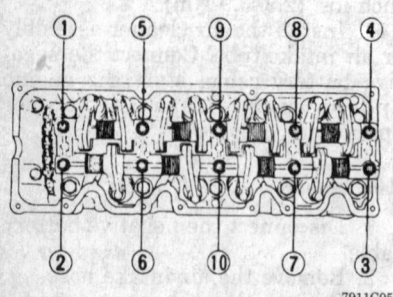

Rocker arm/shaft assembly mounting bolts removal sequence — 2.6L engine

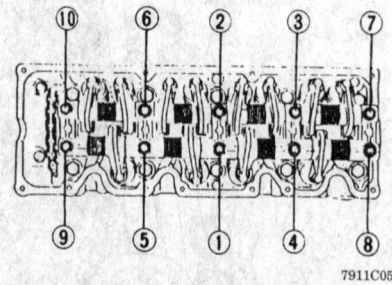

Rocker arm/shaft assembly mounting bolts torque sequence — 2.6L engine

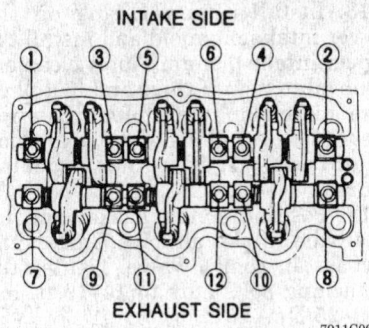

Rocker arm/shaft assembly mounting bolts removal sequence — 3.0L engine

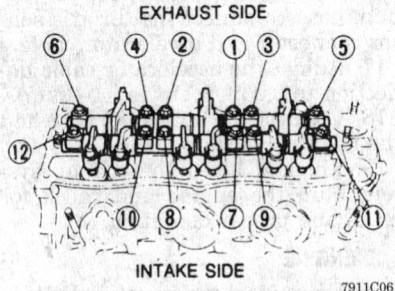

Rocker arm/shaft assembly mounting bolts torque sequence — 3.0L engine

4.0L Engine

1. Disconnect the negative battery cable. Remove the intake shield and air intake tube.

2. If removing the right rocker arm/shaft assembly, proceed as follows:

   a. Remove the alternator and the coil pack.

   b. Remove the retaining bolt from the air conditioning pipe over the upper intake manifold.

   c. Remove the spark plug wires from the clips on the valve cover.

   d. Remove the 2 wiring harnesses from the right rocker arm cover.

   e. Disconnect the vacuum hose at the coupling over the rocker arm cover.

   f. Remove the engine wiring harness clip from the rocker arm cover-to-intake manifold stud. Do not pull on the harness but rather lift up on the clip.

   g. Remove the right rocker arm cover bolts, reinforcement plate and the cover.

3. If removing the right rocker arm/shaft assembly, proceed as follows:

   a. Remove the bolt from the air conditioning pipe over the upper intake manifold, if not already done.

   b. Disconnect the air conditioning compressor clutch connector and remove the wiring harness from the back of the compressor.

   c. Remove the air conditioning compressor bolts, pull up the tube that goes around the back of the engine and reposition the compressor and tube aside.

   d. Tag and disconnect the power brake vacuum hose and other hoses from the vacuum tee on the plenum.

   e. Remove the PCV hose and valve. Remove the wiring harness from the valve cover and position aside.

   f. Tag and disconnect the spark plug wires from the spark plugs and clips on the rocker arm cover.

   g. Remove the engine wiring harness clip from the rocker arm cover-to-intake manifold stud. Do not pull on the harness but rather lift up on the clip.

   h. Remove the retaining bolt from the fuel hose clip to the front of the engine to allow the fuel hoses to be moved enough to gain access to the upper front rocker arm cover bolt.

   i. Remove the bolts, reinforcement plates and left rocker arm cover.

4. Remove the rocker arm shafts by loosening the support bolts 2 turns at a time until the shaft can be removed.

5. If necessary, disassemble the rocker arm/shaft assembly by removing the spring washer and pin from each end of the shaft and sliding the rocker arms, springs and rocker arm shaft supports off the shaft. Note the position of each component to ease reassembly.

6. Inspect all components for wear and replace, as necessary.

To install:

7. Coat the rocker arms and shafts with clean engine oil and reassemble. The oil holes in the shaft must point down when the shaft is installed. This position can be recognized by a notch on the front face of the shaft.

8. Lubricate the pushrod ends and valve stem tips.

9. Install the rocker arm/shaft assembly and draw the shaft support bolts down evenly, 2 turns at a time, until the rocker arm/shaft assembly is fully down. Tighten the shaft support bolts to 46–52 ft. lbs. (62–70 Nm).

10. Clean all gasket mating surfaces.

11. Apply silicone sealant to the intake manifold-to-cylinder head parting seam and an 1/8 in. ball of sealer to the rocker arm cover bolt holes on the exhaust side.

12. Install a new gasket on the rocker arm cover and install the cover, reinforcing plates and bolts. Tighten the bolts to 53–70 inch lbs. (6–8 Nm) working in a criss-cross pattern and starting at the center bolts.

13. Install the remaining components in the reverse order of their removal. Connect the negative battery cable, start the engine and check for leaks and proper operation.

Intake Manifold

REMOVAL AND INSTALLATION

Carbureted Engine

1. Relieve the fuel system pressure and disconnect the negative battery cable. Drain the cooling system.

2. Remove the air cleaner assembly.

3. Disconnect the accelerator cable. Tag and disconnect the necessary electrical connectors and vacuum hoses.

4. Disconnect the coolant hoses and fuel line.

5. Remove the intake manifold mounting nuts and remove the intake manifold.

To install:

6. Clean all gasket mating surfaces.

7. Position a new intake manifold gasket on the cylinder head and install the intake manifold.

8. Install the intake manifold mounting nuts and tighten, in 2–3 steps, to 14–19 ft. lbs. (19–25 Nm) on 2.2L engine or 11–14 ft. lbs. (15–19 Nm) on 2.6L engine. Tighten the nuts at the center of the manifold 1st and work towards the ends.

9. Connect the fuel line and the coolant hoses.

10. Connect the electrical connectors and vacuum hoses. Connect the accelerator cable.

11. Install the air cleaner assembly and connect the negative battery cable.

12. Fill and bleed the cooling system. Run the engine and check for leaks.

Fuel Injected Engine

2.2L AND 2.6L ENGINES

1. Relietry cable. Drain the cooling system.

2. Disconnect the air intake tube and ventilation hose. Remove the air pipe and resonance chamber on 2.6L engine.

3. Disconnect the accelerator cable and coolant hoses. Tag and disconnect the electrical connectors to the solenoid valve, throttle sensor and idle switch.

4. Remove the throttle body.

5. Remove the upper intake manifold brackets.

6. Tag and disconnect the vacuum hoses and PCV hose. Tag and disconnect the intake air thermosensor connector and ground wire.

7. Remove the injector harness bracket and remove the upper intake manifold.

8. Tag and disconnect the vacuum hoses from the lower intake manifold. Disconnect the fuel lines.

9. Remove the fuel supply manifold and the injectors. Remove the injector harness and bracket.

10. Remove the pulsation damper and the intake manifold bracket. Remove the attaching nuts and remove the lower intake manifold.

To install:

11. Clean all gasket mating surfaces.

12. Position a new intake manifold-to-cylinder head gasket and install the lower intake manifold. Tighten the nuts to 14–19 ft. lbs. (19–25 Nm).

13. Install the intake manifold bracket and pulsation damper. Install the injector harness and bracket. Tighten the pulsation damper and injector harness bracket bolts to 69–95 inch lbs. (7.8–11.0 Nm).

14. Install the injectors and the fuel supply manifold. Tighten the fuel supply manifold attaching bolts and tighten to 14–19 ft. lbs. (19–25 Nm).

15. Connect the fuel lines. Connect the vacuum hoses to the lower intake manifold.

16. Position a new gasket and install the upper intake manifold. Tighten the attaching bolts/nuts to 14–19 ft. lbs. (19–25 Nm).

17. Install the injector harness bracket. Connect the ground wire and air thermosensor electrical connector. Connect the PCV hose and the vacuum hoses to the upper intake manifold.

18. Install the upper intake manifold brackets.

19. Position a new gasket and install the throttle body. Tighten the mounting nuts to 14–19 ft. lbs. (19–25 Nm).

20. Connect the electrical connectors at the idle switch, throttle sensor and solenoid valve.

21. Connect the coolant hoses and the accelerator cable. On 2.6L engine, install the air pipe and resonance chamber.

22. Connect the ventilation hose and air intake hose. Connect the negative battery cable.

23. Fill and bleed the cooling system. Run the engine and check for leaks and proper operation.

3.0L ENGINE

1. Relieve the fuel system pressure and disconnect the negative battery cable. Drain the cooling system.

2. Remove the air intake tube from the throttle body. Disconnect the accelerator cable.

3. Disconnect the throttle sensor connector and the coolant hoses. Remove the throttle body.

4. Tag and disconnect the vacuum hoses. Remove the bypass air control valve and the intake air pipe.

5. Remove the extension manifolds. Remove the upper intake plenum with the shutter valve actuator.

6. Remove the fuel supply manifold and the injectors. Disconnect the coolant hoses.

7. Loosen the lower intake manifold nuts, in sequence, in 2 steps and remove the lower intake manifold.

To install:

8. Clean all gasket mating surfaces.

9. Position new lower intake manifold-to-cylinder head gaskets and install the lower intake manifold.

10. Install the intake manifold washers with the white paint mark upward. Install the nuts and tighten, in sequence, in 2 steps to a maximum torque of 14–19 ft. lbs. (19–25 Nm).

11. Install the injectors and the fuel supply manifold. Tighten the attaching bolts to 14–19 ft. lbs. (19–25 Nm).

12. Connect the coolant hoses.

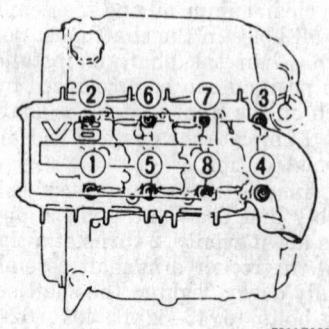

7911C062

Intake manifold mounting nuts removal sequence — 3.0L engine

13. Install a new O-ring on the lower intake manifold and install the upper intake plenum. Apply clean engine oil to new O-rings and install on the extension manifolds. Position new gaskets and install the extension manifolds. Tighten the attaching nuts to 14–19 ft. lbs. (19–25 Nm).

14. Position a new gasket and install the intake air pipe. Install the bypass air control valve. Tighten the attaching bolts/nuts to 14–19 ft. lbs. (19–25 Nm).

15. Position a new gasket and install the throttle body. Tighten the attaching nuts to 14–19 ft. lbs. (19–25 Nm).

16. Connect the coolant and vacuum hoses. Connect the throttle sensor connector and accelerator cable.

17. Adjust the accelerator cable deflection to 0.039–0.118 in. (1–3mm).

18. Connect the air intake tube and the negative battery cable.

19. Fill and bleed the cooling system. Run the engine and check for leaks and proper operation.

4.0L ENGINE

1. Disconnect the negative battery cable and relieve the fuel system pressure.

2. Remove the air cleaner air intake duct from the throttle body.

3. Remove the snow/ice shield and disconnect the throttle cable and bracket assembly.

4. Tag and disconnect the vacuum hoses from the fittings on the upper intake manifold.

5. Tag and disconnect the electrical connectors at the throttle body, upper intake manifold, lower intake manifold and injectors.

6. Disconnect the fuel lines from the fuel supply manifold.

7. Remove the ignition coil and bracket assembly.

8. Remove the mounting nuts and remove the upper intake manifold.

9. Remove the rocker arm covers.

10. Remove the intake manifold attaching bolts and nuts. Tap the manifold lightly with a plastic mallet to break the gasket seal and remove the manifold.

To install:

11. Clean all gasket mating surfaces.

12. Apply silicone sealer to the block and cylinder head mating surfaces at the 4 corners of the lifter valley opening. Install the intake manifold gaskets and again apply sealer to the same location.

13. Position the intake manifold on the 2 guide studs and install the nuts and bolts hand tight. Tighten the bolts, in sequence, in 4 steps, first to

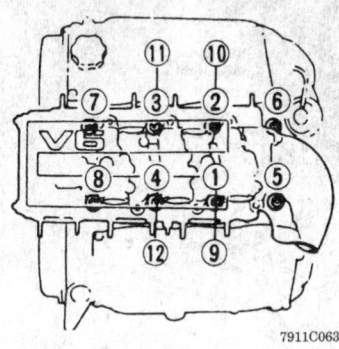

Intake manifold mounting nuts torque
sequence — 3.0L engine

3–6 ft. lbs. (4–8 Nm), then to 6–11 ft.
lbs. (8–15 Nm), then to 11–15 ft. lbs.
(15–21 Nm) and finally to 15–18 ft.
lbs. (21–25 Nm).

14. Apply silicone sealer to the 4 locations where the intake manifold
and the cylinder heads meet. Install
the rocker arm covers with new gaskets and tighten evenly to 3–5 ft. lbs.
(4–7 Nm). Wait 2 minutes and
tighten the bolts again to the same
specification.

15. Install the upper intake manifold and tighten the nuts to 15–18 ft.
lbs. (20–25 Nm).

16. Install the ignition coil and
bracket assembly. Connect the fuel
lines to the fuel supply manifold.

17. Connect the electrical connectors at the throttle body, upper intake manifold, lower intake manifold
and injectors.

18. Connect the vacuum hoses to
the fittings on the upper intake
manifold.

19. Install the throttle cable and
bracket assembly and the snow/ice
shield to the throttle body.

20. Connect the air cleaner air intake duct to the throttle body.

21. Connect the negative battery
cable. Fill and bleed the cooling system. Run the engine and check for
leaks.

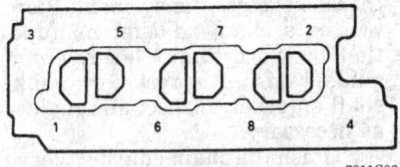

Intake manifold bolts/nuts torque sequence —
4.0L engine

Exhaust Manifold

REMOVAL AND INSTALLATION

1. Disconnect the negative battery
cable. Raise and safely support the
vehicle.

2. Disconnect the exhaust pipe
from the exhaust manifold. If necessary, disconnect the electrical connector from the oxygen sensor.

3. Lower the vehicle.

4. Remove the exhaust manifold
insulator, if equipped.

5. If necessary, remove the engine
oil dipstick and the dipstick tube. On
4.0L engine, if necessary, remove the
power steering pump pressure and
return hoses.

6. On 3.0L engine, disconnect or
remove the exhaust manifold
crossover.

7. If removing the right manifold
on 4.0L engine, remove the heater
hose support bracket and disconnect
the heater hoses.

8. Remove the exhaust manifold
attaching bolts/nuts and remove the
exhaust manifold.

9. Installation is the reverse of the
removal procedure. Tighten the exhaust manifold attaching bolts/nuts
to 19 ft. lbs. (25 Nm).

Timing Chain Front Cover

REMOVAL AND INSTALLATION

2.6L Engine

1. Relieve the fuel system pressure and disconnect the negative battery cable. Drain the cooling system.

2. Remove the air cleaner assembly or disconnect the air intake tube.

3. Remove the cylinder head.

4. Remove the cooling fan and fan
shroud. Remove the accessory drive
belts.

5. Remove the alternator mounting bolts and remove the alternator
and bracket.

6. Remove the power steering
pump mounting bolts and position
the pump and bracket aside.

7. Remove the air conditioning
compressor mounting bolts and position the compressor and bracket
aside.

8. Remove the water pump and
disconnect the bypass pipe, if
equipped.

9. Raise and safely support the vehicle. Remove the splash shield, if
equipped and drain the engine oil.

10. Remove the oil pan and lower
the vehicle.

11. Remove the retaining bolt and
the crankshaft pulley.

12. Remove the attaching bolts and
remove the timing chain cover.

To install:

13. Clean all gasket mating
surfaces.

14. Coat new timing chain cover
gaskets with sealer and install on the
cylinder block. Install the timing
chain cover and tighten the mounting
bolts to 14–19 ft. lbs. (19–25 Nm).

15. Install the crankshaft pulley.
Tighten the retaining bolt to 130–145
ft. lbs. (177–196 Nm).

16. Raise and safely support the vehicle. Install the oil pan and the
splash shield, if equipped.

17. Lower the vehicle and install
the water pump. If equipped, apply
vegetable oil to a new O-ring and install the coolant bypass pipe. Tighten
the attaching bolt to 27–38 ft. lbs.
(37–52 Nm).

18. Install the air conditioning compressor, if equipped, power steering
pump and alternator. Install the accessory drive belts.

19. Install the cooling fan and fan
shroud. Adjust the drive belt tension.

20. Install the cylinder head.

21. Fill the crankcase with the
proper type and quantity of engine
oil.

22. Connect the negative battery
cable. Fill and bleed the cooling
system.

23. Run the engine and check for
leaks and proper operation.

4.0L Engine

1. Disconnect the negative battery
cable and drain the cooling system
and crankcase.

2. Remove the oil pan and the
radiator.

3. Remove the air conditioning
compressor and power steering
bracket, if equipped.

4. Remove the alternator and
drive belt(s). Remove the fan.

5. Remove the water pump and
heater and radiator hoses.

6. Remove the crankshaft pul-
ley/damper assembly and the crankshaft timing sensor.

7. Remove the front cover retaining bolts, noting their positions. If
necessary, tap the cover lightly with
a plastic hammer to break the gasket
seal. Remove the front cover.

To install:

8. Clean all gasket mating surfaces. Apply sealer to the gasket surfaces on the cylinder block and the
back side of the front cover plate. Install the guide sleeves.

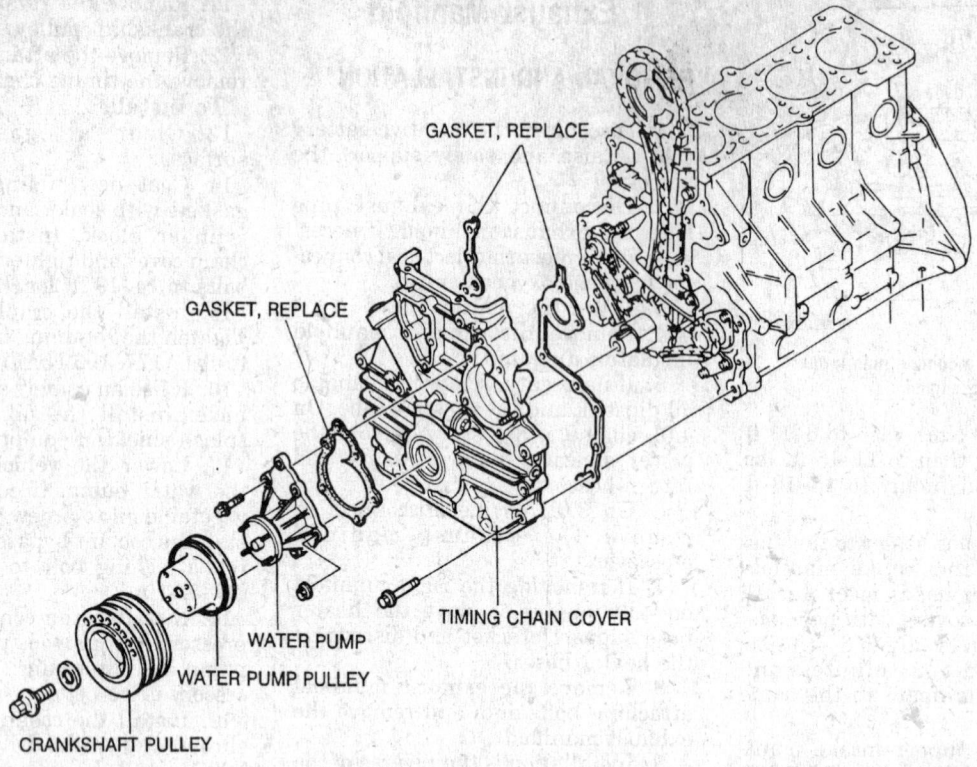

GASKET, REPLACE

GASKET, REPLACE

WATER PUMP

WATER PUMP PULLEY

TIMING CHAIN COVER

CRANKSHAFT PULLEY

7911C065

Timing chain cover installation — 2.6L engine

9. Apply sealer to the front cover gasket surface and position a new gasket on the front cover.

10. Install the front cover with the retaining screws. Note the different bolt lengths. Tighten the bolts to 13–15 ft. lbs. (17–21 Nm).

11. Install the crankshaft timing sensor.

12. Install the crankshaft pulley/damper assembly. Tighten the attaching bolt to 30–37 ft. lbs. (40–50 Nm), then tighten an additional 80–90 degrees.

13. Install the remaining components in the reverse order of their removal. Fill and bleed the cooling system. Run the engine and check for leaks.

Front Cover Oil Seal

REPLACEMENT

1. Disconnect the negative battery cable and drain the cooling system.

2. Remove the cooling fan and fan shroud.

3. Disconnect the upper and lower radiator hoses and remove the radiator, if necessary to provide access.

4. Remove the accessory drive belt(s).

5. Remove the crankshaft pulley.

6. Pry the seal from the front cover, using a suitable prybar. Be careful not to damage the seal housing.

7. Clean the pulley and seal area. Inspect the crankshaft pulley surface for grooving or other damage. Repair or replace, as necessary.

To install:

8. Install a new seal into the cover using a seal installer. The seal must be flush with the edge of the timing chain cover.

9. Apply clean engine oil to the seal lip and the crankshaft pulley. Install the crankshaft pulley and tighten the retaining bolt to 130–145 ft. lbs. (177–196 Nm) or 30–37 ft. lbs. (40–50 Nm), then tighten an additional 80–90 degrees on 4.0L engine.

10. Install the accessory drive belt(s). Install the radiator, if removed.

11. Install the cooling fan and fan shroud. Adjust the drive belt tension.

12. Connect the negative battery cable, start the engine and check for leaks.

Timing Chain and Sprockets

REMOVAL AND INSTALLATION

2.6L Engine

1. Relieve the fuel system pressure and disconnect the negative battery cable. Drain the cooling system and engine oil.

2. Remove the cylinder head, oil pan and timing chain cover.

3. Before replacing any further components, check the following:

a. Check the timing chain tension; if the adjuster sleeve protrudes 13 notches or more, replace the timing chain.

b. Push the chain lever towards the driver's side of the vehicle. If there is excessive movement, there will be a chain adjuster malfunction or worn chain lever, chain guide, camshaft sprocket or crankshaft sprocket. Inspect and replace as necessary.

c. Push the chain adjuster sleeve towards the passenger's side of the vehicle. If it moves back, the chain adjuster ratchet will be faulty and the chain adjuster must be replaced.

4. Remove the crankshaft spacer.

5. Loosen the idler sprocket lock bolt, then remove the balancer shaft chain guides.

6. Remove the idler sprocket assembly, crankshaft balancer chain sprocket and balancer chain.

7. Remove the timing chain adjuster.

8. Remove the camshaft sprocket, timing chain and crankshaft sprocket. Remove the key from the crankshaft.

9. Remove the timing chain lever and chain guide.

10. Inspect all components for damage and/or wear and replace as necessary.

To install:

11. Install the chain guide and tighten the mounting bolts to 61–78 inch lbs. (6.9–8.8 Nm).

12. Install the chain lever and check that it moves smoothly from right to left. Tighten the mounting bolt to 69–95 inch lbs. (7.8–11.0 Nm).

13. Push the chain adjuster sleeve in toward the left and insert a 1.77 in. (45mm) long by 0.08 in. (2mm) diameter pin into the lever hole to hold it in position. Install the chain adjuster and tighten the mounting bolts to 69–95 inch lbs. (7.8–11.0 Nm).

14. Install the key into the crankshaft keyway. Install the crankshaft sprocket.

15. Install the timing chain on the crankshaft sprocket, aligning the 2 white links with the crankshaft sprocket timing mark.

16. Install the camshaft sprocket so the timing mark on the sprocket aligns with the single white link of the timing chain. Secure the timing chain to the sprocket with wire and temporarily rest it between the chain lever and guide.

17. Install the crankshaft balancer sprocket.

18. Set the balancer chain on the idler sprocket assembly so the timing mark on the idler sprocket assembly and the brown link of the balancer chain align.

19. Install the balancer chain so the 5 alignment marks on the chain, sprocket and block align and attach the idler sprocket assembly to the cylinder block. Loosely tighten the idler sprocket assembly lock bolt.

20. Install the right and left lower balancer chain guides and tighten the mounting bolts to 69–95 inch lbs. (7.8–11.0 Nm).

21. Install the upper chain guide and loosely tighten the mounting and adjusting bolts.

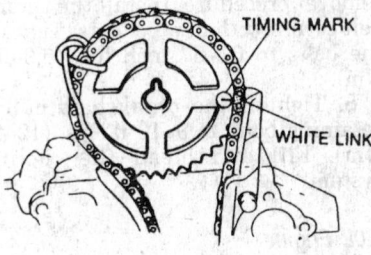

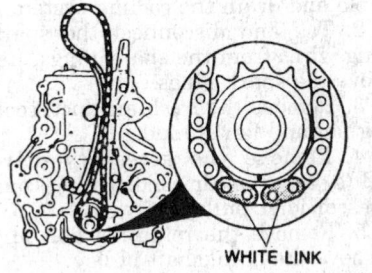

Timing chain and sprockets alignment — 2.6L engine

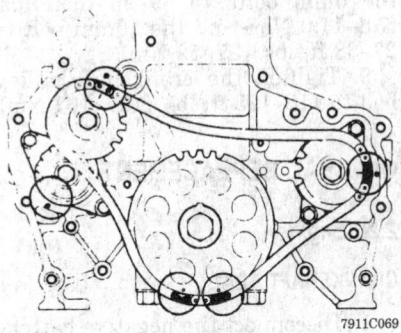

Balancer chain and sprockets alignment — 2.6L engine

22. Tighten the idler sprocket assembly lock bolt to 27–38 ft. lbs. (37–52 Nm) and install the spacer.

23. On 1991 B Series Pick-Up with manual transmission, adjust the balancer chain tension as follows:

a. Tighten the upper chain guide pivot bolt to 69–95 inch lbs. (7.8–11.0 Nm).

b. Loosen the upper chain guide adjusting bolt.

c. Push on the chain guide just above the adjusting slot with a force of approximately 11 lbs., then pull back the guide 0.24 0.012 in.

(6.0 0.3mm) and tighten the bolt to 69–95 inch lbs. (7.8–11.0 Nm).

d. The chain slack at the notch in the guide should be 0.23 in. (5.8mm) when the guide is properly adjusted.

24. On all 1992–95 and 1991 B Series Pick-Up with automatic transmission, adjust the balancer chain tension as follows:

a. Fabricate a piece of wood, 0.118–0.138 in. (3.0–3.5mm) thick and 0.335–0.374 in. (8.5–9.5mm) wide.

b. Insert the piece of wood in the notch in the upper chain guide.

c. Push on the chain guide just above the adjusting slot with a force of 2.9–3.7 lbs. and tighten the adjusting and pivot bolts to 69–95 inch lbs. (7.8–11.0 Nm).

d. Remove the wood from between the chain and chain guide, making sure no wood shavings are left.

e. Measure the chain slack. It should be 0.039–0.059 in. (1.0–1.5mm) at the notch in the guide.

NOTE: If the upper chain guide bottoms on the adjusting bolt during the adjustment procedure, the balancer chain must be replaced.

25. Install the remaining components in the reverse order of removal.

NOTE: Be sure to remove the pin from the timing chain adjuster before installing the service cover.

26. Fill the crankcase with the proper type and quantity of oil. Fill and bleed the cooling system.

27. Run the engine and check for leaks and proper operation. Check the idle speed and ignition timing.

4.0L Engine

1. Disconnect the negative battery cable and drain the cooling system and crankcase.

2. Remove the oil pan and radiator. Remove the accessory drive belt and crankshaft damper.

3. Remove the water pump and timing chain front cover.

4. Remove the camshaft sprocket retaining bolt and the crankshaft sprocket key.

5. Push the timing chain tensioner into the retracted position and install the retaining clip.

6. Remove the crankshaft and camshaft sprockets with the timing chain. Remove the tensioner and guide, as required.

To install:

7. Install the timing chain guide to the cylinder block with the pin of the guide inserted into the oil hole in the block. Install the 2 retaining bolts and tighten to 7–9 ft. lbs. (10–12 Nm).

8. Position the camshaft and crankshaft so the sprocket timing marks will align.

9. Install the sprockets and timing chain together. Install the timing chain tensioner with the clip in place to lock the tensioner in the retracted position.

10. Install the crankshaft key and check the timing marks on the sprockets for correct alignment. Make sure the tensioner side of the timing chain is held inward and the guide side of the chain is straight and tight.

11. Install the camshaft sprocket retaining bolt and tighten to 44–50 ft. lbs. (60–68 Nm). Remove the clip from the tensioner assembly.

12. Install the timing chain front cover and the remaining components in the reverse order of their removal. Fill the crankcase with the proper type and quantity of engine oil. Fill and bleed the cooling system. Run the engine and check for leaks.

Timing Belt Front Cover

REMOVAL AND INSTALLATION

2.2L Engine

1. Disconnect the negative battery cable and drain the cooling system.

2. Remove the cooling fan and fan shroud.

3. Remove the accessory drive belt(s) and the cooling fan pulley and bracket.

4. If carburetor equipped, remove the secondary air pipe assembly.

5. Remove the retaining bolts and the crankshaft pulley.

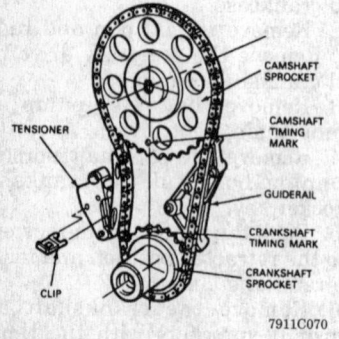

Timing chain and sprockets alignment — 4.0L engine

6. Remove the upper and lower timing belt covers.

7. Installation is the reverse of the removal procedure. Install the timing belt covers with new gaskets. Tighten the bolts to 61–87 inch lbs. (6.9–9.8 Nm).

8. Tighten the crankshaft pulley retaining bolts to 9–13 ft. lbs. (12–17 Nm). Fill and bleed the cooling system.

3.0L Engine

1. Disconnect the negative battery cable and drain the cooling system.

2. Tag and disconnect the spark plug wires from the spark plugs. Remove the spark plugs.

3. Remove the fresh air duct, cooling fan and fan shroud.

4. Remove the accessory drive belt(s) and the air conditioning compressor idler pulley.

5. Remove the retaining bolt and remove the crankshaft pulley.

6. Remove the coolant bypass hose and upper radiator hose.

7. Remove the timing belt covers and gaskets.

8. Installation is the reverse of the removal procedure. Install the timing belt covers with new gaskets. Tighten the 6mm bolts to 69–95 inch lbs. (7.8–11.0 Nm) and the 10mm bolt to 27–38 ft. lbs. (37–52 Nm).

9. Tighten the crankshaft pulley bolt to 116–123 ft. lbs. (157–167 Nm).

OIL SEAL REPLACEMENT

2.2L Engine

CRANKSHAFT SEAL

1. Disconnect the negative battery cable and drain the engine oil.

2. Remove the timing belt cover and timing belt.

3. Remove the crankshaft sprocket retaining bolt and the crankshaft sprocket.

4. Remove the oil seal using a small prybar. Be careful not to damage the seal housing.

To install:

5. Apply clean engine oil to the lip of a new seal and fit the seal into the oil pump body.

6. Tap the seal in place using a seal installer until the seal is flush with the edge of the pump body.

7. Install the crankshaft sprocket and tighten the retaining bolt to 116–123 ft. lbs. (157–167 Nm).

8. Install the timing belt and timing belt cover. Connect the negative battery cable.

9. Fill the engine with the proper type and quantity of engine oil. Run the engine and check for leaks.

CAMSHAFT SEAL

1. Disconnect the negative battery cable and drain the cooling system.

2. Remove the timing belt cover and the timing belt.

3. Remove the camshaft sprocket retaining bolt and remove the sprocket.

4. Disconnect the upper radiator hose and remove the distributor.

5. Remove the front housing.

6. Pry or press the oil seal from the front housing.

To install:

7. Clean all gasket mating surfaces.

8. Apply engine oil to the front housing and a new oil seal. Press the seal into the front housing.

9. Apply engine oil to the seal lip and install the front housing with a new gasket. Tighten the mounting bolts to 14–19 ft. lbs. (19–25 Nm).

10. Connect the upper radiator hose and install the distributor.

11. Install the camshaft sprocket and tighten the retaining bolt to 35–48 ft. lbs. (47–65 Nm).

12. Install the timing belt and the timing belt cover. Connect the negative battery cable.

13. Fill and bleed the cooling system. Run the engine and check for leaks. Check the ignition timing.

3.0L Engine

CRANKSHAFT SEAL

1. Disconnect the negative battery cable. Drain the engine oil and the cooling system.

2. Remove the timing belt cover and timing belt. Remove the crankshaft sprocket.

3. Remove the thermostat assembly.

4. Remove the oil pan and oil pump pickup.

5. Remove the oil pump.

6. Remove the oil seal from the pump using a suitable driver. Be careful not to damage the seal housing.

To install:

7. Coat the lip of a new seal and the seal housing with clean engine oil. Install the seal using a seal installer.

8. Install the oil pump, pickup and oil pan.

9. Install the thermostat assembly.

10. Install the crankshaft sprocket, timing belt and timing belt covers.

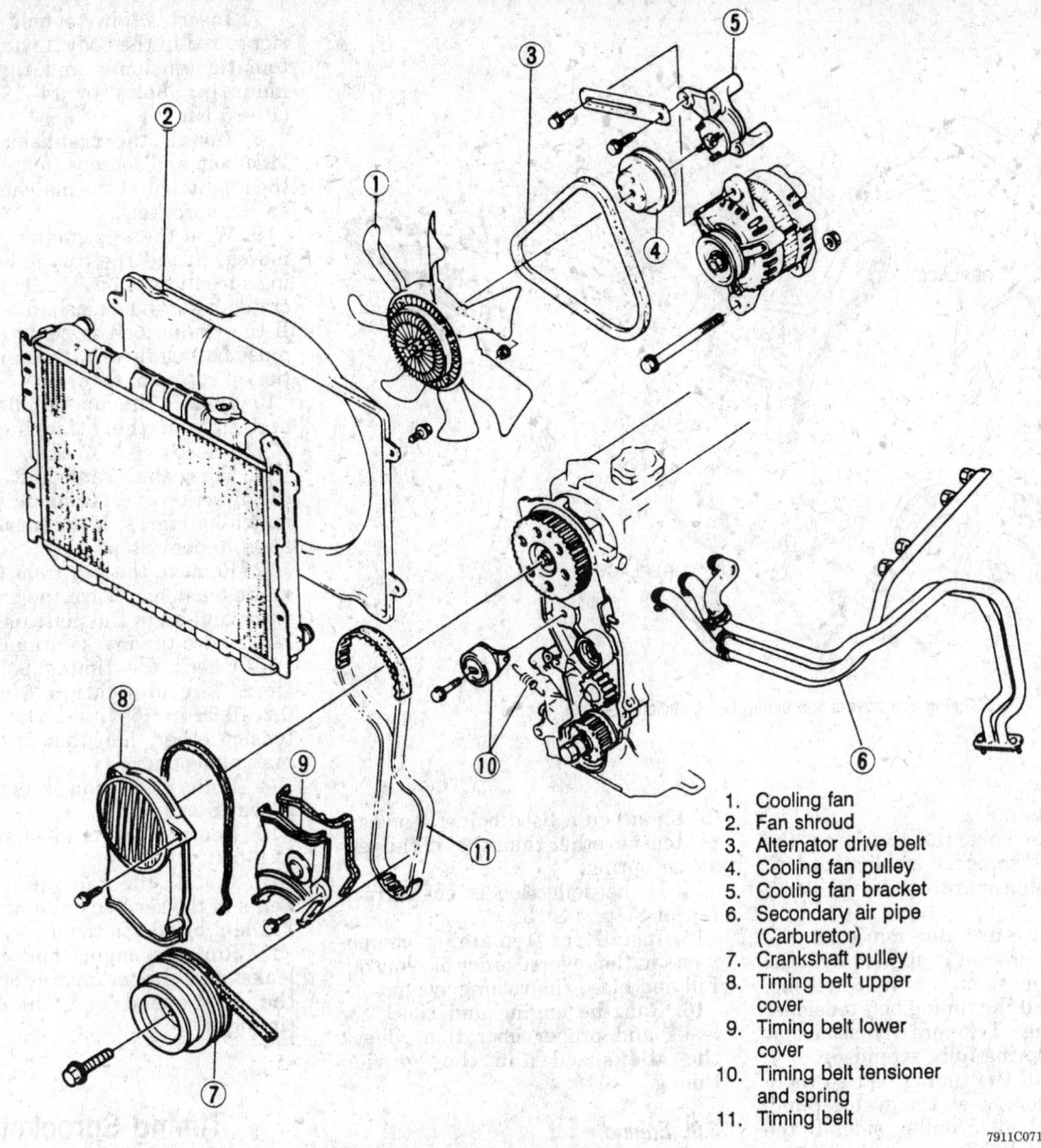

1. Cooling fan
2. Fan shroud
3. Alternator drive belt
4. Cooling fan pulley
5. Cooling fan bracket
6. Secondary air pipe
   (Carburetor)
7. Crankshaft pulley
8. Timing belt upper
   cover
9. Timing belt lower
   cover
10. Timing belt tensioner
    and spring
11. Timing belt

7911C071

Timing belt covers and timing belt installation — 2.2L engine

11. Fill the crankcase with the proper type and quantity of engine oil. Connect the negative battery cable.

12. Fill and bleed the cooling system. Run the engine and check for leaks.

CAMSHAFT SEAL

1. Disconnect the negative battery cable.

2. Remove the timing belt cover and timing belt.

3. Remove the camshaft sprocket retaining bolt and remove the camshaft sprocket.

4. Remove the seal plate and the camshaft seal.

To install:

5. Lubricate the oil seal lip with clean engine oil and install in the cylinder head, using a seal installer.

6. Install the seal plate and tighten the bolts to 69–95 inch lbs. (7.8–11.0 Nm).

7. Install the camshaft sprocket and tighten the retaining bolt to 52–59 ft. lbs. (71–80 Nm).

8. Install the timing belt and timing belt cover. Connect the negative battery cable, start the engine and check for leaks.

Timing Belt and Tensioner

REMOVAL AND INSTALLATION

2.2L Engine

1. Disconnect the negative battery cable and drain the cooling system.

2. Remove the timing belt cover.

3. Turn the crankshaft to align the mark of the camshaft sprocket with the front housing matching mark.

4. Remove the tensioner and spring. Mark the timing belt direction of rotation if it is to be reused.

5. Remove the timing belt.

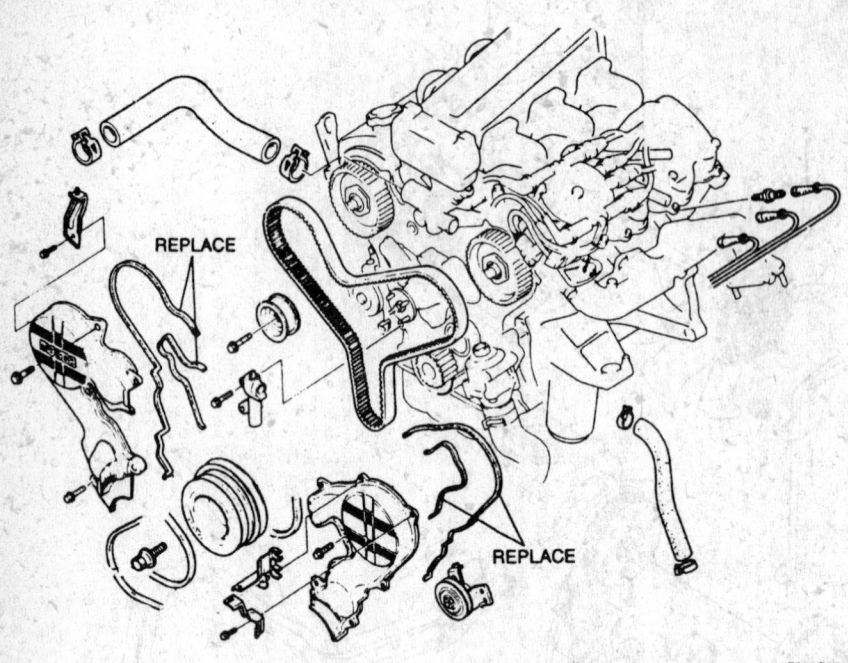

REPLACE

REPLACE

7911C072

Timing belt covers and timing belt installation — 3.0L engine

To install:

6. Make sure the mark on the crankshaft sprocket is aligned with the matching mark on the oil pump body.

7. Make sure the mark on the camshaft sprocket is aligned with the matching mark on the front housing.

8. Install the timing belt tensioner and spring. Temporarily secure it with the spring fully extended.

9. Install the timing belt so there is no looseness at the water pump pulley and idler pulley side. If the timing belt is being reused, it must be installed in the same direction of rotation.

10. Remove the spark plugs to make engine rotation easier.

11. Turn the crankshaft twice clockwise in the direction of rotation. Make sure the matching marks are correctly aligned; if not repeat the installation procedure.

12. Loosen the tensioner lock bolt and apply tension to the belt. Tighten the tensioner lock bolt to 27–38 ft. lbs. (37–52 Nm).

13. Turn the crankshaft twice clockwise in the direction of rotation and align the matching marks. Check the timing belt deflection. The deflection should be 0.31–0.35 in. (8–9mm) on a new belt or 0.35–0.39 in.

(9–10mm) on a used belt. Do not apply tension other than that of the tensioner spring.

14. If the deflection is not correct, repeat Steps 11–13.

15. Install the remaining components in the reverse order of removal. Fill and bleed the cooling system.

16. Run the engine and check for leaks and proper operation. Check the idle speed and the ignition timing.

3.0L Engine

1. Disconnect the negative battery cable and drain the cooling system.

2. Remove the timing belt cover.

3. Remove the upper idler pulley.

4. Turn the crankshaft to align the matching marks on the sprockets. If the timing belt is to be reused, make an arrow on the belt to indicate rotation direction.

5. Remove the timing belt and automatic tensioner.

To install:

6. Set a plane washer at the bottom of the tensioner body to prevent damage to the body plug. Press in the tensioner rod slowly, using a press or a vise.

NOTE: Do not press the tensioner rod more than 2200 lbs.

7. Insert a pin to hold the tensioner rod in the body. Install the automatic tensioner and tighten the mounting bolts to 14–19 ft. lbs. (19–25 Nm).

8. Install the crankshaft pulley lock bolt and loosely tighten. Check the alignment of the matching marks on the sprockets.

9. With the upper idler pulley removed, install the timing belt, making sure there is no slack between the crankshaft and camshaft sprockets. If the timing belt is being reused, it must be installed in the same direction of rotation.

10. Install the upper idler pulley and tighten the attaching bolt to 27–38 ft. lbs. (37–52 Nm).

11. Turn the crankshaft twice in the direction of rotation and align the matching marks. If the marks do not align, repeat Steps 8–11.

12. Remove the pin from the automatic tensioner. Turn the crankshaft twice and align the matching marks. Make sure the marks are aligned.

13. Check the timing belt deflection. The deflection should be 0.20–0.28 in. (5–7mm). Do not apply tension other than that of the automatic tensioner.

14. If the deflection is not correct, repeat Steps 10–13.

15. Remove the crankshaft pulley lock bolt.

16. Install the remaining components in the reverse order of removal. Fill and bleed the cooling system.

17. Run the engine and check for leaks and proper operation. Check the idle speed and the ignition timing.

Timing Sprockets

REMOVAL AND INSTALLATION

1. Disconnect the negative battery cable.

2. Remove the timing belt cover and the timing belt.

3. Remove the sprocket retaining bolt and remove the sprocket.

4. Installation is the reverse of the removal procedure. Tighten the camshaft sprocket bolt to 35–48 ft. lbs. (47–65 Nm) on 2.2L engine or 52–59 ft. lbs. (71–80 Nm) on 3.0L engine. Tighten the crankshaft sprocket bolt to 116–123 ft. lbs. (157–167 Nm) on 2.2L engine.

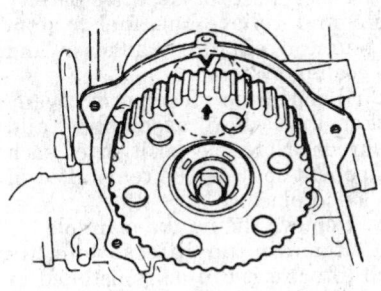

Timing belt sprocket matching marks — 2.2L engine

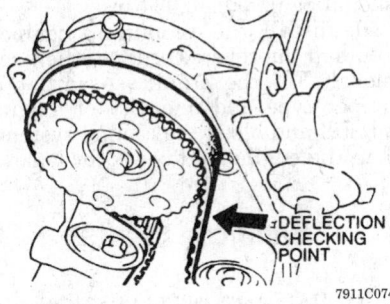

DEFLECTION CHECKING POINT

7911C074

Timing belt deflection checking point — 2.2L engine

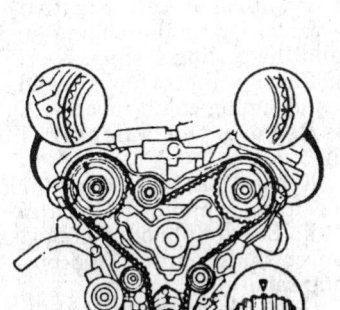

7911C075

Timing belt sprocket matching marks — 3.0L engine

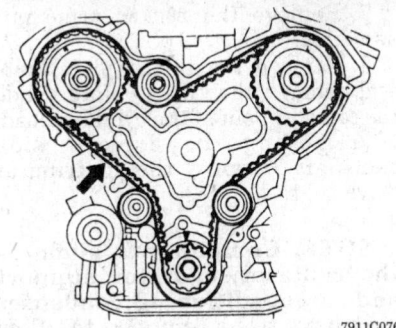

7911C076

Timing belt deflection checking point — 3.0L engine

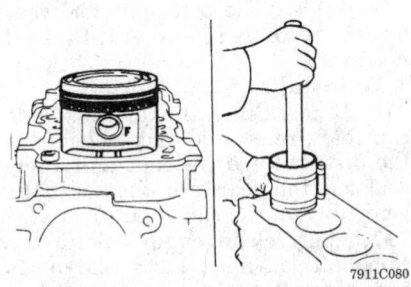

7911C080

Install the piston with the F mark facing the front of the engine — 2.2L and 2.6L engine

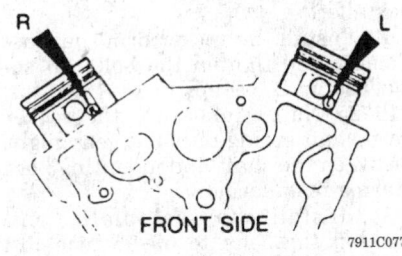

R   L

FRONT SIDE   7911C077

Install the piston with the L mark (left bank) and the R mark (right bank) facing the front of the engine — 3.0L engine

Camshaft

REMOVAL AND INSTALLATION

2.2L Engine

1. Disconnect the negative battery cable and drain the cooling system. Remove the air cleaner assembly or air intake tube.

2. Remove the timing belt cover and the timing belt. Remove the camshaft sprocket.

3. Remove the rocker arm cover and rocker arm/shaft assembly.

4. Disconnect the upper radiator hose from the thermostat housing.

5. Remove the distributor.

6. Remove the front housing.

7. Remove the seal cap from the rear of the cylinder head and remove the camshaft.

8. Check the camshaft for wear and/or damage and replace, as necessary.

To install:

9. Clean all gasket mating surfaces.

10. Apply clean engine oil to the camshaft journals, lobes and bearings. Install the camshaft with the dowel pin facing straight up.

11. Apply silicone sealer to the cylinder head on the front and rear camshaft bearing cap mating surface. Install the rocker arm/shaft assembly and tighten the bolts in sequence, in 2–3 steps to 13–20 ft. lbs. (18–26 Nm).

12. Install a new camshaft seal in the front housing. Apply clean engine oil to the seal lip and install the housing using a new gasket. Tighten the bolts to 14–19 ft. lbs.

13. Apply silicone sealer to the cylinder head and both sides of the front and rear camshaft bearing caps where the caps meet the cylinder head. Install the rocker arm cover and tighten the bolts to 26–35 inch lbs. (2.9–3.9 Nm).

14. Install the camshaft sprocket and tighten the retaining bolt to 35–48 ft. lbs. (47–65 Nm).

15. Install the timing belt and timing belt cover.

16. Install the distributor.

17. Connect the upper radiator hose.

18. Install the remaining components in the reverse order of removal. Fill and bleed the cooling system.

19. Run the engine and check for leaks and proper operation. Check the idle speed and the ignition timing.

2.6L Engine

1. Disconnect the negative battery cable.

2. Remove the air cleaner assembly or air intake tube.

3. Remove the rocker arm cover.

4. Remove the distributor.

5. Remove the rocker arm/shaft assembly.

6. Remove the service cover on the timing chain cover. Push the chain adjuster sleeve toward the left and insert a 1.77 in. (45mm) long by 0.08 in. (2mm) diameter pin into the lever hole to hold it.

7. Remove the bolt and the distributor drive gear. Remove the camshaft sprocket from the camshaft and hold it in position with mechanics wire.

NOTE: Do not allow the timing chain to become disconnected from the camshaft sprocket or engine timing will be disturbed.

8. Remove the camshaft. Inspect the camshaft for wear or damage and replace, if ne. Inspect the camshaft for wear or damage and replace, if necessary.

To install:

9. Install the camshaft, aligning the dowel with the camshaft sprocket.

10. If equipped, coat the circular packing with sealant and install in the end of the cylinder head.

11. Install the rocker arm/shaft assembly and tighten the bolts, in sequence, in 2–3 steps to 14–19 ft. lbs. (19–25 Nm).

12. Install the camshaft sprocket to the camshaft. Install the distributor drive gear and lock bolt. Tighten the lock bolt to 36–43 ft. lbs. (49–58 Nm).

13. Remove the chain adjuster sleeve retaining pin and install the service cover with a new gasket. Tighten the bolts to 69–95 inch lbs. (7.8–11.0 Nm) and the nuts to 61–87 inch lbs. (6.9–9.8 Nm).

14. Apply sealant to the half circle seal(s) and install in the cylinder head. Apply sealer to the tops of the seals and to the cylinder head in the seal area.

15. Coat a new gasket with sealant and install the rocker arm cover. Tighten the bolts to 52–78 inch lbs. (5.9–8.8 Nm).

16. Install the distributor.

17. Install the air cleaner assembly or air intake tube. Connect the negative battery cable.

18. Run the engine and check for leaks and proper operation. Check the idle speed and ignition timing.

3.0L Engine

1. Disconnect the negative battery cable and drain the cooling system.

2. Remove the timing belt covers and timing belt.

3. Remove the rocker arm cover.

4. If removing the driver's side camshaft, remove the distributor and the distributor spacer.

5. Remove the bolt and the camshaft sprocket.

6. Remove the seal plate. Pry out the camshaft seal, being careful not to damage the seal housing.

7. Remove the rocker arm/shaft assembly.

8. Remove the thrust plate bolts and remove the thrust plate. Slide the camshaft out of the cylinder head. If removing the driver's side camshaft, remove the distributor drive gear.

NOTE: Components such as the radiator, radiator support and air conditioning condenser may have to be removed to allow camshaft removal. It may be necessary to remove the cylinder head to remove the camshaft.

9. Inspect the camshaft for wear and/or damage and replace if necessary.

To install:

10. If installing the driver's side camshaft, remove all old sealer from the distributor drive gear and apply sealer to the gear face, then seat the gear fully on the camshaft.

11. Apply clean engine oil to the camshaft journals, lobes and bearings. Install the camshaft and the thrust plate. Tighten the thrust plate to 69–95 inch lbs. (7.8–11.0 Nm).

12. Apply clean engine oil to a new camshaft seal lip and press the seal into the cylinder head, using a seal installer.

13. Install the rocker arm/shaft assembly and tighten the bolts, in sequence, in 2–3 steps to 14–19 ft. lbs. (19–25 Nm). Make sure the rocker arm shaft spring does not get caught between the shaft and mounting boss during installation.

14. Install the seal plates and tighten the bolts to 69–95 inch lbs. (7.8–11.0 Nm).

15. Install the camshaft sprocket and tighten the bolt to 52–59 ft. lbs. (71–80 Nm).

16. If installing the driver's side camshaft, apply clean engine oil to a new O-ring and install on the distributor spacer. Install the spacer and tighten the nuts to 69–95 inch lbs. (7.8–11.0 Nm).

17. Coat a new gasket with silicone sealant and install on the rocker arm cover. Install the cover and tighten the bolts to 30–39 inch lbs. (3.4–4.4 Nm).

18. Install the timing belt and timing belt covers.

19. Install the distributor.

20. Connect the negative battery cable. Fill and bleed the cooling system.

21. Run the engine and check for leaks and proper operation. Check the idle speed and ignition timing.

4.0L Engine

1. Disconnect the negative battery cable and relieve the fuel system pressure. Drain the crankcase and the cooling system.

2. Remove the rocker arm covers, rocker arm shaft assemblies and pushrods. Note the position of each component so it can be reinstalled in the same place.

3. Remove the intake manifold.

4. Remove the lifters. Identify each lifter so it can be reinstalled in the original position.

5. Remove the front timing chain cover and the timing chain and sprockets.

6. Remove the thrust plate bolts and remove the thrust plate. Carefully remove the camshaft, being careful not to damage the journals, lobes or bearings.

To install:

7. Coat the camshaft lobes with grease and the journals with heavy engine oil. Carefully install the camshaft, being careful not to damage the journals, lobes or bearings.

8. Install the thrust plate and the thrust plate retaining bolts. Tighten the bolts to 7–10 ft. lbs. (10–13 Nm).

9. Check the camshaft endplay using a dial indicator. The endplay should be 0.0008–0.004 in.

10. Install the remaining components in the reverse order of their removal. Fill the crankcase with the proper type and quantity of engine oil. Fill and bleed the cooling system. Run the engine and check for leaks.

Balancer Shafts

REMOVAL AND INSTALLATION

2.6L Engine

1. Relieve the fuel system pressure and disconnect the negative battery cable. Drain the engine engine oil and the cooling system.

2. Remove the cylinder head, oil pan and timing chain cover.

3. Remove the balancer shaft chain.

4. Remove the thrust plate lock bolts and remove the balancer shaft(s). Be careful not to damage the balancer shaft journals and bushing during removal.

5. Check the balancer shaft(s) and bushings for wear and/or damage and replace as necessary.

To install:

6. Apply clean engine oil to the balancer shaft journals and install in

the cylinder block, being careful not to damage the bushings and journals.

7. Loosely tighten the thrust plate lock bolts and make sure the balancer shaft(s) rotate smoothly. Tighten the lock bolts to 69–95 inch lbs. (7.8–11.0 Nm).

8. Install the crankshaft balancer sprocket.

9. Set the balancer chain on the idler sprocket assembly so the timing mark on the idler sprocket assembly and the brown link of the balancer chain align.

10. Install the balancer chain so the 5 alignment marks on the chain, sprocket and block align and attach the idler sprocket assembly to the cylinder block. Loosely tighten the idler sprocket assembly lock bolt.

11. Install the right and left lower balancer chain guides and tighten the mounting bolts to 69–95 inch lbs. (7.8–11.0 Nm).

12. Install the upper chain guide and loosely tighten the mounting and adjusting bolts.

13. Tighten the idler sprocket assembly lock bolt to 27–38 ft. lbs. (37–52 Nm) and install the spacer.

14. On 1991 B Series Pick-Up with manual transmission, adjust the balancer chain tension as follows:

   a. Tighten the upper chain guide pivot bolt to 69–95 inch lbs. (7.8–11.0 Nm).

   b. Loosen the upper chain guide adjusting bolt.

   c. Push on the chain guide just above the adjusting slot with a force of approximately 11 lbs., then pull back the guide 0.24 0.012 in. (6.0 0.3mm) and tighten the bolt to 69–95 inch lbs. (7.8–11.0 Nm).

   d. The chain slack at the notch in the guide should be 0.23 in. (5.8mm) when the guide is properly adjusted.

15. On all 1992–95 and 1991 B Series Pick-Up with automatic transmission, adjust the balancer chain tension as follows:

   a. Fabricate a piece of wood, 0.118–0.138 in. (3.0–3.5mm) thick and 0.335–0.374 in. (8.5–9.5mm) wide.

   b. Insert the piece of wood in the notch in the upper chain guide.

   c. Push on the chain guide just above the adjusting slot with a force of 2.9–3.7 lbs. and tighten the adjusting and pivot bolts to 69–95 inch lbs. (7.8–11.0 Nm).

   d. Remove the wood from between the chain and chain guide, making sure no wood shavings are left.

   e. Measure the chain slack. It should be 0.039–0.059 in. (1.0–1.5mm) at the notch in the guide.

NOTE: If the upper chain guide bottoms on the adjusting bolt during the adjustment procedure, the balancer chain must be replaced.

16. Install the remaining components in the reverse order of removal.

NOTE: Be sure to remove the pin from the timing chain adjuster before installing the service cover.

17. Fill the crankcase with the proper type and quantity of oil. Fill and bleed the cooling system.

18. Run the engine and check for leaks and proper operation. Check the idle speed and ignition timing.

Piston and Connecting Rod

POSITIONING

ENGINE LUBRICATION

Oil Pan

REMOVAL AND INSTALLATION

B Series Pick-Up

1. Disconnect the negative battery cable.

2. Raise and support the vehicle safely.

3. Remove the necessary splash shields.

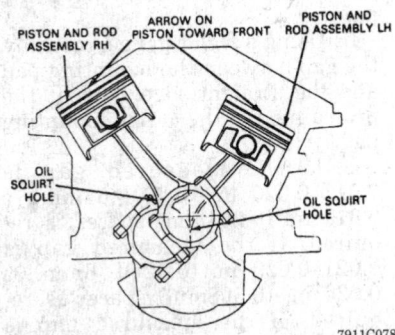

ARROW ON PISTON TOWARD FRONT
PISTON AND ROD ASSEMBLY RH
PISTON AND ROD ASSEMBLY LH
OIL SQUIRT HOLE
OIL SQUIRT HOLE

7911C078

Piston and connecting rod positioning — 4.0L engine

4. Position a drain pan under the oil pan. Remove the drain plug and drain the engine oil.

5. If equipped with 4WD, remove the front differential.

6. Remove the necessary steering linkage.

7. Remove the gusset plates and the clutch housing cover, if equipped.

8. Remove the oil pan mounting bolts and remove the oil pan. If necessary, insert a scraper or other suitable tool between the oil pan and block to separate them. Be careful not to bend the pan.

To install:

9. Clean all gasket mating surfaces and the oil p

10. Apply sealant to the cylinder block and install a new gasket.

11. Install the oil pan with the attaching bolts. Tighten the bolts to 61–104 inch lbs. (6.9–12.0 Nm) on 2.2L engine, or 69–95 inch lbs. (7.8–11.0 Nm) on 2.6L engine.

NOTE: Apply Loctite® to the bolt threads on 2.6L engine before installation.

12. Install the remaining components in the reverse order of removal. Lower the vehicle and connect the negative battery cable.

13. Fill the crankcase with the proper type and quantity of engine oil. Run the engine and check for leaks.

MPV

2.6L ENGINE

1. Disconnect the negative battery cable. Remove the fan shroud.

2. Install engine support tool 49 G017 5A0 or equivalent. Remove the engine mount nuts and lift the engine slightly to gain removal clearance.

3. Raise and safely support the vehicle. Support the transmission with a transmission jack and remove the transmission lower mount.

4. Remove the splash shield.

5. Position a drain pan under the oil pan. Remove the drain plug and drain the engine oil.

6. Remove the gusset plates and stabilizer bar brackets.

7. Remove the oil pan mounting bolts and remove the oil pan. If necessary, insert a scraper or other suitable tool between the oil pan and block to separate them. Be careful not to bend the pan.

To install:

8. Clean all gasket mating surfaces and the oil pan.

9. Apply sealant to the cylinder block and install a new gasket.

10. Install the oil pan with the attaching bolts. Tighten the bolts to 69–95 inch lbs. (7.8–11.0 Nm). Apply Loctite® to the bolt threads before installation.

11. Install the remaining components in the reverse order of removal. Lower the vehicle and remove the engine support tool. Install the engine mount nuts.

12. Connect the negative battery cable. Fill the crankcase with the proper type and quantity of engine oil. Run the engine and check for leaks.

3.0L ENGINE — 2WD

1. Disconnect the negative battery cable.

2. Raise and safely support the vehicle.

3. Remove the splash shield.

4. Position a drain pan under the oil pan. Remove the drain plug and drain the engine oil.

5. Remove the gusset plates.

6. Remove the oil pan mounting bolts and remove the oil pan. If necessary, insert a scraper or other suitable tool between the oil pan and block to separate them. Be careful not to bend the pan.

To install:

7. Clean all gasket mating surfaces and the oil pan.

8. Apply sealant to the cylinder block and install a new gasket.

9. Install the oil pan with the attaching bolts. Tighten to 61–87 inch lbs. (6.9–9.8 Nm).

10. Install the remaining components in the reverse order of removal. Lower the vehicle and connect the negative battery cable.

11. Fill the crankcase with the proper type and quantity of engine oil. Run the engine and check for leaks.

3.0L ENGINE — 4WD

1. Disconnect the negative battery cable. Remove the fresh air duct and fan shroud.

2. Install engine support tool 49 G017 5A0 or equivalent. Remove the engine mount nuts and lift the engine slightly to gain removal clearance.

3. Raise and safely support the vehicle. Support the transmission with a transmission jack and remove the transmission lower mount.

4. Remove the splash shields.

5. Position a drain pan under the oil pan. Remove the drain plug and drain the engine oil.

6. If equipped with automatic transmission, disconnect and reposition the oil cooler hose and tube.

7. Remove the stabilizer bar.

8. Remove the oil pan mounting bolts and remove the oil pan. If necessary, insert a scraper or other suitable tool between the oil pan and block to separate them. Be careful not to bend the pan.

To install:

9. Clean all gasket mating surfaces and the oil pan.

10. Apply sealant to the cylinder block and install a new gasket.

11. Install the oil pan with the attaching bolts. Tighten to 61–87 inch lbs. (6.9–9.8 Nm).

12. Install the remaining components in the reverse order of removal. Lower the vehicle and remove the engine support tool. Install the engine mount nuts.

13. Connect the negative battery cable. Fill the crankcase with the proper type and quantity of engine oil. Run the engine and check for leaks.

Navajo

1. Remove the engine assembly and install on a workstand with the oil pan facing up.

2. Remove the oil pan retaining bolts and remove the pan.

To install:

3. Clean all gasket mating surfaces and the oil pan.

4. Install a new crankshaft rear main bearing cap wedge seal. The seal should fit snugly into the sides of the rear main bearing cap.

5. Position a new oil pan gasket to the engine block and place the oil pan in position on the 4 locating studs. Tighten the retaining nuts and bolts evenly to 5–7 ft. lbs. (7–10 Nm).

6. Measure the gap between the surface of the rear face of the oil pan, at the spacer locations, and the rear face of the engine block as follows:

a. With the oil pan installed on the engine, position a straight-edge flat on the rear of the engine block so it extends over one of the oil pan/transmission bolt mounting pads.

b. Using a feeler gauge, measure the gap between the mounting pad and the straight-edge. Repeat the procedure for the other mounting pad.

c. If the measured gap is 0.011–0.020 in. (0.27–0.51mm), a 0.010 in. (0.254mm) spacer is required. If the measured gap is 0.021–0.029 in. (0.52–0.76mm), a 0.020 in. (0.508mm) spacer is required. If the measured gap is 0.030–0.039 in. (0.77–1.00mm), a 0.030 in. (0.762mm) spacer is required.

d. Install the selected spacers to the mounting pads on the rear of the oil pan before bolting the engine and transmission together.

NOTE: Failure to use the correct spacer can result in improper clearance between the oil pan and transmission, resulting in oil pan damage and/or an oil leak.

7. Remove the engine from the workstand and install in the vehicle.

Oil Pump

REMOVAL AND INSTALLATION

2.2L and 2.6L Engines

1. Disconnect the negative battery cable and drain the cooling system.

2. Raise and safely support the vehicle. Remove the splash shield and drain the engine oil.

3. Remove the oil pan, stiffener and oil pump pickup.

4. Lower the vehicle.

5. On 2.2L engine, remove the timing belt cover, timing belt and crankshaft sprocket. Remove the oil pump mounting bolts and remove the oil pump.

6. On 2.6L engine, remove the cylinder head and the timing chain cover. The oil pump is part of the timing chain cover.

7. If necessary, disassemble the oil pump and check for wear and/or damage. Replace parts as necessary.

8. Pry out the crankshaft oil seal, being careful not to damage the seal housing.

To install:

9. Clean all gasket material.

10. Apply clean engine oil to the lip of a new seal and press the seal into the pump, using a seal installer.

11. On 2.2L engine, proceed as follows:

a. Apply a continuous bead of silicone sealant to the contact surface of the oil pump. Do not allow sealant to get into the oil hole.

b. Lubricate a new O-ring and install into the pump body.

c. Install the oil pump and tighten the **A** bolts to 14–19 ft. lbs. (19–25 Nm) and the **B** bolts to 27–38 ft. lbs. (37–52 Nm).

d. Install the crankshaft sprocket, timing belt and timing belt cover.

12. On 2.6L engine, install the timing chain cover and cylinder head.

13. Raise and safely support the vehicle.

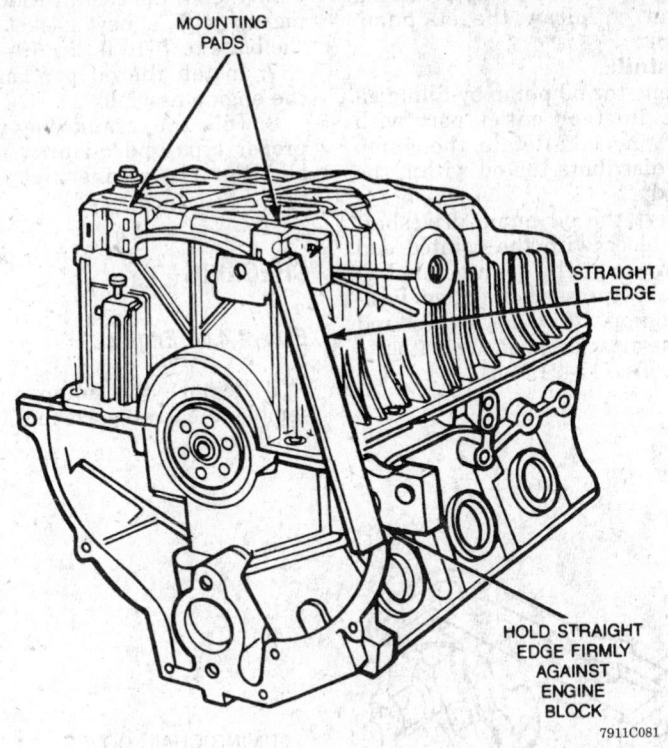

Measuring oil pan-to-transmission gap — Navajo

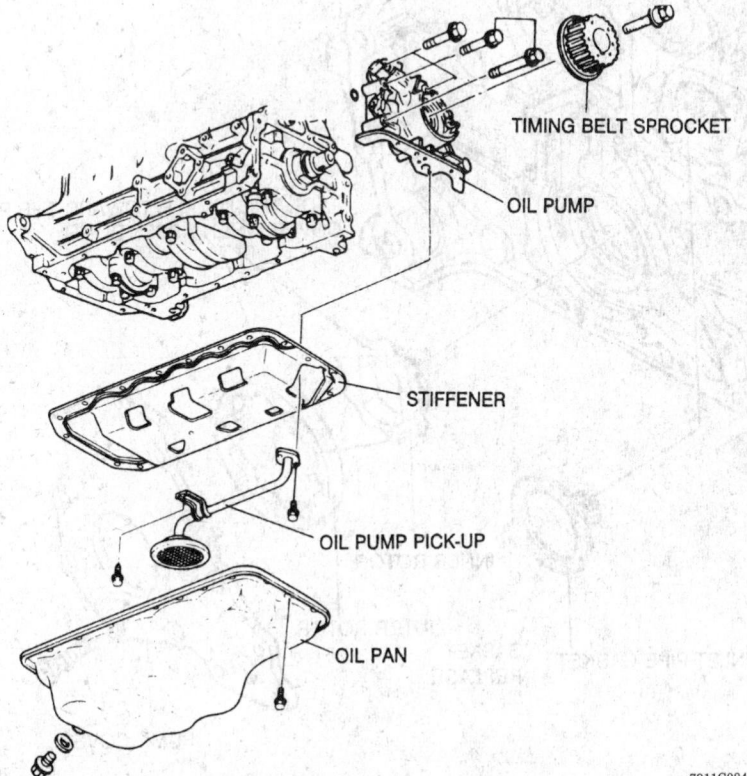

Oil pan and pump removal — 2.2L engine

14. Install the stiffener. Install a new gasket and the oil pump pickup. Tighten the bolts to 69–95 inch lbs. (7.8–11.0 Nm).

15. Install the oil pan and the splash shield.

16. Lower the vehicle and connect the negative battery cable. Fill the engine with the proper type and quantity of engine oil.

17. Run the engine and check for leaks and proper operation.

3.0L Engine

1. Disconnect the negative battery cable and drain the cooling system.

2. Raise and safely support the vehicle. Remove the splash shield and drain the engine oil. Remove the oil pan and oil pump pickup.

3. Lower the vehicle.

4. Remove the timing belt cover, timing belt and crankshaft sprocket.

5. Remove the thermostat assembly.

6. Remove the attaching bolts and remove the oil pump. Disassemble the pump, if necessary and inspect all components for wear and/or damage. Replace components, as necessary.

7. Pry out the crankshaft seal from the pump, being careful not to damage the seal housing.

To install:

8. Clean all gasket mating surfaces.

9. Apply clean engine oil to a new crankshaft seal and install in the pump, using a seal installer.

10. Install a new gasket and the oil pump. Tighten the mounting bolts to 14–19 ft. lbs. (19–25 Nm).

11. Install a new gasket and the thermostat housing. Tighten the mounting bolts to 14–19 ft. lbs. (19–25 Nm).

12. Install the crankshaft sprocket, timing belt and timing belt cover.

13. Raise and safely support the vehicle.

14. Lubricate and install a new O-ring and the oil pump pickup. Tighten the mounting bolts to 69–95 inch lbs. (7.8–11.0 Nm).

15. Install the oil pan and the splash shield. Lower the vehicle.

16. Fill the crankcase with the proper type and quantity of engine oil. Connect the negative battery cable.

17. Fill and bleed the cooling system. Run the engine and check for leaks and proper operation.

4.0L Engine

1. Remove the engine assembly.

2. Remove the oil pan.

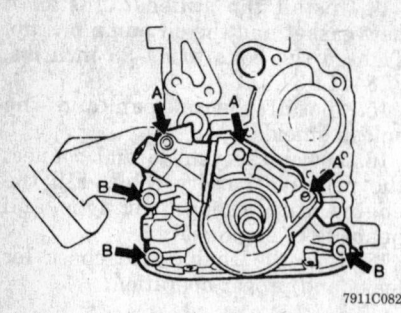

Oil pump mounting bolt identification — 2.2L engine

7911C082

3. Remove the oil pump attaching bolts and withdraw the oil pump driveshaft.

To install:

4. Prime the oil pump by filling either the inlet or outlet port with clean engine oil. Rotate the pump shaft to distribute the oil within the pump body.

5. Insert the oil pump driveshaft into the block with the pointed end facing inward. The pointed end is closest to the pressed-on flange. Position the pump with a new gasket and install the attaching bolts. Tighten to 13–15 ft. lbs. (17–21 Nm).

6. Clean and install the oil pump pickup with a new gasket. Tighten the bolts to 7–10 ft. lbs. (9–13 Nm).

7. Install the oil pan and install the engine assembly.

8. Fill the crankcase with the proper type and quantity of engine oil. Run the engine and check for leaks.

CHECKING

Except 4.0L Engine

1. Remove the pump and disassemble. Clean all parts in solvent and allow to dry.

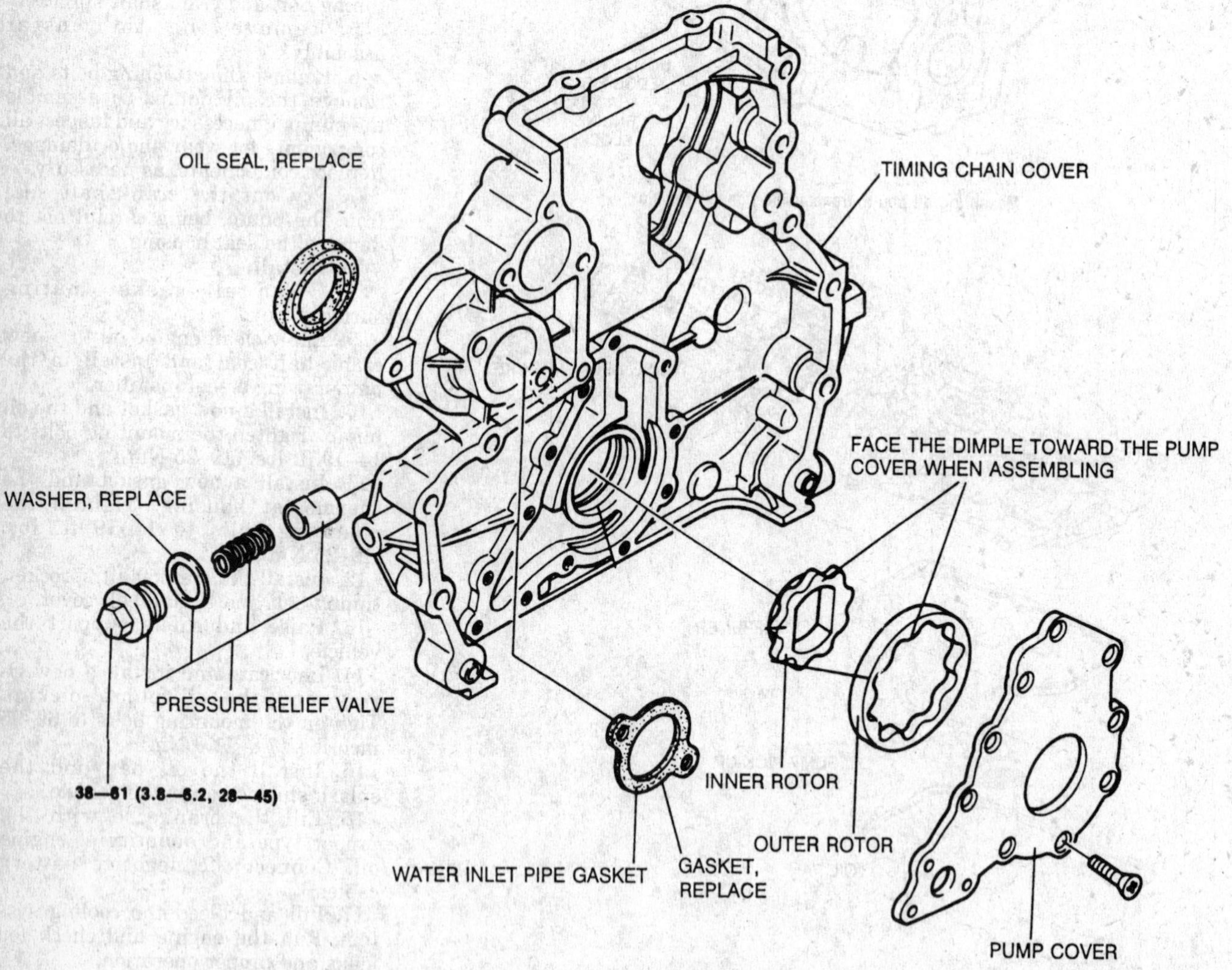

OIL SEAL, REPLACE

TIMING CHAIN COVER

FACE THE DIMPLE TOWARD THE PUMP COVER WHEN ASSEMBLING

WASHER, REPLACE

PRESSURE RELIEF VALVE

38—61 (3.8—6.2, 28—45)

INNER ROTOR

OUTER ROTOR

WATER INLET PIPE GASKET

GASKET, REPLACE

PUMP COVER

7911C083

Oil pump exploded view — 2.6L engine

2. Check for a distorted or damaged oil pump body or cover.

3. Check the relief valve plunger for wear or damage. Check for a weak or broken plunger spring. The spring free length should be 1.827 in. (46.4mm).

4. Check the rotor side clearance as follows: Lay a straightedge across the pump body and, using a feeler gauge, measure between the gear faces and the straight-edge. If the clearance exceeds 0.0039 in. (0.10mm) on 2.2L and 2.6L engines or 0.0051 in. (0.13mm) on 3.0L engine, replace the pump.

5. Check the rotor tooth tip clearance as follows: Insert a feeler gauge between the gears at the gear tip. If the clearance exceeds 0.0071 in. (0.18mm) on 2.2L and 2.6L engines or 0.0094 in. (0.24mm) on 3.0L engine, replace the gears or pump.

6. Check the outer rotor-to-pump body clearance as follows: Insert a feeler gauge between the outer gear and the pump body. If the clearance exceeds 0.0078 in. (0.20mm) on 2.2L and 2.6L engine or 0.0091 in. (0.23mm) on 3.0L engine, replace the pump.

7. Apply clean engine oil to the pump components and reassemble. Apply oil to the lip of a new seal and install in the pump, using a seal installer.

4.0L Engine

1. Remove the pump and disassemble. Thoroughly clean all parts in solvent and dry with compressed air.

2. Check the inside of the pump housing and the inner and outer gears for damage or excessive wear. Check the mating surfaces of the pump cover for wear. Minor scuff marks are normal, but if the cover, gears or housing surfaces are excessively worn, scored or grooved, replace the entire pump.

3. Measure the inner to outer rotor tip clearance. With the rotor assembly removed from the pump and resting on a flat surface, the inner and outer rotor tip clearance must not exceed 0.012 in. (0.30mm) with a feeler gauge inserted ½ in. (13mm) minimum.

4. With the rotor assembly installed in the housing, place a straight-edge over the rotor assembly and the housing. Measure the vertical clearance, the rotor endplay, between the straight-edge and the inner rotor and outer race. Maximum clearance must not exceed 0.005 in. (0.13mm).

5. Inspect the relief valve spring for collapsed or worn condition. Check the spring tension. The tension should be 13.6–14.7 lbs. at 1.39 in.

6. If any part of the oil pump requires replacement, replace the complete pump assembly.

Rear Main Oil Seal

REMOVAL AND INSTALLATION

Except 4.0L Engine

1. Disconnect the negative battery cable. Raise and safely support the vehicle.

2. Drain the engine oil.

3. Remove the transmission. If equipped with manual transmission, remove the clutch assembly.

4. Remove the flywheel.

5. Using a small prybar and a rag, remove the oil seal, being careful not to damage the crankshaft or seal housing.

6. When the seal has been removed, clean the mounting recess.

To install:

7. Coat the new seal and the crankshaft with a light film of clean engine oil.

8. Start the seal into the recess with the seal lip facing forward and install it with a suitable rear oil seal replacer tool. Install the seal until it is flush with the edge of the rear cover.

9. Position the flywheel on the crankshaft flange. Install the flywheel attaching bolts and tighten to 71–76 ft. lbs. (96–103 Nm) on 2.2L engine, 67–72 ft. lbs. (91–98 Nm) on 2.6L engine or 76–81 ft. lbs. (103–110 Nm) on 3.0L engine.

10. If equipped with manual transmission, install the clutch assembly. Install the transmission and lower the vehicle.

11. Fill the crankcase with the proper type and quantity of engine oil. Connect the negative battery cable, start the engine and check for leaks.

4.0L Engine

1. Disconnect the negative battery cable. Raise and safely support the vehicle.

2. Remove the transmission. If equipped with manual transmission, remove the clutch assembly.

3. Remove the flywheel.

4. Using a sharp awl or equivalent, punch 2 holes into the seal on opposite sides of the crank-

shaft and just above the bearing cap-to-cylinder block split line. Install a sheet metal screw in each hole.

5. Using 2 small prybars, pry against both screws and remove the seal. It may be necessary to place small blocks of wood against the cylinder block to provide a fulcrum point for the prybars. Be careful not to damage the crankshaft oil seal surface or seal housing.

6. When the seal has been removed, clean the mounting recess and the oil seal contact surface on the crankshaft.

To install:

7. Coat the new seal and the crankshaft with a light film of engine oil.

8. Start the seal into the recess with the seal lip facing forward and install it with a suitable rear oil seal replacer tool. Keep the tool straight with the centerline of the crankshaft and install the seal until it is fully seated.

9. After removing the tool, inspect the seal to make sure it was not damaged during installation.

10. Position the flywheel on the crankshaft flange. Install the flywheel attaching bolts and tighten to 59 ft. lbs. (80 Nm).

11. If equipped with manual transmission, install the clutch assembly. Install the transmission and lower the vehicle.

12. Connect the negative battery cable, start the engine and check for leaks.

MANUAL TRANSMISSION

Transmission Assembly

REMOVAL AND INSTALLATION

B2200

1. Disconnect the negative battery cable.

2. Remove the gearshift knob and shift console attaching screws. Remove the console.

3. Remove the shift lever to extension housing attaching bolts and remove the shift lever.

4. Raise and support the vehicle safely.

5. Drain the transmission oil.

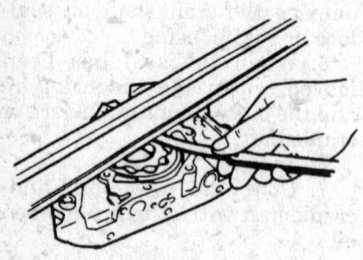

SIDE CLEARANCE

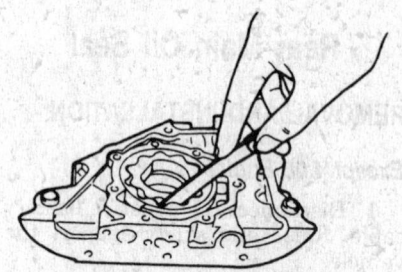

TOOTH TIP CLEARANCE

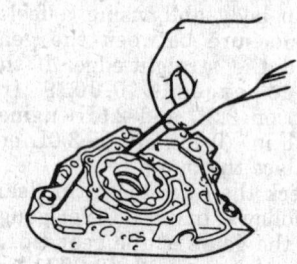

OUTER ROTOR-TO-PUMP BODY CLEARANCE

7911C086

Oil pump checking — 2.2L engine; 2.6L and 3.0L engines similar

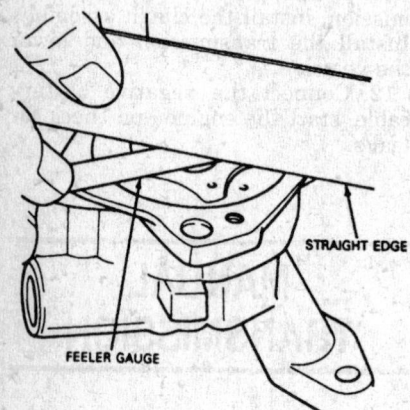

7911C087

Oil pump checking — 4.0L engine

6. Matchmark and remove the driveshaft.

7. Disconnect the speedometer cable from the transmission.

8. Remove the starter motor.

9. Disconnect and the back-up light switch wiring at the transmission.

10. Disconnect the parking brake return spring and parking brake cables.

11. Disconnect the clutch slave cylinder from the transmission.

12. Remove the transmission-to-engine gusset plate.

13. Disconnect the exhaust pipe at the transmission and manifold.

14. Support the transmission and engine separately with jacks.

15. Remove the transmission crossmember.

16. Lower the transmission to gain access to the top bolts and remove the transmission-to-engine bolts.

17. Pull the transmission straight back, away from the engine and remove transmission from the vehicle.

To install:

18. Raise the transmission into position. Install the transmission-to-engine bolts and tighten to 51–65 ft. lbs. (69–88 Nm).

19. Install the transmission crossmember and tighten the cross-

member-to-chassis bolts to 23–34 ft. lbs. (31–46 Nm).

20. Lower the transmission onto the crossmember and remove the jacks. Install the crossmember-to-mount bolts to 12–17 ft. lbs. (16–23 Nm).

21. Connect the exhaust pipe to the manifold and transmission bracket. Tighten the exhaust pipe-to-manifold nuts to 30–41 ft. lbs. (40–55 Nm) and the exhaust pipe-to-transmission bracket bolt to 13–20 ft. lbs. (18–26 Nm).

22. Install the gusset plate and tighten the bolts to 27–38 ft. lbs. (37–52 Nm).

23. Install the clutch slave cylinder and tighten the bolts to 12–17 ft. lbs. (16–23 Nm).

24. Install the parking brake cables and the return spring.

25. Connect the back-up light switch wiring connector and connect the speedometer cable.

26. Install the driveshaft, aligning the marks that were made during removal.

27. Install the starter. Fill the transmission with the proper type and quantity of engine oil.

28. Lower the vehicle. Install the shift lever and tighten the attaching bolts to 69–95 inch lbs. (7.8–11.0 Nm).

29. Install the console and the shift knob.

30. Connect the negative battery cable. Check the transmission for leaks and proper operation.

B2600

NOTE: On 4WD vehicles, the transmission and transfer case are removed as a unit.

1. Disconnect the negative battery cable.

2. Remove the knobs from the transfer case, if equipped and transmission shifters.

3. Remove the console box, if equipped.

4. Remove the insulator plate and shifter boot.

5. Remove the attaching bolts and the shift lever(s).

6. Raise and support the vehicle safely. Drain the transfer case, if equipped and transmission oil.

7. Remove the transmission and transfer case, if equipped, splash shields. Remove the starter.

8. Disconnect and remove the front exhaust pipe.

9. Matchmark and remove the driveshaft(s).

10. Disconnect the speedometer cable, 4WD switch, if equipped and backup light switch wires from the transmission or transmission/transfer case.

11. Remove the slave cylinder without disconnecting the fluid line. Support the slave cylinder aside.

12. Remove the transmission gusset plates and clutch housing cover. Support the transmission and engine with jacks.

13. Raise the transmission or transmission/transfer case and remove the crossmember.

14. Remove the transmission or transmission/transfer case.

To install:

15. Raise the transmission or transmission/transfer case assembly into position. Install the transmission-to-engine bolts and tighten to 51–65 ft. lbs. (69–88 Nm).

16. Install the crossmember and tighten the crossmember-to-chassis bolts to 23–34 ft. lbs. (31–46 Nm).

17. Lower the transmission or transmission/transfer case assembly and remove the jacks. Install the crossmember-to-transmission mount nuts and tighten to 23–34 ft. lbs. (31–46 Nm).

18. Install the gusset plates and clutch housing cover. Install the slave cylinder.

19. Connect the backup light and, if equipped, 4WD switch electrical connectors and the speedometer cable.

20. Install the driveshaft(s), aligning the marks that were made during removal.

21. Install the front exhaust pipe and the starter. Install the splash shields.

22. Fill the transmission and, if equipped, transfer case with the proper type and quantity of fluid. Lower the vehicle.

23. Install the shift lever(s) and tighten the mounting bolts to 25–37 ft. lbs. (34–50 Nm).

24. Install the shifter boot and insulator plate. Tighten the mounting bolts to 25–37 ft. lbs. (34–50 Nm).

25. Install the console box and the shifter knob(s).

26. Connect the negative battery cable. Check the transmission for leaks and proper operation.

MPV

NOTE: On 4WD vehicles, the transmission and transfer case are removed as a unit.

1. Disconnect the negative battery cable.

2. Remove the shift lever knob and boot.

3. Shift the transmission into **N** and unbolt and remove the shift lever.

4. Raise and safely support the vehicle.

5. Drain the transmission and, if equipped, transfer case fluid.

6. Disconnect the speedometer cable and, if equipped, the Hi-Lo shift cable.

7. Disconnect the electrical connectors at the transmission.

8. Matchmark and remove the driveshaft(s). Stuff a rag in the double offset joint to prevent damage to the boot from the driveshaft.

9. On the 2.6L engine, unbolt the support bracket from the transmission, if equipped.

10. Remove the starter.

11. Remove the clutch slave cylinder and hydraulic line bracket. It is not necessary to disconnect the hydraulic line. Remove the bellhousing inspection plate.

12. Remove the front exhaust pipe and heat shield.

13. Remove the gusset plates and support the transmission and engine with jacks.

14. Remove the transmission-to-crossmember bolts. Remove the crossmember.

15. Remove the transmission-to-engine bolts.

16. Pull the transmission straight back on the jack until the mainshaft clears the clutch. Lower the transmission and pull it out from under the vehicle.

To install:

17. Raise the transmission into position.

18. Push the transmission straight forward on the jack until the mainshaft enters the clutch and the transmission engages the locating dowels on the engine.

19. Install the transmission-to-engine bolts and tighten to 27–38 ft. lbs. (37–52 Nm) on 3.0L engine or 51–65 ft. lbs. (69–88 Nm) on 2.6L engine.

20. Install the starter. Install the crossmember and tighten the crossmember-to-chassis bolts to 32–45 ft. lbs. (43–61 Nm).

21. Lower the transmission and remove the jacks. Install the transmission-to-crossmember bolts and tighten the nuts to 23–34 ft. lbs. (31–46 Nm) or bolts/nuts to 32–45 ft. lbs. (43–61 Nm).

22. Install the front exhaust pipe and heat shield.

23. Install the transmission gusset plates. Install the clutch release cylinder and bracket.

24. Connect the speedometer cable and, if equipped, Hi-Lo shift cable. Connect the electrical wiring.

25. Install the driveshaft(s), aligning the marks that were made during removal. Install the bellhousing inspection plate.

26. Fill the transmission and, if equipped, transfer case with the proper type and quantity of fluid. Lower the vehicle.

27. Install the shift lever.

28. Install the shifter knob and boot.

29. Connect the negative battery cable. Check the transmission for leaks and proper operation.

Navajo

1. Disconnect the negative battery cable.

2. Place the gearshift lever in the **N** position.

3. Remove the shifter boot retainer screws and slide the boot up the shift lever shaft. Remove the shift lever attaching bolt(s) and remove the shift lever. Cover the opening in the transmission to prevent dirt from entering.

4. Raise and safely support the vehicle. Drain the transmission fluid.

5. Mark the position of the driveshaft(s) on the flange(s) and remove the driveshaft(s). Plug the

transmission or transfer case opening to prevent fluid leakage.

6. Disconnect the clutch hydraulic fluid line. Plug the line to prevent fluid leakage.

7. Disconnect the speedometer cable from the transmission or transfer case.

8. Disconnect the starter and backup lamp switch wires.

9. Place a jack under the engine, with a wood block to protect the oil pan.

10. Remove the transfer case, if equipped.

11. Remove the starter. Place a transmission jack under the transmission.

12. Remove the bolts attaching the transmission and clutch housing to the engine. Remove the nuts and bolts attaching the transmission mount and damper to the crossmember.

13. Remove the nuts and/or bolts attaching the crossmember to the frame side rails and remove the crossmember.

14. Lower the engine jack. Work the clutch housing off the locating dowels and slide the clutch housing and transmission rearward until the input shaft clears the clutch disc. Remove the transmission.

15. Remove the clutch housing from the transmission, if necessary.

To install:

16. Make sure the machined mating surfaces and the locating dowels on the engine rear plate and the mating face of the clutch housing and locating dowel holes are free of burrs, dirt or paint. Install the clutch housing on the transmission, if removed.

17. Mount the transmission on a transmission jack and raise into position. Start the input shaft into the clutch disc, aligning the splines. Move the transmission forward until the clutch housing seats on the locating dowels.

18. Install the clutch housing-to-engine attaching bolts and tighten to 28–38 ft. lbs. (38–51 Nm). Remove the transmission jack.

19. Install the starter.

20. Raise the engine and install the crossmember, insulator and damper with the attaching nuts and bolts. Install the nuts and bolts attaching the transmission mount to the crossmember.

21. Install the transfer case, if equipped.

22. Remove the plug(s) and install the driveshaft(s), aligning the marks on the flange(s) that were made during the removal procedure.

23. Connect the starter cable and backup lamp switch wires.

24. Connect the hydraulic clutch line and bleed the system.

25. Connect the speedometer cable.

26. Fill the transmission with the proper type and quantity of flu

27. Lower the vehicle and remove the cover from the transmission opening.

28. Install the gearshift lever with the attaching bolt(s). Install the shifter bolts.

29. Connect the negative battery cable. Check the transmission for leaks and proper operation.

CLUTCH

Clutch Assembly

REMOVAL AND INSTALLATION

1. Disconnect the negative battery cable.

2. On Navajo, disconnect the hydraulic clutch master cylinder from the clutch pedal.

3. Raise and safely support the vehicle. Remove the starter.

4. On all except Navajo, remove the slave cylinder, leaving the hydraulic line connected. On Navajo, use coupling disconnect tool T88T–70522–A or equivalent, to slide the white plastic sleeve toward the slave cylinder, then apply a slight tug on the tube to disconnect the hydraulic coupling. Plug the hose.

5. Remove the transmission.

6. Mark the position of the pressure plate on the flywheel so if the pressure plate is reused, it can be reinstalled in the same position.

7. Loosen the pressure plate attaching bolts evenly until the diaphragm spring is expanded. Remove the bolts, pressure plate and clutch disc.

8. Inspect the flywheel for wear, scoring and cracks. Machine or replace, as necessary. Inspect the clutch pilot bearing for wear and free movement. If replacement is necessary, remove using puller tool T58L–101–B or equivalent.

9. Inspect the clutch release bearing for wear and free movement; replace as necessary. On Navajo, remove the release bearing by twisting it until resistance is felt, then turning further will allow the preload

spring to push the bearing assembly off the slave cylinder.

To install:

10. If the pilot bearing was removed, a new one must be installed. Install using a suitable driver. On Navajo, install the pilot bearing with the seal facing the transmission so the adapter is not cock

11. If the flywheel was removed, make sure the mating surfaces of the crank flange and flywheel are clean, and install the flywheel. Tighten the flywheel bolts to 71–76 ft. lbs. (96–103 Nm) on 2.2L engine, 67–72 ft. lbs. (91–98 Nm) on 2.6L engine, 76–81 ft. lbs. (103–110 Nm) on 3.0L engine or 59 ft. lbs. (80 Nm) on 4.0L engine.

12. Position the clutch disc on the flywheel so a suitable alignment tool can enter the pilot bearing and align the disc.

13. Install the pressure plate. If the original pressure plate is being reused, align the marks that were made during the removal procedure. Install the attaching bolts and tighten, in sequence, to 13–20 ft. lbs. (18–26 Nm) on all except 4.0L engine where the torque is 15–24 ft. lbs. (21–32 Nm), then remove the alignment tool.

14. On all except Navajo, lightly lubricate the release bearing fork pivot and release bearing contact surfaces with high temperature grease. Install the fork and the release bearing.

15. On Navajo, lubricate the release bearing bore and bearing carrier with high temperature grease and install over the transmission input shaft and onto the slave cylinder.

16. Install the transmission. If equipped, reuse the aluminum washers under the attaching bolts to prevent galvanic corrosion.

17. On all except Navajo, install the slave cylinder to the transmission. On Navajo, connect the hydraulic coupling by pushing the male coupling into the slave cylinder female coupling.

18. Lower the vehicle. On Navajo, connect the clutch master cylinder to the brake pedal and bleed the clutch system, if necessary.

19. Connect the negative battery cable.

PEDAL HEIGHT/FREE-PLAY ADJUSTMENT

Except Navajo

1. Measure the distance from the top of the clutch pedal pad to the carpet. The distance should be as fol-

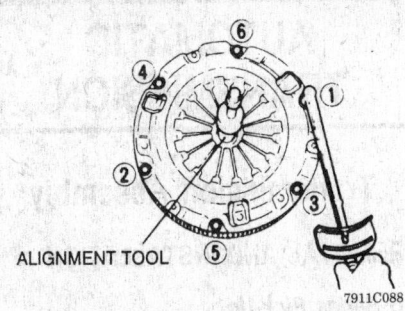

ALIGNMENT TOOL

7911C088

Clutch pressure plate bolt torque sequence —
except 4.0L engine

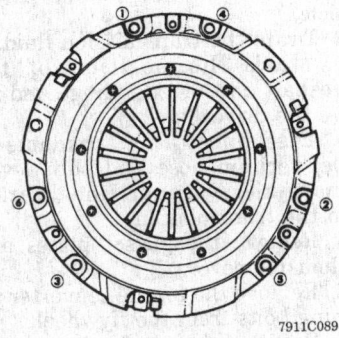

7911C089

Clutch pressure plate bolt torque
sequence — 4.0L engine

lows: B2200: 7.13–7.52 in.
(181–191mm) B2600: 7.52–7.91
in. (191–201mm) MPV: 8.19–8.58
in. (208–218mm)

2. If the distance is not within
specification, loosen the clutch switch
locknut and turn the switch until the
distance is correct. Tighten the lock
nut.

3. Check the free-play by pressing
the pedal by hand until clutch resistance is felt. The free-play should be
0.02–0.12 in. (0.6–3.0mm).

4. If the free-play is not within
specification, loosen the locknut on
the actuator rod and turn the actuator rod until the free-play is correct.

5. Check that the disengagement
height from the upper surface of the
pedal height to the carpet is correct
when the pedal is fully depressed.
The disengagement height should be
as follows: B2200: 2.60 in.
(66mm) B2600: 2.80 in.
(71mm) MPV: 1.38 in. (35mm)

6. Tighten the actuator rod
locknut to 8.7–12.0 ft. lbs. (12–17
Nm).

7. Recheck the pedal height.

Navajo

The hydraulic clutch system provides
automatic adjustment. No adjust-

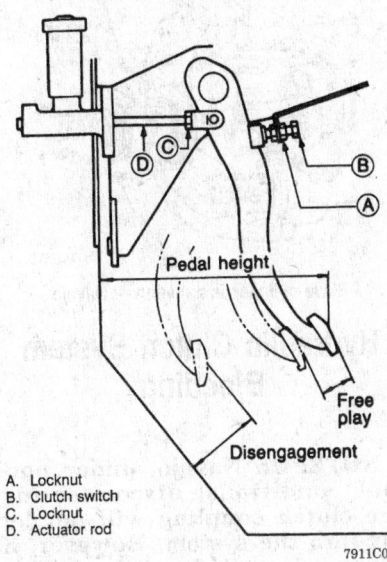

Pedal height

Free
play

Disengagement

A. Locknut
B. Clutch switch
C. Locknut
D. Actuator rod

7911C090

Clutch pedal height/free-play measurement
points — except Navajo

ment of clutch linkage or pedal position is required.

Clutch Master Cylinder

REMOVAL AND INSTALLATION

Except Navajo

1. Disconnect the negative battery
cable. Disconnect and plug the fluid
outlet line at the fitting on the
master cylinder.

2. Working inside the vehicle, remove the nuts mounting the master
cylinder to the firewall.

3. On MPV, remove the fluid reservoir mounting bolts.

4. On B Series Pick-Up, remove
the master cylinder from the firewall.
On MPV, remove the master cylinder
and the remote fluid reservoir as an
assembly.

To install:

5. Start the pedal pushrod into the
master cylinder and position the
master cylinder on the firewall.

6. Working inside the vehicle, install the mounting nuts and tighten
to 12–17 ft. lbs. (16–23 Nm) on B Series Pick-Up or 14–19 ft. lbs. (19–25
Nm) on MPV.

7. On MPV, install the remote
fluid reservoir mounting bolts and

tighten to 69–95 inch lbs. (7.8–11.0
Nm).

8. Connect the fluid outlet line to
the master cylinder fitting.

9. Bleed the hydraulic clutch
system.

10. Check the clutch pedal
height/free-play and adjust if
necessary.

Navajo

1. Disconnect the negative battery
cable.

2. Disconnect the clutch master
cylinder pushrod from the clutch
pedal by prying the retainer bushing
and pushrod off the pedal pin.

3. Remove the switch from the
master cylinder assembly.

4. Remove the screw retaining the
fluid reservoir to the cowl access
cover.

5. Use coupling disconnect tool
T88T–70522–A or equivalent, to slide
the white plastic sleeve toward the
slave cylinder, then apply a slight tug
on the tube to disconnect the hydraulic coupling.

6. Remove the retaining bolts and
the clutch master cylinder.

To install:

7. Install the pushrod through the
hole in the engine compartment.
Make sure it is located on the correct
side of the clutch pedal. Install the
master cylinder and tighten the bolts
to 12 ft. lbs. (16 Nm).

8. Insert the coupling end into the
slave cylinder and install the tube
into the clips.

9. Install the fluid reservoir on the
cowl access cover with the retaining
screw.

10. Replace the retainer bushing in
the clutch master cylinder pushrod if
worn or damaged. Install the retainer
and pushrod on the clutch pedal pin.
Make sure the flange of the bushing
is against the pedal blade. Install the
switch.

11. Connect the negative battery
cable and bleed the clutch hydraulic
system, if necessary.

Clutch Slave Cylinder

REMOVAL AND INSTALLATION

Except Navajo

1. Disconnect the negative battery
cable. Raise and support the vehicle
safely.

2. Disconnect the flexible hose
from the slave cylinder or hydraulic
tube. Plug the hose.

3. Pull off the hose retaining clip, if equipped. Remove the tube from the slave cylinder on MPV.

4. Remove the slave cylinder attaching bolts and the slave cylinder.

5. Installation is the reverse of the removal procedure.

6. Tighten the slave cylinder mounting bolts to 12–17 ft. lbs. (16–23 Nm) on B Series Pick-Up or 14–19 ft. lbs. (19–25 Nm) on MPV.

7. Bleed the clutch hydraulic system.

Navajo

NOTE: Before any vehicle service that requires slave cylinder removal, the clutch master cylinder pushrod must be disconnected from the clutch pedal. If not disconnected, permanent damage to the master cylinder will occur if the clutch pedal is depressed while the slave cylinder is disconnected.

1. Disconnect the negative battery cable.

2. Disconnect the coupling at the transmission using tool T88T–70522–A or equivalent, by sliding the white plastic sleeve toward the slave cylinder while applying a slight tug on the tube.

3. Raise and safely support the vehicle. Remove the transmission and clutch housing.

4. Remove the bolts retaining the slave cylinder to the transmission. Remove the slave cylinder from the transmission input shaft.

5. If necessary, remove the release bearing from the slave cylinder by twisting until resistance is felt, then turning further to allow the preload spring to push the bearing assembly off.

To install:

6. Push the release bearing into place, if removed.

7. Position the slave cylinder over the transmission input shaft with the bleed screw and coupling facing the left side of the transmission.

8. Install the slave cylinder attaching bolts and tighten to 13–19 ft. lbs. (18–26 Nm).

9. Install the transmission.

10. Insert the male coupling into the female coupling on the clutch slave cylinder and make sure the connection is secure.

11. Bleed the clutch hydraulic system, if necessary. Lower the vehicle and connect the negative battery cable.

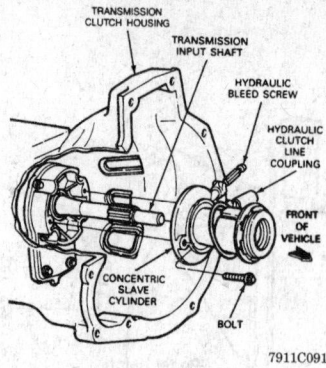

Slave cylinder installation — Navajo

Hydraulic Clutch System Bleeding

NOTE: On Navajo, under normal conditions, disconnecting the clutch coupling will not let air into the system. However, if there appears to be air in the system, indicated by a spongy pedal or insufficient bearing travel, the system must be bled.

1. Clean all dirt and grease from around the reservoir cap.

2. Remove the cap and fill the reservoir with heavy duty brake fluid.

3. Raise and safely support the vehicle, as necessary. Loosen the bleed screw, located in the slave cylinder body, next to the inlet connection.

4. Fluid should now begin to flow from the master cylinder, down the tube and into the slave cylinder.

NOTE: Keep the reservoir full at all times to make sure no additional air is drawn into the system.

5. Bubbles should begin to appear at the bleed screw outlet, indicating air is being expelled. When the slave cylinder is full, a steady stream of fluid will come from the slave cylinder outlet. Tighten the bleed screw.

6. Slowly depress the clutch pedal to the floor and hold. Loosen the bleed screw to allow air and excess fluid to be expelled. Retighten the bleed screw when fluid flow stops.

7. Depress and release the clutch pedal slowly, waiting 2 seconds between each cycle. Repeat 5 times.

8. Check the fluid level in the reservoir and add, if necessary. If evidence of air still exists, repeat Steps 6 and 7.

AUTOMATIC TRANSMISSION

Transmission Assembly

REMOVAL AND INSTALLATION

B Series Pick-Up

2WD

1. Disconnect the negative battery cable. Raise and safely support the vehicle.

2. Drain the transmission fluid.

3. Mark the position of the driveshaft on the axle flange and remove the driveshaft.

4. Disconnect the speedometer cable, vacuum hose and shift lever. Remove the vacuum line bracket from the transmission.

5. Remove the gusset plates and bellhousing cover.

6. Remove the torque converter attaching bolts from the flywheel.

7. Support the transmission and engine with jacks.

8. Remove the transmission mount and mount bracket.

9. Tag and disconnect the electrical connectors from the neutral safety switch, kickdown solenoid and overdrive cancel solenoid.

10. Remove the transmission fluid dipstick and tube.

11. Disconnect and plug the transmission fluid lines.

12. Remove the transmission.

To install:

13. Raise the transmission into position. Install the transmission-to-engine bolts and tighten to 27–38 ft. lbs. (37–52 Nm).

14. Unplug and connect the transmission fluid lines. Tighten the banjo bolts to 17–26 ft. lbs. (24–35 Nm).

15. Install the transmission fluid dipstick and tube.

16. Connect the electrical connectors.

17. Install the transmission mount and tighten the bolts to 7.2–17 ft. lbs. (9.8–23 Nm). Install the mount bracket to the crossmember and tighten the bolts to 23–34 ft. lbs. (31–46 Nm). Install the mount-to-mount bracket bolts to 23–34 ft. lbs. (31–46 Nm).

18. Remove the support jacks.

19. Install the torque converter attaching bolts and tighten to 25–36 ft. lbs. (34–49 Nm).

20. Install the gusset plates and bellhousing cover. Tighten the gusset

plate bolts to 27–38 ft. lbs. (37–52 Nm).

21. Connect the shift lever, vacuum hose and speedometer cable. Attach the vacuum line bracket.

22. Install the driveshaft and lower the vehicle.

23. Connect the negative battery cable. Fill the transmission with the proper type and quantity of fluid.

24. Check the transmission for leaks and proper operation.

4WD

1. Disconnect the negative battery cable.

2. Remove the shifter knob and the console box.

3. Remove the insulator plate and boot. Remove the 4WD shift lever.

4. Raise and safely support the vehicle. Remove the splash shields and drain the transmission fluid.

5. Disconnect and remove the front exhaust pipe.

6. Mark the position of the driveshaft on the flanges and remove the driveshafts.

7. Disconnect the speedometer cable and the 4WD indicator switch connector, if equipped.

8. Disconnect the shift cable and vacuum hose, if equipped.

9. Remove the gusset plate, if equipped.

10. Loosen the front differential mounting bolts and remove the No. 2 crossmember.

11. Remove the torque converter attaching bolts.

12. Support the transmission and engine with jacks. Remove the transmission-to-engine bolts.

13. Disconnect and plug the transmission fluid lines at the transmission. Remove the bracket from the transmission.

14. Remove the rear transmission crossmember.

15. Tag and disconnect the electrical connectors at the transmission.

16. Remove the transmission dipstick and tube.

17. Lower the transmission from the vehicle.

To install:

18. Raise the transmission into position and install the transmission-to-engine bolts. Tighten to 27–38 ft. lbs. (37–52 Nm).

19. Install the dipstick and tube.

20. Connect the electrical connectors.

21. Install the rear transmission crossmember and tighten the transmission-to-chassis bolts to 23–34 ft. lbs. (31–46 Nm).

22. Lower the transmission to the crossmember and install the mount-to-crossmember nuts. Tighten to 23–34 ft. lbs. (31–46 Nm). Remove the jacks.

23. Connect the transmission fluid lines to the transmission and tighten the banjo bolts to 17–26 ft. lbs. (24–35 Nm). Attach the fluid line bracket.

24. Install the torque converter attaching bolts and tighten to 27–40 ft. lbs. (36–54 Nm). Install the bellhousing cover.

25. Install the No. 2 crossmember.

26. Install the gusset plates, if equipped.

27. Connect the shifter cable and the bracket, if equipped. Connect the speedometer cable.

28. Install the driveshafts, aligning the marks that were made during removal.

29. Install the front exhaust pipe and the splash shields. Lower the vehicle.

30. Install the 4WD shift lever and the insulator plate and boot.

31. Install the console box and the shifter knob.

32. Connect the negative battery cable. Fill the transmission with the proper type and quantity of fluid.

33. Check the transmission for leaks and proper operation.

MPV

1. Disconnect the negative battery cable. Raise and safely support the vehicle.

2. Drain the transmission fluid.

3. Disconnect the speedometer cable. Tag and disconnect the necessary electrical connectors.

4. Disconnect and remove the front exhaust pipe and heat shield.

5. Mark the position of the driveshaft on the flanges and remove the driveshaft(s).

6. Disconnect and remove the shift linkage from the transmission.

7. Remove the dipstick and tube.

8. Remove the bellhousing cover and then remove the torque converter bolts.

9. Remove the starter and bracket, if equipped.

10. Remove the exhaust pipe bracket and the gusset plates.

11. Support the engine and transmission with jacks. Remove the transmission crossmember.

12. Disconnect and plug the transmission fluid lines at the transmission. Disconnect the vacuum line, if equipped.

13. Remove the transmission-to-engine bolts and lower the transmission from the vehicle.

To install:

14. Raise the transmission into position. Install the transmission-to-engine bolts. Tighten all bolts to 27–38 ft. lbs. (37–52 Nm) except the 4 larger diameter bolts on the hydraulically controlled transmission which are tightened to 51–65 ft. lbs. (69–88 Nm).

15. Install the transmission crossmember and tighten the crossmember-to-chassis bolts to 32–45 ft. lbs. (43–61 Nm). Tighten the mount-to-crossmember nuts to 23–34 ft. lbs. (31–46 Nm) or bolts/nuts to 32–45 ft. lbs. (43–61 Nm). Remove the jacks.

16. Connect the transmission fluid lines and tighten the banjo bolts to 17–26 ft. lbs. (24–35 Nm). Connect the vacuum line, if equipped.

17. Install the exhaust pipe bracket and gusset plates. Tighten the gusset plate bolts to 27–38 ft. lbs. (37–52 Nm).

18. Install the starter and the bracket, if equipped.

19. Install the torque converter bolts and tighten to 27–40 ft. lbs. (36–54 Nm). Install the bellhousing cover.

20. Install the dipstick and tube. Connect the shift linkage.

21. Install the driveshaft(s), aligning the marks that were made during removal.

22. Install the front exhaust pipe and heat shield.

23. Connect the electrical connectors and speedometer cable. Lower the vehicle.

24. Connect the negative battery cable. Fill the transmission with the proper type and quantity of fluid.

25. Check the transmission for leaks and proper operation.

Navajo

1. Disconnect the negative battery cable. Raise and safely support the vehicle.

2. Position a drain pan under the transmission fluid pan. Pry the lower clips of the transmission heat shield back slightly to allow access to the pan bolts.

3. Starting at the rear of the transmission pan and working toward the front, loosen the attaching bolts and allow the fluid to drain. Remove all the bolts except the 2 at the front to allow the fluid to further drain. After all fluid has drained, reinstall 2 bolts at the rear of the pan to temporarily hold it in place.

4. Disconnect the starter cable and remove the starter.

5. Place a 22mm socket and breaker bar on the crankshaft pulley

attaching bolt. Rotate the pulley clockwise, as viewed from the front, to gain access to each converter attaching nut. Remove the nuts through the starter mounting hole.

6. Mark the position of the driveshaft on the axle flange and remove the driveshaft. Plug the transmission to prevent fluid leakage.

7. Disconnect the speedometer cable from the transmission.

8. Disconnect the shift cable at the transmission manual lever. Remove the kickdown cable from the ball stud lever. Depress the tab on the cable downshift retainer and remove the cable from the bracket.

9. Disconnect the neutral safety switch wires and converter clutch solenoid connector. Disconnect the vacuum line from the vacuum modulator.

10. Position a transmission jack under the transmission and raise it slightly. Remove the engine rear support-to-crossmember bolts.

11. Remove the crossmember-to-frame side support attaching bolts and remove the crossmember insulator and support and damper.

12. Lower the jack under the transmission and allow the transmission to hang. Position a jack to the front of the engine and raise it to gain access to the 2 upper converter housing-to-engine attaching bolts.

13. Disconnect the oil cooler lines at the transmission. Plug the lines and transmission to prevent the entrance of dirt.

14. Remove the lower converter housing-to-engine attaching bolts and remove the transmission filler tube.

15. Secure the transmission to the jack with a safety chain. Remove the 2 upper converter housing-to-engine attaching bolts. Move the transmission to the rear so it disengages from the dowel pins and the converter is disengaged from the flywheel. Lower the transmission from the vehicle.

NOTE: If the transmission is to be removed for an extended period, support the engine with a safety stand and wood block.

To install:

16. Position the converter to the transmission making sure the converter hub is fully engaged in the pump gear. To make sure the converter is fully engaged, push and rotate the converter until 2 "bumps" are felt. Keep pushing and rotating until the distance between the converter pilot and the edge of the converter housing is 7/16–9/16.

17. Place the transmission on a transmission jack and secure with a safety chain. Rotate the converter so the drive studs are in alignment with the holes in the flywheel.

18. Raise the transmission and move it forward into position, being careful not to damage the flywheel and converter pilot.

NOTE: When moving the transmission, do not let the front of the transmission tilt downward. This will cause the converter to move forward and disengage from the pump gear. The converter must rest squarely against the flywheel. This indicates that the converter pilot is not binding in the engine crankshaft.

19. Install 2 converter housing-to-engine attaching bolts at the engine dowel locations and tighten to 28–38 ft. lbs. (38–51 Nm). Install the remaining attaching bolts and tighten to the same specification.

20. Remove the safety chain from the transmission.

21. Insert the filler tube in the stub tube and secure it to the cylinder block with the attaching bolt. Tighten the bolt to 28–38 ft. lbs. (38–51 Nm). If the stub tube is loosened or dislodged, it should be replaced.

22. Install the oil cooler lines in the retaining clip at the cylinder block. Connect the lines to the transmission.

23. Remove the jack supporting the front of the engine.

24. Raise the transmission and position the crossmember, insulator and support and damper to the frame side supports. Install the attaching bolts and nuts and tighten to 65–85 ft. lbs. (88–115 Nm).

25. Lower the transmission and install the rear engine support-to-crossmember nuts. Tighten the nuts to 65–85 ft. lbs. (88–115 Nm). Remove the transmission jack.

26. Install the vacuum hose on the vacuum modulator and attach the line to the clip. Connect the neutral safety switch plug and the converter clutch solenoid connector.

27. Install the flywheel-to-converter nuts and tighten to 20–34 ft. lbs. (27–46 Nm).

28. Install the starter and tighten the attaching bolts to 15–20 ft. lbs. (20–27 Nm). Connect the starter cable.

29. Connect the exhaust pipe to the exhaust manifold, if disconnected for removal.

30. Connect the shift cable to the manual lever and the downshift cable

to the downshift lever. Connect the speedometer cable.

31. Install the driveshaft, aligning the marks on the axle flange. Adjust the manual and downshift linkage, as required.

32. Remove the bolts temporarily holding the transmission fluid pan and remove the pan. Discard the gasket and clean all old gasket material and dirt from the gasket mating surfaces.

33. Install the pan using a new gasket. Tighten the attaching bolts to 8–10 ft. lbs. (11–13.5 Nm).

34. Lower the vehicle and connect the negative battery cable. Fill the transmission with the proper type and quantity of fluid.

35. Run the vehicle and check for leaks and proper operation.

SHIFT LINKAGE ADJUSTMENT

B Series Pick-Up w/Hydraulically Controlled Transmission

1. Move the gearshift lever to the **P** range.

2. Loosen locknuts A and B so they are both at least 0.039 in. (1mm) away from the adjustment lever.

3. Shift the transmission to the **P** range by moving the manual shaft of the transmission.

4. With the link at 90 degrees to the lever, adjust the clearance using a feeler gauge to 0.039 in. (1mm) between the adjustment lever and locknut A.

5. Remove the feeler gauge and tighten locknut B to 69–95 inch lbs. (8–11 Nm).

6. Measure the clearance between the guide plate and the guide pin in the **P** range. There should be 0.039 in. (1mm) clearance in front of the pin and 0.020 in. (0.5mm) clearance behind the pin.

7. Move the gearshift lever to the **N** and **D** ranges and check that the clearance between the guide plate and guide pin is the same in both ranges. If not, readjust locknuts A and B.

B Series Pick-Up w/Electronically Controlled Transmission

1. Disconnect the negative battery cable to deactivate the shift-lock.

2. Remove the selector knob and console.

3. Loosen locknuts A and B and lock bolt C.

4. Shift the transmission manual shaft to the **P** range.

5. Push and hold the selector lever forward with a force of approximately

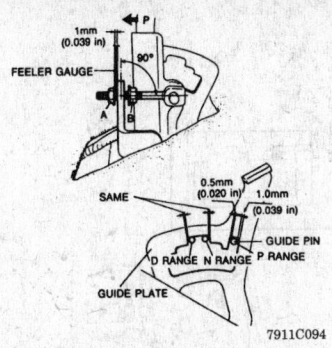

Shift linkage adjustment — B Series Pick-Up with hydraulically controlled transmission

22 lbs., then tighten lock bolt C to 67–95 inch lbs. (8–11 Nm).

6. Turn locknut A by hand until it just touches the spacer, then tighten locknut B to 67–95 inch lbs. (8–11 Nm).

7. Check the lever so the clearance between the guide plate and the guide pin in the **P** range with the pushrod lightly depressed is as specified.

8. Move the selector lever to the **N** and **D** ranges and make sure there is the same clearance between the guide plate and guide pin. If not, re-adjust the lever.

9. Install the console. Clean and apply locking compound to the selector knob screw threads and tighten the screws to 13–26 inch lbs. (1.5–2.9 Nm).

10. Connect the negative battery cable.

MPV

1. Move the selector lever to the **P** range.

2. Remove the column covers.

3. Pull the selector lever rearward, toward the driver and insert a 0.197 in. (5mm) outer diameter pin into the gearshift rod assembly.

4. Remove the air intake tube.

5. Loosen the shift lever and the top lever mounting bolts.

6. Shift the transmission manual shaft to the **P** range position.

7. Adjust the clearance between the lower bracket and the shift lever bushing by sliding the shift lever assembly until there is no clearance.

8. Tighten the shift lever mounting bolts to 12–17 ft. lbs. (16–23 Nm).

9. Make sure the detent ball is positioned in the center of the **P** range detent. If not, loosen bolts A and turn the bracket to adjust the position, then retighten bolts A to 61–87 inch lbs. (6.9–9.8 Nm).

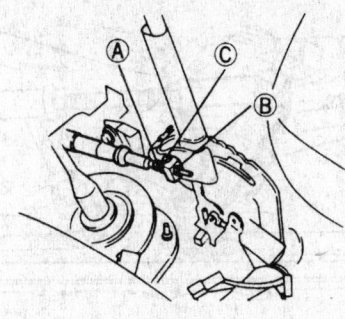

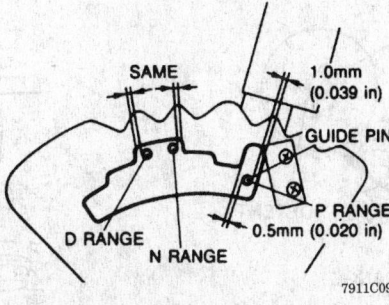

Shift linkage adjustment — B series Pick-Up with electronically controlled transmission

10. Adjust the clearance between the lower bracket and the shift lever bushing by turning the top lever until there is no clearance. Tighten the top lever mounting bolt A then retighten bolt B to 12–17 ft. lbs. (16–23 Nm).

11. Remove the pin from the gear shift rod assembly and install the column covers. Check selector lever operation.

Navajo

1. Raise and safely support the vehicle, as necessary. From inside the vehicle, place the column shift selector lever in the **OD** position. Hang an 8 lb. weight on the selector lever.

2. From below the vehicle, pull down the lock tab on the shift cable and remove the fitting from the transmission manual control lever ball stud.

3. Position the transmission manual control lever in the **OD** position by moving the lever all the way rearward and then moving it 3 detents forward.

4. Connect the cable end fitting to the transmission manual control lever. Push up on the lock tab to lock the cable in the correctly adjusted position.

5. Remove the 8 lb. weight from the column shift selector.

6. After adjustment, check for **P** engagement. Check the column shift selector lever in all detent positions with the engine running to ensure correct adjustment.

KICKDOWN SWITCH ADJUSTMENT

Except Navajo

1. Disconnect the negative battery cable.

2. Disconnect the switch connector, located above the accelerator pedal.

3. Loosen the locknut and back the switch out fully.

4. Depress the accelerator pedal fully and hold it.

5. With the accelerator pedal fully down, turn the kickdown switch clockwise until it turns **ON** (clicking sound heard), then turn the switch ¼ turn further clockwise.

6. Tighten the locknut to 10–13 ft. lbs. (14–18 Nm) and release the accelerator pedal.

7. Reconnect the connector and the negative battery cable.

8. Depress the accelerator pedal fully and verify that the kickdown switch clicks at the fully depressed position.

SELECTOR INDICATOR ADJUSTMENT

Navajo

1. Remove the steering column shroud.

2. With the engine stopped and the parking brake applied, place the transmission selector lever at the steering column in the **OD** position.

3. Secure a 3 lb. weight to the end of the transmission selector lever.

4. On 1991 vehicles, proceed as follows:

   a. Loosen the selector indicator screw on the column casting. Move the selector indicator adjustment until the orange pointer is completely within the letter "D" inside the **OD** graphic.

   b. Tighten the selector indicator screw on the column to 10–15 inch lbs. (1.1–1.7 Nm).

5. Rotate the thumb wheel until the orange pointer is completely within the letter "D" inside the **OD** graphic.

6. Install the steering column shroud.

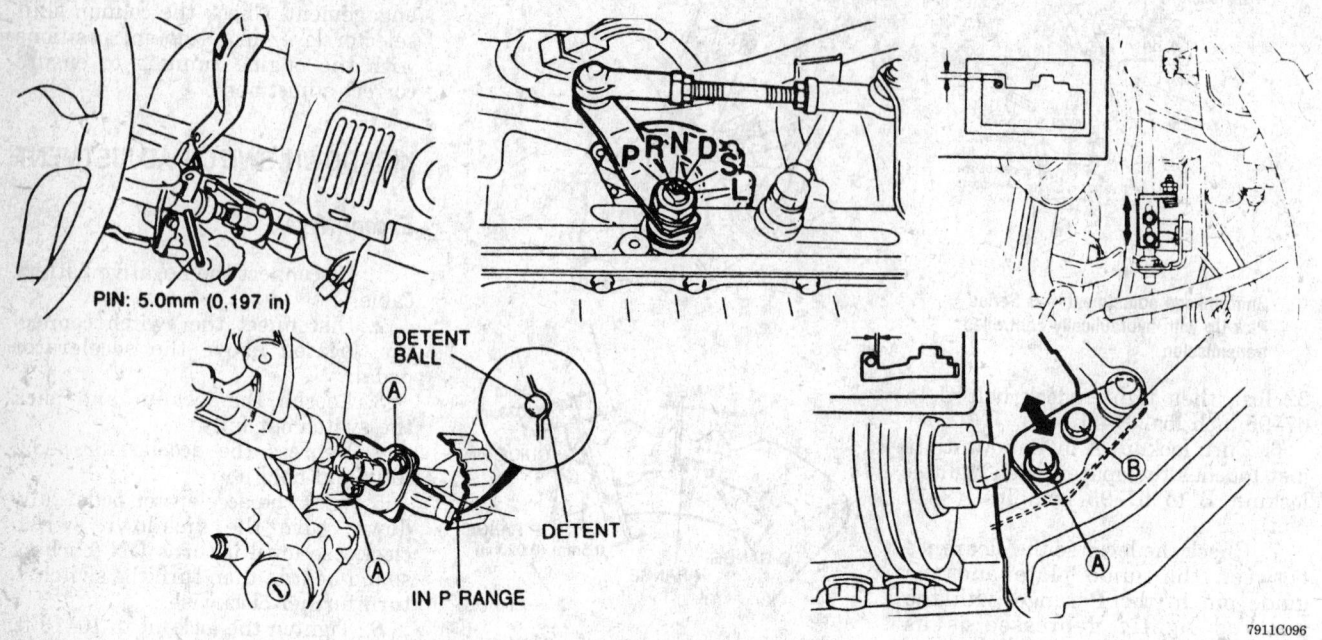

PIN: 5.0mm (0.197 in)

DETENT BALL

DETENT

IN P RANGE

Shift linkage adjustment — MPV

7911C096

TRANSFER CASE

Transfer Case Assembly

REMOVAL AND INSTALLATION

B Series Pick-Up

1. Disconnect the negative battery cable. Raise and safely support the vehicle.
2. Remove the transmission assembly from the vehicle.
3. Place the transmission in a vertical position, converter housing down.
4. Remove the control cover assembly, roll pin and control lever end from the transfer case.
5. Remove the extension housing-to-transfer case attaching bolts. Lift the transfer case off vertically to prevent damaging the control rod.

To install:

6. Install the input sleeve, if removed.
7. Coat the contacting surfaces of the transfer case and extension housing with sealant.
8. Install the transfer case to the extension housing. Apply sealant to

the bolt threads and tighten to 27–35 ft. lbs. (36–47 Nm).
9. Install the control lever end and secure with a new roll pin.
10. Coat the contacting surfaces of the control cover assembly and transfer case with sealant and install the control cover assembly on the transfer case.
11. Apply sealant to the bolt threads and tighten to 16–22 ft. lbs. (22–30 Nm).
12. Install the transmission assembly and lower the vehicle. Connect the negative battery cable.
13. Check the transfer case for leaks and proper operation.

MPV

1. Disconnect the negative battery cable. Raise and safely support the vehicle. Drain the transfer case.
2. Mark the position of the driveshafts on the flanges and remove the driveshafts. Push a rag into the double-offset joint to hold the rear driveshaft straight to prevent damaging the boot.
3. Support the transmission with a jack and remove the transmission lower mount. Remove the upper mount.
4. Remove the front exhaust pipe and heat insulator.

5. Disconnect the speedometer and transfer case shift cable. Tag and disconnect the electrical connectors.
6. Support the transfer case with a jack and remove the transfer case attaching bolts. Remove the transfer case.

To install:

7. Apply silicone sealant to the transfer case flange.
8. Support the transfer case with a jack and install the transfer case. Apply sealant to the bolt threads and tighten to 27–40 ft. lbs. (36–54 Nm).
9. Connect the electrical connectors, transfer case shift cable and speedometer cable. Adjust the transfer case shift cable.
10. Install the exhaust pipe and heat insulator.
11. Install the upper transmission mount.
12. Install the lower transmission mount. Loosely install the center washers and nuts and tighten the outer bolts to 32–45 ft. lbs. (43–61 Nm), then tighten the center nuts to 23–34 ft. lbs. (31–46 Nm).
13. Remove the support jacks.
14. Install the driveshafts, aligning the marks that were made during removal. Remove the rag from the double-offset joint and check the boot for damage.

15. Fill the transfer case with the proper type and quantity of fluid.

16. Lower the vehicle and connect the negative battery cable. Check the transfer case for leaks and proper operation.

Navajo

MECHANICAL SHIFT TYPE

1. Disconnect the negative battery cable. Raise and safely support the vehicle.

2. If equipped, remove the skid plate from the frame. Remove the damper from the transfer case, if equipped.

3. Place a drain pan under the transfer case, remove the drain plug and drain the fluid. Disconnect the 4WD indicator switch wire connector at the transfer case.

4. Disconnect the front driveshaft from the transfer case output shaft yoke and wire the driveshaft out of the way.

5. Disconnect the rear driveshaft from the transfer case output shaft flange and wire the driveshaft out of the way.

6. Disconnect the speedometer driven gear from the transfer case rear cover. Disconnect the vent hose from the control lever.

7. Disconnect the nut from the shift lever and remove the shift lever.

8. Remove the large and small bolts retaining the shifter to the extension housing. Remove the lever assembly and bushing.

9. Support the transfer case with a transmission jack. Remove the 5 bolts retaining the transfer case to the transmission and extension housing.

10. Slide the transfer case rearward off the transmission output shaft and lower the transfer case from the vehicle. Remove the gasket from between the transfer case and extension housing.

To install:

11. Install a new gasket on the front mounting face of the transfer case assembly.

12. Raise the transfer case with the transmission jack so the transmission output shaft aligns with the transfer case input shaft. Slide the transfer case forward onto the transmission output shaft and onto the dowel pin. Install the 5 retaining bolts and tighten, in sequence, to 25–43 ft. lbs. (34–58 Nm).

13. Remove the transmission jack.

14. Install and adjust the shifter. Always tighten the large bolt retaining the shifter to the extension housing before tightening the small bolt.

15. Install the vent assembly so the white marking on the hose is in position in the notch in the shifter. The upper end of the vent hose should be ³/₄ in. above the top of the shifter and positioned just below the floor pan.

16. Connect the speedometer driven gear to the transfer case rear cover. Tighten the screw to 20–25 inch lbs. (2.3–2.8 Nm).

17. Connect the rear driveshaft to the transfer case output shaft flange. Tighten the bolts to 61–87 ft. lbs. (83–118 Nm).

18. Connect the front driveshaft to the transfer case output shaft yoke. Tighten the bolts to 12–16 ft. lbs. (16–22 Nm).

19. Connect the 4WD indicator switch wire connector at the transfer case.

20. Install the drain plug and tighten to 14–22 ft. lbs. (19–30 Nm). Remove the fill plug and fill the transfer case with the proper type of fluid to the bottom of the fill hole. Install the fill plug and tighten to 14–22 ft. lbs. (19–30 Nm).

21. Install the damper to the transfer case, if equipped. Using new damper bolts, tighten to 25–35 ft. lbs. (34–48 Nm).

22. Install the skid plate, if equipped. Tighten the nuts and bolts to 15–20 ft. lbs. (20–27 Nm).

23. Lower the vehicle and connect the negative battery cable.

ELECTRONIC SHIFT TYPE

1. Disconnect the negative battery cable. Raise and safely support the vehicle.

2. If equipped, remove the nuts, bolts and skid plate from the frame. Remove the damper from the transfer case, if equipped.

3. Place a drain pan under the transfer case, remove the drain plug and drain the fluid.

4. Remove the wire connector from the feed wire harness at the rear of the transfer case. First squeeze the locking tabs, then pull the connectors apart.

NOTE: Do not pull directly on the wires or pull outwardly on the locking tabs.

5. Remove the connector for the transfer case motor from the mounting bracket.

6. Disconnect the front driveshaft from the transfer case output shaft yoke and wire the driveshaft out of the way.

7. Disconnect the rear driveshaft from the transfer case output shaft flange and wire the driveshaft out of the way.

8. Disconnect the speedometer driven gear from the transfer case rear cover. Disconnect the vent hose from the mounting bracket.

9. Support the transfer case with a transmission jack. Remove the 5 bolts retaining the transfer case to the transmission and extension housing.

10. Slide the transfer case rearward off the transmission output shaft and lower the transfer case from the vehicle. Remove the gasket from between the transfer case and extension housing.

To install:

11. Install a new gasket on the front mounting face of the transfer case assembly.

12. Raise the transfer case with the transmission jack so the transmission output shaft aligns with the transfer case input shaft. Slide the transfer case forward onto the transmission output shaft and onto the dowel pin. Install the 5 retaining bolts and tighten, in sequence, to 25–43 ft. lbs. (34–58 Nm).

13. Remove the transmission jack.

14. Install the vent hose so the white marking on the hose aligns with the notch in the mounting bracket.

15. Connect the speedometer driven gear to the transfer case rear cover. Tighten the screw to 20–25 inch lbs. (2.3–2.8 Nm).

16. Connect the rear driveshaft to the transfer case output shaft flange. Tighten the bolts to 61–87 ft. lbs. (83–118 Nm).

17. Connect the front driveshaft to the transfer case output shaft yoke. Tighten the bolts to 12–16 ft. lbs. (16–22 Nm).

18. Attach the connector for the transfer case motor to the mounting bracket.

19. Connect the wire connectors on the rear of the transfer case, making sure the retaining tabs lock.

20. Install the drain plug and tighten to 14–22 ft. lbs. (19–30 Nm). Remove the fill plug and fill the transfer case with the proper type of fluid to the bottom of the fill hole. Install the fill plug and tighten to 14–22 ft. lbs. (19–30 Nm).

21. Install the damper to the transfer case, if equipped. Using new damper bolts, tighten to 25–35 ft. lbs. (34–48 Nm).

22. Install the skid plate, if equipped. Tighten the nuts and bolts to 15–20 ft. lbs. (20–27 Nm).

23. Lower the vehicle and connect the negative battery cable.

LINKAGE ADJUSTMENT

MPV

NOTE: Make sure the Hi-Low lever and transfer case are in Low mode while adjusting the Hi-Low lever.

1. Shift the Hi-Low lever to the **4LO** position.
2. Remove the column covers.
3. Pull the column lever rearward, toward the driver and insert a 0.197 in. (5mm) outside diameter pin into the shift rod assembly.
4. Raise and safely support the vehicle.
5. Mark the position of the front driveshaft on the flanges and remove the front driveshaft.
6. Loosen the locknuts of the transfer case side.
7. Remove the air intake tube.
8. Loosen the shift lever mounting bolts and adjust the clearance between the lower bracket and the shift lever bushing by sliding the shift lever assembly until there is no clearance.
9. Tighten the shift lever mounting bolts to 12–17 ft. lbs. (16–23 Nm).
10. Install the air intake tube.
11. While pushing the outer cable toward the transfer case, turn locknut B until it touches Z surface. Tighten locknut A to 18–29 ft. lbs. (25–39 Nm) and locknut C to 25–37 ft. lbs. (34–50 Nm).
12. Install the front driveshaft, aligning the marks that were made during removal.
13. Remove the pin from the shift rod assembly and install the column covers. Check the Hi-Low lever operation.

Navajo

MECHANICAL SHIFT TYPE

1. Raise the shift boot to expose the top surface of the cam plate.
2. Loosen the bolts "A" and "B" on the control lever assembly approximately 1 turn. Move the transfer case shift lever to the **4L** position.
3. Rotate the cam plate clockwise around bolt "A" until the bottom chamfered corner of the neutral lug just contacts the forward right edge of the shift lever, point "C".
4. Hold the cam plate in this position and tighten bolt "A" first to 70–90 ft. lbs. (94–122 Nm), then tighten bolt "B" to 31–42 ft. lbs. (43–56 Nm).
5. Move the transfer case in-cab shift lever to all shift positions to check the positive engagement. There should be clearance, not exceeding

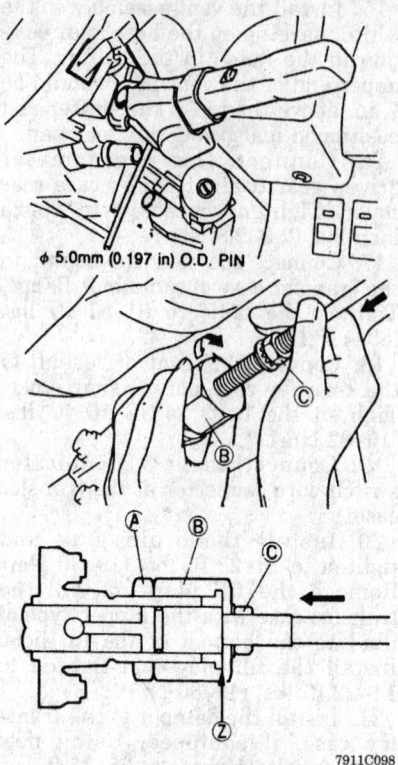

φ 5.0mm (0.197 in) O.D. PIN

Transfer case linkage adjustment — MPV

7911C098

0.13 in. (3.30mm), between the shift lever and cam plate in **2H** front **4H** rear and **4L** shift positions.
6. Install the shift boot assembly.

DRIVE AXLE

Halfshaft

REMOVAL AND INSTALLATION

B Series Pick-Up

1. Raise and safely support the vehicle. Remove the wheel and tire assembly.
2. Remove the drive flange hub.
3. Remove the caliper, mounting support and knuckle arm. Support the caliper aside with rope or mechanics wire; do not let the caliper hang by the brake hose.
4. Disconnect the stabilizer bar and the tie rod end.
5. Remove the lower mount of the shock absorber.
6. Remove the snapring and spacer.
7. Support the lower control arm with a jack.

8. Disconnect the upper and lower ball joints and the knuckle.
9. Lower the lower control arm and remove the knuckle assembly.
10. Remove the splash shield.
11. Using a suitable prybar, pry out the halfshaft from the differential and remove the halfshaft from the vehicle. Be careful not to damage the dust cover or oil seal.
   To install:
12. Install a new clip on the halfshaft. Coat the differential seal with clean transmission fluid.
13. Install the halfshaft in the differential, being careful not to damage the seal. After installation, attempt to pull the halfshaft outward to make sure it does not come out.
14. Install the knuckle and hub to the halfshaft and ball joints. Install the spacer and a new snapring.
15. Install the lower mount of the shock absorber and loosely tighten the boot.
16. Connect the stabilizer bar and tie rod end.
17. Install the caliper assembly, knuckle arm and wheel and tire assembly. Apply sealant to the drive flange and install it.
18. Install the splash shield and lower the vehicle.
19. Tighten the lower shock absorber mount to 41–59 ft. lbs. (55–80 Nm).
20. Check the front end alignment.

MPV

1. Raise and safely support the vehicle. Remove the wheel and tire assembly.
2. Remove and discard the halfshaft locknut.
3. Disconnect the tie rod end from the knuckle.
4. Remove the caliper and brake rotor from the knuckle. Support the caliper aside with rope or mechanics wire; do not let it hang by the brake hose.
5. Remove the nut and bolts and remove the lower ball joint. Remove the bolts and nuts and remove the knuckle/hub assembly from the strut.

NOTE: If the halfshaft is stuck to the hub, install a used locknut so it is flush with the end of the shaft, then tap the nut with a soft mallet.

6. Remove the splash shield.
7. Using a suitable prybar, pry out the halfshaft from the differential and remove the halfshaft from the vehicle. Be careful not to damage the dust cover or oil seal.

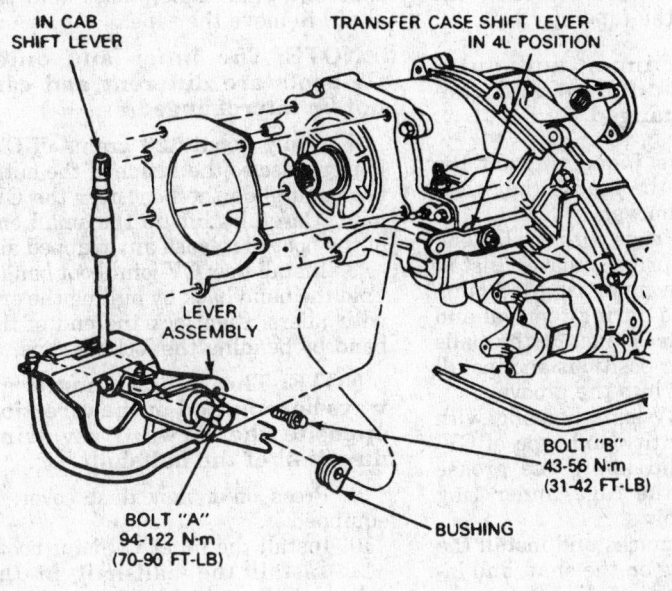

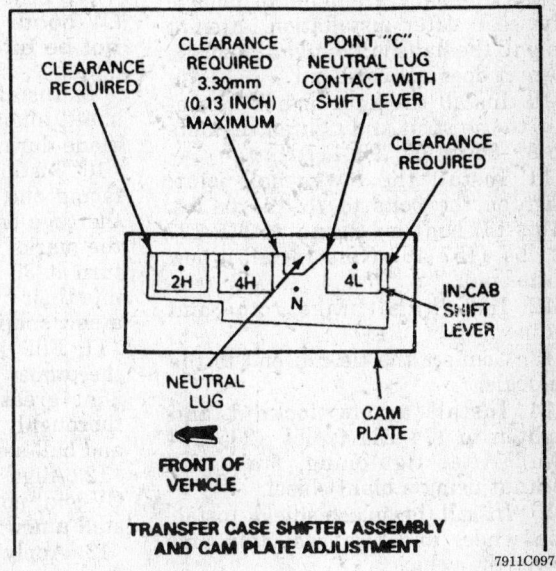

Transfer case linkage adjustment — Navajo with mechanical shift transfer case

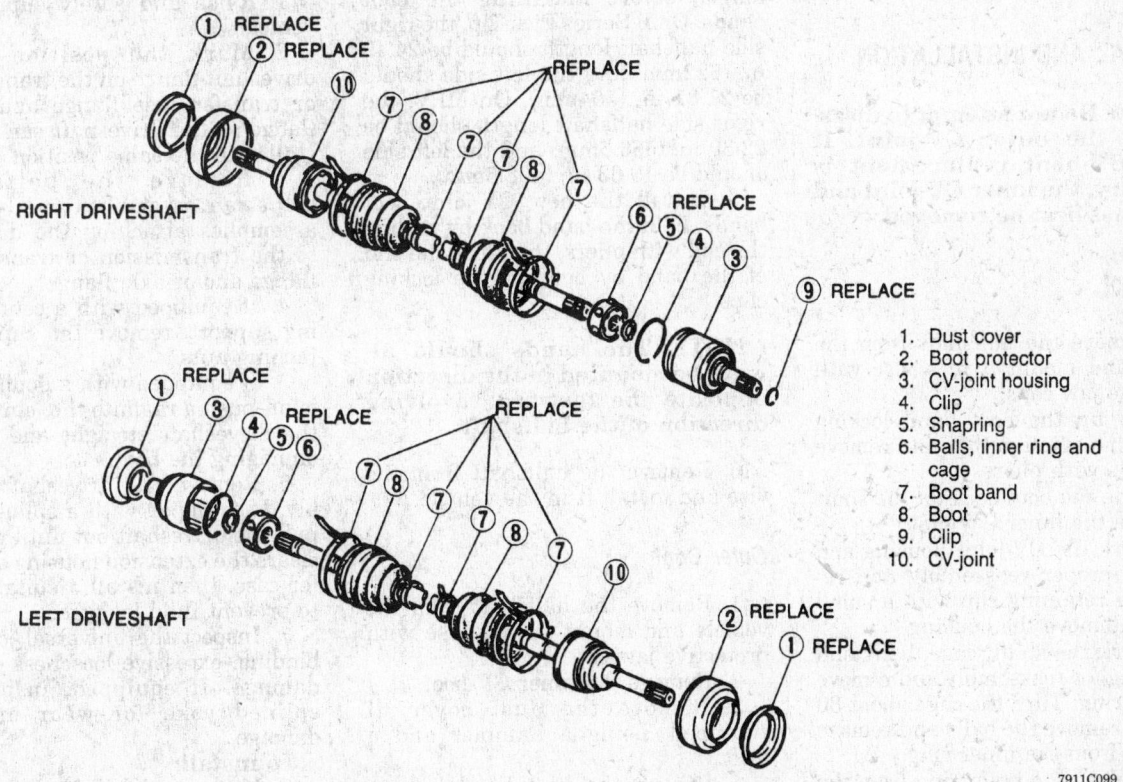

1. Dust cover
2. Boot protector
3. CV-joint housing
4. Clip
5. Snapring
6. Balls, inner ring and cage
7. Boot band
8. Boot
9. Clip
10. CV-joint

Halfshaft assembly exploded view — B Series Pick-Up

To install:

8. Install a new clip on the halfshaft. Coat the differential seal with clean transmission fluid.

9. Install the halfshaft in the differential, being careful not to damage the seal. After installation, attempt to pull the halfshaft outward to make sure it does not come out.

10. Install the knuckle/hub assembly to the strut and tighten the nuts to 69–86 ft. lbs. (93–117 Nm).

11. Install the lower ball joint. Tighten the bolts to 75–101 ft. lbs. (102–137 Nm) and the nut to 115–137 ft. lbs. (157–187 Nm). Install a new cotter pin.

12. Install the brake rotor and caliper.

13. Connect the tie rod end to the knuckle.

14. Install a new locknut and tighten to 174–231 ft. lbs. (235–314 Nm). After tightening, stake the locknut using a blunt chisel.

15. Install the splash shield. Install the wheel and tire assembly and lower the vehicle.

16. Check the front end alignment.

CV-Boot

REMOVAL AND INSTALLATION

NOTE: Do not attempt to disassemble the outer CV-joint. If outer CV-boot replacement is necessary, the inner CV-joint and boot must first be removed.

Inner Boot

1. Remove the halfshaft from the vehicle and mount it in a vise with protective jaw caps.

2. Pry up the boot band locking clips with a small prybar and remove the bands with pliers.

3. Slide the boot back on the shaft to expose the inner CV-joint.

4. Mark the CV-joint housing and cage for proper reassembly and remove the retaining clip with a small prybar. Remove the housing.

5. Mark the shaft, cage, balls and inner ring for reassembly and remove the snapring. Turn the cage about 30 degrees, remove the balls and remove the cage from the inner ring.

6. Remove the inner ring from the shaft with a press or drive it off with a hammer and brass drift.

7. Wrap the shaft splines with tape and remove the inner boot.

To install:

8. Wrap the shaft splines with tape and slide a new boot onto the shaft. Remove the tape.

NOTE: The inner and outer CV-boots are different and cannot be interchanged.

9. Install the inner ring on the shaft, aligning the marks that were made during removal.

10. Install the cage with the big end facing the snapring groove. Install the cage on the inner ring, aligning the marks made during removal and turn it 30 degrees. Install the balls into their proper positions and install a new snapring into the groove.

11. Fill the CV-joint housing with the proper quantity and type of CV-joint grease and apply the grease thoroughly to the cage, inner ring and ball assembly.

12. Align the marks and install the CV-joint housing on the shaft and install a new retaining clip.

13. Apply about 120 grams of CV-joint grease to the inside of the inner boot and slide the boot over the CV-joint. Carefully lift up the small end of the boot to release any trapped air.

14. Set the halfshaft to the required length before installing the boot bands. On B Series Pick-Up, the right side halfshaft length should be 24.49 in. (622mm) and the left side should be 21.81 in. (554mm). On MPV, the right side halfshaft length should be 22.30 in. (566.5mm) and the left side should be 19.63 in. (498.5mm).

15. Install the new CV-joint boot bands. Fold the band back by pulling the end with pliers, then lock the end of the band by bending the locking clips.

NOTE: The bands should always be mounted in the direction opposite the forward revolving direction of the halfshaft.

16. Remove the halfshaft from the vise and install it in the vehicle.

Outer Boot

1. Remove the halfshaft from the vehicle and mount it in a vise with protective jaw caps.

2. Remove the inner CV-boot.

3. Remove the dust cover, if equipped, using a hammer and a drift.

4. Pry up the boot band locking clips with a small prybar and remove the bands with pliers.

5. Slide the outer CV-boot off of the shaft.

To install:

6. Wrap the shaft splines with tape and slide a new boot onto the shaft. Remove the tape.

NOTE: The inner and outer CV-boots are different and cannot be interchanged.

7. Apply about 120 grams of CV-joint grease to the inside of the outer boot and slide the boot over the CV-joint. Carefully lift up the small end of the boot to release any trapped air.

8. Install new CV-joint boot bands. Fold the band back by pulling the end with pliers, then lock the end of the band by bending the locking clips.

NOTE: The bands should always be mounted in the direction opposite the forward revolving direction of the halfshaft.

9. Press on a new dust cover, if equipped.

10. Install the inner CV-joint boot.

11. Install the halfshaft in the vehicle.

Driveshaft and U-Joints

REMOVAL AND INSTALLATION

1. Raise and safely support the vehicle.

2. Mark the position of the driveshaft flange on the transmission or transfer case flange and/or axle flange so the driveshaft can be reinstalled in the same position.

3. Remove the bolts/nuts, bolt/strap assemblies or U-bolt/nut assemblies attaching the driveshaft to the transmission or transfer case flange and/or axle flange.

4. If equipped with a center bearing support, remove the support attaching nuts.

5. If equipped with a double-offset joint, push a rag into the joint to hold the driveshaft straight and prevent damaging the boot.

6. Remove the driveshaft assembly. If equipped with a splined yoke, pull the driveshaft out until the yoke clears the extension housing or transfer case, then install a suitable plug to prevent fluid leakage.

7. Inspect the universal joints for binding, excessive looseness or other damage. If equipped, inspect the splined yoke for wear or other damage.

To install:

8. If equipped with a splined yoke, lubricate the splines with Molybdenum grease and remove the plug from the extension housing or transfer case. During installation, do not

allow the yoke to bottom on the output shaft with excessive force.

9. Raise the driveshaft into position. Install the yoke on the output shaft or align the flange marks that were made during removal, as required.

10. Install the bolts/nuts, bolt/strap assemblies or U-bolt/nut assemblies attaching the driveshaft to the transmission or transfer case flange and/or axle flange. Tighten the bolts/nuts, if equipped, to 36–43 ft. lbs. (49–59 Nm) on all except Navajo. On Navajo, tighten the axle flange bolts to 70–95 ft. lbs. (95–129 Nm) and the transfer case bolts to 12–16 ft. lbs. (17–22 Nm).

11. On Navajo, if equipped with bolt/strap assemblies, tighten the bolts to 10–15 ft. lbs. (14–20 Nm). If equipped with U-bolt/nut assemblies, tighten the nuts to 8–15 ft. lbs. (11–20 Nm).

12. If equipped with a center bearing support, install and tighten the nuts or bolts to 27–39 ft. lbs. (36–53 Nm).

13. Lower the vehicle and road test for proper operation.

Front Axle Shaft, Bearing and Seal

REMOVAL AND INSTALLATION

Navajo

1. Raise and safely support the vehicle. Remove the front wheel and tire assemblies.

2. Remove the disc brake caliper and wire it to the frame. Do not let the caliper hang by the brake hose.

3. Remove the hub locks, wheel bearings and locknuts.

4. Remove the hub, rotor and outer wheel bearing.

5. Remove the grease seal from the rotor with a seal removal tool. Remove the inner wheel bearing.

6. If the wheel bearings are to be replaced, remove the inner and outer bearing races with a suitable puller or a hammer and brass drift.

7. Remove the nuts retaining the spindle to the steering knuckle. Tap the spindle with a plastic hammer to jar the spindle from the knuckle. Remove the splash shield.

8. On the left side of the vehicle, remove the shaft and joint assembly by pulling the assembly out of the carrier. On the right side of the carrier, remove and discard the clamp from the shaft and joint assembly and the stub shaft. Pull the shaft and joint assembly from the splines of the stub shaft.

9. If required, remove the oil seal and needle bearing from the spindle. If necessary, remove the slinger from the shaft by driving it off with a hammer.

To install:

10. If removed, install a new bearing and seal in the spindle and/or press on a new shaft sling

11. On the right side of the carrier, install the rubber boot and new keystone clamps on the stub shaft slip yoke. Slide the right shaft and joint assembly into the slip yoke making sure the splines are fully engaged. Slide the boot over the assembly and crimp the keystone clamp using suitable pliers.

NOTE: Make sure the yoke ears are in phase (in line) during assembly.

12. On the left side of the carrier, slide the shaft and joint assembly through the knuckle and engage the splines on the shaft in the carrier.

13. Install the splash shield and spindle onto the steering knuckle. Install and tighten the spindle nuts to 45 ft. lbs. (61 Nm).

14. If removed, drive the bearing races into the rotor using a suitable driver. Pack the inner and outer wheel bearings and the lip of a new seal with high-temperature wheel bearing grease.

15. Position the inner wheel bearing in the race and install the seal using a seal installer. Install the rotor on the spindle and install the outer wheel bearing in the race.

16. Install the wheel bearing, locknut, thrust bearing, snapring and locking hubs.

17. Install the caliper and the wheel and tire assemblies. Lower the vehicle.

Rear Axle Shaft, Bearing and Seal

REMOVAL AND INSTALLATION

B Series Pick-Up

1. Raise and safely support the vehicle. Remove the wheel and tire assembly.

2. Remove the brake drum and the brake shoes.

3. Disconnect the parking brake cable from the brake backing plate. Disconnect and plug the brake line at the wheel cylinder.

4. Remove the axle shaft/backing plate assembly from the axle housing, then pry out the axle seal. On 4WD vehicles, remove the O-ring from the axle housing.

5. Install the axle shaft in a vise, attached at the axle flange. Remove the lockwasher and the bearing locknut.

NOTE: Left side axles have left-hand thread; remove by turning clockwise.

6. Reposition the axle shaft in the vise so the jaws grip the shaft. Use protective jaw caps on the vise to protect the axle shaft.

7. Use a suitable puller to remove the bearing and bearing housing.

8. Remove the bearing and seal from the hub, then remove the outer bearing race using a suitable drift.

9. Clean all components and inspect for wear and/or damage. Replace as necessary. If bearing replacement is necessary, the bearing races must also be replaced.

To install:

10. Press in a new seal and inner race, using suitable drivers. Liberally apply high temperature wheel bearing grease to the area of the race and seal.

11. Using a suitable seal installer, tap a new seal into the axle housing until it is flush with the end of the housing. Coat the seal lip with grease.

12. Install the spacer on the axle shaft and position the axle shaft and backing plate in a press. Using suitable press tools, press the wheel bearing onto the axle shaft.

NOTE: The standard press-fit force should be 30,379–44,121 lbs. If the force is too high or too low, replace the bearing collar or shaft.

13. Position the axle shaft in a vise, attached at the axle flange. Install the bearing locknut and a new lockwasher. Tighten the locknut to 145–217 ft. lbs. (196–294 Nm). Align the lockwasher craws to the locknut notches and crimp the lockwasher.

14. Install a new O-ring in the axle housing, if removed. Install the axle shaft assembly, being careful not to damage the axle seal. Adjust the wheel bearing play as follows:

    a. There can only be one axle shaft installed in the housing during the adjustment procedure. If only one shaft was removed, remove the other shaft at this time. If both shafts were removed, leave the other shaft out of the axle housing for now.

b. Attach a dial indicator to the backing plate and place the indicator foot on the end of the axle shaft.

c. Check the axle shaft endplay; it should be 0.026–0.037 in. (0.65–0.95mm).

d. If the bearing play is not within specification, remove the axle and install shims which are available from Mazda.

e. If both axle shafts were removed, at this time, remove the axle shaft that has been adjusted and install and adjust the bearing play of the other shaft.

15. Install the backing plate mounting nuts and tighten to 72–87 ft. lbs. (98–118 Nm).

16. Connect the brake line and the parking brake cable.

17. Install the brake shoes and the brake drum.

18. Install the wheel and tire assembly. Bleed the brakes and lower the vehicle.

MPV

1. Raise and safely support the vehicle. Remove the wheel and tire assembly.

2. Remove the brake drum and the brake shoes.

3. Disconnect the parking brake cable from the brake backing plate. Disconnect and plug the brake line at the wheel cylinder.

4. Remove the axle shaft/backing plate assembly using a slide hammer. Remove the axle seal using a prybar.

5. Remove and discard the retaining ring from the axle shaft.

6. Using a suitable grinder, grind a section of the bearing collar until only approximately 0.0197 in. (0.5mm) remains. Be careful not to damage the axle shaft.

7. Cut the bearing collar with a chisel at the ground section and remove the bearing collar.

8. Use a suitable bearing puller to remove the bearing from the shaft. Remove the seal.

To install:

9. Apply grease to the lip of a new seal and install on the backing plate.

10. Install a new bearing over the axle shaft. Make sure there is no oil or grease on the new collar or axle shaft, then install the collar on the axle shaft.

11. Position the axle shaft/backing plate assembly in a press and press the bearing and collar onto the axle shaft.

NOTE: If the press-fit force of the collar is 5952 lbs. (2.7 tons) or less, replace the collar or the axle shaft.

12. Apply grease to the lip of a new axle seal and install the seal into the axle housing, using a suitable seal installer.

13. Install the axle shaft/backing plate assembly in the axle housing, being careful not to damage the axle seal. Install the backing plate nuts and tighten to 72–87 ft. lbs. (98–118 Nm).

14. Connect the brake line and the parking brake cable.

15. Install the brake shoes and the brake drum.

16. Install the wheel and tire assembly. Bleed the brakes and lower the vehicle.

Navajo

1. Raise and safely support the vehicle.

2. Remove the rear wheel and tire assemblies and the brake drums.

3. Clean all dirt from the carrier cover area. Position a drain pan under the carrier, remove the cover and drain the rear axle.

4. Remove the differential pinion shaft lock bolt and pinion shaft.

5. Push the flanged end of the axle shafts toward the center of the vehicle and remove the C-lock from the button end of the axle shaft. Be careful not to lose or damage the rubber O-ring which is in the axle shaft groove under the C-lock.

6. Remove the axle shaft from the housing.

7. Reinstall the pinion shaft and lock bolt to ensure the pinion gears remain in place.

8. Using bearing remover 49 UN01 033 or equivalent and a suitable slide hammer, remove the axle bearing and seal as a unit.

To install:

9. Lubricate a new bearing with rear axle lubricant and install in the housing bore using a suitable driver.

10. Apply grease to the lips of a new axle seal and install, using a seal installer.

11. Remove the pinion shaft lock bolt and pinion shaft.

12. Slide the axle shaft into the axle housing, being careful not to damage the seal or axle bearing. Start the splines into the side gear and push until the button end of the axle shaft can be seen in the differential case.

13. Install the C-lock on the button end of the axle shaft, then pull the shaft outboard until the shaft splines engage and the C-lock seats in the counterbore of the differential side gear. Make sure the O-ring is in the groove at the button end of the axle shaft before installing the C-lock.

14. Slide the pinion shaft through the case and pinion gears, aligning the hole in the shaft with the lock bolt hole. Apply stud and bearing mount compound and install the lock bolt. Tighten to 15–30 ft. lbs. (20–40 Nm).

15. Clean all old sealer from the carrier surface. Apply a bead of RTV sealer ⅛–¼ in. wide. The bead should be continuous and should not pass through or outside the holes.

16. Install a new cover and new bolts. Tighten the bolts to 15–20 ft. lbs. (21–27 Nm). Fill the carrier with the proper type and quantity of fluid.

17. Install the brake drums and the wheel and tire assemblies. Lower the vehicle.

Front Wheel Hub, Knuckle/Spindle and Bearings

REMOVAL AND INSTALLATION

B Series Pick Up

1. Raise and safely support the vehicle. Remove the wheel and tire assembly.

2. Remove the drive flange.

3. Remove the brake caliper. Support the caliper aside with rope or mechanics wire; do not let the caliper hang by the brake hose.

4. Remove the snapring and spacer. Remove the set bolts and bearing set plate.

5. Remove the bearing locknut using a suitable removal tool. Remove the hub and rotor without letting the washer and bearing fall.

6. Remove the dust cover.

7. Disconnect the tie rod end from the knuckle. Disconnect the stabilizer bar and the lower shock mount.

8. Support the lower control arm with a jack. Remove the lower ball joint nut and separate the knuckle from the lower arm using a suitable tool.

9. Remove the upper ball joint nut and separate the knuckle from the lower arm using a suitable tool.

10. Lower the lower control arm and remove the knuckle.

11. Inspect the knuckle, hub and bearings for wear and or damage. Replace components, as necessary.

12. Remove the oil seal and the bearing inner race from the knuckle. Using a suitable drift, remove the bearing outer race by tapping lightly with a hammer.

13. Using a slide hammer, remove the needle bearing from the knuckle.

14. Mark the position of the disc brake rotor on the hub, then remove the bolts and disassemble the rotor and hub.

15. Remove the oil seal and the bearing inner race from the hub. Using a suitable drift, remove the bearing outer race by tapping lightly with a hammer.

To install:

16. Press a new needle bearing into the knuckle using a suitable driver.

17. After installing the inner bearing into the knuckle, press in a new oil seal. Apply wheel bearing grease to the oil seal lip.

18. Press fit the outer side bearing outer race, then the inner side bearing outer race, into the hub using suitable drivers. Press in a new oil seal until it is flush with the hub end surface. Apply wheel bearing grease to the seal lip.

19. Align the matching marks of the hub and brake rotor and tighten the mounting bolts to 40–51 ft. lbs. (54–69 Nm).

20. Liberally apply high temperature wheel bearing grease to the inside of the hub. Install the outer bearing race and washer in the hub.

21. Insert the halfshaft into the knuckle and install the nut for the lower ball joint. Tighten the nut by hand.

22. Raise the lower control arm with the jack until the upper ball joint is connected to the knuckle. Install the nut and tighten by hand.

23. Tighten the upper ball joint nut to 22–38 ft. lbs. (29–51 Nm) and the lower ball joint nut to 87–116 ft. lbs. (118–157 Nm). Install new cotter pins.

24. Connect the tie rod end to the knuckle, tighten the nut to 23–43 ft. lbs. (44–59 Nm) and install a new cotter pin.

25. Install the dust cover to the knuckle and tighten to 14–19 ft. lbs. (19–26 Nm).

26. After loosely installing the lower shock absorber mount, install the stabilizer bar.

27. Install the hub and rotor assembly, then adjust the bearing preload as follows:

a. Tighten the locknut, then turn the hub and rotor 2–3 times to seat the bearing.

b. Loosen the locknut so they can be turned by hand.

c. Attach a suitable pull scale to a wheel lug bolt and measure the frictional forces. The preload is the frictional force plus 1.3–2.6 lbs.

d. Tighten the locknut until the preload is as specified.

e. Install the bearing set plate using 2 bolts. Tighten the bolts to 43–61 inch lbs. (5–7 Nm).

f. Coat the spacer with grease and install it. Install a new snapring.

28. Install the caliper, wheel and tire assembly and drive flange. Lower the vehicle.

29. Tighten the lower shock mount to 41–59 ft. lbs. (55–80 Nm) with the vehicle unloaded.

30. Check the front end alignment.

MPV

1. Raise and safely support the vehicle. Remove the wheel and tire assembly.

2. Remove and discard the locknut from the end of the halfshaft.

3. Remove the brake caliper and disc brake rotor. Support the caliper aside with rope or mechanics wire; do not let the caliper hang by the brake hose.

4. Remove the cotter pin and nut and, using a suitable tool, disconnect the tie rod end from the knuckle.

5. Remove the cotter pin and loosen the lower ball joint nut. Separate the lower arm from the knuckle using a suitable tool.

6. Remove the knuckle-to-strut bolts and nuts and remove the

1. Freewheel hub bolts
2. Snapring
3. Spacer
4. Bearing set plate
5. Locknut
6. Bearing
7. Hub assembly
8. Bearing
9. Oil seal
10. Rotor
11. Dust cover
12. Knuckle
13. Needle bearing
14. Halfshaft

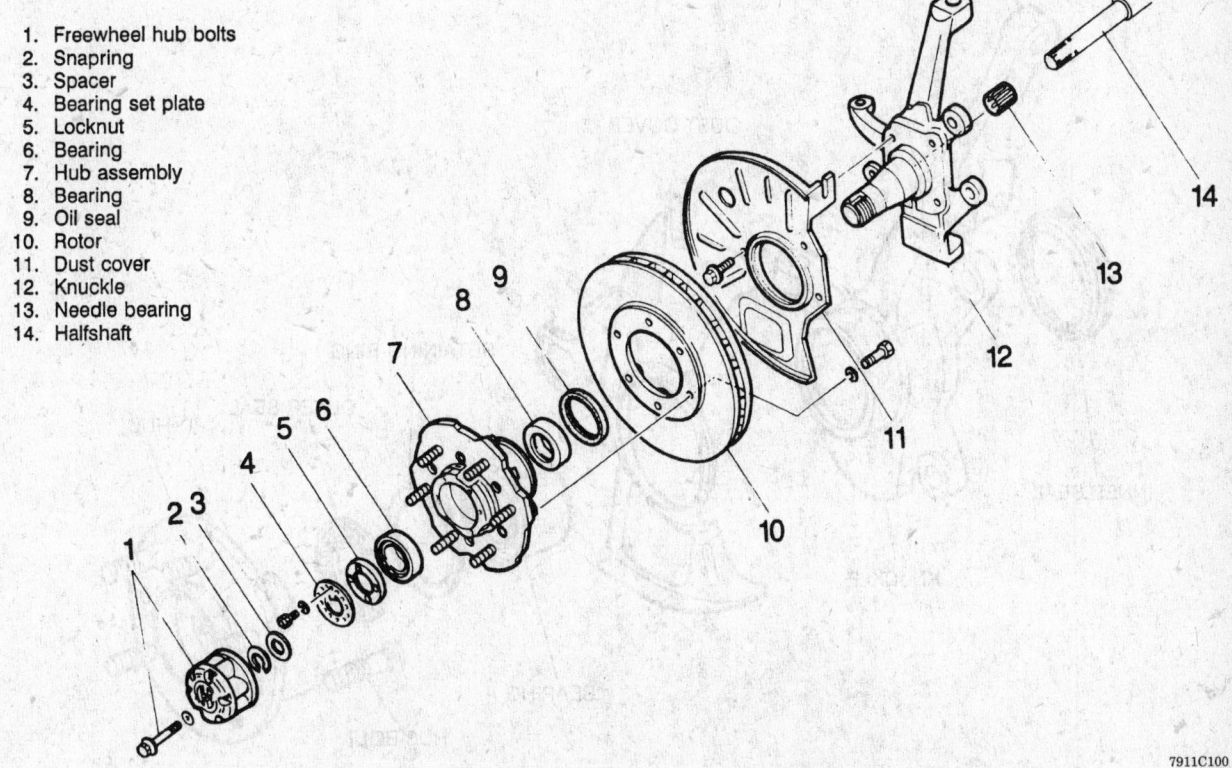

Front hub, knuckle and bearing assembly — 4WD B Series Pick-Up

7911C100

knuckle/hub assembly from the vehicle.

7. Pry out the inner oil seal from the knuckle.

8. Position the knuckle/hub assembly in a press and, using a suitable driver, press the hub from the knuckle.

NOTE: If the inner bearing race remains on the hub, position the hub in a vise, secured by the flange. Move the race away from the hub using a hammer and chisel, then position the hub in a press and press the race off of the hub.

9. Pry out the outer oil seal from the knuckle.

10. Remove the retaining ring and position the knuckle in a press. Using a suitable driver, press the wheel bearing from the knuckle.

11. If necessary, mark the position of the dust shield on the knuckle and remove the dust shield, using a hammer and chisel. Do not reuse the dust cover, if removed.

To install:

12. If the dust cover was removed, mark the new cover in the same place as the old was marked during removal. Align the cover and knuckle

marks and press the cover onto the knuckle.

13. Press a new wheel bearing into the knuckle, using a suitable driver. Install the retaining ring and a new outer seal. Apply grease to the seal lip.

14. Press the hub into the knuckle, using a suitable driver. Install a new inner seal and lubricate the seal lip with grease.

15. Install the knuckle/hub assembly onto the strut, install the bolts and nuts and tighten to 69–86 ft. lbs. (93–117 Nm).

16. Install the lower ball joint into the knuckle and tighten the nut to 115–137 ft. lbs. (157–187 Nm). Install a new cotter pin.

17. Connect the tie rod to the knuckle and install the nut. Tighten to 43–58 ft. lbs. (59–78 Nm) and install a new cotter pin.

18. Install the brake rotor and caliper.

19. Install a new locknut on the end of the halfshaft and tighten to 174–231 ft. lbs. (235–314 Nm). After tightening, stake the nut with a blunt chisel.

20. Install the wheel and tire assembly and lower the vehicle. Check the front end alignment.

Navajo

1. Raise and safely support the vehicle. Remove the front wheel and tire assemblies.

2. Remove the disc brake caliper and wire it to the frame. Do not let the caliper hang by the brake hose.

3. Remove the hub locks, wheel bearings and locknuts.

4. Remove the hub, rotor and outer wheel bearing.

5. Remove the grease seal from the rotor with a seal removal tool. Remove the inner wheel bearing.

6. If the wheel bearings are to be replaced, remove the inner and outer bearing races with a suitable puller or a hammer and brass drift.

7. Remove the nuts retaining the spindle to the steering knuckle. Tap the spindle with a plastic hammer to jar the spindle from the knuckle. Remove the splash shield.

8. On the left side of the vehicle, remove the shaft and joint assembly by pulling the assembly out of the carrier. On the right side of the carrier, remove and discard the clamp from the shaft and joint assembly and the stub shaft. Pull the shaft and joint assembly from the splines of the stub shaft.

9. Place the spindle in a vise on the second step of the spindle. Wrap

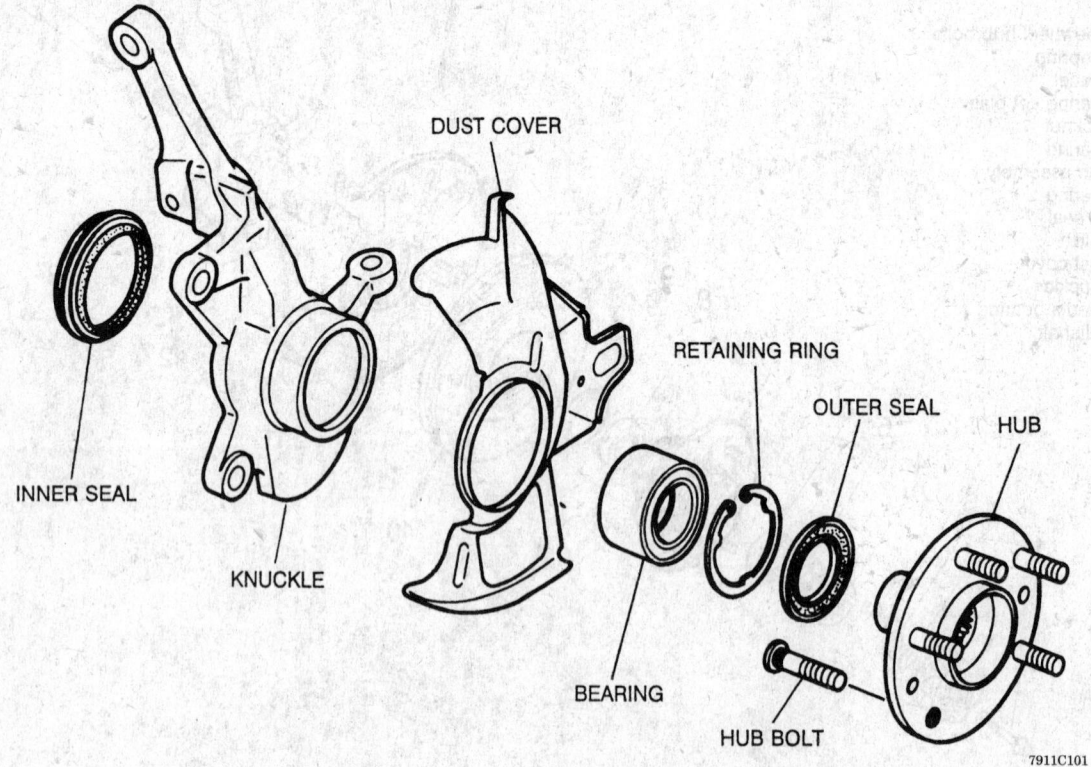

INNER SEAL

KNUCKLE

DUST COVER

RETAINING RING

OUTER SEAL

HUB

BEARING

HUB BOLT

7911C101

Front wheel hub, knuckle and bearing assembly — 4WD MPV

a shop towel around the spindle or use a brass-jawed vise to protect the spindle.

10. Remove the oil seal and needle bearing from the spindle with a slide hammer and seal remover TOOL-1175-AC or equivalent. If necessary, remove the slinger from the shaft by driving off with a hammer.

11. Remove the cotter pin from the tie rod nut and then remove the nut. Tap on the tie rod stud to free it from the steering arm.

12. Remove the upper ball joint snapring and remove the upper ball joint pinch bolt. Loosen the lower ball joint nut to the end of the stud.

13. Strike the inside of the knuckle near the upper and lower ball joints to break the knuckle loose from the ball joint studs.

14. Remove the camber adjuster sleeve. Note the position of the slot in the camber adjuster so it can be reinstalled in the same position during assembly.

15. Remove the lower ball joint nut. Place the knuckle in a vise and remove the snapring from the bottom ball joint socket, if equipped.

16. Assemble C-frame T74P-4635-C and ball joint remover T83T-3050-A or equivalents on the lower ball joint. Turn the forcing screw clockwise until the lower ball joint is removed from the steering knuckle.

17. Assemble the C-frame and ball joint remover on the upper ball joint and remove in the same manner.

NOTE: Always remove the lower ball joint first.

To install:

18. Clean the steering knuckle bore and insert the lower ball joint in the knuckle as straight as possible.

19. Assemble C-frame T74P-4635-C, ball joint installer T83T-3050-A and receiver cup T80T-3010-A3 or equivalents to install the lower ball joint. Turn the forcing screw clockwise until the lower ball joint is firmly seated. Install the snapring on the lower ball joint.

NOTE: The lower ball joint must always be installed first.

20. Assemble the C-frame, ball joint installer and receiver cup to install the upper ball joint. Turn the forcing screw clockwise until the ball joint is firmly seated.

21. Install the camber adjuster into the support arm, making sure the slot is in the original position.

NOTE: The torque sequence in Steps 22 and 23 must be followed exactly when securing the knuckle. Excessive knuckle turning effort may result in reduced steering returnability if this procedure is not followed.

22. Install a new nut on the bottom ball joint stud. Tighten the nut to 90 ft. lbs. (122 Nm) minimum, then tighten to align the next slot in the nut with the hole in the stud. Install a new cotter pin.

23. Install the snapring on the upper ball joint stud. Install the upper ball joint pinch bolt and tighten to 48-65 ft. lbs. (65-88 Nm).

NOTE: The camber adjuster will seat itself into the knuckle at a predetermined position during the tightening sequence. Do not attempt to adjust this position.

24. Clean all dirt and grease from the spindle bearing bore. The bearing bores must be free from nicks and burrs.

25. Place the bearing in the bore with the manufacturers identification facing outward. Drive the bearing into the bore using spindle bearing replacer T80T-4000-S and driver handle T80T-4000-W or equivalents.

26. Install the grease seal in the bearing bore with the lip side of the seal facing towards the tool. Drive the seal in the bore using the same tools as in Step 25. Coat the bearing seal lip with high-temperature lubricant.

27. If removed, press on a new shaft slinger.

28. On the right side of the carrier, install the rubber boot and new keystone clamps on the stub shaft slip yoke. Slide the right shaft and joint assembly into the slip yoke making sure the splines are fully engaged. Slide the boot over the assembly and crimp the keystone clamp using suitable pliers.

NOTE: The Dana model 35 axle does not have a blind spline, therefore pay special attention to make sure the yoke ears are in phase (in line) during assembly.

29. On the left side of the carrier, slide the shaft and joint assembly through the knuckle and engage the splines on the shaft in the carrier.

30. Install the splash shield and spindle onto the steering knuckle. Install and tighten the spindle nuts to 45 ft. lbs. (61 Nm).

31. If removed, drive the bearing races into the rotor using a suitable driver. Pack the inner and outer wheel bearings and the lip of a new seal with high-temperature wheel bearing grease.

32. Position the inner wheel bearing in the race and install the seal using a seal installer. Install the rotor on the spindle and install the outer wheel bearing in the race.

33. Install the wheel bearing, locknut, thrust bearing, snapring and locking hubs.

34. Install the caliper and the wheel and tire assemblies. Lower the vehicle.

Manual Locking Hubs

REMOVAL AND INSTALLATION

Navajo

1. Raise and support the vehicle safely.

2. Remove the lug nuts and remove the wheel and tire assembly.

3. Remove the retainer washers from the lug nut studs and remove the manual locking hub assembly. To remove the internal hub lock assembly from the outer body assembly, remove the outer lock ring seated in the hub body groove. The internal assembly, spring and clutch gear will now slide out of the hub body. Do not remove the screw from the plastic dial.

4. Rebuild the hub assembly in the reverse order of disassembly.

5. Adjust the wheel bearing if necessary. Install the manual locking hub assembly over the spindle and place the retainer washers on the lug nut studs.

6. Install the wheel and tire assembly and lower the vehicle.

ADJUSTMENT

Navajo

1. Raise and safely support the vehicle. Remove the wheel and tire assembly.

2. Remove the retainer washers from the lug nut studs and remove the manual locking hub assembly from the spindle.

3. Remove the snapring from the end of the spindle shaft.

4. Remove the axle shaft spacer.

5. Remove the outer wheel bearing locknut from the spindle using locknut wrench 49 UN01 042 or equivalent. Make sure the tabs on the tool engage the slots in the locknut.

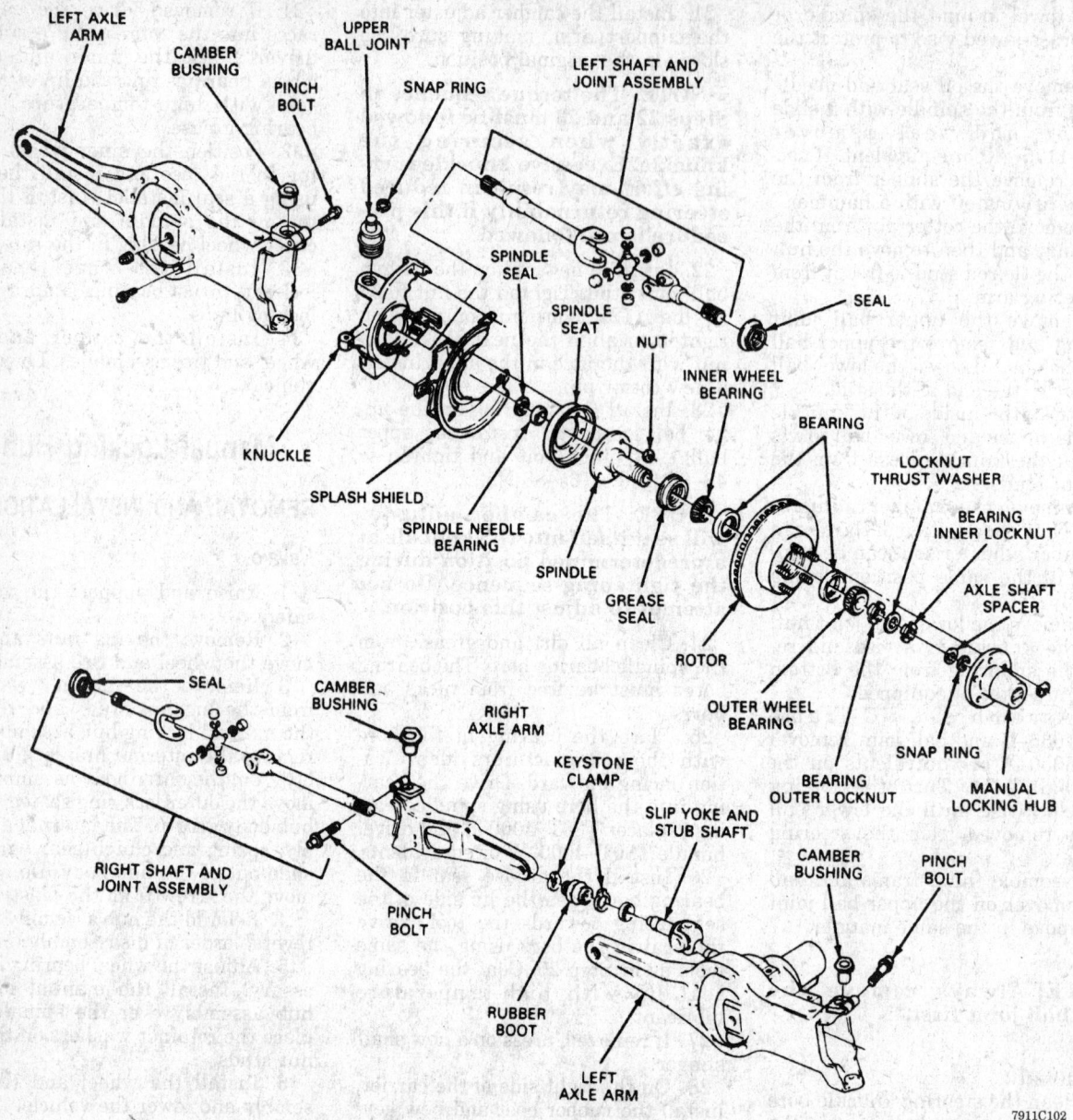

Front axle shaft, hub, knuckle and bearing assembly — Navajo

7911C102

6. Remove the locknut washer from the spindle.

7. Loosen the inner wheel bearing locknut using the locknut wrench. Make sure the tabs on the tool engage the slots in the locknut and the slot in the tool is centered over the locknut pin.

8. Tighten the inner locknut to 35 ft. lbs. (47 Nm) to seat the bearings.

9. Spin the rotor and back off the inner locknut 1/4 turn. Retighten the inner locknut to 16 inch lbs. (1.8 Nm). Install the lockwasher on the spindle. It may be necessary to tighten the inner locknut slightly so the pin on the locknut aligns with the closest hole in the lockwasher.

10. Install the outer wheel bearing locknut using the locknut wrench. Tighten the locknut to 150 ft. lbs. (203 Nm).

11. Install the axle shaft spacer.

12. Clip the snapring onto the end of the spindle. Install the manual hub assembly over the spindle and install the retainer washers.

13. Install the wheel and tire assembly. Check the endplay of the wheel and tire assembly on the spindle. Final endplay should be 0–0.003 in. (0–0.08mm). The maximum torque to rotate the hub should be 25 inch lbs. (2.8 Nm).

14. Lower the vehicle.

Automatic Locking Hubs

REMOVAL AND INSTALLATION

Navajo

1. Raise and support the vehicle safely. Remove the wheel lug nuts and remove the wheel and tire assembly.

2. Remove the retainer washers from the lug nut studs and remove the automatic locking hub assembly from the spindle.

3. Remove the snapring from the end of the spindle shaft.

4. Remove the axle shaft spacer.

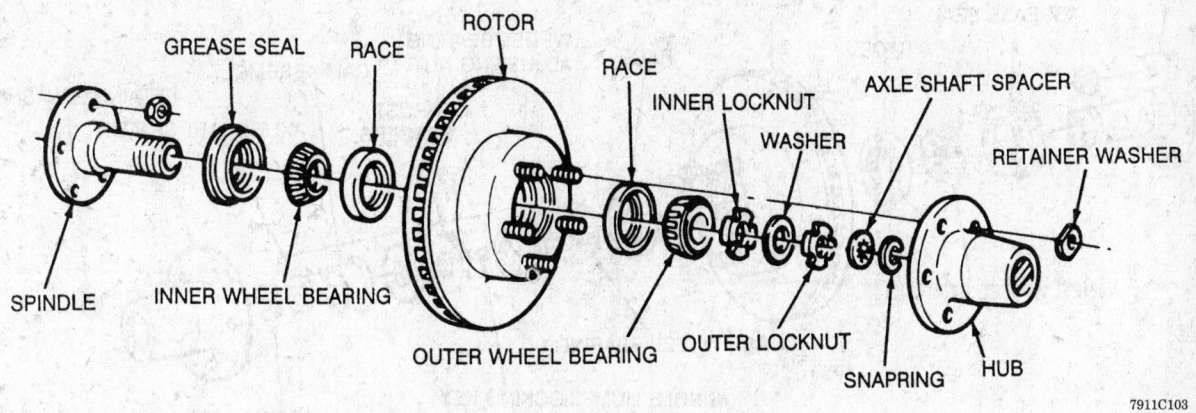

GREASE SEAL | RACE | ROTOR | RACE | INNER LOCKNUT | WASHER | AXLE SHAFT SPACER | RETAINER WASHER

SPINDLE | INNER WHEEL BEARING | OUTER WHEEL BEARING | OUTER LOCKNUT | SNAPRING | HUB

7911C103

Manual locking hub assembly — Navajo

5. Being careful not to damage the plastic moving cam or thrust spacers, pull the cam assembly off the wheel bearing adjusting nut. Remove the 2 plastic thrust spacers from the adjusting nut.

6. Using a magnet, remove the locking key. It may be necessary to rotate the adjusting nut slightly to relieve the pressure against the locking key, before the key can be removed.

NOTE: To prevent damage to the spindle threads, look into the spindle keyway under the adjusting nut and remove the separate locking key before removing the adjusting nut.

7. Loosen the wheel bearing adjusting nut from the spindle using a 2⅜ in. hex socket tool.

8. While rotating the hub and rotor assembly, tighten the wheel bearing adjusting nut to 35 ft. lbs. (47 Nm) to seat the bearings. Spin the rotor and back off the nut ¼ turn.

9. Retighten the adjusting nut to 16 inch lbs. (1.8 Nm) using a torque wrench. Align the closest hole in the wheel bearing adjusting nut with the center of the spindle keyway slot. Advance the nut to the next lug if required. Install the separate locking

key in the spindle keyway under the adjusting nut.

NOTE: Extreme care must be taken when aligning the spindle nut adjustment lug with the center of the spindle keyway slot to prevent damage to the separate locking key.

10. Install the 2 thrust spacers. Push or press the cam assembly onto the locknut by lining up the key in the fixed cam with the spindle keyway.

NOTE: Extreme care must be taken when aligning the fixed cam key with the spindle keyway to prevent damage to the fixed cam.

11. Install the axle shaft spacer.

12. Clip the snapring onto the end of the spindle.

13. Install the automatic locking hub assembly over the spindle by lining up the 3 legs in the hub assembly with the 3 pockets in the cam assembly. Install the retainer washers.

14. Install the wheel and tire assembly. Check the endplay of the wheel and tire assembly on the spindle. Final endplay should be 0–0.003 in. (0–0.08mm). The maximum torque to rotate the hub should be 25 inch lbs. (2.8 Nm).

15. Lower the vehicle.

Freewheel Mechanism

REMOVAL AND INSTALLATION

NOTE: The Remote Freewheel Mechanism on 4WD B Series Pick-Up and the Automatic Freewheel Mechanism on 4WD MPV are used in place of automatic locking hubs.

1. Disconnect the negative battery cable. Raise and safely support the vehicle. Remove the left front wheel and tire assembly.

2. Drain the fluid from the front differential.

3. Remove the left side halfshaft assembly.

4. Tag and disconnect the vacuum hoses and electrical connector from the control box assembly.

5. Remove and discard the snap pin at the control box assembly.

6. Remove the attaching bolts and remove the joint shaft assembly.

7. Remove the attaching bolts and remove the control box assembly.

8. Remove the gear sleeve from the side of the differential, if necessary.

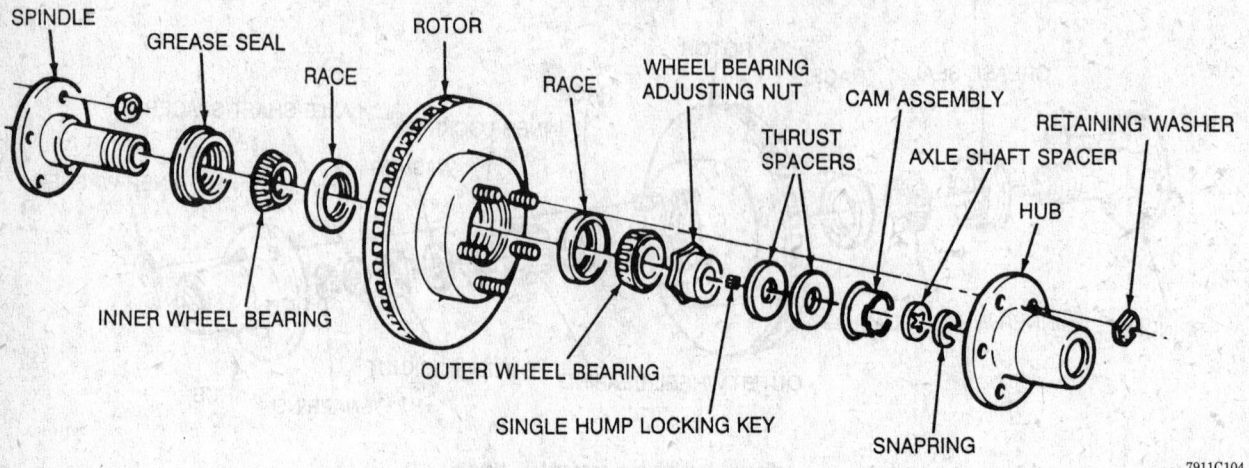

Automatic locking hub assembly — Navajo

9. If necessary, remove the output shaft from the differential using a slide hammer.

To install:

10. If removed, install a new clip on the end of the output shaft and install in the differential. Install the gear sleeve, if removed.

11. Install the control box and tighten the attaching bolts to 17–20 ft. lbs. (23–26 Nm).

12. Install the joint shaft assembly and tighten the attaching bolts to 27–40 ft. lbs. (36–54 Nm). On MPV, install the attaching nut and tighten to 49–72 ft. lbs. (67–97 Nm).

13. Install a new snap pin at the control box assembly.

14. Connect the electrical connector and vacuum hoses at the control box assembly.

15. Install the left side halfshaft assembly.

16. Fill the differential with the proper type and quantity of fluid.

17. Install the wheel and tire assembly and lower the vehicle.

Pinion Seal

REMOVAL AND INSTALLATION

NOTE: This service procedure disturbs the pinion bearing preload and this preload must be carefully reset when assembling.

1. Raise the vehicle and support it safely.

2. Remove the wheels and the brake drums.

3. Mark the driveshaft and the axle companion flange so the driveshaft can be reinstalled in the same position. Remove the driveshaft.

4. Using an inch pound torque wrench on the pinion nut, record the torque required to maintain rotation of the pinion through several revolutions.

5. While holding the companion flange with a suitable tool, remove the pinion nut. Mark the companion flange in relation to the pinion shaft so the flange can be reinstalled in the same position.

6. Using a suitable puller, remove the rear axle companion flange. Use a small prybar to remove the seal from the carrier.

To install:

7. Make sure the splines of the pinion shaft are free of burrs.

8. Apply grease to the lips of the pinion seal and install, using a seal installer.

9. Check the seal surface of the companion flange for scratches, nicks or a groove. Replace the companion flange, as necessary. Apply a small amount of lubricant to the splines. Align the mark on the flange with the mark on the pinion shaft and install the companion flange.

NOTE: The companion flange must never be hammered on or installed with power tools.

10. Install a new nut on the pinion shaft. Hold the companion flange with a suitable tool while tightening the nut.

11. Tighten the pinion nut, rotating the pinion occasionally to ensure proper bearing seating. Take frequent pinion bearing torque preload readings until the original recorded preload reading is obtained.

NOTE: Under no circumstances should the pinion nut be backed off to reduce preload. If reduced preload is required, a new collapsible pinion spacer and pinion nut must be installed.

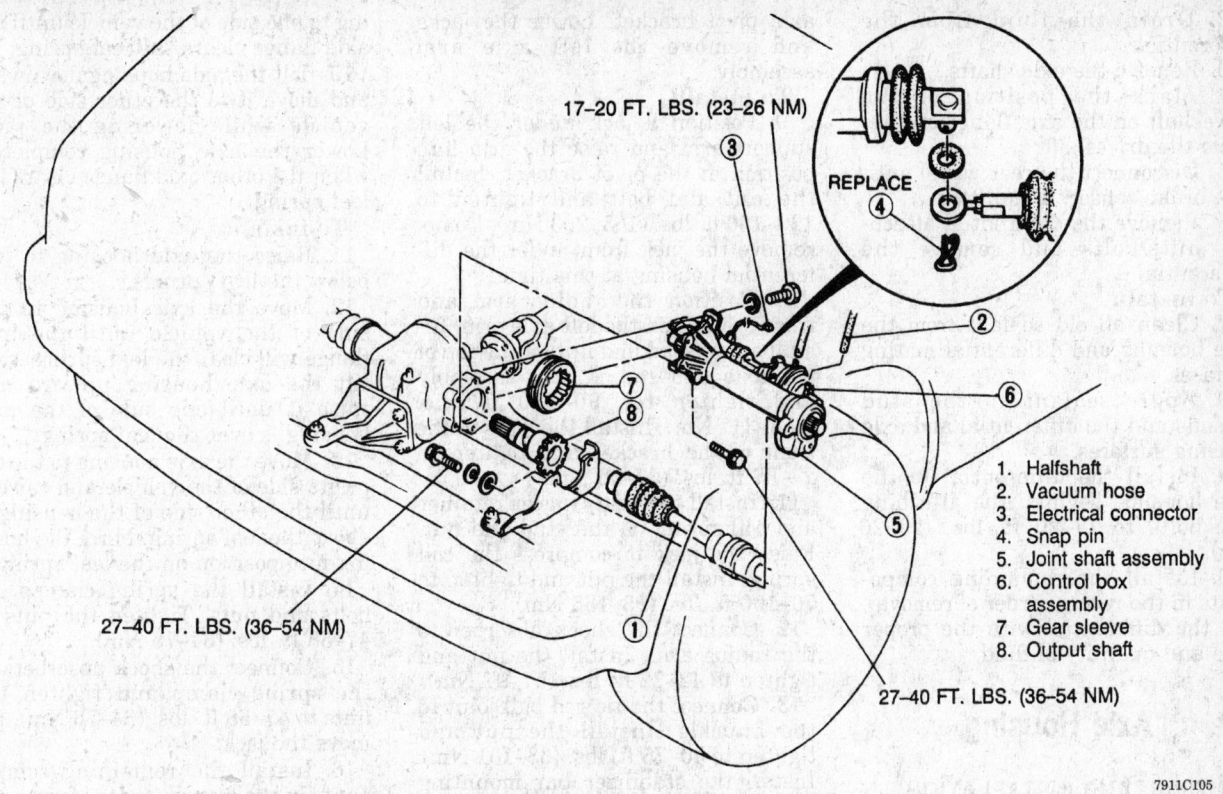

17–20 FT. LBS. (23–26 NM)

REPLACE

27–40 FT. LBS. (36–54 NM)

27–40 FT. LBS. (36–54 NM)

1. Halfshaft
2. Vacuum hose
3. Electrical connector
4. Snap pin
5. Joint shaft assembly
6. Control box assembly
7. Gear sleeve
8. Output shaft

7911C105

Remote Freewheel Mechanism installation — 4WD B Series Pick-Up; MPV similar

12. Install the driveshaft and check the fluid level in the carrier. Lower the vehicle.

Differential Carrier

REMOVAL AND INSTALLATION

Front

B SERIES PICK-UP

NOTE: The differential is removed as a unit with the freewheel mechanism. After removal, the differential can then be separated from the freewheel mechanism, if necessary.

1. Raise and safely support the vehicle. Remove the wheel and tire assemblies.
2. Remove the splash shield and drain the differential fluid.
3. Remove the halfshafts.
4. Mark the position of the driveshaft on the axle flange and remove the driveshaft.
5. Tag and disconnect the vacuum hoses and electrical connector from the freewheel mechanism control box.
6. Support the differential with a jack.
7. Remove the crossmember bolts adjacent to the lower control arm.

Lower the differential/crossmembers assembly from the vehicle.
8. Remove the crossmembers from the differential, if necessary. Remove the freewheel mechanism from the differential, if necessary.
To install:
9. If removed, install the freewheel mechanism.
10. If removed, install the differential to the crossmembers.
11. Raise the differential/crossmembers assembly into position. Install the crossmember mounting bolts and tighten to 69–85 ft. lbs. (93–116 Nm). Remove the jack.
12. Install the remaining components in the reverse order of their removal. Fill the differential with the proper type and quantity of fluid.

MPV

NOTE: The differential is removed as a unit with the freewheel mechanism. After removal, the differential can then be separated from the freewheel mechanism, if necessary.

1. Raise and safely support the vehicle. Remove the wheel and tire assemblies.
2. Remove the splash shield and drain the differential fluid.

3. Remove the halfshafts.
4. Mark the position of the driveshaft on the axle flange and remove the driveshaft.
5. Tag and disconnect the vacuum hoses and electrical connector from the freewheel mechanism control box.
6. Support the differential with a jack.
7. Remove the bolts/nuts attaching the differential/freewheel mechanism assembly in 3 places and lower the assembly from the vehicle.
8. If necessary, separate the freewheel mechanism from the differential.
To install:
9. If removed, install the freewheel mechanism.
10. Raise the differential/freewheel mechanism assembly into position and install the attaching bolts/nuts. Tighten to 49–72 ft. lbs. (67–97 Nm). Remove the jack.
11. Install the remaining components in the reverse order of their removal. Fill the differential with the proper type and quantity of fluid.

Rear

1. Raise and safely support the vehicle. Remove the wheel and tire assemblies.

2. Drain the fluid from the differential.

3. Remove the axle shafts.

4. Mark the position of the driveshaft on the axle flange and remove the driveshaft.

5. Disconnect the rear wheel anti-lock brake sensor, if equipped.

6. Remove the differential attaching nuts/bolts and remove the differential.

To install:

7. Clean all old sealant from the axle housing and differential mating surfaces.

8. Apply sealant to the stud threads and the differential and axle housing surfaces.

9. Install the differential to the axle housing. Tighten the attaching nuts/bolts to 17–20 ft. lbs. (23–26 Nm).

10. Install the remaining components in the reverse order of removal. Fill the differential with the proper type and quantity of fluid.

Axle Housing

REMOVAL AND INSTALLATION

Front

NAVAJO

1. Raise and safely support the vehicle. Remove the front axle shaft and spindle assemblies.

2. Mark the front axle yoke and the driveshaft so they can be reassembled in the same position. Disconnect the driveshaft from the front axle yoke.

3. Remove the cotter pin and nut retaining the steering linkage to the knuckle. Disconnect the linkage from the knuckle.

4. Remove the left stabilizer bar link lower bolt and remove the link from the radius arm bracket.

5. Position a jack under the left axle arm and slightly compress the coil spring. Remove the shock absorber lower nut and disconnect the shock absorber from the radius arm bracket.

6. Remove the nut that retains the lower part of the spring to the axle arm. Slowly lower the jack and remove the coil spring, spacer, seat and stud.

7. Remove the stud and bolts that connect the radius arm bracket and radius arm to the axle arm. Remove the bracket and radius arm.

8. Position another jack under the differential housing. Remove the bolt that connects the left axle arm to the

axle pivot bracket. Lower the jacks and remove the left axle arm assembly.

To install:

9. Position a jack under the left support arm and raise the arm into position in the pivot bracket. Install the nut and bolt and tighten to 120–150 ft. lbs. (163–203 Nm). Do not remove the jack from under the differential housing at this time.

10. Position the radius arm and front bracket on the left axle arm. Install a new stud and nut on the top of the axle and radius arm assembly and tighten to 190–230 ft. lbs. (258–311 Nm). Install the bolts in the front of the bracket and tighten to 27–37 ft. lbs. (37–50 Nm).

11. Install the seat, spacer retainer and coil spring on the stud and nut. Raise the jack to compress the coil spring. Install the nut and tighten to 70–100 ft. lbs. (95–135 Nm).

12. Connect the shock absorber to the radius arm. Install the nut and tighten to 42–72 ft. lbs. (57–97 Nm).

13. Connect the tie rod ball joint to the knuckle. Install the nut and tighten to 50–75 ft. lbs. (68–101 Nm). Install the stabilizer bar mounting bracket and tighten to 203–240 ft. lbs. (275–325 Nm).

14. Connect the front driveshaft to the front axle yoke, aligning the marks that were made during removal. Install the U-bolts and tighten the nuts to 8–15 ft. lbs. (11–20 Nm).

15. Install the spindle and axle shaft assemblies. Lower the vehicle.

Rear

2WD B SERIES PICK-UP

1. Raise and safely support the vehicle. Remove the rear wheel and tire assemblies.

2. Remove the brake drums and brake shoes. Disconnect the parking brake cables from the backing plate.

3. Mark the position of the driveshaft on the axle flange and remove the driveshaft.

4. Disconnect and plug the flexible brake hose at the axle housing or chassis.

5. Disconnect the rear wheel anti-lock brake sensor, if equipped.

6. Support the axle housing with a jack.

7. Disconnect the shock absorbers at the spring clamp.

8. Remove the nuts from the U-bolts and remove the U-bolts and spring clamp.

9. While supporting the axle housing with the jack, move the axle hous-

ing to one side of the vehicle until the axle flange clears the leaf spring.

10. Tilt the axle housing downward and move it to the other side of the vehicle while lowering the jack. Lower the axle housing completely when the other axle flange clears the leaf spring.

To install:

11. Raise the axle housing to just below the leaf springs.

12. Move the axle housing to one side of the vehicle until the axle flange will clear the leaf spring, then tilt the axle housing upward and raise it until one side of the axle housing is over the leaf spring.

13. Move the axle housing to the opposite side of the vehicle and raise it until the other side of the housing is above the leaf spring. Move the housing into position on the leaf springs.

14. Install the spring clamps, U-bolts and nuts. Tighten the nuts to 47–58 ft. lbs. (64–78 Nm).

15. Connect the shock absorbers to the spring clamp and tighten the nuts to 47–58 ft. lbs. (64–78 Nm). Remove the jack.

16. Install the remaining components in the reverse order of removal. Bleed the brake system.

4WD B SERIES PICK-UP

1. Raise and safely support the vehicle. Remove the rear wheel and tire assemblies.

2. Remove the brake drums and brake shoes. Disconnect the parking brake cables from the backing plate.

3. Mark the position of the driveshaft on the axle flange and remove the driveshaft.

4. Disconnect and plug the flexible brake hose at the axle housing or chassis.

5. Disconnect the rear wheel anti-lock brake sensor, if equipped.

6. Support the axle housing with a jack.

7. Disconnect the shock absorbers at the axle housing.

8. Remove the nuts from the U-bolts and remove the U-bolts and spring clamp.

9. Lower the axle housing from the vehicle.

To install:

10. Raise the axle housing into position.

11. Install the U-bolts and spring clamps and tighten to 88–101 ft. lbs. (120–137 Nm).

12. Connect the shock absorbers at the axle housing and tighten the nuts and bolts to 47–58 ft. lbs. (64–78 Nm). Remove the jack.

13. Install the remaining components in the reverse order of removal. Bleed the brake system.

MPV

1. Raise and safely support the vehicle. Remove the rear wheel and tire assemblies.
2. Remove the brake drums and brake shoes. Disconnect the parking brake cables from the backing plate.
3. Mark the position of the driveshaft on the axle flange and remove the driveshaft.
4. Disconnect and plug the flexible brake hose at the axle housing or chassis.
5. Disconnect the rear wheel anti-lock brake sensor, if equipped.
6. Remove the rear stabilizer bar, if equipped.
7. Support the axle housing with a jack.
8. Disconnect the shock absorbers at the axle housing.
9. If equipped, remove the nut and disconnect the height sensor link from the axle housing.
10. Remove the lateral rod.
11. Slowly lower the axle housing until the coil spring tension is relieved. Remove the coil springs.
12. Remove the upper and lower control arms from the axle housing and lower the housing from the vehicle.
   To install:
13. Raise the axle housing into position.
14. Connect the upper and lower control arms to the axle housing and snug the attaching bolts and nuts.
15. Install the coil springs. Install the lateral rod and snug the attaching bolts. Connect the height sensor link, if equipped.
16. Connect the shock absorbers to the axle housing and snug the attaching bolts and nuts. Remove the jack from the axle housing.
17. Install the stabilizer bar and snug the attaching bolts.
18. Connect the rear wheel anti-lock brake sensor, if equipped. Connect the height sensor link to the axle housing, if equipped.
19. Connect the brake hose to the brake line.
20. Install the driveshaft, aligning the marks that were made during the removal procedure.
21. Connect the parking brake cables to the backing plates. Install the brake shoes and the brake drums.
22. Install the wheel and tire assemblies and lower the vehicle.

23. With the vehicle in the normal ride height position, tighten the following to specification:
   • Control arm nuts and bolts: 108–127 ft. lbs. (146–167 Nm).
   • Lateral arm nuts: 108–127 ft. lbs. (146–167 Nm)
   • Shock absorber nuts and bolts: 56–76 ft. lbs. (76–103 Nm)
   • Stabilizer bar bushing mount bolts: 23–38 ft. lbs. (34–51 Nm)

NAVAJO

1. Raise the vehicle and support it safely.
2. Remove the cover and drain the lubricant from the axle.
3. Remove the rear wheel and tire assemblies and remove the rear anti-lock brake system sensor. Remove the axle shafts.
4. Remove the 4 retaining nuts from each backing plate. Wire the backing plates to the underbody.
5. Disconnect the vent hose from the axle housing. Remove the connector from the rear anti-lock sensor.
6. Remove the brake line from the clips that retain the line to the axle housing.
7. Remove the hydraulic brake junction block from the axle housing. Do not open the hydraulic brake system lines.
8. Mark the driveshaft and the axle companion flange so they can be reassembled in the same position. Remove the driveshaft.
9. Support the rear axle housing on a jack, then remove the axle housing U-bolt nuts. Remove the U-bolts and shock absorber plates. Leave the shock absorbers attached to the plates.
10. Remove the stabilizer bar attaching bracket bolts from the axle housing and position the stabilizer bar assembly away from the axle housing.
11. Raise the axle housing off the springs with the jack and move to the right side of the vehicle. Lower the left side of the axle housing below the left spring enough to clear the spring.
12. Remove the axle housing from the vehicle by lowering the axle and moving to the left until the right axle tube clears the right spring.
   To install:
13. Install the axle housing on the transmission jack. Guide the right side of the axle housing over the right spring. Lift the left side of the axle housing over the left spring and position the axle housing on the spring center bolts.
14. Install the stabilizer bar to the axle housing. Tighten the stabilizer

bar bracket bolts to 30–42 ft. lbs. (40–57 Nm).
15. Install the axle housing U-bolts over the axle tube. Position the shock absorber plates under the springs and install the U-bolts through the holes. Install the nuts and tighten to 88–108 ft. lbs. (119–146 Nm).
16. Install the axle vent tube to the axle vent fitting and secure with a clamp. Connect the rear anti-lock sensor connector.
17. Install the brake backing plates on the axle housing flanges. Tighten the nuts to 20–40 ft. lbs. (28–54 Nm).
18. Position the brake lines to the axle housing and secure with the retaining clips. Position the brake junction block to the axle housing and install the retaining screw.
19. Install the axle shafts. Install the driveshaft, aligning the marks that were made during removal. Tighten the attaching bolts to 70–95 ft. lbs. (95–128 Nm).
20. Install the brake drums and the wheel and tire assemblies.
21. Clean all old sealer from the carrier surface. Apply a bead of RTV sealer ⅛–¼ in. wide. The bead should be continuous and should not pass through or outside the holes.
22. Install a new cover and tighten the bolts to 15–20 ft. lbs. (21–27 Nm) in a criss-cross pattern.
23. Add the proper type of lubricant through the filler hole until the lubricant level is ¼–⁹⁄₁₆ in. below the bottom of the filler hole with the axle in the running position.
24. Lower the vehicle.

STEERING

Steering Wheel

REMOVAL AND INSTALLATION

1. Disconnect the negative battery cable.
2. Remove the steering wheel pad from the steering wheel. Pull the pad back and disconnect the horn switch and, if equipped, cruise control wires. Remove the steering wheel pad.
3. Remove the steering wheel attaching bolt or nut. Check to see if the steering wheel and steering shaft have alignment marks or flats. If there are no steering wheel-to-steering column shaft alignment marks or flats, matchmark the steering wheel

and column shaft so they can be reassembled in the same position.

4. Using a suitable puller, remove the steering wheel from the steering column shaft.

NOTE: Do not hammer on the steering wheel or steering shaft or use a knock-off type steering wheel puller, as either will damage the steering column.

To install:

5. Install the steering wheel on the steering column shaft, aligning the marks or flats on the steering wheel with the marks or flats on the steering shaft.

6. On all except Navajo, install the steering wheel attaching nut and tighten to 29–36 ft. lbs. (39–49 Nm). On Navajo, install the steering wheel attaching bolt and tighten to 23–33 ft. lbs. (31–45 Nm).

7. Connect the horn switch and, if equipped, cruise control wires and install the steering wheel pad.

8. Connect the negative battery and check the steering column for proper operation.

Steering Column

REMOVAL AND INSTALLATION

Except Navajo

1. Disconnect the negative battery cable. Place the front wheels in the straight-ahead position.

2. Remove the steering wheel.

3. Remove the steering column covers and remove the combination switch.

4. Remove the necessary dash panels from under the steering column.

5. Disconnect the automatic transmission interlock cable, if equipped. Tag and disconnect the necessary electrical connectors.

6. Disconnect the steering column from the steering linkage by removing the bolt at the intermediate shaft universal joint. Mark the position of the intermediate shaft on the column shaft before disconnecting so they can be reassembled in the same position.

7. Remove the nuts attaching the column at the firewall.

8. Support the steering column and remove the bolts attaching the column to the underside of the dash. Remove the steering column.

9. Installation is the reverse of the removal procedure. Tighten the bolts attaching the steering column to the

underside of the dash to 12–17 ft. lbs. (16–23 Nm). Tighten the nuts attaching the column at the firewall to 14–19 ft. lbs. (19–26 Nm) on B Series Pick-Up or 12–17 ft. lbs. (16–23 Nm) on MPV.

10. Align the marks that were made on the intermediate shaft and steering column shaft during removal. Tighten the intermediate shaft universal joint bolt to 13–18 ft. lbs. (18–25 Nm).

Navajo

1. Disconnect the negative battery cable and apply the parking brake. Place automatic transmission in **N**.

2. Remove the bolt that holds the intermediate shaft to the steering column shaft. Using a prybar, compress the intermediate shaft until it is clear of the steering column shaft.

3. If equipped with automatic transmission, remove the nuts from the studs and remove the shift cable bracket from the steering column bracket. Disconnect the shift cable from the column lever.

4. Remove the steering wheel. If equipped with tilt column, make sure the steering wheel is in the full up position before removal.

5. If equipped with tilt column, remove the tilt lever and remove the column collar by pressing on the collar from the top and bottom while removing the collar.

6. Remove the retaining screws and remove the panel trim cover.

7. Remove the 2 screws from the bottom of the column shroud. Remove the bottom half of the shroud by pulling the shroud down and toward the rear of the vehicle. If equipped with automatic transmission, move the shift lever as required to ease shroud removal. Lift the top half of the shroud from the column.

8. If equipped with automatic transmission, disconnect the selector indicator cable by removing the screw from the column casting and the plastic plug at the end of the cable. To remove the plastic plug from the shift lever socket casting, push on the nose of the plug until the head clears the casting, then pull the plug from the casting.

9. Remove the plastic clip that holds the combination switch wiring to the steering column bracket. Remove the 2 screws from the combination switch and remove the switch from the column, leaving the wiring connectors attached to the switch. Position the switch and wiring aside.

10. Disconnect the key warning buzzer wire from the horn brush

wire. Remove the screw that holds the horn brush connector to the column and remove the connector.

11. Remove the 5 screws that hold the toe plate to the dash panel and loosen the toe plate clamp bolt.

12. Support the column and remove the bolts that hold the breakaway bracket to the pedal support bracket. Pry apart the locking tabs and disconnect the ignition switch wiring harness.

13. Carefully remove the column from the vehicle.

To install:

14. Carefully position the column in the hole in the vehicle floor. Connect the ignition switch wiring harness to the column connect

15. Install the bolts that hold the breakaway bracket to the pedal support bracket, but do not tighten at this ti

16. Tighten the bolts that hold the toe plate to the floor to 8 ft. lbs. (11 Nm), then tighten the breakaway bracket-to-pedal support bracket bolts to 19–27 ft. lbs. (25–36 Nm). Tighten the toe plate clamp to 6–13 ft. lbs. (8–18 Nm).

17. Install the horn brush connector to the column and tighten the retaining screw to 21–29 inch lbs. (2.3–3.3 Nm). Attach the key warning buzzer wire connector to the horn brush wire. Route the wiring to prevent contact with moving parts.

18. Position the combination switch on the column with the attaching screws. Tighten to 18–26 inch lbs. (2–3 Nm). Install the plastic clip that holds the switch wiring to the steering column breakaway bracket.

19. If equipped with automatic transmission, connect the selector indicator cable by pushing the plastic plug at the end of the cable into the shift lever socket casting. When installed, the nose of the plug should be facing the steering wheel and the head of the plug away from the wheel. Install the cable retaining screw in the column and adjust the cable. If the shift lever was removed, install it at this time.

20. Position the top half of the shroud on the column so the screw moldings on the shroud seat in the mounting bores in the column. Place the automatic transmission shift lever in the lowest position to aid assembly.

21. Install the bottom half of the shroud by sliding the guides in the shroud bottom half into the tabs in the shroud top half. Install the shroud retaining screws and tighten to 6–10 inch lbs. (0.7–1.1 Nm).

22. If equipped with tilt column, install the column collar by pressing on the collar from the top and bottom while installing the collar on the column. Install the tilt lever and tighten to 2.2–3.6 ft. lbs. (3–5 Nm).

23. If equipped, place the automatic transmission selector lever in **N**. Install the steering wheel and the lower trim cover panel.

24. If equipped with automatic transmission, install the nuts on the studs and install the shift cable bracket on the steering column bracket. Connect the shift cable to the column lever.

25. Connect the column shaft to the intermediate shaft U-joint and tighten the pinch bolt to 25–35 ft. lbs. (34–47 Nm). The intermediate shaft must be in collapsed state to align, both shafts have a flat side, and then pulled up the column shaft until the bolt holes align. Make sure the intermediate shaft does not contact the plastic retainer at the base of the column. If it does, pull the lower shaft of the column slightly out of the column.

26. Connect the negative battery cable and check the adjustment of the selector indicator cable. Pull the shift lever toward the steering wheel until the **OD** detent in the transmission is felt. Release the shift lever, it should be against the detent wall in the column.

27. Release the parking brake lever and test drive the vehicle.

Manual Steering Gear

ADJUSTMENT

B Series Pick-Up

1. Remove the steering gear and mount the assembly in a vise.
2. Set the toe of a dial indicator at the end of the Pitman arm.
3. Loosen the locknut and turn the adjusting screw until there is no backlash.

NOTE: Adjust the backlash with the steering gear in the center position. Otherwise, the backlash becomes excessively small and the gears may become damaged.

4. Reinstall the steering gear in the vehicle.

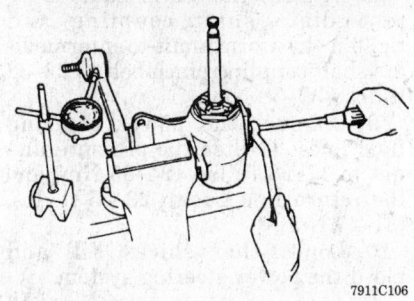

7911C106

Manual steering gear adjustment — B Series Pick-Up

REMOVAL AND INSTALLATION

B Series Pick-Up

1. Position the front wheels in the straight-ahead position. Raise and safely support the vehicle.
2. Remove the pinch bolt securing the worm shaft to the intermediate shaft coupling.
3. Remove the cotter pin and nut from the Pitman arm stud. Use separator tool 49 0118 850C or equivalent, to separate the Pitman arm from the center link.
4. Support the steering gear and remove the steering gear attaching bolts/nuts. Remove the steering gear from the vehicle.
5. Installation is the reverse of the removal procedure. Tighten the steering gear-to-chassis bolts/nuts to 46–69 ft. lbs. (63–93 Nm). Tighten the Pitman arm stud nut to 33–43 ft. lbs. (44–59 Nm) and install a new cotter pin. Tighten the worm shaft-to-intermediate shaft coupling pinch bolt to 22–38 ft. lbs. (30–38 Nm).

Power Steering Gear

ADJUSTMENT

B Series Pick-Up

1. Remove the power steering gear from the vehicle and install in a suitable holding fixture.
2. Position the worm shaft in the center position.
3. Install tool 49 0180 510B or equivalent, on the end of the worm shaft and attach a pull scale to the end of the tool.
4. Set the sector shaft adjusting screw so the preload with the worm

shaft in the center position is 1.3–2.0 lbs.

NOTE: The preload at the center position must be 0.4–0.9 lbs. higher than the preload when the worm shaft is turned 360 degrees to the left and right.

5. If the specified preload is not obtained, disassemble the steering gear and check for dirt and foreign material and that the oil seal is correctly installed. Reassemble the steering gear and adjust the preload.
6. After adjustment, tighten the sector shaft adjusting screw locknut to 25–35 ft. lbs. (34–47 Nm).
7. Reinstall the steering gear in the vehicle.

Navajo

1. Raise and safely support the vehicle. Disconnect the Pitman arm from the sector shaft using puller T64P–3590–F or equivalent.
2. Disconnect the fluid return line at the reservoir and cap the reservoir return line tube. Place the end of the return line in a clean container and turn the steering wheel from stop to stop several times to empty the steering gear. Discard the fluid.
3. Turn the steering wheel to 45 degrees from the right stop.
4. Attach an inch pound torque wrench to steering wheel nut and record the torque required to rotate the shaft slowly approximately 1/8 turn toward center from the 45 degree position.
5. Turn the steering gear back to center and record the torque required to rotate the shaft back and forth across the center position.
6. If the vehicle has less than 5000 miles, resetting is required if total mesh-load over center is not 12–24 inch lbs. If the vehicle has more than 5000 miles or the sector shaft has been replaced, resetting is required if mesh-load over center is less than 10 inch lbs. greater than the torque 45 degrees from the right stop.
7. The set torque specification is measured rocking across center to a value of 9–13 inch lbs. greater than that measured 45 degrees from the right stop.
8. If reset is required, loosen the locknut and turn the sector shaft adjusting screw until the reading is the specified value greater than the torque at 45 degrees from the stop.
9. Tighten the adjusting screw locknut and recheck. Install the Pitman arm and steering wheel cover.

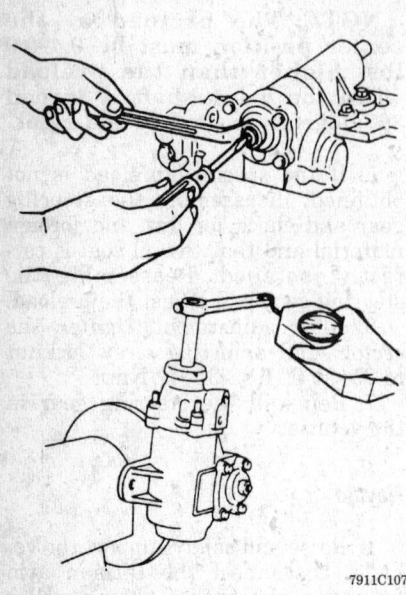

Power steering gear adjustment — B Series Pick-Up

10. Connect the fluid return line to the reservoir and refill the system with fluid. Bleed the system.

REMOVAL AND INSTALLATION

B Series Pick-Up

1. Position the front wheels in the straight-ahead position. Raise and safely support the vehicle.

2. Remove the pinch bolt securing the worm shaft to the intermediate shaft coupling.

3. Disconnect and plug the power steering fluid lines at the steering gear.

4. Remove the cotter pin and nut from the Pitman arm stud. Use separator tool 49 0118 850C or equivalent, to separate the Pitman arm from the center link.

5. Support the steering gear and remove the steering gear attaching bolts/nuts. Remove the steering gear from the vehicle.

To install:

6. Install the steering gear and tighten the steering gear-to-chassis bolts/nuts to 46–69 ft. lbs. (63–93 Nm).

7. Connect the Pitman arm and tighten the Pitman arm stud nut to 33–43 ft. lbs. (44–59 Nm). Install a new cotter pin.

8. Connect the worm shaft-to-intermediate shaft coupling and tighten the worm shaft-to-intermediate shaft coupling pinch bolt to 22–38 ft. lbs. (30–38 Nm).

9. Connect the power steering fluid lines. Tighten the pressure line nut to 17–26 ft. lbs. (24–35 Nm) and the return line nut to 23–35 ft. lbs. (31–47 Nm).

10. Lower the vehicle. Fill and bleed the power steering system.

Navajo

1. Disconnect the pressure and return lines from the steering gear. Plug the lines and the ports in the gear to prevent entry of dirt.

2. Remove the upper and lower steering gear shaft U-joint shield from the flex coupling. Disconnect the flex coupling at the steering gear by removing the bolt.

3. Raise the vehicle and support it safely. Remove the Pitman arm attaching nut and washer. Remove the Pitman arm from the sector shaft using tool T64P–3590–F or equivalent. Be careful not to damage the seals.

4. Support the steering gear and remove the attaching bolts. Work the steering gear free from the flex coupling and remove the gear from the vehicle.

To install:

5. Install the lower U-joint shield onto the steering gear lugs. Slide the upper U-joint shield into place on the steering shaft assembly. Turn the steering wheel so the spokes are in the horizontal position.

6. Center the steering gear input shaft with the indexing flat facing down.

7. Slide the steering gear input shaft into the flex coupling and into place on the frame side rail. Install the attaching bolts and tighten to 50–62 ft. lbs. (68–84 Nm). Tighten the flex coupling bolt to 26–34 ft. lbs. (34–47 Nm).

8. Make sure the wheels are in the straight-ahead position, then install the Pitman arm on the sector shaft. Install the attaching washer and nut and tighten to 170–228 ft. lbs. (230–310 Nm).

9. Connect and tighten the pressure and return lines to the steering gear to 20–30 ft. lbs. (27–40 Nm). Snap the upper and lower steering gear shaft U-joint shields together.

10. Fill and bleed the power steering system.

Power Rack and Pinion

REMOVAL AND INSTALLATION

MPV

2WD

1. Place the front wheels in the straight-ahead position. Raise and safely support the vehicle.

2. Remove the wheel and tire assemblies. Remove the splash shield.

3. Remove the cotter pins and nuts from both tie rod end studs. Use separator tool 49 0727 575 or equivalent, to separate the tie rod ends from the knuckles.

4. Remove the pinch bolt from the intermediate shaft-to-pinion shaft coupling.

5. Disconnect and plug the pressure line from the rack and pinion assembly. Loosen the clamp and disconnect the return line from the rack and pinion assembly. Plug the line.

6. If equipped with automatic transmission, remove the change counter assembly to remove the protector plate mounting bolt indicated by the arrow.

7. Remove the steering bracket mounting bolts and remove the rack and pinion assembly and brackets.

8. If necessary, remove the brackets.

To install:

9. If removed, install the brackets and tighten the mounting bolts, in sequence, to 54–69 ft. lbs. (74–93 Nm).

10. Install the rack and pinion assembly and brackets in the vehicle. Tighten the bracket-to-chassis bolts to 46–69 ft. lbs. (63–93 Nm).

11. If equipped with automatic transmission, install the change counter assembly.

12. Connect the return line and tighten the clamp. Connect the pressure line and tighten the nut to 23–35 ft. lbs. (31–47 Nm).

13. Install the pinch bolt in the intermediate shaft-to-pinion shaft coupling and tighten to 13–20 ft. lbs. (18–26 Nm).

14. Position the tie rod end studs in the knuckles and install the nuts. Tighten the nuts to 43–58 ft. lbs. (59–78 Nm) and install new cotter pins.

15. Install the splash shield and the wheel and tire assemblies. Lower the vehicle and bleed the power steering system.

4WD

1. Place the front wheels in the straight-ahead position. Raise and safely support the vehicle.

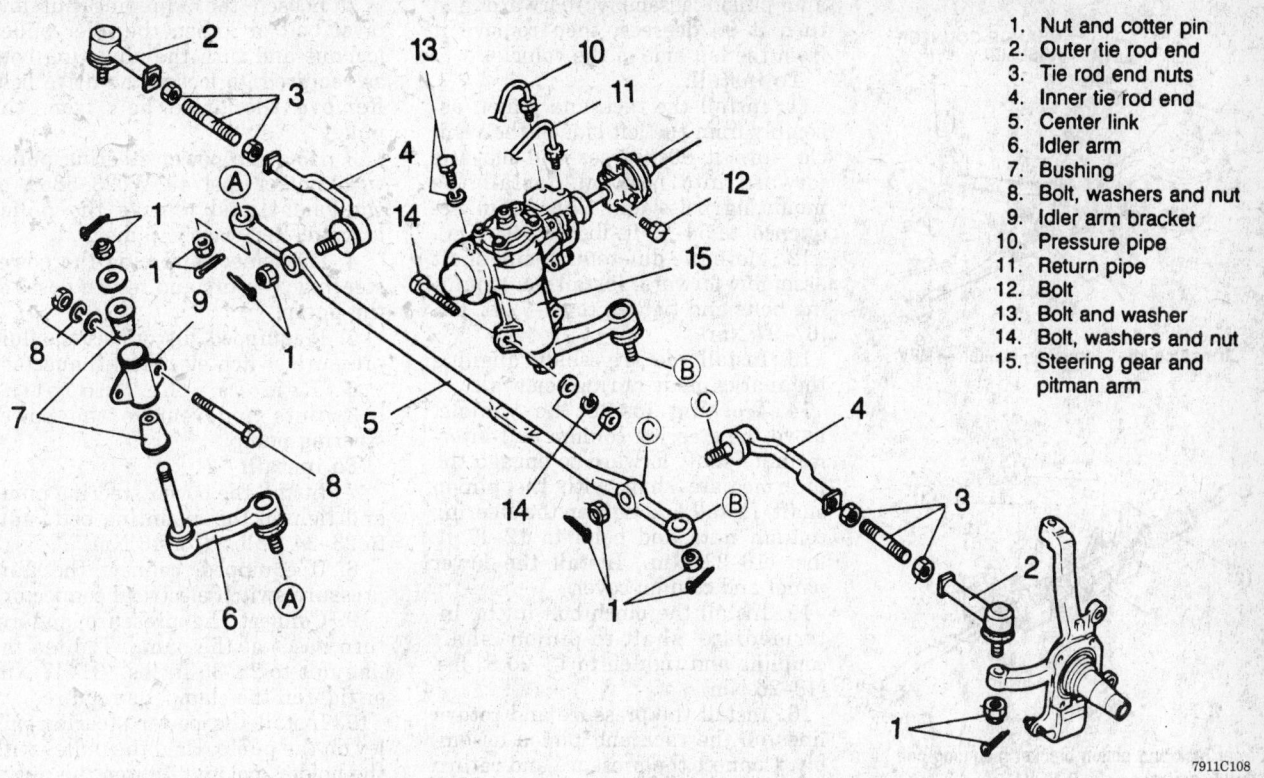

1. Nut and cotter pin
2. Outer tie rod end
3. Tie rod end nuts
4. Inner tie rod end
5. Center link
6. Idler arm
7. Bushing
8. Bolt, washers and nut
9. Idler arm bracket
10. Pressure pipe
11. Return pipe
12. Bolt
13. Bolt and washer
14. Bolt, washers and nut
15. Steering gear and pitman arm

Power steering gear and linkage exploded view — B Series Pick-Up

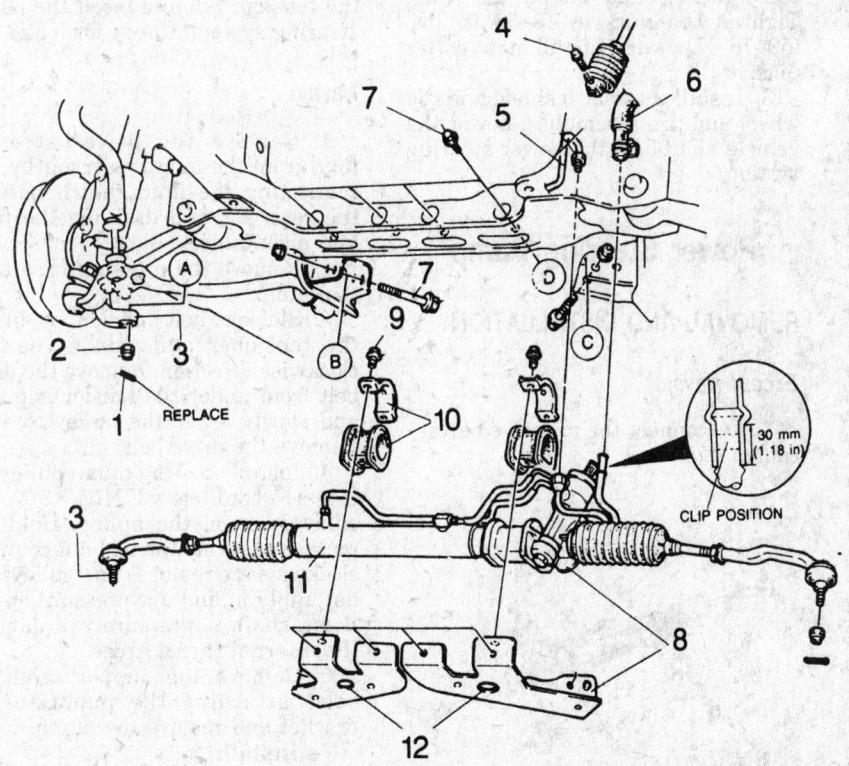

1. Cotter pin
2. Nut
3. Tie rod end and steering knuckle
4. Fixing bolt
5. Pressure pipe
6. Return hose
7. Steering bracket mounting bolts
8. Steering gear and linkage bracket assembly
9. Mounting bracket bolts
10. Mounting bracket and rubbers
11. Steering gear and linkage
12. Steering brackets

Power rack and pinion removal — 2WD MPV

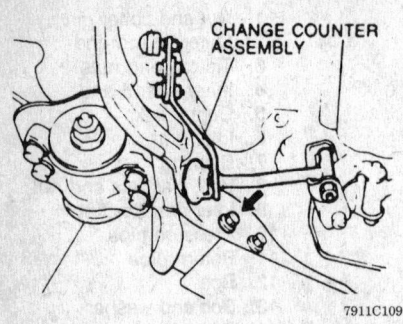

CHANGE COUNTER ASSEMBLY

7911C109

Change counter assembly location — MPV

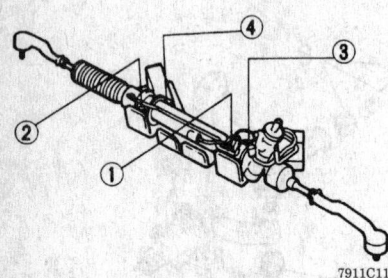

7911C110

Power rack and pinion bracket mounting bolt torque sequence — 2WD MPV

2. Remove the wheel and tire assemblies. Remove the splash shield.

3. Remove the cotter pins and nuts from both tie rod end studs. Use separator tool 49 0727 575 or equivalent, to separate the tie rod ends from the knuckles.

4. Disconnect and plug the pressure and return hoses at the pressure and return lines.

5. Remove the pressure and return lines from the rack and pinion assembly.

6. Remove the pinch bolt from the intermediate shaft-to-pinion shaft coupling.

7. Working inside the vehicle, remove the lower panel and column cover from under the steering column. Remove the steering column mounting bolts and nuts and pull the column and intermediate shaft rearward to separate the intermediate shaft from the pinion shaft.

8. Mark the position of the front driveshaft on the axle flange and remove the front driveshaft.

9. Remove the rack and pinion assembly mounting bracket bolts and the front differential/joint shaft assembly mounting bolts.

10. Slide the differential/joint shaft assembly rearward. Slide the rack

and pinion assembly rearward and turn it 90 degrees, then remove it from the left side of the vehicle.

To install:

11. Install the rack and pinion assembly from the left side of the vehicle, turn it 90 degrees and move it forward into position. Install the mounting bolts and tighten, in sequence, to 54–69 ft. lbs. (74–93 Nm).

12. Move the differential/joint shaft assembly forward, install the mounting bolts and tighten to 49–72 ft. lbs. (67–97 Nm).

13. Install the driveshaft, aligning the marks made during removal.

14. Working inside the vehicle, move the steering column and intermediate shaft forward to engage the intermediate shaft with the pinion shaft. Install and tighten the steering column nuts and bolts to 12–17 ft. lbs. (16–23 Nm). Install the lower panel and column cover.

15. Install the pinch bolt in the intermediate shaft-to-pinion shaft coupling and tighten to 13–20 ft. lbs. (18–26 Nm).

16. Install the pressure and return lines on the rack and pinion assembly. Connect the pressure and return hoses to the lines.

17. Position the tie rod end studs in the knuckles and install the nuts. Tighten the nuts to 43–58 ft. lbs. (59–78 Nm) and install new cotter pins.

18. Install the splash shield and the wheel and tire assemblies. Lower the vehicle and bleed the power steering system.

Power Steering Pump

REMOVAL AND INSTALLATION

Except Navajo

1. Disconnect the negative battery cable.

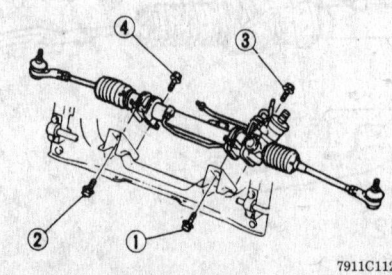

7911C112

Power rack and pinion bracket mounting bolt torque sequence — 4WD MPV

2. Loosen the adjusting nut and pivot bolt or loosen the idler pulley locknut and turn the adjusting bolt, as required, to loosen the drive belt. Remove the drive belt from the pulley.

3. Hold the power steering pulley with holder tool 49 W023 585A or equivalent, and remove the pulley locknut. Remove the pulley.

4. Disconnect and plug the power steering pressure and return hoses at the pump.

5. If equipped, disconnect the fluid pressure switch electrical connector.

6. Remove the mounting bolts/nuts and remove the power steering pump.

To install:

7. Install the power steering pump and tighten the mounting bolts/nuts to 23–34 ft. lbs. (31–46 Nm).

8. If equipped, connect the fluid pressure switch electrical connector.

9. Connect the pressure and return hoses at the pump. Tighten the line nut to 23–35 ft. lbs. (31–47 Nm) or tighten the clamp, as required.

10. Install the power steering pulley on the pump. Hold the pulley with the holder tool and tighten the nut to 36–43 ft. lbs. (49–59 Nm).

11. Install the drive belt and adjust the tension. Fill and bleed the power steering system. Check for leaks.

Navajo

1. Remove the power steering fluid from the pump reservoir by disconnecting the fluid return hose at the reservoir and draining the fluid into a container.

2. Remove the pressure hose from the pump.

3. Slacken belt tension by lifting the tensioner pulley in a counterclockwise direction. Remove the drive belt from under the tensioner pulley and slowly lower the pulley to stop. Remove the drive belt.

4. Install steering pump pulley removal tool 49 UN01 005 or equivalent, on the pulley. Hold the pump and rotate the tool nut counterclockwise to remove the pulley. Do not apply in and out pressure on the pump shaft as pressure will damage the internal thrust areas.

5. Remove the support and the bolts attaching the pump to the bracket and remove the pump.

To install:

6. Install the pump on the bracket and tighten the bolts to 30–40 ft. lbs. (40–55 Nm).

7. Install steering pump pulley replacement tool 49 UN01 006 or

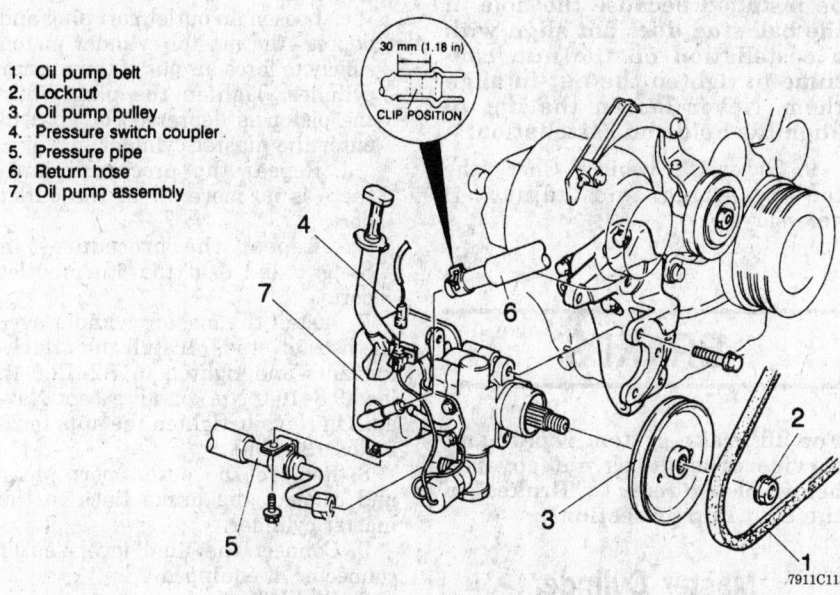

1. Oil pump belt
2. Locknut
3. Oil pump pulley
4. Pressure switch coupler
5. Pressure pipe
6. Return hose
7. Oil pump assembly

30 mm (1.18 in)
CLIP POSITION

7911C113

Power steering pump installation — 3.0L engine

equivalent and press the pulley on the pump shaft.

NOTE: Fore and aft location of the pulley on the pump shaft is critical for correct belt alignment. Make sure the pull-off groove on the pulley is facing front and flush with the end of the shaft 0.010 in. (0.254mm).

8. Position and rotate the drive belt on the engine. While lifting the tensioner pulley in a counterclockwise direction, slide the belt under the tensioner pulley and lower the pulley to the belt.

9. Install the pressure hose to the pump fitting. Connect the return hose to the pump and tighten the clamp.

10. Fill and bleed the power steering system. Check for leaks.

BELT ADJUSTMENT

Except 4.0L Engine

1. Adjust the belt tension by loosening the idler pulley locknut and turning the adjusting bolt until belt tension is as specified.

2. To obtain the correct belt tension, measure the belt deflection at the center of the belt span between the idler pulley and power steering

pump pulley. Deflection specifications are as follows:

- 2.2L Engine — New belt: 0.28–0.31 in. (7–8mm); Used belt: 0.31–0.35 in. (8–9mm)
- 2.6L Engine — New belt: 0.26–0.28 in. (6.6–7.2mm); Used belt: 0.28–0.31 in. (7.2–8.0mm)
- 3.0L Engine — New belt: 0.28–0.30 in. (7.0–7.5mm); Used belt: 0.30–0.32 in. (7.5–8.2mm)

3. After correct belt tension is obtained, tighten the idler pulley locknut or the adjusting nut and pivot bolt, as required.

4.0L Engine

Belt tension is maintained by an automatic tensioner. No adjustment is necessary.

SYSTEM BLEEDING

Except Navajo

1. Raise and safely support the vehicle.
2. Check the fluid level and add, if necessary.
3. Turn the steering wheel fully left and right several times with the engine not running.
4. Recheck the fluid level and add, if necessary.

5. Repeat Steps 3 and 4 until the fluid level stabilizes. Lower the vehicle.
6. Start the engine and let it run at idle. Turn the steering wheel fully left and right several times. If noise is heard in the oil line, air is still present.
7. Put the wheels in the straight-ahead position and turn the engine **OFF**. The fluid level in the pump should not increase; if it does, air is still present. Repeat Step 6.
8. Recheck the fluid level and check for leaks.

NOTE: If air is still present in the system, raise the fluid temperature to 122–176°F (50–80°C) by turning the steering wheel right and left, stop the engine and repeat Step 6 for 5–10 minutes. Air can be completely bled in this manner.

Navajo

1. Fill the power steering fluid reservoir.
2. Disconnect the distributorless ignition module connector.
3. Crank the engine with the starter and continue adding fluid until the level remains constant. Do not prolong cranking as the battery may be drained and the starter damaged.
4. Rotate the steering wheel approximately 30 degrees each side of center while continuing to crank the engine.
5. Recheck the fluid level and fill, as required.
6. Reconnect the distributorless ignition module connector.
7. Start the engine and allow it to run for several minutes.
8. Rotate the steering wheel from stop to stop.
9. Shut off the engine and recheck the fluid level. Add fluid, as required.
10. If air is still trapped in the system, proceed as follows:
    a. Fabricate a purging tool.
    b. Make sure the reservoir fluid level is correct.
    c. Insert the rubber stopper end of the fabricated purging tool tightly into the filler tube.
    d. Connect a suitable length of hose to the purging tool. Connect the other end of the hose to an air conditioner vacuum pump or distributor machine. Do not use engine vacuum.
    e. Start the engine and let it idle for approximately 15 minutes. Turn the steering wheel 1 full cycle every 5 minutes but do not hit the stops. This will assist in removing trapped air.

f. Stop the engine and disconnect the vacuum source. Remove the purging tool.

g. Check the fluid level and install the filler tube dipstick.

Tie Rod Ends

REMOVAL AND INSTALLATION

1. Place the front wheels in the straight-ahead position. Raise and safely support the vehicle.

2. Remove the cotter pin and nut from the tie rod end ball stud. Discard the cotter pin.

3. Separate the tie rod end from the knuckle or center link using separator tool 49 0118 850C or equivalent.

4. On all except Navajo, mark the position of the locknut and tie rod end on the tie rod threads. Grip the tie rod with suitable pliers, loosen the locknut and remove the tie rod end from the tie rod.

5. On Navajo, loosen the bolts on the tie rod adjusting sleeve. Count the number of turns required to remove the tie rod from the tie rod adjusting sleeve and remove the tie rod.

To install:

6. On all except Navajo, thread the replacement tie rod end onto the tie rod to the same location as the one that was removed. Hold the tie rod end with a wrench and tighten the locknut to 51–58 ft. lbs. (69–78 Nm).

7. On Navajo, install the tie rod into the adjusting sleeve the same number of turns required to remove it. With the adjusting sleeve clamps pointed down, tighten the adjusting sleeve nuts to 30–42 ft. lbs. (40–57 Nm).

8. Install the tie rod ball stud into the knuckle or center link. Install the nut and tighten to 33–43 ft. lbs. (44–59 Nm) on B Series Pick-Up, 43–58 ft. lbs. (59–78 Nm) on MPV or

50–75 ft. lbs. (70–100 Nm) on Navajo. Install a new cotter pin.

NOTE: If the cotter pin cannot be installed because the hole in the ball stud does not align with a castellation on the nut, continue to tighten the nut to align them. Never loosen the nut to align the hole and castellation.

9. Lower the vehicle. Check the toe-in setting and adjust, if necessary.

BRAKES

For all brake system repair and service procedures not covered below, please refer to "Brakes" in the Unit Repair section.

Master Cylinder

REMOVAL AND INSTALLATION

1. Disconnect the negative battery cable. Disconnect the fluid level sensor connector, if equipped.

2. Disconnect and plug the brake lines at the master cylinder.

3. Remove the attaching nuts from the power brake booster studs and remove the master cylinder.

To install:

4. On all except Navajo, adjust the power brake booster pushrod clearance as follows:

a. Install clearance adjusting tool 49 F043 001 or equivalent on the rear of the master cylinder. Turn the adjusting bolt until it bottoms in the pushrod hole in the piston.

b. Apply 19.7 in. Hg vacuum to the power brake booster with a vacuum pump.

c. Invert the clearance adjusting tool and place it on the power brake booster. Turn the pushrod locknut until there is no clearance between the tool and the pushrod.

5. On Navajo, measure the distance between the outer end of the booster pushrod and the front face of the booster assembly; the distance should be 0.995 in. Turn the pushrod adjusting screw in or out until the distance is as specified.

6. If a new or dry master cylinder is being installed, bench bleed the master cylinder as follows:

a. Support the master cylinder in a suitable holding fixture. Fill

the reservoir with the proper type of brake fluid.

b. Install plugs in the brake outlet ports.

c. Loosen an outlet port plug and depress the master cylinder piston slowly to force air out of the master cylinder. Tighten the plug while the piston is depressed or air will enter the master cylinder.

d. Repeat the procedure until there is no more air at the outlet port.

e. Repeat the procedures in Steps c and d at the other outlet ports.

7. Install the master cylinder over the booster studs. Install the attaching nuts and tighten to 7.2–12.0 ft. lbs. (9.8–16.0 Nm) on all except Navajo. On Navajo, tighten the nuts to 20 ft. lbs. (27 Nm).

8. Remove the outlet port plugs and connect the brake lines to the master cylinder.

9. Connect the fluid level sensor connector, if equipped.

10. Bleed the brake system.

Proportioning/Bypass Valve

REMOVAL AND INSTALLATION

Except Navajo

1. Disconnect the negative battery cable. Disconnect the pressure differential switch connector, if equipped.

2. Disconnect and plug the brake lines at the valve.

3. Remove the attaching bolt(s) and the valve.

4. Installation is the reverse of the removal procedure.

5. Bleed the brake system.

Power Brake Booster

REMOVAL AND INSTALLATION

Except Navajo

1. Disconnect the negative battery cable.

2. On MPV with 3.0L engine, remove the wiper arms, wiper motor and wiper linkage.

3. Remove the master cylinder.

4. Disconnect the vacuum hose at the booster.

5. Working inside the vehicle, remove the cotter pin and clevis pin from the booster pushrod. Remove the booster attaching nuts.

6. Remove the power brake booster and gasket.

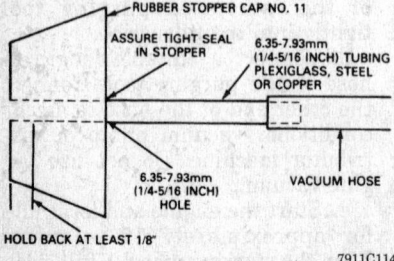

RUBBER STOPPER CAP NO. 11

ASSURE TIGHT SEAL IN STOPPER

6.35-7.93mm (1/4-5/16 INCH) TUBING PLEXIGLASS, STEEL OR COPPER

6.35-7.93mm (1/4-5/16 INCH) HOLE

VACUUM HOSE

HOLD BACK AT LEAST 1/8"

7911C114

Fabricated purging tool dimensions — Navajo

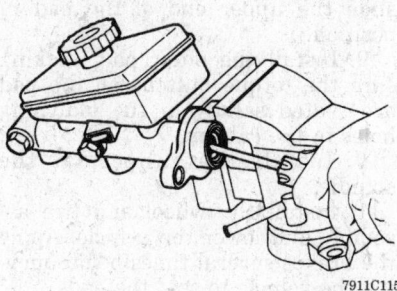

7911C115

Master cylinder bench bleeding

7. Installation is the reverse of the removal procedure. Apply gasket sealant to the booster gasket and grease to the clevis pin prior to installation. Tighten the booster attaching nuts to 12–17 ft. lbs. (16–23 Nm) on B Series Pick-Up or 14–19 ft. lbs. (19–25 Nm) on MPV.

8. Bleed the brake system.

Navajo

1. Disconnect the negative battery cable. Support the master cylinder from the underside with a prop.

2. Disconnect the vacuum hose from the booster check valve and remove the check valve.

3. Remove the master cylinder-to-booster retaining nuts. Pull the master cylinder off the booster and leave it supported by the prop, out of the way enough to allow booster removal.

4. Working inside the cab, remove the hairpin retainer and slide the stoplight switch, valve rod, spacers and bushing off the brake pedal arm. Remove the nuts retaining the booster and remove the booster.

To install:

5. Mount the booster assembly on the engine side of the dash panel by sliding the bracket mounting bolts and valve operating rod in through the holes in the dash panel.

6. Working inside the cab, install the booster mounting nuts and tighten to 13–25 ft. lbs. (18–33 Nm).

7. Before installing the master cylinder, check the distance from the outer end of the vacuum booster assembly pushrod to the front face of the vacuum brake booster assembly. The distance should be 0.995 in. Turn the pushrod adjusting screw in or out, as required, to obtain the proper length.

8. Install the master cylinder and tighten the retaining nuts to 20 ft. lbs. (27 Nm). Remove the prop from under the master cylinder.

9. Install the booster check valve and connect the vacuum hose. Check the hose routing to make sure the hose is not crimped.

10. Working inside the cab, install the bushing and position the switch on the end of the valve rod, then install the switch and rod on the pedal arm along with the spacers and hairpin retainer.

NOTE: Use only the factory supplied hairpin retainer. Do not substitute other types of retainers.

11. Connect the negative battery cable, start the engine and check brake operation.

Brake Caliper

REMOVAL AND INSTALLATION

Except Navajo

1. Raise and safely support the vehicle. Remove the wheel and tire assembly.

2. Remove the banjo bolt and disconnect the brake hose from the caliper. Plug the hose to prevent fluid leakage.

3. On B Series Pick-Up, remove the caliper mounting bolt and pivot the caliper about the mounting pin and off of the brake rotor. Remove the caliper from the pin.

4. On MPV, remove the caliper mounting bolts and remove the caliper.

5. Installation is the reverse of the removal procedure. Lubricate the caliper mounting bolts or bolt and pin prior to installation.

6. Tighten the caliper mounting bolt(s) to 23–30 ft. lbs. (31–41 Nm) on B Series Pick-Up or 61–69 ft. lbs. (83–93 Nm) on MPV. Bleed the brake system.

Navajo

1. Siphon part of the brake fluid out of the master cylinder to avoid overflow when the caliper piston is pressed into the caliper bore.

2. Raise the vehicle and support it safely. Remove the wheel and tire assembly.

3. Position an 8 in. C-clamp on the caliper and tighten the clamp to move the caliper piston into the bore approximately ⅛ in. Avoid clamp contact with the outer shoe spring clip. Remove the clamp.

NOTE: Do not pry the piston away from the rotor.

4. Clean excess dirt from the pin tab area.

5. Using a ¼ in. drive socket, ⅜ in. deep and a light hammer, tap the upper caliper pin towards the outboard side until the pin tabs pass the spindle face.

6. Place one end of a ⁷⁄₁₆ in. diameter punch against the end of the caliper pin and tap the pin out of the caliper slide groove.

7. Repeat Steps 5 and 6 to remove the lower pin.

8. Disconnect and plug the brake hose at the caliper. Remove the caliper from the rotor.

To install:

9. Make sure the caliper mounting surfaces are free of dirt. Lubricate the caliper grooves with disc brake caliper grease and install the caliper.

10. From the caliper outboard side, position the pin between the caliper and spindle grooves. The pin must be positioned so the tabs will be installed against the spindle outer face.

11. Tap the pin on the outboard end with a hammer until the retention tabs on the sides of the pin contact the spindle face.

12. Repeat Steps 10 and 11 for the lower pin.

NOTE: During installation, do not allow the tabs of the caliper pin to be tapped too far into the spindle groove. If this happens, it will be necessary to tap the other end of the caliper pin until the tabs snap in place. The tabs on each end of the pin must be free to catch on the spindle face.

13. Connect the brake hose to the caliper. Bleed the brake system.

14. Install the wheel and tire assembly and lower the vehicle. Check the brake fluid level and check the brakes for proper operation.

Disc Brake Pads

REMOVAL AND INSTALLATION

Except Navajo

1. Siphon part of the brake fluid out of the master cylinder to avoid overflow when the caliper piston is pressed into the caliper bore.

2. Raise the vehicle and support it safely. Remove the wheel and tire assembly.

3. Remove the brake caliper, but do not disconnect the brake hose. Se-

cure the caliper aside with mechanics wire.

NOTE: Do not let the caliper hang by the brake hose.

4. Remove the disc brake pads, shims and guide plates from the mounting support, noting the position of the shims and guide plates for reassembly.

To install:

5. Using a C-clamp or similar tool, bottom out the caliper piston in the caliper bore.

6. Install the new disc pads along with the shims and guide plates in the mounting support. Make sure the shims and guide plates are installed in their original positions.

7. Install the brake caliper.

8. Install the wheel and tire assembly and lower the vehicle. Apply the brakes several times before moving the vehicle to seat the pads.

9. Check the brake fluid level. Check the brakes for proper operation.

Navajo

1. Siphon part of the brake fluid out of the master cylinder to avoid overflow when the caliper piston is pressed into the caliper bore.

2. Raise the vehicle and support it safely. Remove the wheel and tire assembly.

3. Remove the brake caliper, but do not disconnect the brake hose. Secure the caliper aside with mechanics wire.

NOTE: Do not let the caliper hang by the brake hose.

4. Compress the anti-rattle clip and remove the inner brake pad from the caliper.

5. Press each ear of the outer brake pad away from the caliper and slide the torque buttons out of the retention notches.

To install:

6. Bottom out the caliper piston in the caliper bore using an 8 in. C-clamp and a worn out inner brake pad or block of wood to push against the piston. Do not attempt to bottom out the piston with the outer brake pad installed.

7. Place a new anti-rattle clip on the lower end of the inner brake pad. Make sure the tabs on the clip are properly positioned and the clip is fully seated.

8. Position the inner brake pad and anti-rattle clip in the pad abutment with the ant-rattle clip tab against the pad abutment and the loop-type spring away from the rotor.

Compress the anti-rattle clip and slide the upper end of the pad in position.

9. Install the outer pad, making sure the torque buttons on the pad are seated solidly in the matching holes in the caliper.

10. Install the caliper on the spindle.

11. Install the wheel and tire assembly and lower the vehicle. Apply the brakes several times before moving the vehicle to seat the pads.

12. Check the brake fluid level. Check the brakes for proper operation.

Brake Rotor

REMOVAL AND INSTALLATION

1. Raise and safely support the vehicle. Remove the wheel and tire assembly.

2. Remove the caliper and support it aside with mechanics wire; do not let the caliper hang by the brake hose. On all except Navajo, remove the disc brake pads and mounting support.

3. On 2WD B Series Pick-Up and Navajo, remove the dust cap, cotter pin, nut, washer and outer bearing

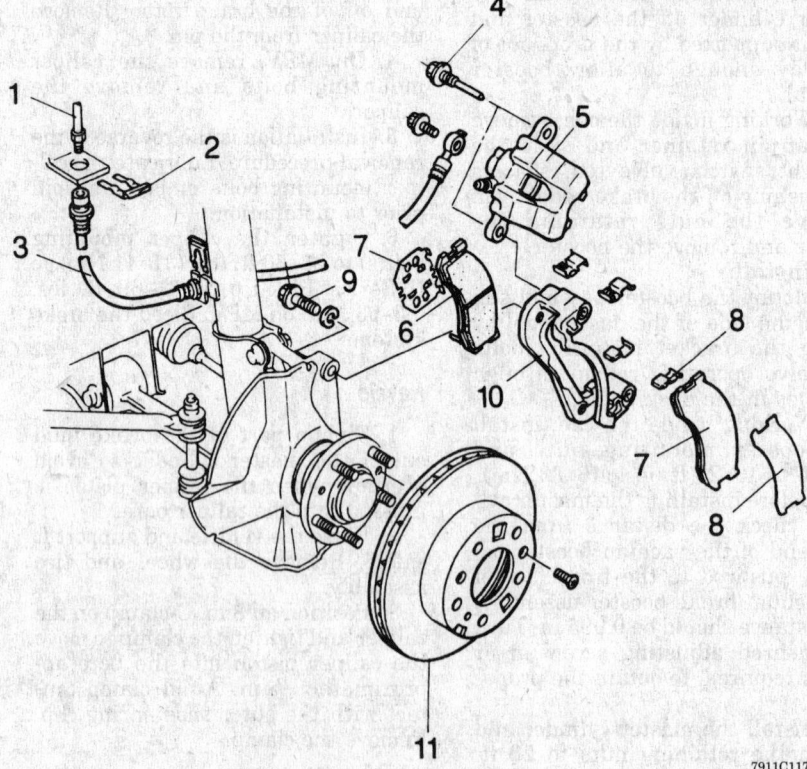

1. Brake pipe
2. Clip
3. Brake hose
4. Lockbolt
5. Brake caliper assembly
6. Disc pad
7. Shim
8. Guide plate
9. Bolt
10. Mounting support
11. Disc plate

Disc brake assembly — MPV

7911C117

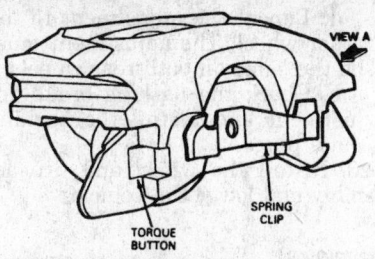

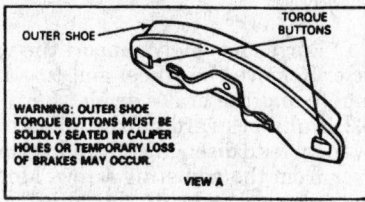

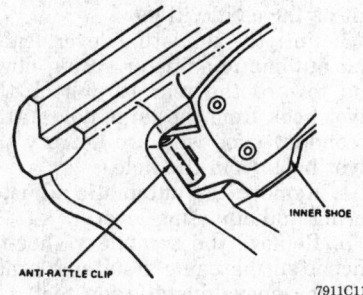

7911C118

Correct disc brake pad installation — Navajo

and remove the rotor from the spindle.

4. On 4WD B Series Pick-Up, remove the locking hub or drive flange, snapring and spacer, set bolts and bearing set plate. Remove the bearing locknut using a suitable puller and remove the hub and rotor assembly, being careful not to let the washer and bearing fall.

5. On 4WD Navajo, remove the locking hub and remove the brake rotor.

6. On MPV, remove the attaching screw and remove the rotor.

7. Inspect the rotor for scoring, wear and runout; machine or replace as necessary.

8. If rotor replacement is necessary on B Series Pick-Up, remove the attaching bolts and separate the rotor from the hub.

9. Install in the reverse order of removal. On B Series Pick-Up, tighten the rotor-to-hub bolts to 40–51 ft. lbs. (54–69 Nm). Adjust the wheel bearings.

Brake Drums

REMOVAL AND INSTALLATION

1. Make sure the parking brake is released.

2. Raise and safely support the vehicle. Remove the wheel and tire assembly.

3. On all except Navajo, remove the retaining screw, if equipped, and remove the brake drum.

4. On Navajo, remove the spring retaining nuts, if equipped, and remove the brake drum.

NOTE: If the brake drum will not come off, insert a narrow prybar through the appropriate hole in the backing plate and disengage the adjusting lever from the adjusting screw. While holding the adjusting lever away from the adjusting screw, insert a brake adjusting tool through the backing plate hole adjacent to the star wheel of the adjusting screw and loosen the adjusting screw.

5. Inspect the brake drum surface for wear, scoring and runout. Machine or replace, as necessary.

6. Installation is the reverse of the removal procedure.

Brake Shoes

REMOVAL AND INSTALLATION

2WD B Series Pick-Up

1. Raise and safely support the vehicle. Remove the wheel and tire assembly and the brake drum.

2. Remove the upper and lower return springs from the brake shoes.

3. Compress the forward brake shoe spring and remove the forward brake shoe, spring and pin.

4. Remove the adjuster screw, pawl lever return spring and pawl lever.

5. Compress the rearward brake shoe spring and remove the rearward brake shoe, spring and pin.

6. Pull back the spring on the parking brake cable and disengage the brake shoe parking brake lever from the cable.

7. Remove the clip and pin and the parking brake lever from the brake shoe, if necessary.

To install:

8. Lubricate the adjusting screw threads, backing plate shoe sliding surfaces and wheel cylinder and anchor sliding points with a small amount of grease.

9. If removed, install the parking brake lever to the rearward brake shoe with the pin and clip. Make sure the clip is secure.

10. Pull back the spring on the parking brake cable and connect the cable end to the parking brake lever.

11. Install the brake shoes to the backing plate and secure with the pins and shoe springs.

12. Turn the adjusting screw all the way in and install the adjusting screw, pawl lever return spring and pawl lever.

13. Install the upper and lower return springs.

14. Install the brake drum.

15. Adjust the brake shoes as follows:

   a. Remove the 2 adjusting hole plugs from the back of the backing plate.

   b. Place a suitable brake adjusting tool against the star wheel of the adjust screw through the hole adjacent to the star wheel and turn the star wheel toward the arrow direction marked on the backing plate until the wheel is locked.

   c. Insert a suitable drift through the other adjusting hole and push the pawl lever of the self-adjuster away from the star wheel, then back off the star wheel about 6–7 notches with the brake adjusting tool, so the drum rotates freely without drag.

   d. Repeat the adjustment to the other wheel. The adjustment must be the same on both rear wheels.

   e. Check the parking brake adjustment and install the backing plate plugs.

16. Install the wheel and tire assembly and lower the vehicle.

MPV and 4WD B Series Pick-Up

1. Raise and safely support the vehicle. Remove the wheel and tire assembly and the brake drum.

2. Remove the upper return springs from the spring anchor and the primary and secondary shoes. Remove the adjuster lever link spring.

3. Remove the primary shoe holding spring assembly and disengage the primary shoe from the lower shoe spring. Remove the primary shoe and the holding spring pin.

NOTE: The primary shoe holding spring is yellow and the secondary shoe holding spring is white. The springs cannot be interchanged.

4. Remove the parking brake lever strut, lower shoe spring and adjuster screw assembly.

5. Remove the secondary shoe holding spring assembly and remove the secondary shoe, adjust lever and

parking brake lever assembly and the holding spring pin.

6. Pull back the spring on the parking brake cable and disengage the brake shoe parking brake lever from the cable.

7. Remove the clip and the parking brake lever from the brake shoe, if necessary.

To install:

8. Lubricate the adjusting screw threads, backing plate shoe sliding surfaces and spring anchor sliding points with a small amount of grease.

9. If removed, install the parking brake lever to the secondary shoe with the clip. Make sure the clip is secure.

10. Pull back the spring on the parking brake cable and connect the cable end to the parking brake lever.

11. Install the secondary shoe with the adjust lever to the backing plate and secure with the pin and holding spring assembly.

12. Install the primary shoe to the backing plate and secure with the pin and holding spring assembly.

13. Install the lower shoe spring to the primary and secondary shoes.

14. Turn the adjuster screw assembly all the way in. Pry the shoes apart slightly and install the adjuster screw assembly between the shoes.

Make sure the star wheel is toward the secondary shoe and is in contact with the adjust lever.

15. Install the parking brake lever strut.

16. Install the adjuster lever link spring between the adjust lever and the spring anchor.

17. Install the primary and secondary shoe upper return springs between the shoes and the spring anchor.

18. Install the brake drum.

19. Adjust the brake shoes as follows:

a. Remove the 2 adjusting hole plugs from the back of the backing plate.

b. Place a suitable brake adjusting tool against the star wheel of the adjust screw through the hole adjacent to the star wheel and turn the star wheel toward the arrow direction marked on the backing plate until the wheel is locked.

c. Insert a suitable drift through the other adjusting hole and push the pawl lever of the self-adjuster away from the star wheel, then back off the star wheel about 8–10 notches on B Series Pick-Up or 13–15 notches on MPV with the brake adjusting tool, so the drum rotates freely without drag.

d. Repeat the adjustment to the other wheel. The adjustment must be the same on both rear wheels.

e. Check the parking brake adjustment and install the backing plate plugs.

20. Install the wheel and tire assembly and lower the vehicle.

Navajo

1. Raise and safely support the vehicle. Remove the wheel and tire assembly and the brake drum.

2. Pull backward on the adjusting lever cable to disengage the adjusting lever from the adjusting screw. Move the outboard side of the adjusting screw upward and back off the pivot nut as far as it will go.

3. Pull the adjusting lever, cable and automatic adjuster spring down and toward the rear to unhook the pivot hook from the large hole in the secondary shoe web. Do not pry the pivot hook from the hole.

4. Remove the automatic adjuster spring and adjusting lever.

5. Remove the secondary shoe-to-anchor spring using a suitable brake spring removal/installation tool. Using the tool, remove the primary shoe-to-anchor spring and unhook the cable anchor. Remove the anchor pin plate, if equipped.

1. Brake drum
2. Parking brake cable
3. Hold spring and sleeve
4. Adjust lever
5. Link
6. Pull-off spring
7. Shoe spring
8. Return spring
9. Return spring
10. Adjuster
11. Primary brake shoe
12. Secondary brake shoe
13. Strut
14. Wheel cylinder assembly

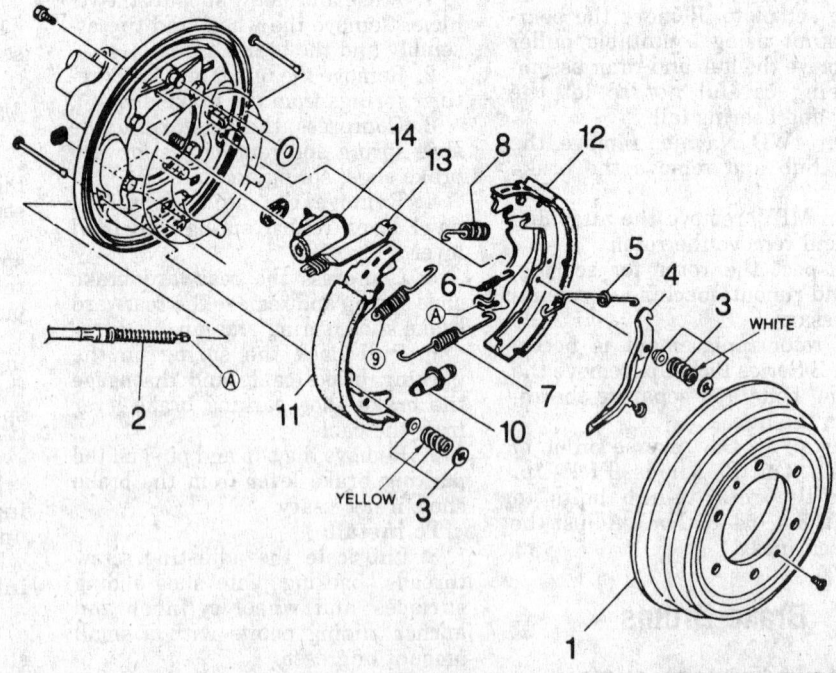

Drum brake assembly — MPV and 4WD B Series Pick-Up

6. Remove the cable guide from the secondary shoe.

7. Remove the shoe hold-down springs, shoes, adjusting screw, pivot nut and socket. Note the color and position of each hold-down spring so they can be reassembled in the same position.

8. Remove the parking brake link and spring. Disconnect the parking brake cable from the parking brake lever.

9. Remove the secondary brake shoe. Remove the retainer clip and spring washer and remove the parking brake lever.

To install:

10. Clean the backing plate ledge pads and sand lightly. Apply a light coating of high temperature lithium grease to the points where the brake shoes touch the backing plate.

11. Install the parking brake lever on the secondary shoe and secure with the spring washer and retaining clip.

12. Position the brake shoes on the backing plate and install the hold-down spring pins, springs and cups. Install the parking brake link, spring and washer. Connect the parking brake cable to the parking brake lever.

13. Install the anchor pin plate, if equipped, and place the cable anchor over the anchor pin with the crimped side toward the backing plate.

14. Install the primary shoe-to-anchor spring using the brake spring removal/installation tool.

15. Install the cable guide on the secondary shoe with the flanged hole fitted into the hole in the secondary shoe. Thread the cable around the cable guide groove.

NOTE: Make sure the cable is positioned in the groove and not between the guide and shoe web.

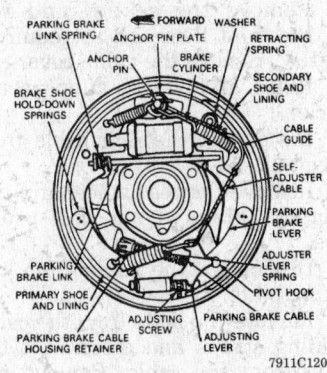

Rear brake shoe assembly — Navajo

16. Install the secondary shoe-to-anchor (long) spring.

NOTE: Make sure the cable end is not cocked or binding on the anchor pin when installed. All parts should be flat on the anchor pin.

17. Apply high temperature lithium grease to the threads and the socket end of the adjusting screw. Turn the adjusting screw into the adjusting pivot nut to the end of the threads and then loosen, ½ turn.

18. Place the adjusting socket on the screw and install the assembly between the shoe ends with the adjusting screw nearest the secondary shoe.

NOTE: Be sure to install the adjusting screw on the same side of the vehicle from which it came. To prevent incorrect installation, the socket end of each adjusting screw is stamped with R or L, to indicate installation on the right or left side of the vehicle. The adjusting pivot nuts have lines machined around the body of the nut, two lines indicating the right side nut and one line indicating the left side nut.

19. Hook the cable hook into the hole in the adjusting lever from the outboard plate side. The adjusting levers are also stamped with an **R** or **L** to indicate right or left side installation.

20. Place the hooked end of the adjuster spring in the large hole in the primary shoe web and connect the loop end of the spring to the adjuster lever hole.

21. Pull the adjuster lever, cable and automatic adjuster spring down toward the rear to engage the pivot hook in the large hole in the secondary shoe web.

22. After installation, check the action of the adjuster by pulling the section of the cable between the cable guide and the adjusting lever toward the secondary shoe web far enough to lift the lever past a tooth on the adjusting screw wheel. The lever should snap into position behind the next tooth and releasing the cable should cause the adjuster spring to return the lever to its original position. This return action will turn the adjusting screw 1 tooth.

23. If pulling the cable does not produce the action described in Step 22 or if lever action is sluggish instead of positive and sharp, check the position of the lever on the adjusting screw toothed wheel. With the brake in a vertical position, anchor at the top,

the lever should contact the adjusting wheel 1 tooth above the center line of the adjusting screw. If the contact point is below the center line, the lever will not lock on the adjusting screw wheel teeth and the screw will not turn as the lever is actuated by the cable.

24. To find the cause of the condition described in Step 23, proceed as follows:

a. Check the cable and fittings. The cable should completely fill or extend slightly beyond the crimped section of the fittings. If this does not happen, the cable assembly may be damaged and should be replaced.

b. Check the cable guide for damage. The cable groove should be parallel to the shoe web and the body of the guide should lie flat against the web. Replace the guide if it shows damage.

c. Check the pivot hook on the lever. The hook surfaces should be square with the body on the lever for proper pivoting. Repair the hook or replace the lever if the hook shows damage.

d. Be sure the adjusting screw socket is properly seated in the notch in the shoe web.

25. Adjust the brake shoes using either a brake adjustment gauge or manually with the drums installed.

26. If using a brake adjustment gauge, proceed as follows:

a. Measure the inside diameter of the brake drum with the gauge.

b. Reverse the tool and adjust the brake shoes until they touch the gauge. The gauge contact points on the shoes must be parallel to the vehicle with the center line through the center of the axle.

c. Install the drum and wheel and tire assembly. Lower the vehicle.

d. Apply the brakes sharply several times while driving the vehicle in reverse. Check brake operation by making several stops while driving forward.

27. If manually adjusting the brakes, proceed as follows:

a. Install the brake drum and wheel and tire assembly.

b. Remove the cover from the adjusting hole at the bottom of the backing plate and turn the adjusting screw, using a suitable brake adjusting tool, to expand the brake shoes until they drag against the brake drum.

c. When the shoes are against the drum, insert a narrow prybar through the brake adjusting hole and disengage the adjusting lever from the adjusting screw. While holding the adjusting lever away from the adjusting screw, loosen the adjusting screw with the brake adjusting tool, until the drum rotates freely without drum.

d. Install the adjusting hole cover and lower the vehicle.

e. Apply the brakes. If the pedal travels more than halfway to the floor, there is too much clearance between the brake shoes and drums. Repeat the adjustment procedure.

Wheel Cylinder

REMOVAL AND INSTALLATION

1. Raise and safely support the vehicle. Remove the wheel and tire assembly, brake drum and brake shoes.
2. Disconnect and plug the brake line at the wheel cylinder.
3. Remove the wheel cylinder retaining bolts and remove the cylinder from the brake backing plate.
4. Installation is the reverse of the removal procedure. Adjust the brakes and bleed the system.

Parking Brake Cable

ADJUSTMENT

Except Navajo

1. Make sure the rear brake shoes are properly adjusted.
2. Start the engine and depress the brake pedal several times while the vehicle is moving in reverse.
3. Stop the engine.
4. On MPV, remove the screw and remove the parking brake lever cover. Remove the adjusting nut clip.
5. On B Series Pick-Up, loosen the locknut at the end of the front cable, near the parking brake lever.
6. Turn the adjusting nut until the parking brake is fully applied when the lever is pulled 7–12 notches on B Series Pick-Up or 5–7 notches on MPV.
7. Tighten the locknut on B Series Pick-Up.
8. Install the adjusting nut clip and the parking brake lever cover on MPV.

Navajo

NOTE: Adjust the drum brakes before adjusting the parking brake. The brake drums must be cold for correct adjustment.

INITIAL ADJUSTMENT

1. Use this procedure when a new tension limiter is install
2. Apply the parking brake pedal to the fully engaged position.
3. Raise and safely support the vehicle, as necessary. Hold the threaded rod end of the right brake cable to keep it from spinning and thread the equalizer nut 2½ in. up the rod.
4. Check to make sure the cinch strap has slipped and there are less than 1⅜ in. remaining.
5. Release the parking brake and check for proper operation.

FIELD ADJUSTMENT

1. Use this procedure to correct a slack system if a new tension limiter is not install
2. Apply the parking brake pedal to the fully engaged position.
3. Raise and safely support the vehicle, as necessary. Grip the threaded rod to keep it from spinning and tighten the equalizer nut 6 full turns past its original position on the threaded rod.
4. Attach a suitable cable tension gauge in front of the equalizer assembly on the front cable and measure the cable tension. The cable tension should be 400–600 lbs. with the parking brake pedal in the last detent position. If tension is low, repeat Steps 2 and 3.
5. Release parking brake and check for rear wheel drag. There should be no brake drag.

REMOVAL AND INSTALLATION

Front

EXCEPT NAVAJO

1. Make sure the parking brake is fully released.
2. Remove the parking brake lever adjusting nut from the forward end of the front cable.
3. Remove the seat(s) and roll back the front floormat, as required. On MPV, remove the cable cover.
4. Raise and safely support the vehicle, as necessary.
5. Disengage the rear cables from the equalizer and remove the spring. Disconnect the front cable from the equalizer.
6. Remove the bolts from the cable retaining straps and remove the cable.
7. Installation is the reverse of the removal procedure. Adjust the parking brake.

NAVAJO

1. Raise and safely support the vehicle.
2. Back off the equalizer nut and remove the cable end of the intermediate cable from the tension limiter.
3. Remove the intermediate cable from the bracket and disconnect the intermediate cable from the front cable.
4. Lower the vehicle. Remove the forward ball end of the parking brake cable from the control assembly clevis.
5. Remove the cable from the control assembly.
6. Using a cord attached to the control lever end of the cable, remove the cable from the vehicle pulling it up into the passenger compartment.

To install:

7. Transfer the cord to the new cable. Position the cable in the vehicle, routing the cable through the dash panel. Remove the cord and secure the cable to the control.
8. Connect the forward ball end of the brake cable to the clevis of the control assembly. Raise and safely support the vehicle.
9. Route the cable through the bracket. Connect the front cable to the intermediate cable.
10. Connect the slug of the front or intermediate cable cable to the tension limiter connector. Adjust the parking brake cable at the equalizer using initial adjustment or field adjustment, as necessary.
11. Rotate both wheels to make sure the parking brakes are not dragging.

Rear

EXCEPT NAVAJO

1. Make sure the parking brake is fully released.
2. Loosen the parking brake lever adjusting nut.
3. Remove the seat(s) and roll back the front floormat, as required. On MPV, remove the cable cover.
4. Raise and safely support the vehicle.
5. Disconnect the rear cable from the equalizer.
6. Remove the rear wheel and tire assembly, brake drum and brake shoes.
7. Disconnect the cable from the backing plate.
8. Remove the bolts from the cable retaining straps and disconnect the spring from the cable. Remove the cable.

9. Installation is the reverse of the removal procedure. Adjust the parking brake.

NAVAJO

1. Release the parking brake control.
2. Raise and safely support the vehicle. Remove the wheel and tire assembly, brake drum and brake shoes.
3. Remove the locknut on the threaded rod at the equalizer. Disconnect the rear parking brake cable from the equalizer.
4. Compress the prongs that retain the cable housing to the frame bracket or crossmember and pull out the cable and housing.
5. Working on the wheel side of the backing plate, compress the prongs on the cable retainer so they can pass through the hole in the brake backing plate.
6. Lift the cable out of the slot in the parking brake lever, attached to the secondary brake shoe, and remove the cable through the brake backing plate hole.
To install:
7. Route the cable through the hole in the backing plate. Insert the cable anchor behind the slot in the parking brake lever. Make sure the cable is securely engaged in the parking brake lever so the cable return spring is holding the cable in the parking brake lever.
8. Push the retainer through the hole in the backing plate so the retainer prongs engage the backing plate.
9. Properly route the cable and insert the front of the cable through the frame bracket or crossmember until the prongs expand. Connect the rear cables to the equalizer.
10. Rotate the equalizer 90 degrees and recouple the threaded rod to the equalizer.
11. Install the brake shoes, brake drum and wheel and tire assembly. Adjust the rear brakes.
12. Adjust the parking brake tension using the initial adjustment or the field adjustment procedure, as necessary.
13. Apply and release the parking brake control several times. Rotate both wheels to make sure the parking brakes are applied and released and not dragging.

BRAKE SYSTEM BLEEDING

1. Clean all dirt from the master cylinder filler cap.

2. If the master cylinder is known or suspected to have air in the bore, it must be bled before any of the wheel cylinders or calipers. Proceed as follows:
   a. Loosen the brake line fitting approximately ¾ turn. Wrap a shop cloth around the tubing below the fitting to absorb escaping brake fluid.
   b. Have an assistant depress the brake pedal slowly through its full travel to force air trapped in the master cylinder to escape at the fitting.
   c. Tighten the fitting and let the pedal return slowly to the fully released position. Do not release the pedal until the fitting is tightened or air will reenter the master cylinder.
   d. Wait 5 seconds and then repeat the operation until all air bubbles disappear.
   e. Repeat Steps a–d on the remaining master cylinder brake line fitting(s).
3. Continue to bleed the brake system by removing the rubber dust cap from the wheel cylinder bleeder fitting at the right-hand rear of the vehicle. Place a box wrench on the bleeder fitting and attach a rubber drain hose to the fitting. The end of the tube should fit snugly around the bleeder fitting. Submerge the other end of the tube in a container partially filled with clean brake fluid and loosen the fitting ¾ turn.
4. Have an assistant push the brake pedal down slowly through its full travel. Close the bleeder fitting and allow the pedal to slowly return to its full release position. Wait 5 seconds and repeat the procedure until no bubbles appear at the submerged end of the bleeder tube. Secure the bleeder fitting and remove the bleeder hose. Install the rubber dust cap on the bleeder fitting.
5. Repeat the procedure in Steps 3 and 4 and bleed the rest of the system in the following sequence: left rear, right front and left front.

NOTE: If equipped with anti-lock brakes, the electro-hydraulic valve must also be bled. It is not necessary to energize the valve to bleed it.

6. Refill the master cylinder reservoir after each wheel cylinder or caliper has been bled and install the master cylinder cover and gasket. When brake bleeding is completed, the fluid level should be filled to the

maximum level indicated on the reservoir.
7. Always make sure the disc brake pistons are returned to their normal positions by depressing the brake pedal several times until normal pedal travel is established. If the pedal feels spongy, repeat the bleeding procedure.

Anti-Lock Brake System Service

PRECAUTIONS

Use caution when disassembling any hydraulic components as the system will contain residual pressure.
   Cover the area around the component to be removed with a shop cloth to catch any brake fluid spray.
   Do not allow brake fluid to come in contact with painted surfaces.

Electronic Control Unit

REMOVAL AND INSTALLATION

1. Disconnect the negative battery cable.
2. On B Series Pick-Up, remove the driver's seat. On MPV, remove the inside trim panel from the left rear of the vehicle. The control unit is located under the dash on Navajo.
3. Disconnect the electrical connector from the control unit.
4. Remove the attaching screws and remove the control unit.
5. Installation is the reverse of the removal procedure. Check the system for proper operation.

Electro-Hydraulic Valve

REMOVAL AND INSTALLATION

1. Disconnect the negative battery cable.
2. Disconnect and plug the 2 brake lines connected to the valve.
3. Disconnect the wiring harness from the valve harness.
4. Remove the screw(s) retaining the valve and remove the valve.
5. Installation is the reverse of the removal procedure. Tighten the valve retaining screw(s) to 14–19 ft. lbs. (19–25 Nm) on all except Navajo where the torque is 11–14 ft. lbs. (15–20 Nm). Bleed the brake system.

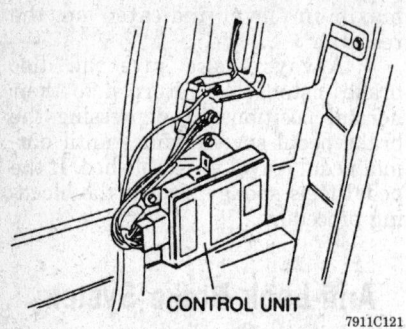

CONTROL UNIT

7911C121

Rear anti-lock brake system electronic control unit location — MPV

Speed Sensor

REMOVAL AND INSTALLATION

Except Navajo

1. Disconnect the negative battery cable.

2. Pull the wiring harness connector off.

3. Remove the sensor hold-down bolt and remove the sensor from the axle housing.

To install:

4. Clean the axle mounting surface. Use care to prevent dirt from entering the axle housing.

5. Inspect the sensor O-ring for damage and replace, if necessary.

6. Check the sensor-to-sensor rotor clearance as follows:

   a. Measure the distance between the sensor attaching surface and the sensor rotor teeth.

   b. Measure the distance between the sensor attaching surface and the sensor pole piece.

   c. Subtract the distance recorded in Step b from the distance recorded in Step a.

   d. The clearance should be 0.020–0.039 in. (0.5–1.0mm) on B2200 or 0.020–0.047 in. (0.5–1.2mm) on B2600 and MPV.

7. If the clearance is less than specified, increase it using adjusting shim P049 27 155 or equivalent, during sensor installation. If the clearance is more than specified, replace the speed sensor.

8. Lubricate the speed sensor with clean engine oil and install the speed sensor. Tighten the attaching bolt to 12–17 ft. lbs. (16–23 Nm).

Navajo

1. Disconnect the negative battery cable.

2. Pull the wiring harness connector off.

3. Remove the sensor hold-down bolt and remove the sensor from the axle housing.

To install:

4. Clean the axle mounting surface. Use care to prevent dirt from entering the axle housing.

5. Inspect and clean the magnetized sensor pole piece to ensure that it is free from loose metal particles which could cause erratic system operation. Inspect the sensor O-ring for damage and replace, if necessary.

6. Lightly lubricate the sensor O-ring with motor oil, align the sensor bolt hole and install. Do not apply force to the plastic sensor connector. The sensor flange should slide to the mounting surface. This will insure the air gap setting is between 0.005–0.045 in. (0.127–1.14mm).

7. Install the hold-down bolt and tighten to 25–30 ft. lbs. (34–40 Nm).

8. Inspect the blue sensor connector seal and replace if missing or damaged. Push the connector on the sensor.

9. Connect the negative battery cable.

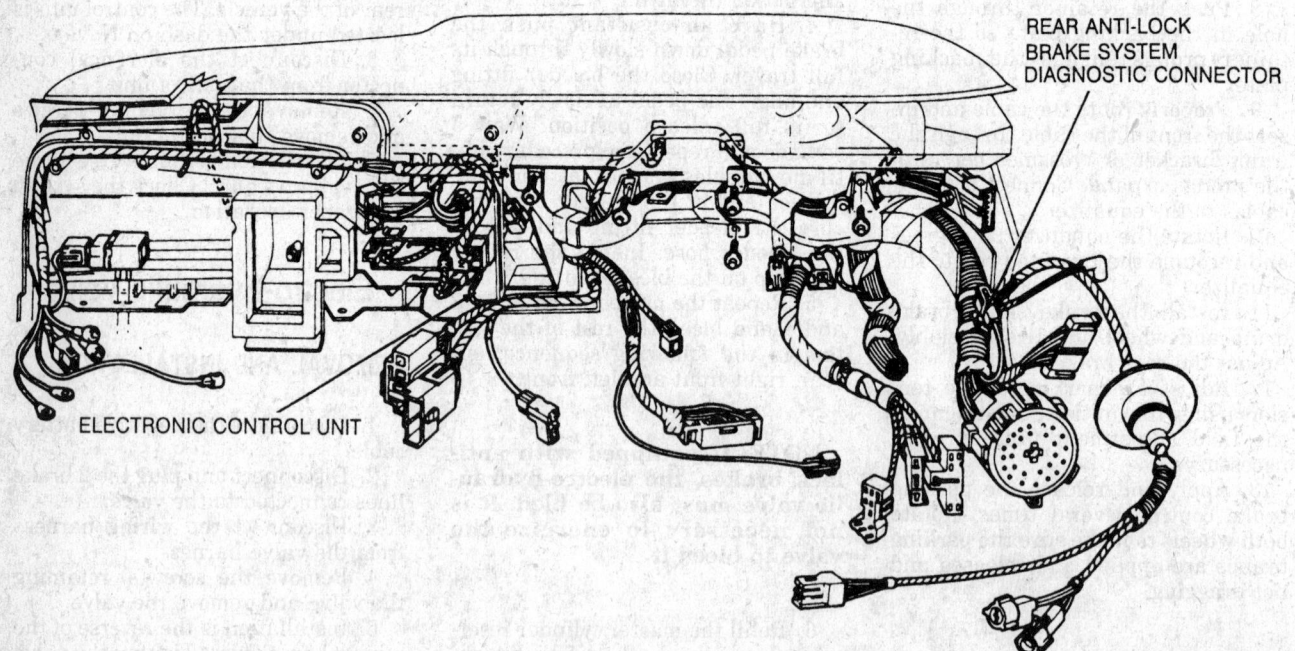

REAR ANTI-LOCK BRAKE SYSTEM DIAGNOSTIC CONNECTOR

ELECTRONIC CONTROL UNIT

7911C122

Rear anti-lock brake system electronic control unit location — Navajo

Sensor Rotor

INSPECTION

1. Remove the speed sensor.
2. View the sensor rotor teeth through the sensor hole. Rotate the rear axle and check the sensor rotor teeth for damage or breakage. Dented or broken teeth could cause the rear anti-lock brake system to function when not required.

REMOVAL AND INSTALLATION

To service the sensor rotor, the differential case must be removed from the axle housing and the sensor rotor pressed off the case.

NOTE: Upon removal, the sensor rotor is to be discarded. It is not to be reused.

FRONT SUSPENSION

Shock Absorbers

REMOVAL AND INSTALLATION

B Series Pick-Up

1. Raise and safely support the vehicle. Remove the wheel and tire assembly.
2. Remove the upper shock absorber nuts, retainer and bushing.
3. Remove the lower shock absorber-to-lower control arm mounting bolt, nut and washer.
4. Slightly compress the shock absorber and remove it from the vehicle. Remove the remaining retainers and bushing from the upper shock absorber stud.
To install:
5. Install the shock absorber and install the mounting bolts, nuts, washers and bushings. Do not tighten at this time.
6. Install the wheel and tire assembly and lower the vehicle.
7. With the vehicle unladen, tighten the upper shock absorber mounting nuts until the stud protrudes 0.28 in. (7mm) above the upper nut. Tighten the lower mounting bolt and nut to 41–59 ft. lbs. (55–80 Nm).
8. Check the front end alignment.

Navajo

1. Raise and safely support the vehicle. Remove the wheel and tire assembly.
2. Remove the shock absorber upper attaching nut, washer and bushing from the upper spring seat.
3. Remove the shock absorber lower mounting nut and washer from the radius arm stud.
4. Slide the shock absorber off the stud. Slightly compress the shock absorber and remove it from the vehicle.
5. Installation is the reverse of the removal procedure. Tighten the lower mounting nut to 39–53 ft. lbs. (53–72 Nm) and the upper mounting nut to 25–35 ft. lbs. (34–48 Nm).

MacPherson Strut

REMOVAL AND INSTALLATION

MPV

1. Raise and safely support the vehicle. Remove the wheel and tire assembly.
2. Support the lower control arm with a jack.
3. Remove the clip attaching the brake hose to the strut and disconnect the hose from the strut.
4. Remove the strut-to-knuckle attaching bolts and nuts.
5. Working in the engine compartment, remove the 4 attaching nuts from the strut tower and remove the strut assembly from the vehicle.
6. Remove the rubber cap from the upper mounting block. Loosen the upper attaching nut, but do not remove it.
7. Install a suitable spring compressor and compress the coil spring.
8. Remove the upper attaching nut and slowly relieve the tension on the coil spring, using the spring compressor. When the spring is no longer under tension, remove the spring compressor.
9. Remove the upper mounting block, upper spring seat, spring seat, coil spring, bump stopper and ring rubber from the strut.
To install:
10. Secure the strut in a vise equipped with protective jaw covers, so the strut will not be damaged.
11. Apply a suitable rubber grease to the ring rubber and install it on the bump stopper. Install the bump stopper on the strut.

12. Attach the spring compressor to the coil spring and compress the spring.
13. Install the compressed spring on the strut and install the spring seat.
14. Install the upper spring seat. The flat of the strut rod must fit correctly into the upper spring seat.
15. Install the upper mounting block. Install and loosely tighten the upper attaching nut.
16. Remove the spring compressor. Make sure the spring is properly seated in the upper and lower spring seats.
17. Secure the upper spring seat in a vise and tighten the upper attaching nut to 47–59 ft. lbs. (64–80 Nm). Install the rubber cap on the upper mounting blots.
18. Install the strut assembly in the strut tower, making sure the white mark on the upper mounting block is in the front-inside direction. Install the attaching nuts and tighten to 22–27 ft. lbs. (29–36 Nm).
19. Install the strut to the knuckle and tighten the attaching bolts and nuts to 69–86 ft. lbs. (93–117 Nm).
20. Position the brake hose on the strut and install the clip. Remove the jack from under the lower control arm.
21. Install the wheel and tire assembly and lower the vehicle. Check the front end alignment.

Coil Springs

REMOVAL AND INSTALLATION

Navajo

1. Raise and safely support the vehicle. Place a jack under the axle.
2. Remove the nut attaching the shock absorber to the radius arm and slide the shock absorber off of the stud.
3. Remove the nut securing the spring to the axle and remove the retainer.
4. Slowly lower the axle to relieve the spring tension. Remove the spring by rotating the upper coil out of the tabs in the upper spring seat.
To install:
5. Install the top of the spring in the upper seat, rotating into position.
6. Raise the axle until the spring is seated in the lower spring seat. Install the lower retainer and tighten the nut to 70–100 ft. lbs. (95–136 Nm).
7. Connect the shock absorber to the radius arm and lower the vehicle.

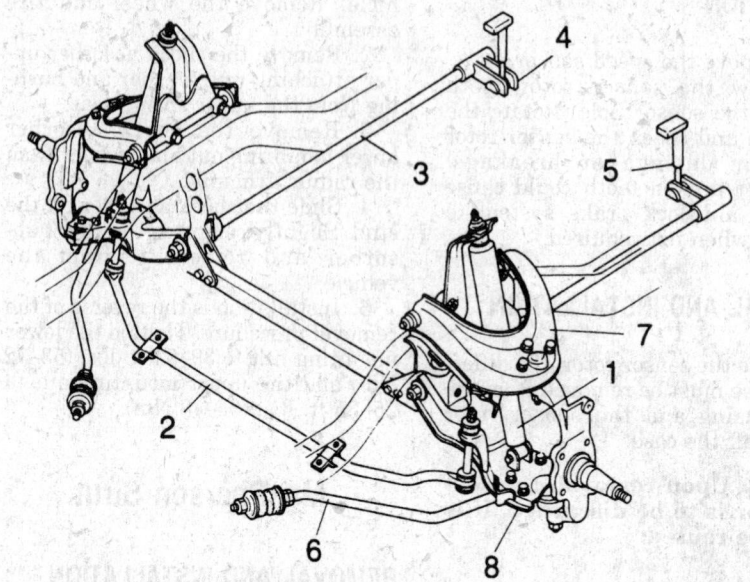

1. Shock absorber
2. Stabilizer
3. Torsion bar
4. Anchor arm
5. Anchor bolt
6. Tension rod
7. Upper arm
8. Lower arm

7911C123

Front suspension assembly — 2WD B Series Pick-Up

1. Shock absorber
2. Stabilizer
3. Torsion bar
4. Anchor arm
5. Anchor bolt
6. Upper arm
7. Lower arm

7911C124

Front suspension assembly — 4WD B Series Pick-Up

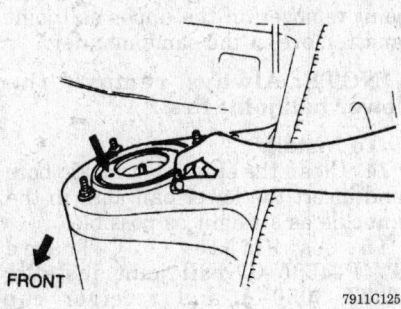

FRONT

7911C125

Strut upper mounting block positioning — MPV

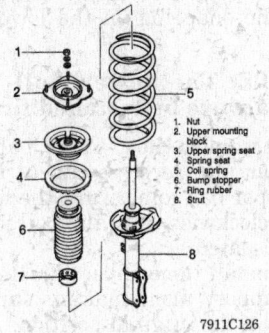

1. Nut
2. Upper mounting block
3. Upper spring seat
4. Spring seat
5. Coil spring
6. Bump stopper
7. Ring rubber
8. Strut

7911C126

MacPherson strut assembly exploded view — MPV

Torsion Bar

REMOVAL AND INSTALLATION

B Series Pick-Up

1. Raise and safely support the vehicle. Remove the wheel and tire assembly.

2. Support the lower control arm with a jack.

3. Remove the cotter pin and nut from the lower ball joint stud. Separate the ball joint from the knuckle using tool 49 0727 575 or equivalent.

4. Remove the bolt, washer and nut attaching the shock absorber to the lower control arm.

5. Mark the position of the anchor bolt and swivel for reference during reassembly and remove the anchor bolt and swivel.

6. Mark the position of the torsion bar on the anchor arm and remove the anchor arm.

7. Mark the position of the torsion bar on the torque plate and remove the torsion bar. If removing the torsion bar from both sides of the vehicle, mark their positions as the torsion bars are not interchangeable.

8. Remove the attaching bolts and remove the torque plate.

9. Check the torsion bar for bending or for looseness between the serrations of the torsion bar and anchor arm and/or torque plate, replace as necessary.

To install:

10. Install the torque plate and tighten the attaching bolts to 55–69 ft. lbs. (75–93 Nm).

11. Coat the serrations of the torsion bar with grease and install in the torque plate, aligning the marks made during the removal procedure. If both torsion bars were removed, make sure the correct torsion bar is being installed.

12. Coat the serrations on the other end of the torsion bar with grease and install the anchor arm onto the torsion bar, aligning the marks made during the removal procedure.

13. Install the anchor bolt and swivel. Tighten the anchor bolt until the marks made during removal are aligned.

14. Connect the shock absorber to the lower control arm and loosely tighten the nut and bolt. Connect the lower ball joint to the knuckle; install the nut and tighten to 87–116 ft. lbs. (118–157 Nm). Install a new cotter pin.

15. Install the wheel and tire assembly and lower the vehicle. With the vehicle unladen, tighten the shock absorber-to-lower control arm bolt and nut to 41–59 ft. lbs. (55–80 Nm).

16. Check vehicle ride height as follows:

    a. Check the front and rear tire pressures and bring to specification.

    b. Measure the distance from the center of each front wheel to the fender brim. The difference must not be greater than 0.39 in. (10mm).

    c. If the difference is not as specified, turn the necessary torsion spring anchor bolt to adjust.

17. Check the front end alignment.

Upper Ball Joints

INSPECTION

B Series Pick-Up

1. Raise and safely support the vehicle. Remove the wheel and tire assembly.

2. Support the lower control arm with a jack.

3. Remove the clip attaching the brake hose to the upper control arm and disconnect the hose from the arm.

4. Remove the cotter pin and nut from the upper ball joint stud. Using tool 49 0727 575 or equivalent, separate the upper ball joint from the knuckle.

5. Install tool 49 0180 510B or equivalent to the ball joint stud and attach a suitable pull scale to the stud.

6. After rocking the ball joint stud back and forth 3–4 times, measure the pull scale reading while the ball joint stud is rotating. The pull scale reading should be 4.4–7.7 lbs.

7. If the pull scale reading is not as specified, replace the upper ball joint.

Navajo

NOTE: Always check and adjust the wheel bearings before ball joint inspection.

1. Raise and safely support the vehicle.

2. Place a jack under the axle beneath the coil spring.

3. Grasp the upper edge of the tire and move the wheel in and out. A $\frac{1}{32}$ in. or greater movement between the upper spindle arm and the upper control arm or upper part of the axle jaw indicates that the upper ball joint must be replaced.

REMOVAL AND INSTALLATION

B Series Pick-Up

1. Raise and safely support the vehicle. Remove the wheel and tire assembly.

2. Support the lower control arm with a jack.

3. Remove the clip attaching the brake hose to the upper control arm and disconnect the hose from the arm.

4. Remove the cotter pin and nut from the upper ball joint stud. Using tool 49 0727 575 or equivalent, separate the upper ball joint from the knuckle.

5. Remove the upper ball joint-to-upper control arm attaching bolts and remove the upper ball joint.

6. Installation is the reverse of the removal procedure. Tighten the upper ball joint-to-upper control arm attaching bolts to 18–25 ft. lbs. (25–33 Nm) and the upper ball joint stud nut to 22–38 ft. lbs. (29–51 Nm). Install a new cotter pin.

7. Check the front end alignment.

Navajo

2WD

1. Raise and safely support the vehicle. Remove the wheel and tire assembly.

2. Remove the brake caliper and support it aside with mechanics wire. Do not let the caliper hang by the brake hose.

3. Remove the dust cap, cotter pin, nut retainer, washer and outer bearing and remove the brake rotor from the spindle. Remove the brake dust shield.

4. Disconnect the steering linkage from the spindle and spindle arm by removing the cotter pin and nut. Remove the tie rod end from the spindle arm.

5. Remove the cotter pin and nut from the lower ball joint stud. Remove the axle clamp bolt from the axle.

6. Remove the camber adjuster from the upper ball joint stud and axle beam.

7. Strike the inside area of the axle to pop the lower ball joint loose from the axle beam. Remove the spindle and ball joint assembly from the axle.

NOTE: Do not use a pickle fork to separate the ball joint from the axle as this will damage the seal and ball joint socket.

To install:

8. Install the spindle assembly in a vise and remove the snapring from the lower ball joint. Remove the lower ball joint from the spindle using C-frame T74P-4635-C or equivalent and a suitable receiver cup to press the ball joint from the spindle.

NOTE: The lower ball joint must be removed first.

9. Repeat the procedure in Step 8 to remove the upper ball joint.

NOTE: Do not heat the ball joints or the spindle to aid in removal.

10. Assemble the C-frame and receiver cup and press in the upper ball joint.

11. Repeat the procedure in Step 10 to install the lower ball joint.

NOTE: Do not heat the ball joints or axle to aid in installation.

12. Install the snapring onto the ball joint.

13. Place the spindle and ball joints into the axle. Install the camber adjuster in the upper spindle over the ball joint stud making sure it is properly aligned.

14. Tighten the lower ball joint stud nut to 104–146 ft. lbs. (141–198 Nm). Continue tightening the castellated nut until it lines up with the hole in the stud, then install the cotter pin.

15. Install the clamp bolt into the axle boss and tighten to 48–65 ft. lbs. (65–88 Nm).

16. Install the remaining components in the reverse order of their removal.

4WD

1. Raise and safely support the vehicle. Remove the front wheel and tire assemblies.

2. Remove the disc brake caliper and wire it to the frame. Do not let the caliper hang by the brake hose.

3. Remove the hub locks, wheel bearings and locknuts.

4. Remove the hub, rotor and outer wheel bearing.

5. Remove the nuts retaining the spindle to the steering knuckle. Tap the spindle with a plastic hammer to jar the spindle from the knuckle. Remove the splash shield.

6. On the left side of the vehicle, remove the shaft and joint assembly by pulling the assembly out of the carrier. On the right side of the carrier, remove and discard the clamp from the shaft and joint assembly and the stub shaft. Pull the shaft and joint assembly from the splines of the stub shaft.

7. Remove the cotter pin from the tie rod nut and then remove the nut. Tap on the tie rod stud to free it from the steering arm.

8. Remove the upper ball joint snapring and remove the upper ball joint pinch bolt. Loosen the lower ball joint nut to the end of the stud.

9. Strike the inside of the knuckle near the upper and lower ball joints to break the knuckle loose from the ball joint studs.

10. Remove the camber adjuster sleeve. Note the position of the slot in the camber adjuster so it can be reinstalled in the same position during assembly.

11. Remove the lower ball joint nut. Place the knuckle in a vise and remove the snapring from the bottom ball joint socket, if equipped.

12. Assemble C-frame T74P-4635-C and ball joint remover T83T-3050-A or equivalents on the lower ball joint. Turn the forcing screw clockwise until the lower ball joint is removed from the steering knuckle.

13. Assemble the C-frame and ball joint remover on the upper ball joint and remove in the same manner.

NOTE: Always remove the lower ball joint first.

To install:

14. Clean the steering knuckle bore and insert the lower ball joint in the knuckle as straight as possible.

15. Assemble C-frame T74P-4635-C, ball joint installer T83T-3050-A and receiver cup T80T-3010-A3 or equivalents to install the lower ball joint. Turn the forcing screw clockwise until the lower ball joint is firmly seated. Install the snapring on the lower ball joint.

NOTE: The lower ball joint must always be installed first.

16. Assemble the C-frame, ball joint installer and receiver cup to install the upper ball joint. Turn the forcing screw clockwise until the ball joint is firmly seated.

17. Install the camber adjuster into the support arm, making sure the slot is in the original position.

NOTE: The torque sequence in Steps 18 and 19 must be followed exactly when securing the knuckle. Excessive knuckle turning effort may result in reduced steering returnability if this procedure is not followed.

18. Install a new nut on the bottom ball joint stud. Tighten the nut to 90 ft. lbs. (122 Nm) minimum, then tighten to align the next slot in the nut with the hole in the stud. Install a new cotter pin.

19. Install the snapring on the upper ball joint stud. Install the upper ball joint pinch bolt and tighten to 48–65 ft. lbs. (65–88 Nm).

NOTE: The camber adjuster will seat itself into the knuckle at a predetermined position during the tightening sequence. Do not attempt to adjust this position.

20. On the right side of the carrier, install the rubber boot and new keystone clamps on the stub shaft slip yoke. Slide the right shaft and joint assembly into the slip yoke making sure the splines are fully engaged. Slide the boot over the assembly and crimp the keystone clamp using suitable pliers.

NOTE: The Dana model 35 axle does not have a blind spline, therefore pay special attention to make sure the yoke ears are in phase (in line) during assembly.

21. On the left side of the carrier, slide the shaft and joint assembly through the knuckle and engage the splines on the shaft in the carrier.

22. Install the splash shield and spindle onto the steering knuckle. Install and tighten the spindle nuts to 45 ft. lbs. (61 Nm).

23. Install the rotor on the spindle and install the outer wheel bearing in the race.

24. Install the wheel bearing, locknut, thrust bearing, snapring and locking hubs.

25. Install the caliper and the wheel and tire assemblies. Lower the vehicle.

Lower Ball Joints

INSPECTION

B Series Pick-Up and 4WD MPV

1. Raise and safely support the vehicle. Remove the wheel and tire assembly.

2. On B Series Pick-Up, support the lower control arm with a jack.

3. On 4WD MPV, disconnect the stabilizer bar from the lower control arm.

4. Remove the cotter pin and nut from the lower ball joint stud. Separate the ball joint from the knuckle using tool 49 0727 575 or equivalent.

5. Install tool 49 0180 510B or equivalent to the ball joint stud and attach a suitable pull scale to the stud.

6. After rocking the ball joint stud back and forth 3–4 times, measure the pull scale reading while the ball joint stud is rotating. The pull scale reading should be 4.4–7.7 lbs.

7. If the pull scale reading is not as specified, replace the lower ball joint.

2WD MPV

1. Raise and safely support the vehicle. Remove the wheel and tire assembly.

2. Remove the brake caliper and support it aside with mechanics wire; do not let it hang by the brake hose.

3. Remove the nuts, bolts, spacer, washers and bushings and remove the compression rod from the lower control arm and chassis and disconnect the stabilizer bar from the lower control arm.

NOTE: The left-hand compression rod nut has left-hand threads.

4. Remove the cotter pin and nut and, using tool 49 0118 850C or equivalent, separate the tie rod end from the knuckle.

5. Remove the bolts and nuts and disconnect the strut from the knuckle.

6. Remove the cotter pin and nut from the lower ball joint stud. Using tool 49 0727 575 or equivalent, separate the lower ball joint from the knuckle.

7. Install tool 49 0180 510B or equivalent to the ball joint stud and attach a suitable pull scale to the stud.

8. After rocking the ball joint stud back and forth 3–4 times, measure the pull scale reading while the ball joint stud is rotating. The pull scale reading should be 39.6 lbs. or less. The rotation torque should be 156.2 inch lbs. (18 Nm) or less.

9. If the pull scale reading or rotation torque is not as specified, replace the entire lower control arm.

Navajo

NOTE: Always check and adjust the wheel bearings before ball joint inspection.

1. Raise and safely support the vehicle.

2. Place a jack under the axle beneath the coil spring.

3. Grasp the lower edge of the tire and move the wheel in and out. A $^1/_{32}$ in. or greater movement between the lower control arm or lower axle and the spindle indicates that the lower ball joint must be replaced.

REMOVAL AND INSTALLATION

B Series Pick-Up

1. Raise and safely support the vehicle. Remove the wheel and tire assembly.

2. Support the lower control arm with a jack.

3. On 2WD vehicles, remove the bolts attaching the tension rod to the lower control arm.

4. Remove the cotter pin and nut from the lower ball joint stud. Separate the ball joint from the knuckle using tool 49 0727 575 or equivalent.

5. Remove the bolts/nuts attaching the lower ball joint to the lower control arm and remove the lower ball joint.

6. Installation is the reverse of the removal procedure. Tighten the lower ball joint-to-lower control arm bolts/nuts to 32–40 ft. lbs. (43–54 Nm) on 2WD vehicles or 41–50 ft. lbs. (55–68 Nm) on 4WD vehicles. Tighten the lower ball joint stud nut to 87–116 ft. lbs. (118–157 Nm) and install a new cotter pin.

7. Check the front end alignment.

MPV

2WD

If the lower ball joint needs replacement, the entire lower control arm must be replaced.

4WD

1. Raise and safely support the vehicle. Remove the wheel and tire assembly.

2. Disconnect the stabilizer bar from the lower control arm.

3. Remove the cotter pin and nut from the lower ball joint stud. Separate the ball joint from the knuckle using tool 49 0727 575 or equivalent.

4. Remove the lower ball joint-to-lower control arm bolts and remove the lower ball joint.

5. Installation is the reverse of the removal procedure. Tighten the lower ball joint-to-lower control arm bolts to 75–101 ft. lbs. (102–137 Nm). Tighten the lower ball joint stud nut to 115–137 ft. lbs. (157–186 Nm) and install a new cotter pin.

Navajo

2WD

1. Raise and safely support the vehicle. Remove the wheel and tire assembly.

2. Remove the brake caliper and support it aside with mechanics wire. Do not let the caliper hang by the brake hose.

3. Remove the dust cap, cotter pin, nut retainer, washer and outer bearing and remove the brake rotor from the spindle. Remove the brake dust shield.

4. Disconnect the steering linkage from the spindle and spindle arm by removing the cotter pin and nut. Remove the tie rod end from the spindle arm.

5. Remove the cotter pin and nut from the lower ball joint stud. Remove the axle clamp bolt from the axle.

6. Remove the camber adjuster from the upper ball joint stud and axle beam.

7. Strike the inside area of the axle to pop the lower ball joint loose from the axle beam. Remove the spin-

dle and ball joint assembly from the axle.

NOTE: Do not use a pickle fork to separate the ball joint from the axle as this will damage the seal and ball joint socket.

To install:

8. Install the spindle assembly in a vise and remove the snapring from the lower ball joint. Remove the lower ball joint from the spindle using C-frame T74P–4635–C or equivalent and a suitable receiver cup to press the ball joint from the spindle.

NOTE: Do not heat the ball joint or the spindle to aid in removal.

9. Assemble the C-frame and receiver cup and press in the lower ball joint.

NOTE: Do not heat the ball joint or axle to aid in installation.

10. Install the snapring onto the ball joint.
11. Place the spindle and ball joints into the axle. Install the camber adjuster in the upper spindle over the ball joint stud making sure it is properly aligned.
12. Tighten the lower ball joint stud nut to 104–146 ft. lbs. (141–198 Nm). Continue tightening the castellated nut until it lines up with the hole in the stud, then install the cotter pin.
13. Install the clamp bolt into the axle boss and tighten to 48–65 ft. lbs. (65–88 Nm).
14. Install the remaining components in the reverse order of their removal.

4WD

1. Raise and safely support the vehicle. Remove the front wheel and tire assemblies.
2. Remove the disc brake caliper and wire it to the frame. Do not let the caliper hang by the brake hose.
3. Remove the hub locks, wheel bearings and locknuts.
4. Remove the hub, rotor and outer wheel bearing.
5. Remove the nuts retaining the spindle to the steering knuckle. Tap the spindle with a plastic hammer to jar the spindle from the knuckle. Remove the splash shield.
6. On the left side of the vehicle, remove the shaft and joint assembly by pulling the assembly out of the carrier. On the right side of the carrier, remove and discard the clamp from the shaft and joint assembly

and the stub shaft. Pull the shaft and joint assembly from the splines of the stub shaft.

7. Remove the cotter pin from the tie rod nut and then remove the nut. Tap on the tie rod stud to free it from the steering arm.
8. Remove the upper ball joint snapring and remove the upper ball joint pinch bolt. Loosen the lower ball joint nut to the end of the stud.
9. Strike the inside of the knuckle near the upper and lower ball joints to break the knuckle loose from the ball joint studs.
10. Remove the camber adjuster sleeve. Note the position of the slot in the camber adjuster so it can be reinstalled in the same position during assembly.
11. Remove the lower ball joint nut. Place the knuckle in a vise and remove the snapring from the bottom ball joint socket, if equipped.
12. Assemble C-frame T74P–4635–C and ball joint remover T83T–3050–A or equivalents on the lower ball joint. Turn the forcing screw clockwise until the lower ball joint is removed from the steering knuckle.

To install:

13. Clean the steering knuckle bore and insert the lower ball joint in the knuckle as straight as possible.
14. Assemble C-frame T74P–4635–C, ball joint installer T83T–3050–A and receiver cup T80T–3010–A3 or equivalents to install the lower ball joint. Turn the forcing screw clockwise until the lower ball joint is firmly seated. Install the snapring on the lower ball joint.
15. Install the camber adjuster into the support arm, making sure the slot is in the original position.

NOTE: The torque sequence in Steps 16 and 17 must be followed exactly when securing the knuckle. Excessive knuckle turning effort may result in reduced steering returnability if this procedure is not followed.

16. Install a new nut on the bottom ball joint stud. Tighten the nut to 90 ft. lbs. (122 Nm) minimum, then tighten to align the next slot in the nut with the hole in the stud. Install a new cotter pin.
17. Install the snapring on the upper ball joint stud. Install the upper

ball joint pinch bolt and tighten to 48–65 ft. lbs. (65–88 Nm).

NOTE: The camber adjuster will seat itself into the knuckle at a predetermined position during the tightening sequence. Do not attempt to adjust this position.

18. On the right side of the carrier, install the rubber boot and new keystone clamps on the stub shaft slip yoke. Slide the right shaft and joint assembly into the slip yoke making sure the splines are fully engaged. Slide the boot over the assembly and crimp the keystone clamp using suitable pliers.

NOTE: The Dana model 35 axle does not have a blind spline, therefore pay special attention to make sure the yoke ears are in phase (in line) during assembly.

19. On the left side of the carrier, slide the shaft and joint assembly through the knuckle and engage the splines on the shaft in the carrier.
20. Install the splash shield and spindle onto the steering knuckle. Install and tighten the spindle nuts to 45 ft. lbs. (61 Nm).
21. Install the rotor on the spindle and install the outer wheel bearing in the race.
22. Install the wheel bearing, locknut, thrust bearing, snapring and locking hubs.
23. Install the caliper and the wheel and tire assemblies. Lower the vehicle.

Upper Control Arms

REMOVAL AND INSTALLATION

B Series Pick-Up

1. Raise and safely support the vehicle. Remove the wheel and tire assembly.
2. Support the lower control arm with a jack.
3. Remove the clip attaching the brake hose to the upper control arm and disconnect the hose from the arm.
4. Remove the cotter pin and nut from the upper ball joint stud. Using tool 49 0727 575 or equivalent, separate the upper ball joint from the knuckle.
5. Remove the bolts and washers retaining the upper control arm shaft to the chassis and remove the upper control arm. Note the position of the alignment shims so they can be reassembled in their original positions.

6. Installation is the reverse of the removal procedure. Tighten the upper control arm shaft-to-chassis bolts to 54–69 ft. lbs. (74–93 Nm) on 2WD vehicles or 69–85 ft. lbs. (93–117 Nm) on 4WD vehicles. Tighten the upper ball joint stud nut to 22–38 ft. lbs. (29–51 Nm) and install a new cotter pin.

Lower Control Arms

REMOVAL AND INSTALLATION

B Series Pick-Up

1. Raise and safely support the vehicle. Remove the wheel and tire assembly.

2. Support the lower control arm with a jack.

3. Remove the cotter pin and nut from the lower ball joint stud. Separate the ball joint from the knuckle using tool 49 0727 575 or equivalent.

4. Remove the bolt, washer and nut attaching the shock absorber to the lower control arm.

5. Remove the torsion bar, anchor arm and torque plate assembly.

6. Remove the bolt(s) and nut(s) attaching the lower control arm to the frame.

7. On 2WD vehicles, remove the bolts attaching the tension rod to the lower control arm.

8. Remove the bolts, bushings, retainers, spacer and nuts connecting the stabilizer bar to the lower control arm.

9. Remove the lower control arm from the vehicle. Remove the lower ball joint, if necessary.

To install:

10. Position the lower control to the frame and install the attaching bolt(s) and nut(s), but do not tighten at this time.

11. Install the torsion bar, anchor arm and torque plate assembly.

12. On 2WD vehicles, install the tension rod bolt and tighten to 69–86 ft. lbs. (93–117 Nm).

13. Attach the stabilizer bar to the control arm with the bolts, bushings, retainers, spacer and nuts. Tighten the nuts so 0.73 in. (18.5mm) of thread is exposed at the end of the bolt.

14. Install the shock absorber to the lower control arm and loosely tighten the mounting bolt and nut.

15. Install the wheel and tire assembly and lower the vehicle.

16. With the vehicle unladen, tighten the lower control arm-to-frame bolt and nut on 2WD vehicles and the front side lower control arm-to-frame bolt and nut on 4WD vehicles to 87–116 ft. lbs. (118–157 Nm). Tighten the rear side lower control arm-to-frame bolt and nut on 4WD vehicles to 116–145 ft. lbs. (157–196 Nm).

17. With the vehicle unladen, tighten the shock absorber-to-lower control arm bolt and nut to 41–59 ft. lbs. (55–80 Nm).

18. Check vehicle ride height as follows:

   a. Check the front and rear tire pressures and bring to specification.

   b. Measure the distance from the center of each front wheel to the fender brim. The difference must not be greater than 0.39 in. (10mm).

   c. If the difference is not as specified, turn the necessary torsion spring anchor bolt to adjust.

19. Check the front end alignment.

MPV

2WD

1. Raise and safely support the vehicle. Remove the wheel and tire assembly.

2. Remove the brake caliper and support it aside with mechanics wire; do not let it hang by the brake hose.

3. Remove the nuts, bolts, spacer, washers and bushings and remove the compression rod from the lower control arm and chassis and disconnect the stabilizer bar from the lower control arm.

4. Remove the cotter pin and nut and, using tool 49 0118 850C or equivalent, separate the tie rod end from the knuckle.

5. Remove the bolts and nuts and disconnect the strut from the knuckle.

6. Remove the cotter pin and nut from the lower ball joint stud. Using tool 49 0727 575 or equivalent, separate the lower ball joint from the knuckle.

7. Remove the mounting bolt and nut and remove the lower control arm from the vehicle.

To install:

8. Position the lower control arm to the chassis and install the bolt and nut, but do not tighten at this time.

9. Install the knuckle to the lower control arm. Tighten the lower ball joint stud nut to 87–116 ft. lbs. (118–157 Nm) and install a new cotter pin.

10. Connect the strut to the knuckle and tighten the attaching bolts and nuts to 69–86 ft. lbs. (93–117 Nm).

11. Connect the tie rod end to the knuckle. Tighten the tie rod end stud nut to 43–58 ft. lbs. (59–78 Nm) and install a new cotter pin.

12. Install the compression rod to the lower control arm and chassis. Tighten the compression rod-to-lower control arm mounting bolts to 76–93 ft. lbs. (103–126 Nm) and the compression rod bushing-to-chassis bolts to 61–76 ft. lbs. (83–103 Nm). Install the compression rod nut but do not tighten at this time.

NOTE: The left-hand compression rod nut has left-hand threads.

13. Connect the stabilizer bar to the control arm with the bolt, washers, bushings, spacer and nuts. Tighten the nuts so 0.24 in. (6mm) of thread is exposed at the end of the bolt.

14. Install the caliper and the wheel and tire assembly. Lower the vehicle.

15. With the vehicle unladen, tighten the lower control arm-to-chassis bolt and nut to 94–108 ft. lbs. (146–172 Nm). Tighten the compression rod nut to 108–127 ft. lbs. (146–172 Nm).

16. Check the front end alignment.

4WD

1. Raise and safely support the vehicle. Remove the wheel and tire assembly.

2. Remove the bolt, retainers, bushings, spacer and nuts and disconnect the stabilizer bar from the lower control arm.

3. Remove the cotter pin and nut from the lower ball joint stud. Separate the ball joint from the knuckle using tool 49 0727 575 or equivalent.

4. Remove the lower control arm-to-chassis nuts and bolts and remove the lower control arm.

To install:

5. Position the lower control arm to the chassis and install the bolts and nuts. Do not tighten at this time.

6. Connect the lower ball joint to the knuckle and tighten the ball joint stud nut to 115–137 ft. lbs. (157–186 Nm). Install a new cotter pin.

7. Install the bolt, retainers, bushings, spacer and nuts and connect the stabilizer bar to the lower control arm. Tighten the nuts so 0.24 in. (6mm) of thread is exposed at the end of the bolt.

8. Install the wheel and tire assembly and lower the vehicle. With the vehicle unladen, tighten the lower control arm-to-chassis nuts and bolts to 101–127 ft. lbs. (137–172 Nm).

9. Check the front end alignment.

Stabilizer Bar

REMOVAL AND INSTALLATION

Except Navajo

1. Raise and safely support the vehicle. Remove the wheel and tire assembly.
2. On MPV, remove the splash shield.
3. Remove the bolt, retainers, bushings, spacer and nuts connecting the stabilizer bar end to the lower control arm.
4. Remove the stabilizer bar bushing bracket bolts and remove the stabilizer bar.

To install:

5. Install the stabilizer bar to the vehicle and loosely tighten the bushing bracket bolts.
6. Install the bolt, retainers, bushings, spacer and nuts connecting the stabilizer bar end to the lower control arm. Tighten the nuts so 0.73 in. (18.5mm) of thread is exposed at the end of the bolt on B Series Pick-Up or 0.24 in. (6mm) of thread is exposed at the end of the bolt on MPV.
7. Install the wheel and tire assembly and lower the vehicle.
8. With the vehicle unladen, tighten the stabilizer bar bushing bracket bolts to 16–20 ft. lbs. (22–26 Nm) on B Series pickup, 37–45 ft. lbs. (50–61 Nm) on 2WD MPV or 14–19 ft. lbs. (19–26 Nm) on 4WD MPV.
9. Check the front end alignment.

Navajo

2WD

1. Raise and safely support the vehicle.
2. Remove the nut and washer and disconnect the stabilizer link assembly from the front I-beam axle.
3. Remove the mounting bolts and remove the stabilizer bar retainers from the stabilizer bar assembly. Remove the stabilizer bar.
4. Installation is the reverse of the removal procedure. Tighten the retainer bolts to 35–50 ft. lbs. (47–68 Nm). Tighten the stabilizer bar link nuts to 30–44 ft. lbs. (40–60 Nm).

4WD

1. Raise and safely support the vehicle.
2. Remove the bolts and retainers from the center and right end of the stabilizer bar.
3. Remove the nut, bolt and washer retaining the stabilizer bar to the stabilizer link.

4. Remove the stabilizer bar and bushings.
5. Installation is the reverse of the removal procedure. Tighten the retainer bolts to 35–50 ft. lbs. (48–68 Nm). Tighten the stabilizer bar link nut to 30–44 ft. lbs. (40–60 Nm).

I-Beam Axle

REMOVAL AND INSTALLATION

2WD Navajo

1. Raise and safely support the vehicle. Remove the front wheel spindle, the front spring and the stabilizer bar, if equipped.
2. Remove the spring lower seat from the radius arm and then remove the bolt and nut that attaches the stabilizer bar bracket, if equipped, and radius arm to the front axle.
3. Remove the axle-to-frame pivot bracket bolt and nut.

To install:

4. Position the axle to the frame pivot bracket and install the bolt and nut finger tight.
5. Position the opposite end of the of the axle to the radius arm, install the attaching bolt from underneath through the bracket, the radius arm and the axle. Install the nut and tighten to 191–220 ft. lbs. (258–298 Nm).
6. Install the spring lower seat on the radius arm so the hole in the seat indexes over the arm-to-axle bolt. Install the front spring.
7. Install the front wheel spindle and stabilizer bar, if equipped.
8. Lower the vehicle and with the weight on the suspension, tighten the axle-to-frame pivot bracket bolts to 120–150 ft. lbs. (163–203 Nm).

Front Wheel Bearings

ADJUSTMENT

2WD Vehicles

B SERIES PICK-UP

1. Raise and safely support the vehicle. Remove the wheel and tire assembly.
2. Remove the brake caliper and suspend it aside with rope or mechanics wire; do not let the caliper hang by the brake hose.
3. Remove the dust cap and cotter pin.
4. Tighten the locknut to 14–22 ft. lbs. (20–29 Nm) and turn the hub and rotor 2–3 times to seat the bearings.

5. Loosen the locknut until it can be turned by hand.
6. Attach a suitable pull scale to a wheel lug bolt and measure the frictional force.
7. Tighten the locknut until the pull scale reading, the initial turning torque, reaches the frictional force plus 1.3–2.4 lbs. Insert the retainer and secure with a new cotter pin.
8. Install the dust cap and the caliper. Install the wheel and tire assembly and lower the vehicle.
9. Before driving the vehicle, pump the brake pedal several times to restore normal brake travel.

NAVAJO

1. Raise the vehicle and support it safely.
2. Remove the wheel cover and the grease cap from the hub. Remove the cotter pin and the retainer. Discard the cotter pin.
3. Loosen the adjusting nut 3 turns.
4. Obtain running clearance between the brake rotor and disc brake pads by rocking the entire wheel and tire assembly in and out several times to push the caliper and brake pads away from the rotor.

NOTE: Do not pry on the caliper piston to obtain clearance.

5. While rotating the wheel, tighten the adjusting nut to 17–25 ft. lbs. (23–34 Nm) to seat the bearings.
6. Back off the adjusting nut ½ turn. Retighten the nut to 18–20 inch lbs. (2.0–2.3 Nm).
7. Install the retainer on the adjusting nut so the castellations line up with the hole in the spindle without moving the nut. Install a new cotter pin.
8. Check the front wheel rotation. If the wheel rotates properly, reinstall the grease cap and the wheel cover. If rotation is noisy or rough, remove, inspect and lubricate the bearings and bearing races.
9. Before driving the vehicle, pump the brake pedal several times to restore normal brake travel.

REMOVAL AND INSTALLATION

2WD Vehicles

1. Raise and safely support the vehicle. Remove the wheel and tire assembly.
2. Remove the brake caliper and support it with mechanics wire. Do not let the caliper hang by the brake hose.

3. Remove the grease cap, cotter pin, retainer, adjusting nut and washer. Discard the cotter pin.

4. Remove the outer bearing and pull the hub and rotor off the spindle. Remove the grease seal using a seal removal tool. Discard the grease seal.

5. Remove the inner bearing from the hub. Remove all traces of old lubricant from the bearings, hub and spindle with solvent and dry thoroughly.

6. Inspect the bearings and bearing races for scratches, pits or cracks. If the bearings and/or races are worn or damaged, remove the races with a brass drift.

To install:

7. If the bearing races were removed, install new races in the hub with suitable installation tools. Make sure the races are properly seated.

8. Using a bearing packer, pack the bearings with high-temperature wheel bearing grease. If a packer is not available, work as much grease as possible between the rollers and cages by hand.

9. Place a small amount of grease within the hub and grease the races. Install the inner bearing. Install a new wheel seal using a seal installer. Apply grease to the lips of the seal.

10. Install the hub and rotor assembly on the spindle. Install the outer bearing, washer and adjusting nut. Adjust the bearings.

11. Install the retainer, a new cotter pin and the grease cap.

12. Install the caliper and the wheel and tire assembly. Lower the vehicle.

13. Before driving the vehicle, pump the brake pedal several times to restore normal brake travel.

REAR SUSPENSION

Shock Absorber

REMOVAL AND INSTALLATION

1. Raise and safely support the vehicle.

2. Place a jack under the rear axle and raise slightly to take the load off the shock absorbers.

3. On MPV equipped with automatic load leveling, disconnect the air line from the shock absorber.

4. Remove the shock absorber lower attaching nut or nut and bolt and swing the lower end free of the mounting bracket or stud on the axle housing or spring plate.

5. Remove the upper attaching bolt and/or nut(s) and remove the shock absorber.

6. Installation is the reverse of the removal procedure. Tighten the lower attaching nut or nut and bolt to 47–58 ft. lbs. (64–78 Nm) on B Series Pick-Up, 56–76 ft. lbs. (76–103 Nm) on MPV or 41–53 ft. lbs. (55–72 Nm) on Navajo.

7. Tighten the upper attaching bolt and/or nut(s) to 47–58 ft. lbs. (64–78 Nm) on B Series Pick-Up, 56–76 ft. lbs. (76–103 Nm) on MPV or 15–21 ft. lbs. (21–29 Nm) on Navajo.

Coil Springs

REMOVAL AND INSTALLATION

MPV

1. Raise and safely support the vehicle. Remove the splash shield.

2. Remove the stabilizer bar.

3. Remove the nut and disconnect the height sensor from the rear axle.

4. Remove the bolt attaching the parking brake cable bracket.

5. Support the rear axle housing with a jack. Raise the jack slightly to take the load off the shock absorbers.

6. Remove the attaching bolts and nuts and disconnect the shock absorbers from the lower axle housing.

7. Slowly lower the axle housing until the spring tension is relieved. Remove the coil springs.

8. Remove the spring seats and bump stopper, if equipped.

To install:

9. Install the upper and lower spring seats and the bump stopper, if removed.

10. Install the coil springs, making sure the larger diameter coil is toward the axle housing.

11. Raise the axle housing enough to connect the shock absorbers. Install the attaching bolts and nuts and tighten to 56–76 ft. lbs. (76–103 Nm). Remove the jack.

12. Install the bolt attaching the parking brake cable bracket and the nut attaching the height sensor.

13. Install the stabilizer bar. Tighten the link bolt nut until 0.28 in. (7mm) of thread is exposed at the top of the link bolt. Do not tighten the stabilizer bar bushing bracket bolts at this time.

14. Lower the vehicle. With the vehicle unladen, tighten the stabilizer bar bushing bracket bolts to 23–38 ft. lbs. (34–51 Nm).

15. Install the splash shield.

Leaf Springs

REMOVAL AND INSTALLATION

B Series Pick-Up

2WD

1. Raise and safely support the vehicle. Remove the wheel and tire assembly.

2. Support the rear axle housing with a jack. Raise the jack slightly to take the load off the shock absorber.

3. Remove the nut and washers and slide the shock absorber off of the spring clamp stud. Lower the jack just enough to relieve the spring tension.

4. Remove the nuts and washers and remove the U-bolts, spring clamp and rubber stopper.

5. Support the leaf spring and remove the attaching bolts and the spring pin at the front of the leaf spring.

6. Remove the nuts and washers and remove the shackle pin and shackle plate at the rear of the leaf spring.

7. Remove the leaf spring and bushings.

8. Installation is the reverse of the removal procedure. Tighten the shackle pin nuts to 43–58 ft. lbs. (59–78 Nm). Tighten the spring pin attaching bolts to 12–17 ft. lbs. (16–23 Nm) and the spring pin nut to 58–72 ft. lbs. (78–98 Nm).

9. Tighten the U-bolt nuts to 47–58 ft. lbs. (64–78 Nm) and the lower shock absorber mounting nut to 47–58 ft. lbs. (64–78 Nm).

4WD

1. Raise and safely support the vehicle. Remove the wheel and tire assembly.

2. Support the rear axle housing with a jack. Raise the jack slightly to take the load off the shock absorber.

3. Remove the nut and bolt and disconnect the shock absorber from the axle housing. Lower the jack just enough to relieve the spring tension.

4. Remove the nuts and washers and remove the U-bolts, spring clamp and rubber stopper.

5. Remove the attaching bolts and the spring pin at the front of the leaf spring.

6. Remove the nuts and washers and remove the shackle pin and shackle plate at the rear of the leaf spring.

7. Lower the axle enough to allow clearance for the leaf spring to be removed. Remove the leaf spring and bushings.

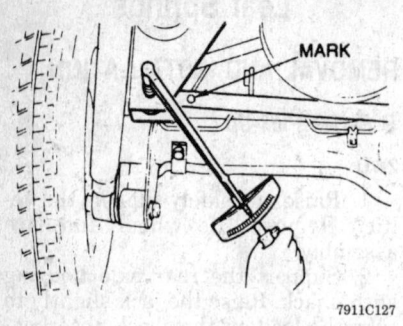

Lateral rod positioning — MPV

7911C127

8. Installation is the reverse of the removal procedure. Tighten the shackle pin nuts to 43–58 ft. lbs. (59–78 Nm). Tighten the spring pin attaching bolts to 12–17 ft. lbs. (16–23 Nm) and the spring pin nut to 58–72 ft. lbs. (78–98 Nm).

9. Tighten the U-bolt nuts to 88–101 ft. lbs. (120–137 Nm) and the lower shock absorber mounting bolt and nut to 47–58 ft. lbs. (64–78 Nm).

Navajo

1. Raise the vehicle and safely support on the frame until the weight is off the rear spring, with the tires still touching the floor.

2. Remove the nuts from the spring U-bolts and drive the U-bolts from the U-bolt plate.

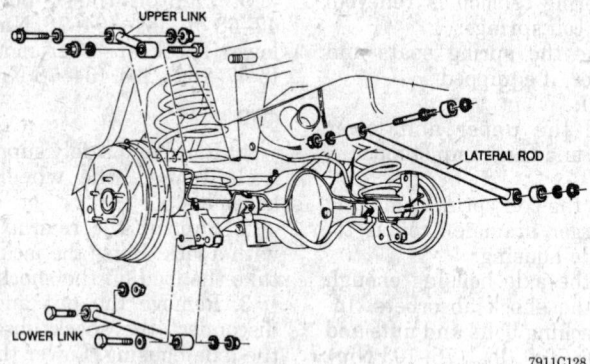

Rear control arm installation — MPV

7911C128

3. Remove the spring-to-bracket nut and bolt at the front of the spring.

4. Remove the shackle upper and lower nuts and bolts at the rear of the spring. Remove the spring and shackle assembly from the rear shackle bracket.

To install:

5. Position the spring in the shackle and install the upper shackle-to-spring bolt and nut with the bolt head facing outboard.

6. Position the front end of the spring in the bracket and install the bolt and nut. Position the shackle in the rear bracket and install the bolt and nut.

7. Position the spring on the bottom of the axle with the spring tie bolt centered in the hole provided in the seat.

8. Install the spring U-bolts, U-bolt plate and nuts and lower the vehicle. Tighten the spring U-bolt nuts to 88–108 ft. lbs. (119–146 Nm), the front spring bolt and nut to 64–91 ft. lbs. (87–123 Nm) and the rear shackle bolts and nuts to 75–115 ft. lbs. (100–155 Nm).

Rear Control Arms

REMOVAL AND INSTALLATION

MPV

LATERAL ROD

1. Raise and safely support the vehicle.

2. Support the axle housing with a jack.

3. Remove the lateral rod-to-chassis stud bolt and nut and the lateral rod-to-axle housing nut.

4. Remove the lateral rod.

5. Installation is the reverse of the removal procedure. Make sure the lateral rod is installed with the identification mark toward the body.

6. Tighten the lateral rod-to-axle housing nut to 108–127 ft. lbs. (146–167 Nm). Tighten the lateral rod-to-chassis stud bolt and nuts to 94–127 ft. lbs. (128–167 Nm).

UPPER CONTROL ARMS

1. Raise and safely support the vehicle.

2. Support the axle housing with a jack.

3. Remove the upper control arm-to-chassis bolt and nut and the upper control arm-to-axle housing bolt and nut.

4. Remove the upper control arm.

5. Installation is the reverse of the removal procedure. Tighten the upper control arm attaching bolts and nuts to 94–127 ft. lbs. (128–167 Nm).

LOWER CONTROL ARMS

1. Raise and safely support the vehicle.

2. Support the axle housing with a jack.

3. Remove the lower control arm-to-chassis bolt and nut and the lower control arm-to-axle housing bolt and nut.

4. Remove the lower control arm.

5. Installation is the reverse of the removal procedure. Tighten the upper control arm attaching bolts and nuts to 101–127 ft. lbs. (137–167 Nm).

SPECIFICATIONS

ENGINE IDENTIFICATION

Year	Model	Engine Displacement Liters (cc)	Engine Series (ID/VIN)	Fuel System	No. of Cylinders	Engine Type
1991	Truck	2.4 (2350)	4G64	MFI	4	SOHC
	Truck	3.0 (2972)	6G72	MFI	4	SOHC
	Montero	2.6 (2555)	G54B	2 BBL	4	SOHC
	Montero	3.0 (2972)	6G72	MFI	6	SOHC
1992	Expo	1.8 (1834)	4G93	MFI	4	SOHC
	Expo	2.4 (2350)	4G64	MFI	4	SOHC
	Truck	2.4 (2350)	4G64	MFI	4	SOHC
	Truck	3.0 (2972)	6G72	MFI	4	SOHC
	Montero	3.0 (2972)	6G72	MFI	6	SOHC
1993	Expo	1.8 (1834)	4G93	MFI	4	SOHC
	Expo	2.4 (2350)	4G64	MFI	4	SOHC
	Truck	2.4 (2350)	4G64	MFI	4	SOHC
	Truck	3.0 (2972)	6G72	MFI	4	SOHC
	Montero	3.0 (2972)	6G72	MFI	6	SOHC
1994-95	Expo	1.8 (1834)	4G93	MFI	4	SOHC
	Expo	2.4 (2350)	4G64	MFI	4	SOHC
	Truck	2.4 (2350)	4G64	MFI	4	SOHC
	Truck	3.0 (2972)	6G72	MFI	4	SOHC
	Montero	3.0 (2972)	6G72	MFI	6	SOHC
	Montero	3.5 (3497)	6G74	MFI	6	DOHC

MFI - Multiport fuel injection
SOHC - Single overhead camshaft
DOHC - Double overhead camshaft

7911dC02

GENERAL ENGINE SPECIFICATIONS

Year	Engine ID/VIN	Engine Displacement Liters (cc)	Fuel System Type	Net Horsepower @ rpm	Net Torque @ rpm (ft. lbs.)	Bore x Stroke (in.)	Compression Ratio	Oil Pressure @ rpm
1991	4G64	2.4 (2350)	MFI	116@5000	136@3500	3.41x3.94	8.5:1	41@2000
	G54B	2.6 (2555)	2 BBL	109@5000	142@3000	3.59x3.86	8.7:1	50-64@2000
	6G72	3.0 (2972)	MFI	151@5000	174@4000	3.59x2.99	8.9:1	30-80@2000
1992	4G93	1.8 (1834)	MFI	113@6000	116@4500	3.19x3.50	9.5:1	41@2000
	4G63	2.0 (1997)	MFI	102@5000	116@4500	3.35x3.46	8.5:1	41@2000
	6G72	3.0 (2972)	MFI	151@5000	174@4000	3.59x2.99	8.9:1	30-80@2000
1993	4G93	1.8 (1834)	MFI	113@6000	116@4500	3.19x3.50	9.5:1	41@2000
	4G64	2.4 (2350)	MFI	116@5000	136@3500	3.41x3.94	8.5:1	41@2000
	6G72	3.0 (2972)	MFI	151@5000	174@4000	3.59x2.99	8.9:1	30-80@2000
1994-95	4G93	1.8 (1834)	MFI	113@6000	116@4500	3.19x3.50	9.5:1	41@2000
	4G64	2.4 (2350)	MFI	136@5500	145@4250	3.41x3.94	9.5:1	41@2000
	6G72	3.0 (2972)	MFI	151@5000	174@4000	3.59x2.99	8.9:1	30-80@2000
	6G74	3.5 (3496)	MFI	215@5500	228@3000	3.66x3.38	9.5:1	30-80@2000

MFI - Multiport fuel injection

7911dC03

GASOLINE ENGINE TUNE-UP SPECIFICATIONS

Year	Engine ID/VIN	Engine Displacement Liters (cc)	Spark Plugs Gap (in.)	Ignition Timing (deg.) MT	AT	Fuel Pump (psi)	Idle Speed (rpm) MT	AT	Valve Clearance In.	Ex.
1991	4G64	2.4 (2350)	0.039-0.043	5B	5B	38	750	750	HYD	HYD
	G54B	2.6 (2555)	0.039-0.043	7B	7B	2.8-4.2	800	800	HYD	HYD
	6G72	3.0 (2972)	0.039-0.043	5B	5B	38	700	700	HYD	HYD
1992	4G93	1.8 (1834)	0.039-0.043	5B	5B	38	750 ①	750 ①	0.008	0.012
	4G63	2.0 (1997)	0.039-0.043	5B	5B	38	750	750	HYD	HYD
	6G72	3.0 (2972)	0.039-0.043	5B	5B	38	700	700	HYD	HYD
1993	4G93	1.8 (1834)	0.039-0.043	5B	5B	38	750 ①	750 ①	0.008	0.012
	4G64	2.4 (2350)	0.039-0.043	5B	5B	38	750	750	HYD	HYD
	6G72	3.0 (2972)	0.039-0.043	5B	5B	38	700	700	HYD	HYD
1994-95	4G93	1.8 (1834)	0.039-0.043	5B	5B	38	750	750	0.008	0.012
	4G64	2.4 (2350)	0.039-0.043	5B	5B	38	750	750	HYD	HYD
	6G72	3.0 (2972)	0.039-0.043	5B	5B	38	700	700	HYD	HYD
	6G74	3.5 (3496)	0.039-0.043	5B	5B	38	700	700	HYD	HYD

NOTE: The Vehicle Emission Control Information label reflects changes made during production. The label figures must be used if they differ from above

B - Before top dead center

HYD - Hydraulic

① California 700 rpm

7911dC04

FIRING ORDERS

NOTE: To avoid confusion, always replace spark plug wires one at a time.

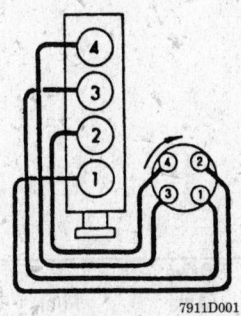

7911D001

2.4L engine
Engine Firing Order: 1 —
3 — 4 — 2
Distributor Rotation:
Clockwise

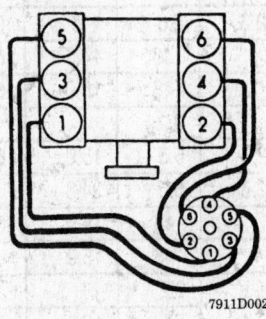

7911D002

3.0L engine
Engine Firing Order: 1 — 2 —
3 — 4 — 5 — 6
Distributor Rotation:
Counterclockwise

CAPACITIES

Year	Model	Engine ID/VIN	Engine Displacement Liters (cc)	Engine Oil with Filter	Transmission (pts.) 4-Spd	5-Spd	Auto.	Transfer Case (pts.)	Drive Axle Front (pts.)	Rear (pts.)	Fuel Tank (gal.)	Cooling System (qts.)
1991	Truck	4G64	2.4 (2350)	4.2	-	4.9	14.8	-	-	3.2	1	4
	Truck	6G72	3.0 (2972)	5.0	5.3	5.3	-	4.7	5.5	5.5	2	8.9
	Montero	G54B	2.6 (2555)	6.0	4.7	4.7	-	4.7	2.3	3.8	15.9	8.4
	Montero	6G72	3.0 (2972)	6.0	5.3	5.3	15.2	4.7	2.3	5.5	3	9.5
1992	Expo	4G93	1.8 (1834)	4.0	-	5	6	1.2	-	1.48	14.5	6.3
	Expo	4G64	2.4 (2350)	4.0	-	4.8	6	1.2	-	1.48	15.9	6.8
	Truck	4G64	2.4 (2350)	4.2	-	4.9	14.8	-	-	3.2	1	4
	Truck	6G72	3.0 (2972)	5.0	5.3	5.3	-	4.7	2.4	5.5	15.9	8.9
	Montero	6G72	3.0 (2972)	5.5	5.3	5.3	15.2	4.8	2.6	5.5	24.3	10.0
1993	Expo	4G93	1.8 (1834)	4.0	-	5	6	1.2	-	1.48	14.5	6.3
	Expo	4G64	2.4 (2350)	4.1	-	4.8	6	1.2	-	1.48	15.9	6.8
	Truck	4G64	2.4 (2350)	4.2	-	4.9	14.8	-	-	3.2	1	4
	Truck	6G72	3.0 (2972)	5.0	-	5.3	-	4.7	2.4	5.5	15.9	8.9
	Montero	6G72	3.0 (2972)	5.5	-	5.3	15.2	4.8	2.6	5.5	24.3	10.0
1994-95	Expo	4G93	1.8 (1834)	4.0	-	5	6	1.2	-	1.48	14.5	6.3
	Expo	4G64	2.4 (2350)	4.1	-	4.8	6	1.2	-	1.48	15.9	6.8
	Truck	4G64	2.4 (2350)	4.2	-	4.9	14.8	-	-	3.2	1	4
	Truck	6G72	3.0 (2972)	5.0	-	5.3	-	4.7	2.4	5.5	15.9	8.9
	Montero	6G72	3.0 (2972)	5.5	-	5.3	15.2	4.8	2.6	5.5	24.3	10.0
	Montero	6G74	3.5 (3497)	5.5	-	5.3	15.2	5.2	2.6	5.5	24.3	10.0

NA - Not Available
1 Std. body: 13.7 gals.
   Long body: 18.2 gals.
2 Std. body: 15.9 gals.
   Long body: 19.8 gals.
3 2 door: 19.8 gals.
   4 door: 14.3 gals.
4 Manual transmission: 6.3 qts.
   Automatic transmission: 6.4 qts.

5 FWD: 3.8 pts.
   AWD: 4.8 pts.
6 FWD: 12.8 pts.
   AWD: 13.8 pts.
7 5 speed: 0.58 pts.
   6 speed: 1.26 pts.

7911dC06

CAMSHAFT SPECIFICATIONS

All measurements given in inches.

Year	Engine ID/VIN	Engine Displacement Liters (cc)	Journal Diameter					Elevation		Bearing Clearance	Camshaft End Play
			1	2	3	4	5	In.	Ex.		
1991	4G64	2.4 (2350)	1.339	1.339	1.339	1.339	1.339	1.669	1.669	0.002-0.004	0.002-0.008
	G54B	2.6 (2555)	1.339	1.339	1.339	1.339	1.339	1.671	1.671	0.002-0.004	0.002-0.008
	6G72	3.0 (2972)	1.339	1.339	1.339	1.339	-	1.624	1.624	0.002-0.004	0.002-0.008
1992	4G93	1.8 (1834)	1.769	1.769	1.769	1.769	1.769	1.487	1.500	0.002-0.004	0.002-0.008
	4G63	2.0 (1997)	1.339	1.339	1.339	1.339	1.339	[1]	[1]	0.002-0.004	0.002-0.008
	6G72	3.0 (2972)	1.339	1.339	1.339	1.339	-	1.374	1.374	0.002-0.004	0.002-0.008
1993	4G93	1.8 (1834)	1.769	1.769	1.769	1.769	1.769	1.487	1.500	0.002-0.004	0.002-0.008
	4G64	2.4 (2350)	1.339	1.339	1.339	1.339	1.339	1.669	1.669	0.002-0.004	0.002-0.008
	6G72	3.0 (2972)	1.339	1.339	1.339	1.339	-	1.374	1.374	0.002-0.004	0.002-0.008
1994-95	4G93	1.8 (1834)	1.769	1.769	1.769	1.769	1.769	1.487	1.500	0.002-0.004	0.002-0.008
	4G64	2.4 (2350)	1.339	1.339	1.339	1.339	1.339	1.669	1.669	0.002-0.004	0.002-0.008
	6G72	3.0 (2972)	1.339	1.339	1.339	1.339	-	1.374	1.374	0.002-0.004	0.002-0.008
	6G74	3.5 (3496)	1.022	1.022	1.022	1.022	-	1.366	1.355	0.002-0.004	0.002-0.008

1 ID mark D: 1.669
   ID mark AR: 1.753

7911dC07

CRANKSHAFT AND CONNECTING ROD SPECIFICATIONS

All measurements are given in inches.

Year	Engine ID/VIN	Engine Displacement Liters (cc)	Crankshaft				Connecting Rod		
			Main Brg. Journal Dia.	Main Brg. Oil Clearance	Shaft End-play	Thrust on No.	Journal Diameter	Oil Clearance	Side Clearance
1991	4G64	2.4 (2350)	2.244	0.001-0.002	0.002-0.007	3	1.772	0.001-0.002	0.004-0.010
	G54B	2.6 (2555)	2.362	0.001-0.002	0.002-0.007	3	2.087	0.001-0.002	0.004-0.010
	6G72	3.0 (2972)	2.362	0.001-0.002	0.002-0.010	3	1.969	0.001-0.002	0.004-0.010
1992	4G93	1.8 (1834)	1.969	0.001-0.002	0.002-0.010	3	1.772	0.001-0.002	0.004-0.010
	4G63	2.0 (1997)	2.244	0.001-0.002	0.002-0.007	3	1.772	0.001-0.002	0.004-0.010
	6G72	3.0 (2972)	2.362	0.001-0.002	0.002-0.010	3	1.969	0.001-0.002	0.004-0.010
1993	4G93	1.8 (1834)	1.969	0.001-0.002	0.002-0.010-	3	1.772	0.001-0.002	0.004-0.010
	4G64	2.4 (2350)	2.244	0.001-0.002	0.002-0.007	3	1.772	0.001-0.002	0.004-0.010
	6G72	3.0 (2972)	2.362	0.001-0.002	0.002-0.010-	3	1.969	0.001-0.002	0.004-0.010
1994-95	4G93	1.8 (1834)	1.969	0.001-0.002	0.002-0.010	3	1.772	0.001-0.002	0.004-0.010
	4G64	2.4 (2350)	2.244	0.001-0.002	0.002-0.007	3	1.772	0.001-0.002	0.004-0.010
	6G72	3.0 (2972)	2.362	0.001-0.002	0.002-0.010	3	1.969	0.001-0.002	0.004-0.010
	6G74	3.5 (3496)	2.520	0.001-0.002	0.002-0.010	3	2.165	0.001-0.002	0.004-0.010

7911dC08

VALVE SPECIFICATIONS

Year	Engine ID/VIN	Engine Displacement Liters (cc)	Seat Angle (deg.)	Face Angle (deg.)	Spring Test Pressure (lbs. @ in.)	Spring Installed Height (in.)	Stem-to-Guide Clearance (in.)		Stem Diameter (in.)	
							Intake	Exhaust	Intake	Exhaust
1991	4G64	2.4 (2350)	44-44.5	45-45.5	73@1.59	1.59	0.001-0.002	0.002-0.004	0.315	0.315
	G54B	2.6 (2555)	44-44.5	45-45.5	73@1.59	1.59	0.001-0.002	0.002-0.004	0.314	0.313
	6G72	3.0 (2972)	44-44.5	45-45.5	74@1.59	1.59	0.001-0.002	0.002-0.004	0.314	0.313
1992	4G93	1.8 (1834)	44-44.5	45-45.5	49@1.74	1.74	0.001-0.002	0.001-0.002	0.236	0.236
	4G63	2.0 (1997)	44-44.5	45-45.5	73@1.59	1.59	0.001-0.002	0.002-0.004	0.315	0.311
	6G72	3.0 (2972)	44-44.5	45-45.5	72.5@1.59	1.59	0.001-0.002	0.002-0.004	0.315	0.311
1993	4G93	1.8 (1834)	44-44.5	45-45.5	49@1.74	1.74	0.001-0.002	0.001-0.002	0.236	0.236
	4G63	2.0 (1997)	44-44.5	45-45.5	73@1.59	1.59	0.001-0.002	0.002-0.004	0.315	0.311
	6G72	3.0 (2972)	44-44.5	45-45.5	72.5@1.59	1.59	0.001-0.002	0.002-0.004	0.315	0.311
1994-95	4G93	1.8 (1834)	44-44.5	45-45.5	49@1.74	1.74	0.001-0.002	0.001-0.002	0.236	0.236
	4G64	2.4 (2350)	44-44.5	45-45.5	73@1.59	1.59	0.001-0.002	0.002-0.004	0.315	0.311
	6G72	3.0 (2972)	44-44.5	45-45.5	72.5@1.59	1.59	0.001-0.002	0.002-0.004	0.315	0.311
	6G74	3.5 (3496)	44-44.5	45-45.5	52.9@1.49	1.49	0.001-0.002	0.002-0.004	0.260	0.256

7911dC09

PISTON AND RING SPECIFICATIONS

All measurements are given in inches.

Year	Engine ID/VIN	Engine Displacement Liters (cc)	Piston Clearance	Ring Gap			Ring Side Clearance		
				Top Compression	Bottom Compression	Oil Control	Top Compression	Bottom Compression	Oil Control
1991	4G63	2.0 (1997)	0.001-0.002	0.010-0.018	0.018-0.024	0.008-0.028	0.001-0.003	0.001-0.003	-
	4G64	2.4 (2350)	0.001-0.002	0.010-0.016	0.010-0.016	0.008-0.028	0.001-0.003	0.001-0.002	-
	G54B	2.6 (2555)	0.001-0.002	0.012-0.018	0.010-0.016	0.008-0.028	0.002-0.004	0.001-0.002	-
	6G72	3.0 (2972)	0.001-0.002	0.012-0.018	0.010-0.016	0.008-0.028	0.001-0.004	0.001-0.002	-
1992	4G93	1.8 (1834)	0.001-0.002	0.010-0.016	0.016-0.022	0.008-0.024	0.001-0.003	0.001-0.002	-
	4G63	2.0 (1997)	0.001-0.002	0.010-0.016	0.008-0.016	0.008-0.028	0.001-0.003	0.001-0.002	-
	6G72	3.0 (2972)	0.001-0.002	0.012-0.018	0.018-0.024	0.008-0.024	0.001-0.003	0.001-0.002	-
1993	4G93	1.8 (1834)	0.001-0.002	0.010-0.016	0.016-0.022	0.008-0.024	0.001-0.003	0.001-0.002	-
	4G63	2.0 (1997)	0.001-0.002	0.010-0.014	0.018-0.028	0.004-0.016	0.001-0.002	0.001-0.002	-
	6G72	3.0 (2972)	0.001-0.002	0.012-0.018	0.018-0.024	0.008-0.024	0.001-0.003	0.001-0.002	-
1994-95	4G93	1.8 (1834)	0.001-0.002	0.010-0.016	0.016-0.022	0.008-0.024	0.001-0.003	0.001-0.002	-
	4G64	2.4 (2350)	0.001-0.002	0.010-0.016	0.018-0.024	0.008-0.024	0.001-0.002	0.001-0.002	-
	6G72	3.0 (2972)	0.001-0.002	0.012-0.018	0.018-0.024	0.008-0.024	0.001-0.003	0.001-0.002	-
	6G74	3.5 (3496)	0.001-0.002	0.012-0.018	0.018-0.024	0.004-0.014	0.001-0.003	0.001-0.002	-

7911dC10

TORQUE SPECIFICATIONS
All readings in ft. lbs.

Year	Engine ID/VIN	Engine Displacement Liters (cc)	Cylinder Head Bolts	Main Bearing Bolts	Rod Bearing Bolts	Crankshaft Damper Bolts	Flywheel Bolts	Manifold Intake	Manifold Exhaust	Spark Plugs	Lug Nut
1991	4G64	2.4 (2350)	65-72	37-39	37-83	80-94	94-101	11-14	11-14	18	87-101
	G54B	2.6 (2555)	65-72 [1]	55-61	33-34	80-94	94-101	11-15	11-15	18	72-87
	6G72	3.0 (2972)	65-72	65-72	37-38	130-137	53-55	11-14	11-16	18	[2]
1992	4G93	1.8 (1834)	[4]	18 [3]	14.5 [3]	134	72	14	[5]	18	65-80
	4G63	2.0 (1997)	80	38	14	87	98	13	13	18	87-101
	6G72	3.0 (2972)	80	57	38	136	54	10	14	18	[2]
1993	4G93	1.8 (1834)	[4]	18 [3]	14.5 [3]	134	72	14	[5]	18	65-80
	4G63	2.0 (1997)	[4]	18 [3]	14.5 [3]	87	98	26	20	18	[2]
	6G72	3.0 (2972)	80	57	38	136	54	10	14	18	[2]
1994-95	4G93	1.8 (1834)	[4]	18 [3]	14.5 [3]	134	72	14	[5]	18	65-80
	4G64	2.4 (2350)	[4]	18 [3]	14.5 [3]	87	98	13	13	18	65-80
	6G72	3.0 (2972)	80	57	38	136	54	10	14	18	[2]
	6G74	3.5 (3496)	80	54	38	136	54	10	14	18	72-87

1 M8 bolts: 11-16 ft. lbs.
2 Truck: 87-101 ft. lbs.
   Galant: 65-80 ft. lbs.
   Montero: 72-87 ft. lbs.
3 Torque to valve plus an additional 1/4 turn
4 Step 1: 54 ft. lbs.
   Step 2: Loosen completely, then torque to 14.5 ft. lbs. plus an additional 1/4 turn
   Step 3: Torque an additional 1/4 turn
5 M10: 22 ft. lbs.
   M8: 13 ft. lbs.

7911dC11

BRAKE SPECIFICATIONS

All measurements in inches unless noted

Year	Model		Master Cylinder Bore	Brake Disc Original Thickness	Brake Disc Minimum Thickness	Brake Disc Maximum Runout	Brake Drum Diameter Original Inside Diameter	Brake Drum Diameter Max. Wear Limit	Brake Drum Diameter Maximum Machine Diameter	Minimum Lining Thickness Front	Minimum Lining Thickness Rear
1991	Montero		0.938	0.866	0.803	0.006	10.00	10.08	-	0.079	0.039
	Truck		0.938	0.866	0.803	0.006	10.00	10.08	-	0.079	0.039
1992	Expo		0.938 [1]	2	6	0.003	4	5	-	0.079	0.079 [8]
	Montero		0.938	3	6	0.003	-	7.80	-	0.079	0.079 [7]
	Truck		0.938	0.866	0.803	0.006	10.00	10.08	-	0.079	0.039
1993	Expo		0.938 [1]	2	6	0.003	4	5	-	0.079	0.079 [8]
	Montero		0.938	3	6	0.003	-	7.80	-	0.079	0.079 [7]
	Truck		0.938	0.866	0.866	0.006	10.00	10.08	-	0.079	0.039
1994-95	Expo		0.938 [1]	2	6	0.003	4	5	-	0.079	0.079 [8]
	Montero		0.938	3	6	0.003	-	7.80	-	0.079	0.079 [7]
	Truck		0.938	0.866	0.803	0.006	10.00	10.08	-	0.079	0.039

1 With ABS: 1.000
2 Front: 0.940; Rear: 0.390
3 Front: 0.940; Rear: 0.710
4 8 inch drum: 8.00
   9 inch drum: 9.00
5 8 inch drum: 8.10
   9 inch drum: 9.10
6 Front: 0.880; Rear: 0.330
7 Drum shoe: 0.177
8 Drum shoe: 0.040

7911dC12

ENGINE ELECTRICAL

NOTE: Disconnecting the negative battery cable on some vehicles may interfere with the functions of the on board computer systems and may require the computer to undergo a relearning process, once the negative battery cable is reconnected.

Distributor

REMOVAL

1. Disconnect the negative battery cable.
2. Disconnect the distributor pickup lead wires and vacuum hose(s), if equipped.
3. Unfasten the distributor cap retaining clips or screws and lift off the distributor cap with all ignition wires connected. Remove the coil wire if necessary.
4. Matchmark the rotor to the distributor housing and the distributor housing to the engine.

NOTE: Do not crank the engine during this procedure. If the engine is cranked, the matchmark must be disregarded.

5. Remove the retaining nut and remove the distributor from the engine.

INSTALLATION

Timing Not Disturbed

1. Install a new distributor housing O-ring.
2. Install the distributor in the engine so the rotor is aligned with the matchmark on the housing and the housing is aligned with the matchmark on the engine. Make sure the distributor is fully seated and that the distributor shaft is fully engaged.
3. Install the retaining nut finger-tight only. Connect the vacuum hose(s), if removed.
4. Connect the distributor pickup electrical harness.
5. Install the distributor cap and secure.
6. Connect the negative battery cable.
7. Check the ignition timing and adjust as required. Tighten the retaining nut.

Timing Disturbed

1. Install a new distributor housing O-ring.
2. Position the engine so No. 1 piston is on TDC of compression stroke and the timing mark on the vibration damper is aligned with **0** on the timing indicator.
3. Install the distributor so the rotor is aligned with the No. 1 ignition wire on the distributor cap. Take note that the distributor shaft is fully engaged and the housing is fully seated.

NOTE: There are distributor cap runners inside the cap on vehicles with 3.0L engine. Make sure the rotor is pointing to where the No. 1 runner originates inside the cap and not where the No. 1 ignition wire plugs into the cap.

4. Install the retaining nut finger-tight only. Connect the vacuum hose(s), if removed.
5. Connect the distributor electrical harness.
6. Install the distributor cap and secure.
7. Connect the negative battery cable.
8. Adjust the ignition timing and tighten the retaining nut.

Ignition Timing

ADJUSTMENT

1. Start the engine, set the parking brake and run the engine until normal operating temperature is reached. Keep all lights and accessories **OFF** and the transmission in neutral.
2. Without disconnecting the connector, insert a paper clip into the tachometer terminal. Connect the red lead of a tachometer to the paper clip and connect the black lead to a

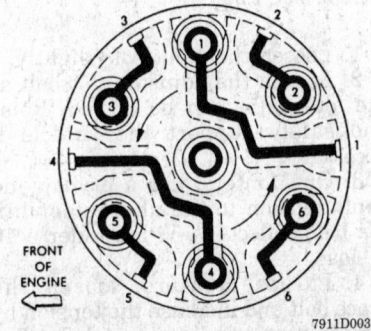

FRONT OF ENGINE

7911D003

Distributor cap terminal routing viewed from the top of the cap — 3.0L engine

ground. Set the idle speed to specifications.
3. Turn the engine **OFF**. Remove the water-proof cover from the ignition timing adjusting connector. Connect a jumper wire from the ignition timing adjusting terminal to ground.
4. Connect a conventional timing light to No. 1 cylinder spark plug wire. Start the engine and allow to idle.
5. Aim the timing light at the timing scale.
6. Loosen the distributor nut to allow for distributor rotation.
7. Turn the distributor in the proper direction until the specified timing is reached. Tighten the retainer nut and recheck the timing. Turn the engine **OFF**.
8. Remove the jumper wire from the ignition timing adjusting terminal and install the water-proof cover.
9. Start the engine and check the actual ignition timing. This reading should be 3 degrees more than basic timing for the 2.4L engine or 10 degrees more than basic timing for the 3.0L engine.

NOTE: The actual timing may fluctuate according to the control mode of the engine control unit; this is a normal condition.

10. Turn the engine **OFF** and remove all test equipment.

Alternator

PRECAUTIONS

Several precautions must be observed when working with the alternator to avoid damaging the unit.

If the battery is removed for any reason, make sure it is reconnected with the correct polarity. Reversing the battery connections may result in damage to the one-way rectifiers.

When utilizing a booster battery as a starting aid, always connect the positive to positive terminals and the negative terminal from the booster battery to a good engine ground on the vehicle being started.

Never use a fast charger as a booster to start vehicles.

Disconnect the battery cables when charging the battery with a fast charger.

Never attempt to polarize the alternator.

Do not use test lights of more than 12 volts when checking diode continuity.

Do not short across or ground any of the alternator terminals.

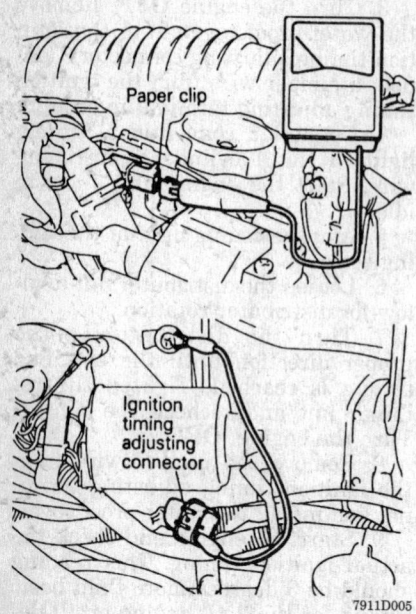

Tachometer terminal and ignition timing adjusting connector — Montero with 3.0L engine

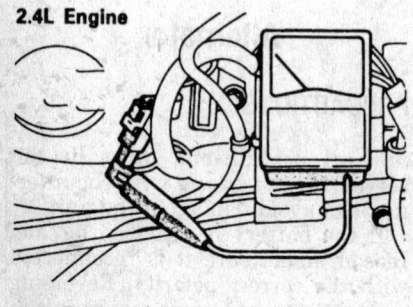

2.4L Engine

3.0L Engine

Tachometer connector — Pick-Up

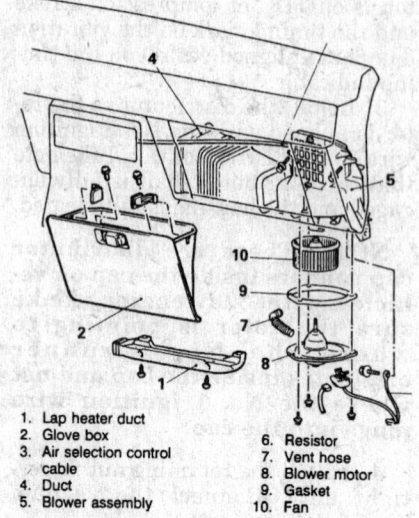

1. Lap heater duct
2. Glove box
3. Air selection control cable
4. Duct
5. Blower assembly

6. Resistor
7. Vent hose
8. Blower motor
9. Gasket
10. Fan

Blower motor removal and installation

The polarity of the battery, alternator and regulator must be matched and considered before making any electrical connections within the system.

Never separate the alternator on an open circuit. Make sure all connections within the circuit are clean and tight.

Disconnect the battery ground terminal when performing any service on electrical components.

Disconnect the battery if arc welding is to be done on the vehicle.

BELT TENSION ADJUSTMENT

Except 3.0L Engine

1. Loosen the pivot bolt slightly.
2. Loosen the adjuster slot bolt so the alternator can be moved. Raise and safely support the vehicle if necessary.
3. On Montero, use a prybar and apply tension to the alternator until the belt deflects ¼–½ in. under a 10 lb. load.
4. Except Montero, loosen the brace bolt and increase the tension by tightening the adjustment bolt. The belt should deflect ¼–½ in. under a 10 lb. load.

5. Torque the adjuster strap bolt to 10 ft. lbs. (15 Nm). and the pivot bolt to 16 ft. lbs. (23 Nm).

3.0L Engine

1. Loosen the tensioner pulley locknut.
2. Turn the adjusting bolt until the belt deflects ¼–½ in. under a 10 lb. load.
3. Tighten the locknut.

REMOVAL AND INSTALLATION

1. Disconnect the negative battery cable.
2. Remove the alternator cover, if equipped.
3. Remove the alternator belt.
4. Remove the alternator brace bolt(s), nut(s) and applicable spacers.
5. Remove the alternator from the mounting bracket, label and disconnect all wires from the rear of the unit.

To install:
6. Connect all wiring to their proper terminals on the rear of the alternator.
7. Position the alternator in the mounting bracket.
8. Install the alternator brace bolt(s), nut(s) and applicable spacers.
9. Install the alternator belt and adjust the tension as required.
10. Install the alternator cover, if equipped.
11. Connect the negative battery cable and check the alternator for proper operation.

Starter

REMOVAL AND INSTALLATION

1. Disconnect the negative battery cable.
2. Raise the vehicle and support safely.
3. Remove the starter cover, if equipped.
4. Label and disconnect the wiring to the starter motor.
5. Remove the starter mounting bolts.
6. Remove the starter motor from the vehicle.
7. The installation is the reverse of the removal procedure. Torque the mounting bolts to 20–25 ft. lbs. (27–34 Nm).

CHASSIS ELECTRICAL

Heater Blower Motor

REMOVAL AND INSTALLATION

Without Air Conditioning

1. Disconnect the negative battery cable.
2. Remove the lap heater duct and the left side defroster duct.
3. Remove the blower motor retaining screws and remove the motor assembly from the housing. Remove the fan from the motor.
To install:
4. Inspect the gasket on housing for cracking or breaks and repair as required.
5. Install the fan to the blower motor shaft. Install the blower motor assembly to the housing and install the retaining screws.
6. Install the heater housing as follows:
   a. Install the vent hose and resistor block.
   b. Install the assembled blower housing to the vehicle.
   c. Position the ground wire and install the grounding screw.
   d. Connect the resistor block connector.
   e. Install the air distribution duct to the blower housing.
   f. Move the air selection control lever to the recirculation position. Pull the air selection control damper lever up and connect the air selection control cable to the end of the air selection damper lever. Secure the cable with the retaining clip.
7. Install the lap heater duct, glove box and stopper.
8. Connect the negative battery cable and check the blower motor for proper operation.

With Air Conditioning

1. Disconnect negative battery cable.
2. Remove the lap duct and the left side defroster duct.
3. If necessary remove the heater/air conditioner housing as follows:
   a. Open the glove box lid and release the glove box stoppers. Remove the glove box from its hinge.
   b. Use a small prying device to remove the clip that holds the air selection control cable in place.
   c. Disconnect the air selection control cable from the end of the air selection damper lever.
   d. Remove the air distribution duct from the left side of the blower assembly.
   e. Disconnect the connector to the resistor block and remove the grounding screw.
   f. Remove the blower housing attaching bolts and remove the housing from the vehicle.
   g. Remove the resistor block and vent hose from the housing.
4. Remove the blower motor retaining screws and remove the motor assembly from the housing. Remove the fan from the motor.
To install:
5. Inspect the gasket on housing for cracking or breaks and repair as required.
6. Install the fan to the blower motor shaft. Install the blower motor assembly to the housing and install the retaining screws.
7. Install the heater housing as follows:
   a. Install the vent hose and resistor block.
   b. Install the assembled blower housing to the vehicle.
   c. Position the ground wire and install the grounding screw.
   d. Connect the resistor block connector.
   e. Install the air distribution duct to the blower housing.
   f. Move the air selection control lever to the recirculation position. Pull the air selection control damper lever up and connect the air selection control cable to the end of the air selection damper lever. Secure the cable with the retaining clip.
8. Install the lap duct and the left side defroster duct.
9. Install the glove box and stopper.
10. Connect the negative battery cable and check the blower motor for proper operation.

Windshield Wiper Motor

REMOVAL AND INSTALLATION

1. Disconnect the negative battery cable.
2. Disconnect the wiper motor electrical connector.
3. Remove the wiper motor retaining bolts and pull the motor out far enough to gain access to the wiper linkage.
4. Matchmark position of wiper linkage to aid installation. Pry the wiper linkage from the motor output shaft.
5. Remove the wiper motor from the vehicle.
6. The installation is the reverse of the removal procedure.

Rear Wiper Motor

REMOVAL AND INSTALLATION

1. Disconnect the negative battery cable.
2. Tilt the wiper mount nut cover up and remove the nut. Remove the wiper arm from the shaft. Remove the wiper pivot nut and washer, if equipped.
3. Remove rear door inside trim panel and waterproof film.
4. Disconnect the wiper motor wire connector.
5. Remove the wiper motor retaining bolts and remove the motor from the vehicle.
6. The installation is the reverse of the removal procedure.

Windshield Wiper Switch

The front windshield wiper switch is part of the combination switch located on the steering column. The rear wiper switch is located on the instrument panel.

REMOVAL AND INSTALLATION

Rear Wiper Switch

1. Disconnect negative battery cable.
2. Insert a trim stick between the instrument panel and the switch housing. Pry switch out from the panel.
3. Disconnect electrical connector and service the wiper/washer switch, as required.
To install:
4. Connect the electrical connector and snap switch assembly into instrument panel.
5. Connect the negative battery cable and check operation of the wiper/washer assembly.

Instrument Cluster

REMOVAL AND INSTALLATION

1. Disconnect the negative battery cable.

2. On Pick-Up, remove the hazard flasher switch and the matching cover on the other side of the column.

3. If equipped with tilt steering, operate the tilt lever to bring the steering down. Remove the hood attaching screws and remove the hood.

4. Remove the 4 cluster attaching screws and pull out cluster. Disconnect the speedometer cable from the back of the cluster by pushing the stopper of the plug on the speedometer cable side of the connection.

5. Disconnect the electrical connectors at the cluster and remove the cluster from the vehicle.

6. To remove the speedometer, remove the cover and glass. Remove the retaining screws and remove the speedometer or gauges as required.

To install:

7. Install the speedometer or gauges into the cluster and install the retaining screws. Install the cover and the glass.

8. Connect the electrical connectors and the speedometer to the cluster and install in vehicle. Install 4 retaining screws.

9. Install the instrument cluster hood and retaining screws.

10. Install the hazard flasher switch and the matching cover on the other side of the column.

11. Connect the negative battery cable.

Radio

REMOVAL AND INSTALLATION

1. Disconnect negative battery cable.

2. Remove the heater control lever knob.

3. Remove the center console or trim panel. On Pick-up, after removal of the attaching screws, use a trim stick to remove the upper side of the center panel.

4. Remove the electrical connector of the center panel wiring harness.

5. Remove the radio front trim panel. Remove the retaining screws and the radio from the vehicle.

6. The installation is the reverse of the removal procedure.

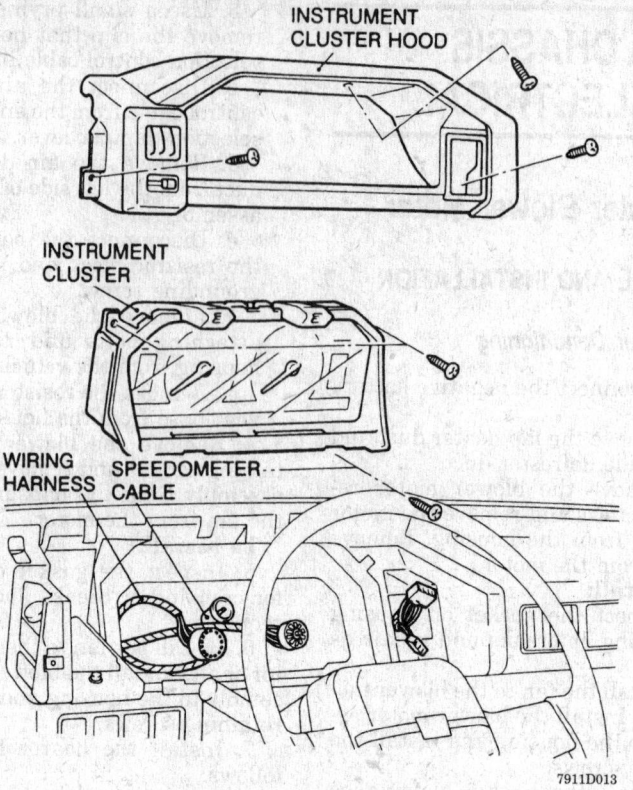

Instrument cluster mounting

Combination Switch

REMOVAL AND INSTALLATION

NOTE: The combination switch incorporates the windshield, headlight, dimmer and turn signal switches into a single switch assembly.

1. Disconnect negative battery cable.

2. Remove the steering wheel pad. Matchmark and remove the steering wheel.

3. If equipped with tilt steering column, put the column in lowest position.

4. Remove the upper and lower column covers.

5. Remove the wiring harness band and disconnect the harness connectors.

6. Remove the combination switch mounting screws and remove the switch.

7. The installation is the reverse of the removal procedure. Torque the steering wheel nut to 30 ft. lbs. (41 Nm).

Ignition Lock/Switch

REMOVAL AND INSTALLATION

1. Disconnect the negative battery cable.

2. If equipped with tilt steering column, put the column in its lowest position.

3. Remove the upper and lower column covers.

4. Remove the wiring harness band and disconnect the ignition switch harness.

5. Remove the ignition switch retaining screw, if equipped and remove the switch from the lock.

6. Using a hacksaw blade, cut a groove into the head of the ignition switch break-off bolts and remove.

7. Remove the assembly from the steering column.

To install:

8. With the key inserted in the switch, install the switch and lock assembly to the steering column. Tighten the screws gradually making sure the key does not bind.

9. If using break-off bolts, tighten them until the heads break off.

10. Connect the switch harness and check the assembly for proper operation.

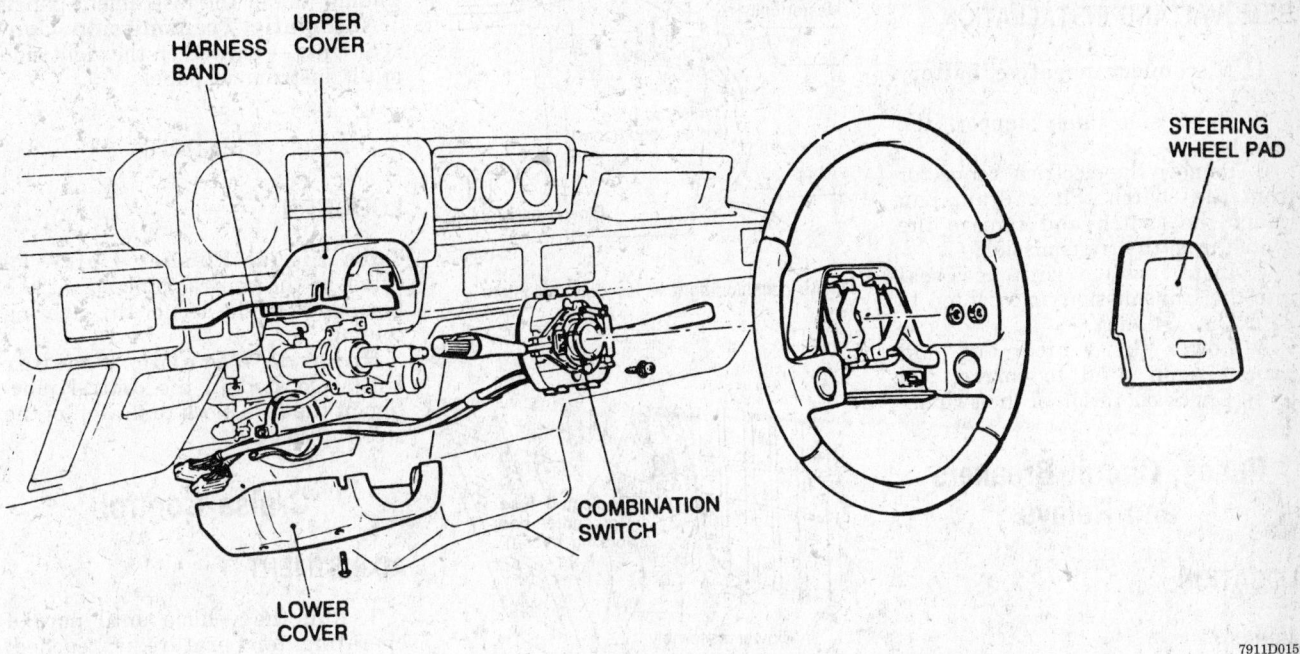

Combination switch mounting

HARNESS BAND

UPPER COVER

STEERING WHEEL PAD

COMBINATION SWITCH

LOWER COVER

7911D015

11. Install upper and lower column covers.

Stoplight Switch

ADJUSTMENT

1. Disconnect negative battery cable.
2. Loosen locknut on the brake light switch.
3. Turn the switch in the mounting bracket until the plunger does not contact the brake pedal.

4. Rotate the switch outward ½–1 revolution and secure the locknut. Reconnect the negative battery cable and check switch operation.

REMOVAL AND INSTALLATION

1. Disconnect the negative battery cable.
2. Unplug the connector to the switch.
3. Remove the locknut and remove the switch from its mount bracket.
4. The installation is the reverse of the removal procedure.

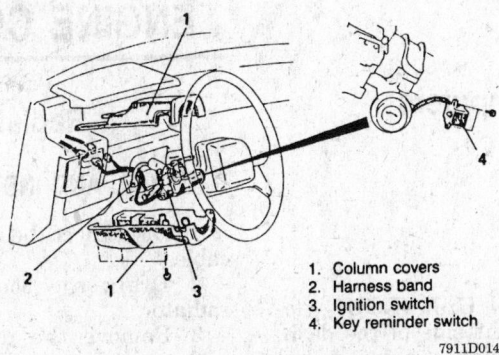

1. Column covers
2. Harness band
3. Ignition switch
4. Key reminder switch

7911D014

Ignition switch and lock assembly

Clutch Switch

ADJUSTMENT

1. Disconnect negative battery cable.
2. Loosen the locknut on the switch. Adjust the switch so the distance from the floorboard to the top of the pedal pad in a released position is 7½ in. for Montero or 6½ in. for Pick-Up.
3. Reconnect the negative battery cable. Check operation of cruise control system making sure the cruise control is not operational when the pedal is depressed.

REMOVAL AND INSTALLATION

1. Disconnect the negative battery cable.
2. Unplug the connector to the switch.
3. Remove the locknut and the switch from its mounting bracket.
4. The installation is the reverse of the removal procedure.

Neutral Safety Switch

REMOVAL AND INSTALLATION

1. Disconnect negative battery cable.
2. Raise and safely support the vehicle.
3. Remove the electrical connector from the switch. Place drain pan under the switch and remove the switch from the transmission.
4. Install switch with new seal into the transmission and tighten to 25 ft. lbs. (34 Nm).
5. Lower the vehicle and add transmission fluid to correct the level. Check operation of the switch.

Fuses, Circuit Breakers and Relays

LOCATION

Montero

Main Fuse Block/Relay Box — located on the left side of the instrument panel, covered by a removable access panel. There are also several dedicated fuses located throughout the vehicle.

Air Conditioning Compressor Power Relay — located in the engine compartment on the left fender.

Engine Control Relay — located on the right side of the instrument panel, mounted under a removable trim panel.

Pick-Up

Fuse Block/Relay Box — located to the left of the steering column, covered by a removable access panel or mounted to the left kick-panel.

Air Conditioning Compressor Power Relay — located in the engine compartment on the left fender.

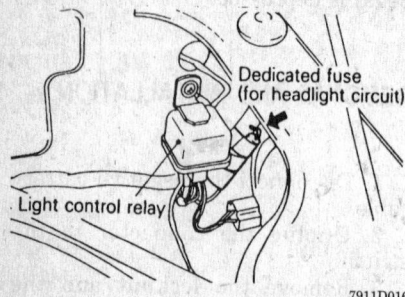

Headlight circuit fuse location — Montero

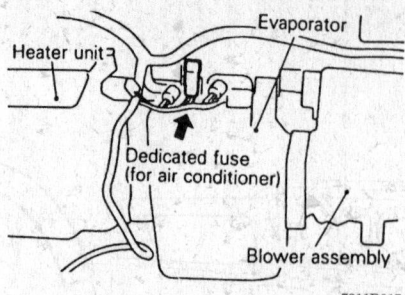

Air conditioning fuse location — Montero

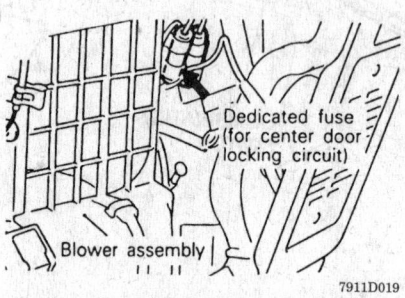

Electric door locks fuse location — Montero

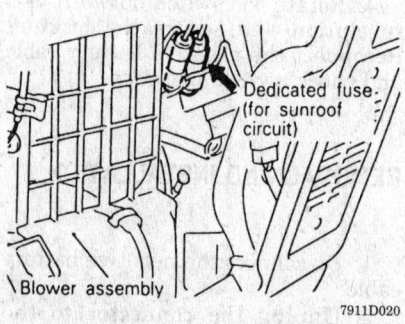

Sunroof fuse location — Montero

Engine Control Relay — located on the right side of the instrument panel behind a removable trim panel.

Computers

LOCATION

Engine Control Unit (ECU) — located on the right side of the dash, under a removable trim panel mounted to the kick-panel.

Auto-Cruise Control Unit — located on the left side of the steering column, under the instrument panel.

Automatic Transmission Control Unit — located on the right side of the instrument panel.

Flashers

LOCATION

Turn Signal Flasher Unit — located in the central junction at the fuse block to the left of the steering column.

Hazard Warning Flasher Unit — located in the central junction at the fuse block to the left of the steering column.

Cruise Control

ADJUSTMENT

1. Run the vehicle until normal operating temperature is reached. Adjust the idle speed to match the emission sticker on the vehicle.
2. Turn the engine **OFF** and remove the actuator cover.
3. Loosen the locknuts on the accelerator cables and let the inner cables sag.
4. While keeping lever "P" and the stopper in contact with each other, turn the adjusting nut to lengthen the outer cable. Adjust until the lever "P" begins to operate. Turn the nut back ½ turn and secure the locknut. Accelerator cable "A" free-play should be 0–0.04 in. (0–1mm).
5. To adjust accelerator cable "B", alter the adjusting nuts to meet the preferred free-play setting of 0–0.08 in. (0–2mm).
6. Install actuator cover and check operation of cruise control.

ENGINE COOLING

Radiator

REMOVAL AND INSTALLATION

1. Disconnect the negative battery cable.
2. Drain the coolant from the radiator.
3. Remove the upper hose and coolant reserve tank hose from the radiator.

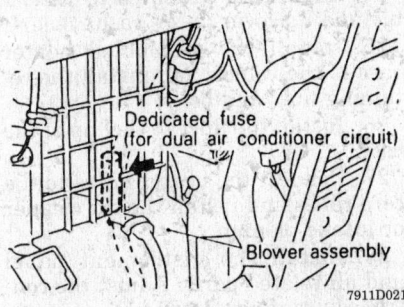

Dedicated fuse (for dual air conditioner circuit)

Blower assembly

7911D021

Dual air conditioning fuse location — Montero

4. Remove the shroud assembly from the radiator.

5. Raise the vehicle and support safely.

6. Remove the lower hose from the radiator.

7. Disconnect and plug the automatic transmission cooler hoses, if equipped. Lower the vehicle.

8. Remove the mounting screws and carefully lift the radiator out of the engine compartment.

To install:

9. Lower the radiator into position and install the mounting screws.

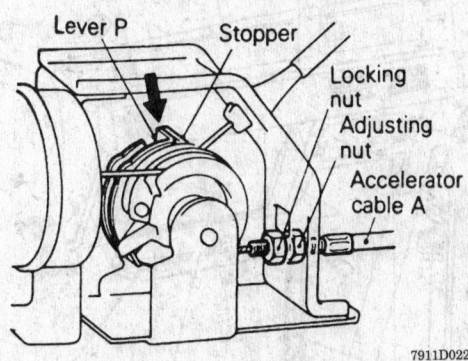

Lever P   Stopper

Locking nut
Adjusting nut
Accelerator cable A

7911D022

Accelerator cable "A" adjustment — with cruise control

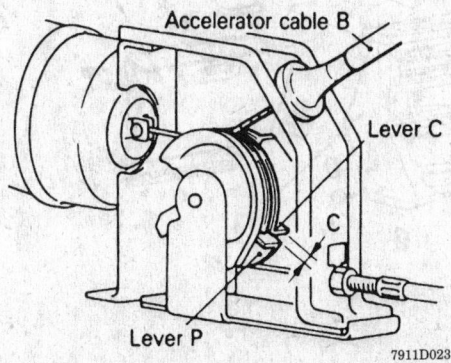

Accelerator cable B

Lever C

C

Lever P

7911D023

Accelerator cable "B" adjustment — with cruise control

10. Raise the vehicle and support safely. Connect the automatic transmission cooler hoses, if removed.

11. Connect the lower radiator hose. Lower the vehicle.

12. Install the shroud assembly.

13. Connect the upper hose and coolant reserve tank hose.

14. Fill the cooling system.

15. Connect the negative battery cable and run the vehicle until the thermostat opens. Fill the radiator completely and check the automatic transmission fluid level, if equipped.

16. Check for leaks. Once vehicle has cooled, recheck coolant level.

Electric Cooling Fan

An electric condenser cooling fan is used on the Montero. The condenser on Pick-Up is cooled by the belt-driven fan.

TESTING

1. Disconnect negative battery cable.

— **CAUTION** —

Make sure the key is in the OFF position when checking the electric cooling fan. If not the fan could turn on at any time, causing serious personal injury.

2. Disconnect the electrical connector from the condenser fan.

3. Connect the green-with-black-tracer wire to 12 volt supply and ground the black wire.

4. Make sure the fan runs smoothly, without abnormal noise or vibration.

5. Connect the negative battery cable.

REMOVAL AND INSTALLATION

1. Disconnect negative battery cable.

2. Open the hood and remove the grille. It is fastened with 3 screws along the top and 3 nuts along the bottom.

3. Disconnect the electrical connector from the fan.

4. Remove the mounting screws and the fan through the grille opening.

5. The installation is the reverse of the removal procedure.

6. Connect the negative battery cable and check the fan for proper operation.

Heater Core

REMOVAL AND INSTALLATION

Montero

1. Disconnect the negative battery cable.

2. Set the temperature control to the extreme right position and drain the cooling system.

3. Remove the air filter assembly. Remove the heater hose clamps and disconnect the heater hoses.

4. Remove the lap heater ducts and the hood release cable bracket.

5. Remove the side demister grills by carefully prying them from the instrument panel.

6. Remove the glove box and center console assembly. Remove the center reinforcement.

7. Remove the steering wheel.

8. Remove the instrument cluster.

9. Remove the oil pressure gauge, inclinometer and voltmeter pod cover and remove the gauge assembly.

10. Label and disconnect the recirculation/fresh air door control cable.

11. Label and disconnect the mode selection control cable.

12. Label and disconnect the water valve control cable.

13. Remove the fuse box retaining screw and position the fuse box aside.

14. Remove the instrument panel retaining nuts and bolts and carefully remove the instrument panel from the vehicle.

15. Remove the air cleaner or air intake plenum, as required.

16. Remove the duct from the top of the heater case.

17. Remove the retaining nuts and bolts and remove the heater case from the vehicle.

18. Remove the water valve cover and carefully remove the water valve from the case.

19. Remove the foot/defroster selection link from the mode selection lever.

20. Move the lever up to a position which will not interfere with the removal of the heater core.

21. Remove the heater core from the heater case. If the mode lever is in the way, remove it.

To install:

22. Install the heater core to the heater case. Install the mode lever, if removed.

23. Install the foot/defroster selection link to the mode selection lever.

24. Install the water valve assembly and its cover to the case.

25. Install the assembled heater case to the vehicle and install the retaining nuts and bolts.

26. Install the duct to the top of the case.

27. Connect the heater hoses to the core tubes and install the air cleaner or intake plenum.

28. Install the instrument panel and all related parts. Adjust the control cables if necessary.

29. Fill the system with coolant.

30. Connect the negative battery cable, run the vehicle until the thermostat opens and fill the radiator completely.

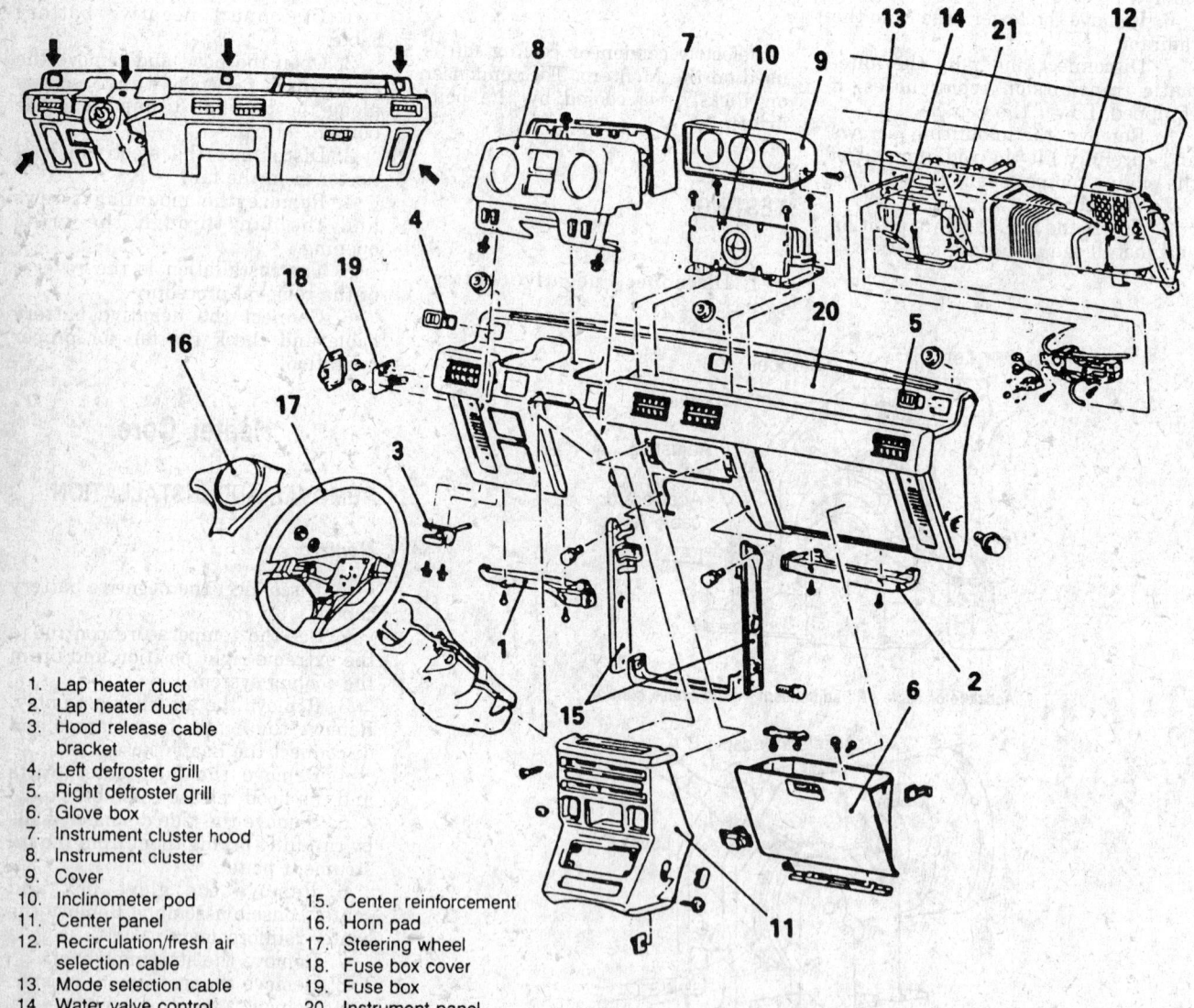

1. Lap heater duct
2. Lap heater duct
3. Hood release cable bracket
4. Left defroster grill
5. Right defroster grill
6. Glove box
7. Instrument cluster hood
8. Instrument cluster
9. Cover
10. Inclinometer pod
11. Center panel
12. Recirculation/fresh air selection cable
13. Mode selection cable
14. Water valve control cable
15. Center reinforcement
16. Horn pad
17. Steering wheel
18. Fuse box cover
19. Fuse box
20. Instrument panel
21. Heater case

7911D024

Instrument panel assembly — Montero

31. Check for leaks. Once cooled, recheck the coolant lever.

32. Check the entire climate control system and all gauges for proper operation.

Pick-Up

1. Disconnect the negative battery cable. Position the heater controls in the extreme right position.

2. Drain the coolant. Disconnect the heater hoses from the core tubes.

3. Remove the hazard flasher switch and the matching cover on the other side of the column. Remove the instrument cluster.

4. Remove the fuse box cover and remove the fuse box retaining screws. Position the fuse box aside.

5. Remove the glove box assembly.

6. Remove the defroster ducts.

7. Label and disconnect the air, mode and temperature control cables from the heater case.

8. Remove the front speaker grilles.

9. Remove the parcel box or clock, as equipped.

10. Remove the nut cover from the top center of the instrument panel.

11. Remove the center cover.

12. Remove the shift knob and floor console assembly, if equipped.

13. Move the tilt steering column down as far as it will go.

14. Remove the instrument panel retaining nuts and bolts and carefully remove the instrument panel from the vehicle.

15. Remove the duct from the top center of the heater case.

16. Remove the defroster duct from the the left side of the case.

17. Remove the center reinforcement braces.

18. Remove the mounting nuts and remove the heater case from the vehicle.

19. Remove the hose cover, joint hose clamp and the plate from the case.

20. Remove the heater core from the case.

To install:

21. Install the heater core to the heater case.

22. Install the plate, joint hose clamp and hose cover.

23. Install the assembled heater case to the vehicle. Connect the heater hoses to the core tubes.

24. Install the center reinforcement braces.

25. Install the defroster and center ducts to the case.

26. Install the instrument panel and all related parts. Adjust the control cables if necessary.

27. Fill the system with coolant.

28. Connect the negative battery cable, run the vehicle until the thermostat opens and fill the radiator completely.

29. Check for leaks. Once cooled, recheck the coolant level.

30. Check the entire climate control system and all gauges for proper operation.

Water Pump

REMOVAL AND INSTALLATION

2.4L Engine

1. Disconnect the negative battery cable.

2. Drain the cooling system.

3. Release the fuel pressure if equipped with fuel injection.

4. Remove the upper radiator shroud.

5. Remove all accessory belts. Remove the air conditioning compressor tensioner pulley, if equipped.

6. Remove the cooling fan assembly along with the water pump pulley.

7. Disconnect the radiator hose from the water pump.

8. Remove the crankshaft pulley(s).

9. Remove the timing belt covers. If the same timing belt will be reused, mark the direction of the timing belt's rotation, for installation in the same direction. Make sure the engine is positioned so the No. 1 cylinder is at the TDC of its compression stroke and the sprockets timing marks are aligned with the engine's timing mark indicators. Remove the timing belt.

10. The water pump bolts are different lengths, note their positions before removing. Remove the water pump mounting bolts and remove the pump from the block and the water pipe connection. Remove the O-ring from the water pipe connection.

To install:

11. Clean and dry the mating surfaces of the block and water pump. Install a new O-ring to the water pipe connection. Coat the new O-ring with water to aid in installation.

12. Install the water pump with a new gasket to the block and torque the bolts (except the alternator bracket bolt) to 10 ft. lbs. (13 Nm). Torque the aforementioned bolt to 17 ft. lbs. (23 Nm).

13. Install the timing belt(s) and covers.

14. Install the crankshaft pulley(s).

15. Connect the radiator hose to the water pump.

16. Install the water pump pulley and cooling fan assembly.

17. Install the air conditioning compressor tensioner pulley, if equipped, and install and adjust the accessory belts.

18. Install the upper radiator shroud.

19. Fill the system with coolant.

20. Connect the negative battery cable, run the vehicle until the thermostat opens and fill the radiator completely.

21. Check for leaks. Once cooled, recheck the coolant level.

22. Install the seat underframe and all related parts, if removed.

3.0L Engine

1. Disconnect the negative battery cable.

2. Drain the cooling system. Remove the upper radiator hose and the upper radiator shroud.

3. If equipped with electric fuel pump, loosen the fuel filler cap to release fuel tank pressure and release the fuel pressure in the supply lines.

4. Remove the cooling fan assembly along with the water pump pulley. Remove all belts.

5. Remove the power steering pump from the bracket and remove the bracket.

6. Remove the tensioner pulley bracket, the air conditioning compressor and its bracket.

7. Remove the cooling fan bracket assembly. Remove the crankshaft pulley and flange.

8. Remove the timing belt covers. If the same timing belt will be reused, mark the direction of the timing belt's rotation, for installation in the same direction. Make sure the engine is positioned so the No. 1 cylinder is at the TDC of its compression stroke and the sprockets timing marks are aligned with the engine's timing mark indicators.

9. Loosen the timing belt tensioner bolt and remove the belt. Position the tensioner as far away from the center of the engine as possible and tighten the bolt. Remove the water pump mounting bolts, separate the pump from the water inlet pipe and remove the pump from the engine. Remove the water inlet fitting from the pump.

To install:

10. Clean and dry the mating surfaces of the block and water pump. Install a new O-ring to the water inlet pipe. Install the pump and water inlet fitting with new gaskets to the

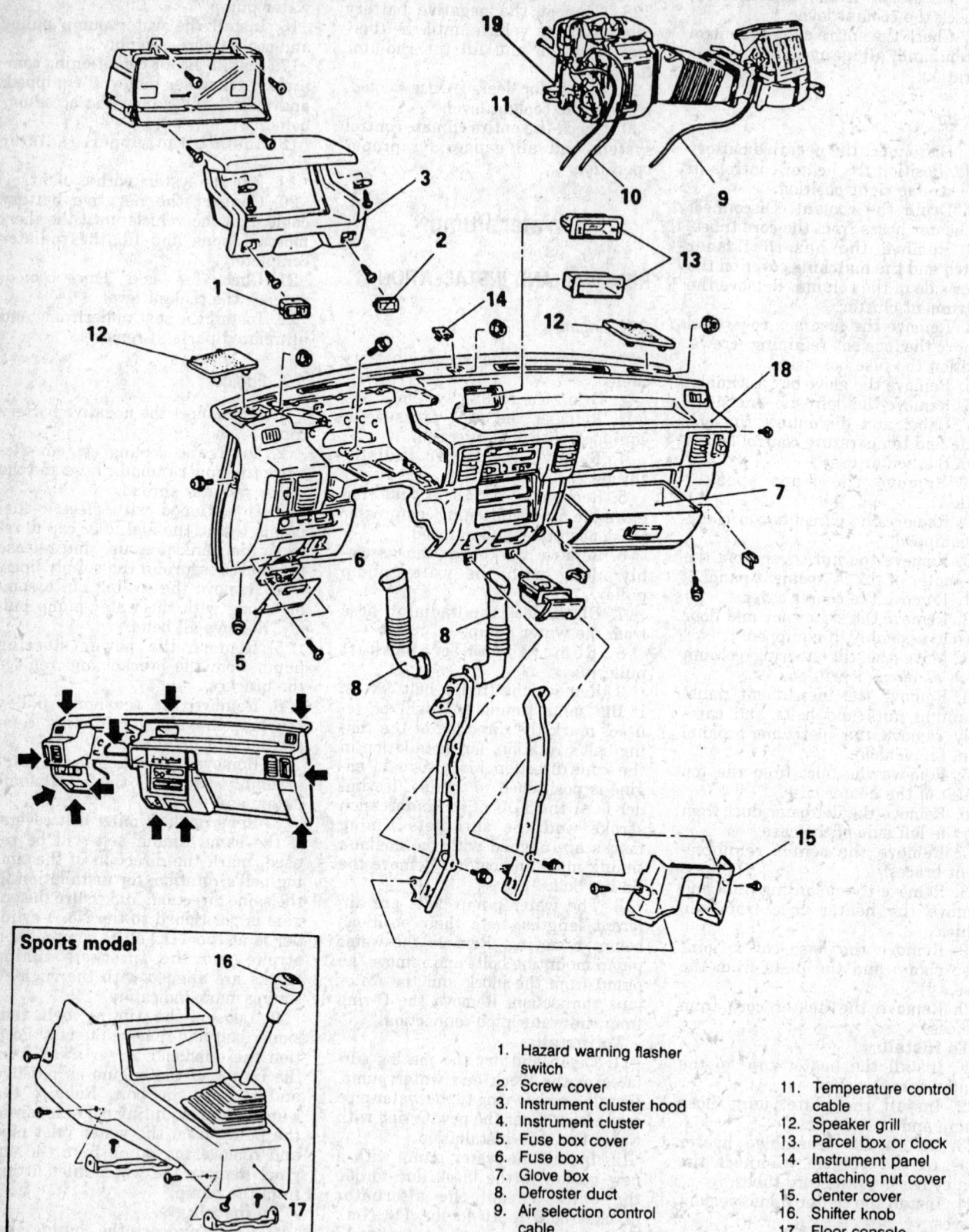

Sports model

1. Hazard warning flasher switch
2. Screw cover
3. Instrument cluster hood
4. Instrument cluster
5. Fuse box cover
6. Fuse box
7. Glove box
8. Defroster duct
9. Air selection control cable
10. Mode selection control cable
11. Temperature control cable
12. Speaker grill
13. Parcel box or clock
14. Instrument panel attaching nut cover
15. Center cover
16. Shifter knob
17. Floor console
18. Instrunment panel
19. Heater case

7911D025

Instrument panel assembly — Pick-Up

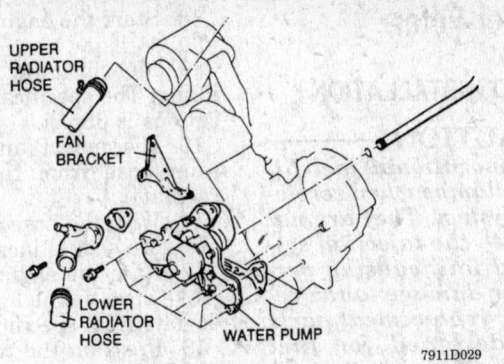

7911D029

Water pump assembly — 3.0L engine

engine and water pipe. Torque the water pump mounting bolts to 20 ft. lbs. (27 Nm).

11. If not already done, position both camshafts so the marks line up with those on the alternator bracket and inner timing cover. Rotate the crankshaft so the timing mark aligns with the mark on the oil pump.

12. Install the timing belt on the crankshaft sprocket and while keeping the belt tight on the tension side (right side), install the belt on the front camshaft sprocket.

13. Install the belt on the water pump pulley, then the rear camshaft sprocket and the tensioner.

14. Rotate the front camshaft counterclockwise to tension the belt between the front camshaft and the crankshaft. If the timing marks came out of line, repeat the procedure.

15. Install the crankshaft sprocket flange.

16. Loosen the tensioner bolt and allow the spring to tension the belt.

17. Turn the crankshaft 2 full turns in the clockwise direction only until the timing marks align again. Now that the belt is properly tensioned, torque the tensioner lock bolt to 21 ft. lbs. (29 Nm).

18. Install the timing belt covers and all related parts.

19. Fill the cooling system.

20. Connect the negative battery cable, run the vehicle until the thermostat opens and fill the radiator completely.

21. Check for leaks. Once cooled, recheck the coolant level.

Thermostat

REMOVAL AND INSTALLATION

1. Disconnect the negative battery cable. Drain the coolant down to thermostat level or below.

2. Disconnect the engine coolant temperature switch connector, if equipped.

3. Remove the thermostat housing.

4. Remove the thermostat and gasket from the housing.

5. Clean the housing mating surfaces and install a new gasket.

6. The installation is the reverse of the removal procedure.

Cooling System Bleeding

All engines are equipped with self-bleeding thermostats. Cooling system bleeding is not necessary in any vehicles when servicing the cooling system.

FUEL SYSTEM

Fuel System Service Precaution

Safety is the most important factor when performing not only fuel system maintenance but any type of maintenance. Failure to conduct maintenance and repairs in a safe manner may result in serious personal injury or death. Maintenance and testing of the vehicle's fuel system components can be accomplished safely and effectively by adhering to the following rules and guidelines.

To avoid the possibility of fire and personal injury, always disconnect the negative battery cable unless the repair or test procedure requires that battery voltage be applied.

Always relieve the fuel system pressure prior to disconnecting any fuel system component (injector, fuel rail, pressure regulator, etc.), fitting or fuel line connection. Exercise extreme caution whenever relieving fuel system pressure to avoid exposing skin, face and eyes to fuel spray. Please be advised that fuel under pressure may penetrate the skin or any part of the body that it contacts.

Always place a shop towel or cloth around the fitting or connection prior to loosening to absorb any excess fuel due to spillage. Ensure that all fuel spillage (should it occur) is quickly removed from engine surfaces. Ensure that all fuel soaked cloths or towels are deposited into a suitable waste container.

Always keep a dry chemical (Class B) fire extinguisher near the work area.

Do not allow fuel spray or fuel vapors to come into contact with a spark or open flame.

Always use a backup wrench when loosening and tightening fuel line connection fittings. This will prevent unnecessary stress and torsion to fuel line piping. Always follow the proper torque specifications.

Always replace worn fuel fitting O-rings with new. Do not substitute fuel hose or equivalent where fuel pipe is installed.

RELIEVING FUEL SYSTEM PRESSURE

1. Disconnect the fuel pump harness connector.

2. Start the engine and allow the engine to run out of fuel.

3. Once the engine has stalled, turn the key to the **OFF** position and connect the electrical connector.

4. Disconnect the negative battery cable pressure cannot build up until work has been completed.

Fuel Tank

REMOVAL AND INSTALLATION

1. If equipped with an electric fuel pump, release fuel pressure as follows:

   a. Disconnect the fuel pump harness connector.

   b. Start the engine and allow the engine to run out of fuel.

   c. Once engine has stalled, turn the key to the **OFF** position.

   d. Disconnect the negative battery cable so pressure cannot build up until work has been completed.

2. Raise the vehicle and support safely.

3. Using the proper equipment, drain the fuel tank.

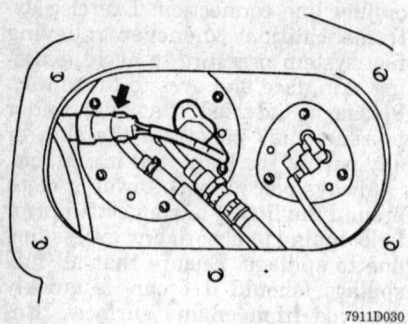

Fuel pump harness connector at the rear of the fuel tank — Montero

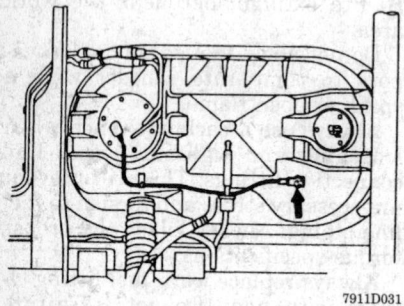

Fuel pump harness connector at the rear of the fuel tank — Pick-Up

4. Remove the side skirt panel, if equipped.

5. Disconnect the fuel gauge unit connector.

6. Disconnect the main hose connector.

7. Disconnect the return hose connector and the vapor hose connector.

8. Remove the filler hose connector and the breather hose connection.

9. Place a transmission jack or equivalent, under the center of the tank and apply slight pressure. Remove the tank mounting nuts.

10. Lower the tank and disconnect any lines still connected.

11. Remove the tank from the vehicle.

To install:

12. Raise the tank into position and connect all harnesses and hoses.

13. Install the fuel tank mounting nuts and torque to 18–22 ft. lbs. (25–30 Nm).

14. Reconnect the fuel gauge unit connector.

15. Install the side skirt panel, if equipped.

16. Refill the fuel tank and start the engine. Check fuel system for leaks.

Fuel Filter

REMOVAL AND INSTALLATION

—— **CAUTION** ——
Do not use conventional fuel filters, hoses or clamps when servicing this fuel system. They are not compatible with the injection system and could fail, causing personal injury or damage to the vehicle. Use only replacement parts specifically designed for fuel injection.

1. Relieve the fuel pressure.
2. Disconnect the negative battery cable.
3. Raise the vehicle and support safely.
4. Remove the fuel filter protector plate, if equipped.
5. Disconnect the main and high pressure lines. Remove any other fuel hose that are damaged or worn.
6. Remove the filter mounting bolts and remove the filter from the vehicle.

To install:

7. Install new filter onto vehicle using new gaskets. Replace any hoses that are worn or damaged.
8. Install the protector plate, if equipped.
9. Connect the negative battery cable, check for leaks and road test the vehicle.

Electric Fuel Pump

PRESSURE TESTING

1. Relieve the fuel pressure.
2. Disconnect the negative battery cable.
3. Cover the high pressure fuel hose hose with a clean shop towel to prevent fuel spray from residual pressure in the line. Disconnect the high pressure fuel hose at the delivery pipe.
4. Connect the proper fuel pressure gauge and accompanying special adaptor tools to the delivery pipe.
5. If not already done, place the key in the **OFF** position. Connect the negative battery cable.
6. Connect a jumper wire from the fuel pump activation terminal to the positive battery post. This will pressurize the system so the fuel pump installation assembly can be inspected for leaks. If a leak is found, repair it before proceeding.
7. Disconnect the jumper wire to stop the fuel pump.

8. Start the engine and allow it to idle.
9. Measure the pressure during idling. The specification for all applications is 38 psi.
10. Disconnect and plug the vacuum hose from the fuel pressure regulator.
11. With the hose disconnected, the pressure should increase to 50 psi.
12. Race the engine a few times to make sure the fuel pressure does not deviate from specifications.
13. Press on the return hose while racing the engine to make sure there is pressure in the hose. Reconnect the vacuum hose.
14. Stop the engine and allow pressure to remain in the system. There should be no decrease in pressure for at least 2 minutes.
15. Release the fuel pressure.
16. Remove the fuel pressure measuring equipment.
17. Connect the high pressure fuel hose to the delivery pipe using new O-rings where necessary.
18. Connect the jumper wire from the fuel pump activation terminal to the positive battery post, inspect the system for leaks.
19. Road test the vehicle.

REMOVAL AND INSTALLATION

Montero

1. Relieve the fuel pressure.
2. Disconnect the negative battery cable.
3. Remove the pump and sending unit access panel.
4. Cover the high pressure fuel hose with a clean shop towel to prevent fuel spray from residual pressure in the line. Disconnect the fuel hose and pipe from the pump.
5. Remove the fuel pump mounting screws and remove the pump from the tank.
6. The installation is the reverse of the removal procedure.

Pick-Up

1. Relieve the fuel pressure.
2. Disconnect the negative battery cable.
3. Raise the vehicle and support safely. Remove protective plates, if equipped.
4. Using the proper equipment, drain the fuel tank.
5. Remove the fuel tank from the vehicle.
6. Remove the pump retaining screws and remove the pump from the tank.

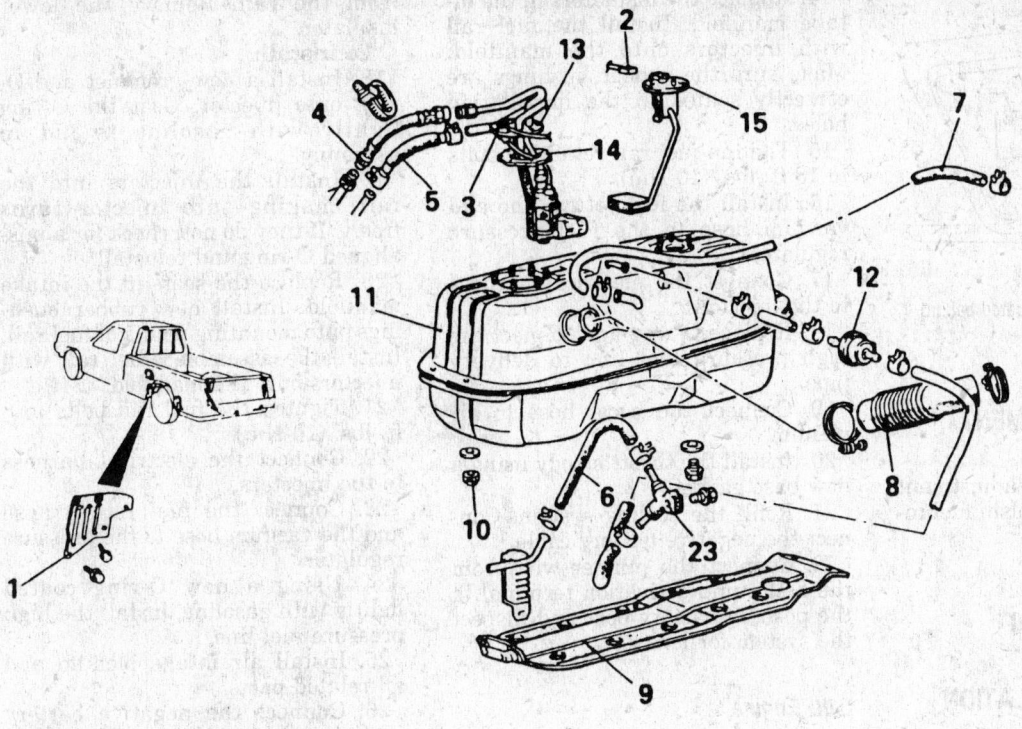

1. Side skirt stay
2. Sending unit connector
3. Fuel pump connector
4. Main hose
5. Return hose
6. Vapor hose
7. Breather hose
8. Filler hose
9. Fuel tank protector (4WD)
10. Tank mounting nut
11. Fuel tank
12. Two-way overfill limiter valve
13. Fuel pump assembly
14. Gasket
15. Sending unit

7911D038

Fuel tank assembly — Pick-Up

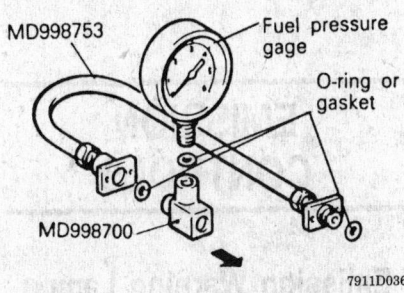

Fuel pressure testing equipment. The hose is not used on Pick-Up with 2.4L engine.

7911D036

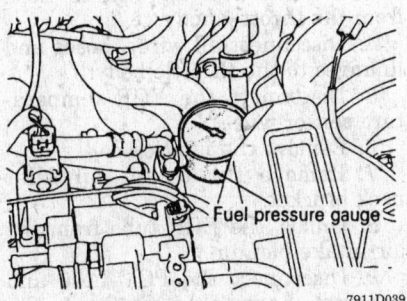

Fuel pressure gauge installation — Pick-Up with 2.4L engine

7911D039

To install:

7. Clean the seal area of the tank. Install a new gasket.

8. Install the pump in the same position as originally installed.

9. Install the retaining screws and torque them to 15 inch lbs.

10. Install the fuel tank and all related parts.

11. Connect the negative battery cable. Connect the jumper wire from the fuel pump activation terminal to the positive battery post and inspect the system for leaks.

12. Road test the vehicle.

Fuel Injection

IDLE SPEED ADJUSTMENT

The idle speed is automatically regulated by the idle speed control system which receives data from various sensors and switches in the system and adjusts the engine idle to a predetermined speed. Idle speed specifications can be found on the Vehicle Emission Control Information (VECI) label located in the engine compartment.

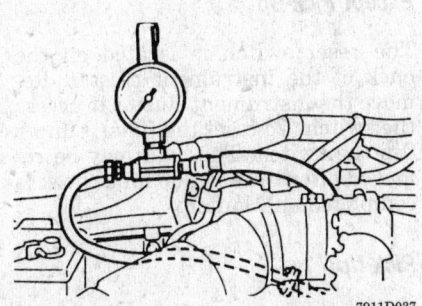

7911D037

Fuel pressure gauge installation — 3.0L engine

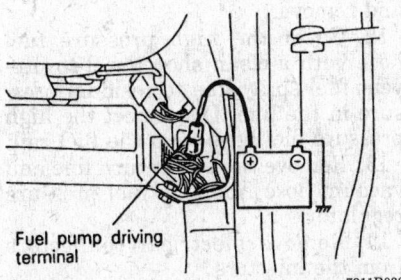

7911D033

Fuel pump activation terminal — Montero

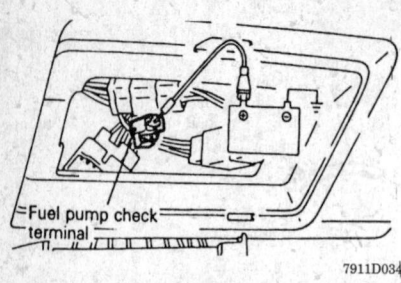

Fuel pump activation terminal located behind the fuse box — Pick-Up

IDLE MIXTURE ADJUSTMENT

There is no idle mixture adjustment provided with any Mitsubishi fuel injection system.

Fuel Injector

REMOVAL AND INSTALLATION

2.4L Engine

1. Relieve the fuel pressure.
2. Disconnect negative battery cable. Remove the front seat underframe as required.
3. Drain the cooling system.
4. Matchmark and remove all linkage and vacuum hoses from the throttle body and remove the throttle body from the intake manifold.
5. Remove the boost hose from opposite end of air plenum.
6. Cover the high pressure fuel hose with a clean shop towel to prevent fuel spray from residual pressure in the line. Disconnect the high pressure fuel hose from the fuel rail.
7. Remove the fuel return line and vacuum hose from the fuel pressure regulator.
8. Disconnect the electrical connectors from the injectors.
9. Remove the fuel rail retaining bolts.
10. Lift the rail with injectors attached up and away from the engine.
11. Remove the injector from the rail by pulling straight out away from rail. Remove the lower insulator.
   To install:
12. Install a new grommet and O-rings onto the injector. Coat the O-rings lightly with gasoline.
13. Install the injector to the rail, making sure injector turns freely. If it does not turn, check for a damaged or misaligned O-ring and reinstall.

14. Replace the insulators in the intake manifold. Install the fuel rail with injectors onto the manifold. Make sure the rubber bushings are correctly seated in the installation holes.
15. Tighten fuel rail retaining bolts to 18 ft. lbs. (10 Nm).
16. Install the fuel return line and vacuum hose to the fuel pressure regulator.
17. Connect the electrical harness to the injectors.
18. Replace O-ring and connect the high pressure fuel line to delivery pipe.
19. Connect the boost hose to air plenum.
20. Install the throttle body using a new base gasket.
21. Refill the cooling system. Connect the negative battery cable.
22. Connect the jumper wire from the fuel pump activation terminal to the positive battery post and inspect the system for leaks.

3.0L Engine

1. Relieve the fuel pressure.
2. Disconnect negative battery cable.
3. Remove the air intake hose from the throttle body.
4. Disconnect all wires, hoses and linkages to the throttle body.
5. Disconnect the EGR temperature sensor wire.
6. Remove the ignition coil.
7. Remove the engine oil filler neck bracket.
8. Unbolt the EGR tube from the air intake plenum.
9. Disconnect the PCV hose and vacuum hose cluster from the plenum.
10. Remove the plenum to engine brackets.
11. Unbolt the air intake plenum assembly from the intake manifold and remove.
12. Cover the high pressure fuel hose with a clean shop towel to prevent fuel spray due to residual pressure in the line. Disconnect the high pressure fuel hose from the fuel rail.
13. Remove the fuel return line and vacuum hose from the fuel pressure regulator.
14. Remove electrical connectors from the injectors.
15. Remove the fuel rail retaining bolts.
16. Lift the rail with injectors attached up and away from the engine.
17. Remove the injectors from the fuel rail pulling straight away

from the rail. Remove the lower insulator.
   To install:
18. Install a new grommet and O-ring onto injector. Coat the O-ring lightly with gasoline to aid in assembly.
19. Install the injectors into the rail, making sure injector turns freely. If they do not, check for a misaligned O-ring and reinstall.
20. Replace the seats in the intake manifold. Install new rubber bushings onto mounting points of fuel rail. Install the assembled fuel rail with injectors onto the manifold.
21. Tighten the fuel rail bolts to 8 ft. lbs. (10 Nm).
22. Connect the electrical harness to the injectors.
23. Connect the fuel return hose and the vacuum hose to the pressure regulator.
24. Using a new O-ring coated lightly with gasoline, install the high pressure fuel line.
25. Install air intake plenum and all related parts.
26. Connect the negative battery cable.
27. Connect the jumper wire from the fuel pump activation terminal to the positive battery post inspect the system for leaks.

EMISSION CONTROLS

Emission Warning Lamps

RESETTING

Except Pick-Up

The reset switch is located on the back of the instrument cluster. Remove the instrument cluster to access the switch. To reset the timer, simply flip the switch. The bulb may be removed after the 150,000 mile check is completed on Montero.

Pick-Up

Remove the glass in from of the instrument cluster to access the reset switch. To reset the timer, simply flip the switch. The bulb may be removed after the 120,000 mile check is completed.

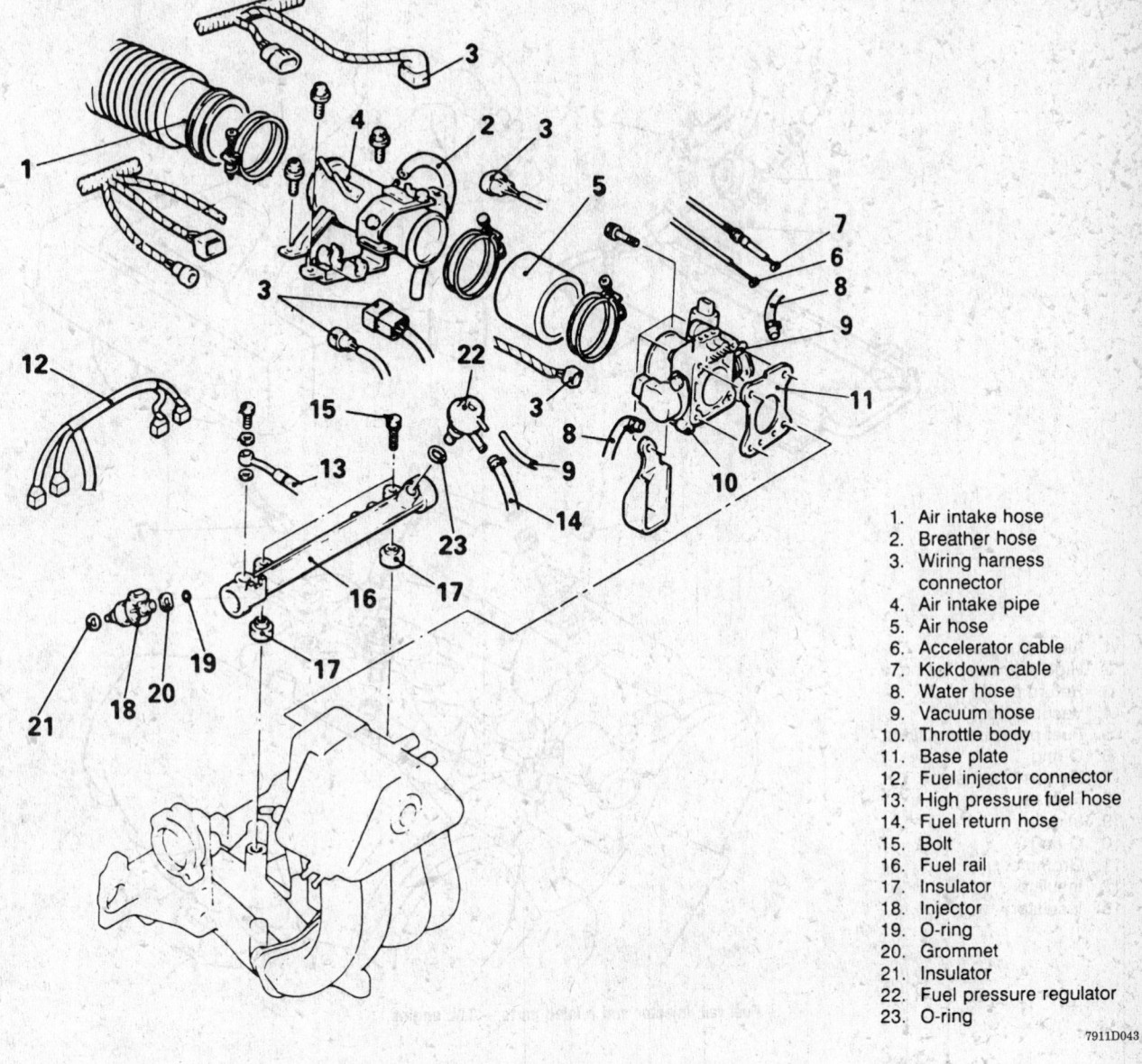

1. Air intake hose
2. Breather hose
3. Wiring harness connector
4. Air intake pipe
5. Air hose
6. Accelerator cable
7. Kickdown cable
8. Water hose
9. Vacuum hose
10. Throttle body
11. Base plate
12. Fuel injector connector
13. High pressure fuel hose
14. Fuel return hose
15. Bolt
16. Fuel rail
17. Insulator
18. Injector
19. O-ring
20. Grommet
21. Insulator
22. Fuel pressure regulator
23. O-ring

7911D043

Throttle body, fuel rail, injector and related parts — 2.4L engine

ENGINE MECHANICAL

NOTE: Disconnecting the negative battery cable on some vehicles may interfere with the functions of the on board computer systems and may require the computer to undergo a relearning process, once the negative battery cable is reconnected.

Engine

REMOVAL AND INSTALLATION

1. Relieve the fuel pressure if equipped with fuel injection. Disconnect the negative battery cable from the battery and from the engine.

2. Matchmark and remove the hood. Remove the oil dipstick.

3. Raise the vehicle and support safely. Remove the engine under cover. Drain the engine oil and coolant.

4. Remove the starter. Remove the lower radiator hose.

5. Remove the exhaust pipe from the exhaust manifold(s).

6. If equipped with a manual transmission, remove the transmission and all related parts.

7. If equipped with an automatic transmission, remove the inspection plate, matchmark the flexplate to the converter, remove the torque converter bolts and push the torque converter backwards as far as it will go. Remove the lower bell housing bolts. Lower the vehicle.

8. Remove all ductwork and air intake hoses. Disconnect all linkages and cables from the throttle body.

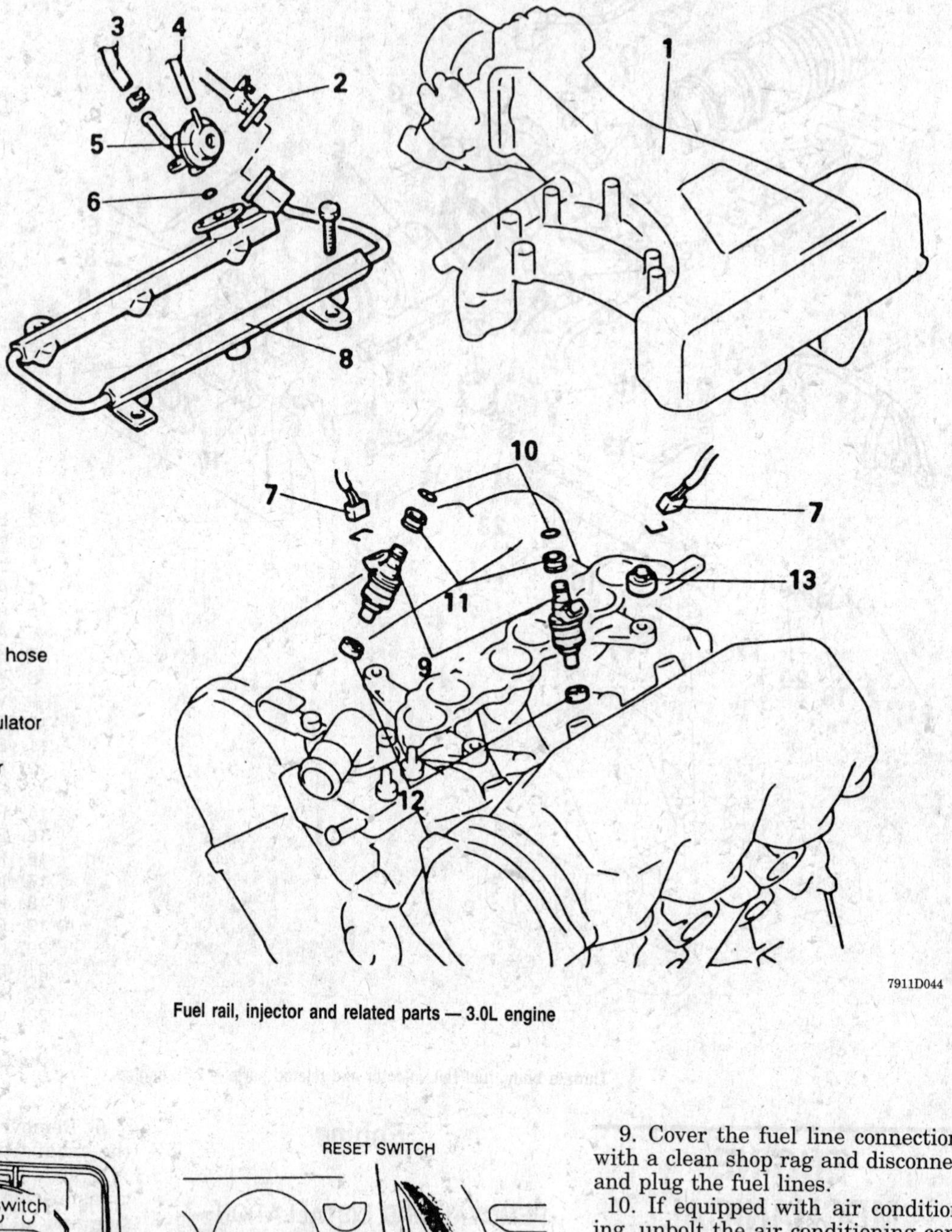

1. Air intake plenum
2. High pressure fuel hose
3. Return hose
4. Vacuum hose
5. Fuel pressure regulator
6. O-ring
7. Harness connector
8. Fuel rail
9. Injector
10. O-ring
11. Grommet
12. Insulator
13. Insulator

7911D044

Fuel rail, injector and related parts — 3.0L engine

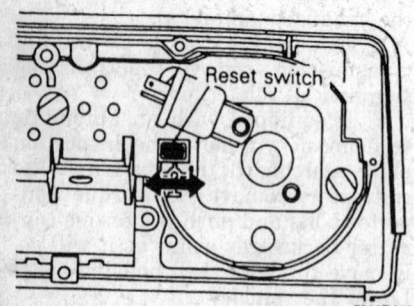

7911D045

Emissions warning light reset switch — Montero

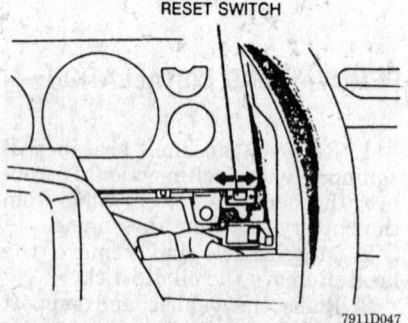

7911D047

Emissions warning light reset switch — Pick-Up

9. Cover the fuel line connections with a clean shop rag and disconnect and plug the fuel lines.

10. If equipped with air conditioning, unbolt the air conditioning compressor from the engine and position aside. It is not necessary to remove the lines from the compressor.

11. Remove the radiator and shroud. Remove the fan and all related parts. Disconnect the heater hoses.

12. Unbolt the power steering pump from its brackets and position it to the side. Do not remove the hoses from the pump.

13. Remove the alternator. Remove the ignition coil and power transistor assembly, if equipped.

14. Label and disconnect all remaining electrical connectors, vacuum hoses and check for any other items preventing engine removed.

15. Attach an engine removal device to the engine support eyes on the engine.

16. If equipped with an automatic transmission, support the transmission with a floor jack or equivalent. Remove the remaining bell housing bolts.

17. Remove the engine mount nuts and remove the engine from the vehicle.

To install:

18. Lower the engine into position and install the engine mount nuts. Torque the nuts to 20 ft. lbs. (27 Nm). Install the upper bell housing bolts. Remove the engine removal device. Install the oil dipstick.

19. Raise the vehicle and support safely.

20. Install the remaining bell housing bolts.

21. If equipped with a manual transmission, install the transmission and all related parts.

22. If equipped with an automatic transmission, align the torque con-

verter and flexplate and install the bolts. Install the inspection plate and starter.

23. Install the exhaust pipe to the exhaust manifold(s) using new gaskets. Install the lower radiator hose. Lower the vehicle.

24. Connect the heater hoses.

25. Make sure the negative battery cable is not connected to the battery. Connect the engine side of the negative cable to the engine. Install the alternator, power steering pump and all brackets.

26. Install the air conditioning compressor.

27. Connect all linkages and cables to the throttle body.

28. Install the ignition coil and power transistor assembly, if equipped. Connect all electrical connectors and vacuum hoses that were disconnected during the engine removal procedure.

29. Install the fan and all related parts. Adjust all belt tensions, as required.

30. Install the radiator, shroud and upper hose.

31. Install the air cleaner assembly, ducts and air intake hoses.

32. Fill the engine with the specified amount of oil and fill the radiator with coolant.

33. Connect the negative battery cable and connect the jumper wire from the fuel pump activation terminal, if equipped, to the positive battery post to inspect the system for leaks.

34. Check the automatic transmission fluid level, if equipped. Set all adjustments to specifications.

35. Install and align the hood.

Engine Mounts

REMOVAL AND INSTALLATION

Front Mount

1. Disconnect negative battery cable.

2. Install engine support fixture in place. Raise and safely support the vehicle.

3. Remove the engine front support insulator.

4. Remove the front insulator stopper.

5. Remove the heat protector.

6. Unbolt the rear crossmember or mount as required.

7. Raise engine with support fixture far enough to remove mounts, remove remaining bolts and mounts. Transfer insulator and stopper to new mount.

1. Throttle position sensor connector
2. Ignition coil connector
3. Power transistor connector
4. EGR temperature sensor connector
5. Coolant temperature sending unit connector
6. Coolant temperature sensor connector
7. Thermo switch connector (automatic transmission only)
8. Oxygen sensor connector
9. Alternator connector
10. Oil pressure sending unit connector
11. Coolant temperature switch connector (automatic transmission only)
12. Ground cable
13. Emission control vacuum hose
14. Brake booster vacuum hose
15. Ground cable
16. I.S.C. connector
17. Motor position sensor connector
18. Engine controller wiring harness
19. Heat shield
20. Engine mount bolt

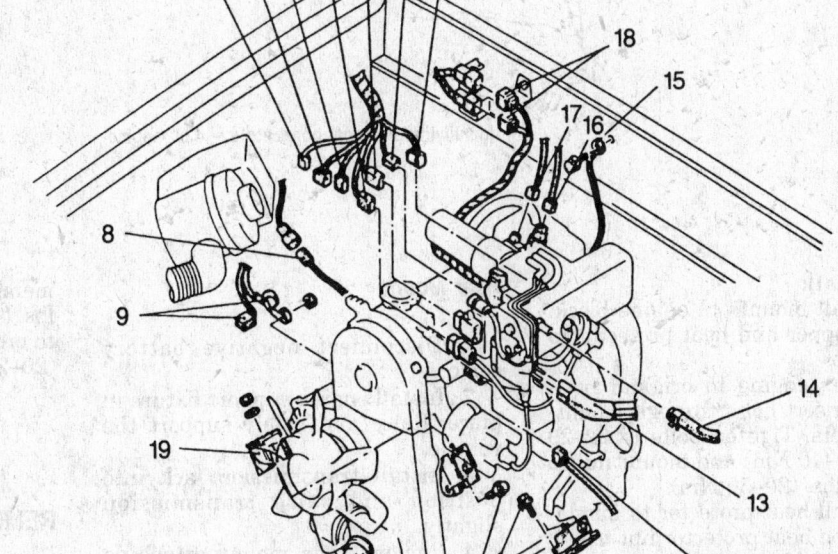

7911D048

Identifying electrical connectors — Pick-Up with 2.4L engine

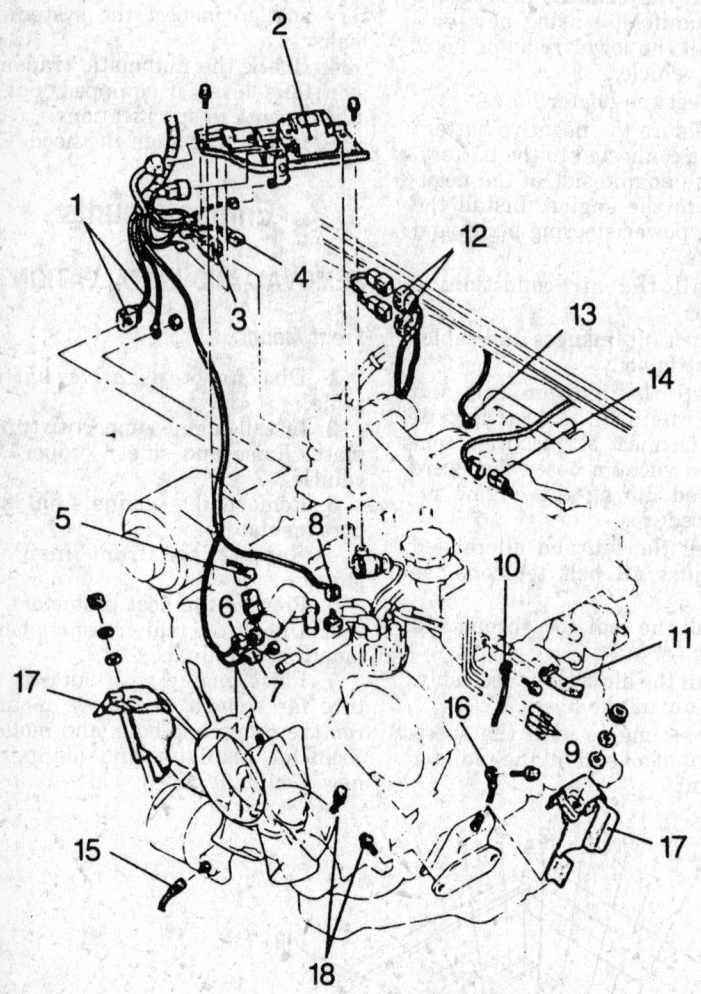

1. Alternator connector
2. Ignition coil and power transistor assembly
3. I.S.C. connector
4. Throttle position sensor connector
5. Coolant temperature switch connector (automatic transmission only)
6. Coolant temperature sensor connector
7. Thermo switch connector (automatic transmission only)
8. Coolant temperature sending unit connector
9. Emission control vacuum hose
10. Ground cable
11. Brake booster vacuum hose
12. Engine controller wiring harness
13. Ground cable
14. EGR temperature sensor connector
15. Oil pressure sending unit connector
16. Oil pressure sending unit connector
17. Ground cable
18. Heat shield

7911D049

Identifying electrical connectors — 3.0L engine

To install:

8. Install mounts to engine block. Install stopper and heat protector to the mount.

9. Lower engine to original position and insert bolts through mounting brackets. Tighten bolts to 22–29 ft. lbs. (30–40 Nm) and mount nut to 14–22 ft. lbs. (20–30 Nm).

10. Install heat protector to insulator. Tighten heat protector nut to 6–9 ft. lbs. (8–12 Nm).

11. Lower the vehicle and remove the engine support fixture. Reconnect the negative battery cable.

Rear Mount

1. Disconnect negative battery cable.

2. Install engine support fixture in place. Raise and safely support the vehicle.

3. Install transmission jack into position and raise transmission slightly.

4. Remove rear mount attaching bolts and crossmember attaching bolts. Remove crossmember from the vehicle and transfer mounting bracket, support insulator plate assembly to new mount.

5. Installation is the reverse of the removal procedure. Torque cross-member retaining bolts to 29–40 ft. lbs. (40–55 Nm). Torque the insulator to crossmember nuts to 14–18 ft. lbs. (20–25 Nm).

Cylinder Head

REMOVAL AND INSTALLATION

2.4L Engine

1. Rotate the engine so No. 1 cylinder is at TDC. If equipped with fuel injection, relieve the fuel pressure.

2. Disconnect negative battery cable. Drain the cooling system.

3. Remove the upper radiator hose and disconnect the heater hoses. Remove the dipstick bracket bolt.

4. Remove the air cleaner assembly or air intake hose.

5. Disconnect all linkages and cables from the throttle body. Disconnect and plug the fuel lines to the fuel rail. Remove the valve cover.

6. On the 2.4L engines, perform the following:

   a. Without disconnecting the lines, unbolt the power steering pump from its brackets and position it to the side, if equipped.

   b. Remove the timing belt upper cover.

   c. Align the timing mark, if not already aligned. Secure the timing belt to the sprocket with wire tie.

   d. Remove the camshaft bolt.

   e. Remove the sprocket from the camshaft and allow it to rest on the lower cover. If this is not possible, tie it to a fabricated device so the belt remains taut and the timing is not lost.

7. Disconnect and label all vacuum lines, hoses and wiring connectors from the manifolds, throttle body and cylinder head.

8. Raise the vehicle and support safely.

9. Remove the exhaust pipe from the exhaust manifold. Lower the vehicle.

10. Remove the cylinder head from the engine.

11. Clean the cylinder head gasket mating surfaces.

To install:

12. Install the camshaft gear or sprocket.

13. Install the timing belt cover, if equipped.

14. Install the power steering pump, if removed.

15. Connect the heater hoses.

16. Install the dipstick bracket bolt. Install the valve cover with a new gasket.

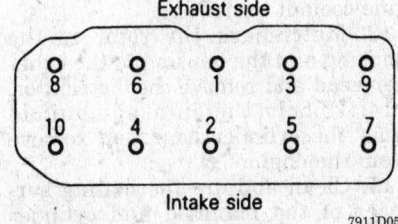

⟸ Timing belt side

Exhaust side

8 6 1 3 9

10 4 2 5 7

Intake side

7911D052

Cylinder head bolt torque sequence — 2.4L engine

17. Connect and plug the fuel lines to the fuel rail. Connect all linkages and cables to the throttle body.

18. Install the air cleaner assembly or air intake hose.

19. Connect the upper radiator hose.

20. Fill the radiator with coolant.

21. Connect the negative battery cable and connect the jumper wire from the fuel pump activation terminal to the positive battery post to inspect the system for leaks, if equipped.

22. Set all adjustments to specifications.

23. Install the seat underframe, if removed.

3.0L Engine

1. Relieve the fuel pressure. Disconnect the negative battery cable. Drain the cooling system. Disconnect the upper radiator hose.

2. Remove the drive belts, air conditioning compressor and power steering pump from the mounts and position them to the side.

3. Using a ½ in. drive breaker bar, insert it into the square hole of the serpentine drive belt tensioner, rotate it counterclockwise (to reduce the belt tension) and remove the belt.

4. Remove the alternator.

5. Remove the crankshaft pulley and the torsional damper.

6. To remove the timing belt, perform the following procedures:

   a. Remove the covers. Rotate the crankshaft to position No. 1 cylinder on the TDC of its compression stroke; the crankshaft sprocket timing mark should align with the oil pan timing indicator and the camshaft sprockets timing marks (triangles) should align with the rear timing belt covers timing marks.

   b. Mark the timing belt in the direction of rotation for reinstallation purposes.

   c. Loosen the timing belt tensioner and remove the timing belt.

NOTE: When removing the timing belt from the camshaft sprocket, make sure the belt does not slip off of the other camshaft sprocket. Support the belt so it can not slip off of the crankshaft sprocket and opposite side camshaft sprocket.

7. Remove the air intake hose.

8. Label and disconnect the spark plug wires and vacuum hoses.

9. Remove the valve cover.

10. If removing the left cylinder head, matchmark the distributor ro-

tor to the distributor housing and the housing to distributor extension locations. Remove the distributor and the distributor extension. Also, remove the EGR pipe.

11. Remove the air intake plenum and intake manifold assembly.

12. Remove the exhaust manifold.

13. Remove the cylinder head bolts starting from the outside and working inward.

14. Remove the cylinder head from the engine.

15. Clean the gasket mounting surfaces.

To install:

16. Install the new cylinder head gaskets over the dowels on the engine block.

17. Install the cylinder heads on the engine and torque the cylinder head bolts in sequence using 3 even steps, to 70 ft. lbs. (95 Nm).

18. Install the intake and exhaust manifolds.

19. Install the EGR pipe with a new gasket, if removed.

20. Install the distributor and extension, if removed.

21. Install the timing belt and all related items. When installing the timing belt over the camshaft sprocket, use care not to allow the belt to slip off the opposite camshaft sprocket.

22. Make sure the timing belt is installed on the camshaft sprocket in the same position as when removed.

23. Install the alternator and power steering pump.

24. Install the air conditioning compressor and belts.

25. Connect the upper radiator hose.

26. Fill the radiator with coolant.

27. Connect the negative battery cable and connect the jumper wire from the fuel pump activation terminal to the positive battery post to inspect the system for leaks, if equipped.

28. Set all adjustments to specifications.

Valve Lifters

REMOVAL AND INSTALLATION

2.4L Engine

1. Disconnect negative battery cable.

2. Remove the valve cover.

3. Have a helper hold the rear of the camshaft down during removal. Then install the rear cap loosely to hold the camshaft in position.

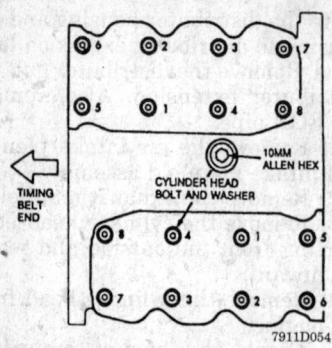

Cylinder head bolt torque sequence — 3.0L engine

4. Loosen the camshaft cap bolts but do not remove them from the caps. Remove the caps, arms, shafts and bolts all as an assembly. Install tool MD998443-01 or equivalent, on the end of all rocker arms to assure the hydraulic lifters will not fall out of the rocker arms on removal of the assembly.

5. Disassemble the unit keeping all parts in the order of removal and repair as required.

NOTE: The rocker arms have identification marks on them. Arms with 1–3 on them should only be used on cylinders 1 or 3. Arms with 2–4 on them should only be used on cylinders 2 and 4.

6. The installation is the reverse of the removal procedure. Make sure the arrows on the caps are all pointing to the front of the engine. Torque the cap bolts first to 85 inch lbs. (10 Nm), then to 175 inch lbs. (18 Nm) in the following order: No. 3 cap, No. 2 cap, No. 4 cap, Front cap and Rear cap.

3.0L Engine

1. Disconnect the negative battery cable. Remove the air cleaner assembly.

2. Remove the valve cover.

3. Using the auto lash adjuster retainer tools MD998443 or equivalent, install them on the rocker arms to keep the lash adjusters from falling out.

4. On the right side cylinder head, remove the distributor extension.

5. Have a helper hold the rear end of the camshaft down. If the rear of the camshaft cannot be held down, the belt will dislodge and the valve timing will be lost. Loosen the camshaft cap bolts but do not remove them from the caps. Remove the caps, arms, shafts and bolts all as an assembly.

6. Disassemble the unit keeping all parts in order and repair as required.

7. The installation is the reverse of the removal procedure. Apply a drop of sealant to the rear edge of the rear cap. Torque the cap bolts first to 85 inch lbs. (19 Nm), then to 180 inch lbs. (19 Nm) in the following order: No. 3 cap, No. 2 cap, No. 1 cap and No. 4 cap.

Rocker Arms and Shafts

REMOVAL AND INSTALLATION

2.4L Engine

1. Disconnect the negative battery cable.

2. Remove the valve cover.

3. Have a helper hold the rear of the camshaft down. Then install the rear cap loosely to hold the camshaft in position. If the rear of the camshaft cannot be held down, the belt will dislodge and the valve timing will be lost.

4. Loosen the camshaft cap bolts but do not remove them from the caps. Remove the caps, arms, shafts and bolts all as an assembly. Install tool MD998443-01 or equivalent, on the end of all rocker arms to assure the hydraulic lifters will not fall out of the rocker arms on removal of the assembly.

5. Disassemble the unit keeping all parts in the order of removal and repair as required.

NOTE: The rocker arms have identification marks on them. Arms with 1–3 on them should only be used on cylinders 1 or 3. Arms with 2–4 on them should only be used on cylinders 2 and 4.

6. The installation is the reverse of the removal procedure. Make sure the arrows on the caps are all pointing to the front of the engine. Torque the cap bolts first to 85 inch lbs. (10 Nm), then to 175 inch lbs. (18 Nm) in the following order: No. 3 cap, No. 2 cap, No. 4 cap, Front cap and Rear cap.

3.0L Engine

1. Disconnect the negative battery cable. Remove the air cleaner assembly.

2. Remove the valve cover.

3. Using the auto lash adjuster retainer tools MD998443 or equivalent, install them on the rocker arms to keep the lash adjusters from falling out.

4. On the right side cylinder head, remove the distributor extension.

5. Have a helper hold the rear end of the camshaft down. If the rear of the camshaft cannot be held down, the belt will dislodge and the valve timing will be lost. Loosen the camshaft cap bolts but do not remove them from the caps. Remove the caps, arms, shafts and bolts all as an assembly.

6. Disassemble the unit keeping all parts in order and repair as required.

7. The installation is the reverse of the removal procedure. Apply a drop of sealant to the rear edge of the rear cap. Torque the cap bolts first to 85 inch lbs. (19 Nm), then to 180 inch lbs. (19 Nm) in the following order: No. 3 cap, No. 2 cap, No. 1 cap and No. 4 cap.

Intake Manifold

REMOVAL AND INSTALLATION

2.4L Engine

1. Relieve the fuel pressure.

2. Disconnect the negative battery cable.

3. Drain the engine coolant. Disconnect the upper radiator hose from the thermostat housing.

4. Remove the air intake hoses and the air intake pipe.

5. Disconnect all wires, hoses and linkages to the throttle body.

6. Remove the ignition coil.

7. Disconnect the brake booster hose and vacuum hose cluster from the air intake plenum.

8. Unbolt the air intake plenum from the intake manifold and remove the plenum from the engine.

9. Cover the fuel line with a clean shop rag and disconnect the fuel lines from the fuel rail. Keep the line covered or plugged.

10. Remove the fuel rail assembly with injectors.

11. Disconnect the heater hose from the manifold.

12. Disconnect the wires to the engine coolant switches.

13. Matchmark the rotor to the housing and the housing to the cylinder head and remove the distributor.

14. Unbolt the intake manifold from the cylinder head and remove from the engine.

15. Clean and dry the mating surfaces of the manifold and cylinder head.

To install:

16. Using a new gasket, install the intake manifold to the head. Starting

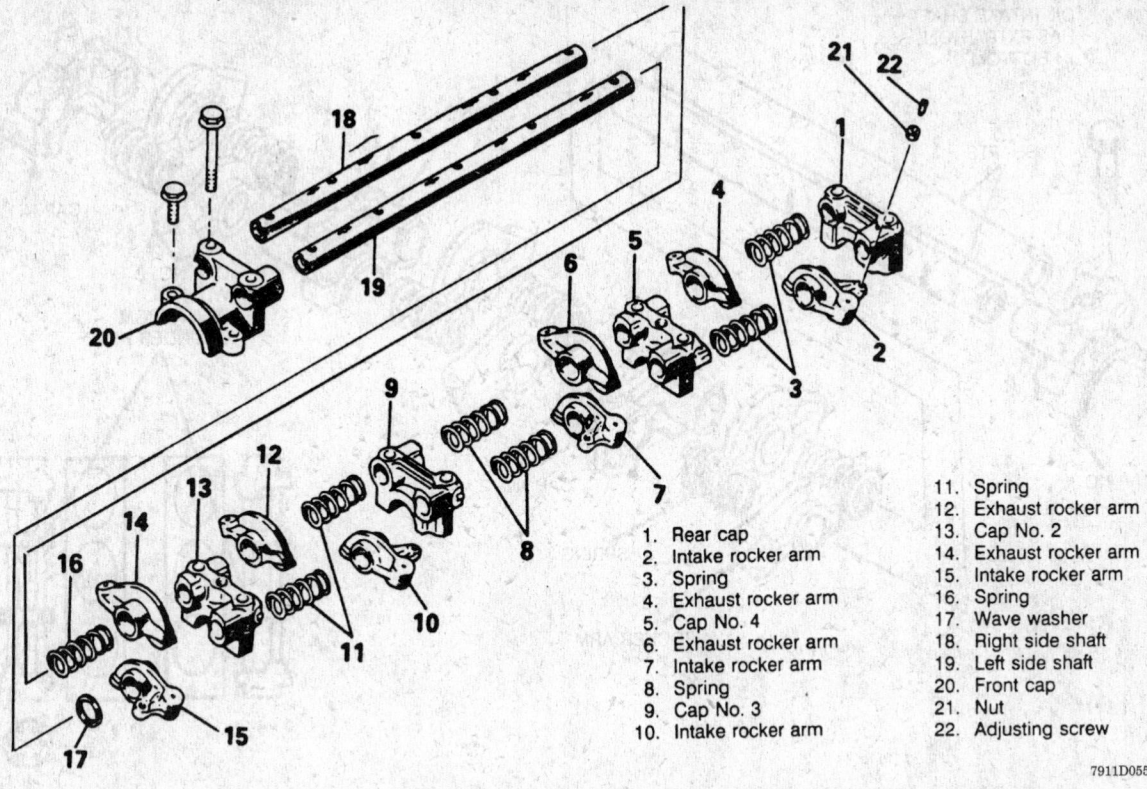

1. Rear cap
2. Intake rocker arm
3. Spring
4. Exhaust rocker arm
5. Cap No. 4
6. Exhaust rocker arm
7. Intake rocker arm
8. Spring
9. Cap No. 3
10. Intake rocker arm
11. Spring
12. Exhaust rocker arm
13. Cap No. 2
14. Exhaust rocker arm
15. Intake rocker arm
16. Spring
17. Wave washer
18. Right side shaft
19. Left side shaft
20. Front cap
21. Nut
22. Adjusting screw

7911D055

Rocker arms/shafts assembly — 2.4L engine. Note that 2.4L rocker arms are not equipped with jet valve extensions.

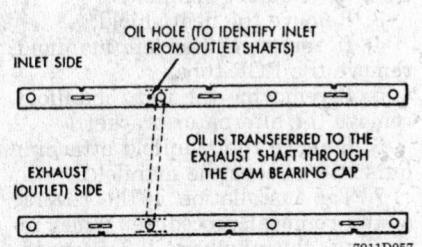

Identifying rocker shafts — 3.0L engine

from the middle and working outward, torque the retaining nuts to 12 ft. lbs. (16 Nm).

17. Connect the wires to the engine coolant switches. Install the distributor with matchmarks aligned.

18. Install the fuel rail assembly to the manifold and connect the fuel line using a new O-ring.

19. Connect the heater hose to the manifold.

20. Install the air intake plenum with a new gasket. Torque the retaining bolts to 12 ft. lbs. (16 Nm).

21. Connect the vacuum hoses cluster, brake booster hose and all wires,

hoses and linkages to the throttle body.

22. Install the ignition coil.

23. Install the air intake pipe and hoses.

24. Connect the upper radiator hose.

25. Fill the radiator with coolant.

26. Connect the negative battery cable and connect the jumper wire from the fuel pump activation terminal to the positive battery post to inspect the system for leaks.

27. Set all adjustments to specifications.

3.0L Engine

1. Relieve the fuel pressure.

2. Disconnect the negative battery cable.

3. Drain the engine coolant. Disconnect the upper radiator hose from the thermostat housing.

4. Remove the air intake hose from the throttle body.

5. Disconnect all wires, hoses and linkages to the throttle body.

6. Disconnect the EGR temperature sensor wire.

7. Remove the ignition coil.

8. Remove the engine oil filler neck bracket.

9. Unbolt the EGR tube from the air intake plenum.

10. Disconnect the PCV hose and vacuum hose cluster from the plenum.

11. Remove the plenum to engine brackets.

12. Unbolt the air intake plenum assembly from the intake manifold and remove.

13. Cover the fuel lines with a clean shop rag and disconnect the fuel lines from the fuel rail. Keep the lines covered.

14. Remove the fuel rail with injectors in place.

15. Disconnect the bypass hose and the upper radiator hose from the thermostat housing. Disconnect the wires to the coolant temperature switches.

16. Remove the intake manifold retaining nuts and remove the manifold from the cylinder heads.

17. Remove the gaskets and thoroughly clean and dry the mating surfaces of the manifold and heads.

To install:

18. Position the manifold over the studs and install the retaining nuts. Torque to 12 ft. lbs. (16 Nm), starting from the center and working outward.

19. Connect the hoses and connect the wires to the coolant switches.

20. Install the fuel rail assembly and connect the fuel hoses.

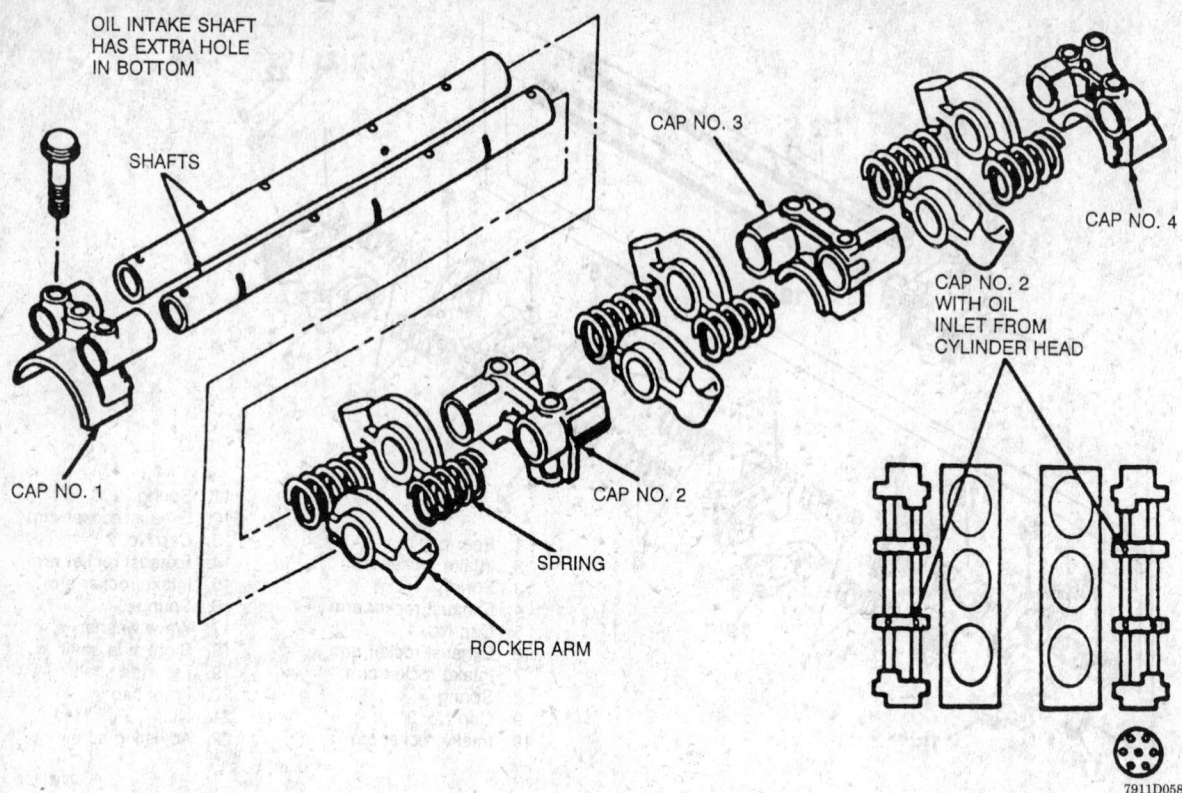

OIL INTAKE SHAFT HAS EXTRA HOLE IN BOTTOM

SHAFTS

CAP NO. 3

CAP NO. 4

CAP NO. 2 WITH OIL INLET FROM CYLINDER HEAD

CAP NO. 1

CAP NO. 2

SPRING

ROCKER ARM

7911D058

Rocker arms/shafts assembly — 3.0L engine

21. Using a new gasket, install the air intake plenum to the intake manifold. Torque the nuts and bolts to 12 ft. lbs. (16 Nm). Install the plenum to engine brackets.

22. Connect the PCV hose and vacuum hose cluster to the plenum.

23. Connect the EGR tube.

24. Install the engine oil filler neck bracket.

25. Install the ignition coil assembly.

26. Connect the EGR temperature sensor wire.

27. Connect all wires, hoses and linkages to the throttle body.

28. Install the air intake hose to the throttle body.

29. Connect the upper radiator hose to the thermostat housing.

30. Fill the radiator with coolant.

31. Connect the negative battery cable and connect the jumper wire from the fuel pump activation terminal to the positive battery post to inspect the system for leaks.

32. Set all adjustments to specifications.

Exhaust Manifold

REMOVAL AND INSTALLATION

2.4L Engine

1. Disconnect the negative battery cable.

2. Remove the heat cowl from the exhaust manifold.

3. Remove the aspirator valve assembly, if equipped.

4. Raise the vehicle and support safely. Disconnect the exhaust pipe from the manifold. Lower the vehicle.

5. Disconnect the oxygen sensor connector or ground cable, if equipped.

6. Remove the manifold mounting nuts and remove the manifold and gasket from the engine.

7. The installation is the reverse of the removal procedure. Torque the manifold mounting nuts to 13 ft. lbs. (18 Nm) starting from the middle and working outward.

8. Start the engine and check for exhaust leaks.

3.0L Engine

1. Disconnect the negative battery cable. Raise and safely support the vehicle.

2. Disconnect the exhaust pipe from the exhaust manifolds.

3. Remove the heat shield.

4. If removing the left manifold, remove the EGR tube.

5. If removing the right manifold, remove the alternator bracket.

6. Remove the manifold attaching nuts and remove the manifold.

7. The installation is the reverse of the removal procedure. When installing, the numbers 1–3–5 on the gaskets are used with the right side cylinders and 2–4–6 are on the gasket for the left side cylinders. Torque the manifold nuts to 14 ft. lbs. (19 Nm).

8. Start the engine and check for exhaust leaks.

Timing Belt Front Cover

REMOVAL AND INSTALLATION

2.4L Engine

1. Disconnect the negative battery cable. Remove the spark plug wires from the tree on the upper cover.

2. Drain the cooling system. Remove the shroud, fan, belts and radiator as required.

3. Remove the power steering pump, alternator, air conditioning

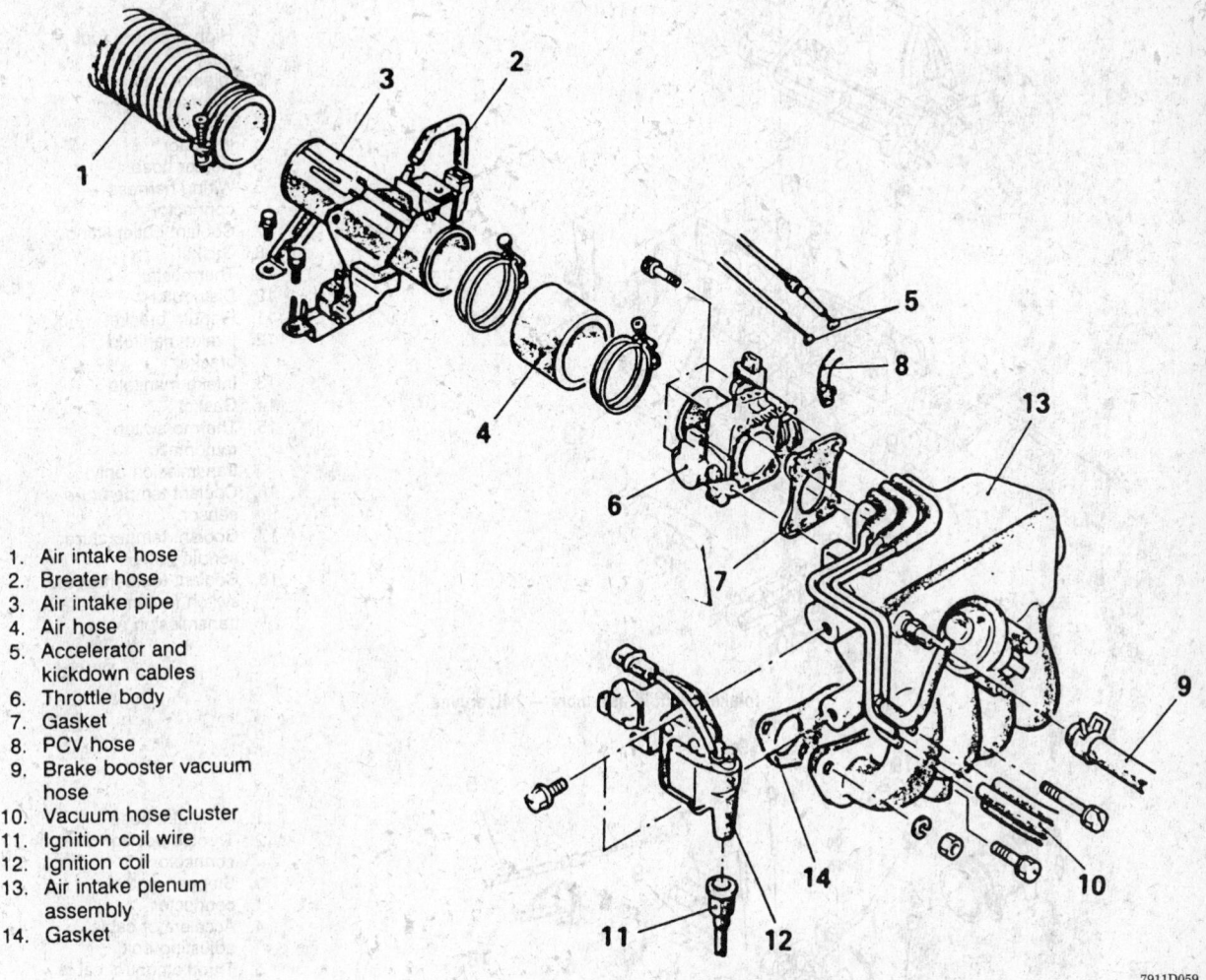

1. Air intake hose
2. Breater hose
3. Air intake pipe
4. Air hose
5. Accelerator and kickdown cables
6. Throttle body
7. Gasket
8. PCV hose
9. Brake booster vacuum hose
10. Vacuum hose cluster
11. Ignition coil wire
12. Ignition coil
13. Air intake plenum assembly
14. Gasket

7911D059

Air intake plenum assembly — 2.4L engine

compressor, tension pulley and accompanying brackets, as required.

4. Remove the upper front timing belt cover.

5. Remove the water pump pulley and the crankshaft pulley(s).

6. Remove the lower timing belt cover to engine screws and remove the cover.

7. The installation is the reverse of the removal procedure. Make sure the packing is positioned in the inner grooves of the covers properly when installing.

3.0L Engine

1. Disconnect the negative battery cable.

2. Drain the cooling system. Remove the drive belts.

3. Remove the upper radiator shroud.

4. Remove the fan and fan pulley.

5. Without disconnecting the lines, remove the power steering pump from its bracket and position it to the side. Remove the pump brackets.

6. Remove the belt tensioner pulley bracket.

7. Without releasing the refrigerant remove the air conditioning compressor from its bracket and position it to the side. Remove the bracket.

8. Remove the cooling fan bracket.

9. Remove the crankshaft pulley bolt and the pulley from the crankshaft.

10. Remove the timing belt cover bolts and the upper and lower covers from the engine.

To install:

11. Install the upper and lower covers to the engine and secure with the retaining screws. Make sure the packing is positioned in the inner grooves of the covers properly when installing.

12. Install the crankshaft pulley and bolts. Torque bolt to 110 ft. lbs. (150 Nm).

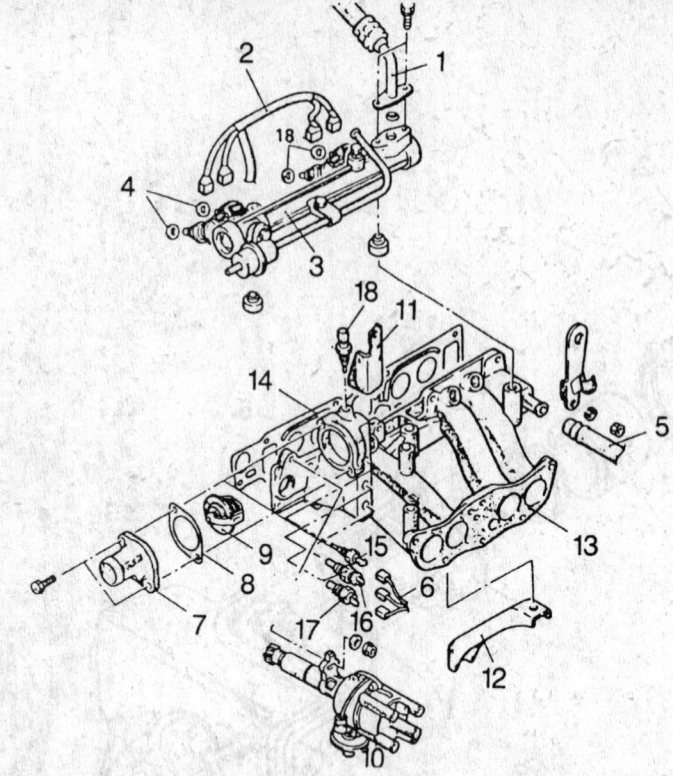

1. High pressure fuel hose
2. Injector harness connector
3. Fuel rail
4. Insulator
5. Heater hose
6. Wiring harness connector
7. Coolant outlet fitting
8. Gasket
9. Thermostat
10. Distributor
11. Plenum bracket
12. Intake manifold bracket
13. Intake manifold
14. Gasket
15. Thermo switch (automatic transmission only)
16. Coolant temperature sensor
17. Coolant temperature sending unit
18. Coolant temperature switch (automatic transmission only)

7911D060

Intake manifold assembly — 2.4L engine

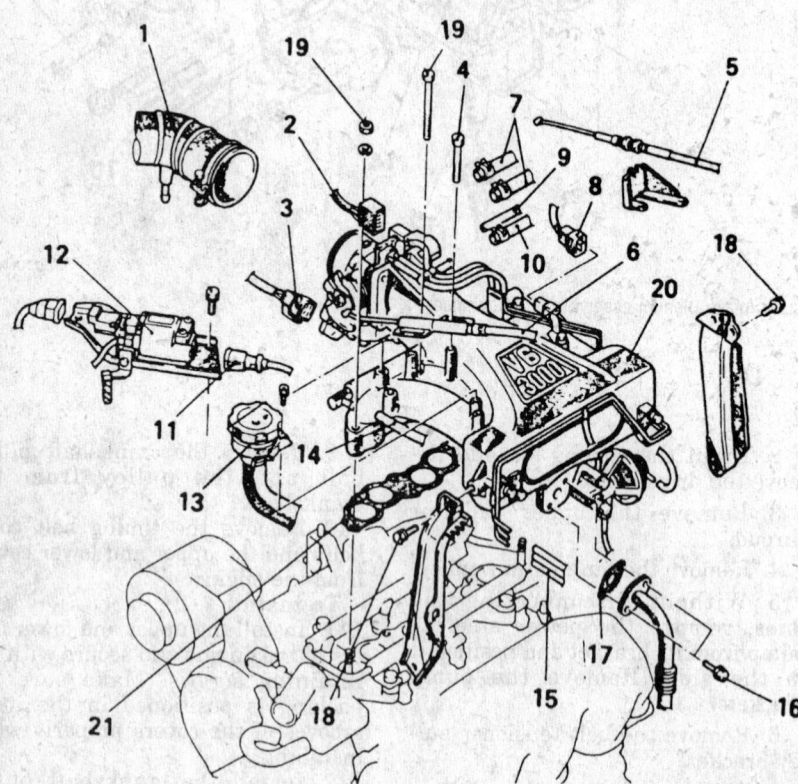

1. Air intake hose
2. Throttle position sensor connector
3. Stepper motor connector
4. Accelerator cable adjusting bolt
5. Throttle control cable (automatic transmission only)
6. Accelerator cable
7. Coolant hoses
8. EGR temperature connector
9. Vacuum hose
10. Brake booster vacuum hose
11. Ignition coil wire
12. Ignition coil
13. Oil filler neck bracket
14. PCV hose
15. Vacuum hose cluster
16. EGR pipe attaching bolt
17. Gasket
18. Bracket bolt
19. Attaching bolt and nut
20. Air intake plenum and throttle body assembly
21. Gasket

7911D061

Air intake plenum assembly — 3.0L engine

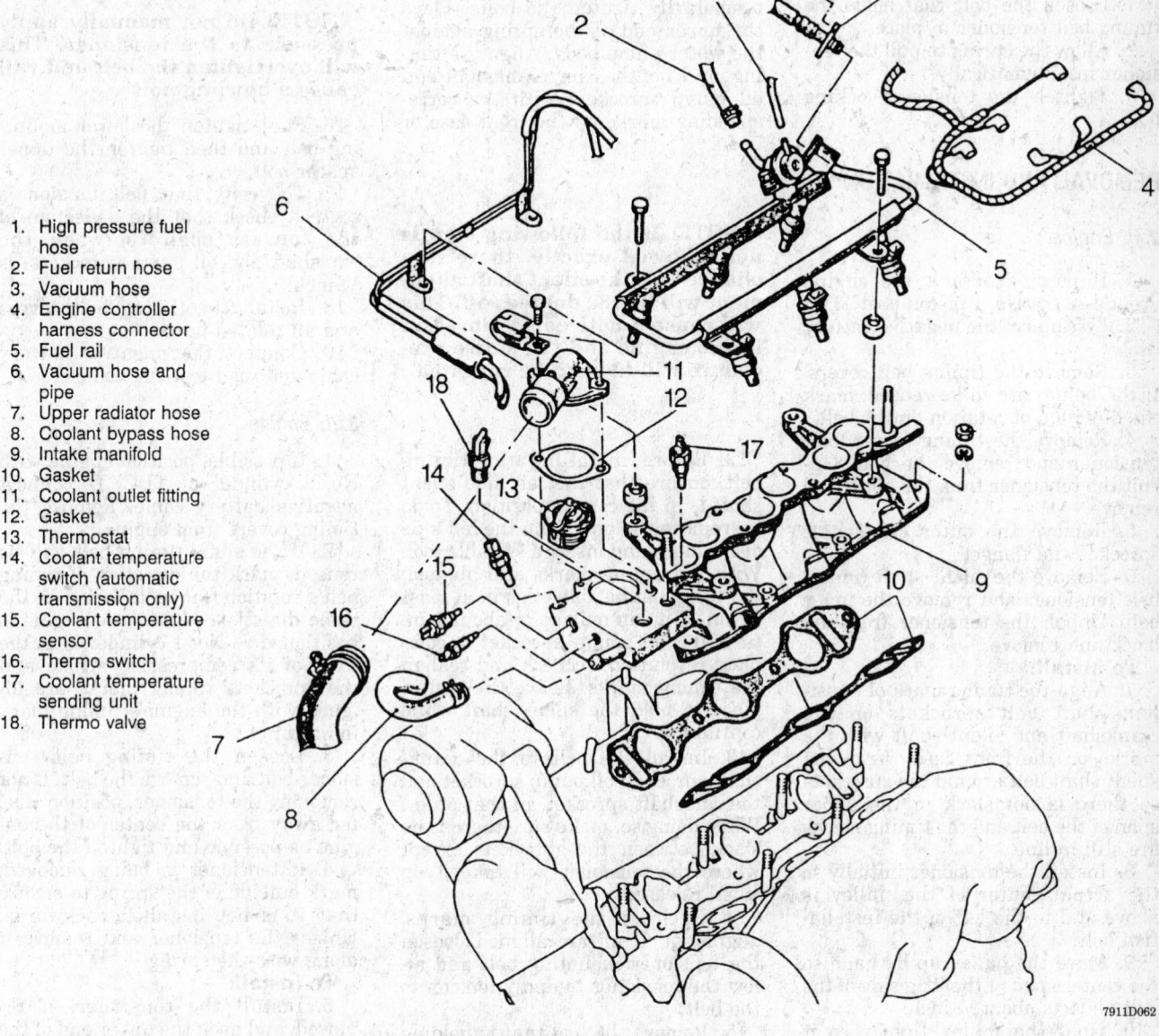

1. High pressure fuel hose
2. Fuel return hose
3. Vacuum hose
4. Engine controller harness connector
5. Fuel rail
6. Vacuum hose and pipe
7. Upper radiator hose
8. Coolant bypass hose
9. Intake manifold
10. Gasket
11. Coolant outlet fitting
12. Gasket
13. Thermostat
14. Coolant temperature switch (automatic transmission only)
15. Coolant temperature sensor
16. Thermo switch
17. Coolant temperature sending unit
18. Thermo valve

Intake manifold assembly — 3.0L engine

13. Install the air conditioning bracket and compressor to the engine. Install the belt tensioner.

14. Install the power steering pump into position. Install the fan pulley and fan.

15. Install the fan shroud to the radiator. Fill the cooling system, reconnect the negative battery cable and check for fluid leaks.

Timing Belt and Tensioner

ADJUSTMENT

2.4L Engine

1. Disconnect the negative battery cable.

2. Remove the timing belt cover.

3. Adjust the silent shaft (inner) belt first. Loosen the pulley center bolt so the pulley may be moved.

4. Move the pulley up by hand so the center span of the long side of the belt deflects ¼ inch.

5. Hold the pulley tight so the pulley itself does not rotate when the bolt is tightened. Tighten the bolt to

15 ft. lbs. (20 Nm). If the pulley has moved, the belt will be too tight.

6. Check the timing (outer) belt tension.

7. To adjust the timing (outer) belt, first loosen the tensioner pulley bolts.

8. Allow the spring to take up any slack. Check that the the deflection of the longest span (between the camshaft and oil pump sprockets) is ½ inch. Do not overtighten the belt or it will howl.

9. First tighten the lower pulley bolts to 35 ft. lbs. (47 Nm) and then the upper bolt to the same value.

10. Install the covers and all related parts.

3.0L Engine

1. Loosen the bolt that holds the timing belt tensioner in place.
2. Allow the spring to pull the tensioner in automatically.
3. Tighten the tensioner locking bolt.

REMOVAL AND INSTALLATION

2.4L Engine

1. If possible, crank the engine around so the No. 1 piston is at TDC.
2. Disconnect the negative battery cable.
3. Remove the timing belt covers. If the belt(s) are to be reused, mark the direction of rotation on the belt.
4. Remove the timing (outer) belt tensioner and remove the belt. Unbolt the tensioner from the block and remove.
5. Remove the outer crankshaft sprocket and flange.
6. Remove the silent shaft (inner) belt tensioner and remove the inner belt. Unbolt the tensioner from the block and remove.

To install:

7. Align the timing mark of the silent shaft belt sprockets on the crankshaft and silent shaft with the marks on the front case. Wrap the silent shaft belt around the sprockets so there is not slack in the upper span of the belt and the timing marks are still in line.
8. Install the tensioner initially so the actual center of the pulley is above and to the left of the installation bolt.
9. Move the pulley up by hand so the center span of the long side of the belt deflects about ¼ inch.
10. Hold the pulley tightly so it does not rotate when the bolt is tightened. Tighten the bolt to 15 ft. lbs. (20 Nm). If the pulley has moved, the belt will be too tight.

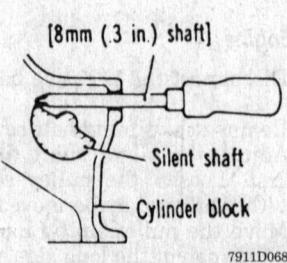

TIMING MARKS

TIMING MARKS

7911D067

Silent shaft belt installation — 2.4L engine

11. Install the timing belt tensioner fully toward the water pump and temporarily tighten the bolts. Place the upper end of the spring against the water pump body. Align the timing marks of the cam, crankshaft and oil pump sprockets with the corresponding marks on the front case or head.

NOTE: If the following step is not followed exactly, there is a chance that the silent shaft alignment will be 180 degrees off. This will cause a noticeable vibration in the engine and the entire procedure will have to be repeated.

12. Before installing the timing belt, ensure that the left side silent shaft is in the correct position. To do so, remove the plug from the left side of the block and insert a suitable tool. With the timing marks still aligned, the tool must be able to go in at least 2⅓ in. If it can only go in about 1 in. turn the oil pump sprocket 1 complete revolution, recheck and realign the timing marks. Leave the tool in place to hold the silent shaft while continuing.
13. Install the belt to the crankshaft sprocket, oil pump sprocket and the camshaft sprocket, in that order. While doing so, make sure there is no slack between the sprockets except where the tensioner will take it up when released.
14. Recheck the timing marks' alignment. If all are aligned, loosen the tensioner mounting bolt and allow the tensioner to apply tension to the belt.
15. Remove the tool that is holding the silent shaft in place and turn the crankshaft clockwise a distance equal to 2 teeth of the camshaft sprocket. This will allow the tensioner to auto-

[8mm (.3 in.) shaft]

Silent shaft

Cylinder block

7911D068

Checking the left side silent shaft for proper positioning

matically tension the belt the proper amount.

NOTE: Do not manually apply pressure to the tensioner. This will overtighten the belt and will cause a howling noise.

16. First tighten the lower mounting bolt and then tighten the upper spacer bolt.
17. To verify that belt tension is correct, check that the deflection of the longest span (between the camshaft and oil pump sprockets) is ½ inch.
18. Install the timing belt covers and all related parts.
19. Connect the negative battery cable and road test the vehicle.

3.0L Engine

1. If possible, position engine with No. 1 cylinder at TDC. Disconnect negative battery cable. Remove the timing covers from engine.
2. If the same timing belt will be reused, mark the direction of timing belt's rotation, for installation in the same direction. Make sure engine is positioned so No. 1 cylinder is at the TDC of its compression stroke and the sprockets timing marks are aligned with the engine's timing mark indicators.
3. Loosen the timing belt tensioner bolt and remove the belt. If not removing the tensioner, position it as far away from the center of the engine as possible and tighten the bolt.
4. If tensioner is being removed, mark outside of the spring to ensure that it is not installed backwards. Unbolt the tensioner and remove it along with the spring.

To install:

5. Install the tensioner, if removed, and hook the upper end of the spring to the water pump pin. Install the lower end of the spring to the tensioner in exactly the same position as originally installed. If not already done, position both camshafts so the timing marks line up with those on the alternator bracket (rear bank) and inner timing cover (front bank). Rotate the crankshaft so the timing mark aligns with the mark on the oil pump.
6. Install the timing belt on the crankshaft sprocket and while keeping the belt tight on the tension side (right side), install the belt on the front camshaft sprocket.
7. Install the belt on the water pump pulley, then the rear camshaft sprocket and the tensioner.
8. Rotate the front camshaft counterclockwise to tension the belt be-

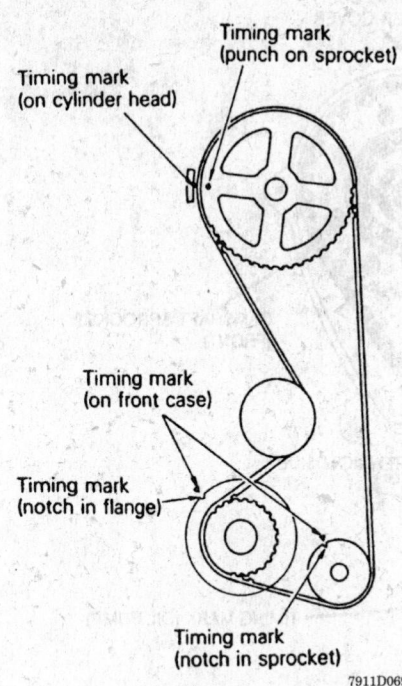

Timing mark (punch on sprocket)

Timing mark (on cylinder head)

Timing mark (on front case)

Timing mark (notch in flange)

Timing mark (notch in sprocket)

7911D069

Timing belt installation — 2.4L engine

tween the front camshaft and the crankshaft. If the timing marks came out of line, repeat the procedure.

9. Install the crankshaft sprocket flange.

10. Loosen the tensioner bolt and allow the spring to tension the belt.

11. Turn the crankshaft 2 full turns in the clockwise direction until the timing marks align. Now that the belt is properly tensioned, torque the tensioner lock bolt to 21 ft. lbs. (29 Nm).

12. Install the timing belt covers and all related parts.

13. Connect the negative battery cable and road test the vehicle.

Timing Sprockets

REMOVAL AND INSTALLATION

1. Disconnect the negative battery cable.

2. Remove the timing belt(s).

3. To remove the camshaft sprocket, hold the sprocket with a holding tool and remove the retaining bolt and washer.

4. To remove the crankshaft sprocket, remove the bolt, if equipped, and remove the sprocket from the crankshaft.

5. To remove a silent shafts sprocket, remove the retaining bolt and pry the sprocket from its mounting shaft.

6. The installation is the reverse of the removal procedure. Torque the sprocket retaining bolt to specifications.

Camshaft

REMOVAL AND INSTALLATION

2.4L Engine

1. Disconnect the negative battery cable. Remove the valve cover and the upper timing belt cover.

2. Matchmark the rotor to the distributor housing and remove the distributor.

3. Secure the belt to the sprocket and hold the sprocket up with a fabricated device to keep the belt taut. If timing is lost, it will have to be reset. Remove the camshaft sprocket from the camshaft.

4. Remove the camshaft cap bolts evenly and gradually. Install the auto lash adjuster retainers to the rocker arms.

5. Remove the caps, shafts, rocker arms and bolts together as an assembly.

6. Remove the camshaft with the front seal from the engine.

To install:

7. Install a new roll pin to the camshaft. Lubricate the camshaft and install with the front seal in place. Install the camshaft so the hole in the sprocket will line up with the roll pin.

8. Install the caps, shafts and arms assembly. Tighten the camshaft bearing cap bolts in the following order to 85 inch lbs. (10 Nm): No. 3, No. 2, No. 4, front cap, rear cap. Repeat the sequence increasing the torque to 175 inch lbs. (19 Nm).

9. Install the sprocket to the camshaft, engaging the roll pin. Torque the bolt to 70 ft. lbs. (95 Nm). Install the distributor.

10. Install the valve cover and all related parts.

3.0L Engine

1. Disconnect the negative battery cable. Remove the valve cover.

2. Remove the timing belt and remove the sprocket from the camshaft.

3. Install auto lash adjuster retainers MD998443 or equivalent on the rocker arms.

4. If removing the left side camshaft, remove the distributor and the distributor extension.

5. Remove the camshaft bearing caps but do not remove the bolts from the caps. Remove the rocker arms, rocker shafts and bearing caps, as an assembly.

6. Remove the camshaft from the cylinder head.

7. Inspect the bearing journals on the camshaft, cylinder head and bearing caps.

To install:

8. Lubricate the camshaft journals and camshaft with clean engine oil and install the camshaft in the cylinder head.

9. Align the camshaft bearing caps with the arrow mark (depending on cylinder numbers) and in numerical order.

10. Apply sealer at the ends of the bearing caps and install the assembly.

11. Torque the bearing cap bolts, in the following sequence: No. 3, No. 2, No. 1 and No. 4 to 85 inch lbs. (10 Nm).

12. Repeat the sequence increasing the torque to 175 inch lbs. (18 Nm).

13. Install the distributor, if removed.

14. Install the sprocket, timing belt and all related parts.

15. Install the valve cover.

Silent Shafts

REMOVAL AND INSTALLATION

2.4L Engine

1. Disconnect the negative battery cable.

2. Remove the timing belts.

3. Remove the timing belt sprockets.

4. Remove the front case.

5. Remove the oil pump components.

6. Remove the shafts from the block.

To install:

7. Coat the silent shafts with clean engine oil and install into the block.

8. Install the drive and driven gears into place so the timing marks are mated with each other. Coat gears with clean engine oil.

9. Attach the special tool to the end of the crankshaft and coat the outer surface with engine oil.

10. Install the front case assembly with a new front case gasket to the block. Install retaining bolts loosely.

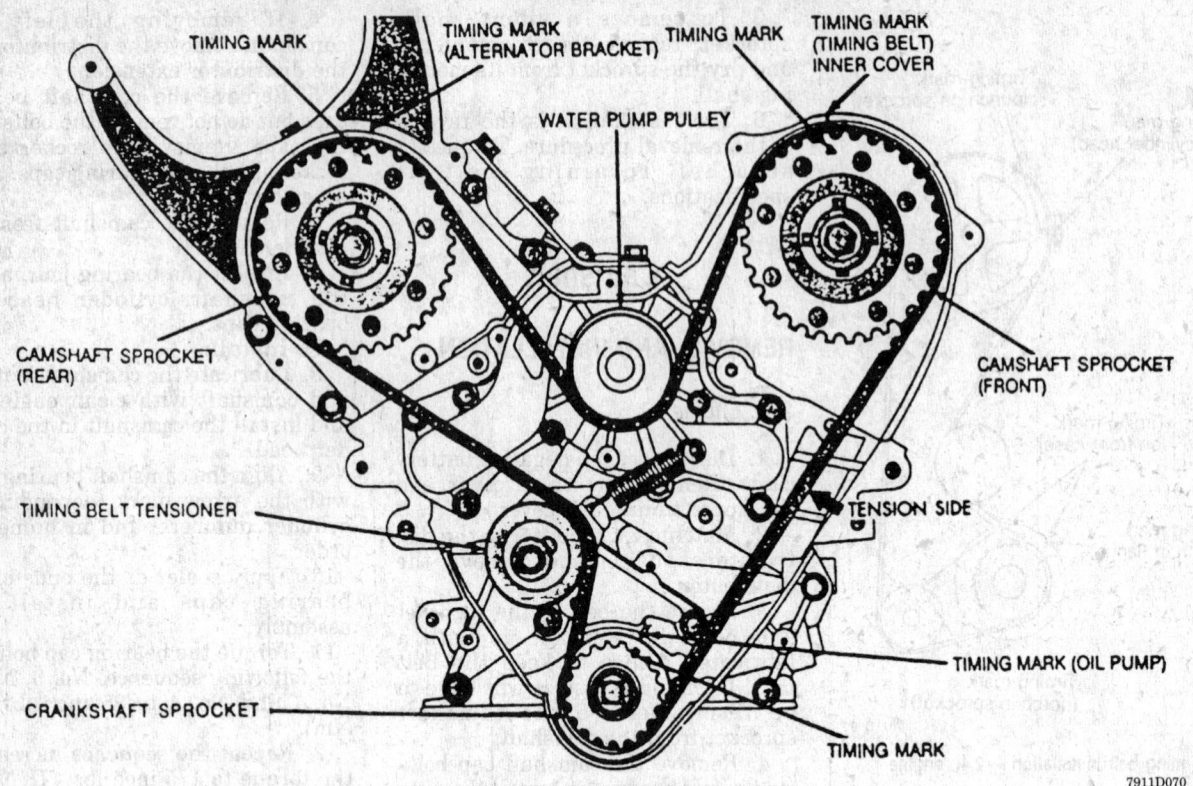

Timing belt installation — 3.0L engine

7911D070

11. Install the oil filter bracket together with the oil filter bracket gasket and install retaining bolts with washers. Torque bolts to 15 ft. lbs. (22 Nm).

12. Install the timing belts, covers and related parts.

13. Connect the negative battery cable, start engine and check for leaks.

Piston and Connecting Rod

POSITIONING

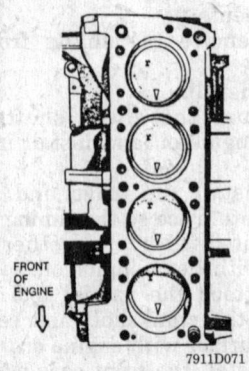

FRONT OF ENGINE

7911D071

Piston positioning — 2.4L engine

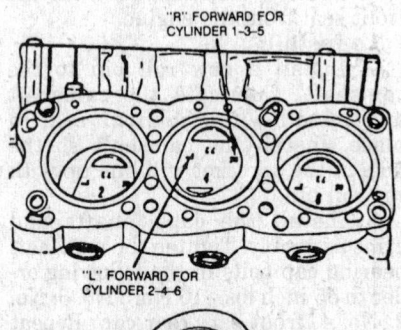

"R" FORWARD FOR CYLINDER 1-3-5

"L" FORWARD FOR CYLINDER 2-4-6

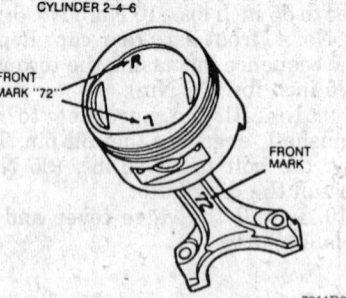

FRONT MARK "72"

R

L

FRONT MARK

7911D072

Piston positioning — 3.0L engine

ENGINE LUBRICATION

Oil Pan

REMOVAL AND INSTALLATION

1. Disconnect the negative battery cable.

2. Raise the vehicle and support safely.

3. Remove the skid plate(s), engine undercover, air guide plate and cross shaft plate as required.

4. Remove the front exhaust pipe and steering linkage components as required.

5. On some vehicles, unbolt the motor mounts and raise the engine safely using the proper equipment.

6. Drain the engine oil from the pan.

7. Disconnect the fluid level sensor wire, if equipped. Remove the attaching bolts and remove the pan.

8. The installation is the reverse of the removal procedure.

Oil Pump

REMOVAL AND INSTALLATION

2.4L Engine

1. Disconnect the negative battery cable. Remove the timing belt covers, timing belts and sprockets.

2. Raise the vehicle and support safely. Drain the oil and remove the oil filter. Remove the oil pan and gasket. Remove the oil pump pickup and gasket.

3. Remove the oil pressure relief plunger plug and gasket. Remove the spring and plunger from the oil filter bracket.

4. Remove the 4 bracket mounting bolts and remove the oil filter mount and gasket.

5. Using special tool MD998162, remove the cap and gasket that covers the oil pump driven gear shaft. This is located on the right side of the front case at the front of the engine, just above the protruding drive gear shaft.

6. Using a long socket, remove the retaining bolt from the oil pump driven gear located behind the plug removed earlier.

7. Remove the front case mounting bolts and remove the case from the block.

8. Remove the case gasket from the block.

To install:

9. Prime the pump by pouring fresh oil into the pump intake and turning the driveshaft until oil comes out the pressure port. Repeat a few times until no air bubbles are present. Replace all seals on the case assembly.

10. Install a special seal guide to the crankshaft MD998285 or equivalent so the smaller diameter faces outward. Coat the outer diameter of the seal with clean engine oil.

11. Install a new front case gasket and install the front case by carefully positioning the crankshaft seal over the seal guide and lining up all bolt holes. Install and tighten the bolts to 17 ft. lbs. (23 Nm).

12. Remove the plug from the left side of the block. Hold the left side silent shaft by inserting a tool in the plug hole and torque the driven gear bolt to 26 ft. lbs. (35 Nm). Using a new O-ring, install the plug cover.

13. Install the oil filter mounting bracket gasket. Install the mounting bracket and bolts tightening the oil filter mounting bracket bolts to 12 ft. lbs. (16 Nm).

14. Clean or replace the oil pickup screen and install with a new gasket.

15. Install the oil pan using a new gasket.

16. Install the timing sprockets, belts and covers.

17. Fill the engine with the proper amount of engine oil.

18. Connect the negative battery cable and check for proper oil pressure and leaks.

3.0L Engine

1. Disconnect the negative battery cable. Remove the dipstick.

2. Raise the vehicle and support safely. Remove the timing belt, drain the engine oil and remove the oil pan from the engine. Remove the oil pickup.

3. Remove the oil pump mounting bolts and remove the pump from the front of the engine. Note the different length bolts and their position in the pump for installation.

To install:

4. Clean the gasket mounting surfaces of the pump and engine block.

5. Prime the pump by pouring fresh oil into the inlet and turning the rotors or by packing pump with petroleum jelly. Using a new gasket, install the oil pump on the engine and torque all bolts to 11 ft. lbs. (15 Nm).

6. Install the balancer and crankshaft sprockets.

7. Clean out the oil pickup or replace as required. Replace the oil pickup gasket ring and install the pickup to the pump.

8. Install the timing belt, oil pan and all related parts.

9. Install the dipstick. Fill the engine with the proper amount of oil.

10. Connect the negative battery cable and check the oil pressure.

CHECKING

2.4L Engine

1. Remove the oil pump cover from the front case.

2. Check the case and gear cover for stepped wear from the gears.

3. Measure the drive gear tip clearance between the teeth and the case. The specification is 0.006–0.010 in. (0.16–0.25mm).

4. Check the driven gear tip clearance in the similar manner. The specification is 0.005–0.010 in. (0.13–0.25mm).

5. Place a straight-edge across the gears resting on the opposite sides of the case. If a 0.010 in. (0.25mm) feeler gauge can be inserted under the straight-edge, replace the assembly.

6. Check the relief plunger for freedom of movement and check the spring for deformation and rust.

7. If the gears were removed from the body, install them with the mating marks aligned. If not aligned properly, the silent shaft will be out of time.

8. Install the gear cover to the case using a new gasket. Torque the bolts to 12 ft. lbs. (16 Nm).

3.0L Engine

1. Remove the rear cover.

2. Remove the pump rotors and inspect the case for excessive wear.

3. Measure the diameter of the inner rotor hub that sits in the case. Measure the inside diameter of the inner rotor hub bore. Subtract the first measurement from the second; if the result is over 0.006 in. (0.15mm), replace the oil pump assembly.

4. Measure the clearance between the outer rotor and the case. The specification is 0.004–0.007 in. (0.10–0.18mm).

5. Check the side clearance of the rotors using a feeler gauge and a straight-edge placed across the case. The specification is 0.0015–0.0035 in. (0.04–0.09mm).

6. Check the relief plunger and spring for damage and replace as required.

7. Install the rear pump cover to the case.

Rear Main Bearing Oil Seal

REMOVAL AND INSTALLATION

1. Disconnect the negative battery cable.

2. Support the weight of the engine using a jack stand with a block of wood to protect the oil pan. Remove the transmission from the vehicle.

3. If equipped with manual transmission, remove the clutch cover retainer bolts in a cross fashion gradually. Remove the clutch cover and disc. Remove the retaining bolt, flywheel and ball bearing from the engine.

4. If equipped with automatic transmission, remove the adapter plate, driveplate, crankshaft adapter and bushing.

5. Remove the rear engine plate and lower bell housing cover.

6. Remove the rear oil seal case and gasket from the back of the engine.

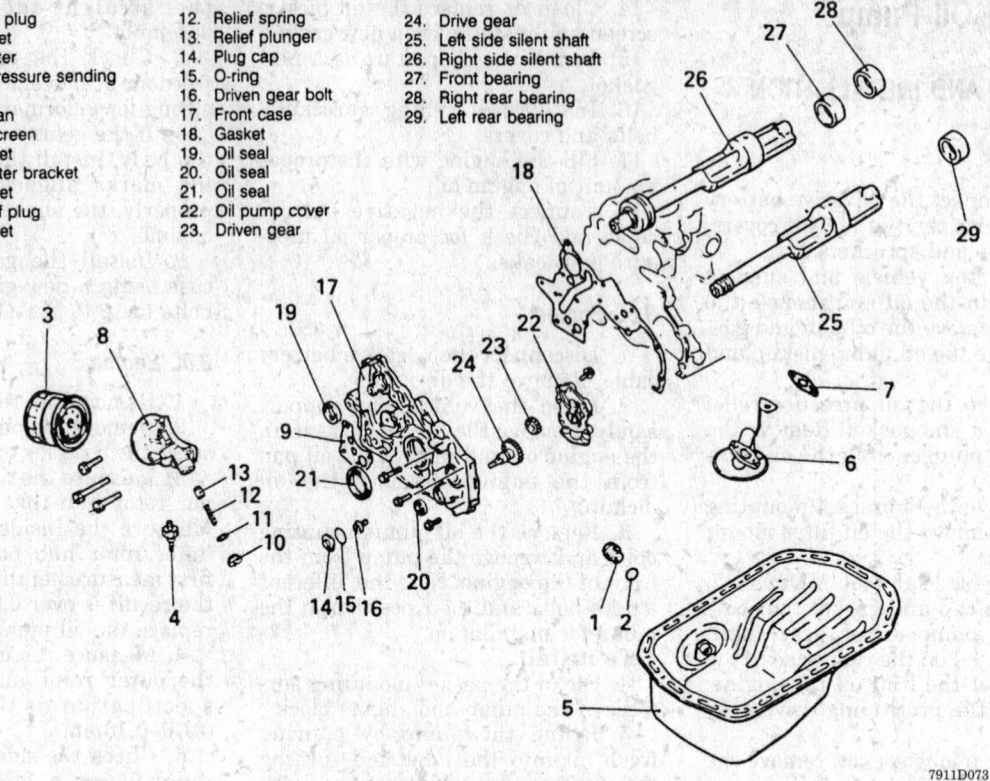

1. Drain plug
2. Gasket
3. Oil filter
4. Oil pressure sending unit
5. Oil pan
6. Oil screen
7. Gasket
8. Oil filter bracket
9. Gasket
10. Relief plug
11. Gasket
12. Relief spring
13. Relief plunger
14. Plug cap
15. O-ring
16. Driven gear bolt
17. Front case
18. Gasket
19. Oil seal
20. Oil seal
21. Oil seal
22. Oil pump cover
23. Driven gear
24. Drive gear
25. Left side silent shaft
26. Right side silent shaft
27. Front bearing
28. Right rear bearing
29. Left rear bearing

Engine lubrication components — 2.4L engine

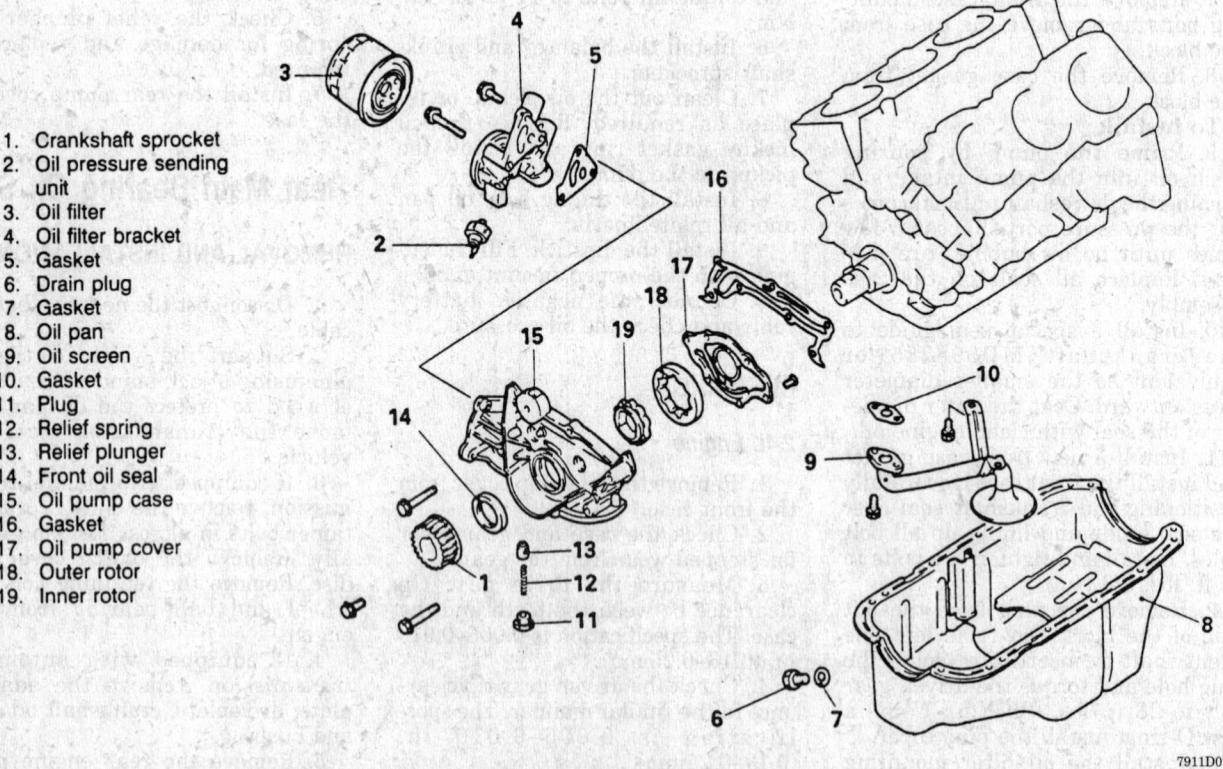

1. Crankshaft sprocket
2. Oil pressure sending unit
3. Oil filter
4. Oil filter bracket
5. Gasket
6. Drain plug
7. Gasket
8. Oil pan
9. Oil screen
10. Gasket
11. Plug
12. Relief spring
13. Relief plunger
14. Front oil seal
15. Oil pump case
16. Gasket
17. Oil pump cover
18. Outer rotor
19. Inner rotor

Engine lubrication components — 3.0L engine

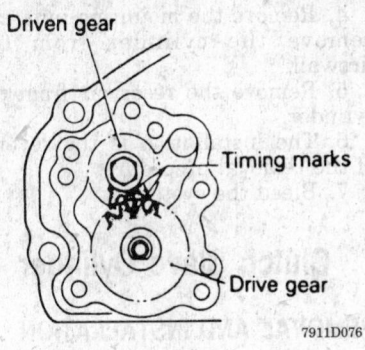

Aligning the oil pump gear timing marks — 2.4L engine

7. Remove the oil separator if remaining on the engine block.

To install:

8. Install the new oil seal into the seal case using seal installer.

9. Press the oil separator into the seal case making sure the separator oil hole is on the very bottom of the case. In the correct position, the hole is on the adjacent to the oil pan mounting surface.

10. Clean and dry the seal case to block sealing surfaces. Install seal case to the rear of the block using a new gasket.

11. Install the rear engine plate and the lower bell housing cover.

12. If equipped with automatic transmission, install the adapter plate, driveplate, crankshaft adapter and bushing.

13. If equipped with manual transmission, install the retaining bolt, flywheel and ball bearing to the engine.

14. Install the transmission into the vehicle. Lower the vehicle and remove the engine support. Connect the negative battery cable and check the oil level, adding as required.

MANUAL TRANSMISSION

Transmission Assembly

REMOVAL AND INSTALLATION

1. Disconnect the negative battery cable.

2. Place the shifter(s) in the neutral position.

3. Unscrew the shift knob from the control lever. Remove the re-tainer screws from the dust cover retaining plate and slide plate and boot off of the lever.

4. Remove the retainer screws from the stopper plate and remove the lever assembly from the transmission. Cover the opening with a clean towel to prevent dirt from entering the transmission.

5. Raise the vehicle and support safely. Remove the skid plate, if equipped.

6. Drain the transmission fluid. If equipped with 4WD, drain the oil from the transfer case.

7. Matchmark and remove the driveshaft(s) from the vehicle.

8. Disconnect the speedometer cable from the transmission or transfer case.

9. Disconnect the clutch cable connection.

10. Disconnect the reverse light switch harness connector.

11. Remove the starter and the bell housing cover.

12. Support the weight of the engine using a jack stand with a block of wood to protect the oil pan.

13. Support the transmission with a transmission jack.

14. Remove the transfer case bracket, if equipped.

15. Remove the rear crossmember.

16. Remove the transmission to bell housing bolts.

17. Slide the transmission backwards until the input shaft clears the clutch disc. Remove the transmission from the vehicle.

To install:

18. Lubricate the pilot bushing and input shaft splines very lightly with high temperature lubricant.

19. Mount the transmission securely on a transmission jack and lift it in place until the input shaft is centered in the bell housing opening. Roll the transmission forward until the input shaft splines fully engage with the clutch disc.

20. Install the transmission to bell housing bolts. Torque the bolts to 35 ft. lbs. (47 Nm).

21. Lift the transmission using the transmission jack and install the rear crossmember into position. Install the transfer case bracket, if equipped. Torque the frame bolts to 50 ft. lbs. (68 Nm). Remove the transmission and engine support fixtures.

22. Install the starter and torque the mounting bolt to 20 ft. lbs. (27 Nm). Install the bell housing cover.

23. Connect the reverse light switch and clip all wiring to the transmission case.

24. Connect the speedometer cable.

25. Install the driveshaft(s) making sure to align matchmarks.

26. Connect the clutch cable, using a new cotter pin.

27. Fill the transmission and transfer case with the proper amount of SAE 80W or 75W/85W hypoid gear oil with an API classification of GL–4 or higher.

28. Install the skip plate, if equipped. Lower the vehicle.

29. Install the shift lever assembly, boot and console.

30. Connect the negative battery cable and check the transmission and transfer case for proper operation. Check operation of the reverse lights.

LINKAGE ADJUSTMENT

Since this transmission uses a direct engage shift mechanism, there are no provisions for adjustment.

CLUTCH

Clutch Assembly

REMOVAL AND INSTALLATION

1. Disconnect the negative battery cable.

2. Place the shifter(s) in the neutral position.

3. Unscrew the shift knob from the control lever. Remove the retainer screws from the dust cover retaining plate and slide plate and boot off of the lever.

4. Remove the retainer screws from the stopper plate and remove the lever assembly from the transmission. Cover the opening with a clean towel to prevent dirt from entering the transmission.

5. Raise the vehicle and support safely. Remove the skid plate, if equipped.

6. Drain the transmission fluid. If equipped with 4WD, drain the oil from the transfer case.

7. Matchmark and remove the driveshaft(s) from the vehicle.

8. Disconnect the speedometer cable from the transmission or transfer case.

9. Disconnect the clutch cable connection.

10. Disconnect the reverse light switch harness connector.

11. Remove the starter and the bell housing cover.

12. Support the weight of the engine using a jack stand with a block of wood to protect the oil pan.

13. Support the transmission with a transmission jack.

14. Remove the transfer case bracket, if equipped.

15. Remove the rear crossmember.

16. Remove the transmission to bell housing bolts.

17. Slide the transmission backwards until the input shaft clears the clutch disc. Remove the transmission from the vehicle.

18. Remove the clutch cover retainer bolts in a cross fashion gradually. Remove the clutch cover and disc.

To install:

19. Raise the clutch cover and disc into place and use a clutch aligning tool or spare input shaft to center the disc. Apply Loctite® to the threads and tighten all of the bolts finger tight.

20. Tighten cover bolts gradually, evenly and to the proper torque to avoid distorting the cover. Torque the bolts to 15 ft. lbs. (20 Nm).

21. Lubricate the pilot bushing and input shaft splines very lightly with high temperature lubricant.

22. Mount the transmission securely on a transmission jack and raise in place until the input shaft is centered in the bell housing opening. Roll the transmission forward until the input shaft splines fully engage with the clutch disc.

23. Install the transmission to bell housing bolts. Torque the bolts to 35 ft. lbs. (47 Nm).

24. Lift the transmission using the transmission jack and install the rear crossmember into position. Install the transfer case bracket, if equipped. Torque the frame bolts to 50 ft. lbs. (68 Nm). Remove the transmission and engine support fixtures.

25. Install the starter and torque the mounting bolt to 20 ft. lbs. (27 Nm). Install the bell housing cover.

26. Connect the reverse light switch and clip all wiring to the transmission case.

27. Connect the speedometer cable.

28. Install the driveshaft(s) making sure to align matchmarks.

29. Connect the clutch cable, using a new cotter pin.

30. Fill the transmission and transfer case with the proper amount of SAE 80W or 75W/85W hypoid gear oil with an API classification of GL–4 or higher.

31. Install the skip plate, if equipped. Lower the vehicle.

32. Install the shift lever assembly and boot.

33. Connect the negative battery cable and check the transmission and transfer case for proper operation. Check operation of the reverse lights.

Pedal Height/Free-Play Adjustment

1. Measure the distance from the face of the pedal pad to the floorboard. This distance should be about 6½–7 in.

2. If the pedal height is not correct, adjust the pedal stopper to contact the pedal at correct height.

Clutch Cable

ADJUSTMENT

1. Check the pedal's free-play. The specification is 1 inch.

2. Adjust the clutch cable by turning the star shaped adjusting wheel on the firewall. Turning the wheel counterclockwise will increase the amount of free-play in the pedal or vice-versa.

REMOVAL AND INSTALLATION

1. Disconnect the negative battery cable.

2. Turn the cable adjusting wheel counterclockwise to provide enough play to remove the cable end from the clutch lever inside the vehicle. Remove the cable from the lever.

3. Raise the vehicle and support safely.

4. Remove the cotter pin from the lever on the transmission.

5. Remove the clutch cable from the vehicle.

6. The installation is the reverse of the removal procedure. Make sure the insulator is positioning properly.

Clutch Master Cylinder

REMOVAL AND INSTALLATION

1. Disconnect the negative battery cable. Remove as much fluid as possible from the clutch master cylinder reservoir.

2. Remove the cotter pin and remove the clevis pin from the clutch pedal.

3. Disconnect and plug the fluid line from the clutch master cylinder.

4. Remove the mounting nuts and remove the cylinder from the firewall.

5. Remove the reservoir from the cylinder.

6. The installation is the reverse of the removal procedure.

7. Bleed the system.

Clutch Slave Cylinder

REMOVAL AND INSTALLATION

1. Disconnect the negative battery cable.

2. Raise the vehicle and support safely.

3. Remove the eye-bolt and washer from the slave cylinder.

4. Remove the mounting bolts and remove the slave cylinder from the transmission case.

5. Replace the eye-bolt washer.

6. The installation is the reverse of the removal procedure.

7. Bleed the system.

Hydraulic Clutch System Bleeding

───── **CAUTION** ─────

When bleeding, keep the facial area well away from the slave cylinder and protect all painted surfaces from fluid contact. Brake fluid will damage painted surfaces and could cause physical injury.

1. Fill the clutch master cylinder with fresh DOT 3 brake fluid.

2. Have a helper sit in the vehicle. Raise the vehicle and support safely.

3. Remove the bleeder screw cap.

4. If the system is empty, the most efficient way to get fluid down to the cylinder is to loosen the bleeder about ½–¾ turn, place a finger firmly over the bleeder and have the helper pump the brakes slowly until fluid pressure is felt at the bleeder. Once fluid is at the bleeder, close before the pedal is released.

NOTE: If the pedal is pump rapidly, the fluid will churn and create small air bubbles, which are difficult and time consuming to remove from the system. These air bubbles will eventually congregate and will result in a spongy pedal.

5. Once fluid has been pumped to the slave cylinder, open the bleeder screw, have the helper depress the clutch pedal, lock the bleeder and

have the helper release the pedal. Wait 15 seconds and repeat the procedure (including the 15 second wait) until no air bubbles flow from the bleeder. Remember to close the bleeder before the pedal is released. If the bleeder is left open when the pedal is released, air will be induced into the system.

6. If a helper is not available, connect a small hose to the bleeder, submerge the other end in a clean container of fresh brake fluid placed in a position that is visible from the driver's seat. Pump the pedal until no air comes out of the tube.

AUTOMATIC TRANSMISSION

Transmission Assembly

REMOVAL AND INSTALLATION

1. Disconnect the negative battery cable.
2. If equipped with 4WD, disconnect the 4WD indicator light switch connector and the ground cable at the transfer case.
3. Disconnect the pulse generator connector.
4. Raise and safely support the vehicle.
5. Remove the skid plate, if equipped. Drain the transmission and transfer case, as equipped.
6. Matchmark and remove the driveshaft(s).
7. Disconnect the speedometer cable from the transmission or transfer case.
8. Disconnect the shifter linkage or cable.
9. Unplug all transmission electrical connectors.
10. Remove the exhaust pipe from the vehicle. Remove the exhaust bracket from the transmission case.
11. Remove the filler neck and dipstick.
12. Remove the torque converter inspection plate. Matchmark the flexplate to the torque converter and remove the torque converter bolts.
13. Remove the starter assembly.
14. Disconnect and plug the oil cooler lines.
15. Using a transmission jack, support the transmission. Remove the retaining bolts at the crossmember.

16. Remove the crossmember.
17. Lower the transmission down slightly and unbolt the 4WD shifter from the transfer case, if equipped.
18. Remove the bell housing bolts and mounting brackets.
19. Pull the transmission assembly rearward to clear the aligning dowels and remove from the vehicle.
   To install:
20. Install the transmission assembly to the engine using dowels as guides. Install the bell housing bolts and torque to 35 ft. lbs. (47 Nm).
21. Install the 4WD shifter, if equipped. Raise the assembly up into position and install the rear crossmember and mounting hardware. Torque the crossmember to frame bolts to 50 ft. lbs. (68 Nm).
22. Align the flexplate to torque converter. Apply Loctite® to the threads and install the torque converter bolts. Torque the bolts to 25 ft. lbs. (34 Nm). Install the inspection plate.
23. Install the filler tube with a new O-ring and the dipstick.
24. Install the exhaust and the support brackets.
25. Connect all switch connectors that were removed.
26. Connect the shifter linkage or cable and the throttle cable.
27. Align and install the driveshaft(s).
28. Install the ground wire and the 4WD indicator light switch connector to the transfer case, if equipped.
29. Fill the transfer case with hypoid gear oil with an API classification of GL–4 or higher.
30. Connect the throttle cable.
31. Install the skid plate, if equipped. Lower the vehicle.
32. Install the transfer shifter boot, if equipped.
33. Fill the transmission with the proper amount of Dexron® II.
34. Connect the negative battery cable, start the engine and run through all gears. Add fluid until the transmission is properly filled.
35. Check the operation of the neutral safety switch and the reverse lights.
36. Road test the vehicle and check for leaks.

SHIFT LINKAGE ADJUSTMENT

1. Position the shifter in the **N** detent.
2. Loosen the nut or bolt on the shifter linkage or cable.
3. Make sure the lever on the transmission is in the **N** detent and the needle in the indicator is also in

the **N** position. Jiggle the selector rod to settle the assembly in position. If the rod is equipped with a notch, it should be at the 6 o'clock position.
4. Tighten the adjusting bolt or nut and check for proper assembly.
5. Check the operation of the neutral safely switch.
6. Liberally lubricate all pivoting points within the system.

THROTTLE CABLE ADJUSTMENT

1. Depress the accelerator fully and make sure the throttle valve opens all the way. Adjust as required.
2. Measure the distance between the end of the rubber boot and the stopper on the cable at wide open throttle.
3. The specification is 0–0.04 in. (0–1mm). Adjust the cable itself.
4. Road test the vehicle and check for proper shift points.

NEUTRAL SAFETY SWITCH ADJUSTMENT

NOTE: Some vehicles are not equipped with an adjustable switch.

1. Adjust the shifter linkage.
2. Raise the vehicle and support safely. Place the shifter in the **N** detent. Loosen the mounting bolts.
3. Align the lever with the positioning boss.
4. Hold in position and tighten the bolts.
5. Check the switch and the reverse lights for proper operation.

TRANSFER CASE

Transfer Case Assembly

REMOVAL AND INSTALLATION

Automatic Transmission

1. Disconnect the negative battery cable.
2. Remove the transmission and transfer case assembly from the vehicle.
3. Remove the plug from the right side of the transfer case under the control housing.
4. Remove the select spring and plunger from the housing bore.
5. Remove the control lever housing assembly, cover and gasket.

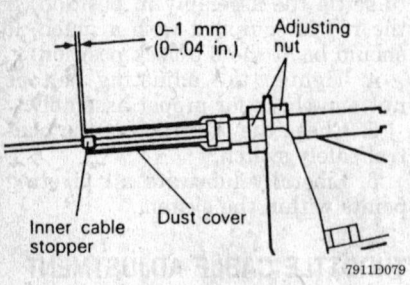

Throttle cable adjustment

7911D079

0–1 mm (0–.04 in.)
Adjusting nut
Inner cable stopper
Dust cover

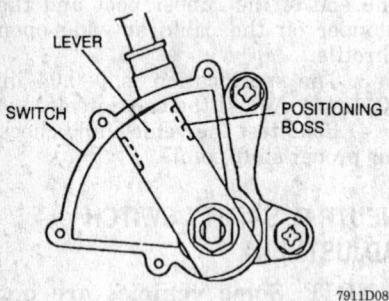

Neutral safety switch adjustment

7911D080

LEVER
SWITCH
POSITIONING BOSS

6. Remove the transfer case to adapter attaching nuts. Pull transfer case from the adapter.

To install:

7. Install the transfer case to adapter and secure using retainer nuts. Torque nuts to 30 ft. lbs. (42 Nm).

8. Install the control lever housing assembly, cover and gasket.

9. Install the select spring and plunger into the housing and install plug in bore.

10. Install the transfer case and transmission into the vehicle, connect the negative battery cable and road test for proper operation.

Manual Transmission

1. Disconnect the negative battery cable.

2. Remove the transmission and transfer case assembly from the vehicle.

3. Remove the plug from the right side of the transfer case under the control housing.

4. Remove the select spring and plunger from the housing bore.

5. Remove the spring pin that retains the shift changer to the control shaft using a pin punch.

6. Remove the transfer case to transmission attaching nuts and remove from the transmission.

To install:

7. Install the transfer case to transmission and tighten attaching nuts to 30 ft. lbs. (42 Nm).

8. Install the shift changer to the control shaft and install new roll pin.

9. Install select spring, plunger and plug into housing bore.

10. Install the transfer case and transmission into vehicle. Connect negative battery cable and road test for proper operation.

Linkage Adjustment

Since this transfer case uses a directly engaging shift mechanism, there are no provisions for adjustment.

DRIVE AXLE

Front Halfshaft

REMOVAL AND INSTALLATION

Right Halfshaft

OUTER

1. Place the free-wheeling hub in the free condition by placing the transfer lever in the **2H** position and moving in reverse for about 6 or 7 feet.

2. Disconnect negative battery cable.

3. Raise and safely support the vehicle. Remove the skid plate, if equipped.

4. Remove the tire and wheel assembly.

5. Remove the hub cover with the use of an oil filter wrench. Install a protective cloth between the wrench and the cover to avoid damage to the cover.

6. Remove the snapring from the inside of the hub. Remove the shim.

7. Remove the front brake caliper and brake pads from the vehicle. Do not allow the caliper to hang from the brake hose.

8. Separate the tie rod from the steering knuckle.

9. Separate the upper and lower ball joints from the steering knuckle.

10. Remove the front hub/knuckle assembly with the inner and outer bearings intact.

11. Remove the halfshaft to axle housing retaining nuts and remove the halfshaft from the vehicle.

To install:

12. Install the halfshaft and the retaining nuts and tighten to 43 ft. lbs. (60 Nm).

13. Install the front hub/knuckle and bearing assembly.

14. Install the upper ball joint to the knuckle and torque retaining nut to 130 ft. lbs. (180 Nm). Install the lower ball joint to knuckle and torque retaining nut to 65 ft. lbs. (90 Nm). Install new cotter pins.

15. Install the tie rod end to the steering knuckle and torque to 33 ft. lbs. (45 Nm). Install new cotter pin.

16. Install the shim and snapring to the axle shaft. Install the front hub cover.

17. Install front brake caliper assembly.

18. Install the tire and wheel assembly. Install skid plate, if removed.

INNER

1. Disconnect negative battery cable.

2. Raise and safely support the vehicle.

3. Remove the skid plate, if equipped.

4. Remove the right outer halfshaft.

5. Remove the lower shock absorber mounting bolts.

6. Install slide hammer to inner shaft flange and pull from housing. Press the bearing from the axle, as required.

To install:

7. Press new bearing and seal on axle, as required. Install new circlip to inner halfshaft and install into housing. Drive the axle into position.

8. Install the lower shock absorber mounting bolts.

9. Install the right outer halfshaft.

10. Install the skid plate, if equipped.

Left Halfshaft

1. Place the free-wheeling hub in the free condition by placing the transfer lever in the **2H** position and moving in reverse for about 6 or 7 feet.

2. Disconnect negative battery cable.

3. Raise and safely support the vehicle. Remove the skid plate, if equipped.

4. Remove the tire and wheel assembly.

5. Remove the hub cover with the use of an oil filter wrench. Install a protective cloth between the wrench

1. Oil filler plug
2. Drain plug
3. Plug
4. Gasket
5. Selector spring
6. Selector plunger
7. Roll pin
8. Wire clamp
9. Wire clamp
10. Shift changer
11. Transfer case assembly
12. Gasket
13. Plug
14. Spring
15. Steel ball
16. Plug
17. Neutral return spring
18. Plunger
19. Plunger

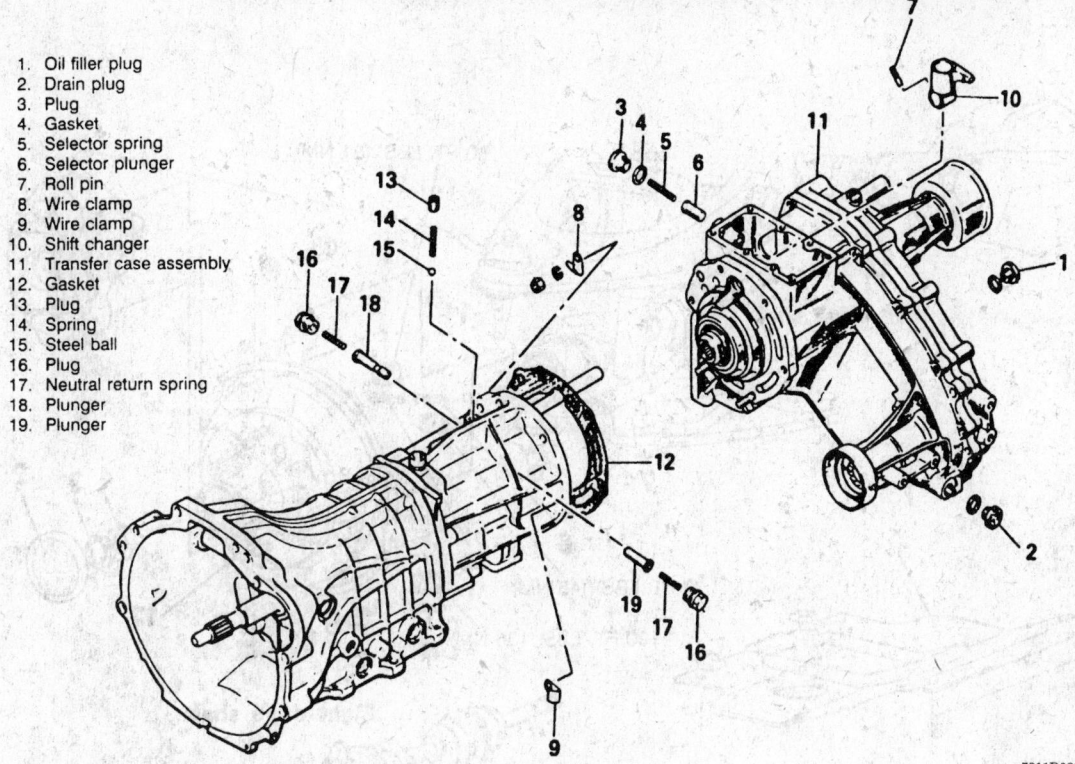

Transfer case assembly

7911D081

and the cover to avoid damage to the cover.

6. Remove the snapring from the inside of the hub. Remove the shim.

7. Remove the front brake caliper and brake pads from the vehicle. Do not allow the caliper to hang from the brake hose.

8. Separate the tie rod from the steering knuckle.

9. Separate the upper and lower ball joints from the steering knuckle.

10. Remove the front hub/knuckle assembly with the inner and outer bearings intact.

11. Pull the left halfshaft out from the differential carrier. Use care not to damage the oil seal with the splines of the shaft.

To install:

12. Replace circlip on the end of the shaft. Install the halfshaft into the front differential case and drive into position using a plastic hammer.

13. Install the front hub/knuckle and bearing assembly.

14. Install the upper ball joint to the knuckle and torque retaining nut to 130 ft. lbs. (180 Nm). Install the lower ball joint to knuckle and torque retaining nut to 65 ft. lbs. (90 Nm). Install new cotter pin.

15. Install the tie rod end to the steering knuckle and torque to 33 ft. lbs. (45 Nm). Install new cotter pin.

16. Install the shim and snapring to the axle shaft. Install the front hub cover.

17. Install front brake caliper assembly.

18. Install the tire and wheel assembly. Install skid plate, if removed.

CV-Boot

REMOVAL AND INSTALLATION

1. Disconnect negative battery cable.

2. Raise and safely support the vehicle.

3. Remove the skid plate, if equipped.

4. Remove the left halfshaft.

5. Remove the boot bands on the inner Double Offset Joint (DOJ). Remove the circlip.

6. Remove the DOJ outer race from the shaft.

7. Remove the dust cover from the DOJ outer race.

8. Remove the balls from the DOJ cage prying from the inside of the cage outward.

9. Rotate the cage while pushing toward the Birfield Joint (BJ). The cage will drop down to expose a snapring on the halfshaft. Remove the snapring.

10. Remove the DOJ cage and inner race from the halfshaft.

11. Wrap tape over the threads of the halfshaft. Remove the boot from the shaft.

12. Remove the dust cover from the BJ and the halfshaft.

13. Remove the boot bands. Remove the BJ boot from the DOJ end of the driveshaft.

NOTE: Do not disassemble the Birfield Joint (BJ).

To install:

14. Install the BJ boot and bands over the DOJ end of the driveshaft. Install the DOJ boot and bands onto the halfshaft. Apply ½ of supplied grease into the BJ boot and install the bands onto the boot.

15. Install the DOJ cage onto the halfshaft with the smaller diameter side of the cage facing the installed BJ boot. Install the snapring.

16. Apply remaining grease to the DOJ inner race, the DOJ cage and the balls. Insert the balls into the cage from the outside pushing in toward the shaft.

17. Install the outer race over the DOJ cage assembly and install the retaining circlip.

18. Install the dust cover into place and install the boot bands.

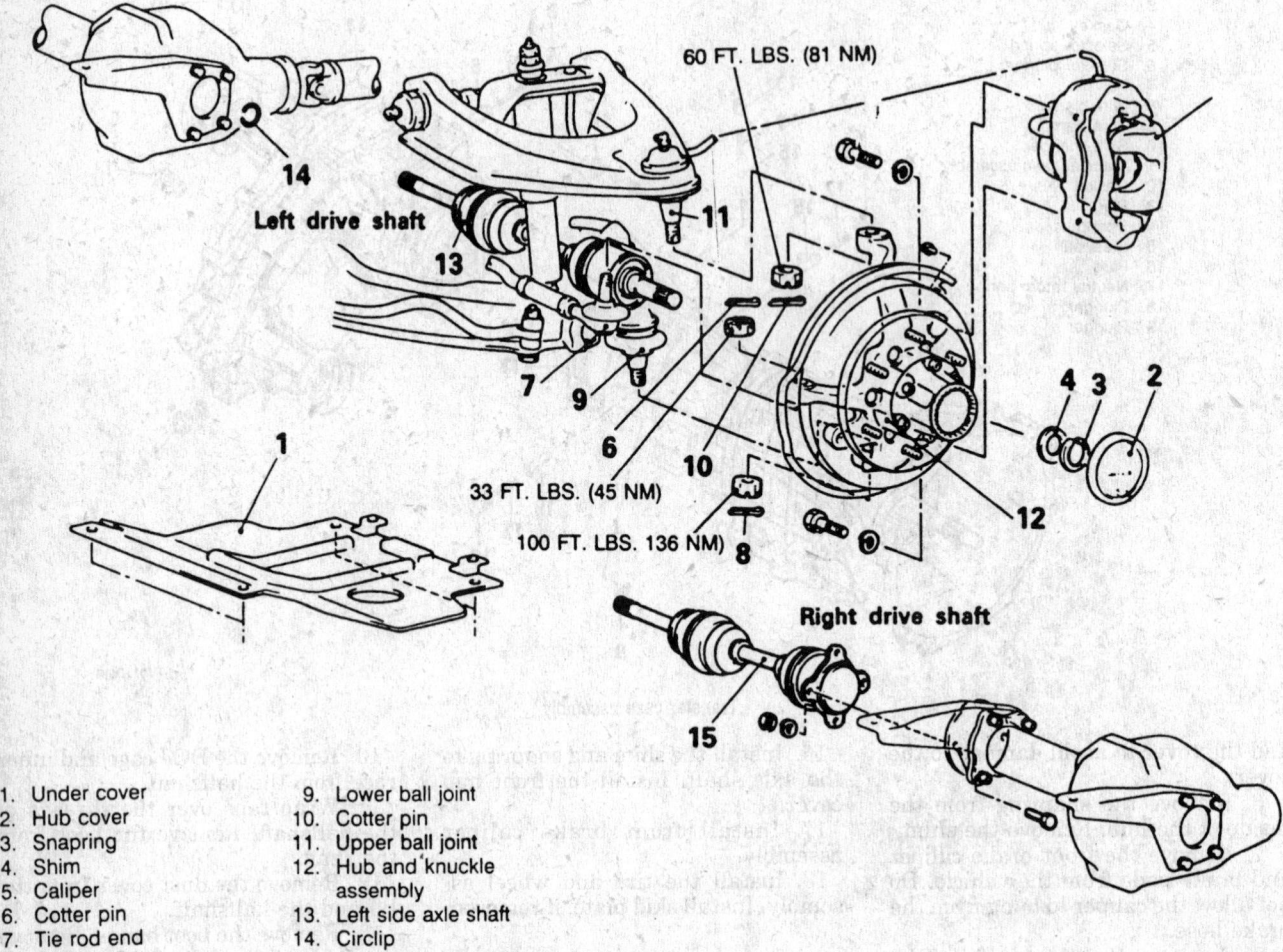

60 FT. LBS. (81 NM)

14

Left drive shaft

13

11

7

9

6

10

1

33 FT. LBS. (45 NM)

100 FT. LBS. 136 NM)

8

4 **3** **2**

12

Right drive shaft

15

1. Under cover
2. Hub cover
3. Snapring
4. Shim
5. Caliper
6. Cotter pin
7. Tie rod end
8. Cotter pin
9. Lower ball joint
10. Cotter pin
11. Upper ball joint
12. Hub and knuckle assembly
13. Left side axle shaft
14. Circlip
15. Right side axle shaft

7911D082

Front axle shafts and related parts

19. Install the halfshaft into the vehicle.
20. Install the skid plate, if equipped.

Driveshaft and U-Joints

REMOVAL AND INSTALLATION

1. If equipped with 4WD, place the free-wheeling hub in the free condition by placing the transfer lever in the **2H** position and moving in reverse for about 6 or 7 feet.
2. Disconnect negative battery cable.

3. Raise and safely support the vehicle.
4. Remove the skid plate, if equipped.
5. Matchmark the flange yoke and the differential companion flange. Do not damage the oil seals when marking positions.
6. Drain the transfer case. Position a drain pan under the transmission tailshaft, as required. Fluid will leak from unit on removal of the driveshaft.
7. Remove the retaining bolts and nuts from the flange(s) and remove the shaft from the vehicle.

8. Remove the U-Joint from the driveshaft as follows:
   a. Make mating marks on the yokes of the universal joint that is to be disassembled.
   b. Remove the snaprings from the yoke. When disassembling, note the position of the snaprings so they may be reinstalled in the same positions.
   c. Remove the grease fitting from the journal.
   d. Remove the journal bearings from the propeller shaft yoke using tool MB990840–01.
   e. Remove the journal and the flange yoke or sleeve yoke from the shaft.

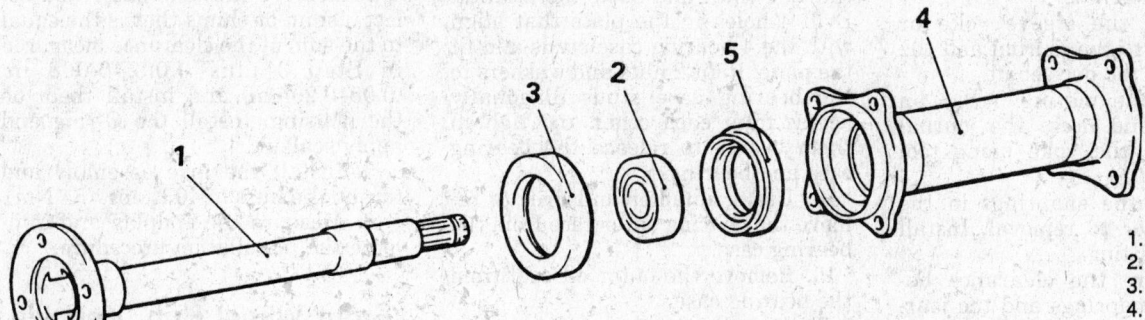

1. Inner shaft
2. Bearing
3. Dust cover
4. Housing tube
5. Dust seal

7911D083

Inner axle shaft and tube

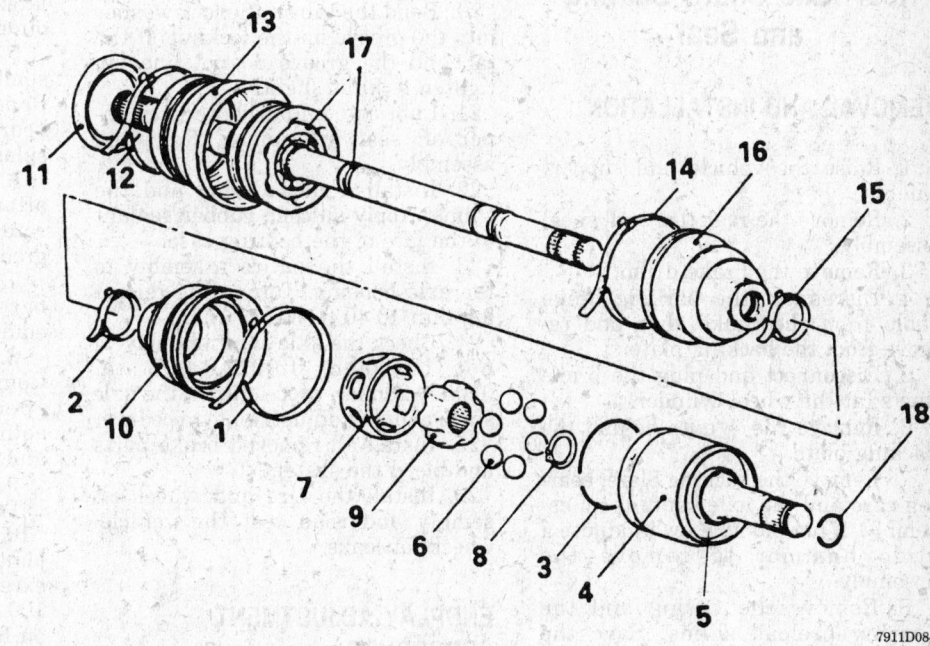

1. Boot band
2. Boot band
3. Circlip
4. DOJ outer race
5. Dust cover
6. Balls
7. DOJ cage
8. Snapring
9. DOJ inner race
10. DOJ boot
11. Dust cover
12. Boot protector band
13. Boot protector
14. Boot band
15. Boot band
16. BJ boot
17. Drive shaft and BJ
18. Circlip

7911D084

Front halfshaft assembly — 4WD Vehicle

9. Inspect the propeller shaft yokes for wear, cracks or damage. Check the sleeve yoke and the flange yoke for wear, cracks or damage. Replace as required.

To install:

10. Install the U-Joint on the driveshaft as follows:

a. Install the sleeve yoke or flange yoke to the journal and the assembly to the driveshaft.

b. Align the mating marks on the yokes and press the journal bearings to the yoke using tool MB990840–01.

c. Install the snaprings in the position prior to removal. Install the grease fitting.

d. Measure the clearance between the snaprings and the journal bearing. If the clearance is larger than 0.0024 in. (0.06mm), it will be necessary to replace the snapring.

11. With the mating marks aligned, install the driveshaft into the transfer case/transmission and the companion flange.

12. Clean the mounting nuts and bolts and install to the flange(s). Torque to 43 ft. lbs. (60 Nm).

13. Install the skid plate, if equipped.

Rear Axle Shaft, Bearing and Seal

REMOVAL AND INSTALLATION

1. Raise the vehicle and support safely.

2. Remove the rear tire and wheel assembly.

3. Remove the brake drum.

4. Disconnect the parking brake cable from the brake shoe and remove from the backing plate.

5. Disconnect and plug the brake line(s) at the wheel cylinder.

6. Remove the 4 nuts behind the backing plate.

7. Remove the backing plate, bearing case and the axle shaft as an assembly. If not possible by hand, use a slide hammer to remove the assembly.

8. Remove the O-ring and the bearing preload shims. Save the preload shims for reassembly.

9. Remove the oil seal from the axle tube with a hooked slide hammer.

10. To remove the axle shaft bearing, remove the notched locknut with tool MB990785–01 or a brass drift.

11. Remove the lock washer and flat washer.

12. Screw the locknut onto the axle shaft about 3 turns.

13. If tool MB990787–01 is not available, it will be necessary to fabricate a metal plate that fits over the axle shaft and butts the locknut. Drill 4 holes in the plate that align with the 4 bearing case studs and fit the plate. Refit 2 nuts and washers to the bearing case studs and fit the plate. Refit 2 nuts and washers to the bearing case studs diagonally across from each other and tighten them evenly to release the bearing case and bearing.

14. Use a hammer and drift to remove the bearing outer race from the bearing case.

15. Remove the outer oil seal from the bearing case.

To install:

16. Apply grease to the outer surface on the bearing outer race and the lip of the outer oil seal. Drive into the bearing case.

17. Slide the bearing case and bearing over the rear axle shaft. Apply grease on the bearing rollers and install the inner race by pressing into place. Be careful not to damage the dust cover.

18. Pack the bearing with grease.

19. Install the washer, the crowned lock washer and the locknut in that order and torque the locknut to 130–159 ft. lbs. (176–220 Nm).

20. Bend the tab on the lock washer into the groove on the locknut. If the tab and the groove do not line up, tighten locknut slightly.

21. Lubricate and drive the new inner oil seal into place. Refit the assembly.

22. Install a new O-ring and the shims. Apply silicone rubber sealant to the face of the bearing case.

23. Install the entire assembly to the axle housing. Torque the retaining nuts to 40 ft. lbs. (54 Nm).

24. Check the axle shaft endplay. If not between 0.002–0.008 in. (0.05–0.20mm), proceed with the axle shaft endplay adjustment procedure.

25. Install all removed brake parts and bleed the system.

26. Install the tire and wheel assembly and road test the vehicle. Check for leaks.

ENDPLAY ADJUSTMENT PROCEDURE

1. Begin with the left side rear axle assembly and insert a 0.04 in. (1mm) shim between the bearing case and the axle shaft housing. Torque the nuts to specification.

2. Install the right side axle assembly into the right side housing without its shim and O-ring. Torque the 4 nuts to about 50 inch. lbs.

3. Using a feeler gauge, measure the gap between the bearing case and the axle housing face.

4. Remove the axle shaft and select a shim or shims that is the equal to the sum of the clearance measured in Step 3 plus 0.002–0.008 in. (0.05–0.20mm) and install them on the housing. Install the O-ring and apply sealant.

5. Install the axle assembly and torque the nuts to 40 ft. lbs. (54 Nm).

6. Measure the endplay and complete the installation procedure.

Front Wheel Hub, Knuckle and Bearings

REMOVAL AND INSTALLATION

2WD Vehicle

1. Raise the vehicle and support safely. Remove the tire and wheel assembly.

2. Remove the brake caliper, pads and adaptor and position them out of the way.

3. Remove the grease cap.

4. Remove the cotter pin, castellated nut and washer. Remove the outer bearing.

5. Remove the front hub/rotor assembly from the steering knuckle. Remove the grease seal and inner bearing from the hub. Remove the splash shield.

6. Remove the nuts and bolts that attach the hub to the rotor and separate them. Clean out all of the old grease from the inside of the hub.

7. Remove the shock and stabilizer bar from the lower control arm, if equipped.

8. Support the lower control arm. Compress the coil spring with the special spring compressor, if equipped. Separate the ball joints and tie rod end from the knuckle.

9. Remove the steering knuckle.

To install:

10. Install the knuckle to the ball joint studs. Torque all ball joint nuts, except the upper nut on Pick-Up to 100 ft. lbs. (136 Nm). Torque the nut on Pick-Up to 60 ft. lbs. (82 Nm). Install new cotter pins. Remove the spring compressor, if used. Install the shoes.

11. Connect the stabilizer bar to the lower control arm. Tighten the nut until the bushing is the same diameter as the washer.

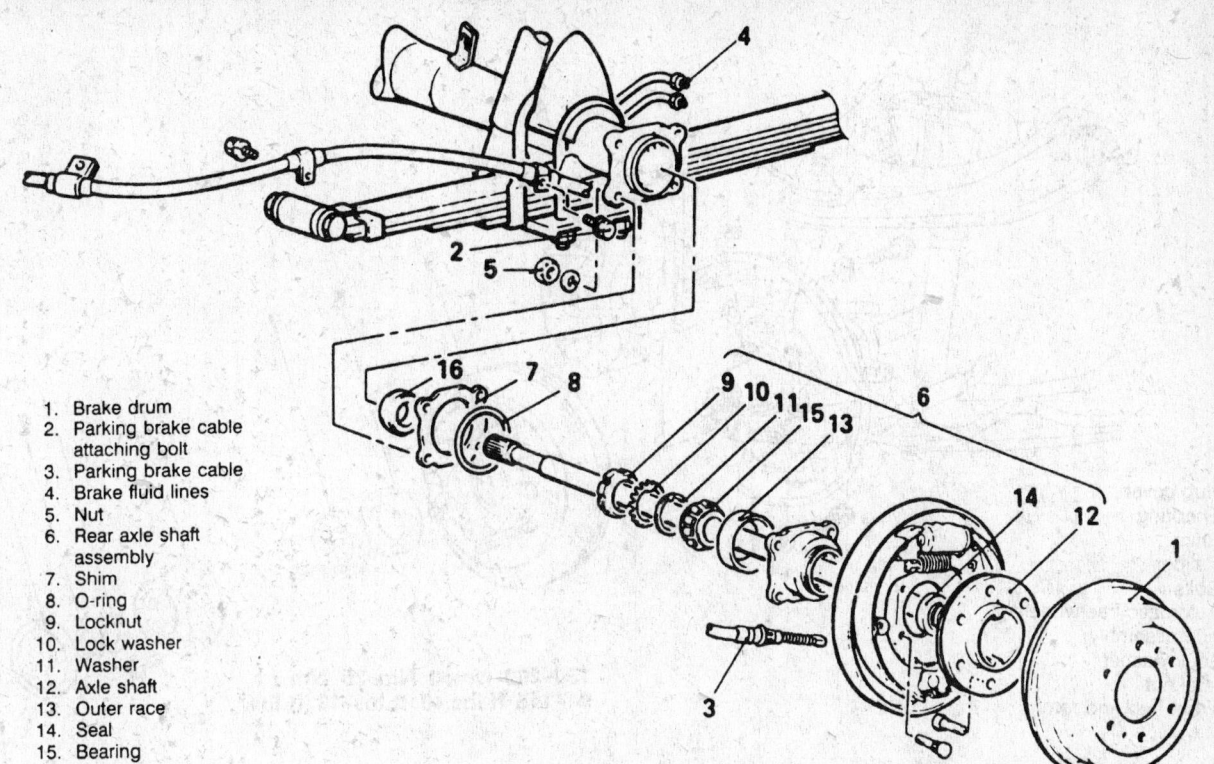

1. Brake drum
2. Parking brake cable attaching bolt
3. Parking brake cable
4. Brake fluid lines
5. Nut
6. Rear axle shaft assembly
7. Shim
8. O-ring
9. Locknut
10. Lock washer
11. Washer
12. Axle shaft
13. Outer race
14. Seal
15. Bearing
16. Seal

7911D085

Rear axle shaft assembly

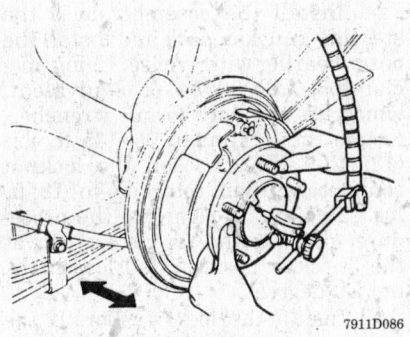

7911D086

Measuring axle shaft endplay

12. Connect the tie rod end to the knuckle. Torque the nut to 30 ft. lbs. (41 Nm). Install a new cotter pin.

13. Assemble the rotor and front hub. Torque the nuts to 40 ft. lbs. (54 Nm).

14. Apply wheel bearing grease to the inside of the front hub.

15. Pack the inner bearing and install into hub.

16. Install a new oil seal into the hub so it is flush with the hub end face.

17. Install the assembly onto the steering knuckle, pack and install the outer bearing.

18. Install the washer and castellated nut. Spin the rotor and torque the nut to 22 ft. lbs. (30 Nm), loosen completely and retighten to 6 ft. lbs. (8 Nm) while turning the wheel.

19. Install a new cotter pin and install the grease cap.

20. Install the brake hardware and bleed the system, if opened.

21. Install the tire and wheel assembly and road test the vehicle.

Automatic Locking Hubs, Knuckle and Bearings

REMOVAL AND INSTALLATION

1. Place the free-wheeling hub in the free condition by placing the transfer lever in the **2H** position and moving in reverse for about 6 or 7 feet.

2. Disconnect negative battery cable.

3. Raise and safely support the vehicle. Remove the skid plate, if equipped.

4. Remove the tire and wheel assembly.

5. Remove the hub cover with the use of an oil filter wrench. Install a protective cloth between the wrench and the cover to avoid damage to the cover.

6. Remove the snapring from the inside of the hub. Remove the shim.

7. Remove the front brake caliper and brake pads from the vehicle. Do not allow the caliper to hang from the brake hose.

8. Using hexagonal wrench, remove the automatic free wheeling hub assembly.

9. Remove the retaining screws from lock washer and remove lock washer.

10. Remove the outer bearing and front hub assembly from the steering knuckle.

11. Remove the inner oil seal and inner bearing from the hub assembly.

12. To separate the hub from the brake disc, matchmark the brake disc to the front hub. Remove the retaining bolts and separate the hub from the disc.

13. Remove the dust cover from the spindle. Separate the tie rod from the steering knuckle.

14. Separate the upper and lower ball joints from the steering knuckle.

15. Remove the knuckle from the axle shaft.

16. Remove the knuckle oil seal and spacer. If damaged, drive out the needle bearing through the spindle end of the knuckle.

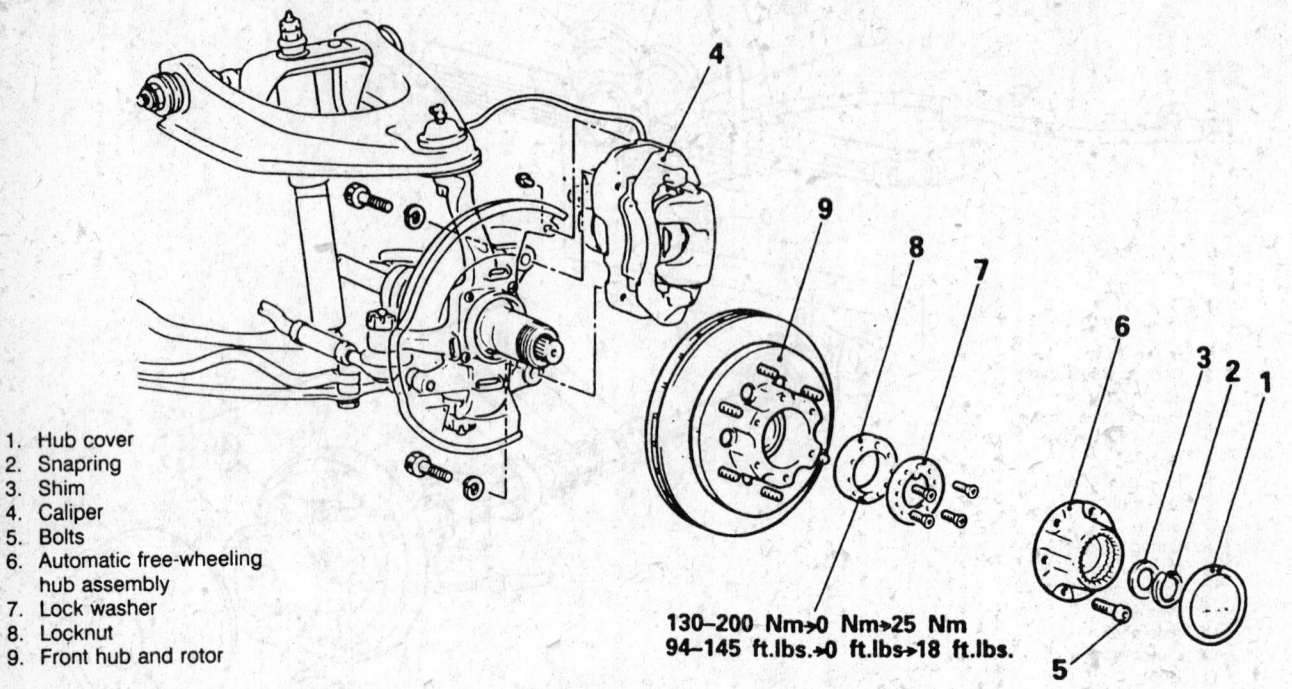

1. Hub cover
2. Snapring
3. Shim
4. Caliper
5. Bolts
6. Automatic free-wheeling hub assembly
7. Lock washer
8. Locknut
9. Front hub and rotor

130–200 Nm→0 Nm→25 Nm
94–145 ft.lbs.→0 ft.lbs→18 ft.lbs.

7911D087

Front hub and automatic hub assembly — 4WD vehicle

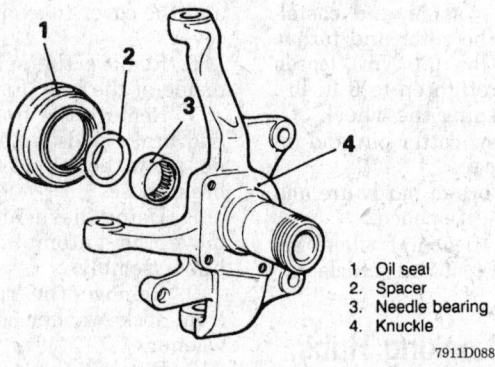

1. Oil seal
2. Spacer
3. Needle bearing
4. Knuckle

7911D088

Steering knuckle, needle bearing, spacer and seal — 4WD vehicle

To install:

17. Use Mitsubishi tool MB990985 to press the new needle bearing into the knuckle. Lubricate the rollers and install the spacer with the chamfered side facing the center of the vehicle. Install a new seal until it is flush with the knuckle end face and lubricate the l

18. Install the knuckle to the ball joint studs. Torque all ball joint nuts except the upper nut on Pick-Up to 100 ft. lbs. (136 Nm). Torque the nut on Pick-Up to 60 ft. lbs. (82 Nm). Install new cotter pins.

19. Connect the tie rod end to the knuckle. Torque the nut to 30 ft. lbs. (41 Nm). Install a new cotter pin.

20. Clean out all of the old grease from inside the front hub.

21. Bolt the rotor and front hub together. Torque the nuts to 40 ft. lbs. (54 Nm). Install new races and apply wheel bearing grease to the inside of the front hub.

22. Pack the inner bearing and install to the race.

23. Install new oil seal into the front hub until it is flush with the front hub end face.

24. Install the assembly onto the steering knuckle, pack and install the outer bearing with grease. Using special tool MB990954 or equivalent, which fits standard torque wrenches, torque the locknut to 95–145 ft. lbs. (130–200 Nm). Loosen the locknut completely, then retorque to 18 ft. lbs. (25 Nm). To complete the procedure, position the torque wrench at the 3 o'clock position and loosen the nut 30 degrees.

25. Install the lock washer. If the lock washer and locknut holes do not align, loosening the nut slightly.

26. Before installing the automatic hub assembly, measure the turning force of the front hub. If the measured value does not meet specifications of 2.5–11.5 inch lbs. and retorque the locknut. Also check the hub for an maximum endplay of 0.002 in. (0.05mm).

27. Apply even coating of sealant on the freewheeling hub body and front hub contact surface. Align the key of the brake and the keyway of the spindle and loosely install the automatic free wheeling hub assembly. The mounting surfaces of the automatic hub and the front hub must be perfectly flush before the mounting bolts are torqued.

28. Torque the automatic hub mounting bolts to 40 ft. lbs. (54 Nm).

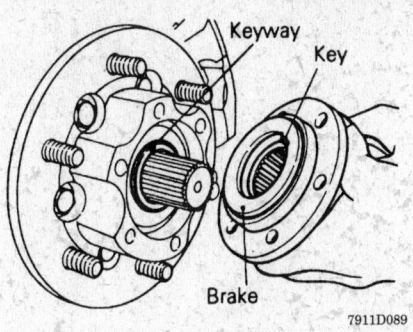

Aligning the key in the automatic hub with the keyway in the spindle

29. Check the front hub turning resistance again. If the difference between the reading in Step 26 and this reading is more than 8.7 inch lbs., repair or replace the automatic hub, as required.

30. Install the shim and snapring. Rotate the axle shaft forward and backward and stop at a position midway between 2 heavy spots where there is a heavy feeling.

31. Set a dial indicator so the pin is resting on the end of the axle shaft. The endplay specification is 0.008–0.020 in. (0.2–0.5mm). If not within specifications, adjust by adding or removing shims.

32. Install the hub cover.

33. Install the brake hardware and install the wheel and tire assembly.

34. Road test the vehicle.

Pinion Seal

REMOVAL AND INSTALLATION

1. Raise the vehicle and support safely.

2. Matchmark and remove the driveshaft.

3. Check the turning torque of the pinion before proceeding. It should be 3.5–4.5 inch lbs. (0.4–0.5 Nm). This is the torque that must be reached during installation of the pinion nut.

4. Using a pinion flange holding tool, remove the pinion nut and washer.

5. Remove the companion flange from the drive pinion.

6. Pry the pinion seal out of the differential carrier.

7. Clean and inspect the sealing surface of the housing.

To install:

8. Using a seal driver, drive the new seal into the housing until the flange on the seal is flush with the carrier.

9. With the seal installed, the pinion bearing preload must be set.

10. Tighten the pinion nut, while holding the flange, until the turning torque is the same as before removal of the nut. The final pinion nut torque must be 137–181 ft. lbs. (190–250 Nm).

11. Align the matchmarks and install the drive shaft.

12. Check the level of the differential lubricant.

Differential Carrier

REMOVAL AND INSTALLATION

Front Differential

1. Raise the vehicle and support safely. Remove the skid plate.

2. Remove both side outer and right side inner axle shafts.

3. Drain the front differential and remove the cover.

4. Remove the bearing cap retaining bolts, matchmark and remove the caps.

5. Pry the differential case out of the housing.

6. Label any loose parts as removed from the assembly.

To install:

7. Install the differential case to the housing.

8. Install the caps aligning the matchmarks made previously. Torque the retaining bolts evenly and gradually to 45 ft. lbs. (61 Nm).

9. Clean the sealing surfaces of the cover and the differential housing. Reseal the cover and install to the housing.

10. Install the axle shafts.

11. Level the vehicle and fill the differential with Hypoid gear oil with an API classification of at least GL–5, until oil level reaches hole.

12. Install the skid plate and lower the vehicle.

13. Perform a road test and check the differential for leaks.

Rear Differential

1. Raise the vehicle and support safely. Drain the rear differential.

2. Remove the rear tire and wheel assemblies and brake drums.

3. Remove the axle shafts.

4. Matchmark and remove the driveshaft.

5. Remove the differential carrier retaining nuts and remove the carrier from the axle housing.

6. The installation is the reverse of the removal procedure. Torque the carrier retaining nuts to 22 ft. lbs. (27 Nm).

Front Differential Housing

REMOVAL AND INSTALLATION

1. Raise the vehicle and support safely. Remove the skid plate and drain the differential.

2. Remove the hubs and knuckles from the vehicle.

3. Remove the outer and inner axle shafts.

4. Matchmark and remove the front driveshaft.

5. Support the differential housing with a jack.

6. Remove the differential mounting brackets.

7. Remove the front suspension crossmember.

8. Lower the differential housing and remove from the vehicle.

To install:

9. Mount the housing safely on jack and raise into position.

10. Lubricate the bushings and install the crossmember and housing brackets.

11. Torque all crossmember mounting nuts and bolt to 80 ft. lbs. (109 Nm) and the housing bracket mounting bolts to 65 ft. lbs. (88 Nm).

12. Replace the circlips and install the axle shafts.

13. Install the knuckle and hub assemblies.

14. Install the front driveshaft.

15. Level the vehicle and fill the differential with Hypoid gear oil with an API classification of at least GL–5, until oil level is up to the fill hole.

16. Install the under cover and lower the vehicle.

17. Perform a road test and check the differential for leaks.

Rear Axle Housing

REMOVAL AND INSTALLATION

Except Montero With Coil Springs

1. Raise the vehicle and support safely. Drain the differential

2. Remove the tire and wheel assemblies. Remove the brake drums.

3. Remove the parking brake cable attaching bolts, disconnect the cables from the shoes and unclip them from the backing plates.

4. Disconnect the brake hose at the T-fitting.

5. Remove the load sensing proportioning valve spring support, if equipped.

6. Disconnect the breather hose, if equipped.

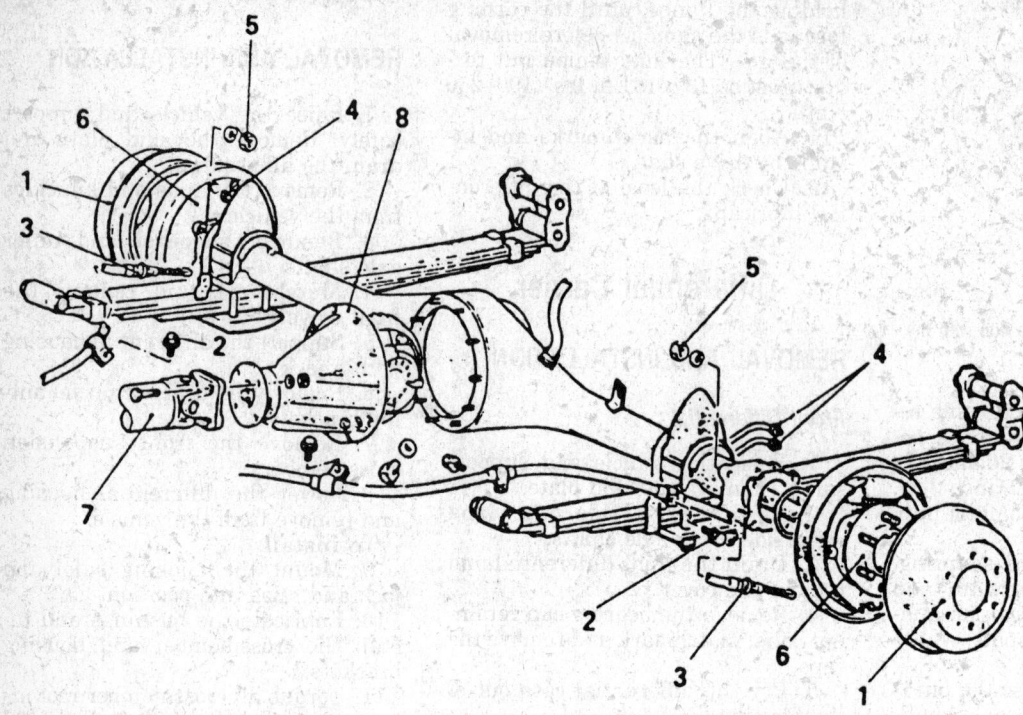

1. Brake drum
2. Parking brake cable attaching bolt
3. Parking brake cable
4. Brake fluid lines
5. Nut
6. Rear axle shaft
7. Driveshaft
8. Differential carrier

7911D090

Differential carrier removal and installation

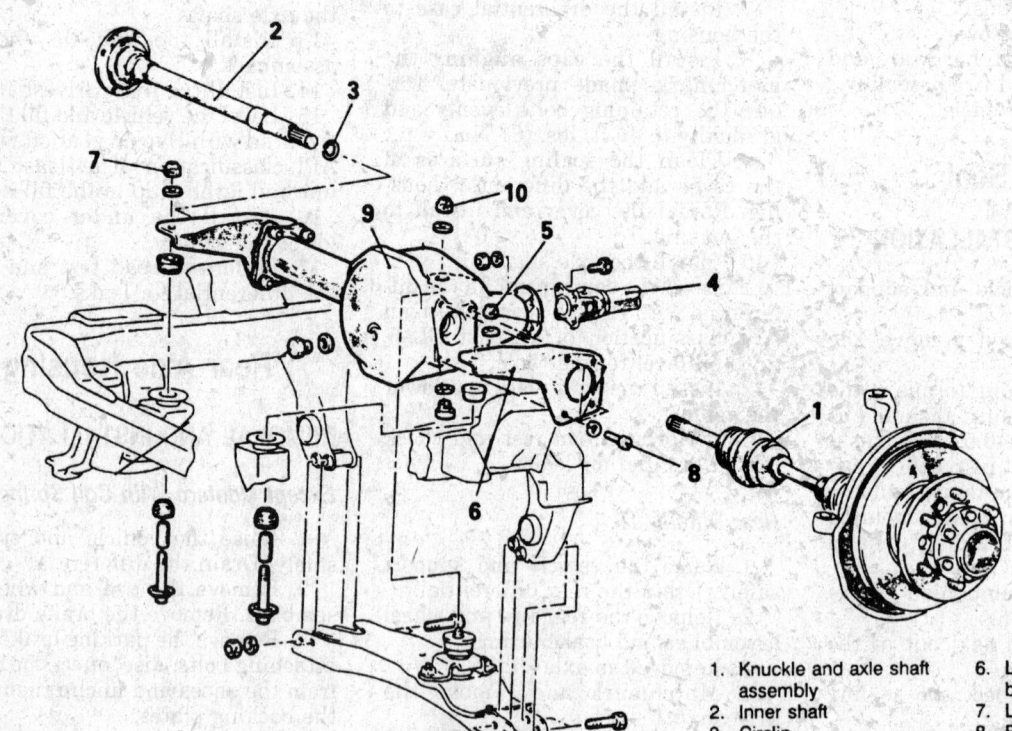

1. Knuckle and axle shaft assembly
2. Inner shaft
3. Circlip
4. Front driveshaft
5. Locknut
6. Left side mounting bracket
7. Locknut
8. Bolt
9. Crossmember
10. Locknut

7911D091

Front axle removal and installation

7. Matchmark and remove the driveshaft.

8. Place jack under the center of the differential housing and unbolt the shocks from their lower mounts.

9. Remove the U-bolts, shackle assemblies and remove the leaf springs.

10. Lower the axle and remove from vehicle.

To install:

11. Raise the axle assembly into position and install the leaf springs. Make sure the shackle nuts are on the inside of the shackles.

12. Install the lower shock mounting bolts.

13. Matchmark and install the driveshaft.

14. Install the breather hose, if equipped.

15. Connect the brake hose.

16. Connect the parking brake cables and install the retaining bolts.

17. Install the load sensing spring support and spring, if equipped.

18. Level the vehicle and fill the differential using Hypoid gear oil with an API classification of at least GL–5, until level reaches the fill hole.

NOTE: If equipped with a limited slip differential, add the proper amount of limited slip friction modifier additive before filling the differential with gear oil.

19. Bleed the rear brakes. Install the brake drums and tire and wheel assemblies.

20. Lower the vehicle so full weight of the vehicle is on the ground. Unload any excess weight that may be weighing down the rear of the vehicle. Adjust the load sensing proportioning valve lever so the distance from the proportioning valve lever to the spring support is 7 in.

21. Road test the vehicle and check for leaks.

Montero With Coil Springs

1. Raise the vehicle and support safely. Drain the differential.

2. Remove the tire and wheel assemblies. Remove the brake drums.

3. Remove the parking brake cable attaching bolts, disconnect the cables from the shoes and unclip from backing plates.

4. Disconnect the brake hose at the T-fitting.

5. Disconnect the breather hose.

6. Matchmark and remove the driveshaft.

7. Remove the rear stabilizer bar attaching bolts, links and bushings.

8. Place jack under the center of the differential housing and remove the rear trailing arm, if equipped.

9. Remove the lateral rod.

10. Unbolt the shocks from their lower mounts.

11. Lower the axle housing enough to remove the coil springs and the stabilizer bar.

12. Lower the axle assembly fully and remove from vehicle.

To install:

13. With the coil springs and stabilizer bar in place, raise the axle assembly into place and install the lower shock mounting bolts.

14. Install the lateral rod and tighten nuts loosely.

15. Assemble the trailing arm with its front mounting spacers, bushings and nuts. Make sure the washer's concave side faces away from the bushings. Install the trailing arm and torque the rear mount nuts to 90 ft. lbs. (122 Nm). Do not tighten the front mounting nuts yet.

16. Install the stabilizer bar and tighten the mounting nuts until the diameter of the bushing is the same as the diameter of the washers.

17. Install the driveshaft and breather hose.

18. Connect the parking brake cables and install the retaining bolts.

19. Level the vehicle and fill the differential with Hypoid gear oil with an API classification of at least GL–5, until it spills out the fill hole.

NOTE: If equipped with a limited slip differential, add the proper amount of limited slip friction modifier additive before filling the differential with gear oil.

20. Bleed the rear brakes. Install the brake drums and tire and wheel assemblies.

21. Lower vehicle so full weight of vehicle is on the ground. Torque the lateral rod mounting nuts to 90 ft. lbs. (122 Nm) and the front trailing arm mounting nuts to 100 ft. lbs. (150 Nm).

22. Road test the vehicle and check for leaks.

STEERING

Steering Wheel

REMOVAL AND INSTALLATION

1. Disconnect the negative battery cable.

2. Remove the center pad.

3. Remove the steering wheel jam nut. Matchmark the steering wheel to the shaft.

4. Using steering wheel puller, pull the steering wheel off of the shaft.

5. The installation is the reverse of the removal procedure.

Steering Column and Shaft

REMOVAL AND INSTALLATION

1. Disconnect negative battery cable.

2. Remove the steering wheel from the vehicle.

3. Remove the upper and lower steering column covers.

4. Disconnect the electrical connectors at the base of the steering column.

5. Remove the gear indicator cable connection at the top of the steering column, if equipped.

6. Remove the brake pedal return spring from the lower portion of the steering column, if equipped.

7. Remove the bolt in the steering shaft coupler above the steering gear.

8. Remove the column retainer bolts at the floor panel and under the instrument panel. Remove the steering column and shaft from the vehicle.

To install:

9. Install the steering column into the vehicle positioning the steering shaft over the bevel gear or steering gear shaft.

10. Install the under dash retaining bolts loosely.

11. Install the bolt at the lower steering coupler and torque to 25 ft. lbs. (35 Nm).

12. Tighten the bolts under the instrument panel to 14 ft. lbs. (20 Nm). Install the bolts in the column retainer at the floor panel tightening to 4 ft. lbs. (6 Nm).

13. Reconnect the electrical connectors at the base of the steering column.

14. Connect the gear indicator cable, if removed.

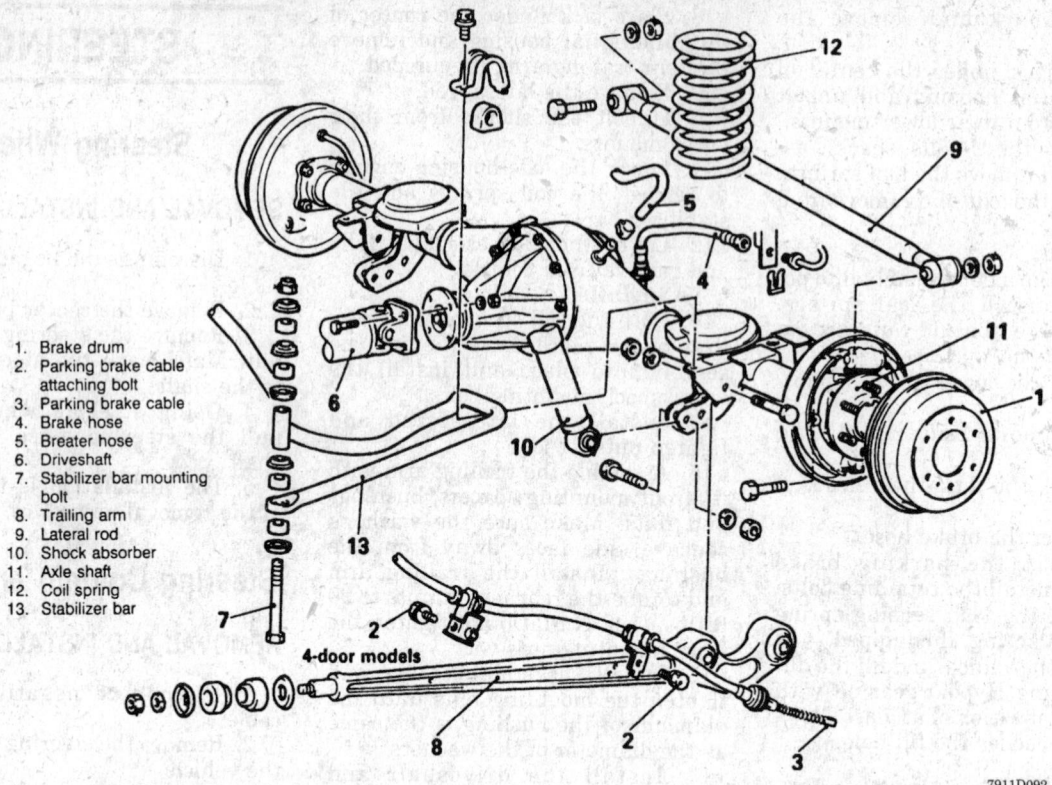

1. Brake drum
2. Parking brake cable attaching bolt
3. Parking brake cable
4. Brake hose
5. Breater hose
6. Driveshaft
7. Stabilizer bar mounting bolt
8. Trailing arm
9. Lateral rod
10. Shock absorber
11. Axle shaft
12. Coil spring
13. Stabilizer bar

4-door models

7911D092

Rear axle assembly — Montero with coil springs

15. Install the upper and lower steering column covers.

16. Align and install the steering wheel tightening the jam nut to 36 ft. lbs. (50 Nm).

17. Install the horn pad. Reconnect the negative battery cable and check operation of all related electrical components.

Manual Steering Gear

REMOVAL AND INSTALLATION

1. Disconnect negative battery cable.

2. Remove the cotter pin and nut on the pitman arm. Separate the relay rod from the pitman arm.

3. Remove the bolt from the steering shaft to steering gear coupling.

4. Remove the steering gear mounting bolts and nuts, slide the gear from the steering shaft and remove the gear from the vehicle. Matchmark and remove the pitman arm, as required.

5. The installation is the reverse of the removal procedure. Torque the gear mounting nuts to 40 ft. lbs. (55 Nm). Torque the relay rod retainer nut to 33 ft. lbs. (45 Nm) and install new cotter pin.

ADJUSTMENT

1. Disconnect negative battery cable.

2. Remove the steering gear from the vehicle.

3. Turn the cross shaft 36 degrees. Position the mainshaft in the straight ahead position.

4. Using torque wrench, measure the starting torque of the mainshaft. The recommended value is 7.7–8.6 inch lbs. (0.9–1.0 Nm). If measured value is not within this value, Use tool MB990914 to screw adjusting cover to correct reading.

Manual Rack and Pinion

REMOVAL AND INSTALLATION

1. Disconnect the negative battery cable.

2. Raise the vehicle and support safely. Remove the under cover.

3. Remove the front tire and wheel assemblies.

4. Remove the cotter pins, castellated nuts and tie rod ends from the steering knuckles.

5. Disconnect the intermediate shaft from the steering gear yoke.

6. Remove the gear housing clamps and remove the gear from the vehicle. Transfer the mounting insulators to the new gear.

7. The installation is the reverse of the removal procedure. Torque the steering gear mounting bolt to 60 ft. lbs. (81 Nm).

8. Align the front end.

ADJUSTMENT

1. With the rack removed from the vehicle, total pinion torque can be checked and adjusted. Remove the boots before attempting to adjust, they will create an additional drag.

2. Attach special tool MB990228–01 to the pinion shaft and rotate the pinion with an inch lb. torque wrench. Turn the pinion 180 degrees to the right and left to measure.

3. The specification is 5–10 inch lbs.

4. If not within specifications, loosen the large locknut below the shaft and tighten the adjust cover until the proper turning torque is reached.

5. If the proper torque is not obtained after tightening the adjust cover 60 degrees, disassemble the unit and replace worn parts.

Power Steering Gear

REMOVAL AND INSTALLATION

1. Disconnect the negative battery cable. Raise the vehicle and support safely.

2. Fold back the dust boot which covers the steering shaft joint, if equipped. Remove the pinch bolt that holds the steering column shaft to the steering gear main shaft.

3. Remove the cotter pin, castellated nut the steering linkage from the pitman arm.

4. Disconnect and plug the fluid lines from the steering gear.

5. Remove the steering gear mounting nuts and remove the gear from the vehicle.

6. Matchmark and remove the pitman arm from the pitman shaft.

7. The installation is the reverse of the removal procedure. Torque the pitman arm nut to 100 ft. lbs. (136 Nm) and the gear mounting nuts to 28 ft. lbs. (38 Nm) on 2WD vehicles or 45 ft. lbs. (61 Nm) on 4WD vehicles.

8. Fill and bleed the system.

ADJUSTMENT

1. Attach special tool MB990228–01 to the mainshaft and rotate the pinion with an inch lb. torque wrench to measure the starting torque.

2. The specification is 4–11 inch lbs.

3. If not within specifications, loosen the adjusting screw locknut and adjust the locknut until the proper starting torque is reached.

4. Tighten the adjusting bolt locknut.

Power Rack and Pinion Steering Gear

REMOVAL AND INSTALLATION

1. Disconnect the negative battery cable.

2. Raise the vehicle and support safely. Remove the under cover.

3. Remove the front tire and wheel assemblies.

4. Remove the cotter pins, castellated nuts and tie rod ends from the steering knuckles.

5. Disconnect the intermediate shaft from the steering gear yoke.

6. Disconnect and plug the fluid lines from the fittings on the rack.

7. Remove the gear housing clamps and remove the gear from the vehicle.

To install:

8. Transfer the mounting insulators to the new gear and replace the O-rings on the the fluid lines.

9. Install the gear and the gear housing clamps in place and torque the steering gear mounting bolts to 60 ft. lbs. (81 Nm).

10. Connect the fluid lines to the fittings on the rack. Connect the intermediate shaft to the steering gear yoke.

11. Install the tie rod ends of the rack to the steering knuckles. Install the castellated nuts on tie rod end studs and install new cotter pins.

12. Install the tire and wheel assemblies.

13. Install the under cover, if equipped.

14. Lower the vehicle. Fill and bleed the power steering system.

15. Align the front end.

ADJUSTMENT

1. With the rack removed from the vehicle, total pinion torque can be checked and adjusted. Remove the boots before attempting to adjust, they will create an additional drag.

2. Attach special tool MB990228–01 to the pinion shaft and rotate the pinion with an inch lb. torque wrench. Turn the pinion 180 degrees to the right and left to measure.

3. The specification is 6–12 inch lbs.

4. If not within specifications, loosen the large locknut below the shaft and tighten the adjust cover until the proper turning torque is reached.

5. If the proper torque is not obtained after tightening the adjust cover 60 degrees, disassemble the unit and replace worn parts.

Power Steering Pump

REMOVAL AND INSTALLATION

1. Disconnect the negative battery cable.

2. Position a drain pan under the power steering pump.

3. Disconnect the fluid lines from the pump and plug them.

4. Remove the front bracket attaching bolts and remove the belt from the pulley.

5. Remove the pump from the vehicle.

6. The installation is the reverse of the removal procedure.

7. Adjust the belt tension, as required.

1. Pinch bolt
2. Pressure hose
3. Return hose
4. Cotter pin
5. Castellated nut
6. Steering linkage
7. Locknuts
8. Steering shaft
9. Power steering gear

7911D093

Power steering gear and related parts — Montero and Pick-Up

8. Fill and bleed the system.

BELT ADJUSTMENT

Except 3.0L Engine

1. Loosen the pump mounting bolts.
2. Using a pry bar, move the pump away from the engine.
3. Adjust the tension until the belt deflects about 1/4–1/2 in. under a 10 lb. load, tighten the mounting bolts.

3.0L Engine

1. Loosen the tensioner pulley locknut.
2. Tighten the adjuster bolt until the belt deflects about 1/4–1/2 in. under a 10 lb. load
3. Tighten the tensioner lock bolt.

System Bleeding

1. Fill the reservoir with Dexron® II.
2. Raise the front end of the vehicle.
3. Disconnect the ignition coil wire.
4. Simultaneously crank the engine and turn the steering wheel from lock to lock. Repeat this several times.

NOTE: If the bleeding procedure is done with the engine running, high speed rotation of the pump will churn the fluid and create air bubbles filling the system with air. Bleed the system while cranking the engine only.

5. Lower the front end.
6. Connect one end of a tube to the breather plug on the steering box and place the other a container.
7. Start the engine and allow it to idle.
8. Loosen the breather plug and turn the steering wheel from lock to lock continuously until no more air bubbles appear in the fluid coming out the tube.

NOTE: Do not hold the steering wheel all the way against the stop for more than 5 seconds.

9. After the bleeding is done, tighten the breather plug and refill the reservoir.

Tie Rod Ends

REMOVAL AND INSTALLATION

1. Raise the vehicle and support safely.
2. Remove the cotter pin and nut from the tie rod end stud.
3. Using a puller, remove the tie rod from the steering knuckle or center link.
4. Loosen the sleeve clamp nut, if equipped, and unthread the tie rod end.
5. The installation is the reverse of the removal procedure. Torque the stud nuts to 33 ft. lbs. (45 Nm) and install a new cotter pin.
6. Lubricate the front end.
7. Align the front end.

BRAKES

For all brake system repair and service procedure not detained below, please refer to "Brakes" in the Unit Repair section.

Master Cylinder

REMOVAL AND INSTALLATION

1. Disconnect the negative battery cable. Disconnect the fluid level sensor connector.
2. Disconnect and plug the brake lines from the master cylinder.
3. Remove the nuts attaching the master cylinder to the power booster.
4. Remove the master cylinder from the mounting studs.
5. Remove the fluid reservoir from the cylinder.
To install:
6. Bench bleed the master cylinder.
7. Install master cylinder on power booster studs and install the retaining nuts.
8. Install the brake lines loosely to the master cylinder.
9. Have a helper press down on the brake pedal, holding pedal to the floor. Tighten hydraulic lines at master cylinder.
10. Connect the fluid level sensor connector and refill reservoir as required.

Proportioning Valve

REMOVAL AND INSTALLATION

Montero

1. Disconnect the negative battery cable.

2. Raise the vehicle and support safely.
3. Identify and disconnect the brake lines from the proportioning valve.
4. Disconnect the wires to the valve, if any.
5. Remove the proportioning valve from the vehicle.
6. The installation is the reverse of the removal procedure.
7. Bleed the brakes in the following order:
   a. Right rear
   b. Left rear
   c. Right front
   d. Left front

Load Sensing Proportioning Valve

REMOVAL AND INSTALLATION

Pick-Up

1. Disconnect the negative battery cable.
2. Raise the vehicle and support safely.
3. Identify and disconnect the brake lines from the valve.
4. Remove the mounting bolts and remove the valve leaving the spring hanging from the lever.
To install:
5. Install the valve to the vehicle engaging the spring.
6. Connect the brake lines.
7. Bleed the brakes in the following order:
   a. Right rear
   b. Load sensing proportioning valve
   c. Left rear
   d. Right front
   e. Left front
8. Lower the vehicle so the full weight of the vehicle is on the ground. Unload any excess weight that is weighing down the rear of the vehicle. Adjust the load sensing proportioning valve lever so the distance from the proportioning valve lever to the spring support is about 7 in.

Power Brake Booster

REMOVAL AND INSTALLATION

1. Disconnect the negative battery cable. Disconnect the vacuum hose from the booster.
2. Remove the nuts attaching master cylinder to the booster and move the master cylinder to the side.

7911D095

Adjusting the load sensing proportioning valve

3. From inside of the vehicle, remove the clevis pin that secures the booster pushrod to the brake pedal.

4. Remove the nuts that attach the booster to the dash panel and remove from the vehicle.

5. The installation is the reverse of the removal procedure.

Brake Caliper

REMOVAL AND INSTALLATION

1. Raise the vehicle and support safely.

2. Remove the tire and wheel assembly. Disconnect the brake line from the caliper.

3. Label and remove the 2 mounting bolts.

4. Lift the caliper off of the adaptor.

5. The installation is the reverse of the removal procedure. Make sure all anti-rattle and anti-squeal clips and pads are in place. Torque the guide pin bolt to 35 ft. lbs. (47 Nm) and the lock pin bolt to 30 ft. lbs. (41 Nm).

6. Bleed the brakes and road test the vehicle.

Disc Brake Pads

REMOVAL AND INSTALLATION

1. Remove ½ of the fluid from the master cylinder reservoir.

2. Raise the vehicle and support safely. Remove the tire and wheel assemblies.

3. Remove the calipers.

4. Remove the pads from the adaptor.

To install:

5. Use a large C-clamp to compress the piston into the caliper bore.

6. Install the pads to the adaptor with anti-rattle and anti-squeal clips in place.

7. Install the caliper and secure to the adapter.

8. Refill the master cylinder.

9. Bleed the brakes if any brake lines were opened.

Brake Rotor

REMOVAL AND INSTALLATION

1. Raise the vehicle and support safely.

2. Remove the tire and wheel assembly.

3. Remove the caliper and brake pads.

4. Remove the caliper adaptor.

5. If equipped with 4WD, remove the automatic hub, the shim, lock washer and locknut.

6. If equipped with 2WD, remove the dust cap, cotter pin, nut lock, nut, washer and outer bearing.

7. Remove the front hub and rotor assembly from the spindle.

8. To separate the hub from the rotor, remove the bolts and split the rotor from the hub.

To install:

9. Install the rotor to the hub and torque the attaching nuts and bolts to 40 ft. lbs. (54 Nm).

10. Install the hub and rotor assembly to the spindle and install the retaining nuts and bearings as removed. Adjust bearing preload.

11. Install the caliper adaptor. Torque the bolts to 70 ft. lbs. (95 Nm).

12. Install the brake pads and caliper, making sure all anti-rattle and anti-squeal clips are in place.

13. Bleed the brakes if any brake line was opened.

14. Install the tire and wheel assembly and road test the vehicle.

Brake Drums

REMOVAL AND INSTALLATION

1. Raise the vehicle and support safely.

2. Remove the tire and wheel assembly.

3. Remove the drum retaining screw with an impact driver if equipped.

4. Remove the drum. If difficult to remove the drum, remove the adjuster access cover in the backing plate and rotate the star wheel to loosen the adjustment on the rear brake shoes.

5. The installation is the reverse of the removal procedure.

Brake Shoes

REMOVAL AND INSTALLATION

1. Raise the vehicle and support safely.

2. Remove the tire and wheel assemblies and the brake drums.

3. Remove the upper return spring along with the adjuster.

4. Remove the adjuster spring.

5. Remove the lower retaining spring.

6. Remove the hold-down springs.

7. Remove the shoes, disengaging the parking brake lever.

To install:

8. Clean and dry the backing plate. Lubricate the bosses, anchor contacts and wheel cylinder piston grooves where the top of the shoe fits lightly with lithium based grease.

9. Lubricate the star wheel shaft threads with anti-seize lubricant and transfer all parts to their proper locations on the new shoes. Make sure the longer end of the upper return spring is installed toward the shoe with the parking brake lever.

10. Spread the shoes apart, engage the parking brake lever and position them on the backing plate making sure the wheel cylinder pins engage the webs of the brake shoes.

11. Install the lower retaining spring and the hold-down springs.

12. Adjust the star wheel until the brake shoes are in contact with the brake drum when installed.

13. Remove any grease from the linings and install the brake drums.

14. Install the tire and wheel assemblies and complete the brake adjustment.

Wheel Cylinder

REMOVAL AND INSTALLATION

1. Raise the vehicle and support safely.

2. Remove the wheel, drum and brake shoes.

3. Remove the hydraulic brake line from the wheel cylinder.

4. Remove the wheel cylinder mounting bolts and remove the cylinder from the backing plate.

To install:

5. Position the wheel cylinder on the backing plate and install the retaining bolts. Torque the bolts to 14 ft. lbs. (19 Nm).

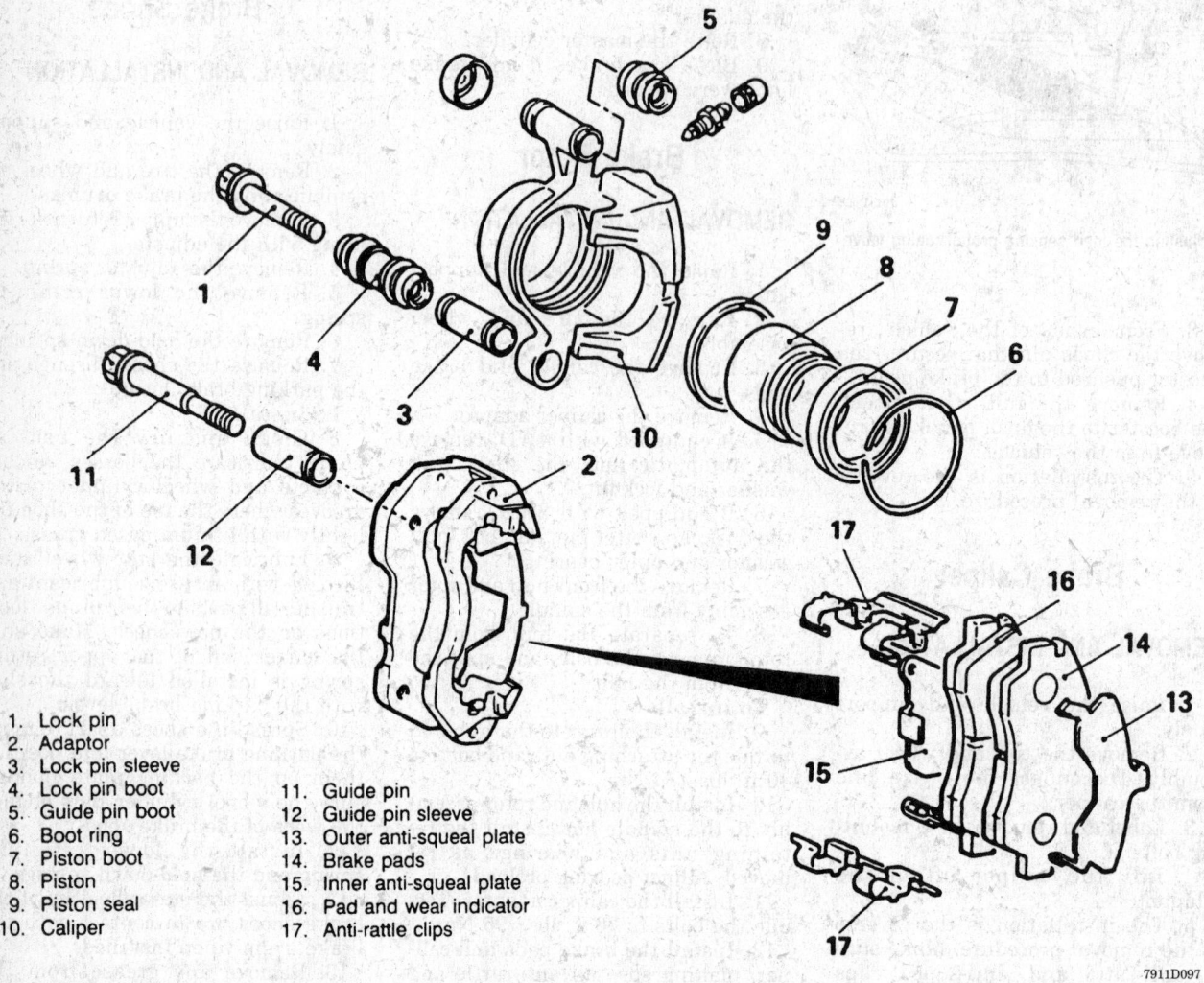

1. Lock pin
2. Adaptor
3. Lock pin sleeve
4. Lock pin boot
5. Guide pin boot
6. Boot ring
7. Piston boot
8. Piston
9. Piston seal
10. Caliper
11. Guide pin
12. Guide pin sleeve
13. Outer anti-squeal plate
14. Brake pads
15. Inner anti-squeal plate
16. Pad and wear indicator
17. Anti-rattle clips

7911D097

Front brakes — Montero and Pick-Up

6. Connect the brake line to the wheel cylinder.

7. Lubricate the backing plate and install the brake linings.

8. Install the brake drums, tire and wheel assemblies and adjust the brake linings.

9. Bleed the brakes system and re-fill the master cylinder reservoir.

Front Parking Brake Cable

REMOVAL AND INSTALLATION

1. Raise the vehicle and support safely. Remove the front cable adjusting nut.

2. Remove the brake lever cover, if floor mounted.

3. Remove the pin securing the cable to the equalizer and slide the cable out of the bracket.

4. Remove the pin or retainer attaching the cable to the handle assembly. Disengage the cable from the handle.

5. Remove the cable grommet from the floor pan and remove the cable from the vehicle.

6. The installation is the reverse of the removal procedure.

Rear Parking Brake Cable

REMOVAL AND INSTALLATION

1. Release the parking brakes fully.

2. Raise the vehicle and support safely.

3. Loosen the adjusting nut from the front cable.

4. Remove the brake drums. Remove the shoes, if necessary. Disconnect the cable from the lever.

5. Compress the cable retainer tabs to remove the cable from the backing plate.

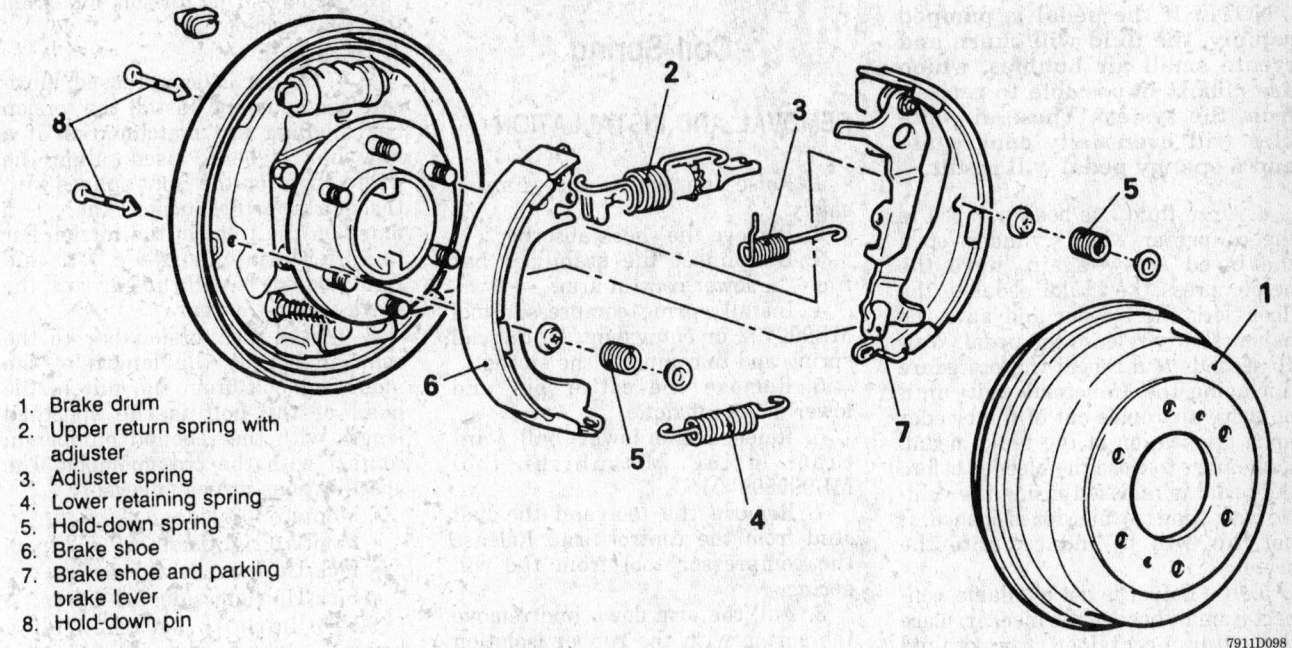

1. Brake drum
2. Upper return spring with adjuster
3. Adjuster spring
4. Lower retaining spring
5. Hold-down spring
6. Brake shoe
7. Brake shoe and parking brake lever
8. Hold-down pin

7911D098

Rear brake shoes and hardware

6. Remove the cable from the equalizer and remove the retaining bolts from the frame.

7. The installation is the reverse of the removal procedure.

ADJUSTMENT

1. Release the parking brakes fully.

2. Raise the vehicle and support safely.

3. Adjust the rear brake linings.

4. Loosen the adjusting nut until there is slack in all the cables.

5. Rotate the rear wheels and tighten the cable adjusting nut until there is a slight drag at the wheels.

6. Continue to rotate the rear wheels and loosen the nut until all drag is eliminated.

7. Back off the nut an additional 2 turns.

8. Apply and release the parking brake several times. Upon the last release, verify that there is no drag on the wheels. The pedal should depress 4–7 notches.

9. To check operation, make sure the parking brake holds on an incline.

Brake System Bleeding

If using a pressure bleeder, follow the instructions furnished with the unit and choose the correct adaptor for the application. Do not substitute an adapter that "almost fits" as it will not work and could be dangerous.

MASTER CYLINDER

1. If the master cylinder is off the vehicle it can be bench bled.

2. Connect short piece(s) of brake line to the outlet fitting(s), bend them until the free end is below the fluid level in the master cylinder reservoirs.

3. Fill the reservoir with fresh brake fluid. Pump the piston slowly until no more air bubbles appear in the reservoir(s).

4. Disconnect the lines, refill the master cylinder and securely install the cylinder cap.

5. If the master cylinder is on the vehicle, it can still be bled, using a flare nut wrench.

6. Open the brake line(s) slightly with the flare nut wrench while pressure is applied to the brake pedal by a helper inside the vehicle.

7. Be sure to tighten the line before the brake pedal is released.

8. Repeat the process with both lines until no air bubbles come out.

CALIPERS AND WHEEL CYLINDERS

1. Fill the master cylinder with fresh brake fluid. Check the level often during the procedure.

2. Starting with the right rear wheel, remove the protective cap from the bleeder and place where it will not be lost. Clean the bleed screw.

— **CAUTION** —

When bleeding the brakes, keep face away from the brake area. Spewing fluid may cause facial and/or visual damage. Do not allow brake fluid to spill on the car's finish; it will remove the paint.

3. If the system is empty, the most efficient way to get fluid down to the wheel is to loosen the bleeder about $1/2$–$3/4$ turn, place a finger firmly over the bleeder and have a helper pump the brakes slowly until fluid comes out the bleeder. Once fluid is at the

bleeder, close it before the pedal is released inside the vehicle.

NOTE: If the pedal is pumped rapidly, the fluid will churn and create small air bubbles, which are almost impossible to remove from the system. These air bubbles will eventually congregate and a spongy pedal will result.

4. Once fluid has been pumped to the caliper or wheel cylinder, open the bleed screw again, have the helper press the brake pedal to the floor, lock the bleeder and have the helper slowly release the pedal. Wait 15 seconds and repeat the procedure (including the 15 second wait) until no more air comes out of the bleeder upon application of the brake pedal. Remember to close the bleeder before the pedal is released inside the vehicle each time the bleeder is opened. If not, air will be induced into the system.

5. If a helper is not available, connect a small hose to the bleeder, place the end in a container of brake fluid and proceed to pump the pedal from inside the vehicle until no more air comes out the bleeder. The hose will prevent air from entering the system.

6. Repeat the procedure on remaining wheel cylinders in order:
   a. Left rear
   b. Right front
   c. Left front

7. Hydraulic brake systems must be totally flushed if the fluid becomes contaminated with water, dirt or other corrosive chemicals. To flush, bleed the entire system until all fluid has been replaced with new fluid.

8. Install the bleeder cap(s) on the bleeder to keep dirt out. Always road test the vehicle after brake work of any kind is done.

FRONT SUSPENSION

Shock Absorbers

REMOVAL AND INSTALLATION

1. Raise and safely support the vehicle. Remove the upper shock nut, washer and bushing.
2. Raise the vehicle fully and support safely.
3. Remove the lower mounting bolt(s) and discard the shock.

4. The installation is the reverse of the removal procedure.

Coil Spring

REMOVAL AND INSTALLATION

1. Raise the vehicle and support safely.
2. Remove the shock absorber.
3. Disconnect the stabilizer bar from the lower control arm.
4. Install spring compressor tool MB990792 or equivalent, to the coil spring and to compress the spring.
5. Remove the cotter pin and lower ball joint nut.
6. Release the lower ball joint taper using Mitsubishi tool MB990809–01.
7. Remove the tool and the ball stud from the control arm. Release the compressor tool from the coil spring.
8. Pull the arm down and remove the spring with the rubber isolation pad from the vehicle.

To install:

9. Install the spring with the rubber isolator. Install the compressor tool and compress spring so the lower ball joint can be inserted through the knuckle.
10. Torque the lower ball joint nut to 100 ft. lbs. (136 Nm). Install a new cotter pin. Remove the spring compressor.
11. Connect the sway bar to the lower control arm, if equipped.
12. Install the shock absorber.

Torsion Bar

REMOVAL AND INSTALLATION

4WD Vehicle

1. Raise the vehicle and support safely. Remove skid plate, if equipped.
2. Support the lower control arm at a point away from the torsion bar mounting point to the arm.
3. Fold the dust covers back and slide them away from the ends of the bar.
4. If the bars are to be reused, matchmark the torsion bar at both ends to the anchor and identify left from right.
5. Paint or measure the distance of the exposed threads of the rear mounting bolt down to the nut to aid in adjustment when installing. Re-

move the rear anchor arm mounting nut and bolt.
6. Remove the torsion bar from the front anchor arm.

To install:

7. If the bar is being reused, lubricate the ends and install the torsion bar aligning the matchmarks. If a new bar is being used, align the white stripe on the front splines with the mark on the anchor. There is a mark on the front of the torsion bar to differentiate between left and right. Do not install the bar with the mark facing the rear.
8. Install the torsion bar to the rear anchor so the length of the mounting bolt from the nut to the head of the bolt is the specified length with the rebound bumper in contact with the crossmember. The specifications are as follows:
   • Montero left side: 5.3–5.6 in.
   • Montero right side: 4.9–5.2 in.
   • Pick-Up left side: 5.5–5.8 in.
   • Pick-Up right side: 5.3–5.6 in.
9. To initially set the riding height, tighten the rear anchor mounting nut to the same point at which it was removed if the old bar is being reused. If a new bar has been installed, tighten the nut so the exposed length of the bolt threads is the specification:
   • Montero left side: 2.4 in.
   • Montero right side: 2.8 in.
   • Pick-Up left side: 3.9 in.
   • Pick-Up right side: 3.4 in.
10. Fill the dust covers with grease and fold them back into position.
11. Adjust the torsion to the correct riding height.

Adjusting the Torsion Bar

1. Lower the vehicle so its full weight is on the ground. If there is any excess weight in the vehicle, unload it.
2. To set the riding height, measure the distance from the rebound bumper to its stopper bracket on the frame. Adjust the torsion bar nut until that distance is within the specification: — —Montero—2.8 in.— —Pick-Up—3.1 in.
3. If, after adjusting the riding height, the exposed portion of the adjusting bolt is protruding up or down far enough to interfere with any front suspension component or is hanging down too low, the bar must be removed and repositioned.
4. Road test the vehicle and remeasure the riding height. Readjust if necessary.

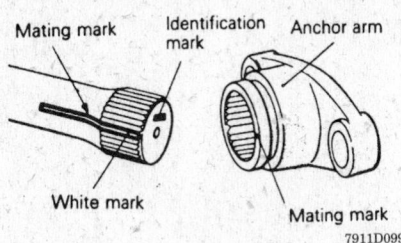

Aligning the front of the torsion bar with the front anchor

7911D099

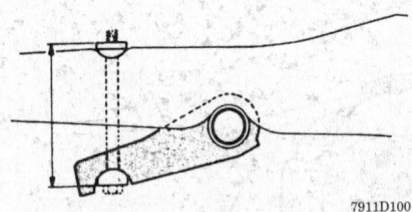

Measure this distance when installing the torsion bar to the rear anchor

7911D100

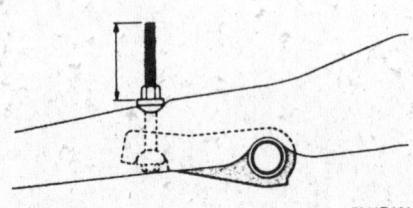

Measure this distance to initially set the riding height

7911D101

Upper Ball Joint

INSPECTION

With the control arm removed, check the starting torque required to turn the ball stud. The specification for all vehicles is 7–30 inch lbs.

REMOVAL AND INSTALLATION

NOTE: The upper ball joint and upper control arm must be replaced as an assembly on Pick-Up.

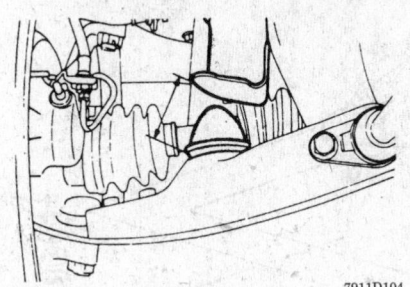

7911D104

Measure this distance to set the riding height

1. Raise the vehicle and support safely. Release the tension on the torsion bar and remove the upper control arm from the vehicle.
2. Remove the ring and boot from the ball joint. Remove the snapring.
3. Remove the ball joint from the control arm using tool sets MB990800 and MB990799.

To install:
4. Align the mating mark on the upper ball joint with the mark on the arm.
5. Press the ball joint in using the same tools that were used to remove the ball joint.
6. Install the snapring. If the snapring is loose, install a new one.
7. Fill the boot with grease and install the boot and ring.
8. Install the upper control arm.
9. Lubricate the ball joint with a grease gun.
10. Adjust the riding height, if equipped with a torsion bar and align the front end.

Lower Ball Joint

INSPECTION

With the control arm removed, check the up and down endplay of the ball stud. If it exceeds 0.02 in. (0.5mm), the ball joint should be replaced.

REMOVAL AND INSTALLATION

1. Raise the vehicle and support safely. Remove the lower control arm.
2. Remove the ball joint retaining nuts and bolts and remove the ball joint from the arm.
3. The installation is the reverse of the removal procedure. Torque the ball joint retaining nuts and bolts to 50 ft. lbs. (68 Nm) on 4WD vehicles or 28 ft. lbs. (38 Nm) on 2WD vehicles. Torque the ball stud nut to 100 ft.

lbs. (136 Nm) and install a new cotter pin.
4. Lubricate the ball joint with a grease gun.
5. Adjust the riding height, if equipped with a torsion bar.
6. Align the front suspension.

Upper Control Arm

REMOVAL AND INSTALLATION

2WD Pick-Up

1. Raise the vehicle and support safely. Remove the tire and wheel assembly.
2. Remove the shock absorber.
3. Install Mitsubishi spring compressor tool MB990792 or equivalent, to the coil spring and compress the spring.
4. Remove the cotter pin and upper ball joint nut.
5. Suspend the rotor assembly so there is not excessive pull on the brake hose.
6. Release the upper ball joint taper using Mitsubishi tool MB990809–01 or equivalent.
7. Remove the tool and remove the ball stud from the knuckle.
8. Loosen the pivot bar retaining nuts and bolts, identify and remove the alignment shims, remove the nuts and bolts and remove the arm from the vehicle.

To install:
9. Install the arm to the frame rail bracket, install the shims in their original locations and install the retaining nuts and bolts. Torque the nuts initially to 40 ft. lbs. (54 Nm).
10. Torque the ball joint nut to 60 ft. lbs. (81 Nm). Install a new cotter pin. Remove the spring compressor.
11. Install the shock absorber.
12. Align the front end. When all settings are at specifications, torque the pivot bar retaining bolts to 80 ft. lbs. (109 Nm).

Montero and 4WD Pick-Up

1. Raise the vehicle and support safely. Remove the skid plate.
2. Remove the shock absorber.
3. Turn the torsion bar adjustment nut counterclockwise to relieve all tension from the torsion bar. Disconnect the brake hose from the brake line and remove the hose from the bracket.
4. Remove the cotter pin from the upper ball stud.
5. Release the upper ball joint taper using Mitsubishi tool MB990809–01 or equivalent. Remove

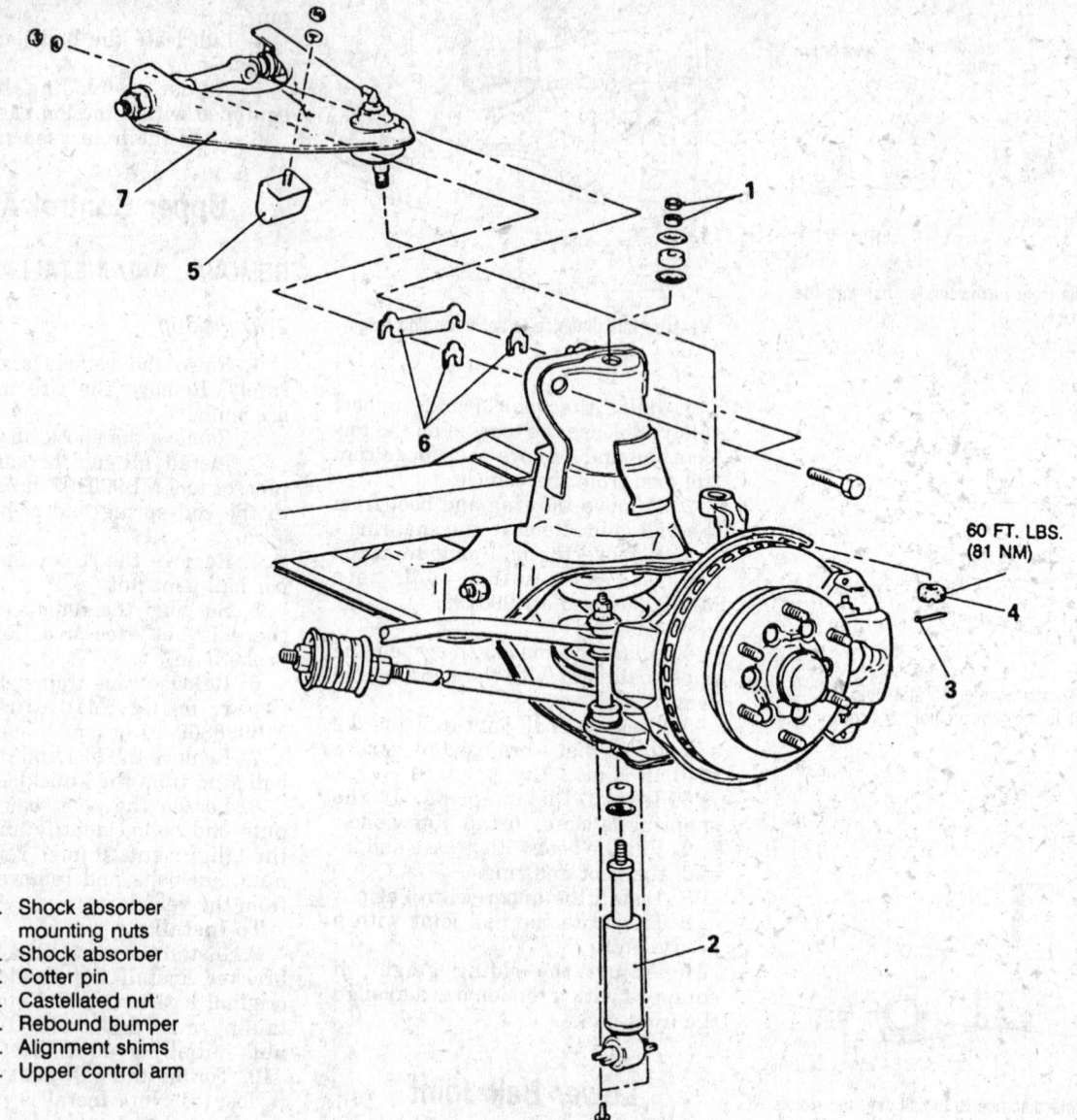

1. Shock absorber
   mounting nuts
2. Shock absorber
3. Cotter pin
4. Castellated nut
5. Rebound bumper
6. Alignment shims
7. Upper control arm

60 FT. LBS.
(81 NM)

7911D106

Front suspension components — 2WD Pick-Up

the tool. Remove the ball stud from the steering knuckle.

6. Loosen the pivot bar retaining nuts and bolts, identify and remove the alignment shims, remove the nuts and bolts.

7. Remove the control arm from the vehicle.

To install:

8. Position the arm at the frame rail bracket.

9. Install the shims in their original locations and install the retaining nuts and bolts. Torque the nuts initially to 40 ft. lbs. (54 Nm).

10. Insert the upper ball stud in the steering knuckle arm bore and install

the nut. Torque the nut to 60 ft. lbs. (81 Nm) and install a new cotter pin. Install the shock absorber and attach the brake hose.

11. Turn the torsion bar adjustment nut clockwise to apply a load on the bar.

12. Lower the vehicle.

13. Set the riding height and align the front end.

Lower Control Arm

REMOVAL AND INSTALLATION

2WD Pick-Up

1. Raise the vehicle and support safely.

2. Remove the shock absorber.

3. Disconnect the sway bar and strut bar from the lower control arm.

4. Install Mitsubishi spring compressor tool MB990792 or equivalent, to the coil spring and compress the spring.

5. Remove the cotter pin and lower ball joint nut.

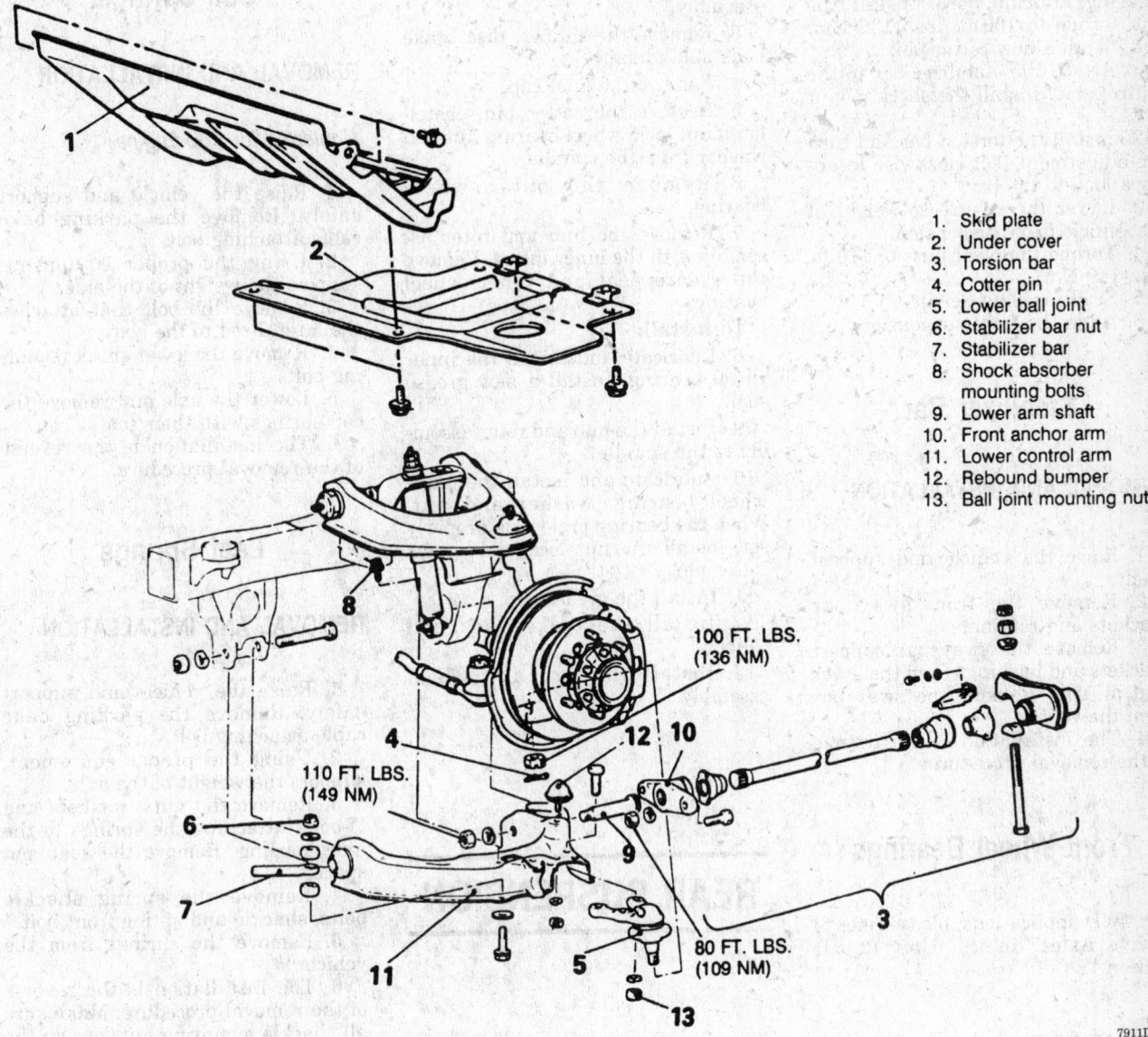

1. Skid plate
2. Under cover
3. Torsion bar
4. Cotter pin
5. Lower ball joint
6. Stabilizer bar nut
7. Stabilizer bar
8. Shock absorber mounting bolts
9. Lower arm shaft
10. Front anchor arm
11. Lower control arm
12. Rebound bumper
13. Ball joint mounting nuts

100 FT. LBS. (136 NM)

110 FT. LBS. (149 NM)

80 FT. LBS. (109 NM)

7911D107

Front suspension components — 4WD Vehicle

6. Release the lower ball joint taper using Mitsubishi tool MB990809–01 or equivalent.

7. Remove the tool and remove the ball stud from the knuckle. Remove the spring compressor.

8. Pull the arm down and remove the spring with the rubber isolation pad from the vehicle. Remove the lower arm shaft mounting nuts and remove the arm from the vehicle.

To install:

9. Install the arm to the crossmember finger tight. Install the spring with the rubber isolators. Install the compressor tool and compress spring so lower ball joint can be inserted through the knuckle.

10. Torque the lower ball joint nut to 100 ft. lbs. (136 Nm). Install a new cotter pin. Remove the spring compressor.

11. Connect the sway bar and strut bar to the lower control arm.

12. Install the shock absorber.

13. Lower the vehicle completely. With the weight of the vehicle on suspension torque the lower control arm to crossmember mounting nut to 50 ft. lbs. (68 Nm).

14. Align the front suspension.

Montero and 4WD Pick-Up

1. Raise the vehicle and support safely.

2. Remove the skid plate.

3. Remove the torsion bar, anchors and lower arm shaft.

4. Remove the shock absorber lower attaching bolt.

5. Disconnect the stabilizer bar from the lower control arm.

6. Remove the cotter pin and the nut from the lower ball stud. Separate the lower ball stud from the steering knuckle using a puller.

7. Remove the pivot bolts and remove the arm from the vehicle.

To install:

8. Install the new control arm to the vehicle.

9. Install the pivot bolts, but do not torque.

10. Insert the ball stud into the steering knuckle bore. Install the nut, torque to 100 ft. lbs. (136 Nm) and install a new cotter pin.

11. Attach the stabilizer bar to the control arm. Install the shock mount bolts.

12. Install the torsion bar and turn the adjustment bolt clockwise to apply a load to the bar.

13. Lower the vehicle so weight of the vehicle is on suspension.

14. Torque the pivot nuts to 110 ft. lbs. (149 Nm).

15. Set the riding height.

16. Align the front suspension.

Stabilizer Bar

REMOVAL AND INSTALLATION

1. Raise the vehicle and support safely.

2. Remove the front sway bar brackets and retainers.

3. Remove the sway bar support brackets and bushings from the lower control arm. Remove the sway bar from the vehicle.

4. The installation is the reverse of the removal procedure.

Front Wheel Bearings

For 4WD applications, please refer to "Drive Axles" in the Unit Repair section.

ADJUSTMENT

1. Tighten the wheel bearing nut to 22 ft. lbs. (30 Nm) while turning the rotor.

2. Loosen the wheel bearing adjusting nut completely.

3. Tighten the nut to 6 ft. lbs. (8 Nm).

4. Check the wheel bearing endplay. The specification is 0.001–0.003 in.

5. Install the nut lock and cotter pin.

REMOVAL AND INSTALLATION

1. Raise the vehicle and support safely.

2. Remove the tire and wheel assembly.

3. Remove the caliper, disc brake pads and adaptor.

4. Remove the dust cap.

5. Remove the cotter pin, castellated nut lock, wheel bearing nut and washer from the spindle.

6. Remove the outer wheel bearing.

7. Remove the hub and rotor assembly with the inner intact. Remove the grease seal and inner wheel bearing.

To install:

8. Lubricate and install the inner wheel bearing. Install a new grease seal.

9. Install the hub and rotor assembly to the spindle.

10. Lubricate and install the outer wheel bearing, washer and nut. When the bearing preload is properly set, install the nut lock and a new cotter pin.

11. Install the grease cap.

12. Install the brake pads and caliper.

13. Install the tire and wheel assembly.

REAR SUSPENSION

Shock Absorber

REMOVAL AND INSTALLATION

1. Raise the vehicle and support safely. If equipped with rear coil springs, support the rear axle using the proper equipment.

2. Remove the bolts that attach the shock to the frame or bracket.

3. Remove the shock from the vehicle.

4. The installation is the reverse of the removal procedure.

Coil Springs

REMOVAL AND INSTALLATION

Montero With 3.0L Engine

1. Raise the vehicle and support safely. Remove the parking bake cable attaching bolt.

2. Using the proper equipment, support the weight of the axle.

3. Remove the bolt that attaches the lateral rod to the body.

4. Remove the lower shock mounting bolts.

5. Lower the axle and remove the coil springs with their seats.

6. The installation is the reverse of the removal procedure.

Leaf Springs

REMOVAL AND INSTALLATION

1. Raise the vehicle and support safely. Remove the parking bake cable attaching bolt.

2. Using the proper equipment, support the weight of the axle.

3. Remove the nuts, washers and U-bolts attaching the springs to the axle housing. Remove the seat and spacer.

4. Remove the spring shackle bolts, shackle and spring front bolt.

5. Remove the springs from the vehicle.

6. The installation is the reverse of the removal procedure. Make sure all shackle mounting nuts are on the inside of the springs. Torque the front spring mount nut to 116 ft. lbs. (160 Nm) and the rear shackle nuts to 43 ft. lbs. (60 Nm). Torque U-bolt nuts to 87 ft. lbs. (120 Nm).

Rear Wheel Bearings

For RWD and 4WD applications, please refer to "Drive Axle" section in this repair manual.

Rear Axle Assembly

For RWD and 4WD applications, please refer to "Drive Axle" section of this manual.

Nissan **11**

Pathfinder • Pickup • Quest

SPECIFICATIONS

ENGINE IDENTIFICATION

Year	Model	Engine Displacement Liters (cc)	Engine Series (ID/VIN)	Fuel System	No. of Cylinders	Engine Type
1991	Pick-up	2.4 (2389)	KA24E	MFI	4	SOHC
	Pick-up	3.0 (2960)	VG30E	MFI	6	SOHC
	Pathfinder	3.0 (2960)	VG30E	MFI	6	SOHC
1992	Pick-up	2.4 (2389)	KA24E	MFI	4	SOHC
	Pick-up	3.0 (2960)	VG30E	MFI	6	SOHC
	Pathfinder	3.0 (2960)	VG30E	MFI	6	SOHC
1993	Pick-up	2.4 (2389)	KA24E	MFI	4	SOHC
	Pick-up	3.0 (2960)	VG30E	MFI	6	SOHC
	Quest	3.0 (2960)	VG30E	MFI	6	SOHC
	Pathfinder	3.0 (2960)	VG30E	MFI	6	SOHC
1994-95	Pick-up	2.4 (2389)	KA24E	MFI	4	SOHC
	Pick-up	3.0 (2960)	VG30E	MFI	6	SOHC
	Quest	3.0 (2960)	VG30E	MFI	6	SOHC
	Pathfinder	3.0 (2960)	VG30E	MFI	6	SOHC

MFI - Multiport fuel injection
SOHC - Single overhead camshaft

7911eC02

GENERAL ENGINE SPECIFICATIONS

Year	Engine ID/VIN	Engine Displacement Liters (cc)	Fuel System Type	Net Horsepower @ rpm	Net Torque @ rpm (ft. lbs.)	Bore x Stroke (in.)	Compression Ratio	Oil Pressure @ rpm
1991	KA24E	2.4 (2389)	MFI	134@5200	154@3600	3.50x3.78	8.6:1	60@3000
	VG30E	3.0 (2960)	MFI	160@5200	182@2800	3.43x3.27	9.0:1	53@3200
1992	KA24E	2.4 (2389)	MFI	134@5200	154@3600	3.50x3.78	8.6:1	60@3000
	VG30E	3.0 (2960)	MFI	153@4800	180@4000	3.43x3.27	9.0:1	53@3200
1993	KA24E	2.4 (2389)	MFI	134@5200	154@3600	3.50x3.78	8.6:1	60@3000
	VG30E	3.0 (2960)	MFI	153@4800	180@4000	3.43x3.27	9.0:1	53@3200
1994-95	KA24E	2.4 (2389)	MFI	134@5200	154@3600	3.50x3.78	8.6:1	60@3000
	VG30E	3.0 (2960)	MFI	1	2	3.43x3.27	9.0:1	53@3200

MFI - Mutliport fuel injection

1 Quest: 151@4800

   Pick-up and Pathfinder: 153@4800

2 Quest: 174@4400

   Pick-up and Pathfinder: 180@4000

7911eC03

GASOLINE ENGINE TUNE-UP SPECIFICATIONS

Year	Engine ID/VIN	Engine Displacement Liters (cc)	Spark Plugs Gap (in.)	Ignition Timing (deg.) MT	AT	Fuel Pump (psi)	Idle Speed (rpm) MT	AT	Valve Clearance In.	Ex.
1991	KA24E	2.4 (2389)	0.033	10B	10B	33 ①	800	800 ②	HYD	HYD
	VG30E	3.0 (2960)	0.041	15B	15B	33 ①	750	750 ②	HYD	HYD
1992	KA24E	2.4 (2389)	0.033	10B	10B	33 ①	750	750 ②	HYD	HYD
	VG30E	3.0 (2960)	0.041	15B	15B	33 ①	750	750 ②	HYD	HYD
1993	KA24E	2.4 (2389)	0.033	10B	10B	33 ①	800	800 ②	HYD	HYD
	VG30E	3.0 (2960)	0.041	15B	15B	33 ①	750	750 ②	HYD	HYD
1994-95	KA24E	2.4 (2389)	0.033	10B	10B	33 ①	800	800 ②	HYD	HYD
	VG30E	3.0 (2960)	0.041	15B	15B	33 ①	750	750 ②	HYD	HYD

NOTE: The Vehicle Emission Control Information label often reflects changes made during production. The label must be used if different from above.

B - Before top dead center

HYD - Hydraulic

1 System pressure at idle with vacuum hose connected; should increase to 43 psi when disconnected

2 Automatic transmission in neutral

7911eC04

CAPACITIES

Year	Model	Engine ID/VIN	Engine Displacement Liters (cc)	Engine Crankcas with Filter	Transmission (pts.) 4-Spd	5-Spd	Auto.	Transfer Case (pts.)	Drive Axle Front (pts.)	Rear (pts.)	Fuel Tank (gal.)	Cooling System (qts.)
1991	Pick-up	KA24E	2.4 (2389)	1	-	2	3	-	4	5	16.0	6
	Pick-up	VG30E	3.0 (2960)	7	-	8	3	-	9	5.8	21.0 ⑩	11
	Pathfinder	VG30E	3.0 (2960)	7	-	8	3	-	9	5.8	21.0	11
1992	Pick-up	KA24E	2.4 (2389)	1	-	2	3	-	4	5	16.0	6
	Pick-up	VG30E	3.0 (2960)	7	-	8	3	-	9	5.8	21.0 ⑩	11
	Pathfinder	VG30E	3.0 (2960)	7	-	8	3	-	9	5.8	21.0	11
1993	Pick-up	KA24E	2.4 (2389)	1	-	2	3	-	6	5	16.0	8
	Pick-up	VG30E	3.0 (2960)	7	-	8	3	-	11	5.8	21.0 ⑩	11
	Pathfinder	VG30E	3.0 (2960)	7	-	8	3	-	11	5.8	20.0	11
	Quest	VG30E	3.0 (2960)	4.0	-	-	20.0	-	-	-	20.0	2
1994-95	Pick-up	KA24E	2.4 (2389)	1	-	2	3	-	6	5	16.0	6
	Pick-up	VG30E	3.0 (2960)	7	-	8	3	-	11.0	5.8	21.0 ⑩	11
	Pathfinder	VG30E	3.0 (2960)	7	-	8	3	-	11.0	5.8	20.0	11
	Quest	VG30E	3.0 (2960)	4.0	-	-	20.0	-	-	-	20.0	11

1 2WD: 4.1 qts.; 4WD: 3.5 qts.

2 2WD: 4.25 pts.; 4WD: 8.50 pts.

3 2WD: 16.75 pts.; 4WD: 18.0 pts.

4 Front differential: 2.75 pts.
   Transfer case: 4.62 pts.

5 2WD: 3.12 pts.; 4WD: 2.75 pts.

6 2WD: 8.6 qts.; 4WD: 9.5 qts.

7 2WD: 4.25 qts.; 4WD: 3.60 qts.

8 2WD: 5.1 pts.; 4WD: 7.6 pts.

9 Front differential: 3.12 pts.
   Transfer case: 4.62 pts.

10 SE models: 16 gals.

11 2WD: 11.4 qts.; 4WD: 12.4 qts.

7911eC06

CAMSHAFT SPECIFICATIONS
All measurements given in inches.

Year	Engine ID/VIN	Engine Displacement Liters (cc)	Journal Diameter					Elevation		Bearing Clearance	Camshaft End Play
			1	2	3	4	5	In.	Ex.		
1991	KA24E	2.4 (2389)	1.2967-1.2974	1.2967-1.2974	1.2967-1.2974	1.2967-1.2974	1.2967-1.2974	1.765-1.773	1.765-1.773	1	0.002-0.005
	VG30E	3.0 (2960)	2	2	2	2	2	1.557-1.564	1.557-1.564	1	0.001-0.002
1992	KA24E	2.4 (2389)	1.2967-1.2974	1.2967-1.2974	1.2967-1.2974	1.2967-1.2974	1.2967-1.2974	1.765-1.773	1.765-1.773	1	0.002-0.005
	VG30E	3.0 (2960)	2	2	2	2	2	1.557-1.564	1.557-1.564	1	0.001-0.002
1993	KA24E	2.4 (2389)	1.2967-1.2974	1.2967-1.2974	1.2967-1.2974	1.2967-1.2974	1.2967-1.2974	1.765-1.773	1.765-1.773	1	0.002-0.005
	VG30E	3.0 (2960)	2	2	2	2	2	1.557-1.564	1.557-1.564	1	0.001-0.002
1994-95	KA24E	2.4 (2389)	1.2967-1.2974	1.2967-1.2974	1.2967-1.2974	1.2967-1.2974	1.2967-1.2974	1.765-1.773	1.765-1.773	1	0.002-0.005
	VG30E	3.0 (2960)	2	2	2	2	2	3	3	1	0.001-0.002

1 Wear limit: 0.008
   Left camshaft:
2 Rear Journal: 1.6732-1.6742
   Middle three Journals: 1.8504-1.8514
   Front Journal: 1.8898-1.8907
   Right camshaft:
   Rear Journal: 1.6732-1.6742
   Other Journals: 1.8504-1.8514
3 Pick-up and Pathfinder: 1.557-1.564
   Quest: 1.533-1.541

7911eC07

CRANKSHAFT AND CONNECTING ROD SPECIFICATIONS

All measurements are given in inches.

Year	Engine ID/VIN	Engine Displacement Liters (cc)	Crankshaft				Connecting Rod		
			Main Brg. Journal Dia.	Main Brg. Oil Clearance	Shaft End-play	Thrust on No.	Journal Diameter	Oil Clearance	Side Clearance
1991	KA24E	2.4 (2389)	1	0.0008-0.0019 [2]	0.002-0.007	3	2.3603-2.3612	0.0004-0.0014 [2]	0.008-0.016
	VG30E	3.0 (2960)	3	0.0011-0.0022 [4]	0.002-0.007	4	1.9667-1.9675	0.0006-0.0021 [4]	0.008-0.014
1992	KA24E	2.4 (2389)	1	0.0008-0.0019 [2]	0.002-0.007	3	2.3603-2.3612	0.0004-0.0014 [2]	0.008-0.016
	VG30E	3.0 (2960)	3	0.0011-0.0022 [4]	0.002-0.007	4	1.9667-1.9675	0.0006-0.0021 [4]	0.008-0.014
1993	KA24E	2.4 (2389)	1	0.0008-0.0019 [2]	0.002-0.007	3	2.3603-2.3612	0.0004-0.0014 [2]	0.008-0.016
	VG30E	3.0 (2960)	3	0.0011-0.0022 [4]	0.002-0.007	4	1.9667-1.9675	0.0006-0.0021 [4]	0.008-0.014
1994-95	KA24E	2.4 (2389)	1	0.0008-0.0019 [2]	0.002-0.007	3	2.3603-2.3612	0.0004-0.0014 [2]	0.008-0.016
	VG30E	3.0 (2960)	3	0.0011-0.0022 [4]	0.002-0.007	4	1.9667-1.9675	0.0006-0.0021 [4]	0.008-0.014

[1] Grade 0: 2.3609-2.3612
   Grade 1: 2.3606-2.3609
   Grade 2: 2.3603-2.3606
[2] Wear limit: 0.004

[3] Grade 0: 2.4790-2.4793
   Grade 1: 2.4787-2.4790
   Grade 2: 2.4784-2.4787
[4] Wear limit: 0.0035

7911eC08

VALVE SPECIFICATIONS

Year	Engine ID/VIN	Engine Displacement Liters (cc)	Seat Angle (deg.)	Face Angle (deg.)	Spring Test Pressure (lbs. @ in.)	Spring Installed Height (in.)	Stem-to-Guide Clearance (in.)		Stem Diameter (in.)	
							Intake	Exhaust	Intake	Exhaust
1991	KA24E	2.4 (2389)	45	45.5	1	NA	0.0008-0.0021	0.0016-0.0028	0.2742-0.2748	0.3129-0.3134
	VG30E	3.0 (2960)	45	45.25-45.75	2	3	0.0008-0.0021	0.0016-0.0029	0.2742-0.2748	0.3136-0.3138
1992	KA24E	2.4 (2389)	45	45.5	1	NA	0.0008-0.0021	0.0016-0.0028	0.2742-0.2748	0.3129-0.3134
	VG30E	3.0 (2960)	45	45.25-45.75	2	3	0.0008-0.0021	0.0016-0.0029	0.2742-0.2748	0.3136-0.3138
1993	KA24E	2.4 (2389)	45	45.5	1	NA	0.0008-0.0021	0.0016-0.0028	0.2742-0.2748	0.3129-0.3134
	VG30E	3.0 (2960)	45	45.25-45.75	2	NA	0.0008-0.0021	0.0016-0.0029	0.2742-0.2748	0.3136-0.3138
1994-95	KA24E	2.4 (2389)	45	45.5	106@1.026	NA	0.0008-0.0021	0.0016-0.0028	0.2742-0.2748	0.3129-0.3134
	VG30E	3.0 (2960)	45	45.25-45.75	2	NA	0.0008-0.0021	0.0016-0.0029	0.2742-0.2748	0.3136-0.3138

NA - Not Available

1 Intake:
Inner: 63.9 @ 1.28
Outer: 135.2 @ 1.48
Exhaust:
Inner: 74 @ 1.15
Outer: 144 @ 1.34

2 Inner: 57.3 @ 0.984
Outer: 117.7 @ 1.181

3 Inner: 1.378; Outer: 1.575

7911eC09

PISTON AND RING SPECIFICATIONS

All measurements are given in inches.

Year	Engine ID/VIN	Engine Displacement Liters (cc)	Piston Clearance	Ring Gap			Ring Side Clearance		
				Top Compression	Bottom Compression	Oil Control	Top Compression	Bottom Compression	Oil Control
1991	KA24E	2.4 (2389)	0.0008-0.0016	0.011-0.020	1	0.008-0.027	0.002-0.003	0.001-0.003	NA
	VG30E	3.0 (2960)	0.0006-0.0014	0.008-0.017	0.007-0.017	0.008-0.030	0.002-0.003	0.001-0.003	0.001-0.007
1992	KA24E	2.4 (2389)	0.0008-0.0016	0.011-0.020	1	0.008-0.027	0.002-0.003	0.001-0.003	NA
	VG30E	3.0 (2960)	0.0006-0.0014	0.008-0.017	0.007-0.017	0.008-0.030	0.002-0.003	0.001-0.003	0.001-0.007
1993	KA24E	2.4 (2389)	0.0008-0.0016	0.011-0.020	0.018-0.027	0.008-0.027	0.002-0.003	0.001-0.003	NA
	VG30E	3.0 (2960)	0.0006-0.0014	0.008-0.017	0.007-0.017	0.008-0.030	0.002-0.003	0.001-0.003	0.001-0.007
1994-95	KA24E	2.4 (2389)	0.0008-0.0016	0.011-0.020	0.018-0.027	0.008-0.027	0.002-0.003	0.001-0.003	0.003-0.005
	VG30E	3.0 (2960)	0.0006-0.0014	0.008-0.017	0.007-0.017	0.008-0.030	0.002-0.003	0.001-0.003	0.001-0.007

NA - Not Available

1 Grade 1 (R or T): 0.018-0.027
  Grade 2 (N): 0.022-0.028

7911eC10

TORQUE SPECIFICATIONS
All readings in ft. lbs.

Year	Engine ID/VIN	Engine Displacement Liters (cc)	Cylinder Head Bolts	Main Bearing Bolts	Rod Bearing	Crankshaft Damper Bolts	Flywheel Bolts	Manifold		Spark Plugs	Lug Nut
								Intake	Exhaust		
1991	KA24E	2.4 (2389)	1	34-38	2	105-112	3	14	14	18	4
	VG30E	3.0 (2960)	5	67-74	6	90-98	61-69	7	15	18	4
1992	KA24E	2.4 (2389)	1	34-38	2	105-112	3	14	14	18	4
	VG30E	3.0 (2960)	5	67-74	2	90-98	72-80	7	15	18	4
1993	KA24E	2.4 (2389)	1	34-38	2	105-112	3	14	14	18	4
	VG30E	3.0 (2960)	5	67-74	2	90-98	72-80	7	15	18	4
1994-95	KA24E	2.4 (2389)	1	34-38	2	105-112	3	14	14	18	4
	VG30E	3.0 (2960)	5	67-74	2	90-98	72-80	7	15	18	4

1 Step 1: 22 ft. lbs.
  Step 2: 58 ft. lbs.
  Step 3: Loosen completely then retorque to 22 ft. lbs.
  Step 4: 58 ft. lbs. or an additional 80-85 degrees
2 12 ft. lbs. plus an additional 60-65 degrees
3 Manual transmission: 105-112 ft. lbs.
  Automatic transmission: 69-76 ft. lbs.
4 Pick-up and Pathfinder with single wheel: 87-108 ft. lbs.
  Dual wheel: 166-203 ft. lbs.

5 Step 1: 22 ft. lbs.
  Step 2: 43 ft. lbs.
  Step 3: Loosen completely then retorque to 22 ft. lbs.
  Step 4: 40-47 ft. lbs. or an additional 60-65 degrees
6 Step 1: 12 ft. lbs.
  Step 2: 30 ft. lbs. or an additional 60-65 degrees
7 Nuts: 18 ft. lbs., in two steps
  Bolts: 13 ft. lbs., in two steps

7911eC11

BRAKE SPECIFICATIONS

All measurements in inches unless noted

Year	Model	Master Cylinder Bore	Brake Disc			Brake Drum Diameter			Minimum Lining Thickness	
			Original Thickness	Minimum Thickness	Maximum Runout	Original Inside Diameter	Max. Wear Limit	Maximum Machine Diameter	Front	Rear
1991	Pick-up	3	NA	4	0.003	NA	-	6	0.079	1
	Pathfinder	3	NA	5	0.003	NA	-	7	0.079	1
1992	Pick-up	3	NA	4	0.003	NA	-	6	0.079	1
	Pathfinder	3	NA	5	0.003	NA	-	7	0.079	1
1993	Pick-up	1.000	NA	4	0.003	NA	-	6	0.079	1
	Pathfinder	1.000	NA	5	0.003	NA	-	8	0.079	1
	Quest	1.000	NA	0.945	0.003	NA	-	9.900	0.079	0.079
1994-95	Pick-up	1.000	NA	4	0.003	NA	-	6	0.079	2
	Pathfinder	1.000	NA	5	0.003	NA	-	8	0.079	2
	Quest	1.000	NA	0.945	0.003	NA	-	9.900	0.079	0.079

NA - Not Available
1 Disc brake: 0.079
  Drum brake: 0.059
2 Disc brake: 0.079
  Drum brake: 0.059
3 With ABS: 1.000
  Without ABS: 0.9375
4 2WD, KA24E engine:
  Front: 0.787; Rear: 0.630
  VG30E engine:
  Front: 0.945; Rear: 0.630

5 Front: 0.945
  Rear: 0.630
6 2WD: 10.30
  4WD: 11.67
7 Rear disc parking brake drum: 7.52
8 Rear drum brake: 10.30
  Rear disc parking brake drum: 7.52

7911eC12

FIRING ORDERS

NOTE: To avoid confusion, always replace spark plug wires one at a time.

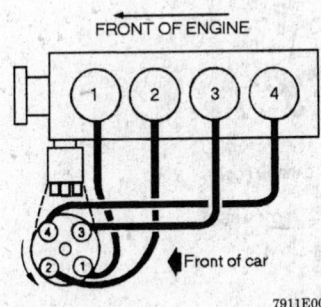

7911E002

2.4L (KA24E) Engine
Engine Firing Order: 1 — 3 — 4 — 2
Distributor Rotation: Counterclockwise

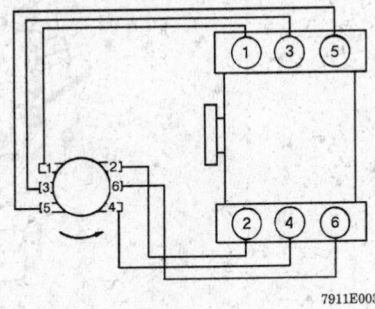

7911E003

3.0L (VG30E) Engines
Engine Firing Order: 1 — 2 — 3 — 4 — 5 — 6
Distributor Rotation: Counterclockwise

ENGINE ELECTRICAL

NOTE: Disconnecting the negative battery cable on some vehicles may interfere with the functions of the on board computer systems and may require the computer to undergo a relearning process when the battery cable is reconnected. This usually requires only a few minutes of driving.

Distributor

REMOVAL

1. Disconnect the negative battery cable.
2. Remove the distributor cap without disconnecting the spark plug wires. Unplug the distributor connector.
3. Rotate the crankshaft to TDC of No. 1 cylinder, if possible. Mark the position of the rotor to the distributor body and to the engine.
4. Remove the hold-down bolt and lift the distributor out of the engine.

INSTALLATION

Timing Not Disturbed

1. If necessary, install a new O-ring on the distributor body. Lightly oil the O-ring and make sure the distributor mounting boss is clean.
2. If the crankshaft was not rotated with the distributor removed, align the rotor with the mark on the distributor body and insert the distributor into place. Some distributors have alignment marks on the shaft and housing. Align the shaft mark with the protruding mark on the housing.
3. The rotor may turn as the distributor is pushed in; remove it and turn the rotor back or forward 1 tooth and try again. When installing the distributor, it is important to make sure the rotor points to the mark made on the engine.
4. Check and adjust ignition timing as required.

Timing Disturbed

1. If the crankshaft was turned or if no alignment marks were made when the distributor was removed,

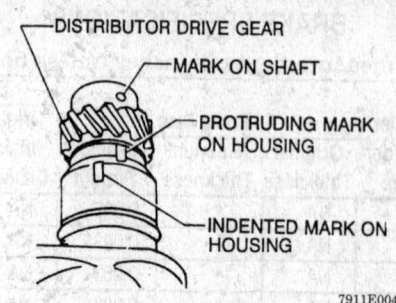

- DISTRIBUTOR DRIVE GEAR
- MARK ON SHAFT
- PROTRUDING MARK ON HOUSING
- INDENTED MARK ON HOUSING

7911E004

Some distributors have alignment marks on the shaft and body — 3.0L engine shown

make sure the engine is at TDC of No. 1 cylinder.

  a. Remove the spark plug from No. 1 cylinder.

  b. Rotate the crankshaft while holding a thumb over the spark plug hole to determine that the piston is coming up on the compression stroke.

  c. Align the red TDC mark on the crankshaft pulley with the pointer on the cover.

2. Check the O-ring on the distributor body and replace as necessary. Lightly lubricate the O-ring and make sure the mounting boss is clean.

3. Align the marks on the distributor shaft and body and install the distributor. The rotor may turn as the distributor is pushed in; remove it and turn the rotor back or forward 1 tooth and try again. When installing the distributor, it is important to make sure the rotor points to the mark made on the engine.

4. Check and adjust ignition timing as required.

Ignition Timing

ADJUSTMENT

The ignition timing is controlled by the ECU, but a basic setting is required if the distributor has been removed. Always check and adjust ignition timing and idle speed together.

Locate the timing marks on the crankshaft pulley and the front of the engine. The timing marks are in 5 degree increments and the TDC mark is always painted red.

1. Connect a tachometer and a timing light to the engine, according to the manufacturer's instructions. Make sure the wires do not interfere with the fan.

2. Make sure all lights and accessories are switched OFF and run the

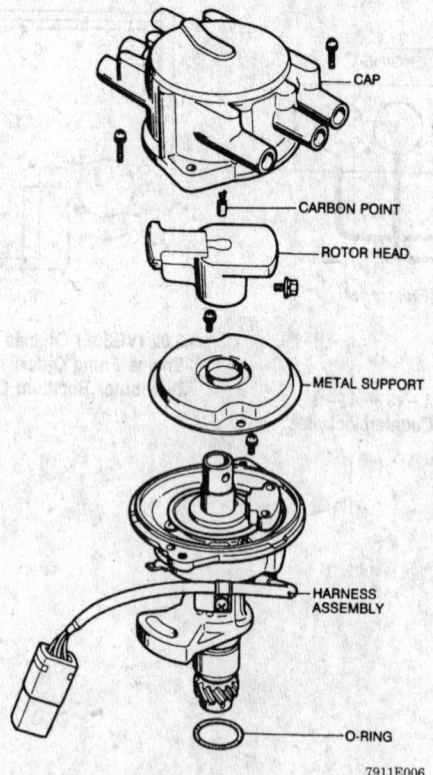

- CAP
- CARBON POINT
- ROTOR HEAD
- METAL SUPPORT
- HARNESS ASSEMBLY
- O-RING

7911E006

Exploded view of the distributor — 3.0L engine

engine to warm it to normal operating temperature.

3. With the transmission out of gear, run the engine at 2000 rpm for about 2 minutes, then check idle speed and ignition timing. If timing and idle speed are different from specification, look for another engine problem such as a vacuum leak or bad electrical connection. If no other obvious problem is found, a full diagnostic test should be run.

4. Check idle speed; on engines with throttle body fuel injection, ignition timing is not adjustable.

5. On models with automatic transmission, idle speed should be 750 rpm in **N**.

6. To adjust timing on 3.0L engines with multi-point fuel injection:

a. If timing is not correct, disconnect the wiring for the air control valve at the back of the intake manifold. Idle speed should be 700 rpm. The adjusting screw is on the same body assembly as the air control valve.

b. Reconnect the wiring, idle speed should be about 750 rpm. Do not change it with the wiring connected.

c. Check and adjust timing as required to 15 degrees BTDC. Adjust by rotating the distributor. Idle speed should remain at about 750 rpm.

7. To adjust timing on the 2.4L engine with multi-point fuel injection:

a. Check idle speed at about 800 rpm. If not correct, disconnect the throttle sensor wiring and adjust the speed to less than 800 rpm.

b. Check and adjust ignition timing to 10 degrees.

8. Run the engine at 2000 rpm for about 2 minutes and check idle speed and timing again. If idle speed or timing are not correct, other problems exist with the engine control system and a full diagnostic test must be run.

Alternator

PRECAUTIONS

Several precautions must be observed with alternator equipped vehicles to avoid damage to the unit.

If the battery is removed for any reason, make sure it is reconnected with the correct polarity. Reversing the battery connections may result in damage to the one-way rectifiers.

When utilizing a booster battery as a starting aid, always connect the positive to positive terminals and the

negative terminal from the booster battery to a good engine ground on the vehicle being started.

Never use a fast charger as a booster to start vehicles.

Disconnect the battery cables when charging the battery with a fast charger.

Never attempt to polarize the alternator.

Do not use test lights of more than 12 volts when checking diode continuity.

Do not short across or ground any of the alternator terminals.

The polarity of the battery, alternator and regulator must be matched and considered before making any electrical connections within the system.

Never separate the alternator on an open circuit. Make sure all connections within the circuit are clean and tight.

Disconnect the battery ground terminal when performing any service on electrical components.

Disconnect the battery if arc welding is to be done on the vehicle.

BELT TENSION ADJUSTMENT

Belt tension can be checked with a gauge made for the purpose. If a tension gauge is not available, belt deflection can be measured in inches. Press on the belt with 22 lbs. (10 kg) of force at a point halfway between the alternator and water pump pulleys. Deflection should be 0.200–0.310 in. (5–8mm) on the 3.0L engine or 0.310–0.470 in. (8–12mm) on the 2.4L engine.

1. Loosen the alternator's pivot bolt and the tension adjuster lock bolt.

2. Turn the adjuster bolt as required until the tension is correct.

3. Tighten the bolts and recheck the tension. If new belts have been installed, run the engine for a few minutes, then check tension again. The ideal adjustment is toward the looser end of the specification, belts adjusted too tight will cause bearing failure.

REMOVAL AND INSTALLATION

1. Disconnect the negative battery cable.

2. Label and disconnect the wiring from the alternator.

3. Loosen the drive belt tension and remove the belt from the pulley.

4. Remove the alternator-to-bracket bolts and the alternator from the vehicle.

5. To install, reverse the removal procedures. Adjust the drive belt tension and torque the lower bracket bolt to 35 ft. lbs. (47 Nm), the adjuster lock bolt to 10 ft. lbs. (13 Nm).

6. Connect all wiring before connecting the battery cable. Terminal **E** is always alternator ground.

Voltage Regulator

The voltage regulator on all models is an internal part of the alternator and cannot be removed separately. The regulator can be replaced when rebuilding the alternator.

Starter

REMOVAL AND INSTALLATION

1. If necessary, raise and safely support the vehicle.

2. Disconnect the negative battery cable.

3. Disconnect the electrical connectors from the starter, taking note of the positions for reinstallation purposes.

4. Remove the starter-to-engine bolts and the starter from the vehicle.

5. To install, reverse the removal procedures. Torque the starter mounting bolts to 29–36 ft. lbs. (39–49 Nm).

CHASSIS ELECTRICAL

Heater Blower Motor

REMOVAL AND INSTALLATION

The blower motor is accessible from under the right side of the instrument pan.

1. Disconnect the negative battery cable.

2. Disconnect the electrical connector from the blower motor.

3. Remove the blower motor-to-heater unit screws and the blower motor from the unit. It may be necessary to remove the glove box or package tray.

4. To install, reverse the removal procedures.

Windshield Wiper Motor

REMOVAL AND INSTALLATION

Front Wiper

1. Disconnect the negative battery cable.
2. Remove the wiper blades and arms as an assembly from the pivots. The arms are retained to the pivots by nuts; remove the nuts and pull the arms straight off.
3. Remove the cowl top grille screws (from the front edge) and pull the grille forward to disengage the rear tabs.
4. Remove the wiper motor arm-to-connecting rod stop ring.
5. From under the instrument panel, disconnect the electrical connector from the wiper motor harness.
6. Remove the wiper motor-to-cowl screws and the wiper motor from the vehicle.
   To install:
7. Install the motor and secure the bolts. Install the cowl grille.
8. Before installing the wiper arms, be sure the motor is in the PARK position. To do this, connect the battery and wiper motor wiring and turn the ignition switch **ON**. Turn the wiper switch **ON** and cycle the motor 3–4 times, then turn the wiper switch **OFF**. The motor should stop in the correct PARK position.
9. When installing the arms, the blades should be the correct distance above the lower windshield molding.

Rear Wiper

1. Disconnect the negative battery cable.
2. From the rear door, remove the wiper blade/arm as an assembly from the pivot. The arm is retained to the pivots by a nut; remove the nut and pull the arm straight off.
3. From inside the rear door, remove wiper motor cover plate.
4. Remove the wiper motor arm-to-connecting rod stop ring.
5. Disconnect the wiring and remove the screws to remove the motor from the door.
6. Installation is the reverse of removal. Before installing the wiper arm, be sure the motor is in the PARK position. To do this, connect the battery and wiper motor wiring and turn the ignition switch **ON**. Turn the wiper switch **ON** and cycle the motor 3–4 times, then turn the wiper switch **OFF**. The motor should stop in the correct PARK position.

Windshield Wiper Switch

The windshield wiper switch is a part of the combination switch located on the steering column. The switch can be removed without removing the steering wheel.

REMOVAL AND INSTALLATION

1. Disconnect the negative battery cable.
2. Remove the steering column covers.
3. Disconnect the windshield wiper switch electrical connector.
4. Remove the windshield wiper switch-to-combination switch screws and the windshield wiper switch from the steering column.
5. To install, reverse the removal procedures.

Instrument Cluster

REMOVAL AND INSTALLATION

1. Disconnect the negative battery cable.
2. Remove the instrument cluster bezel screws and the bezel. The bezel is also secured with clips; pull the bezel straight out after removing the screws.
3. Remove the instrument cluster-to-dash screws and pull the cluster assembly out far enough to disconnect the wiring and speedometer cable.
4. Installation is the reverse of removal.

Headlight and Turn Signal Switch

The headlight, turn signal and dimmer switches make up the combination switch located on the steering column.

REMOVAL AND INSTALLATION

1. Disconnect the negative battery cable.
2. Remove the steering column covers.
3. Disconnect the switch electrical connector.
4. Remove the switch-to-combination switch screws and remove the switch from the steering column.
5. To install, reverse the removal procedures.

Combination Switch

The combination switch includes switches for headlights and dimmer, windshield washer and wiper, turn signal, horn contacts and cruise control, if equipped. The steering wheel must be removed to remove the combination switch as an assembly. The 2 switch stalks can be removed separately as described above.

REMOVAL AND INSTALLATION

1. Disconnect the negative battery cable.
2. Remove the steering column covers.
3. Remove the horn pad and the steering wheel nut. Use a puller to remove the steering wheel.
4. Disconnect the wiring harness from the clip which retains it to the lower instrument panel.
5. Disconnect the electrical connectors from the combination switch.
6. Loosen the combination switch-to-steering column screw and remove the switch assembly.
7. To install, align the hole in the steering column with the protrusion on the switch base and reverse the removal procedures.

Ignition Lock/Switch

REMOVAL AND INSTALLATION

1. Disconnect the negative battery cable.
2. From the upper steering column, remove the shell cover screws and the covers.
3. Disconnect the electrical connector from the rear of the ignition switch.
4. Drill out the 2 self shear screws from the ignition lock holder.
5. Remove the screws and the ignition switch.
6. To install, reverse the removal procedures. Torque the shear-type screws until the heads shear.

Stoplight Switch

The stoplight switch is attached to a bracket at the top of the brake pedal.

REMOVAL AND INSTALLATION

1. Disconnect the negative battery cable.

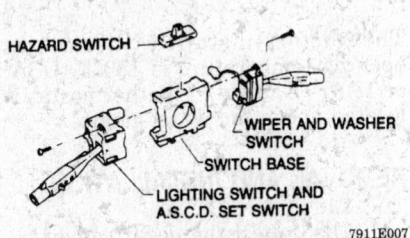

HAZARD SWITCH

WIPER AND WASHER SWITCH

SWITCH BASE

LIGHTING SWITCH AND A.S.C.D. SET SWITCH

7911E007

Combination switch assembly — the steering wheel must be removed to remove the base

2. From under the dash, disconnect the electrical connector from the stoplight switch.

3. Loosen and remove the locknut from the stoplight switch.

4. Unscrew the stoplight switch from the brake pedal bracket.

5. To install, reverse the removal procedures. Adjust the switch, torque the stoplight switch locknut to 11 ft. lbs. (15 Nm) and adjust the pedal height and free-play.

ADJUSTMENT

1. Start the engine and press the brake pedal by hand until pushrod resistance is felt.

2. Use a feeler gauge to measure stoplight switch-to-brake pedal gap: it should be 0.012–0.039 in. (0.3–1.0mm).

3. If necessary, disconnect the electrical connector from the stoplight switch, loosen the switch locknut and adjust the gap. Torque the lock nut to 11 ft. lbs. (15 Nm).

Clutch Cruise Control Switch

If equipped with cruise control, the clutch switch is attached to a bracket mounted to the upper portion of the clutch pedal bracket. This switch replaces the stopper bolt and is used to set pedal height.

REMOVAL AND INSTALLATION

1. Disconnect the negative battery cable.

2. Disconnect the electrical connector from the clutch switch.

3. Loosen the locknut and unscrew the switch from the bracket.

4. To install, screw the clutch switch into the bracket until the correct pedal height is established and

torque the locknut to 11 ft. lbs. (15 Nm). 2.4L engine — 9.29–9.69 in. (236–246mm) 3.0L engine — 8.94–9.33 in. (227–237mm).

NOTE: The pedal height is the distance from the floor board to the pedal pad with the carpet removed.

Clutch Interlock Switch

The clutch interlock switch is attached to a bracket mounted to the cowl.

REMOVAL AND INSTALLATION

1. Disconnect the negative battery cable.

2. Disconnect the electrical connector from the clutch interlock switch.

3. Loosen the locknut and unscrew the switch from the bracket.

4. To install, use a feeler gauge and screw the clutch interlock switch into the bracket.

5. Fully depress the clutch pedal and adjust the gap between the clutch pedal stopper bracket and the threaded end of the clutch interlock switch. The gap should be 0.012–0.039 in. (0.3–1.0mm).

6. Torque the locknut to 11 ft. lbs. (15 Nm), reconnect the wiring and make sure the started operates only when the clutch pedal is fully depressed.

Fuses, Circuit Breakers and Relays

LOCATION

The main fuse block is behind a panel to the right or the steering column. The turn signal flasher is to the left

of the brake pedal. A series relays in mounted under the hood on the right fender. All relays are color coded according to internal circuitry and should be replaced with relays of the same color.

Computers

LOCATION

The main engine ECU is under the passenger's seat along with the automatic transmission control unit. The power window amplifier is behind the passenger's side kick panel. The door lock timer and cruise control ECU are under the driver's seat.

Turn Signal Flasher

LOCATION

The turn signal flasher is to the left of the brake pedal.

ENGINE COOLING

Radiator

REMOVAL AND INSTALLATION

1. Disconnect the negative battery cable.

2. Drain the engine cooling system and disconnect the radiator hoses. If equipped with an automatic transmission, disconnect and plug the transmission oil cooler lines at the radiator.

3. There is a lower section that can be removed from the fan shroud. Remove this section, unbolt the 2 up-

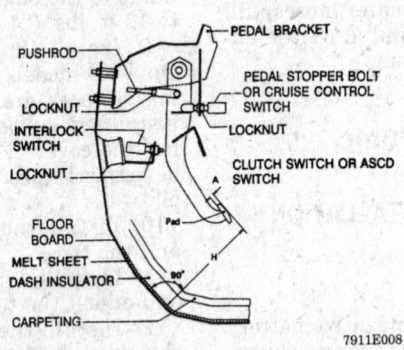

PEDAL BRACKET

PUSHROD

PEDAL STOPPER BOLT OR CRUISE CONTROL SWITCH

LOCKNUT

LOCKNUT

INTERLOCK SWITCH

CLUTCH SWITCH OR ASCD SWITCH

LOCKNUT

Pad

FLOOR BOARD

MELT SHEET

DASH INSULATOR

CARPETING

90°

7911E008

Clutch pedal switch adjustment; vehicles with cruise control use the switch to set pedal height

per radiator brackets and lift the radiator and shroud out together.

4. Installation is the reverse of removal. Make sure the lower radiator mounting rubbers are in place before installing radiator.

5. If equipped with an automatic transmission, check and/or refill the transmission.

Heater Core

REMOVAL AND INSTALLATION

1. The dashboard must be removed. Disconnect the negative battery cable. On vehicles with theft protected radios, obtain the security code.

2. Set the temperature control to full hot and drain the cooling system.

3. Disconnect the heater hoses at the engine compartment.

4. Remove the ash tray and remove the screws holding the center console face cover. Remove the cover, heater controls, radio and center vent.

5. Disconnect the control cables and wiring as needed to remove the dashboard.

6. Disconnect the ducts and remove the heater unit. Remove the clips or screws and split the case to remove the heater core. Note how the air flow control doors fit into the case.

7. Install the heater core and assemble the case halves. Use a new gasket and make sure the air flow doors work properly.

8. Install the heater unit and use a new gasket to seal it to the cooling unit or blower fan housing. Connect the ducts.

9. Install the instrument panel.

10. Install the heater control assembly, center vent and radio.

11. Adjust the heater controls and air flow doors as required.

12. Connect the heater hoses, fill the cooling system and start the engine to test the system.

Water Pump

REMOVAL AND INSTALLATION

2.4L Engine

1. Disconnect the negative battery cable and drain the cooling system. Don't forget the block drain.

2. Remove the upper radiator hose to provide working room and loosen

the alternator to remove the drive belt from the pulleys.

3. Remove the lower section of the fan shroud and remove the screws to lift the shroud from the engine. Hold the pulley and remove the nuts to remove the fan and pulley from the water pump.

4. Remove the bolts and remove the water pump from the engine.

To install:

5. Make sure all gasket surfaces are clean and use a new gasket or silicone sealer when installing the pump to the engine. Torque the 6mm bolts to 7 ft. lbs. (10 Nm) and the 8mm bolts to 12 ft. lbs. (16 Nm).

6. Install the fan clutch, fan and pulley and torque the nuts or bolts to 7 ft. lbs. (10 Nm).

7. Install the fan shroud and drive belts and fill the cooling system to check for leaks.

3.0L Engine

1. The timing belt cover must be removed. Disconnect the negative battery cable and drain the cooling system. Don't forget the block drain.

2. Remove the radiator hoses and, on automatic transmission, disconnect and plug the fluid cooling lines.

3. Remove the lower section of the fan shroud and remove the screws to lift the shroud from the engine. Remove the bracket bolts and lift radiator out of the vehicle.

4. Remove all the accessory drive belts.

5. Hold the pulley and remove the nuts to remove the fan and pulley from the water pump.

6. Remove the timing belt covers.

7. Remove the bolts to remove the water pump from the engine.

To install:

8. Make sure all gasket surfaces are clean and use a new gasket or silicone sealer when installing the pump to the engine. Torque the bolts to 15 ft. lbs. (21 Nm).

9. Install the timing belt covers. On 4WD models, make sure the sealing surfaces are clean and carefully install the rubber seal when installing the cover. The timing belt must be properly protected from dirt and oil.

10. Install the pulley, fan clutch and the fan.

11. Install the accessory drive belts and adjust the tension.

12. Install the radiator and fan shroud and connect the cooling system hoses.

13. Fill the system and check for leaks.

Thermostat

The factory-installed thermostat opening temperature is 180°F (USA) or 190°F (Canada). The thermostat is located above the water pump.

REMOVAL AND INSTALLATION

1. Disconnect the negative battery cable.

2. Drain the engine coolant to a level below the thermostat housing.

3. Disconnect the coolant hose from the thermostat water outlet.

4. Remove the water outlet-to-thermostat housing bolts, gasket and thermostat.

NOTE: The thermostat spring must face the inside of the engine.

5. Clean the gasket mounting surfaces.

NOTE: If the thermostat is equipped with an air bleed or jiggle valve, be sure to position it in the upward direction.

6. To install, use a new gasket or sealant and reverse the removal procedures. Torque the thermostat housing bolts to 6 ft. lbs. (8 Nm) on the 2.4L engine or 15 ft. lbs. (21 Nm) on all except 2.4L engine.

Cooling System Bleeding

1. Set the heater temperature control to **HOT** and open the air relief plug. Pour coolant into the radiator until it comes out the relief plug, then close the plug.

2. Fill the reservoir and run the engine to warm it to operating temperature.

3. When the engine is cool again, check the coolant level in the reservoir.

FUEL SYSTEM

Fuel System Service Precautions

Keep a Class B dry chemical fire extinguisher available.

Always relieve the fuel system pressure before disconnecting any fitting.

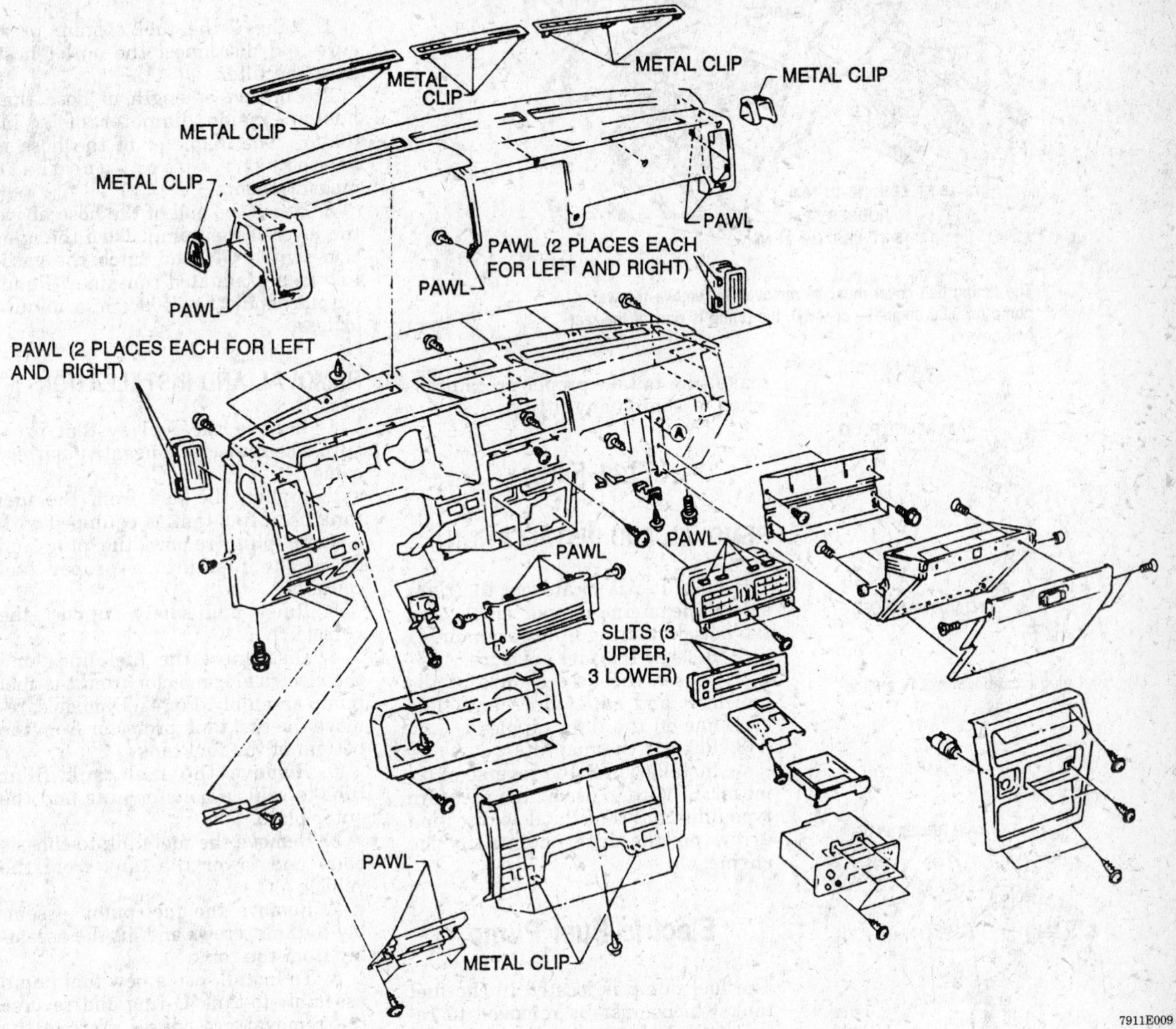

METAL CLIP

METAL CLIP

METAL CLIP

METAL CLIP

METAL CLIP

PAWL

PAWL (2 PLACES EACH FOR LEFT AND RIGHT)

PAWL

PAWL (2 PLACES EACH FOR LEFT AND RIGHT)

PAWL

PAWL

SLITS (3 UPPER, 3 LOWER)

PAWL

METAL CLIP

7911E009

Dashboard assembly must be removed to remove the heater core

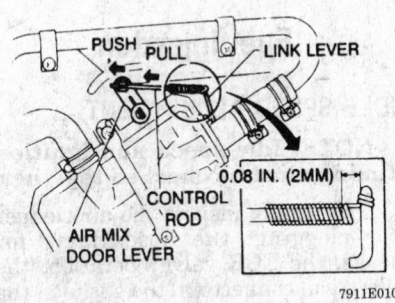

PUSH PULL LINK LEVER

0.08 IN. (2MM)

CONTROL ROD

AIR MIX DOOR LEVER

7911E010

Water valve and air mix door control rod adjustment

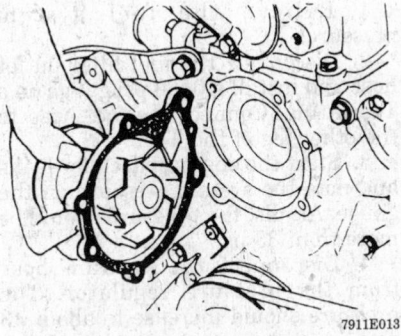

7911E013

Water pump replacement on 2.4L engine

Make sure the work area is well ventilated to minimize the possibility of explosion.

Turn off any source of ignition, such as a heater or welding equipment before beginning work on a fuel system.

Always use new O-rings or gaskets and do not replace the metal fuel pipes with hose.

Always us a back-up wrench when opening or closing a fuel line. Wrap a rag around the fitting to catch any fuel spilled.

Deposit of fuel soaked rags or clothing in a proper safety container.

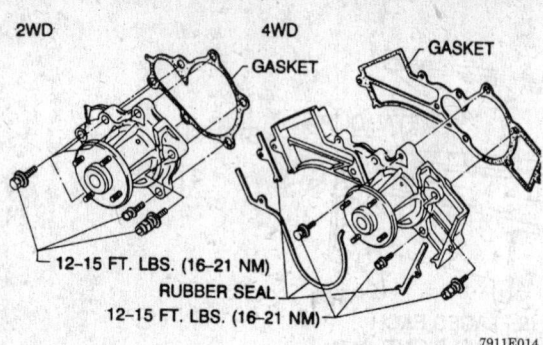

12-15 FT. LBS. (16-21 NM)
RUBBER SEAL
12-15 FT. LBS. (16-21 NM)

7911E014

The timing belt cover must be removed to remove the water pump on 3.0L engine — on 4WD, the pump is part of the rear belt cover

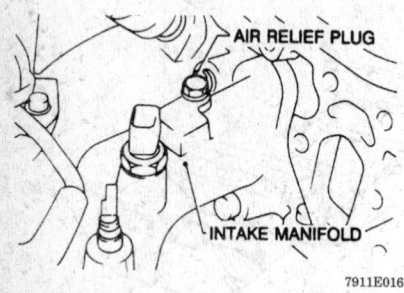

Air relief plug location with 2.4L engine

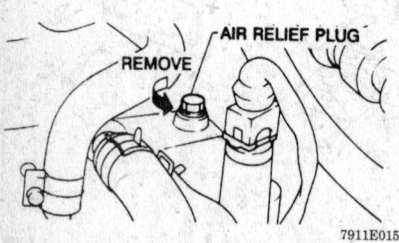

Air relief plug on 3.0L engine

RELIEVING FUEL SYSTEM PRESSURE

1. Locate the fuel pump fuse or relay and remove it.
2. Start the engine and allow it to run.
3. After the engine stalls, crank it 2–3 times to make sure the pressure is released.
4. Turn the ignition switch **OFF** and replace the fuse. There may still be some pressure in the system, make sure to take proper precautions when loosening any fittings.

Fuel Filter

REMOVAL AND INSTALLATION

The fuel filter is located in the right side of the engine compartment, near the power steering fluid reservoir.
1. Release the fuel pressure.
2. Loosen the hose clamps at the fuel inlet and outlet lines and slide each line off the filter nipples.
3. Remove the fuel filter.
4. Installation is the reverse of removal. Be sure to use a high pressure type filter and that the flow direction arrow on the filter points to the engine.

Electric Fuel Pump

The fuel pump is located in the fuel tank which must be removed to remove the fuel pump.

SYSTEM PRESSURE TEST

1. Relieve the fuel system pressure.
2. Disconnect the fuel filter outlet hose and install a 0–60 psi gauge on a Tee fitting. Connect the fuel hose to the other leg of the Tee.
3. Start the engine, check for leaks and note the system pressure on the gauge. At idle, the pressure should be more than 33 psi.
4. Disconnect the vacuum hose from the pressure regulator. The pressure should increase to about 43 psi.
5. If the pressure is not correct, check for a system leak or a faulty injector or pressure regulator before removing the pump.

PUMP DELIVERY TEST

1. Relieve the fuel system pressure and disconnect the outlet hose from the filter.
2. Connect a length of hose that has an inside diameter of 1/4 in. (6mm). The diameter of the hose is important for accurate measurements.
3. Raise the end of the hose above the level of the pump. Turn the ignition switch **ON** and catch the gasoline in a graduated container. Pump output should be 1400cc in a minute or less.

REMOVAL AND INSTALLATION

1. Relieve the fuel system pressure. Disconnect the negative battery cable.
2. Siphon the fuel from the fuel tank or, if fuel tank is equipped with a drain plug, remove the plug and drain the fuel into a proper fuel container.
3. Raise and safely support the vehicle.
4. Disconnect the fuel lines and the electrical connector from the fuel pump assembly. For 4WD models, remove the fuel tank protector from the bottom of the fuel tank.
5. Remove the fuel tank filler tube-to-vehicle bolts or nuts and the outer plate.
6. Remove the fuel tank-to-chassis bolts and lower the tank from the vehicle.
7. Remove the fuel pump assembly-to-tank screws and lift the assembly from the tank.
8. To install, use a new fuel pump assembly-to-tank O-ring and reverse the removal procedures. Torque the fuel pump assembly-to-tank screws to 24 inch lbs. (3 Nm) and the fuel tank protectors-to-chassis bolts to 26 ft. lbs. (35 Nm).

Fuel Injection

IDLE SPEED ADJUSTMENT

NOTE: Idle speed and ignition timing must be checked together.

1. Visually inspect the air cleaner for clogging, the hoses/ducts for leaks, the EGR valve operation, the electrical connectors, the gaskets, the throttle valve and throttle sensor operation and the AIV hose.
2. Start the engine and allow it to reach normal operating temperature.
3. Operate the engine under no-load for 2 minutes at about 2000 rpm.

Make sure all accessories and lights are **OFF**. Race the engine 2–3 times, then let it run at idle speed for 1 minute.

4. To adjust idle speed on the 3.0L engine with multi-point fuel injection:

    a. Disconnect the wiring for the Auxiliary Air Control (AAC) valve at the back of the intake manifold. Idle speed should drop to about 700 rpm. The adjusting screw is on the same body assembly as the AAC.

    b. Reconnect the wiring, idle speed should be about 750 rpm. Do not change it with the wiring connected. If idle speed is not correct, look for another problem in the engine control system.

    c. Check and adjust timing as required to 15 degrees BTDC by rotating the distributor. Idle speed should remain at about 750 rpm.

5. To adjust timing on the 2.4L engine with multi-point fuel injection:

    a. Check idle speed at about 800 rpm. If it is not correct, disconnect the throttle sensor wiring and adjust the speed to less than 800 rpm. The adjustment is on the throttle body Idle Air Adjusting (IAA) unit that has small engine coolant hoses connected to it.

    b. Check and adjust ignition timing to 15 degrees BTDC.

6. Run the engine at 2000 rpm for about 2 minutes and check idle speed and timing again. If idle speed or timing are not correct, other problems exist with the engine control system and a full diagnostic test must be run.

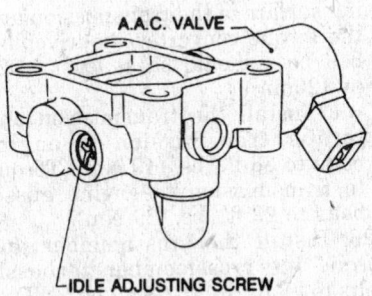

7911E018

Idle speed adjustment on engines with multi-point fuel injection is on the Idle Air Adjusting unit — 3.0L engine shown

Fuel Injector

REMOVAL AND INSTALLATION

2.4L Engine

The engine is equipped with 4 fuel injectors, with one located at each cylinder.

1. Relieve the fuel pressure.
2. Label and disconnect the electrical connectors from the fuel injectors.
3. Disconnect the fuel hoses from the rail.
4. Remove the fuel rail with injectors attached from the intake manifold.
5. Separate the fuel injectors from the fuel rail.

To install:

6. Replace the fuel injector O-rings.
7. Install the fuel injectors to the fuel rail.
8. Lubricate the fuel injector O-rings with automatic transmission fluid and press them, with the fuel rail, into the intake manifold.
9. Install the fuel rail-to-intake manifold bolts.
10. Connect the fuel hoses to the rail.
11. Connect the electrical connectors to the fuel injectors.
12. Turn the ignition switch **ON** and check for fuel leaks at the fuel rail.

3.0L Engine

The engine is equipped with 6 fuel injectors, with one located at each cylinder.

1. Relieve the fuel pressure.
2. Remove the air cleaner from the throttle body.
3. Label and disconnect the vacuum hoses, electrical connectors and throttle cable from the throttle body/upper intake manifold assembly.
4. Remove the upper intake manifold.
5. Disconnect the electrical connectors from the fuel injectors.
6. Disconnect the supply and return hoses from the fuel rail.
7. The injectors can be removed separately or as an assembly with the rail. Do not use the old O-rings or insulators when installing the injectors.

To install:

8. Replace the fuel injector O-rings.

9. Lubricate the fuel injector O-rings with automatic transmission fluid and press them into the fuel rail assembly.
10. Connect the electrical connectors to the fuel injectors.
11. Using a new gasket, install the upper intake manifold and torque the bolts to 16 ft. lbs. (22 Nm).
12. Connect the electrical connectors, the vacuum lines and the accelerator cable.
13. Install the air cleaner to the throttle body.
14. Turn the ignition switch **ON** and check for fuel leaks at the fuel rail.

EMISSION CONTROLS

Emission Warning Lamp

The check engine light is located on the instrument panel of California vehicles only and indicates an emission performance malfunction.

RESETTING

1. Turn the ignition switch **ON**.
2. If the check engine light turns **ON**, perform the self-diagnosis procedures to determine the malfunction.
3. Turn the ignition switch **OFF**.
4. Locate and repair the malfunction.

NOTE: When the malfunction is repaired and the fault code memory cleared, the check engine light will stay OFF.

ENGINE MECHANICAL

NOTE: Disconnecting the negative battery cable on some vehicles may interfere with the functions of the on board computer systems and may require the computer to undergo a relearning process when the battery cable is reconnected. This usually requires only a few minutes of driving.

Engine

REMOVAL AND INSTALLATION

On vehicles equipped with the 3.0L engine, the transmission must be removed before removing the engine. If equipped with the 2.4L engine, the engine and transmission are removed together. On vehicles with 4WD, the front torsion bar must be removed to remove the transfer case.

1. Relieve the fuel system pressure.

2. Disconnect the battery cables and remove the battery.

3. Using a scribing tool, mark the location of the hood hinges on the body and remove the hood.

4. Drain the engine oil and the cooling system, including the block drains. Dispose of old fluids properly.

5. Remove the air cleaner. Wrap a shop rag around the fuel filter outlet and disconnect the hose.

6. Raise and safely support the vehicle. If equipped, remove the splash pan from under the engine.

7. To remove the radiator:

   a. Remove the upper and lower radiator hoses.

   b. If equipped with an automatic transmission, disconnect and plug the transmission oil cooler lines from the radiator.

   c. Remove the lower radiator shroud.

   d. Remove the bracket bolts and lift the radiator and shroud out together.

8. If equipped with air conditioning, loosen the belt tension and remove the belt. Disconnect the wiring, remove the compressor and secure it out of the way. Do not disconnect the pressure hoses.

9. If equipped with power steering, remove the drive belt and the power steering pump and secure it out of the way. Do not disconnect the pressure hoses.

10. Label and disconnect all wiring and vacuum hoses.

11. Disconnect the heater hoses from the engine and disconnect the throttle cable.

12. Remove the starter.

13. Matchmark the driveshaft flange at the rear pinion flange and remove the driveshaft. Plug the extension housing opening to prevent the oil from draining out.

14. If equipped with 4WD, matchmark both front driveshaft flanges so the shaft can be installed in the same position. Remove the driveshaft.

15. Disconnect the exhaust pipe from the manifold(s) and from the catalytic converter and remove the pipe.

16. Disconnect the speedometer cable and the wiring from the transmission.

17. If equipped an automatic transmission:

   a. Disconnect the selector lever and throttle cables from the transmission.

   b. Remove the dipstick tube and disconnect the cooler lines.

   c. Remove the torque converter housing dust cover. Matchmark the converter with the driveplate for reassembly; these are balanced together at the factory. Remove the torque converter-to-driveplate (flywheel) bolts. Use a wrench on the crankshaft pulley bolt to rotate the crankshaft to expose the hidden torque converter bolts.

18. On vehicles with a manual transmission:

   a. Remove the shifter knob and boot and remove the snapring to lift the shift lever out of the transmission. Stuff a rag in the opening to keep dirt out of the transmission.

   b. Without disconnecting the hydraulic hose, remove the clutch slave cylinder from the transmission and secure it aside.

19. If equipped with 4WD:

   a. Working under the vehicle, measure and note the length of the threads on the torsion bar adjustment. At the front of the bar, pull the boot back and matchmark the bar to the mounting plate. The spline on the bar must be re-installed in the same position on the plate.

   b. Remove the locknut and adjustment nut from both torsion bars. Remove the 3 nuts at the mounting plate and remove the bars. Mark the bars left and right side for proper installation.

   c. Remove the transfer case shift lever assembly from the transfer case.

   d. If necessary, the transfer case can be removed at this point so the jack will be available to remove the transmission.

   e. If equipped with a 3.0L engine, remove the gusset securing the engine to the transmission.

20. Using a chain hoist, attach it to the engine and lift the engine slightly to take the weight off the mounts. Using an appropriate transmission jack, properly support the transmission and remove the transmission mount and crossmember.

21. Remove the transmission-to-engine bolts and move the transmission back away from the engine. On automatic transmissions, secure the torque converter so it does not fall out. Lower the transmission from the vehicle.

NOTE: When removing the engine mounts, do not loosen the 4 mount cover nuts. The mount is fluid filled and will not function properly if the fluid leaks out.

22. Check to make sure all wires and hoses have been disconnected. Remove the front engine mount bolts and carefully lift the engine out.

To install:

23. Carefully guide the engine into place and start the mount bolts. Tighten the bolts temporarily.

24. On manual transmission:

   a. Lightly grease the input shaft splines. On 4WD, apply a silicone sealant to the engine block or rear plate to seal the engine to the transmission.

   b. Fit the transmission into place and start all the engine-to-transmission bolts. Make sure the input shaft fits properly into the clutch disc and pilot bearing.

   c. Torque the 2.36 in. (60mm) and 2.56 in. (65mm) engine-to-transmission bolts to 36 ft. lbs. (49 Nm).

   d. Torque the remaining bolts to 18 ft. lbs. (25 Nm) on the 2.4L cylinder engine, 29 ft. lbs. (39 Nm) on the 3.0L engine.

25. On automatic transmission:

   a. Use a dial indicator to check the driveplate runout while turning the crankshaft. Maximum allowable runout is 0.020 in. (0.5mm); if beyond specifications, replace the driveplate.

   b. Measure and adjust how far the torque converter is recessed into the transmission housing. The distance between the front mounting surface of the transmission and the torque converter-to-driveplate bolt boss should be at least 1.024 in. (26mm).

   c. Install the transmission and torque transmission-to-engine bolts to 36 ft. lbs. (49 Nm). Torque to transmission-to-engine gusset bolts to 29 ft. lbs. (39 Nm).

26. Install the crossmember and torque the crossmember-to-chassis bolts to 38 ft. lbs. (52 Nm) on 2WD or 31 ft. lbs. (42 Nm) on 4WD.

27. Install the rear transmission mount bolts, loosen the engine mount bolts, then torque all the mount bolts

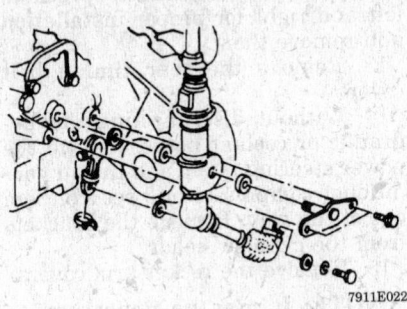

Removing the transfer case shift lever on 4WD

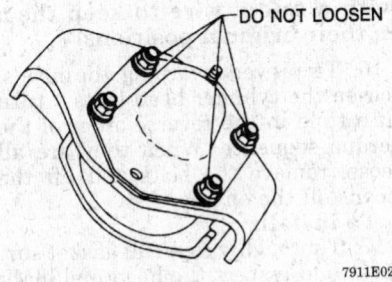

DO NOT LOOSEN

Do not remove these engine mount nuts

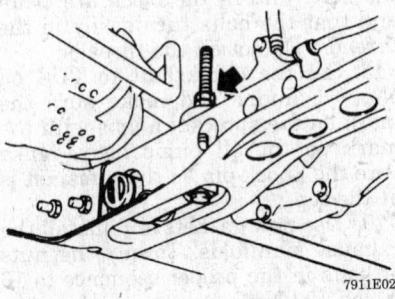

Measure thread length before removing torsion bar

to 31 ft. lbs. (42 Nm), starting at the rear.

28. On automatic transmission, align the matchmarks on the driveplate and torque converter, install the bolts and torque to 36 ft. lbs. (49 Nm). Turn the crankshaft after tightening the bolts to make sure there is no binding at the driveplate.

29. If the torsion bars were removed, install them in their original location. Make sure the splines are in their original position and set the adjustment to its original position.

30. If the transfer case was removed, apply silicone sealant to seal the case to the transmission. Install the transfer case and torque the bolts to 30 ft. lbs. (41 Nm). Install the shift lever.

31. When installing the driveshafts, be sure to align the matchmarks. Torque the bolts on the front driveshaft (4WD) to 33 ft. lbs. (44 Nm). On single piece rear driveshafts, torque the bolts to 65 ft. lbs. (88 Nm). On 2 piece driveshafts, torque the bolts to 33 ft. lbs. (44 Nm). Torque the center bearing bracket bolts to 16 ft. lbs. (22 Nm).

32. When installing the exhaust system, use new gaskets and torque the flange bolts to 27 ft. lbs. (36 Nm).

33. Install the remaining components in order of removal and connect the wiring and hoses.

Cylinder Head

REMOVAL AND INSTALLATION

2.4L Engine

1. Relieve the fuel system pressure. Disconnect the negative battery cable.

2. Remove the air cleaner. Disconnect the accelerator cable from the throttle body.

3. Drain the engine coolant, including the block drain.

4. Label and disconnect all wiring and hoses as required.

5. Remove the intake and exhaust manifolds.

6. Remove the steering pump, alternator and/or air conditioner compressor as required and remove the brackets as required. When removing the steering pump or air conditioner compressor, disconnect the drive belt, remove the unit pump and move it aside; do not disconnect the pressure hoses.

7. Remove the distributor cap and rotor and disconnect the wires from the spark plugs.

8. Remove the valve cover.

9. Rotate the crankshaft until the No. 1 cylinder is on the TDC of its compression stroke. Make sure the TDC mark on the crankshaft pulley is aligned with the pointer. The silver link on the timing chain should be aligned with the mark on the camshaft sprocket and the knock pin on the camshaft will be at the top.

10. To remove the camshaft sprocket:

    a. Fabricate a wooden wedge tool to hold the timing chain in place.

    b. Remove the camshaft sprocket bolt and the camshaft sprocket. Hold the sprocket up to keep tension on the chain.

    c. Install the wedge tool and rest the chain on the tool with the driver's side of the chain pulled snug against the crankshaft sprocket. If the chain falls off the crankshaft sprocket or if the crankshaft is turned, the front cover must be removed to properly set the valve timing.

11. Remove the cylinder head–to–timing chain cover bolts.

12. Carefully loosen the cylinder head bolts one turn at a time in the reverse order of the tightening sequence. When all bolts are loose, remove the bolts and remove the cylinder head.

To install:

13. Thoroughly clean all gasket surfaces and inspect the head and block for damage to the surfaces. Before installing, check the cylinder head for warping. The limit is 0.006 in. (0.15mm). Make sure the threads on the bolts and in the block are clean and that the bolts turn easily in the threads. Do not oil the threads.

14. Make sure the camshaft knock pin is at the top; both valves for No. 1 cylinder will be closed.

15. Install the new gasket and carefully install the cylinder head. Torque the cylinder head bolts in the sequence shown in 5 steps: Step 1 — 22 ft. lbs. (29 Nm) Step 2 — 58 ft. lbs. (78 Nm) Step 3 — loosen all bolts Step 4 — 22 ft. lbs. (29 Nm) Step 5 — 54–61 ft. lbs. (74–83 Nm)

16. Correctly position the camshaft sprocket into the timing chain with the silver link aligned with the mark on the sprocket. Install the sprocket onto the camshaft, install the sprocket bolt but do not fully torque it yet.

17. Carefully turn the crankshaft 2 full turns and make sure the timing marks still line up. If not, remove the camshaft sprocket and try again. Torque the sprocket bolt to 87–116 ft. lbs. (118–157 Nm).

18. Adjust the valves.

19. Use a silicone sealer when installing the rubber covers at each end of the camshaft. Install the valve cover with a new gasket and torque the bolts in a circular pattern to 7 ft. lbs. (10 Nm). Loosen all the bolts and torque them again in the same pattern.

20. Complete the installation of the remaining components. Change the oil before running the engine.

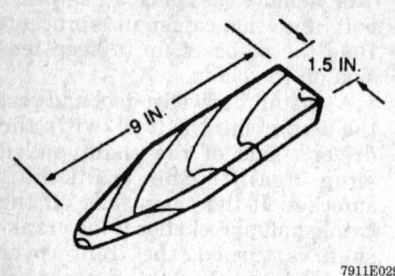

Make a wooden wedge to support the timing chain when removing 2.4L engine cylinder head

7911E029

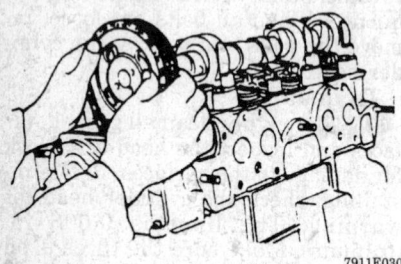

Remove the camshaft sprocket with chain attached — 2.4L engine

7911E030

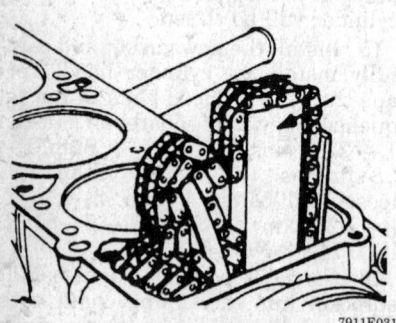

Support the chain with the wedge to keep it engaged with the crankshaft sprocket

7911E031

3.0L Engine

1. Relieve the fuel system pressure. Disconnect the negative battery cable.

2. Remove the air cleaner. Disconnect the accelerator cable from the throttle body.

3. Drain the engine coolant, including the block drain.

4. Label and disconnect all wiring and hoses as required.

5. Remove the distributor and spark plug wires as an assembly.

6. Remove the timing belt covers and the timing belt.

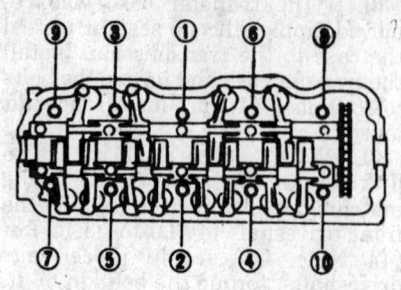

Cylinder head bolt torque sequence — 2.4L engine

7911E033

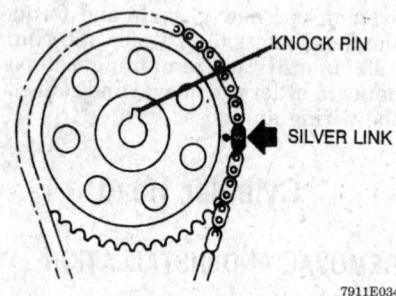

KNOCK PIN

SILVER LINK

7911E034

To install the camshaft sprocket, the knock pin will be at the top and the silver link will align with the mark

7. Remove the upper intake manifold section (5 bolts).

8. Label and disconnect the wiring to the fuel injectors and disconnect the fuel supply and return hoses.

9. Remove the injectors and rail as an assembly. Place the assembly where it will stay clean.

10. Loosen the intake manifold bolts 1 turn at a time in the reverse order of the torque sequence. This is important to prevent warping the manifold.

11. Remove the bolts and lift the manifold off.

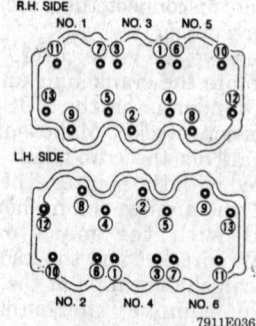

R.H. SIDE
NO. 1   NO. 3   NO. 5

L.H. SIDE

NO. 2   NO. 4   NO. 6

7911E036

Cylinder head bolt torque sequence — 3.0L engine

12. Mark the camshaft sprockets left and right for proper installation and remove them.

13. Remove the rear timing belt cover.

14. Without disconnecting the hydraulic or coolant hoses, remove the power steering pump and the air conditioner compressor and secure them out of the way. Remove the brackets from the cylinder heads.

15. Remove the rocker arm covers.

NOTE: It may be necessary to remove the rocker shafts and valve lifter guide to provide access to the cylinder head bolts. Before removing the valve lifter guide, secure the valve lifters with a safety wire to keep them in their original positions.

16. To prevent warping the heads, loosen the cylinder head bolts 1 turn at a time in the reverse order of the torque sequence. When they are all loose, remove the bolts and lift the heads off the engine.

To install:

17. Thoroughly clean all gasket surfaces and inspect the head and block for damage to the surfaces. Before installing, check the cylinder head for warping. The limit is 0.004 in. (0.10mm). Make sure the threads on the bolts and in the block are clean and that the bolts turn easily in the threads. Do not oil the threads.

18. Set the crankshaft to TDC on No. 1 cylinder and make sure the mark on the sprocket aligns with the mark on the oil pump body. Make sure the knock pin on the camshaft is at the top.

19. Use new gaskets and install the exhaust manifolds. Torque the nuts or bolts in the proper sequence to 16 ft. lbs. (22 Nm).

20. Apply sealant to the block cooling system drain plugs and install the plugs.

21. Make sure the new head gaskets are properly fitted and install the cylinder heads. When installing the bolts, the long bolts go into positions 4, 5, 12 and 13; the flat side of the washer goes towards the head.

22. Torque the cylinder head bolts in the proper sequence in five steps: Step 1 — 22 ft. lbs. (30 Nm) Step 2 — 43 ft. lbs. (58 Nm) Step 3 — loosen all bolts Step 4 — 22 ft. lbs. (30 Nm) Step 5 — 40–47 ft. lbs. (54–64 Nm) or; Step 5 alternate — 22 ft. lbs. (30 Nm) plus 65 degrees.

23. If the lifter guide and rocker arms were removed, install them now and tighten the bolts 1 turn at a time

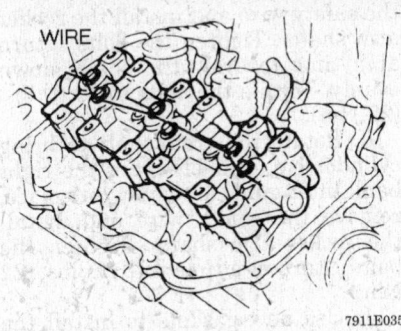

Secure the lifters with wire before removing the guide assembly — 3.0L engine

to draw the shafts down evenly against the valve springs without bending the shafts. Torque the bolts to 16 ft. lbs. (22 Nm).

24. Install the rocker arm covers and torque the bolts to 25 inch lbs. (3 Nm).

25. Install the rear timing belt cover and the camshaft sprockets. Make sure the sprockets are on the correct side and torque the sprocket bolts to 65 ft. lbs. (88 Nm).

26. Make sure the crankshaft and camshafts are properly positioned to install the timing belt. Be careful if it is necessary to turn either shaft; this is not a free wheeling engine and the valves will contact the pistons if the crankshaft is turned without the timing belt in place.

27. Install the timing belt, set the tension and turn the crankshaft 2 full turns to make sure the timing marks still align properly.

28. Use a new gasket to install the intake manifold and torque the nuts bolts in the proper sequence in 3 steps. Torque the nuts to 20 ft. lbs. (27 Nm) and the bolts to 14 ft. lbs. (20 Nm).

29. Connect the exhaust pipes to the manifolds and torque the bolts to 20 ft. lbs. (27 Nm).

30. Install the remaining components using new gaskets, O-rings or seals as required. Adjust belt tensions and change the oil before starting the engine.

31. When the engine is first started, the hydraulic valve lifters may be noisy. Run the engine for 10–20 minutes at about 1000 rpm. If the noise has not subsided, the lifter will probably never pump up and must be replaced.

Rocker Arms and Valve Lifters

REMOVAL AND INSTALLATION

2.4L Engine

The hydraulic lifters are built into the rocker arms. Do not allow the arms to lay on their side or they will become air bound. Keep the rocker arms upright or lay them in a pan of new engine oil. On all models, the same bolts that hold the rocker arm assembly also hold the camshaft bearing caps. To avoid damage to the bearing surfaces, the camshaft sprocket must be removed.

1. Relieve the fuel system pressure and disconnect the negative battery cable.

2. Remove the rocker arm cover and turn the crankshaft to align the timing marks at TDC on No. 1 cylinder.

3. Use a wire tie or wire to secure the timing chain to the camshaft sprocket.

4. Hold the camshaft sprocket to loosen the bolt and remove the sprocket. Secure the sprocket so the chain does not fall off the crankshaft sprocket.

5. Loosen each rocker shaft bolt 1 turn at a time to prevent bending the shafts.

6. When all the bolts are loose, remove the rocker arm shafts with the bolts still in the shafts. This will hold the assembly together.

7. If the rocker arms are to be removed from the shafts, mark them so they can be returned into their original position. Remove the bolts from the shaft assembly and remove the parts. Keep the rocker arms upright or lay them in a pan of new engine oil. Note the punch marks on the front of each shaft that tell which shaft is for the intake side and which is for the exhaust side. This is important for correct rocker arm oiling.

To install:

8. Lubricate the shafts with engine oil and assemble them with the punch marks facing up. Use the bolts to hold the assembly together. Make sure the camshaft and the bearing surfaces are in good condition and lubricate with engine oil. Make sure the pin on the camshaft sprocket end is up.

9. Install the rocker arm shafts and tighten the bolts in the proper sequence 1 turn at a time to draw the shafts down evenly against the valve springs without bending the shafts. Torque the bolts to 16 ft. lbs. (22 Nm).

10. Install the camshaft sprocket and remove the tie securing the chain. Install the sprocket bolt but do not torque it yet. Rotate the crankshaft 2 full turns to make sure the timing marks line up. When the valve timing is correct, torque the sprocket bolt to 116 ft. lbs. (157 Nm).

11. If required, adjust the valves.

12. Use a silicone sealer on the rubber end plugs and install the rocker arm cover with a new gasket. Torque the bolts to 7 ft. lbs. (10 Nm) in a crisscross pattern starting at the middle. Install the remaining components.

13. When the engine is first started, the hydraulic valve lifters may be noisy. Run the engine for 10–20 minutes at about 1000 rpm. If the noise has not subsided, the lifter will probably never pump up and must be replaced.

3.0L Engine

1. Relieve the fuel system pressure and disconnect the negative battery cable.

2. Remove the rocker arm covers.

3. Turn the crankshaft to align the timing marks at TDC on No. 1 cylinder. Remove the distributor cap and matchmark the position of the rotor to the distributor body and to the engine. Remove the distributor.

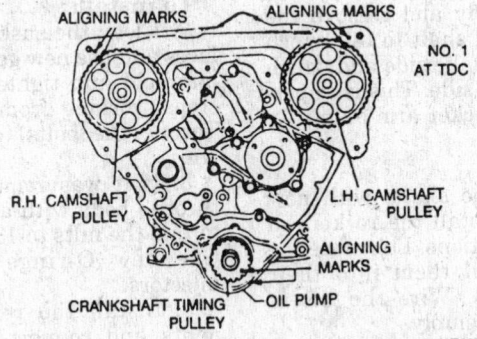

View of the camshaft sprocket timing marks — 3.0L engine

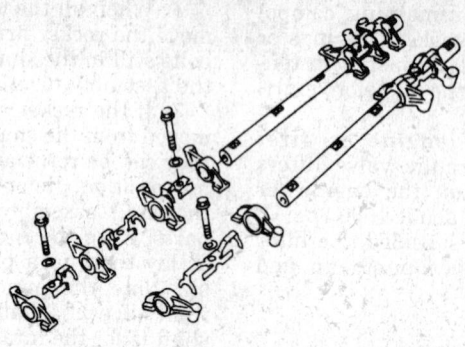

Rocker arm assembly on 2.4L engine with hydraulic lash adjusters

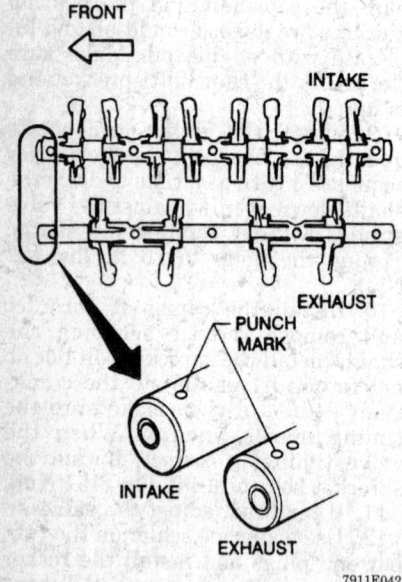

Make sure the rocker arm shafts are returned to their correct position — 2.4L engine shown

4. Loosen each rocker shaft bolt 1 turn at a time to prevent bending the shafts.

5. When all the bolts are loose, remove the rocker arm shafts with the bolts still in the shafts. This will hold the assembly together.

6. If the lifters are to be removed:

a. Secure the valve lifters in the guide assembly with safety wire to keep them in their original positions, then remove the entire assembly.

b. Before removing a lifter from the guide assembly, tag the lifters to make sure they are returned to

their original position. Do not disassemble a lifter.

c. Keep the lifters upright to prevent air from getting in or lay them down in a pan of new engine oil.

d. Check the lifter for signs of wear or damage. Measure the outside diameter of the lifter and the inside diameter of the bore it came from. The clearance should be 0.0017–0.0026 in. (0.043–0.066mm).

7. If the rocker arms are to be removed from the shafts, mark them so they can be returned into their original position. Remove the bolts from the shaft assembly and remove the rockers. Tag each shaft to tell which shaft is for the intake side and which is for the exhaust side. This is important for correct rocker arm oiling.

To install:

8. Lubricate the shafts with new engine oil and install the rockers in their original positions. Lubricate the lifters and install them into their original positions. Wire the lifters into the guide assembly.

9. Make sure the engine is at TDC on No. 1 cylinder. Install the left bank lifter guide assembly, remove

Rocker arm shaft tightening sequence — 2.4L engine

the safety wire and install the rocker arm shafts. Tighten the bolts 1 turn at a time to draw the shafts down evenly. Torque the bolts to 16 ft. lbs. (22 Nm).

10. Rotate the crankshaft to bring cylinder No. 4 to TDC. Set the right bank lifter guide assembly into place, remove the safety wire and install the rocker arm shafts. Tighten the bolts 1 turn at a time to 16 ft. lbs. (22 Nm).

11. Use new gaskets to install the rocker arm covers and torque the bolts to 24 inch lbs. (3 Nm). Install the remaining components.

12. When the engine is first started, the hydraulic valve lifters may be noisy. Run the engine for 10–20 minutes at about 1000 rpm. If the noise has not subsided, the lifter will probably never pump up and must be replaced.

Intake Manifold

REMOVAL AND INSTALLATION

2.4L Engine

1. Release the fuel pressure. Disconnect the negative battery cab

2. Remove the air cleaner assembly together with all of the attending hoses. Remove the EGR tube.

3. Label and disconnect all wiring and hoses as required.

4. Drain the engine coolant to a level below the thermostat housing, then, disconnect the upper coolant hose from the thermostat housing.

5. Disconnect the throttle linkage from the throttle body. The throttle body and/or fuel injectors can be removed from the manifold at this point or the entire assembly can be removed.

6. Remove the manifold bracket, if equipped, and remove the bolts and the intake manifold.

To install:

7. Clean the gasket mounting surfaces and use new gaskets. Install the manifold and tighten the bolts in 2 steps working from the center out. Torque the bolts to 15 ft. lbs. (21 Nm).

8. If it was removed, install the throttle body with a new gasket and torque the nuts to 13 ft. lbs. (18 Nm). Use new O-rings to install the injectors.

9. Install the remaining components and connect the wiring and hoses. Fill the cooling system and run the engine to check ignition timing and idle speed.

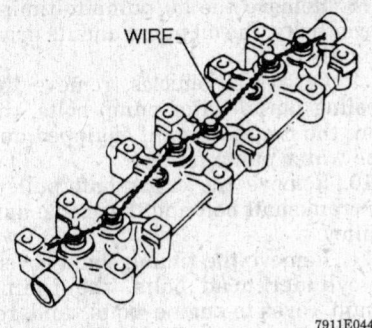

WIRE

7911E044

Hold the lifters in place with wire before removing the guide assembly — 3.0L engine

ROCKER SHAFT DIRECTION

EXHAUST

R.H. CYLINDER HEAD FRONT   L.H. CYLINDER HEAD FRONT

INTAKE

7911E045

Make sure rocker arm shafts are installed in their original position — 3.0L engine

⑤ ① ③ ⑦

⑧ ④ ② ⑥

7911E051

Intake manifold bolt torque sequence — 3.0L engine

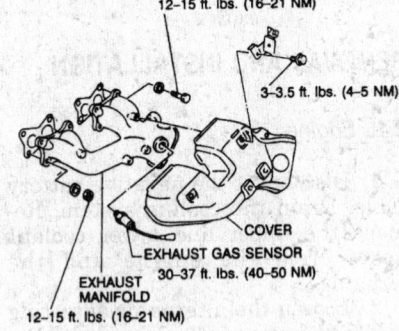

12–15 ft. lbs. (16–21 NM)

3–3.5 ft. lbs. (4–5 NM)

COVER

EXHAUST GAS SENSOR

30–37 ft. lbs. (40–50 NM)

EXHAUST MANIFOLD

12–15 ft. lbs. (16–21 NM)

7911E052

Exhaust manifold — 2.4L engine

THROTTLE BODY

MANIFOLD BRACKET

⑦

⑤ ② ④

③ ①

⑥

7911E046

Intake manifold assembly on 2.4L engine

3.0L Engine

1. Release the fuel system pressure and disconnect the negative battery cable.

2. Drain the cooling system to a level below the intake manifold.

3. Remove the air duct from the throttle body. Disconnect the accelerator linkage from the throttle body.

4. Remove the upper radiator hose from the water outlet housing and the exhaust tube from the EGR valve. If necessary, remove the EGR valve-to-intake manifold nuts and the EGR valve.

5. Label and disconnect the wiring and hoses as required.

6. Remove the 5 intake manifold collector-to-intake manifold bolts and the lift the collector off the engine.

7. Remove the fuel rail and the injectors as an assembly from the intake manifold.

8. To prevent warping, loosen the intake manifold nuts and bolts 1 or 2 turns at a time in the reverse of the torque sequence. Remove the manifold.

To install:

9. Clean the gasket surfaces and install new gaskets.

10. Install the intake manifold and torque the nuts and bolts in the proper sequence in the following steps: Step 1: all to 43 inch lbs. (5 Nm). Step 2: bolts to 14 ft. lbs. (20 Nm), nuts to 20 ft. lbs. (27 Nm). Step 3: repeat Step 2.

11. Use new O-rings and install the fuel injectors and rail assembly. Connect the wiring.

12. Use a new gasket and install the intake manifold collector. Torque the bolts to 12 ft. lbs. (16 Nm).

13. Install the remaining components and connect the wiring and hoses. Refill the cooling system and run the engine to check ignition timing and idle speed.

Exhaust Manifold

REMOVAL AND INSTALLATION

2.4L Engine

1. Disconnect the negative battery cable.

2. If equipped, remove the hot air duct from the exhaust manifold cover.

3. Disconnect the spark plug wires from the exhaust side of the engine; if necessary, remove the spark plugs from the exhaust side of the engine.

4. If necessary, raise and safely support the vehicle.

5. If equipped, remove the air induction tubes from the exhaust manifold. Remove the EGR tube from the exhaust manifold.

6. Remove the hot air cover and the exhaust pipe from the exhaust manifold.

7. Remove the exhaust manifold-to-engine nuts and the manifold from the engine.

To install:

8. Clean the gasket mounting surfaces and install new gaskets.

9. Install the manifold and torque the nuts/bolts to 15 ft. lbs. (20 Nm), working from the center to the ends, in 2 steps.

10. Use new gaskets and connect the exhaust pipe to the manifold. Torque the nuts to 27 ft. lbs. (36 Nm).

11. Install the remaining components and run the engine to check for leaks.

3.0L Engine

LEFT SIDE

1. Disconnect the negative battery cable.

2. Remove the hot air tube from the exhaust manifold cover. Remove the exhaust manifold cover-to-exhaust manifold bolts and cover.

3. Remove the EGR and the AIR tubes from the exhaust manifold, if equipped.

NOTE: If the alternator is in the way, remove the drive belt and the alternator.

4. Raise and safely support the vehicle.

5. Remove the exhaust pipe-to-exhaust manifold nuts and separate the exhaust pipe from the manifold.

6. Remove the exhaust manifold-to-cylinder head bolts and the manifold from the engine.

7. Clean the gasket mounting surfaces.

8. To install, use new gaskets and reverse the removal procedures. Torque the exhaust manifold-to-cylinder head nuts to 16 ft. lbs. (22 Nm) and the exhaust pipe-to-exhaust manifold bolts to 20 ft. lbs. (27 Nm).

RIGHT SIDE

1. Disconnect the negative battery cable.

2. Remove the upper/lower exhaust manifold cover-to-exhaust manifold bolts and covers.

3. Remove the AIR tube from the exhaust manifold, if equipped.

4. Raise and safely support the vehicle.

5. Remove the exhaust pipe-to-exhaust manifold bolts and separate the exhaust pipe from the manifold.

6. Remove the exhaust manifold-to-cylinder head bolts and the manifold from the engine.

7. Clean the gasket mounting surfaces.

8. To install, use new gaskets and reverse the removal procedures. Torque the exhaust manifold-to-cylinder head nuts to 16 ft. lbs. (22 Nm) and the exhaust pipe-to-exhaust manifold bolts to 20 ft. lbs. (27 Nm).

Timing Chain Front Cover

REMOVAL AND INSTALLATION

2.4L Engine

1. Disconnect the negative battery cable. Drain the cooling system. Remove the upper and lower coolant hoses from the engine and the radiator.

2. Loosen the alternator adjusting bolt and remove the drive belt. Remove the alternator bracket-to-engine bolts and move the alternator aside.

3. If equipped with air conditioning, remove the drive belt. If necessary, remove the air conditioner compressor and move it aside without disconnecting the coolant hoses.

4. If equipped with power steering, remove the drive belt.

5. Rotate the crankshaft to position the No. 1 cylinder on TDC of its compression stroke.

6. Drain the oil and remove the oil pan.

7. Remove the distributor cap. Matchmark the rotor to the distributor housing and the distributor housing to the timing chain cover. Remove the distributor hold-down bolt and the distributor.

Exhaust manifolds — 3.0L engine

8. Remove the oil pump-to-timing cover bolts, the oil pump and its drive spindle.

9. On RWD vehicles, remove the cooling fan-to-water pump bolts, the fan, the fan coupling if equipped and the water pump pulley.

10. Remove the crankshaft pulley-to-crankshaft bolt and the crankshaft pulley.

11. Remove the timing chain cover-to-cylinder head bolts, the timing chain cover-to-engine bolts, and remove the cover. Clean the gasket mounting surfaces.

NOTE: Whenever the timing cover is removed, replace the oil seal.

12. Install a new oil seal and use new gaskets and silicone sealant as necessary to install the front cover.

13. Install the oil pan.

14. When installing the oil pump, place the new gasket over the shaft and make sure the drive spindle mark aligns with the hole in the pump body. This will align the shaft properly for the distributor.

15. Install the distributor and make sure the rotor aligns with the matchmark.

16. Install the crankshaft pulley and torque the bolt to 116 ft. lbs. (157 Nm).

17. Install the remaining components and refill the cooling system. Run the engine to check ignition timing and idle speed.

Front Cover Oil Seal

REPLACEMENT

2.4L Engine

1. Disconnect the negative battery cable.

2. Remove the radiator shroud as required and remove crankshaft pulley.

3. Carefully pry the front oil seal out of the timing cover without damaging the crankshaft.

4. Lubricate the new seal with light grease. Using an oil seal installation tool, drive the new oil seal into the timing cover until it seats. Clean all oil and grease away from the seal and crankshaft area.

5. To complete the installation, reverse the removal procedures. Torque the crankshaft pulley-to-crankshaft bolt to 116 ft. lbs. (157 Nm). Start the engine, allow it reach normal operating temperatures and check for leaks.

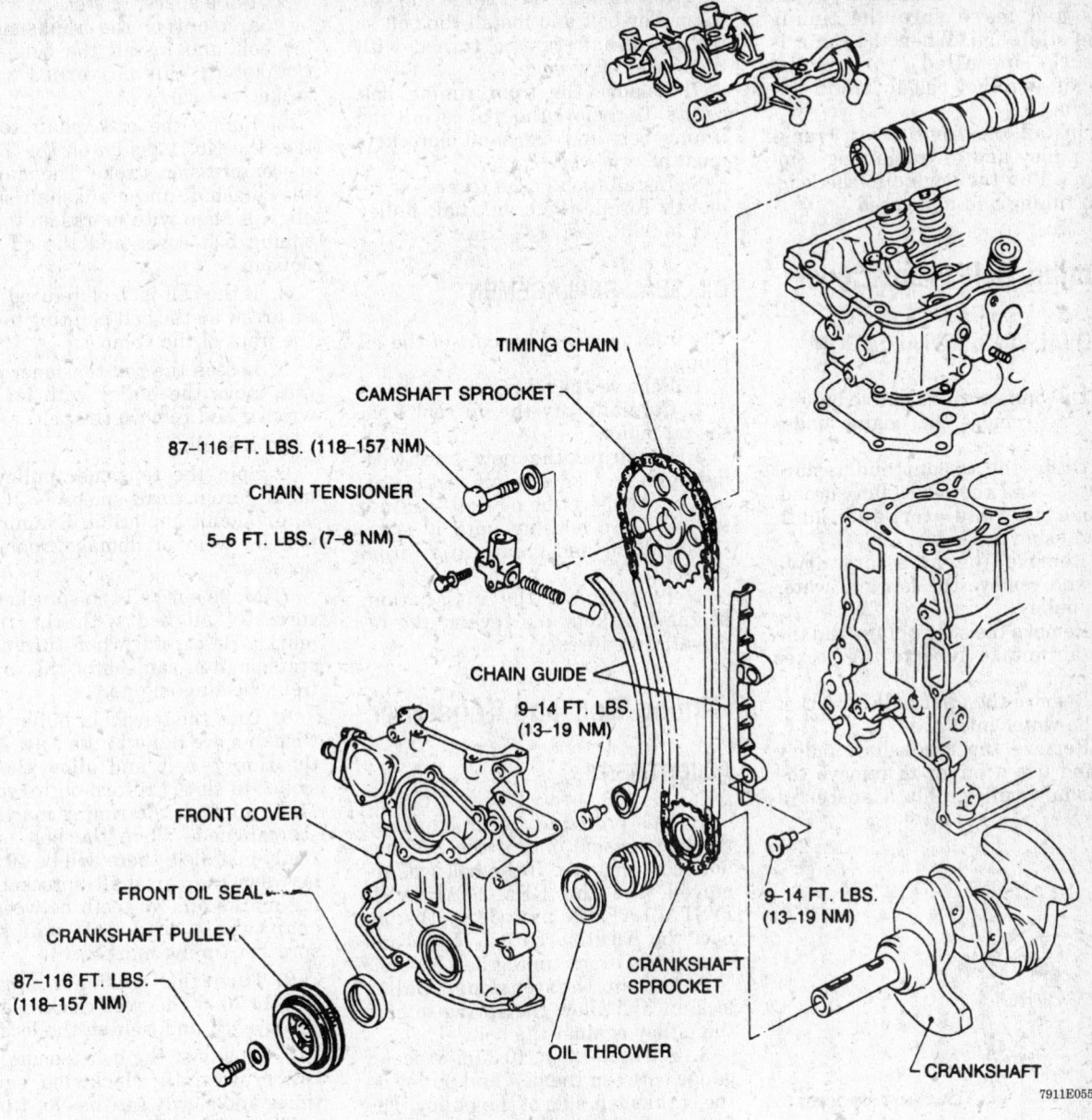

TIMING CHAIN

CAMSHAFT SPROCKET

87–116 FT. LBS. (118–157 NM)

CHAIN TENSIONER

5–6 FT. LBS. (7–8 NM)

CHAIN GUIDE

9–14 FT. LBS. (13–19 NM)

FRONT COVER

FRONT OIL SEAL

CRANKSHAFT PULLEY

87–116 FT. LBS. (118–157 NM)

OIL THROWER

CRANKSHAFT SPROCKET

9–14 FT. LBS. (13–19 NM)

CRANKSHAFT

7911E055

Timing chain assembly — 2.4L engine

Timing Chain and Sprockets

REMOVAL AND INSTALLATION

2.4L Engine

1. Disconnect the negative battery cable.

2. Rotate the crankshaft to align the timing marks on the crankshaft pulley at TDC of No. 1 cylinder. Remove the timing chain cover and the rocker arm cover.

3. Make sure the No. 1 piston is at TDC of its compression stroke; the No. 1 camshaft lobes will both be down. The timing marks on the camshaft sprocket and crankshaft sprocket should align with the silver links on the timing chain. If the chain has no silver links, paint alignment marks on the chain.

4. Remove the chain tensioner.

NOTE: When the chain tensioner bolts are removed, the tensioner will come apart. Hold on to the piston and do not drop any of the parts into the oil pan. There is no need to remove the chain guide unless it is being replaced.

5. Hold the camshaft sprocket from turning, remove the bolt and re- move the sprocket along with the chain.

6. Inspect the timing chain for cracked links, wear and/or damage; if necessary, replace the chain. Inspect the guides and sprockets for wear or damage and replace as necessary. If replacing a sprocket, always replace the chain.

7. Install the timing chain and camshaft sprocket, making sure to align all the timing marks. Install the camshaft sprocket bolt but do not torque it yet.

8. Install the chain tensioner and torque the bolts to 72 inch lbs. (8 Nm).

9. Rotate the crankshaft 2 full turns and make sure the timing marks still align. When the chain is correctly installed, torque the camshaft sprocket bolt to 116 ft. lbs. (157 Nm).

10. Install the front cover, crankshaft pulley and all remaining components. Run the engine to check ignition timing and idle speed.

Timing Belt Front Cover

REMOVAL AND INSTALLATION

1. Disconnect the negative battery cable and remove the engine under cover.

2. Drain the coolant and remove the hoses and the lower fan shroud. Remove the radiator and main shroud as an assembly.

3. Remove the accessory drive belts and remove the fan and water pump pulley.

4. Remove the spark plugs and the fresh air intake tube to the rocker arm cover.

5. Remove the idler pulley bracket and the water inlet hose.

6. Remove the crankshaft pulley bolt and use a puller to remove the crankshaft pulley. Put a spacer (a

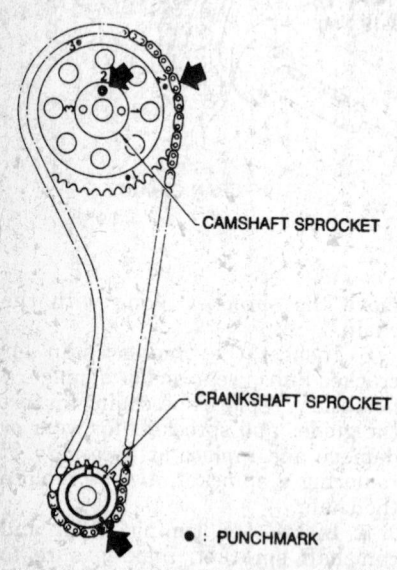

CAMSHAFT SPROCKET

CRANKSHAFT SPROCKET

● : PUNCHMARK

7911E057

Timing chain and sprocket marks on 2.4L engine; later models may have only 1 timing mark on the camshaft sprocket

stack of washers or a large nut) on the pulley bolt and install the bolt so the crankshaft can be turned with the socket or wrench.

7. Remove the front timing belt covers. To remove the rear covers, the timing belt and camshaft sprockets must be removed.

8. Installation is the reverse of removal. Torque the crankshaft pulley bolt to 98 ft. lbs. (132 Nm).

OIL SEAL REPLACEMENT

The front oil seal is a part of the oil pump.

1. Remove the oil pump.

2. Carefully pry the oil seal from the oil pump.

3. Lubricate the new seal with light grease. Using an oil seal installation tool, drive the new oil seal into the oil pump housing until it seats. Clean all oil and grease away from the seal.

4. To complete the installation, use new gaskets and reverse the removal procedures.

Timing Belt and Tensioner

ADJUSTMENT

1. This procedure is for adjusting the belt tension only if the belt has not been removed. If the belt was removed, see the REMOVAL AND INSTALLATION procedure. Disconnect the negative battery cable and remove the front timing belt covers.

2. Loosen the tensioner pulley locknut and allow the spring to hold the pulley against the belt.

3. Set a 0.014 in. (0.35mm) feeler gauge between the belt and pulley on the crankshaft side of the pulley. The feeler gauge should be at least ½ in. (12.7mm) wide.

4. Rotate the crankshaft clockwise to make the feeler gauge roll up between the tensioner pulley and the belt. Make sure the gauge is centered on the tensioner pulley.

5. Push in on the belt halfway between the tensioner pulley and the camshaft sprocket with a force of 22 lbs. (98 N) and torque the locknut to 43 ft. lbs. (58 Nm).

6. Rotate the crankshaft to remove the feeler gauge. Install the covers.

REMOVAL AND INSTALLATION

1. Disconnect the negative battery cable and remove the timing belt cover.

2. Put a spacer (a stack of washers or a large nut) on the crankshaft pulley bolt and install the bolt so the crankshaft can be turned with a socket wrench.

3. Rotate the crankshaft to position the No. 1 piston on the TDC of its compression stroke. The marks on the camshaft and crankshaft sprockets will align with marks on the rear timing belt cover and the oil pump housing.

4. If the belt is to be re-used, paint an arrow on the belt pointing towards the front of the vehicle.

5. Loosen the belt tensioner pulley nut, move the pulley with an Allen wrench and remove the belt.

To install:

6. Spin the tensioner pulley and make sure it turns smoothly. If there is any doubt, replace it. Examine the belt for wear or damage, replace as necessary.

7. Make sure all the sprockets are correctly aligned with the timing marks. Be careful when turning the crankshaft or camshafts, this is not a free wheeling engine.

8. Turn the tensioner pulley clockwise to move it out of the way, install the timing belt and allow the tensioner to slowly return on its spring. Make sure all the timing marks are still aligned. When the belt is correctly installed, there will be 40 teeth between the camshaft sprocket timing marks and 43 teeth between the crankshaft and left camshaft sprocket timing marks.

9. Turn the tensioner approximately 70–80 degrees clockwise with the wrench and tighten the locknut.

10. To adjust the belt tension, turn the crankshaft clockwise several times and slowly set the No. 1 piston to TDC of the compression stroke.

11. Set a 0.014 in. (0.35mm) feeler gauge between the belt and pulley on the crankshaft side of the pulley. The feeler gauge should be at least ½ in. (12.7mm) wide.

12. Rotate the crankshaft clockwise to make the feeler gauge roll up between the tensioner pulley and the belt. Make sure the gauge is centered on the tensioner pulley.

13. Loosen the tensioner pulley locknut, push in on the belt halfway between the tensioner pulley and the camshaft sprocket with a force of 22 lbs. (98 N) and torque the locknut to 43 ft. lbs. (58 Nm).

14. Rotate the crankshaft to remove the feeler gauge. Install the covers.

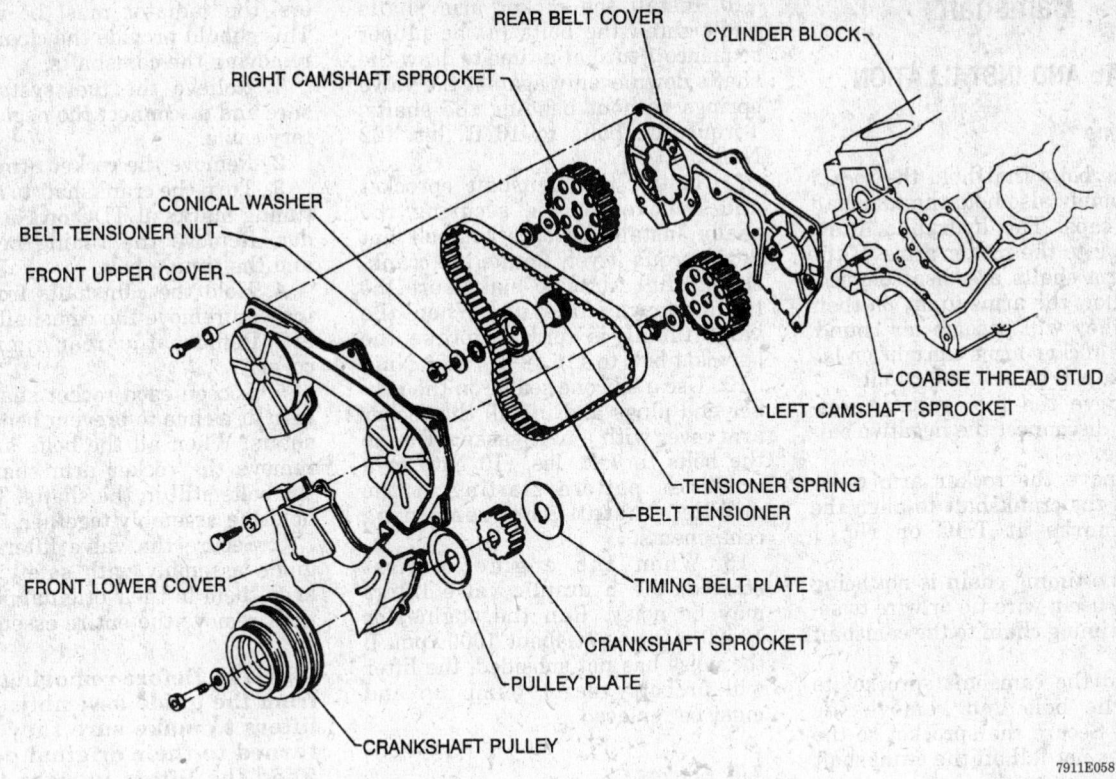

Timing belt and cover assembly on 3.0L engine

7911E058

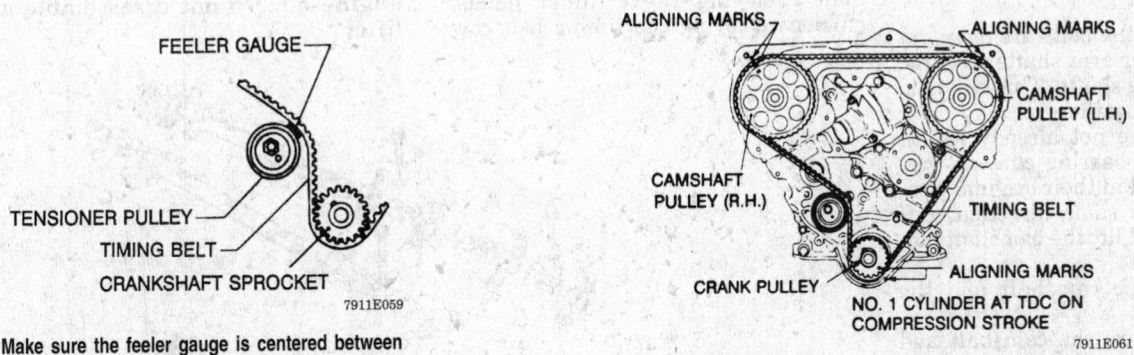

Timing belt and sprockets on 3.0L engine

7911E061

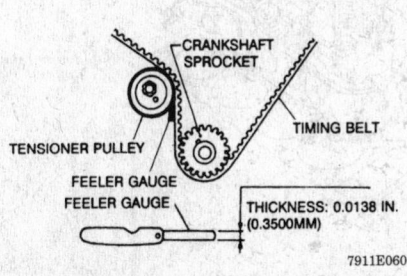

Make sure the feeler gauge is centered between the tensioner pulley and the belt — 3.0L engine

7911E059

THICKNESS: 0.0138 IN. (0.3500MM)

Place the feeler gauge under the tensioner pulley — 3.0L engine

7911E060

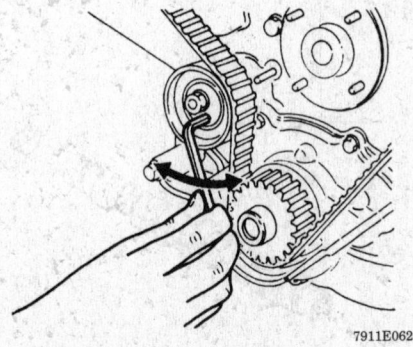

7911E062

Move the tensioner pulley with an Allen wrench

Timing Sprockets

REMOVAL AND INSTALLATION

1. Disconnect the negative battery cable.

2. Remove the timing belt cover and the timing belt.

3. Use an appropriate tool to prevent the camshafts from turning and remove the sprocket bolts. Pull the sprockets straight off.

4. Installation is the reverse of removal. Torque the camshaft sprocket bolts to 65 ft. lbs. (88 Nm). Install the timing belt.

Camshaft

REMOVAL AND INSTALLATION

2.4L Engine

The same bolts that hold the rocker arm assembly also hold the camshaft bearing caps. The hydraulic lifters are built into the rocker arms. If the rocker arm shafts are disassembled, do not allow the arms to lay on their side or they will become air bound. Keep the rocker arms upright or lay them in a pan of new engine oil.

1. Relieve the fuel system pressure and disconnect the negative battery cable.

2. Remove the rocker arm cover and turn the crankshaft to align the timing marks at TDC on No. 1 cylinder.

3. If the timing chain is not being removed, use a wire tie or wire to secure the timing chain to the camshaft sprocket.

4. Hold the camshaft sprocket to loosen the bolt and remove the sprocket. Secure the sprocket so the chain does not fall off the crankshaft sprocket.

5. Loosen each rocker shaft bolt 1 turn at a time to prevent bending the shafts.

6. When all the bolts are loose, remove the rocker arm shafts with the bolts still in the shafts. This will hold the assembly together.

7. If they are not already identified, mark the bearing caps so they can be installed in their original position facing the same direction. Lift the caps off and lift the camshaft out.

To install:

8. Inspect the camshaft and the bearings:

   a. Make sure the camshaft and the bearing surfaces are in good condition.

   b. Install the bearing caps without the camshaft, torque the rocker arm shaft bolts to specification and measure the inside diameter of the bearing circle.

   c. Measure the diameter of the camshaft bearings.

   d. The difference between the measurements is the camshaft journal clearance; it should be no more than 0.0047 in. (0.12mm)

   e. Install the camshaft without the rocker arms and torque the bolts to specification. The camshaft end-play should be no more than 0.008 in. (0.2mm).

9. Lubricate the camshaft with engine oil and set it in place. Make sure the pin on the sprocket end is up.

10. Install the rocker arm shafts and tighten the bolts in the proper sequence 1 turn at a time to draw the shafts down evenly against the valve springs without bending the shafts. Torque the bolts to 16 ft. lbs. (22 Nm).

11. Install the camshaft sprocket and remove the tie securing the chain. Install the sprocket bolt but don't torque it yet. Rotate the crankshaft 2 full turns to make sure the timing marks line up. When the valve timing is correct, torque the sprocket bolt to 116 ft. lbs. (157 Nm).

12. Use a silicone sealer on the rubber end plugs and install the rocker arm cover with a new gasket. Torque the bolts to 7 ft. lbs. (10 Nm) in a crisscross pattern starting at the middle. Install the remaining components.

13. When the engine is first started, the hydraulic valve lifters may be noisy. Run the engine for 10–20 minutes at about 1000 rpm. If the noise has not subsided, the lifter will probably never pump up and must be replaced.

3.0L Engine

The camshafts can be removed without removing the cylinder heads. When removing the timing belt covers, the radiator must be removed. This should provide the clearance for removing the camshafts.

1. Relieve the fuel system pressure and disconnect the negative battery cable.

2. Remove the rocker arm covers.

3. Turn the crankshaft to align the timing marks at TDC on No. 1 cylinder. Remove the timing belt cover and the timing belt.

4. Hold the camshafts from turning and remove the camshaft sprockets. Remove the rear timing belt cover.

5. Loosen each rocker shaft bolt 1 turn at a time to prevent bending the shafts. When all the bolts are loose, remove the rocker arm shafts with the bolts still in the shafts. This will hold the assembly together.

6. Secure the valve lifters in the guide assembly with safety wire to keep them in their original positions, then remove the entire assembly.

NOTE: Before removing a lifter from the guide assembly, tag the lifters to make sure they are returned to their original position. Keep the lifters upright to keep them from becoming air bound or lay them down in a pan of new engine oil. Do not disassemble a lifter.

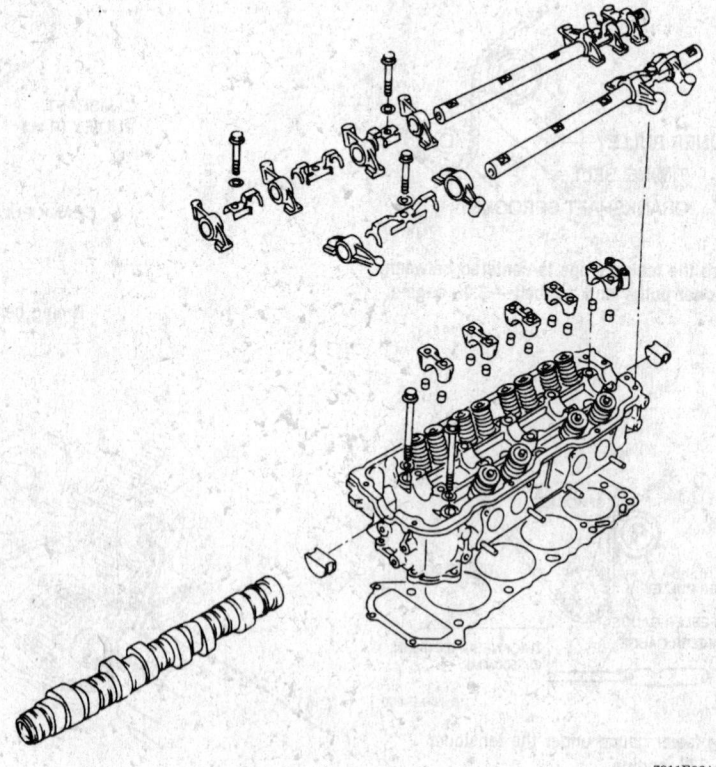

7911E064

Camshaft removal on 2.4L engine

7. Remove the plates from the front and rear of the cylinder heads and pry the oil seals out. Remove the bolt at the rear of the camshafts and remove the locating plates. Carefully withdraw the camshafts out towards the front.

To install:

8. Inspect the camshaft and the bearing surfaces:

 a. Make sure the camshaft and the bearing surfaces are in good condition.

 b. Measure the inside diameter of the bearing circle.

 c. Measure the diameter of the camshaft bearings.

 d. The difference between the measurements is the camshaft journal clearance; it should be no more than 0.0059 in. (0.15mm)

 e. To check endplay, install the camshaft and the locating plates and torque the bolts to 65 ft. lbs. (88 Nm). The camshaft endplay should be no more than 0.0024 in. (0.06mm).

9. Lubricate the camshaft with engine oil and carefully set it in place. Install the locating plate at the rear and torque the bolt to 65 ft. lbs. (88 Nm). Turn the camshaft so the pin on the sprocket end is up.

10. Install the rear camshaft cover plate with a new gasket.

11. Lubricate a new camshaft front oil seal with grease and use an appropriate seal installation tool to carefully drive the new seal into place. Make sure the seal seats in the cylinder head.

12. With the rocker arm assemblies removed, all the valves will be closed. The rear timing belt cover, sprockets and timing belt can be installed without risk of damage to valves or pistons. Adjust timing belt tension according to correct procedure.

13. Make sure the engine is at TDC on No. 1 cylinder. Install the left bank lifter guide assembly, remove the safety wire and install the rocker arm shafts. Tighten the bolts 1 turn at a time to draw the shafts down evenly. Torque the bolts to 16 ft. lbs. (22 Nm).

14. Rotate the crankshaft to bring cylinder No. 4 to TDC. Set the right bank lifter guide assembly into place, remove the safety wire and install the rocker arm shafts. Tighten the bolts 1 turn at a time to 16 ft. lbs. (22 Nm).

15. Use new gaskets to install the rocker arm covers and torque the bolts to 24 inch lbs. (3 Nm). Install the remaining components.

16. When the engine is first started, the hydraulic valve lifters may be noisy. Run the engine for 10–20 minutes at about 1000 rpm. If the noise has not subsided, the lifter will probably never pump up and must be replaced.

Piston and Connecting Rod

POSITIONING

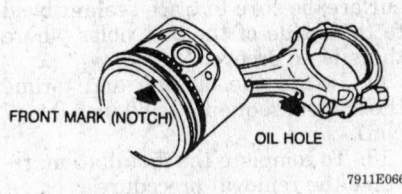

Piston and rod orientation on 2.4L engine

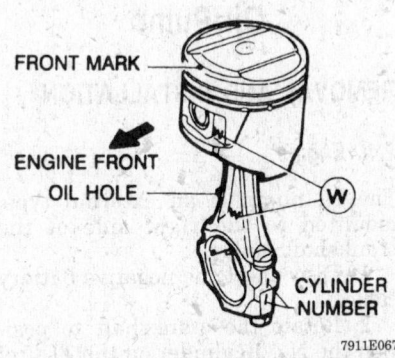

Piston and rod orientation on 3.0L engine

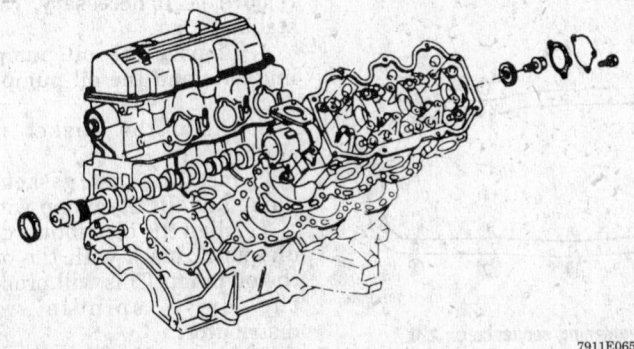

Camshaft removal on 3.0L engine can be done without removing the cylinder head

ENGINE LUBRICATION

Oil Pan

REMOVAL AND INSTALLATION

2.4L Engine

1. Raise and safely support the vehicle.

2. Remove the engine undercover and drain the engine oil.

3. On 4WD models, perform the following procedures:

 a. Remove the bolt from the front differential carrier member.

 b. Position a floor jack under the front differential carrier and remove the mounting bolts.

 c. Remove the transmission-to-rear engine mount bracket nuts.

 d. Remove the engine mount nuts and bolts.

 e. Attach an engine hoist and raise the engine slightly.

4. On 2WD models, remove the front crossmember.

5. Remove the oil pan-to-engine bolts. Insert a seal cutter tool between the cylinder block and the oil pan and tap it around the circumference of the pan with a hammer. Remove the oil pan.

NOTE: Be careful not to drive the seal cutter into the oil pump or rear oil seal retainer as damage may occur.

6. Clean the gasket mounting surfaces.

7. Apply a continuous 1/8 in. bead of silicone sealant to the oil pan mounting surface; be sure to trace sealant bead to the inside of the bolt holes where there is no groove.

8. Install the oil pan and torque the bolts in sequence to 60 inch lbs. (7 Nm).

9. To complete the installation, reverse the removal procedures.

10. Wait at least 30 minutes and refill the crankcase. Start the engine and allow it to reach normal operating temperatures and check for leaks.

3.0L Engine

1. Raise and safely support the vehicle.

2. Remove the undercover and drain the engine oil.

3. On 2WD models, remove the stabilizer bar bracket bolts.

4. On 4WD models, remove the front driveshaft and disconnect the halfshafts at the transfer case. Position a floor jack under the front differential carrier and remove the mounting bolts.

5. On 2WD models, remove the front crossmember.

6. Remove the idler arm and the starter motor.

7. On 4WD models, remove the transmission-to-rear engine mount bracket nuts and the engine mount nuts/bolts.

8. Remove the engine gussets.

9. On 4WD models, attach a hoist to the engine and raise the engine slightly.

10. Remove the oil pan-to-engine bolts in the correct sequence to avoid warping the pan. Insert a seal cutter tool between the cylinder block and the oil pan and tap the tool around the circumference with a hammer to remove the oil pan.

NOTE: Be careful not to drive the seal cutter into the oil pump or rear oil seal retainer for damage may occur.

11. Clean the gasket mounting surfaces.

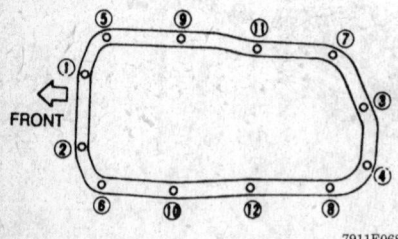

Oil pan bolt tightening sequence on 2.4L engine

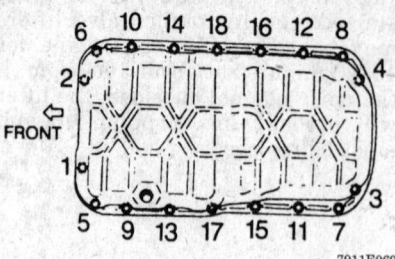

Oil pan bolt loosening sequence on 3.0L engine — tighten the bolts in the reverse of this sequence

To install:

12. Apply silicone sealant to the oil pump and oil seal retainer gasket.

13. Apply a continuous ⅛ in. bead of sealant to the oil pan mounting surface; be sure to trace sealant bead to the inside of the bolt holes where there is no groove.

14. Install the oil pan and torque the bolts in sequence to 60 inch lbs. (7 Nm).

15. To complete the installation, reverse the removal procedures.

16. Wait at least 30 minutes and refill the crankcase. Start the engine and allow it to reach normal operating temperatures and check for leaks.

Oil Pump

REMOVAL AND INSTALLATION

2.4L Engine

The oil pump is an external type, mounted to the right side of the crankshaft pulley.

1. Disconnect the negative battery cable.

2. Rotate the crankshaft to position the No. 1 cylinder on the TDC of the compression stroke.

3. If equipped with a splash pan, remove it. If necessary, remove the stabilizer bar.

4. Remove the oil pump-to-housing bolts and the oil pump from the engine.

5. Clean the gasket mounting surfaces.

6. Install a new gasket onto the pump and fill the pump with oil.

7. Align the distributor drive spindle punch mark with the oil hole on the oil pump. This will properly align the drive spindle with the distributor.

8. Insert the oil pump into the housing until the drive spindle tang fits into the distributor shaft notch.

Torque the oil pump-to-engine housing bolts to 11 ft. lbs. (15 Nm).

9. Start the engine to check the ignition timing and to check for leaks.

3.0L Engine

The oil pump is mounted at the front of the engine behind the crankshaft pulley.

1. Disconnect the negative battery cable.

2. Raise and safely support the vehicle. Drain the cooling system and the crankcase.

3. Remove the oil pan and the timing belt.

4. Remove the crankshaft timing sprocket using a wheel puller and the timing belt plate.

5. Remove the oil pump strainer and the pickup tube from the oil pump.

6. Remove the oil pump-to-engine bolts and the oil pump from the engine.

7. Clean the gasket mounting surfaces.

NOTE: Whenever the oil pump is removed, replace the oil seal.

8. To install, use new gaskets or silicone sealant. Pack the oil pump cavity with petroleum jelly and reverse the removal procedures. Torque as follows: Oil pump-to-engine: 6mm bolts — 60 inch lbs. (7 Nm) 8mm bolts — 12 ft. lbs. (16 Nm) Pickup tube-to-oil pump bolts — 15 ft. lbs. (21 Nm)

CHECKING

To check the oil pump clearances, the oil pump must be removed from the engine and disassembled. If the parts do not meet specifications, replace the oil pump as an assembly.

2.4L Engine

Using a feeler gauge, check the following clearances:

Inner rotor tip-to-outer rotor — 0.0047 in. (0.12mm) max. Outer rotor-to-housing — 0.0059–0.0083 in. (0.15–0.21mm)

Side clearance (with gasket) — 0.0016–0.0031 in. (0.04–0.07mm)

3.0L Engine

Using a feeler gauge, check the following clearances:

Pump body-to-outer gear — 0.0043–0.0079 in. (0.11–0.20mm)

Inner gear-to-crescent — 0.0047–0.0091 in. (0.12–0.23mm)

Outer gear-to-crescent — 0.0083–0.0126 in. (0.22–0.33mm)

When installing the oil pump on 2.4L engine, align the mark on the shaft with the oil hole

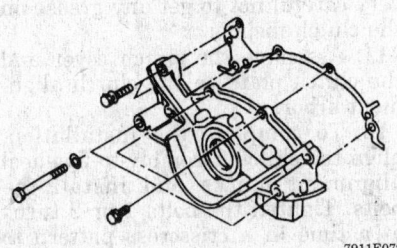

Oil pump on 3.0L engine is on the front of the engine block

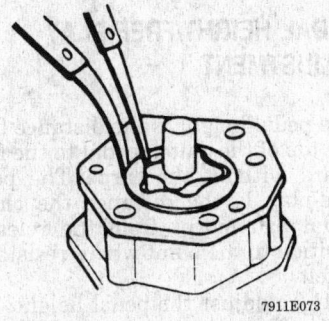

Inspecting oil pump rotor clearance — 2.4L engine

Housing-to-inner gear — 0.0020–0.0035 in. (0.05–0.09mm)
Housing-to-outer gear — 0.0020–0.0043 in. (0.05–0.11mm)

Rear Main Bearing Oil Seal

REMOVAL AND INSTALLATION

1. Disconnect the negative battery cable.

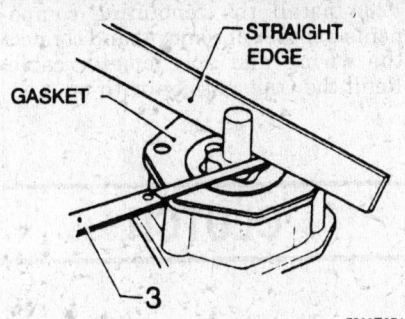

Inspecting oil pump side clearance — 2.4L engine

2. Raise and safely support the vehicle. Remove the starter.
3. Remove the transmission from the vehicle.
4. If equipped with a manual transmission, remove the clutch-to-flywheel bolts and the clutch assembly from the vehicle.
5. Remove the flywheel-to-crankshaft bolts and the flywheel from the engine.
6. Remove the rear oil seal retainer-to-engine bolts, the rear oil seal retainer-to-oil pan bolts and the retainer.
7. Carefully pry the rear oil seal from the retainer; be careful not to damage the mounting surfaces. Clean the oil seal mounting surfaces.
8. Using an appropriate oil seal installation tool, lubricate the new oil seal lips with engine oil and drive the the seal into the retainer until it seats.
9. To complete the installation, reverse the removal procedures. Start the engine and check for leaks.

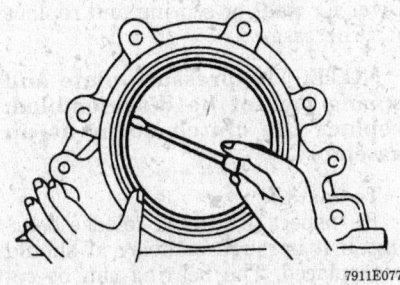

Removing the rear oil seal from the retainer; the transmission and flywheel must be removed

MANUAL TRANSMISSION

Transmission Assembly

REMOVAL AND INSTALLATION

1. Disconnect the negative battery cable.
2. Remove the shifter knob and boot and remove the snapring to lift the shift lever out of the transmission. Stuff a rag in the opening to keep dirt out of the transmission.
3. Raise and safely support the vehicle. If equipped, remove the splash pan or skid plate.
4. Remove the starter and drain the oil from the transmission.
5. Matchmark the driveshaft flange at the rear pinion flange and remove the driveshaft. Plug the extension housing opening to prevent dirt from getting in.
6. If equipped with 4WD, matchmark both front driveshaft flanges so the shaft can be installed in the same position. Remove the driveshaft.
7. Disconnect the exhaust pipe from the manifolds and from the catalytic converter and remove the pipe.
8. Disconnect the speedometer cable and the wiring from the transmission.
9. Without disconnecting the hydraulic hose, remove the clutch slave cylinder from the transmission and secure it aside.
10. If equipped with 4WD, the torsion bars must be removed.

   a. Working under the vehicle, measure and record the length of the threads on the torsion bar adjustment. At the front of the bar, pull the boot back and matchmark the bar to the mounting plate. The spline on the bar must be re-installed in the same position on the plate.

   b. Remove the locknut and adjustment nut from both torsion bars. Remove the 3 nuts at the mounting plate and remove the bars. Mark the bars left and right side for proper installation.

   c. Remove the transfer case shift lever assembly from the transfer case.

   d. If necessary, the transfer case can be removed at this point so the jack will be available to remove the transmission.

e. If equipped with a 3.0L engine, remove the gusset securing the engine to the transmission.

11. Using a chain hoist, attach it to the engine and lift the engine slightly to take the weight off the mounts. Using an appropriate transmission jack, properly support the transmission and remove the transmission mount and crossmember.

12. Remove the transmission-to-engine bolts and move the transmission back away from the engine. Lower the transmission carefully from the vehicle.

To install:

13. Lightly grease the input shaft splines. On 4WD, apply a silicone sealant to the engine block or rear plate to seal the engine to the transmission.

14. Fit the transmission into place and start all the engine-to-transmission bolts. Make sure the input shaft fits properly into the clutch disc and pilot bearing.

15. Torque the 2.36 in. (60mm) and 2.56 in. (65mm) engine-to-transmission bolts to 36 ft. lbs. (49 Nm).

16. Torque the remaining bolts to 18 ft. lbs. (25 Nm) on the 2.4L engine or 29 ft. lbs. (39 Nm) on the 3.0L engine.

17. Install the crossmember and torque the crossmember-to-chassis bolts to 38 ft. lbs. (52 Nm) on 2WD or 31 ft. lbs. (42 Nm) on 4WD.

18. Install the rear transmission mount bolts, loosen the engine mount bolts, then torque all the mount bolts to 31 ft. lbs. (42 Nm), starting at the rear.

19. If the torsion bars were removed, install them in their original location. Make sure the splines are in their original position and set the adjustment to its original position.

20. If the transfer case was removed, apply silicone sealant to seal the case to the transmission. Install the transfer case and torque the bolts to 30 ft. lbs. (41 Nm). Install the shift lever.

21. When installing the driveshafts, be sure to align the matchmarks. Torque the bolts on the front driveshaft (4WD) to 33 ft. lbs. (44 Nm). On single piece rear driveshafts, torque the bolts to 65 ft. lbs. (88 Nm). On 2 piece driveshafts, torque the bolts to 33 ft. lbs. (44 Nm). Torque the center bearing bracket bolts to 16 ft. lbs. (22 Nm).

22. When installing the exhaust system, use new gaskets and torque the flange bolts to 27 ft. lbs. (36 Nm).

23. Install the remaining components in order of removal and connect the wiring and speedometer cable. Refill the transmission with oil.

CLUTCH

Clutch Assembly

REMOVAL AND INSTALLATION

1. Disconnect the negative battery cable. Raise and safely support the vehicle.
2. Remove the transmission or the transaxle.
3. Using a piece of chalk, paint or a center punch, mark the clutch assembly-to-flywheel relationship so it can be reassembled in the same position from which it is removed.
4. Using a clutch aligning tool, insert it into the clutch disc hub.
5. Loosen the clutch cover-to-flywheel bolts, a turn at a time in an alternating sequence, until the spring tension is relieved to avoid distorting or bending the clutch cover. Remove the clutch assembly.
6. Inspect the flywheel for scoring, roughness or signs of overheating. Light scoring may be cleaned up with emery cloth, but any deep grooves or overheating (blue marks) warrant replacement or refacing of the flywheel. If the clutch facings or flywheel are oily, inspect the transmission/transaxle front cover oil seal, the pilot bushing and engine rear seals, etc. for leakage; replace any leaking seals before replacing the clutch.
7. If the crankshaft pilot bushing is worn, replace it. Install it using a soft hammer. The factory supplied part does not have to be oiled, but check the procedure if using an aftermarket part. Inspect the clutch cover for wear or scoring and replace it, if necessary.

NOTE: The pressure plate and spring cannot be disassembled; replace the clutch cover as an assembly.

To install:

8. Inspect the clutch release bearing. If it is rough or noisy, it should be replaced. The bearing can be removed from the sleeve with a puller; this requires a press to install the new bearing. After installation, coat the sleeve groove, the release lever contact surfaces, the pivot pin/sleeve and the release bearing-to-transmission/transaxle contact surfaces with a light coat of grease. Be careful not to use too much grease, which will run at high temperatures and get onto the clutch facings. Reinstall the release bearing on the lever.

9. Apply a thin coat of grease to the pressure plate wire ring, diaphragm spring, clutch cover grooves and the pressure plate drive bosses.

10. Apply a thin coat of Lubriplate® to the splines in the driven plate. Slide the clutch disc onto the splines and move it back and forth several times. Remove the disc and wipe off the excess lubricant. Be very careful not to get any grease on the clutch facings.

11. Assemble the clutch cover and the clutch plate on the clutch alignment arbor.

12. To complete the installation, align the clutch assembly-to-flywheel alignment marks and install the bolts. Tighten the bolts 1 or 2 turns at a time in a crisscross pattern to avoid distorting the cover. Torque the bolts to 22 ft. lbs. (30 Nm).

13. Install the transmission/transaxle and adjust the pedal height as necessary.

PEDAL HEIGHT/FREE-PLAY ADJUSTMENT

The pedal height is the distance from the top of the clutch pedal to the floor board without the carpet. The pedal free-play is the distance the clutch pedal pad moves from the released position to the point where resistance is felt.

1. To adjust the pedal height:
   a. From under the dash, loosen the pedal stopper locknut.
   b. Turn the pedal stopper until the specified pedal height is obtained: 9.29–9.69 in. (236–246mm) with the 2.4L engine or 8.94–9.33 in. (227–236mm) with the 3.0L engine.
   c. After adjustment, torque the pedal stopper locknut to 16 ft. lbs. (22 Nm).
2. To adjust the pedal free-play, perform the following procedures:
   a. Loosen the clutch pedal, pushrod locknut.
   b. Using a ruler, measure the clutch pedal free-play.
   c. Turn the clutch pedal pushrod to set the free-play at 0.040–0.120 in. (1–3mm).
   d. After adjustment, torque the locknut to 9 ft. lbs. (12 Nm).

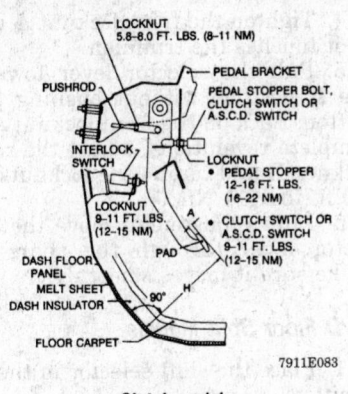

LOCKNUT
5.8–8.0 FT. LBS. (8–11 NM)

PUSHROD

PEDAL BRACKET

PEDAL STOPPER BOLT,
CLUTCH SWITCH OR
A.S.C.D. SWITCH

INTERLOCK
SWITCH

LOCKNUT
• PEDAL STOPPER
12–16 FT. LBS.
(16–22 NM)

LOCKNUT
9–11 FT. LBS.
(12–15 NM)

• CLUTCH SWITCH OR
A.S.C.D. SWITCH
9–11 FT. LBS.
(12–15 NM)

PAD

DASH FLOOR
PANEL

MELT SHEET

DASH INSULATOR

90°

FLOOR CARPET

7911E083

Clutch pedal

Clutch Master Cylinder

REMOVAL AND INSTALLATION

The master cylinder is attached to a bracket located under the dash.

1. Disconnect the negative battery cable.
2. From under the dash, remove the clevis pin snap pin and pull the clevis pin from the clutch pedal.
3. Disconnect the clutch pedal arm from the pushrod clevis. Remove the dust cover (boot) from the master cylinder body and pushrod. It will not go through the cowl without tearing.
4. Disconnect and plug the hydraulic line from the clutch master cylinder.

NOTE: Take precautions to keep brake fluid from coming in contact with any painted surfaces.

5. Remove the clutch master cylinder.
6. Installation is the reverse of removal. Torque the clutch master cylinder-to-cowl bolts/nuts to 9 ft. lbs. 12 Nm) on all others.
7. Bleed the clutch hydraulic system.

Clutch Slave Cylinder

REMOVAL AND INSTALLATION

1. If necessary, raise and safely support the vehicle.
2. Remove the slave cylinder-to-clutch housing bolts and the pushrod from the shift fork.
3. Disconnect and plug the hydraulic hose from the slave cylinder, then, remove the cylinder from the vehicle.
4. To install, reverse the removal procedures. Torque the slave cylinder-to-clutch housing bolts to 30 ft. lbs. (40 Nm).
5. Bleed the clutch hydraulic system.

Hydraulic Clutch System Bleeding

1. Check and refill the clutch fluid reservoir to the full mark. During the bleeding process, continue to check and replenish the reservoir to prevent the fluid level from getting lower than ½ full.
2. Connect a clear vinyl hose to the bleeder screw on the slave cylinder. Immerse the other end of the hose in a clear jar ½ filled with brake fluid.
3. Have an assistant pump the clutch pedal several times and hold it down. Loosen the bleeder screw slowly.
4. Tighten the bleeder screw and release the clutch pedal gradually. Repeat this operation until the air bubbles disappear from the brake fluid being expelled out through the bleeder screw.
5. When the air is completely removed, securely tighten the bleeder screw and replace the dust cap.
6. Check and refill the master cylinder reservoir as necessary.

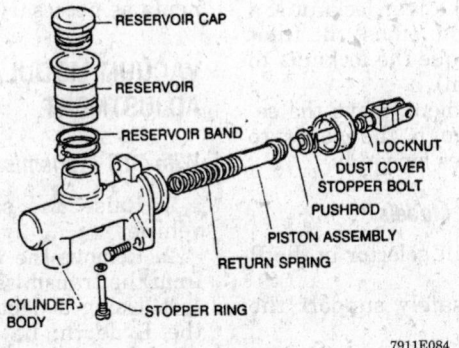

RESERVOIR CAP

RESERVOIR

RESERVOIR BAND

LOCKNUT

DUST COVER

STOPPER BOLT

PUSHROD

PISTON ASSEMBLY

RETURN SPRING

CYLINDER
BODY

STOPPER RING

7911E084

Clutch master cylinder assembly

7. Depress the clutch pedal several times to check the operation of the clutch and check for leaks.

AUTOMATIC TRANSMISSION

Transmission Assembly

REMOVAL AND INSTALLATION

1. Disconnect the negative battery cable.
2. Raise and safely support the vehicle. If equipped, remove the splash pan or skid plate.
3. Remove the starter.
4. Matchmark the driveshaft flange at the rear pinion flange and remove the driveshaft. Plug the extension housing opening to prevent fluid from leaking out.
5. If equipped with 4WD, matchmark both front driveshaft flanges so the shaft can be installed in the same position. Remove the driveshaft.
6. Disconnect the exhaust pipe from the manifolds and from the catalytic converter and remove the pipe.
7. Disconnect the speedometer cable and the wiring from the transmission.
8. Disconnect the selector lever and throttle cables from the transmission.
9. Remove the dipstick tube and disconnect the cooling lines from the transmission.
10. Remove the torque converter housing dust cover. Matchmark the torque converter with the driveplate for reassembly; these are balanced together at the factory. Remove the torque converter-to-driveplate (flywheel) bolts. Use a wrench on the crankshaft pulley bolt to rotate the crankshaft to expose the hidden torque converter bolts.
11. If equipped with 4WD, the torsion bars must be removed.

   a. Working under the vehicle, measure and record the length of the threads on the torsion bar adjustment. At the front of the bar, pull the boot back and matchmark the bar to the mounting plate. The spline on the bar must be re-installed in the same position on the plate.

   b. Remove the locknut and adjustment nut from both torsion bars. Remove the 3 nuts at the

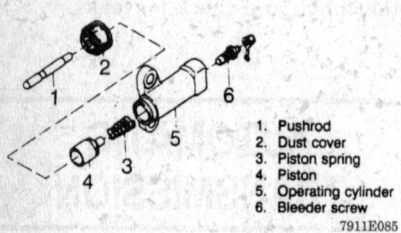

1. Pushrod
2. Dust cover
3. Piston spring
4. Piston
5. Operating cylinder
6. Bleeder screw

7911E085

Clutch slave cylinder

mounting plate and remove the bars. Mark the bars left and right side for proper installation.

c. Remove the transfer case shift lever assembly from the transfer case.

d. If necessary, the transfer case can be removed at this point.

e. If equipped with a 3.0L engine, remove the gusset securing the engine to the transmission.

12. Using a chain hoist, attach it to the engine and lift the engine slightly to take the weight off the mounts. Using an appropriate transmission jack, properly support the transmission and remove the transmission mount and crossmember.

13. Remove the transmission–to–engine bolts and move the transmission back away from the engine. Lower the transmission carefully from the vehicle.

To install:

14. Use a dial indicator to check the driveplate runout while turning the crankshaft. Maximum allowable runout is 0.020 in. (0.5mm); if beyond specification, replace the driveplate.

15. Measure and adjust how far the torque converter is recessed into the transmission housing. The distance between the front mounting surface of the transmission and the torque converter-to-driveplate bolt boss should be at least 1.024 in. (26mm).

16. Install the transmission and torque the 4 upper transmission–to–engine bolts to 36 ft. lbs. (49 Nm). Torque to transmission–to–engine gusset bolts to 29 ft. lbs. (39 Nm). Torque the remaining bolts to 18 ft. lbs. (25 Nm) on the 2.4L engine or 29 ft. lbs. (39 Nm) on the 3.0L engine.

17. Install the crossmember and torque the crossmember-to-chassis bolts to 38 ft. lbs. (52 Nm) on 2WD or 31 ft. lbs. (42 Nm) on 4WD.

18. Install the rear transmission mount bolts, loosen the engine mount

bolts, then torque all the mount bolts to 31 ft. lbs. (42 Nm), starting at the rear.

19. Align the matchmarks on the driveplate and torque converter, install the bolts and torque to 36 ft. lbs. (49 Nm). Turn the crankshaft after tightening the bolts to make sure there is no binding at the driveplate.

20. If the torsion bars were removed, install them in their original location. Make sure the splines are in their original position and set the adjustment to its original position.

21. If the transfer case was removed, apply silicone sealant to seal the case to the transmission. Install the transfer case and torque the bolts to 30 ft. lbs. (41 Nm). Install the shift lever.

22. When installing the driveshafts, be sure to align the matchmarks. Torque the bolts on the front driveshaft (4WD) to 33 ft. lbs. (44 Nm). On single piece rear driveshafts, torque the bolts to 65 ft. lbs. (88 Nm). On 2 piece driveshafts, torque the bolts to 33 ft. lbs. (44 Nm). Torque the center bearing bracket bolts to 16 ft. lbs. (22 Nm).

23. When installing the exhaust system, use new gaskets and torque the flange bolts to 27 ft. lbs. (36 Nm).

24. Install the remaining components in order of removal and connect the wiring, cooling lines and speedometer cable. Refill the transmission with fluid and adjust as required.

SHIFT LINKAGE ADJUSTMENT

2WD Floor Shift Models

1. Place the shift selector in the **P** position.
2. Raise and safely support the vehicle.
3. From under the vehicle, loosen the shift lever locknuts.
4. Tighten the rear locknut **X** until it touches the trunnion.
5. Pull the selector lever toward the **R** position without pushing the button. Back off the rear locknut **X** a complete revolution, adjust the front locknut **Y** and torque the locknuts to 8.0 ft. lbs. (11 Nm).
6. After adjustment, move the selector lever through the gears to make sure it moves smoothly.

2WD Column Shift Models

1. Place the shift selector in the **P** position.
2. Raise and safely support the vehicle.
3. From under the vehicle, loosen the shift lever locknuts.

4. Tighten the front locknut **A** until it touches the trunnion.
5. Pull the selector lever toward the **R** position without pushing the button. Back off the front locknut **A** 2 complete revolutions, adjust the rear locknut **B** and torque the locknuts to 8.0 ft. lbs. (11 Nm).
6. After adjustment, move the selector lever through the gears to make sure it moves smoothly.

4WD Floor Shift Models

1. Place the shift selector in the **P** position.
2. Raise and safely support the vehicle.
3. Remove the console cover.
4. Loosen the turn buckle locknuts.
5. Tighten the turn buckle until it aligns with the inner cable.
6. Pull the selector lever toward the **R** position without pushing the button. Back off the turn buckle a complete revolution, torque the locknuts to 48 inch lbs. (5 Nm).
7. After adjustment, move the selector lever through the gears to make sure it moves smoothly.

KICKDOWN SWITCH ADJUSTMENT

With 71B Transmission

A kickdown switch is located inside the vehicle at the upper post of the accelerator pedal. Its purpose is to provide transmission downshifting when the accelerator pedal is fully depressed; a click can be heard just before the pedal bottoms out.

With the ignition switch in the **ON** position and the engine **OFF**, when the accelerator pedal is depressed fully, the kick-down switch contacts should be closed and the downshift solenoid activated, emitting a clicking sound. If the components fail to operate in this manner, check for continuity at the switch and then at the solenoid. Replace either of the components as necessary.

VACUUM MODULATOR ADJUSTMENT

With 71B Transmission

1. Raise and safely support the vehicle.
2. Remove the vacuum modulator from the transmission.
3. Using a depth gauge, measure the **L** depth; be sure the vacuum throttle valve is pushed into the valve body as far as possible.

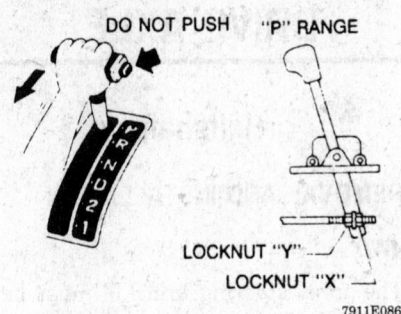

Automatic with 2WD and floor shift

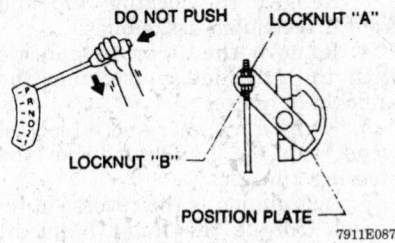

Automatic with 2WD and column shift

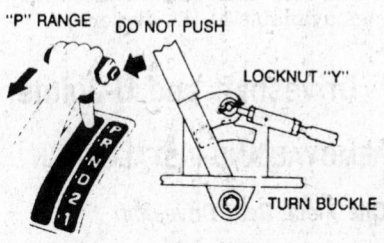

Automatic with 4WD and floor shift

4. Select the correct length rod and install it into the vacuum modulator.

5. Using a new O-ring, install the modulator into the transmission.

THROTTLE LINKAGE ADJUSTMENT

With RL4R01A Transmission

1. Press the lock plate and move the adjusting tube in the direction **T**.
2. Release the lock plate.

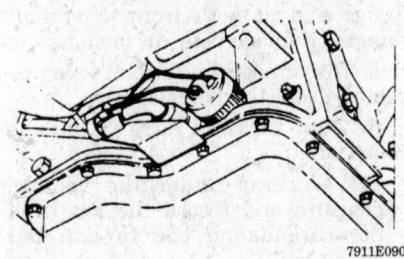

Downshift solenoid on with 71B transmission

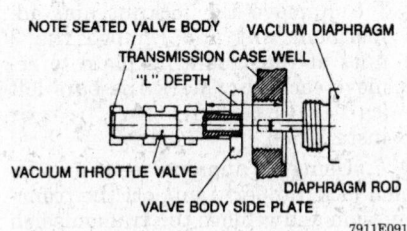

Vacuum diaphragm with 71B transmission

3. Quickly move the throttle drum from P-to-P.

4. Ensure the throttle wire stroke **L** is within specified range between full throttle and idle; the throttle wire stroke **L** should be 1.50–1.65 in. (38–42mm).

NOTE: Adjust the throttle wire stroke when the throttle wire/accelerator wire is installed. Place marks on the throttle wire to facilitate measuring the wire stroke.

5. If the throttle wire stroke is not adjusted, the following problems may arise:

a. When full-open position **P** of the throttle drum is closer to the direction **T**, the kickdown range will greatly increase.

b. When the full-open position **P** of the throttle drum is closer to the direction **U**, the kickdown range will not occur.

NEUTRAL SAFETY SWITCH ADJUSTMENT

The neutral safety switch is located on the transmission shift selector lever. The switch operates the back-up lights and controls the operation of the starter. The starter should only operate when the transmission is in **P** or **N**.

With 71B Transmission

1. Unscrew the securing nut of the shift selector lever and the switch-to-transmission screws.

2. Position the shift selector to the **N** position (in vertical detent position). Move the switch slightly aside so the screw hole will be aligned with the pin hole of the shift selector lever.

3. Using a 0.080 in. (2mm) diameter alignment pin, place it in the alignment holes of the neutral start switch and the shift selector lever.

NOTE: A No. 47 drill bit will substitute for the pin gauge.

4. Secure the switch body with the screws and pull out the pin.

NOTE: If the neutral safety switch does not perform satisfactorily after adjustment, replace it with a new one.

With "R" Series Transmission

1. Unscrew the securing nut of the shift selector lever and the switch-to-transmission screws.

Measured depth "L" mm (in)	Rod length mm (in)	Part number
Under 25.55 (1.0059)	29.0 (1.142)	31932-X0103
25.65 - 26.05 (1.0098 - 1.0256)	29.5 (1.161)	31932-X0104
26.15 - 26.55 (1.0295 - 1.0453)	30.0 (1.181)	31932-X0100
26.65 - 27.05 (1.0492 - 1.0650)	30.5 (1.201)	31932-X0102
Over 27.15 (1.0689)	31.0 (1.220)	31932-X0101

7911E092

Vacuum diaphragm rod selection chart — 71B transmission

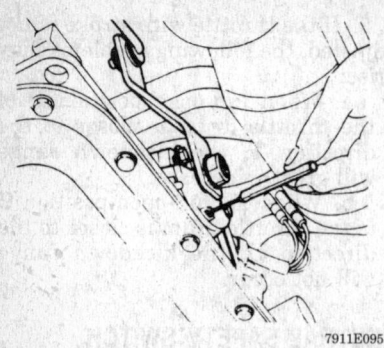

7911E095

Neutral safety switch alignment with 71B transmission

2. Position the shift selector to the **N** position (in vertical detent).

3. Using a 0.16 in. (4mm) diameter alignment pin, place it in the alignment holes of the neutral start switch and the shift selector lever.

4. Secure the switch body with the screws and pull out the pin.

NOTE: If the neutral safety switch does not perform satisfactorily after adjustment, replace it with a new one.

TRANSFER CASE

Transfer Case

REMOVAL AND INSTALLATION

1. Disconnect the negative battery cable.

2. Raise and safely support the vehicle. If equipped, remove the splash pan or skid plate.

3. Remove the starter. Drain the oil from both the transmission and the transfer case.

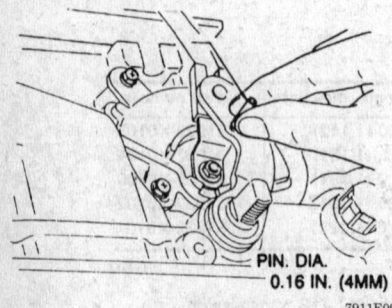

PIN. DIA.
0.16 IN. (4MM)

7911E096

Neutral safety switch alignment with "R" series transmission

4. Matchmark the driveshaft flange at the rear differential pinion flange and at both front driveshaft flanges. Remove both driveshafts.

5. Disconnect the selector lever assembly from the transfer case.

6. The torsion bars must be removed:

    a. Working under the vehicle, measure and record the length of the threads on the torsion bar adjustment.

    b. At the front of the bar, pull the boot back and matchmark the bar to the mounting plate. The spline on the bar must be re-installed in the same position on the plate.

    c. Remove the locknut and adjustment nut and remove the 3 nuts at the mounting plate to remove each bar. Mark the bars left and right side for proper installation.

7. Using an appropriate transmission jack, properly support the transmission and remove the transmission mount and crossmember.

8. Remove the transfer case–to–transmission bolts and move the unit back away from the transmission.

To install:

9. Clean the mating surfaces and apply a bead of silicone sealant to the transfer case mounting flange.

10. Carefully fit the case into place and start all the mounting bolts. Torque the bolts to 30 ft. lbs. (41 Nm).

11. Install the crossmember and torque the bolts to 58 ft. lbs. (78 Nm). Install the mount bolts and torque to 38 ft. lbs. (52 Nm).

12. Install the driveshafts and make sure to align the matchmarks:

    a. On the front driveshaft, torque the bolts to 33 ft. lbs. (44 Nm).

    b. On 2 piece rear driveshafts, torque the flange bolts to 33 ft. lbs. (44 Nm) and the center bearing bracket bolts to 16 ft. lbs. (22 Nm).

    c. On single piece rear driveshafts, torque the flange bolts to 65 ft. lbs. (88 Nm).

13. Install the selector lever assembly.

14. Install the torsion bars in their original location. Make sure the splines are in their original position and set the adjustment to its original position.

15. Install the remaining components and fill the transfer case and transmission with oil. Check and adjust front suspension height.

DRIVE AXLE

Halfshaft

REMOVAL AND INSTALLATION

4WD

The front steering knuckle must be removed to remove the halfshaft.

1. Raise and safely support the vehicle.

2. Have an assistant depress the brake pedal and remove the halfshaft-to-differential flange bolts.

3. Remove the locking hub and front drive clutch assemblies.

4. Remove the steering knuckle with the halfshaft and clamp the knuckle in a vise.

5. Using a hammer and a block of wood, tap the halfshaft from the steering knuckle.

6. Installation is the reverse of removal. Torque the inner halfshaft drive flange bolts to 33 ft. lbs. (44 Nm).

7. Measure halfshaft end-play with a dial indicator against the end of the shaft; it should be 0.004–0.012 in. (0.1–0.3mm). Endplay can be adjusted with different thickness snaprings available at the dealer.

Driveshaft and U-Joints

REMOVAL AND INSTALLATION

One Piece Rear Driveshaft

1. Raise and safely support the vehicle.

2. Matchmark the driveshaft flange to the pinion flange on the differential.

3. Remove the flange bolts, lower the driveshaft and pull it from the transmission.

4. Using a clean rag, plug the rear of the transmission to keep the oil from leaking out.

5. To install, insert the sleeve yoke into the transmission, align the matchmarks and fit the driveshaft into place. Torque the driveshaft flange nuts/bolts to 65 ft. lbs. (88 Nm).

Two Piece Rear Driveshaft

1. Raise and safely support the vehicle.

2. Matchmark the driveshaft flange to the pinion flange on the dif-

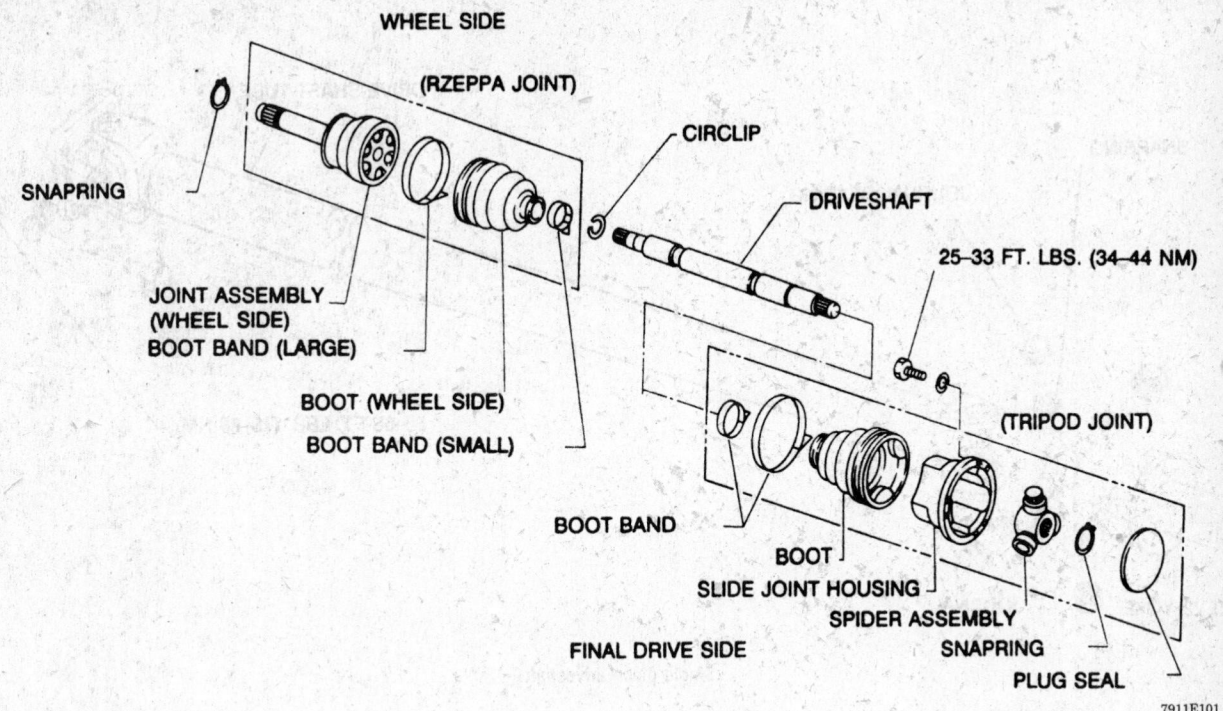

WHEEL SIDE

(RZEPPA JOINT)

CIRCLIP

DRIVESHAFT

25–33 FT. LBS. (34–44 NM)

SNAPRING

JOINT ASSEMBLY
(WHEEL SIDE)

BOOT BAND (LARGE)

BOOT (WHEEL SIDE)

BOOT BAND (SMALL)

(TRIPOD JOINT)

BOOT BAND

BOOT

SLIDE JOINT HOUSING

SPIDER ASSEMBLY

SNAPRING

PLUG SEAL

FINAL DRIVE SIDE

7911E101

Front halfshaft on 4WD with 2.4L engine

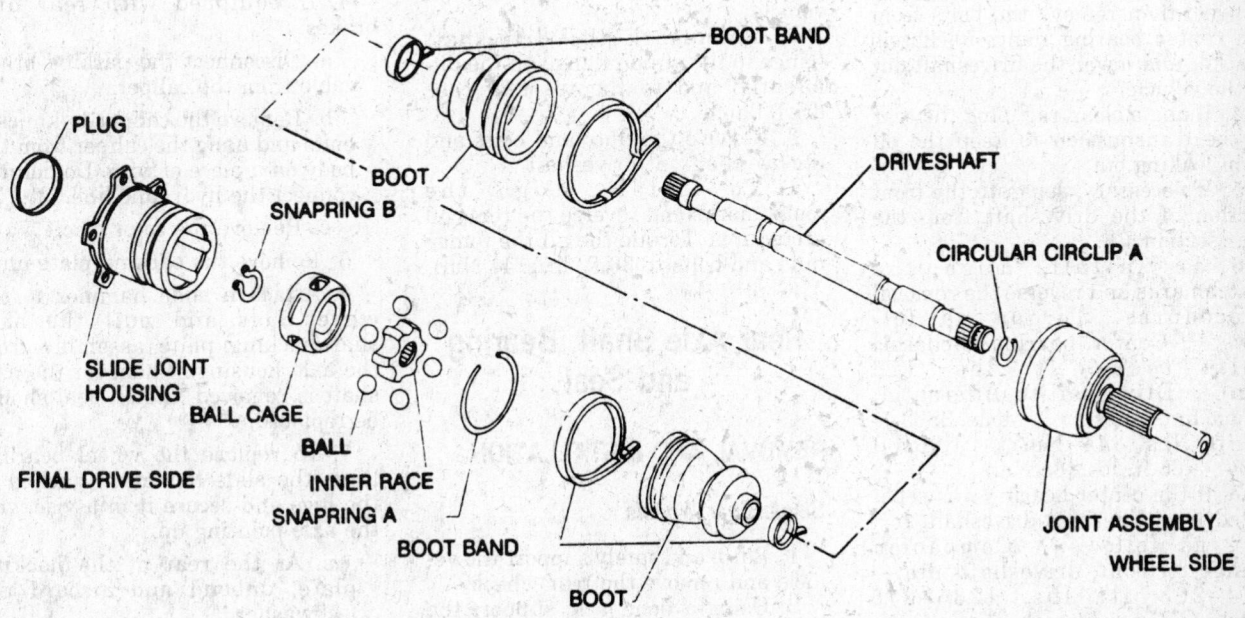

BOOT BAND

PLUG

BOOT

SNAPRING B

DRIVESHAFT

CIRCULAR CIRCLIP A

SLIDE JOINT
HOUSING

BALL CAGE

BALL

FINAL DRIVE SIDE

INNER RACE

SNAPRING A

BOOT BAND

JOINT ASSEMBLY

WHEEL SIDE

BOOT

7911E100

Front halfshaft on 4WD with 3.0L engine

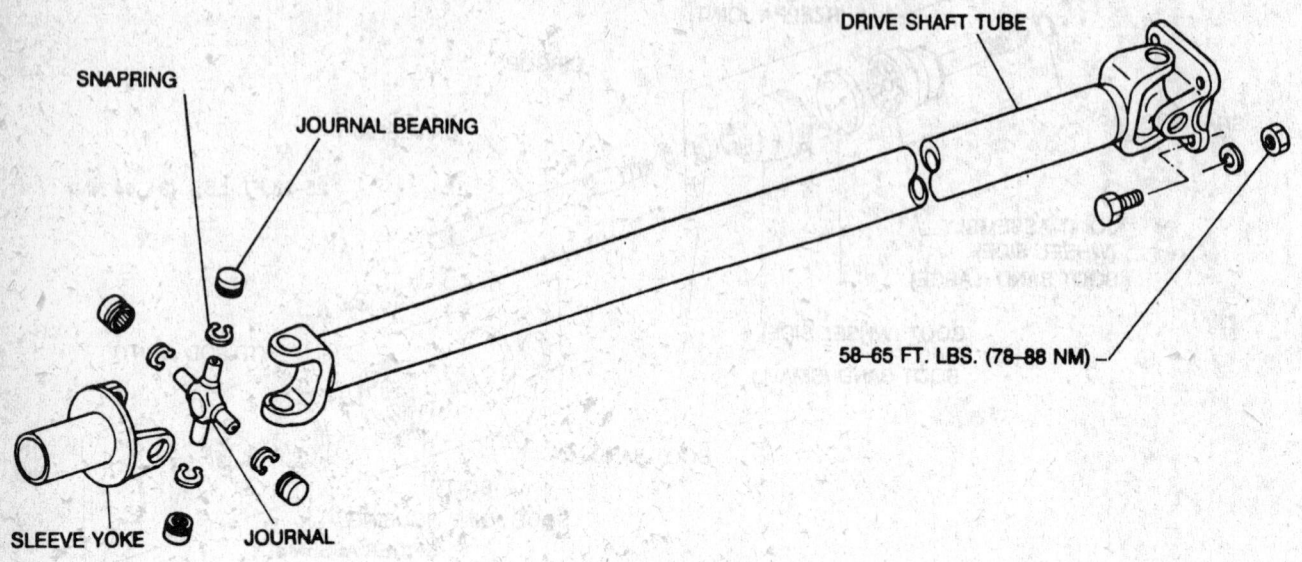

SNAPRING

JOURNAL BEARING

DRIVE SHAFT TUBE

58–65 FT. LBS. (78–88 NM)

SLEEVE YOKE

JOURNAL

Single piece driveshaft

7911E105

ferential and to the transmission drive flange.

3. Remove the bolts from both flanges, then remove the bolts from the center bearing flange-to-chassis bracket and lower the driveshaft out of the vehicle.

4. Using a clean rag, plug the rear of the transmission to keep the oil from leaking out.

5. If necessary, separate the front section of the driveshaft from the rear section.

6. To install, align the matchmarks and reverse the removal procedures. Torque as follows: Center bearing-to-chassis bolts to 16 ft. lbs. (22 Nm) Driveshaft-to-differential flange nuts/bolts Model 3S63 — 33 ft. lbs. (44 Nm) Model 3S80 — 65 ft. lbs. (88 Nm)

7. If the center bearing was separated from the front driveshaft, torque as follows: Companion flange-to-front driveshaft nut — 174–203 ft. lbs. (235–275 Nm) Rear driveshaft-to-center bearing flange nuts/bolts Model 3S63 — 17–24 ft. lbs. (24–32 Nm) Model 3S71H — 29–33 ft. lbs. (39–44 Nm) Model 3S80 — 58–65 ft. lbs. (78–88 Nm)

4WD Front Driveshaft

1. Raise and safely support the vehicle.

2. Matchmark the driveshaft flange to the pinion flange on the differential and to the transfer case drive flange.

3. Remove the nuts and bolts and remove the front driveshaft.

4. To install, align the matchmarks and reverse the removal procedures. Torque the all the flange nuts and bolts to 33 ft. lbs. (44 Nm).

Rear Axle Shaft, Bearing and Seal

REMOVAL AND INSTALLATION

Single Rear Wheels

1. Raise and safely support the vehicle and remove the rear wheels.

2. Using a floor jack, support the differential.

3. If equipped with rear drum brakes:

  a. Remove the brake drum.

  b. Disconnect the parking brake cable from the brake shoes.

  c. Disconnect and plug the hydraulic line from wheel cylinder.

  d. Remove the brake shoe assembly.

4. If equipped with rear disc brakes:

  a. Disconnect the parking brake cable from the caliper.

  b. Remove the caliper-to-knuckle bolts and hang the caliper from the body on a piece of wire. Do not disconnect the hydraulic line.

  c. Remove the rotor disc.

5. Remove the backing plate nuts.

6. Attach a slide hammer to the wheel lugs and pull the axle shaft/backing plate assembly from the axle housing. Whenever the axle shaft is removed, the oil seal should be replaced.

7. To replace the wheel bearing, leave the slide hammer attached to the lugs and secure it in a vise with the axle pointing up.

  a. At the rear of the backing plate, unbend and discard the lockwasher.

  b. Using a brass drift and a hammer, loosen and remove the locknut.

  c. Using a shop press, press the axle shaft out of the bearing.

  d. Press the bearing out of the bearing housing.

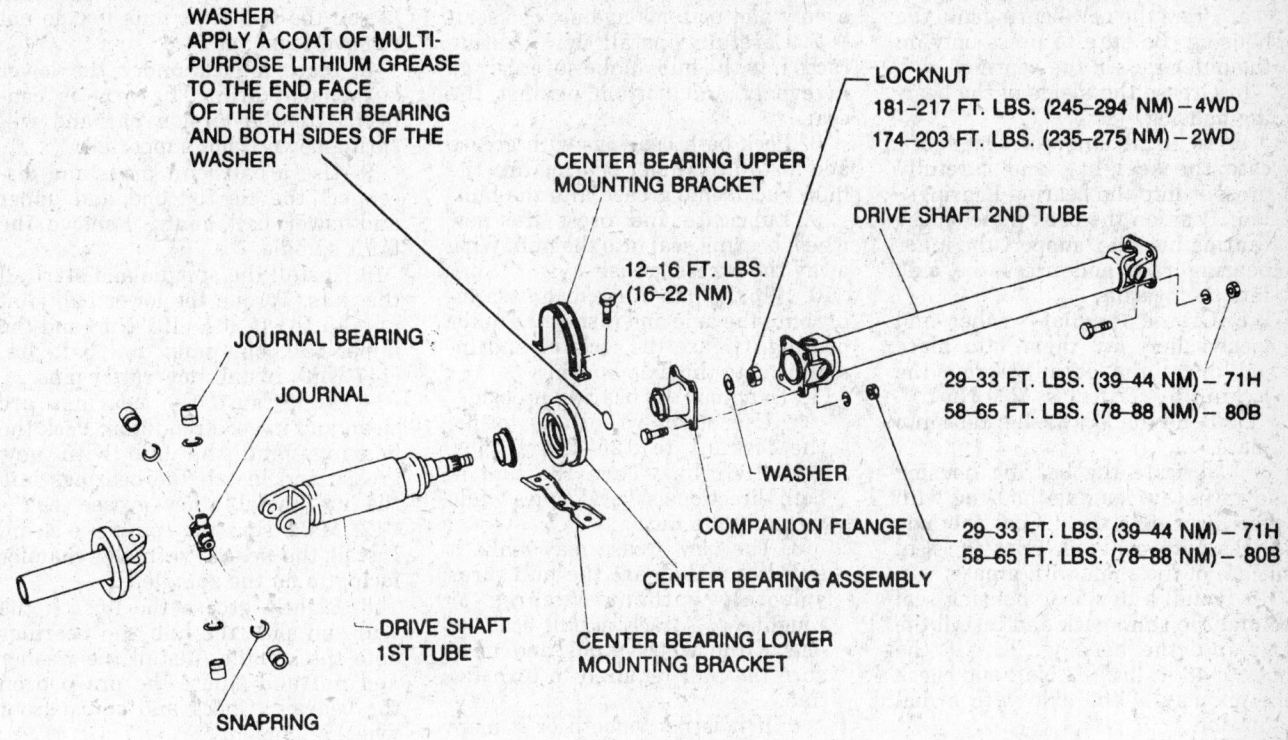

WASHER
APPLY A COAT OF MULTI-
PURPOSE LITHIUM GREASE
TO THE END FACE
OF THE CENTER BEARING
AND BOTH SIDES OF THE
WASHER

LOCKNUT
181–217 FT. LBS. (245–294 NM) – 4WD
174–203 FT. LBS. (235–275 NM) – 2WD

CENTER BEARING UPPER
MOUNTING BRACKET

DRIVE SHAFT 2ND TUBE

12–16 FT. LBS.
(16–22 NM)

JOURNAL BEARING

JOURNAL

29–33 FT. LBS. (39–44 NM) – 71H
58–65 FT. LBS. (78–88 NM) – 80B

WASHER

COMPANION FLANGE

29–33 FT. LBS. (39–44 NM) – 71H
58–65 FT. LBS. (78–88 NM) – 80B

CENTER BEARING ASSEMBLY

DRIVE SHAFT
1ST TUBE

CENTER BEARING LOWER
MOUNTING BRACKET

SNAPRING

7911E106

Two piece driveshaft

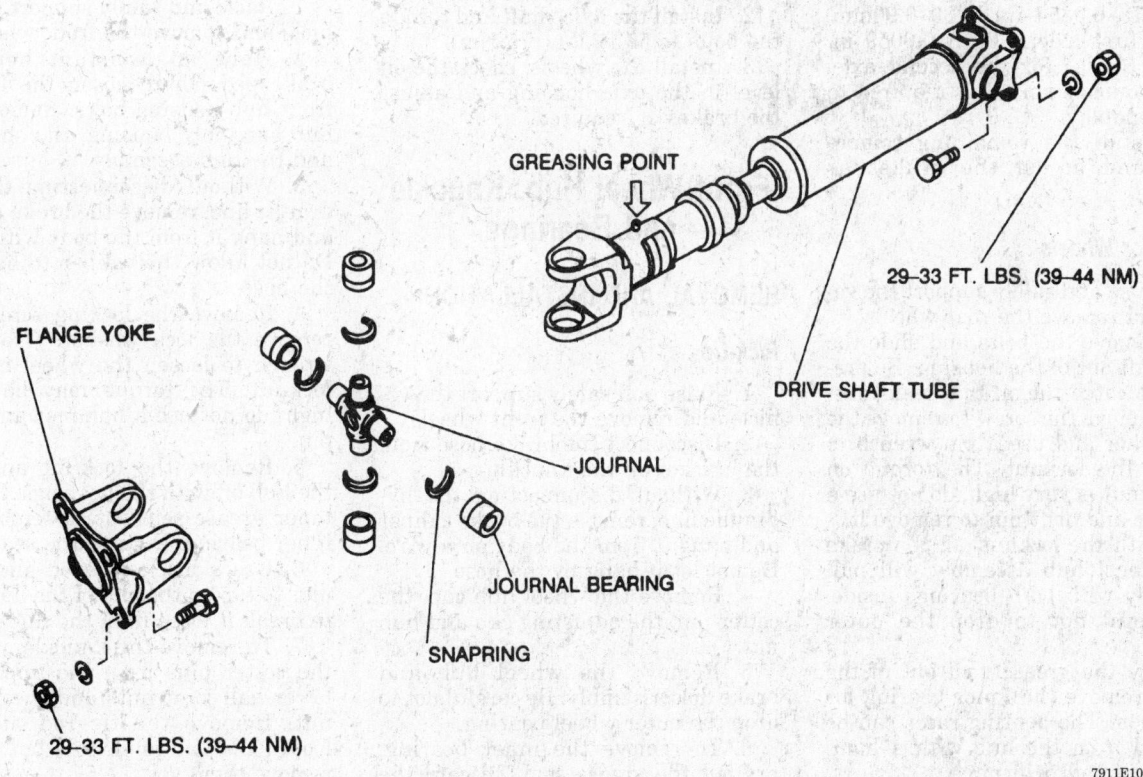

GREASING POINT

29–33 FT. LBS. (39–44 NM)

DRIVE SHAFT TUBE

FLANGE YOKE

JOURNAL

JOURNAL BEARING

SNAPRING

29–33 FT. LBS. (39–44 NM)

7911E107

Front driveshaft used on 4WD

8. To install the wheel bearing:

a. Press the new bearing into the housing. Be sure to press only on the outer race of the bearing.

b. Grease the inside of the bearing housing.

c. To install a new oil seal, lubricate the seal lips and carefully press it into the bearing housing.

d. Position the back plate on the bearing housing, support the inner bearing race and press the axle into the bearing.

e. Grease the flat washer and lockwasher, lay them into place and install the locknut. Torque the locknut to 217 ft. lbs. (294 Nm).

f. Bend the lockwasher tabs into place.

9. Lubricate the bearing housing and recess in the axle housing with wheel bearing grease. Coat the axle splines with gear oil. Coat the seal surface of the shaft with grease.

10. Install a new axle housing seal behind the shim pack and install the axle into the housing. Torque the nuts to 46 ft. lbs. (63 Nm) and check the endplay of the axle with a dial indicator.

11. The axle endplay should be 0.0008–0.0059 in. (0.02–0.15mm) when servicing only one axle. When servicing both sides, endplay should be 0.0118–0.0354 in. (0.30–0.90mm) for the first axle, 0.0008–0.0059 in. (0.02–0.15mm) for the second axle. Add or remove shims as required to adjust endplay.

12. Install the remaining components and adjust the brakes as required.

Dual Rear Wheels

1. Raise and safely support the vehicle and remove the rear wheels.

2. Remove the bolts and slide the axle shaft out of the housing. Be prepared to catch the oil that leaks out.

3. Remove the screw to remove the lockwasher and use a pin wrench to remove the locknut. The torque on the locknut is very high, do not use a hammer and drift pin to remove it.

4. With the locknut off, the brake drum/wheel hub assembly will pull off easily with both bearings inside. Be careful not to drop the outer bearing.

5. Pry the grease seal out of the hub to remove the inner bearing for inspection. The bearing races can be removed from the hub with a hammer and a soft drift pin.

6. To replace the oil seal, pry the old seal out of the axle housing. Lubricate and carefully install the new seal with an appropriate seal instal-

lation tool. Make sure it goes in evenly and bottoms against the seat.

7. Carefully install new bearing races into the hub. Make sure they go in evenly and bottom against the seat.

8. Pack both bearings with grease and install the inner bearing into the hub. Pack some grease into the hub.

9. Lubricate and press the new wheel bearing seal into the hub. Wipe away the excess grease.

10. Slip the hub/brake drum assembly onto the axle and install the outer bearing. Grease the locknut and install it onto the axle housing.

11. To adjust the bearing preload:

a. Use a pin wrench to torque the locknut to 125–145 ft. lbs. (167–196 Nm). Turn the hub in both directions several times while torquing the nut.

b. The new grease may make it stiff but make sure the hub turns smoothly without catching or roughness. Attach a pull scale to one of the wheel studs and measure the pull required to turn the hub.

c. If it is not smooth or if more than 4.7 lbs. of pull is required to turn the hub, remove the hub and look for improperly installed bearing races or dirt in the bearings.

12. Install the axle shaft and torque the bolts to 55 ft. lbs. (75 Nm).

13. Install the wheels, check the oil level in the axle housing and adjust the brakes as required.

Front Wheel Hub, Knuckle and Bearings

REMOVAL AND INSTALLATION

Pick-Up

1. Raise and safely support the vehicle and remove the front wheels.

2. Disconnect the brake hose from the bracket on the knuckle.

3. Without disconnecting the hydraulic line, remove the brake caliper and hang it from the body on a wire. Do not let it hang by the hose.

4. Remove the wheel hub cup, the cotter pin, the adjusting cap and hub nut.

5. Remove the wheel hub and brake disc assembly. Be careful not to drop the outer wheel bearing.

6. To remove the inner bearing, pry out the grease seal. Discard the seal.

7. To remove the knuckle, remove the cotter pins from the upper and lower ball joint nuts and the tie rod

nut. Remove the tie rod nut and loosen the ball joint nuts but do not remove them yet.

8. Place a jack under the lower suspension arm. The arm is connected to the torsion bar and will spring down if not supported.

9. Use a ball joint press and disconnect the tie rod end and upper and lower ball joints. Remove the front spindle.

10. Install the spindle and start all the nuts. Torque the lower ball joint nuts to 141 ft. lbs. (191 Nm) and the upper ball joint nuts to 108 ft. lbs. (147 Nm). Install new cotter pins.

11. Make sure the bearings are clean and in good condition. Pack the bearings and the hub with new grease and install the bearings into the hub. Install a new grease seal.

12. Make sure the spindle is clean. Install the spacer with the chamfer facing in on the spindle.

13. Lightly grease the lips on the seal and slide the hub and bearings onto the spindle. Install the washer and nut and adjust the pre-load on the bearing. Check and adjust front wheel alignment.

Pathfinder

1. Raise and safely support the vehicle and remove the front wheels.

2. Have an assistant hold the brake pedal and loosen the locking front hub housing bolts. Remove the hub assembly housing, the snapring and the hub assembly.

3. Without disconnecting the hydraulic line, remove the brake caliper and hang it from the body with wire. Do not allow the caliper to hang by the hose.

4. Remove the locking screw and remove the lock washer. Use a pin wrench to loosen the wheel bearing locknut. The torque may be fairly high, do not use a hammer and drift pin.

5. Remove the locknut and pull the hub off with the bearings. Pry the inner grease seal out to remove the inner bearing. Discard the seal.

6. Use a block of wood and hammer to tap on the end of the halfshaft to break it loose from the hub spline.

7. To remove the knuckle, remove the cotter pins from the upper and lower ball joint nuts and the tie rod nut. Remove the tie rod nut and loosen the ball joint nuts but do not remove them yet.

8. Place a jack under the lower suspension arm. The arm is connected to the torsion bar and will spring down if not supported.

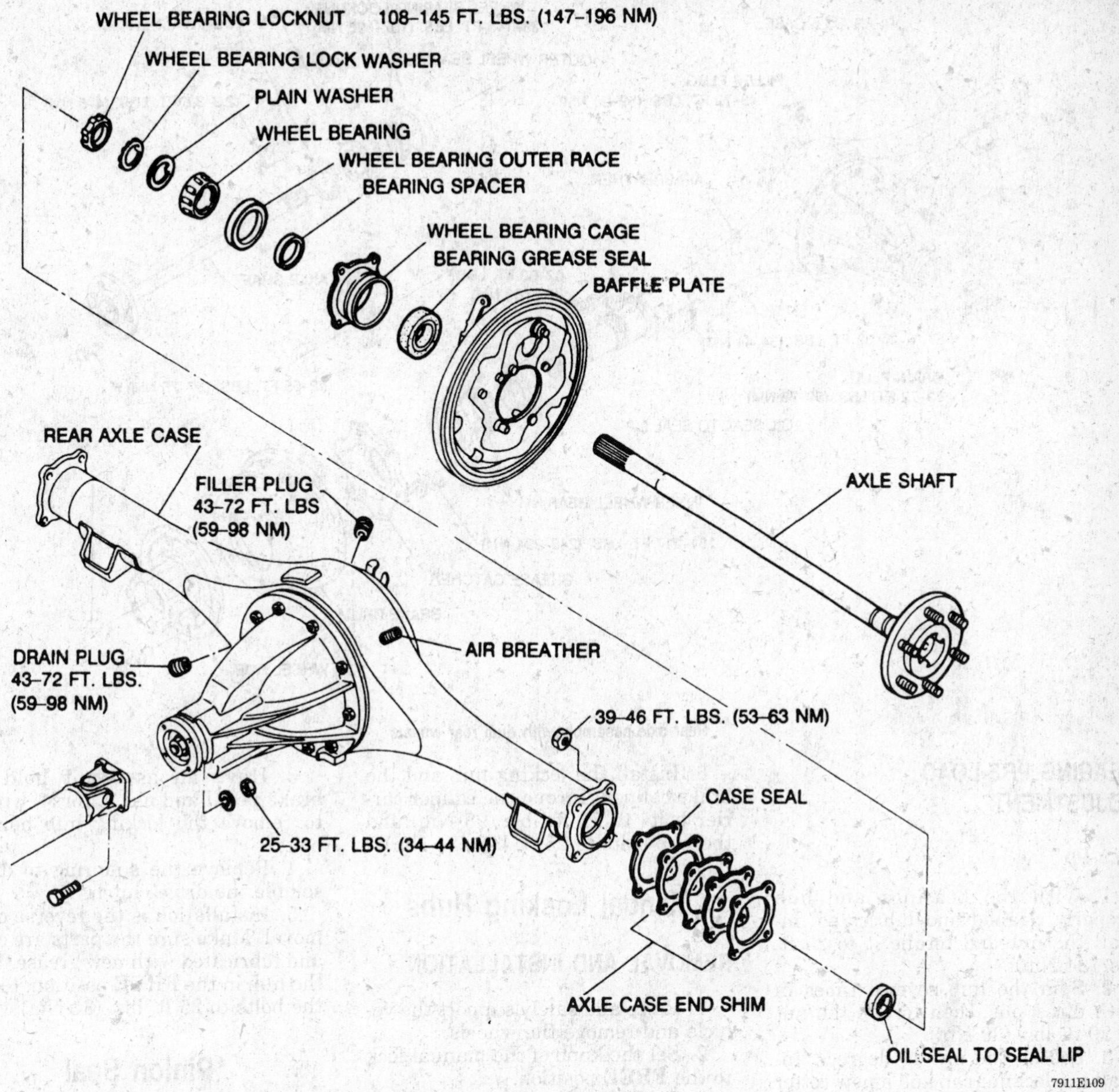

WHEEL BEARING LOCKNUT   108–145 FT. LBS. (147–196 NM)

WHEEL BEARING LOCK WASHER

PLAIN WASHER

WHEEL BEARING

WHEEL BEARING OUTER RACE

BEARING SPACER

WHEEL BEARING CAGE

BEARING GREASE SEAL

BAFFLE PLATE

REAR AXLE CASE

FILLER PLUG
43–72 FT. LBS.
(59–98 NM)

AXLE SHAFT

DRAIN PLUG
43–72 FT. LBS.
(59–98 NM)

AIR BREATHER

39–46 FT. LBS. (53–63 NM)

CASE SEAL

25–33 FT. LBS. (34–44 NM)

AXLE CASE END SHIM

OIL SEAL TO SEAL LIP

7911E109

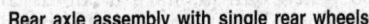

Rear axle assembly with single rear wheels

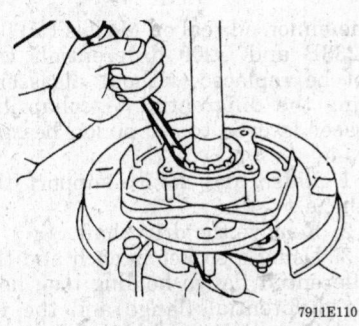

7911E110

Unbend the lockwasher and remove the locknut, then press the bearing off, with single rear wheels

9. Use a ball joint press and disconnect the tie rod end and upper and lower ball joints. Remove the front spindle.

10. Clean all parts of grease and check the condition of the bearings. If bearings are to be replaced, the inner races can be removed from the hub with a hammer and soft drift pin. Be careful not to damage the hub.

11. Install the spindle and start all the nuts. Torque the lower ball joint nuts to 141 ft. lbs. (191 Nm) and the upper ball joint nuts to 108 ft. lbs. (147 Nm). Install new cotter pins.

12. Connect the tie rods and torque the nuts to 72 ft. lbs. (98 Nm). Install new cotter pins.

13. Carefully install the new inner races with the drift pin, making sure they seat in the hub.

14. Pack the bearings with new grease and pack grease into the hub. Install the inner bearing and press a new inner seal into the hub.

15. Slip the hub assembly onto the spindle and install the outer bearing. Grease the locknut, thread it into place and set the bearing pre-load. Check and adjust front wheel alignment.

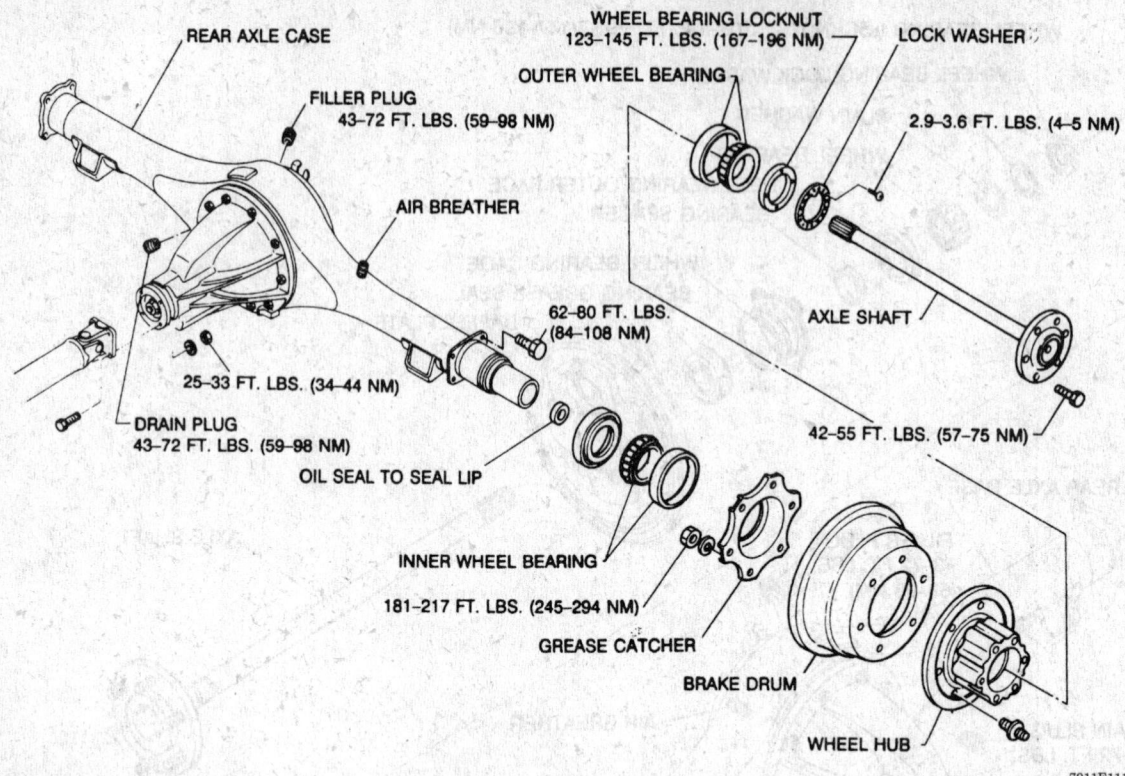

REAR AXLE CASE

FILLER PLUG
43–72 FT. LBS. (59–98 NM)

WHEEL BEARING LOCKNUT
123–145 FT. LBS. (167–196 NM)

OUTER WHEEL BEARING

LOCK WASHER

2.9–3.6 FT. LBS. (4–5 NM)

AIR BREATHER

62–80 FT. LBS.
(84–108 NM)

AXLE SHAFT

25–33 FT. LBS. (34–44 NM)

DRAIN PLUG
43–72 FT. LBS. (59–98 NM)

OIL SEAL TO SEAL LIP

42–55 FT. LBS. (57–75 NM)

INNER WHEEL BEARING

181–217 FT. LBS. (245–294 NM)

GREASE CATCHER

BRAKE DRUM

WHEEL HUB

7911E111

Rear axle assembly with dual rear wheels

BEARING PRE-LOAD ADJUSTMENT

Pick-Up

1. With the bearings and hub properly cleaned and lubricated, install the nut and torque it to 25 ft. lbs. (34 Nm).
2. Spin the hub several times in both directions, then torque the nut to 29 ft. lbs. (39 Nm).
3. Loosen the nut 45 degrees. Install the locknut cap and a new cotter pin.

Pathfinder

1. Use a pin wrench to torque the locknut to 58–72 ft. lbs. (78–98 Nm). Turn the hub in both directions several times while torquing the nut.
2. Loosen the locknut, then torque again to 13 inch lbs. (1.5 Nm).
3. Turn the hub several times and check the nut torque again.
4. Install the lock washer. When installing the screw, make sure the locknut turns no more than 30 degrees in either direction.
5. When bearing pre-load is properly set, there will be no endplay in the hub and it will require no more than 4.7 lbs. of pull at the wheel stud to turn the hub.

6. Install the locking hub and the brake caliper. Torque the caliper carrier bolts to 72 ft. lbs. (98 Nm) and the hub bolts to 25 ft. lbs. (34 Nm).

Manual Locking Hubs

REMOVAL AND INSTALLATION

1. Raise and safely support the vehicle and remove the wheels.
2. Set the knob of the manual lock to the **FREE** position.
3. Have an assistant hold the brake pedal and use a Torx® wrench to remove the locking hub housing bolts.
4. Remove the snapring to disassemble the drive clutch.
5. Installation is the reverse of removal. Make sure the parts are clean and lubricated with new grease. With the hub in the **FREE** position, torque the bolts to 25 ft. lbs. (34 Nm).

Automatic Locking Hubs

REMOVAL AND INSTALLATION

1. Raise and safely support the vehicle and remove the wheels.
2. Set the knob of the manual lock to the **FREE** position.

3. Have an assistant hold the brake pedal and use a Torx® wrench to remove the locking hub housing bolts.
4. Remove the snapring to disassemble the drive clutch.
5. Installation is the reverse of removal. Make sure the parts are clean and lubricated with new grease. With the hub in the **FREE** position, torque the bolts to 25 ft. lbs. (34 Nm).

Pinion Seal

REMOVAL AND INSTALLATION

The pinion oil seal on Models H190A, H233B and C200 differentials can not be replaced without disassembling the differential. A collapsible spacer is used to set pinion bearing pre-lo

1. Raise and safely support the vehicle.
2. Remove the driveshaft.
3. Using a socket wrench and the differential flange holding tool, hold the differential flange and the remove the differential pinion nut.
4. Using a wheel puller tool, pull the pinion flange from the differential.
5. Using a small prybar, pry the oil seal from the differential.

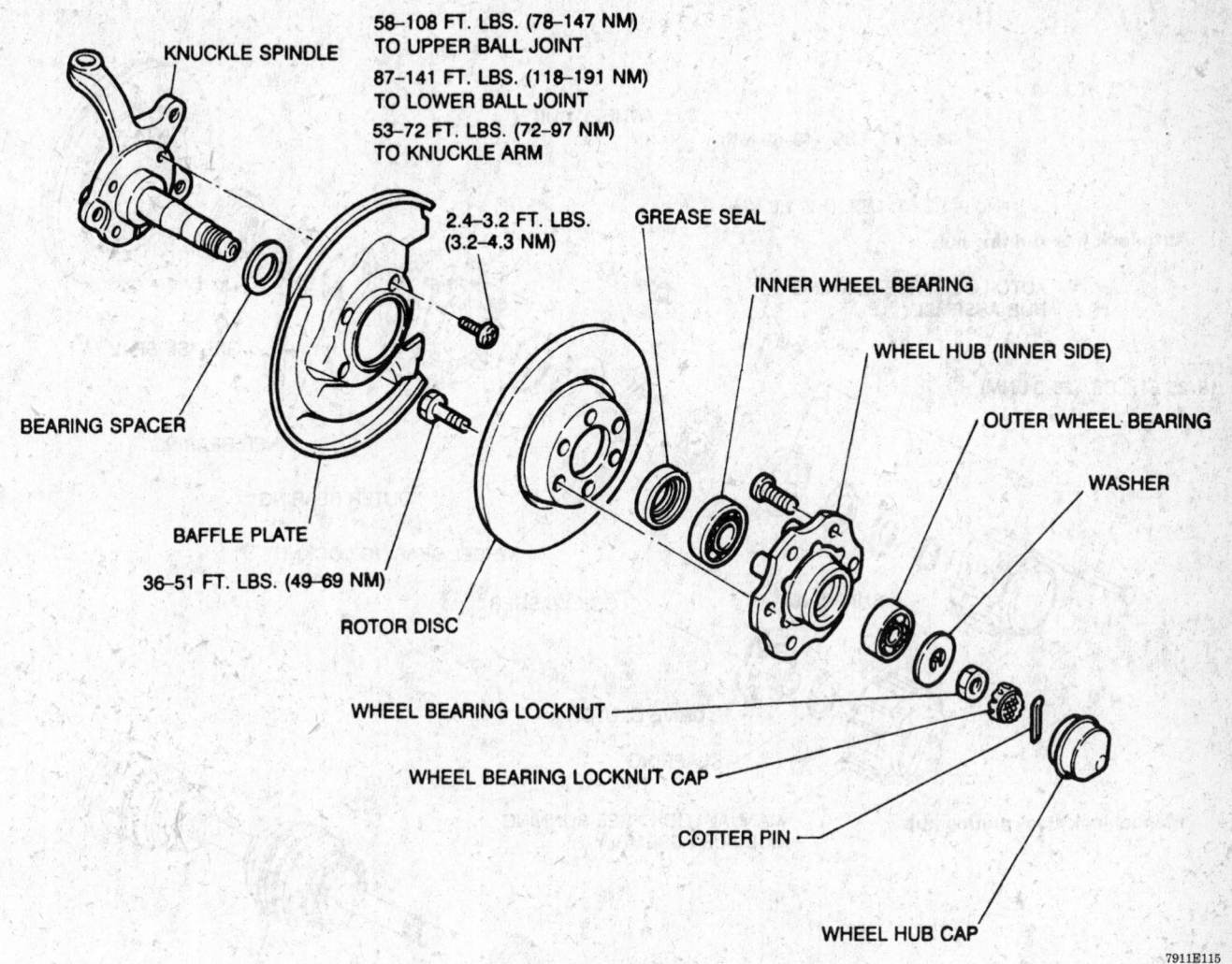

KNUCKLE SPINDLE

58–108 FT. LBS. (78–147 NM)
TO UPPER BALL JOINT

87–141 FT. LBS. (118–191 NM)
TO LOWER BALL JOINT

53–72 FT. LBS. (72–97 NM)
TO KNUCKLE ARM

2.4–3.2 FT. LBS.
(3.2–4.3 NM)

GREASE SEAL

INNER WHEEL BEARING

WHEEL HUB (INNER SIDE)

OUTER WHEEL BEARING

WASHER

BEARING SPACER

BAFFLE PLATE

36–51 FT. LBS. (49–69 NM)

ROTOR DISC

WHEEL BEARING LOCKNUT

WHEEL BEARING LOCKNUT CAP

COTTER PIN

WHEEL HUB CAP

7911E115

Front wheel bearing, with 2WD

6. Using the oil seal driver tool (Model R180A and H190A differentials), lubricate the new oil seal lips with multi-purpose grease and drive the new seal into the differential housing until it is flush with the end of the housing.

7. Using a soft hammer, tap the pinion flange onto the pinion shaft.

8. Using a socket wrench and the differential flange holding tool (Model H233B), hold the differential flange and torque the pinion flange nut as follows: Model R180 — 123–145 ft. lbs. (166–196 Nm) Model R190A — 94–217 ft. lbs. (127–294 Nm) Model

H233B — 145–181 ft. lbs. (196–245 Nm)

9. Check the oil level in the differential and install the driveshaft and wheels.

Differential Carrier

REMOVAL AND INSTALLATION

4WD

FRONT

1. Raise and safely support the vehicle. Drain the differential.

2. Matchmark and disconnect the front halfshafts from the front differential.

3. Matchmark the front driveshaft to the flanges and remove the driveshaft.

4. Attach a chain hoist to the engine, remove the front engine mount bolts and raise the engine slightly.

5. Remove the differential crossmember-to-chassis bolts and lower the differential with the crossmember as an assembly.

6. Fit the differential without the crossmember into place and start all the bolts. Torque all nuts and bolts to

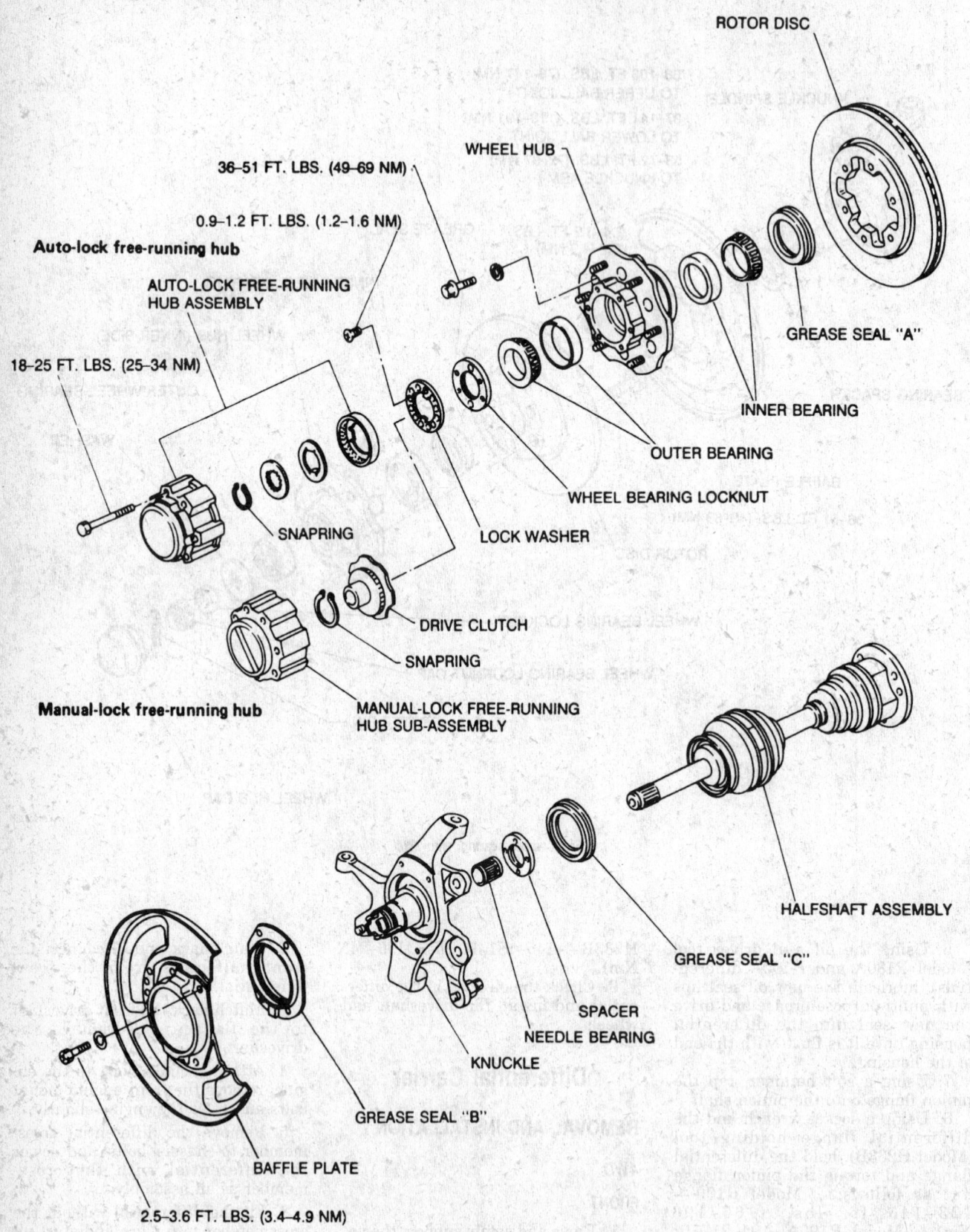

ROTOR DISC

WHEEL HUB

36–51 FT. LBS. (49–69 NM)

0.9–1.2 FT. LBS. (1.2–1.6 NM)

Auto-lock free-running hub

AUTO-LOCK FREE-RUNNING HUB ASSEMBLY

GREASE SEAL "A"

18–25 FT. LBS. (25–34 NM)

INNER BEARING

OUTER BEARING

WHEEL BEARING LOCKNUT

SNAPRING

LOCK WASHER

DRIVE CLUTCH

SNAPRING

Manual-lock free-running hub

MANUAL-LOCK FREE-RUNNING HUB SUB-ASSEMBLY

HALFSHAFT ASSEMBLY

GREASE SEAL "C"

SPACER

NEEDLE BEARING

KNUCKLE

GREASE SEAL "B"

BAFFLE PLATE

2.5–3.6 FT. LBS. (3.4–4.9 NM)

7911E116

Front wheel bearing and hub assembly on Pathfinder with 2WD and Pick-Up and Pathfinder with 4WD

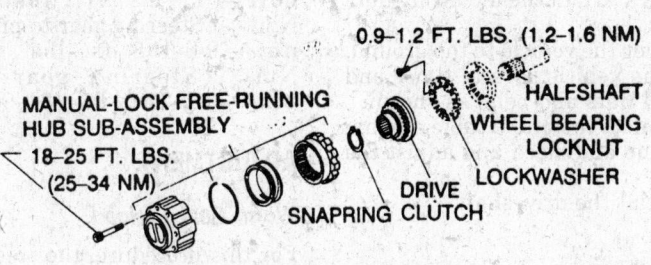

0.9–1.2 FT. LBS. (1.2–1.6 NM)

MANUAL-LOCK FREE-RUNNING HUB SUB-ASSEMBLY
18–25 FT. LBS. (25–34 NM)

HALFSHAFT

WHEEL BEARING LOCKNUT

LOCKWASHER

SNAPRING DRIVE CLUTCH

7911E121

Manual locking hubs used on 4WD

64 ft. lbs. (87 Nm) in the following sequence to avoid excess vibration:

a. First tighten the differential mounts to the frame.

b. Tighten the 2 long differential mount bolts.

c. Install the crossmember and tighten the differential mount bolts.

d. Tighten the crossmember mount bolts.

7. Install the front driveshaft and halfshafts, making sure to align the matchmarks.

REAR

1. Raise and safely support the vehicle.

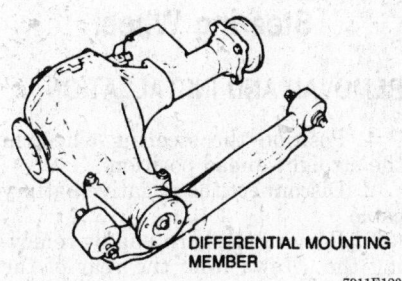

DIFFERENTIAL MOUNTING MEMBER

7911E123

When removing the front differential from, remove the crossmember

2. Matchmark the driveshaft to the differential flange and remove the driveshaft.

3. Drain all fluid from the differential carrier and remove the axle shafts.

4. Remove the differential carrier-to-axle housing bolts and remove the carrier.

5. Installation is the reverse of removal. Be sure the gasket is correctly installed and torque the bolts to 18 ft. lbs. (25 Nm).

6. Install the remaining components and fill the differential with oil.

Axle Housing

REMOVAL AND INSTALLATION

Pick-Up

1. Block the front wheels.

2. Raise and safely support the vehicle. Using a floor jack, position it under the differential and support its weight.

3. Remove the rear wheel/tire assemblies.

4. Matchmark the driveshaft to the differential flange and remove the driveshaft.

5. If equipped with drum brakes, remove the brake drum and discon-

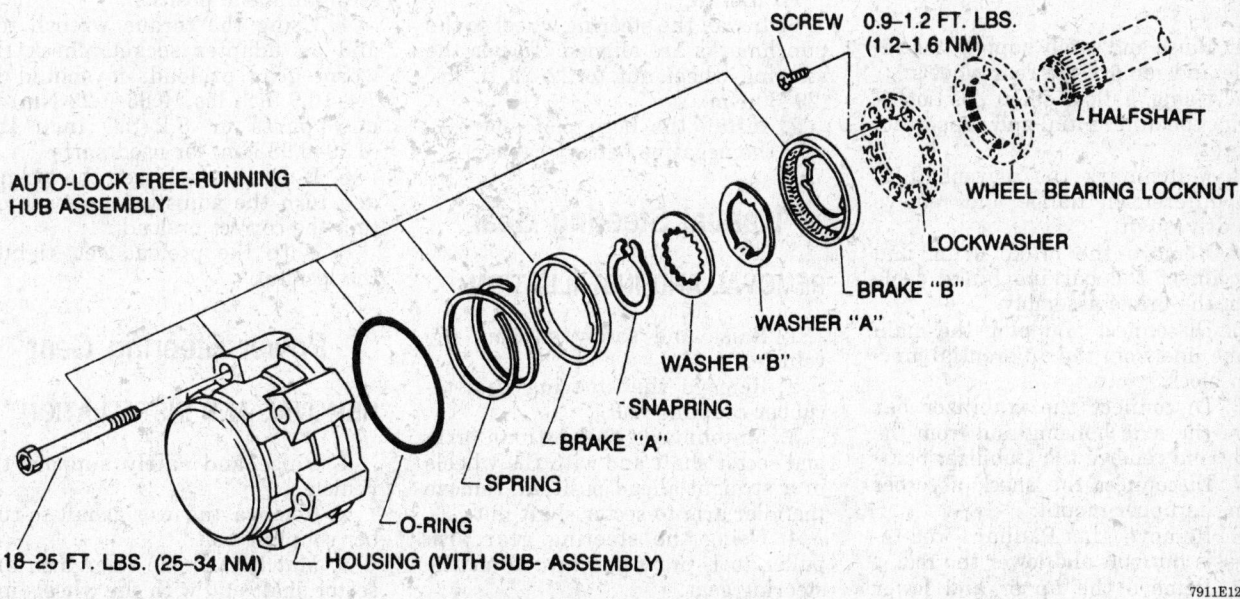

SCREW 0.9–1.2 FT. LBS. (1.2–1.6 NM)

HALFSHAFT

WHEEL BEARING LOCKNUT

LOCKWASHER

BRAKE "B"

AUTO-LOCK FREE-RUNNING HUB ASSEMBLY

WASHER "A"

WASHER "B"

SNAPRING

BRAKE "A"

SPRING

O-RING

18–25 FT. LBS. (25–34 NM)   HOUSING (WITH SUB- ASSEMBLY)

7911E122

Automatic locking hubs used on 4WD

nect the parking brake cable from the brake assembly. If equipped with disc brakes, disconnect the parking brake cable from the caliper.

6. Disconnect and plug the brake hydraulic lines from the wheel cylinders or calipers. Disconnect the brake line from the retaining clips.

7. Disconnect the shock absorber from the lower mount.

8. On vehicles with 2WD, remove all 4 leaf spring mount bolts and remove the axle housing and leaf springs together. Remove the U-bolts to remove the springs from the axle housing.

9. On vehicles with 4WD, remove the nuts from the U-bolts and lower the axle housing away from the leaf springs.

10. Raise the axle housing into position and start all the mounting bolts and nuts.

11. On 2WD vehicles, torque the rear leaf spring mount nuts to 50 ft. lbs. (68 Nm), then the front mount nuts to 72 ft. lbs. (98 Nm).

12. On 4WD vehicles, torque the U-bolt nuts to 72 ft. lbs. (98 Nm).

13. Connect the shock absorbers to the lower mounts and torque the nut to 30 ft. lbs. (40 Nm).

14. Connect the brake parking cable and hydraulic line and bleed and adjust the brakes.

15. Install the driveshaft with the matchmarks aligned.

Pathfinder

1. Raise and safely support the vehicle and remove the rear wheels.

2. Using a floor jack, position it under the differential and support its weight.

3. Matchmark the driveshaft to the differential flange and remove the driveshaft.

4. Remove the brake drum and disconnect the parking brake cable from the brake assembly.

5. Disconnect and plug the main brake line from the differential junction block.

6. Disconnect the stabilizer bar from the axle housing and from the body and remove the stabilizer bar.

7. Disconnect the shock absorber from its upper mount.

8. Remove the Panhard rod-to-chassis nut/bolt and lower the rod.

9. Remove the upper and lower links-to-chassis nuts/bolts, then separate the links from the chassis supports.

10. Lower the axle housing and remove the coil springs.

11. Raise the axle housing into position and start all the mounting nuts and bolts.

12. Lower the vehicle to the ground, bounce the vehicle several times and torque all nuts and bolts as shown.

13. Connect the brake cable and hydraulic line and bleed and adjust the brakes.

14. Install the driveshaft.

STEERING

Steering Wheel

REMOVAL AND INSTALLATION

1. Position the steering wheel in the straight-ahead position.

2. Disconnect the negative battery cable.

3. Remove the horn pad by removing the screws from the rear of the steering wheel crossbar.

4. Matchmark the top of the steering column shaft and the steering wheel flange.

5. Remove the attaching nut and remove the steering wheel with a puller.

NOTE: Do not strike the shaft with a hammer, the steering column may collapse.

To install:

6. Install the steering wheel so the punchmarks are aligned. Torque the steering wheel nut to 22–29 ft. lbs. (29–39 Nm).

7. Install the horn pad and connect the negative battery cable.

Manual Steering Gear

REMOVAL AND INSTALLATION

1. Raise and safely support the vehicle.

2. Remove the steering gear-to-rubber coupling bolt.

3. Matchmark the pitman arm and sector shaft and with the wheels in a straight-ahead position, remove the idler arm-to-sector shaft nut.

4. Using the steering gear arm puller tool, press the arm from the steering gear.

5. Remove the steering gear-to-chassis bolts and the steering gear from the vehicle.

6. To install, reverse the removal procedures. Torque as follows: Steering gear-to-coupling bolt — 17–22 ft. lbs. (24–29 Nm) Steering gear-to-pitman arm nut — 94–108 ft. lbs. (127–147 Nm) Steering gear-to-frame bolts — 62–71 ft. lbs. (84–96 Nm)

ADJUSTMENT

Worm Gear Preload

For this procedure, the steering gear must be removed from the vehicle and placed in a vise.

1. Using the locknut wrench tool, loosen the locknut.

2. Rotate the worm shaft a few times, in both directions, to settle the worm bearing and check the preload.

3. Using the adjusting plug wrench, the torque wrench and an adapter socket, check the worm bearing preload; it should be 1.7–5.2 inch lbs. (0.20–0.59 Nm).

4. To adjust the worm gear preload, turn the adjusting plug with the special pin wrench and recheck the preload.

5. With the worm gear preload set, hold the adjusting plug and tighten the locknut. Check preload again.

Steering Gear Preload

1. Loosen the adjusting screw locknut.

2. Rotate the worm shaft a few times in both directions to settle the worm bearing and check the preload.

3. Set the worm gear in the straight ahead position.

4. Using the torque wrench tool and an adapter socket, check the worm gear preload; it should be 7.4–10.9 inch lbs. (0.83–1.23 Nm) for new parts or 5.2–8.7 inch lbs. (0.59–0.98 Nm) for used parts.

5. If necessary, loosen the locknut and turn the adjusting screw to obtain the correct preload.

6. With the preload set, tighten the locknut.

Power Steering Gear

REMOVAL AND INSTALLATION

1. Raise and safely support the vehicle.

2. Remove the wormshaft-to-rubber coupling bolt.

3. Matchmark the idler arm and sector shaft and with the wheels in a straight-ahead position, remove the idler arm-to-sector shaft nut.

4. Disconnect the fluid lines from the gear and cap the lines and openings in the gear.

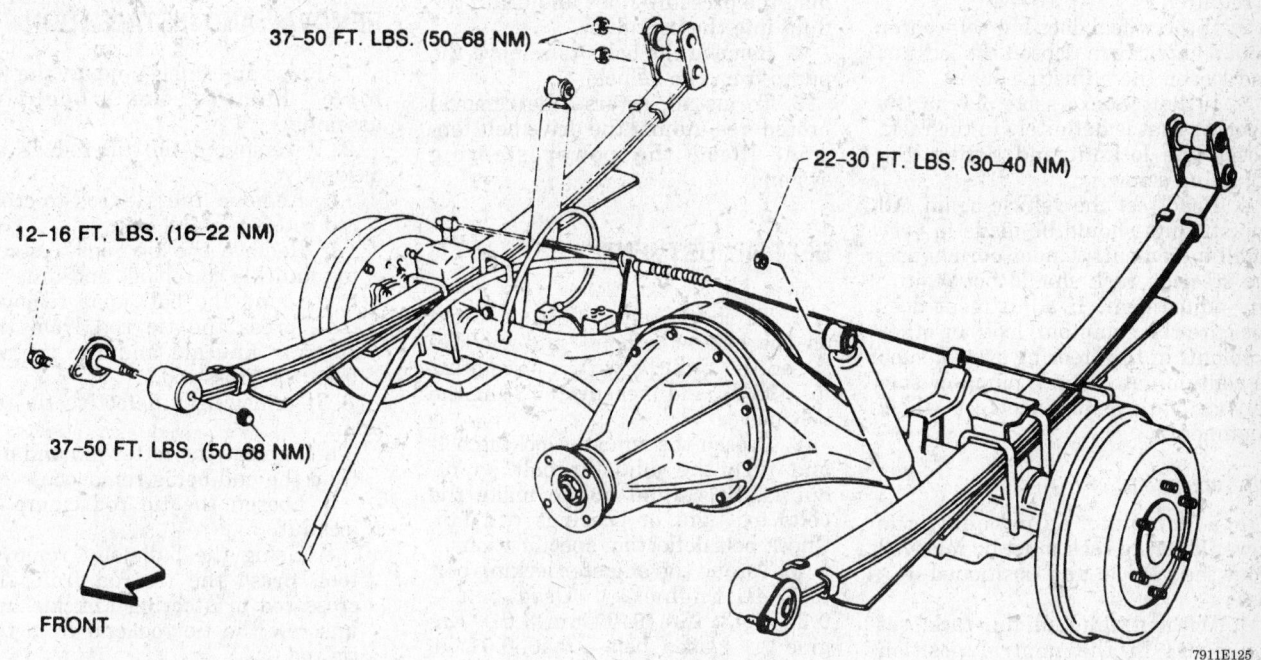

37–50 FT. LBS. (50–68 NM)

22–30 FT. LBS. (30–40 NM)

12–16 FT. LBS. (16–22 NM)

37–50 FT. LBS. (50–68 NM)

FRONT

7911E125

Rear axle housing on Pick-Up with 4WD; with 2WD, the leaf springs must be removed to remove the axle housing

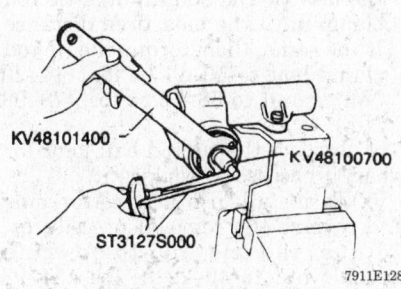

KV48101400

KV48100700

ST3127S000

7911E128

Use the special wrench to adjust worm gear preload

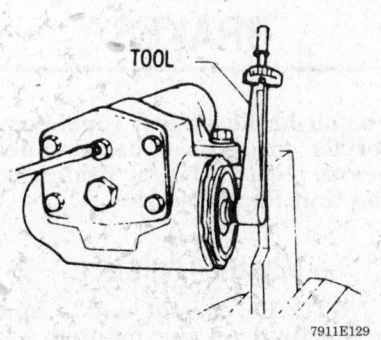

TOOL

7911E129

Adjusting steering gear preload

5. Using the steering gear arm puller, press the gear arm from the steering knuckle.

6. Remove the steering gear-to-chassis bolts and the steering gear from the vehicle.

7. To install, reverse the removal procedures. Torque as follows: Steering gear coupling bolt — 17–22 ft. lbs. (49–51 Nm) Steering gear-to-pitman arm nut — 101–130 ft. lbs. (137–177 Nm) Steering gear-to-frame bolts — 62–71 ft. lbs. (84–96 Nm).

8. Refill the power steering pump reservoir and bleed the system.

ADJUSTMENT

1. Remove the power steering gear and position it in a vise.

2. Loosen the adjusting screw locknut.

3. Set the worm gear in the straight-ahead position.

4. Using the torque wrench and an adapter socket, check the turning torque; it should be 0.9–3.5 inch lbs. (0.1–0.4 Nm).

5. If necessary, use a screwdriver, then, turn the adjusting screw to obtain the correct preload.

6. With the preload set, tighten the adjusting screw nut.

Power Steering Rack

REMOVAL AND INSTALLATION

1. Raise and safely support the vehicle and remove the front wheels.

2. Remove the cotter pins and nuts and use a ball joint press to separate the tie rod ends from the steering knuckles.

3. Remove the rubber coupling pinch bolt.

4. Disconnect and plug the hydraulic lines coming from the pump.

5. Remove the bracket bolts and remove the rack from the chassis.

6. Installation is the reverse of removal. Torque the bracket bolts to 72 ft. lbs. (98 Nm). Torque the coupling pinch bolt to 22 ft. lbs. (29 Nm).

7. Connect the tie rod ends and torque the nuts to 29 ft. lbs. (39 Nm). Tighten as required to install new cotter pins.

8. Use new O-rings to connect the hydraulic lines. Fill and bleed the system.

ADJUSTMENT

On Vehicle

1. Drive the vehicle on a flat road and turn the steering wheel about 20

degrees. If the wheel returns to center when released, no adjustment is required.

2. If the wheel does not self-center from a slight turn, loosen the locknut and loosen the adjusting screw.

3. If there is excessive play in the steering that is definitely in the rack, loosen the locknut and tighten the adjusting screw.

4. Road test the vehicle again. All adjustments should be made in very small increments. Under normal use, the steering rack should not require any adjustment. If adjustment does not cure the symptom, look for other problems in the steering system such as contaminated fluid, pump or suspension failure or incorrect wheel alignment.

Off Vehicle

For a complete adjustment, the power steering rack must be removed from the vehicle and positioned in a vise.

1. Without fluid in the rack, set the gears in the neutral position (wheels straight-ahead).

2. Lubricate the adjusting screw with locking sealant and screw it in.

3. Lightly, tighten the locknut.

4. Torque the adjusting screw to 43–52 inch lbs. (4.9–5.9 Nm).

5. Loosen the adjusting screw and retorque it to 0.43–1.74 inch lbs. (0.05–0.20 Nm).

6. Move the rack over its entire stroke several times.

7. Using an inch lb. torque wrench, measure the pinion rotating torque within the range of 180 degrees from the neutral position.

8. Loosen the adjusting screw and retorque it to 43–52 inch lbs. (4.9–5.9 Nm).

9. Loosen the adjusting screw 40–60 degrees.

10. While securing the adjusting screw in position, torque the locknut to 29–43 ft. lbs. (39–59 Nm).

11. Using a spring gauge, connect it to the tie rod end, pull the tie rod to check the frictional sliding force; it should be 27.6–37.5 lbs. (122.6–166.7 N) at neutral point or 27.6–41.9 lbs. (122.6–186.3 N) other than neutral point.

Power Steering Pump

REMOVAL AND INSTALLATION

1. Disconnect the negative battery cable.

2. Remove the drive belt from the power steering pump.

3. Place a container under the power steering pump, disconnect and plug the pressure lines and drain the fluid into the container.

4. Remove the bolts to remove the pump from the vehicle.

5. To install, reverse the removal procedures. Adjust the drive belt tension. Bleed the power steering system.

BELT ADJUSTMENT

1. To check belt deflection, press on the belt at a point mid-way between the pulleys with a force of 22 lbs. (98 N) and measure how far the belt moves.

2. Loosen the adjuster locking bolt and turn the adjuster bolt as required. Be careful not to make the belts too tight or bearings will fail. Check belt deflection specifications.

3. Torque the adjuster locking bolt to: 2.4L Engine   Used belt — 0.35–0.43 in. (9–11mm)3.0L Engine   Used belt — 0.43–0.51 in. (11–13mm)   New belt — 0.35–0.43 in. (9–11mm).

SYSTEM BLEEDING

1. Raise and support the vehicle safely.

2. Check and add fluid to the reservoir, if necessary.

3. Start the engine. Turn the steering wheel quickly (all the way), right and left, just touching the stops; turn the steering wheel at least 10 times.

NOTE: When bleeding the system, make sure the temperature of the fluid reaches 140–176°F (60–80°C).

4. Stop the engine, check the fluid level, add as required.

5. Start and run the engine for 3–5 seconds.

6. Stop the engine, check the fluid level, add as required.

7. Start the engine. Turn the steering wheel (all the way) right and left, just touching the stops; turn the steering wheel at least 10 times.

8. Stop the engine, check the fluid level, add as required.

9. Repeat the steps until all of the air is bled from the system.

10. If the air cannot be bleed from the system, turn and hold the steering wheel at each stop for at least 5 seconds but never more than 15 seconds.

Tie Rod Ends

REMOVAL AND INSTALLATION

1. Raise and safely support the vehicle. Remove the wheel/tire assembly.

2. If removing the tie rod as an assembly:

   a. Remove the tie rod-to-cross rod cotter pin and nut.

   b. Remove the tie rods-to-steering knuckle cotter pin and nut.

   c. Using the ball joint remover tool, press the tie rod from the steering knuckle and the tie rod from the cross rod.

3. If removing a defective tie rod end:

   a. Remove the cotter pin and nut from the end being removed.

   b. Loosen the tie rod clamp or locknut.

   c. Using the ball joint remover tool, press the tie rod from the cross rod or steering knuckle and unscrew the tie rod end from the tie rod.

   d. Measure the tie rod end-to-tie rod clamp distance.

   e. Unscrew the tie rod end from the tie rod.

   f. Using a new tie rod end, screw the new tie rod end into the tie rod clamp until the measured distance is the same, then, torque the tie rod clamp bolt to 10–14 ft. lbs. (14–20 Nm) or nut to 58–72 ft. lbs. (78–98 Nm).

4. Inspect the tie rod ball joint for wear; if necessary, replace it.

5. To install, use new cotter pins and reverse the removal procedures. Torque the tie rod-to-steering knuckle nut to 40–72 ft. lbs. (54–98 Nm) and the tie rod-to-cross rod nut to 40–72 ft. lbs. (54–98 Nm). Check and/or adjust the front-end alignment.

BRAKES

For all brake system repair and service procedures not detailed below, please refer to "Brakes" in the Unit Repair section.

Master Cylinder

NOTE: Be careful not to spill brake fluid on the painted surfaces of the vehicle; it will damage the paint.

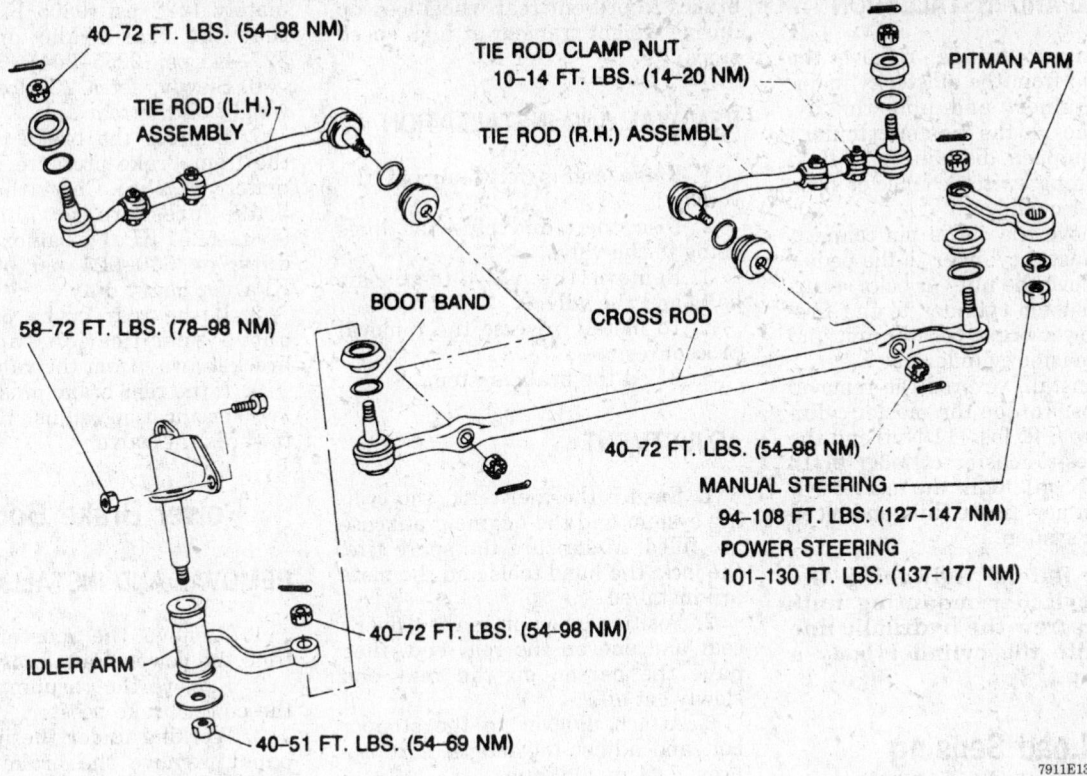

40–72 FT. LBS. (54–98 NM)

TIE ROD CLAMP NUT
10–14 FT. LBS. (14–20 NM)

PITMAN ARM

TIE ROD (L.H.) ASSEMBLY

TIE ROD (R.H.) ASSEMBLY

58–72 FT. LBS. (78–98 NM)

BOOT BAND

CROSS ROD

40–72 FT. LBS. (54–98 NM)

MANUAL STEERING
94–108 FT. LBS. (127–147 NM)

POWER STEERING
101–130 FT. LBS. (137–177 NM)

40–72 FT. LBS. (54–98 NM)

IDLER ARM

40–51 FT. LBS. (54–69 NM)

7911E131

Steering linkage on 2WD

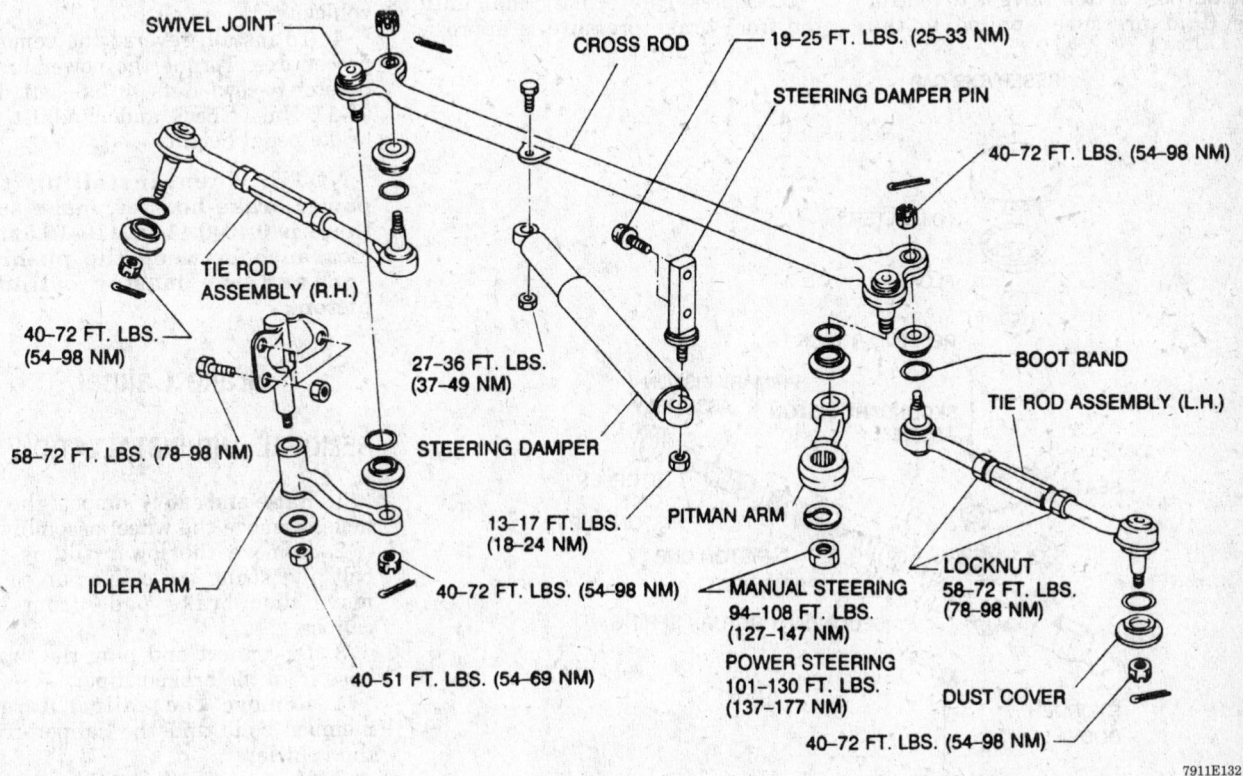

SWIVEL JOINT

CROSS ROD

19–25 FT. LBS. (25–33 NM)

STEERING DAMPER PIN

40–72 FT. LBS. (54–98 NM)

TIE ROD ASSEMBLY (R.H.)

40–72 FT. LBS.
(54–98 NM)

27–36 FT. LBS.
(37–49 NM)

BOOT BAND

TIE ROD ASSEMBLY (L.H.)

58–72 FT. LBS. (78–98 NM)

STEERING DAMPER

13–17 FT. LBS.
(18–24 NM)

PITMAN ARM

LOCKNUT
58–72 FT. LBS.
(78–98 NM)

IDLER ARM

40–72 FT. LBS. (54–98 NM)

MANUAL STEERING
94–108 FT. LBS.
(127–147 NM)

POWER STEERING
101–130 FT. LBS.
(137–177 NM)

DUST COVER

40–51 FT. LBS. (54–69 NM)

40–72 FT. LBS. (54–98 NM)

7911E132

Steering linkage on 4WD

REMOVAL AND INSTALLATION

1. Using a syringe, remove the brake fluid from the master cylinder.
2. Disconnect and plug the hydraulic lines at the master cylinder.
3. If equipped, disconnect the fluid level warning switch connector from the master cylinder.
4. Remove the clevis pin connecting the master cylinder to the pedal.
5. Remove the nuts or bolts securing the master cylinder to the firewall or power brake booster unit and pull the master cylinder out.
6. To install, reverse the removal procedures. Torque the master cylinder nuts to 8 ft. lbs. (11 Nm) and the brake lines-to-master cylinder to 13 ft. lbs. (18 Nm). Refill the master cylinder with new brake fluid and bleed the brake system.

NOTE: Before tightening the master cylinder mounting nuts or bolts, screw the hydraulic line fitting into the cylinder body a few turns.

Load Sensing Proportioning Valve

The purpose of this valve is to control the fluid pressure applied to the brakes to prevent rear wheel lock-up during weight transfer at high speed stops.

REMOVAL AND INSTALLATION

1. Raise and safely support the vehicle.
2. Disconnect and plug the lines going to the valve.
3. Remove the valve-to-chassis bolts and the valve.
4. To install, reverse the removal procedures.
5. Bleed the brake system.

ADJUSTMENT

1. Ensure the fuel tank, the cooling system and the engine crankcase are filled. Make sure the spare tire, the jack, the hand tools and the mats are installed.
2. Position a person in the driver's seat and one on the rear end, then have the person on the rear end slowly get off.
3. Attach a lever to the stopper bolt and adjust the length **L** to approx. 7.44 in. (189mm).
4. Install pressure gauges at the front and rear brakes.
5. Depress the brake pedal until the front brake pressure is approxi-

mately 1422 psi (9805 KPa). Check that the rear brake pressure is 327–441 psi (2255–3041 KPa).
6. Slowly, set a 220 lbs. (100 kg) weight on the rear axle.
7. Depress the brake pedal until the front brake pressure is approximately 1422 psi. Check that the rear brake pressure is 711–995 psi (4902–6861 KPa) for all except heavy duty, or 640–924 psi (4413–6371 KPa) for heavy duty.
8. If the rear brake pressure is above specification, adjust the bracket away from the valve.
9. If the rear brake pressure is below specification, adjust the bracket towards the valve.

Power Brake Booster

REMOVAL AND INSTALLATION

1. Remove the master cylinder from the power brake booster.
2. Remove the vacuum hose form the power brake booster.
3. Working under the instrument panel, remove the brake pedal-to-brake booster rod clevis pin. Remove the power brake booster mounting bolts and the booster from the vehicle.
4. To install, reverse the removal procedures. Torque the power brake booster-to-cowl nuts to 5.8–8 ft. lbs. (8–11 Nm). Check and/or adjust the brake pedal height.

NOTE: When installing the power brake booster, make sure there is 0.40–0.41 in. (10–10.5mm) clearance between the pushrod end and the master cylinder piston.

Brake Caliper

REMOVAL AND INSTALLATION

1. Raise and safely support the vehicle. Remove the wheel assembly.
2. Remove the lower sliding pin bolt and swing the caliper up to remove disc brake pads from the caliper.
3. Disconnect and plug the brake hose from the brake caliper.
4. Remove the caliper torque member bolts and the caliper from the vehicle.
5. Make sure the caliper moves freely on the sliding pins. Clean, repair or replace as necessary. Use a small amount of light grease to lubricate the pins.

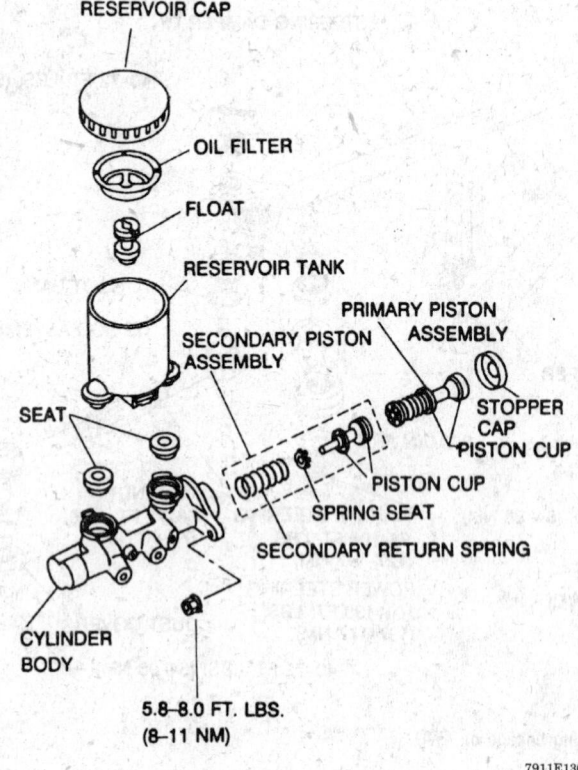

RESERVOIR CAP
OIL FILTER
FLOAT
RESERVOIR TANK
PRIMARY PISTON ASSEMBLY
SECONDARY PISTON ASSEMBLY
STOPPER CAP
PISTON CUP
SEAT
PISTON CUP
SPRING SEAT
SECONDARY RETURN SPRING
CYLINDER BODY
5.8–8.0 FT. LBS. (8–11 NM)

7911E136

Master cylinder assembly

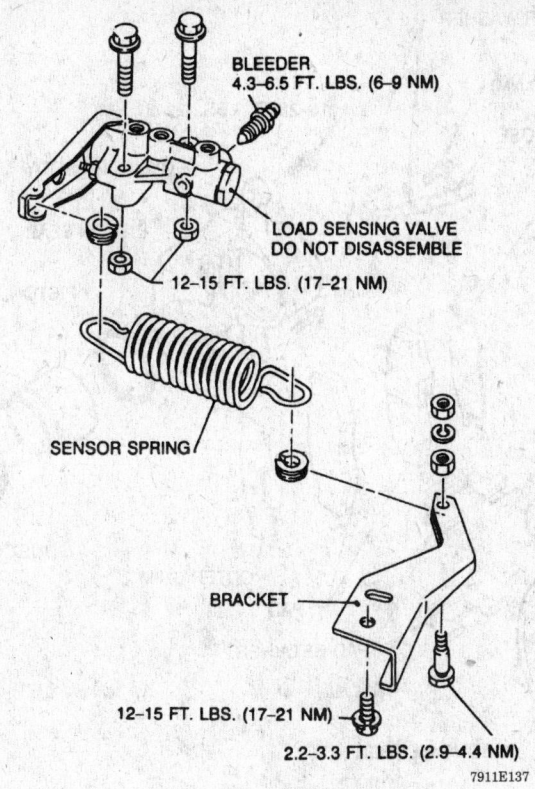

BLEEDER
4.3–6.5 FT. LBS. (6–9 NM)

LOAD SENSING VALVE
DO NOT DISASSEMBLE

12–15 FT. LBS. (17–21 NM)

SENSOR SPRING

BRACKET

12–15 FT. LBS. (17–21 NM)

2.2–3.3 FT. LBS. (2.9–4.4 NM)

7911E137

Load sensing brake valve

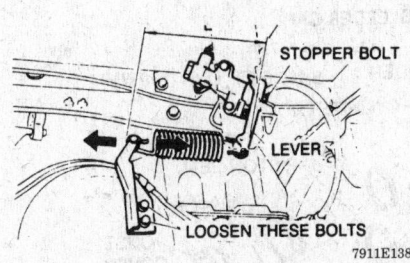

STOPPER BOLT

LEVER

LOOSEN THESE BOLTS

7911E138

Adjust load sensing valve spring length

6. To install, reverse the removal procedures. Torque the sliding pin bolts to 31 ft. lbs. (42 Nm). If the torque member was removed, torque the bolts to 72 ft. lbs. (97 Nm) on front calipers, 38 ft. lbs. (52 Nm) on rear calipers.

7. Bleed the brake system.

Disc Brake Pads

REMOVAL AND INSTALLATION

1. Raise and safely support the vehicle. Remove the wheels and the brake fluid reservoir cap.

2. Remove the bottom sliding pin bolt and swing the caliper upward.

3. Remove the pad retainers, the inner/outer retainers and the brake pads.

4. Using a medium C-clamp and a block of wood, place the wood against the caliper piston(s) and use the C-clamp press the piston into the caliper. Some brake fluid may be expelled from the reservoir.

5. Inspect the caliper for signs of fluid leakage; if necessary, replace or rebuild the caliper.

6. Make sure the caliper moves freely on the sliding pins. Clean, repair or replace as necessary. Use a small amount of light grease to lubricate the pins.

7. To install, use new brake pads and reverse the removal procedures. Torque the sliding pin bolts to 31 ft. lbs. (42 Nm).

Front Brake Rotor

REMOVAL AND INSTALLATION

2WD

1. Raise and safely support the vehicle. Remove the wheel assembly.

2. Without disconnecting the hydraulic hose, remove the and torque

member bolts and hang the caliper from the body with wire. Do not let the caliper hang by the hose.

3. Remove the wheel bearing grease cup, cotter pin, adjusting nut cap, hub nut, thrust washer and wheel bearing.

4. Pull the wheel hub/brake rotor assembly from the wheel spindle.

5. From the rear of the brake rotor, remove the rotor-to-wheel hub bolts and the rotor.

6. Inspect the rotor for cracks, wear and/or other damage; if necessary, replace it.

To install:

7. Install the rotor to the hub and torque the bolts to 51 ft. lbs. (69 Nm).

8. Install the hub and wheel bearings and adjust the bearing pre-load according to the proper procedure.

9. Install the caliper and pump the brake pedal to adjust the brakes.

4WD

1. Raise and support the vehicle safely under the axle case. Remove the wheels and brake calipers.

2. Block the wheels or hold the brake pedal and remove the Torx® screws to remove the locking hubs.

3. Remove the snapring from the end of the halfshaft and remove the locking hub parts.

4. From the halfshaft, remove the thrust washer, the snapring, the lockwasher screw and the lockwasher.

5. Use the proper pin wrench to loosen and remove the wheel bearing locknut.

6. Pull the wheel hub/rotor assembly from the spindle.

7. Remove the bolts to remove the rotor from the hub.

To install:

8. Install the rotor to the hub and torque the bolts to 51 ft. lbs. (69 Nm).

9. Install the hub and bearings and adjust the wheel bearing pre-load according to the proper procedure.

10. Install the brake caliper.

11. Install the locking hub assembly and torque the Torx® bolts to 25 ft. lbs. (34 Nm).

Rear Brake Rotor

REMOVAL AND INSTALLATION

1. Raise and safely support the vehicle.

2. Remove the wheel assembly.

3. Without disconnecting the hydraulic hose, remove the and torque member bolts and hang the caliper

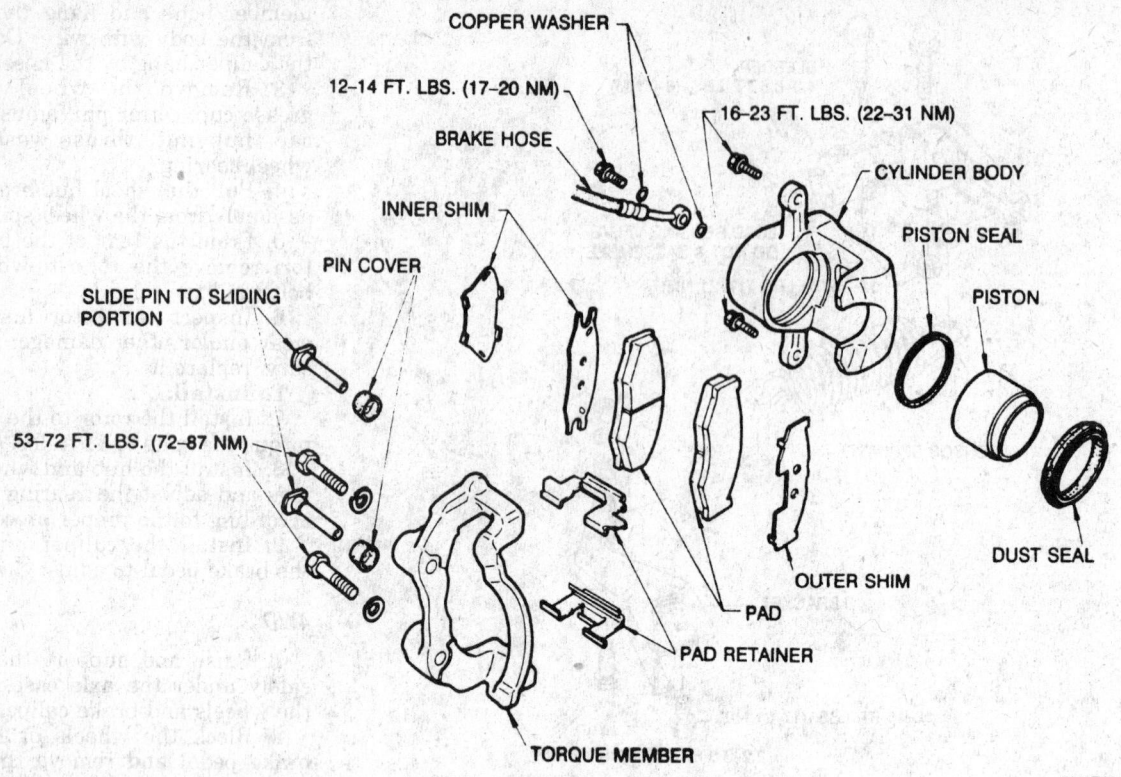

COPPER WASHER

12–14 FT. LBS. (17–20 NM)

BRAKE HOSE

16–23 FT. LBS. (22–31 NM)

CYLINDER BODY

PISTON SEAL

PISTON

INNER SHIM

PIN COVER

SLIDE PIN TO SLIDING PORTION

53–72 FT. LBS. (72–87 NM)

DUST SEAL

OUTER SHIM

PAD

PAD RETAINER

TORQUE MEMBER

7911E142

Single piston front caliper

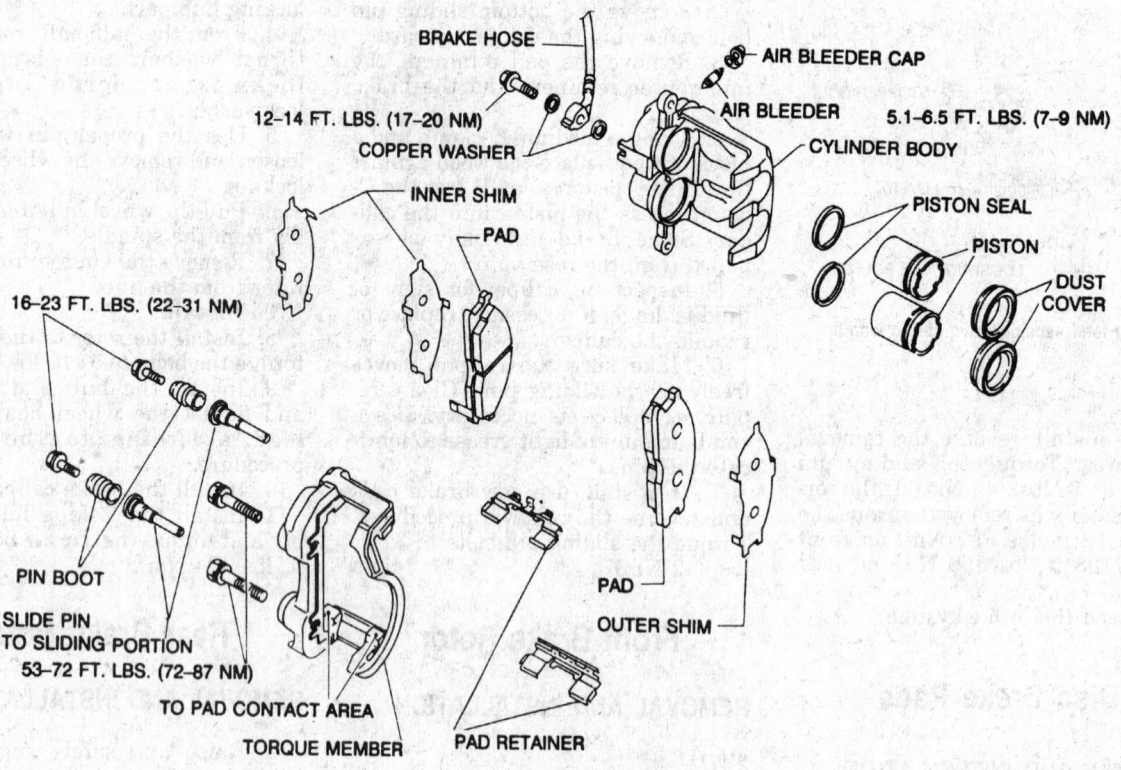

BRAKE HOSE

AIR BLEEDER CAP

12–14 FT. LBS. (17–20 NM)

COPPER WASHER

AIR BLEEDER

5.1–6.5 FT. LBS. (7–9 NM)

CYLINDER BODY

INNER SHIM

PAD

PISTON SEAL

PISTON

DUST COVER

16–23 FT. LBS. (22–31 NM)

PIN BOOT

SLIDE PIN TO SLIDING PORTION

53–72 FT. LBS. (72–87 NM)

TO PAD CONTACT AREA

TORQUE MEMBER

PAD

OUTER SHIM

PAD RETAINER

7911E144

Dual piston front caliper

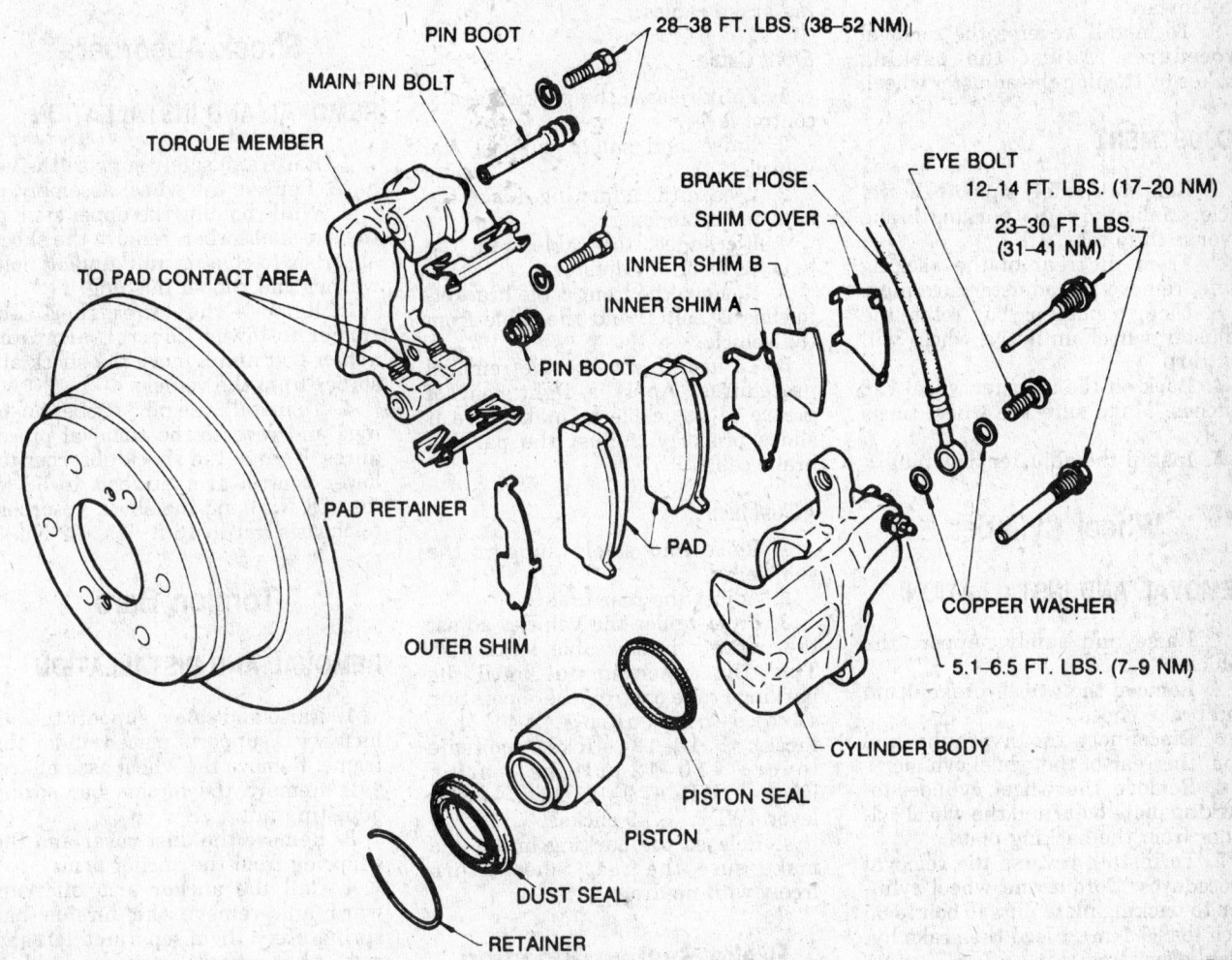

28–38 FT. LBS. (38–52 NM)

PIN BOOT

MAIN PIN BOLT

TORQUE MEMBER

BRAKE HOSE

EYE BOLT
12–14 FT. LBS. (17–20 NM)

SHIM COVER

23–30 FT. LBS.
(31–41 NM)

TO PAD CONTACT AREA

INNER SHIM B

INNER SHIM A

PIN BOOT

PAD RETAINER

PAD

COPPER WASHER

OUTER SHIM

5.1–6.5 FT. LBS. (7–9 NM)

CYLINDER BODY

PISTON SEAL

PISTON

DUST SEAL

RETAINER

7911E145

Single piston rear caliper

from the body with wire. Do not let the caliper hang by the hose.

4. Remove the rotor from the hub.

5. To install, reverse the removal procedures.

Brake Drums

REMOVAL AND INSTALLATION

1. Raise and safely support the vehicle.

2. Remove the wheel assembly.

3. Pull the brake drum from the wheel hub. It may be necessary to back off the brake adjustment.

4. Inspect and/or replace the brake drum.

5. To install, reverse the removal procedures.

Brake Shoes

REMOVAL AND INSTALLATION

1. Fully release the parking brake.

2. Raise and safely support the vehicle. Remove the wheel assembly.

3. If equipped with rear disc brakes and drum type parking brakes, remove the brake caliper. Without disconnecting the hydraulic

hose, remove the torque member bolts and hang the caliper from the body with wire. Do not let the caliper hang by the hose.

4. Remove the brake drum. If the drum is difficult to remove, insert 2 — 8mm x 1.25 screws in the disc/drum holes, tighten the screws to press the drum from the hub.

5. When removing the brake shoes and springs, be sure to remove only one side at a time so the other side is available as a reference. Remove the brake shoe retainers and the springs.

6. Separate the parking brake cable from the from the parking brake lever.

7. Using brake grease, lubricate the shoe adjuster and the backing plate contact points.

8. Turn the shoe adjuster all the way inward.

9. To install, reverse the removal procedures. Adjust the parking brakes by turning the adjuster wheel.

ADJUSTMENT

1. Raise and safely support the vehicle. Make sure the parking brake lever is fully released.

2. From the rear of the backing plate, remove the adjuster hole plug.

3. Using a small prybar, rotate the adjuster wheel until the wheel will not turn.

4. Back off the adjuster wheel 7–8 notches. Make sure the wheel turns freely.

5. Install the adjuster hole plug.

Wheel Cylinder

REMOVAL AND INSTALLATION

1. Raise and safely support the vehicle.

2. Remove the wheel, brake drum and brake shoes.

3. Disconnect the hydraulic line from the rear of the wheel cylinder.

4. Remove the wheel cylinder-to-backing plate bolts and the wheel cylinder from the backing plate.

5. To install, reverse the removal procedures. Torque the wheel cylinder-to-backing plate nuts to bolt to 65 inch lbs. (7 Nm). Bleed the brake hydraulic system.

Parking Brake Cable

REMOVAL AND INSTALLATION

Rear Cable

1. Fully release the parking brake handle.

2. Raise and safely support the vehicle.

3. Loosen the adjusting nut at the adjuster cable lever.

4. Disconnect the cable from the balance lever or adjuster.

5. Disconnect the rear parking brake cable(s) from the parking brake toggle levers of the rear service brake assemblies.

6. Remove the rear parking brake cable brackets-to-chassis bracket screws.

7. Remove parking brake cable(s) from the vehicle.

8. To install, reverse the removal procedures. Apply a light coat of grease to the cables to make sure they slide properly. Adjust the parking brake cables.

Front Cable

1. Fully release the parking brake control lever.

2. Raise and safely support the vehicle.

3. Loosen the adjusting nut at the adjuster cable lever.

4. Disconnect the cable from the balance lever or adjuster.

5. Remove the front cable bracket-to-chassis bolt(s) and the cable from the vehicle.

6. To install, reverse the removal procedures. Apply a light coat of grease to the cable to make sure it slides properly. Adjust the parking brake cables.

Adjustment

1. Raise and safely support the vehicle.

2. Adjust the rear brakes.

3. From under the vehicle, adjust the parking brake cable locknut(s). Turn the adjusting nut until the parking brake control lever operating stroke is (using 44 lbs. force): 10–12 clicks — console lever 10–12 clicks — stick lever — 2WD 9–11 clicks — stick lever 4WD 7–9 clicks.

4. Release the parking brake and make sure the rear wheels turn freely with no drag.

Brake System Bleeding

1. Raise and safely support the vehicle.

2. Make sure the brake fluid reservoir is full. Keep checking the level during the procedure, do not allow the level to fall too low.

3. Have an assistant pump the brake pedal, then hold pressure on the pedal.

4. Connect a tube to the bleeder at the right rear wheel and put the other end in a container. Open the bleeder until the pressure is released, then close it before releasing pressure on the pedal.

5. Repeat the procedure until fluid flows from the bleeder with no air bubbles. If the bubbles do not stop, a problem is indicated in the master cylinder.

6. Repeat the procedure on the left rear, right front, then left front brakes in order.

7. Fill the fluid reservoir.

FRONT SUSPENSION

Shock Absorbers

REMOVAL AND INSTALLATION

1. Raise and safely support the vehicle. Remove the wheel assembly.

2. While holding the upper stem of the shock absorber, remove the shock absorber-to-chassis nut and/or bolt, washer and rubber bushing.

3. Remove the lower shock absorber-to-lower control arm nut and/or bolt and remove the shock absorber from the vehicle.

4. To install, use new rubber bushings and reverse the removal procedures. Torque the shock absorber-to-lower control arm nut/bolt to 58 ft. lbs. (78 Nm) and the shock absorber-to-chassis nut to 16 ft. lbs. (22 Nm).

Torsion Bars

REMOVAL AND INSTALLATION

1. Raise and safely support the vehicle with supports placed under the frame. Remove the wheel assemblies.

2. Remove the torsion bar spring adjusting nut.

3. Remove the dust cover and the snapring from the anchor arm.

4. Pull the anchor arm off rearward and remove the torsion bar spring. Keep them separated left and right, they are not interchangeable.

5. Remove the torque arm.

To install:

6. Check the torsion bars for wear, cracks or other damage; replace them if necessary.

7. Install the torque arm on the lower link (control arm) and torque the bolts to 50 ft. lbs. (68 Nm).

8. Install the snapring and dust cove on the torsion bar.

9. Coat the splines on the inner end of the torsion bar with chassis lube and install it into the torque arm. The torsion bars are marked **L** and **R** and are not interchangeable. Adjust the torsion bars.

ADJUSTMENT

2WD Models

1. Position a floor jack under the lower suspension arm and raise it so the clearance between the arm and the rebound bumper is 0.

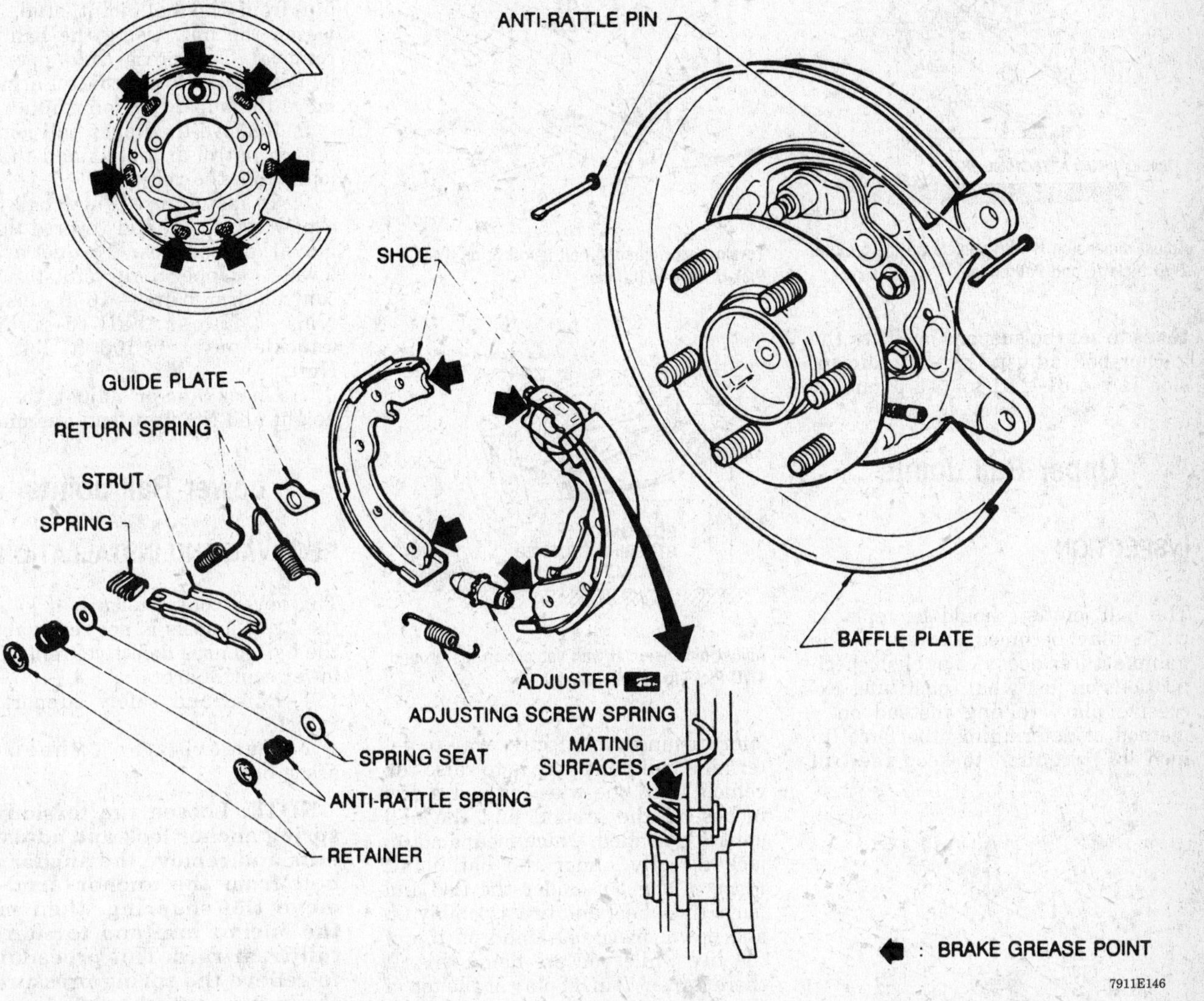

ANTI-RATTLE PIN

SHOE

GUIDE PLATE

RETURN SPRING

STRUT

SPRING

ADJUSTER

ADJUSTING SCREW SPRING

MATING SURFACES

SPRING SEAT

ANTI-RATTLE SPRING

RETAINER

BAFFLE PLATE

◄ : BRAKE GREASE POINT

7911E146

Rear brake shoe assembly on vehicles with rear disc brakes and drum parking brakes; lightly lubricate the backing plate in the areas indicated with arrows

2. Install the anchor arm so the dimension **L** is 0.24–0.71 in. (6–18mm).

3. Install the snapring to the anchor arm and dust cover. Make sure the snapring is properly installed in the groove of the anchor arm.

4. Tighten the anchor arm adjusting nut until dimension **L** is 1.38 in. (35mm) for heavy duty, cab/chassis and std models or 1.93 in. (49mm) for all other models.

5. Lower the vehicle so it is resting on the wheels and bounce it several times to set the suspension. Turn the anchor bolt adjusting nut so dimension **H** is 4.25–4.65 in. (108–118mm).

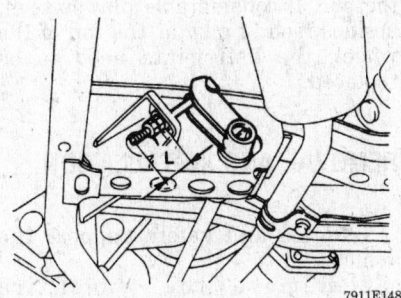

7911E148

Adjusting torsion bar anchor arm bolt length L on 2WD Pick-Up and Pathfinder

4WD Models

1. Position a floor jack under the lower suspension arm and raise it so the clearance between the arm and the rebound bumper is 0.

2. Install the anchor arm so the dimension **G** is 01.97–2.36 in. (50–60mm).

3. Install the snapring to the anchor arm and dust cover. Make sure the snapring is properly installed in the groove of the anchor arm.

4. Tighten the anchor arm adjusting nut until dimension **L** is 3.03 in. (77mm).

5. Lower the vehicle so it is resting on the wheels and bounce it several

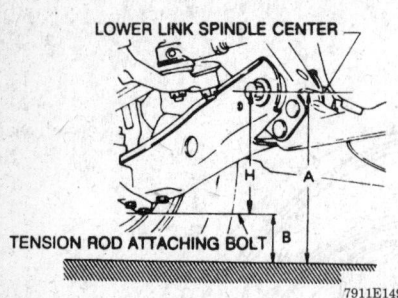

Adjust dimension H with vehicle on the ground; 2WD Pick-Up and Pathfinder

times to set the suspension. Turn the anchor bolt adjusting nut so dimension **H** is 1.61–2.01 in. (41–51mm).

Upper Ball Joints

INSPECTION

The ball joint(s) should be replaced when play becomes excessive. The manufacturer does not publish specifications on just what constitutes excessive play, relying instead on a method of determining the force (in inch lbs.) required to keep the ball

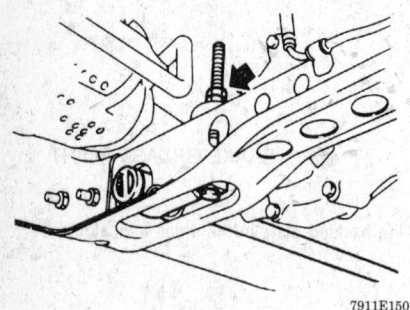

Adjusting torsion bar on 4WD Pick-Up and Pathfinder

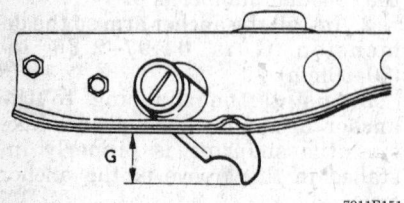

Anchor arm length G on 4WD Pick-Up and Pathfinder

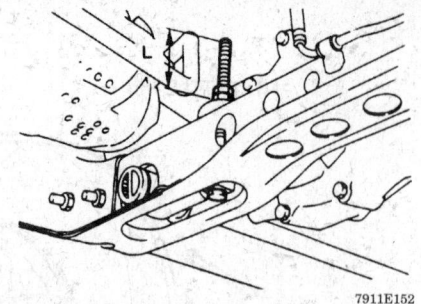

Torsion bar adjusting bolt length L on 4WD Pick-Up and Pathfinder

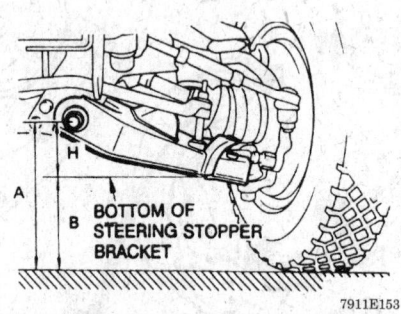

Adjust dimension H with vehicle on the ground; 4WD Pick-Up and Pathfinder

joint turning. An effective way to determine ball joint play is to raise the vehicle until the wheel is just a few inches off the ground and the ball joint is unloaded, which means not to jack directly under the ball joint. Place a long bar under the tire and move the wheel and tire assembly up and down; place one hand on top of the tire while you are doing this. If there is over ¼ in. of play at the top of the tire, the ball joint is probably bad. This assumes that the wheel bearings are in good shape and properly adjusted. As a double check, have someone watch the ball joint while you move the tire up and down with the bar. If considerable play is seen, besides feeling play at the top of the wheel, the ball joints need to be replaced.

REMOVAL AND INSTALLATION

1. Raise and safely support the vehicle.
2. Remove the wheel/tire assembly.

NOTE: It may be necessary to loosen the torsion bar anchor lock and adjusting nuts to relieve spring tension.

3. Place a floor jack under the steering knuckle and support it.
4. Remove and discard the cotter pin from the ball joint stud, then loosen the nut. Using the ball joint removal tool, press the upper ball joint from the lower control arm. Remove the upper ball joint nut.
5. Remove the upper ball joint-to-upper control arm bolts and the ball joint from the vehicle.
6. To install, use a new ball joint, a new cotter pin and reverse the removal procedures. Torque as follows: Upper ball joint-to-upper control arm bolts — 16 ft. lbs. (22 Nm) Upper ball-to-steering knuckle nut 108 ft. lbs. (146 Nm).
7. Check and/or adjust the ride height and the front end alignment.

Lower Ball Joints

REMOVAL AND INSTALLATION

The lower control arm ball joint on the 2WD models is not removable; if the ball joint is defective, replace the lower control arm.

1. Raise and safely support the vehicle.
2. Remove the wheel/tire assembly.

NOTE: Loosen the torsion bar spring anchor lock and adjusting nuts and remove the anchor arm bolt from the anchor arm. Remove the snapring, then move the anchor arm and torsion bar fully rearward. This procedure is to relieve the spring pressure on the lower control arm.

3. If equipped, it may be necessary to disconnect the sway bar from the lower arm.
4. Disconnect the tension rod from the lower arm.
5. Remove the cotter pin from the ball joint stud and loosen the nut.
6. Using the ball joint separator tool, press the ball joint from the steering knuckle.
7. Remove the lower ball joint-to-lower control arm bolts and the ball joint.
8. To install, use a new cotter pin and reverse the removal procedures. Torque as follows: Ball joint-to-control arm bolts 4WD — 45 ft. lbs. (61 Nm) Ball joint-to-steering knuckle nut 141 ft. lbs. (190 Nm).
9. Check and/or adjust the torsion bar ride height assembly and the front end alignment.

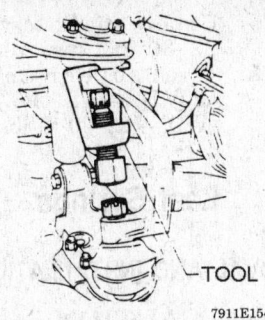

7911E154

Use a ball joint press to disconnect the ball joints and tie rod ends

Upper Control Arms

REMOVAL AND INSTALLATION

1. Raise and safely support the vehicle.
2. Remove the wheels.
3. Remove the upper shock absorber-to-chassis nut and compress the shock absorber.

NOTE: If may be necessary to loosen the torsion bar anchor lock and adjusting nuts to relieve the torsion bar tension.

4. Remove the upper ball joint-to-upper control arm bolts.
5. Using a floor jack, raise the lower control arm.
6. Remove the upper control arm-to-chassis bolts and the upper control arm from the vehicle.

NOTE: If shims are used, be sure to keep them in order for reinstallation purposes.

7. Inspect the ball joint, replace as required.
8. To install, replace the shims, if used, in their original locations and reverse the removal procedures. Torque the upper control arm-to-chassis bolts to 108 ft. lbs. (146 Nm), the ball joint-to-upper control arm bolts to 15 ft. lbs. (20 Nm) and the upper shock absorber-to-chassis nut to 16 ft. lbs. (21 Nm).
9. Lower the vehicle and adjust the ride height. Check and/or adjust the front end alignment.

Lower Control Arms

REMOVAL AND INSTALLATION

2WD Pick-Up

1. Raise and safely support the vehicle and remove the front wheels.

2. Remove the shock absorber.
3. Remove the torsion bar.
4. Disconnect the stabilizer bar linkage from the lower control arm.
5. Disconnect the tension rod-to-lower control arm bolts.
6. Remove the cotter pin from the ball joint and use a ball joint press to separate the ball joints from the control arm.
7. Remove the lower control arm-to-chassis nut/bolt, tap the pivot shaft from the bushing. Push down on the tension rod and remove the lower control arm.

To install:

8. Install the control arm and the pivot bolt and torque the bolt to 108 ft. lbs. (147 Nm).
9. Connect the lower ball joint and sway bar. Torque the ball joint nut to 141 ft. lbs. (191 Nm) and tighten as required to install a new cotter pin.
10. Connect the tension rod and torque the bolts to 47 ft. lbs. (64 Nm). Install the stabilizer bar linkage and torque to 16 ft. lbs. (22 Nm).
11. Install the torsion bar and shock absorber. Adjust the torsion bar height and front wheel alignment.

4WD Pick-Up and Pathfinder

1. Raise and safely support the vehicle and remove the front wheels.
2. Remove the torsion bar.
3. Remove the shock absorber.
4. Disconnect the stabilizer bar linkage.
5. Remove the cotter pin and lower ball joint nut and use a ball joint press to separate the ball joint from the lower control arm.
6. Remove the lower arm spindle bolt and the bushing nut and remove the lower control arm.

To install:

7. Fit the arm into place and start the bushing nut. Torque this nut when the vehicle is on the wheels. Torque the spindle bolt to 108 ft. lbs. (147 Nm).
8. Connect the lower ball joint and torque the nut to 141 ft. lbs. (191 Nm) and tighten as required to install a new cotter pin.
9. Connect the stabilizer bar linkage and install the shock absorber. Torque the stabilizer bar linkage when the vehicle is on the wheels to 16 ft. lbs. (22 Nm).
10. Install the torsion bar and adjust the bar and front wheel alignment.

Stabilizer Bar

REMOVAL AND INSTALLATION

1. Raise and safely support the vehicle.
2. From both sides of the vehicle, remove the stabilizer bar-to-lower control arm connecting rod nut, bushings and tube.
3. Remove the bracket bolts and brackets and remove the stabilizer bar.
4. Installation is the reverse of removal. Make sure the brackets are to the outside of the white painted marks on the stabilizer bar. Torque the bracket bolts to 16 ft. lbs. (22 Nm) and the connecting rod nut to 22 ft. lbs. (30 Nm).

Front Wheel Bearings

For the 4WD models, please refer to the Drive Axle section.

REMOVAL AND INSTALLATION

2WD Pick-Up

1. Raise and safely support the vehicle and remove the front wheels.
2. Disconnect the brake hose from the bracket on the knuckle.
3. Without disconnecting the hydraulic line, remove the brake caliper and hang it from the body on a wire. Do not let it hang by the hose.
4. Remove the wheel hub cup, the cotter pin, the adjusting cap and hub nut.
5. Remove the wheel hub and brake disc assembly. Be careful not to drop the outer wheel bearing.
6. To remove the inner bearing, pry out the grease seal. Discard the seal.

To install:

7. Make sure the bearings are clean and in good condition. Replace as required. Pack the bearings and the hub with new grease and install the bearings into the hub. Install a new grease seal.
8. Make sure the spindle is clean. Install the spacer with the chamfer facing in on the spindle.
9. Lightly grease the lips on the seal and slide the hub and bearings onto the spindle. Install the washer and nut.
10. To adjust the pre-load on the bearing:
    a. Torque the wheel bearing nut to 25 ft. lbs. (34 Nm).

b. Spin the hub several times in both directions, then torque the nut to 29 ft. lbs. (39 Nm).

c. Loosen the nut 45 degrees. Install the locknut cap and a new cotter pin.

11. Install the brake caliper and torque the caliper carrier bolts to 72 ft. lbs. (98 Nm).

2WD Pathfinder

1. Raise and safely support the vehicle and remove the front wheels.

2. Without disconnecting the hydraulic line, remove the brake caliper and hang it from the body with wire. Do not allow the caliper to hang by the hose.

3. Remove the hub cap and locking screw and remove the lock washer. Use a pin wrench to loosen the wheel bearing locknut. The torque may be fairly high, do not use a hammer and drift pin.

4. Remove the locknut and pull the hub off with the bearings. Pry the inner grease seal out to remove the inner bearing. Discard the seal.

To install:

5. Clean all parts of grease and check the condition of the bearings. If bearings are to be replaced, the inner races can be removed from the hub with a hammer and soft drift pin. Be careful not to damage the hub.

6. Carefully install the new inner races with the drift pin, making sure they seat in the hub.

7. Pack the bearings with new grease and pack grease into the hub. Install the inner bearing and press a new inner seal into the hub.

8. Slip the hub assembly onto the spindle and install the outer bearing. Grease the locknut and thread it into place.

9. To adjust bearing pre-load:

a. Use a pin wrench to torque the locknut to 58–72 ft. lbs. (78–98 Nm). Turn the hub in both directions several times while torquing the nut.

b. Loosen the locknut, then torque again to 13 inch lbs. (1.5 Nm).

c. Turn the hub several times and check the nut torque again.

d. Install the lock washer. When installing the screw, make sure the locknut turns no more than 30 degrees in either direction.

e. When bearing pre-load is properly set, there will be no end-play in the hub and it will require no more than 4.7 lbs. of pull at the wheel stud to turn the hub.

10. Install the brake caliper and torque the caliper carrier bolts to 72 ft. lbs. (98 Nm).

REAR SUSPENSION

Shock Absorbers

REMOVAL AND INSTALLATION

1. Raise and safely support the vehicle.

2. Remove the upper shock-to-vehicle nut, the lower shock-to-vehicle nut and the shock from the vehicle.

NOTE: The weight of the vehicle must be on the rear wheels before tightening the shock absorber attaching nuts.

3. To install, reverse the removal procedures. Torque as follows: Upper shock-to-vehicle nut 22–30 ft. lbs. Lower shock absorber-to-axle nut 2WD — 12–16 ft. lbs. 4WD — 22–30 ft. lbs.

Leaf Springs

REMOVAL AND INSTALLATION

—— CAUTION ——

The leaf springs are under a considerable amount of tension. Be very careful when removing or installing them; they can exert enough force to cause serious injuries.

1. Raise and safely support the vehicle. Using a floor jack, support the axle housing.

2. Disconnect the shock absorbers at their lower end.

3. Remove the axle housing-to-spring pad U-bolt nuts and the spring pad.

4. Raise the axle housing to remove the weight off the springs.

5. Remove the spring shackle nuts, drive out the shackle pins and remove the spring from the vehicle.

NOTE: The weight of the vehicle must be on the rear wheels before torquing the front pin, shackle and shock absorber nuts.

6. To install, reverse the removal procedures. Torque as follows: Front pin and shackle nuts — 94 ft. lbs. (127 Nm) U-bolt nuts — 72 ft. lbs. (98 Nm) Shock absorber lower end nut 2WD — 16 ft. lbs. (22 Nm) 4WD — 30 ft. lbs. (41 Nm)

SPECIFICATIONS

ENGINE IDENTIFICATION

Year	Model		Engine Displacement Liters (cc)	Engine Series (ID/VIN)	Fuel System	No. of Cylinders	Engine Type
1991	Samurai		1.3 (1298)	5	EFI	4	SOHC
	Sidekick		1.6 (1590)	0	EFI	4	SOHC
1992	Samurai		1.3 (1298)	5	EFI	4	SOHC
	Sidekick		1.6 (1590)	0	EFI	4	SOHC
	Sidekick	1	1.6 (1590)	0	MFI	4	SOHC
1993	Samurai		1.3 (1298)	5	EFI	4	SOHC
	Sidekick		1.6 (1590)	0	EFI	4	SOHC
	Sidekick	1	1.6 (1590)	0	MFI	4	SOHC
1994-95	Samurai		1.3 (1298)	5	EFI	4	SOHC
	Sidekick		1.6 (1590)	0	EFI	4	SOHC
	Sidekick	1	1.6 (1590)	0	MFI	4	SOHC

EFI - Electronic fuel injection
MFI - Multiport fuel injection
SOHC - Single overhead camshaft
1 4 door

7911fC02

GENERAL ENGINE SPECIFICATIONS

Year	Engine ID/VIN		Engine Displacement Liters (cc)	Fuel System Type	Net Horsepower @ rpm	Net Torque @ rpm (ft. lbs.)	Bore x Stroke (in.)	Compression Ratio	Oil Pressure @ rpm
1991	5		1.3 (1298)	EFI	66@6000	76@3500	2.91x2.97	9.5:1	43-60@3000
	0		1.6 (1590)	EFI	80@5400	94@3000	2.95x3.54	8.9:1	51-63@3000
1992	5		1.3 (1298)	EFI	66@6000	76@3500	2.91x2.97	9..5:1	43-60@3000
	0		1.6 (1590)	EFI	80@5400	94@3000	2.95x3.54	8.9:1	51-63@3000
	0	1	1.6 (1590)	MFI	995@5600	98@4000	2.95x3.54	9.5:1	47-61@3000
1993	5		1.3 (1298)	EFI	66@6000	76@3500	2.91x2.97	9.5:1	43-60@3000
	0		1.6 (1590)	EFI	80@5400	94@3000	2.95x3.54	8.9:1	51-63@3000
	0	1	1.6 (1590)	MFI	95@5600	98@4000	2.95x3.54	9.5:1	47-61@3000
1994-95	5		1.3 (1298)	EFI	66@6000	76@3500	2.91x2.97	9.5:1	43-60@3000
	0		1.6 (1590)	EFI	80@5400	94@3000	2.95x3.54	8.9:1	51-63@3000
	0	1	1.6 (1590)	MFI	95@5600	98@4000	2.95x3.54	9.5:1	47-61@3000

EFI - Electronic fuel injection
MFI - Multiport fuel injection
  Sidekick 4 door

7911fC03

GASOLINE ENGINE TUNE-UP SPECIFICATIONS

Year	Engine ID/VIN	Engine Displacement Liters (cc)	Spark Plugs Gap (in.)	Ignition Timing (deg.)		Fuel Pump (psi)	Idle Speed (rpm)		Valve Clearance	
				MT	AT		MT	AT	In.	Ex.
1991	5	1.3 (1298)	0.029	8B	NA	34-40	800	NA	0.009-0.011	0.0102-0.012
	0	1.6 (1590)	0.029	8B	8B	34-40	800	800	0.009-0.011	0.0102-0.012
1992	5	1.3 (1298)	0.029	8B	NA	34-40	800	NA	0.009-0.011	0.0102-0.012
	0	1.6 (1590)	0.029	8B	8B	34-40	800	800	0.009-0.011	0.0102-0.012
1993	5	1.3 (1298)	0.029	8B	NA	34-40	800	NA	0.009-0.011	0.0102-0.012
	0	1.6 (1590)	0.029	8B	8B	34-40	800	800	0.009-0.011	0.0102-0.012
1994-95	5	1.3 (1298)	0.029	8B	NA	34-40	800	NA	0.009-0.011	0.0102-0.012
	0	1.6 (1590)	0.029	8B	8B	34-40	800	800	0.009-0.011	0.0102-0.012

NOTE: The Vehicle Emission Control Information label reflects changes made during production. The label must be used if it differs from above

B - Before top dead center

NA - Not Available

HYD - Hydraulic

7911fC04

FIRING ORDERS

NOTE: To avoid confusion, always replace spark plug wires one at a time.

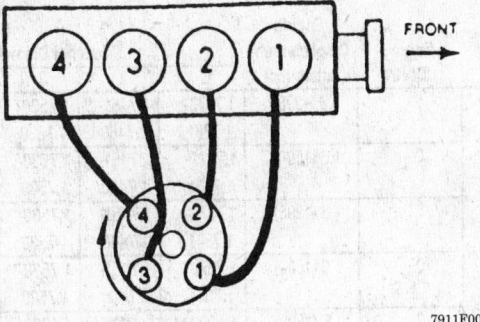

7911F001

Engine Firing Order: 1 — 3 — 4 — 2
Distributor Rotation: Clockwise

CAPACITIES

Year	Model	Engine ID/VIN	Engine Displacement Liters (cc)	Engine Crankcase with Filter	Transmission (pts.)			Transfer Case (pts.)	Drive Axle		Fuel Tank (gal.)	Cooling System (qts.)
					4-Spd	5-Spd	Auto.		Front (pts.)	Rear (pts.)		
1991	Samurai	5	1.3 (1298)	3.7	-	2.7	-	1	4.2	3.2	10.6	5.1
	Sidekick	0	1.6 (1590)	4.5	-	3.2	10.8	4	2.1	4.6	11.1	1
1992	Sammurai	5	1.3 (1298)	3.7	-	2.7	-	1	4.2	3.2	10.6	5.1
	Sidekick	0	1.6 (1590)	4.5	-	3.2	10.8	4	2.1	4.6	11.1	1
1993	Samurai	5	1.3 (1298)	3.7	-	2.7	-	1	4.2	3.2	10.6	5.1
	Sidekick	0	1.6 (1590)	4.5	-	3.2	10.8	4	2.1	4.6	11.1	1
1994-95	Samurai	5	1.3 (1298)	3.7	-	2.7	-	1	4.2	3.2	10.6	5.1
	Sidekick	0	1.6 (1590)	4.5	-	3.2	10.8	4	2.1	4.6	11.1	1

1 Automatic transmission: 5.5 qts.
Manual transmission: 5.6 qts.

7911fC06

CAMSHAFT SPECIFICATIONS

All measurements given in inches.

Year	Engine ID/VIN	Engine Displacement Liters (cc)	Journal Diameter					Elevation		Bearing Clearance	Camshaft End Play
			1	2	3	4	5	In.	Ex.		
1991	5	1.3 (1298)	1.7372-1.7381	1.7451-1.7460	1.7530-1.7539	1.7609-1.7618	1.7687-1.7697	1.4763-1.4724	1.4763-1.4724	0.0020-0.0036	0.0039
	0	1.6 (1590)	1.7372-1.7381	1.7451-1.7460	1.7530-1.7539	1.7609-1.7618	1.7687-1.7697	1.4763-1.4724	1.4763-1.4724	0.0020-0.0036	0.0039
1992	5	1.3 (1298)	1.7372-1.7381	1.7451-1.7460	1.7530-1.7539	1.7609-1.7618	1.7687-1.7697	1.4763-1.4724	1.4763-1.4724	0.0020-0.0036	0.0039
	0	1.6 (1590)	1.7372-1.7381	1.7451-1.7460	1.7530-1.7539	1.7609-1.7618	1.7687-1.7697	1.4763-1.4724	1.4763-1.4724	0.0020-0.0036	0.0039
	0 1	1.6 (1590)	1.1000-1.1008	1.1000-1.1008	1.1000-1.1008	1.1000-1.1008	1.1000-1.1008	1.4551-1.4557	1.4328-1.4334	0.0016-0.0032	0.0039
1993	5	1.3 (1298)	1.7372-1.7381	1.7451-1.7460	1.7530-1.7539	1.7609-1.7618	1.7687-1.7697	1.4763-1.4724	1.4763-1.4724	0.0020-0.0036	0.0039
	0	1.6 (1590)	1.7372-1.7381	1.7451-1.7460	1.7530-1.7539	1.7609-1.7618	1.7687-1.7697	1.4763-1.4724	1.4763-1.4724	0.0020-0.0036	0.0039
	0 1	1.6 (1590)	1.1000-1.1008	1.1000-1.1008	1.1000-1.1008	1.1000-1.1008	1.1000-1.1008	1.4551-1.4557	1.4328-1.4334	0.0016-0.0032	0.0039
1994	5	1.3 (1298)	1.7372-1.7381	1.7451-1.7460	1.7530-1.7539	1.7609-1.7618	1.7687-1.7697	1.4763-1.4724	1.4763-1.4724	0.0020-0.0036	0.0039
	0	1.6 (1590)	1.7372-1.7381	1.7451-1.7460	1.7530-1.7539	1.7609-1.7618	1.7687-1.7697	1.4763-1.4724	1.4763-1.4724	0.0020-0.0036	0.0039
	0 1	1.6 (1590)	1.1000-1.1008	1.1000-1.1008	1.1000-1.1008	1.1000-1.1008	1.1000-1.1008	1.4241-1.4303	1.4314-1.4376	0.0016-0.0032	0.0039

1 Sidekick 16 valve engine

7911fC07

CRANKSHAFT AND CONNECTING ROD SPECIFICATIONS

All measurements are given in inches.

Year	Engine ID/VIN	Engine Displacement Liters (cc)	Crankshaft				Connecting Rod		
			Main Brg. Journal Dia.	Main Brg. Oil Clearance	Shaft End-play	Thrust on No.	Journal Diameter	Oil Clearance	Side Clearance
1991	5	1.3 (1298)	1	0.0008-0.0016	0.0044-0.0122	3	1.7710-1.7716	0.0012-0.0019	0.0039-0.0078
	0	1.6 (1590)	2	0.0012-0.0023	0.0100-0.0149	3	1.7316-1.7323	0.0008-0.0031	0.0039-0.0137
1992	5	1.3 (1298)	1	0.0008-0.0016	0.0044-0.0122	3	1.7710-1.7716	0.0012-0.0019	0.0039-0.0078
	0	1.6 (1590)	2	0.0008-0.0023	0.0044-0.0149	3	1.7316-1.7323	0.0008-0.0031	0.0039-0.0137
1993	5	1.3 (1298)	1	0.0008-0.0016	0.0044-0.0122	3	1.7710-1.7716	0.0012-0.0019	0.0039-0.0078
	0	1.6 (1590)	2	0.0008-0.0023	0.0044-0.0149	3	1.7316-1.7323	0.0008-0.0031	0.0039-0.0137
1994-95	5	1.3 (1298)	1	0.0008-0.0016	0.0044-0.0122	3	1.7710-1.7716	0.0012-0.0019	0.0039-0.0078
	0	1.6 (1590)	2	0.0008-0.0023	0.0044-0.0149	3	1.7316-1.7323	0.0008-0.0031	0.0039-0.0137

1 No. 1: 1.7714-1.7716
No. 2: 1.7712-1.7714
No. 3: 1.7710-1.7712
2 No. 1: 2.0470-2.0472
No. 2: 2.0468-2.0470
No. 3: 2.0465-2.0468

7911fC08

VALVE SPECIFICATIONS

Year	Engine ID/VIN	Engine Displacement Liters (cc)	Seat Angle (deg.)	Face Angle (deg.)	Spring Test Pressure (lbs. @ in.)	Spring Installed Height (in.)	Stem-to-Guide Clearance (in.)		Stem Diameter (in.)	
							Intake	Exhaust	Intake	Exhaust
1991	5	1.3 (1298)	45	45	55-64@ 1.63	1.9409	0.0008-0.0019	0.0014-0.0025	2.742-0.2748	0.2737-0.2742
	0	1.6 (1590)	45	45	55-64@ 1.63	1.9074	0.0008-0.0019	0.0014-0.0025	2.742-0.2748	0.2737-0.2742
1992	5	1.3 (1298)	45	45	55-64@ 1.63	1.9409	0.0008-0.0019	0.0014-0.0025	0.2742-0.2748	0.2737-0.2742
	0	1.6 (1590)	45	45	55-64@ 1.63	1.9074	0.0008-0.0019	0.0014-0.0025	0.2742-0.2748	0.2737-0.2742
	0	1.6 (1590) [1]	45	45	24-28@ 1.24	1.4500	0.0008-0.0018	0.0018-0.0028	0.2152-0.2157	0.2142-0.2148
1993	5	1.3 (1298)	45	45	55-64@ 1.63	1.9409	0.0008-0.0019	0.0014-0.0025	0.2742-0.2748	0.2737-0.2742
	0	1.6 (1590)	45	45	55-64@ 1.63	1.9866	0.0008-0.0019	0.0014-0.0025	0.2742-0.2748	0.2737-0.2742
	0	1.6 (1590) [1]	45	45	24-28@ 1.24	1.4500	0.0008-0.0018	0.0018-0.0028	0.2152-0.2157	0.2142-0.2148
1994-95	5	1.3 (1298)	45	45	55-64@ 1.63	1.9409	0.0008-0.0019	0.0014-0.0025	0.2742-0.2748	0.2737-0.2742
	0	1.6 (1590)	45	45	55-64@ 1.63	1.9866	0.0008-0.0019	0.0014-0.0025	0.2742-0.2748	0.2737-0.2742
	0	1.6 (1590) [1]	45	45	24-28@ 1.24	1.4500	0.0008-0.0018	0.0018-0.0028	0.2152-0.2157	0.2142-0.2148

[1] Sidekick 16 valve engine

7911fC09

PISTON AND RING SPECIFICATIONS
All measurements are given in inches.

Year	Engine ID/VIN	Engine Displacement Liters (cc)	Piston Clearance	Ring Gap			Ring Side Clearance		
				Top Compression	Bottom Compression	Oil Control	Top Compression	Bottom Compression	Oil Control
1991	5	1.3 (1298)	0.0008-0.0015	0.0079-0.0129	0.0079-0.0137	0.0079-0.0275	0.0012-0.0027	0.0008-0.0023	NA
	0	1.6 (1590)	0.0008-0.0015	0.0079-0.0129	0.0079-0.0137	0.0079-0.0275	0.0012-0.0027	0.0008-0.0023	NA
1992	5	1.3 (1298)	0.0008-0.0015	0.0079-0.0129	0.0079-0.0137	0.0079-0.0275	0.0012-0.0027	0.0008-0.0023	NA
	0	1.6 (1590)	0.0008-0.0015	0.0079-0.0137	0.0079-0.0137	0.0079-0.0275	0.0012-0.0027	0.0008-0.0023	NA
1993	5	1.3 (1298)	0.0008-0.0015	0.0079-0.0129	0.0079-0.0137	0.0079-0.0275	0.0012-0.0027	0.0008-0.0023	NA
	0	1.6 (1590)	0.0008-0.0015	0.0079-0.0137	0.0079-0.0137	0.0079-0.0275	0.0012-0.0027	0.0008-0.0023	NA
1994-95	5	1.3 (1298)	0.0008-0.0015	0.0079-0.0129	0.0079-0.0137	0.0079-0.0275	0.0012-0.0027	0.0008-0.0023	NA
	0	1.6 (1590)	0.0008-0.0015	0.0079-0.0137	0.0079-0.0137	0.0079-0.0275	0.0012-0.0027	0.0008-0.0023	NA

NA - Not Available

7911fC10

TORQUE SPECIFICATIONS
All readings in ft. lbs.

Year	Engine ID/VIN	Engine Displacement Liters (cc)	Cylinder Head Bolts	Main Bearing Bolts	Rod Bearing Bolts	Crankshaft Damper Bolts	Flywheel Bolts	Manifold		Spark Plugs	Lug Nut
								Intake	Exhaust		
1991	5	1.3 (1298)	51-54	36-41	24-26	76-83 [1]	41-47	13-20	13-20	14-21	36-57
	0	1.6 (1590)	51-54	36-41	24-26	76-83 [1]	55-58	13-20	13-20	14-21	58-80
1992	5	1.3 (1298)	51-54	36-41	24-26	76-83 [1]	41-47	13-20	13-20	14-21	36-57
	0	1.6 (1590)	51-54	36-41	24-26	76-83 [1]	55-58	13-20	13-20	14-21	58-80
1993	5	1.3 (1298)	51-54	36-41	24-26	76-83 [1]	41-47	13-20	13-20	14-21	36-57
	0	1.6 (1590)	51-54	36-41	24-26	76-83 [1]	55-58	13-20	13-20	14-21	58-80
1994-95	5	1.3 (1298)	51-54	36-41	24-26	76-83 [1]	41-47	13-20	13-20	14-21	36-57
	0	1.6 (1590)	48-51	36-41	24-26	76-83 [1]	55-58	13-20	13-20	14-21	58-80

1 Specification shown is for Crankshaft timing sprocket nut

7911fC11

BRAKE SPECIFICATIONS

All measurements in inches unless noted

Year	Model		Master Cylinder Bore	Brake Disc Original Thickness	Brake Disc Minimum Thickness	Brake Disc Maximum Runout	Brake Drum Diameter Original Inside Diameter	Brake Drum Diameter Max. Wear Limit	Brake Drum Diameter Maximum Machine Diameter	Minimum Lining Thickness Front	Minimum Lining Thickness Rear
1991	Samurai		NA	0.394	0.334	0.006	8.66	8.74	8.74	0.236	0.120
	Sidekick		NA	0.394	0.315	0.006	8.66	8.74	8.74	0.315	0.120
1992	Samurai		NA	0.394	0.334	0.006	8.66	8.74	8.74	0.236	0.120
	Sidekick		NA	0.394	0.315	0.006	8.66	8.74	8.74	0.315	0.120
	Sidekick 1		NA	0.669	0.591	0.006	10.00	10.07	10.07	0.315	0.120
1993	Samurai		NA	0.394	0.334	0.006	8.66	8.74	8.74	0.236	0.120
	Sidekick		NA	0.394	0.315	0.006	8.66	8.74	8.74	0.315	0.120
	Sidekick 1		NA	0.669	0.591	0.006	10.00	10.07	10.07	0.315	0.120
1994	Samurai		NA	0.394	0.334	0.006	8.66	8.74	8.74	0.236	0.120
	Sidekick		NA	0.394	0.315	0.006	8.66	8.74	8.74	0.315	0.120
	Sidekick 1		NA	0.669	0.591	0.006	10.00	10.07	10.07	0.315	0.120

NA - Not Available

1 4 door model

7911fC12

WHEEL ALIGNMENT

Year	Model		Caster Range (deg.)	Caster Preferred Setting (deg.)	Camber Range (deg.)	Camber Preferred Setting (deg.)	Toe-in (in.)	Steering Axis Inclination (deg.)
1991	Samurai		2 1/2P-4 1/2P	3 1/2P	1/4P-1 3/4P	1P	5/32P-15/32P	9
	Sidekick		1/2P-2 1/2P	1 1/2P	1/2N-1 1/2P	1/2P	5/16P-1/4P	NA
1992	Samurai		2 1/2P-4 1/2P	3 1/2P	1/4P-1 3/4P	1P	5/32P-15/32P	9
	Sidekick		1/2P-2 1/2P	1 1/2P	1/2N-1 1/2P	1/2P	5/16P-1/4P	NA
1993	Samurai		2 1/2P-4 1/2P	3 1/2P	1/4P-1 3/4P	1P	5/32P-15/32P	9
	Sidekick		1/2P-2 1/2P	1 1/2P	1/2N-1 1/2P	1/2P	5/16P-1/4P	NA
1994-95	Samurai		2 1/2P-4 1/2P	3 1/2P	1/4P-1 3/4P	1P	5/32P-15/32P	9
	Sidekick		1/2P-2 1/2P	1 1/2P	1/2N-1 1/2P	1/2P	5/16P-1/4P	NA

NA - Not Available

7911fC13

ENGINE ELECTRICAL

NOTE: Disconnecting the battery cable on some vehicles may interfere with the functions of the on board computer systems and may require the computer to undergo a relearning process, once the negative battery cable is disconnected.

Distributor

REMOVAL

1. Disconnect the negative battery terminal at the battery.
2. Disconnect the electrical connections and the vacuum hoses.

NOTE: Do not bend or twist the spark plug wires to avoid internal damage. Grip the the wire boot when removing or installing the wires.

3. Remove the distributor cap.
4. Mark the rotor position on the distributor housing and the distributor housing position on the engine.
5. Remove the distributor flange bolt and remove the distributor.

NOTE: Do not crank the engine with the distributor removed.

INSTALLATION

Timing Not Disturbed

1. Install the distributor and align the marks on the distributor housing and on the engine.
2. Tighten the flange bolt and install the distributor cap.
3. Connect the electrical connections and the vacuum hoses.
4. Connect the negative battery cable and set the timing to specification.

Timing Disturbed

1. Rotate the crankshaft in a clockwise position until the specified timing mark on the flywheel aligns with the timing matchmark on the engine.

NOTE: After aligning the 2 marks, remove the cylinder head cover to visually check that the rocker arms are not riding on the camshaft cams at No. 1 cylinder. If the arms are found to be riding on the cams, turn the crankshaft **360 degrees to realign the 2 marks.**

2. Position the rotor to the No. 1 cylinder position and install the distributor.
3. Connect the vacuum and electrical connections.
4. Connect the negative battery cable and set the timing to specification.

Ignition Timing

ADJUSTMENT

1. Start the engine and warm up to normal operating temperature. Prior to any adjustment, be sure all the electrical accessories are **OFF**.
2. After warming up, make sure the idle speed is within the proper specification.
3. Remove cap from monitor coupler next to battery and connect terminals **C** and **D** with a jumper wire. Connect the timing light to the No. 1 cylinder spark plug wire.
4. With the engine running at the specified idle speed, direct the timing light to the crankshaft pulley. If the specified timing mark on the timing tab is aligned with the timing notch on the crankshaft pulley the ignition is properly timed.
5. If the timing is out of adjustment, loosen the distributor flange bolt and turn the distributor housing to advance or retard the timing. Turn the distributor counterclockwise to advance the timing and clockwise to retard the timing.
6. After the adjustment, tighten the flange bolt and recheck the timing.

Alternator

BELT TENSION ADJUSTMENT

1. Disconnect the negative battery cable.
2. Loosen the alternator bolt and pivot bolts.
3. Adjust the belt tension to 0.24 – 0.32 in. of deflection using a belt tension gauge and measuring at the midpoint between the 2 pulleys.

REMOVAL AND INSTALLATION

1. Disconnect the negative battery cable.
2. Disconnect the wire coupler and white lead wire from the alternator.

3. Remove the radiator shroud from under the radiator.
4. Remove the alternator mounting bolt and alternator drive belt adjusting bolt.
5. Remove the alternator from the vehicle.
To install:
6. Install the alternator and tighten the mounting bolt and drive belt adjusting bolt. Adjust the drive belt tension.
7. Install the shroud under the radiator and connect the wire coupler to the alternator.
8. Connect the negative battery cable.

Starter

REMOVAL AND INSTALLATION

1. Disconnect the negative battery cable.
2. Raise and support the vehicle safely.
3. Disconnect the lead wire and battery cable from the starter motor.
4. Support the starter and remove the 2 mounting bolts.
5. Remove the starter.
To install:
6. Install the starter and tighten the 2 mounting bolts to 22 ft. lbs..
7. Connect the lead wire and battery cable to the starter.
8. Lower the vehicle and connect the negative battery cable.

CHASSIS ELECTRICAL

Heater Blower Motor

REMOVAL AND INSTALLATION

Samurai

1. Disconnect the negative battery cable. Drain the cooling system.
2. Disconnect the inlet and outlet heater hoses from the heater core.
3. Remove the horn pad and the steering wheel retaining nut and remove the steering wheel by using the proper puller.
4. Disconnect and tag the radio and cigar lighter wires. Remove the radio from the vehicle.
5. Remove the ash tray and mounting plate.

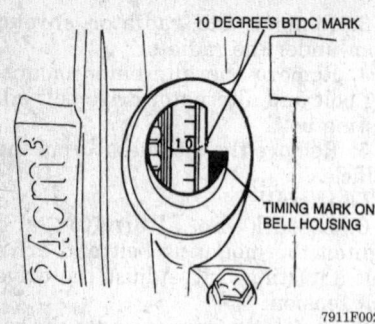

Timing mark alignment — Samurai{?nl}The timing marks are visible by removing the rubber plug in the bell housing

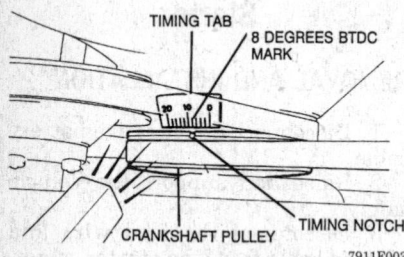

Timing marks — Sidekick

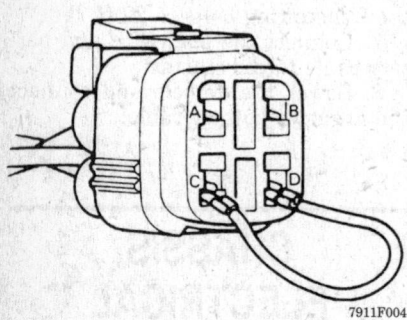

Installed jumper wire at monitor coupler

6. Disconnect the hood release cable from the release lever.

7. Disconnect and tag the heater control cables and wires at the controls.

8. Remove the heater control lever knobs and facing plate. Loosen the lever case screws.

9. Remove the defroster and side ventilator hoses.

10. Disconnect the lead wires and speedometer cable from the speedometer and remove the lead wires from the heater controls.

11. Disconnect the wiring harness clamps from the instrument panel.

12. Loosen the instrument panel mounting screws and remove the instrument panel.

NOTE: When removing the heater lever case which is fitted in the steering column holder, be very careful not to damage it.

13. Loosen the front door opening stop screws and remove the steering column bracket.

14. Disconnect and tag the blower motor and resistor connections at the coupler.

15. Loosen the heater case securing nut on the engine side.

16. Remove the heater assembly from the vehicle.

17. Remove the blower motor from the case.

To install:

18. Install the blower motor in the heater case and install the assembly in the vehicle.

19. Tighten the heater case securing nut on the engine side.

20. Install the blower motor and resistor connections at the coupler.

21. Tighten the front door opening stop screws and install the steering column holder.

22. Tighten the instrument panel mounting screws and replace the instrument panel.

23. Reconnect the wiring harness clamps to the instrument panel.

24. Reconnect the lead wires and speedometer cable to the speedometer.

25. Install the defroster and side ventilator hoses.

26. Install the heater control knobs and plate, and tighten the lever case screws.

27. Reconnect the heater control cables at the controls.

28. Reconnect the hood release cable to the release lever.

29. Install the ash tray and mounting plate.

30. Reconnect the radio and cigar lighter wires. Install the radio in the vehicle.

31. Install the horn pad and steering wheel retaining nut and install the steering wheel.

32. Reconnect the inlet and outlet heater hoses to heater core.

33. Refill the cooling system with the proper coolant. Install the negative battery cable.

Sidekick

1. Disconnect the negative battery cable.

2. Remove the glove box assembly.

3. Disconnect the blower motor and resistor wire connectors.

4. Disconnect the fresh air control cable from the blower motor case.

5. Loosen but do not remove the blower housing fastener bolts.

6. Remove the 3 blower motor mounting screws.

7. Remove the blower motor.

To install:

8. Install the blower motor and install the 3 mounting screws.

9. Tighten the blower housing fastener bolts.

10. Connect the fresh air control cable to the blower motor case.

11. Connect the blower motor and resistor wire connectors.

12. Install the glove box assembly and connect the negative battery cable.

Windshield Wiper Motor

REMOVAL AND INSTALLATION

Samurai

SOFT TOP

1. Disconnect the negative battery cable.

2. Remove the wiper linkage to wiper mounting nut.

3. Disconnect the wire connector from the wiper motor.

4. Remove the 3 wiper mounting bolts and remove the wiper motor.

To install:

5. Install the wiper motor and install the 3 mounting bolts.

6. Connect the wire connector and linkage to the wiper motor and tighten the linkage mounting nut.

7. Connect the negative battery cable.

HARD TOP

NOTE: On some models, the windshield frame may have to be removed.

1. Disconnect the negative battery cable.

2. Remove the wiper linkage to wiper mounting nut.

3. Disconnect the wire connector from the wiper motor.

4. Remove the 3 wiper mounting bolts and remove the wiper motor.

To install:

5. Install the 3 wiper motor mounting bolts and install the motor.

6. Install the linkage and mounting nut and connect the negative battery cable.

Sidekick

1. Disconnect the negative battery cable.

2. Remove the right and left cowl grilles.

3. Remove the wiper linkage to wiper mounting nut.

4. Disconnect the wire connector from the wiper motor.

5. Remove the 4 wiper mounting bolts and remove the wiper motor.

To install:

6. Install the 4 wiper mounting bolts and the wiper motor.

7. Connect the wiper motor wire connector and wiper linkage.

8. Install the right and left cowl grilles and connect the negative battery cable.

Instrument Cluster

REMOVAL AND INSTALLATION

1. Disconnect the negative battery cable.

2. Remove the lower instrument panel cover.

3. Loosen, but do not remove the 2 upper and 4 lower steering column mounting bolts. Lower and support the steering column, if necessary.

4. Remove the outer instrument cluster cover.

5. Remove the 4 cluster mounting screws and slide the cluster outwards.

6. Disconnect the lead wires and speedometer cable from the speedometer.

7. Remove the speedometer.

8. Disconnect and tag the wire connector from the rear of the instrument cluster.

9. Remove the instrument cluster.

To install:

10. Install the instrument cluster and connect the wire connector to the rear.

11. Install the speedometer and connect the speedometer cable and lead wires.

12. Install the 4 cluster mounting screws.

13. Install the outer instrument cluster cover.

14. Secure the steering column and the column mounting bolts.

15. Install the lower instrument cover panel and connect the negative battery cable.

Radio

REMOVAL AND INSTALLATION

1. Disconnect the negative battery cable.

2. Loosen the retaining bracket at the rear of the radio.

3. Remove the pins from the 4 corners of the radio front.

4. Disconnect the electrical connections and antenna lead from the radio.

5. Remove the radio from the vehicle.

To install:

6. Install the radio and connect the electrical connections and antenna lead.

7. Install the 4 pins to the corners of the front of the radio.

8. Tighten the bolts connecting the retaining bracket to the rear of the radio.

9. Connect the negative battery cable.

Combination Switch

The combination switch incorporates the turn signal, windshield wiper, dimmer and headlight switches into 1 switch.

REMOVAL AND INSTALLATION

1. Disconnect the negative battery cable.

2. Remove the horn button and remove the steering wheel retaining nut and remove the steering wheel.

3. Remove the upper and lower column cover screws and remove the covers.

4. Disconnect the lead wires from the combination switch at the connector.

5. Remove the combination switch assembly screws.

6. Remove the combination switch from the steering column.

To install:

7. Install the combination switch to the steering column and install the assembly screws.

8. Connect the lead wires at the connector to the combination switch.

9. Install the lower and upper column covers and reinstall the steering wheel.

10. Install the horn button and connect the negative battery cable.

Ignition Switch

REMOVAL AND INSTALLATION

1. Disconnect the negative battery cable.

2. Remove the steering wheel from the vehicle.

3. Disconnect the wire connector at the ignition switch.

4. With the ignition switch in the **OFF** position, remove the mounting bolts and remove the switch.

To install:

5. With the ignition switch in the **OFF** position, install the switch and the mounting bolts.

6. Connect the wire connector at the ignition switch.

7. Install the steering wheel and horn pad and connect the negative battery cable.

Ignition Lock

REMOVAL AND INSTALLATION

1. Disconnect the negative battery cable.

2. Remove the steering wheel.

3. Disconnect the lead wires to the ignition and combination switches.

4. Disconnect the steering joint by removing the joint bolt.

5. Remove the steering column fastening bolts.

6. Remove the steering column assembly.

7. Using a center punch, loosen and remove the steering lock mounting bolts.

8. Turn the key to **ACC** or **ON** position and remove the steering lock assembly from the steering column.

To install:

9. To install the switch position the oblong hole of the steering shaft in the center of the hole in the steering column.

10. Turn the ignition key to the **ACC** or **ON** position and install the steering lock assembly onto the column. Turn the ignition key to the **LOCK** position and remove it. Align the hub on the lock with oblong hole of the steering shaft.

11. Rotate the shaft to assure that the steering shaft is locked.

12. Tighten the 2 new steering lock mounting bolts.

13. Turn the ignition key to the **ACC** or **ON** position and check to be sure the steering shaft rotates smoothly and check the lock mechanism operation.

14. Install the steering column assembly by attaching the joint and joint bolt.

15. Install the steering column attaching bolts.

16. Connect the wires to the ignition and combination switches.

17. Install the steering wheel and connect the negative battery cable.

Stoplight Switch

REMOVAL AND INSTALLATION

1. Disconnect the negative battery cable. Remove the under dash trim panel, if equipped.
2. Push the brake pedal down and remove the stoplight locknut.
3. Disconnect the wire connector at the switch.
4. Remove the switch from the bracket.
To install:
5. Install the new switch on the bracket.
6. Adjust the switch so there is 0.02–0.04 in. clearance between the switch and the brake pedal.
7. Install the locknut and tighten to 7.5–10.5 ft. lbs.
8. Install the negative battery cable.

Clutch Switch

REMOVAL AND INSTALLATION

Sidekick

1. Disconnect the negative battery cable. Remove the under dash trim panel, if equipped.
2. Disconnect the start switch connector beside the clutch pedal bracket.
3. Remove the locknut and start switch.
To install:
4. Installation is the reverse of the removal procedure.

Fuses and Relays

LOCATION

The fuse box and turn signal relay are located under the driver side of the dashboard.

Computers

LOCATION

Sidekick

The Electronic Control Module computer is located under the dash on the driver side of the vehicle.

Cruise Control

ADJUSTMENT

Sidekick

1. Remove the top cover from the actuator assembly.
2. Loosen the lock nuts on the cable.
3. Adjust the cable so the play range is between 0.04–0.08 in. when the actuator lever is fully closed.
4. Tighten the locknut to 4 ft. lbs. and replace the top cover.

ENGINE COOLING

Radiator

REMOVAL AND INSTALLATION

1. Disconnect the negative battery cable.
2. Drain the cooling system.
3. Disconnect and plug the transmission cooler lines, if equipped.
4. Loosen the water pump drive belt tension.
5. Remove the cooling fan and radiator shroud.
6. Disconnect the water hoses to the radiator.
7. Remove the radiator retaining bolts and remove the radiator.
To install:
8. Install the radiator mounting bolts and install the radiator.
9. Reconnect the water hoses to the radiator.
10. Install the cooling fan and radiator shroud.
11. Tighten the water pump drive belt to the proper tension.

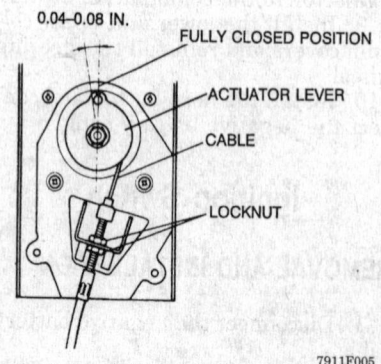

0.04–0.08 IN.

FULLY CLOSED POSITION

ACTUATOR LEVER

CABLE

LOCKNUT

7911F005

View of the cruise control adjustment

12. Reconnect the transmission cooler lines, if necessary, to the radiator.
13. Refill the cooling system.
14. Reconnect the negative battery cable.

Heater Core

REMOVAL AND INSTALLATION

Samurai

1. Disconnect the negative battery cable. Drain the cooling system.
2. Disconnect the heater hoses from the heater core.
3. Disconnect the radio and cigar lighter lead wires, and remove the radio from the vehicle.
4. Remove the ash tray and mounting plate.
5. Disconnect the hood release cable from the release lever.
6. Disconnect and tag the heater control cables and wires at the controls.
7. Remove the heater control lever knobs and facing plate, and loosen the lever case screws.
8. Remove the defroster and side ventilator hoses.
9. Disconnect the lead wires and speedometer cables from the speedometer.
10. Disconnect the wiring harness from the instrument panel.
11. Loosen the instrument panel mounting screws and remove the instrument panel.
12. Loosen the front door opening stop screws and remove the steering column holder.
13. Disconnect and tag the blower motor and resistor connections at the coupler.
14. Loosen the heater securing nut on the engine side.
15. Remove the heater assembly from the vehicle.
16. Remove the clips holding the heater case together, separate the case and remove the heater core.
To install:
17. Install the heater core into the heater case and install the clips holding the case together.
18. Install the heater assembly in the vehicle and tighten the securing nut on the engine side.
19. Reconnect the blower motor and resistor connections at the coupler.
20. Install the steering column holder and tighten the front door opening stop screws.
21. Install the instrument panel and tighten panel mounting screws.

22. Reconnect the wiring harness to the instrument panel.

23. Install the defroster and side ventilator hoses.

24. Tighten the heater control case screws and install the control knobs and facing plate.

25. Reconnect the heater control cable and wires to the heater control.

26. Reconnect the hood release cable the release lever.

27. Install the ash tray and mounting plate.

28. Install radio and connect the cigar lighter and radio wires.

29. Reconnect the heater hoses to the heater core.

30. Refill the cooling system to the proper with the proper coolant. Install the negative cable.

Sidekick

1. Disconnect the negative battery cable. Drain the cooling system.

2. Disconnect the 2 water hoses from the heater assembly.

3. Remove the steering wheel. Disconnect the lead wires from the ignition and combination switches.

4. Disconnect the wiring harness and remove the instrument panel and related parts.

5. Disconnect all the wiring connections and cables from the heater controls.

6. Disconnect the defroster duct and speedometer cable retaining bracket from the heater case.

7. Remove the heater assembly and remove the heater core.

To install:

8. Install the heater core and install the heater assembly.

9. Reconnect the defroster duct and speedometer cable retaining bracket to the heater case.

10. Reconnect the wiring connections and cables to the heater controls.

11. Reconnect the wiring harness and install the instrument panel.

12. Connect the lead wires to the ignition and combination switches and install the steering wheel.

13. Connect the water hoses to the heater assembly.

14. Refill the cooling system with the proper coolant and replace the negative battery cable.

Water Pump

REMOVAL AND INSTALLATION

1. Disconnect the negative battery cable.

2. Drain the cooling system.

3. Loosen the drive belt tension and remove the drive belt. If equipped, loosen the air conditioning belt tension and remove the air conditioning belt.

4. Remove the radiator fan shroud mounting bolts and radiator fan mounting bolts. Remove the radiator shroud, fan and water pump pulley from the vehicle.

5. Remove the crankshaft pulley bolts and remove the crankshaft pulley.

NOTE: The crankshaft pulley bolt can be removed without removing the center crankshaft bolt.

6. Remove the timing belt cover mounting bolts and remove the cover.

7. Loosen the timing belt tensioner adjusting bolt and pivot nut. Hold the tensioner to loosen the timing belt and remove the belt from the camshaft pulley.

8. Remove the timing belt tensioner mounting bolts and remove the tensioner plate and spring.

9. Remove the water pump mounting bolts and remove the water pump assembly.

10. On Sidekick, it is necessary to remove the dipstick tube and the alternator bracket from the vehicle.

To install:

11. Clean and inspect the surface of the engine before installation.

12. Using a new gasket, install the new water pump on the engine. Torque the mounting bolts to 8 ft. lbs.

13. On Sidekick, install the dipstick tube and the alternator bracket to the vehicle.

14. Install the rubber seals between the water pump to cylinder head and water pump to oil pump.

15. Install the timing belt tensioner plate, tensioner and spring.

16. Align the marks on the timing belt and the camshaft sprocket. Install the timing belt in the same position on the camshaft sprocket as when removed.

17. Adjust the timing belt to be free of any slack. Torque the tensioner bolts to 7.5–9.0 ft. lbs.

18. Install the crankshaft and water pump pulleys. Torque the crankshaft and water pulley bolts to 7.5–9.0 ft. lbs.

19. Install the timing belt cover, cooling fan/clutch, shroud and d rive belt. Adjust the belt tension.

20. As required, adjust the intake and exhaust valve lash.

21. Refill the cooling system with the proper coolant and replace the negative battery cable.

Thermostat

REMOVAL AND INSTALLATION

1. Disconnect the negative battery cable.

2. Drain the cooling system.

3. Disconnect the thermostat cap from the intake manifold and remove the thermostat.

To install:

4. Clean and inspect the surfaces of the housing and the engine.

5. Install the new thermostat with the spring facing towards the engine.

6. Install a new gasket and the thermostat cap to the intake manifold.

7. Refill the cooling system with the proper coolant and replace the negative battery cable.

FUEL SYSTEM

Fuel System Service Precaution

When working with the fuel system, certain precautions should be taken; always work in a well ventilated area, keep a dry chemical (Class B) fire extinguisher near the work area. Always disconnect the negative battery cable and do not make any repairs to the fuel system until all the necessary steps for repair have been reviewed.

RELIEVING FUEL SYSTEM PRESSURE

The fuel pressure must be relieved before performing any service on the fuel system.

1. Disconnect the negative battery cable.

2. Remove the fuel filler cap from the fuel filler neck to release the fuel vapor pressure in the fuel tank. Reinstall the fuel cap.

3. Raise and support the vehicle safely.

4. Place an appropriate container under the fuel filter.

5. Cover the plug bolt on the fuel filter inlet union bolt with a rag and loosen the plug bolt slowly to release the fuel pressure gradually.

6. When the pressure has been released, tighten the plug bolt to 7.5 ft. lbs. so the fuel does not leak.

7. Lower the vehicle.

8. Reconnect the negative battery cable.

Fuel Filter

REMOVAL AND INSTALLATION

1. Disconnect the negative battery cable.

2. Remove the fuel filler cap to release the fuel vapor pressure in the fuel tank. After releasing pressure, reinstall filler cap.

3. Raise and support the vehicle safely.

4. Release the fuel pressure.

5. Disconnect the inlet and outlet hoses from the fuel filter.

6. Remove the fuel filter from the chassis frame.

To install:

7. Install the new fuel filter to the frame and connect the inlet and outlet hoses.

8. Lower the vehicle and connect the negative battery cable.

Mechanical Fuel Pump

REMOVAL AND INSTALLATION

1.3L Engine

1. Disconnect the negative battery cable.

2. Remove the fuel filler cap to release the fuel vapor pressure in the fuel tank.

3. Disconnect and tag the fuel inlet, outlet and return hoses from the fuel pump.

4. Remove the fuel pump mounting bolts and remove the fuel pump.

5. Remove the fuel pump mounting rod from the engine and lubricate it with engine oil before installation.

To install:

6. Clean the engine mounting surface and install the fuel pump using a new gasket and tighten to 10 ft. lbs.

7. Connect the inlet, outlet and return hoses to the fuel pump.

8. Reconnect the negative battery cable. Run the engine and check for leaks when finished.

Electric Fuel Pump

PRESSURE TESTING

1.6L Engine

1. Remove the fuel filler cap to release the fuel vapor pressure in the fuel tank. Reinstall the cap.

2. Raise and support the vehicle safely.

3. Release the fuel pressure in the fuel feed line.

4. Remove the plug bolt on the fuel filter union bolt and connect the proper fuel pressure gauge to the fuel filter inlet union bolt.

NOTE: If the pressure in the fuel tank is not released prior to system service, the fuel in the fuel tank may be forced out through the fuel hoses during disconnection.

5. Start the engine and warm up to the proper operating temperature.

6. Measure the fuel pressure under each of the following conditions.

   a. At a specified idle speed or with the fuel pump operating and the engine stopped the fuel pressure should read: 34.1–39.8 psi.

   b. Within 1 minute after the fuel pump has stopped the fuel pressure should read: 21.3 psi.

7. Release the fuel pressure and remove the fuel pressure gauge.

8. Install the plug bolt to the fuel filter inlet bolt. Use a new gasket.

9. Start the engine and check for leaks.

10. Lower the vehicle.

REMOVAL AND INSTALLATION

1.6L Engine

The fuel pump is located in the fuel tank. The fuel pump and the fuel gauge sending unit are located on the upper part of the fuel tank.

1. Disconnect the negative battery cable.

2. Remove the rear bumper assembly.

3. Disconnect the fuel gauge sending unit and fuel pump electrical connectors from the fuel tank.

4. Relieve the fuel system pressure.

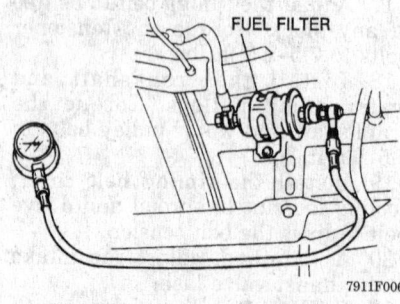

Checking fuel pump pressure

7911F006

5. Disconnect the fuel tank filler hose cover, filler hose, and fuel tank inlet valve (breather hose).

6. Disconnect the inlet pipe from the fuel filter and remove the fuel vapor and return hoses.

7. If necessary, drain the fuel from the tank using a hand operated pump.

8. Disconnect the fuel tank protector and remove the fuel tank and cover from the vehicle.

9. Remove the fuel pump from the fuel tank.

To install:

10. Install the fuel pump to the fuel tank using a new gasket.

11. Install the fuel tank and cover to the vehicle and connect the fuel tank protector.

12. Install the fuel vapor and return hoses and connect the inlet pipe to the fuel filter.

13. Reconnect the fuel tank inlet valve (breather hose), filler hose, and fuel tank filler hose cover.

14. Connect the fuel pump and fuel gauge sending sending unit electrical connectors to the fuel tank.

15. Install the bumper assembly and reconnect the negative battery cable.

16. If necessary, refill the fuel tank with any fuel that had been drained.

17. Start the engine and check for leaks when finished.

Fuel Injection

IDLE SPEED ADJUSTMENT

NOTE: Before starting the engine, place the gear shift lever in neutral for manual transmission or P for automatic transmission. Set the parking brake and block the drive wheels.

1. Warm the engine to the normal operating temperature.

2. Race the engine until the engine speed exceeds 1500 rpm and then let it slow down to idle speed.

3. Check and see if the idle speed is within 750–850 rpm.

Fuel Injector

REMOVAL AND INSTALLATION

1.6L Engine

1. Disconnect the negative battery cable.

2. Release the fuel pressure.

3. Remove the the air intake case from the throttle body.

4. Remove the fuel feed pipe clamp from the intake manifold and disconnect the fuel feed pipe from the throttle body.

5. Remove the injector cover.

6. Disconnect the injector coupler, release its wire harness from the clamp and remove its grommet from the throttle body.

7. Place a cloth over the injector and a hand on top of it. Using an air gun, blow low pressure compressed air into the fuel inlet port of the throttle body and the injector can be removed.

NOTE: Be precise about the pressure of the compressed air. Using excessively high pressure may force the injector to jump out and may cause damage, not only to the injector itself but also to other parts.

To install:

8. Apply thin coating of gasoline to O-rings and then install the fuel injector to the throttle body.

9. Connect fuel injector wire connector and install injector cover.

10. Connect fuel feed pipe to throttle body.

11. Connect the negative battery cable and turn the ignition **ON** for 3 seconds and then **OFF** until fuel pressure is felt at the return hose.

12. Install air intake case assembly.

13. Secure all wires and check for any leaks in the system.

EMISSION CONTROLS

Emission Warning Lamps

The CHECK ENGINE light automatically comes on at 50,000, 80,000 and 100,000 miles.

RESETTING

The lamp reset switch is located on the left side of the dashboard, mounted on the steering column support. The lamp can be reset by moving the switch upwards and then downwards. If the switch remains on after being reset, check the system.

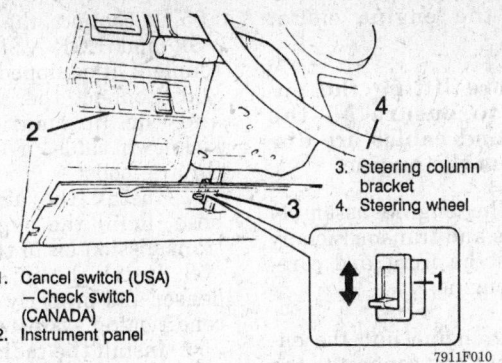

1. Cancel switch (USA) — Check switch (CANADA)
2. Instrument panel
3. Steering column bracket
4. Steering wheel

7911F010

Resetting the check engine light

ENGINE MECHANICAL

NOTE: Disconnecting the battery cable on some vehicles may interfere with the functions of the on board computer systems and may require the computer to undergo a relearning process, once the negative battery cable is disconnected.

Engine

REMOVAL AND INSTALLATION

1. Disconnect both cables at the battery.

2. Remove the hood from the vehicle and drain the cooling system.

3. Remove the radiator reservoir tank, fan shroud, cooling fan and radiator.

4. Properly discharge the air conditioning system and remove the air conditioning condenser, if equipped.

5. Remove the air cleaner outlet hose.

6. Disconnect the accelerator and the automatic transmission kickdown cable from the throttle body, if equipped.

7. Disconnect and tag the throttle opener VSV and EGR VSV wires at the coupler.

8. Disconnect and tag the water temperature, oil pressure, air temperature and ground cable wires at the intake manifold.

9. Disconnect and tag the injector, throttle position sensor and idle speed control solenoid valve wires at their couplers, if equipped.

10. Disconnect the PTC heater wires at the coupler for automatic transmission vehicles, if equipped.

11. Disconnect and tag the wires at the starter and alternator.

12. Disconnect and tag the oxygen sensor and distributor wires at their couplers. Remove the coil wire.

13. Remove the starter motor and disconnect the ground wires from the distributor.

14. Remove the fuel tank filler cap to relieve the pressure. Reinstall the cap.

15. Relieve the fuel pressure in the fuel feed line. Disconnect and tag the fuel feed and return hoses.

16. Remove the gear shift lever mounting bolts and remove the shifter.

17. Disconnect the canister purge hose and remove the pressure sensor hose from the fuel filter, if equipped.

18. Disconnect the brake booster hose from the intake manifold.

19. Remove the vacuum hose from the intake manifold, if equipped with automatic transmission.

20. Disconnect the heater hoses from the intake manifold and from the heater core outlet.

21. Raise and safely support the vehicle.

22. Drain the engine oil and disconnect the exhaust pipe from the exhaust manifold and muffler.

23. Disconnect the clutch cable, if equipped with manual transmission.

24. Drain the automatic transmission fluid, if equipped.

25. Disconnect the clutch (torque converter) housing lower plate.

26. Remove the lock driveplate, using the special tool, if equipped with automatic transmission.

27. Lower the vehicle.

28. Remove the nuts and bolts between the cylinder block and transmission.

29. Support the transmission, using an appropriate stand or jack.

30. Support the engine from the top using a chain type hoist.

31. Remove the engine mount assemblies.

NOTE: Before lifting the engine, check to ensure all the hoses, wires and cables are disconnected from the engine.

32. Remove the engine assembly from the chassis and transmission by sliding towards the front and carefully hoist the engine.

To install:

33. Install the engine into the engine compartment and connect to the transmission.

34. Install the engine mounts with the chassis side mounting brackets. Tighten to 41 ft. lbs.

35. Remove the lifting device.

36. Install the nuts and bolts fastening the cylinder block and transmission.

37. Raise and safely support the vehicle.

38. Install the lock driveplate, if equipped.

39. Connect the clutch (torque converter) housing lower plate.

40. Connect the automatic transmission lines, if equipped.

41. Reconnect the clutch cable to the bracket, if equipped with manual transmission.

42. Connect the exhaust pipe to the exhaust manifold and muffler.

43. Lower the vehicle.

44. Reconnect the heater hoses to the outlet pipe and the intake manifold.

45. Install the vacuum hose for the automatic transmission to the intake manifold, if equipped.

46. Reconnect the brake booster hose to the intake manifold.

47. Install the pressure sensor hose to the fuel filter and connect the canister purge hose, if equipped.

48. Install the fuel return and feed hoses.

49. Install the starter motor and connect the ground wire to the distributor.

50. Install the coil wire and connect the oxygen sensor and distributor wires to their couplers.

51. Reconnect the wires at the alternator and starter motor.

52. Install the PTC heater wires at the coupler, if equipped with automatic transmission.

53. Install the idle speed control solenoid valve, throttle position sensor and injector wires at their couplers, if equipped.

54. Connect the ground cable, air temperature, oil pressure and water temperature wires to the intake manifold.

55. Reconnect the throttle opener VSR and EGR VSR wires to their couplers, if equipped.

56. Connect the accelerator cable and the automatic transmission kickdown cable, if equipped, to the throttle body.

57. Install the air cleaner outlet hose. Refill the engine oil and the transmission oil to the proper level.

58. Install the air conditioning condenser and properly recharge the air conditioning system, if equipped.

59. Install the radiator, cooling fan, fan shroud and radiator reservoir tank.

60. Install the hood and refill the cooling system with the proper coolant.

61. Reconnect the positive and negative battery cables.

62. Adjust the accelerator, clutch and the automatic transmission kickdown cable, if equipped.

63. Before starting the engine, check to see if all the parts disassembled are back in place securely.

64. Start the engine and check the timing. Check for any oil leaks.

Cylinder Head

REMOVAL AND INSTALLATION

NOTE: When removing the cylinder head, remove the pools from the top of the cylinder head before loosening the head bolts to avoid possible contamination.

1. Disconnect the negative battery cable. Drain the cooling system.

2. Disconnect the air intake case.

3. Disconnect the brake booster hose from the intake manifold and the air valve water hose from the throttle body.

4. Disconnect the accelerator and the automatic transmission kickdown cable, if equipped, from the throttle body.

5. Disconnect and tag the throttle opener VSV and EGR VSV wires at the coupler.

6. Disconnect and tag the water temperature, oil pressure, air temperature sensor and ground cable wires at the intake manifold.

7. Disconnect and tag the injector, throttle position sensor and idle speed control solenoid valve wires at their couplers, if equipped.

8. Disconnect the PTC heater wires at the coupler, if equipped with automatic transmission.

9. Disconnect and tag the oxygen sensor and distributor wires at their couplers. Remove the coil wire.

10. Disconnect the ground wires from the distributor assembly.

11. Remove the fuel tank filler cap to relieve the pressure. Reinstall the cap.

12. Release the fuel pressure in the fuel feed line.

13. Disconnect the fuel feed and return hoses.

14. Disconnect the canister purge hose and remove the pressure sensor hose from the fuel filter, if equipped.

15. Remove the vacuum hose for the automatic transmission from the intake manifold, if equipped.

16. Disconnect the radiator cooling fan, fan shroud, water pump drive belt and water pump pulley.

17. Disconnect the crankshaft pulley, timing belt cover and the timing belt.

18. Raise and safely support the vehicle.

19. Disconnect the exhaust pipe from the exhaust manifold and lower the vehicle.

20. Disconnect the air conditioner compressor adjusting arm, if equipped.

21. Remove the cylinder head cover mounting bolts and remove the cylinder head cover.

22. Loosen all the valve adjusting screw locknuts, turn the adjusting screws back all the way to allow all the valves to close.

23. Remove the intake manifold from the cylinder head.

24. Remove the distributor and distributor housing from the cylinder head.

25. Remove the cylinder head mounting bolts and remove the cylinder head from the engine.

26. Remove any oil and water in the cylinder bores and on top of the pistons.

27. Clean and inspect the sealing surfaces of the cylinder head and the engine block.

To install:

28. Install the cylinder head using a new gasket.

29. Install the cylinder head mounting bolts and tighten in the proper sequence to 46–54 ft. lbs.

30. Install the distributor and distributor housing to the cylinder head.

31. Install the intake manifold to the cylinder head.

32. Tighten all the valve adjusting screw nuts.

33. Adjust the valve lash.

34. Install the cylinder head cover and connect the air conditioning compressor adjusting arm, if equipped.

35. Raise and safely support the vehicle.

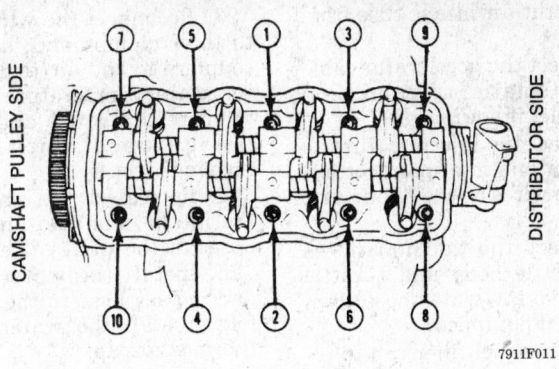

CAMSHAFT PULLEY SIDE

DISTRIBUTOR SIDE

7911F011

Cylinder head bolt tightening sequence

36. Connect the exhaust pipe to the exhaust manifold and safely lower the vehicle.

37. Reconnect the timing belt, timing belt cover and crankshaft pulley.

38. Install the water pump pulley, water pump drive belt, fan shroud and the radiator cooling fan.

39. Install the vacuum hose for the automatic transmission to the intake manifold, if equipped.

40. Install the pressure sensor hose to the fuel filter and connect the canister purge hose, if equipped.

41. Install the fuel return and feed hoses.

42. Connect the ground wires to the distributor assembly.

43. Install the coil wire. Connect the oxygen sensor and distributor wires to their couplers.

44. Connect the PTC wires to the coupler.

45. Connect the idle speed control valve, throttle position sensor and injector wires to their couplers, if equipped.

46. Reconnect the ground cable, air temperature, oil pressure and water temperature sensor wires to the intake manifold.

47. Install the throttle opener VSV and EGR VSV wires to their couplers.

48. Connect the accelerator and the automatic transmission kickdown cables, if equipped, to the throttle body.

49. Install the air valve water hose to the throttle body and connect the brake booster hose to intake manifold.

50. Reconnect the air intake case.

51. Refill the cooling system with the proper coolant and connect the negative battery.

52. Check for any water and oil leaks when finished.

Valve Lash

ADJUSTMENT

Valve lash can be adjusted with the engine hot or cold. Specifications are provided for both adjustments.

1. Remove the air intake case and cylinder head cover.

2. On the 1.3L engine, remove the rubber plug from the transmission case to gain access to the timing marks.

3. Turn the crankshaft clockwise so:

a. The **T** mark punched on the flywheel is aligned with the matchmark on the transmission on the 1.3L engine.

b. The **V** mark on the crankshaft pulley is aligned with the **0** mark on the timing belt cover on the 1.6L engine.

4. Remove the distributor cap and confirm that the rotor is facing the No. 1 firing position. If the rotor is out of place, turn the crankshaft 360 degrees.

5. With the engine in this position, check the valve lash at valves 1, 2, 5 and 7.

NOTE: The valves are adjusted by loosening the locknut on the valve adjuster and turning the adjusting screw to obtain the proper clearance.

6. Once the proper clearance is obtained, the locknut must be torqued to 11–13 ft. lbs. while holding the adjusting screw. Check the clearance after the locknut is torqued. Clearance should be: Intake: 0.0051–0.0067 in. (0.13–0.17mm.) cold or 0.009–0.011 in. (0.23–0.27mm.) hot. Exhaust: 0.0063–0.0079 in. (0.16–0.20mm.) cold or 0.0102–0.0118 in. (0.26–0.30mm.) hot.

7. Rotate the crankshaft 360 degrees, and check the valve lash at valves 3, 4, 6 and 8.

8. After adjusting and checking all the valves, install the cylinder head cover, distributor cover and intake hose.

Rocker Arms/Shafts

REMOVAL AND INSTALLATION

NOTE: The rocker arm shafts are not identical and must be kept in the proper order for installation. If the shafts get mixed up before installation, the intake rocker shaft has a 14mm stepped end and the exhaust rocker shaft has a 13mm stepped end. The stepped end of the intake rocker shaft faces the front of the engine and the stepped end of the exhaust rocker shaft faces the rear of the engine.

1. Disconnect the negative battery cable. Drain the cooling system.

2. Remove the front grille for Sidekick.

3. Remove the radiator cooling fan and fan shroud. Properly discharge the air conditioning system and remove the air conditioning flexible suction hose, if equipped.

4. Remove the radiator for Sidekick.

5. Remove the water pump drive belt and the water pump pulley.

6. Remove the timing belt outside cover, timing belt and the tensioner.

7. Disconnect the air intake case and remove the cylinder head cover.

8. Remove the camshaft timing belt pulley and timing belt inside cover. Insert the proper size rod into the hole in the camshaft to lock the camshaft and loosen the pulley nut.

9. Loosen all the valve lash adjusting screw locknuts and turn the adjusting screws out all the way.

10. Remove the rocker arm shaft screws.

11. Remove the intake and exhaust rocker arm shafts, rocker arms and springs.

12. Keep all the valve train parts in the order that they were removed.

To install:

13. Apply engine oil to the rocker arms, springs and shafts. Install the shafts in the correct direction, into the cylinder head by placing the rocker arms and springs on the shafts as they are installed. With the rocker arms, springs and shafts installed, torque the rocker shaft mounting screws to 7.0–8.5 ft. lbs.

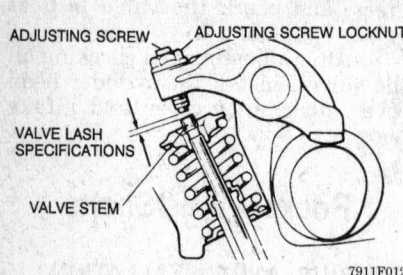

Valve lash adjustment screw and locknut

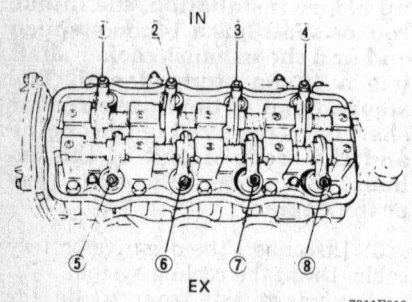

Valve adjustment sequence

14. Install the timing belt pulley and inside cover by locking the camshaft with the proper size rod.

15. With the camshaft locked, tighten the camshaft pulley bolt to 41–46 ft. lbs.

16. The remainder of the installation is the reverse of the removal procedure.

Intake Manifold

REMOVAL AND INSTALLATION

1. Disconnect the negative battery cable. Drain the cooling system.

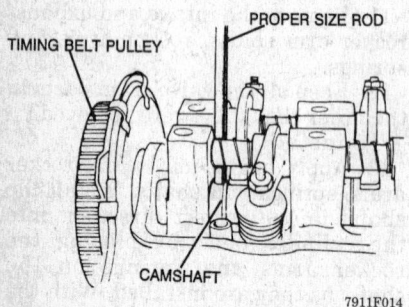

Locking the camshaft in position

2. Remove the air intake case from the manifold.

3. Disconnect the accelerator cable and the automatic transmission kickdown cable, if equipped.

4. Disconnect the injector, throttle position sensor and idle speed control solenoid valve wires at their couplers, if equipped.

5. Disconnect the vacuum hoses from the throttle body and throttle opener. Remove the water hose from the air valve, if equipped.

6. Remove the fuel filler cap to release the fuel pressure in the fuel tank. Reinstall the cap.

7. Release the fuel pressure in the fuel line.

8. Disconnect the fuel line from the throttle body and intake manifold.

9. Remove the fuel return hose.

10. Disconnect the throttle body from the intake manifold and remove the PCV hose from the cylinder head cover.

11. Disconnect the pressure sensor hose from the fuel filter, if equipped, and remove the brake booster hose from the intake manifold.

12. Disconnect the vacuum hose for the automatic transmission from the intake manifold, if equipped.

13. Remove the VSV (for throttle opener) hose from the intake manifold.

14. Disconnect the water hose from the thermostat cap. Remove the heater inlet and water bypass hose from the intake manifold.

15. Disconnect the hoses at the EGR valve and remove the ground wire from the intake manifold.

16. Disconnect the wire couplers from the air temperature sensor, water temperature sensor, water temperature gauge and PTC heater, if equipped with automatic transmission.

17. Disconnect the wire harnesses from their clamps.

18. Remove the intake manifold mounting bolts and remove the intake manifold from the cylinder head.

19. Remove the PCV valve, EGR valve, fuel filter, thermostat from the intake manifold.

20. Clean and inspect the sealing surfaces of the intake manifold and the cylinder head.

To install:

21. Install the thermostat, fuel filter, EGR valve and PCV valve to the intake manifold.

22. Using a new gasket, install the intake manifold and tighten the mounting bolts to 13–20 ft. lbs.

23. Reconnect the wiring harnesses to their clamps and fasten the wire couplers to the air temperature sensor, water temperature sensor, water temperature gauge and PTC heater, if equipped with automatic transmission.

24. Reconnect the ground wire to the intake manifold and replace the hoses at the EGR valve.

25. Install the water heater inlet and bypass hose to the intake manifold. Install the water hose to the thermostat cap.

26. Install the VSV (for throttle opener) hose to the intake manifold.

27. Connect the automatic transmission vacuum hose to the intake manifold, if equipped.

28. Install the brake booster hose to the intake manifold and connect the pressure sensor to the fuel filter, if equipped.

29. Reconnect the fuel line to the throttle body.

30. Install the water hose to the air valve. Connect the vacuum hoses to the throttle body and throttle opener, if equipped.

31. Reconnect the injector, throttle position sensor and idle speed control wires to their couplers, if equipped.

32. Connect the accelerator cable and the automatic transmission kickdown cable, if equipped, to the throttle body.

33. Install the air intake case to the manifold.

34. Refill the cooling system with the proper coolant. Connect the negative battery cable.

35. Check for vacuum, water and oil leaks when finished.

Exhaust Manifold

REMOVAL AND INSTALLATION

1. Disconnect the negative cable.

2. Raise and safely support the vehicle.

3. Disconnect the exhaust pipe from the exhaust manifold and lower the vehicle.

4. Disconnect the oxygen sensor lead wire at the coupler and remove the air intake case bracket for 1.6L engine.

5. Disconnect the exhaust manifold upper and lower covers or heat shields from the exhaust manifold.

6. Remove the exhaust manifold mounting bolts and remove the exhaust manifold from the cylinder head.

To install:

7. Clean and inspect the sealing surfaces of the exhaust manifold and the cylinder head.

8. Using new gaskets, install the exhaust manifold to the cylinder head and tighten the mounting bolts to 13–20 ft. lbs.

9. The remainder of the installation is the reverse of the removal procedure.

10. Check for exhaust leaks when finished.

Timing Belt Front Cover

REMOVAL AND INSTALLATION

1. Disconnect the negative battery cable.

2. Remove the radiator cooling fan and fan shroud.

3. If equipped, disconnect the air conditioning compressor drive belt, properly discharge the air conditioning system and remove the air conditioning compressor flexible suction hose.

4. Loosen the alternator mounting bolts and remove the water pump drive belt and pulley.

5. Remove the crankshaft mounting bolts and remove the crankshaft pulley.

NOTE: The crankshaft drive belt pulley can be removed without loosening the center crankshaft bolt.

6. Disconnect the timing belt cover mounting bolts and remove the timing belt cover.

To install:

7. Clean and inspect all mounting surfaces.

8. Install the timing belt cover and tighten the bolts to 7.0–9.0 ft. lbs.

9. Install the crankshaft pulley. Install the 5 mounting bolts and tighten to 7.0–9.0 ft. lbs.

10. Install the water pump drive belt pulley and install the drive belt.

11. Adjust the belt tension and tighten the alternator mounting bolts.

12. Install the air conditioning flexible suction hose and replace the air conditioning compressor drive belt. Adjust the belt tension and properly recharge the air conditioning system, if equipped.

13. Install the radiator cooling fan and fan shroud.

14. Connect the negative battery cable.

OIL SEAL REPLACEMENT

1. Disconnect the negative battery cable.

2. Remove the the timing belt and the crankshaft sprocket.

3. Insert a suitable tool between the crankshaft and the oil seal and pull the seal outwards to remove it.

NOTE: Use care when removing or installing the oil seal, not to damage the crankshaft or the oil pump sealing surfaces.

4. Clean and inspect the surfaces of the crankshaft and the oil pump assembly.

5. Install the crankshaft sleeve, using the proper tool, onto the crankshaft.

6. Install the new seal over the crankshaft and into the oil pump, making sure the oil seal lip is not upturned.

7. Install the crankshaft sprocket and the timing belt. Check the timing.

8. Connect the negative battery cable.

Timing Belt and Tensioner

ADJUSTMENT

1. Disconnect the negative battery cable.

2. Remove the radiator cooling fan and fan shroud.

3. If equipped, disconnect the air conditioning compressor drive belt, properly discharge the air conditioning system and remove the air conditioning compressor flexible suction hose.

4. Loosen the alternator mounting bolts and remove the water pump drive belt and pulley.

5. Remove the crankshaft mounting bolts and disconnect the crankshaft pulley.

NOTE: The crankshaft drive belt pulley can be removed without loosening the center crankshaft bolt.

6. Disconnect the timing belt cover mounting bolts and remove the timing belt cover. Loosen but do not remove the tensioner bolt.

7. Disconnect the air intake case from the intake manifold.

8. Remove the cylinder head cover and loosen all the valve adjusting screws to permit free rotation of the camshaft.

9. Turn the camshaft pulley clockwise and align the timing marks.

10. Turn the crankshaft clockwise, using a 17mm wrench to crank the timing belt pulley bolt.

11. Align the punch mark on the timing belt pulley with the arrow mark on the oil pump.

12. With the 4 marks aligned, remove any slack from the drive side of the belt. Tighten the tensioner bolt to 17.5–21.5 ft. lbs.

13. To allow the belt to be free of any slack, turn the crankshaft clockwise 2 full rotations. Confirm that the 4 marks are aligned.

14. Install the timing cover and tighten the bolts to 7.0–8.5 ft. lbs.

15. Adjust the valve lash and install the cylinder head cover, using a new gasket.

16. Connect the air intake case to the throttle body.

17. Install the crankshaft pulley and install the 5 mounting bolts. Tighten to 7.0–8.5 ft. lbs.

18. Install the water pump pulley, water pump drive belt and tighten the alternator mounting bolts.

19. Install the air conditioning compressor flexible suction hose and the air compressor belt. Properly recharge the air conditioning system, if equipped.

20. Install the radiator cooling fan and fan shroud.

21. Connect the negative battery cable.

22. Properly recharge the air conditioning system, if equipped. Run the engine and check for any leaks.

REMOVAL AND INSTALLATION

1. Disconnect the negative battery cable.

2. Remove the radiator cooling fan and fan shroud.

3. If equipped, disconnect the air conditioning compressor drive belt, properly discharge the air conditioning system and remove the air conditioning compressor flexible suction hose.

4. Loosen the alternator mounting bolts and remove the water pump drive belt and pulley.

5. Remove the crankshaft mounting bolts and disconnect the crankshaft pulley.

NOTE: The crankshaft drive belt pulley can be removed without loosening the center crankshaft bolt.

6. Disconnect the timing belt cover mounting bolts and remove the timing belt cover. Loosen but do not remove the tensioner bolt.

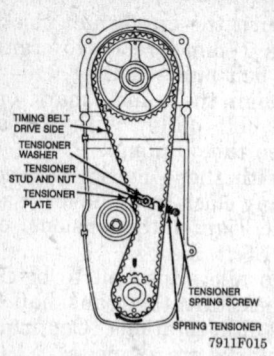

Exploded view of the timing belt and the tensioner assembly

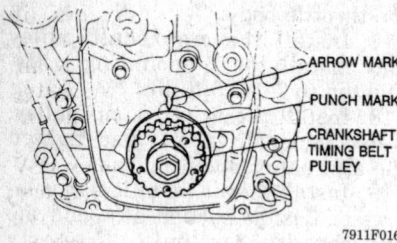

Timing marks on the crankshaft pulley

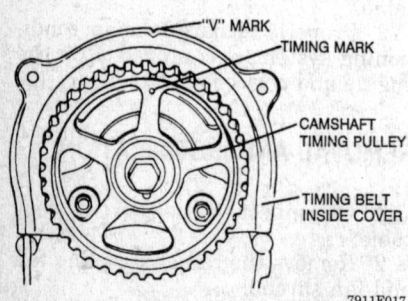

Timing marks on the camshaft pulley

7. Disconnect the air intake case from the intake manifold.

8. Loosen the timing belt tensioner adjusting bolt and pivot nut. Hold pressure on the tensioner to loosen the timing belt and remove the timing belt from the camshaft and crankshaft pulleys.

9. Remove the timing belt tensioner, tensioner plate and tensioner spring.

To install:

10. Install the timing belt tensioner, plate and spring. Hand

tighten the tensioner bolt and stud only at this time.

11. Remove the cylinder head cover and loosen all the valve adjusting screws to permit free rotation of the camshaft.

12. Turn the camshaft pulley clockwise and align the timing marks.

13. Turn the crankshaft clockwise, using a 17mm wrench to crank the timing belt pulley bolt.

14. Align the punch mark on the timing belt pulley with the arrow mark on the oil pump.

15. With the 4 marks aligned, remove any slack from the drive side of the belt. Tighten the tensioner bolt to 17.5–21.5 ft. lbs.

16. To allow the belt to be free of any slack, turn the crankshaft clockwise 2 full rotations. Confirm that the 4 marks are aligned.

17. Install the timing cover and tighten the bolts to 7.0–8.5 ft. lbs.

18. Adjust the valve lash and install the cylinder head cover, using a new gasket.

19. Connect the air intake case to the intake manifold.

20. Install the crankshaft pulley and install the mounting bolts. Tighten to 7.0–8.5 ft. lbs.

21. Install the water pump pulley, water pump drive belt and tighten the alternator mounting bolts.

22. If equipped, install the air conditioning compressor flexible suction hose and install the air conditioning compressor belt.

23. Install the radiator cooling fan and fan shroud.

24. Connect the negative battery cable.

25. If equipped, properly recharge the air conditioning system. Run the engine and check for any leaks.

Timing Sprockets

REMOVAL AND INSTALLATION

1. Disconnect the negative battery cable.

2. Remove the radiator cooling fan/clutch and fan shroud.

3. If equipped, disconnect the air conditioning compressor drive belt, properly discharge the air conditioning system and remove the air conditioning compressor flexible suction hose.

4. Loosen the alternator mounting bolts and remove the water pump drive belt and pulley.

5. Remove the crankshaft mounting bolts and disconnect the crankshaft pulley.

NOTE: The crankshaft drive belt pulley can be removed without loosening the center crankshaft bolt.

6. Disconnect the timing belt cover mounting bolts and remove the timing belt cover. Loosen, but do not remove the tensioner bolt.

7. Disconnect the air intake case from the throttle body.

8. Remove the cylinder head cover and loosen all the valve adjusting screws to permit rotation of the camshaft.

9. Remove the camshaft timing belt pulley by inserting a proper size rod into the hole in the camshaft to lock the camshaft.

10. Remove the camshaft sprocket mounting bolt, sprocket and sprocket pin.

To install:

11. Install the camshaft sprocket pin in the sprocket and replace the sprocket and the bolt on the camshaft.

12. With the camshaft locked, tighten the sprocket bolt to 41–46 ft. lbs.

13. Remove the crankshaft sprocket by using a gear stopper to hold the flywheel (driveplate for automatic transmission vehicles). Remove the crankshaft timing belt pulley bolt, sprocket and key.

14. With the crankshaft locked, install the crankshaft timing belt sprocket and key. Install the crankshaft pulley bolt and tighten to 47–54 ft. lbs.

15. Align the camshaft and crankshaft timing marks, install the timing belt and tighten the timing belt tensioner.

16. Adjust the valve lash and install the cylinder head cover.

17. Reconnect the air intake hose to the throttle body.

18. If equipped, install the air conditioning compressor flexible suction hose and install the air conditioning com pressor belt.

19. Install the water pump pulley and install the water pump drive belt.

20. Install the radiator cooling fan and fan shroud.

21. Install the water pump pulley and install the water pump drive belt.

22. Connect the negative battery cable.

23. If equipped, properly recharge the air conditioning system. Run the engine and check for any leaks.

Camshaft

REMOVAL AND INSTALLATION

1. Disconnect the negative battery cable. Drain the cooling system.

2. Disconnect the air intake case.

3. Disconnect the brake booster hose from the intake manifold and the air valve water hose from the throttle body.

4. Disconnect the accelerator and the automatic transmission kickdown cable, if equipped, from the throttle body.

5. Disconnect and tag the throttle opener VSV and EGR VSV wires at the coupler.

6. Disconnect and tag the water temperature, oil pressure, air temperature sensor and ground cable wires at the intake manifold.

7. Disconnect and tag the injector, throttle position sensor and idle speed control solenoid valve wires at their couplers, if equipped.

8. Disconnect the PTC heater wires at the coupler, if equipped with automatic transmission.

9. Disconnect and tag the oxygen sensor and distributor wires at their couplers. Remove the coil wire.

10. Disconnect the ground wires from the distributor assembly.

11. Remove the fuel tank filler cap to relieve the pressure. Reinstall the cap.

12. Release the fuel pressure in the fuel feed line.

13. Disconnect the fuel feed and return hoses.

14. Disconnect the canister purge hose and remove the pressure sensor hose from the fuel filter, if equipped.

15. Remove the vacuum hose for the automatic transmission from the intake manifold, if equipped.

16. Disconnect the radiator cooling fan, fan shroud, water pump drive belt and water pump pulley.

17. Disconnect the crankshaft pulley, timing belt cover and the timing belt.

18. Raise and safely support the vehicle.

19. Disconnect the exhaust pipe from the exhaust manifold and lower the vehicle.

20. Disconnect the air conditioner compressor adjusting arm, if equipped.

21. Remove the cylinder head cover mounting bolts and remove the cylinder head cover.

22. Loosen all the valve adjusting screw locknuts, turn the adjusting screws back all the way to allow all the valves to close.

23. Remove the intake manifold from the cylinder head.

24. Remove the distributor and distributor housing from the cylinder head.

25. Remove the cylinder head mounting bolts and remove the cylinder head from the engine.

26. Remove any oil and water in the cylinder bores and on top of the pistons.

27. Clean and inspect the sealing surfaces of the cylinder head and the engine block.

28. Disconnect the timing belt gear from the camshaft and remove the rocker arms, springs and rocker arm shafts.

29. Keep all the valve train parts in the order that they were removed.

30. Remove the camshaft from the rear of the cylinder head.

To install:

31. To install, lubricate the lobes and journals of the camshaft and the oil seal on the cylinder head with engine oil.

32. Install the camshaft to the cylinder head from the transmission side.

33. Install the cylinder head using a new gasket.

34. Install the cylinder head mounting bolts and tighten to 46–54 ft. lbs.

35. Install the distributor and distributor housing to the cylinder head.

36. Install the intake manifold to the cylinder head.

37. Tighten all the valve adjusting screw nuts.

38. Adjust the valve lash.

39. Install the cylinder head cover and connect the air conditioning compressor adjusting arm, if equipped.

40. Raise and safely support the vehicle.

41. Connect the exhaust pipe to the exhaust manifold and safely lower the vehicle.

42. Reconnect the timing belt, timing belt cover and crankshaft pulley.

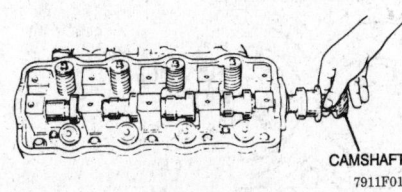

CAMSHAFT
7911F018

Camshaft removal

43. Install the water pump pulley, water pump drive belt, fan shroud and the radiator cooling fan.

44. Install the vacuum hose for the automatic transmission to the intake manifold, if equipped.

45. Install the pressure sensor hose to the fuel filter and connect the canister purge hose, if equipped.

46. Install the fuel return and feed hoses.

47. Connect the ground wires to the distributor assembly.

48. Install the coil wire. Connect the oxygen sensor and distributor wires to their couplers.

49. Connect the PTC wires to the coupler.

50. Connect the idle speed control valve, throttle position sensor and injector wires to their couplers, if equipped.

51. Reconnect the ground cable, air temperature, oil pressure and water temperature sensor wires to the intake manifold.

52. Install the throttle opener VSV and EGR VSV wires to their couplers.

53. Connect the accelerator and the automatic transmission kickdown cables, if equipped, to the throttle body.

54. Install the air valve water hose to the throttle body and connect the brake booster hose to intake manifold.

55. Reconnect the air intake case.

56. Refill the cooling system with the proper coolant and connect the negative battery.

57. Check for any water and oil leaks when finished.

Piston and Connecting Rod

POSITIONING

ENGINE LUBRICATION

Oil Pan

REMOVAL AND INSTALLATION

1. Raise and safely support the vehicle.

2. On Sidekick, drain and remove the front differential assembly from the chassis.

3. Drain the engine oil.

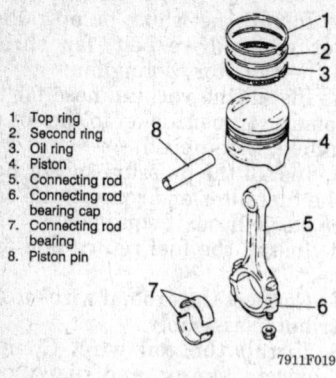

1. Top ring
2. Second ring
3. Oil ring
4. Piston
5. Connecting rod
6. Connecting rod bearing cap
7. Connecting rod bearing
8. Piston pin

7911F019

Exploded view of the piston and rod assembly

4. Remove the clutch housing lower plate or the torque converter housing lower plate on the Sidekick.

5. Remove the oil pan mounting bolts and remove the pan.

To install:

6. Clean and inspect the sealing surfaces on the oil pan and the engine block.

7. Using new gaskets, install the oil pan and tighten the oil pan bolts to 7.0–8.5 ft. lbs.

NOTE: Tightening should begin at the center moving outward on both sides.

8. Install the oil drain plug and tighten to 22.0–28.5 ft. lbs.

9. Install the clutch housing lower plate or the torque converter on the Sidekick.

10. Install the front differential assembly to the chassis and fill with differential oil on the Sidekick.

11. Lower the vehicle.

12. Refill the engine with engine oil. Run the engine and check for leaks.

Oil Pump

REMOVAL AND INSTALLATION

1. Disconnect the negative battery cable.

2. Remove the radiator cooling fan, fan shroud, water pump pulley and water pump drive belt.

3. Remove the timing belt outside cover, timing belt and timing belt tensioner.

4. Disconnect the alternator and remove the bracket, if necessary.

5. If equipped, properly discharge the air conditioning system and remove the air compressor and bracket.

6. Disconnect the crankshaft timing belt pulley and the timing belt guide.

NOTE: To lock the crankshaft, engage a special tool (gear stopper) with the flywheel ring gear (driveplate ring gear for automatic transmission vehicles). With the crankshaft locked, remove the crankshaft timing belt pulley bolt.

7. Raise and safely support the vehicle. Drain the engine oil.

8. Remove the clutch (torque converter) housing lower plate.

9. Remove the oil pan mounting bolts and remove the oil pan. Disconnect the oil pump strainer.

10. Remove the oil pump mounting bolts and remove the oil pump.

To install:

11. Clean and inspect the sealing surfaces of the oil pump and the engine block.

12. Using a new gasket, install the oil pump to the engine.

13. Install the No. 1 and No. 2 mounting bolts and tighten to 7.0–8.5 ft. lbs.

NOTE: To prevent the oil seal lip from being damaged when installing the oil pump to the

crankshaft, use a proper seal guide tool when installing. After installing the oil pump, check to be sure the oil lip is not up-turned, then remove the special tool. The edge of the oil pump gasket might bulge out. If it does, cut off bulge with knife.

14. With the crankshaft locked, install the crankshaft timing belt guide and the timing belt pulley.

15. Install the clutch (torque converter) housing lower plate.

16. If equipped, install the air conditioning bracket, the air conditioning compressor and properly recharge the air conditioning system.

17. Install the alternator and the alternator bracket, if necessary.

18. Reinstall the timing belt tensioner, timing belt and timing belt outside cover.

19. Install the water pump drive belt, water pump pulley, radiator cooling fan and fan shroud.

20. Refill the engine with engine oil and connect the negative battery cable.

21. Run the engine and check for any leaks.

CHECKING

With the oil pump removed from the engine, certain clearances must be checked on the oil pump, if it is being reused.

1. The radial clearance is the clearance between the oil pump outer rotor and the oil pump case. The maximum clearance is 0.0122 in.

2. The side clearance is measured with a straightedge across the mounting surface of the oil pump. The measurement is taken between the oil pump gears and the straightedge. The maximum clearance is 0.0059 in.

Rear Main Bearing Oil Seal

REMOVAL AND INSTALLATION

1. Raise and safely support the vehicle.

2. Support the engine and remove the transmission from the vehicle.

3. Disconnect the clutch and flywheel, for manual transmission vehicles or the driveplate, for automatic transmission vehicles.

4. Using a suitable tool, remove the seal by pulling it outwards. Use care not to damage the sealing surface of the crankshaft.

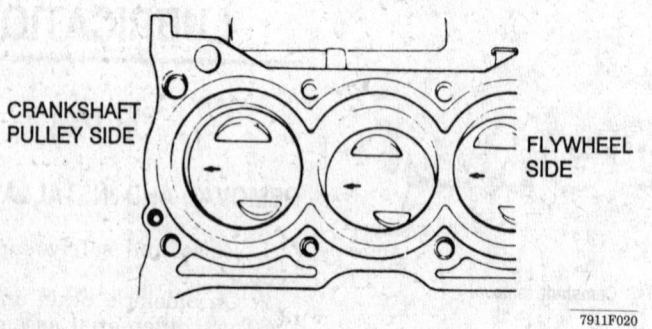

CRANKSHAFT PULLEY SIDE

FLYWHEEL SIDE

7911F020

Piston arrow marks to the cylinder head

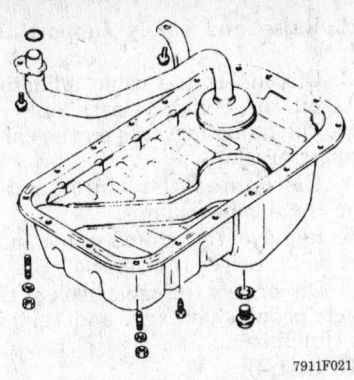

7911F021

Oil pump installation — bolt location

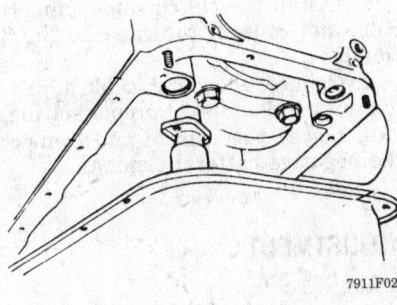

7911F022

Oil pump installation

To install:

5. Clean and inspect the sealing surfaces of the crankshaft and seal housing.

6. Using a seal driver, install the new seal into the seal housing with the lip of the seal facing the engine.

7. Install the flywheel and clutch, for manual transmission vehicles or the driveplate, for automatic transmission vehicles.

8. Install the transmission in the vehicle.

9. Lower the vehicle and check all the fluids. Run the engine and check for any leaks.

MANUAL TRANSMISSION

Transmission Assembly

REMOVAL AND INSTALLATION

1. Disconnect the negative battery cable.

2. Remove the console cover, remove the 4 gear shift boot mounting bolts and slide the boot upwards on the gear shifter.

3. On Samurai, loosen the 3 gear shift lever mounting bolts and remove the gear shift lever.

4. On the Sidekick remove the boot clamp and remove the second boot from the shift lever case.

5. Push the gear shift lever control case down with fingers, turn it counterclockwise and remove the shift control lever.

6. Remove the transfer case shift control lever the same way.

7. Disconnect the breather hose and clamp at the rear of the cylinder head.

8. Remove the clamp at the rear of the intake manifold to free up the wiring harness. Disconnect the harness coupler.

9. Raise and safely support the vehicle. Drain the oil from the transmission and the transfer case.

10. Disconnect the starter lead wires and mounting bolts and remove the starter.

11. Remove the fuel line clamp on the transmission. Disconnect the bolts fastening the engine to the transmission.

12. Disconnect the flange bolts from the front driveshaft and remove and mark the shaft.

13. Disconnect the flange bolts from the rear driveshaft and remove and mark the shaft.

14. Disconnect the clutch cable and remove the clutch housing lower plate.

15. Disconnect the center exhaust pipe and remove the nuts from the joint with the engine.

16. Disconnect the speedometer cable from the transfer case.

17. Position a transmission jack and remove the engine rear mounting member from the vehicle. Move the transmission and transfer case rearwards and lower.

18. Disconnect the wiring harness and breather hose at the transmission.

19. Separate the gearshift lever case and transfer case from the transmission. Remove the transmission from the vehicle.

To install:

20. Install the gear shift lever case and transfer case to the transmission.

21. Install the transmission wiring harness and breather hose.

22. Raise the transmission and transfer case on the transmission jack and place under the vehicle.

23. Install the engine rear mounting member and tighten the bolts. Tighten the Samurai mounting bolts to 19 ft. lbs. or the Sidekick mounting bolts to 37 ft. lbs.

24. Connect the speedometer cable to the transfer case and joint with the engine.

25. Remove the transmission jack and install the center exhaust pipe.

26. Install the clutch housing lower plate and connect the clutch cable.

27. Install the front and rear driveshaft and install the flange bolts. Tighten the Samurai flange bolts to 20 ft. lbs. or the Sidekick flange bolts to 37 ft. lbs.

28. Connect the bolts attaching the engine to the transmission and install the fuel line clamp on the transmission.

29. Install the starter and install the mounting bolts and the lead wires to the starter.

30. Lower the vehicle. Refill the transmission and transfer case with the recommended gear oil.

31. Connect the wiring coupler and install the clamp holding the wiring harness at the rear of the intake manifold.

32. Install the breather hose and clamp at the rear of the cylinder head.

33. Install the transfer case and the gear shift control levers. Install the gear shift lever boots and the console cover.

34. Connect the negative battery cable. Run the engine and check for any leaks.

CLUTCH

Clutch Assembly

REMOVAL AND INSTALLATION

1. Disconnect the negative battery cable.

2. Raise and safely support the vehicle.

3. Support the engine and remove the transmission from the vehicle.

4. Support the pressure plate and remove the 6 pressure plate to flywheel mounting bolts.

5. Remove the pressure plate and the clutch disc from the flywheel.

6. Inspect the condition of the flywheel, pressure plate and clutch disc and install as necessary.

7. Remove the flywheel mounting bolts and remove the flywheel.

To install:

8. Install the new flywheel and install the mounting bolts, tighten to 41–47 ft. lbs.

9. Inspect the condition of the clutch release bearing and the input shaft bearing and install if necessary.

10. Align the clutch disc to the flywheel using a clutch disc alignment tool.

11. Install the pressure plate and evenly tighten the pressure plate bolts to 13–20 ft. lbs.

NOTE: Before assembling, make sure the clutch disc and the pressure plate are clean and dry.

12. With the engine supported, install the transmission in the vehicle.

13. Lower the vehicle. Connect the negative battery cable.

14. Check and adjust the clutch pedal height. Check the clutch operation when finished.

PEDAL HEIGHT/FREE-PLAY ADJUSTMENT

The only adjustment possible is the clutch pedal free-play. The free-play is adjusted by moving the joint nut on the release bearing arm. Clutch linkage free-play should be between 0.8–1.1 in. for the Samurai or 0.6–1.1 in. for the Sidekick.

Clutch Cable

REMOVAL AND INSTALLATION

1. Disconnect the negative battery cable.

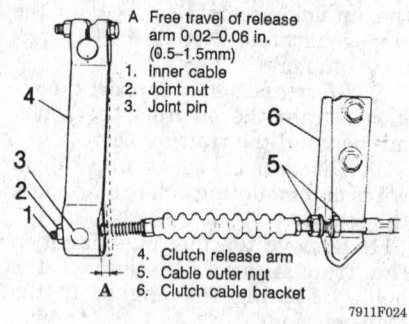

A Free travel of release arm 0.02–0.06 in. (0.5–1.5mm)
1. Inner cable
2. Joint nut
3. Joint pin
4. Clutch release arm
5. Cable outer nut
6. Clutch cable bracket

7911F024

Clutch cable free travel adjustment

2. Raise and safely support the vehicle.

3. Disconnect the cable adjusting nuts on the clutch release and remove the adjusting nuts on the cable support bracket.

4. Disconnect the clutch cable from the 3 cable clamps.

5. Remove the clutch cable support bolts. Lower the vehicle.

6. Disconnect the cable hook at the clutch pedal shaft arm and remove the clutch cable.

To install:

7. Install the cable after first applying grease to the cable end hook and the joint pin.

8. Install the clutch cable support bolts and connect the cable to the 3 clamps.

9. Connect the clutch cable adjusting nuts and adjust to proper setting.

10. Lower the vehicle and connect the negative battery terminal.

ADJUSTMENT

Adjust the clutch pedal height by adjusting the bolt located on the pedal bracket so the clutch pedal exceeds the height of the brake pedal by 0.2 in. (5mm). Tighten the locknut.

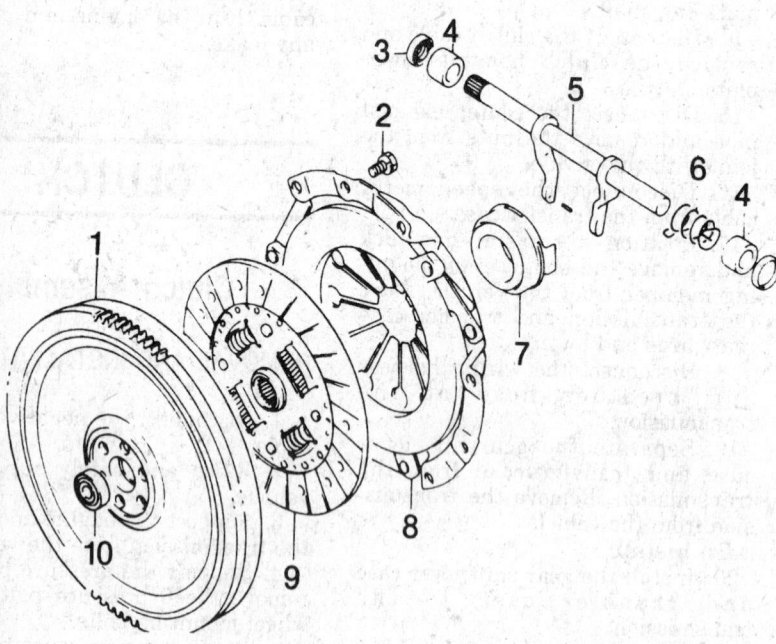

1. Flywheel
2. Cover bolt
3. Release shaft seal
4. Release shaft bushing
5. Clutch release shaft
6. Release return spring
7. Clutch release bearing
8. Clutch cover
9. Clutch disc
10. Input shaft bearing

7911F023

Exploded view of the clutch assembly

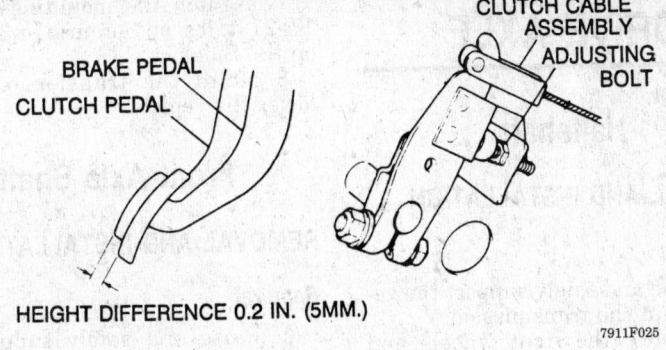

HEIGHT DIFFERENCE 0.2 IN. (5MM.)

7911F025

Clutch pedal height adjustment

AUTOMATIC TRANSMISSION

Transmission Assembly

REMOVAL AND INSTALLATION

1. Disconnect the negative battery cable.
2. Disconnect the transmission shift control lever and remove the transfer case shift control lever knob.
3. Disconnect the breather hose from the clamp at the rear of the cylinder head.
4. Disconnect the wiring harness clamp at the rear end of the intake manifold to free up the harness.
5. Disconnect the wiring harness coupler and remove the detent cable at the throttle body, if equipped.
6. Remove the vacuum modulator hose at the intake manifold.
7. Raise and safely support the vehicle.
8. Disconnect the starter lead wires and mounting bolts, remove the starter motor.
9. Drain the oil from the transmission and transfer case.
10. Disconnect the flange bolts from the front driveshaft, remove and mark the shaft.
11. Disconnect the flange bolts from the rear driveshaft, remove and mark the shaft.
12. Disconnect the select cable from the transmission and the speedometer cable from the transfer cases.
13. Remove the torque converter housing lower plate and disconnect and plug the oil cooler lines.
14. Disconnect the transfer case skid plate and remove the kickdown cable from the transmission.

15. Remove the exhaust bracket at the catalytic converter and at the transmission. Disconnect the center exhaust pipe, if necessary.
16. Remove the transmission to engine retaining bolts and nuts.
17. Support the transmission using a transmission jack or equivalent. Disconnect the transmission crossmember mounting bolts and remove the crossmember.
18. Disconnect the lead wires and breather hose at the transmission and lower the transmission with the transfer case from the vehicle. Remove the torque converter-to-flywheel bolts.
19. Disconnect the transmission-to-transfer case bolts and separate the transmission from the transfer case.

To install:
20. Connect the transfer case to the transmission and tighten the bolts to 20 ft. lbs. Install the torque converter to flywheel bolts.
21. Raise the transmission, connect the lead wires and breather to the transmission.
22. Install the transmission to engine retaining bolts and tighten to 62 ft. lbs.
23. Install the transmission cross member and tighten the bolts to 62 ft. lbs.
24. Connect the exhaust bracket to the catalytic converter and the transmission.
25. Reconnect the center exhaust pipe, if necessary.
26. Connect the transfer case skid plate and install the kickdown cable to the transmission.
27. Install the torque converter housing lower plate and connect the transmission cooler lines.
28. Reconnect the select cable to the transmission and the speedometer cable to the transfer case.
29. Install the front and rear driveshafts and tighten the flange bolts to 37 ft. lbs.

30. Install the starter and connect the lead wires to the starter.
31. Lower the vehicle and connect the vacuum modulator hose to the intake manifold.
32. Connect the wiring coupler at the rear of the intake manifold and install the wiring harness clamp in the proper position.
33. Connect the breather hose to the clamp at the rear of the cylinder head.
34. Install the transmission shift control lever and install the transfer case shift control lever knob.
35. Connect the negative battery cable and refill the transmission. Run the engine and check for leaks.

SHIFT LINKAGE ADJUSTMENT

1. Adjust the shift cable assembly by moving the shift selector to the N position and placing a pin in the selector to hold that position.
2. Put the manual select lever in L position and set the cable to the cable bracket with the E-clip.
3. Adjust the manual select lever back to the N position and tighten the cable end locknut to 5 ft. lbs.

NOTE: When the manual selector is moved to the N position, there should be a little clearance between the lever and the adjusting nut.

THROTTLE LINKAGE ADJUSTMENT

Adjust the cable by loosening the cable locknut and adjust the cable to specifications. The cable endplay should be 0.12–0.20 in. when the throttle valve is in the idle position.

DETENT CABLE ADJUSTMENT

1. Make sure accelerator play is within specifications.
2. Loosen the kickdown cable nut and adjusting nut.
3. With the accelerator pedal fully depressed and the cable pulled towards the firewall, adjust the locknut-to-bracket clearance to 0.039 in. by turning the locknut.

NOTE: When adjusting the clearance make sure the adjusting nut does not rub against the bracket.

4. Release the accelerator pedal and adjust the locknut-to-bracket clearance by tightening the adjusting nut. Tighten the locknut securely.

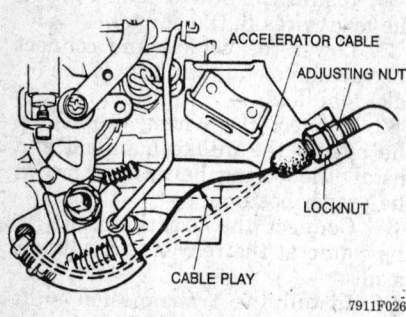

ACCELERATOR CABLE

ADJUSTING NUT

LOCKNUT

CABLE PLAY

7911F026

Accelerator cable adjustment

TRANSFER CASE

Transfer Case Assembly

REMOVAL AND INSTALLATION

1. Disconnect the negative battery cable.
2. Raise and safely support the vehicle.
3. Support the transmission and transfer case using a suitable transmission jack.
4. Disconnect the speedometer cable from the transfer case.
5. Remove the transmission and transfer case from the vehicle.
6. Remove the 10 transmission-to-transfer case mounting bolts and separate the transfer case from the transmission.
To install:
7. Install the 10 transmission-to-transfer case mounting bolts and tighten the bolts to 20 ft. lbs.
8. Install the transmission and transfer case into the vehicle and tighten the mounting bolts to 20 ft. lbs. on the Samurai or 37 ft. lbs. on the Sidekick.
9. Connect the speedometer cable to the transfer case.
10. Lower the vehicle and connect the negative battery cable.

DRIVE AXLE

Halfshaft

REMOVAL AND INSTALLATION

Sidekick

1. Raise and safely support the vehicle. Drain the transmission.
2. Remove the front wheels and disconnect the locking hubs.
3. Disconnect the halfshaft snaprings and remove the stabilizer bar links.
4. Disconnect the tie rod ends and remove the brake caliper and brake disc. Support the brake caliper.
5. Disconnect the wheel hub and remove the steering knuckle.
6. Disconnect the differential side joint snapring and remove the right side halfshaft by prying the joint away from the differential assembly.
7. Disconnect the left side halfshaft bolts and remove the halfshaft from the left inner axle flange.
To install:
8. Install the left side halfshaft and the halfshaft bolts and tighten to 37 ft. lbs.
9. Install the right side halfshaft and install the snapring.
10. Install the steering knuckle and wheel hub assembly.
11. Install the brake caliper and connect the tie rod ends.
12. Install the left side driveshaft snapring and connect the stabilizer bar links.
13. Install the front wheels and lower the vehicle.
14. Refill the transmission with the proper fluid.

Driveshaft and U-Joints

REMOVAL AND INSTALLATION

1. Raise and safely support the vehicle.
2. Matchmark the driveshafts to the yokes on the transfer case and the differential.
3. Drain the transfer case oil, when servicing the front driveshaft shaft.
4. Support the driveshaft and remove the attaching bolts.
5. Remove the driveshaft from the vehicle.
To install:
6. Install the driveshaft, by aligning the matchmarks to the vehicle.

7. Tighten the mounting bolts to 17–21 ft. lbs. on Samurai or 37 ft. lbs. on the Sidekick.
8. Refill the transfer case and lower the vehicle.

Front Axle Shaft

REMOVAL AND INSTALLATION

Samurai

1. Raise and safely support the vehicle.
2. Drain the oil in the front differential.
3. Remove the front wheels and disconnect the brake caliper. Support the brake caliper.
4. Disconnect the tie rod end from the steering knuckle. The tie rod end removal may require the use of a puller.
5. Remove the 8 oil seal cover mounting bolts and disconnect the felt pad, oil seal and the retainer from the steering knuckle.
6. Mark the upper and lower kingpins. Remove the 4 mounting bolts and disconnect the kingpins from the steering knuckle.
7. Remove the axle shaft from the housing with the steering knuckle attached.
To install:
8. Transfer the steering knuckle to the new axle.
9. Align the marks on the upper and lower kingpins and install the kingpins and tighten to 14–21 ft. lbs.
10. Install the steering knuckle oil seal and install the 8 mounting bolts and tighten to 6–8 ft. lbs.
11. Connect the tie rod end to the steering knuckle and tighten the bolts to 22–39 ft. lbs. Install the front brake caliper assembly.
12. Install the front wheels and refill the differential with the proper fluid.
13. Lower the vehicle.

Front Axle Shaft, Bearing and Seal

REMOVAL AND INSTALLATION

Samurai

1. Raise and safely support the vehicle. Drain the front differential assembly.
2. Disconnect the front wheel hub and remove the bearing.
3. Support the hub and drive out the seal in the hub.

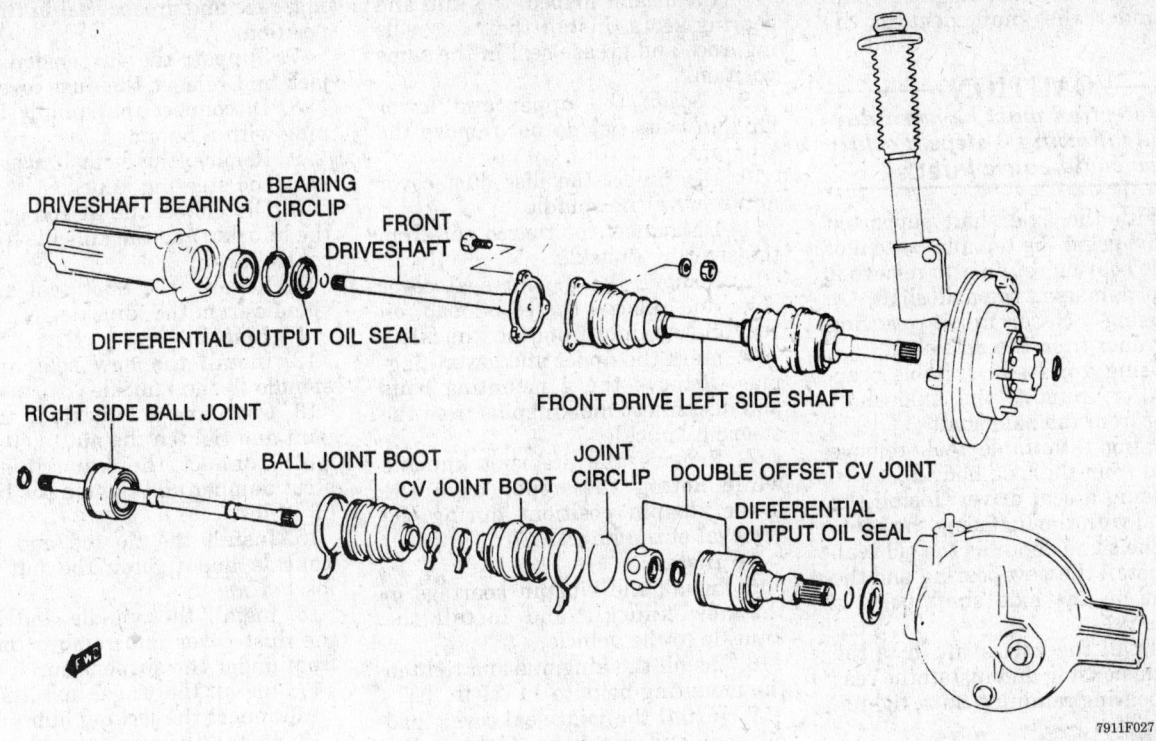

DRIVESHAFT BEARING   BEARING CIRCLIP   FRONT DRIVESHAFT
DIFFERENTIAL OUTPUT OIL SEAL
FRONT DRIVE LEFT SIDE SHAFT
RIGHT SIDE BALL JOINT   BALL JOINT BOOT   CV JOINT BOOT   JOINT CIRCLIP   DOUBLE OFFSET CV JOINT   DIFFERENTIAL OUTPUT OIL SEAL

7911F027

Exploded view of the halfshaft assembly

4. Using a seal driver, install the new seal in the hub until it is flush with the hub face. Apply a thin film of oil to the lip of the seal before installation to the vehicle.

5. Install the hub and the bearing to the vehicle.

6. Refill the front differential assembly and safely lower the vehicle.

Rear Axle Shaft, Bearing and Seal

REMOVAL AND INSTALLATION

Samurai

1. Raise and safely support the vehicle. Drain the rear differential assembly.

2. Make sure the rear brake is released.

3. Remove the rear wheels and remove the rear brake drums from the vehicle.

4. Disconnect the parking brake cables from the levers. Remove the parking brake lever stop plates.

5. Disconnect and plug the brake lines to the wheel cylinders.

6. Remove the backing plate mounting bolts.

7. Remove the rear axles with the backing plates attached using a slide hammer type puller.

8. Using a suitable tool, remove the axle seal from the housing.

9. If the axle, axle bearing or backing plate is being replaced, support the axle in a vise with additional support under the shaft next to the bearing.

--- CAUTION ---
Eye protection must be worn during the following 3 steps. Failure to do so could cause injury.

10. With the axle shaft supported properly, grind the top and bottom of the axle bearing retainer to remove it without damaging the axle shaft.

11. Using a chisel, finish removing the retainer from the axle shaft.

12. Using a press or suitable bearing puller, remove the axle shaft bearing from the axle shaft.

13. Remove the backing plate from the axle shaft.
To install:
14. Using a seal driver, install the new seal with the lip facing the housing to the same depth as the old seal.

15. Install the backing plate on the axle shaft and using a press, install the bearing and the retainer on the axle shaft.

16. Install the axle shaft in the housing.

17. Install the backing plate mounting bolts and connect the brake lines to the wheel cylinders.

18. Install the parking brake lever plates and connect the brake cables to the parking brake lever.

19. Install the rear brake drums and install the rear wheels.

20. Adjust the brakes and bleed the brake system.

21. Refill the rear differential and safely lower the vehicle.

Sidekick

1. Raise and safely support the vehicle.

2. Remove the rear wheels and remove the rear brake drums from the vehicle.

3. Drain the gear oil from the rear axle housing.

4. Remove the rear wheel bearing retainer nuts from the rear axle housing.

5. Using a suitable tool, remove the axle shaft from the housing.

NOTE: Do not remove the backing plate with the axle. This may cause damage to the inner seal.

6. If the axle, axle bearing or backing plate is being replaced, support

the axle in a vise with additional support under the shaft next to the bearing.

— CAUTION —
Eye protection must be worn during the following 3 steps. Failure to do so could cause injury.

7. With the axle shaft supported properly, grind the top and bottom of the axle bearing retainer to remove it without damaging the axle shaft.

8. Using a chisel, finish removing the retainer from the axle shaft.

9. Using a press or suitable bearing puller, remove the axle shaft bearing from the axle shaft.

10. Using a suitable tool, remove the seal from the axle housing.

11. Using a seal driver, install the new seal with the lip facing the housing to the same depth as the old seal.

12. Install the new bearing and the retainer on the axle shaft using a suitable press.

13. Install the axle shaft in to the rear axle housing and install the rear wheel bearing retaining nuts, tighten to 17 ft. lbs.

14. Install the rear brake drums and install the rear tires on the vehicle.

15. Refill the rear axle housing with the proper gear oil and safely lower the vehicle.

Front Wheel Hub, Knuckle and Bearings

REMOVAL AND INSTALLATION

Samurai

1. Raise and safely support the vehicle. Remove the front wheels.

2. Disconnect the locking hub. Remove the caliper mounting bolts and move the caliper out of position with the brake line attached.

NOTE: Do not allow the caliper to hang on the brake hose. Support it by the mounting bracket.

3. Install 2 (8mm) bolts into the threaded holes and tighten evenly. This will remove the rotor from the hub assembly.

4. Remove the front axle shaft cap and the circlip. Remove the drive flange from the steering knuckle.

5. Straighten the bent lock washer and remove the hub nut and washer.

6. Remove the front wheel hub and bearing from the spindle.

7. Remove the oil seal and race from the wheel hub.

8. Clean and inspect the hub and bearing seats. Install the new bearing, race and grease seal in the same position.

9. Loosen the upper and lower kingpin bolts but do not remove the kingpins.

10. Disconnect the disc dust cover and remove the spindle.

11. Disconnect the tie rod end from the steering knuckle.

12. Remove the 8 joint seal cover bolts and remove the cover, pad, oil seal and retainer from the knuckle.

13. Mark the upper and lower kingpins. Remove the 4 mounting bolts and disconnect the kingpins from the steering knuckle.

14. Remove the steering knuckle while noting the upper from the lower kingpin positions during the removal of the knuckle.

To install:

15. Install the kingpin bearings in the new knuckle and install the knuckle to the vehicle.

16. Install the kingpins and tighten the mounting bolts to 14–21 ft. lbs.

17. Install the joint seal cover, pad oil seal and retainer. Tighten the joint seal cover bolts to 6.0–8.5 ft. lbs.

18. Connect the tie rod end to the steering knuckle.

19. Install the spindle and connect disc dust cover.

20. Install the front wheel hub and bearing to the spindle.

21. Install the front hub nut and lock washer and install the drive flange to the steering knuckle.

22. Install the front axle cap and circlip. Connect the rotor to the hub assembly.

23. Place the brake caliper into position and install the caliper mounting bolts.

24. Reconnect the locking hub assembly and install the front wheels.

25. Lower the vehicle.

Sidekick

1. Disconnect the negative battery cable. Raise and safely support the vehicle.

2. Remove the wheels and disconnect the locking hub assembly.

3. Remove the caliper mounting bracket and position the caliper out of the way. Support the caliper.

4. Disconnect the front wheel bearing lock plate and washer and remove the wheel hub complete with bearings and seals.

5. Remove the oil seal and race from the wheel hub.

6. Clean and inspect the hub and bearing seats. Install the new bearing, race and grease seal in the same position.

7. Support the suspension with a jack and remove the dust cover.

8. Disconnect the spindle by tapping with a hammer.

9. Remove the strut bracket bolts from the steering knuckle.

10. Disconnect the tie rod end from the knuckle and the knuckle from the control arm.

11. Remove the dust seal and the spindle from the knuckle.

To install:

12. Install the new seal and the spindle to the knuckle.

13. Connect the knuckle to the ball joint and tighten the nut to 40 ft. lbs.

14. Connect the knuckle to the strut damper and tighten the bolts to 66 ft. lbs.

15. Install the tie rod end to the knuckle and tighten the nut 30 ft. lbs.

16. Install the spindle and install the dust cover and remove the jack from under the suspension.

17. Install the wheel hub assembly and connect the locking hub.

18. Install the wheels and lower the vehicle.

19. Connect the negative battery cable.

Manual Locking Hubs

REMOVAL AND INSTALLATION

1. Disconnect the negative battery cable and remove the locking hub cover.

2. Raise and support the vehicle safely.

3. Remove the locking hub body assembly.

NOTE: The manual locking hubs must not be packed with grease.

To install:

4. Install a new O-ring to the locking hub assembly.

5. Install the locking hub body assembly to the wheel hub flange and tighten the hub body bolts to 18 ft. lbs.

6. Install a new gasket in the manual locking hub cover.

7. Install the locking hub cover and tighten the bolts to 106 inch lbs.

8. Connect the negative battery cable and lower the vehicle.

NOTE: The O mark on the hub knob must be in the FREE position.

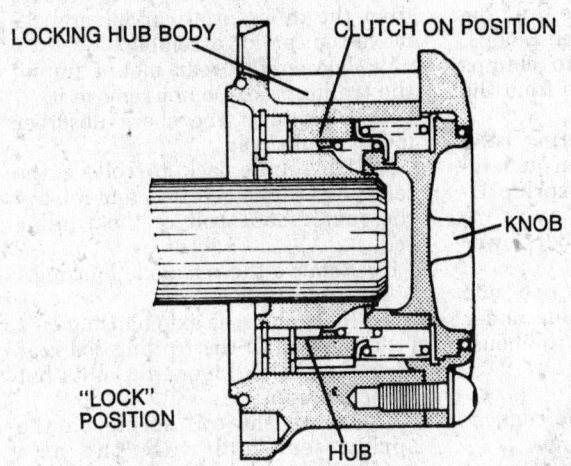
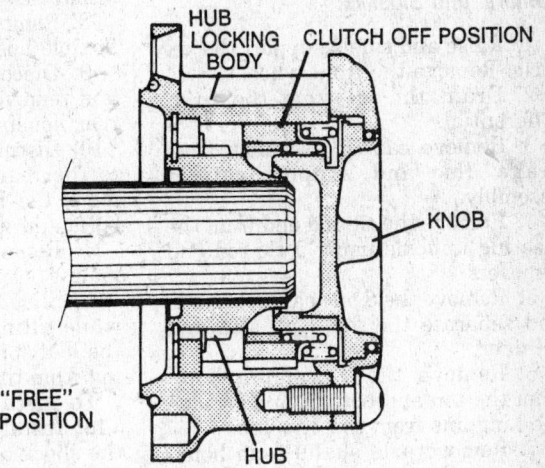

Wheel hub "LOCK" and "FREE" positions

Automatic Locking Hubs

REMOVAL AND INSTALLATION

1. Disconnect the negative battery cable and remove the automatic locking hub cover.
2. Disconnect the automatic hub body assembly.
To install:
3. Install a new O-ring to the automatic locking hub body assembly.
4. Connect the automatic locking hub body assembly to the wheel hub flange. Tighten the hub body assembly bolts to 18 ft. lbs.
5. Install a new gasket in the automatic locking hub cover and install the hub cover.
6. Connect the negative battery cable.

Pinion Seal

REMOVAL AND INSTALLATION

1. Raise and safely support the vehicle.
2. Matchmark and remove the driveshaft.
3. Check the turning torque of the pinion before proceeding.

4. Using a pinion flange holding tool, remove the pinion nut and the washer.
5. Remove the pinion flange from the differential carrier and pry out the pinion seal.
To install:
6. Clean and inspect the sealing surface of the carrier.
7. Using a seal driver, install the new seal into the carrier until the flange of the seal is flush with the carrier.

NOTE: Tightening the flange nut will preload the pinion bearings. Exceeding the preload specifications will compress the collapsible spacer too far and require the spacer to be replaced.

8. Install the pinion flange and using the pinion flange holding tool, install the pinion nut and washer. Tighten the pinion nut to the same torque as before.
9. Align the matchmarks and install the driveshaft.
10. Check the level of the differential fluid when finished.

Differential Carrier

REMOVAL AND INSTALLATION

1. Raise and safely support the vehicle. Drain the oil from the rear differential.
2. Remove the left and right axle shafts.
3. Disconnect the driveshaft.
4. Remove the 4 mounting bolts to the upper rear suspension arm.
5. Support the differential carrier with a proper jack.
6. Remove the differential case nuts and lower the differential assembly from the rear housing.
7. Clean and inspect the sealing surfaces of the carrier and the housing.
To install:
8. Using a liquid sealant on the carrier, install the carrier in the housing and tighten the nuts to 16 ft. lbs.
9. Install the 4 mounting bolts to the upper rear suspension arm and tighten to 41 ft. lbs.
10. Connect the rear driveshaft and install the left and rear axles.
11. Refill the rear differential to the proper level with SAE 75W–90W API GL5 Hypoid Gear Oil.
12. Lower the vehicle.

Front Axle Housing

REMOVAL AND INSTALLATION

Samurai and Sidekick

1. Raise and safely support the vehicle. Remove the front wheels.
2. Drain the oil from the front differential.
3. Remove caliper assembly from brake disc and support caliper assembly.
4. Remove the tie rod end from the steering knuckle using a tie rod end remover.
5. Remove the 8 oil seal cover bolts and separate the felt pad, seal and retainer.
6. Remove the 4 kingpin bolts from the top and bottom and remove the kingpins from the knuckle.
7. Remove axle shafts from housing by pulling outward gradually.
8. Remove the bolts between the flange yoke and companion flange and disconnect the front drive shaft.
9. Remove the 8 U-joint nuts and remove the housing assembly.
To install:
10. Install the housing and torque the housing U-bolt nuts to 43 ft. lbs.
11. Connect the front drive shaft and install the bolts.
12. Install the front axle shafts and install the kingpins. Apply Loctite° to the kingpin bolts.
13. Install oil seal, felt pad and retainer and tighten the 8 cover bolts.
14. Install the tie rod end to the steering knuckle.
15. Install the front caliper to the disc and install the front wheels.
16. Refill the differential with the proper lubricant and bleed the brake system when finished.

Rear Axle Housing

REMOVAL AND INSTALLATION

Samurai

1. Raise and safely support the vehicle. Drain the rear differential assembly.
2. Make sure the parking brake is in the released position.
3. Remove the rear wheels and remove the rear brake drums from the vehicle.
4. Disconnect the parking brake cables from the levers. Remove the parking brake lever stop plates.
5. Disconnect and plug the brake lines to the wheel cylinders.

6. Remove the backing plate mounting bolts and remove the axle shafts with the backing plates attached.
7. Disconnect the driveshaft and remove from the transmission.
8. Remove the brake line from the flexible hose and remove the E-clip.
9. Disconnect the brake clamps and remove the brake lines from the rear housing.
10. Disconnect the mounting bolts to the rear suspension arm and remove the housing to leaf spring U-bolts and nuts.
11. Remove shock absorber lower mount bolt.
12. Slide the housing to one side while tilting the opposite side under the leaf spring and remove the housing from the vehicle.
To install:
13. Install the housing and connect the shock absorber lower bolt.
14. Connect the U-bolts to the leaf spring and tighten the bolts to 50 ft. lbs.
15. Connect brake lines to the rear housing and install the driveshaft.
16. Install the axle shafts with backing plates attached and tighten the backing plate bolts to 19 ft. lbs.
17. Connect the brake lines to the rear wheel cylinders.
18. Install the parking brake lever stop plates and connect the parking brake cables.
19. Install the rear brake drums and install the rear wheels.
20. Lower the vehicle and refill the rear differential with the proper fluid.

Sidekick

1. Raise and safely support the vehicle. Drain the rear differential assembly.
2. Make sure the parking brake is in the released position.
3. Remove the rear wheels and remove the rear brake drums from the vehicle.
4. Disconnect the parking brake cables from the levers. Remove the parking brake lever stop plates.
5. Disconnect and plug the brake lines to the wheel cylinders.
6. Remove the rear wheel bearing retainer nuts and remove the axle shafts from the vehicle. Do not remove the axle shafts with the backing plates attached.
7. Remove the rear axle carrier assembly.
8. Disconnect the brake line from the flexible hose and remove the E-clip. Remove the brake lines from the rear housing.

9. Remove the breather hose from the axle housing and disconnect the rear driveshaft.
10. Support the rear axle housing with a jack.
11. Remove the ball joint bracket from the differential carrier and remove the carrier assembly.
12. Loosen the rear mount nut of the trailing rod. Do not remove it.
13. Disconnect the shock absorber lower mount bolt.
14. Lower the jack to relieve the tension of the coil springs and remove the rear mount bolt of the trailing rod.
15. Remove the rear axle housing.
To install:
16. Place the rear axle housing on a jack and install the trailing rod rear mounting bolts. Mount the nuts but do not tighten.
17. Install the coil spring on the spring seat and raise the axle housing.
18. Install the shock absorber lower mounting bolts. Do not tighten.
19. Install the differential carrier assembly and install the rear upper ball joint bracket onto the carrier assembly and tighten to 37 ft. lbs.
20. Install the rear driveshaft and remove the jack from under the axle housing.
21. Install the breather hose and brake lines to axle housing. Tighten them securely.
22. Connect the flexible brake hose to the bracket on the axle housing and secure with the E-clip.
23. Tighten the trailing rod nuts and the shock absorber nuts to 66 ft. lbs.
24. Install the brake line to the flexible hose and install the axle shafts.
25. Install the brake line to the wheel cylinders and install the brake drums.
26. Install the wheels and refill the rear differential assembly.
27. Bleed the brake system and lower the vehicle.

STEERING

Steering Wheel

REMOVAL AND INSTALLATION

1. Disconnect the negative battery cable.

2. Disconnect the horn button and remove the steering wheel shaft nut.

3. Make matchmarks on the steering wheel and the shaft to use as a guide during reinstallation.

4. Remove the steering wheel, using a suitable steering wheel puller.

To install:

5. Install the steering wheel onto the shaft, aligning the matchmarks.

6. Install and tighten the shaft nut to 24 ft. lbs.

7. Install the horn button and connect the negative battery cable.

Manual Steering Gear

REMOVAL AND INSTALLATION

Samurai

1. Remove the steering shaft coupler bolt and disconnect the coupler from the steering box. As required, raise and support the vehicle safely.

2. Remove the radiator under cover and disconnect the ball stud of the drag rod. Remove the steering damper from the pitman arm.

3. Support the steering gear and remove the mounting bolts.

4. Remove the steering gear from the vehicle.

To install:

5. Install the steering gear in the vehicle and tighten the nuts to 55 ft. lbs.

6. Install the steering damper to the pitman arm tighten the nut to 30 ft. lbs.

7. Install the radiator under cover and connect the drag rod.

8. Connect the steering coupler to the steering box and tighten the bolt to 18 ft. lbs. Lower the vehicle.

Sidekick

1. As required, raise and support the vehicle safely. Disconnect the steering lower shaft mounting bolts.

2. Disconnect the center link end from the pitman arm.

3. Remove the 3 steering gear box mounting bolts.

4. Disconnect the steering lower shaft joint and remove the steering gear.

To install:

5. Install the steering gear box by connecting to the lower shaft joint.

NOTE: Align the flat part of the steering gear worm shaft with the bolt hole of the lower shaft joint.

6. Install the steering gear box mounting bolts and tighten to 60 ft. lbs.

7. Attach the center link to the pitman arm and tighten the nut to 35 ft. lbs.

8. Connect the lower shaft mounting bolts and tighten to 18 ft. lbs.

ADJUSTMENT

1. Check the worm shaft to make sure it is free from thrust play.

2. Place the pitman arm in a position that is nearly parallel with the worm shaft.

3. With the pitman arm in this position, the front wheels should be in a straight forward position.

4. Measure the worm shaft starting torque from its straight foward position. The torque should be 0.4–0.7 ft. lbs.

5. Adjust the worm shaft adjusting bolt to specifications.

Power Steering Gear

REMOVAL AND INSTALLATION

1. Disconnect the negative battery cable.

2. Remove the coolant reservoir tank from the radiator.

3. Disconnect the steering column lower shaft from the gear box.

4. Raise and safely support the vehicle.

5. Remove the center link nut and lock washer and disconnect the center link from the pitman arm, using a pitman arm puller.

6. Lower the vehicle. Remove the pressure hose from the power steering gear assembly and plug.

7. Disconnect the return hose and plug. Remove the 3 power steering gear mounting bolts.

8. Remove the power steering gear. Disconnect the pitman arm

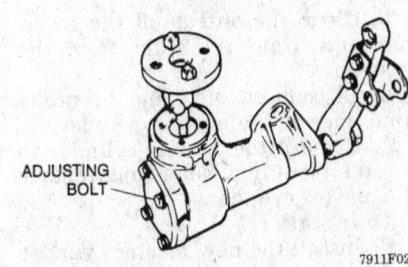

ADJUSTING BOLT

7911F029

Adjusting the steering gear assembly

from the gear assembly, note the alignment marks.

To install:

9. Align the matchmarks on the pitman arm and the power steering gear sector shaft. Install the pitman arm to the gear assembly and tighten the nut to 95 ft. lbs.

10. Install the power steering gear assembly on the vehicle and tighten the mounting bolts to 72 ft. lbs.

11. Connect the power steering pressure and return hoses.

12. Raise and safely support the vehicle.

13. Install the center link to the pitman arm and tighten the nut to 40 ft. lbs. Lower the vehicle.

14. Connect the steering column lower shaft to the gear assembly and tighten the bolts to 29 ft. lbs.

15. Install the coolant reservoir tank to the radiator. Refill the power steering pump.

16. Connect the negative battery cable. Run the engine and operate the power steering. Recheck the fluid level and for any leaks.

ADJUSTMENT

1. Check the worm shaft to make sure it is free from thrust play.

2. Place the pitman arm in a position that is nearly parallel with the worm shaft.

3. With the pitman arm in this position, the front wheels are in a straight-forward state.

4. Measure the worm shaft starting torque from its straight-forward state. The torque should be 0.5 ft. lbs.

5. Adjust the worm shaft adjusting bolt to specifications.

Power steering Pump

REMOVAL AND INSTALLATION

NOTE: Before disconnecting the power steering pressure and return line at the pump assembly, make sure any dirt or grease is removed.

1. Disconnect the negative battery cable. Remove the coolant reservoir tank from the radiator.

2. Loosen the air conditioning compressor adjusting and pivot bolts, if equipped.

3. Loosen the power steering pump adjusting and mounting bolts, if not equipped with air conditioning.

4. Remove the power steering belt.

5. Disconnect the power steering pressure and return hose and plug.

6. Disconnect the power steering pressure switch lead wire at the switch terminal.

7. Remove the engine oil filter.

8. Remove the power steering pump mounting and adjusting bolts.

9. Remove the power steering pump.

To install:

10. Install the power steering pump and install the pump mounting bolts. Do not tighten.

11. Install the power steering pump pressure switch lead wire to the switch terminal.

12. Install the power steering pressure and return hoses.

13. Install the power steering belt and tighten the power steering pump mounting bolts to 21 ft. lbs.

14. Tighten the air conditioning mounting bolts to 21 ft. lbs., if equipped.

15. Install the coolant reservoir tank to the radiator and connect the negative battery cable. Refill the power steering pump.

16. Install the oil filter and fill the crankcase to the proper level.

17. Run the engine and operate the power steering. Recheck the fluid level and for any leaks.

BELT ADJUSTMENT

1. To adjust the power steering belt tension, loosen the adjusting bolt of the air conditioning compressor, if equipped, or that of the power steering pump for vehicles without air conditioning.

2. Adjust the belt tension to 0.30 in. deflection, using the proper belt tension gauge and measuring at the midpoint between the 2 pulleys.

3. Tighten the proper adjusting and mounting bolts to the specified torque.

SYSTEM BLEEDING

1. Raise and support the vehicle safely.

2. Fill the power steering reservoir to the specified level.

3. Run the engine for 3–5 minutes, stop the engine and add fluid if necessary to reach specified level.

4. With the engine stopped, turn the steering wheel to the left and to the right as far as it turns. Repeat a few times and refill the reservoir.

5. With the engine running at idle speed, bleed air from the system by loosening the bleeder valve at the gear assembly.

6. Repeat the stop to stop turn of the steering wheel until all the foam is gone.

7. Tighten the bleed valve securely. Recheck the fluid level in the reservoir.

NOTE: When air bleeding is not complete, it is indicated by a foaming fluid on the level indicator or a humming noise from the power steering pump.

Tie Rod Ends

REMOVAL AND INSTALLATION

1. Raise and safely support the vehicle and remove the wheels.

2. Remove the tie rod end from the steering knuckle, using a suitable tie rod end remover tool.

3. Mark the tie rod end locknut position on the tie rod thread.

4. Loosen the locknut and remove the tie rod end from the tie rod.

To install:

5. Install the tie rod end locknut and the tie rod end to the tie rod. Align the locknut with the mark on the tie rod thread and tighten the locknut to 48 ft. lbs.

6. Connect the tie rod end to the steering knuckle. Tighten the castle nut until the holes of the split pin are aligned but only within the specified torque 33 ft. lbs.

BRAKES

Master Cylinder

REMOVAL AND INSTALLATION

1. Disconnect the negative battery cable. Remove the air cleaner case.

2. Disconnect the reservoir lead wire.

3. Clean the outside of the reservoir and drain the fluid from the reservoir.

4. Disconnect and plug the brake fluid lines at the master cylinder.

5. Remove the master cylinder to booster mounting bolts and remove the master cylinder.

To install:

6. Install the new master cylinder and tighten the mounting bolts to 12 ft. lbs.

7. Install the hydraulic brake lines to the master cylinder and tighten

and tighten the flare nuts to 13 ft. lbs.

8. Install the reservoir lead wire, if equipped.

9. Fill the reservoir with the specified brake fluid. Install the air cleaner case.

10. Bleed the air from the brake hydraulic system and check the brake pedal play.

Proportioning Valve

REMOVAL AND INSTALLATION

1. Raise and safely support the vehicle.

2. Disconnect and plug the hydraulic brake lines from the proportioning valve assembly.

3. Remove the proportioning valve from the vehicle body.

NOTE: The proportioning valve should be removed with the spring attached.

To install:

4. Install the proportioning valve to the vehicle body and tighten the mounting bolts to 20 ft. lbs.

5. Connect the hydraulic brake lines to the proportioning valve and tighten the flare nuts to 13 ft. lbs.

6. Fill the brake reservoir to the proper level and bleed the brake hydraulic system. Lower the vehicle.

NOTE: Bleed the air from the proportioning valve bleeder valve.

Power Brake Booster

REMOVAL AND INSTALLATION

1. Disconnect the negative battery cable. Remove the air cleaner case.

2. Drain the fluid from the brake reservoir and remove the master cylinder.

3. Disconnect the vacuum hose from the booster. Remove the pushrod clevis pin and cotter pin from the brake pedal arm.

4. Disconnect the brake booster mounting nuts and remove the brake booster from the vehicle.

5. Disconnect the pedal attachment from the booster.

To install:

6. Connect the pedal attachment to the booster and install the booster to the vehicle. Tighten the mounting nuts to 7.5–11.5 ft. lbs.

7. Install the pushrod clevis pin and cotter pin to the brake pedal arm.

8. Connect the vacuum hose to the booster and install the master cylinder to the booster.

9. Install the air cleaner case and connect the negative battery cable.

10. Fill the brake reservoir and bleed the brake hydraulic system.

Brake Caliper

REMOVAL AND INSTALLATION

1. Disconnect the negative battery cable. Raise and safely support the vehicle.

2. Remove the wheels. Disconnect and plug the brake line.

3. Remove the caliper mounting bolts and remove the caliper from the vehicle.

To install:

4. Install the caliper on the vehicle. Tighten the mounting bolts to 65 ft. lbs.

5. Connect the hydraulic brake line, using 2 new washers. Replace the front wheels.

6. Lower the vehicle. Connect the negative battery cable.

7. Fill the brake reservoir and bleed the hydraulic brake system.

Disc Brake Pads

REMOVAL AND INSTALLATION

1. Disconnect the negative battery cable. Raise and safely support the vehicle.

2. Remove the wheels.

3. Disconnect the brake caliper.

NOTE: Do not allow the caliper to hang from the brake hose. Support it by the mounting bracket.

4. Remove the disc pads from the caliper. Disconnect the anti-rattle springs.

To install:

5. Connect the anti-rattle springs and install the brake pads on to caliper assembly.

6. Connect the brake caliper and install the front wheels.

7. Lower the vehicle. Connect the negative battery cable.

Brake Rotor

REMOVAL AND INSTALLATION

1. Disconnect the negative battery cable.

2. Raise and safely support the vehicle. Remove the front wheels.

3. Remove the brake caliper mounting bracket, with the caliper and the brake line attached, move it out of the way and support it.

NOTE: Do not allow the caliper to hang from the brake hose. Support it by the mounting bracket.

4. Install 2 bolts into the threaded holes in the brake rotor and tighten them evenly. This will remove the rotor from the hub assembly.

To install:

5. Install the new brake rotor on the hub and install the caliper assembly.

6. Install the front wheels and tighten to 65 ft. lbs.

7. Lower the vehicle and connect the negative battery terminal.

Brake Drums

REMOVAL AND INSTALLATION

1. Disconnect the negative battery cable. Raise and safely support the vehicle.

2. Make sure the parking brake is released.

3. Remove the wheels and the rear drum nuts from the vehicle.

4. Remove the brake drum, using a slide hammer.

To install:

5. Install the brake drum, tighten the brake drum nuts to 58 ft. lbs.

6. Install the rear wheels and lower the vehicle. Connect the negative battery cable.

Brake Shoes

REMOVAL AND INSTALLATION

1. Disconnect the negative battery cable. Raise and safely support the vehicle.

2. Remove the rear wheels from the vehicle. Remove the rear brake drums.

3. Remove the brake shoe hold-down springs.

4. Disconnect the parking brake cable from the parking brake shoe lever and remove the brake shoes.

To install:

5. Install the brake shoes and install the parking brake lever and cable to the assembly.

6. Install the brake shoe hold-down springs and install the rear brake drums.

7. Install the rear wheels, lower the vehicle and connect the negative battery terminal.

Wheel Cylinder

REMOVAL AND INSTALLATION

1. Disconnect the negative battery cable. Raise and safely support the vehicle.

2. Remove the rear wheels and brake drums from the vehicle.

3. Remove the rear brake shoes and disconnect the brake line from the rear of the wheel cylinder. Plug the brake lines.

4. Remove the 2 rear wheel cylinder mounting bolts. Remove the rear wheel cylinder.

To install:

5. Install the wheel cylinder and tighten the mounting bolts to 7 ft. lbs.

6. Connect the brake line to the wheel cylinder and install the rear brake shoes.

7. Install the rear brake drums and the rear wheels.

8. Lower the vehicle. Connect the negative battery cable.

9. Bleed the brake system and check for any leaks when finished.

Parking Brake Cable

REMOVAL AND INSTALLATION

1. Disconnect the parking brake cable from the parking lever.

2. Raise and safely support the vehicle. Remove the rear wheels and brake drums from the vehicle.

3. Remove the rear brake shoes and disconnect the park brake cable from the park brake shoe lever.

4. Remove the cable from the brake backing plate by squeezing the park brake cable stop ring.

To install:

5. Install the cable to the backing plate and to the brake shoe lever.

6. Install the brake shoes and install the rear brake drums.

7. Install the rear wheels and lower the vehicle. Connect the cable to the parking brake lever.

ADJUSTMENT

1. Adjust the parking brake lever by loosening or tightening the self locking nut at the park brake lever.

2. The proper adjustment is when the park brake lever is within 7–9 notches, when the lever is pulled up at 44 lbs.

3. Check the rear drum for dragging after adjustment.

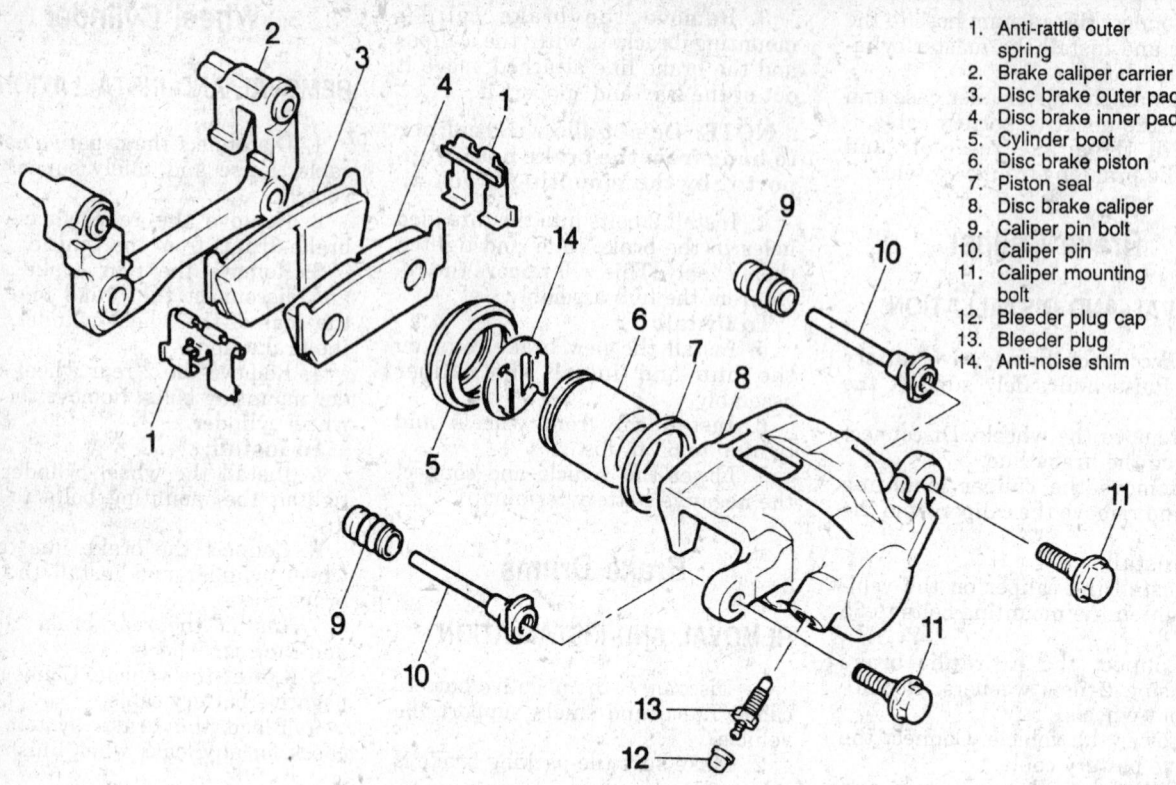

1. Anti-rattle outer spring
2. Brake caliper carrier
3. Disc brake outer pad
4. Disc brake inner pad
5. Cylinder boot
6. Disc brake piston
7. Piston seal
8. Disc brake caliper
9. Caliper pin bolt
10. Caliper pin
11. Caliper mounting bolt
12. Bleeder plug cap
13. Bleeder plug
14. Anti-noise shim

7911F030

Exploded view of the front disc brake assembly

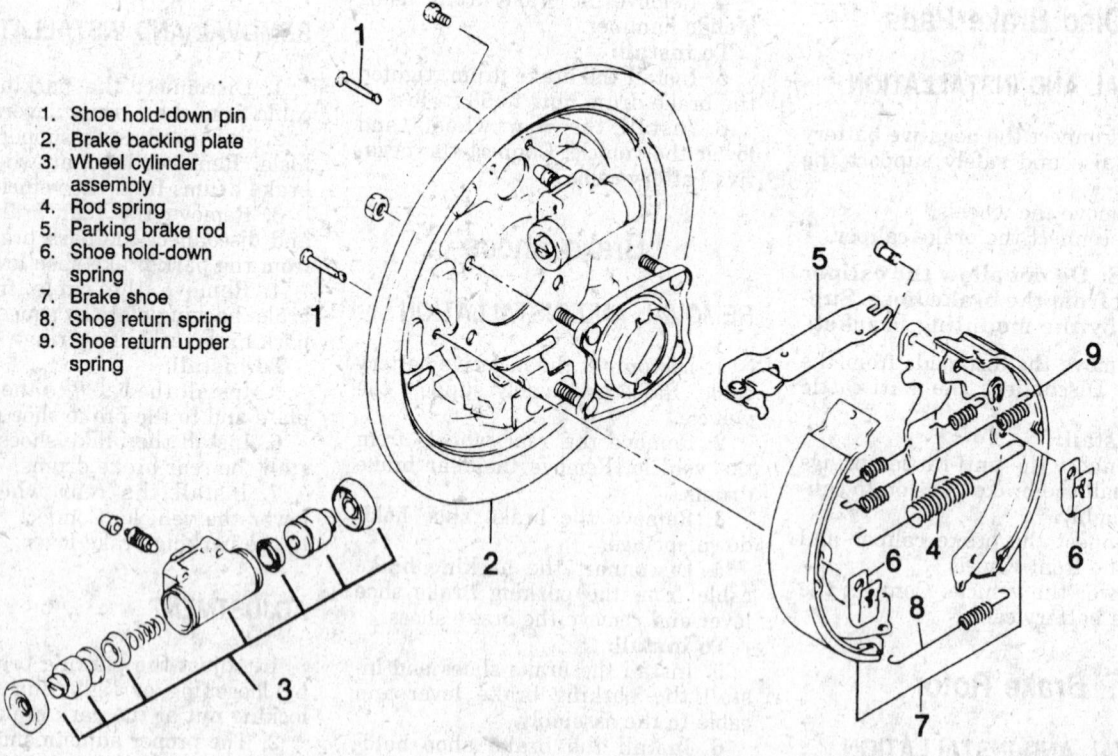

1. Shoe hold-down pin
2. Brake backing plate
3. Wheel cylinder assembly
4. Rod spring
5. Parking brake rod
6. Shoe hold-down spring
7. Brake shoe
8. Shoe return spring
9. Shoe return upper spring

7911F031

Exploded view of the rear brake drum assembly

FRONT SUSPENSION

Shock Absorbers

REMOVAL AND INSTALLATION

1. Raise and safely support the vehicle.
2. Support the axle assembly and remove the upper shock absorber mounting nut.
3. Remove the lower shock absorber mounting nut and remove the shock absorber.
   To install:
4. Install the shock absorber and install the lower mounting nut to the bolt.
5. Torque the upper mounting nut to 20 ft. lbs. and the lower nut to 33 ft. lbs. and lower the vehicle.

MacPherson Strut

REMOVAL AND INSTALLATION

1. Disconnect the negative battery cable.
2. Raise and safely support the vehicle. Allow the front suspension to hang free.
3. Remove the front wheel. Disconnect the E-clip mounting the brake hose and remove the brake hose from the strut bracket.
4. Remove the strut bracket to steering knuckle bolts.
5. Lower the vehicle. Remove the strut support nuts, while holding the strut by hand. Remove the strut assembly.
   To install:
6. Install the strut assembly and tighten the strut support nuts to 18 ft. lbs.
7. Raise and safely support the vehicle.
8. Install the strut bracket to steering knuckle bolts and tighten to 66 ft. lbs.
9. Connect the brake hose to the strut bracket using the E-clip.
10. Install the front wheels and lower the vehicle.
11. Connect the negative battery cable.

Coil Springs

REMOVAL AND INSTALLATION

Sidekick

1. Disconnect the negative battery cable.
2. Raise and safely support the vehicle. Allow the front suspension to hang free.
3. Remove the front wheel and the locking hub assembly.
4. Remove the front axle circlip and washer.
5. Remove the brake caliper mounting bracket, with the caliper and the brake line attached, move it out of the way and support it.
6. Remove the brake disc and disconnect the stabilizer link from the control arm.
7. Disconnect the tie rod end and support the lower control arm using a jack.
8. Disconnect the strut bracket and remove the ball joint castle nut.
9. Remove the steering knuckle and the wheel hub assembly while lowering the jack.
10. Remove the coil spring from the vehicle.
    To install:
11. Install the coil spring in the vehicle and install the wheel hub and steering knuckle while raising the jack.
12. Connect the strut bracket and install the ball joint castle nut.
13. Connect the tie rod end and connect the stabilizer link to the control arm.
14. Install the brake disc and connect the caliper assembly.
15. Install the front axle circlip and washer and install the locking hub assembly.
16. Install the front wheels, lower the vehicle and connect the negative battery cable.

Leaf Springs

REMOVAL AND INSTALLATION

Samurai

1. Disconnect the negative battery cable.
2. Raise and safely support the vehicle. Allow the front suspension to hang free.
3. Remove the stabilizer bar pivot bolt.
4. Support the front axle assembly with an adjustable stand.
5. Remove the leaf spring to spring plate mounting U-bolts.
6. Remove the shackle pin and the nut from the front of the leaf spring.
7. Disconnect the leaf spring bolt and remove the leaf spring.

NOTE: Removal of the leaf spring causes the axle housing to hang. Support it with a safety stand to prevent it from damaging the U-joint of the driveshaft.

To install:
8. Install the leaf spring and leaf spring bolts and tighten to 57 ft. lbs.
9. Install the shackle pin and nut to the front of the leaf spring and install the spring plate and U-bolts.
10. Install the stabilizer bar pivot bolt.
11. Lower the vehicle and connect the negative battery terminal.

Lower Ball Joints

INSPECTION

Sidekick

1. Inspect for the smoothness of the rotation.
2. Check the ball stud for damage.
3. Inspect the dust shield for damage.

REMOVAL AND INSTALLATION

Sidekick

1. Disconnect the negative battery cable. Remove the coil spring from the vehicle.
2. Remove the control arm mounting bolts.
3. Remove the control arm.
4. Disconnect the ball joint from the control arm and remove the ball joint.
   To install:
5. Install the ball joint to the control arm and tighten the bolts to 63 ft. lbs.
6. Install the control arm to the chassis and tighten the bolts to 74 ft. lbs.
7. Install the coil spring and connect the negative battery cable.

Lower Control Arms

REMOVAL AND INSTALLATION

Sidekick

1. Disconnect the negative battery cable. Remove the coil spring from the vehicle.

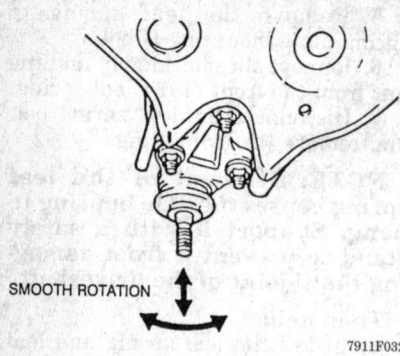

SMOOTH ROTATION

7911F032

Checking ball joint play

2. Remove the control arm mounting bolts.

3. Remove the control arm.

To install:

4. Install the control arm to the chassis and tighten the bolts to 74 ft. lbs.

5. Install the coil spring and connect the negative battery cable.

Stabilizer Bar

REMOVAL AND INSTALLATION

1. Disconnect the negative battery cable. Raise and safely support the vehicle.

2. Disconnect the left and the right stabilizer ball joints from the front control arms on the Sidekick.

3. Disconnect the stabilizer bar pivot bolts on the Samurai.

4. Remove the stabilizer bar mount bushing bracket bolts and nuts.

5. Remove the stabilizer bar.

6. Remove the stabilizer links on the Sidekick.

To install:

7. Install the new stabilizer bar, using new bushings.

8. Torque the stabilizer bar pivot bolts to 51–65 ft. lbs. and the mounting bracket nuts to 13–20 ft. lbs. for the Samurai. Torque the stabilizer link nuts to 21 ft. lbs. and the stabilizer bar bracket bolts and nuts to 37 ft. lbs. for the Sidekick.

9. Lower the vehicle and connect the negative battery cable.

Kingpin and Bushings

REMOVAL AND INSTALLATION

Samurai

1. Raise and safely support the vehicle.

2. Remove the steering knuckle from the vehicle.

NOTE: When the steering knuckle is pulled, the lower kingpin bearing sometimes falls off.

3. Remove the upper and the lower kingpins, mark them. Check the number of shims on each side.

To install:

4. Install the new kingpin bearings in the steering knuckle holding them in with grease.

5. Install the steering knuckle on the axle assembly.

6. Install the new kingpins in the steering knuckle, shim them correctly and torque the bolts to 17 ft. lbs.

NOTE: The correct procedure for installing the kingpins is to check the turning torque of the spindle while pulling it outwards from the tie rod end hole. A spring type gauge is required for this procedure. The correct force should be 2.20–3.96 lbs. of force required to turn the spindle without the oil seal being installed. Use additional shims, if necessary, to correct the turning torque.

7. With the turning torque of the spindle correct and the oil seal installed, install the front wheels and lower the vehicle.

REAR SUSPENSION

Shock Absorbers

REMOVAL AND INSTALLATION

1. Raise and safely support the vehicle. Support the rear axle housing.

2. Remove the upper and lower shock absorber mounting bolts.

3. Remove the rear shock absorber.

To install:

4. Install the the rear shock absorber.

5. On the Sidekick, install the upper mounting bolts and tighten to 21 ft. lbs. Install the lower mounting bolts and tighten to 63 ft. lbs.

6. On Samurai, tighten the upper and the lower mounting bolts to 25 ft. lbs.

Coil Springs

REMOVAL AND INSTALLATION

Sidekick

1. Raise and safely support the vehicle. Remove the rear wheels.

2. Support the rear axle housing, using a floor jack.

3. Remove the shock absorber lower mounting bolt and disconnect the parking brake cable hangers.

4. Lower the rear axle housing so the coil spring can be removed.

5. Remove the coil spring from the vehicle.

To install:

6. Install the coil spring to the spring seat and raise the axle housing.

7. Install the lower shock absorber mounting bolt but do not tighten.

8. Connect the parking brake cable hangers and install the rear wheels.

9. Lower the vehicle and tighten the lower shock absorber nut to 64 ft. lbs.

Leaf Springs

REMOVAL AND INSTALLATION

Samurai

1. Raise and safely support the vehicle. Remove the rear wheels.

2. Safely support the axle housing separately and disconnect the shocks.

3. Disconnect the stabilizer bar from the shackle plate under the leaf spring.

NOTE: Do not let the axle housing hang on the brake hoses or lines.

4. Remove the rear axle housing U-bolt nuts and remove the bolts.

5. Raise the rear axle housing to release spring tension and remove the shackle plate.

6. Support the leaf spring and disconnect the rear leaf spring mounting bolts.

7. Remove the leaf spring assembly.

To install:

8. Install the leaf spring and shackle assembly and tighten the mounting bolts to 56 ft. lbs.

9. Install the U-bolt nuts and tighten to 50 ft. lbs.

10. Connect the stabilizer bar under the shackle plate. Install the shocks and tighten to 30 ft. lbs.

11. Install the rear wheels and lower the vehicle.

TOYOTA **13**

Land Cruiser • Pickup • Previa • 4Runner

SPECIFICATIONS

ENGINE IDENTIFICATION

Year	Model	Engine Displacement Liters (cc)	Engine Series (ID/VIN)	Fuel System	No. of Cylinders	Engine Type
1991	Previa	2.4 (2393)	2TZ-FE	EFI	4	DOHC
	4Runner	2.4 (2366)	22R-E	EFI	4	SOHC
	4Runner	3.0 (2959)	3VZ-E	EFI	6	SOHC
	Pick-up	2.4 (2366)	22R-E	EFI	4	SOHC
	Pick-up	3.0 (2959)	3VZ-E	EFI	6	SOHC
	Land Cruiser	4.0 (3956)	3F-E	EFI	6	OHV
1992	Previa	2.4 (2393)	2TZ-FE	EFI	4	DOHC
	4Runner	2.4 (2366)	22R-E	EFI	4	SOHC
	4Runner	3.0 (2959)	3VZ-E	EFI	6	SOHC
	Pick-up	2.4 (2366)	22R-E	EFI	4	SOHC
	Pick-up	3.0 (2959)	3VZ-E	EFI	6	SOHC
	Land Cruiser	4.0 (3956)	3F-E	EFI	6	OHV
1993	Previa	2.4 (2393)	2TZ-FE	EFI	4	DOHC
	4Runner	2.4 (2366)	22R-E	EFI	4	SOHC
	4Runner	3.0 (2959)	3VZ-E	EFI	6	SOHC
	Pick-up	2.4 (2366)	22R-E	EFI	4	SOHC
	Pick-up	3.0 (2959)	3VZ-E	EFI	6	SOHC
	Land Cruiser	4.5 (4477)	1FZ-FE	EFI	6	DOHC
	T100	3.0 (2959)	3VZ-E	EFI	6	SOHC
1994-95	Previa	2.4 (2393)	2TZ-FE	EFI	4	DOHC
	4Runner	2.4 (2366)	22R-E	EFI	4	SOHC
	4Runner	3.0 (2959)	3VZ-E	EFI	6	SOHC
	Pick-up	2.4 (2366)	22R-E	EFI	4	SOHC
	Pick-up	3.0 (2959)	3VZ-E	EFI	6	SOHC
	Land Cruiser	4.5 (4477)	1FZ-FE	EFI	6	DOHC
	T100	3.0 (2959)	3VZ-E	EFI	6	SOHC

EFI - Electronic fuel injection
SOHC - Single overhead camshaft
DOHC - Double overhead camshaft
OHV - Overhead valve

7911gc02

GENERAL ENGINE SPECIFICATIONS

Year	Engine ID/VIN	Engine Displacement Liters (cc)	Fuel System Type	Net Horsepower @ rpm	Net Torque @ rpm (ft. lbs.)	Bore x Stroke (in.)	Com-pression Ratio	Oil Pressure @ rpm
1991	2TZ-FE	2.4 (2393)	EFI	138@5000	154@4000	3.74x3.39	9.1:1	36-71@3000
	22R-E	2.4 (2366)	EFI	116@4800	140@2800	3.62x3.50	9.3:1	36-71@3000
	3VZ-E	3.0 (2959)	EFI	150@4800	180@3400	3.44x3.23	9.0:1	36-71@4000
	3F-E	4.0 (3956)	EFI	154@4000	220@3000	3.70x3.74	8.1:1	36-71@4000
1992	2TZ-FE	2.4 (2393)	EFI	138@5000	154@4000	3.74x3.39	9.1:1	36-71@3000
	22R-E	2.4 (2366)	EFI	116@4800	140@2800	3.62x3.50	9.3:1	36-71@3000
	3VZ-E	3.0 (2959)	EFI	150@4800	180@3400	3.44x3.23	9.0:1	36-71@4000
	3F-E	4.0 (3956)	EFI	154@4000	220@3000	3.70x3.74	8.1:1	36-71@4000
1993	2TZ-FE	2.4 (2393)	EFI	138@5000	154@4000	3.74x3.39	9.1:1	36-71@3000
	22R-E	2.4 (2366)	EFI	116@4800	140@2800	3.62x3.50	9.3:1	36-71@3000
	3VZ-E	3.0 (2959)	EFI	150@4800	180@3400	3.44x3.23	9.0:1	36-71@4000
	1FZ-FE	4.5 (4477)	EFI	212@4600	275@3200	3.94x3.74	9.0:1	36-71@3000
1994-95	2TZ-FE	2.4 (2393)	EFI	138@5000	154@4000	3.74x3.39	9.1:1	36-71@3000
	22R-E	2.4 (2366)	EFI	116@4800	140@2800	3.62x3.50	9.3:1	36-71@3000
	3VZ-E	3.0 (2959)	EFI	150@4800	180@3400	3.44x3.23	9.0:1	36-71@4000
	1FZ-FE	4.5 (4477)	EFI	212@4600	275@3200	3.94x3.64	9.0:1	36-71@3000

EFI - Electronic fuel injection

7911gc03

GASOLINE ENGINE TUNE-UP SPECIFICATIONS

Year	Engine ID/VIN	Engine Displacement Liters (cc)	Spark Plugs Gap (in.)	Ignition Timing (deg.) MT	Ignition Timing (deg.) AT	Fuel Pump (psi)	Idle Speed (rpm) MT	Idle Speed (rpm) AT	Valve Clearance In.	Valve Clearance Ex.
1991	22R-E	2.4 (2366)	0.031	5B	5B	38-44	750	850	0.008	0.012
	2TZ-FE	2.4 (2393)	0.043	5B	5B	38-44	700	750	3	4
	3VZ-E	3.0 (2959)	0.031	10B	10B	38-44	800	800	1	2
	3F-E	4.0 (3956)	0.031	-	7B	37-46	650	650	0.008	0.014
1992	22R-E	2.4 (2366)	0.031	5B	5B	38-44	750	850	0.008	0.012
	2TZ-FE	2.4 (2393)	0.043	5B	5B	38-44	700	750	3	4
	3VZ-E	3.0 (2959)	0.031	10B	10B	38-44	800	800	1	2
	3F-E	4.0 (3956)	0.031	-	7B	37-46	650	650	0.008	0.014
1993	22R-E	2.4 (2366)	0.031	5B	5B	38-44	750	850	0.008	0.012
	2TZ-FE	2.4 (2393)	0.043	5B	5B	38-44	700	750	3	4
	3VZ-E	3.0 (2959)	0.041	10B	10B	38-44	800	800	1	2
	1FZ-FE	4.5 (4477)	0.031	-	3B	38-44	-	600-700	0.006-0.010	0.010-0.014
1994-95	22R-E	2.4 (2366)	0.031	5B	5B	38-44	750	850	0.008	0.012
	2TZ-FE	2.4 (2393)	0.043	5B	5B	38-44	700	750	3	4
	3VZ-E	3.0 (2959)	0.031	10B	10B	38-44	800	800	1	2
	1FZ-FE	4.5 (4477)	0.031	-	3B	38-44	-	600-700	0.006-0.010	0.010-0.014

NOTE: The Vehicle Emission Control Information label reflects changes made during production. The label must be used if the data differs from above.

B - Before top dead center

1 Intake: 0.007-0.011 (cold)

2 Exhaust: 0.009-0.013 (cold)

3 Intake: 0.006-0.010 (cold)

4 Exhaust: 0.010-0.014 (cold)

7911gc04

FIRING ORDERS

NOTE: To avoid confusion, always replace spark plug wires one at a time.

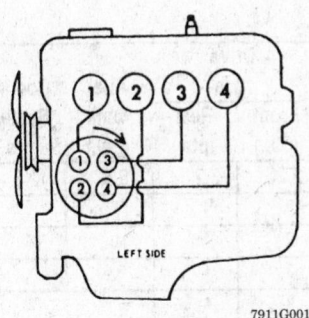

7911G001

22R — E and 2TZ — FE Engines
Engine Firing Order: 1 — 3 — 4 — 2
Distributor Rotation: Clockwise

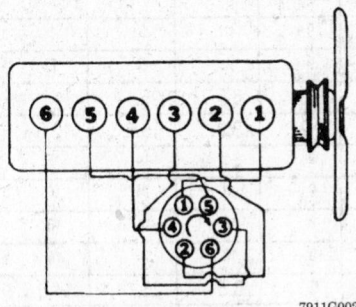

7911G002

3F — E Engine
Engine Firing Order: 1-5-3-6-2-4
Distributor Rotation: Clockwise

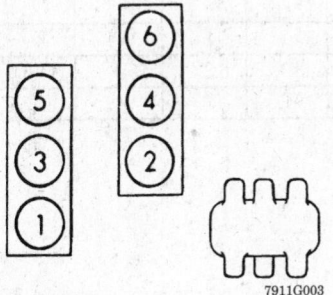

7911G003

3VZ — E Engine
Engine Firing Order: 1 — 2 — 3 — 4 —
5 — 6
Distributor Rotation: Counterclockwise

CAPACITIES

Year	Model	Engine ID/VIN	Engine Displacement Liters (cc)	Engine Crankcase with Filter	Transmission (pts.)			Transfer Case (pts.)	Drive Axle		Fuel Tank (gal.)	Cooling System (qts.)
					4-Spd	5-Spd	Auto.		Front (pts.)	Rear (pts.)		
1991	Previa	2TZ-FE	2.4 (2393)	6.1	-	10	5.0	3	2.2	3.2	19.8	12.3
	Pick-up	22R-E	2.4 (2366)	4.5	-	6	7	8	9	4	11	1
	Pick-up	3VZ-E	3.0 (2959)	4.8	-	6	7	8	9	4	11	2
	4Runner	22R-E	2.4 (2366)	4.5	-	6	7	8	9	4	5	1
	4Runner	3VZ-E	3.0 (2959)	4.8	-	6	7	8	9	4	5	2
	Land Cruiser	3F-E	4.0 (3956)	8.2	-	-	12.6	4	6.0	6.8	25.1	3
1992	Previa	2TZ-FE	2.4 (2393)	6.1	-	10	5.0	3	2.2	3.2	19.8	12.3
	Pick-up	22R-E	2.4 (2366)	4.5	-	6	7	8	9	4	11	1
	Pick-up	3VZ-E	3.0 (2959)	4.8	-	6	7	8	9	4	11	2
	4Runner	22R-E	2.4 (2366)	4.5	-	6	7	8	9	4	5	1
	4Runner	3VZ-E	3.0 (2959)	4.8	-	6	7	8	9	4	5	2
	Land Cruiser	3F-E	4.0 (3956)	8.2	-	-	12.6	4	6.0	6.8	25.1	3
1993	Previa	2TZ-FE	2.4 (2393)	6.1	-	10	5.0	3	2.2	3.2	19.8	12.3
	Pick-up	22R-E	2.4 (2366)	4.5	-	12	7	19	9	13	14	15
	Pick-up	3VZ-E	3.0 (2959)	4.8	-	12	7	19	9	13	14	16
	4Runner	22R-E	2.4 (2366)	4.5	-	12	7	19	9	13	14	15
	4Runner	3VZ-E	3.0 (2959)	4.8	-	12	7	19	9	13	14	16
	Land Cruiser	1FZ-FE	4.5 (4477)	7.8	-	-	12.6	17	18	6.8	25.1	14.8
	T100	3VZ-E	3.0 (2959)	4.8	-	6.4	7	2	9	13	14	16
1994-95	Previa	2TZ-FE	2.4 (2393)	6.1	-	10	5.0	3	2.2	3.2	19.8	12.3
	Pick-up	22R-E	2.4 (2366)	4.5	-	12	7	19	9	13	14	15
	Pick-up	3VZ-E	3.0 (2959)	4.8	-	12	7	19	9	13	14	16
	4Runner	22R-E	2.4 (2366)	4.5	-	12	7	19	9	13	14	15
	4Runner	3VZ-E	3.0 (2959)	4.8	-	12	7	19	9	13	14	16
	Land Cruiser	1FZ-FE	4.5 (4477)	7.8	-	-	12.6	17	18	6.8	25.1	14.8
	T100	3VZ-E	3.0 (2959)	4.8	-	6.4	7	2	9	13	14	16

1 Except 4WD automatic tranmission: 8.9
   4WD automatic transmission: 9.6
2 Manual transmission: 11.1
   Automatic transmission: 10.9
3 With rear heater: 20.6
   Without rear heater: 18.5
4 2WD with 7.5" differential: 2.8
   2WD with 8.0" differential: 3.8
   4WD: 4.6
5 Standard: 17.2; Optional: 19.3
6 2WD with W56: 5.0
   4WD with W56: 6.4
   4WD with G58: 8.2
   4WD with R150F: 6.4
7 2WD with A340E: 3.4
   4WD with A340H: 9.6
   4WD with A340F: 4.2
8 Counter gear type: 2.4
   Planetary gear type: 2.4
   A340H: 1.8
9 Standard: 3.4; ADD: 4.0

10 2WD: 4.6; 4WD: 5.4
11 Short bed: 13.7
    Long bed: 17.2
    Long bed 4WD: 19.3
12 G58: 8.2
    R150F: 6.4
13 2WD: 3.8; 4WD: 4.6
14 With standard tires: 17.2
    With optional 31x10.5 tires: 18.8
15 With rear heater: 9.2
    Without rear heater: 8.9
16 2WD with manual transmission: 11.0
    2WD with automatic transmission: 10.8
    4WD with manual transmission: 11.1
    4WD with automatic transmission: 10.9
17 With ABS: 2.8
    Without ABS: 3.6
18 With differential lock: 5.6
    Without differential lock: 5.8
19 Except 3VZ-E AT (VF1A type): 2.4
    3VZ-E AT (A340H): 1.6

7911gC06

CAMSHAFT SPECIFICATIONS

All measurements given in inches.

Year	Engine ID/VIN	Engine Displacement Liters (cc)	Journal Diameter					Elevation		Bearing Clearance	Camshaft End Play
			1	2	3	4	5	In.	Ex.		
1991	2TZ-FE	2.4 (2164)	1.0614-1.0620	1.0614-1.0620	1.0614-1.0620	1.0614-1.0620	NA	1.7839-1.7878	1.7740-1.7779	0.0010-0.0031	0.0016-0.0047
	22R-E	2.4 (2366)	1.2984-1.2992	1.2984-1.2992	1.2984-1.2992	1.2984-1.2992	NA	1.6783-1.6819	1.6807-1.6842	0.0004-0.0024	0.0031-0.0071
	3VZ-E	3.0 (2959)	1.0610-1.0616	1.0610-1.0616	1.0610-1.0616	1.0610-1.0616	NA	1.5102-1.5142	1.5059-1.5098	0.0010-0.0030	0.0079-0.0103
	3F-E	4.0 (3956)	1.8880-1.8888	1.8289-1.8297	1.7699-1.7707	1.7108-1.7116	NA	1.5102-1.5142	1.5059-1.5098	0.0010-0.0030	0.0079-0.0103
1992	2TZ-FE	2.4 (2393)	1.0614-1.0620	1.0614-1.0620	1.0614-1.0620	1.0614-1.0620	NA	1.7839-1.7878	1.7740-1.7779	0.0010-0.0031	0.0016-0.0047
	22R-E	2.4 (2366)	1.2984-1.2992	1.2984-1.2992	1.2984-1.2992	1.2984-1.2992	NA	1.6783-1.6819	1.6807-1.6842	0.0004-0.0024	0.0031-0.0071
	3VZ-E	3.0 (2959)	1.0610-1.0616	1.0610-1.0616	1.0610-1.0616	1.0610-1.0616	NA	1.6598-1.6638	1.6520-1.6559	0.0014-0.0028	0.0013-0.0031
	3F-E	4.0 (3956)	1.8880-1.8888	1.8289-1.8297	1.7699-1.7707	1.7108-1.7116	NA	1.5102-1.5142	1.5059-1.5098	0.0010-0.0030	0.0079-0.0103
1993	2TZ-FE	2.4 (2393)	1.0614-1.0620	1.0614-1.0620	1.0614-1.0620	1.0614-1.0620	NA	1.7839-1.7878	1.7740-1.7779	0.0010-0.0031	0.0016-0.0047
	22R-E	2.4 (2366)	1.2984-1.2992	1.2984-1.2992	1.2984-1.2992	1.2984-1.2992	NA	1.6783-1.6819	1.6807-1.6842	0.0004-0.0024	0.0031-0.0071
	3VZ-E	3.0 (2959)	1.0610-1.0616	1.0610-1.0616	1.0610-1.0616	1.0610-1.0616	NA	1.6598-1.6638	1.6520-1.6559	0.0014-0.0028	0.0013-0.0031
	1FZ-FE	4.5 (4477)	1.0614-1.0620	1.0614-1.0620	1.0614-1.0620	1.0614-1.0620	NA	1.9925-1.9965	1.9925-1.9965	0.0010-0.0024	0.0012-0.0031
1994-95	2TZ-FE	2.4 (2393)	1.0614-1.0620	1.0614-1.0620	1.0614-1.0620	1.0614-1.0620	NA	1.7839-1.7878	1.7740-1.7779	0.0010-0.0031	0.0016-0.0047
	22R-E	2.4 (2366)	1.2984-1.2992	1.2984-1.2992	1.2984-1.2992	1.2984-1.2992	NA	1.6783-1.6819	1.6807-1.6842	0.0004-0.0024	0.0031-0.0071
	3VZ-E	3.0 (2959)	1.0610-1.0616	1.0610-1.0616	1.0610-1.0616	1.0610-1.0616	NA	1.6598-1.6638	0.6520-1.6559	0.0014-0.0028	0.0013-0.0031
	1FZ-FE	4.5 (4477)	1.0614-1.0620	1.0614-1.0620	1.0614-1.0620	1.0614-1.0620	NA	1.9925-1.9965	1.9925-1.9965	0.0010-0.0024	0.0012-0.0031

NA - Not Available

7911gC07

CRANKSHAFT AND CONNECTING ROD SPECIFICATIONS

All measurements are given in inches.

Year	Engine ID/VIN	Engine Displacement Liters (cc)	Crankshaft				Connecting Rod		
			Main Brg. Journal Dia.	Main Brg. Oil Clearance	Shaft End-play	Thrust on No.	Journal Diameter	Oil Clearance	Side Clearance
1991	2TZ-FE	2.4 (2393)	2.3617-2.3622	0.0009-0.0019	0.0008-0.0087	3	2	0.0012-0.0023	0.0063-0.0121
	22R-E	2.4 (2366)	2.3616-2.3622	0.0010-0.0022	0.0008-0.0087	3	2.0861-2.0866	0.0010-0.0022	0.0008-0.0087
	3VZ-E	3.0 (2959)	2.5195-2.5197	0.0009-0.0017	0.0008-0.0098	3	2.1648-2.1654	0.0009-0.0021	0.0059-0.0130
	3F-E	4.0 (3956)	1	0.0008-0.0017	0.0024-0.0063	3	2.1252-2.1260	0.0008-0.0024	0.0043-0.0091
1992	2TZ-FE	2.4 (2393)	2.3617-2.3622	0.0009-0.0019	0.0008-0.0087	3	2	0.0012-0.0023	0.0063-0.0123
	22R-E	2.4 (2366)	2.3616-2.3622	0.0010-0.0022	0.0008-0.0087	3	2.0861-2.0866	0.0010-0.0022	0.0008-0.0087
	3VZ-E	3.0 (2959)	2.5195-2.5197	0.0009-0.0017	0.0008-0.0098	3	2.1648-2.1654	0.0009-0.0021	0.0059-0.0130
	3F-E	4.0 (3956)	1	0.0008-0.0017	0.0024-0.0063	3	2.1252-2.1260	0.0008-0.0024	0.0043-0.0091
1993	2TZ-FE	2.4 (2393)	2.3617-2.3622	0.0009-0.0019	0.0008-0.0087	3	2	0.0012-0.0023	0.0063-0.0123
	22R-E	2.4 (2366)	2.3616-2.3622	0.0010-0.0022	0.0008-0.0087	3	2.0861-2.0866	0.0010-0.0022	0.0008-0.0087
	3VZ-E	3.0 (2959)	2.5195-2.5197	0.0009-0.0017	0.0008-0.0098	3	2.1648-2.1654	0.0009-0.0021	0.0059-0.0130
	1FZ-FE	4.5 (4477)	2.7158-2.7165	0.0016-0.0032	0.0008-0.0087	3	3	0.0013-0.0020	0.0063-0.0103
1994-95	2TZ-FE	2.4 (2393)	2.3617-2.3622	0.0009-0.0019	0.0008-0.0087	3	2	0.0012-0.0023	0.0063-0.0123
	22R-E	2.4 (2366)	2.3616-2.3622	0.0010-0.0022	0.0008-0.0087	3	2.0861-2.0866	0.0010-0.0022	0.0008-0.0087
	3VZ-E	3.0 (2959)	2.5195-2.5197	0.0009-0.0017	0.0008-0.0098	3	2.1648-2.1654	0.0009-0.0021	0.0059-0.0130
	1FZ-FE	4.5 (4477)	2.7158-2.7165	0.0016-0.0032	0.0008-0.0087	3	3	0.0013-0.0020	0.0063-0.0103

1 No. 1: 2.6367-2.6376
  No. 2: 2.6957-2.6967
  No. 3: 2.7548-2.7557
  No. 4: 2.8139-2.8148

2 No. 1: 2.2047-2.2050
  No. 2: 2.2051-2.2053
  No. 3: 2.2054-2.2057

3 There are five sizes of standard connecting rod bearings, marked 2, 3, 4, 5 and 6 accordingly.
Replace with one having the same number (number located on outside of bearing end)
If the number of bearing cannot be determined, select correct bearing by adding together the numbers
imprinted on connecting rod and crankshaft, then selecting the bearing with the same number as the total
EXAMPLE: Connecting rod 3 + Crankshaft 1= 4. Use bearing 4.

7911gC08

VALVE SPECIFICATIONS

Year	Engine ID/VIN	Engine Displacement Liters (cc)	Seat Angle (deg.)	Face Angle (deg.)	Spring Test Pressure (lbs. @ in.)	Spring Installed Height (in.)	Stem-to-Guide Clearance (in.)		Stem Diameter (in.)	
							Intake	Exhaust	Intake	Exhaust
1991	2TZ-FE	2.4 (2393)	45 [1]	44.5	57-63	1.406	0.0010-0.0024	0.0012-0.0026	0.2350-0.2356	0.2348-0.2354
	22R-E	2.4 (2366)	45 [1]	44.5	66.1	1.594	0.0010-0.0024	0.0012-0.0026	0.3138-0.3144	0.3136-0.3142
	3VZ-E	3.0 (2959)	45 [1]	44.5	54-57	1.575	0.0010-0.0024	0.0012-0.0026	0.3138-0.3144	0.3136-0.3142
	3F-E	4.0 (3956)	45	44.5	71.6	1.693	0.0012-0.0024	0.0016-0.0028	0.3140	0.3137
1992	2TZ-FE	2.4 (2393)	45 [1]	44.5	57-63	1.406	0.0010-0.0024	0.0012-0.0026	0.2350-0.2356	0.2348-0.2354
	22R-E	2.4 (2366)	45 [1]	44.5	66.1	1.594	0.0010-0.0024	0.0012-0.0026	0.3138-0.3144	0.3136-0.3142
	3VZ-E	3.0 (2959)	45 [1]	44.5	54-57	1.575	0.0010-0.0024	0.0012-0.0026	0.3138-0.3144	0.3136-0.3142
	3F-E	4.0 (3956)	45	44.5	71.6	1.693	0.0012-0.0024	0.0016-0.0028	0.3140	0.3137
1993	2TZ-FE	2.4 (2393)	45 [1]	44.5	57-63	1.406	0.0010-0.0024	0.0012-0.0026	0.2350-0.2356	0.2348-0.2354
	22R-E	2.4 (2366)	45 [1]	44.5	66.1	1.594	0.0010-0.0024	0.0012-0.0026	0.3138-0.3144	0.3136-0.3142
	3VZ-E	3.0 (2959)	45 [1]	44.5	54-57	1.575	0.0010-0.0024	0.0012-0.0026	0.3138-0.3144	0.3136-0.3142
	1FZ-FE	4.5 (4477)	45	44.5	53.4	1.437	0.0010-0.0024	0.0012-0.0026	0.2744-0.2750	0.2742-0.2748
1994-95	2TZ-FE	2.4 (2393)	45 [1]	44.5	57-63	1.406	0.0010-0.0024	0.0012-0.0026	0.2350-0.2356	0.2348-0.2354
	22R-E	2.4 (2366)	45 [1]	44.5	66.1	1.594	0.0010-0.0024	0.0012-0.0026	0.3138-0.3144	0.3136-0.3142
	3VZ-E	3.0 (2959)	45 [1]	44.5	54-57	1.575	0.0010-0.0024	0.0012-0.0026	0.3138-0.3144	0.3136-0.3142
	1FZ-FE	4.5 (4477)	45	44.5	53.4	1.437	0.0010-0.0024	0.0012-0.0026	0.2744-0.2750	0.2742-0.2748

[1] Blend seat with 30 and 60 degree cutters to center the 45 degree portion on valve face

7911gC09

PISTON AND RING SPECIFICATIONS

All measurements are given in inches.

Year	Engine ID/VIN	Engine Displacement Liters (cc)	Piston Clearance	Ring Gap			Ring Side Clearance		
				Top Compression	Bottom Compression	Oil Control	Top Compression	Bottom Compression	Oil Control
1991	2TZ-FE	2.4 (2393)	0.0012-0.0020	0.0118-0.0169	0.0177-0.0236	0.0051-0.0150	0.0008-0.0028	0.0012-0.0028	SNUG
	22R-E	2.4 (2366)	0.0008-0.0016	0.0098-0.0185	0.0236-0.0323	0.0079-0.0224	0.0012-0.0028	0.0012-0.0028	SNUG
	3VZ-E	3.0 (2959)	0.0031-0.0039	0.0091-0.0327	0.0150-0.0366	0.0059-0.0354	0.0012-0.0028	0.0012-0.0028	SNUG
	3F-E	4.0 (3956)	0.0011-0.0019	0.0079-0.0165	0.0197-0.0283	0.0079-0.0323	0.0012-0.0028	0.0020-0.0035	SNUG
1992	2TZ-FE	2.4 (2393)	0.0012-0.0020	0.0118-0.0169	0.0177-0.0236	0.0051-0.0150	0.0008-0.0028	0.0012-0.0028	SNUG
	22R-E	2.4 (2366)	0.0008-0.0016	0.0098-0.0185	0.0236-0.0323	0.0079-0.0224	0.0012-0.0028	0.0012-0.0028	SNUG
	3VZ-E	3.0 (2959)	0.0031-0.0039	0.0091-0.0327	0.0150-0.0366	0.0059-0.0354	0.0012-0.0028	0.0012-0.0028	0.0012-0.0028
	3F-E	4.0 (3956)	0.0011-0.0019	0.0079-0.0165	0.0197-0.0283	0.0079-0.0323	0.0012-0.0028	0.0020-0.0035	SNUG
1993	2TZ-FE	2.4 (2393)	0.0012-0.0020	0.0118-0.0169	0.0177-0.0236	0.0051-0.0150	0.0008-0.0028	0.0012-0.0028	SNUG
	22R-E	2.4 (2366)	0.0008-0.0016	0.0098-0.0185	0.0236-0.0323	0.0079-0.0224	0.0012-0.0028	0.0012-0.0028	SNUG
	3VZ-E	3.0 (2959)	0.0031-0.0039	0.0091-0.0327	0.0150-0.0366	0.0059-0.0354	0.0012-0.0028	0.0012-0.0028	SNUG
	1FZ-FE	4.5 (4477)	0.0016-0.0024	0.0118-0.0205	0.0177-0.0264	0.0059-0.0205	0.0016-0.0031	0.0012-0.0028	SNUG
1994-95	2TZ-FE	2.4 (2393)	0.0012-0.0020	0.0118-0.0169	0.0177-0.0236	0.0051-0.0150	0.0008-0.0028	0.0012-0.0028	SNUG
	22R-E	2.4 (2366)	0.0008-0.0016	0.0098-0.0185	0.0236-0.0323	0.0079-0.0224	0.0012-0.0028	0.0012-0.0028	SNUG
	3VZ-E	3.0 (2959)	0.0031-0.0039	0.0091-0.0327	0.0150-0.0366	0.0059-0.0354	0.0012-0.0028	0.0012-0.0028	0.0012-0.0028
	1FZ-FE	4.5 (4477)	0.0016-0.0024	0.0118-0.0205	0.0177-0.0264	0.0059-0.0205	0.0016-0.0031	0.0012-0.0028	SNUG

7911gC10

TORQUE SPECIFICATIONS
All readings in ft. lbs.

Year	Engine ID/VIN	Engine Displacement Liters (cc)	Cylinder Head Bolts	Main Bearing Bolts	Rod Bearing Bolts	Crankshaft Damper Bolts	Flywheel Bolts	Manifold Intake	Manifold Exhaust	Spark Plugs	Lug Nut
1991	2TZ-FE	2.4 (2393)	4	5	6	192	7	15	36	11-15	-
	22R-E	2.4 (2366)	53-63	69-83	40-47	120-130	73-86	13-19	26-36	11-15	-
	3VZ-E	3.0 (2959)	3	43-47	16-20	176-186	63-67	11-15	252-33	11-15	-
	3F-E	4.0 (3956)	87-93	1	40-46	247-259	60-68	2	2	11-15	-
1992	2TZ-FE	2.4 (2393)	4	5	6	192	7	15	36	11-15	-
	22R-E	2.4 (2366)	53-63	69-83	40-47	120-130	73-86	13-19	26-36	11-15	-
	3VZ-E	3.0 (2959)	3	43-47	16-20	176-186	63-67	11-15	25-33	11-15	-
	3F-E	4.0 (3956)	87-93	1	40-46	247-259	60-68	2	2	11-15	-
1993	2TZ-FE	2.4 (2393)	4	5	6	192	7	15	36	11-15	-
	22R-E	2.4 (2366)	53-63	69-83	40-47	120-130	73-86	13-19	26-36	11-15	-
	3VZ-E	3.0 (2959)	3	43-47	16-20	176-186	63-67	11-15	25-33	11-15	-
	1FZ-FE	4.5 (4477)	8	9	10	304	-	15	29	15	-
1994-95	2TZ-FE	2.4 (2393)	4	5	6	192	7	15	36	11-15	-
	22R-E	2.4 (2366)	53-63	69-83	40-47	120-130	73-86	13-19	26-36	11-15	-
	3VZ-E	3.0 (2959)	3	43-47	16-20	176-186	63-67	11-15	25-33	11-15	-
	1FZ-FE	4.5 (4477)	8	9	10	304	-	15	29	15	-

1 17mm bolts: 85 ft. lbs.
   19mm bolts: 99 ft. lbs.
2 14mm bolts: 37 ft. lbs.
   17mm bolts: 51 ft. lbs.
3 Step 1: 27 ft. lbs.
   Step 2: 33 ft. lbs.
   Step 3: 90 degree turn
   Step 4: 90 degree turn
4 Step 1: 29 ft. lbs.
   Step 2: 90 degree turn
   Step 3: 90 degree turn
5 Step 1: 20 ft. lbs.
   Step 2: 35 ft. lbs.
   Step 3: 58 ft. lbs.

6 Step 1: 22 ft. lbs.
   Step 2: 90 degree turn
7 Manual transmission: 65 ft. lbs.
   Automatic transmission: 54 ft. lbs.
8 Step 1: 27 ft. lbs.
   Step 2: 90 degree turn
   Step 3: 90 degree turn
9 Step 1: 54 ft. lbs.
   Step 2: 90 degree turn
10 Step 1: 35 ft. lbs.
   Step 2: 90 degree turn

7911gC11

BRAKE SPECIFICATIONS
All measurements in inches unless noted

Year	Model	Master Cylinder Bore	Brake Disc			Brake Drum Diameter			Minimum Lining Thickness	
			Original Thickness	Minimum Thickness	Maximum Runout	Original Inside Diameter	Max. Wear Limit	Maximum Machine Diameter	Front	Rear
1991	Previa	NA	0.866 [2]	0.827 [3]	0.0028 [1]	10.00	-	10.079 [4]	0.039	0.039
	Pick-up	NA	[5]	[6]	[7]	[8]	-	[9]	0.039	0.039
	4Runner	NA	0.787	0.709	0.0035	11.61	-	11.69	0.039	0.039
	Land Cruiser	NA	0.984	0.906	0.0059	11.61	-	11.69	0.059	0.059
1992	Previa	NA	[10]	[11]	0.0028	10.00	-	10.079	0.039	0.039
	Pick-up	NA	[12]	[13]	[14]	[15]	-	[16]	0.039	0.039
	4Runner	NA	0.984	0.906	0.0035	11.61	-	11.69	0.059	0.039
	Land Cruiser	NA	0.984	0.906	0.0059	11.61	-	11.69	0.059	0.059
1993	Previa	NA	[10]	[11]	0.0028	10.00	-	10.079	0.039	0.039
	Pick-up	NA	[12]	[13]	[14]	[15]	-	[16]	0.039	0.039
	4Runner	NA	0.984	0.906	0.0035	11.61	-	11.69	0.059	0.039
	Land Cruiser	NA	1.260	1.181	0.0059	11.61	-	11.69	0.059	[17]
1994-95	Previa	NA	[10]	[11]	0.0028	10.00	-	10.079	0.039	0.039
	Pick-up	NA	[12]	[13]	0.0035	[15]	-	[16]	[18]	0.039
	4Runner	NA	0.984	0.906	0.0035	11.61	-	11.69	0.059	0.039
	Land Cruiser	NA	1.260	1.181	0.0059	11.61	-	11.69	0.059	[17]

NA - Not Available
1 Rear: 0.0039
2 Rear: 0.709
3 Rear: 0.669
4 Rear disc (parking brake): 7.52
5 2WD (PD60 type disc): 0.984
 2WD (PD66 type disc): 1.181
 2WD (FS17, 18 type disc): 0.866
 4WD (S12+12 type disc): 0.787
6 2WD (PD60 type disc): 0.906
 2WD (PD66 type disc): 1.102
 2WD (FS17, 18 type disc): 0.787
 4WD (S12+12 type disc): 0.709
7 PD60, FS17, FS18, S12+12 type discs: 0.0035
8 2WD: 10.00; 4WD: 11.61
9 2WD: 10.08; 4WD: 11.69
10 Front with rear drum brake: 0.984
 Front with rear disc brake: 0.866
 Rear disc brake: 0.709

11 Front with rear drum brake: 0.906
 Front with rear disc brake: 0.787
 Rear disc brake: 0.669
12 2WD (PD60 type disc): 0.984
 2WD (PD66 type disc): 1.181
 2WD (FS17, 18 type disc): 0.866
 4WD (S12+12 type disc): 0.787
13 2WD (PD60 type disc): 0.787
 2WD (PD66 type disc): 1.102
 2WD (FS17, 18 type disc): 0.787
 4WD (S12+12 type disc): 0.709
14 PD60, FS17, FS18, S12+12 type discs: 0.0035
15 2WD: 10.00; 4WD: 11.61
16 2WD: 10.08; 4WD: 11.69
17 Brake shoe lining: 0.059
 Disc pad lining: 0.039
18 2WD: 0.059; 4WD: 0.039

7911gC12

WHEEL ALIGNMENT

Year	Model		Caster Range (deg.)	Caster Preferred Setting (deg.)	Camber Range (deg.)	Camber Preferred Setting (deg.)	Toe-in (in.)	Steering Axis Inclination (deg.)
1991	Pick-up 2WD	1,2	0-1 1/2P	3/4P	1/4N-1 1/4P	1/2P	1/32N-5/32N	10
	Pick-up 2WD	1,3	1/16N-1 7/16P	11/16P	5/16N-1 3/16P	7/16P	1/32N-5/32N	10
	Pick-up 2WD	1,4	7/32P-1 23/32P	31/32P	5/16N-1 3/16P	7/16P	0-3/16N	10 1/32
	Pick-up 2WD	1,5	1/2P-2P	1 1/4P	3/8N-1 1/8P	3/8P	1/32N-7/32N	10 3/32
	Pick-up 2WD	1,6	3/16N-1 5/16P	9/16P	1/4N-1 1/4P	1/2P	1/8N-5/16N	10
	Pick-up 2WD	1,7	3/16N-1 5/16P	9/16P	1/4N-1 1/4P	1/2P	3/32N-9/32N	10
	Pick-up 2WD	1,8	1P-2 1/2P	1 3/4P	1/4N-1 1/4P	1/2P	3/32N-9/32N	10
	Pick-up 2WD	1,9	7/16P-1 15/16P	1 3/16P	3/8N-1 1/8P	3/8P	1/32N-7/32N	10 3/32
	Pick-up 2WD	1,10	7/16P-1 15/16P	1 3/16P	5/16N-1 3/16P	7/16P	1/32N-7/32N	10 1/16
	Pick-up 2WD	1,11	1P-2 1/2P	1 3/4P	1/4N-1 1/4P	1/2P	1/8N-5/16N	10
	Pick-up 4WD		2P-3P	2 1/2P	1/4P-1 1/4P	3/4P	0-1/16N	11 13/16
	4-Runner		2P-3P	2 1/2P	1/4P-1 1/4P	3/4P	1/32N-3/32N	11 13/16
	Land Cruiser	1,12	11/16P-2 11/16P	1 11/16P	1/4P-1 3/4P	1P	0-3/16N	13
	Land Cruiser	1,13	2P-4P	3P	1/4P-1 3/4P	1P	0-3/16N	13
	Previa 2WD		4 3/4P-6 1/4P	5 1/2P	21/32N-27/32P	3/32P	0-3/16N	10 19/32
	Previa 4WD		4 19/32P-6 3/32P	5 11/32P	1/2N-1P	1/4P	1/32N-7/32N	10 11/32
1992	Pick-up 2WD	1,2	0-1 1/2P	3/4P	1/4N-1 1/4P	1/2P	1/32N-5/32N	10
	Pick-up 2WD	1,3	1/16N-1 71/6P	11/16P	5/16N-1 3/16P	7/16P	1/32N-5/32N	10
	Pick-up 2WD	1,4	7/32P-1 23/32P	31/32P	5/16N-1 3/16P	7/16P	0-3/16N	10 1/32
	Pick-up 2WD	1,5	1/2P-2P	1 1/4P	3/8N-1 1/8P	3/8P	1/32N-7/32N	10 1/32
	Pick-up 2WD	1,6	3/16N-1 5/16P	9/16P	1/4N-1 1/4P	1/2P	1/8N-5/16N	10
	Pick-up 2WD	1,7	3/16N-1 5/16P	9/16P	1/4N-1 1/4P	1/2P	3/32N-9/32N	10
	Pick-up 2WD	1,8	1P-2 1/2P	1 3/4P	1/4N-1 1/4P	1/2P	3/32N-9/32N	10
	Pick-up 2WD	1,9	7/16P-1 15/16P	1 3/16P	3/8N-1 1/8P	3/8P	1/32N-7/32N	10 3/32
	Pick-up 2WD	1,10	7/16P-1 15/16P	1 3/16P	5/16N-1 3/16P	7/16P	1/32N-7/32N	10 1/16
	Pick-up 2WD	1,11	1P-2 1/2P	1 3/4P	1/4N-1 1/4P	1/2P	1/8N-5/16N	10
	Pick-up 4WD		2P-3P	2 1/2P	1/4P-1 1/4P	3/4P	0-1/16N	11 13/16
	4-Runner		2P-3P	2 1/2P	1/4P-1 1/4P	3/4P	1/32N-3./32N	11 13/16
	Land Cruiser	1,12	11/16P-2 11/16P	1 11/16P	1/4P-1 3/4P	1P	0-3/16N	13
	Land Cruiser	1,13	2P-4P	3P	1/4P-1 3/4P	1P	0-3/16N	13
	Previa 2WD		4 3/4P-6 1/4P	5 1/2P	21/32N-27/32P	3/32P	0-3/16N	10 19/32
	Previa 4WD		4 19/32P-6 3/32P	5 11/32P	1/2N-1P	1/4P	1/32N-7/32N	10 19/32
1993	Pick-up 2WD	1,14	0-1 1/2P	3/4P	1/4N-1 1/4P	1/2P	0.03N-0.13P	9 1/4
	Pick-up 2WD	1,15	0-1 1/2P	3/4P	1/4N-1 1/4P	1/2P	0.02N-0.14P	9 1/4
	Pick-up 2WD	1,16	1/4P-1 3/4P	1P	1/4N-1 1/4P	1/2P	0-0.16P	9 1/4
	Pick-up 2WD	1,17	1/2P-2P	1 1/4P	1/4N-1 1/4P	1/3P	0.05N-0.20P	9 1/4
	Pick-up 2WD	1,18	1/4N-1 1/4P	1/2P	1/4N-1 1/4P	1/2P	0.14N-0.30P	9 1/4
	Pick-up 2WD	1,19	1P-2 1/4P	1 3/4P	1/4N-1 1/4P	1/2P	0.14N-0.30P	9 1/4
	Pick-up 2WD	1,20	1/2P-2P	1 1/4P	1/4N-1 1/4P	1/3P	0.05N-0.20P	9 1/4
	Pick-up 2WD	1,21	1P-2 1/2P	1 3/4P	1/4N-1 1/4P	1/2P	0.014N-0.30P	9 1/4
	Pick-up 4WD		1 3/4P-3 1/4P	2 1/2P	0-1 1/2P	3/4P	0.04N-0.012P	11
	4-Runner		1 3/4P-3 1/4P	2 1/2P	0-1 1/2P	3/4P	0.04N-0.012P	11
	Land Cruiser		2P-4P	3P	1/4P-1 3/4P	1P	0-016P	12 1/4
	Previa		4 3/4P-6 1/4P	5 1/2P	3/4N-3/4P	0	0-016P	9 3/4
	T100 2WD	1,22	1 3/4P-3 1/4P	2 1/2P	1/4N-1 1/4P	1/2P	0.04N-0.20P	9 1/4
	T100 2WD	1,23	2 1/4P-3 3/4P	3P	0-1 1/2P	1/2P	0.27N-0.43P	8 3/4

7911gC13

WHEEL ALIGNMENT

Year	Model			Caster		Camber		Toe-in (in.)	Steering Axis Inclination (deg.)
				Range (deg.)	Preferred Setting (deg.)	Range (deg.)	Preferred Setting (deg.)		
1993	T100 4WD			3/4P-2 1/4P	1 1/2P	0-1 1/2P	3/4P	0.04N-0.20P	11
1994-95	Pick-up 2WD	1,24		0-1 1/2P	3/4P	1/4N-1 1/4P	1/2P	0.03N-0.13P	9 1/4
	Pick-up 2WD	1,20		1/2P-2P	1 1/4P	1/4N-1 1/4P	1/3P	0.05N-0.20P	9 1/4
	Pick-up 4WD			1 3/4P-3 1/4P	2 1/2P	0-1 1/2P	3/4P	0.04N-0.12P	11
	4-Runner			1 3/4P-3 1/4P	2 1/2P	0-1 1/2P	3/4P	0.04N-0.12P	11
	Land Cruiser			2P-4P	3P	1/4P-1 3/4P	1P	0-0.16P	12 1/4
	Previa 2WD			4 3/4P-6 1/4P	5 1/2P	3/4N-3/4P	0	0-0.16P	9 3/4
	Previa 4WD			4 3/4P-6 1/4P	5 1/2P	1/2N-1P	1/4P	0.04N-0.20P	9 3/4
	T100 2WD	1,22		1 34P-3 1/4P	2 1/2P	1/4N-1 1/4P	1/2P	0.04N-0.20P	9 3/4
	T100 2WD	1,23		2 1/4P-3 3/4P	3P	0-1 1/2P	1/2P	0.27N-0.43P	8 3/4
	T100 4WD			3/4P-2 1/4P	1 1/2P	0-1 1/2P	3/4P	0.04N-0.20P	11

1 NOTE: Front end alignment specifications are given according to Vehicle Identification Numbers. Before
   using this information, verify that the Vehicle Identification Number is correct for the alignment data that is used
2 RN80L-TRSD, RN80L-TRMD
3 RN80L-TRMR
4 RN85L-TRMD, RN85L-TRSD
5 RN90L-CRSD, RN90L-CRMD
6 VZN85L-THMD
7 VZN85L-THSD
8 VZN85L-TWMR, VZN85L-TWSR
9 VZN90L-CRMD, VZN90L-CRSD
10 VZN90L-CRMG, VZN90L-CRPG
11 VZN95L-TWMR, VZN95L-TWSR
12 With 10.5R tire

13 Without 10.5R tire
14 RN80L-TRSDEA, RN80L-TRSDEK, RN80L-TRMDEA, RN80L-TRMDEK
15 RN80L-TRMREA, RN80L-TRMREK
16 RN85L-TRMDEA, RN85L-TRMDEK, RN85L-TRSDEA, RN85L-TRSDEK
17 RN90L-CRSDEA, RN90L-CRSDEK, RN90L-CRMDEA, RN90L-CRMDEK
18 VZN85L-THMDEA, VZN85L-THSDEA
19 VZN85L-TWMREA6, VZN85L-TWSREA6
20 RN90L-CRSDEA, RN90L-CRMDEA, RN90L-CRSDEK, RN90L-CRMDEK,
   VZN90L-CRMDEA, VZN90L-CRMDEK, VZN90L-CRSDEA, VZN90L-CRSDEK,
   VZN90L-CRMGEA, VZN90L-CRPGEA
21 VZN95L-TWMREA6, VZN95L-TWSREA6, VZN95L-TWSREK6
22 1/2 ton
23 1 ton
24 RN80L-TRSDEA, RN80L-TRMDEA, RN80L-TRMREA, RN80L-TRMREK

7911g13a

ENGINE ELECTRICAL

NOTE: Disconnecting the battery cable on some vehicles may interfere with the functions of the on board computer systems and may require the computer to undergo a relearning process, once the negative battery cable is disconnected.

Distributor

REMOVAL

3F–E, 22R–E and 3VZ–E Engines

1. Disconnect the negative battery cable. Label and disconnect the high tension cables from the spark plugs. Remove the high tension cable from the coil.
2. Remove the primary wire or the electrical connector and the vacuum line, if equipped, from the distributor. Remove the distributor cap spring clips or screws, then the cap.
3. Using a piece of chalk, match-mark the rotor-to-distributor housing and the distributor-to-engine block. This will aid in correct positioning of the distributor during installation.
4. Remove the distributor hold-down clamp bolt and the distributor from the engine.

NOTE: It is easier to install the distributor if the engine timing is not disturbed while it is removed.

2TZ–FE Engine

1. Disconnect the negative battery cable.
2. Label and disconnect the spark plug wires.
3. Disconnect the distributor connector and ventilation hoses.
4. Remove the cap and packing.
5. Set the No. 1 cylinder to TDC of the compression stroke. Install the service bolt and nut into the equipment driveshaft to turn the crankshaft pulley. Turn the crankshaft 1 turn if the rotor is not facing No. 1 spark plug wire.
6. Remove the distributor hold-down and pull the distributor out of the engine.

INSTALLATION

Timing Not Disturbed

ALL ENGINES

1. Insert the distributor into the engine block by aligning the matchmarks made during removal.
2. Engage the distributor drive with the oil pump drive shaft.
3. Install the distributor hold-down clamp, the cap, the high tension wire, the primary wire or the electrical connector and the vacuum line(s).
4. Install the spark plugs cables.
5. Connect the negative battery cable.

Timing Disturbed

3F–E, 22R–E AND 3VZ–E ENGINES

If the engine has been cranked, dismantled or the timing otherwise lost, proceed as follows:
1. Determine the Top Dead Center (TDC) of the No. 1 cylinder's compression stroke by removing the spark plug from the No. 1 cylinder and placing a finger or a compression gauge over the spark plug hole.

NOTE: Using a wrench, turn the crankshaft until the compression pressure starts to build up. Continue cranking the engine until the timing marks indicate TDC (0°).

2. Turn the crankshaft to align the timing marks on the 22R–E engines to 5° BTDC or on the 3VZ–E engines to 0 degree TDC.
3. Temporarily install the rotor on the distributor shaft so the rotor is pointing toward the No. 1 terminal of the distributor cap.
4. Using a small prybar, align the slot on the distributor drive (oil pump driveshaft) with the key on the bottom of the distributor shaft.
5. Install the distributor in the block by rotating it slightly (no more than a gear tooth in either direction) until the driven gear meshes with the drive.

NOTE: Oil the distributor drive gear and the oil pump driveshaft end before installation.

6. Temporarily tighten the lock bolt.
7. Remove the rotor, then install the dust cover, the rotor and the distributor cap.

8. Install the primary wire or the electrical connector and the vacuum line(s).
9. Install the No. 1 cylinder spark plug. Connect the cables to the spark plugs in the proper order. Install the high tension wire on the coil.
10. Start the engine and adjust the ignition timing.

2TZ–FE ENGINE

1. Remove the No. 1 spark plug, place a finger over the opening and rotate the equipment driveshaft, using a turning tool, in the clockwise direction, until pressure is felt, then replace the spark plug.

NOTE: Make sure the slit in the exhaust camshaft is in the proper position.

2. Remove the service bolt and nut.
3. Align the cut-out portion of the coupling with the groove on the housing.
4. Install the distributor and align the center of the flange with the bolt hole on the cylinder head.
5. Install the hold-down bolt loosely.
6. Install the seal packing, distributor cap and connect the wiring.
7. Adjust the timing to specifications and torque the hold-down bolt to 14 ft. lbs. (19 Nm).

Ignition Timing

ADJUSTMENT

Except 2TZ–FE Engine

NOTE: The timing mark locations differ between the engines used in the Pick-Up and 4Runner for 22R–E and 3VZ–E engines, and Land Cruiser for 3F–E engine. On the 22R–E, and 3VZ–E engines, the timing marks are located on the crankshaft pulley (painted notch) and the timing cover (plate). On the 3F–E engine, the timing marks are located on the flywheel (ball) and the bell housing (pointer).

1. Set the parking brake and block the wheels.
2. Clean off the timing marks and mark them with chalk or paint. The crankshaft may have to rotated to find the marks.
3. Warm the engine to operating temperatures. Connect a tachometer

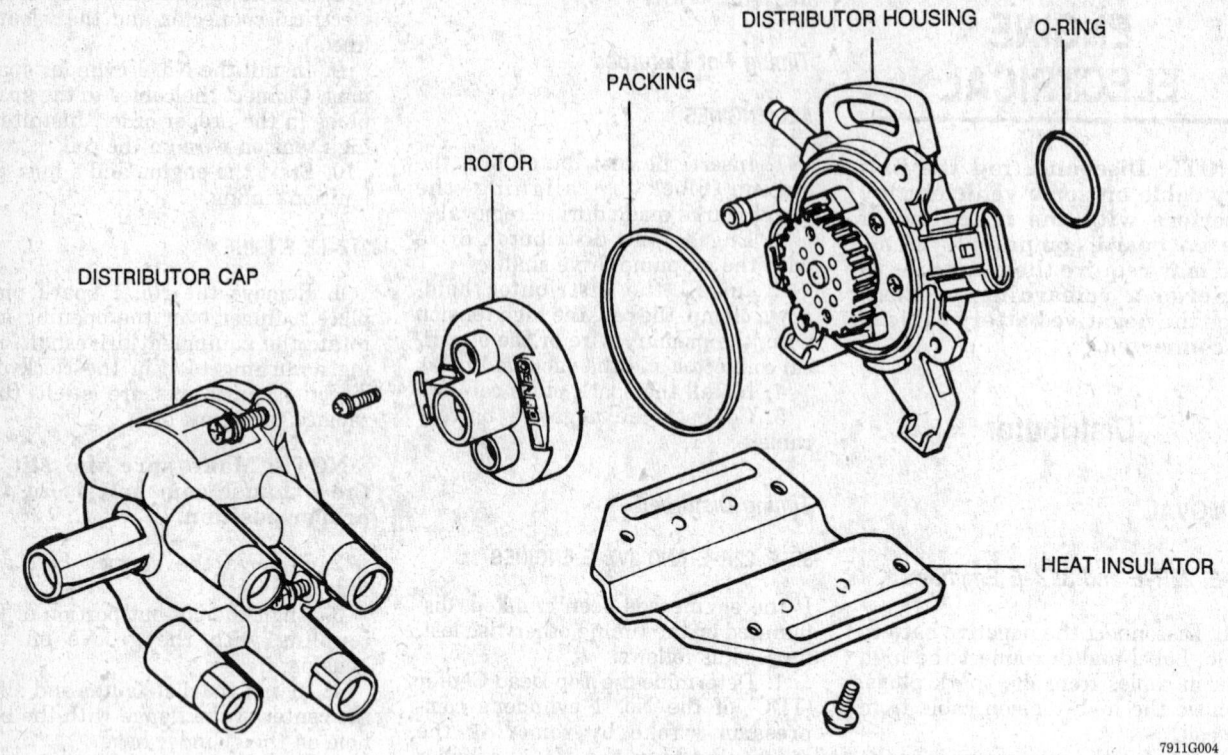

Distributor assembly — 2TZ — FE engine

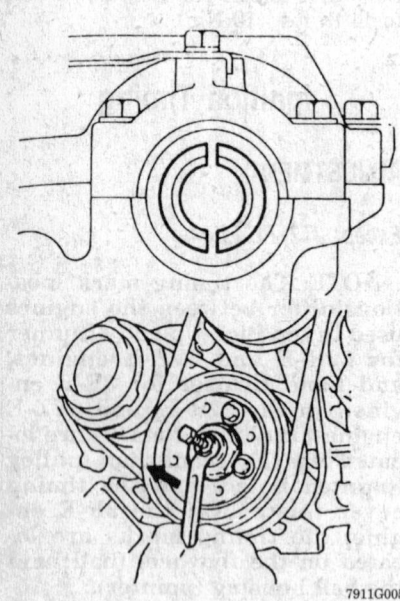

Exhaust camshaft positioning — 2TZ — FE engine

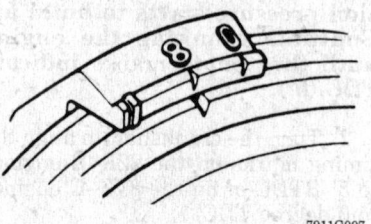

Timing mark location — 22R — E engine

to the engine, then, check and/or adjust the engine idle speed.

NOTE: On the 22R–E, and 3VZ–E engines, connect the positive (+) tachometer terminal either to the negative (–) ignition coil terminal or to the yellow service connector. On the 3VZ–E engines, use a service wire to short the engine check connector.

4. Turn off the engine and connect a timing light according to the manufacturer's directions.

5. On the 22R–E engines, disconnect and short the **T** and the **E₁** connector of the engine check harness (near the front of the vehicle). On all other engines, disconnect and plug the vacuum hose(s) from the distributor vacuum unit.

NOTE: If equipped with a High Altitude Compensation (HAC) system there are 2 vacuum hoses which connect to the distributor. Both must be disconnected and plugged. These systems require an extra step in the timing procedure.

6. Be sure the timing light wires are clear of the fan and pulleys, then start the engine.

7. Allow the engine to run at the specified idle speed with the shift selector in neutral for manual transmission or **D** for automatic transmission.

8. Point the timing light at the marks. With the engine at the specified idle, the marks should align.

9. If the timing is incorrect, loosen the bolt at the base of the distributor just enough so the distributor can be turned. Hold the distributor by its base and turn it slightly to advance or retard the timing as required. Once the marks are seen to align properly, tighten the bolt.

10. After tightening the distributor bolt or adjusting the octane selector, recheck the timing. Turn off the en-

gine, then, disconnect the timing light and connect the vacuum line(s) at the distributor or the electrical **T** and **E₁** connector, except on engines with HAC.

11. On engines with HAC after setting the initial timing, reconnect the vacuum hoses at the distributor. Recheck the timing.

12. If the advance is still low, pinch the hose between the HAC valve and the 3 way connector; it should now be to specifications. If not, the HAC valve should be checked for proper operation.

2TZ–FE Engine

NOTE: On the 2TZ–FE engine the timing mark is on the equipment driveshaft U-joint and front cover.

Ground terminals **TE1** and **E1** of the check connector using tool 0984318020 or equivalent. The check connector is next to the emergency brake lever.

1. Connect a timing light terminal to terminal **30** of the starter and test probe to the No. 1 spark plug wire (light blue).

2. Start the engine and warm up. Slowly turn the distributor until the timing mark on the crankshaft pulley is aligned with the **5** mark. Tighten the distributor bolt and recheck timing.

3. Remove the grounding tool from the check connector and timing light.

Alternator

PRECAUTIONS

Several precautions must be observed with alternator equipped vehicles to avoid damage to the unit.

If the battery is removed for any reason, make sure it is reconnected with the correct polarity. Reversing

Timing mark location — 3VZ — E engine

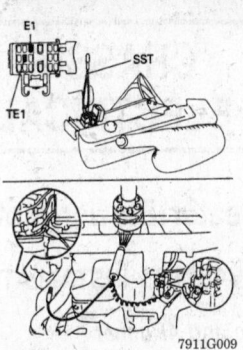

7911G009

Grounding the check connector and timing location — 2TZ — FE engine

the battery connections may result in damage to the one-way rectifiers.

When utilizing a booster battery as a starting aid, always connect the positive to positive terminals and the negative terminal from the booster battery to a good engine ground on the vehicle being started.

Never use a fast charger as a booster to start vehicles.

Disconnect the battery cables when charging the battery with a fast charger.

Never attempt to polarize the alternator.

Do not use test lights of more than 12 volts when checking diode continuity.

Do not short across or ground any of the alternator terminals.

The polarity of the battery, alternator and regulator must be matched and considered before making any electrical connections within the system.

Never separate the alternator on an open circuit. Make sure all connections within the circuit are clean and tight.

Disconnect the battery ground terminal when performing any service on electrical components.

Disconnect the battery if arc welding is to be done on the vehicle.

BELT TENSION ADJUSTMENT

Inspection and adjustment to the alternator drive belt should be performed every 30,000 miles or if the alternator has been removed.

1. Inspect the drive belt to see that it is not cracked or worn. Be sure its surfaces are free of grease or oil.

2. If not using a belt tension gauge, push down on the belt halfway between the fan and the alternator pulleys, (or crankshaft pulley) with thumb pressure; belt deflection should be ³⁄₈–¹⁄₂ in.

3. If using the an appropriate belt tension gauge, position it in the middle of the drive belt and check the belt tension; a new belt should be 100–150 lbs. (135–202 N) for Pick-Up and 4Runner, a used belt should be 60–100 lbs. (81–136 N) for Pick-Up and 4Runner.

4. If the belt tension requires adjustment, loosen the adjusting link bolt and move the alternator until the proper belt tension is obtained.

5. Tighten the adjusting link bolt.

REMOVAL AND INSTALLATION

NOTE: On some engines the alternator is mounted very low. On these engines it may be necessary to remove the gravel shield and work from under the vehicle in order to gain access to the alternator.

Except 2TZ–FE Engine

1. Disconnect the negative battery cable.

2. Remove the air cleaner, if necessary, to gain access to the alternator.

3. On the 22R–E engines, drain the engine coolant. If necessary, remove the under engine cover.

4. If equipped with power steering, remove the water inlet pipe bolts and the water inlet hose from the engine.

NOTE: If equipped with air conditioning, it may be necessary to remove the No. 2 fan shroud.

5. Remove the nut or the wiring connector and the wire(s) from the alternator.

6. Remove the adjusting lock, the pivot and the adjusting bolt(s), then the drive belt from the alternator.

7. Remove the alternator attaching bolt and then withdraw the alternator from its bracket.

8. To install, reverse the removal procedures. Rotate the drive belt 8 revolutions for new belt or 5 revolutions for used belt. Adjust the drive belt tension.

9. Refill the cooling system, if it was drained.

10. Connect the negative battery cable.

2TZ–FE Engine

NOTE: To service many engine components, the engine service hole covers may have to be removed. Remove the front seats, scuff plate and seat legs to access the service hole covers.

1. Disconnect the negative battery cable.

2. Raise the vehicle and safely support it.

3. Disconnect the alternator connectors.

4. Loosen the lock bolt, adjusting bolt and pivot bolt. Remove the drive belt.

5. Remove the pivot and lock bolts. Remove the alternator.

To install:

6. Install the alternator and bolts.

7. Install the belt and adjust. Do not pry against the housing.

8. Connect the alternator wiring and battery cable.

9. Lower the vehicle and check operation.

Starter

REMOVAL AND INSTALLATION

Except 2TZ–FE Engine

1. If necessary, raise and support the vehicle safely.

2. Disconnect the negative battery terminal from the battery.

NOTE: On some 22R–E engines equipped with an automatic transmission, it may be necessary to remove the transmission oil filler tube.

3. Disconnect the wiring connectors and the wiring from the starter.

4. Remove the starter-to-engine bolts and the starter from the engine.

5. To install, reverse the removal procedures.

2TZ–FE Engine

1. Disconnect the negative battery cable.

2. Raise the vehicle and safely support it.

3. Disconnect the starter connectors.

4. Remove the front driveshaft for 4WD only.

5. Remove the clutch release cylinder (manual transmission only).

6. Remove the 3 bolts and starter for 2WD. Remove the 4 bolts, 2 nuts, center support bracket and starter for 4WD.

To install:

7. Install the starter and torque the long bolts to 41 ft. lbs. (54 Nm) and short bolts to 30 ft. lbs. (41 Nm).

8. Connect the starter connectors.

9. Install the front driveshaft and clutch release cylinder, if removed.

10. Connect the battery cable and lower the vehicle.

CHASSIS ELECTRICAL

Heater Blower Motor

REMOVAL AND INSTALLATION

Pick-Up and 4Runner

1. Disconnect the negative battery cable.

2. Disconnect the electrical connector from motor.

3. Remove the blower motor-to-case screws and lift the motor from the case.

4. To install, reverse the removal procedures. Make sure the seal around the motor flange is in good condition.

Land Cruiser

1. Disconnect the negative battery cable. Disconnect the electrical connector from the blower motor.

2. Disconnect the flexible tube from the side of the blower motor.

3. Remove the blower motor fasteners and lower the blower motor out of the air inlet duct.

4. To install, reverse the removal procedures. During installation, be sure to position the motor so the flexible tube can be attached to the motor.

Previa

1. Disconnect the negative battery cable and open the hood.

2. Remove the air duct and disconnect the motor connectors.

3. Remove the blower motor from the housing.

4. Installation is the reverse of removal.

Windshield Wiper Motor

REMOVAL AND INSTALLATION

Pick-Up and 4Runner

FRONT

Disconnect the negative battery cable. Disconnect the wiring from the wiper motor. Remove the motor from the fire wall.

1. Remove the nut, then, pry the wiper link from the crank arm.

2. Remove the motor.

3. To install, reverse the removal procedures and inspect he operation.

REAR

1. Disconnect the negative battery cable. At the rear of the vehicle, remove the wiper motor cover panel.

2. Remove the wiper arm from the wiper motor.

3. Disconnect the electrical connector from the wiper motor.

4. Remove the wiper motor-to-door bolts and the motor from the vehicle.

5. To install, reverse the removal procedures and inspect the operation.

Land Cruiser and Previa

NOTE: On these vehicles, the wiper motor is removed with the linkage assembly.

1. Disconnect the negative battery cable. Remove the wiper arm retaining nuts, then, the wiper arm/blade assemblies.

2. Remove both wiper arm pivot covers and the pivot-to-cowl attaching screws.

3. Remove the service hole covers from the cowl area of the engine compartment.

4. Disconnect the wiring from the wiper motor.

5. From the engine compartment, remove the wiper motor plate-to-cowl screws. Withdraw the wiper motor and the linkage from the cowl panel as an assembly.

6. Pry the linkage from of the wiper motor.

7. To install, reverse the removal procedures.

Windshield Wiper Linkage

REMOVAL AND INSTALLATION

Pick-Up, 4Runner and Previa

1. Disconnect the negative battery cable. Remove the wiper motor.

2. Remove the wiper arms by removing their retaining nuts and working them off their shafts.

3. Remove the wiper shafts nuts/spacers and push the shafts down into the body cavity. Pull the linkage out of the cavity through the wiper motor hole.

4. To install, reverse the removal procedures.

Land Cruiser

1. Disconnect the negative battery cable. Remove the wiper arm assemblies.

2. Remove the end plate from the pivot housing.

3. Remove the wiper motor with the linkage cable.

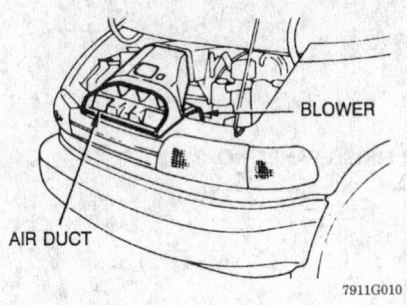

Blower motor location — Previa

7911G010

4. Separate the wiper motor and the transmission.

5. Remove the linkage cable.

6. To install, reverse the removal procedures.

Windshield Wiper Switch

REMOVAL AND INSTALLATION

Front

1. Disconnect negative battery terminal from the battery.

2. Remove the upper and the lower steering column shrouds.

3. Disconnect the combination switch electrical connector.

4. Remove the terminal from the horn contact.

5. To remove the windshield/wiper switch wires from the electrical connector, place a small prybar into the end of the connector, pry up on the retaining tab and pull the wire(s) from the connector.

6. Remove the windshield/wiper switch-to-combination switch screw and the switch.

7. Installation is the reverse of the removal procedure. To install, place the wire(s) into the electrical connector's slots, place a suitable tool behind the wire terminal and push the wire into the connector until the retaining tab locks it into place.

Rear

If equipped with a rear wiper switch, it will be located in the center of the dash.

1. Disconnect the negative battery cable. Using a small prybar, pry the rear wiper switch from the center of the dash.

2. Disconnect the electrical connector from the rear of the switch.

3. To install, reverse the removal procedures.

Instrument Cluster

REMOVAL AND INSTALLATION

Pick-Up and 4Runner

1. Disconnect the negative battery terminal from the battery.

2. Remove the upper and lower steering column covers.

3. Remove the instrument trim panel screws and the panel.

4. Disconnect the speedometer cable from the speedometer.

5. Remove the instrument panel screws and pull the panel forward. Disconnect the electrical connectors from the back of the panel and remove the panel.

6. To install, reverse the removal procedures.

Land Cruiser

1. Disconnect the negative battery terminal from the battery.

2. Disconnect the speedometer cable. Remove the instrument panel screws.

3. Loosen the steering column clamp by removing the attaching bolts.

4. Pull out the instrument panel and the speedometer, disconnect the electrical connectors and remove the panel.

5. To install, reverse the removal procedures.

Previa

1. Disconnect the negative battery cable.

2. Remove the cluster finish panel by removing the 2 screws and prying on the left side with a prybar. Be careful not to damage the plastic components.

3. Remove the 4 screws and disconnect the cable from the automatic transmission indicator. Remove the cable from the roller and cluster.

4. Disconnect the electrical connectors and remove from the vehicle.

5. To install, reverse the removal procedures.

Speedometer Cable

REMOVAL AND INSTALLATION

1. Disconnect the negative battery cable. Remove the instrument cluster and disconnect the cable from the speedometer.

2. Disconnect the other end of the speedometer cable from the transmission extension housing and pull the cable from its jacket at the transmission end.

NOTE: If the cable is being replaced because it is broken, be sure to remove both pieces of the broken cable.

3. Using graphite, lubricate the new speedometer cable and insert it into the cable jacket at the lower end.

4. Connect the speedometer cable to the transmission, then, to the instrument cluster.

5. To complete the installation, reverse the removal procedures.

Radio

REMOVAL AND INSTALLATION

1. Disconnect the negative battery cable.

2. Remove the screws retaining the lower center finish panel to the dash board.

3. Remove the radio retaining screws.

4. Remove the radio by pulling it out of the instrument panel.

5. Disconnect the antenna wire and the multi-plug connector.

6. Remove it from the vehicle.

7. Installation is the reverse of the removal procedure.

Headlight/Dimmer Switch

REMOVAL AND INSTALLATION

1. Disconnect the negative battery terminal from the battery.

2. Remove the upper and lower steering column covers.

3. Disconnect the electrical connector from the combination switch.

4. Remove the headlight switch-to-combination switch screws and the headlight switch from the combination switch.

5. To install, reverse the removal procedures.

Turn Signal Switch

REMOVAL AND INSTALLATION

1. Disconnect negative battery terminal.

2. Remove the upper and the lower steering column shrouds.

3. Disconnect the combination switch electrical connector.

4. At the left-rear of the combination switch, remove the mounting screws and the turn signal switch.

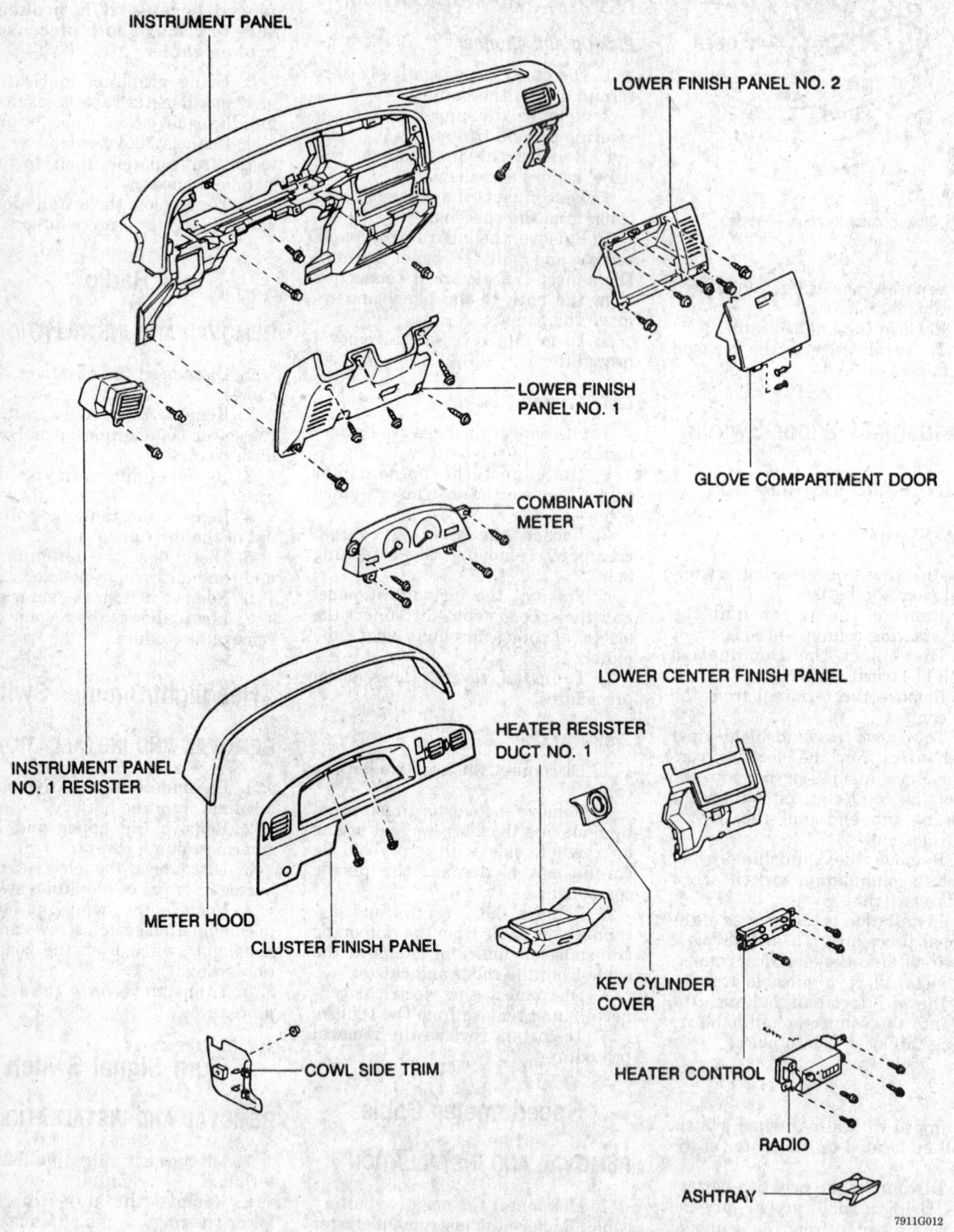

INSTRUMENT PANEL

LOWER FINISH PANEL NO. 2

LOWER FINISH PANEL NO. 1

GLOVE COMPARTMENT DOOR

COMBINATION METER

LOWER CENTER FINISH PANEL

HEATER RESISTER DUCT NO. 1

INSTRUMENT PANEL NO. 1 RESISTER

METER HOOD

CLUSTER FINISH PANEL

KEY CYLINDER COVER

COWL SIDE TRIM

HEATER CONTROL

RADIO

ASHTRAY

7911G012

Typical instrument cluster and panel assembly — Pick-Up and 4Runner

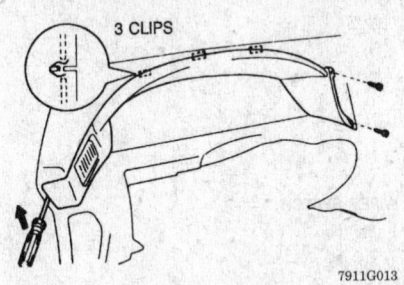

Instrument cluster finish panel removal — Previa

7911G013

5. If necessary to the remove the turn signal switch wires from the electrical connector, place a small prybar into the end of the connector, pry up on the retaining tab and pull the wire(s) from the connector.

6. To install, place the wire(s) into the electrical connector's slots, place a prybar behind the wire terminal and push the wire into the connector until the retaining tab locks it into place.

7. To complete the installation, reverse the removal procedures.

Combination Switch

REMOVAL AND INSTALLATION

The combination switch is composed of the turn signal, the headlight control, the dimmer, the hazard, the wiper and the washer switches.

1. Disconnect the negative battery cable. Remove the steering wheel.

2. Remove the upper and lower steering column shroud screws and the shrouds.

3. Remove the combination switch screws and the switch from the column.

4. Disconnect the electrical connector from the combination switch. To remove the wires from the electrical connector, perform the following procedures.

   a. Using a small prybar, insert it into the open end between the locking lugs and the terminal.

   b. Pry the locking lugs upward and pull the terminal out from the rear.

   c. To install the terminals, simply push them into the connector until they lock securely in place.

5. To complete the installation, reverse the removal procedures.

Ignition Lock/Switch

REMOVAL AND INSTALLATION

The ignition lock/switch is located behind the combination switch on the steering column.

1. Disconnect the negative battery terminal.

2. Remove the upper and lower steering column covers.

3. Disconnect the ignition switch from the electrical connector.

4. Using the key in the ignition switch, turn it to the **ACC** position.

5. Using a thin rod, place it into the hole of the cylinder lock housing. Pushing down on the thin rod, pull out the cylinder lock.

6. Remove the unlock warning switch-to-combination switch screws and the unlock warning switch.

7. Remove the ignition switch-to-combination switch screw and the ignition switch.

8. To install, push the ignition switch into the housing and install the screw. Using the key, install cylinder lock into the housing until the retaining tab locks it in place.

9. To complete the installation, reverse the removal procedures.

Stoplight Switch

REMOVAL AND INSTALLATION

1. Disconnect the negative battery cable.

2. Remove the electrical connector from the switch at the brake pedal.

3. Remove the mounting nut and remove the switch from the bracket.

4. Installation is the reverse of the removal procedure.

Clutch Switch

REMOVAL AND INSTALLATION

1. Disconnect the negative battery cable.

2. Remove the electrical connector from the switch at the clutch pedal.

3. Remove the mounting nut and remove the switch from the bracket.

4. Installation is the reverse of the removal procedure.

Fuses, Circuit Breakers and Relays

LOCATION

There are 3 fuse boxes on the Pick-Up, 4Runner and Land Cruiser. One is located in the engine compartment, 1 at the drivers side kick panel and 1 behind the glove box. The junction/relay and fuse block is located behind the heater controls, under the center instrument panel hood. Each fuse box has the fuse numbers and circuits protected on the lid of the box.

Computer

LOCATION

The ECU is located at the right kick panel for the Pick-Up and 4Runner, behind the glove compartment for the Land Cruiser, behind the trim panel or under the driver's seat for the Previa.

Flashers

The turn signal/hazard flasher is located under the instrument panel near the steering column for the Pick-Up, 4Runner and Previa. The flasher unit is located in the relay block at the left kick panel for the Land Cruiser.

Cruise Control

ADJUSTMENT

Pick-Up, 4Runner and Land Cruiser

1. Inspect that the control cable freeplay is less than 0.39 in. (10mm).

2. Connect the positive lead from the battery to terminals **1** and **2** and negative lead to terminal **3** of the actuator.

3. Slowly apply vacuum from 0–11.81 in. Hg. (0–300mm Hg), check that the control cable can be pulled smoothly and the cable stroke is at least 1.42 in. (36mm).

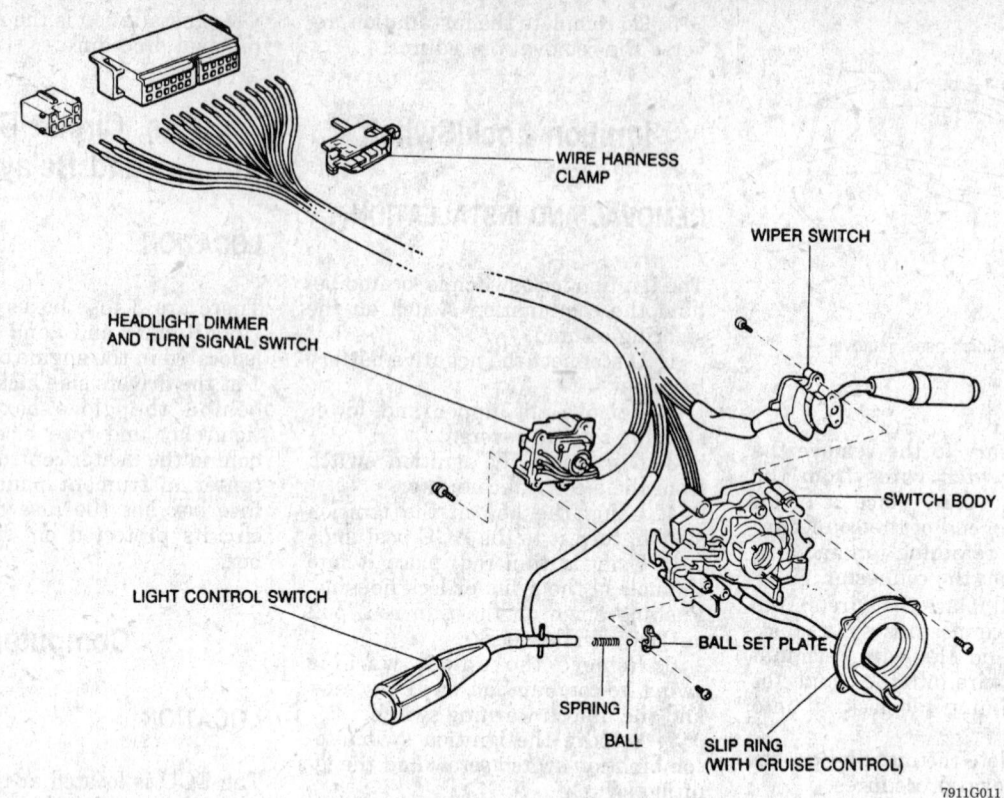

Typical combination switch assembly

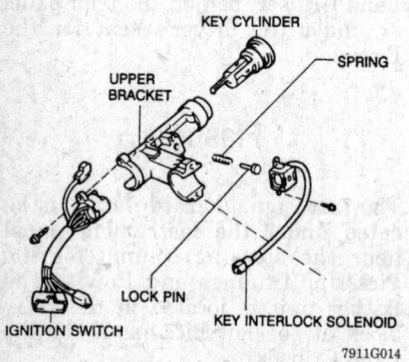

Ignition lock assembly

ENGINE COOLING

Radiator

REMOVAL AND INSTALLATION

1. Disconnect the negative battery cable. Drain the cooling system.
2. Unfasten the hose clamps and disconnect the hoses from the radiator. On Land Cruiser equipped with air conditioning, properly discharge the system.

3. Disconnect the transmission cooling lines, if equipped with an automatic transmission. Disconnect the air conditioning air intake duct on Previa vehicles.
4. Remove the fan shrouds, if equipped.
5. Remove the grille assembly and remove the hood lock from the radiator support. On Land Cruiser with air conditioning, remove the condenser to radiator bolts.
6. Remove the coolant recovery tank. Unbolt the radiator and remove it from the vehicle.

To install:

7. Install the radiator into position.
8. On Land Cruiser with air conditioning, install the condenser-to-radiator bolts.
9. Install the hood lock assembly and install the grille.
10. Connect the radiator hoses and the transmission cooler lines.
11. Connect the air conditioning intake duct on Previa vehicles.
12. Install the coolant recovery bottle. Refill the cooling system to the correct level.
13. Run the engine and check for leaks.

Heater Core

REMOVAL AND INSTALLATION

NOTE: If equipped with air conditioning, the heater and the air conditioner are completely separate units. Be certain when working under the dashboard that only the heater hoses are disconnected.

— CAUTION —
The air conditioning hoses are under pressure; if disconnected, the escaping refrigerant will freeze any surface with which it comes in contact, including skin and eyes.

Pick-Up, 4Runner and Previa

1. Disconnect the negative battery terminal from the battery.
2. Drain the cooling system.
3. Remove the glove box, the defroster hoses, the air damper, the air duct and the 2 side defroster ducts.
4. Remove the control unit from the instrument panel.
5. Disconnect the heater hoses from the core tubes.
6. Remove the retaining bolts and lift out the heater unit. At this point, the core may be pulled from the case.

7. To install, reverse the removal procedures. Refill the cooling system.

Land Cruiser

FRONT HEATER

NOTE: The entire heater unit must be removed to gain access to the heater core. This procedure requires almost complete disassembly of the instrument panel and lowering of the steering column.

1. Disconnect the negative battery terminal. Remove the glove box and the glove box door.
2. Remove the lower heater ducts. Remove the large heater duct from the passenger side of the heater unit.
3. Remove the ductwork from behind the instrument panel. Remove the radio.
4. Disconnect the wiring connector from the right-side inner portion of the glove opening.
5. Remove the instrument panel pad. Remove the hood release lever. Disconnect the hand throttle control cable.
6. Remove the retaining screw from the left-side of the fuse block.
7. Remove the steering column-to-instrument panel attaching nuts and carefully lower the steering column. Tag and disconnect the wiring as necessary in order to lower the column assembly.
8. Disconnect the electrical connector from the rheostat located to the left of the steering column opening.
9. Remove the center dual outlet duct which is attached to the upper portion of the heater unit.
10. Remove the lower instrument panel.
11. Tag and disconnect the hoses from the heater unit. Remove the heater unit-to-firewall fasteners and the heater unit.
12. Remove the heater core-to-heater unit pipe clamps and the heater core retaining clamp, then, withdraw the heater core from the heater unit.
13. To install, reverse the removal procedures. Torque the steering column-to-instrument panel fasteners to 14–15 ft. lbs. (18–20 Nm). Refill the cooling system.

REAR HEATER

1. Turn off the water valve and disconnect both hoses from the rear heater core.
2. Disconnect the wiring from the rear heater.

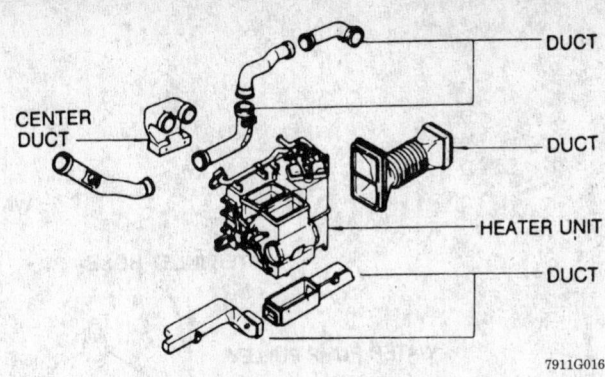

Typical heater unit and ducts — Land Cruiser

3. Remove the mounting bolts and lift out the core.
4. To install, reverse the removal procedures. Refill the cooling system.

Water Pump

REMOVAL AND INSTALLATION

22R–E Engine

1. Disconnect the negative battery cable. Drain the cooling system.
2. If equipped, remove the fan shroud bolts and the shroud.
3. Loosen the alternator adjusting link bolt and remove the drive belt, then, swing the alternator toward the engine.
4. If equipped with an air pump, air conditioning compressor or power steering pump drive belts, it may be necessary to loosen the adjusting bolt, remove the drive belt(s) and move the component(s) aside.
5. Remove the fan from the fluid coupling, the fluid coupling and pulley from the water pump, then, the water pump-to-engine bolts and the pump.
6. Clean the gasket mounting surfaces.
7. To install, use a new gasket, sealant and reverse the removal procedures. Adjust the drive belt(s) tension. Refill the cooling system.

3F–E Engine

1. Disconnect the negative battery cable.
2. Drain the engine coolant.
3. Remove the accessory drive belt. Loosen the power steering pump mount, idler pulley and adjusting bolts.
4. Disconnect the overflow tank hose.
5. Disconnect the radiator inlet hose and remove the fan shroud.
6. Remove the fan, fluid coupling and the water pump pulley.

7. Remove the alternator. Disconnect the hoses from the water pump.
8. Remove the water pump, power steering idler pulley and bracket as an assembly.
9. Installation is the reverse of the removal procedure. Torque the water pump mounting bolts to 27 ft. lbs. (37 Nm).

3VZ–E Engine

1. Disconnect the negative battery cable.
2. Remove the timing belt assembly.
3. Remove the thermostat.
4. Remove the idler pulley.
5. Remove the water pump mounting bolts and remove the water pump.
To install:
6. Clean the mounting surface.
7. Use new seal packing on the water pump and install it in position on the engine.
8. Torque the bolts marked **A** to 13 ft. lbs. (18 Nm) and the bolts marked **B** to 14 ft. lbs. (20 Nm).
9. Install the thermostat and the idler pulley.
10. Install the timing belt assembly.
11. Connect the negative battery cable and refill the cooling system.

2TZ–FE Engine

1. Disconnect the negative battery cable and drain the engine coolant.
2. Raise the vehicle and support safely.
3. Disconnect the heater hose and radiator outlet hoses.
4. Remove the oil filter bracket.
5. Remove the pump retaining bolts and pump from the timing cover.
To install:
6. Install the water pump with new O-rings.
7. Torque the bolts to 21 ft. lbs. (28 Nm).

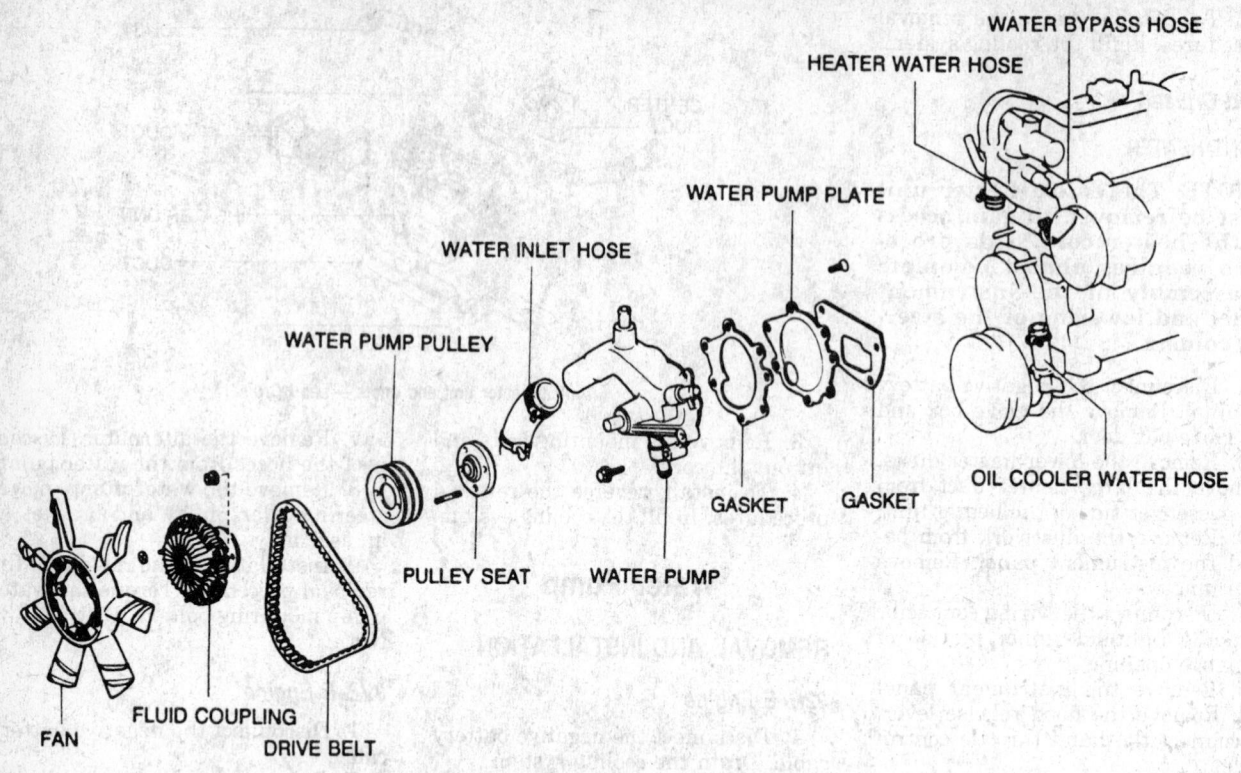

Water pump assembly — 3F — E engine

7911G018

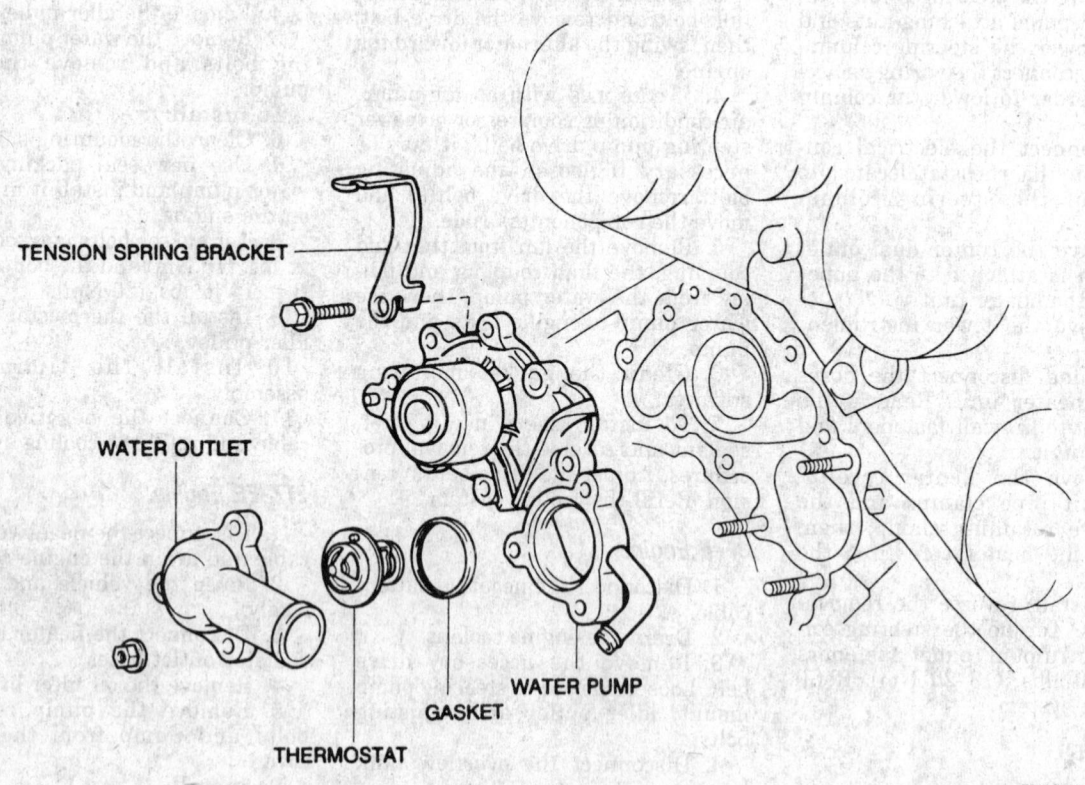

Water pump assembly — 3VZ — E engine

7911G019

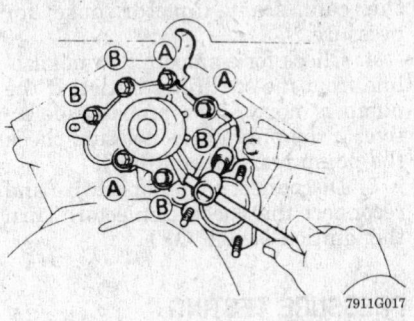

Water pump mounting bolt locations — 3VZ — E engine

8. Reconnect the coolant hoses and oil filter bracket.

9. Refill the engine coolant and check for leaks.

Thermostat

REMOVAL AND INSTALLATION

22R–E Engine

1. Disconnect the negative battery cable. Partially drain the cooling system to a level below the thermostat.

NOTE: Unless the upper radiator hose is positioned over one of the thermostat housing (water outlet) bolts, it is not necessary to detach the hose.

2. Remove the mounting bolts, the water outlet and the thermostat from the intake manifold.

3. Clean the gasket mounting surfaces.

4. To install, use a new gasket, sealant and reverse the removal procedures. Refill the cooling system.

NOTE: When installing a new thermostat, be sure the thermostat is positioned with the spring down.

5. Bleed the cooling system.

3F–E Engine

1. Disconnect the negative battery cable.

2. Drain the cooling system.

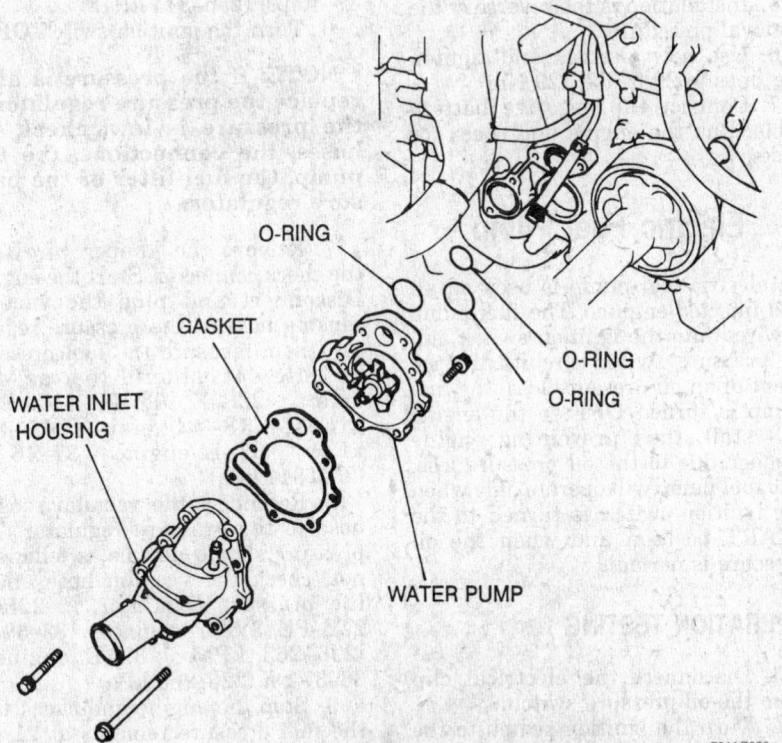

O-RING

GASKET

WATER INLET HOUSING

O-RING

O-RING

WATER PUMP

Water pump assembly — 2TZ — FE engine

3. Disconnect the cold start injector wire and the BVSV vacuum lines.

4. Remove the thermostat housing bolts and remove the housing.

5. Remove the thermostat housing.

6. Installation is the reverse of removal. Torque the housing bolts to 13 ft. lbs. (18 Nm).

7. Refill the cooling system. Bleed the cooling system.

3VZ–E Engine

1. Disconnect the negative battery cable.

2. Drain the coolant.

3. Remove the radiator outlet hose from the housing.

4. Remove the thermostat housing and thermostat from the engine.

5. Installation is the reverse of the removal procedure. Use a new gasket for installation.

6. Torque the housing bolts to 14 ft. lbs. (20 Nm). Refill and bleed the cooling system.

2TZ–FE Engine

1. Raise the vehicle and support safely. Drain the engine coolant.

2. Disconnect the radiator outlet hose.

3. Remove the retaining bolts, water inlet and thermostat.

4. Installation is the reverse of removal. Torque the retaining bolts to 14 ft. lbs. (20 Nm). Refill the engine with coolant.

Cooling System Bleeding

After working on the cooling system, even to replace the thermostat, it must be bled. Air trapped in the system will prevent proper filling and leave the radiator coolant level low, causing a risk of overheating.

1. To bleed the system, start with the system cool, the radiator cap off and the radiator filled to about an inch below the filler neck.

2. Start the engine and run it at slightly above normal idle speed. This will insure adequate circulation. If air bubbles appear and the coolant level drops, fill the system with an antifreeze/water mixture to bring the level back to the proper level.

3. Run the engine this way until the thermostat opens. When this happens, coolant will move abruptly across the top of the radiator and the temperature of the radiator will suddenly rise.

4. At this point, air is often expelled and the level may drop quite a bit. Keep refilling the system until

the level is near the top of the radiator and remains constant.

5. If the vehicle has an overflow tank, fill the radiator right up to the filler neck. Replace the radiator filler cap.

FUEL SYSTEM

Fuel System Service Precaution

When working with the fuel system certain precautions should be taken; always work in a well ventilated area, keep a dry chemical (Class B) fire extinguisher near the work area. Always disconnect the negative battery cable and do not make any repairs to the fuel system until all the necessary steps for repair have been reviewed.

RELIEVING FUEL SYSTEM PRESSURE

1. Disconnect the negative battery terminal from the battery.
2. Allow the system enough time to bleed off the fuel pressure through the fuel return line.
3. Before disconnecting any fuel line component, place a rag under the item to catch any excess fuel.
4. After installation, install the negative battery terminal, turn **ON** the ignition switch and check for fuel leaks.

Fuel Tank

REMOVAL AND INSTALLATION

1. Disconnect the negative battery cable. Relieve the fuel pressure.
2. This procedure should be done with the tank less than ¼ full. Place a suitable drain pan under fuel tank and remove the drain plug. If the tank is more than ¼ full, use an approved siphon to remove the fuel from the tank.
3. Disconnect all hoses, wiring and tubes that can be accessed at this point.
4. Place a jack under the tank and remove the tank protector.
5. Remove the tank straps or retaining bolts.

6. Lower the tank far enough to disconnect the remaining hoses, wiring and tubes. Be careful not to damage the nylon fuel lines when removing the tank.
7. Drain the remaining fuel from the tank.
   To install:
8. Raise the tank and connect the hoses, wiring and tubes.
9. Install the retaining bolts or straps and torque to 29 ft. lbs. (39 Nm).
10. Connect the remaining hoses or wiring.
11. Refill the fuel tank and check for leaks.
12. Check the battery cable and check operation.

Fuel Filter

REMOVAL AND INSTALLATION

The fuel filter is located in the engine compartment, at the inlet line to the fuel rail.
1. Disconnect the negative battery cable.
2. Relieve the fuel system pressure.
3. Disconnect and plug the inlet and outlet lines from the filter.
4. Remove the filter retaining bolts and remove the filter.
5. Installation is the reverse of removal procedure.
6. Use new O-rings and tighten the lines to 22 ft. lbs. (29 Nm).
7. Connect the negative battery cable. Run the engine and check for leaks.

Electric Fuel Pump

An electric fuel pump is used on all fuel injected engines. The fuel pump is wired into the ignition switch and oil pressure switch circuits. In the event of an oil pressure loss, the fuel pump is turned **OFF** so the engine will stall, thus preventing engine damage due to the oil pressure loss. The fuel pump will operate only when the ignition switch is turned to the **START** position and when the oil pressure is normal.

OPERATION TESTING

1. Disconnect the electrical clip from the oil pressure switch.
2. Turn the ignition switch to the **ON** position; do not start the engine.
3. Short the **Fp** and the **+B** terminals of the check connector. Check

the cold start injector hose for pressure.
4. Check for a smooth flow of gasoline from the fuel filter outlet. If the pump is noisy, it is probably defective. If the pump does not run, check the pump resistor and relay.
5. Disconnect the jumper wire and reconnect the check connector. Turn the ignition switch **OFF**.

PRESSURE TESTING

1. Disconnect the negative battery terminal from the battery and the wiring connector from the cold start injector.
2. Place a container or a shop towel near the end of the delivery tube.
3. Slowly loosen the cold start injector union bolt, then, remove the bolt and the gaskets. Drain the fuel line.
4. Using pressure gauge tool 09268–45011 or equivalent, connect it in line with the cold start injector. Reconnect the battery cable.
5. Short the **Fp** and **+B** terminals of the check connector wire. Turn the ignition switch to the **ON** position and measure the fuel pump pressure. It should be as follows: 22R–E, 2TZ–FE, 3VZ–E engines— 38–44 psi (265–304 kPa) 3F–E engine— 37–46 psi (255–314 kPa)
6. Turn the ignition switch **OFF**.

NOTE: If the pressure is high, replace the pressure regulator; if the pressure is low, check the hoses, the connections, the fuel pump, the fuel filter or the pressure regulator.

7. Remove the jumper wire from the check connector. Start the engine. Disconnect and plug the vacuum sensing hose at the pressure regulator, then, measure the fuel pressure at idle. It should be as follows: 22R–E, 2TZ–FE, 3VZ–E engines — 38–44 psi (265–304 kPa) 3F–E engine — 37–46 psi (255–314 kPa)
8. Reconnect the vacuum sensing hose to the pressure regulator. The pressure should now be as follows; if not, check the vacuum hose and/or the pressure regulator. 22R–E, 2TZ–FE, 3VZ–E engines— 33–38 psi (226–265 kPa) 3F–E engine— 33–37 psi (226–265 kPa)
9. Stop the engine and check that the fuel pressure remains at 21 psi. for 5 minutes. If not, check the fuel pump, the pressure regulator and/or the injectors.

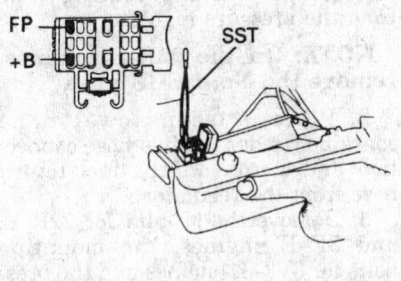

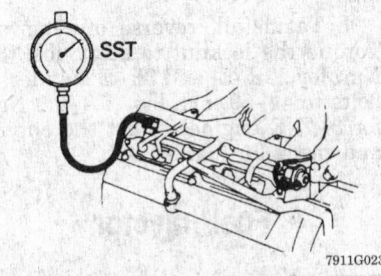

7911G023

Testing the fuel pump pressure

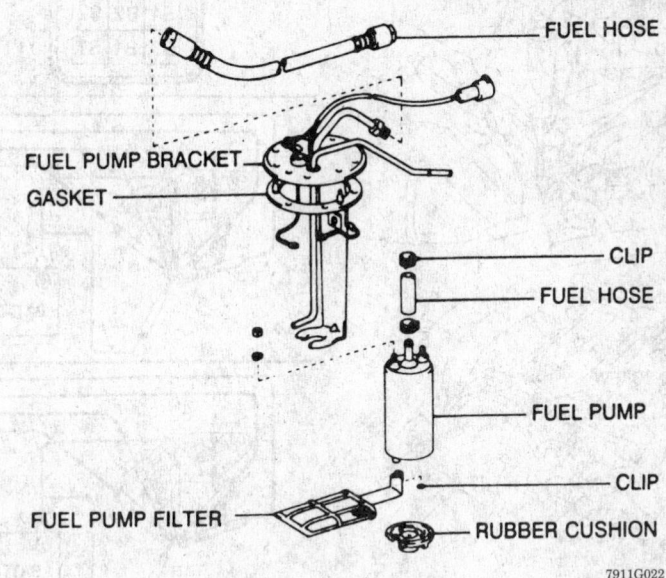

7911G022

Typical electric fuel pump

REMOVAL AND INSTALLATION

1. Disconnect the negative battery cable. Drain the fuel tank.

2. Disconnect the electrical connector and the fuel lines from the fuel tank.

3. Remove the inlet tube and mounting bolts/straps, then, the fuel tank from the vehicle.

4. Remove the access plate-to-fuel tank bolts, then, pull out the plate/fuel pump assembly.

5. Disconnect the electrical connectors from the fuel pump. Pull the bracket from the lower-side of the fuel pump, then, remove the fuel pump from the fuel hose.

6. Remove the rubber cushion, the clip and the fuel filter from the bottom of the fuel pump.

To install:

7. Install the fuel pump and use new gaskets.

8. Install the fuel tank and connect all electrical and fuel connections.

9. Torque the fuel pump bracket-to-fuel tank to 43 inch lbs. (5.3 Nm). Refill the fuel tank and check for leaks.

Fuel Injection

IDLE SPEED ADJUSTMENT

22R–E and 3VZ–E Engines

The engines are equipped with a computer activated, electronic fuel injection system. Prior to adjusting the idle speed, make sure the air cleaner is installed. All vacuum hoses are connected. All pipes and hoses in the air intake system are connected and in good condition. All fuel injection system wiring is connected and in good condition. The engine is at normal operating temperature. All accessories are **OFF**. Transmission selector lever in **N**.

1. Connect the tachometer positive (+) lead to the coil's (−) negative terminal or to the igniter's service connector, if provided.

2. Run the engine at 2500 rpm for 2 minutes.

3. Run the engine at idle and turn the idle speed adjusting screw to obtain the correct idle speed.

4. Disconnect and remove the tachometer.

3F–E and 2TZ–FE Engines

These engines are equipped with electronic fuel injection. The idle speed is controlled by an Idle Speed Control (ISC) valve. The idle speed is preset at the factory and requires no adjustment. There are no idle speed adjusting screws.

3F–E engine: check that there is a clicking sound immediately after stopping the engine. Check the valve using an ohmmeter. Measure the resistance between the terminals.

• Terminals **B1–S1** or **S3** — 10–30 ohms
• Terminals **B2–S2** or **S4** — 10–30 ohms

1. If not within specifications, replace the ISC valve.

2. 2TZ–FE engine: using an ohmmeter, measure the resistance between terminals **B+** and the other 2. The resistance should be 18.8–22.8 ohms.

3. If the valve check OK, remove the valve and clean the mounting area with carburetor cleaner. Install new gaskets and valve

Idle Mixture Adjustment

The idle mixture adjustment is preset at the factory and controlled by the electronic control unit. There is no adjustment necessary or possible.

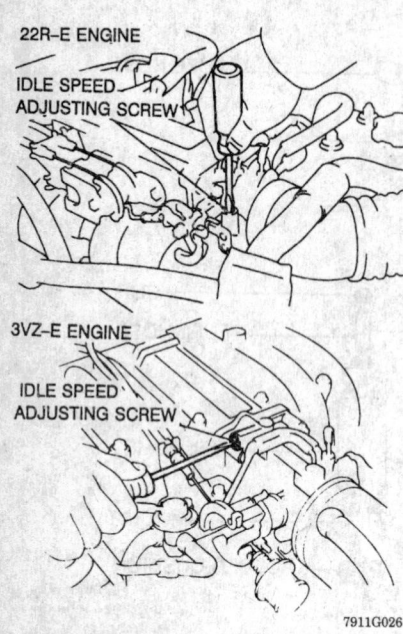

idle speed adjusting screw — 22R — E and 3VZ — E engines

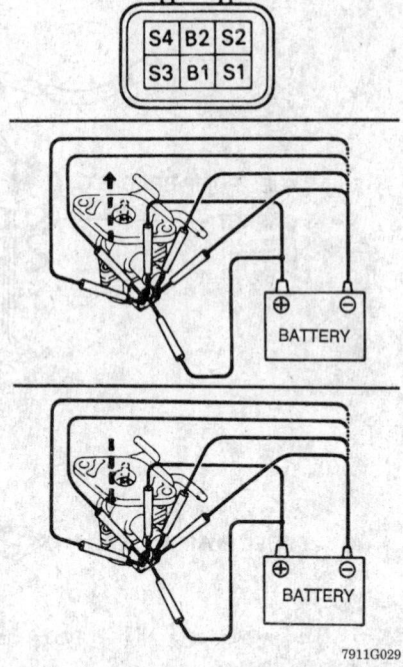

Checking ISC valve — 3F — E engine

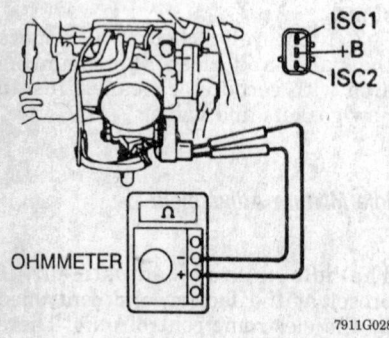

Checking Idle Speed Control (ISC) valve — 3F — E engine

Checking ISC valve — 2TZ — FE engine

OHMMETER

ISC1
+B
ISC2

Cold Start Injector

The EFI engines have a cold start injector located in the intake air chamber which aids in cold weather starting.

REMOVAL AND INSTALLATION

1. Disconnect the negative battery cable and the cold start injector wire.
2. Place a shop towel or a container under the fuel delivery pipe and drain the fuel from the pipe.
3. Disconnect the fuel pipe from the cold start injector.
4. Remove the mounting bolts and the cold start injector from the intake air chamber.
5. To install, use new gaskets and reverse the removal procedures. Torque the injector bolts to 44–60 inch lbs. (5.4–6.8 Nm).

Fuel Pressure Regulator

The fuel pressure regulator is located on the fuel delivery pipe of the fuel system; it maintains a constant fuel pressure in the injection system.

REMOVAL AND INSTALLATION

1. Relieve the fuel pressure. Disconnect the vacuum sensing hose from the pressure regulator.

NOTE: On the 22R–E engines, remove the No. 1 EGR pipe.

2. Place a shop towel or a container under the fuel hose connection and disconnect the fuel return hose from the regulator.
3. Remove the locknut for 22R–E, and 3F–E engines, the mounting bolts for 3VZ–E engines and the pressure regulator from the fuel delivery pipe.
4. To install, reverse of removal. Torque the locknut to 22 ft. lbs. (30 Nm) for 22R–E, and 3F–E engines or bolts to 44–60 inch lbs. (5.4–6.8 Nm) for 3VZ–E engines. Start the engine and check for fuel leaks.

Fuel Injector

TESTING

Except 2TZ–FE Engine

Each injector may be tested for operation while on the engine, in 2 ways.
1. Listen for a clicking at the injector.
2. Using an ohmmeter, check the continuity at each injector's terminal; the resistance should be 1.5–3.0 ohms.

REMOVAL AND INSTALLATION

Except 2TZ–FE Engine

1. Disconnect the negative battery terminal from the battery and the ground strap from the rear side of the engine.
2. Disconnect the accelerator wire. If equipped with an automatic transmission, disconnect the throttle cable from the bracket and the clamp.
3. Disconnect the No. 1 and No. 2 PCV hoses.
4. Disconnect the following items:
   a. Brake power booster hose
   b. Air control valve hoses
   c. Vacuum Switching Valve (VSV)
   d. Evaporative emission control hose
   e. EGR vacuum hose and modulator
   f. Pressure regulator hose for 2WD
   g. Fuel pressure-up (VSV) and hose
   h. No. 1 and No. 2 air valve hose from the throttle body

i. No. 2 and No. 3 water bypass hoses from the throttle body

   j. Cold start injector wire

   k. Throttle position wire

5. Remove the following items:

   a. Cold start injector-to-plenum chamber bolt.

   b. No. 1 EGR pipe-to-plenum chamber bolts.

   c. Manifold stay-to-plenum chamber bolts.

   d. Fuel hose clamp, 4 bolts, 2 nuts and the bond strap.

   e. The plenum chamber with the throttle body and gaskets.

6. Disconnect the fuel return hose.

7. Disconnect the following wires:

   a. Auxiliary air valve wire

   b. Knock sensor wire

   c. Oil pressure sender gauge/switch

   d. Starter wire (terminal 50)

   e. Transmission wires

   f. Air conditioning compressor wires

   g. Injector wires

   h. Water temperature sender gauge wire

   i. Overdrive temperature switch wire (air conditioning)

   j. Oxygen sensor and igniter wire

   k. Vacuum Switching Valve (VSV) wire (air conditioning)

   l. Cold start injector time switch wire

   m. Water temperature sensor wire

8. Disconnect the fuel hose from the delivery pipe with the pulsation damper and gaskets.

9. Remove the injectors from the engine. Take care in handling the injectors.

NOTE: Injector performance tests are possible but special tools are required. If these tools are unavailable, use the test procedures above.

10. To install, use new O-rings and reverse the removal procedures. Torque the hold-down bolts to 14 ft. lbs. (20 Nm). Check for fuel leakage.

NOTE: Each injector should have 4 insulators. Prior to installation, coat the O-rings with clean gasoline. Prior to tightening the hold-down bolts, make sure the injector rotates smoothly in its bore. If not, the O-rings are twisted.

2TZ–FE Engine

1. Relieve the fuel pressure and disconnect the negative battery cable.

2. Remove the right engine service hole.

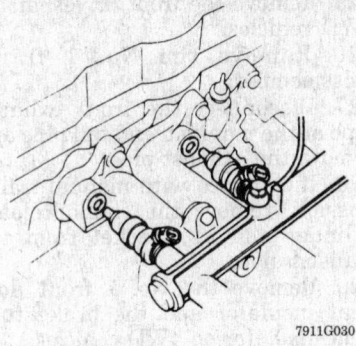

7911G030

Removing the fuel delivery pipe

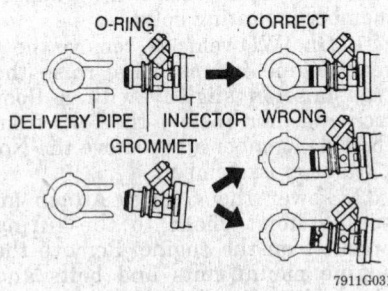

7911G031

Injector installation into delivery pipe

3. Disconnect the PCV hose, engine wiring harness-to-injectors, vacuum hoses and fuel rail pipes.

4. Remove the fuel rail and spacers.

5. Remove the 4 bolts from the injector covers.

6. Using an injector remover tool 0926874010 and remove the injector from the fuel rail.

To install:

7. Lubricate the O-rings with spindle oil or gasoline. Install new O-rings.

8. Push the injector into the fuel rail. Check the connectors are along the center line of the fuel rail.

9. Install the insulator onto each injector.

10. Install the injector covers and fuel rail onto the engine.

11. Torque the fuel rail retaining bolts to 14 ft. lbs. (20 Nm).

12. Install the remaining components, connect the battery cable and check for leaks.

ENGINE MECHANICAL

NOTE: Disconnecting the negative battery cable on some vehicles may interfere with the functions of the on board computer systems and may require the computer to undergo a relearning process, one the negative battery cable is reconnected.

Engine Assembly

REMOVAL AND INSTALLATION

22R–E Engine

1. Disconnect the negative battery cable.

2. Remove the engine undercover.

3. Disconnect the windshield washer hose and then remove the hood. Scribe matchmarks around the hinges for easy installation.

4. Drain the engine oil. Drain the engine coolant from the radiator and the cylinder block.

5. Drain the automatic transmission fluid on models so equipped.

6. Disconnect the air cleaner hose and then remove the air cleaner.

7. Remove the radiator and shroud.

8. Remove the coupling fan.

9. Disconnect the heater hoses at the engine.

10. If equipped with automatic transmissions, disconnect the accelerator and throttle cables at their bracket.

11. Disconnect the following:

   a. No. 1 and No. 2 PCV hoses

   b. Brake booster hose

   c. Air control valve hoses

   d. EVAP hose at the canister

   e. Actuator hose on vehicles with cruise control

   f. Vacuum modulator hose at the EGR valve

   g. Air valve hoses at the throttle body and chamber

   h. Two water bypass hoses at the throttle body

   i. Air control valve hose at the actuator

   j. Pressure regulator hose at the chamber

   k. Cold start injector pipe

   l. BVSV hose.

12. Tag and disconnect the cold start injector wire and the throttle position sensor wire.

13. Remove the EGR valve from the throttle chamber.

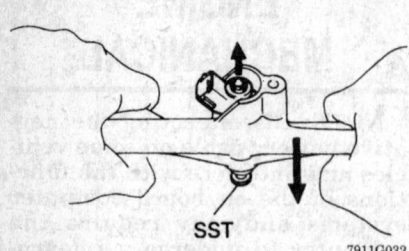

SST

7911G032

Injector removal — 2TZ — FE engine

14. Disconnect the throttle chamber at the stay. Remove the chamber-to-intake manifold mounting bolts and lift off the throttle chamber.

15. Tag and disconnect the following wires:

   a. Cold start injector time switch wire

   b. Water temperature sensor wire

   c. If equipped with air conditioning: VSV and air conditioning compressor wires

   d. OD temperature switch wire (with automatic transmission)

   e. Injector wires

   f. Knock sensor connector

   g. Air valve wire

   h. Oil pressure switch wire

   i. Starter wire.

16. Remove the power steering pump from its bracket, if equipped. Disconnect the ground strap from the bracket.

17. If equipped with air conditioning, loosen the drive belt and remove the air conditioning compressor. Position it aside with the refrigerant lines attached.

18. Disconnect the engine ground straps at the rear and right side of the engine.

19. If equipped with a manual transmission, remove the shift lever from inside the vehicles.

20. Raise and safely support the vehicle. Drain the engine oil. Remove the rear driveshaft.

21. If equipped with automatic transmission, disconnect the manual shift linkage at the neutral start switch. On 4WD vehicles with automatic transmission, disconnect the transfer shift linkage.

22. Disconnect the speedometer cable. Be sure not to lose the felt dust protector and washers.

23. Remove the transfer case undercover on 4WD vehicles.

24. Remove the stabilizer bar on 4WD vehicles.

25. Remove the front driveshaft on 4WD vehicles.

26. Remove the No. 1 frame crossmember.

27. Disconnect the front exhaust pipe at the manifold and tail pipe and remove the exhaust pipe.

28. If equipped with manual transmission, remove the clutch release cylinder and its bracket from the transmission.

29. Remove the No. 1 front floor heat insulator and the brake tube heat insulator on 4WD vehicles.

30. On 2WD vehicles, remove the rear engine mount bolts, raise the transmission slightly with a floor jack and then remove the support member mounting bolts.

31. On 4WD vehicles, remove the 4 rear engine mount bolts, raise the transmission slightly with a floor jack and then remove the bolts from the side member and remove the No. 2 frame crossmember.

32. Lower the vehicle. Attach an engine hoist chain to the lifting brackets on the engine. Remove the engine mount nuts and bolts and slowly lift the engine/transmission out of the truck.

To install:

33. Slowly lower the engine assembly into the engine compartment.

34. Raise the transmission onto the crossmember with a floor jack.

35. Align the holes in the engine mounts and the frame, install the bolts and then remove the engine hoist chain.

36. On 2WD vehicles, raise the transmission slightly and align the rear engine mount with the support member and tighten the bolts to 9 ft. lbs. (13 Nm). Lower the transmission until it rests on the extension housing and then tighten the bracket mounting bolts to 19 ft. lbs. (25 Nm).

37. On 4WD vehicles, raise the transmission slightly and tighten the No. 2 frame crossmember-to-side frame bolts to 70 ft. lbs. (95 Nm). Lower the transmission and tighten the rear engine mount bolts to 9 ft. lbs. (13 Nm).

38. On 4WD vehicles, install the brake tube and front floor heat insulators.

39. Install the clutch release cylinder and its bracket to the manual transmission. Tighten the bracket bolts to 29 ft. lbs. (39 Nm) and the cylinder bolts to 9 ft. lbs. (13 Nm).

40. Reconnect the exhaust pipe. Install the No. 1 frame crossmember.

41. On 4WD vehicles, install the front driveshaft, stabilizer bar and the transfer case undercover.

42. Connect the speedometer cable. Connect the transfer shift linkage on 4WD vehicles with automatic transmission.

43. Connect the manual shift linkage to the neutral start switch (automatic transmission only).

44. Install the rear driveshaft. Install the shift lever for manual transmission only.

45. Connect the engine ground straps. Install the air conditioning compressor.

46. Install the power steering pump and connect the ground strap.

47. Connect all of the following wires:

   a. Cold start injector time switch wire

   b. Water temperature sensor wire

   c. If equipped with air conditioning: VSV and air conditioning compressor wires

   d. OD temperature switch wire for automatic transmission

   e. Injector wires

   f. Knock sensor connector

   g. Air valve wire

   h. Oil pressure switch wire

   i. Starter wire.

48. Connect all of the following parts:

   a. No. 1 and No. 2 PCV hoses

   b. Brake booster hose

   c. Air control valve hoses

   d. EVAP hose at the canister

   e. Actuator hose on vehicles with cruise control

   f. Vacuum modulator hose at the EGR valve

   g. Air valve hoses at the throttle body and chamber

   h. Two water bypass hoses at the throttle body

   i. Air control valve hose at the actuator

   j. Pressure regulator hose at the chamber

   k. Cold start injector pipe

   l. BVSV hose.

49. Connect the accelerator and throttle cables to the bracket for automatic transmission only.

50. Connect the heater hoses and install the coupling fan. Install the radiator and shroud.

51. Install the air cleaner. Refill the engine with oil and the radiator with coolant. Install the engine undercover.

52. Install and adjust the hood.

53. Connect the battery cable, start the engine and road test the vehicle.

3F–E Engine

1. Disconnect the negative battery cable. Drain the engine coolant.

2. Scribe matchmarks around the hood hinges and then remove the hood.

3. Remove the battery and its tray.

4. Disconnect the accelerator and throttle cables.

5. Remove the air intake hose, air flow meter and air cleaner assembly.

6. Remove the coolant reservoir tank.

7. Remove the radiator.

8. Tag and disconnect the following wires and connectors:
   a. Oil pressure connector
   b. High tension cord at the coil
   c. Neutral start switch and transfer connectors near the starter
   d. Front differential lock connector
   e. Starter wire and connector
   f. Starter ground strap
   g. O$_2$ sensor connectors
   h. Alternator wire and connector
   i. Cooling fan connector
   j. Check connector.

9. Disconnect the following hoses:
   a. Heater hoses
   b. Fuel hoses
   c. Transfer case hose
   d. Brake booster hose
   e. Air injection hoses
   f. Distributor hose
   g. Emission control hoses

10. Remove the glove box, pull out the 4 connectors and then pull the EFI wiring harness from the cowl.

11. Unbolt the power steering pump and position it aside with the hoses still connected.

12. Do the same with the air conditioning compressor.

13. Raise the vehicle and remove the transfer case undercover. Drain the engine oil.

14. Remove the front and rear driveshafts.

15. Disconnect the speedometer cable.

16. Disconnect the engine ground strap.

17. Disconnect the 2 vacuum hoses at the diaphragm cylinder under the transfer case.

18. Remove the clip and pin and then disconnect the shift rod at the transfer case. Remove the nut, disconnect the washers and the shift lever at the shift rod.

19. Disconnect the transmission control rod.

20. Disconnect the exhaust pipe at the manifold.

21. With a floor jack under the transmission, remove the bolts and nuts that attach the frame cross-member and then remove the cross-member. Lower the vehicle.

22. Attach an engine hoist to the lifting brackets on the engine. Remove the engine mount nuts and bolts and slowly lift the engine/transmission out of the vehicle.

To install:

23. Slowly lower the engine assembly into the engine compartment.

24. Raise the transmission onto the crossmember with a floor jack.

25. Align the holes in the engine mounts and the frame, install the bolts and then remove the engine hoist chain.

26. Raise the transmission slightly and tighten the frame crossmember-to-chassis bolts to 29 ft. lbs. (39 Nm). Tighten the 2 nuts to 43 ft. lbs. (59 Nm).

27. Install the exhaust pipe with a new gasket and tighten the nuts to 46 ft. lbs. (62 Nm).

28. Connect the transmission control rod. Connect the transfer case shift lever.

29. Connect the engine ground strap. Connect the speedometer cable.

30. Install the front and rear driveshafts. Tighten the nuts to 65 ft. lbs. (88 Nm).

31. Install the transfer case undercover. Install the air conditioning compressor. Install the power steering pump and tighten the pulley nut to 35 ft. lbs. (47 Nm).

32. Connect the EFI wiring harness at the ECU. Connect all of the following hoses:
   a. Heater hoses
   b. Fuel hoses
   c. Transfer case hose
   d. Brake booster hose
   e. Air injection hoses
   f. Distributor hose
   g. Emission control hoses.

33. Connect all of the following the wires and connectors:
   a. Oil pressure connector
   b. High tension cord at the coil
   c. Neutral start switch and transfer connectors near the starter
   d. Front differential lock connector
   e. Starter wire and connector
   f. Starter ground strap
   g. Oxygen sensor connectors
   h. Alternator wire and connector
   i. Cooling fan connector
   j. Check connector.

34. Install the radiator and the coolant reservoir tank.

35. Install the air intake hose, air flow meter and air cleaner. Connect the accelerator and throttle cables.

36. Refill the engine with oil and the radiator with coolant. Install the engine undercover. Install the battery.

37. Install and adjust the hood.

38. Install the battery, start the engine and road test the vehicle.

3VZ–E Engine

1. Disconnect the battery cables and remove the battery.

2. Remove the engine undercover.

3. Disconnect the windshield washer hose and then remove the hood. Scribe matchmarks around the hinges for easy installation.

4. Drain the engine coolant from the radiator and the cylinder block.

5. Raise and safely support the vehicle. Drain the engine oil. Drain the automatic transmission fluid, if equipped.

6. Lower the vehicle. Disconnect the air cleaner hose and then remove the air cleaner.

7. Remove the radiator.

8. Remove all drive belts and then remove the fluid coupling and fan pulley.

9. Tag and disconnect the following wires and connectors:
   a. Left side and rear ground straps
   b. Alternator connector and wire
   c. Igniter connector
   d. Oil pressure switch connector
   e. ECU connectors
   f. VSV connectors
   g. Starter relay connector for manual transmission only
   h. Solenoid resistor connector
   i. Check connector
   j. Air conditioning compressor connector

10. Tag and disconnect the following hoses:
   a. Power steering hoses at the gas filter and air pipe
   b. Brake booster hose
   c. Cruise control vacuum hose, if equipped
   d. Charcoal canister hose at the canister
   e. VSV vacuum hoses.

11. Disconnect the accelerator, throttle and cruise control cables where applicable.

12. Unbolt the power steering pump and position it aside with the hydraulic lines still connected.

13. Properly discharge the air conditioning system. Remove the air conditioning compressor if equipped.

14. Disconnect the clutch release cylinder hose for manual transmission only.

15. Disconnect the 2 heater hoses.

16. Disconnect and plug the fuel inlet and outlet lines.

17. Remove the shift levers for manual transmission only.

18. Raise and safely support the vehicle. Remove the rear driveshaft.

19. Disconnect the manual shift linkage for automatic transmission only.

20. Disconnect the speedometer cable, don't lose the felt dust protector and washers.

21. Remove the transfer case undercover. Remove the stabilizer bar.

22. Remove the front driveshaft. Remove the front exhaust pipe.

23. Remove the No. 1 front floor heat insulator and the brake tube heat insulator.

24. Remove the rear engine mount bolts, raise the transmission slightly with a floor jack and then remove the 4 bolts from the side member and remove the No. 2 frame crossmember. Lower the vehicle.

25. Attach an engine hoist chain to the lifting brackets on the engine. Remove the engine mount nuts and bolts and slowly lift the engine/transmission out of the vehicle.

To install:

26. Slowly lower the engine assembly into the engine compartment.

27. Raise the transmission onto the crossmember with a floor jack.

28. Align the holes in the engine mounts and the frame, install the bolts and then remove the engine hoist chain.

29. Raise the transmission slightly and tighten the No. 2 frame crossmember-to-side frame bolts to 70 ft. lbs. (95 Nm). Lower the transmission and tighten the 4 rear engine mount bolts to 9 ft. lbs. (13 Nm).

30. Install the brake tube and front floor heat insulators. Reconnect the exhaust pipe.

31. Install the No. 1 frame crossmember. Install the front driveshaft, stabilizer bar and the transfer case undercover.

32. Connect the speedometer cable. Connect the manual shift linkage for automatic transmission only.

33. Install the rear driveshaft. Install the shift levers for manual transmission only.

34. Install the fuel inlet and outlet lines. Connect the heater hoses.

35. Connect the clutch release cylinder hose. Install the air conditioning compressor.

36. Install the power steering pump and connect the ground strap.

37. Connect the throttle, cruise control and accelerator cables.

38. Connect the following hoses:
   a. Power steering hoses at the gas filter and air pipe
   b. Brake booster hose
   c. Cruise control vacuum hose, if equipped
   d. Charcoal canister hose at the canister
   e. VSV vacuum hoses.

39. Install the fan pulley, belt guide, fluid coupling and drive belt. Connect the following wires:
   a. Left side and rear ground straps
   b. Alternator connector and wire
   c. Igniter connector
   d. Oil pressure switch connector
   e. ECU connectors
   f. VSV connectors
   g. Starter relay connector for manual transmission only
   h. Solenoid resistor connector
   i. Check connector
   j. Air conditioning compressor connector

40. Install the air conditioning belt. Install the power steering pump and connect the ground strap.

41. Install the radiator and shroud. Install the air cleaner. Refill the engine with oil and the radiator with coolant.

42. Install the engine undercover. Install the battery. Install and adjust the hood.

43. Install the battery, start the truck and road test it.

2TZ–FE Engine

1. Disconnect the negative battery cable and drain the engine coolant and oil.

2. Raise the vehicle and support safely. Remove the engine under covers.

3. Remove the accessory drive belt and service bolts and nut at the front end of the driveshaft. Matchmark and disconnect the equipment driveshaft from the crankshaft pulley.

4. Label and disconnect all hoses, electrical connectors, vacuum lines and cables from the engine and position aside.

5. Disconnect the front driveshaft for 4WD only.

6. Remove the air intake connector and disconnect the engine wiring from the floor pan.

7. Remove the exhaust pipe from the manifold and disconnect the oxygen sensor connector.

8. Remove the rear driveshaft.

9. Place a engine/transmission jack under the engine and remove the engine and transmission mount bolts.

10. With the vehicle on a hoist, lower the engine with the transmission from the bottom of the vehicle.

To install:

11. Raise the engine/transmission into the vehicle and torque the mount bolts and nuts to 27 ft. lbs. (34 Nm).

12. Connect all electrical connectors, vacuum hoses, coolant hoses and cables to the engine.

13. Install the exhaust pipe and torque to 32 ft. lbs. (43 Nm).

14. Install the front (4WD) and rear driveshafts and torque to 14 ft. lbs. (21 Nm).

15. Install the equipment drive shaft to the pulley and torque the bolts to 38 ft. lbs. (54 Nm). Install the belt and adjust.

16. Refill the engine with oil and coolant.

17. Connect the battery cable, start the engine and check for leaks.

Engine Mounts

REMOVAL AND INSTALLATION

1. Raise the vehicle and support safely.

2. Remove the engine-to-mount bolt or nut.

3. Position a suitable jack under the engine block.

4. Raise the engine far enough to take the weight off of the mount. Wedge a piece of wood between the engine and chassis in case the engine jack slips.

5. Remove the mount-to-body bolts and remove the mount.

To install:

6. Install the mount and torque the bolts to 34 ft. lbs. (46 Nm).

7. Remove the wood and lower the jack to engage the mount.

8. Torque the mount nut to 25 ft. lbs. (34 Nm).

9. Lower the vehicle and check for proper operation.

Cylinder Head

REMOVAL AND INSTALLATION

22R–E Engine

1. Relieve the fuel system pressure. Disconnect the negative battery cable.

2. Drain the coolant from the radiator and the cylinder block. Raise and safely support the vehicle. Drain the engine oil.

3. Disconnect and remove the air cleaner hose on the 22R–E engine.

4. Disconnect the oxygen sensor wire. Remove the nuts attaching the manifold to the exhaust pipe and then separate them.

5. Remove the oil dipstick. Remove the distributor with the spark plug leads attached.

6. Disconnect the upper radiator hose and the heater hoses where they attach to the engine and then position them aside.

7. Disconnect the actuator cable, the accelerator cable and the throttle cable for the automatic transmission at their bracket.

8. Tag and disconnect the following:

    a. Both PCV vacuum hoses

    b. Brake booster hose

    c. Actuator hose, if equipped with cruise control

    d. Air control valve hoses

    e. Air control valve.

9. Tag and disconnect the EGR vacuum modulator hoses and then remove the modulator itself along with the bracket.

10. Tag and disconnect the following:

    a. Green and brown BVSV hoses

    b. Vacuum advance hoses

    c. The 2 air valve hoses; one at the throttle body, the other at the air chamber

    d. Air control valve hose, if equipped with air conditioning

    e. Pressure regulator hose at the air chamber

    f. Cold start injector pipe and wire

    g. Throttle position sensor wire.

11. Remove the bolt holding the EGR valve to the air chamber. Disconnect the chamber from the stay. Remove the air chamber-to-intake manifold bolts and then lift off the chamber with the throttle body.

12. Disconnect the fuel return hose.

13. Tag and disconnect the following:

    a. Water temperature sender gauge wire

    b. Temperature sensor wire

    c. Start injection time switch wire

    d. Fuel injector wires.

14. Remove the pulsation damper. Remove the bolt holding the fuel hose to the delivery pipe and then disconnect and remove the fuel hose.

15. Disconnect the wire and hose and then remove the air valve from the intake manifold.

16. Disconnect the bypass hose at the intake manifold on the 22R–E engine.

17. If equipped with power steering, remove the pump and position it

aside without disconnecting the hydraulic lines.

18. Remove the 4 nuts and then remove the cylinder head cover.

19. Remove the rubber camshaft seals. Turn the crankshaft until the No. 1 piston is at TDC of its compression stroke. Matchmark the timing sprocket to the timing chain and then remove the semi-circular plug. Using a 19mm wrench, remove the camshaft sprocket bolt. Slide the distributor drive gear and spacer off the camshaft and wire the cam sprocket in place.

20. Remove the timing chain cover bolt in front of the cylinder head.

NOTE: This must be done before the cylinder head bolts are removed.

21. Remove the cylinder head bolts gradually, in 2–3 stages, in the correct order.

22. Using prybars applied evenly at the front and rear of the rocker arm assembly, pry the assembly off of its mounting dowels.

23. Lift the cylinder head off of its mounting dowels.

To install:

24. Apply liquid sealer to the front corners of the block and install the head gasket.

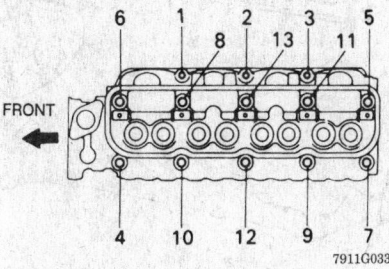

Cylinder head bolt removal sequence — 22R – E engine

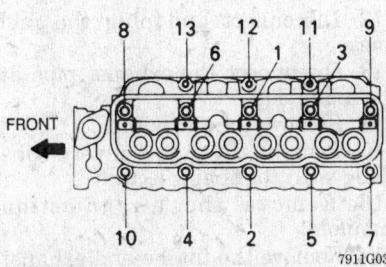

Cylinder head bolt installation sequence — 22R – E engine

25. Lower the head over the locating dowels. Do not attempt to slide it into place.

26. Rotate the camshaft so the sprocket aligning pin is at the top. Remove the wire and hold the cam sprocket. Manually rotate the engine so the sprocket hole is also at the top. Wire the sprocket in place.

27. Install the rocker arm assembly over its positioning dowels.

28. Tighten the cylinder head bolts evenly, in 3 stages and in the correct order.

29. Install the timing chain cover bolt and tighten it to 7–11 ft. lbs. (11–15 Nm).

30. Remove the wire and fit the sprocket over the camshaft dowel. If the chain won't allow the sprocket to reach, rotate the crankshaft back and forth, while lifting up on the chain and sprocket.

31. Install the distributor drive gear and tighten the crankshaft bolt to 51–65 ft. lbs. (68–88 Nm).

32. Set the No. 1 piston at TDC of its compression stroke and adjust the valves.

33. After completing valve adjustment, rotate the crankshaft 352°, so the 8 BTDC mark on the pulley aligns with the pointer.

34. Install the distributor.

35. Install the spark plugs and leads.

36. Make sure the oil drain plug is installed. Fill the engine with oil after installing the rubber cam seals. Pour the oil over the distributor drive gear and the valve rockers.

37. Install the rocker cover and tighten the bolts to 8–11 ft. lbs. (11–15 Nm).

38. Connect all the vacuum hoses and electrical leads which were removed during disassembly. Install the spark plug lead supports. Fill the cooling system. Install the air cleaner.

39. Tighten the exhaust pipe-to-manifold flange bolts to 25–33 ft. lbs. (34–45 Nm).

40. Reconnect the battery. Start the engine and allow it to reach normal operating temperature. Check and adjust the timing and valve clearance. Adjust the idle speed and mixture. Road test the vehicle.

3F–E Engine

1. Drain the coolant.

2. Disconnect the negative battery cable.

3. Scribe matchmarks around the hood hinges and then remove the hood.

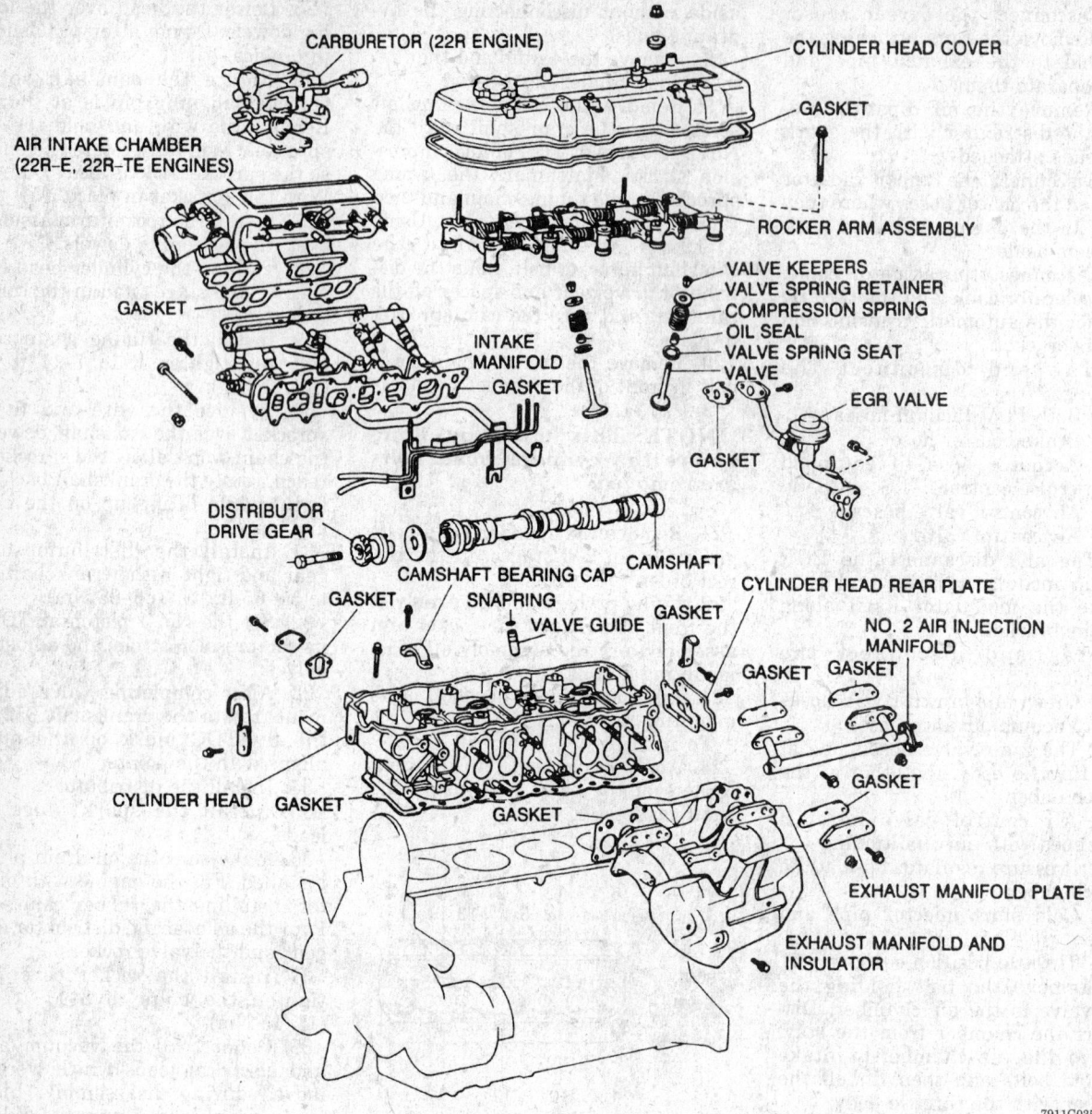

CARBURETOR (22R ENGINE)

CYLINDER HEAD COVER

GASKET

AIR INTAKE CHAMBER
(22R–E, 22R–TE ENGINES)

ROCKER ARM ASSEMBLY

VALVE KEEPERS
VALVE SPRING RETAINER
COMPRESSION SPRING
OIL SEAL
VALVE SPRING SEAT
VALVE

GASKET

INTAKE
MANIFOLD
GASKET

EGR VALVE

GASKET

DISTRIBUTOR
DRIVE GEAR

CAMSHAFT BEARING CAP   CAMSHAFT
SNAPRING

CYLINDER HEAD REAR PLATE

GASKET   VALVE GUIDE   GASKET

NO. 2 AIR INJECTION
MANIFOLD

GASKET   GASKET

GASKET

CYLINDER HEAD   GASKET

GASKET

EXHAUST MANIFOLD PLATE

EXHAUST MANIFOLD AND
INSULATOR

7911G035

Exploded view of the cylinder head assembly — 22R — E engine

4. Disconnect the accelerator and throttle cables.

5. Remove the air intake hose, air flow meter and air cleaner cap.

6. Unbolt the power steering pump and position it aside without disconnecting the hydraulic lines.

7. Unbolt the air conditioning compressor and position it aside without disconnecting the refrigerant lines.

8. Remove the power steering pump and air conditioning compressor brackets.

9. Disconnect the high tension leads from the spark plugs and the coil.

10. Disconnect and remove the heater water (oil cooler) pipe.

11. Disconnect the upper radiator hose.

12. Disconnect and plug the fuel lines.

13. Disconnect the exhaust pipe at the manifold.

14. Remove the air pump.

15. Remove the fuel delivery pipe along with the fuel injectors.

16. Remove the air injection manifold.

17. Remove the intake and exhaust manifolds.

18. Disconnect the water bypass hose at the water outlet and then remove the outlet.

19. Remove the spark plugs.

20. Remove the cylinder head cover and its gasket.

21. Loosen the bolts and nuts that attach the rocker shaft assembly in several stages and then remove the rocker shaft.

22. Remove the pushrods.

23. Remove the cylinder head bolts in the reverse of the tightening sequence. Remove the air pump bracket and engine hanger.

24. Lift the cylinder head off of its mounting dowels.

To install:

25. Install the cylinder head on the cylinder block using a new gasket.

26. Lightly coat the threads of the cylinder head bolts with engine oil and then install them into the head. Tighten in several stages, to the correct torque.

27. Install the pushrods in the order that they were removed.

28. Position the rocker shaft assembly on the cylinder head and align the rocker arm adjusting screws with the heads of the pushrods. Tighten the mounting bolts with a 12mm head to 17 ft. lbs. (24 Nm); tighten the bolts with a 14mm head to 25 ft. lbs. (33 Nm).

29. Adjust the valve clearance and install the spark plugs. Install the cylinder head cover and tighten the cap nuts to 78 inch lbs. (8.8 Nm).

30. Install the water outlet and connect the bypass hose. Tighten the bolts to 18 ft. lbs. (25 Nm).

31. Install the intake and exhaust manifolds using a new gasket. Make sure the front mark on the gasket is towards the front of the engine.

32. Install the heat insulators and the manifold stay.

33. Install the air injection manifold and tighten the union nuts and clamp bolts to 15 ft. lbs. (21 Nm).

34. Install the fuel injector/delivery pipe assembly.

35. Install the air pump and connect the air hose.

36. Connect the exhaust pipe to the manifold. Use a new gasket and tighten the bolts to 46 ft. lbs. (62 Nm).

37. Connect the fuel lines and the upper radiator hose.

38. Install the heater water pipe.

39. Connect the high tension cords.

40. Install the air conditioning compressor and the power steering pump. Remember to adjust the belt tension later.

41. Install the air intake hose, the air flow meter and the air cleaner cap.

42. Connect and adjust the accelerator and throttle cables.

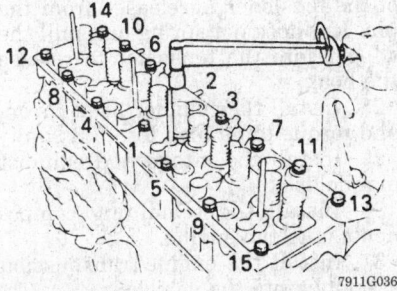

Cylinder head bolt installation sequence — 3F – E engine

43. Connect the battery cable, fill the engine with coolant, start the engine and check for any leaks. Road test the vehicle.

3VZ–E Engine

1. Disconnect the negative battery cable.

2. Remove the air cleaner hose and case.

3. Drain the engine coolant.

4. Remove the radiator.

5. Unbolt the power steering pump and position it aside with the hoses still attached.

6. Remove all drive belts and then remove the fluid coupling and fan pulley.

7. Tag and disconnect all wires and connectors that will interfere with cylinder head removal.

8. Disconnect the following hoses:
   a. Power steering air hoses
   b. Brake booster hose
   c. Cruise control vacuum hose
   d. Charcoal canister has at the canister
   e. VSV vacuum hose.

9. Disconnect the accelerator, throttle and cruise control cables.

10. Disconnect the clutch release cylinder hose for manual transmission only.

11. Disconnect the heater hoses and the fuel lines.

12. Remove the left side scuff plate and disconnect the O_2 sensor and then remove the front exhaust pipe.

13. Remove the timing belt as detailed later in this section.

14. Remove the distributor with the spark plug leads attached; position it aside.

15. Remove the air intake chamber.

16. Disconnect the connectors and then remove the engine wire.

17. Remove the Nos. 2 and 3 fuel pipes.

18. Remove the No. 4 timing belt cover.

19. Remove the No. 2 idler pulley and the No. 3 timing belt cover.

20. Disconnect the hose and remove the water bypass outlet.

21. Remove the intake manifold.

22. Remove the exhaust crossover pipe.

23. For the right side, remove the following:
   a. Remove the reed valve with the No. 1 air injection manifold.
   b. Remove the water bypass pipe mounting bolt.
   c. Remove the cylinder head cover.
   d. Remove the camshaft.
   e. Loosen the cylinder head bolts in several stages, in the opposite

order of the tightening sequence. Remove the air pump bracket and engine hanger.
   f. Lift the cylinder head off of its mounting dowels, do not pry it off.

24. For the left side, remove the following:
   a. Remove the alternator.
   b. Remove the oil dipstick guide tube.
   c. Remove the cylinder head cover.
   d. Remove the camshaft.
   e. Loosen the cylinder head bolts in several stages, in the opposite order of the tightening sequence. Remove the air pump bracket and engine hanger.
   f. Lift the cylinder head off of its mounting dowels, do not pry it off.

To install:

25. Install the cylinder head on the cylinder block using a new gasket.

26. Lightly coat the threads of the cylinder head bolts with engine oil and then install them into the head. Tighten them in several stages, in the correct order. After the initial tightening, mark the front side the the top of the bolt with paint. Tighten the bolts an additional 90° (¼ turn) and check that the mark is now facing the side of the head. Tighten the bolts an additional 90° and check that the mark is now facing the rear of the head. Install the bolt (A) and tighten it to 27 ft. lbs. (37 Nm).

27. Install the camshaft.

28. Install the alternator and the water bypass pipe mounting bolt.

29. Install the reed valve with the No. 1 injection manifold.

30. Install the oil dipstick tube.

31. Install the crossover pipe and tighten it to 29 ft. lbs. (39 Nm).

32. Connect the oxygen sensor wire.

33. Install the intake manifold with new gaskets and tighten the mounting bolts to 29 ft. lbs. (39 Nm).

34. Install the water bypass outlet and tighten the bolts to 13 ft. lbs. (18 Nm).

35. Install the fuel delivery pipes and injectors.

36. Install the No. 2 idler pulley. Install the Nos. 3 and 4 timing belt covers and tighten the bolts to 74 inch lbs. (8.3 Nm).

37. Install the fuel pipes and tighten the union bolts to 22 ft. lbs. (29 Nm).

38. Install the timing belt. Install the cylinder head covers.

39. Install the air intake chamber and tighten the nuts and bolts to 13 ft. lbs. (18 Nm).

40. Install the EGR valve and connect all hoses and lines. Install the

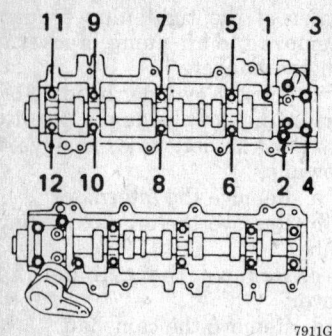

Camshaft bearing cap loosening sequence — 3VZ — E engine

7911G039

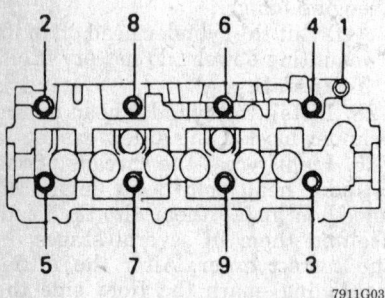

Cylinder head bolt removal sequence — 3VZ — E engine

7911G037

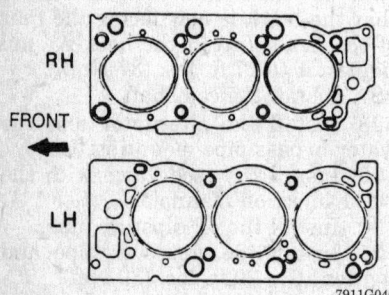

Cylinder head gasket installation — 3VZ — E engine

7911G040

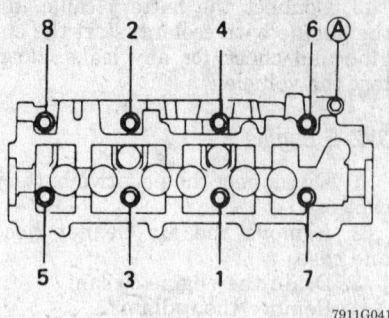

Cylinder head bolt tightening sequence — 3VZ — E engine

7911G041

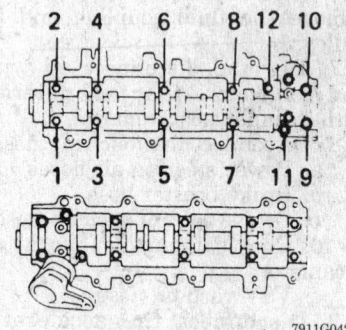

Camshaft bearing cap bolt tightening sequence — 3VZ — E engine

7911G042

2TZ–FE Engine

1. Disconnect the negative battery cable.

2. Remove the engine/transmission assembly from the vehicle.

3. Remove the engine wiring from the engine and move aside.

4. Remove the No. 2 valve cover.

5. Mark the spark plug wires and disconnect. Matchmark the distributor, rotor and cylinder head. Remove the distributor and disconnect the wiring.

6. Remove the EGR valve.

7. Remove the fuel rail and water outlet.

8. Remove the intake and exhaust manifold assembly.

9. Remove the exhaust manifold heat insulator and oil return pipe.

10. Remove the No. 1 valve cover.

11. Place matchmarks on the timing sprocket and chain. Hold the camshaft with a wrench and remove the sprocket bolt.

12. Remove the chain tensioner and gasket.

13. Remove the No. 6 camshaft bearing cap.

14. Set the knock pin hole of the exhaust camshaft at the 5–30 degree BTDC. Uniformly loosen the

camshaft bearing caps and remove the exhaust camshaft from the head.

15. Set the knock pin hole of the intake camshaft at the 75–100 degree BTDC. Uniformly loosen the camshaft bearing caps and remove the exhaust camshaft from the head.

16. Remove the 2 bolts in front of the head before removing the cylinder head retaining bolts.

17. Using a 12 sided socket wrench, remove the 10 cylinder head retaining bolts in sequence and remove the head from the engine.

To install:

18. Clean the gasket mating surfaces and check for warpage.

19. Install the head gasket and install the cylinder head.

20. Oil the bolts and torque in 3 steps, in sequence. If any of the bolts break, deform or do not meet the torque specification replace them.

21. Install and torque the 2 front bolts to 15 ft. lbs. (21 Nm).

22. Grease to all camshaft journals and caps.

23. Place the intake camshaft at 75–100° BTDC. Install the bearing caps with the marking arrows facing forward. Uniformly torque the bearing cap bolts to 12 ft. lbs. (16 Nm).

24. Apply sealant to the bearing cap next to the timing chain sprocket. Place the intake camshaft at 5–30° BTDC. Install the bearing caps with the marking arrows facing forward. Uniformly torque the bearing cap bolts to 12 ft. lbs. (16 Nm). Make sure the exhaust and intake camshaft gear alignment marks are facing each other. The one gear has 2 dots and the other has 1 dot.

25. Install the timing chain sprocket and torque the bolt to 54 ft. lbs. (74 Nm).

26. Release the ratchet pawl and fully push the plunger and apply the hook on the tensioner so it cannot spring out.

27. Install the tensioner and torque the bolts to 15 ft. lbs. (21 Nm). Turn the crankshaft to the left so the hook of the tensioner is released from the pin. If it does not spring out, pull the slipper into the tensioner to release the hook.

28. Install the cylinder head covers and torque to 69 inch lbs. (7.8 Nm).

29. Install the intake and exhaust manifolds.

30. Install the remaining components onto the engine.

31. Install the engine/transmission assembly into the vehicle.

32. Refill the engine coolant and oil. Connect the battery cable and check for leaks.

distributor and the front exhaust pipe.

41. Connect the fuel lines and heater hoses. Connect the clutch release cylinder hose.

42. Install the power steering pump. Connect and adjust all cables, hoses and wires previously removed.

43. Install the fan pulley, fluid coupling and drive belts. Install the radiator.

44. Install the air cleaner hose, refill the engine with coolant and connect the battery cable.

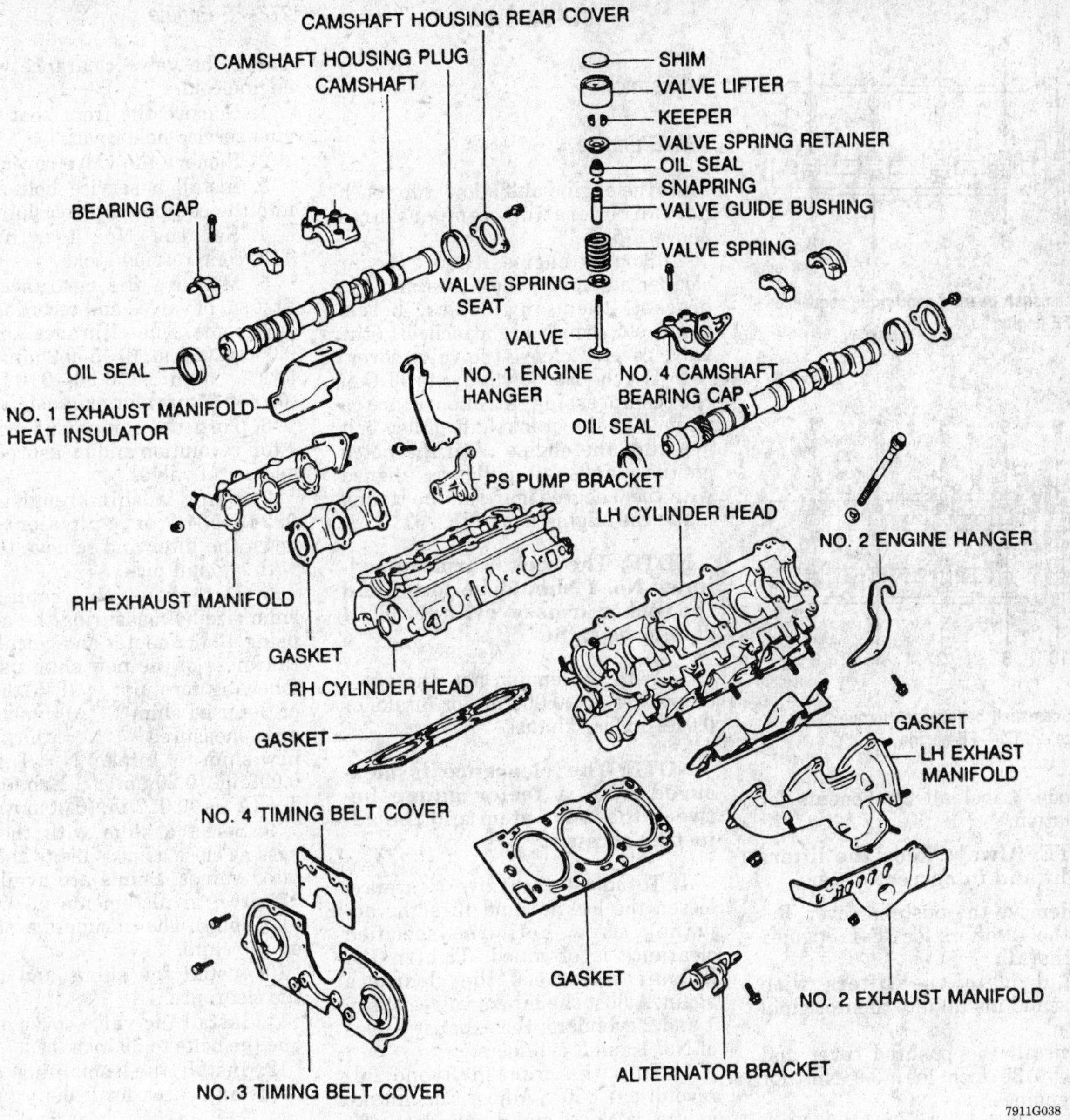

CAMSHAFT HOUSING REAR COVER

CAMSHAFT HOUSING PLUG

CAMSHAFT

SHIM

VALVE LIFTER

KEEPER

VALVE SPRING RETAINER

OIL SEAL

SNAPRING

VALVE GUIDE BUSHING

BEARING CAP

VALVE SPRING

VALVE SPRING SEAT

VALVE

OIL SEAL

NO. 1 ENGINE HANGER

NO. 4 CAMSHAFT BEARING CAP

OIL SEAL

NO. 1 EXHAUST MANIFOLD HEAT INSULATOR

PS PUMP BRACKET

RH EXHAUST MANIFOLD

LH CYLINDER HEAD

NO. 2 ENGINE HANGER

GASKET

RH CYLINDER HEAD

GASKET

GASKET

LH EXHAST MANIFOLD

NO. 4 TIMING BELT COVER

GASKET

NO. 2 EXHAUST MANIFOLD

ALTERNATOR BRACKET

NO. 3 TIMING BELT COVER

7911G038

Exploded view of the cylinder head assembly — 3VZ — E engine

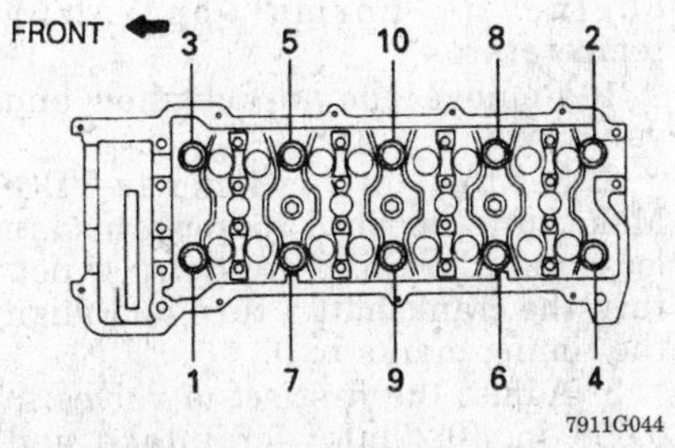

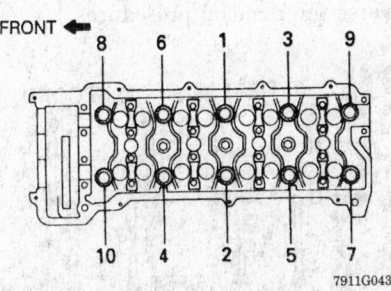

FRONT 3 5 10 8 2

1 7 9 6 4

7911G044

Cylinder head loosening sequence — 2TZ — FE engine

FRONT 8 6 1 3 9

10 4 2 5 7

7911G043

Cylinder head torque sequence — 2TZ — FE engine

Valve Lifters

REMOVAL AND INSTALLATION

Always replace the camshaft and lifters as a set. If not replacing, label all components for exact reinstallation.

3F–E Engine

1. Disconnect the negative battery cable.

2. Remove the spark plugs and tubes.

3. Remove the valve cover.

4. Uniformly loosen and remove the rocker arm shaft bolts and nuts. Remove the rocker shaft and

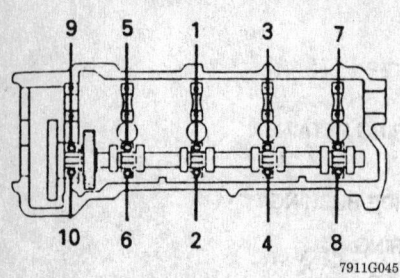

Intake camshaft bearing cap torque sequence — 2TZ — FE engine

7911G045

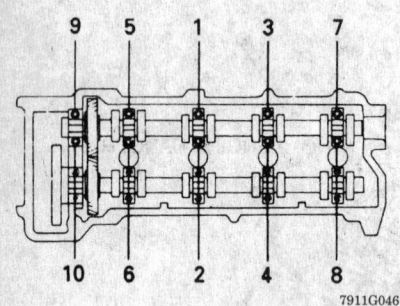

Exhaust camshaft bearing cap torque sequence — 2TZ — FE engine

7911G046

pushrods. Label all components for installation.

NOTE: Always keep the lifters upright and in correct order.

5. Remove the pushrod cover. Remove the 12 lifters for 3F–E engine.
To install:
6. Lubricate the lifters with prelube and install into their original location.
7. Install the pushrod cover and torque to 35 inch lbs. (3.9 Nm) for 3F–E engines.
8. Install the rocker arms and torque to 17 ft. lbs. (24 Nm) and 25 ft. lbs. (33 Nm) for the bolt/nut combination.
9. Install the rocker arm cover and torque to 69 inch lbs. (7.8 Nm).
10. Install the remaining components and check for leaks.

Valve Lash

Adjustment

22R–E Engine

Start the engine and allow it to reach normal operating temperatures above 175° F.
1. Stop the engine. Remove the air cleaner assembly, the hoses and the bracket, then any cables, hoses, wires, etc., which are attached to the valve cover. Remove the valve cover.
2. Set the No. 1 cylinder to TDC of the compression stroke. Place a wrench on the crankshaft pulley bolt and turn the engine until the notch on the crankshaft pulley is aligned with the 0 degree mark on the timing plate; the engine is at TDC.

NOTE: The rocker arms on cylinder No. 1 should be loose and the rocker arms on cylinder No. 4 should be tight.

3. With the engine hot, the valve clearances are 0.008 in. for intake or 0.012 in. for exhaust.

NOTE: The clearance is measured with a feeler gauge between the valve stem and the adjusting screw.

4. To adjust the valve clearance, loosen the locknut and turn the adjusting screw until the specified clearance is obtained. Tighten the locknut and check the clearance again. Adjust the intake valves of No. 1 and 2 cylinders; the exhaust valves of No. 1 and 3 cylinders.
5. Turn the crankshaft one full revolution, 360°. Adjust the intake valves of No. 3 and 4 cylinders; the exhaust valves of No. 2 and 4 cylinders.
6. To install the components, reverse the removal procedures.

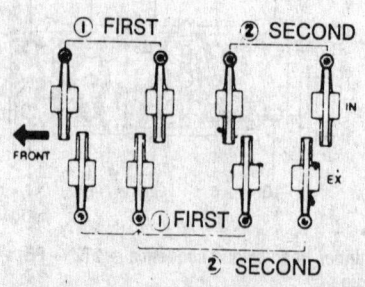

Valve adjustment sequence — 22R — E engine

7911G047

2TZ–FE Engine

Check the valve clearance with the engine cold.
1. Remove the front seat and engine service hole cover.
2. Remove the valve cover.
3. Install a service bolt and nut into the equipment driveshaft.
4. Set the No. 1 cylinder to TDC/compression stoke.
5. Measure the clearance of the first set of valves and record the measurements. The clearance should be 0.006–0.010 in. (0.15–0.25mm) for intake and 0.010–0.014 in. (0.25–0.35mm) for exhaust.
6. Turn the equipment driveshaft 1 full revolution and measure the second set of valves.
7. Using a shim removing tool 0924855010 or equivalent, press down the lifter and remove the shim with a small pick.
8. Determine the replacement shim size by measuring the old shim using a micrometer and calculate the thickness of the new shim using the following formula: T = thickness of the used shim A = valve clearance measured N = thickness of new shim Intake: $N = T + (A - 0.008$ in. $(0.20$mm$))$ Exhaust: $N = T + (A - 0.012$ in. $(0.30$mm$))$
9. Select a shim with the thickness as close as possible to the calculated values. Shims are available in 17 sizes, in increments of 0.002 in. (0.050mm). The thickness is stamped on the shim.
10. Install the shims and recheck the clearance.
11. Install the valve cover and torque the bolts to 69 inch lbs. (7.8 Nm).
12. Install the remaining components and check for leaks.

3F–E Engine

Check the valve clearance with the engine at normal operating temperature.
1. Remove the air cleaner and valve cover.
2. Set the No. 1 cylinder to TDC. Make sure the No. 1 rocker arms are loose and the No. 6 are tight. If not, turn the crankshaft 1 turn and align the timing marks to 0.
3. Adjust the first set of valves to 0.008 in. (0.20mm) for intake and 0.014 in. (0.35mm) for exhaust valves.
4. After tightening the lock nut, recheck the valve clearance.
5. Install the valve cover and air cleaner.

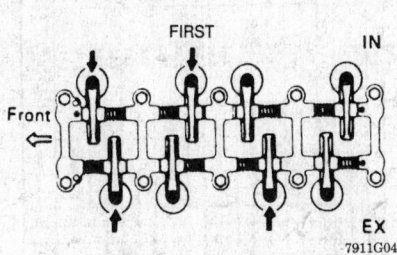

First step of the valve adjustment procedure — 22R — E engine

7911G048

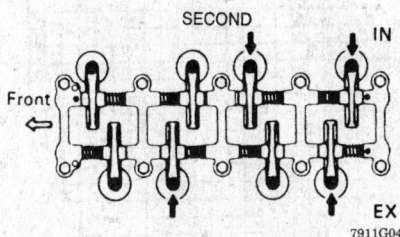

Second step of the valve adjustment procedure — 22R — E engine

7911G049

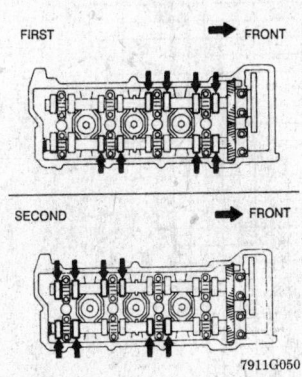

Valve clearance check — 2TZ — FE engine

7911G050

3VZ–E Engine

Check the valve clearance with the engine cold.

1. Remove the valve cover.

2. Set the No. 1 cylinder to TDC/compression.

3. Measure the clearance of the lifters with the camshaft lobe at the base circle and record the measurements. The clearance should be as follows: 0.007–0.011 in. (0.18–0.28mm) intake; 0.009–0.013 in. (0.22–0.32mm) exhaust

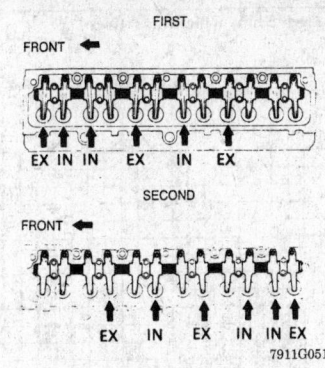

Adjusting valve clearance — 3F — E engine

7911G051

4. Using a shim removing tool 0924855010 or equivalent, press down the lifter and remove the shim with a small pick.

5. Determine the replacement shim size by measuring the old shim using a micrometer and calculate the thickness of the new shim using the following formula: T = thickness of the used shim A = valve clearance measured N = thickness of new shim Intake: $N = T + (A - 0.0091$ in. (0.23mm); Exhaust: $N = T + (A - 0.0126$ in. (0.32mm)

6. Select a shim with the thickness as close as possible to the calculated values. Shims are available in 25 sizes, in increments of 0.002 in. (0.050mm). The thickness is stamped on the shim.

7. Install the shims and recheck the clearance.

8. Install the valve cover and torque the bolts to 69 inch lbs. (7.8 Nm).

9. Install the remaining components and check for leaks.

Rocker Arm Shafts

REMOVAL AND INSTALLATION

22R–E Engine

1. Remove the valve cover.

2. Remove the timing chain sprocket and secure the sprocket and chain to the engine with wire to ensure correct valve timing.

3. Uniformly loosen the 10 shaft bolts in the opposite sequence of tightening.

4. Remove the rocker arm shaft assembly from the cylinder head. Keep the assembly together for installation.

To install:

5. Install the rocker arm shaft assembly and torque the bolts in 3 steps, in sequence to 58 ft. lbs. (78 Nm).

6. Install the chain cover bolt and torque to 9 ft. lbs. (13 Nm).

7. Install timing chain sprocket, distributor drive gear and thrust plate. Torque the bolt to 58 ft. lbs. (78 Nm).

8. Adjust the valve clearance.

9. Install the valve cover and torque to 69 ft. lbs. (7.8 Nm).

10. Install the remaining components and check for leaks.

3F–E Engine

1. Remove the valve cover.

2. Uniformly loosen the shaft bolts starting from the ends and working inward.

3. Remove the rocker arm shaft assembly from the cylinder head. Keep the assembly together for installation.

To install:

4. Install the rocker arm shaft assembly and torque the bolts in 3 steps, start in the middle and move to the ends and torque in sequence to 17 ft. lbs. (24 Nm) and 25 ft. lbs. (33 Nm) for the bolt and nut combination.

5. Adjust the valve clearance.

6. Install the valve cover and torque to 69 ft. lbs. (7.8 Nm).

7. Install the remaining components and check for leaks.

Intake Manifold

REMOVAL AND INSTALLATION

22R–E Engine

1. Relieve the fuel system pressure. Disconnect the negative battery cable.

2. Drain the cooling system.

3. Disconnect the air intake hose from both the air cleaner assembly on one end and the air intake chamber on the other.

4. Tag and disconnect all vacuum lines attached to the intake chamber and manifold.

5. Tag and disconnect the wires to the cold start injector, throttle position sensor and the water hoses from the throttle body.

6. Remove the EGR valve from the intake chamber.

7. Tag and disconnect the actuator cable, accelerator cable and throttle valve cable, if equipped, from the cable bracket on the intake chamber.

8. Unbolt the air intake chamber from the intake manifold and remove the chamber with the throttle body attached.

9. Disconnect the fuel hose from the fuel delivery pipe.

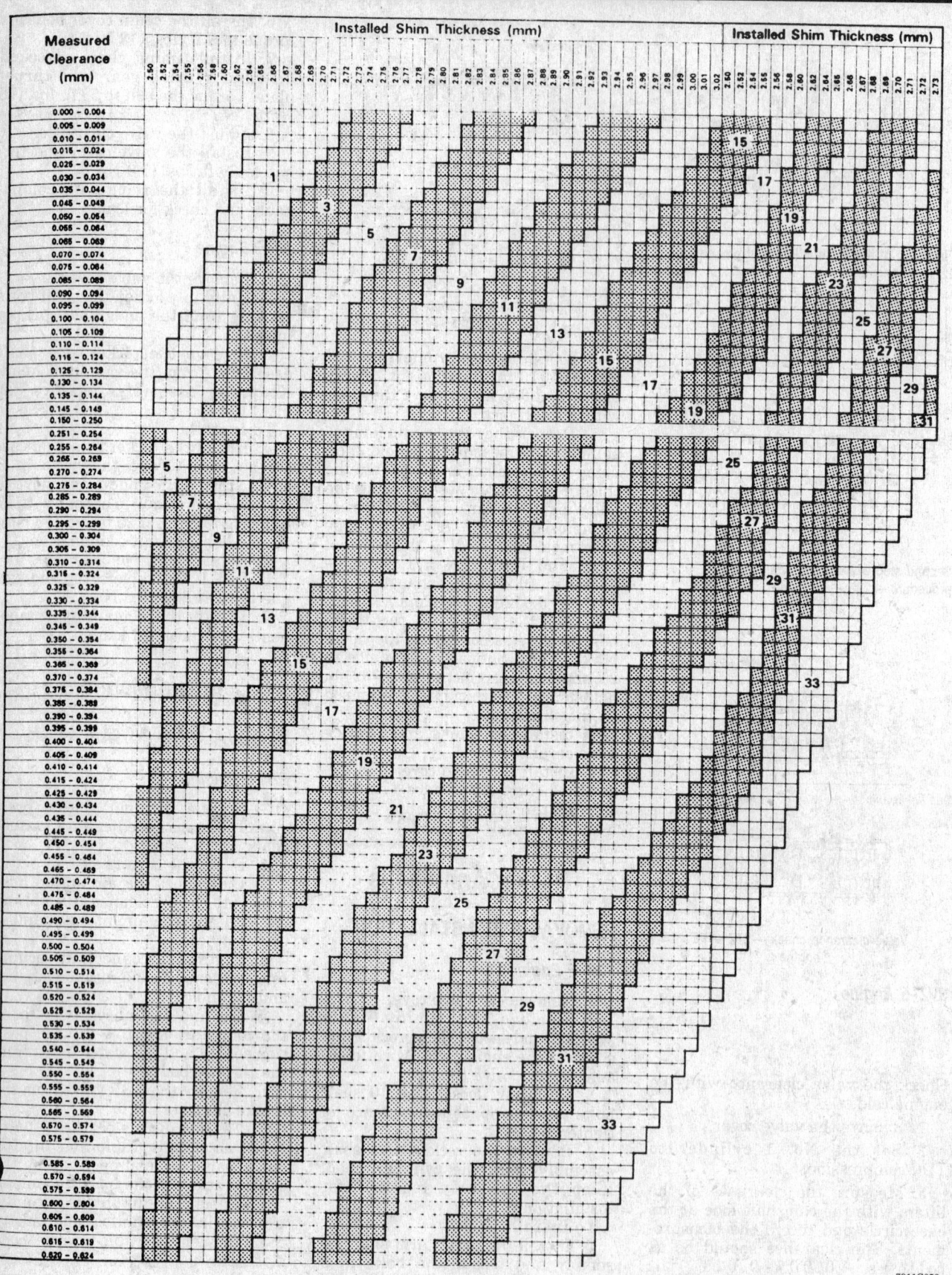

Intake shim selection chart — 2TZ — FE engine

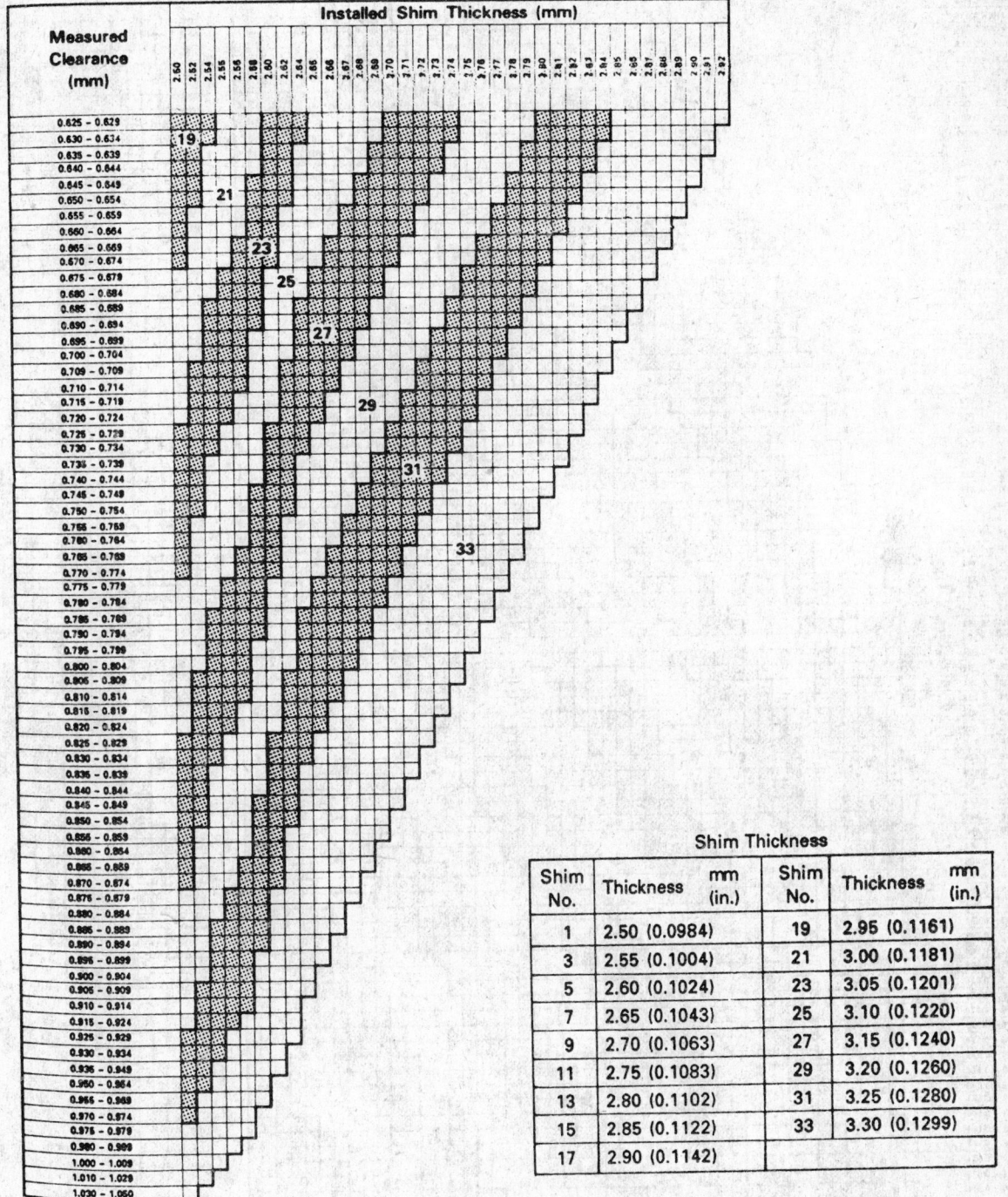

Intake shim selection chart (cont) — 2TZ — FE engine

Shim Thickness

Shim No.	Thickness mm (in.)	Shim No.	Thickness mm (in.)
1	2.50 (0.0984)	19	2.95 (0.1161)
3	2.55 (0.1004)	21	3.00 (0.1181)
5	2.60 (0.1024)	23	3.05 (0.1201)
7	2.65 (0.1043)	25	3.10 (0.1220)
9	2.70 (0.1063)	27	3.15 (0.1240)
11	2.75 (0.1083)	29	3.20 (0.1260)
13	2.80 (0.1102)	31	3.25 (0.1280)
15	2.85 (0.1122)	33	3.30 (0.1299)
17	2.90 (0.1142)		

7911G053

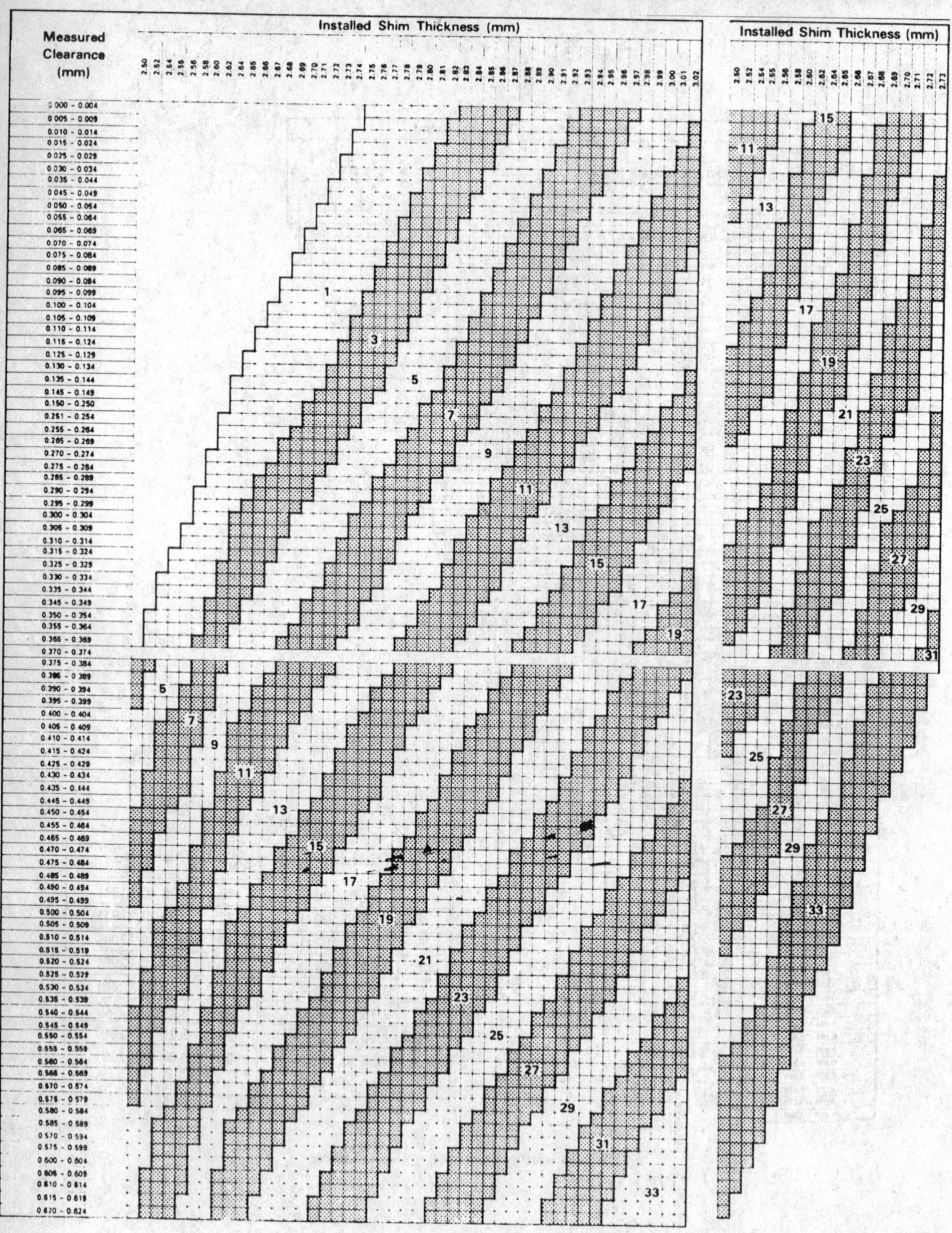

Exhaust shim selection chart (cont) — 2TZ — FE engine

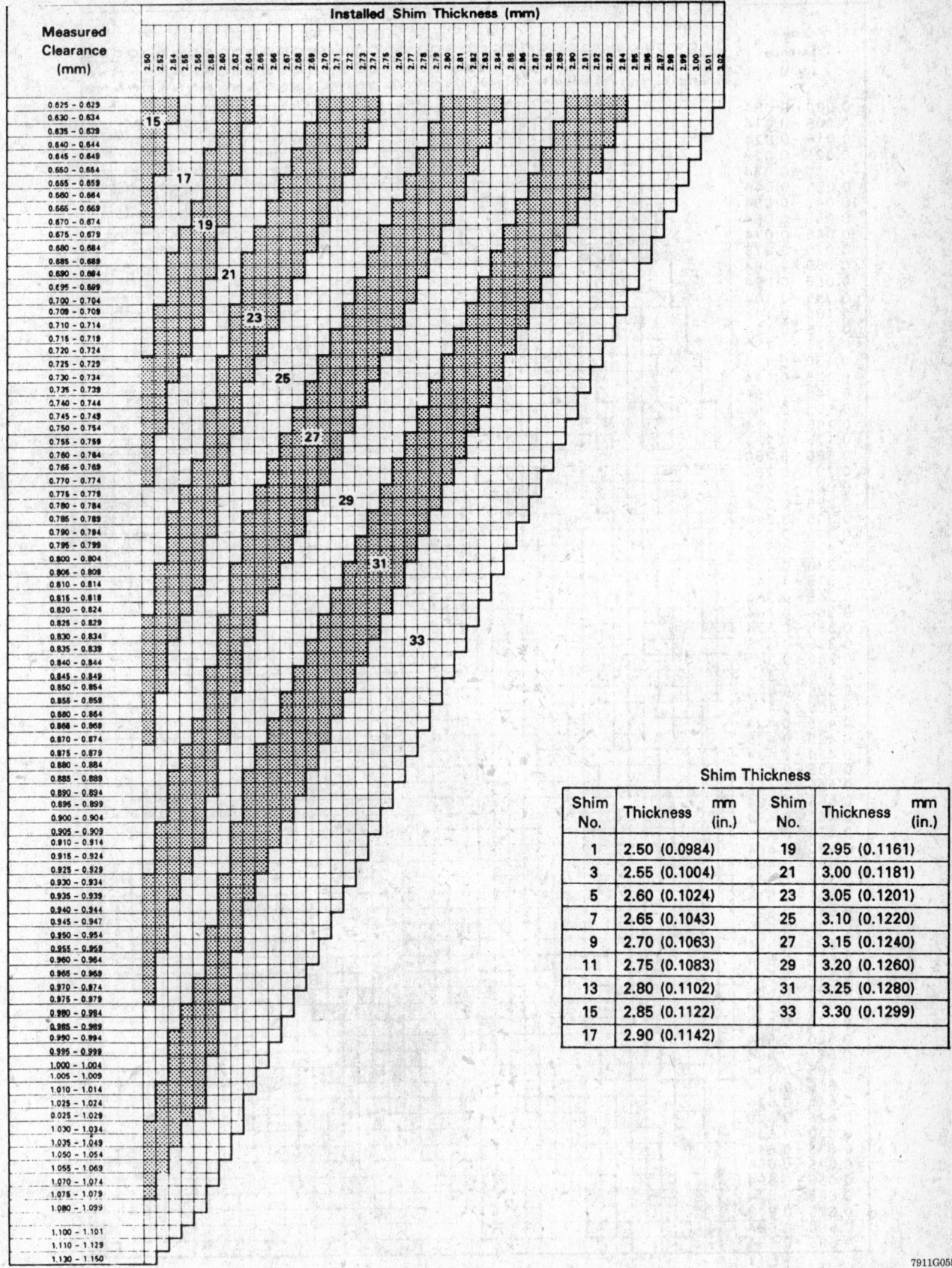

Shim Thickness				
Shim No.	Thickness mm (in.)		Shim No.	Thickness mm (in.)
1	2.50 (0.0984)		19	2.95 (0.1161)
3	2.55 (0.1004)		21	3.00 (0.1181)
5	2.60 (0.1024)		23	3.05 (0.1201)
7	2.65 (0.1043)		25	3.10 (0.1220)
9	2.70 (0.1063)		27	3.15 (0.1240)
11	2.75 (0.1083)		29	3.20 (0.1260)
13	2.80 (0.1102)		31	3.25 (0.1280)
15	2.85 (0.1122)		33	3.30 (0.1299)
17	2.90 (0.1142)			

Exhaust shim selection chart (cont) — 2TZ — FE engine

7911G055

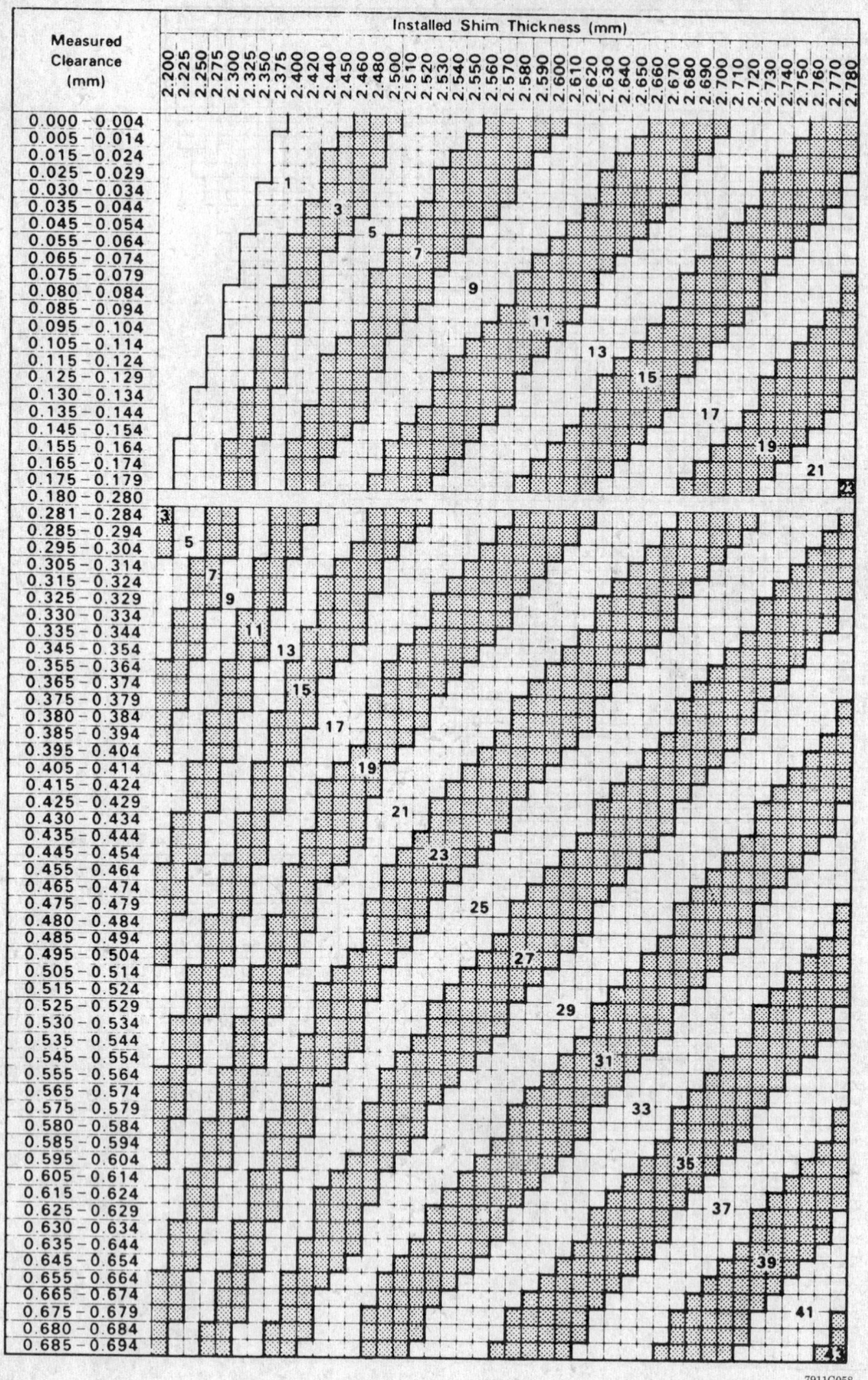

Intake shim selection chart — 3VZ — E engine

7911G058

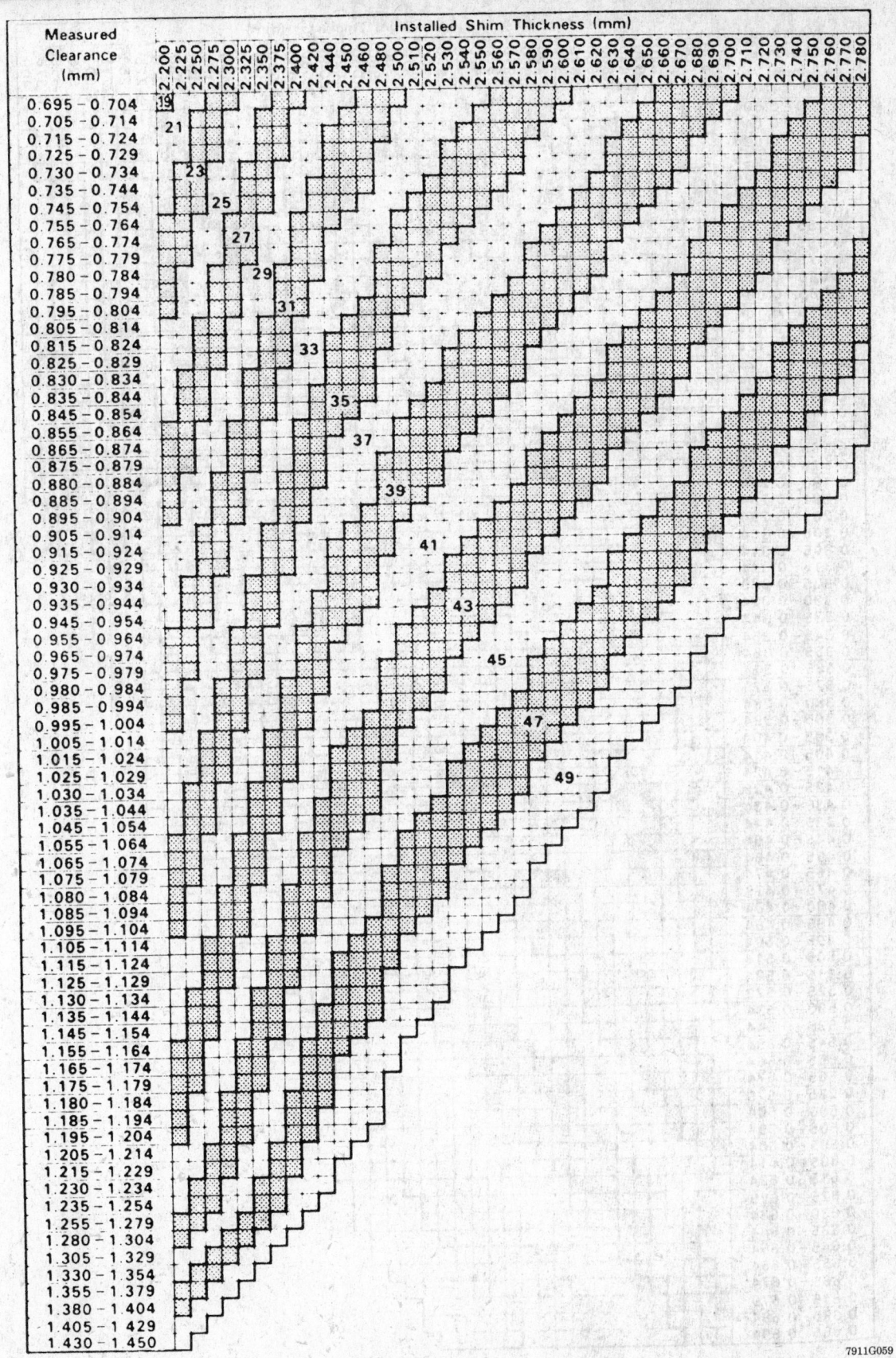

Intake shim selection chart (cont) — 3VZ — E engine

7911G059

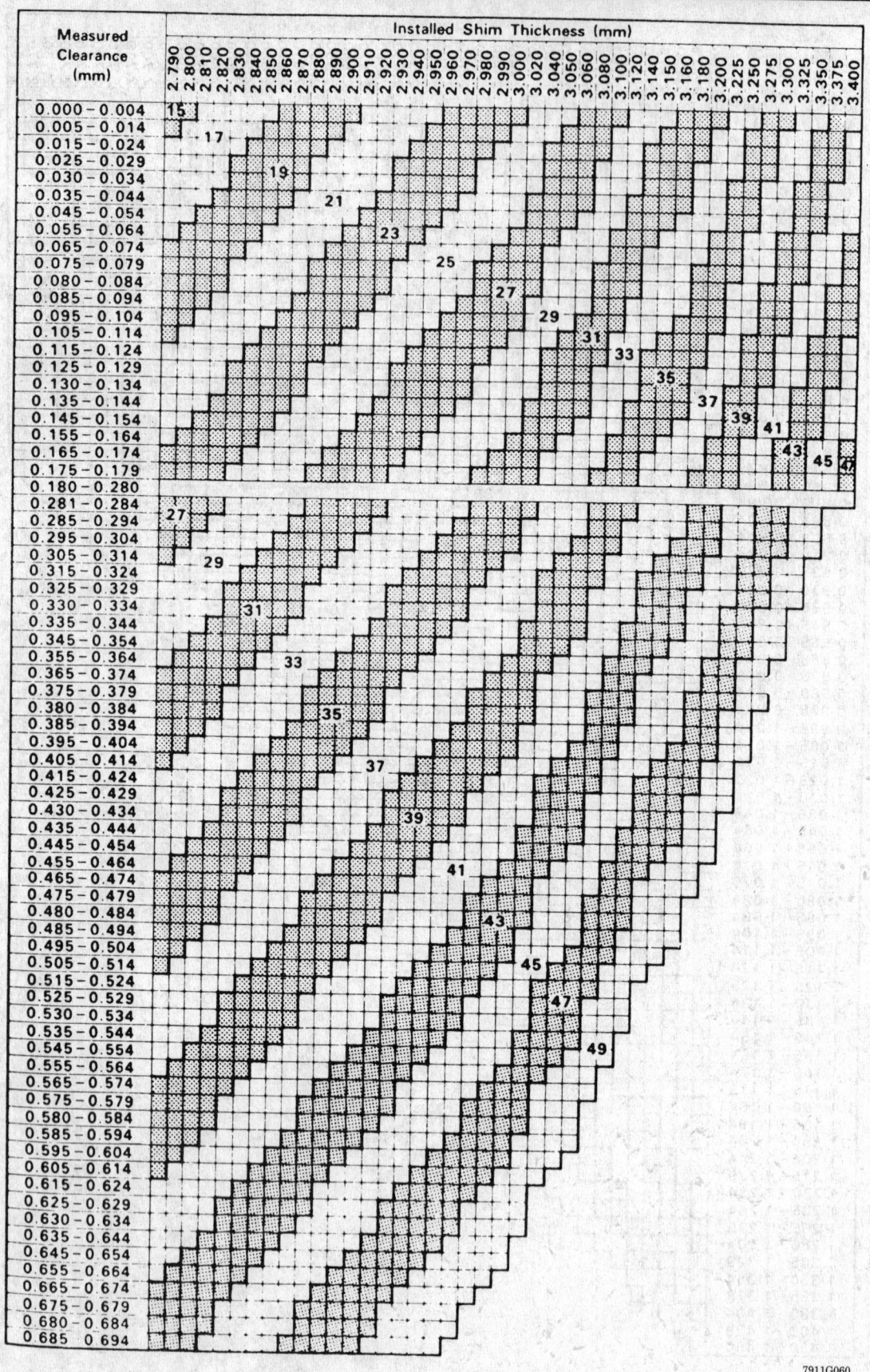

Intake shim selection chart (cont) — 3VZ — E engine

7911G060

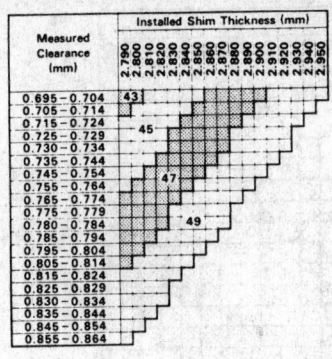

Measured Clearance (mm)	Installed Shim Thickness (mm) 2.790 2.800 2.810 2.820 2.830 2.840 2.850 2.860 2.870 2.880 2.890 2.900 2.910 2.920 2.930 2.940 2.950
0.695 – 0.704	43
0.705 – 0.714	
0.715 – 0.724	45
0.725 – 0.729	
0.730 – 0.734	
0.735 – 0.744	
0.745 – 0.754	47
0.755 – 0.764	
0.765 – 0.774	
0.775 – 0.779	49
0.780 – 0.784	
0.785 – 0.794	
0.795 – 0.804	
0.805 – 0.814	
0.815 – 0.824	
0.825 – 0.829	
0.830 – 0.834	
0.835 – 0.844	
0.845 – 0.854	
0.855 – 0.864	

Shim Thickness

Shim No.	Thickness mm (in.)	Shim No.	Thickness mm (in.)
01	2.20 (0.0866)	27	2.85 (0.1122)
03	2.25 (0.0886)	29	2.90 (0.1142)
05	2.30 (0.0906)	31	2.95 (0.1161)
07	2.35 (0.0925)	33	3.00 (0.1181)
09	2.40 (0.0945)	35	3.05 (0.1201)
11	2.45 (0.0965)	37	3.10 (0.1220)
13	2.50 (0.0984)	39	3.15 (0.1240)
15	2.55 (0.1004)	41	3.20 (0.1260)
17	2.60 (0.1024)	43	3.25 (0.1280)
19	2.65 (0.1043)	45	3.30 (0.1299)
21	2.70 (0.1063)	47	3.35 (0.1319)
23	2.75 (0.1083)	49	3.40 (0.1339)
25	2.80 (0.1102)		

7911G061

Intake shim selection chart (cont) — 3VZ — E engine

10. Tag and disconnect the air valve hose from the intake manifold.

11. Make sure all hoses, lines and wires are tagged for later installation and disconnected from the intake manifold. Unbolt the manifold from the cylinder head, removing the delivery pipe and injection nozzle with the manifold.

To install:

12. Clean the gasket mating surfaces and check for warpage.

13. Install the gasket and manifold. Torque the bolts to 13 ft. lbs. (18 Nm) starting from the middle and move outward.

14. Install the remaining components and check for leaks.

3VZ–E Engine

1. Relieve the fuel system pressure. Disconnect the negative battery cable.

2. Drain the cooling system.

3. Disconnect the air intake hose from both the air cleaner assembly on one end and the air intake chamber on the other.

4. Tag and disconnect all vacuum lines attached to the intake chamber and manifold.

5. Disconnect the throttle position sensor connector at the air chamber.

Disconnect the PCV hose at the union.

6. Disconnect the No. 4 water bypass hose at the manifold. Remove the No. 5 bypass hose at the water bypass pipe.

7. Disconnect the cold start injector and the vacuum hose at the fuel filter.

8. Remove the union bolt and gaskets, then remove the cold start injector tube.

9. Disconnect the EGR gas temperature sensor and the EGR vacuum hoses from the air pipe and the vacuum modulator.

10. Remove the EGR valve.

11. Disconnect the No. 1 air hose at the reed valve.

12. Remove the air intake chamber and then remove the engine wire.

13. Remove the union bolts and then remove the No. 2 and 3 fuel pipes.

14. Remove the No. 4 timing belt cover. Remove the the No. 2 idler pulley and the No. 3 timing belt cover.

15. Remove the fuel delivery pipes with their injectors.

16. Remove the water bypass outlet and then remove the intake manifold.

To install:

17. Install the intake manifold with new gaskets and tighten the mounting bolts to 29 ft. lbs. (39 Nm).

18. Install the water bypass outlet and tighten the 2 bolts to 13 ft. lbs. (18 Nm).

19. Install the fuel delivery pipes and injectors.

20. Install the No. 2 idler pulley. Install the No. 3 and 4 timing belt covers and tighten the bolts to 74 inch lbs. (8.3 Nm).

21. Install the fuel pipes and tighten the union bolts to 22 ft. lbs. (29 Nm).

22. Install the engine wire.

23. Install the air intake chamber and tighten the nuts and bolts to 13 ft. lbs. (18 Nm).

24. Install the EGR valve and connect all hoses and lines.

25. Install the air cleaner hose, refill the engine with coolant and connect the battery cable.

2TZ–FE Engine

1. Disconnect the negative battery cable.

2. Remove the engine from the vehicle.

3. Disconnect the fuel pipes and remove the fuel rail.

4. Label and disconnect all hoses, wiring and cables from the intake manifold.

5. Remove the water outlet and disconnect the PCV hose.

6. Remove the intake manifold stays.

7. Remove the retaining bolts, intake manifold and gasket.

To install:

8. Clean the gasket mating surfaces and check for warpage.

9. Install the gasket, manifold and bolts. Torque the bolts to 15 ft. lbs. (21 Nm), starting from the inside and work outward.

10. Install the manifold stays and torque the bolts to 27 ft. lbs. (37 Nm).

11. Connect the water bypass pipe and fuel rail.

12. Reconnect all wiring, hoses and cables to the manifold.

13. Install the engine into the vehicle.

14. Install the remaining components and check for leaks.

Exhaust Manifold

REMOVAL AND INSTALLATION

22R–E and 3VZ–E Engines

1. Tag and disconnect the spark plug leads.

2. Position the spark plug wires aside.

3. Use a 14mm wrench to remove the manifold securing nuts.

4. Remove the manifold(s), complete with air injection tubes and the inner portion of the heat stove.

To install:

5. Separate the inner portion of the heat stove from the manifold.

6. When installing the manifold(s), torque the retaining nuts to 29–36 ft. lbs. (41–49 Nm), working from the inside out, and in several stages. Install the distributor and set the timing. Tighten the exhaust pipe flange nuts to 25–32 ft. lbs. (34–45 Nm).

7. Install the remaining components and check for leaks.

Combinaton Manifold

REMOVAL AND INSTALLATION

3F–E Engine

1. Relieve the fuel system pressure on the 3F–E engine. Disconnect the negative battery cable. Remove the air cleaner assembly, complete with hoses.

2. Disconnect the accelerator and choke linkages, as well as the fuel

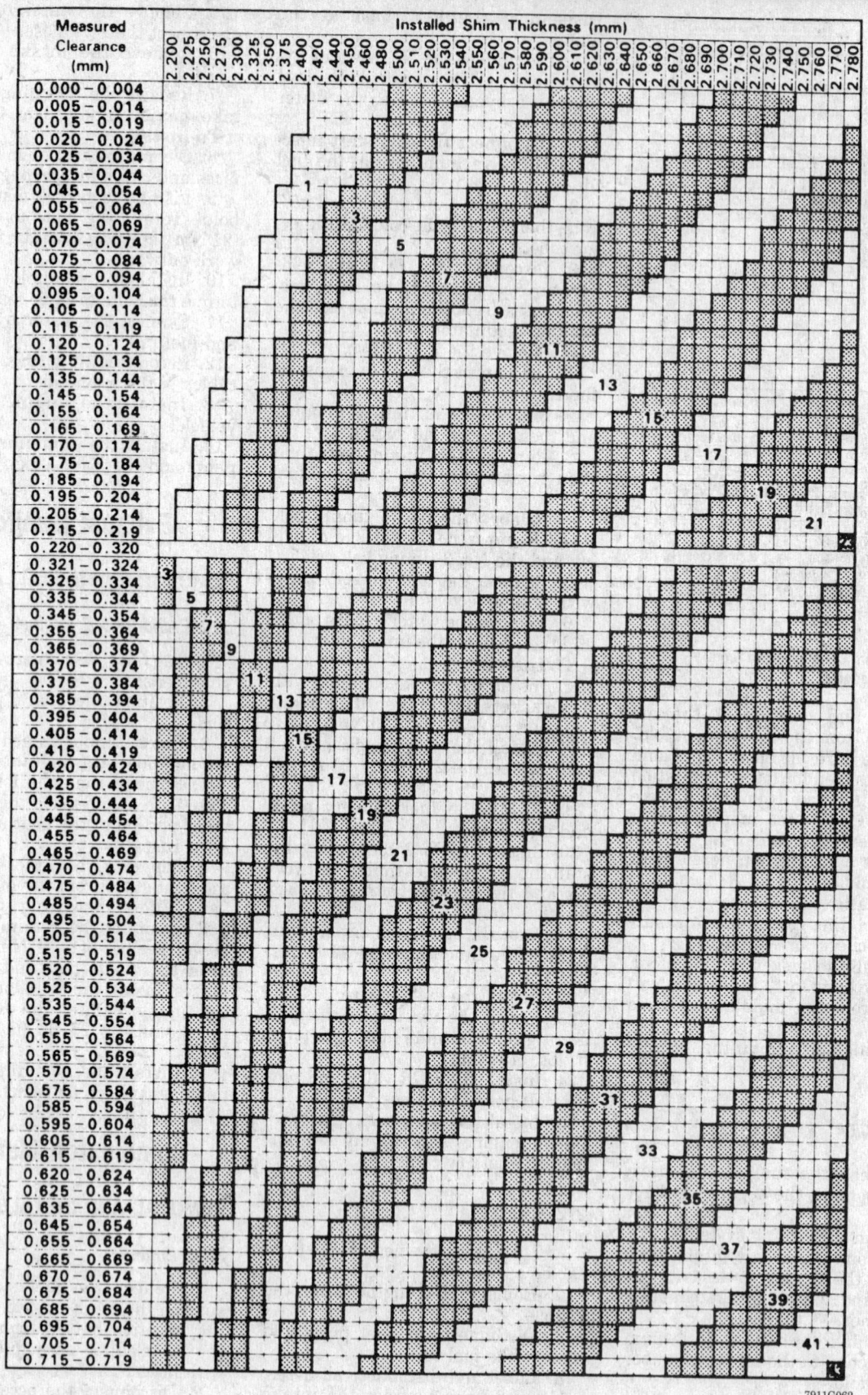

Exhaust shim selection chart — 3VZ — E engine

7911G062

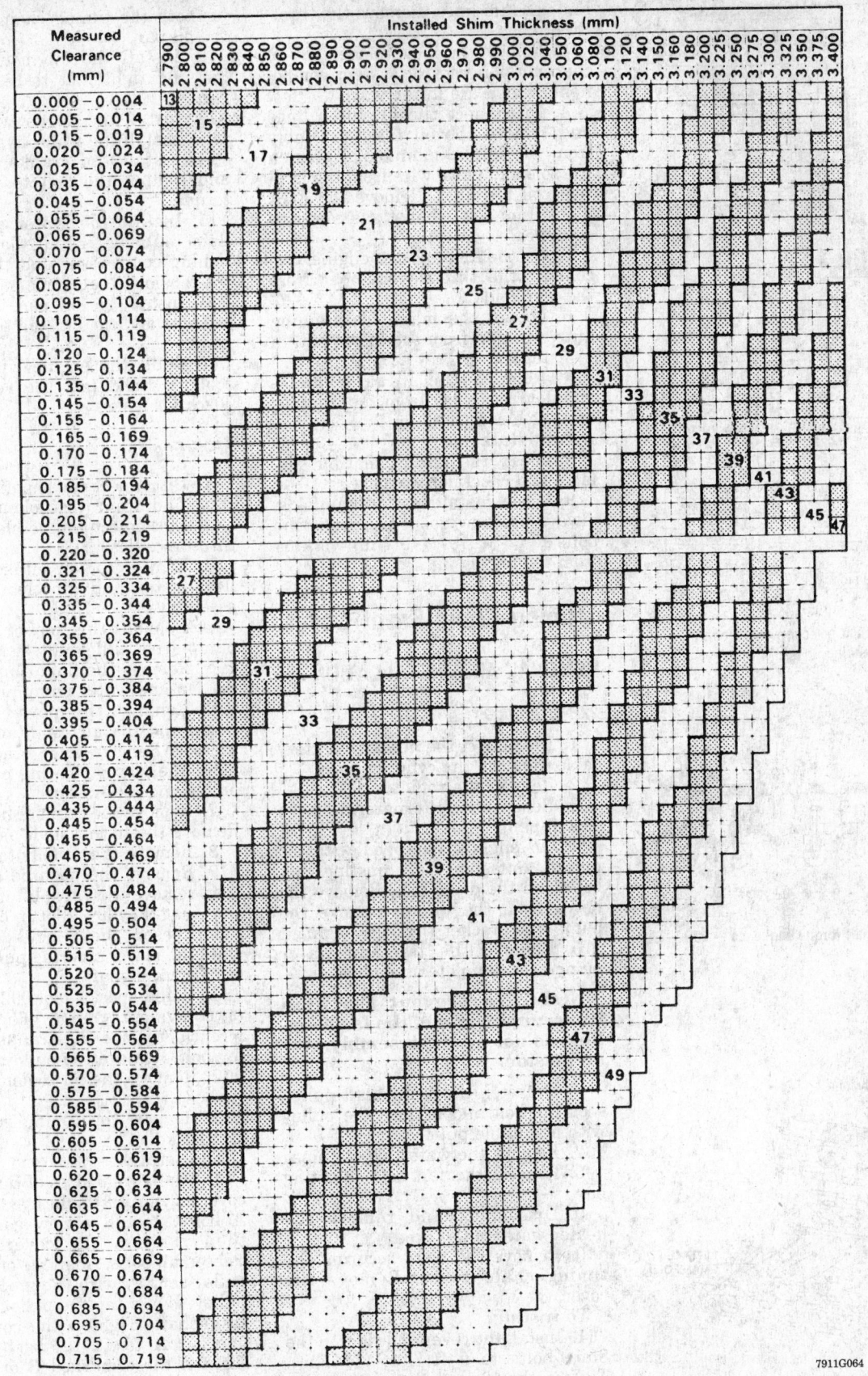

Exhaust shim selection chart (cont) — 3VZ — E engine

7911G064

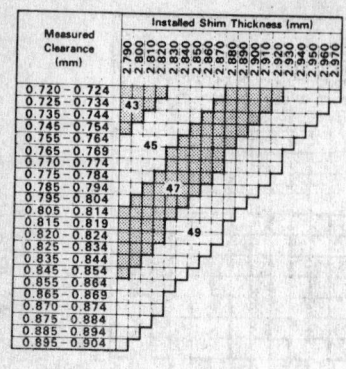

Measured Clearance (mm)	Installed Shim Thickness (mm)
	2.790 / 2.800 / 2.810 / 2.820 / 2.830 / 2.840 / 2.850 / 2.860 / 2.870 / 2.880 / 2.890 / 2.900 / 2.910 / 2.920 / 2.930 / 2.940 / 2.950 / 2.960 / 2.970
0.720 – 0.724	
0.725 – 0.734	43
0.735 – 0.744	
0.745 – 0.754	
0.755 – 0.764	45
0.765 – 0.769	
0.770 – 0.774	
0.775 – 0.784	
0.785 – 0.794	47
0.795 – 0.804	
0.805 – 0.814	
0.815 – 0.819	
0.820 – 0.824	49
0.825 – 0.834	
0.835 – 0.844	
0.845 – 0.854	
0.855 – 0.864	
0.865 – 0.869	
0.870 – 0.874	
0.875 – 0.884	
0.885 – 0.894	
0.895 – 0.904	

7911G065

Exhaust shim selection chart (cont) — 3VZ — E engine

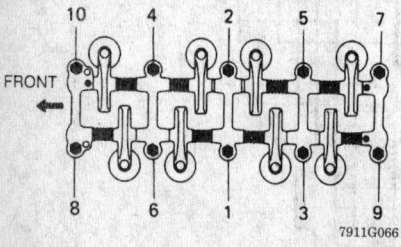

Shim Thickness

Shim No.	Thickness mm (in.)	Shim No.	Thickness mm (in.)
01	2.20 (0.0866)	27	2.85 (0.1122)
03	2.25 (0.0886)	29	2.90 (0.1142)
05	2.30 (0.0906)	31	2.95 (0.1161)
07	2.35 (0.0925)	33	3.00 (0.1181)
09	2.40 (0.0945)	35	3.05 (0.1201)
11	2.45 (0.0965)	37	3.10 (0.1220)
13	2.50 (0.0984)	39	3.15 (0.1240)
15	2.55 (0.1004)	41	3.20 (0.1260)
17	2.60 (0.1024)	43	3.25 (0.1280)
19	2.65 (0.1043)	45	3.30 (0.1299)
21	2.70 (0.1063)	47	3.35 (0.1319)
23	2.75 (0.1083)	49	3.40 (0.1339)
25	2.80 (0.1102)		

7911G066

Rocker arm shaft torque sequence — 22R — E engine

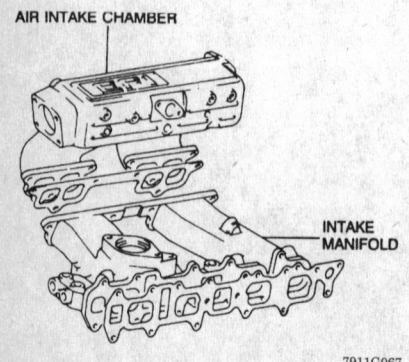

AIR INTAKE CHAMBER

INTAKE MANIFOLD

7911G067

Intake manifold assembly — 22R — E engine

and vacuum lines. Remove the throttle linkage.

3. Remove or move aside, any of the emission control system components which are in the way.

4. Disconnect the oil filter lines and remove the oil filter assembly from the intake manifold. Unfasten the solenoid valve wire from the ignition coil terminal. Remove the EGR pipes from the exhaust gas cooler, if equipped.

5. On the fuel injected engine, disconnect the throttle chamber from the manifold.

6. Loosen the manifold retaining nuts, working from the inside out, in 2–3 stages.

7. Remove the intake/exhaust manifold assembly from the cylinder head as a complete unit.

To install:

8. Clean the gasket mating surfaces and check for warpage.

9. When installing the manifolds, always use new gaskets. Torque the bolts to 36 ft. lbs. (49 Nm) working from the inside out.

Timing Chain Front Cover

REMOVAL AND INSTALLATION

22R–E Engine

1. Disconnect the negative battery cable. Remove the cylinder head.
2. Remove the radiator.
3. Remove the alternator. Remove the oil pan.
4. On engines equipped with air pumps, unfasten the adjusting link bolts and the drive belt. Remove the hoses from the pump; remove the pump and bracket from the engine.
5. Remove the fan and water pump as a complete assembly.

NOTE: To prevent the fluid from running out of the fan coupling, do not tip the assembly over on its side.

6. Unfasten the crankshaft pulley securing bolt and remove the pulley with a suitable puller.
7. Remove the water bypass pipe.
8. Remove the fan belt adjusting bar.
9. Disconnect and remove the heater water outlet pipe.
10. Remove the bolts securing the timing chain cover. Remove the cover.

To install:

11. Install the cover and tighten the 8mm bolts to 9 ft. lbs. (13 Nm). Tighten the 10mm bolts to 29 ft. lbs. (39 Nm). Apply sealer to the gaskets for both the timing chain cover and the oil pan.

12. Install the fan belt adjusting bar and tighten it to 9 ft. lbs. (13 Nm).
13. Install the heater water outlet pipe. Install the water bypass pipe.
14. Install the crankshaft pulley and tighten the bolt to the proper torque.
15. Install the water pump and fluid coupling. Install the air conditioning compressor and then adjust the tension on all drive belts.
16. Install the oil pan. Install the radiator and then install the cylinder head.
17. Refill the engine with oil and coolant. Road test the vehicle and check for leaks.

3F–E Engine

1. Disconnect the negative battery cable and drain the coolant.
2. Disconnect the accelerator and throttle cables.
3. Remove the air intake hose, air flow meter and air cleaner as an assembly.
4. Loosen the power steering pump drive pulley nut.
5. Remove the fluid coupling with the fan and water pump pulley.
6. Remove the power steering pump and the air conditioning compressor. Remove their brackets. Remove the power steering pump idler pulley and its bracket.
7. Remove the cylinder head cover. Remove the rocker shaft assembly.
8. Remove the distributor.
9. Remove the pushrod cover and then remove the valve lifters. Be certain that they are kept in order.
10. Loosen the 6 bolts and then slide the power steering pump pulley off the crankshaft.
11. Using special tool 09213–58011 or equivalent, and a 46mm socket wrench, remove the crankshaft pulley bolt. Remove the pulley.
12. Remove the oil cooler pipe and its hose.
13. Remove the timing gear cover and gasket.

To install:

14. There are 3 sizes of timing gear cover bolts. Apply adhesive to the two **A** bolts. Install a new gasket and then position the cover. Finger tighten all bolts. Align the crankshaft pulley set key with the groove of the pulley; gently tap the pulley onto the crankshaft. Tighten the cover bolts marked **A** to 18 ft. lbs. (25 Nm). Tighten those marked **B** or **C** to 43 inch lbs. (5 Nm). Tighten the pulley bolt to 253 ft. lbs. (343 Nm).

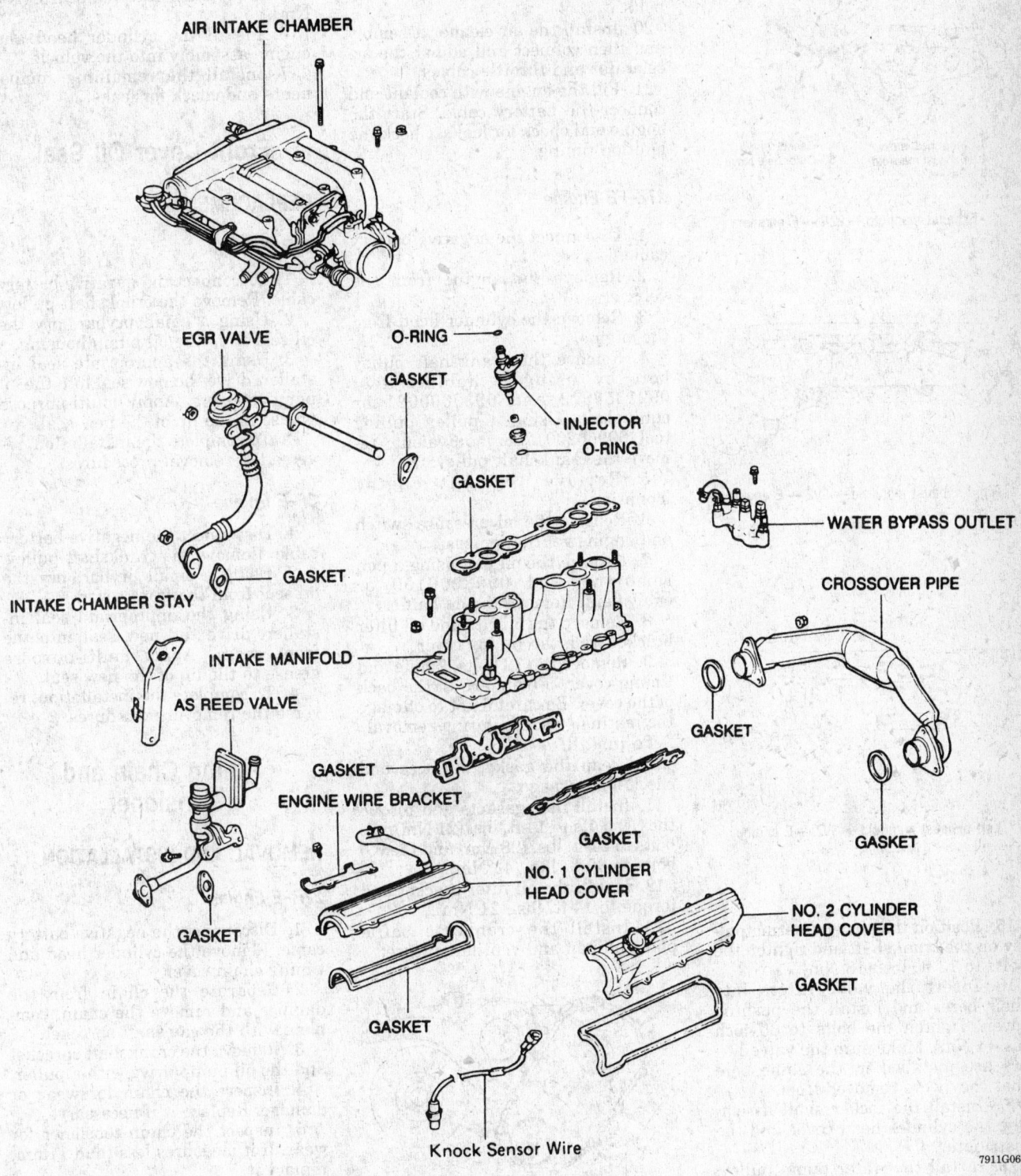

AIR INTAKE CHAMBER

EGR VALVE

O-RING

GASKET

INJECTOR

O-RING

GASKET

WATER BYPASS OUTLET

CROSSOVER PIPE

GASKET

INTAKE CHAMBER STAY

INTAKE MANIFOLD

AS REED VALVE

GASKET

GASKET

ENGINE WIRE BRACKET

GASKET

GASKET

NO. 1 CYLINDER HEAD COVER

NO. 2 CYLINDER HEAD COVER

GASKET

GASKET

GASKET

Knock Sensor Wire

7911G068

Intake manifold assembly — 3VZ — E engine

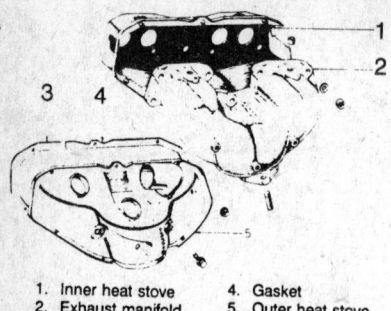

1. Inner heat stove
2. Exhaust manifold
3. Gasket
4. Gasket
5. Outer heat stove

7911G069

Exhaust manifold — 22R — E engine

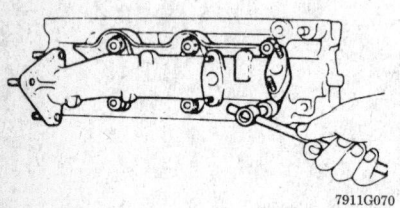

7911G070

Right exhaust manifold — 3VZ — E engine

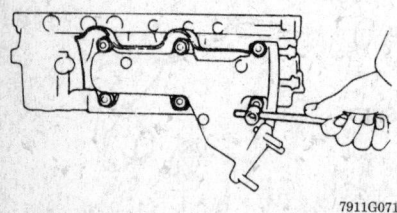

7911G071

Left exhaust manifold — 3VZ — E engine

15. Position the power steering pulley on the crankshaft and tighten the bolts to 13 ft. lbs. (18 Nm).

16. Insert the valve lifters into their bores and install the pushrod cover. Tighten the bolts to 35 inch lbs. (4 Nm). Make sure the valve lifters are installed in the same bore that they were removed from.

17. Install the rocker shaft assembly, the cylinder head cover and the distributor.

18. Install the water pump pulley, fluid coupling and fan.

19. Install the power steering pump idler pulley and bracket. Install the power steering pump and air conditioning compressor. Adjust the drive belts.

20. Install the air cleaner assembly and then connect and adjust the accelerator and throttle cables.

21. Fill the engine with coolant and connect the battery cable. Start the engine and check for leaks. Check the ignition timing.

2TZ–FE Engine

1. Disconnect the negative battery cable.

2. Remove the engine from the vehicle.

3. Remove the cylinder head from the engine.

4. Remove the crankshaft pulley bolt by using a holding tool 0921358012 and 0933000021 or equivalent. Using a pulley pulling tool 0995020017 or equivalent, remove the crankshaft pulley.

5. Remove the left engine mounting.

6. Remove the oil pressure switch and engine ventilation case.

7. Remove the oil pan using a pan removing tool 0903200100 or equivalent. Remove the oil baffle.

8. Remove the 3 bolts and oil filter bracket from the timing cover.

9. Remove the 12 bolts, 2 nuts and timing cover. 3 bolts are in the back of the cover. Be careful not to damage the mating surfaces during removal.

To install:

10. Clean the gasket surfaces and check for warpage.

11. Install new gaskets and torque the (A) bolts to 14 ft. lbs. (21 Nm), (B) bolts to 21 ft. lbs. (28 Nm) and the (C) bolts to 32 ft. lbs. (43 Nm).

12. Install the oil filter bracket and torque to 14 ft. lbs. (21 Nm).

13. Install the crankcase baffle plate, oil pan and ventilation case.

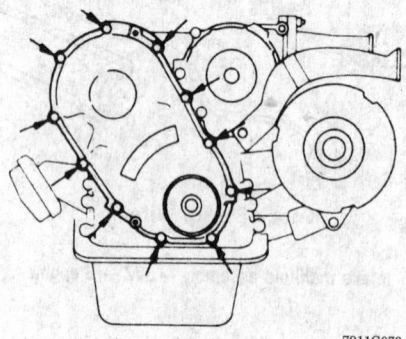

7911G073

Timing chain cover removal — 3F — E engine

14. Install the oil pressure switch and left engine mount.

15. Install the crankshaft pulley and torque the bolt to 192 ft. lbs. (260 Nm).

16. Install the cylinder head and engine assembly into the vehicle.

17. Install the remaining components and check for leaks.

Front Cover Oil Seal

REPLACEMENT

22R–E Engine

1. Disconnect the negative battery cable. Remove the crankshaft pulley.

2. Using a small prybar, pry the oil seal from the oil pump housing.

3. Using the appropriate seal installer, drive the new seal into the oil pump housing. Apply multi-purpose grease to the lip of the new seal.

4. To complete the installation, reverse the removal procedures.

3F–E Engine

1. Disconnect the negative battery cable. Remove the crankshaft pulley.

2. Using a small prybar, pry the oil seal from the front cover.

3. Using the appropriate seal installer, drive the new seal into the front cover. Apply multi-purpose grease to the lip of the new seal.

4. To complete the installation, reverse the removal procedures.

Timing Chain and Tensioner

REMOVAL AND INSTALLATION

22R–E Engine

1. Disconnect the negative battery cable. Remove the cylinder head and timing chain cover.

2. Separate the chain from the damper and remove the chain, complete with the camshaft sprocket.

3. Remove the crankshaft sprocket and the oil pump drive with a puller.

4. Inspect the chain for wear or damage. Replace it if necessary.

5. Inspect the chain tensioner for wear. If it measures less than 11mm, replace it.

6. Check the dampers for wear. If their measurements are below the following specifications, replace them. The specification for the upper damper is 5.0mm and the lower damper is 4.5mm.

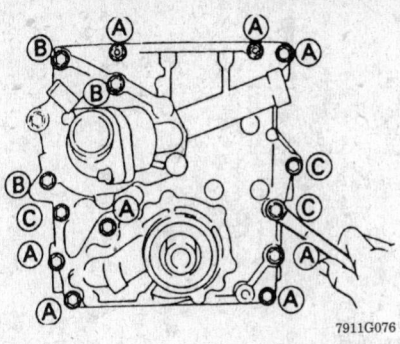

7911G076

Timing chain cover torque sequence — 2TZ — FE engine

To install:

7. Rotate the crankshaft until its key is at TDC. Slide the sprocket in place over the key.

8. Place the chain over the sprocket so its single bright link aligns with the mark on the camshaft sprocket.

9. Install the cam sprocket so the timing mark falls between the two bright links on the chain.

10. Fit the oil pump drive spline over the crankshaft key.

11. Install the timing cover gasket on the front of the block.

12. Rotate the camshaft sprocket counterclockwise to remove the slack from the chain.

13. Install the timing chain cover and cylinder head.

2TZ–FE Engine

1. Remove the timing chain cover. There are 3 bolts in back of the cover.

2. Remove the cylinder head assembly.

3. Remove the chain slipper, damper and oil nozzle.

4. Remove the oil pump drive chain and idle gear.

5. Remove the crankshaft sprocket using a puller 0921336020 or equivalent.

To install:

6. Install the crankshaft sprocket using an installer tool 0960806040 or equivalent.

7. Install the idle sprocket, chain and tensioner. Torque the bolts to 14 ft. lbs. (21 Nm).

8. Check the spring is operating normally against the chain guide by pressing on the chain with a finger and then release. With the guide against the chain, torque the tensioner bolt to 14 ft. lbs. (20 Nm).

9. Install the oil nozzle and torque to 13 ft. lbs. (18 Nm).

10. Install the timing chain damper and slipper. Torque the bolts to 20 ft. lbs. (26 Nm).

11. Place the timing chain on the camshaft sprocket so the timing mark is between the bright chain links and at 12 o'clock.

12. Place the timing chain on the crankshaft sprocket with the single bright link indicated aligned with the timing mark on the crankshaft sprocket at 6 o'clock.

13. Turn the camshaft sprocket counterclockwise to take the slack out of the chain.

14. Tie the timing chain with a cord.

15. Install the timing chain cover and cylinder head.

16. Install the remaining components and check for leaks.

Timing Gears

REMOVAL AND INSTALLATION

3F–E Engine

1. Disconnect the negative battery cable. Remove the cylinder head and the front cover from the engine.

2. Remove the oil slinger from the crankshaft. Remove the camshaft thrust plate retaining bolts, by working through the holes provided in the camshaft timing gear.

3. Remove the camshaft through the front of the cylinder block. Support the camshaft while removing it, so the bearings or the lobes do not become damaged.

NOTE: The timing gear is a press-fit and cannot be removed without removing the camshaft.

4. Inspect the crankshaft timing gear. Replace it if it has worn or damaged teeth.

5. Remove the sliding key, then, pull the crankshaft timing gear from the crankshaft with a gear puller.

6. Use a large piece of pipe to drive the timing gear onto the crankshaft. Lightly and evenly tap the end of the pipe until the gear is in its original position.

To install:

7. Apply a coat of engine oil to the camshaft journals and bearings, then, insert the camshaft into the block.

8. Align the mating marks on the timing gears. Slip the camshaft into position. Torque the camshaft thrust plate bolts to 14.5 ft. lbs. (20 Nm).

9. Using a feeler gauge, check gear backlash, inserted between the crankshaft and the camshaft timing gears. The maximum backlash should be 0.002–0.005 in.; if it ex-

ceeds this, replace one or both of the gears, as required.

10. Using a dial indicator, check the gear run-out. Maximum run-out, for both gears, is 0.008 in.; if not, replace the gear.

11. Install the oil nozzle, if removed, by screwing it in place with a screwdriver and punching it in 2 places, to secure it.

NOTE: Be sure the oil hole in the nozzle is pointed toward the timing gear before securing it.

12. To complete the installation, use new gaskets, sealant and reverse the removal procedures.

Timing Belt Front Cover

REMOVAL AND INSTALLATION

3VZ–E Engine

1. Disconnect the negative battery cable and drain the coolant.

2. Remove the radiator and shroud.

3. Remove the power steering belt and pump.

4. Remove the spark plugs.

5. Disconnect the No. 2 and 3 air hoses at the air pipe.

6. Disconnect the No. 1 water by-pass hose at the air pipe and then remove the water outlet.

7. Remove the air conditioning belt. Remove the alternator drive belt, fluid coupling, guide and fan pulley.

8. Disconnect the high tension cords and their clamps at the No. 2 (upper) timing belt cover and then remove the cover and its gaskets.

9. Rotate the crankshaft pulley until the groove on its lip is aligned with the **0** on the No. 1 (lower) timing belt cover, this should set the No. 1 cylinder at TDC of its compression stroke. The matchmarks on the camshaft timing pulleys must be in alignment with those on the No. 3 (upper rear) timing cover. If not, rotate the engine 360° (1 complete revolution).

10. Remove the crankshaft pulley using a puller.

11. Remove the fan pulley bracket and then remove the No. 1 timing belt cover.

To install:

12. Install the No. 1 cover with the 2 gaskets and tighten the bolts to 48 inch lbs. (5.4 Nm).

13. Install the fan pulley bracket and tighten it to 30 ft. lbs. (41 Nm).

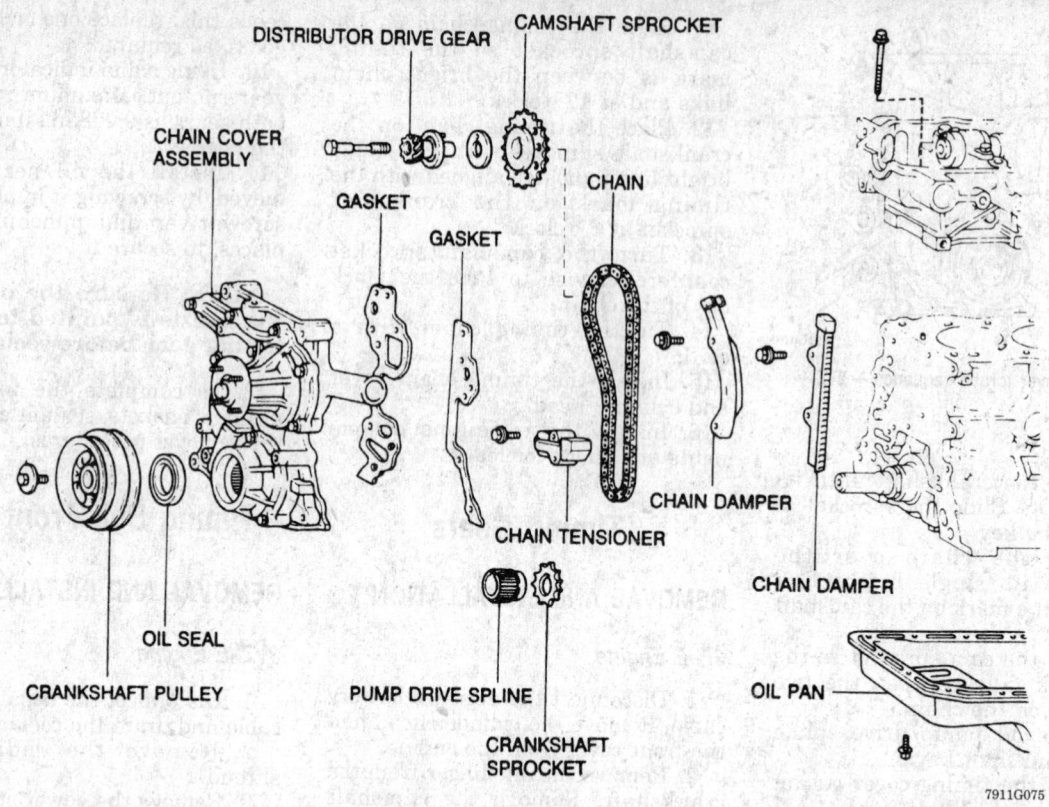

Timing chain and sprockets — 22R — E engine

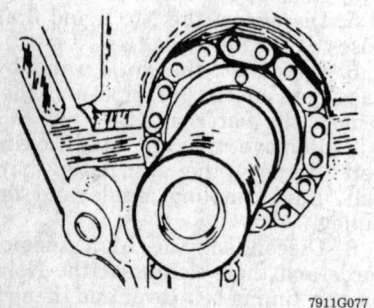

Aligning the crankshaft gear with the single bright link of the timing chain — 22R — E engine

Aligning the camshaft sprocket mark between the 2 bright links of the timing chain — 22R — E engine

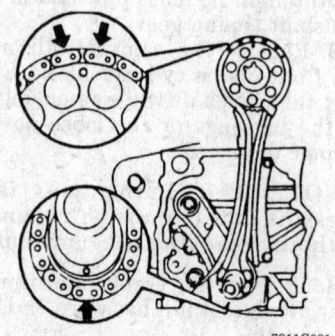

Timing chain alignment marks — 2TZ — FE engine

14. Install the No. 2 cover and tighten the bolts 48 inch lbs. (5.4 Nm).

15. Position the crankshaft pulley so the groove in the pulley is aligned with the Woodruff key in the crankshaft. Tighten the bolt to 181 ft. lbs. (245 Nm).

16. Install the fan pulley, guide, fluid coupling and alternator drive belt. Adjust the belt tension.

17. Install the power steering pump and belt. Install the air conditioning belt. Adjust the belt tension.

18. Install the water outlet and connect the bypass hose. Connect the No. 2 and 3 air hoses.

19. Install the spark plugs. Install the radiator, fill with coolant and road test the vehicle. Check for leaks and check the ignition timing.

Timing Belt and Tensioner

REMOVAL AND INSTALLATION

3VZ–E Engine

1. Disconnect the negative battery cable. Remove the timing belt covers.

2. Draw a directional arrow on the timing belt and matchmark the belt to each of the pulleys. Remove the timing belt guide and then remove the tension spring.

3. Loosen the idler pulley bolt and shift it left as far as it will go. Tighten the set bolt and relieve the tension on the timing belt. Remove the belt.

4. Remove the crankshaft and camshaft sprocket timing pulleys. Remove the No. 1 idler pulley.

To install:

5. Align the groove in the crankshaft pulley with the key on the crankshaft and press the pulley onto the shaft.

6. Install the idler pulley. Align the groove on the pulley with the cavity of the oil pump and then force it to

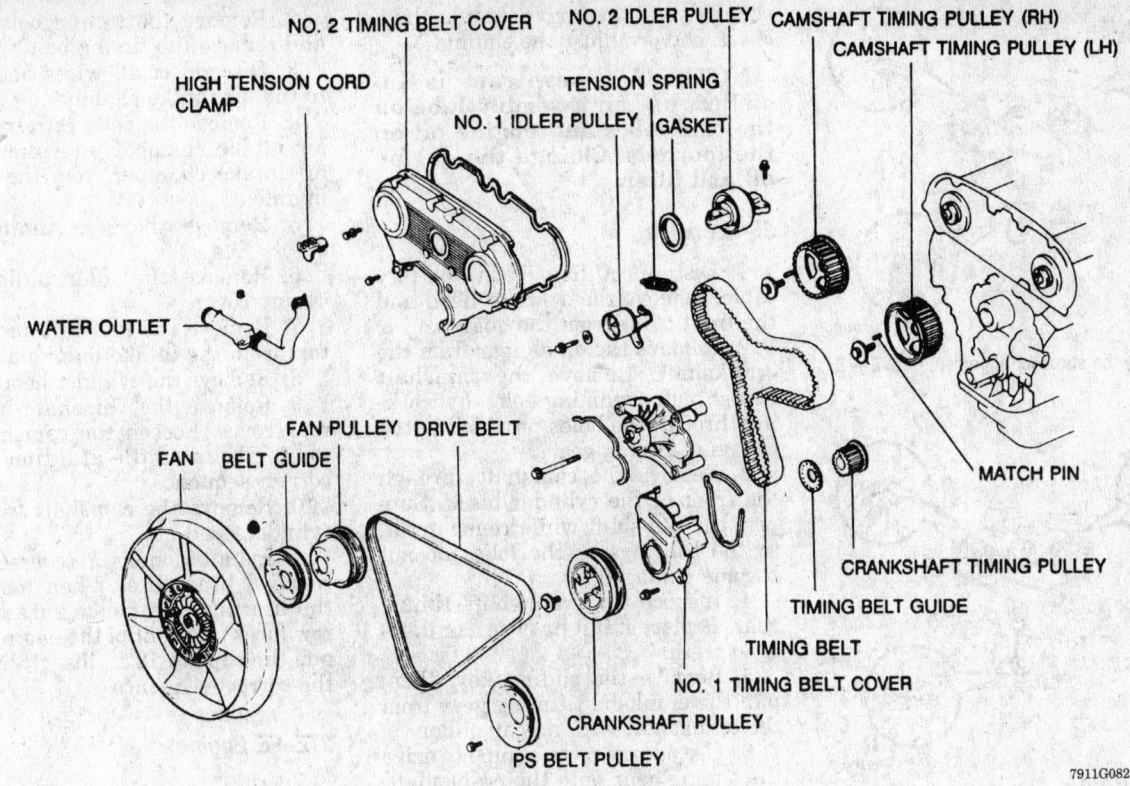

Timing belt components — 3VZ — E engine

the left as far as it will go. Temporarily tighten it to 27 ft. lbs. (37 Nm).

7. Position the camshaft pulleys on the camshafts so the match holes in each pulley are in alignment with those on the No. 3 (upper rear) timing cover. Align the pulley matchmark with the one on the cover.

NOTE: Do not install the match pin. Check that the bolt head is not touching the pulley.

8. Install the timing belt around the timing pulleys. If reusing the old belt, make sure the arrow and matchmarks all line up with those made earlier on the pulleys.

9. Move the idler pulley to the right as far as it will go. Install the tension spring and then loosen the pulley bolt until the pulley moves lightly with the tension spring force.

10. Check the valve timing and belt tension by turning the crankshaft 2 complete revolutions clockwise. Check that each pulley aligns with its timing marks. Retighten the idler pulley bolt to 27 ft. lbs. (37 Nm).

11. Remove the camshaft timing pulley bolts and align the match pin hole with the match pin hole in the camshaft. Install the pin and bolt and tighten to 80 ft. lbs. (108 Nm).

12. Remove the crankshaft timing pulley bolt and position the belt guide

over the crankshaft pulley so the cupped side is out.

13. Install the timing covers. Connect the negative battery cable.

Oil Seal Replacement

Remove the crankshaft pulley, timing belt cover and pry the seal from the retainer. Be careful not to damage the crankshaft. Install the seal with a seal installer 0930937010 or equivalent.

Aligning the timing marks with the rear cover — 3VZ — E engine

Camshaft

REMOVAL AND INSTALLATION

22R–E

1. Disconnect the negative battery cable. Remove the cylinder head cover.

2. Remove the rocker arm assembly from the cylinder head.

NOTE: It may be necessary to use a small prybar to lift the rocker arm assembly from the cylinder head.

3. Using a feeler gauge, measure the thrust bearing clearance at the front of the camshaft; the standard clearance is 0.003–0.007 in., it should not exceed 0.0098 in.

4. Remove the camshaft bearing caps and lift out the camshaft. Keep the bearings in order so they may be installed in their original position.

5. Check the camshaft journal caps for damage. Clean all of the bearing surfaces, including the caps, cam journal and the cylinder head.

To install:

6. With the camshaft in place on the cylinder head, lay small strips of plastigage® on each of the camshaft journals (at the tops of the journals, facing front-to-rear).

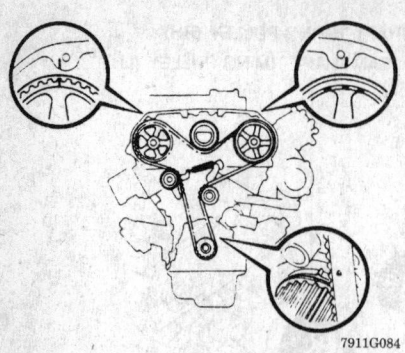

Rechecking the sprocket alignment — 3VZ — E engine

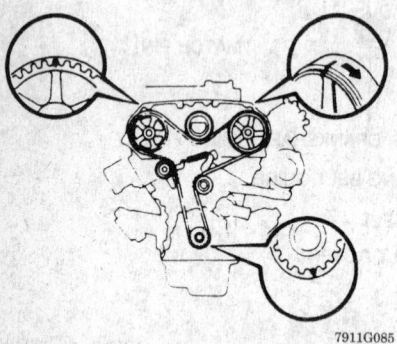

Installing the timing belt — 3VZ — E engine

7. Reinstall the journal caps in their original locations, arrows facing forward, and torque the caps to 13–16 ft. lbs. (18–22 Nm).

8. Remove the journal caps and gauge the width of the plastigage® against the chart on the plastigage® package. Maximum journal clearance is 0.004 in. If the journal clearance is greater than specified, measure the cam journal diameters with a micrometer. If the diameter of any cam journal is less than specified, obtain a new camshaft and recheck the journal clearance. If the clearance is still excessive, the cylinder head must be replaced.

9. To complete the installation, use new gaskets, sealant and reverse the removal procedures. Refill the cooling system. Torque the camshaft bearing cap bolts to 14 ft. lbs. (19 Nm), the cylinder head-to-engine block to 58 ft. lbs. (79 Nm), the timing chain cover-to-cylinder head bolt to 9 ft. lbs. (12 Nm), the camshaft sprocket-to-camshaft bolt to 58 ft. lbs. (79 Nm), the intake manifold bolts to 14 ft. lbs. (19 Nm), the exhaust manifold bolts to 33 ft. lbs. (45 Nm) and the rocker arm cover to 7–12 ft. lbs. (10–15 Nm). Replace the cooling system fluid and the engine oil. Adjust

the valves, the drive belts, then, check and/or adjust the timing.

NOTE: If a new cam is installed, use an assembly lube on the cam lobes and engine oil on the journals. Change the engine oil and filter.

3F–E Engine

1. Disconnect the negative battery cable. Remove the cylinder head and the front cover from the engine.

2. Remove the oil slinger from the crankshaft. Remove the camshaft thrust plate retaining bolts, by working through the holes provided in the camshaft timing gear.

3. Remove the camshaft through the front of the cylinder block. Support the camshaft while removing it, so the bearings or the lobes do not become damaged.

4. Inspect the crankshaft timing gear. Replace it if it has worn or damaged teeth.

5. Remove the sliding key, then, pull the crankshaft timing gear from the crankshaft with a gear puller.

6. Use a large piece of pipe to drive the timing gear onto the crankshaft. Lightly and evenly tap the end of the pipe until the gear is in its original position.

To install:

7. Apply a coat of engine oil to the camshaft journals and bearings, then, insert the camshaft into the block.

8. Align the mating marks on the timing gears. Slip the camshaft into position. Torque the camshaft thrust plate bolts to 14.5 ft. lbs. (20 Nm).

9. Using a feeler gauge, check the gear backlash, inserted between the crankshaft and the camshaft timing gears. The maximum backlash should be 0.002–0.005 in.; if it exceeds this, replace one or both of the gears, as required.

10. Using a dial indicator, check the gear run-out. Maximum run-out, for both gears, is 0.008 in.; if not, replace the gear.

11. Install the oil nozzle, if removed, by screwing it in place and punching it in 2 places, to secure it.

NOTE: Be sure the oil hole in the nozzle is pointed toward the timing gear before securing it.

12. To complete the installation, use new gaskets, sealant and reverse the removal procedures.

3VZ–E Engine

1. Disconnect the negative battery cable.

2. Remove the timing belt covers and remove the timing belt.

3. Disconnect all wires and hoses to the air intake chamber.

4. Remove the bolts retraining the air intake chamber and remove the air intake chamber from the intake manifold.

5. Remove the rear timing belt cover.

6. Remove the idler pulley and timing cover.

7. Remove the fuel rail and injectors from the intake manifold.

8. Remove the cylinder head cover.

9. Remove the camshaft housing rear cover. Loosen the camshaft retaining bolts a little at a time in the correct sequence.

10. Remove the camshaft from the cylinder head.

11. Installation is the reverse of the removal procedure. When installing the bearing caps, make sure the arrow faces the front of the engine. Torque the caps to 12 ft. lbs. (16 Nm) in the correct sequence.

2TZ–FE Engine

1. Disconnect the negative battery cable.

2. Remove the engine from the vehicle.

3. Remove the valve cover.

4. Place the matchmarks on the camshaft sprocket and chain facing 12 o'clock with the engine upright.

5. Hold the camshaft with a wrench and remove the sprocket bolt.

6. Remove the chain tensioner, timing chain and sprocket. Hold to the side with a wire to maintain camshaft timing.

7. Remove the No. 6 camshaft bearing cap.

8. To remove the exhaust camshaft:

   a. Set the knock pin hole of the camshaft at 5–30° BTDC.

   b. Secure the camshaft sub-gear to the main gear with a service bolt.

   c. Uniformly loosen and remove the bearing caps No. 1, No. 2, No. 3 and No. 5., in that order. Leave No. 4 tight.

   d. Loosen and remove the No. 4 bearing cap.

9. To remove the intake camshaft:

   a. Set the knock pin hole of the camshaft at 75–100° BTDC.

   b. Uniformly loosen and remove the bearing caps No. 1, No. 2, No. 4 and No. 5., in that order. Leave No. 3 tight.

   c. Loosen and remove the No. 3 bearing cap.

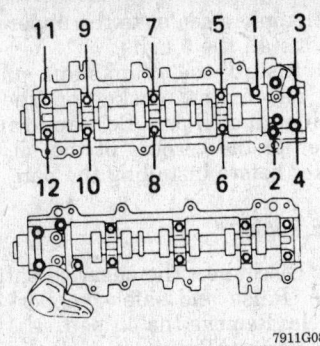

7911G086

Removing the camshaft bearing bolts —
3VZ — E engine

To install:

10. To install the intake camshaft:

a. Set the knock pin hole of the camshaft at 75–100° BTDC.

b. Apply prelube to the camshaft journals and lobes.

c. Tighten caps No. 1 and No. 3 to draw the camshaft to the cylinder head.

d. Uniformly torque the bearing caps in several steps to ensure the camshaft does not bend.

e. Torque the bolts to 12 ft. lbs. (16 Nm).

11. To install the exhaust camshaft:

a. Set the knock pin hole of the camshaft at 5–30° BTDC.

b. Apply prelube to the camshaft journals and lobes.

c. Tighten caps No. 2 and No. 4 to draw the camshaft to the cylinder head.

d. Uniformly torque the bearing caps in several steps to ensure the camshaft does not bend.

e. Torque the bolts to 12 ft. lbs. (16 Nm).

12. Apply sealer to the bottom of No. 6 bearing cap and install. Torque the cap to 12 ft. lbs. (16 Nm).

13. Install the camshaft sprocket and chain. Torque the bolt to 54 ft. lbs. (74 Nm).

14. Release the chain tensioner ratchet pawl. Fully push in the plunger and apply the hook to the pin so the plunger can not spring out and install the tensioner. Torque the bolts to 15 ft. lbs. (21 Nm).

15. Set the tensioner by pulling back on the slipper to release the hook.

16. Check the valve clearances.

17. Install the remaining components and check for leaks.

Piston and Connecting Rod

Positioning

ENGINE LUBRICATION

Oil Pan

REMOVAL AND INSTALLATION

Pick-Up and 4Runner

1. Raise the hood and disconnect the negative battery cable.

2. Raise and safely support the vehicle.

3. Drain the engine oil.

4. Remove the steering relay rod and the tie rods from the idler arm, pitman arm, and steering knuckles.

5. Remove the engine stiffening plates.

6. Remove the splash pans from under the engine.

7. Position a floor jack under the transmission and raise the engine/transmission assembly slightly.

8. Remove the front motor mount attaching bolts.

9. Remove the oil pan bolts and remove the oil pan.

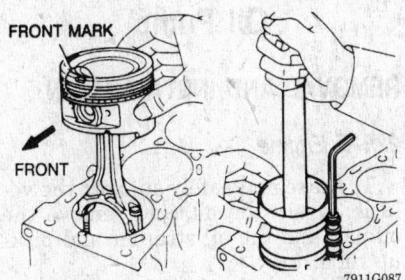

7911G087

Piston and connecting rod positioning — 22R — E and 3F — E engines

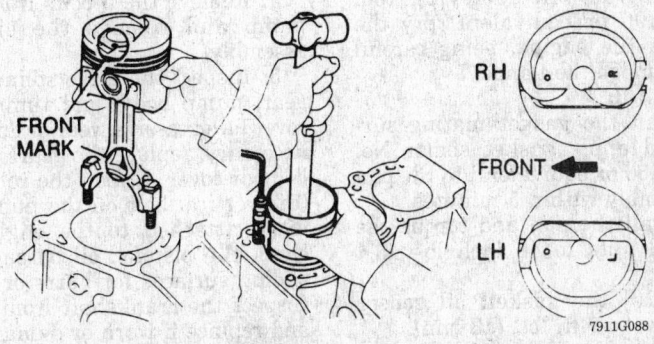

7911G088

Piston and connecting rod positioning — 3VZ — E engine; 2TZ — FE similar

10. Scrape the cylinder block and oil pan mating surfaces clean of any old sealing material. Apply gasket sealer to the oil pan when installing a new gasket. If equipped with the 22R–E engine, apply 5mm bead of gasket sealer; if equipped with the 3F–E or 3VZ–E engine, use 3mm bead. The parts should be installed within 5 minutes of applying the sealer.

11. The oil pan bolts should be tightened to 9 ft. lbs. (12 Nm) on the 22R–E engines; 6 ft. lbs. (10 Nm) (bolt), 52 inch lbs. (70 Nm) for the 3F–E or 3V–ZE engines. Tighten the bolts in a circular pattern, starting in the middle of the pan and working out towards the ends.

12. Lower the engine and tighten the motor mount bolts. Install the splash shields and stiffening plates.

13. Install any steering arms removed in Step 4 and then lower the vehicle; tighten all suspension components and the motor mounts to their final torque with the vehicle resting on the ground.

14. Fill the engine with oil, road test the vehicle and check for leaks.

Land Cruiser

1. Raise and safely support the vehicle. Remove the engine skid plates.

2. Remove the flywheel side cover and skid plate.

3. Disconnect the front driveshaft from the engine.

4. Drain the engine oil.

5. Remove the bolts which secure the oil pan. Remove the pan and its gasket.

6. Scrape away any old gasket material and then apply gasket sealer to the cylinder block mating surface and the No. 1 and No. 4 main bearing caps.

7. Install the oil pan and tighten the bolts to 69 inch lbs. (7.8 Nm). Always use a new pan gasket.

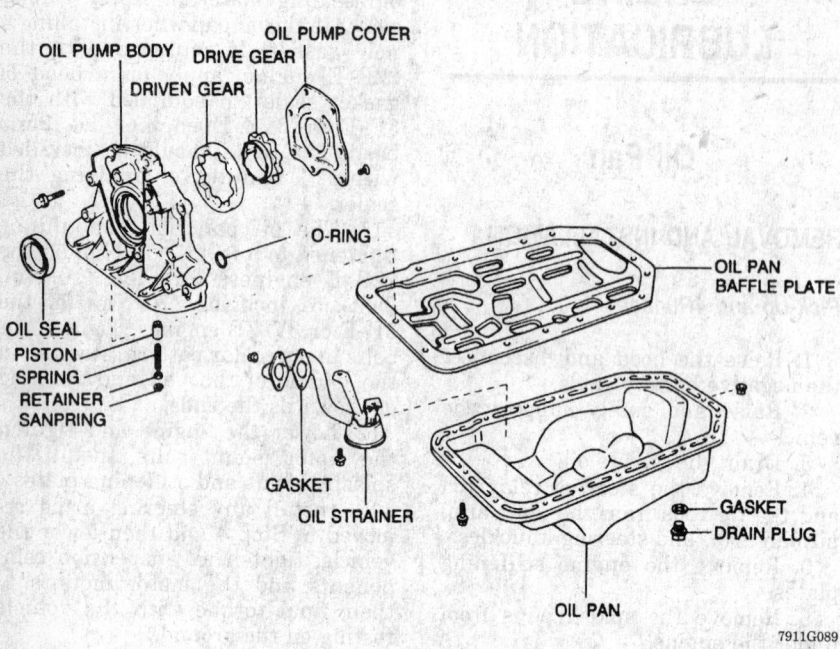

OIL PUMP BODY DRIVE GEAR OIL PUMP COVER
DRIVEN GEAR
O-RING
OIL SEAL
PISTON
SPRING
RETAINER
SANPRING
GASKET
OIL STRAINER
OIL PAN BAFFLE PLATE
GASKET
DRAIN PLUG
OIL PAN

7911G089

Oil pan and pump assembly — 3VZ — E engine

8. Connect the driveshaft, skid plate and flywheel side cover.

9. Lower the vehicle, fill the engine with oil and check for any leaks.

Previa

This engine has 2 oil pans. If the crankshaft is going to be serviced, the side crankcase pan has to be removed. If the oil pump sump is going to be serviced, the bottom oil pan has to be removed.

1. Drain the engine oil and disconnect the battery cable.

2. Remove the oil level sensor and gasket. Be careful not to drop the sensor when removing.

3. Remove the 14 bolts and 2 nuts.

4. Using a pan removing tool 0903200100 or equivalent, pry the pan from the engine, being careful not to damage the flange.

To install:

5. Clean the gasket mating surfaces and apply gasket sealer No. 0882600080 or equivalent, to the pan and assembly within 5 minutes.

6. Install the pan and torque the bolts and nuts to 48 inch lbs. (5.4 Nm).

7. Install the gasket, oil sensor and torque to 9 ft. lbs. (13 Nm).

8. Install the remaining components.

9. Refill the engine with oil and check for leaks.

Oil Pump

REMOVAL AND INSTALLATION

22R–E Engine

1. Raise and safely support the vehicle. Drain the oil, and remove the oil pan and the oil strainer and pick-up tube.

2. Remove the drive belts from the crankshaft pulley.

3. Remove the crankshaft bolt, and remove the pulley with a gear puller.

4. Remove the 5 bolts from the oil pump and remove the oil pump assembly.

5. Inspect the drive spline, driven gear, pump body, and timing chain cover for excessive wear or damage. If necessary, replace the gears or pump body or cover. Unbolt the relief valve (the vertical bolt on the pump body when attached to the engine) and check the pistons, oil passages, and sliding surfaces for burrs or scoring. Inspect the crankshaft front oil seal and replace if worn or damaged.

6. When installing, use a new O-ring if necessary.

7. Apply a sealer to the upper bolt and install the 5 bolts.

8. Install the crankshaft pulley and use a new gasket on the oil strainer and oil pan. Be sure to apply sealer to the corners of the oil pan gasket before installing the pan.

3F–E Engine

1. Disconnect the negative battery cable. Raise and safely support the vehicle. Remove the oil pan.

2. Remove the oil strainer and unfasten the union nuts on the oil pump pipe.

3. Remove the lock wire and the oil pump retaining bolt and pipe from the engine.

4. Remove the oil pump cover and inspect the following parts for nicks, scoring, grooving, etc.:
   a. pump cover
   b. drive and driven gears
   c. pump body

5. Replace either the damaged parts or the complete pump if damage is excessive.

To install:

6. Install the oil pump so the slot in the oil pump shaft is in alignment with the protrusion on the governor shaft of the distributor. Tighten the mounting bolts to 13 ft. lbs. (18 Nm).

7. Install the outlet pipe and tighten the union bolt to 33 ft. lbs. (44 Nm); use new gaskets.

8. Install the oil pan, fill the engine with oil and check for leaks.

NOTE: Be sure to check all of the gaskets and replace if necessary.

3VZ–E Engine

1. Disconnect the negative battery cable.

2. Remove the timing belt.

3. Raise and safely support the vehicle. Remove the engine under cover.

4. Remove the front differential.

5. Drain the oil.

6. Remove the crankshaft timing pulley.

7. Raise the engine slightly and remove the oil pan.

8. Remove the oil strainer. Insert a drift between the cylinder block and the oil pan baffle plate, cut off the sealer and remove the baffle plate.

NOTE: When removing the baffle plate with the drift, do not damage the baffle plate flange.

9. Remove the oil pump and O-ring.

To install:

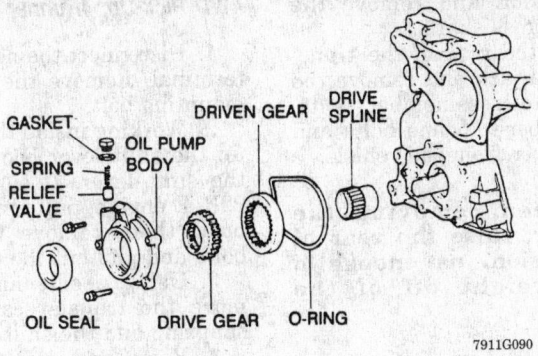

Oil pump assembly — 22R — E engine

Apply sealer to the oil pump mating surface running the bead on the inside of the bolts holes. Position a new O-ring in the groove in the cylinder block and install the pump so the spline teeth of the drive gear engage the large teeth on the crankshaft. Tighten the mounting bolts to 14 ft. lbs. (20 Nm).

10. Remove any old sealer and install the baffle plate with new sealer.

11. Install the oil strainer and tighten the bolts to 61 inch lbs. (7 Nm).

12. Install the oil pan, crankshaft pulley and the timing belt.

13. Install the front differential and the undercovers.

14. Fill the engine with oil and check for leaks.

2TZ–FE Engine

1. Remove the oil level sensor. Be careful not to drop the sensor when removing.

2. Remove the pan retaining bolts and pry the pan from the engine using a prying tool 0903200100 or equivalent. Be careful not to damage the flange.

To install:

3. Clean the gasket mating surfaces with a scraper and solvent.

4. Apply a bead of sealer 0882600080 or equivalent to the pan surface and install within 5 minutes.

5. Torque the pan bolts and nuts to 48 inch lbs. (5.4 Nm).

6. Install the oil sensor and torque the bolts to 9 ft. lbs. (13 Nm).

7. Install the remaining components and refill the with engine oil.

Rear Main Bearing Oil Seal

REMOVAL AND INSTALLATION

22R–E Engine

1. Raise and safely support the vehicle. Remove the transmission and the clutch assembly, if equipped. Remove the transfer case, if equipped.

2. Remove the flywheel or the flexplate from the crankshaft. Remove the cover plate from the rear of the engine.

3. Remove oil pan-to-oil seal retaining plate bolts, the oil seal retaining plate-to-engine bolts and oil seal retaining plate.

4. Carefully pry or drive the old seal from the retaining plate. Be careful not to damage the retaining plate.

5. Using an oil seal driver tool, drive the new seal into the oil seal retaining plate, until the surface is flush.

6. Lubricate the lips of the seal with multi-purpose grease.

7. Clean the gasket mounting surfaces.

8. To install, use new gaskets and reverse the removal procedures. Adjust the clutch.

3F–E Engine

1. Raise and safely support the vehicle. Remove the transfer case, the transmission and the clutch assembly.

2. Remove the flywheel from the crankshaft.

3. Using a small prybar, carefully pry the oil seal from the rear of the crankshaft.

4. Lubricate the lips of the seal with multipurpose grease.

5. Using an oil seal driver tool, drive the new seal into the rear of the crankshaft.

6. Clean the gasket mounting surfaces.

7. To install, reverse the removal procedures. Adjust the clutch.

3VZ–E Engine

1. Raise and safely support the vehicle. Remove the transmission and the clutch assembly, if equipped. Remove the transfer case, if equipped.

2. Remove the flywheel or the flexplate from the crankshaft. Remove the cover plate from the rear of the engine.

3. To replace the oil seal with the retaining plate removed:

a. Remove oil pan-to-oil seal retaining plate bolts, the oil seal retaining plate-to-engine bolts and oil seal retaining plate.

b. Carefully pry or drive the old seal from the retaining plate. Be careful not to damage the retaining plate.

c. Using an oil seal driver tool, drive the new seal into the oil seal retaining plate, until the surface is flush.

d. Lubricate the lips of the seal with multipurpose grease.

4. To replace the oil seal with the retaining plate installed:

a. Cut off the oil seal lip.

b. Using a small prybar, pry the oil seal from the retaining plate.

c. Apply multi-purpose grease to the new oil seal.

d. Using an oil seal driver tool, drive the new seal into the oil seal retaining plate until the surface is flush.

5. Clean the gasket mounting surfaces.

6. To complete the installation, reverse the removal procedures. Adjust the clutch.

2TZ–FE Engine

1. Remove the transmission from the vehicle.

2. Separate the transmission from the engine.

3. Remove the flywheel or flexplate.

4. Using a knife, cut off the lip of the oil seal.

5. Using a suitable prybar, pry the seal out of the seal carrier. Be careful not to damage the crankshaft.

To install:

6. Clean the seal mating surfaces with solvent.

7. Apply grease to the seal lip.

8. Using an installer tool 0922356010 or equivalent, install the seal. Make sure the seal is seated properly. If not, the seal will leak and the engine will have to be removed again.

9. Install the engine/transmission assembly.

10. Install the remaining components and check for leaks.

MANUAL TRANSMISSION

Transmission Assembly

REMOVAL AND INSTALLATION

2WD Pick-Up

Disconnect the negative battery terminal.

1. Perform the following:

a. Remove the center floor console, if equipped.

b. Remove the shift lever handle, then the floor mat or carpet along with the shift lever boot in order to gain access to the shift lever.

c. Using an shift lever removal tool, remove the shift lever.

NOTE: On the Pick-Up, remove the boot and the shift lever from inside the vehicle.

2. Raise and safely support the vehicle. Drain the transmission fluid.

3. Make matchmarks on the driveshaft flange and the differential pinion flange to indicate their relationships; these marks must be aligned during installation.

4. Remove the driveshaft flange bolts and the center support bearing-to-frame bolts, if equipped with a 2-piece driveshaft. Lower the driveshaft out of the vehicle. Using an appropriate tool, insert it into the end of the transmission to prevent oil leakage.

5. Disconnect the back-up lamp switch electrical connector and the speedometer cable from the transmission, then tie the cable aside.

6. Disconnect the wiring at the starter. Remove the starter mounting bolts and lower the starter out of the vehicle.

7. Remove the exhaust pipe clamp and the exhaust pipe.

8. If the hydraulic line from the clutch release cylinder is clamped to the frame, remove the clamp retaining bolt. Remove the release cylinder mounting bolts and the fork spring, if equipped. Tie the release cylinder aside.

NOTE: It is not necessary to disconnect the hydraulic line from the release cylinder.

9. On column shift vehicles, disconnect the shift selector linkage at

the transmission and remove the transmission cross shafts.

10. Support the rear of the transmission with a jack and remove the transmission-to-crossmember bolts, the crossmember-to-frame bolts and the crossmember from the vehicle.

NOTE: When removing the crossmember, raise the rear of the transmission, just enough to take the weight off of the crossmember.

11. Place a support under the engine with a wooden block (¾ in. thick) between the support and the engine oil pan.

NOTE: The wooden block and support should be no more than about ¼ in. away from the engine so when the engine is lowered, damage will not occur to any underhood components. If possible, shim the support so the wooden block touches the engine.

12. Remove the transmission-to-engine bolts, draw the transmission rearward and down, away from the engine.
To install:
13. Raise the transmission into position under the vehicle.
14. Install the transmission-to-engine bolts.
15. Torque transmission-to-engine bolts to 53 ft. lbs. (70 Nm), the stiffener plate bolts to 27 ft. lbs. (38 Nm), the transmission mount/bracket bolts to 19 ft. lbs. (28 Nm), the rear engine mount bracket-to-crossmember bolts to 9 ft. lbs. (12 Nm).
16. Connect the exhaust pipes and brackets. Install the starter and the clutch release cylinder.
17. Tighten the exhaust pipe-to-manifold bolts to 29 ft. lbs. (40 Nm), the upper exhaust pipe bracket-to-clutch housing bolts to 27 ft. lbs. (38 Nm), the lower exhaust pipe bracket-to-clutch housing bolts to 51 ft. lbs. (68 Nm), the lower starter bolt/release cylinder tube bracket bolt to 29 ft. lbs. (41 Nm), the clutch release cylinder bolts to 9 ft. lbs. (12 Nm).
18. Connect the remaining linkages and connect the driveshaft, aligning the matchmarks made during removal.
19. Refill the transmission to the correct level.
20. Install the shift lever and the center console.
21. Connect the negative battery cable.

4WD Pick-Up, 4Runner

1. Disconnect the negative battery terminal. Remove the starter upper mounting bolt.

2. Working inside the vehicle, pull up the shift lever boot and pull out the shift lever. If equipped with a 22R–E engine, pull up the shift lever boot, then, remove the mounting bolts and pull out the shift lever.

3. Using needle nose pliers, remove the transfer case shift lever snapring and the shift lever.

4. Raise and safely support the vehicle.

5. Drain the lubricant from both the transmission and the transfer case.

6. Make matchmarks on the driveshaft flanges and the differential pinion flanges to indicate their relationships. These marks must be aligned during installation.

7. Remove the driveshaft mounting bolts and remove the front driveshaft assembly.

NOTE: Do not disassemble the front driveshaft to remove it.

8. Using a piece of chalk, place matchmarks on the rear driveshaft and the slip yoke to indicate their relationships; these marks must be aligned during installation.

9. Remove the mounting bolts from the rearward flange of the rear driveshaft. Lower the driveshaft out of the vehicle. Remove the mounting bolts from the slip yoke flange, then, remove the flange and yoke assembly.

10. Unbolt the clutch release cylinder and tie it aside.

NOTE: It is not necessary to disconnect the hydraulic line from the clutch release cylinder.

11. Disconnect the starter motor electrical connectors. Remove the starter bolts and lower the starter from the vehicle.

12. At the transfer case, disconnect the speedometer cable (tie it aside), the back-up light switch connector and the 4WD indicator switch connector.

13. Disconnect the exhaust pipe clamp and the exhaust pipe from the transmission housing.

14. Remove the clutch release cylinder and the tube bracket, then, move the cylinder aside.

NOTE: When removing the clutch release cylinder, do not disassemble the hydraulic line from the cylinder.

15. Remove the crossmember-to-transfer case mounting bolts. Using a jack, raise the transmission and transfer case assembly off of the crossmember. Remove the cross-member-to-frame attaching bolts and remove the crossmember.

16. Place a support under the engine oil pan, with a wooden block (3/4 in. thick) between the support and the engine oil pan.

NOTE: The wooden block and support should be no more than about 1/4 in. away from the engine so when the engine is lowered, damage will not occur to any underhood components. If possible, shim the support so the wooden block touches the engine.

17. Lower the jack until the engine rests on the support.

18. Remove the exhaust pipe bracket and the stiffener plate bolts.

19. Remove the transmission-to-engine bolts, draw the transmission/transfer case assembly rearward and down away from the engine.

20. Remove the transmission-to-transfer case adapter bolts and pull the transfer case from the transmission.

To install:

21. Raise the transmission into position under the vehicle.

22. Install the transmission-to-engine bolts.

23. Torque transmission-to-engine bolts to 53 ft. lbs. (73 Nm), the stiffener plate bolts to 27 ft. lbs. (38 Nm), the transmission mount/bracket bolts to 19 ft. lbs. (27 Nm), the rear engine mount bracket-to-crossmember bolts to 9 ft. lbs. (12 Nm).

24. Connect the exhaust pipes and brackets. Install the starter and the clutch release cylinder.

25. Tighten the exhaust pipe-to-manifold bolts to 29 ft. lbs. (40 Nm), the upper exhaust pipe bracket-to-clutch housing bolts to 27 ft. lbs. (38 Nm), the lower exhaust pipe bracket-to-clutch housing bolts to 51 ft. lbs. (69 Nm), the lower starter bolt/release cylinder tube bracket bolt to 29 ft. lbs. (40 Nm), the clutch release cylinder bolts to 9 ft. lbs. (12 Nm).

26. Connect the remaining linkages and connect the driveshaft, aligning the matchmarks made during removal.

27. Refill the transmission to the correct level.

28. Install the shift lever and the center console.

29. Connect the negative battery cable.

Previa

1. Disconnect the negative battery cable and drain the transmission fluid.

2. Raise the vehicle and support safely.

3. Remove the starter motor. Matchmark the driveshafts-to-flange and remove the front (4WD) and rear driveshafts.

4. Remove the clutch release cylinder, hose and bracket.

5. Remove the exhaust pipe and bracket.

6. Disconnect the control cables/bracket and speed sensor connector.

7. Remove the engine-to-transmission stiffener plate.

8. Place a suitable transmission jack under the transmission.

9. Remove the engine rear mounting bolts and raise the rear side of the engine.

10. Remove the engine-to-transmission bolts, pull the transmission toward the rear and remove.

To install:

11. Align the input shaft with the clutch disc and push the transmission fully into position.

12. Install the transmission bolts and torque to 53 ft. lbs. (72 Nm).

13. Install the rear engine mounts and stiffener plate. Torque the bolt to 27 ft. lbs. (37 Nm).

14. Connect the speed sensor and control cables.

15. Install the exhaust pipe and torque the bracket to 37 ft. lbs. (51 Nm).

16. Install the clutch release cylinder, starter and driveshafts. Torque the starter to 41 ft. lbs. (56 Nm) and driveshaft bolts to 20 ft. lbs. (25 Nm).

17. Lower the vehicle.

18. Connect the battery cable and refill with transmission fluid.

CLUTCH

Clutch Assembly

REMOVAL AND INSTALLATION

1. Raise and safely support the vehicle. Remove the transmission from the vehicle.

2. Make matchmarks on the clutch cover and flywheel, indicating their relationship.

3. Loosen the clutch cover-to-flywheel retaining bolts a turn at a time. The pressure on the clutch disc must be released gradually.

4. Remove the clutch cover-to-flywheel bolts. Remove the clutch cover and the clutch disc.

5. If the clutch release bearing is to be replaced, perform the following:

   a. Remove the bearing retaining clip(s), the bearing and hub.

   b. Remove the release fork and the boot.

   c. The bearing is press fitted to the hub.

   d. Clean all parts and lightly grease the input shaft splines and all of the contact points.

   e. Install the bearing/hub assembly, the fork, the boot and the retaining clip(s) in their original locations.

To install:

6. Inspect the flywheel surface for cracks, heat scoring (blue marks) and warpage. Replace or resurface the flywheel, if any damage is present.

NOTE: Before installing any new parts, make sure they are clean. During installation, do not get grease or oil on any of the components, as this will shorten clutch life considerably.

7. Using an clutch alignment tool, position the clutch disc against the flywheel. The raised center section of the disc faces the transmission.

8. Install the clutch cover over the disc and install the bolts loosely. Align the pressure plate-to-flywheel matchmarks. If a new or rebuilt clutch cover assembly is installed, use the matchmark on the old cover assembly as a reference. Torque the pressure plate-to-flywheel bolts to 14 ft. lbs. (19 Nm), using a criss-cross pattern.

9. Install the transmission into the vehicle.

PEDAL HEIGHT/FREE-PLAY ADJUSTMENT

The pedal height measurement is gauged from the angled section of the floorboard to the center of the clutch pedal pad. The correct pedal height is 6.12 for Pick-Up and 4Runner or 6.46 in. for Previa. If necessary, adjust the pedal height by loosening the locknut and turning the pedal stop bolt which is located above the pedal towards the driver's seat. Tighten the locknut after the adjustment.

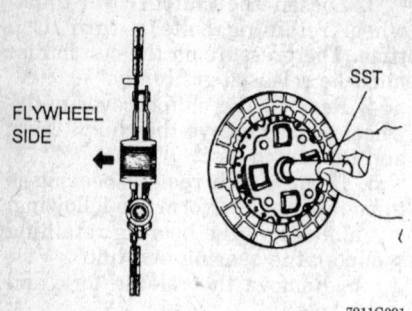

7911G091

Clutch disc installation

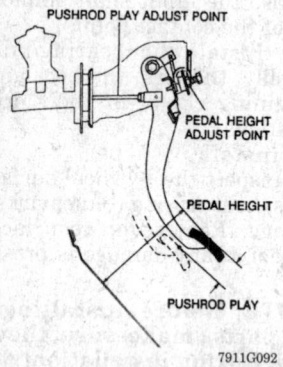

7911G092

Clutch pedal adjustment points

Clutch Master Cylinder

REMOVAL AND INSTALLATION

Pick-Up and 4Runner

1. Disconnect the negative battery cable. Disconnect the master cylinder pushrod pin from the top of the clutch pedal.
2. Remove the hydraulic line from the master cylinder, being careful not to damage the compression fitting.
3. Remove the master cylinder-to-cowl nuts/ bolts.
4. To install, reverse the removal procedures. Partially tighten the hydraulic line before tightening the master cylinder mounting nut(s). Torque the nuts/bolts to 9 ft. lbs. (12 Nm). Bleed the clutch system. Adjust the push rod play clearance.

Previa

1. Disconnect the negative battery cable.
2. Remove the instrument panel lower finish panel and steering column cover.
3. Remove the clip and clevis pin.
4. Disconnect the reservoir hose and clutch union line.

5. Remove the mounting bolts and pull out the master cylinder.
6. Installation is the reverse of removal. Torque the mounting bolts to 9 ft. lbs. (12 Nm).
7. Bleed the clutch hydraulic system.
8. Connect the battery cable and check operation.

Adjustment

PUSHROD

The pedal pushrod play is the distance between the clutch master cylinder piston and the pedal pushrod located above the pedal towards the firewall. Since it is nearly impossible to measure this distance at the source, it must be measured at the pedal pad, preferably with a dial indicator gauge. The pushrod play specification is: 0.040–0.200 for Land Cruiser and Previa or 0.039–0.197 for Pick-Ups and 4Runner. If necessary, adjust the pedal play by loosening the pedal pushrod locknut and turning the pushrod. Tighten the locknut after the adjustment.

FREE-PLAY

The free-play measurement is the total travel of the clutch pedal from the fully released position to where resistance is felt as the pedal is pushed downward. The free-play specification is: 0.20–0.59 in. for all vehicles.

Clutch Slave Cylinder

REMOVAL AND INSTALLATION

1. Raise and safely support the vehicle.
2. If equipped, remove the tension spring on the clutch fork.
3. Remove the hydraulic line from the release cylinder. Be careful not to damage the fitting.
4. Turn the release cylinder pushrod in sufficiently to gain clearance from the fork.
5. Remove the mounting bolts and withdraw the cylinder.
6. To install, reverse the removal procedures. Bleed the clutch system. Adjust the fork tip clearance.

Hydraulic Clutch System Bleeding

1. Fill the master cylinder reservoir with brake fluid.
2. Remove the cap and loosen the bleeder screw on the clutch release cylinder. Cover the hole with a finger.

3. Pump the clutch pedal several times. Take the finger off the hole while the pedal is being depressed so the air in the system can be released. Put a finger back on the hole and release the pedal.
4. When fluid pressure can be felt tighten the bleeder screw.
5. Place a short length of hose over the bleeder screw and the other end in a jar half full of clean brake fluid.
6. Depress the clutch pedal and loosen the bleeder screw. Allow the fluid to flow into the jar.
7. Tighten the plug, then release the clutch pedal.
8. Repeat this procedure until no air bubbles are visible in the bleeder tube.
9. When there are no more air bubbles in the system, tighten the plug fully with the pedal depressed. Replace the plastic cap.
10. Refill the master cylinder to the correct level with brake fluid. Check the system for leaks.

AUTOMATIC TRANSMISSION

Transmission Assembly

REMOVAL AND INSTALLATION

1. Disconnect the negative battery terminal from the battery. On the Pick-Up and 4Runner, remove the air cleaner assembly.
2. Disconnect the transmission throttle cable from the throttle body.
3. Raise and safely support the vehicle. Drain the transmission fluid.
4. Disconnect the wiring connectors (near the starter) for the neutral start switch and the back-up light switch. If equipped, disconnect the solenoid (overdrive) switch wiring at the same location. Disconnect the oil level gauge for Previa.
5. Disconnect the starter wiring at the starter. Remove the mounting bolts and the starter from the engine.
6. Make matchmarks on the rear driveshaft flange and the differential pinion flange. These marks must be aligned during installation.
7. Unbolt the rear driveshaft flange. If the vehicle has a 2 piece driveshaft, remove the center bearing bracket-to-frame bolts. Remove the driveshaft from the vehicle.

8. Disconnect the speedometer cable (tie it aside) and the shift linkage from the transmission.

9. Disconnect the transmission oil cooler lines at the transmission.

10. Disconnect the exhaust pipe clamp and remove the oil filler tube.

11. Support the transmission, using a jack with a wooden block placed between the jack and the transmission pan. Raise the transmission, just enough to take the weight off of the rear mount.

12. On the Pick-Up and 4Runner, remove the rear engine mount with the bracket and the engine under cover, to gain access to the engine crankshaft pulley. Remove the stiffener plates on 2WD Previa.

13. Place a wooden block (or blocks) between the engine oil pan and the front frame crossmember.

14. Slowly, lower the transmission until the engine rests on the wooden block.

15. Remove the rubber plug(s) from the service holes located at the rear of the engine in order to gain access to the torque convertor bolts.

16. Rotate the crankshaft (to remove the torque convertor bolts) to access the bolts through the service holes.

17. Obtain a bolt of the same dimensions as the torque convertor bolts. Cut the head off of the bolt and hacksaw a slot in the bolt opposite the threaded end.

NOTE: This modified bolt is used as a guide pin. Two guides pins are needed to properly install the transmission.

18. Thread the guide pin into one of the torque convertor bolt holes. The guide pin will help keep the convertor with the transmission.

19. Remove the stiffener plates from the transmission.

20. Remove the transmission-to-engine bolts, then carefully move the transmission rearward by prying on the guide pin through the service hole.

21. Pull the transmission rearward and lower it (front end down) out of the vehicle.

To install:

22. Apply a coat of multi-purpose grease to the torque convertor stub shaft and the corresponding pilot hole in the flywheel.

23. Install the torque convertor into the front of the transmission. Push inward on the torque convertor while rotating it to completely couple the torque convertor to the transmission.

24. To make sure the convertor is properly installed, measure the distance between the torque convertor mounting lugs and the front mounting face of the transmission. The proper distance is 0.080 in.

25. Install guide pins into 2 opposite mounting lugs of the torque converter.

26. Raise the transmission to the engine, align the transmission with the engine alignment dowels and position the convertor guide pins into the mounting holes of the flywheel.

27. Install and tighten the transmission-to-engine mounting bolts. Torque the bolts to 47 ft. lbs. (63 Nm).

28. Remove the convertor guide pins and install the convertor mounting bolts. Rotate the crankshaft as necessary to gain access to the guide pins and bolts through the service holes. Evenly, tighten the convertor mounting bolts to 13 ft. lbs. (17 Nm). Install the rubber plugs into the access holes.

29. Install the engine undercover. Raise the transmission slightly and remove the wood block(s) from under the engine oil pan.

30. Install the transmission crossmember. Torque the crossmember-to-frame bolts to 26–36 ft. lbs. (34–48 Nm).

31. Lower the transmission onto the crossmember and install the transmission mounting bolts. Torque the bolts to 19 ft. lbs. (26 Nm).

32. Install the oil filler tube and connect the exhaust pipe clamp.

33. Connect the oil cooler lines to the transmission and torque the fittings to 25 ft. lbs. (34 Nm).

34. To complete the installation, reverse the removal procedures. Adjust the transmission throttle cable. Refill the transmission with Dexron®II fluid. Road test the vehicle and check for leaks.

SHIFT LINKAGE ADJUSTMENT

1. Loosen the adjustment nut on the transmission shift cable.

2. Push the manual lever of the transmission fully rearward except for Previa, or downward for Previa.

3. Move the manual lever back 2 notches, which is the **N** position except for Previa or 3 notches for Previa.

4. Set the gearshift selector lever in the **N** position.

5. Apply a slight amount of forward pressure on the selector lever and tighten the shift cable adjustment nut.

THROTTLE LINKAGE ADJUSTMENT

1. Remove the air cleaner assembly.

2. Push the accelerator to the floor and check that the throttle valve opens fully; if not, adjust the accelerator link, so it does.

3. Push back the rubber boot from the throttle cable which runs down to the transmission. Loosen the throttle cable adjustment nuts so the cable housing can be adjusted.

4. Fully open the throttle by pressing the accelerator all the way to the floor.

5. Adjust the cable housing so, with the throttle wide open, the distance between the outer cable end rubber cap to the inner cable stopper is 0–0.04 in. (0–1mm).

6. Tighten the nuts and double check the adjustment. Install the rubber boot and the air cleaner.

NEUTRAL START SWITCH ADJUSTMENT

The neutral safety switch prevents the vehicle from starting unless the gearshift selector is in either the **P** or **N** positions. If the vehicle will start in these positions, adjustment of the switch is required.

1. Loosen the neutral start switch bolt.

2. Place the selector lever in the **N** position.

3. Disconnect the wires from the neutral start switch.

4. Connect an ohmmeter between the terminals of the switch.

5. Adjust the switch until there is continuity between the N and B terminals.

6. Reconnect the wires. Torque the bolt to 48 inch lbs. (65 Nm).

TRANSFER CASE

Transfer Case Assembly

REMOVAL AND INSTALLATION

The transfer case and transmission are connected together. It is recommended by the manufacturer that they be removed from the vehicle as an assembly and then separated for repairs.

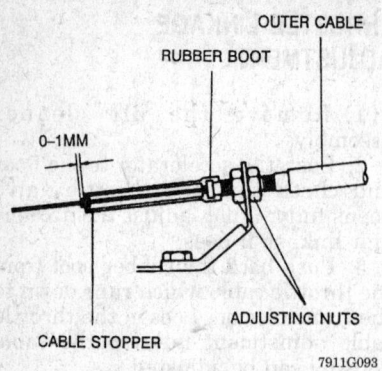

OUTER CABLE
RUBBER BOOT
0–1MM
CABLE STOPPER
ADJUSTING NUTS
7911G093

Throttle linkage adjustment

DRIVE AXLE

Front Halfshaft

REMOVAL AND INSTALLATION

4WD Pick-Up, 4Runner

Remove the 4WD hub (with the flange) from the axle hub.

1. Raise and safely support the vehicle. Remove the wheel and tire assembly.
2. Disconnect and plug the brake line from the caliper. Remove the caliper from the axle hub.
3. Using a drift punch and a hammer, drive the lock washer tabs away from the locknut.
4. Remove the locknut from the halfshaft. Remove the lock washer, the adjusting nut, the thrust washer, the outer bearing and the axle hub/disc assembly from the vehicle.
5. Remove the knuckle spindle bolts, the dust seal and the dust cover. Using a brass bar and a hammer, tap the steering spindle from the steering knuckle.
6. Turn the halfshaft until a flat spot on the outer shaft is in the upper position, then pull the halfshaft from the steering knuckle.
7. Using a slide hammer, pull the oil seal from the axle housing.
8. Using a clean shop towel, wipe the grease from inside the steering knuckle housing and the halfshaft.

To install:
9. Using an oil seal installation tool, drive a new oil seal into the axle housing until it seats.
10. Install the halfshaft into the axle housing.
11. Using multi-purpose grease, fill the steering knuckle cavity to about ¾ full.

12. To complete the installation, use new seals/gaskets and reverse the removal procedures.
13. Torque the steering spindle-to-steering knuckle bolts to 38 ft. lbs. (52 Nm), the axle hub adjusting nut to 18 ft. lbs. (25 Nm), the axle hub locknut to 33 ft. lbs. (44 Nm), the free wheel/locking hub nuts to 23 ft. lbs. (33 Nm) and the brake caliper to 65 ft. lbs. (88 Nm).

NOTE: To install the wheel bearings with the axle hub, torque the adjusting nut to 43 ft. lbs. (58 Nm), turn the axle hub (back and forth, several times), loosen the nut and retorque the adjusting nut to 18 ft. lbs. (24 Nm).

14. Install the wheel and tire assembly. Lower the vehicle.

4WD Previa

1. Raise and safely support the vehicle. Remove the wheel and tire assembly.
2. Using a drift punch and a hammer, drive the lock washer tabs away from the locknut.
3. Remove the locknut from the halfshaft. Remove the lock washer, the adjusting nut and the thrust washer.
4. Using a tie rod remover 0962810011 or equivalent, remove the tie rod end from the knuckle.
5. Remove the 6 bolts from the inner shaft joint.
6. Disconnect the knuckle from the lower ball joint by removing the 2 bolts.
7. Remove the halfshaft by pulling the knuckle outward and remove the halfshaft from the wheel hub. If the outer shaft will not come out of the hub, soak the splines with penetrating lube, install the nut and tap on the halfshaft with a soft hammer. Be careful not to damage the shaft threads.

To install:

NOTE: Coat the halfshaft splines with anti-seize compound to prevent spline seizure. This will help for future halfshaft removal.

8. Connect the lower ball joint and torque the bolts to 94 ft. lbs. (127 Nm).
9. Install the inner halfshaft joint-to-drive axle and torque the bolts to 51 ft. lbs. (61 Nm).
10. Install the tie rod end, torque the nut to 36 ft. lbs. (49 Nm) and install a new cotter pin.
11. Install the halfshaft nut and torque to 76 ft. lbs. (103 Nm).

12. Install the remaining components and check operation.

CV-Boot

REMOVAL AND INSTALLATION

Inner Boot

1. Remove the halfshaft from the vehicle.
2. Clean the halfshaft with soap and water. Remove the 2 inner boot clamps and slide the inner boot off the the joint.
3. Place matchmarks on the inner joint outer race and shaft. Remove the outer race from the shaft. Some joints have a large snapring on the inner side of the race that has to be removed.
4. Remove the snapring from the end of the inner joint. Pull the inner joint and boot from the shaft splines.
To install:
5. Install the boot, inner joint and snapring.
6. Fill the boot and joint with special CV-boot grease. The grease usually comes with replacement boot kits.
7. Slide the boot into position and install the clamps.

Outer Boot

1. Remove the inner joint and boot from the shaft.
2. Tape the inner splines to protect the boots.
3. Remove the 2 outer boot clamps and slide the outer boot off the the joint.
4. Installation is the reverse of removal. Clean the outer joint from all old grease and repack with special CV-boot grease.

Driveshaft and U-Joints

REMOVAL AND INSTALLATION

4WD Vehicles, Except Long Bed Pick-Up

REAR

1. Raise and safely support the vehicle and place a drain pan under the transmission.
2. Paint a mating mark on the halves of the rear universal joint flange.
3. Remove the bolts which hold the rear flange together.
4. Remove the splined end of the driveshaft from the transmission.

The Previa rear shaft is bolted at both ends.

NOTE: Plug the end of the transmission with a rag or dummy flange to avoid losing transmission oil.

5. Remove the driveshaft from under the vehicle.

6. To install, reverse the removal procedures. Grease the splined end of the shaft before installing. Torque bolts to 31 ft. lbs. (42 Nm) for Previa or 54 ft. lbs. (75 Nm) for Pick-up and 4Runner.

All 4WD Vehicles

FRONT

1. Raise and safely support the vehicle.
2. Matchmark the driveshaft flange at the front axle housing and the transfer case.
3. On Previa, remove the center support bracket.
4. On Pick-Up and 4Runner, remove the flange dust cover.
5. Remove the bolts retaining the driveshaft and remove the driveshaft from the vehicle.
6. Install the driveshaft by aligning the matchmarks made during removal.
7. Torque the retaining bolts to 54–58 ft. lbs. (74–78 Nm). Torque the center support bracket bolts to 35 ft. lbs. (48 Nm). Make sure the drain hole is facing downward.

2WD Long Bed Pick-Up

REAR

1. Raise and safely support the vehicle.
2. Paint mating marks on all 6 flange halves.
3. Remove the bolts attaching the rear universal joint flange to the drive pinion flange.
4. Drop the rear section of the shaft slightly and pull the unit out of the center bearing sleeve yoke.
5. Remove the center bearing support from the crossmember.
6. Unbolt the driveshaft flange from the rear of the transmission and remove driveshaft along with center bearing support.
7. To install, align the matchmarks and reverse the removal procedures. Torque the flange bolts to 54 ft. lbs. (75 Nm).

Front Axle Shaft

REMOVAL AND INSTALLATION

Land Cruiser

1. Raise and safely support the vehicle. Remove the wheel and tire assembly.
2. Plug the brake master cylinder reservoir to prevent brake fluid leakage from the disconnected brake flexible hose.
3. Remove the outer axle shaft flange cap for automatic locking hub, the hub cover bolts and the cover for free wheel locking hub and the shaft snaping from the axle hub.
4. Remove the outer axle shaft flange for automatic locking hub or the hub ring-to-axle hub bolts, then alternately, screw 2 service bolts into the shaft flange or hub ring and remove the shaft flange or the hub ring with it's gasket.
5. Remove the caliper and disc.
6. Straighten the lock washer and remove the front wheel bearing adjusting nuts with front wheel adjusting nut wrench or similar tool.
7. Remove the front axle hub together with its claw washer, bearings and oil seal.
8. Remove the clip and disconnect the brake flexible hose from brake tube.
9. Cut and remove the lock wire.
10. Using a soft mallet, lightly, tap the steering knuckle spindle and remove the spindle with it's gasket.

NOTE: When removing the steering knuckle spindle, if equipped with the ball joint type axle shaft joint, be prepared for the disconnection of the outer axle shaft from the joint. Prevent the shaft joint ball from falling from the joint.

11. If equipped with the ball type axle shaft joint, slide the inner front axle shaft out of the axle housing. If equipped with the Birfield constant velocity joint type of axle shaft joint, remove the entire axle shaft assembly from the axle housing.
12. Using a bearing puller, remove the bushing from inside of knuckle spindle and the axle housing oil seal. Using a metal tube as a seating tool, drive oil seal into the axle housing and the new bushing into the knuckle spindle.

NOTE: If equipped with the ball joint type axle joint, install the inner axle with its proper spacer in position until the splines are fully meshed with the differential. If equipped with the Birfield constant velocity joint axle joint, install the axle into the housing and rotate the axle shaft until its splines mesh with the differential. Fill the steering knuckle about ¾ full with grease and place the joint ball on the inner shaft end.

13. To complete the installation, reverse the removal procedures. Adjust the wheel bearing preload.

Front Axle Bearing

REMOVAL AND INSTALLATION

Except 4WD Previa

1. Raise and safely support the vehicle. Remove the 4WD hubs.
2. Using a small prybar, pry the grease seal from the rear of the disc/hub assembly, then remove the inner bearing from the assembly.
3. Using a shop cloth, wipe the grease from inside the disc/hub assembly.
4. Using a brass drift, drive the outer bearing races from each side of the disc/hub assembly.
5. Using solvent, clean all of the parts and blow dry with compressed air.

To install:

6. Using a bearing installation tool, drive the outer races into the disc/hub assembly until they seat against the shoulder.
7. Using multi-purpose grease, coat the area between the races and pack the bearings.
8. Place the inner bearing into the rear of the disc/hub assembly. Using a bearing installation tool, drive a new grease seal into the rear of the disc/hub assembly until it is flush with the housing.
9. Install the disc/hub assembly onto the axle shaft, the outer bearing, the thrust washer and the adjusting nut.
10. To adjust the bearing preload, perform the following:
   a. Torque the adjusting nut to 43 ft. lbs. (58 Nm).
   b. Turn the disc/hub assembly 2–3 times, from the left to the right.
   c. Loosen the adjusting nut until it can be turned by hand.
   d. Retorque the adjusting nut to 18 ft. lbs. (25 Nm).
   e. Install the lock washer and the locknut. Torque the locknut to 33 ft. lbs. (44 Nm).

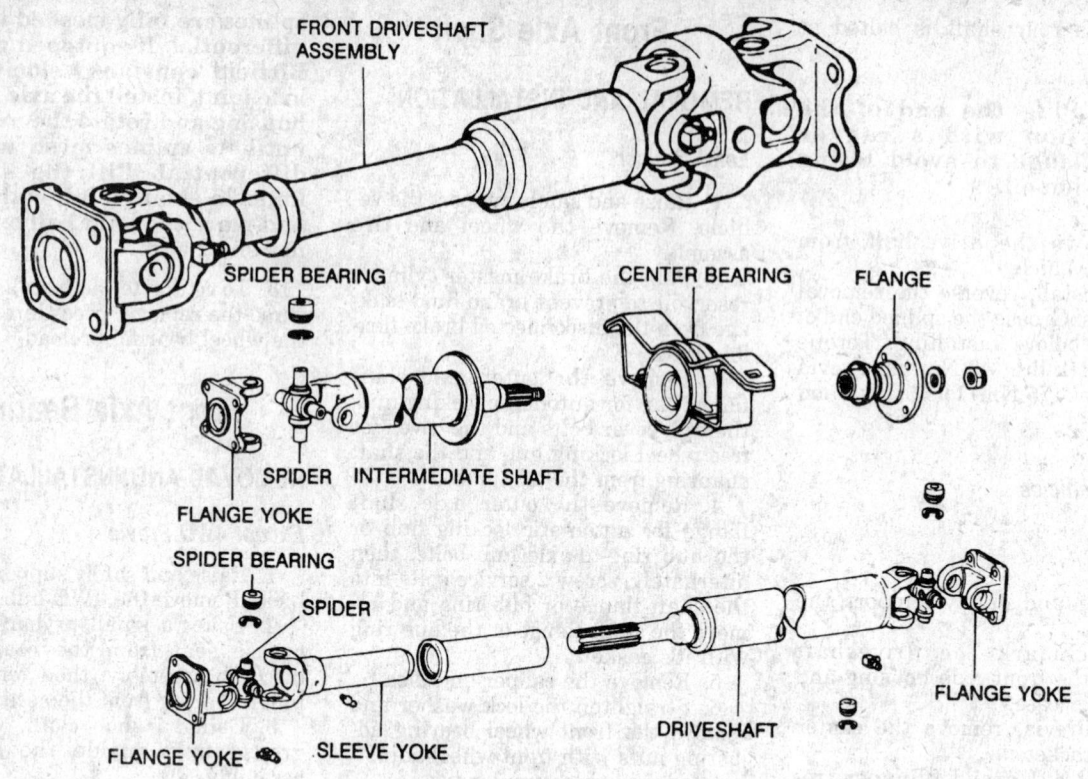

FRONT DRIVESHAFT ASSEMBLY

SPIDER BEARING

CENTER BEARING

FLANGE

SPIDER   INTERMEDIATE SHAFT

FLANGE YOKE

SPIDER BEARING

SPIDER

FLANGE YOKE

FLANGE YOKE   SLEEVE YOKE

DRIVESHAFT

FLANGE YOKE

7911G094

Driveshaft assemblies — Pick-Up and 4Runner

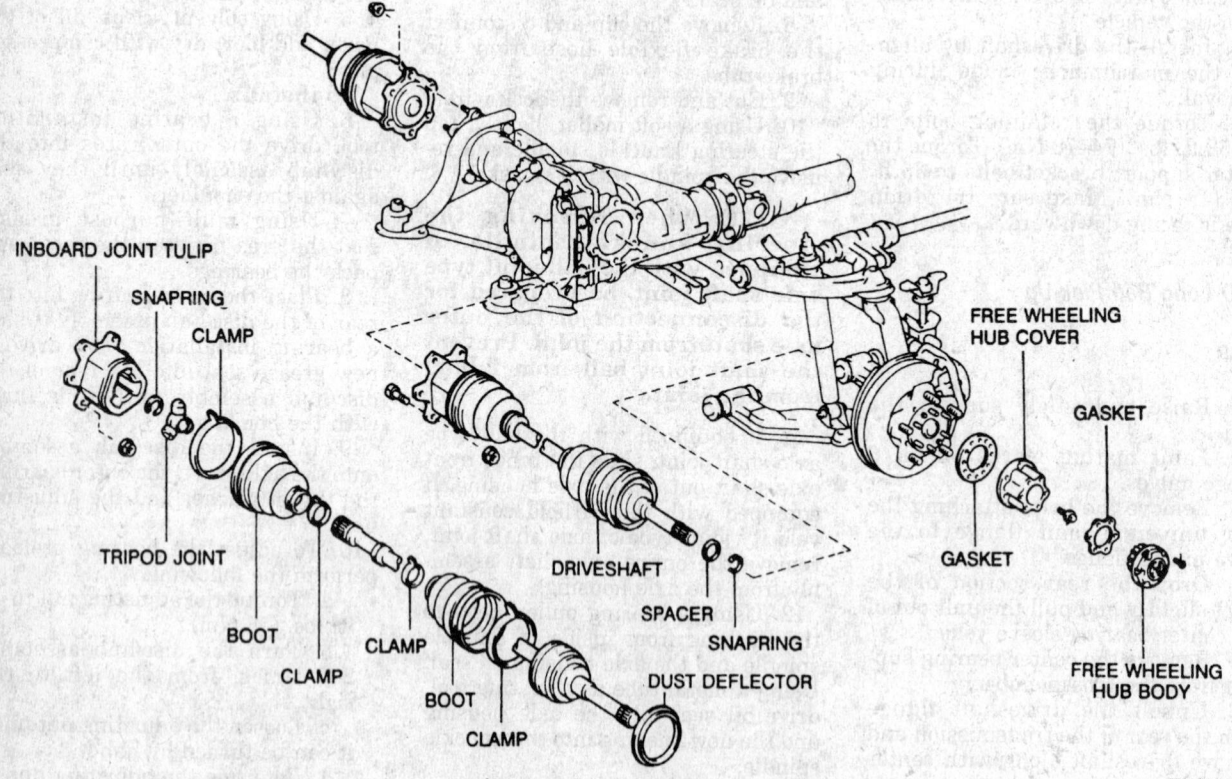

INBOARD JOINT TULIP

SNAPRING

CLAMP

FREE WHEELING HUB COVER

GASKET

TRIPOD JOINT

DRIVESHAFT

GASKET

BOOT

CLAMP

SPACER

CLAMP

BOOT

SNAPRING

FREE WHEELING HUB BODY

CLAMP

DUST DEFLECTOR

7911G095

Front drive axle assembly — Pick-Up and 4Runner

f. Check that the bearing has no play.

g. Using a spring gauge, connect it to a wheel stud, the gauge should be held horizontal, then measure the rotating force, it should be 6–12 lbs. (8–15 Nm).

11. Lower the vehicle.

4WD Previa

1. Remove the steering knuckle from the vehicle. Place the assembly in a vise.

2. Using a hub remover tool (slide hammer) 0952000031 or equivalent, remove the wheel hub from the knuckle.

3. Remove the bearing from the hub using a press and arbor tool 0955010012 or equivalent.

4. Remove the dust protector and oil seal.

5. Remove the bearing snapring from the knuckle and press the inner bearing from the knuckle.

To install:

6. Using a press and arbor tool 0960810010 or equivalent, press the bearing into the knuckle.

7. Install the outer oil seal and dust cover.

8. Press the axle hub onto the knuckle.

9. Install the knuckle assembly onto the vehicle.

Rear Axle Shaft and Bearings

REMOVAL AND INSTALLATION

Pick-Up and 4Runner

1. Loosen the rear wheel lug nuts, then raise and safely support the vehicle. Remove the wheel and tire assembly.

2. Place a pan under the axle, remove the plug and drain the axle housing.

3. For 2WD vehicles, remove the clip/clamp-to-frame bolts and disconnect the parking brake cable from the equalizer. For 4WD vehicles, remove the pin and disconnect the rear parking brake cable from the bell crank.

4. Remove the brake drum securing screw and the drum.

5. Disconnect the brake line from the wheel cylinder and plug it, being careful not to damage the fitting.

6. Remove the brake backing plate-to-axle housing nuts and pull the backing plate with the axle from the axle housing.

NOTE: When removing the axle shaft, be careful not to damage the oil seal.

7. Using a pair of snapring pliers, remove the snapring from the axle shaft.

8. Slip tool 09521–25011 or equivalent, over the axle shaft and fasten it to the backing plate. Using 2 metal blocks and a press, press the axle from the backing plate assembly.

9. If necessary to remove the bearing from backing plate, perform the following:

a. Remove the brake spring, the retracting spring clamp bolt, the lower springs, the shoe strut, the brake shoes and the parking brake lever.

b. Using a slide hammer puller, pull the outer oil seal from the backing plate.

c. Press the bearing from the backing plate.

d. Using the proper installation tools, press the new bearing into the backing plate.

e. Using the proper seal installation tool, press the new oil seal into the backing plate.

f. Reassemble the brake components to the backing plate.

10. Using a slide hammer, pull the oil seal from the axle housing.

To install:

11. Using the installation tool and a hammer, drive a new oil seal into the axle housing.

12. Using a press, press the axle shaft into the backing plate and the bearing retainer. Using snapring pliers, install the snapring onto the axle shaft.

13. Clean the gasket mounting surfaces.

14. To complete the installation, reverse the removal procedures. Torque the backing plate-to-axle housing nuts to 51 ft. lbs. (69 Nm). Adjust the brake shoe clearance and bleed the brake system. Refill the axle housing with SAE 90W GL5 gear oil.

Previa

1. Loosen the rear wheel nuts. Raise and safely support the vehicle. Remove the wheel and tire assembly.

2. Remove the brake drum or caliper and rotor.

3. Working through the hole in the axle flange, remove the backing plate-to-axle housing bolts.

4. Using a slide hammer puller, pull the axle shaft from the housing.

5. Using a grinder, grind down the inner bearing retainer on the axle shaft. Using a chisel and a hammer, cut off the retainer and remove it from the shaft.

6. Using a arbor press, press the bearing from the axle shaft.

7. Using a slide hammer puller, pull the oil seal from the axle housing.

To install:

8. Lubricate the new oil seal with multi-purpose grease. Using the proper seal installation tool and a hammer, drive the new oil seal into the axle housing to a depth of 0.236 in.

9. To install, use new gaskets and reverse the removal procedures. Torque the axle retainer-to-housing bolts to 48 ft. lbs. (65 Nm).

Land Cruiser

SEMI-FLOATING TYPE

1. Loosen the rear wheel nuts. Raise and safely support the vehicle. Remove the wheel and tire assembly.

2. Place a pan under the axle, remove the plug and drain the oil from the differential.

3. Remove the brake drum and related parts, as follows:

a. Remove the cover from the back of the differential housing.

b. Remove the pin from the differential pinion shaft.

c. Withdraw the pinion shaft and it's spacer from the case.

d. Use a mallet to tap the rear axle shaft toward the differential, then remove the C-lock from the axle shaft.

e. Withdraw the axle shaft from the housing.

4. Using a bearing puller, remove axle bearing and oil seal together from the axle housing. Using a metal tube and a hammer, drive the bearing and the seal into the housing until they seat.

NOTE: Do not mix the parts of the left and right axle shaft assemblies.

5. To complete the installation, reverse the removal procedures. Refill the axle housing with SAE 90W GL5 gear oil.

6. After installing the axle shaft, C-lock, spacer and pinion shaft, measure the clearance between the axle shaft and the pinion shaft spacer with a feeler gauge. The clearance should fall between 0.0024–0.0181 in. If the clearance is not within specifications, use one of the following spacers to adjust it:

a. 1.172–1.173 in.

b. 1.188–1.189 in.
c. 1.204–1.205 in.

FULL FLOATING TYPE

1. Loosen the rear wheel nuts. Raise and safely support the vehicle. Remove the wheel and tire assembly.
2. Place a pan under the axle, remove the plug and drain the oil from the differential.
3. Remove the rear axle shaft plate nuts.
4. Remove the cone washers from the mounting studs by tapping the slits of the washers with a tapered punch.
5. Install bolts into the 2 unused holes of the axle shaft plate.
6. Tighten the bolts to draw the axle shaft assembly out of the housing.
7. To install, use a new gasket, sealant and reverse the removal procedures. Torque the axle shaft nuts to 21–25 ft. lbs. (27–34 Nm).

Front Wheel Hub, Knuckle and Bearing

REMOVAL AND INSTALLATION

2WD Pick-Up

1. Raise and safely support the vehicle. Remove the wheel and tire assembly.
2. Remove the brake caliper. Do not disconnect the brake hose from the caliper. Suspend it on a wire.
3. Remove axle hub dust cap, the cotter pin, the nut lock, the adjusting nut, the thrust washer and the outer bearing, then pull the hub/disc assembly from the axle spindle.
4. Remove the backing plate cotter pins and the mounting nuts or bolts, then the backing plate.
5. Remove steering knuckle arm from the back of the steering knuckle.
6. Remove the nuts, the retainers and the bushings, then the shock absorber from the lower control arm.
7. Support the lower arm with a jack and raise to put pressure on spring.

NOTE: Be careful not to unbalance vehicle support stands when jacking up lower arm.

8. Remove cotter pins, then the upper and lower ball joint nuts. Using a ball joint removal tool, separate the ball joints from the steering knuckle.

9. Remove the steering knuckle from the vehicle.

NOTE: Whenever the hub/disc assembly is removed from the vehicle, it is good practice to replace the grease seal.

10. To install, reverse the removal procedures. Torque the upper ball joint nut to 80 ft. lbs. (109 Nm) for Pick-Up, the lower ball joint nut to 105 ft. lbs. (142 Nm) for Pick-Up or 76 ft. lbs. (109 Nm), the steering knuckle arm-to-steering knuckle bolts to 80 ft. lbs. (109 Nm) for Pick-Up, the shock absorber-to-lower control arm nuts to 19 ft. lbs. (27 Nm) and the backing plate-to-steering knuckle bolts to 80 ft. lbs. (109 Nm) for Pick-Up. Adjust the wheel bearing.

4WD Vehicles

1. Remove the front axle shaft assembly from the vehicle.
2. Remove the oil seal retainer and the oil seal set from the rear of the steering knuckle.
3. At the drag link end of the steering knuckle arm, remove the cotter pin. Using the proper tool, remove the plug from the drag link, then disconnect the drag link from the steering knuckle arm.

4. Remove the tie rod-to-steering knuckle, cotter pin and nut. Using the proper ball joint removal tool, separate the tie rod from the steering knuckle arm.
5. Remove the steering knuckle arm-to-steering knuckle (top) nuts and the steering knuckle-to-bearing cap (bottom) nuts. Using a tapered punch, tap the cone washers slits and remove the washers.

NOTE: Do not mix or lose the upper and lower bearing cap shims.

6. Using a bearing removal tool (without a collar), press the steering knuckle arm with the shims from the steering knuckle.
7. Using a bearing removal tool (without a collar), press the bearing cap with the shims from the steering knuckle.
8. Remove the steering knuckle from the vehicle.
To install:
9. To install the steering knuckle, use a suitable tool to support the upper inner bearing. Using a hammer, tap the steering knuckle arm into the bearing inner race.
10. Using the proper tool, support the lower bearing inner race. Using a

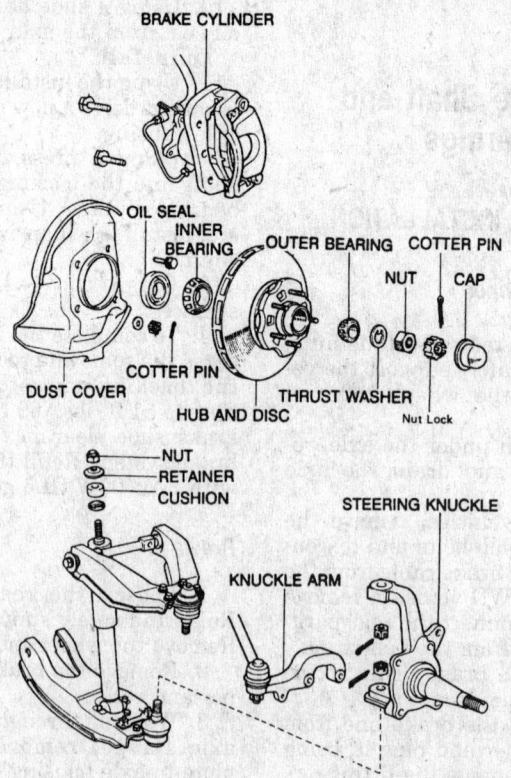

Front hub and steering knuckle — typical 2WD vehicle

hammer, tap the bearing cap into the bearing inner race.

NOTE: When installing the drag link-to-steering knuckle arm, torque the plug all the way, then loosen it 1⅓ turns and secure it with the cotter pin.

11. To install, use gaskets, seals, pack the steering knuckle with multi-purpose grease and reverse the removal procedures for the Land Cruiser.

    a. Torque the steering knuckle arm-to-steering knuckle nuts to 71 ft. lbs. (95 Nm).

    b. Torque the bearing cap-to-steering knuckle nuts to 71 ft. lbs. (95 Nm).

    c. Torque the tie rod-to-steering knuckle arm nut to 67 ft. lbs. (91 Nm).

    d. Torque the axle spindle-to-steering knuckle bolts to 38 ft. lbs. (51 Nm).

12. To install, reverse the removal procedures for the Pick-Up, 4Runner.

    a. Torque the upper ball joint nut to 80 ft. lbs. (109 Nm) for Pick-Up and 4Runner.

    b. Torque the lower ball joint nut to 105 ft. lbs. (142 Nm) for Pick-Up and 4Runner.

    c. Torque the steering knuckle arm-to-steering knuckle bolts to 80 ft. lbs. (109 Nm) for Pick-Up and 4Runner.

    d. Torque the shock absorber-to-lower control arm nuts to 19 ft. lbs. (27 Nm).

    e. Torque the backing plate-to-steering knuckle bolts to 80 ft. lbs. (109 Nm) for Pick-Up and 4Runner.

13. Adjust the wheel bearing preload.

NOTE: To test the knuckle bearing preload, attach a spring scale to the tie rod end hole (at a right angle) in the steering knuckle arm. The force required to move the knuckle from side to side should be 6.6–13 lbs. (10–17 N) except for Land Cruiser or 4–5 lbs. (5–7 N) for Land Cruiser. If the preload is not correct, adjust by replacing shims.

Previa

1. Loosen the halfshaft nut. Raise the vehicle and support safely.

2. Remove the front wheels and disconnect the ABS sensor, if equipped.

3. Remove the front caliper and rotor. Secure the caliper to the vehicle with wire.

4. Remove the the halfshaft nut for 4WD.

5. Remove the MacPherson strut-to-knuckle bolts and lower ball joint bolts.

6. Separate the tie rod end using a remover tool 0962810011 or equivalent.

7. Make sure the halfshaft splines are loose and remove the halfshaft from the knuckle.

8. Remove the steering knuckle/hub assembly.

9. 2WD only, remove the grease cap. Using a chisel, release the nut stake and nut.

10. Using a hub remover tool (slide hammer) 0952000031 or equivalent, remove the wheel hub from the knuckle.

11. Remove the bearing from the hub using a press and arbor tool 0955010012 or equivalent.

12. Remove the dust protector and oil seal.

13. Remove the bearing snapring from the knuckle and press the inner bearing from the knuckle.

To install:

14. Using a press and arbor tool 0960810010 or equivalent, press the bearing into the knuckle.

15. Install the outer oil seal and dust cover.

16. Press the axle hub onto the knuckle.

17. 2WD only, install the lock nut and torque to 147 ft. lbs. (199 Nm). Stake the nut and install the dust cover.

18. Install the knuckle assembly onto the vehicle. Torque the tie rod ends to 36 ft. lbs. (49 Nm), strut bolts to 231 ft. lbs. (314 Nm), lower ball joint bolts to 94 ft. lbs. (147 Nm) and 4WD halfshaft nut to 137 ft. lbs. (186 Nm).

Locking Hubs

REMOVAL AND INSTALLATION

If equipped with free-wheeling hubs, turn the hub control handle to the **FREE** position.

1. Remove the hub cover bolts and pull off the cover.

2. If equipped with automatic locking hubs, remove the axle bolt with the washer.

3. Using snapring pliers, remove the snapring from the axle shaft.

4. Remove the hub body mounting nuts.

5. Remove the cone washers from the hub body mounting studs by tapping on the washer slits with a tapered punch.

6. Remove the hub body from the axle hub.

7. Apply multi-purpose grease to the inner hub splines.

8. To install, use new gaskets and reverse the removal procedures. Torque the hub body-to-axle hub nuts to 23 ft. lbs. (31 Nm), the plate washer/bolt to 13 ft. lbs. (17 Nm) for auto. locking hub and the hub cover-to-hub body bolts to 7 ft. lbs. (10 Nm).

NOTE: To install the snapring onto the axle shaft, install a bolt into the axle shaft, pull it out and install the snapring.

Pinion Seal

REMOVAL AND INSTALLATION

1. Raise and safely support the vehicle.

2. Matchmark and remove the driveshaft.

3. Remove the companion flange from the differential.

4. Using puller 09308–10010 or equivalent, remove the oil seal from the housing.

5. Remove the oil slinger.

6. Remove the bearing and spacer.

To install:

7. Install the bearing spacer and bearing.

NOTE: Lubricate the seal lips with multi-purpose grease before installing it.

8. Install the oil slinger and using seal installer 09554–30011 or equivalent, install the oil seal.

9. Drive the seal into place, to a depth of 0.59 in below the housing lip for 7.5 in. axles and 0.39 in. below the lip for 8 in. axles.

10. Install the companion flange and install the driveshaft.

11. Lower the vehicle.

Differential Carrier

REMOVAL AND INSTALLATION

Rear and Land Cruiser Front

1. Raise and safely support the vehicle. Drain the lubricant from the differential.

2. Remove the axle shafts from the axle housing.

3. Matchmark the driveshaft flange to the differential flange. Remove the mounting bolts and sepa-

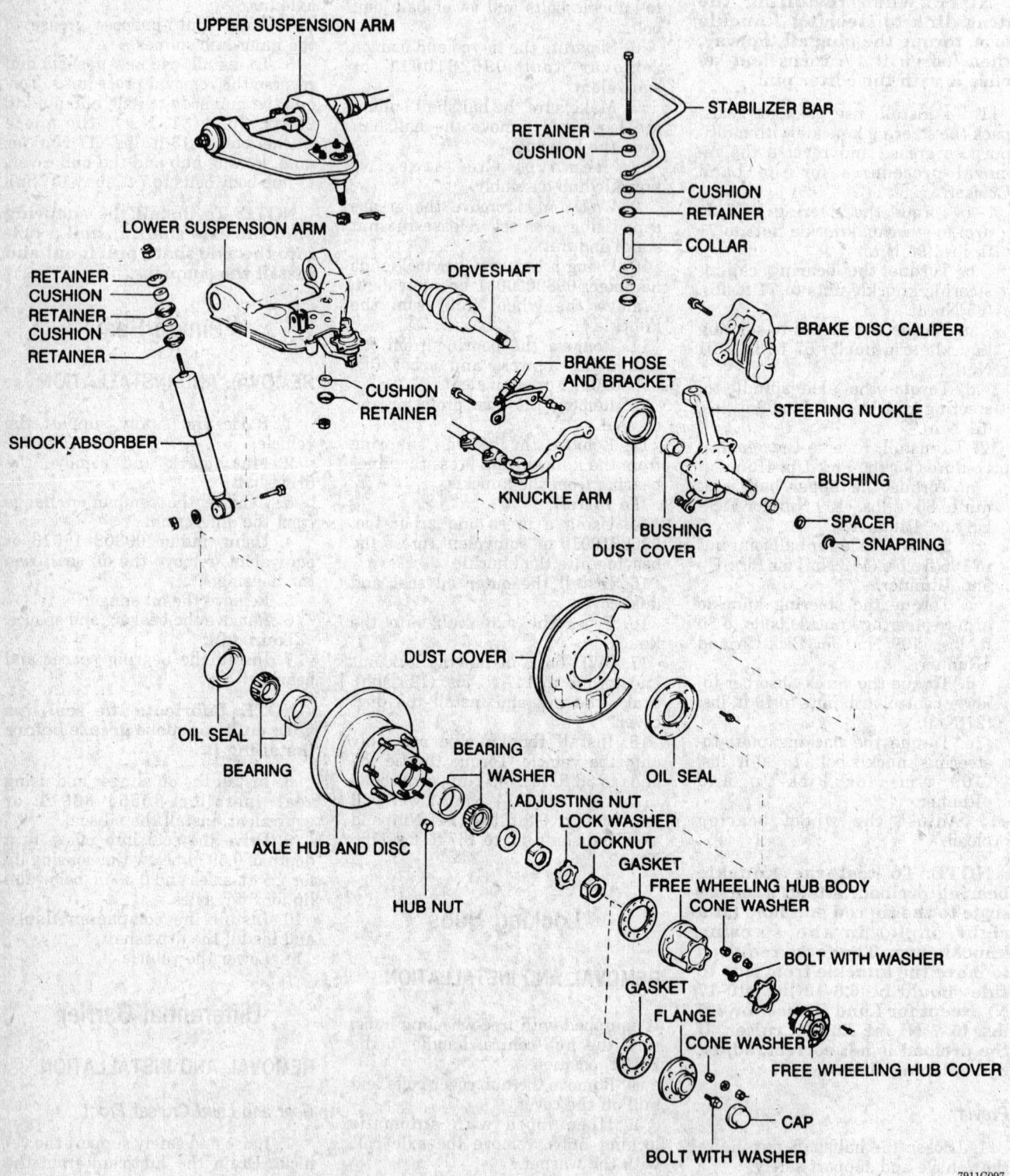

UPPER SUSPENSION ARM

STABILIZER BAR

RETAINER
CUSHION

CUSHION
RETAINER
COLLAR

LOWER SUSPENSION ARM

DRVESHAFT

BRAKE DISC CALIPER

RETAINER
CUSHION
RETAINER
CUSHION
RETAINER

BRAKE HOSE
AND BRACKET

STEERING NUCKLE

CUSHION
RETAINER

SHOCK ABSORBER

BUSHING

SPACER
SNAPRING

KNUCKLE ARM

BUSHING
DUST COVER

DUST COVER

OIL SEAL

BEARING

BEARING
WASHER
ADJUSTING NUT
LOCK WASHER
LOCKNUT
GASKET

OIL SEAL

AXLE HUB AND DISC

FREE WHEELING HUB BODY
CONE WASHER

HUB NUT

BOLT WITH WASHER

GASKET
FLANGE
CONE WASHER

FREE WHEELING HUB COVER

CAP

BOLT WITH WASHER

Front hub and steering knuckle — typical 4WD vehicle

7911G097

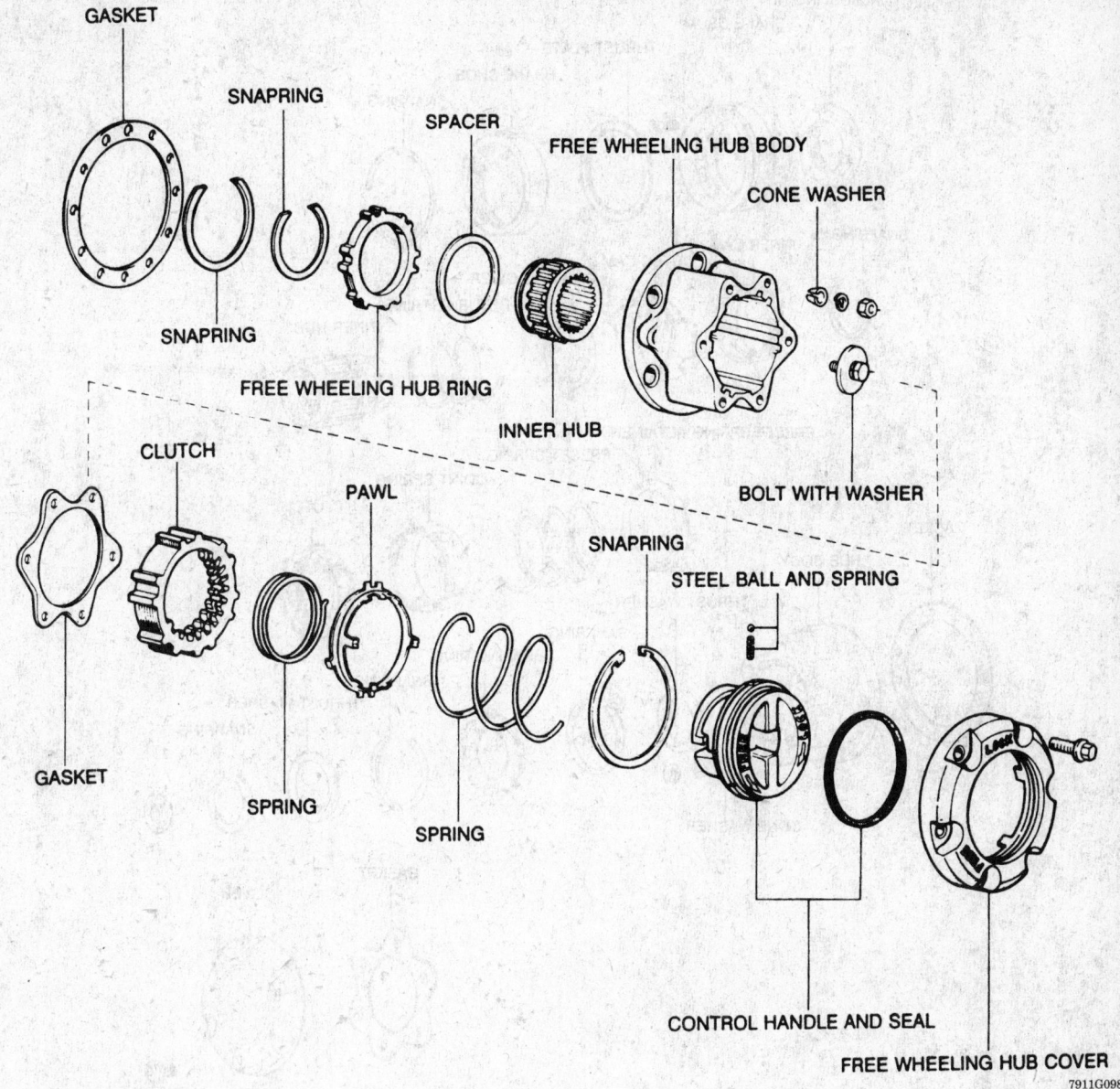

GASKET

SNAPRING

SPACER

FREE WHEELING HUB BODY

CONE WASHER

SNAPRING

FREE WHEELING HUB RING

INNER HUB

BOLT WITH WASHER

CLUTCH

PAWL

SNAPRING

STEEL BALL AND SPRING

GASKET

SPRING

SPRING

CONTROL HANDLE AND SEAL

FREE WHEELING HUB COVER

7911G098

Manual 4WD hub assembly

rate the driveshaft from the differential.

4. Remove the differential retaining nuts and pull the assembly out of the differential housing.

5. Place matchmarks on the bearing caps, carrier housing and adjusting nuts.

6. Remove the 4 bolts and bearing caps. Lift the carrier assembly with the side bearings out of the housing. Tag the parts to show the location for reassembly.

To install:

7. Place the bearing outer races on their respective bearings. Install the carrier into the differential housing.

8. Install the bearing adjusting nuts, bearing caps and bolts to the original position. Torque the cap bolts to 58 ft. lbs. (78 Nm).

9. Use new gaskets and reverse the removal procedures. Torque the differential-to-axle nuts to 23 ft. lbs. (31 Nm) and the driveshaft flange-to-differential flange nuts/bolts to 31 ft. lbs. (43 Nm). Refill the axle with 80W–90 gear oil to a level of ¼ in. below the fill hole.

NOTE: Before installing the carrier, apply a thin coat of liquid or silicone sealer to the carrier housing gasket and to the carrier side face of each carrier retaining nut.

Front

1. Raise and safely support the vehicle. Drain the lubricant from the differential.

2. Remove the front axle shafts or halfshafts from the axle housing.

3. Matchmark the front driveshaft flange to the differential flange. Remove the mounting bolts and separate the driveshaft from the differential.

4. Remove the differential carrier cover.

5. Using needle nose pliers or equivalent, remove the snapring from the side gear shafts. Pull the side gear shafts from the housing. A slide

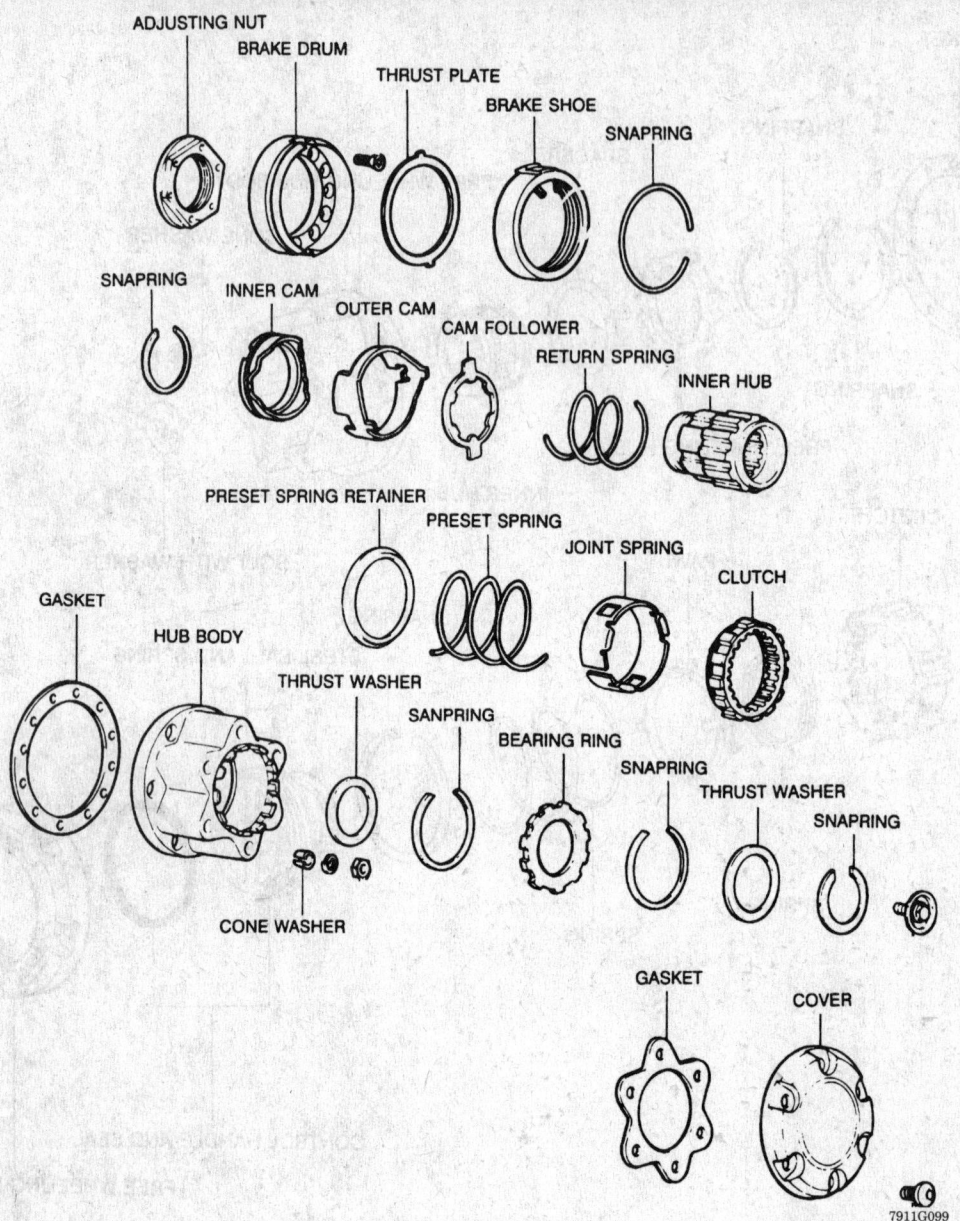

ADJUSTING NUT · BRAKE DRUM · THRUST PLATE · BRAKE SHOE · SNAPRING · SNAPRING · INNER CAM · OUTER CAM · CAM FOLLOWER · RETURN SPRING · INNER HUB · PRESET SPRING RETAINER · PRESET SPRING · JOINT SPRING · CLUTCH · GASKET · HUB BODY · THRUST WASHER · SANPRING · BEARING RING · SNAPRING · THRUST WASHER · SNAPRING · CONE WASHER · GASKET · COVER

7911G099

Automatic 4WD hub assembly

hammer may have to be used to remove the side shafts.

6. Remove the A.D.D actuator from the left side of the differential assembly. Remove the intermediate shaft using a puller 0935020015 or equivalent, for A.D.D equipped vehicles.

7. Place matchmarks on the bearing caps, carrier housing and adjusting nuts.

8. Remove the 4 bolts and bearing caps. Using carrier removing tool 0950422011 or equivalent, and a hammer. Dislodge the carrier assembly and lift the carrier assembly with the side bearings out of the housing.

Tag the parts to show the location for reassembly.

To install:

9. Install the carrier, bearing races and adjusting shims into the housing.

10. Install the bearing caps and torque the bolts to 58 ft. lbs. (78 Nm).

11. Install the side gear shaft oil seals and side shafts.

12. Install new snaprings to the end of the shafts.

13. Apply sealer to the A.D.D actuator and install. Torque the bolts to 15 ft. lbs. (21 Nm) for A.D.D only.

14. Apply sealer to the differential cover and torque the bolts to 34 ft. lbs. (47 Nm).

15. Torque the differential-to-axle nuts to 19 ft. lbs. (27 Nm) and the front driveshaft flange-to-differential flange nuts/bolts to 54 ft. lbs. (74 Nm). Refill the axle with 80W–90 gear oil to a level of ¼ in. below the fill hole.

Axle Housing

REMOVAL AND INSTALLATION

Front

4WD VEHICLES, EXCEPT PREVIA

1. Raise and safely support the vehicle.

2. Matchmark and remove the front driveshaft.

3. Disconnect the axle shafts from the axle assembly.

4. Disconnect vacuum hoses, if equipped with automatic locking hubs.

5. Disconnect the 4WD indicator. Remove the front differential mounting bolt.

6. Support the axle housing with an suitable jack and remove the rear mounting bolts.

To install:

7. Install the differential in position under the vehicle.

8. Install the rear mounting bolts and torque to 123 ft. lbs. (167 Nm).

9. Install the front mounting bolt and torque to 108 ft. lbs. (147 Nm).

10. Connect vacuum hoses and the 4WD indicator.

11. Install the axle shafts and the driveshaft, aligning matchmarks made during removal.

12. Refill the axle with the correct oil and lower the vehicle.

4WD PREVIA

1. Raise the vehicle and safely support it and drain the axle housing.

2. Disconnect the driveshaft and halfshafts. Hang the shafts aside with wire. Be careful not to damage the rubber shaft boots.

3. Remove the left engine undercover and differential support protectors.

4. Place a jack under the axle housing and remove the 3 support bolts and 6 cushions. Lower the jack and remove the assembly.

To install:

5. Raise the assembly and install the collars, cushions and support bolts. Torque the bolts to 54 ft. lbs. (73 Nm).

6. Install the support protectors and torque the bolts to 9 ft. lbs. (12 Nm).

7. Install the under cover.

8. Install the halfshafts and torque the bolts to 51 ft. lbs. (69 Nm).

9. Install the front driveshaft and torque the bolts to 27 ft. lbs. (37 Nm).

10. Refill the axle housing with hypoid gear oil.

Rear

EXCEPT PREVIA

1. Raise and safely support the vehicle.

2. Remove the tire and wheel assemblies.

3. Support the axle housing with a suitable jack.

4. Disconnect the shock absorber lower bolts.

5. Disconnect the stabilizer bar and lateral rod. Disconnect the brake lines from the axle housing.

6. Remove the leaf spring U-bolts and carefully lower the axle housing from the vehicle.

7. Installation is the reverse of the removal procedure. Torque the shock absorber lower bolts to 19 ft. lbs (25 Nm) on 2WD vehicles or to 47 ft. lbs (64 Nm) on 4WD vehicles. Tighten the U-bolt nuts to 90 ft. lbs (123 Nm).

PREVIA

1. Raise and safely support the vehicle.

2. Remove the tire and wheel assemblies.

3. Support the axle housing with a suitable jack.

4. Disconnect the shock absorber lower bolts.

5. Disconnect the stabilizer bar and lateral rod. Also, disconnect the upper and lower control arms.

6. Remove the brake lines from the axle housing.

7. Slowly lower the axle housing from the vehicle.

8. Installation is the reverse of the removal procedure. Observe the following torques:

a. Shock absorber bottom bolts — 27 ft. lbs (37 Nm) for 2WD, 94 ft. lbs. (127 Nm) for 4WD vehicles.

b. Upper control arm bolts — 105 ft. lbs. (142 Nm).

c. Lower control arm bolts — 105 ft. lbs. (142 Nm).

d. Stabilizer bar bolts — 19 ft. lbs. (25 Nm).

STEERING

Steering Wheel

REMOVAL AND INSTALLATION

1. Disconnect the negative battery cable.

2. Position the wheels in a straight ahead position.

3. Remove the steering wheel center cover, some vehicles use a screw to retain the cover.

4. Disconnect the horn wire. Matchmark the wheel and the shaft.

5. Using an appropriate wheel puller tool 0960920011 or equivalent, remove the steering wheel.

6. Installation is the reverse of the removal procedure. Tighten the

steering wheel nut to 25 ft. lbs. (34 Nm).

Steering Column

REMOVAL AND INSTALLATION

1. Disconnect the negative battery cable.

2. Remove the instrument panel and steering column finish panels.

3. Disconnect all electrical connectors from the column.

4. Remove the lower joint protectors, if equipped.

5. Remove the pinch bolt and nut from the intermediate shaft or flex joint. Disconnect the intermediate shaft from the column shaft or disconnect the flex joint from the steering gear.

6. Remove the column bracket-to-instrument panel nuts.

7. Remove the floor boot retainers and pull the boot away from the floor.

8. Remove the column from the vehicle. Do not hammer on the shaft.

To install:

9. Install the column into the vehicle. Make sure the plastic retainers are properly aligned. Torque the column-to-instrument panel nuts to 18 ft. lbs. (25 Nm).

10. Install the floor boot and retainers. Make sure the boot is sealed from water. Use sealer between the floor and boot.

11. Connect the intermediate shaft to the column shaft or connect the flex joint to the steering gear. Install the pinch bolt and nut to the intermediate shaft or flex joint. Torque the bolt to 26 ft. lbs. (35 Nm).

12. Install the lower joint protectors, if equipped.

13. Connect all electrical connectors to the column.

14. Install the instrument panel and steering column finish panels.

15. Connect the negative battery cable and check operation.

Manual Steering Gear

ADJUSTMENT

Pick-Up and 4Runner

1. Install a special socket 0961600010 or equivalent, and inch lbs. torque wrench on the worm shaft.

2. Turn the adjusting screw while measuring the preload. It should be 6.9–9.5 inch lbs. (0.8–1.1 Nm).

3. Install the lock nut and torque to 34 ft. lbs. (46 Nm).

REMOVAL AND INSTALLATION

2WD Pick-Up

1. Raise and safely support the vehicle. Remove the pitman arm-to-relay rod cotter pin and nut. Separate the relay rod from the pitman arm.
2. Matchmark the flexible steering coupling-to-steering gear, then remove the lock bolt and separate the steering coupling from the steering gear.
3. Remove the steering gear housing mounting bolts and the gear housing.
4. To install, reverse the removal procedures. Torque the housing-to-frame bolts to 48 ft. lbs. (62 Nm), the pitman arm-to-relay rod nut 67 ft. lbs. (93 Nm) and the steering gear-to-coupling yoke to 15–20 ft. lbs. (20–27 Nm).

4WD Pick-Up and 4Runner

1. Raise and safely support the vehicle. Remove the stone shield from the gear housing, if equipped.
2. Matchmark the intermediate shaft-to-steering gear and disconnect them.
3. Remove the cotter pin and plug from the drag link.
4. Disconnect the drag link from the pitman arm.
5. Remove the pitman arm nut. Using a puller tool, separate the pitman arm from the steering gear.
6. Remove the steering gear housing-to-frame bolts and the gear housing.
7. To install, reverse the removal procedures. Torque the steering gear-to-frame bolts to 42 ft. lbs. (56 Nm), the steering gear-to-intermediate bolts to 29 ft. lbs. (40 Nm), the pitman arm-to-steering gear nut to 127 ft. lbs. (170 Nm).

NOTE: When installing the drag link to the pitman arm, tighten the plug completely and loosen it 1⅓ turns.

Land Cruiser

1. Raise and safely support the vehicle. Remove the worm yokes from the worm and the main shaft.
2. Remove the intermediate shaft assembly.
3. Remove the pitman arm from the sector shaft.
4. Remove the steering gear-to-frame bolts and the steering gear from the vehicle.

5. To install, reverse the removal procedures. Torque the pitman arm to 119–141 ft. lbs. (163–190 Nm).

NOTE: The intermediate shaft must be installed with the wheels in a straight-ahead position and the steering wheel straight-ahead.

Power Steering Gear

ADJUSTMENT

Pick-Up and 4Runner

1. Install a special socket 0961600010 or equivalent, and inch lbs. torque wrench on the worm shaft.
2. Turn the adjusting screw while measuring the preload. It should be 2.6–4.8 inch lbs. (0.3–0.5 Nm).
3. Install the lock nut and torque to 34 ft. lbs. (46 Nm).

Previa

1. Loosen the lock nut and torque the rack guide spring cap to 18 ft. lbs. (25 Nm).
2. Return the cap 30°.
3. Turn the control valve shaft right and left 1 or 2 times. Loosen the spring cap until the rack guide compression spring is not functioning.
4. Using an inch lbs. torque wrench on the worm shaft, tighten the rack guide cap until the preload is within 6.1–11.3 inch lbs. (0.7–1.3 Nm).
5. Torque the lock nut to 41 ft. lbs. (56 Nm).

Land Cruiser

1. Turn the worm shaft in both directions to determine the exact center.
2. Install special socket 0961600010 or equivalent, and an inch lbs. torque wrench to the worm shaft.
3. Turn the adjusting screw while measuring the preload. It should be 6.5–9.5 inch lbs. (0.7–1.1 Nm).
4. Install a new seal and torque the locknut to 34 ft. lbs. (46 Nm).

REMOVAL AND INSTALLATION

2WD Pick-Up

1. Raise and safely support the vehicle. Disconnect and plug the pressure line clamp bolts at the steering gear.
2. Matchmark the intermediate shaft-to-steering gear, then, remove the coupling bolt and separate the in-

termediate shaft from the steering gear.
3. Remove the pitman arm-to-steering gear and the pitman arm-to-relay rod nuts.
4. Using a puller tool, separate the pitman arm from the relay rod and the pitman arm from the steering gear.
5. Remove the steering gear-to-frame bolts and the steering gear from the vehicle.
6. To install, reverse the removal procedures. Torque the steering gear-to-frame bolts to 48 ft. lbs. (66 Nm), the pitman arm-to-steering gear nut to 90 ft. lbs. (122 Nm), the pitman arm-to-relay rod nut to 67 ft. lbs. (92 Nm), the intermediate shaft-to-steering gear bolt to 19 ft. lbs. and the pressure line nuts to 33 ft. lbs. (45 Nm). Bleed the power steering system.

4WD Pick-Up and 4Runner

1. Remove the battery and the engine lower gravel shield.
2. Raise and safely support the vehicle. Disconnect and plug the pressure lines at the steering gear.
3. Remove the steering gear stone shield.
4. Matchmark the intermediate shaft-to-steering gear, then remove coupling bolt and the intermediate shaft from the steering gear.
5. Remove the pitman arm-to-steering gear nut. Using a puller tool, separate the pitman arm from the steering gear.
6. Remove the gear housing-to-frame bolts and the steering gear from the vehicle.
7. To install, reverse the removal procedures. Torque the steering gear-to-frame bolts to 42 ft. lbs. (57 Nm), the pitman arm-to-steering gear nut to 127 ft. lbs. (172 Nm), the intermediate shaft-to-steering gear bolt to 29 ft. lbs. (41 Nm) and the pressure line union nuts to 33 ft. lbs. (44 Nm). Bleed the power steering system.

Land Cruiser

1. Raise and safely support the vehicle. Disconnect the pressure lines from the steering gear.
2. Remove the intermediate shaft-to-steering gear bolt and the steering column-to-firewall bolts.
3. Loosen the steering column-to-dash bolts. Remove the pitman arm-to-steering gear nut.
4. Using a puller, separate the relay rod from the pitman shaft and the pitman arm from the steering gear.
5. Pull the steering column towards the passenger compartment to

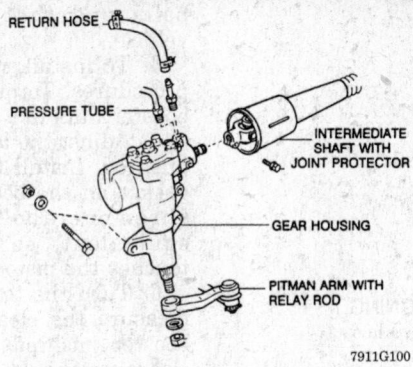

RETURN HOSE

PRESSURE TUBE

INTERMEDIATE SHAFT WITH JOINT PROTECTOR

GEAR HOUSING

PITMAN ARM WITH RELAY ROD

7911G100

Power steering gear assembly — all models

uncouple the steering shaft from the steering gear.

6. Remove the steering gear-to-frame bolts and the steering gear from the vehicle.

7. To install, reverse the removal procedures. Torque the steering gear-to-frame bolts to 40–63 ft. lbs. (54–86 Nm) , the pitman arm-to-steering gear nut to 120–141 ft. lbs. (163–190 Nm), the intermediate shaft-to-steering gear bolt to 22–32 ft. lbs. (29–43 Nm), the pressure hose fitting to 29–36 ft. lbs. (40–48 Nm) and the return hose fitting to 24–30 ft. lbs. (35–41 Nm). Bleed the power steering system.

NOTE: During installation of the hydraulic lines, position each line clear of any surrounding components, then tighten the fittings.

Previa

1. Raise the vehicle and support safely.

2. Disconnect the tie rod ends using a separator tool 0961112010 or equivalent.

3. Place matchmarks on the universal joint and shaft. Remove the lower and upper joint bolts and slide the joint upward to disconnect.

4. Disconnect the pressure and return pipes using flarenut wrenches.

5. 4WD only: remove the front differential assembly, equipment driveshaft housing insulator and drive housing stay.

6. Remove the bracket bolts and grommets. Turn the housing toward the back side and slide the housing to the right side. Put the left tie rod end in the body panel. Pull the housing out through the opening in the left lower side of the body.

To install:

7. Install the gear housing and torque the mounting bolts to 56 ft. lbs. (76 Nm).

8. 4WD only: install the front differential assembly, equipment driveshaft housing insulator and drive housing stay.

9. Connect the pressure and return pipes using flarenut wrenches. Torque the fittings to 33 ft. lbs. (44 Nm).

10. Install the lower and upper joint bolts. Torque the bolts to 26 ft. lbs. (35 Nm).

11. Connect the tie rod ends, torque to 36 ft. lbs. (49 Nm) and install a new cotter pin.

12. Align the front end and lower the vehicle.

Power Steering Pump

REMOVAL AND INSTALLATION

Pick-Up, Land Cruiser and 4Runner

NOTE: Disconnect the air hoses from the air control valve and the high tension wires from the distributor.

1. Disconnect the negative battery cable. Loosen the power steering pump pulley nut.

NOTE: Use the drive belt as a brake to keep the pulley from rotating.

2. Place a container under the pump. Disconnect the return line and the pressure tube, then drain the fluid into the container.

3. Loosen the idler pulley nut and the adjusting bolt, then remove the drive belt.

4. Remove the drive pulley and the Woodruff key from the pump shaft.

5. Remove the mounting bolts and the power steering pump from the vehicle.

6. To install, reverse the removal procedures. Torque the pump pulley mounting bolt to 29 ft. lbs. (40 Nm),

the pump pulley nut to 32 ft. lbs. (42 Nm) and the pressure hoses to 33 ft. lbs. (45 Nm). Adjust the drive belt tension. Bleed the power steering system.

Belt Adjustment

Use a belt tension gauge and make sure the tension is 105 lbs. (140 N) for new or 85 lbs. (115 N) for an old belt. Loosen the air pump mounting and adjust the pump until the proper tension is obtained. Tighten the mounting brackets and check operation.

System Bleeding

1. Raise and safely support the vehicle.

2. Fill the pump reservoir with power steering fluid.

3. With the engine running, rotate the steering wheel from lock to lock several times. Add fluid as necessary.

NOTE: Perform the bleeding procedure until all of the air is bled from the system.

4. The fluid level should not have risen more than 0.2 in.; if it does, check the pump.

Tie Rod Ends

REMOVAL AND INSTALLATION

1. Raise and safely support the vehicle.

2. Remove the wheel and tire assembly. Remove the cotter pin and nut.

3. Using a tie rod end puller, disconnect the tie rod from the relay rod.

4. Using a tie rod end puller remove the tie rod from the steering knuckle.

5. Remove the tie rod end from the vehicle.

6. Installation is the reverse of the removal procedure. Tighten the clamp nuts to 19 ft. lbs. (25 Nm) and the knuckle-to-arm nuts to 67 ft. lbs. (90 Nm). Always install a new cotter pin.

BRAKES

For all brake system repair and service procedures not detailed below, please refer to ""Brakes" in the Unit Repair section.

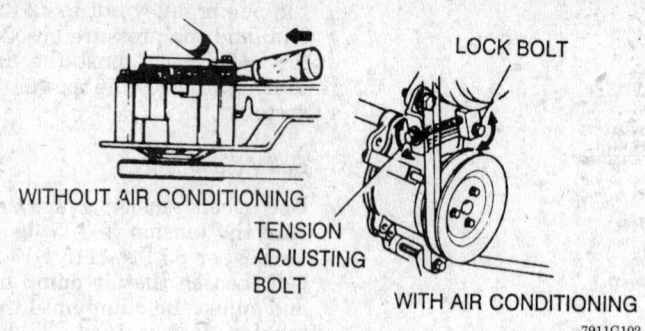

WITHOUT AIR CONDITIONING

LOCK BOLT

TENSION
ADJUSTING
BOLT

WITH AIR CONDITIONING

7911G102

Drive belt adjustment

Master Cylinder

REMOVAL AND INSTALLATION

Pick-Up, 4Runner, Land Cruiser and Previa

1. Disconnect the negative battery cable. Using a syringe, remove the brake fluid from the master cylinder.
2. Disconnect and plug the hydraulic lines at the master cylinder.
3. If equipped, disconnect the level warning switch connector from the master cylinder.
4. Remove the master cylinder-to-power booster nuts and the master cylinder assembly from the power brake unit.
5. To install, reverse the removal procedures. Torque the master cylinder mounting bolts to 9 ft. lbs. (12 Nm) and the brake lines-to-master cylinder to 11 ft. lbs. (15 Nm). Refill the master cylinder with new brake fluid and bleed the brake system.

Load Sensing Proportioning Valve

REMOVAL AND INSTALLATION

1. Raise and safely support the vehicle, so it is level.
2. Disconnect the No. 2 shackle from the bracket.
3. Disconnect and plug the brake lines from the load sensing valve.
4. Remove the load sensing valve bracket from the frame.
5. To install, reverse the removal procedures. Torque the load sensing valve-to-frame bolts to 14 ft. lbs. (18 Nm) and the brake tubes to 11 ft. lbs. (15 Nm). Bleed the brake system.

ADJUSTMENT

Except Previa

1. Adjust the load sensing valve and the rear axle load.
2. Check and/or adjust the length of the No. 2 shackle for distance from the center of the No. 2 shackle-to-shackle bracket bolt to the center of the No. 1 shackle-to-spring bolt, 3.07 in. for 2WD Pick-Up, and Land Cruiser, 4.72 in. for 4WD Pick-Up, 4Runner and 1.18 in. for Previa.
3. Shortening the **A** length raises the pressure to the rear brakes and lengthening lowers the pressure. 1 turn will adjust the pressure to the rear brakes 28–51 psi (196–353 kPa).

Power Brake Booster

REMOVAL AND INSTALLATION

1. Disconnect the negative battery cable. Separate the master cylinder from the power brake booster.
2. Remove the vacuum hose form the power brake booster.
3. Working under the instrument panel, remove the brake pedal-to-brake booster rod clevis pin. Remove the power brake booster mounting

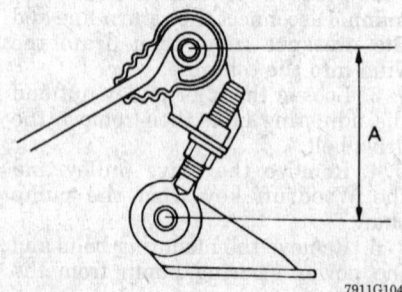

7911G104

Load sensing proportioning valve adjustment — Except Previa

bolts and the booster from the vehicle.
4. To install, reverse the removal procedures. Torque the power brake booster nuts to 9 ft. lbs. (12 Nm).
5. Adjust the length of the booster pushrod. Install the master cylinder gasket on the booster. Set the measurement tool 0973700010 or equivalent, so the pin slightly touches the piston.
6. Turn the tool upside down and measure the clearance between the booster pushrod and pin head. The clearance should be **0**.

NOTE: When installing a new booster, make sure there is a little clearance between the pushrod end and the master cylinder piston.

Brake Caliper

REMOVAL AND INSTALLATION

1. Raise and safely support the vehicle.
2. Remove the wheel and tire assembly.
3. Remove the 2 wire clips at the ends of the brake pad pins.
4. Pull out the pads and anti-rattle springs.
5. Lift out the anti squeal shims.
6. Plug the vent hole on the master cylinder cap, to prevent fluid leakage. Disconnect and plug the brake line from the caliper.
7. Remove the 2 caliper mounting bolts and remove the caliper.
   To install:
8. Position the caliper and install the mounting bolts. Tighten the caliper mounting bolts to 18 ft. lbs. (25 Nm) on Previa, 29 ft. lbs. (41 Nm) on 2WD vehicles or to 90 ft. lbs. (122 Nm) on 4WD vehicles.
9. Install the brake pads. Install the wheel and tire assembly and lower the vehicle.
10. Road test the vehicle.

Disc Brake Pads

REMOVAL AND INSTALLATION

2WD Vehicle

1. Raise and safely support the vehicle.
2. Remove the wheel and tire assembly.
3. Remove the bottom caliper retaining bolt and loosen the top bolt.

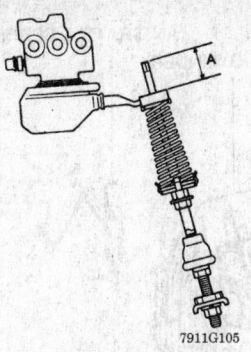

Load sensing proportioning valve adjustment — Previa

4. Pivot the caliper upward and suspend it with wire, do not disconnect the brake line.

5. Remove the anti-squeal springs, brake pads, anti-squeal shims, wear indicator plates and the 4 pad support plates.

6. Installation is the reverse of the removal procedure. Compress the caliper piston into the bore using a C-clamp. Torque the caliper bolts to 29 ft. lbs. (39 Nm).

4WD Vehicle

1. Raise and safely support the vehicle.

2. Remove the wheel and tire assembly.

3. Remove the brake pad retaining clips, 2 locating pins, anti-rattle spring, brake pads and anti-squeal shims.

4. Installation is the reverse of the removal procedure. Compress the caliper piston into the bore using a C-clamp. Torque the caliper bolt to 27 ft. lbs. (36 Nm).

Brake Rotor

REMOVAL AND INSTALLATION

1. Raise and safely support the vehicle.

2. Remove the wheels and tires.

3. Remove the caliper and brake pads. Hang from the suspension with a wire.

4. Remove the brake torque plate.

5. On 2WD vehicles, remove the grease cap, cotter pin and nut from the hub, except Previa.

6. On 4WD vehicles, remove the bolt retaining the locking hub assembly and lift it off the hub. Remove the nuts and washer from the inside of the hub.

7. Remove the rotor from the vehicle. Be careful not to drop the bearings from the hub.

To install:

8. Install the rotor. Be careful not to drop the bearings from the hub.

9. On 4WD vehicles, install the hub bolts.

10. On 2WD vehicles, install the grease cap, cotter pin and nut to the hub. Adjust the bearing preload.

11. Install the brake torque plate and torque the bolts to 77 ft. lbs. (104 Nm).

12. Install the caliper and brake pads.

13. Install the wheels and tires.

14. Bleed the brake system.

Brake Drums

REMOVAL AND INSTALLATION

1. Raise and safely support the vehicle.

2. Remove the wheel and tire assemblies.

3. Remove the brake drum retaining screws.

4. Remove the drum from the vehicle. If the drum is difficult to remove, release the brake adjusters from behind the drum.

5. Install the drum on the vehicle and install the retaining screws.

6. Install the wheel and tire assemblies.

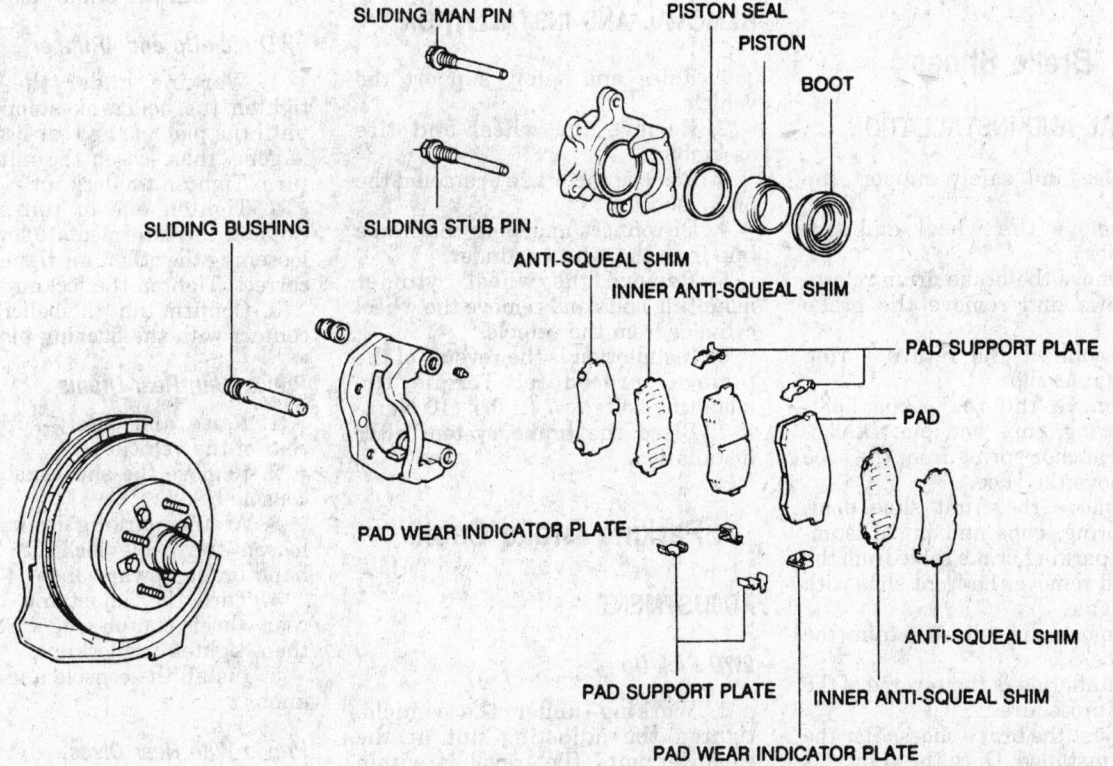

Front brake assembly — typical 2WD vehicle

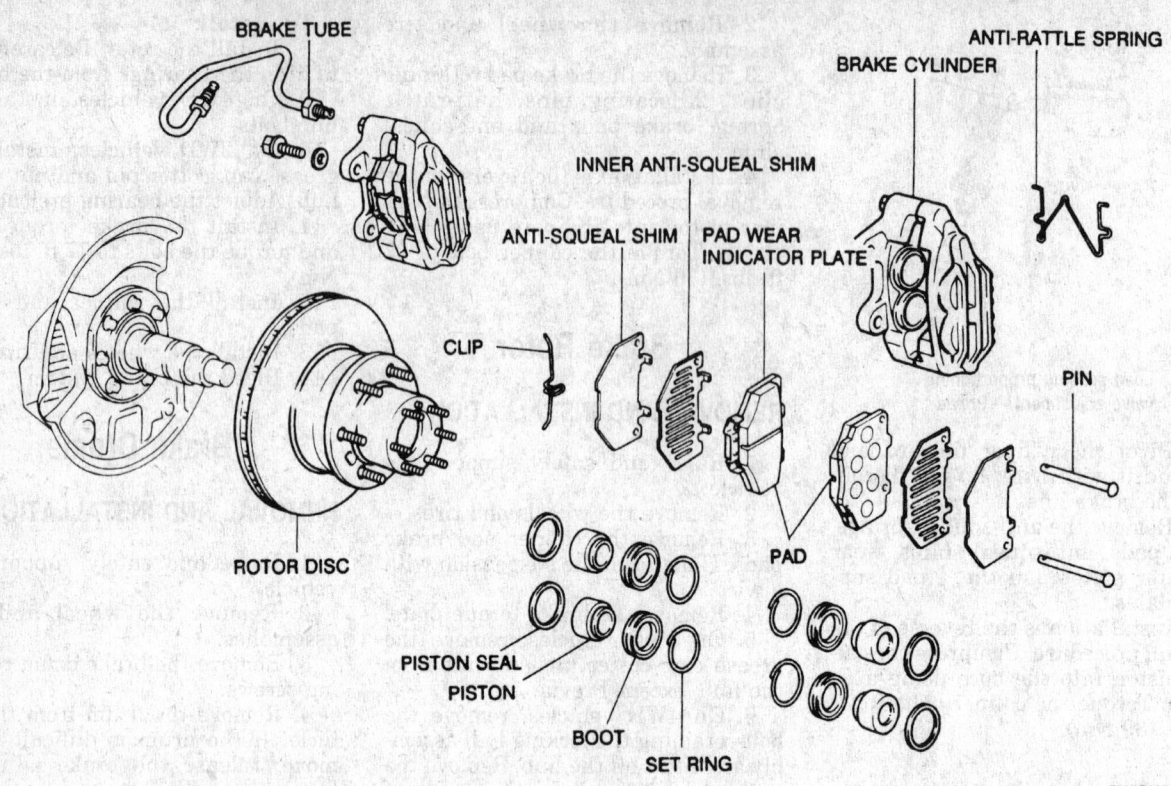

BRAKE TUBE

ANTI-RATTLE SPRING

BRAKE CYLINDER

INNER ANTI-SQUEAL SHIM

ANTI-SQUEAL SHIM

PAD WEAR INDICATOR PLATE

CLIP

PIN

ROTOR DISC

PAD

PISTON SEAL

PISTON

BOOT

SET RING

7911G107

Front brake assembly — typical 4WD vehicle

7. Lower the vehicle. Check the operation of the brakes and adjust as needed.

Brake Shoes

REMOVAL AND INSTALLATION

1. Raise and safely support the vehicle.
2. Remove the wheel and tire assemblies.
3. Remove the brake drum retaining screws and remove the brake drum.
4. Disconnect the return spring from the rear shoe.
5. Remove the rear shoe holddown spring, cups and pin. Disconnect the anchor spring from the shoe and remove the shoe.
6. Remove the front shoe holddown spring, cups and pin. Disconnect the parking brake cable from the lever and remove the front shoe with the adjuster.
7. Remove the adjuster from the front shoe.
8. Installation is the reverse of removal procedure.
9. Adjust the brake shoes after the drum is installed. Once the vehicle is lowered, pump the brakes several times to seat the shoes.

Wheel Cylinder

REMOVAL AND INSTALLATION

1. Raise and safely support the vehicle.
2. Remove the wheel and tire assemblies.
3. Remove the brake drum and the brake shoes.
4. Disconnect and plug the brake line from the wheel cylinder.
5. Remove the wheel cylinder mounting bolts and remove the wheel cylinder from the vehicle.
6. Installation is the reverse of removal procedure. Torque the mounting bolts to 7 ft. lbs. (10 Nm).
7. Bleed the brake system after installation.

Parking Brake Cable

ADJUSTMENT

2WD Pick-Up

1. Working under the vehicle, tighten the adjusting nut at the equalizer until the travel is within limits and there is no drag at the rear shoes.

2. Apply the parking brake several times and again check that there is no drag with the brake released.

4WD Pick-Up and 4Runner

1. Working under the vehicle, tighten the bellcrank stopper screw until the play at the rear brake links is gone, then loosen the nut one full turn. Tighten the locknut.
2. Tighten one of the adjusting nuts on the intermediate lever while loosening the other, until the travel is correct. Tighten the locknuts.
3. Confirm that the bellcrank is in contact with the backing plate.

Previa With Rear Drums

1. Raise and safely support the rear of the vehicle.
2. Remove the shift knob and the console box.
3. At the parking brake handle, loosen the cable locknut. Pull the hand brake upward about 4–5 clicks.
4. Turn the adjust nut until the rear wheels can no longer be turned, then, tighten the locknut.
5. Install the console and the shift knob.

Previa With Rear Discs

1. Raise the vehicle and support safely. Remove the rear wheel.

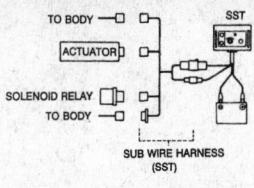

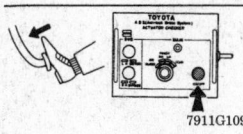

7911G109

Brake system bleeding when the power steering system is disconnected at the actuator — Pick-Up and 4Runner with ABS

2. Unfasten the double nuts at the top end of the shock absorber. Remove the cushions and the cushion retainers.

3. Remove the shock absorber-to-axle housing bolt.

4. Compress the shock absorber and remove it from the vehicle.

5. To install, reverse the removal procedures. Torque the shock absorber-to-suspension arm nut/bolt to 101 ft. lbs. (137 Nm) for Pick-Up and 4Runner and the shock absorber-to-body nuts to 19 ft. lbs. (27 Nm).

Land Cruiser

1. Raise and safely support the vehicle. Remove the wheel and tire assembly.

2. Remove mounting bolts from the top and the bottom of the shock and remove shock.

3. To install, reverse the removal procedures.

MacPherson Strut

REMOVAL AND INSTALLATION

Previa

1. Raise the vehicle and support safely.

2. Remove the halfshaft cotter pin and nut for 4WD only.

3. Disconnect the brake hose from the strut by removing the clips.

4. Loosen the 2 strut-to-knuckle bolts, do not remove.

5. Remove the stabilizer bar from the strut.

6. Remove the 3 upper strut mount nuts and remove the strut from the vehicle.

To install:

7. Install the strut and torque the upper nuts to 47 ft. lbs. (64 Nm).

8. Install the strut-to-knuckle bolts and torque to 231 ft. lbs. (314 Nm).

9. Install the stabilizer bar and torque the bolt to 76 ft. lbs. (103 Nm).

10. Install the halfshaft nut and torque to 136 ft. lbs. (187 Nm). Install a new cotter pin.

11. Install the remaining components and align the front end.

Coil Springs

REMOVAL AND INSTALLATION

Land Cruiser

1. Raise the vehicle and support the body and frame.

2. Raise the front axle assembly.

3. Hold the shock piston rod with a wrench and remove the upper mounting nut.

4. Disconnect the stabilizer bar with the cushion and bracket.

5. Lower the jack and install a spring compressor tool 0972730020 or equivalent. Compress the spring to take the load off the the upper mount.

6. Remove the upper spring retaining nuts and remove the spring.

To install:

7. Compress the coil spring and position into the vehicle. Align the spring end with the lower seat. Torque the upper mount nuts to 30 ft. lbs. (40 Nm).

8. Remove the spring compressor and raise the jack.

9. Connect the stabilizer bar and shock absorber. Torque the stabilizer and shock nut to 51 ft. lbs. (69 Nm).

Torsion Bars

REMOVAL AND INSTALLATION

2WD Pick-Up

The vehicles are equipped with torsion bar front springs.

NOTE: Great care must be taken to make sure springs are not mixed after removal. It is strongly suggested that before removal, each spring be marked with paint, showing front and rear of spring and from which side of the vehicle it was taken. If the springs are installed backwards or on the wrong sides of the vehicle, they could fracture. If replacing the springs, it is not necessary to mark them.

1. Raise and safely support the front of the vehicle.

2. Slide the boot from the rear of torsion bar spring, then paint an alignment mark from the torsion bar spring onto the anchor arm and the torque arm. There are right and left identification marks on the rear end of the torsion bar springs.

3. On the rear torsion bar spring holder, there is a long bolt that passes through the arm of the holder and up through the frame crossmember. Remove the locking nut only from this bolt.

4. Using a small ruler, measure the length from the bottom of the remaining nut to the threaded tip of the bolt and record this measurement.

5. Place a jack under the rear torsion bar spring holder arm and raise the arm to remove the spring pressure from the long bolt. Remove the adjusting nut from the long bolt.

6. Slowly lower jack.

7. Remove the long bolt, the spacers, the anchor arm and the torsion bar spring. The torsion bar should be easily pulled out of the anchor and the torque arms.

NOTE: Inspect all parts for wear damage or cracks. Check the boots for rips and wear. Inspect the splined ends of the torsion bar spring and the splined holes in the rear holder and the front torque arm for damage. Replace as necessary.

To install:

8. Coat the splined ends of the torsion bar with multi-purpose grease.

9. If installing the old torsion bars, perform the following:

a. Slide the front of the torsion bar spring into the torque arm, making sure the alignment marks are matched.

b. Slide the anchor arm onto the rear of the torsion bar spring, making sure the alignment marks are matched. Install the long bolt and it's spacers.

c. Tighten the adjusting nut so it is the same length as it was before removal.

NOTE: Do not install the locknut.

10. When installing a new torsion bar spring, perform the following:

a. Raise the front of the vehicle, replace the wheel and tire assembly, place a wooden block (7½ in. high) under the front tire. Lower the jack until the clearance between the spring bumper (on the

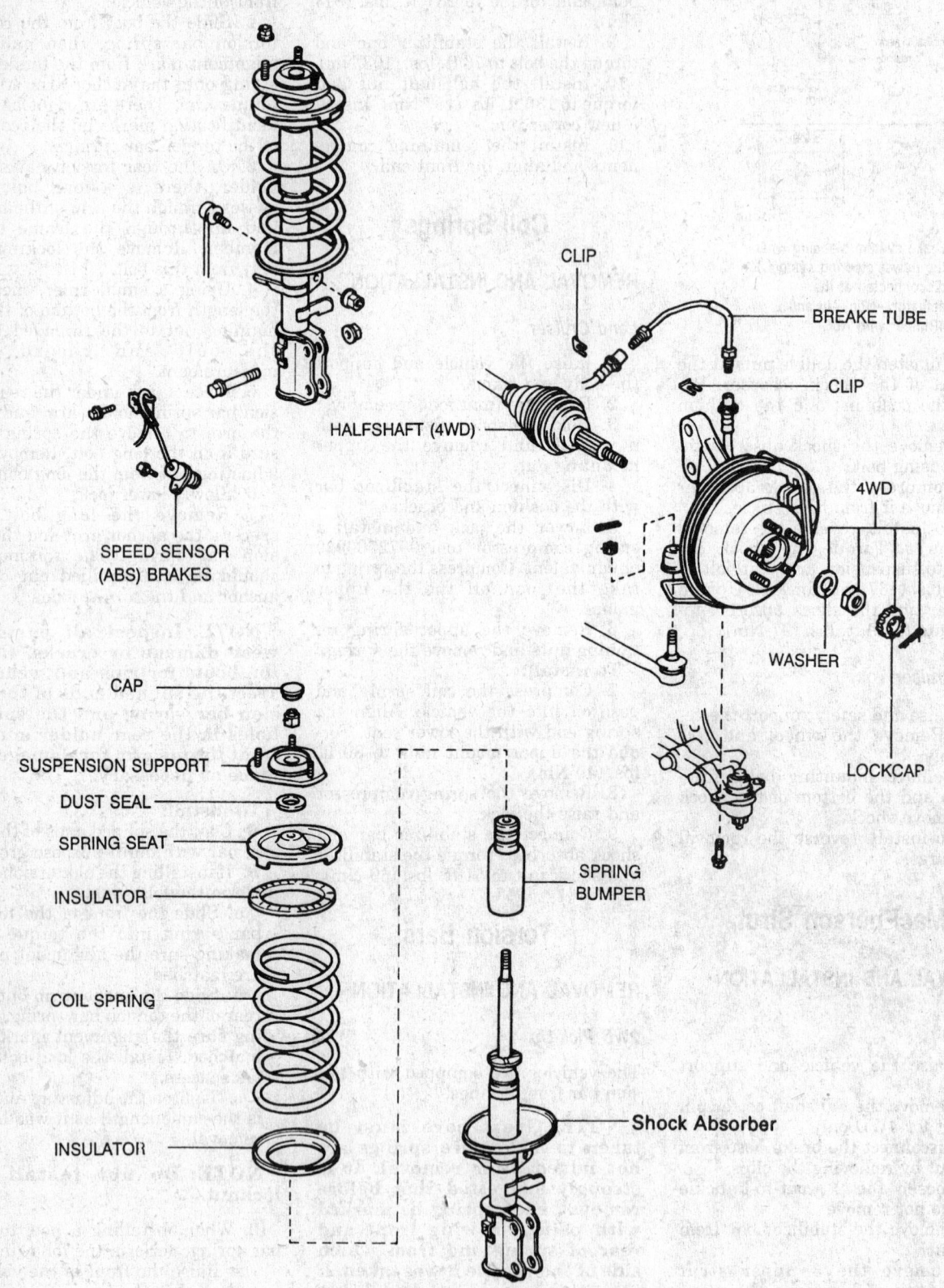

CLIP

BREAKE TUBE

CLIP

HALFSHAFT (4WD)

SPEED SENSOR
(ABS) BRAKES

4WD

WASHER

LOCKCAP

CAP

SUSPENSION SUPPORT

DUST SEAL

SPRING SEAT

INSULATOR

SPRING
BUMPER

COIL SPRING

INSULATOR

Shock Absorber

7911G110

Front suspension components — Previa

lower control arm) and the frame is ½ in.

b. Slide the front of the torsion bar spring into the torque arm.

c. Install the anchor arm into the rear of the torsion bar spring, then the long bolt and the spacers. the distance from the top of the upper spacer to the tip of the threaded end of bolt is 0.310–1.100 in. for ½ ton vehicles or 0.430–1.220 in. for ¾ ton vehicles.

NOTE: Make sure the bolt and bottom spacer are snuggly in the holder arm while measuring.

d. Remove the wooden block and lower the vehicle until it rests on the jackstands.

e. Install and tighten the adjusting nut until the distance from the bottom of the nut to the tip of the threaded end of the bolt is 2.7–3.5 in.

NOTE: Do not install the locknut.

11. Apply multi-purpose grease to the boot lips, then refit the boots to the torque and the anchor arms.

12. Lower the vehicle to the floor and bounce it several times to settle the suspension. With the wheels on the ground, measure the distance from the ground to the center of the lower control arm-to-frame shaft. Adjust the vehicle height using the adjusting nut on the anchor arm. The height should be approximately 10.31 in.

NOTE: If, after achieving the correct vehicle height, the distance from the bottom of the adjusting nut to the top of the threaded end of the long bolt is not within 2.7–3.5 in., change the position of the anchor arm-to-tension bar spring spline and reassemble.

13. Install and torque the locknut on the long bolt to 61 ft. lbs. (83 Nm).

NOTE: Make sure the adjusting nut does not move when tightening locknut.

4WD Pick-Up and 4Runner

These vehicles are equipped with torsion bar front springs.

NOTE: Great care must be taken to make sure springs are not mixed after removal. It is strongly suggested that before removal, each spring be marked with paint, showing front and rear of spring and from which

side of the vehicle it was taken. If the springs are installed backwards or on the wrong sides of the vehicle, they could fracture. If replacing the springs, it is not necessary to mark them.

1. Raise and safely support the vehicle.

2. Using a piece of chalk, remove the boots. Matchmark the torsion bar spring, the anchor arm and the torque arm.

3. Remove the locknut.

4. Measure the protruding length of the adjusting arm bolt, from the nut to the end of the bolt.

NOTE: The adjusting arm bolt measurement is used as a reference to establish the chassis ground clearance.

5. Remove the adjusting nut, the anchor arm and the torsion bar spring.

NOTE: When installing the torsion bar springs, be sure to check the left/right indicating marks on the rear end of the springs; be careful not to interchange the springs.

To install:

6. Using molybdenum disulphide lithium base grease, apply a coat to the torsion bar spring splines.

7. If installing a used torsion bar spring, perform the following procedures:

a. Align the matchmarks, install the torsion bar spring to the torque arm.

b. Align the matchmarks and install the anchor arm to the torsion bar spring.

c. Tighten the adjusting nut until the bolt protrusion is the same as it was before.

8. If installing a new torsion bar spring, perform the following procedures:

a. Make sure the upper and lower arms rebound.

b. Install the boots onto the torsion bar spring.

c. Install one end of the torsion bar spring to the torque arm.

d. Install the torsion bar spring onto the opposite end of the anchor arm.

e. Finger tighten the adjusting nut until the adjusting bolt protrudes about 1.570 in.

f. Tighten the adjusting nut until the adjusting bolt protrudes about 3.430 in.

g. Install the wheel(s) and remove the jackstands. Bounce the

front of the vehicle to stabilize the suspension.

9. To adjust the ground clearance, turn the adjusting nut until the center of the cam plate nut, located of the front end of the lower suspension arm, about 11.220 in. above the ground.

10. After adjusting the ground clearance, torque the locknut to 61 ft. lbs. (83 Nm), then, install the boots.

Upper Ball Joints

INSPECTION

1. Raise the vehicle and place jackstands under the lower control arms and check for excess play.

2. Move the suspension arm up and down. Maximum vertical play should be 0.

3. If the ball joints are within specifications and a looseness problem still exists, check the other suspension parts for wheel bearings, tie rods and etc.

4. The bottom of the tire should not move more than 0.200 in. when the tire is pushed and pulled inward and outward. The tire should not move more than 0.090 in. up and down.

5. If the play is greater than these figures, replace the ball joint.

REMOVAL AND INSTALLATION

1. Raise and safely support the vehicle. Remove the wheel and tire assembly.

2. Support the lower control arm with a floor jack.

3. Remove the brake caliper and support it aside, with a wire.

4. Using a ball joint removal tool, separate the tie rod end from the knuckle arm.

5. Remove the ball joint-to-control arm mounting bolts and separate the joint from the arm.

6. To install, reverse the removal procedures. Torque the ball joint-to-upper control arm bolts 20 ft. lbs. (27 Nm) for 2WD Pick-Up, 25 ft. lbs. (34 Nm) for 4WD Pick-Up, the ball joint-to-lower control arm bolts to 51 ft. lbs. (69 Nm) for Pick-Up, and the lower ball joint-to steering knuckle nut to 25 ft. lbs. (34 Nm) for 4WD Pick-Up.

NOTE: Be sure to grease the ball joints before moving the vehicle.

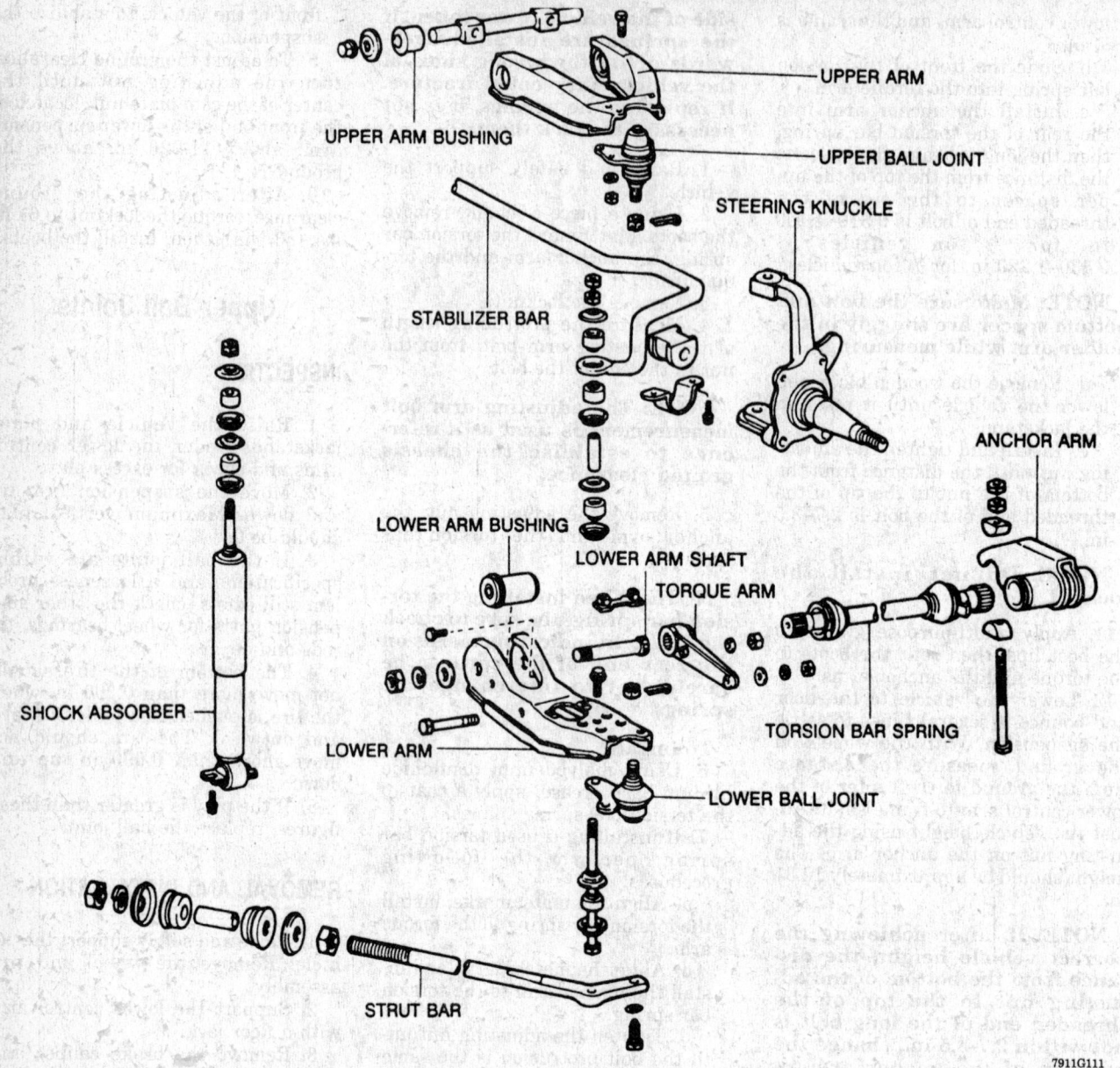

UPPER ARM

UPPER ARM BUSHING

UPPER BALL JOINT

STEERING KNUCKLE

STABILIZER BAR

ANCHOR ARM

LOWER ARM BUSHING

LOWER ARM SHAFT

TORQUE ARM

SHOCK ABSORBER

LOWER ARM

TORSION BAR SPRING

LOWER BALL JOINT

STRUT BAR

7911G111

Front suspension components — 2WD vehicle

Lower Ball Joint

INSPECTION

1. Raise the vehicle and place jackstands under the lower control arms and check for excess play.

2. Move the suspension arm up and down. Maximum vertical play should be 0.

3. If the ball joints are within specifications and a looseness problem still exists, check the other suspension parts for wheel bearings, tie rods and etc.

4. The bottom of the tire should not move more than 0.200 in. when the tire is pushed and pulled inward and outward. The tire should not move more than 0.090 in. up and down.

5. If the play is greater than these figures, replace the ball joint.

REMOVAL AND INSTALLATION

1. Raise and safely support the vehicle. Remove the wheel and tire assembly.

2. Support the lower control arm with a floor jack.

3. Remove the brake caliper and support it aside, with a wire.

4. Using a ball joint removal tool, separate the tie rod end from the knuckle arm.

5. Using a ball joint removal tool, separate the upper ball joint from the steering knuckle.

6. Remove the ball joint-to-control arm mounting bolts and separate the joint from the arm.

7. To install, reverse the removal procedures. Torque the ball joint-to-upper control arm bolts 20 ft. lbs. (27 Nm) for 2WD Pick-Up, 25 ft. lbs. (34 Nm) for 4WD Pick-Up, the upper ball joint-to-steering knuckle nut to 80 ft.

lbs. (122 Nm) for 2WD Pick-Up, 105 ft. lbs. (140 Nm) for 4WD Pick-Up.

NOTE: Be sure to grease the ball joints before moving the vehicle.

Knuckle Joint Bearings

REMOVAL AND INSTALLATION

Land Cruiser

1. Raise the vehicle and support safely. Remove the front wheels.
2. Remove the caliper, axle hub and rotor.
3. Remove the spindle mounting bolts, dust seal and cover.
4. Using a brass drift, tap the spindle off of the steering knuckle. Tap around the side to dislodge.
5. Position 1 flat part of the outer shaft upward and pull out the axle shaft.
6. Using a tie rod remover 0961122012 or equivalent, remove the tie rod end.
7. Remove the oil seal and retainer from the rear of the knuckle.
8. Remove the upper and lower knuckle arm and bearing caps. Tap with a brass drift on the cone washers and remove them from the arm or bearing cap.
9. Using a bearing remover tool 0960660020 or equivalent, push the bearing cap and shims from the steering knuckle. Label each bearing set upper and lower for installation.
10. Remove the outer race using a brass drift.
    To install:
11. Using a race installer tool 0960560010 or equivalent, drive in the new bearing outer races.
12. Coat the knuckle bearings with molybdenum lithium grease.
13. Mount the bearings on tool 0963460013 or equivalent.
14. Add preload to the bearing by tightening the upper nut on the bearings installation tool. Preload should be 6.6–13.2 lbs. (29–59 N).
15. Measure distance **A** and **B**. The difference between the 2 measurements is required to maintain the correct bearing preload (shim thickness).
16. Install the oil seal into the axle housing using tool 0951860010 or equivalent.
17. Install the felt dust seal, rubber seal and steel ring to the knuckle.
18. Pack the bearings with molybdenum lithium grease.

19. Install the bearings into position on the knuckle and axle housing. Insert the knuckle on the axle housing.
20. Using a support tool 0960660020 or equivalent, to support the bearings. Install the knuckle over the shims that were originally used or selected in the adjustment operation.
21. Using a hammer, tap the knuckle arm into the bearing inner races.
22. Install the bearing cap over the shims and tap with a hammer.
23. Install the cone washers and torque the nuts to 71 ft. lbs. (96 Nm).
24. Install the tie rod end and torque to 67 ft. lbs. (91 Nm). Install a new cotter pin.
25. Install the seal set to the knuckle.
26. Install the axle shaft and pack with grease.
27. Install the spindle dust cover and axle hub.
28. Install the remaining components and check operation.

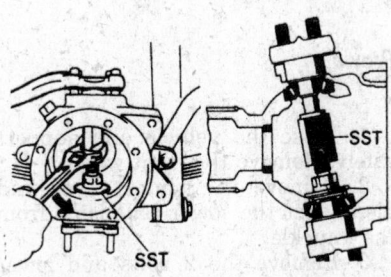

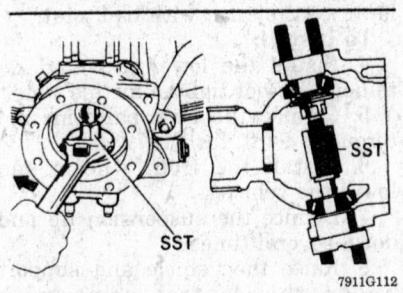

Knuckle joint bearings — Land Cruiser

7911G112

Upper Control Arm

REMOVAL AND INSTALLATION

Pick-Up and 4Runner

1. Raise and safely support the vehicle. Remove the wheel and tire assembly.
2. Using a floor jack, support the lower control arm.
3. Remove the upper ball joint-to-upper control arm nuts/bolts, then, disconnect the upper control arm.
4. Remove the upper control arm-to-chassis bolts and camber adjusting shims and the upper control arm.

NOTE: When removing the camber adjusting shims, be sure to record their location and thickness of shims, so they may be reinstalled in their original positions.

5. To install, reverse the removal procedures. Torque the upper control arm-to-chassis bolts to 72 ft. lbs. (96 Nm) and the upper control arm-to-upper ball joint nuts/bolts to 20 ft. lbs. (27 Nm). Check and/or adjust the front wheel alignment.

Lower Control Arm

REMOVAL AND INSTALLATION

2WD Pick-Up

1. Raise and safely support the vehicle. Remove the torsion bar spring.
2. Remove the shock absorber, the stabilizer bar and the strut bar from the lower arm.
3. Remove the shock absorber from the lower arm.
4. From the lower ball joint, remove the cotter pin and the nut. Using a ball joint removal tool, press the ball joint from the lower control arm.

NOTE: If the lower ball joint is not to be replaced, simply unbolt it from the lower control arm. It is not necessary to separate the ball joint from the steering knuckle.

5. Remove the lower control arm shaft nut. Remove the spring torque arm from the other side of the lower control arm, then, remove the lower arm shaft bolt and the lower arm.
6. To install, reverse the removal procedures.
7. Tighten the bolt(s) holding the lower control arm to the frame but do not torque them until the vehicle is on the ground.

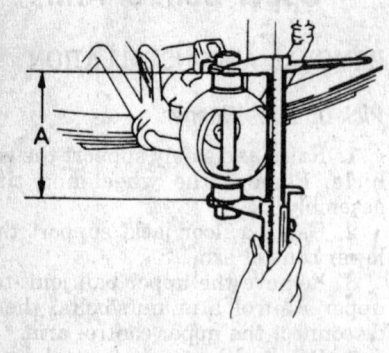

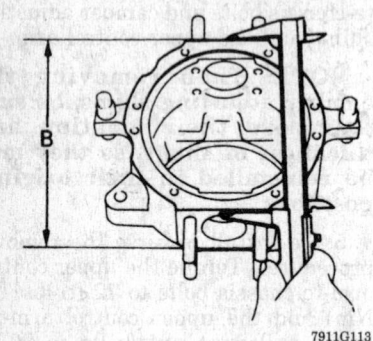

7911G113

Knuckle bearing preload — Land Cruiser

8. Torque the ball joint-to-lower control arm nuts/bolts to 51 ft. lbs. (69 Nm), the strut bar-to-lower control arm bolts to 70 ft. lbs. (95 Nm), the stabilizer bar-to-lower control arm bolts to 9 ft. lbs. (12 Nm), the lower shock absorber bolt to 13 ft. lbs. (18 Nm), upper shock absorber bolt to 18 ft. lbs. (25 Nm) and the lower arm mounting nuts to 166 ft. lbs. (220 Nm).

9. Check and/or adjust the front end alignment.

NOTE: Do not torque the control arm bolts fully until the vehicle is lowered and bounced several times; if the bolts are tightened with the control arm(s) hanging, excessive bushing wear will result.

4WD Pick-Up and 4Runner

1. Raise and safely support the vehicle. Remove the shock absorber.
2. Disconnect the stabilizer bar from the lower suspension arm.
3. Remove the lower ball joint-to-lower control arm bolts, then, separate the control arm from the ball joint.
4. Using a piece of chalk, place matchmarks on the front/rear adjusting cams.

5. Remove the nuts and adjusting cams and the lower control arms.
6. To install, reverse the removal procedures. Torque the lower ball joint-to-lower control arm bolts to 20 ft. lbs. (27 Nm), the stabilizer bar-to-lower control arm bolts to 19 ft. lbs. (26 Nm), the shock absorber-to-lower control arm nut/bolt to 101 ft. lbs. (137 Nm).
7. Lower the vehicle to the ground, bounce it a few times, align the matchmarks and torque the adjusting cam nuts to 203 ft. lbs. (270 Nm). Check and/or adjust the front wheel alignment.

Land Cruiser

1. Raise the vehicle and support safely.
2. Support the frame and body. Support the axle assembly with a floor jack.
3. Remove the lower control arm-to-frame bolt.
4. Remove the 2 bolts and nuts from the axle housing and remove the control arm.
To install:
5. Install the control and bolts loosely. Do not tighten at this time.
6. Install the front wheels and lower the vehicle. Bounce the front end to stabilize the front suspension.
7. Torque the retaining bolts to 127 ft. lbs. (171 Nm).

Previa

1. Raise the vehicle and support safely. Remove the front wheel.
2. Remove the under cover and disconnect the lower ball joint from the knuckle.
3. Remove the 2 bolts and lower arm bracket, nut and arm shaft and lower control arm with ball joint.
To install:
4. Install the lower arm and retainers. Do not tighten at this time.
5. Connect the lower ball joint and torque to 94 ft. lbs. (127 Nm).
6. Install the front wheels and lower the vehicle.
7. Bounce the suspension up and down several times.
8. Raise the vehicle and support under the control arms with jackstands.
9. Torque the lower arm retainers to 121 ft. lbs. (164 Nm).
10. Install the remaining components and lower the vehicle.

Sway Bar

REMOVAL AND INSTALLATION

1. Raise the vehicle and support safely.
2. Remove the under covers.
3. Disconnect the sway bar links-to-control arms.
4. Disconnect the sway bar-to-frame bolts and remove the sway bar.
5. Installation is the reverse of removal. torque the frame bracket bolts to 14 ft. lbs. (19 Nm) and the link nuts to 76 ft. lbs. (103 Nm).

Front Wheel Bearings

REMOVAL AND INSTALLATION

NOTE: This procedure applies to RWD vehicles only. Refer to drive axle instructions for 4WD vehicles.

Pick-Up

1. Raise the vehicle and support safely. Remove the front wheel.
2. Remove the brake caliper and suspend with a wire.
3. Remove the bearing cap, cotter pin, nut and outer bearing. Do not drop the bearing.
4. Remove the rotor/hub assembly.
5. Pry the inner grease seal from the hub and remove the inner bearing.
6. Clean the grease out of the hub using shop rags and compressed air. Do not use solvent to remove grease.
7. Using a brass drift and hammer, drive out the bearing outer races. Do not use old races with new bearings.
NOTE: Grind the outer circumference of the old bearing race with a grinder. Use this race to press in the new race.

8. Install the outer race into the hub using a press. If the new race gets damaged, replace with a new one.
9. Pack the bearings with high temperature bearing grease and install inner bearing.
10. Install the grease seal with tool 0960804100 or equivalent.
11. Install the rotor, outer bearing, washer, nut and nut lock.
12. Torque the nut to 25 ft. lbs. (34 Nm). Loosen the nut until it can be turned by hand.
13. Using a spring tension gauge attached to a lug bolt.
14. Tighten the nut until the bearing preload is 0.9–2.2 lbs. (3.9–9.8 N)

for dual wheel vehicles or 1.3–4.0 lbs. (5.9–17.7 N) for single wheel vehicles.

15. Install nut lock, new cotter pin and grease cap.

16. Install the remaining components and check operation.

2WD Previa

1. Remove the steering knuckle from the vehicle. Place the assembly in a vise.

2. Remove the grease cap and lock nut. Remove the spacer or speed sensor rotor.

3. Using a hub remover tool (slide hammer) 0952000031 or equivalent, remove the wheel hub from the knuckle.

4. Remove the bearing from the hub using a press and arbor tool 0955010012 or equivalent.

5. Remove the dust protector and oil seal.

6. Remove the bearing snapring from the knuckle and press the inner bearing from the knuckle.

To install:

7. Using a press and arbor tool 0960810010 or equivalent, press the bearing into the knuckle.

8. Install the outer oil seal and dust cover.

9. Press the axle hub onto the knuckle.

10. Install the new nut and torque to 147 ft. lbs. (199 Nm). Stack the nut using a chisel. Install the grease cap.

11. Install the knuckle assembly onto the vehicle.

REAR SUSPENSION

Shock Absorbers

REMOVAL AND INSTALLATION

Pick-Up, 4Runner and Land Cruiser

1. Raise and safely support the vehicle.

2. Remove the upper shock absorber retaining bolts from the upper frame member.

3. Remove the lower end bolt of the shock absorber from the spring seat.

4. Remove the shock absorber from the vehicle.

NOTE: Inspect the shock for wear, leaks or other signs of damage.

5. To install, reverse the removal procedures. Torque the upper bolt to 19 ft. lbs. (25 Nm) for 2WD vehicle or 47 ft. lbs. (63 Nm) for 4WD vehicle and the lower bolt to 19 ft. lbs. (25 Nm) for 2WD vehicle or 47 ft. lbs. (63 Nm) for 4WD vehicle.

Previa

1. Raise the vehicle and support safely.

2. Support the axle housing with a jack.

3. Disconnect the shock from the lower control arm. Hold the shaft with a suitable tool and remove the nut.

4. Remove the upper mount bolt and shock absorber.

To install:

5. Install the shock and torque the upper mount to 27 ft. lbs. (37 Nm) and tighten the lower mount nut until the shaft protrudes 0.0059 in. (1.5mm).

6. Lower the vehicle and check operation.

Coil Spring

REMOVAL AND INSTALLATION

Previa

1. Raise and safely support the vehicle. Support the axle housing with a floor jack. Remove the wheel and tire assembly.

2. Remove the shock absorber-to-axle housing bolt.

3. Remove the stabilizer-to-axle housing bar bushing bracket bolts.

4. Remove the lateral control arm-to-axle housing nut and disconnect the lateral control arm.

5. Lower the floor jack, then remove the coil spring(s) and the insulators.

NOTE: While lowering the axle housing, be careful not to snag the brake line of the parking brake cable.

6. To install, reverse the removal procedures. Torque the shock absorber bolt to 27 ft. lbs. (34 Nm), the lateral control arm-to-axle housing nut to 43 ft. lbs. (60 Nm) and the stabilizer-to-axle housing bolts to 27 ft. lbs. (34 Nm).

NOTE: Before tightening the lateral control arm and the stabilizer nuts/bolts, bounce the vehicle to stabilize the suspension.

Leaf Springs

REMOVAL AND INSTALLATION

1. Raise and safely support the vehicle. Support the axle housing with a floor jack. Remove the wheel and tire assembly.

2. Lower the floor jack to take the tension off of the spring. Remove the shock absorber mounting nuts/bolts and the shock absorber.

3. Remove the cotter pins and the nuts from the lower end of the stabilizer link. Detach the link from the axle housing.

4. Remove the spring-to-axle housing U-bolt nuts, the spring bumper and the U-bolt.

5. At the front of the spring, remove the hanger pin bolt. Disconnect the spring from the bracket.

6. Remove the spring shackle retaining nuts and the spring shackle inner plate, then carefully pry out the spring shackle with a pry bar.

7. Remove the spring from the vehicle.

To install:

8. To install, perform the following procedure:

    a. Install the rubber bushings in the eye of the spring.

    b. Align the eye of the spring with the spring hanger bracket and drive the pin through the bracket holes and rubber bushings.

NOTE: Use soapy water as lubricant, if necessary, to aid in pin installation. Never use oil or grease.

    c. Finger-tighten the spring hanger nuts/bolts.

    d. Install the rubber bushings in the spring eye at the opposite end of the spring.

    e. Raise the free end of the spring. Install the spring shackle through the bushings and the bracket.

    f. Install the shackle inner plate and finger-tighten the retaining nuts.

    g. Center the bolt head in the hole which is provided in the spring seat on the axle housing.

    h. Fit the U-bolts over the axle housing. Install the lower spring seat for 2WD vehicle or spring bumper for 4WD vehicle and the nuts.

9. To complete the installation, reverse the removal procedures. Torque the U-bolt nuts to 90 ft. lbs. (122 Nm), the hanger pin-to-frame nut to 67 ft. lbs. (90 Nm), the shackle pin nuts to 67 ft. lbs. (90 Nm), the shock

absorber bolts to 47 ft. lbs. (65 Nm) for 4WD vehicle.

NOTE: When installing the U-bolts, tighten the nuts so the length of the bolts are equal.

Rear Control Arms

REMOVAL AND INSTALLATION

Previa

1. Raise and safely support the vehicle. Place a floor jack under the axle housing to support it.

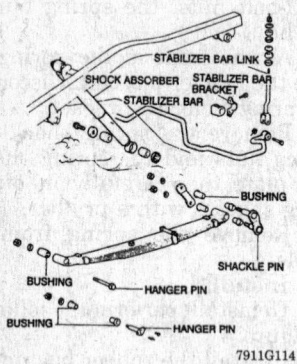

7911G114

Rear suspension — except Previa

2. Remove the upper control arm-to-body bolt, the upper control arm-to-axle housing bolt and the upper control arm from the vehicle.

3. Disconnect the brake line from the lower control arm.

4. Remove the lower control arm-to-body bolt, the lower control arm-to-axle housing bolt and the lower control arm from the vehicle.

To install:

5. Install the upper control arm to the body and to the axle housing with the nuts. Do not tighten the nuts.

6. Install the lower control arm to the body and to the axle housing with the nuts. Do not tighten the nuts.

7. Remove the jack and the supports from under the vehicle. Bounce the vehicle to stabilize the suspension.

8. Using the floor jack under the axle housing, raise the vehicle. Place jackstands under the frame but do not let them touch the frame.

9. To complete the installation, torque the upper control arm-to-body bolt to 105 ft. lbs. (140 Nm), the upper control arm-to-axle housing bolt to 105 ft. lbs. (140 Nm), the lower control arm-to-body bolt to 130 ft. lbs. (176 Nm) and the lower control arm-to-axle housing bolt to 105 ft. lbs. (140 Nm).

Lateral Control Rod

REMOVAL AND INSTALLATION

Previa

1. Raise and safely support the vehicle. Place a floor jack under the axle housing and support it.

2. Remove the lateral control rod-to-axle housing nut.

3. Remove the lateral control rod-to-body nut and the control rod from the vehicle.

To install:

4. Raise the axle housing until the frame is just free of the jack.

5. Install the lateral control rod-to-body with the nut. Do not tighten the nut.

6. Install the lateral control rod-to-axle housing in the following order: washer, bushing, spacer, lateral control rod, bushing, washer and nut. Do not tighten the nut.

7. Remove the jack, lower the vehicle to the floor and bounce it to stabilize the suspension.

8. Using the floor jack under the axle housing, raise the vehicle. Torque the lateral control rod-to-body nut to 81 ft. lbs. (110 Nm) and the lateral control rod-to-axle housing nut to 43 ft. lbs. (58 Nm).

DOMESTIC CHARGING SYSTEMS

Electrical Diagnosis

To satisfy the growing trend toward organized engine diagnosis and tune-up, the following gauge and meter hook-ups, as well as diagnosis procedures are covered. The most sophisticated tune-up and diagnostic facilities are no more than a complex of the basic gauges and meters in common, everyday use. Therefore, to understand gauge and meter hook-ups, their applications and procedures, is to be equipped with the know how to perform the most exacting diagnosis.

Test Instruments

OHMMETER

An ohmmeter is used to measure electrical resistance in a unit or circuit. The ohmmeter has a self contained power supply. In use, it is connected across (or in parallel with) the terminals of the unit being tested.

AMMETER

An ammeter is used to measure the amount of electricity flowing through a unit, or circuit. Ammeters are always connected in series with the unit or circuit being tested.

VOLTMETER

A voltmeter is used to measure voltage pushing the current through a unit, or circuit. The meter is connected across the terminals of the unit being tested.

Alternator Testing

Diagnosis

The first step in diagnosing troubles of the charging system, is to identify the source of failure. Does the fault lie in the alternator or the regulator? The next move depends upon preference or necessity, either repair or replace the defective unit.

If the system is equipped with an external voltage regulator, it is easy to separate an alternator, electrically, from the regulator. Alternator output is controlled by the amount of current supplied to the field circuit of the system.

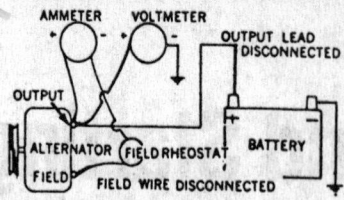

Checking field circuit current draw—as a precaution, a field rheostat should be used to control the amount of current allowed to pass through the circuit during isolation test

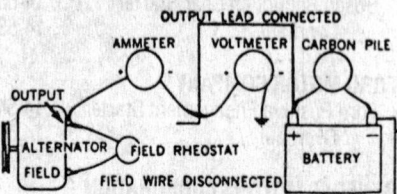

Checking current output of the charging system—if an overcharge of 10–15 amps is indicated, check for a faulty regulator

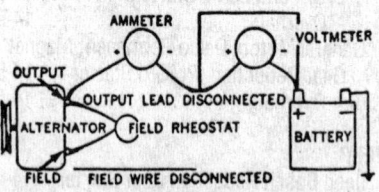

Checking charging system resistance to determine the amount of voltage drop between the alternator output terminal wire and the battery

An alternator is capable of producing substantial current at idle speed. Higher maximum output is also a possibility. This presents a potential danger when testing. As a precaution, a field rheostat should be used in the field circuit when making the following isolation test. The field rheostat permits positive control of the amount of current allowed to pass through the field circuit during the isolation test. Unregulated alternator capacity could ruin the unit.

Most manufacturers of precision gauges offer special test connectors, in sets, that will adapt to the leads and connections of any charging system.

Alternator Test Plans

The following is a procedure pattern for testing the various alternators and their control systems.

There are certain precautionary measures that apply to alternator tests in general. These items are listed in detail to avoid repetition when testing each make of alternator and to encourage a habit of good test procedure.

1. Check alternator drive belt for condition and tension.
2. Disconnect the battery cables.

Check physical, chemical and electrical condition of battery.

3. Be absolutely sure of polarity before connecting any battery in the circuit. Reversed polarity will ruin the diodes.
4. Never use a battery charger to start the engine.
5. Disconnect both battery cables when making a battery recharge hook-up.
6. Be sure of polarity hook-up when using a booster battery for starting.
7. Never ground the alternator output or battery terminal.
8. Never ground the field circuit between alternator and regulator.
9. Never run any alternator on an open circuit with the field energized.
10. Never try to polarize an alternator, unless directed by the manufacturer.
11. Do not attempt to motor an alternator.
12. When making engine idle speed adjustments, always consider potential load factors that influence engine rpm. To compensate for electrical load, switch **ON** the lights, radio, heater, air conditioner, etc.

Low or No Charging

1. Blown fuse
2. Broken or loose fan belt
3. Voltage regulator not working
4. Brushes sticking
5. Slip ring dirty
6. Open circuit
7. Faulty wiring connections
8. Faulty diode rectifier
9. High resistance in charging circuit
10. Grounded stator
11. May be open rectifiers (check all 3 phases)
12. If rectifiers are found or open, check capacitor

Noisy Unit

1. Damaged rotor bearings
2. Poor alignment of unit
3. Broken or loose belt
4. Open diode rectifiers.

CHRYSLER CORPORATION

Chrysler 60 Amp, 78 Amp and 114 Amp Alternator with External Regulator

The 60 and 78 amp alternators are

equipped with 6 built-in silicon rectifiers, while the 114 amp alternator is equipped with 12 built-in silicon rectifiers.

ON VEHICLE SERVICE

System Operation

NOTE: If the current indicator is to give an accurate reading, the battery cables must be of the same gauge and length as the original equipment.

1. With the engine running and all electrical systems **OFF**, place a current indicator over the positive battery cable.

2. If a charge of about 5 amps is recorded, the charging system is working. If a draw of about 5 amps is recorded the system is not working. The needle moves toward the battery when a charge condition is indicated and away from the battery when a draw condition is indicated. If a draw is indicated, proceed to the next testing procedure. If an overcharge of 10–15 amps is indicated, check for a faulty regulator.

Ignition Switch to Regulator Circuit Check

1. Disconnect the regulator wires at the regulator.

2. Turn the key **ON** but do not start the engine.

3. Using a voltmeter or test light check for voltage across the I and F terminals. If there is current present the circuit is good. If there is no current check for faulty connections, a faulty ballast resistor, a faulty ammeter, broken wires or a faulty ground at the alternator or voltage regulator. Also, check for voltage from the I wire to ground, current should be present. Check for voltage from the F terminal to ground, current should not be present.

Isolation Check

This test determines whether the regulator or alternator is faulty if everything else in the circuit was ok.

1. Disconnect, at the alternator, the wire that runs between 1 of the alternator field connections and the voltage regulator.

2. Run a jumper wire from the disconnected alternator terminal to ground.

3. Connect a voltmeter to the battery. The positive voltmeter lead connects to the positive battery terminal and the negative lead goes to the negative terminal. Record the reading.

4. Make sure that all electrical systems are turned **OFF**. Start the engine. Do not race the engine.

5. Gradually increase engine speed to 1500–2000 rpm. There should be an increase of 1–2 volts on the voltmeter. If this is true the alternator is good and the voltage regulator should be repaired. If there is no voltage increase the alternator is faulty.

Charging Circuit Resistance Check

The purpose of this test is to determine the amount of voltage drop between the alternator output terminal wire and the battery.

1. Disconnect the battery ground cable and the BAT lead at the alternator output terminal.

2. Connect an ammeter with a scale to 100 amps in series between the alternator BAT terminal and the disconnected BAT wire.

3. Connect the positive lead of a voltmeter to the disconnected BAT wire. Connect the negative lead of the voltmeter to the negative post of the battery.

4. Disconnect the green colored regulator field wire from the alternator. Connect a jumper lead from the alternator field terminal to ground.

5. Connect a tachometer to the engine and reconnect the battery ground cable.

6. Connect a variable carbon pile rheostat to the battery cables. Be sure

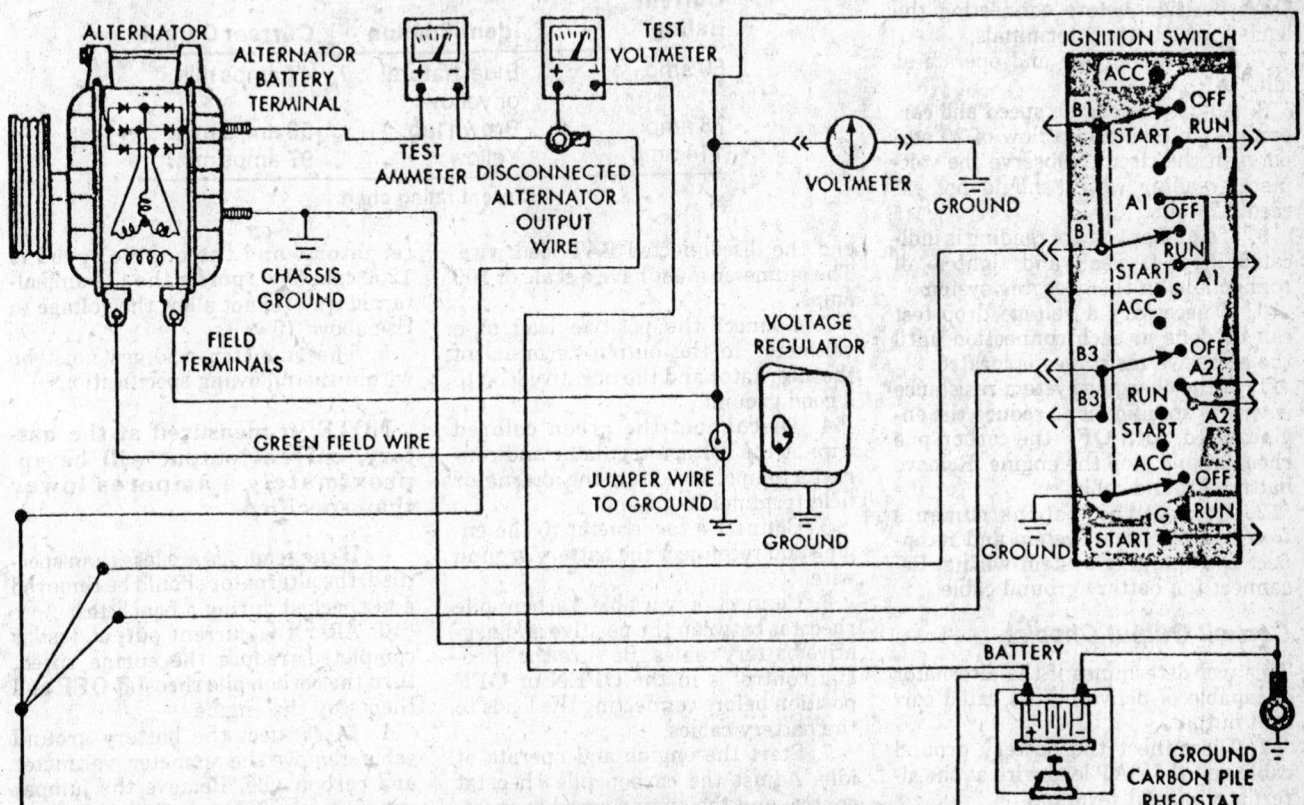

Chrysler 60, 78 and 114 amp alternators with external voltage regulator–charging system resistance test–adjust engine speed and carbon pile to maintain 20 amps, voltmeter reading should not exceed 0.5 volts

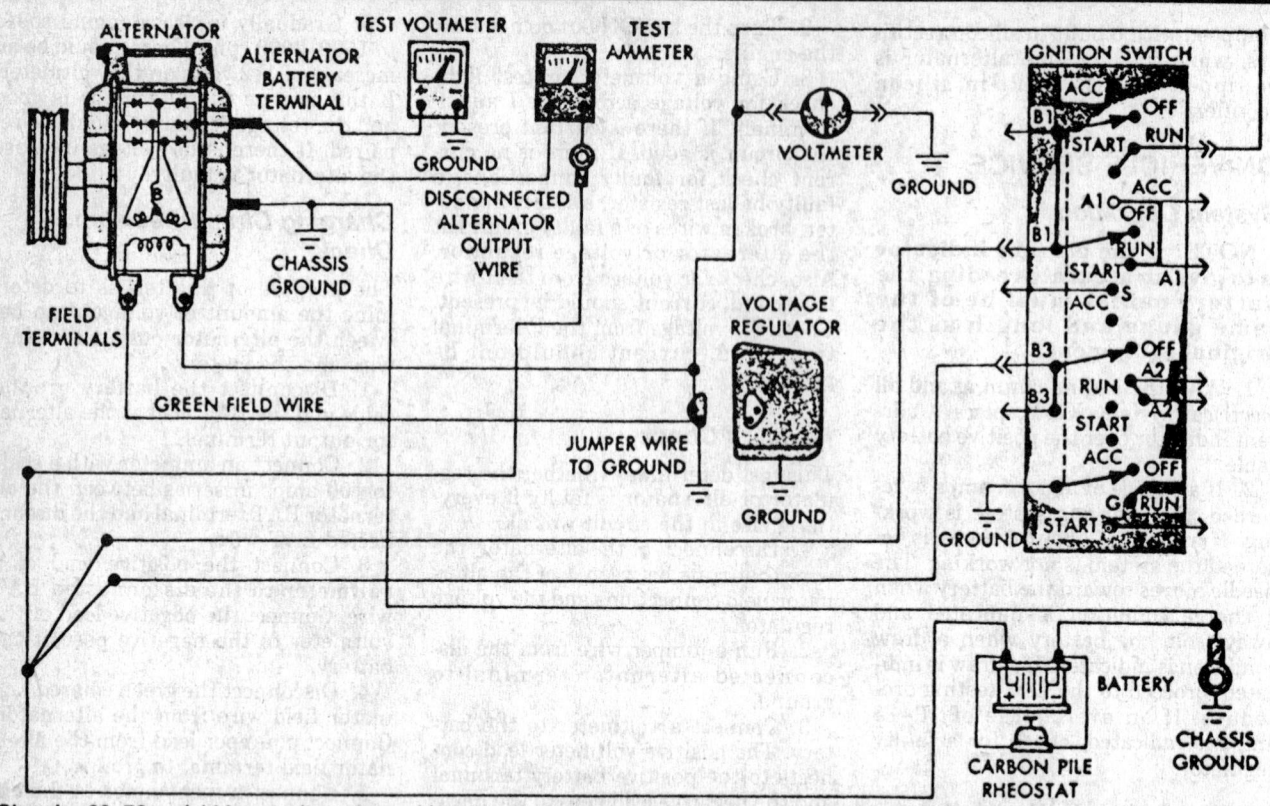

Chrysler 60, 78 and 144 amp alternators with external voltage regulator—current output test—adjust engine speed and carbon pile increments until an engine speed of 1250 rpm is reached, voltmeter reading should not exceed 15 volts

the carbon pile is in the **OPEN** or **OFF** position before connecting the leads to the battery terminals.

7. Start the engine and operate at idle.

8. Adjust the engine speed and carbon pile to maintain a flow of 20 amperes in the circuit. Observe the voltmeter reading which should not exceed 0.7 volts.

9. If a higher voltage reading is indicated inspect, clean and tighten all connections in the charging system.

10. If necessary a voltage drop test can be done at each connection until the excessive resistance is located.

11. If the charging system resistance is within specifications reduce the engine speed, turn **OFF** the carbon pile rheostat and stop the engine. Remove battery ground cable.

12. Remove the test instruments from the electrical system and reconnect the charging system wiring. Reconnect the battery ground cable.

Current Output Check

This test determines if the alternator is capable of delivering its rated current output.

1. Disconnect the battery ground cable and the BAT lead wire at the alternator output terminal.

2. Connect an ammeter in series between the alternator output terminal

Current Rating	Identification	Current Output
60 amp	Blue, natural or yellow	47 amps min.
78 amp	Brown tag	58 amps min.
114 amp	Yellow	97 amps min.

Current rating chart

and the disconnected BAT lead wire. The ammeter must have a scale of 100 amps.

3. Connect the positive lead of a voltmeter to the output terminal of the alternator and the negative lead to a good ground.

4. Disconnect the green colored wire at the voltage regulator and connect a jumper wire from the alternator field terminal to ground.

5. Connect a tachometer to the engine and reconnect the battery ground wire.

6. Connect a variable carbon pile rheostat between the positive and negative battery cables. Be sure the rheostat control is in the **OPEN** or **OFF** position before connecting the leads to the battery cables.

7. Start the engine and operate at idle. Adjust the carbon pile rheostat control and the engine speed in increments until the voltmeter reading is 15 volts (13 volts for the 114 amp al-

ternators) and the engine speed is 1250 rpm (900 rpm for the 114 amp alternators). Do not allow the voltage to rise above 16 volts.

8. The ammeter readings must be within the following specifications.

NOTE: If measured at the battery, current output will be approximately 5 amperes lower than specified.

9. If the readings are less than specified, the alternator should be removed and checked during a bench test.

10. After the current output test is completed, reduce the engine speed, turn the carbon pile rheostat **OFF** and then stop the engine.

11. Disconnect the battery ground cable, remove the ammeter, voltmeter and carbon pile. Remove the jumper wire from the field terminal and reconnect the green colored wire to the alternator field terminal.

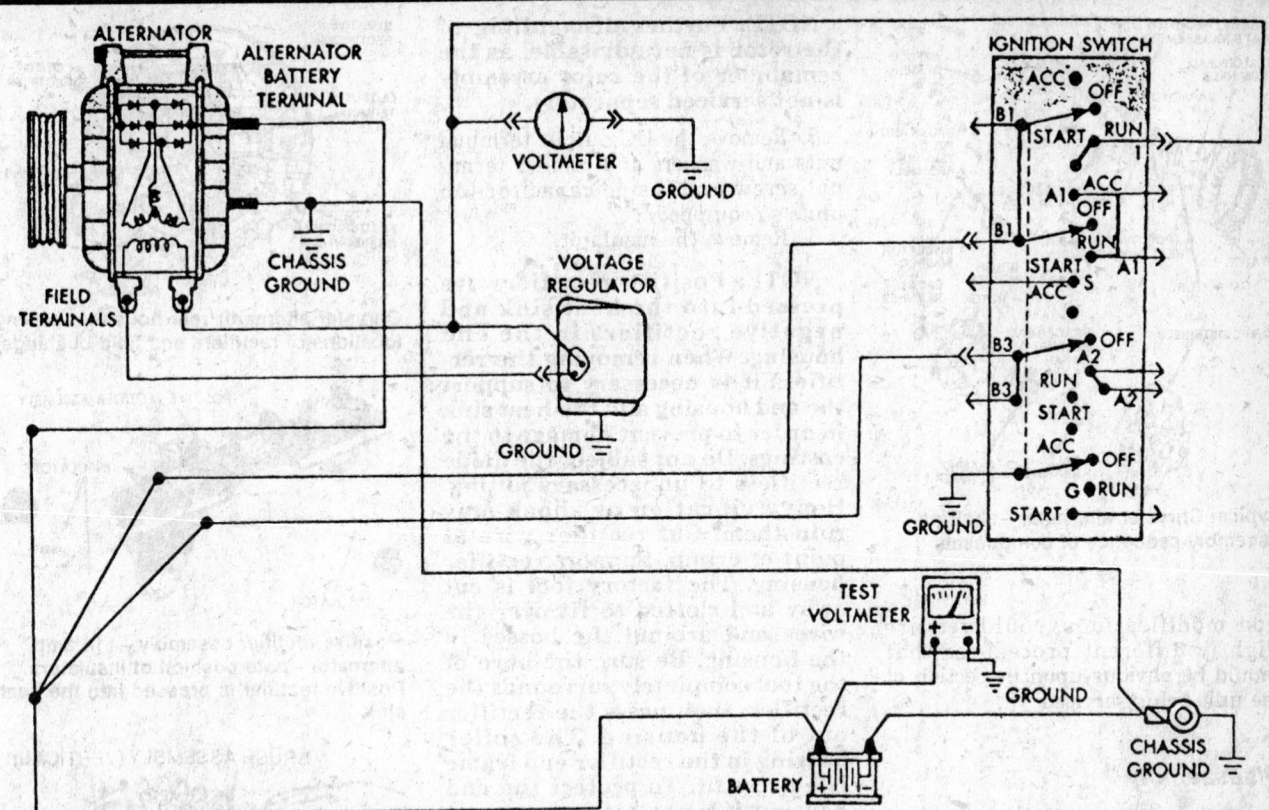

Chrysler 60, 78 and 114 amp alternators with external voltage regulator—voltage regulator test—idle engine at 1250 rpm with all lights and accessories off, voltmeter readings should equal shown in table

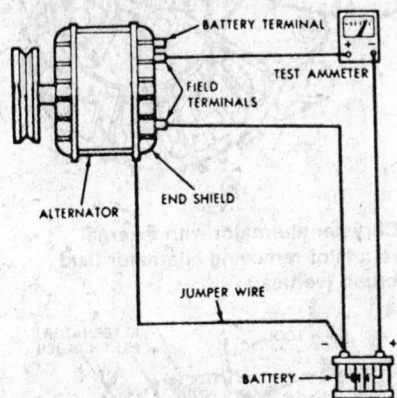

Chrysler alternator with external regulator—rotor field coil current draw test. Connect ammeter as shown, rotate alternator pulley slowly by hand and observe the field coil draw reading.

12. Reconnect the battery cable, if no further testing is to be done to the charging circuit.

Rotor Field Coil Draw Check

1. If on the vehicles remove the drive belt and wiring connections from the alternator.

2. Connect a jumper wire from the negative terminal of the battery to 1 of the field terminals of the alternator.

3. Connect the test ammeter positive lead to the other field terminal of

the alternator and the negative ammeter lead to the positive battery terminal.

4. Connect a jumper wire between the alternator end shield and the battery negative terminal.

5. Slowly rotate the alternator pulley by hand and observe the ammeter reading.

6. The field coil draw should be 4.5–6.5 amperes at 12 volts. (4.75–6.0 amperes at 12 volts 114 amp alternators).

7. A low rotor coil draw is an indication of high resistance in the field coil circuit (brushes, slip-rings or rotor coil). A higher rotor coil draw indicates possible shorted rotor coil or grounded rotor. No reading indicates an open rotor or defective brushes.

8. Remove the test equipment and jumper leads.

Electronic Voltage Regulator Check

1. Make sure battery terminals are clean and battery is charged.

2. Connect the positive lead of a test voltmeter to ignition terminal No. 1 of the ballast resistor.

3. Connect the negative voltmeter lead to a good body ground.

4. Start engine and allow it to idle at 1250 rpm, all lights and accessories turned **OFF**. Voltage should be as in-

dicated in the figure of the voltage regulator test.

5. If the voltage is below specification check the following. Voltage regulator ground check, voltage drop between regulator cover and ground. Harness wiring, disconnect regulator plug (ignition switch **OFF**), then turn **ON** ignition switch and check for battery voltage at the terminals having the red and green leads. Wiring harness must be disconnected from the regulator when checking individual leads. If no voltage is present in either lead the problem is in the wiring or alternator field.

6. If Step 5 tests showed no malfunctions, install a new regulator and repeat Step 4.

7. If voltage is above specifications (Step 4), or fluctuates, check the following. Ground between regulator and body, between body and engine. Ignition switch circuit between switch and regulator.

8. If voltage is still more than ½ volt above specifications install a new regulator and repeat Step 4.

ALTERNATOR OVERHAUL

Alternator disassembly, repair and assembly procedures are basically the same for all Chrysler alternators. Certain variations in design, or produc-

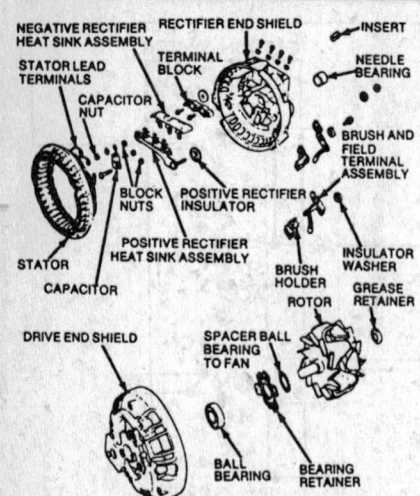

Typical Chrysler alternator—showing assembly sequence of components

NOTE: Further dismantling of the rotor is not advisable, as the remainder of the rotor assembly is not serviced separately.

7. Remove the DC output terminal nuts and washers and remove terminal screw and inside capacitor (on units so equipped).

8. Remove the insulator.

NOTE: Positive rectifiers are pressed into the heat sink and negative rectifiers in the end housing. When removing the rectifiers it is necessary to support the end housing and the heat sink in order to prevent damage to the castings. Do not subject the diode rectifiers to unnecessary jolting. Heavy vibration or shock may ruin them. Cut rectifier wire at point of crimp. Support rectifier housing. The factory tool is cut away and slotted to fit over the wires and around the bosses in the housing. Be sure the bore of the tool completely surrounds the rectifier, then press the rectifier out of the housing. The roller bearing in the rectifier end frame is a press fit. To protect the end housing, it is necessary to support the housing with a tool when pressing out the bearing.

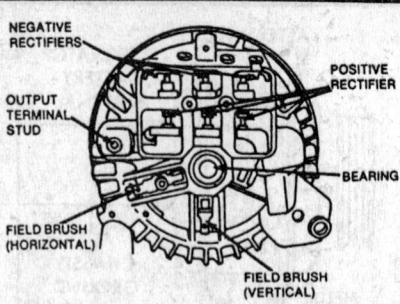

Chrysler alternator rear housing showing locations of rectifiers and field bushings

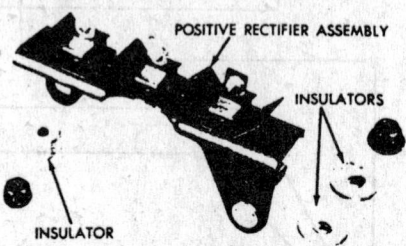

Positive rectifier assembly—114 amp alternator—note position of insulators. Positive rectifier is pressed into the heat sink

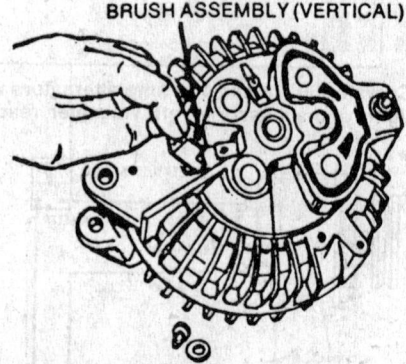

Chrysler alternator with external regulator removing alternator field brush (vertical)

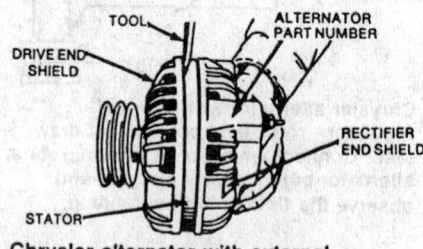

Chrysler alternator with external regulator—separating alternator drive end shield from stator, using a prybar

tion modifications, could require slightly different procedures that should be obvious upon inspection of the unit being serviced.

Disassembly

To prevent damage to the brush assemblies (114 amp), they should be removed before proceeding with the disassembly of the alternator. The brushes are mounted in a plastic holder that positions the brushes vertically against the slip-rings.

1. Remove the retaining screw, flat washer, nylon washer and field terminal and carefully lift the plastic holder containing the spring and brush assembly from the end housing.

2. The ground brush (60 amp) is positioned horizontally against the slip-ring and is retained in the holder that is integral with the end housing. Remove the retaining screw and lift the clip, spring and brush assembly from the end housing. The stator is laminated so don't burr the stator or end housings.

3. Remove the through bolts and pry between the stator and drive end housing with a suitable tool. Carefully separate the drive end housing, pulley and rotor assembly from the stator and rectifier housing assembly.

4. The pulley is an interference fit on the rotor shaft. Remove with a puller and special adapters.

5. Remove the nuts and washers and, while supporting the end frame, tap the rotor shaft with a plastic hammer and separate the rotor and end housing.

6. The drive end ball bearing is an interference fit with the rotor shaft. Remove the bearing with puller and adapters.

Inspection

RECTIFIERS OPEN IN ALL 3 PHASES

Testing with Ohmmeter

Disassemble the alternator and separate the wires at the Y-connection of the stator.

There are 6 diode rectifiers mounted in the back of the alternator (60 amp). Three of them are marked with a plus (+) and 3 are marked with a minus (−). These marks indicate diode case polarity. The 114 amp alternator has 12 silicon diodes; 6 positive and 6 negative.

To test, set ohmmeter to its lowest range. If case is marked positive (+), place positive meter probe to case and negative probe to the diode lead. Meter should read between 4–10 ohms. Now, reverse leads of ohmmeter, connecting negative meter probe to positive case and positive meter probe to wire of rectifier. Set meter on a high range. Meter needle should move very little, if any (infinite reading). Do this to all positive diode rectifiers.

The diode rectifiers with minus (−) marks on their cases are checked the same way as above. Only now the negative ohmmeter probe is connected to the case for a reading of 4–10 ohms. Reverse leads as above for the other part to test. If a reading of 4–10 ohms

is obtained in 1 direction and no reading (infinity) is read on the ohmmeter in the other direction, diode rectifiers are good. If either infinity or a low resistance is obtained in both directions on a rectifier it must be replaced. If meter reads more than 10 ohms when ohmmeter positive probe is connected to positive on diode and negative probe

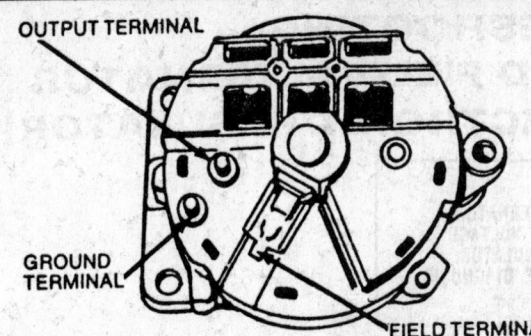

OUTPUT TERMINAL

GROUND TERMINAL

FIELD TERMINAL

View of the rear housing terminal location—Chrysler alternator 114 amp

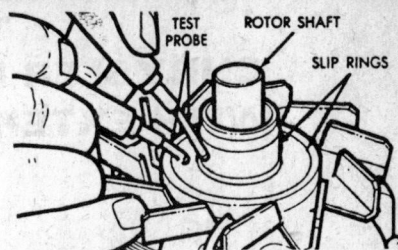

TEST PROBE   ROTOR SHAFT

SLIP RINGS

Chrysler alternator with external regulator testing the rotor for short circuit or open circuits using a 110 volt AC test bulb

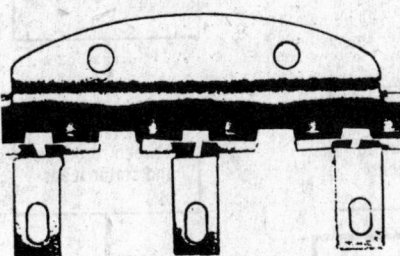

POSITIVE RECTIFIER ASSEMBLY (65 AMP)

POSITIVE RECTIFIER ASSEMBLY (65 AMP) (EXCEPT

Chrysler alternator with external regulator—positive and negative rectifier identification—note the different types are not interchangeable

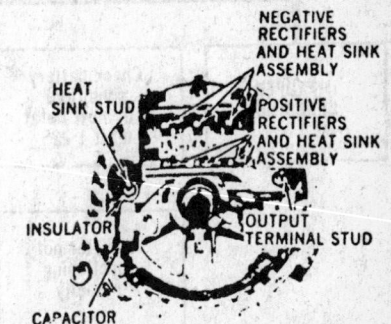

NEGATIVE RECTIFIERS AND HEAT SINK ASSEMBLY

HEAT SINK STUD

POSITIVE RECTIFIERS AND HEAT SINK ASSEMBLY

INSULATOR

OUTPUT TERMINAL STUD

CAPACITOR

Chrysler alternator with external regulator—location of negative and positive rectifiers check part number of rectifier to be sure correct rectifier is being used

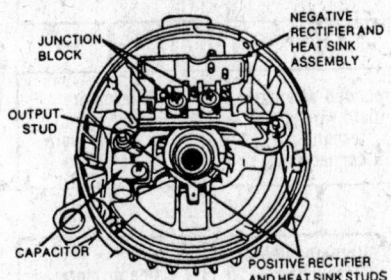

Negative rectifier assembly—114 amp alternator. Negative rectifier is pressed into the end housing

JUNCTION BLOCK

NEGATIVE RECTIFIER AND HEAT SINK ASSEMBLY

OUTPUT STUD

CAPACITOR

POSITIVE RECTIFIER AND HEAT SINK STUDS

Chrysler alternator with external regulator—heat sink and rectifier assembly removal—note location of studs

to negative diode, replace diode rectifier.

NOTE: With this test it is necessary to determine the polarity of the ohmmeter probes. This can be done by connecting the ohmmeter to a DC voltmeter. The voltme-

ter will read up scale when the positive probe of the ohmmeter is connected to the positive side of the voltmeter and the negative probe of the ohmmeter is connected to the negative side of the voltmeter.

Alternate Method with Test Lamp

Be sure that the lead from the center of the diode rectifiers is disconnected. To test rectifiers with positive cases, touch the positive probe of tester to case and the negative probe to lead wire of rectifier. Bulb should light if rectifier is good. If bulb does not light, replace rectifier. Now reverse tester probe connections to rectifier. Bulb should not light. If bulb does light, replace rectifier. For testing minus (−) marked cases follow the above procedure except that now bulb should light with negative probe of tester touching rectifier case and positive probe touching lead wire. Rectifier is good if the bulb lights when tester probes are connected one way and does not light when tester connections are reversed. Rectifier must be replaced if the bulb does not light either way. Also, replace rectifier if bulb lights both ways.

NOTE: The usual cause of an open diode or rectifier is a defective capacitor or a battery that has been installed in reverse polarity. If the battery is installed properly and the diodes are open, test the capacitor.

FIELD COIL DRAW TEST

1. Connect a jumper between one FLD terminal and the positive terminal of a fully charged 12 volt battery.
2. Connect the positive lead of a test ammeter to the other field (Fld) terminal and the negative test lead to the negative battery terminal.
3. Slowly rotate the rotor by hand and observe the ammeter. The proper field coil draw is 2.3–2.7 amps at 12 volts.

NOTE: Field coil draw for the 114 ampere alternators should be 4.75–6.0 amperes at 12 volts.

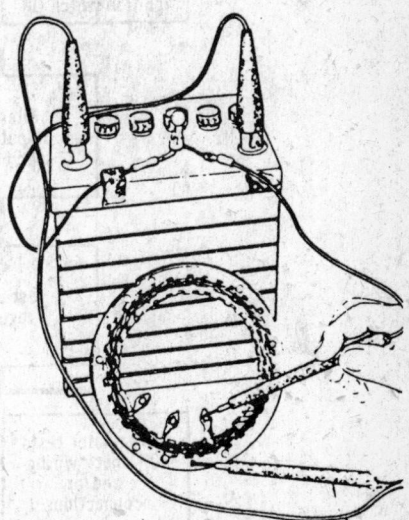

Chrysler alternator with external regulator—testing the stator for grounding using a 110 volt test lamp if lamp lights stator is grounded and must be replaced

FIELD CIRCUIT GROUND TEST

1. Touch a test lead of a 110 volt AC test bulb to 1 of the alternator brush (field) terminals and the other test lead to the end shield.
2. If the lamp lights, remove the field brush assemblies and separate the end housing by removing the through bolts.

TROUBLESHOOTING
CHRYSLER ISOLATED FIELD ALTERNATOR
(WITH EXTERNAL ELECTRONIC REGULATOR)

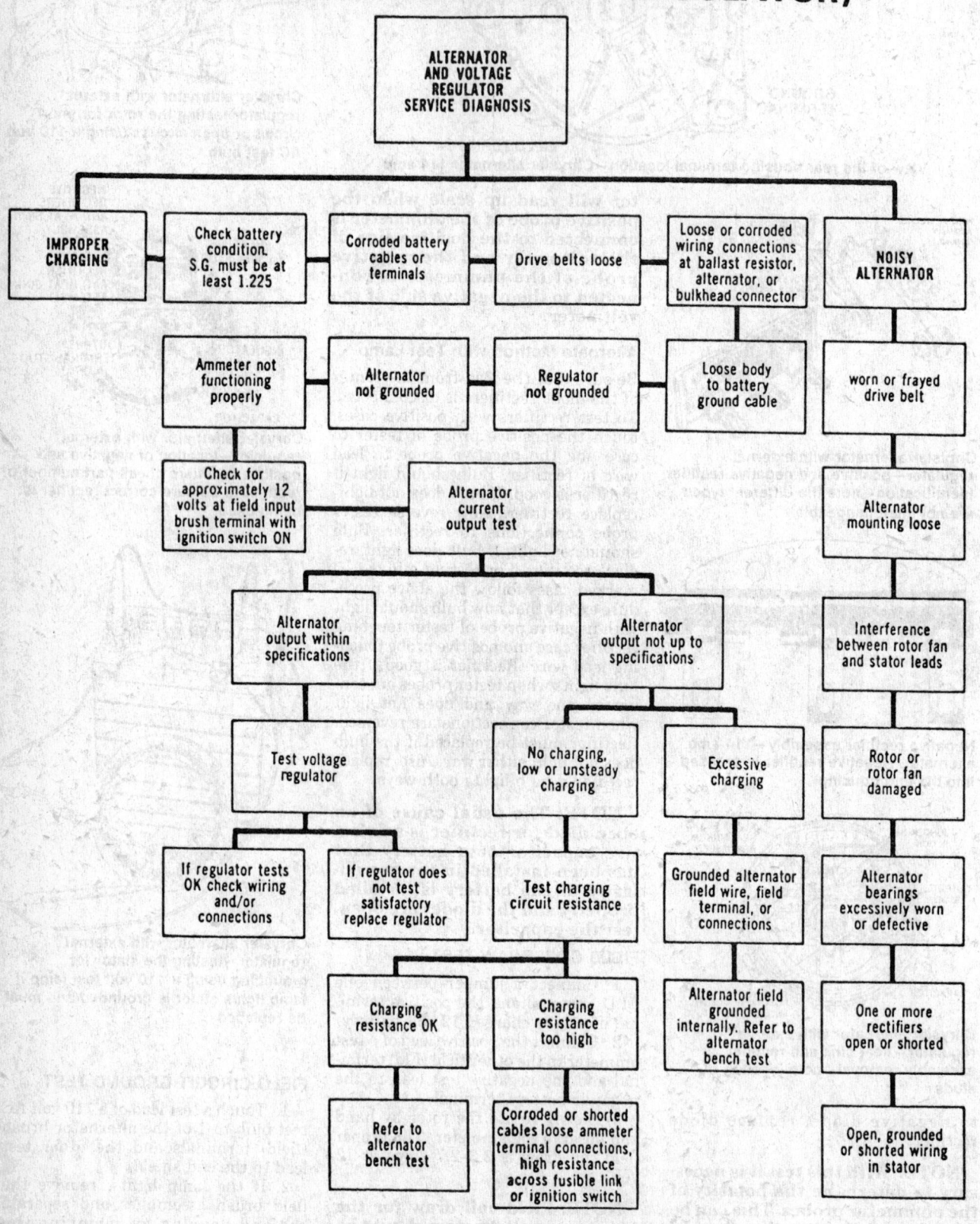

3. Place one test lead on a slip-ring and the other on the end shield.

4. If the lamp lights, the rotor assembly is grounded internally and must be replaced.

5. If the lamp does not light, the cause of the problem was a grounded brush.

GROUNDED STATOR

1. Disconnect the diode rectifiers from the stator leads.

2. Test from stator leads to stator core, using a 110 volt test lamp. Test lamp should not light. If it does, the stator is grounded and must be replaced.

LOW OUTPUT TEST

About 50% output accompanied with a growl-hum caused by a shorted phase or a shorted rectifier. If the rectifiers are found to be within specifications replace the stator assembly.

If the rectifier tests satisfactorily, inspect the stator connections before replacing the stator.

Assembly

1. Support the heat sink or rectifier end housing on circular plate.

2. Check rectifier identification to be sure the correct rectifier is being used. The part numbers are stamped on the case of the rectifier. They are also marked red for positive and black for negative.

3. Start the new rectifier into the casting and press it in squarely. Do not start rectifier with a hammer or it will be ruined.

4. Crimp the new rectifier wire to the wires disconnected at removal or solder using a heat sink with rosin core solder.

5. Support the end housing on tool so that the notch in the support tool will clear the raised section of the heat sink, press the bearing into position with tool SP–3381, or equivalent. New bearings are prelubricated, additional lubrication is not required.

6. Insert the drive end bearing in the drive end housing and install the bearing plate, washers and nuts to hold the bearing in place.

7. Position the bearing and drive end housing on the rotor shaft and, while supporting the base of the rotor shaft, press the bearing and housing in position on the rotor shaft with an arbor press and arbor tool. Be careful that there is no cocking of the bearing at installation; or damage will result. Press the bearing on the rotor shaft until the bearing contacts the shoulder on the rotor shaft.

8. Install pulley on rotor shaft. Shaft of rotor must be supported so that all pressing force is on the pulley hub and rotor shaft. Do not exceed 6800 lbs. pressure. Pulley hub should just contact bearing inner race.

9. Some alternators will be found to have the capacitor mounted internally. Be sure the heat sink insulator is in place.

10. Install the output terminal screw with the capacitor attached through the heat sink and end housing.

11. Install insulating washers, lockwashers and locknuts.

12. Make sure the heat sink and insulator are in place and tighten the locknut.

13. Position the stator on the rectifier end housing. Be sure that all of the rectifier connectors and phase leads are free of interference with the rotor fan blades and that the capacitor (internally mounted) lead has clearance.

14. Position the rotor assembly in the rectifier end housing. Align the through bolt holes in the stator with both end housings.

15. Enter stator shaft in the rectifier end housing bearing, compress stator and both end housings manually and install through bolts, washers and nuts.

16. Install the insulated brush and terminal attaching screw.

17. Install the ground screw and attaching screw.

18. Rotate pulley slowly to be sure the rotor fan blades do not hit the rectifier and stator connectors.

Chrysler 40/90 Amp and 50/120 Amp Alternator with External Electronic Regulator

The 40/90 alternator is used as standard equipment. A 50/120 alternator is optional.

ON VEHICLE SERVICE

Charging Circuit Resistance Test

1. Be sure that the battery is fully charged.

2. Turn **OFF** ignition switch.

3. Disconnect negative battery cable.

4. Disconnect BAT terminal wire from alternator output BAT terminal post.

5. Connect a 0–100 amps minimum range scale DC test ammeter in series between alternator BAT terminal and disconnected BAT terminal wire. Connect ammeter positive lead wire to alternator BAT terminal and negative ammeter lead to disconnected alternator BAT terminal wire.

6. Connect a 0–18 volt minimum range scale test voltmeter between disconnected alternator BAT terminal wire and positive battery cable. Connect voltmeter positive lead to disconnected alternator BAT terminal wire and negative voltmeter lead to battery positive cable.

7. Disconnect wiring harness connector from electronic voltage regulator on vehicle.

8. Connect a jumper wire from wiring harness connector green wire (outside terminal), to ground. Do not connect blue J2 lead of wiring connector to ground.

9. Connect an engine tachometer and reconnect negative battery cable. Connect a variable carbon pile rheostat to battery terminals. Be sure carbon pile is in **OPEN** or **OFF** position before connecting leads.

10. Start engine. Immediately after starting reduce engine speed to idle.

11. Adjust engine speed and carbon pile to maintain 20 amperes flowing in circuit. Observe voltmeter reading. Voltmeter reading should not exceed 0.5 volts.

NOTE: If a higher voltage drop is indicated, inspect, clean and tighten all connections in the charging circuit. A voltage drop test may be performed at each connection to locate connection with excessive resistance. If charging circuit resistance tested satisfactorily, reduce engine speed, turn OFF carbon pile and turn OFF ignition switch.

12. Disconnect negative battery cable.

13. Remove test ammeter, test voltmeter, variable carbon pile rheostat and engine tachometer.

14. Remove jumper wire connected between electronic voltage regulator wiring harness connector green wire terminal and ground.

15. Connect wiring harness connector to electronic voltage regulator.

16. Connect BAT terminal wire to alternator output BAT terminal.

17. Connect negative battery cable.

Current Output Test

1. Be sure that the battery is fully charged.

2. Turn **OFF** ignition switch.

3. Disconnect negative battery cable.

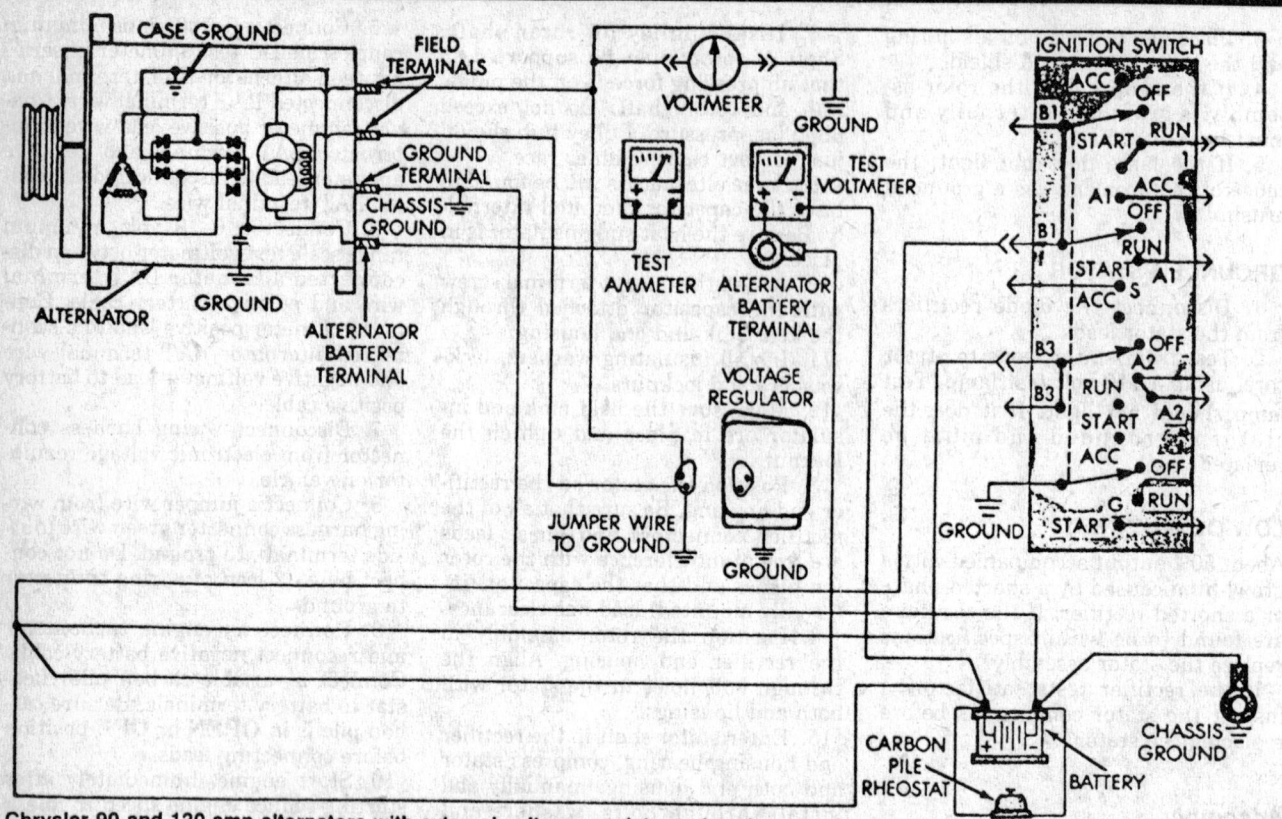

Chrysler 90 and 120 amp alternators with external voltage regulator—charging system resistance test—adjust engine speed and carbon pile to maintain 20 amps, voltmeter reading should not exceed 0.5 volts

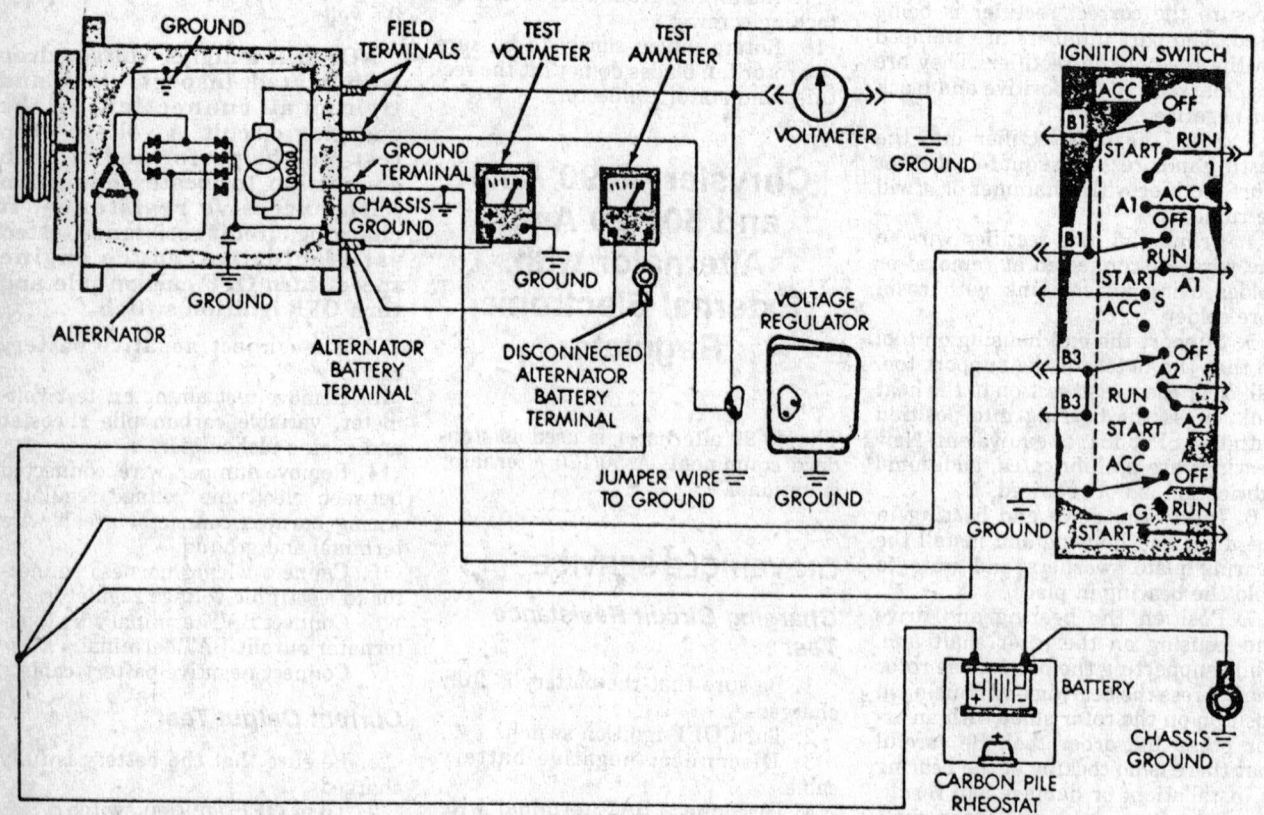

Chrysler 90 and 120 amp alternators with external voltage regulator—current output test—adjust engine speed and carbon pile increments until an engine speed of 1250 rpm is reached, voltmeter reading should not exceed 15 volts

4. Disconnect the output wire from the alternator BAT terminal. Connect a 0-100 ammeter positive lead wire to alternator BAT terminal and negative ammeter lead to disconnected alternator BAT terminal wire.

5. Connect a 0–18 volt minimum range scale test voltmeter between alternator BAT terminal post and ground. Connect voltmeter positive lead to alternator BAT terminal post. Connect negative lead of test voltmeter to a good ground.

6. Disconnect wiring harness connector from electronic voltage regulator on vehicle.

7. Connect a jumper wire from wiring harness connector green wire (outside terminal) to ground. Do not connect blue J2 lead of wiring connector to ground.

8. Connect an engine tachometer and reconnect negative battery cable. Connect a variable carbon pile rheostat between battery terminals. Be sure the carbon pile is in **OPEN** or **OFF** position before connecting leads.

9. Start the engine. Immediately after starting reduce engine speed to idle.

10. Adjust carbon pile and engine speed in increments until a speed of 1250 rpm and voltmeter reading of 15 volts is obtained. Do not allow voltage meter to read above 16 volts.

11. Ammeter reading must be within the proper limits. If reading is less than specified alternator should be removed from vehicle and bench tested.

12. After current output test is completed reduce engine speed, turn **OFF** carbon pile and turn **OFF** ignition switch.

13. Disconnect negative battery cable.

14. Remove test ammeter, test voltmeter, tachometer and variable carbon pile rheostat.

15. Remove jumper wire connected between electronic voltage regulator wiring harness connector green wire terminal and ground.

16. Connect wiring harness connector to electronic voltage regulator.

17. Connect BAT terminal wire to alternator output BAT terminal.

18. Connect negative battery cable.

Voltage Regulator Test

1. Be sure that the battery is fully charged.

2. Turn **OFF** ignition switch.

3. Connect a 0–18 volts minimum range scale test voltmeter between vehicle battery and ground. Connect positive lead of voltmeter to positive battery cable terminal. Connect negative lead of voltmeter to a good vehicle body ground.

4. Connect a tachometer to engine.

5. Start engine and adjust engine speed to 1250 rpm with all lights and accessories turned **OFF**.

6. Check voltmeter, regulator is working properly if voltage readings are in accordance with the voltage chart.

7. If voltage is below limits or is fluctuating, proceed as follows. Check for a good voltage regulator ground. Voltage regulator ground is obtained through regulator case to mounting screws and to sheet metal of vehicle. This is ground circuit that is to be checked for opens.

8. Turn **OFF** ignition switch and disconnect voltage regulator wiring harness connector. Be sure terminals of connector have not spread open to cause an open or intermittant connection.

9. Do not start engine or distort terminals with voltmeter probe: turn **ON** ignition switch and check for battery voltage at voltage regulator wiring harness connector terminals. Both blue and green terminals should read battery voltage. Turn **OFF** ignition switch.

10. If satisfactory, replace the regulator and repeat the test.

11. If the voltage is above limits specification proceed as follows. Turn **OFF** ignition switch and disconnect voltage regulator wiring harness connector. Be sure terminals in connector have not spread open.

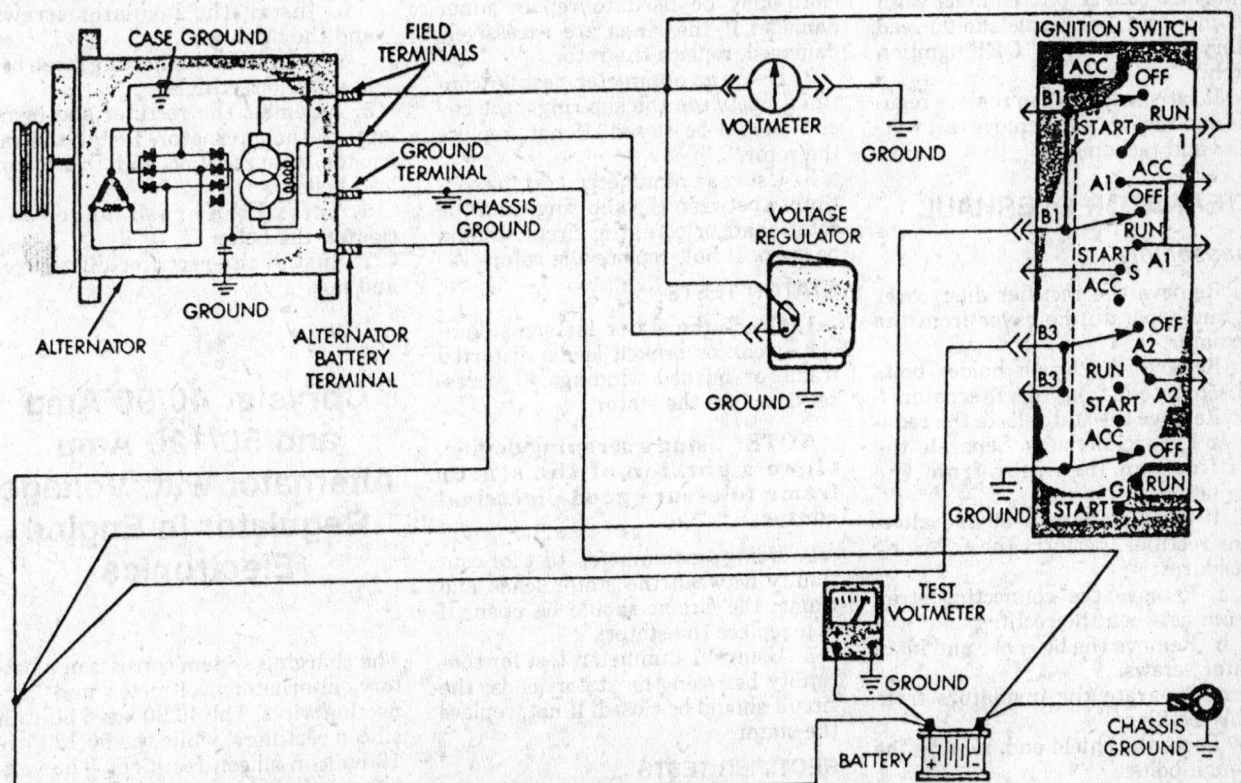

Chrysler 90 and 120 amp alternators with external voltage regulator—voltage regulator test—idle engine at 1250 rpm with all lights and accessories off, voltmeter readings should equal shown in table

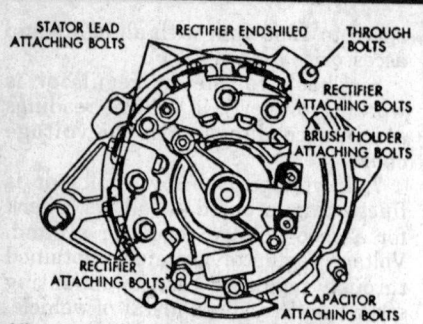

View of the rectifier end shield assembly bolts—Chrysler 90 and 120 amp alternator

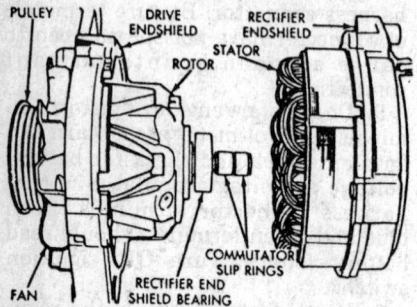

View of the stator and drive end shield assemblies—Chrysler 90 and 120 amp alternator

12. Do not start engine or distort terminals with voltmeter probe. Turn **ON** ignition switch and check for battery voltage at voltage regulator wiring harness connector terminals. Both blue and green terminals should read battery voltage. Turn **OFF** ignition switch.

13. If satisfactory, then replace regulator and repeat test. Remove test voltmeter and tachometer.

ALTERNATOR OVERHAUL

Disassembly

1. Remove the rectifier dust cover nut and separate the cover from the alternator.
2. Remove the brush holder bolts and separate it from the alternator.
3. Remove the stator lead, the rectifier and capacitor bolts. Separate the rectifier and insulator from the alternator.
4. If disassembling a 50/120 alternator rectifier, perform the following procedures:
   a. Remove the connecting strap from between the rectifiers.
   b. Remove the buss bar and insulator screws.
   c. Separate the insulators from the rectifier.
5. From the shield end, remove the through bolts.
6. From the drive end shield, separate the stator and rectifier end shield.

7. Separate the stator from the rectifier end shield.
8. Remove the drive pulley nut, the washer, the pulley and the fan.
9. Press the rotor shaft from the drive end shield.

NOTE: If the bearing is defective, replace the drive end shield as an assembly.

10. If necessary to remove the rectifier end bearing from the rotor assembly, use a bearing puller to press the bearing from the rotor shaft.

Inspection

BRUSH HOLDER TESTS

1. Make sure the brushes move smoothly and return fully to the stops when released; if not, replace the brush holder assembly.
2. Using an ohmmeter, test inner brush-to-field terminals continuity; a field terminal should be open and the other closed. If not, replace the brush holder.
3. Using an ohmmeter, test outer brush-to-field terminals continuity; 1 field terminal should be open and the other closed. If not, replace the brush holder.

ROTOR TESTS

1. Check the field slip rings for excessive wear or roughness; fine emery cloth may be used to repair minor damage. If the rings are excessively damaged, replace the rotor.
2. Using an ohmmeter, test for continuity between the slip rings; the circuit should be closed. If not, replace the rotor.
3. Using an ohmmeter, test for continuity between the slip rings and the rotor shaft or core; the circuit should be open. If not, replace the rotor.

STATOR TESTS

1. Check the stator for signs damage—weak or broken leads, distorted frame or burned windings; if necessary, replace the stator.

NOTE: Using a scraping device, clean a portion of the stator frame to assure good electrical contact.

2. Using an ohmmeter, test for continuity between the stator leads and frame; the circuit should be open. If not, replace the stator.
3. Using an ohmmeter, test for continuity between the stator leads; the circuit should be closed. If not, replace the stator.

RECTIFIER TESTS

1. Separate the positive diode leads from the negative diode leads.

2. Using an ohmmeter, test for continuity between the positive heat sink to each of the positive diode leads. Reverse the test probes and repeat the test; the diodes should show continuity in 1 direction. If not, replace the rectifier assembly.
3. Using an ohmmeter, test for continuity between the negative heat sink to each of the negative diode leads. Reverse the test probes and repeat the test; the diodes should show continuity in 1 direction. If not, replace the rectifier assembly.

Assembly

1. If the rectifier end bearing was removed, support the rotor assembly in a holding fixture and drive the bearing onto the rotor shaft with a bearing driver.
2. Press the drive end shield onto the rotor shaft and install the fan, the pulley, the washer and the drive pulley nut.
3. Assemble the stator to the rectifier end shield and the stator to the drive end shield. Install the through bolts.
4. If assembling a 50/120 alternator rectifier, perform the following procedures:
   a. Install the insulators on the rectifier.
   b. Install the insulator screws and the buss bar.
   c. Attach the connecting strap between the rectifiers.
5. Assemble the rectifier and insulator to the alternator. Install the capacitor, the rectifier and the stator lead bolts.
6. Install the brush holder and tighten the bolts.
7. Install the rectifier dust cover and nut.

Chrysler 40/90 Amp and 50/120 Amp Alternator with Voltage Regulator in Engine Electronics

The charging system consists of a battery, alternator, voltmeter and connecting wires. The 40/90 has 6 built-in silicon rectifiers, while the 50/120 has 12 built-in silicon rectifiers. The voltage regulator is built into the power and logic modules. The rectifiers convert AC current into DC current. Cur-

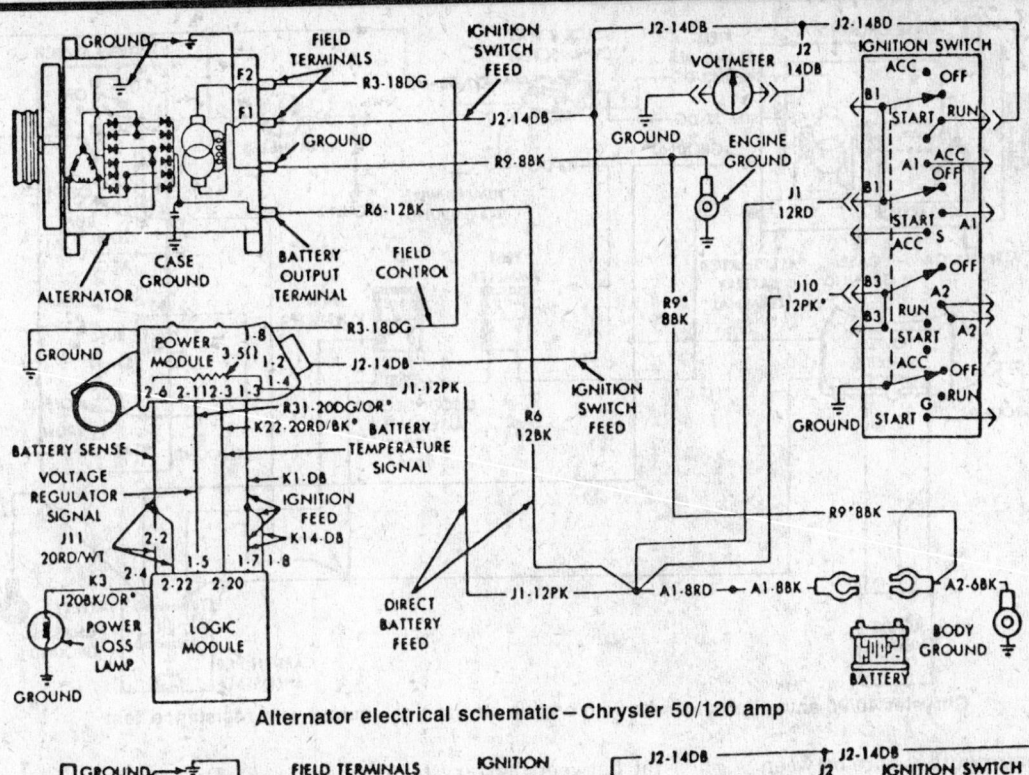

Alternator electrical schematic – Chrysler 50/120 amp

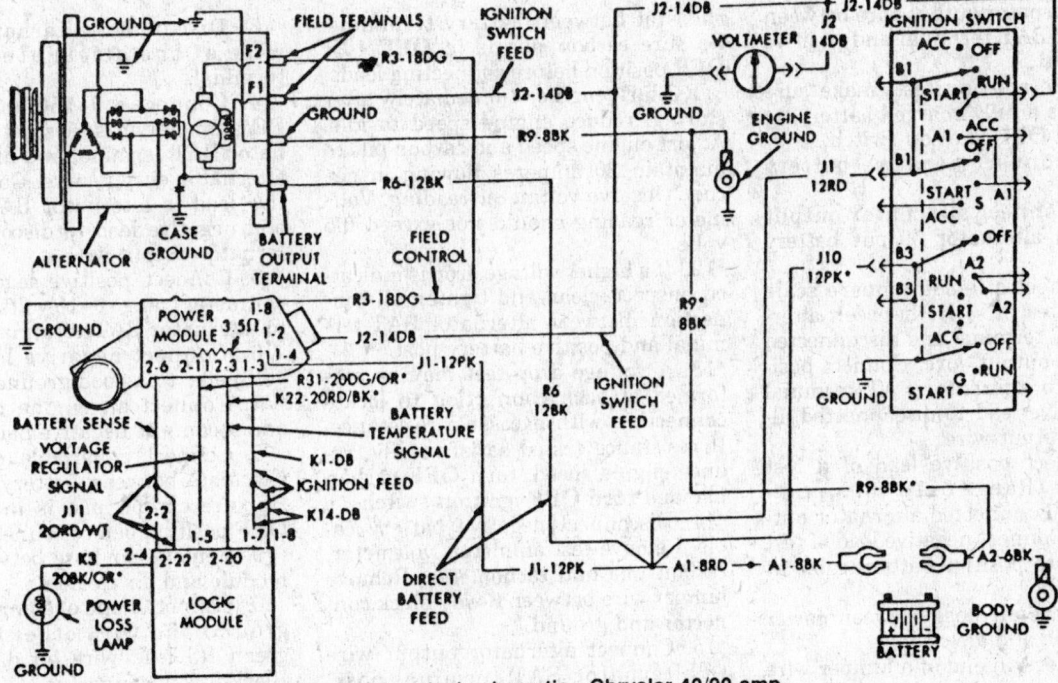

Alternator electrical schematic – Chrysler 40/90 amp

rent at the alternator battery terminal is DC. The alternator's main components are rotor, stator, capacitor, rectifiers, end shields, brushes, bearings, poly-vee drive pulley and fan.

The electronic voltage regulator is contained within engine electronics power module and logic module. It is a device that regulates vehicle electrical system voltage by limiting output voltage that is generated by the alternator. This is accomplished by controlling

amount of current that is allowed to pass through alternator field winding. The alternator field is turned **ON** by a driver in power module which is controlled by a predriver in the logic module. The logic module looks at battery temperature to determine control voltage. The field is then driven at a duty cycle proportional to the difference between battery voltage and desired control voltage. One important feature of the electronic regulator is the ability of

its control circuit to vary regulated system voltage up or down as temperature changes. This provides varying charging conditions for battery throughout seasons of the year.

ON VEHICLE SERVICE

Resistance Test

Alternator output wire resistance test will show amount of voltage drop

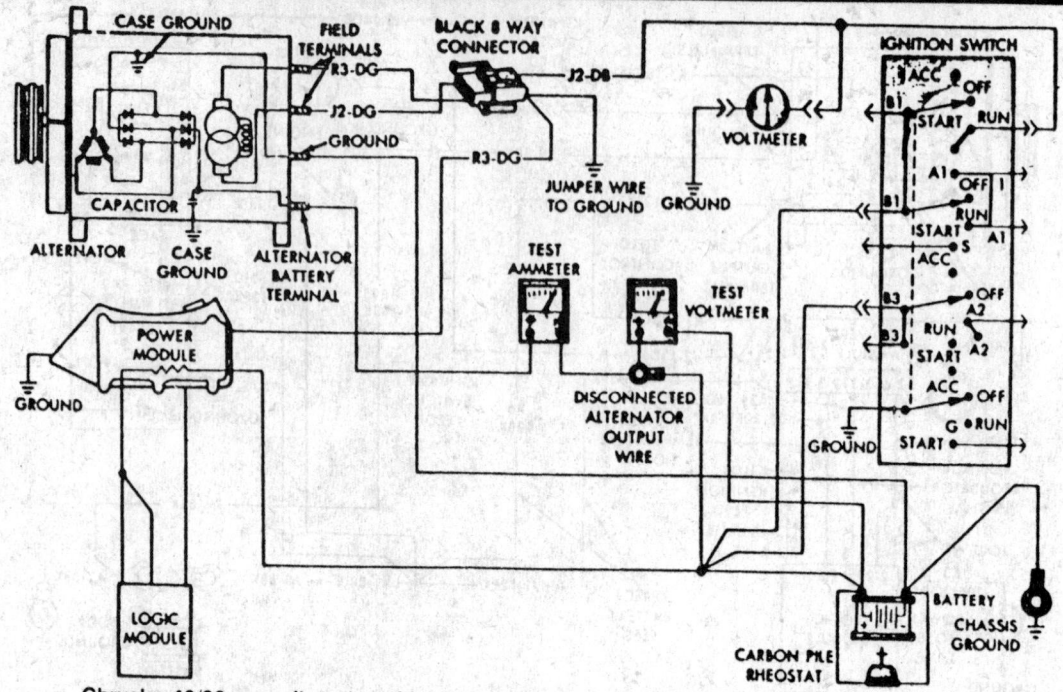

Chrysler 40/90 amp alternator with with engine electronics—output wire resistance test

across alternator output wire between alternator BAT terminal and positive battery post.

1. Before starting test, make sure vehicle has a fully charged battery.

2. Turn **OFF** ignition switch.

3. Disconnect negative battery cable.

4. Disconnect alternator output wire from alternator output battery terminal.

5. Connect a 0–150 ampere scale DC ammeter in series between alternator BAT terminal and disconnected alternator output wire. Connect positive lead to alternator BAT terminal and negative lead to disconnected alternator output wire.

6. Connect positive lead of a test voltmeter (Range 0–18 volts minimum) to disconnected alternator output wire. Connect negative lead of test voltmeter to positive battery cable at positive post.

7. Remove air hose between power and module and air cleaner.

8. Connect an end of a jumper wire to ground and with other end probe green R3 lead wire on dash side of black 8-way connector.

NOTE: Do not connect the blue J2 lead of the 8-way connector to ground. Both R3 and J2 leads are green on the alternator side of the 8-way connector. At the dash end of the connector, R3 is green and J2 is blue.

9. Connect an engine tachometer and reconnect negative battery cable.

10. Connect a variable carbon pile

rheostat between battery terminals. Be sure carbon pile is in **OPEN** or **OFF** position before connecting leads.

11. Start engine. Immediately after staring, reduce engine speed to idle. Adjust engine speed and carbon pile to maintain 20 amperes flowing in circuit. Observe voltmeter reading. Voltmeter reading should not exceed 0.5 volts.

12. If a higher voltage drop is indicated, inspect, clean and tighten all connections between alternator BAT terminal and positive battery post.

13. A voltage drop test may be performed at each connection to locate connection with excessive resistance. If resistance tested satisfactorily, reduce engine speed, turn **OFF** carbon pile and turn **OFF** ignition switch.

14. Disconnect negative battery cable. Remove test ammeter, voltmeter, carbon pile and tachometer. Remove jumper wire between 8-way black connector and ground.

15. Connect alternator output wire to alternator BAT terminal post. Tighten 45–75 inch lbs. Reconnect negative battery cable. Reconnect hose between power module and air cleaner.

Current Output Test

Current output test determines whether or not alternator is capable of delivering its rated current output.

1. Before starting any tests, make sure vehicle has a fully charged battery.

2. Disconnect negative battery cable.

3. Disconnect alternator output wire at the alternator battery terminal.

4. Connect a 0–150 ampere scale DC ammeter in series between alternator BAT terminal and disconnected alternator output wire. Connect positive lead to alternator BAT terminal and negative lead to disconnected alternator output wire.

5. Connect positive lead of a test voltmeter (range 0–18 volts minimum) to alternator BAT terminal.

6. Connect negative lead of test voltmeter to a good ground.

7. Connect an engine tachometer and reconnect negative battery cable.

8. Connect a variable carbon pile rheostate between battery terminals. Be sure carbon pile is in **OPEN** or **OFF** position before connecting leads.

9. Remove air hose between power module and air cleaner.

10. Connect 1 end of a jumper wire to ground and with other end probe green R3 lead wire on dash side of Black 8-way connector.

NOTE: Do not connect the blue J2 lead of the 8-way connector to ground. Both R3 and J2 leads are green on the alternator side of the 8-way connector. At the dash end of the connector, R3 is green and J2 is blue.

11. Start engine. Adjust carbon pile and engine speed in increments until a speed of 1250 rpm and voltmeter reading of 15 volts is obtained. Do not allow the voltage meter to read above 16 volts.

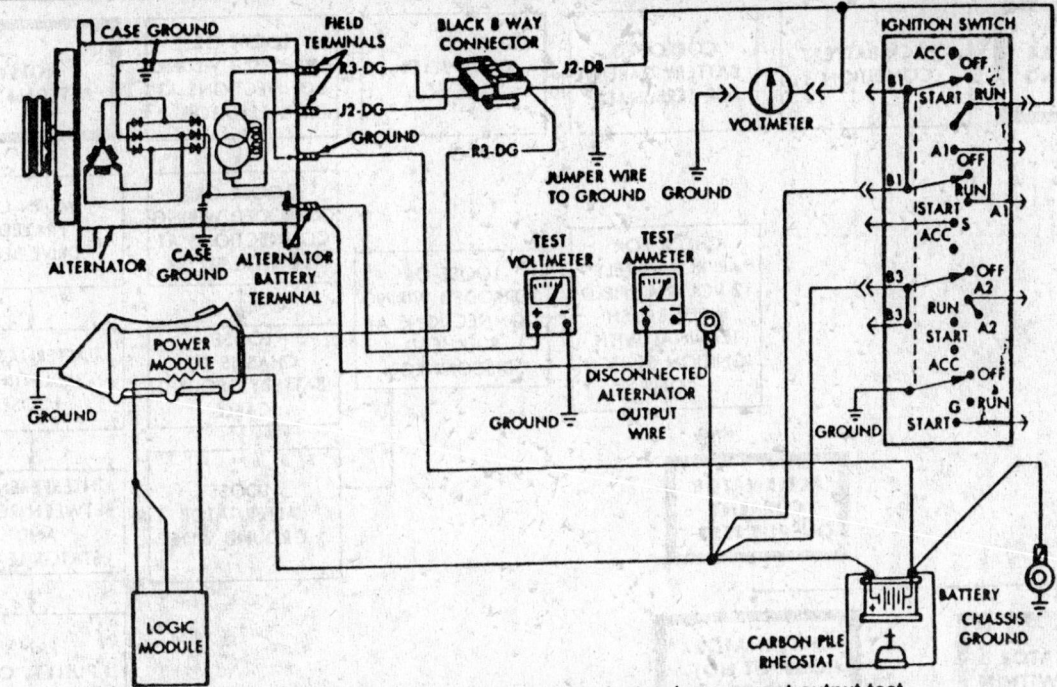

Chrysler 40/90 amp alternator with engine electronics—current output test

12. The ammeter reading must be within the proper limits.

13. If reading is less than specified and alternator output wire resistance is not excessive alternator should be removed from vehicle and bench tested.

14. After current output test is completed, reduce engine speed, turn **OFF** carbon pile and ignition switch. Disconnect negative battery cable.

15. Remove test ammeter, voltmeter, tachometer and carbon pile. Remove jumper wire between 8-way black connector and ground. Disconnect alternator output wire to alternator BAT terminal post.

16. Reconnect negative battery cable. Reconnect air hose between power module and air cleaner.

Voltage Regulator Test

On board diagnostic fault codes play a major role in case of a charging system failure.

Fault codes are 2 digit numbers that identify which circuit is at fault. In most cases, they do not identify which component in a circuit is at fault. Therefore, a fault code is only a result, not necessarily a reason for the problem. It is important that the test procedure be followed in order to understand the fault codes of the on-board diagnostic system.

DIAGNOSTIC READOUT BOX OPERATION

The diagnostic readout box is used to put the on-board diagnostic system in 3 different modes of testing as called

for in the driveability test procedure, only one of which is used in charging system diagnosis.

DIAGNOSTIC MODE

1. Connect diagnostic readout box C-4805 to the mating connector located in the wiring harness by right front shock tower.

2. Place read/hold switch on readout box in **READ** position.

3. Turn ignition switch **ON/OFF**, **ON/OFF**, on within 5 seconds.

4. Record all codes, displaying of codes may be stopped by moving read/hold button to **HOLD** position. Returning to **READ** position will continue displaying of codes.

5. If for some reason diagnostic readout box is not available, logic module can show fault codes by means of flashing power loss lamp on instrument cluster.

HOW TO USE POWER LOSS OR POWER LIMIT LAMP FOR CODES

To activate this function, turn ignition key **ON/OFF/ON/OFF/ON** within 5 seconds. The power loss lamp will turn **ON** for 2 seconds as a bulb check. Immediately following this it will display a fault code by flashing on and off. There is a short pause between flashes and a longer pause between digits. All codes displayed are 2 digit numbers with a 4 second pause between codes. An example of a code is as follows.

1. Lamp on for 2 seconds then turns off.

2. Lamp flashes 4 times, pauses and flashes once.

3. Lamp pauses for 4 seconds, flashes 4 times, pauses and flashes 7 times.

4. The 2 codes are 41 and 47. Any number of codes can be displayed as long as they are in memory. The lamp will flash until all of them are displayed.

CHARGING SYSTEM FAULT CODES

Perform test procedure categories using the following guide lines.

1. Each category is made up of many tests. Always start at the first test of a category. Starting at any other test will only give incorrect results.

2. Each test may have many steps. Only perform steps indicated under action required. It is not necessary to perform all steps in a test.

3. At the end of each test (not step) reconnect all wires and turn the engine **OFF** and reinstall any components that were removed for testing.

4. The vehicle being tested must have a fully charged battery.

DIAGNOSTIC TESTING WITH READOUT BOX

Test 1

CHECKING BATTERY SENSING CIRCUIT CODE 16

This test will check for direct battery feed to logic module. Circuit is also memory feed to logic module. Code 16 with lower battery voltage will turn **ON** power loss lamp.

1. Turn the ignition switch **OFF**.

2. Disconnect the (black on EFI,

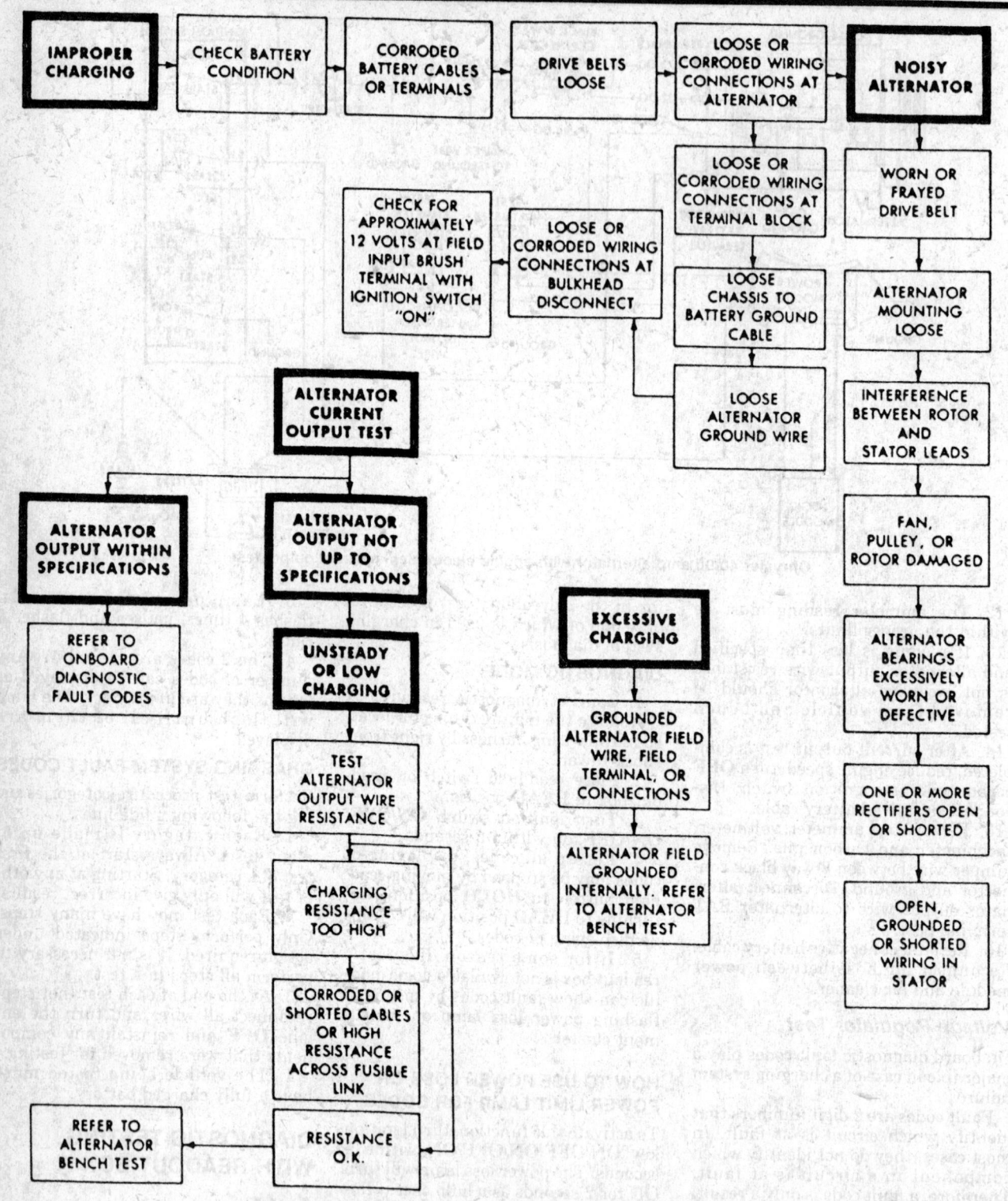

Alternator fault code chart—Chrysler 40/90 amp and 50/120 amp

blue on turbocharged engines) connector from the logic module.

3. Connect a voltmeter to cavity No. 22 of logic module connector and ground.

4. Voltmeter should read within 1 volt of battery voltage. Voltage okay,

replace logic module. Before replacing logic module, make sure the terminal in cavity No. 22 is not crushed so that it cannot touch logic module pin.

5. Zero volts, repair wire of cavity No. 22 for an open circuit to the wiring harness splice.

Test 2
FAULT CODES 41 AND 46
CHECKING CHARGING SYSTEM
Step A

1. Disconnect the power module 10-way connector.

2. Connect a voltmeter between cavity No. 8 of 10-way connector and ground.

3. Turn ignition switch to **RUN** position.

4. Voltmeter should read within 1 volt of battery voltage. Not within 0–1 volts, repair alternator field circuit for short to ground. Voltage okay, perform Step B.

Step B

1. Turn the ignition switch **OFF**.

2. Reconnect the power module 10-way connector.

3. Disconnect the power module 12-way connector.

4. Connect a voltmeter between F2 terminal on alternator and ground.

5. Turn the ignition switch to the **RUN** position.

6. Voltage should read within 1 volt of battery voltage. Not within 0–1 volt, replace power module. Voltage okay, perform Step C.

Step C

1. Turn the ignition switch **OFF**.

2. With power module 12-way connector disconnected.

3. Disconnect the logic module (white on EFI, red on turbocharged engines) connector.

4. Connect an ohmmeter between cavity No. 11 of power module 12-way connector and ground.

5. Ommeter should not show continuity. No continuity, replace logic module. Continuity repair wire of cavity No. 11 for short circuit to ground.

Test 3

CHECKING CODES 41 AND 47

Step A

1. Connect a voltmeter between battery positive and ground.

2. Connect an end of jumper wire to a good engine ground.

3. Start the engine and note reading of voltmeter.

4. Very quickly touch other end of jumper wire to F2 terminal on alternator and watch voltmeter.

5. Voltmeter should show an increase in voltage. Voltage increases, this indicates alternator is operating correctly. Move on to Step B, for field circuit check. Voltage does not increase, this indicates alternator is not operating. If this is the case, perform Step E which checks for voltage to alternator field.

Step B

1. Connect a voltmeter between cavity No. 2 of logic module (black on EFI, blue on turbocharged engines) connector and ground.

2. Connect an end of a jumper wire

to cavity No. 5 of the logic module white connector.

3. Very quickly touch other end of jumper wire to logic module mounting stud and watch voltmeter.

4. Voltmeter should show an increase in voltage. If voltage increases, this indicates all components of system, except logic module, are operating correctly.

5. Before replacing the logic module, be sure that the terminal in cavity No. 5 is not crushed so that it cannot touch the logic module pin. If terminal in cavity 5 is not damaged, replace logic module. If no increase is indicated, move on to Step C.

Step C

1. Turn the engine **OFF**.

2. Disconnect logic module (white on EFI, red on turbocharged engines) connector.

3. Connect a voltmeter between cavity No. 5 of logic module connector and ground.

4. Turn the ignition switch to **RUN** position. Voltmeter should read within 1 volt of battery voltage. Zero volts, disconnect power module 12-way and connect an ohmmeter between cavity 5 of logic module (white on EFI, red on turbocharged engines) connector and cavity 11 of power module. If open, repair wire or connector. If meter shows continuity replace power module. If voltage shown is within 1 volt of battery voltage, go on to Step D.

Step D

1. Turn ignition switch **OFF**.

2. Disconnect 10-way connector from power module.

3. Connect a voltmeter between cavity No. 8 of 10-way connector and ground.

4. Turn ignition switch to **RUN** position. Voltmeter should read within 1 volt of battery voltage. If voltage shown is within 1 volt of battery voltage, replace power module.

5. Zero volts, turn ignition switch **OFF** and place an ohmmeter between cavity 8 of power module 10-way connector and F2 terminal of alternator. If open, repair wire or connector. If meter shows continuity, proceed to Step E.

Step E

1. Turn ignition switch to **OFF** position.

2. Connect a voltmeter between F1 terminal of alternator and ground.

3. Turn ignition switch to **RUN** position.

4. Voltmeter should read within 1 volt of battery voltage. If voltage shown is within 1 volt of battery voltage, alternator is not functioning

properly and must be removed form vehicle and repaired.

5. If no voltage is shown, this indicates an open circuit and the wire from the F1 terminal to ignition switch must be repaired.

Test 4

CHECKING CODE 44

Step A

1. Turn the ignition switch **OFF**.

2. Disconnect the logic module (black connector on EFI, blue connector on turbocharged engines).

3. Connect an ohmmeter between cavity No. 20 of logic module (black on EFI, blue on turbocharged engines) connector and ground.

4. Ohmmeter should show resistance, amount of resistance should be 8–29K ohms. Correct resistance, replace logic module. If 0 resistance, perform Step B. Open circuit, perform Step C.

Step B

1. Ohmmter connected between cavity No. 20 of logic module (black on EFI, blue on turbocharged engines) connector and ground.

2. Disconnect power module 12-way connector.

3. Ohmmeter should show an open circuit. Open circuit, replace power module. If 0 resistance, repair wire of cavity No. 20 and cavity No. 3 of power module 12-way connector.

Step C

1. Disconnect power module 12-way connector.

2. Connect an ohmmeter between pin 3 of power module 12-way and ground.

3. Ohmmeter should show resistance, amount of resistance should be between 8–29K ohms.

4. Correct resistance, repair wire in cavity No. 20 of logic module (black on EFI, blue on turbocharged engines) and cavity No. 3 of power module 12-way connector. Open circuit, replace power module.

ALTERNATOR OVERHAUL

Disassembly

1. Remove the dust cover mounting nut. Remove the dust cover.

2. Remove the brush holder assembly mounting screws. Remove the brush holder assembly.

3. Remove the stator to rectifier mounting screws. Remove the stator-to-rectifier assembly mounting screws. Remove the rectifier insulator. Remove the capacitor mounting screw. Remove the rectifier assembly.

4. Remove the through bolts. Care-

fully pry between the stator and the drive end shield, using a suitable tool and separate the end shields. The stator is laminated, do not burr the stator or the end shield.

5. Position the drive end of the alternator over the bosses of the holding fixture. Do not position the rotor plastic termination plate over the fixture boss or damage to the assembly will result.

6. Bolt the drive end of the assembly to shield fixture. Loosen the pulley mounting nut. Remove the pulley mounting nut. Remove the pulley washer.

7. Remove the poly-vee pulley. Remove the fan. Remove the front bearing spacer. Press the rotor assembly out of the drive end shield.

8. Remove the inner bearing spacer. Position the alternator bearing puller tool under the rear rotor bearing. Tighten the right puller bolt a ½ turn. Tighten the left puller bolt a ½ turn. Continue tightening the tool a ½ turn on each bolt until the rear rotor bearing is free. Remove the rear rotor bearing assembly from the rotor.

9. Position the rotor assembly in the holding fixture. Position the rear rotor bearing onto the rotor shaft.

10. Drive the rear rotor bearing onto the rotor until it bottoms. The rear rotor position is critical and must be installed using special tools C-4885 and C-4894.

11. Remove the front bearing retaining screws. Press the front bearing out of the drive end shield.

12. Carefully remove the stator from the rectifier end shield.

Inspection

ROTOR ASSEMBLY TEST

Check the outside circumference of slip-ring for dirtiness and roughness. Clean or polish with fine sandpaper, if required. A badly roughened slip-ring or a worn down slip-ring should be replaced.

Slip-rings are not serviced as a separate item. They are serviced with the rotor assembly.

ROTOR FIELD COILS FOR OPENS AND SHORTS TEST

To check for an open rotor field coil connect an ohmmeter between slip-rings. Ohmmeter readings should be between 1.5–2 ohms on rotor field coils at room ambient conditions. Resistance between 2.5–3.0 ohms would result from alternator rotor field coils that have been operated on vehicle at higher engine compartment temperatures. Readings about 3.5 ohms would indicate high resistance rotor field coils and further testing or replacement may be required.

To check for a shorted rotor field coil, connect an ohmmeter between both slip-rings. If the reading is below 1.5 ohms, the rotor field coil is shorted.

ROTOR FIELD COIL FOR GROUND TEST

To check for a grounded rotor field coil, connect an ohmmeter from each slip-ring to rotor shaft. Ohmmeter should be set for infinite reading when probes are apart and 0 when probes are shorted. The ohmmeter should read infinite. If the reading is 0 or low in value, rotor is grounded.

STATOR ASSEMBLY TEST

Stator Coil for Ground Test

1. Remove varnish from a spot on the stator frame.
2. Press an ohmmeter test probe firmly onto cleaned spot on frame. Be sure varnish has been removed from stator so that spot is bare.
3. Press the other ohmmeter test probe firmly to each of the 3 phase (stator) lead terminals 1 at a time. If ohmmeter reads 0 or low in value stator lead is grounded.
4. Replace stator if stator tested grounded.

Stator for Open or Short Circuit Test

The stator windings are delta wound. Therefore, they cannot be tested for opens or shorts with an ohmmeter. They can only be tested for these items with test equipment not common to automotive service test equipment. If stator is not grounded and all other electrical circuits and components of alternator test okay, it can be suspected that stator could possibly be open or shorted and must be replaced.

Rectifier Assemblies Test

When testing rectifiers with an ohmmeter, disconnect the 3 phase stator lead terminals from rectifier assembly. Pry stator lead terminals away from rectifier assembly.

Positive Rectifier Test

With an ohmmeter check for continuity between each positive (+) rectifier strap and positive (+) heat sink. Reverse test probes and retest. There should be continuity in one direction only. If there is continuity in both directions, rectifier is short circuited. If there is no continuity in either direction, rectifier is open. If rectifier is shorted or open, replace rectifier assembly.

Negative Rectifier Test

With an ohmmeter, check for continuity between each negative (−) rectifier strap and negative (−) sink. Reverse test probes and retest. There should be continuity in one direction only. If there is continuity in both directions, rectifier is short circuited. If there is no continuity in either direction, rectifier is open. If rectifier is shorted or open, replace rectifier assembly. When installing a new rectifier assembly, apply 3 dabs (0.1 grams each) of heat sink compound to bottom of negative rectifier prior to mounting rectifier assembly to rectifier end shield.

Brushes and Brush Springs Continuity Test

When testing brushes and brush springs make sure that brushes move smoothly in brush holder. Sticking brushes require replacement of brush holder assembly.

Inner Brush Circuit Test

With an ohmmeter, touch a test probe to inner brush and another probe to field terminal. If there is no continuity, replace the brush assembly.

Outer Brush Circuit Test

With an ohmmeter, touch a test probe to outer brush and other probe to field terminal. If there is no continuity, replace brush assembly.

Cleaning Alternator Parts

Do not immerse stator field coil assembly, rotor assembly or rectifier assembly in cleaning solvent, as solvent will damage these parts.

Assembly

1. Be sure to repair or replace defective components as required.
2. To the front of the rotor, install the inner bearing spacer and press the drive end shield onto the rotor.

NOTE: The front drive end shield bearing must be replaced anytime the rotor or drive end shield is removed, for the front bearing is a press fit and may be damaged upon removal.

3. Position the drive end shield into a holding fixture so the fixture bosses do not contact the rotor plastic termination plate and tighten the holding bolt.
4. At the front of the drive end shield, install the pulley spacer (flat side up), the fan, the pulley, the washer and the pulley nut. Torque the pulley nut to 80–105 ft. lbs. (108–125 Nm).
5. Remove the drive end shield and rotor assembly from the holding fixture.

6. To assemble the rear end housing for the 40/90, perform the following procedures:

a. Apply joint compound to the rectifier end shield surface, under the rectifier mounting position.

b. Position the rectifier assembly to the rectifier drive end housing.

c. Install the rectifier insulator, the insulator mounting screws and torque the screws to 36–46 inch lbs. (4–6 Nm).

d. Install the capacitor terminal over the rectifier assembly battery terminal stud and torque the mounting screw to 36–48 inch lbs. (4–6 Nm).

e. Install the alternator battery terminal nut and torque to 30–50 inch lbs. (3–6 Nm).

f. Install the stator-to-rectifier screws and torque to 12–18 inch lbs. (1–2 Nm).

g. Slide the brushes into their cavity, Install the brush holder assembly and torque the screws to 12–18 inch lbs. (1–2 Nm).

h. Install the dust cover and torque the nut to 12–18 inch lbs. (1–2 Nm).

7. To assemble the rear end housing for the 50/120, perform the following procedures:

a. Apply joint compound to the rectifier end shield surface, under the rectifier mounting position.

b. Position the rectifier assemblies No. 1 and No. 2 to the rectifier drive end housing.

c. Install the rectifier insulators, the insulator mounting screws and torque to 36–46 inch lbs. (4–6 Nm).

d. Install the capacitor terminal over the rectifier assembly No. 2 battery terminal stud and torque the mounting screw to 36–48 inch lbs. (4–6 Nm).

e. Install the alternator battery terminal nut and torque to 30–50 inch lbs. (3–6 Nm).

f. Position a jumper strap between the rectifier assemblies No. 1 and No. 2.; torque the nut to 36–48 inch lbs. (4–6 Nm) and the screw to 15–35 inch lbs. (2–4 Nm).

g. Position the 3 buss bars to the rectifier assemblies and torque the screws to 12–18 inch lbs. (1–2 Nm).

h. Slide the brushes into their cavity, Install the brush holder assembly and torque the screws to 12–18 inch lbs. (1–2 Nm).

i. Install the dust cover and torque the nut to 12–18 inch lbs. (1–2 Nm).

8. Assemble the rear end housing to the drive end shield and rotor assemlby. Install the through bolts and torque to 48–72 inch lbs. (5–8 Nm).

9. Install the rear bearing oil seal.

Bosch 35/75, 40/90, 75 HS, 90 HS and 90 RS Amp Alternators

The alternators are alike, except, the 35/75 has 6 built-in silicon rectifiers and the 40/90 has 12 (1986–87) or 8 (1988–90) built-in silicon rectifiers.

The voltage regulator is built into the power and logic modules for 1986–88 or the Single Module Engine Controller (SMEC) for 1989–90.

ON VEHICLE SERVICE
Charging Circuit Resistance Test

1. Be sure that the battery is fully charged.

2. Disconnect negative battery cable.

3. Disconnect BAT lead at alternator output terminal.

4. Connect a 0–150 ampere scale DC ammeter in series between alternator output BAT terminal and disconnected BAT terminal wire. Connect positive lead to alternator output BAT terminal and negative lead to disconnected alternator BAT lead.

5. Connect positive lead of a test voltmeter (range 0–18 volts minimum) to alternator BAT terminal. Connect negative lead of test voltmeter to battery positive post.

6. Remove air hose between power module (1986–88) or single module engine controller (1989–90) and air cleaner.

7. Connect an end of a jumper wire to ground and with other end, probe the green R3 lead wire of black 8-way connector (1986–88) or back of the alternator (1989–90). Do not connect blue J2 lead of 8-way wiring connector (1986–88) or back of the alternator (1989–90) to ground.

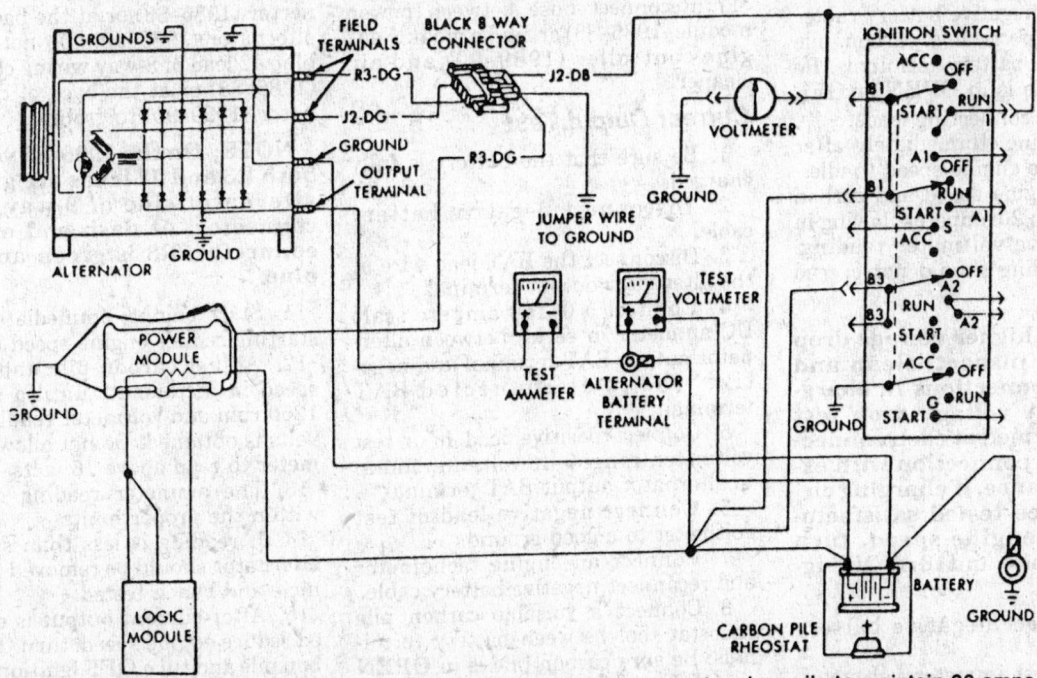

Bosch 40/90 amp alternator – charging resistance test – adjust engine speed and carbon pile to maintain 20 amps flowing in circuit. Voltmeter reading should not exceed 0.5 volts

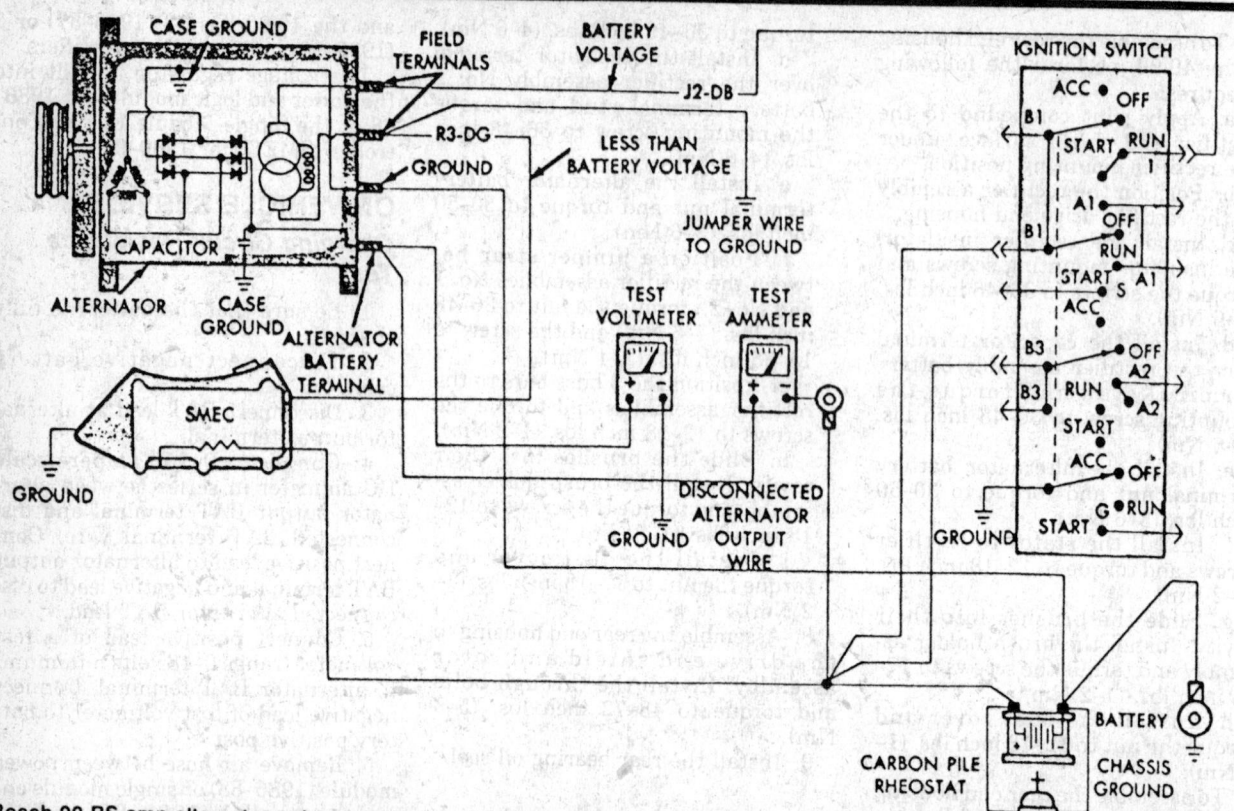

Bosch 90 RS amp alternator—charging system resistance test—adjust engine speed and carbon pile to maintain 20 amps, voltmeter reading should not exceed 0.5 volts

NOTE: On the 1986–88 vehicles, both R3 and J2 leads are green on alternator side of 8-way wiring connector. At dash end of 8-way connector, R3 is green and J2 is blue.

8. Connect an engine tachometer and reconnect negative battery cable.

9. Connect a variable carbon pile rheostat to the battery terminals. Be sure carbon pile is in **OPEN** or **OFF** position before connecting leads.

10. Start engine. Immediately after starting, reduce engine speed to idle.

11. Adjust engine speed and carbon pile to maintain 20 amperes flowing in circuit. Observe voltmeter reading. Voltmeter reading should not exceed 0.5 volts.

NOTE: If a higher voltage drop is indicated, inspect, clean and tighten all connections in charging circuit. A voltage drop test may be performed at each connection to locate connection with excessive resistance. If charging circuit resistance tested satisfactorily, reduce engine speed, turn OFF carbon pile and turn OFF ignition switch.

12. Disconnect negative battery cable.

13. Remove test ammeter, voltmeter, carbon pile and tachometer.

14. Remove jumper wire between 8-way black connector (1986–88) or the back of the alternator (1989–90) and ground.

15. Connect BAT lead to alternator output BAT terminal post.

16. Reconnect negative battery cable.

17. Reconnect hose between power module (1986–88) or single module engine controller (1989–90) and air cleaner.

Current Output Test

1. Be sure that the battery is fully charged.

2. Disconnect negative battery cable.

3. Disconnect the BAT lead wire at the alternator output terminal.

4. Connect a 0–150 ampere scale DC ammeter in series between alternator output BAT terminal and negative lead to disconnected BAT terminal.

5. Connect positive lead of a test voltmeter (range 0–18 volts minimum) to alternator output BAT terminal.

6. Connect negative lead of test voltmeter to a good ground.

7. Connect an engine tachometer and reconnect negative battery cable.

8. Connect a variable carbon pile rheostat tool between battery terminals. Be sure carbon pile is in **OPEN** or **OFF** position before connecting leads.

9. Remove air hose between power module (1986–88) or the single module engine controller (1989–90) and air cleaner.

10. Connect 1 end of a jumper wire to ground and with other end, probe the green R3 lead wire of black 8-way connector (1986–88) or at the back of the alternator (1989–90). Do not connect blue J2 lead of 8-way wiring connector (1986–88) or at the back of the alternator (1989–90) to ground.

NOTE: On the 1986–88 vehicles, both R3 and J2 leads are green on alternator side of 8-way wiring connector. At dash end of 8-way connector, R3 is green and J2 is blue.

11. Start engine. Immediately after starting reduce engine speed to idle.

12. Adjust carbon pile and engine speed in increments until a speed of 1250 rpm and voltmeter reading of 15 volts is obtained. Do not allow voltmeter to read above 16 volts.

13. The ammeter reading must be within the proper limits.

14. If reading is less than specified, alternator should be removed from vehicle and bench tested.

15. After current output is completed reduce engine speed, turn **OFF** carbon pile and turn **OFF** ignition switch.

16. Disconnect negative battery cable.

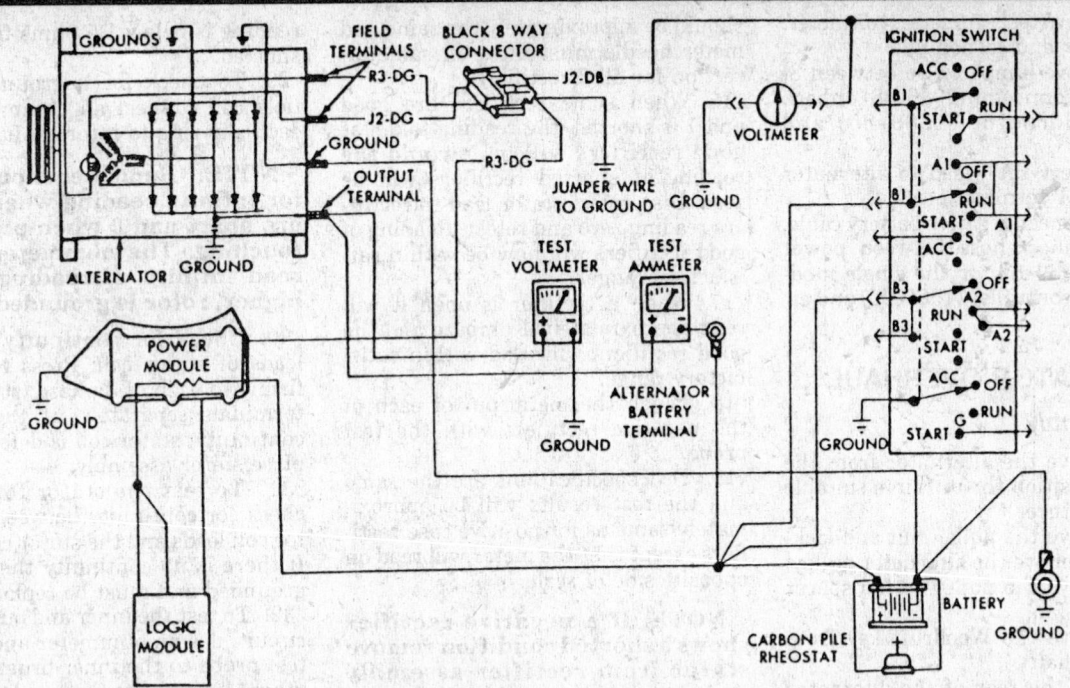

Bosch 40/90 amp alternator—current output test—adjust carbon pile and engine speed until a speed of 1250 rpm and voltmeter reading of 15 volts is obtained. The ammeter reading must be within proper limits

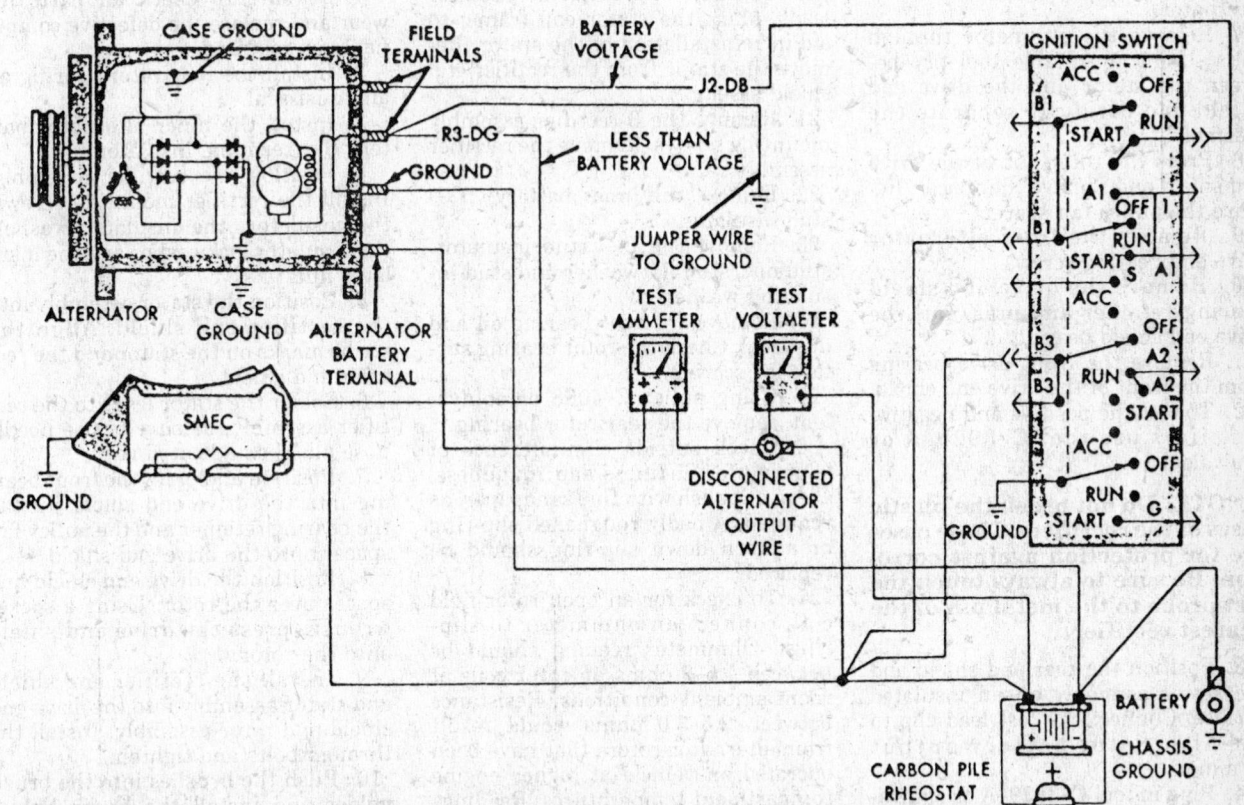

Bosch 90 RS amp alternator—current output test—adjust engine speed and carbon pile increments until an engine speed of 1250 rpm is reached, voltmeter reading should not exceed 15 volts

17. Remove test ammeter, voltmeter, tachometer and carbon pile.

18. Remove jumper wire between 8-way black connector (1986–88) or back of the atlernator (1989–90) and ground.

19. Connect BAT lead to alternator output BAT terminal post.

20. Reconnect negative battery cable.

21. Reconnect hose between power module (1986–88) or the single module engine controller (1989–90) and air cleaner.

ALTERNATOR OVERHAUL

Disassembly

1. Remove the alternator from the vehicle. Position the unit in a suitable holding fixture.

2. Remove the pulley nut and lockwasher. Remove the alternator pulley.

3. Remove the pulley to fan spacer and pulley fan.

4. Remove the Woodruff key from the rotor shaft.

5. From the rear of the alternator disconnect the electrical terminal from the capacitor. Remove the capacitor retaining screw and the capacitor.

6. Remove the brush holder retaining screw and remove the brush holder from its mounting on the rear of the alternator.

7. Remove the alternator through bolts. Using a suitable tool pry between the stator and the drive end shield and carefully separate the assembly.

8. Press the rotor out of the drive end shield and remove the spacer. Remove the pulley fan spacer.

9. Remove the front alternator drive end bearing screws.

10. Remove the drive end shield bearing retainer and press out the drive end shield bearing.

11. Remove the front drive bearing from the front of the drive end shield.

12. To test the positive and negative rectifiers use tool C–3929–A or equivalent.

NOTE: Do not break the plastic cases of the rectifiers. These cases are for protection against corrosion. Be sure to always touch the test probe to the metal pin of the nearest rectifier.

13. Position the rear end shield and the stator assembly on an insulated surface. Connect the test lead clip to the alternator battery output terminal.

14. Plug in tool C–3829–A or equivalent. Touch the metal pin of each of the positive rectifiers with the test probe.

15. Reading for satisfactory rectifiers will be 1¾ amperes or more. Reading

should be approximately the same and meter needle must move in same direction for all 3 rectifiers.

16. When some rectifiers are good and 1 is shorted, the reading taken at good rectifiers will be low and the reading at shorted rectifiers will be zero. Disconnect stator lead to rectifiers reading zero and retest. Reading of good rectifiers will now be within satisfactory range.

17. When a rectifier is open it will read approximately 1 ampere and the good rectifiers will read within satisfactory range.

18. Touch the metal pin of each of the negative rectifiers with the test probe.

19. Test specifications are the same and the test results will be approximately same as for positive case rectifiers except that the meter will read on opposite side of scale.

NOTE: If a negative rectifier shows a shorted condition remove stator from rectifier assembly and retest. It is possible that a stator winding could be grounded to stator laminations or to an rectifier end shield, which would indicate a shorted negative rectifier.

20. Unsolder the stator to rectifier leads. Mark the stator coil frame, to aid in reinstallation of the stator. Remove the stator from the rectifier end shield assembly.

21. Remove the 3 rectifier assembly mounting screws. Remove the rectifier assembly.

22. Remove the inner battery (B+) stud insulator.

23. Remove the D+ stud insulator, stud nut, stud flatwasher and stud insulating washer.

24. Remove the rear bearing oil and dust seal. Check the rotor bearing surface for scoring.

25. Using puller C–4068 or equivalent, remove the rear rotor bearing.

26. Check outside circumference of slip-ring for dirtiness and roughness. Clean or polish with fine sandpaper, as required. A badly roughened slip-ring or a worn down slip-ring should be replaced.

27. To check for an open rotor field coil, connect an ohmmeter to slip-rings. Ohmmeter reading should be between 1.5–2 ohms on rotor coils at room ambient conditions. Resistance between 2.5–3.0 ohms would result from alternator rotors that have been operated on vehicle at higher engine compartment temperatures. Readings above 3.5 ohms would indicate high resistance rotor coils and further testing or replacement may be required.

28. To check for a shorted field coil connect an ohmmeter to slip-rings. If

reading is below 1.5 ohms field coil is shorted.

29. To check for a grounded rotor field coil connect an ohmmeter from each slip-ring to rotor shaft.

NOTE: Ohmmeter should be set for infinite reading when probes are apart and 0 when probes are touching. The ohmmeter should read infinite. If reading is 0 or higher, rotor is grounded.

30. Check for continuity between leads of stator coil. Press test probe firmly to each of 3 phase (stator) lead terminals separately. If there is no continuity, stator coil is defective. Replace stator assembly.

31. To test the stator for ground check for continuity between the stator coil leads and the stator coil frame. If there is no continuity the stator is grounded and must be replaced.

32. To test the inner and outer brush circuit, use an ohmmeter and touch 1 test probe to the inner brush and the other test probe to the brush terminal. If continuity does not exist replace the brush assembly. Repeat the same procedure for the outer brush.

Assembly

1. Be sure to check all parts for wear and replace the defective components as required.

2. Install the rear rotor bearing oil and dust seal.

3. Install the inner alternator battery B+ terminal insulator.

4. Position the rectifier assembly. Install the rectifier mounting screws, the insulator, the insulator washer, the insulator lockwasher and the insulator nut.

5. Position the stator assembly into the rectifier end shield. Align the scribe marks on the stator and the rectifier end shield.

6. Solder the stator leads to the rectifier assembly; be sure to use needle nose pliers as a heat sink.

7. Position and press the front bearing into the drive end shield. Install the bearing retainer and the pulley fan spacer onto the drive end shield.

8. Position the drive end shield and spacer over the rotor. Using a socket wrench, press the drive end shield onto the rotor.

9. Install the rectifier end shield and stator assembly into the drive end shield and rotor assembly. Install the through bolts and tighten.

10. Push the brushes into the brush holder and install the brush holder onto the alternator assembly.

11. Install the capacitor and terminal plug onto the alternator assembly.

12. Install the Woodruff key into the shaft and the fan over the shaft.

13. Install the drive pulley-to-fan spacer, the pulley, the lockwasher and the nut over the shaft. Secure the pulley and tighten the nut.

Mitsubishi 75 Amp and 90 Amp Alternator with Internal Regulator

This integrated circuit alternator has 15 built-in rectifiers, that convert A.C. current into D.C. current at the alternator battery terminal. The main components of the alternator are the rotor, stator, rectifiers, end shields, pulley, fan and capacitor. The electronic voltage regulator is very compact and is built into the rectifier end shield of the alternator.

ON VEHICLE SERVICE

Charging Circuit Resistance Test

1. With the ignition switch in the **OFF** position, disconnect the negative battery cable and the output wire from the alternator battery cable.

2. Using a 0–100 amp DC ammeter, connect it in series between the alter-

nator **BAT** terminal and the disconnected alternator output wire; connect the positive (+) wire to the alternator **BAT** terminal and the negative (−) wire to the disconnected alternator output wire.

3. Using a 0–18 volt voltmeter, connect its positive (+) lead to the disconnected alternator output wire and its negative (−) lead to the positive (+) battery post.

4. Install the tachometer to the engine.

5. Using a variable carbon pile rheostat, connect it between the battery posts; be sure the pile is in the **OPEN** or **OFF** position before connecting the leads.

6. Start the engine and allow it to idle.

7. Adjust the engine speed and the carbon pile to maintain 20 amps flowing in the circuit; the voltmeter should not exceed 0.5 volts.

8. If a higher voltage drop is indicated, clean and tighten the alternator output wire circuit connections.

9. If the resistance is satisfactory, reduce the engine speed, turn **OFF** the carbon pile and the ignition switch.

10. Remove the test equipment. Reconnect the alternator output wire and the negative battery cable.

Voltage Regulator Test

1. With the ignition switch in the **OFF** position, disconnect the positive cable from the battery and place a knife switch on the battery post and connect the cable to the knife switch.

2. Install the leads from an ammeter to the knife switch connectors and open the switch-to-battery current.

3. Connect the leads of a voltmeter between the **L** terminal of the alternator and a good ground.

4. The voltage reading should be zero volts. Should voltage be present, a defective alternator or wiring is indicated.

5. If no voltage is present, turn the ignition switch **ON** position. The voltage present should be lower than battery voltage, by about 1 volt or less. If the voltage reading is higher or at battery voltage, a defective alternator is indicated.

6. Connect a tachometer to the engine and close the knife switch, mounted on the battery post. Start the engine. Do not apply any starting current through the ammeter when starting the engine. The ammeter can be ruined.

7. After the engine is operating, open the knife switch and increase the

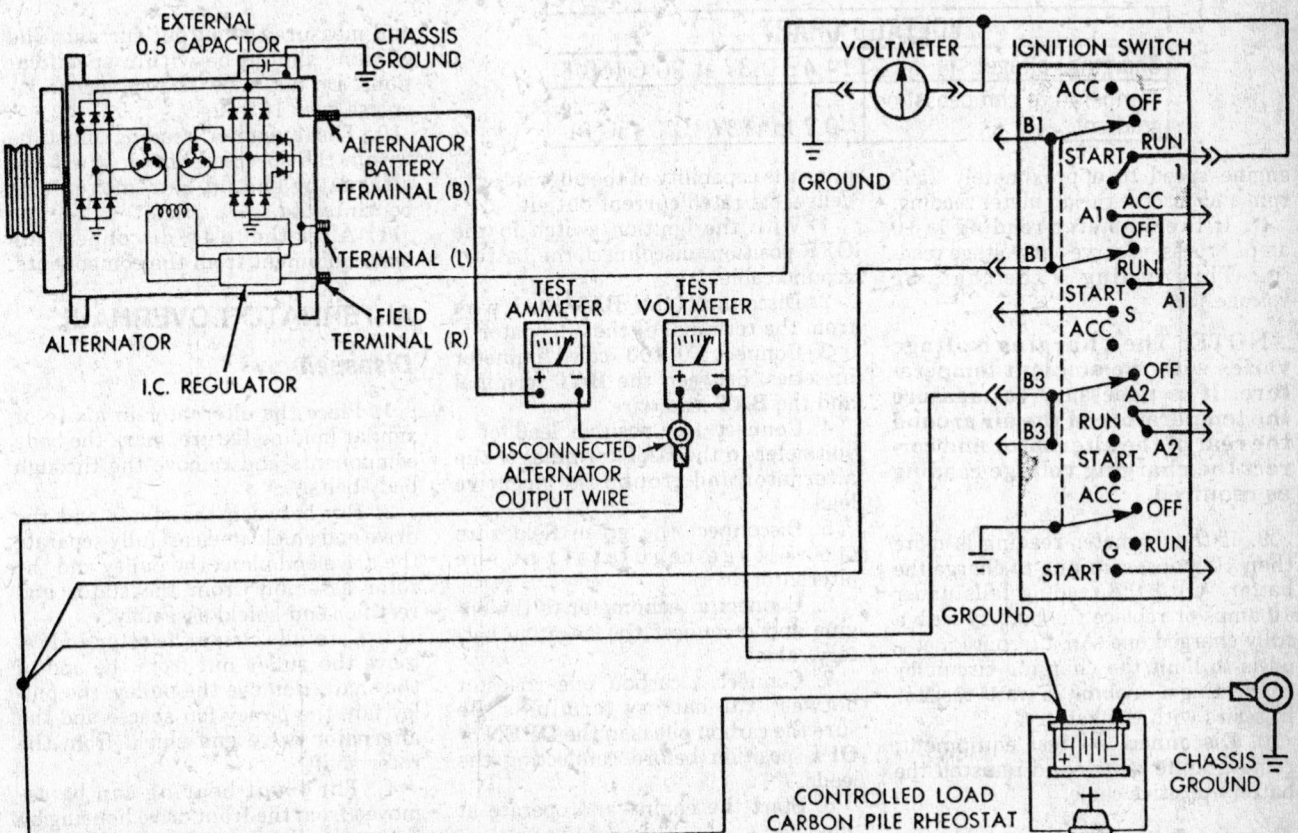

Mitsubishi 75 and 90 amp alternators with internal voltage regulator—charging system resistance test—adjust engine speed and carbon pile to maintain 20 amps, voltmeter reading should not exceed 0.5 volts

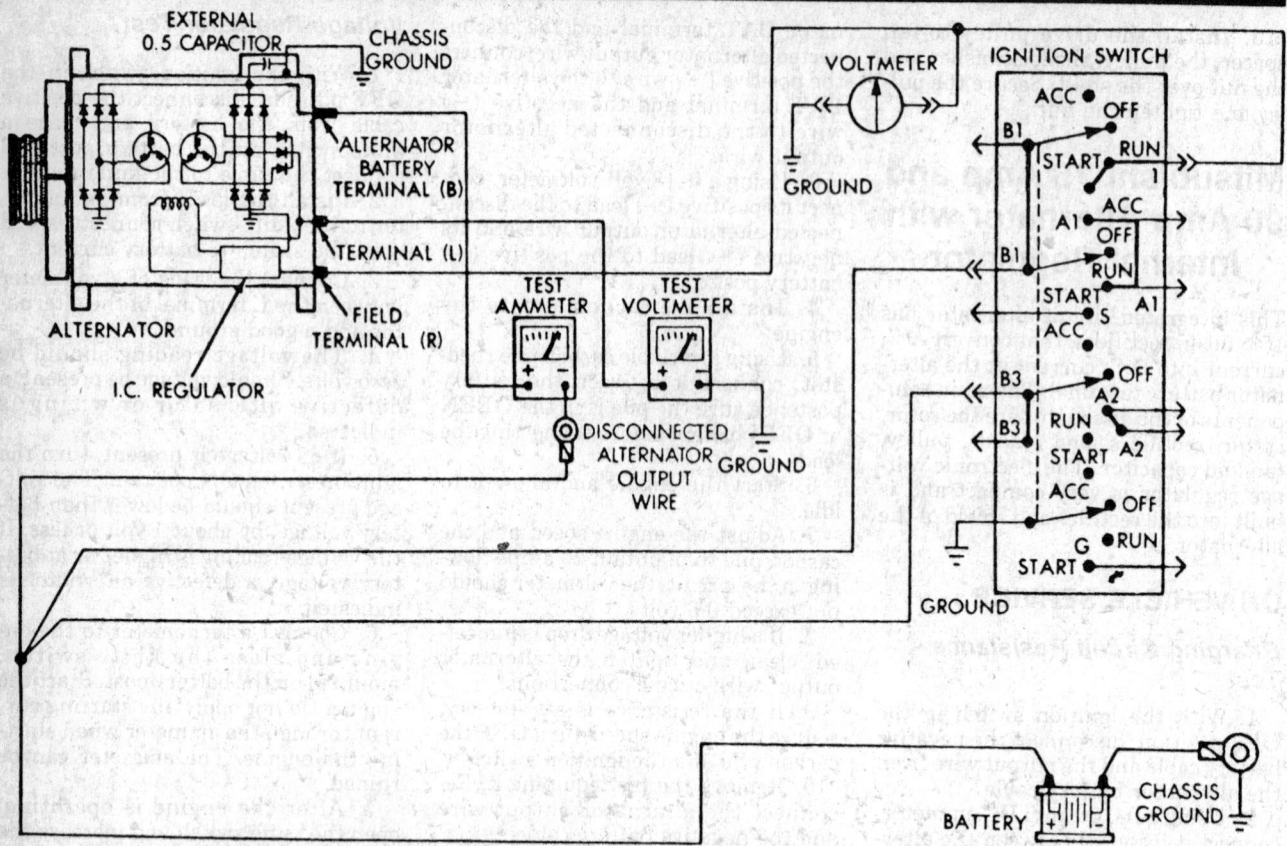

Mitsubishi 75 and 90 amp alternators with external voltage regulator — voltage regulator test

VOLTAGE CHART	
Charging voltage	14.4 ± 0.3V at 20°C (68°F)
Temperature compensation gradient	−0.7 to .13V/10°C (50°F)

engine speed to approximately 2500 rpm and observe the ammeter reading.

8. If the ammeter reading is 10 amps or less, observe the voltage reading. This reading is the charging voltage.

NOTE: The charging voltage varies with the ambient temperature. It is necessary to measure the temperature of the air around the rear of the alternator and correct the charging voltage reading as required.

9. If the ammeter reading is more than 10 amps, continue to charge the battery until the reading falls under 10 amps or replace the battery with a fully charged one. An alternate method is to limit the charging circuit by connecting a ¼ ohm (25 watt) resistor in series with the battery.

10. Disconnect all test equipment, remove knife switch and reinstall the battery positive cable.

Current Output Test

The purpose of this test is to deter-

mine the capability of the alternator to deliver its rated current output.

1. With the ignition switch in the **OFF** position, disconnect the battery ground cable.

2. Disconnect the **BAT** lead wire from the terminal of the alternator.

3. Connect a 0–100 scaled ammeter in series, between the **BAT** terminal and the **BAT** lead wire.

4. Connect the positive lead of a voltmeter to the **BAT** terminal of the alternator and ground the negative lead.

5. Disconnect the green field wire (to voltage regulator) at the alternator.

6. Connect a tachometer to the engine and reconnect the negative battery cable.

7. Connect a carbon pile rheostat between the battery terminals. Be sure the carbon pile is in the **OPEN** or **OFF** position before connecting the leads.

8. Start the engine and operate at idle.

9. Adjust the carbon pile and accelerate the engine to the specified speed

and measure the output current. The current should be within specifications. Do not allow the voltage to increase over 16 volts.

10. The ammeter reading must be within the specified limits. If not the alternator should be removed and bench tested.

11. After the tests, disconnect the test equipment from the components.

ALTERNATOR OVERHAUL

Disassembly

1. Place the alternator in a vise or similar holding fixture, mark the body components and remove the through body bolts.

2. Pry between the stator and the drive end shield and carefully separate the drive end plate, the pulley and the rotor assembly from the stator and rectifier end shield assembly.

3. Carefully clamp the rotor and remove the pulley nut from the end of the shaft. Remove the pulley, the pulley fan, the pulley fan spacer and the alternator drive end shield from the rotor shaft.

4. The front bearing can be removed from the front drive housing by the removal of the dust seals, front and rear, the bearing retainer screws, the retainer, exposing the bearing so

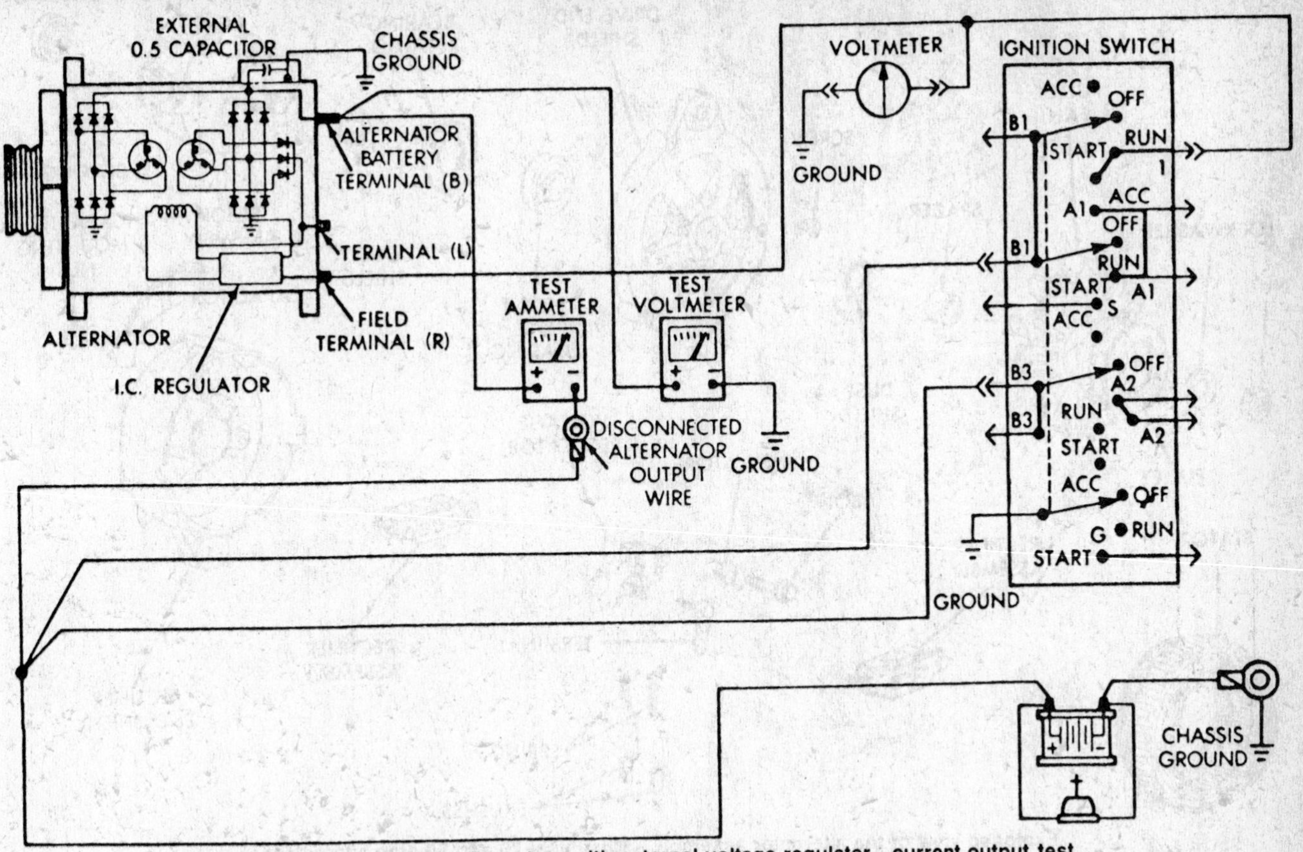

Mitsubishi 75 and 90 amp alternators with external voltage regulator—current output test

75 AMP CURRENT OUTPUT CHART	
Output current (Hot or Cold) at Engine RPM	44-51A at 13.0 Volts and 750 RPM
	64-72A at 13.0 Volts and 1000 RPM
	74-78A at 13.0 Volts and 2000 RPM

90 AMP CURRENT OUTPUT CHART	
Output current (Hot or Cold) at Engine RPM	58-71A at 13.0 Volts and 750 RPM
	80-91A at 13.0 Volts and 1000 RPM
	95-102A at 13.0 Volts and 2000 RPM

that it can be tapped from the drive housing.

5. To remove the stator assembly. The 6 stator leads must be unsoldered from the rectifiers, as per the manufacturer's recommendation.

6. Remove the rectifiers from the stator end shield housing.

7. Remove the brush holder and regulator retaining screw.

8. Remove the Bat terminal retaining nut and remove the capacitor from the terminal.

9. Remove the regulator and rectifier assembly. Unsolder 1 rectifier-to-regulator assembly and remove the other rectifier assembly by sliding the battery stud out of the regulator.

10. Inspect the rotor bearing surface for scores and make the necessary off vehicle test on the electrical components.

Inspection

ROTOR ASSEMBLY

1. Check the outside circumference of the slip ring for dirtiness and roughness. Clean or polish with fine sandpaper, if required. A badly roughened slip ring or a slip ring worn down beyond the service limit should be replaced.

2. Check for continuity between the field coil and slip ring. If there is no continuity, the field coil is defective. Replace the rotor assembly.

3. Check for continuity between the slip ring and shaft (or core). If there is continuity, the coil or slip ring is grounded. Replace the rotor assembly.

STATOR ASSEMBLY

Check for continuity between the leads of the stator coil. If there is no

continuity the stator coil is defective. Replace the stator assembly.

RECTIFIER ASSEMBLY

Positive (+) Heat Sink Assembly

Check for continuity between the posi-

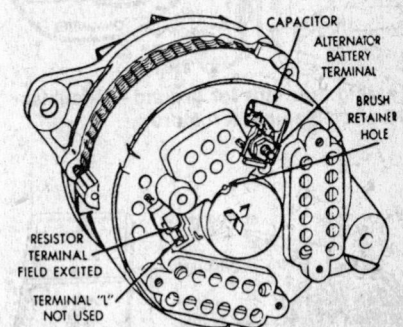

View of the alternator terminals— Mitsubishi 75 and 90 amp alternators

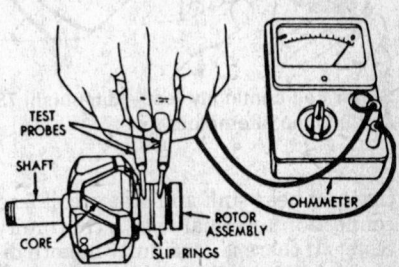

Testing the rotor field coils—Mitsubishi 75 and 90 amp alternators

Exploded view of the alternator assembly—Mitsubishi 75 and 90 amp alternators

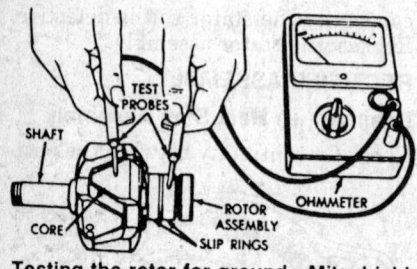

Testing the rotor for ground—Mitsubishi 75 and 90 amp alternators

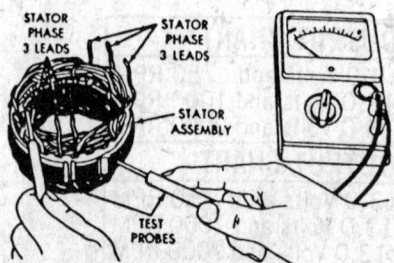

Stator coil ground test—Mitsubishi 75 and 90 amp alternators

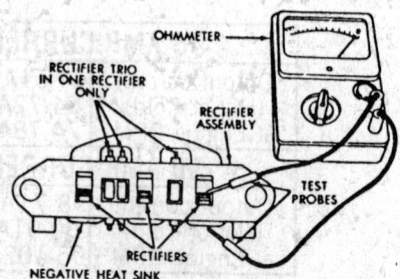

Testing the negative rectifiers—Mitsubishi 75 and 90 amp alternators

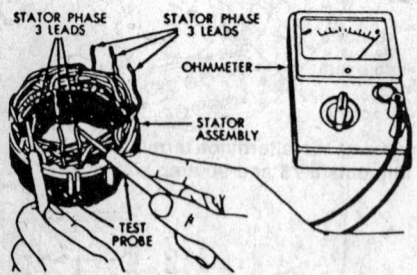

Stator coil continuity test—Mitsubishi 75 and 90 amp alternators

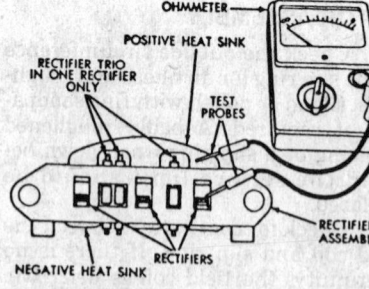

Testing the positive rectifiers—Mitsubishi 75 and 90 amp alternators

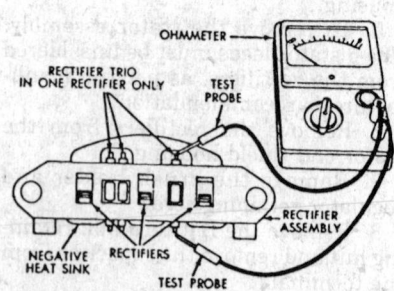

Testing the rectifier trio—Mitsubishi 75 and 90 amp alternators

tive (+) heat sink and stator coil lead connection terminal with a continuity tester. If there is continuity in both directions the diode is short circuited. Replace the rectifier assembly.

Negative (−) Heat Sink Assembly

Check for continuity between the negative (−) heat sink and stator coil lead connection terminal. If there is continuity in both directions the diode is

short circuited. Replace the rectifier assembly.

RECTIFIER TRIO TEST

Using a circuit tester check the 3 diodes for continuity in both directions.

If there is either continuity or an open circuit in both directions the diode is defective. Replace the rectifier assembly.

Assembly

1. The assembly of the alternator is the reverse of the removal procedure. Certain steps must be performed as the alternator is assembled.

2. Install the seals in the front and in the rear of the front bearing with the angled lip away from the bearing.

3. Push the brushes into the brush holder and insert a wire to hold them in the raised position. Install the rotor and remove the holding wire.

Nippondenso 75, 90 and 120 Amp Alternators

The alternators are the same, except, the 75 amp alternator is equipped with 3 sets of diodes and the 90 amp and 120 amp alternators are equipped with 4 sets.

ALTERNATOR OVERHAUL

Disassembly

1. Remove the B+ insulator nut and insulator. Remove the rear cover nuts and cover.

2. Remove the brush holder screws and the brush holder.

3. Remove the field block screws and the block.

4. Remove the rectifier and stator terminal screws and the rectifier.

5. Remove the stator terminal rubber insulators and the end shield through stud nuts.

6. With the drive pulley facing downward, tap rectifier end shield upward and separate the 2 end shields.

7. Remove the drive pulley nut and the drive pulley.

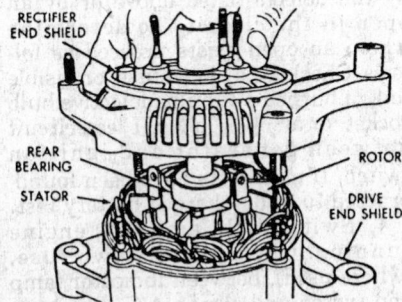

Separating the end shields— Nippondenso 75, 90 and 120 amp alternators

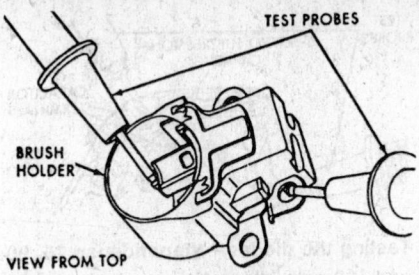

Testing the inner brush continuity— Nippondenso 75, 90 and 120 amp alternators

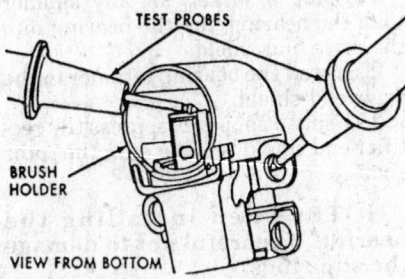

Testing the outer brush continuity— Nippondenso 75, 90 and 120 amp alternators

NOTE: Do not handle the slip rings of the rotor with the bare hands for oil or grease may restrict contact.

8. Pull the rotor assembly from the drive end shield.

9. Using a wheel puller, press the rectifier end shield bearing from the rotor shaft.

NOTE: When removing the bearing, be careful not to damage the slip rings.

10. From the drive end shield, remove the bearing retainer screws and the retainer.

11. Using a socket and a light hammer, place the drive end shield on a work surface and tap the bearing from the shield.

Inspection
BRUSH HOLDER TESTS

1. Make sure the brushes move smoothly and return fully to the stops when released; if not, replace the brush holder assembly.

2. Using an ohmmeter, test inner brush-to-field terminals continuity; 1 field terminal should be open and the other closed. If not, replace the brush holder.

3. Using an ohmmeter, test outer brush-to-field terminals continuity; a field terminal should be open and the other closed. If not, replace the brush holder.

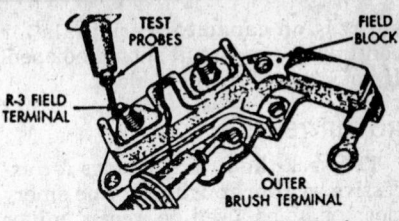

Testing the field block R-3 terminal— Nippondenso 75, 90 and 120 amp alternators

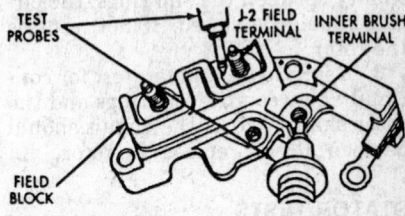

Testing the field block J-2 terminal— Nippondenso 75, 90 and 120 amp alternators

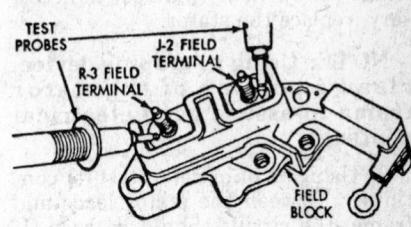

Testing the field block R-3 to J-2 terminals—Nippondenso 75, 90 and 120 amp alternators

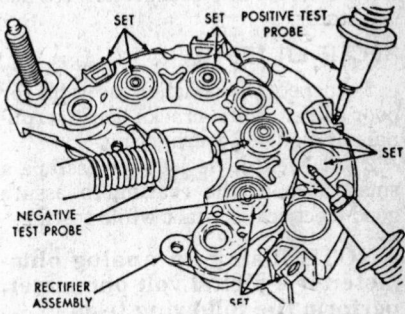

Testing the field block radio capacitor— Nippondenso 75, 90 and 120 amp alternators

FIELD BLOCK TESTS

1. Using an ohmmeter, test the outer brush terminal to R-3 field terminal for continuity; the circuit should be closed. If not, replace the field block.

2. Using an ohmmeter, test the inner brush terminal to J-2 field terminal for continuity; the circuit should be closed. If not, replace the field block.

3. Using an ohmmeter, test the R-3 field terminal to the J-2 terminal for continuity; the circuit should be open. If not, replace the field block.

4. Turn the field block over. Using an ohmmeter, test the radio

supression capacitor terminals for continuity; the circuit should be open. If not, replace the field block.

ROTOR TESTS

1. Check the field slip rings for excessive wear or roughness; fine emery cloth may be used to repair minor damage. If the rings are excessively damaged, replace the rotor.

2. Using an ohmmeter, test for continuity between the slip rings; the circuit should be closed. If not, replace the rotor.

3. Using an ohmmeter, test for continuity between the slip rings and the rotor shaft or core; the circuit should be open. If not, replace the rotor.

STATOR TESTS

1. Check the stator for signs damage—weak or broken leads, distorted frame or burned windings; if necessary, replace the stator.

NOTE: Using a scraping device, clean a portion of the stator frame to assure good electrical contact.

2. Using an ohmmeter, test for continuity between the stator leads and frame; the circuit should be open. If not, replace the stator.

3. Using an ohmmeter, test for continuity between the stator leads; the circuit should be closed. If not, replace the stator.

RECTIFIER TESTS

1. Inspect the rectifier assembly for poor solder joints, cracks, loose terminals or signs of overheating.

2. At each diode location, scrape a small area of the coating to assure good electrical contact while testing.

NOTE: Using an analog ohmmeter or a digital volt ohmmeter, perform the following tests.

3. Position the negative test probe on the stator terminal and the positive test probe on a negative diode; the resistance should be 7–11 ohms or 0.4–0.6 volts. Position the positive test probe on a positive diode; there should be no continuity. Perform this procedure on each terminal to diode set. If failure is detected, replace the rectifier.

4. Position the positive test probe on the stator terminal and the negative test probe on a positive diode; the resistance should be 7–11 ohms or 0.4–0.6 volts. Position the negative test probe on a negative diode; there should be no continuity. Perform this procedure on each terminal to diode set. If failure is detected, replace the rectifier.

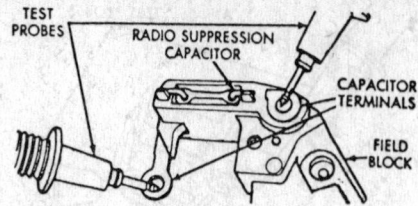

Testing the diodes—Nippondenso 75, 90 and 120 amp alternators

Assembly

1. Using a socket, slightly smaller than the bearing, tap the bearing into the drive end shield.

2. Install the bearing retainer in the drive end shield.

3. Using a shop press, press the rectifier end shield bearing onto the rotor shaft.

NOTE: When installing the bearing, be careful not to damage the slip rings.

4. Install the rotor assembly into the drive end shield.

5. Install the drive pulley and nut.

6. Using solvent, clean the slip rings on the rotor shaft.

7. Align the rectifier end shield and assemble it to the drive end shield.

8. Install the through stud nuts and the stator terminal insulators.

9. Position the rectifier and install the screws.

10. Position the field block and install the screws.

11. Retract the brushes into the brush holder, slide the assembly over the commutator rings and install the screws.

12. Install the rear cover and nuts.

13. Align the B+ insulator guide boss into the rear cover hole and install the nut.

GENERAL MOTORS CORPORATION

SI Alternators

Delcotron alternators are available with different idle outputs and rated amp outputs.

All alternators incorporate a solid state voltage regulator which is mounted inside the alternator. The construction and operation of each alternator is basically the same. The Delcotron alternator consists of a forward and rear end frame assembly, a rotor, a stator, brushes, slip-rings and diodes. The rotor is supported in the drive end frame by ball bearings and in the slip-ring end frame by roller bearings. The bearings do not require periodic lubrication.

There are 2 brushes which carry current through the slip-rings to the field coil. The field coil is mounted on the rotor. The stator windings are assembled on the inside of a laminated core that is part of the alternator frame. The rectifier bridge which is connected to the stator windings contains 6 diodes, 3 of which are negative and 3 of which are positive. The positive and negative diodes are moulded into the assembly. The rectifier bridge changes stator AC voltage into DC voltage which appears at the output BAT terminal.

The blocking action of the diodes prevents the battery from discharging, back through the alternator. The need for a cutout relay is eliminated because of this blocking action. The alternator field current is supplied through a diode trio, which is connected to the stator windings. A capacitor is mounted in the end frame to protect the rectifier bridge and the 6 diodes from high voltage and radio interference. Periodic alternator adjustment or maintenance is not required. The voltage regulator is preset and needs no adjustment.

ON VEHICLE SERVICE

Indicator Lamp Operation Test

1. Check the indicator lamp for normal operation. If the indicator lamp operates properly, refer to the undercharged battery test. If the indicator lamp does not operate properly, proceed accordingly.

2. Switch **OFF**, lamp **ON**. Unplug the connector from the generator No. 1 and No. 2 terminals. If the lamp stays **ON**, there is a short between these 2 leads. If the lamp goes out, replace the rectifier bridge.

3. Switch **ON**, lamp **OFF**, engine stopped. This condition can be caused by the defects listed above or by an open in the circuit. To determine where an open exists proceed as follows. Check for a blown fuse, or fusible link, a burned out bulb, defective bulb socket, or an open in No. 1 lead circuit between generator and ignition switch. If no defects have been found, proceed to undercharged battery test.

4. Switch **ON**, lamp **ON**, engine running. Check for a blown fuse, (where used), between indicator lamp and switch and also in A/C circuit.

Undercharged Battery Test

1. Be sure that the undercharged

Delcotron Alternator Availability Chart

Alternator Type	Rated Amp Output
10SI	37, 42, 63
12SI	56, 66, 78, 94
15SI	70, 85
17SI	—
27SI	65, 80, 100

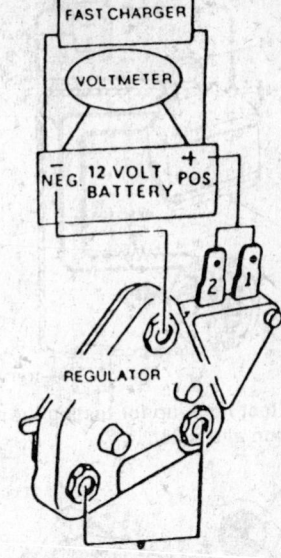

Delcotron alternator—voltage test

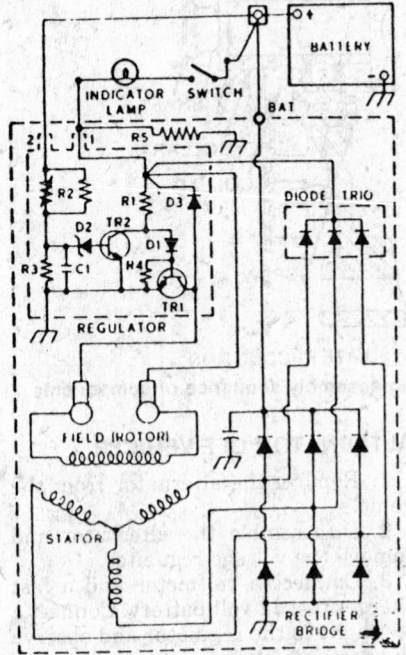

Typical Delcotron charging system wiring schematic

battery condition has not been caused by accessories that have been left **ON** for an extended period of time.

2. Check the alternator belt for proper belt tension. Inspect the battery for physical defects replace as required.

3. Inspect the wiring for defects. Check all connections for proper contact and cleanliness, including the slip connectors at the generator and baulkhead connections.

4. With ignition switch **ON** and all wiring harness leads connected, connect a voltmeter from the generator BAT terminal to ground, from the generator No. 1 terminal to ground and from the generator No. 2 terminal to ground. A zero reading indicates an open circuit between voltmeter connection and battery.

5. Delcotron alternators have a built in feature, which prevents overcharge and accessory damage by preventing the alternator from turning on if there is an open circuit in the wiring harness connected to the No. 2 alternator terminal.

6. If Steps 1–5 check out okay, check the alternator as follows. Disconnect negative battery cable. Connect an ammeter or alternator tester in the circuit at the BAT terminal of the alternator. Reconnect negative battery cable.

7. Turn **ON** radio, windshield wipers, lights high beam and blower motor on high speed. Connect a carbon pile across the battery (or use alternator tester). Operate engine about 2000 rpm and adjust carbon pile as required, to obtain maximum current output. If ampere output is within 10 amperes of rated output as stamped on generator frame, alternator is not defective. Recheck Steps 1–5.

8. If ampere output is not within 10 percent of rated output, determine if test hole is accessible. Ground the field winding by inserting a suitable tool into the test hole. Tab is within ¾ in. of casting surface. Do not force suit-

able tool deeper than 1 in. into end frame to avoid damaging alternator.

9. Operate engine at moderate speed as required and adjust carbon pile as required to obtain maximum current output.

10. If output is within 10 amperes of rated output, check field winding, diode trio and rectifier bridge. Test regulator with an approved regulator tester.

11. If output is not within 10 amperes of rated output, check the field winding, diode trio, rectifier bridge and stator. If test hole is not accessible, disassemble alternator and repair as required.

Overcharged Battery Test

1. Check the condition of the battery before any testing is done.

2. If an obvious overcharging condition exists, remove the alternator from the vehicle and check the field wind-

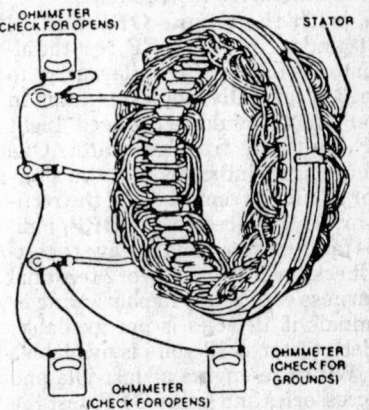

Testing Delcotron alternator stator—use an ohmmeter to check for opens or grounds

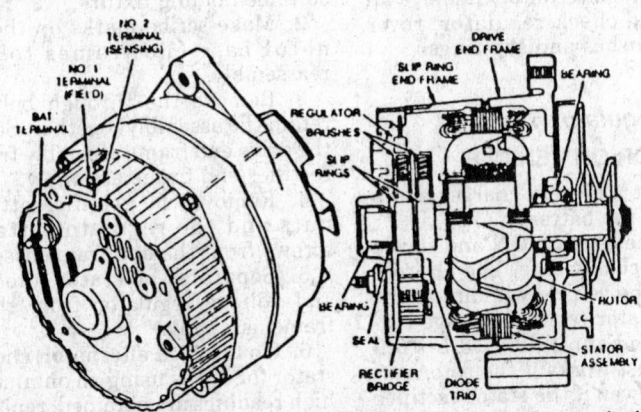

Delcotron alternator—showing location of components—typical

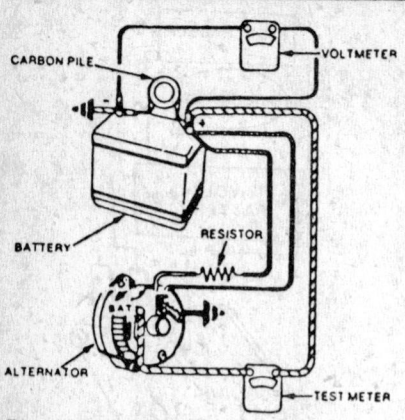

Bench test hook-up for testing the Delcotron alternator

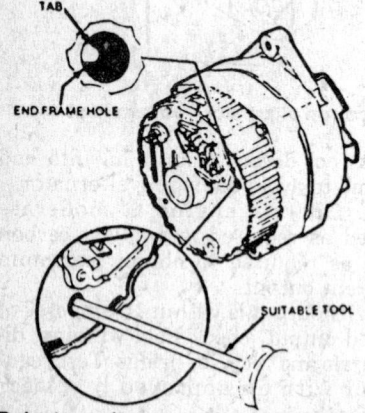

Delcotron alternator — if test hole is accessible, ground the field winding by inserting a suitable tool into the test hole (max. or 1 in.)

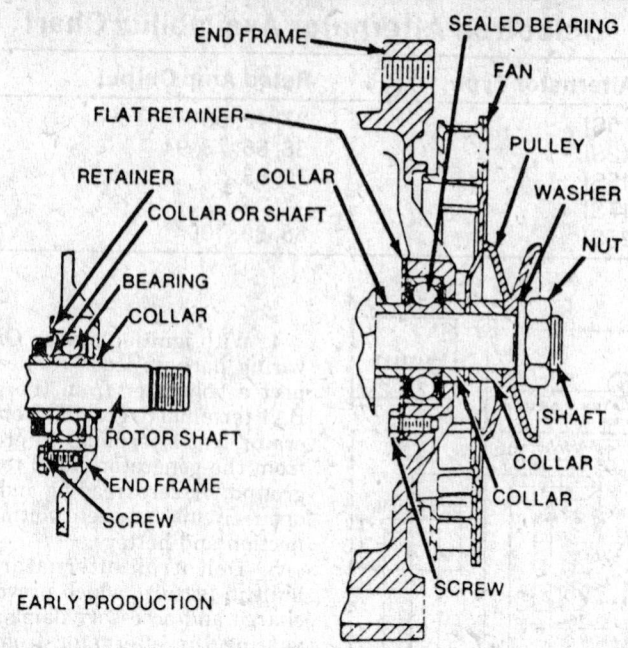

EARLY PRODUCTION

LATE PRODUCTION

Delcotron alternator — exploded view showing assembly sequence of components

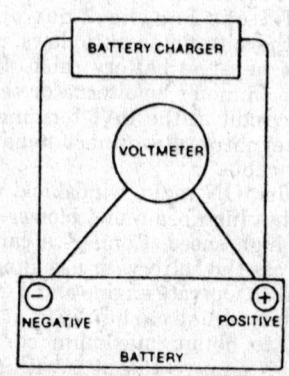

Delcotron alternator — voltage regulator test (on vehicle) voltage regulator setting should be 13.5-16.0 volts

ings for grounds or shorts. If defective, replace the rotor. Test the regulator.

Alternator Diagnostic Tester (J-26290)

This special diagnostic tester is designed to determine if the alternator should be removed from the vehicle.

1. Install tester J-26290 according to manufacturers instructions.

2. With the engine **OFF** and all lights and accessories **OFF**, test the alternator as follows. Light flashes, go to Step 3. Light **ON**, indicates fault in tester which should be replaced. Light **OFF**, pull plug from generator. One flashing light, indicates that the alternator should be removed and the rectifier bridge replaced. Light **OFF**, indicates faulty tester or no voltage to tester. Check for 12 volts at No. 2 terminal of harness connector. Repair wiring or terminals if 12 volts is not available. Replace tester if 12 volts is available.

3. With the engine at fast idle and all accessories and lights **OFF**, test the alternator as follows. Light **OFF** indicates that the charging system good, do not remove alternator. Light **ON**

indicates a component failure within the alternator. Remove alternator and check diode trio, rectifier bridge and stator. Light flashing indicates a problem within the alternator. Remove alternator and check regulator, rotor field coil, brushes and slip-rings.

Voltage Regulator Test
ALTERNATOR ON VEHICLE

1. Connect a battery charger and a voltmeter to the battery.

2. Turn the ignition **ON** and slowly increase the charge rate. The alternator light in the vehicle will dim at the voltage regulator setting. Voltage regulator setting should be 13.5-16.0 volts. This test works if the rotor setting is good, even if the stator rectifier bridge or diode trio is bad.

ALTERNATOR OFF VEHICLE

1. Remove the alternator from the vehicle.

2. Disassemble the alternator and remove the voltage regualtor.

3. Connect a voltmeter and a fast charger to a 12 volt battery. Connect a test light to the regulator and observe the battery polarity.

4. The test light should light.

5. Turn **ON** the fast charger and slowly increase the charge rate. Observe the voltmeter, the light should go out at the voltage regualtor setting. The voltage regulator setting specification is 13.5-16.0 volts.

ALTERNATOR OVERHAUL

Disassembly

1. Remove the alternator from the vehicle. Position the assembly in a suitable holding fixture.

2. Make scribe marks on the alternator case end frames to aid in reassembly.

3. Remove the through bolts that retain the assembly together. Separate the drive end frame assembly from the rectifier end frame assembly.

4. Remove the rectifier attaching nuts and the regulator attaching screws from the end frame assembly.

5. Separate the stator, diode trio and voltage regulator from the end frame assembly.

6. On the 10SI alternator, check the stator for opens using an ohmmeter. If high readings are obtained, replace the stator.

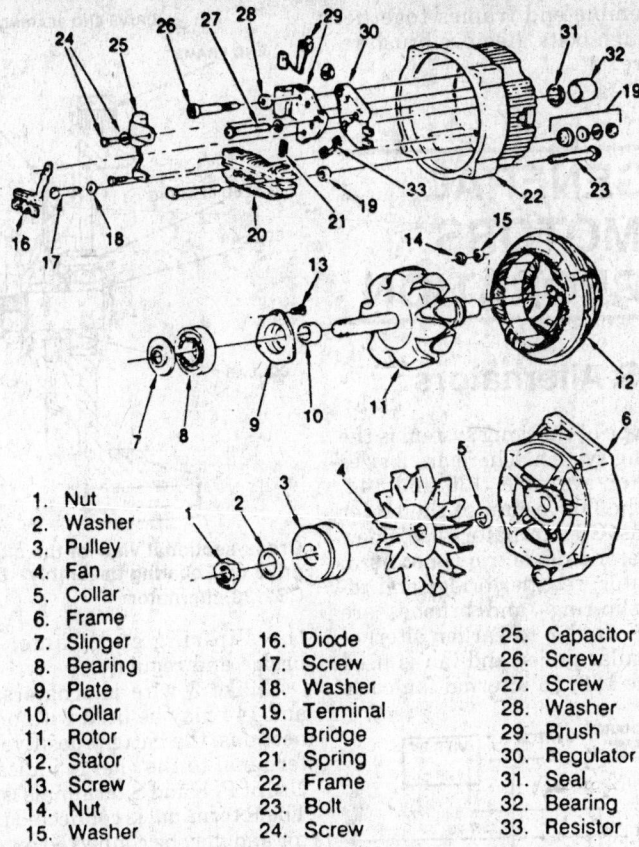

1. Nut
2. Washer
3. Pulley
4. Fan
5. Collar
6. Frame
7. Slinger
8. Bearing
9. Plate
10. Collar
11. Rotor
12. Stator
13. Screw
14. Nut
15. Washer
16. Diode
17. Screw
18. Washer
19. Terminal
20. Bridge
21. Spring
22. Frame
23. Bolt
24. Screw
25. Capacitor
26. Screw
27. Screw
28. Washer
29. Brush
30. Regulator
31. Seal
32. Bearing
33. Resistor

Delcotron alternator drive end bearing and related components

7. Check the stator for grounds using an ohmmeter. If readings are low, replace the stator.

8. Using an ohmmeter check the rotor for grounds. The ohmmeter reading should be very high. If not, replace the rotor.

9. Using an ohmmeter, check the rotor for opens. If the ohmmeter reading is not 2.4–3.5 ohms replace the rotor.

10. To check the diode trio connect the ohmmeter to the diode trio and then reverse the lead connections. The ohmmeter should read high and low if not replace the diode trio. Repeat the same test between the single connector and each of the other connectors.

11. Check rectifier bridge with ohmmeter connected from grounded heat sink to flat metal on terminal. Reverse leads. If both readings are the same replace rectifier bridge.

12. Repeat test between grounded heat sink and other 2 flat metal clips.

13. Repeat test between insulated heat sink and 3 flat metal clips.

14. Clean or replace the alternator brushes as required. Position the brushes in the brush holder and retain them in place using the brush retainer wire or equivalent.

15. To remove the rotor and drive end bearing, remove the shaft nut, washer and pulley, fan and collar. Push the rotor from the housing.

16. Remove the retainer plate from inside the drive end frame. Push the bearing out. Clean or replace parts as required.

Inspection

ALTERNATOR BENCH TEST

1. Remove the alternator from the vehicle. Position the unit in a suitable test stand.

2. Connect the alternator in series, but leave the carbon pile disconnected.

NOTE: Ground polarity of the battery must be the same as the alternator. Be sure to use a fully charged battery and a 10 ohm resistor rated at 6 watts or more between the alternator No. 1 terminal and the battery.

3. Increase the alternator speed slowly and observe the voltage.

4. If the voltage is uncontrolled with speed and increases above 15.5 volts, test regulator with an approved regulator tester and check field winding. If voltage is below 15.5 volts, connect the carbon pile.

5. Operate the alternator at moderate speed as required and adjust the

carbon pile as required to obtain maximum current output.

6. If output is within 10 amperes of rated output as stamped on alternator frame, alternator is good. If output is not within 10 amperes of rated output, keep battery loaded with carbon pile and ground alternator field.

7. Operate alternator at moderate speed and adjust carbon pile as required to obtain maximum output. If output is within 10 amperes of rated output, test regulator with an approved regulator tester and check field winding.

8. If output is not within 10 amperes of rated output, check the field winding, diode trio, rectifier bridge and stator.

Assembly

1. Press against the outer bearing race to push the bearing in. On early production alternators it will be necessary to fill the bearing cavity with lu-

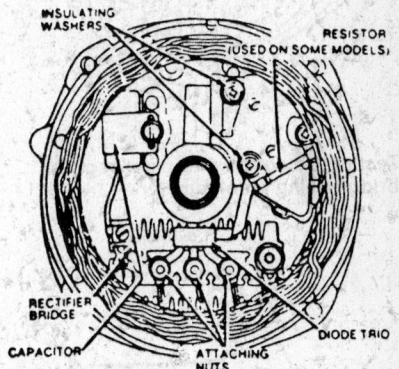

Delcotron alternator end frame— showing location of related components

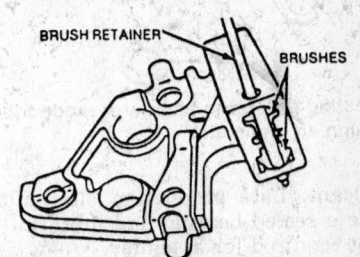

Delcotron alternator brush installation

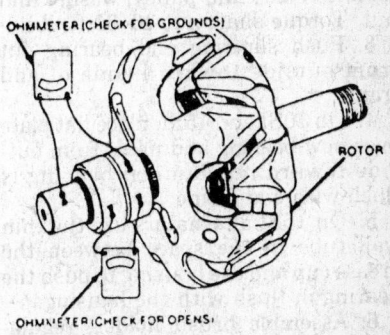

Delcotron alternator—rotor test use an ohmmeter to check for opens or grounds

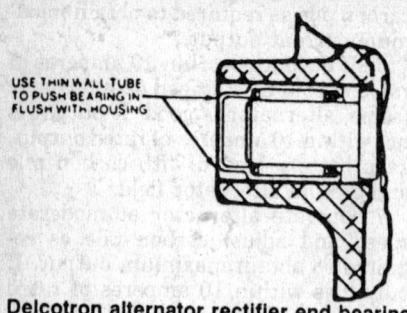

Delcotron alternator rectifier end bearing Installation

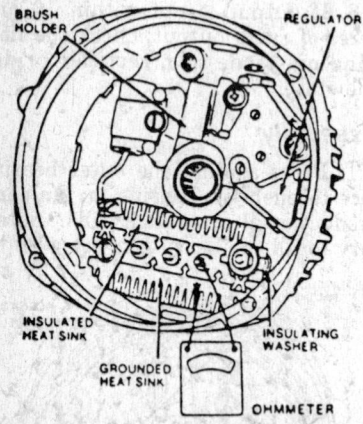

Testing Delcotron alternator rectifier bridge using an ohmmeter

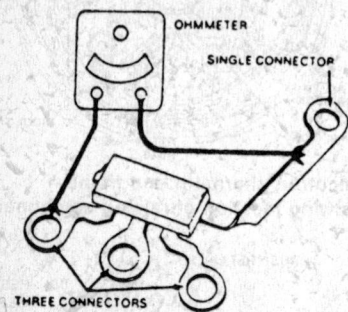

Testing Delcotron alternator diode trio using an ohmmeter

bricant. Late production alternators use a sealed bearing and lubricant is not required for assembly.

2. Press rotor into end frame. Assemble collar, fan, pulley, washer and nut. Torque shaft nut 40–60 ft. lbs.

3. Push slip-ring end bearing out from outside toward inside of end frame.

4. On 10SI and 15SI, place flat plate over new bearing and press from outside toward inside until bearing is flush with end frame.

5. On 15SI alternators use the thin wall tube in the space between the grease cup and the housing to push the bearing in flush with the housing.

6. Assemble brush holder, regulator, resistor, diode trio, rectifier bridge and stator to slip-ring end frame.

7. Assemble end frames together with through bolts. Remove brush retainer wire.

GENERAL MOTORS CORPORATION

CS Alternators

Another type of charging system is the CS charging system. There are 2 sizes of alternator available, 130 and 144, denoting the OD in mm of the stator laminations. CS alternators use a new type regulator a diode trio is not used. A delta stator, rectifier bridge and rotor with slip-rings and brushes are electrically similar to earlier alternators. A regular pulley and fan is used and, on the 130, an internal fan cools

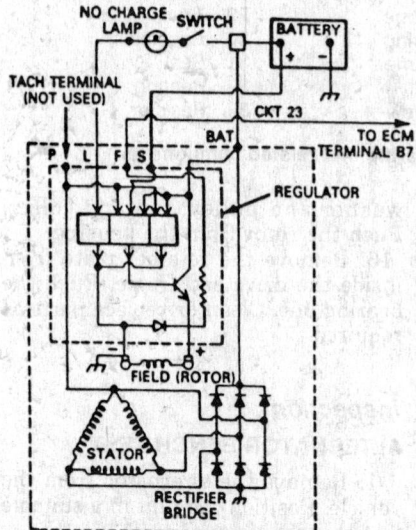

Schematic of the Delcotron CS series charging system

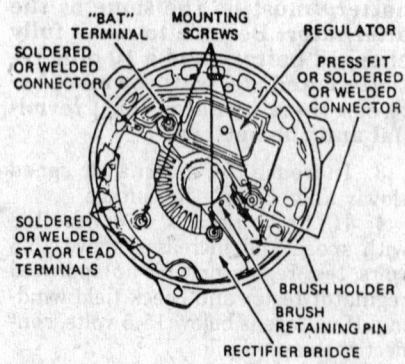

Location of the components—CS 130 alternator

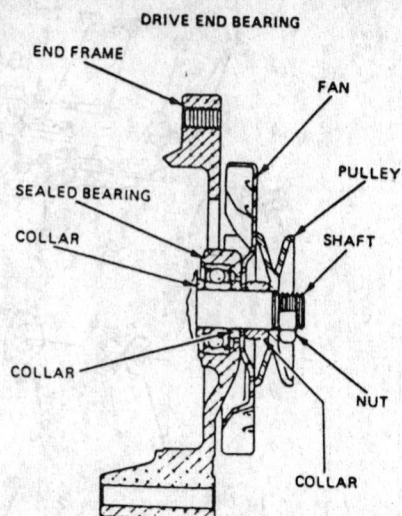

Cross-sectional view of the alternator drive end bearing assembly—Delcotron CS 130 alternator

the slip-ring end frame, rectifier bridge and regulator.

Unlike 3 wire alternators, the 130 and 144 may be used with only 2 connections, the battery positive and an L terminal to the charge indicator bulb. Use of P, F and S terminals is optional. The P terminal is connected to the stator and may be connected externally to a tachometer or other device. The F terminal is connected internally to field positive and may be used as a fault indicator. The S terminal may be connected externally to a voltage, such as battery voltage, to sense voltage to be controlled.

As on other charging systems, the charge indicator lights when the switch is closed and goes out when the engine is running. If the charge indicator is ON with the engine running, a charging system defect is indicated. For all kinds of defects, the indicator will glow at full brilliance, not half lit. Also, the charge indicator will be ON with the engine running if system voltage is too high or too low. The regulator voltage setting varies with temperature and limits system voltage by controlling rotor field current.

This regulator switches rotor field current ON and OFF at a fixed frequency of about 400 cycles per second. By varying the ON/OFF time, correct average field current for proper system voltage control is obtained. At high speeds, the ON time may be 10% and the OFF time 90%. At low speeds, with high electrical loads, ON/OFF time may be 90% and 10% respectively. No periodic maintenance on the generator is required.

ON VEHICLE SERVICE

When operating normally, the indica-

TROUBLESHOOTING
GM DELCOTRON ALTERNATOR
(WITH INTERNAL CONTROL REGULATOR)

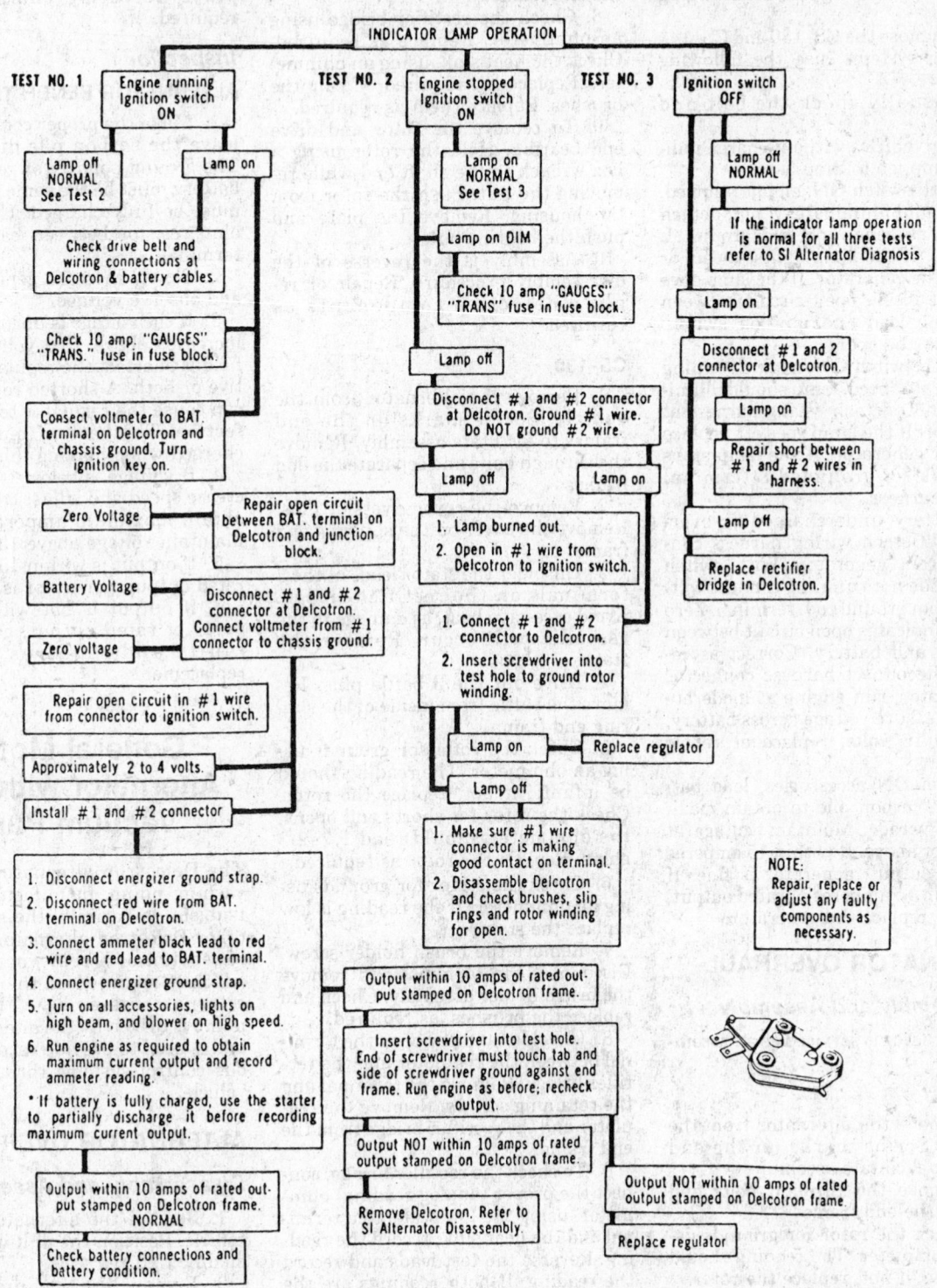

INDICATOR LAMP OPERATION

TEST NO. 1

Engine running Ignition switch ON

Lamp off NORMAL See Test 2

Lamp on

Check drive belt and wiring connections at Delcotron & battery cables.

Check 10 amp. "GAUGES" "TRANS." fuse in fuse block.

Connect voltmeter to BAT. terminal on Delcotron and chassis ground. Turn ignition key on.

Zero Voltage → Repair open circuit between BAT. terminal on Delcotron and junction block.

Battery Voltage → Disconnect #1 and #2 connector at Delcotron. Connect voltmeter from #1 connector to chassis ground.

Zero voltage

Repair open circuit in #1 wire from connector to ignition switch.

Approximately 2 to 4 volts.

Install #1 and #2 connector

1. Disconnect energizer ground strap.
2. Disconnect red wire from BAT. terminal on Delcotron.
3. Connect ammeter black lead to red wire and red lead to BAT. terminal.
4. Connect energizer ground strap.
5. Turn on all accessories, lights on high beam, and blower on high speed.
6. Run engine as required to obtain maximum current output and record ammeter reading.*

* If battery is fully charged, use the starter to partially discharge it before recording maximum current output.

Output within 10 amps of rated output stamped on Delcotron frame. NORMAL

Check battery connections and battery condition.

TEST NO. 2

Engine stopped Ignition switch ON

Lamp on NORMAL See Test 3

Lamp on DIM

Check 10 amp "GAUGES" "TRANS" fuse in fuse block.

Lamp off

Disconnect #1 and #2 connector at Delcotron. Ground #1 wire. Do NOT ground #2 wire.

Lamp off | Lamp on

1. Lamp burned out.
2. Open in #1 wire from Delcotron to ignition switch.

1. Connect #1 and #2 connector to Delcotron.
2. Insert screwdriver into test hole to ground rotor winding.

Lamp on → Replace regulator

Lamp off

1. Make sure #1 wire connector is making good contact on terminal.
2. Disassemble Delcotron and check brushes, slip rings and rotor winding for open.

Output within 10 amps of rated output stamped on Delcotron frame.

Insert screwdriver into test hole. End of screwdriver must touch tab and side of screwdriver ground against end frame. Run engine as before; recheck output.

Output NOT within 10 amps of rated output stamped on Delcotron frame.

Remove Delcotron. Refer to SI Alternator Disassembly.

TEST NO. 3

Ignition switch OFF

Lamp off NORMAL

If the indicator lamp operation is normal for all three tests refer to SI Alternator Diagnosis

Lamp on

Disconnect #1 and 2 connector at Delcotron.

Lamp on

Repair short between #1 and #2 wires in harness.

Lamp off

Replace rectifier bridge in Delcotron.

NOTE: Repair, replace or adjust any faulty components as necessary.

Output NOT within 10 amps of rated output stamped on Delcotron frame.

Replace regulator

tor lamp will illuminate when the ignition switch is turned **ON** and go out when the engine starts. If the lamp operates abnormally, or if an undercharged or overcharged battery condition occurs, the following procedure may be used to diagnose the charging system.

To diagnose the CS–130 and CS–144 charging systems, use the following procedure.

1. Visually check the belt and wiring.

2. For vehicles without charge indicator lamp, go to Step 5.

3. With switch **ON**, engine stopped, lamp should illuminate. If not, detach harness at generator and ground L terminal. If the lamp lights, repair or replace the generator. If the lamp does not light locate open circuit between grounding lead and ignition switch. Lamp may be open.

4. With switch **ON**, engine running at moderate speed, lamp should illuminate. If not, detach wiring harness at generator. If the lamp goes off, replace or repair generator. If the lamp stays on, check for grounded L terminal wire in harness.

5. Battery undercharged or overcharged. Detach wiring harness connector from generator. With switch **ON**, engine not running, connect voltmeter from ground to L terminal. Zero reading indicates open circuit between terminal and battery. Correct as required. Reconnect harness connector to generator, run engine at moderate speed. Measure voltage across battery. If above 16 volts, replace or repair generator.

6. Turn **ON** accessories, load battery with carbon pile to obtain maximum amperage. Maintain voltage at 13 volts or above. If within 15 amperes of rated output, generator is fine. If not within 15 amperes of rated output, repair or replace the generator.

ALTERNATOR OVERHAUL

Disassembly and Assembly

The alternator is serviced as an assembly only.

CS–144

1. Remove the alternator from the vehicle. Scribe marks on the end frames to facilitate assembly.

2. Remove the through bolts and separate the end frames.

3. Check the rotor for grounds using an ohmmeter. The reading should be infinite, if not, replace the rotor.

4. Check the rotor for shorts and opens. Replace the rotor as required.

5. Remove the attaching nuts and the stator from the end frame.

6. Check the stator for grounds us-

ing an ohmmeter. If the reading is low replace the stator.

7. Unsolder the connections, remove the retaining screws and connector from the end frame. Separate the regulator and the brush holder from the end frame.

8. Check the rectifier bridge using an ohmmeter. Replace as required. Check the heat sink, using an ohmmeter. Replace as required. Clean the brushes. Replace them as required.

9. To remove the rotor and drive end bearing, Hold the rotor using a hex wrench in the shaft end while removing the nut. Push the rotor from the housing. Remove the plate and push the bearing out.

10. Assembly is the reverse of the disassembly procedure. Repair or replace defective components as required.

CS–130

1. Remove the alternator from the vehicle. Scribe marks on the end frames to facilitate assembly. Remove the through bolts and separate the end frames.

2. Remove the cover rivets or pins. Remove the cover on the slip-ring end frame.

3. Unsolder the stator leads at the 3 terminals on the rectifier bridge. Avoid excessive heat, as damage to the assembly will occur. Remove the stator.

4. Drive out the 3 baffle pins. Remove the baffle from inside of the slip-ring end frame.

5. Check the rotor for grounds using an ohmmeter. The reading should be infinite, if not, replace the rotor. Check the rotor for shorts and opens, the ohmmeter should read 1.7–2.3 ohms. Replace the rotor as required.

6. Check the stator for grounds using an ohmmeter. If the reading is low replace the stator.

7. Remove the brush holder screw. Disconnect the terminal and remove the brush holder assembly. Check and replace the brushes, as required.

8. Unsolder and pry open the terminal between the regulator and the rectifier bridge. Remove the terminal and the retaining screws. Remove the regulator and the rectifier bridge from the end frame.

9. To check the rectifier bridge, connect the proper (analog reading) ohmmeter, using the low scale, to a terminal and the heat sink, record the reading. Reverse the test leads and record the reading. If both readings are the same replace the rectifier bridge. Check the other diodes in the same manner.

10. To remove the rotor and drive end bearing, Hold the rotor using a

hex wrench in the shaft end while removing the nut. Push the rotor from the housing. Remove the plate and push the bearing out.

11. Assembly is the reverse of the disassembly procedure. Repair or replace defective components as required.

Inspection

ALTERNATOR BENCH TEST

1. Make the proper connections but leave the carbon pile disconnected. The ground polarity of generator and battery must be the same. The battery must be fully charged. Use a 30–500 ohm resistor between battery and L terminal.

2. Slowly increase generator speed and observe voltage.

3. If the voltage is uncontrolled and increases above 16.0 volts, the rotor field is shorted, the regulator is defective or both. A shorted rotor field coil can cause the regulator to become defective. The battery must be fully charged when making this test.

4. If voltage is below 16.0 volts, increase speed and adjust carbon pile to obtain maximum amperage output. Maintain voltage above 13.0 volts.

5. If output is within 15 amperes of rated output, generator is good.

6. If output is not within 15 amperes of rated output, generator is faulty and requires repair or replacement.

General Motors Alternator with Rear Vacuum Pump

The Delcotron alternator with rear vacuum pump, manufactured by Mitsubishi, is basically the same as the other Delcotron alternators, with the exception of a rear vacuum pump which is mounted on the back of the alternator assembly. The vacuum pump is driven by the alternator shaft and is used to provide vacuum to various control systems throughtout the vehicle.

ALTERNATOR OVERHAUL

Disassembly and Assembly

1. Remove the alternator from the vehicle. Position the unit in a suitable holding fixture.

2. Remove the vacuum pump retaining bolts. Remove the vacuum pump from the rear of the alternator while holding the center plate.

3. Remove the brush cover retaining bolts and brushes. Wrap the pump

drive shaft spline with tape in order to protect the rear seal from damage.

4. Inspect the vacuum pump for wear and damage, replace defective components as required. Measure the length of the vanes, replace if not within specification (0.511–0.531 in.) Measure the inside diameter of the housing and replace if not within specification (2.440–2.441 in.).

5. Examine the check valve for damage. Apply light pressure to the valve and make sure the valve operates properly. Replace as required.

6. Check the inner face of the rear cover on the vacuum pump for oil leakage. Check the inner face of the oil seal for wear and damage. Replace the oil seal in the rear end housing of the vacuum pump as required.

7. Remove the alternator through bolts which hold the unit together. Matchmark the assembly to aid in reassembly. Separate the front end housing from the stator and rear end housing.

8. Remove the pulley nut, fan and front end housing from the rotor.

9. Remove the front bearing retainer screws. Remove the front bearing retainer and the bearing from the front end housing.

10. Remove the bolt and nuts retaining the stator, diodes and brush holder to the rear end housing. Note the position of the insulating washers for reassembly.

11. Separate the rear end housing from the stator and diode assembly.

12. Remove the diodes from the stator by melting the solder from the terminals. Be sure to protect the diodes while melting the solder.

13. Remove the solder from the voltage regulator holder plate terminal. Remove the voltage regulator.

14. Check the slip-ring surfaces of the rotor for wear and damage, repair or replace as required.

15. Measure the outside diameter of the rotor slip-rings. If ring diameter is not 1.18–1.24 in., replace the rotor.

16. Connect the ohmmeter test leads to each slip-ring. Resistance should be 4.2 ohms at 68°F. If continuity does not exist the coil is open and the rotor must be replaced.

17. Connect the ohmmeter to either slip-ring and the rotor core. If continuity exists the coil is grounded and the rotor must be replaced.

18. Check the front and rear rotor bearings for wear and damage. Replace defective parts as required.

19. Check for continuity across the stator coils. If continuity does not exist in any 1 stator coil replace the stator assembly.

20. Check for continuity across any of the stator coils and the stator core.

If continuity exists, 1 of the stator coils is grounded and the stator must be replaced.

21. Coil resistance should be 0.05 ohms at 68°F and should be measured from the coil lead to terminal N.

22. Inspect the alternator brush assembly for wear and damage. Replace defective components as required.

23. Check for continuity of positive diodes between each stator coil terminal and the battery terminal of rectifier assembly. Reverse the ohmmeter leads and recheck for continuity.

24. If continuity exists in both polarity directions or does not exist in both directions diode is defective and must be replaced.

25. Check for continuity of negative diodes between each stator lead and E terminal or rectifier assembly. Reverse ohmmeter leads and recheck for continuity. Continuity should exist in 1 direction only.

26. Assemble a test circuit using the following components: One 10 ohm 3 watt resistor (R_1) one 0–300 ohm 3 watt variable resistor (R_2), two 12 volt batteries $(BAT_1$ and $BAT_2)$ and one 0–30 volt DC voltmeter.

27. Adjust variable resistor (R_2) until voltage at V_4 reads the same as voltage at V_3 (this should be all the way to 1 end of travel or 0 ohms).

28. Connect the test circuit to the integrated circuit regulator terminals. Measure voltage at V_1 and V_2. Voltage should measure 10–13 volts at V_1 and 0–2 volts at V_2.

29. Disconnect terminal S from circuit and measure voltage at V_3. Voltage at V_3 should be 20–26 volts. Reconnect terminal S.

30. Measure voltage at V_4 while increasing resistance at R_2 from 0 ohms. V_4 voltmeter reading should increase from 2 volts to 10–13 volts. Stop increasing R_2 when voltage reaches 10–13 volts.

31. If increase at V_4 is interrupted at any point up to 10–13 volts, while increasing resistance at R_2, regulator is defective.

32. Measure voltage at V_4 with R_2 at same setting as previous step that produced 10–13 volt reading at V_2. If V_4 not within 14–14.6 volts, regulator is defective.

33. Disconnect wire at terminal S. Connect it to terminal B. Repeat Step 30. If V_2 does not vary or V_4 is not within 14.5–16.6 volts, regulator is defective.

34. To assemble the alternator reverse the disassembly procedure. Be sure to check all parts for wear and damage. Replace defective components as required.

35. Insert the brushes into the brush holder and insert a wire to retain them

in place. Install the rotor and remove the retaining wire.

FORD MOTOR COMPANY

Rear Terminal Alternator

The Ford charging system is a negative ground system. It includes an alternator, electronic regulator, a charge indicator or an ammeter and a storage battery.

CHARGING SYSTEM OPERATION

NOTE: If the current indicator is to give an accurate reading, the battery cables must be of the same gauge and length as the original equipment.

1. With the engine running and all electrical systems turned **OFF**, position a current indicator over the positive battery cable.

2. If a charge of about 5 amps is recorded, the charging system is working. If a draw of about 5 amps is recorded, the system is not working. The needle moves toward the battery when a charge condition is indicated and away from the battery when a draw condition is indicated. If a draw is indicated continue to the next testing procedure. If an overcharge of 10–15 amps is indicated check for a faulty regulator or a bad ground at the regulator or the alternator.

Ignition Switch to Regulator Circuit Test

1. Disconnect the regulator wiring harness from the regulator.

2. Turn **ON** the key. Using a test light or voltmeter check for voltage between the I wire and ground. Check for voltage between the A wire and ground. If voltage is present at this part of the system the circuit is OK. If there is no voltage at the I wire check for a burned out charge indicator bulb, a burned-out resistor, or a break or short in the wiring. If there is no voltage present at the A wire check for a faulty connection at the starter relay or a break or short in the wire.

Isolation Test

This test determines whether the regulator or the alternator is faulty after

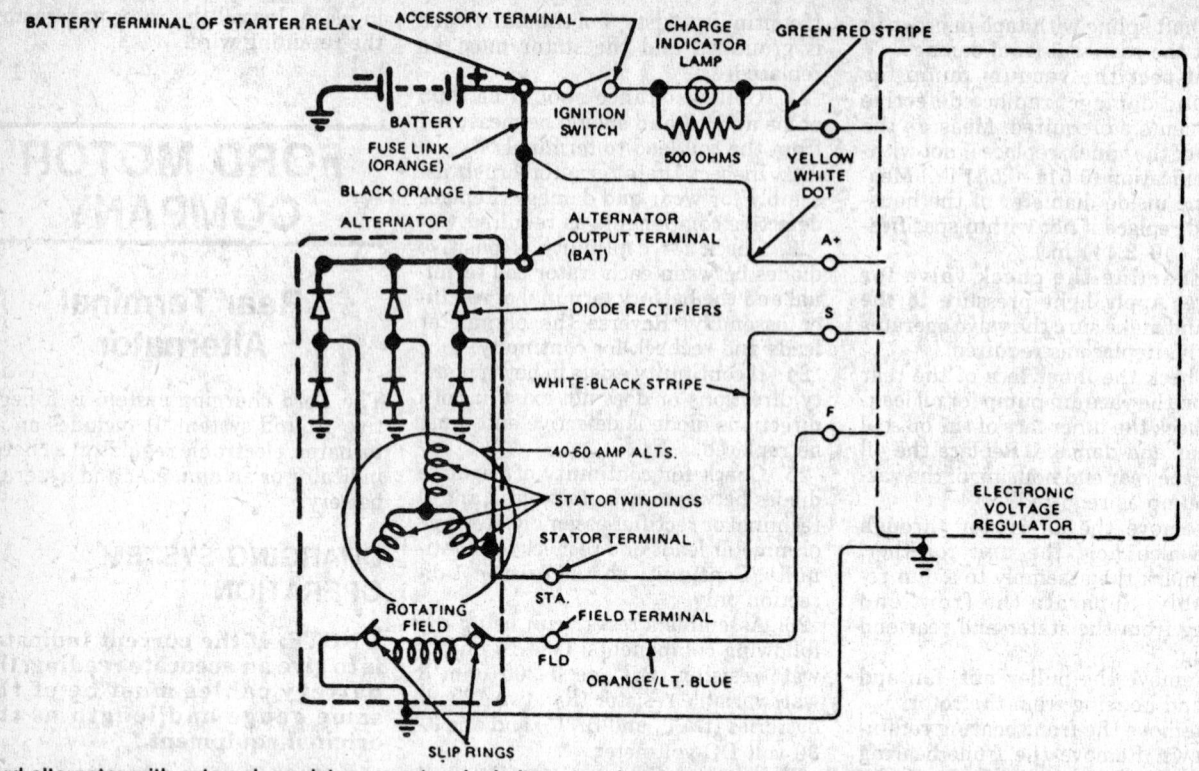

Ford alternator with external regulator—rear terminal alternator charging system schematic with indicator light—typical side terminal alternator

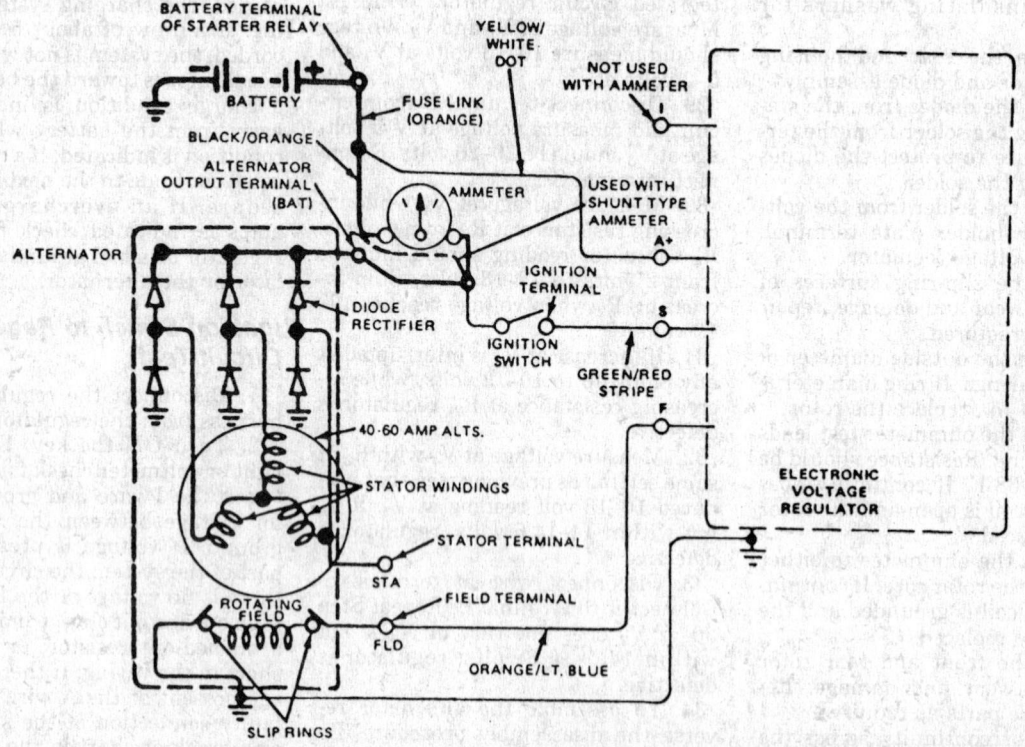

Ford alternator with external regulator—rear terminal alternator charging system schematic with ammeter—typical side terminal alternator

the rest of the circuit is found to be in good working order.

1. Disconnect the regulator wiring harness from the regulator.

2. Connect a jumper wire from the A wire to the F wire in the wiring harness plug.

3. Connect a voltmeter to the battery. The positive voltmeter lead goes to the positive terminal and the negative lead to the negative terminal. Record the reading on the voltmeter.

4. Turn **OFF** all of the electrical systems and start the engine. Do not race the engine.

5. Gradually increase engine speed 1500–2000 rpm. The voltmeter reading should increase above the previously recorded battery voltage reading by at least 1–2 volts. If there is no increase the alternator is not working correctly. If there is an increase the voltage regulator needs to be replaced.

ALTERNATOR OVERHAUL

Disassembly

1. Matchmark both end housings for assembly.

2. Remove the housing through bolts.

3. Separate the front housing and rotor from the stator and rear housing.

4. Remove the nuts from the rectifier to rear housing mounting studs and remove rear housing.

5. Remove the brush holder mounting screws and the holder, brushes, springs, insulator and terminal.

6. If replacement is necessary press the bearing from the rear end housing while supporting the housing on the inner boss.

7. If rectifiers are to be replaced carefully unsolder the leads from the terminals. Use only a 100 watt soldering iron. Leave the soldering iron in contact with the diode terminals only long enough to remove the wires. Use pliers as temporary heat sinks in order to protect the diodes.

8. There are various types of rectifi-

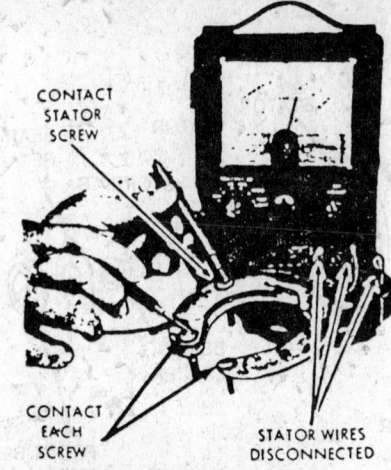

Testing diodes with stator wires connected—Ford alternator with external regulator

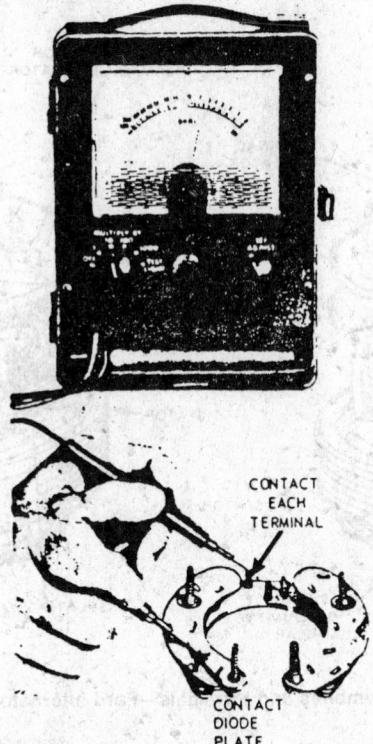

Testing diodes—Ford alternator with external regulator

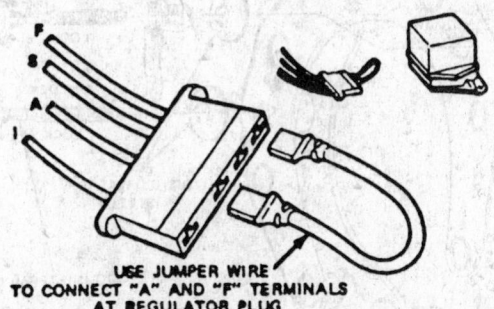

USE JUMPER WIRE TO CONNECT "A" AND "F" TERMINALS AT REGULATOR PLUG

Using a jumper wire at the regulator plug to test the alternator for normal output amps and for field circuit wiring continuity—Ford alternator with external regulator

er assembly circuit boards installed in production. One type has the circuit board spaced away from the diode plates and the diodes are exposed. Another type consists of a single circuit board with integral diodes; and still another has integral diodes with an additional booster diode plate containing 2 diodes.

9. This last type is used only on the 8 diode (61 amp) alternator. To disassemble use the following procedures. Exposed diodes remove the screws from the rectifier by rotating bolt heads ¼ turn clockwise to unlock and unscrewing. Integral diodes, press out the stator terminal screw, making sure not to twist it while doing this. Do not remove grounded screw. Booster diodes, press out the stator terminal screw about ¼ in., remove the nut from the end of the screw and lift screw from circuit board. Be sure not to twist it as it comes out.

10. Remove the drive pulley and fan. On alternator pulleys with threaded holes in the outer end of the pulley use a standard puller for removal.

11. Remove the front bearing retainer screws and the front housing. If the bearing is to be replaced press it from the housing.

Brush Replacement

1. Remove the brush holder and cover assembly from the rear housing.

2. Remove the terminal bolts from the brush holder and cover assembly. Remove the brush assemblies.

3. Position the new brush terminals on the terminal bolts and assemble the terminals, bolts, brush holder washers and nuts. The insulating washer mounts under the FLD terminal nut. The entire brush and cover assembly also is available for service.

4. Depress the brush springs in the brush holder cavities and insert the brushes on top of the springs. Hold the brushes in position by inserting a stiff wire in the brush holder as shown. Position the brush leads as shown.

5. Install the brush holder and cover assembly into the rear housing. Remove the brush retracting wire and put a dab of silicone cement over the hole.

Inspection

1. The rotor, stator, diode rectifier assemblies and bearings are not to be cleaned with solvent. These parts are to be wiped off with a clean cloth. Cleaning solvent may cause damage to the electrical parts or contaminate the bearing internal lubricant. Wash all other parts in solvent and dry them.

2. Rotate the front bearing on the driveshaft. Check for any scraping noise, looseness or roughness that in-

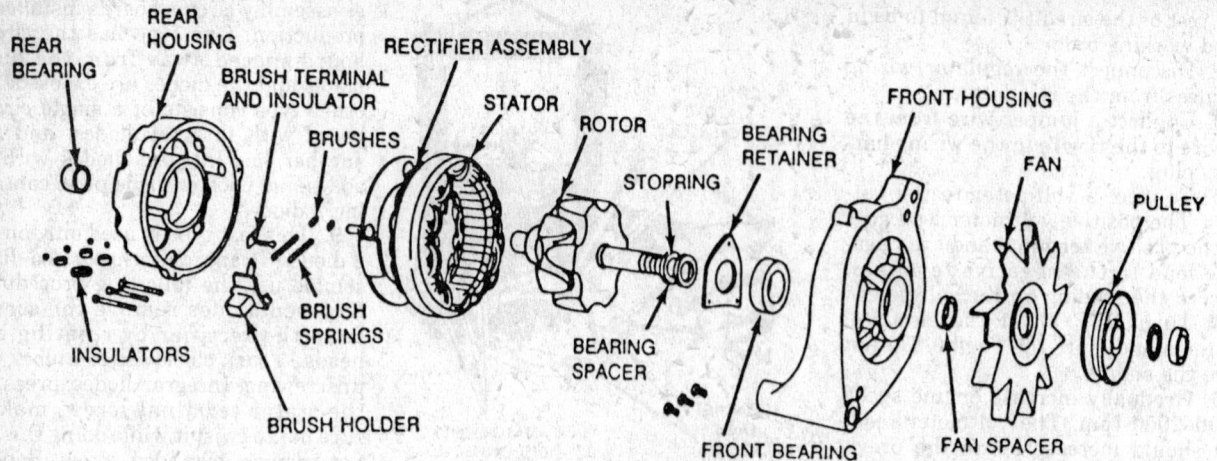

Exploded view of the rear terminal alternator with an external voltage regulator

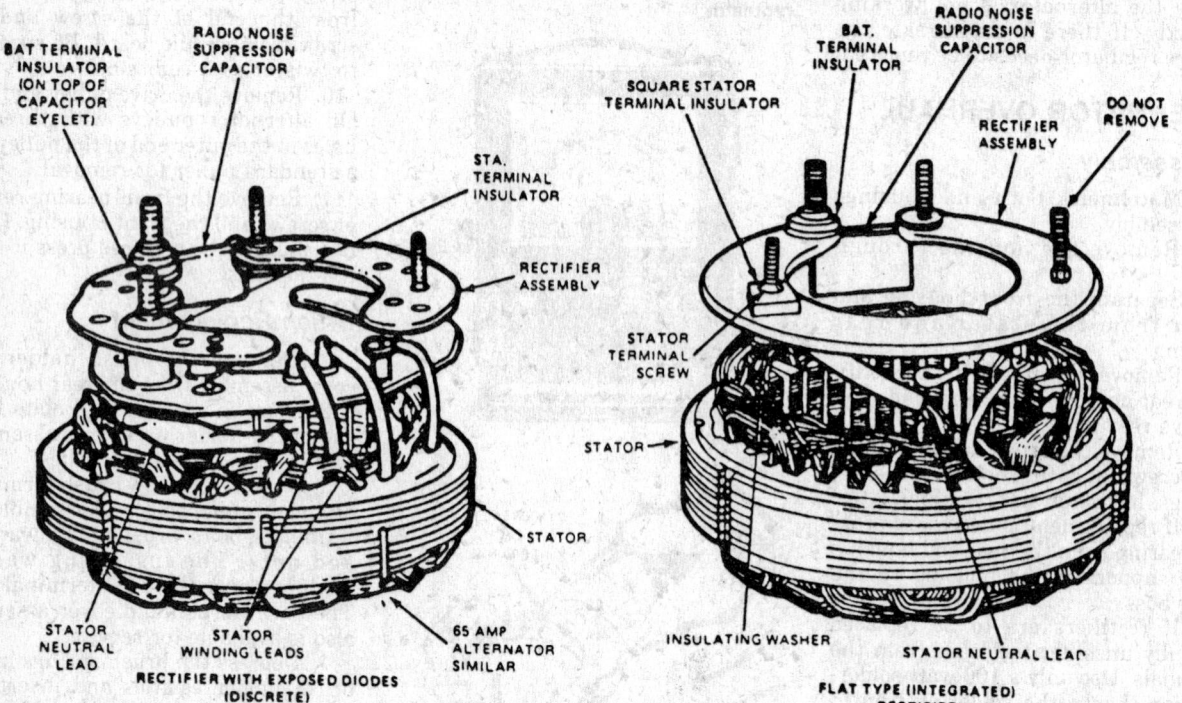

View of stator and rectifier assemblies and terminals—Ford alternator with external regulator

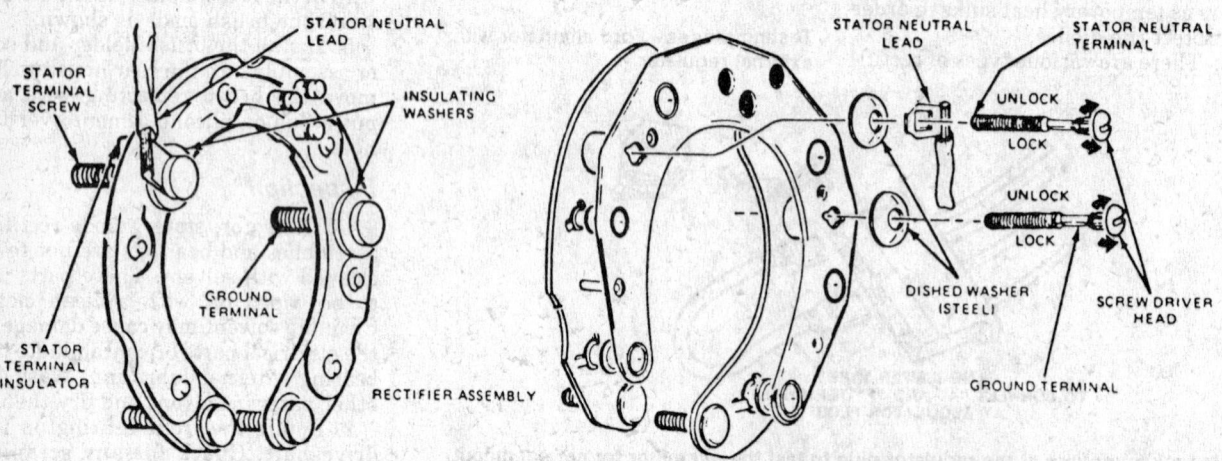

Rectifier assemblies and related components—Ford alternator with external regulator

FORD AUTOLITE/MOTORCRAFT ALTERNATOR
(WITH EXTERNAL REGULATOR)

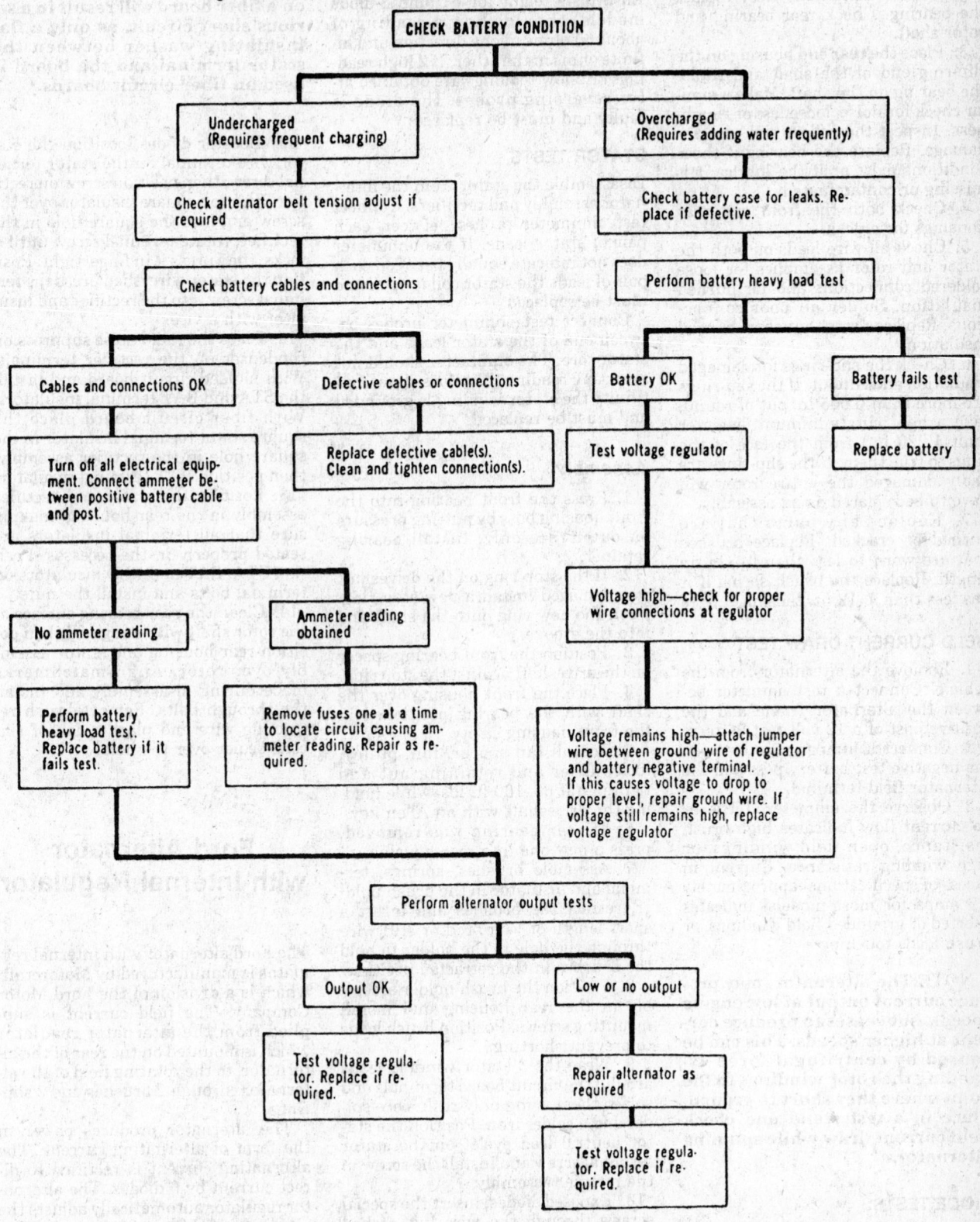

dicates that the bearing is excessively worn. As the bearing is being rotated look for excessive lubricant leakage. If any of these conditions exist replace the bearing. Check rear bearing and rotor shaft.

3. Place the rear end housing on the slip-ring end of the shaft and rotate the bearing on the shaft. Make a similar check for noise, looseness or roughness. Inspect the rollers and cage for damage. Replace the bearing if these conditions exist or if the lubricant is missing or contaminated.

4. Check both the front and rear housings for cracks.

5. Check all wire leads on both the stator and rotor assemblies for loose soldered connections and for burned insulation. Solder all poor connections. Replace parts that show burned insulation.

6. Check the slip-rings for damaged insulation and runout. If the slip-rings are more than 0.005 in. out of round, take a light cut (minimum diameter limit 1.220 in.) from the face of the rings to true them. If the slip-rings are badly damaged the entire rotor will have to be replaced as an assembly.

7. Replace any parts that are burned or cracked. Replace brushes that are worn to less than $5/16$ in. in length. Replace the brush spring if it has less than 7–12 oz. tension.

FIELD CURRENT DRAW TEST

1. Remove the alternator from the vehicle. Connect a test ammeter between the alternator frame and the positive post of a 12 volt test battery.

2. Connect a jumper wire between the negative test battery post and the alternator field terminal.

3. Observe the ammeter. Little or no current flow indicates high brush resistance, open field windings, or high winding resistance. Current in excess of specifications (approximately 2.9 amps for most models) indicates shorted or grounded field windings, or brush leads touching.

NOTE: The alternator, may produce current output at low engine speeds, but ceases to produce current at higher speeds. This can be caused by centrifugal force expanding the rotor windings to the point where they short to ground. Place in a test stand and check field current draw while spinning alternator.

DIODE TESTS

Disassemble the alternator. Disconnect diode assembly from stator and make tests. To test one set of diodes contact 1 ohmmeter probe to the diode plate and contact each of the 3 stator lead terminals with the other probe. Reverse the probes and repeat the test. All 6 tests (eight for 61 amp 8-diode models) should show a reading of about 60 ohms in one direction and infinite ohms in the other. If 2 high readings or 2 low readings are obtained after reversing probes, the diode is faulty and must be replaced.

STATOR TESTS

Disassemble the stator from the alternator assembly and rectifiers. Connect test ohmmeter probes between each pair of stator leads. If the ohmmeter does not indicate equally between each pair of leads the stator coil is open and must be replaced.

Connect test ohmmeter probes between one of the stator leads and the stator core. The ohmmeter should not show any reading. If it does show continuity the stator winding is grounded and must be replaced.

Assembly

1. Press the front bearing into the front housing boss by putting pressure on outer race only. Install bearing retainer.

2. If the stop ring on the driveshaft was damaged install a new stop ring. Push the new ring onto the shaft and into the groove.

3. Position the front bearing spacer on the driveshaft against the stop ring.

4. Place the front housing over the shaft with the bearing positioned in the front housing cavity.

5. Install fan spacer, fan, pulley, lockwasher and retaining nut and tighten nut 60–100 ft. lbs. while holding the driveshaft with an Allen key.

6. If rear bearing was removed, press a new one into rear housing.

7. Assemble brushes, springs, terminal and insulator in the brush holder, retract the brushes and insert a short length of $1/8$ in. rod or stiff wire through the hole in the holder to hold the brushes in the retracted position.

8. Position the brush holder assembly in the rear housing and install mounting screws. Position brush leads to prevent shorting.

9. Wrap the 3 stator winding leads around the circuit board terminals and solder them using only rosin core solder and a solder iron. Position the stator neutral lead eyelet on the stator terminal screw and install the screw in the rectifier assembly.

10. Exposed diodes, insert the special screws through the wire lug, dished washers and circuit board. Turn ¼ turn counterclockwise to lock in place. Integral diodes, insert the screws straight through the holes.

NOTE: The dished washers are to be used on the molded circuit boards only. Using these washers on a fiber board will result in a serious short circuit, as only a flat insulating washer between the stator terminal and the board is used on fiber circuit boards.

11. Booster diodes, position the stator wire terminal on the stator terminal screw, then position screw on rectifier. Position square insulator over the screw and into the square hole in the rectifier, rotate terminal screw until it locks, then press it in fingertight. Position the stator wire, then press the terminal screw into the rectifier and insulator with a vise.

12. Place the radio noise suppression condenser on the rectifier terminals. With molded circuit board and install the STA and BAT terminal insulators. With fiber circuit board place the square stator terminal insulator in the square hole in the rectifier assembly, then position BAT terminal insulator.

13. Position the stator and rectifier assembly in the rear housing, making sure that all terminal insulators are seated properly in the recesses. Position STA, BAT and FLD insulators on terminal bolts and install the nuts.

14. Clean the rear bearing surface of the rotor shaft with a rag and then position rear housing and stator assembly over rotor. Align matchmarks made during disassembly and install the through bolts. Remove brush retracting wire and place a dab of silicone sealer over the hole.

Ford Alternator with Internal Regulator

The Ford alternator with internal regulator is manufactured by Motorcraft, which is a division of the Ford Motor Company. The field current is supplied from the alternator regulator which is mounted on the rear of the alternator, to the rotating field of the alternator through 2 brushes and 2 slip-rings.

The alternator produces power in the form of alternating current. The alternating current is rectified to direct current by 6 diodes. The alternator regulator automatically adjusts the alternator field current to maintain the alternator output voltage within prescribed limits to correctly charge the battery. The alternator is self current limiting.

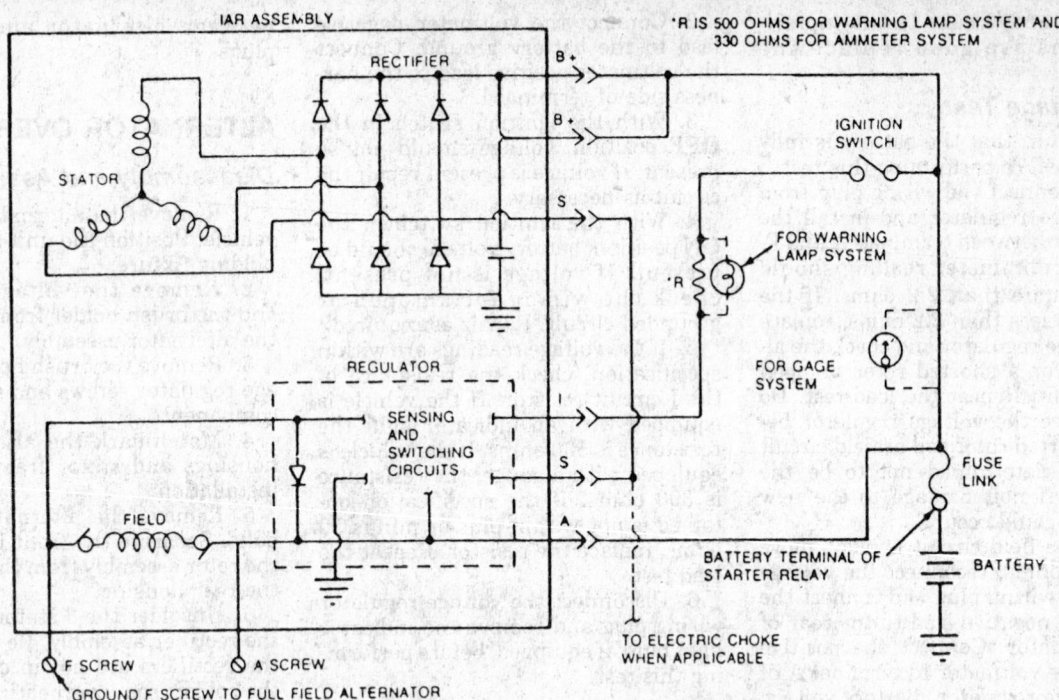

Schematic of the charging system—Ford alternator with internal regulator

The regulator voltage control circuit is turned on when the ignition switch is **ON** and voltage is applied to the regulator I terminal through a resistor in the I circuit. When the ignition switch is **OFF** the control circuit is turned off and no field current flows to the alternator.

On warning lamp equipped vehicles, the warning lamp is connected across the terminals of a 500 ohm resistor at the instrument cluster. Current passes through the warning lamp when the ignition switch is in the **RUN** position and there is no voltage at terminal S. When voltage at S rises to a preset value the regulator switching circuits stop the flow of current into terminal I and the lamp turns off.

System voltage is sensed and alternator field current is drawn through terminal A. The regulator switching circuits will turn the warning lamp on, indicating a system fault, if terminal A voltage is excessively high or low, or if the terminal S voltage signal is abnormal. A fusible link is included in the charging system wiring on all models. The fusible link is used to prevent damage to the wiring harness and alternator if the wiring harness should become grounded, or if a booster battery is connected to the charging system with the wrong polarity.

ON VEHICLE SERVICE

NOTE: The following diagnostic tests are made with the alternator installed in the vehicle. Be
sure that the battery is fully charged before any testing is done.

Battery Voltage Test

1. Connect a voltmeter to the positive and negative battery terminals.
2. Record the battery voltage. If battery voltage is not within specification, correct as required.

Load Test

1. Be sure that the battery is fully charged before performing this test.
2. Connect the tachometer to the engine.
3. Start the engine. Turn the heater/air condition switch to the **HIGH** blower position. Turn **ON** the headlights with the high beams. Increase engine speed to 2000 rpm.
4. Voltmeter should indicate a minimum of a 0.5 volt increase over battery voltage.
5. If the system is working the above readings will be obtained. Be sure not to ground the A terminal of the voltage regulator.

No Load Test

1. Be sure that the battery is fully charged before performing this test.
2. Connect a tachometer to the engine. Start the engine and increase the engine speed to 1500 rpm with no electrical load.
3. The voltmeter reading should be taken when the voltmeter needle stops moving. The voltmeter reading should

be 1–2 volts above the voltage of the battery.
4. If the voltage increased properly proceed with another test.

High Voltage Test

1. Be sure that the battery is fully charged before performing this test.
2. Turn the ignition switch to the **ON** position. Connect the voltmeter negative lead to the rear of the alternator housing.
3. Connect voltmeter positive lead to alternator output terminal and record voltage. Connect voltmeter positive lead to the A terminal of regulator. Compare voltage difference recorded at alternator output terminal.
4. If voltage difference is greater than 0.5 volt, repair or replace wiring circuit to A terminal.
5. If high voltage condition still exists check ground connections at regulator to alternator, alternator to engine, firewall to engine and engine to battery.
6. If high voltage condition still exists connect voltmeter negative lead to rear of alternator housing. With ignition to **OFF** position connect voltmeter positive lead to A terminal of regulator and record reading. Connect voltmeter positive lead to F terminal of regulator.
7. Check if different voltage is present at A and F terminals. Different voltage readings indicate a defective regulator, grounded brush leads or grounded rotor coil.
8. If same voltage is present at both

terminals and circuits tested in previous steps are good replace the regulator.

Low Voltage Test

1. Be sure that the battery is fully charged before performing this test.

2. Disconnect the wiring plug from the voltage regulator and install the ohmmeter between terminals A and F.

3. The ohmmeter reading should indicate more than 2.2 ohms. If the reading is less than 2.2 ohms, replace the voltage regulator and check the alternator for a shorted rotor or open field circuit. Repeat the load test. Do not replace the voltage regulator before a shorted rotor coil or field circuit has been determined not to be the problem. If not, damage to the new regulator could occur.

4. If the field circuit is okay, more than 2.2 ohms, reconnect the voltage regulator wiring plug and connect the voltmeter negative lead to the rear of the alternator. Connect the positive lead of the voltmeter to terminal A of the voltage regulator. Battery voltage should be present, if so go on. If not, repair wiring in circuit A.

5. With the ignition switch in the **OFF** position connect the positive lead of the voltmeter to the F terminal of the voltage regulator.

6. If battery voltage is present, go on. If not, replace the voltage regulator. Repeat the load test.

7. Turn the ignition switch to the **ON** position. The voltmeter should indicate 1.5 volts or less. If the reading is more than 1.5 volts, perform the regulator I circuit test. Repair the voltage regulator as required. Repeat the load test.

8. If the voltmeter reading is 1.5 volts or less, disconnect the alternator wiring plug and connect a 12 gauge jumper wire between the alternator plug terminal and the wiring harness connector.

9. Connect the positive lead of the voltmeter to 1 of the B terminals. Repeat the load test.

10. If the reading is 0.5 volt above battery voltage, repair the wiring harness from the alternator to the starter relay.

11. If the voltmeter reading is less than 0.5 volt above battery voltage, connect a jumper wire from the rear of the alternator housing to terminal F of the voltage regulator.

12. Repeat the load test. If the voltmeter reading is more than 0.5 volt, replace the voltage regulator. If the reading is less than 0.5 volt, repair the alternator.

Voltage Regulator Circuit I Test

1. Disconnect the voltage regulator wiring plug harness.

2. Connect the voltmeter negative lead to the battery ground. Connect the voltmeter positive lead to the harness side of terminal I.

3. With the ignition switch in the **OFF** position, voltage should not be present. If voltage is present repair the circuit as necessary.

4. With the ignition switch in the **ON** position, battery voltage should be present. If voltage is not present, check the wiring for an open or grounded circuit. Repair as required.

5. If the voltage readings are within specification, check the resistance of the I circuit resistor. If the vehicle is equipped with an indicator light the resistance is 500 ohms. If the vehicle is equipped with a gauge, the resistance is 300 ohms. If the specification obtained is not within plus or minus 50 ohms, replace the resistor. Repeat the load test.

6. Disconnect the voltage regulator wiring plug and remove the indicator light bulb, if equipped, before performing this test.

Field Circuit Drain Test

1. Connect the negative lead of the voltmeter to the rear of the alternator housing. Turn the ignition switch to the **OFF** position. Connect the positive lead of the voltmeter to the F terminal of the voltage regulator.

2. Battery voltage should be present. If no voltage is present, proceed.

3. If voltage is less than battery voltage, check I circuit. Disconnect regulator wiring harness. Connect voltmeter positive lead to S terminal of wiring plug. If voltage is present, proceed to Step 4. If no voltage is present replace regulator.

4. Disconnect wiring plug from alternator. Check S terminal for voltage. If no voltage is present, replace alternator rectifier assembly. If voltage is still present, replace or repair wiring

between alternator and regulator plugs.

ALTERNATOR OVERHAUL

Disassembly and Assembly

1. Remove the alternator from the vehicle. Position the unit in a suitable holding fixture.

2. Remove the voltage regulator and the brush holder from the rear of the alternator assembly.

3. Remove the brush holder-to-voltage regulator screws and separate the components.

4. Matchmark the alternator end housings and stator frame to aid in installation.

5. Remove the alternator through bolts. Separate the front housing and the rotor assembly from the stator and the rear housing.

6. Unsolder the 3 stator leads from the rectifier assembly. Be careful that the rectifiers are not in contact with the solder iron, overheating them will cause damage.

7. Remove the rectifier assembly from the rear of the alternator housing. Press the rear alternator housing bearing from the rear housing.

8. From the front housing of the alternator remove the drive pulley nut from the rotor shaft.

9. Remove the lockwasher, drive pulley, fan and fan spacer from the rotor shaft.

10. Remove the rotor from the front housing. Remove the front bearing spacer from the rotor shaft. Do not remove the rotor stop ring unless it must be replaced.

11. Remove the front housing bearing retainer and bearing.

12. Assembly of the alternator is the reverse of the disassembly procedure. Be sure to clean and check all parts for wear and defects. Repair or replace defective components as required.

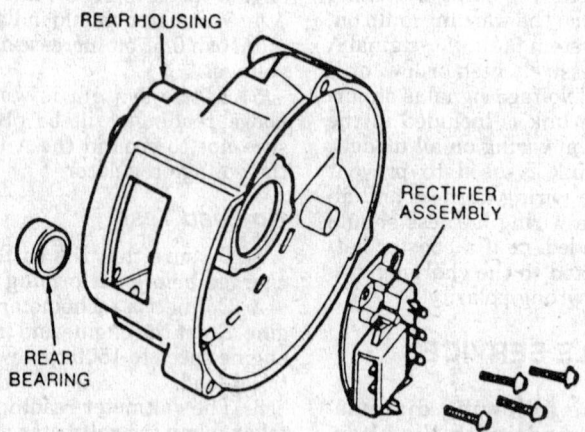

Ford alternator with internal regulator rear housing and related components.

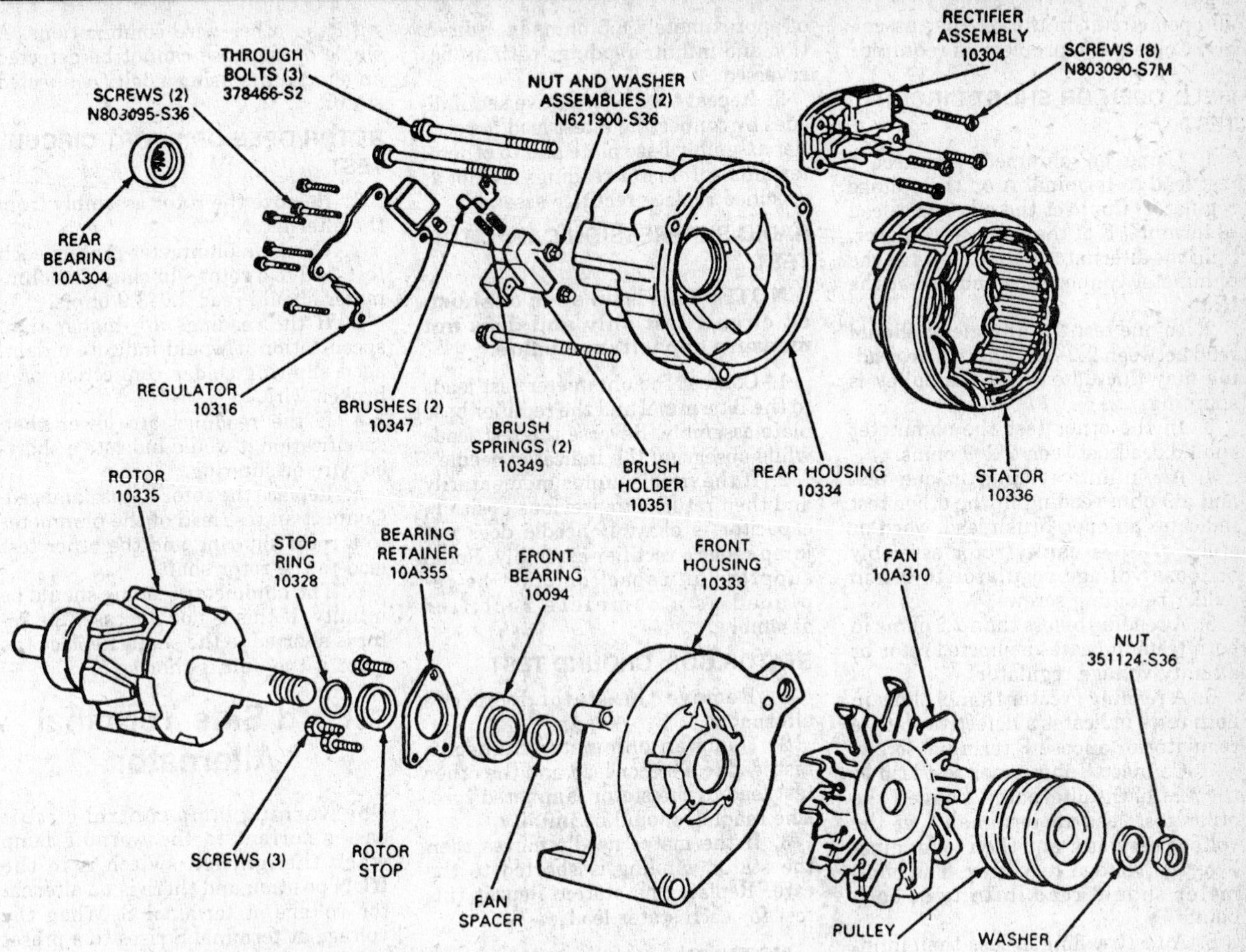

RECTIFIER ASSEMBLY 10304
SCREWS (8) N803090-S7M
THROUGH BOLTS (3) 378466-S2
SCREWS (2) N803095-S36
NUT AND WASHER ASSEMBLIES (2) N621900-S36
REAR BEARING 10A304
REGULATOR 10316
BRUSHES (2) 10347
BRUSH SPRINGS (2) 10349
BRUSH HOLDER 10351
REAR HOUSING 10334
STATOR 10336
ROTOR 10335
STOP RING 10328
BEARING RETAINER 10A355
FRONT BEARING 10094
FRONT HOUSING 10333
FAN 10A310
NUT 351124-S36
SCREWS (3)
ROTOR STOP
FAN SPACER
PULLEY
WASHER

Exploded view of the Ford alternator with internal regulator

VOLTAGE REGULATOR ASSEMBLY
BRUSH HOLDER
BRUSH SPRINGS
BRUSHES

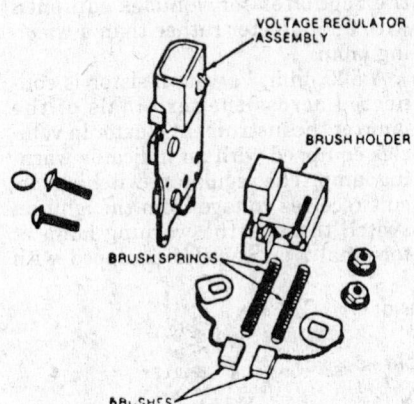

View of the voltage regulator and brush holder – Ford alternator with internal regulator

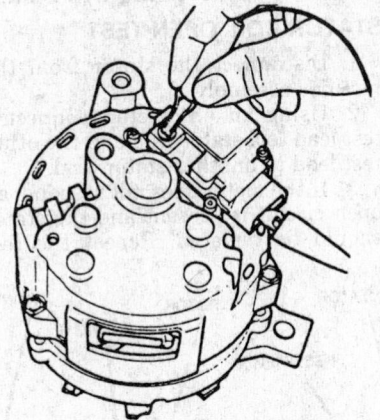

Ford alternator with external regulator

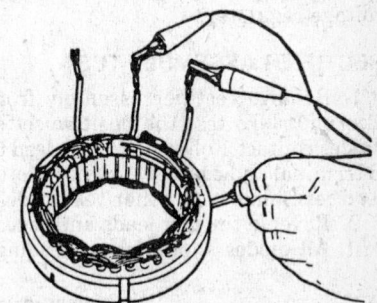

Stator coil open test, using an ohmmeter. If meter does not respond, an open is present and the stator should be replaced – Ford alternator with internal regulator

Inspection

NOTE: In performing the following tests, digital meters cannot be used.

STATOR GROUND TEST

1. Using an ohmmeter connect 1 test lead to the B terminal and the other test lead to the S terminal.

2. Reverse the test leads and repeat the test. The ohmmeter should read about 6.5 ohms in 1 direction and infinity when the test probes are reversed.

3. A reading in both directions indicates a faulty positive diode or a shorted radio suppression capacitor.

4. Perform the same test using the S terminal and the alternator rear housing.

5. Readings in both directions indicate a faulty negative diode, grounded stator winding, grounded stator terminal or a shorted radio suppression capacitor.

6. If the ohmmeter needle does not move in 1 direction, or high resistance in the other direction exists, there is

an open circuit in the rectifier assembly. Correct the problem as required.

FIELD OPEN OR SHORT CIRCUIT TEST

1. Using an ohmmeter connect 1 test lead to terminal A on the voltage regulator. Connect the other test lead to terminal F of the voltage regulator. Spin the alternator pulley. Reverse the ohmmeter connections and repeat the test.
2. In one test the ohmmeter should read between 2.2–100 ohms. The reading may fluctuate while the pulley is spinning.
3. In the other test the ohmmeter should read between 2.2–9 ohms.
4. An infinite reading in one test and a 9 ohm reading in the other test indicates an open brush lead, worn or stuck brushes, faulty rotor assembly or loose voltage regulator to brush holder retaining screws.
5. A reading of less than 2.2 ohms in both tests indicates a shorted rotor or a faulty voltage regulator.
6. A reading greater than 9 ohms in both tests indicates a defective voltage regulator or a loose F terminal screw.
7. Connect 1 ohmmeter test lead to the rear of the alternator. Connect the other test lead to terminal A of the voltage regulator and then to terminal F of the voltage regulator. The ohmmeter should read infinity at both points.
8. A test reading of less than infinity at both points indicates a grounded brush lead, grounded rotor or a faulty voltage regulator.

RECTIFIER ASSEMBLY TEST

1. Remove rectifier assembly from alternator. To test the positive set of diodes contact 1 ohmmeter test lead to B terminal and contact each of 3 stator lead terminals with other test lead.
2. Reverse the test leads and repeat test. All diodes should show readings

of approximately 6.5 ohms in 1 direction and infinite readings with probes reversed.
3. Repeat test for negative set of diodes by connecting a test lead to rectifier assembly base plate and to other 3 terminals. If meter readings are not as specified replace rectifier assembly.

RADIO SUPPRESSION CAPACITOR TEST

NOTE: This is an open or shorted circuit test only and does not measure capacitance value.

1. Contact the ohmmeter test leads to the B terminal and the rectifier base plate assembly. Reverse the test leads while observing the indicator needle.
2. If the needle jumps momentarily and then returns to previous position, capacitor is okay. If needle does not jump replace rectifier assembly. Radio suppression capacitor must be replaced as a complete rectifier assembly.

STATOR COIL GROUND TEST

1. Remove the stator from the alternator.
2. Using an ohmmeter connect a test lead to a stator lead and the other test lead to the stator laminated core. The reading should be infinity.
3. If the meter needle moves then the stator winding is shorted to the core. Replace the stator. Repeat this test for each stator lead.

NOTE: Do not touch the metal test leads or the stator leads, an incorrect test reading will result.

STATOR COIL OPEN TEST

1. Disconnect the stator from the rectifier assembly.
2. Using an ohmmeter, connect a test lead to a stator lead and the other test lead to another stator lead.
3. If the meter does not respond, an open circuit is present and the stator should be replaced. Repeat the test

with the other wire combinations. A single open phase cannot be detected on alternators using a delta connected stator.

ROTOR OPEN OR SHORT CIRCUIT TEST

1. Remove the rotor assembly from the alternator.
2. Using an ohmmeter contact each test lead to a rotor slip-ring. The ohmmeter should read 2.0–3.9 ohms.
3. If the readings are higher than specification it would indicate a damaged slip-ring solder connection or a broken wire.
4. If the readings are lower than specification it would indicate a shorted wire or slip-ring.
5. Replace the rotor if it is damaged. Connect an test lead of the ohmmeter to a rotor slip-ring and the other test lead to the rotor shaft.
6. The ohmmeter reading should be infinity. If this is not the case, the rotor is shorted to the shaft. Replace the rotor if the unit is shorted.

Ford Side Terminal Alternator

The warning lamp control circuit passes current to the warning lamp when the ignition switch is in the **RUN** position and there is no alternator voltage at terminal S. When the voltage at terminal S rises to a preset value, current is cut off to the warning lamp. This circuit is not included in the regulator for vehicles equipped with an ammeter rather than a warning lamp.

A 500 ohm, ¼ watt resistor is connected across the terminals of the lamp at the instrument cluster in vehicles equipped with an indicator warning lamp. The regulator switching circuit receives voltage from the ignition switch through the warning lamp at terminal I on vehicles equipped with

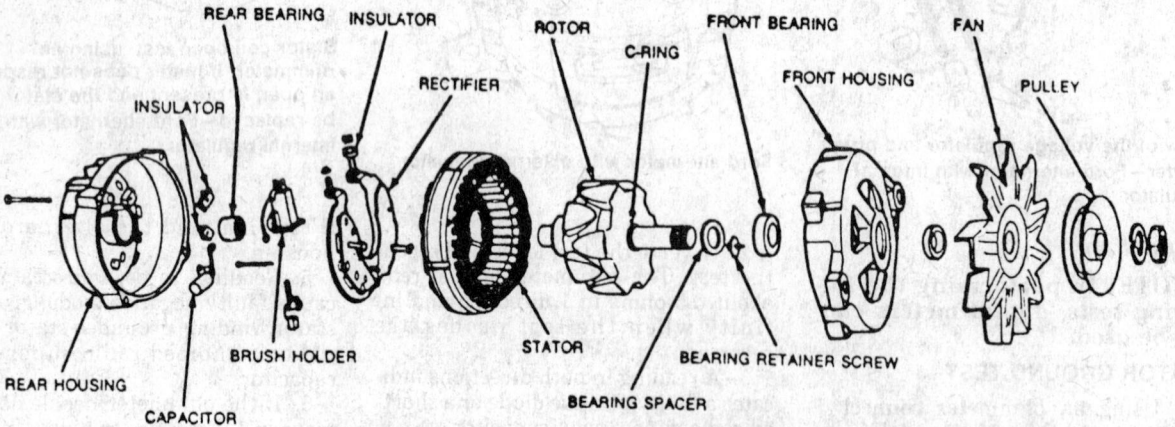

Exploded view of the Ford side terminal alternator

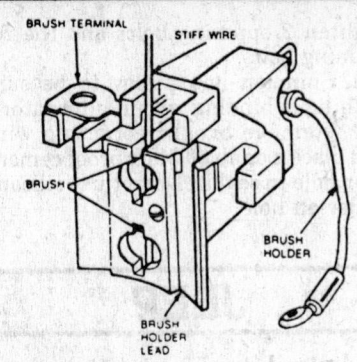

Ford side terminal alternator brush holder assembly

an indicator warning lamp or through terminal S on vehicles equipped with an ammeter. With an input voltage present, the switching circuit turns on the voltage control circuit which, in turn, adjusts field current to control alternator output voltage.

Fuse links are included in the charging system wiring on all models. This fuse link is used to prevent damage to the wiring harness and alternator if the wiring harness should become grounded or if a booster battery is connected to the charging system with the wrong polarity.

ALTERNATOR OVERHAUL

Disassembly

1. Mark both end housings and stator with a scribe mark for assembly.
2. Remove housing through bolts and separate front housing and rotor from rear housing and stator. Slots are provided in front housing to aid in disassembly. Do not separate rear housing from stator at this time.
3. Remove drive pulley nut, lockwasher, pulley, fan and fan spacer from rotor shaft.
4. Pull rotor and shaft from front housing and remove spacer from rotor shaft.
5. Remove the screws retaining bearing to front housing. If bearing is damaged or has lost lubricant, remove bearing from housing. To remove bearing, support housing close to bearing boss and press bearing from housing.
6. Unsolder and disengage 3 stator leads from rectifier. Work quickly to prevent overheating rectifier.
7. Lift stator from rear housing.
8. Unsolder and disengage brush holder lead from rectifier. Work quickly to prevent overheating rectifier.
9. Remove screw attaching capacitor lead to rectifier.
10. Remove the screws attaching rectifier to rear housing.
11. Remove the terminal nuts and

insulator from outside housing. Remove rectifier from housing.
12. Remove the screws attaching brush holder to housing. Remove brushes and holder.
13. Remove any sealing compound from rear housing and brush holder.
14. Remove the screw attaching capacitor to rear housing and remove capacitor.

Inspection

RECTIFIER SHORT GROUNDED AND STATOR GROUNDED TEST

NOTE: These tests are performed with an ohmmeter. Digital meters cannot be used to perform rectifier tests

1. Connect an ohmmeter probe to alternator BAT terminal (red insulator) and other probe to STA terminal (rear blade terminal). Then reverse ohmmeter probes and repeat test. Normally, there will be no needle movement in one direction, indicating rectifier diodes are being checked in reverse current direction and are not shorted. A low reading with probes reversed indicates that rectifier positive diodes are being checked in forward current direction. Using referenced tester, low reading should be about 6 ohms, but may vary if another type of test is used. A reading in both directions indicates a damaged positive diode, a grounded positive diode plate, or a grounded BAT terminal.
2. Perform same test using STA and GND (ground) terminals of alternator. A reading in both directions indicates either a damaged negative diode, a grounded positive diode plate or a grounded BAT terminal.
3. If there is no needle movement with probes in 1 direction and no needle movement or high resistance (significantly over 6 ohms) in opposite direction, a faulty connection exists in stator circuit inside alternator.

FIELD OPEN OR SHORT CIRCUIT TEST

1. Using an ohmmeter, contact the alternator field terminal with 1 probe and ground terminal with other probe. Then, spin alternator pulley. Ohmmeter reading should be between 2.4–100 ohms and should fluctuate while pulley is turning.
2. An infinite reading (no meter movement) indicates a grounded brush lead, worn or stuck brushes or a worn or damaged rotor assembly.
3. An ohmmeter reading less than 2.4 ohms indicates a grounded brush assembly, a grounded field terminal or a worn or damaged rotor.

DIODE TEST

1. Remove the rectifier assembly from the alternator stator. To test a set of diodes, contact one probe to a terminal screw and contact each of 3 stator lead terminals with other probe. Reverse probes and repeat test. All diodes should show a low reading of about 6 ohms in one direction and an infinite reading (no needle movement) with probes reversed. Low reading may vary with type of ohmmeter used.
2. Repeat preceding tests for other set of diodes by contacting the other terminal screw and 3 stator lead terminals.
3. If meter readings are not as specified, replace rectifier assembly.

STATOR COIL GROUNDED TEST

1. Connect ohmmeter probes to a stator lead and to stator laminated core. Ensure that probe makes a good electrical connection with stator core. The meter should show an infinite reading (no meter movement).
2. If meter does not indicate an infinite reading (needle moves), stator winding is shorted to core and must be replaced.
3. Repeat this test for each stator lead. Do not touch the metal probes or stator leads with the hands. Such contact will result in an incorrect reading.

STATOR COIL OPEN TEST

NOTE: A single open phase will not be diagnosed by this test on a 100 amp alternator that has a delta connected stator.

1. Connect ohmmeter probe to a stator phase lead and touch other probe to another stator lead. Check meter reading.
2. Repeat this test with the other 2 stator lead combinations. If no meter movement occurs (infinite resistance) on a lead paired with either of the other phase leads, that phase is open and the stator should be replaced.

ROTOR OPEN OR SHORT CIRCUIT TEST

1. Contact each ohmmeter probe to a rotor slip-ring. The meter reading should be 2.3–2.5 ohms.
2. A higher reading indicates a damaged slip-ring solder connection or a broken wire.
3. A lower reading indicates a shorted wire or slip-ring. Replace rotor if it is damaged and cannot be serviced.
4. Contact an ohmmeter probe to a slip-ring and the other probe to rotor shaft. Meter reading should be infinite (no deflection).
5. A reading other than infinite indicates rotor is shorted to shaft. Inspect slip-ring soldered terminals to assure they are not bent and not

touching rotor shaft, or that excess solder is not grounding rotor coil connections to shaft. Replace the rotor if it is shorted and cannot be serviced.

Assembly

1. If bearing replacement is necessary, support rear housing close to bearing boss and press bearing out of housing.

2. Wipe rotor, stator and bearings with a clean cloth. Do not clean these parts with solvent.

3. Rotate front bearing on drive end of rotor shaft. Check for any scraping noise, looseness or roughness. Look for excessive lubricant leakage. If any of these conditions exist, replace bearing.

4. Inspect rotor shaft rear bearing surface for roughness or sever chatter marks. Replace rotor assembly if shaft is not smooth.

5. Place rear bearing on slip-ring end of rotor shaft and rotate bearing. Make the same check for noise, looseness, or roughness as was made for front bearing. Inspect rollers and cage for damage. Replace bearing if these conditions exist, or if lubricant is lost or contaminated.

6. Check pulley and fan for excessive looseness on rotor shaft. Replace any pulley that is loose or bent out of shape.

7. Check both front and rear housing for cracks, particularly in webbed areas and at mounting ear. Replace damaged or cracked housing.

8. Check all wire leads on both stator and rotor assemblies for loose or broken soldered connections and for burned insulation. Resolder poor connections. Replace parts that show signs of burned insulation.

9. Check slip-rings for nicks and surface roughness. Nicks and scratches may be removed by turning down slip-rings. Do not go beyond minimum diameter of 1.220 in. If rings are badly damaged, replace rotor assembly.

10. Replace brushes if they are worn shorter than ¼ in.

12. If front housing bearing is being replaced, press new bearing in housing. Apply pressure on bearing outer race only. Install bearing retaining screws and tighten to 25–40 inch lbs.

13. Place inner spacer on rotor shaft and insert rotor shaft into front housing and bearing.

14. Install fan spacer, fan, pulley, lockwasher and nut on rotor shaft. Use the proper tool to tighten pulley nut.

15. If rear bearing is being replaced, press a new bearing in from inside housing until rear bearing face is flush with boss outer surface.

16. Position brush terminal on brush holder. Install springs and brushes in brush holder and insert a piece of stiff wire to hold brushes in place.

17. Brushes and springs are serviced as part of brush holder assembly. Position brush holder in rear housing and install attaching screws. Brush retaining wire must stick out enough to be grabbed and pulled from housing assembly.

18. Waterproof glue sealer may have to be pushed out of pin hole in housing. Push brush holder toward brush holder attaching screws. Reseal crack between brush holder and brush cavity in rear housing with caulking cord or equivalent body sealer. Do not use silicone base sealer for this application.

19. Position capacitor to rear housing and install attaching screw. Place 2 rectifier insulators on bosses inside housing.

20. Place insulator on BAT (large) terminal of rectifier and position rectifier in rear housing. Place outside insulator on BAT terminal and install nuts on BAT and Grd terminals fingertight. Install, but do not tighten, the rectifier attaching screws.

21. Tighten the BAT terminal nuts to 35–50 inch lbs. and GRD terminal nuts to 25–35 inch lbs. on outside of rear housing. Then, tighten rectifier attaching screws to 40–50 inch lbs.

22. Position capacitor lead to rectifier and install attaching screw.

23. Press brush holder lead on rectifier pin and solder securely. Work quickly to prevent overheating of rectifier.

24. Position stator in rear housing and align scribe marks. Press 3 stator leads on rectifier pins and solder securely using resin core electrical solder. Work quickly to prevent overheating rectifier.

25. Position rotor and front housing into stator and rear housing. Align scribe marks and install through bolts.

Tighten 2 opposing bolts and the remaining bolts.

26. Spin fan and pulley to be sure nothing is binding within alternator.

27. Remove brush retracting wire and place a daub of waterproof cement over hole to seal it. Do not use silicone sealer on hole.

JEEP

Delcotron SI-Series

The charging system is an integrated AC generating system containing a built in voltage regulator. The regulator is mounted inside the slip ring end frame. All regulator components are enclosed in an epoxy molding and the regulator cannot be adjusted.

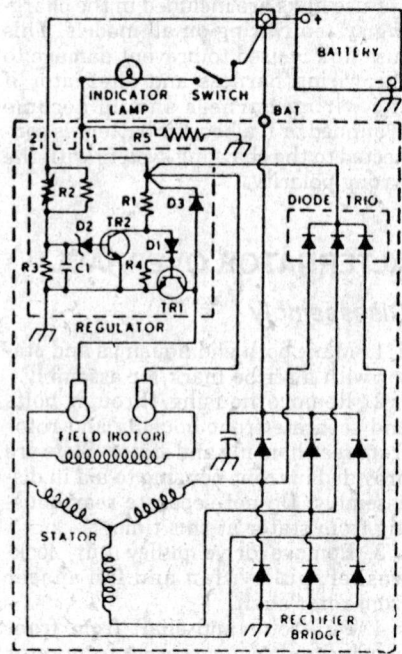

Schematic of the Delcotron charging system—Jeep

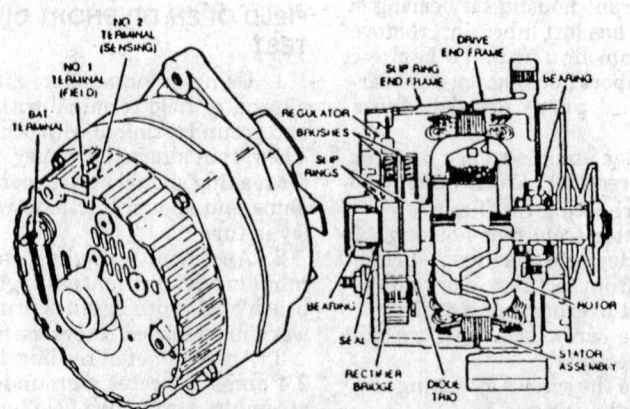

Cross-sectional view of the Delcotron alternator—Jeep

ALTERNATOR OVERHAUL

Disassembly

1. Remove the alternator from the vehicle. Position the assembly in a suitable holding fixture.

2. Make scribe marks on the alternator case end frames to aid in reassembly.

3. Remove the assembly through bolts and separate the drive end frame assembly from the rectifier end frame assembly.

4. Remove the rectifier nuts and the regulator screws from the end frame assembly.

5. Separate the stator, diode trio and voltage regulator from the end frame assembly.

6. Check the stator for opens using an ohmmeter. If high readings are obtained, replace the stator.

7. Check the stator for grounds, using an ohmmeter. If readings are low, replace the stator.

8. Using an ohmmeter, check the rotor for grounds. The ohmmeter reading should be very high. If not, replace the rotor.

9. Using an ohmmeter, check the rotor for opens. If the ohmmeter reading is not 2.4–3.5 ohms, replace the rotor.

10. To check the diode trio connect the ohmmeter to the diode trio and then reverse the lead connections. The ohmmeter should read high and low. If not, replace the diode trio. Repeat the same test between the single connector and each of the other connectors.

11. Check rectifier bridge with ohmmeter connected from grounded heat sink to flat metal on terminal. Reverse leads. If both readings are equal then replace rectifier bridge.

12. Repeat test between grounded heat sink and other 2 flat metal clips.

13. Repeat test between insulated heat sink and 3 flat metal clips.

14. Clean or replace the alternator brushes as required. Position the brushes in the brush holder and retain them in place using the brush retainer wire or equivalent.

15. To remove the rotor and drive end bearing, remove the shaft nut, washer and pulley, fan and collar. Push the rotor from the housing.

16. Remove the retainer plate from inside the drive end frame. Push the bearing out. Clean or replace parts as required.

Inspection

INDICATOR LAMP OPERATION TEST

1. Check the indicator lamp for normal operation. If the indicator lamp operates properly, refer to the under-

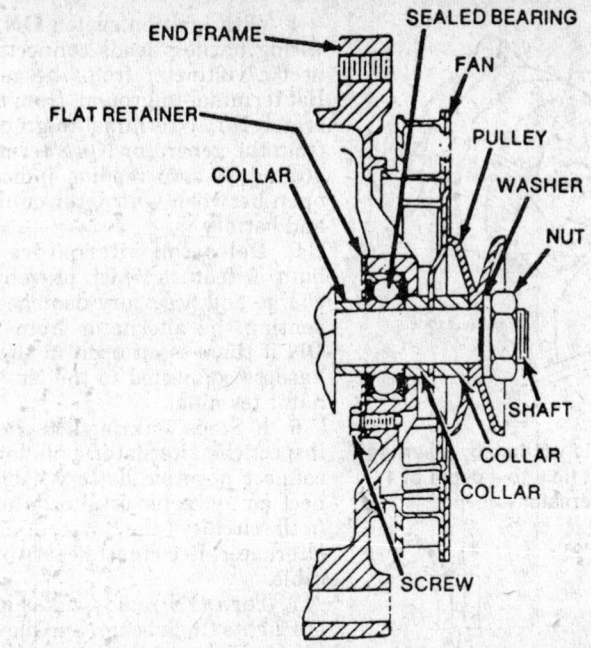

Exploded view of the end frame assembly — Delcotron alternator — Jeep

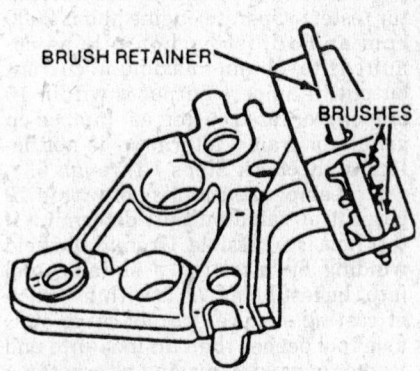

Installing the brushes into the holder — Delcotron alternator — Jeep

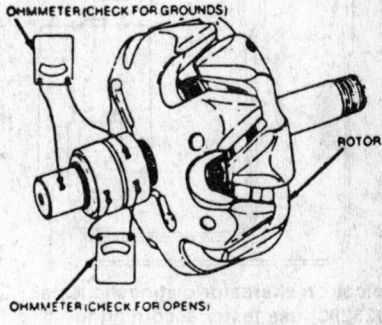

Using an ohmmeter to check the rotor for opens or grounds — Delcotron alternator — Jeep

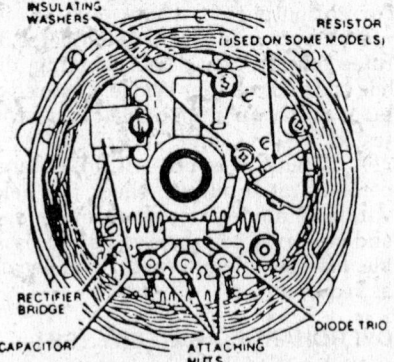

Location of the related end frame components — Delcotron alternator — Jeep

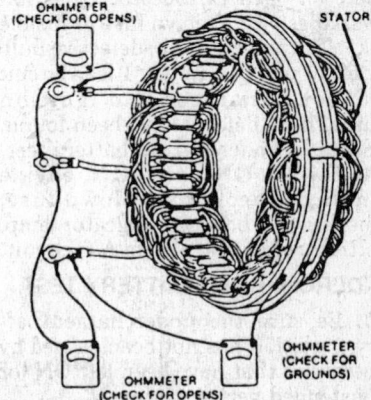

Using an ohmmeter to check the stator for opens or grounds — Delcotron alternator — Jeep

charged battery test. If the indicator lamp does not operate properly, proceed with the following tests.

2. Switch **OFF**, lamp **ON**. Unplug the connector from the generator No. 1 and No. 2 terminals. If the lamp stays **ON**, there is a short between these 2 leads. If the lamp goes out, replace the rectifier bridge.

3. Switch **ON**, lamp **OFF**, engine stopped. This condition can be caused

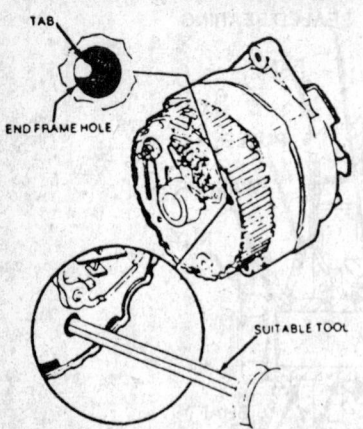

Grounding the field winding by inserting a tool into the test hole to a depth of 1 in. – Delcotron alternator – Jeep

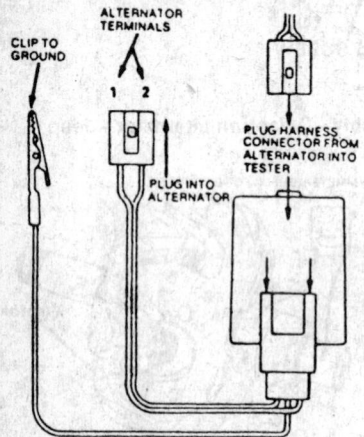

Delcotron alternator diagnostic tester (J-26290) use tester according to manufacturers instructions

by the defects listed above or by an open in the circuit. To determine where an open exists, proceed as follows. Check for a blown fuse or fusible link, a burned out bulb, defective bulb socket or an open in No. 1 lead circuit between generator and ignition switch. If no defects have been found, proceed to undercharged battery test.

4. Switch **ON**, lamp **ON**, engine running. Check for a blown fuse, (where used), between indicator lamp, ignition switch and also in A/C circuit.

UNDERCHARGED BATTERY TEST

1. Be sure the undercharged battery condition has not been caused by accessories that have been left **ON** for an extended period of time.

2. Check the alternator belt for proper belt tension. Inspect the battery for physical defects. Replace as required.

3. Inspect the wiring for defects. Check all connections for proper contact and cleanliness, including the slip connectors at the generator and bulkhead connections.

4. With ignition switch **ON** and all wiring harness leads connected, connect a voltmeter from the generator Bat terminal-to-ground, from the generator No. 1 terminal-to-ground and from the generator No. 2 terminal-to-ground. A zero reading indicates an open between voltmeter connection and battery.

5. Delcotron alternators have a built in feature which prevents overcharge and accessory damage by preventing the alternator from turning **ON** if there is an open in the wiring harness connected to the No. 2 alternator terminal.

6. If Steps 1 through 5 check out, inspect the alternator as follows. Disconnect negative battery cable. Connect an ammeter or alternator tester in the circuit at the Bat terminal of the alternator. Reconnect negative battery cable.

7. Turn **ON** radio, windshield wipers, lights (high beam) and blower motor on high speed. Connect a carbon pile across the battery (or use alternator tester). Operate engine about 2000 rpm and adjust carbon pile as required, to obtain maximum current output. If ampere output is within 10 amperes of rated output as stamped on generator frame, alternator is not defective. Recheck Steps 1 through 5.

8. If ampere output is not within 10 percent of rated output, determine if test hole is accessible. Ground the field winding by inserting a suitable tool into the test hole. Tab is within ¾ inch of casting surface. Do not force suitable tool deeper than an inch into end frame to avoid damaging alternator.

9. Operate engine at moderate speed as required and adjust carbon pile as required to obtain maximum current output.

10. If output is within 10 amperes of rated output, check field winding, diode trio and rectifier bridge. Test regulator with an approved regulator tested.

11. If output is not within 10 amperes of rated output, check the field winding, diode trio, rectifier bridge and stator. If test hole is not accessible, disassemble alternator and repair as required.

OVERCHARGED BATTERY TEST

1. Check the condition of the battery before any testing is done.

2. If an obvious overcharging condition exists, remove the alternator from the vehicle and check the field windings for grounds or shorts. If defective, replace the rotor. Test the regulator.

ALTERNATOR DIAGNOSTIC TESTER (J-26290)

This special diagnostic tester is de-signed to determine if the alternator should be removed from the vehicle.

1. Install tester J-26290 according to manufacturers instructions.

2. With the engine **OFF** and all lights and accessories **OFF**, test the alternator as follows. Light flashes, go to Step 3. Light **ON**, indicates fault in tester which should be replaced. Light **OFF**, pull plug from generator. A flashing light, indicates the alternator should be removed and the rectifier bridge replaced. Light **OFF**, indicates faulty tester or no voltage to tester. Check for 12 volts at No. 2 terminal of harness connector. Repair wiring or terminals if 12 volts is not available. Replace tester if 12 volts is available.

3. With the engine at fast idle and all accessories and lights **OFF**, test the alternator as follows. Light **OFF** indicates the charging system good, do not remove alternator. Light **ON** indicates a component failure within the alternator. Remove alternator and check diode trio, rectifier bridge and stator. Light flashing indicates a problem within the alternator. Remove alternator and check regulator, rotor field coil, brushes and slip rings.

VOLTAGE REGULATOR TEST (ALTERNATOR ON VEHICLE)

1. Connect a battery charger and a voltmeter to the battery.

2. Turn the ignition switch **ON** and slowly increase the charge rate. The alternator light in the vehicle will dim at the voltage regulator setting. Voltage regulator setting should be 13.5–16.0 volts. This test is performed to determine if the voltage regulator setting is within specifications. This test is accu-

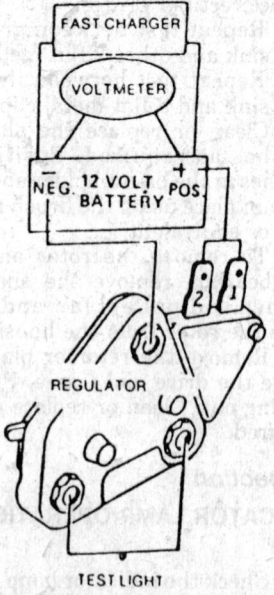

Testing the voltage regulator – Delcotron alternator – Jeep

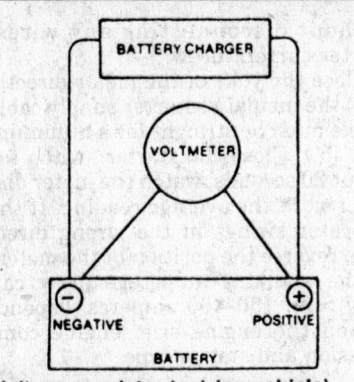

Voltage regulator test (on vehicle)
voltage regulator setting should be
13.5–16.0 volts – Delcotron alternator –
Jeep

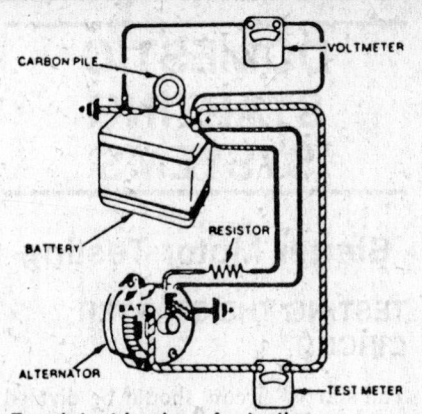

Bench test hook-up for testing –
Delcotron alternator – Jeep

rate even if the stator rectifier bridge
or diode trio is faulty.

VOLTAGE REGULATOR TEST (ALTERNATOR OFF VEHICLE)

1. Remove the alternator from the vehicle.

2. Disassemble the alternator and remove the voltage regualtor.

3. Connect a voltmeter and a fast charger to a 12 volt battery. Connect a test light to the regulator and observe the battery polarity.

4. The test light should light.

5. Turn **ON** the fast charger and slowly increase the charge rate. Observe the voltmeter, the light should go out at the voltage regulator setting. The voltage regulator setting specification is 13.5–16.0 volts.

1. Rotor
2. Front bearing retainer plate
3. Inner collar
4. Bearing
5. Front housing
6. Inner collar
7. Fan
8. Pulley
9. Lockwasher
10. Pulley nut
11. Terminal assembly
12. Bridge rectifier
13. Regulator
14. Brush assembly
15. Screw
16. Stator
17. Insulating washer
18. Capacitor
19. Diode trio
20. Rear housing
21. Through bolt
22. Bearing and seal assembly

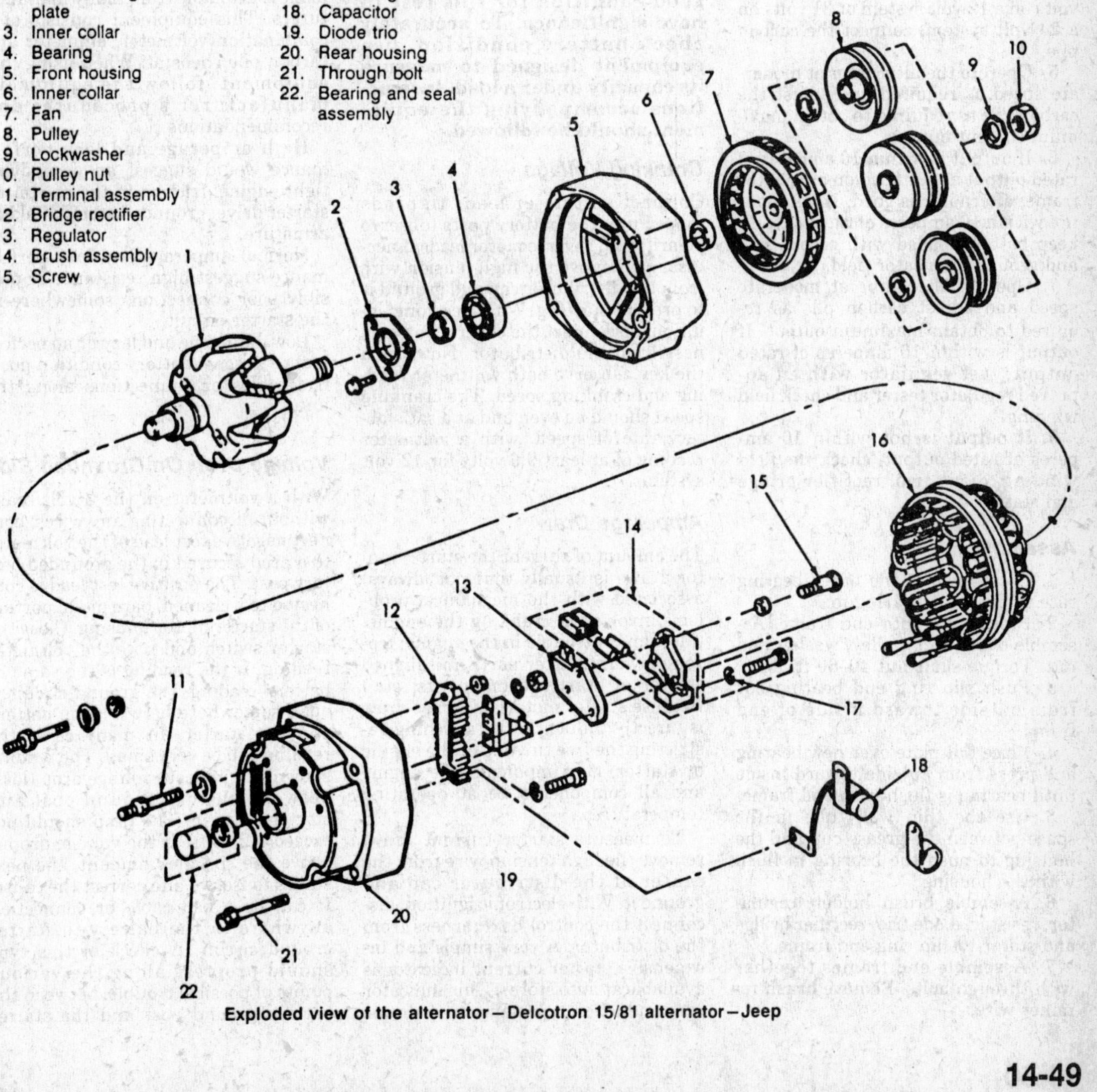

Exploded view of the alternator – Delcotron 15/81 alternator – Jeep

ALTERNATOR BENCH TEST

1. Remove the alternator from the vehicle. Position the unit in a suitable test stand.

2. Connect the alternator in series, but leave the carbon pile disconnected.

NOTE: Ground polarity of the battery must be the same as the alternator. Be sure to use a fully charged battery and a 10 ohm resistor rated at 6 watts or more between the alternator No. 1 terminal and the battery.

3. Increase the alternator speed slowly and observe the voltage.

4. If the voltage is uncontrolled with speed and increases above 15.5 volts on a 12 volt system or 31 volts on a 24 volt system, test regulator with an approved regulator tester and check field winding. If voltage is below 15.5 volts on a 12 volt system or 31 volts on a 24 volt system, connect the carbon pile.

5. Operate the alternator at moderate speed as required and adjust the carbon pile as required to obtain maximum current output.

6. If output is within 10 amperes of rated output as stamped on alternator frame, alternator is good. If output is not within 10 amperes of rated output, keep battery loaded with carbon pile and ground alternator field.

7. Operate alternator at moderate speed and adjust carbon pile as required to obtain maximum output. If output is within 10 amperes of rated output, test regulator with an approved regulator tester and check field winding.

9. If output is not within 10 amperes of rated output, check the field winding, diode trio, rectifier bridge and stator.

Assembly

1. Press against the outer bearing race to push the bearing in.

2. Press rotor into end frame. Assemble collar, fan, pulley, washer and nut. Torque shaft nut 40–60 ft. lbs.

3. Push slip ring end bearing out from outside toward inside of end frame.

4. Place flat plate over new bearing and press from outside toward inside until bearing is flush with end frame.

5. Use the thin wall tube in the space between the grease cup and the housing to push the bearing in flush with the housing.

6. Assemble brush holder, regulator, resistor, diode trio, rectifier bridge and stator-to-slip ring end frame.

7. Assemble end frames together with through bolts. Remove brush retainer wire.

DOMESTIC STARTING SYSTEMS

Starter Motor Testing

TESTING THE STARTER CIRCUIT

The starter circuit should be divided and tested in 4 separate phases:
1. Cranking voltage check
2. Amperage draw
3. Voltage drop on grounded side
4. Voltage drop on battery side

NOTE: The battery must be in good condition for this test to have significance. To accurately check battery condition, use equipment designed to measure its capacity under a load. Instructions accompanying the equipment should be followed.

Cranking Voltage

Connect voltmeter leads to prods tapped into the battery posts (observe polarity and reverse meter leads if necessary). Remove the high tension wire from the distributor cap and ground it to prevent starting. With electronic ignition, disconnect the control box harness from the distributor. Now, turn the key. Observe both voltmeter reading and cranking speed. The cranking speed should be even and at a satisfactory rate of speed, with a voltmeter reading of at least 9.6 volts for 12 volt systems.

Amperage Draw

The amount of current the starter motor draws is usually (but not always) associated with the mechanical problems involved in cranking the engine. (Mechanical trouble in the engine, frozen or worn starter parts, misaligned starter or starter components, etc.) Because starter motor amperage draw is directly influenced by anything restricting the free turning of the engine or starter, it is important the engine and all components be at operating temperatures.

To measure starter current draw, remove the high tension wire from the center of the distributor cap and ground it. With electronic ignition, disconnect the control box harness from the distributor. A very simple and inexpensive starter current indicator is available at auto stores. This indicator is an induction type gauge and shows, without disconnecting any wires, starter current draw.

Place the yoke of the meter directly over the insulated starter supply cable (cable must be straight for a minimum of 2 in.). Close the starter switch for about 20 seconds, watch the meter dial and record the average reading. If the indicator swings in the wrong direction, reverse the position of the meter.

The cranking amperage draw can vary from 150–400 amperes, depending on the engine size, engine compression and starter type.

NOTE: When starter specifications are not available, average starter draw amperage can be derived from testing a like starter unit, known to be operating satisfactorily.

More accurate but complex equipment is available from many manufacturers. This equipment consists of a combination voltmeter, ammeter and carbon pile rheostat. When using this equipment, follow the equipment manufacturer's procedures and recommendations.

High amperage and lazy performance would suggest an excessively tight engine, friction in the starter or starter drive, grounded starter field or armature.

Normal amperage and lazy performance suggest high resistance or possibly poor connections somewhere in the starter circuit.

Low amperage and lazy or no performance suggest battery condition poor, bad cables or connections along the line.

Voltage Drop On Grounded Side

With a voltmeter on the 3 volt scale, without disconnecting any wires, connect negative test lead of the voltmeter to a prod secured in the grounded battery post. The positive test lead is connected to a cleaned, bare metal portion of the starter motor housing. Close the starter switch and note the voltmeter reading. If the reading is the same as battery reading, the ground circuit is open somewhere between the battery and the starter. In many cases the reading will be very small. The reading shown will indicate voltage drop (loss) between battery ground post and starter housing. The drop should not exceed 0.2 volts. If the voltage drop is above the specified amount, the next step is to isolate and correct the cause. It can be a bad cable or connection anywhere in the battery-to-starter ground circuit. A check of this type should progress along the various points of possible trouble, between the battery ground post and the starter motor housing, until the trouble spot has been located.

Voltage Drop On Battery Side

Bad starter cranking may result from poor connections or faulty components of the battery or hot phase of the starter motor circuit. To check this phase of the circuit, without disconnecting any wires, connect a lead of a voltmeter to a prod secured in the hot post of the battery and the other voltmeter lead to the field terminal of the starting motor. The meter should be set to the 16–20 volt scale. Before closing the starter switch, the voltmeter reading will be that of the battery. After closing the starter switch, change the selector on the voltmeter to the 3 volt scale. With a jumper wire between the relay battery terminal and the relay starter switch terminal, crank the engine. If the starting motor cranks the engine, the relay (solenoid) is operating.

While the engine is being cranked, watch the voltmeter. It should not register more than 0.5 volts. If more than this, check each part of the circuit for voltage drop to isolate the trouble, (high resistance).

Without disturbing the voltmeter-to-battery hook-up, move the free voltmeter lead to the battery terminal of the relay (solenoid) and crank the engine. The voltmeter should show no more than 0.1 volts.

If this reading is correct, move the same voltmeter lead to the starting motor terminal of the relay (solenoid).

While the engine is being cranked, the voltmeter should show no more than 0.3 volts. If it does, the trouble lies in the relay.

If the reading is correct, the trouble is in the cable or connections between the relay and the starting motor.

DIAGNOSIS

Starter Won't Start Engine

1. Dead battery.
2. Open starter circuit, such as:
   a. Broken or loose battery cables
   b. Inoperative starter motor solenoid
   c. Broken or loose wire from starter switch-to-solenoid
   d. Poor solenoid or starter ground
   e. Faulty starter switch
3. Defective starter internal circuit, such as:
   a. Dirty or burned commutator
   b. Stuck, worn or broken brushes
   c. Open or shorted armature
   d. Open or grounded fields
4. Starter motor mechanical faults, such as:
   a. Jammed armature end bearings
   b. Faulty bearings, allowing armature to rub fields
   c. Bent shaft
   d. Broken starter housing
   e. Faulty starter worm or drive mechanism
   f. Faulty starter drive or flywheel driven gear

5. Engine hard or impossible to crank such as:
   a. Hydrostatic lock, water in combustion chamber
   b. Crankshaft seizing in bearings
   c. Piston or ring seizing
   d. Bent or broken connecting rod
   e. Seizing of connecting rod bearing
   f. Flywheel jammed or broken
6. Starter spins free, won't engage such as:
   a. Sticking or broken drive mechanism

CHRYSLER CORPORATION

Chrysler Reduction Gear Starter

OVERHAUL

Disassembly and Assembly

1. Support assembly in a soft jawed vise; be careful not to distort or damage the die cast aluminum.
2. Remove the through bolts and the end housing.
3. To disassemble the starter motor, perform the following procedures:
   a. Carefully pull the armature up and out of the gear housing and the starter frame and field assembly.

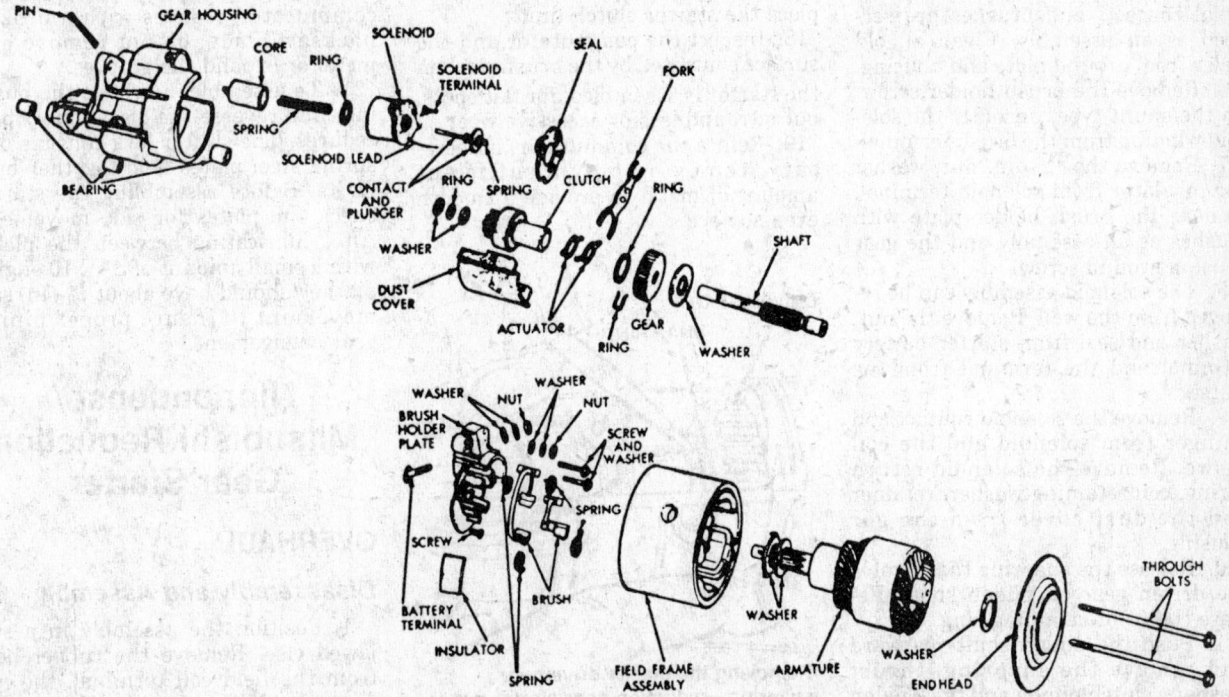

Exploded view of the reduction gear starter motor—Chrysler

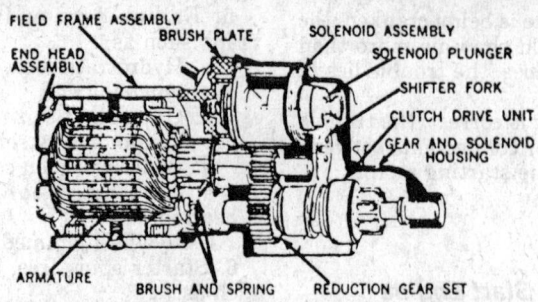

Cross-sectional view of the reduction gear starter motor — Chrysler

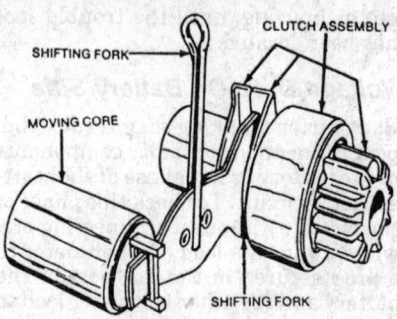

Shift fork and clutch arrangement — reduction gear starter motor — Chrysler

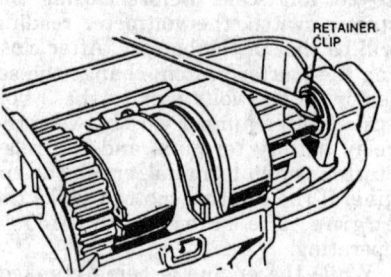

Removing the starter drive gear retaining ring — reduction gear starter motor — Chrysler

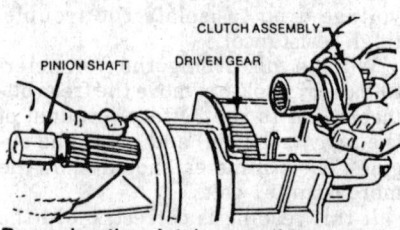

Removing the clutch assembly — reduction gear starter motor — Chrysler

b. Remove the steel and fiber thrust washer.

c. Carefully pull the frame and field assembly up enough to expose the terminal screw and the solder connection of the shunt field at the brush terminal.

d. Place 2 wood blocks between the starter frame and starter gear housing to facilitate removal of the terminal screw.

e. Unsolder the shunt field wire at the brush terminal.

NOTE: The starting motors have the wire of the shunt field coil soldered to the brush terminal. A pair of brushes are connected to this terminal. Another pair of brushes are attached to the series field coils by means of a terminal screw.

4. Support the brush terminal with a finger behind terminal and remove the screw. Unsolder the shunt field coil lead from the brush terminal and housing.

5. The brush holder plate (with terminal contact) and brushes are serviced as an assembly. Clean all old sealer from around plate and housing.

6. Remove the brush holder screw. On the shunt type, unsolder the solenoid winding from the brush terminal.

7. Remove the $\frac{11}{32}$ in. nut, washer and insulator from solenoid terminal. Remove the brush holder plate with brushes as an assembly and the gear housing ground screw.

8. The solenoid assembly can be removed from the well. Remove the nut, washer and seal from starter battery terminal and the terminal from the plate.

9. Remove the solenoid contact and plunger from solenoid and the coil sleeve. Remove the solenoid return spring, coil retaining washer, retainer and the dust cover from the gear housing.

10. Release the snapring that locates the driven gear on pinion shaft. Release the front retaining ring.

11. Push the pinion shaft rearward and remove the snapring, thrust washers, clutch/pinion and the 2 nylon shift fork actuators.

12. Remove the driven gear and friction washer. Pull the shifting fork forward and remove the moving core.

13. Remove the fork retaining pin and shifting fork assembly. The gear housing with bushings is serviced as an assembly.

14. Brushes that are worn to ½ the length of new brushes or oil soaked, should be replaced.

15. When resoldering the shunt field and solenoid lead, make a strong low resistance connection using a high temperature solder and resin flux; do not use acid or acid core solder. Be careful not to break the shunt field wire units when removing and installing the brushes.

16. Do not immerse the starter clutch unit in a cleaning solvent. The outside of the clutch and pinion must be cleaned with a cloth so the lubricant is not washed from inside the clutch.

17. Rotate the pinion. The pinion gear should rotate smoothly and in 1 direction only. If the starter clutch unit does not function properly or the pinion is worn, chipped or burred, replace the starter clutch unit.

18. Inspect the commutator and the surface contacted by the brushes when the starter is assembled, for flat spots, out of roundness or excessive wear.

19. Reface the commutator, if necessary, removing only a sufficient amount of metal to provide a smooth even surface.

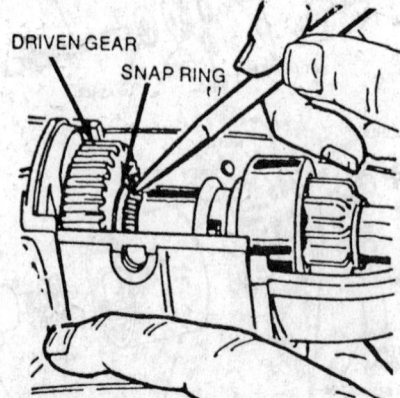

Removing the starter drive gear snapring — reduction gear starter motor — Chrysler

20. Using light pressure scrape the commutator grooves with a broken hacksaw blade; do not remove any metal or expand the grooves.

21. To assemble, lubricate the bushings and reverse the disassembly procedures. The shifter fork consists of 2 spring steel plates held together by 2 rivets. Before assembling the starter check the plates for side movement. After lubricating between the plates with a small amount of SAE 10 engine oil they should have about $\frac{1}{16}$ in. side movement to insure proper pinion gear engagement.

Nippondenso/ Mitsubishi Reduction Gear Starter

OVERHAUL

Disassembly and Assembly

1. Position the assembly in a soft jawed vise. Remove the rubber boot from the field coil terminal, the nut from the field coil terminal stud and the field coil terminal from the stud.

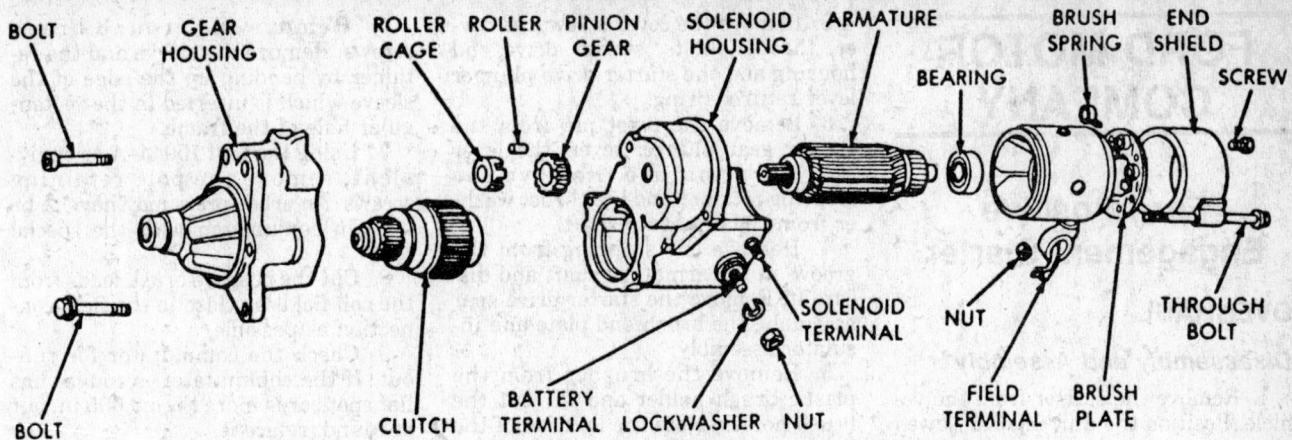

BOLT · GEAR HOUSING · ROLLER CAGE · ROLLER · PINION GEAR · SOLENOID HOUSING · ARMATURE · BEARING · BRUSH SPRING · END SHIELD · SCREW

BOLT · CLUTCH · BATTERY TERMINAL · SOLENOID TERMINAL · LOCKWASHER · NUT · NUT · FIELD TERMINAL · BRUSH PLATE · THROUGH BOLT

Exploded view of the Nippondenso reduction gear starter motor—Chrysler

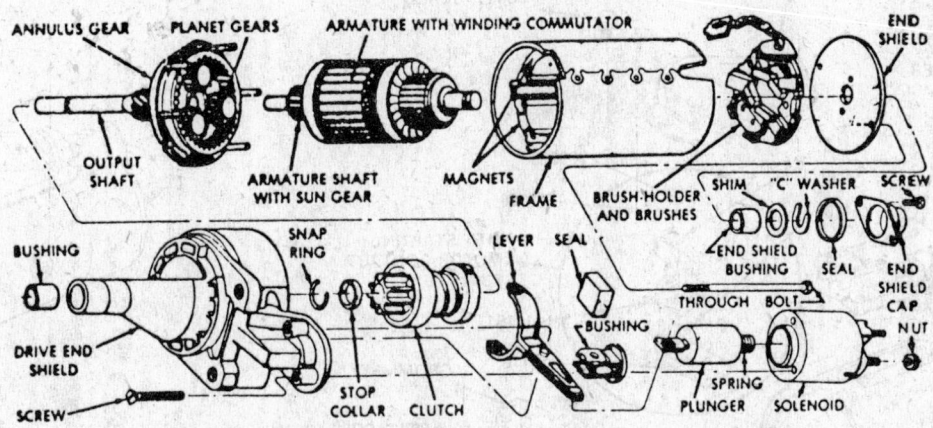

ANNULUS GEAR · PLANET GEARS · ARMATURE WITH WINDING COMMUTATOR · END SHIELD

OUTPUT SHAFT · ARMATURE SHAFT WITH SUN GEAR · MAGNETS · FRAME · BRUSH HOLDER AND BRUSHES · SHIM · "C" WASHER · SCREW

BUSHING · SNAP RING · LEVER · SEAL · END SHIELD BUSHING · SEAL · END SHIELD CAP

DRIVE END SHIELD · BUSHING · THROUGH BOLT · NUT

SCREW · STOP COLLAR · CLUTCH · SPRING · PLUNGER · SOLENOID

Exploded view of the Bosch reduction gear starter—Chrysler

2. Remove the through bolts and the splash shield. Remove the end shield screws from the brush plate and the end shield.

3. Slide the brushes from the holders. Pry the retaining springs back for access and remove the brush plate.

4. Slide the armature from the starter housing. Remove the starter housing from the gear housing and the solenoid terminal cover.

5. Remove the solenoid terminal nut/washer, the battery terminal nut/washer and the solenoid terminal assembly from the terminal posts.

6. Remove the solenoid terminal and the battery terminal from the insulator.

7. Remove the solenoid cover screws, the solenoid cover, the seal, the solenoid plunger from the housing and the plunger spring.

8. Remove the gear housing-to-solenoid screws and separate the gear housing from the solenoid housing.

9. Remove the reduction gear and clutch assembly from the gear housing.

10. Remove the reduction gear, pin-ion gear, retainer and roller assembly from the gear housing.

11. Inspect and clean all parts, as required. Repair or replace defective parts as required. Brushes that are worn less than ½ the length of new brushes or oil soaked should be replaced.

12. To assemble, reverse the disassembly procedures.

Bosch Reduction Starter

OVERHAUL

Disassembly and Assembly

1. Position the assembly in a soft jawed vise. Remove the field terminal nut, the terminal and the washer.

2. Remove the solenoid-to-starter screws. Work the solenoid from the shift fork and remove the solenoid from the starter.

3. Remove the starter end shield bushing cap screws, the starter end shield bushing cap, the end shield bushing and C-washer.

4. Remove the starter end shield bushing washer and seal.

5. Remove the starter through bolts, the starter end shield and the brush plate.

6. Slide the field frame from the starter and over the armature. Remove the armature assembly from the drive end housing.

7. Remove the rubber seal from the drive end housing. Remove the starter drive gear train.

8. Remove the dust plate. Press the stop collar from the snapring. Using snapring pliers, loosen the snapring.

9. Remove the output shaft snapring, the clutch stoping collar and the clutch assembly from the starter.

10. Remove the clutch shift lever bushing, the clutch shift lever and the C-clip retainer.

11. Remove the retaining washer, the sun and the planetary gears from the annulus gear.

12. To assemble, lubricate the necessary parts and reverse the disassembly procedures. Replace all defective components as required.

FORD MOTOR COMPANY

Ford Positive Engagement Starter

OVERHAUL

Disassembly and Assembly

1. Remove the starter from the vehicle. Position the unit in a soft jawed vise.

2. Remove the cover screw, the cover, through bolts, starter drive end housing and the starter drive plunger lever return spring.

3. Remove the pivot pin from the starter gear plunger lever, the lever and the armature. Remove the stopring retainer and the thrust washer from the armature shaft.

4. Remove the stopring from the groove in the armature shaft and discard it. Remove the starter drive gear assembly, the brush end plate and insulator assembly.

5. Remove the brushes from the plastic brush holder and lift out the brush holder. Note the location of the holder in relation to the end terminal.

6. Remove the ground brush screws. Remove the sleeve and the retainer by bending up the edge of the sleeve which is inserted in the rectangular hole of the frame.

7. Using the tool 10044–A or equivalent, remove the pole retaining screws. An arbor press may have to be used in conjunction with the special tool.

8. Cut the positive brush leads from the coil fields as close to the field connection as possible.

9. Check the commutator for runout. If the commutator is rough, has flat spots or is more than 0.005 in. out of round, reface it.

10. Inspect the armature shaft and

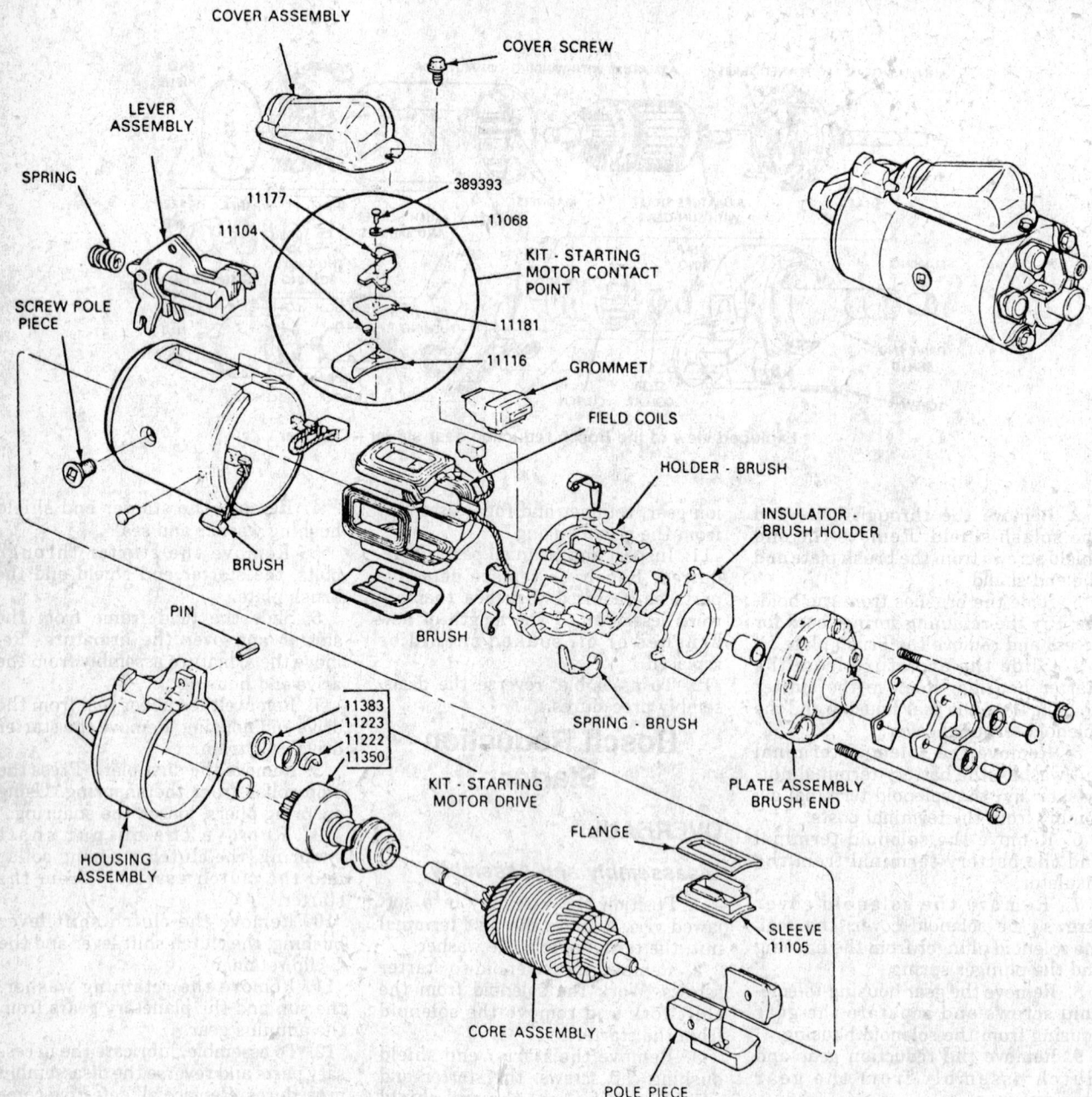

Exploded view of the Ford positive engagement starter — 4 inch plunger pole

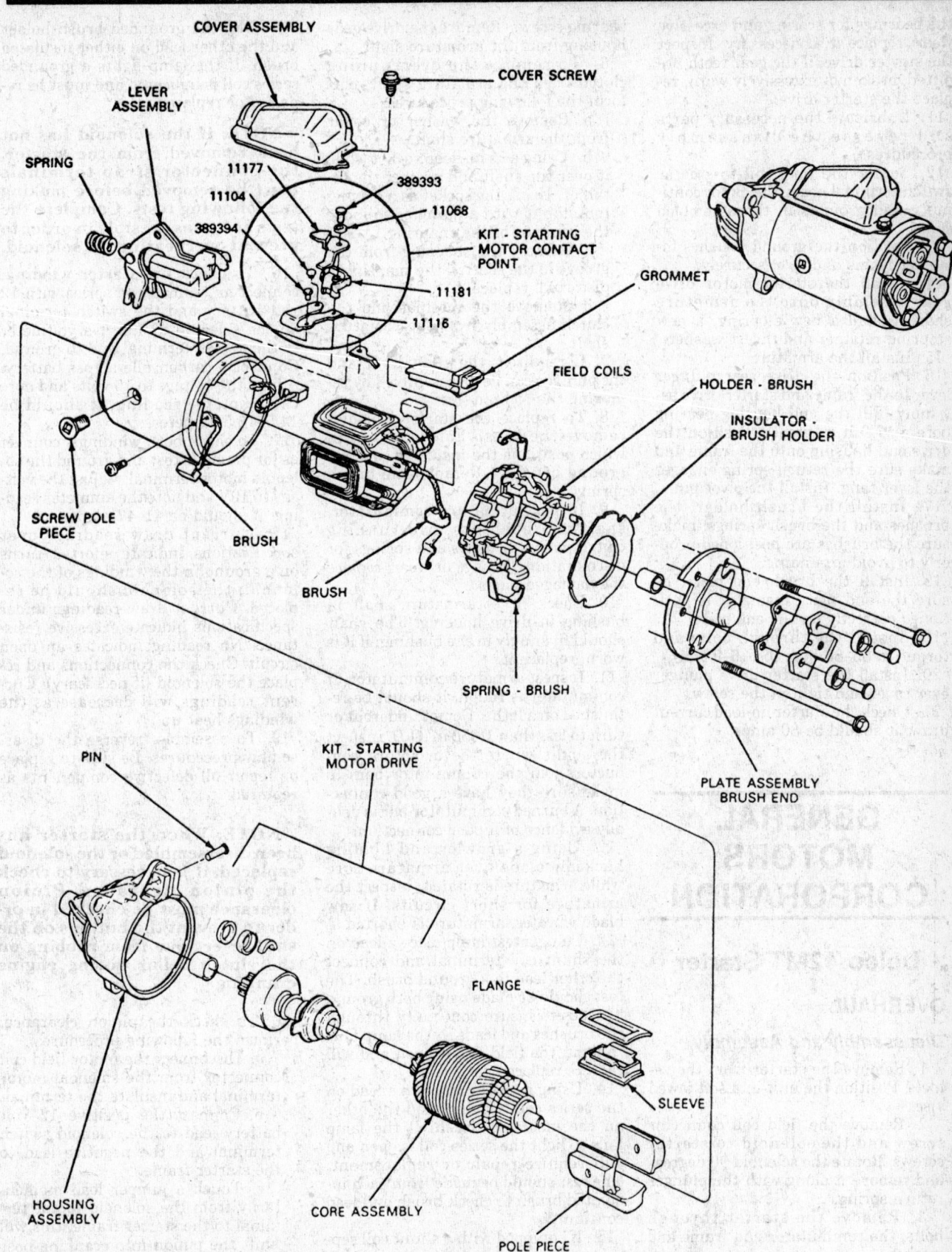

COVER ASSEMBLY

COVER SCREW

LEVER ASSEMBLY

SPRING

11177
11104
389394
389393
11068

KIT - STARTING MOTOR CONTACT POINT

11181
11116

GROMMET

FIELD COILS

HOLDER - BRUSH

INSULATOR - BRUSH HOLDER

SCREW POLE PIECE

BRUSH

BRUSH

SPRING - BRUSH

PLATE ASSEMBLY BRUSH END

PIN

KIT - STARTING MOTOR DRIVE

FLANGE

SLEEVE

HOUSING ASSEMBLY

CORE ASSEMBLY

POLE PIECE

Exploded view of the Ford positive engagement starter — 4½ inch plunger pole

the bearings for scoring and excessive wear; replace it, if necessary. Inspect the starter drive; if the gear teeth are pitted, broken or excessively worn, replace the starter drive.

11. Lubricate the necessary parts and reverse the disassembly procedures.

12. Solder the field coil-to-starter switch terminal posts. Check for continuity and grounds in the assembled coils.

13. Position the ground brushes-to-starter frame and rivet securely.

14. Install the starter motor drive gear assembly onto the armature shaft. Install a new stopring, a new stopring retainer and thrust washer.

15. Install the armature.

16. Position the drive gear plunger lever to the frame and starter drive assembly. Fill the end housing bearing bore a ¼ full or grease. Position the drive end housing onto the frame and make sure the return spring engages the lever tang. Install the pivot pin.

17. Install the brush holder, the brushes and the brush springs; make sure the brushes are positioned properly to avoid grounding.

18. Install the brush end plate; be sure the end plate insulator is positioned correctly on the end plate.

19. Install the through bolts and torque to 55–80 inch lbs. (6–9 Nm).

20. Install the starter drive plunger ever cover and tighten the screw.

21. Check the starter no-load current draw; it should be 80 amps.

GENERAL MOTORS CORPORATION

Delco 42MT Starter

OVERHAUL

Disassembly and Assembly

1. Remove the starter from the vehicle. Position the unit in a soft jawed vise.

2. Remove the field coil connector screw and the solenoid-to-starter screws. Rotate the solenoid 90 degrees and remove it along with the plunger return spring.

3. Remove the starter through bolts, the commutator end frame and washer.

4. Remove the field frame assembly from the drive gear housing.

5. If equipped, remove the center

bearing screws. Remove the drive gear housing from the armature shaft.

6. To remove the overrunning clutch from the armature shaft, perform the following procedures:

   a. Remove the washer or collar from the armature shaft.

   b. Using a ⅝ in. deep socket, slide it over the shaft and against the retainer. Using the socket as a driving tool, tap it with a hammer to move the retainer of the snapring.

   c. Remove the snapring from the groove in the shaft; if the snapring is distorted, replace it.

   d. Remove the retainer and the clutch assembly from the armature shaft.

7. If required, the shaft lever and the plunger can be disassembled by removing the roll pin.

8. To replace the starter brushes, remove the brush holder pivot pin which positions the insulated and the ground brushes. Remove the brush spring.

9. Inspect armature commutator, shaft and bushings, overrunning clutch pinion, brushes and springs for discoloration, damage or wear; replace the damaged parts.

10. Check fit of armature shaft in bushing in drive housing. The shaft should fit snugly in the bushing; if it is worn, replace it.

11. Inspect armature commutator. If commutator is rough, it should be refinished on a lathe. Do not undercut or turn to less than 1.650 in. O.D. Inspect the points where the armature conductors join the commutator bars to make sure they have a good connection. A burned commutator bar is usually evidence of a poor connection.

12. Using a growler and holding hacksaw blade over armature core while armature is rotated, inspect the armature for short circuits. If saw blade vibrates, armature is shorted.

13. Using a test lamp place a lead on the shunt coil terminal and connect the other lead to a ground brush. The test should be made using both ground brushes to insure continuity through the brushes and leads. If the lamp fails to light, the field coil is open and will require replacement.

14. Using a test lamp place a lead on the series coil terminal and the other on the insulated brush. If the lamp fails to light the series coil is open and will require repair or replacement. The test should be made from each insulated brush to check brush and lead continuity.

15. If equipped with a shunt coil separate the series and shunt coil strap terminals during the test. Do not allow the strap terminals to touch case or other ground. Using a test lamp place

a lead on the grounded brush holder and the other lead on either insulated brush. If the lamp lights a grounded series coil is indicated and must be repaired or replaced.

NOTE: If the solenoid has not been removed from the starter, the connector strap terminals must be removed before making the following tests. Complete the tests as fast as possible in order to prevent overheating the solenoid.

16. To check the starter winding, connect an ammeter in series with 12 volt battery and the switch terminal on the solenoid. Connect a voltmeter to the switch terminal and to ground. Connect a carbon pile across battery. Adjust the voltage to 10 volts and note the ammeter reading; it should be 14.5–16.5 amperes.

17. To check both windings, connect as for previous test and ground the solenoid motor terminal. Adjust the voltage to 10V and note the ammeter reading; it should be 41–47 amperes.

18. Current draw readings over specifications indicate shorted turns on a ground in the windings of the solenoid; the solenoid should be replaced. Current draw readings under specifications indicate excessive resistance. No reading indicates an open circuit. Check the connections and replace the solenoid (if necessary). Current readings will decrease as the windings heat up.

19. To assemble, reverse the disassembly procedures. Be sure to replace or repair all defective components as required.

NOTE: When the starter has been disassembled or the solenoid replaced, it is necessary to check the pinion clearance. Pinion clearance must be checked in order to prevent the buttons on the shift lever yoke from rubbing on the clutch collar during engine cranking.

20. To check the pinion clearance, perform the following procedures:

   a. Disconnect the motor field coil connector from the solenoid motor terminal and insulate the terminal.

   b. Connect the positive 12 volt battery lead to the solenoid switch terminal and the negative lead to the starter frame.

   c. Touch a jumper lead momentarily from the solenoid motor terminal to the starter frame; this will shift the pinion into cranking position and remain there until the battery is disconnected.

   d. Using a feeler gauge, push the pinion back as far as possible and

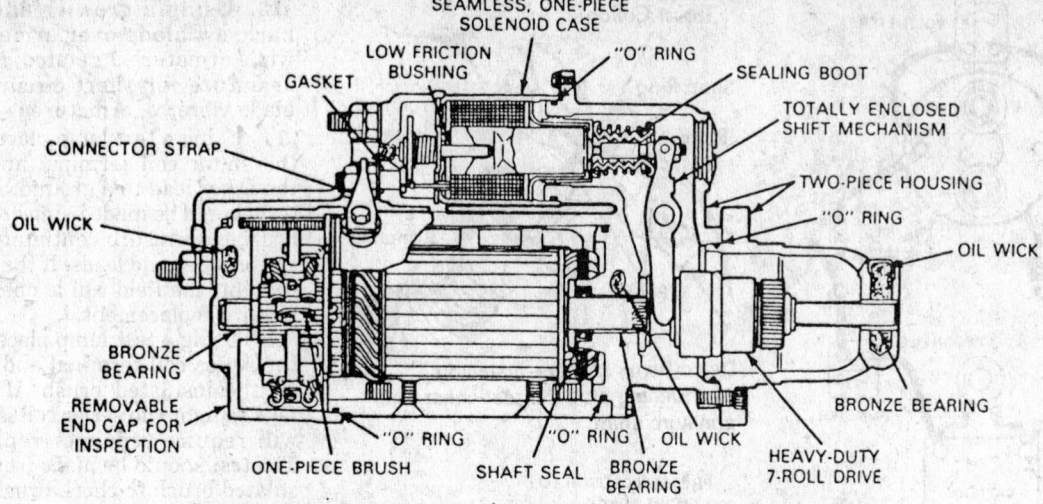

Cross-sectional view of the Delco 42 MT starter—Ford diesel engine

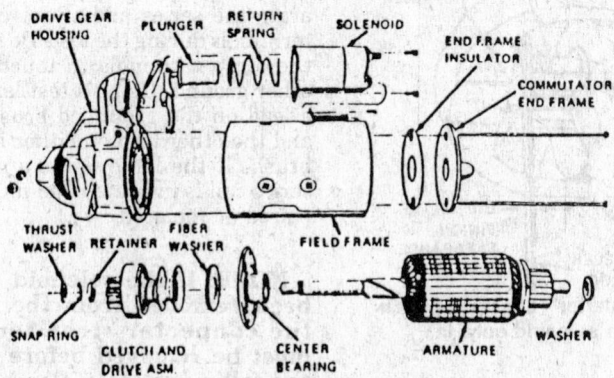

Exploded view of a typical Delco-Remy starter motor

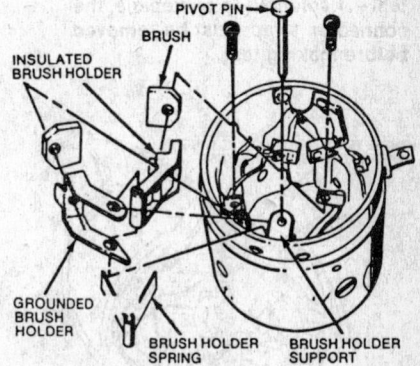

Delco-Remy starter brush replacement—all except 5MT starter

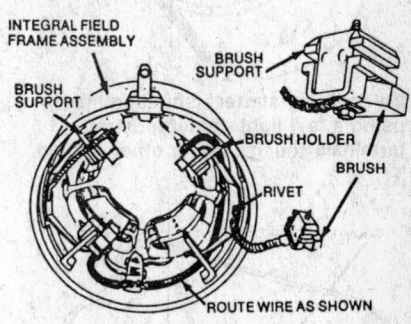

Delco-Remy 5MT starter-brush replacement

check the clearance; the clearance should be 0.010–0.140 in.

e. Pinion clearance adjustment is not provided on the starter motor. If the clearance does not fall within limits check for improper installation and replace all worn parts.

Delco 5MT, 10MT, 27MT and 28MT Starters

NOTE: In 1989, the identification for the 5MT starter was converted to the SD200/SD250 starters and the 10MT was converted to the SD300 starter. The 28MT is used with the diesel engine.

OVERHAUL

Disassembly and Assembly

1. Remove the starter from the vehicle. Position the unit in a soft jawed vise.

2. Remove the field coil connector screw and the solenoid-to-starter screws. Rotate the solenoid 90 degrees and remove it along with the plunger return spring.

3. Remove the starter through bolts, the commutator end frame and washer.

4. Remove the field frame assembly from the drive gear housing.

5. If equipped, remove the center bearing screws. Remove the drive gear housing from the armature shaft.

6. To remove the overrunning clutch from the armature shaft, perform the following procedures:

a. Remove the washer or collar from the armature shaft.

b. Using a ⅝ in. deep socket, slide it over the shaft and against the retainer. Using the socket as a driving tool, tap it with a hammer to move the retainer of the snapring.

c. Remove the snapring from the groove in the shaft; if the snapring is distorted, replace it.

d. Remove the retainer and the clutch assembly from the armature shaft.

7. If required, the shaft lever and the plunger can be disassembled by removing the roll pin.

8. To replace the starter brushes, remove the brush holder pivot pin which positions the insulated and the ground brushes. Remove the brush spring.

9. On 5MT starters, to replace the brushes remove the screw from the brush holder and separate the brushes from the holder.

10. Inspect armature commutator, shaft and bushings, overrunning clutch pinion, brushes and springs for discoloration, damage or wear; replace the damaged parts.

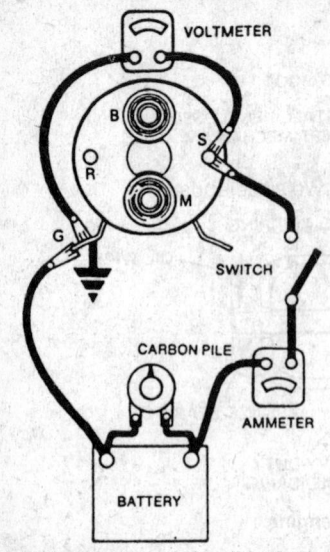

Delco-Remy starter—solenoid winding test—if solenoid is on vehicle, the connector strap must be removed before making test

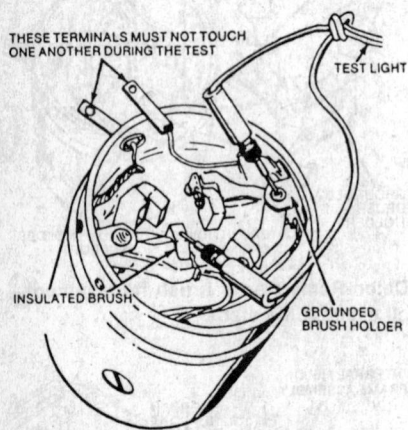

Delco-Remy starter—shunt coil test—using a test light do not let strap terminals touch case or other ground

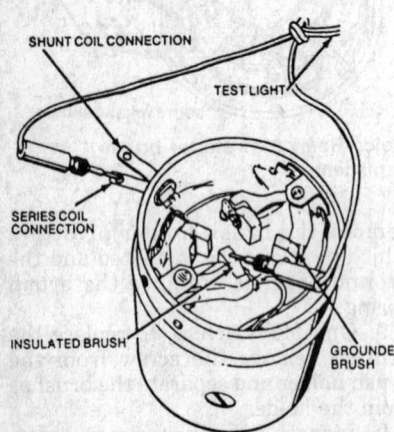

Delco-Remy starter—coil test—using a test light if lamp fails to light, the series coil is open and must be repaired or replaced

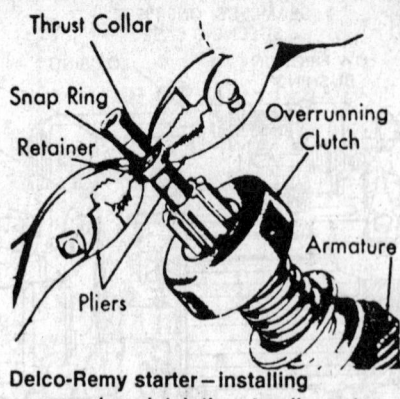

Delco-Remy starter—installing overrunning clutch thrust collar onto armature shaft

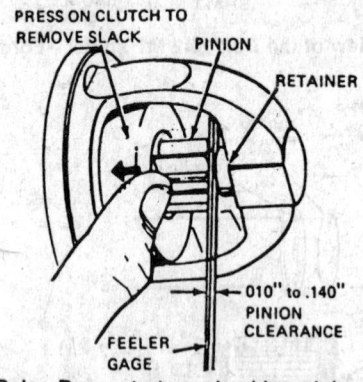

Delco-Remy starter—checking pinion clearance with solenoid only, in operation

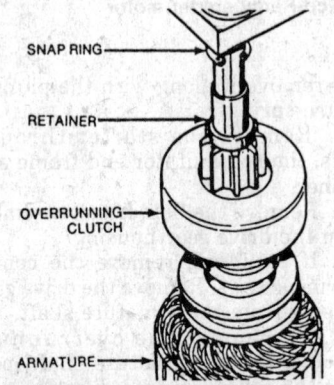

Delco-Remy starter—installing overrunning clutch thrust collar snapring onto armature shaft

11. Check fit of armature shaft in bushing in drive housing. The shaft should fit snugly in the bushing; if it is worn, replace it.

12. Inspect armature commutator. If commutator is rough, it should be refinished on a lathe. Do not undercut or turn to less than 1.650 in. O.D. Inspect the points where the armature conductors join the commutator bars to make sure they have a good connection. A burned commutator bar is usually evidence of a poor connection.

13. Using a growler and holding hacksaw blade over armature core while armature is rotated, inspect the armature for short circuits. If saw blade vibrates, armature is shorted.

14. Using a test lamp place a lead on the shunt coil terminal and connect the other lead to a ground brush. The test should be made using both ground brushes to insure continuity through the brushes and leads. If the lamp fails to light, the field coil is open and will require replacement.

15. Using a test lamp place a lead on the series coil terminal and the other on the insulated brush. If the lamp fails to light the series coil is open and will require repair or replacement. The test should be made from each insulated brush to check brush and lead continuity.

16. If equipped with a shunt coil separate the series and shunt coil strap terminals during the test. Do not allow the strap terminals to touch case or other ground. Using a test lamp place a lead on the grounded brush holder and the other lead on either insulated brush. If the lamp lights a grounded series coil is indicated and must be repaired or replaced.

NOTE: If the solenoid has not been removed from the starter, the connector strap terminals must be removed before making the following tests. Complete the tests as fast as possible in order to prevent overheating the solenoid.

17. To check the starter winding, connect an ammeter in series with 12V battery and the switch terminal on the solenoid. Connect a voltmeter to the switch terminal and to ground. Connect a carbon pile across battery. Adjust the voltage to 10V and note the ammeter reading; it should be 14.5–16.5 amperes.

18. To check both windings, connect as for previous test and ground the solenoid motor terminal. Adjust the voltage to 10V and note the ammeter reading; it should be 41–47 amperes.

19. Current draw readings over specifications indicate shorted turns on a ground in the windings of the solenoid; the solenoid should be replaced. Current draw readings under specifications indicate excessive resistance. No reading indicates an open circuit. Check the connections and replace the solenoid (if necessary). Current readings will decrease as the windings heat up.

20. To assemble, reverse the disassembly procedures. Be sure to replace or repair all defective components as required.

NOTE: When the starter has been disassembled or the solenoid replaced, it is necessary to check the pinion clearance. Pinion clearance must be checked in order to prevent the buttons on the shift lever yoke from rubbing on the clutch collar during engine cranking.

21. To check the pinion clearance, perform the following procedures:

a. Disconnect the motor field coil connector from the solenoid motor terminal and insulate the terminal.

b. Connect the positive 12 volt battery lead to the solenoid switch terminal and the negative lead to the starter frame.

c. Touch a jumper lead momentarily from the solenoid motor terminal to the starter frame; this will shift the pinion into cranking position and remain there until the battery is disconnected.

d. Using a feeler gauge, push the pinion back as far as possible and check the clearance; the clearance should be 0.010–0.140 in.

e. Pinion clearance adjustment is not provided on the starter motor. If the clearance does not fall within limits check for improper installation and replace all worn parts.

Delco Permanent Magnet Gear Reduction (PMGR) Starter

OVERHAUL

Disassembly and Assembly

1. Remove the starter from the ve-hicle. Position the unit in a soft jawed vise.

2. Remove the field coil screw, the field frame through bolts and separate the field frame assembly from the drive gear assembly. Separate the armature and the commutator end frame from the field frame.

3. Remove the solenoid screws and the solenoid from the drive housing.

4. Remove the retaining ring, shift lever shaft and housing through bolts. Separate the drive assembly, drive housing and gear assembly.

5. To remove the overrunning clutch from the armature shaft, perform the following procedures:

a. Remove the washer or collar from the armature shaft.

b. Using a 5/8 in. deep socket, slide it over the shaft and against the retainer. Use the socket as a driving tool, tap the socket with a hammer to move the retainer off of the snapring.

c. Remove the snapring from the groove in the shaft; if the snapring is distorted, replace it.

d. Remove the retainer and the clutch assembly from the armature shaft.

6. To replace the starter brushes, remove the brush holder pivot pin which positions the insulated and the ground brushes. Remove the brush spring.

7. Inspect armature commutator, shaft and bushings, overrunning clutch pinion, brushes and springs for discoloration, damage or wear; replace the damaged parts (if necessary). Check the armature shaft fit in drive housing bushing; the shaft should fit snugly in the bushing. If the bushing is worn, it should be replaced.

8. Inspect armature commutator. If commutator is rough, it should be re-finished on a lathe; do not undercut or turn to less than 1.650 in. O.D. Inspect the points where the armature conductors join the commutator bars to make sure they have a good connection. A burned commutator is usually evidence of a poor connection.

9. Using a growler and holding hacksaw blade over armature core while armature is rotated, check the armature for short circuits; if the saw blade vibrates, the armature is shorted.

10. Using a test lamp, place a lead on the shunt coil terminal and the other lead to a ground brush. The test should be made from both ground brushes to insure continuity through both brushes and leads. If the lamp fails to light, the field coil is open and will require replacement.

11. Using a test lamp, place a lead on the series coil terminal and the other lead on the insulated brush. If the lamp fails to light, the series coil is open and will require repair or replacement. The test should be made from each insulated brush to check brush and lead continuity.

12. If equipped with a shunt coil, separate the series and shunt coil strap terminals during this test; do not allow the strap terminals to touch the case or other ground. Using a test lamp, place a lead on the grounded brush holder and the other lead on either insulated brush. If the lamp lights, a grounded series coil is indicated and must be repaired or replaced.

NOTE: If the solenoid has not been removed from the starter, the connector strap terminals must be removed before making the following tests. Complete the tests as fast as possible in order to prevent overheating the solenoid.

9. Clutch
12. Shift lever
13. Plunger
16. Solenoid
19. Bearing
22. Armature
31. TIG welds
32. Return spring
33. Planetary gear reduction assembly
34. Permanent magnet field
35. Brushes
36. Welded connections
37. Pinion stop

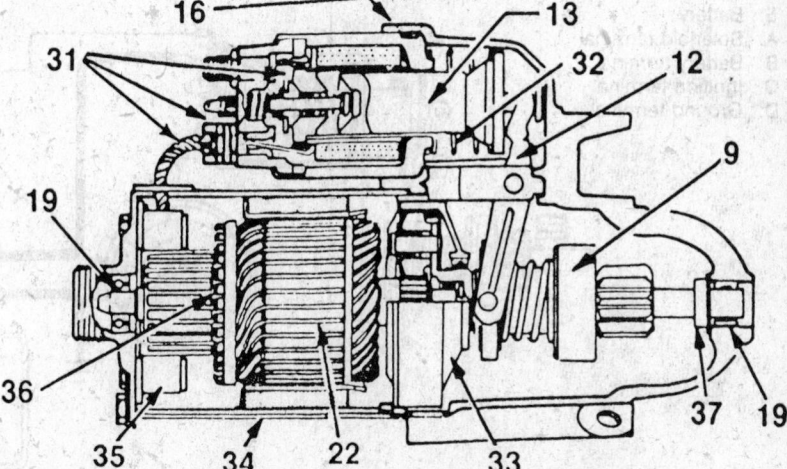

Cross-sectional view of the Permanent Magnet Gear Reduction (PMGR) starter

13. To check the starter winding, connect an ammeter in series with a 12 volt battery, the switch terminal and to ground. Connect a carbon pile across the battery. Adjust the voltage to 10 volts and note the ammeter reading; it should be 14.5–16.5 amperes.

14. To check both windings, connect as for previous test. Ground the solenoid motor terminal, adjust the voltage to 10 volts and note the ammeter reading; it should be 41–47 amperes.

15. Current draw readings above specifications indicate shorted turns or a ground in the windings of the solenoid; the solenoid should be replaced. Current draw readings under specifications indicate excessive resistance. No reading indicates an open circuit. Check the connections and replace solenoid (if necessary). Current readings will decrease as windings heat up.

16. The roller bearing in the drive housing and the roller bearings in the gear housing must be replaced (if they are dry); do not lubricate or reuse the bearings.

17. To replace the gear housing bearing, use a tube or solid cylinder that just fits inside the housing to push bearing toward the armature side. In the opposite direction, use the tube or cylinder to press bearing flush with housing.

18. To replace the gear housing driveshaft bearing, use a tube or collar that just fits inside the housing and press bearing out; press against the open end of bearing. To install a new bearing, press against the closed end, using a thin wall tube or collar that fits in space between bearing and housing.

Do not press against the flat end of the bearing; this will bend the thin metal of the bearing. As required, replace the drive housing bearing.

19. To assemble, reverse the disassembly procedures. Be sure to replace or repair all defective components as required.

NOTE: When the starter has been disassembled or the solenoid replaced, it is necessary to check the pinion clearance. The pinion clearance must be checked in order to prevent the buttons on the shift lever yoke from rubbing on the clutch collar during engine cranking.

20. To check the pinion clearance, perform the following procedures:

   a. Disconnect the motor field coil connector from the solenoid motor terminal and insulate the terminal.

   b. Connect the positive (+) 12 volt battery lead to the solenoid switch terminal and the other to the starter frame.

   c. Touch a jumper lead momentarily from the solenoid motor terminal to the starter frame; this will shift the pinion into cranking position and retain it until the battery is disconnected.

   d. Using a feeler gauge, push the pinion back as far as possible, to take up any movement, and check the clearance; the clearance should be 0.010–0.140 in.

   e. Means for adjusting pinion clearance is not provided on the starter motor. If the clearance does not fall within limits, check for im-

proper installation and replace worn parts.

JEEP

Bosch Reduction Gear Starter

OVERHAUL

Disassembly

NOTE: When performing disassembly procedures, do not stike the thin wall stator frame with a hammer or any other instrument. Do not clamp the thin wall stator frame in the jaws of a vise. This may result in damage to the permanent magnets and the stator housing. The starter may be clamped by the mounting flange only.

1. Position the starter motor assembly in a suitable holding fixture, ensuring that it is secured by the mounting flange only. Remove the field terminal nut. Remove the field coil wire (Terminal 45) from the solenoid switch terminal post. Remove the field washer.

2. Remove the solenoid switch mounting screws. Disengage the solenoid switch from the fork lever and withdraw the solenoid switch with the

1. Starter
2. Solenoid
3. Relay
4. Ignition switch
5. Battery
A. Solenoid terminal
B. Battery terminal
C. Ignition terminal
D. Ground terminal

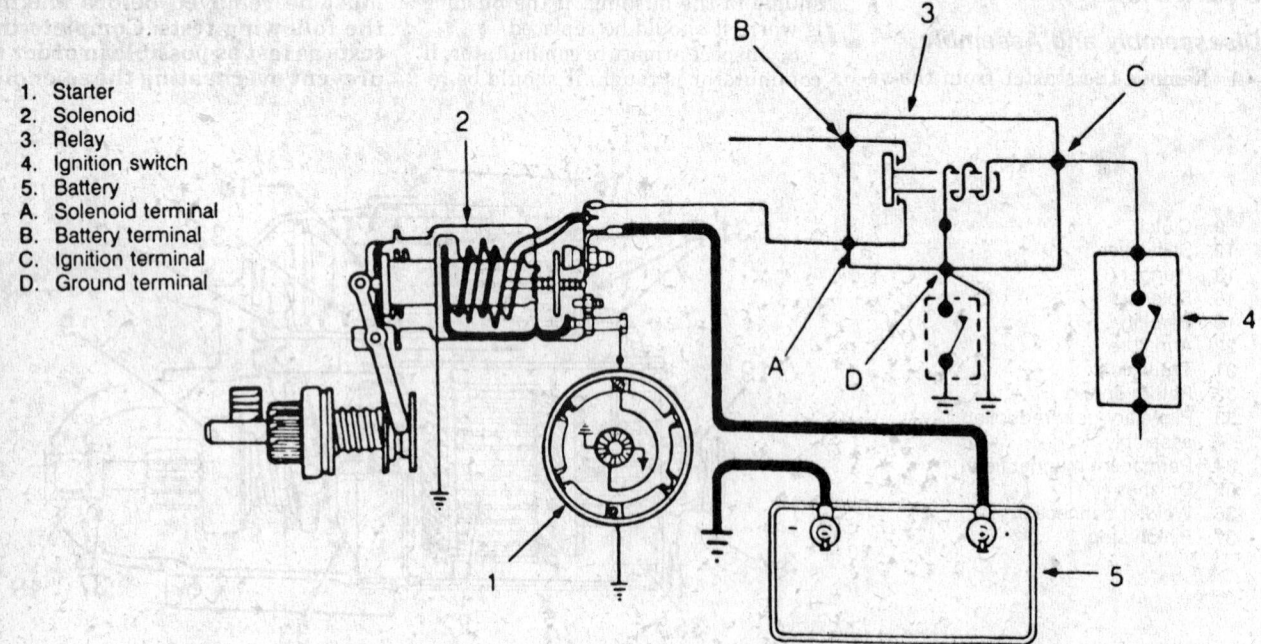

Schematic of the Bosch starter motor circuit–Jeep

1. Bushing
2. Screw
3. Shield
4. **Solenoid switch**
5. Retainer
6. Stop ring
7. Bushing
8. Overrunning clutch drive
9. Fork
10. Bearing pedestal
11. Sealing rubber
12. Planetary gear system
13. Armature
14. Stator frame
15. Brush frame
16. Gasket
17. Commutator end shield
18. Bushing
19. Seal ring
20. Shim
21. Shim
22. Retaining washer
23. Closure cap
24. Hexagon screw
25. Screw
26. Coverplate

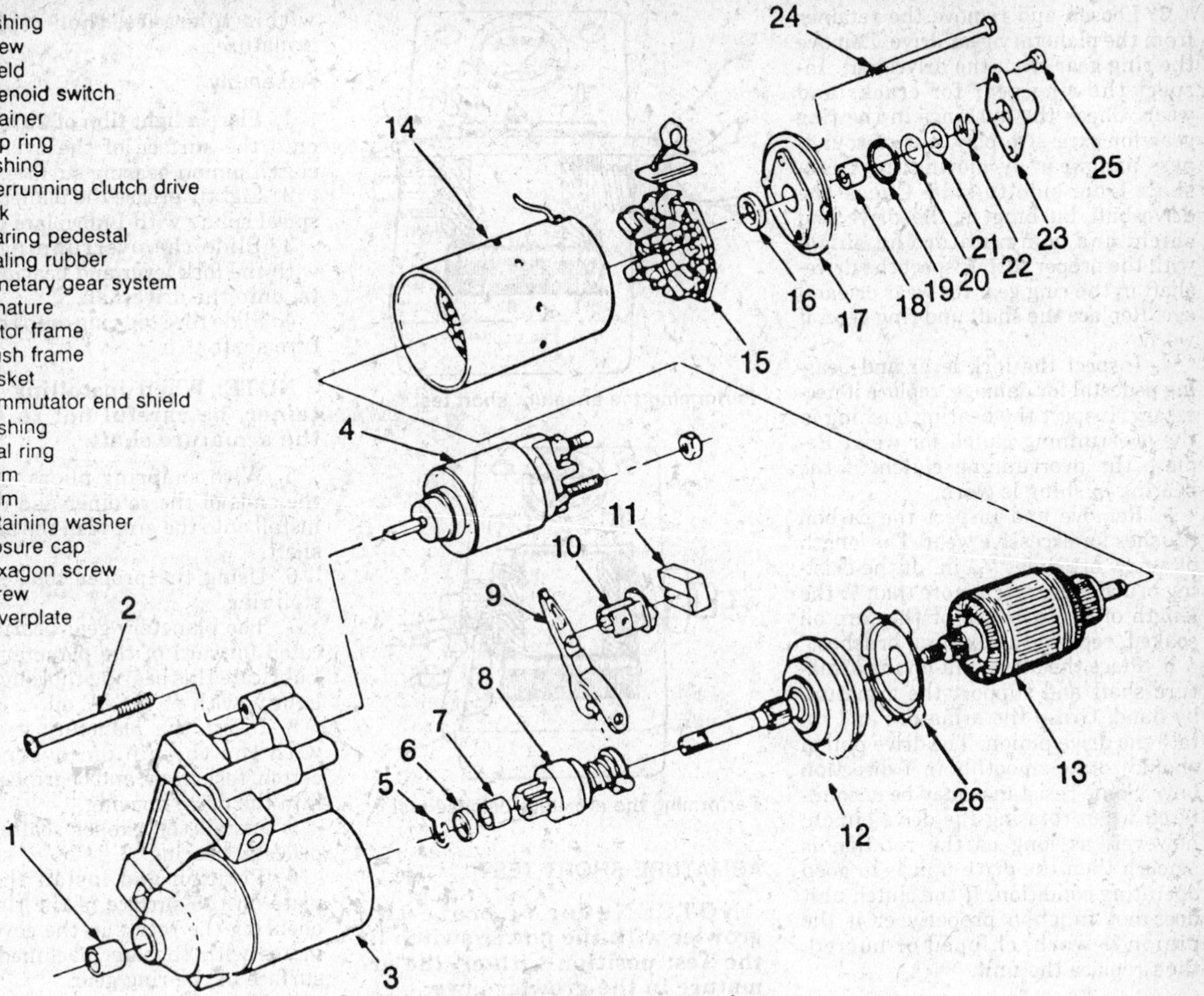

Exploded view of the Bosch starter—Jeep

armature and return spring from the housing.

3. Loosen the closure cap screws, but do not remove them.

4. Remove the stator frame-to-shield housing hexagon screws, the stator frame with the cover plate, armature and commutator end shield.

5. Remove the cover plate from the surface of the drive end bearing ring gear. Carefully press the armature with the commutator end shield out of the stator frame. At the same time, push out terminal 45 with the sealing rubber.

6. Remove the closure cap screws and remove the closure cap from the commutator end shield. Remove the retaining washer and shims from the armature shaft and remove the commutator end shield.

7. Using the proper tool, remove the brush plate from the armature shaft.

8. Remove the sealing rubber from the bearing pedestal. Remove the planetary gear system with the overruning clutch and fork assembly from the drive end shield. Mount both assem-

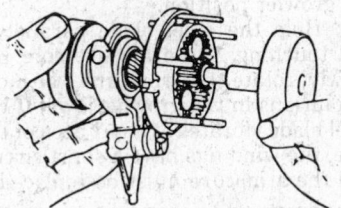

Removing the overrunning clutch and planetary gear system—Bosch starter—Jeep

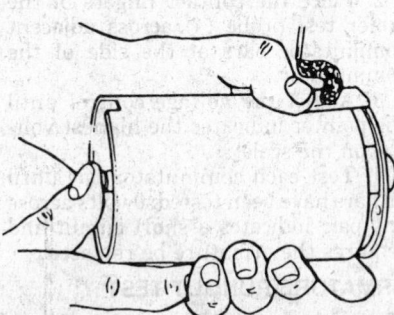

Removing the armature—Bosch starter—Jeep

blies onto a suitable mounting base in the verticle position.

9. Using the proper tool, drive the stop ring down the input shaft.

10. With snapring pliers, separate the ends of the retainer far enough to allow removal from the driveshaft without damage. If necessary, carefully remove any burrs or nicks from the ddriveshaft, otherwise damage to the drive bushing may occur.

Inspection

1. The armature windings, overrunning clutch, and relay must be cleaned with compressed air and a clean dry rag. Other parts, such as screws and the armature shaft may be cleaned with cleaning solvent. Inspect all parts, seals, bushings, and gaskets for wear and damage. Replace all worn parts as required.

2. Inspect the stator frame and permanent magnets for damage. Do not remove the magnets from the stator frame. If the stator frame or magnets are damaged, then replace the stator frame.

3. Loosen and remove the retainer from the planetary gear drive. Remove the ring gear from the driveshaft. Inspect the ring gear for cracks and wear. Check the bushings in the ring gear for excessive play, out of roundness and wear by moving the driveshaft from side-to-side. Check the driveshaft bushing in the drive end shield and commutator end shield with the proper tool. Inspect the driveshaft in the ring gear for wear or damage. Replace the shaft and ring gear, if worn.

4. Inspect the fork lever and bearing pedestal for damage, replace if necessary. Inspect the bearing bushing in the overrunning clutch for wear. Replace the overrunning clutch, if the bearing bushing is worn.

5. Remove and inspect the carbon brushes for excessive wear. The length of a new brush is $\frac{11}{16}$ in.. If the existing brushes are worn more than $\frac{1}{2}$ the length of a new brush or they are oil soaked, replace the existing brushes.

6. Place the drive unit on the armature shaft and support the armature by hand. Grasp the armature and rotate the drive pinion. The drive pinion should rotate smoothly in 1 direction only. Some resistance may be encountered when rotating the drive pinion; however, as long as the rotation is smooth then the drive unit is in good operating condition. If the clutch unit does not function properly or if the pinion is worn, chipped or burred, then replace the unit.

ARMATURE GROUND TEST

1. Position the armature in the growler jaws and turn the power switch-to-**TEST** position.

2. Touch a test probe-to-armature core (1). Touch the other test probe-to-commutator bar (2), 1 at a time and observe the test lamp. The test lamp should not light. If the test lamp lights on any bar, the armature has a short circuit-to-ground and must be replaced.

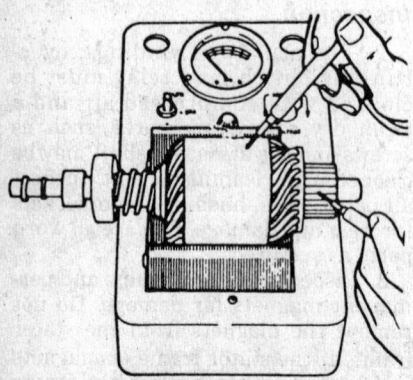

Performing the armature ground test

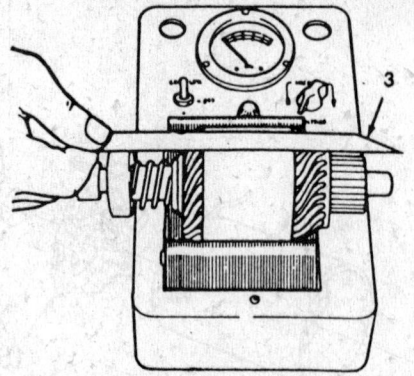

Performing the armature short test

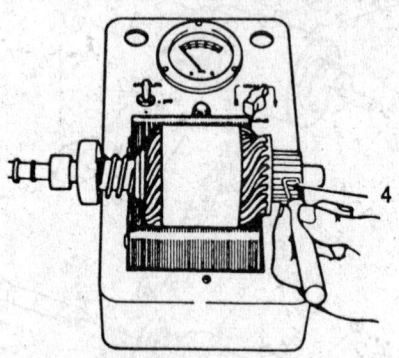

Performing the armature balance test

ARMATURE SHORT TEST

NOTE: Never operate the growler with the power switch in the Test position without the armature in the growler jaws.

1. Place the armature in the growler jaws and turn the power switch to the growler position.

2. Hold the steel blade parallel to and touching, the armature core (3). Slowly, rotate the armature 1 or more revolutions in the growler jaws. If the steel blade vibrates at any 1 area of the core, the windings have a short circuit and the armature must be replaced.

ARMATURE BALANCE TEST

1. Place the armature in the growler jaws and turn the power switch to the growler position.

2. Place the contact fingers of the meter test probe (4) across adjacent commutator bars at the side of the commutator.

3. Adjust the voltage control until the pointer indicates the highest voltage on the scale.

4. Test each commutator bar until all bars have been tested; 0 volts across any pair indicates a short circuit and requires the armature be replaced.

ARMATURE RUNOUT TEST

Check that the armature runout is within the range shown on the specifications chart. If the runout is not within tolerance, then replace the armature.

Assembly

1. Place a light film of 20W SAE oil onto the surface of the overrunning clutch pinion bearing surface.

2. Lightly grease the planetary gear spiral spline with Lubriplate grease.

3. Slide the overrunning clutch with the fork lever and bearing pedestal onto the driveshaft.

4. Slide the stop ring onto the armature shaft

NOTE: When installing the retainer, be careful not to scratch the armature shaft.

5. With snapring pliers, separate the ends of the retainer and carefully install onto the groove in the armature shaft.

6. Using the proper tool, seat the stop ring.

7. The planetary gear bearing is located forward of the planetary gears. Lubricate this bearing thoroughly, but lightly, with 20W SAE oil.

8. Insert the planetary gear train with the the pinion, overrunning clutch, fork lever, and bearing pedestal into the drive housing.

9. Install the rubber seal onto the pedestal bearing.

10. Position and install the cover plate on the surface of the ring gear, ensuring the recess in the cover plate mates with the lug machined in the surface of the ring gear.

11. Using the proper tool, position the brush plate onto the end of the commutator shaft. Slide the brush

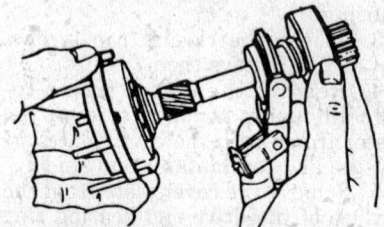

Assembling the overrunning clutch and planetary gear system

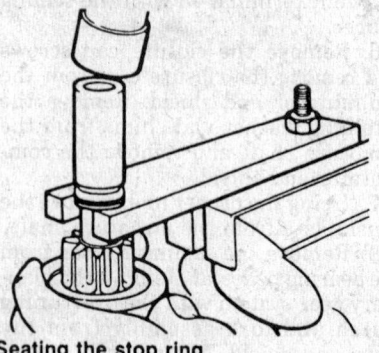

Seating the stop ring

plate over the commutator, ensuring the brush holders are properly seated in the anchor point. After the are properly seated, remove the tool.

12. Place a light coat of 20W SAE oil onto the felt ring gasket and install the gasket on the commutator end of the armature shaft.

13. Place a light coat of Lithium-based lubricant onto the commutator end shield bushing and install the bushing onto the armature shaft.

14. Slide the commutator end shield onto the armature shaft.

15. Set the armature endplay by installing a shim and the retaining washer. After installing the shim, check the endplay with the appropriate feeler gauges. The endplay should be within the range of 0.002–0.016 in. Should additional shims be required, 3 sizes are available: 0.004 in., 0.047 in. and 0.055 in..

16. After setting the armature endplay, place a light coat of lubricant onto the retaining washer.

17. Position the seal ring against the commutator end shield and place the closure cap into installation position.

18. Install the closure cap screws far enough to hold the closure cap into place, but do not tighten at this time.

19. Support the stator frame by hand and slide the armature with the brush plate and commutator end shield carefully into the stator frame, ensuring that a sufficient gap is left to allow installation of the rubber seal.

20. Install the rubber seal onto terminal 45 and slide the rubber seal into the groove of the stator frame.

21. With moderate force, rotate the pinion gear until the armature spline meshes evenly with the planetary gears.

NOTE: When positioning the stator frame ensure the groove in the stator frame is aligned with and fits into the sealing rubber of the bearing pedestal.

22. Rotate the commutator end shield until the groove on the commutator end shield is aligned with the sealing rubber on the bearing pedestal.

23. When the proper alignment is achieved, install the stator frame-to-drive shield screws and torque to 2.0–2.6 ft. lbs.

24. Torque the closure cap screws to 1.0–1.5 ft. lbs..

25. Engage the solenoid switch armature with the fork lever and insert the armature return spring.

26. Install and tighten the solenoid switch housing screws.

27. Install the field washer and terminal wire (Terminal 45) onto the field terminal post. Install the field terminal nut.

Motorcraft Positive Engagement Starter Motor

The Motorcraft starting system, that is used with all Jeep 6 and 8 cylinder engines, consists of a lightweight positive engagement starter motor, a starter motor solenoid, an ignition/start switch, circuits protected by fusible links and the battery. Vehicles equipped with an automatic transmission also have a neutral safety switch to prevent operation of the starter if the selector lever is not in the **N** or **P** position. The Motorcraft starter motor has a moveable pole shoe and appropriate linkage to engage the drive mechanism. Inside the drive assembly, an overrunning clutch prevents the starter motor from being driven by the ring gear.

OVERHAUL

Disassembly

1. Remove the cover screw, cover, and through bolts.

2. Withdraw the pivot pin that retains the starter gear plunger lever. Remove the plunger lever, starter drive end housing and lever return spring.

3. Remove the stop ring retainer. Remove and discard the starter drive gear-to-armature shaft stop ring. Remove the starter drive gear assembly.

4. Remove the brush end plate and insulator assembly.

5. Remove the brushes from the plastic brush holder and lift out the brush holder. Note location of brush holder with respect to the end terminal.

6. Remove the ground brushes-to-frame screws or rivets.

7. Locate the field coil that operates the starter drive gear actuating lever. Bend the edges on the retaining sleeve of this field coil and remove the sleeve and retainer.

8. Position the starter frame in an arbor press and remove the coil screws. Cut the field coil connection at the switch post lead and remove the small diameter ground wire from the upper tab riveted to the frame. Remove the pole shoes and the coils from the frame.

9. Cut the positive brush leads from the the field coils as close to the field connection point as possible.

Inspection

1. Use a clean dry rag or compressed air to clean the field coils, armature, commutator, armature shaft, brush end plate and drive end housing.

Wash all other parts in solvent and dry with a clean rag.

2. Inspect the armature windings for broken or burned insulation and unsoldered or open connections.

3. Check the plastic brush holder for cracks or broken mounting pads. Replace the brushes if worn to ¼ in. in length or if oil soaked.

4. Inspect the armature shaft and bushings for excessive wear and scoring. If necessary, lightly polish damaged surfaces.

5. Examine the wear pattern on the starter drive teeth. To eliminate premature starter and ring gear failure, the pinion teeth must penetrate to a depth greater than ½ the ring gear tooth depth.

6. Examine the starter drive gear for milled, pitted or broken teeth. Replace the starter drive if necessary.

7. Inspect the overrunning clutch by grasping and rotating the pinion gear. The pinion gear should rotate smoothly and freely in the clockwise direction and lock in the counterclockwise direction. Replace if necessary.

HOLD-IN COIL WINDING RESISTANCE TEST

1. Insert a piece of paper (1) between the contact points to insulate them.

2. With an ohmmeter, measure the resistance between the **S** terminal (2) and the starter motor frame. This will

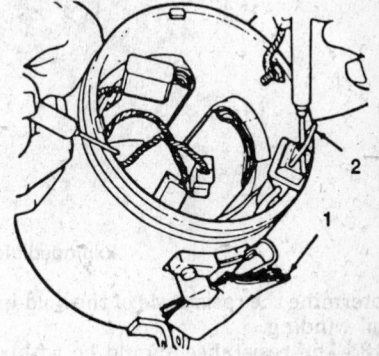

Performing the hold-in coil winding resistance test

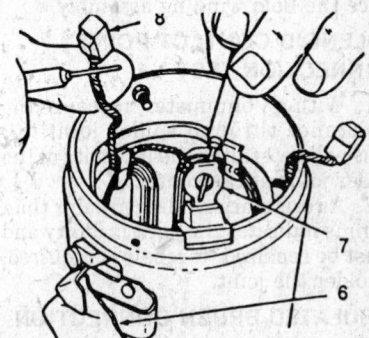

Performing the solenoid contact point connection test

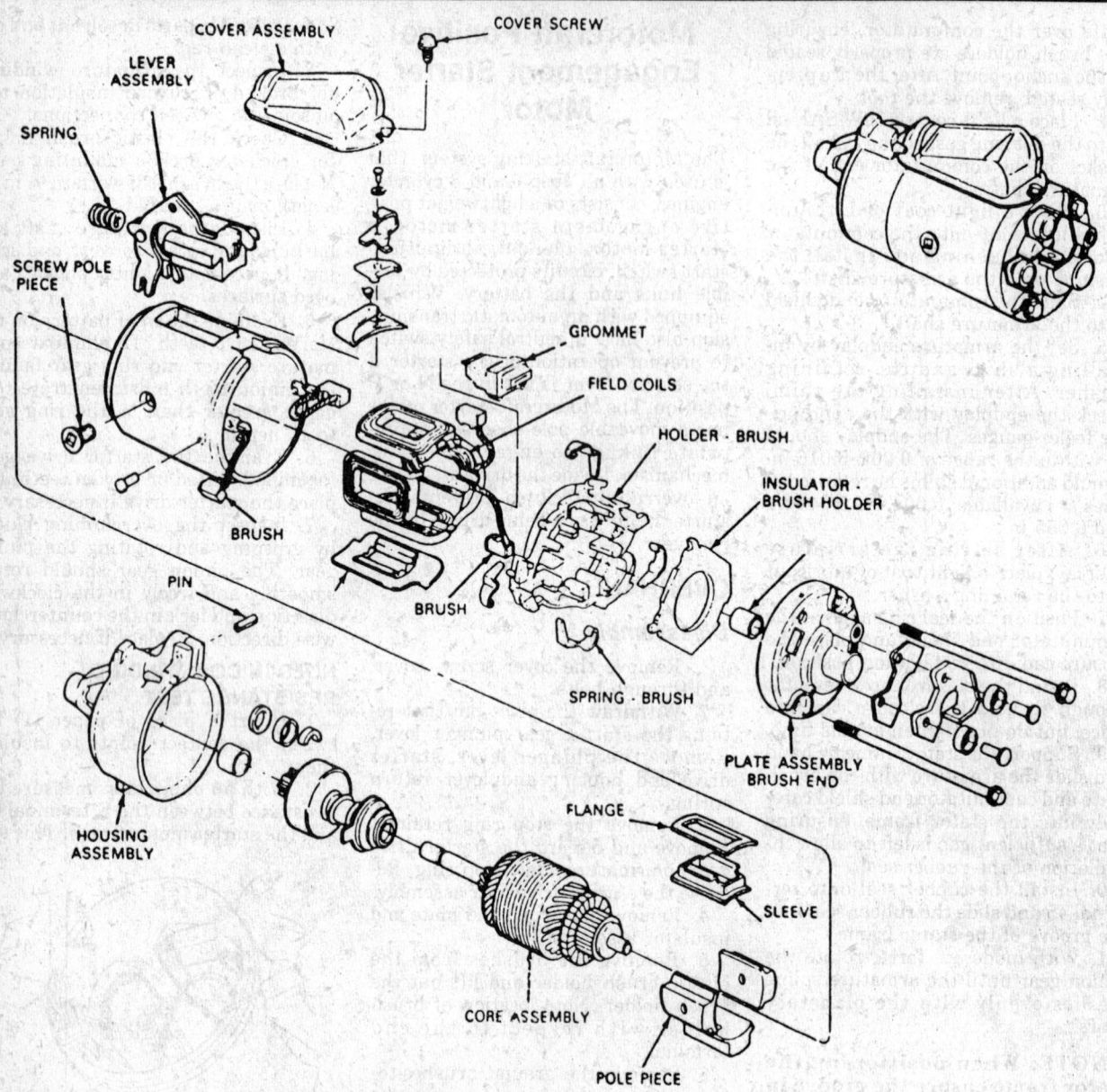

Exploded view of the Motorcraft positive engagement starter motor—Jeep

determine the resistance of the hold-in coil winding.

3. The resistance should be within the range of 2.0–3.5 ohms. If the resistance is not within this range, replace the field winding assembly.

SOLENOID CONTACT POINT CONNECTION TEST

1. With an ohmmeter, measure the resistance through solder joint (3). This will determine the integrity of the solder joint at the contacts.

2. A resistance reading greater than 0 ohms indicates the joint is faulty and must be repaired. If repair is required, resolder the joint.

INSULATED BRUSH CONNECTION TEST

1. With an ohmmeter, measure the

resistance through the solder joint by touching the test probes to the brush and to the copper test bar. This will determine the integrity of the solder joint between the the insulated brush braided wire and the field windings.

2. A resistance reading greater than 0 ohms indicates the joint is faulty and must be repaired. If repair is required, resolder the joint.

FIELD WINDING TERMINAL TO BRUSH CONTINUITY TEST

1. Insert a piece of paper between the contact points to insulate them.

2. Touch the test probes to the field winding terminal and to the insulated brush. This will determine the integrity of all the field winding solder joints.

3. A resistance reading greater than 0 ohms indicates that 1 or more of the

field winding solder joints is faulty. Test each solder joint to identify the faulty joint(s). If repair is required, resolder the joint(s).

TERMINAL BRACKET INSULATION TEST

1. With an ohmmeter, measure the resistance between the bracket and the cap. This will determine if the terminal bracket is properly insulated from the end cap.

2. If the resistance is less than infinite, then the end cap is faulty and must be replaced.

ARMATURE GROUND TEST

1. Position the armature in the growler jaws and turn the power switch to the test position.

2. Touch a test probe-to-armature

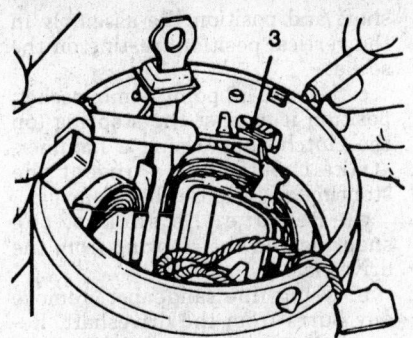

Performing the insulated brush connection test

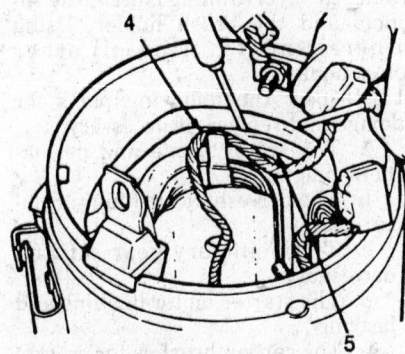

Performing the field winding terminal-to-brush continuity test

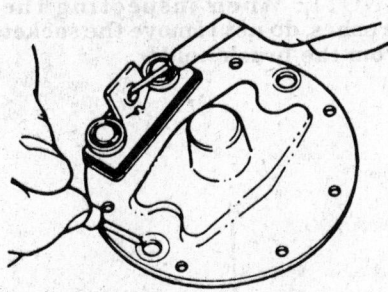

Performing the terminal bracket insulation test

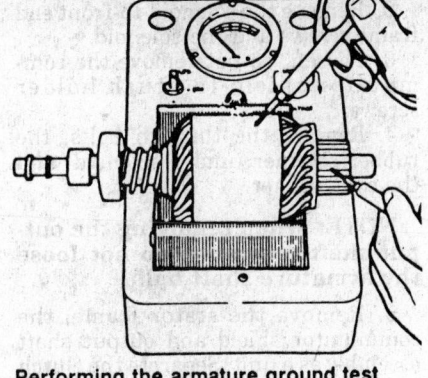

Performing the armature ground test

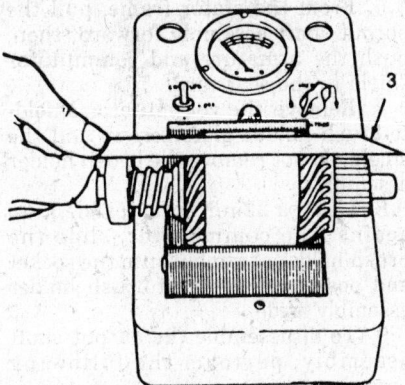

Performing the armature short test

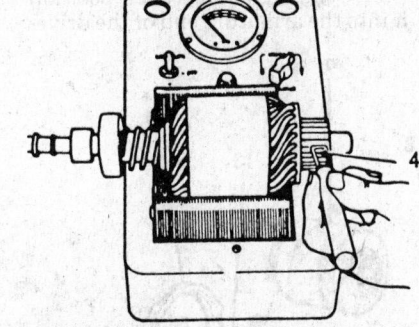

Performing the armature balance test

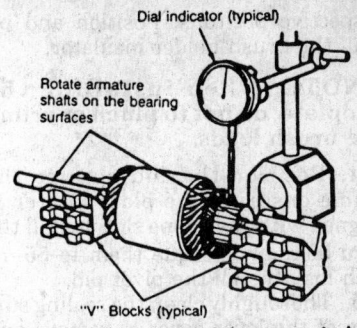

Performing the commutator runout check

core and the other test probe to each commutator bar, 1 at a time and observe the test lamp. The test lamp should not light. If the test lamp lights on any bar, the armature has a short circuit to ground and must be replaced.

ARMATURE SHORT TEST

NOTE: Never operate the growler with the power switch in the test position without the armature in the growler jaws.

1. Place the armature in the growler jaws and turn the power switch to the growler position.
2. Hold the steel blade parallel to, and touching, the armature core (3). Slowly, rotate the armature 1 or more revolutions in the growler jaws. If the steel blade vibrates at any 1 area of the

core, the windings have a short circuit and the armature must be replaced.

ARMATURE BALANCE TEST

1. Place the armature in the growler jaws and turn the power switch to the growler position.
2. Place the contact fingers of the meter test probe (4) across adjacent commutator bars at the side of the commutator.
3. Adjust the voltage control until the pointer indicates the highest voltage on the scale.
4. Test each commutator bar until all bars have been tested; 0 volts across any pair indicates a short circuit and requires the armature be replaced.

ARMATURE RUNOUT TEST

1. Lightly polish the commutator

with commutator cloth prior to measuring runout. Do not use emory cloth.
2. Measure the commutator runnout. If the commutator is greater than 0.005 in. out-of-round or has insulation protruding from between the bars, turn it down on a lathe.

Assembly

1. Position 3 coils and pole pieces into the stator frame interior and support by hand. Install the pole piece screws.
2. Tighten the pole piece screws evenly and at regular intervals, tap the frame lightly with a soft-faced mallet. This will facilitate the alignment of the screws and the pole pieces.
3. Repeat procedures until the screws are securely fastened and the pole pieces are in proper alignment.
4. Install the remaining coil and retainer and bend the tabs to secure the coils to the frame.
5. Solder the field coils and solenoid wire to the starter terminal.
6. Check for continuity and grounds in the assembled coils.
7. Ground the coil, that is located in the area of the retaining sleeve, by positioning the small diameter wire leading from the coil under the copper tab, held by the rivet attaching the contact to the frame.
8. Attach the ground brushes to the starter frame with the screws or rivets.
9. Apply a thin coating of Lubriplate or equivalent, to the armature shaft splines. Install the starter motor drive gear assembly to the armature shaft and install a new retaining stop ring. Install a new stop retainer.
10. Install the armature assembly into the starter frame.
11. Fill the drive end housing bearing bore 1/4 in. with lubricant. Position the starter drive gear plunger lever to the frame and starter drive assembly.
12. Position the starter drive plunger lever return spring and the drive end housing to the frame.
13. Install the brush holder and insert the brushes and springs into their

respective positons. Position and install the brush holder insulator.

NOTE: When installing the endplate, do not to pinch or crimp the brush leads.

14. Position the end plate to the frame ensuring the plate locater is aligned with the frame slot. Install the thru-bolts and torque them to 55–75 inch lbs.. Install the pivot pin.

15. Thoroughly clean the sealing surface of the lever cover to remove any traces of existing gasket material. Apply rubber gasket compound or equivalent, to the sealing surface of the lever cover. Position the lever cover on the frame and secure with the attaching screw.

16. Check the starter no-load current draw.

Mitsubishi Reduction Gear Starter

OVERHAUL

Disassembly and Assembly

NOTE: Do not place the stator frame in a vise or strike it with a hammer for damage to the permanent magnets could occur.

1. Disconnect the coil wire from the solenoid.

2. Remove the solenoid-to-front end frame screws and the solenoid.

3. Loosen, do not remove the commutator shield-to-brush holder screws.

4. Remove the through bolts, the rubber retainer (under solenoid) and the coin washer.

NOTE: When removing the output shaft assembly, do not loose the armature shaft ball.

5. Remove the stator frame, the commutator shield and output shaft assembly as a unit. Separate the clutch fork from the output shaft assembly.

6. From the stator frame, pull the output shaft assembly forward, then, push the armature and commutator shield to the rearward.

7. Remove the commutator shield-to-brush holder plate screws and the shield; do not remove the brush holder assembly.

8. Using a 22mm socket, slide it up against the commutator, slide the brush holder assembly onto the socket and position the socket/brush holder assembly aside.

9. To disassemble the output shaft assembly, perform the following procedures:

   a. Remove the rubber packing ring and the gears.

   b. Using a 17mm socket, position it into the armature end of the drive-shaft and position the assembly in the vertical position, resting on the socket.

   c. Using a 12 point 14mm socket, position it against the stopring (on the clutch end). Using a hammer, strike the socket to unseat the stopring and expose the snapring.

   d. Remove the socket, the snapring and the stopring from the driveshaft.

   e. Using fine sandpaper, remove any burrs from the driveshaft. Remove the overruning clutch.

10. Using compressed air or dry cloths, clean the armature, the stator frame, the overrunning clutch, the solenoid and the brush holder. Using mineral spirits, clean all other components.

11. Inspect the following parts for damage and replace, if necessary:

   a. The stator frame and permanent magnets

   b. The driveshaft bushing (armature side)

   c. The planetary gear set and driveshaft

   d. The starter motor bushing and bearing

   e. The carbon brushes for cracks, distortion and wear below 0.354 in.

NOTE: When inspecting the brushes, do not remove the socket from the brush holder.

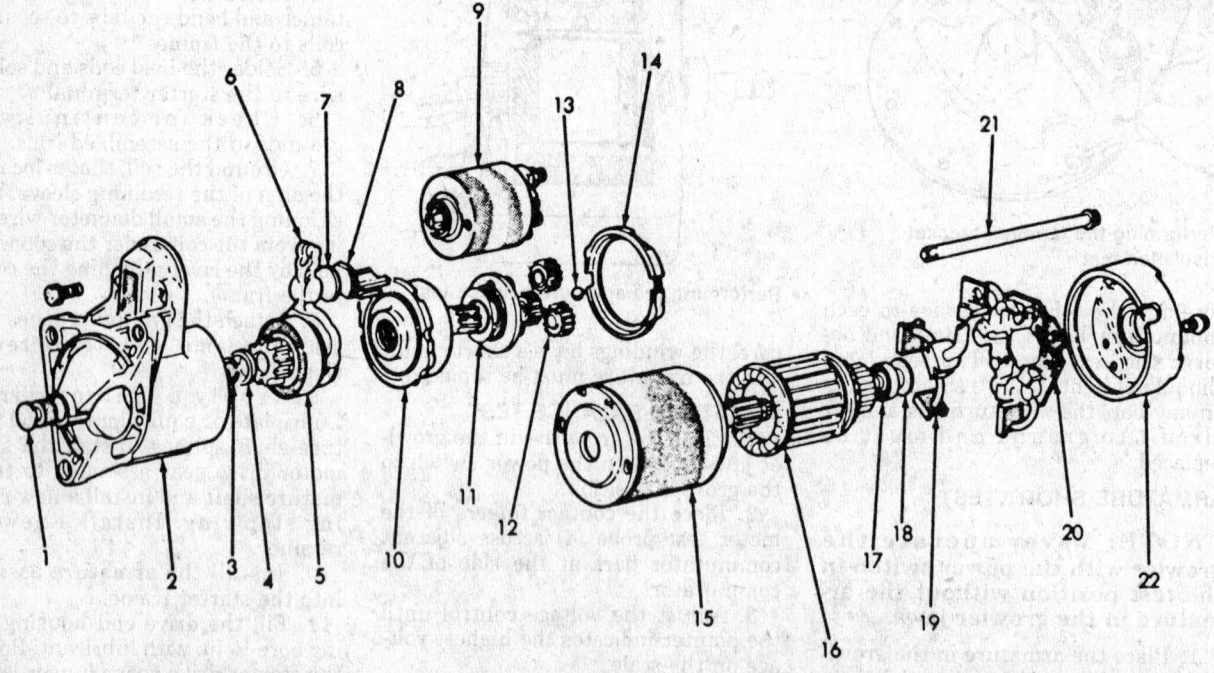

1. Bushing	5. Overrunning clutch	10. Planetary annulus gear	14. Seal ring	18. Washer
2. Overrunning clutch housing	6. Clutch yoke	11. Planetary carrier and pinion shaft	15. Armature frame and magnet assembly	19. Carbon brushes
3. Snapring	7. Yoke washer	12. Planetary gears	16. Armature	20. Brush holder
4. Stop ring	8. Retainer	13. Armature shaft ball	17. Bearing	21. Armature frame bolts
	9. Solenoid			22. End cover

Exploded view of the Mitsubishi reduction gear starter—Jeep

12. Using a growler and a hacksaw blade (placed on top of the armature), rotate the armature and check it for a shorted condition. If the hacksaw blade vibrates, a short exists; replace the armature.

13. Using a test light, place a lead on the armature's core and the other on each commutator segment, inspect the armature for a grounded condition. If a ground exists, the test light will turn **ON**; replace the armature.

14. Using a test light, place the leads on the adjacent commutator segments. If the test light turns **ON** between any 2 segments, the armature is shorted and must be replaced.

15. Inspect the commutator out-of-round, if it is more than 0.001 in., reface it on a lathe.

16. Using motor oil, lubricate the driveshaft and the overrunning clutch bushing. Using Lubriplate®, lubricate the overrunning clutch spiral cut splines.

17. Install the overrunning clutch on the driveshaft/planetary gear assembly, followed by the stopring and the snapring; sure to seat the snapring in the shaft groove and crimp the it with a pair of pliers.

18. Using a battery terminal puller, attach it to the driveshaft tip and press the stopring over the snapring.

NOTE: When installing the stopring, be careful not to scratch the driveshaft.

19. Install the clutch fork, with the assembled planetary gear set, lubricated with lithium grease, into the front end housing; make sure the locating lugs are properly seated in the front end housing.

20. Install the coin washer and the rubber fork retainer.

21. Install the rubber backing ring by placing the largest rubber lug at the top.

22. Install the brush holder onto the armature's commutator; make sure the brushes and brush holders are seated in the holder. Inspect the flex washer and install the commutator shield onto the armature. Install the brush holder screws but do not tighten them.

23. Install the armature assembly into the stator frame and seat the wire grommet into the frame.

24. Be sure the armature spline gear is seated in the planetary gear seat with the armature shaft ball in place. Seat the armature shaft in the shaft bushing bore; rotate the stator frame to align the tabs on the drive housing frame.

25. Install the through bolts and torque to 28 inch lbs. Torque the brush holder screws to 18 inch lbs.

26. To complete the assembly, reverse the disassembly procedures.

IMPORT CHARGING SYSTEMS

Test Instruments

OHMMETER

An ohmmeter is used to measure electrical resistance in a unit or circuit. The ohmmeter has a self-contained power supply. In use, it is connected across (or in parallel with) the terminals of the unit being tested.

AMMETER

An ammeter is used to measure current (amount of electricity) flowing through a unit or circuit. Ammeters are always connected in the line (in series) with the unit or circuit being tested.

VOLTMETER

A voltmeter is used to measure voltage (electrical pressure) pushing the current through a unit or circuit. The meter is connected across the terminals of the unit being tested.

Alternator Testing

Diagnosis

The first step in diagnosing troubles of the charging system, is to identify the source of failure. Does the fault lie in the alternator or the regulator. The next move depends upon preference or necessity; either repair or replace the offending unit.

Alternator output is controlled by the amount of current supplied to the field circuit of the system.

The alternator is capable of produc-

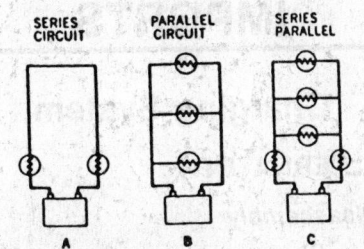

Basic electrical circuits

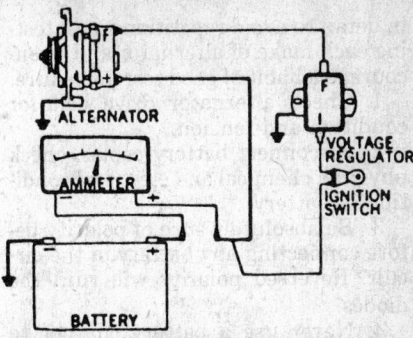

Ammeter connected to test wire—circuit equipped with external voltage regulator

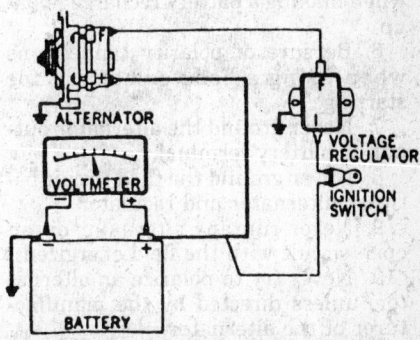

Voltmeter connected in parallel circuit—circuit equipped with external voltage regulator

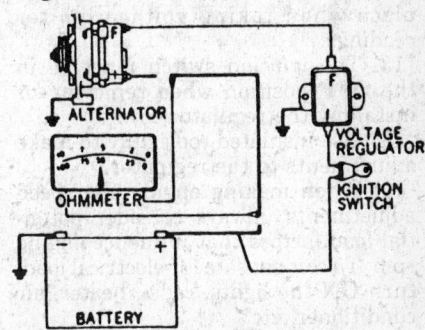

Ohmmeter connected to test wire resistance—circuit equipped with external voltage regulator

ing substantial current at idle speed. Higher maximum output is also a possibility. This presents a potential danger when testing. As a precaution, a field rheostat should be used in the field circuit when making the following isolation test. The field rheostat permits positive control of the amount of current allowed to pass through the field circuit during the isolation test. Unregulated alternator capacity could ruin the unit.

NOTE: Most manufacturers of precision gauges offer special test connectors, in sets, that will adapt to the leads and connections of any charging system.

There are certain precautionary measures that apply to alternator tests in general. These items are listed in detail to avoid repetition when testing each make of alternator and to encourage a habit of good test procedure.

1. Check alternator drive belt for condition and tension.

2. Disconnect battery cables, check physical, chemical and electrical condition of battery.

3. Be absolutely sure of polarity before connecting any battery in the circuit. Reversed polarity will ruin the diodes.

4. Never use a battery charger to start the engine.

5. Disconnect both battery cables when making a battery recharge hook-up.

6. Be sure of polarity connections when using a booster battery for starting.

7. Never ground the alternator output or battery terminal.

8. Never ground the field circuit between alternator and regulator.

9. Never run any alternator on an open circuit with the field energized.

10. Never try to polarize an alternator, unless directed by the manufacturer of the alternator.

11. Do not attempt to motor an alternator.

12. The regulator cover must be in place when taking voltage limiter readings.

13. The ignition switch must be in the **OFF** position when removing or installing the regulator cover.

14. Use insulated tools only to make adjustments to the regulator.

15. When making engine idle speed adjustments, always consider potential load factors that influence engine rpm. To compensate for electrical load, turn **ON** the lights, radio, heater, air conditioner, etc.

Diagnosis of Charging System
LOW OR NO CHARGING

1. Blown fuse.
2. Broken or loose fan belt.
3. Voltage regulator not working.
4. Brushes sticking.
5. sliping dirty.
6. Open circuit.
7. Bad wiring connections.
8. Bad diode rectifier.
9. High resistance in charging circuit.
10. Voltage regulator needs adjusting.
11. Grounded stator.
12. Open rectifiers (check all 3 phases).
13. If rectifiers are found blown or open, check capacitor.

NOISY UNIT

1. Damaged rotor bearings.
2. Poor alignment of unit.
3. Broken or loose belt.

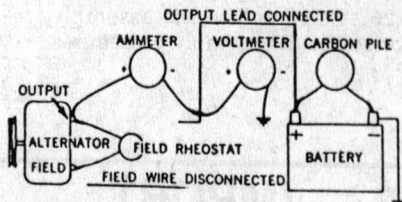

Checking the current output of the charging system

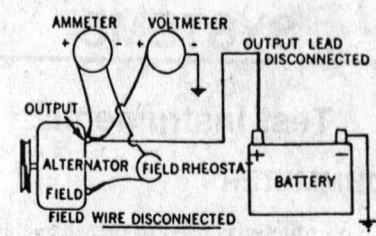

Checking the field current draw

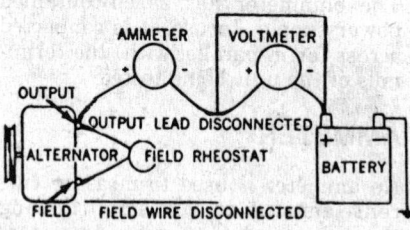

Checking the charging system resistance

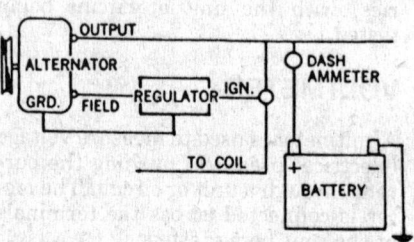

Alternator system with ammeter in the circuit

4. Open diode rectifiers.

REGULATOR POINTS BURNED OR STUCK

1. Regulator set too high.
2. Poor ground connections.
3. Shorted generator field.
4. Regulator air gap incorrect.

CHRYSLER IMPORTS

Charging System

ALTERNATOR

Disassembly

1. Place a soldering iron on the bearing box for approximately 3-4

minutes to heat it, then pull out the bolts and insert a flat tip tool between the stator and front bracket and separate them.

NOTE: The bearing box must be heated or the bearing cannot be pulled out.

2. Separate the front and rear sections, being careful not to lose the stopper spring that fits around the circumference of the rear bearing.

3. Remove the pulley nut, then disassemble the pulley, rotor and front bracket.

4. The rear bearing can be removed by using a bearing puller.

5. Remove the nut of the **B** terminal and the insulation bushing. Remove the rectifier retaining screws and the brush holder retaining screw. Separate the rear bracket and stator.

6. Remove the IC regulator.

7. Remove the solder from the rectifier and stator leads.

NOTE: Do not use the soldering iron for more than 5 seconds as the rectifier may be damaged if overheated.

8. The brush may be removed by removing the solder from the pigtail.

ROTOR

Testing

1. Using an ohmmeter, check for continuity at the slip end rings. If there is no continuity, replace the rotor.

2. Using an ohmmeter, make an insulation test. Check for continuity between the slipring and the rotor core. If continuity exists, replace the rotor.

3. Measure the slipring outer diameter for wear. Mimimum diameter is 1.18 in. (30mm).

STATOR

Testing

1. Using an ohmmeter, make a continuity test between the stator lead wires. If there is no continuity, replace the stator.

2. Using an ohmmeter, make an insulation test between the stator core and the lead wire. If the continuity exists, replace the stator.

DIODE

Testing

1. Using and ohmmeter, perform a continuity test on diodes in both directions.

2. Replace as necessary.

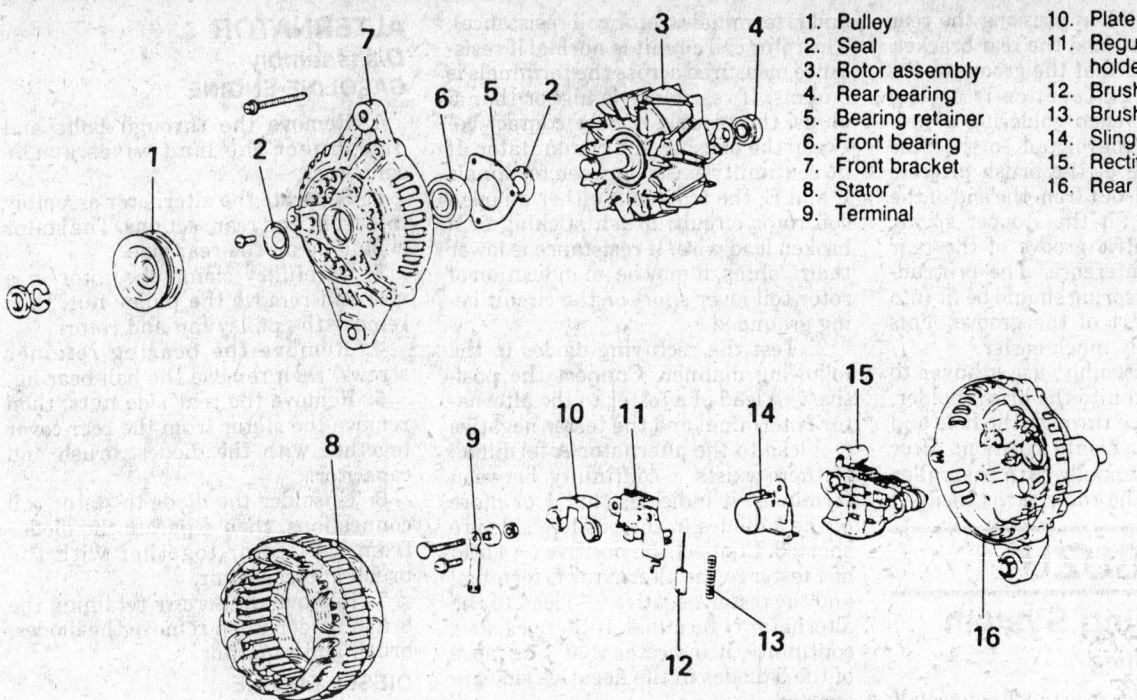

1. Pulley
2. Seal
3. Rotor assembly
4. Rear bearing
5. Bearing retainer
6. Front bearing
7. Front bracket
8. Stator
9. Terminal
10. Plate
11. Regulator and brush holder
12. Brush
13. Brush spring
14. Slinger
15. Rectifier assembly
16. Rear bracket

Exploded view of the Chrysler alternator—2.6L engine

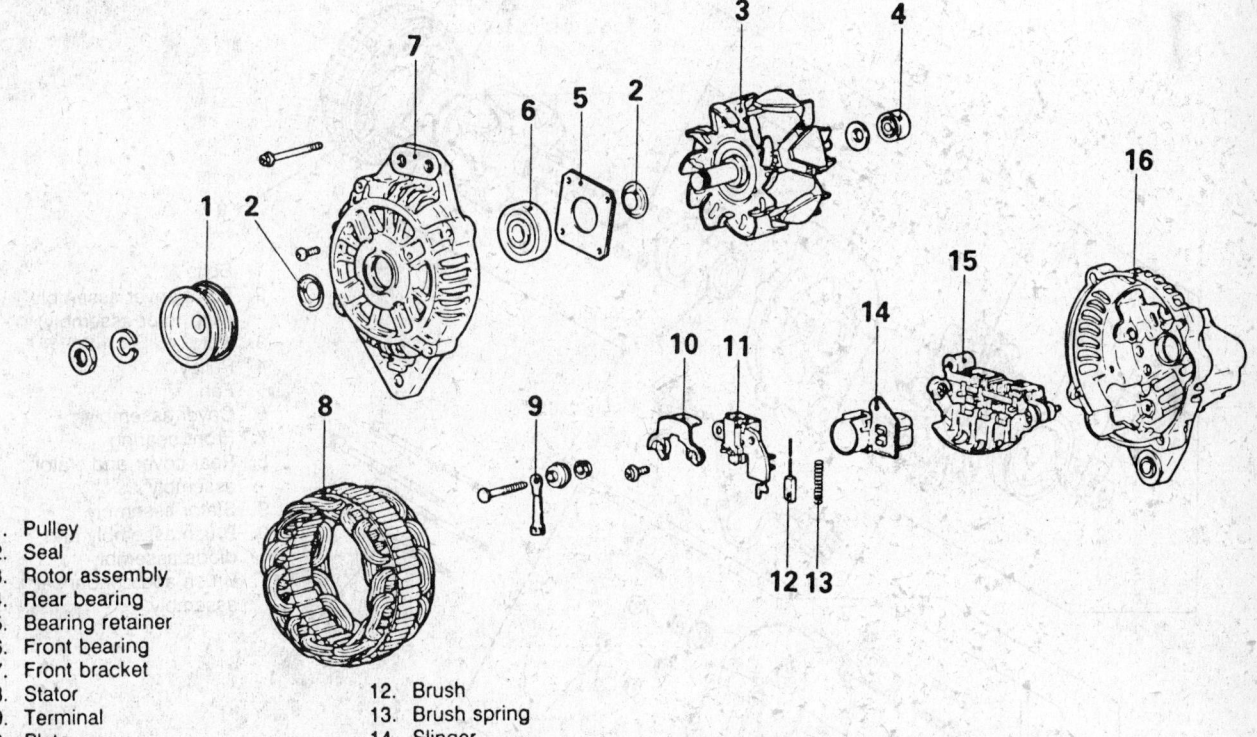

1. Pulley
2. Seal
3. Rotor assembly
4. Rear bearing
5. Front bearing
6. Front bearing
7. Front bracket
8. Stator
9. Terminal
10. Plate
11. Regulator and brush holder

12. Brush
13. Brush spring
14. Slinger
15. Rectifier assembly
16. Rear bracket

Exploded view of the Chrysler alternator—3.0L engine

BRUSHES

Testing

1. Check for smooth movement of the brush and clean the brush holder if necessary.

2. Check for brush wear by looking at the wear limit line on the brush and replace if necessary.

Assembly

1. Assembly of the alternator is the reverse of disassembly with the following instructions:

2. When installing the front bearing, use a socket which exactly fits the outer race of the bearing, then use a hand press or vise and press the bear-

ing in evenly. When pressing the rear bearing on, first heat the rear bracket, then press it so that the groove at the bearing circumference is at the slipring side. When soldering a new brush, solder the pigtail so that the wear limit line of the brush projects 0.079–0.118 in. out from the end of the brush holder. Fit the stopper spring into the eccentric groove of the rear bearing circumference. The protruding part of the spring should be fit into the deepest part of the groove. This makes assembly much easier.

3. Before assembly, use a finger to push the brush into the brush holder, then pass a wire through the hole and secure the brush into position. After reassembly, manually turn the pulley to make sure the rotor turns easily.

ISUZU

Charging System

Troubleshooting

1. Measure the resistance between **F** and **E** terminals (rotor coil resistance): The rotor coil circuit is normal if resistance measured across the terminals is 5 ohms. If resistance is higher than 5 ohms, the trouble is poor contact between the brushes and commutator. If no continuity exists between terminals **F** and **E**, the trouble is either an open coil rotor circuit, brush sticking or a broken lead wire. If resistance is lower than 5 ohms, it may be an indication of rotor coil layer short or the circuit being grounded.

2. Test the rectifying diodes in the following manner: Connect the positive (+) lead of a tester to the alternator **N** terminal and the tester negative (−) lead to the alternator **A** terminal. If there exists a continuity between terminals, it indicates that 1 or more of the 3 diodes in the positive side are shorted. Connect the positive (+) lead of a tester to the alternator **E** terminal and the tester negative (−) lead to the alternator **N** terminal. If there exists a continuity, it indicates that 1 or more of the 3 diodes in the negative side are shorted.

ALTERNATOR
Disassembly
GASOLINE ENGINE

1. Remove the through bolts and disconnect the lead wires at the connector.

2. Separate the alternator assembly into front and rear sections. The stator should be on the rear side.

3. Carefulley clamp the rotor in a vise and remove the pulley nut, then remove the pulley fan and rotor.

4. Remove the bearing retainer screws, then remove the ball bearing.

5. Remove the rear side nuts, then remove the stator from the rear cover together with the diodes, brush and capacitor.

6. Unsolder the diode-to-stator coil connections, then separate the diodes from the stator together with the brush and capacitor.

7. Remove the screws retaining the brush holder, then remove the diodes, brush and capacitor.

DIESEL ENGINE

1. If so equipped, remove the vacu-

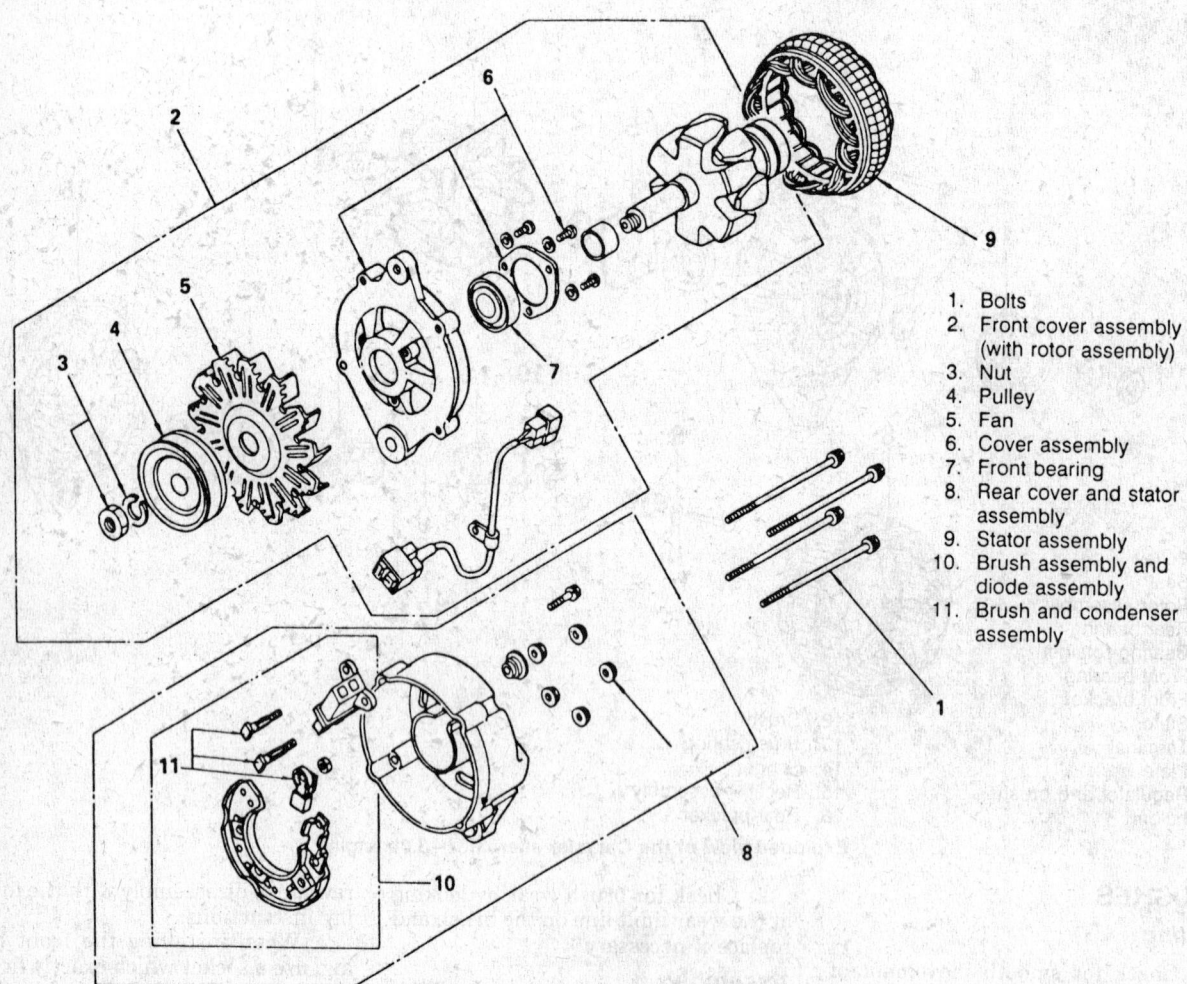

1. Bolts
2. Front cover assembly (with rotor assembly)
3. Nut
4. Pulley
5. Fan
6. Cover assembly
7. Front bearing
8. Rear cover and stator assembly
9. Stator assembly
10. Brush assembly and diode assembly
11. Brush and condenser assembly

Exploded view of the Isuzu alternator—2.0L gas engine

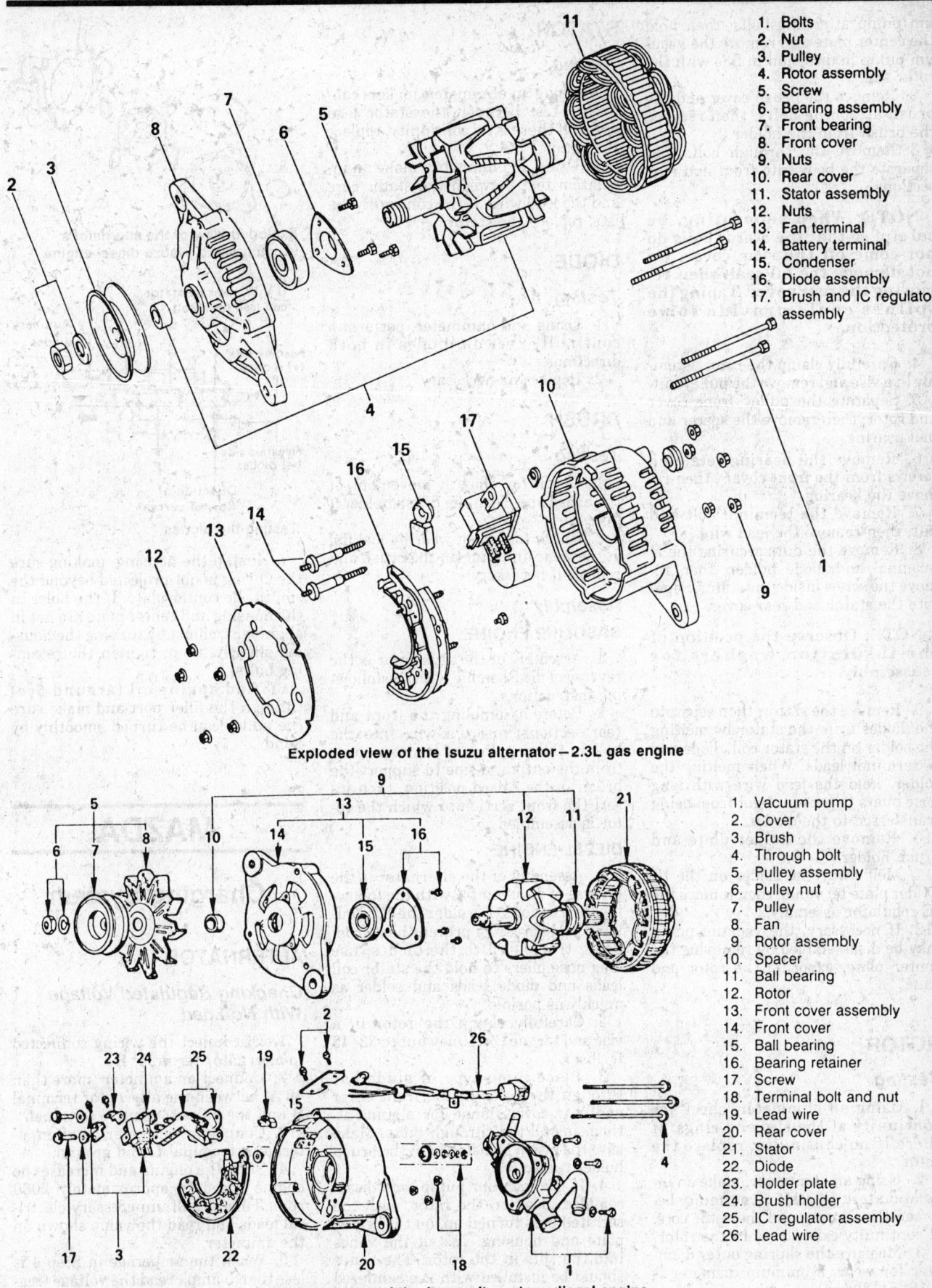

1. Bolts
2. Nut
3. Pulley
4. Rotor assembly
5. Screw
6. Bearing assembly
7. Front bearing
8. Front cover
9. Nuts
10. Rear cover
11. Stator assembly
12. Nuts
13. Fan terminal
14. Battery terminal
15. Condenser
16. Diode assembly
17. Brush and IC regulator assembly

Exploded view of the Isuzu alternator — 2.3L gas engine

1. Vacuum pump
2. Cover
3. Brush
4. Through bolt
5. Pulley assembly
6. Pulley nut
7. Pulley
8. Fan
9. Rotor assembly
10. Spacer
11. Ball bearing
12. Rotor
13. Front cover assembly
14. Front cover
15. Ball bearing
16. Bearing retainer
17. Screw
18. Terminal bolt and nut
19. Lead wire
20. Rear cover
21. Stator
22. Diode
23. Holder plate
24. Brush holder
25. IC regulator assembly
26. Lead wire

Exploded view of the Isuzu alternator — diesel engine

um pump attaching bolts, then hold the center plate and remove the vacuum pump in direction in line with the rotor shaft.

2. Remove the brush cover and the brush attaching bolts, then remove the brush from the holder.

3. Remove the through bolts and separate the body into front and rear sections.

NOTE: When separating, be careful so that the stator coils do not come off the rear cover. Do not damage the oil seal when removing the rear cover. Taping the splines could provide some protection.

4. Carefully clamp the rotor assembly in a vise and remove the pulley nut.

5. Separate the pulley front cover and rotor, then remove the spacer and ball bearing.

6. Remove the bearing retaining screws from the front cover, then remove the bearing.

7. Remove the terminal bolt and nut, then remove the lead wire.

8. Remove the nuts securing the **B** terminal and diode holder, then remove the screw inside the stator. Separate the stator and rear cover.

NOTE: Observe the position of the insulation washers for reassembly.

9. Remove the stator, then separate the diodes from the stator by melting the solder on the stator coil, diode and **N** terminal leads. When melting the solder, hold the lead wire with long nose pliers to prevent heat from being transferred to the diodes.

10. Remove the holder plate and brush holder.

11. Melt away the solder on the IC holder plate terminal, then remove the IC regulator assembly.

12. If necessary, the vacuum pump may be disassembled by removing the center plate, exposing the rotor and vane.

ROTOR

Testing

1. Using an ohmmeter, check for continuity at the slip end rings. If there is no continuity, replace the rotor.

2. Using an ohmmeter, make an insulation test. Check for continuity between the slipring and the rotor core. If continuity exists, replace the rotor.

3. Measure the slipring outer diameter for wear. Mimimum diameter is 1.18 in. (30mm).

STATOR

Testing

1. Using an ohmmeter, make a continuity test between the stator lead wires. If there is no continuity, replace the stator.

2. Using an ohmmeter, make an insulation test between the stator core and the lead wire. If the continuity exists, replace the stator.

DIODE

Testing

1. Using and ohmmeter, perform a continuity test on diodes in both directions.

2. Replace as necessary.

BRUSH

Testing

1. Check for smooth movement of the brush and clean the brush holder if necessary.

2. Check for brush wear by looking at the wear limit line on the brush and replace if necessary.

Assembly

GASOLINE ENGINE

1. Assembly of the alternator is the reverse of disassembly with the following instructions.

2. Before assembling the front and rear sections. insert a wire into the hole in the rear face of the rear cover from the outboard side to support the brush in the raised position, then insert the front section to which the rotor is assembled

DIESEL ENGINE

1. Assembly of the alternator is the reverse of disassembly with the following instructions: Resolder the IC regulator lead wires. To prevent heat from being transfered to the diodes, use long nose pliers to hold the stator coil leads and diode leads and solder as quickly as posible.

2. Carefully clamp the rotor in a vise and torque the pulley nut to 33–43 ft. lbs.

3. Place some type of guide bar through the holes in the front cover and rear cover flange for alignment, then, install the through bolts. Make sure the brush is installed in the brush holder correctly.

4. If the vacuum pump was disassembled, position the rotor, with the serrated boss turned up, on the center plate and housing. Install the vanes into the slits in the rotor. The vanes should be installed with the camfered side turned outward.

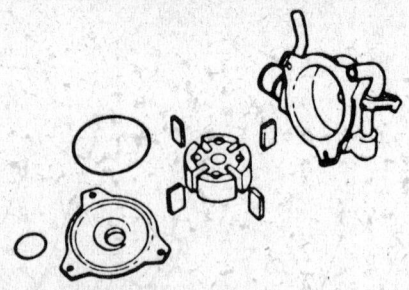

Exploded view of the alternator's vacuum pump—Isuzu diesel engine

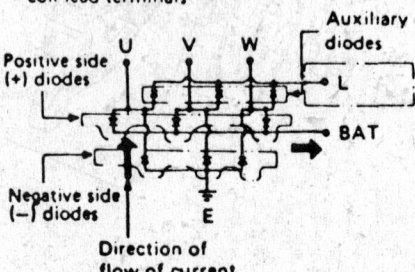

U.V.W. indicate stator coil lead terminals

Testing the diodes

5. Install the housing, making sure the O-ring is not projected beyond the slot in the center plate. If the holes in the housing and center plate are not in alignment, adjust by turning the housing slightly, then, tighten the retaining bolts.

6. Add engine oil (around 5cc) through the filler port and make sure the pulley can be turned smoothly by hand.

MAZDA

Charging System

ALTERNATOR

Checking Regulated Voltage With No Load

1. Disconnect the wiring connected to alternator terminal **B**.

2. Connect an ammeter (more than 40 A) between the alternator terminal **B** and the battery positive terminal.

3. Connect a voltmeter between alternator terminal **L** and ground.

4. Start the engine and increase the engine speed to approximately 2000 rpm. Turn off all unnecessary electrical loads and read the value shown on the ammeter.

5. When the amperage in Step 4 is less than 5 amp., read the voltage (regulated voltage) of terminal **L**. The reg-

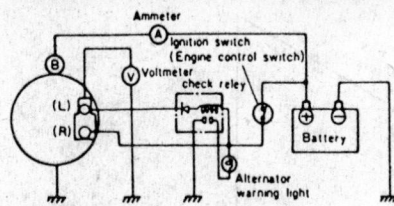

Checking the charging system

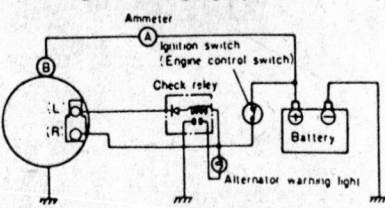

Checking the regulated voltage with no-load

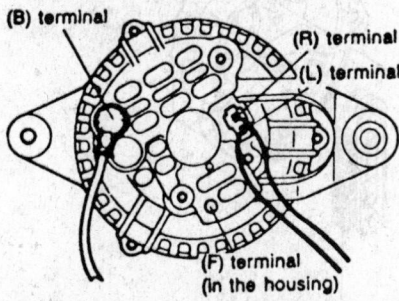

Location of the Mazda atlernator terminals

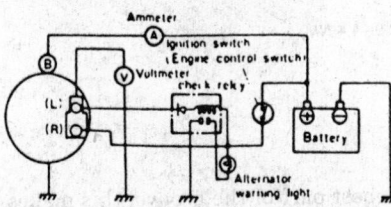

Checking the current output

ulated voltage should be 14.1–14.7 V.

Checking Output Current

1. Disconnect the wiring connected to the alternator terminal **B**.

2. Connect an ammeter (more than 40 amp) between the alternator terminal **B** and the battery positive terminal.

3. Start the engine and increase the engine speed to more than 2500 rpm and read the maximum value shown on the ammeter. Apply all electrical loads. If the value shown on the ammeter is more than 90% of the rated output, the alternator is normal.

Disassembly

1986–88

1. Place a soldering iron on the bearing box for approximately 3–4 minutes to heat it, then pull out the bolts and insert a flat tip tool between the stator and front bracket and separate them.

NOTE: The bearing box must be heated or the bearing cannot be pulled out.

2. Separate the front and rear sections, being careful not to lose the stopper spring that fits around the circumference of the rear bearing.

3. Remove the pulley nut, then disassemble the pulley, rotor and front bracket.

4. The rear bearing can be removed by using a bearing puller.

5. Remove the nut of the **B** terminal and the insulation bushing. Remove the rectifier retaining screws and the

brush holder retaining screw and then separate the rear bracket and stator.

6. Remove the IC regulator.

7. Remove the solder from the rectifier and stator leads.

NOTE: Do not use the soldering iron for more than 5 seconds as the rectifier may be damaged if overheated.

8. The brush may be removed by removing the solder from the pigtail.

1989–90

1. Remove the alternator through bolts and separate the rear case/stator assembly from the front case/rotor assembly.

2. Secure the rotor and remove the pulley nut, the washer and the pulley.

3. Pull the rotor from the front case.

4. Remove the front bearing retainer-to-front case bolts, the retainer and the bearing.

5. From the rear case, remove the stator, the brush holder/rectifier assembly-to-rear case nuts/washers, the brush holder assembly and the rectifier.

ROTOR

Testing

1. Using an ohmmeter, check for continuity at the slip end rings. If there is no continuity, replace the rotor.

2. Using an ohmmeter, make a ground test. Check for continuity between the slipring and the rotor core. If continuity exists, replace the rotor.

3. Measure the slipring outer diam-

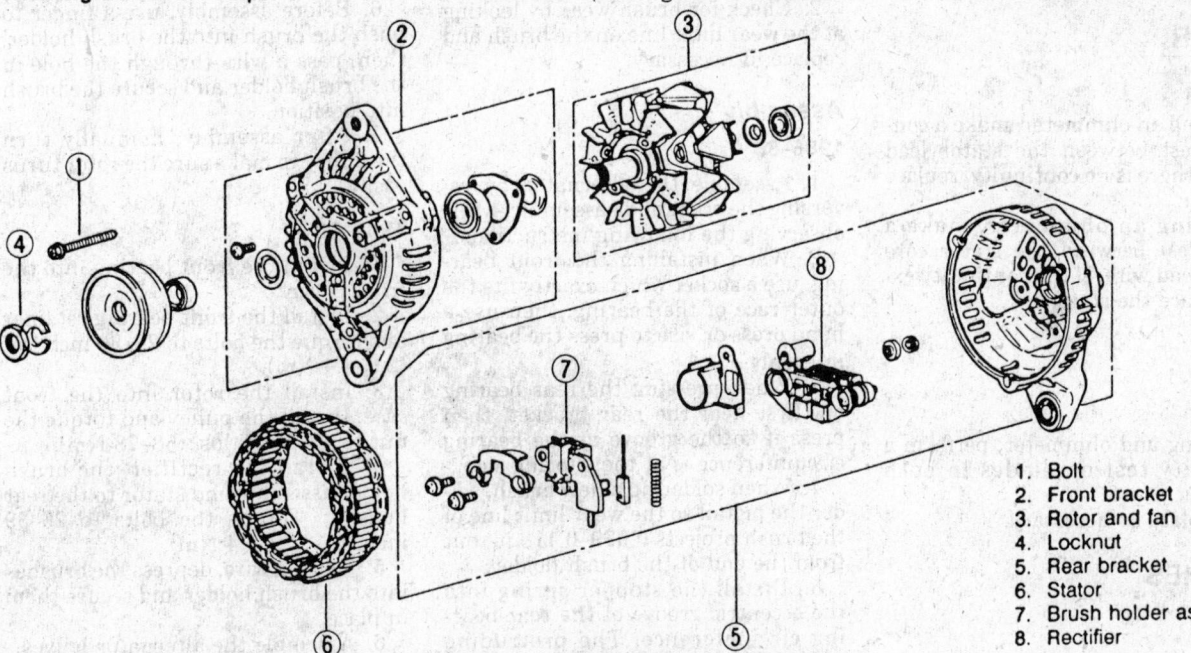

1. Bolt
2. Front bracket
3. Rotor and fan
4. Locknut
5. Rear bracket
6. Stator
7. Brush holder assembly
8. Rectifier

Exploded view of the 1986-88 Mazda alternator

1. Pulley
2. Front cover
3. Rotor
4. Stator
5. Brush holder
6. Rectifier
7. Rear bracket

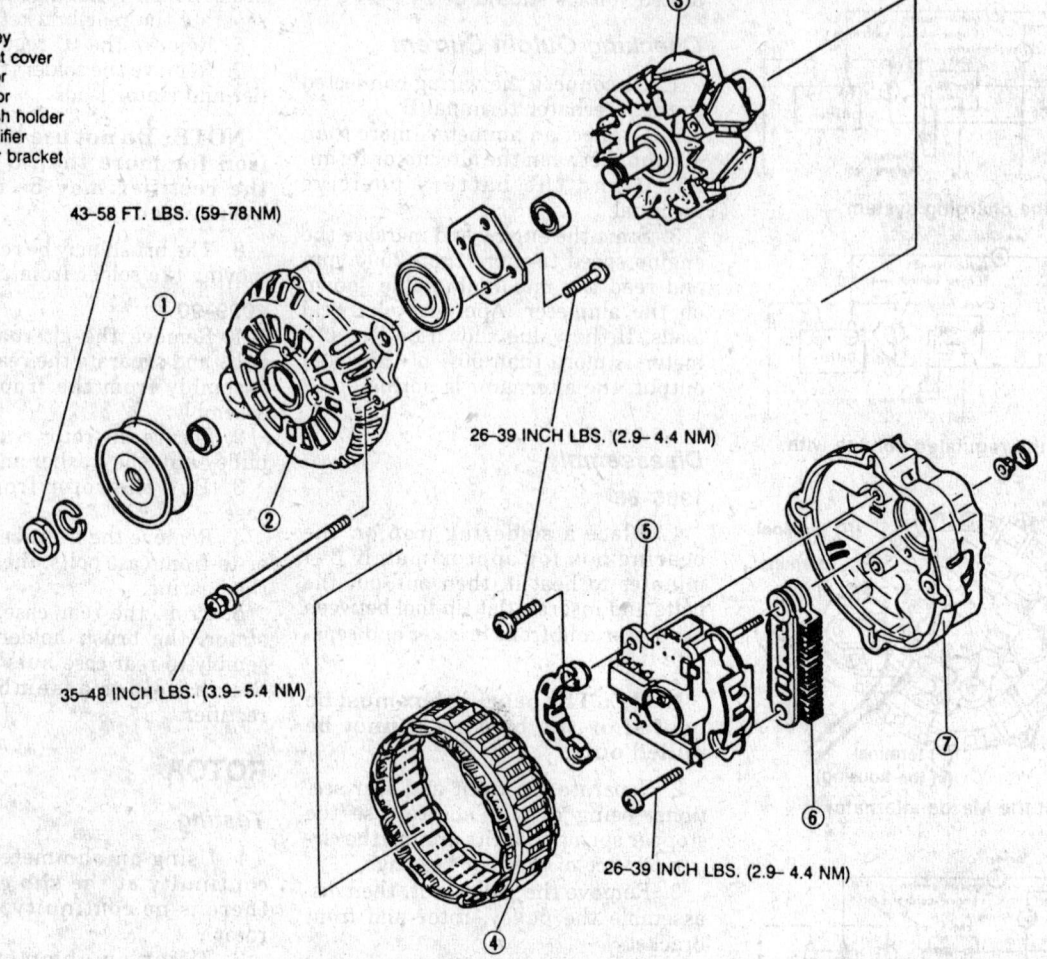

43–58 FT. LBS. (59–78 NM)

26–39 INCH LBS. (2.9–4.4 NM)

35–48 INCH LBS. (3.9–5.4 NM)

26–39 INCH LBS. (2.9–4.4 NM)

Exploded view of the 1989–90 Mazda alternator

eter for wear. Mimimum diameter is 1.18 in. (30mm).

STATOR

Testing

1. Using an ohmmeter, make a continuity test between the stator lead wires. If there is no continuity, replace the stator.

2. Using an ohmmeter, make a ground test between the stator core and the lead wire. If the continuity exists, replace the stator.

DIODE

Testing

1. Using and ohmmeter, perform a continuity test on diodes in both directions.

2. Replace as necessary.

BRUSHES

Testing

1. Check for smooth movement of

the brush and clean the brush holder, if necessary.

2. Check for brush wear by looking at the wear limit line on the brush and replace, if necessary.

Assembly

1986–88

1. Assemble the alternator by reversing the order of disassembly, while observing the following instructions:

2. When installing the front bearing, use a socket which exactly fits the outer race of the bearing, then use a hand press or vise to press the bearing in evenly.

3. When pressing the rear bearing on, first heat the rear bracket, then press it so the groove at the bearing circumference is at the sliping side.

4. When soldering a new brush, solder the pigtail so the wear limit line of the brush projects 0.079–0.118 in. out from the end of the brush holder.

5. Install the stopper spring into the eccentric groove of the rear bearing circumference. The protruding part of the spring should be fit into the

deepest part of the groove; this makes assembly much easier.

6. Before assembly, use a finger to push the brush into the brush holder, then, pass a wire through the hole in the brush holder and secure the brush into position.

7. After assembly, manually turn the pulley to make sure the rotor turns easily.

1989–90

1. Install the front bearing into the front cover.

2. Install the front bearing retainer and torque the bolts to 26–39 inch lbs. (2.9–4.4 Nm).

3. Install the rotor into the front case. Install the pulley and torque the nut to 43–58 ft. lbs. (58–78 Nm).

4. Install the rectifier, the brush holder assembly and stator to the rear housing. Torque the bolts to 26–39 inch lbs. (2.9–4.4 Nm).

5. Using a wire, depress the brushes into the brush holder and secure them in place.

6. Assemble the alternator halves.

7. Install the through bolts and

torque to 35–48 ft. lbs. (3.9–5.4 Nm).

8. Remove the wire securing the brushes.

9. Manually, turn the pulley to make sure the rotor turns easily.

MITSUBISHI

Charging System

ALTERNATOR

Troubleshooting (On Vehicle)

1. Place the ignition switch in the **OFF** position.

2. Disconnect the battery ground cable.

3. Disconnect the cable from terminal **B** of the alternator and connect an ammeter between the terminal **B** and the cable.

4. Connect a voltmeter between terminal **B** (+) and ground (−).

5. Set the engine tachometer.

6. Connect the battery ground cable to the battery. The voltmeter should indicate the battery voltage.

7. Start the engine.

8. Turn **ON** the lamps, accelerate the engine to the speed specified and measure the output current. Check it against the specifications.

Disassembly

1. Remove alternator from vehicle.

2. Remove the through bolts from the alternator body.

3. Insert an appropriate pry tool between the front bracket and stator. Pry the front bracket away from the stator. Remove the front bracket along with the rotor.

NOTE: If the tool is inserted too deeply, the stator coil might be damaged.

4. Hold the rotor in a vise and remove the pulley nut. Then remove the pulley, fan, spacer and seal. Remove

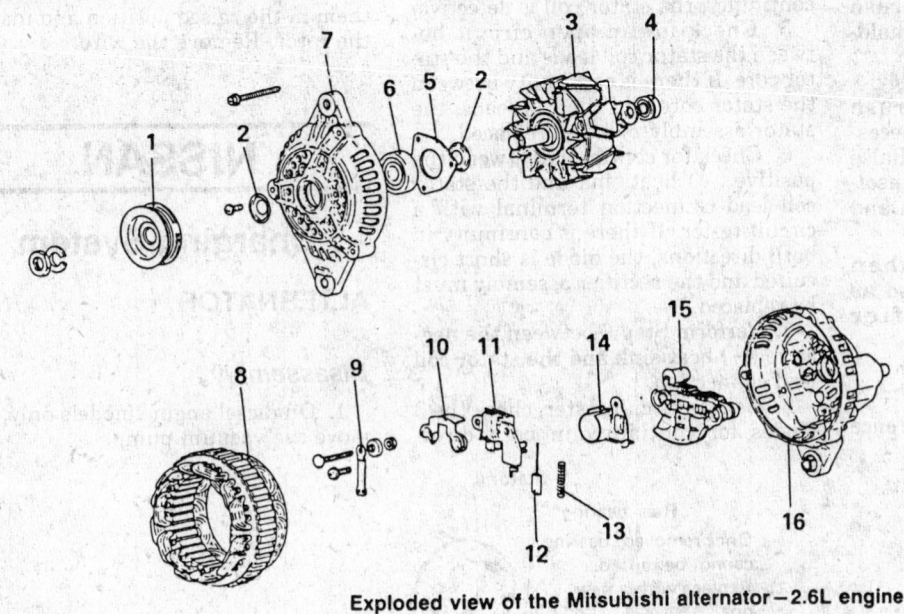
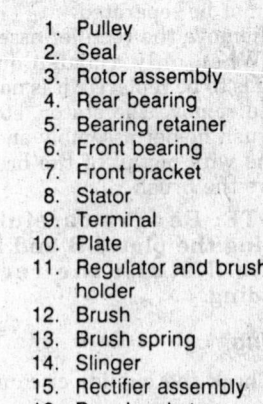

1. Pulley
2. Seal
3. Rotor assembly
4. Rear bearing
5. Bearing retainer
6. Front bearing
7. Front bracket
8. Stator
9. Terminal
10. Plate
11. Regulator and brush holder
12. Brush
13. Brush spring
14. Slinger
15. Rectifier assembly
16. Rear bracket

Exploded view of the Mitsubishi alternator—2.6L engine

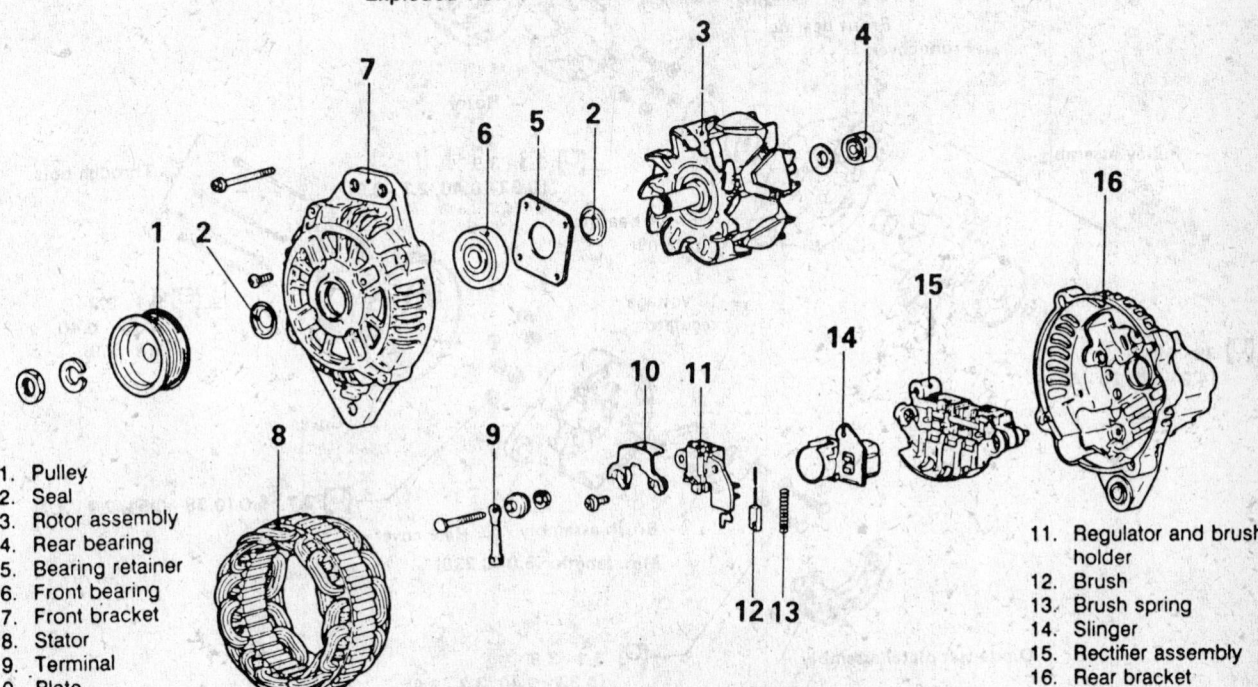

1. Pulley
2. Seal
3. Rotor assembly
4. Rear bearing
5. Bearing retainer
6. Front bearing
7. Front bracket
8. Stator
9. Terminal
10. Plate

11. Regulator and brush holder
12. Brush
13. Brush spring
14. Slinger
15. Rectifier assembly
16. Rear bracket

Exploded view of the Mitsubishi alternator—3.0L engine

the rotor from the front bracket and remove the seal.

5. Unsolder the rectifier from the stator coil lead wires and remove the stator assembly.

NOTE: Make sure the solder is removed quickly (in less than 5 seconds). If a diode is heated to more than 150°C, it might be damaged.

6. Remove the condenser from terminal **B**.

7. Unsolder the plates **B** and **L** from the rectifier assembly.

8. Remove the mounting screw and terminal **B** bolt and remove the electronic voltage regulator and brush holder. The regulator and brush holder cannot be separated.

9. Remove the rectifier assembly.

10. When only a brush or brush spring is to be replaced, it is not necessary to remove the stator, etc. Raise the brush holder assembly and unsolder the wire pigtail of the brush and remove the brush.

NOTE: Be very careful when bending the plates B and L so as not to disturb the rectifier moulding.

Testing

1. Check the outside circumference of the slipring for dirtiness and roughness. Clean or polish with armature paper, if required. A badly damaged slipring or a slipring worn down beyond the service limit should also be replaced. The service limit for the slipring outside diameter is 1.268 in.

2. Check for continuity between the field coil and slipring. If there is not continuity, the field coil is defective and the rotor must be replaced.

3. Check for continuity between the slipring and the shaft (or core). If there is continuity, the rotor assembly must be replaced.

4. Check for continuity between the leads of the stator coil. If there is no continuity, the stator coil is defective.

5. Check for an open circuit between the stator coil leads and the stator core. If there is continuity between the stator core and the coil leads, the stator assembly must be replaced.

6. Check for continuity between the positive (+) heat sink and the stator coil lead connection terminal with a circuit tester. If there is continuity in both directions, the diode is short circuited and the rectifier assembly must be replaced.

7. Perform Step 6 between the negative (−) heat sink and the stator coil lead connection.

8. Using a circuit tester, check the 3 diodes for continuity in both directions. If there is either continuity or an open circuit in both directions, the diode is defective and must be replaced.

9. Measure the length of the brush. If it is worn below 0.315 in., it must be replaced.

Assembly

1. Assembly is the reverse of disassembly with the following instructions:

2. Be sure to install both the front and rear seals on the front bearing. To install the rotor assembly in the rear bracket, push the brushes into the brush holder, insert a wire to hold them in the raised position and install the rotor. Remove the wire.

NISSAN

Charging System

ALTERNATOR

Disassembly

1. On diesel engine models only, remove the vacuum pump.

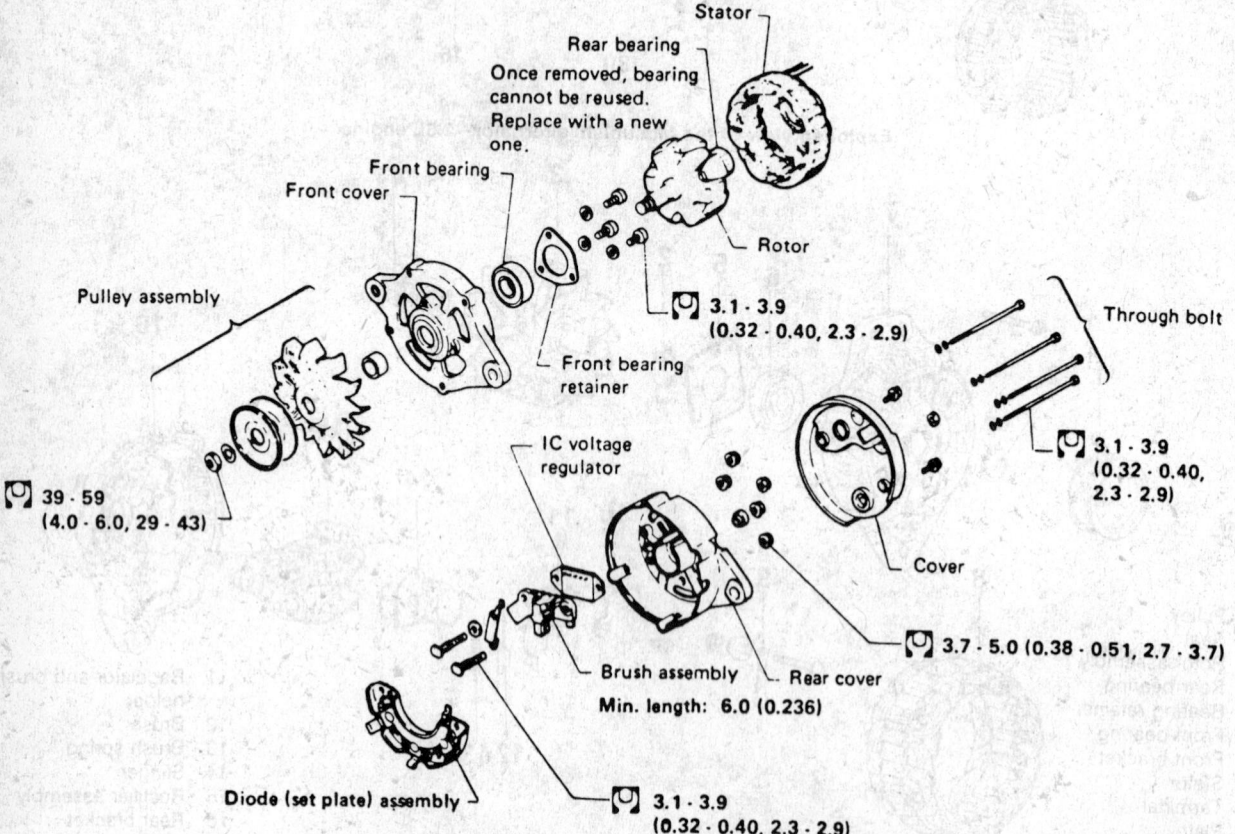

Exploded view of the 1986–89 Nissan alternator—gas engines

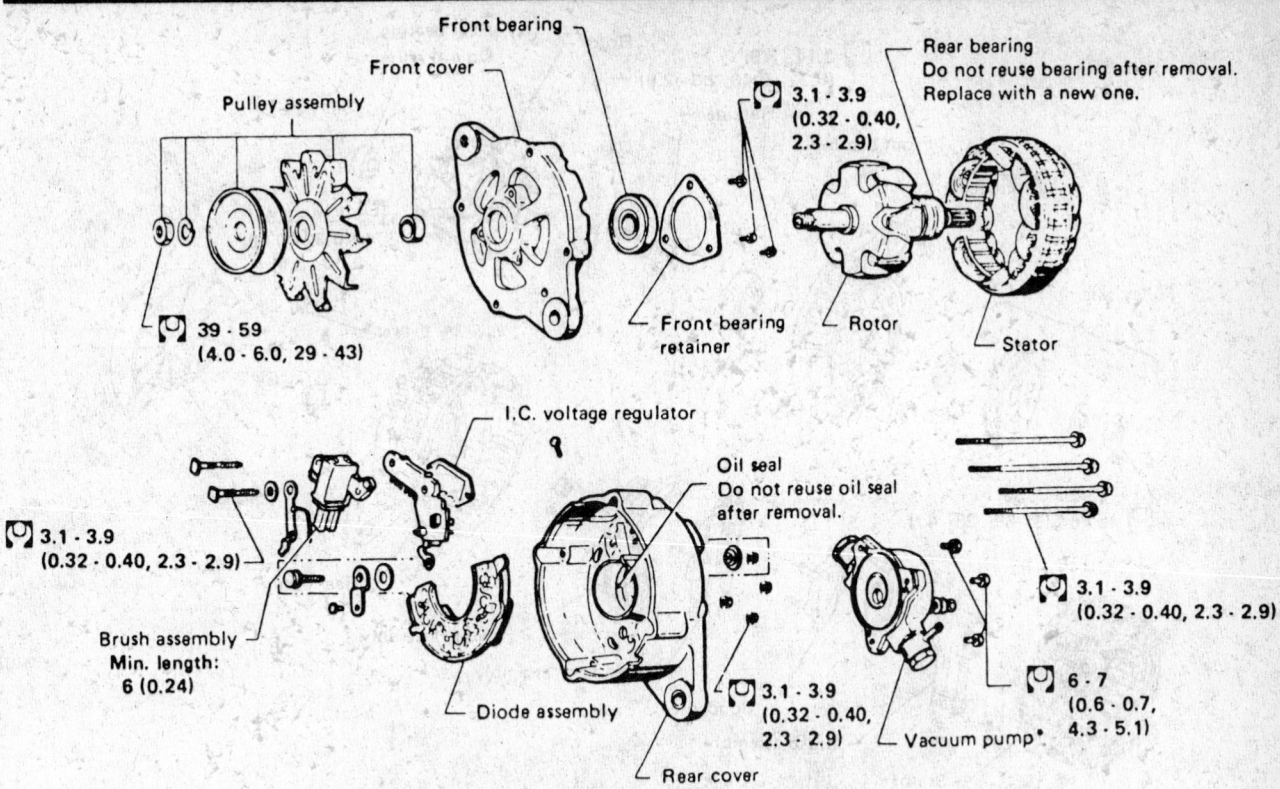

Exploded view of the Nissan alternator—diesel engine

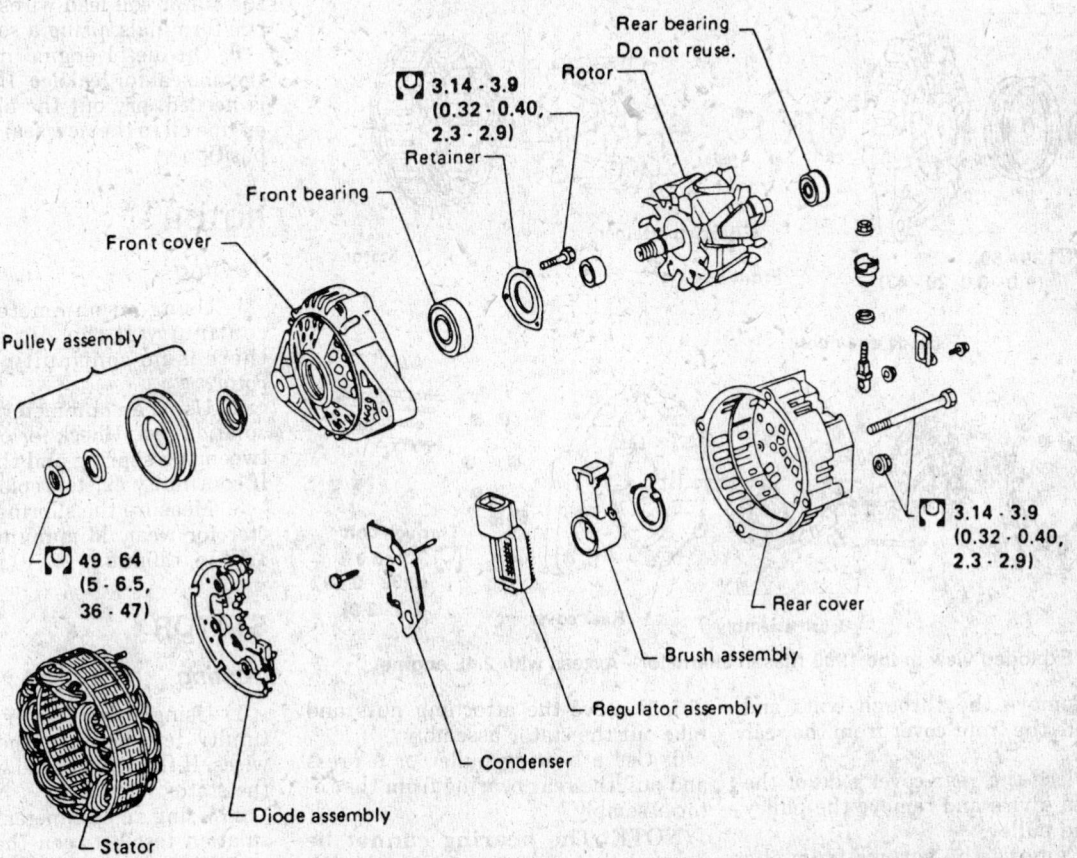

Exploded view of the 1990 Nissan alternator—Pick-Up and Pathfinder with 2.4L engine

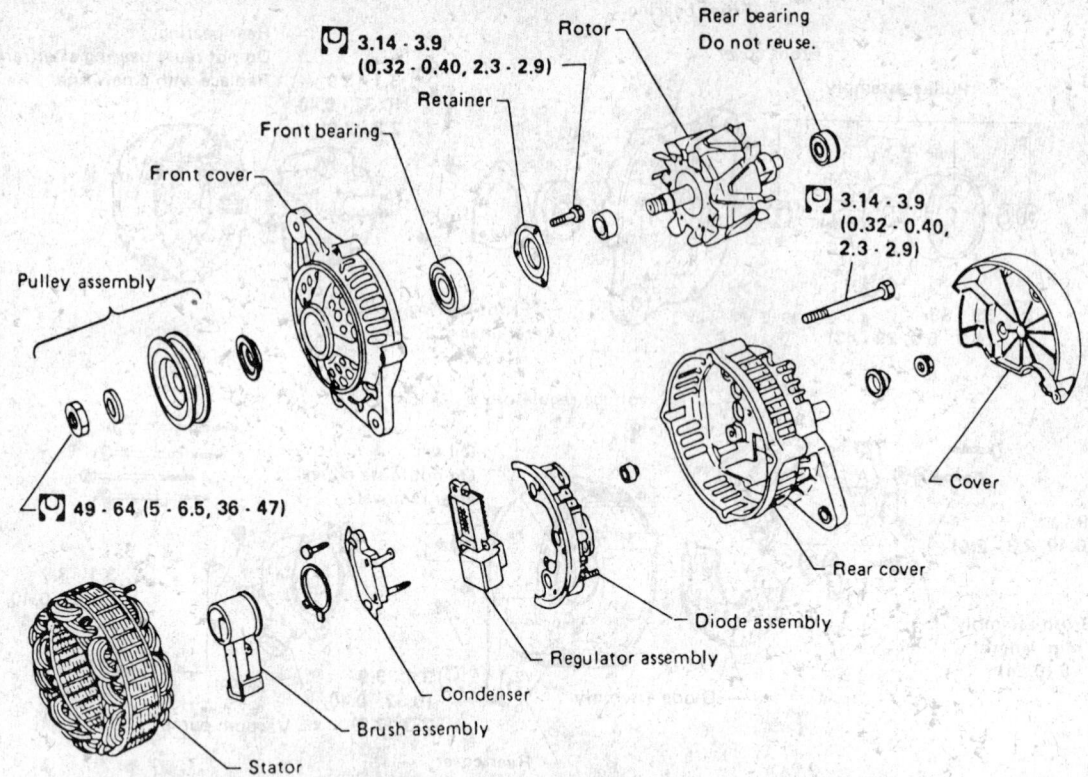

3.14 - 3.9
(0.32 - 0.40, 2.3 - 2.9)

Rotor

Rear bearing
Do not reuse.

Retainer

Front bearing

Front cover

3.14 - 3.9
(0.32 - 0.40,
2.3 - 2.9)

Pulley assembly

49 - 64 (5 - 6.5, 36 - 47)

Cover

Rear cover

Diode assembly

Regulator assembly

Condenser

Brush assembly

Stator

Exploded view of the 1990 Nissan alternator—Pick-Up and Pathfinder with 3.0L engine

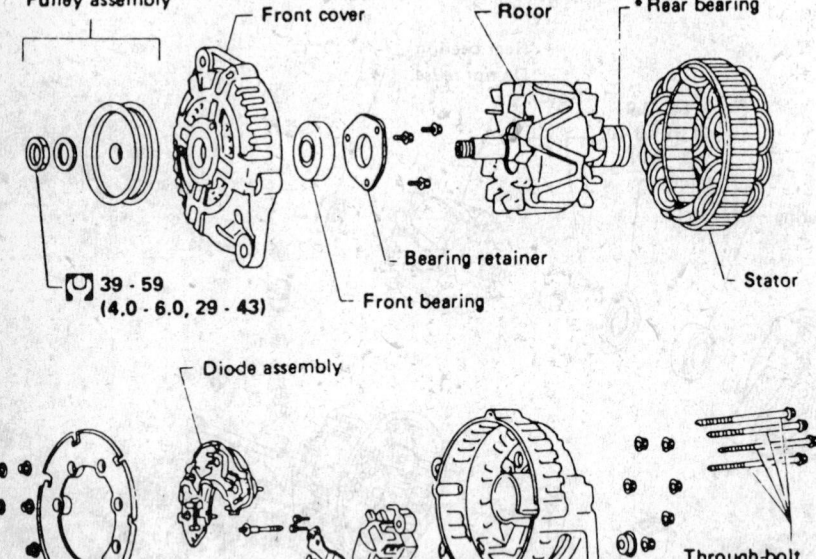

Pulley assembly

Front cover

Rotor

Rear bearing

39 - 59
(4.0 - 6.0, 29 - 43)

Bearing retainer

Front bearing

Stator

Diode assembly

Through-bolt

3.1 - 3.9
(0.32 - 0.40,
2.3 - 2.9)

Brush assembly

Rear cover

Exploded view of the 1990 Nissan alternator—Axxess with 2.4L engine

2. Remove the through bolts and separate the front cover from the rear cover.

3. Place the rear cover side of the rotor in a vise and remove the pulley nut and pulley.

4. Remove the screws from the bearing retainer.

5. Remove the attaching nuts and take out the stator assembly.

6. Use a bearing puller or a press and pull the rear bearing from the rotor assembly.

NOTE: The bearing cannot be reused and must be replaced with a new one.

7. To remove the stator, disconnect the stator coil lead wires from the diode terminals, using a soldering iron.

8. On diesel engine models, check the oil seal for leakage. If replacement is needed, pry out the old seal, apply engine oil to the new seal and install in position.

ROTOR

Testing

1. Using an ohmmeter, check for continuity at the slip end rings. If there is no continuity, replace the rotor.

2. Using an ohmmeter, make an insulation test. Check for continuity between the slipring and the rotor core. If continuity exists, replace the rotor.

3. Measure the slipring outer diameter for wear. Mimimum diameter is 1.18 in. (30mm).

STATOR

Testing

1. Using an ohmmeter, make a continuity test between the stator lead wires. If there is no lead wires, replace the stator.

2. Using an ohmmeter. make an insulation test between the stator core and the lead wire. If the continuity exists, replace the stator.

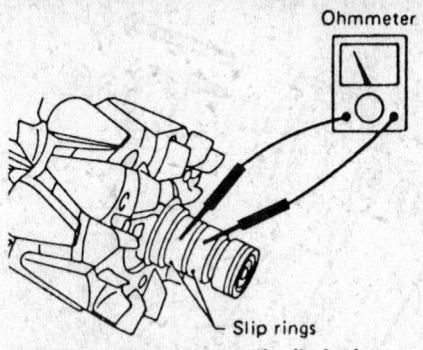

Performing the rotor continuity test

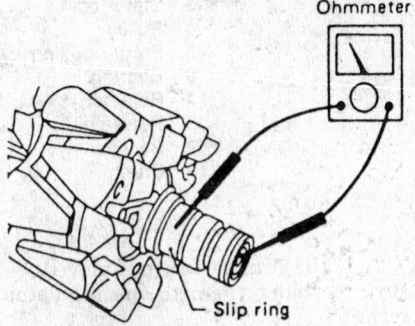

Performing the rotor ground test

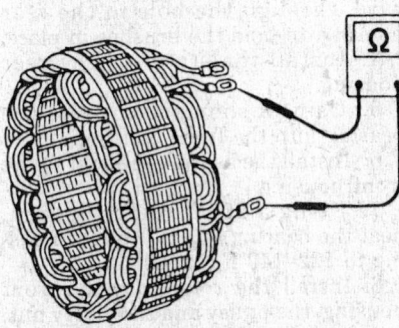

Performing the stator continuity test

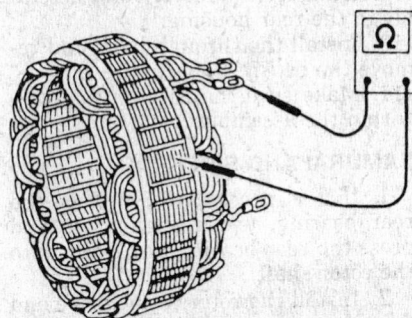

Performing the stator ground test

DIODE

Testing

1. Using an ohmmeter, perform a continuity test on diodes in both directions.
2. Replace diodes as necessary.

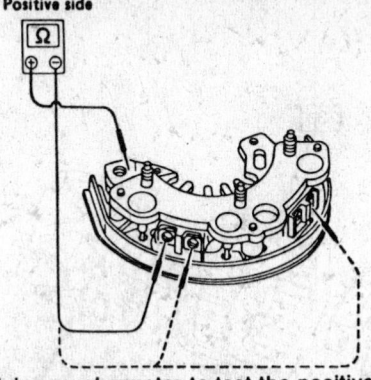

Using an ohmmeter to test the positive side diodes

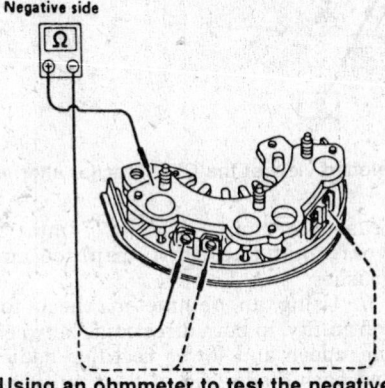

Using an ohmmeter to test the negative side diodes

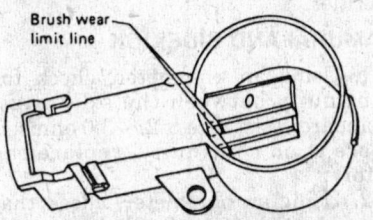

View of the brush wear limit lines

BRUSH

Testing

1. Check for smooth movement of the brush and clean the brush holder if necessary.
2. Check for brush wear at the wear limit line on the brush and replace if necessary.

Assembly

1. Assembly of the alternator is the reverse of disassembly with the following instructions: Solder each stator coil lead wire to the diode assembly terminal as quickly as possible. When soldering the brush lead wire, position the brush so that it extends 0.43 in. from the brush holder and wrap the coil lead wire at least 1.5 times around the terminal groove. Solder the outside of the terminal.
2. Tighten the pulley nut to 29–43 ft. lbs.

3. Before installing the front and rear sides of the alternator, push the brush up and retain the brush by inserting a wire from the outside into a lift hole. After installing the front and rear sides of the alternator, pull the brush lift by pushing towards the center.

NOTE: Do not pull brush lift by pushing towards the outside of the cover as it will damage the slipring as it slides.

SUZUKI/GEO

Charging System

ALTERNATOR

Disassembly

TRACKER

1. Place alignment matchmarks on the cases.
2. Remove the through bolts from the generator.
3. Using a 200W soldering iron, position it against the bearing box at the rear of the alternator housing and heat the bearing box to a temperature of 122–140°F (50–60°C).
4. Using a medium prybar, position it between the stator core and the front housing, then, separate the front housing from the rear housing.
5. Position the rotor/front housing into a soft jawed vise.
6. Remove the pulley nut and the pulley.
7. Remove the front housing from the rotor.
8. Remove the front bearing by performing the following procedures:
   a. Remove the front bearing retainer screws and the retainer.
   b. Using a shop press, press the bearing from the front housing.
9. Remove the stator-to-rear housing screws, the battery terminal nut and the stator.
10. If necessary to remove the brushes from the brush holder, use a soldering iron to unsolder the brush wire(s).
11. Remove the rectifier and condensor assembly from the rear housing.

SAMURAI

1. Remove the nut and terminal insulator, then, the nuts and end cover.
2. Remove the screws, the brush holder and the brush holder cover.
3. Remove the screws and the IC regulator.

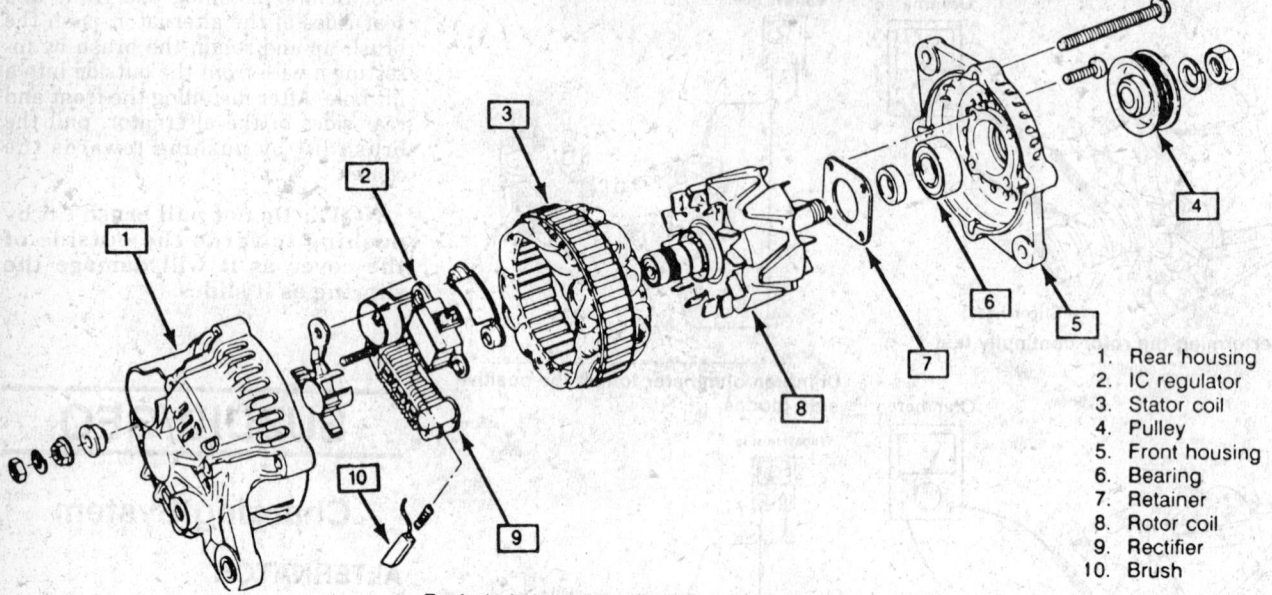

Exploded view of the GEO Tracker alternator

1. Rear housing
2. IC regulator
3. Stator coil
4. Pulley
5. Front housing
6. Bearing
7. Retainer
8. Rotor coil
9. Rectifier
10. Brush

4. Remove the rectifier holder screws and the holder.

5. Remove the terminal insulator.

6. Loosen the alternator pulley nut and the pulley.

NOTE: To prevent damage to the rotor shaft, do not loosen the pulley nut more than ½ turn.

7. Remove the nuts from the rear end frame. Using a puller, remove the rear end frame.

8. Remove the rotor from the drive end frame.

9. If necessary, remove the front bearing by removing the screws from the bearing retainer.

10. If necessary, remove the rear bearing by using a puller. Remove the rear bearing with the cover from the rotor shaft.

Testing

TRACKER

1. Using an ohmmeter, check the resistance between the rotor sliprings; if there is no continuity, replace the rotor.

2. Using an ohmmeter, check the resistance between the rotor slipring and the rotor; if there is continuity, replace the rotor.

3. Inspect the rotor sliprings for scoring or roughness; if rough or scored, replace the rotor.

4. Using an ohmmeter, check for continuity between the leads; if there is no continuity, replace the stator.

5. Using an ohmmeter, check for continuity between the leads and the stator core; if there is continuity, replace the stator.

6. Inspect the brushes for wear; the brush wear limit is 0.55 in. (14mm), if worn past the limit, replace the brushes.

7. Using an ohmmeter, check for continuity, in both directions, between the upper and lower rectifier bodies and each diode lead; if continuity exists in both directions, replace the rectifier.

SAMURAI AND SIDEKICK

1. Using an ohmmeter, check for continuity between the ssliprings. Standard resistance is 2.8–3.0 ohms; if there is no continuity, replace the rotor.

2. Using an ohmmeter, check that there is no continuity between the slipring and rotor; if there is continuity, replace the rotor.

3. Using an ohmmeter, check all leads for continuity. If there is no continuity, replace the drive end frame assembly.

4. Using an ohmmeter, check for continuity between the coil leads and the drive end frame; if there is continuity, replace the drive end frame assembly.

5. Measure the exposed brush length and replace, if necesary. Minimum length is 0.200 in. Also check that the brush moves smoothly in the brush holder.

6. Inspect the front and rear bearings for roughness and replace, if necessary.

Assembly

TRACKER

1. Install the rectifier and condensor assembly into the rear housing.

2. If the brushes were removed, be sure to solder them to the regulator terminal.

3. Using a stiff wire, depress the brushes into the holder and insert the wire, through the hole in the rear housing, to hold the brushes in place.

4. Install the stator to the rear housing.

5. Using a shop press, press the bearing into the front housing.

6. Install the bearing retainer to the front housing.

7. Using a 200W soldering iron, heat the bearing box of the rear housing to 122–140°F (50–60°C).

8. Install the rotor into the front housing, the pulley and the pulley nut; torque the nut to 48 ft. lbs. (65 Nm).

9. Align the matchmarks and assemble the front housing/rotor assembly to the rear housing.

10. Install the through bolts and remove the brush support wire.

11. Make sure the rotor turn freely within the assembly.

SAMURAI AND SIDEKICK

1. If it is necessary to replace the rear bearing, use a shop press and press the rear bearing and cover onto the rotor shaft.

2. Install the rotor to the drive end frame.

3. Using a plastic hammer, lightly tap the rear end frame on the drive end frame and install the nuts.

4. Install the alternator pulley and torque to 37–47 ft. lbs.

5. Install the terminal insulators on the lead wires.

6. Install the rectifier holder with the screws.

7. If it is necessary to install a new

1. Pulley nut
2. Pulley
3. Drive end frame
4. Stator
5. Stud bolt
6. Drive end bearing
7. Bearing retainer
8. Rotor
9. End housing bearing
10. Bearing cover
11. Wave washer
12. Rear end frame
13. Frame bolt
14. Rectifier
15. Insulator
16. Regulator
17. Brush
18. Brush holder
19. Rear end cover

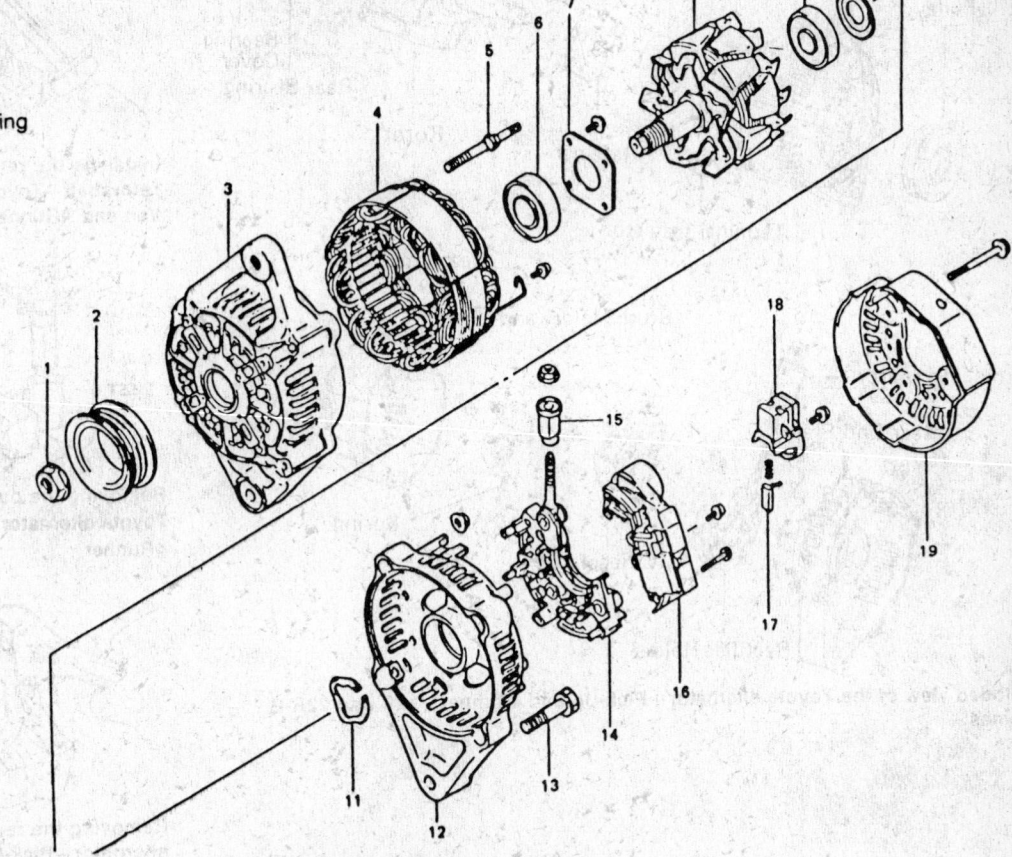

Exploded view of the Suzuki alternator

brush, unsolder and remove the brush and spring. Put the new brush wire through the spring and insert it into the brush holder. Solder the wire to the brush holder and cut off any excess.

8. Install the brush holder with the IC regulator and the screws to the IC regulator. Install the retaining screws and the brush holder cover to the rear end frame.

9. Install the end cover with the retaining nuts and install the terminal insulator and nut.

10. Make sure the rotor rotates smoothly.

TOYOTA

Charging System

ALTERNATOR

Disassembly
EXCEPT LAND CRUISER
1. Remove the nut and terminal in-

sulator. Remove the nuts and end cover.

2. Remove the screws, the brush holder and brush holder cover.

3. Remove the screws and the IC regulator.

4. Remove the rectifier holder screws and the holder.

5. Remove the terminal insulator(s).

6. To remove the pulley and nut, use tool 09820–63010 to perform the following procedure:

a. Hold the tool at point **A** with a torque wrench and tighten at point **B** clockwise to 29 ft. lbs. (39 Nm).

b. Confirm that the tool **A** is secured to the pulley shaft.

c. Grip tool **C** in a vise and install the alternator into the special tool **C**.

d. Loosen the pulley nut by turning tool **A** counter-clockwise.

NOTE: To prevent damage to the rotor shaft, do not loosen the pulley nut more than ½ turn.

e. Turn the tool at point **B** and remove all the tool components.

f. Remove the pulley nut and the pulley.

7. Remove the nuts from the rear end frame. Using a puller, tool 09286–46011, remove the rear end frame.

8. Remove the rotor from the drive end frame.

9. If necessary, remove the front bearing by removing the screws from the bearing retainer.

10. If necessary, remove the rear bearing by using a puller, tool 09820–00020. Remove the rear bearing with the cover from the rotor shaft.

LAND CRUISER

1. Remove the through bolts.

2. Using a prybar, pry the drive end frame/rotor assembly from the stator.

3. Mount the rotor assembly into a soft jawed vise.

4. Remove the pulley nut, the spring washer, the pulley and fan from the rotor.

5. Remove the spacer collar, drive end frame, the spacer ring and snapring.

6. From the rectifier end frame, remove the nuts, the condenser and the terminal insulators.

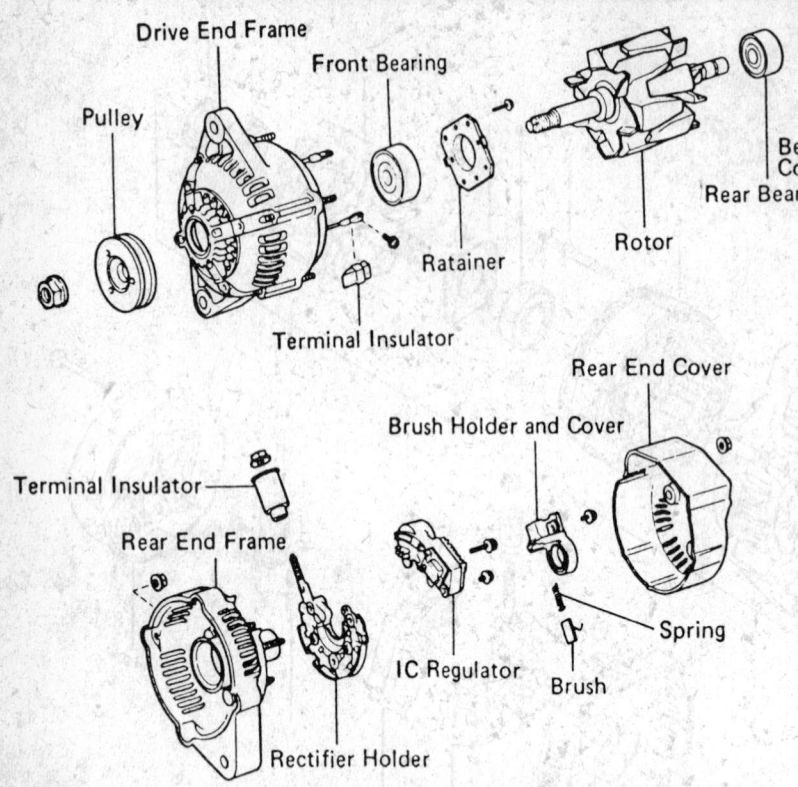

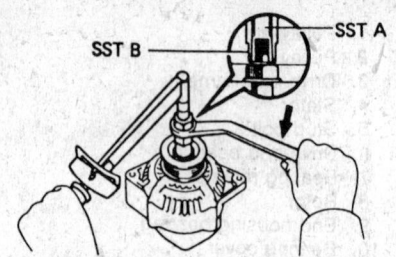

Installing the removal tool to the rotorshaft—Toyota alternator—Pick-Up, Van and 4Runner

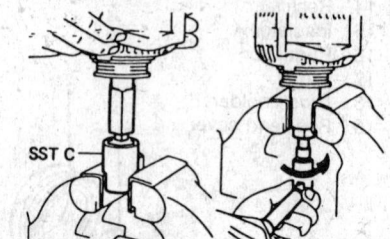

Removing the pulley nut and pulley—Toyota alternator—Pick-Up, Van and 4Runner

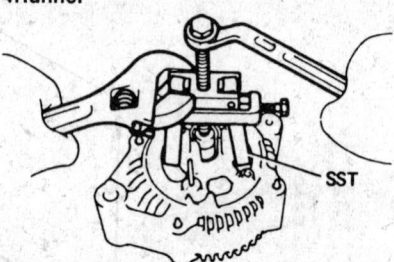

Removing the rear frame—Toyota alternator—Pick-Up, Van and 4Runner

Exploded view of the Toyota alternator—Pick-Up and 4Runner—22R and 22R-E engines

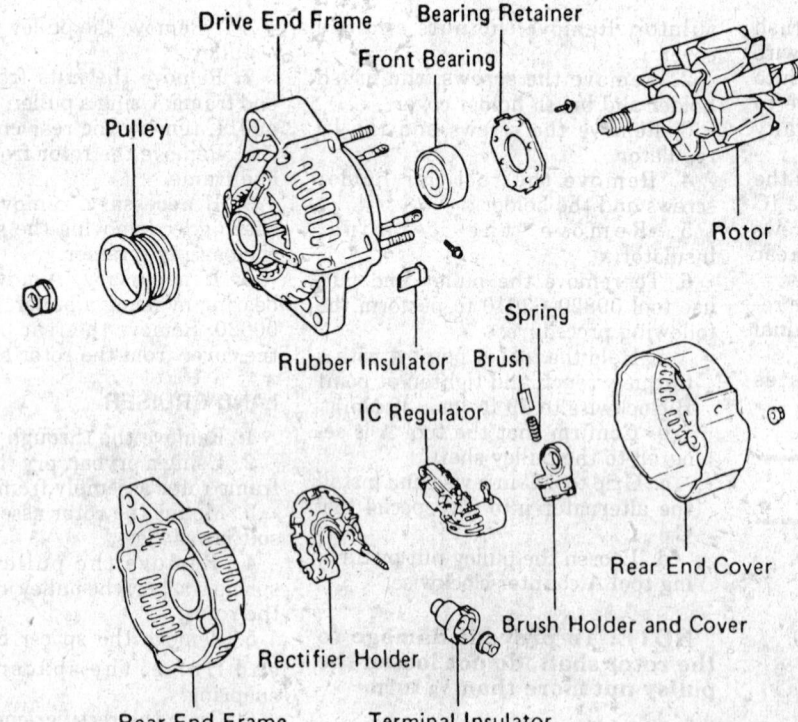

Exploded view of the Toyota alternator—Pick-Up and 4Runner—3VZ-E engine

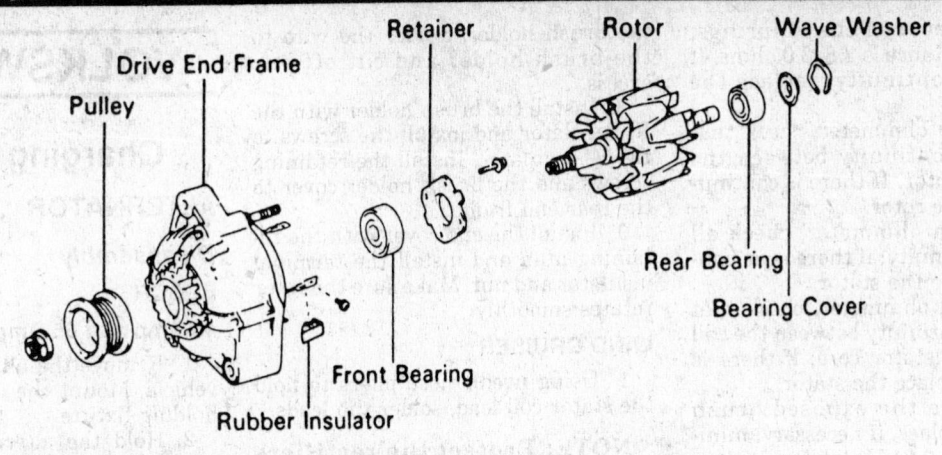

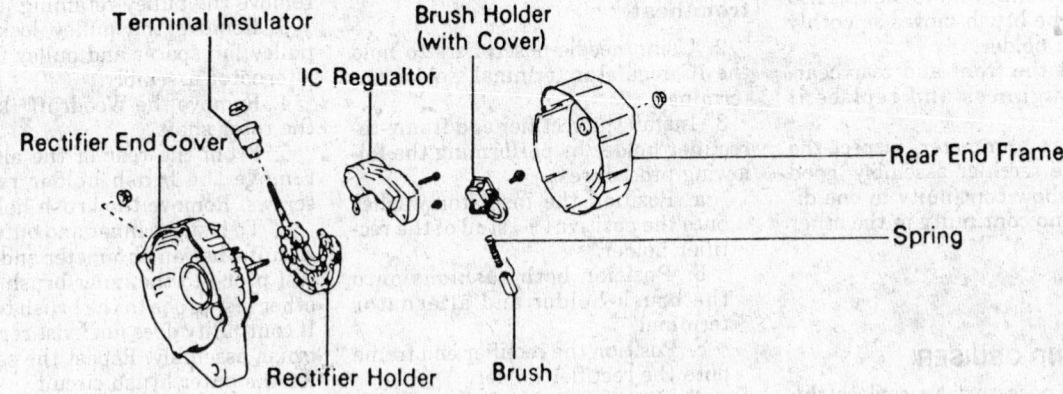

Exploded view of the Toyota alternator – Van

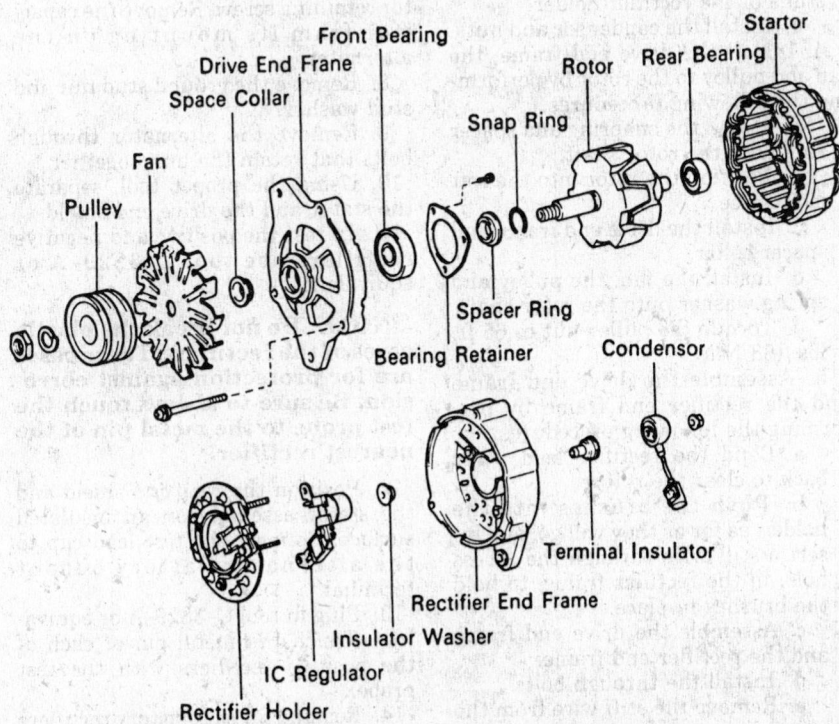

Exploded view of the Toyota alternator – Land Crusier

7. Remove the rectifier end frame and the insulator washer from the rectifier holder stud.

8. Using needle-nose pliers to hold the IC regulator terminal, unsolder the terminals.

NOTE: Protect the rectifiers from heat.

9. Using needle-nose pliers to hold the stator coil lead, unsolder the leads.

Testing
EXCEPT LAND CRUISER

1. Using an ohmmeter, check for continuity between the ssliprings; standard resistance is 2.8–3.0 ohms. If there is no continuity, replace the rotor.

2. Using an ohmmeter, check that there is no continuity between the slipring and rotor. If there is continuity, replace the rotor.

3. Using an ohmmeter, check all leads for continuity. If there is no continuity, replace the drive end frame assembly.

4. Using an ohmmeter, check that there is no continuity between the coil leads and the drive end frame. If there is continuity, replace the drive end frame assembly.

5. Measure the exposed brush length and replace, if necessary; minimum length is 0.059 in. (1.5mm). Also check that the brush moves smoothly in the brush holder.

6. Inspect the front and rear bearings for roughness and replace if necessary.

LAND CRUISER

1. Using an ohmmeter, check for

continuity between the ssliprings; standard resistance is 2.8–3.0 ohms. If there is no continuity, replace the rotor.

2. Using an ohmmeter, check that there is no continuity between the slipring and rotor. If there is continuity, replace the rotor.

3. Using an ohmmeter, check all leads for continuity; if there is no continuity, replace the stator.

4. Using an ohmmeter, check that there is no continuity between the coil leads and the stator core; if there is continuity, replace the stator.

5. Measure the exposed brush length and replace, if necessary; minimum length is 0.217 in. (5.5mm). Also check that the brush moves smoothly in the brush holder.

6. Inspect the front and rear bearings for roughness and replace if necessary.

7. Using an ohmmeter, inspect the diodes of the rectifier assembly; good diodes will show continuity in one direction and no continuity in the other direction.

Assembly
EXCEPT LAND CRUISER

1. If it is necessary to replace the rear bearing, use bearing installation and press the rear bearing and cover onto the rotor shaft.

2. Install the rotor to the drive end frame.

3. Using a plastic hammer, lightly tap the rear end frame on the drive end frame and install the nuts.

4. Install the pulley in the following manner:

   a. Using the pulley onto the rotor shaft and tighten the pulley nut by hand.

   b. Using tool 09820–63010, hold it at point **A** with a torque wrench and tighten tool **B** clockwise to 29 ft. lbs. (39 Nm).

   c. Confirm tool **A** is secured to the pulley shaft.

   d. Grip special tool **C** in a vise and then install the alternator to tool **C**.

   e. To torque the pulley nut, turn tool **A** and tighten to 81 ft. lbs. (110 Nm).

   f. Turn tool **B** and remove all the tool components.

5. Install the terminal insulators on the lead wires.

6. Install the rectifier holder with the screws.

7. If it is necessary to install a new brush, unsolder and remove the brush and spring. Put the new brush wire through the spring and insert it into the brush holder. Solder the wire to the brush holder and cut off any excess.

8. Install the brush holder with the IC regulator and install the screws to the IC regulator. Install the retaining screws and the brush holder cover to the rear end frame.

9. Install the end cover with the retaining nuts and install the terminal insulator and nut. Make sure the rotor rotates smoothly.

LAND CRUISER

1. Using needle-nose pliers to hold the stator coil lead, solder the leads.

NOTE: Protect the rectifiers from heat.

2. Using needle-nose pliers to hold the IC regulator terminal, solder the terminals.

3. Install the rectifier end frame-to-rectifier holder by performing the following procedures:

   a. Position the insulator washer onto the positive (+) stud of the rectifier holder.

   b. Position both cushions onto the brush holder and alternator terminal.

   c. Position the rectifier end frame onto the rectifier holder.

   d. Position both terminal insulators onto the positive (+) studs of the rectifier holder.

   e. Install the condensor and nuts.

4. Install the drive end frame, the fan and pulley to the rotor by performing the following procedures:

   a. Install the snapring and spacer ring onto the rotor shaft.

   b. Position the rotor into the soft jawed vise.

   c. Install the drive end frame and spacer collar.

   d. Install the fan, the pulley and spring washer onto the rotor shaft.

   e. Torque the pulley nut to 65 ft. lbs. (88 Nm).

5. Assemble the drive end frame and the rectifier end frame by performing the following procedures:

   a. Bend the rectifier lead wires back to clear the rotor.

   b. Push the brushes into the holder, as far as they will go, and insert a stiff wire through the access hole, in the rectifier frame, to hold the brushes in place.

   c. Assemble the drive end frame and the rectifier end frame.

   d. Install the through bolts.

   e. Remove the stiff wire from the access hole.

   f. Turn the rotor to make sure it turns freely.

   g. Seal the access hole.

VOLKSWAGEN

Charging System
ALTERNATOR

Disassembly
BOSCH
45 Amp and 65 Amp

1. Remove the alternator from the vehicle. Mount the unit in a suitable holding fixture.

2. Hold the alternator pulley and remove the pulley retaining nut.

3. Remove the pulley lockwasher, pulley fan spacer and pulley from the alternator assembly.

4. Remove the Woodruff® key from the rotor shaft.

5. From the rear of the alternator remove the brush holder retaining screws. Remove the brush holder.

6. To test the inner and outer brush circuits, use an ohmmeter and touch 1 test probe to the inner brush and the other test probe to the brush terminal. If continuity does not exist replace the brush assembly. Repeat the same test for the outer brush circuit.

7. Disconnect the capacitor electrical connection and remove the capacitor retaining screw. Remove the capacitor from its mounting on the alternator.

8. Remove the ground stud nut and stud washer.

9. Remove the alternator through bolts that retain the unit together.

10. Using the proper tool, separate the stator and the drive end shield.

11. To test the positive and negative rectifiers use tool C–3929–A or equivalent.

NOTE: Do not break the plastic cases of the rectifiers. These cases are for protection against corrosion. Be sure to always touch the test probe to the metal pin of the nearest rectifier.

12. Position the rear end shield and the stator assembly on an insulated surface. Connect the test lead clip to the alternator battery output terminal.

13. Plug in tool C–3829–A or equivalent. Touch the metal pin of each of the positive rectifiers with the test probe.

14. Reading for satisfactory rectifiers will be 1¾ amperes or more. Reading should be approximately the same and meter needle must move in the same direction for all 3 rectifiers.

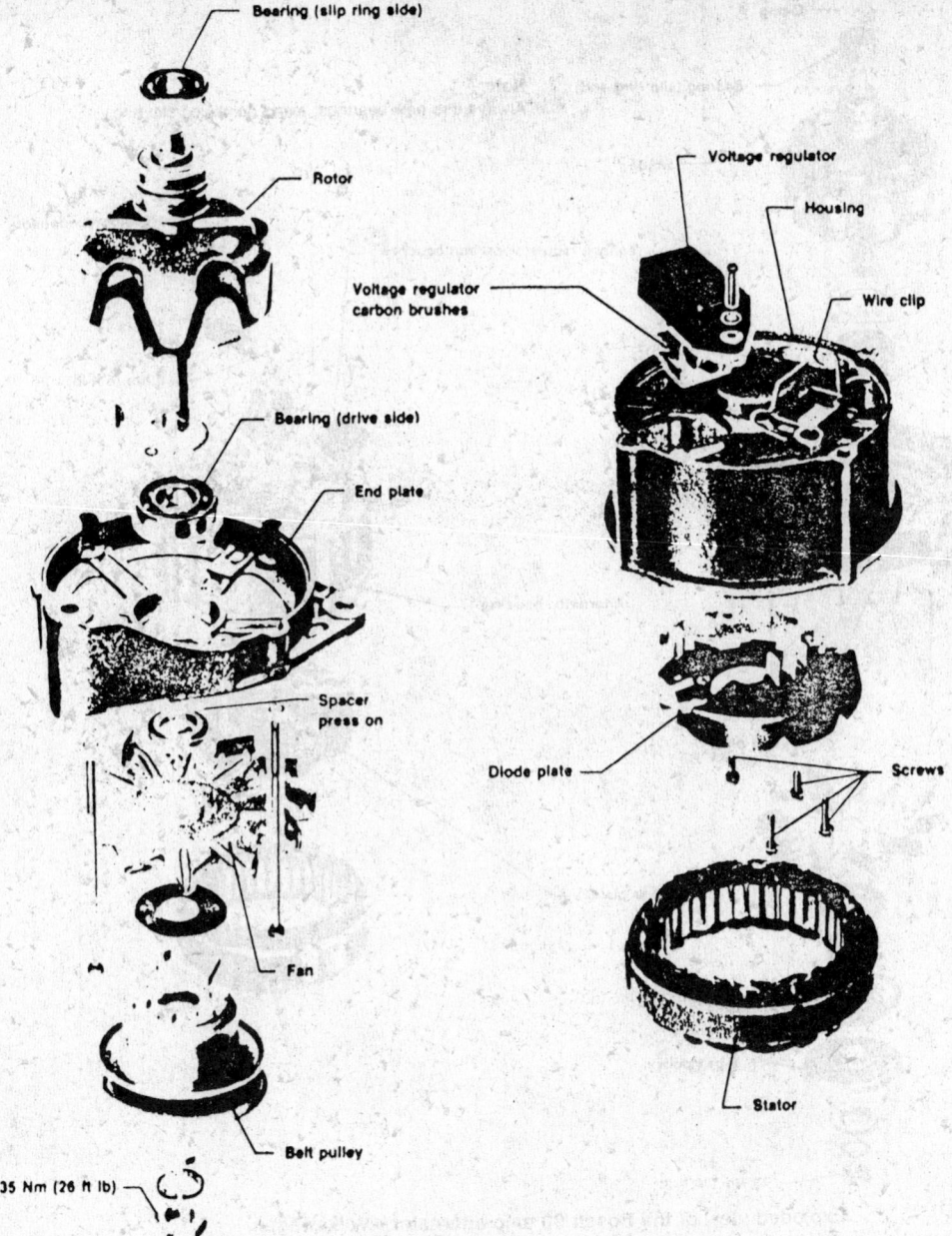

Exploded view of the Bosch 45 amp and 65 amp alternators—Volkswagen

15. When some rectifiers are good and 1 is shorted, the reading taken at good rectifiers will be low and the reading at shorted rectifiers will be 0. Disconnect stator lead to rectifiers reading 0 and retest. Reading of good rectifiers will now be within satisfactory range.

16. When a rectifier is open, it will read approximately 1 ampere and good rectifiers will read within satisfactory range.

17. To test the negative rectifiers, connect the test clip of tool C–3829–A to the rectifier end housing.

18. Touch the metal pin of each of the negative rectifiers with the test probe.

19. Test specifications are the same and test results will be approximately same as for positive case rectifiers except that the meter will read on opposite side of scale.

NOTE: If a negative rectifier shows a shorted condition, remove stator from rectifier assembly and retest. It is possible that a stator winding could be grounded to stator laminations or to an rectifier end shield which would indicate a shorted negative rectifier.

20. Remove the battery (B+) stud nut, stud lockwasher, stud flatwasher and stud insulator.

21. Remove the rectifier assembly retaining screws. Remove the stator as-

sembly along with the rectifier unit. Unsolder the stator to rectifier leads.

22. Check for continuity between stator coil leads. Press test probe firmly to each of 3 phase (stator) lead terminals separately. If there is no continuity, stator coil is defective. Replace stator assembly.

23. To test stator for ground, check for continuity between stator coil leads and stator coil frame. If there is continuity stator is grounded. Replace stator assembly.

24. Remove the rear bearing oil and dust seal. Check the rotor bearing surface for wear and scoring. Replace as required.

25. Remove the inner battery (B+) stud insulator.

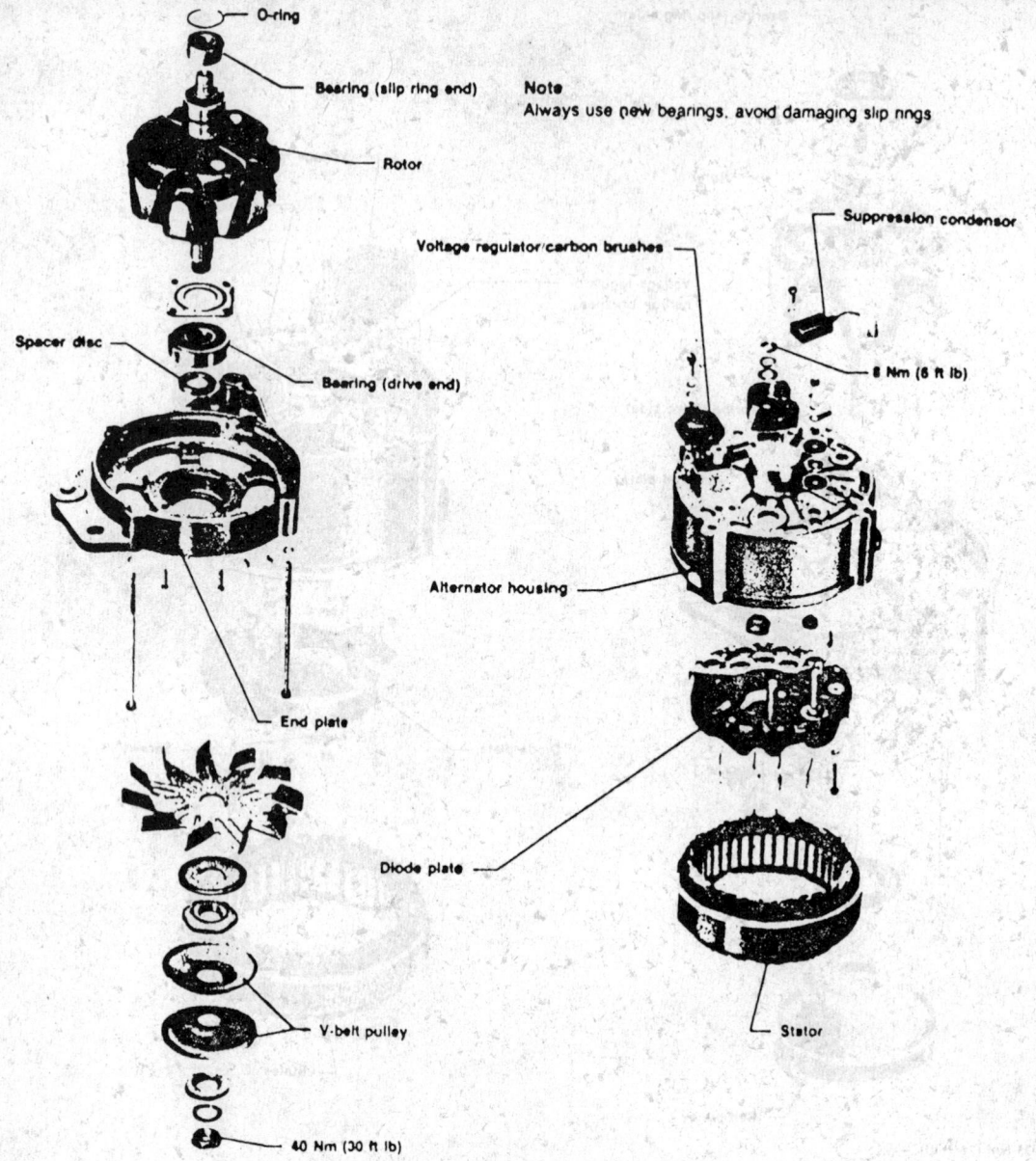

Exploded view of the Bosch 90 amp alternator—Volkswagen

26. Press the rotor out of the drive end shield and remove the spacer.

27. Check outside circumference of slipring for dirtiness and roughness. Clean or polish with fine sandpaper, if required. A badly roughened slipring or a worn down slipring should be replaced.

28. Check for continuity between field coil and sliprings. If there is no continuity, field coil is defective. Replace rotor assembly.

29. Check for continuity between sliprings and shaft (or core). If there is continuity, it means that coil or slipring is grounded. Replace rotor assembly.

30. Using a puller remove the rotor bearing.

31. Remove the front bearing from the drive end shield by removing the front bearing retaining screws.

32. Press out the drive end shield bearing. Remove the front drive bearing from the front drive end shield.

90 Amp

1. Remove the alternator from the vehicle. Position the unit in a suitable holding fixture.

2. Remove the pulley nut and lockwasher. Remove the alternator pulley.

3. Remove the pulley to fan spacer and pulley fan.

4. Remove the Woodruff key from the rotor shaft.

5. From the rear of the alternator disconnect the electrical terminal from the capacitor. Remove the capacitor retaining screw and the capacitor.

6. Remove the brush holder retaining screw and remove the brush holder from its mounting on the rear of the alternator.

7. Remove the alternator through bolts. Using a suitable tool pry between the stator and the drive end shield and carefully separate the assembly.

8. Press the rotor out of the drive end shield and remove the spacer. Remove the pulley fan spacer.

9. Remove the front alternator drive end bearing screws.

10. Remove the drive end shield bearing retainer and press out the drive end shield bearing.

11. Remove the front drive bearing from the front of the drive end shield.

12. To test the positive and negative

rectifiers use tool C-3929-A or equivalent.

NOTE: Do not break the plastic cases of the rectifiers. These cases are for protection against corrosion. Be sure to always touch the test probe to the metal pin of the nearest rectifier.

13. Position the rear end shield and the stator assembly on an insulated surface. Connect the test lead clip to the alternator battery output terminal.

14. Plug in tool C-3829-A or equivalent. Touch the metal pin of each of the positive rectifiers with the test probe.

15. Reading for satisfactory rectifiers will be 1¾ amperes or more. Reading should be approximately the same and meter needle must move in same direction for all 3 rectifiers.

16. When some rectifiers are good and 1 is shorted, the reading taken at good rectifiers will be low and the reading at shorted rectifiers will be 0. Disconnect stator lead to rectifiers reading 0 and retest. Reading of good rectifiers will now be within satisfactory range.

17. When a rectifier is open it will read approximately 1 ampere and the good rectifiers will read within satisfactory range.

18. Touch the metal pin of each of the negative rectifiers with the test probe.

19. Test specifications are the same and the test results will be approximately same as for positive case rectifiers except that the meter will read on opposite side of scale.

NOTE: If a negative rectifier shows a shorted condition remove stator from rectifier assembly and retest. It is possible that a stator winding could be grounded to stator laminations or to an rectifier end shield, which would indicate a shorted negative rectifier.

20. Unsolder the stator to rectifier leads. Mark the stator coil frame, to aid in reinstallation of the stator. Remove the stator from the rectifier end shield assembly.

21. Remove the 3 rectifier assembly mounting screws. Remove the rectifier assembly.

22. Remove the inner battery (B+) stud insulator.

23. Remove the D+ stud insulator, stud nut, stud flatwasher and stud insulating washer.

24. Remove the rear bearing oil and dust seal. Check the rotor bearing surface for scoring.

25. Using puller C-4068 or equivalent, remove the rear rotor bearing.

26. Check outside circumference of slipring for dirtiness and roughness. Clean or polish with fine sandpaper, as required. A badly roughened slipring or a worn down slipring should be replaced.

27. To check for an open rotor field coil, connect an ohmmeter to sliprings. Ohmmeter reading should be between 1.5-2.0 ohms on rotor coils at room ambient conditions. Resistance between 2.5-3.0 ohms would result from alternator rotors that have been operated on vehicle at higher engine compartment temperatures. Readings above 3.5 ohms would indicate high resistance rotor coils and further testing or replacement may be required.

28. To check for a shorted field coil connect an ohmmeter to sliprings. If reading is below 1.5 ohms field coil is shorted.

29. To check for a grounded rotor field coil connect an ohmmeter from each slipring to rotor shaft.

NOTE: Ohmmeter should be set for infinite reading when probes are apart and 0 when probes are touching. The ohmmeter should read infinite. If reading is 0 or higher, rotor is grounded.

30. Check for continuity between leads of stator coil. Press test probe firmly to each of 3 phase (stator) lead terminals separately. If there is no continuity, stator coil is defective. Replace stator assembly.

31. To test the stator for ground check for continuity between the stator coil leads and the stator coil frame. If there is no continuity the stator is grounded and must be replaced.

32. To test the inner and outer brush circuit, use an ohmmeter and touch 1 test probe to the inner brush and the other test probe to the brush terminal. If continuity does not exist replace the brush assembly. Repeat the same procedure for the outer brush.

Assembly

45 Amp and 65 Amp

1. Be sure to check all parts for wear and replace the defective components as required.

2. Install the rear rotor bearing oil and dust seal.

3. Install the inner alternator battery terminal insulator.

4. Solder the stator leads to the rectifier assembly; be sure to use needle nose pliers as a heat sink.

5. Position the stator and rectifier assembly. Install the rectifier mounting screws, both terminal insulators, the insulator washer, the insulator lockwasher and the insulator nut.

6. Position and press the front bearing into the drive end shield. Install the bearing retainer.

7. Position the drive end shield and spacer over the rotor. Press the drive end shield onto the rotor.

8. Install the rectifier end shield over the drive end shield. Install the through bolts and tighten.

9. Install the capacitor and terminal plug onto the alternator assembly.

10. Push the brushes into the brush holder and install the brush holder onto the alternator assembly.

11. Install the Woodruff key into the shaft and the fan over the shaft.

12. Install the drive pulley-to-fan spacer, the pulley, the lockwasher and the nut over the shaft. Secure the pulley and tighten the nut.

90 Amp

1. Be sure to check all parts for wear and replace the defective components as required.

2. Install the rear rotor bearing oil and dust seal.

3. Install the inner alternator battery B+ terminal insulator.

4. Position the rectifier assembly. Install the rectifier mounting screws, the insulator, the insulator washer, the insulator lockwasher and the insulator nut.

5. Position the stator assembly into the rectifier end shield. Align the scribe marks on the stator and the rectifier end shield.

6. Solder the stator leads to the rectifier assembly; be sure to use needle nose pliers as a heat sink.

7. Position and press the front bearing into the drive end shield. Install the bearing retainer and the pulley fan spacer onto the drive end shield.

8. Position the drive end shield and spacer over the rotor. Using a socket wrench, press the drive end shield onto the rotor.

9. Install the rectifier end shield and stator assembly into the drive end shield and rotor assembly. Install the through bolts and tighten.

10. Push the brushes into the brush holder and install the brush holder onto the alternator assembly.

11. Install the capacitor and terminal plug onto the alternator assembly.

12. Install the Woodruff key into the shaft and the fan over the shaft.

13. Install the drive pulley-to-fan spacer, the pulley, the lockwasher and the nut over the shaft. Secure the pulley and tighten the nut.

SWITCHES AND SOLENOIDS

Magnetic Switches

Magnetic switches serve only to make contact for the starter motor. Usually, such switches are located on the inner fender panel, although they are found mounted on the starter in a few cases.

MAGNETIC SWITCHES WITH TWO CONTROL TERMINALS

On this type of magnetic switch current is supplied from the ignition switch or transmission neutral button to 1 of the magnetic switch control terminals. The other control terminal is connected to the transmission neutral safety switch (on the transmission) where it is grounded.

MAGNETIC SWITCHES WITH IGNITION RESISTOR BYPASS TERMINALS

All normally use a magnetic switch with a single control terminal. The second terminal is an ignition resistor bypass terminal.

SOLENOIDS WITHOUT RELAYS

This type of starter solenoid is always mounted on the starter. Makes electrical contact for the starter and pulls the starter and drive clutch into mesh with the flywheel. The Chrysler reduction gear starter has this solenoid embodied in the starter housing.

There is only 1 control terminal on the solenoid and the ignition bypass terminal is usually marked **R** or **IGN**, if it is used.

SOLENOIDS WITH SEPARATE RELAYS

The solenoid itself is always mounted on the starter. In addition to making contact for the starter, it also pulls the starter drive clutch gear into mesh with the flywheel. A single control terminal is used on the solenoid itself. The relay is usually found mounted to the inner fender panel or on the firewall.

SOLENOIDS WITH BUILT-IN RELAYS

These units are always mounted on the starter and are connected, through linkage, to the starter drive clutch. The relay portion is built into and integral with the front end of the solenoid assembly.

NEUTRAL SAFETY SWITCHES

The purpose of the neutral safety switch is to prevent the starter from cranking the engine except when the transmission is in neutral or park.

On some vehicles, the neutral safety switch is located on the transmission. It serves to ground the solenoid or magnetic switch, whichever is used.

On other vehicles, the neutral safety switch is located on the steering column, where it contacts the shift mechanism within the steering column or on the shift linkage when a console is used.

Some manual transmission models have a clutch linkage safety switch to prevent starter operation unless the clutch pedal is depressed.

On most vehicles, the neutral safety switch and the backup light switch are combined into a single switch mechanism.

Troubleshooting Neutral Safety Switches Quick Test

If the starter fails to function and the neutral safety switch is to be checked, a jumper can be placed across its terminals. If the starter then functions, the safety switch is defective.

In the case of neutral safety switches with 1 wire, the wire must be grounded for testing purposes. If the starter works with the wire grounded, the switch is defective.

NEUTRAL SAFETY SWITCH/ BACK-UP LIGHT SWITCH

When the neutral safety switch is built in combination with the back-up light switch, the quickest way to determine which terminals are for the back-up lights is to take a test lamp/jumper wire and cross from a hot wire to a neutral wire. The wires which light the back-up lamps should be ignored when testing the neutral safety switch. Once the back-up light wires have been located, jump the other pair of wires to test the neutral safety switch. If the starter functions only when the jumper is placed across both wires, the neutral safety switch is defective or requires adjustment.

IMPORT STARTING SYSTEMS

STARTER MOTOR TESTING

The starter circuit should be divided and tested in 4 separate phases:
1. Cranking voltage check
2. Amperage draw
3. Voltage drop on grounded side
4. Voltage drop on battery side

NOTE: The battery must be in good condition for this test to have significance. To accurately check battery condition, use equipment designed to measure its capacity under a load. Instructions accompanying the equipment should be followed.

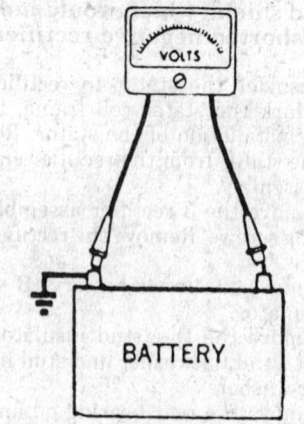

Voltmeter connected to battery for cranking voltage test

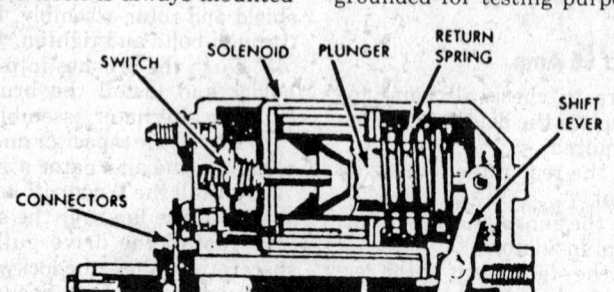

Sectional view of the starter solenoid mounted on a starter motor

Cranking Voltage

Connect voltmeter leads to prods tapped into the battery posts (observe polarity and reverse meter leads if necessary). Remove the high tension wire from the distributor cap and ground it to prevent engine starting. With electronic ignition, disconnect the control box harness from the distributor. Turn the key to the start position. Observe both voltmeter reading and cranking speed. The cranking speed should be even and at a satisfactory rate of speed, with a voltmeter reading of at least 9.6 volts for 12 volt systems.

Amperage Draw

The amount of current the starter motor draws is usually (but not always) associated with the mechanical problems involved in cranking the engine. (Mechanical trouble in the engine, frozen or worn starter parts, misaligned starter or starter components, etc.) Because starter motor amperage draw is directly influenced by anything restricting the free turning of the engine or starter, it is important that the engine and all components be at operating temperatures.

To measure starter current draw, remove the high tension wire from the center of the distributor cap and ground it. With electronic ignition, disconnect the control box harness from the distributor. A very simple and inexpensive starter current indicator is available. This indicator is an induction type gauge and shows, without disconnecting any wires, starter current draw.

Place the yoke of the meter directly over the insulated starter supply cable (cable must be straight for a minimum of 2 in.). Close the starter switch for about 20 seconds, watch the meter dial and record the average reading. If the indicator swings in the wrong direction, reverse the position of the meter.

The cranking amperage draw can vary from 150–400 amperes, depending on the engine size, engine compression and starter type.

NOTE: When starter specifications are not available, average starter draw amperage can be derived from testing a like starter unit, known to be operating satisfactorily.

More accurate equipment is available from many manufacturers. This equipment consists of a combination voltmeter, ammeter and carbon pile rheostat. When using this equipment, follow the equipment manufacturer's procedures and recommendations.

High amperage and lazy performance would suggest an excessively

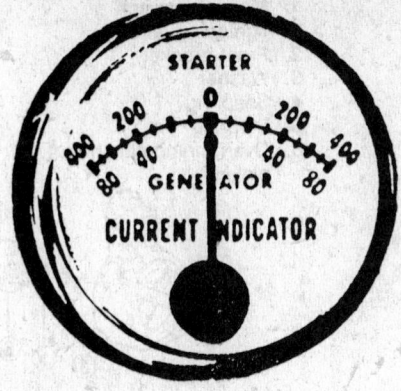

Starter current indicator

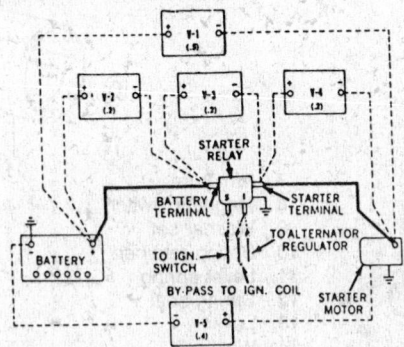

Starter cable resistance tests

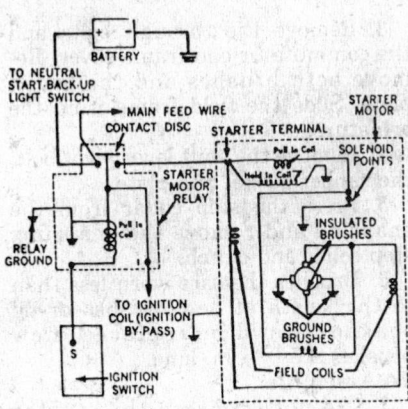

Positive engagement starter circuits

tight engine, friction in the starter or starter drive, grounded starter field or armature.

Normal amperage and lazy performance suggest high resistance or possibly poor connections somewhere in the starter circuit.

Low amperage and lazy or no performance suggest battery condition poor, bad cables or connections along the line.

Voltage Drop on Grounded Side

With a voltmeter on the 3 volt scale and without disconnecting any wires, connect negative test lead of the voltmeter to a prod secured in the ground-

ed battery post. The positive test lead is connected to a cleaned, bare metal portion of the starter motor housing. Close the starter switch and note the voltmeter reading. If the reading is the same as battery reading, the ground circuit is open somewhere between the battery and the starter. In many cases, the reading will be very small. The reading shown will indicate voltage drop (loss) between battery ground post and starter housing. The drop should not exceed 0.2 volts. If the voltage drop is above the specified amount, the next step is to isolate and correct the cause. It can be a bad cable or connection anywhere in the battery-to-starter ground circuit. A check of this type should progress along the various points of possible trouble between the battery ground post and the starter motor housing until the trouble spot has been located.

Voltage Drop on Battery Side

Bad starter cranking may result from poor connections or faulty components of the battery or hot phase of the starter motor circuit. To check this phase of the circuit, without disconnecting any wires, connect a lead of a voltmeter to a prod secured in the hot post of the battery and the other voltmeter lead to the field terminal of the starting motor. The meter should be set to the 16–20 volt scale. Before closing the starter switch, the voltmeter reading will be that of the battery. After closing the starter switch, change the selector on the voltmeter to the 3 volt scale. With a jumper wire between the relay battery terminal and the relay starter switch terminal, crank the engine. If the starting motor cranks the engine, the relay (solenoid) is operating.

While the engine is being cranked, watch the voltmeter. It should not register more than 0.5 volts. If more than this, check each part of the circuit for voltage drop to isolate the trouble, (high resistance).

Without disturbing the voltmeter-to-battery hook-up, move the free voltmeter lead to the battery terminal of the relay (solenoid) and crank the engine. The voltmeter should show no more than 0.1 volts.

If this reading is correct, move the same voltmeter lead to the starting motor terminal of the relay (solenoid). While the engine is being cranked, the voltmeter should show no more than 0.3 volts. If it does, the trouble lies in the relay.

If the reading is correct, the trouble is in the cable or connections between the relay and the starting motor.

DIAGNOSIS

Starter won't Crank Engine

1. Dead battery.
2. Open starter circuit, such as:
   a. Broken or loose battery cables.
   b. Inoperative starter motor solenoid.
   c. Broken or loose wire from starter switch to solenoid.
   d. Poor solenoid or starter ground.
   e. Bad starter switch.
3. Defective starter internal circuit, such as:
   a. Dirty or burnt commutator.
   b. Stuck, worn or broken brushes.
   c. Open or shorted armature.
   d. Open or grounded fields.
4. Starter motor mechanical faults, such as:
   a. Jammed armature end bearings.
   b. Bad bearing, allowing armature to rub fields.
   c. Bent shaft.
   d. Broken starter housing.
   e. Bad starter worm or drive mechanism.
   f. Bad starter drive or flywheel driven gear.
5. Engine hard or impossible to crank such as:
   a. Hydrostatic lock caused by water or other liquid in combustion chamber.
   b. Crankshaft seizing in bearings.
   c. Piston or ring seizing.
   d. Bent or broken connecting rod.
   e. Seizing of connecting rod bearing.
   f. Flywheel jammed or broken.

Starter Spins Free, won't Engage

Sticking or broken drive mechanism.

CHRYSLER IMPORTS

Starter System

DIRECT DRIVE TYPE

Disassembly and Assembly

MITSUBISHI

1986

1. Position the assembly in the soft jawed vise. Disconnect the field coil wire from the solenoid terminal.
2. Remove the solenoid-to-starter screws and the solenoid.

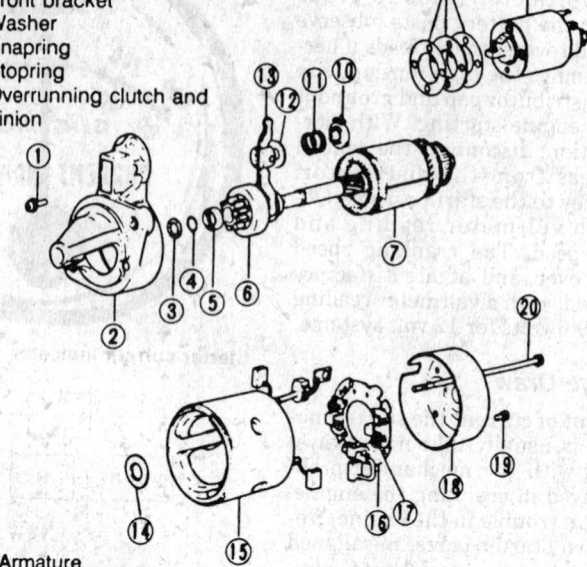

1. Magnetic switch tightening screw
2. Front bracket
3. Washer
4. Snapring
5. Stopring
6. Overrunning clutch and pinion

7. Armature
8. Magnetic switch
9. Washer set
10. Spring retainer
11. Lever spring
12. Spring seat
13. Lever
14. Washer

15. Yoke
16. Brush holder assembly
17. Brush spring
18. Rear bracket
19. Brush holder tightening screw
20. Through bolt

Exploded view of the Mitsubishi direct drive starter–1986 Chrysler

3. Remove the through bolts and the commutator end frame cover. Remove both brushes and the brush plate. Slide the field frame from the armature.
4. Remove the shift lever pivot bolt, the rubber gasket and plate.
5. Press the stop collar from the snapring and remove the snapring, stop collar and clutch.
6. Brushes that are worn less than ½ the length of new brushes or oil soaked should be replaced; new brushes are $^{11}/_{16}$ in. long.

To Assemble:

7. Do not immerse the starter clutch unit in cleaning solvent; lubricant will be washed from inside the clutch.
8. Place the drive unit on the armature shaft and while holding the armature, rotate the pinion. The drive pinion should rotate smoothly in 1 direction only. The pinion may not rotate easily but as long as it rotates smoothly it is in good condition. If the clutch unit does not function properly or if the pinion is worn, chipped or burred replace the unit.
9. To assemble, lubricate the bushings/splines and reverse the disassembly procedures.
10. Install the clutch, stop collar, lock ring and shaft fork on the armature.
11. Install the armature assembly and shift fork in the drive end housing.
12. To complete the installation, reverse the removal procedures.

FORD

1987–90

1. Position the starter in a soft jawed vice.
2. Remove the field strap-to-solenoid nut and the field strap.
3. Remove the solenoid-to-drive end housing screws. Remove the solenoid from the housing by guiding it away from the drive end housing and the plunger.
4. Disconnect the plunger from the drive yoke. If shims are present between the solenoid and the starter, save them for reinstallation; the shims determine the starter pinion depth clearance.
5. Remove the starter housing through bolts. Separate the starter housing from the drive end housing.
6. Remove the rear cover-to-field frame screws, separate the strap grommet from the rear cover and remove the rear cover.
7. Using a small pry bar, lift the retaining springs and remove the brushes from their channels.

NOTE: Before removing the brush plate, remove the brushes from the plate; this will prevent possible damage to the brushes.

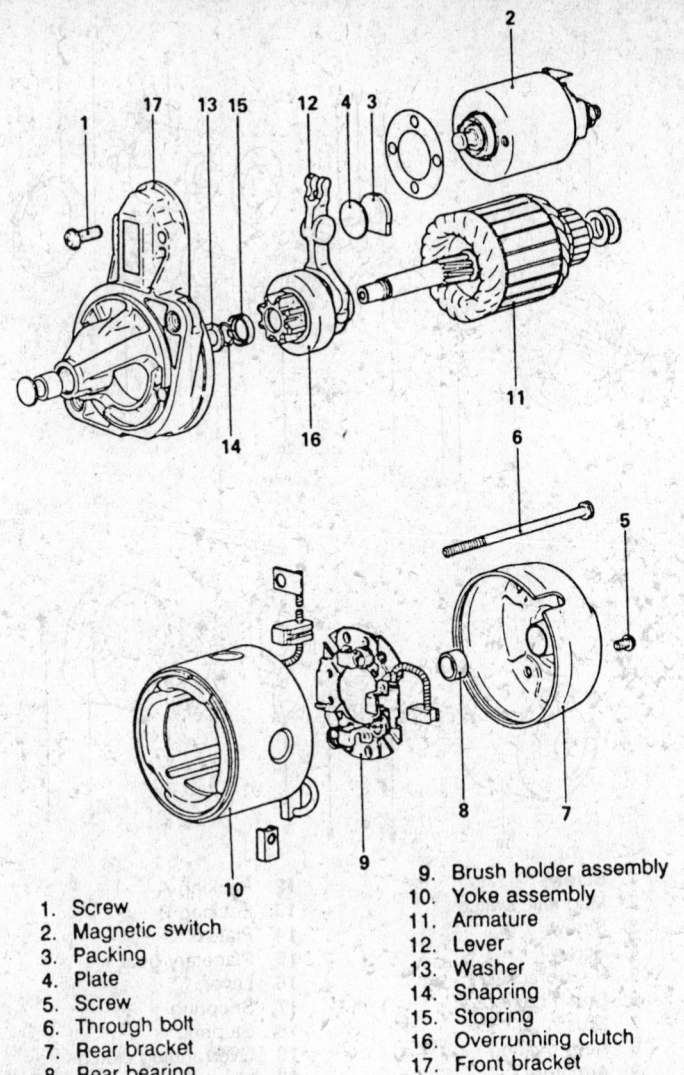

1. Screw
2. Magnetic switch
3. Packing
4. Plate
5. Screw
6. Through bolt
7. Rear bracket
8. Rear bearing
9. Brush holder assembly
10. Yoke assembly
11. Armature
12. Lever
13. Washer
14. Snapring
15. Stopring
16. Overrunning clutch
17. Front bracket

Exploded view of the Ford direct drive starter – 1987–90 Chrysler

8. Note the position of the yoke and separate it from the drive pinion. Remove the armature and the drive pinion from the drive housing.

9. From the drive housing, remove the yoke, seal and washer.

NOTE: When removing the drive pinion, do not clamp it in a vise, for damage to the internal parts may occur.

10. Using a deep socket or equivalent, drive the armature collar towards the armature to expose the snapring. Remove the snapring from the armature's groove and slide the drive pinion from the armature.

11. Using an ohmmeter, check the each commutator-to-armature core for grounds; there should be no movement of the ohmmeter indicator. If the ohmmeter indicates a ground, replace the armature.

12. Inspect the commutator burn spots, scored surface and/or dirt. Us-ing a set of V-blocks and a dial indicator, check the commutator runout. If the runout is greater than 0.002 in., refinish the commutator or replace the armature.

13. Using a micrometer, check the commutator's outer diameter; if it is less than 1.220 in., replace the armature.

NOTE: Never use emery cloth to clean the commutator face.

14. Inspect the depth of the insulating material between the commutator segments; it should be greater than 0.008 in. If necessary to undercut the insulating material, use a broken hacksaw blade and scrap the material to a depth of 0.020–0.031 in.

15. If the armature core shows signs of scuffing, the bushings are probably worn and need replacement.

16. Inspect the field coil for corrosion, insulation burnt/bare spots and/or deterioration; if necessary, replace the field coil housing assembly.

17. Using an ohmmeter, check the field strap connector-to-brushes for continuity; if there is no continuity, replace the field coil housing assembly.

18. Using an ohmmeter, check for continuity between the field strap connector and the field coil housing; if there is continuity, replace the field coil housing assembly. When performing this test, be certain the brushes and wires are not touching the housing.

19. Measure the brush lengths for wear, if they are near or beyond 0.453 in., replace the brushes.

20. To inspect the drive pinion, perform the following procedures:

a. Inspect the drive pinion teeth for excessive wear or milling. If either condition exists, the drive pinion and flywheel (manual) or flexplate (automatic) must be replaced.

b. To check the one-way clutch, try to turn the drive pinion in both directions; it should turn freely one-way and lock up the other way.

To Assemble:

21. Lubricate the armature splines with Lubriplate® 777 or equivalent. Install the drive pinion and the locking collar on the armature. Install the snapring and pull the collar over the snapring to secure it.

22. Using Lubriplate® 777 or equivalent, lubricate the shift fork and install it into the drive end housing. Engage the armature assembly into the drive end housing and couple the shift fork with the drive pinion.

23. Position the drive end housing into a soft jawed vise (nose down) and install the plug and seal into the housing recess.

24. Lower the field coil housing over the armature and seat it onto the drive end housing; position the housing so the field strap is on the solenoid side.

25. Install the washers onto the armature. Load the brushes into the brush plate holders. With the brushes pull all the way back in the holders, position the brush springs on the brush sides.

26. Install the brush plate over the commutator. Push the brushes toward the commutator until the springs snap onto the brush ends. Make sure the brush wires do not contact any metal parts.

27. Seat the field strap grommet and the rear cover. Install the through bolts and torque them to 55–75 ft. lbs. Install the solenoid and shims, if equipped.

28. To check the pinion depth, perform the following procedures:

a. If the field strap was connected to the solenoid terminal, disconnect it.

b. Using a 12 volt battery, attach the negative (−) terminal to the solenoid's M-terminal and the positive (+) terminal to the solenoid's S-terminal; this will energize the solenoid.

NOTE: When energizing the solenoid, do not engage it for more than 20 seconds. Between each engagement, allow it to cool for at least 3 minutes.

c. Using a feeler gauge and the solenoid energized, check the drive pinion-to-collar gap; it should be 0.020–0.080 in. If necessary, add or subtract shims between the solenoid and drive end housing until the desired depth is achieved.

REDUCTION DRIVE TYPE
Disassembly and Assembly
MITSUBISHI

NOTE: Do not place the stator frame in a vise or strike it with a hammer for damage to the permanent magnets could occur.

1. Disconnect the coil wire from the solenoid.
2. Remove the solenoid-to-front end frame screws and the solenoid.
3. Loosen, do not remove the commutator shield-to-brush holder screws.
4. Remove the through bolts, the rubber retainer (under solenoid) and the coin washer.

NOTE: When removing the output shaft assembly, do not loose the armature shaft ball.

5. Remove the stator frame, the commutator shield and output shaft assembly as a unit. Separate the clutch fork from the output shaft assembly.
6. From the stator frame, pull the output shaft assembly forward, then, push the armature and commutator shield to the rearward.
7. Remove the commutator shield-to-brush holder plate screws and the shield; do not remove the brush holder assembly.
8. Using a 22mm socket, slide it up against the commutator, slide the brush holder assembly onto the socket and position the socket/brush holder assembly aside.
9. To disassemble the output shaft assembly, perform the following procedures:
   a. Remove the rubber packing ring and the gears.
   b. Using a 17mm socket, position it into the armature end of the driveshaft and position the assembly in the vertical position, resting on the socket.

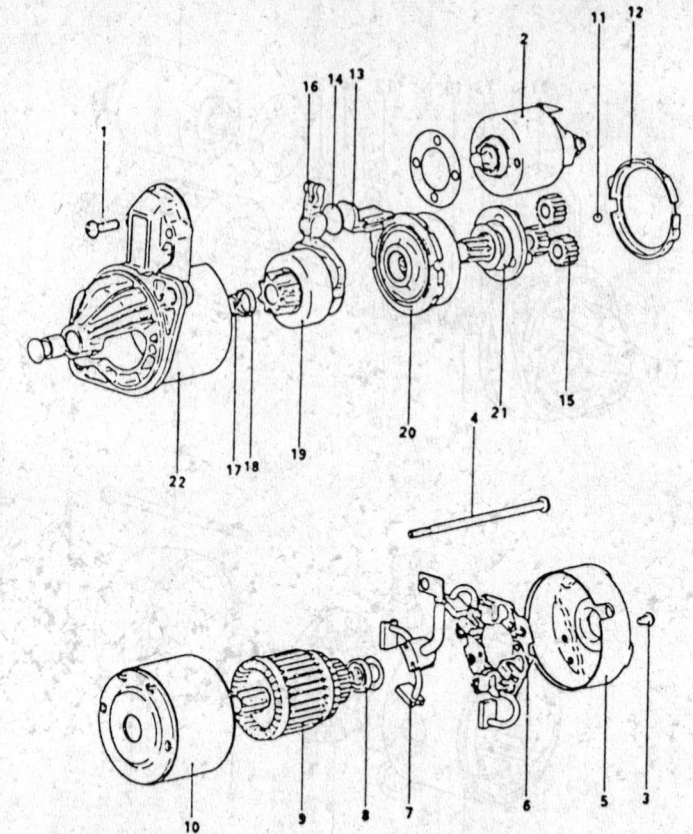

1.	Screw	12.	Packing A
2.	Magnetic switch	13.	Packing B
3.	Screw	14.	Plate
4.	Screw	15.	Planetary gear
5.	Rear bracket	16.	Lever
6.	Brush holder	17.	Snapring
7.	Brush	18.	Stopring
8.	Rear bearing	19.	Overrunning clutch
9.	Armature	20.	Internal gear
10.	Yoke assembly	21.	Planetary gear holder
11.	Ball	22.	Front bracket

Exploded view of the Mitsubishi reduction drive starter – Chrysler

c. Using a 12 point 14mm socket, position it against the stopring (on the clutch end). Using a hammer, strike the socket to unseat the stopring and expose the snapring.
d. Remove the socket, the snapring and the stopring from the driveshaft.
e. Using fine sandpaper, remove any burrs from the driveshaft. Remove the overruning clutch.
10. Using compressed air or dry cloths, clean the armature, the stator frame, the overrunning clutch, the solenoid and the brush holder. Using mineral spirits, clean all other components.
11. Inspect the following parts for damage and replace, if necessary:
   a. The stator frame and permanent magnets
   b. The driveshaft bushing (armature side)

c. The planetary gear set and driveshaft
d. The starter motor bushing and bearing
e. The carbon brushes for cracks, distortion and wear below 0.354 in.

NOTE: When inspecting the brushes, do not remove the socket from the brush holder.

12. Using a growler and a hacksaw blade (placed on top of the armature), rotate the armature and check it for a shorted condition. If the hacksaw blade vibrates, a short exists; replace the armature.
13. Using a test light, place a lead on the armature's core and the other on each commutator segment, inspect the armature for a grounded condition. If a ground exists, the test light will turn **ON**; replace the armature.
14. Using a test light, place the leads

on the adjacent commutator segments. If the test light turns **ON** between any 2 segments, the armature is shorted and must be replaced.

15. Inspect the commutator out-of-round, if it is more than 0.001 in., reface it on a lathe.

To Assemble:

16. Using motor oil, lubricate the driveshaft and the overrunning clutch bushing. Using Lubriplate®, lubricate the overrunning clutch spiral cut splines.

17. Install the overrunning clutch on the driveshaft/planetary gear assembly, followed by the stopring and the snapring; sure to seat the snapring in the shaft groove and crimp the it with a pair of pliers.

18. Using a battery terminal puller, attach it to the driveshaft tip and press the stopring over the snapring.

NOTE: **When installing the stopring, be careful not to scratch the driveshaft.**

19. Install the clutch fork, with the assembled planetary gear set, lubricated with lithium grease, into the front end housing; make sure the locating lugs are properly seated in the front end housing.

20. Install the coin washer and the rubber fork retainer.

21. Install the rubber backing ring by placing the largest rubber lug at the top.

22. Install the brush holder onto the armature's commutator; make sure the brushes and brush holders are seated in the holder. Inspect the flex washer and install the commutator shield onto the armature. Install the brush holder screws but do not tighten them.

23. Install the armature assembly into the stator frame and seat the wire grommet into the frame.

24. Be sure the armature spline gear is seated in the planetary gear seat with the armature shaft ball in place. Seat the armature shaft in the shaft bushing bore; rotate the stator frame to align the tabs on the drive housing frame.

25. Install the through bolts and torque to 28 inch lbs. Torque the brush holder screws to 18 inch lbs.

26. To complete the assembly, reverse the disassembly procedures.

ISUZU

Starting System

NON-REDUCTION TYPE

Disassembly and Assembly

1. Position the starter in a soft

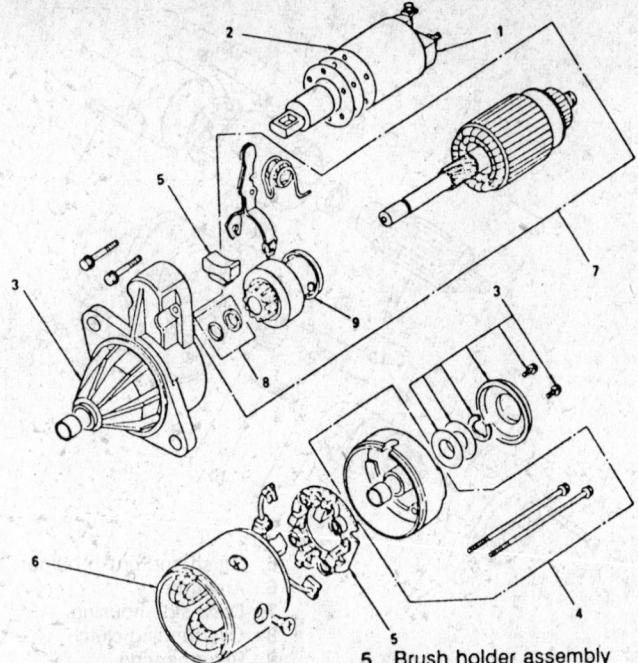

1. Magnetic switch lead
2. Solenoid switch assembly
3. Dust cover and snapring
4. Rear cover assembly
5. Brush holder assembly
6. Field coil assembly
7. Armature assembly with shift lever
8. Pinion stop clip
9. Pinion assembly

Exploded view of the Isuzu non-reduction drive starter — 2.0L engine

jawed vise. Disconnect the field coil wire from the solenoid terminal.

2. Remove the solenoid mounting screws and work the solenoid from the shift fork. Remove the bearing cover, the armature shaft lock, the washer, the spring and the seal.

3. Remove the commutator end frame cover through bolts, the cover, the brushes and the brush plate.

4. Slide the field frame from over the armature. Remove the shift lever pivot bolt, the rubber gasket and the metal plate.

5. Remove the armature assembly and the shift lever from the drive end housing. Press the stop collar from the snapring, then remove the snapring, the stop collar and the clutch assembly.

6. If the brushes are worn more than ½ the length of new brushes or are oil-soaked, should be replaced; the new brushes are 0.630 in. long.

7. Do not immerse the starter clutch unit in cleaning solvent as the solvent will wash the lubricant from the clutch. Place the drive unit on the armature shaft, then, while holding the armature, rotate the pinion.

NOTE: **The drive pinion should rotate smoothly in one direction only. The pinion may not rotate easily but as long as it rotates smoothly it is in good condition. If the clutch unit does not function**

properly or if the pinion is worn, chipped or burred, replace the unit.

To Assemble:

8. Lubricate the armature shaft and splines with lubricant.

9. Install the clutch, the stop collar, the lock ring and the shift fork onto the armature.

10. Install the armature assembly into the drive end housing.

11. Install the field frame housing over the armature and onto the drive end housing.

12. Position the brushes into the brush holder and position the brush holder over the armature's commutator.

13. Install the commutator end housing and the through bolts.

14. Install the cap end gasket, the armature brake spring, the armature plate and the commutator end cap.

REDUCTION GEAR TYPE

Disassembly and Assembly

1. Position the starter in a soft jawed vise. Remove the nut and disconnect the motor wire from the magnetic switch terminal.

2. Remove the through bolts and

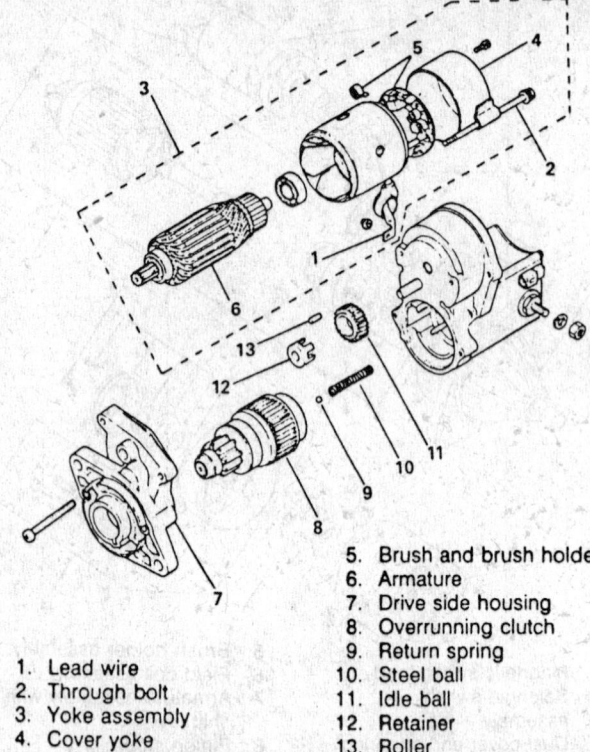

5. Brush and brush holder
6. Armature
7. Drive side housing
8. Overrunning clutch
9. Return spring
10. Steel ball
11. Idle ball
12. Retainer
13. Roller

1. Lead wire
2. Through bolt
3. Yoke assembly
4. Cover yoke

Exploded view of the Isuzu reduction drive starter—2.3L engine

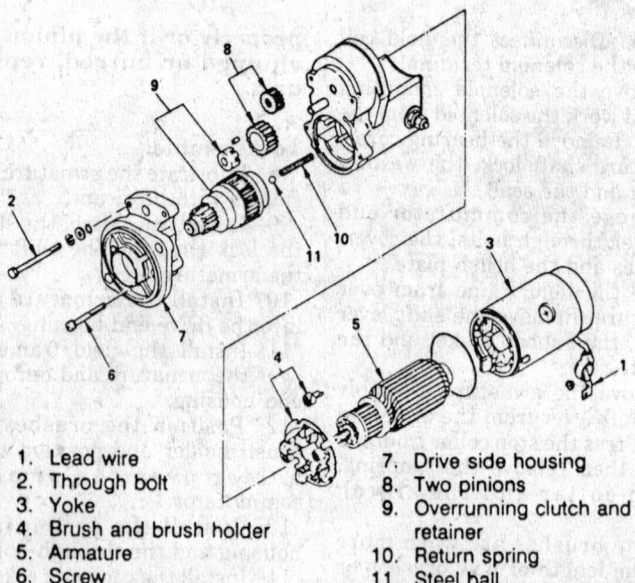

1. Lead wire
2. Through bolt
3. Yoke
4. Brush and brush holder
5. Armature
6. Screw

7. Drive side housing
8. Two pinions
9. Overrunning clutch and retainer
10. Return spring
11. Steel ball

Exploded view of the Isuzu reduction drive starter—2.2L diesel engine

pull the field frame (with the armature) from the magnetic switch assembly.

3. Remove the starter housing to magnetic switch assembly bolts and separate the housing from the assembly. Remove the pinion gear, the pinion retainer/bearings and the clutch assembly.

4. Using a magnetic finger, remove the spring and the steel ball from the

hole in the clutch assembly shaft. Remove the field frame end cover.

5. Using a small pry bar, separate the brush springs, then remove the brushes from the brush holder and pull the brush holder from the field frame.

6. Remove the armature from the field frame. Using an ohmmeter, make sure there is no continuity between the commutator and the armature coil

core. If there is continuity, replace the armature.

7. Using an ohmmeter, check for continuity between the commutator segments. If there is no continuity between any of the segments, replace the armature.

8. If the commutator is dirty, burnt or the runout exceeds 0.002 in., use a lathe to reface the surface; do not machine the diameter to less than 1.140 in. diameter.

9. Using an ohmmeter, make sure there is continuity between the lead wire and the brush lead of the field coil. If there is no continuity, replace the field frame.

10. Using an ohmmeter, make sure there is no continuity between the field coil and the field frame. If there is continuity, replace the field frame.

11. If the brush length is less than 0.394 in., replace the brush and dress with emery cloth.

12. Check the gear teeth for wear or damage, if damaged, replace them. Turn the clutch assembly pinion clockwise and make sure it rotates freely, try to turn the pinion counterclockwise and make sure it locks. If the pinion does not respond correctly, replace it.

13. While applying inward force on the bearings, turn each by hand; if resistance or sticking is noticed, replace the bearings. To replace the bearings, use the tool 09286–46011 or equivalent, to pull the bearing(s) from the armature shaft. Using the tool 09285–76010 or equivalent, and an arbor press, press the new bearing(s) onto the armature shaft.

14. Using an ohmmeter, check for continuity between the grounded terminal and the insulated terminal, then between the grounded terminal and the housing. If there is no continuity in either case, replace the magnetic switch assembly.

To Assemble:

15. Lubricate the gears, the shafts and bearings with high temperature grease.

16. Install the over-running clutch, the pinion and the retainers with the rollers into the housing.

17. Install the spring to the magnetic switch and assembly the housing and the magnetic switch.

18. Install the brushes into the brush holder; make sure the positive brush leads are not grounded.

19. Install the rear end frame to the yoke, engage the tab with the wire grommet and install the cover.

20. Install the yoke to the magnetic switch, engage the tab on the yoke with the magnetic switch notch.

21. To complete the installation, reverse the removal procedures.

MAZDA

Starting System

NON-REDUCTION TYPE

Disassembly and Assembly

1. Remove the starter from the engine.

2. Disconnect the field strap from the solenoid.

3. Remove the screws attaching the solenoid to the drive end housing. Disengage the solenoid plunger hook from the shift fork and remove the solenoid.

4. Remove the shift fork pivot bolt, nut and lockwasher.

5. Remove the through bolts and separate the drive end housing from the starter frame. At the same time, disengage the shift fork from the drive assembly.

6. Remove the brush end bearing cover-to-brush end cover screws.

7. Remove the C-washer, washer and spring from the brush end of the armature shaft.

8. Pull the brush end cover from the starter frame.

9. Slide the armature from the starter frame and brushes.

10. Slide the drive stop-ring retainer toward the armature and remove the stop-ring. Slide the retainer and drive assembly off the armature shaft.

11. Remove the field brushes from the brush holder and separate the brush holder from the starter frame.

To Assemble:

12. Position the drive assembly on the armature shaft.

13. Position the drive stop-ring retainer on the armature shaft and install the drive stop-ring. Slide the stop-ring retainer over the stop-ring to secure the stop-ring on the shaft.

14. Position the armature in the starter frame. Install the brush holder on the armature and starter frame. Install the brushes in the brush holder.

15. Install the drive end housing on the armature shaft and starter housing. Engage the shift fork with the starter drive assembly as the drive end housing is moved toward the starter frame.

16. Install the brush end cover on the starter frame making sure that the rear tabs of the brush holder are aligned with the through-bolt holes.

17. Install the through-bolts.

18. Install the rubber washer, spring, washer and C-washer on the armature shaft at the brush end. Install the brush end bearing cover on the brush end cover and install the attaching screws. If the brush end cover is not properly positioned, the bearing cover screws cannot be installed.

19. Align the shift fork with the pivot bolt hole and install the pivot bolt, lockwasher and nut. Tighten the nut securely.

20. Position the solenoid on the drive end housing. Be sure that the solenoid plunger hook is engaged with the shift fork.

21. Install both solenoid retaining screws and washers.

22. Apply 12 volts to the solenoid **S** terminal (ground the **M** terminal) and check the clearance between the starter drive and the stop-ring retainer. The clearance should be 0.080–0.200 in. If not, the solenoid plunger is not properly adjusted. The clearance can be adjusted by inserting an adjusting shim between the solenoid body and drive end housing.

23. Install the field strap and tighten the nut.

24. Install the starter. Check the operation of the starter.

COAXIAL REDUCTION TYPE

Disassembly and Assembly

1. Position the starter in a soft jawed vise. Disconnect the field coil wire from the solenoid terminal.

2. Remove the solenoid mounting screws and work the solenoid from the shift fork. Remove the bearing cover, the armature shaft lock, the washer, the spring and the seal.

3. Remove the commutator end frame cover through bolts, the cover, the brushes and the brush plate.

4. Slide the field frame from over the armature. Remove the shift lever pivot bolt, the rubber gasket and the metal plate.

5. Remove the armature assembly and the shift lever from the drive end housing. Press the stop collar from the snapring, then remove the snapring, the stop collar and the clutch assembly.

6. If the brushes are worn more than ½ the length of new brushes or are oil-soaked, should be replaced; the new brushes are 0.630 in. long.

7. Do not immerse the starter clutch unit in cleaning solvent as the solvent will wash the lubricant from the clutch. Place the drive unit on the armature shaft, then, while holding the armature, rotate the pinion.

NOTE: The drive pinion should rotate smoothly in one direction only. The pinion may not rotate easily but as long as it rotates smoothly it is in good condition. If the clutch unit does not function properly or if the pinion is worn, chipped or burred, replace the unit.

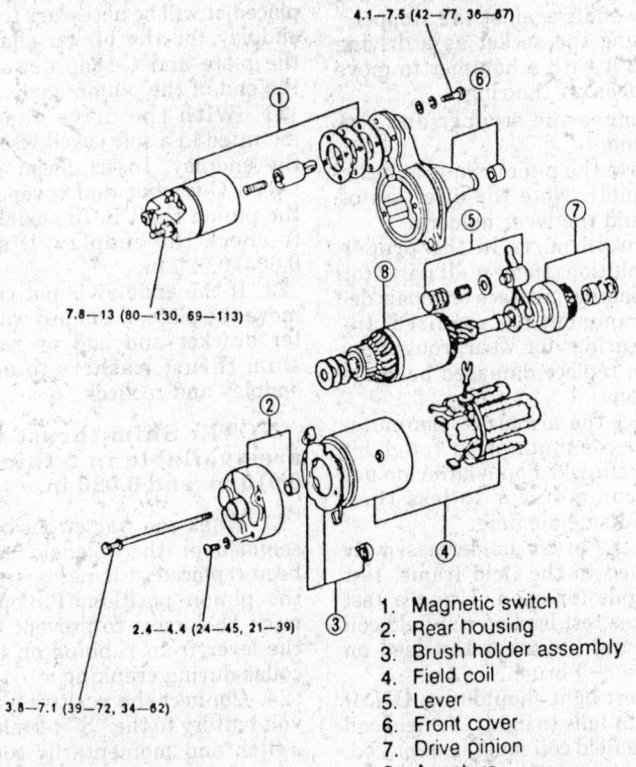

4.1—7.5 (42—77, 36—67)

7.8—13 (80—130, 69—113)

2.4—4.4 (24—45, 21—39)

3.8—7.1 (39—72, 34—62)

1. Magnetic switch
2. Rear housing
3. Brush holder assembly
4. Field coil
5. Lever
6. Front cover
7. Drive pinion
8. Armature

Exploded view of the Mazda non-reduction drive starter

1. Magnetic switch
2. Rear housing
3. Front cover
4. Drive pinion
5. Internal gear
6. Gear shaft
7. Planetary gear
8. Magnet coil
9. Armature
10. Brush holder assembly

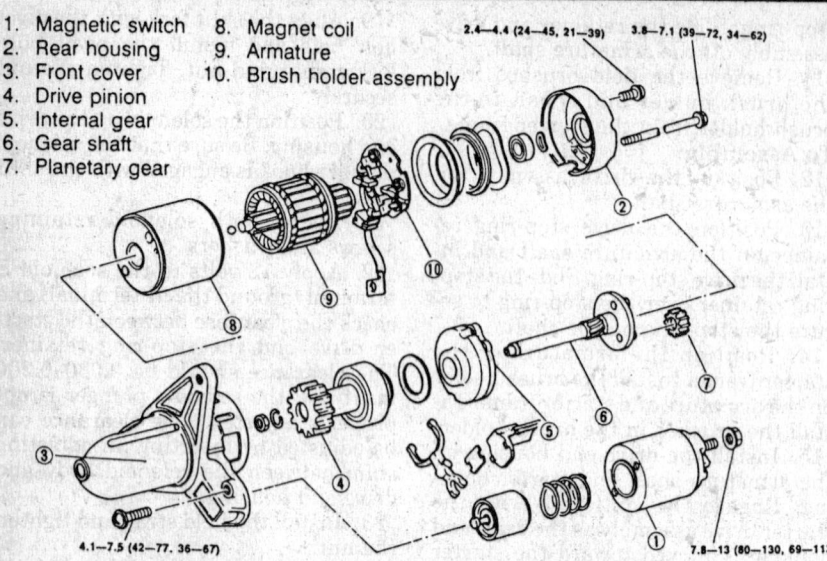

Exploded view of the Mazda coaxial reduction drive starter

To Assemble:

8. Lubricate the armature shaft and splines with lubricant.

9. Install the clutch, the stop collar, the lock ring and the shift fork onto the armature.

10. Install the armature assembly into the drive end housing.

11. Install the field frame housing over the armature and onto the drive end housing.

12. Position the brushes into the brush holder and position the brush holder over the armature's commutator.

13. Install the commutator end housing and the through bolts.

14. Install the cap end gasket, the armature brake spring, the armature plate and the commutator end cap.

NON-COAXIAL REDUCTION TYPE

Disassembly and Assembly

1. Remove the starter from the vehicle. Position the unit in a soft jawed vise.

2. Remove the field connector nut, the solenoid switch screws and the solenoid.

3. If equipped with shims between the solenoid and the drive end housing, retain these for installation purposes.

4. Remove the starter through bolts and the brush holder bolts. Remove the commutator end frame from the armature and bearing assembly. Remove the field frame assembly and the armature from the center housing.

5. Pry back each brush spring so that each brush can be backed away from the armature about ¼ in. Release the spring to hold the brushes in

the backed out position, then remove the armature from the field frame and brush holder.

6. Remove the shaft cover-to-center housing screws and the shaft cover. Remove the C-shaped washer and plate. Remove the center housing bolts, the center housing shim and thrust washers.

7. Remove the reduction gear, the spring holder and the lever springs.

8. To remove the drive pinion, perform the following procedures:

a. Using a ⅝ in. socket, slide it over the shaft against the stopper.

b. Using the socket as a driving tool, tap it with a hammer to move the stopper off the ring.

c. Remove the stopper and the drive pinion.

9. Remove the pinion shaft and the lever assembly. Note the direction of the lever and the lever holders.

10. Clean all parts in the proper cleaning solution. Inspect all parts for wear and damage; replace or repair defective components as required. Inspect all bearings for wear, roughness or dryness; replace damaged bearings with new ones.

11. Inspect the armature commutator. If the commutator is rough, it should be refinished on a lathe; do not turn the commutator to less than 1.480 in. outside diameter.

12. With the brush holder assembly still attached to the field frame, test the field coils for open. Using a test lamp, place a test lead on the field coil connector and the other test lead on the positive (+) brush.

13. The test light should turn **ON**. If the test light fails to light, the field coil is open; the field coil must be replaced. Repeat the test on the other positive brush.

14. To test the field coil for ground, place a test light lead on the field coil connector and the other lead on the field frame; the test light should stay **OFF**. If the test light turns **ON**, the field coils are grounded to the field frame assembly; the field frame must be replaced.

To Assemble:

15. To replace the brushes, remove the positive brushes from the brush holder, the brush holder and the negative brush assembly from the field frame.

16. Cut the old brush leads off of their mountings as close to brush connection point as possible. Solder the new brushes as required. Careful installation of the positive side is necessary to prevent grounding of the brush connection point having no insulation.

17. Reinstall the positive and negative brushes in the brush holder assembly and position the assembly in the starter housing.

18. In order to replace the drive end bearing, it will be necessary to press the bearing out of the drive end housing using a press. Replace the armature commutator end bearing and the armature drive end bearing.

19. To assemble, reverse the disassembly procedures. Be sure to check all parts for wear and damage; repair or replace defective parts.

20. If either the drive end housing, pinion shaft, reduction gear, shim washers or center housing were replaced, it will be necessary to check the endplay for the pinion shaft. Install the plate and C-shaped washer onto the end of the pinion shaft.

21. With the drive end housing mounted in a soft jawed vise, measure the endplay. Insert feeler gauge between C-washer and cover plate, pry the pinion shaft in the axial direction to check the endplay; it should be 0.004–0.020 in.

22. If the endplay is not correct, remove the plate, C-shaped washer, center bracket and add or remove the shim thrust washers to adjust the endplay and recheck.

NOTE: Shim thrust washers are available in 2 thicknesses 0.010 in. and 0.020 in.

23. When the starter has been disassembled or the solenoid switch has been replaced, it is necessary to check the pinion position. Pinion position must be correct to prevent the top of the lever from rubbing on the clutch collar during cranking.

24. Connect the positive lead of a 12 volt battery to the "S" terminal on the switch and momentarily connect the other to the starter frame. This will shift the pinion into cranking position

① 7.8—13 (80—130, 69—113)

2.4—4.4 (24—45, 21—39)   3.8—7.1 (39—72, 34—62)

4.1—7.5 (42—77, 36—67)

② ⑨

1. Magnetic switch
2. Rear housing
3. Drive pinion
4. Front cover
5. Reduction gear
6. Center bracket
7. Armature
8. Field coil
9. Brush holder assembly

Exploded view of the Mazda non-coaxial reduction drive starter

and will retain it until the battery is disconnected. Do not leave engaged more than 30 seconds at a time.

25. Using a dial indicator (with pinion engaged), push the pinion shaft back by hand and measure the amount of pinion shaft movement; the clearance should be 0.020–0.080 in.

26. If the amount does not fall within limits, adjust it by adding or removing the shims which are located between the switch and the front bracket; adding shims decreases the amount of the movement. Solenoid switch shims are available in 2 thicknesses 0.020 in. and 0.010 in.

MITSUBISHI

Starting System

DIRECT DRIVE TYPE

Disassembly and Assembly

1. Remove the wire connecting the starter solenoid to the starter.

2. Remove the starter solenoid-to-starter drive housing screws and the solenoid.

3. Remove both long through bolts from the rear of the starter and separate the armature yoke from the armature.

4. Carefully remove the armature and the starter drive engagement lever from the front bracket, noting the way they are positioned along with the attendant spring and spring retainer.

5. Loosen both screws and remove the rear bracket.

6. Tap the stopper ring at the end of the drive gear engagement shaft in towards the drive gear to expose the snapring. Remove the snapring.

7. Pull the stopper, drive gear and overrunning clutch from the end of the shaft.

8. Inspect the pinion and spline teeth for wear or damage. If the engagement teeth are damaged, visually check the flywheel ring gear through the starter hole to insure that it is not damaged. It will be necessary to turn the engine over by hand to completely inspect the ring gear.

9. Check the brushes for wear.

Their service limit length is 0.453 in. Replace if necessary.

To Assemble:

10. Install the spring retainer and spring on the armature shaft.

11. Install the overrunning clutch assembly on the armature shaft.

12. Fit the stopper ring with its open side facing out on the shaft.

13. Install a new snapring and, using a gear puller, pull the stopper ring into place over the snapring.

14. Fit the small washer on the front end of the armature shaft.

15. Fit the engagement lever into the overrunning clutch and refit the armature into the front housing.

16. Fit the engagement lever spring and spring retainer into place and slide the armature yoke over the armature. Position the yoke with the spring retainer cut-out space in line with the spring retainer.

NOTE: Make sure the brushes are seated on the commutator.

17. Replace the rear bracket and retainer screws.

18. Install the through bolts in the end of the yoke.

19. Refit the starter solenoid, fitting the plunger over the engagement lever. Install the screws and connect the wire running from the starter yoke to the starter solenoid.

REDUCTION DRIVE TYPE

1. Remove the wire connecting the starter solenoid to the starter.
2. Remove the solenoid screws, pull it out, unhook and disengage it from the engagement lever.
3. Remove the through bolts in the end of the starter and the bracket screws. Pull off the rear bracket.

NOTE: Since the conical spring washer is contained in the rear bracket, be sure to take it out.

4. Remove the yoke and brush holder assembly while pulling the brush upward.
5. Pull the armature assembly out of the mounting bracket.
6. In the side of the mounting bracket that the armature fits into, there is a small dust cap held by screws. Remove it and remove the snapring and washer under it.
7. Remove the remaining bolts in the mounting bracket and split the reduction case.

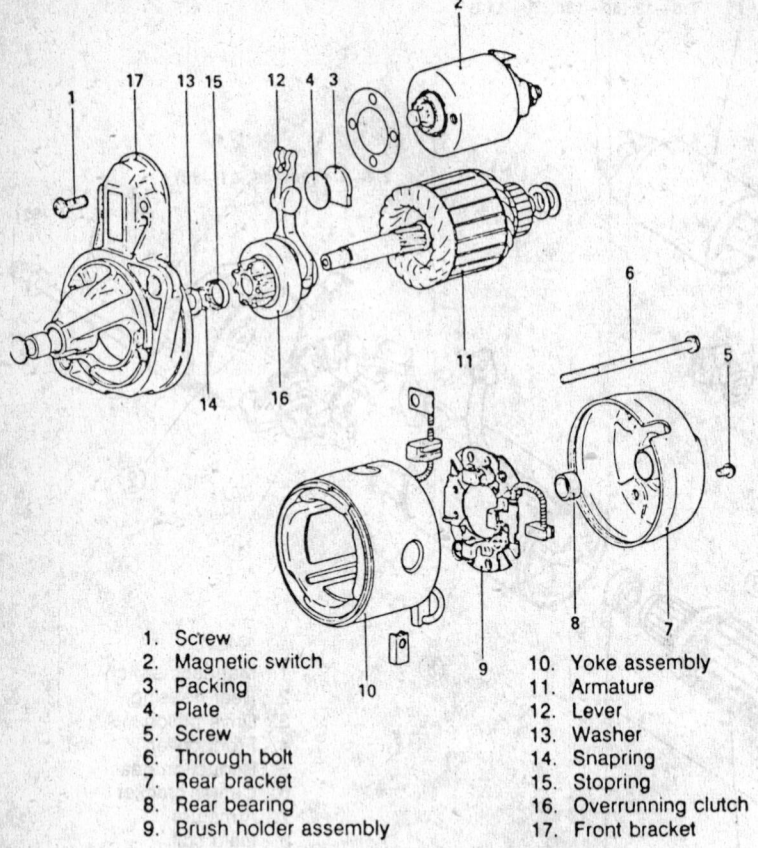

1. Screw
2. Magnetic switch
3. Packing
4. Plate
5. Screw
6. Through bolt
7. Rear bracket
8. Rear bearing
9. Brush holder assembly
10. Yoke assembly
11. Armature
12. Lever
13. Washer
14. Snapring
15. Stopring
16. Overrunning clutch
17. Front bracket

Exploded view of the Mitsubishi direct drive starter

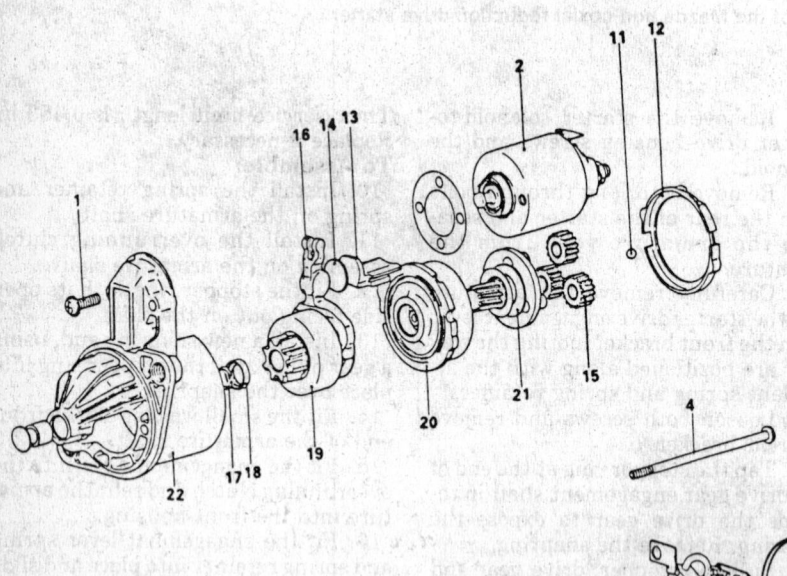

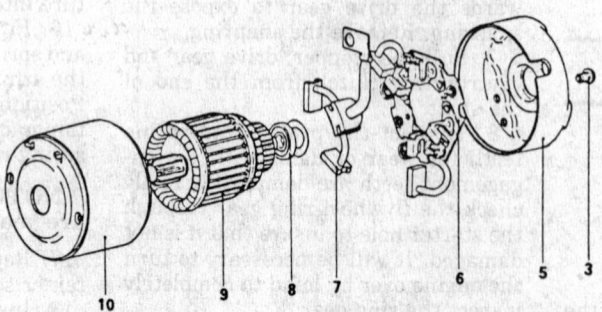

1. Screw
2. Magnetic switch
3. Screw
4. Screw
5. Rear bracket
6. Brush holder
7. Brush
8. Rear bearing
9. Armature
10. Yoke assembly
11. Ball
12. Packing A
13. Packing B
14. Plate
15. Planetary gear
16. Lever
17. Snapring
18. Stopring
19. Overrunning clutch
20. Internal gear
21. Planetary gear holder
22. Front bracket

Exploded view of the Mitsubishi reduction drive starter

NOTE: **Several washers will come out when the case is split. These adjust the endplay for the pinion shaft. Do not lose them.**

8. Remove the reduction gear, lever and lever spring from the front bracket.

9. Using a brass drift or deep socket, knock the stopper ring on the end of the shaft in toward the pinion. Remove the snapring. Remove the stopper, pinion and pinion shaft assembly.

10. Remove the ball bearings at both ends of the armature.

NOTE: **The ball bearings are pressed in the front bracket and are not replaceable. Replace them together with the bracket.**

11. Inspect the pinion and spline teeth for wear or damage. If the engagement teeth are damaged, visually check the flywheel ring gear through the starter hole to insure that it is not damaged also. It will be necessary to turn the engine over by hand to completely inspect the ring gear.

12. Check the brushes for wear. Their service limit length is 0.453 in. Replace if necessary.

13. Assembly is the reverse of disassembly. Be sure to replace all adjusting and thrust washers. When replacing the rear bracket, fit the conical spring pinion washer with its convex side fac-

ing out. Make sure the brushes seat on the commutator.

NISSAN

Starting System

NON-REDUCTION GEAR TYPE

Disassembly and Assembly
MITSUBISHI

1. Position the assembly in the soft jawed vise. Disconnect the field coil wire from the solenoid terminal.

2. Remove the solenoid-to-starter screws and the solenoid from the shifting fork.

3. Remove the through bolts and the commutator end frame cover. Remove the 2 brushes and the brush plate. Slide the field frame from the armature.

4. Remove the shift lever pivot bolt, the rubber gasket and plate.

5. Remove the snapring, the clutch assembly and the drive end housing from the armature.

6. Press the stop collar from the snapring and remove the snapring, stop collar and clutch.

7. Brushes that are worn less than ½ the length of new brushes or oil soaked should be replaced; new brushes are $^{11}/_{16}$ in. long.

To Assemble:

8. Do not immerse the starter clutch unit in cleaning solvent; lubricant will be washed from inside the clutch.

9. Place the drive unit on the armature shaft and while holding the armature, rotate the pinion. The drive pinion should rotate smoothly in one direction only. The pinion may not rotate easily but as long as it rotates smoothly it is in good condition. If the clutch unit does not function properly or if the pinion is worn, chipped or burred replace the unit.

10. To assemble, lubricate the bushings/splines and reverse the disassembly procedures.

11. Install the clutch, stop collar, lock ring and shaft fork on the armature.

12. Install the armature assembly and shift fork in the drive end housing. Check the endplay, it should be 0.002–0.021 in.

HITACHI

1. Support the starter in a soft jawed vise.

2. Remove the rear mounting bracket-to-starter nuts, the bracket and the plastic cap.

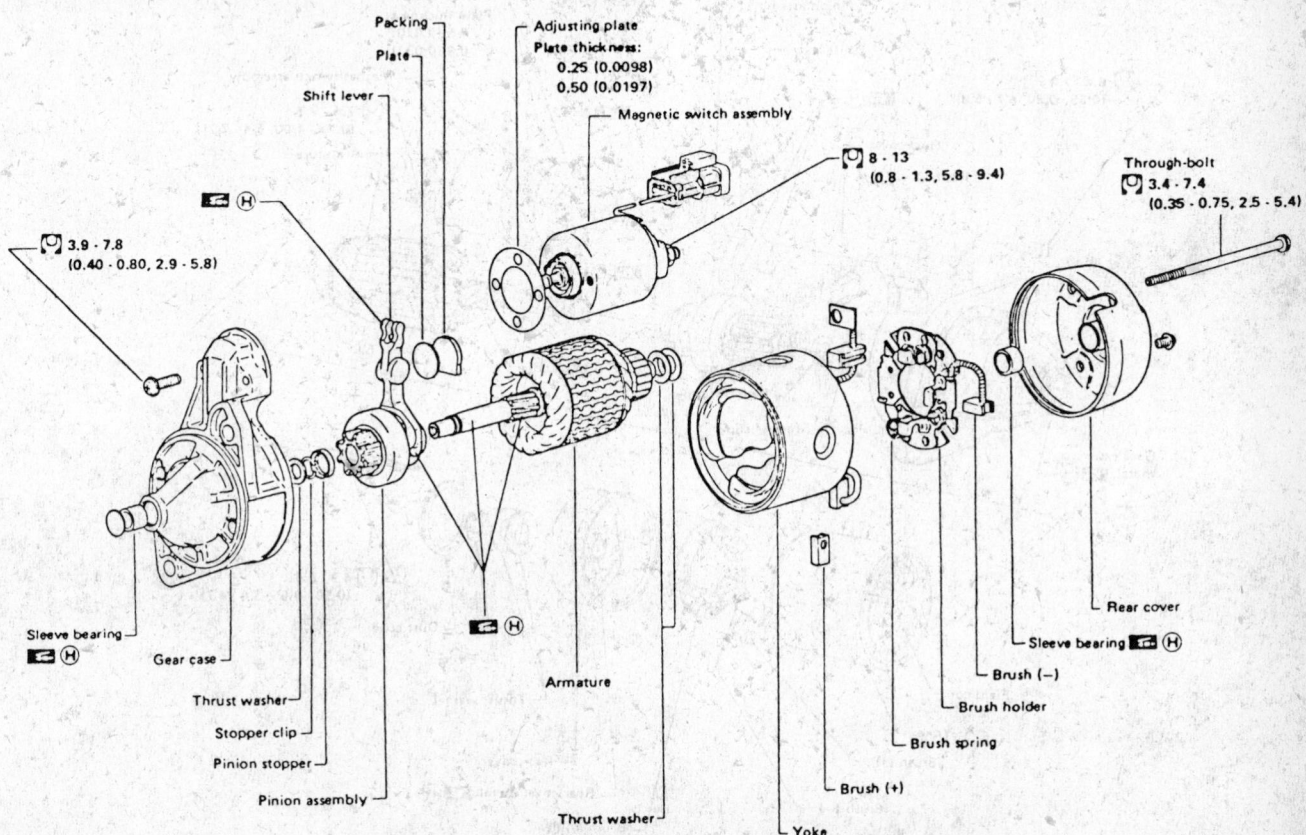

Exploded view of the Mitsubishi non-reduction drive starter – Nissan

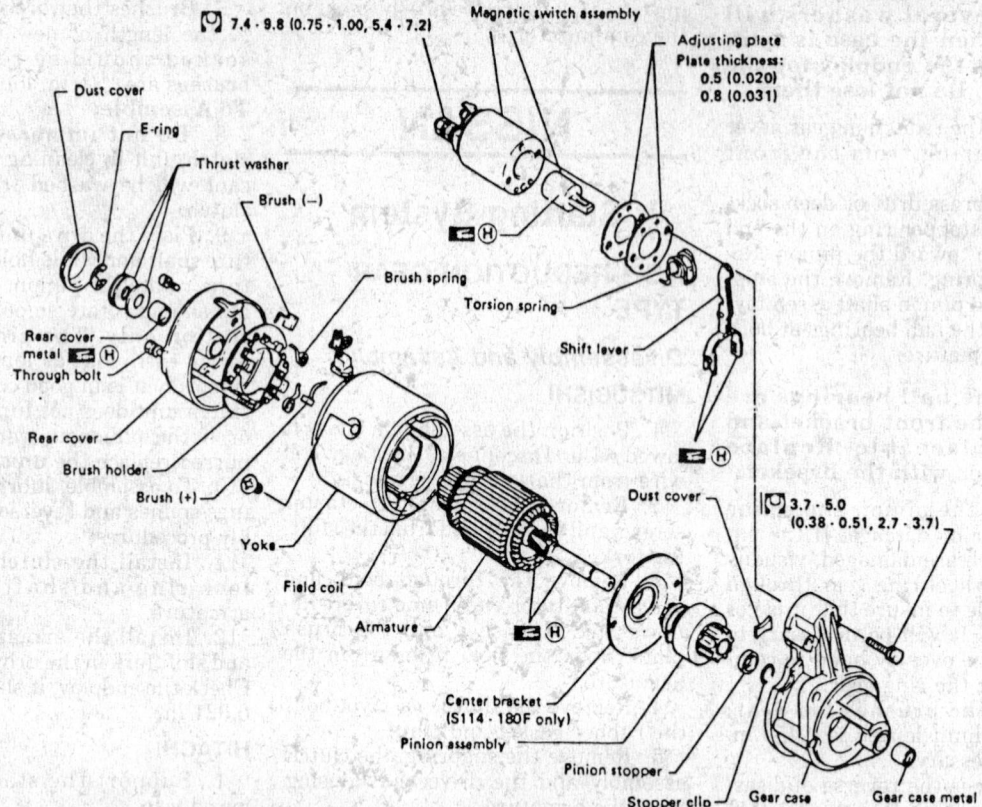

Exploded view of the Hitachi non-reduction drive starter with center bracket — Nissan

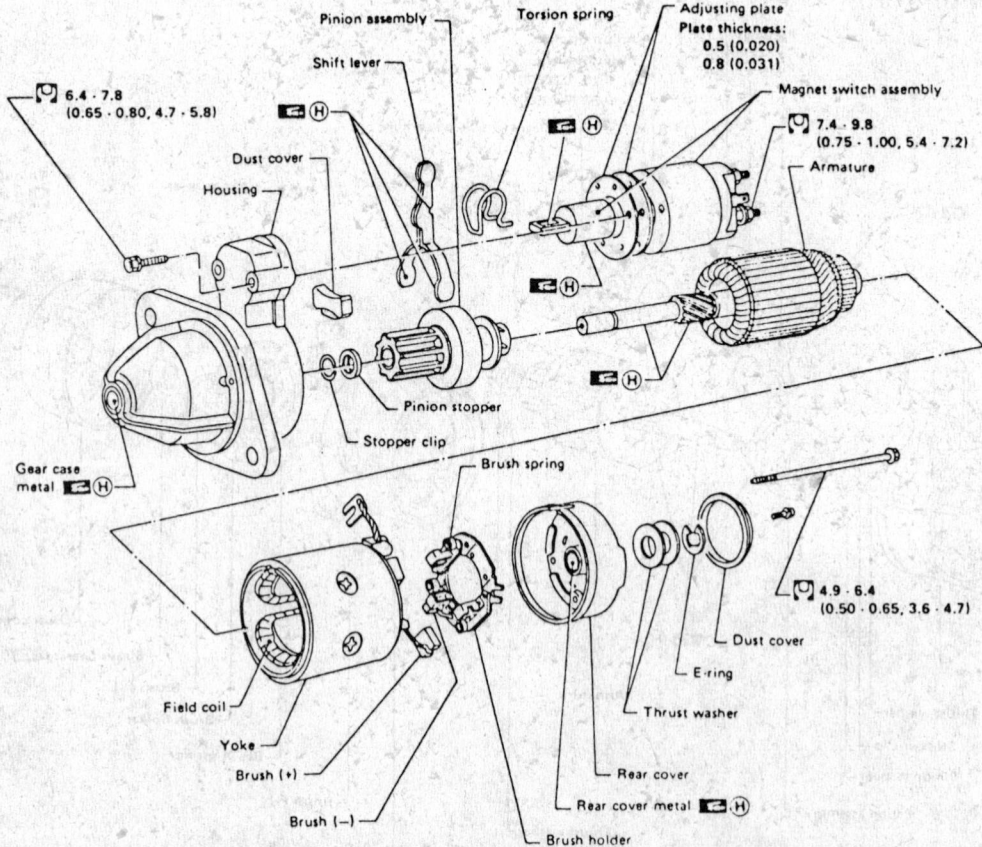

Exploded view of the Hitachi non-reduction drive starter without center bracket — Nissan

3. Disconnect the field wire from the solenoid.

4. Remove the solenoid-to-starter nuts and the solenoid.

5. From the end cover, remove the nuts, the through bolts and the end cover.

6. Using a pin punch at the drive end housing, drive the yoke axle pin from the housing.

7. At the positive brushes, move the brush spring clips to the side of the brushes and pull the brushes away from the armature. Carefully pull the field housing from the drive end housing.

8. Remove the pinion yoke and armature from the drive end housing.

9. Using a deep socket which fits over the armature shaft, tap the stop collar (driving it toward the armature) to expose the snapring. Remove the snapring from the groove and slide it from the shaft.

NOTE: When removing the snapring, be careful not to bend or distort it.

10. Remove the stop collar, the drive pinion and the support plate.

11. Using compressed air or a brush, clean the drive pinion, the drive end frame, the armature, the field coils

and the starter frame; all other parts can be cleaned in solvent.

12. Inspect the condition of the starter parts, perform the following procedures:

a. Check for broken wires or badly soldered connections.

b. Replace any bushings which are scored or badly worn.

c. If the armature's commutator more than 0.005 in. out of round, reface it on a lathe.

NOTE: Never use emery cloth to clean a commutator.

d. The drive pinion should be free of excessive wear or damage.

e. If the brushes are cracked, broken, distorted or worn to less than 0.314 in., replace them.

f. Using a growler and a hacksaw blade (placed on top of the armature), rotate the armature and check it for a shorted condition. If the hacksaw blade vibrates, a short exists; replace the armature.

g. Using a test light, place a lead on the armature's core and the other on each commutator segment, inspect the armature for a grounded condition. If a ground exists, the test light will turn **ON**; replace the armature.

h. Using a test light, check for

continuity between the positive brushes; if no continuity exists, replace the winding. Repeat this test for the negative windings.

i. Using a test light, place a lead on the coil housing and the other on each coil lead, make sure there is NO continuity; if continuity exists, replace the winding(s).

To Assemble:

13. Lubricate the necessary parts and place the support plate onto the armature, followed by the drive pinion and the stop collar. Carefully slide the snapring into the armature groove. Slide the stop collar over the snapring until it locks.

14. Position the armature and pinion yoke into the drive end housing. Install the pinion yoke axle pin.

15. Install the coil housing over the armature and onto the drive end housing. Position the brush holder onto the coil housing. Install the brushes and secure the brush springs.

16. Install the solenoid to the drive end housing. Position the field wire grommet to the coil housing and connect field wire to the solenoid.

17. Install the end cover to the coil housing and the install the through bolts/nuts. Install the armature brake assembly, the plastic cover and the mounting bracket.

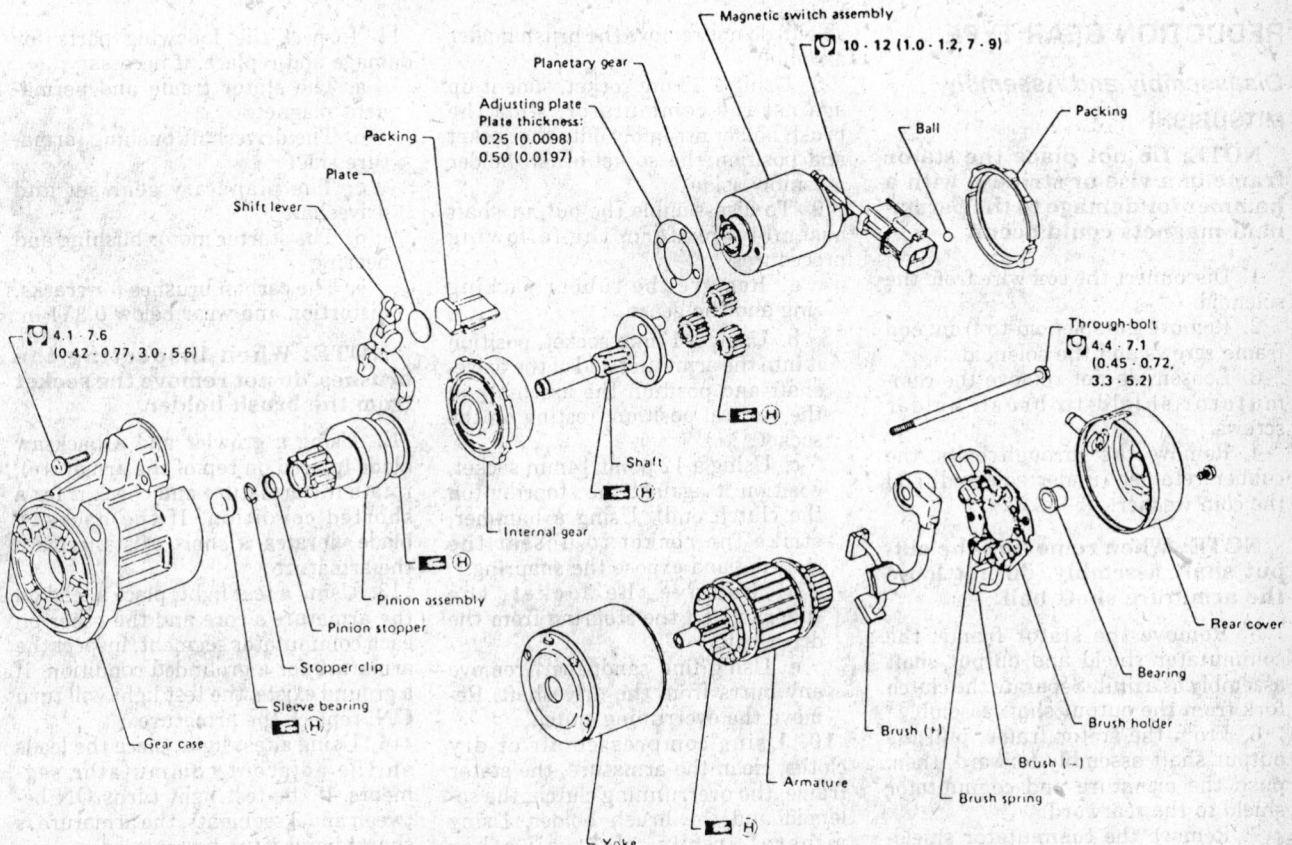

Exploded view of the Mitsubishi reduction drive starter—Nissan Pick-Up and Pathfinder

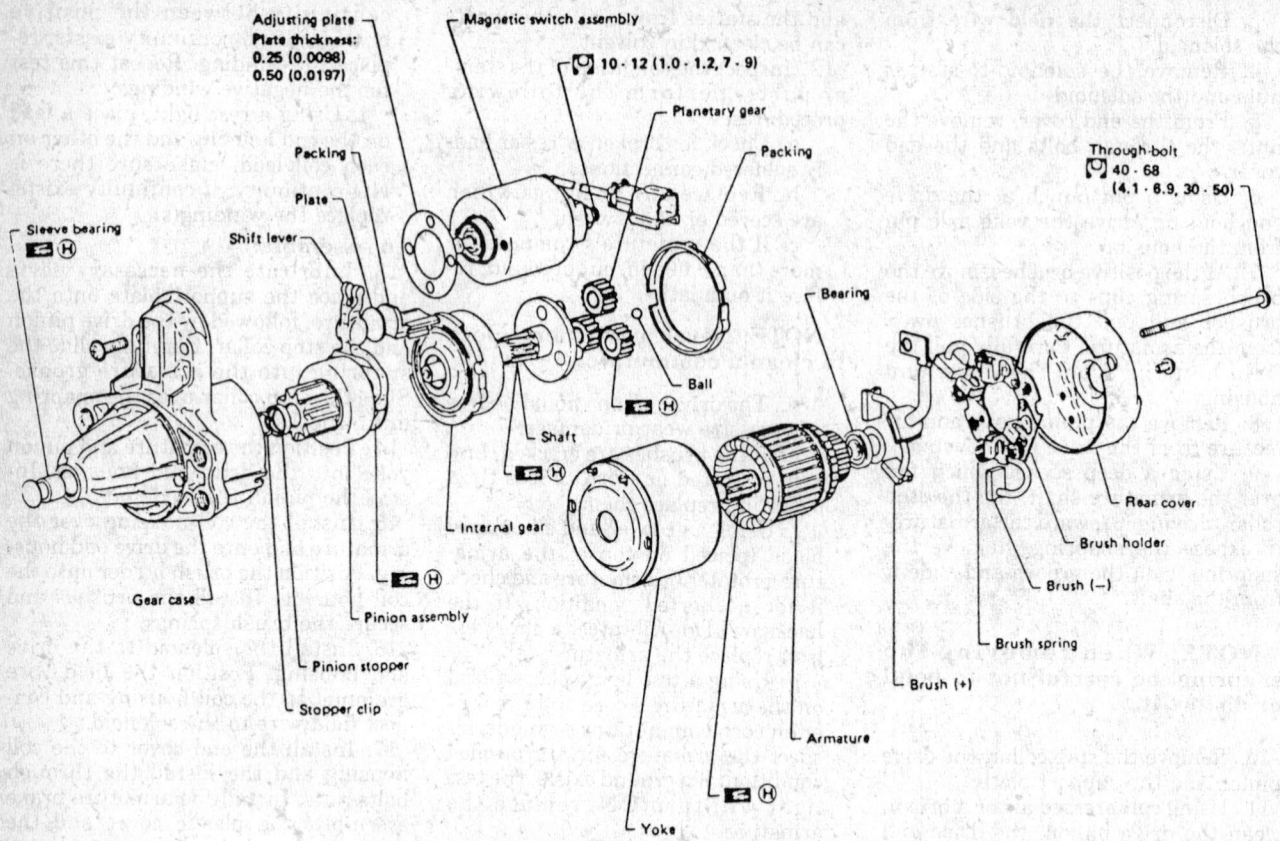

Exploded view of the Mitsubishi reduction drive starter—Nissan Axxess

REDUCTION GEAR TYPE

Disassembly and Assembly

MITSUBISHI

NOTE: Do not place the stator frame in a vise or strike it with a hammer for damage to the permanent magnets could occur.

1. Disconnect the coil wire from the solenoid.
2. Remove the solenoid-to-front end frame screws and the solenoid.
3. Loosen, do not remove the commutator shield-to-brush holder screws.
4. Remove the through bolts, the rubber retainer (under solenoid) and the coin washer.

NOTE: When removing the output shaft assembly, do not loose the armature shaft ball.

5. Remove the stator frame, the commutator shield and output shaft assembly as a unit. Separate the clutch fork from the output shaft assembly.
6. From the stator frame, pull the output shaft assembly forward, then, push the armature and commutator shield to the rearward.
7. Remove the commutator shield-to-brush holder plate screws and the

shield; do not remove the brush holder assembly.
8. Using a 22mm socket, slide it up against the commutator, slide the brush holder assembly onto the socket and position the socket/brush holder assembly aside.
9. To disassemble the output shaft assembly, perform the following procedures:
   a. Remove the rubber packing ring and the gears.
   b. Using a 17mm socket, position it into the armature end of the driveshaft and position the assembly in the vertical position, resting on the socket.
   c. Using a 12 point 14mm socket, position it against the stopring (on the clutch end). Using a hammer, strike the socket to unseat the stopring and expose the snapring.
   d. Remove the socket, the snapring and the stopring from the driveshaft.
   e. Using fine sandpaper, remove any burrs from the driveshaft. Remove the overruning clutch.
10. Using compressed air or dry cloths, clean the armature, the stator frame, the overrunning clutch, the solenoid and the brush holder. Using mineral spirits, clean all other components.

11. Inspect the following parts for damage and replace, if necessary:
   a. The stator frame and permanent magnets
   b. The driveshaft bushing (armature side)
   c. The planetary gear set and driveshaft
   d. The starter motor bushing and bearing
   e. The carbon brushes for cracks, distortion and wear below 0.354 in.

NOTE: When inspecting the brushes, do not remove the socket from the brush holder.

12. Using a growler and a hacksaw blade (placed on top of the armature), rotate the armature and check it for a shorted condition. If the hacksaw blade vibrates, a short exists; replace the armature.
13. Using a test light, place a lead on the armature's core and the other on each commutator segment, inspect the armature for a grounded condition. If a ground exists, the test light will turn **ON**; replace the armature.
14. Using a test light, place the leads on the adjacent commutator segments. If the test light turns **ON** between any 2 segments, the armature is shorted and must be replaced.
15. Inspect the commutator out-of-

round, if it is more than 0.001 in., reface it on a lathe.

To Assemble:

16. Using motor oil, lubricate the driveshaft and the overrunning clutch bushing. Using Lubriplate®, lubricate the overrunning clutch spiral cut splines.

17. Install the overrunning clutch on the driveshaft/planetary gear assembly, followed by the stopring and the snapring; sure to seat the snapring in the shaft groove and crimp the it with a pair of pliers.

18. Using a battery terminal puller, attach it to the driveshaft tip and press the stopring over the snapring.

NOTE: When installing the stopring, be careful not to scratch the driveshaft.

19. Install the clutch fork, with the assembled planetary gear set, lubricated with lithium grease, into the front end housing; make sure the locating lugs are properly seated in the front end housing.

20. Install the coin washer and the rubber fork retainer.

21. Install the rubber backing ring by placing the largest rubber lug at the top.

22. Install the brush holder onto the armature's commutator; make sure the brushes and brush holders are seated in the holder. Inspect the flex washer and install the commutator shield onto the armature. Install the brush holder screws but do not tighten them.

23. Install the armature assembly into the stator frame and seat the wire grommet into the frame.

24. Be sure the armature spline gear is seated in the planetary gear seat with the armature shaft ball in place. Seat the armature shaft in the shaft bushing bore; rotate the stator frame to align the tabs on the drive housing frame.

25. Install the through bolts and torque to 28 inch lbs. Torque the brush holder screws to 18 inch lbs.

26. To complete the assembly, reverse the disassembly procedures.

HITACHI

With Planetary Gears

1. Position the assembly in a soft jawed vise. Remove the field terminal nut, the terminal and the washer.

2. Remove the solenoid-to-starter screws. Work the solenoid from the shift fork and remove the solenoid from the starter.

3. Remove the starter end shield bushing cap screws, the starter end shield bushing cap, the end shield bushing and C-washer.

4. Remove the starter end shield bushing washer and seal.

5. Remove the starter through bolts, the starter end shield and the brush plate.

6. Slide the field frame from the starter and over the armature. Remove the armature assembly from the drive end housing.

7. Remove the rubber seal from the drive end housing. Remove the starter drive gear train.

8. Remove the dust plate. Press the stop collar from the snapring. Using snapring pliers, loosen the snapring.

9. Remove the output shaft snapring, the clutch stopring collar and the clutch assembly from the starter.

10. Remove the clutch shift lever bushing, the clutch shift lever and the C-clip retainer.

11. Remove the retaining washer, the sun and the planetary gears from the annulus gear.

12. To assemble, lubricate the necessary parts and reverse the disassembly procedures. Replace all defective components as required.

Without Planetary Gears

1. Remove the starter from the vehicle. Position the unit in a soft jawed vise.

2. Remove the field coil screw, the

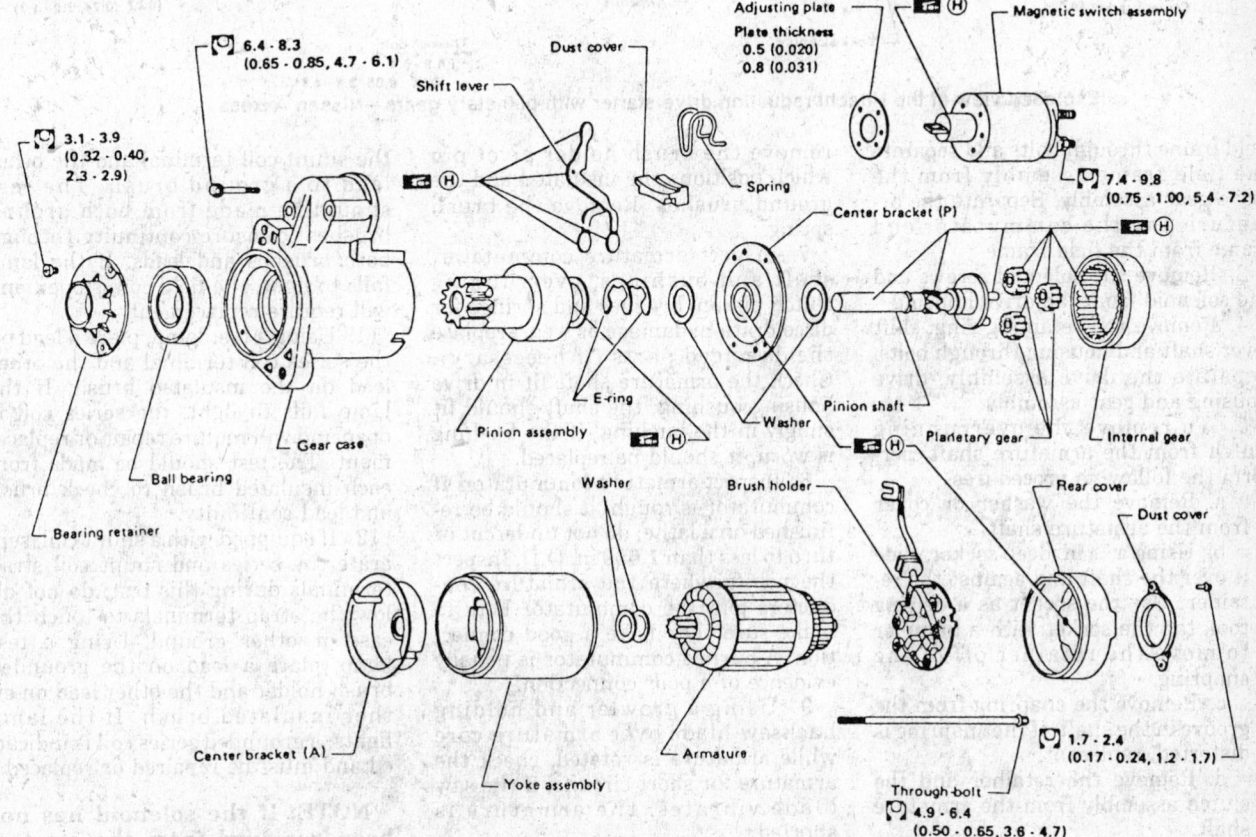

Exploded view of the Hitachi reduction drive starter with planetary gears—Nissan Pick-Up and Pathfinder

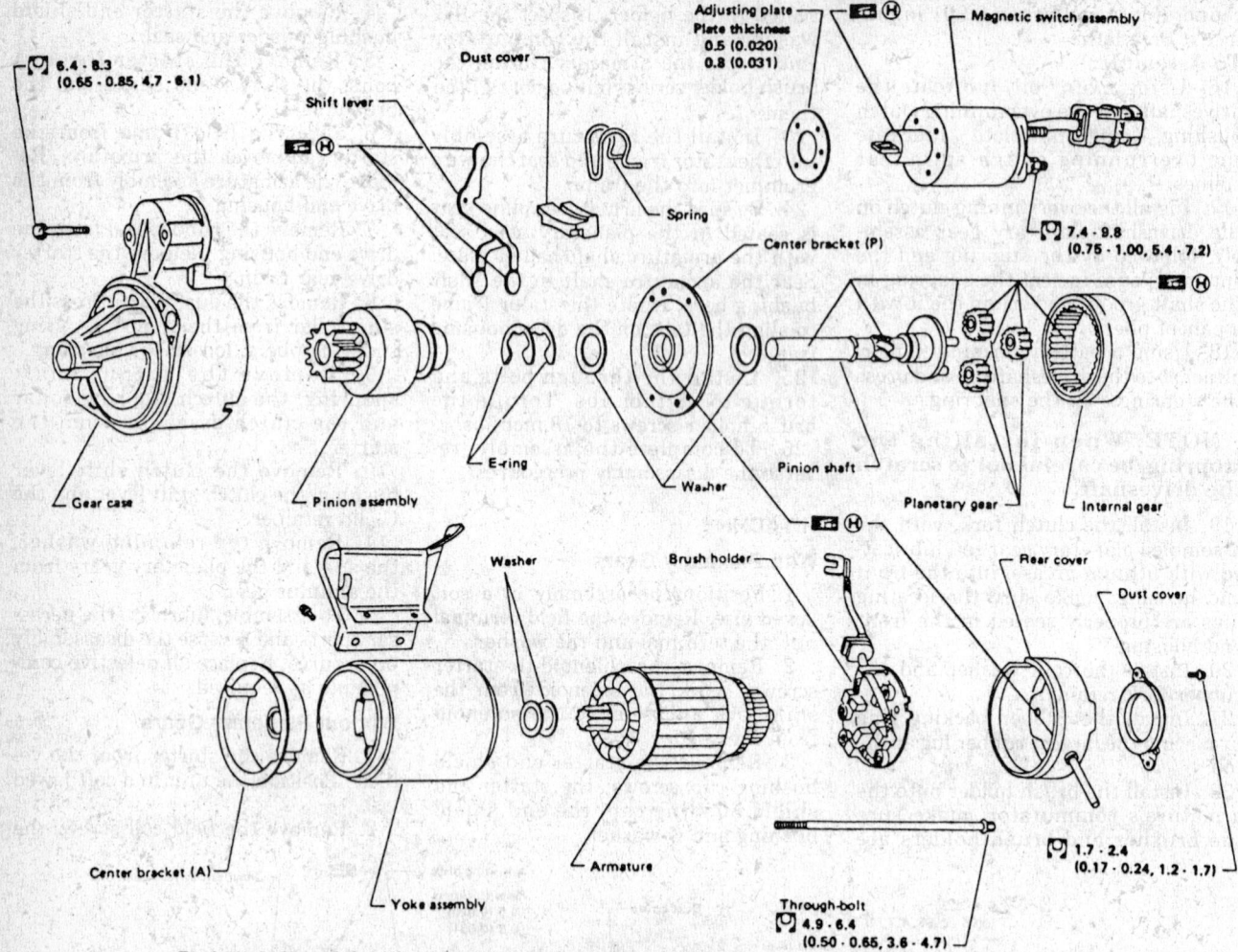

Exploded view of the Hitachi reduction drive starter with planetary gears—Nissan Axxess

field frame through bolts and separate the field frame assembly from the drive gear assembly. Separate the armature and the commutator end frame from the field frame.

3. Remove the solenoid screws and the solenoid from the drive housing.

4. Remove the retaining ring, shift lever shaft and housing through bolts. Separate the drive assembly, drive housing and gear assembly.

5. To remove the overrunning clutch from the armature shaft, perform the following procedures:

a. Remove the washer or collar from the armature shaft.

b. Using a ⅝ in. deep socket, slide it over the shaft and against the retainer. Use the socket as a driving tool, tap the socket with a hammer to move the retainer off of the snapring.

c. Remove the snapring from the groove in the shaft; if the snapring is distorted, replace it.

d. Remove the retainer and the clutch assembly from the armature shaft.

6. To replace the starter brushes, remove the brush holder pivot pin which positions the insulated and the ground brushes. Remove the brush spring.

7. Inspect armature commutator, shaft and bushings, overrunning clutch pinion, brushes and springs for discoloration, damage or wear; replace the damaged parts (if necessary). Check the armature shaft fit in drive housing bushing; the shaft should fit snugly in the bushing. If the bushing is worn, it should be replaced.

8. Inspect armature commutator. If commutator is rough, it should be refinished on a lathe; do not undercut or turn to less than 1.650 in. O.D. Inspect the points where the armature conductors join the commutator bars to make sure they have a good connection. A burned commutator is usually evidence of a poor connection.

9. Using a growler and holding hacksaw blade over armature core while armature is rotated, check the armature for short circuits; if the saw blade vibrates, the armature is shorted.

10. Using a test lamp, place a lead on the shunt coil terminal and the other lead to a ground brush. The test should be made from both ground brushes to insure continuity through both brushes and leads. If the lamp fails to light, the field coil is open and will require replacement.

11. Using a test lamp, place a lead on the series coil terminal and the other lead on the insulated brush. If the lamp fails to light, the series coil is open and will require repair or replacement. The test should be made from each insulated brush to check brush and lead continuity.

12. If equipped with a shunt coil, separate the series and shunt coil strap terminals during this test; do not allow the strap terminals to touch the case or other ground. Using a test lamp, place a lead on the grounded brush holder and the other lead on either insulated brush. If the lamp lights, a grounded series coil is indicated and must be repaired or replaced.

NOTE: If the solenoid has not been removed from the starter, the connector strap terminals

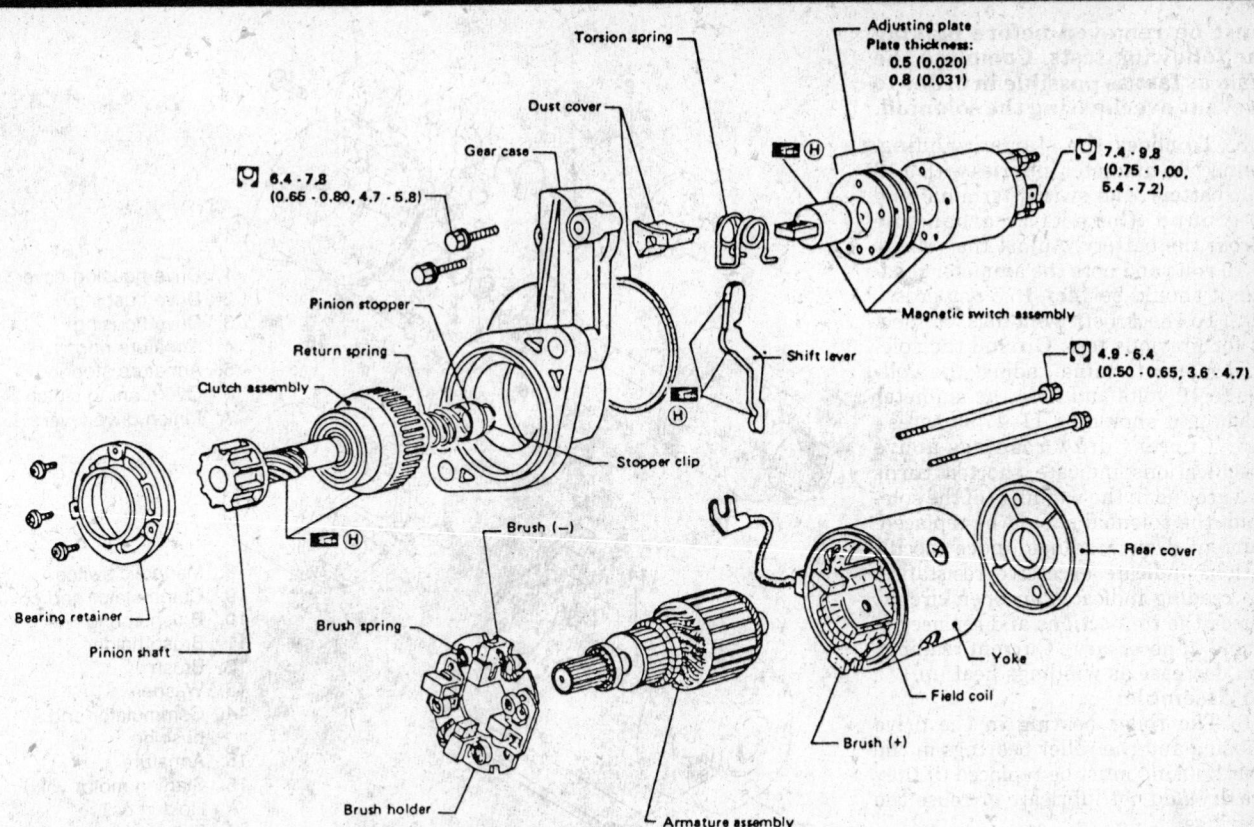

Exploded view of the Hitachi reduction drive starter with front mounted brush holder—Nissan

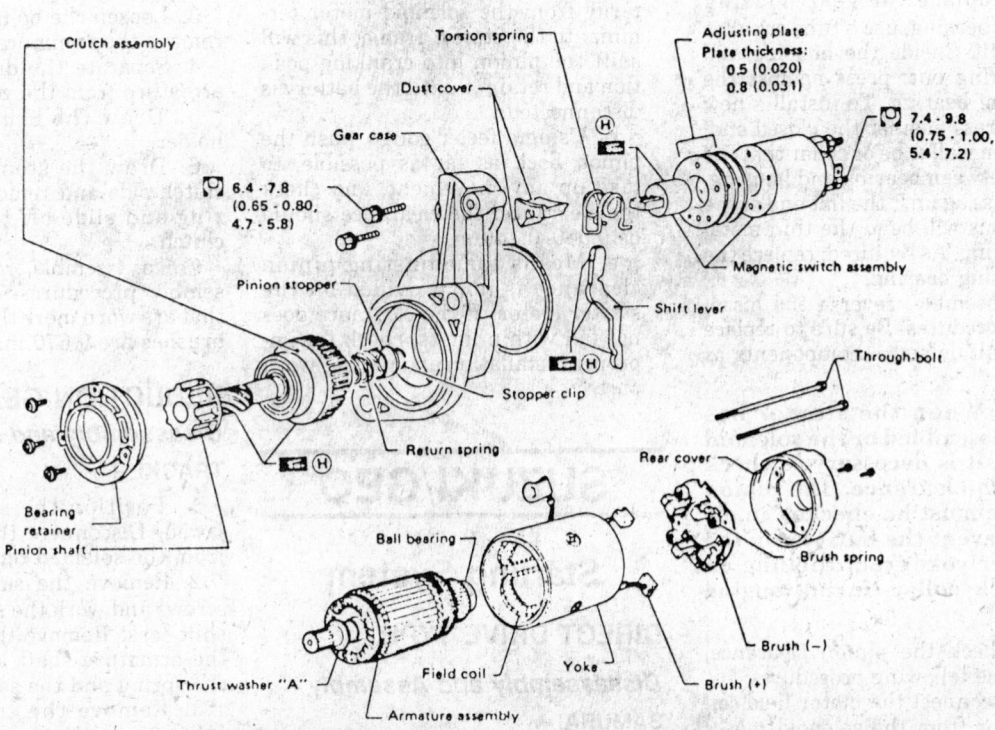

Exploded view of the Hitachi reduction drive starter with rear mounted brush holder—Nissan

must be removed before making the following tests. Complete the tests as fast as possible in order to prevent overheating the solenoid.

13. To check the starter winding, connect an ammeter in series with a 12 volt battery, the switch terminal and to ground. Connect a carbon pile across the battery. Adjust the voltage to 10 volts and note the ammeter reading; it should be 14.5–16.5 amperes.

14. To check both windings, connect as for previous test. Ground the solenoid motor terminal, adjust the voltage to 10 volts and note the ammeter reading; it should be 41–47 amperes.

15. Current draw readings above specifications indicate shorted turns or a ground in the windings of the solenoid; the solenoid should be replaced. Current draw readings under specifications indicate excessive resistance. No reading indicates an open circuit. Check the connections and replace solenoid (if necessary). Current readings will decrease as windings heat up.

To Assemble:

16. The roller bearing in the drive housing and the roller bearings in the gear housing must be replaced (if they are dry); do not lubricate or reuse the bearings.

17. To replace the gear housing bearing, use a tube or solid cylinder that just fits inside the housing to push bearing toward the armature side. In the opposite direction, use the tube or cylinder to press bearing flush with housing.

18. To replace the gear housing driveshaft bearing, use a tube or collar that just fits inside the housing and press bearing out; press against the open end of bearing. To install a new bearing, press against the closed end, using a thin wall tube or collar that fits in space between bearing and housing. Do not press against the flat end of the bearing; this will bend the thin metal of the bearing. As required, replace the drive housing bearing.

19. To assemble, reverse the disassembly procedures. Be sure to replace or repair all defective components as required.

NOTE: When the starter has been disassembled or the solenoid replaced, it is necessary to check the pinion clearance. The pinion clearance must be checked in order to prevent the buttons on the shift lever yoke from rubbing on the clutch collar during engine cranking.

20. To check the pinion clearance, perform the following procedures:

   a. Disconnect the motor field coil connector from the solenoid motor terminal and insulate the terminal.

   b. Connect the positive (+) 12 volt battery lead to the solenoid switch terminal and the other to the starter frame.

   c. Touch a jumper lead momentarily from the solenoid motor terminal to the starter frame; this will shift the pinion into cranking position and retain it until the battery is disconnected.

   d. Using a feeler gauge, push the pinion back as far as possible, to take up any movement, and check the clearance; the clearance should be 0.010–0.140 in.

   e. Means for adjusting pinion clearance is not provided on the starter motor. If the clearance does not fall within limits, check for improper installation and replace worn parts.

SUZUKI/GEO

Starting System

DIRECT DRIVE TYPE

Disassembly and Assembly

SAMURAI

1. Remove the nut securing the end

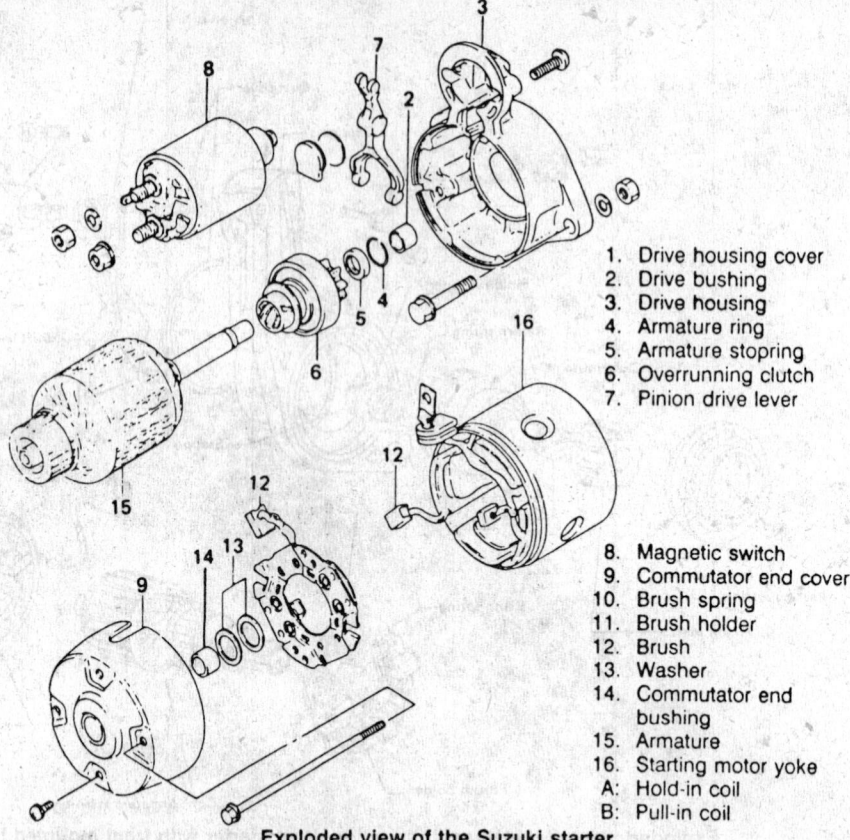

1. Drive housing cover
2. Drive bushing
3. Drive housing
4. Armature ring
5. Armature stopring
6. Overrunning clutch
7. Pinion drive lever
8. Magnetic switch
9. Commutator end cover
10. Brush spring
11. Brush holder
12. Brush
13. Washer
14. Commutator end bushing
15. Armature
16. Starting motor yoke
A: Hold-in coil
B: Pull-in coil

Exploded view of the Suzuki starter

of the field coil lead to the terminal on the head of the magnetic switch.

2. Remove the magnetic switch mounting screws and the switch from the motor body.

3. Loosen the bolts and screws and remove the commutator end cover.

4. Separate the drive housing and armature from the yoke.

5. Draw the brushes out of the holder.

6. Draw the stopring toward the clutch side and remove the armature ring and slide off the overrunning clutch.

7. To assemble, reverse the disassembly procedures. Replace brushes that are worn more than 0.450 in. New brushes are 0.670 in.

REDUCTION GEAR TYPE

Disassembly and Assembly

TRACKER

1. Position the starter in a soft jawed. Disconnect the field coil wire from the solenoid terminal.

2. Remove the solenoid mounting screws and work the solenoid from the shift fork. Remove the bearing cover, the armature shaft lock, the washer, the spring and the seal.

3. Remove the commutator end frame cover through bolts, the cover, the brushes and the brush plate.

Brakes 15

DOMESTIC BRAKE SYSTEM SERVICE

HYDRAULIC BRAKE SYSTEM TROUBLE DIAGNOSIS

Condition	Possible Cause	Correction
Insufficient brakes	1. Improper brake adjustment. 2. Worn lining. 3. Sticking brakes. 4. Brake valve pressure low. 5. Master cylinder low on brake fluid.	1. Adjust brakes. 2. Replace brake lining and adjust brakes. 3. Lubricate brake pivots and support platforms. 4. Inspect for leaks and obstructed brake lines. 5. Fill master cylinder and inspect for leaks.
Brakes apply slowly	1. Improper brake adjustment or lack of lubrication. 2. Excessive leakage with brakes applied. 3. Restriction in brake line or hose.	1. Adjust brakes and lubricate linkage. 2. Inspect all fittings and lines for leaks and repair as necessary. 3. Clean or replace brake line or hose.
Spongy pedal	1. Air in hydraulic system. 2. Swollen rubber parts due to contaminated brake fluid. 3. Improper brake shoe adjustment. 4. Brake fluid with low boiling point. 5. Brake drums ground excessively.	1. Fill and bleed hydraulic system. 2. Clean hydraulic system and recondition wheel cylinders and master cylinder. 3. Adjust brakes. 4. Flush hydraulic system and refill with proper brake fluid. 5. Replace brake drums.
Erratic brakes	1. Linings soaked with grease or brake fluid. 2. Primary and secondary shoes mounted in wrong position.	1. Correct the leak and replace brake lining. 2. Match the primary and secondary shoes and mount in proper position.
Chattering brakes	1. Improper adjustment of brake shoes. 2. Loose front wheel bearings. 3. Hard spots in brake drums. 4. Out-of-round brake drums. 5. Grease or brake fluid on lining.	1. Adjust brakes. 2. Clean, pack and adjust wheel bearings. 3. Grind or replace brake drums. 4. Grind or replace brake drums. 5. Correct leak and replace brake lining.
Squealing brakes	1. Incorrect lining. 2. Distorted brakedrum. 3. Bent brake support plate. 4. Bent brake shoes. 5. Foreign material embedded in brake lining. 6. Dust or dirt in brake drum. 7. Shoes dragging on support plate. 8. Loose support plate. 9. Loose anchor bolts. 10. Loose lining on brake shoes or improperly ground lining.	1. Install correct lining. 2. Grind or replace brake drum. 3. Replace brake support plate. 4. Replace brake shoes. 5. Replace brake shoes. 6. Use compressed air and blow out drums and support plate and shoes. 7. Sand support plate platforms and lubricate. 8. Tighten support plate attaching nuts. 9. Tighten anchor bolts. 10. Replace brake shoes and cam-grind lining.
Brakes fading	1. Improper brake adjustment. 2. Improper brake lining. 3. Improper type of brake fluid. 4. Brake drums ground excessively.	1. Adjust brakes correctly. 2. Replace brake lining. 3. Drain, flush and refill hydraulic system. 4. Replace brake drums.

HYDRAULIC BRAKE SYSTEM TROUBLE DIAGNOSIS

Condition	Possible Cause	Correction
Dragging brakes	1. Improper brake adjustment. 2. Distorted cylinder cups. 3. Brake shoe seized on anchor bolt. 4. Broken brake shoe return spring. 5. Loose anchor bolt. 6. Distorted brake shoe. 7. Loose wheel bearings. 8. Obstruction in brake line. 9. Swollen cups in wheel cylinder or master cylinder. 10. Master cylinder linkage improperly adjusted.	1. Correct adjust brakes. 2. Recondition or replace cylinder. 3. Clean and lubricate anchor bolt. 4. Replace brake shoe return spring. 5. Adjust and tighten anchor bolt. 6. Replace defective brake shoes. 7. Lubricate and adjust wheel bearings. 8. Clean or replace brake line. 9. Recondition wheel or master cylinder. 10. Correctly adjust master cylinder linkage.
Hard pedal	1. Incorrect brake lining. 2. Incorrect brake adjustment. 3. Frozen brake pedal linkage. 4. Restricted brake line or hose.	1. Install matched brake lining. 2. Adjust brakes and check fluid. 3. Free up and lubricate brake linkage. 4. Clean out or replace brake line hose.
Wheel locks	1. Loose or torn brake lining. 2. Incorrect wheel bearing adjustment. 3. Wheel cylinder cups sticking. 4. Saturated brake lining.	1. Replace brake lining. 2. Clean, pack and adjust wheel bearings. 3. Recondition or replace the wheel cylinder. 4. Reline front, rear or all four brakes.
Brakes fade (high speed)	1. Improper brake adjustment. 2. Distorted or out of round brake drums. 3. Overheated brake drums. 4. Incorrect brake fluid (low boiling temperature). 5. Saturated brake lining.	1. Adjust brakes and check fluid. 2. Grind or replace the drums. 3. Inspect for dragging brakes. 4. Drain, flush and refill and bleed the hydraulic brake system. 5. Reline brakes as necessary.

General Information

Servicing the hydraulic brake system is chiefly a matter of adjustments, replacement of worn or damaged parts and correcting the damage caused by grit, dirt or contaminated brake fluid. Always make sure the brake system is clean and tightly sealed when a brake job is completed and that only approved heavy duty brake fluid is used.

Approved heavy duty brake fluid keeps the correct consistency throughout the widest temperature range, will not affect rubber parts, helps protect metal parts and assures long, trouble free brake operation.

Never use brake fluid from a container that has been used for any other liquid. Mineral oil, alcohol, antifreeze or cleaning solvents, even in very small quantities, will contaminate brake fluid. Contaminated brake fluid will cause piston cups and the valve(s) in the master cylinder to swell and deteriorate.

Brake fluid will also absorb moisture from the air. Over time, brake fluid stored for long periods, and fluid inside the brake system will be affected by this moisture. Rust, corrosion and pitting of system components result.

Some fleet managers change the brake fluid in their vehicles every year or two to avoid brake system problems caused by moisture.

Use extreme care when using brake fluid. It will damage painted surfaces.

The hydraulic braking system consists of a master cylinder, sometimes a power booster depending on application, hydraulic line and hoses, control valves and calipers and/or wheel cylinders. Newer models incorporate a computer controlled wheel antilock system. When the brake pedal is depressed, the master cylinder forces brake fluid to the calipers and/or cylinders. Sliding rubber seals contain the fluid and prevent leakage.

Return springs in the master cylinder help the brake pedal return to the unapplied position. Check valves, in most cases regulate the return flow of the fluid to the master cylinder. Other valves, such as the metering valve, proportioning valve, or combination valve, regulate the flow of fluid to the caliper/wheel cylinder, to achieve efficient braking.

Dual braking systems were introduced on many light trucks and vans during the late 1960's. The main difference is the use of a tandem master cylinder which is essentially two master cylinders in one. Two separate pistons share one bore and two fluid reservoirs are built into one housing. Dual brake lines split the calipers and/or wheel cylinders into two groups, each actuated by its own, separate master cylinder piston. In the event of failure of one of the dual systems, the other should provide enough braking to safely stop the vehicle. The development of dual braking systems is an improvement over the older single systems where a leak anywhere would allow the fluid to escape resulting in loss of braking.

The dual system usually includes some type of warning light on the instrument panel, activated by a pressure differential valve. The valve reacts to loss of hydraulic pressure that might result from failure on either side of the system.

Vehicles are generally equipped with either a front/rear wheel split or a diagonally split system. On front/rear systems, the front wheels are connected to one circuit while the rear wheels are connected to the other circuit. These systems are the most popular ever since front disc brakes were introduced. Since a greater amount of brake fluid is moved in a disc brake system, one chamber of the the master cylinder feeds the front discs, while the other chamber feeds the rear wheel cylinders.

Diagonally split systems have diagonally opposite wheels connected to each circuit.

Because of the differences in pressure and the amount of fluid required to operate disc brakes and drum brakes, a control valve, often called a proportioning valve or combination valve is installed between the master cylinder and the rest of the system. This valve has several sections and functions.

The metering, or hold off section of the valve limits the pressure to the front disc brakes until a predetermined front input pressure is reached, enough to overcome the rear shoe retractor springs. The is generally no restriction to the inlet pressures below about 3 psi to allow for pressure equalization during the no apply periods.

Another section of the valve proportions, or measures out the outlet pressure to the rear brakes after a predetermined rear input pressure has been reached. This is done to prevent rear wheel lockup on vehicles with light rear wheel loads.

Yet another section of the valve is

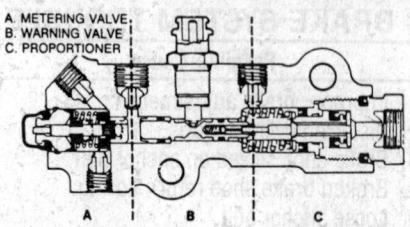

A. METERING VALVE.
B. WARNING VALVE.
C. PROPORTIONER

Brake proportioning valve – typical

designed to constantly compare front and rear brake pressure from the master cylinder and will turn on a warning light in the event of a front or rear system malfunction. On some systems, after repairs are made, and the system properly bled, the valve will center itself and the warning lamp will shut off. On other systems, the switch will latch so the warning light will stay on until a repair is made, and the switch manually reset, generally be one or two firm brake applications.

These valves are also designed with a bypass feature which assures full system pressure to the rear brakes in the event of a front brake system malfunction, and full front pressure is retained in the event of a rear malfunction.

These valves are not to be disassembled for repair. Replace defective valves.

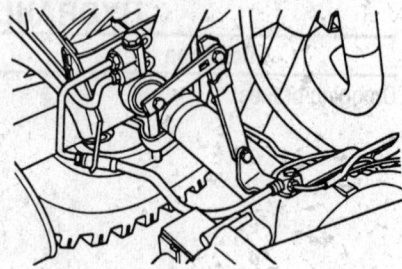

General Motors Corporation height sensing valve – typical

Most light trucks have some sort of height sensing valve. The vehicle braking force is distributed to the front and rear wheels as determined be either a light or heavy payload. The valve is usually mounted on the frame and a linkage connects the valve to a bracket mounted on the axle.

Adding suspension accessories or other equipment (such as load leveling kits, lift kits, extra springs, etc.) or making modifications that will change the distance between the axle and frame without changing the load will provide a false reading to the sensing valve. This could result in unsatisfactory brake performance which could result in an accident and possibly personal injury.

GM CORPORATION BRAKE DIAGNOSIS

DIAGNOSIS OF THE BRAKE SYSTEM

Problem	Possible Cause	Correction
Uneven brake action (brakes pull)	1. Incorrect tire pressure.	1. Inflate evenly on both sides to specifications.
	2. Front end out of alignment.	2. Check and align to specifications.
	3. Loose suspension parts.	3. Check all the suspension mountings.
	4. Worn out brake linings.	4. Replace with lining of the correct material.
	5. Incorrect lining material.	5. Replace with linings of the correct material.
	6. Malfunctioning caliper assembly.	6. Check for frozen or sluggish pistons and the lubrication of the retainer bolts. Caliper should slide.
	7. Loose calipers.	7. Check and torque.
	8. Contaminated brake linings.	8. Repair as necessary. Replace the linings in complete axle sets.
	9. Malfunctioning rear brakes.	9. Check for inoperative self adjusters. Weak return springs. Leaking wheel cylinders.
	10. Leaking wheel or piston cylinder seal.	10. Repair as necessary.
	11. Restricted brake tubes or hoses.	11. Check for collapsed rubber hoses or damaged lines. Repair as necessary.
	12. Unmatched tires on the same axle.	12. Same style tires with about the same tread should be used on the same axle.
Slow pedal return	Compensating (peripheral) holes in the quick take-up valve are clogged.	Replace the master cylinder.

DIAGNOSIS OF THE BRAKE SYSTEM

Problem	Possible Cause	Correction
Brakes squeak	1. Worn out linings.	1. Replace the linings.
	2. Glazed brake linings.	2. Replace the linings.
	3. Heat spotted rotors or drums.	3. Check per instructions. If within specifications, machine the rotor or drum.
	4. Weak or incorrect brake shoe retention springs.	4. Replace with new retention springs.
	5. Contaminated brake linings.	5. Repair as necessary. Replace the linings in complete axle sets.
	6. Incorrect lining material.	6. Replace with linings of correct material.
	7. Brake assembly attachments missing or loose.	7. Repair as necessary.
	8. Excessive brake lining dust.	8. Clean the dust from the brake assembly.
Brake pedal pulsates	1. Excessive rotor lateral runout.	1. Check per instructions. If within specifications, machine the rotor.
	2. Rear drums out of round.	2. Check per instructions. If within specifications, machine the drum.
	3. Heat spotted rotors or drums.	3. Check per instructions. If within specifications, machine the drum.
	4. Incorrect wheel bearing adjustments.	4. Repair as necessary.
	5. Out of balance wheel assembly.	5. Repair as necessary.
	6. Brake assembly attachments missing or loose.	6. Repair as necessary.
Excessive pedal effort	1. Leaking vacuum system.	1. Repair as necessary.
	2. Malfunctioning power brake unit.	2. Repair as necessary.
	3. Worn out linings.	3. Replace the linings.
	4. Malfunctioning proportioning valve.	4. Replace the combination valve.
	5. Incorrect lining material.	5. Replace with linings of the correct materials.
	6. Incorrect wheel cylinder.	6. Replace with the correct size wheel cylinder.
	7. Center orifice in quick take-up valve clogged.	7. Replace the master cylinder.
Brakes drag	1. Malfunctioning caliper assembly.	1. Check for frozen or sluggish pistons and the lubrication of the retainer bolts. Caliper should slide.
	2. Contaminated or improper brake fluid.	2. Repair as necessary.
	3. Improperly adjusted parking brake.	3. Adjust as necessary.
	4. Restricted brake tube or hoses.	4. Check for collapsed rubber hoses or damaged lines. Repair as necessary.
	5. Malfunctioning proportioning valve.	5. Replace the combination valve.
	6. Malfunctioning self adjusters.	6. Repair as necessary.
	7. Malfunctioning master cylinder.	7. Repair as necessary.
	8. Improperly adjusted master cylinder pushrod.	8. Adjust pushrod length.

Brake warning light comes on	1. Air in the brake system.	1. Check the fluid level. Check for leaks in the lines, wheel cylinders, or master cylinder. Bleed the system.
	2. Malfunctioning master cylinder.	2. Check for malfunctioning or leaking metering valve. Repair as necessary.
	3. Contaminated or improper brake fluid.	3. Repair as necessary.
	4. Parking brake on or not fully released.	4. Check the parking brake. Repair as necessary.
	5. Worn out brake lining.	5. Replace the linings.
	6. Incorrect wheel bearing adjustment.	6. Repair as necessary.
	7. Malfunctioning self adjusters.	7. Repair as necessary.
	8. Brake assembly attachments missing or loose.	8. Replace or repair as necessary.
	9. Improperly adjusted master cylinder pushrod.	9. Adjust the pushrod length.
	10. RWAL system malfunction.	**10. Refer to REAR WHEEL ANTILOCK BRAKE SYSTEM**
Excessive pedal travel	1. Fluid level low in the master cylinder reservoir.	1. Fill the reservoir with approved brake fluid. Check for leaks and air in the system. Check the warning light.
	2. Air in the brake system.	2. Check for leaks in the lines, wheel cylinders, or master cylinder. Bleed the system.
	3. Malfunctioning self adjusters.	3. Repair as necessary.
	4. Master cylinder.	4. Replace or repair as necessary.
	5. Incorrect wheel bearing adjustment.	5. Repair as necessary.
	6. Improperly adjusted master cylinder pushrod.	6. Adjust the master cylinder pushrod.
	7. Fluid bypassing quick take-up valve to the reservoir.	7. Replace the master cylinder.
	8. Leaking brake line or connection.	8. Repair as necessary.
	9. Leaking wheel cylinder or caliper.	9. Repair as necessary.

General Diagnosis

LOW PEDAL

Normal brake lining wear reduces pedal reserve. Low pedal reserve may also be caused by the lack of brake fluid in the master cylinder. This means a brake adjustment is required. Check fluid level in the master cylinder and add as required.

FLUID LOSS

If the master cylinde00requires constant addition of brake fluid, fluid may be leaking past the piston cups in the master cylinder, calipers or wheel cylinders, the brake lines or hoses, or through loose connections. Brake hoses often have copper sealing washers which may need to be replaced. Replace worn or damaged components, refill and bleed the system.

FLUID CONTAMINATION

To determine if contamination exists in the brake fluid as indicated by swollen, deteriorated rubber parts like wheel cylinder cups, the following tests can be made.

Place a small amount of the drained brake fluid into a clear glass bottle. Separation of the fluid into distinct layers will indicate mineral oil content. Be safe and discard the old brake fluid that has been bled from the system since it may contain dirt and other contamination and should not be reused.

BRAKE ADJUSTMENT

Self adjusting brakes usually do not require manual adjustment. When installing new brake shoes, it may be advisable to make the initial adjustment manually to speed up adjusting time.

AUTOMATIC ADJUSTER CHECK

On most vehicles, drum brakes appear only on the rear axle. Brake adjustments are automatic and are made during reverse brake applications.

Raise and safely support the vehicle, have a helper in the driver's seat to apply brakes. Remove the plug from the adjustment slot to observe the star wheel (some models are adjusted through the face of the brake drum others through the backing plate). Tighten the adjusting star wheel screw until the wheel can just be turned by hand. The brake drag should be equal at both wheels. Back off the adjusting star wheel about 30-35 notches. The brakes should have no drag after the adjuster has been backed off about 15 notches. If heavy drag is present, the parking brake likely needs service and/ or adjustment.

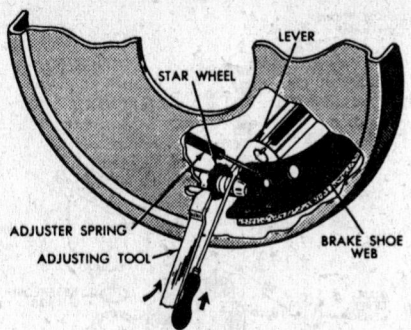

Brake adjustment—typical

NOTE: It will be necessary to carefully insert a small, stiff wire or thin tool to hold the automatic adjustment lever away from the star wheel to allow backing off the adjustment. Install adjusting hole cover in backing plate. Check parking brake adjustment.

HYDRAULIC LINE REPAIR

Steel tubing is used for the hydraulic lines, and special flexible hoses connect moving parts such as front calipers which turn with the steering, and the connection between the body and rear axle.

When replacing steel lines or flexible brake hose, use only exact replacements, in both size and quality.

NOTE: Never use copper tubing for hydraulic brake lines because copper is subject to fatigue cracking and corrosion which could result in brake failure. Use specially made steel brake lines only. Tighten all connections securely.

After replacement, bleed the brake system at each wheel and at the booster, if equipped with a bleeder screw.

Flexible hoses should be inspected for any signs of road damage, cracks or chafing any of which requires immediate hose replacement. Hoses feeding front brake calipers are often equipped with a copper washer for a seal. This should also be replaced when hoses are renewed. After hose installation, make sure the hose is not twisted. Many hoses have a painted stripe on them to help the technician determine if the hose is twisted.

If a length of steel brake line tubing must be replaced and the end flared, a special double flaring tool must be used. Always inspect newly formed flares for defects. Double lap flaring tools must be used. Single flares cannot hold the high pressures in the brake system and tend to crack. When bending brake tubing to fit the frame or rear axle contours, be careful not to

crack or kink the tube. Always clean the inside of even new brake tube with clean isopropyl alcohol.

GENERAL MOTORS CORPORATION

Master Cylinder Service

In addition to standard master cylinder functions, a quick take-up feature in included on most models. This provides a large volume of fluid to the wheels at low pressure with the initial brake application. This large volume of fluid is needed to overcome the clearance created by the seals retracting the pistons into the front calipers and the spring retraction of the rear brake shoes.

NOTE: General Motors Corporation does not recommend honing the bores of either cast iron or composite master cylinders. When the brake master cylinder is overhauled, it is recommended that the cylinder body be replaced rather than attempting to clean up a scuffed cylinder by honing the bore. The master cylinder has a hard, highly polished surface which is produced by diamond boring followed by ball or roller burnishing under heavy pressure. Honing will destroy this surface which will cause rapid wear of rubber cups. Do not use kerosene, gasoline or other solvents for cleaning or flushing master cylinder and components. The use of these solvents or any other with a trace of mineral oil, will damage rubber parts.

DELCO CAST IRON MASTER CYLINDER

Disassembly and Assembly

1. Remove cover, rubber diaphragm and drain reservoir.
2. Remove snapring and primary piston assembly.
3. Plug rear port and apply low pressure air to front port. Secondary piston will pop out. Use shop cloth to catch piston. Use caution. The piston may come out with considerable force. Remove seals from secondary piston.
4. If tube seats must be replaced,

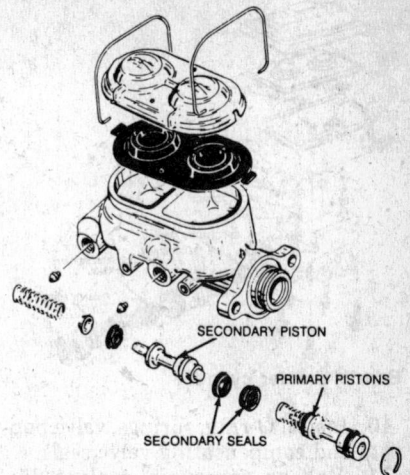

GM cast iron master cylinder

thread in self-tapping screw into seat, remove with locking pliers.
5. Clean all metal parts in denatured alcohol, the rubber parts in brake fluid.
6. A stained or discolored bore may be cleaned with fine crocus cloth. Do not attempt to hone the bore.
7. Lube all seals and the bore with clean brake fluid. Install seals on pistons and assemble.
8. Push primary piston with smooth rounded end tool and install snapring.
9. If tube seats are replaced, seat with spare brake tube nut.
10. Install diaphragm into cover and install cover.

BENDIX MASTER CYLINDER

Disassembly and Assembly

1. Remove cover, rubber diaphragm and drain reservoir.
2. Locate and remove the reservoir to body bolts and remove reservoir.
3. Remove the seals, poppet valves and springs from cylinder body.
4. Remove snapring and primary piston.
5. Plug rear port and apply low pressure air to front port. Secondary piston will pop out. Use shop cloth to catch piston. Use caution. The piston may come out with considerable force. Remove seals from secondary piston.
6. Clean all metal parts in denatured alcohol, the rubber parts in brake fluid.
7. A stained or discolored bore may be cleaned with fine crocus cloth. Do not attempt to hone the bore.
8. Lube all seals and the bore with clean brake fluid. Install seals on pistons and assemble.
9. Push primary piston with smooth rounded end tool and install snapring.

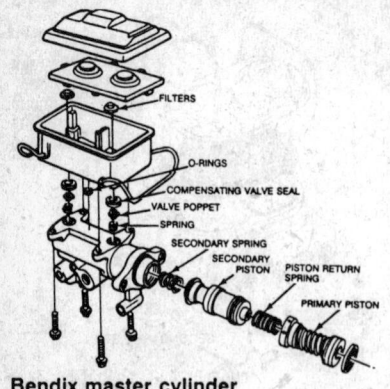

Bendix master cylinder

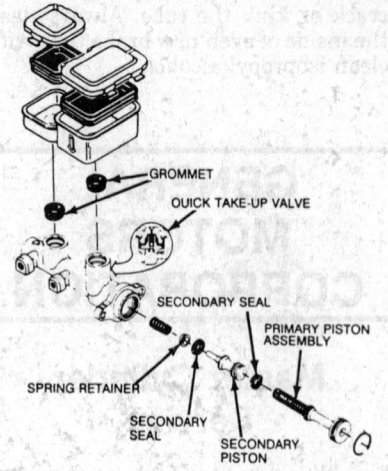

GM composite master cylinder

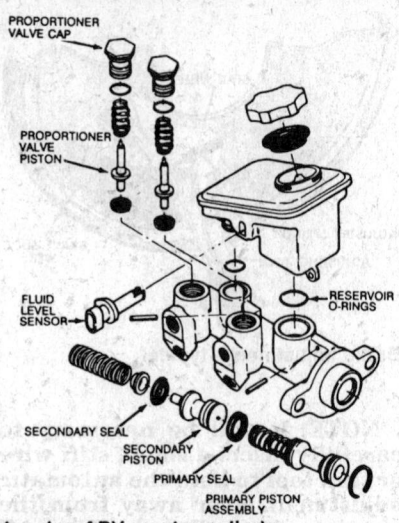

Lumina APV master cylinder

10. Install O-ring, springs, valve poppets and compensating valve seals.

11. Bolt on reservoir, replace diaphragm and cover.

DELCO COMPOSITE MASTER CYLINDER

Disassembly and Assembly

EXCEPT LUMINA APV, SILHOUETTE AND TRANS SPORT

1. Remove cover, rubber diaphragm and drain reservoir.

2. Clamp mounting flange of cylinder in vise and pry reservoir off. Remove reservoir grommet.

3. Remove snapring and primary piston.

4. Plug rear port and apply low pressure air to front port. Secondary piston will pop out. Use shop cloth to catch piston. Use caution. The piston may come out with considerable force. Remove seals from secondary piston.

5. Clean all metal parts in denatured alcohol, the rubber parts in brake fluid.

6. A stained or discolored bore may be cleaned with fine crocus cloth. Do not attempt to hone the bore.

7. Lube all seals and the bore with clean brake fluid. Install seals on pistons and assemble.

8. Push primary piston with smooth rounded end tool and install snapring.

9. Install grommets and press reservoir on with rocking motion. Install diaphragm and cover.

LUMINA APV, SILHOUETTE AND TRANS SPORT

1. Remove cover, rubber diaphragm and drain reservoir.

2. Remove fluid level sensor switch using needle nose pliers to compress switch locking tabs at the inboard side of master cylinder.

3. Remove proportioner valve assemblies by unscrewing cap assemblies, removing the O-rings and

springs. Remove pistons with needle nose pliers. Use care. Do not scratch piston stems.

3. Remove snapring and primary piston.

4. Apply low pressure air into upper outlet port at blind end of bore. Plug all other ports. Secondary piston will pop out. Use shop cloth to catch piston. Use caution. The piston may come out with considerable force. Remove seals from secondary piston.

5. To remove reservoir, clamp flange (never the body) of master cylinder in vise. Drive out spring pins with 1/8 in. punch. Pull reservoir straight up and remove O-rings from grooves in reservoir.

6. Clean all metal parts in denatured alcohol, the rubber parts in brake fluid.

7. Inspect the bore for scoring or corrosion. If noted, replace master cylinder. No abrasives should be used in bore. Do not attempt to hone the bore.

8. Install new seals on reservoir, press into cylinder body and install spring pins.

9. Lube all seals and the bore with clean brake fluid. Install seals on pistons and assemble.

10. Push primary piston with smooth rounded end tool and install snapring.

11. Install proportioner valve assemblies and fluid level sensor, making sure locking tabs snap into place.

12. Install diaphragm and cover.

Brake Booster Service

Three types of brake boosters are used: a single diaphragm, a tandem diaphragm and a hydraulic booster (Hydro-Boost).

The vacuum boosters may have a vacuum switch to activate a brake warning light in case of low booster vacuum or vacuum pump malfunc-

tion. Under normal operating conditions, with brakes released, a vacuum suspended booster operates with vacuum on both sides of its diaphragm (or both diaphragms in a tandem diaphragm unit). When the brakes are applied, air at atmospheric pressure is admitted to one side of the diaphragm (or both diaphragms in a tandem diaphragm unit) to provide the power assist.

The hydraulic brake booster (Hydro-Boost) uses a hydraulic pump to power the system and a pneumatic accumulator as a reserve system. In this system no special fluids are used. However, care must be taken to use the correct fluids. The master cylinder and brake system operate on standard brake fluid. The hydraulic pump, which is driven by pressure from the power steering pump, uses power steering fluid.

VACUUM DIAPHRAGM BRAKE BOOSTERS

Disassembly and Assembly

SINGLE DIAPHRAGM BOOSTER

NOTE: A special brake booster disassembly tool is recommended for this procedure.

1. With the booster unit off the vehicle, and the master cylinder removed from the booster, take off the boot and silencer disc from the booster's mounting (rear) side. Remove the vacuum check valve, grommet and the front housing seal from the face side.

2. Scribe a mark across the front and rear housings to aid assembly.

3. Clamp the base in a vise with the power section facing up. Separate the front and rear housings by pressing down (using the special disassembly

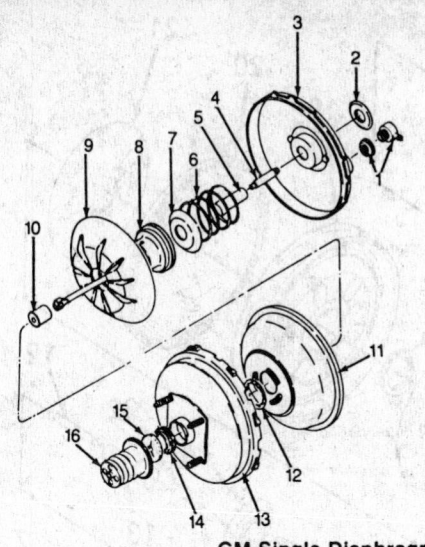

1. Check valve and grommet
2. Front housing seal
3. Front housing
4. Piston rod
5. Reaction retainer
6. Return springs
7. Reaction body retainer
8. Power piston and pushrod assembly
9. Diaphragm support
10. filter
11. Diaphragm
12. Diaphragm retainer
13. Rear housing
14. Power piston bearing
15. Silencer disc
16. Boot

GM Single Diaphragm Booster

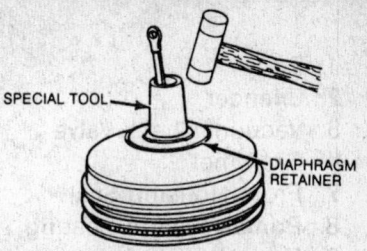

Installing diaphragm retainer with pipe driver

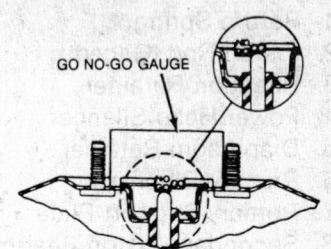

Using special GO NO-GO gauge to check pushrod

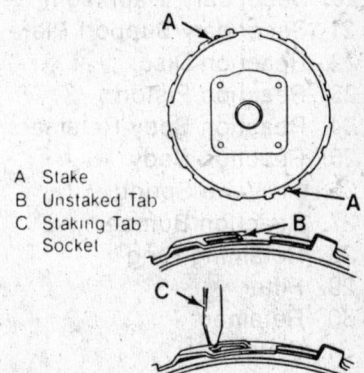

A Stake
B Unstaked Tab
C Staking Tab Socket

Staking the booster tabs

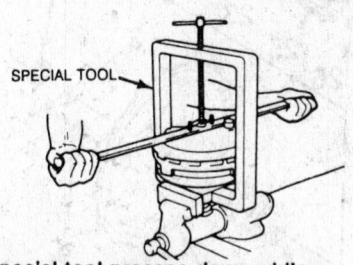

Special tool presses down while housing is turned

tool, if available) and rotating the housing counterclockwise to the unlocked position. Loosen carefully as it is spring loaded.

4. Remove the large return spring and the power piston which will likely still have the pedal pushrod attached.

5. Remove the power piston bearing from the rear housing.

6. Remove the reaction body retainer, the master cylinder piston rod, and reaction retainer.

7. Remove the filter with an awl or similar tool.

8. Separate the power piston and pedal pushrod. To do this, grasp the outside edge of the diaphragm support and diaphragm. Hold the pushrod down against a hard surface and use slight force to dislodge the diaphragm retainer.

9. Remove the diaphragm from the diaphragm support.

10. Inspect all parts for corrosion, nicks, cracks, cuts, scoring, distortion or excessive wear. Replace as required. Clean all parts in denatured alcohol, but do not immerse the power piston and pushrod assembly in alcohol. Dry with compressed air.

11. Lubricate the inside diameter of the diaphragm lip with a thin layer of silicone grease and install the diaphragm into its support.

12. Install the diaphragm and support onto the power piston and push-

rod assembly. Use a new retainer and install with a pipe-like driver.

13. Install filter, reaction retainer, piston rod, and reaction body retainer.

14. Install the reaction body retainer disc. Lubricate the inside and outside of the power piston bearing with silicone grease and install in rear housing.

15. Install power piston group into rear housing, put in the return spring and assemble front housing to rear housing. Align mating marks made before disassembly.

16. Apply pressure in clockwise direction to lock housings together. Stake the housing at two tabs 180 degrees apart. Do not stake a tab that has been previously staked.

17. Lubricate the check valve grommet with silicone grease and install it and the check valve in the front cover.

18. Install the front housing seal, silencer disc, and boot.

19. The piston rod should be checked with a special GO, NO-GO gauge. If not within limits, replace the rod.

TANDEM DIAPHRAGM BOOSTER

NOTE: A special brake booster disassembly tool is recommended for this procedure.

1. With the booster unit off the vehicle, and the master cylinder removed from the booster, take off the boot and silencer disc from the booster's mounting (rear) side. Remove the vacuum check valve, grommet and the front housing seal from the face side.

2. Scribe a mark across the front and rear housings to aid assembly.

3. Clamp the base in a vise with the power section facing up. Separate the front and rear housings by pressing down (using the special disassembly tool, if available) and rotating the housing counterclockwise to the un-

locked position. Loosen carefully as it is spring loaded.

4. Remove the large return spring and the power piston which will likely still have the pedal pushrod attached.

5. Remove the power piston bearing from the rear housing.

6. Remove the piston rod, reaction retainer and power head silencer ring.

7. Remove the power piston assembly and pushrod. Grasp the assembly at the outside edge of the housing divider and both diaphragms. Hold the pushrod down against a hard surface. Tap to dislodge diaphragm retainer.

8. Remove the primary diaphragm (rear) and primary support plate from the housing divider and separate the primary diaphragm from its support plate.

9. Remove the secondary diaphragm (front) and secondary support plate from the housing divider. Remove the secondary piston bearing from the center of the housing divider and separate the secondary diaphragm from its support plate.

10. Remove the reaction plate retainer, reaction body, reaction disc, and reaction piston from the reaction body.

1. Boot
2. Silencer
3. Vacuum Check Valve
4. Grommet
7. Front Housing Seal
8. Primary Piston Bearing
9. Rear Housing
10. Front Housing
11. Return Spring
12. Piston Rod (Gaged)
13. Reaction Retainer
14. Power Head Silencer
15. Diaphragm Retainer
16. Primary Diaphragm
17. Primary Support Plate
18. Secondary Piston Bearing
19. Housing Divider
20. Secondary Diaphragm
21. Secondary Support Plate
22. Reaction Disc
23. Reaction Piston
24. Reaction Body Retainer
25. Reaction Body
26. Air Valve Spring
27. Reaction Bumper
28. Retaining Ring
29. Filter
30. Retainer
31. O-Ring
32. Air Valve Push Rod Assembly
33. Power Piston

Tandem diaphragm vacuum booster components

11. Remove the air valve spring and reaction bumper from the end of the pushrod. Carefully remove the retaining ring from the pushrod, and remove the pushrod by inserting a suitable tool through the eyelet end and pulling straight out. Considerable force will be required.

12. Remove the filter, retainer and O-ring from the pushrod assembly.

13. Inpect all parts for corrosion, nicks, cracks, cuts, scoring, distortion or excessive wear. Replace as required. Clean all parts in denatured alcohol, but do not immerse the power piston and pushrod assembly in alcohol. Dry with compressed air.

14. Using silicone grease, lubricate the O-ring for the pushrod and install the pushrod into the power piston, then install the retainer and seat.

15. Install the filter over the pushrod eyelet and into the power piston and install the retaining ring.

16. Assemble the reaction bumber, air valve spring, reaction piston and reaction disc into the reaction body. Install the reaction body retainer.

17. Using silicone grease, lubricate the inside diameter of both the primary and secondary diaphragm as well as the secondary piston bearing.

18. Assemble the secondary diaphragm to its support plate and slide the assembly over the power piston/pushrod assembly.

19. Install the secondary piston bearing into the housing divider. The flat surface of the bearing goes on the same side as the six raised lugs on the divider. Install this assembly over the power piston/pushrod assembly.

20. Assemble the primary diaphragm to its support plate, by folding the diaphragm up, away from the support plate and slide the assembly over the power piston/pushrod assembly. Fold the primary diaphragm back into position and pull the outside edge of the diaphragm over the formed flange of the housing divider.

NOTE: Check that the beads on the secondary diaphragm are seated evenly around the complete circumference.

21. Install a new diaphragm retainer and install with a pipe-like driver.

22. Install the silencer ring, reaction retainer and piston rod.

23. Lubricate the inside and outside diameters of the primary piston bearing with silicone grease and install into the rear housing.

24. Install the power piston assembly into the rear housing.

25. Install the return spring and assemble front housing to rear housing. Align mating marks made before disassembly.

26. Apply pressure in clockwise direction to lock housings together. Stake the housing at two tabs 180 degrees apart. Do not stake a tab that has been previously staked.

27. Lubricate the check valve grommet with silicone grease and install it and the check valve in the front cover.

28. Install the front housing seal, silencer disc, and boot.

29. The piston rod should be checked with a special GO, NO-GO gauge. If not within limits, replace the rod.

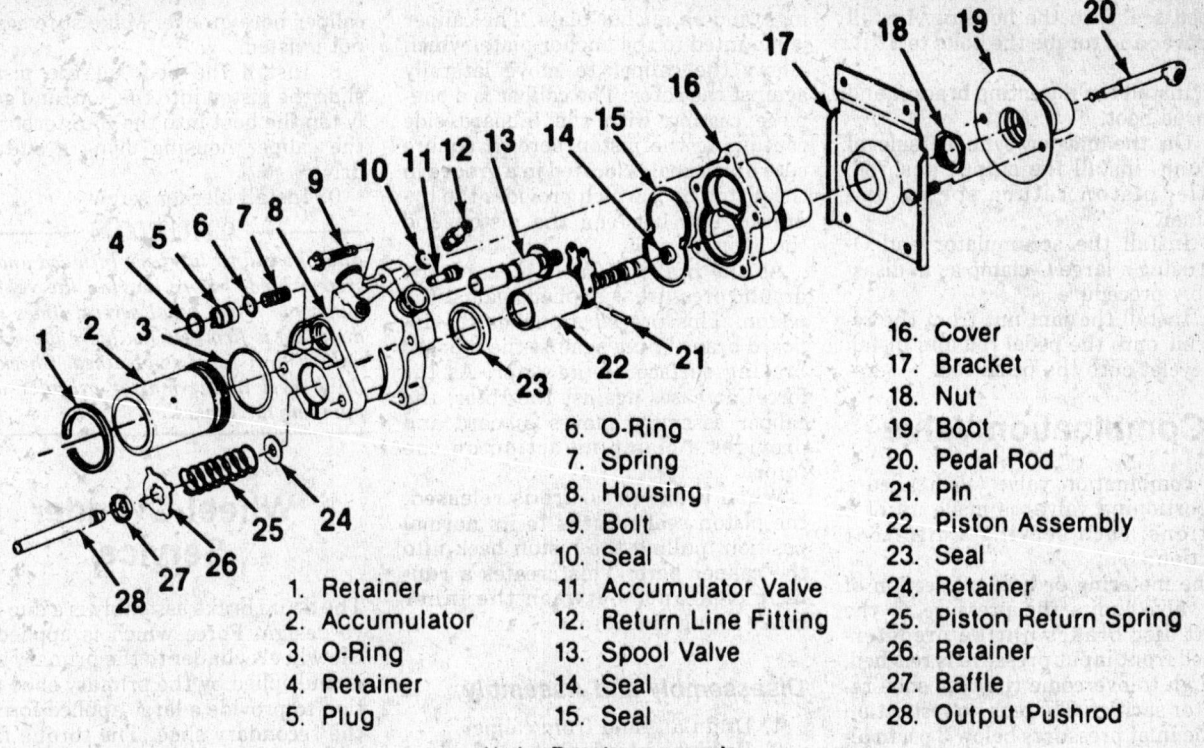

Hydro-Boost components

6. O-Ring
7. Spring
8. Housing
9. Bolt
10. Seal
11. Accumulator Valve
12. Return Line Fitting
13. Spool Valve
14. Seal
15. Seal

1. Retainer
2. Accumulator
3. O-Ring
4. Retainer
5. Plug

16. Cover
17. Bracket
18. Nut
19. Boot
20. Pedal Rod
21. Pin
22. Piston Assembly
23. Seal
24. Retainer
25. Piston Return Spring
26. Retainer
27. Baffle
28. Output Pushrod

HYDRO-BOOST

The Bendix Hydro-Boost uses the hydraulic pressure supplied by the power steering pump to provide a power assist to brake application. It has identifying information stamped into the housing near the inlet line. When servicing this unit, there are some special tools that are recommended for these procedures.

Disassembly and Assembly

--- CAUTION ---

This system uses an accumulator that contains compressed gas. Do not apply heat to the accumulator. Use caution or personal injury may result. Do not attempt to repair a defective accumulator. Always replace with a new unit. Dispose of an inoperative accumulator by drilling a 1/16 in. diameter hole through the end of the accumulator can opposite the O-ring.

1. Remove the accumulator. Note that an adapter (a piece of strap metal with a hole drilled to accommodate a mounting stud) is used along with a C-clamp. Depress the accumulator with the C-clamp, insert a punch into the hole on the housing and remove the snapring retainer. Release the C-clamp, remove the accumulator and O-ring.

2. Remove the retainer from the small port above the mounting flange and remove the plug, O-ring and spring.

3. Remove the star-shaped retainer from the mounting flange bore and remove the output pushrod, baffle, piston return spring and inner retainer.

4. Remove the pedal rod boot and mounting bracket, if installed. On some installations it may be necessary to saw off the eyelet of the pedal rod. Make sure exact replacements or the replacement parts kit is available before cutting the original part.

5. Remove the cover bolts and separate the cover from the housing. Remove the seals (a large figure-eight shaped seal and a smaller seal on the end of the piston assembly).

6. Remove the piston assembly and the O-ring seal on the housing side of the piston.

7. Remove the spool valve.

8. Remove the accumulator valve. It may be necessary to fish this small valve out with a thin wire hook.

9. Remove the return line fitting and seal.

10. Clean all parts in power steering fluid. Inspect the spool valve for corrosion, nicks and scoring. If found, replace the complete booster. Discoloration of the spool or bore is not harmful and is no cause for replacement. Check all components for damage or wear, especially the tube seat in the housing. This seat can be removed with a Easy-Out type remover, and its replacement installed by tapping gently into place.

11. Lubricate all seals and friction

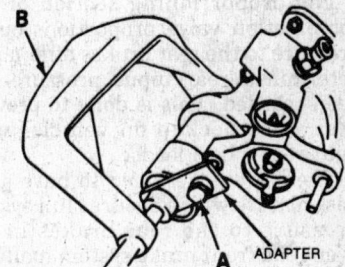

A. Nut
B. C-clamp
Removing the accumulator

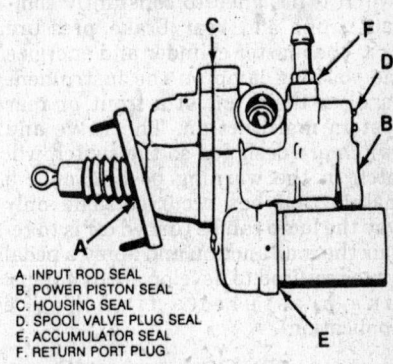

A. INPUT ROD SEAL
B. POWER PISTON SEAL
C. HOUSING SEAL
D. SPOOL VALVE PLUG SEAL
E. ACCUMULATOR SEAL
F. RETURN PORT PLUG
Hydro-Boost seal leak areas

metal parts with power steering fluid.

12. Install the return line fitting and seal.

13. Install the accumulator valve and spool valve.

14. Install the seal and piston assembly, place the small seal on the piston assembly end and the figure eight

shaped seal into the housing. Install the cover and torque the bolts to 22 ft. lbs.

15. Install the mounting bracket and pushrod boot.

16. On the master cylinder side of the unit, install the output pushrod, baffle, piston return spring and retainer.

17. Install the accumulator and O-ring using a large C-clamp as in disassembly procedure.

18. Install the jam nut from the repair kit onto the pedal rod and install the eyelet onto the pedal rod.

Combination Valve

The combination valve (also called a proportioning valve) is made up of 3 sections, each serving a different function.

The metering or hold-off section of the valve limits the pressure to the front disc brakes until a predetermined front input pressure is reached, enough to overcome the rear shoe retractor springs. There is no restriction to the inlet pressures below 3 psi to allow for pressure equalization during no-apply periods.

The proportioning section of the combination valve proportions outlet pressure to the rear brakes after a predetermined rear input pressure has been reached. This is done to prevent rear wheel lockup on vehicles with light rear wheel loads.

The valve is designed to have a by-pass feature which ensures full system pressure to the rear brakes in the event of a front brake system malfunction. Full front pressure is retained in the event of rear malfunction.

The pressure-differential warning switch is designed to constantly compare front and rear brake pressure from the master cylinder and energize the warning lamp on the instrument panel in the event of a front or rear system malfunction. The valve and switch are designed so the switch will latch in the warning postion once a malfunction has occurred. The only way the lamp can be turned off is to repair the malfunction and apply a pedal force required to develop about 450 psi line pressure (a firm brake application).

Valve Overhaul

The combination valve is not repairable and must be replaced as a complete assembly.

Caliper Service

The disc brake assembly consists of a caliper and piston assembly, rotor, lin-ings, and an anchor plate. The caliper is mounted to the anchor plate, which allows the caliper to move laterally against the rotor. The caliper is a one-piece casting with the inboard side containing the piston bore. A square cut rubber seal is located in a groove in the piston bore which provides the hydraulic seal between the piston and the cylinder wall.

As the brake pedal is pressed, hydraulic pressure is applied against the piston. This pressure pushes the inboard brake lining against the inboard braking surface of the rotor. As the force increases against the rotor, the caliper assembly moves inboard and provides a clamping action on the rotor.

When brake pressure is released, the piston seal returns to its normal position, pulling the piston back into the caliper bore. This creates a running clearance between the inner brake lining and rotor.

Disassembly and Assembly

1. Drain all fluid from caliper.
2. Pad interior of caliper with clean shop towels and use just enough compressed air to ease the piston out of the bore.

CAUTION

Do not place fingers in front of piston to try to catch piston or protect it when applying compressed air. This could result in serious injury. Use just enough air to ease the piston out. If piston is blown out, even with padding, it may be damaged.

3. Remove boot, being careful not to scratch bore.
4. Remove square cut seal from caliper bore groove.
5. Remove bleeder valve.
6. Clean all parts with denatured alcohol, blow dry with compressed air. Inspect all parts for scoring, corrosion or damage to chrome plating on piston. Replace if damaged. Fine crocus cloth can be used to polish out any light corrosion.
7. Lubricate the new piston seal, caliper bore and piston with clean brake fluid and install the seal in the

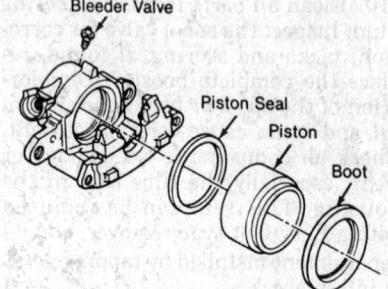

Caliper components – typical

caliper bore groove. Make sure seal is not twisted.

8. Install the boot on the piston, slide the piston into the bore and gently tap the boot into the counterbore of the caliper housing using a suitable driver.

9. Install bleeder screw.

CAUTION

After the caliper has been installed and the system bled, before moving the vehicle, pump the brake pedal several times until the pedal is firm. Do not move the vehicle until a firm pedal is obtained. Check the fluid level in the master cylinder after pumping the brakes.

Wheel Cylinder Service

The drum brake assembly is a duo-servo design. Force which is applied by the wheel cylinder to the primary shoe is multiplied by the primary shoe friction to provide a large applied force to the secondary shoe. The torque from the brake shoes is transferred to the anchor pin and through the backing plate, the the axle flange. Brake adjustments are automatic and are made during reverse brake applications.

Wheel cylinders may need reconditioning or replacement whenever the brake shoes are replaced or when required to correct a leak condition. In some cases, the wheel cylinders can be disassembled without removing them from the backing plate. On others, the cylinder must be removed before being disassembled.

Leaks which coat the boot and the cylinder with fluid, or result in a dropped reservoir fluid level, or dampen and stain the brake linings are dangerous. Such leaks can cause the brakes to grab or fail and should be immediately corrected. A leakage, not immediately apparent, can be detected by pulling back the cylinder boot. A small amount of fluid seepage dampening the interior of the boot is normal. However, a dripping boot is not. Unless other conditions causing a brake to pull, grab or drag become obvious, the wheel cylinder is suspect and should and should be included in general reconditioning.

Cylinder binding may be caused by rust, deposits, grime, or swollen cups due to fluid contamination, or by a cup wedged into an excessive piston clearance.

Hydraulic system parts should not be allowed to come into contact with oil or grease, neither should those be handled with greasy hands. Even a trace of any petroleum based product

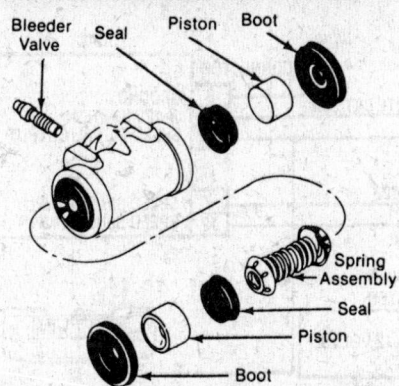

Bleeder Valve — Seal — Piston — Boot — Spring Assembly — Seal — Piston — Boot

Wheel cylinder components — typical

is sufficient to cause damage to the rubber parts.

Disassembly and Assembly

1. Remove the brake bleeder screw.
2. Remove the dust boots, allow any brake fluid to drain out.
3. Remove the pistons, seals and inner spring.
4. Inspect the bore for scoring and corrosion. The inside of the bore may be cleaned with fine crocus cloth. If bore is scored, replace cylinder. Clean the cylinder with brake fluid.
5. Lubricate the seals with brake fluid and install. Cup lips should always face inward. Install the spring assembly and seals.
6. Carefully install the pistons and dust boots.
7. Install the bleeder screw.

Anti-Lock Braking Systems

Despite advances in brake design over the years, even the best systems in use

can still lock up during certain road conditions, such as wet road surfaces. When the brakes lock up, the driver can lose control of the vehicle, because a locked wheel cannot absorb any cornering or lateral forces, and steering is lost. It is impossible to brake to a maximum and at the same time steer the vehicle when the front wheels are locked. If the back wheels are locked the vehicle will become unstable and start to slide.

While many different ways have been tried over the years to solve this problem, mechanical sensors could not provide sufficient information about wheel rotation speed and mechanical control units could not operate the brakes fast enough to prevent brake lockup.

The growth of the electronics industry has allowed small computers (microprocessors) to be reduced in both size and cost. Coupled with fast reacting electronic sensors, anti-lock braking has become more reliable with widespread application.

REAR WHEEL ANTI-LOCK BRAKE SYSTEM

General Motors Corporation's system is called Rear Wheel Antilock System (RWAL) and its application is on selected 2WD models. It is designed to reduce the occurrence of the rear wheel lockup during a severe brake application.

The system functions by regulating the rear hydraulic brake line pressure. The pressure regulation is accomplished by a control valve which is located under the master cylinder. The control valve is made up of 2 valves, a dump valve which releases pressure

into an accumulator, and an isolation valve which maintains rear brake pressure. The valve is controlled by a microcomputer which is part of the Electronic Control Unit (ECU). The ECU is mounted next to the master cylinder. In a severe brake application as pressure is applied to the brake pedal, the ECU is designed to permit the valve to do one of three functions or a combination of all three. The ECU will allow the valve to either maintain the same amount of hydraulic pressure, release hydraulic pressure through the dump valve into the accumulator, or increase the pressure by pulsing the isolation valve.

The ECU operates by receiving signals from the speed sensor which is located in the transmission and the brake lamp switch. The speed sensor sends its signal to the digital ratio adapter which is part of the instrument cluster. If the axle ratio or tire size is changed, it will be necessary to recalibrate the digital ratio adapter.

The RWAL system is connected to the existing brake warning lamp located on the dash. An indication of the RWAL operation is a bulb check performed each time the ignition is turned ON. The warning lamp will remain on for about two seconds. A RWAL system malfunction is indicated by a brake warning lamp.

To aid in diagnosing problems, trouble codes are produced by the system. Trouble codes are available at the ALDL (Assembly Line Diagnostic Link) the twelve terminal connector wired to the Electronic Control Module and located under the instrument panel in the passenger compartment. These codes are read by jumping terminal A (which is ground) of the ALDL

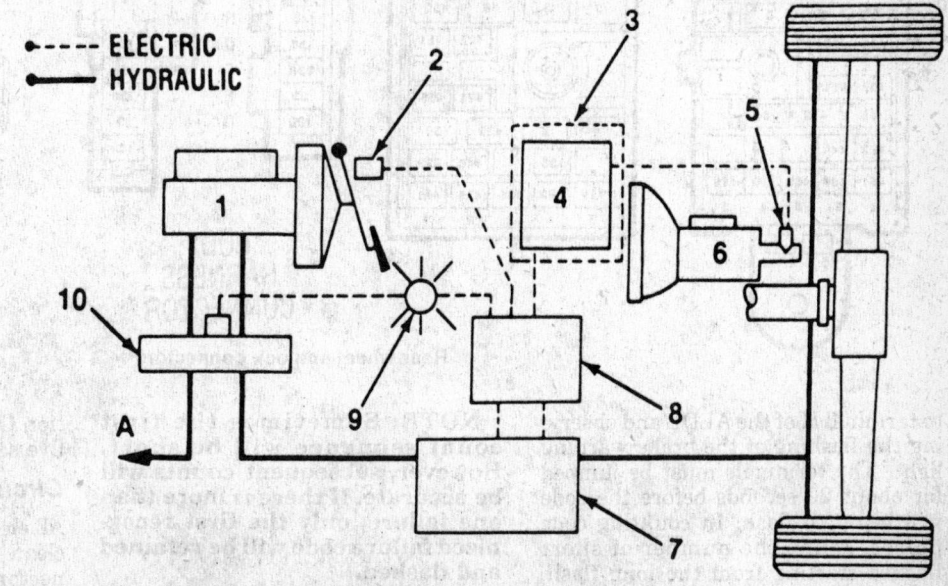

- - - - ELECTRIC
—— HYDRAULIC

A. To Front Brakes
1. Master Cylinder
2. Brake Light Switch
3. Instrument Cluster
4. Digital Ratio Adapter
5. Speed Sensor
6. Transmission
7. Isolation/Dump Valve
8. RWAL Control Module
9. Brake Warning Light
10. Combination Valve

General Motors Rear Wheel Antilock Brake System (RWAL)

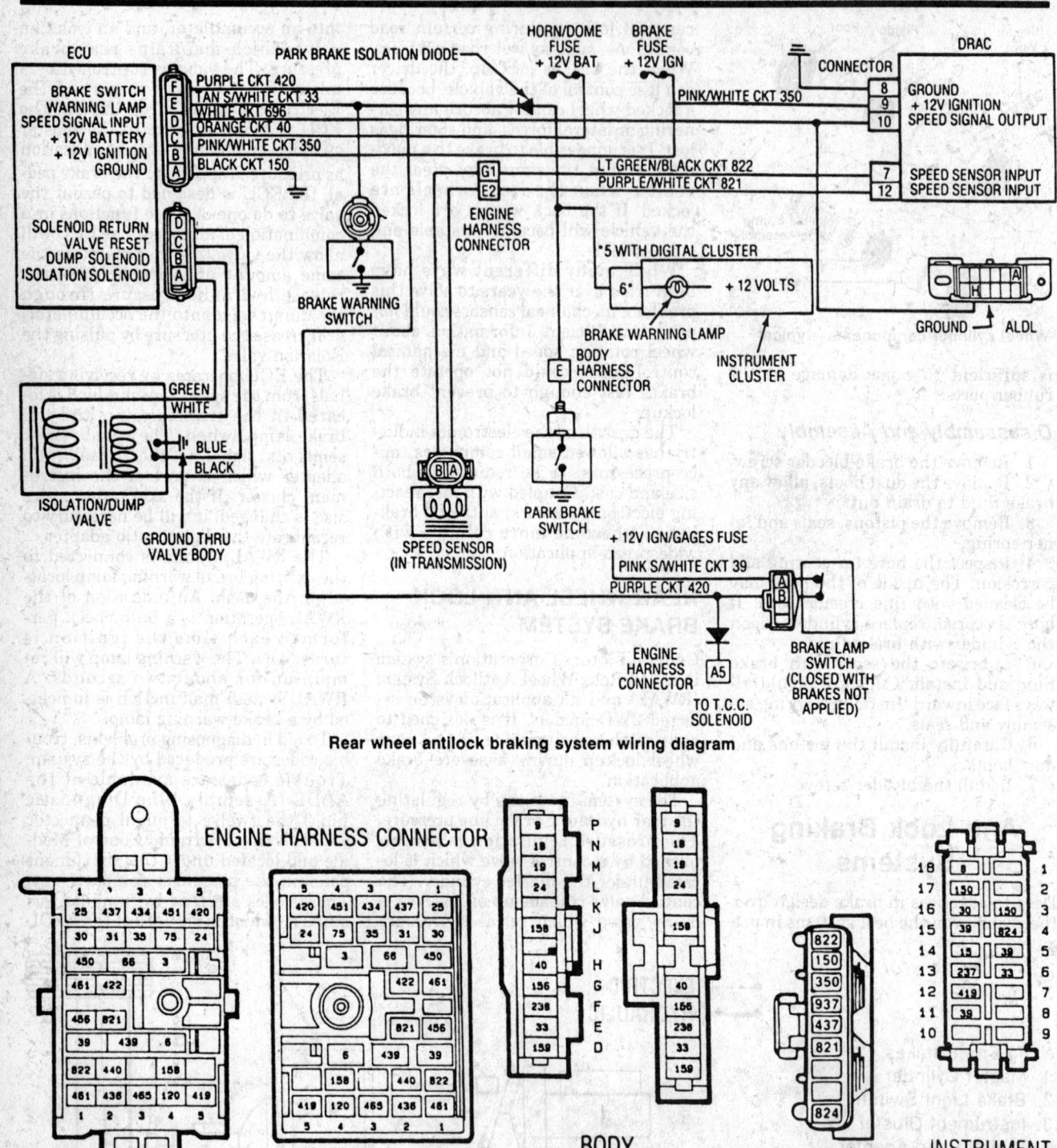

Rear wheel antilock braking system wiring diagram

ENGINE HARNESS CONNECTOR

BODY HARNESS CONNECTOR

D.R.A.C.

INSTRUMENT CLUSTER CONNECTOR

Rear wheel antilock connectors

to terminal H of the ALDL and observing the flashing of the brake warning light. The terminals must be jumped for about 20 seconds before the code will begin to flash. In counting code pulses, count the number of short flashes starting from the long flash. Include the long flash as a count.

NOTE: Sometimes the first count sequence will be short. However, subsequent counts will be accurate. If there is more than one failure, only the first recognized failure code will be retained and flashed.

To clear trouble codes, with the igni-

tion **OFF**, remove the brake fuse, wait five seconds, and then reinstall fuse.

Circuit Maintenance and Repair

All electrical connections must be kept clean and tight. Make sure that connectors are properly seated and all of the sealing rings on weather-proof

REAR WHEEL ANTILOCK BRAKE SYSTEM DIAGNOSTIC CODES

CODE	SYSTEM PROBLEM
CODE 1	Electronic control unit malfunction
CODE 2	Open isolation valve or faulty ECU
CODE 3	Open dump valve or faulty ECU
CODE 4	Grounded antilock valve switch
CODE 5	Excessive actuations of dump valve during an antilock stop
CODE 6	Erratic speed signal
CODE 7	Shorted isolation valve or faulty ECU
CODE 8	Shorted dump valve or faulty ECU
CODE 9	Open circuit to the speed signal
CODE 10	Brake lamp switch circuit
CODE 11	Electronic control unit malfunction
CODE 12	Electronic control unit malfunction
CODE 13	Electronic control unit malfunction
CODE 14	Electronic control unit malfunction
CODE 15	Electronic control unit malfunction

Trouble codes are available at the ALDL (Assembly Line Diagnostic Link), the twelve terminal connector wired to the Electronic Control Module and located under the instrument panel in the passenger compartment. These codes are ready by jumping terminal A (which is ground) of the ALDL to terminal H of the ALDL and observing the flashing of the brake warning light.

connections are in place. With the low current and voltage levels found in some circuits, it is important that all connections be the best possible. Special tools are required for servicing GM's Weather-Pack connectors. This special tool is required to remove the pin and sleeve terminals. If removal is attempted with an ordinary pick, there is a good chance that the terminal will be bent or deformed. These terminals cannot be straightened once they are bent.

Use care when probing the connections or replacing terminals in them. It is possible to short between opposite terminals. If this happens to the wrong terminal part, it is possible to damage certain components. Always use jumper wires between connectors for circuit checking. Never probe through Weather-Pack seals.

When diagnosing for possible open circuits, it is often difficult to locate them by sight because oxidation or terminal misalignment are hidden by the connectors. Merely wiggling a connector on a sensor or in the wiring harness may correct the open circuit condition. This should always be considered when an open circuit is indicated while troubleshooting. Intermittent problems may also be caused by oxidized or loose connections.

Removal and Installation
RWAL ELECTRONIC CONTROL UNIT

The RWAL Electronic Control Unit (ECU) is not serviceable. It should be replaced when diagnosis shows it to be malfunctioning.

NOTE: Do not touch the electrical connections and pins or allow them to come into contact with brake fluid as this will damage the RWAL ECU.

1. Disconnect the electrical connectors.
2. Remove the RWAL ECU by prying the tab at the rear of the ECU and pulling it forward toward the front of the vehicle.
3. To install, simply slide the RWAL ECU into its bracket on the master cylinder until the tab locks into the hole.
4. Install the electrical connectors. If brake fluid has gotten on the connectors, clean then with water followed by isopropyl alcohol.

ISOLATION/DUMP VALVE

The Isolation/Dump valve is not serviceable. It should be replaced when diagnosis shows it to be malfunctioning.

NOTE: Do not touch the electrical connections and pins or allow them to come into contact with brake fluid as this will damage the RWAL ECU.

1. Disconnect the brake line fittings and remove the bolts holding the valve to the bracket.
2. Disconnect the electrical connector from the RWAL ECU. Do not allow the valve to hang by the pigtail.
3. Remove the valve from the vehicle.
4. When installing replacement valve, torque bolts to 21 ft. lbs, brake line fittings to 18 ft. lbs.
5. Install the electrical connectors.

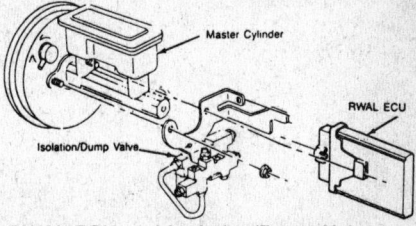

RWAL ECU and Isolation/Dump Valve

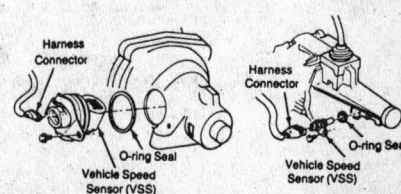

Transmission speed sensor installation

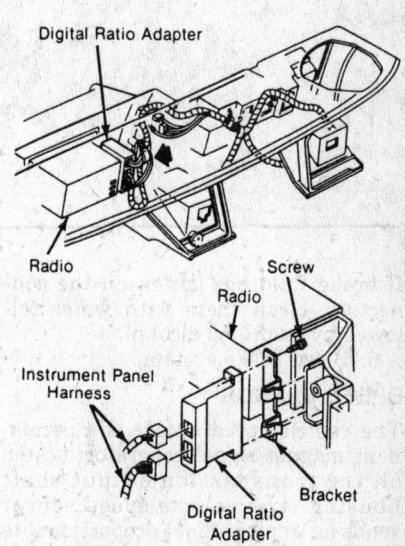

Digital ratio adapter controller mounting

DIAGNOSIS OF THE VEHICLE SPEED SENSOR
AND DIGITAL RATIO ADAPTER CONTROLLER

Problem	Possible Cause	Correction
Speedometer and odometer are inaccurate	Incorrect digital ratio adapter.	Check for the correct digital ratio adapter.
Speedometer and odometer do not operate properly	1. Inoperative digital ratio adapter.	1. Disconnect the digital ratio adapter, and place the ignition in run. Check for voltage between the pink/black wire in the harness and a good chassis ground. If the voltage is less than the battery voltage, check for an open or short in the pink/black wire.
	2. Poor ground path from the digital ratio adapter.	2. Check for voltage between the pink/black wire in the harness and the black/white wire. If the voltage is less than battery voltage, check for an open or short in the black/white wire.
	3. No signal from the vehicle speed sensor.	3. Raise and support the vehicle, start the engine, and place the transmission in drive. Check for AC voltage that changes with the engine rpm between the purple/white wire, and the light green/black wire at the digital ratio adapter. If there is not AC voltage at these wires, check for opens in the purple/white wire and the light green wire. If there are not shorts or opens, replace the vehicle speed sensor.
	4. Inoperative digital ratio adapter (speedometer output).	4. Raise and support the vehicle, start the engine, and place the transmission in drive. Check for AC voltage that changes with the engine rpm between the light blue/black and the black/white wires at the digital ratio adapter connector (connector attached) if AC voltage varies with RPM, replace the digital ratio adapter.
	5. Inoperative digital ratio adapter (cruise output).	5. Raise and support the vehicle, start the engine, and place the transmission in drive. Check for AC voltage that changes with the engine rpm between the yellow and the black/white wires at the digital ratio adapter connector (connector attached) if AC voltage varies with rpm, replace the digital ratio adapter.
	6. Inoperative instrument cluster.	6. Refer to DIAGNOSIS OF THE ELECTRONIC INSTRUMENT CLUSTER.

If brake fluid has gotten on the connectors, clean them with water followed by isopropyl alcohol.

6. Bleed brake system.

SPEED SENSOR

The vehicle speed sensor is a permanent magnet signal generator located on the transmission output shaft housing. The vehicle speed sensor sends an analog signal proportional to the propeller shaft speed. This signal goes to the Digital Ratio Adapter Controller (DRAC) which is used to change the speed sensor signal to a digital signal for the electronic instrument cluster.

The speed sensor is not serviceable. It should be replaced when necessary. The sensor is located in the left rear of the transmission on 2WD models and on the transfer case of 4WD models.

The resistance of the speed sensor should be 900-2000 ohms.

1. Remove the electrical connector.
2. Most applications will use a bolt retainer which is removed. Pull the speed sensor out of transmission housing. Have container ready to catch transmission fluid.
3. Always use new O-ring seal when installing speed sensor. Coat the seal with a thin film of transmission fluid.
4. Install retainer bolt if used and connect electrical harness.

NOTE: The speedometer must recalibrated when rear axle ratio or tire size is changed.

FOUR WHEEL ANTI-LOCK BRAKE SYSTEM

In 1990, selected G.M. models with All Wheel Drive (AWD) could be equipped with a Four Wheel Antilock System (4WAL). It too is designed to to reduce wheel lockup during severe braking. Like the Rear Wheel Antilock system, it works by regulating hyrdaulic brake line pressure. An Electro-Hydraulic Control Unit valve (EHCU), located under the master cylinder, is made up of two types of valves. Isolation valves maintain pressure to each front wheel separately and to the rear wheels combined. Dump valves dump pressure to each front wheel separately and to the rear wheels combined. The valves are controlled by a microcomputer which is part of the EHCU valve. Under severe braking, the EHCU valve will allow the valves to either maintain the same hydraulic pressure, release hydraulic pressure through the dump valves into the accumulator, or increase pressure.

The EHCU valve operates by receiving signals from the speed sensors, located at each wheel and the brake lamp switch. The speed sensors are connected directly to the EHCU valve through the 8 pin connector.

The 4WAL system is connected to the ANTILOCK warning lamp in the instrument panel. An indication of 4WAL operation and a bulb check is performed each time the ignition is turned ON. The warning lamp will remain on for about 2 seconds. A 4WAL system malfunction is indicated by the ANTILOCK warning light.

To aid in diagnosing problems, trouble codes are produced by the system. Trouble codes are available at the Assembly Line Diagnostic Link (ALDL) the twelve terminal connector wired to the Electronic Control Module and located under the instrument panel in the passenger compartment. These codes are read by jumping terminal A of the ALDL to terminal H of the ALDL and observing the flashing of the ANTILOCK. The terminals must be jumped for a few seconds before the code will begin to flash. The ANTILOCK light will flash in a manner similar to the SERVICE ENGINE SOON light for the fuel and emissions system.

In counting code pulses, count the number of short flashes starting from the long flash. Include the long flash as a count.

TERMINAL IDENTIFICATION

A	GROUND	
B	DIAGNOSTIC TERMINAL	
C	A.I.R. (IF USED)	
E	SERIAL DATA (V6/V8)	

F	T.C.C. (IF USED)	
G	FUEL PUMP (CK)	
H	BRAKE SENSE SPEED INPUT (CK)	
M	SERIAL DATA (L4)	

ALDL Connector

G.M. FOUR WHEEL ANTILOCK BRAKE SYSTEM DIAGNOSTIC CODES

CODE	SYSTEM PROBLEM
CODE 21	Right front speed sensor or circuit open
CODE 22	Missing right front speed signal (set with vehicle in motion)
CODE 23	Erratic right front speed sensor
CODE 25	Left front speed sensor or circuit open
CODE 26	Missing left front speed signal (set with vehicle in motion)
CODE 27	Erratic left front speed sensor
CODE 28	Erratic speed sensor signal (two drop-outs above 20 mph)
CODE 29	Simultaneous drop-out of all four sensors (at speeds above 8 mph)
CODE 31	Right rear speed sensor or circuit open
CODE 32	Missing right rear speed signal (set with vehicle in motion)
CODE 33	Erratic right rear speed sensor
CODE 35	Left rear speed sensor or circuit open
CODE 36	Missing left rear speed signal (set with vehicle in motion)
CODE 37	Erratic left rear speed sensor
CODE 38	Wheel speed error (set with vehicle in motion)
CODES 41 thru 66 and 71 thru 74	4WAL control unit
CODE 67	Open motor circuit or shorted ECU output
CODE 68	Locked motor or shorted motor circuit
CODES 68, 43, 44, 47, 48, 53 and 54	Loss of power ground
CODE 81	Brake switch circuit shorted or open
CODE 85	Open antilock warning lamp
CODE 86	Shorted antilock warning lamp
CODE 88	Shorted brake warning lamp

FOUR WHEEL ANTILOCK WIRING DIAGRAM

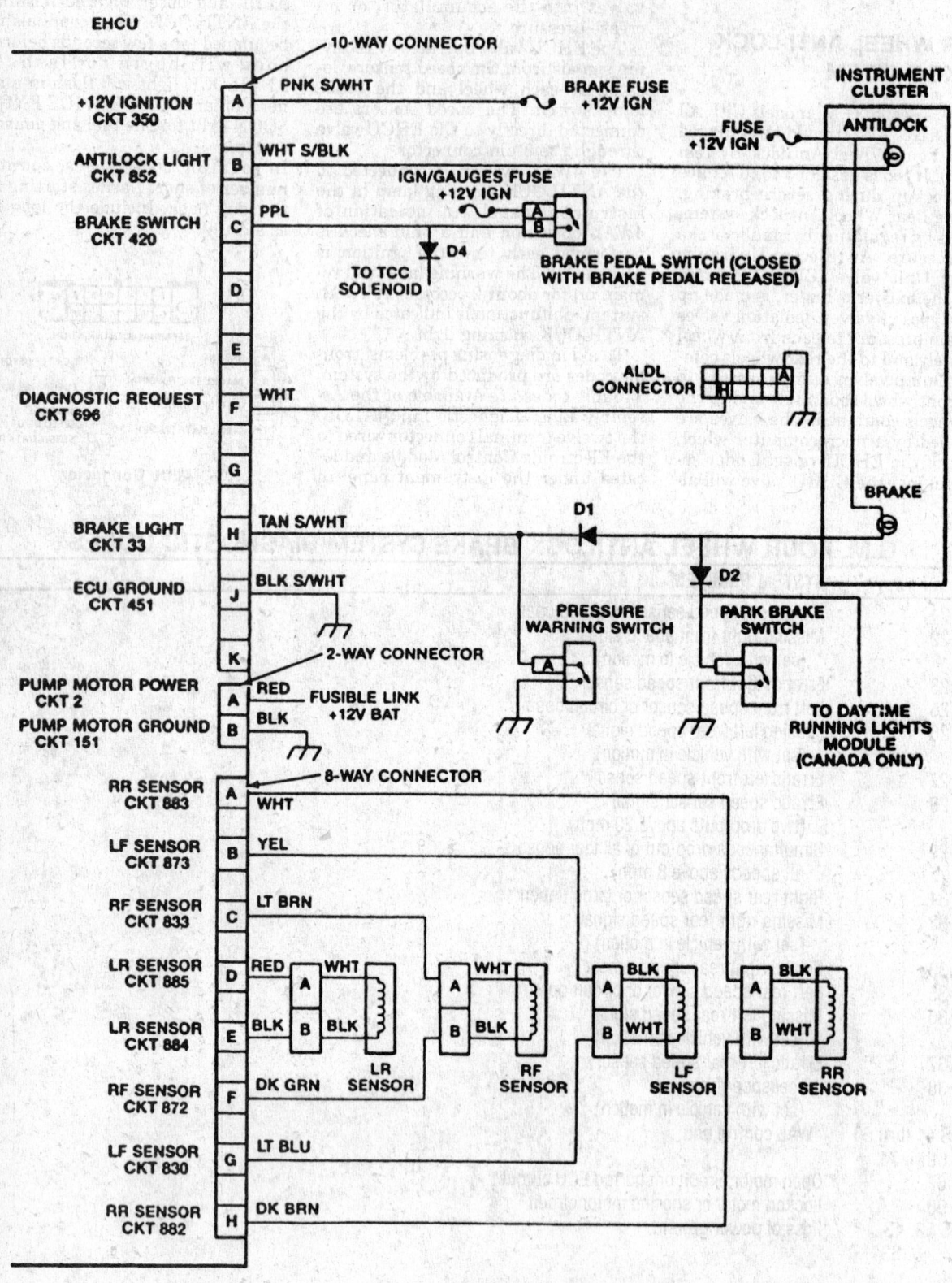

FOUR WHEEL ANTILOCK BRAKE SYSTEM

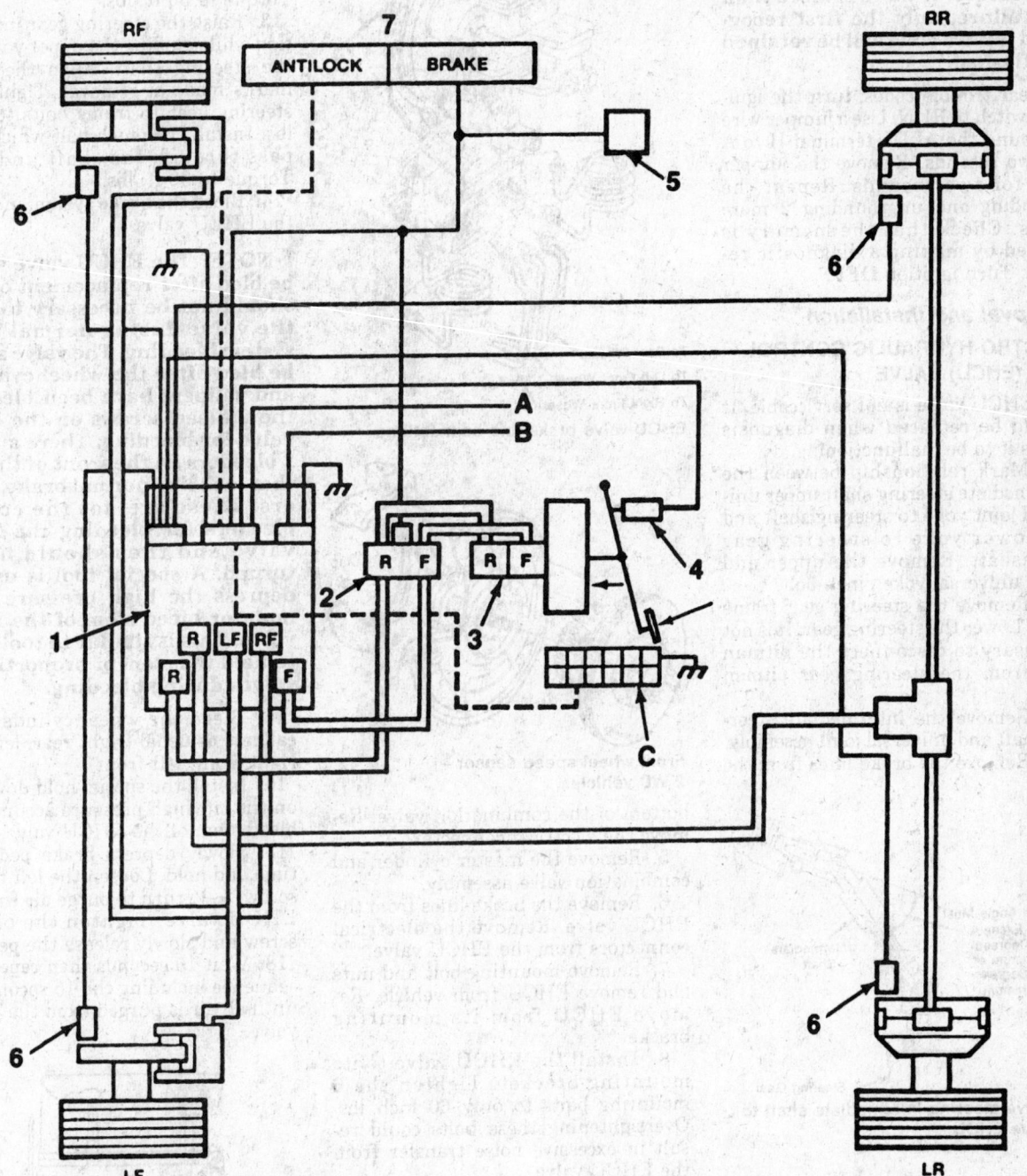

A. To Ignition Switch (B+)
B. To Battery (B+)
C. ALDL
1. 4WAL EHCU Valve
2. Combination Valve
3. Master Cylinder
4. Brake Pedal Switch
5. Park Brake Switch
6. Wheel Speed Sensors
7. Warning Lamps

NOTE: Sometimes the first count sequence will be short. However, subsequent counts will be accurate. If there is more than one failure, only the first recognized failure code will be retained and flashed.

To clear trouble codes, turn the ignition switch to RUN. Use a jumper wire to ground the ALDL terminal H to A for two seconds. Remove the jumper wire for two seconds. Repeat the grounding and ungrounding 2 more times. Check that the memory is cleared by making a diagnostic request. Turn ignition **OFF**.

Removal and Installation

ELECTRO-HYDRAULIC CONTROL UNIT (EHCU) VALVE

The EHCU valve is not serviceable. It should be replaced when diagnosis shows it to be malfunctioning.

1. Mark relationship between the intermediate steering shaft upper universal joint yoke to steering shaft and the lower yoke to steering gear wormshaft. Remove the upper and lower universal yoke pinch bolt.

2. Remove the steering gear frame bolts. Lower the steering gear. It is not necessary to disconnect the pitman arm from the steering gear pitman shaft.

3. Remove the intermediate steering shaft and universal joint assembly.

4. Remove the brake lines from the

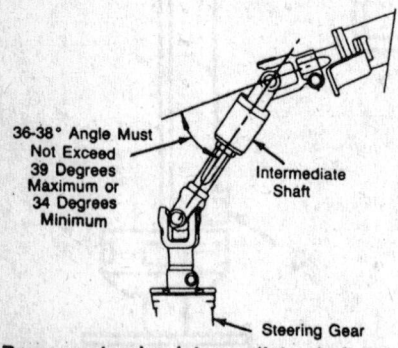

36-38° Angle Must Not Exceed 39 Degrees Maximum or 34 Degrees Minimum
Intermediate Shaft
Steering Gear

Remove steering intermediate shaft to remove EHCU

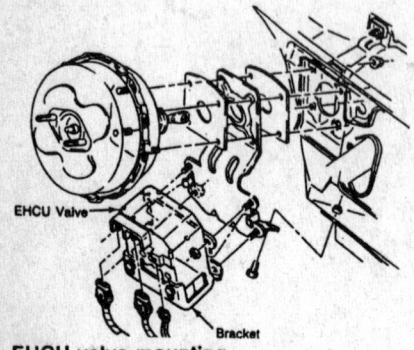

EHCU Valve
Bracket

EHCU valve mounting

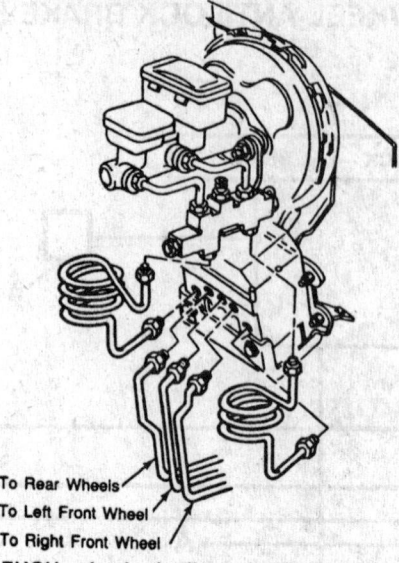

To Rear Wheels
To Left Front Wheel
To Right Front Wheel

EHCU valve brake line connections

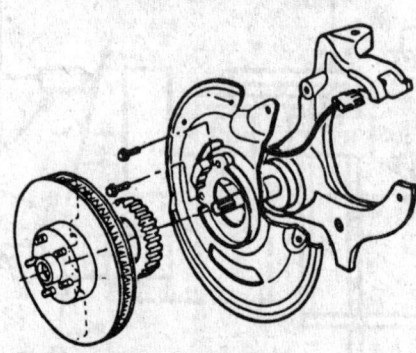

Front wheel speed sensor— 2 WD vehicles

bottom of the combination valve. Remove the electrical connector.

5. Remove the master cylinder and combination valve assembly.

6. Remove the brake lines from the EHCU valve. Remove the electrical connectors from the EHCU valve.

7. Remove mounting bolt and nuts and remove EHCU from vehicle. Remove EHCU from its mounting bracket.

8. Install the EHCU valve to its mounting bracket. Tighten the 6 mounting bolts to only 60 inch lbs. Overtightening these bolts could result in excessive noise transfer from the EHCU valve.

9. Install the EHCU valve and bracket assembly into the vehicle and install the fasteners. Torque the bolt to 33 ft. lbs., the nuts to 20 ft. lbs.

10. Install the electrical connectors and the brake lines to the EHCU valve.

11. Install the master cylinder and combination valve assembly. Connect the electrical connector and the brake lines to the combination valve.

12. Install the steering intermediate shaft. Start by placing the lower yoke onto the steering gear wormshaft.

Align the match marks made at removal. Install the pinch bolt which must pass through the shaft undercut. Torque to 30 ft. lbs.

13. Raise the steering gear into position while guiding the upper yoke onto the steering shaft. Align the match marks made at removal. Tighten the steering gear to frame bolts to 55 ft. lbs. Install the pinch bolt which must pass through the shaft undercut. Torque to 30 ft. lbs.

14. Bleed the brake system including the EHCU valve.

NOTE: The EHCU valve should be bled after replacement only. It should not be necessary to bleed the valve during normal brake system bleeding. The valve should be bled after the wheel cylinders and calipers have been bled. Use the 2 bleed screws on the EHCU valve for bleeding. There are also 2 bleeders on the front of the unit that look like normal brake bleeders. These are not the correct bleeders for bleeding the EHCU valve and they should not be turned. A special tool is used to depress the high pressure accumulator bleed stem of the EHCU valve. This is similar to tools used to hold the stem of proportioning valves during bleeding.

15. Bleed the wheel cylinders and calipers as usual (right rear, left rear, right front, left front).

16. Install the special hold-down tool on the left high pressure accumulator bleed stem of the EHCU valve.

17. Slowly depress brake pedal one time and hold. Loosen the left bleeder screw ¼–½ turn to purge air from the EHCU valve. Tighten the bleeder screw and slowly release the pedal.

18. Wait 15 seconds then repeat this sequence including the 15 second wait until all air is purged from the EHCU valve.

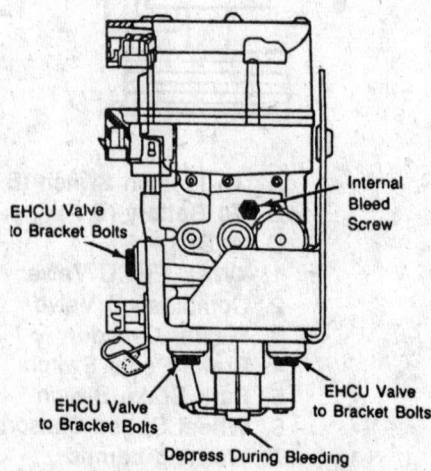

Internal Bleed Screw
EHCU Valve to Bracket Bolts
EHCU Valve to Bracket Bolts
EHCU Valve to Bracket Bolts
Depress During Bleeding

EHCU valve bleeders

19. Repeat these steps on the right side of the EHCU valve.

20. Remove valve depressor tool.

FRONT WHEEL SPEED SENSOR

2 WD Vehicles

1. Raise and safely support vehicle. Remove wheel and tire assembly.

2. Remove brake caliper.

3. Remove hub and bearing assembly.

4. Remove rotor assembly.

5. Disconnect the sensor wire connector and unclip from upper control arm.

6. Remove bolts holding speed sensor/splash shield assembly and remove unit from vehicle.

7. Install speed sensor/splash shield assembly and torque bolts to 11 ft. lbs.

8. Clip wire to upper control arm and reconnect electrical connection.

9. Apply recommended lubricant to the spindle and inside of hub. Install hub and outer wheel bearing and washer with a generous amount of lubricant. Tighten nut to 12 ft. lbs. while turning the hub forward to seat the bearing. Back off the nut to the just loose position. Hand tighten nut, then back off until the hole in the spindle lines up with a slot on the nut. The nut should not be backed off more than ½ a flat. Install new cotter pin.

10. Install brake rotor.

11. Install caliper. Bushings should be lubricated with silicone grease. Make sure the brake line is not twisted.

12. Install wheel and tire assembly. Lower vehicle.

4WD Vehicles

1. Raise and safely support vehicle. Remove wheel and tire assembly.

2. Remove brake caliper.

3. Remove brake rotor.

4. Remove hub and bearing assembly. On all-wheel drive models, remove the drive axle nut and washer. Use a puller to ease the hub from the drive axle.

5. Disconnect the sensor wire connector and unclip from upper control arm.

6. Remove bolts holding speed sensor/splash shield assembly and remove unit from vehicle.

7. Install speed sensor/splash shield assembly to the steering knuckle and torque the bolts to 11 ft. lbs.

8. Clip wire to upper control arm and reconnect electrical connection.

9. Install hub and bearing assembly, align threaded holes and tighten bolts to 66 ft. lbs.

10. Install outer axle washer and nut. Torque to 175 ft. lb.

11. Install brake rotor and caliper.

12. Install wheel and tire. Lower vehicle.

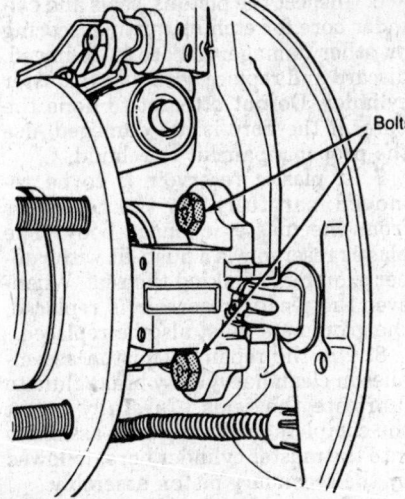

Rear wheel speed sensor

REAR WHEEL SPEED SENSOR

4WD Vehicles

1. Raise and safely support vehicle. Remove wheel and tire assembly.

2. Remove brake drum.

3. Remove the primary brake shoe.

4. Disconnect the sensor electric wire connector and wire from its axle clips.

5. Remove the retainer bolts and take out speed sensor by pulling the wire through the hole in the backing plate.

6. Install the speed sensor and bolts. Torque to 26 ft. lbs.

7. Install sensor wire to axle clips and hook up connector.

8. Install primary brake shoe, brake drum and wheel and tire assembly. Lower vehicle.

FORD MOTOR COMPANY

Master Cylinder Service

Ford Motor Company uses a dual master cylinder which contains a double hydraulic cylinder with 2 fluid reservoirs and primary and secondary hydraulic pistons. The rear wheel brakes are connected to the secondary outlet port and are actuated by the secondary piston piston assembly. The front wheel brakes are connected to the primary outlet port (nearest the dash panel) and are actuated by the primary piston assembly. Both primary and secondary pistons function together. On most models, the master cylinder is assisted by a vacuum booster. Both single and double diaphragm Bendix models are used. Hydraulic rear drum brakes with automatic adjusters and self-adjusting front disc brakes are standard on all models.

Two types of master cylinders were used by Ford Motor Company; 1) a cast iron master cylinder with built-in reservoir and separate proportioning valve and, 2) a master cylinder with a pressed on plastic see-through reservoir with the proportioning valve built into or integral with the master cylinder. In the event of a front brake system malfunction the proportioning valve with a bypass feature allows full hydraulic pressure to the rear brake system.

A Fluid Level Indicator (FLI) is built into the reservoir. It is serviced as part of the reservoir assembly. This master cylinder is found from 1987.

During normal operation, the fluid level in the master cylinder will rise during brake operation and fall during release. It is also expected that the fluid level will decrease with brake pad wear. In addition, a trace of brake fluid on the booster shell below the master cylinder mounting flange will often be found as a result of the normal lubricating action of the master cylinder bore and seal. All of these conditions are considered normal and are not indications the master cylinder needs service.

Disassembly and Assembly

CAST IRON MASTER CYLINDER

1. Clean outside of master cylinder, remove cover and diaphragm. Drain and discard remaining brake fluid.

2. Depress the primary piston and remove snapring at rear of the master cylinder bore.

3. Remove the primary piston and inspect for damage.

4. Remove the secondary piston assembly by directing a little compressed air into the outlet at the blind end of the bore while plugging the other outlet port.

5. Inspect the pistons, seals and cylinder bore for etching, pitting, scoring or other damage. If the bore is damaged, discard and replace with a new master cylinder. Do not attempt to hone the bore. If the bore is not damaged, use the proper repair kit to rebuild.

6. Clean the master cylinder body with isopropyl alcohol.

7. Dip the repair kit piston assemblies in clean heavy duty brake fluid to lubricate the seals. Carefully insert the complete secondary piston assembly into the master cylinder bore, followed by the primary piston assembly.

8. Depress the primary piston and

install snapring. On manual brake vehicle, install the push rod retainer onto the push rod and install into primary piston. Make sure retainer is properly seated and holding the push-rod securely.

9. Install cover and gasket and secure with retainer.

PLASTIC RESERVOIR MASTER CYLINDER

1. Clean the outside of the master cylinder and remove the cap and gasket. Drain and discard remaining brake fluid.

2. Remove the proportioning valve from the master cylinder.

3. Remove the stop-bolt from the bottom of the master cylinder assembly.

4. Depress the secondary piston, remove the snapring from the bore and remove the secondary piston assembly.

5. Remove the primary piston assembly by directing a little compressed air into the outlet at the blind end of the bore while plugging the other outlet port.

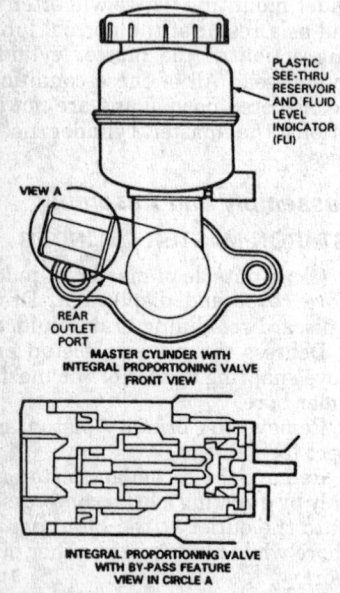

Ford master cylinder with see-through reservoir and integral proportioning valve

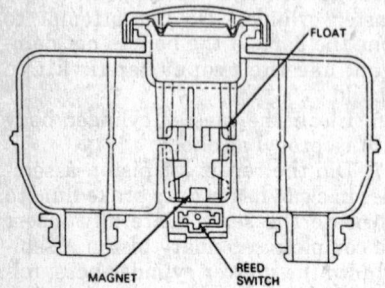

Plastic reservoir with fluid level indicator

6. Inspect the pistons, seals and cylinder bore for etching, pitting, scoring or other damage. If bore is damaged, discard and replace with a new master cylinder. Do not attempt to hone the bore. If the bore is not damaged, use the proper repair kit to rebuild.

7. If plastic reservoir is to be removed, carefully pry the reservoir from the master cylinder body. The plastic reservoir is a push-fit with rubber grommets making the seal. Whenever the plastic reservoir is replaced, the grommets must also be replaced.

8. Dip the repair kit piston assemblies in clean heavy duty brake fluid to lubricate the seals. Carefully insert the complete primary piston assembly into the master cylinder bore, followed by the secondary piston assembly.

8. Depress the secondary piston and install snapring. Install the stop bolt in the bottom of the master cylinder bore.

9. Install the proportioning valve assembly in the master cylinder.

10. If the reservoir was removed, lubricate new retainer grommets with brake fluid and insert the grommets into the master cylinder body. Press the plastic reservoir into the grommets. The reservoir should snap into place indicating that it is secure. Make sure the fluid level indicator socket is facing the correct direction for the vehicle on which it is to be installed.

11. Install plastic cap on the master cylinder.

Brake Booster Service

Three types of brake boosters are used, a single diaphragm, a tandem diaphragm and a hydraulic booster (Hydro-Boost).

Under normal operating conditions, with brakes released, a vacuum suspended booster operates with vacuum on both sides of its diaphragm (or both diaphragms in a tandem diaphragm unit). When the brakes are applied, air at atmospheric pressure is admitted to one side of the diaphragm (or both diaphragms in a tandem diaphragm unit) to provide the power assist.

The hydraulic brake booster (Hydro-Boost) uses the power steering pump to power the system and a pneumatic accumulator as a reserve system. In this system no special fluids are used. However, care must be taken to use the correct fluids. The master cylinder and brake system operate on standard brake fluid. The hydraulic pump uses power steering fluid.

VACUUM DIAPHRAGM BRAKE BOOSTER

Both the Bendix single diaphragm and

Bendix tandem diaphragm booster are not repairable and the booster must be replaced as a unit. The booster check valve is the only component which can be serviced on the booster assembly.

The Bendix Hydro-Boost uses the hydraulic pressure supplied by the power steering pump to provide a power assist to brake application. It has identifying information stamped into the housing near the inlet

The Hydro-Boost power brake booster is not to be disassembled and is to be serviced as a unit.

Brake Booster Vacuum Pump

On gasoline engine vehicles, engine vacuum is used to power the vacuum booster brake. On diesel engines, there is not enough vacuum available for vacuum booster brake operation.

On diesel engine vehicles, vacuum is supplied from a pump usually located on the top right side of the engine. It is usually driven by a single belt off the alternator.

Diesel engine equipped vehicles will also usually have a low-vacuum indicator switch which activates the BRAKE warning lamp when vacuum gets below a certain level. The switch senses vacuum through a fitting in the vacuum manifold (sometimes called a vacuum tree). Note that the BRAKE light will glow until vacuum builds up to normal level.

The low-vacuum switch for E Series is located to the left side fender panel. On F 250 – F 350 Series the switch is located on the right side of the engine compartment adjacent to the vacuum pump.

The vacuum pump is not to be disassembled. It is only serviced as a unit. The pulley is serviced as a separate item.

Caliper Service

Ford Motor Company uses several types of brake calipers, a Heavy Duty (HD) Pin Rail caliper, a Light Duty (LD) sliding brake caliper and a Light Duty (LD) Pin Rail caliper.

The HD caliper has 2 pistons on the same side of the rotor. Pin Rail type calipers slide on 2 pins that also attach the caliper to the spindle (or anchor plate on some F-Super Duty rear disc brakes). The caliper is a one-piece casting with the inboard side containing the piston bore(s). A square cut rubber seal is located in a groove in the piston bore which provides the hydraulic seal between the piston and the cylinder wall.

As the brake pedal is pressed, hy-

BRACKET
ALTERNATOR
ALTERNATOR ADJUSTMENT ARM
ALTERNATOR SUPPORT BRACKET
"J" NUT
ADJUSTMENT BOLT
PIVOT BOLT
ALTERNATOR MOUNTING BRACKET
VACUUM PUMP MOUNTING BOLTS
VACUUM PUMP
VACUUM PUMP PULLEY
VACUUM PUMP ADJUSTMENT PLATE

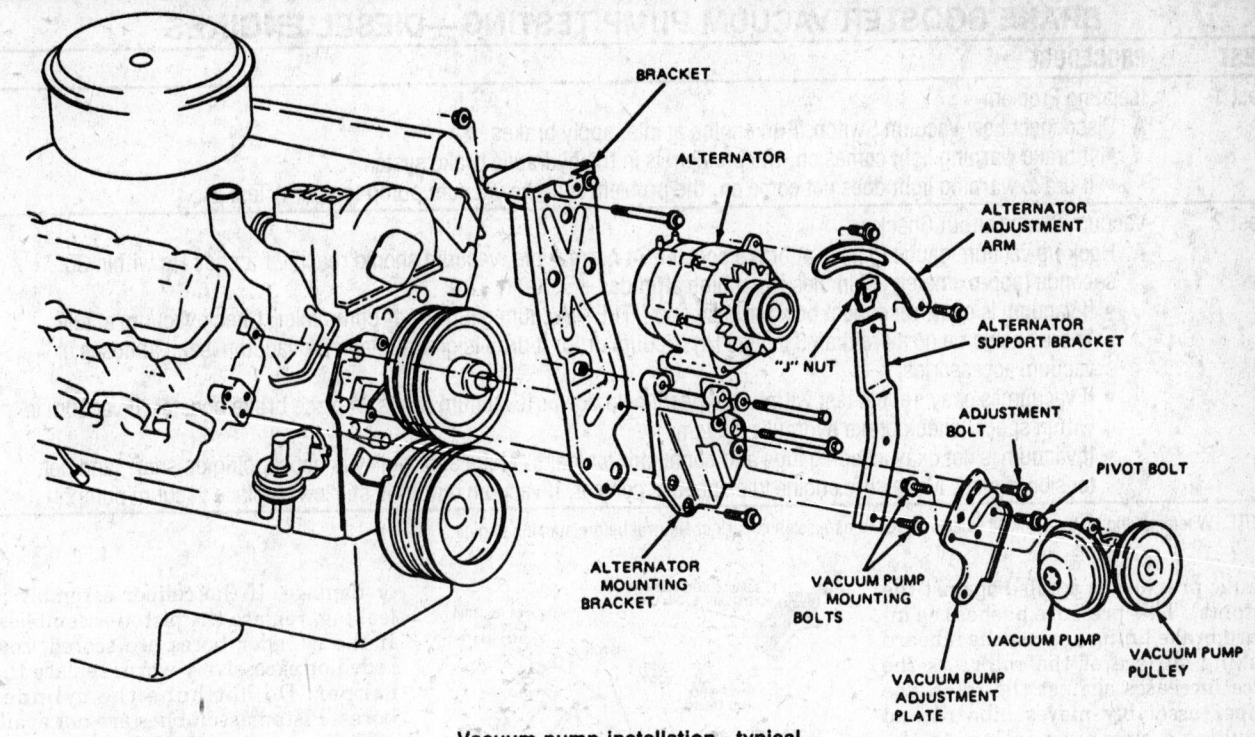

Vacuum pump installation – typical

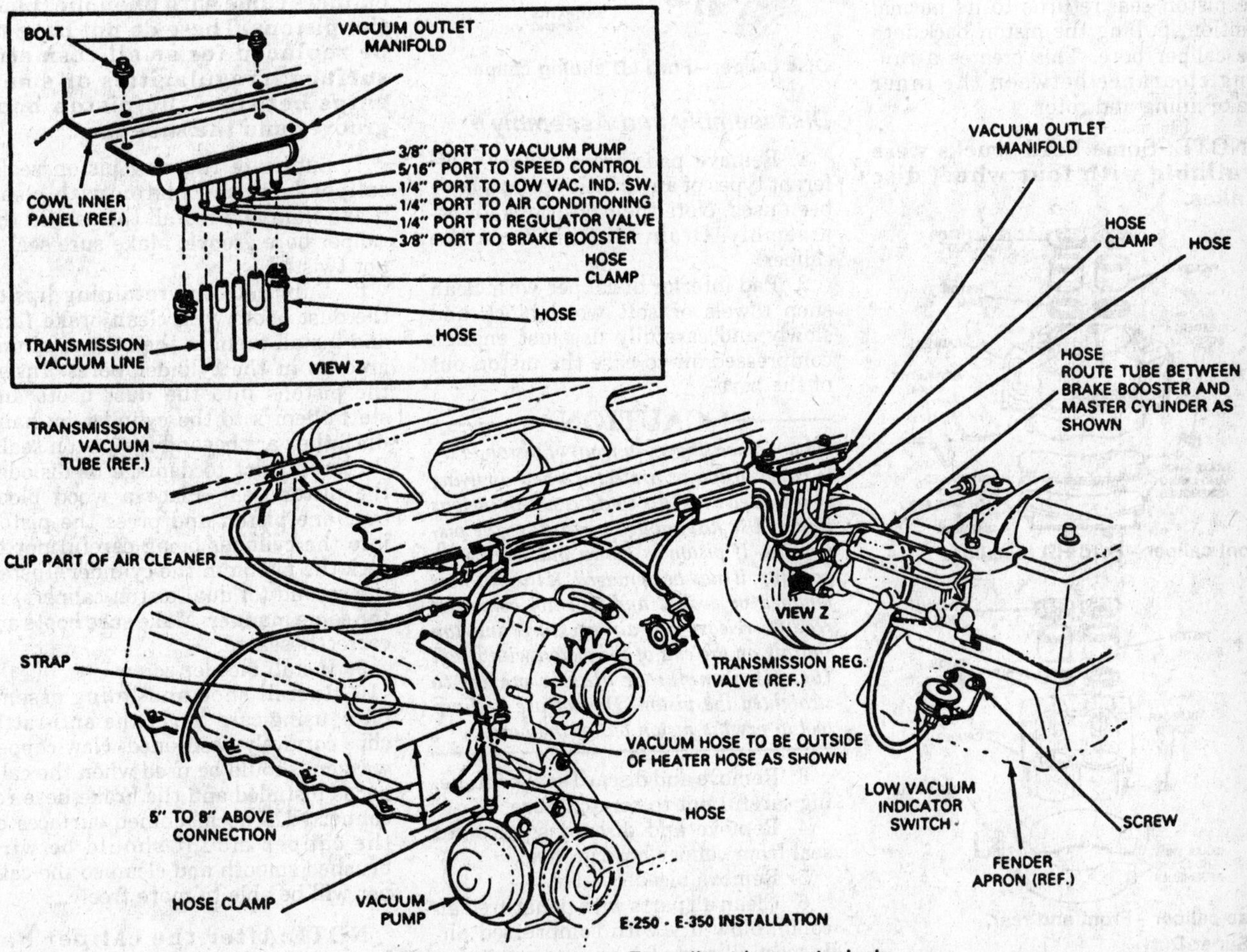

BOLT
VACUUM OUTLET MANIFOLD
3/8" PORT TO VACUUM PUMP
5/16" PORT TO SPEED CONTROL
1/4" PORT TO LOW VAC. IND. SW.
1/4" PORT TO AIR CONDITIONING
1/4" PORT TO REGULATOR VALVE
3/8" PORT TO BRAKE BOOSTER
COWL INNER PANEL (REF.)
HOSE CLAMP
HOSE
HOSE
TRANSMISSION VACUUM LINE
VIEW Z
VACUUM OUTLET MANIFOLD
HOSE CLAMP
HOSE
HOSE
ROUTE TUBE BETWEEN BRAKE BOOSTER AND MASTER CYLINDER AS SHOWN
TRANSMISSION VACUUM TUBE (REF.)
CLIP PART OF AIR CLEANER
VIEW Z
TRANSMISSION REG. VALVE (REF.)
STRAP
VACUUM HOSE TO BE OUTSIDE OF HEATER HOSE AS SHOWN
5" TO 8" ABOVE CONNECTION
HOSE
LOW VACUUM INDICATOR SWITCH
SCREW
FENDER APRON (REF.)
HOSE CLAMP
VACUUM PUMP
E-250 - E-350 INSTALLATION

Vacuum pump layout – Econoline shown – typical

BRAKE BOOSTER VACUUM PUMP TESTING—DIESEL ENGINES

TEST	PROCEDURE
Test 1	Isolating Problem A. Disconnect Low Vacuum Switch. Run engine at idle, apply brakes. • If brake warning light comes on, the problem is in the hydraulic brake system. • If brake warning light does not come on, the problem is in the vacuum pump (perform Test 2).
Test 2	Vacuum Pump Output Check A. Hook up vacuum gauge to hose at brake booster. At normal idle, vacuum should reach 21 inches Hg within 30 seconds (approximately 16 inches Hg at high altitudes—5,000 ft.). • If vacuum is okay, reconnect booster base and "TEE" vacuum gauge near pump inlet. Check vacuum at idle. There should be no more than 3 inches Hg vacuum drop. If drop is greater, look for vacuum leaks at hoses or vacuum accessories. • If vacuum is okay, repeat test with brake pedal held down. If vacuum drops, replace brake booster. If vacuum is within specs, check brake hydraulic system. • If vacuum is not okay, check gauge and connector for leaks. Make sure pulley is not slipping on shaft, and belt tension is okay. Make sure engine idle speed is correct. If vacuum output is still low, replace vacuum pump.

NOTE: When making tests, block all wheels, place transmission in Park or Neutral before starting engine.

draulic pressure is applied against the piston(s). This pressure pushes the inboard brake lining against the inboard braking surface of the rotor. As the force increases against the rotor, the caliper assembly moves inboard and provides a clamping action on the rotor.

When brake pressure is released, the piston seal returns to its normal position, pulling the piston back into the caliper bore. This creates a running clearance between the inner brake lining and rotor.

NOTE: Some Ford trucks were available with four wheel disc brakes.

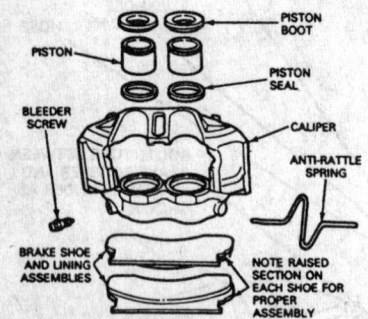

Front caliper—Ford HD rail slider

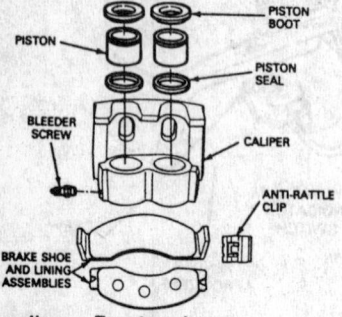

Disc caliper—Front and rear, F-Super Duty

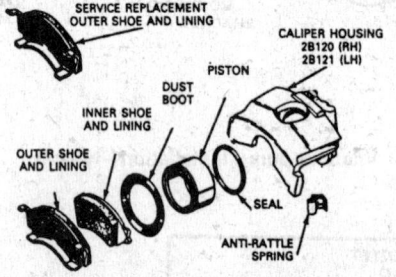

Disc caliper—Ford LD sliding caliper

Disassembly and Assembly

1. Remove pads from caliper. Different types of anti-rattle springs have been used. Note their positions for reassembly. Drain all fluid from the caliper.

2. Pad interior of caliper with clean shop towels or soft wood block and slowly and carefully use just enough compressed air to ease the piston out of the bore.

CAUTION

Do not place fingers in front of piston to try to catch piston or protect it when applying compressed air. This could result in serious injury. Use just enough air to ease the piston out. If piston is blown out, even with padding, it may be damaged. If the piston is jammed or cocked and will not come out readily, release the air pressure and tap sharply on the end of the piston with a soft (brass) hammer or plastic mallet to straighten the piston. Do not use a sharp tool to pry the piston out of the bore.

3. Remove and discard seal boot, being careful not to scratch bore.

4. Remove and discard square cut seal from caliper bore groove.

5. Remove bleeder valve.

6. Clean all parts with denatured alcohol, blow dry with compressed air. Inspect all parts for scoring, corrosion

or damage. If the caliper assembly is leaking, replace the piston assemblies. If the cylinder bores are scored, corroded or excessively worn, replace the caliper. Do not hone the cylinder bores. Piston assemblies are not available for oversize bores.

NOTE: Some versions of Ford calipers came with phenolic (plastic) pistons. These do not have to be replaced for small cosmetic surface irregularities or small chips between the piston boot grooves and the shoe face.

7. Lubricate the new piston seals, caliper bore and piston with clean brake fluid and install the seal in the caliper bore groove. Make sure seal is not twisted.

8. Lubricate the retaining lips of the dust boots with clean brake fluid and install them in the boot retaining grooves in the cylinder bores. Insert the pistons into the dust boots and start them into the cylinder by hand until they are beyond the piston seals. Be careful not to damage or dislodge the piston seal. Place a wood block over one piston and press the piston into the cylinder being careful not to cock the piston in the cylinder. Install the second (of dual piston calipers) in the same manner. Make sure boots are correctly seated.

9. Install bleeder screw.

10. Install shoe and lining assemblies, using care to get the anti-rattle clips correctly positioned. New copper washers should be used when the caliper is installed and the brake hose reconnected. The machined surfaces of the caliper mount should be wire brushed smooth and clean so the caliper will be able to more freely.

NOTE: After the caliper has been installed and the system

bled, before moving the vehicle, pump the brake pedal several times until the pedal is firm. Do not move the vehicle until a firm pedal is obtained. Check the fluid level in the master cylinder after pumping the brakes.

Wheel Cylinder Service

The rear brakes are drum type with internal expanding shoes. The rear brakes are of the single anchor type, mounted to the same anchor and actuated by one wheel cylinder. The wheel cylinder has 2 pistons. Brake adjustments are automatic and are made during reverse brake applications.

Wheel cylinders may need reconditioning or replacement whenever the brake shoes are replaced or when required to correct a leak condition. Leaks which coat the boot and the cylinder with fluid, or result in a dropped reservoir fluid level, or dampen and stain the brake linings are dangerous. Such leaks can cause the brakes to grab or fail and should be immediately corrected. A leakage, not immediately apparent, can be detected by pulling back the cylinder boot. A small amount of fluid seepage dampening the interior of the boot is normal. However, a dripping boot is not. Unless other conditions causing a brake to pull, grab or drag become obvious, the wheel cylinder is suspect and should and should be included in general reconditioning.

Cylinder binding may be caused by rust, deposits, grime, or swollen cups due to fluid contamination, or by a cup wedged into an excessive piston clearance.

Hydraulic system parts should not be allowed to come into contact with oil or grease, neither should those be handled with greasy hands. Even a trace of any petroleum based product is sufficient to cause damage to the rubber parts.

Disassembly and Assembly

1. Remove the brake bleeder screw.
2. Remove the dust boots, allow any brake fluid to drain out.
3. Remove the pistons, seals and inner spring.
4. Inspect the bore for scoring and corrosion. The inside of the bore may be cleaned with fine crocus cloth. If bore is scored, replace cylinder. Clean the cylinder with brake fluid.
5. Lubricate the seals with brake fluid and install. Cup lips should always face inward. Install the spring assembly and seals.

6. Carefully install the pistons and dust boots.
7. Install the bleeder screw.

Anti-Lock Braking Systems

Despite advances in brake design over the years, even the best systems in use can still lock up during certain road conditions, such as wet road surfaces. When the brakes lock up, the driver can lose control of the vehicle, because a locked wheel cannot absorb any cornering or lateral forces, and steering is lost. It is impossible to brake to a maximum and at the same time steer the vehicle when the front wheels are locked. If the back wheels are locked the vehicle will become unstable and start to slide.

While many different ways have been tried over the years to solve this problem, mechanical sensors could not provide sufficient information about wheel rotation speed and mechanical control units could not operate the brakes fast enough to prevent brake lockup.

The growth of the electronics industry has allowed small computers (microprocessors) to be reduced in both size and cost. Coupled with fast reacting electronic sensors, anti-lock braking has become more reliable with widespread application.

REAR WHEEL ANTI-LOCK BRAKE SYSTEM

Ford Motor Company's Rear Antilock Brake System (RABS) continually monitors rear wheel speed with a sensor mounted on the rear axle. When the teeth on an exciter ring, mounted inside the rear axle on the differential gear case, pass the sensor pole piece, an AC voltage is induced in the sensor circuit with a frequency proportional to the average rear wheel speed. In the event of an impending lockup condition during braking, the RABS modulates hydraulic pressure to the rear brakes. This inhibits rear wheel lockup.

When the brake pedal is applied, the RABS module senses the drop in rear wheel speed. If the rate of deceleration is too great, indicating the wheel lockup is going to occur, the RABS module activates the electro-hydraulic valve causing isolation valve to close. With the isolation valve closed, the rear wheel cylinders are isolated from the master cylinder and the rear brake pressure cannot increase. If the rate of deceleration is still too great, the RABS module will energize the dump solenoid with a series of rapid pulses to

bleed off rear wheel cylinder fluid into an accumulator built into the RABS valve. This will reduce the rear wheel cylinder pressure and allow the rear wheels to spin back up to vehicle speed. Continuing under RABS module control, the dump and isolation solenoids will be pulsed in a manner that will keep the rear wheels rotating while still maintaining high levels of deceleration during braking.

At the end of the stop, when the operator releases the brake pedal, the isolation valve de-energizes and any fluid in the accumulator is returned to the master cylinder. Normal brake operation is resumed.

System Self Test

The RABS module performs system tests and self-tests during startup and normal operation. The valve, sensor, and fluid level circuits are monitored for proper operation. If a fault is found the RABS will be deactivated and the REAR ANTILOCK light will be illuminated. Most faults will cause the light to stay illuminated until the ignition is turned **OFF**. While the light is illuminated a diagnostic flashout code may be obtained. However, there are certain faults (those associated with the fluid level switch or loss of power to the module) that will cause the system to be deactivated and the REAR ANTILOCK light to be illuminated but will not provide a diagnostic flashout code will be available.

Warning Lights

The RABS uses both the BRAKE and REAR ANTILOCK instrument panel warning lights to alert the driver to a system malfunction. Both lights must be working properly to assist in problem diagnosis. The red BRAKE warning light is used to indicate a low fluid level condition, parking brake applied condition or, for vehicles equipped with diesel engines, a low vacuum condition. To check this light, turn the key to **START**. The light should glow in this position. If it fails to glow, service of the electrical system is required.

NOTE: If the red light continues to glow after the key is in the RUN position, repair the brake system as required. If the brake system checks out OK, troubleshooting will be required to diagnose the RABS problem.

The yellow REAR ANTILOCK warning light is used to indicate an RABS malfunction and a deactivation of the RABS. To check this light, turn the key to **ON** or **START**. The light should perform a self-check, glowing for about 2 seconds. If the light fails to glow or continues to glow after two seconds, troubleshooting of the warning lights is required.

A diode/resistor is located on the main trunk of the instrument panel wiring harness where the RABS module connector pigtail intersects the main trunk. The diode/resistor isolates the RABS module from the parking brake switch and the low vacuum switch (diesel engines). If the diode/resistor did not prevent voltage from reaching the RABS module, the yellow REAR ANTILOCK lamp would turn on and the system would be shut down whenever the parking brake was applied or the low vacuum switch was closed.

Flashout Codes

Whenever the yellow REAR ANTILOCK light comes on during normal operation, a flashout code may be obtained to aid in problem diagnosis. If the vehicle is shut off before the code is read, the code will be lost. In some cases the code may reappear when the vehicle is restarted. In other cases, the vehicle may have to be driven to reproduce the problem and, if the problem was associated with an intermittent condition, it may be difficult to reproduce. Therefore, whenever possible, it is recommended that the code be read before the vehicle is shut off.

NOTE: Place blocks behind the rear wheels and in front of the front wheels to prevent the vehicle from moving while the flashout code is being taken. If the BRAKE light is also on, due to a grounding of the fluid level cir-cuit (perhaps low brake fluid), no flashout code will be flashed and the REAR ANTILOCK light will remain on steadily. If there is more than one system fault only the first recognized flashout code may be obtained.

Obtaining the Flashout Code

A flashout code may be obtained only when the yellow REAR ANTILOCK light is ON. No code will be flashed if the system is operating properly.

Before obtaining the flashout code, drive the vehicle to a level area, and place the shift lever in **P** for automatic transmissions and neutral for manual transmissions.

Notice whether the red BRAKE light is on or not (for future reference) and then apply the parking brake. Keep the ignition ON so that the code will not be lost.

NOTE: Place blocks behind the rear wheels and in front of the front wheels to prevent the vehicle from moving while the flashout code is being taken.

To obtain the flashout code, locate the RABS diagnostic connector (with the black/orange wire) and attach a jumper wire to it. Momentarily ground it to the chassis. When the ground is made and then broken the REAR ANTILOCK light should begin to flash.

NOTE: If the red BRAKE light was on (as noticed before the parking brake was applied) the problem may be with the low fluid level circuit and, in this case, no flashout code will be flashed and the light will remain on steadily.

The code consists of a number of short flashes and ends with a long flash. Count the short flashes and include the following long flash in the count to obtain the proper code number. For example, three short flashes followed by one long flash indicates Flashout Code Four. The code will continue to repeat itself until the key is turned off. It is recommended that the code be verified by reading it several times. In addition, the first code flashed may be too short because it may have been started in the middle. It should be ignored.

COMPONENT LOCATION

F SERIES AND BRONCO

The RABS consists of the following components:

1. RABS module located in the cab to the right of the brake pedal under the upper dash panel.

2. RABS valve (dual solenoid electro-hydraulic) is located on the left frame rail just behind the number one crossmember.

3. RABS speed sensor located on the rear axle housing and the exciter ring is located inside on the gear carrier.

4. Yellow REAR ANTILOCK warning light is in the instrument cluster.

5. RABS diagnostic connector is located in the cab and clipped on the main instrument panel wiring harness about six inches from the firewall near the parling brake pedal.

6. Diode/resistor element is located

FORD REAR WHEEL ANTILOCK BRAKE SYSTEM DIAGNOSTIC CODES

Codes Are Yellow REAR ANTILOCK Light Flashing—Count Flashes

CODE	SYSTEM PROBLEM
CODE 1	This code is not used and should not occur
CODE 2	Open isolate circuit
CODE 3	Open dump circuit
CODE 4	Check RABS valve
CODE 5	Check RABS valve/4 × 4 indicator switch (4WD)
CODE 6	Check sensor for loose connections, chips, bad exciter ring
CODE 7	No isolate valve self test
CODE 8	No dump valve self test
CODE 9	Check sensor wiring for high resistance
CODE 10	Check sensor wiring for low resistance (shorted)
CODE 11	Check stop lamp switch
CODE 12	Check for low fluid level, bad brake light wiring
CODE 13	RABS module failure—replace
CODE 14	RABS module failure—replace
CODE 15	RABS module failure—replace
CODE 16	This code is not used and should not occur

NOTE: CODES 1 and 16 are not used. When checking resistance in the antilock brake system, always disconnect the battery. Improper resistance readings will occur with the vehicle battery connected.

Component Location F-150—F-350, Bronco

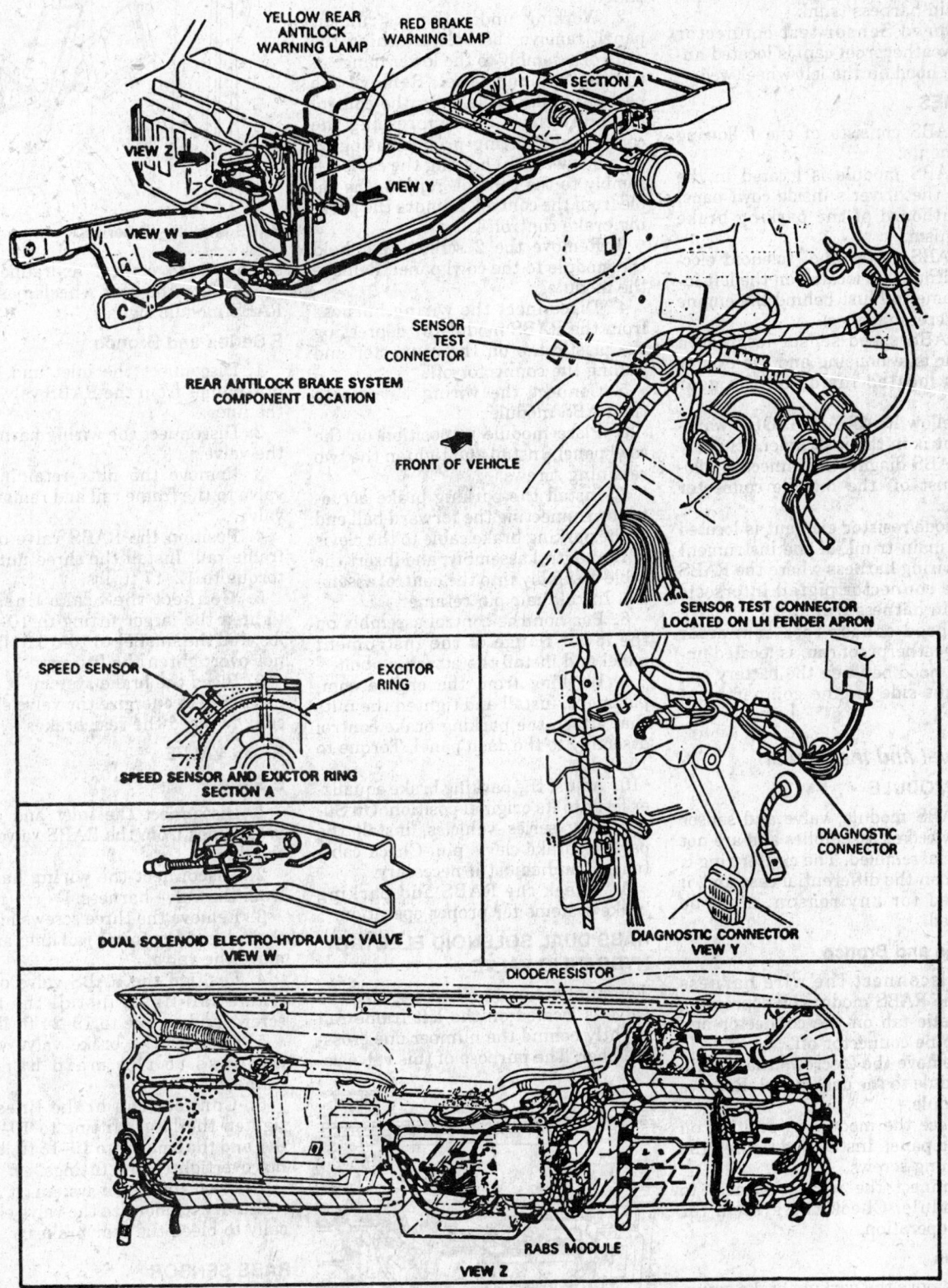

YELLOW REAR ANTILOCK WARNING LAMP

RED BRAKE WARNING LAMP

SECTION A

VIEW Z

VIEW Y

VIEW W

SENSOR TEST CONNECTOR

REAR ANTILOCK BRAKE SYSTEM COMPONENT LOCATION

FRONT OF VEHICLE

SENSOR TEST CONNECTOR LOCATED ON LH FENDER APRON

SPEED SENSOR

EXCITOR RING

SPEED SENSOR AND EXICTOR RING SECTION A

DIAGNOSTIC CONNECTOR

DUAL SOLENOID ELECTRO-HYDRAULIC VALVE VIEW W

DIAGNOSTIC CONNECTOR VIEW Y

DIODE/RESISTOR

RABS MODULE

VIEW Z

RABS Component Location—F Series, Bronco shown—typical

on the main trunk of the instrument panel wiring harness where the RABS module connector pigtail intersects the main harness trunk.

7. Speed Sensor test connector (with weatherproof cap) is located under the hood on the left wheel well.

E SERIES

The RABS consists of the following components:

1. RABS module is located in the cab on the driver's inside cowl panel just outboard of the parking brake mechanism.

2. RABS valve (dual solenoid electro-hydraulic) is located on the left inside frame rail just behind the engine mount crossmember.

3. RABS speed sensor located on the rear axle housing and the exciter ring is located inside on the gear carrier.

4. Yellow REAR ANTILOCK warning light is in the instrument cluster.

5. RABS diagnostic connector is located just off the module connector harness.

6. Diode/resistor element is located on the main trunk of the instrument panel wiring harness where the RABS module connector pigtail intersects the main harness trunk.

7. Speed sensor test connector (with weatherproof cap) is located under the hood between the battery and the right side engine compartment wall.

Removal and Installation

RABS MODULE

The RABS module, valve and sensor are serviced as assemblies and are not to be disassembled. The exciter ring is pressed on the differential case and, if removed for any reason, must be discarded.

F Series and Bronco

1. Disconnect the wire harness from the RABS module by depressing the plastic tab on the connector and pulling the connector off.

2. Remove the 2 screws that retain the module to the dash panel. Remove the module.

3. Place the module in position on the dash panel. Install and tighten the 2 retaining screws.

4. Connect the wiring harness to the module. Check the system for proper operation.

E Series

1. Remove the parking brake actuator assembly. Start by loosening the adjusting nut at the equalizer. On F-Super Duty series vehicles remove the clevis pin at the parking brake. Working from the engine compartment, re-

move the nuts attaching the parking brake control assembly to the dash panel.

2. Working under the instrument panel, remove the bolt attaching the control assembly to the lower flange of the instrument panel. Remove the parking brake cable from the control assembly clevis by compressing the conduit end fitting prongs (using ½ inch box wrench) holding the cable assembly to the control, remove the cable from the control. Remove the parking brake control.

3. Remove the 2 screws that hold the module to the cowl panel. Remove the module.

4. Disconnect the wiring harness from the RABS module by depressing the plastic tab on the connector and pulling the connector off.

5. Connect the wiring harness to the RABS module.

6. Place module in position on the cowl panel. Install and tighten the two retaining screws.

7. Install the parking brake actuator by connecting the forward ball end of the parking brake cable to the clevis of the control assembly, and insert the cable assembly into the control assembly. Install hair pin retainer.

8. Position the control assembly on the lower flange of the instrument panel and install the attaching bolt.

9. Working from the engine compartment, install and tighten the nuts that attach the parking brake control assembly to the dash panel. Torque to 15 ft. lbs.

10. Adjust the parking brake equalizer lever to its original position. On Super Duty series vehicles, install the parking brake clevis pin. Check cable tension and adjust if necessary.

11. Check the RABS and parking brake systems for proper operation.

RABS DUAL SOLENOID ELECTRO-HYDRAULIC VALVE

The dual solenoid electro-hydraulic valve is located on the left frame rail slightly behind the number one crossmember. The purpose of this valve as-

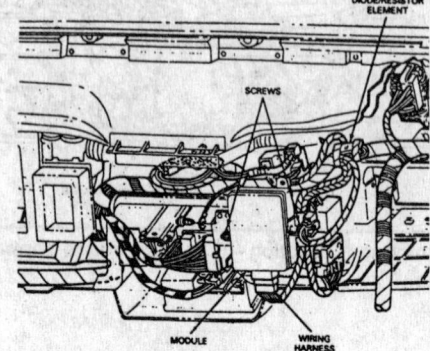

RABS Module—F Series, Bronco

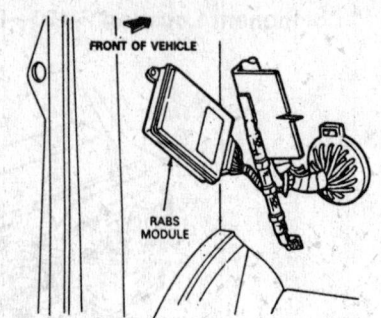

RABS Module—E Series

sembly is to control hydraulic fluid pressure to the rear wheels based on RABS module signals.

F Series and Bronco

1. Disconnect the inlet and outlet brake lines from the RABS valve. Cap the lines.

2. Disconnect the wiring harness to the valve.

3. Remove the nuts retaining the valve to the frame rail and remove the valve.

4. Position the RABS valve on the frame rail. Install the three nuts and torque to 12–17 ft. lbs.

5. Connect the brake lines and tighten the larger fitting to 10–17 ft. lbs. and the smaller to 10–15 ft. lbs. Do not overtighten the fittings.

6. Bleed the brake system. It is not necessary to energize the valve electrically to bleed the rear brakes.

RABS VALVE

E Series

1. Disconnect the inlet and outlet brake lines from the RABS valve. Cap the lines.

2. Disconnect the wiring harness from the valve harness.

3. Remove the three screws holding the valve to the frame rail liner and remove the valve.

4. Position the RABS valve on the frame rail liner, install the three screws and torque to 19–24 ft. lbs.

5. Connect the brake valve wiring harness to the main harness connector.

6. Connect the brake lines and tighten the larger fitting to 10–17 ft. lbs. and the smaller to 10–15 ft. lbs. Do not overtighten the fittings.

7. Bleed the brake system. It is not necessary to energize the valve electrically to bleed the rear brakes.

RABS SENSOR

1. The RABS sensor is located on the rear axle housing. Remove the wiring connector.

2. Remove the sensor hold-down bolt and remove the sensor from the axle housing.

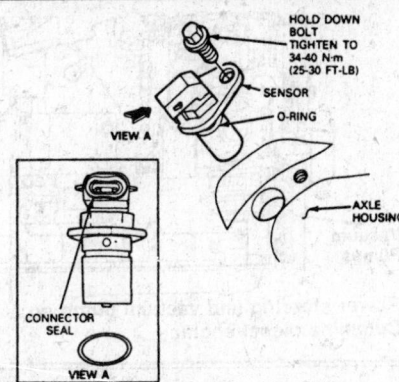

Speed sensor mounts to rear axle housing

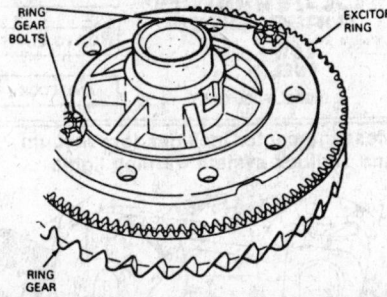

Exciter ring for speed sensor is pressed on gear carrier

3. When installing, clean the axle mounting surface. Use care to keep dirt from entering the axle housing.

4. Inspect and clean the magnetized sensor pole piece to ensure that it is free from loose metal particles which could cause erratic system operation. Inspect the sensor O-ring for damage and replace if necessary.

5. Lightly lubricate the sensor O-ring with motor oil, align the sensor bolt hole and install. Do not apply force to the plastic sensor connector. The sensor flange should slide to the mounting surface. This will insure the air gap setting is between 0.005–0.045 in.

6. Inspect the blue sensor connector seal and replace it if missing or damaged. Push the connector on the sensor.

NOTE: The clearance between the sensor and the excitor ring should be no greater than 0.050 inches. To measure, remove the sensor from the axle carrier, and measure the height of the pole piece from the mounting face of the flange. Sensor pole should be 1.07–1.08 in. Measure the depth from axle housing mounting face to the top of the exciter teeth. Subtract the two measurements to get the sensor gap.

EXCITER RING

The exciter ring is pressed onto the

differential case. To inspect it, remove the sensor from the axle housing.

View the exciter ring teeth through the sensor hole. Rotate the rear axle and check the exciter ring teeth for damage or breakage. Dented or broken teeth could cause the RABS system to function when not required. To replace the exciter ring, the differential case must be removed from the axle housing and the exciter ring pressed off the case. Upon removal, the exciter ring must be discarded. It is not to be reused.

FUSES

Three replaceable fuses are involved with RABS. The fuses are located in the fuse box. A 20 amp fuse protects the total RABS. A 15 amp fuse protects the red BRAKE and yellow REAR ANTILOCK warning lights. Another 15 amp fuse protects the four-way stop lamp cluster.

CHRYSLER CORPORATION

Master Cylinder Service

Chrysler Corporation uses a dual master cylinder which contains a double hydraulic cylinder with two fluid reservoirs and primary and secondary hydraulic pistons. On most models, the master cylinder is assisted by a vacuum booster. The unit is mounted on a 90 degree bracket on some models, directly to the dashboard on others. Hydraulic rear drum brakes with automatic adjusters and self-adjusting front disc brakes are standard on all models.

The front outlet tube of the master cylinder is connected to the hydraulic system control valve and then to the rear brakes. This is referred the secondary system. The rear outlet tube is connected to the control valve and to the front brakes. This system is referred to as the primary system. No residual pressure valves are used in the master cylinder outlets.

During normal operation, the fluid level in the master cylinder will rise during brake operation and fall during release. It is also expected that the fluid level will decrease with brake pad wear. In addition, a trace of brake fluid on the booster shell below the master cylinder mounting flange will often be found as a result of the normal lubricating action of the master cylinder

bore and seal. All of these conditions are considered normal and are not indications the master cylinder needs service.

Two types of master cylinders were used by Chrysler Corporation; 1) a cast iron master cylinder with built-in reservoir and separate proportioning valve and, 2) a master cylinder with a pressed on plastic reservoir. The body of the two-piece master cylinder is made of anodized aluminum and the reservoir is made of nylon. Both compartments of the reservoir are interconnected to permit equalization of the fluid level. However, a sufficient quantity of fluid is retained in the reservoir of the unaffected system to permit operation of that half of the master cylinder even if the other half of the reservoir is drained due to a hydraulic leak.

Use extra care when servicing aluminum master cylinders not to cross thread brake line fittings and do not overtighten any threaded connection.

In the event of a front brake system malfunction the proportioning valve with a bypass feature allows full hydraulic pressure to the rear brake system.

Disassembly and Assembly

The manufacturer does not recommend that either the cast iron or the aluminum master cylinder be overhauled, but replaced only. Do not attempt to hone the bore of the aluminum master cylinder or the hard anodized surface will be removed.

On the aluminum master cylinders, the plastic reservoir may be replaced using the following procedure:

1. Clean the outside of the master cylinder and remove the caps. Drain and discard remaining brake fluid.

2. Hold the master cylinder in a soft-jaw vise and grasp the plastic reservoir. Firmly rock reservoir from side to side to and remove it from the master cylinder housing. Don't pry the reservoir off with a tool that could damage the plastic body.

3. Remove and discard the reservoir grommets from the master cylinder body.

4. To install, lubricate the new

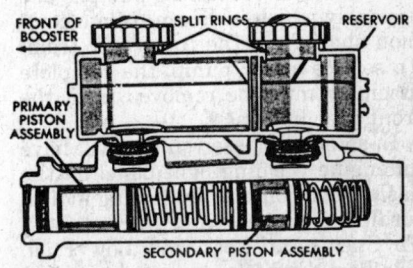

Chrysler aluminum master cylinder

Removing plastic reservoir from master cylinder housing

grommets with brake fluid and install in the master cylinder body.

5. Place the reservoir in position. Make sure it is positioned properly and that the lettering can be read from the left side of the master cylinder. Install by rocking the reservoir while pressing down. Seat the reservoir until the bottom of the reservoir touches the top of the grommets.

Brake Booster Service

Chrysler Corporation models covered here are equipped with vacuum boosters.

Under normal operating conditions, with brakes released, a vacuum suspended booster operates with vacuum on both sides of its diaphragm (or both diaphragms in a tandem diaphragm unit). When the brakes are applied, air at atmospheric pressure is admitted to one side of the diaphragm (or both diaphragms in a tandem diaphragm unit) to provide the power assist.

Brake boosters used by Chrysler Corporation are not repairable and if defective, the booster must be replaced as a unit, including the check valve. Do not remove the check valve.

Brake Booster Vacuum Pump

On gasoline engine vehicles, engine vacuum is used to power the vacuum booster brake. On diesel engines, there is not enough vacuum available for vacuum booster brake operation.

On diesel engine vehicles, vacuum is supplied from a pump. On Chrysler Corporation vehicles equipped with Cummins Turbo Diesels, the vacuum pump and power steering pump is one assembly which is driven from a common shaft from the front gear train. To service either pump, the complete assembly must be removed from the front gear housing.

In addition, these vehicles will have an engine warning light panel on the dash which contains 5 warning indicator lights: Brake, Water In Fuel, Wait To Start, Anti-Lock and Low Fuel. The Brake light is connected to a sen-

sor that monitors vacuum in the brake booster system. The Brake light when lit indicates low vacuum. If this light comes on, the brake system must be serviced.

NOTE: The brake light will also be activated when the parking brake is on or there is a hydraulic brake failure.

The vacuum pump provides vacuum for the brake booster and dash controllers. The vacuum sensor is mounted under the left hood hinge in the engine compartment. Vacuum is supplied by a hose Teed off the check valve in the brake booster. The sensor will activate the brake light on the warning light panel in 10 seconds or less when the vacuum drops to 8.5 in. Hg or less in the brake system.

Removal and Installation

1. Remove the vacuum pump and power steering pump assembly by removing the 2 bolts which should need an 18mm wrench.
2. Clean the gasket from the engine rear cover.
3. Install the pump assembly with a new gasket. Torque the bolts to 57 ft. lbs.

Disassembly and Assembly

1. Before starting pump disassembly, make a pin 2 in. long and 0.312 in. diameter. Use an 8mm bolt or a piece of drill rod. This pin must be hard so make it from a 10.9 grade metric bolt or at least an SAE Grade 8 capscrew.
2. Insert the pin into the pump shaft and screw a M14-2mm threaded capscrew in against the pin.
3. Tighten the capscrew against the pin to draw off the gear/eccentric/bearing assembly off of the power steering pump shaft.
4. Inspect the gear for excessive wear or damage. If the gear/eccentric is not damaged, do not separate them. If the gear is damaged, use a press to press the eccentric out of the gear. Use a flat plate over the new gear and press it onto the eccentric until it bottoms.
5. Inspect the bearing, turning it by hand. If the bearing looks good, and feels smooth to turn, do not separate them. If the bearing/eccentric is bad, press the eccentric out of the bearing, using the appropriate size socket or tool that will press only on the inner race of the bearing. Press it on until it bottoms.
6. With a 15mm wrench, remove the 4 nuts and separate the power steering pump from the vacuum pump housing. Remove the two short pushrods from the housing. Inspect the pushrods. They should move smoothly, but should not move side to side.

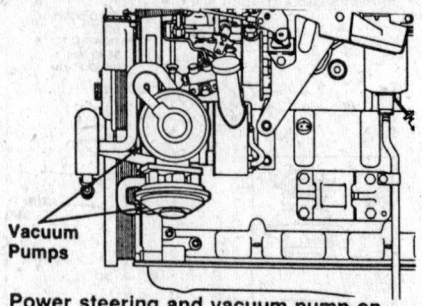

Power steering and vacuum pump on Cummins diesel engine

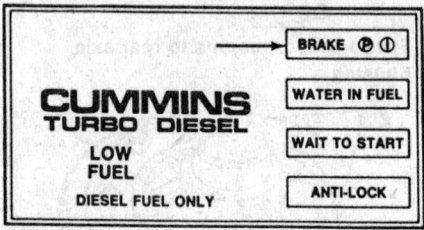

Message center includes low vacuum and antilock system warning lights

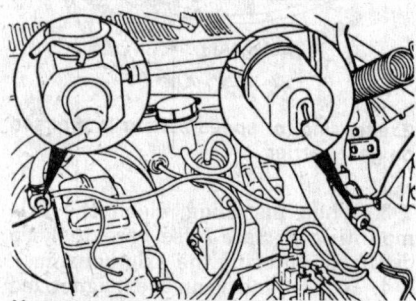

Vacuum sensor under the left hood hinge is supplied by line from booster check valve

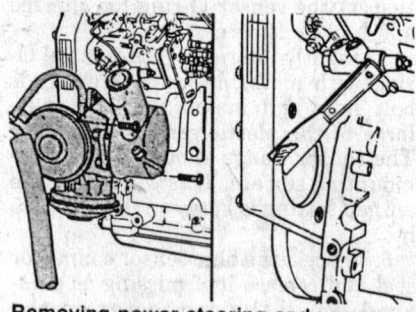

Removing power steering and vacuum pump assembly from engine's gear drive

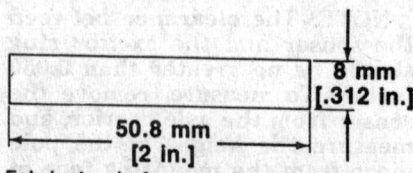

Fabricate pin for pressing off drive gear

7. With a 10mm wrench, remove the diaphragm assemblies from the housing. If the diaphragms are bad, they must be replaced as a unit. Remove and discard the O-rings from the center bore and push rod bores.
8. Clean all parts and dry with compressed air.

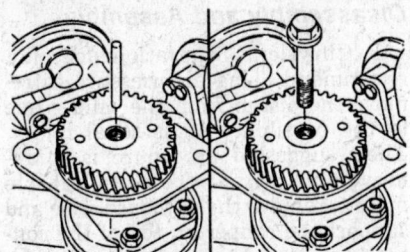

Insert pin, tighten capscrew against pin to draw gear assembly from pump shaft

9. Install the two spacers on the studs on the back of the power steering pump.

10. Install the vacuum pump housing onto the power steering pump using a new O-ring lubricated with engine oil. Install the four retaining nuts and with a 15mm wrench, torque to 18 ft. lbs.

11. Install the gear/eccentric/bearing assembly by pulling it onto the power steering shaft with a ⅜–18 thread capscrew with a flat washer threaded into the power steering pump shaft. Pull it on until it bottoms. Remove capscrew and washer.

NOTE: This is a press fit and will require a minimum of SAE Grade 8 capscrew for thread strength.

12. Lubricate the pushrods with engine oil and install then in the housing. Install new O-rings, also lubricated with engine oil.

13. With a 10mm wrench, install the

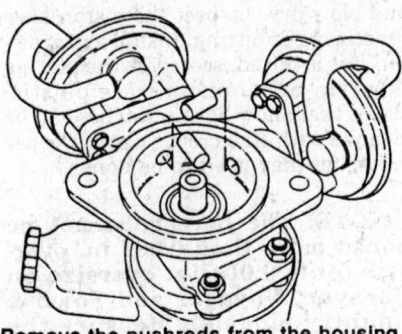

Remove the pushrods from the housing

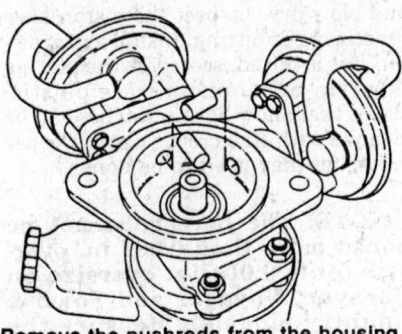

Remove vacuum diaphragm assemblies, then O-rings

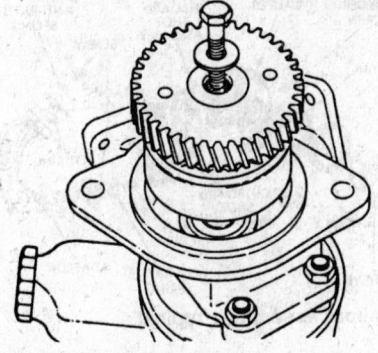

Use ⅜-18 Grade 8 capscrew to draw assembly onto pump shaft

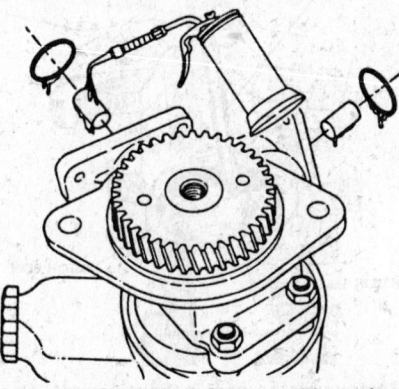

Lube pushrods and O-rings with engine oil at assembly

diaphragm assemblies. Torque to 7 ft. lbs.

14. Install the assembly to the engine with new gaskets. Torque bolts to 57 ft. lbs.

Combination Valve

All models have a hydraulic system control valve in the brake system. The valve is usually mounted on the frame rail below the master cylinder. The control valve assembly in B-150 and B-250 models combines a brake warning switch with a hold-off and proportioning valve assembly. A brake warning switch and hold-off valve assembly are combined and used on B-350 models. Hold-off and Proportioning valves are used because of different braking characteristics between disc and drum brakes.

A height sensing proportioning valve regulates the front to rear braking balance based upon vehicle load conditions. The valve senses vehicle loads through variations in rear suspension height. With a light load on the rear axle, the valve reduces hydraulic pressure to the rear brakes. As the load increases, more hydraulic pressure is released to the rear brakes.

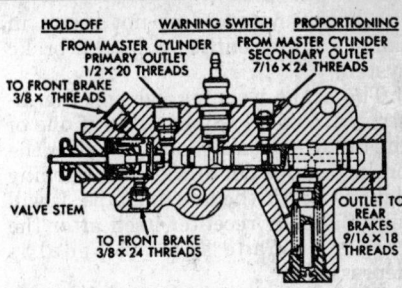

Combination warning switch/hold-off/ proportioning valve assembly – Typical

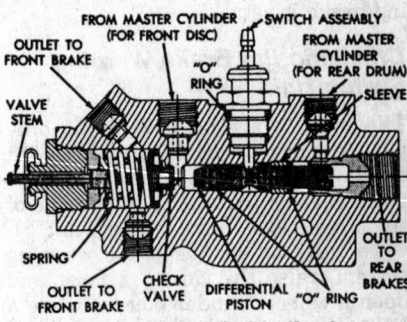

Brake warning switch/hold-off valve

HOLD-OFF VALVE

The hold-off valve section of the combination valve holds off pressure to the front disc brakes to allow the rear drum brake shoes to overcome the return springs and begin to contact the drums. This valve keeps the output pressure to the front brakes in the 3–30 psi range until the hold-off pressure is reached (117 psi) and then blends back to give full output pressure to the front brakes under heavy brake applications. This feature helps keep the front brakes from locking under light pedal applications when driving on icy surfaces. The hold-off valve has no effect on front brake pressure during hard stops.

Checking the Hold-Off Valve

A visual check will show that the valve stem extends slightly when the brakes are applied and retracts when the brakes are released.

In case of a hold-off valve malfunction, remove the valve and install a new combination valve assembly.

BRAKE WARNING SWITCH

The hydraulic brake system is split. The front brakes are part of one system and the rear brakes are part of the other. Both systems are routed through, but hydraulically separated by the pressure differential switch. The function of this switch is to alert the driver to a malfunction in one of the hydraulic systems. Since the brake system is split, a failure in one part of the brake system does not result in failure of the entire hydraulic brake system.

The brake warning light on the instrument panel will come on if one of the brake systems should fail after the brake pedal is depressed. The warning light switch is the latching type. It will automatically recenter itself after the repair is made and the pedal is depressed.

The instrument panel bulb can be checked each time the ignition switch is turned to **ON, START** or the parking brake is set.

Checking the Brake Warning Switch Unit

The brake warning light is lit only when the parking brake is applied with the ignition key turned **ON**. The same light will also illuminate should one of the two service brake systems fail.

To test the service brake warning system, raise the car on a hoist and open a wheel cylinder bleeder while a helper depresses the brake pedal and watches the warning light. If the light fails to light, check for a burned out bulb, disconnected socket, or a broken or disconnected wire at the switch. If the bulb is not burned out and the wire continuity is uninterupted check the service brake warning switch operation with a test lamp between the switch terminal and a voltage source.

If the light still fails to light, disconnect the brake tubes from the valve assembly and install a new valve assembly. If a new is installed, bleed the system. The warning switch is not serviced separately. Do not remove the switch or attempt to repair.

After repairing and bleeding the brake system applying the brakes with moderate force will hydraulically recenter the valve's piston and automatically turn off the warning light. Do not disassemble to reset the piston.

Caliper Service

Chrysler Corporation calipers are one-piece castings with the inboard side containing the piston bore. A square cut rubber seal is located in a groove in the piston bore which provides the hydraulic seal between the piston and the cylinder wall.

As the brake pedal is pressed, hydraulic pressure is applied against the piston. This pressure pushes the inboard brake lining against the inboard braking surface of the rotor. As the force increases against the rotor, the caliper assembly moves inboard and provides a clamping action on the rotor.

When brake pressure is released,

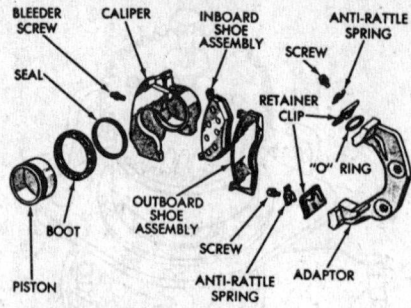

Caliper assembly – typical

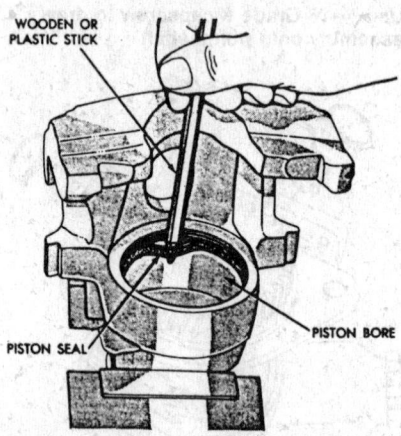

Remove piston seal without scratching bore

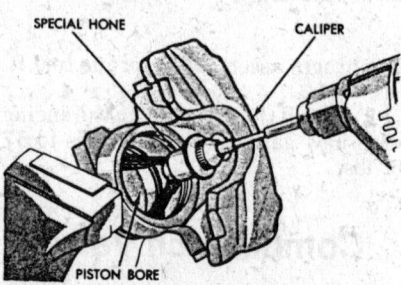

Clean up bore with light honing

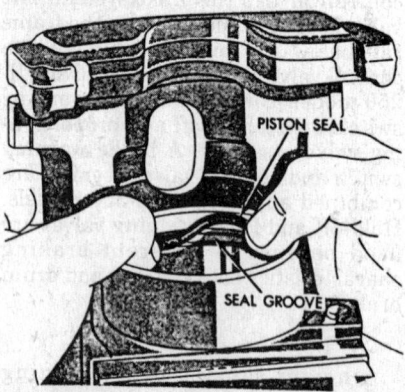

Lubricate piston seal with brake fluid and install

the piston seal returns to its normal position, pulling the piston back into the caliper bore. This creates a running clearance between the inner brake lining and rotor.

Disassembly and Assembly

1. Chrysler Corporation does not recommend using air pressure to remove the piston from the caliper due to the possibility of personal injury. Their suggested procedure is to remove the caliper from its vehicle mount, remove the outboard shoe and support the caliper on top of the control arm on shop towels to absorb brake fluid. Carefully depress the brake pedal and allow hydraulic pressure to push the piston out of the bore.

NOTE: The brake pedal will fall away when the piston has passed the bore opening. Prop the brake pedal to any position below the first inch of pedal travel to prevent loss of brake fluid. If both front caliper pistons are to be removed, disconnect the flexible brake line at the frame bracket after removing the first piston. Plug the brake line to remove the piston from the opposite caliper.

2. Remove the flexible brake line from the caliper.

3. Mount the caliper in a soft-jaw vise. Do not use too much pressure or the bore will be distorted and the piston will bind.

4. Remove the dust boot and, using a small, pointed wooden stick, work the piston seal out of its groove in the caliper bore. Do not use any metal tool which might nick or scratch the bore. Discard the seal.

5. Clean all parts well using alcohol and blow dry. Inspect the piston bore for scoring or pitting. Install a new piston if it is pitted, scored or the plating is worn. Some pistons are plastic. Bores that show light scratches can be cleaned with fine crocus cloth. Deeper scratches may need to be hones.

NOTE: The bore must not be honed more than 0.002 in. oversize (only 0.001 in. oversize on Caravan, Voyager and Town & Country models). Measure the bore accurately before honing. If the bore does not clean up within this specification, a new housing must be installed. Black stains on the piston are caused by the piston seal and will do no harm.

6. Clean all parts well, dip the new piston seal in brake fluid and install in the bore. Position the seal at one area at a time and, using fingers, gently work the seal into the groove.

7. Coat the new piston boot with brake fluid leaving a generous amount inside the boot. Position the boot on the piston and install the piston into the bore, pushing it past the piston seal until it bottoms in the bore.

8. Position the dust boot and with a circular driver, seal into the caliper.

9. Install the flexible brake hose with new seals.

Wheel Cylinder Service

The rear brakes are drum type with internal expanding shoes. The rear brakes are of the single anchor type, mounted to the same anchor and actuated by 1 wheel cylinder. The wheel cylinder has 2 pistons. Brake adjustments are automatic and are made during reverse brake applications.

Wheel cylinders may need reconditioning or replacement whenever the brake shoes are replaced or when required to correct a leak condition. Leaks which coat the boot and the cylinder with fluid, or result in a dropped reservoir fluid level, or dampen and stain the brake linings are dangerous. Such leaks can cause the brakes to grab or fail and should be immediately corrected. A leakage, not immediately apparent, can be detected by pulling back the cylinder boot. A small amount of fluid seepage dampening the interior of the boot is normal. However, a dripping boot is not. Unless other conditions causing a brake to pull, grab or drag become obvious, the wheel cylinder is suspect and should and should be included in general reconditioning.

Cylinder binding may be caused by rust, deposits, grime, or swollen cups due to fluid contamination, or by a cup wedged into an excessive piston clearance.

Hydraulic system parts should not be allowed to come into contact with oil or grease, neither should those be handled with greasy hands. Even a trace of any petroleum based product is sufficient to cause damage to the rubber parts.

Disassembly and Assembly

1. Remove the brake bleeder screw.
2. Remove the dust boots, allow any brake fluid to drain out.
3. Remove the pistons, seals and inner spring.
4. Inspect the bore for scoring and corrosion. The inside of the bore may be cleaned with fine crocus cloth. If bore is scored, replace cylinder. Black stains on the cylinder walls are caused by the piston cups and will not impair operation of the cylinder. Clean the cylinder with alcohol or brake fluid.
5. Lubricate the seals with brake fluid and install. Cup lips should always face inward. Install the spring assembly and seals.

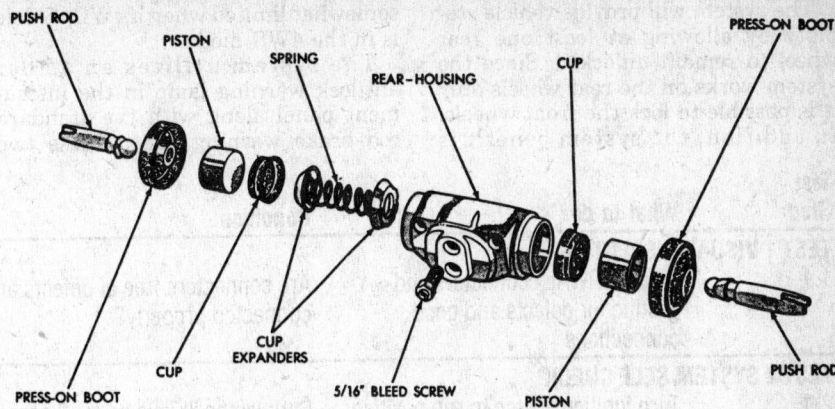

Rear wheel cylinder for 11 and 12 inch brakes—typical

6. Carefully install the pistons and dust boots.
7. Install the bleeder screw.

Anti-Lock Braking Systems

Despite advances in brake design over the years, even the best systems in use can still lock up during certain road conditions, such as wet road surfaces. When the brakes lock up, the driver can lose control of the vehicle, because a locked wheel cannot absorb any cornering or lateral forces, and steering is lost. It is impossible to brake to a maximum and at the same time steer the vehicle when the front wheels are locked. If the back wheels are locked the vehicle will become unstable and start to slide.

While many different ways have been tried over the years to solve this problem, mechanical sensors could not provide sufficient information about wheel rotation speed and mechanical control units could not operate the brakes fast enough to prevent brake lockup.

The growth of the electronics industry has allowed small computers (microprocessors) to be reduced in both size and cost. Coupled with fast reacting electronic sensors, anti-lock braking has become more reliable with widespread application.

REAR WHEEL ANTI-LOCK BRAKE SYSTEM

Chrysler Corporation's Rear Wheel Antilock brake system (RWAL) is designed to prevent rear wheel lockup under heavy braking conditions. Anti-lock braking allows a vehicle to stop without locking the wheels and therefore maintains directional stability.

The RWAL system uses a standard master cylinder and booster arrangement with a vertical split hydraulic circuit. An electronic control module,

rear wheel speed sensor, and a dual solenoid control valve (hydraulic valve) are the major components added to this system. No hydraulic pumps are used. Braking pressure comes directly from pushing on the brake pedal.

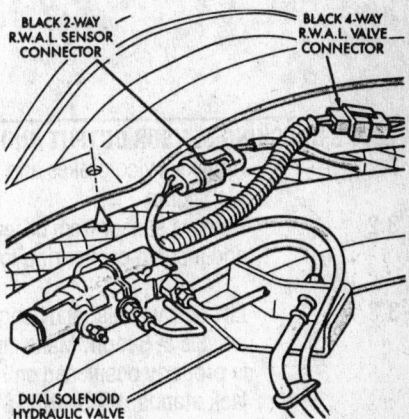

RWAL connection locations—typical

RWAL electronic brake control module—typical

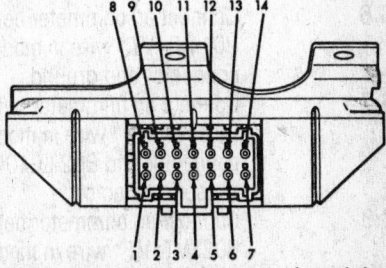

RWAL electronic brake control module pin locations

The system will provide vehicle stability by allowing at least one rear wheel to remain unlocked. Since the system works on the rear wheels only it is possible to lock the front wheels. In addition, the system benefit is somewhat limited when a 4WD vehicle is in the 4WD mode.

The system utilizes an amber antilock warning lamp in the instrument panel along with the standard red brake warning light. These two lights work together to notify the driver the driver that the system is working correctly. These lamps are also used to blink fault codes for system diagnosis.

Test Step	What to do	Condition	Yes	No
TEST 1 VISUAL INSPECTION				
1.1	Inspect RWAL connectors and ground for defects and good connections	Are connectors free of defects and connected properly?	Go to Test 2	Repair or connect terminals as required
TEST 2 SYSTEM SELF CHECK				
2.1	Turn ignition switch to run position	Both lamps illuminate for 2 seconds then go out as system performs self check	Go to Test 3	Choose another condition
		Antilock and brake lamp stay on	Go to Test 6	Choose another condition
		Antilock lamp off and will not self check, brake lamp checked OK	Go to Test 7	Choose another condition
		Brake lamp on, antilock lamp off and does self check	Go to Test 8	Choose another condition
		Antilock and brake lamps flashing	Go to Test 9	Choose another condition
		Brake lamp off, antilock lamp on, antilock lamp does self check	Go to Test 10	Choose another condition
TEST 3 CHECKING SENSOR OUTPUT AND PHYSICAL CONDITION				
3.1	Apply service brakes and check stop lights	Stop lamps illuminate	Go to Test 3.2	Repair stop lamp circuit
3.2	Remove sensor from differential and inspect exciter ring for damage	Exciter ring in good condition	Reinstall sensor. Go to Test 3.3	Replace exciter ring and retest system
3.3	Lift rear wheels, start engine, run wheels at 5 mph. Make sure vehicle is properly positioned on hoist or jack stands. **WARNING: STAY CLEAR OF ROTATING WHEELS.** Using a voltmeter set on 2 volt AC scale and connect between B01 PK and B02 LG/OR* wires of the sensor connector	Is voltage 650 MV (RMS) or greater?	Go to Test 5	Go to Test 3.4
3.4		Has sensor been replaced?	Go to Test 3.5	Replace sensor and go to Test 3.3
3.5	Disconnect 14 way module connector. Disconnect sensor connector and connect an ohmmeter between B01 RD/VT* wire in module connector and B01 PK wire in sensor connector	Is there continuity?	Go to Test 3.6	Repair open circuit
3.6	Connect an ohmmeter between B01 RD/VT* wire in module connector and ground	Is there continuity?	Repair circuit for a short to ground	Go to Test 3.7
3.7	Connect an ohmmeter between B02 WT/VT* wire in module connector and B02 LG/OR wire in sensor connector	Is there continuity?	Repair circuit for an open circuit	Go to Test 3.8
3:8	Connect an ohmmeter between B02 WT/VT* wire in module connector and ground	Is there a short to ground?	Repair circuit for short to ground	Go to Test 4

Test Step	What to do	Condition	Yes	No
TEST 4 CHECKING SENSOR GAP				
4.1	Remove sensor from differential. Measure height of sensor pole piece from mounting face of sensor (should be 1.07″–1.08″). Measure top of exciter ring teeth from sensor mounting face on differential (should be 1.085″–1.12″). Subtract measurements as shown to obtain sensor gap. Gap must be a minimum of 0.005″ and a maximum of 0.05″	Was gap within specifications?	Go to Test 5	Go to Test 4.2
4.2	Look at sensor measurement from Test 4.1 (should be 1.07″–1.08″)	Was sensor measurement within specifications?	Repair differential and retest system	Replace sensor and retest system
TEST 5 CHECKING FOR BRAKE MECHANICAL PROBLEMS				
5.1	Check rear brakes for mechanical problems such as grabbing, locking or pulling	Are the rear brakes functioning properly?	Replace module and retest system	Repair mechanical problem and retest system
TEST 6 CHECKING THE DIAGNOSTIC CONNECTOR GROUND				
6.1	Locate the black 2-way diagnostic connector below RWAL module. Connect a jumper wire between the diagnostic connector and ground	Is there a flashout code?	Go to Test 11	Go to 6.2
6.2	Turn ignition off. Disconnect 14-way connector from module and connect an ohmmeter between the BK* in the 14-way connector and the BK* in the 2-way diagnostic connector	Is there continuity?	Go to Test 6.3	Repair open circuit and retest system
6.3	Check brake fluid level in master cylinder reservoir	Is brake fluid level correct?	Go to Test 6.4	Find and repair leak and retest system
6.4	Reconnect 14-way module connector. Disconnect connector from the pressure differential switch. Turn ignition to run position	Do both antilock lamp and brake lamp stay on?	Go to Test 6.5	Check brake system for air in lines or mechanical damage
6.5		Does the vehicle have a diesel engine?	Go to Test 6.6	Go to Test 6.7
6.6	Disconnect harness connector from vacuum warning switch. Turn the ignition switch to the run position	Do both lamps stay on?	Go to Test 6.7	Check complete vehicle vacuum system, repair as required and retest system
6.7	Disconnect 14-way module connector. Turn ignition switch to the run position	Are both antilock lamp and brake lamp off?	Go to Test 6.8	Choose another condition
		Antilock lamp on, brake lamp off	Repair AT1 orange wire for short to ground between module and antilock lamp	Repair short to ground in differential switch sensor wiring, B01 PK and B02 LG/OR
6.8	Remove and inspect antilock fuse	Is fuse open?	Check and repair all circuits fuse is protecting. Replace fuse	Go to Test 6.9

Test Step	What to do	Condition	Yes	No
TEST 6 CHECKING THE DIAGNOSTIC CONNECTOR GROUND				
6.9	Connect a voltmeter between pin 3 RD/YL wire of 14-way module connector and ground	Is voltage greater than 9 volts?	Go to Test 6.10	Repair the D1 RD/YL wire for an open
6.10	Remove and inspect the stop lamp fuse.	Is the fuse open?	Check and repair all circuits fuse is protecting for shorts and replace fuse	Go to Test 6.11
6.11	Connect a voltmeter between pin 9 of the 14-way module connector and ground	Is voltage greater than 9 volts?	Replace module and retest system	Repair the D3B PK/DB* wire for open circuit.
TEST 7 CHECKING MODULE GROUND AND POWER				
7.1	Make sure module connector is fully-plugged into module	Is connector plugged in?	Go to Test 7.2	Plug connector in and retest system
7.2	Disconnect battery and 14-way module connector. With an ohmmeter set on 200 ohm scale, check resistance between pin 10 BK/LG wire on the 14-way harness connector and ground	Is resistance less than 1 ohm?	Go to Test 7.3	Repair H40 BK/LG for an open or damaged circuit
7.3	Remove and inspect antilock lamp fuse	Is fuse open?	Check and repair all circuits fuse is protecting for shorts and replace fuse	Go to Test 7.4
7.4	Connect battery and turn ignition to run position. With a voltmeter set on 20 volt DC scale, check voltage between Pin 2 Orange wire of the 14-way module connector and ground	Is voltage less than 9 volts?	Go to Test 7.5	Replace the module and retest system
7.5	Check antilock bulb	Is bulb open?	Replace bulb and retest system	Repair AT1 orange wire for open circuit between pin 2 and fuse
TEST 8 CHECKING PARKING SYSTEM AND MODULE				
8.1	Turn ignition to run position. Pull lever to release parking brake	Does the brake lamp go off?	Disregard failure, retest antilock and brake lamp for 2 second self check	Go to Test 8.2
8.2	Pull park brake release lever with one hand and pull pedal up with other hand	Did brake lamp go off	Repair park brake mechanism or switch	Go to Test 8.3
8.3	Disconnect black 1-way park brake switch connector	Did brake lamp go off	Adjust or replace park brake switch and retest system	Go to Test 8.4
8.4	Disconnect 14-way module connector	Did brake lamp go off?	Replace module and retest system	Repair P5 BK/GY wire for a short to ground

Test Step	What to do	Condition	Yes	No
TEST 9 CHECKING FOR INTERMITTENT PROBLEMS				
9.1	Disconnect 14-way module connector. With a voltmeter set on 20 volt DC scale, check voltage between pin 3 RD/YL wire on the 14-way module connector and ground. Turn ignition to run position and shake the instrument panel harness	Is voltage steady at 9 volts?	Go to Test 9.2	Repair D1 RD/YL for open circuit
9.2	Disconnect battery, set ohmmeter on 200 scale and connect between pin 12 BK wire of the 14-way module connector and ground, then shake the instrument panel harness	Is resistance 100K or greater and steady?	Go to Test 9.3	Repair DK/BK/* wire for a short to ground
9.3	With ohmmeter set on 200 ohm scale, connect between Pin 10 BK/LG wire of the 14-way module connector and ground, then shake instrument panel harness.	Is resistance steady at 1 ohm?	Replace module and retest system	Repair open circuit in H40 BK/LG
TEST 10 CHECKING FOR OPEN OR DISCONNECTED PARK BRAKE SWITCH CONNECTOR				
10.1	Make sure the P5 BK/GY wire at the park brake switch is connected	Is P5 wire connected to park brake switch?	Go to Test 10.2	Connect and retest system
10.2	Disconnect park brake switch connector. Disconnect message center black 6-way connector. Connect an ohmmeter between P5 BK/BY wire in both connectors	Is there continuity?	Go to Test 10.3	Repair P5 BK/GY wire for open circuit
10.3	Inspect instrument cluster printed circuit board for damage	Is printed circuit board damaged?	Replace printed circuit board	Replace brake warning lamp bulb
TEST 11 FLASHCODES				
11.1	Connect a jumper wire between the diagnostic connector and ground. Count the flashes including the long flash that starts the flash code count. Choose the proper condition	Antilock lamp and brake lamp flash 1 time	Go to Test 12	
		Antilock lamp and brake lamp flash 2 times.	G to Test 13	
		Antilock lamp and brake lamp flash 3 times.	G to Test 14	
		Antilock lamp and brake lamp flash 4 times.	G to Test 15	
		Antilock lamp and brake lamp flash 5 times.	G to Test 16	
		Antilock lamp and brake lamp flash 6 times.	G to Test 17	
		Antilock lamp and brake lamp flash 7 times.	G to Test 18	
		Antilock lamp and brake lamp flash 8 times.	G to Test 19	
		Antilock lamp and brake lamp flash 9 times.	G to Test 20	
		Antilock lamp and brake lamp flash 10 times.	G to Test 21	
		Antilock lamp and brake lamp flash 11 times.	G to Test 22	

Test Step	What to do	Condition	Yes	No
TEST 11 FLASHCODES				
		Antilock lamp and brake lamp flash 12 times.	G to Test 23	
		Antilock lamp and brake lamp flash 13 times.	G to Test 24	
		Antilock lamp and brake lamp flash 14 times.	G to Test 25	
		Antilock lamp and brake lamp flash 15 times.	G to Test 26	
		Antilock lamp and brake lamp flash 16 times.	G to Test 27	
TEST 12 ONE FLASH				
12.1	One flash code should not occur. Perform flashcode procedure several times	Are you still getting code 1?	Go to Test 5	Go to Test 11
TEST 13 TWO FLASHES				
13.1	Disconnect battery and 14-way module connector. Set an ohmmeter on 200 ohm scale and connect between pin 1 LG wire of the 14-way module connector and ground	Does the circuit have over 6 ohms?	Go to test 13.2	Replace the module and retest system
13.2	Disconnect valve harness connector from valve connector. Connect an ohmmeter between B09 GY/WT of harness connector and ground	Is resistance greater than 1 ohm?	Repair B09 GY/WT wire for an open circuit or high resistance. Check for contaminated or loose connector pins and retest system	Go to Test 13.3
13.3	Connect an ohmmeter between IS1 LG/* and B09 GY/WT wires in the 4-way black valve connector	Does circuit have over 6 ohms?	Replace antilock valve and retest system	Repair the IS1 LG/* for open circuit from valve to computer module and retest system
TEST 14 THREE FLASHES				
14.4	Disconnect battery. Remove 14-way module harness connector from module. Set ohmmeter on 200 ohm scale and connect to pin 8 DS1 WT/BR and ground	Does circuit have over 3 ohms?	Go to Test 14.2	Replace module and retest system
14.2	Disconnect the 4-way valve harness connector. Connect an ohmmeter between DS1 WT and B09 GY/NT wires in valve connector	Does circuit have over 3 ohms?	Replace antilock valve and retest system	Repair DS1 WT wire for open between module connector and valve connector

Test Step	What to do	Condition	Yes	No
TEST 15 FOUR FLASHES				
15.1	Disconnect the 4-way valve harness connector from valve connector. Set ohmmeter on 20K scale and connect between VS1 LB wire in valve body and ground	Is resistance greater than 10K ohms?	Go to Test 15.2	Replace the antilock valve and retest system
15.2	Connect an ohmmeter between VS1 LB and B09 GY/WY wires in the valve connector	Is resistance greater than 10K ohms?	Go to Test 15.3	Replace the antilock valve and retest system
15.3	Disconnect battery. Disconnect the 14-way module harness connector from module. Set ohmmeter on 200K scale and connect to pin 11, VS1 LB of 14-way connector and ground	Is resistance greater than 100K ohms?	Replace module and retest system	Repair VS1 LB wire for a short to ground between valve and module. Retest system
TEST 16 FIVE FLASHES				
16.1	Did the failure occur in 2 wheel drive mode?		Go to Test 16.2	Go to Test 16.3
16.2	Disconnect 14-way module connector from module to deactivate antilock system. Drive the vehicle in 2 wheel drive mode and make normal and safe stops to determine the condition of the rear brakes	Are the brakes functioning normally?	Replace the antilock valve and retest system	Repair rear brakes and retest system
16.3	Disconnect 14-way module connector from module. Turn ignition key to run position. Shift into 4 wheel drive. Set a voltmeter to 20 vdc and connect between pin 4, X4 LG/BR wire and ground	Is voltage greater than 1 volt?	Repair X4 wire for an open or 4 wheel drive indicator switch. Retest system	Replace antilock valve and retest system
TEST 17 SIX FLASHES				
17.1	Recheck flashcode after driving vehicle	Antilock light and brake light flash 6 times	Go to Test 17.2	Go to Test 11
17.2	Disconnect battery. Disconnect 14-way module connector. Set ohmmeter on 200 ohm scale and connect between pin 13, B02 WT/VT and pin 14, B01 RD/VT of harness connector. Shake antilock wiring harness from differential to module	Is resistance constant at 1000–2000 ohms?	Go to Test 17.3	Repair circuit. Retest system
17.3	Remove sensor from the differential and inspect for build up of metal chips on sensor pole piece	Are metal chips present?	Drain and clean differential. Check exciter ring for broken or chipped teeth. Retest system	Go to Test 17.4
17.4	Look into sensor hole in differential and rotate exciter ring and check for damage (missing or bent teeth)	Is exciter ring intact?	Go to 17.5	Replace exciter ring. Retest system
17.5	Reinstall sensor. Disconnect 2-way RWAL sensor connector. With a voltmeter on 2 volt scale, connect between B01 PK and B2 LG/OR wires of sensor connector. Raise rear wheels off floor and run at 5 mph. **WARNING: STAY CLEAR OF ROTATING WHEELS**	Is voltage greater than 650 MV and steady?	Replace module. Retest system	Replace sensor. Recheck sensor output. Retest system

Test Step	What to do	Condition	Yes	No
TEST 18 SEVEN FLASHES				
18.1	Disconnect 4-way valve harness connector from valve connector. Connect an ohmmeter between IS1 LG/* and B09 GY/WT wire in valve connector	Is resistance less than 3 ohms?	Replace antilock valve. Retest system	Go to Test 18.2
18.2	Disconnect battery. Disconnect 4-way valve harness connector from valve connector. Disconnect 14-way module harness connector from module. Set ohmmeter on 20K ohms scale and connect between Pin 1, IS1 LG/* wire in harness connector and ground	Is resistance greater than 20K ohms?	Replace module. Retest system	Repair IS1 LG/* for a short between antilock valve and module. Retest system
TEST 19 EIGHT FLASHES				
19.1	Disconnect 4-way valve harness from valve connector. Set ohmmeter on 200 ohm scale and connect between DS1 WT and B09 GY/WT wires in valve connector	Is resistance less than 1 ohm?	Replace antilock valve. Retest system	Go to Test 19.2
19.2	Disconnect battery. Disconnect 4-way valve connector. Disconnect 14-way module connector. Set ohmmeter on 20K ohm scale and connect between pin 8, DS1 WT/BR and ground	Is resistance greater than 20K ohms?	Replace module	Repair DS1 WT/BR for a short to ground between antilock valve and module. Retest system
TEST 20 NINE FLASHES				
20.1	Disconnect 2-way sensor harness connector from sensor on differential housing. Set ohmmeter on 20K scale and connect to B01 PK and B02 LG/OR wire on sensor	Is resistance greater than 2500 ohms?	Replace sensor. Recheck resistance. **Make sure seal is in place between sensor and connector.** Retest system	Go to Test 20.2
20.2	Reconnect sensor harness **making sure seal is in place.** Disconnect battery. Disconnect 14-way module connector. Connect an ohmmeter between pin 13, B02 WT/VT and pin 14, B01 RD/VT wires in module harness connector	Is resistance greater than 2500 ohms?	Repair B02 WT/VT and B01 RD/VT for open circuits between the module and sensor. Retest system	Replace computer module. When reconnecting sensor, **make sure seal is in place.** Retest system
TEST 21 TEN FLASHES				
21.1	Set ohmmeter on 20K ohm scale. Disconnect 2-way sensor connector from sensor on differential. Connect an ohmmeter between B01 PK and B02 LG/OR wires on sensor	Is resistance less then 1000 ohms?	Replace sensor. Recheck resistance. **Make sure seal is in place between sensor and connector.** Retest system	Go to Test 21.2

Test Step	What to do	Condition	Yes	No
TEST 21 TEN FLASHES				
21.2	Disconnect battery. Disconnect 14-way module. Connect an ohmmeter between pin 14, B01 RD/VT and ground	Is resistance greater than 20K ohms?	Go to Test 21.3	Repair B01 RD/VT circuit between the module and sensor. When reconnecting sensor, **make sure seal is in place.** Retest system
21.3	Connect an ohmmeter between pin 13, B02 WT/VT and pin 14 B01 RD/VT wire in module harness connector	Is resistance greater than 20K ohms?	Replace module. Retest system	Repair B01 RD/VT and B02 WT/VT circuit. When reconnecting sensor, **make sure seal is in place.** Retest system
TEST 22 ELEVEN FLASHES				
22.1	Recheck flash code after driving vehicle at 35 mph or greater.	Antilock light and brake light flash 11 times	Go to Test 22.1	Go to Test 11
22.2	Apply vehicle service brakes and check vehicle stop lights	Are stop lights operating correctly?	Go to Test 22.3	Repair the stop lamp circuit. Retest system
22.3	Turn ignition switch off. Disconnect 14-way module connector from module. Connect a voltmeter between pin 7, D4 WT of harness connector and ground while stepping on brake pedal	Is voltage less than 9 volts?	Repair D4 WT for open circuit between stop light switch and module. Retest system	Check 4-way flasher, directional wiring, and feedback through stop light circuit. Also check for proper operation of cruise control. Recheck antilock and brake lights for proper 2 second bulb check
TEST 23 TWELVE FLASHES				
23.1	This code should not occur. Read flashcodes several times	Are 12 flashes still present?	Replace module. Retest system	Go to Test 11
TEST 24 THIRTEEN FLASHES				
24.1	Read flashcode	Are 13 flashes present?	Replace module. Retest system	Go to Test 11
TEST 25 FOURTEEN FLASHES				
25.1	Read flashcode	Are 14 flashes present?	Replace module. Retest system	Go to Test 11
TEST 26 FIFTEEN FLASHES				
26.1	Read flashcode	Are 15 flashes present?	Replace module. Retest system	Go to Test 11
TEST 27 SIXTEEN OR MORE FLASHES				
27.1	Read flashcode	Are 16 or more flashes present?	Replace module. Retest system	Go to Test 11

Major Components

The system continually monitors rear wheel speed with a sensor mounted on the rear axle. A toothed exciter ring is press fit onto the differential case next to the differential ring gear and provides a signal for the sensor. When the teeth on the exciter ring pass the sensor pole piece, an AC voltage is induced in the sensor circuit with a frequency proportional to the average rear wheel speed. In the event of an impending lockup condition during braking, the RWAL modulates hydraulic pressure to the rear brakes. This inhibits rear wheel lockup.

The electronic brake control module is located behind the glove box on most models. The control module monitors the rear wheel speed and controls the dual solenoid valve.

The control module determines if the rear wheels are decelerating too quickly and sends a signal to the dual solenoid hydraulic valve to prevent rear wheel lockup. The control module also performs a system self check every time the ignition key is turned ON.

The dual solenoid hydraulic valve is located between the rear brakes and the proportioning valve and is attached on the left frame rail near the rear axle.

Under normal conditions, the valve will allow brake fluid to flow freely between the master cylinder and the rear brakes. Once antilock braking begins the control module will trigger the valve to either isolate or reduce pressure to the rear wheels.

The electronic control module has the capability of generating and storing fault codes. Only one code can be stored and shown at any one time. Also, if a fault code is generated the electronic control module will retain the code even after a key OFF condition.

If a problem is detected the electronic control module will illuminate the amber antilock brake warning lamp and set a fault code. When a fault code is set the red brake warning lamp will also be lit. To determine what the fault code is, momentarily ground the RWAL diagnostic connector and count the flashes of the amber antilock warning lamp. The initial flash will be a long flash followed by a number of short flashes. The long flash indicates the beginning of the fault number sequence and the short flashes are a continuation of that sequence. Count the long flash with the short flashes to have an accurate fault code count.

To clear a fault code disconnect the control module connector from the module or disconnect the battery for at least five seconds. During system re-test, wait 30 seconds to make sure the fault code does not reappear.

Removal and Installation
RWAL SPEED SENSOR

1. Raise and safely support vehicle.
2. Remove sensor hold down-bolt.
3. Remove sensor shield and sensor from differential by pulling sensor out of differential.
4. Disconnect wiring from sensor.
5. To install, connect wiring to sensor. Be sure the seal is in place between the sensor and wiring connector.
6. Install the sensor into the differential housing with a new O-ring.
7. Install sensor shield.
8. Install sensor hold down bolt. Torque to 170–230 inch lbs.
9. Lower vehicle.

ELECTRONIC BRAKE CONTROL MODULE

1. Remove the right side sill plate.
2. Remove the right side cowl cover.
3. Remove the three screws that attach electronic brake control module to side cowl.
4. Disconnect the wiring from the module.
5. When installing, connect the wiring first, then install the three mounting screws.
6. Install the right side cowl cover and sill plate.

DUAL SOLENOID HYDRAULIC VALVE

1. Raise and safely support vehicle.
2. Remove the brake lines from the valve.
3. Remove the 2 nuts that hold the valve to the frame.
4. Remove the valve from the frame and disconnect the wiring.
5. When installing, connect the wiring first, then install the valve. Torque the hold-down nuts to 16–25 ft. lbs.
6. Bleed the solenoid valve and the rear brakes.
7. Lower hoist.

JEEP

Master Cylinder Service

Jeep uses a dual master cylinder which contains a double hydraulic cylinder with 2 fluid reservoirs and primary and secondary hydraulic pistons. The master cylinder is assisted by a vacuum booster. Hydraulic rear drum brakes with automatic adjusters and self-adjusting front disc brakes are standard on all models.

The front outlet tube of the master cylinder is connected to the hydraulic system control valve and then to the rear brakes. This is referred the secondary system. The rear outlet tube is connected to the control valve and to the front brakes. This system is referred to as the primary system. No residual pressure valves are used in the master cylinder outlets.

During normal operation, the fluid level in the master cylinder will rise during brake operation and fall during release. It is also expected that the fluid level will decrease with brake pad wear. In addition, a trace of brake fluid on the booster shell below the master cylinder mounting flange will often be found as a result of the normal lubricating action of the master cylinder bore and seal. All of these conditions are considered normal and are not indications the master cylinder needs service.

Two types of master cylinders were used; 1) a cast iron master cylinder with built-in reservoir and separate proportioning valve and, 2) a master cylinder with a pressed on plastic reservoir. The body of the two-piece master cylinder is made of anodized aluminum and the reservoir is made of nylon. The two compartments of the reservoir are interconnected to permit equalization of the fluid level. However, a sufficient quantity of fluid is retained in the reservoir of the unaffected system to permit operation of that half of the master cylinder even if the other half of the reservoir is drained due to a hydraulic leak. There is another type of master cylinder used on vehicles equipped with anti-lock brakes. See anti-lock section.

Use extra care when servicing aluminum master cylinders not to cross thread brake line fittings and do not overtighten any threaded connection.

In the event of a front brake system malfunction the proportioning valve with a bypass feature allows full hydraulic pressure to the rear brake system.

Disassembly and Assembly

While the manufacturer does give overhaul procedures for the master cylinder, the manufacturer does not permit that either the cast iron or aluminum master cylinder be honed in an attempt to restore the surface. Replace the cylinder if the bore is corroded or if doubt exists about cylinder bore condition.

CAST IRON MASTER CYLINDER— EXCEPT ANTI-LOCK

1. Remove the cover and drain fluid. Examine cover seal for damage.

2. Mount cylinder in vise and press in primary piston. Remove snapring.

3. Remove the primary piston and discard. It is serviced only as an assembly.

4. Apply a small amount of compressed air to secondary outlet while covering small ports at bottom of rear reservoir to ease secondary piston from bore.

5. Disassemble secondary piston. Discard seals, spring and retainer.

6. Inspect bore carefully, clean and dry with compressed air. Do not hone bore if scored. Replace the assembly.

7. Coat the cylinder bore and new piston assemblies with clean brake fluid. Assemble the secondary piston, making sure the seals are facing the proper direction. Assemble the retainer and return spring and insert assembly into the cylinder bore.

8. Install the primary piston into the bore, push in and install the snapring.

9. Bench-bleed the master cylinder before installation.

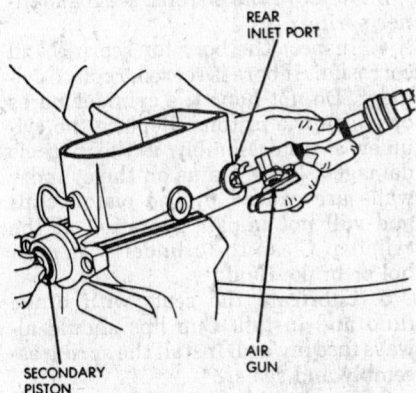

Use air pressure to remove secondary piston

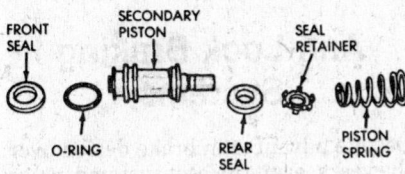

Secondary piston components

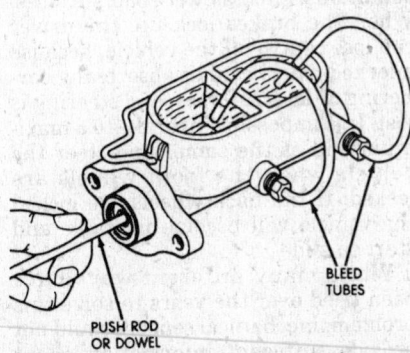

Bleeding master cylinder – typical

ALUMINUM MASTER CYLINDER – EXCEPT ANTI-LOCK

1. Remove the reservoir cover, seal and drain fluid.

2. Mount the cylinder in a vise. Clamp vise jaws on one of the mounting ears so cylinder bore not be distorted.

3. Remove reservoir with a rocking motion. Remove the seal grommets.

4. Remove the cylinder snapring.

5. Remove the primary piston and discard. It is serviced only as an assembly.

6. Apply a small amount of compressed air to secondary outlet to ease secondary piston from bore.

7. Disassemble secondary piston. Discard seals, spring and retainer.

8. Inspect the cylinder check valve. Remove the valve snapring and replace the valve if necessary. The tube seats can also be replaced. Remove by forced a hardened self-tapping screw, then pry upwards using the proper tools to remove the seat.

9. Inspect bore carefully, clean and dry with compressed air. Do not hone bore if scored. Replace the assembly.

10. Coat the cylinder bore and new piston assemblies with clean brake fluid. Assemble the secondary piston, making sure the seals are facing the proper direction. Assemble the retainer and return spring and insert assembly into the cylinder bore.

11. Install the primary piston into the bore, push in and install the snapring.

12. Install new reservoir grommets into the cylinder body.

13. Turn reservoir upside down and place on flat surface. Press the master cylinder onto the reservoir with a rocking motion. Be sure it is fully seated.

14. Bench bleed the master cylinder before installation.

Brake Booster Service

Jeep models covered are equipped with vacuum boosters, except those equipped with the anti-lock brake system. Anti-lock master cylinder/power brake booster units are not repairable and must be replaced as a complete assembly.

Under normal operating conditions, with brakes released, a vacuum suspended booster operates with vacuum on both sides of its diaphragm (or both diaphragms in a tandem diaphragm unit). When the brakes are applied, air at atmospheric pressure is admitted to one side of the diaphragm (or both diaphragms in a tandem diaphragm unit) to provide the power assist.

Brake boosters used by Jeep are not repairable and if defective, the booster

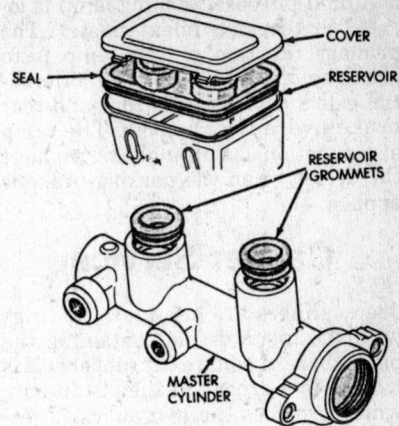

Aluminum master cylinder

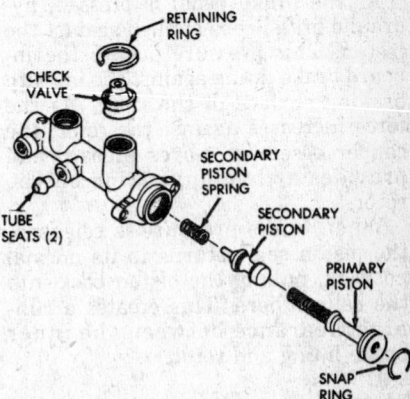

Aluminum master cylinder components

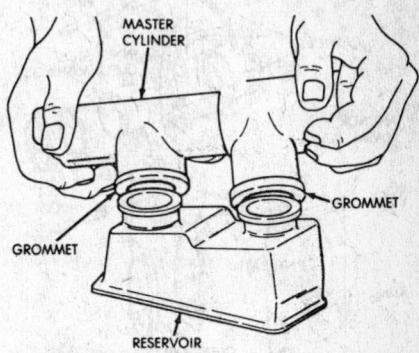

Installing the reservoir to the master cylinder body

must be replaced as a unit. The brake boosters used on different models of Jeeps are different sizes and are not interchangeable.

Combination Valve

All models have a hydraulic system control valve in the brake system. A combination proportioning valve/pressure differential switch is used on all models. Comanche series pickup trucks are equipped with a rear brake height sensing proportioning valve. The valve adjusts front-rear brake proportioning to maintain brake balance when the vehicle is loaded.

A dual purpose warning lamp is located in the instrument cluster. The primary function of this lamp is to alert the driver if a pressure differential exists between the front and rear brake hydraulic systems. The lamp also functions as an indicator to alert the driver when the parking brake is applied.

Caliper Service

Jeep calipers are one-piece castings with the inboard side containing the piston bore. A square cut rubber seal is located in a groove in the piston bore which provides the hydraulic seal between the piston and the cylinder wall.

As the brake pedal is pressed, hydraulic pressure is applied against the piston. This pressure pushes the inboard brake lining against the inboard braking surface of the rotor. As the force increases against the rotor, the caliper assembly moves inboard and provides a clamping action on the rotor.

When brake pressure is released, the piston seal returns to its normal position, pulling the piston back into the caliper bore. This creates a running clearance between the inner brake lining and rotor.

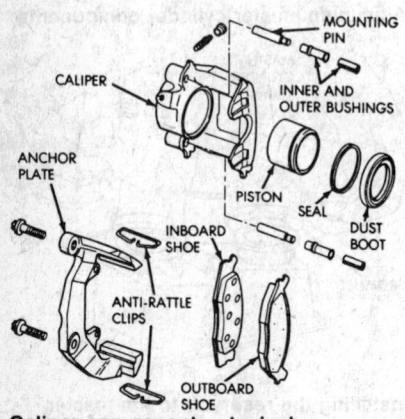

Caliper components – typical

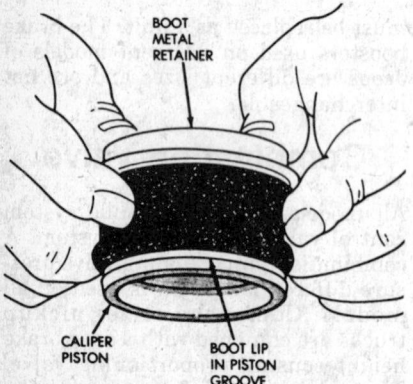

Install dust boot on piston, then fold into place

Disassembly and Assembly

1. Place shop towels under the caliper piston to protect it during removal. Apply a little compressed air to slowly ease the piston out of the bore.

2. Remove the dust boot and, using a small, pointed wooden stick, work the piston seal out of its groove in the caliper bore. Do not use a screwdriver or metal tool which might nick or scratch the bore. Discard the seal.

3. Clean all parts well using alcohol and blow dry. Inspect the piston bore for scoring or pitting. Install a new piston if it is pitted, scored or the plating is worn. Jeep does not recommend honing calipers or using any type of abrasives on the basis that abrasives will ruin the piston plating and cause it to corrode and bind. Replace the piston if damaged in any way.

4. Clean all parts well, dip the new piston seal in brake fluid and install in the bore. Position the seal at one area at a time and, using fingers, gently work the seal into the groove.

5. Slide the metal retainer part of the new dust boot over the open end of the piston. Pull the retainer rearward until the boot lip seats in the groove at the end of the piston. Push the metal retainer part of the boot forward until flush with the rim at the open end of the piston. Then snap boot folds in place. Finally, install the piston in the caliper bore with a twisting motion being careful not to unseal the piston seal.

6. Seal the metal retainer part of the dust boot in the caliper with a circular driver.

7. Install new mounting bushings in the caliper as required. Install bleeder screw if removed.

8. Install the flexible brake hose with new seals.

Wheel Cylinder Service

The rear brakes are drum type with internal expanding shoes. The rear brakes are of the single anchor type, mounted to the same anchor and actuated by one wheel cylinder. The wheel cylinder has two pistons. Brake adjustments are automatic and are made during reverse brake applications.

Wheel cylinders may need reconditioning or replacement whenever the brake shoes are replaced or when required to correct a leak condition. Leaks which coat the boot and the cylinder with fluid, or result in a dropped reservoir fluid level, or dampen and stain the brake linings are dangerous. Such leaks can cause the brakes to grab or fail and should be immediately corrected. A leakage, not immediately apparent, can be detected by pulling back the cylinder boot. A small amount of fluid seepage dampening the interior of the boot is normal. However, a dripping boot is not. Unless other conditions causing a brake to pull, grab or drag become obvious, the wheel cylinder is suspect and should and should be included in general reconditioning.

Cylinder binding may be caused by rust, deposits, grime, or swollen cups due to fluid contamination, or by a cup wedged into an excessive piston clearance.

Hydraulic system parts should not be allowed to come into contact with oil or grease, neither should those be handled with greasy hands. Even a trace of any petroleum based product is sufficient to cause damage to the rubber parts.

Disassembly and Assembly

1. Remove the brake bleeder screw.
2. Remove the dust boots, allow any brake fluid to drain out.
3. Remove the pistons, seals and inner spring.
4. Inspect the bore for scoring and corrosion. If bore is scored, replace cylinder. Do not hone the cylinder bores or polish the pistons. Replace the cylinder as an assembly if the bore is damaged. Black stains on the cylinder walls are caused by the piston cups and will not impair operation of the cylinder. Clean the cylinder with alcohol or brake fluid.
5. Lubricate the seals with brake fluid and install. Cup lips should always face inward. Install the spring assembly and seals.
6. Carefully install the pistons and dust boots.
7. Install the bleeder screw.

Anti-Lock Braking Systems

Despite advances in brake design over the years, even the best systems in use can still lock up during certain road conditions, such as wet road surfaces. When the brakes lock up, the driver can lose control of the vehicle, because a locked wheel cannot absorb any cornering or lateral forces, and steering is lost. It is impossible to brake to a maximum and at the same time steer the vehicle when the front wheels are locked. If the back wheels are locked the vehicle will become unstable and start to slide.

While many different ways have been tried over the years to solve this problem, mechanical sensors could not provide sufficient information about

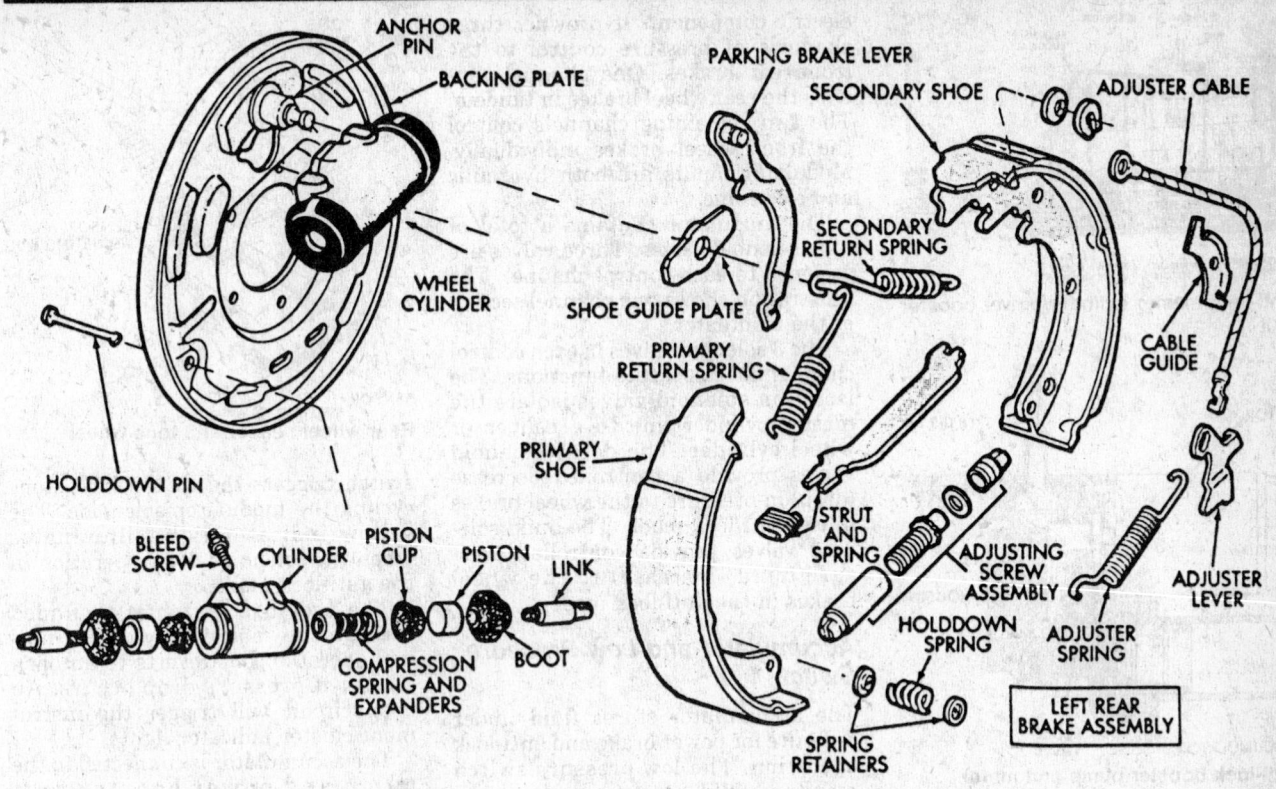

Rear drum brakes—Comanchee, Cherokee, Wrangler shown—typical

wheel rotation speed and mechanical control units could not operate the brakes fast enough to prevent brake lockup.

The growth of the electronics industry has allowed small computers (microprocessors) to be reduced in both size and cost. Coupled with fast reacting electronic sensors, anti-lock braking has become more reliable with widespread application.

Jeep's antilock brake system is available on 70 Series (Cherokee/Wagoneer Sport trucks) models with Select-Track four wheel drive. It is an electronically operated, power assisted, all wheel brake control system. The system is designed to retard wheel lockup during periods of high wheel slip when braking. Retarding wheel lockup is accomplished by modulating fluid pressures to the wheel brake units.

The 70 Series anti-lock system is a 3 channel design. The front wheel brakes are controlled individually and the rear wheel brakes in tandem.

System pressure is modulated according to wheel speed, degree of wheel slip and rate of deceleration. A sensor at each wheel converts wheel speed into electronic signals. The signals are transmitted to the brake system control unit for processing and determination of deceleration rate and wheel slip.

Basic system components include wheel sensors, fluid level and pressure switches, a pressure modulator, an accumulator, an electric booster pump, a master cylinder/power boost unit and an electronic control unit. Two instrument cluster indicator lights (one red, one yellow) are used to signal system condition and operating status.

The anti-lock electronic control system is separate from other electrical circuits in the vehicle. A specially programmed ECU is used for operational control.

The accumulator tank and the small accumulator on the booster pump both contain high pressure gas charges which assist in maintaining boost pressure. Do not puncture or attempt to disassemble either of these components at any time.

When servicing the anti-lock system, keep system components clean. Do not allow any dirt or foreign material to enter the system. Clean the reservoir cap and exterior thoroughly before removing the cap to add fluid. Dirt or foreign material in the system could result in poor brake system performance and possible component failure.

The manufacturer recommends Mopar brake fluid or equivalent meeting DOT 3 standards only. Never use reclaimed fluid or fluid from an open container that has been allowed to stand for any length of time.

CAUTION

The normal working pressure of the anti-lock boost system is 1650–2050 psi. System pressure must be pumped down before any pressure lines are loosened or disconnected. Failure to do so could result in personal injury. To reduce system pressure, turn the ignition key OFF. They apply the brakes 45–50 times (until pedal is firm) to reduce fluid pressure in the accumulator, booster, pump and lines. Wear safety goggles when disconnecting fluid lines.

MAJOR COMPONENTS

Master Cylinder/Power Booster Unit

The master cylinder and power booster pistons are located in a single, cast aluminum, cylinder body. A fluid reservoir is attached to the body with rubber seals.

The fluid reservoir is internally separated into 3 sections by bulkheads. A common fluid fill is used for the 3 sections.

The power booster is a demand type component. Power assist occurs only when the brakes are applied. Power assist is from high pressure brake fluid supplied by the electric motor driven booster pump. The pump is connected to the system accumulator. The accumulator is connected to the booster unit in the master cylinder.

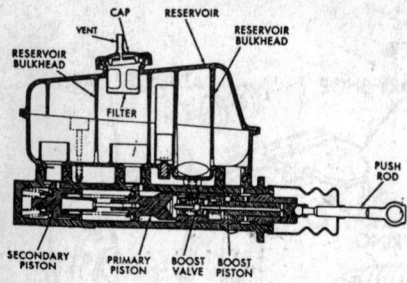

Anti-lock master cylinder/power booster unit

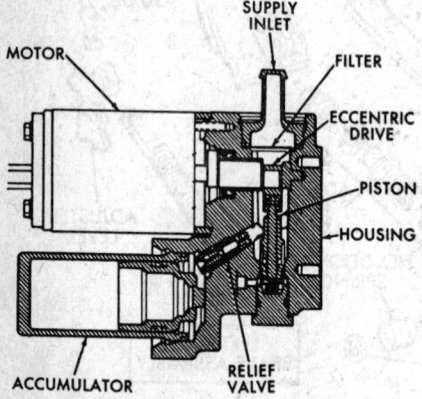

Anti-lock booster pump and motor assembly

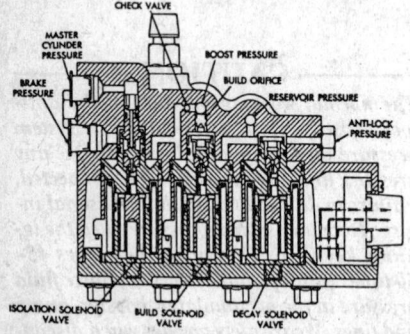

Pressure modulator channel – typical

Pump and Motor Assembly

The booster pump is powered by an electric motor. The motor and pump are combined in a common housing. The pump piston operates off an eccentric drive. An internal relief valve and a pressure switch control pump output. The housing contains a small accumulator which operates in tandem with the main accumulator.

The pump supplies fluid boost pressure for both standard and anti-lock brake operation. Pump operating pressure range is 1650–2050 psi. The pump motor is equipped with a thermal fuse which stops the pump if operating temperatures approach overheat range. The fuse does not reset once it is tripped.

Pressure Modulator

The pressure modulator is a hydro-electric component. It provides three channels of pressure control to the front/rear brakes. One channel controls the rear wheel brakes in tandem. The two remaining channels control the front wheel brakes individually. Modulator inputs are both hydraulic and electronic.

The modulator contains a total of nine solenoid valves. Three valves are assigned to each control channel. The illustration shows one channel section of the modulator.

The 3 solenoid valves in each control channel have separate functions. The isolation solenoid valves isolate the master cylinder line to a caliper or wheel cylinder. The decay solenoid valves provide a controlled decrease (drop) in pressure to the wheel brakes in the anti-lock mode. The build solenoid valves provide controlled pressure build (increase) to the wheel brakes in the anti-lock mode.

Accumulator and Low Pressure Switch

The accumulator stores fluid under pressure for power brake and anti-lock operation. The low pressure switch monitors fluid pressure and is connected to the ECU.

If pressure falls below a minimum value of approximately 1050 psi, the

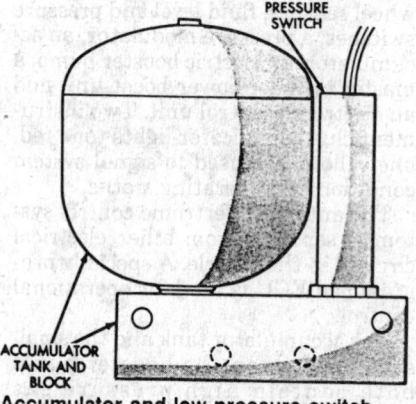

Accumulator and low pressure switch

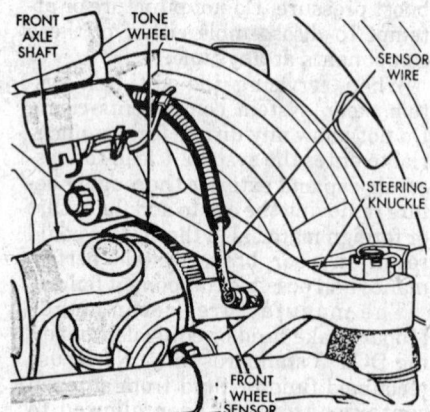

Front wheel sensor and tone wheel

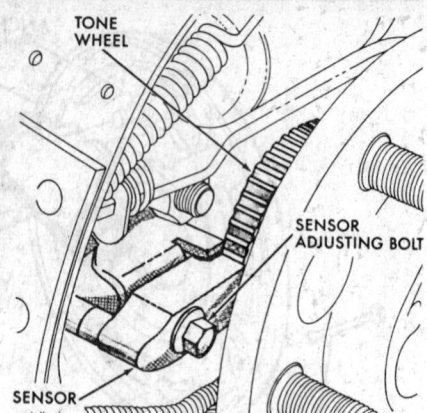

Rear wheel sensor and tone wheel

switch triggers the ECU which stops cycling the modulator solenoids. The yellow indicator light illuminates when the solenoids cease operation in the anti-lock mode.

The pressure switch is grounded through the vehicle body during normal operation but reverts to an open circuit if pressure drop occurs. An open circuit will trigger the instrument cluster indicator lights.

The accumulator is connected to the pump and power booster unit respectively.

Wheel Sensors

A sensor is used at each wheel. The sensors convert wheel speed into an electronic signal which is transmitted to the anti-lock ECU. A toothed-type tone wheel serves as the trigger mechanism for each sensor. The tone wheels are mounted at the outboard ends of the front and rear axle shafts.

Boost Pressure Differential Switch

The boost pressure differential switch is mounted in the pressure modulator. The switch checks the pressure differential between modulated boost pres-

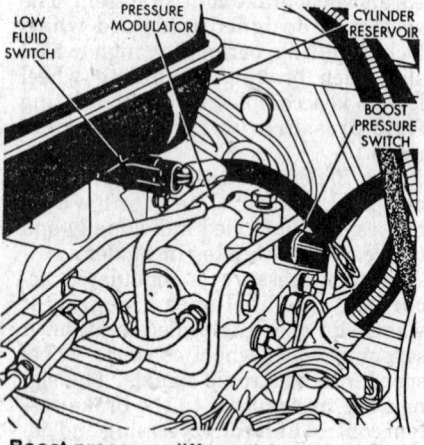

Boost pressure differential and low fluid switch locations

sure and the master cylinder primary system pressure.

The switch is in circuit with the ECU and instrument panel indicator lights. The switch is open when pressure differential is normal. The switch will ground (through the vehicle body) if a pressure differential problem is detected. Once grounded, the switch signals the ECU to illuminate the indicator lights.

Fluid Level Switch

A fluid level switch is located in the master cylinder reservoir. The switch activates the red indicator light if the fluid level falls below the required level. The yellow light also comes on if the vehicle speed is above approximately 2.5 mph.

Electronic Control Unit

A separate electronic control unit (ECU) is used to monitor and control the entire anti-lock system. The ECU is attached to a bracket located under the rear seat. The power up voltage source for the ECU is through the ignition switch in the ON or RUN position.

The anti-lock ECU is separate from the other vehicle electronic systems. It contains a self-diagnostic program

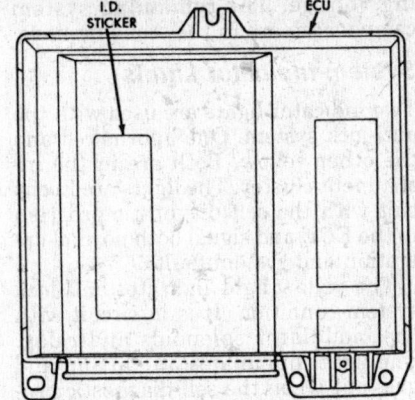

Anti-lock electronic control unit

which triggers the indicator lights when a system fault is detected. Faults are stored in a diagnostic program memory. Faults remain in memory until cleared. However, if the battery is disconnected, stored faults are erased.

The ECU is also equipped with a mercury switch. The switch monitors the degree of vehicle deceleration to determine what type of surface the vehicle is on. The switch provides input to the ECU for improved operation in the 4WD mode on low traction (slippery) road surfaces.

NOTE: Proper mounting angle of the ECU is critical to correct and accurate operation of the mercury switch.

System Relays

There are 3 system relays. The yellow indicator light and modulator power relays are located on the driver side inner fender panel. The relay wires are in the engine compartment wire harness.

The pump/motor relay is part of the pump motor harness and is located at the passenger side of the engine compartment.

The modulator power relay is connected to the pressure modulator solenoids and the ECU. When the system is powered up, the ECU supplies the operating voltage (12 volts) to the solenoids through the relay.

The indicator light relay is connected to the modulator solenoid relay and the indicator light. The relay turns the yellow light on when the modulator power relay is off. The yellow light is turned off when the modulator power relay is energized.

The pump motor relay starts/stops the pump motor when signaled by the pump switch.

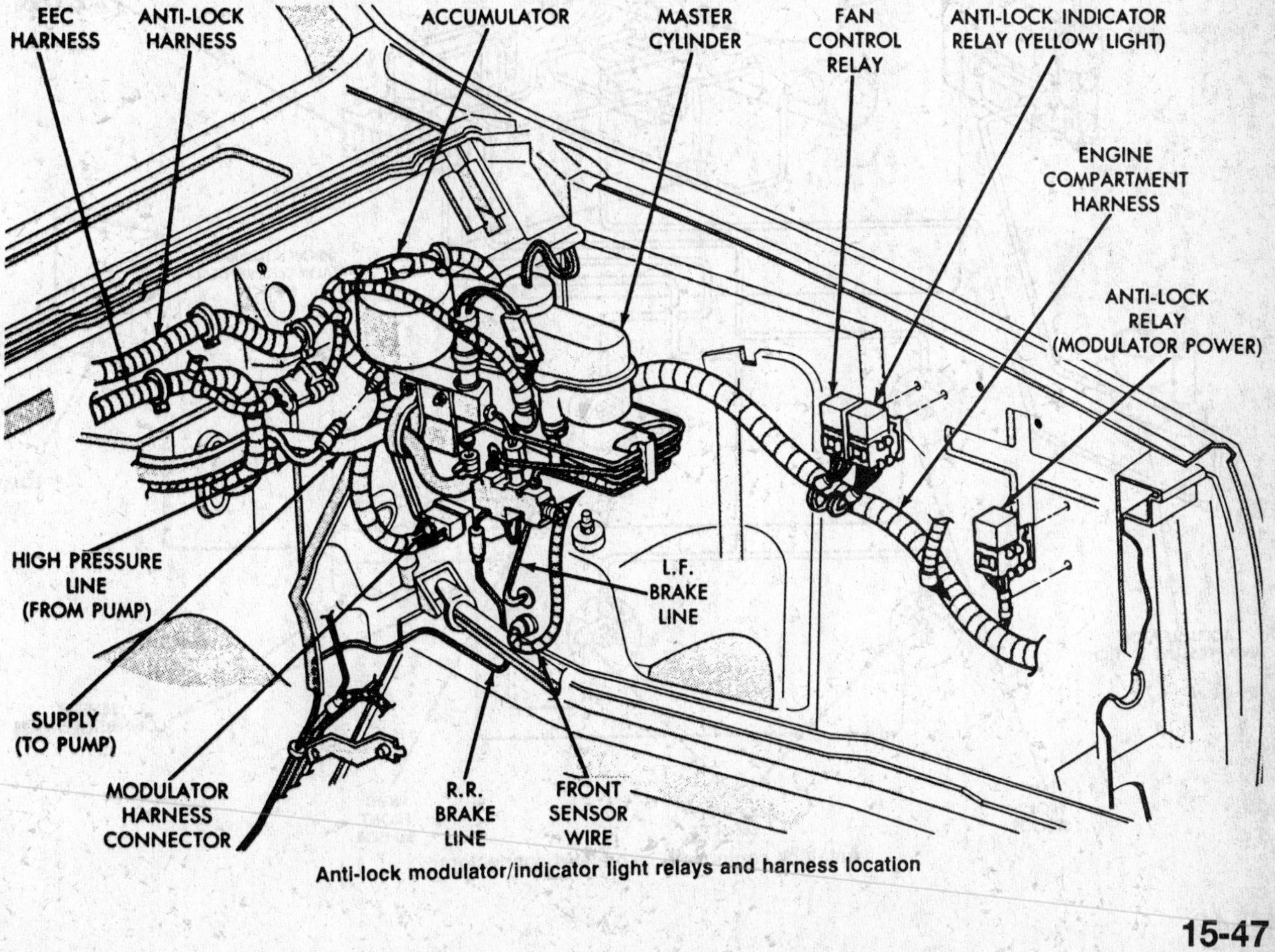

Anti-lock modulator/indicator light relays and harness location

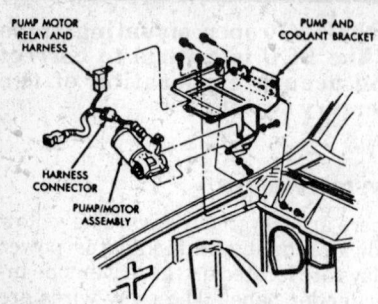

Pump motor relay and harness

Ignition Switch

The anti-lock ECU and indicator lights are in the standby mode with the ignition switch in the **OFF** or **AC-CESSORY** position. No operating voltage is supplied.

In the **ON** and **RUN** positions, the switch supplies the ECU, pump motor and indicator lights with a 12 volt power supply.

In the **START** position, only the indicator lights are supplied with operating voltage. The remaining system components are in the standby mode.

System Indicator Lights

Two indicator lights are used with the anti-lock system. One light is red and the other yellow. Both are in the instrument cluster. The lights are in circuit with the self diagnostic program in the ECU and signal both normal operation and system faults.

The yellow light indicates anti-lock system condition. It is in circuit with the modulator solenoids and relay. The light illuminates at start-up and goes out when the self-diagnostic program determines system operation is normal.

If a fault occurs, the yellow light remains on until the fault is either corrected, the battery is disconnected, or the ignition switch is cycled (turned **OFF** and then **ON**). Cycling the ignition switch may not turn the light off after some faults.

The yellow light illuminates in tandem with the red warning light to indicate certain types of faults. The pressure modulator solenoids are in circuit with the yellow indicator light. When the yellow light is on (system fault occurred), the solenoids are disabled.

The red light serves as the system warning light (low fluid, parking brake on, system pressure differential, etc.). The light illuminates in tandem with the yellow anti-lock light when certain system faults occur.

There are time delays built into indicator light illumination. These delays are provided as a means of identifying some system faults.

Proportioning Valve

The combination front/rear brake pressure switch and proportioning valve is connected between the master cylinder and modulator. The switch and valve operate normally in the standard braking mode. In the anti-lock mode, the proportioning valve is

BOOSTER PUMP AND MOTOR

MASTER CYLINDER/POWER BOOSTER

BOOST PRESSURE SWITCH (IN MODULATOR)

PRESSURE MODULATOR

PROPORTIONING VALVE/DIFFERENTIAL SWITCH

ECU

ACCUMULATOR AND PRESSURE SWITCH

SENSOR CONNECTORS/WIRES

LEFT FRONT SENSOR

REAR WHEEL SENSORS

RIGHT FRONT SENSOR

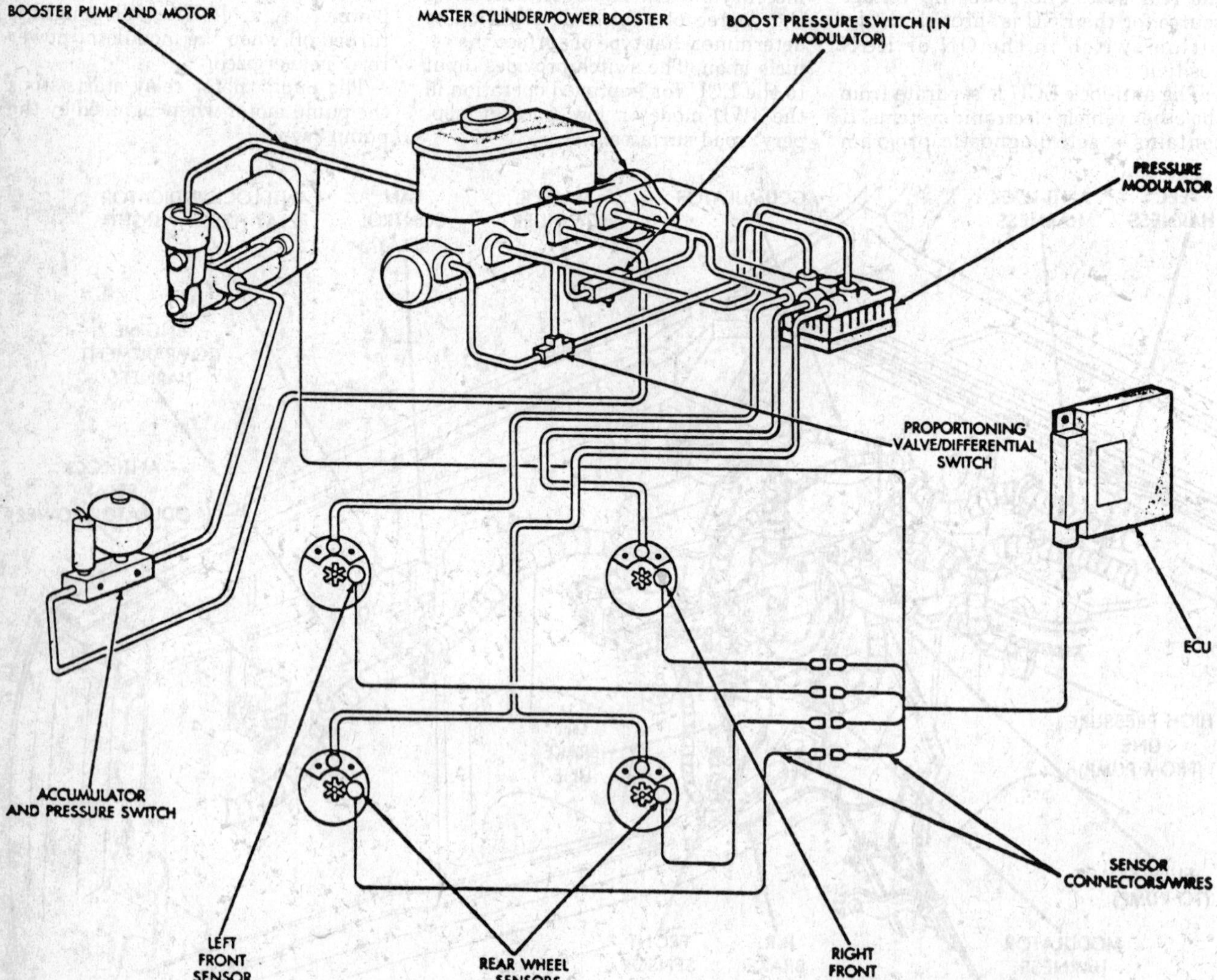

Anti-lock component layout and connectors

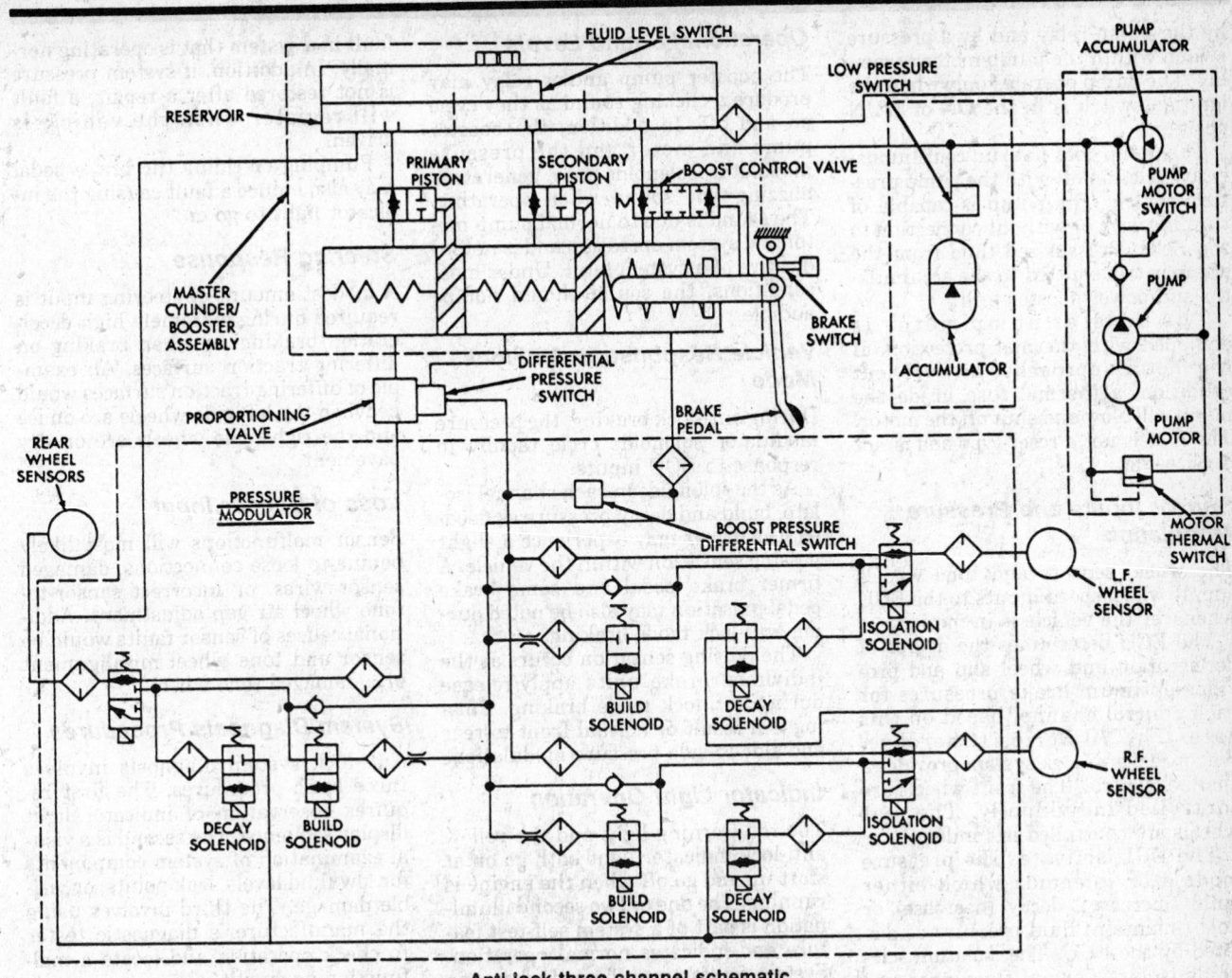

Anti-lock three channel schematic

isolated to enable brake pressure modulation during anti-lock system operation. The pressure differential switch is only activated by a difference in pressure between the front and rear (primary and secondary) brake circuits.

Parking Brake Switch

The switch is connected to the ECU low fluid circuit. When the switch is activated, the red indicator light illuminates. If vehicle speed is above approximately 2.5 mph, the yellow light will also illuminate.

SYSTEM OPERATION

Component Connections

The booster pump and accumulator provide the fluid pressure needed for power assist. The accumulator is connected to the pump by a high pressure feed line. A second high pressure line connects the accumulator to the booster section of the master cylinder. A low

pressure supply line connects the reservoir to the booster pump.

Fluid from the master cylinder is channeled to the calipers and wheel cylinders through the pressure modulator. The 3 solenoid valves (isolation—decay—build) in each modulator control channel are contained within the modulator body.

The wheel sensors are connected directly to the ECU. The sensor triggering devices (tone wheels), are mounted on the axle shafts.

System Power Up

The anti-lock system is in standby mode with the ignition switch in **OFF** or **ACCESSORY** position. When the switch is moved to the **START** position, voltage through the switch activates the indicator lights only. The ECU and other system components are still the standby mode.

The indicator lights illuminate as part of the self test feature and remain on until the switch is in the **RUN** position.

In the **ON** and **RUN** position, 12 volts are supplied through the ignition

switch to power up the ECU and system components.

When the vehicle is motionless (no wheel speed inputs) and the ignition switch is in the **ON** or **RUN** position, the modulator solenoids are activated and briefly exercised. This serves two purposes: It checks solenoid function as part of the self diagnostic feature and ensures proper solenoid operation after periods of inactivity.

Boost Pressure and Fluid Supply

The system main fluid supply is contained in the master cylinder reservoir and the accumulator. Additional fluid is also contained in the booster pump accumulator. The pump and main accumulator provide the reserve fluid pressure needed for power brake assist.

Fluid stored in the accumulator is at normal working pressure of 1650–2050 psi. The accumulator contains enough pressurized fluid for 25–30 power assisted brake applications if a pump fault should occur.

Pump motor operation is controlled

by the pump relay and by a pressure switch within the pump/motor assembly. The pump operates only when the ignition switch is in the **ON** or **RUN** position.

The pump does not run continuously. It cycles on/off with the pump pressure switch. The pump is capable of running with or without connection to the ECU. Pressurized fluid from the pump is transmitted to the accumulator and power booster unit.

The booster pump motor is equipped with thermal protection. If the motor approaches an overheat condition, a thermal fuse inside the pump will blow and shut off the motor. The fuse is not a reset-type and is not serviceable.

Sensor Inputs and Pressure Modulation

The wheel sensors and tone wheels supply wheel speed inputs to the ECU whenever the vehicle is in motion.

The ECU determines the degree of deceleration and wheel slip and provides optimum brake pressures for each control channel based on this data. The 70 Series (Cherokee/Wagoneer) anti-lock system provides 3 channel control. The front wheels are controlled individually. The rear wheels are controlled in tandem.

The ECU activates the pressure modulator solenoids which either build (increase), decay (decrease) or hold (maintain) fluid pressure as dictated by the ECU. The isolation solenoids isolate normal fluid pressure from the master cylinder to the rear wheels or to the left or right front wheel as needed. Wheel brake isolation occurs prior to build/decay solenoid operation.

Solenoid operation is not entirely constant in the anti-lock mode. Operation occurs in brief, rapid cycles front-to-rear and side-to-side. Rapid changes in input data will produce equally rapid changes in pressure modulation. The isolation, decay, and build functions are capable of function changes and cycle times measured in milliseconds.

SERVICE DIAGNOSIS

Wheel/Tire Size and Input Signals

Anti-lock system operation is dependent on signals from the wheel sensors. The vehicle wheels and tires must all be the same size and type in order to generate accurate signals. Variations in wheel/tire size will produce inaccurate input signals resulting in incorrect pressure modulation.

Operational Sound Levels

The booster pump and/or relay may produce a clicking sound as they cycle on and off. In addition, the booster pump and motor and the pressure modulator solenoids may generate a buzzing-type sound when operating. The sound is due to normal pump motor and system operation and is not indicative of a system fault. Under most conditions, the sound should not be audible.

Vehicle Response in Anti-lock Mode

During anti-lock braking, the pressure modulator solenoids cycle rapidly in response to ECU inputs.

As the solenoids in each channel isolate, build and decay pressure as needed, the driver may experience a slight pulsing sensation within the vehicle. A firmer brake pedal and some brake pedal pulsation may also be noted during anti-lock mode braking.

The pulsing sensation occurs as the individual brake units apply/release during anti-lock mode braking. Pulsing is a result of normal front-to-rear and side-to-side pressure modulation.

Indicator Light Operation

The red warning light and the yellow anti-lock indicator light both go on at start-up and go off when the engine is running. The one or two second illumination is part of a system self-test feature and indicates normal operation. System faults are indicated when one or both of the lights illuminate after initial start-up system check.

Driver Induced System Faults

Some driving or operational situations can induce faults in an anti-lock system that is actually operating correctly.

Induced faults are not true faults; a component malfunction as not actually occurred. Instead, they are a result of driver actions recognized by the diagnostic program as improper operation.

Improper parking brake use can induce faults in the self-diagnostic program. If a driver attempts to move the vehicle with the parking brakes applied, a system fault will register. Or, if the parking brake is applied while the vehicle is in motion, a system fault will also register. One or both indicator lights will illuminate in either situation. The red light illuminates for parking brake faults. The yellow light illuminates if a fault is sensed.

Faults can be induced in the system through excessive wheel spin. Wheel spin due to low traction surfaces or high speed acceleration can induce a

fault in a system that is operating normally. In addition, if system pressure is not restored after a repair, a fault will register when the vehicle is driven.

Pumping or riding the brake pedal may also induce a fault causing the indicator light to go on.

Steering Response

A modest amount of steering input is required during extremely high deceleration braking, or when braking on differing traction surfaces. An example of differing traction surfaces would be when the left side wheels are on ice and the right side wheels are on dry pavement.

Loss of Sensor Input

Sensor malfunctions will most likely be due to loose connections, damaged sensor wires, or incorrect sensor-to-tone wheel air gap adjustment. Additional causes of sensor faults would be sensor and tone wheel misalignment or a damaged tone wheel.

System Diagnosis Procedures

Anti-lock system diagnosis involves three basic procedures. The first requires observation of indicator light display sequence. The second is a visual examination of system components for low fluid levels, leak points, or visible damage. The third involves using the manufacturer's diagnostic tester to check operation and locate a malfunctioning circuit.

The two indicator lights will illuminate separately, simultaneously, or with varying delays depending on the fault. The lights signal low fluid levels, pressure drops and other hydro-electrical faults. The service diagnosis charts for indicator light display can be used when a fault is detected.

COMPONENT SERVICEABILITY

The master cylinder/power booster unit, pressure modulator, accumulator, pump/motor and proportioning valve are not repairable components. If diagnosis indicates a malfunction has occurred, these components are to be replaced as a complete assembly.

The fluid level switch in the master cylinder and the boost pressure switch in the modulator are also not serviceable. The switches cannot be removed from their respective components.

The electrical harnesses, pump bracket, pump high pressure and supply hose, system relays and wheel sensors can be serviced individually.

The tone wheels are permanently attached to the axle shafts and are not

SERVICE DIAGNOSIS

SYSTEM FAULT	POSSIBLE CAUSE	INDICATOR LIGHT DISPLAY
LOW FLUID	SYSTEM LEAK. ACCUMULATOR CHARGE LOW OR LOST.	RED LIGHT ON. YELLOW LIGHT ON WITHIN 1/2 SECOND WHEN VEHICLE SPEED EXCEEDS 2.5 MPH.
PARKING BRAKES APPLIED	PARKING BRAKES NOT RELEASED BEFORE DRIVING VEHICLE.	RED LIGHT ON. YELLOW LIGHT ON IF VEHICLE SPEED EXCEEDS 2.5 MPH.
PRESSURE DROP AT ACCUMULATOR	ACCUMLATOR GAS CHARGE LOST. SYSTEM LEAK. PUMP/MOTOR MALFUNCTION. PROLONGED STOP ON ICY SURFACE WITH TRANSMISSION IN GEAR.	YELLOW LIGHT ON. RED LIGHT WILL ALSO COME ON WITHIN 20 SECONDS.
DIFFERENTIAL PRESSURE SWITCH (IN PROPORTIONING VALVE) ACTUATED	SYSTEM LEAK. MASTER CYLINDER MALFUNCTION (SECONDARY PISTON). AIR IN SYSTEM.	RED LIGHT ON. YELLOW LIGHT COMES ON AT VEHICLE SPEED OF 3 MPH.
PRESSURE DROP AT BOOST PRESSURE SWITCH AND PRESSURE DIFFERENTIAL SWITCH	MASTER CYLINDER MALFUNCTION (PRIMARY PISTON). SYSTEM LEAK. AIR IN SYSTEM.	RED LIGHT ON. YELLOW LIGHT COMES ON AT VEHICLE SPEED OF 3 MPH.
WHEEL SENSOR FAULT (FRONT ONLY)	SENSOR-TO-TONE WHEEL SPACING INCORRECT (SPACE TOO LARGE). DAMAGED SENSOR WIRE, SENSOR, OR TONE WHEEL. SENSOR AND TONE WHEEL MISALIGNED. SENSOR DISCONNECTED.	YELLOW LIGHT ON. (AFTER 15 MPH)
WHEEL SENSOR FAULT (FRONT OR REAR ONE OR TWO MISSING SIGNALS)	DAMAGED SENSOR, WIRE, OR CONNECTOR. SENSOR DISCONNECTED. EXCESSIVE WHEEL SPIN. MISALIGNED OR DAMAGED TONE WHEEL. OPEN SENSOR OR WIRE.	YELLOW LIGHT ON AT 15 MPH IF FAULT OCCURRED BEFORE VEHICLE DRIVE-OFF. OR, ORANGE LIGHT ON AT 8 MPH IF FAULT OCCURRED AFTER VEHICLE DRIVE-OFF.
EXCESSIVE DECAY SOLENOID OPERATION	MODULATOR/SOLENOID FAULT. WHEEL SENSOR FAULT. EXTREMELY LOW AMBIENT TEMPERATURES. VEHICLE ON ICE COVERED SURFACE. TIRES HYDROPLANING ON WATER COVERED ROAD SURFACE.	YELLOW LIGHT ON WITHIN 1-2 SECONDS.
PRESSURE MODULATOR SOLENOID FAULT	SOLENOID SHORTED OR OPEN. DECAY AND BUILD SOLENOID ON AT SAME TIME. OPEN/SHORT IN MODULATOR HARNESS.	YELLOW LIGHT ON.
PUMP/MOTOR RUN-ON	EXCESSIVE RUN TIME. RELAY SHORTED, MOTOR SWITCH SHORTED.	RED LIGHT ON IF PUMP RUNS MORE THAN 4 MINUTES WITH NO BRAKE.
PUMP/MOTOR INOPERATIVE	PUMP RELAY FAULT. NO VOLTAGE TO MOTOR. DAMAGED PUMP OR MOTOR. PUMP GAS CHARGE LOST.	YELLOW LIGHT ON. RED LIGHT ON AFTER 20 SECONDS.
LOW VOLTAGE	SYSTEM VOLTAGE BELOW 9V. SHORT, OPEN IN FEED WIRES OR RELAY. FUSE BAD. POOR GROUND. LOOSE, DISCONNECTED WIRE IN SYSTEM. BATTERY LOW OR DISCHARGED	YELLOW LIGHT ON.
NO BRAKE SIGNAL	SYSTEM LEAK. MASTER CYLINDER MALFUNCTION. PUMP/MOTOR MALFUNCTION. ACCUMULATOR OR MODULATOR FAULT.	RED LIGHT ON DURING BRAKING.
RELAY FAULT	RELAY SHORTED OR OPEN.	YELLOW LIGHT ON.
ECU SELF DIAGNOSTIC FEATURE INOPERATIVE (SOLENOIDS NOT TEST-EXERCISED AT START-UP)	IGNITION SWITCH IN OFF POSITION. PARKING BRAKES ON (NOT RELEASED AT DRIVE-OFF). SYSTEM COMPONENT HAS MALFUNCTIONED. LOW FLUID LEVEL/LEAK IN SYSTEM.	YELLOW LIGHT ON.

Jeep Anti-lock diagnosis chart

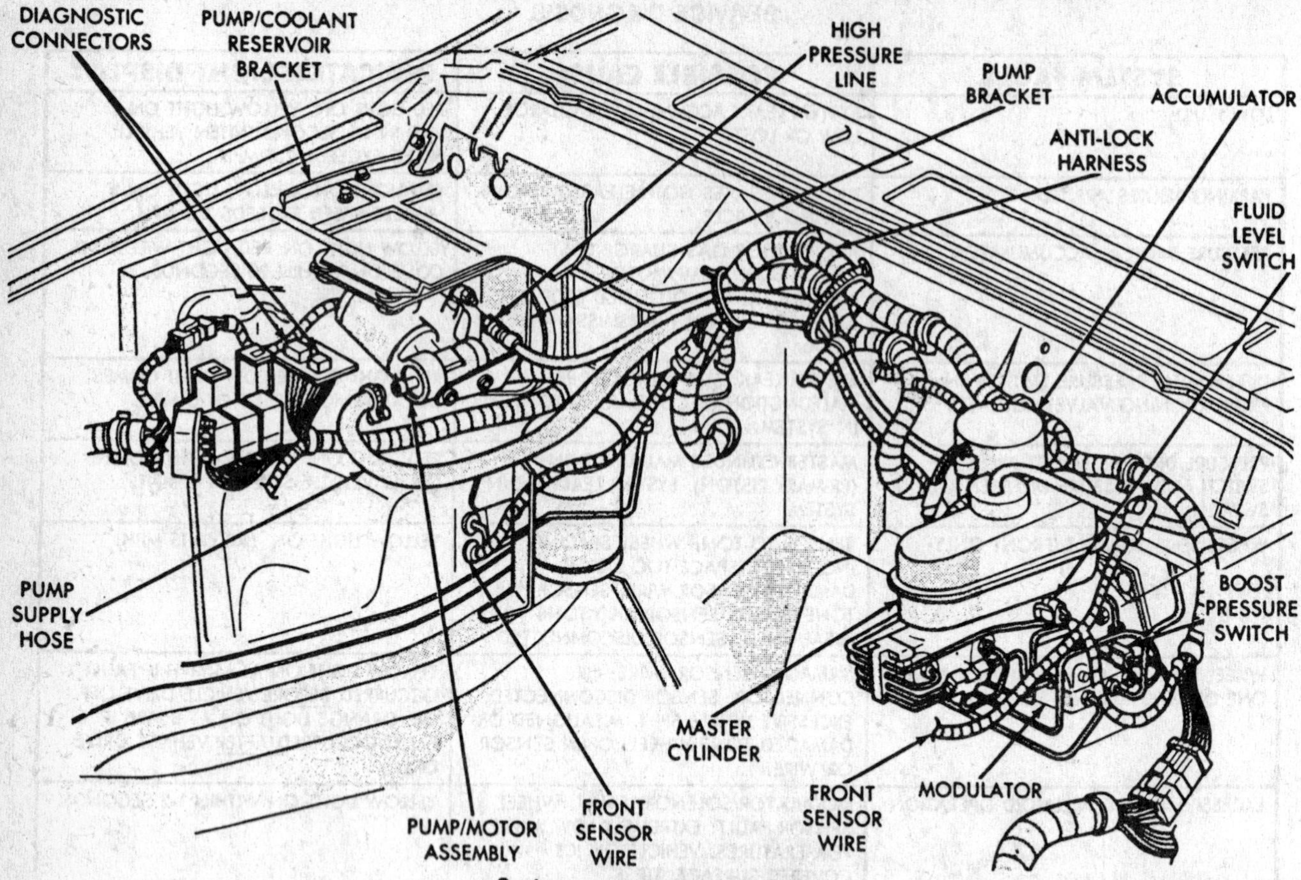

DIAGNOSTIC CONNECTORS

PUMP/COOLANT RESERVOIR BRACKET

HIGH PRESSURE LINE

PUMP BRACKET

ACCUMULATOR

ANTI-LOCK HARNESS

FLUID LEVEL SWITCH

PUMP SUPPLY HOSE

BOOST PRESSURE SWITCH

MASTER CYLINDER

FRONT SENSOR WIRE

PUMP/MOTOR ASSEMBLY

FRONT SENSOR WIRE

MODULATOR

System components and location

replaceable. If a tone wheel becomes damaged, it will be necessary to replace the tone wheel and axle shaft as an assembly.

The wheel brake components such as calipers, brakeshoes, wheel cylinders, rotors and drums are all serviced the same as standard brake system components.

Wheel Sensors

The wheel sensors have a polyethylene spacer strip attached to the sensor contact face. When installing a sensor, be very sure this spacer strip actually touches the tone wheel. The strips are made in the exact thickness needed for correct sensor-to-tone wheel spacing (air gap). If the spacing (air gap) is too great, the sensors will not transmit accurate speed signals to the ECU. If the spacer strip is missing from an original, or not provided on a replacement sensor, the correct sensor-to-tone wheel air gap will have to be established with a brass feeler gauge. A brass feeler gauge must be used to avoid disrupting sensor polarity

If the sensor strip is missing from a reuseable sensor, set the sensor-to-tone wheel air gap as follows:

Set front sensor air gap to 0.013–0.019 in.

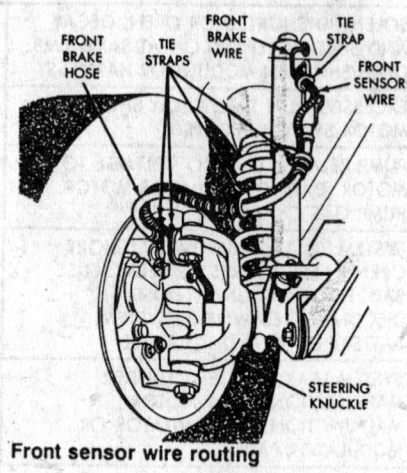

FRONT BRAKE HOSE

TIE STRAPS

FRONT BRAKE WIRE

TIE STRAP

FRONT SENSOR WIRE

STEERING KNUCKLE

Front sensor wire routing

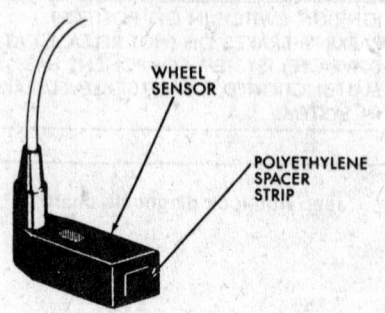

WHEEL SENSOR

POLYETHYLENE SPACER STRIP

Sensor spacer strip

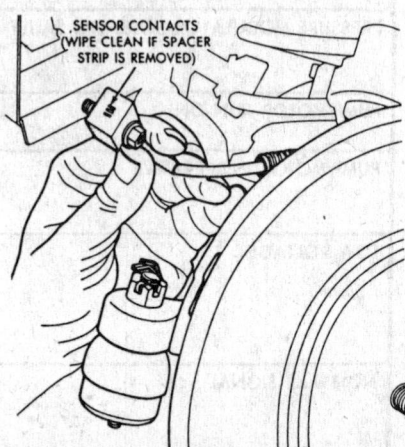

SENSOR CONTACTS (WIPE CLEAN IF SPACER STRIP IS REMOVED)

Sensor contacts (spacer strip removed)

Set rear sensor air gap to 0.030–0.036 in.

Removal and Installation

FRONT WHEEL SENSOR

1. Raise and safely support vehicle and turn wheel outward for easier access tosensor.

2. Note the sensor wire routing for installation reference. Cut the tie straps holding wire to steering knuckle and brake line.

3. If sensor is covered with heavy

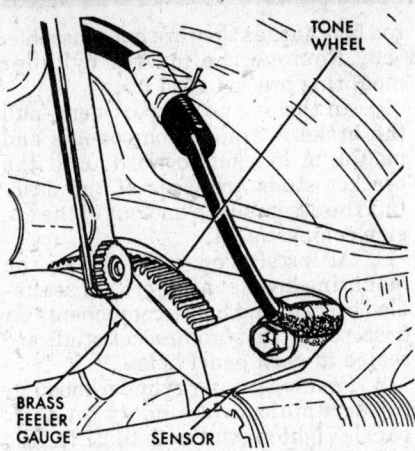

Adjusting sensor-to-wheel air gap

accumulations of dirt or mud, clean the sensor and surrounding area before proceeding. This is necessary to avoid possible damage when removing the sensor.

4. Remove the sensor attaching screw and remove the sensor from the steering knuckle.

5. Unseat the grommet holding the sensor wire in the wheel house panel.

6. In the engine compartment, disconnect the sensor wire connector at the anti-lock harness plug. Remove sensor and wire.

7. Before installing a new or reinstalling an original front wheel sensor, note the condition of the spacer strip. If the strip is securely attached and in good condition, a spacing (air gap) adjustment will not be needed. However, if the strip is loose or damaged, the correct air gap will have to be established with a brass feeler gauge.

8. To install a new sensor, route the wire through the wheelhouse grommet hole and connect the sensor to the harness plug. Seat the grommet.

9. Position the sensor on the steering knuckle and install the bolt finger-tight.

10. If the sensor strip was in good condition and securely attached, lightly press the sensor against the tone wheel and tighten the bolt to 11 ft. lbs.

11. If the sensor spacer strip is missing, loose or damaged and the sensor contacts are exposed, perform the following steps:

    a. Remove the spacer strip if completely loose or torn. Wipe sensor contacts clean with a shop towel.

    b. Set sensor-to-tone wheel air gap to 0.013 – 0.019 in. with a brass feeler gauge.

    c. Tighten the sensor bolt to 11 ft. lbs. and recheck spacing.

12. Secure the sensor wire to the brake line and steering knuckle with new tie straps.

REAR WHEEL SENSOR REMOVAL

1. Raise and fold the rear seat forward for access to the rear sensor connectors. They are located near the ECU. Separate both sensor connections.

2. Push the sensor grommets and sensor wires through the floorpan.

3. Raise and safely support the vehicle and remove the wheel and brake drum.

4. Cut and remove the tie straps securing the sensor wires to the axle and rear brake hose.

5. Unseat the sensor backing plate grommet, remove the sensor attach bolt and remove the sensor by pulling the wire through the grommet hole in the backing plate.

6. Before installing a new or reinstalling an original rear wheel sensor, note the condition of the spacer strip. If the strip is securely attached and in good condition, a spacing (air gap) adjustment will not be needed. However, if the strip is loose or damaged, the correct air gap will have to be established with a brass feeler gauge.

7. If the sensor spacer strip is missing, loose or damaged and the sensor contacts are exposed, perform the following steps:

    a. Remove the spacer strip if completely loose or torn. Wipe sensor contacts clean with a shop towel.

    b. Set sensor-to-tone wheel air gap to 0.030 – 0.036 in. with a brass feeler gauge.

    c. Tighten the sensor bolt to 11 ft. lbs. and recheck spacing.

8. Route the sensor wires to the rear seat area, feed the wires through the access holes and seat the grommets in the floorpan.

9. Secure the sensor wire with wire ties to the rear brake hose and axle. Make sure the wire is clear or rotating components.

10. Install the brake drum and wheel then lower vehicle.

11. Connect the sensors to the harness connects, reposition carpet and fold down rear seat.

BOOSTER PUMP AND MOTOR

———— CAUTION ————

The normal working pressure of the anti-lock boost system is 1650–2050 psi. System pressure must be pumped down before any pressure lines are loosened or disconnected. Failure to do so could result in personal injury. To reduce system pressure, turn the ignition key OFF. Then apply the brakes 45–50 times (until pedal is firm) to reduce fluid pressure in the accumulator, booster, pump and lines. Wear safety goggles when disconnecting fluid lines.

1. Turn the ignition switch to the **OFF** position and apply the brakes

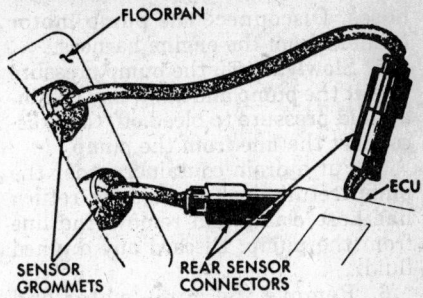

Rear sensor connectors

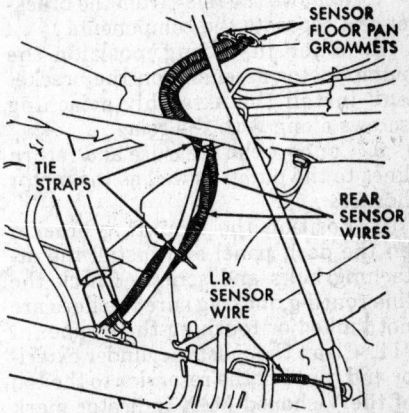

Rear sensor wire routing and attachment

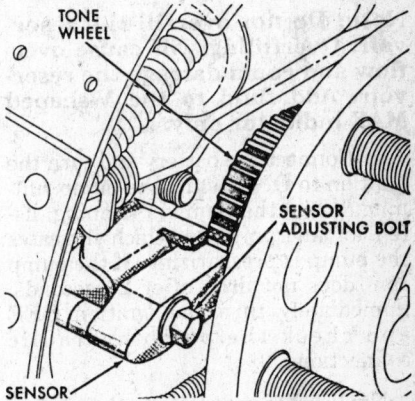

Rear sensor installation

(pump pedal) 45–50 times (until pedal is firm on initial apply) to reduce system fluid pressure.

NOTE: When the reserve pressure is depleted, the reservoir fluid level will rise above the MAX fill mark but will not overflow unless the reservoir was overfilled to begin with.

2. Disconnect the battery negative cable.

3. Remove the coolant pressure bottle retaining strap and move the reservoir aside. It is not necessary to disconnect the bottle hoses, just move the bottle aside for working clearance.

4. Remove the bolts holding the two-piece mounting bracket to the dash and inner fender panels. Rotate the bracket and pump/motor assembly to one side for access to the wires and

hoses. Disconnect the pump motor harness from the engine harness.

5. Slowly loosen the pump pressure line at the pump and allow any residual fluid pressure to bleed off, then disconnect the line from the pump.

5. Put a drain container under the pump return line, loosen the return line hose clamp and remove the line from the pump. Discard any drained fluid.

6. Remove the pump/motor and bracket as an assembly.

7. Remove the relay from the bracket and separate the components.

8. When installing, position the pump/motor assembly on the bracket and install the assembly attaching screws along with the relay.

9. Connect the pressure and return lines to the pump as well as the motor harness.

10. Position the mounting bracket on the dash panel and install the attaching bolts and screws. Check the line routing, making sure the lines are not kinked or touching the engine.

11. Clean the master cylinder exterior and cap. Fill the reservoir to the top of the V-shaped MAX indicator mark with DOT 3 rated brake fluid.

Note: Do not overfill the reservoir. Overfilling will cause overflow and could damage the reservoir. Add fluid to the V-shaped MAX indicator only.

12. Connect the battery and turn the ignition to **ON** to start the pump running. While the pump is running, listen for an rpm drop which indicates the pump is pressurizing. If the pump rpm does not drop after 20 seconds, immediately turn the ignition **OFF** and check the pump hydraulic connections.

NOTE: Do not allow the pump to run if it does not pressurize. If the pump rpm does not drop after 20 seconds run time, turn the ignition OFF immediately to avoid pump damage.

13. Add fluid to master cylinder reservoir if necessary.

MASTER CYLINDER

The master cylinder, modulator and accumulator are serviced as an assembly only. Do not attempt to disassemble or repair these components.

─── **CAUTION** ───

CAUTION: The normal working pressure of the anti-lock boost system is 1650–2050 psi. System pressure must be pumped down before any pressure lines are loosened or disconnected. Failure to do so could result in personal injury. To reduce system pressure, turn the ignition key OFF. Then apply

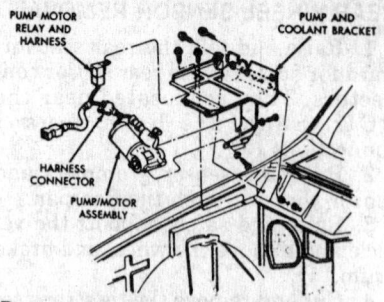

Pump motor harness and relay

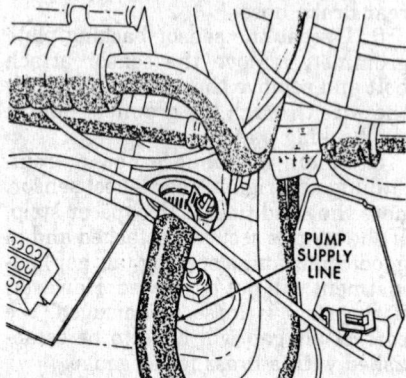

Pump pressure and supply line connections

the brakes 45–50 times (until pedal is firm) to reduce fluid pressure in the accumulator, booster, pump and lines. Wear safety goggles when disconnecting fluid lines.

1. Pump system pressure down by turning the ignition **OFF** and applying the brake pedal 45–50 times until the pedal becomes firm.

2. Disconnect the battery. Remove the windshield washer fluid reservoir attaching screrws, disconnect the hoses and wires and remove the reservoir.

3. Remove the air cleaner assembly, disconnect the ECU harness wire connectors at the pressure modulator and disconnect the wires at the proportioning valve differential switch.

4. Disconnect the high pressure line at the accumulator block. Cap the line to keep out dirt.

5. Disconnect the supply line at the reservoir. Cap the line to keep out dirt. Discard any fluid that drains from the line.

6. Disconnect the low pressure switch wires, and the wires at the modulator boost pressure and fluid level switches.

7. Disconnect the front brakelines at the outboard side of the pressure modulator.

8. In the passenger compartment, remove the instrument panel lower trim cover for access to the brake pedal. Disconnect the brakelight switch wires, remove the master cylinder pushrod bolt and disconnect the pushrod from the pedal. Discard the push-

rod bolt nuts as they are not reuseable.

9. Remove the master cylinder mounting bracket stud nuts.

10. In the engine compartment, pull the brake hydraulic components and mounting bracket forward until the bracket studs are clear of the dash. Lift the assembly up and out of the engine compartment.

11. At installation, place pad on the mounting bracket and position assembled bracket and brake components on dash panel. Be sure bracket studs are seated in dash panel holes.

12. Position master cylinder/modulator/accumulator assembly on dash panel. Tighten stud nuts to 27 ft. lbs.

13. Connect the harness wires to the modulator, low pressure switch, differential switch and the low fluid and boost pressure switches.

14. In the passenger compartment, install the nuts of the mounting studs and torque to 31 ft. lbs.

15. Align the brake pedal, brakelight switch, master cylinder push rod and install the pushrod bolt.

NOTE: The pushrod bolt must be installed correctly to avoid interference with the dash bracket. The bolt must be installed with the bolt head at the left side of the pedal.

16. Install new nuts on the push rod bolt. Tighten the inner lock nut to 25 ft. lbs. and the outer jam nut to only 75 in. lbs. Connect the brake light switch wires and install the lower trim panel.

17. In the engine compartment, connect the brake lines to the proportioning valve and connect the pressure and return lines to the accumulator.

18. Install the air cleaner assembly, connect the hoses and wires to the washer reservoir and install the reservoir to the fender panel.

19. Clean the master cylinder exterior and cap. Fill the reservoir to the top of the V-shaped MAX indicator mark with DOT 3 rated brake fluid.

20. Connect the battery and turn the ignition to ON to start the pump running. While the pump is running, listen for an rpm drop which indicates the pump is pressurizing. If the pump rpm does not drop after 20 seconds, immediately turn the ignition **OFF** and check the pump hydraulic connections.

NOTE: Do not allow the pump to run if it does not pressurize. If the pump rpm does not drop after 20 seconds run time, turn the ignition OFF immediately to avoid pump damage.

21. Add fluid to master cylinder reservoir if necessary.
22. Bleed the brake system.

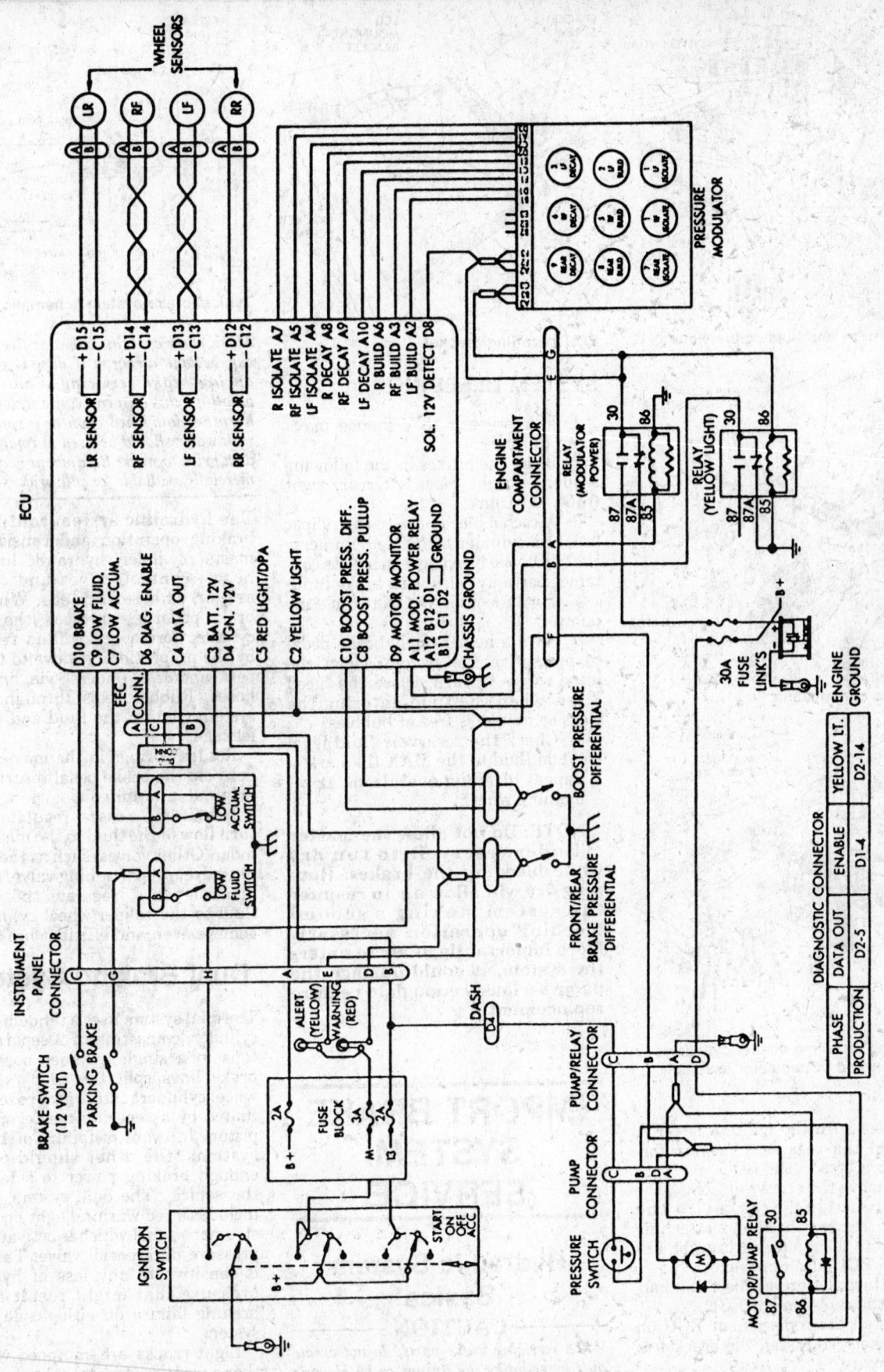

Anti-lock ECU wiring diagram

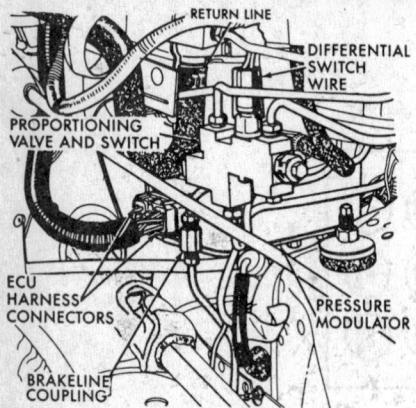

ECU harness and fluid connections

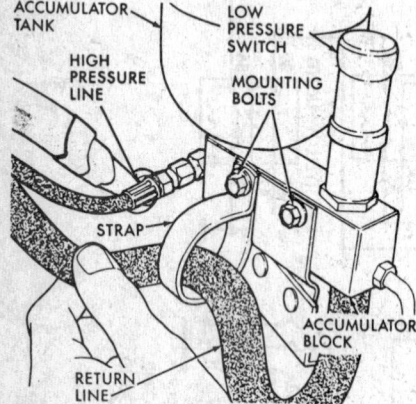

Fluid line connections

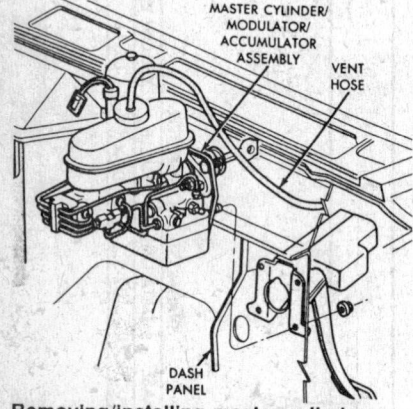

Removing/installing master cylinder, modulator and accumulator assembly

ECU

1. Make sure the ignition is **OFF**, then fold the rear seat cushion forward for access to the ECU.
2. Remove the screws attaching the ECU mounting bracket to the floorpan, remove the screws attaching the ECU to the mounting bracket, unplug the ECU and remove.
3. At installation, connect the harness of the replacement ECU.
4. Install the replacement ECU on the bracket and install the bracket to the floorpan.
5. Fold down rear seat.

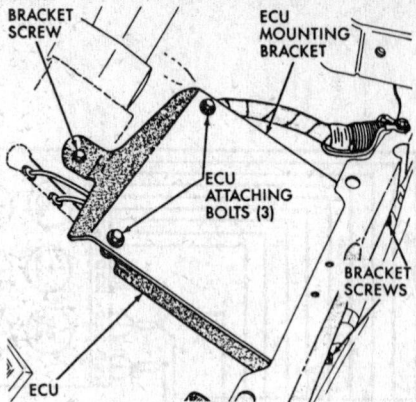

ECU mounting bracket

SYSTEM BLEEDING

1. Fill reservoir to V-shaped maximum fill mark.
2. Bleed the brakes in the following sequence: right rear, left rear, right front, left front.
3. Attach a bleed hose to the caliper or wheel cylinder being bled, immerse the end of the bleed hose in a glass container partially filled with brake fluid.
4. Turn the ignition **ON** to cycle the pump.
5. Have a helper apply brake pedal to pressurize the system. Open the bleed screw ½ turn. Close the bleed screw when the fluid entering the glass container is free of bubbles.
6. Check the reservoir fluid level and add fluid to the **MAX** fill mark.
7. Repeat bleeding operations at remaining wheels.

NOTE: Do not allow the master cylinder reservoir to run dry while bleeding the brakes. Running dry will allow air to re-enter the system making a second bleeding operation necessary. More importantly, if air re-enters the system, it could damage the pump seriously enough to require replacement.

IMPORT BRAKE SYSTEM SERVICE

Hydraulic System Basics
—— CAUTION ——

When servicing brake parts, do not create dust by grinding the linings or by blowing them clean with compressed air. Many

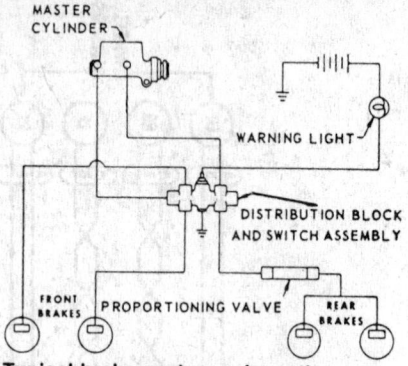

Typical brake system schematic

brake parts contains asbestos fibers which can become airborne if dust is generated during brake servicing. Continuously breathing this dust can cause serious bodily harm. A dampened cloth or a spray bottle with water should be used to remove brake dust prior to work. Equipment is also commercially available for safe brake servicing.

The hydraulic system controls the braking operation and consists of a master cylinder, hydraulic lines and hoses, control valves and calipers and/or wheel cylinders. When the brake pedal is depressed, the master cylinder forces brake fluid regulated by the proportioning valve to the calipers and/or cylinders, via lines and hoses. Rubber seals throughout the system contain the fluid and prevent leakage.

Return springs in the master cylinder help the brake pedal return to the original unapplied position. Check valves (in most cases) regulate the return flow of the fluid to the master cylinder. Other valves, such as the metering valve, proportioning valve, or combination valve, regulate the flow of fluid to the caliper/wheel cylinder, to achieve even and efficient braking.

Dual Braking Systems

The dual system uses a tandem master cylinder, consisting of 2 separate pistons in a single cylinder bore. Dual brake lines split the calipers and/or wheel cylinders into 2 groups, each actuated by a separate master cylinder piston. In event of failure of 1 of the systems, the other should provide enough braking power to safely stop the vehicle. The dual system usually includes a red warning light on the instrument panel which is activated by a pressure differential valve. The valve is sensitive to any loss of hydraulic pressure that might result from a braking failure on either side of the system.

Light trucks are equipped with either a front/rear wheel split or a diagonally split system. On front/rear systems, the front wheels are connected to one circuit while the rear wheels are connected to the other circuit. Diagonally split systems have diagonally opposite wheels connected to each circuit.

Brake Fluid

Approved DOT3 heavy duty type brake fluid retains the correct consistency throughout the widest range of temperature variation, will not affect rubber cups or seals, helps protect the metal parts of the brake system against failure and has a high boiling point to assure long trouble free brake operation when properly maintained.

Never use brake fluid from a container that has been used for any other liquid. Mineral oil, alcohol, antifreeze, or cleaning solvents, even in very small quantities, will contaminate brake fluid. Contaminated brake fluid will cause rubber parts within the hydraulic system to swell or deteriorate. Always return the cap to the fluid container after using the brake fluid. This will prevent the fluid from absorbing moisture from the surrounding air, which will lower its boiling point. Do not use brake fluid that has been sitting in an unopened container for an extending amount of time.

Hydraulic Line Repair

Steel tubing is used for the hydraulic lines between any 2 parts that do not move independant of one another, like the master cylinder and the front brake tube connector, or along the axle tubes between the brake tee and the rear brake cylinders. Brake hoses provide a flexible connection between the tube and a part that moves independantly, like the hose that connects the brake tube to the front brake calipers or the tube that runs along the body of the vehicle conneted by hose down to the axle.

BRAKE TUBING

If a section of the brake tube becomes damaged, the entire section should be replaced with tubing of the same type, size, shape, and length. Copper tubing should not be used in the hydraulic system. When bending brake tubing to fit the frame or axle contours, be careful not to kink or crack the tube.

All brake tubing should be double flared to provide good leak proof connections. Always clean the inside of a new brake tube with clean isopropyl alcohol before installing.

BRAKE HOSE

All flexible brake hoses should be carefully inspected often. They should be replaced if they show any signs of softening, cracking, swelling or other damage.

When installing a new brake hose, make sure it is not twisted and is positioned to avoid contact with other vehicle components under any suspension condition.

Hydraulic Control Valves

PRESSURE DIFFERENTIAL VALVE

The valve activates a panel warning lamp in event of pressure loss failure. As pressure fails in one split system, the other system's normal pressure causes a piston in the switch to compress a spring and move until an electrical circuit is completed lighting the dash lamp. Normally, the spring balanced piston automatically recenters when the brake pedal is released, thus flashing the warning lamp only during brake application and will re-center automatically after repairs are successfully completed and the system is properly bled.

METERING VALVE

Often used on vehicles equipped with front disc and rear drum brakes, the metering valve improves braking balance during light brake applications by preventing application of the front disc brakes unitl the pressure to the rear brakes overcomes the tension of the rear brake shoe return springs. Thus, when the front brake pads contact the rotor the rear brakes shoes move outward to contact the brake drum at approximately the same time.

The metering valve should be inspected whenever the brakes are serviced. A slight amount of moisture inside the boot does not indicate a defective valve, however a great deal of fluid indicates a faulty valve and replacement is recommended. Make sure to install the brake lines in the correct ports when installing a new valve, crossed lines will cause the hydraulic system to malfunction.

PROPORTIONING VALVE

The proportioning valve is used to transmit full inpout pressure to the rear brakes up to a certain point (the split point). Beyond the split point, it reduces the amount of pressure to the rear brakes according to a certain ratio, which is built into the valve. On light pedal applications, equal braking pressure is transmitted to to the front and rear brakes. During heavier brake applications, however, the pessure delivered to the rear brakes is lower than that at the front brakes to prevent rear wheel lockup and skidding.

Whenever the brakes are serviced, the valve should be inspected. To check valve operation, install hydraulic gauges ahead of and behind the valve and determine that it has an operative transition point above which rear brake pressure is proportioned. If the valve is leaking replacement is required.

COMBINATION VALVE

A valve combining two or three functions (metering, proportioning, and/or brake warning) may be used. The combination valve is usually mounted under the hood close to the master cylinder, where the brake lines can be easily routed to the front and rear wheels. The combination valve is a non-serviceable unit, and if found to be malfunctioning, must be replaced as a unit.

Disc Brakes

CALIPERS

Caliper disc brakes can be divided into 3 types: dual piston floating caliper, single piston floating caliper and single piston sliding caliper. On the floating caliper types, the inner pad is hydraulically pushed into contact with the disc, while the reaction force generated is used to pull the outer pad into contact with the other side of the disc, made possible by allowing the caliper to move slightly along the axle's center line. Two pistons may be used to accomodate more severe braking requirements on heavier vehicles. All disc brake systems are self-adjusting and have no provision for amnual adjustment.

With the sliding caliper type, the caliper assembly slides along the smooth and lubricated surface of a key or stopper (nomenclature varies between manufacturers) which is held stationary by cotter pins or retaining bolts. The caliper is held against the key or stopper with 1 or 2 support springs.

One of the recent developments in brake equipment materials includes the use of aluminum calipers for the purpose of saving weight. A vast array of anti-rattle springs, clips and anti-squeal pads and shims are used by manufacturers to prevent the pads from making any undesirable noises.

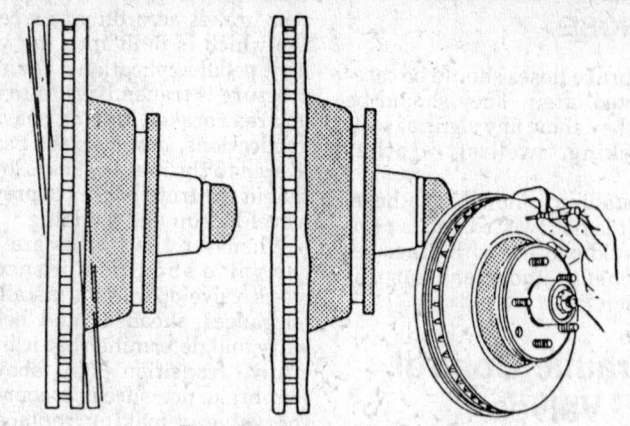

Excessive runout of parallelism

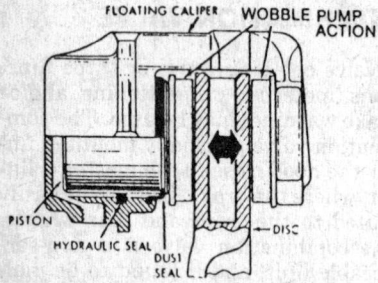

Wobble pump action resulting from a warped rotor

In addition, most brake pads are comprised of semi-metallic materials for improved stopping power. Unfortunately, these pads often produce an annoying brake squeal as they are used, but the squeal itself should not reduce the stopping ability of the pads. If the pad kit includes plates or shims, use them when servicing the disc brakes. Aftermarket spreads and sprays are also available to quiet noisy pads. It is always a good idea to replace calipers in pairs.

ROTORS

Rotors provide the surface upon which the brake pads stop the vehicle. It is important to inspect the rotor carefully. Manufacturers differ some on permissable runout (warpage) of the rotor, but excessive runout can usually be felt as a pulsation at the brake pedal regardless of the specification. A wobble pump effect is created when the rotor is not perfectly flat and the pad hits a high spot forcing fluid back into the master cylinder. This alternating pressure causes a pulsation which is felt at the pedal when the brakes are applied. An excessive amount of runout can cause the pads to be out of adjustment because they cannot hold themselves against the rotor at all times. It is always a good idea to replace rotors in pairs.

To check the actual runout of the ro-

tor, fist tighten the spindle nut to remove all endplay. Fasten a dial indicator on the suspension on a convenient location so the pin contacts the rotor surface about 1 in. from its outer edge. Set the dial at 0 and check the runout while turning the rotor 1 full revolution. If the rotor is warpe beyond limits, the rotor should be replace. Rotors the are not warped beyond the specification can be resurfaced, as long as the final thickness is not below the minumum thickness requirement of the rotor.

Lateral Runout: The wobbly movement of the rotor from side to side as it rotates. Excessive lateral runout causes the rotor faces to knock bask the pads and can cause brake chatter, shudder or vibration.

Parallelism (lack of): This refers to the amount of variation in the thickness of the rotor. Excessive variation of parallelism can cause a pulsating pedal, front end vibrations or grabbing condiions; a condition comparable to an out-of-round brake drum. Check parallelism with a micrometer. Measure the thickness of the rotor a 8 or more equally spaced locations on the rotor, equally distant from the outer edge, preferably at the mid-point of the braking surface. A lack of parallelism results if the difference between the highest and lowest measurement is beyond the parallelism specification.

Surface or Micro-inch finish, flatness and smoothness: These terms all refer to the degree of perfection of the flat surfaces on each side fo the rotor. Visually inspecting the rotor, the machined surface should have a fine ground polish with a swirling pattern to reduce brake squeal.

DISC BRAKE PADS

The brake pads are the tools that the vehicle uses to stop, so they should be inspected very carefully. Pads should be replaced only in sets (both sides of

the same axle) if any of the pads is damaged, cracked, burnt, or separated from the plate that it is glued or riveted to. Replace the pads if any of their lining is worn down to $\frac{1}{16}$ in. thickness. If the lining is allowed to wear beyond this point, severe and costly damage to the rotor may result. Note that individual state inspection guidelines take precedence over these general recommendation.

Floating caliper type disc brake pads may wear at an angle; the measurement should always be taken at the narrowest end of the taper. Tapered lining should be replaced if the amount of tapering exceeds $\frac{1}{8}$ in. from end to end and the cause of the taper should be investigated and repaired, if possible.

To prevent potentially costly paint damage, remove some brake fluid from the master fluid reservoir and install the reservoir cover before servicing the pads. When installing new pads, the caliper piston is forced back into its bore and fluid is pushed back into the master cylinder. The fluid could spill and remove paint from any painted surface that it comes in contact with.

When the caliper is removed from its mouning adaptor, do not allow it to dangle by the brake hose. This can damage the hose and create a dangerous situation. Always rest the caliper on a suspensiion member of suspent it from the frame with a wire or rope.

Drum Brakes

WHEEL CYLINDERS

The wheel cylinder performs in response to the master cylinder. It receives fluid from the hydraulic tube through its inlet port. As the pressure increases, the wheel cylinder cups and pistons are forced apart. As a result, the hydraulic pressure is converted into mechanical force acting on the brake shoes. The variation in wheel cylinder size (diameter) is one of the factors controlling the distribution of braking force in a vehicle. Bleeder screws are provided to remove air or vapor trapped in the system.

Wheel cylinders may need reconditioning or replacement whenever the brake shoes are replaced or when required to correct a leak condition. On many designs, the wheel cylinders can be diassembled without removing them from the backing plate. On some designs, however, the cylinder is mounted in an indention in the backing plate or a cylinder piston stop is welded to the backing plate. When servicing brakes of this type, the cylinder must be removed from the backing plate before being disassembled.

Leaks which coat the boot and the cylinder with fluid, or result in a dropped reservoir fluid level, or dampen and stain the brake linings are dangerous. Such leaks can cause the brakes to grab or fail and should be immediately corrected. A leakage, not immediately apparent, can be detected by pulling back the cylinder boot. A small amount of fluid seepage dampening the interior of the boot is normal, however a dripping boot is not. Unless other conditions causing a brake to pull, grab, or drag becomes obvious, the wheel cylinder is a suspect and should be included in general reconditioning. It is always a good idea to service wheel cylinders in pairs.

Cylinder binding may be caused by rust, deposits, grime, or swollen cups due to fluid contamination, or by a cup wedged into an excessive piston clearance. If the clearance between the pistons and the bore wall exceeds allowable values, a condition called heel drag may exist. It can result in rapid cup wear and can cause the pistons to retract very slowly when the brakes are released. A ring of a hard, crystal like substance is sometimes noticed in the cylinder bore where the piston stops after the brakes are released.

BRAKE DRUMS

The condition of the brake drum surface is equally as important as the surface of the lining. All drum braking surfaces should be clean, smooth, free of hark spots, heat damage, scoring and forign material embedded in the braking surface. The drum should not be out-of-round, bell mouthed or barrel shaped. Drums should be checked with a drum micrometer before resurfacing to see if it within over-size limits. If the drum is within safe limints, it should be machined to true the drum surface and to remove any contamination in the surface from previous linings and road matter. Too much metal removed from the drum is unsafe and may result in brake fade due to the thin drum not being able to absorb the heat generated, vibration from ensuing drum distortion and generally unsafe conditions.

Brake drum runout should not exceed 0.005 in. Drums machined to more than 0.060 in. oversize are unsafe and should be replaced. It is always good practice to replace drums on both whels at the same time to ensure even braking.

If the drums are in good condition, smooth up any slight scoring by sanding with coarse sand paper, then polishing with emery cloth. If deep scores or grooves are present, machine the drums to restore them to good operating condition.

DRUM BRAKE SHOES

The brake shoes should also be inspected very carefully. Shoes should be replaced only in sets (both sides of the same axle) if any of the shoes are bent, damaged, cracked, burnt, or separated from the plate that it is glued or riveted to. Replace the shoes if any of their lining is worn down to $1/16$ in. thickness. If the lining is allowed to wear beyond this point, severe and costly damage to the drum may result. Note that individual state inspection guidelines take precedence over these general recommendation.

Rear brakes with weak return springs may create a bell-mouthed condition which causes the shoes to wear at an angle; if so the measurement should always be taken at the narrowest end of the taper. Tapered linings should be replaced if the amount of tapering is excessive and the cause of the taper should be investigated and repaired, if possible.

Disassemble and assemble one side at a time to prevent improper assembly. This will also provide a model that can be refered if the installation doen not look right. Always complete the brake shoe adjustment with the wheels installed and adjust the parking brake cable last.

STAR AND SCREW ADJUSTER

Star and screw self-adjusters are used on most late-model vehicles. This system requires manual adjustment only when the shoes have been replaced, when the star wheel has been disturbed or when the star wheel is not operating properly. As the star wheel is turned, it expands or contracts the shoes accordingly.

In most cases, the brakes can be initially adjusted by removing the drum, measuring the internal diameter of the drum and setting the shoes to slightly less than that measurement. Normally, the brakes will self-adjust when the vehicle is backed up and the brakes are firmly applied.

CHRYSLER IMPORTS/ MITSUBISHI

NOTE: When cleaning brake system components, use only brake fluid or denatured isopropyl alcohol. Never use a mineral- based solvent such as gasoline or paint thinner; these fluids will leave a residue that may swell and deteriorate rubber parts within the system. All alcohol must be removed from the system when the work is done because alcohol mixed with brake fluid lowers its boiling point.

Do not hone any aluminum bores. There is a hard anodized coating on the aluminum which will be removed if honed. This will allow the aluminum to wear quickly and will damage the component.

Master Cylinder Service

Disassembly and Assembly

1. Remove the brake fluid reservoir retainer screw, if equipped and remove the reservoir and seals or remove the reservoir unions and seals.
2. Push the primary piston in and remove the stopper bolt and gasket.
3. Hold the pistons in and remove the snapring.

NOTE: Do not disassemble either piston assembly as the individual parts are not serviceable.

4. Let the primary piston spring out and remove the primary piston assembly.
5. Remove the secondary piston assembly. This may be accomplished by carefully blowing low pressure compressed air through the secondary side outlet port if it does not come out easily.
6. Remove the brake line connector block and washers.
To assemble:
7. Coat the master cylinder bore and secondary and primary piston seals liberally with brake fluid.
8. Install both assemblies to the master cylinder.
9. Push the pistons in and install a new snapring.
10. Hold the pistons in and install the stopper bolt with a new gasket.
11. Install the brake fluid reservoir or unions with new seals. Install the reservoir retaining screw, if equipped.
12. Install the connector block with new washers.

Brake Booster Service

The brake boosters installed on these vehicles are not serviceable. If the unit is defective, replace the entire assembly.

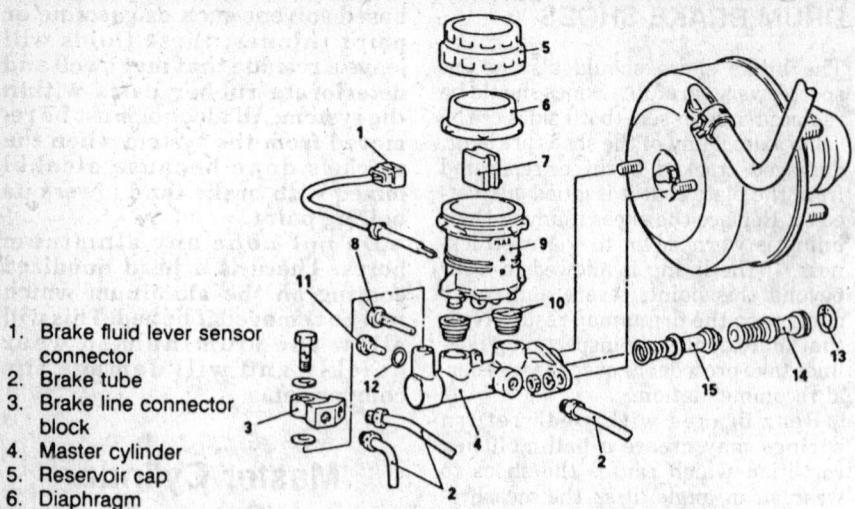

1. Brake fluid lever sensor connector
2. Brake tube
3. Brake line connector block
4. Master cylinder
5. Reservoir cap
6. Diaphragm
7. Float
8. Reservoir retainer screw
9. Reservoir
10. Reservoir seals
11. Piston stopper bolt
12. Gasket
13. Snapring
14. Primary piston assembly
15. Secondary assembly

Chrysler/Mitsubishi master cylinder

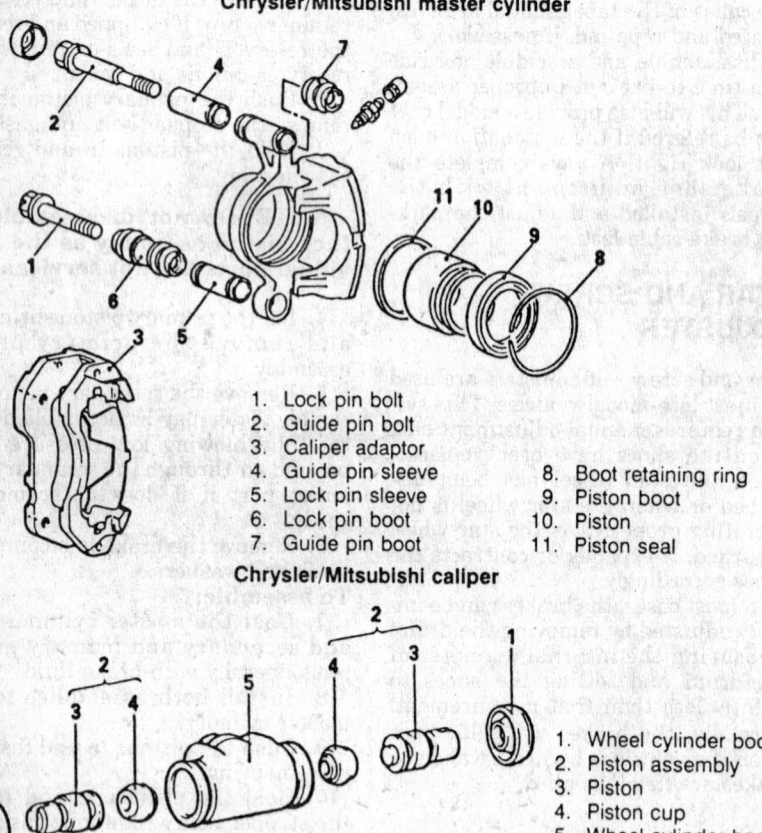

1. Lock pin bolt
2. Guide pin bolt
3. Caliper adaptor
4. Guide pin sleeve
5. Lock pin sleeve
6. Lock pin boot
7. Guide pin boot
8. Boot retaining ring
9. Piston boot
10. Piston
11. Piston seal

Chrysler/Mitsubishi caliper

1. Wheel cylinder boot
2. Piston assembly
3. Piston
4. Piston cup
5. Wheel cylinder body

Chrysler/Mitsubishi wheel cylinder

Caliper Service

Disassembly and Assembly

1. Remove the bleeder screw.
2. Drain the brake fluid from the caliper.
3. Remove the dust boot ring and the dust boot.
4. Position a shop rag opposite the top of the piston. Apply low pressure compressed air to the fluid inlet and carefully blow the piston out of the bore.

— CAUTION —

Do not put fingers where the piston will land when it is blown out of the caliper. The force of the piston can crush fingers and cause personal injury.

5. Remove the piston seal from inside the bore.

To assemble:

6. Blow out all fluid passages with low pressure compressed air.
7. Slight corrosion or rust can be removed with commutator paper or crocus cloth. Replace pistons that are pitted, scored, peeling or otherwise damaged.
8. Lubricate the caliper bore with brake fluid. Lubricate the seal and install it in the groove.
9. Lubricate the piston and install it to the bore until it bottoms. If the piston will not go in by hand, it is probably cocked. Remove it and reinstall.
10. Apply lubricant to the dust boot mounting groove and install the boot. Install the retaining ring.
11. Apply rust penetrant to the bleeder screw threads and install to the caliper. Tighten it until it is just snug against its seat.

Wheel Cylinder Service

Disassembly and Assembly

1. Remove the bleeder screw, if equipped.
2. Remove both rubber boots.
3. Remove the piston assemblies, each consisting of the piston and cup.
4. Remove the piston cups from the pistons, being careful not to scratch the piston.
5. Remove the return spring.
6. Hone the cylinder bore if it is not damaged, to provide a smooth sealing surface. Wash and dry the bore after honing.

To assemble:

7. Lubricate the cylinder bore with brake fluid.
8. Lubricate the new piston cups and install them to the pistons, with their lips facing upward. Install the assemblies to the cylinder.
9. Apply lubricant to the boot mounting groove and install the boot.
10. Apply rust penetrant to the bleeder screw threads and install to the wheel cylinder. Tighten it until it is just snug against its seat.

ISUZU

NOTE: When cleaning brake system components, use only brake fluid or denatured isopropyl alcohol. Never use a mineral-based solvent such as gasoline or paint thinner; these fluids will leave a residue that may swell and deteriorate rubber parts within

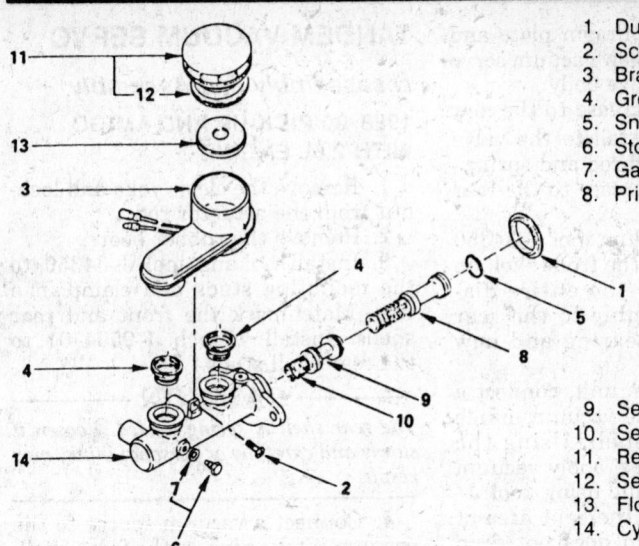

1. Dust seal
2. Screw
3. Brake fluid reservoir
4. Grommet
5. Snapring
6. Stopper bolt
7. Gasket
8. Primary piston assembly

9. Secondary assembly
10. Secondary piston spring
11. Reservoir cover
12. Seal
13. Float magnet
14. Cylinder body

Isuzu master cylinder

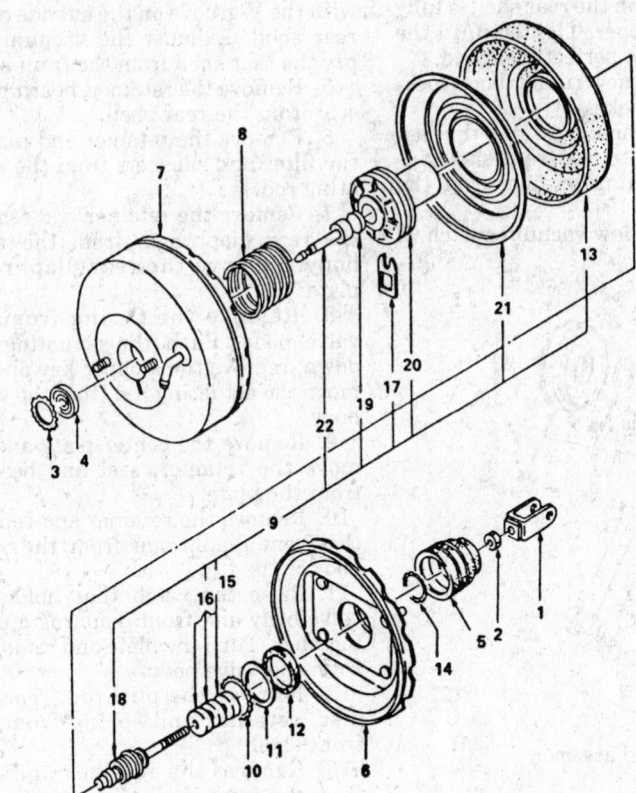

1. Clevis yoke
2. Locknut
3. Servo retainer
4. Seal
5. Rubber boot
6. Rear shell assembly
7. Front shell assembly
8. Servo spring
9. Diaphragm plate assembly
10. Servo retainer
11. Servo bearing

12. Valve body seal
13. Vacuum servo diaphragm
14. Servo retainer
15. Vacuum filter
16. Vacuum silencer
17. Actuating rod stopper key
18. Actuating rod
19. Reaction disc
20. Valve body
21. Diaphragm plate
22. Pushrod assembly

Isuzu single vacuum servo booster

the system. All alcohol must be removed from the system when the work is done because alcohol mixed with brake fluid lowers its boiling point.

Do not hone any aluminum bores. There is a hard anodized coating on the aluminum which will be removed if honed. This will allow the aluminum to wear quickly and will damage the component.

Master Cylinder Service

Disassembly and Assembly

1. Remove the brake fluid reservoir retainer screw.
2. Remove the reservoir and seals.
3. Remove the dust seal from the end of the master cylinder.
4. Push the primary piston in and remove the stopper bolt and gasket.
5. Hold the pistons in and remove the snapring.

NOTE: Do not disassemble either piston assembly as the individual parts are not serviceable.

6. Let the primary piston spring out and remove the primary piston assembly and end washer, if equipped.
7. Remove the secondary piston assembly. This may be accomplished by carefully blowing low pressure compressed air through the secondary side outlet port if it does not come out easily.

To assemble:

8. Coat the master cylinder bore and secondary and primary piston seals liberally with brake fluid. Install both assemblies to the master cylinder.
9. Push the pistons in and install a new snapring.
10. Hold the pistons in and install the stopper bolt with a new gasket.
11. Install the brake fluid reservoir with new seals. Install the reservoir retaining screw.

Brake Booster Service

SINGLE VACUUM SERVO

Disassembly and Assembly

EXCEPT 1988–90 PICK-UP AND AMIGO WITH 2.6L ENGINE AND 1988–90 TROOPER II

1. Remove the clevis yoke and locknut from the actuator rod. Remove the low vacuum switch (diesel only).
2. Pry out the servo retainer and seal from the front of the shell.
3. Remove the rubber boot.

4. Matchmark the front and rear shells. Using wrench J–9504–01 and holder J–34250, separate the 2 shells slowly. Remove the servo spring.

5. Remove the entire diaphragm plate assembly.

6. Pry out the servo retainer, bearing and valve body seal from the rear shell.

7. Using a knife, cut the vacuum servo diaphragm and remove. Make sure not to damage the valve body when cutting.

8. Remove the servo retainer from the valve body.

9. Remove the vacuum filter and silencers from the valve body.

10. Push in the actuating rod and remove the rod stopper key. Remove the actuating rod assembly.

11. Remove the reaction disc, valve body, diaphragm plate and pushrod assembly.

To assemble:

12. Install the new actuating rod assembly to the valve body. Push the rod in and install the rod stopper key.

13. Install the new vacuum silencers and filter.

14. Install the new retainer to the rear shell.

15. Install the diaphragm plate and carefully attach the new vacuum servo diaphragm to the valve body.

16. Apply silicone grease to the new reaction disc and install to the valve body. Install the pushrod and spring.

17. Apply silicone grease to the rear shell seal and install.

18. Mount the holding tool J–34350 in a vise and mount the front shell on the tool. Assemble the entire diaphragm plate assembly to the rear shell. Install the bearing and new retainer.

19. To assemble the unit, connect a vacuum source to the vacuum intake pipe on the front shell. Using the matchmarks as guides, apply vacuum and assemble the unit using tool J–9504–01. Make sure the bent area of the of the diaphragm does not drop against the front and rear shell engaging areas.

20. Rotate the rear shell clockwise until the notch on the rear shell is fully against the stopper. Tighten until the matchmarks are perfectly aligned.

21. Install the new rubber boot, locknut and clevis yoke.

22. Apply silicone grease to the last seal and install to the front shell. Install the remaining retainer over the seal.

23. Install the low vacuum switch, if equipped.

TANDEM VACUUM SERVO

Disassembly and Assembly

1988–90 PICK-UP AND AMIGO WITH 2.6L ENGINE

1. Remove the clevis yoke and locknut from the actuator rod.

2. Remove the rubber boot.

3. Install holding tool J–34350 to the mounting studs and clamp in a vise. Matchmark the front and rear shells. Install wrench J–9504–01 to the rear shell studs.

4. Connect a vacuum source to the vacuum intake pipe on the front shell. Rotate the lever counterclockwise slightly (about $\frac{1}{20}$ turn). The front shell caulking tab must be aligned with the V-groove on the outside of the rear shell. Exhaust the vacuum and pry the rear shell from the front shell.

5. Remove the retainer, bearing and seal from the rear shell.

6. Remove the retainer and remove the filter and silencers from the actuating rod.

7. Remove the retainer and remove the rear diaphragm from the valve body. Remove the rear diaphragm plate.

8. Remove the C-ring from the valve body. Push the actuating rod down, remove the stopper key and remove the actuating rod from the valve body.

9. Remove the center plate and remove the retainer, seal and bearing from the plate.

10. Remove the retainer and remove the front diaphragm from the valve body.

11. Raise the catch that holds the valve body and front diaphragm plate together. Turn the plate and remove it from the valve body.

12. Remove the pushrod, reaction disc, retainer and spring from the front shell.

13. Remove the retainer and seal from the front shell.

To assemble:

14. Apply silicone grease to the seal and install to the front shell and install the retainer over the seal. Lubricate the diaphragm sliding surfaces on the inside of the shell with the same grease.

15. Lubricate the pushrod stem and install to the hub reaction disc.

16. Lubricate and install a new O-ring to the valve body. Set the diaphragm plate on the valve body and rotate it to install. Install the new front diaphragm and retainer.

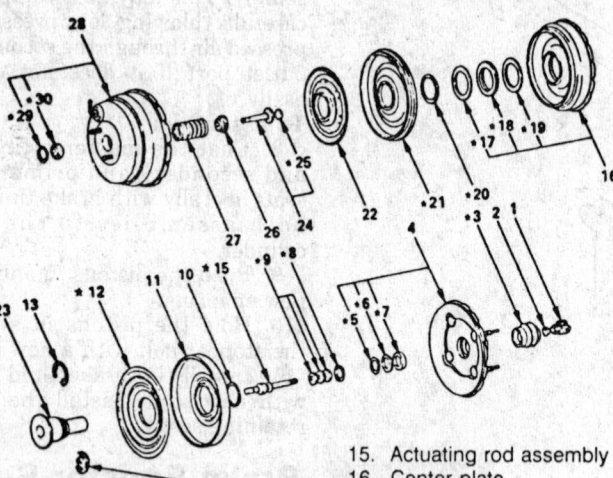

1. Clevis yoke	15. Actuating rod assembly
2. Locknut	16. Center plate
3. Rubber boot	17. Retainer
4. Rear shell	18. Seal
5. Retainer	19. Bearing
6. Bearing	20. Retainer
7. Valve body seal	21. Front diaphragm
8. Retainer	22. Front diaphragm plate
9. Filter and silencer	23. Valve body and O-ring
10. Retainer	24. Pushrod
11. Rear diaphragm	25. Reaction disc
12. Rear diaphragm plate	26. Retainer
13. C-ring	27. Spring
14. Stopper key	28. Front shell
	29. Retainer
	30. Plate and seal

Isuzu tandam vacuum servo — Pick-Up and Amigo with 2.6L engine

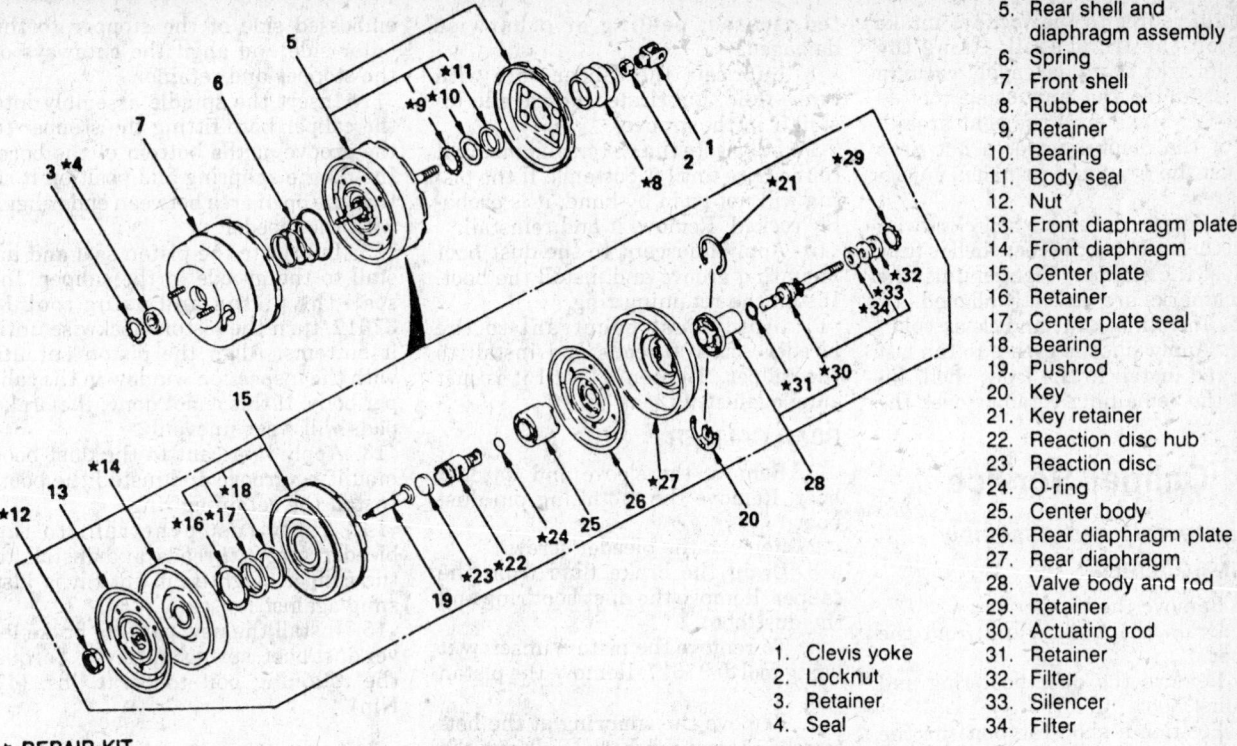

5. Rear shell and
   diaphragm assembly
6. Spring
7. Front shell
8. Rubber boot
9. Retainer
10. Bearing
11. Body seal
12. Nut
13. Front diaphragm plate
14. Front diaphragm
15. Center plate
16. Retainer
17. Center plate seal
18. Bearing
19. Pushrod
20. Key
21. Key retainer
22. Reaction disc hub
23. Reaction disc
24. O-ring
25. Center body
26. Rear diaphragm plate
27. Rear diaphragm
28. Valve body and rod
29. Retainer
30. Actuating rod
31. Retainer
32. Filter
33. Silencer
34. Filter

1. Clevis yoke
2. Locknut
3. Retainer
4. Seal

★ REPAIR KIT

Isuzu tandam vacuum servo — Trooper II

17. Lubricate and install the bearing, seal and retainer to the center plate.

18. Use installer J–38595 to install the center plate assembly to the valve body.

19. Align the arrow mark on the rear diaphragm plate with the mark on the valve body and install the plate with a tilting motion. Install a new retainer to the front diaphragm.

20. Lubricate and install the actuating rod to the valve body and install the C-ring. Push the actuating rod and install the stopper key.

21. Install the first filter, the silencer then the second filter to the valve body and install the new retainer.

22. Lubricate and install the seal, bearing and retainer to the rear shell.

23. Mount the holding tool J–34350 in a vise and mount the front shell on the tool. Assemble the entire center plate assembly to the rear shell.

24. To assemble the unit, connect a vacuum source to the vacuum intake pipe on the front shell. Using the matchmarks as guides, apply vacuum and assemble the unit using tool J–9504–01. Make sure the bent area of the of the diaphragm does not drop against the front and rear shell engaging areas.

25. Rotate the rear shell clockwise until the notch on the rear shell is fully against the stopper. Tighten until the matchmarks are perfectly aligned.

26. Install the new rubber boot, locknut and clevis yoke.

27. Adjust the pushrod length to 0.7 in. (18.2mm) from the base of the front shell.

1988–90 TROOPER/TROOPER II

1. Remove the clevis yoke and locknut from the actuator rod.

2. Pry out the servo retainer and seal from the front of the shell.

— **CAUTION** —

The rear shell is spring loaded. Loosen it slowly and carefully or personal injury may result.

3. Matchmark the front and rear shells. Using wrench J–9504–01 and holder J–34250 (holes elongated to fit), separate the 2 shells slowly. Remove the servo spring.

4. Remove the rubber boot.

5. Remove the retainer, bearing and seal from the rear shell.

6. To disassemble the diaphragm assembly, remove the large nut with a 1½ in. wrench.

7. Remove the front diaphragm plate, front diaphragm and center plate.

8. Remove the retainer, center plate seal and bearing from the center plate.

9. Remove the pushrod and carefully pry the key out of the valve body. Remove the key retainer.

10. Rotate the reaction disc hub 90 degrees and release it from the center body. Remove the reaction disc and O-ring.

11. Remove the center body and dia-phragm from the rear diaphragm plate.

12. Remove the valve body assembly. Remove the retainer and remove the actuating rod.

13. Remove the retainer and remove the filters and silencer from the rod.
To assemble:

14. Install the valve body to the rear diaphragm plate. Install the retainer with the cutouts to the valve body lugs and install the actuating rod.

15. Install the new rear diaphragm.

16. Install the center body and new lubricated O-ring. Set the retainer against the outer pawl of the reaction and rotate the reaction disc 90 degrees to install.

17. Install the key retainer, push in the actuating rod and install the stopper key.

18. Install the new filter, silencer, second filter and retainer.

19. Install the reaction disc (with the stepped side inward) and pushrod.

20. Lubricate and install the bearing, center plate seal and retainer to the center plate.

21. Install the front diaphragm to its plate and install 1½ in. nut.

22. Lubricate and install the seal, bearing and retainer to the rear shell. Install the new rubber boot.

23. Mount the holding tool J–34350 in a vise and mount the front shell on the tool. Assemble the entire diaphragm assembly, spring and the rear shell.

24. To assemble the unit, connect a

vacuum source to the vacuum intake pipe on the front shell. Using the matchmarks as guides, apply vacuum and assemble the unit using tool J-9504-01. Make sure the bent area of the of the diaphragm does not drop against the front and rear shell engaging areas.

25. Rotate the rear shell clockwise until the notch on the rear shell is fully against the stopper. Tighten until the matchmarks are perfectly aligned.

26. Install locknut and clevis yoke.

27. Apply silicone grease to the last seal and install to the front shell. Install the remaining retainer over the seal.

Caliper Service

Disassembly and Assembly

FRONT CALIPER

1. Remove the bleeder screw.
2. Drain the brake fluid from the caliper.
3. Remove the dust boot ring and the dust boot.
4. Position a shop rag opposite the top of the piston. Apply low pressure compressed air to the fluid inlet and carefully blow the piston out of the bore.

--- CAUTION ---

Do not put fingers where the piston will land when it is blown out of the caliper. The force of the piston can crush fingers and cause personal injury.

5. Remove the piston seal from inside the bore.

To assemble:

6. Blow out all fluid passages with low pressure compressed air.
7. Slight corrosion or rust can be removed with commutator paper or crocus cloth. Replace pistons that are pit-

ted, scored, peeling or otherwise damaged.

8. Lubricate the caliper bore with brake fluid. Lubricate the seal and install it in the groove.
9. Lubricate the piston and install it to the bore until it bottoms. If the piston will not go in by hand, it is probably cocked. Remove it and reinstall.
10. Apply lubricant to the dust boot mounting groove and install the boot. Install the retaining ring.
11. Apply rust penetrant to the bleeder screw threads and install to the caliper. Tighten it until it is just snug against its seat.

REAR CALIPER

1. Remove the sleeve and its dust boot. Remove the mounting pin dust boot.
2. Remove the bleeder screw.
3. Drain the brake fluid from the caliper. Remove the dust boot ring and the dust boot.
4. To remove the piston, unscrew it using tool J-37617. Remove the piston seal.
5. Remove the snapring at the bottom of the caliper bore. Pull out the pad adjusting spindle along with the retainer, spring, spring plate and stopper. Remove the O-ring from the spindle.
6. Remove the parking brake link from the bottom of the caliper bore.
7. Remove the parking brake lever, spring and dust boot.
8. Remove the return spring and dust boot.

To assemble:

9. Apply brake system compatible grease to the parking brake link and install.
10. Lubricate and install the new O-ring to the pad adjusting spindle. Attach the stopper, spring plate, spring and retainer to the spindle. Turn the

embossed side of the stopper to the outer side and align the cutaways of the stopper and retainer.

11. Insert the spindle assembly into the caliper bore fitting the stopper to the groove at the bottom of the bore. Install the snapring and position it so that the open area between ends aligns with the bleeder.
12. Lubricate the piston seal and install to the groove in the caliper. Install the piston and using tool J-37617, turn the piston clockwise until it bottoms. Align the piston cutouts with the inspection window in the caliper body. If this is not done, the brake pads will wear unevenly.
13. Apply lubricant to the dust boot mounting groove and install the boot. Install the retaining ring.
14. Apply rust penetrant to the bleeder screw threads and install to the caliper. Tighten it until it is just snug against its seat.
15. Install the new parking brake lever dust boot, spring and lever. Torque the retaining bolt to 35 ft. lbs. (47 Nm).

Wheel Cylinder Service

Disassembly and Assembly

1. Remove the bleeder screw.
2. Remove both rubber boots.
3. Remove the pistons from the cylinder.
4. Remove the piston cups.
5. Remove the return spring.
6. Hone the cylinder bore if it is not damaged, to provide a smooth sealing surface. Wash and dry the bore after honing.

To assemble:

7. Lubricate the cylinder bore with brake fluid. Install the return spring.
8. Lubricate the new piston cups and install them to the cylinder. Lubricate and install the pistons.
9. Apply lubricant to the boot mounting groove and install the boot.
10. Apply rust penetrant to the bleeder screw threads and install to the wheel cylinder. Tighten it until it is just snug against its seat.

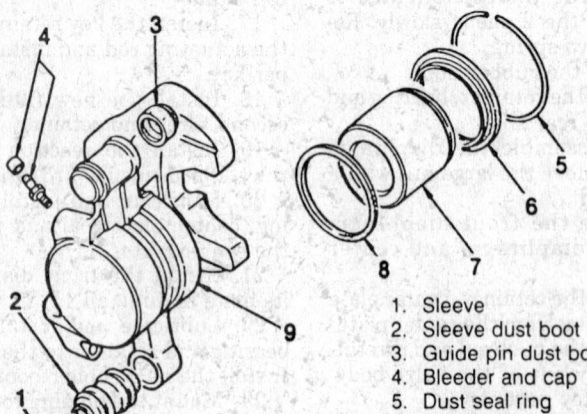

1. Sleeve
2. Sleeve dust boot
3. Guide pin dust boot
4. Bleeder and cap
5. Dust seal ring
6. Dust seal
7. Piston
8. Piston seal
9. Caliper body

Isuzu front caliper

MAZDA

NOTE: When cleaning brake system components, use only brake fluid or denatured isopropyl alcohol. Never use a mineral-based solvent such as gasoline or paint thinner; these fluids will leave a residue that may swell and deteriorate rubber parts within

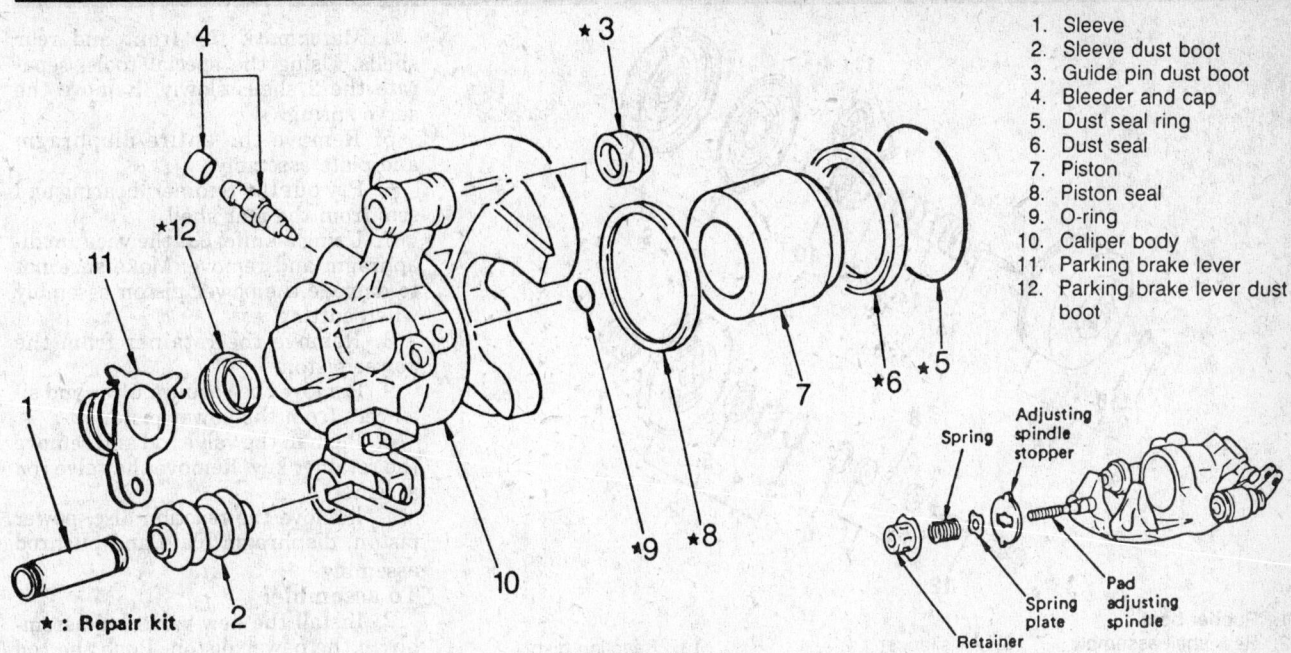

1. Sleeve
2. Sleeve dust boot
3. Guide pin dust boot
4. Bleeder and cap
5. Dust seal ring
6. Dust seal
7. Piston
8. Piston seal
9. O-ring
10. Caliper body
11. Parking brake lever
12. Parking brake lever dust boot

★ : Repair kit

Isuzu rear caliper

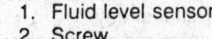

the system. All alcohol must be removed from the system when the work is done because alcohol mixed with brake fluid lowers its boiling point.

Do not hone any aluminum bores. There is a hard anodized coating on the aluminum which will be removed if honed. This will allow the aluminum to wear quickly and will damage the component.

Master Cylinder Service

Disassembly and Assembly

1. Remove the brake fluid reservoir retainer screw.
2. Remove the reservoir and seals.
3. Remove the dust seal from the end of the master cylinder.
4. Push the primary piston in and remove the stopper bolt and gasket.
5. Hold the pistons in and remove the snapring.
6. Let the primary piston spring out and remove the primary piston assembly and end washer, if equipped.
7. Remove the secondary piston assembly. This may be accomplished by carefully blowing low pressure compressed air through the secondary side outlet port if it does not come out easily.

To assemble:

8. Assemble the secondary piston assembly, if it was disassembled. Coat the master cylinder bore and secondary and primary piston seals liberally with brake fluid. Install both assemblies to the master cylinder.

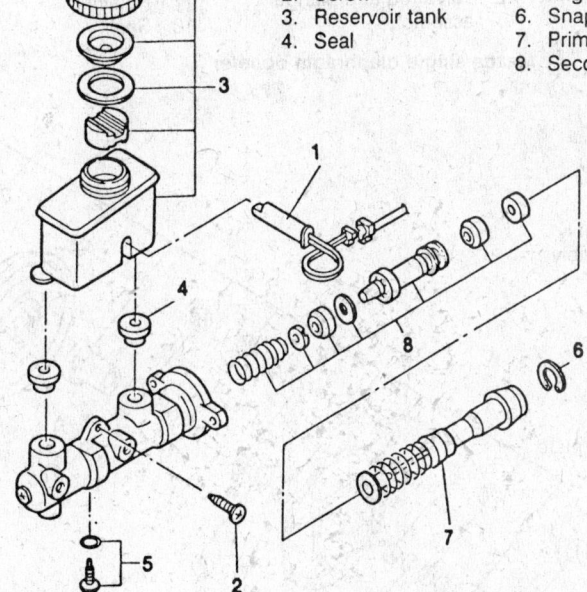

1. Fluid level sensor
2. Screw
3. Reservoir tank
4. Seal
5. Stopper screw and O-ring
6. Snapring
7. Primary piston assembly
8. Secondary assembly

Mazda master cylinder

9. Push the pistons in and install a new snapring.
10. Hold the pistons in and install the stopper bolt with a new gasket.
11. Install the brake fluid reservoir with new seals. Install the reservoir retaining screw.

Brake Booster Service

The brake booster on the MPV is not serviceable. If the unit is defective, replace the entire assembly.

SINGLE DIAPHRAGM

Disassembly and Assembly

2WD PICK-UP

1. Remove the clevis yoke and locknut from the actuator rod.
2. Pry out the servo retainer and seal from the front of the shell.
3. Remove the rubber boot.

— **CAUTION** —

The rear shell is spring loaded. Loosen it slowly and carefully or personal injury may result.

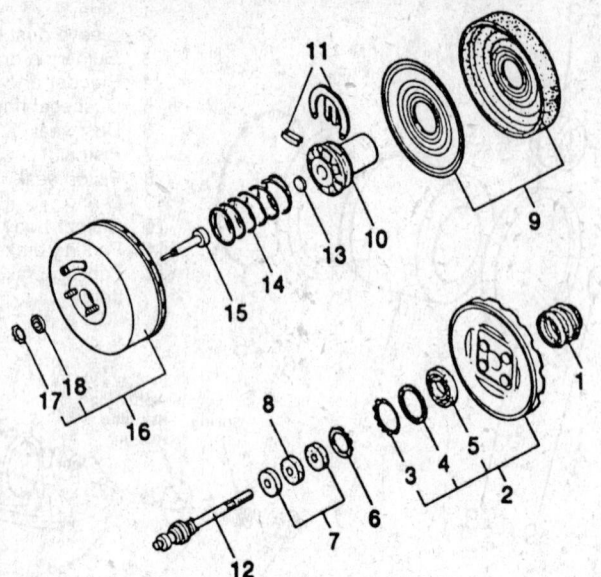

4. Matchmark the front and rear shells. Using the special tools, separate the 2 shells slowly. Remove the servo spring.

5. Remove the entire diaphragm and plate assembly.

6. Pry out the retainer, bearing and seal from the rear shell.

7. Using a knife, cut the vacuum diaphragm and remove. Make sure not to damage the power piston assembly when cutting.

8. Remove the retainer from the power piston.

9. Remove the vacuum filter and silencers from the power piston.

10. Push in the valve rod and remove the retainer key. Remove the valve rod assembly.

11. Remove the reaction disc, power piston, diaphragm plate and pushrod assembly.

To assemble:

12. Install the new valve rod assembly to the power piston. Push the rod in and install the retainer key.

13. Install the new vacuum silencers and filter.

14. Install the new retainer to the rear shell.

15. Install the diaphragm plate and

1. Rubber boot
2. Rear shell assembly
3. Retainer
4. Bearing
5. Dust seal
6. Retainer
7. Air filter
8. Air silencer
9. Diaphragm and plate
10. Power piston assembly
11. Retainer key
12. Valve rod and plunger assembly
13. Reaction disc
14. Spring
15. Pushrod
16. Front shell assembly
17. Retainer
18. Seal

Mazda single diaphragm booster

1. Clevis
2. Nut
3. Rear shell assembly
4. Front shell assembly
5. Boot
6. Rear shell
7. Retainer
8. Bearing
9. Seal
10. Locknut
11. Front diaphragm
12. Front diaphragm plate
13. Center plate
14. Retainer
15. Seal
16. Bearing
17. Stopper and key
18. Retainer
19. Retainer
20. Air filter
21. Air silencer
22. Valve rod and plunger assembly
23. Valve body
24. O-ring
25. Reaction disc hub
26. O-ring
27. Reaction disc
28. Rear diaphragm
29. Rear diaphragm plate
30. Spring
31. Pushrod
32. Front shell
33. Retainer
34. Seal

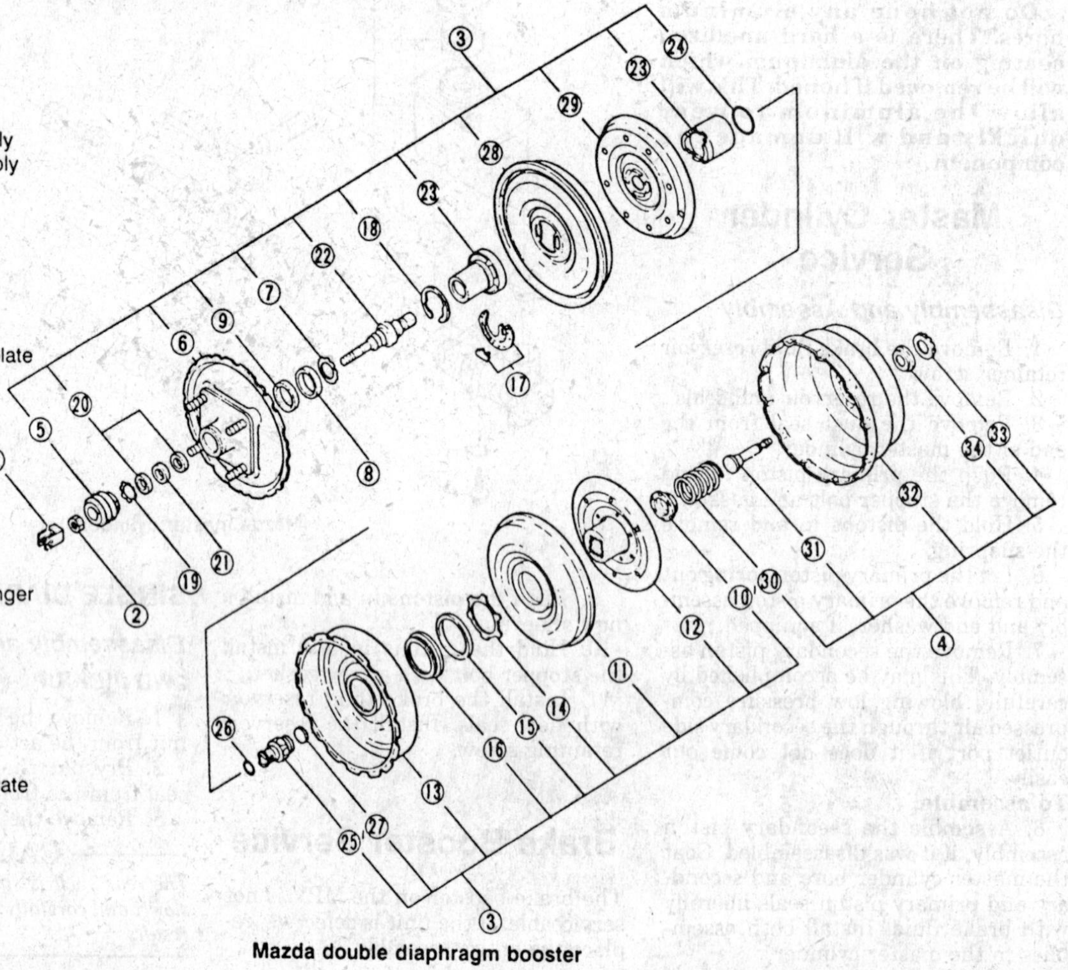

Mazda double diaphragm booster

carefully attach the new vacuum diaphragm to the power piston.

16. Apply silicone grease to the new reaction disc and install to the power piston. Install the pushrod and spring.

17. Apply silicone grease to the rear shell seal and install.

18. Mount the front shell in a vise. Assemble the entire diaphragm plate assembly to the rear shell. Install the bearing and new retainer.

19. To assemble the unit, connect a vacuum source to the vacuum intake pipe on the front shell. Using the matchmarks as guides, apply vacuum and assemble the unit using the special tool. Make sure the bent area of the of the diaphragm does not drop against the front and rear shell engaging areas.

20. Rotate the rear shell clockwise until the notch on the rear shell is fully against the stopper. Tighten until the matchmarks are perfectly aligned.

21. Install the new rubber boot, locknut and clevis yoke.

22. Apply silicone grease to the last seal and install to the front shell. Install the remaining retainer over the seal.

DOUBLE DIAPHRAGM

Disassembly and Assembly

4WD PICK-UP

1. Remove the clevis yoke and locknut from the actuator rod.

2. Pry out the servo retainer and seal from the front of the shell.

— CAUTION —

The rear shell is spring loaded. Loosen it slowly and carefully or personal injury may result.

3. Matchmark the front and rear shells. Using the special tools, separate the shells. Remove the rubber boot.

4. Clamp holding tool 49–U043–002 in a vise and mount the rear shell assembly to the tool. Uncrimp the reaction disc hub. Remove the locknut using wrench 49–U043–001.

5. Remove the retainer, bearing and seal from the rear shell.

6. Remove the front diaphragm plate, front diaphragm and center plate.

7. Remove the retainer, center plate seal and bearing from the center plate.

8. Remove the pushrod and carefully pry the key out of the valve body.

9. Remove the key retainer.

10. Rotate the reaction disc hub 90 degrees and release it from the center body. Remove the reaction disc and O-ring.

11. Remove the center body and dia-

phragm from the rear diaphragm plate.

12. Remove the valve body assembly. Remove the retainer and remove the actuating rod.

13. Remove the retainer and remove the filters and silencer from the rod.

To assemble:

14. Install the valve body to the rear diaphragm plate. Install the retainer with the cutouts to the valve body lugs and install the actuating rod.

15. Install the new rear diaphragm.

16. Install the center body and new lubricated O-ring. Set the retainer against the outer pawl of the reaction and rotate the reaction disc 90 degrees to install.

17. Install the key retainer, push in the actuating rod and install the stopper key.

18. Install the new filter, silencer, second filter and retainer.

19. Install the reaction disc (with the stepped side inward) and pushrod.

20. Lubricate and install the bearing, center plate seal and retainer to the center plate.

21. Install the front diaphragm to its plate and install locknut using the special wrench. Crimp the reaction disc hub in the 2 notched grooves of the locknut.

22. Lubricate and install the seal, bearing and retainer to the rear shell. Install the new rubber boot.

23. Align the notches of the rear shell and center plate. Make sure the bent area of the of the diaphragm does not drop against the front and rear shell engaging areas.

24. Push down and rotate the rear shell clockwise until the notch on the rear shell is fully against the stopper. Tighten until the matchmarks are perfectly aligned.

25. Install locknut and clevis yoke.

26. Apply silicone grease to the last seal and install to the front shell. Install the remaining retainer over the seal.

Caliper Service

Disassembly and Assembly

1. Remove the bleeder screw.

2. Drain the brake fluid from the caliper.

3. Remove the dust boot.

4. Position a shop rag opposite the top of the piston. Apply low pressure compressed air to the fluid inlet and carefully blow the piston out of the bore.

— CAUTION —

Do not put fingers where the piston will land when it is blown out of the caliper. The force of the piston can crush fingers and cause personal injury.

5. Remove the piston seal from inside the bore.

To assemble:

6. Blow out all fluid passages with low pressure compressed air.

7. Slight corrosion or rust can be removed with commutator paper or crocus cloth. Replace pistons that are pitted, scored, peeling or otherwise damaged.

8. Lubricate the caliper bore with brake fluid. Lubricate the seal and install it in the groove.

9. Lubricate the piston and install it to the bore until it bottoms. If the piston will not go in by hand, it is probably cocked. Remove it and reinstall.

10. Apply lubricant to the dust boot mounting groove and install the boot.

11. Apply rust penetrant to the bleeder screw threads and install to the caliper. Tighten it until it is just snug against its seat.

Wheel Cylinder Service

Disassembly and Assembly

1. On the left side, remove the bleeder screw and check ball.

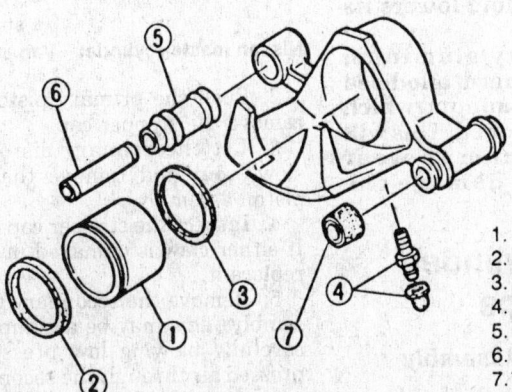

1. Piston
2. Dust seal
3. Piston seal
4. Bleeder and cap
5. Pin boot
6. Pin
7. Bushing

Mazda caliper

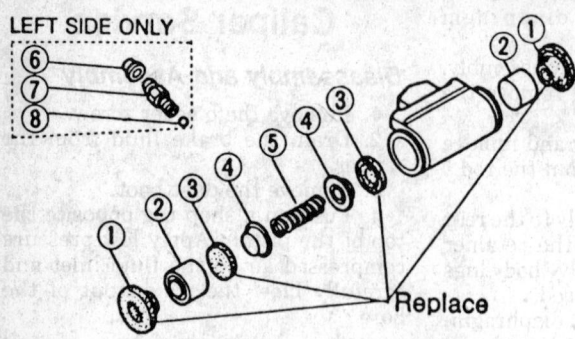

LEFT SIDE ONLY

1. Dust boots
2. Pistons
3. Piston rubber cups
4. Piston cups
5. Spring
6. Rubber cap
7. Bleeder
8. Steel ball

Replace

Mazda wheel cylinder

2. Remove both dust boots.
3. Remove the pistons.
4. Remove the piston cups.
5. Remove the return spring.
6. Hone the cylinder bore if it is not damaged, to provide a smooth sealing surface. Wash and dry the bore after honing.

To assemble:

7. Lubricate the cylinder bore with brake fluid. Install the return spring.
8. Lubricate the new piston cups and install them to the cylinder. Lubricate and install the pistons.
9. Apply lubricant to the boot mounting groove and install the boot.
10. Apply rust penetrant to the bleeder screw threads and install it with the check ball to the wheel cylinder. Tighten until just snug.

NISSAN

NOTE: When cleaning brake system components, use only brake fluid or denatured isopropyl alcohol. Never use a mineral-based solvent such as gasoline or paint thinner; these fluids will leave a residue that may swell and deteriorate rubber parts within the system. All alcohol must be removed from the system when the work is done because alcohol mixed with brake fluid lowers its boiling point.

Do not hone any aluminum bores. There is a hard anodized coating on the aluminum which will be removed if honed. This will allow the aluminum to wear quickly and will damage the component.

Master Cylinder Service

Disassembly and Assembly

1. Remove the brake fluid reservoir and seals or hoses.

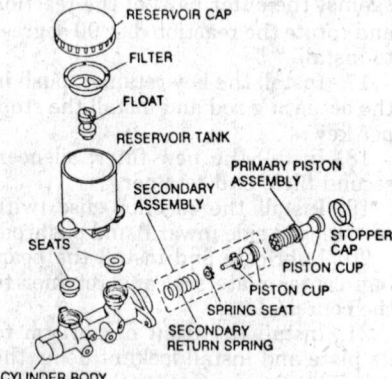

RESERVOIR CAP
FILTER
FLOAT
RESERVOIR TANK
PRIMARY PISTON ASSEMBLY
SECONDARY ASSEMBLY
STOPPER CAP
PISTON CUP
PISTON CUP
SPRING SEAT
SECONDARY RETURN SPRING
SEATS
CYLINDER BODY

Nissan master cylinder—Pick-Up and Pathfinder

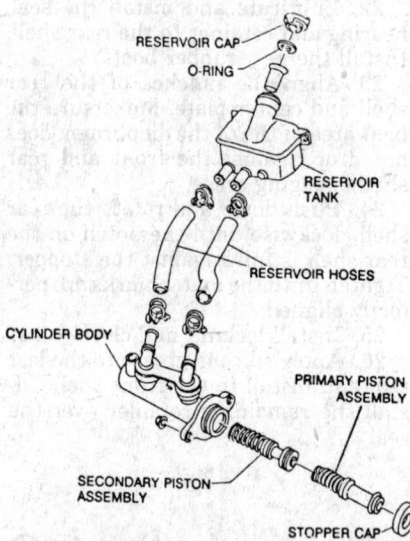

RESERVOIR CAP
O-RING
RESERVOIR TANK
RESERVOIR HOSES
CYLINDER BODY
PRIMARY PISTON ASSEMBLY
SECONDARY PISTON ASSEMBLY
STOPPER CAP

Nissan master cylinder—Van and Axxess

2. Push the primary piston in and remove the stopper cap.
3. Let the primary piston spring slowly out and remove the primary piston assembly.
4. Inspect the stopper cap carefully. If either claw is damaged in any way, replace it.
5. Remove the secondary piston assembly. This may be accomplished by carefully blowing low pressure compressed air through the secondary side outlet port if it does not come out easily. Disassemble the secondary piston

on Pick-Up and Pathfinder only. The secondary pistons on Axxess and Van are not serviceable.

To assemble:

6. Assemble the secondary piston assembly, if it was disassembled. Coat the master cylinder bore and secondary and primary piston seals liberally with brake fluid. Install both assemblies to the master cylinder.
7. Push the pistons in and install the stopper cap.
8. Install the fluid reservoir and seals, if equipped.

Brake Booster Service

The brake boosters installed on these vehicles are not serviceable. If the unit is defective, replace the entire assembly.

Caliper Service

SINGLE PISTON TYPE

Disassembly and Assembly

1. Remove the bleeder screw.
2. Drain the brake fluid from the caliper.
3. Remove the dust seal ring (rear caliper only) and the dust seal.
4. Position a shop rag opposite the top of the piston. Apply low pressure compressed air to the fluid inlet and carefully blow the piston out of the bore.

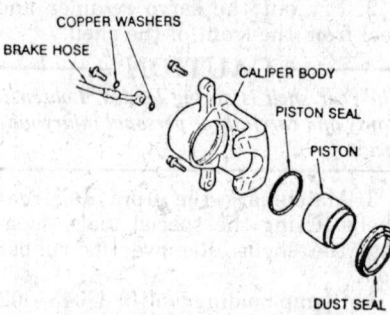

COPPER WASHERS
BRAKE HOSE
CALIPER BODY
PISTON SEAL
PISTON
DUST SEAL

Nissan single piston front caliper

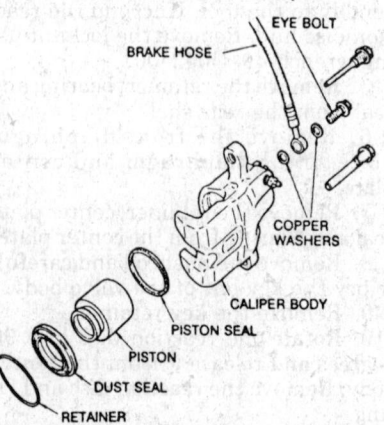

EYE BOLT
BRAKE HOSE
COPPER WASHERS
CALIPER BODY
PISTON SEAL
PISTON
DUST SEAL
RETAINER

Nissan rear caliper

CAUTION

Do not put fingers where the piston will land when it is blown out of the caliper. The force of the piston can crush fingers and cause personal injury.

5. Remove the piston seal from inside the bore.

To assemble:

6. Blow out all fluid passages with low pressure compressed air.

7. Slight corrosion or rust can be removed with commutator paper or crocus cloth. Replace pistons that are pitted, scored, peeling or otherwise damaged.

8. Lubricate the caliper bore with brake fluid. Lubricate the seal and install it in the groove.

9. Lubricate the piston and install it to the bore until it bottoms. If the piston will not go in by hand, it is probably cocked. Remove it and reinstall.

10. Apply lubricant to the dust seal mounting groove and install the seal. Install the retaining ring, if equipped.

11. Apply rust penetrant to the bleeder screw threads and install to the caliper. Tighten it until it is just snug against its seat.

DOUBLE PISTON TYPE

Disassembly and Assembly

1. Remove the bleeder screw.
2. Drain the brake fluid from the caliper.
3. Remove the dust covers.
4. Position a shop rag opposite the top of the piston. Apply low pressure compressed air to the fluid inlet and carefully blow the pistons out of their bores. If both pistons do not come out, install the extracted piston back into its bore just enough to seal; then hold it place with a suitable spacer. Blow out the remaining piston.

CAUTION

Do not put fingers where the piston will land when it is blown out of the caliper. The force of the piston can crush fingers and cause personal injury.

5. Remove the piston seals from inside the bore.

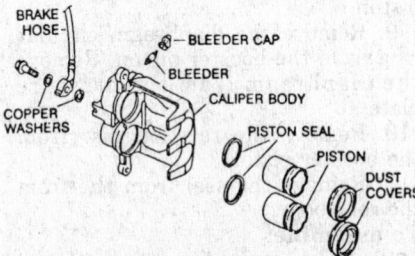

Nissan double piston caliper

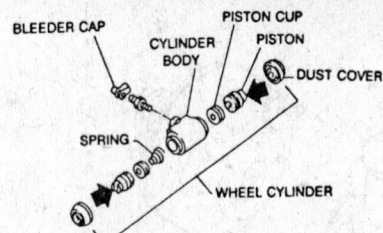

Nissan wheel cylinder

To assemble:

6. Blow out all fluid passages with low pressure compressed air.

7. Slight corrosion or rust can be removed with commutator paper or crocus cloth. Replace pistons that are pitted, scored, peeling or otherwise damaged.

8. Lubricate the caliper bore with brake fluid. Lubricate the seals and install in the grooves.

9. Lubricate the pistons and install them to the bores until they bottom. If either piston will not go in by hand, it is probably cocked. Remove it and reinstall.

10. Apply lubricant to the dust seal mounting grooves and install the seals.

11. Apply rust penetrant to the bleeder screw threads and install to the caliper. Tighten it until it is just snug against its seat.

Wheel Cylinder Service

Disassembly and Assembly

1. Remove the bleeder screw.
2. Remove both dust covers.
3. Remove the pistons.
4. Remove the piston cups.
5. Remover the return spring.
6. Hone the cylinder bore if it is not damaged, to provide a smooth sealing surface. Wash and dry the bore after honing.

To assemble:

7. Lubricate the cylinder bore with brake fluid. Install the return spring.

8. Lubricate the new piston cups and install them to the cylinder. Lubricate and install the pistons.

9. Apply lubricant to the dust cover mounting groove and install the cover.

10. Apply rust penetrant to the bleeder screw threads and install to the wheel cylinder. Tighten it until it is just snug against its seat.

SUZUKI/GEO

NOTE: When cleaning brake system components, use only brake fluid or denatured isopropyl alcohol. Never use a mineral-based solvent such as gasoline or paint thinner; these fluids will leave a residue that may swell and deteriorate rubber parts within the system. All alcohol must be removed from the system when the work is done because alcohol mixed with brake fluid lowers its boiling point.

Do not hone any aluminum bores. There is a hard anodized coating on the aluminum which will be removed if honed. This will allow the aluminum to wear quickly and will damage the component.

Master Cylinder Service

Disassembly and Assembly

1. Remove the brake fluid reservoir screw of roll pin and remove the reservoir. Remove the seals.

2. Push the primary piston in and remove the snapring.

3. Let the primary piston spring slowly out and remove piston stopper, cups and the primary piston assembly.

4. Remove the secondary piston stopper bolt.

5. Remove the secondary piston assembly. This may be accomplished by carefully blowing low pressure compressed air through the secondary side outlet port if it does not come out easily. Disassemble the secondary piston.

To assemble:

6. Assemble the secondary piston assembly. Coat the master cylinder bore and secondary and primary piston seals liberally with brake fluid. Install both assemblies to the master cylinder.

7. Install the cups and stoppers. Install the snapring.

8. Push the pistons in and install the stopper bolt.

9. Install the fluid reservoir and seals.

Brake Booster Service

SINGLE DIAPHRAGM TYPE

SAMURAI

1. Remove the piston rod from the booster.

2. Remove the pushrod clevis and locknut.

3. Mount the booster on the holding tool set 09950-88210. Do not tighten the 2 outer nuts to more than 30 inch lbs. or the booster may become deformed.

4. Turn the special tool bolt clock-

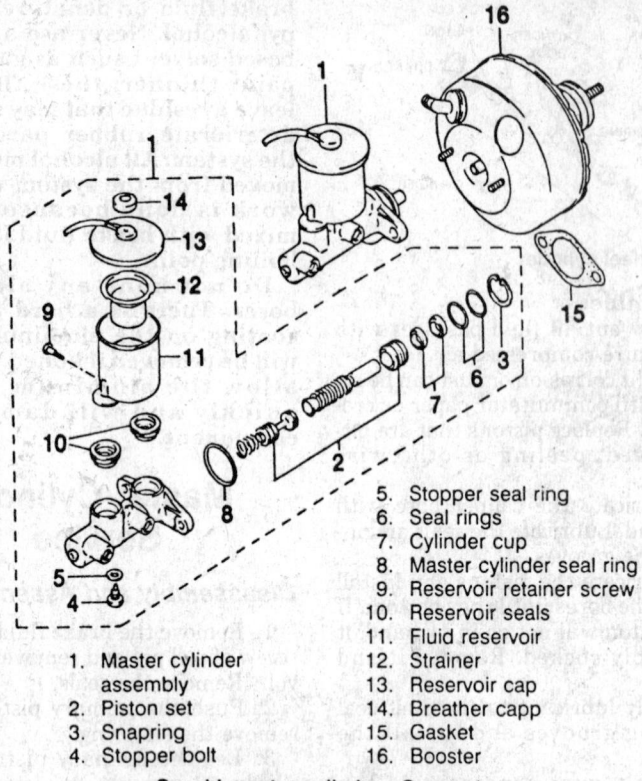

1. Master cylinder assembly
2. Piston set
3. Snapring
4. Stopper bolt
5. Stopper seal ring
6. Seal rings
7. Cylinder cup
8. Master cylinder seal ring
9. Reservoir retainer screw
10. Reservoir seals
11. Fluid reservoir
12. Strainer
13. Reservoir cap
14. Breather capp
15. Gasket
16. Booster

Suzuki master cylinder—Samarai

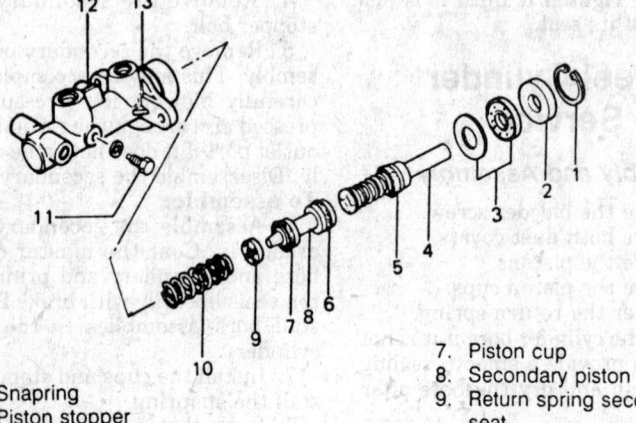

1. Snapring
2. Piston stopper
3. Cylinder cup and plate
4. Primary piston
5. Piston cup
6. Secondary piston pressure cap
7. Piston cup
8. Secondary piston
9. Return spring secondary seat
10. Secondary piston return spring
11. Stopper bolt
12. Cylinder body
13. Seal

Suzuki/Geo master cylinder—Sidekick and Tracker

wise until the projection and depression in the bodies match. Matchmark the bodies at that location.

5. Remove the booster from the tool and separate the bodies.

CAUTION

The rear body is spring loaded. Loosen it slowly and carefully or personal injury may result.

6. Remove the boot, air cleaner elements and air cleaner separator from the rear body.

7. Turn the booster piston counterclockwise and remove from the rear body.

8. Push the air valve down and remove the stopper key. Remove the air valve assembly from the booster piston.

9. Remove the diaphragm from the pressure plate.

10. Remove the reaction disc from the booster piston.

11. Remove the seal from the from the rear body.

To assemble:

12. Lubricate and install a new seal to the rear body.

13. Lubricate and install the booster air valve to the booster piston.

14. Compress the valve and install the stopper key.

15. Install the new diaphragm to the pressure plate. Make sure the diaphragm is seated securely in its groove in the pressure plate.

16. Lubricate the face and install the reaction disc to the booster piston.

17. Install the booster piston to the rear body by turning it clockwise into place.

18. Install the new air cleaner separator and elements to the air valve rod.

19. Install the boot to the rear body.

20. Place the front body on the special tool and place the return spring on it large end up.

21. Place the assembled rear body on the return spring and compress. Turn the special bolt counterclockwise until the projection and depression line up.

22. Install the piston rod into the booster piston.

23. Install the locknut and clevis.

SIDEKICK AND TRACKER

1. Remove the pushrod clevis.

2. Remove the locknut.

3. Mount the booster on the holding tool set 09950–88210. Do not tighten the 2 outer nuts to more than 30 inch lbs. or the booster may become deformed.

4. Turn the special tool bolt clockwise until the projection and depression in the bodies match. Matchmark the bodies at that location.

5. Remove the booster from the tool and separate the bodies.

CAUTION

The rear body is spring loaded. Loosen it slowly and carefully or personal injury may result.

6. Remove the piston rod, boot, air cleaner elements and air cleaner separator from the rear body.

7. Remove the valve stopper key cushion.

8. Push the air valve down and remove the stopper key. Remove the air valve assembly from the booster piston.

9. Remove the diaphragm circular ring from the booster piston. Remove the diaphragm from the pressure plate.

10. Remove the reaction disc from the booster piston.

11. Remove the seal from the from the rear body.

To assemble:

12. Lubricate and install a new seal to the rear body.

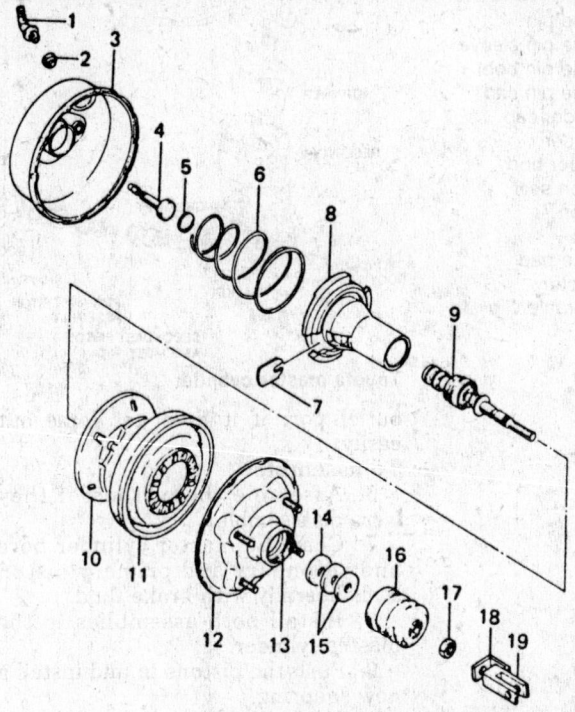

1.	Vacuum check valve		
2.	Grommet		
3.	Front body		
4.	Piston rod		
5.	Reaction disc		
6.	Booster piston return spring		
7.	Valve stopper key		
8.	Booster piston		
9.	Booster air valve assembly		
10.	Pressure plate		
11.	Diaphragm		
12.	Rear body		
13.	Seal		
14.	Air cleaner separator		
15.	Air cleaner element		
16.	Rubber boot		
17.	Nut		
18.	Bracket		
19.	Pushrod clevis		

Suzuki single diaphragm booster — Samarai

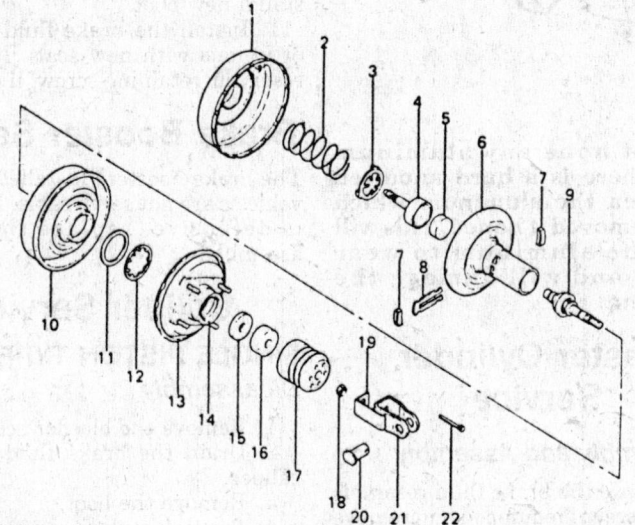

1.	Front body	8.	Valve stopper key	15.	Air cleaner separator
2.	Booster piston return spring	9.	Booster air valve assembly	16.	Air cleaner element
3.	Piston rod retainer	10.	Diaphragm	17.	Rubber boot
4.	Piston rod	11.	Diaphragm retainer	18.	Nut
5.	Reaction disc	12.	Retainer	19.	Bracket
6.	Booster piston	13.	Rear body	20.	Clevis pin
7.	Key cushion	14.	Seal	21.	Pushrod clevis
				22.	Cotter pin

Suzuki/Geo single diaphragm booster — Sidekick and Tracker

13. Install the retainer to the diaphragm. Install the diaphragm to the booster piston and install the new diaphragm circular ring.

14. Lubricate and install the booster air valve to the booster piston. Compress the valve and install the stopper key. Install the cushion to the notch in the stopper.

15. Install the booster piston to the rear body.

16. Install the new air cleaner separator and elements to the air valve rod.

17. Install the boot to the rear body.

18. Lubricate the face and install the reaction disc to the booster piston.

19. Place the front body on the special tool and place the return spring,

rod retainer and piston rod on it.

20. Place the assembled rear body on the return spring and compress. Turn the special bolt counterclockwise until the projection and depression line up.

21. Install the piston rod into the booster piston.

22. Install the locknut and clevis.

Caliper Service

Disassembly and Assembly

1. Remove the bleeder screw.

2. Drain the brake fluid from the caliper.

3. Remove the cylinder boot.

4. Position a shop rag opposite the top of the piston. Apply low pressure compressed air to the fluid inlet and carefully blow the piston out of the bore.

— CAUTION —

Do not put fingers where the piston will land when it is blown out of the caliper. The force of the piston can crush fingers and cause personal injury.

5. Remove the piston seal from inside the bore.

To assemble:

6. Blow out all fluid passages with low pressure compressed air.

7. Slight corrosion or rust can be removed with commutator paper or crocus cloth. Replace pistons that are pitted, scored, peeling or otherwise damaged.

8. Lubricate the caliper bore with brake fluid. Lubricate the seal and install it in the groove.

9. Lubricate the piston and install it to the bore until it bottoms. If the piston will not go in by hand, it is probably cocked. Remove it and reinstall.

10. Apply lubricant to the cylinder boot mounting groove and install.

11. Apply rust penetrant to the bleeder screw threads and install to the caliper. Tighten it until it is just snug against its seat.

Wheel Cylinder Service

Disassembly and Assembly

1. Remove the bleeder screw, if equipped.

2. Remove both rubber boots.

3. Remove the piston assemblies, each consisting of the piston and cup.

4. Remove the piston cups from the pistons, being careful not to scratch the piston.

5. Remove the return spring.

6. Hone the cylinder bore if it is not damaged, to provide a smooth sealing

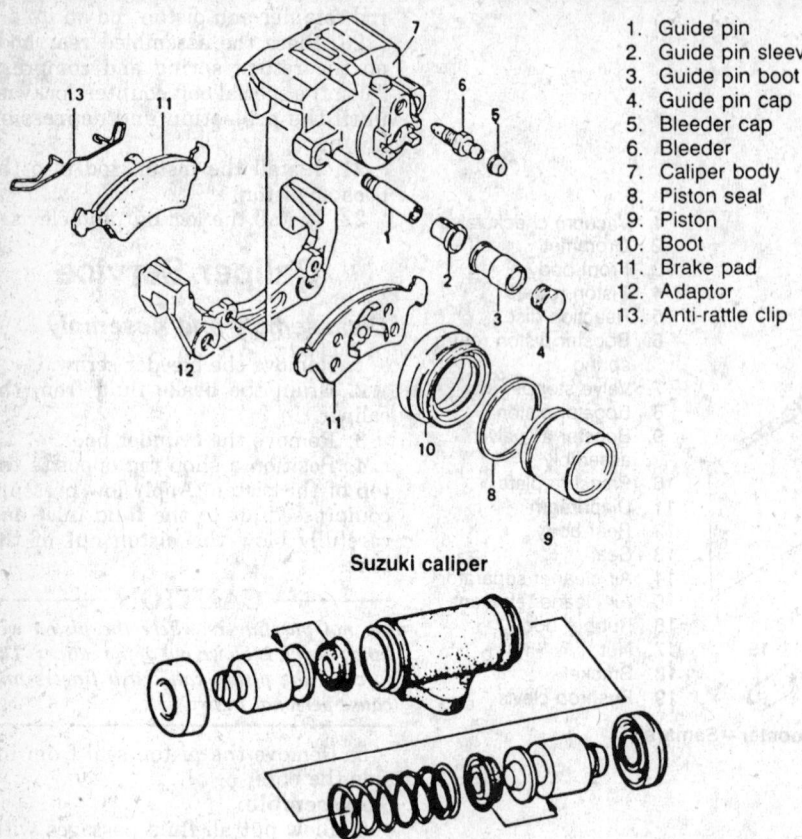

1. Guide pin
2. Guide pin sleeve
3. Guide pin boot
4. Guide pin cap
5. Bleeder cap
6. Bleeder
7. Caliper body
8. Piston seal
9. Piston
10. Boot
11. Brake pad
12. Adaptor
13. Anti-rattle clip

Suzuki caliper

Suzuki wheel cylinder

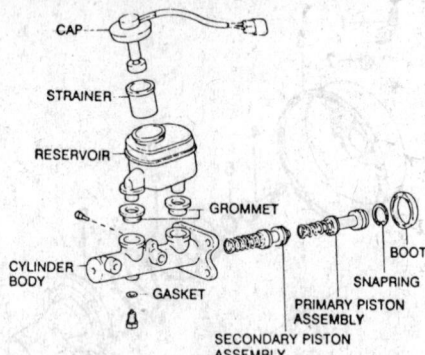

Toyota master cylinder

surface. Wash and dry the bore after honing.

To assemble:

7. Lubricate the cylinder bore with brake fluid.

8. Lubricate the new piston cups and install them to the pistons, with their lips facing upward. Install the assemblies to the cylinder.

9. Apply lubricant to the boot mounting groove and install the boot.

10. Apply rust penetrant to the bleeder screw threads and install to the wheel cylinder. Tighten until it is just snug against its seat.

TOYOTA

NOTE: When cleaning brake system components, use only brake fluid or denatured isopropyl alcohol. Never use a mineral-based solvent such as gasoline or paint thinner; these fluids will leave a residue that may swell and deteriorate rubber parts within the system. All alcohol must be removed from the system when the work is done because alcohol mixed with brake fluid lowers its boiling point.

Do not hone any aluminum bores. There is a hard anodized coating on the aluminum which will be removed if honed. This will allow the aluminum to wear quickly and will damage the component.

Master Cylinder Service

Disassembly and Assembly

1. Remove the brake fluid reservoir retainer screw, if equipped and remove the reservoir and seals or remove the reservoir unions and seals.

2. Remove the boot. Push the primary piston in and remove the stopper bolt and gasket.

3. Hold the pistons in and remove the snapring.

NOTE: Do not disassemble either piston assembly on the Pick-Up and 4Runner. The individual parts are not serviceable.

4. Let the primary piston spring out and remove the primary piston assembly.

5. Remove the secondary piston assembly. This may be accomplished by carefully blowing low pressure compressed air through the secondary side

outlet port if it does not come out easily.

To assemble:

6. Assemble the pistons, if they were disassembled.

7. Coat the master cylinder bore and secondary and primary piston seals liberally with brake fluid.

8. Install both assemblies to the master cylinder.

9. Push the pistons in and install a new snapring.

10. Hold the pistons in and install the stopper bolt with a new gasket. Install a new boot.

11. Install the brake fluid reservoir or unions with new seals. Install the reservoir retaining screw, if equipped.

Brake Booster Service

The brake boosters installed on these vehicles are not serviceable. If the unit is defective, replace the entire assembly.

Caliper Service

SINGLE PISTON TYPE

Disassembly

1. Remove the bleeder screw.
2. Drain the brake fluid from the caliper.
3. Remove the boot.
4. Position a shop rag opposite the top of the piston. Apply low pressure compressed air to the fluid inlet and carefully blow the piston out of the bore.

— CAUTION —

Do not put fingers where the piston will land when it is blown out of the caliper. The force of the piston can crush fingers and cause personal injury.

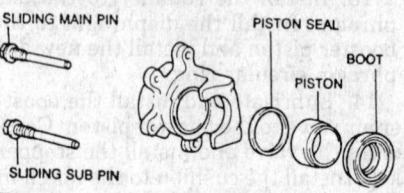

Toyota single piston caliper

EMISSION CONTROLS

Emission control devices are designed to eliminate the chemical compounds that escape from the engine crankcase, from the exhaust and from evaporation of fuel out of the tank and carburetor. With the growing use of on-board computers, it has become possible for car manufacturers to meet strict Federal emission standards by using electronic engine controls to monitor operating conditions and adjust engine calibrations for the best possible performance and economy with minimum emissions.

Engine calibration has a big effect on emissions out the tailpipe. The calibration consists of spark timing, fuel mixture, choke setting, idle speed and spark plug gap. Calibrations are not a service problem as long as the engine is adjusted to the factory specifications, which are found on a sticker in the engine compartment. Engines must be adjusted to these factory specifications or emissions will be high. Additionally, emission control systems have become such an integral part of the overall engine design that best engine performance is dependent on best emission control system performance. This is especially true for computer controlled systems.

NOTE: Any attempt to disconnect or bypass any OEM emission device is a violation of federal law.

The latest emission control systems use electronic instead of vacuum devices and are much more sensitive to malfunctions in any component. Following is a description of each group of controls and how they work to reduce emissions. Due to the complex nature of modern electronic engine control systems, comprehensive diagnosis and testing procedures fall outside the confines of this repair manual. For complete information on diagnosis, testing and repair procedures concerning all modern engine and emission control systems, please refer to *Chilton's Guide To Electronic Engine Controls*.

Emission Service Indicators

RESET PROCEDURES

Indicator lights or flags will periodically appear on or near instrument cluster to alert the driver that various emission control components need to be serviced or replaced. Most of the reminder lights are triggered at preset mileages programmed into either mechanical or electronic counter or odometer switches. The mechanical counter switches are normally operated by the speedometer cable, while the electronic counter switches are pulsed by a speed sensor usually located in the speedometer assembly. After servicing the indicated emission system (EGR valve, Oxygen sensor, etc.). the service indicator device will have to be reset to eliminate the light or flag. Follow the appropriate procedure outlined below for each year and model listed.

American Motors

The reset switch is located under the hood on the left side of the firewall, between the upper and lower speedometer cables. There is a rest screw on the unit that must be rotated one-quarter turn to the detent position.

Chrysler Corporation

1980 MODELS

Chrysler models use either electronic or mechanical service counters. The electronic switch is located under the instrument panel somewhere near the lower left instrument cluster. It is usually covered with a green plastic case.

To reset the mechanical switch, remove it from the mounting bracket and then remove the plastic case. Insert a small screwdriver or rod into the hole in the switch body to close the contacts and turn off the indicator light. To reset the mechanical switch, first locate the unit between the upper and lower speedometer cables. Turn the screw on the upper side of the switch to reset.

General Motors

1980 CADILLAC

To reset the switch, remove the lower steering column cover and locate the reset cable at the lower left side of the speedometer cluster. Pull the cable lightly to reset the switch, then replace the column cover. Do not pull hard on the cable or damage to the cable and switch will occur.

1980 AND LATER MODELS

An emission indicator flag will appear in the odometer window when service is necessary on some 1980 and later GM models. The flags are marked SENSOR, EMISSION, or

FLAG WINDOW IN SPEEDOMETER FACE

RESETTING FLAG WITH DOWNWARD MOVEMENT

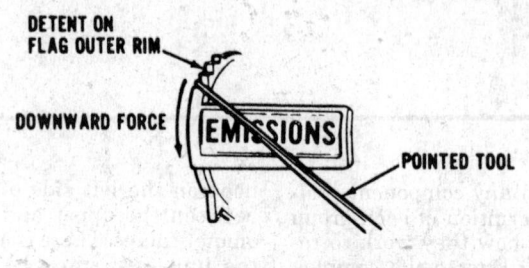

FLAG IN RESET POSITION

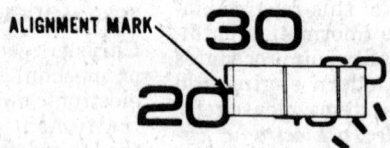

Resetting Emissions flag on 1980 GM models—typical

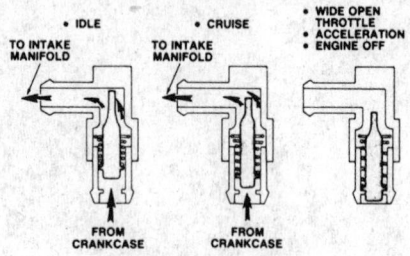

PCV valve operation

CATALYST, depending on the device that is schedule for regular maintenance. To reset the flag, first remove the instrument panel trim plate and instrument cluster cover lens. There are reset notches on the driver's side of the indicator flag.

Insert a long, pointed probe diagonally into the detents on the upper left side and rotate the flag downward until an alignment mark becomes visible in the left side of the odometer window. Once the flag has been lowered, replace the cluster lens and trim plate.

CRANKCASE CONTROLS

PCV System

Ventilation of a crankcase is necessary because of the compression blow-by past and piston rings. This blow-by is mostly unburned gasoline. If allowed to stay in the crankcase, it dilutes the oil and increases engine wear. The PCV system uses engine vacuum to draw out the crankcase fumes. The crankcase or the rocker arm cover is connected by a hose to engine vacuum at the intake manifold or carburetor. When the engine is running, the crankcase fumes are drawn into the engine and burned in the combustion chamber. Fresh air enters the crankcase through the oil filler cap on the open system. When the oil filler cap is connected to the air cleaner, it is known as a closed system.

At wide open throttle, there is little vacuum in the engine, so the PCV system doesn't pull any fumes out of the crankcase. Because the hose connection from the crankcase to the intake manifold acts like a vacuum leak, there has to to be some kind of control to limit the air flow. The PCV valve is the control. It can be an actual valve, with an internal plunger, or a simple orifice without any moving parts. In the plunger types, a spring moves the plunger against engine vacuum, allowing less flow at high vacuum and more flow at low vacuum. In the event of a backfire, the plunger moves to close the PCV valve and prevent a possible crankcase explosion.

Fresh air enters the air cleaner and goes through a hose to the crankcase or rocker cover. The fumes exit the crankcase and enter the intake manifold, either through a hose or some other type of connection, usually with a PCV valve controlling the flow. Most systems use some kind of PCV filter, usually mounted at the end of the hose in the air cleaner. The filter keeps dust from entering the crankcase and also prevents oil fumes from ruining the air cleaner element.

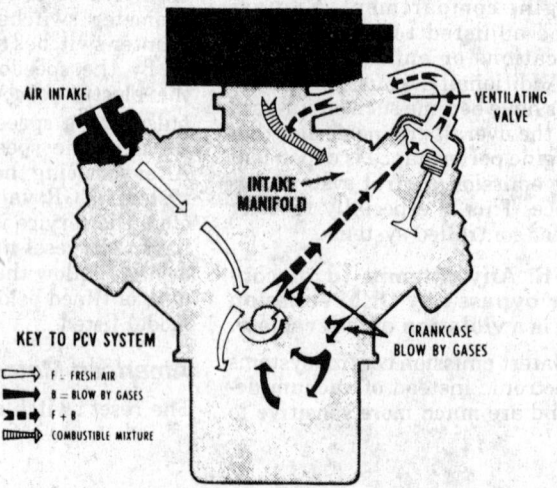

KEY TO PCV SYSTEM

F : FRESH AIR
B : BLOW BY GASES
F + B
COMBUSTIBLE MIXTURE

Typical open crankcase ventilation system

GM diesel V8 engines are equipped with one of two different crankcase ventilation systems. The first system uses a crankcase vacuum regulator valve to meter the flow of crankcase gases back into the engine. The regulator limits crankcase vacuum as the gases are drawn from the valve covers through the regulator, and into the air crossover. This sytem is used on 1981 and later non-California models. Other models use a crankcase flow control valve to meter the blow-by gases back into the engine. On these models, a ventilation hose runs from each valve cover and connects the flow control valve, which is screwed into the back of the air crossover.

TESTING PCV SYSTEMS

NOTE: Do not attempt to test the crankcase controls on GM V8 diesels. Instead, clean the valve cover filter assemblies and vent pipes and check rubber fittings every 15,000 miles, and replace or clean the breather cap assembly and ventilation regulator valve (if equipped) every 30,000 miles.

Checking crankcase vacuum is the most effective way to test any PCV sytem. If there is a vacuum in the crankcase, then the major part of the system has to be working. Inspect the system to find out where the fresh air enters the engine. This is usually through a hose attached to the air cleaner, but is may be through the oil filler cap on some models. If the fresh air entry is separate from the oil filler cap, simply remove the cap.

On all models, use a piece of paper or a PCV tester to measure the crankcase vacuum at the oil filler cap, with the cap removed and the engine idling in Park or Neutral. It may take a few seconds for the vacuum to build up enough to suck the piece of paper against the oil filler hole. If the vacuum does not build up, check to be sure you have plugged the fresh air entry. An alternate method on some cars is to use the piece of paper or PCV tester on the end of the fresh air entry hose. When you do it that way, the oil filler cap must be the solid type and you must leave it in place.

If there is no crankcase vacuum, pull the PCV valve from the crankcase and hold your finger over the end of it. You should feel full manifold vacuum with the engine idling. If not, the valve is plugged or there is an obstruction in a hose or passageway. On some designs the valve may be screwed into its mounting, with a hose leading to the rocker cover or crankcase. If the valve has good suc-

tion, but there is no crankcase vacuum, check the hose to be sure it is open. PCV valves that are restricted or plugged must be replaced, unless they are the type that will come apart for cleaning. Lack of crankcase vacuum can also be caused by vacuum leaks at rocker cover, oil pan, or other engine gaskets. Usually, tightening the bolts will stop the leak.

In some extreme cases, usually on high mileage engines, the PCV system is in good shape, but the blow-by past the rings is so much that the system can't handle it, and the engine will blow smoke out the oil filler hole. Switching to a PCV valve with a higher flow may temporarily correct the problem. But the only good solution is to overhaul the engine. If the motor oil is contaminated with gasoline, the PCV system will pick up the unburned vapors, add them to the intake mixture and cause the engine to run excessively rich. After checking crankcase vacuum, always check the condition of the fresh air filter and hose, to be sure they are clean and not clogged.

NOTE: The PCV system operation is not computer controlled, but if inoperative it will directly affect the operation of any computerized emission system. If poor performance is a problem, check the PCV system first.

FUEL EVAPORATION CONTROLS

Charcoal Canister Vapor System

Evaporation controls are made up of hoses which allow the tank and carburetor vapors to go to a canister filled with charcoal. When the engine is

running, a hose to the intake manifold or carburetor base allows engine vacuum to pull fresh air through the canister, drawing the vapors into the engine where they are burned. Fresh air enters the canister through a filter, which keeps the charcoal clean.

When the engine is running, air must enter the tank to replace the fuel that is used up and prevent a vacuum. On all makes of canister storage models, air enters the tank through the filter in the canister, but air can also enter the tank through the pressure-vacuum tank cap.

All evaporation control systems use some sort of vapor separator at the fuel tank to prevent liquid fuel from traveling along the vent line to the canister. The early models had very elaborate separators mounted separately from the tank, but now they are simpler and usually attached to the top of the tank. The only periodic servicing required on evaporation controls is replacement of the canister filter on those models on which it is replaceable.

NOTE: If the vent lines become blocked, it is possible for some evaporation control systems to pull liquid fuel from the tank into the charcoal canister. If any charcoal canister is found to be fuel-soaked it should be replaced and all hoses checked for obstructions.

EXHAUST CONTROLS

Thermostatic Air Cleaner (TAC)

Fresh air supplied to the air cleaner comes either from the normal snorkle, or from a tube connected to an exhaust manifold stove. A door in the snorkle regulates the source of incoming air so that a warm engine always

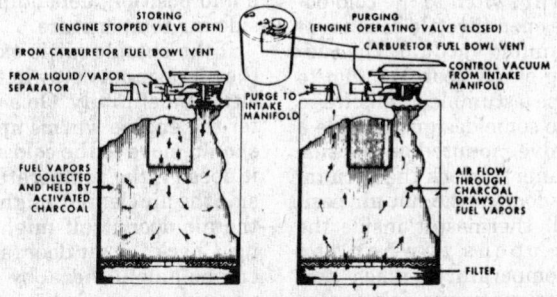

Vapor storage canister operation—typical

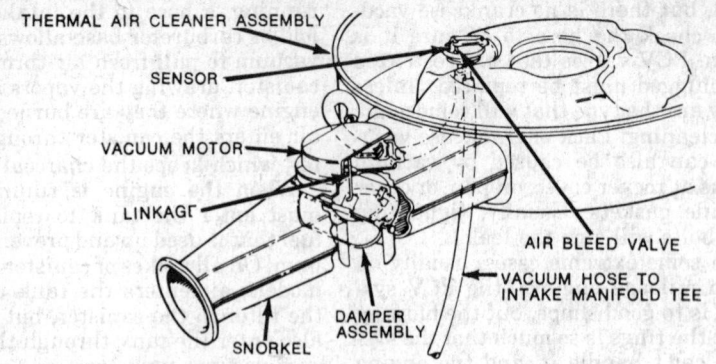

THERMAL AIR CLEANER ASSEMBLY
SENSOR
VACUUM MOTOR
LINKAGE
AIR BLEED VALVE
VACUUM HOSE TO INTAKE MANIFOLD TEE
DAMPER ASSEMBLY
SNORKEL

Vacuum controlled thermostatic air cleaner.

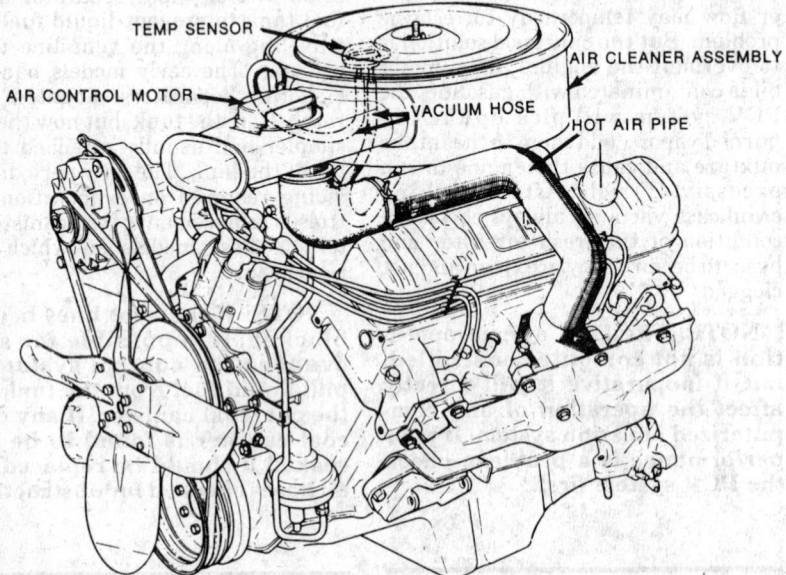

TEMP SENSOR
AIR CONTROL MOTOR
VACUUM HOSE
AIR CLEANER ASSEMBLY
HOT AIR PIPE

A typical heated air cleaner system, with the hot air pipe connected to the left exhaust manifold

takes in warm air, approximately 100°F. The snorkel door may be controlled in any number of ways, but most are vacuum operated. The vacuum operated designs use a thermostatic bimetal switch inside the air cleaner that bleeds off vacuum as the engine warms up and regulates the position of the air door.

Vacuum operated air doors are all designed so that the air cleaner takes in cold air when there is no vacuum. This means that an air door in the hot air position will switch to the cold position at wide open throttle because of the loss of manifold vacuum. The sudden switching of the door from hot to cold may cause a stumble or misfire in the engine, so some designs include a modulator valve mounted on the side of the air cleaner to block the vacuum and hold the door in the hot air position. A small thermostat inside the modulator opens it when the underhood temperatures reach normal. Other designs used a delay valve that allows the air door to move to the

cold position slowly, to prevent stumble.

TESTING TAC OPERATION

To test the vacuum type of heated air cleaner, inspect the air door with the engine off. It should be in the cold air position. Start the engine. If the engine is cold, the air door should move to the hot air position. As the engine warms up, the air door should move to a mid position, depending on the outside air temperature.

If the outside air is extremely cold, the air door may stay in the hot air position indefinitely. On a warm day, after the engine warms up, the air door should move to the cold air position. If it doesn't, the temperature sensor inside the air cleaner might be faulty, or the air door itself might be hanging up. Check the air door (a small mirror can be helpful here) by using a hand vacuum pump, or by running a hose from manifold vacuum to the vacuum

motor. Connect and disconnect the hose to see if the air door moves freely. If the air door is free, check out the hoses for leaks or blockage. If the hoses are okay, the trouble must be in the temperature sensor, and it should be replaced.

Both General Motors and Ford use a modulator in the air cleaner vacuum line on some engines. The modulator mounts on the side of the air cleaner and has two hose connections, one to the air cleaner temperature sensor, and the other to the vacuum motor. Below 50–80°F. the modulator is a one-way check valve, which allows vacuum to move the air door to the hot air position, but traps the vacuum so the door will not jump back to the cold air position during acceleration. This prevent a stumble.

After the module warms up the check valve unseats so that the vacuum can pass freely in either direction, and the air door then operates normally. The connections from the modulator are important. The connection in the center (usually the larger diameter) goes to the vacuum motor, and the connection on the edge goes to the vacuum source, which is the temperature sensor.

To test the modulator on a cold engine, apply enough vacuum to the edge port to move the air door to the hot position. Then remove the hose from the port, and the air door should stay in the hot position. Make the same test when the engine is warmed up, and the air door should move to the cold position when you pull off the hose.

Exhaust Gas Recirculation (EGR)

Gasoline Engines

NOx (oxides of nitrogen) is a tailpipe emission caused by the oxidation of nitrogen in the combustion chamber. When the peak combustion temperatures go over 2500°F, NOx is formed in excessive amounts. To keep the combustion temperatures down, exhaust gas is recirculated by allowing intake manifold vacuum to draw exhaust gas into the intake manifold. The lower combustion temperatures also help control spark knock (ping).

An EGR valve is used to control the flow of exhaust gas into the intake manifold. All EGR valves look similar and are operated by vacuum. When the vacuum is off, the valve is closed. Several different types of controls are used to turn the vacuum to the EGR valve on and off. Most of them have to do with engine temperature, as de-

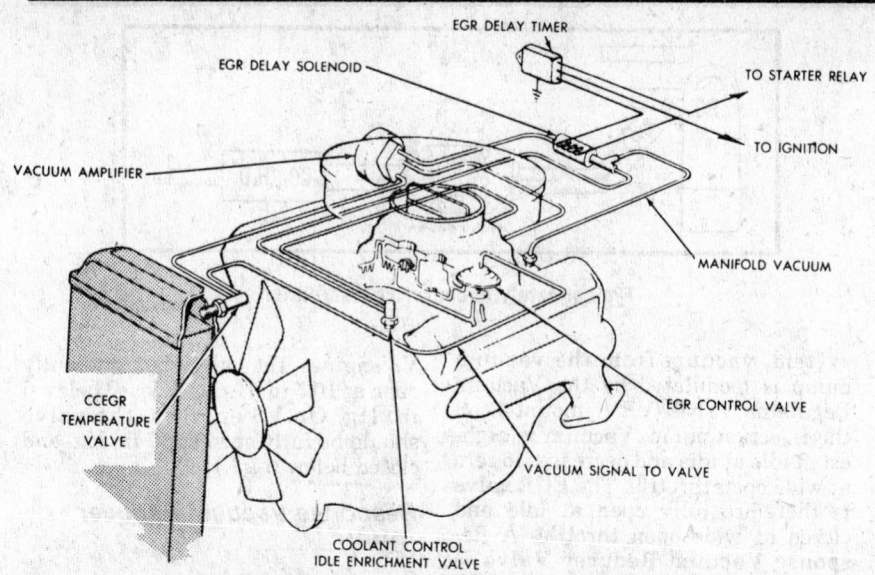

Venturi vacuum exhaust gas recirculation

scribed later. On computerized control systems, EGR operation is regulated by the electronic control unit.

NOTE: All EGR systems are designed to cut off exhaust recirculation when the engine is cold, at idle, or under hard acceleration. If the EGR valve is stuck open, the engine won't idle.

TESTING EGR SYSTEM

— CAUTION —

The EGR valve gets hot during normal operation. Take normal precautions to avoid accidental burns.

Testing of EGR systems should verify that when the engine is at normal operating temperature the EGR valve is closed at idle, open above idle, and that the exhaust gas is actually recirculating. If the EGR valve sticks open at idle, the engine will run very rough, or may not even start. If this happens the valve should be removed and cleaned or replaced. To check for valve operating above idle, check with a mirror to see if the diaphragm or stem moves when the engine is at fast idle in Park or Neutral. If the diaphragm does not move when the throttle is opened, there is either a problem with vacuum, or the valve is stuck closed. With a vacuum gauge hooked up to the EGR port, you should see vacuum on the gauge when the throttle is opened. EGR valves should not leak when tested with a hand vacuum pump. If they do they must be replaced.

NOTE: The EGR valve should open when about 3–5 in. Hg. is ap-plied with a hand vacuum pump. **Back pressure operated EGR valves cannot be vacuum tested.**

To find out if the exhaust gas is actually recirculating, use a hand vacuum pump to open the EGR valve with the engine idling. If the engine runs rough or dies, the exhaust gas is recir-

culating. If the engine does not run rough, make a second test of 2500 rpm. Opening the EGR valve at that rpm should cause a change in engine speed. If it does, the exhaust gas is recirculating. To make the 2500 rpm test, remove and plug the hose from the EGR port. Attach the suction hose to the EGR valve before running the engine at 2500 rpm. Simply pulling off the EGR hose at 2500 rpm is not a valid test, because the extra air entering the engine through the hose could cause a speed change by itself.

If the exhaust is not recirculating, it means that a passageway or the valve itself is clogged up. The only way to fix it is to clean out the clogging as best you can, replace the clogged part, or replace the EGR valve. Many EGR valves have a back pressure sensor built into the valve. This sensor is a pressure operated bleed that disables the EGR valve and keeps it closed when there is no exhaust pressure. This type of valve cannot be tested with a hand vacuum pump with the engine off because the bleed is open. The only practical way to test these new valves is by substitution of a known good valve. If a valve is not available, the suspect valve can be removed, and the mounting holes temporarily taped shut. If this cor-

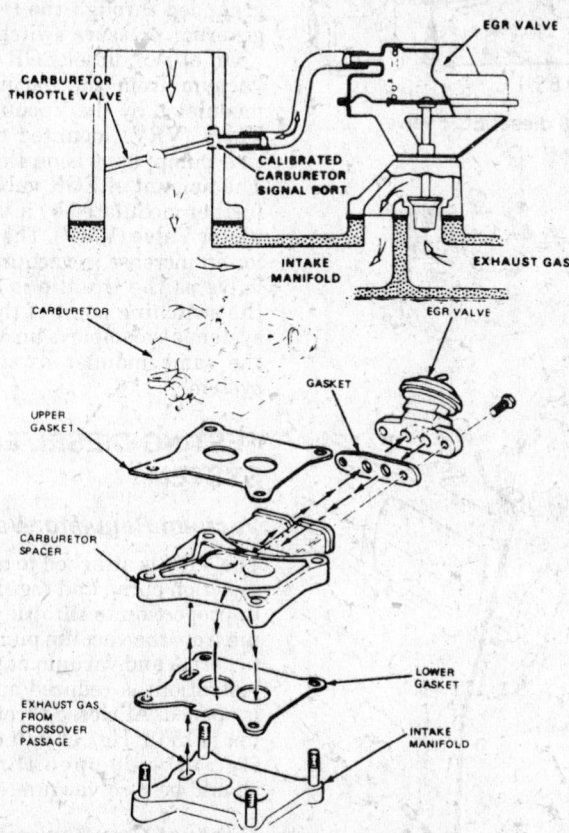

Most cars use an EGR system with a valve and a ported vacuum signal, as shown here. Some cars use the venturi vacuum with a separate amplifier to operate the valve.

rects the problem, then a new valve should be installed.

Diesel EGR Systems

GM V6 and V8 Engines

GM has equipped its V8 and V6 diesel engines with EGR systems. The diesel EGR systems work in the same basic manner as gasoline engine EGR systems: exhaust gases are introduced into the combustion chambers to reduce combustion temperatures, and thus lower the formation of nitrogen oxides (NOx). There are two systems used on the V8 diesels. One is used on the B (large body) type station wagons, and one system is used on all other cars.

On the B–body station wagon EGR

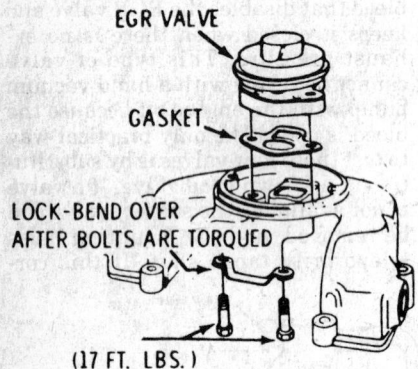

GM V6 diesel EGR valve

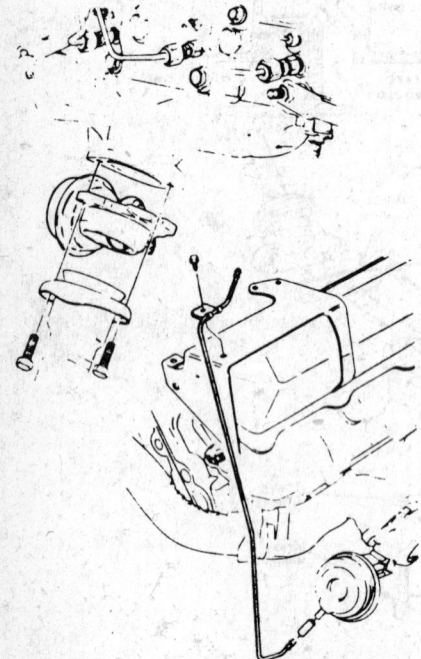

GM V6 diesel Exhaust Pressure Regulator Valve (EPR)

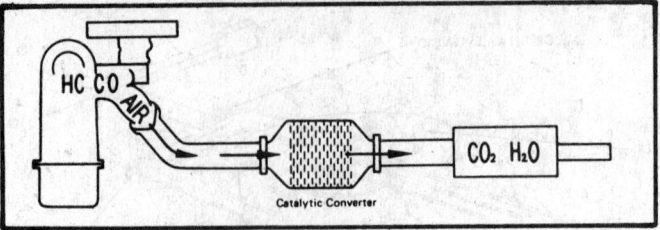

Typical catalytic converter installation

system, vacuum from the vacuum pump is modulated by the Vacuum Regulator Valve (VRV) mounted on the injection pump. Vacuum is highest at idle at idle and decreases to zero at wide open throttle. The EGR valve is therefore fully open at idle and closed at wide open throttle. A Response Vacuum Reducer Valve is used between the VRV and the EGR valve to allow the EGR valve to change position quickly as throttle position is changed.

On all other V8 diesel engines, the EGR system is the same as used on the B-body wagon, except a solenoid is added to the system that shuts off vacuum to the EGR valve when the Torque Converter Clutch is engaged. This solenoid is fed 12V from the TCC switch portion of the VRV and is grounded through the transmission's governor pressure switch.

On all V6 diesel EGR systems, the vacuum from the vacuum pump is modulated by the Vacuum Regulator Valve (VRV) mounted on the injection pump, as it is on the V8 diesels. The amount of EGR valve opening is further modulated by a Vacuum Modulator Valve (VMV). The VMV allows for an increase in vacuum to the EGR valve as the throttle is closed, up to the switching point of the VMV. The system also employs an VRV valve in the same manner as the V8 diesel system.

TESTING DIESEL EGR SYSTEM

Vacuum Regulator Valve (VRV)

The VRV is attached to the side of the injection pump and regulates vacuum in proportion to throttle angle. Vacuum from the vacuum pump is supplied to port A and vacuum at port B (see illustration) is reduced as the throttle is opened. At closed throttle the vacuum is 15 in. Hg.; at half throttle, 6 in. Hg.; at wide open throttle there should be zero vacuum.

Exhaust Gas Recirculation (EGR) VALVE

Apply vacuum to the vacuum port. On

V8 engines, the valve should be fully open at 10.5 in. Hg. and closed below 6 in. Hg. On V6 engines, the valve should be fully open at 12 in. Hg. and closed below 6 in. Hg.

Response Vacuum Reducer (RVR)

Connect a vacuum gauge to the port marked "To EGR valve or TCC solenoid". Connect a hand operated vacuum pump to the VRV port. Draw 15 in. of vacuum on the pump and the reading on the vacuum gauge should be .75 in. Hg. lower than the vacuum pump reading on all except High Altitude V8 engines. On High Altitude V8 engines ONLY, the reading should be 2.5 in. Hg. lower.

Exhaust Pressure Regulator Valve (V6 Diesels)

Apply vacuum to the vacuum port of the valve. The valve should be fully closed at 12 in. Hg. and open below 6 in. Hg.

Vacuum Modulator Valve (VMV)

To test the VMV, block the drive wheels, and apply the parking brake. With the shift lever in Park, start the engine and run at a slow idle. Connect a vacuum gauge to the hose that connects to the port marked "MAN". There should be at least 14 in. Hg. of vacuum. If not, check the vacuum pump, VRV, RVR, solenoid, and all connecting hoses. Reconnect the hose to the "MAN" port. Connect a vacuum gauge to the "DIST" port on the VMV. The vacuum reading should be 12 in. Hg. except on High Altitude cars, which should be 9 in. Hg.

Catalytic Converters

Two main types of converters are used on today's vehicles. The first is an oxidation type converter containing two precious (noble) metals, platinum and palladium to effectively catalyze the oxidation of the hydrocarbons (HC) and carbon monoxide (CO). The second type converter used is considered a three-way catalyst, containing plat-

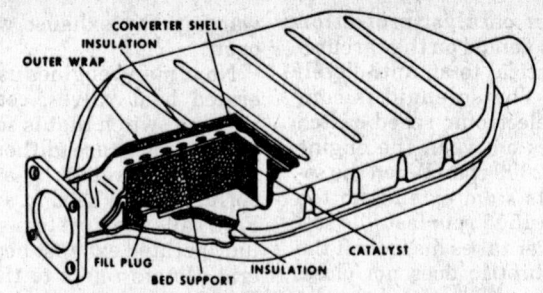

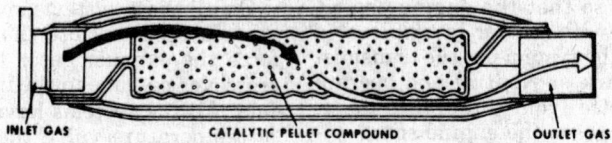

Pellet type catalytic converter

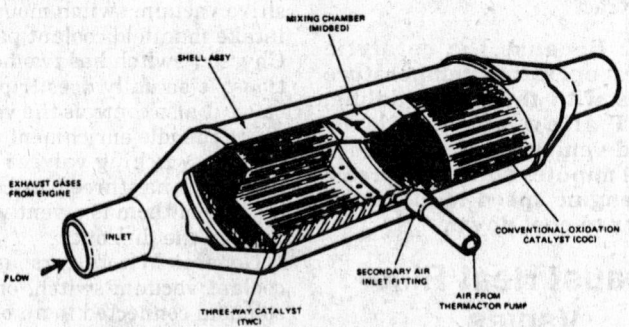

Sectional view of three-way catalytic converter

inum and rhoduim in the front part of the converter to reduce the oxides of nitrogen (NOx), while platinum and palladium are used in the rear section to oxidize the hydrocarbons (HC) and carbon monoxide (CO), as was done in the two-way converters.

Oxidizing Catalytic Converters

These converters do not operate unless there is sufficient oxygen in the exhaust stream. It is extremely important that the proper amount of oxygen is supplied at all times. This is accomplished by a secondary air source, provided by either an air pump system or a pulse air type system. The catalytic converter system is protected by several devices that block out the secondary air supply when the engine is laboring under any abnormal hot or cold operating situation, preventing converter overheating and burnout. Converter temperatures are normally between 900 and 1500°F, with peak temperatures around 1800°F, so the converterss must be hot to properly perform their functions. Should the converter be

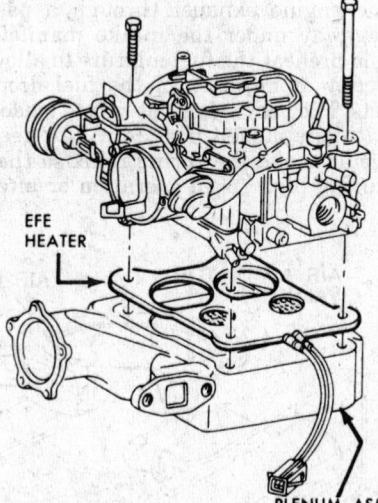

1981 G.M. electric early fuel evaporative heater

supplied too rich a mixture of hydrocarbons (HC), such as would result from a misfiring spark plug or stuck choke valve, along with an oversupply of fresh air, the converter temperature would increase sharply, causing a burnout of the catalyst material.

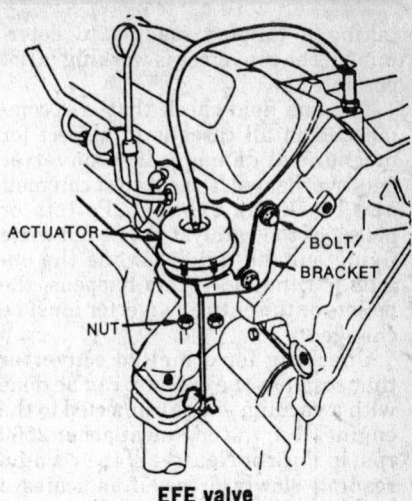

EFE valve

NOTE: Some computer controlled systems use an air management valve to increase converter efficiency by routing air to the exhaust system under certain conditions.

Three-Way Catalytic Converters

The three-way catalytic converters use a combination of catalysts which produce two different chemical reactions, oxidation and reduction. By adding fresh air to the unburned hydrocarbons (HC) and carbon monoxide (CO) within the converter, the oxidizing or combustion process takes place.

Just the reverse process is required to lower the oxides of nitrogen (NOx) emissions. The oxides of nitrogen (NOx) already contains excessive oxygen and the process of separating the excess oxygen from the nitrogen is called a reducing reaction. This reducing or reduction process is done in the front section of the converter while the oxidizing process is accomplished in the rear section. A fresh air connector is located on the center of the converter shell to add fresh air from the air system as required.

To enable the three-way converter to operate properly, the engine air/fuel ratio must be held within a tight range, called a "Stoichiometric" range. This is accomplished with the use of the closed loop, feedback fuel management systems incorporating the latest electronic controls.

TESTING

There is no way to test a catalytic converter in the field to see if it is actually working. Tailpipe readings may be used to set carburetor idle mixtures, when the car maker requires it, but taking a tailpipe reading to determine if the converter is working is not possible.

The one field check that is recommended in all cases is to inspect for mechanical damage. If a converter gets overheated, the catalyst can melt and block the exhaust. Pellets or pieces of the catalyst may even come flying out the tailpipe while the engine is running. If this happens, the pellets or the entire converter must be changed.

Checking for a melted converter that restricts the exhaust can be done with a vacuum gauge connected to the engine. Run the engine at about 2500 rpm in Park or Neutral. If the vacuum reading slowly drops, it indicates a buildup of pressure in the exhaust.

The use of leaded fuel will slowly destroy the efficiency of the catalyst. If used long enough, leaded fuel can even cause catalyst plugging to the point where the engine will not run. If you know that a car has been run on several tanks or leaded fuel, then you can be sure that the catalyst is ruined. The only thing you can do is change the catalyst or install a new converter.

NOTE: Do not change the catalyst if the car has been run on only one tank or less of leaded fuel. Switching back to unleaded will allow the catalyst to recover and be almost as efficient as it was.

CONVERTER OVERHEAT PROTECTION

Some cars have overheat protection systems for the converter. Ford Motor Co. sometimes uses a heat sensitive switch mounted in the floorpan above the converter. The switch turns a vacuum to the air pump bypass valve. When the vacuum is shut off the bypass valve dumps the pump air into the atmosphere, diverting the air away from the exhaust system. Without the air in the exhaust, the converter's catalytic heat reaction slows and the system cools down.

Chrysler Corporation cars use an overheat protection system that holds the throttle open to prevent high speed closed throttle deceleration. Any engine decelerating on closed throttle is usually running rich, because the high vacuum pulls so much fuel out of the carburetor bowl through the idle circuit. This rich mixture can cause the catalytic reaction to speed up, increasing the heat generated to dangerous levels. Holding the throttle open slightly while decelerating allows more air into the engine and eliminates the problem.

The Chrysler catalyst protection system uses a solenoid on the carburetor that is identical to an anti-dieseling solenoid. The solenoid is controlled by an electronic speed switch and only comes on when the engine speed is above 2000 rpm. When the solenoid is on, its stem extends to the equivalent of a 1500 rpm fast idle setting. If the driver takes his foot off the throttle, the throttle does not close, but rests against the extended solenoid stem. The solenoid goes off below 2000 rpm so that the engine doesn't run away with the car in traffic.

To test the system put the transmission in Park or Neutral and operate the throttle from under the hood. Slowly increase the engine speed until it is above 2000 rpm. The solenoid stem should extend. As the speed drops below 2000 rpm, the stem should retract

NOTE: Because the catalytic converter operating temperature increases with the engine idling, DO NOT allow any catalyst-equipped vehicle to idle for more than five minutes without increasing the engine speed to allow the converter to cool down.

Exhaust Heat Riser Valves

Exhaust heat riser valves have been used for many years to force part of the engine exhaust through a passageway under the intake manifold and preheat the fuel mixture to allow better atomization of the fuel droplets. The heat valve was spring loaded into the closed position, but heat would make the spring relax so that during high speed operation or after

warmup the exhaust would push it open.

Now, many engines use vacuum-operated heat valves, controlled by a vacuum switch that is sensitive to engine temperature (although the above system, controlled by a thermostatic spring, is still used in some engines). Ford calls their system simply a vacuum operated exhaust heat valve. General Motors refers to theirs as Early Fuel Evaporation, and Chrysler calls theirs a Power Heat Control Valve.

On all these systems, manifold vacuum is used to close the valve, and force the exhaust gases through the crossover passage in the intake manifold. All the systems have some kind of temperature valve that shuts the vacuum off when the engine warms up. Both Chrysler and Ford products use a simple coolant temperature-sensitive vacuum switch mounted on the intake manifold coolant passage. The Chyrsler switch has two hose connections. It actually does triple duty because it also controls the vacuum supply to the idle enrichment system and the air switching valve. Ford's vacuum switch has three hose connections, but one of them is a vent with a filter to keep the dirt out.

General Motors cars use either a coolant vacuum switch, or a vacuum solenoid connected to an oil temperature switch. The coolant vacuum switch has two hose connections and a vent when it controls the heat valve only. When it is tied into other emission control systems, it can have as many as five hose connections, and a vent. Many General Motors cars also have a check valve in the hose so that vacuum will be trapped in the heat valve actuator when the engine is accelerated. This keeps the heat valve

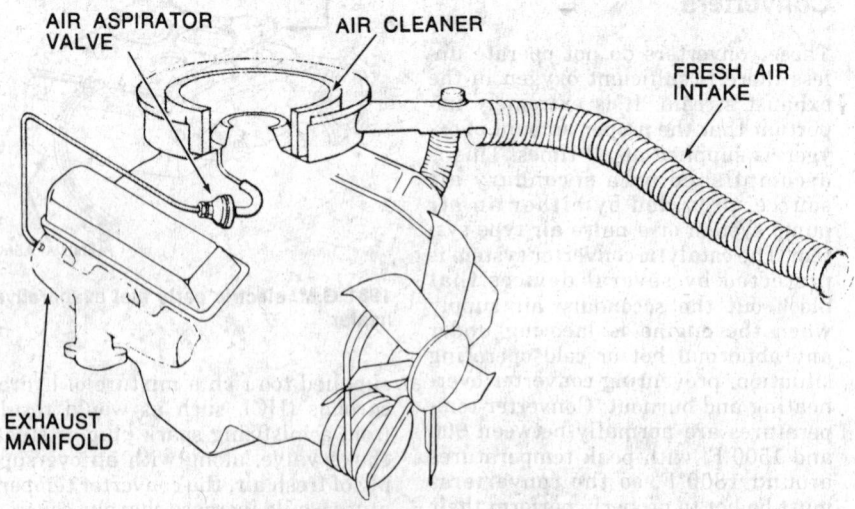

AIR ASPIRATOR VALVE   AIR CLEANER   FRESH AIR INTAKE

EXHAUST MANIFOLD

Chrysler Air Aspirator system

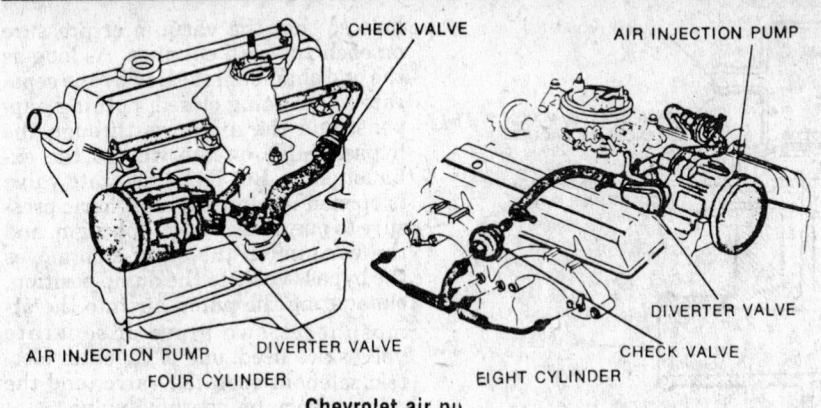

Chevrolet air pu

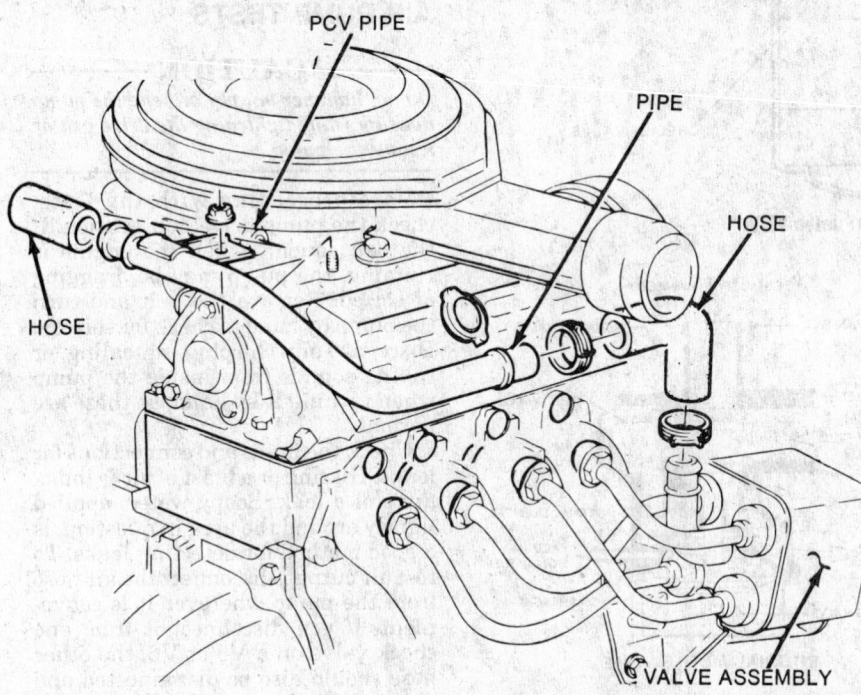

Pulse Air pipe and hose

TESTING EFE HEATER

To check the resistance of the heater, turn the ignition OFF, disconnect the heater electrical connector, using a ohmmeter, measure the resistance across the two terminals of the heater connector. If resistance is under 2 ohms, the heater is good. If not, replace the heater.

Air Injection Systems

On these systems, a belt-driven air pump supplies air to small tubes positioned in the exhaust port near each exhaust valve. The air mixes with any unburned hydrocarbons in the exhaust and the hydrocarbons burn up in the exhaust system. On late model engines, air may not be pumped to every exhaust port, and some engines have only a single air injection fitting on the exhaust pipe near its connection to the exhaust manifold. Air injection systems are frequently used on engines with catalytic converters so that the converter gets enough air to keep the reaction going.

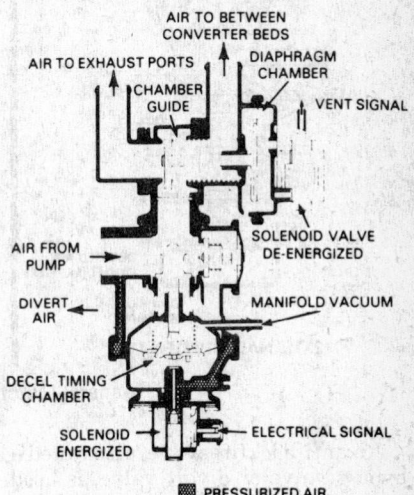

AIR SWITCHING VALVE CROSS SECTION
(AIR TO CONVERTER OPERATION)

Typical GM air switching valve (ASV) assembly

Plumbing on air injection systems varies considerably. At first, all the plumbing was external, with individual tubes inserted into each exhaust port either through the cylinder head or the exhaust manifold. Now many engines have internal passageways to duct the air to the exhaust port. A check valve is used between the pump and the exhaust port nozzle to keep hot exhaust gases from traveling up the plumbing and destroying the pump. Some V8 and V6 engines use two check valves.

in the closed position and prevents a rattle.

TESTING

Testing the vacuum operated heat riser valve is a matter of making sure it closes and opens freely. You can move it by hand to see if it works, on a warm engine. On a cold engine, the valve should be closed, and disconnecting the hose should allow it to open (engine idling). On a cold engine, there should be vacuum at the vacuum actuator, and on a warm engine the vacuum should be shut off.

GM Early Fuel Evaporation (EFE) System

The electrically operated EFE system

used on some 1981 and later GM engines performs the same function as the vacuum operated heat riser on other engines, which is to preheat the engine induction system during cold driveway. Rapid heating is desirable because it provides quick fuel evaporation and more uniform fuel distribution to aid cold driveability.

The electrically heated EFE system has a ceramic heater grid located underneath the primary bore(s) of the carburetor which is part of the carburetor insulator. When the ignition is turned on and engine coolant temperature is low, voltage is applied to the EFE relay, which in turn transfers the voltage to the EFE heater in the ceramic grid. When temperature increases, a thermal valve switch de-energizes the relay and the heater is turned off.

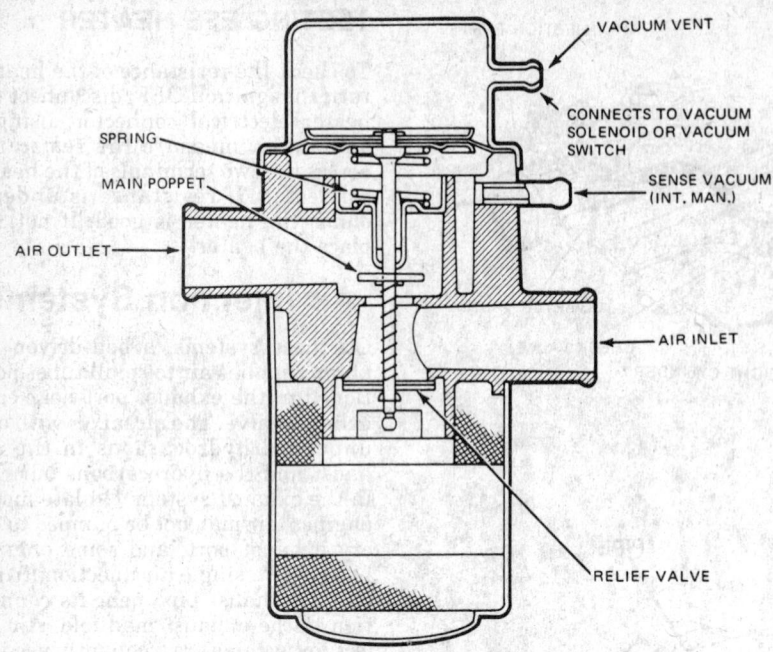

Vacuum differential valve VDV

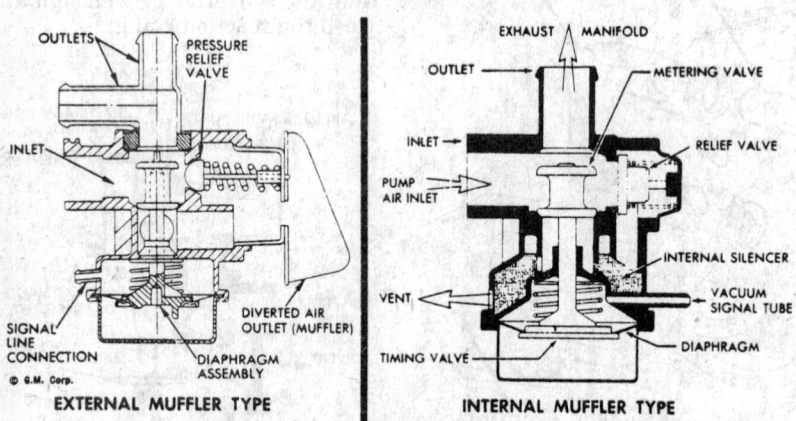

EXTERNAL MUFFLER TYPE   **INTERNAL MUFFLER TYPE**

General Motors Diverter Valves

An anti-backfire valve, also called a bypass valve or divert valve, is used between the pump and the check valve. Usually, the diverter valve is mounted on the pump or near it. A small sensing hose connects the diverter valve to intake manifold vacuum. When the vacuum rises during deceleration, the diverter valve opens and vents the pump air into the atmosphere. This prevents an over-rich fuel mixture in the exhaust system from exploding or backfiring out the tailpipe. Some systems have a delay valve, similar to a spark delay valve, in the sensing hose. This delays for a few seconds the drop in vacuum when the throttle closes, so that the air is not dumped every time the driver takes his foot off the throttle in traffic.

Temperature controls are also used in the sensing hose hookup. Usually, the temperature valve shuts the vacuum off when the engine is cold, so that the pump air doesn't go to the engine exhaust ports until the engine warms up. Some cars have a temperature sensor mounted under the car above the catalytic converter. If the converter overheats, the sensor turns off a solenoid which shuts off the air to the diverter valve. The diverter valve then goes to the dump position, shutting off the air to the exhaust to keep the converter from melting or burning up.

Ford Motor Company 4-cylinder, V6, and some inline 6 engines use a unique air bypass valve, with two small sensing hoes connected to it. Each of the hoses connects to one side of a diaphragm in the valve. The hose on the body of the valve connects to manifold vacuum, and the hose closer to the end connects to a separate on-off valve. The diaphragm has a small

hole so that the vacuum or pressure on each side will equalize. As long as the end chamber is sealed by the separate valve being closed, nothing happens, and the air flows through the bypass valve on the way to the exhaust ports. But if the separate valve is opened it admits atmospheric pressure to one side of the diaphragm, and the vacuum on the other side moves the bypass valve to the dump position, exhausting the pump air into the atmosphere. Two types of separate valves are used, one of them an electric solenoid operated valve, and the other a vacuum-operated valve.

AIR PUMP TESTS

CAUTION

Do not hammer on, pry or bend the pump housing while tightening the drive belt or testing the pump.

Before proceeding with the tests, check the pump drive belt tension. If the belt squeals when the engine is running, the pump may be dragging or seized. Remove the belt and turn the pump by hand to check for seizure. Disregard any chirping, squealing, or rolling sounds from inside the pump when turning it by hand, as these are normal.

Check the hoses and connections for leaks. Hissing or a blast of air is indicative of a leak. Soapy water, applied lightly around the area in question, is a good method for detecting leaks. To test air output, disconnect the air hose from the pump wherever it is convenient. If you disconnect it from one check valve on a V8 or V6, the other hose should also be disconnected and plugged for the test. Run the engine at idle and feel the blast of air from the hose with your hand. Increase the engine speed to 1500 rpm and feel the blast of air again. If the blast increases and is steady, the pump is okay.

Pump Noise Diagnosis

The air pump is normally noisy. As engine speed increases, the noise of the pump will rise in pitch. The rolling sound the pump bearings make is normal. However, if this sound becomes objectionable at certain speeds, the pump is defective and will have to be replaced. A continual hissing sound from the air pump pressure relief valve at idle indicates a defective relief valve. Replace the relief valve.

If the pump rear bearing fails, a continual knocking sound will be hard. Since the rear bearing is not separately replaceable, the pump will have to be replaced as an assembly.

DIVERTER (ANTI-BACKFIRE) VALVE TEST

Detach the hose, which runs from the bypass valve to the check valve. Connect a tachometer to the engine. With the engine running at normal idle speed, check to see that air is flowing from the bypass valve hose connection. Increase the engine speed to 1500-2000 rpm and allow the throttle to snap shut. The flow of air from the bypass valve at the check valve hose connection should stop momentarily and air should then flow from the exhaust port on the valve body or the silencer assembly.

Let the throttle snap shut several times. If the flow of air is not diverted into the atmosphere from the valve exhaust port or if it fails to stop flowing from the hose connection, check the vacuum lines and connections. If these are tight, either the bypass valve or one of the accessory valves in the small sensing hose is defective and must be replaced. A leaking diaphragm will cause the air to flow out both the hose connection and the exhaust port at the same time. If this happens, replace the valve.

NOTE: Late model systems should stop flowing at idle, as described earlier. If not, the bypass valve or accessory valve is defective.

CHECK VALVE TEST

Remove the hose from the check valve. With the engine running at 1500 rpm in Park or Neutral, hold the back of your hand near the check valve to test for exhaust gas leakage. If the valve leaks, it must be replaced.

NOTE: Vibration and flutter of the valve at idle is a normal condition caused by exhaust pulsations. It does not mean that the valve is defective.

VACUUM DIFFERENTIAL VALVE TEST

Disconnect the small sensing hose at the bypass valve and connect a vacuum gauge to the hose. With the engine idling in Park or Neutral, the gauge should read full manifold vacuum. Run the engine at a steady 2500 rpm in Park or Neutral, and release the throttle. As the engine decelerates, the vacuum gauge should drop close to zero, then return to full manifold vacuum as the engine speed drops to idle. If not, the VDV is defective and must be replaced.

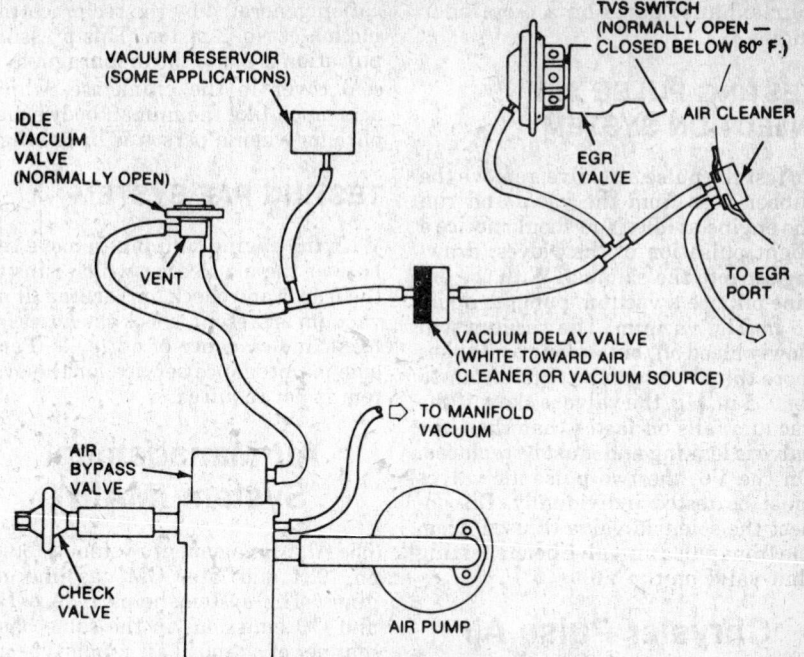

Air pump system using a timed air by-pass valve vacuum vent

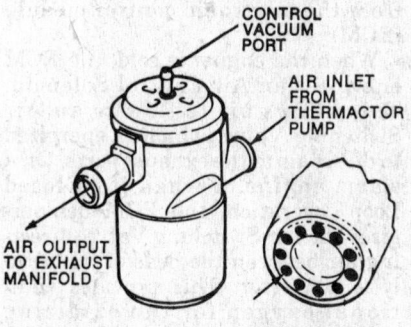

Catalyst cars use a different air bypass valve, with small hose connecting to the end.

NOTE: The small hose nozzle should be connected to manifold vacuum.

Chrysler Air Aspirator System (AAS)

Chrysler Corporation cars which use this system have done away with the air pump. The complete air aspirator system consists of a hose from the clean side of the air cleaner, the aspirator valve mounted on top of the engine, and a tube connecting the valve with the exhaust manifold. The suction in the exhaust draws in air through the air cleaner and this extra air helps the catalytic converter burn up the pollutants. The aspirator valve is similar to the check valve used with all air pump systems. It keeps the exhaust from flowing back into the air cleaner, but allows clean air to go into the exhaust.

TESTING AAS VALVE

Testing the air aspirator valve is done by disconnecting the hose from the air cleaner and checking for slight suction at idle with a piece of paper over the end of the valve. Speeding the engine up slightly will show if the valve is leaking. Exhaust should not come out of the valve. Vibration of the valve diaphragm is normal due to exhaust impulses.

GM Pulse Air Injection System

This system is similar to Chrysler's Air Aspirator. A hose from the clean side of the air cleaner connects to the pulse air valve. Tubes connect the pulse air valve to each cylinder's exhaust port. Suction in the exhaust draws fresh air from the air cleaner into the exhaust, and the air helps the catalytic converter burn up the pollutants. The pulse air valve consists of four or six check valves built into a housing. It allows each exhaust port to suck in fresh air independently of the other ports. The check valves only open when there is suction in the exhaust. If there is any back pressure, the check valves close to prevent exhaust flow back into the air cleaner. On some applications the pulse air valve is connected to only three of the four exhaust ports on a 4-cylinder engine.

TESTING PULSE AIR INJECTION SYSTEM

To test the pulse air valve, remove the rubber hose from the valve and run the engine at idle. You should notice a slight pulsation of the valves, drawing air into the exhaust. With the engine off, use a vacuum pump to apply 15 in. Hg. vacuum. The vacuum will slowly bleed off, but as long as it takes more than two seconds to fall from 15 in. to 5 in. Hg. the valve is okay. If the vacuum falls off faster than that, the valve is leaking and must be replaced. On the V6, the two pulse air valves must be tested individually. Disconnect the solenoid valve (if used) from the front pulse air valve before testing that valve on the V6.

Chrysler Pulse Air Feeder (PAF) System

The PAF system supplies secondary air into the exhaust system between the front and rear catalytic converters, which promotes oxidation of exhaust emissions in the rear catalytic converter. The system consists of a pulse air feeder, which contains two reed valve assemblies, a hose which links the pulse air feeder to the air cleaner, and a tube which runs from the feeder to the exhaust system. At the bottom of the feeder there are two tubes, one which runs into the oil sump and one which connects to No. 3 cylinder crankcase above the oil level. The main reed valve is actuated by a diaphragm in the feeder which, in turn, is activated by the pressure pul-

sation generated by the reciprocating motion of No. 3 piston. This pressure pulsation is fed to the diaphragm by a seal cover in the crankcase, which acts much like the human body's diaphragm when a person is breathing.

TESTING PAF SYSTEM

With the engine running, remove the hose at the air cleaner which runs to the feeder and check for vacuum. If no vacuum is present, check the hoses for leaks and evidence of oil leaks. Periodic maintenance service for the system is not required.

Air Management System (MAIR)

The Air Management System is found on 1981 and later GM gasoline engines. The system helps reduce HC and CO emissions in the same basic manner of a typical air pump-type air injection system, except that the MAIR system is controlled by signals from the electronic control module (ECM).

When the engine is cold, the ECM energizes an Air Control Solenoid. This allows air to flow to an Air Switching Valve, which is energized to direct air to the exhaust ports. On a warm engine or when in "Closed Loop" operation, the ECM de-energizes the Air Switching Valve, directing air between the beds of the catalytic converter. This provides additional oxygen for the oxidizing catalyst to decrease the HC and CO levels. If the Air Control Valve detects a rapid increase in manifold vacuum (deceleration, etc.), certain operating modes (wide open throttle, etc.),

or the ECM self-diagnostic system detects any problem in the MAIR system as a whole, air is diverted (divert mode) to the air cleaner or directly into the atmosphere.

The air flow and control hoses transmit pressurized air to the catalytic converter or to the exhaust ports through internal (intake manifold) passages or external piping. The check valves prevent backflow of exhaust gas into the air distribution system. The valve prevents backflow when the air pump "bypasses" at high speed and loads, or in case the air pump malfunctions.

NOTE: Due to the complex nature of modern electronic engine control systems, comprehensive diagnosis and testing procedures fall outside the confines of this repair manual. For complete information on diagnosis, testing and repair procedures concerning all modern engine and emission control systems, please refer to *Chilton's Guide To Electronic Engine Controls.*

Ford Pulse Air (Thermactor II) System

Some Ford engines are equipped with an air injection system which does not use an air pump. Instead, natural pulses present in the exhaust system are used to pull the air into the system through the pulse air valves. The pulse valve is connected to the exhaust manifold by a tube and to the air cleaner or silencer with a hose. Make sure air can flow freely through the air cleaner or silencer to the check valve.

Electronic Engine Controls

ENGINE ELECTRONICS

In the ladder part of the 1960's, Robert Bosch introduce the first true electronically controlled engine with an on-board computer. Today, almost every car produced has some kind of electronic control. The once mechanically controlled engine functions are all but extinct.

The first system, Bosch D-Jetronic, is comprised of electrically energized fuel injectors in which the injection time is controlled by an electronic control unit (ECU). The early system delivered a basic quantity of fuel and varied from this point depending upon engine load, engine speed and engine temperature.

Since the early days of ECU, the controls have become more complex, with a much greater amount of computer memory and even the ability to learn.

In this section the topics will include different types of electronically controlled fuel induction, spark control, the sensors and switches that provide the ECU with information, other non-engine related controls that the ECU might supply and some ECU self-diagnostics.

The most common fuel induction system with an ECU is electronic fuel injection. In this system fuel can be delivered many different ways. One of which is the single point injection (SPI) were one or two injectors are mounted on a throttle body assembly. Fuel is delivered constantly through the injector(s), but in varying quanti-

ties. The SPI system very much resembles a carbureted system. SPI is more commonly known as throttle body injection (TBI). Another fuel injection system is multi-point injection (MPI). This system supplies one injector for each cylinder, usually positioned in the intake manifold, just above the intake valve. In MPI, fuel can be injected in two ways. One is to energize a group of injectors, thus atomizing fuel in the intake manifold and storing it for a short time until the intake valve opens. The second way is to sequentially energize each cylinder's injector as the intake valve is opened. This injection is the more efficient, effective and more complex system.

Another fuel induction system utilizing an ECU is the feedback carburetor (FBC). A conventional carburetor is still used but it has a more precise air/fuel mixture control which is achieved through an integral mixture control solenoid. The solenoid is energized on and off by the ECU to maintain mixture demand. The ECU calculates air/fuel mixture demand changes by the data it receives through remote sensors. The most important sensor (and makes the system possible) is an oxygen (O_2) sensor (which will be discussed later in this section). The ECU monitors the exhaust gases for rich/lean conditions by way of the O_2 sensor and, in turn, controls the air/fuel mixture by increasing or decreasing the duty cycles (on and off) to the mixture control solenoid for an optimum 14.7:1 air/fuel ratio.

ECU Self-Diagnostics

The ECU can detect a malfuction or

abnornality in the sensors or in the ECU itself and display a warning light on the instrument panel when it does. When this occurs, the ECU stores a trouble code for future system diagnosis. If the problem is sever enough to where it inhibits closed loop operation, the ECU will assume a backup system. This fail-safe circuit is pre-programmed into the ECU for minimal driveability operation so the vehicle can be driven to a nearby service facility. The trouble codes are usually a two digit numbers identified by the number of diagnostic LED or check engine light flashes. The trouble codes assist the service technician in isolating a faulty circuit or component within the system.

Electronic Data Sensors

The engine control system consists of various data sensors. Although data sensor names and applications vary from system to system, the most common input sensors/switches are:

- oxygen (O_2) sensor
- coolant temperature sensor
- manifold air pressure (MAP) sensor
- vehicle speed sensor (VSS)
- throttle position sensor (TPS)
- engine speed reference or distributor reference (rpm)
- air flow sensor
- air intake temperature sensor
- crankshaft sensor
- detonation (knock) sensor
- throttle body temperature sensor
- throttle idle switch
- transmission or drive switch
- a/c compressor clutch switch

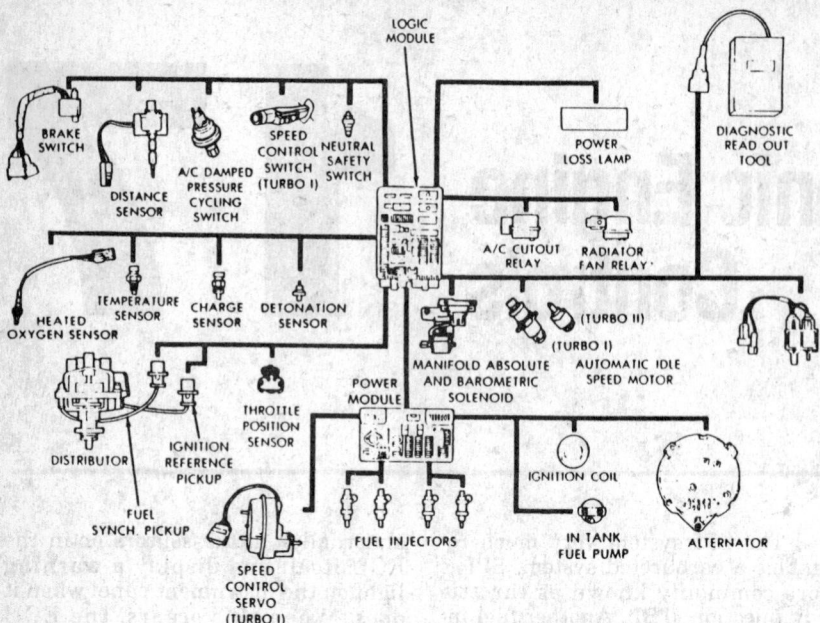

Electronic engine control components

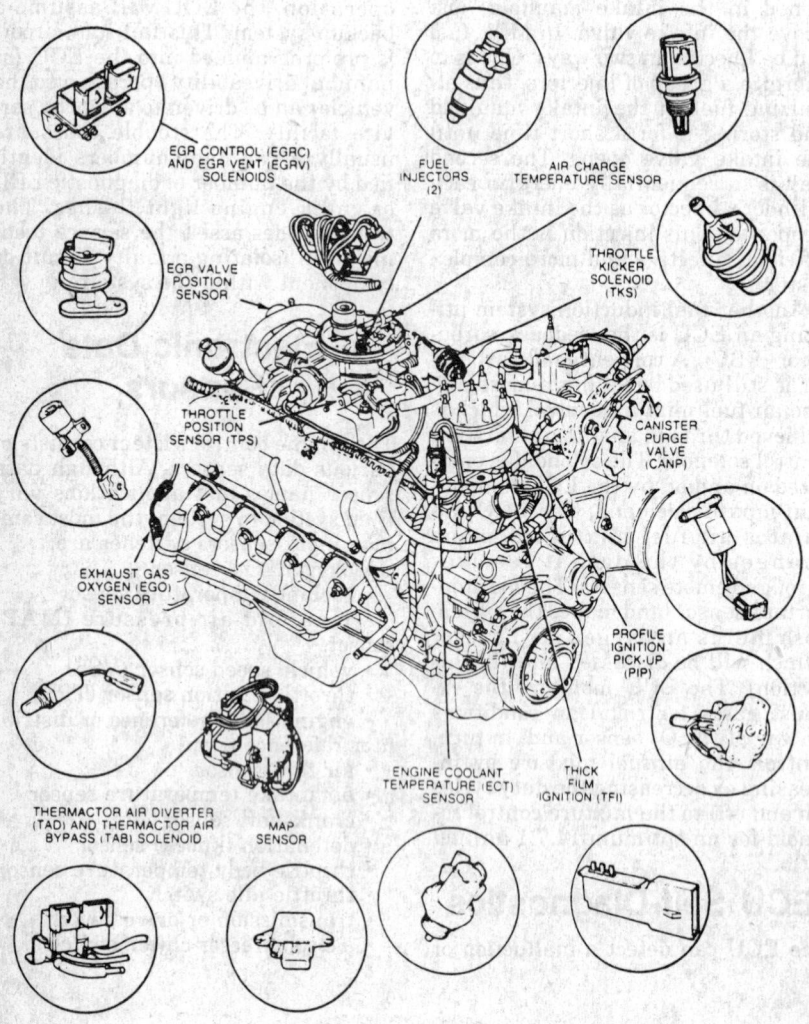

Electronic engine control component locations

- power steering pump switch
- altitude or barometric pressure sensor
- wide open throttle switch

Electronically Controlled Devices

Some of the output devices that the ECU may control vary from system to system, but the most common output or ECU controlled devices are:

- fuel injector(s)
- air/fuel mixture solenoid
- fuel pump relay
- a/c compressor clutch relay
- idle air control (IAC) valve
- idle speed control (ISC) motor
- ignition spark/timing
- canister purge solenoid
- torque converter clutch solenoid (automatic transmission)
- air management system (air induction)
- idle-up or throttle kicker solenoid
- alternator field control (charging system)
- turbocharger boost wastegate
- cooling fan relay

Component Description

THROTTLE BODY

The throttle body, in most fuel injected systems, is usually an aluminum housing that consists of one or two throttle blades which are attached to a throttle shaft. The housing has a throttle position sensor (TPS) sensor, idle air control motor and, in some cases, throttle body temperature sensor. On SPI systems, the housing also has an injector(s) and (in some cases) a fuel pressure regulator. The throttle body throttle blade controls the amount of air that enters the engine as well as the amount of vacuum.

ELECTRONIC CONTROL UNIT (ECU)

The ECU monitors and controls all engine control functions. The ECU consists of input and output devices, a central processing unit, a power supply and various memory banks. The input and output devices of the ECU convert electrical signals received by the data sensors and switches to the digital signal that are used by the central processing unit. The central processing unit receives digital signals that are used to perform all mathematical computations and logic

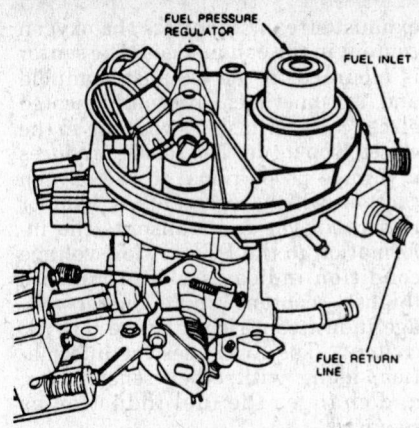

Throttle body—TBI

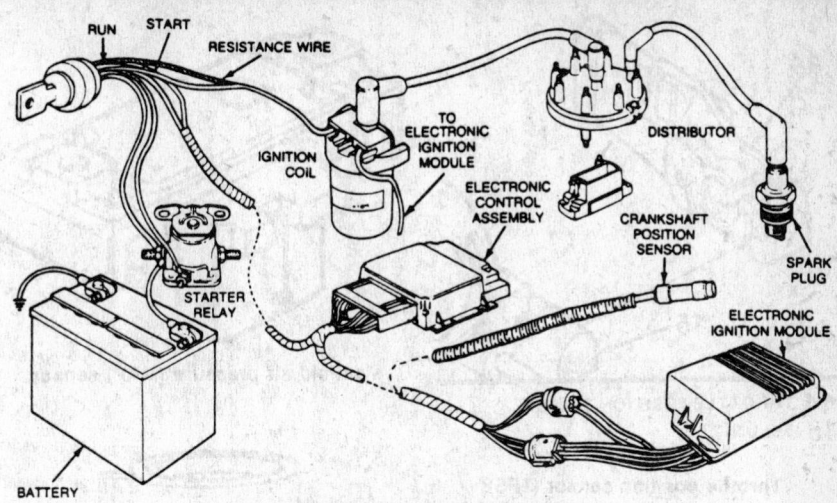

Electronic ignition system using ECU

functions necessary to deliver proper air/fuel mixture. The central processing unit is also responsible for calculating spark timing information. The main source of power that allows the ECU to function is generated from the battery of the vehicle and transported through the ignition system. The memory bank of the ECU is programmed with exact information that is used by the ECU during the open loop mode. This data is also used when a sensor of other component fails, allowing the vehicle to be driven to a repair facility.

CALIBRATION ASSSEMBLY OR PROM (PROGRAMMABLE READ ONLY MEMORY)

Some vehicle manufactures use one ECU for several different model vehicles. This interchangeable ECU is possible through the use of a calibration assembly or prom. Information about the vehicle's engine, transmission, body and drive axle ratio are programmed and permanently stored into the assembly. If the battery supply should become disconnected from the ECU, the data stored into the assembly is not lost.

ELECTRONIC SPARK CONTROL (ESC)

The vehicles equipped with an ESC have the ability to change the ignition timing under any and all operating conditions. Data from various remote sensors (coolant temperature, throttle position, rpm, etc.) is transmitted to the ESC. The ESC computes the information and triggers the ignition spark at precisely the right instant. Some ESC systems (ie.,turbocharged engines) use a detonation (knock) sensor which senses pre-ignition and

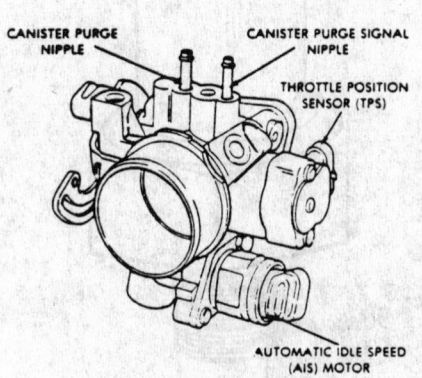

Throttle body—MFI

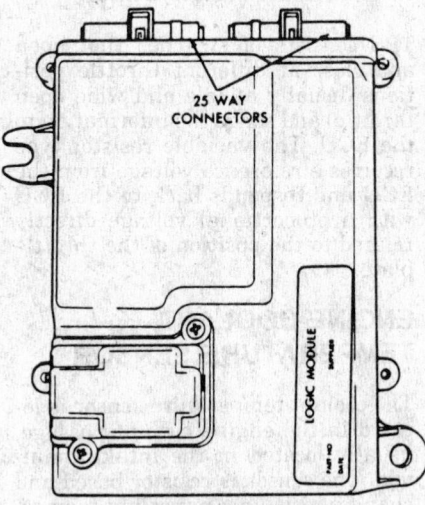

Electronic control unit

transmits the information to the ESC. The ESC modifies spark advance and boost pressure in order to eliminate knock.

MASS AIR FLOW SENSOR

The mass air flow (MAF) sensor is

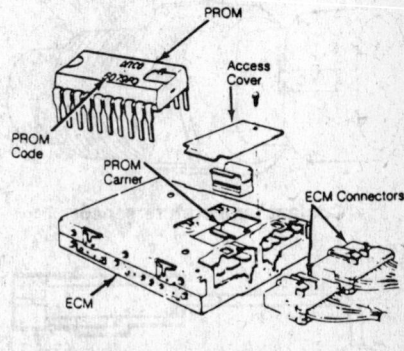

Electronic control unit prom

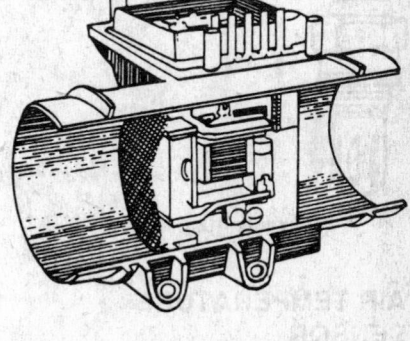

Mass air flow (MAF) sensor

only incorporated in some Multi-point fuel injection systems. The MAF sensor is a very complex device which measures the air mass of the engine intake. Because the air mass is always changing with temperature, humidity and altitude, the fuel delivery rate must be adjusted to compensate for these changes so that a precise fuel mixture can be maintained.

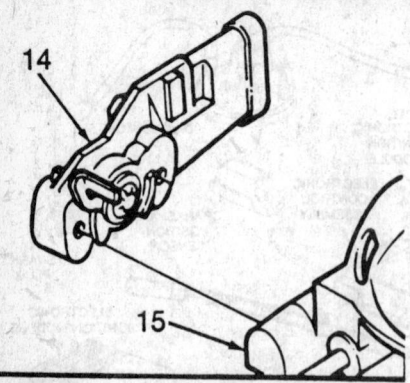

14 THROTTLE POSITION SENSOR
15 TBI UNIT

Throttle position sensor (TPS)

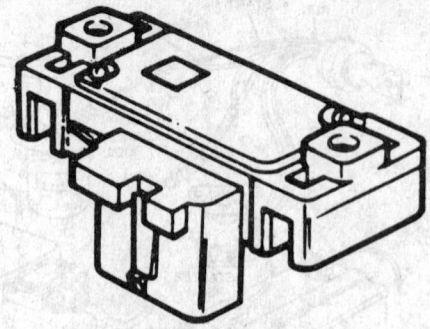

Manifold air pressure (MAP) sensor

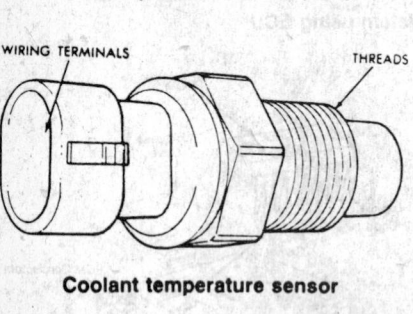

Coolant temperature sensor

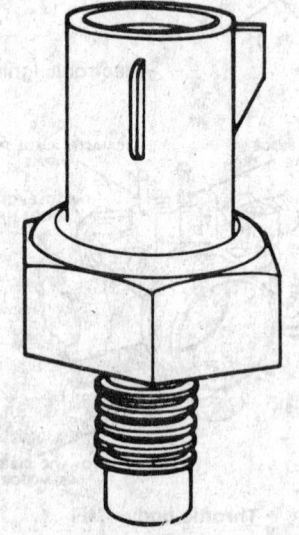

Oxygen (O₂) sensor

Detonation (knock) sensor

exhaust stream, monitors the oxygen content in the exhaust gas. The sensor is mounted in the exhaust manifold and is sometimes internally heated electrically for faster switching to the closed loop mode. The sensor produces a voltage proportional to the oxygen content which represents a lean or rich condition and transmits the information to the ECU. A low voltage condition indicates a lean mixture (high O_2 content) and a higher voltage indicates a rich mixture (low O_2 content). The ECU uses the information, along with other sensor data, and changes the fuel induction as required.

CYLINDER HEAD TEMPERATURE SENSOR

The cylinder head temperature sensor monitors the temperature of the cylinder head and transmits the information to the ECU. The sensor is located in the cylinder head and is a temperature sensitive resistive unit known as a thermistor.

VEHICLE SPEED SENSOR (VSS)

The VSS provides vehicle speed data to the ECU in the form of pulse signals. There are many different types of VSS, some using a reed switch installed in the speed meter unit and others using a optical type. In the optical type a light emitting diode (LED) is used to transmit light and photo diode receives the light. A shutter device, which is usually in-line with the speedometer cable, allows the LED light to reach the photo diode in vehicle speed related pulses. The reed switch type relies on a reed switch that opens and closes by way of a rotating magnet. The magnet rotates proportionally with the vehicle speed.

MANIFOLD AIR PRESSURE (MAP) SENSOR

The MAP sensor is a device that monitors manifold absolute pressure. The sensor is mounted remotely and senses vacuum through a connecting hose. The MAP sensor has a reference voltage from the ECU and transmits remaining voltage to the ECU to calculate engine load. The ECU uses this data along with other data to determine fuel demands.

DETONATION (KNOCK) SENSOR

The detonation sensor generates a

TPS consists of switches that open and close at different throttle positions (usually at idle and wide open throttle) and sends the information to the ECU. The variable resistor type receives a reference voltage from the ECU and responds back to the ECU with a proportional voltage directly related to the position of the throttle plate.

ENGINE COOLANT TEMPERATURE SENSOR

The coolant temperature sensor is located in the engine coolant passage, usually located in the intake manifold. The sensor is resistor based and changes resistance as coolant temperature changes. The sensor uses a reference voltage and the output voltage is sent to the ECU. The ECU calculates engine warm up and provides an optimum fuel enrichment when the engine is cold.

OXYGEN (O₂) SENSOR

The O_2 sensor, which is placed in the

AIR TEMPERATURE SENSOR

The air temperature sensor is located in the air stream of the air flow meter. The sensor supplies incoming air temperature information to the ECU. The ECU uses this data, along with other data, to regulate fuel injection rate.

THROTTLE POSITION SENSOR (TPS)

The TPS can be either a switch (or a combination of switches) or a variable resistor which is much more accurate in throttle position. The switch type

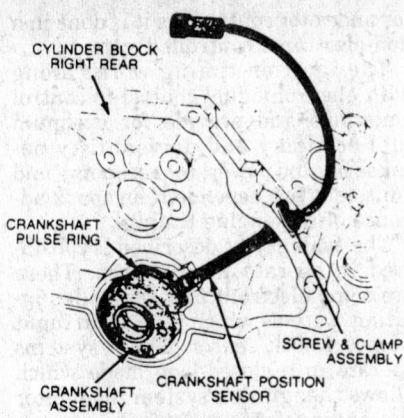

Crankshaft position sensor

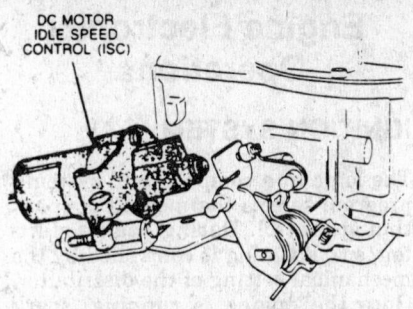

Idle speed control (ISC) motor

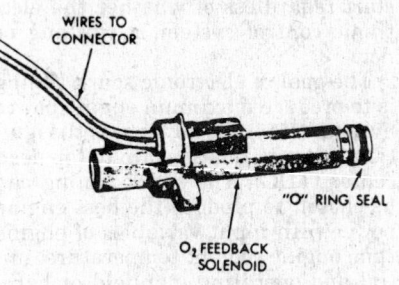

Air/fuel mixture solenoid

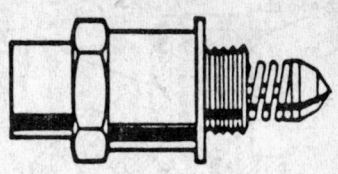

DUAL TAPER VALVE

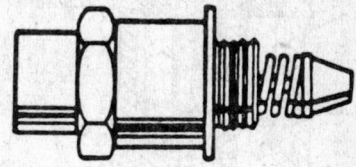

BLUNT PINTLE

Idle air control (IAC) valves

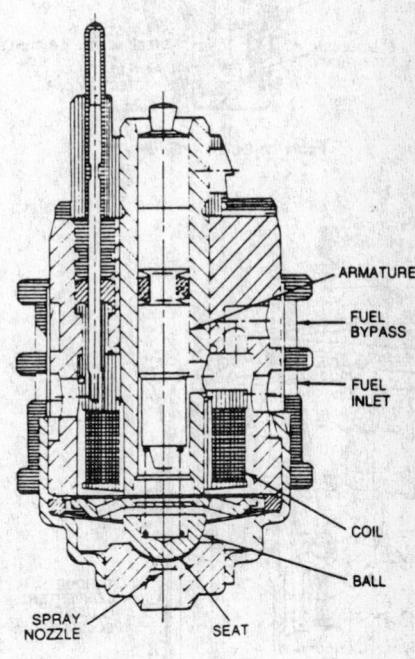

Fuel Injector – TBI type

signal when pre-ignition (knock) occurs in one or more combustion chambers. The sensor is made of a material that is sensitive to oscillation that the engine knock produces and sends signals to the ECU. The ECU, in turn, delays the ignition signal which retards the ignition timing and continues to do this until the engine knock ceases.

CRANKSHAFT (REFERENCE MARK) SENSOR

The crankshaft sensor may be located at either the rear of the engine, at the flywheel or at the front of the engine, near the crankshaft pulley. The sensor detects crankshaft position in relation to top dead center and transmits the signals to ECU.

IDLE SPEED CONTROL (ISC) MOTOR

The ISC is sometimes included on a feedback carburetor system and mounted to the side of the carburetor. The motor driven ISC would maintain a steady idle by way of the ECU. When an added load is put on the engine (air conditioning or when vehicle is in drive) the ECU could increase the idle via the ISC by extending a plunger which would open the throttle valve.

AIR/FUEL MIXTURE SOLENOID

The air/fuel mixture solenoid on feedback carburetor operates in conjunction with the fixed metering jets and/or the manually adjustable idle speed mixture screw. The ECU energizes and de-energizes the solenoid in the closed loop mode. The solenoid usually controls a fixed air bleed and/or fuel discharge port.

IDLE AIR CONTROL (IAC)

The IAC in a fuel injection system controls the air flow around the throttle plate by extending and retracting a bypass valve in the bypass port. The ECU controls the valve by sending voltage pulses called counts or steps to increase or decrease the bypass air flow, thus increasing and decreasing the idle speed.

FUEL INJECTOR

Throttle Body Type

The fuel injector is an electric solenoid controlled by the ECU. The ECU controls the injector by varying voltage pulse widths. When electrical current is supplied to the injector a spring loaded ball is lifted from its seat. This allows fuel to flow through spray orifices and deflects off the sharp edge of the injector nozzle. This action causes the fuel to form a 45° cone shaped spray pattern before entering the air stream in the throttle body.

Multiport Type

The fuel injector is an electric solenoid controlled by the ECU. The ECU controls the injector by varying voltage pulse widths. When electrical current is supplied to the injector, the armature and pintle move a short distance against a spring, opening a small orifice. Fuel is supplied to the inlet of the injector by the fuel pump, then passes through the injector,

around the pintle and out the orifice. Since the fuel is under high pressure, a fine spray is developed in the shape of a hollow cone. The injector, through this spraying action, atomizes the fuel and distributes it into the air entering the combustion chamber.

TORQUE CONVERTER CLUTCH (TCC) SOLENOID

The TCC solenoid is used on some automatic transmission, which allows for better fuel economy. When certain engine and vehicle speeds have been met, the ECU energizes the solenoid. This allows transmission fluid to flow into passages in the torque converter,

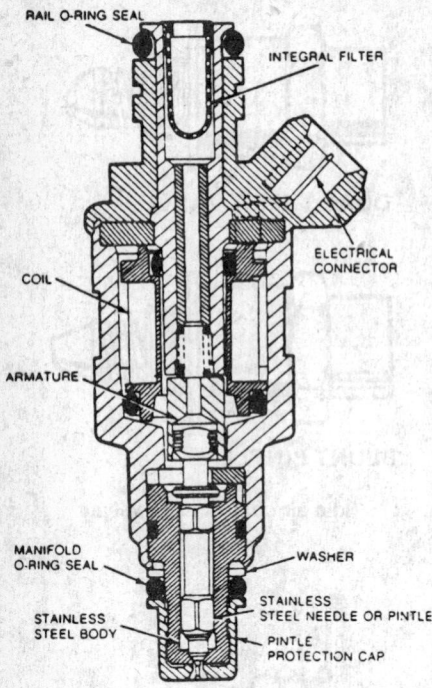

Fuel Injector — MFI type

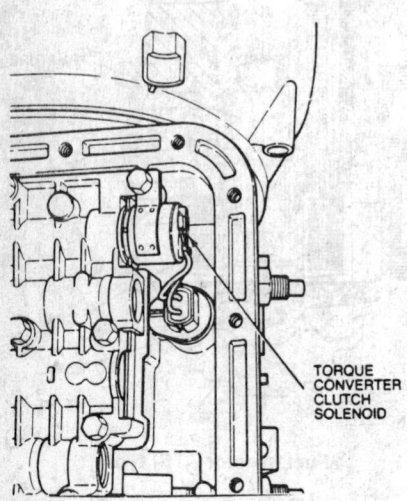

Torque converter clutch (TCC) solenoid

which causes the converter to lock up. This lockup is similar to a direct connection made possible in a manual transmission.

FUEL PUMP RELAY

The fuel is supplied under pressure, usually by an electric fuel pump. The ECU controls the fuel pump relay, which controls the fuel pump operation. When the ignition is switch ON, the fuel pump relay is energized and the fuel pump is activated. The pump primes the fuel system with fuel to a pre-determined pressure.

Engine Electronic Operations

IGNITION SYSTEM

The logic in a computerized system's program selects the method of spark timing control. During engine starting, spark timing is controlled by the mechanical setting of the distributor. Once the engine is running, spark timing is turned over to the ECU. This scheme ensures that the car will start regardless of whether the electronic control system is working or not.

The goal of electronic spark timing is to produce maximum engine power by adjustment the advance of the ignition firing in relationship to top dead center (TDC). The spark timing can be chosen to produce the best engine power with input variables of engine rpm, engine coolant temperature, initial and operating manifold or barometric pressure.

The total spark advance is determined by computing the information received from the various engine sensors which affect spark timing. The processor will then adjust the timing according to information that has been calibrated in it. The processor has programmed into it specific information on:

Warm-Up Spark Advance — this is used when the engine is cold, since a greater amount of advance is required while the engine warms up.

Special Spark Advance — to improve fuel economy during steady driving conditions.

Spark Advance Due to Barometric Pressure — this is used when barometric pressure exceeds a preset calibrated amount.

All of this information is then added together and the initial mechanical advance (if equipped) is subtracted to determine the final spark advance.

The processor receives a timing pulse from a sensor which indicates crankshaft position for top dead center and engines rpm. The processor makes a decision based upon this information and the information that was calibrated into it. at that time, the computer sends a pulse to the ignition actuator circuit which opens the ignition coil primary circuit to generate a secondary voltage pulse to fire the spark plugs. In some cases, the circuitry to open the primary of the coil may be in the computerized controller. The spark selection is performed mechanically by the distributor and rotor contacts as it is done in a non-electronic controlled system.

The ignition timing works along with electronic fuel control to control emissions and provide for optimum fuel economy and driveability because engine power, fuel economy and emissions are dependent on spark advance of the engine timing.

The system just described is considered to operate in open-loop. There are some electronically controlled ignition systems which receive an input from a knock sensor. These systems operate in a closed-loop mode which allows the ignition system to monitor the engine for mechanical changes, such as engine knock.

Engine knock is a condition where the air/fuel mixture in the cylinder does not burn normally. the pressure rise during this burning is so rapid compared to normal combustion that it is accompanied by an audible "knock".

Through some low level knock is acceptable, it is important to avoid excessive knock. To control engine knock, a knock sensor is installed in the engine or intake manifold. This helps to detect excessive engine knock.

The knock sensor is a tuned accelerometer and produces an output voltage depending on the amount of engine vibration occurring in a certain frequency band. When the processor receives a signal from the knock sensor, it retards the spark advance until the knocking stops and then starts increasing it again. This cycle is repeated as long as engine knock occurs.

FUEL CONTROL

In order for the processor to control fuel, it requires a sensor or sensors to monitor the state of the engine, and one or more actuators to do the actual controlling. The sensors measure: exhaust gas oxygen, manifold or barometric absolute pressure, engine rpm and speed, inlet air and coolant temperatures. Actuators are energized to control the air/fuel ratio.

The primary purpose of this control system is to maintain air/fuel ratio at or near 14.7:1 ratio. This is accomplished in two modes (during normal engine operation) open and closed loop. The electronic fuel control system can operate in closed loop only when certain conditions are satisfied. Open loop mode is employed whenever these conditions are not satisfied. However, for either mode, the exhaust emissions will satisfy federal requirements if the average air/fuel ratio is held within the tolerance limits.

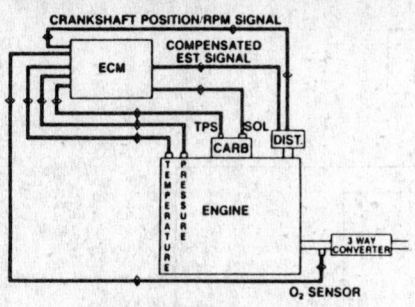

Electronic spark timing system

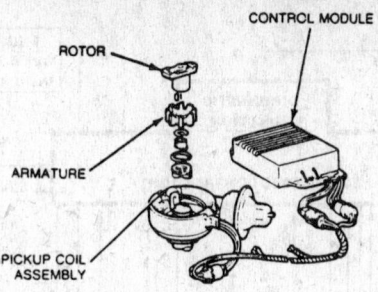

Solid state ignition system

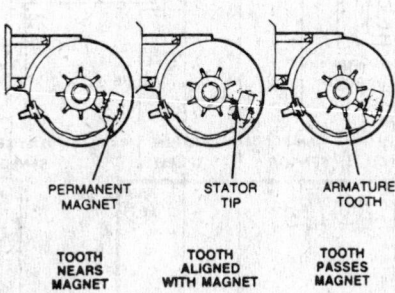

Distributor pick-up coil and armature assembly

In addition to open and closed loop control modes, a practical fuel control system has other operating modes depending on engine conditions. These handle such conditions as starting, rapid acceleration or heavy load, sudden deceleration, idling, etc.

An automotive engine has various operating modes as the operating conditions change. Preprogrammed into the processor, control logic determines the operating mode from the engine conditions that exist. From these engine conditions, the system determines which operating modes are to be performed.

There are seven different engine operating modes which affect fuel control: engine crank, engine warmup, open loop, closed loop control, hard acceleration, deceleration and idle. The program for mode control logic determines the engine operating mode by reading various sensors.

When the ignition switch is initially switched on, the mode control logic automatically selects an engine-start control scheme which provides the low air/fuel ratio required for starting the engine. Once the engine rpm rises above the cranking value, the controller identifies the engine-started mode and passes control to the program for the engine warm-up mode. This operating mode keeps the air/fuel ratio low to prevent engine stall during cool weather until engine coolant temperature rises above a preset value.

When the coolant temperature rises, the mode control logic directs the system to operate in the open loop control mode until a certain time has elapsed and the exhaust gas sensor warms up enough to provide accurate readings. This condition is detected by monitoring the exhaust gas sensor's output for voltage readings above a certain minimum air/fuel mixture voltage set point. when the sensor has indicated a rich mixture a certain number of times (depending on calibration), and after the engine has been in open loop for a specific time, the control mode logic selects the closed loop mode for the system.

The engine remains in the closed loop mode until either the exhaust gas sensor cools and fails to switch (from rich to lean) for a certain length of time, or a hard acceleration or deceleration occurs. If the sensor cools, the control mode logic selects the open loop mode again.

During hard acceleration of heavy engine loads, the control mode logic chooses a scheme which provides a rich air/fuel mixture for the duration of the acceleration or heavy load. This scheme provides maximum power, but poor emissions control and poor fuel economy. After the need for enrichment has passed, control is returned to either open or closed loop depending on the control mode logic selection conditions that exist at that time.

During periods of deceleration, the air/fuel ratio is increased to reduce emissions of HC and CO due to unburned fuel. When idle conditions are present, control mode logic passes system control to the idle speed control mode. In this mode, the engine speed is controlled to reduce engine roughness and stalling which might occur because the idle load has changed due to air conditioner compressor operation, alternator operation, or gearshift positioning from PARK or NEUTRAL to DRIVE.

Engine Crank

While the engine is being cranked, the fuel control system must provide an intake air/fuel ratio anywhere from 2:1 to 12:1, depending on engine temperature. Low temperatures affect the carburetor's ability to atomize or mix the incoming air and fuel. At low temperature, the fuel tends to form into large droplets. The larger fuel droplets tend to increase the apparent air/fuel ratio because the amount of usable fuel in the air is reduced, therefore, the system must provide a decreased air/fuel ratio to provide the engine with a more combustible air/fuel mixture. The engine temperature is read by the processor through an analog to digital converter from a temperature sensor in the engine water coolant passage. The processor's calibration determines what the proper air/fuel ratio must be at that temperature. The air/fuel is determined and controlled as in the open loop mode.

Engine Warm-up

While the engine is warming up, an enriched air/fuel ratio is still needed to keep it running smoothly, but the required air/fuel ratio changes as the temperature increases. Therefore, the fuel control system will stay in the open loop mode, but the air/fuel ratio commands continue to be altered due to the temperature changes. The emphasis in this control mode is on rapid and smooth engine warm-up. Fuel economy and emission control are still a secondary concern. The controller determines the warm-up time period based on the coolant temperature when the warm-up mode was selected. Naturally, an initially cold engine requires a longer warm-up time than a warm engine. The time allowed by the controller timer is chosen according to the calibration of the processor.

OPEN LOOP CONTROL

Open loop fuel control is used when the engine has not reached a preset operating condition. This condition is sensed by various sensors located in and around the engine, and include engine coolant temperature, air charge temperature, engine time on, etc. After all these preset conditions are met, the system will go into closed loop. During certain operating conditions, such as a wide open throttle condition the system will go back into open loop.

CLOSED LOOP CONTROL

Closed loop fuel control is selected when the engine is warm and the exhaust gas oxygen sensor exceeds its minimum operating temperature. The intake air/fuel ratio is controlled in a closed loop by measuring the ex-

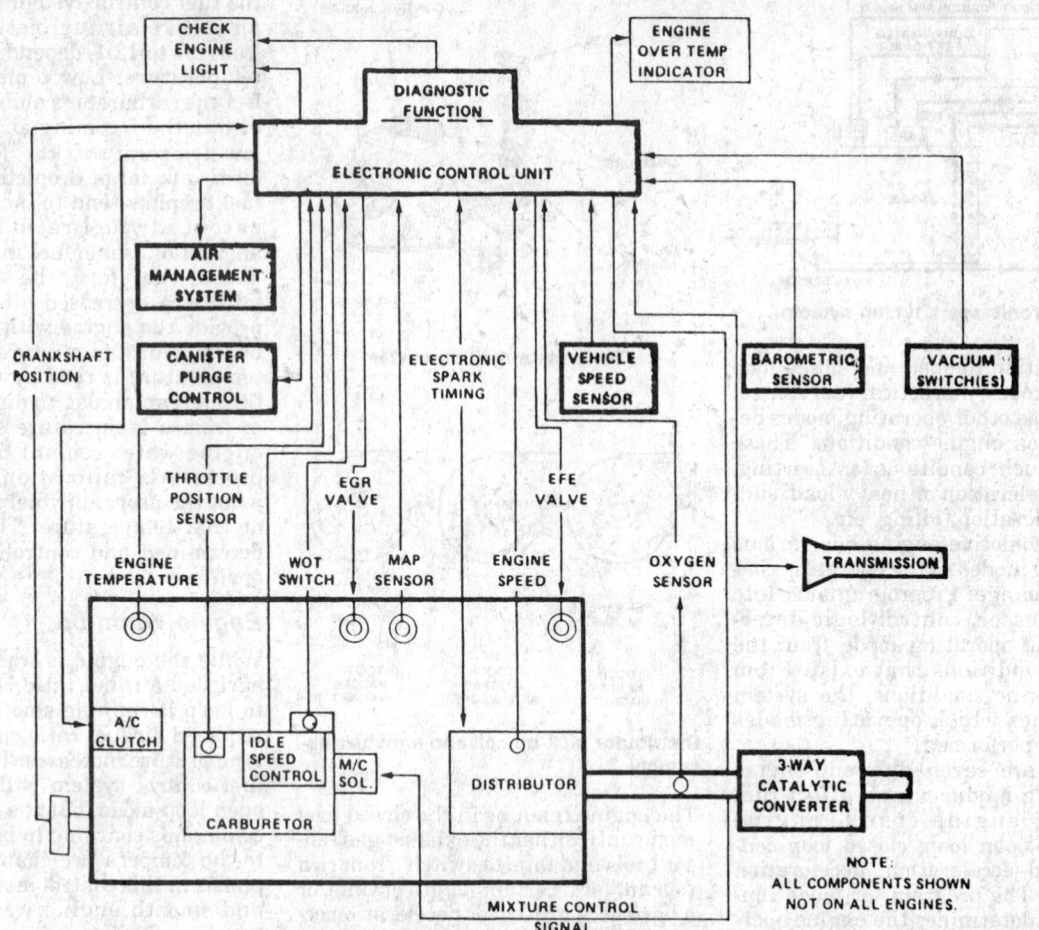

Typical electronic feedback carburetor system

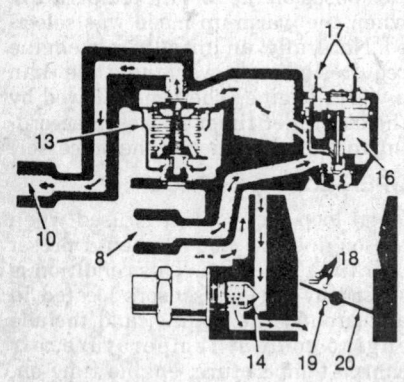

8 FUEL SUPPLY
10 FUEL RETURN
13 PRESSURE REGULATOR (PART OF FUEL METER COVER)
14 IDLE AIR CONTROL (IAC) VALVE (SHOWN OPEN)
16 FUEL INJECTOR
17 FUEL INJECTOR TERMINALS
18 PORTED VACUUM SOURCES*
19 MANIFOLD VACUUM SOURCE*
20 THROTTLE VALVE

*May Be Different on some Models.

Throttle body injection air and fuel flow

haust gas at the exhaust manifold and altering the input fuel flow rate or the air entering the main metering systems (depending on the type of fuel system used).

ACCELERATION ENRICHMENT (OPEN LOOP)

During periods of heavy engine load, such as wide open acceleration, fuel control is adjusted to provide an enriched ratio to maximize engine power while neglecting fuel economy and emission.

The computer detects this condition by reading the throttle position sensor voltage or the MAP sensor. Low intake manifold vacuum or throttle position corresponds to heavy engine loads. The fuel control system controller responds by increasing the amount of fuel to enter the intake manifold or to decrease the amount of air n the main metering system. This enrichment allows the engine to operate with a power greater than that allowed when emissions and fuel economy are controlled within specifications.

DECELERATION AND IDLE SPEED CONTROL (OPEN LOOP)

During periods of light engine load and high rpm, such as during closed throttle deceleration, coasting or engine idle, the engine requires a very lean air/fuel ratio to reduce excess emissions of HC and CO. Deceleration is indicated by a sudden increase in manifold vacuum and throttle position, indicating a closed throttle. When these conditions are detected by the processor, it computes a change in the amount of fuel required or amount of air entering the main or idle speed passages (depending on type of fuel system used). On certain engine engine applications which electronic fuel injection, the fuel may even be turned completely off during closed throttle deceleration.

Idle speed control is used to prevent engine stall during idle. The goal is to allow the engine to idle at as low an rpm as possible, yet keep the engine from running rough and stalling when power takeoff accessories such as air conditioning compressors are turned on.

Engine Rebuilding

18

GENERAL INFORMATION

This section describes in detail, the procedures involved in rebuilding a typical gasoline engine. A rebuilt engine can be expected to give many miles of dependable service only if the proper reconditioning procedures are performed and clearances are kept within the manufacturer's recommended specifications.

The following systems of the gasoline engine should be checked to determine to what degree the rebuilding should be accomplished.

Engine Oil Pressure

The engine oil pressure developed should be compared to the manufacturer's recommended pressure, which is necessary to provide lubricating oil to the engine oil circuits. If the pressure is below specifications, the cause must be located and repaired.

The following wear points should be considered during this determination.

Oil Pump—Check Gear clearances (a new oil pump is a good investment).

Main Bearings and Journals—Check clearances, taper and roundness.

Connecting Rod Bearings and Journals—Check clearances, taper and roundness.

Camshaft and Bearings—Check clearances, taper and roundness.

Rocker Arms, Rocker Arm Shafts—Check arm and shaft wear, ball and seat wear.

Tappets—Check clearances between tappet and bore and for excessive leakdown of hydraulic tappets.

Leakage of oil pressure along external or internal oil galleries or gaskets.

Dilution of the oil by gasoline leakage through failures of the carburetor or mechanical fuel pump.

External damage to the oil pan, causing blockage or movement of the oil pick-up tube, resulting in loss of oil pick-up to the pump.

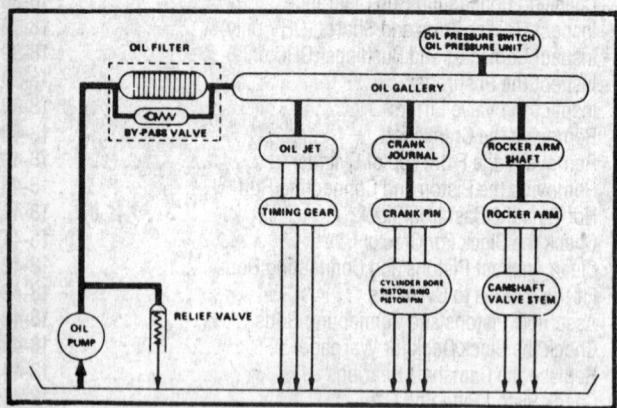

Engine oil flow diagram—typical

Compression

Compression in an engine is determined by the correct fit and sealing efficiency of the piston and rings against the cylinder walls, the quality of the seal between the valve and its seat and the seal between the cylinder head, head gasket and block. Here are some important check points.

Valve, seat and face—Machine face and seat to original specifications.

Valve guides—The reconditioned seat and face won't hold up long if the valve stem clearance isn't within specifications.

Valve seals—Oil reaching the valve seat will become a solid when it combines with the heat in the combustion chamber. This solid (carbon) will build up and eventually keep the valve from seating.

Cylinder walls—Check for taper, out-of-round and hone to proper cross hatch pattern.

Pistons—Check all dimensions. A poorly fitted piston will shorten the life of the new rings.

COMPRESSION PRESSURE COMPARISON CHART

Minimum pressure is 75% of maximum pressure

Maximum PSI	Minimum PSI	Maximum PSI	Minimum PSI	Maximum PSI	Minimum PSI
134	101	174	131	214	160
136	102	176	132	216	162
138	104	178	133	218	163
140	105	180	135	220	165
142	107	182	136	222	166
144	108	184	138	224	168
146	110	186	140	226	169
148	111	188	141	228	171
150	113	190	142	230	172
152	114	192	144	232	174
154	115	194	145	234	175
156	117	196	147	236	177
158	118	198	148	238	178
160	120	200	150	240	180
162	121	202	151	242	181
164	123	204	153	244	183
166	124	206	154	246	184
168	126	208	156	248	186
170	127	210	157	250	187
172	129	212	158		

NOTE: To Determine if the engine compression is satisfactory, most engine manufacturers require that a complete compression test be done, with the lowest cylinder pressure reading not being less than 75% of the highest reading. Look for uniformity of compression between cylinders, rather than specific compression pressures of an engine.

Cooling System

Maintaining engine temperatures within the specified range is critical to the life of a rebuilt engine. Until new parts mate properly with each other, excessive heat can cause permanent damage or substantially reduce the service life of the reconditioned engine. If the engine is operated at temperatures below normal, the oil may not properly lubricate all of the parts. Some parts that should be checked during the rebuilding process are:

Coolant passages—Should be free of rust and corrosion deposits.

Core plugs—All plugs should be replaced during the rebuilding process.

Hoses—Should be free of cracks, hard spots and oil softened spots.

Thermostat—Check for opening and closing at the specified temperature.

Radiator—Check for leaks and rust or corrosion deposits.

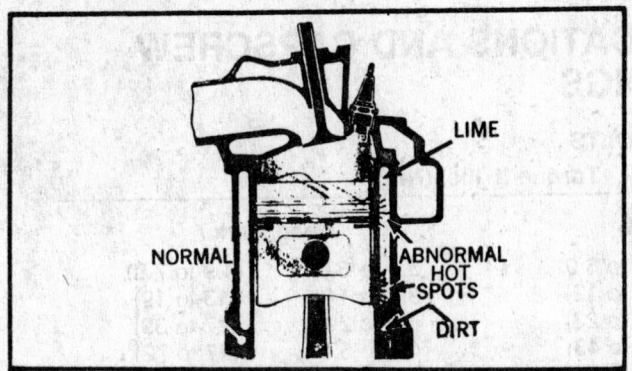

Illustration of cooling system passages being restricted, causing internal engine problems

Pressure cap—Should hold specified pressure, also check the gasket and vent valve operation.

Engine Noises

Engine noises are not only annoying, but indicate conditions inside the engine that can limit the service life of the engine or shut it down completely. Generally, noises are caused by too much clearance between parts or loss of oil supply. Engine noises can be caused by any of the following parts.
- Main bearings
- Connecting rod bearings
- Piston and/or rings

STANDARD TORQUE SPECIFICATIONS AND CAPSCREW MARKINGS

Newton-Meter has been designated as the world standard for measuring torque and will gradually replace the foot-pound and kilogram-meter torque measuring standard. Torquing tools are still being manufactured with foot-pounds and kilogram-meter scales, along with the new Newton-Meter standard. To assist the repairman, foot-pounds, kilogram-meter and Newton-Meter are listed in the following charts and should be followed as applicable.

U.S. BOLTS

SAE Grade Number	1 or 2			5			6 or 7			8		
Capscrew Head Markings Manufacturer's marks may vary. Three-line markings on heads below indicate SAE Grade 5.												
Usage	Used Frequently			Used Frequently			Used at Times			Used at Times		
Quality of Material	Indeterminate			Minimum Commercial			Medium Commercial			Best Commercial		
Capacity Body Size	Torque			Torque			Torque			Torque		
(inches)–(thread)	Ft-Lb	kgm	Nm	Ft-Lb	kgm	Nm	Ft-Lb	kgm	Nm	Ft-Lb	kgm	Nm
1/4–20	5	0.6915	6.7791	8	1.1064	10.8465	10	1.3630	13.5582	12	1.6596	16.2698
–28	6	0.8298	8.1349	10	1.3830	13.5582				14	1.9362	18.9815
5/16–18	11	1.5213	14.9140	17	2.3511	23.0489	19	2.6277	25.7605	24	3.3192	32.5396
–24	13	1.7979	17.6256	19	2.6277	25.7605				27	3.7341	36.6071
3/8–16	18	2.4894	24.4047	31	4.2873	42.0304	34	4.7022	46.0978	44	6.0852	59.6560
–24	20	2.7660	27.1164	35	4.8405	47.4536				49	6.7767	66.4351
7/16–14	28	3.8132	37.9629	49	6.7767	66.4351	55	7.6065	74.5700	70	9.6810	94.9073
–20	30	4.1490	40.6745	55	7.6065	74.5700				78	10.7874	105.7538
1/2–13	39	5.3937	52.8769	75	10.3725	101.6863	85	11.7555	115.2445	105	14.5215	142.3609
–20	41	5.6703	55.5885	85	11.7555	115.2445				120	16.5860	162.6960
9/16–12	51	7.0533	69.1467	110	15.2130	149.1380	120	16.5960	162.6960	155	21.4365	210.1490
–18	55	7.6065	74.5700	120	16.5960	162.6960				170	23.5110	230.4860
5/8–11	83	11.4789	112.5329	150	20.7450	203.3700	167	23.0961	226.4186	210	29.0430	284.7180
–18	95	13.1385	128.8027	170	23.5110	230.4860				240	33.1920	325.3920
3/4–10	105	14.5215	142.3609	270	37.3410	366.0660	280	38.7240	379.6240	375	51.8625	508.4250
–16	115	15.9045	155.9170	295	40.7985	399.9610				420	58.0860	568.4360
7/8–9	160	22.1280	216.9280	395	54.6285	535.5410	440	60.8520	596.5520	605	83.6715	820.2590
–14	175	24.2025	237.2650	435	60.1605	589.7730				675	93.3525	915.1650
1–8	236	32.5005	318.6130	590	81.5970	799.9220	660	91.2780	894.8280	910	125.8530	1233.7780
–14	250	34.5750	338.9500	660	91.2780	849.8280				990	136.9170	1342.2420

STANDARD TORQUE SPECIFICATIONS AND CAPSCREW MARKINGS

METRIC BOLTS

Description				
Thread for general purposes			Torque ft-lbs. (Nm)	
(size x pitch (mm))	Head Mark 4		Head Mark 7	
6 x 1.0	2.2 to 2.9	(3.0 to 3.9)	3.6 to 5.8	(4.9 to 7.8)
8 x 1.25	5.8 to 8.7	(7.9 to 12)	9.4 to 14	(13 to 19)
10 x 1.25	12 to 17	(16 to 23)	20 to 29	(27 to 39)
12 x 1.25	21 to 32	(29 to 43)	35 to 53	(47 to 72)
14 x 1.5	35 to 52	(48 to 70)	57 to 85	(77 to 110)
16 x 1.5	51 to 77	(67 to 100)	90 to 120	(130 to 160)
18 x 1.5	74 to 110	(100 to 150)	130 to 170	(180 to 230)
20 x 1.5	110 to 140	(150 to 190)	190 to 240	(160 to 320)
22 x 1.5	150 to 190	(200 to 260)	250 to 320	(340 to 430)
24 x 1.5	190 to 240	(260 to 320)	310 to 410	(420 to 550)

CAUTION: Bolts threaded into aluminum require much less torque.

Tools

The tools required for the basic rebuilding procedure should, with minor exception, be those included in a mechanic's tool kit. Accurate torque wrench, micrometers and dial indicators should readily be available to the repairman. Special tools are available from the major tool suppliers. The services of a competent automotive machine shop must also be available.

Precautions

When assembling the engine, any parts that will be in frictional contact must be pre-lubricated, to provide protection on initial start-up.

Any product specifically formulated for this purpose may be used. Where semipermanent locked but removable installation of bolts or nuts is desired, threads should be cleaned and coated with a liquid locking compound. Studs may be permanently installed using a stud mounting compound. Bolts and nuts with no torque specification should be tightened according to size (see chart).

Aluminum has become increasingly popular for use in engines, due to its low weight and excellent heat transfer characteristics. The following precautions should be observed when handling aluminum engine parts.

Never hot-tank aluminum parts.

Remove all aluminum parts (identification tags, etc.) from engine parts before hot-tanking (otherwise they will be removed during the process).

Always coat threads lightly with engine oil or anti-seize compounds before installation, to prevent seizure.

Heli-coil

Never over-torque bolts or spark plugs in aluminum threads. Should stripping occur, threads can be restored according to the following procedure, using Heli-Coil thread inserts.

Tap drill the hole with the stripped threads to the specified size, using the Specified tap.

NOTE: Heli-Coil tap sizes refer to the size thread being replaced, rather than the actual tap size.

Tap the hole for the Heli-Coil. Place the insert on the proper installation tool (see Heli-Coil chart with kit). Apply pressure on the insert while winding it clockwise into the hole, until the top of the insert is one turn below the surface. Remove the in-stallation tool and break the installation tang from the bottom of the insert by moving it up and down. If the Heli-Coil must be removed, tap the removal tool firmly into the hole, so that it engages the top thread and turn the tool counterclockwise to extract the insert.

Broken Bolts or Studs

Snapped bolts or studs may be removed, using a stud extractor (unthreaded) or locking pliers (threaded). Penetrating oil will often aid in breaking frozen threads. In cases where the stud or bolt is flush with, or below the surface, proceed as follows.

Drill a hole in the broken stud or bolt, approximately 1/2 its diameter. Select a screw extractor of the proper size and tap it into the stud or bolt. Turn the extractor counterclockwise to remove the stud or bolt.

Locating Metal Flaws and Cracks

Magnaflux and Zyglo are inspection techniques used to locate material flaws, such as stress cracks. Magnafluxing coats the part with fine magnetic particles and subjects the part to a magnetic field. Cracks cause breaks in the magnetic field,

Magnaflux indication of cracks

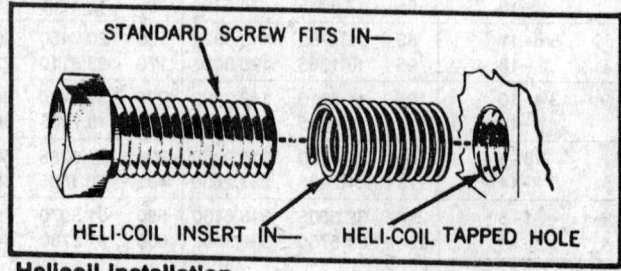

STANDARD SCREW FITS IN—

HELI-COIL INSERT IN—   HELI-COIL TAPPED HOLE

Helicoil Installation

which are outlined by the particles. Since Magnaflux is a magnetic process, it is applicable only to ferrous materials. The Zyglo process coats the material with a fluorescent dye penetrant and then subjects it to blacklight inspection, under which cracks glow brightly. Parts made of any material may be tested using Zyglo. While Magnaflux and Zyglo are excellent for general inspection and locating hidden defects, specific checks of suspected cracks may be made at lower cost and more readily using spot check dye. The dye is sprayed onto the suspected area, wiped off and the area is then sprayed with a developer. Cracks then will show up brightly. Spot check dyes will only indicate surface cracks; therefore, structural cracks below the surface may escape detection. When questionable, the part should be tested using Magnaflux, Zyglo or their equivalent.

REBUILDING GASOLINE ENGINES

The section is divided into two parts. The first, Cylinder Head Reconditioning, assumes that the cylinder head is removed from the engine, all manifolds are removed and the cylinder head is on a workbench. The camshaft should be removed from overhead cam cylinder heads. The second section, Cylinder Block Reconditioning, covers the block, pistons, connecting rods and crankshaft. It is assumed that the engine is mounted on a work stand and the cylinder head and all accessories are removed.

In many cases, a choice of methods is provided. The choice of method for a procedure is at the discretion of the user.

Many makes and types of special tools are available to the rebuilder for the express purpose of making a specific rebuilding operation easier and quicker. It is the choice of the rebuilder as to the tool desired and obtained.

Cylinder Head Reconditioning

IDENTIFY THE VALVES

Invert the cylinder head and clean the carbon from the valve heads. Number the valve heads from front to rear with touch-up paint or a felt tip marking pencil. Upon removal of the valves from the cylinder head, place them in a holder, made from cardboard, wood or metal, in their respective order.

REMOVE THE ROCKER ARMS

Remove the rocker arms and shaft or balls and nuts, if not done during the cylinder head removal. Wire the sets of rockers, balls and nuts together and identify according to the corresponding valve.

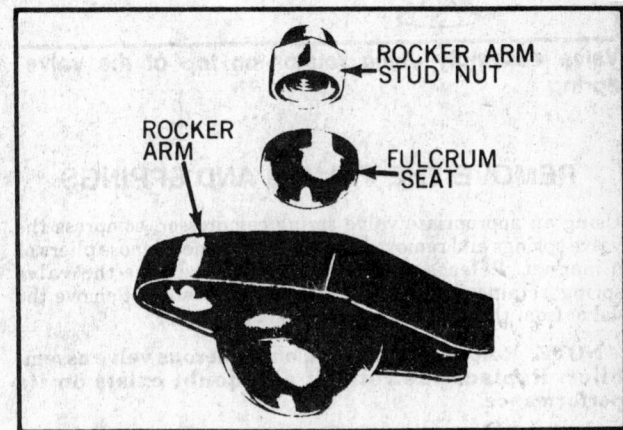

Individual rocker arm assembly

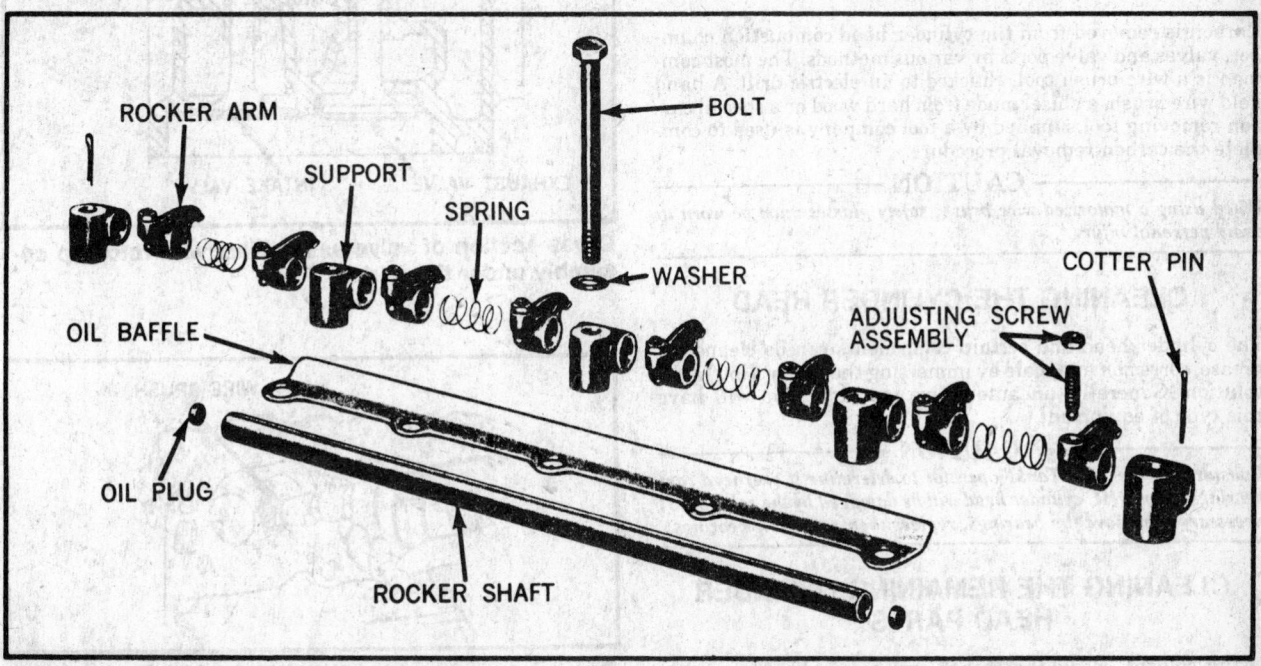

Rocker arm and shaft assembly—typical

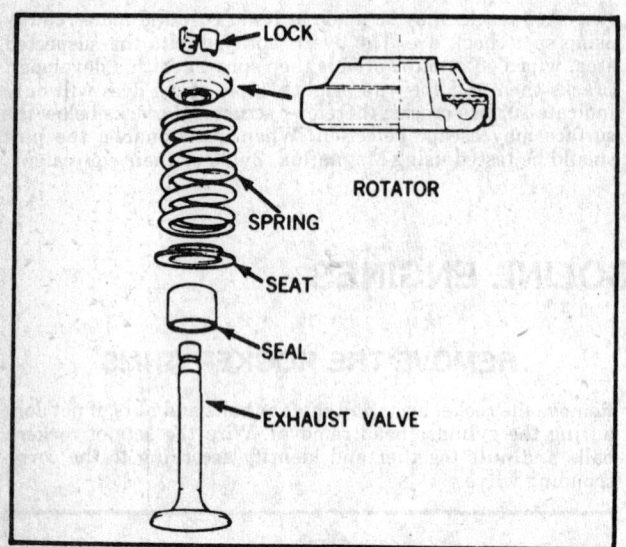

Valve assembly using rotator on top of the valve spring

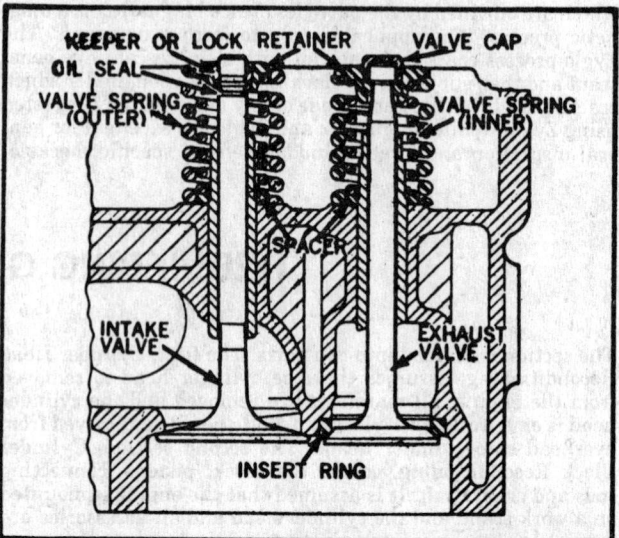

Cross section of valve assemblies with valve rotor cap on exhaust valve stem

REMOVE THE VALVES AND SPRINGS

Using an appropriate valve spring compressor, compress the valve springs and remove the keepers with needlenose pliers or a magnet. Release the compressor and remove the valve spring, retainer and oil seal from the valve stem. Remove the valve from the cylinder head and keep in order.

NOTE: Rotor units are used on numerous valve assemblies. Replace the rotor if any doubt exists on its performance.

DE-CARBON THE CYLINDER HEAD AND VALVES

Carbon is removed from the cylinder head combustion chamber, valves and valve ports by various methods. The most common is a wire brush tool, chucked to an electric drill. A hand held wire brush, a chisel made from hard wood or a special carbon removing tool supplied by a tool company is used to complete the carbon removal procedure.

——————— CAUTION ———————
When using a motorized wire brush, safety glasses must be worn to avoid personal injury.

CLEANING THE CYLINDER HEAD

The cylinder head and certain components can be cleaned of grease, corrosion and scale by immersing them in a "Hot Tank" solution. Generally, an automotive machine shop will have this type of equipment.

——————— CAUTION ———————
Consult with the "Hot Tank" operator to determine if overhead cam bearings of an OHC cylinder head will be damaged by the solution. If necessary to remove the bearings, replace them with new bearings.

CLEANING THE REMAINING CYLINDER HEAD PARTS

Using solvent, clean the rocker arm assemblies, (or rocker

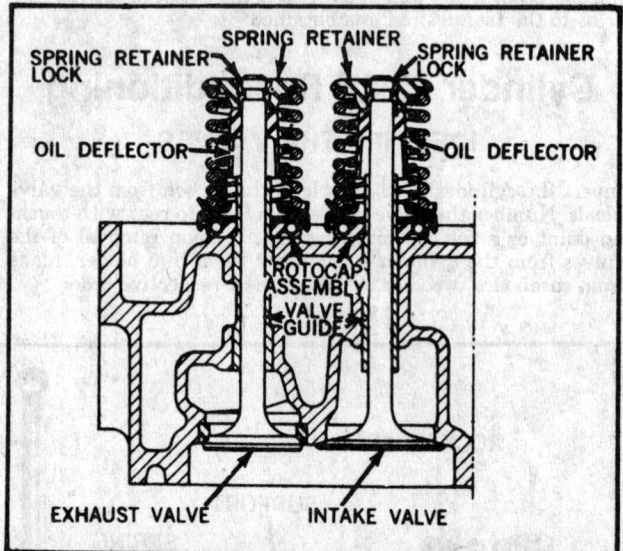

Cross section of valve assemblies with roto-cap assembly under the spring

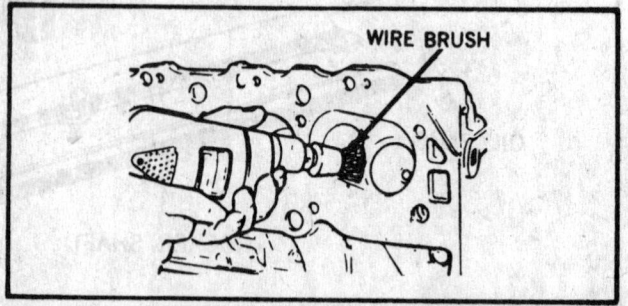

Removing carbon from the cylinder head

balls and nuts) springs, spring retainers, keepers, all bolts and nuts, push rods and rocker arm cover.

CHECK THE CYLINDER HEAD FOR WARPAGE

Place a straightedge across the gasket surface of the cylinder head. Using feeler gauges, determine the clearance at the cen-

Cleaning valve guides with wire type cleaner

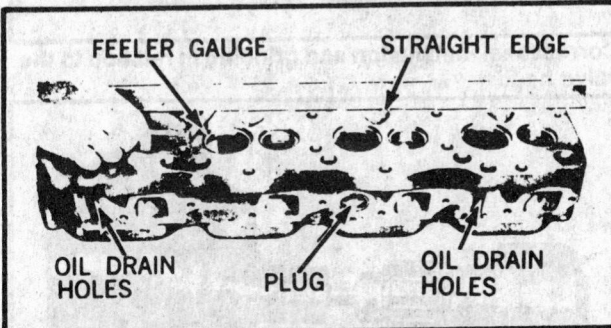

Checking cylinder head surface for flatness with feeler gauge and straight edge

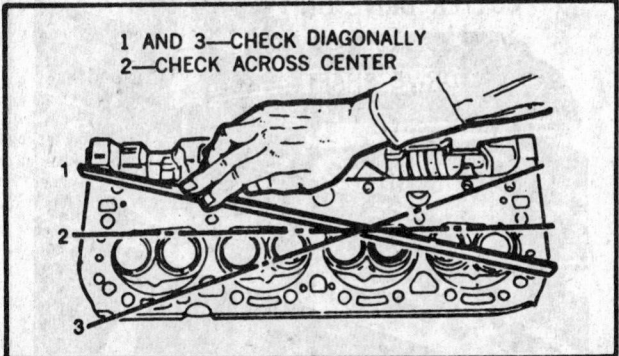

Location of straight edge to check cylinder head surface for flatness

ter of the straightedge. Measure across both diagonals, along the longitudinal centerline and across the cylinder head at several points. If warpage exceeds 0.003 in. in a 6 in. span, or 0.006 in. over the total length, the cylinder head must be resurfaced.

NOTE: If warpage exceeds the manufacturer's maximum tolerance for material removal, the cylinder head must be replaced.

When milling the cylinder heads of V-type engines, the intake manifold mounting position is altered and must be corrected by milling the manifold flange a proportionate amount.

CHECK THE VALVE STEM TO GUIDE CLEARANCE

Clean the valve stem with a solvent to remove all gum and varnish. Clean the valve guides with a solvent and/or a wire type expanding valve guide cleaner tool. Insert the proper valve into its guide and hold the valve head to the valve seat tightly. Mount a dial indicator on the spring side of the cylinder head so that the dial indicator foot is against the valve stem, protruding from the guide, at a 90° angle. Move the valve off its seat and measure the valve guide to stem clearance by moving the stem back and forth to actuate the dial indicator. Measure the

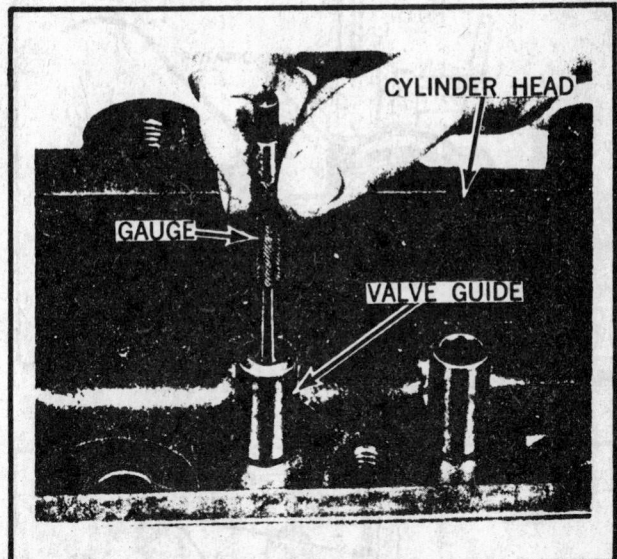

Measuring valve guide with small hole gauge

Measuring valve guide wear at valve stem

valve stems using a micrometer and compare to specifications to determine if the valve stem or valve guide is responsible for the excessive wear clearance.

An alternate method of checking the valve stem to guide clearance is to mount a dial indicator on the combustion side of the cylinder head, with the foot of the indicator to contact the side of the valve head. The valve head is moved away from its seat a predetermined distance, either by a special collar tool, placed on the valve stem between the head of the valve and the guide, or by measuring the height of the valve head above the seat with the use of a scale. The valve head is moved back and forth to actuate the dial indicator. Measure the valve stems, using a micrometer and compare to specifications.

Determination of wear from either the valve guide or valve stem can be made. Other types of measuring methods are available to the rebuilder. Go and No-Go gauge, inside caliper type small hole gauges or shim stock can be used to determine the wear of the guides.

Careful inspection will detect bellmouthing or elliptical wear of the guides, normally at the port end of the guide.

REPLACING VALVE SEAT INSERTS

Most exhaust and some intake valve seats are of the insert type and can be replaced if found to be loose, burned or cracked.

The valve seat insert can be removed by either pulling it from its counterbore with a puller, or by drilling a small hole into the seat insert on two sides and cracking it with a chisel. Care must be exercised to avoid drilling into the cylinder head. The insert counterbore in the cylinder head, should be machined prior to the insert installation, with special emphasis on having the bottom of the counterbore square to insure proper seating of the valve insert. Most inserts are supplied in standard, 0.015 in. and 0.030 in. oversizes.

After installation of the valve seat insert, grinding by a refacing machine should be made to insure the seat is angled to specification and in proper relationship to the valve guide.

REPLACING VALVE GUIDES
INTEGRAL GUIDE TYPE

These types of cylinder heads do not have removable guides but

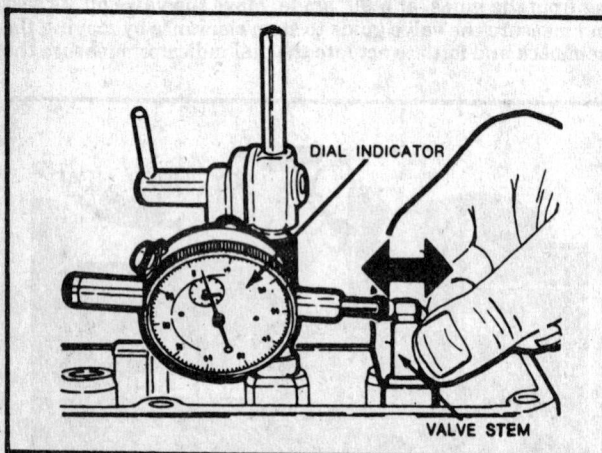

Measuring valve guide wear at valve stem

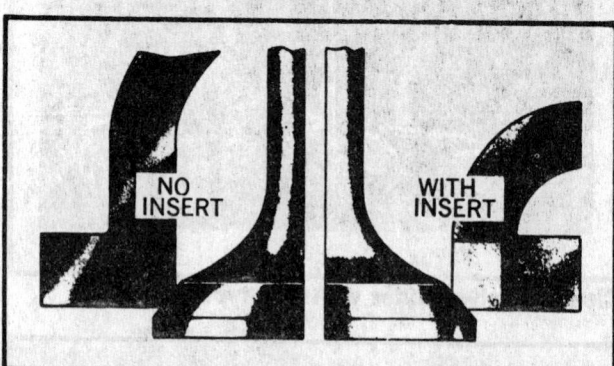

Correct seal installation and grinding in relation to the valve head

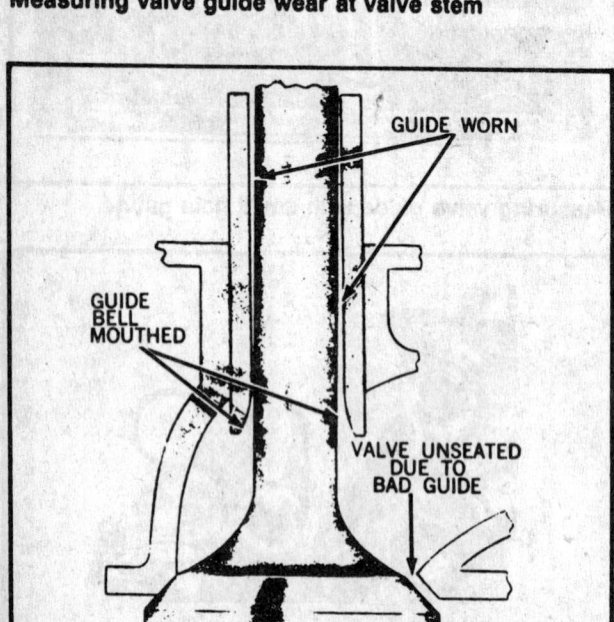

Effects of worn and bellmouthed guides on valve head seating

Machining cylinder head for valve seat insert installation

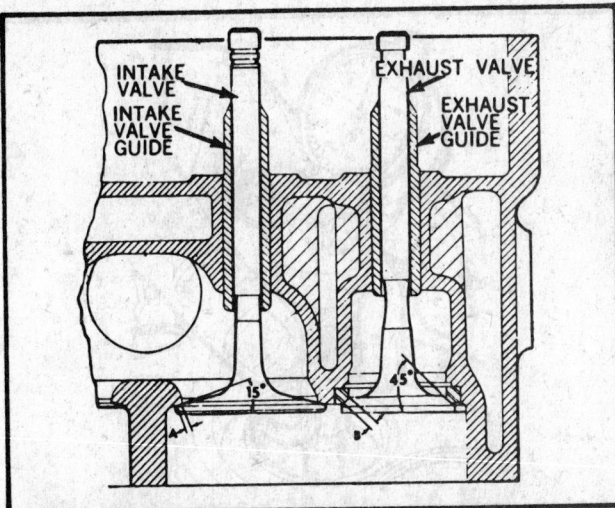

Cross section of valves showing exhaust valve seat insert and intake valve seat without insert. Method of measuring seat angle is illustrated

have the guide holes bored directly in the cylinder head material. When the clearances become excessive between the valve and guide, the guides can be reamed to an oversize dimension and oversize valves used. The "knurling" process may be used to recondition the inside of the guide surface, if the valve to guide clearance is not excessive.

A machining operation can be used to drill out the non-replaceable guide holes and have a standard type guide installed. This operation should be done by an automotive machine shop, equipped with the special boring machine.

REPLACEABLE GUIDE TYPE

Depending on the type of cylinder head, valve guides may be pressed, hammered, or shrunk in. In cases where the guides are shrunk into the head, replacement should be left to an equipped machine shop. In other cases, the guides are replaced as follows: Press or tap the valve guides out of the head using a stepped drift. Determine the height above the boss that the guide must extend and obtain a stack of washers, their I.D. similar to the guides O.D., of that height. Place the stack of washers on the guide and insert the guide into the boss.

NOTE: Valve guides are often tapered or beveled for installation.

Using the stepped installation tool, press or tap the guides into position. Ream the guides according to the size of the valve stem.

RESURFACING (GRINDING) THE VALVE FACE

Using a valve grinder, resurface the valves according to specifications.

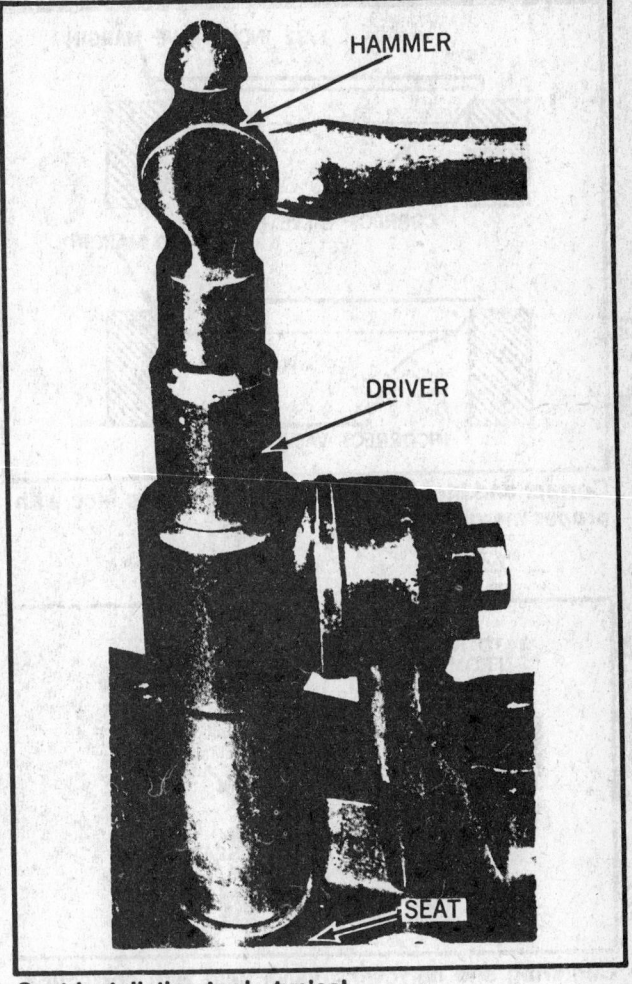

Seat installation tool—typical

━━━━━━━━━━ CAUTION ━━━━━━━━━━

Valve face angle is not always identical to valve seat angle.

A minimum margin of $3/64$ in. should remain after grinding the valve. The valve stem tip should also be squared and resurfaced, by placing the stem in the V-block of the grinder and turning it while pressing lightly against the grinding wheel.

RESURFACING THE VALVE SEATS USING A GRINDER

Select a pilot of the correct size and a coarse stone of the correct seat angle. Lubricate the pilot if necessary and install the tool

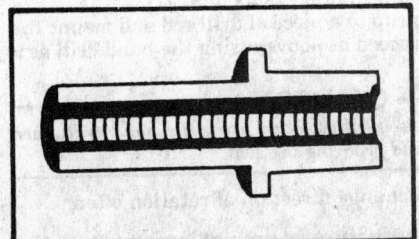

Cross section of knurled valve guide

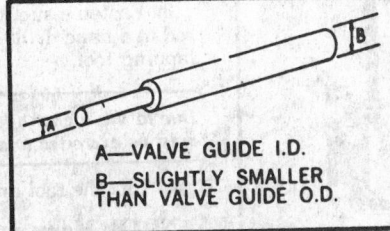

A—VALVE GUIDE I.D.
B—SLIGHTLY SMALLER THAN VALVE GUIDE O.D.

Valve guide removal tool

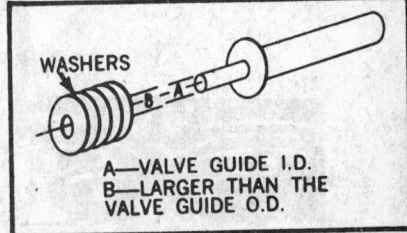

WASHERS
A—VALVE GUIDE I.D.
B—LARGER THAN THE VALVE GUIDE O.D.

Valve guide installation tool with washer stack

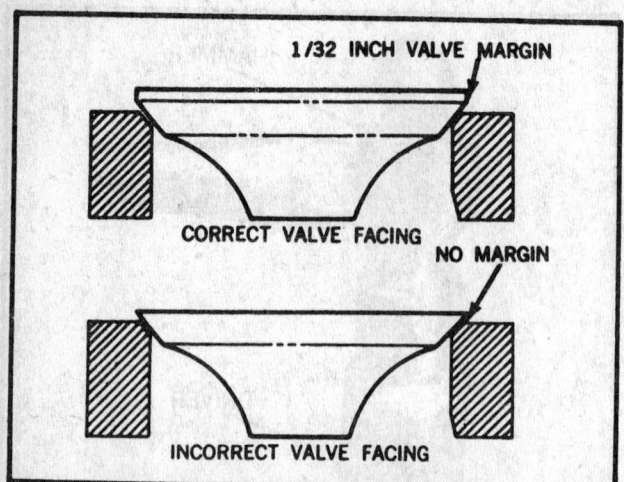

Correct and incorrect grinding of the valve face with proper margin indicated

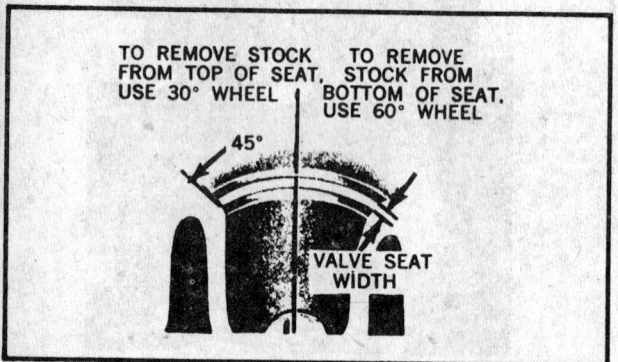

Centering and narrowing valve seat with correction stone

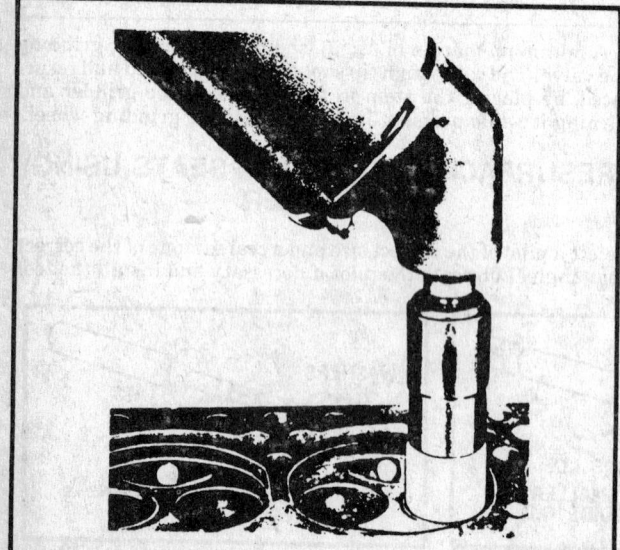

Grinding valve seats

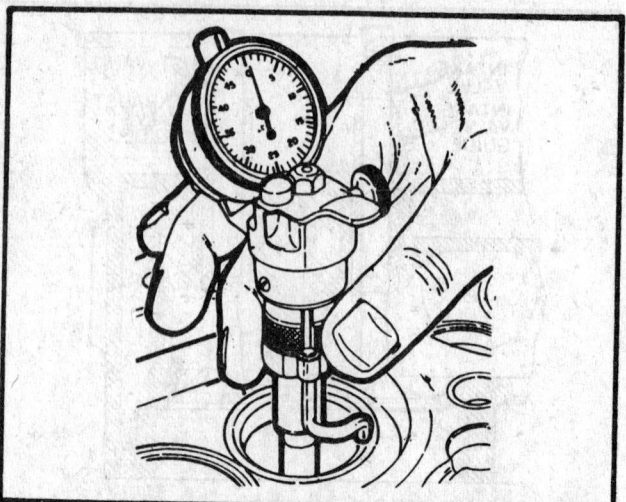

Checking valve seat run-out with a dial gauge

in the valve guide. Move the stone on and off the seat at approximately two cycles per second, until all flaws are removed from the seat. Install a fine stone and finish the seat. Center and narrow the seat using correction stones. Intake seat width $-\frac{1}{16}-\frac{5}{64}$ in. Exhaust seat width $-\frac{3}{64}-\frac{1}{16}$ in.

CHECKING THE VALVE SEAT CONCENTRICITY

1. Coat the valve face with Prussian blue dye, install the valve and rotate it on the valve sat. If the entire seat becomes coated and the valve is known to be concentric, the set is concentric.

2. Install a dial gauge pilot into the guide and rest the arm on the valve seat. Zero the gauge and rotate the arm around the seat. Runout should not exceed 0.002 in.

LAPPING THE VALVES

NOTE: Valve lapping is done to ensure efficient sealing of resurfaced valves and seats. Valve lapping alone is not recommended for use as a resurface procedure.

1. Invert the cylinder head, lightly lubricate the valve stems and install the valves in the head as numbered. Coat valve seats with fine grinding compound and attach the lapping tool suction cup to a valve head.

NOTE: Moisten the suction cup.

2. Rotate the tool between the palms, changing position and lifting the tool often to prevent grooving. Lap the valve until a smooth polished seat is evident. Remove the valve and tool and rinse away all traces of grinding compound.

3. Fasten a suction cup to a piece of drill rod and mount the rod in a hand drill. Proceed as above, using the hand drill as a lapping tool.

─── CAUTION ───
Due to the higher speeds involved when using the hand drill, care must be exercised to avoid grooving the seat.

4. Lift the tool and change direction of rotation often.

NOTE: Many manufacturers do not recommend the lapping of valves to the seats after each has been reground. However, for the rebuilder to be certain a perfect

Hand lapping the valve to the seat

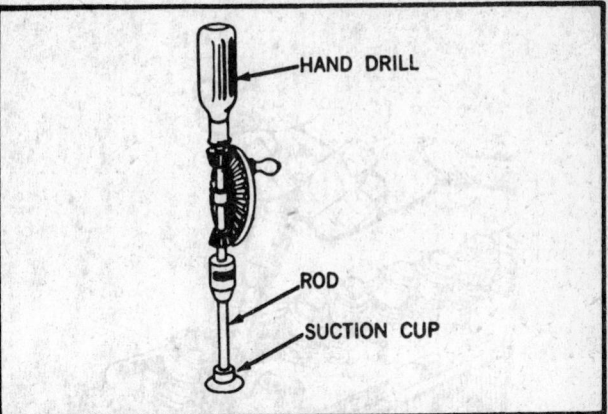

Mechanical valve lapping tool

seal exists between the valve and the seat, lapping is suggested.

CHECK THE VALVE SPRINGS

Test the spring pressure at the installed and compressed (installed height minus valve lift) height using a valve spring tester. Springs used on small displacement engines (up to 3 liters) should be ± 1 lb. of all other springs in either position. A tolerance of ± 5 lbs. is permissable on larger engines.

INSTALL VALVE STEM SEALS

Due to the pressure differential that exists at the ends of the intake valve guides (atmospheric pressure above, manifold vacuum below), oil is drawn through the valve guides into the intake port. This has been alleviated somewhat since the addition of positive crankcase ventilation, which lowers the pressure above the guides. Several types of valve stem seals are available to reduce blow-by. Certain seals simply slip over the stem and guide boss, while others require that the boss be machined. Recently, Teflon guide seals have become popular. Consult a parts supplier or machinist concerning availability and suggested usages.

NOTE: When installing seals, ensure that a small amount of oil is able to pass the seal to lubricate the valve guides; otherwise, excessive wear may result.

INSTALL THE VALVES

Lubricate the valve stems and install the valves in the cylinder head as numbered. Lubricate and position the seals (if used, see above) and the valve springs. Install the spring retainers, compress the springs and insert the keys using needlenose pliers or a tool designed for this purpose.

NOTE: Retain the keys with wheel bearing grease during installation.

CHECK VALVE SPRING INSTALLED HEIGHT

Measure the distance between the spring pad and the lower edge of the spring retainer and compare to specifications. If the installed height is incorrect, add shim washers between the spring pad and the spring.

───── **CAUTION** ─────
Use only washers designed for this purpose.

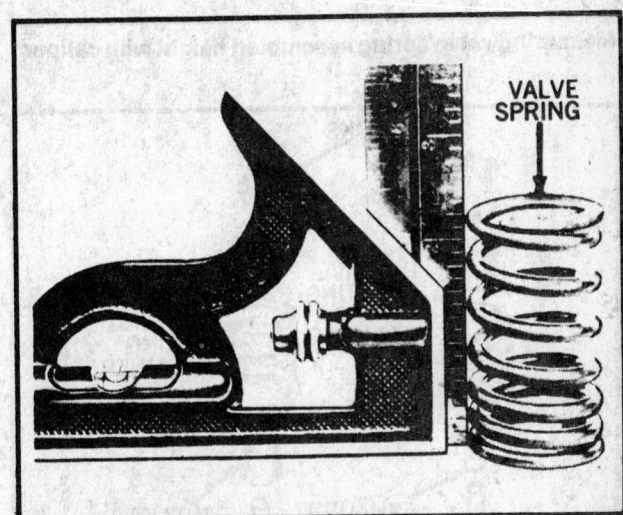

Checking valve spring for free length and squareness

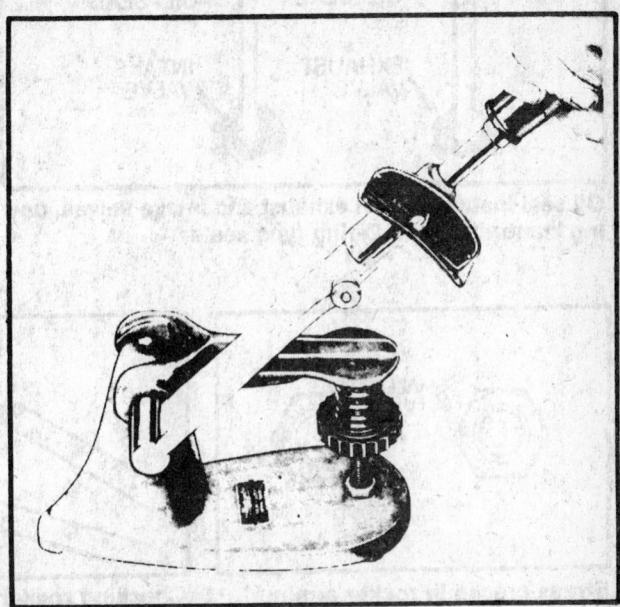

Testing valve spring compressed height

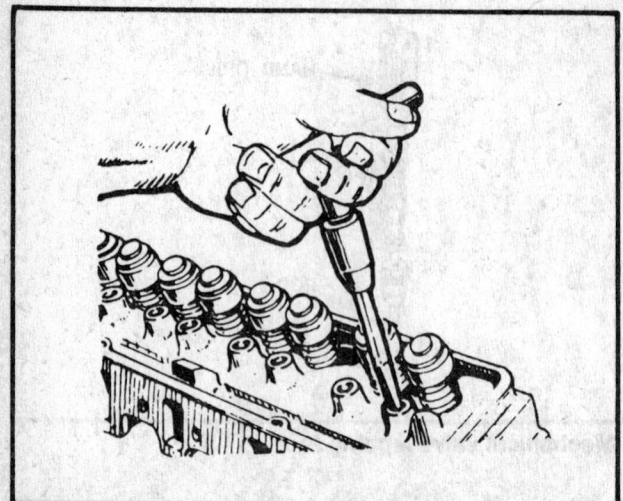

Measuring valve spring assembled height with caliper

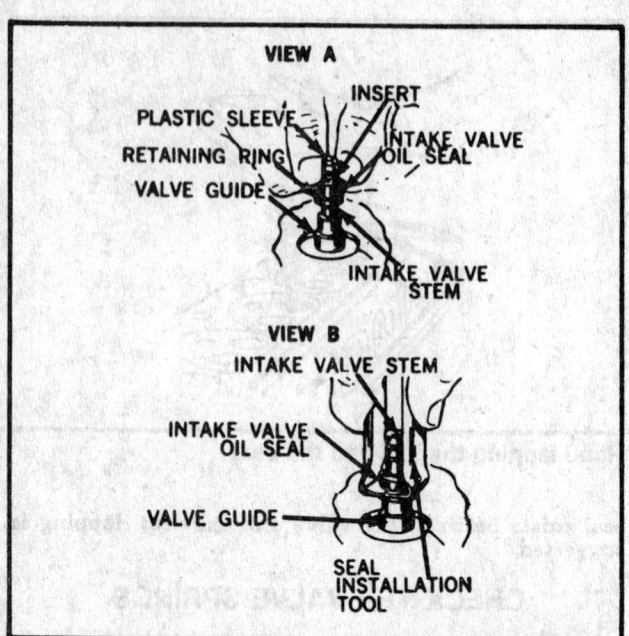

VIEW A

PLASTIC SLEEVE
INSERT
RETAINING RING
INTAKE VALVE
OIL SEAL
VALVE GUIDE
INTAKE VALVE
STEM

VIEW B

INTAKE VALVE STEM
INTAKE VALVE
OIL SEAL
VALVE GUIDE
SEAL
INSTALLATION
TOOL

Installing intake valve oil seals, Perfect Circle type, using plastic sleeve and special installation tool

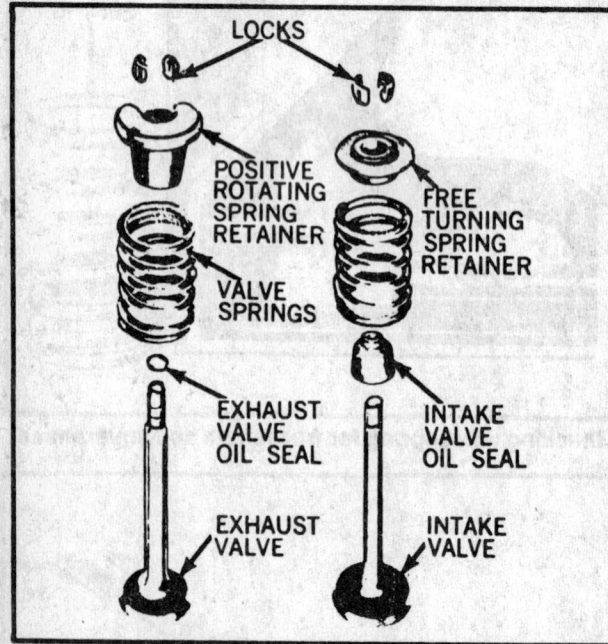

LOCKS

POSITIVE
ROTATING
SPRING
RETAINER
FREE
TURNING
SPRING
RETAINER
VALVE
SPRINGS
EXHAUST
VALVE
OIL SEAL
INTAKE
VALVE
OIL SEAL
EXHAUST
VALVE
INTAKE
VALVE

Oil seal installation on exhaust and intake valves, using "umbrella" and O-ring type seals

INSPECT THE ROCKER ARMS, BALLS, STUDS and NUTS

Visually inspect the rocker arms, balls, studs and nuts for cracks, galling, burning, scoring, or wear. If all parts are intact, liberally lubricate the rocker arms and balls and install them on the cylinder head. If wear is noted on the rocker arm at the point of valve contact, grind it smooth and square, removing as little material as possible. Replace the rocker arm if excessively worn. If a rocker stud shows signs of wear, it must be replaced. If a rocker nut shows stress cracks, replace it.

INSPECT THE ROCKER SHAFT(S) AND ROCKER ARMS

Remove rocker arms, springs and washers from rocker shaft.

NOTE: Lay out parts in the order in which they are removed.

Inspect rocker arms for pitting or wear on the valve contact point, or excessive bushing wear. Bushings need only be replaced if wear is excessive, because the rocker arm normally contacts the shaft at one point only. Grind the valve contact

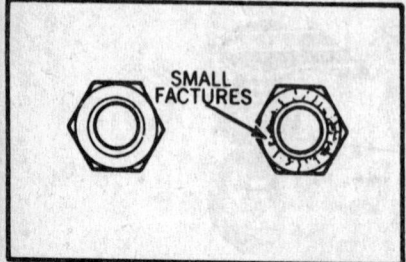

SMALL
FACTURES

Stress cracks in rocker arm nut

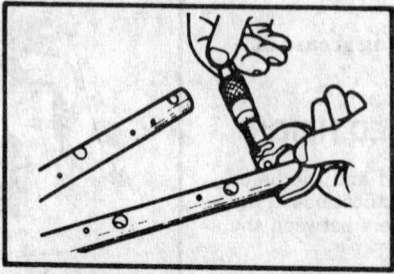

Checking rocker arm shaft O.D. with mi-crometer

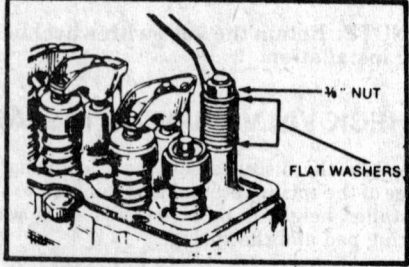

¾" NUT

FLAT WASHERS

Checking rocker arms

Reaming the stud bore for oversize rocker studs

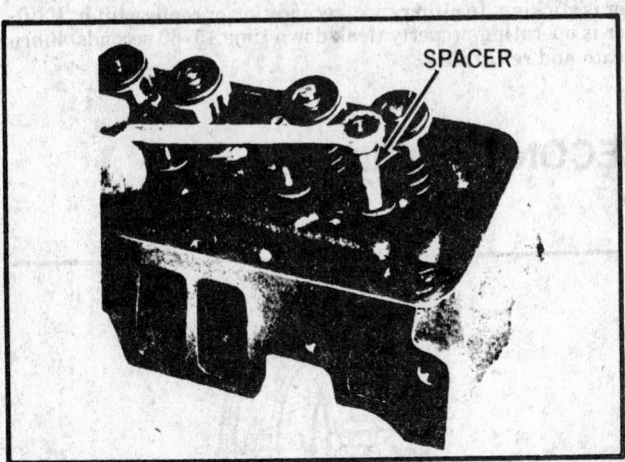

Removing a pressed in rocker stud

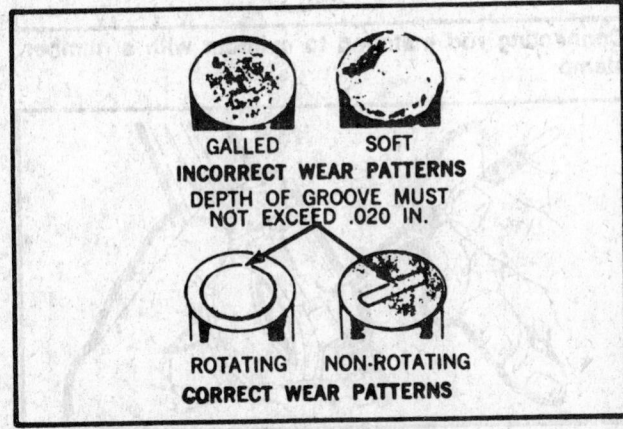

GALLED   SOFT
INCORRECT WEAR PATTERNS
DEPTH OF GROOVE MUST
NOT EXCEED .020 IN.

ROTATING   NON-ROTATING
CORRECT WEAR PATTERNS

Wear patterns on base of lifter bodies

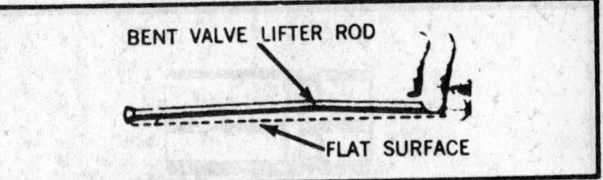

BENT VALVE LIFTER ROD

FLAT SURFACE

Checking for bent push rod

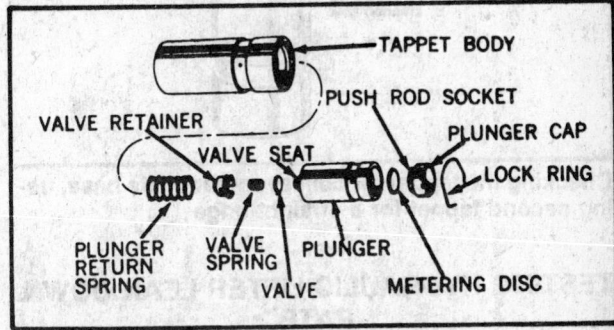

TAPPET BODY
PUSH ROD SOCKET
VALVE RETAINER
PLUNGER CAP
VALVE SEAT
LOCK RING
PLUNGER
RETURN
SPRING
VALVE
SPRING
VALVE
PLUNGER
METERING DISC

Exploded view of hydraulic lifter—typical

point of rocker arm smooth if necessary, removing as little material as possible. If excessive material must be removed to smooth and square the arm, it should be replaced. Clean out all oil holes and passages in rocker shaft. If shaft is grooved or worn, replace it. Lubricate and assemble the rocker shaft.

REPLACING ROCKER STUDS

In order to remove a threaded stud, lock two nuts on the stud and unscrew the stud using the lower nut. Coat the lower threads of the new stud with Loctite and install.

Two alternative methods are available for replacing pressed in studs. Remove the damaged stud using a stack of washers and a nut or use a stud puller. In the first, the boss is reamed 0.005–0.006 in. oversize and an oversize stud pressed in. Control the stud extension over the boss using washers, in the same manner as valve guides. Before installing the stud, coat it with white lead and grease. To retain the stud more positively, drill a hole through the stud and boss and install a roll pin. In the second method, the boss is tapped and a threaded stud installed. Retain the stud using a locking compound.

INSPECT THE PUSHRODS

Remove the pushrods and if hollow, clean out the oil passages using fine wire. Roll each pushrod over a piece of clean glass. If a distinct clicking sound is heard as the pushrod rolls, the rod is bent and must be replaced.

The length of all pushrods must be equal. Measure the length of the pushrods, compare to specifications and replace as necessary.

INSPECT THE VALVE LIFTERS
MECHANICAL OR HYDRAULIC

Remove lifters from their bores and remove gum and varnish, using solvent. Clean walls of lifter bores. Check lifters for concave wear as illustrated. If face is worn concave, replace lifter and carefully inspect the camshaft. Lightly lubricate lifter and insert it into its bore. If play is excessive, an oversize lifter must be installed (where possible). Consult a machinist concerning feasibility. If play is satisfactory, remove, lubricate and reinstall the lifter.

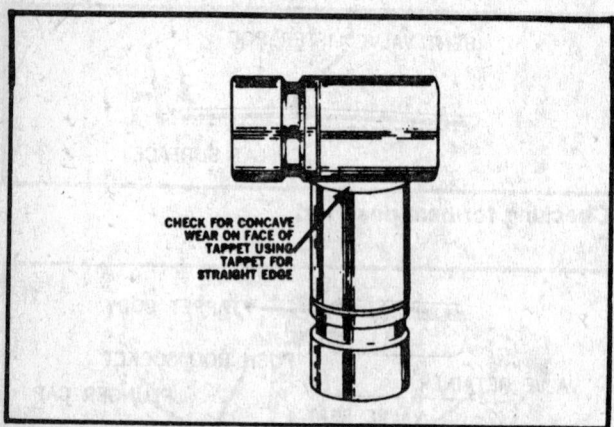

Checking the tappet for concave wear on its base, using second tappet for a straight edge

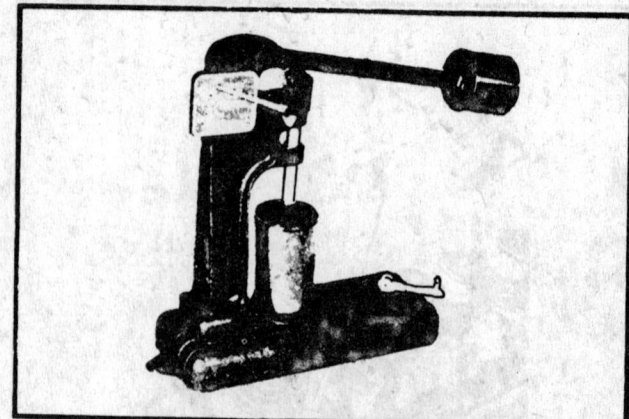

Hydraulic lifter leakdown tester—typical

TESTING HYDRAULIC LIFTER LEAK DOWN RATE

Special testers are available for the checking of the hydraulic lifter leak down rate. Special instructions accompanying the testers should be followed by the rebuilder. If the tester is not available, the following alternate method can be used.

Submerge lifter in a container of kerosene. Chuck a used pushrod or its equivalent into a drill press. Position container of kerosene so pushrod acts on the lifter plunger. Pump lifter with the drill press, until resistance increases. Pump several more times to bleed any air out of lifter. Apply very firm, constant pressure to the lifter and observe rate at which fluid bleeds out of lifter. If the fluid bleeds very quickly (less than 15 seconds), lifter is defective. If the time exceeds 60 seconds, lifter is sticking. In either case, recondition or replace lifter. If lifter is operating properly (leak down time 15–60 seconds) lubricate and reinstall.

ENGINE BLOCK RECONDITIONING

MARKING MAIN AND CONNECTING ROD CAPS

Using a punch, mark the corresponding main bearing caps and saddles according to position (i.e., one punch on the front main cap and saddle, two on the second, three on the third, etc.). Using number stamps, identify the corresponding connecting rods and caps, according to cylinder (if no numbers are present). Remove the main and connecting rod caps and place sleeves of plastic tubing over the connecting rod bolts, to protect the journals as the crankshaft is removed.

REMOVE THE RIDGE

In order to facilitate removal of the piston and connecting rod, the ridge at the top of the cylinder (unworn area; see illustration) must be removed. Place the piston at the bottom of the bore and cover it with a rag. Cut the ridge away using a ridge reamer, exercising extreme care to avoid cutting too deeply. Remove the rag and remove cuttings that remain at the piston.

─── CAUTION ───

If the ridge is not removed and new rings are installed, damage to rings will result.

REMOVING THE PISTON AND CONNECTING ROD

Invert the engine and push the pistons and connecting rods out of the cylinders. If necessary, tap the connecting rod boss with a wooden hammer handle, to force the piston out.

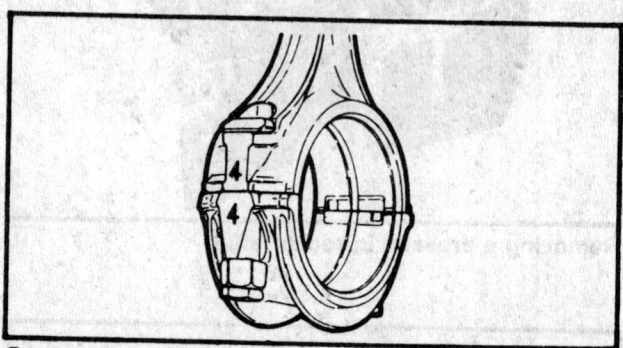

Connecting rod matched to cylinder with a number stamp

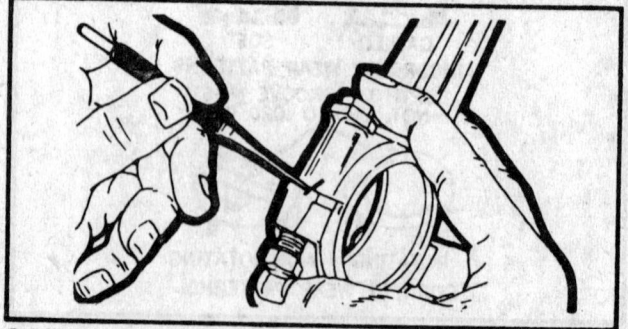

Scribe connecting rod matchmarks

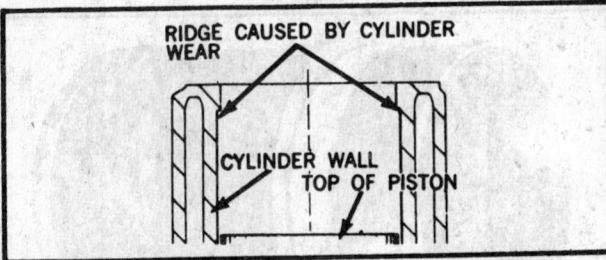

Cylinder bore ridge

Removing the piston and connecting rod assembly

--- **CAUTION** ---
Do not attempt to force the piston past the uncut cylinder ridge.

REMOVE THE OIL GALLERY PLUGS

Threaded plugs should be removed using an appropriate (usually square) wrench. To remove soft, pressed in plugs, drill a hole in the plug and thread in a sheet metal screw. Pull the plug out by the screw using pliers.

REMOVING FREEZE PLUGS

Drill a hole in the center of the freeze plugs and pry them out using a drift or special puller.

CHECK THE BORE DIAMETER AND SURFACE

Visually inspect the cylinder bores for roughness, scoring, or scuffing. If evident, the cylinder bore must be bored or honed oversize to eliminate imperfections and the smallest possible oversize piston used. The new pistons should be given to the machinist with the block, so that the cylinders can be bored or honed exactly to the piston size (plus clearance). If no flaws are evident, measure the bore diameter using a telescope gauge and micrometer, or dial gauge, parallel and perpendicular to the engine centerline, at the top (below the ridge) and bottom of the bore. Subtract the bottom measurements from the top to determine taper and the parallel to the centerline measurements from the perpendicular measurements to determine eccentricity. If the measurements are not within specifications, the cylinder must be bored or honed and an oversize piston installed. If the measurements are within specifications the cylinder may be used as is, with only finish honing.

CYLINDER SLEEVE LINERS
DRY CYLINDER LINERS

Various engines are fitted with dry type cylinder liners at the

Location of oil gallery plugs, core plugs and camshaft bearing bore plug—V8 engine (some engines may be equipped with a balance shaft)

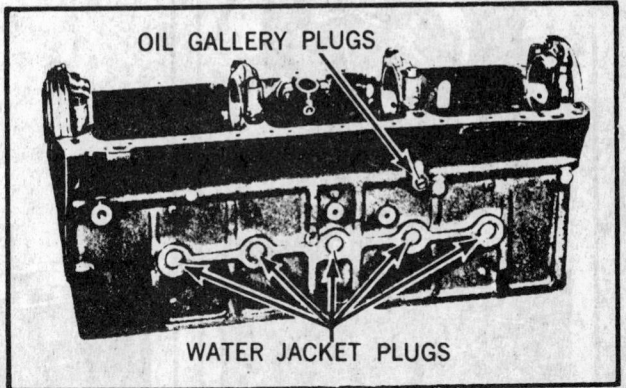

Location of oil galley and water jacket plugs—6 cylinder engine typical

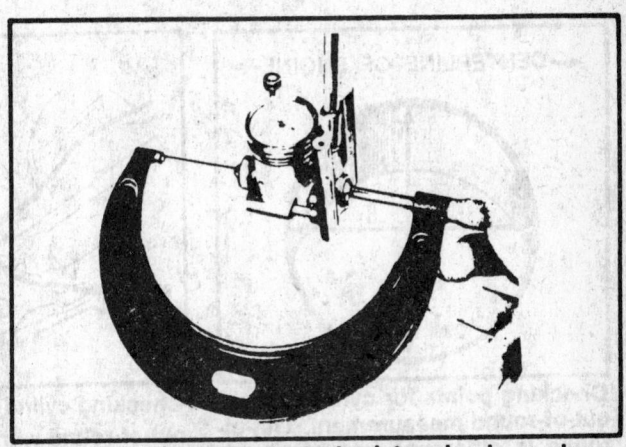

Measuring telescope gauge to determine bore size

time of manufacture. This type of liner can be replaced with the use of special pulling tools at the time of engine overhaul.

When the cylinder bore is part of the block assembly and if only one or two cylinder bores are damaged, sleeves can be installed in the damaged bores to avoid reboring all cylinders to an oversize condition.

The services of a competent automotive machine shop should be used for the boring of the cylinders and the installation of the liners.

WET CYLINDER LINERS

Removable cylinder liners are used in varied engines that can be lifted from the engine block without the use of pullers or of a press. Soft metal rings are used at the base of the liners to seal between machined surfaces on the engine block and the cylinder liner, to prevent coolant from entering the engine lubricating system.

The cylinder head gasket is used to seal the top of the cylinder liner and the cylinder head. Should the cylinder head be re-

Wet cylinder liner—typical

Cylinder reboring machine

Cylinder honing tool

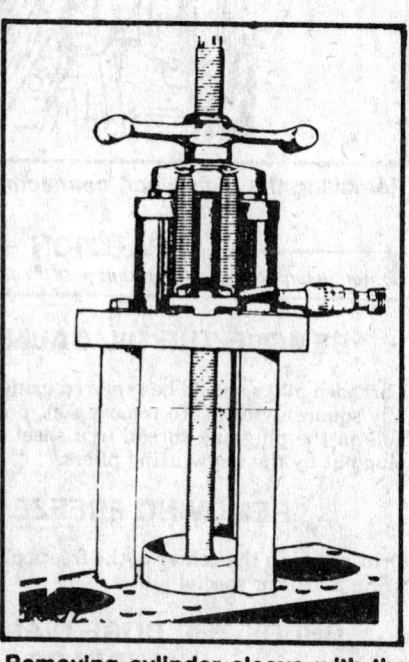

Removing cylinder sleeve with the use of hydraulic tool

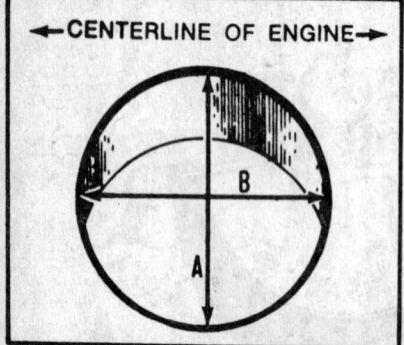

Checking points for cylinder bore out-of-round measurement. Out-of-round is difference between measurement A and B

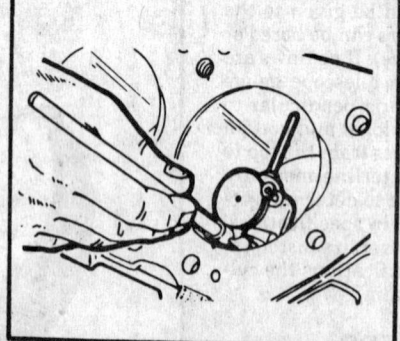

Checking cylinder bore taper and out-of-round with a dial indicator cylinder bore gauge

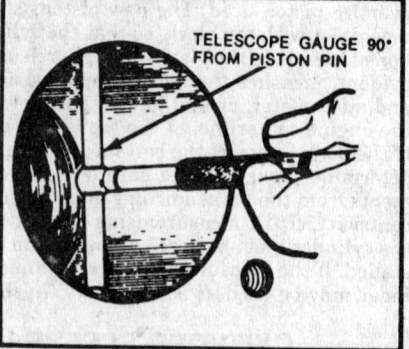

Measuring culinder bore with tele scope

moved or the engine overhauled, a projection of approximately 0.002–0.006 in. (depending upon engine manufacturer's specifications) should exist at the liner top, above the surface of the engine block. If the projection is not present, new sealing rings should be installed at the bottom of the cylinder liners to prevent coolant leakage or compression loss.

Installation of pistons, rings and liners are installed as sets to control the weight and balance of the components.

CHECK THE CYLINDER BLOCK BEARING ALIGNMENT

Remove the upper bearing inserts. Place a straightedge in the bearing saddles along the centerline of the crankshaft. If clear-

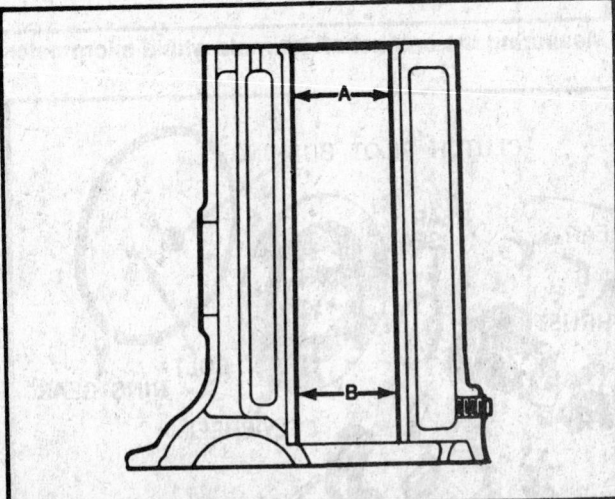

Checking points for cylinder bore taper measurement. Taper is difference between measurement A and B

Checking main bearing saddle alignment

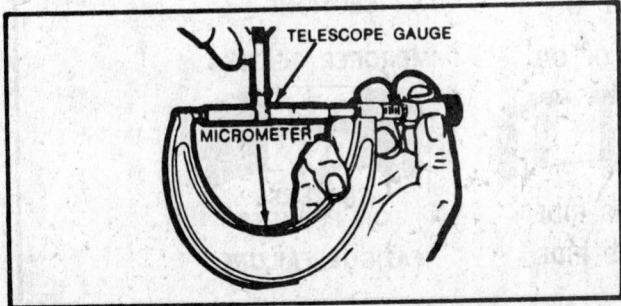

Measuring cylinder gauge to determine bore size

ance exists between the straightedge and the center saddle, the block must be linebored.

HOT-TANK THE BLOCK

Have the block hot-tanked to remove grease, corrosion and scale from the water jackets.

NOTE: Consult the operator to determine whether the camshaft bearings will be damaged during the hot-tank process.

SERVICE THE CRANKSHAFT

Ensure that all oil holes and passages in the crankshaft are open and free of sludge. If necessary, have the crankshaft ground to the largest possible undersize.

Have the crankshaft magnafluxed, to locate stress cracks. Consult a machinist concerning additional service procedures, such as surface hardening (e.g., Nitriding, Tuftriding) to improve wear characteristics, cross drilling and chamfering the oil holes to improve lubrication and balancing.

Measure the main bearing journals at each end twice (90° apart) using a micrometer, to determine diameter, journal taper and eccentricity. If journals are within tolerances, reinstall bearing caps at their specified torque. Using a telescope gauge and micrometer, measure bearing I.D. parallel to piston axis and at 30° on each side of piston axis. Subtract journal O.D. from bearing I.D. to determine oil clearance. If crankshaft journals appear defective, or do not meet tolerances, there is no need to measure bearings; for the crankshaft will require grinding and/or undersize bearings will be required. If bearing appears defective, cause for failure should be determined prior to replacement.

Refer to the failure diagnosis section to help you determine the cause of the failure.

CHECK THE BLOCK FOR CRACKS

Visually inspect the block for cracks or chips. The most common locations are as follows:
1. Adjacent to freeze plugs
2. Between the cylinders and water jackets
3. Adjacent to the main bearing saddles
4. At the entrance bottom of the cylinders

Check only suspected cracks using spot check dye (see introduction). If a crack is located, consult a machinist concerning possible repairs.

Magnaflux the block to locate hidden cracks. If cracks are located, consult a machinist about feasibility of repair.

NOTE: Engine blocks that are porous or have sand holes, can be repaired with metallic plastic where coolant or oil pressure does not exist. Do not attempt to repair cracked blocks with the metallic plastic.

CHECK THE BLOCK DECK FOR WARPAGE

Using a straightedge and feeler gauges, check the block deck for warpage in the same manner that the cylinder head is checked (see Cylinder Head Reconditioning). If warpage exceeds specifications, have the deck resurfaced.

NOTE: In certain cases a specification for total material removal (Cylinder head and block deck) is provided. This specification must not be exceeded.

CHECK THE DECK HEIGHT

The deck height is the distance from the crankshaft centerline to the block deck. To measure, invert the engine and install the crankshaft, retaining it with the center main cap. Measure the distance from the crankshaft journal to the block deck, parallel to the cylinder centerline. Measure the diameter of the end (front and rear) main journals, parallel to the centerline of the cylinders, divide the diameter in half and subtract it from the previous measurement. The results of the front and rear measurements should be identical. If the difference exceeds 0.005 in., the deck height should be corrected.

NOTE: Block deck height and warpage should be corrected together.

INSTALL THE OIL GALLERY PLUGS AND FREEZE PLUGS

Coat freeze plugs with sealer and tap into position using a piece of pipe, slightly smaller than the plug, as a driver. To ensure retention, stake the edges of the plugs. Coat threaded oil

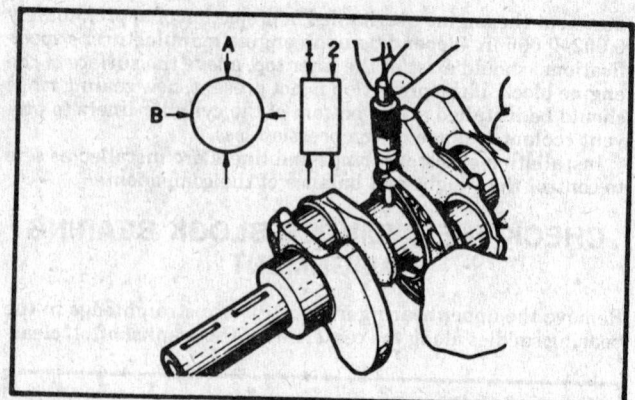

Measuring the crankshaft journals with a micrometer

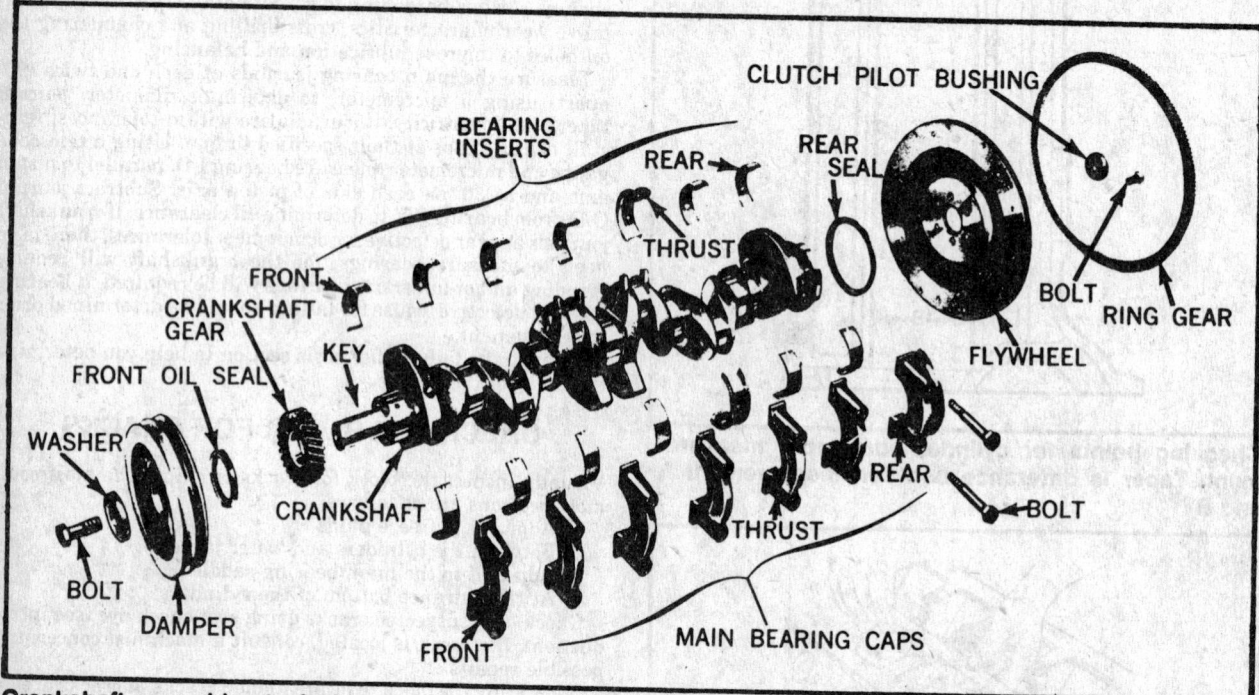

Crankshaft assembly—typical

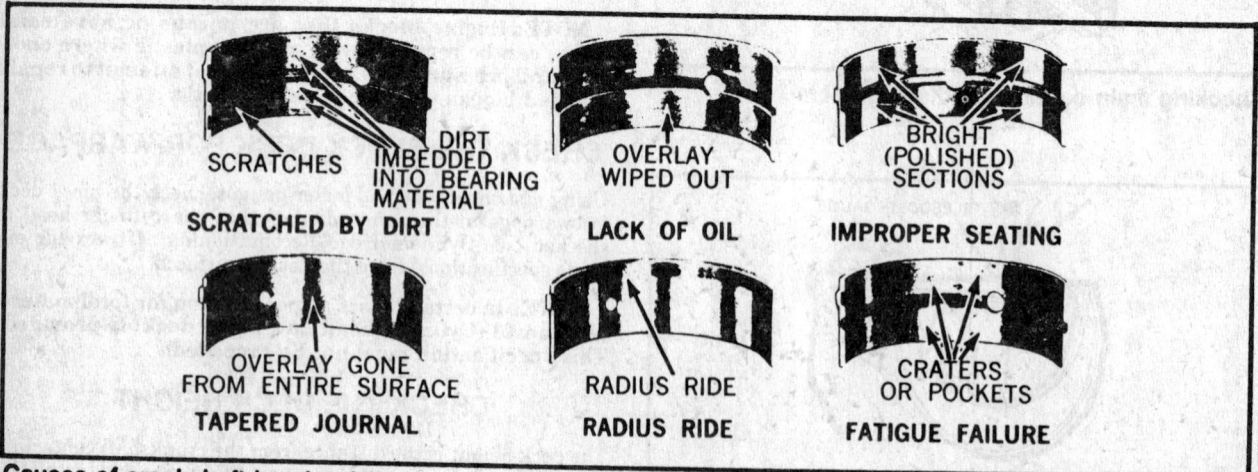

Causes of crankshaft bearing failures

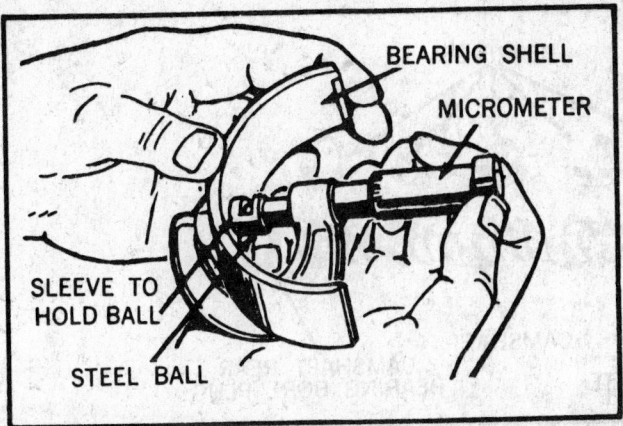

Masuring bearing insert thickness with special ball

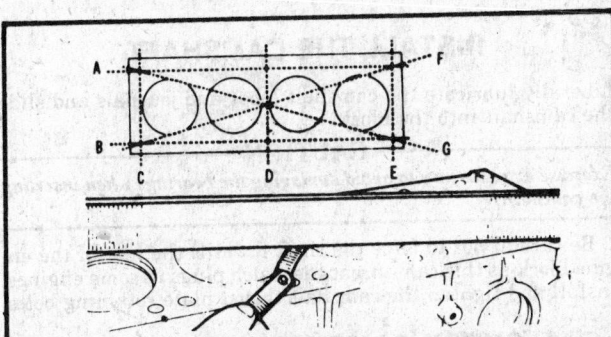

Checking the cylinder block for distortion

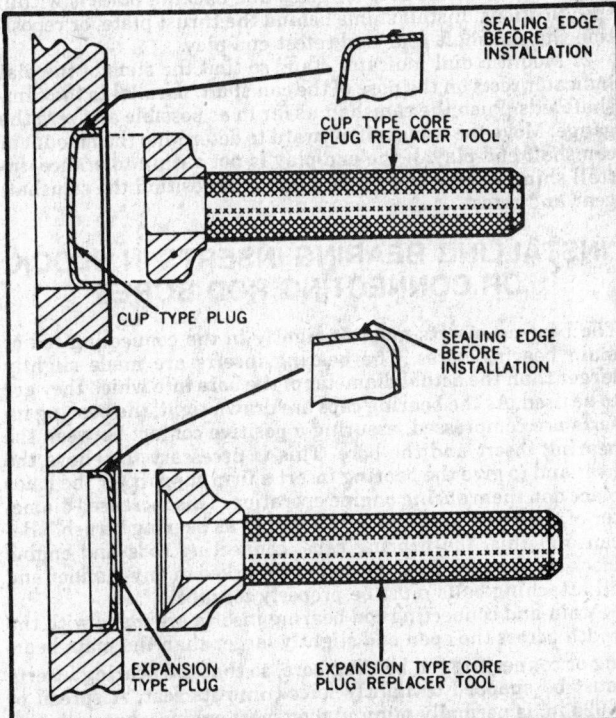

Installation of cup type and expansion type core plugs with special tool

Checking the camshaft for straightness

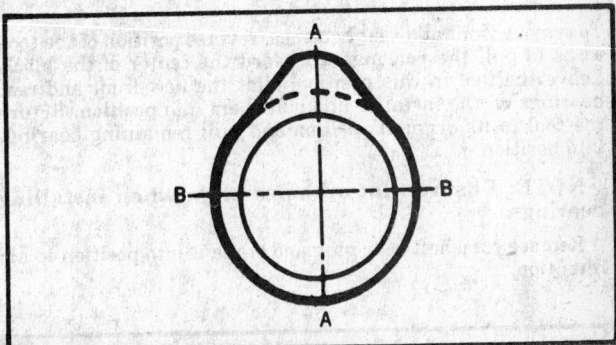

Camshaft lobe measurement—lift is difference between A and B measurement

gallery plugs with sealer and install. Drive replacement soft plugs into block using a large drift as a driver.

Rather than reinstalling lead plugs, drill and tap the holes and install threaded plugs, where possible.

CLEAN AND INSPECT THE CAMSHAFT

Degrease the camshaft, using solvent and clean out all oil holes. Visually inspect cam lobes and bearing journals for excessive wear. If a lobe is questionable, check all lobes as indicated below. If a journal or lobe is worn, the camshaft must be reground or replaced.

NOTE: If a journal is worn, there is a good chance that the bushings are worn.

If lobes and journals appear intact, place the front and rear journals in V-blocks and rest a dial indicator on the center journal. Rotate the camshaft to check straightness. If deviation exceeds 0.001 in., replace the camshaft.

Check the camshaft lobes with a micrometer, by measuring the lobes from the nose to base and again at 90°. The lift is determined by subtracting the second measurement from the first. If all exhaust lobes and all intake lobes are not identical with specs, the camshaft must be reground or replaced.

REPLACE THE CAMSHAFT BEARINGS

If excessive wear is indicated or if the engine is being completely rebuilt, camshaft bearings should be replaced as follows: Drive the camshaft rear plug from the block. Assemble the removal puller with its shoulder on the bearing to be removed. Gradually tighten the puller nut until bearing is removed. Remove remaining bearings, leaving the front and rear for last.

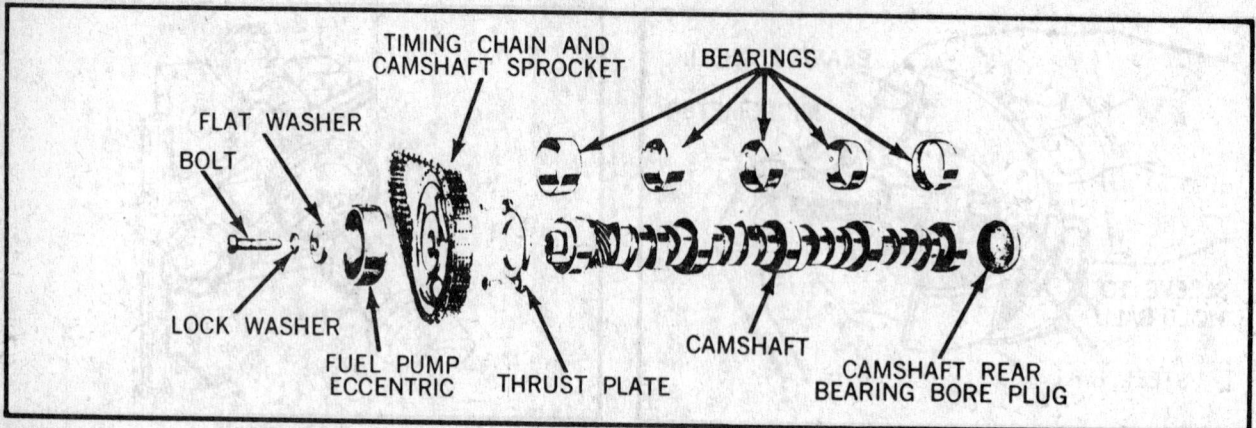

Chain driven camshaft assembly—typical

To remove front and rear bearings, reverse position of the tool, so as to pull the bearings in toward the center of the block. Leave the tool in this position, pilot the new front and rear bearings on the installer and pull them into position. Return the tool to its original position and pull remaining bearings into position.

NOTE: Ensure that oil holes align when installing bearings.

Replace camshaft rear plug and stake it into position to aid retention.

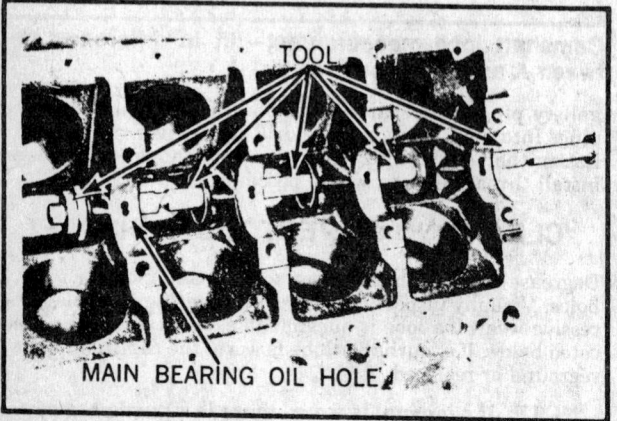

Removing and Installing cam bearings with special puller tool—typical

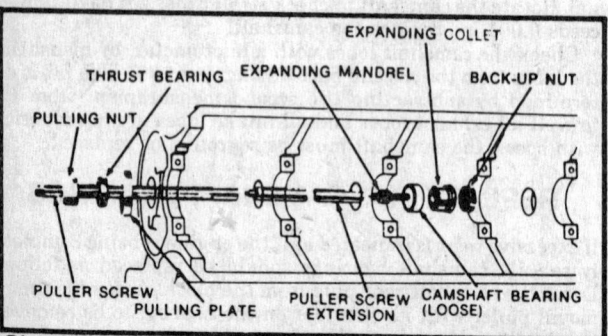

Checking camshaft alignment with Vee blocks and dial indicator

INSTALL THE CAMSHAFT

Liberally lubricate the camshaft lobes and journals and slide the camshaft into the block.

───────── CAUTION ─────────
Exercise extreme care to avoid damaging the bearings when inserting the camshaft.
───────────────────────────

Be careful not to force the shaft towards the rear of the engine block as this can unseat the welch plugs in some engines. Install and tighten the camshaft thrust plate retaining bolts.

CHECK CAMSHAFT END-PLAY

1. Using feeler gauges, determine whether the clearance between the camshaft boss (or gear) and backing plate is within specifications. Install shims behind the thrust plate, or reposition the camshaft gear and retest end-play.
2. Mount a dial indicator stand so that the stem of the dial indicator rests on the nose of the camshaft, parallel to the camshaft axis. Push the camshaft as far in as possible and zero the gauge. Move the camshaft outward to determine the amount of camshaft end-play. If the end-play is not within tolerance, install shims behind the thrust plate or reposition the camshaft gear and retest.

INSTALLING BEARING INSERTS IN BLOCK OR CONNECTING ROD BORES

The bearing inserts must fit tightly in the connecting rod or main bearing bores. The bearing inserts are made slightly larger than the actual diameter of the bore into which they are to be used. As the bearing caps are drawn tight, the bearing inserts are compressed, assuring a positive contact between the bearing insert and the bore. This is necessary to relieve the heat and to give the bearing insert a firm support for the loads place don them during engine operation. This increased diameter of the bearing insert is referred to as bearing "crush". Because of this, the bearing caps, connecting rods and engine block must not be filed, lapped or reworked in any manner and all attaching bolts must be properly torqued.

Main and connecting rod bearing inserts are made with the width across the open end slightly larger than the main bearing or connecting rod bearing bore, so that the bearing inserts must be snapped or lightly forced into its seat. A spread of 0.025 in. is normally minimum on most engines, but will vary from engine to engine. (Some bearing kits will have instructions for the proper installation of the inserts.)

To adjust the bearing spread of the thick wall bearings, such

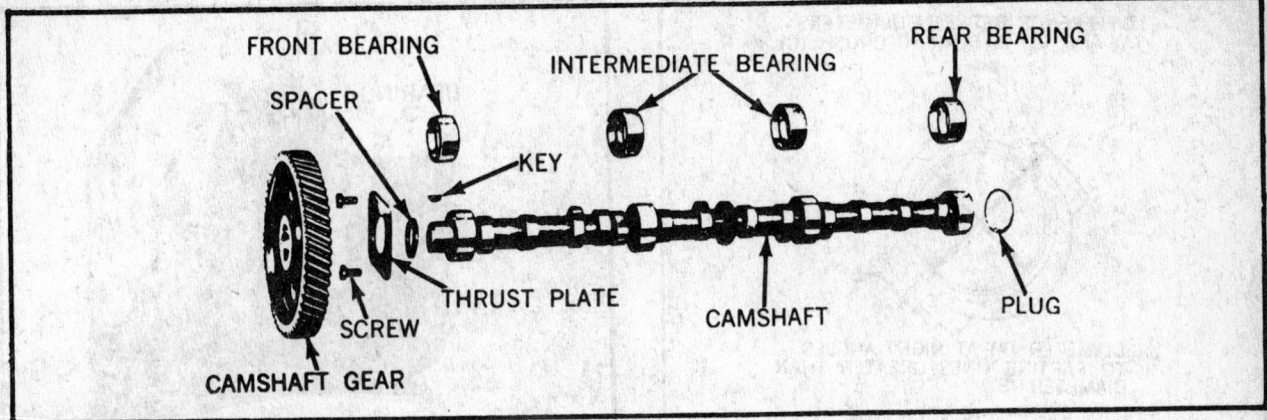

Gear driven camshaft assembly—typical

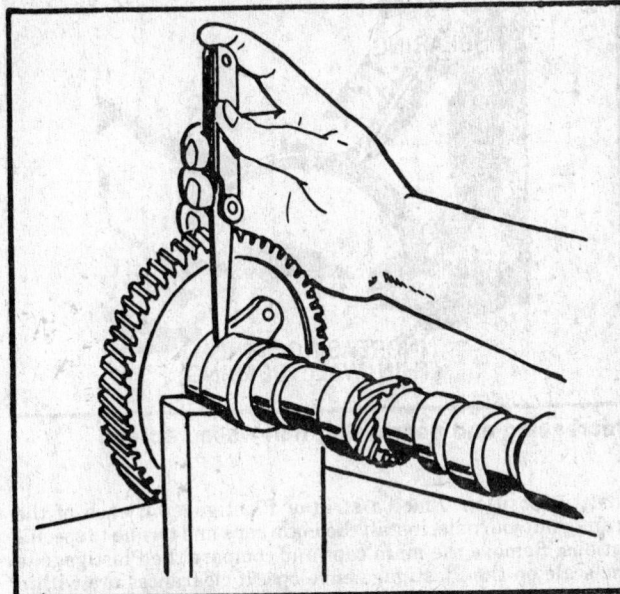

Checking clearance between timing gear and thrust plate with feeler gauge

1. PUSH CAM TO REAR OF ENGINE
2. SET DIAL ON ZERO
3. PULL CAM FORWARD AND RELEASE

Checking camshaft end-play with dial indicator

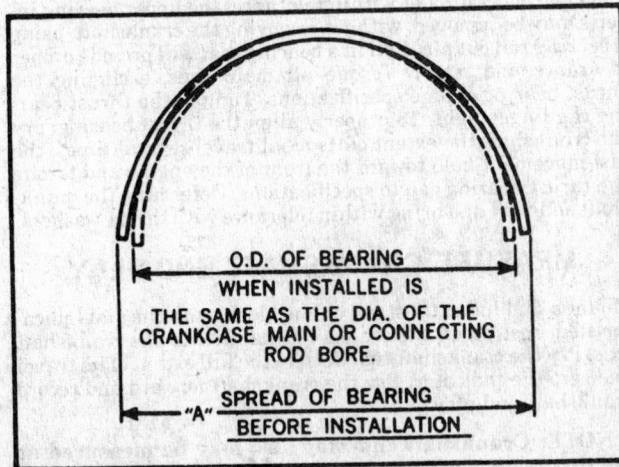

O.D. OF BEARING
WHEN INSTALLED IS
THE SAME AS THE DIA. OF THE
CRANKCASE MAIN OR CONNECTING
ROD BORE.

"A" SPREAD OF BEARING
BEFORE INSTALLATION

Illustration of bearing insert spread

as main bearing inserts, place one end of the bearing insert on a wood block and strike the other end with a soft mallet to decrease the spread. To increase the spread, place the bearing insert ends on a wood block and strike the back of the insert with a soft mallet, squarely and lightly. The bearing spread on the thin walled bearing inserts, such as connecting rod bearing inserts, can be adjusted by hand, either spreading with the thumbs and forefingers of both hands, or by squeezing the bearing insert by the palm of the hand to decrease the spread. Check the spread distance often during the adjustment procedure.

INSTALL THE REAR MAIN SEAL

Position the block with the bearing saddles facing upward. Lay the rear main seal in its groove and press it lightly into its seat. Place a piece of pipe the same diameter as the crankshaft journal into the saddle and firmly seat the seal. Hold the pipe in position and trim the ends of the seal flush if required.

INSTALL THE CRANKSHAFT

Thoroughly clean the main bearing saddles and caps. Place the upper halves of the bearing inserts on the saddles and press into position.

NOTE: Ensure that the oil holes align.

Press the corresponding bearing inserts into the main bearing caps. Lubricate the upper main bearings and lay the crank-

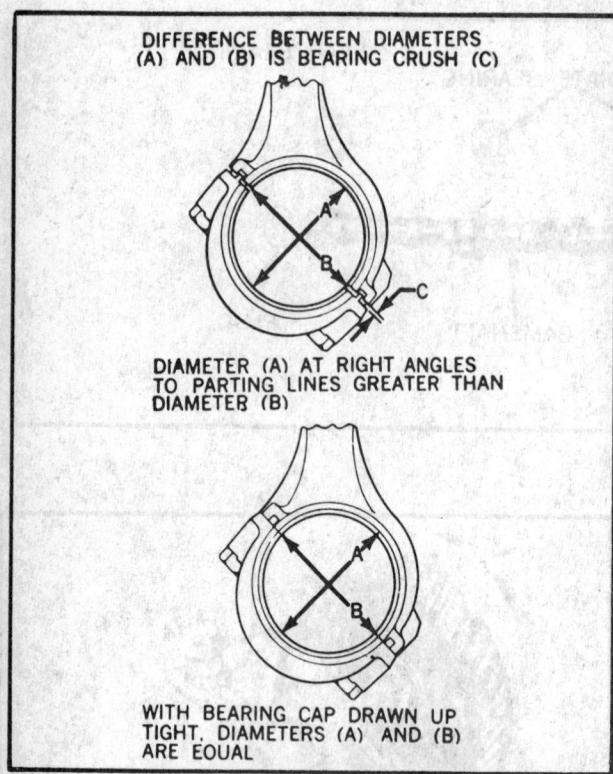

DIFFERENCE BETWEEN DIAMETERS (A) AND (B) IS BEARING CRUSH (C)

DIAMETER (A) AT RIGHT ANGLES TO PARTING LINES GREATER THAN DIAMETER (B)

WITH BEARING CAP DRAWN UP TIGHT, DIAMETERS (A) AND (B) ARE EQUAL

Bearing crush in connecting rod bore

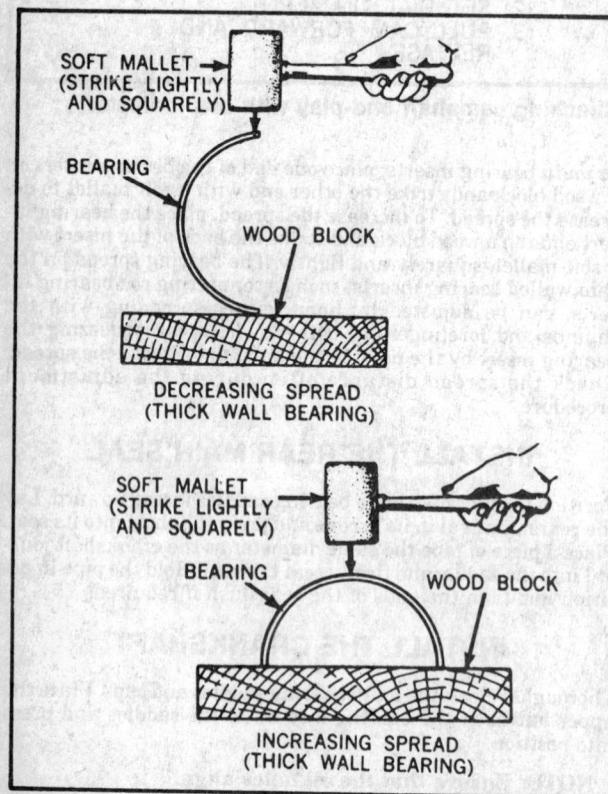

SOFT MALLET (STRIKE LIGHTLY AND SQUARELY)

BEARING

WOOD BLOCK

DECREASING SPREAD (THICK WALL BEARING)

SOFT MALLET (STRIKE LIGHTLY AND SQUARELY)

BEARING

WOOD BLOCK

INCREASING SPREAD (THICK WALL BEARING)

Increasing and decreasing thick-walled bearing spread

BEARING

DECREASING SPREAD (THIN WALL BEARING)

BEARING

INCREASING SPREAD (THIN WALL BEARING)

Increasing and decreasing thin-walled spread

shaft in position. Place a strip of Plastigage on each of the crankshaft journals, install the main caps and torque to specifications. Remove the main caps and compare the Plastigage to the scale on the Plastigage envelope. If clearances are within tolerances, remove the Plastigage, turn the crankshaft 90°, wipe off all oil and retest. If all clearances are correct, remove all Plastigage, thoroughly lubricate the main caps and bearing journals and install the main caps.

If clearances are not within tolerance, the upper bearing inserts may be removed, without removing the crankshaft, using a bearing roll out pin. Roll in a bearing that will provide proper clearance and retest. Torque all main caps, excluding the thrust bearing cap, to specifications. Tighten the thrust bearing cap finger tight. To properly align the thrust bearing, pry the crankshaft the extent of its axial travel several times, the last movement held toward the front of the engine and torque the thrust bearing cap to specifications. Determine the crankshaft end-play and bring within tolerance with thrust washers.

MEASURE CRANKSHAFT END-PLAY

Mount a dial indicator stand on the block, with the dial indicator stem resting on the crankshaft parallel to the crankshaft axis. Pry the crankshaft rearward to the full extent of its travel and zero the indicator. Pry the crankshaft forward and record crankshaft end-play.

NOTE: Crankshaft end-play also may be measured at the thrust bearing, using feeler gauges.

Seating the rear main bearing seal

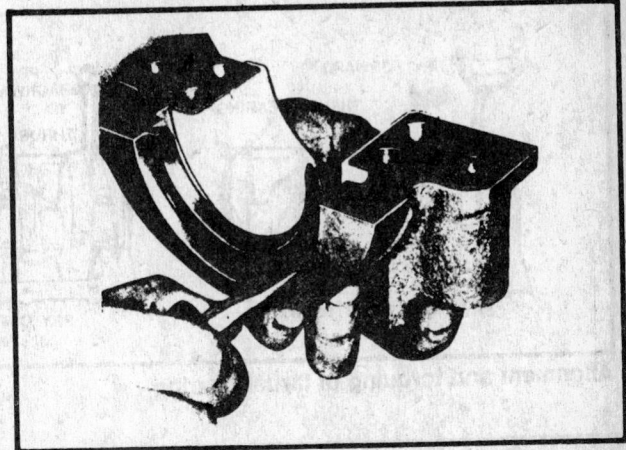

Removing the rear main bearing oil seal from the bearing cap—preformed seal type

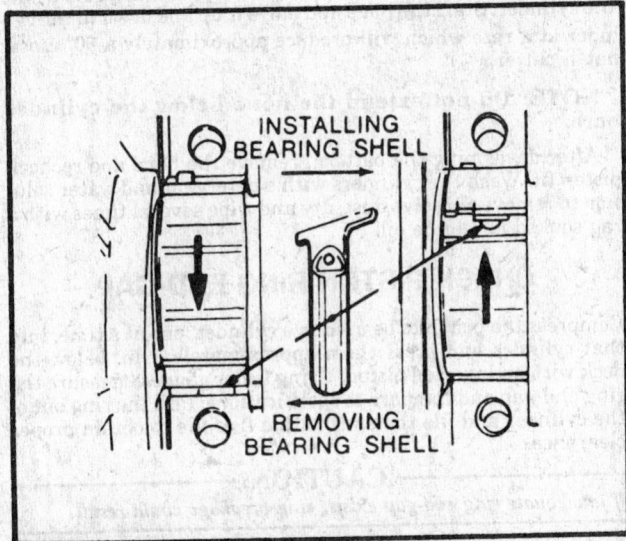

Main bearing insert identification—typical

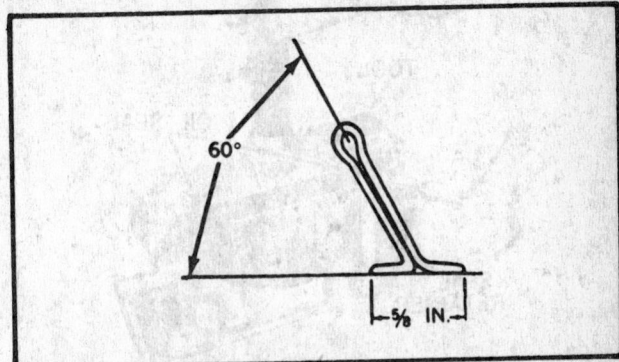

Bearing insert roll-out pin made from cotter pin

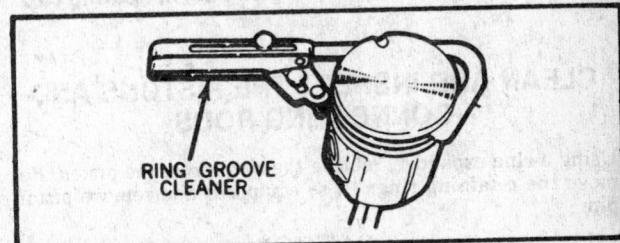

Cleaning piston ring grooves—typical

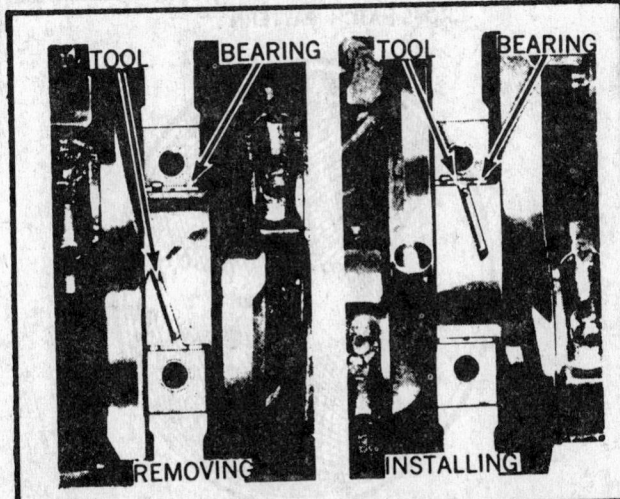

Removing and installing main bearing inserts with roll-out pins (cross section of crankshaft journal)

Checking crankshaft end-play—typical

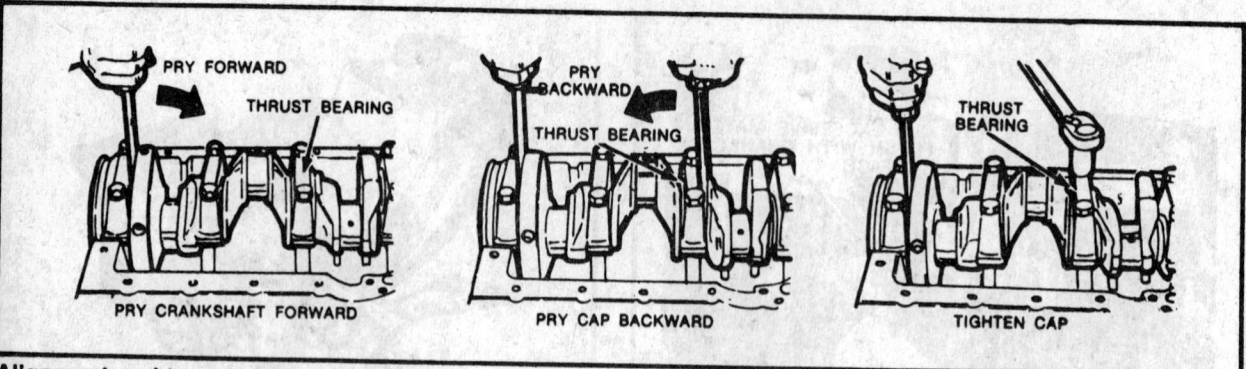

Alignment and torquing of thrust bearing

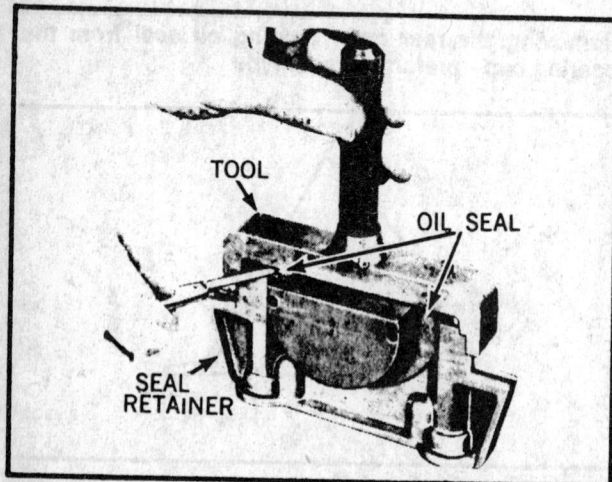

Installing the rear main bearing oil seal in bearing cap

CLEAN AND INSPECT THE PISTONS AND CONNECTING RODS

Using a ring expander, remove the rings from the piston. Remove the retaining rings (if so equipped) and remove piston pin.

--- CAUTION ---

If the piston pin must be pressed out, determine the proper method and use the proper tools; otherwise the piston will distort.

Clean the ring grooves using an appropriate tool, exercising care to avoid cutting too deeply. Thoroughly clean all carbon and varnish from the piston with solvent.

--- CAUTION ---

Do not use a wire brush or caustic solvent on pistons.

Inspect the pistons for scuffing, scoring, cracks, pitting, or excessive ring groove wear. If wear is evident, the piston must be replaced. Check the connecting rod length by measuring the rod from the inside of the large end to the inside of the small end using calipers. All connecting rods should be equal length. Replace any rod that differs from the others in the engine. Have the connecting rod alignment checked in an alignment fixture by a machinist. Replace any twisted or bent rods.

Magnaflux the connecting rods to locate stress cracks. If cracks are found, replace the connecting rod.

FINISH HONE THE CYLINDERS

Chuck a flexible drive hone into a power drill and insert it into the cylinder. Start the hone and move it up and down in the cylinder at a rate which will produce approximately a 60° cross-hatch pattern.

NOTE: Do not extend the hone below the cylinder bore.

After developing the pattern, remove the hone and recheck piston fit. Wash the cylinders with a detergent and water solution to remove abrasive dust, dry and wipe several times with a rag soaked in engine oil.

CHECK PISTON RING END-GAP

Compress the piston to be used in a cylinder, one at a time, into that cylinder and press them approximately 1 in. below the deck with an inverted piston. Using feeler gauges, measure the ring end-gap and compare to specifications. Pull the ring out of the cylinder and file the ends with a fine file to obtain proper clearance.

--- CAUTION ---

If inadequate ring end-gap exists, ring breakage could result.

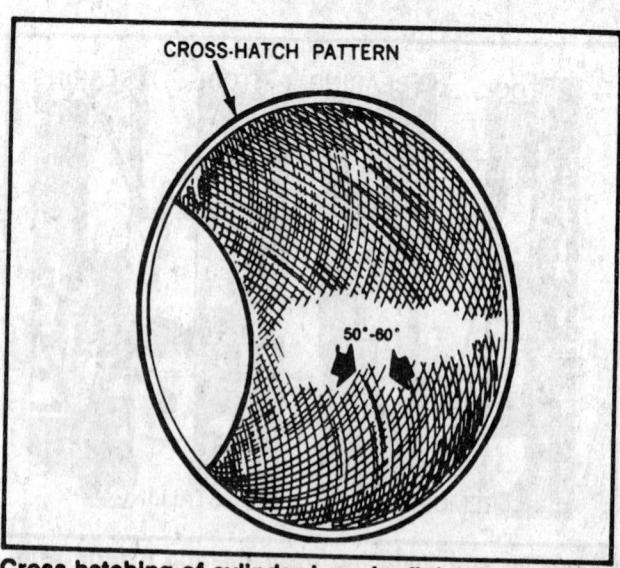

Cross hatching of cylinder bore by finish honing

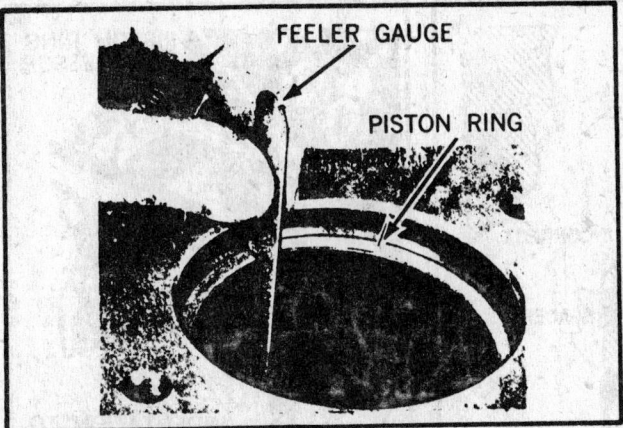

Checking ring gap in cylinder bore

FIT THE PISTONS TO THE CYLINDERS

Using a telescope gauge and micrometer, or a dial gauge, measure the cylinder bore diameter perpendicular to the piston pin, 2 ½ in. below the deck. Measure the piston perpendicular to its pin on the skirt. The difference between the two measurements is the piston clearance. If the clearance is within specifications or slightly below (after boring or honing), finish honing is all that is required. If the clearance is excessive, try to obtain a slightly larger piston to bring clearance within specifications. Where this is not possible, obtain the first oversize piston and hone (or if necessary, bore) the cylinder to size.

ASSEMBLE THE PISTONS AND CONNECTING RODS

Inspect piston pin, connecting rod, small end bushing and piston bore for galling, scoring or excessive wear. If evident, replace defective part(s). Measure the I.D. of the piston boss and connecting rod small end and the O.D. of the piston pin. If within specifications, assemble piston pin and rod.

—— CAUTION ——
CAUTION: If piston pin must be pressed in, determine the proper method and use the proper tools; otherwise the piston will distort.

Install the lock rings; ensure that they seat properly. If the parts are not within specifications, determine the service method for the type of engine. In some cases, piston and pin are serviced as an assembly when either is defective. Others specify reaming the piston and connecting rods for an oversize pin. If the connecting rod bushing is worn, it may in many cases be replaced. Reaming the piston and replacing the rod bushing are machine shop operations.

INSTALL THE PISTON RINGS

Inspect the ring grooves in the piston for excessive wear or taper. If necessary, recut the groove(s) for use with an overwidth ring or a standard ring and spacer. If the groove is worn uniformly, overwidth rings or standard rings and spacers may be installed without recutting. Roll the outside of the ring around the groove to check for burrs or deposits. If any are found, remove with a fine file. Hold the ring in the groove and measure side clearance. If necessary, correct as indicated above.

NOTE: Always install any additional spacers above the piston ring.

The ring grooves must be deep enough to allow the ring to seat below the lands. In many cases, a "go-no-go" depth gauge will be provided with the piston rings. Shallow grooves may be corrected by recutting, while deep grooves require some type of filler or expander behind the piston. Consult the piston ring supplier concerning the suggested method. Install the rings on the piston, lowest ring first, using a ring expander.

NOTE: Position the ring markings as specified by the manufacturer.

INSTALL THE PISTONS

Press the upper connecting rod bearing halves into the connecting rods and the lower halves into the connecting rod caps. Position the piston ring gaps according to specifications and lubricate the pistons. Install a ring compressor on the piston and press two long (8 in.) pieces of plastic tubing over the rod bolts. Using the plastic tubes as a guide, press the piston into the bore and onto the crankshaft with a wooden hammer handle. After seating the rod on the crankshaft journal, remove the tubes and install the cap nuts finger tight. Install the remain-

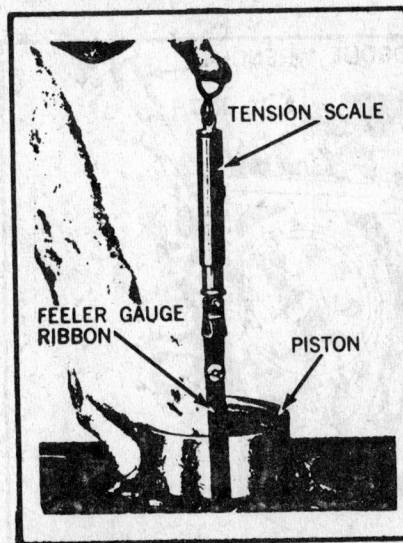

Checking piston to cylinder bore clearance

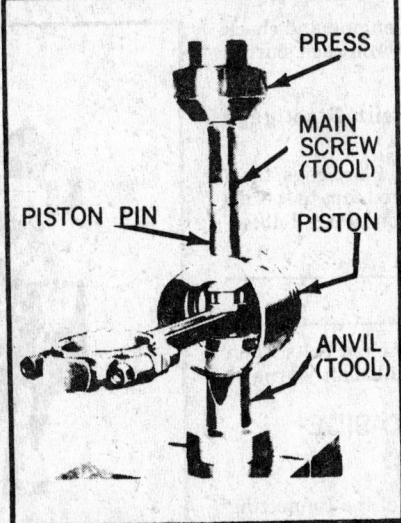

Removing or installing piston pin

Installing rings on piston with the use of an expander tool

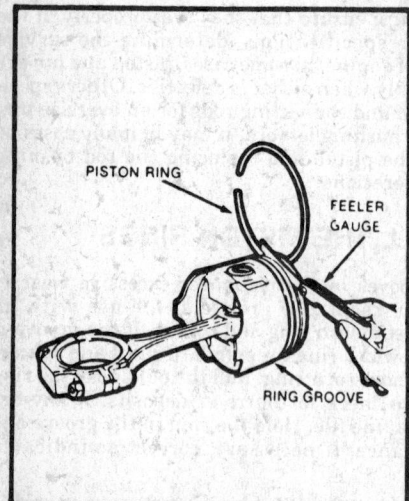

Checking ring to groove side clearance

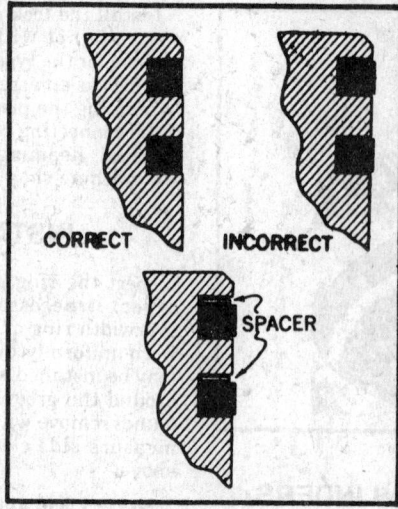

Correct ring spacer installation

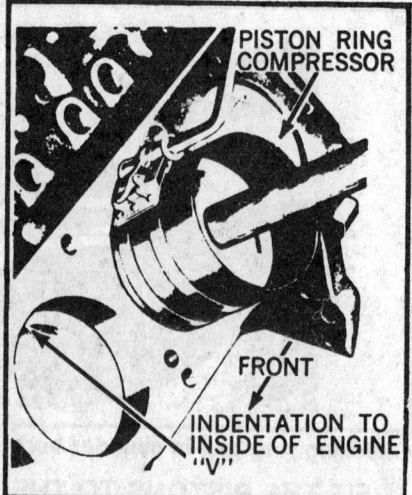

Installing piston assembly with straight sided ring compressor tool

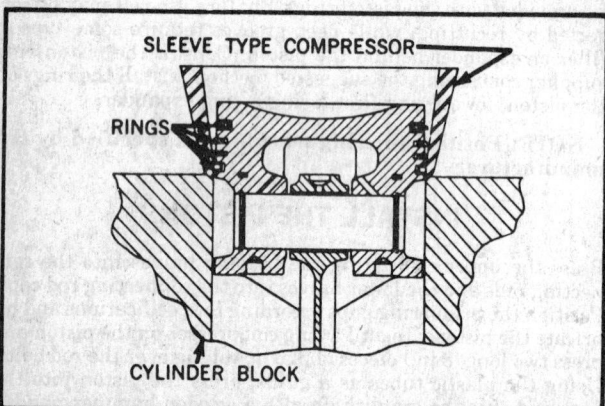

Installing piston assembly with a tapered sleeve type ring compressor tool

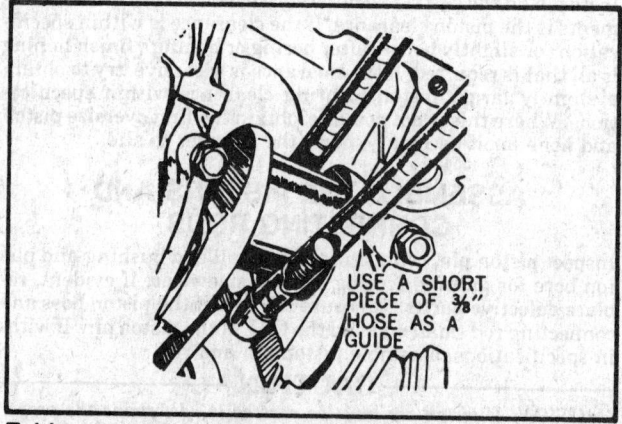

Tubing used as a guide during piston-connecting rod installation

ing pistons in the same manner. Invert the engine and check the bearing clearance at two points (90° apart) on each journal with Plastigage.

NOTE: Do not turn the crankshaft with Plastigage installed.

If clearance is within tolerances, remove all Plastigage, thoroughly lubricate the journals and torque the rod caps to specifications. If clearance is not within specifications, install different thickness bearing inserts and recheck.

─────── CAUTION ───────
Never shim or file the connecting rods or caps.
───────────────────────

Always install plastic tube sleeves over the rod bolts when the caps are not installed, to protect the crankshaft journals.

CHECK CONNECTING ROD SIDE CLEARANCE

Determine the clearance between the sides of the connecting rods and the crankshaft, using feeler gauges. If clearance is below the minimum tolerance, the rod may be machined to pro-

Using torque wrench to measure timing chain deflection

vide adequate clearance. If clearance is excessive, substitute an unworn rod and recheck. If clearance is still outside specifications, the crankshaft must be welded and reground, or replaced.

INSPECT THE TIMING CHAIN

Visually inspect the timing chain for broken or loose links and replace the chain if any are found. If the chain will flex sideways, it must be replaced.

NOTE: If the original timing chain is to be reused, install it in its original position.

INSPECT THE TIMING CHAIN DEFLECTION

Different methods are used by the engine manufacturers to measure the timing chain deflection and to determine the condition of the chain. Three such methods are as follows:

1. Rotate the crankshaft in a counterclockwise direction to remove the slack on the left side of the chain. Make a reference mark on the block and measure from the mark to the outside of the chain, halfway between the camshaft and crankshaft sprockets. Rotate the crankshaft in the opposite direction and remove the slack from the right side of the chain. Force the chain outward on the left side and measure from the original reference mark to the chain. The difference between the first and second measurements is the amount of chain deflection. The allowable deflection can be from $\frac{1}{4}$–$\frac{1}{2}$ in., depending upon the manufacturer.

2. The second method of chain deflection measurement is to block the crankshaft to prevent movement. Using a torque wrench and socket on the camshaft sprocket bolt and placing a scale even with the edge of a chain link, apply 30 ft. lbs. (w/cylinder head on block) or 15 ft. lbs. (w/cylinder head off block) in the direction of engine rotation and obtain a reference point on the scale to link. Apply 30 ft. lbs. (w/cylinder head on block) or 15 ft. lbs. (w/cylinder head off block) in the opposite direction and measure the chain movement on the scale. The measurement should not exceed $\frac{1}{8}$ in.

3. A third method of measuring timing chain deflection is to rotate the crankshaft clockwise until the No. 1 piston is on its firing stroke at TDC. The damper timing mark should point to TDC on the timing degree indicator. Remove the valve cover to expose the rocker arms of the companion or opposite cylinder to the No. 1 piston. Install a dial indicator on the push rod or rocker arm of the exhaust valve of this opposite or companion cylinder in a manner to register the push rod or rocker arm upward movement. Zero the dial indicator and slowly turn the crankshaft counterclockwise until the slightest movement is recorded on the dial indicator. Stop and observe the damper timing mark for the number of degrees of travel from TDC. If the reading on the timing degree indicator exceeds 6–8 degrees, replace the timing chain and sprockets.

OVERHEAD CAM TIMING CHAINS

Timing chains for the overhead cam engines are difficult to examine for looseness due to the chain tensioners used to maintain a controlled tension on the chain while in use. The chains are usually checked off the engine by measurement of a predetermined number of links with the chain stretched tight. Should the measurement of the links exceed the manufacturer's specifications, the chain should be replaced.

Another method is to wrap the chain around the sprockets and measure each chain/sprocket diameter. Should the manufacturer's specifications be exceeded, the chain and sprocket should be replaced. Where no specifications exist, the repairperson must make a professional determination to either replace the chain and sprockets or not.

Checking connecting rod end clearance with feeler gauge blade

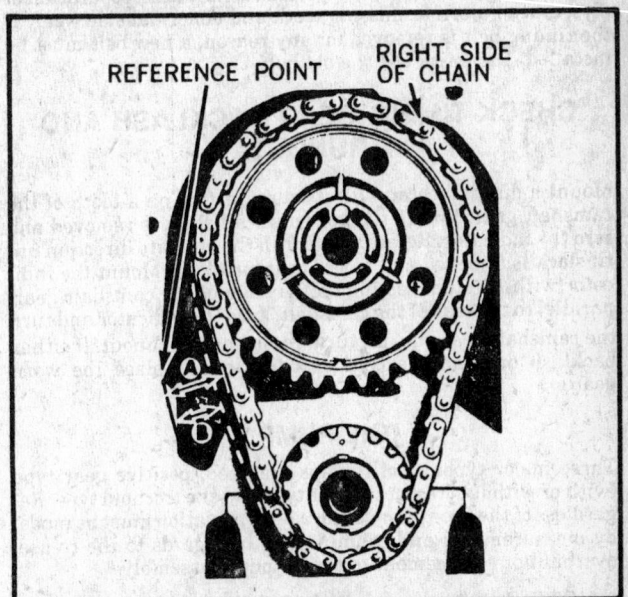

Checking timing chain deflection using point on engine blocks as reference point

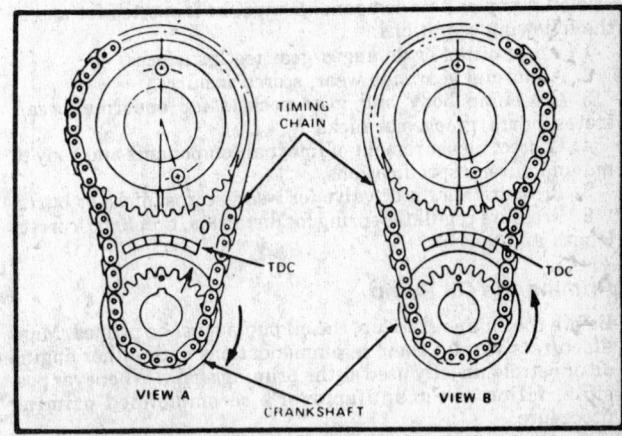

Checking timing chain deflection with timing indicator scale

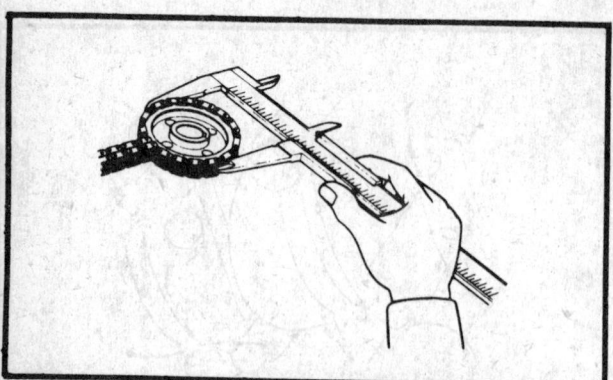

Example of timing chain/sprocket measurement

TIMING BELT INSPECTION

The timing belt should be inspected for hardness, separation of plys, cracks, worn or missing teeth and abnormal side wear. If the timing belt is removed for any reason, a new belt must be installed. Do not re-use the old belt.

CHECK TIMING GEAR BACKLASH AND RUNOUT

Mount a dial indicator with its stem resting on a tooth of the camshaft gear. Rotate the gear until all slack is removed and zero the indicator. Rotate the gear in the opposite direction until slack is removed and record gear backlash. Mount the indicator with its stem resting on the edge of the camshaft gear, parallel to the axis of the camshaft. Zero the indicator and turn the camshaft gear one full turn, recording the runout. If either backlash or runout exceed specifications, replace the worn gear(s).

OIL PUMP

Three major types of oil pumps are used, positive gear type (with or without crescent), rotor type and the trochoid type. Regardless of the type pump used, a determination must be made, by measurements and examination, in regards to the re-use, overhaul or replacement of the oil pump assembly.

NOTE: It is recommended to replace the oil pump assembly with a new unit when a major engine overhaul is done.

Inspection

The oil pump and its components must be inspected for any of the following conditions.
1. Worn, pitted or damaged gear teeth surfaces.
2. Abnormal gear side wear, scores or burrs.
3. Oil pump body and gear pockets for abnormal wear, scores, burrs, grooves or nicks.
4. Correct measurement of internal components and body to manufacturer's specifications.
5. Pressure regulator valve for wear, scores, nicks or burrs.
6. Pressure regulator spring for distortion, breakage, correct length and tension.

Priming the Oil Pump

Before the engine start-up, the oil pump must be primed. Manufacturers vary in their recommendations with either engine oil or petroleum jelly used as the priming agent. Whenever possible, follow the manufacturer's recommended priming procedure.

Completing the Rebuilding Process

Following the above procedures, complete the rebuilding process as follows:

Fill the oil pump with oil, or petroleum jelly, to prevent cavitating (sucking air) on initial engine start up. Install the oil pump and the pickup tube on the engine. Coat the oil pan gasket as necessary and install the gasket and the oil pan. Mount the flywheel and the crankshaft vibrational damper or pulley on the crankshaft.

NOTE: Always use new bolts when installing the flywheel.

Inspect the clutch shaft pilot bushing in the crankshaft. If the bushing is excessively worn, remove it with an expanding puller and a slide hammer and tap a new bushing into place.

Position the engine, cylinder head side up. Lubricate the lifters and install them into their bores. Install the cylinder head and torque it as specified in the car section. Insert the pushrods (where applicable) and install the rocker shaft(s) (if so equipped) or position the rocker arms on the pushrods. If solid lifters are utilized, adjust the valves to the "cold" specifications.

Mount the intake and exhaust manifolds, the carburetor(s), the distributor and spark plugs. Adjust the point gap and the static ignition timing. Mount all accessories and install the engine in the car. Fill the radiator with coolant and the crankcase with high quality engine oil.

BREAK-IN PROCEDURE

Before starting the engine, be sure all coolant hoses are attached and tight, the coolant level is correct, a new oil filter is installed and the crankcase filled with the proper level of oil.

The oil pump should be primed and if possible, the engine lubrication system should be charged with a pressure tank. Adjust the tappets (if required), the timing and carburetor as accurately as possible.

Start the engine and adjust the throttle to an approximate engine speed of 1000 to 2000 rpm, until the engine reaches normal operating temperature, normally within 20–30 minutes.

——— CAUTION ———
Do not leave the vehicle unattended during the warm-up period. Observe the engine operation and check for any oil or coolant leaks. Stop the engine immediately if a problem exists, to avoid engine damage.

After the engine has "run-in", lower the idle speed and stop the engine. Retorque the cylinder head bolts as required.

NOTE: Engines with aluminum heads or blocks must be allowed to cool to room temperature before any bolts are retorqued.

After rechecking the coolant and oil levels, make any further adjustments as necessary.

Follow the manufacturer's recommended driving break-in procedure or as a general rule, the following procedures may be used.

Drive the vehicle on the highway and accelerate from 30–50 mph, approximately 10–15 times, traffic flow permitting, to properly seat the piston rings to the cylinder walls. If traffic flow does not permit this procedure, accelerate the engine rapidly during shifting through the intermediate gears. The vehicle should be put in light duty service for the first 50 miles and sustained high speed should be avoided during the first 100 miles. Most important: Do not lug the engine. (Lugging exists when the engine does not respond to further opening of the throttle.)

Engine Rebuilding
Import Trucks

This section describes, in detail, the procedures involved in rebuilding a typical engine. The procedures are basically identical to those used in rebuilding engines of nearly all design and configurations.

The section is divided into two parts. The first, Cylinder Head Reconditioning, assumes that the cylinder head is removed from the engine, all manifolds are removed, and the cylinder head is on a workbench. The camshaft should be removed from overhead cam cylinder heads. The second section, Cylinder Block Reconditioning, covers the block, pistons, connecting rods and crankshaft. It is assumed that the engine is mounted on a work stand, and the cylinder head and all accessories are removed.

Procedures are identified as follows:

Unmarked—Basic procedures that must be performed in order to successfully complete the rebuilding process.

Starred (*)—Procedures that should be performed to ensure maximum performance and engine life.

Double starred (**)—Procedures that may be performed to increase engine performance and reliability.

In many cases, a choice of methods is also provided. Methods are identified in the same manner as procedures. The choice of method for a procedure is at the discretion of the user.

The tools required for the basic rebuilding procedure should, with minor exceptions, be those included in a mechanic's tool kit. An accurate torque wrench, and a dial indicator (reading in thousandths) mounted on a universal base should be available. Special tools, where required, all are readily available from the major tool suppliers. The services of a competent automotive machine shop must also be readily available.

When assembling the engine, any parts that will be in frictional contact must be prelubricated, to provide protection on initial start-up. Any product specifically formulated for this purpose may be used. NOTE: *Do not use engine oil.* Where semi-permanent (locked but removable) installation of bolts or nuts is desired, threads should be cleaned and coated with Loctite® or a similar product (non-hardening).

Aluminum has become increasingly popular for use in engines, due to its low weight and excellent heat transfer characteristics. The following precautions must be observed when handling aluminum engine parts:
—Never hot-tank aluminum parts.
—Remove all aluminum parts (identification tags, etc.) from engine parts before hot-tanking (otherwise they will be removed during the process).
—Always coat threads lightly with engine oil or anti-seize compounds before installation, to prevent seizure.
—Never over-torque bolts or spark plugs in aluminum threads. Should stripping occur, threads can be restored using any of a number of thread repair kits available (see next section).

Magnaflux and Zyglo are inspection techniques used to locate material flaws, such as stress cracks. Magnafluxing coats the part with fine magnetic particles, and subjects the part to a magnetic field. Cracks cause breaks in the magnetic field, which are outlined by the particles. Since Magnaflux is a magnetic process, it is applicable only to ferrous materials. The Zyglo process coats the material with a fluorescent dye penetrant, and then subjects it to blacklight inspection, under which cracks glow brightly. Parts made of any material may be tested using Zyglo. While Magnaflux and Zyglo are excellent for general inspection, and locating hidden defects, specific checks of suspected cracks may be made at lower cost and more readily using spot check dye. The dye is sprayed onto the suspected area, wiped off, and the area is then sprayed with a developer. Cracks then will show up brightly. Spot check dyes will only indicate surface cracks; therefore, structural cracks below the surface may escape detection. When questionable, the part should be tested using Magnaflux or Zyglo.

REPAIRING DAMAGED THREADS

Several methods of repairing damaged threads are available. Heli-Coil® (shown here), Keenserts® and Microdot® are among the most widely used. All involve basically

the same principle—drilling out stripped threads, tapping the hole and installing a prewound insert— making welding, plugging and oversize fasteners unnecessary.

Two types of thread repair inserts are usually supplied—a standard type for most Inch Coarse, Inch Fine, Metric Coarse and Metric Fine thread sizes and a spark plug type to fit most spark plug port sizes. Consult the individual manufacturer's catalog to determine exact applications. Typical thread repair kits will contain a selection of prewound threaded inserts, a tap (corresponding to the outside diameter threads of the insert) and an installation tool. Most manufacturers also supply blister-packed thread repair inserts separately and a master kit with a variety of taps and inserts plus installation tools.

Before effecting a repair to a threaded hole, remove any snapped, broken or damaged bolts or studs. Penetrating oil can be used to free frozen threads; the offending item can be removed with locking pliers or with a screw or stud extractor. After the hole is clear, the thread can be repaired as follows.

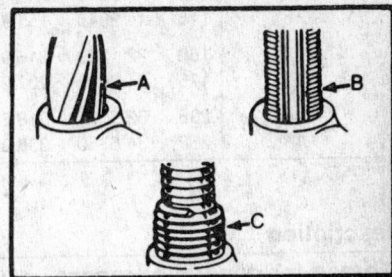

A. Drill out the damaged threads with the specified drill. Drill completely through the hole or to the bottom of a blind hole.

B. With the tap supplied tap the hole to receive the threaded insert. Keep the tap well oiled and back it out frequently to avoid clogging the threads.

C. Screw the threaded insert onto the installation tool until the tang engages the slot. Screw the insert into the tapped hole until it is 1/4–1/2 turn below the top surface. After installation, break the tang off with a hammer and punch.

STANDARD TORQUE SPECIFICATIONS AND CAPSCREW MARKINGS

Newton-Meter has been designated as the world standard for measuring torque and will gradually replace the foot-pound and kilogram-meter torque measuring standard. Torquing tools are still being manufactured with foot-pounds and kilogram-meter scales, along with the new Newton-Meter standard. To assist the repairman, foot-pounds, kilogram-meter and Newton-Meter are listed in the following charts, and should be followed as applicable.

U.S. BOLTS

SAE Grade Number	1 or 2			5			6 or 7			8		
Capscrew Head Markings — Manufacturer's marks may vary. Three-line markings on heads below indicate SAE Grade 5.												
Usage	Used Frequently			Used Frequently			Used at Times			Used at Times		
Quality of Material	Indeterminate			Minimum Commercial			Medium Commercial			Best Commercial		
Capacity Body Size	Torque			Torque			Torque			Torque		
(inches)–(thread)	Ft-Lb	kgm	Nm	Ft-Lb	kgm	Nm	Ft-Lb	kgm	Nm	Ft-Lb	kgm	Nm
1/4–20	5	0.6915	6.7791	8	1.1064	10.8465	10	1.3630	13.5582	12	1.6596	16.2698
–28	6	0.8298	8.1349	10	1.3830	13.5582				14	1.9362	18.9815
5/16–18	11	1.5213	14.9140	17	2.3511	23.0489	19	2.6277	25.7605	24	3.3192	32.5396
–24	13	1.7979	17.6256	19	2.6277	25.7605				27	3.7341	36.6071
3/8–16	18	2.4894	24.4047	31	4.2873	42.0304	34	4.7022	46.0978	44	6.0852	59.6560
–24	20	2.7660	27.1164	35	4.8405	47.4536				49	6.7767	66.4351
7/16–14	28	3.8132	37.9629	49	6.7767	66.4351	55	7.6065	74.5700	70	9.6810	94.9073
–20	30	4.1490	40.6745	55	7.6065	74.5700				78	10.7874	105.7538
1/2–13	39	5.3937	52.8769	75	10.3725	101.6863	85	11.7555	115.2445	105	14.5215	142.3609
–20	41	5.6703	55.5885	85	11.7555	115.2445				120	16.5860	162.6960
9/16–12	51	7.0533	69.1467	110	15.2130	149.1380	120	16.5960	162.6960	155	21.4365	210.1490
–18	55	7.6065	74.5700	120	16.5960	162.6960				170	23.5110	230.4860
5/8–11	83	11.4789	112.5329	150	20.7450	203.3700	167	23.0961	226.4186	210	29.0430	284.7180
–18	95	13.1385	128.8027	170	23.5110	230.4860				240	33.1920	325.3920
3/4–10	105	14.5215	142.3609	270	37.3410	366.0660	280	38.7240	379.6240	375	51.8625	508.4250
–16	115	15.9045	155.9170	295	40.7985	399.9610				420	58.0860	568.4360
7/8–9	160	22.1280	216.9280	395	54.6285	535.5410	440	60.8520	596.5520	605	83.6715	820.2590
–14	175	24.2025	237.2650	435	60.1605	589.7730				675	93.3525	915.1650
1–8	236	32.5005	318.6130	590	81.5970	799.9220	660	91.2780	894.8280	910	125.8530	1233.7780
–14	250	34.5750	338.9500	660	91.2780	849.8280				990	136.9170	1342.2420

METRIC BOLTS

Description — Thread for general purposes (size x pitch (mm))	Torque ft-lbs. (Nm)			
	Head Mark 4		Head Mark 7	
6 x 1.0	2.2 to 2.9	(3.0 to 3.9)	3.6 to 5.8	(4.9 to 7.8)
8 x 1.25	5.8 to 8.7	(7.9 to 12)	9.4 to 14	(13 to 19)
10 x 1.25	12 to 17	(16 to 23)	20 to 29	(27 to 39)
12 x 1.25	21 to 32	(29 to 43)	35 to 53	(47 to 72)
14 x 1.5	35 to 52	(48 to 70)	57 to 85	(77 to 110)
16 x 1.5	51 to 77	(67 to 100)	90 to 120	(130 to 160)
18 x 1.5	74 to 110	(100 to 150)	130 to 170	(180 to 230)
20 x 1.5	110 to 140	(150 to 190)	190 to 240	(160 to 320)
22 x 1.5	150 to 190	(200 to 260)	250 to 320	(340 to 430)
24 x 1.5	190 to 240	(260 to 320)	310 to 410	(420 to 550)

CAUTION: Bolts threaded into aluminum require much less torque

NOTE: This engine rebuilding section is a guide to accepted rebuilding procedures. Typical examples of standard rebuilding procedures are illustrated.

CYLINDER HEAD RECONDITIONING

Procedure	Method
Identify the valves:	Invert the cylinder head, and number the valve faces front to rear, using a permanent felt-tip marker.
Remove the rocker arms (OHV engines only):	Remove the rocker arms with shaft(s) or balls and nuts. Wire the sets of rockers, balls and nuts together, and identify according to the corresponding valve.
Remove the camshaft (OHC engines only):	See the engine service procedures earlier in this book for details concerning specific engines.
Remove the valves and springs:	Using an appropriate valve spring compressor (depending on the configuration of the cylinder head), compress the valve springs. Lift out the keepers with needlenose pliers, release the compressor, and remove the valve, spring, and spring retainer.
Remove glow plugs and fuel injectors (Diesel engines only):	Label and remove all fuel injectors and glow plugs from the head. Glow plugs unscrew. See the appropriate car section for injector removal. Inspect glow plugs for bulges, cracks or signs of melting. Clean injector tips with a steel brush, then inspect for evidence of melting.
**Remove pre-combustion chamber inserts (Diesel engines only): Removing pre-combustion chamber with a drift (© G.M. Corp.)	**Remove the pre-combustion chambers using a hammer and a thin, blunt brass drift, inserted through the injector hole (or glow plug hole, whichever is more convenient). If chamber is to be reused, carefully remove all carbon from it. NOTE: *Remove chamber only if being replaced, if a glow plug tip has broken off and must be removed, or if chamber is obviously damaged or loose.*
Check the valve stem-to-guide clearance: Checking the valve stem-to-guide clearance	Clean the valve stem with lacquer thinner or a similar solvent to remove all gum and varnish. Clean the valve guides using solvent and an expanding wire-type valve guide cleaner. Mount a dial indicator so that the stem is at 90° to the valve stem, as close to the valve guide as possible. Move the valve off its seat, and measure the valve guide-to-stem clearance by rocking the stem back and forth to actuate the dial indicator. Measure the valve stems using a micrometer, and compare to specifications, to determine whether stem or guide wear is responsible for excessive clearance.

CYLINDER HEAD RECONDITIONING

Procedure	Method
De-carbon the cylinder head and valves:	Chip carbon away from the valve heads, combustion chambers, and ports, using a chisel made of hardwood. Remove the remaining deposits with a stiff wire brush. NOTE: *Ensure that the deposits are actually removed, rather than burnished.*

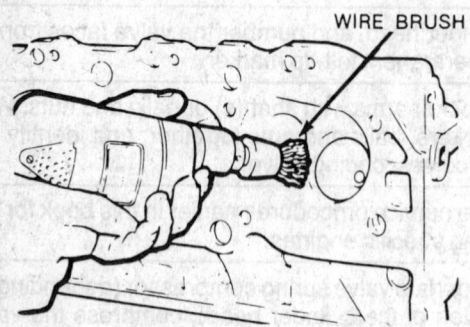

WIRE BRUSH

Removing carbon from the cylinder head

Procedure	Method
Hot-tank the cylinder head (cast iron heads only): CAUTION: *Do not hot-tank aluminum parts.*	Have the cylinder head hot-tanked to remove grease, corrosion, and scale from the water passages. NOTE: *In the case of overhead cam cylinder heads, consult the operator to determine whether the camshaft bearings will be damaged by the caustic solution.*
Degrease the remaining cylinder head parts:	Using solvent (i.e., Gunk), clean the rockers, rocker shaft(s) (where applicable), rocker balls and nuts, springs, spring retainers, and keepers. Do not remove the protective coating from the springs.
Check the cylinder head for warpage:	Place a straight-edge across the gasket surface of the cylinder head. Using feeler gauges, determine the clearance at the center of the straight-edge. Measure across both diagonals, along the longitudinal centerline, and across the cylinder head at several points. If warpage exceeds .003' in a 6' span, or .006' over the total length, the cylinder head must be resurfaced. NOTE: *If warpage exceeds the manufacturer's maximum tolerance for material removal, the cylinder head must be replaced.* When milling the cylinder heads of V-type engines, the intake manifold mounting position is altered, and must be corrected by milling the manifold flange a proportionate amount.

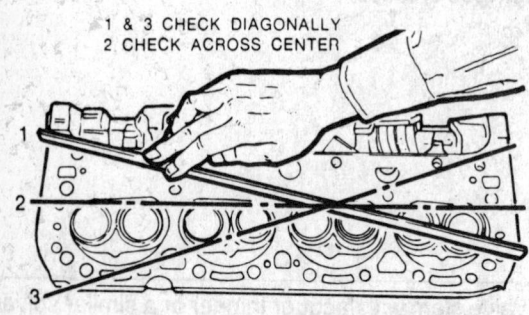

1 & 3 CHECK DIAGONALLY
2 CHECK ACROSS CENTER

Checking cylinder head for warpage

Procedure	Method
**Porting and gasket matching:	**Coat the manifold flanges of the cylinder head with Prussian blue dye. Glue intake and exhaust gaskets to the cylinder head in their installed position using rubber cement and scribe the outline of the ports on the manifold flanges. Remove the gaskets. Using a small cutter in a hand-held power tool gradually taper the walls of the port out to the scribed outline of the gasket. Further enlargement of the ports should include the removal of sharp edges and radiusing of sharp corners. Do not alter the valve guides. NOTE: *The most efficient port configuration is determined only by extensive testing. Therefore, it is best to consult someone experienced with the head in question to determine the optimum alterations.*

CYLINDER HEAD RECONDITIONING

Procedure	Method

*Knurling the valve guides:

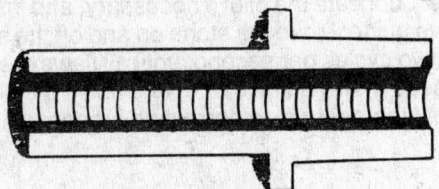

Cut-away view of a knurled valve guide

*Valve guides which are not excessively worn or distorted may, in some cases, be knurled rather than replaced. Knurling is a process in which metal is displaced and raised, thereby reducing clearance. Knurling also provides excellent oil control. The possibility of knurling rather than replacing valve guides should be discussed with a machinist.

Replacing the valve guides:
NOTE: *Valve guides should only be replaced if damaged or if an oversize valve stem is not available.*

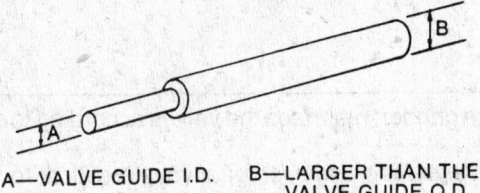

A—VALVE GUIDE I.D. B—LARGER THAN THE VALVE GUIDE O.D.
Valve guide removal tool

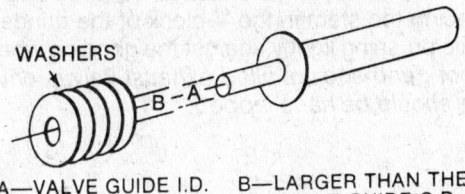

A—VALVE GUIDE I.D. B—LARGER THAN THE VALVE GUIDE O.D.
Valve guide installation tool (with washers used for installation)

Depending on the type of cylinder head, valve guides may be pressed, hammered, or shrunk in. In cases where the guides are shrunk into the head, replacement should be left to an equipped machine shop. In other cases, the guides are replaced as follows: Press or tap the valve guides out of the head using a stepped drift (see illustration). Determine the height above the boss that the guide must extend, and obtain a stack of washers, their I.D. similar to the guide's O.D., of that height. Place the stack of washers on the guide, and insert the guide into the boss.
NOTE: *Valve guides are often tapered or beveled for installation.*
Using the stepped installation tool (see illustration), press or tap the guides into position. Ream the guides according to the size of the valve stem.

Replacing valve seat inserts:

Replacement of valve seat inserts which are worn beyond resurfacing or broken, if feasible, must be done by a machine shop.

Resurfacing the valve seats using reamers:

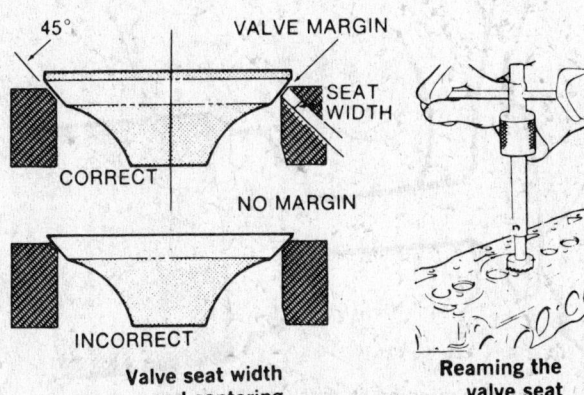

Valve seat width and centering

Reaming the valve seat

Select a reamer of the correct seat angle, slightly larger than the diameter of the valve seat, and assemble it with a pilot of the correct size. Install the pilot into the valve guide, and using steady pressure, turn the reamer clockwise.
CAUTION: *Do not turn the reamer counterclockwise.*
Remove only as much material as necessary to clean the seat. Check the concentricity of the seat (see below). If the dye method is not used, coat the valve face with Prussian blue dye, install and rotate it on the valve seat. Using the dye marked area as a centering guide, center and narrow the valve seat to specifications with correction cutters.
NOTE: *When no specifications are available, minimum seat width for exhaust valves should be 5/64", intake valves 1/16".*
After making correction cuts, check the position of the valve seat on the valve face using Prussian blue dye.
NOTE: *Do not cut induction hardened seats; they must be ground.*

CYLINDER HEAD RECONDITIONING

Procedure	Method

*Resurfacing the valve seats using a grinder:

*Select a pilot of the correct size, and a coarse stone of the correct seat angle. Lubricate the pilot if necessary, and install the tool in the valve guide. Move the stone on and off the seat at approximately two cycles per second, until all flaws are removed from the seat. Install a fine stone, and finish the seat. Center and narrow the seat using correction stones, as described above.

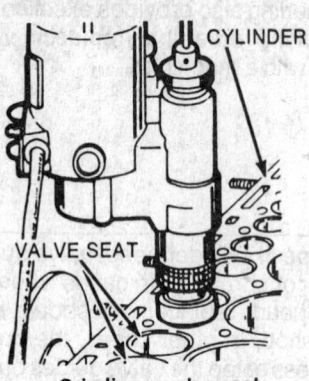

CYLINDER

VALVE SEAT

Grinding a valve seat

Resurfacing (grinding) the valve face:

Using a valve grinder, resurface the valves according to specifications.
CAUTION: *Valve face angle is not always identical to valve seat angle.*
A minimum margin of 1/32" should remain after grinding the valve. The valve stem top should also be squared and resurfaced, by placing the stem in the V-block of the grinder, and turning it while pressing lightly against the grinding wheel.
NOTE: *Do not grind sodium filled exhaust valves on a machine. These should be hand lapped.*

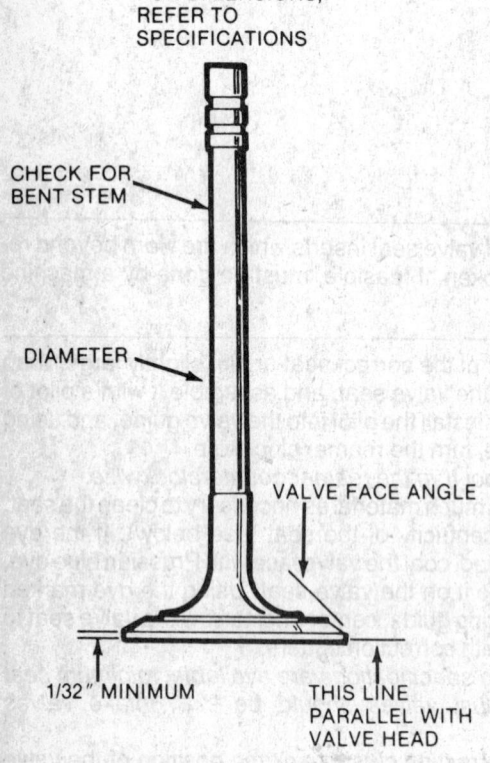

FOR DIMENSIONS, REFER TO SPECIFICATIONS

CHECK FOR BENT STEM

DIAMETER

VALVE FACE ANGLE

1/32" MINIMUM

THIS LINE PARALLEL WITH VALVE HEAD

Critical valve dimensions

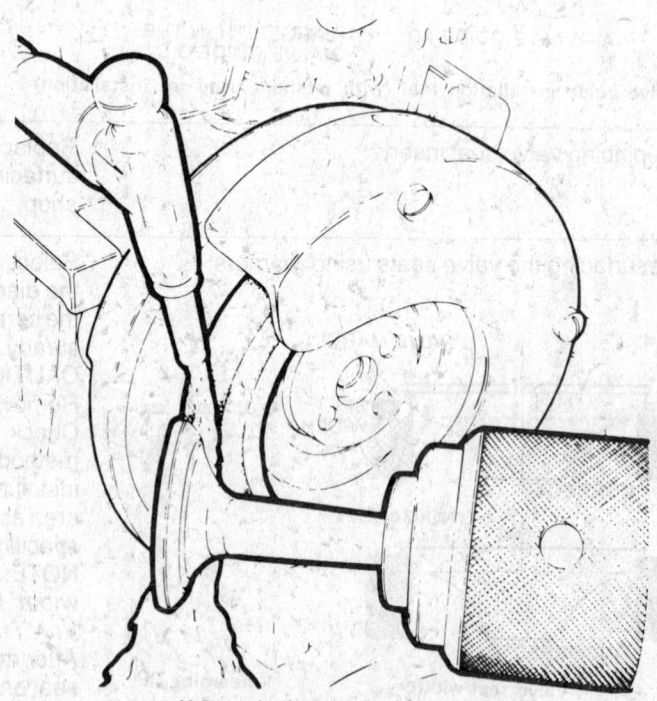

Valve grinding by machine

CYLINDER HEAD RECONDITIONING

Procedure	Method

Checking the valve seat concentricity:

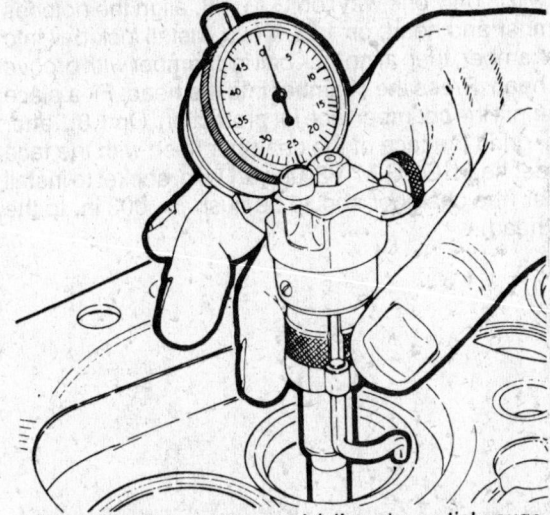

Checking valve seat concentricity using a dial gauge

Coat the valve face with Prussian blue dye, install the valve, and rotate it on the valve seat. If the entire seat becomes coated, and the valve is known to be concentric, the seat is concentric.
*Install the dial gauge pilot into the guide, and rest the arm on the valve seat. Zero the gauge, and rotate the arm around the seat. Run-out should not exceed .002".

*Lapping the valves:
NOTE: *Valve lapping is done to ensure efficient sealing of resurfaced valves and seats.*

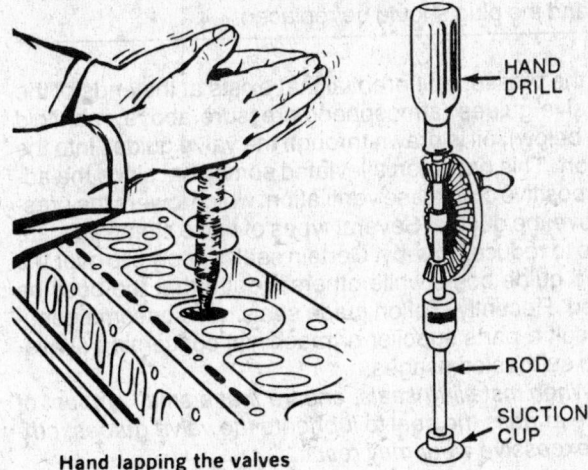

Hand lapping the valves

Home made mechanical valve lapping tool

*Invert the cylinder head, lightly lubricate the valve stems, and install the valves in the head as numbered. Coat valve seats with fine grinding compound, and attach the lapping tool suction cup to a valve head.
NOTE: *Moisten the suction cup.*
Rotate the tool between the palms, changing position and lifting the tool often to prevent grooving. Lap the valve until a smooth, polished seat is evident. Remove the valve and tool, and rinse away all traces of grinding compound.
**Fasten a suction cup to a piece of drill rod, and mount the rod in a hand drill. Proceed as above, using the hand drill as a lapping tool.
CAUTION: *Due to the higher speeds involved when using the hand drill, care must be exercised to avoid grooving the seat.* Lift the tool and change direction of rotation often.

Check the valve springs:

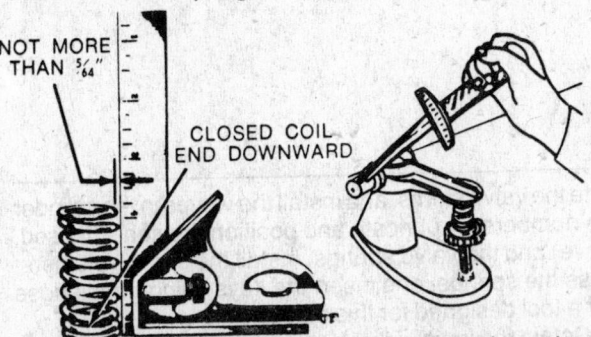

Checking valve spring free length and squareness

Measuring valve spring test pressure

Place the spring on a flat surface next to a square. Measure the height of the spring, and rotate it against the edge of the square to measure distortion. If spring height varies (by comparison) by more than 1/16" or if distortion exceeds 1/16", replace the spring.
**In addition to evaluating the spring as above, test the spring pressure at the installed and compressed (installed height minus valve lift) height using a valve spring tester. Springs used on small displacement engines (up to 3 liters) should be ∓ 1 lb. of all other springs in either position. A tolerance of ∓ 5 lbs. is permissible on larger engines.

CYLINDER HEAD RECONDITIONING

Procedure	Method

Install pre-combustion chambers (Diesel engines only)

Pre-combustion chambers are press-fit into the head. The chambers will fit only one way: on G.M. V8, align the notches in the chamber and head; on 1.8L 4 cyl., install lock ball into groove in chamber, then align lock ball in chamber with groove in cylinder head. Press the chamber into the head. Fit a piece of metal against the chamber face for protection. On 1.8L, after installation, grind the face of the chamber flush with the face of the cylinder head. On G.M. V8, use a 1¼ in. socket to install the chamber (the chamber should be flush ± .003 in. to the face of the head).

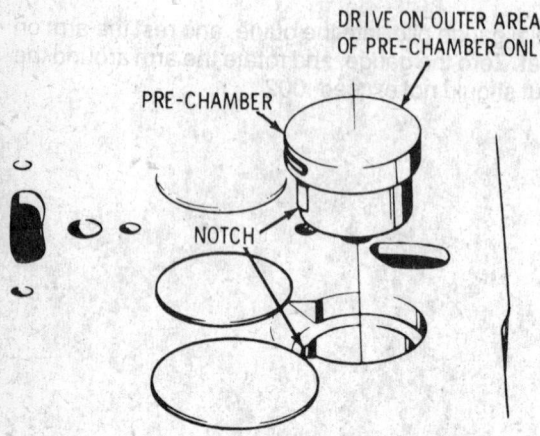

DRIVE ON OUTER AREA OF PRE-CHAMBER ONLY

PRE-CHAMBER

NOTCH

Align the notches to install the pre-combustion chamber

Install fuel injectors and glow plugs (Diesel engines)

Before installing glow plugs, check for continuity across plug terminals and body. If no continuity exists, the heater wire is broken and the plug should be replaced.

*Install valve stem seals:

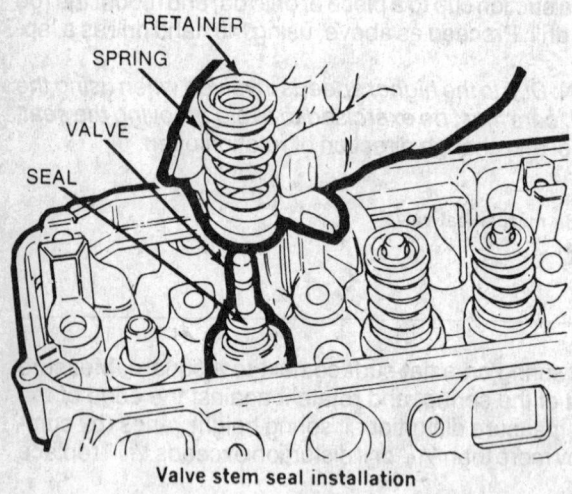

RETAINER

SPRING

VALVE

SEAL

Valve stem seal installation

*Due to the pressure differential that exists at the ends of the intake valve guides (atmospheric pressure above, manifold vacuum below), oil is drawn through the valve guides into the intake port. This has been alleviated somewhat since the addition of positive crankcase ventilation, which lowers the pressure above the guides. Several types of valve stem seals are available to reduce blow-by. Certain seals simply slip over the stem and guide boss, while others require that the boss be machined. Recently, Teflon guide seals have become popular. Consult a parts supplier or machinist concerning availability and suggested usages.

NOTE: *When installing seals, ensure that a small amount of oil is able to pass the seal to lubricate the valve guides; otherwise, excessive wear may result.*

Install the valves:

Lubricate the valve stems, and install the valves in the cylinder head as numbered. Lubricate and position the seals (if used, see above) and the valve springs. Install the spring retainers, compress the springs, and insert the keys using needlenose pliers or a tool designed for this purpose.

NOTE: *Retain the keys with wheel bearing grease during installation.*

CYLINDER HEAD RECONDITIONING

Procedure	Method

Check valve spring installed height:

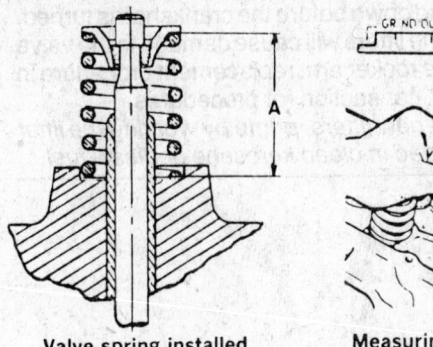

Valve spring installed
height dimension

Measuring valve spring
installed height

Measure the distance between the spring pad and the lower edge of the spring retainer, and compare to specifications. If the installed height is incorrect, add shim washers between the spring pad and the spring.
CAUTION: *Use only washers designed for this purpose.*

Install the camshaft (OHC engines only) and check end play:

See the engine service procedures earlier in this book for details concerning specific engines.

Inspect the rocker arms, balls, studs, and nuts (OHV engines only):

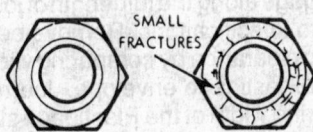

Stress cracks in the rocker nuts

Visually inspect the rocker arms, balls, studs, and nuts for cracks, galling, burning, scoring or wear. If all parts are intact, liberally lubricate the rocker arms and balls, and install them on the cylinder head. If wear is noted on a rocker arm at the point of valve contact, grind it smooth and square, removing as little material as possible. Replace the rocker arm if excessively worn. If a rocker stud shows signs of wear, it must be replaced (see below). If a rocker nut shows stress cracks, replace it. If an exhaust ball is galled or burned, substitute the intake ball from the same cylinder (if it is intact), and install a new intake ball.
NOTE: *Avoid using new rocker balls on exhaust valves.*

Replacing rocker studs (OHV engines only):

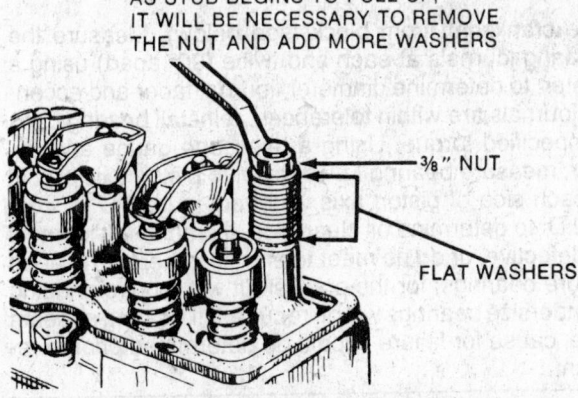

AS STUB BEGINS TO PULL UP,
IT WILL BE NECESSARY TO REMOVE
THE NUT AND ADD MORE WASHERS

⅜" NUT

FLAT WASHERS

Extracting a pressed-in rocker stud

In order to remove a threaded stud, lock two nuts on the stud, and unscrew the stud using the lower nut. Coat the lower threads of the new stud with Loctite®, and install.
Two alternative methods are available for replacing pressed in studs. Remove the damaged stud using a stack of washers and a nut (see illustration). In the first, the boss is reamed .005—.006" oversize, and an oversize stud pressed in. Control the stud extension over the boss using washers, in the same manner as valve guides. Before installing the stud, coat it with white lead and grease. To retain the stud more positively drill a hole through the stud and boss, and install a roll pin. In the second method, the boss is tapped, and a threaded stud installed. Retain the stud using Loctite® Stud and Bearing Mount.

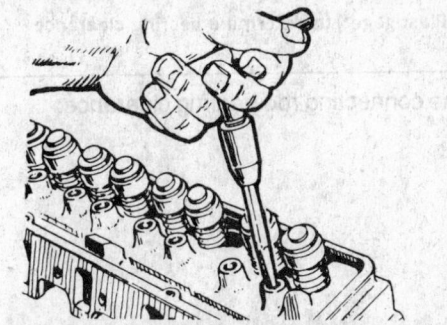

Reaming the stud bore for oversize rocker studs

CYLINDER HEAD RECONDITIONING

Procedure	Method
Bleed the hydraulic lifters (diesel engines only):	After the cylinder heads are installed on G.M. V8 diesels, the valve lifters must be bled down before the crankshaft is turned. Failure to bleed down the lifters will cause damage to the valve train. See diesel engine rocker arm replacement procedure in Oldsmobile 88, 98, etc. car section for procedures. NOTE: *When installing new lifters, prime by working the lifter plunger while submerged in clean kerosene or diesel fuel.*

CYLINDER BLOCK RECONDITIONING

Procedure	Method
Checking the main bearing clearance: 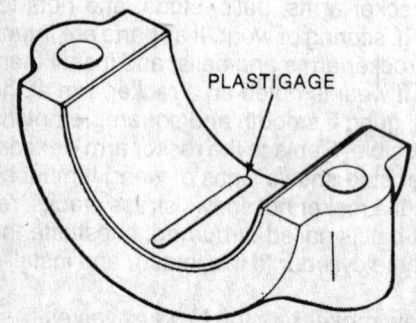 Plastigage® installed on the lower bearing shell 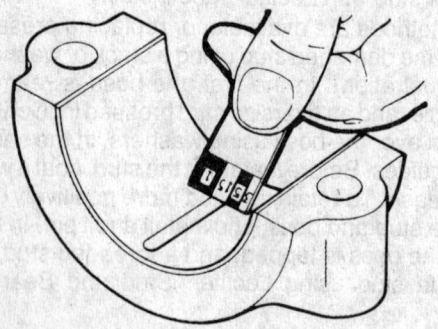 Measuring Plastigage® to determine bearing clearance	Invert engine, and remove cap from the bearing to be checked. Using a clean, dry rag, thoroughly clean all oil from crankshaft journal and bearing insert. NOTE: *Plastigage is soluble in oil; therefore, oil on the journal or bearing could result in erroneous readings.* Place a piece of Plastigage along the full length of journal, reinstall cap, and torque to specifications. Remove bearing cap, and determine bearing clearance by comparing width of Plastigage to the scale on Plastigage envelope. Journal taper is determined by comparing width of the Plastigage strip near its ends. Rotate crankshaft 90° and retest, to determine journal eccentricity. NOTE: *Do not rotate crankshaft with Plastigage installed.* If bearing insert and journal appear intact, and are within tolerances, no further main bearing service is required. If bearing or journal appear defective, cause of failure should be determined before replacement. *Remove crankshaft from block (see below). Measure the main bearing journals at each end twice (90° apart) using a micrometer, to determine diameter, journal taper and eccentricity. If journals are within tolerances, reinstall bearing caps at their specified torque. Using a telescope gauge and micrometer, measure bearing I.D. parallel to piston axis and at 30° on each side of piston axis. Subtract journal O.D. from bearing I.D. to determine oil clearance. If crankshaft journals appear defective, or do no meet tolerances, there is no need to measure bearings; for the crankshaft will require grinding and/or undersize bearings will be required. If bearing appears defective, cause for failure should be determined prior to replacement.
Checking the connecting rod bearing clearance:	Connecting rod bearing clearance is checked in the same manner as main bearing clearance, using Plastigage. Before removing the crankshaft, connecting rod side clearance also should be measured and recorded. *Checking connecting rod bearing clearance, using a micrometer, is identical to checking main bearing clearance. If no other service is required, the piston and rod assemblies need not be removed.

CYLINDER HEAD RECONDITIONING

Procedure	Method

Inspect the rocker shaft(s) and rocker arms (OHV engines only):

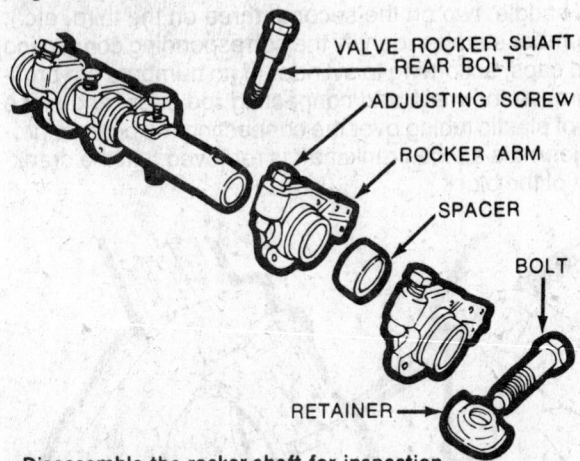

- VALVE ROCKER SHAFT REAR BOLT
- ADJUSTING SCREW
- ROCKER ARM
- SPACER
- BOLT
- RETAINER

Disassemble the rocker shaft for inspection

Remove rocker arms, springs and washers from rocker shaft. NOTE: *Lay out parts in the order as they are removed.*
Inspect rocker arms for pitting or wear on the valve contact point, or excessive bushing wear. Bushings need only be replaced if wear is excessive, because the rocker arm normally contacts the shaft at one point only. Grind the valve contact point of rocker arm smooth if necessary, removing as little material as possible. If excessive material must be removed to smooth and square the arm, it should be replaced. Clean out all oil holes and passages in rocker shaft. If shaft is grooved or worn, replace it. Lubricate and assemble the rocker shaft.

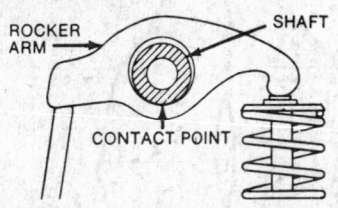

- ROCKER ARM
- SHAFT
- CONTACT POINT

Rocker arm-to-rocker shaft contact area

Inspect the camshaft bushings and the camshaft (OHC engines):

See next section.

Inspect the pushrods (OHV engines only):

Remove the pushrods, and, if hollow, clean out the oil passages using fine wire. Roll each pushrod over a piece of clean glass. If a distinct clicking sound is heard as the pushrod rolls, the rod is bent, and must be replaced.

*The length of all pushrods must be equal. Measure the length of the pushrods, compare to specifications, and replace as necessary.

Inspect the valve lifters (OHV engines only):

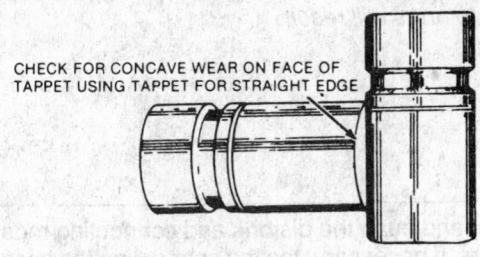

CHECK FOR CONCAVE WEAR ON FACE OF TAPPET USING TAPPET FOR STRAIGHT EDGE

Checking the lifter face

Remove lifters from their bores, and remove gum and varnish, using solvent. Clean walls of lifter bores. Check lifters for concave wear as illustrated. If face is worn concave, replace lifter, and carefully inspect the camshaft. Lightly lubricate lifter and insert it into its bore. If play is excessive, an oversize lifter must be installed (where possible). Consult a machinist concerning feasibility. If play is satisfactory, remove, lubricate, and reinstall the lifter.
NOTE: *1981 and later G.M. diesel V8 valve lifters have roller cam followers. Check these for smooth operation and wear. The roller should rotate freely, but without excessive play. Check the rollers for missing or broken needle bearings. If the roller is pitted or rough, check the camshaft lobe for wear.*

***Testing hydraulic lifter leak down (OHV gasoline engines only):**

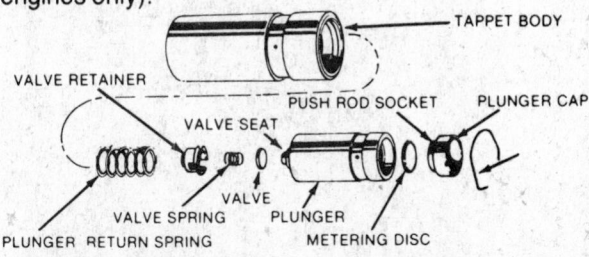

- TAPPET BODY
- VALVE RETAINER
- PUSH ROD SOCKET
- PLUNGER CAP
- VALVE SEAT
- VALVE
- VALVE SPRING
- PLUNGER
- PLUNGER RETURN SPRING
- METERING DISC

Typical exploded view of hydraulic valve lifter

Submerge lifter in a container of kerosene. Chuck a used pushrod or its equivalent into a drill press. Position container of kerosene so pushrod acts on the lifter plunger. Pump lifter with the drill press, until resistance increases. Pump several more times to bleed any air out of lifter. Apply very firm, constant pressure to the lifter, and observe rate at which fluid bleeds out of lifter. If the fluid bleeds very quickly (less than 15 seconds), lifter is defective. If the time exceeds 60 seconds, lifter is sticking. In either case, recondition or replace lifter. If lifter is operating properly (leak down time 15–60 seconds), lubricate and install it.

CYLINDER BLOCK RECONDITIONING

Procedure	Method

Removing the crankshaft:

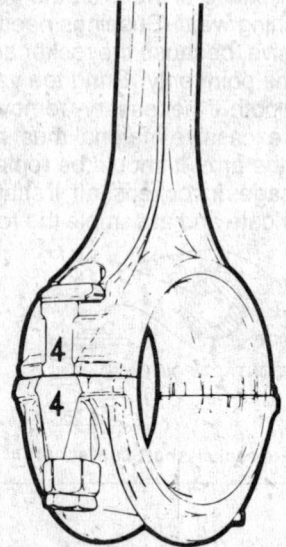

Connecting rod matched to cylinder with a number stamp

Using a punch, mark the corresponding main bearing caps and saddles according to position (i.e., one punch on the front main cap and saddle, two on the second, three on the third, etc.). Using number stamps, identify the corresponding connecting rods and caps, according to cylinder (if no numbers are present). Remove the main and connecting rod caps, and place sleeves of plastic tubing over the connecting rod bolts, to protect the journals as the crankshaft is removed. Lift the crankshaft out of the block.

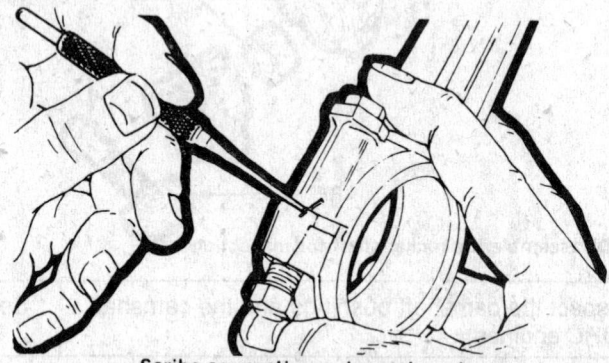

Scribe connecting rod matchmarks

Remove the ridge from the top of the cylinder:

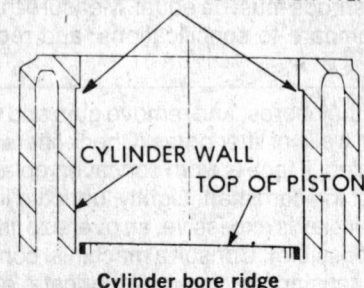

RIDGE CAUSED BY CYLINDER WEAR

CYLINDER WALL
TOP OF PISTON

Cylinder bore ridge

In order to facilitate removal of the piston and connecting rod, the ridge at the top of the cylinder (unworn area; see illustration) must be removed. Place the piston at the bottom of the bore, and cover it with a rag. Cut the ridge away using a ridge reamer, exercising extreme care to avoid cutting to deeply. Remove the rag, and remove cuttings that remain on the piston.
CAUTION: *If the ridge is not removed, and new rings are installed, damage to rings will result.*

Removing the piston and connecting rod:

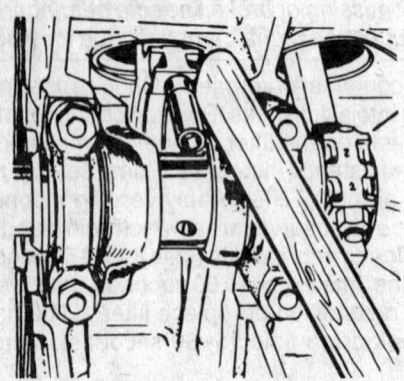

Removing the piston

Invert the engine, and push the pistons and connecting rods out of the cylinders. If necessary, tap the connecting rod boss with a wooden hammer handle, to force the piston out.
CAUTION: *Do not attempt to force the piston past the cylinder ridge (see above).*

CYLINDER BLOCK RECONDITIONING

Procedure	Method
Service the crankshaft:	Ensure that all oil holes and passages in the crankshaft are open and free of sludge. If necessary, have the crankshaft ground to the largest possible undersize. **Have the crankshaft Magnafluxed, to locate stress cracks. Consult a machinist concerning additional service procedures, such as surface hardening (e.g., nitriding, Tuftriding) to improve wear characteristics, cross drilling and chamfering the oil holes to improve lubrication, and balancing.
Removing freeze plugs:	Drill a small hole in the middle of the freeze plugs. Thread a large sheet metal screw into the hole and remove the plug with a slide hammer.
Remove the oil gallery plugs:	Threaded plugs should be removed using an appropriate (usually square) wrench. To remove soft, pressed in plugs, drill a hole in the plug, and thread in a sheet metal screw. Pull the plug out by the screw using pliers.
Hot-tank the block: NOTE: *Do not hot-tank aluminum parts.*	Have the block hot-tanked to remove grease, corrosion, and scale from the water jackets. NOTE: *Consult the operator to determine whether the camshaft bearings will be damaged during the hot-tank process.*

Check the block for cracks:	Visually inspect the block for cracks or chips. The most common locations are as follows:
	    Adjacent to freeze plugs.
	    Between the cylinders and water jackets.
	    Adjacent to the main bearing saddles.
	    At the extreme bottom of the cylinders.
	Check only suspected cracks using spot check dye (see introduction). If a crack is located, consult a machinist concerning possible repairs.
	
	**Magnaflux the block to locate hidden cracks. If cracks are located, consult a machinist about feasibility of repair.

Install the oil gallery plugs and freeze plugs:	Coat freeze plugs with sealer and tap into position using a piece of pipe, slightly smaller than the plug, as a driver. To ensure retention, stake the edges of the plugs. Coat threaded oil gallery plugs with sealer and install. Drive replacement soft plugs into block using a large drift as a driver. *Rather than reinstalling lead plugs, drill and tap the holes, and install threaded plugs.
*Check the deck height:	*The deck height is the distance from the crankshaft centerline to the block deck. To measure, invert the engine, and install the crankshaft, retaining it with the center main cap. Measure the distance from the crankshaft journal to the block deck, parallel to the cylinder centerline. Measure the diameter of the end (front and rear) main journals, parallel to the centerline of the cylinders, divide the diameter in half, and subtract it from the previous measurement. The results of the front and rear measurements should be identical. If the difference exceeds .005″, the deck height should be corrected. NOTE: *Block deck height and warpage should be corrected at the same time.*

CYLINDER BLOCK RECONDITIONING

Procedure	Method

Clean and inspect the pistons and connecting rods:

Using a ring expander, remove the rings from the piston. Remove the retaining rings (if so equipped) and remove piston pin.

NOTE: *If the piston pin must be pressed out, determine the proper method and use the proper tools; otherwise the piston will distort.*

Clean the ring grooves using an appropriate tool, exercising care to avoid cutting too deeply. Thoroughly clean all carbon and varnish from the piston with solvent.

CAUTION: *Do not use a wire brush or caustic solvent on pistons.*

Inspect the pistons for scuffing, scoring, cracks, pitting, or excessive ring groove wear. If wear is evident, the piston must be replaced. Check the connecting rod length by measuring the rod from the inside of the large end to the inside of the small end using calipers (see illustration). All connecting rods should be equal length. Replace any rod that differs from the others in the engine.

*Have the connecting rod alignment checked in an alignment fixture by a machinist. Replace any twisted or bent rods.

*Magnaflux the connecting rods to locate stress cracks. If cracks are found, replace the connecting rod.

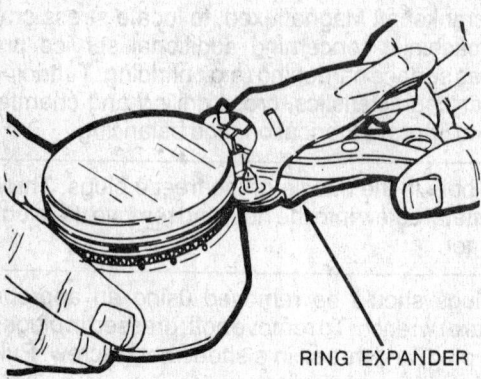

RING EXPANDER

Removing the piston rings

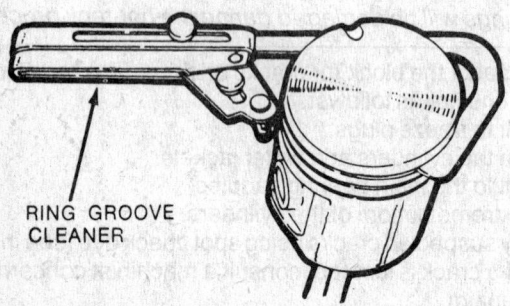

RING GROOVE CLEANER

Cleaning the piston ring grooves

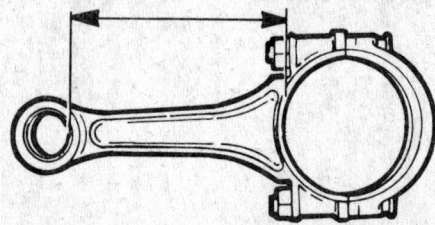

Check the connecting rod length (arrow)

Fit the pistons to the cylinders:

Using a telescope gauge and micrometer, or a dial gauge, measure the cylinder bore diameter perpendicular to the piston pin, 2½° below the deck. Measure the piston perpendicular to its pin on the skirt. The difference between the two measurements is the piston clearance. If the clearance is within specifications or slightly below (after boring or honing), finish honing is all that is required. If the clearance is excessive, try to obtain a slightly larger piston to bring clearance within specifications. Where this is not possible, obtain the first oversize piston, and hone (or if necessary, bore) the cylinder to size.

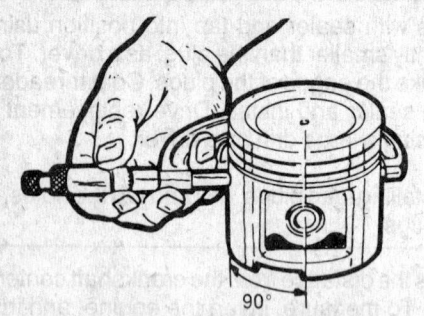

90°

Measuring the piston prior to fitting

Assemble the pistons and connecting rods:

Inspect piston pin, connecting rod small end bushing, and piston bore for galling, scoring, or excessive wear. If evident, replace defective part(s). Measure the I.D. of the piston boss and connecting rod small end, and the O.D. of the piston pin. I within specifications, assemble piston pin and rod.

CAUTION: *If piston pin must be pressed in, determine the proper method and use the proper tools; otherwise the piston will distort.*

CYLINDER BLOCK RECONDITIONING

Procedure	Method

Check the block deck for warpage:

Using a straightedge and feeler gauges, check the block deck for warpage in the same manner that the cylinder head is checked (see Cylinder Head Reconditioning). If warpage exceeds specifications, have the deck resurfaced.
NOTE: *In certain cases a specification for total material removal (Cylinder head and block deck) is provided. This specification must not be exceeded.*

Check the bore diameter and surface:

Measuring the cylinder bore with a dial gauge

Visually inspect the cylinder bores for roughness, scoring, or scuffing. If evident, the cylinder bore must be bored or honed oversize to eliminate imperfections, and the smallest possible oversize piston used. The new pistons should be given to the machinist with the block, so that the cylinders can be bored or honed exactly to the piston size (plus clearance). If no flaws are evident, measure the bore diameter using a telescope gauge and micrometer, or dial guage, parallel and perpendicular to the engine centerline, at the top (below the ridge) and bottom of the bore. Subtract the bottom measurements from the top to determine taper, and the parallel to the centerline measurements from the perpendicular measurements to determine eccentricity. If the measurements are not within specifications, the cylinder must be bored or honed, and an oversize piston installed. If the measurements are within specifications the cylinder may be used as is, with only finish honing (see below).
NOTE: *Prior to boring, check the block deck warpage, height and bearing alignment.*
CAUTION: *The 4 cyl. 140 G.M. engine cylinder walls are impregnated with silicone. Boring or honing can be done only by a shop with the proper equipment.*

TELESCOPE GAUGE 90° FROM PISTON PIN

Measuring cylinder bore with a telescope gauge

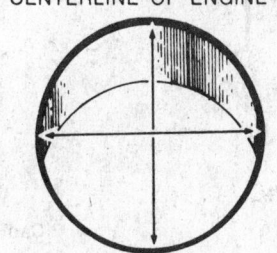

◄— CENTERLINE OF ENGINE —►

A—AT RIGHT ANGLE TO CENTERLINE OF ENGINE
B—PARALLEL TO CENTERLINE OF ENGINE
Cylinder bore measuring points

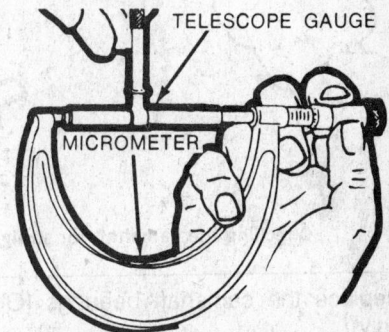

TELESCOPE GAUGE

MICROMETER

Determining cylinder bore by measuring telescope gauge with a micrometer

Check the cylinder block bearing alignment:

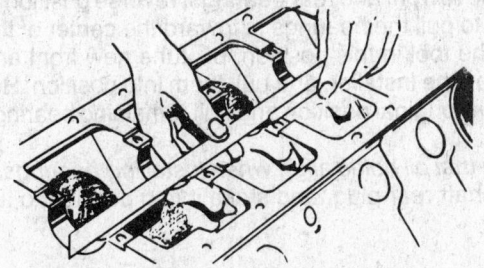

Checking main bearing saddle alignment

Remove the upper bearing inserts. Place a straightedge in the bearing saddles along the centerline of the crankshaft. If clearance exists between the straightedge and the center saddle, the block must be alignbored.

CYLINDER BLOCK RECONDITIONING

Procedure	Method

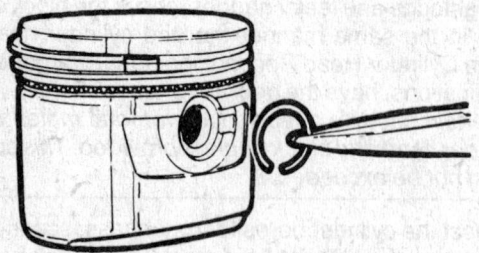

Installing piston pin lock rings

Install the lock rings; ensure that they seat properly. If the parts are not within specifications, determine the service method for the type of engine. In some cases, piston and pin are serviced as an assembly when either is defective. Others specify reaming the piston and connecting rods for an oversize pin. If the connecting rod bushing is worn, it may in many cases be replaced. Reaming the piston and replacing the rod bushing are machine shop operations.

Clean and inspect the camshaft:

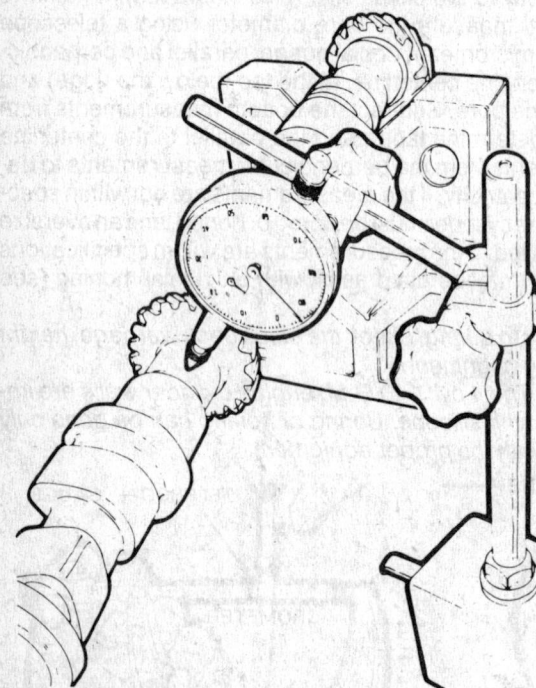

Checking the camshaft for straightness

Degrease the camshaft, using solvent, and clean out all oil holes. Visually inspect cam lobes and bearing journals for excessive wear. If a lobe is questionable, check all lobes as indicated below. If a journal or lobe is worn, the camshaft must be reground or replaced.

NOTE: *If a journal is worn, there is a good chance that the bushings are worn.*

If lobes and journals appear intact, place the front and rear journals in V-blocks, and rest a dial indicator on the center journal. Rotate the camshaft to check straightness. If deviation exceeds .001°, replace the camshaft.

*Check the camshaft lobes with a micrometer, by measuring the lobes from the nose to base and again at 90° (see illustration). The lift is determined by subtracting the second measurement from the first. If all exhaust lobes and all intake lobes are not identical, the camshaft must be reground or replaced.

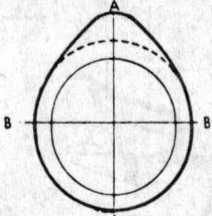

Camshaft lobe measurement

Replace the camshaft bearings (OHV engines only):

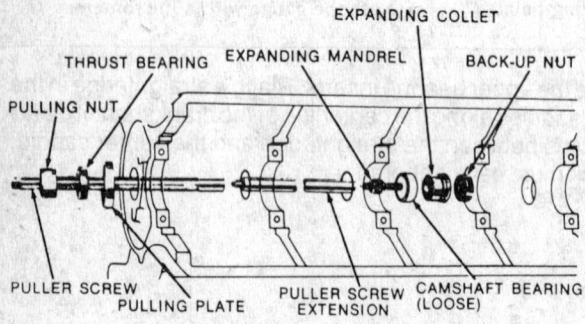

Camshaft removal and installation tool (typical)

If excessive wear is indicated, or if the engine is being completely rebuilt, camshaft bearings should be replaced as follows: Drive the camshaft rear plug from the block. Assemble the removal puller with its shoulder on the bearing to be removed. Gradually tighten the puller nut until bearing is removed. Remove remaining bearings, leaving the front and rear for last. To remove front and rear bearings, reverse position of the tool, so as to pull the bearings in toward the center of the block. Leave the tool in this position, pilot the new front and rear bearings on the installer, and pull them into position: Return the tool to its original position and pull remaining bearings into postion.

NOTE: *Ensure that oil holes align when installing bearings.* Replace camshaft rear plug, and stake it into position to aid retention.

CYLINDER BLOCK RECONDITIONING

Procedure	Method

Finish hone the cylinders:

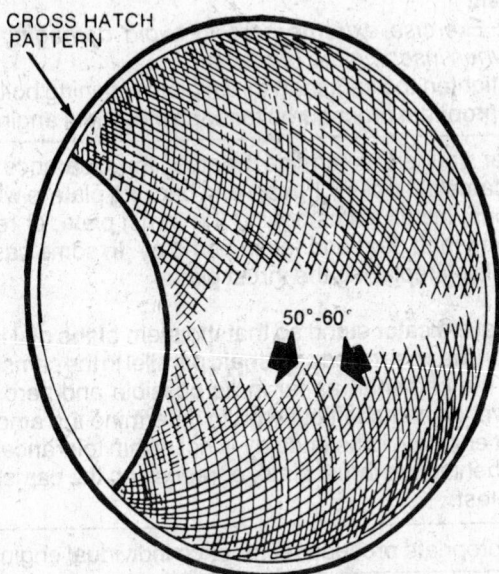

CROSS HATCH PATTERN

50°-60°

Chuck a flexible drive hone into a power drill, and insert it into the cylinder. Start the hone, and move it up and down the cylinder at a rate which will produce approximately a 60° cross-hatch pattern (see illustration).

NOTE: *Do not extend the hone below the cylinder bore.*

After developing the pattern, remove the hone and recheck piston fit. Wash the cylinders with a detergent and water solution to remove abrasive dust, dry, and wipe several times with a rag soaked in engine oil.

Check piston ring end-gap:

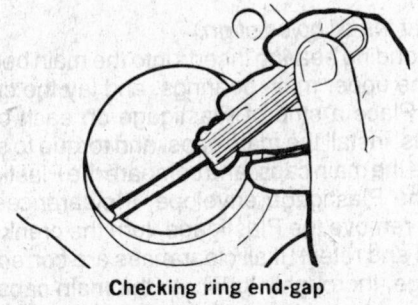

Checking ring end-gap

Compress the piston rings to be used in a cylinder, one at a time, into that cylinder, and press them approximately 1″ below the deck with an inverted piston. Using feeler gauges, measure the ring end-gap, and compare to specifications. Pull the ring out of the cylinder and file the ends with a fine file to obtain proper clearance.

CAUTION: *If inadequate ring end-gap is utilized, ring breakage will result.*

Install the piston rings:

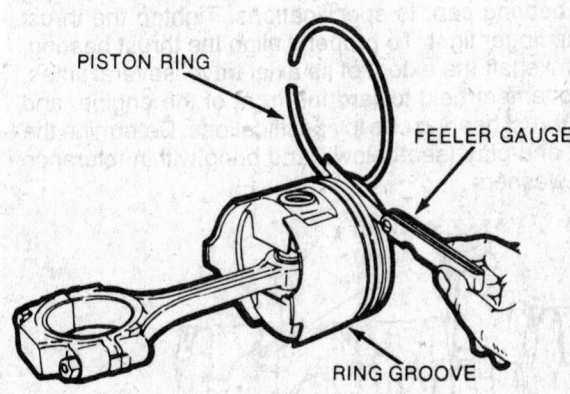

PISTON RING

FEELER GAUGE

RING GROOVE

Checking ring side clearance

Inspect the ring grooves in the piston for excessive wear or taper. If necessary, recut the groove(s) for use with an over-width ring or a standard ring and spacer. If the groove is worn uniformly, overwidth rings, or standard rings and spacers may be installed without recutting. Roll the outside of the ring around the groove to check for burrs or deposits. If any are found, remove with a fine file. Hold the ring in the groove, and measure side clearance. If necessary, correct as indicated above.

NOTE: *Always install any additional spacers above the piston ring.*

The ring groove must be deep enough to allow the ring to seat below the lands (see illustration). In many cases, a "go-no-go" depth gauge will be provided with the piston rings. Shallow grooves may be corrected by recutting, while deep grooves require some type of filler or expander behind the piston. Consult the piston ring supplier concerning the suggested method. Install the rings on the piston, lowest ring first, using a ring expander.

NOTE: *Position the ring markings as specified by the manufacturer (see car section).*

CYLINDER BLOCK RECONDITIONING

Procedure	Method
Install the camshaft (OHV engines only):	Liberally lubricate the camshaft lobes and journals, and install the camshaft. CAUTION: *Exercise extreme care to avoid damaging the bearings when inserting the camshaft.* Install and tighten the camshaft thrust plate retaining bolts. See the appropriate procedures for each individual engine.

Check camshaft end-play (OHV engines only):

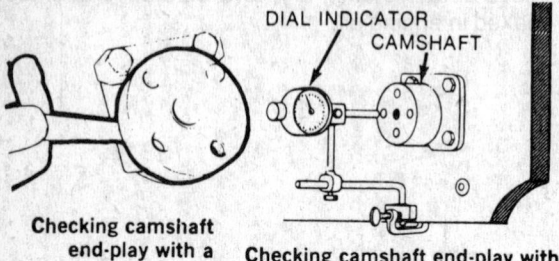

DIAL INDICATOR
CAMSHAFT

Checking camshaft
end-play with a
feeler gauge

Checking camshaft end-play with a
dial indicator

Using feeler gauges, determine whether the clearance between the camshaft boss (or gear) and backing plate is within specifications. Install shims behind the thrust plate, or reposition the camshaft gear and retest end-play. In some cases, adjustment is by replacing the thrust plate.

*Mount a dial indicator stand so that the stem of the dial indicator rests on the nose of the camshaft, parallel to the camshaft axis. Push the camshaft as far in as possible and zero the gauge. Move the camshaft outward to determine the amount of camshaft endplay. If the endplay is not within tolerance, install shims behind the thrust plate, or reposition the camshaft gear and retest.

Procedure	Method
Install the rear main seal (where applicable):	See the appropriate procedures for each individual engine.

Install the crankshaft:

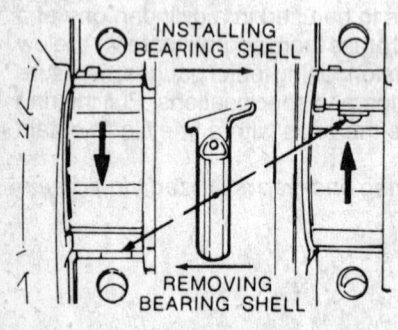

INSTALLING
BEARING SHELL

REMOVING
BEARING SHELL

Removal and installation of upper
bearing insert using a roll-out pin

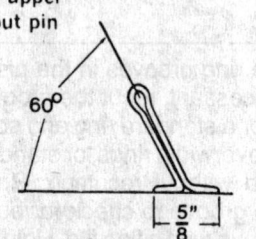

60°

5"/8

Home-made bearing roll-out pin

Thoroughly clean the main bearing saddles and caps. Place the upper halves of the bearing inserts on the saddles and press into position.

NOTE: *Ensure that the oil holes align.*

Press the corresponding bearing inserts into the main bearing caps. Lubricate the upper main bearings, and lay the crankshaft in position. Place a strip of Plastigage on each of the crankshaft journals, install the main caps, and torque to specifications. Remove the main caps, and compare the Plastigage to the scale on the Plastigage envelope. If clearances are within tolerances, remove the Plastigage, turn the crankshaft 90°, wipe off all oil and retest. If all clearances are correct, remove all Plastigage, thoroughly lubricate the main caps and bearing journals, and install the main caps. If clearances are not within tolerance, the upper bearing inserts may be removed, without removing the crankshaft, using a bearing roll out pin (see illustration). Roll in a bearing that will provide proper clearance, and retest. Torque all main caps, excluding the thrust bearing cap, to specifications. Tighten the thrust bearing cap finger tight. To properly align the thrust bearing, pry the crankshaft the extent of its axial travel several times, the last movement held toward the front of the engine, and torque the thrust bearing cap to specifications. Determine the crankshaft end-play* (see below), and bring within tolerance with thrust washers.

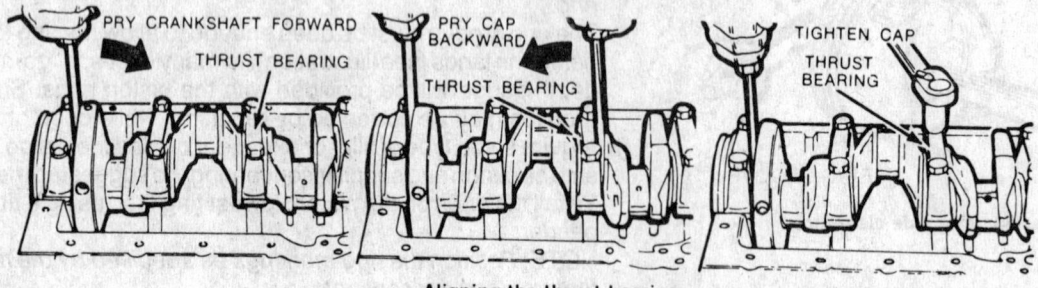

PRY CRANKSHAFT FORWARD
THRUST BEARING

PRY CAP
BACKWARD
THRUST BEARING

TIGHTEN CAP
THRUST
BEARING

Aligning the thrust bearing

CYLINDER BLOCK RECONDITIONING

Procedure	Method

Measure crankshaft end-play:

Mount a dial indicator stand on the front of the block, with the dial indicator stem resting on the nose of the crankshaft, parallel to the crankshaft axis. Pry the crankshaft the extent of its travel rearward, and zero the indicator. Pry the crankshaft forward and record crankshaft end-play.

NOTE: *Crankshaft end-play also may be measured at the thrust bearing, using feeler gauges* (see illustration).

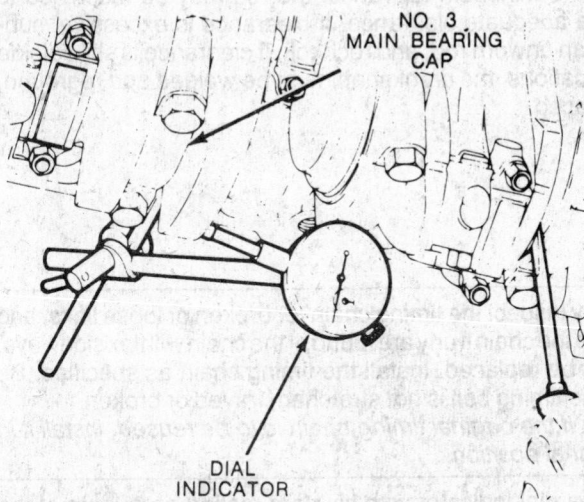

NO. 3 MAIN BEARING CAP

DIAL INDICATOR

Checking crankshaft end-play with a dial indicator

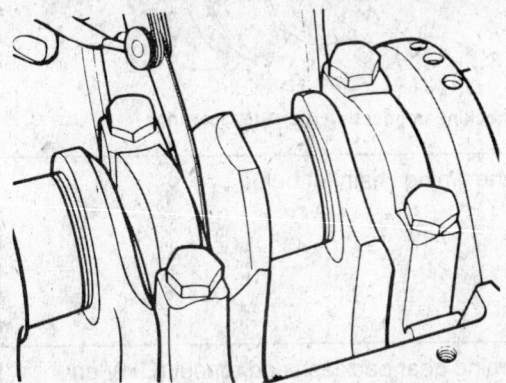

Checking crankshaft end-play with a feeler gauge

Install the pistons:

Press the upper connecting rod bearing halves into the connecting rods, and the lower halves into the connecting rod caps. Position the piston ring gaps according to specifications (see car section), and lubricate the pistons. Install a ring compressor on a piston, and press two long (8″) pieces of plastic tubing over the rod bolts. Using the tubes as a guide, press the pistons into the bores and onto the crankshaft with a wooden hammer handle. After seating the rod on the crankshaft journal, remove the tubes and install the cap finger tight. Install the remaining pistons in the same manner. Invert the engine and check the bearing clearance at two points (90° apart) on each journal with Plastigage.

NOTE: *Do not turn the crankshaft with Plastigage installed.*

If clearance is within tolerances, remove *all* Plastigage, thoroughly lubricate the journals, and torque the rod caps to specifications. If clearance is not within specifications, install different thickness bearing inserts and recheck.

CAUTION: *Never shim or file the connecting rods or caps.* Always install plastic tube sleeves over the rod bolts when the caps are not installed, to protect the crankshaft journals.

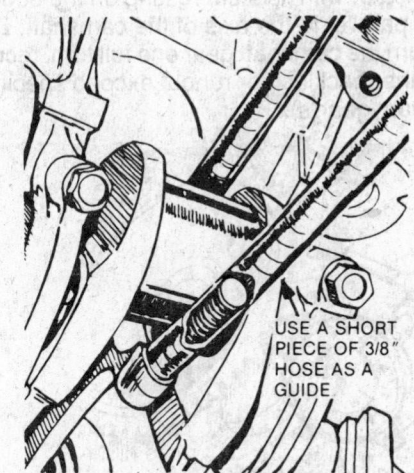

USE A SHORT PIECE OF 3/8″ HOSE AS A GUIDE

Tubing used to protect crankshaft journals and cylinder walls during piston installation

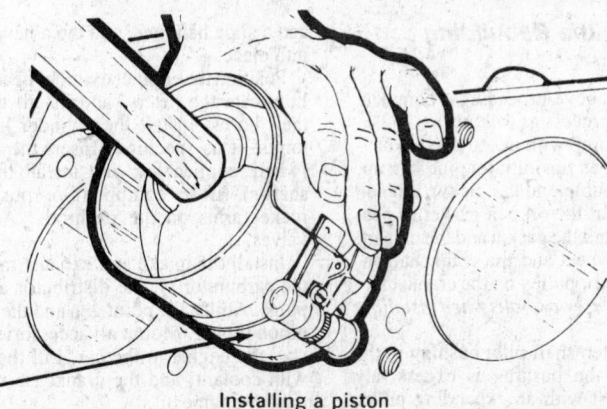

RING COMPRESSOR

Installing a piston

CYLINDER BLOCK RECONDITIONING

Procedure	Method
Check connecting rod side clearance: 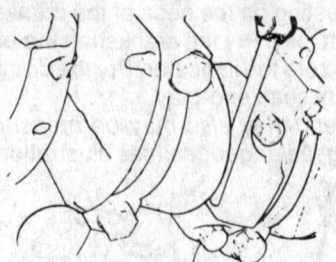 Checking connecting rod side clearance	Determine the clearance between the sides of the connecting rods and the crankshaft, using feeler gauges. If clearance is below the minimum tolerance, the rod may be machined to provide adequate clearance. If clearance is excessive, substitute an unworn rod, and recheck. If clearance is still outside specifications, the crankshaft must be welded and reground, or replaced.
Inspect the timing chain (or belt):	Visually inspect the timing chain for broken or loose links, and replace the chain if any are found. If the chain will flex sideways, it must be replaced. Install the timing chain as specified. Be sure the timing belt is not stretched, frayed or broken. NOTE: *If the original timing chain is to be reused, install it in its original position.*
Check timing gear backlash and runout (OHV engines): 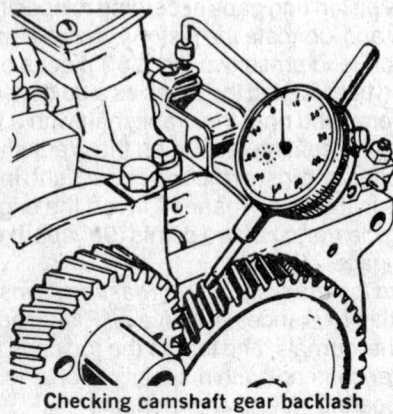 Checking camshaft gear backlash	Mount a dial indicator with its stem resting on a tooth of the camshaft gear (as illustrated). Rotate the gear until all slack is removed, and zero the indicator. Rotate the gear in the opposite direction until slack is removed, and record gear backlash. Mount the indicator with its stem resting on the edge of the camshaft gear, parallel to the axis of the camshaft. Zero the indicator, and turn the camshaft gear one full turn, recording the runout. If either backlash or runout exceed specifications, replace the worn gear(s).  Checking camshaft gear runout

Completing the Rebuilding Process

Following the above procedures, complete the rebuilding process as follows:

Fill the oil pump with oil, to prevent cavitating (sucking air) on initial engine start up. Install the oil pump and the pickup tube on the engine. Coat the oil pan gasket as necessary, and install the gasket and the oil pan. Mount the flywheel and the crankshaft vibration damper or pulley on the crankshaft. NOTE: *Always use new bolts when installing the flywheel.*

Inspect the clutch shaft pilot bushing in the crankshaft. If the bushing is excessively worn, remove it with an expanding puller and a slide hammer, and tap a new bushing into place.

Position the engine, cylinder head side up. Lubricate the lifters, and install them into their bores. Install the cylinder head, and torque it as specified. Insert the pushrods (where applicable), and install the rocker shaft(s) (if so equipped) or position the rocker arms on the pushrods. Adjust the valves.

Install the intake and exhaust manifolds, the carburetor(s), the distributor and spark plugs. Adjust the point gap and the static ignition timing. Mount all accessories and install the engine in the car. Fill the radiator with coolant, and the crankcase with high quality engine oil.

Break-in Procedure

Start the engine, and allow it to run at low speed for a few minutes, while checking for leaks. Stop the engine, check the oil level, and fill as necessary. Restart the engine, and fill the cooling system to capacity. Check the point dwell angle and adjust the ignition timing and the valves. Run the engine at low to medium speed (800–2500 rpm) for approximately ½ hour, and retorque the cylinder head bolts. Road test the car, and check again for leaks.

Follow the manufacturer's recommended engine break-in procedure and maintenance schedule for new engines.

Diesel Maintenance 19

NOTE: Most procedures associated with diesel engined cars are similar to gas engined cars, although many parts of the diesel engine are unique compared to their gas engine counterparts. Standard maintenance and service procedures are given here while component removal, installation and adjustment procedures unique to diesel engines can be found in the appropriate section.

HOW THE DIESEL ENGINE WORKS

Four-stroke diesels require four piston strokes for the complete cycle of actions, exactly like a gasoline engine. The difference lies in how the fuel mixture is ignited. A diesel engine does not rely on a conventional spark ignition to ignite the fuel mixture for the power stroke. Instead, a diesel relies on the heat produced by compressing air in the combustion chamber to ignite the fuel and produce a power stroke. This is known as a compression-ignition engine. No fuel enters the cylinder on the intake stroke, only air. At the end of the compression stroke, fuel is sprayed into the precombustion chamber (prechamber). The mixture ignites and spreads out into the main combustion chamber, forcing the piston downward (power stroke). The fuel/air mixture ignites because of the very high combustion chamber temperatures generated by the extraordinarily high compression ratios used in diesel engines. Typically, the compression ratios used in automotive diesels run anywhere from 16:1 to 23:1 A typical spark-ignition engine has a ratio of about 8:1. This is why a spark-ignition engine which continues to run after you have shut off the engine is said to be "dieseling". It is running on combustion chamber heat alone.

Designing an engine to ignite on its own combustion chamber heat poses certain problems. For instance, although a diesel engine has no need for a coil, spark plugs, or a distributor, it does need what are known as "glow plugs". These superficially resemble spark plugs, but are only used to warm the combustion chambers when the engine is cold. Without these plugs, cold starting would be impossible, due to the enormously high compression ratios and the characteristics of the diesel fuel itself.

All diesel engines use fuel injection, be-

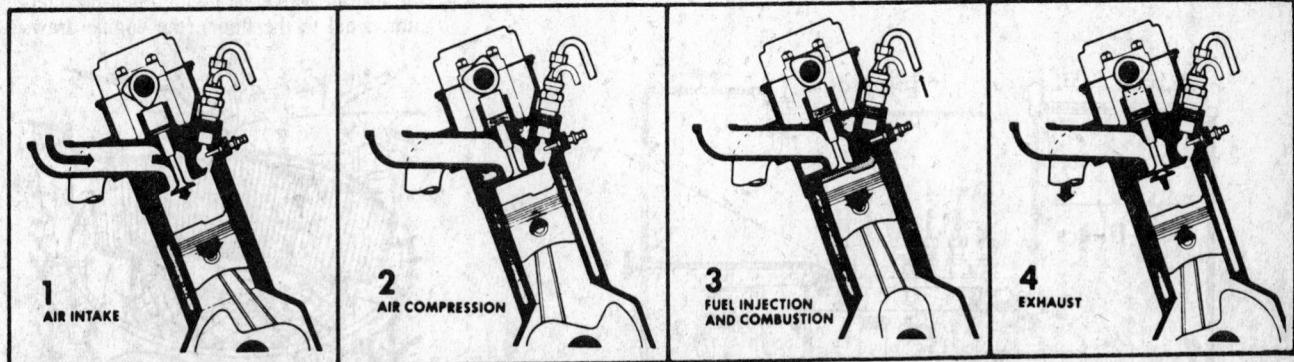

4-stroke diesel engine cycle. At *air intake* (1), rotation of the crankshaft drives a toothed belt that turns the camshaft, opening the intake valve. As the piston moves down, a vacuum is created, sucking fresh air into the cylinder, past the open intake valve. *Air compression* (2): As the piston moves up, both valves are closed, and the air is compressed about 23 times smaller than its original volume. The compressed air reaches a temperature of about 1,650°F., far above the temperature needed to ignite diesel fuel. *Fuel injection and compression* (3): As the piston reaches the top of the stroke, the air temperature is at its maximum. A fine mist of fuel is sprayed into the prechamber, where it ignites, and the flame front spreads rapidly into the combustion chamber. The piston is forced downward by the pressure (about 500 psi) of expanding gases. *Exhaust* (4): As the energy of combustion is spent and the piston begins to move upward again, the exhaust valve opens, and burnt gases are forced out past the open valve. As the piston starts down, the exhaust valve closes, the intake valve opens, and the air intake stroke begins again.

Increasingly, modern diesel engines are being equipped with turbochargers, exhaust gas–driven devices that force more air into the engine to increase power output

Maintenance and Service Procedures

Maintenance procedures for the diesel engine generally fall into three categories:

1. Fuel system
2. Starting system
3. Engine mechanical systems

Of these, the fuel system is usually the most likely source of engine troubles, and should be high on the list for regular maintenance attention.

FUEL SYSTEM

The typical diesel engine fuel system consists of fuel tank, fuel feed and return lines, mechanical fuel injection pump, fuel injectors and lines, and a large capacity fuel filter. On some models, the engine may also be equipped with a small, low pressure fuel pump which feeds the injection pump.

In addition to these, the air intake system (air cleaner, inlet manifold) should be checked over regularly to insure unrestricted air flow into the cylinders.

In operation, fuel is sucked out of the fuel tank by the injection pump (or its feed pump) and fed by the injection pump to the injectors in the cylinder head at a very high pressure. Before the fuel is allowed to enter the main injection pump, it passes through a specially built fuel filter which traps solid particles (and water on some models) in the fuel. Fuel that is not used is pumped back to the fuel tank through the fuel return lines. This recirculated fuel helps cool the injection pump.

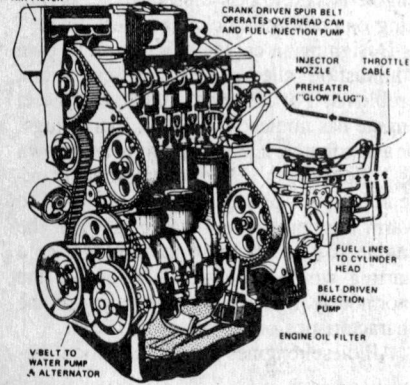

Cutaway view of typical 4-cylinder diesel engine.

cause unlike spark-ignited engines, the fuel cannot be drawn through the intake tract and into the cylinders. The introduction of fuel into a diesel engine must be precisely timed so that each cylinder "fires" at the proper moment. Also, the fuel injection pressure (at the cylinder) must be great enough to overcome the high compression pressures, and properly atomize the fuel without the aid of a moving air mass (as in a carbureted gas engine). It is not uncommon for diesel engine fuel injection pressures to be set at 1500–1700 psi.

Diesel engines share many of their basic mechanical components with gasoline engines, though the cylinder block, head(s), crankshaft, connecting rods, pistons, etc., are manufactured to be much stronger for use in diesel engines. The additional strength of the components is necessary due to the very high cylinder pressure generated within the diesel engine.

Air Cleaner

On a gasoline engine, the volume of air taken in by the engine is controlled by throttle valves. When the throttle valves are closed (engine idling), air intake is restricted. When the throttle valves are wide open (accelerator pedal to the floor), the engine draws

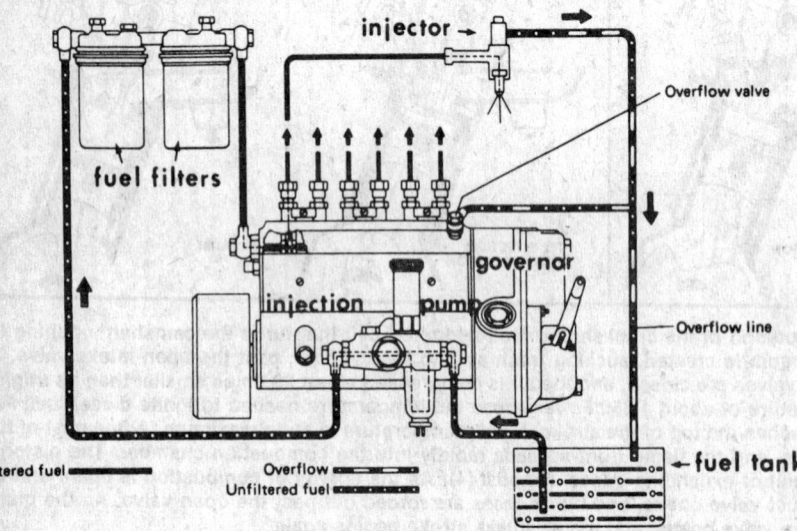

Typical diesel engine fuel system schematic

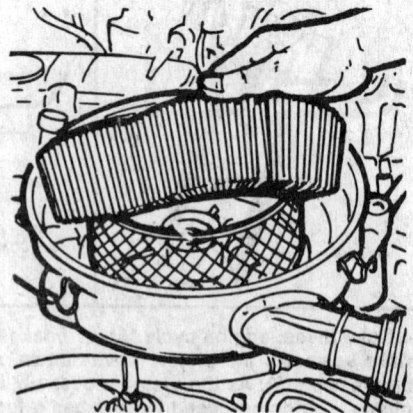

Because a greater quantity of air passes through the diesel engine, air filter maintenance is particularly important. Most diesel air filters on passenger cars are similar to their counterparts on gasoline engines.

in the maximum amount of air it possibly can. This applies to both carbureted and fuel injected gasoline engines.

The speed (rpm) of a diesel engine is controlled by the quantity of fuel which is injected into the engine; no air metering restrictions (throttle valves) are used. Because of this, diesel engines ingest as much air as they possibly can under all conditions. A much greater volume of air passes through the air cleaner of a diesel per mile, therefore, diesel air filters must either be larger or the filter replacement intervals more frequent than those of a similarly sized gasoline engine.

One word of caution: never remove the air cleaner on a diesel with the engine running, and never run the engine with the air cleaner removed. The volume of air drawn through the inlet manifold is very great, and, because the inlet manifold is unobstructed, anything drawn into the inlet manifold (air cleaner wing nut, etc.) goes straight to the combustion chambers, where it can cause major engine damage.

Fuel Filter

The diesel engine fuel filter is usually larger than the filter used on gasoline engines. The extra capacity is needed to trap the suspended particles in diesel fuel, which is generally "dirtier" than gasoline.

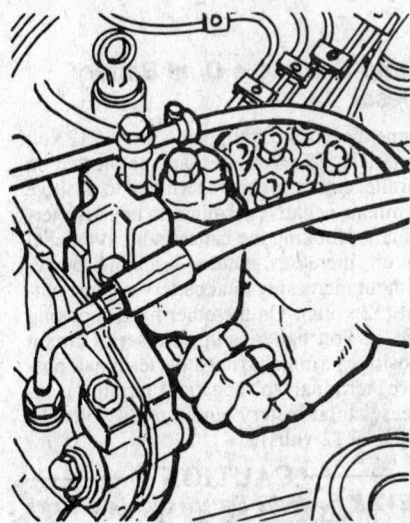

Many diesel engines use a spin-on type primary fuel filter.

On some engines, the fuel filter looks like a second engine oil filter, and is removed and installed in the same manner as the canister-type oil filter.

The fuel filter must be changed according to the manufacturer's suggested interval. See the owner's manual for information.

After installing the fuel filter start the engine and check for leaks. Run the engine for about two minutes, then stop the engine for the same amount of time to allow any air trapped in the injection system to bleed off.

Many diesels also have a small, in-tank filter which is usually maintenance-free.

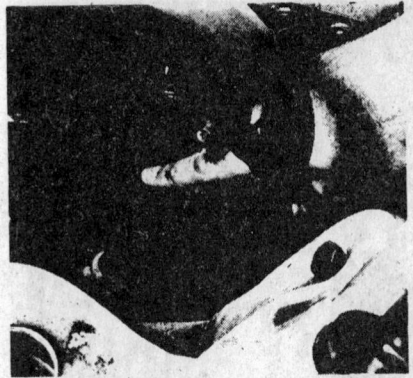

A smaller, in-line secondary filter is used on many engines.

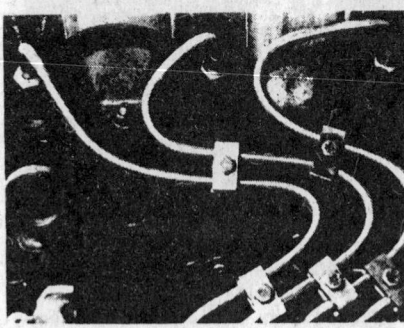

Check the tightness of the clamps securing the injector lines. Note that the injector lines are all the same length.

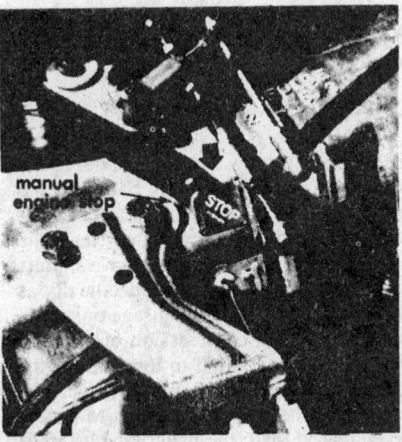

Mercedes-Benz diesel engines use this stop switch, which shuts off fuel delivery

Water In Fuel

Diesel fuel is a hydrophilic fluid, that is, it naturally attracts water. Since diesel fuel and water do not mix, the water remains floating beneath the fuel at the bottom of the tank. This water must be removed every now and then, or it will be sucked into the fuel circuit and pass through the injection system, causing corrosion and possible component failure (injection pumps can cost up to $1,000). Water in the fuel system will also cause the engine to run poorly, if at all.

Most diesel fuel tanks are equipped with a separator which can isolate from 1 to 3 gallons of water from the fuel.

Many diesels are also equipped with "Water in Fuel" lights in the dashboard which warn of the presence of water in the fuel tank. These warning systems can be installed on models not so equipped.

On some diesels, there is a water catcher in the bottom of the fuel filter which can easily be bled off. In addition, there are several bolt-on water filters on the market which attach to the fuel line under the hood and separate water from the fuel. Depending on which kind you buy, draining water from the system is simply a matter of opening the petcock at the bottom of the filter and letting the water drain out, or, if money is no object, a separator is available on which water is drained from the filter simply by activating a switch on the dashboard.

Removing Water from the Fuel Tank

Treat diesel fuel with the same respect you would gasoline, and after the procedure, properly dispose of the fuel.

1. Remove the fuel tank cap.
2. Connect a pump or siphon hose to the ¼ in. fuel return hose (smaller of the two fuel hoses) above the rear axle, or under the hood near the fuel pump (on the passenger's side of the engine, near the front).
3. Siphon until all water is removed from the tank. Do not use your mouth to create siphon vacuum, EVER! The best method is to siphon the water into a large capacity see-through container. The water will collect at the bottom of the container.
4. When all water has been removed from the tank, be sure to reinstall the fuel return hose and fuel cap.

NOTE: If the entire fuel system (not just the tank) is contaminated by water, the vehicle must be stopped immediately and the fuel system must be purged. This includes draining and removing the fuel tank, blowing low pressure compressed air backwards through the fuel feed and return lines, and bleeding the water out of all injection components. This job should be referred to a qualified technician.

Cold Weather Fuel System Maintenance

———— **CAUTION** ————
NEVER use "starting aids" (e.g.—ether) to help start a diesel engine—serious engine damage will result.

As will be explained later under "Fuel Recommendations", diesel fuel tends to become "cloudy", or thicker, as the temperature drops. The thicker the diesel fuel becomes, the slower it flows through the fuel system, until finally it stops flowing altogether somewhere near the bottom of the thermometer.

One way to fight sluggish fuel flow is to use winterized blends of diesel fuel, straight No. 1 diesel fuel or add cold weather additives to the fuel to improve flow in cold weather.

NOTE: Consult your owners manual for recommendations and be sure to use a fuel conditioner compatible with water separators.

Another way is to install an aftermarket fuel system pre-heater. These are generally canisters which connect into the fuel line and use coolant from the engine cooling system to heat the fuel before it reaches the injection pump. The one drawback with this system is the engine must be started before the pre-heater begins to work. Also available are electric fuel warmers. These preheat the fuel going into the filter and can be used in conjunction with the coolant-type fuel heater.

Cold weather additives and fuel conditioners can help improve cold weather flow of diesel fuel.

Some manufacturers offer an optional electric diesel fuel heater and engine block heaters. The fuel heater is thermostatically controlled to heat the fuel before it enters the fuel filter when fuel temperature is 20°F or lower. The fuel heater works only when the ignition key is in the RUN position. On these models, the fuel tank filter has a by-pass valve which allows fuel to flow to the heater when the tank filter is covered with fuel wax. The engine block heater is equipped with an electrical cord wrapped up in the engine compartment. The cord

Some diesel engines come equipped with a built-in heating system to keep the engine warm in cold temperatures.
Most OEM heaters work from 110-volt house current.

Some aftermarket diesel fuel warmers are thermostatically controlled heat exchangers that use engine coolant to keep diesel fuel above its "cloud point," the temperature at which it gels and forms wax that can clog a fuel system.

plugs into regular 110 volt household current. The block heater can be used, according to the type of oil in the crankcase, up to eight hours or overnight to warm up the block.

STARTING SYSTEM

The diesel starting system includes one (sometimes two) heavy duty batteries, the starter, and the glow plug circuit. In addition to the heavy duty battery(ies), the majority of diesel engines also have starters and battery cables designed specifically as heavy duty items for diesel usage only. Because of the high compression of any diesel, the torque required to turn the engine is much greater than a gasoline engine. The starter must be powerful enough to handle the increased load; the battery cables must be thick enough to withstand the heat generated by the starter load.

For battery maintenance, see the regular "Maintenance" section. Jump starting procedures for a dual battery car are given below. Starter maintenance is included in the appropriate car section.

The glow plug circuit is used on the diesel to initially start the engine. When the ignition switch is turned to the ON position, a light will come on in the instrument panel signalling that the glow plugs are preheating the combustion chambers. After a certain interval (depending on how cold the engine is), the light will go off. This signals that the starter may be engaged and the engine started. If the glow plug circuit mal-

functions, especially in cold weather, the engine will be almost impossible to start.

Glow Plug Testing

To test each individual glow plug, disconnect the busbar and/or wire connector from the glow plug and connect a test light between the glow plug terminal and the positive battery terminal. If the test light lights, the glow plug is working. Replace individual glow plugs which do not work.

NOTE: Some diesel engines are equipped with either "slow glow" or "fast glow" glow plugs. Do not attempt to interchange any parts of these two glow plug systems.

To test the glow plug circuit, connect a test light to the terminal of one of the glow plugs (glow plug wiring still attached) and turn the ignition to the heating position. The test light should light for a short while. If not, the glow plug circuit is malfunctioning and must be diagnosed and repaired.

NOTE: Perform this operation on a cold engine only.

Jump-Starting a Dual Battery Diesel

Some diesels are equipped with two 12 volt batteries. The batteries are connected in parallel circuit (positive terminal to positive terminal, negative terminal to negative terminal). Hooking the batteries up in parallel circuit increases battery cranking power without increasing total battery voltage output (12 volts). On the other hand, hooking two 12 volt batteries up in a series circuit (positive terminal to negative terminal, positive terminal to negative terminal) increases total battery output to 24 volts (12 volts + 12 volts).

—————— CAUTION ——————
NEVER hook the batteries up in a series circuit; SEVERE electrical system damage will result.

In the event that a dual battery diesel must be jumped started, use the following procedure.

1. Open the hood and locate the batteries.
2. Position the donor car so that the jumper cables will reach from its battery (must be 12 volt, negative ground) to the appropriate battery in the diesel. Do not allow the cars to touch.
3. Shut off all electrical equipment on both vehicles. Turn off the engine of the donor car, set the parking brakes on both vehicles and block the wheels. Also, make sure both vehicles are in Neutral (manual

transmission models) or Park (automatic transmission models).

4. Using the jumper cables, connect the positive (+) terminal of the donor car battery to the positive terminal of one (not both) of the diesel batteries.

5. Using the second jumper cable, connect the negative (−) terminal of the donor battery to a solid, stationary, metallic point on the diesel (alternator bracket, engine block, etc.). Be very careful to keep the jumper cables away from moving parts (cooling fan, alternator belt, etc.) on both vehicles.

6. Start the engine of the donor car and run it at moderate speed.

7. Start the engine of the diesel.

8. When the diesel starts, disconnect the battery cables in the reverse order of attachment.

ENGINE MECHANICAL SYSTEMS

Included are engine lubrication and engine compression.

Although diesel engines are very low in carbon monoxide (CO) and hydrocarbon (HC) emissions, "particulate" emission output is very high from diesel engines. This is evident from the black smoke emitted by diesels, which is most noticeable during hard acceleration or high engine loads. The particulates are made up of mostly soot (carbon) and sulpher particles. The majority of these particulates are released into the atmosphere. However, some of the particulate matter, because it is produced within the engines cylinders, is left inside the engine and gradually contaminates the engine oil. This contamination makes the oil corrosive, due to the sulpher, and abrasive, due to the carbon. Serious engine damage will result if these contaminants continue to accumulate in the oil. Engine oil and filters of diesel engines must be changed more frequently than those of gasoline engines, due to the increased rate at which the contaminants form in the diesel. Consult the "Maintenance" section for oil and filter change procedures. The manufacturer's recommended oil change interval will be given in the owner's manual. An explanation of diesel engine oils is given at the end of this section.

As explained earlier, very high cylinder compression is the key to the operation of the diesel engine. The normal compression of most gasoline engines will rarely exceed 180 psi; whereas with diesel engines, compression pressures of 350–400 psi are commonplace.

───── CAUTION ─────

DO NOT attempt to check the compression of a diesel engine with a standard compression gauge—personal injury could result. A special, high pressure compression gauge is needed to safely check the compression of any diesel.

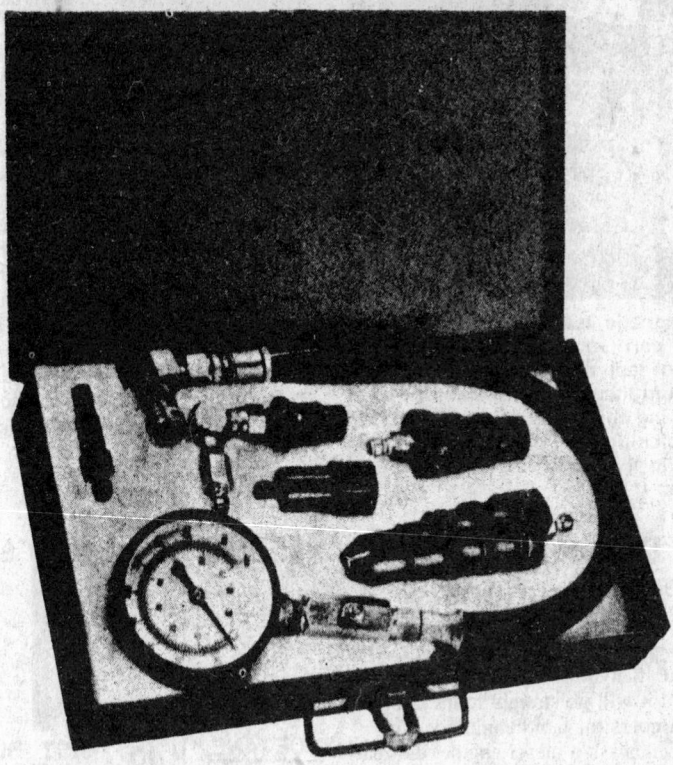

A diesel compression tester kit with adaptors (Courtesy S & G Tools).

Compression Test

1. Remove the air cleaner.
2. Disconnect the wire from the fuel shutoff solenoid terminal of the injection pump.
3. Disconnect the wires from the glow plugs and remove all glow plugs.
4. Screw compression gauge into the glow plug hole in the cylinder being checked.
5. Crank the engine, allowing six "puffs" for each cylinder.

The lowest reading cylinder should not be less than 70% of the highest, and no cylinder should be less than 275 pounds.

Idle Speed Adjustments

Idle speed adjustment procedures for individual diesel engines are given in the car section. Consult the following section for procedures to measure idle speed

Connecting a Tachometer to a Diesel Engine

As mentioned earlier, the diesel engine does not require an electrical ignition system. Because of this, problems arise when attempts are made to connect a tachometer to the engine for the purpose of idle adjustments, etc. The average gasoline engine tachometer senses the ignition spark pulses and converts them into a readable engine rpm signal. This type of tachometer is use-

less on the diesel engine, because of the diesel's compression ignition system.

There are several magnetic and photoelectric tachometers available from various tool manufacturers which were designed specifically for use with the diesel engine. These units can run into a little more money than the average do-it-yourselfer may be willing to spend, in which case any adjustments requiring the monitoring of engine rpm should be performed by a competent service technician.

The newest equipment for measuring idle speed on a diesel engine includes (clockwise from lower left) a digital diesel tach display, photomagnetic pick-up with display input, magnetic swivel base (holder), DC power source for the display unit and a roll of magnetic tape.

The magnetic tape is attached to any moving part (such as the balancer). The pieces of tape must be at least 6 inches apart. Aim the photomagnetic pick-up at the moving object and adjust the position of the pick-up until the "on-target" light is lit. Flip the switch to TACH and read the rpm.

Diesel Engine Precautions

- Never run the engine with the air cleaner removed: if anything is sucked into the inlet manifold it will go straight to the combustion chambers, or jam behind a valve.
- Never wash a diesel engine: the reaction of a warm fuel injection pump to cold (or even warm) water can ruin the pump.
- Never operate a diesel engine with one or more fuel injectors removed unless fully familiar with injector testing procedures: some diesel injection pumps spray fuel at up to 1400 psi—enough pressure to allow the fuel to penetrate your skin.
- Do not skip engine oil and filter changes.
- Strictly follow the manufacturer's oil and fuel recommendations as given in the owner's manual.
- Do not use home heating oil as fuel for your diesel.
- Do not use "starting aids" (e.g.—ether) in the automotive diesel engine, as these "aids" can cause severe internal engine damage.
- Do not run a diesel engine with the "Water in Fuel" warning light on in the dashboard.
- If removing water from the fuel tank yourself, use the same caution you would use when working around gasoline engine fuel components.
- Do not allow diesel fuel to come in contact with rubber hoses or components on the engine, as it can damage them.

Fuel and Oil Recommendations

FUEL

Fuel makers produce two grades of diesel fuel, No. 1 and No. 2, for use in automotive diesel engines. Generally speaking, No. 2 fuel is recommended over No. 1 for driving in temperatures above 20°F. In fact, in many areas, No. 2 diesel is the only fuel available. By comparison, No. 2 diesel fuel is less volatile than No. 1 fuel, and gives better fuel economy. No. 2 fuel is also a better injection pump lubricant.

Two important characteristics of diesel fuel are its cetane number and its viscosity.

The cetane number of a diesel fuel refers to the ease with which a diesel fuel ignites. High cetane numbers mean that the fuel will ignite with relative ease or that it ignites well at low temperatures. Naturally, the lower the cetane number, the higher the temperature must be to ignite the fuel. Most commercial fuels have cetane numbers that range from 35 to 65. No. 1 diesel fuel generally has a higher cetane rating than No. 2 fuel.

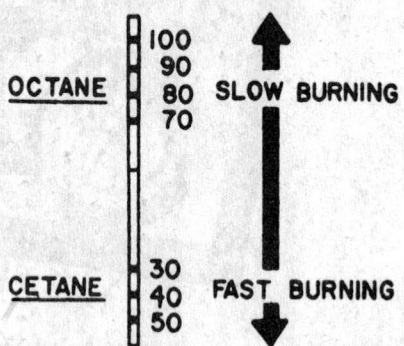

Cetane (diesel engine) versus octane (gasoline engine) ratings. The higher the cetane number, the faster the fuel burns

Viscosity is the ability of a liquid, in this case diesel fuel, to flow. Using straight No. 2 diesel fuel below 20°F can cause problems, because this fuel tends to become cloudy, meaning wax crystals begin forming in the fuel. In extreme cold weather, No. 2 fuel can stop flowing altogether. In either case, fuel flow is restricted, which can result in a "no start" condition or poor engine performance. Fuel manufacturers often "winterize" No. 2 diesel fuel by using various fuel additives and blends (No. 1 diesel fuel, kerosene, etc.) to lower its winter-time viscosity. Generally speaking, though, No. 1 diesel fuel is more satisfactory in extremely cold weather.

NOTE: No. 1 and No. 2 diesel fuels will mix and burn with no ill effects, although the engine manufacturer will undoubtedly recommend one or the other. Consult the owner's manual for information.

Depending on local climate, most fuel manufacturers make winterized No. 2 fuel available seasonally.

Many automobile manufacturers publish pamphlets giving the locations of diesel fuel stations nationwide. Contact the local dealer for information.

Do not substitute home heating oil for automotive diesel fuel. While in some cases, home heating oil refinement levels equal those of diesel fuel, many times they are far below diesel engine requirements. The result of using "dirty" home heating oil will be a clogged fuel system, in which case the entire system may have to be dismantled and cleaned.

One more word on diesel fuels. Don't thin diesel fuel with gasoline in cold weather. The lighter gasoline, which is more explosive, will cause rough running at the very least, and may cause extensive engine damage if enough is used.

OIL

Diesel engines require different engine oil from those used in gasoline engines. Besides doing the things gasoline engine oil does, diesel oil must also deal with increased engine heat and the diesel blow-by gases, which create sulphuric acid, a high corrosive.

Under the American Petroleum Institute (API) classifications, gasoline engine oil codes begin with an "S", and diesel engine oil codes begin with a "C". This first letter designation is followed by a second letter code which explains what type of service (heavy, moderate, light) the oil is meant for. For example, the top of a typical oil can will include: "API SERVICES SC, SD, SE, CA, CB, CC". This means the oil in the can is a good, moderate duty engine oil when used in a diesel engine.

It should be noted here that the further

COMPARISON OF #1 AND #2 DIESEL FUEL

Requirement	1-D	2-D
Flash Point, °F minimum	100	125
Cetane Number, minimum	40	40
Viscosity at 100°F, Centistokes		
Minimum	1.4	2.0
Maximum	2.5	4.3
Water and Sediment, % by volume maximum	Trace	0.05
Sulfur, % by weight maximum	0.5	0.5
Ash, % by weight maximum	0.01	0.01

Flash Point: The temperature at which diesel fuel ignites when exposed to a flame *in the open air.*

Cetane Number: See text

down the alphabet the second letter of the API classification is, the greater the oil's protective qualities are (CD is the severest duty diesel engine oil, CA is the lightest duty oil, etc.). The same is true for gasoline engine oil classifications (SF is the severest duty gasoline engine oil, SA is the lightest duty oil, etc.).

Many diesel manufacturers recommend an oil with both gasoline and diesel engine API classifications. Consult the owner's manual for specifications.

The top of the oil can will also contain an SAE (Society of Automotive Engineers) designation, which gives the oil's viscosity. A typical designation will be: SAE 10W-30, which means the oil is a "winter" viscosity oil, meaning it will flow and give protection at low temperatures.

On the diesel engine, oil viscosity is critical, because the diesel is much harder to start (due to its higher compression) than a gasoline engine. Obviously, if you fill the crankcase with a very heavy oil during winter (SAE 20W-50, for example), the starter is going to require a lot of current from the battery to turn the engine. And, since batteries don't function well in cold weather in the first place, you may find yourself stranded some morning. Consult the owner's manual for recommended oil specifications for the climate you live in.

LUBE OIL ANALYSIS

From an oil sample a laboratory can diagnose many potential engine problems—from piston wear to impending bearing failure. What's more, the laboratory can spot them quicker, and with greater accuracy. Just as easily, the lab can give the diesel a clean bill of health, saving the car owner unnecessary servicing and other routine preventive maintenance, costly in time and money.

There's nothing new about engine lube oil analysis. Thousands of the nation's trucks and buses regularly have their engine's lube oil analyzed by laboratories specializing in this type of work. What is new is the availability of lube oil analysis to individual vehicle owners rather than, as before, almost exclusively to companies operating fleets of diesel equipment.

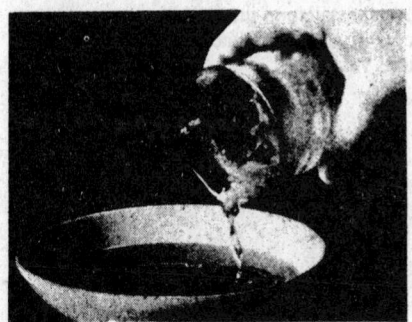

Lube oil analysis can be a valuable indicator of internal engine condition.

Here's how lube oil analysis works. You write one of the several laboratories that offer individual diesel vehicle owners lube analysis service. By return mail you'll receive an oil sampling kit. It will probably contain a two-ounce plastic oil sampling container with a screw-on plastic top. Instructions tell you how to take the sample. Usually, a lab-bound sample of diesel lube oil may be taken in any of three ways, but always right after the engine has been shut off, so that the sampled oil is as close as possible to normal engine operating temperature. That's important to assure that the lab's test will be accurate. Oil samples can be taken during normal oil changes, when lube oil is drained anyway. Between oil changes, a sample can be drawn from the engine through the dipstick tube (where you normally check the oil's level). In drawing an oil sample from the dipstick tube, a small suction bulb fitted with a length of disposable tubing is used. The tubing is merely inserted into the dipstick tube, the suction bulb depressed, and the oil sample drawn. The third method of sampling is by loosening the drain plug on the engine's bypass oil filter (if your diesel has one). A little oil is caught in the lube sampling container. In all cases, extreme cleanliness is a must, so as not to contaminate the sample with dirt, grease, or other substances not actually found inside the engine. For example, using a rag that contains solvents, metal filings, or other impurities can contaminate the oil sample, leading to false and even alarming lab reports. A bit of technique is required: In taking a sample of lube oil during a routine oil drain, about half of the crankcase's lube oil should be allowed to drain out before the sample is taken. The sample taken, the date, make and model of the engine, its mileage, mileage since last oil change, and sometimes oil type are noted on the container's label, and the container is mailed to the laboratory.

Shortly, you'll receive the lab's report, which, based on a number of tests, including spectrochemical analysis (using a spectrometer, which can detect the presence of virtually all basic elements and contaminants), tells what's in the oil in what quantities and analyzes both the probable source of what was found and whether it indicates trouble. For one example, the finding of more than trace amounts of copper in an oil sample may strongly point to excessive bearing wear in a particular diesel whose bearings contain copper. Some analyses report on as many as eighteen basic elements that may be found in a diesel's lube oil sample, and in the report's "recommendation" may pinpoint their probable source—as, "indicates piston ring wear." Also indicated is the presence of such contaminants as water, solids (the products of oxidation and engine blow-by), and fuel dilution. Noted, too, is the lubricity of the sample—whether, or not, in the lab's opinion, it is still doing its internal engine lubricating job.

NOTE: Never use lube analysis and a lab's report of "good oil" to extend, beyond the manufacturer's recommendation, the mileage period between oil changes. Follow the manufacturers recommendations.

The more frequently an engine is lube-sampled, the more accurate and meaningful the lab's reports. Infrequent samplings, although they can spot sudden, unusual changes in internal engine condition, may fail to show the gradual deterioration of engine parts. Ideally, you should have the laboratory analyze a lube sample every other oil change. For most automobile diesels, that's every 6,000 miles. Analysis costs from $7 to $11 per sample. Drive an average 18,000 miles a year and you'd change your diesel's oil three times. In that time, you'd submit three samples to the lab at an annual lube analysis cost of $21 to $33.

Aftermarket Fuel System Accessories

Due to reasons described previously, most diesel engine problems can be attributed to either fuel contamination or cold weather fuel performance characteristics. Diesel-engined vehicle manufacturers have designed and installed various systems to combat these problems, but ultimately, their best efforts are limited by cost.

Inconvenience is a major concern to diesel owners. If water accumulates (in substantial quantities) in the diesel fuel system, the fuel and water must be siphoned from the fuel tank and purged from the remainder of the fuel system. It goes without saying that this operation is a messy, time-consuming process. Even if the vehicle is equipped with a water/fuel separator having a drain valve, the owner must manually open the valve from either under the hood or beneath the vehicle.

Although the fuel filter installed by the manufacturer offers adequate performance when maintained properly, the addition of another, separate diesel fuel filter is a wise improvement.

If you live in an extremely cold climate, you've probably experienced cold starting problems due to fuel "waxing", plugged filters, "gelled" fuel, etc. If your vehicle is not factory-equipped with the optional fuel line or cylinder block heaters, these

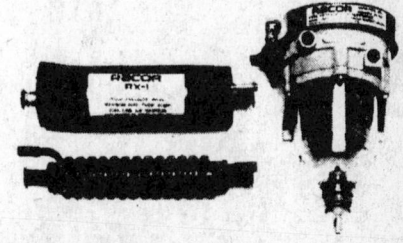

Aftermarket fuel filter/water separator and fuel line heater

heaters can be purchased from the aftermarket (retail auto parts manufacturers). The installation of either of these items can improve cold-starting dramatically.

WATER/FUEL SEPARATORS

Centrifugal Action

Sometimes referred to as a "cyclonic" water/fuel separator, this device uses baffles which spin the fuel as it comes through the separator inlet. Since water is heavier than diesel fuel, the water will spin away from the fuel, sink to the bottom of the separator, and collect in the sediment bowl.

This type of separator is most efficient in dealing with large water droplets. If the water is in emulsion with the fuel, that is, if the water is equally dispersed through the fuel in very small droplets, some of the water will remain with the fuel to travel through the fuel system.

Coalescing Action

In this type of separator, the fuel must pass through a coalescent filtering media before proceeding through the fuel system. The idea behind the coalescent media is to trap even the smallest droplets of water on the media. As the small droplets combine into larger, heavier droplets, gravity acts on the droplets to pull them downward, off of the media and into the sediment bowl.

FUEL FILTER/SEPARATOR COMBINATION UNITS

Most separators of either the centrifugal or coalescent types are available with disposeable fuel filtering elements which are built into the separator unit. If your car already has a large, disposeable filter, it would probably be more cost-effective to stay with a separator only, and to change the factory-equipped filter at the recommended intervals. Should your vehicle have a fairly small filter, and/or an inconveniently located water drain (or none at all), choose the filter/separator combination. The filter/separator offers both increased fuel filtering ability and efficient water separation.

Convenience Add-Ons

Available with many separators and filter/separators are items such as dash-mounted water-in-fuel indicator lamps, audible water-in-fuel alarms, and dash-controlled water ejection systems. A properly chosen system would warn you of water in the fuel, and allow you to eject the water by simply "flipping" a dash-mounted switch.

Installing a Separator

Clear installation instructions and the necessary installation parts will be provided with the separator kit. Follow those instructions exactly. A general list of suggestions follows:

1. Fuel additives should not be used unless approved by the separator manufacturer.

2. Do not install a separator within 4″ of any exhaust system component.

3. If plastic fittings are supplied with the kit, do not replace them with metal fittings. Also, use extreme caution when tightening the fittings, especially those made of plastic.

4. Use a fuel-proof sealer on all fitting threads, only if the threads are not factory-coated with sealer.

5. Use only fuel-proof hoses for the installation.

6. Do not eliminate the original equipment fuel filter, even if a filter/separator is installed.

7. For new car warranty purposes, a filter/separator should be located BEFORE the original equipment filter. The fuel must pass through the original filter last, before entering the fuel injection pump.

8. If any type of fuel line heater is installed, it is best to position the heater between the fuel tank and the separator inlet.

9. To ease the job of the separator, the separator should be installed between the fuel transfer pump and the tank (unless the separator manufacturer specifies otherwise). Fuel and water which have been churned through the fuel transfer pump will be more difficult to separate.

10. Be sure that any wiring (for warning lamps, water ejection, etc.) is routed and connected properly. If the wiring must pass through a drilled hole, be sure to use a rubber grommet between the drilled component(s) and the wire to prevent damage to the wire.

FUEL LINE HEATERS

Two popular types of fuel line heaters are available for diesel passenger cars. Both types raise the temperature of the fuel to prevent "waxing" and "gelling" of the fuel in the lines during cold weather operation. One type uses engine coolant as a heating source. In order for this type to heat the fuel, the engine must first be started and allowed to run until the coolant temperature increases. Though this type of heater will usually increase fuel mileage, it offers no aid in starting ability.

The other type of heater uses a 12V DC electric heating element. This type is recommended, due to its ability to warm the fuel BEFORE the engine is started. This type of heater will also usually increase the overall fuel mileage.

Installation

Follow the manufacturer's instructions exactly. Also, see suggestions 5, 8, and 9 under "Separator Installation".

CYLINDER BLOCK HEATERS

A cylinder block heater electrically (usually 110V house current) heats the engine coolant, which in turn warms the cylinder block, heads, and engine oil. In this case, the warmth is not used to alter the characteristics of the fuel. Block heaters offer two main advantages when starting a diesel in cold weather:

1. The reduced viscosity (thinning) of the engine oil from the warmth allows the engine to be "turned over" easier (and faster) by the starter. Less strain is imposed on the starting system.

2. Because the diesel relies on the heat of compression to ignite the fuel, the increase in the base combustion chamber temperature results in a higher tempearture during compression. This allows the fuel to ignite easier than if just the glow plugs were used.

Installation

Most cylinder block heaters replace one of the existing freeze (or expansion) plugs of the cylinder block. Follow the manufacturers installation instructions exactly. Also, refer to the manufacturers recommendations for usage.

Carburetors 20

DOMESTIC CARBURETOR SERVICE

Carburetor Identification

All carburetors are identified by code numbers, either stamped on the attaching flange side, the main body or on a metal tag retained by a bowl cover screw. This identification number is important in order to obtain the correct carburetor replacement or parts and to properly adjust the carburetor when matched to a specific engine.

Special Tools

An angle degree tool is recommended by Rochester Products Division for use to confirm adjustments to the choke valve and related linkages on late model 2 and 4 barrel carburetors in place of the plug type gauges. Decimal and degree conversion charts are provided for use with the angle degree tool. To use the angle gauge, rotate the degree scale until zero (0) is opposite the pointer. With the choke valve completely closed, place the gauge magnet squarely on top of the choke valve and rotate the bubble until it is centered. Make the necessary adjustments to have the choke valve at the specified degree angle opening as read from the degree angle tool. The carburetor may be off the engine for adjustments, but make sure the carburetor is held firmly during the use of the angle gauge.

A variety of other special adjustment tools may be necessary during the overhaul of different carburetors covered in this section. Most carburetor overhaul kits contain the float level gauges and specifications necessary for complete rebuilding, and if specifications differ from those given in the following charts, use the values listed in the overhaul instructions with a specific kit. Before beginning any overhaul procedures, read through each section to make sure all required special tools are on hand in order to complete the repair.

Carburetor Overhaul Tips

When the carburetor is disassembled, wash all parts (except diaphragms, electric choke units, pump plunger, and any other plastic, leather, fiber, or rubber parts) in clean carburetor solvent. Do not leave parts in the solvent any longer than is necessary to sufficiently loosen the deposits. Excessive cleaning may remove the special finish from the float bowl and choke valve bodies, leaving these parts unfit for service. Rinse all parts in clean solvent and blow them dry with compressed air or allow them to air dry. Wipe clean all cork, plastic, leather, and fiber parts with a clean, lint-free cloth.

Blow out all passages and jets with compressed air and be sure that there are no restrictions or blockages. Never use wire or similar tools to clean jets, fuel passages, or air bleeds. Clean all jets and valves separately to avoid acci-

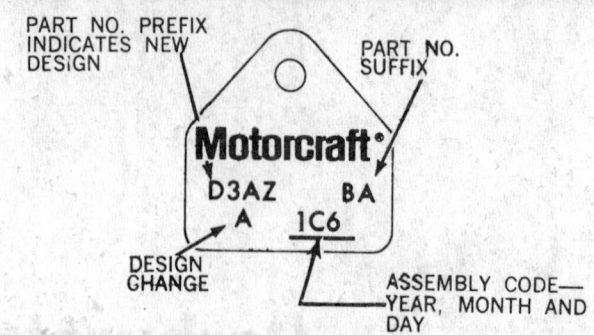

Motorcraft carburetors for Ford usage—typical

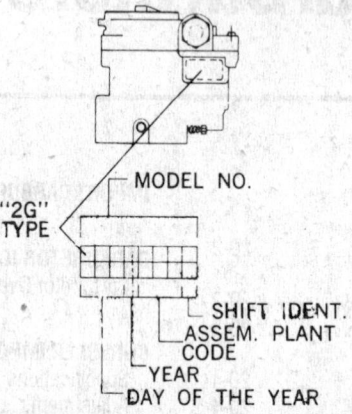

Rochester two barrel models—typical

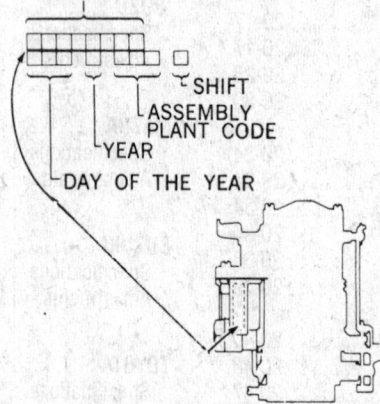

Rochester one barrel models—typical

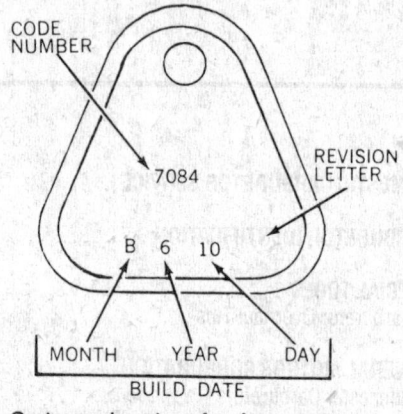

Carter carburetors for Jeep usage—typical

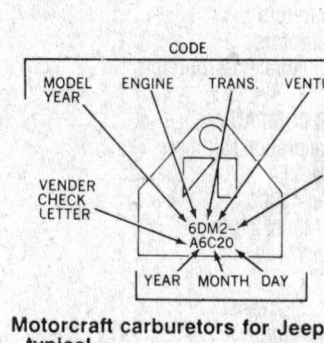

Motorcraft carburetors for Jeep usage—typical

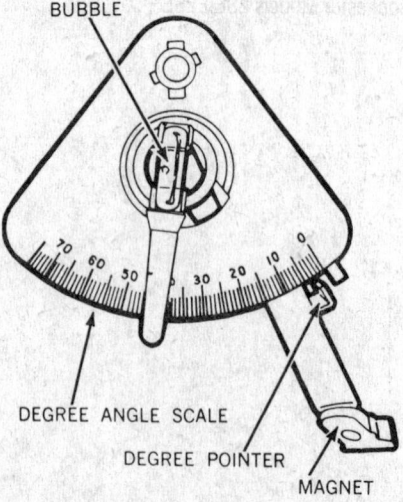

Typical degree angle tool

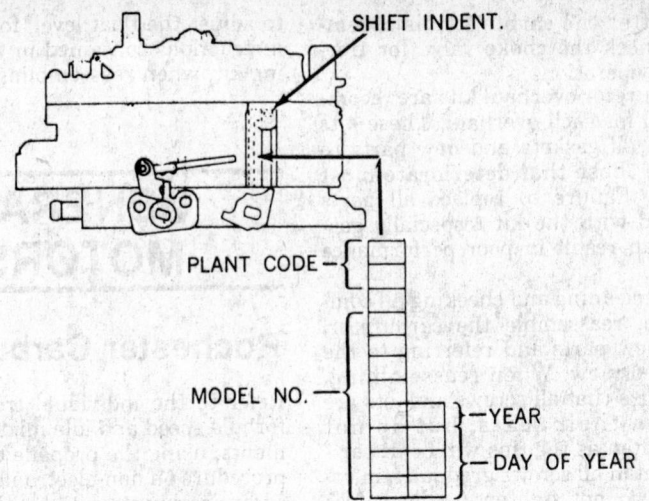

SHIFT INDENT.

PLANT CODE

MODEL NO.

— YEAR

— DAY OF YEAR

Rochester four barrel models – typical

dental interchange. Check all parts for wear or damage. If wear or damage is found, replace the defective parts. Especially check the following:

1. Check the float needle and seat for wear. If wear is found, replace the complete assembly.

2. Check the float hinge pin for wear and the float(s) for dents or distortion. Replace the float if fuel has leaked into it.

3. Check the throttle and choke shaft bores for wear or an out-of-round condition. Damage or wear to the throttle arm, shaft, or shaft bore will often require replacement of the throttle body. These parts require a close tolerance of fit. Wear may allow air leakage, which could affect starting and idling.

NOTE: Throttle shafts and bushings are not included in over-

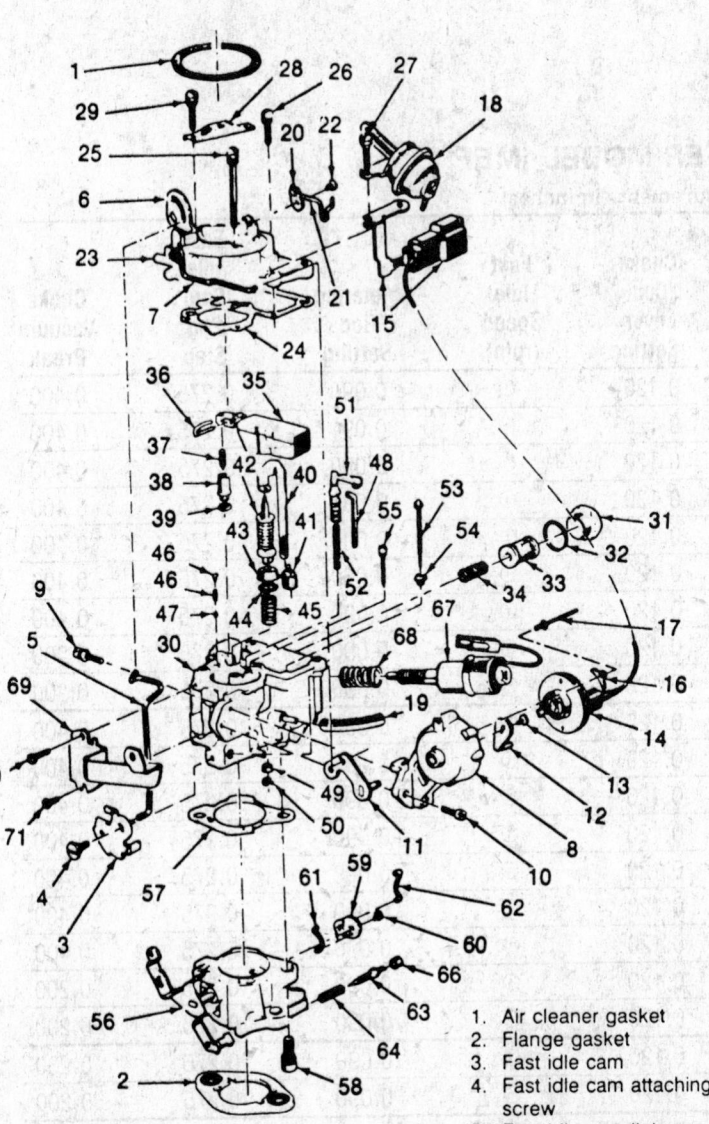

6. Choke shaft, lever and link assembly
7. Choke link
8. Choke housing and bearing assembly
9. Choke housing attaching screw
10. Choke housing attaching screw
11. Choke shaft and lever assembly
12. Choke stat lever
13. Stat lever attaching screw
14. Electric choke cover and stat assembly
15. Connector and bracket assembly
16. Choke cover retainer
17. Choke cover attaching rivet
18. Bowl side vacuum breat assembly
19. Vacuum Break Hose
20. Vacuum break lever and link assembly
21. Vacuum break link
22. Lever attaching screw
23. Air horn assembly
24. Air horn to float bowl gasket
25. Air horn to float bowl (long) screw assembly
26. Air horn to float bowl screw assembly
27. Air horn to float bowl (countersunk) screw
28. Air cleaner bracket
29. Air cleaner bracket attaching screw assembly
30. Float bowl assembly
31. Fuel inlet nut
32. Fuel inlet nut gasket
33. Fuel inlet filter

34. Fuel filter spring
35. Float
36. Float hinge pin
37. Float needle
38. Float needle seat
39. Float needle gasket
40. Pump rod
41. Pump rod seal
42. Pump assembly
43. Pump plunger B cup
44. Pump plunger spring
45. Pump return spring
46. Pump discharge spring guide
47. Pump discharge ball
48. Power piston rod
49. Power piston rod seal
50. Power piston rod seal retainer
51. Power valve piston assembly
52. Power piston spring
53. Metering rod and spring assembly
54. Main metering jet
55. Idle tube assembly
56. Throttle body assembly
57. Float bowl to throttle body gasket
58. Float bowl to throttle body screw assembly
59. Pump and power rod lever
60. Pump lever attaching screw
61. Power rod link
62. Pump link
63. Idle misture needle
64. Idle misture needle spring
65. Idle misture needle limiter
66. Idle misture needle plug
67. Idle stop solenoid
68. Idle stop solenoid spring
69. Throttle return spring anchor bracket
70. Bracket attaching (countersunk) screw
71. Bracket attaching screw

1. Air cleaner gasket
2. Flange gasket
3. Fast idle cam
4. Fast idle cam attaching screw
5. Fast idle cam link

Rochester 1MEF carburetor – exploded view

haul kits. They can be purchased separately.

4. Inspect the idle mixture adjusting needles for burrs or grooves. Any such condition requires replacement of the needle, since you will not be able to obtain a satisfactory idle.

5. Test the accelerator pump check valves. They should pass air one way but not the other. Test for proper seating by blowing and sucking on the valve. Replace the valve if necessary. If the valve is satisfactory, wash the valve again to remove breath moisture.

6. Check the bowl cover for warped surfaces with a straight edge.

7. Closely inspect the valves and seats for wear and damage, replacing as necessary.

8. After the carburetor is assembled, check the choke valve for freedom of operation.

Carburetor overhaul kits are recommended for each overhaul. These kits contain all gaskets and new parts to replace those that deteriorate most rapidly. Failure to replace all parts supplied with the kit (especially gaskets) can result in poor performance later.

After cleaning and checking all components, reassemble the carburetor, using new parts and referring to the exploded view. When reassembling, make sure that all screws and jets are tight in their seats, but do not overtighten as the tips will be distorted. Tighten all screws gradually, in rotation. Do not tighten needle valves into their seats. Uneven jetting will result. Always use new gaskets. Be sure

to adjust the float level, following the instructions contained in the rebuilding kit, when reassembling.

GENERAL MOTORS

Rochester Carburetors

Refer to the individual truck section for idle speed and idle mixture adjustments, using the propane enrichment procedure on non-electronic controlled engine carburetors and with the use of a dwellmeter on the electronic controlled carburetor equipped engines.

ROCHESTER MODEL 1MEF

(All measurements in inches)

Year	Carburetor Number	Float Level	Choke Unloader Setting	Choke Coil Lever Setting	Fast Idle Speed (rpm)	Metering Rod Setting	Fast Idle Cam 2nd Step	Choke Vacuum Break
1986	17081009	$^{11}/_{32}$	0.520	0.120	①	0.090	0.275	0.400
	17084329	$^{11}/_{32}$	0.520	0.120	①	0.090	0.275	0.400
	17085009	$^{11}/_{32}$	0.520	0.120	①	0.090	0.275	0.400
	17085036	$^{11}/_{32}$	0.520	0.120	①	0.090	0.275	0.400
	17085044	$^{11}/_{32}$	0.520	0.120	①	0.090	0.275	0.400
	17085045	$^{11}/_{32}$	0.520	0.120	①	0.090	0.275	0.400
	17086096	$^{11}/_{32}$	0.520	0.120	①	0.090	0.275	0.400
	17086101	$^{11}/_{32}$	0.520	0.120	①	0.090	0.275	0.200
	17086102	$^{11}/_{32}$	0.520	0.120	①	0.090	0.275	0.200
1987	17081009	$^{11}/_{32}$	0.520	0.120	①	0.090	0.275	0.400
	17084329	$^{11}/_{32}$	0.520	0.120	①	0.090	0.275	0.400
	17085009	$^{11}/_{32}$	0.520	0.120	①	0.090	0.275	0.400
	17085036	$^{11}/_{32}$	0.520	0.120	①	0.090	0.275	0.400
	17085044	$^{11}/_{32}$	0.520	0.120	①	0.090	0.275	0.400
	17085045	$^{11}/_{32}$	0.520	0.120	①	0.090	0.275	0.400
	17086096	$^{11}/_{32}$	0.520	0.120	①	0.090	0.275	0.400
	17086101	$^{11}/_{32}$	0.520	0.120	①	0.090	0.275	0.200
	17086102	$^{11}/_{32}$	0.520	0.120	①	0.090	0.275	0.200
1988	17086096	$^{11}/_{32}$	0.520	0.120	①	0.090	0.275	0.200
	17086101	$^{11}/_{32}$	0.520	0.120	①	0.090	0.275	0.200
1989	17086101	$^{11}/_{32}$	0.520	0.120	①	0.090	0.275	0.200

① See emission label under hood

MODEL 1MEF

Rochester 1MEF float level adjustment

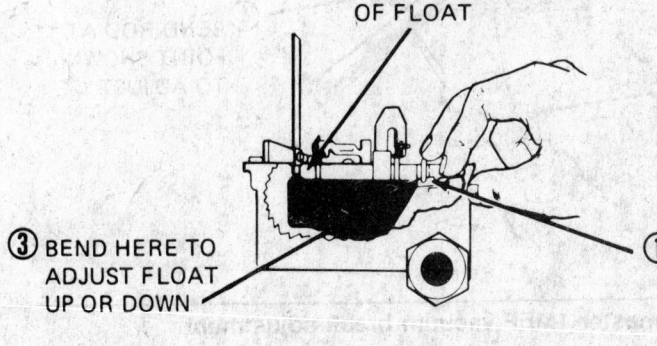

② GAUGE FROM TOP OF CASTING TO TOP OF INDEX POINT AT TOE OF FLOAT

③ BEND HERE TO ADJUST FLOAT UP OR DOWN

① HOLD FLOAT RETAINING PIN FIRMLY IN PLACE — PUSH DOWN ON END OF FLOAT ARM, AGAINST TOP OF FLOAT NEEDLE

Rochester 1MEF metering rod adjustment

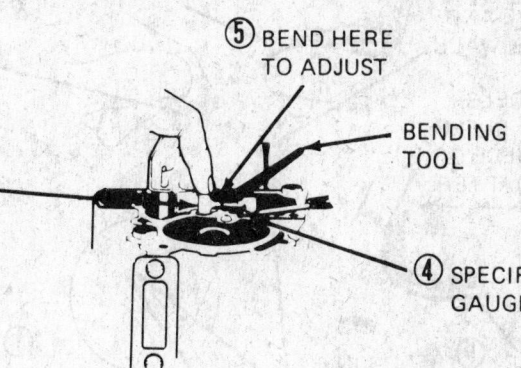

③ HOLD POWER PISTON DOWN AND SWING METERING ROD HOLDER OVER FLAT SURFACE (GASKET REMOVED) OF BOWL CASTING NEXT TO CARBURETOR BORE

⑤ BEND HERE TO ADJUST

BENDING TOOL

① REMOVE METERING ROD BY HOLDING THROTTLE VALVE WIDE OPEN. PUSH DOWNWARD ON METERING ROD AGAINST SPRING TENSION, THEN SLIDE METERING ROD OUT OF SLOT IN HOLDER AND REMOVE FROM MAIN METERING JET.

④ SPECIFIED PLUG GAUGE — SLIDE FIT

② BACK OUT IDLE STOP SOLENOID — HOLD THROTTLE VALVE COMPLETELY CLOSED

Rochester 1MEF choke coil lever adjustment

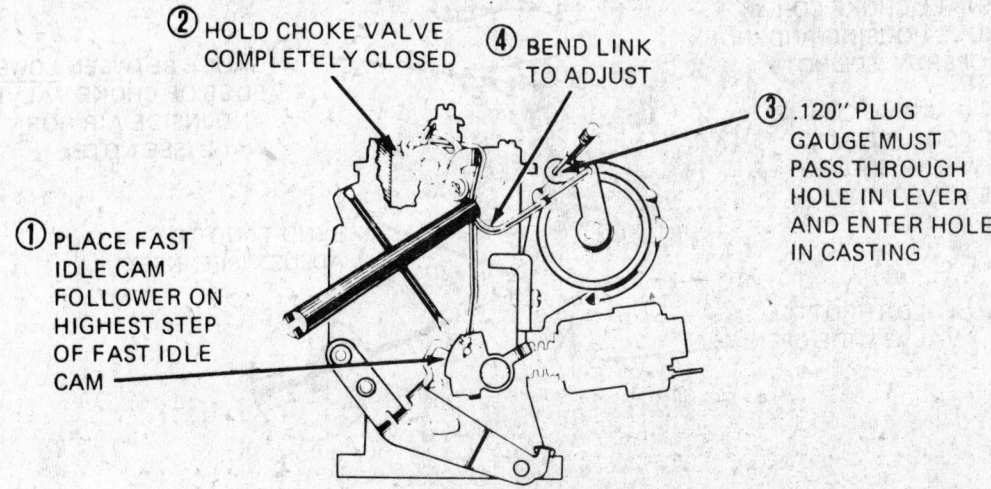

② HOLD CHOKE VALVE COMPLETELY CLOSED

④ BEND LINK TO ADJUST

③ .120" PLUG GAUGE MUST PASS THROUGH HOLE IN LEVER AND ENTER HOLE IN CASTING

① PLACE FAST IDLE CAM FOLLOWER ON HIGHEST STEP OF FAST IDLE CAM

Rochester 1MEF fast idle cam adjustment

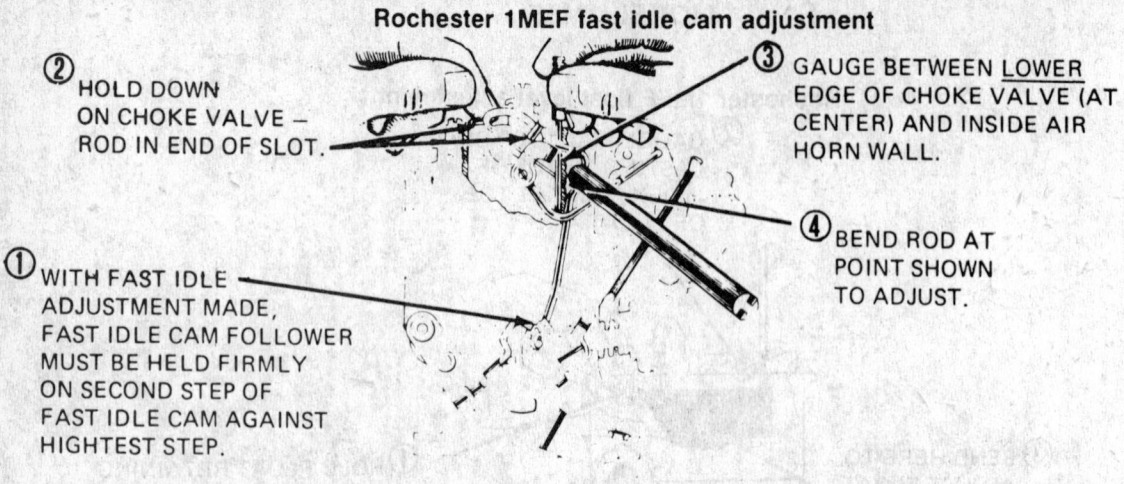

② HOLD DOWN ON CHOKE VALVE – ROD IN END OF SLOT.

③ GAUGE BETWEEN LOWER EDGE OF CHOKE VALVE (AT CENTER) AND INSIDE AIR HORN WALL.

④ BEND ROD AT POINT SHOWN TO ADJUST.

① WITH FAST IDLE ADJUSTMENT MADE, FAST IDLE CAM FOLLOWER MUST BE HELD FIRMLY ON SECOND STEP OF FAST IDLE CAM AGAINST HIGHTEST STEP.

Rochester 1MEF vacuum break adjustment

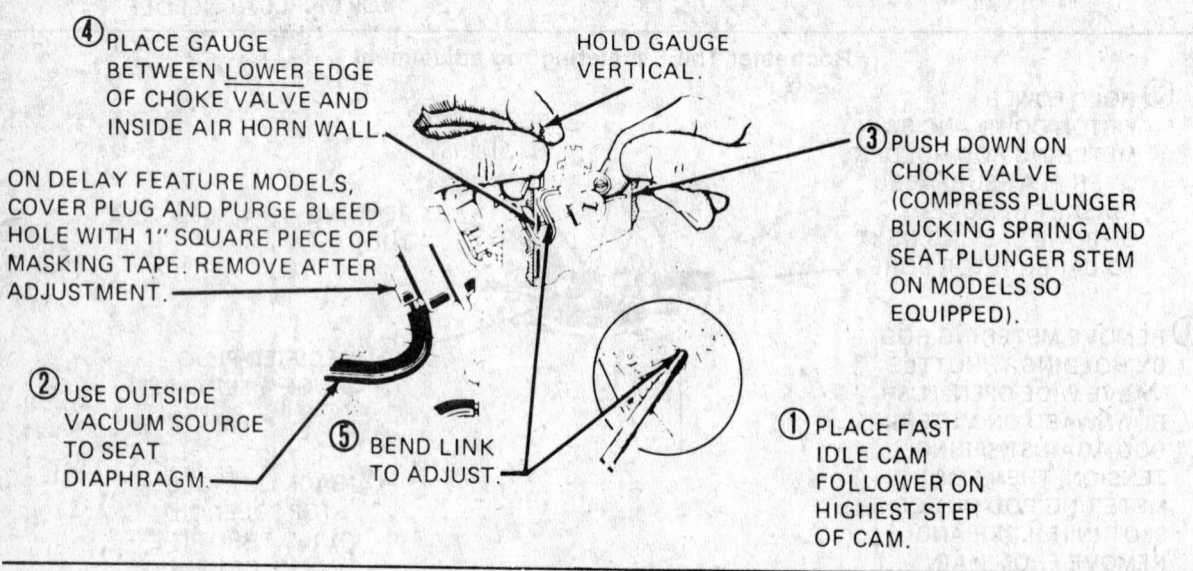

④ PLACE GAUGE BETWEEN LOWER EDGE OF CHOKE VALVE AND INSIDE AIR HORN WALL.

ON DELAY FEATURE MODELS, COVER PLUG AND PURGE BLEED HOLE WITH 1" SQUARE PIECE OF MASKING TAPE. REMOVE AFTER ADJUSTMENT.

HOLD GAUGE VERTICAL.

③ PUSH DOWN ON CHOKE VALVE (COMPRESS PLUNGER BUCKING SPRING AND SEAT PLUNGER STEM ON MODELS SO EQUIPPED).

② USE OUTSIDE VACUUM SOURCE TO SEAT DIAPHRAGM.

⑤ BEND LINK TO ADJUST.

① PLACE FAST IDLE CAM FOLLOWER ON HIGHEST STEP OF CAM.

Rochester 1MEF unloader adjustment

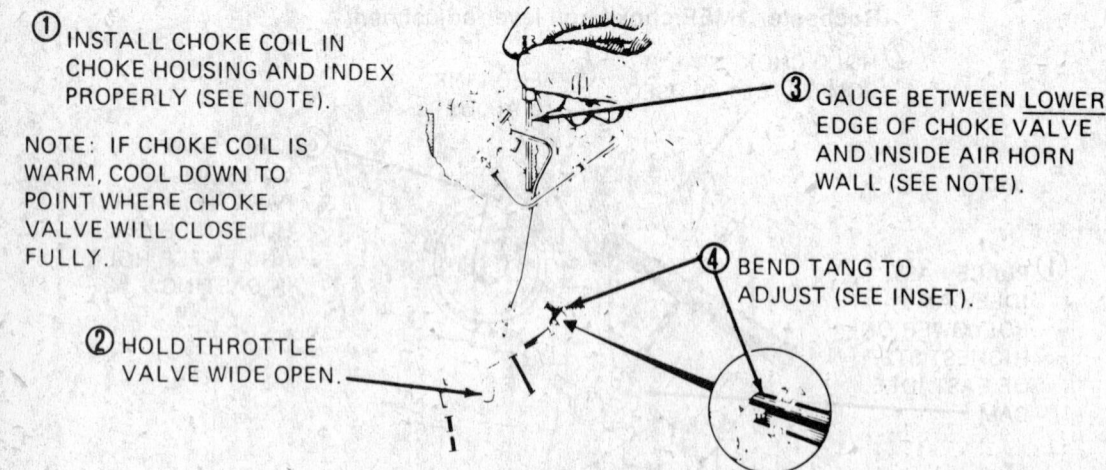

① INSTALL CHOKE COIL IN CHOKE HOUSING AND INDEX PROPERLY (SEE NOTE).

NOTE: IF CHOKE COIL IS WARM, COOL DOWN TO POINT WHERE CHOKE VALVE WILL CLOSE FULLY.

③ GAUGE BETWEEN LOWER EDGE OF CHOKE VALVE AND INSIDE AIR HORN WALL. (SEE NOTE).

④ BEND TANG TO ADJUST (SEE INSET).

② HOLD THROTTLE VALVE WIDE OPEN.

ROCHESTER MODEL M4ME, M4MED, M4MEF QUADRAJET
General Motors Corporation
(All measurements in inches or degrees)

Year	Carburetor Number	Float Level	Pump Rod Hole	Pump Rod Setting	Choke Rod ① Setting	Air Valve Rod	Vacuum Break Front	Vacuum Break Rear	Air Valve Turns	Choke Unloader	Propane Enrichment (rpm)
1986	17084500	12/32	inner	9/32	37°	0.025	23°	30°	1	40°	②
	17084501	12/32	inner	9/32	37°	0.025	23°	30°	1	40°	②
	17084502	12/32	inner	9/32	46°	0.025	24°	30°	7/8	40°	②
	17085000	12/32	inner	9/32	46°	0.025	24°	30°	7/8	40°	②
	17085001	12/32	inner	9/32	46°	0.025	23°	30°	1	40°	②
	17085003	13/32	inner	9/32	46°	0.025	23°	—	7/8	35°	②
	17085004	13/32	inner	9/32	46°	0.025	23°	—	7/8	35°	②
	17085205	13/32	inner	9/32	20°	0.025	26°	38°	7/8	39°	②
	17085206	13/32	inner	9/32	46°	0.025	—	26°	7/8	39°	20
	17085208	13/32	inner	9/32	20°	0.025	26°	38°	7/8	39°	10
	17085209	13/32	outer	3/8	20°	0.025	26°	36°	7/8	39°	50
	17085210	13/32	inner	9/32	20°	0.025	26°	38°	7/8	39°	10
	17085211	13/32	outer	3/8	20°	0.025	26°	36°	7/8	39°	50
	17085212	13/32	inner	9/32	46°	0.025	23°	—	7/8	35°	②
	17085213	13/32	inner	9/32	46°	0.025	23°	—	7/8	35°	②
	17085215	13/32	inner	9/32	46°	0.025	—	26°	7/8	32°	②
	17085216	13/32	inner	9/32	20°	0.025	26°	38°	7/8	39°	②
	17085217	13/32	inner	9/32	20°	0.025	26°	36°	1/2	39°	②
	17085219	13/32	inner	9/32	20°	0.025	26°	36°	1/2	39°	②
	17085220	13/32	outer	3/8	20°	0.025	—	26°	7/8	32°	75
	17085221	13/32	outer	3/8	20°	0.025	—	26°	7/8	32°	75
	17085222	13/32	inner	9/32	20°	0.025	26°	36°	1/2	39°	20
	17085223	13/32	outer	3/8	20°	0.025	26°	36°	1/2	39°	50
	17085224	13/32	inner	9/32	20°	0.025	26°	36°	1/2	39°	20
	17085225	13/32	outer	3/8	20°	0.025	26°	36°	1/2	39°	50
	17085226	13/32	inner	9/32	20°	0.025	—	24°	7/8	32°	20
	17085227	13/32	inner	9/32	20°	0.025	—	24°	7/8	32°	20
	17085228	13/32	inner	9/32	46°	0.025	—	24°	7/8	39°	30
	17085229	13/32	inner	9/32	46°	0.025	—	24°	7/8	39°	30
	17085230	13/32	inner	9/32	20°	0.025	—	26°	7/8	32°	20
	17085231	13/32	inner	9/32	20°	0.025	—	26°	7/8	32°	40
	17085235	13/32	inner	9/32	46°	0.025	—	26°	7/8	39°	80
	17085238	13/32	outer	3/8	20°	0.025	—	26°	7/8	32°	75
	17085239	13/32	outer	3/8	20°	0.025	—	26°	7/8	32°	75
	17085283	13/32	inner	9/32	20°	0.025	—	24°	7/8	32°	20
	17085284	13/32	inner	9/32	20°	0.025	—	26°	7/8	32°	20

ROCHESTER MODEL M4ME, M4MED, M4MEF QUADRAJET
General Motors Corporation
(All measurements in inches or degrees)

Year	Carburetor Number	Float Level	Pump Rod Hole	Pump Rod Setting	Choke Rod ① Setting	Air Valve Rod	Vacuum Break Front	Vacuum Break Rear	Air Valve Turns	Choke Unloader	Propane Enrichment (rpm)
1986	17085285	13/32	inner	9/32	20°	0.025	—	24°	7/8	32°	20
	17085290	13/32	inner	9/32	46°	0.025	—	24°	7/8	39°	30
	17085291	13/32	outer	3/8	46°	0.025	—	26°	7/8	39°	100
	17085292	13/32	inner	9/32	46°	0.025	—	24°	7/8	39°	30
	17085293	13/32	outer	3/8	46°	0.025	—	26°	7/8	39°	100
	17085294	13/32	inner	9/32	46°	0.025	—	26°	7/8	39°	②
	17085298	13/32	inner	9/32	46°	0.025	—	26°	7/8	39°	②
1987	17084500	12/32	inner	9/32	37°	0.025	23°	30°	1	40°	②
	17084501	12/32	inner	9/32	37°	0.025	23°	30°	1	40°	②
	17084502	12/32	inner	9/32	46°	0.025	24°	30°	7/8	40°	②
	17085000	12/32	inner	9/32	46°	0.025	24°	30°	7/8	40°	②
	17085001	12/32	inner	9/32	46°	0.025	23°	30°	1	40°	②
	17085003	13/32	inner	9/32	46°	0.025	23°	—	7/8	35°	②
	17085004	13/32	inner	9/32	46°	0.025	23°	—	7/8	35°	②
	17085205	13/32	inner	9/32	20°	0.025	26°	38°	7/8	39°	②
	17085206	13/32	inner	9/32	46°	0.025	—	26°	7/8	39°	20
	17085208	13/32	inner	9/32	20°	0.025	26°	38°	7/8	39°	10
	17085209	13/32	outer	3/8	20°	0.025	26°	36°	7/8	39°	50
	17085210	13/32	inner	9/32	20°	0.025	26°	38°	7/8	39°	10
	17085211	13/32	outer	3/8	20°	0.025	26°	36°	7/8	39°	50
	17085212	13/32	inner	9/32	46°	0.025	23°	—	7/8	35°	②
	17085213	13/32	inner	9/32	46°	0.025	23°	—	7/8	35°	②
	17085215	13/32	inner	9/32	46°	0.025	—	26°	7/8	32°	②
	17085216	13/32	inner	9/32	20°	0.025	26°	38°	7/8	39°	②
	17085217	13/32	inner	9/32	20°	0.025	26°	36°	1/2	39°	②
	17085219	13/32	inner	9/32	20°	0.025	26°	36°	1/2	39°	②
	17085220	13/32	outer	3/8	20°	0.025	—	26°	7/8	32°	75
	17085221	13/32	outer	3/8	20°	0.025	—	26°	7/8	32°	75
	17085222	13/32	inner	9/32	20°	0.025	26°	36°	1/2	39°	20
	17085223	13/32	outer	3/8	20°	0.025	26°	36°	1/2	39°	50
	17085224	13/32	inner	9/32	20°	0.025	26°	36°	1/2	39°	20
	17085225	13/32	outer	3/8	20°	0.025	26°	36°	1/2	39°	50
	17085226	13/32	inner	9/32	20°	0.025	—	24°	7/8	32°	20
	17085227	13/32	inner	9/32	20°	0.025	—	24°	7/8	32°	20

ROCHESTER MODEL M4ME, M4MED, M4MEF QUADRAJET
General Motors Corporation
(All measurements in inches or degrees)

Year	Carburetor Number	Float Level	Pump Rod Hole	Pump Rod Setting	Choke Rod ① Setting	Air Valve Rod	Vacuum Break Front	Vacuum Break Rear	Air Valve Turns	Choke Unloader	Propane Enrichment (rpm)
1987	17085228	13/32	inner	9/32	46°	0.025	—	24°	7/8	39°	30
	17085229	13/32	inner	9/32	46°	0.025	—	24°	7/8	39°	30
	17085230	13/32	inner	9/32	20°	0.025	—	26°	7/8	32°	20
	17085231	13/32	inner	9/32	20°	0.025	—	26°	7/8	32°	40
	17085235	13/32	inner	9/32	46°	0.025	—	26°	7/8	39°	80
	17085238	13/32	outer	3/8	20°	0.025	—	26°	7/8	32°	75
	17085239	13/32	outer	3/8	20°	0.025	—	26°	7/8	32°	75
	17085283	13/32	inner	9/32	20°	0.025	—	24°	7/8	32°	20
	17085284	13/32	inner	9/32	20°	0.025	—	26°	7/8	32°	20
	17085285	13/32	inner	9/32	20°	0.025	—	24°	7/8	32°	20
	17085290	13/32	inner	9/32	46°	0.025	—	24°	7/8	39°	30
	17085291	13/32	outer	3/8	46°	0.025	—	26°	7/8	39°	100
	17085292	13/32	inner	9/32	46°	0.025	—	24°	7/8	39°	30
	17085293	13/32	outer	3/8	46°	0.025	—	26°	7/8	39°	100
	17085294	13/32	inner	9/32	46°	0.025	—	26°	7/8	39°	②
	17085298	13/32	inner	9/32	46°	0.025	—	26°	7/8	39°	②
1988	17085004	13/32	inner	9/32	46°	0.025	23°	—	7/8	35°	—
	17085212	13/32	inner	9/32	46°	0.025	23°	—	7/8	35°	—
	17088040	13/32	inner	9/32	46°	0.025	27°	—	7/8	35°	—
	17088041	13/32	inner	9/32	46°	0.025	27°	—	7/8	35°	—
1989	17085004	13/32	inner	9/32	46°	0.025	23°	—	7/8	35°	—
	17085212	13/32	inner	9/32	46°	0.025	23°	—	7/8	35°	—

NOTE: Specified angle for use with angle degree tool. Choke coil lever setting is 0.120 in. for all carburetors.
① Second step of fast idle cam
② See underhood specification sticker

MODELS M4ME, M4MED AND M4MEF

GENERAL MOTORS—
ROCHESTER CARBURETOR
SPECIFICATIONS

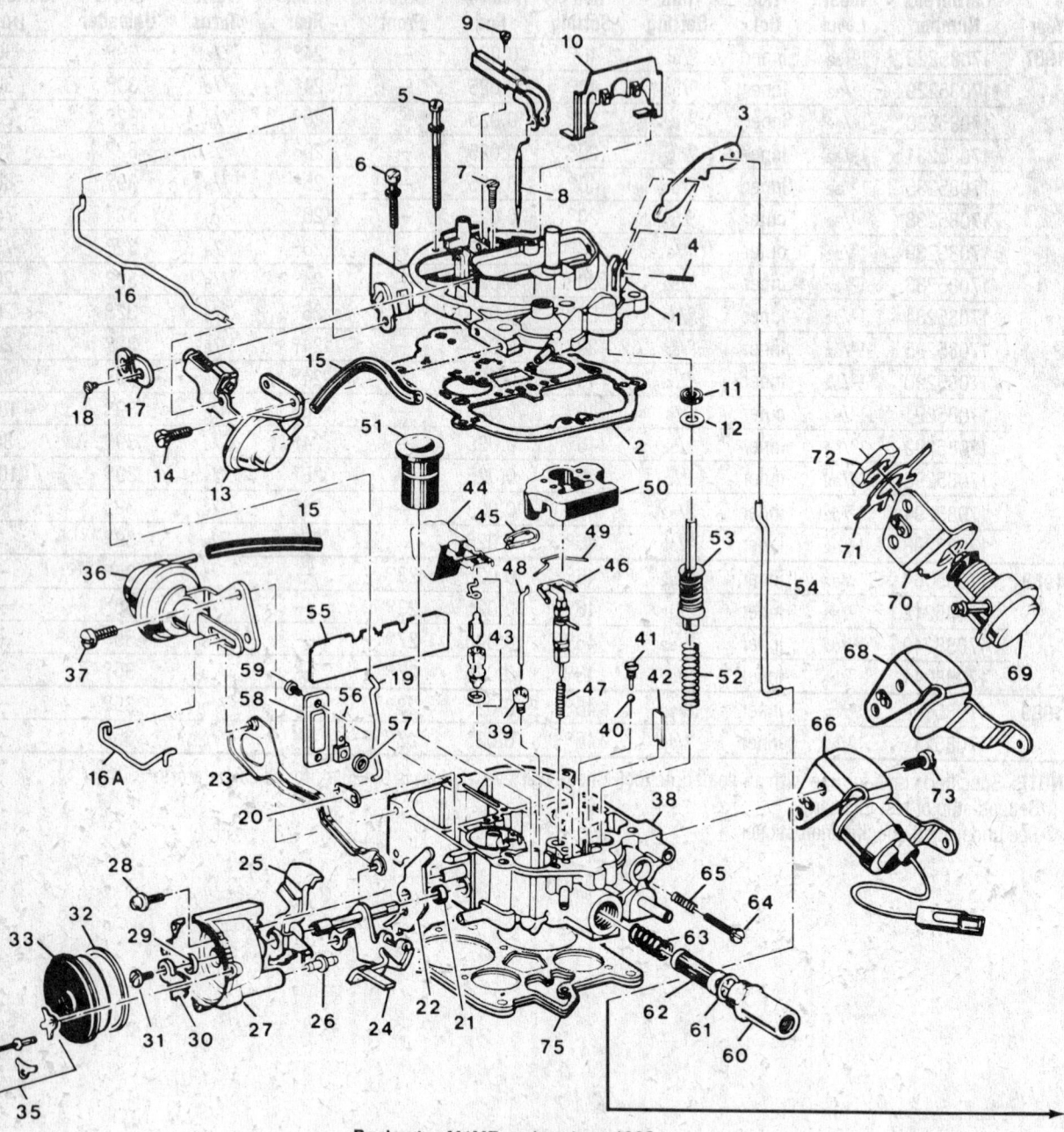

Rochester M4ME carbureter — 1986

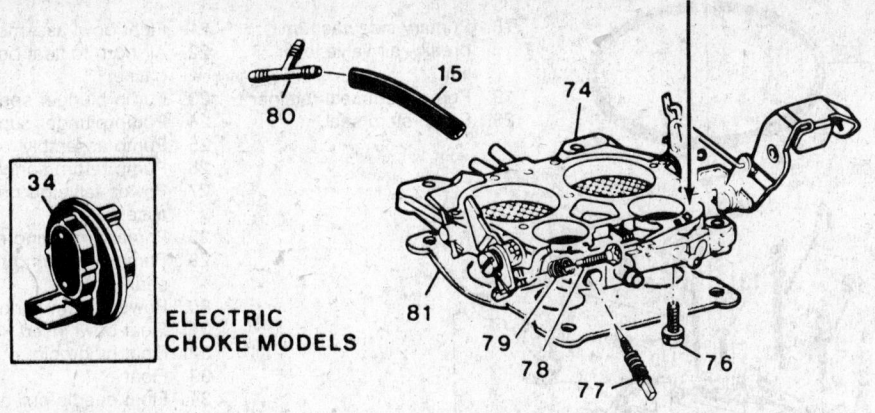

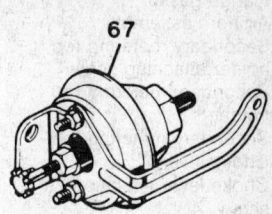

ELECTRIC
CHOKE MODELS

1. Air horn assembly
2. Air horn gasket
3. Pump actuating lever
4. Pump lever hinge roll pin
5. Air horn long (2) screw
6. Air horn short screw
7. Air horn countersunk (2) screw
8. Secondary (2) metering rod
9. Secondary metering rod holder and screw
10. Secondary air baffle
11. Pump plunger seal
12. Pump seal retainer
13. Front vacuum break control and bracket
14. Control attaching (2) screw
15. Vacuum hose
16. Air vavle rod
16a. Air valve rod
17. Choke rod (upper) lever
18. Choke lever screw
19. Choke rod

20. Choke rod (lower) lever
21. Intermediate choke shaft seal
22. Secondary lockout lever
23. Rear vacuum break link
24. Int. choke shaft and lever
25. Fast idle cam
26. Choke housing to bow (hot air choke) seal
27. Choke housing kit
28. Choke housing to bowl screw
29. Intermediate choke shaft (hot air choke) seal
30. Choke coil lever
31. Choke coil lever screw
32. Stat cover (hot air choke) gasket
33. Stat cover and coil assembly (hot air choke)
34. Stat cover and coil assembly (electric choke)
35. Stat cover attaching kit
36. Rear vacuum break assembly

37. Vacuum break attaching (2) screw
40. Pump discharge ball
41. Pump discharge ball retainer
42. Pump well baffle
43. Needle and seat assembly
44. Float assembly
45. Float assembly hinge pin
46. Power piston assembly
47. Power piston spring
48. Primary Metering (2) rod
49. Metering rod retainer spring
50. Float bowl insert
51. Bowl cavity insert
52. Pump return spring
53. Pump assembly
54. Pump rod
55. Secondary bores baffle
56. Idle compensator assembly
57. Idle compensator seal
58. Idle compensator cover
59. Idle compensator cover (2) screw

60. Fuel inlet filter nut
61. Filter nut gasket
62. Fuel inlet filter
63. Fuel filter spring
64. Idle stop srew
65. Idle stop screw spring
66. Idle speed solenoid and bracket assembly
67. Idle load compensator and bracket assembly
68. Throttle return spring bracket
69. Throttle lever actuator
70. Throttle lever actuator bracket
71. Actuator nut washer
72. Actuator attaching nut
73. Bracket attaching (2) screw
74. Throttle body assembly
75. Throttle body gasket
76. Throttle body (3) screw
77. Idle mixture needle and spring assembly (2)
78. Fast idle adjusting screw
79. Fast idle screw spring
80. Vacuum hose tee
81. Flange gasket

1. Air cleaner gasket
2. Flange gasket
3. Air horn assembly
4. Secondary metering rod holder attaching screw
5. Secondary metering rod holder
6. Secondary metering rod
7. Choke lever
8. Choke lever attaching screw
9. Pump lever
10. Pump lever hinge pin

18. Primary side vacuum break – air valve lever link
19. Pump stem seal retainer
20. Pump stem seal

21. Float bowl assembly
22. Air horn to float bowl gasket
23. Pump plunger spring
24. Pump plunger cup
25. Pump assembly
26. Pump return spring
27. Power valve piston assembly
28. Primary metering rod
29. Primary metering rod spring
30. Power piston spring
31. Float bowl insert
32. Float hinge pin
33. Float
34. Float needle pull clip
35. Float needle
36. Float needle seat
37. Float needle seat gasket
38. Primary metering jet
39. Pump discharge (retainer) plug
40. Pump discharge ball
41. Pump well baffle
42. Choke cover attaching rivet
43. Choke cover retainer
44. Electric choke cover and stat assembly
45. Choke housing assembly
46. Choke housing to float bowl screw and washer assembly
47. Choke stat lever attaching screw
48. Choke stat lever
49. Intermediate choke shaft – lever and link assembly
50. Fast idle cam assembly
51. Intermediate choke lever
52. Choke lever
53. Secondary throttle lockout lever
54. Intermediate choke shaft seal
55. Fuel inlet nut
56. Fuel inlet nut gasket
57. Fuel inlet filter
58. Fuel filter spring
59. Throttle stop screw
60. Throttle stop srew spring
61. Throttle body assembly
62. Float bowl to throttle body gasket
63. Float bowl to throttle body screw assembly
64. Pump link
65. Idle mixture needle
66. Idle mixture needle spring
67. Idle mixture needle plug
68. Fast idle adjusting screw
69. Fast idle adjusting screw spring
70. Solenoid and bracket assembly
71. Bracket attaching screw

11. Air horn to throttle body screw assembly
12. Air horn to float bowl screw assembly
13. Air horn to float bowl (countersunk) screw
14. Air horn baffle
15. Primary side (front) vacuum break assembly
16. Primary side (front) vacuum break assembly attaching screw
17. Primary side (front) vacuum break hose

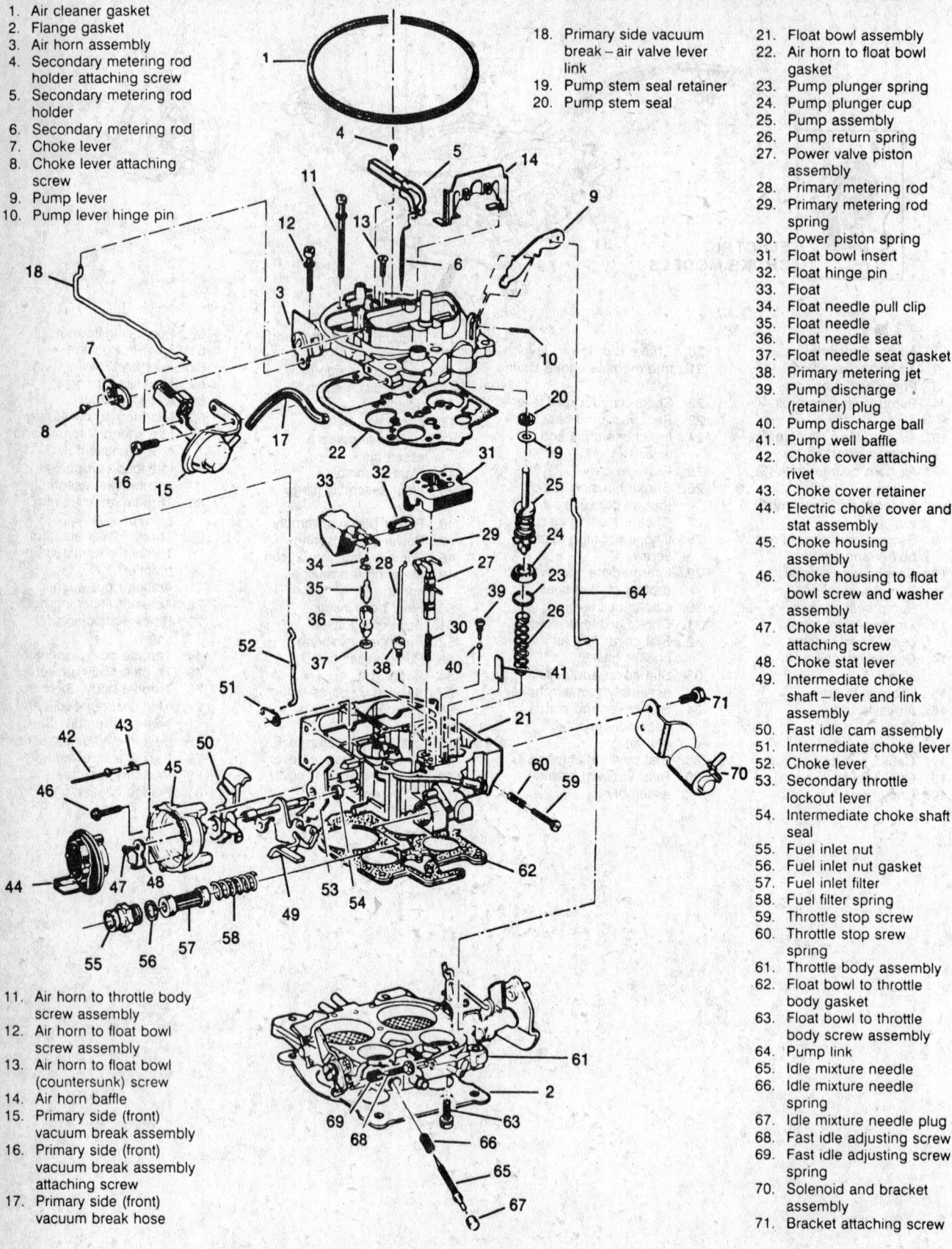

Rochester M4MED and M4MEF carburetor exploded view – 1987–89

1. Air cleaner gasket
2. Flange gasket
3. Air horn assembly
4. Secondary metering rod holder attaching screw
5. Secondary metering rod holder
6. Secondary metering rod
7. Choke lever
8. Choke lever attaching screw

16. Primary side (front) vacuum break assembly attaching screw
17. Primary side (front) vacuum break hose
18. Primary side vacuum break – air valve lever link
19. Pump stem seal retainer
20. Pump stem seal
21. Float bowl assembly

38. Float needle seat gasket
39. Primary metering jet
40. Pump discharge (retainer) plug
41. Pump discharge ball
42. Pump well baffle
43. Secondary side (rear) vacuum break hose
44. Secondary side (rear) vacuum break tee
45. Secondary side (rear) vacuum break assembly
46. Secondary side (rear) vacuum break assembly attaching screw
47. Secondary side (rear) vacuum break to choke link
48. Secondary side vacuum break – air valve lever link
49. Choke cover attaching rivet
50. Choke cover retainer
51. Electric choke cover and stat assembly
52. Choke housing assembly
53. Choke housing to float bowl screw and washer assembly
54. Choke stat lever attaching screw
55. Choke stat lever
56. Intermediate choke shaft – lever and link assembly
57. Fast idle cam assembly
58. Intermediate choke lever
59. Choke lever
60. Secondary throttle lockout lever
61. Intermediate choke shaft seal
62. Fuel inlet nut
63. Fuel inlet nut gasket
64. Fuel inlet filter
65. Fuel filter spring
66. Throttle stop screw
67. Throttle stop srew spring
68. Throttle body assembly
69. Float bowl to throttle body gasket
70. Float bowl to throttle body screw assembly
71. Pump link
72. Idle mixture needle
73. Idle mixture needle spring
74. Idle mixture needle plug
75. Fast idle adjusting screw
76. Fast idle adjusting screw spring
77. Solenoid and bracket assembly
78. Bracket attaching screw
79. Throttle kicker assembly
80. Throttle kicker bracket
81. Throttle kicker assembly attaching nut
82. Tab locking washer

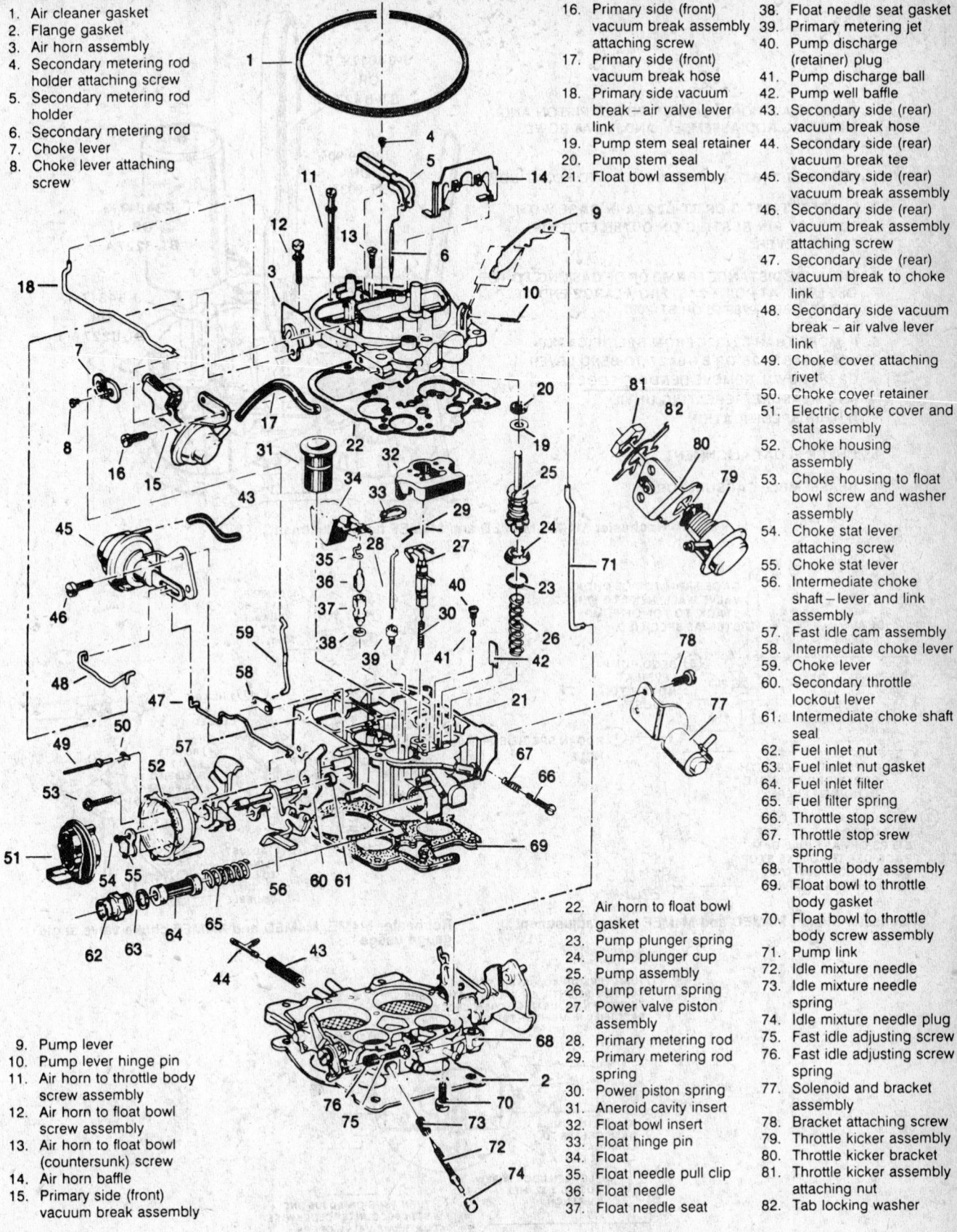

9. Pump lever
10. Pump lever hinge pin
11. Air horn to throttle body screw assembly
12. Air horn to float bowl screw assembly
13. Air horn to float bowl (countersunk) screw
14. Air horn baffle
15. Primary side (front) vacuum break assembly

22. Air horn to float bowl gasket
23. Pump plunger spring
24. Pump plunger cup
25. Pump assembly
26. Pump return spring
27. Power valve piston assembly
28. Primary metering rod
29. Primary metering rod spring
30. Power piston spring
31. Aneroid cavity insert
32. Float bowl insert
33. Float hinge pin
34. Float
35. Float needle pull clip
36. Float needle
37. Float needle seat

Rochester M4ME carburetor exploded view – 1987 California models

1. REMOVE AIR HORN, GASKET, POWER PISTON AND METERING ROD ASSEMBLY, AND FLOAT BOWL INSERT.

2. ATTACH J-34817-1 OR BT-8227A-1 TO FLOAT BOWL.

3. PLACE J-34817-3 OR BT-8227A IN BASE WITH CONTACT PIN RESTING ON OUTER EDGE OF FLOAT LEVER.

4. MEASURE DISTANCE FROM TOP OF CASTING TO TOP OF FLOAT, AT POINT 3/16" FROM LARGE END OF FLOAT. USE J-9789-90 OR BT-8037.

5. IF MORE THAN ±2/32" FROM SPECIFICATION, USE J-34817-25 OR BT-8427 TO BEND LEVER UP OR DOWN. REMOVE BENDING TOOL AND MEASURE, REPEATING UNTIL WITHIN SPECIFICATION.

6. CHECK FLOAT ALIGNMENT.

7. REASSEMBLE CARBURETOR.

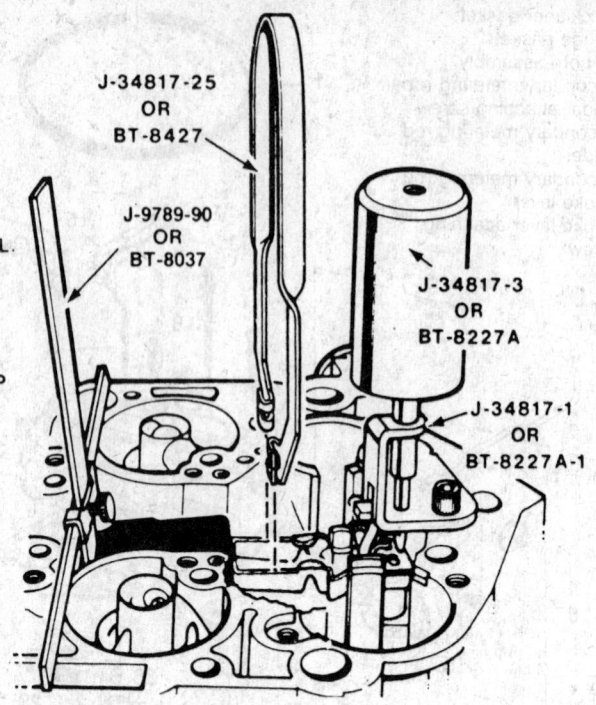

Rochester M4ME, M4MED and M4MEF float adjustment

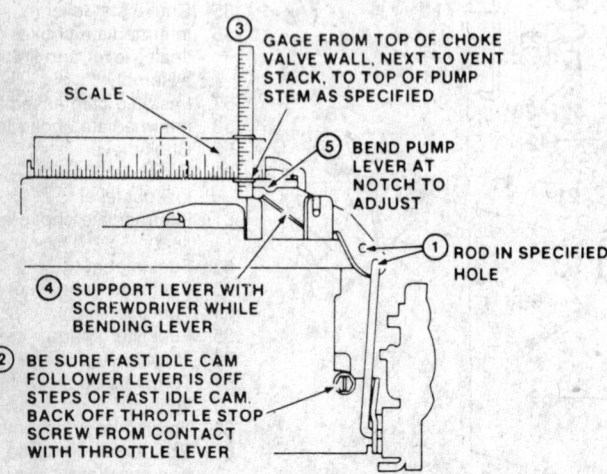

Rochester M4ME, M4MED and M4MEF pump adjustment

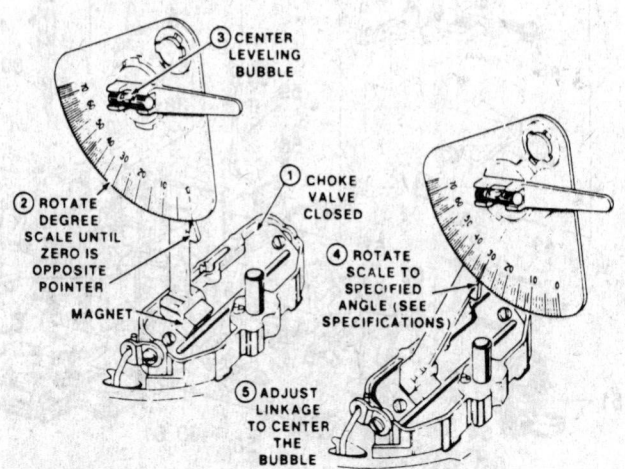

Rochester M4ME, M4MED and M4MEF choke valve angle gauge usage

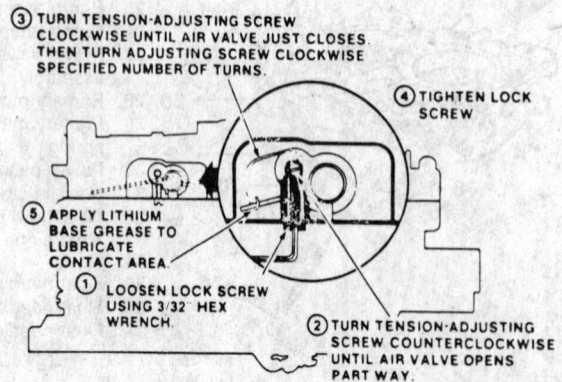

Rochester M4ME, M4MED and M4MEF air valve spring adjustment

① ATTACH RUBBER BAND TO GREEN TANG OF INTERMEDIATE CHOKE SHAFT

② OPEN THROTTLE TO ALLOW CHOKE VALVE TO CLOSE

③ SET UP ANGLE GAGE AND SET ANGLE TO SPECIFICATION

④ ON QUADRAJET, HOLD SECONDARY LOCKOUT LEVER AWAY FROM PIN

⑤ HOLD THROTTLE LEVER IN WIDE OPEN POSITION

⑥ ADJUST BY BENDING TANG OF FAST IDLE LEVER UNTIL BUBBLE IS CENTERED

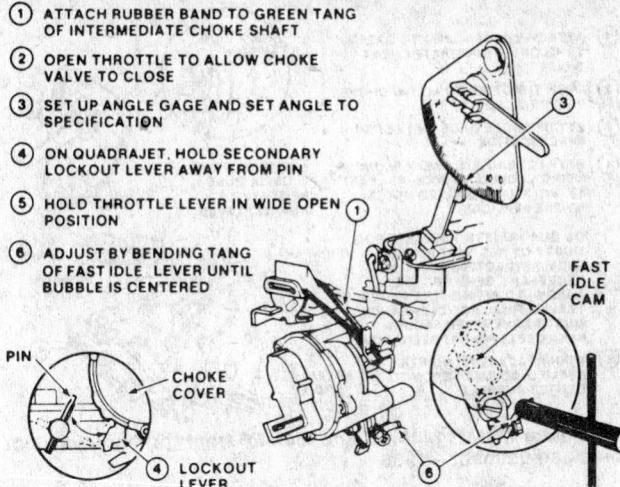

PIN
CHOKE COVER
LOCKOUT LEVER
FAST IDLE CAM

Rochester M4ME, M4MED and M4MEF unloader adjustment

① ATTACH RUBBER BAND TO GREEN TANG OF INTERMEDIATE CHOKE SHAFT.

② OPEN THROTTLE TO ALLOW CHOKE VALVE TO CLOSE.

③ SET UP ANGLE GAGE AND SET ANGLE TO SPECIFICATION.

④ RETRACT VACUUM BREAK PLUNGER, USING VACUUM SOURCE, AT LEAST 18" HG. PLUG AIR BLEED HOLES WHERE APPLICABLE.

④A ON QUADRAJETS, AIR VALVE ROD MUST NOT RESTRICT PLUNGER FROM RETRACTING FULLY. IF NECESSARY, BEND ROD HERE TO PERMIT FULL PLUNGER TRAVEL. WHERE APPLICABLE, PLUNGER STEM MUST BE EXTENDED FULLY TO COMPRESS PLUNGER BUCKING SPRING.

⑤ TO CENTER BUBBLE, EITHER:
A. ADJUST WITH 1/8" HEX WRENCH (VACUUM STILL APPLIED)
-OR-
B. SUPPORT AT "S" AND BEND VACUUM BREAK ROD (VACUUM STILL APPLIED)

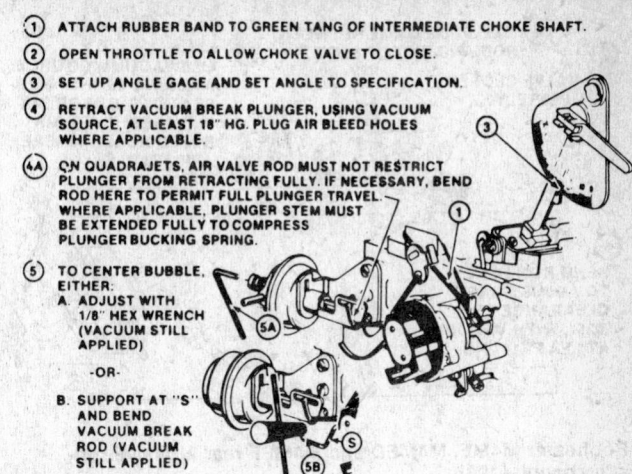

Rochester M4ME, M4MED and M4MEF rear (secondary) vacuum break adjustment

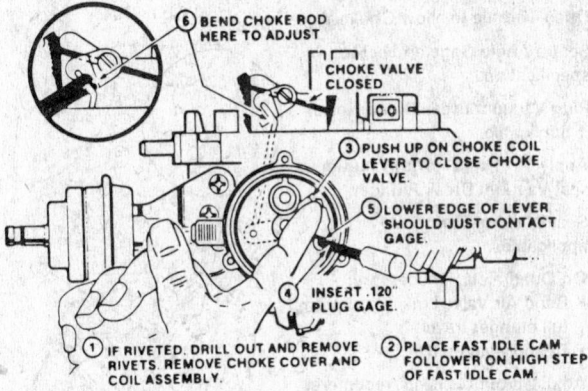

⑥ BEND CHOKE ROD HERE TO ADJUST

CHOKE VALVE CLOSED

③ PUSH UP ON CHOKE COIL LEVER TO CLOSE CHOKE VALVE.

⑤ LOWER EDGE OF LEVER SHOULD JUST CONTACT GAGE.

④ INSERT .120" PLUG GAGE.

① IF RIVETED, DRILL OUT AND REMOVE RIVETS. REMOVE CHOKE COVER AND COIL ASSEMBLY.

② PLACE FAST IDLE CAM FOLLOWER ON HIGH STEP OF FAST IDLE CAM.

Rochester M4ME, M4MED and M4MEF coil lever adjustment

PLUGGING AIR BLEED HOLES

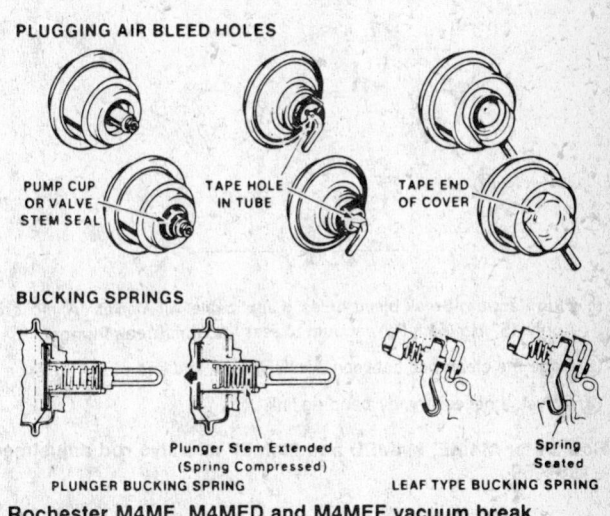

PUMP CUP OR VALVE STEM SEAL
TAPE HOLE IN TUBE
TAPE END OF COVER

BUCKING SPRINGS

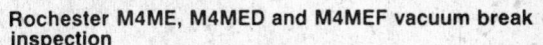

PLUNGER BUCKING SPRING
Plunger Stem Extended (Spring Compressed)
LEAF TYPE BUCKING SPRING
Spring Seated

Rochester M4ME, M4MED and M4MEF vacuum break inspection

③ .025" PLUG GAGE BETWEEN ROD AND END OF SLOT IN LEVER

② AIR VALVE CLOSED COMPLETELY

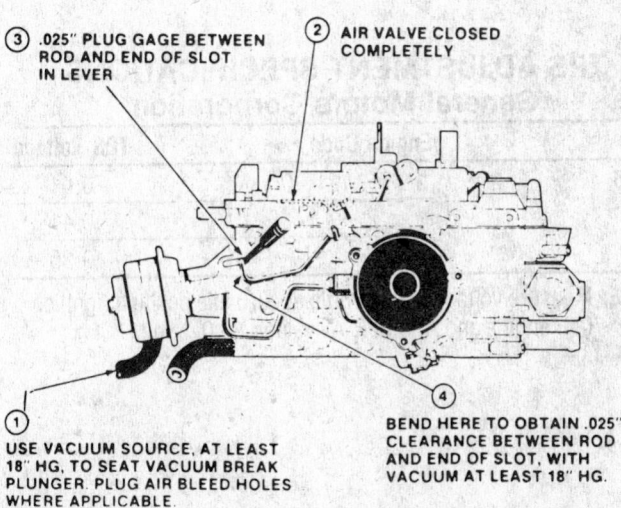

① USE VACUUM SOURCE, AT LEAST 18" HG, TO SEAT VACUUM BREAK PLUNGER. PLUG AIR BLEED HOLES WHERE APPLICABLE.

④ BEND HERE TO OBTAIN .025" CLEARANCE BETWEEN ROD AND END OF SLOT, WITH VACUUM AT LEAST 18" HG.

Rochester M4ME, M4MED and M4MEF rear air valve rod adjustment—1986

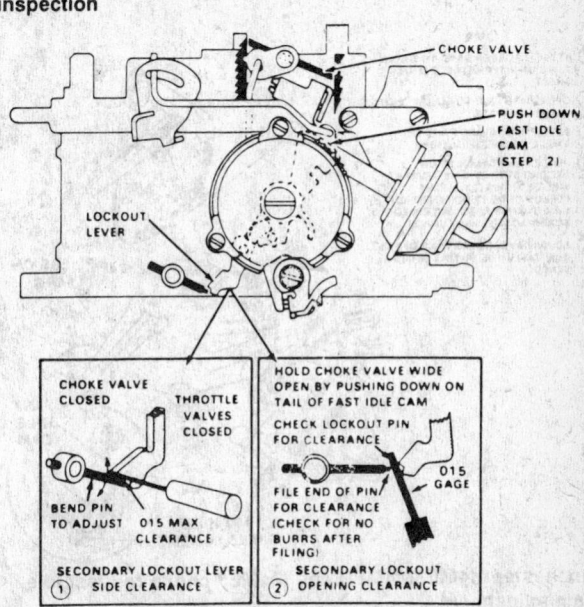

CHOKE VALVE
PUSH DOWN FAST IDLE CAM (STEP 2)
LOCKOUT LEVER

CHOKE VALVE CLOSED
THROTTLE VALVES CLOSED
BEND PIN TO ADJUST
.015 MAX CLEARANCE
SECONDARY LOCKOUT LEVER SIDE CLEARANCE ①

HOLD CHOKE VALVE WIDE OPEN BY PUSHING DOWN ON TAIL OF FAST IDLE CAM
CHECK LOCKOUT PIN FOR CLEARANCE
FILE END OF PIN FOR CLEARANCE (CHECK FOR NO BURRS AFTER FILING)
.015 GAGE
SECONDARY LOCKOUT OPENING CLEARANCE ②

Rochester M4ME, M4MED and M4MEF secondary lockout adjustment

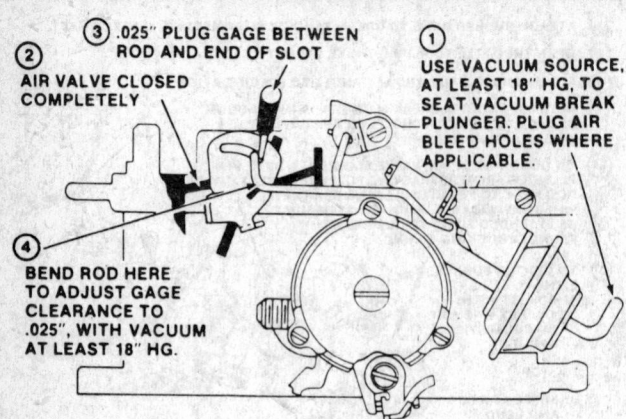

③ .025" PLUG GAGE BETWEEN ROD AND END OF SLOT

② AIR VALVE CLOSED COMPLETELY

① USE VACUUM SOURCE, AT LEAST 18" HG, TO SEAT VACUUM BREAK PLUNGER. PLUG AIR BLEED HOLES WHERE APPLICABLE.

④ BEND ROD HERE TO ADJUST GAGE CLEARANCE TO .025", WITH VACUUM AT LEAST 18" HG.

Rochester M4ME, M4MED and M4MEF rear air valve rod adjustment — 1986

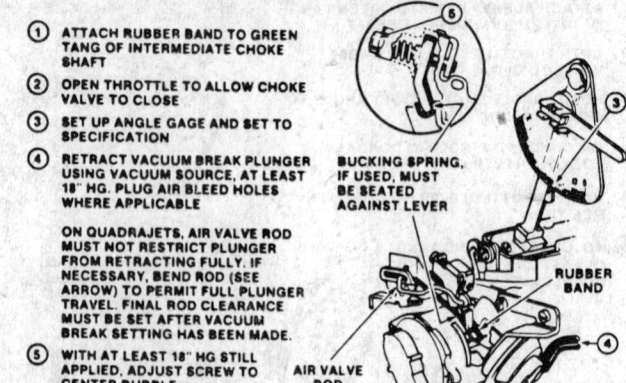

① ATTACH RUBBER BAND TO GREEN TANG OF INTERMEDIATE CHOKE SHAFT

② OPEN THROTTLE TO ALLOW CHOKE VALVE TO CLOSE

③ SET UP ANGLE GAGE AND SET TO SPECIFICATION

④ RETRACT VACUUM BREAK PLUNGER USING VACUUM SOURCE, AT LEAST 18" HG. PLUG AIR BLEED HOLES WHERE APPLICABLE

ON QUADRAJETS, AIR VALVE ROD MUST NOT RESTRICT PLUNGER FROM RETRACTING FULLY. IF NECESSARY, BEND ROD (SEE ARROW) TO PERMIT FULL PLUNGER TRAVEL. FINAL ROD CLEARANCE MUST BE SET AFTER VACUUM BREAK SETTING HAS BEEN MADE.

⑤ WITH AT LEAST 18" HG STILL APPLIED, ADJUST SCREW TO CENTER BUBBLE

BUCKING SPRING, IF USED, MUST BE SEATED AGAINST LEVER

RUBBER BAND

AIR VALVE ROD

Rochester M4ME, M4MED and M4MEF front (primary) vacuum break adjustment — 1986

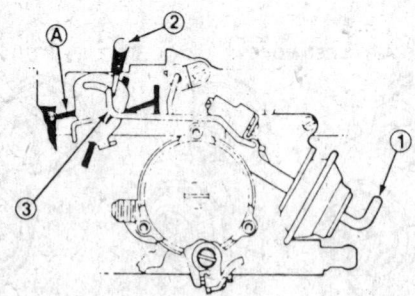

① Plug Vacuum Break bleed holes, if applicable. Air Valves Ⓐ closed. Apply 15" Hg (51 k Pa) vacuum to seat Vacuum Break Plunger.

② Gage the clearance between Air Valve Link and end of slot in lever.

③ Adjust, if necessary, by bending link.

Rochester M4ME, M4MED and M4MEF air valve rod adjustment — 1987–89

① Attach rubber band to Vacuum Break Lever of Intermediate Choke Shaft.

② Open Throttle to allow Choke Valve to close.

③ Set up Angle Gage and set to specification.

④ Plug Vacuum Break Bleed Holes, if applicable.

Apply 15" Hg (51 k Pa) vacuum to seat Vacuum Break Plunger.

Seat Bucking Spring Ⓐ, if applicable.

On Quadrajets, if necessary:
- Bend Air Valve Link Ⓑ to permit full plunger travel.
- Reapply vacuum.

⑤ Adjust, if bubble is not recentered, by turning screw.

J-26701-A or BT-7704

Rochester M4ME, M4MED and M4MEF front (primary) vacuum break adjustment — 1987–89

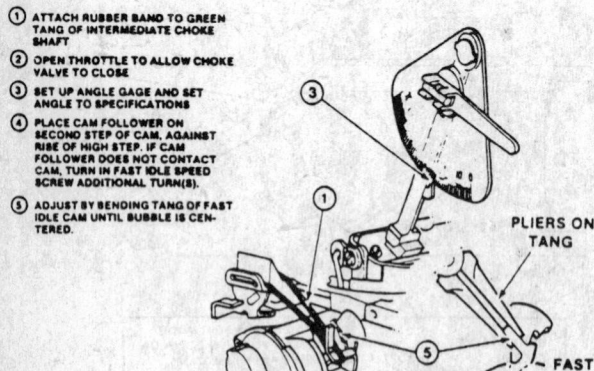

① ATTACH RUBBER BAND TO GREEN TANG OF INTERMEDIATE CHOKE SHAFT

② OPEN THROTTLE TO ALLOW CHOKE VALVE TO CLOSE

③ SET UP ANGLE GAGE AND SET ANGLE TO SPECIFICATIONS

④ PLACE CAM FOLLOWER ON SECOND STEP OF CAM, AGAINST RISE OF HIGH STEP. IF CAM FOLLOWER DOES NOT CONTACT CAM, TURN IN FAST IDLE SPEED SCREW ADDITIONAL TURN(S).

⑤ ADJUST BY BENDING TANG OF FAST IDLE CAM UNTIL BUBBLE IS CENTERED.

PLIERS ON TANG

FAST IDLE CAM

FAST IDLE SPEED SCREW

Rochester M4ME, M4MED and M4MEF choke rod fast idle cam adjustment

TPS ADJUSTMENT SPECIFICATIONS
General Motors Corporation

Year	Engine Code	TPS Voltage
1986	F	0.41
	L	0.41
	N	0.25

NOTE: Measure voltage with throttle at curb idle position, ignition ON, engine and A/C OFF. All values ± 0.1 volt.

CHRYSLER CORPORATION

Holley Carburetors

HOLLEY MODEL 1945
Chrysler Corporation
(All measurements in inches)

Year	Carburetor Number	Dry Float Level	Choke Unloader	Pump Stroke	Pump Rod Hole	Fast Idle Cam Position	Fast Idle Speed	Initial Choke Opening
1986	R40102A	①	0.250	1.70	1	0.080	1800	0.130
	R40244A	①	0.250	1.61	2	0.090	1600	0.130
	R40159	①	0.250	1.61	2	0.080	1600	0.130
	R40160	①	0.250	1.61	2	0.090	1600	0.130
1987	R40102A	①	0.250	1.70	1	0.080	1800	0.130
	R40244A	①	0.250	1.61	2	0.090	1600	0.130
	R40159	①	0.250	1.61	2	0.080	1600	0.130
	R40160	①	0.250	1.61	2	0.090	1600	0.130

NOTE: Choke setting is fixed and nonadjustable
① Flush with top of bowl over gasket, carb inverted

MODEL 1945

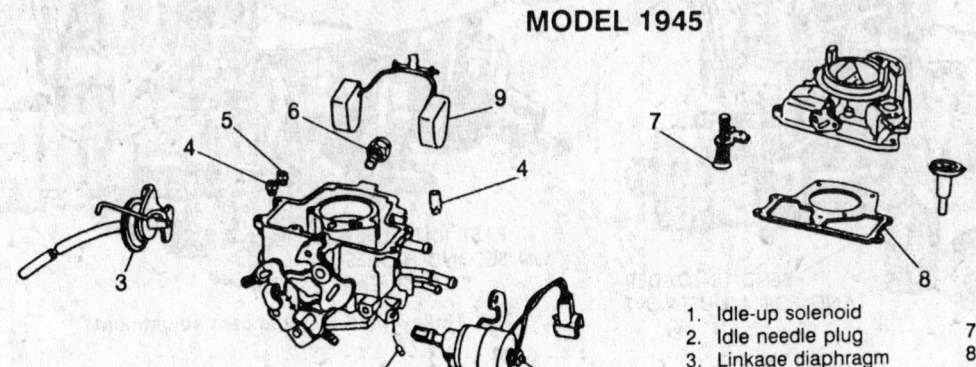

1. Idle-up solenoid
2. Idle needle plug
3. Linkage diaphragm
4. Power valve
5. Main metering jet
6. Needle and seat
7. Piston pump cup
8. Air horn gasket
9. Float
10. Idle solenoid adjusting screw

Holley 1945 carburetor — exploded view

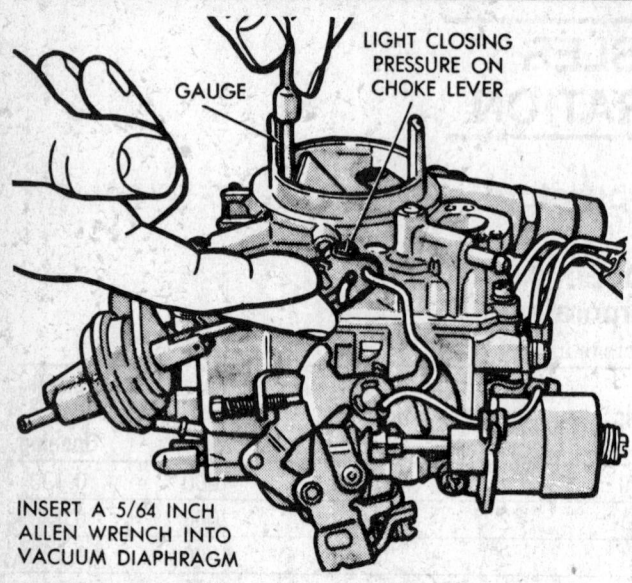

INSERT A 5/64 INCH
ALLEN WRENCH INTO
VACUUM DIAPHRAGM

GAUGE

LIGHT CLOSING
PRESSURE ON
CHOKE LEVER

Holley 1945 initial choke valve setting (vacuum kick)

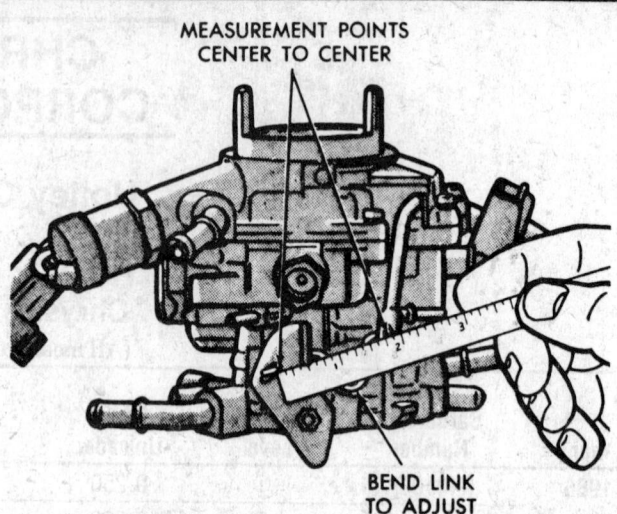

MEASUREMENT POINTS
CENTER TO CENTER

BEND LINK
TO ADJUST

Holley 1945 accelerator pump piston stroke adjustment

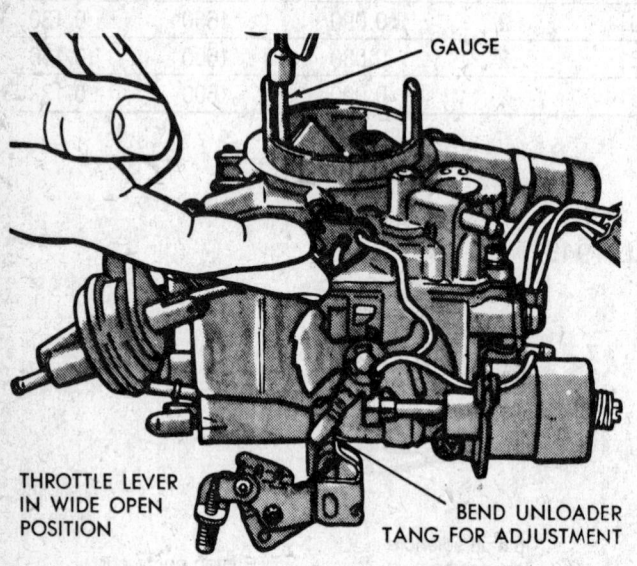

GAUGE

THROTTLE LEVER
IN WIDE OPEN
POSITION

BEND UNLOADER
TANG FOR ADJUSTMENT

Holley 1945 choke valve unloader adjustment

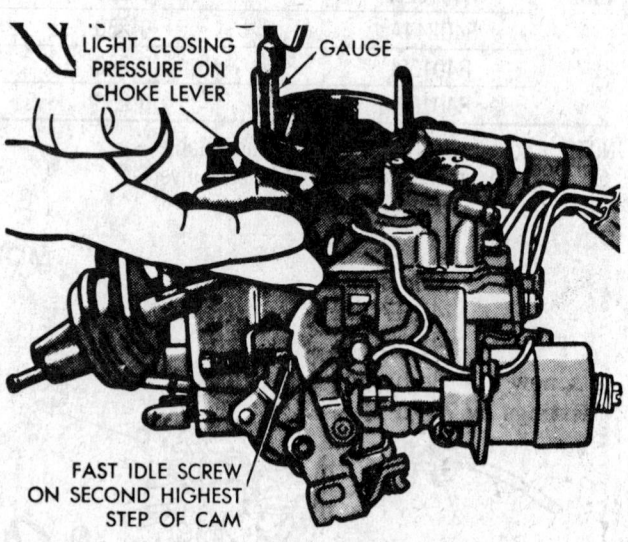

LIGHT CLOSING
PRESSURE ON
CHOKE LEVER

GAUGE

FAST IDLE SCREW
ON SECOND HIGHEST
STEP OF CAM

Holley 1945 fast idle cam adjustment

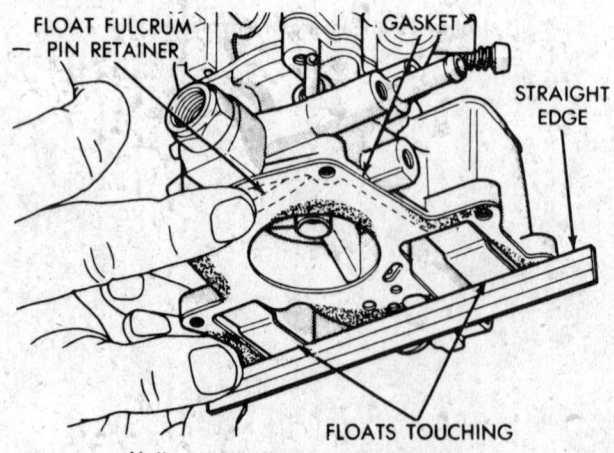

FLOAT FULCRUM
— PIN RETAINER

GASKET

STRAIGHT
EDGE

FLOATS TOUCHING

Holley 1945 float level adjustment

HOLLEY MODEL 2280
Chrysler Corporation
(All measurements in inches)

Year	Carburetor Number	Float Level ①	Choke Vacuum Kick	Fast Idle Cam	Fast Idle (rpm)	Choke Unloader	Bowl Vent Valve
1986	R40172-1A	9/32	0.140	0.052	②	0.250	0.035
	R40216A	9/32	0.140	0.070	②	0.250	0.035
	R40214A	9/32	0.140	0.070	②	0.250	0.035
1987	R40172-1A	9/32	0.140	0.052	②	0.200	0.035
	R40214A	9/32	0.140	0.070	②	0.250	0.035
	R40216A	9/32	0.140	0.070	②	0.250	0.035
	R40221A	9/32	0.130	0.070	②	0.150	0.035
	R40222A	9/32	0.130	0.070	②	0.150	0.035

① Measured from surface of fuel bowl to the toe of each float
② Refer to underhood specification sticker

MODELS 2280

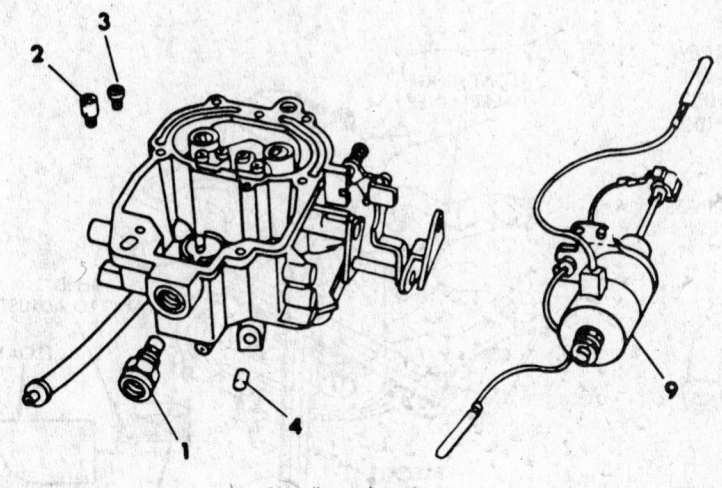
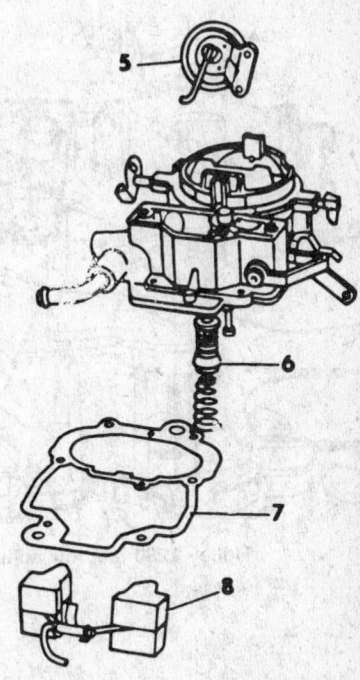

1. Needle and seat
2. Power valve
3. Main metering jet
4. Idle needle plug
5. Choke diaphragm
6. Pump cup
7. Air horn gasket
8. Float
9. Idle-stop solenoid

Holley 2280 carburetor exploded view

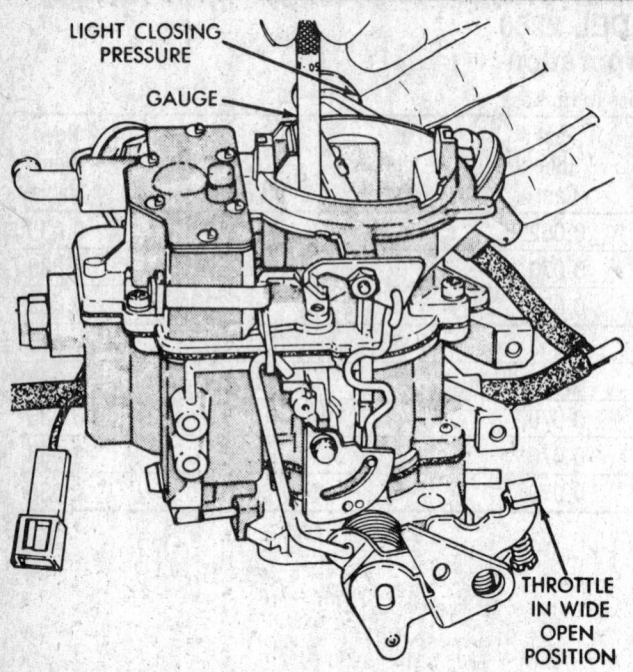

Holley 2280 choke inloader adjustment

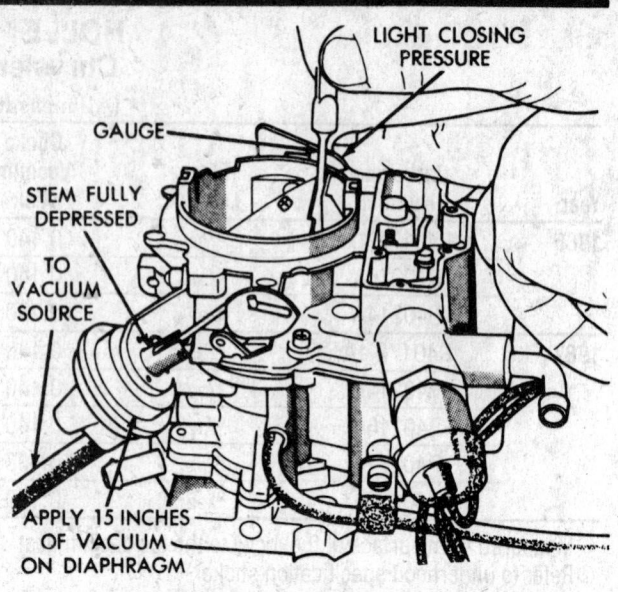

Holley 2280 vacuum kick adjustment

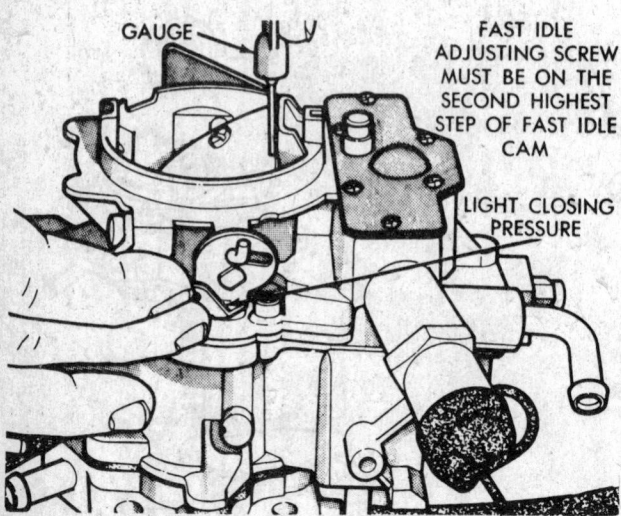

Holley 2280 fast idle adjustment

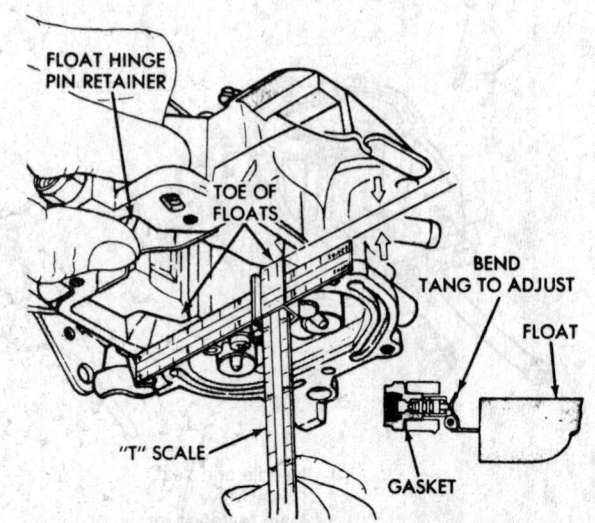

Holley 2280 float level adjustment

HOLLEY MODEL 6145
Chrysler Corporation
(All measurements in inches)

Year	Carburetor Number	Float Setting	Choke Vacuum Kick	Choke Unloader Adjustment	Fast Idle Cam Position	Fast Idle (rpm)	Pump Piston Stroke
1986	R40161	①	0.150	0.250	0.060	②	1.75
	R40162	①	0.150	0.250	0.070	②	1.75
1987	R40161	①	0.150	0.250	0.060	②	1.75
	R40162	①	0.150	0.250	0.070	②	1.75

①With bowl inverted, float lungs just touch a straightedge run along gasket surface
②1986–87: Refer to underhood specification sticker

MODEL 6145

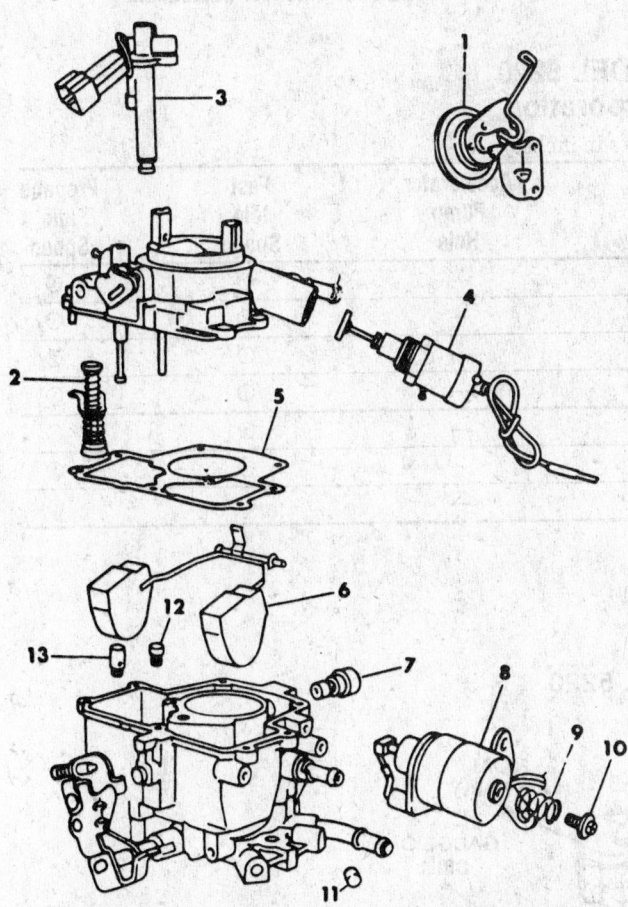

1. Choke Diaphragm
2. Piston pump cup
3. Duty-cycle solenoid
4. Switch vent solenoid
5. Air horn gasket
6. Float
7. Needle and Seat
8. Speed-up solenoid
9. Solenoid adjusting screw spring
10. Solenoid adjusting screw
11. Idle mixture plug
12. Main jet
13. Power valve

Holley 6145 carburetor – exploded view

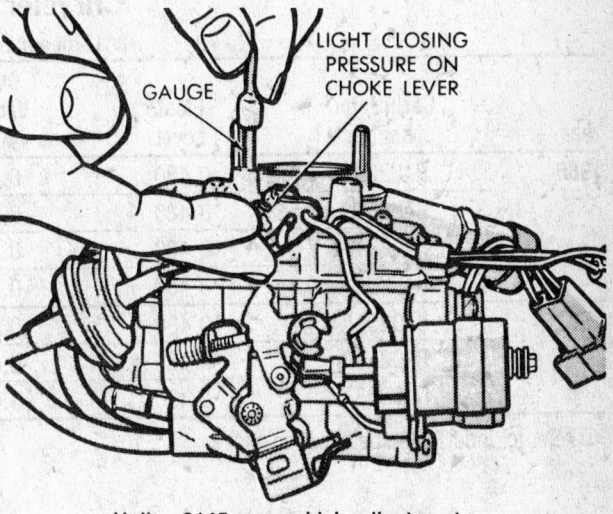

Holley 6145 vacum kick adjustment

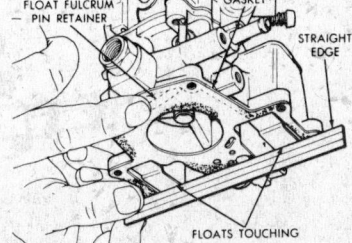

Holley 6145 float adjustment

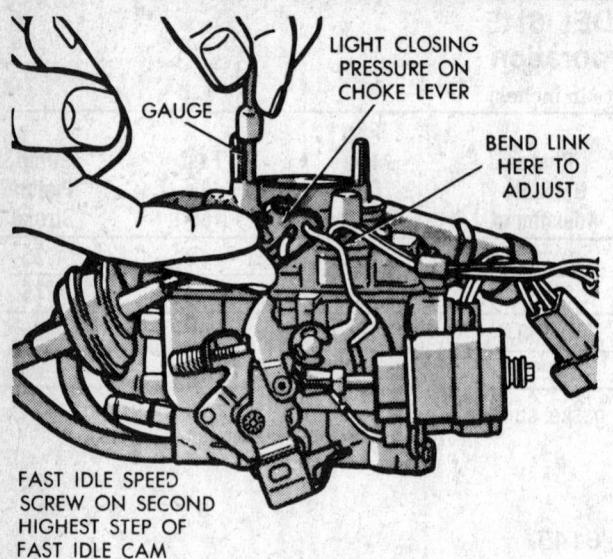

GAUGE

LIGHT CLOSING
PRESSURE ON
CHOKE LEVER

BEND LINK
HERE TO
ADJUST

FAST IDLE SPEED
SCREW ON SECOND
HIGHEST STEP OF
FAST IDLE CAM

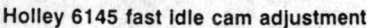

Holley 6145 fast idle cam adjustment

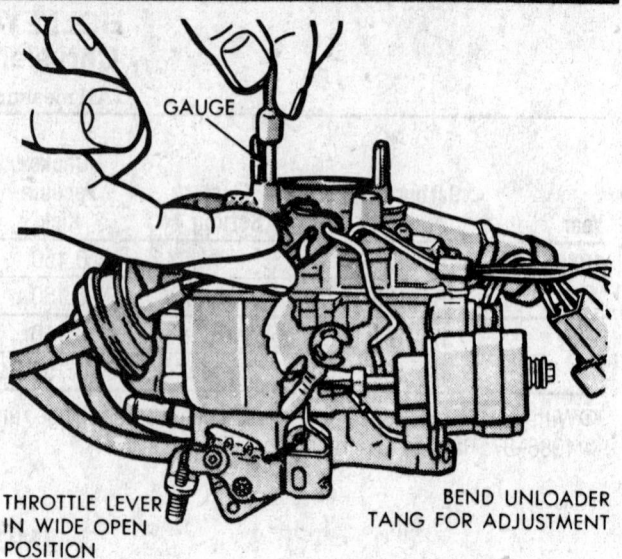

GAUGE

THROTTLE LEVER
IN WIDE OPEN
POSITION

BEND UNLOADER
TANG FOR ADJUSTMENT

Holley 6145 choke unloader adjustment

HOLLEY MODEL 5220
Chrysler Corporation
(All measurements in inches)

Year	Carburetor Number	Float Level	Choke Vacuum Kick	Accelerator Pump Hole	Fast Idle Speed	Propane Idle Speed
1986	R40229A	0.480	0.130	—	①	①
	R40230A	0.480	0.130	—	①	①
	R40231A	0.480	0.130	—	①	①
	R40232A	0.480	0.130	—	①	①
1987	R40234A	0.480	0.095	—	①	①
	R40240A	0.480	0.095	—	①	①
	R40303A	0.480	0.095	—	①	①

① Refer to underhood specification sticker

MODEL 5220

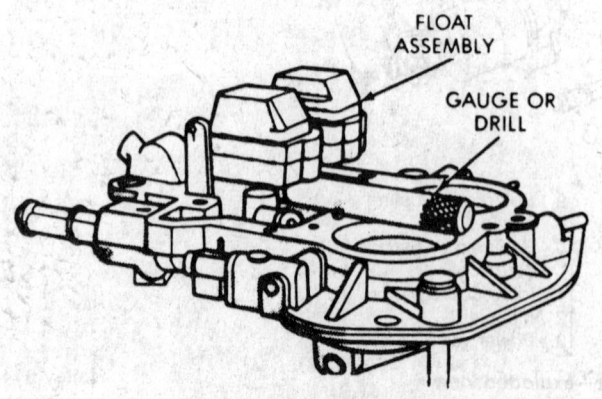

FLOAT
ASSEMBLY

GAUGE OR
DRILL

Holley 5220/6520 dry float setting

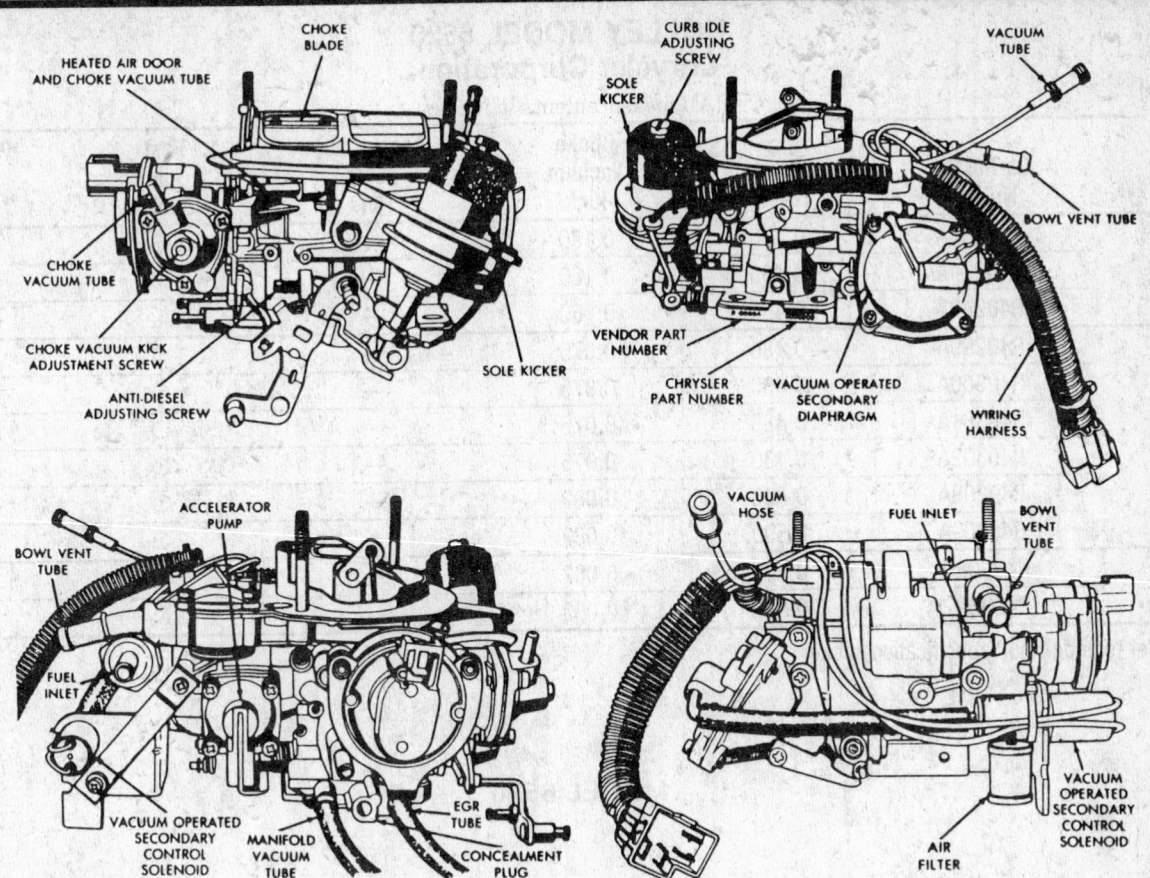

HEATED AIR DOOR
AND CHOKE VACUUM TUBE

CHOKE BLADE

CHOKE
VACUUM TUBE

CHOKE VACUUM KICK
ADJUSTMENT SCREW

ANTI-DIESEL
ADJUSTING SCREW

SOLE KICKER

CURB IDLE
ADJUSTING
SCREW

SOLE
KICKER

VACUUM
TUBE

BOWL VENT TUBE

VENDOR PART
NUMBER

CHRYSLER
PART NUMBER

VACUUM OPERATED
SECONDARY
DIAPHRAGM

WIRING
HARNESS

BOWL VENT
TUBE

ACCELERATOR
PUMP

FUEL
INLET

VACUUM OPERATED
SECONDARY
CONTROL
SOLENOID

MANIFOLD
VACUUM TUBE

EGR
TUBE

CONCEALMENT
PLUG

VACUUM
HOSE

FUEL INLET

BOWL VENT
TUBE

AIR
FILTER

VACUUM
OPERATED
SECONDARY
CONTROL
SOLENOID

Holley 5220 carburetor side view

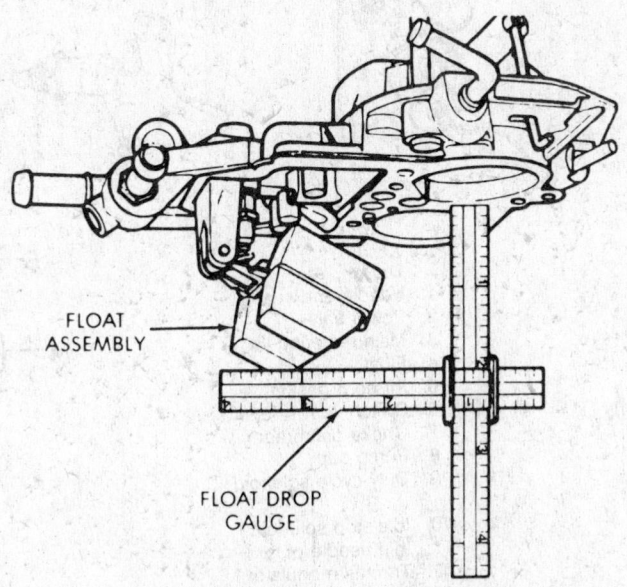

FLOAT
ASSEMBLY

FLOAT DROP
GAUGE

Holley 5220/6520 float drop measurement

HOLLEY MODEL 6520
Chrysler Corporation
(All measurements in inches)

Year	Carburetor Number	Float Level	Choke Vacuum Kick	Accelerator Pump Hole	Fast Idle Speed	Propane Idle Speed
1986	R40233A	0.480	0.160	—	①	①
	R40234A	0.480	0.160	—	①	①
	R40240A	0.480	0.160	—	①	①
1987	R40299A	0.480	0.075	—	①	①
	R40300A	0.480	0.075	—	①	①
	R40301A	0.480	0.075	—	①	①
	R40302A	0.480	0.075	—	①	①
	R40308A	0.500	0.082	—	①	①
	R40309A	0.500	0.082	—	①	①
1988	R40308-1	0.500	0.082	—	①	①
	R40309-1	0.500	0.082	—	①	①

① Refer to underhood specification sticker

MODEL 6520

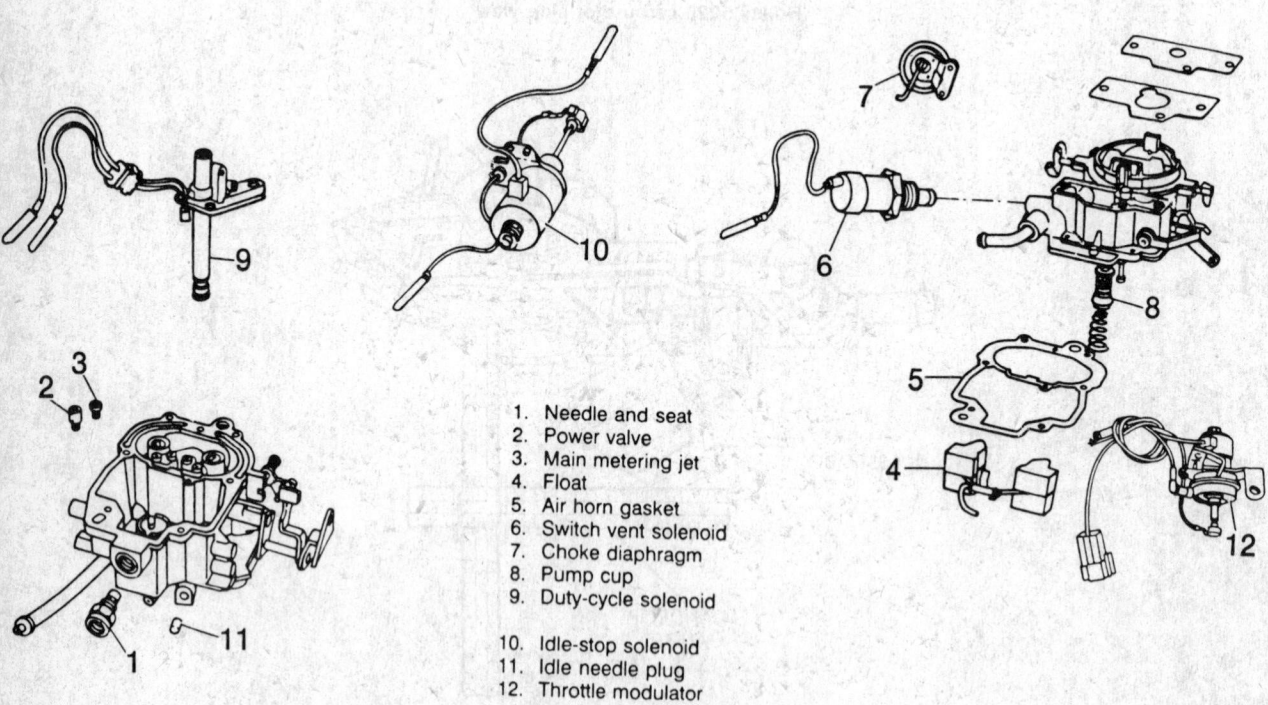

1. Needle and seat
2. Power valve
3. Main metering jet
4. Float
5. Air horn gasket
6. Switch vent solenoid
7. Choke diaphragm
8. Pump cup
9. Duty-cycle solenoid
10. Idle-stop solenoid
11. Idle needle plug
12. Throttle modulator

Holley 6520 carburetor exploded view

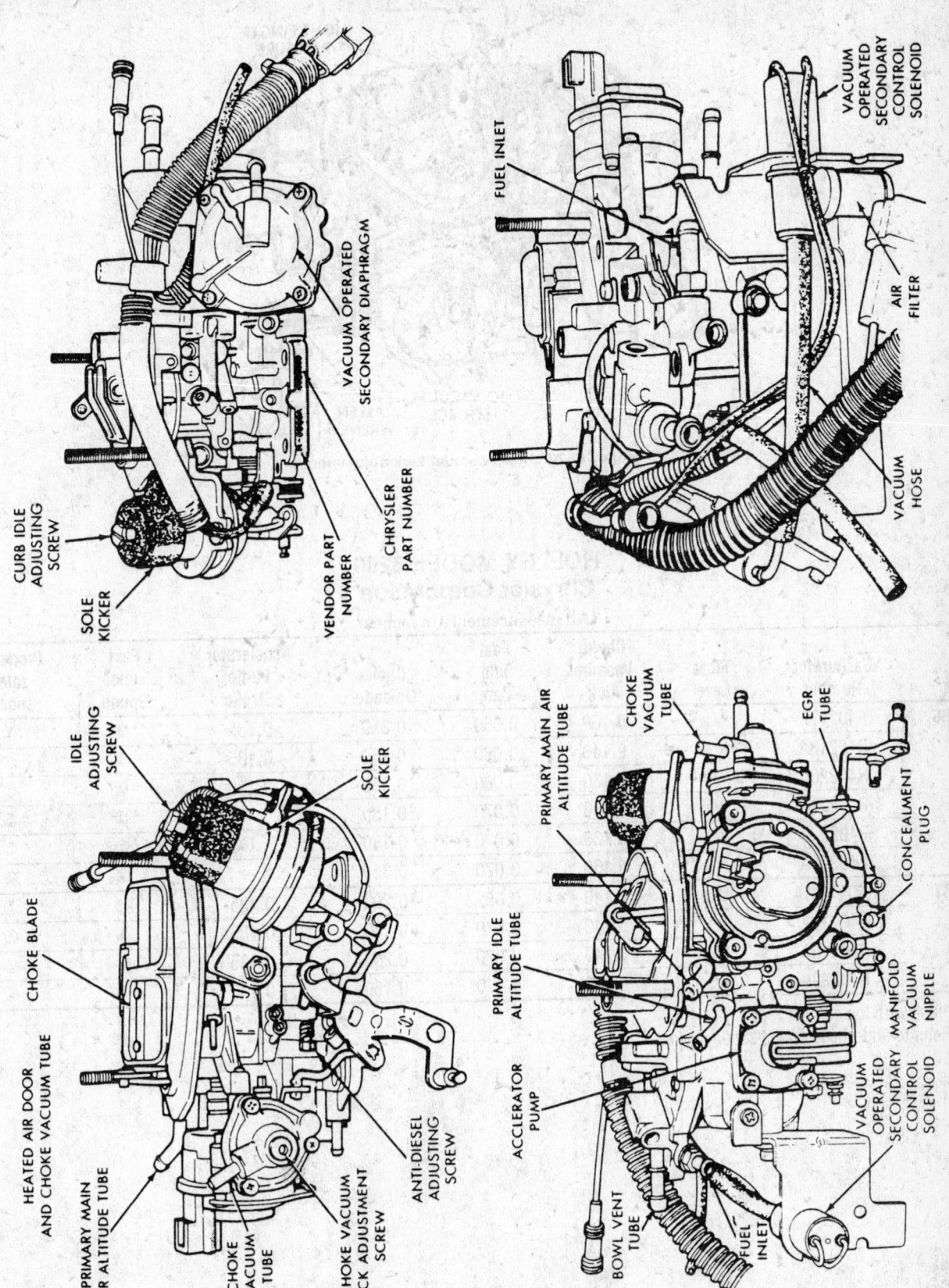

CURB IDLE ADJUSTING SCREW

SOLE KICKER

VENDOR PART NUMBER

CHRYSLER PART NUMBER

VACUUM OPERATED SECONDARY DIAPHRAGM

FUEL INLET

VACUUM OPERATED SECONDARY CONTROL SOLENOID

AIR FILTER

VACUUM HOSE

Holley 6520 carburetor side view

IDLE ADJUSTING SCREW

SOLE KICKER

CHOKE BLADE

HEATED AIR DOOR AND CHOKE VACUUM TUBE

PRIMARY MAIN AIR ALTITUDE TUBE

CHOKE VACUUM TUBE

CHOKE VACUUM KICK ADJUSTMENT SCREW

ANTI-DIESEL ADJUSTING SCREW

PRIMARY MAIN AIR ALTITUDE TUBE

CHOKE VACUUM TUBE

EGR TUBE

PRIMARY IDLE ALTITUDE TUBE

CONCEALMENT PLUG

MANIFOLD VACUUM NIPPLE

ACCLERATOR PUMP

VACUUM OPERATED SECONDARY CONTROL SOLENOID

BOWL VENT TUBE

FUEL INLET

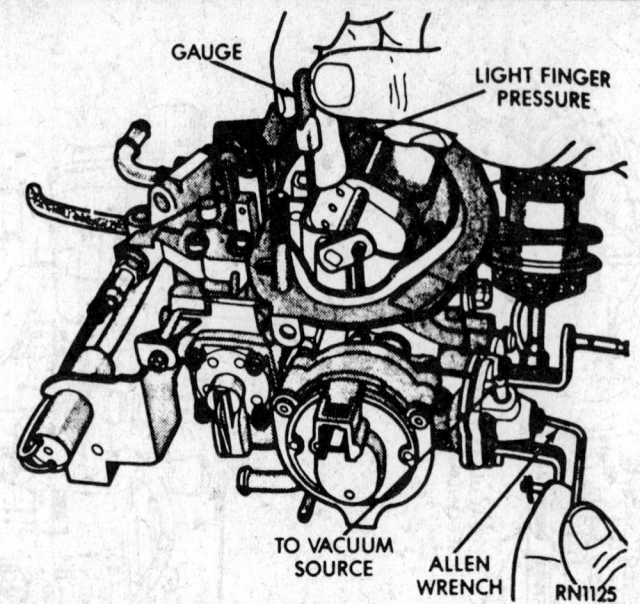

GAUGE

LIGHT FINGER PRESSURE

TO VACUUM SOURCE

ALLEN WRENCH

RN1125

Holley 6520 choke vacuum kick adjustment

HOLLEY MODEL 6280
Chrysler Corporation
(All measurements in inches)

Year	Carburetor Number	Float Level	Choke Vacuum Kick	Fast Idle Cam	Choke Unloader	Accelerator Pump Stroke	Fast Idle Speed	Propane Idle Speed
1986	R40217A	9/32	0.160	0.060	0.350	0.135	②	②
	R40218A	9/32	0.140	0.060	0.250	0.135	②	②
	R40220A	9/32	0.130	0.060	0.350	0.135	②	②
	R40221A	9/32	0.130	0.070	0.150	①	②	②
	R40222A	9/32	0.130	0.070	0.150	①	②	②
	R40294A	9/32	0.160	0.070	0.350	—	②	②
1987	R40172-1A	9/32	0.140	0.052	0.200	0.135	②	②
	R40214A	9/32	0.140	0.070	0.250	0.135	②	②
	R40216A	9/32	0.140	0.070	0.250	0.135	②	②
	R40222A	9/32	0.130	0.070	0.150	0.135	②	②

① Flush with top of bowl vent
② Refer to underhood specification sticker

MODEL 6280

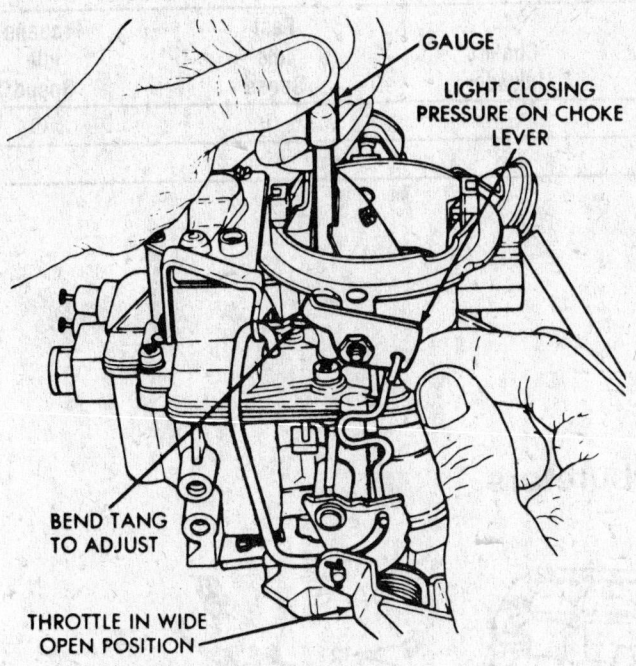

GAUGE

LIGHT CLOSING PRESSURE ON CHOKE LEVER

BEND TANG TO ADJUST

THROTTLE IN WIDE OPEN POSITION

Holley 6280 choke unloader adjustment

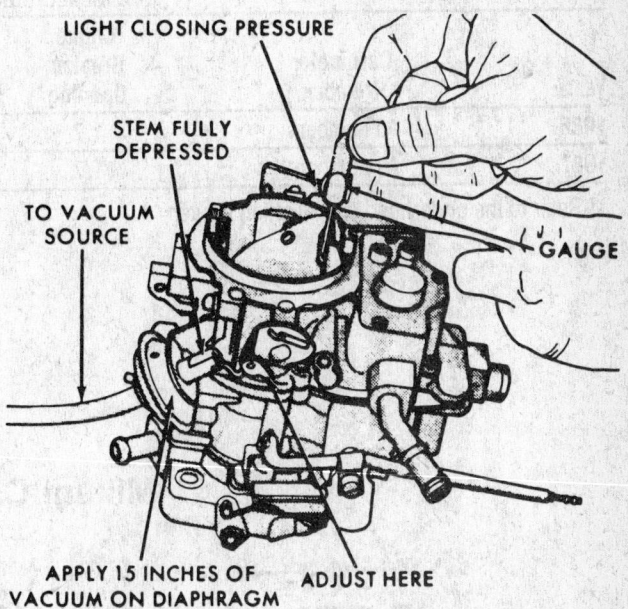

LIGHT CLOSING PRESSURE

STEM FULLY DEPRESSED

TO VACUUM SOURCE

GAUGE

APPLY 15 INCHES OF VACUUM ON DIAPHRAGM

ADJUST HERE

Holley 6280 choke vacuum kick adjustment

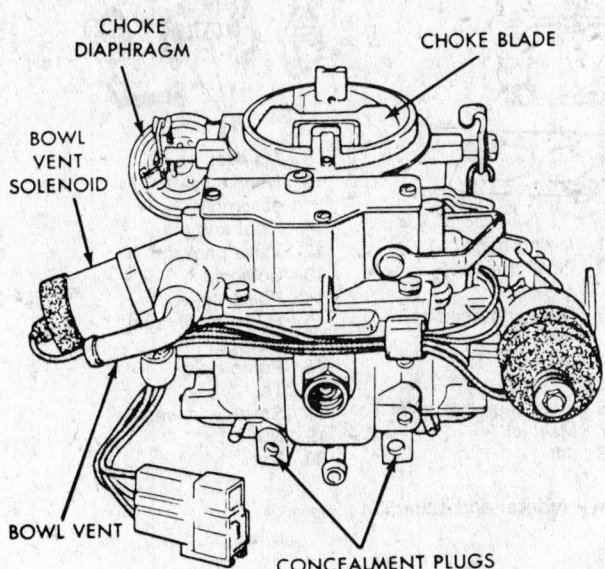

CHOKE DIAPHRAGM

CHOKE BLADE

BOWL VENT SOLENOID

BOWL VENT

CONCEALMENT PLUGS

Holley 6280 carburetor—exploded view

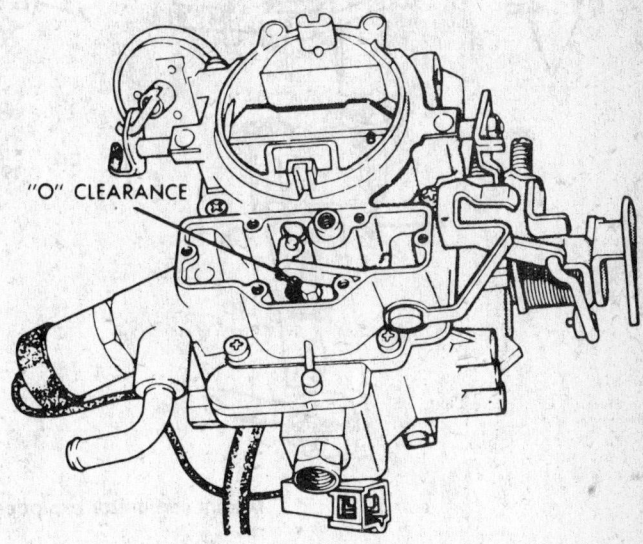

"O" CLEARANCE

Holley 6280 accelerator pump stroke measurement

MIKUNI
Chrysler Corporation
(All measurements in millimeters)

Year	Carburetor Number	Choke Breaker Opening	Choke Unloader	Fast Idle Speed	Propane Idle Speed
1986	All numbers	1.7	1.3	①	①
1987	All numbers	1.7	1.3	①	①

① Refer to the underhood specification sticker

Mikuni Carburetors

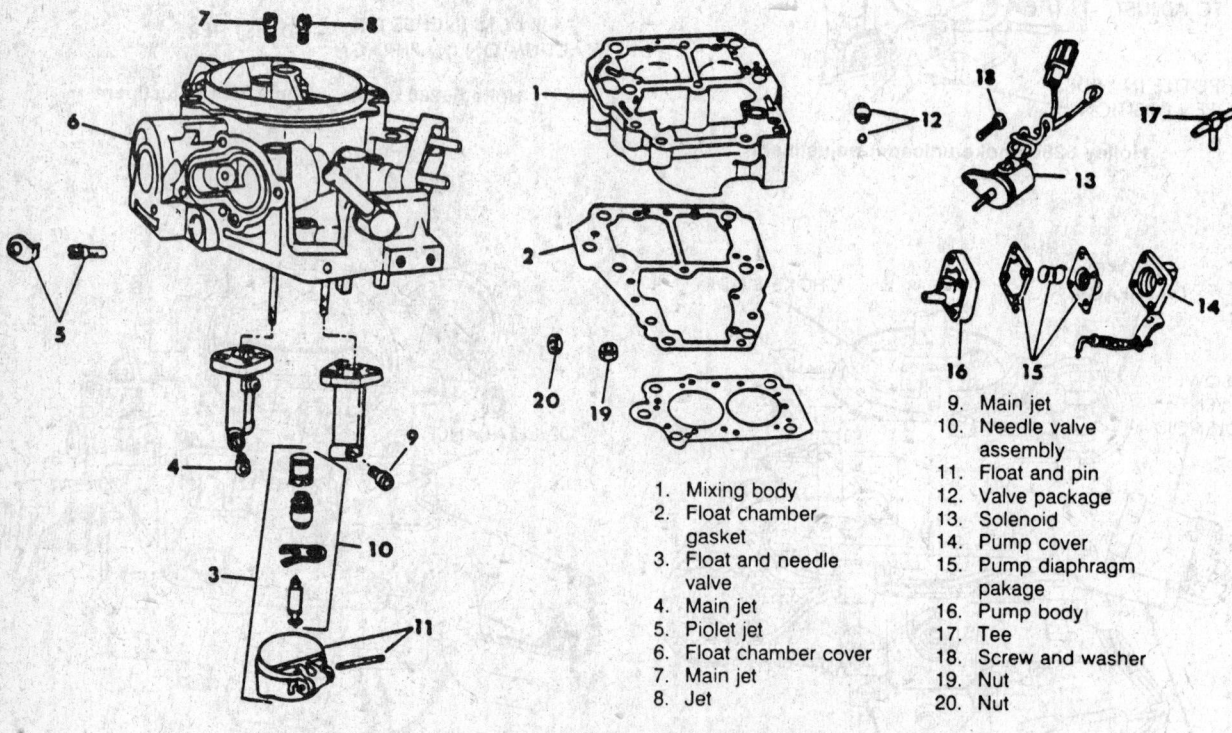

1. Mixing body
2. Float chamber gasket
3. Float and needle valve
4. Main jet
5. Piolet jet
6. Float chamber cover
7. Main jet
8. Jet
9. Main jet
10. Needle valve assembly
11. Float and pin
12. Valve package
13. Solenoid
14. Pump cover
15. Pump diaphragm pakage
16. Pump body
17. Tee
18. Screw and washer
19. Nut
20. Nut

Mikuni carburetor exploded view—federal and canadian

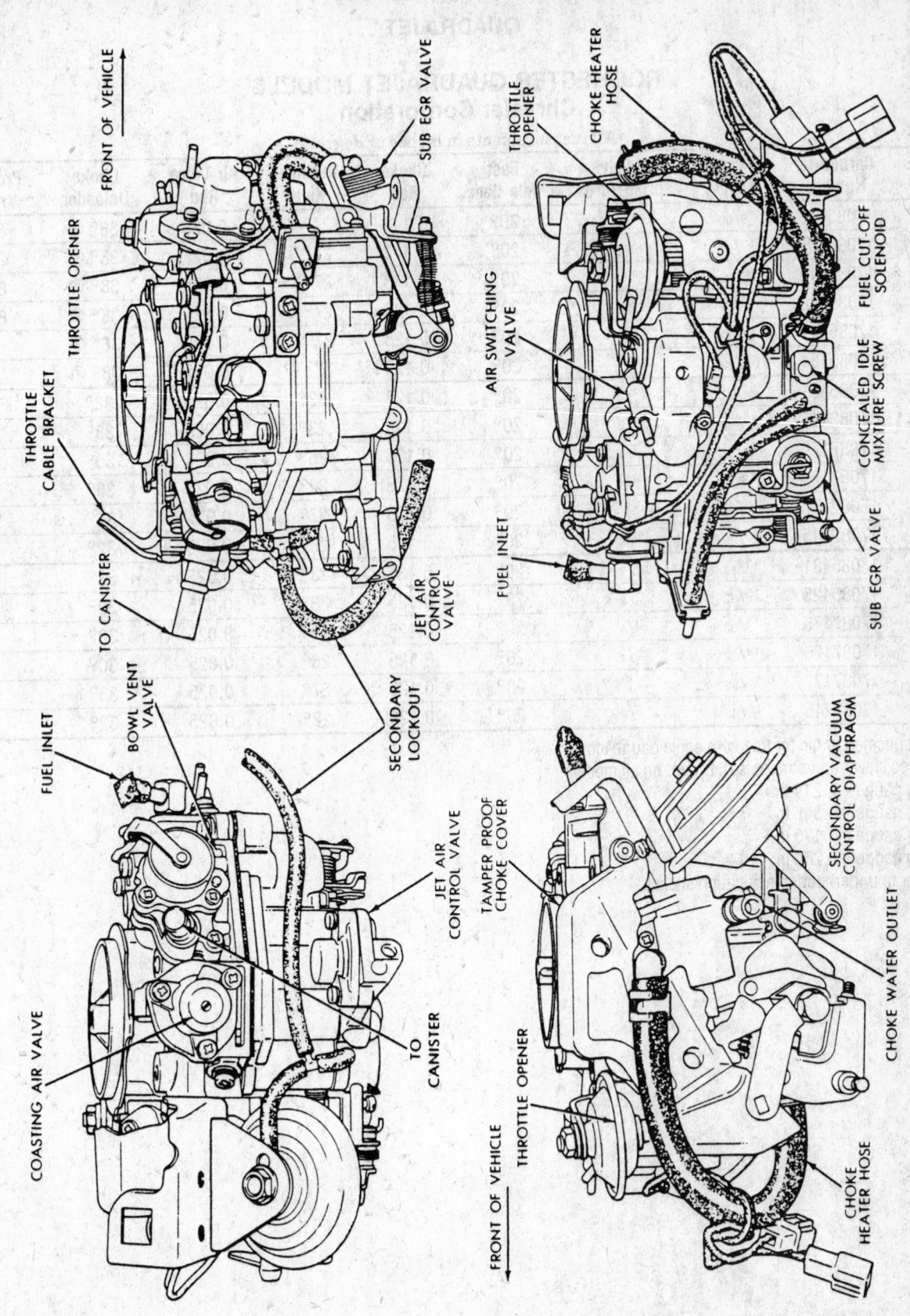

Mikuni carburetor side view—California and high altitude

Rochester Carburetors
QUADRAJET

ROCHESTER QUADRAJET MODELS
Chrysler Corporation
(All measurements in inches or degrees)

Year	Carburetor Number	Float Level	Air Valve Spring Turns	Fast Idle Cam	Choke Rod	Vacuum Kick	Air Valve Rod	Choke Unloader	Propane rpm
1986	17085408	13/32	1/2	20°	0.143	27° ①	0.025	38° ②	800
	17085409	13/32	5/8	20°	0.143	27° ①	0.025	38° ②	750
	17085415	13/32	1/2	20°	0.143	27° ①	0.025	38° ②	800
	17085416	13/32	3/4	20°	0.143	27° ①	0.025	38° ②	800
	17085417	13/32	3/4	20°	0.125	27° ③	0.025	38° ④	⑤
	17085431	13/32	1/2	20°	0.125	27° ③	0.025	38° ④	⑤
1987	17085431	13/32	1/2	20°	0.125	23°	0.025	32°	⑤
	17086425	15/32	1/2	20°	0.125	23°	0.025	38°	⑤
	17087175	13/32	3/4	20°	0.125	26°	0.025	30°	⑤
	17087176	13/32	3/4	20°	0.125	26°	0.025	30°	⑤
	17087177	13/32	1	20°	0.125	27°	0.025	33°	⑤
	17087245	15/32	5/8	20°	0.125	23°	0.025	32°	⑤
1988	17085431	13/32	1/2	20°	0.125	23°	0.025	32°	⑤
	17086425	15/32	1/2	20°	0.125	23°	0.025	32°	⑤
	17087175	13/32	3/4	20°	0.125	26°	0.025	38°	⑤
	17087176	13/32	3/4	20°	0.125	26°	0.025	30°	⑤
	17087177	13/32	1	20°	0.125	27°	0.025	33°	⑤
	17087245	15/32	5/8	20°	0.125	23°	0.025	32°	⑤

Note: specified angle for use with angle gauge tool
Choke coil lever adjustment is 0.120 in. on all models
① Plug gauge—0.214 in.
② Plug gauge—0.345 in.
③ Plug gauge—0.170 in.
④ Plug gauge—0.260 in.
⑤ Refer to underhood specification sticker

1. Air cleaner gasket
2. Flange gasket
3. Air horn assembly
4. Secondary metering rod holder attaching screw
5. Secondary metering rod holder
6. Secondary metering rod
7. Choke lever
8. Choke lever attaching screw
9. Pump lever
10. Pump lever hinge pin
11. Air horn to throttle body screw assembly
12. Air horn to float bowl screw assembly

13. Air horn to float bowl (countersunk) screw
14. Air horn baffle
15. Pump stem seal retainer
16. Pump stem seal
17. Air horn to float bowl gasket
18. Pump assembly
19. Pump return spring
20. Power valve piston assembly
21. Primary metering rod
22. Primary metering rod (M2M, M4M only) spring
23. Power piston spring

24. Float bowl insert
25. Float hinge pin
26. Float
27. Float needle pull clip
28. Float needle
29. Float needle seat
30. Float needle seat gasket
31. Primary metering jet
32. Pump discharge (retainer) plug

33. Pump discharge ball
34. Pump well baffle
35. Throttle body to "T" hose
36. Secondary side vacuum break (choke diaphragm) "T" vacuum hose
37. "T" to secondary side vacuum break (choke diaphragm) hose
38. Secondary side (rear) vacuum break assembly (choke diaphragm)
39. Secondary side (rear) vacuum break assembly (choke diaphragm) attaching screw
40. Secondary side (rear) vacuum break (choke diaphragm) to choke link
41. Secondary side vacuum break (choke diaphragm) air valve lever link
42. Choke cover attaching rivet
43. Choke cover retainer
44. Electric choke cover and stat assembly
45. Choke housing assembly
46. Choke housing to float bowl screw and washer assembly
47. Choke stat lever attaching screw
48. Choke stat lever
49. Intermediate choke shaft lever and link assembly
50. Fast idle cam assembly
51. Intermediate choke lever
52. Choke link
53. Secondary throttle lockout lever
54. Intermediate choke shaft seal
55. Fuel inlet nut
56. Fuel inlet nut gasket
57. Fule inlet filter
58. Fuel filter spring
59. Throttle stop screw
60. Throttle stop srew spring
61. Throttle body assembly
62. Float bowl to throttle body gasket
63. Float bowl to throttle body screw assembly
64. Pump link
65. Idle mixture needle
66. Idle mixture needle spring
67. Idle mixture needle plug
68. Fast idle adjusting screw
69. Fast idle adjusting screw spring
70. Hose
71. Solenoid and bracket assembly
72. Bracket attach screw

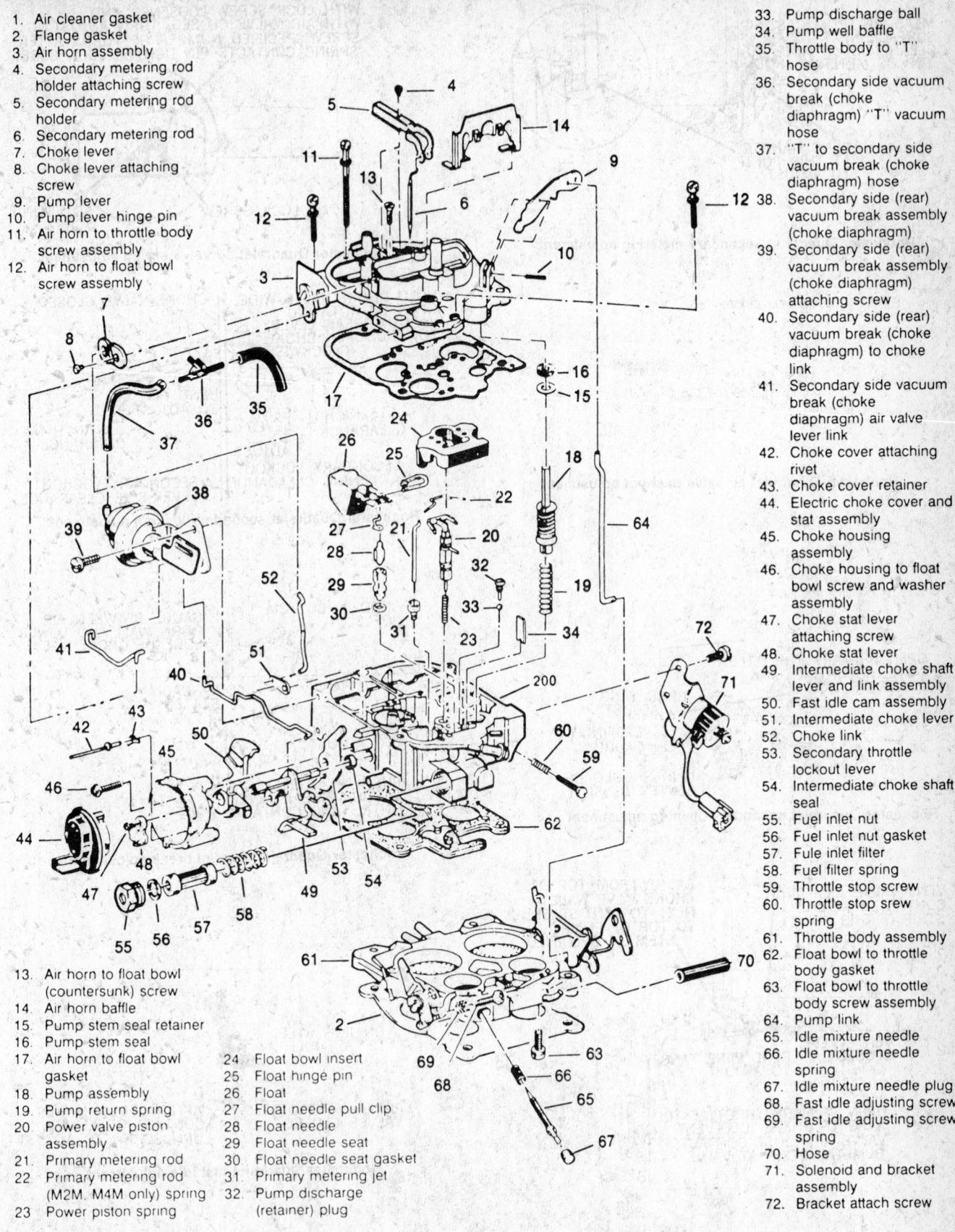

Rochester Quadrajet carburetor exploded view

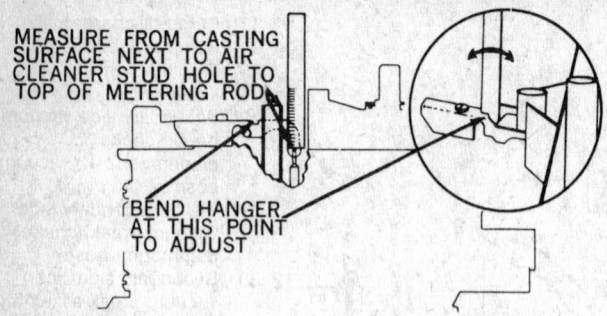

Rochester Quadrajet secondary metering adjustment

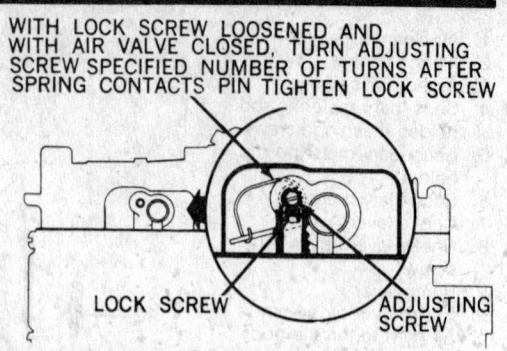

Rochester Quadrajet air valve spring adjustment

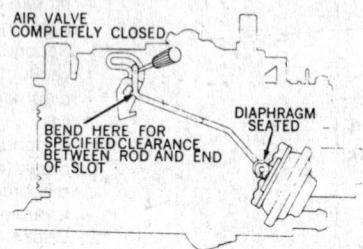

Rochester Quadrajet air valve dashpot adjustment

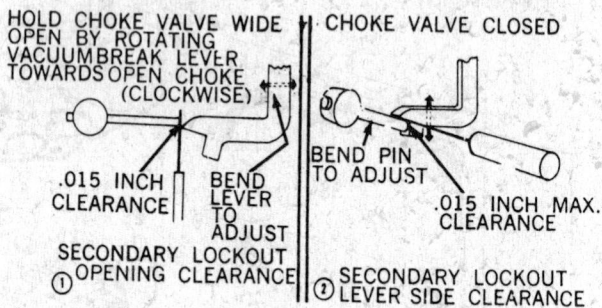

Rochester Quadrajet secondary lockout adjustment

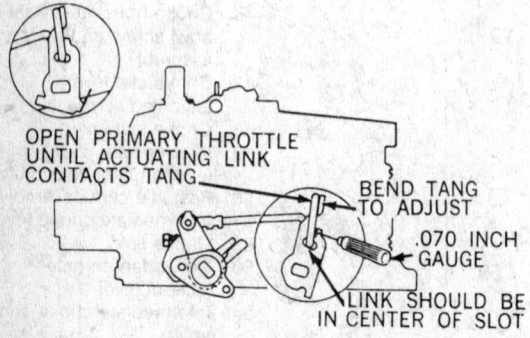

Rochester Quadrajet secondary opening adjustment

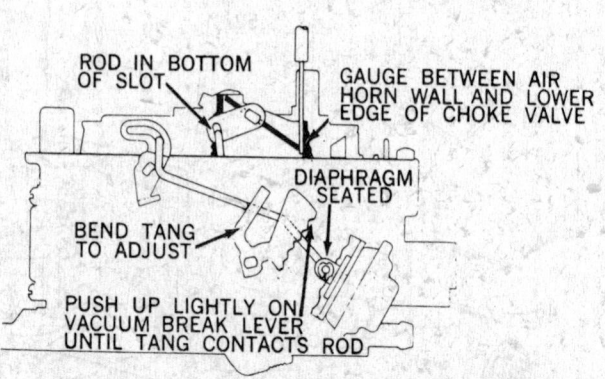

Rochester Quadrajet vacuum break adjustment

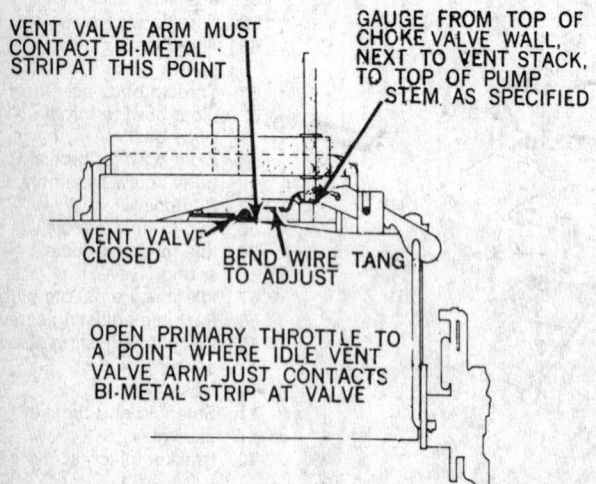

Rochester Quadrajet idle vent adjustment

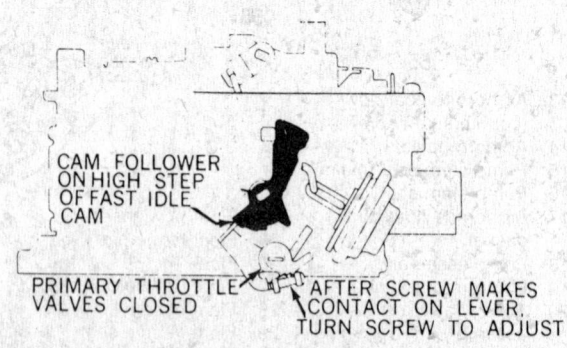

Rochester Quadrajet fast idle adjustment

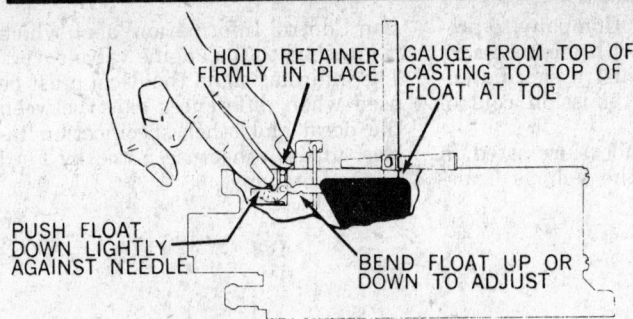

Rochester Quadrajet float level adjustment

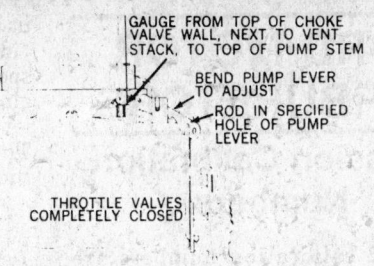

Rochester Quadrajet pump rod adjustment

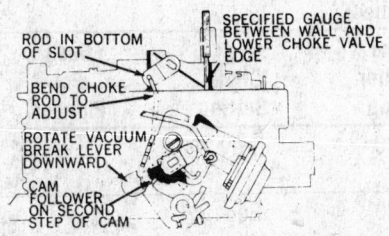

Rochester Quadrajet choke rod adjustment

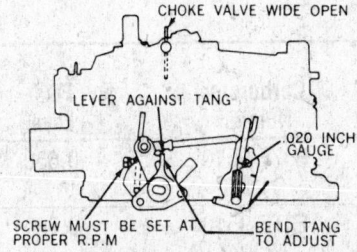

Rochester Quadrajet secondary closing adjustment

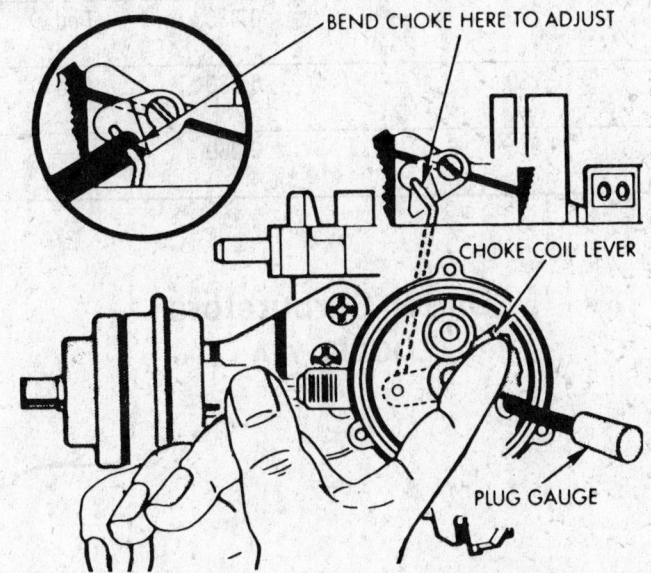

Rochester Quadrajet choke coil lever adjustment

FORD MOTOR CARBURETORS

Emission Calibration Numbers

Emission calibration numbers are used by Ford Motor Company to provide the technician with the necessary specifications to adjust a specific engine to the proper emission control levels.

The calibration numbers are listed on the lower right of the Vehicle Emission Control Information label, which is attached to the engine valve cover. The information on the decal must be used when differences exist between the decal and other specification tables, unless otherwise noted by Ford Motor Company.

CARTER MODEL YFA
Ford Motor Company
(All measurements in inches)

Year	Carburetor Number	Float Level	Float Drop	Choke Unloader Setting	Choke Setting	Dash Pot Plunger	Initial Choke Opening
1986	E57E-9510 DA, DB	0.650	—	0.270	Gray ①	—	0.320
	E5TE-9510 AA, TA, UA, VA, RA, SA, JA, BA, MA, CA, GA	0.780	—	0.330	Red ①	—	0.360
	E5TE-9510 FA	0.780	—	0.330	Red ①	—	0.340
	E5TE-9510 HA	0.780	—	0.330	Red ①	—	0.320
	D5TE-9510 AGB	0.375		0.280	—	—	0.230

① Choke cap index plate color

Carter Carburetors
MODEL YFA

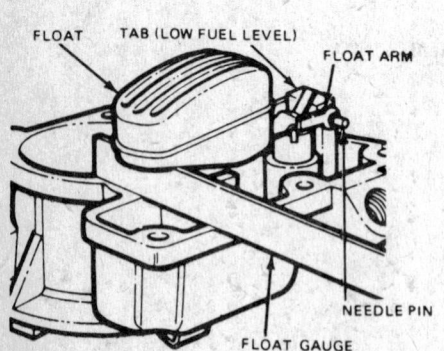

Carter YFA float drop measurement

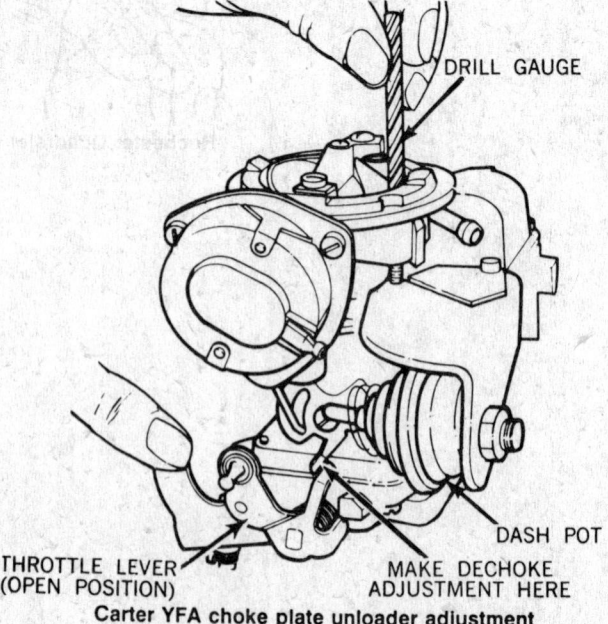

Carter YFA choke plate unloader adjustment

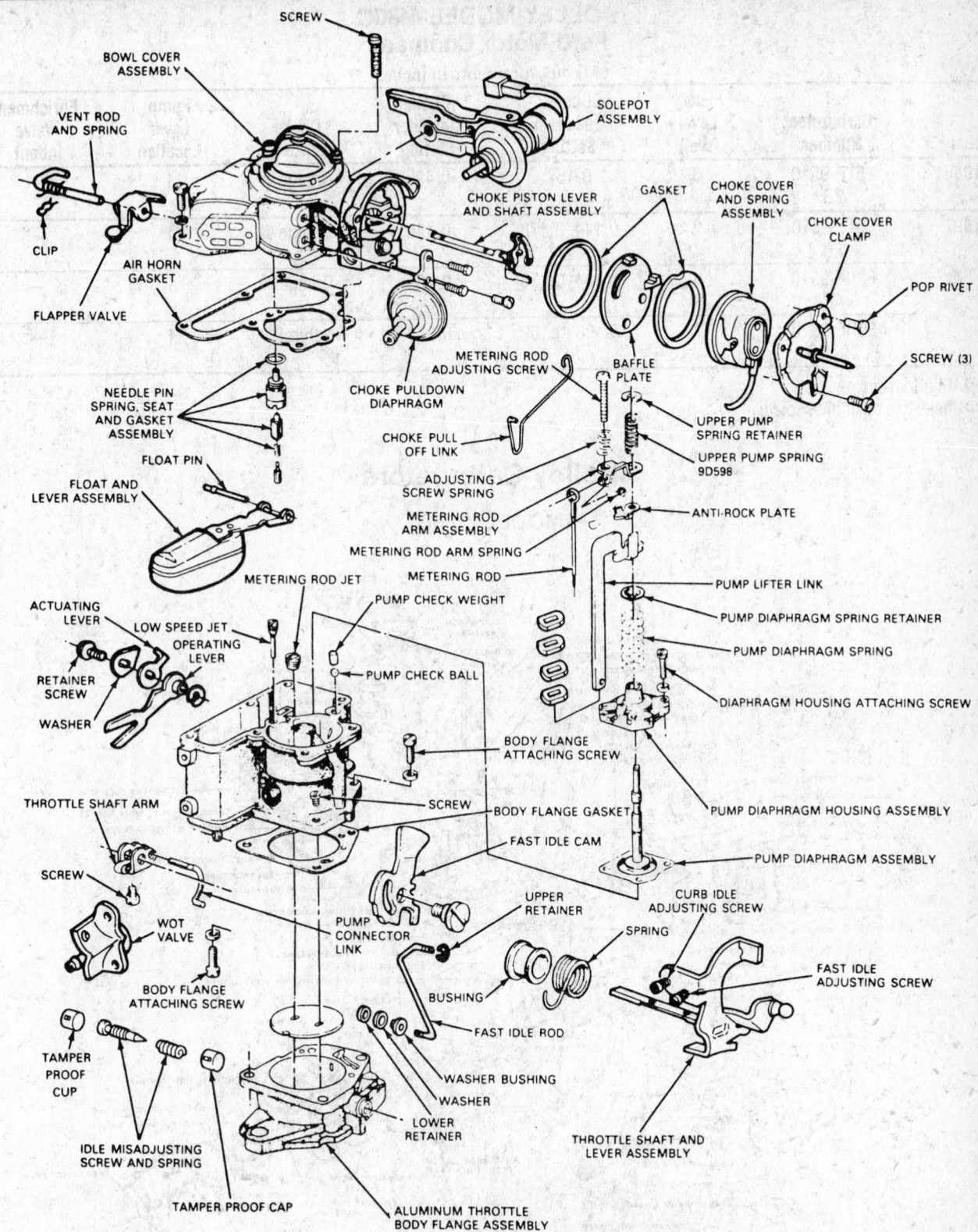

SCREW

BOWL COVER ASSEMBLY

VENT ROD AND SPRING

SOLEPOT ASSEMBLY

CLIP

AIR HORN GASKET

FLAPPER VALVE

CHOKE PISTON LEVER AND SHAFT ASSEMBLY

GASKET

CHOKE COVER AND SPRING ASSEMBLY

CHOKE COVER CLAMP

POP RIVET

SCREW (3)

NEEDLE PIN SPRING, SEAT AND GASKET ASSEMBLY

METERING ROD ADJUSTING SCREW

CHOKE PULLDOWN DIAPHRAGM

CHOKE PULL OFF LINK

BAFFLE PLATE

UPPER PUMP SPRING RETAINER

UPPER PUMP SPRING 9D598

FLOAT PIN

ADJUSTING SCREW SPRING

FLOAT AND LEVER ASSEMBLY

METERING ROD ARM ASSEMBLY

ANTI-ROCK PLATE

METERING ROD ARM SPRING

METERING ROD

METERING ROD JET

PUMP CHECK WEIGHT

PUMP LIFTER LINK

ACTUATING LEVER

LOW SPEED JET

OPERATING LEVER

PUMP DIAPHRAGM SPRING RETAINER

RETAINER SCREW

PUMP CHECK BALL

PUMP DIAPHRAGM SPRING

WASHER

DIAPHRAGM HOUSING ATTACHING SCREW

BODY FLANGE ATTACHING SCREW

THROTTLE SHAFT ARM

SCREW

BODY FLANGE GASKET

PUMP DIAPHRAGM HOUSING ASSEMBLY

FAST IDLE CAM

PUMP DIAPHRAGM ASSEMBLY

WOT VALVE

PUMP CONNECTOR LINK

UPPER RETAINER

CURB IDLE ADJUSTING SCREW

BODY FLANGE ATTACHING SCREW

SPRING

FAST IDLE ADJUSTING SCREW

TAMPER PROOF CUP

BUSHING

FAST IDLE ROD

IDLE MISADJUSTING SCREW AND SPRING

WASHER BUSHING

WASHER

LOWER RETAINER

THROTTLE SHAFT AND LEVER ASSEMBLY

TAMPER PROOF CAP

ALUMINUM THROTTLE BODY FLANGE ASSEMBLY

Carter YFA carburetor exploded view

HOLLEY MODEL 4180C
Ford Motor Company
(All measurements in inches)

Year	Carburetor Number	Fuel Level (Wet)	Choke Pulldown Setting	Choke Unloader Setting	Choke Setting	Pump Lever Location	Enrichment Valve Indent.
1986	E5TE-9510 ZB	①	0.157	0.425	Orange ②	1	—
1987	E5TE-9510 ZB	①	0.144–0.170	0.425	Orange ②	1	—
	E6HE-9510 AC	①	0.144–0.170	0.425	Orange ②	1	—
	E6HE-9510 GA, GB	①	0.138–0.162	0.300	Natural ②	1	—

①At bottom of sight plug
②Choke cap color index plate

Holley Carburetors

MODEL 4180C

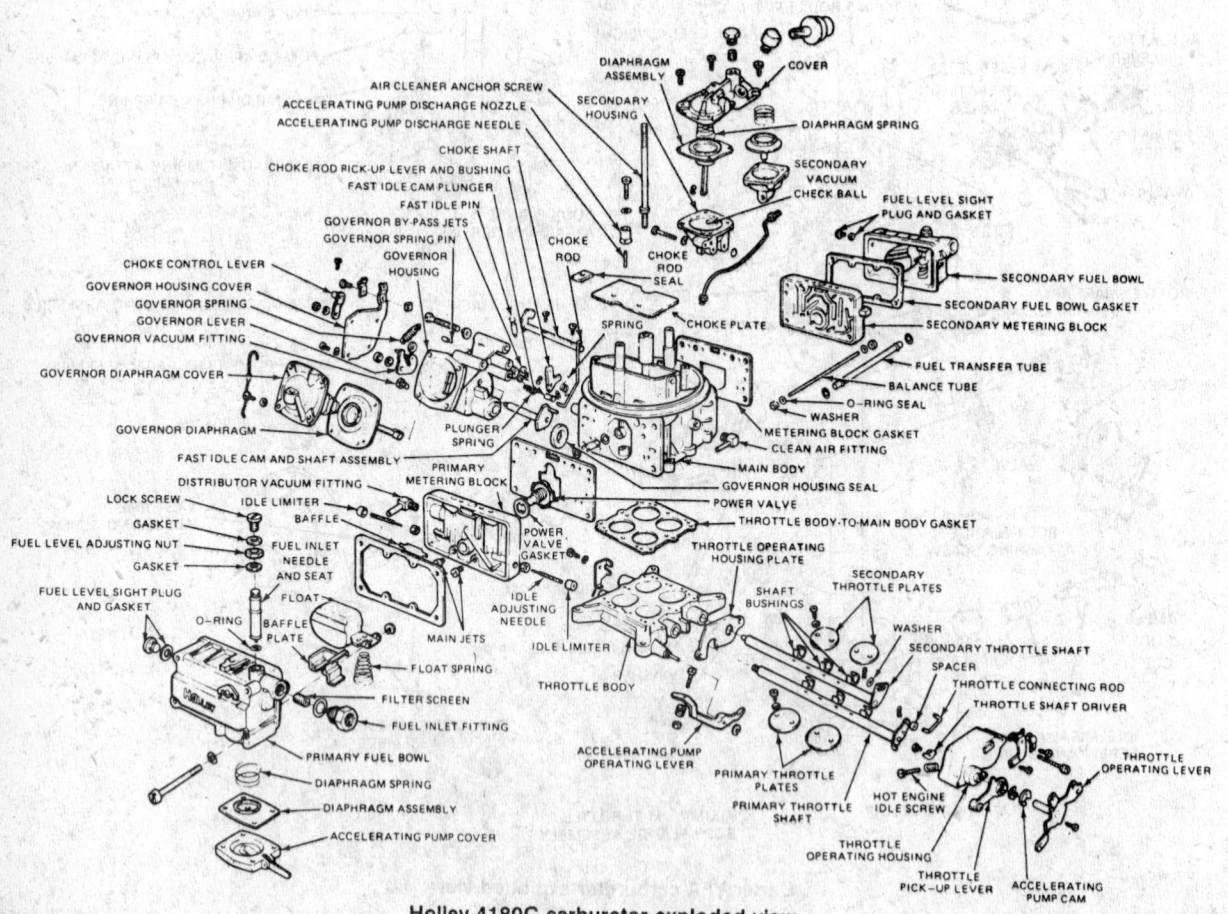

Holley 4180C carburetor exploded view

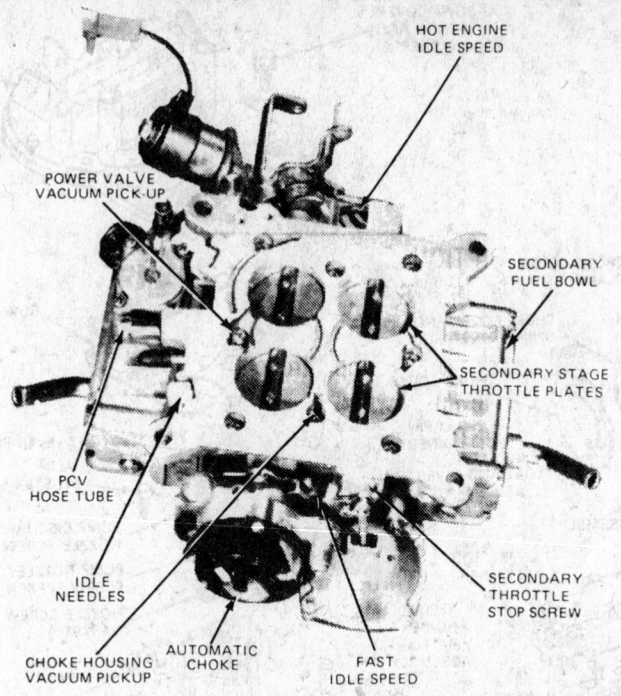

PACKER VALVE
VACUUM PICK-UP

HOT ENGINE
IDLE SPEED

SECONDARY
FUEL BOWL

SECONDARY STAGE
THROTTLE PLATES

PCV
HOSE TUBE

IDLE
NEEDLES

CHOKE HOUSING
VACUUM PICKUP

AUTOMATIC
CHOKE

FAST
IDLE SPEED

SECONDARY
THROTTLE
STOP SCREW

Holley 4180C carburetor bottom view

MOTORCRAFT MODEL 2150
Ford Motor Company

(All measurements in inches)

Year	Carburetor Number	Float Level (Dry)	Choke Unloader Setting	Choke Setting	Accelerator Pump Rod Location	Fuel Level (Wet)	Choke Pulldown Setting (Min.)
1986	E69E-9510 CA, DA	$1/16$	0.250	V notch	4	0.810	0.126–0.146
	AA, BA	$1/16$	0.250	V notch	4	0.810	0.126–0.146

Motorcraft Carburetors

MODEL 2150

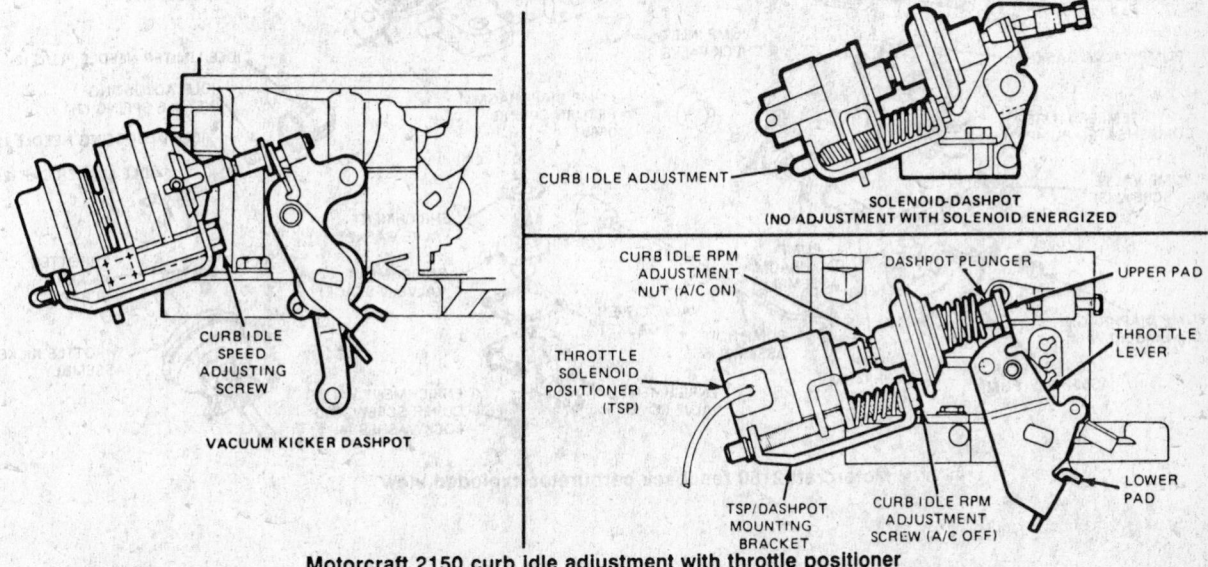

CURB IDLE ADJUSTMENT

SOLENOID-DASHPOT
(NO ADJUSTMENT WITH SOLENOID ENERGIZED

CURB IDLE
SPEED
ADJUSTING
SCREW

VACUUM KICKER DASHPOT

CURB IDLE RPM
ADJUSTMENT
NUT (A/C ON)

DASHPOT PLUNGER

UPPER PAD

THROTTLE
SOLENOID
POSITIONER
(TSP)

THROTTLE
LEVER

LOWER
PAD

TSP/DASHPOT
MOUNTING
BRACKET

CURB IDLE RPM
ADJUSTMENT
SCREW (A/C OFF)

Motorcraft 2150 curb idle adjustment with throttle positioner

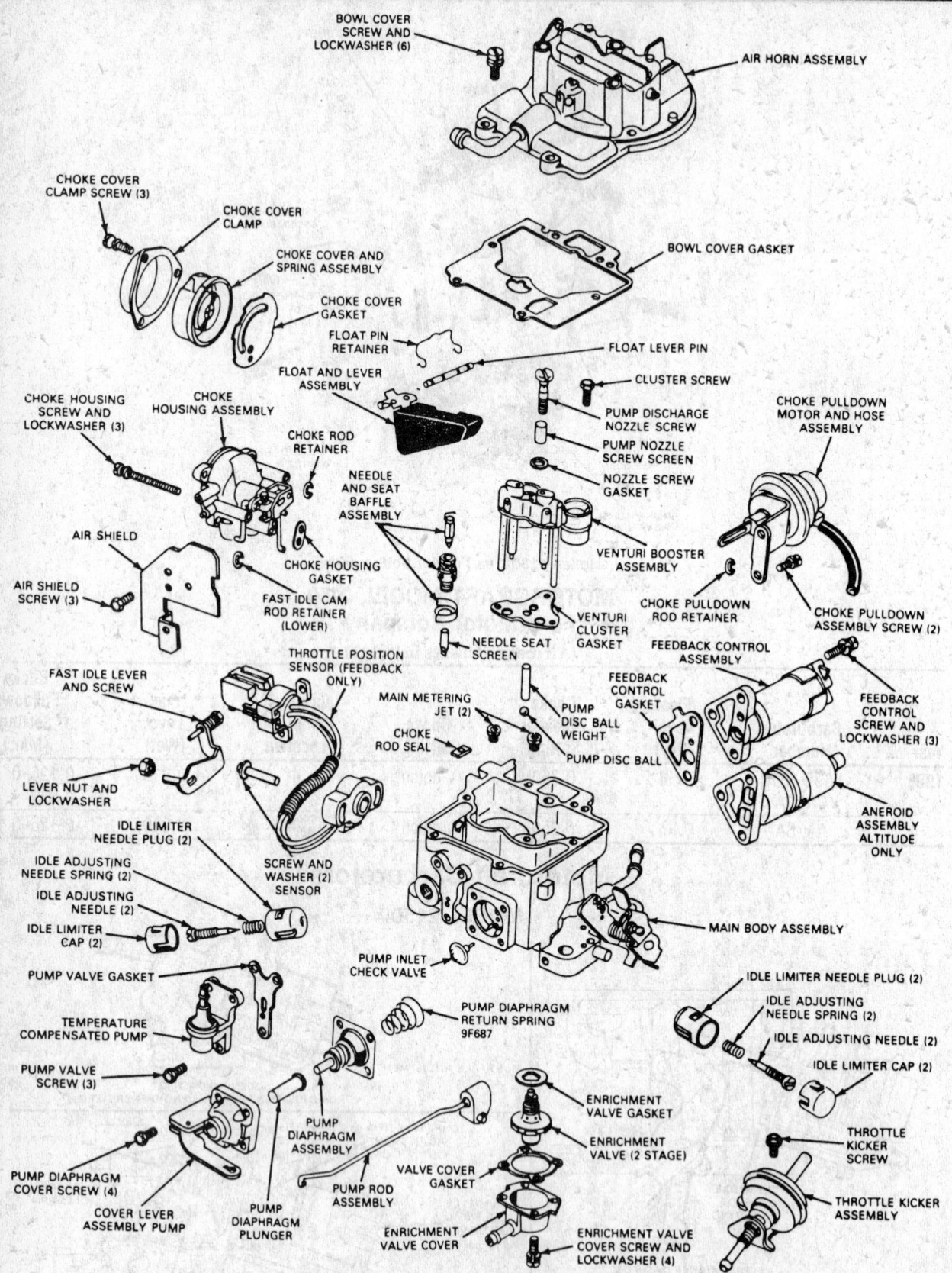

Motorcraft 2150 feedback carburetor exploded view

Motorcraft 2150 accelerator pump stroke hole location

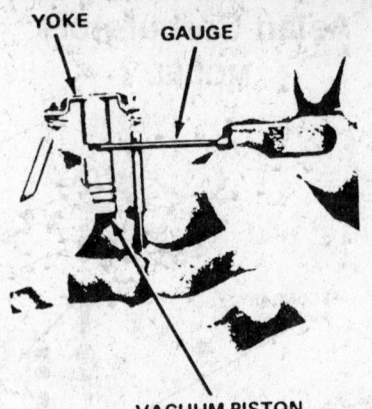

Motorcraft 2150 metering rod vacuum piston adjustment to a clearance of 0.120 In.

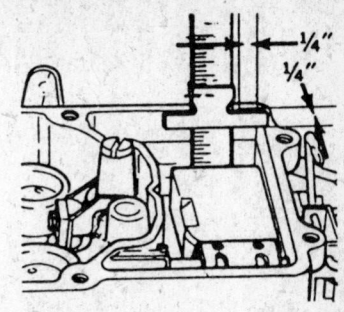

Motorcraft 2150 wet float level adjustment

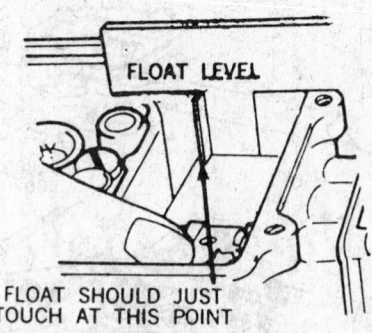

Motorcraft 2150 dry float level adjustment

ASIAN MODEL Y
Ford Motor Company

(All measurements in millimeters and degrees)

Year	Carburetor Number	Choke Pulldown Setting	Fast Idle Cam Setting	Dechoke Setting	Float Setting (Dry)	Choke Cap Setting	Fast Idle
1987	E77E-9510-AA	18°	14.5° ①	195 sec.	47.1 mm	20°C	3200 ②
1988	E87E-9510-AA	18°	14.5° ①	195 sec.	47.1 mm	20°C	3200 ②

① On step #1
② 3100 rpm for vehicles with less than 100 miles.

Asian Carburetors
MODEL Y

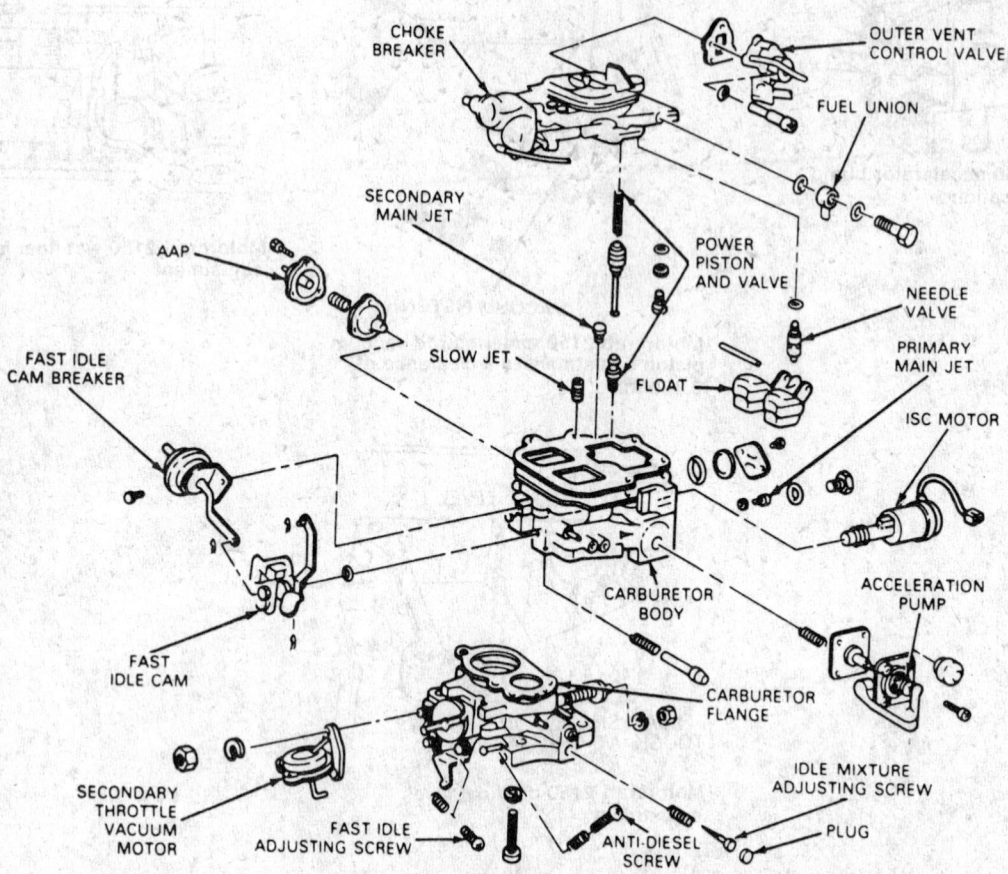

Aisan model Y carburetor — exploded view

JEEP

Carter Carburetors

MODEL BBD

Float Adjustment

1. Remove the air horn.
2. Apply light finger pressure to the vertical float tab to exert gentle pressure against the inlet needle.
3. Lay a straight edge across the float bowl and measure the gap between the straight edge and the top of the float at its highest point. The gap should be 0.250 in.
4. To adjust, remove the float and bend the lower tab. Replace the float and check the gap.

Fast Idle Cam Adjustment

1. Loosen the choke housing cover and turn it ¼ turn right. Tighten one screw.
2. Slightly open the throttle and place the fast idle screw on the second cam step.
3. Measure the distance between the choke plate and the air horn wall. The distance should be 0.095 in.
4. If adjustment is necessary, bend the fast idle cam link down to increase and up to decrease the gap.
5. Return the choke cover cap to the original setting.

Initial Choke Valve Clearance

1. Position the fast idle screw on the top step of the fast idle cam.
2. Using a vacuum pump, seat the choke vacuum break.
3. Apply light closing pressure in the choke plate to position the plate as far closed as possible without forcing it.
4. Measure the distance between the air horn wall and the choke plate,

it should be 0.140 in. If it is not, bend the choke vacuum break link until it is.

Unloader Adjustment

1. With the throttle held fully open, apply pressure on the choke valve toward the closed position and measure the clearance between the lower edge of the choke valve and the air horn wall.
2. The measurement should be 0.280 in. Adjust by bending the tang on the throttle lever which contacts the fast idle cam. Bend toward the cam to increase the clearance.

NOTE: Do not bend the unloader down so that it binds or interferes with any other component.

Vacuum (Step Up) Piston Gap

1. Turn the adjusting screw, mounted on top of the unit, so that the gap between the metering rod lifter lower edge, and the top of the vacuum piston, is 0.035 in.

CARTER MODELS
MODEL BBD

CARTER MODEL BBD-2
Jeep

(All measurements in inches)

Year	Carburetor Number	Float Level	Step-up Piston Gap	Initial Choke Clearance	Fast Idle Cam Setting	Choke Cover Setting	Choke Unloader (Min.)	Fast Idle Speed (rpm) ①
1986	8383	0.250	0.035	0.140	0.095	1 Rich	0.280	1850
	8384	0.250	0.035	0.140	0.095	1 Rich	0.280	1700
1987	8383	0.250	0.035	0.140	0.095	1 Rich	0.280	1850
	8384	0.250	0.035	0.140	0.095	1 Rich	0.280	1700
1988	8383	0.250	0.035	0.140	0.095	1 Rich	0.280	1850
	8384	0.250	0.035	0.140	0.095	1 Rich	0.280	1700
1989	8383	0.250	0.035	0.140	0.095	1 Rich	0.280	1850
	8384	0.250	0.035	0.140	0.095	1 Rich	0.280	1700

① On second step of fast idle cam with TCS solenoid and EGR disconnected.

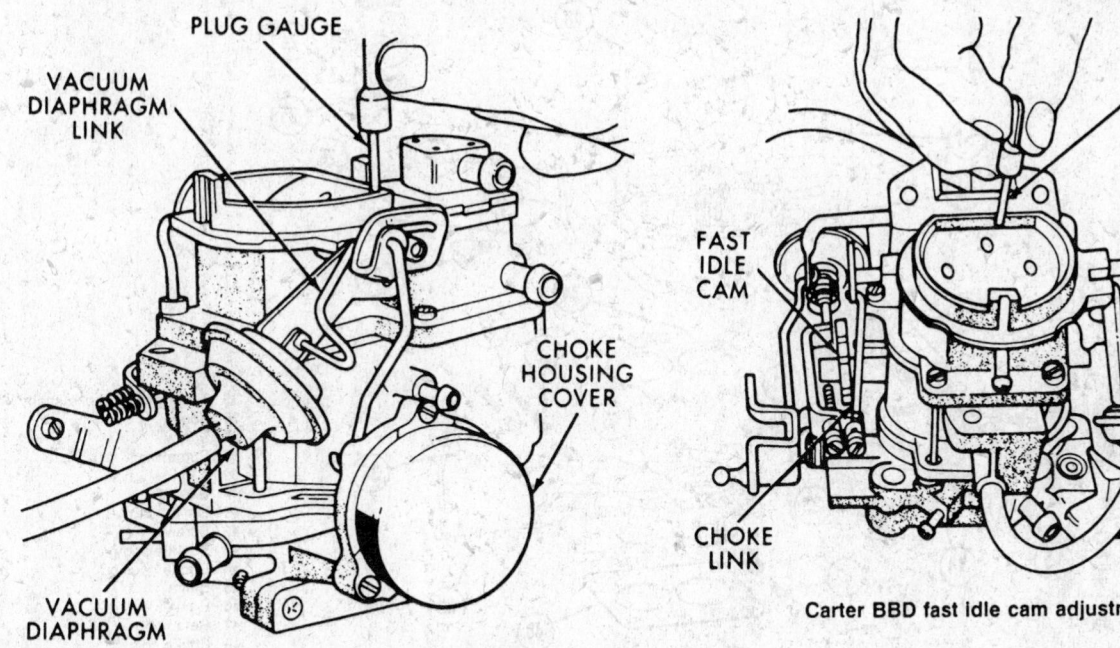

Carter BBD initial choke valve adjustment

Carter BBD fast idle cam adjustment

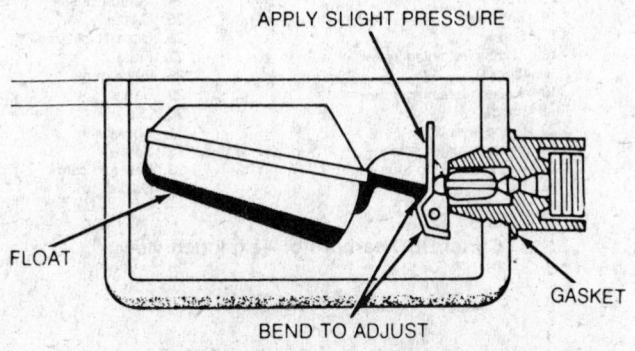

Carter BBD float level adjustment

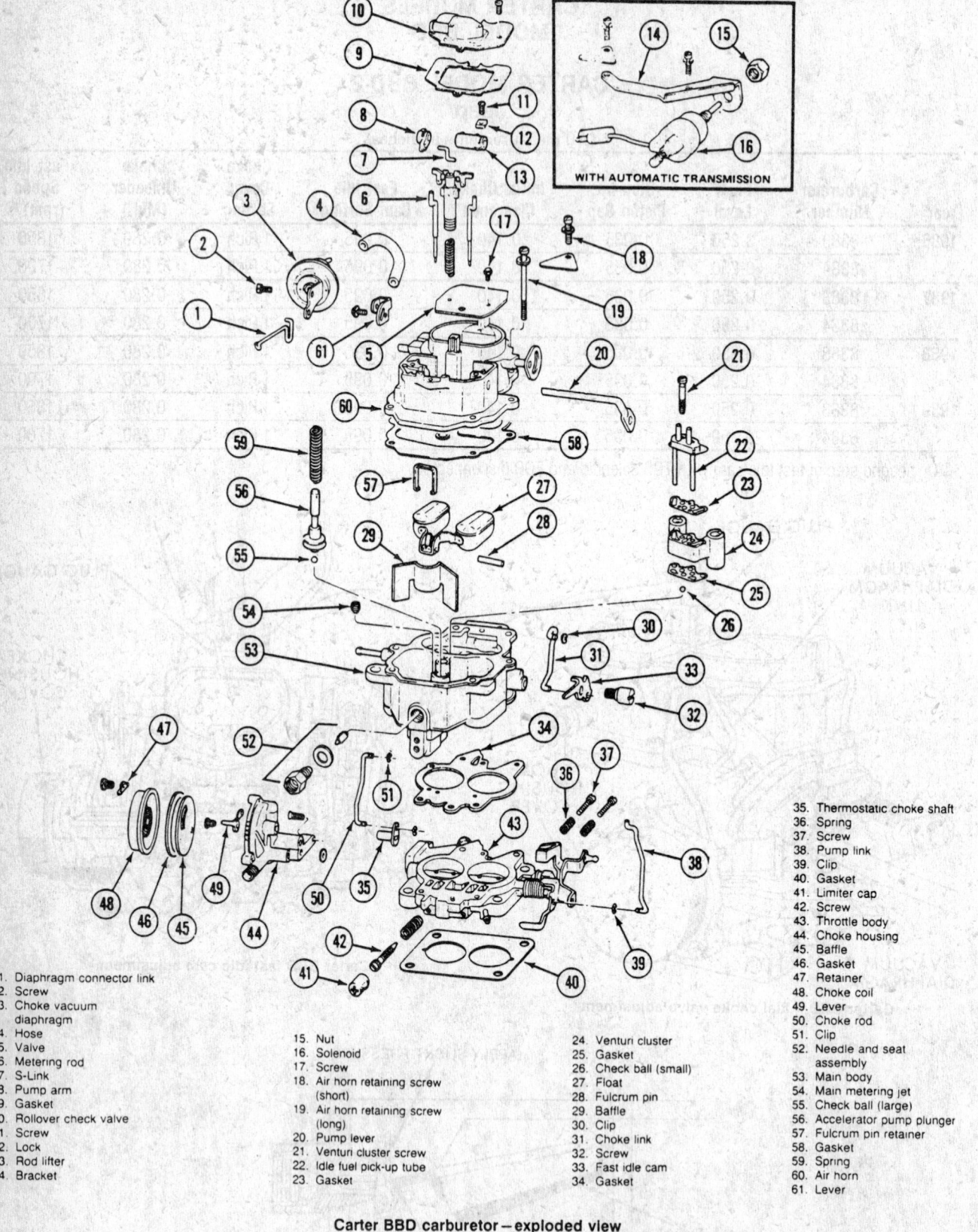

WITH AUTOMATIC TRANSMISSION

Carter BBD carburetor—exploded view

1. Diaphragm connector link
2. Screw
3. Choke vacuum diaphragm
4. Hose
5. Valve
6. Metering rod
7. S-Link
8. Pump arm
9. Gasket
10. Rollover check valve
11. Screw
12. Lock
13. Rod lifter
14. Bracket

15. Nut
16. Solenoid
17. Screw
18. Air horn retaining screw (short)
19. Air horn retaining screw (long)
20. Pump lever
21. Venturi cluster screw
22. Idle fuel pick-up tube
23. Gasket

24. Venturi cluster
25. Gasket
26. Check ball (small)
27. Float
28. Fulcrum pin
29. Baffle
30. Clip
31. Choke link
32. Screw
33. Fast idle cam
34. Gasket

35. Thermostatic choke shaft
36. Spring
37. Screw
38. Pump link
39. Clip
40. Gasket
41. Limiter cap
42. Screw
43. Throttle body
44. Choke housing
45. Baffle
46. Gasket
47. Retainer
48. Choke coil
49. Lever
50. Choke rod
51. Clip
52. Needle and seat assembly
53. Main body
54. Main metering jet
55. Check ball (large)
56. Accelerator pump plunger
57. Fulcrum pin retainer
58. Gasket
59. Spring
60. Air horn
61. Lever

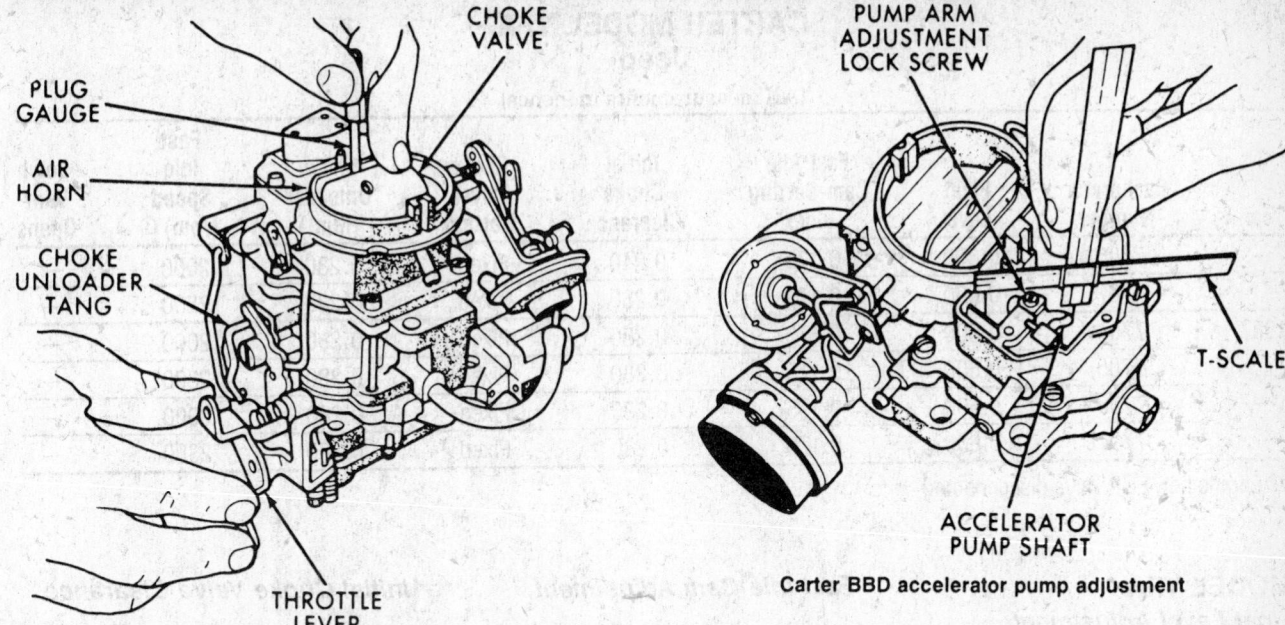

Carter BBD choke unloader adjustment

Carter BBD accelerator pump adjustment

2. Counting the number of turns involved, turn the curb idle adjustment screw counterclockwise, until the throttle valves are completely closed.

3. Fully depress the vacuum piston, while exerting moderate pressure on the metering rod lifter tab. In this position, tighten the rod lifter lock screw.

4. Release the piston and rod lifter.

NOTE: The accelerator pump should now be adjusted.

5. Return the curb idle adjustment screw to its original position.

Accelerator Pump

1. Counting the number of turns involved, turn the curb idle adjustment screw counterclockwise, until the throttle valves are completely closed.

2. Open the choke valve so that the fast idle cam will allow the throttle valves to seat in their bores.

3. Turn the curb idle adjustment screw clockwise, so that it just barely touches the stop, then, turn it 2 full turns further.

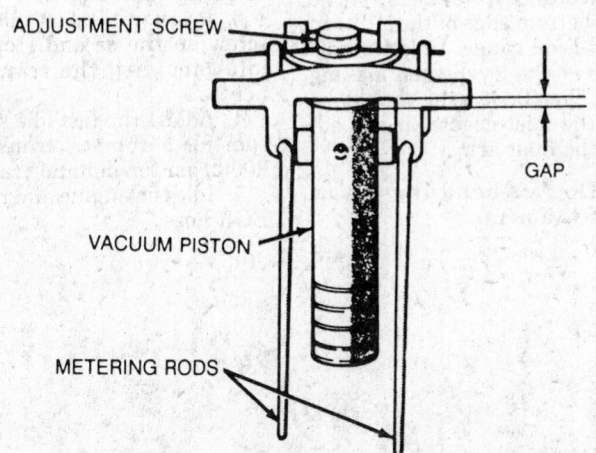

Carter BBD piston gap adjustment

4. Measure the distance between the surface of the air horn and the top of the accelerator pump shaft with a T-scale. The distance should be 0.520 in.

5. If the dimension is not correct,

loosen the pump arm adjusting screw and rotate the sleeve to adjust the pump travel. Tighten the lock screw.

6. Return the curb idle screw to its original position.

CARTER MODEL YFA
Jeep

(All measurements in inches)

Year	Carburetor Number	Float Level	Fast Idle Cam Setting Index	Initial Choke Clearance	Choke Cover Setting	Choke Unloader (Min.)	Fast Idle Speed (rpm) ①	Bowl Vent Opens
1986	7704	0.600	0.175	0.280	Fixed	0.280	2000	—
	7705	0.600	0.175	0.280	Fixed	0.280	2300	—

CARTER MODEL YFA
Jeep
(All measurements in inches)

Year	Carburetor Number	Float Level	Fast Idle Cam Setting Index	Initial Choke Clearance	Choke Cover Setting	Choke Unloader (Min.)	Fast Idle Speed (rpm) ①	Bowl Vent Opens
	7706	0.600	0.175	0.240	Fixed	0.280	2000	—
	7707	0.600	0.175	0.280	Fixed	0.280	2300	—
1987	7704	0.600	0.175	0.280	Fixed	0.280	2000	—
	7705	0.600	0.175	0.280	Fixed	0.280	2300	—
	7706	0.600	0.175	0.280	Fixed	0.280	2000	—
	7707	0.600	0.175	0.280	Fixed	0.280	2300	—

① Engine hot, EGR valve disconnected

MODEL YFA
Float Level Adjustment

1. Remove the top of the carburetor and the gasket.

2. Invert the carburetor top and check the clearance from the top of the float to the bottom edge of the air horn with a float level gauge. Hold the carburetor top at eye level when making the check. The float arm should be resting on the inlet needle pin. To adjust, bend the float arm.

NOTE: Do Not bend the tab at the end of the arm.

Fast Idle Cam Adjustment

1. Run the engine to normal operating temperature. Connect a tachometer.

2. Disconnect and plug the EGR valve vacuum hose.

3. Position the fast idle adjustment screw on the second stop of the fast idle cam with the transmission in neutral.

4. Adjust the fast idle speed to 2300 rpm for automatic transmission and 2000 rpm for manual transmission.

5. Idle the engine and reconnect the EGR hose.

Initial Choke Valve Clearance

1. Position the fast idle screw on the top step of the fast idle cam.

2. Using a vacuum pump, seat the choke vacuum break.

3. Apply light closing pressure in the choke plate to position the plate as far closed as possible without forcing it.

4. Measure the distance between the air horn wall and the choke plate, it should be 0.280 in. If it is not, bend the choke vacuum break link until it is.

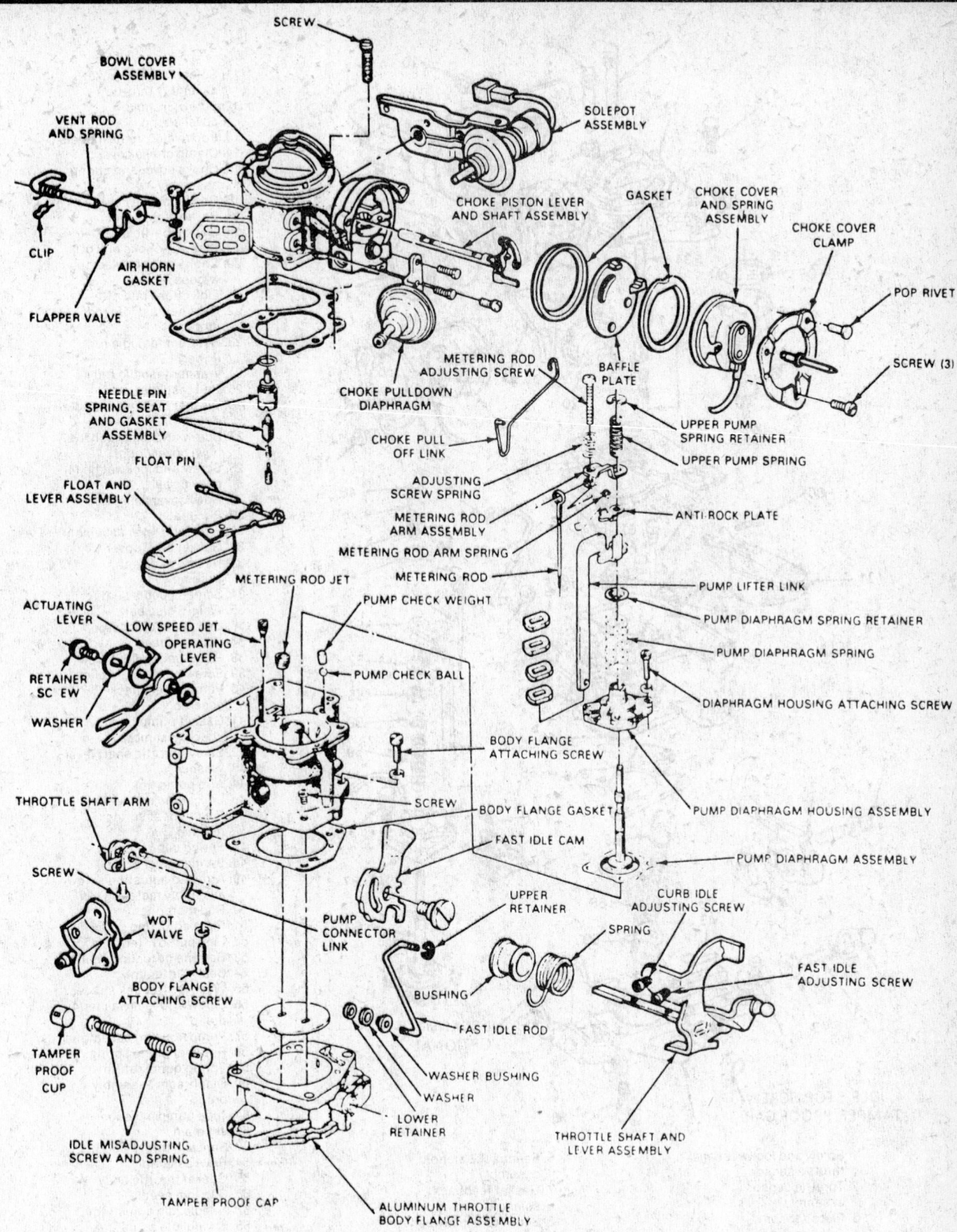

SCREW

BOWL COVER ASSEMBLY

VENT ROD AND SPRING

CLIP

FLAPPER VALVE

AIR HORN GASKET

SOLEPOT ASSEMBLY

CHOKE PISTON LEVER AND SHAFT ASSEMBLY

GASKET

CHOKE COVER AND SPRING ASSEMBLY

CHOKE COVER CLAMP

POP RIVET

SCREW (3)

NEEDLE PIN SPRING, SEAT AND GASKET ASSEMBLY

METERING ROD ADJUSTING SCREW

CHOKE PULLDOWN DIAPHRAGM

BAFFLE PLATE

UPPER PUMP SPRING RETAINER

UPPER PUMP SPRING

CHOKE PULL OFF LINK

FLOAT PIN

FLOAT AND LEVER ASSEMBLY

ADJUSTING SCREW SPRING

METERING ROD ARM ASSEMBLY

METERING ROD ARM SPRING

METERING ROD

ANTI-ROCK PLATE

PUMP LIFTER LINK

PUMP DIAPHRAGM SPRING RETAINER

PUMP DIAPHRAGM SPRING

DIAPHRAGM HOUSING ATTACHING SCREW

ACTUATING LEVER

LOW SPEED JET OPERATING LEVER

METERING ROD JET

PUMP CHECK WEIGHT

PUMP CHECK BALL

RETAINER SCREW

WASHER

BODY FLANGE ATTACHING SCREW

THROTTLE SHAFT ARM

SCREW

SCREW

BODY FLANGE GASKET

FAST IDLE CAM

PUMP DIAPHRAGM HOUSING ASSEMBLY

PUMP DIAPHRAGM ASSEMBLY

WOT VALVE

BODY FLANGE ATTACHING SCREW

PUMP CONNECTOR LINK

UPPER RETAINER

SPRING

CURB IDLE ADJUSTING SCREW

FAST IDLE ADJUSTING SCREW

TAMPER PROOF CUP

BUSHING

FAST IDLE ROD

WASHER BUSHING

WASHER

LOWER RETAINER

THROTTLE SHAFT AND LEVER ASSEMBLY

IDLE MISADJUSTING SCREW AND SPRING

TAMPER PROOF CAP

ALUMINUM THROTTLE BODY FLANGE ASSEMBLY

Carter YFA non-feedback carburetor—exploded view

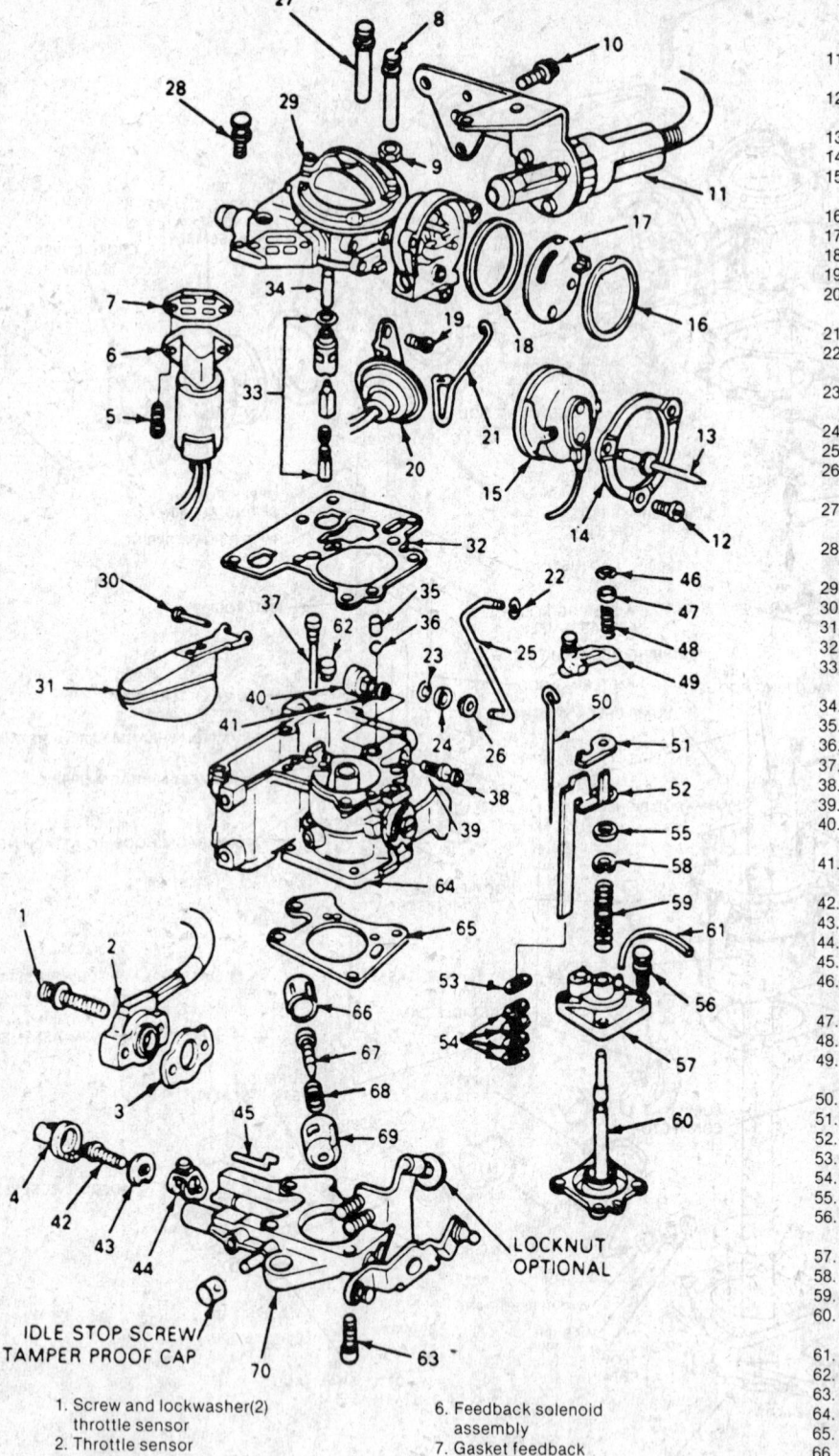

11. Solenoid and bracket assembly throttle
12. Screw(3) choke cover clamp
13. Pop rivet(2) some models
14. Clamp choke cover
15. Choke cover and spring assembly
16. Gasket choke cover
17. Baffle plate cover
18. Gasket baffle plate
19. Screw(2) choke pull off
20. Choke pull off assembly w/hose
21. Link choke pull off
22. Retainer fast idle rod (upper)
23. Retainer fast idle rod (lower)
24. Washer fast idle rod
25. Rod fast idle
26. Washer bushing fast idle rod
27. Screw and lockwasher(2) bowl cover (long)
28. Screw and lockwasher(4) bowl cover
29. Bowl cover assembly
30. Pin float
31. Float and lever assembly
32. Gasket bowl cover
33. Needle and seat assembly
34. Screen needle seat
35. Weight disc ball
36. Ball pump discharge
37. Jet low speed
38. Plug pump relief screw
39. Screw pump relief check
40. Pump relief check assembly
41. Gasket pump relief check assembly
42. Screw throttle shaft lever
43. Washer
44. Arm pump link
45. Link pump connector
46. E-clip upper spring retainer
47. Spring clip
48. Spring upper pump
49. Arm and adjusting screw assembly metering rod
50. Rod metering
51. Plate adjusting screw
52. Link pump lifter
53. Retainer lifter link seal
54. Seal(4) lifter link
55. Washer lifter link spacer
56. Screw and lockwasher(4) pump
57. Pump housing assembly
58. Retainer pump spring
59. Spring pump return
60. Diaphragm assembly pump
61. Tube pump passage
62. Jet main
63. Screw(4) throttle body
64. Bowl assembly
65. Gasket throttle body
66. Cap idle needle
67. Needle idle adjusting
68. Spring idle adjusting needle
69. Clip idle needle
70. Throttle body assembly

IDLE STOP SCREW/
TAMPER PROOF CAP

LOCKNUT
OPTIONAL

1. Screw and lockwasher(2) throttle sensor
2. Throttle sensor assembly
3. Plate sensor
4. Drive coupler sensor
5. Screw(2) feedback solenoid
6. Feedback solenoid assembly
7. Gasket feedback solenoid
8. Screw and lockwasher solenoid bracket
9. Locknut bracket screw
10. Screw(3) bracket

Carter YFA feedback carburetor — exploded view

MOTORCRAFT MODEL 2150
Jeep
(All measurements in inches)

Year	Carburetor Number	Float Level (Dry)	Fuel Level (Wet)	Initial Choke Valve Clearance	Fast Idle Cam Setting ②	Choke Cover Setting	Choke Unloader Valve Clearance	Fast Idle Speed ①	Bowl Vent Clearance	Rod Pump Location Hole
1986	4RHA2	0.575	0.930	0.136	0.086	2 Rich	0.350	1600	0.120	—
	5RHA2	0.328	0.930	0.118	0.076	Y	0.420	1600	—	—
1987	4RHA2	0.575	0.930	0.136	0.086	2 Rich	0.350	1600	0.120	—
	5RHA2	0.328	0.930	0.118	0.076	Y	0.420	1600	—	—
1988	4RHA2	0.575	0.930	0.136	0.086	2 Rich	0.350	1600	0.120	—
	5RHA2	0.328	0.930	0.118	0.076	Y	0.420	1600	—	—
1989	5RHA2	0.328	0.930	0.118	0.076	Y	0.420	1600	—	—

① TCS solenoid and EGR disconnected, fast idle screw on 2nd cam step
② Measured between choke valve and air horn fast idle screw on 2nd cam step

Motorcraft Carburetors

MODEL 2150

Dry Float Adjustment

1. With the air horn assembly and the gasket removed raise the float by pressing down on the float tab until the fuel inlet needle is lightly seated.

2. Using a T-scale, measure the distance from the fuel bowl machined surface to either corner of the float ⅛ in. (3mm) from the free end.

3. To adjust bend the float tab and hold the fuel inlet needle off its eat in order to prevent damage to the seat and the tip of the needle.

Wet Float Adjustment

——————— **CAUTION** ———————
Exercise extreme care when performing this adjustment as fuel vapors and liquid fuel are present and could cause personal injury if ignited.

1. Place the vehicle on a flat, level surface and run the engine to normal operating temperature. Turn off the engine and remove the air cleaner.

2. Remove the air horn attaching screws, but leave the air horn in place.

3. Start the engine and let it idle for one minute. Shut off the engine and remove the air horn and gasket.

4. Use a T-scale to measure the vertical distance between the machined surface of the carburetor body and the fuel level in the bowl. Make this measurement as near the center of the bowl as possible. The proper distance is ¼ in. To adjust, bend the float tab.

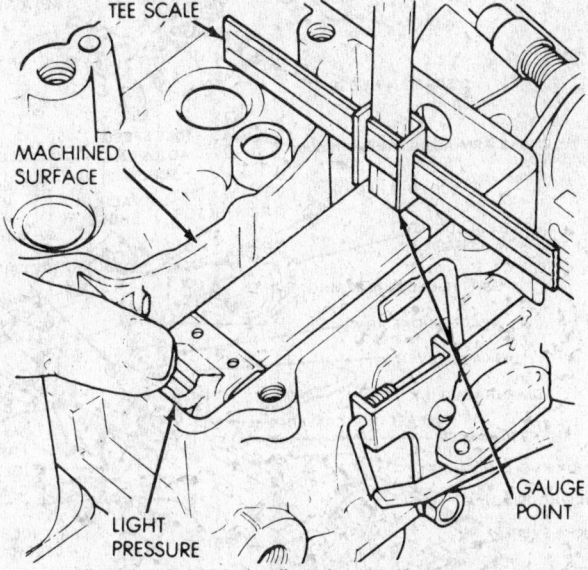

Motorcraft 2150 dry float level adjustment

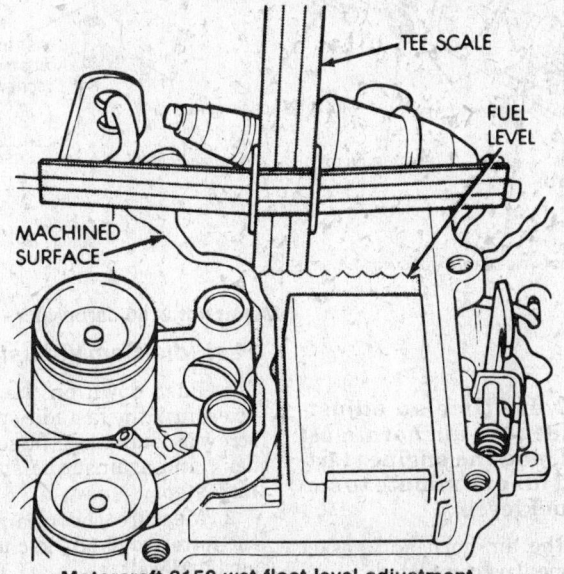

Motorcraft 2150 wet float level adjustment

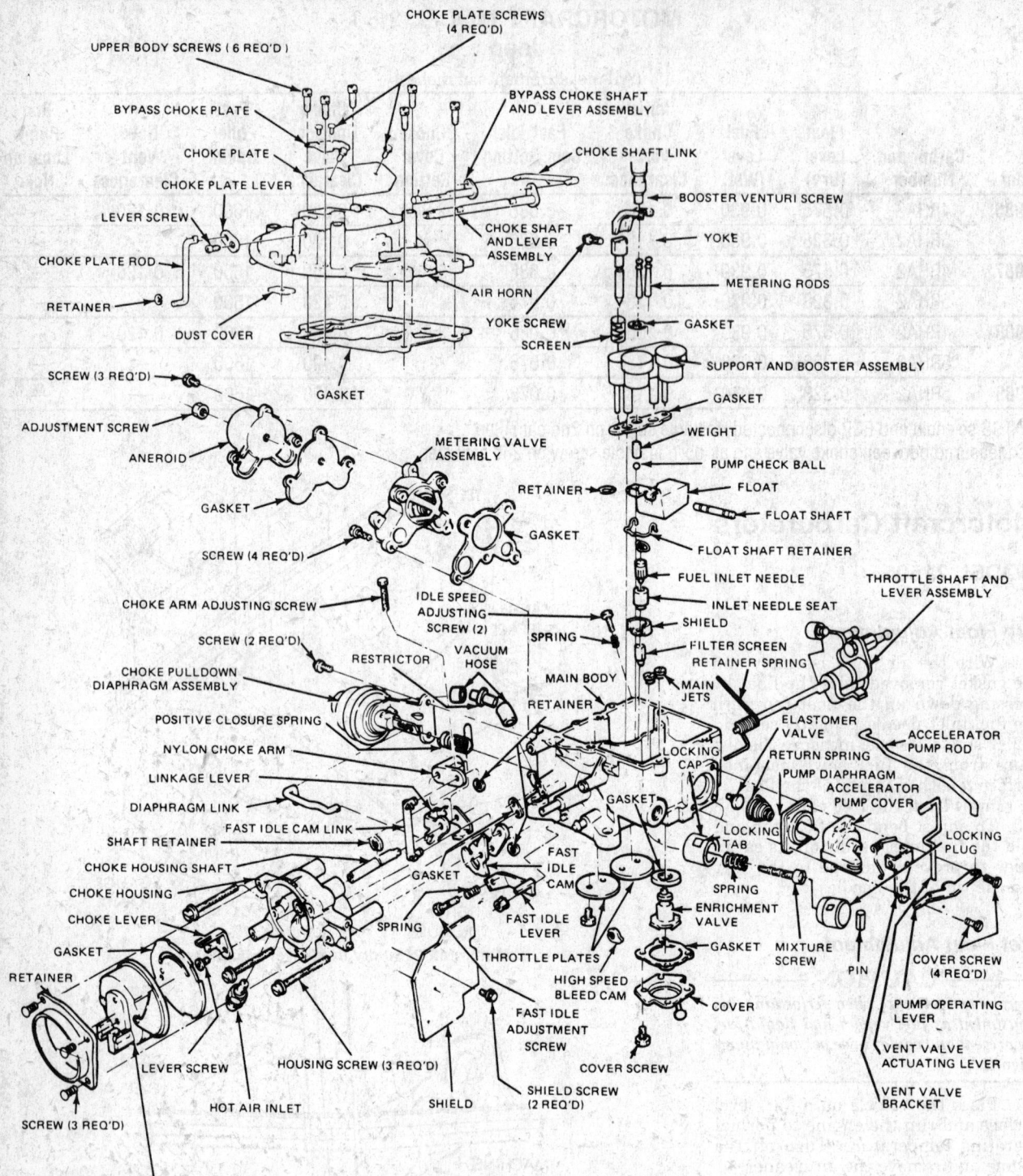

Motorcraft 2150 carburetor — exploded view

NOTE: Every time an adjustment is made, the air horn must be replaced, and the engine started and idled for one minute to stabilize the fuel level.

5. Install the air horn and gasket when adjustment is completed.

Fast Idle Cam Adjustment

1. Push down on the fast idle cam lever until the fast idle speed adjusting screw is contacting the second step (index), and against the shoulder of the high step.

2. Measure the clearance between the lower edge of the choke valve and air horn wall.

3. Adjust by turning the fast idle cam lever screw to obtain the proper specification.

Initial Choke Valve Clearance

1. Remove the choke cover shield.

2. Loosen the choke cover coil retaining screws. If rivets are used grind away and replace with screws.

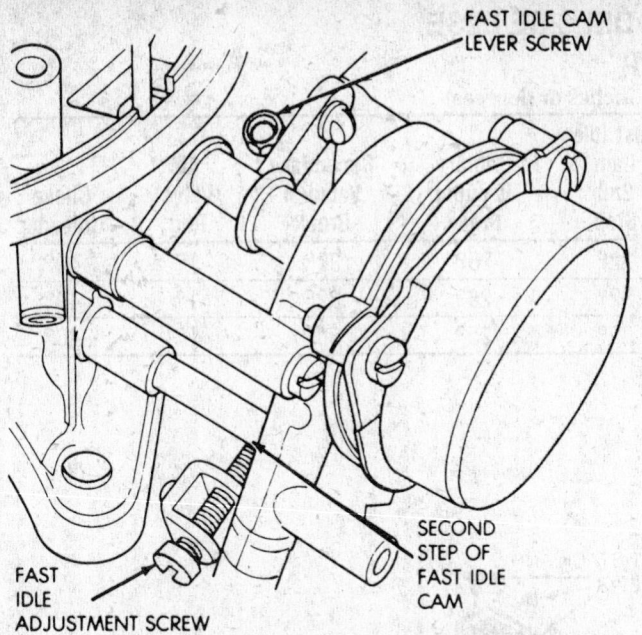

Motorcraft 2150 fast idle cam adjustment

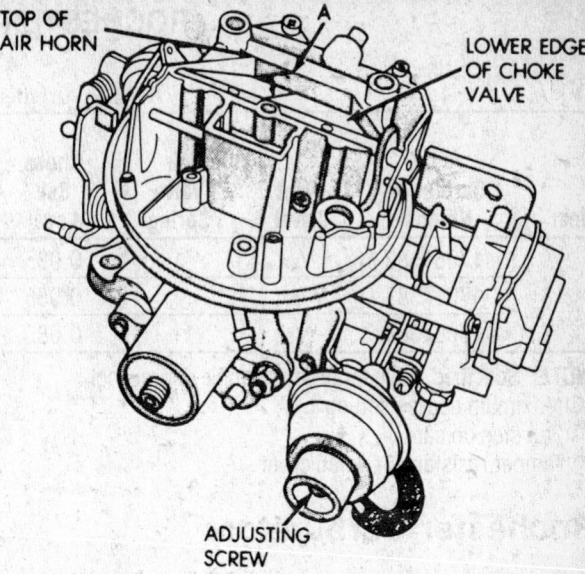

Motorcraft 2150 initial choke valve clearance adjustment

3. Open the throttle and rotate the choke cover until the choke valve is held close.

4. Tighten 1 cover retaining screw.

5. Close the throttle with the fast idle screw adjustment screw on the top step of the cam.

6. Use a hand held vacuum pump and apply vacuum to hold the choke diaphragm aganst the setscrew.

7. Measure the clearance between the lower edge of the choke valve and the top of the air horn.

8. Adjust the clearance by turning the screw at the rear of the choke vacuum diaphragm.

9. Make a fast idle cam adjustment.

Choke Setting

The automatic choke setting is made by loosening the choke cover in the desired direction as indicated by an arrow on the face of the cover. the original setting will be satisfactory for most driving conditions. However, if the engine stumbles or stalls on acceleration during warmup, the choke may be set richer or leaner no more than two graduations from the original setting.

Unloader Adjustment

1. With the throttle held fully open, apply pressure on the choke valve toward the closed position and measure the clearance between the lower edge of the choke valve and the air horn wall.

2. Adjust by bending the tang on the throttle lever which contacts the fast idle cam. Bend toward the cam to increase the clearance.

NOTE: Do not bend the unloader down so that it binds or interferes with any other component.

3. A clearance of 0.070 in. (1.8mm) must be between the unloader tang and the edge of the fast idle cam. Final unloader adjustment must always be done on the vehicle. The throttle should be fully opened by depressing the accelerator pedal to the floor. This is to assure that full throttle is obtained.

ROCHESTER MODEL 2SE/E2SE
Jeep

(All measurements in inches or degrees)

Year	Carburetor Number	Float Level	Air Valve Spring	Choke Coil Level	Fast Idle Cam 2nd Step	Primary Vacuum Break	Secondary Vacuum Break	Air Valve Rod	Choke Unloader
1986	17085380	$5/32$	1	0.085	22°	26°	32°	1°	40°
	17085381	$5/32$	1	0.085	22°	26°	32°	1°	40°
	17085382	$5/32$	1	0.085	22°	26°	32°	1°	40°
	17085383	$5/32$	1	0.085	22°	26°	32°	1°	40°
	17085384	$1/8$	1	0.085	22°	25°	30°	1°	40°
1987	17084580	$5/32$	1	0.085	22°	26°	32°	1°	40°
	17084581	$5/32$	1	0.085	22°	26°	32°	1°	40°

ROCHESTER MODEL 2SE/E2SE
Jeep
(All measurements in inches or degrees)

Year	Carburetor Number	Float Level	Air Valve Spring	Choke Coil Level	Fast Idle Cam 2nd Step	Primary Vacuum Break	Secondary Vacuum Break	Air Valve Rod	Choke Unloader
	17084582	5/32	1	0.085	22°	26°	32°	1°	40°
	17084583	5/32	1	0.085	22°	26°	32°	1°	40°
	17084384	1/8	1	0.085	22°	25°	30°	1°	40°

NOTE: Specified angle for use with angle degree tool
① Maximum degree setting
② 2nd step on cam
③ Tamper resistant—riveted cover

Rochester Carburetors

MODELS 2SE/E2SE

Float and Fuel Level Adjustment

1. Start the engine and run it to normal operating temperature.
2. Remove the vent stack screws and the vent stack.
3. Remove the air horn screw adjacent to the vent stack.
4. With the engine idling and the choke fully opened, carefully insert float gauge J-9789-136 for E2SE carburetors and tool J-9789-138 for 2SE carburetors, into the air horn screw hole and vent hole. Allow the gauge to rest freely on the float.

NOTE: Do not press down on the float.

5. With the gauge at eye level, observe the mark that aligns with the top of the casting at the vent hole. The float level should be within 0.06 in. (1.5mm) of the specifications. If not, remove the air horn and adjust the float as follows:
   a. Hold the retainer pin firmly in place and push the float down, lightly, against the inlet needle.
   b. Using an adjustable T-scale, at a point 3/16 in. (5mm) from the end of the float, at the toe, measure the distance from the float bowl top surface (gasket removed) to the top of the float at the toe. If not within specification, remove the float and bend the arm.

Fast Idle Cam Adjustment

1. Make sure the choke coil adjustment is correct and that the fast idle speed is correct.
2. Obtain a choke angle gauge, tool No. J-26701-A. Rotate the degree scale to the zero degree mark opposite the pointer.

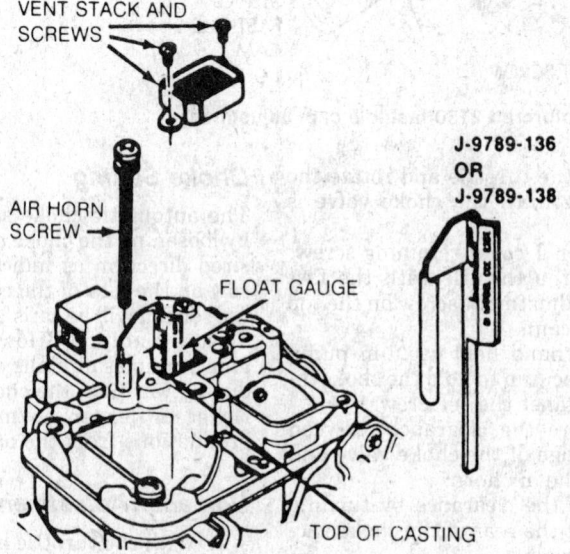

Rochester 2SE and E2SE float level measurement

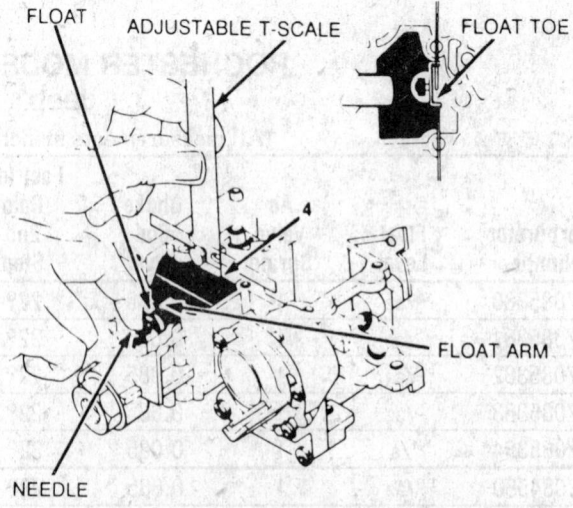

Rochester 2SE and E2SE float level adjustment

1. Screw—air horn (long) (2)
2. Screw—air horn (large)
3. Screw—air horn (short) (3)
4. Screw—air horn (medium)
5. Vent stack assembly
6. Screw—hot idle compensator (2)
7. Hot idle compensator
8. Gasket—hot idle compensator
9. Air horn assembly
10. Gasket—air horn
11. Retainer—pump link
12. Seal—pump stem
13. Retainer—stem seal
14. Vacuum break and bracket assembly— primary
15. Screw—vacuum break attaching
16. Bushing—air valve— link
17. Retainer—air valve link
18. Hose—vacuum break—primary
19. Link—air valve
20. Link—fast idle cam
21. Intermediate choke shaft/lever/link assembly
22. Bushing— intermediate choke shaft link
23. Retainer— intermediate choke shaft link
24. Vacuum break and bracket assembly— secondary
25. Choke cover and coil assembly
26. Screw—choke lever
27. Choke lever and contact assembly
28. Choke housing
29. Screw—choke housing (2)
30. Stat cover retainer kit
31. Screw—vacuum break attaching (2)

32. Float bowl assembly
33. Nut—fuel inlet
34. Gasket—fuel inlet nut
35. Filter—fuel inlet
36. Spring—fuel filter
37. Float assembly
38. Hinge pin—float
39. Insert—float bowl
40. Needle and seat assembly
41. Spring—pump return
42. Pump—assembly
43. Jet—main metering
44. Rod—main metering assembly
45. Ball—pump discharge
46. Spring—pump discharge

47. Retainer—pump discharge spring
48. Power piston assembly
49. Spring—power piston
50. Gasket—throttle body
51. Throttle body assembly
52. Pump rod
53. Clip—cam screw
54. Screw—cam
55. Spring—throttle stop screw
56. Screw—throttle stop
57. Idle needle and spring
58. Screw—throttle body attaching (4)
59. Nut—idle solenoid
60. Retainer—idle solenoid
61. Idle solenoid

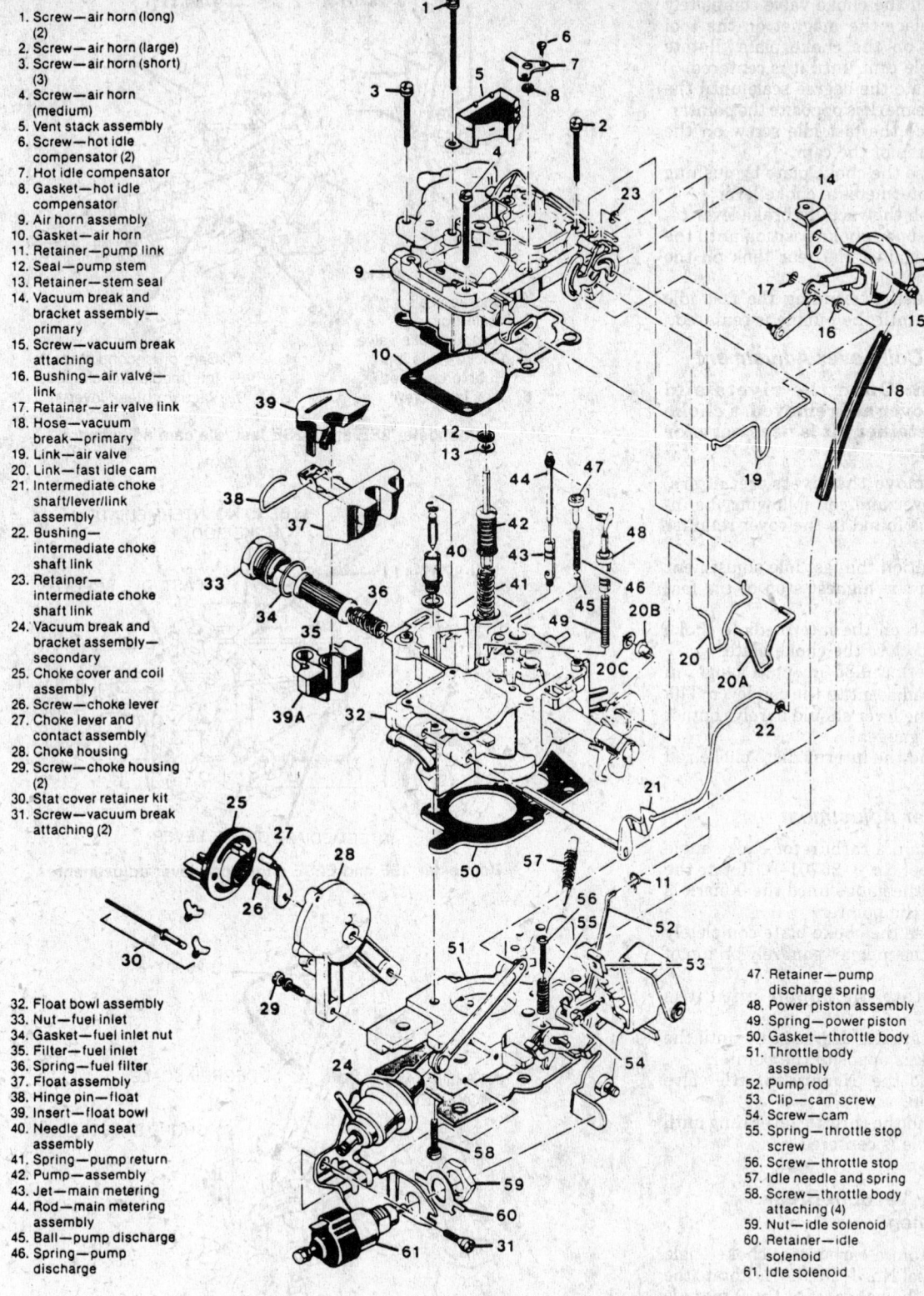

Rochester 2SE and E2SE carburetor—exploded view

3. With the choke valve completely closed, place the magnet on the tool squarely on the choke plate. Rotate the bubble unit until it is centered.

4. Rotate the degree scale until the 22 degree mark is opposite the pointer.

5. Place the fast idle screw on the second step of the cam.

6. Close the choke plate by pushing on the intermediate choke lever.

7. Push the vacuum brake lever toward the open choke position until the lever is against the rear tank on the choke lever.

8. Adjust by bending the fast idle cam rod until the bubble is centered.

Choke Coil Lever Adjustment

NOTE: Once the rivets and choke cover are removed, a choke cover retainer kit is necessary for assembly.

1. Remove the rivets, retainers, choke cover and coil following the instructions found in the cover retainer kit.

2. Position the fast idle adjustment screw on the highest stop of the fast idle cam.

3. Push on the intermediate choke lever and close the choke plate.

4. Insert a 0.85 in. plug gauge, in the hole adjacent to the coil lever. The edge of the lever should barely contact the plug gauge.

5. Bend the intermediate choke rod to adjust.

Unloader Adjustment

1. Obtain a carburetor choke angle gauge, tool No. J–26701–A. Rotate the scale on the gauge until the 0 mark is opposite the pointer.

2. Close the choke plate completely and set the magnet squarely on top of it.

3. Rotate the bubble until it is centered.

4. Rotate the degree scale until the 40° mark is opposite the pointer.

5. Hold the primary throttle valve wide open.

6. Bend the throttle lever tang until the bubble is centered.

Primary Vacuum Break Adjustment

1. Obtain a carburetor choke angle gauge, tool No. J–26701–A. Rotate the scale on the gauge until the 0 mark is opposite the pointer.

2. Rotate the degree scale until the correct specification mark is opposite the pointer.

3. Place tape over the vacuum bleed in the diaphragm.

4. Set the choke vacuum diaphragm using a hand vacuum pump.

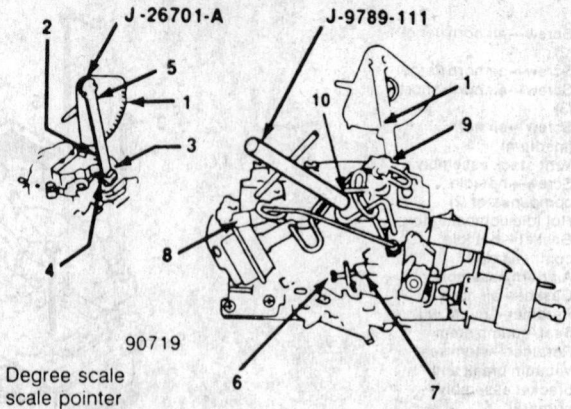

90719

1. Degree scale
2. scale pointer
3. Closed choke valve
4. Choke valve top
5. Bubble centered
6. Fast idle screw
7. Cam on second step
8. Intermediate choke lever
9. Vacuum break lever

Rochester 2SE and E2SE fast idle cam adjustment

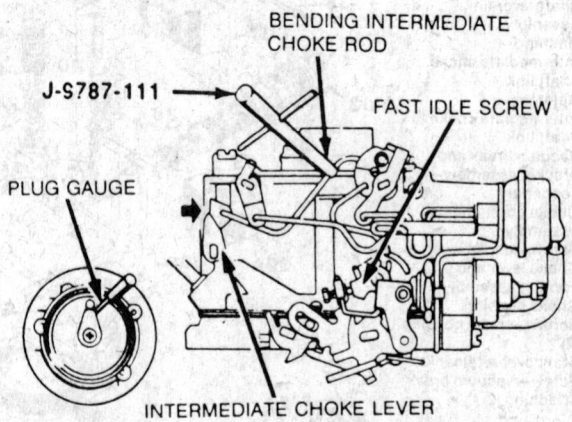

Rochester 2SE and E2SE choke coil lever adjustment

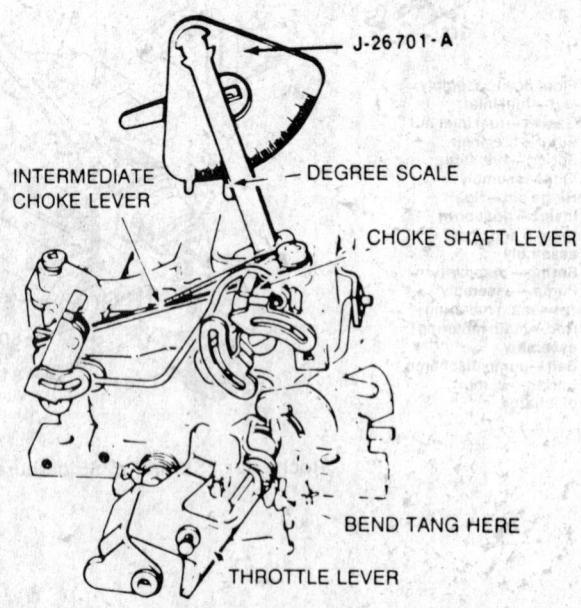

Rochester 2SE and E2SE choke unloader adjustment

5. Hold the choke valve at its closed position by pushing on the choke lever.

6. Adjust to the proper specification by bending the primary vacuum break rod.

Secondary Vacuum Break Adjustment

1. Obtain a carburetor choke angle gauge, tool No. J–26701–A. Rotate the scale on the gauge until the 0 mark is opposite the pointer.

2. Rotate the degree scale until the correct specification mark is opposite the pointer.

3. Place tape over the vacuum bleed in the diaphragm.

4. Seat the choke vacuum diaphragm using a hand vacuum pump.

5. Hold the choke valve at its closed position by pushing on the choke lever.

6. Adjust to the proper specification by bending the secondary vacuum break rod.

Air Valve Spring Adjustment

1. If necessary, remove the intermediate choke rod to gain acess to the lock screw. Loosen the lock screw.

2. Turn the tension adjusting screw clockwise until the air valve opens slightly.

3. Turn the adjusting screw counterclockwise until the air valve just closes.

Continue counterclockwise to the specified number of turns.

4. Tighten the lock screw.

Air Valve Rod Adjustment

1. Obtain a carburetor choke angle gauge, tool No.J–26701–A. Rotate the scale on the gauge until the 0 mark is opposite the pointer.

2. Rotate the degree scale until the correct specification mark is opposite the pointer.

3. Seat the choke vacuum diaphragm using a hand vacuum pump.

4. Place tape over the vacuum bleed in the diaphragm.

5. Apply light pressure to the air valve shaft in the direction to open the valve to ensure all slack is removed between the air valve link and the plunger slot.

6. Bend the air valve link with tool J–97789–111 until the bubble is centered.

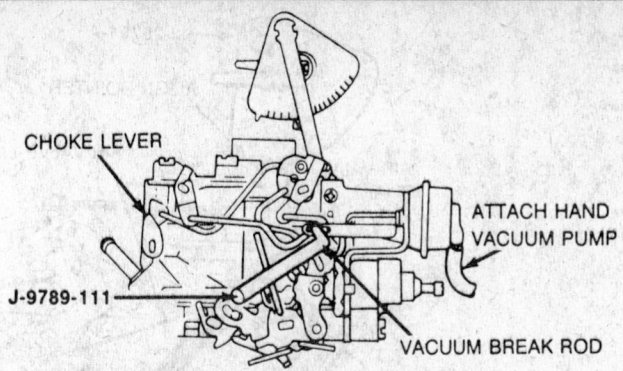

Rochester 2SE and E2SE primary vacuum break adjustment

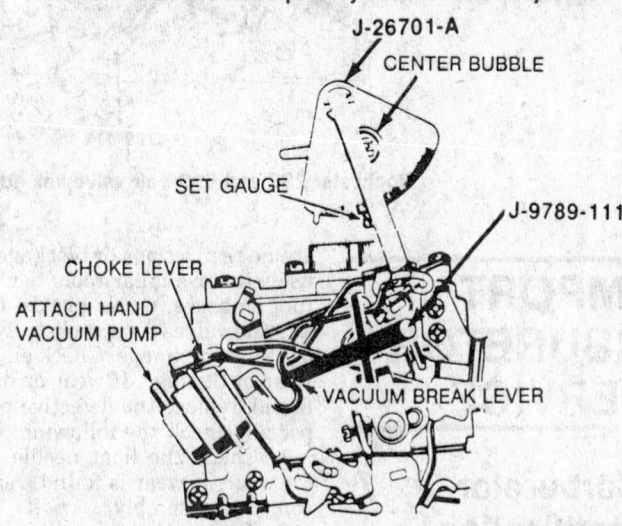

Rochester 2SE and E2SE secondary vacuum break adjustment

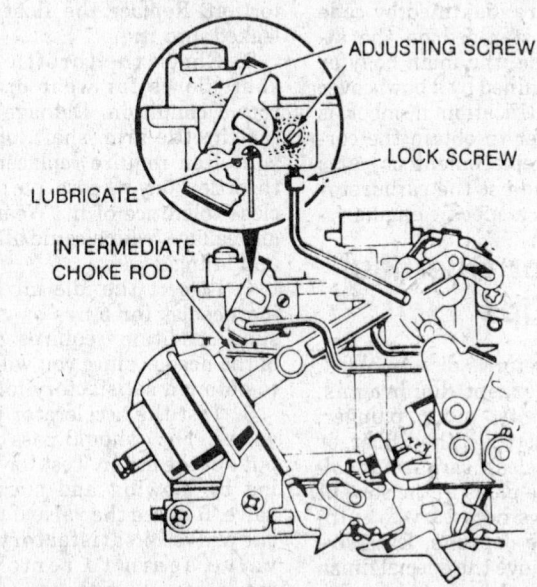

Rochester 2SE and E2SE air valve spring adjustment adjustment

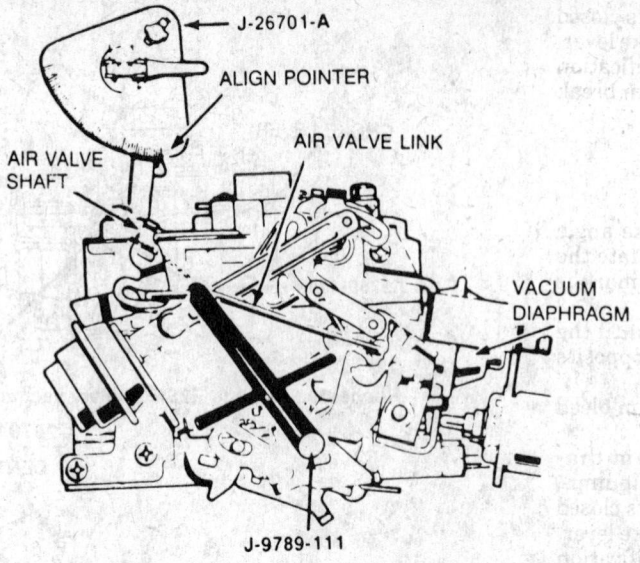

J-9789-111

Rochester 2SE and E2SE air valve link adjustment adjustment

IMPORT CARBURETOR SERVICE

Carburetor Identification

All carburetors are identified by code numbers, either stamped on the attaching flange side, the main body or on a metal tag retained by a bowl cover screw. This identification number is important in order to obtain the correct carburetor replacement or parts and to properly adjust the carburetor when matched to a specific engine.

Carburetor Overhaul Tips

When the carburetor is disassembled, wash all parts (except diaphragms, electric choke units, pump plunger, and any other plastic, leather, fiber, or rubber parts) in clean carburetor solvent. Do not leave parts in the solvent any longer than is necessary to sufficiently loosen the deposits. Excessive cleaning may remove the special finish from the float bowl and choke valve bodies, leaving these parts unfit for service. Rinse all parts in clean solvent and blow them dry with compressed air or allow them to air dry. Wipe clean all cork, plastic, leather, and fiber parts with a clean, lint-free cloth.

Blow out all passages and jets with compressed air and be sure that there are no restrictions or blockages. Never use wire or similar tools to clean jets, fuel passages, or air bleeds. Clean all jets and valves separately to avoid accidental interchange. Check all parts for wear or damage. If wear or damage is found, replace the defective parts. Especially check the following:

1. Check the float needle and seat for wear. If wear is found, replace the complete assembly.

2. Check the float hinge pin for wear and the float(s) for dents or distortion. Replace the float if fuel has leaked into it.

3. Check the throttle and choke shaft bores for wear or an out-of-round condition. Damage or wear to the throttle arm, shaft, or shaft bore will often require replacement of the throttle body. These parts require a close tolerance of fit. Wear may allow air leakage, which could affect starting and idling.

4. Inspect the idle mixture adjusting needles for burrs or grooves. Any such condition requires replacement of the needle, since you will not be able to obtain a satisfactory idle.

5. Test the accelerator pump check valves. They should pass air one way but not the other. Test for proper seating by blowing and sucking on the valve. Replace the valve if necessary. If the valve is satisfactory, wash the valve again to remove breath moisture.

6. Check the bowl cover for warped surfaces with a straight edge.

7. Closely inspect the valves and seats for wear and damage, replacing as necessary.

8. After the carburetor is assembled, check the choke valve for freedom of operation.

Carburetor overhaul kits are recommended for each overhaul. These kits contain all gaskets and new parts to replace those that deteriorate most rapidly. Failure to replace all parts supplied with the kit (especially gaskets) can result in poor performance later.

After cleaning and checking all components, reassemble the carburetor, using new parts and referring to the exploded view. When reassembling, make sure that all screws and jets are tight in their seats, but do not overtighten as the tips will be distorted. Tighten all screws gradually, in rotation. Do not tighten needle valves into their seats. Uneven jetting will result. Always use new gaskets. Be sure to adjust the float level, following the instructions contained in the rebuilding kit, when reassembling.

CHRYSLER IMPORTS AND MITSUBISHI

Float Level Adjustment

1. Invert the float chamber cover assembly without a gasket.

2. Position a float gauge and measure the distance from the bottom of the float to the surface of the float chamber cover.

3. If not within specification the shim under the needle seat must be changed.

CHRYSLER IMPORTS/MITSUBISHI CARBURETORS
(All measurements in inches)

Year	Carburetor Number	Fast Idle Opening	Float Level	Choke Breaker Opening	Throttle Position Sensor (volts)	Dashpot (rpm)
1986	DIDTA-209	0.028	.0394	—	0.25	—
	DIDTA-210	0.031	.0394	—	0.25	—
	DIDTF-205	0.025	.0394	—	0.25	—
	DIDTF-206	0.028	.0394	—	0.25	—
	DIDTF-207	0.028	.0394	—	0.25	—
	DIDTF-208	0.031	.0394	—	0.25	—
1987	DIDEF-400	—	—	①	0.25	2000
	DIDEF-401	—	—	①	0.25	1500
	DIDEF-402	—	—	①	0.25	2000
	DIDEF-403	—	—	①	0.25	1500
	DIDEF-404	—	—	②	0.25	2000
	DIDEF-405	—	—	②	0.25	1500
	DIDEF-406	—	—	②	0.25	2000
	DIDEF-407	—	—	②	0.25	1500
	DIDEF-410	—	—	②	0.25	2000
	DIDEF-411	—	—	②	0.25	1500
	DIDEF-412	—	—	②	0.25	2000
	DIDEF-413	—	—	②	0.25	1500
	DIDEF-420	—	—	①	0.25	2000
	DIDEF-421	—	—	①	0.25	1500
	DIDEF-422	—	—	②	0.25	2000
	DIDEF-432	—	—	②	0.25	1500
1988	DIDEF-400	—	—	①	0.25	2000
	DIDEF-401	—	—	①	0.25	1500
	DIDEF-402	—	—	①	0.25	2000
	DIDEF-403	—	—	①	0.25	1500
	DIDEF-420	—	—	②	0.25	2000
	DIDEF-421	—	—	②	0.25	1500
	DIDEF-429	—	—	②	0.25	2000
	DIDEF-430	—	—	②	0.25	1500
	DIDEF-431	—	—	③	0.25	2000
	DIDEF-432	—	—	③	0.25	1500
	DIDEF-435	—	—	②	0.25	2000
	DIDEF-436	—	—	②	0.25	1500
	DIDEF-437	—	—	②	0.25	2000
	DIDEF-438	—	—	②	0.25	1500
	DIDEF-441	—	—	③	0.25	2000
	DIDEF-442	—	—	③	0.25	1500
	DIDEF-443	—	—	③	0.25	2000
	DIDEF-444	—	—	③	0.25	1500
1989	DIDEF-400	—	—	①	0.25	2000
	DIDEF-401	—	—	④	0.25	1500
	DIDEF-402	—	—	①	0.25	2000
	DIDEF-403	—	—	①	0.25	1500

CHRYSLER IMPORTS/MITSUBISHI CARBURETORS

(All measurements in inches)

Year	Carburetor Number	Fast Idle Opening	Float Level	Choke Breaker Opening	Throttle Position Sensor (volts)	Dashpot (rpm)
	DIDEF-420	—	—	①	0.25	2000
	DIDEF-428	—	—	①	0.25	1500
	DIDEF-429	—	—	②	0.25	2000
	DIDEF-430	—	—	②	0.25	1500
	DIDEF-431	—	—	③	0.25	2000
	DIDEF-435	—	—	②	0.25	2000
	DIDEF-436	—	—	②	0.25	1500
	DIDEF-437	—	—	②	0.25	2000
	DIDEF-438	—	—	②	0.25	1500
	DIDEF-441	—	—	③	0.25	2000
	DIDEF-443	—	—	③	0.25	2000

① 1st stage: 0.087–0.094　② 1st stage: 0.098–0.106　③ 1st stage: 0.091–0.098　④ 1st stage: 0.079–0.087
　 2nd stage: 0.114–0.122　　 2nd stage: 0.126–0.133　　 2nd stage: 0.118–0.126　　 2nd stage: 0.114–0.122

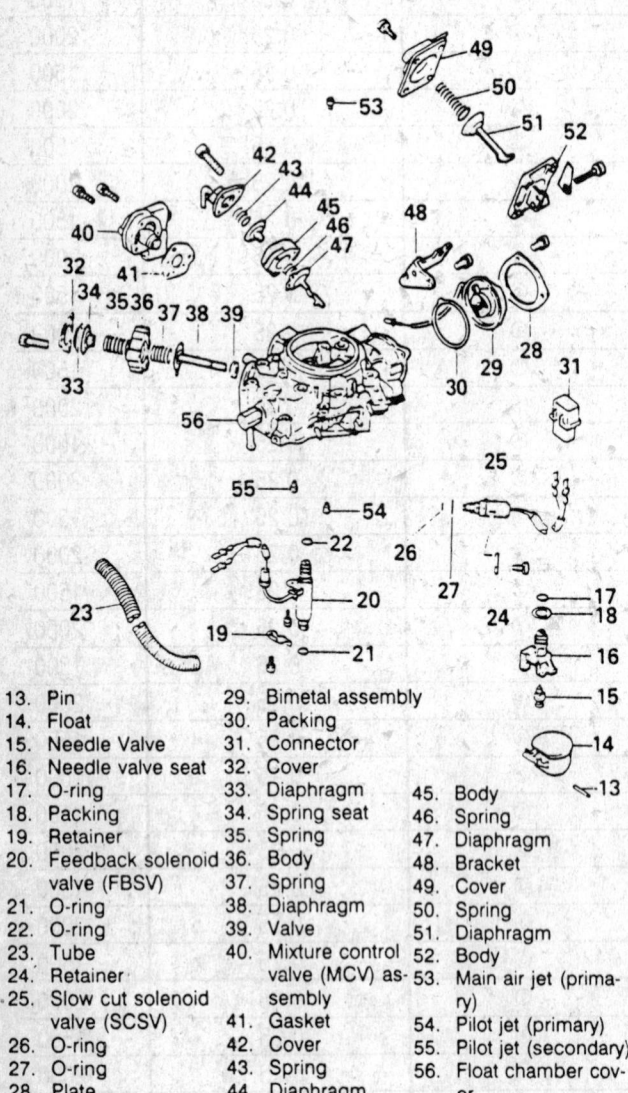

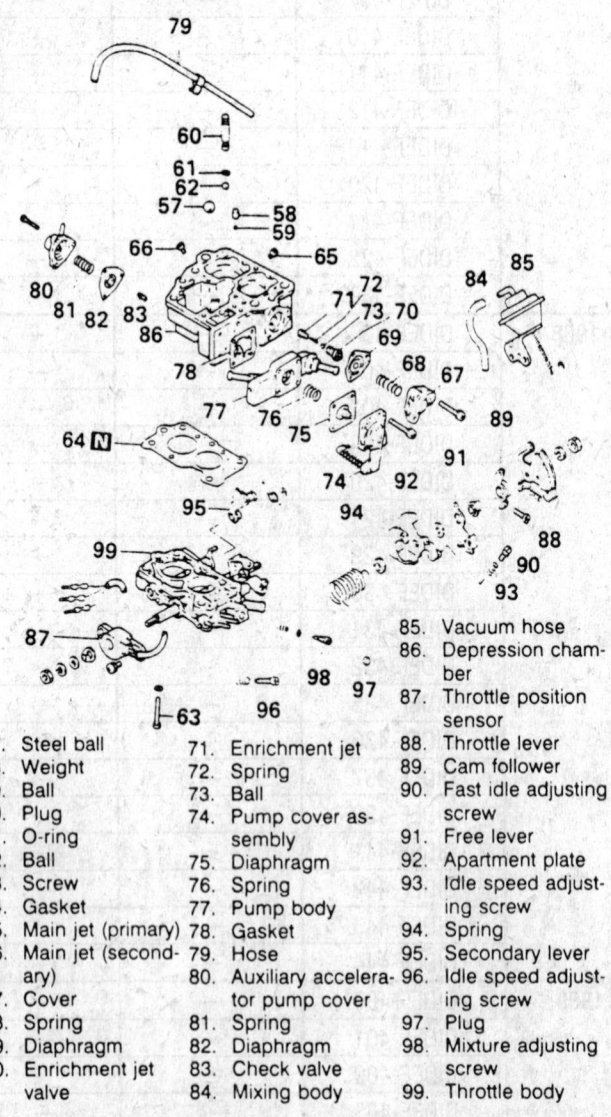

13. Pin
14. Float
15. Needle Valve
16. Needle valve seat
17. O-ring
18. Packing
19. Retainer
20. Feedback solenoid valve (FBSV)
21. O-ring
22. O-ring
23. Tube
24. Retainer
25. Slow cut solenoid valve (SCSV)
26. O-ring
27. O-ring
28. Plate

29. Bimetal assembly
30. Packing
31. Connector
32. Cover
33. Diaphragm
34. Spring seat
35. Spring
36. Body
37. Spring
38. Diaphragm
39. Valve
40. Mixture control valve (MCV) assembly
41. Gasket
42. Cover
43. Spring
44. Diaphragm

45. Body
46. Spring
47. Diaphragm
48. Bracket
49. Cover
50. Spring
51. Diaphragm
52. Body
53. Main air jet (primary)
54. Pilot jet (primary)
55. Pilot jet (secondary)
56. Float chamber cover

57. Steel ball
58. Weight
59. Ball
60. Plug
61. O-ring
62. Ball
63. Screw
64. Gasket
65. Main jet (primary)
66. Main jet (secondary)
67. Cover
68. Spring
69. Diaphragm
70. Enrichment jet valve

71. Enrichment jet
72. Spring
73. Ball
74. Pump cover assembly
75. Diaphragm
76. Spring
77. Pump body
78. Gasket
79. Hose
80. Auxiliary accelerator pump cover
81. Spring
82. Diaphragm
83. Check valve
84. Mixing body

85. Vacuum hose
86. Depression chamber
87. Throttle position sensor
88. Throttle lever
89. Cam follower
90. Fast idle adjusting screw
91. Free lever
92. Apartment plate
93. Idle speed adjusting screw
94. Spring
95. Secondary lever
96. Idle speed adjusting screw
97. Plug
98. Mixture adjusting screw
99. Throttle body

Chrysler/Mitsubishi carburetor upper half exploded view

Chrysler/Mitsubishi carburetor lower half exploded view

Chrysler/Mitsubishi float level adjustment

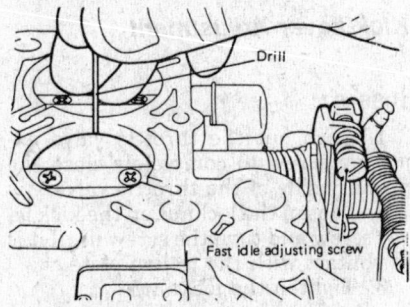

Chrysler/Mitsubishi fast idle opening adjustment

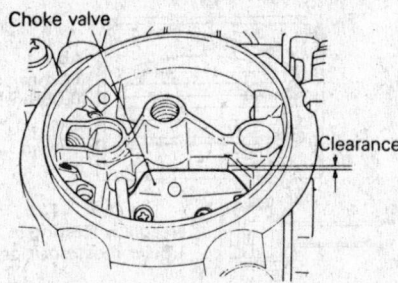

Chrysler/Mitsubishi choke breaker opening measurement

4. Adding or removing a shim will change the float level by 3 times the thickness of the shim.

Fast Idle Opening Adjustment

1986

1. With the carburetor at room temperature, insert the diameter drill specified and adjust the fast idle opening by adjusting the fast idle screw.
2. Use a drill diameter of 0.028 in. on vehicles with manual transmission and 0.031 in. with automatic transmission.

Choke Breaker Opening

1. Disconnect the vacuum hose (yellow stripe) from the choke breaker.
2. With the engine idling, close the choke valve lightly until the choke valve stops.
3. Measure the choke valve to choke bore clearance to see if it is within the 1st stage specification. If necessary,

remove the bimetal assembly and adjust the rod end opening to be within the 1st stage specification.

NOTE: When removing the bimetal assembly, put a mark on the electric choke body.

4. Reconnect the vacuum hose (yellow stripe) from the choke breaker and remeasure the choke valve to choke bore clearance.
5. If the clearance is not within the 2nd stage specification, adjust by the adjusting screw.

Throttle Position Sensor Adjustment

1986

1. Warm the engine to normal operating temperature and make sure the fast idle cam is released.
2. Stop the engine, then turn the No. 1 and 2 idle speed adjusting screws counterclockwise enough to close the throttle valve completely. Record the number of turns.
3. Connect a digital type voltmeter between the TPS connectors 2 and 3.

NOTE: Do not disconnect the TPS connector and body harness.

4. Switch the ignition key on and adjust the TPS adjustment screw so that the TPS output is 0.25 volts.
5. Turn the ignition key off.
6. Close the No. 1 and 2 idle speed adjusting screws the same number of turns.

1987–89

Montero and Pick-up

1. Loosen the accelerator cable.
2. Turn the No. 1 and 2 idle speed adjusting screws counterclockwise enough to close the throttle valve completely. Record the number of turns. At this time, the fast idle control should have been released (the lever not resting on the fast idle cam).
3. Disconnect the carburetor connectors.
4. Connect the special test harness tool MD998474 between the disconnected connectors.
5. Connect a digital voltmeter between the No. 2 red sensor output terminal and the No. 8 blue sensor (ground) of the carburetors connectors.
6. Turn the ignition switch to **ON** but do not start the engine.
7. Measure the output voltage of the TPS. The standard value is 0.25 volts.
8. If adjustment is necessary, loosen the TPS attaching screw and adjust by turning the TPS to the standard value.

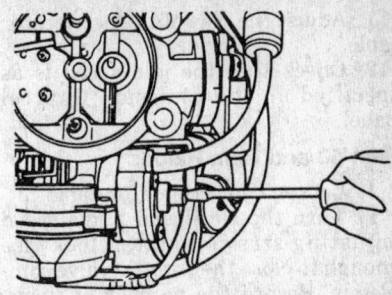

Chrysler/Mitsubishi TPS adjusting screw —1986

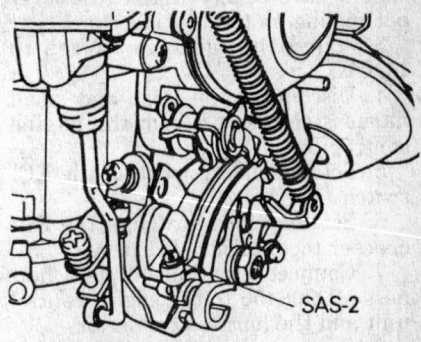

Chrysler/Mitsubishi No. 1 and No. 2 idle speed adjusting screws—1987–89

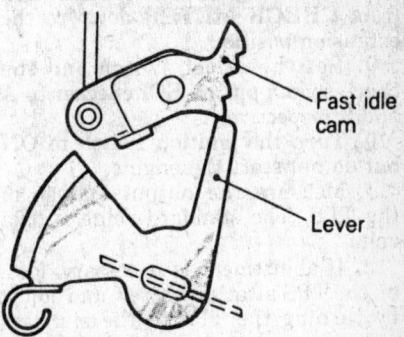

Chrysler/Mitsubishi fast idle cam and lever—1987–89

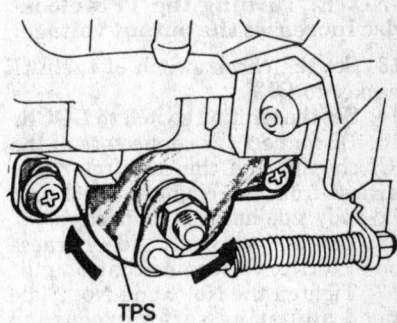

Chrysler/Mitsubishi TPS adjustment—1987–89

NOTE: Turning the TPS clockwise increases the output voltage.

9. Tighten the No. 1 and 2 idle speed adjusting screws recorded earlier.
10. Remove the voltmeter and the special tool and connect the carburetor's connector.

11. Adjust the play of the accelerator cable.

12. Check that the idle speed is as specified on the Emission Control Label.

Ram 50 and Ram Raider

1. Loosen the accelerator cable.

2. Turn the No. 1 and 2 idle speed adjusting screws counterclockwise enough to close the throttle valve completely. Record the number of turns. At this time, the fast idle control should have been released (the lever not resting on the fast idle cam).

3. Turn the ignition switch to **LOCK**.

4. Disconnect the large and small harness connector from the engine control unit.

5. Set the check switch of the ECI switch **OFF**.

6. Set the select switch of the ECI checker toool MD998451 to **A**.

7. Connect a carburetor test harness MD998456 to the engine control unit and the harness connectors.

8. Connect a volt meter to the extension terminals of the ECI checker, and then change the extension switch from **CHECK METER** down to the extension position.

9. Set the select switch and the check switch on the ECI checker to **A** and **3** respectively.

10. Turn the ignition switch to **ON** but do not start the engine.

11. Measure the output voltage of the TPS. The standard velue is 0.25 volts.

12. If adjustment is necessary, loosen the TPS attaching screw and adjust by turning the TPS to the standard value.

NOTE: Turning the TPS clockwise increases the output voltage.

13. Set the check switch of the ECI checker to **OFF**.

14. Set the ignition switch to **LOCK**.

15. Disconnect the connectors of the ECI checker and the carburetor test harness from the engine control unit and body side harness connectors.

16. Connect the body side harness connectors to the engine control unit.

17. Tighten the No. 1 and No. 2 idle speed adjusting screws recorded earlier.

18. Adjust the play of the accelerator cable.

19. Check that the idle speed is as specified on the emission control label.

Dashpot Adjustment

1. Make sure the curb idle speed is correct, the engine temperature is at normal operating temperature, lights and accessories off, manual transmis-

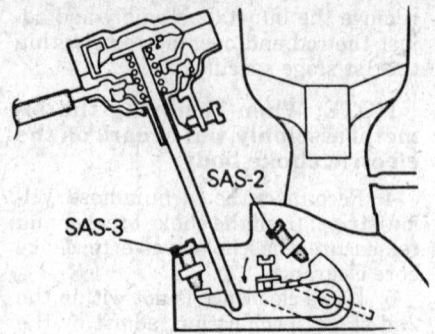

Chrysler/Mitsubishi No. 2 and No. 3 idle speed adjustment screws—1987–89

sion in neutral and automatic transmission in **PARK**.

2. Start the engine and run at idle.

3. Close the throttle valve until the No. 2 idle speed adjusting screw contacts the free lever. Check the engine speed at that moment. If the engine speed is not as specified, adjust the dashpot by turning the No. 3 idle speed adjustment screw.

4. Release the free lever and verify that the engine returns to the idle speed slowly.

ISUZU

Stromberg Carburetor

Float Level Adjustment

1986–90

The fuel level is normal if it is within the mark on the window. If the level is outside the line, adjust by bending the float seat. The needle valve should have an effective stroke of about 0.059 in. The float stroke may be measured as follows:

1. Hold the carburetor with the bottom side up and fully raise the float.

2. Measure the distance between the valve stem (resting at its bottom position) and the float seat.

3. Normal clearance is 0.059 in. Bend the float stopper as necessary to adjust.

Primary Throttle Valve Adjustment

1986–87

The primary throttle valve is opened by means of the fast idle adjusting screw to an angle of 16 degrees when the choke valve is completely closed. The primary throttle valve opening angle may be checked as follows:

1. Close the choke valve completely and measure the clearance between the center of the throttle valve and the wall of the throttle valve chamber. Standard clearance is 0.050–0.059 in.

2. Adjust the throttle valve opening angle with the fast idle adjusting screw.

NOTE: Be sure to turn the throttle stop screw all the way in before measuring the clearance.

Kick Lever Adjustment

1986–87

1. Turn out the throttle valve adjusting screw to completely close the primary side of the throttle valve.

2. Loosen the locknut on the kick lever screw and turn the screw until it is in contact with the return plate.

3. Tighten the locknut.

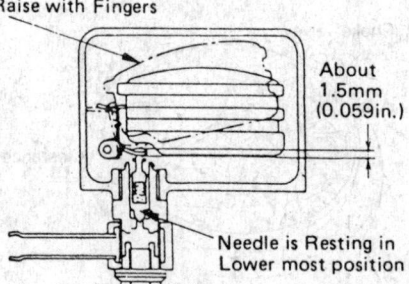

Isuzu (Stromberg) float level adjustment

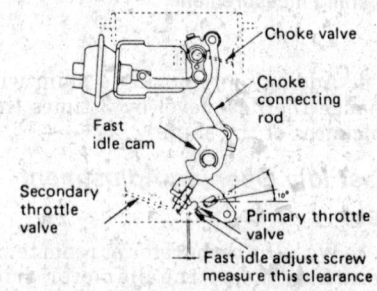

Isuzu (Stromberg) primary throttle valve adjustment—1986–87

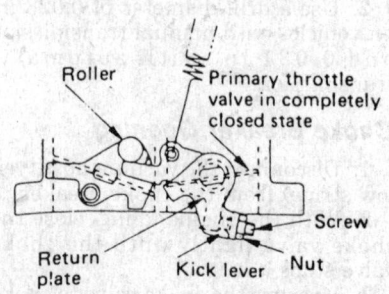

Isuzu (Stromberg) kick lever adjustment—1986–87

ISUZU CARBURETORS
Stromberg Models
(All measurements in inches)

		Float Level	Primary Throttle Valve	Choke Valve Opening	Unloader	Primary and Secondary Throttle Valve
1986	DCH340-227	0.059	0.050–0.059	—	—	—
	DCH340-228	0.059	0.050–0.059	—	—	—
	DFP340-3	0.059	0.050–0.059	—	—	—
	DFP340-4	0.059	0.050–0.059	—	—	—
	DCR384	0.059	0.050–0.059	—	—	—
	DFP384	0.059	0.050–0.059	—	—	—
1987	DCH340-227	0.059	0.050–0.059	—	—	—
	DCH340-228	0.059	0.050–0.059	—	—	—
	DFP340-3	0.059	0.050–0.059	—	—	—
	DFP340-4	0.059	0.050–0.059	—	—	—
	DCR384	0.059	0.050–0.059	—	—	—
	DFP384	0.059	0.050–0.059	—	—	—
1988	DFP384-205	0.059	—	0.031–0.050	0.105–0.050	0.272–0.331
1989	DFP384-205	0.059	—	0.031–0.050	0.105–0.050	0.272–0.331
1990	DFP384-205	0.059	—	0.031–0.050	0.105–0.129	0.272–0.331

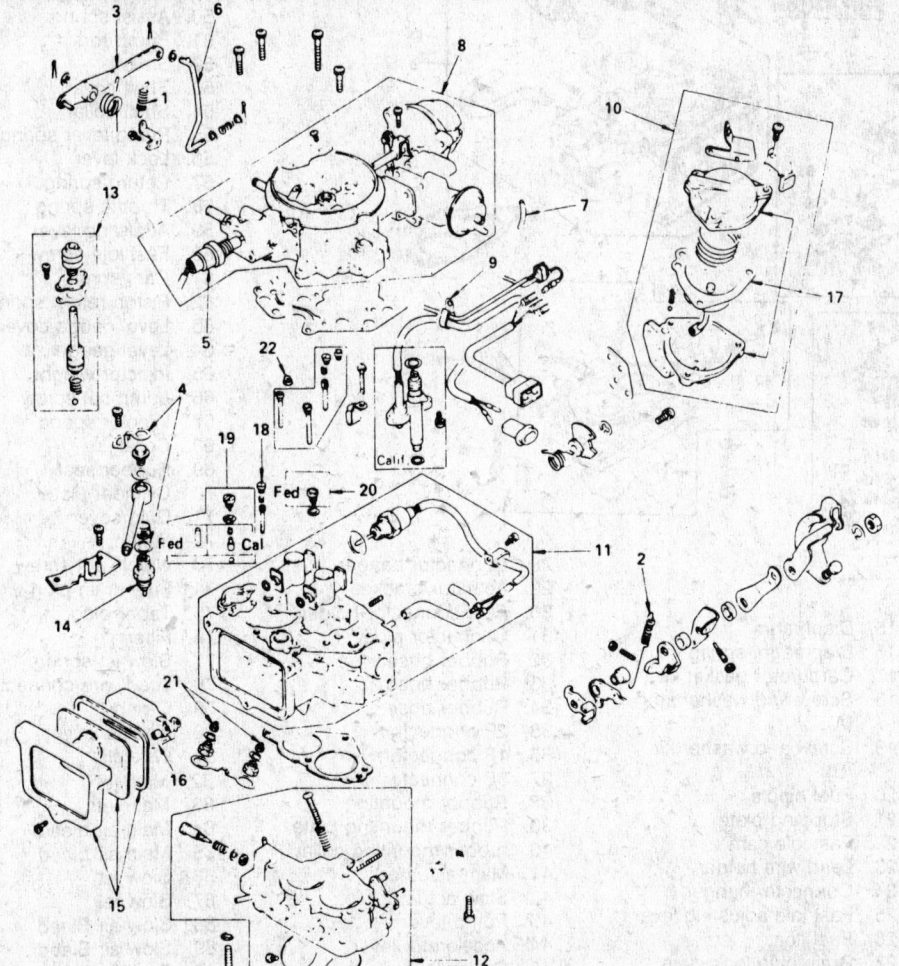

1. Pump lever spring
2. Throttle return spring
3. Pump lever
4. Fuel nipple and strainer
5. Vent valve switch
6. Pump rod
7. Vacuum hose
8. Choke chamber assembly
9. Clip
10. Diaphragm assembly
11. Float chamber assembly
12. Throttle chamber assembly
13. Accelerator pump plunger assembly
14. Float needle assembly
15. Level gauge cover and level gauge
16. Float
17. Diaphragm chamber
18. Jets
19. Injector weight plug, injector weight and check ball
20. Power jet
21. Main jet plugs and primary
22. Primary show air bleed

Isuzu (Stromberg) carburetor disassembled view—1986–90 models DCR384/DFP384

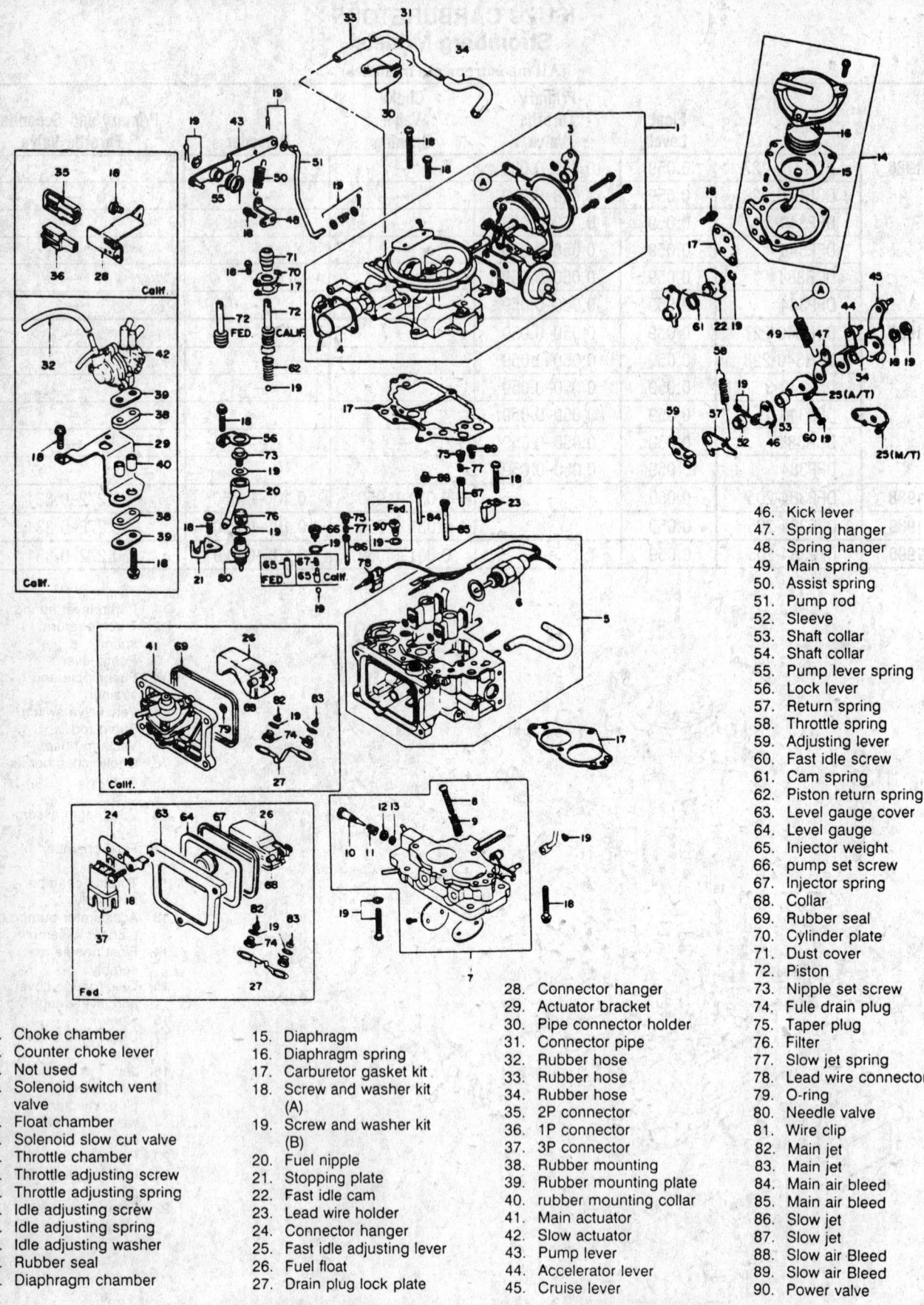

Isuzu (Stromberg) carburetor disassembled view — 1986–90 model DCH340

1. Choke chamber
2. Counter choke lever
3. Not used
4. Solenoid switch vent valve
5. Float chamber
6. Solenoid slow cut valve
7. Throttle chamber
8. Throttle adjusting screw
9. Throttle adjusting spring
10. Idle adjusting screw
11. Idle adjusting spring
12. Idle adjusting washer
13. Rubber seal
14. Diaphragm chamber

15. Diaphragm
16. Diaphragm spring
17. Carburetor gasket kit
18. Screw and washer kit (A)
19. Screw and washer kit (B)
20. Fuel nipple
21. Stopping plate
22. Fast idle cam
23. Lead wire holder
24. Connector hanger
25. Fast idle adjusting lever
26. Fuel float
27. Drain plug lock plate

28. Connector hanger
29. Actuator bracket
30. Pipe connector holder
31. Connector pipe
32. Rubber hose
33. Rubber hose
34. Rubber hose
35. 2P connector
36. 1P connector
37. 3P connector
38. Rubber mounting
39. Rubber mounting plate
40. rubber mounting collar
41. Main actuator
42. Slow actuator
43. Pump lever
44. Accelerator lever
45. Cruise lever

46. Kick lever
47. Spring hanger
48. Spring hanger
49. Main spring
50. Assist spring
51. Pump rod
52. Sleeve
53. Shaft collar
54. Shaft collar
55. Pump lever spring
56. Lock lever
57. Return spring
58. Throttle spring
59. Adjusting lever
60. Fast idle screw
61. Cam spring
62. Piston return spring
63. Level gauge cover
64. Level gauge
65. Injector weight
66. pump set screw
67. Injector spring
68. Collar
69. Rubber seal
70. Cylinder plate
71. Dust cover
72. Piston
73. Nipple set screw
74. Fule drain plug
75. Taper plug
76. Filter
77. Slow jet spring
78. Lead wire connector
79. O-ring
80. Needle valve
81. Wire clip
82. Main jet
83. Main jet
84. Main air bleed
85. Main air bleed
86. Slow jet
87. Slow jet
88. Slow air Bleed
89. Slow air Bleed
90. Power valve

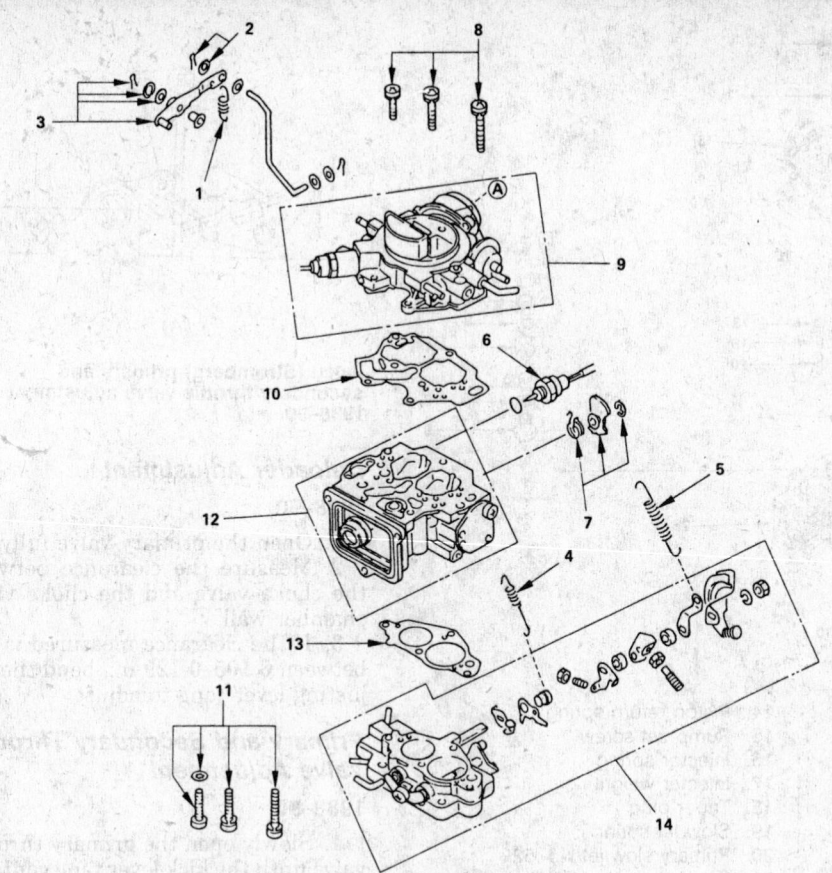

1. Assist spring
2. Pump rod split pin with washer
3. Pump lever and split pin with washer
4. Return spring
5. Main spring
6. Slow cut solenoid valve
7. Fast idler cam and spring
8. Choke chamber screw and washer
9. Choke chamber assembly
10. Choke and float chamber gasket
11. Throttle chamber screw and washer
12. Float chamber assembly
13. Float and throttle chamber assembly
14. Throttle chamber assembly

Isuzu (Stromberg) carburetor disassembled view of major components – 1988–90

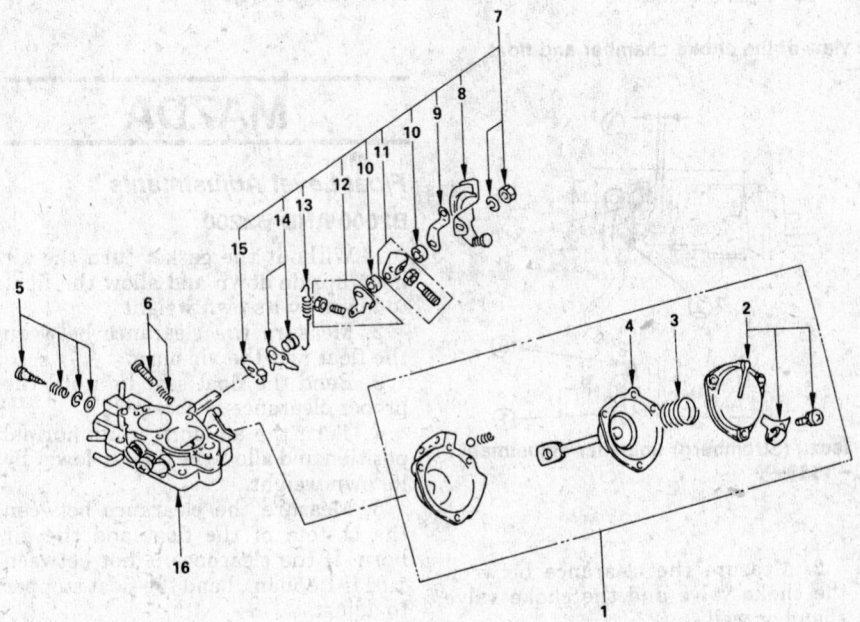

1. Diaphragm chamber assembly
2. Diaphragm chamber cover
3. Diaphragm spring
4. Diaphragm
5. Idler adjusting screw
6. Throttle adjusting screw
7. Throttle shaft nut and washer
8. Throttle lever
9. Spring hanger
10. Shaft collar
11. Fast idle adjusting lever and screw
12. Kick lever
13. Return spring
14. Return plate and sleeve
15. Adjusting lever
16. Throttle chamber

Isuzu (Stromberg) carburetor disassembled view of the throttle chamber – 1988–90

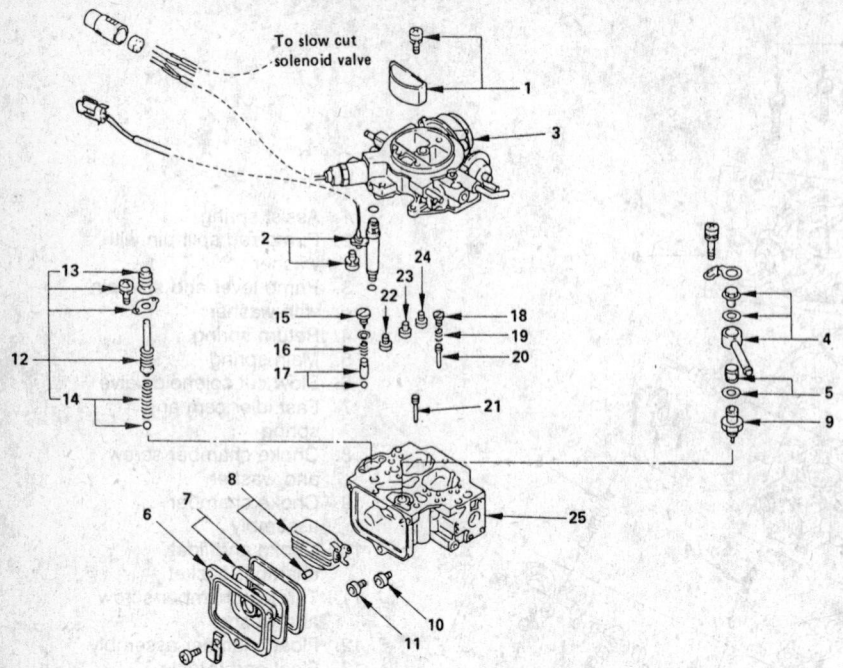

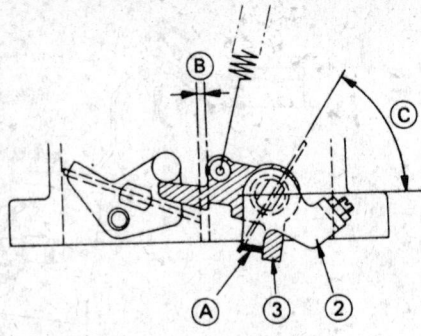

Isuzu (Stromberg) primary and secondary throttle valve adjustment — 1988–90

1. Vent cover
2. Duty solenoid valve
3. Choke chamber
4. Fuel nipple
5. Fuel filter
6. Level gauge cover
7. Level gauge and rubber seal
8. Float and collar
9. Needle valve
10. Secondary main jet No. 170
11. Primary main jet No. 88
12. Piston
13. Pump cover
14. Piston return spring
15. Pump set screw
16. Injector spring
17. Injector weight
18. Taper plug
19. Slow jet spring
20. Primary slow jet No. 52
21. Secondary slow jet No. 100
22. Primary main air bleed No. 100
23. Secondary main air bleed No. 60
24. Primary slow air bleed
25. Float chamber

Isuzu (Stromberg) carburetor disassembled view of the choke chamber and float chamber — 1988–90

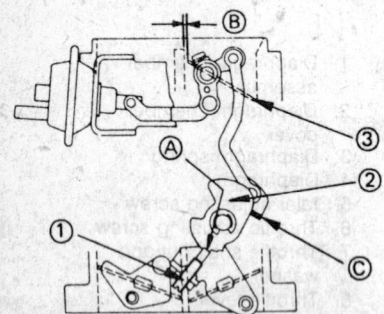

Isuzu (Stromberg) choke valve opening — 1988–90

Isuzu (Stromberg) unloader adjustment — 1988–90

Choke Valve Opening Adjustment

1988–90

1. Move the fast idle screw tip against the 2nd step of the fast idle cam.

2. Measure the clearance between the choke valve and the choke valve chamber wall.

3. If the clearance measured is not between 0.031–0.050 in., bend the counter lever tang to adjust.

Unloader Adjustment

1988–90

1. Open the primary valve fully.

2. Measure the clearance between the choke valve and the choke valve chamber wall.

3. If the clearance measured is not between 0.105–0.129 in., bend the adjusting lever tang to adjust.

Primary and Secondary Throttle Valve Adjustment

1988–90

1. Slowly open the primary throttle valve until the kick lever tang contacts the return plate.

2. Measure the clearance between the choke valve and the choke valve chamber wall.

3. If the clearance measured is not between 0.272–0.331 in., bend the kick lever tang to adjust.

MAZDA

Float Level Adjustments

B2000 AND B2200

1. Without the gasket turn the air horn upside down and allow the float to lower by its own weight.

2. Measure the clearance between the float and the air horn.

3. Bend the float seat lip until the proper clearance is obtained.

4. Turn the air horn to its normal position and allow the float to lower by its own weight.

5. Measure the clearance between the bottom of the float and the air horn. If the clearance is not between 1.811–1.850 in., bend the float stopper to adjust.

B2600

1. Invert the float chamber cover assembly.

MAZDA CARBURETORS
(All measurements in inches)

Year	Model	Fast Idle Cam	Fast Idle Opening	Float Level	Choke Valve Opening	Choke Diaphragm	Choke Unloader	Dashpot (rpm)	Secondary Throttle Valve
1986	B-2000	0.029– 0.044	—	0.453– 0.492	0.023– 0.039	0.066– 0.084	0.108– 0.142	—	0.289– 0.325
1987	B-2200	0.033– 0.041	—	①	0.024– 0.039	0.067– 0.085	0.110– 0.143		0.289– 0.325
	B-2600	—	②	0.748– 0.827	—	—	—	③	—
1988	B-2200	0.033– 0.041	—	①	0.024– 0.045	0.067– 0.085	0.110– 0.143		0.289– 0.325
	B-2600	—	0.026– 0.030	0.748– 0.827	—	—	—	③	—
1989	B-2200	0.033– 0.041	—	①	0.024– 0.045	0.067– 0.085	0.110– 0.143		0.289– 0.325
1990	B-2200	0.033– 0.041	—	①	0.024– 0.045	0.067– 0.085	0.110– 0.143		0.289– 0.325

① Manual transmission: 0.457–0.496
  Automatic transmission: 0.421–0.461
② Manual transmission: 0.028
  Automatic transmission: 0.031
③ Manual transmission: 1400–1600 rpm
  Automatic transmission: 1900–2100 rpm

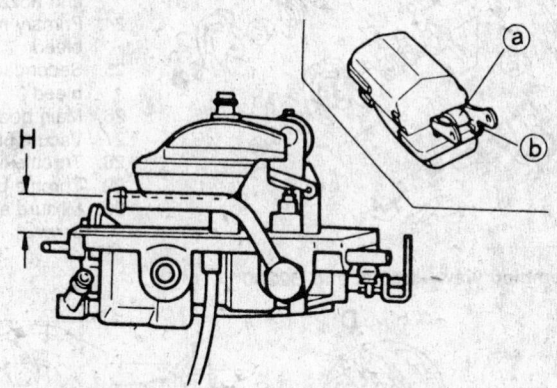

Mazda float level adjustment with the air horn upside down—B2000 and B2200

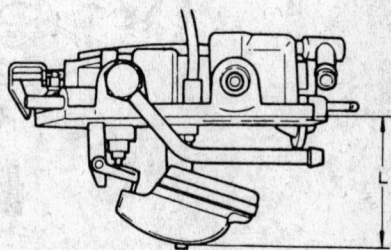

Mazda float level adjustment with the air horn at its normal position—B2000 and B2200

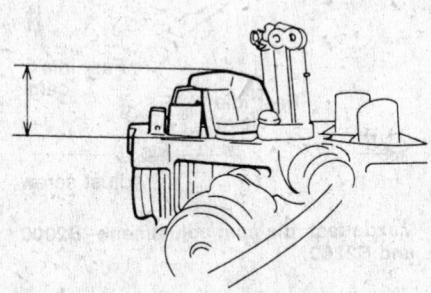

Mazda float level measurement—B2600

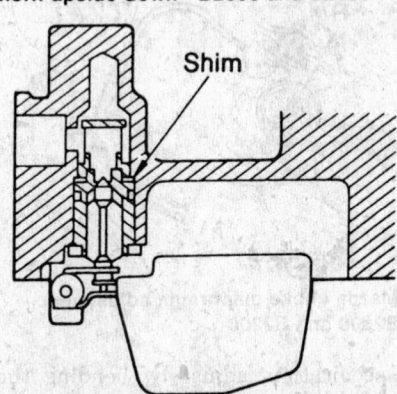

Mazda valve seat shim location—B2600

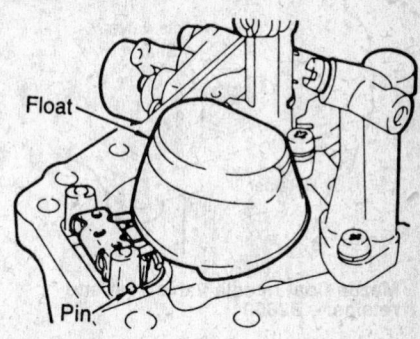

Mazda float and pin—B2600

a. Pull out the float lever pin and remove the float.

b. Remove the needle valve and retainer.

c. Use pliers and remove the needle valve seat.

d. Insert the necessary shims for adjustment and reinstall the needle valve, float and float chamber cover.

2. Measure the clearance between the bottom of the float to the surface of the float chamber cover without the gasket in place.

3. Add or remove needle valve seat shims to adjust the fuel level to specifications as follows:

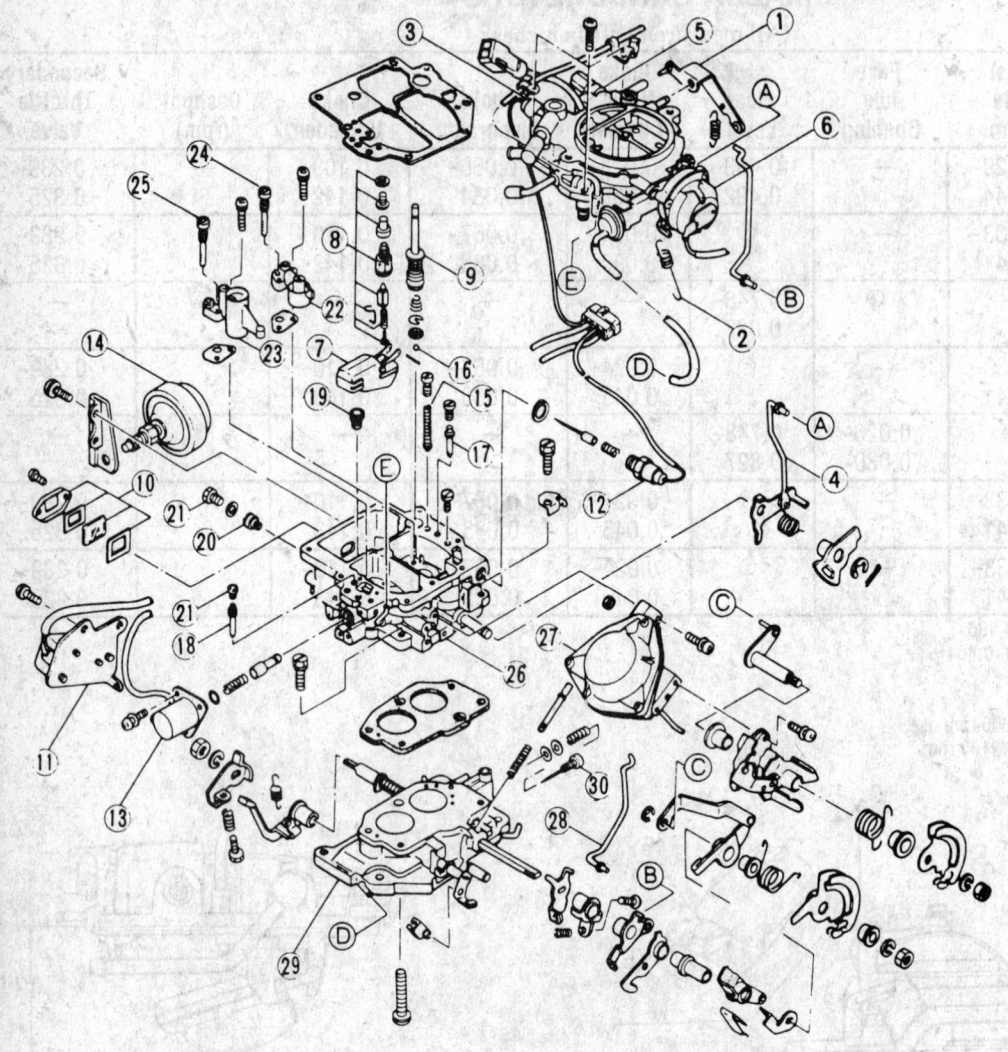

1. Accelerator pump connecting rod
2. Connecting spring
3. Air vent solenoid valve
4. Chock rod
5. Air horn
6. Automatic choke assembly
7. Float
8. Needle valve asssembly
9. Accelerator pump plunger
10. Fuel bowl sight glass
11. Idle switch
12. Slow fuel cut solenoid valve
13. Coasting richer solenoid valve
14. Dashpot
15. Accelerator pump outlet check ball and spring
16. Accelerator pump inlet check ball
17. Primary slow jet
18. Secondary slow jet
19. Primary main jet
20. Secondary main jet
21. Plug
22. Primary venturi and nozzle
23. Secondary venturi and nozzle
24. Primary main air bleed
25. Secondary main air bleed
26. Main body
27. Vacuum Diaphragm
28. Throttle link
29. Throttle body
30. Mixture adjusting screw

Mazda carburetor disassembled view—B2000 and B2200

Mazda float needle valve seat and retainer—B2600

Mazda choke diaphragm adjustment—B2000 and B2200

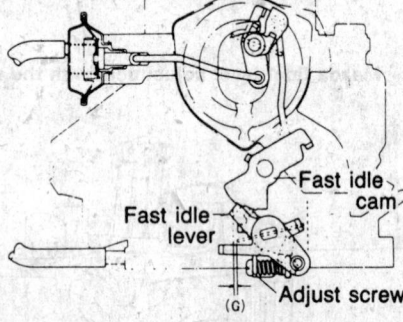

Mazda fast idle cam adjustment—B2000 and B2200

Choke Diaphragm Adjustment

B2000 AND B2200

1. Using a vacuum pump, apply a vacuum of about 15.7 in.Hg to the choke diaphragm vacuum tube.

2. Push the choke valve lightly to close it and measure the clearance.

3. If the clearance is not within specification, adjust by bending the choke lever.

Fast Idle Cam Adjustment

B2000 AND B2200

1. Set the fast idle cam to the second position.

2. Adjust the throttle valve clearance by turning the adjusting screw.

NOTE: The clearance will become larger as the screw is turned clockwise.

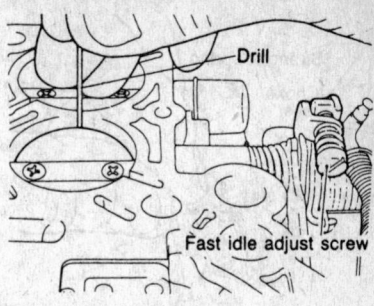

Mazda fast idle opening adjustment— B2600

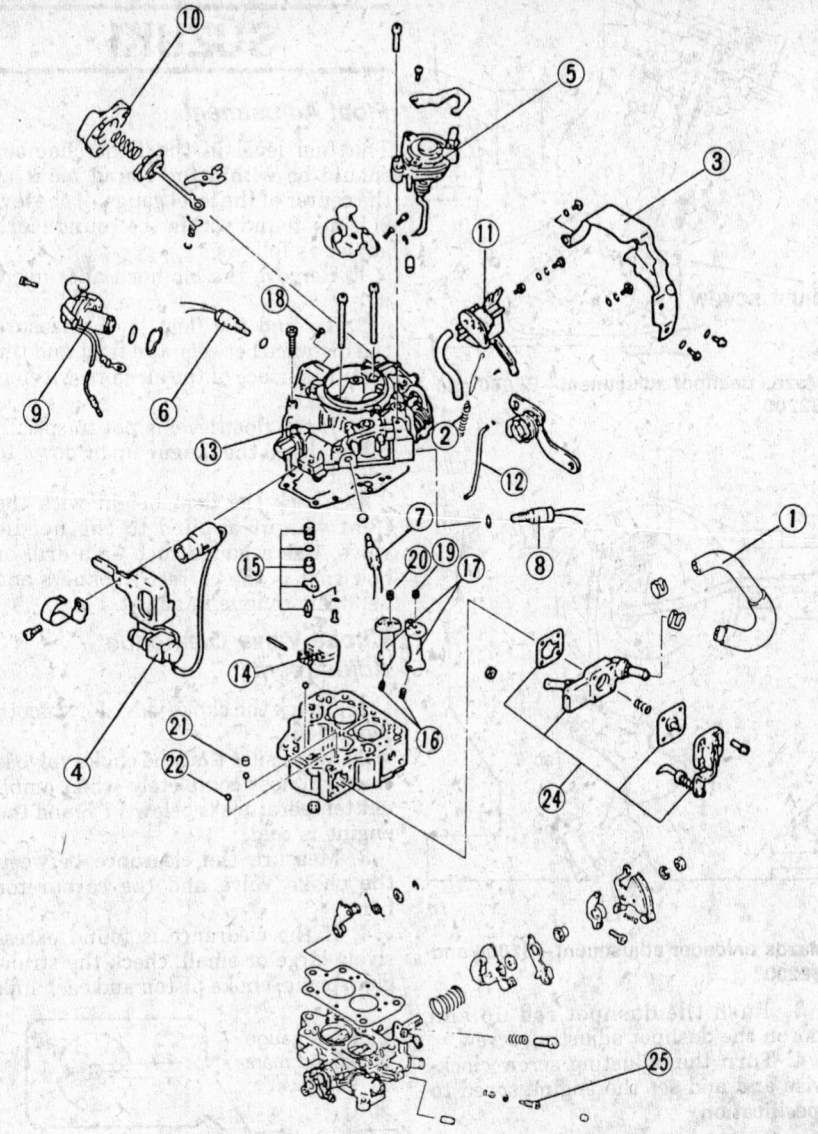

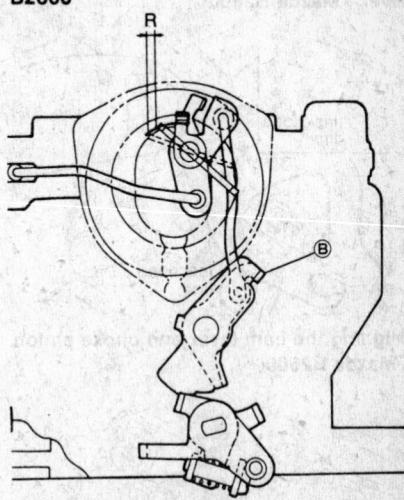

Mazda choke valve clearance adjustment —B2000 and B2200

1. Water hose
2. Throttle return spring
3. Choke cover
4. Throttle sensor
5. Dash pot
6. Jet mixture solenoid valve
7. Enrichment solenoid valve
8. Slow fuel cut solenoid valve
9. Air vent solenoid valve
10. Choke breaker
11. Secondary diaphragm
12. Choke rod
13. Float chamber cover
14. Float
15. Needle valve
16. Main jet
17. Jet block
18. Secondary slow jet
19. Primary slow jet
20. Jet mixture jet
21. Outlet check weight and ball
22. Inlet check ball
23. Steel ball
24. Accelerator pump
25. Mixture adjust screw

Mazda carburetor disassembled view—B2600

Fast Idle Opening Adjustment

B2600

1. With the carburetor at normal room temperature, insert the specified drill between the choke valve and the choke valve chamber wall.

2. Adjust the fast idle opening by using the fast idle adjusting screw.

Choke Valve Clearance Adjustment

B2000 AND B2200

1. Set the fast idle cam select to the second position.

2. Make sure the choke valve clearance is within specification.

3. Adjust the choke valve clearance by bending the starting arm. If a large adjustment is required, bend the choke rod.

Choke Valve Setting Adjustment

B2600

1. Remove the choke cover. It may be necessary to grind off the rivet heads and replace with locking screws.

2. Remove the 2 lock screws and remove the choke valve pinion assembly.

3. Install the strangler spring to the choke lever.

4. Assemble, aligning the inscribed line or black painted line on the teeth of the choke pinion with the inscribed line on the cam lever.

5. Loosely tighten the new lock screws.

6. Set the choke valve by moving the pinion arm up or down, align the punch mark on the float chamber cover with the center of the 3 inscribed lines and secure the pinion arm with the lock screws.

Dashpot Adjustment

B2600

1. Warm the engine to normal operating temperature.

2. Connect a tachometer to the engine.

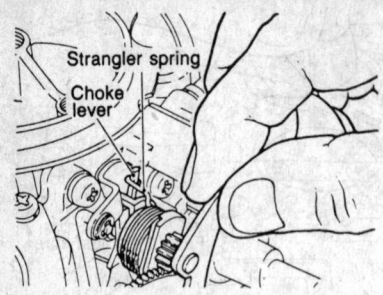

Fit the strangler spring to the choke lever—Mazda B2600

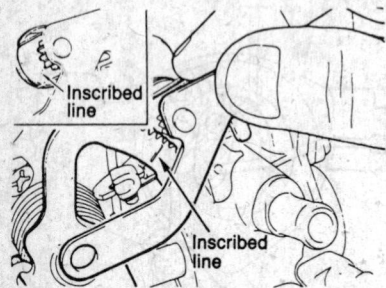

Aligning the cam lever and choke pinion—Mazda B2600

Pinion arm lock screws—Mazda B2600

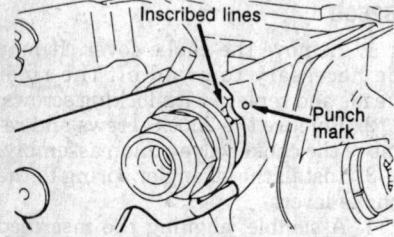

Align the center line with the punch mark—Mazda B2600

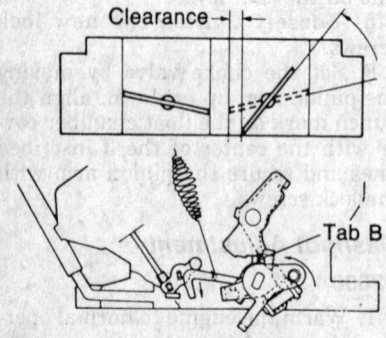

Mazda secondary throttle valve adjustment—B2000 and B2200

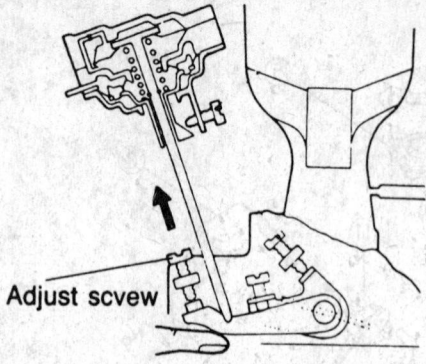

Mazda dashpot adjustment—B2000 and B2200

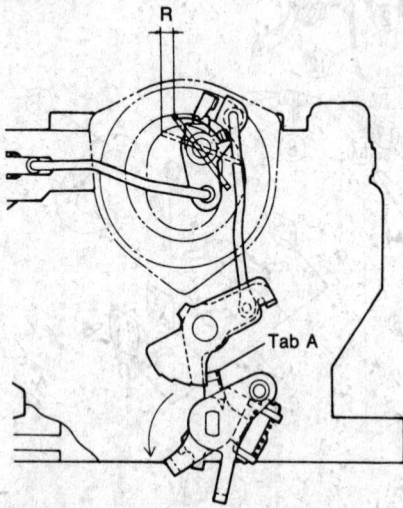

Mazda unloader adjustment—B2000 and B2200

3. Push the dashpot rod up and loosen the dashpot adjusting screw.

4. Turn the adjusting screw clockwise and and set the engine speed to specification.

Unloader Adjustment

B2000 AND B2200

1. Open the primary throttle valve all the way.

2. Measure the choke valve clearance and if not within specification adjust by bending the tab.

Secondary Throttle Valve Adjustment

B2000 AND B2200

1. The secondary throttle valve should start to open when the primary throttle valve opens (50–52 degrees), and should be completely open at the same time that the primary throttle valve is fully open.

2. Check the clearance between the primary throttle valve and the wall of the throttle bore when the secondary throttle valve starts to open.

3. If the clearance is not within specification, bend the tab to adjust.

SUZUKI

Float Adjustment

The fuel level in the float chamber should be within the round mark at the center of the level gauge. If the level is not found within the round mark adjust as follows:

1. Remove the air horn and invert it.

2. To find the float level, measure the distance between the float and the mating surface of the air horn without the gasket.

3. If the float level is not to specification, bend the tongue up or down to adjust.

4. Check the float height with the float weight applied to the needle valve. Use a gauge such as a drill or bolt that is the correct thickness and bend the tongue to adjust.

Choke Valve Clearance Adjustment

1. Check the choke valve for smooth movement.

2. Make sure that the choke valve is closed almost completely when ambient temperature is below 77°F and the engine is cold.

3. Measure the clearance between the choke valve and the carburetor bore.

4. If the clearance is found excessively large or small, check the strangler spring, choke piston and each link

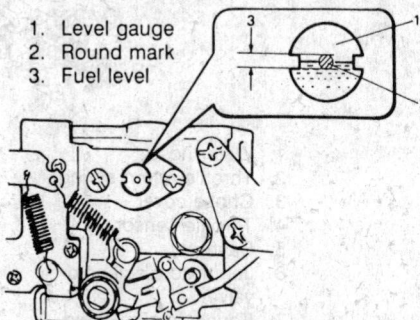

1. Level gauge
2. Round mark
3. Fuel level

Suzuki fuel level adjustment

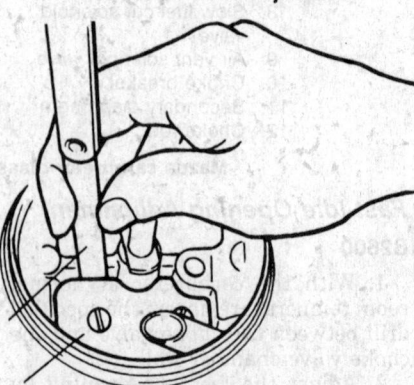

Suzuki choke valve to carburetor bore measurement

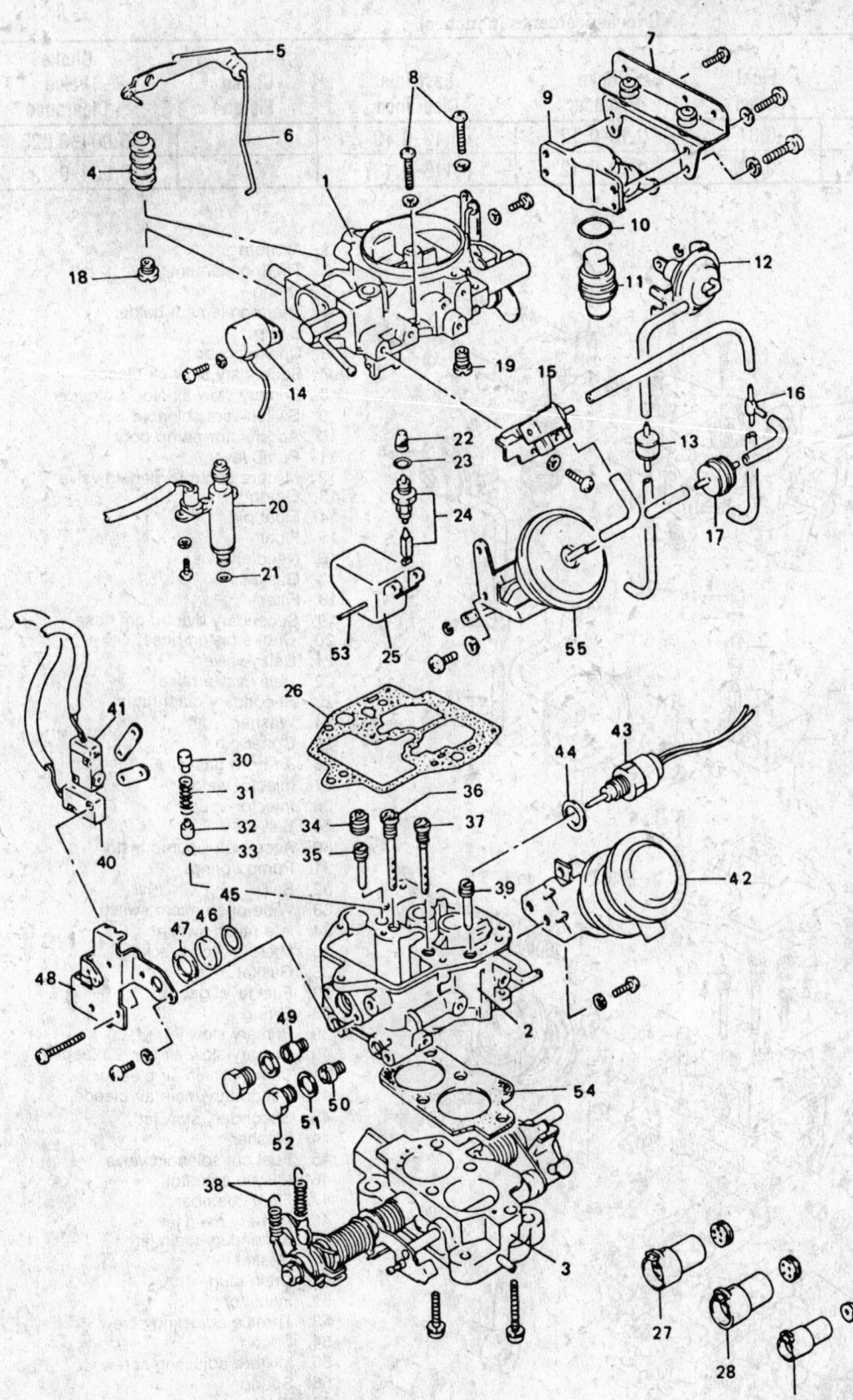

1. Air Horn
2. Float chamber
3. Throttle chamber
4. Pump boot
5. Pump lever
6. Pump rod
7. Bracket
8. Screw
9. Thermo element holder
10. Seal
11. Thermo element
12. Choke piston
13. Delay valve
14. Switch vent solenoid
15. Vacuum switching valve
16. 3-way joint
17. Vacuum transmitting valve
18. Primary slow air No. 1 bleeder
19. Secondary slow air bleeder
20. Mixture control solenoid
21. Solenoid valve seal
22. Needle valve filter
23. Needle valve gasket
24. Needle valve
25. Float
26. Air horn gasket
27. Connector (5 terminal)
28. Connector (4 terminal)
29. Connector (1 terminal)
30. Injector weight
31. Injector spring
32. Injector weight
33. Ball
34. Primary slow air No. 2 bleeder
35. Primary slow jet
36. Primary main air bleeder
37. Secondary main air bleeder
38. Spring
39. Secondary slow jet
40. Idle micro jet
41. Wide open micro switch
42. Idle up actuator
43. Solenoid valve
44. Washer
45. Level gauge seal
46. Level gauge
47. Level gauge gasket
48. Micro switch bracket
49. Primary main jet
50. Secondary main jet
51. Drain plug gasket
52. Drain plug
53. Float pin
54. Insulator
55. Secondary actuator (diaphragm)

Suzuki (Samurai) carburetor exploded view

SUZUKI CARBURETORS
(All measurements in inches)

Year	Model	Float Level	Choke Unloader	Fast Idle Clearance	Choke Piston	Choke Valve Clearance
1986–90	Samurai	0.31	0.10–0.12	0.10–0.12	①	0.004–0.023 ②
	Sidekick	0.31	0.10–0.12	0.118–0.137	①	0.004–0.023 ②

① See procedure in text
② At 75°F (25°C)

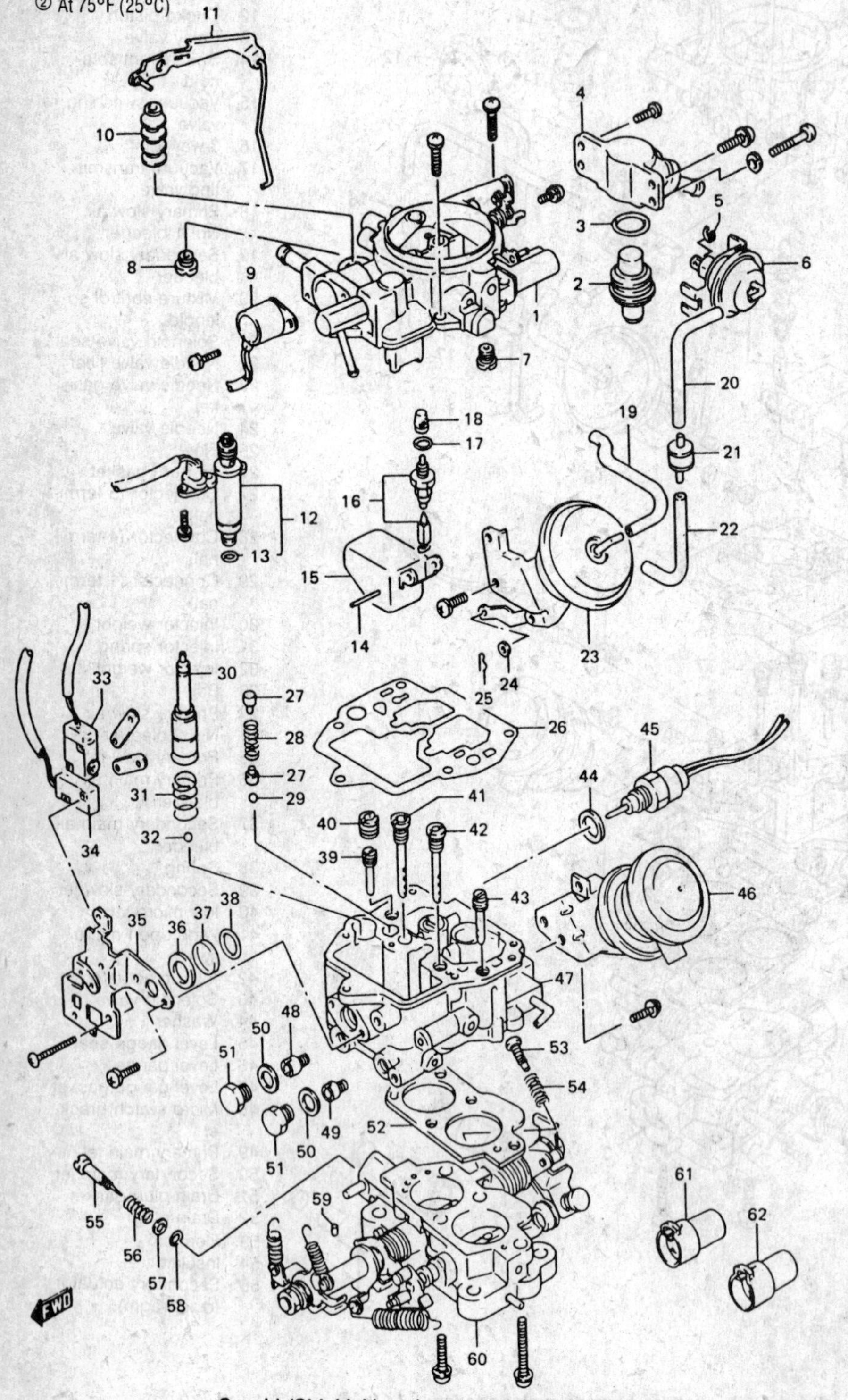

1. Air horn
2. Thermo element
3. O-ring
4. Thermo element holder
5. E ring
6. Choke piston
7. Secondary slow air bleeder
8. Primary slow air Noi. 1 bleeder
9. Switch vent solenoid
10. Accelerator pump boot
11. Pump lever
12. Mixture control solenoid valve
13. O-ring
14. Float pin
15. Float
16. Needle valve
17. Gasket
18. Filter
19. Secondary diaphragm hose
20. Choke piston hose
21. Delay valve
22. Delay valve hose
23. Secondary diaphragm
24. Washer
25. Cotter pin
26. Air horn gasket
27. Injector weight
28. Injector spring
29. Ball
30. Accelerator pump piston
31. Pump spring
32. Ball
33. Wide open micro switch
34. Idle micro switch
35. Micro switch bracket
36. Gasket
37. Fuel level gauge
38. O-ring
39. Primary slow jet
40. Primary slow air No. 2 bleeder
41. Primary main air bleeder
42. Secondary main air bleeder
43. Secondary slow jet
44. Washer
45. Fuel cut solenoid valve
46. Idle up actuator
47. Float chamber
48. Primary main jet
49. Secondary main jet
50. Gasket
51. Drain plug
52. Insulator
53. Throttle adjusting screw
54. Spring
55. Mixture adjusting screw
56. Spring
57. Washer
58. O-ring
59. Tamper pin
60. Throttle chamber
61. 4P coupler
62. 6p coupler

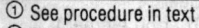

Suzuki (Sidekick) carburetor exploded view

in the choke system for smooth operation. Lubricate as necessary.

5. If clearance is still out of specification, remove the carburetor from the intake manifold and remove the idle up actuator from the carburetor.

6. Turn the fast idle cam counterclockwise and insert a pin into the holes in the cam and bracket to lock the cam.

7. Using pliers, bend the choke lever up or down. Bending up causes the choke valve to close and down to open.

8. Start the engine and warm it up fully.

9. Stop the engine and check to see if the choke valve is fully open.

10. If the choke valve doesn't open fully, the wax element or its link system is defective. Replace defective parts as necessary.

Choke Piston Adjustment

1. Disconnect the choke piston hose at the throttle chamber.

2. While pushing down lightly on the choke valve to its closed position, apply vacuum to the choke piston hose with a vacuum gauge and measure the choke valve to carburetor bore clearance. It should be 0.04–0.05 in. for the Samarai and 0.070–0.078 in. for the Sidekick.

3. With vacuum applied, use a suitable tool and move the choke piston rod in towards the choke piston and check to see if the choke valve to carburetor bore clearance is 0.13–0.14 in. for the Samarai and 0.142–0.163 in. on the Sidekick.

Fast Idle Cam Adjustment

1. Make sure the ambient temperature is between 71–81°F and that the mark on the cam and the center of the cam follower are in alignment.

2. Disconnect the vacuum hose from the Three Way Solenoid Valve (TWSV) on the Samurai, or the Vacuum Switching Valve (VSV) on the

Sidekick and connect a vacuum pump gauge to its hose.

3. While applying 15 in.Hg of vacuum to the actuator, check the clearance between the actuator rod and the idle up adjusting screw.

4. Adjust as needed using the fast idle adjusting screw.

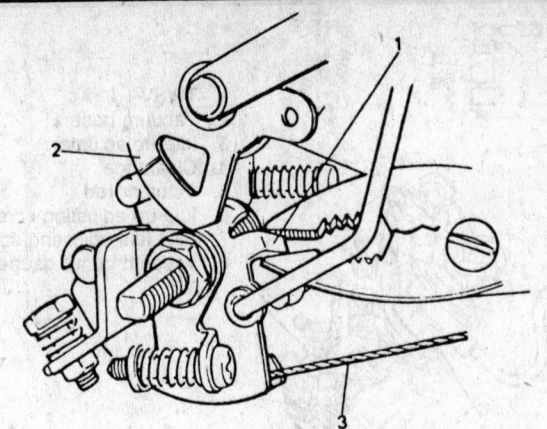

1. Unloader arm
2. Fast idle cam
3. Accelerator cable

Suzuki unloader lever arm

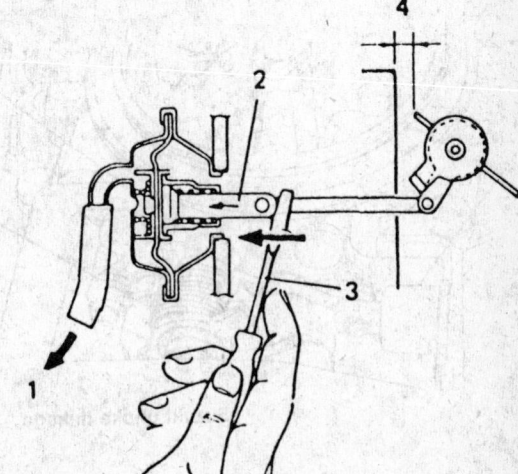

1. Apply vacuum
2. Choke piston rod
3. Suitable tool
4. Choke valve to carburetor bore clearance

Moving the choke piston rod—Suzuki

Unloader Adjustment

1. Make sure the engine is cool and the choke valve is fully closed.

1. VSV
2. Vacuum hose
3. Idle up actuator
4. Actuator rod
5. Idle up adjusting screw
6. Fast idle adjusting screw
7. Vaccum pump gauge
8. Clearance

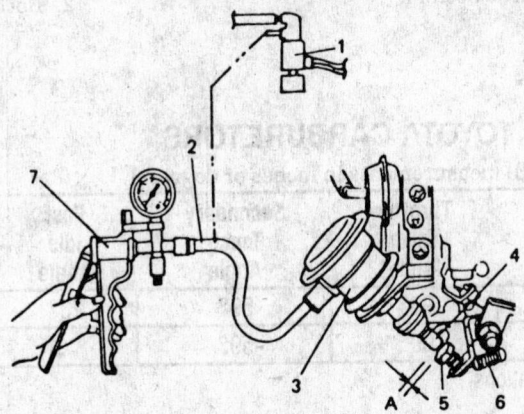

Applying vacuum to the VSV—Sidekick

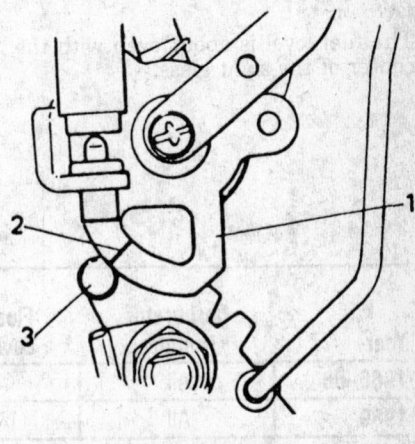

1. Fast idle cam
2. Mark on cam
3. Cam follower

Suzuki cam and cam follower mark

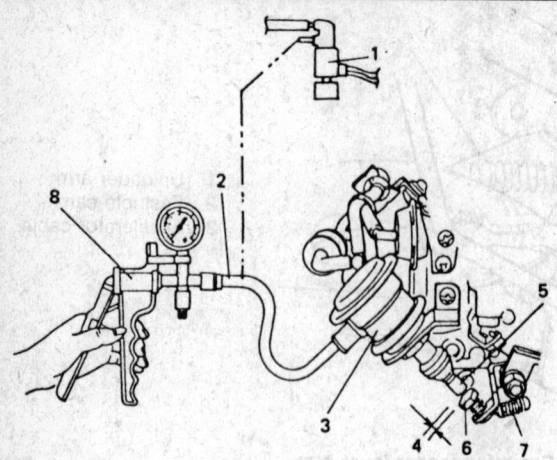

1. TWSV
2. Vacuum hose
3. Idle up actuator
4. Clearance
5. Actuator rod
6. Idle up adjusting screw
7. Fast idle adjusting screw
8. Vaccum pump gauge

Applying vacuum to the TWSV—Samural

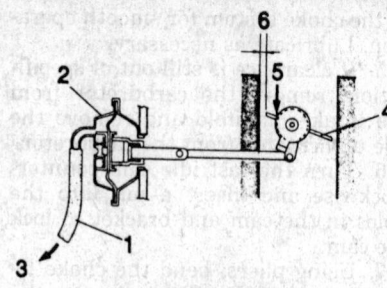

1. Choke piston hose
2. Choke piston
3. Apply vacuum
4. Choke valve
5. Push here lightly
6. Choke valve to carburetor bore clearance

Suzuki choke piston measurement

1. Choke piston
2. Choke valve shaft
3. Strangler spring

Suzuki choke linkage

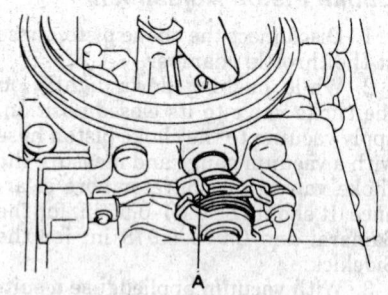

Suzuki choke lever adjusting tab

2. Fully open the throttle valve and measure the choke valve to carburetor bore clearance.

3. If necessary, bend the unloader arm to adjust.

TOYOTA

Fuel Level

The fuel level is about even with the center of the sight glass.

Float Level Adjustment

1. Allow the float to hang down by its own weight.

2. Measure the clearance between the float top and the air horn without the gasket.

3. The float level (raised position), should be 0.386 in. If necessary, bend the upper portion of the float tab to adjust to specification.

4. Lift up the float and measure the distance between the air horn and the float bottom.

5. The float level (lowered position), should be 0.189 in. If necessary, bend the lower portion of the float tab to adjust to specification.

Throttle Valve Opening (Angle) Adjustment

1. Measure the full opening angle of the of the primary and secondary throttle valves.

2. Both the primary and secondary

TOYOTA CARBURETORS

(All measurements in inches or degrees)

Year	Carburetor Number	Float Level	Throttle Valve Angle	Secondary Touch Angle	Fast Idle Angle	Choke Unloader Angle	Idle-Up Angle
1986–88	All	②	90°	59°	23°	45°	16.5°
1989	All	②	90°	59°	①	45°	16.5°

NOTE: Use angle degree tool for angle specifications
① Federal and Canada: 24.5 degrees
   California: 23.0 degrees
② Raised position: 0.386 in.
   Lowered position: 1.89 in.

throttle valve openings should be 90 degrees from the horizontal plane.

3. Adjust by bending the respective throttle arm levers of either valve.

Secondary Touch Angle Measurement

1. Measure the primary throttle valve opening at the same time the second throttle valve just starts to open.

2. The standard angle is 59 degrees from the horizontal plane.

3. Adjustment is not necessary, this is a rebuilding reference only.

Fast Idle Angle Adjustment

1. Set the throttle shaft lever to the first step of the fast idle cam.

2. Measure the primary throttle valve angle with the choke valve fully closed.

3. Adjust by turning the fast idle adjusting screw.

Choke Unloader Adjustment

1. Measure the choke valve angle with the primary throttle valve fully opened.

2. Adjust by bending the primary throttle arm.

Idle-Up Angle Adjustment

1. Apply vacuum to the idle-up diaphragm.

2. Measure the throttle valve opening angle.

3. Adjust the angle by turning the idle adjusting screw.

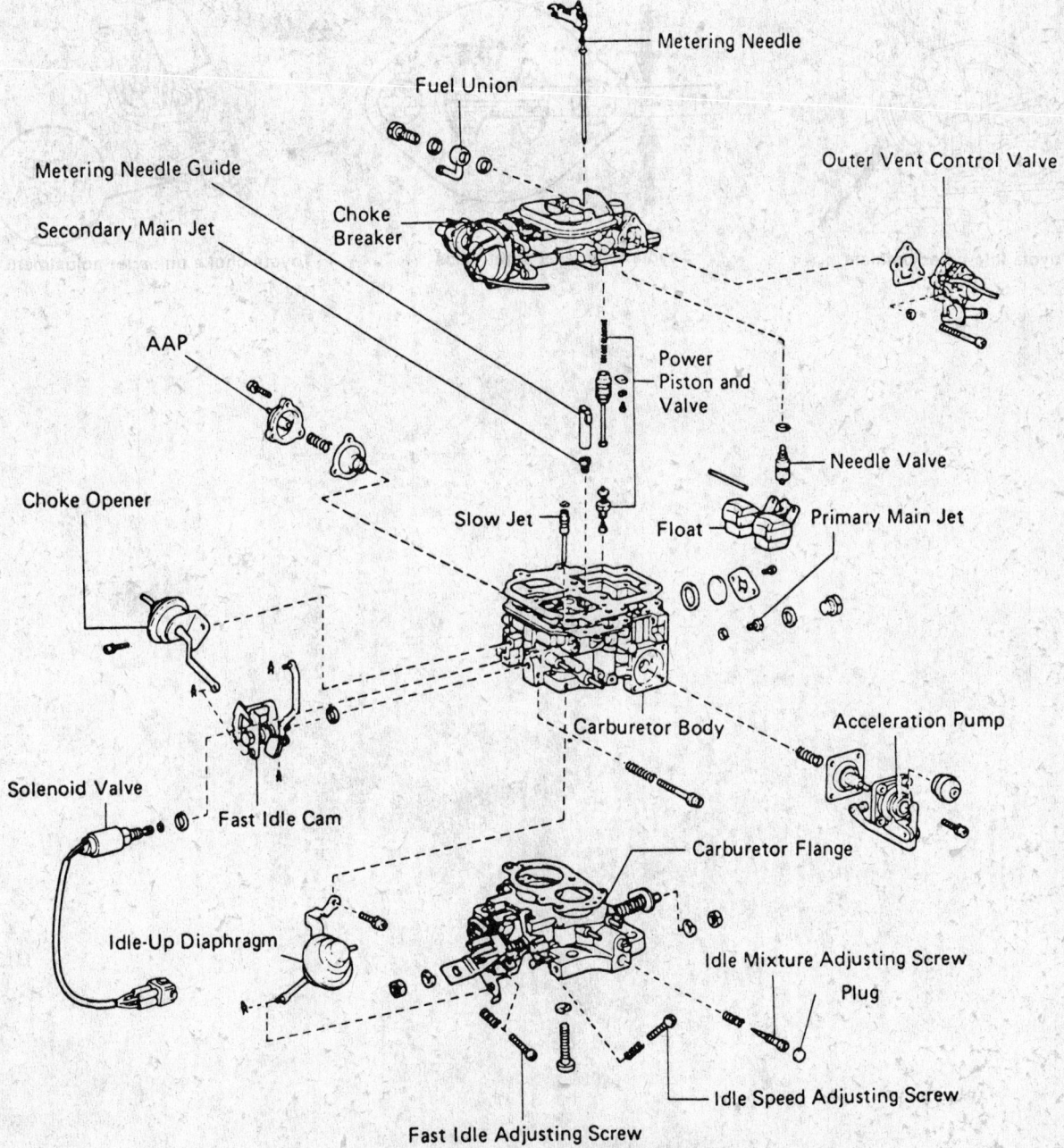

Toyota carburetor exploded view

Dashpot Adjustment

1. Open the throttle valve until the throttle lever separates from the dashpot end.

2. Release the throttle valve gradually, and check the dashpot touch angle when the throttle lever touches the dashpot end.

3. If the dashpot touch angle is not 24.5 degrees from the horizontal plane, unlock the locknut and adjust the angle by turning the dashpot diaphragm.

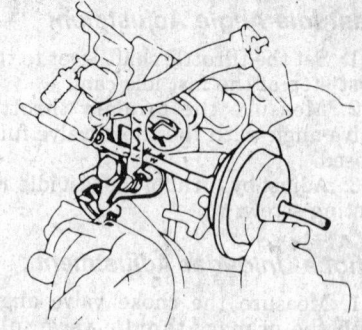

Toyota fast idle angle adjustment

Toyota secondary touch angle adjustment

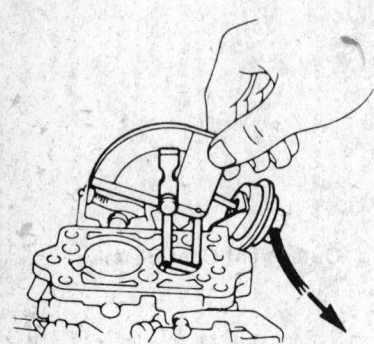

Toyota idle-up adjustment

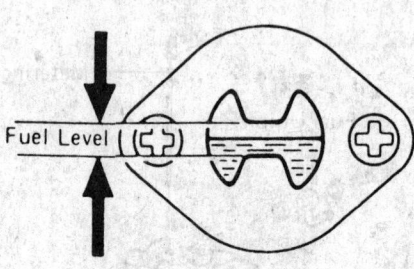

Fuel Level

Toyota float level sight glass

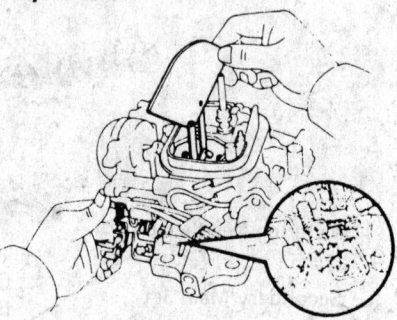

Toyota choke unloader adjustment

Wheel Alignment 21

FRONT END TROUBLE DIAGNOSIS

Condition	Cause	Correction
Hard steering	1) Steering linkage needs lubrication 2) Low or uneven front tire pressure 3) Incorrect front wheel alignment (manual steering)	1) Lubricate the linkage 2) Inflate tires to the recommended pressure 3) Check and align the front suspension
Poor directional stability	1) Steering linkage need lubrication 2) Low or uneven tire pressure 3) Loose wheel bearings 4) Incorrect front wheel alignment (caster) 5) Broken springs 6) Malfunctioning shock absorber 7) Broken stabilizer bar or a missing link	1) Lubricate the linkage 2) Inflate tires to the recommended pressure 3) Adjust or replace the wheel bearings 4) Check and align the front suspension 5) Replace the springs 6) Check and replace the shock absorber 7) Replace the stabilizer bar or link
Front wheel shimmy (smooth road shake)	1) Tire and wheel are out of balance or out of round 2) Worn or loose wheel bearings 3) Worn ball joints 4) Malfunctioning shock absorber	1) Balance the tires, check run-out 2) Adjust the wheel bearings 3) Replace the ball joints 4) Check and replace the shock absorber
Vehicle pulls to one side (no braking action)	1) Low or uneven tire pressure 2) Front or rear brakes dragging 3) Broken or sagging front spring 4) Incorrect front wheel alignment (camber)	1) Inflate the tires to the recommended pressure 2) Adjust the brakes 3) Replace the spring 4) Check and align the front suspension
Noise in the front end	1) Steering linkage needs lubrication 2) Loose shock absorber or worn bushings 3) Worn control arm bushings 4) Worn or loose wheel bearings 5) Loose stabilizer bar 6) Loose wheel nuts 7) Spring is improperly positioned 8) Loose suspension bolts	1) Lubricate at the recommended intervals 2) Tighten the bolts or replace the shock absorber 3) Replace the bushings or control arm 4) Adjust or replace the wheel bearings 5) Tighten all the stabilizer bar attachments 6) Tighten the wheel nuts 7) Reposition the spring 8) Tighten to specifications or replace
Wheel tramp	1) Tire and the wheel are out of balance 2) Tire and the wheel are out of round 3) Blister or bump on the tire 4) Improper shock absorber action	1) Balance the wheels 2) Replace the tire 3) Replace the tire 4) Replace the shock absorber
Excessive or uneven tire wear	1) Underinflated or overinflated tires 2) Improper toe-in 3) Wheels are out of balance 4) Hard driving 5) Overloading the vehicle	1) Inflate the tire to the recommended pressure 2) Adjust toe-in setting 3) Balance the wheels 4) Follow proper driving techniques 5) Do not exceed the maximum recommended payload rating
Scuffed tires	1) Toe-in is incorrect 2) Excessive speed on turns 3) Tires are improperly inflated 4) Suspension arm is bent or twisted	1) Adjust toe-in setting 2) Follow proper driving techniques 3) Inflate the tires to the recommended pressure 4) Replace the suspension arm
Noisy shocks	1) Loose mounting	1) Check all mounting torques (bolt and/or nut)
Excessive road shock	1) Tire air pressure is too high 2) Loose wheel bearings	1) Deflate to correct pressure 2) Adjust bearings

FRONT END TROUBLE DIAGNOSIS

Condition	Cause	Correction
Excessive road shock	3) Camber adjustment is incorrect (negative camber contributes to road shock)	3) Adjust camber
	4) Weak or broken front spring	4) Replace the spring
	5) Loose suspension components	5) Inspect, repair, and adjust as necessary
Leaky shocks	1) Seals are worn out	1) Replace shocks
Weak shocks	1) Shocks are worn out	1) Replace shocks
	2) Loss of shock fluid	2) Replace shocks
Vibration and shimmy	1) Seal damage resulting in loss of lubricant, corrosion, excessive wear	1) Replace damaged parts as necessary
	2) Tires and wheels, or brake drums, are out of balance	2) Balance the tires and wheels, turn the brake drums
	3) Bent wheel or tire is out of round	3) Replace the wheel and remount, or replace the tire
	4) Wheel stud nuts torqued unevenly	4) Retorque the wheel stud nuts
	5) Loose steering linkage	5) Adjust, torque or repair as necessary
	6) Wheel is loose on the hub	6) Inspect the wheel bolt for damage Replace the wheel if needed. Replace all wheel studs
	7) Driveline universal joints are rough or defective (may be confused with steering vibration)	7) Repair driveline
	8) Malfunctioning shock absorbers	8) Replace the shock absorbers
Cupped tires	1) Front shock absorbers are defective	1) Replace the shock absorbers
	2) Worn ball joints	2) Replace the ball joints
	3) Wheel bearings are incorrectly adjusted or worn	3) Adjust or replace the wheel bearings (also replace the races)
	4) Wheel and tire is out of balance	4) Balance the wheel and tire
	5) Excessive tire or wheel runout	5) Check and compensate for runout

GENERAL INFORMATION

Wheel Alignment

For a vehicle to have safe steering control with a minimum of tire wear, certain established rules must be followed. These rules fix the values of planes, angles and radii relative to each other and to vehicle and tire dimensions. Some factors are built in, with no provision for adjustment; others are adjustable within limits. The entire system depends upon all value factors, separately and combined. It is therefore difficult to change some of the established settings without influencing others.

This system is called steering geometry or wheel alignment and requires a complete check of all the factors involved. Definitions of these factors and the effect each has on the vehicle are as follows.

Steering Wheel Position

Always check steering wheel alignment in conjunction with and at the same time as toe-in. In fact, the steering wheel spoke position, with the vehicle on a straight section of highway, may be the first indication of front end misalignment.

If the vehicle has been wrecked, or indicates any evidence of steering gear or linkage disturbance, the pitman arm should be disconnected from the sector shaft. The steering wheel (or gear) should be turned from extreme right to extreme left to determine the halfway point in its turning scope. This will be the spot on the gear that is in action during straight ahead driving and in which position the steering gear should be adjusted. With the steering wheel in the straight ahead position and the steering gear adjusted to zero lash status, reconnect the pitman arm.

CAMBER ANGLE

Camber is the amount that the front wheels are inclined outward or inward at the top. Camber is spoken of, and measured, in degrees from the perpendicular. The purpose of the camber angle is to take some of the load off the spindle outboard bearing.

CASTER ANGLE

Caster is the amount that the king pin (or in the case of vehicles without king pins, the knuckle support pivots) is tilted towards the back or front of the vehicle. Caster is usually spoken of, and measured, in degrees. Positive caster means that the top of the king-pin is tilted toward the back of the vehicle. Positive caster is indicated by the sign +.

Negative caster is exactly the opposite; the top of the king pin is tilted toward the front of the vehicle. This is generally indicated by the sign −. Negative caster is sometimes referred to as reverse caster.

The effect of positive caster is to cause the vehicle to steer in the direction in which it tends to go. Positive caster in the front wheels may cause the vehicle to steer down off a crowned road or steer in the direction of a cross wind. For this reason, a number of our modern vehicles are arranged with negative caster so that the opposite is true, which is that the vehicle tends to steer up a crowned road and into a cross wind.

Caster angle specifications are based on the vehicle load limits, which will usually result in a level frame. Since load requirements may vary, the frame does not always remain level and must be considered when determining the correct caster angle.

Because of their naturally straight running characteristics, front wheel drive vehicles are not overly sensitive to caster, therefore caster is not adjustable.

To measure the caster angle, the vehicle should be on a smooth and level surface. Place a bubble protractor on the frame rail and measure the degree of frame tilt and in what direction, either front or rear.

Two methods of determining caster angles are used. The first method is to determine the caster angle from the wheel with alignment equipment, and the second method is to obtain the desired caster angle from the specification charts. The frame angle is then added to or subtracted from the caster angles as necessary.

METHOD ONE

1. Determine the frame angle.
    a. If the frame is high at rear, than the frame angle is negative.
    b. If the frame is low at rear, than the frame angle is positive.
2. Determine the caster angle at the wheel with the alignment checking equipment.
3. Add or subtract frame angle to determine caster angle.
    a. Negative frame angle is added to positive caster angle.
    b. Positive frame angle is subtracted from positive caster angle.
    c. Negative frame angle is subtracted from negative caster angle.
    d. Positive frame angle is added to negative caster angle.
4. Determine the correct caster angle and the specified caster angle. Correct the vehicle caster, as required.

METHOD TWO

1. Measure the frame angle.
    a. If the front of the frame is down than the frame angle is positive.
    b. If the front of the frame is up than the frame angle is negative.
2. From the specifications, determine the specified or desired caster setting.
3. Add or subtract the frame angle from the specified caster setting.
    a. Positive frame angle is subtracted from the specified setting.

    b. Negative frame angle is added to the specified caster setting.
4. Using wheel alignment equipment, obtain the measured caster angle from the wheel and determine the corrected specified setting.

KINGPIN INCLINATION ANGLE

In addition to the caster angle, the kingpins, if equipped, (or knuckle support pivots) are also inclined toward each other at the top. This angle is known as kingpin inclination and is usually spoken of, and measured, in degrees.

The effect of kingpin inclination is to cause the wheels to steer in a straight line, regardless of outside forces such as crowned roads, cross winds, etc., which may tend to make the vehicle steer at a tangent. As the spindle is moved from extreme right to extreme left it apparently rises and falls. The spindle reaches its highest position when the wheels are in the straight ahead position. In actual operation, the spindle cannot rise and fall because the wheel is in constant contact with the ground.

Therefore, the vehicle itself will rise at the extreme right turn and come to its lowest point at the straight ahead position, and again rise for an extreme left turn. The weight of the vehicle will tend to cause the wheels to come to the straight ahead position, which is the lowest position of the vehicle itself.

KINGPIN INCLUDED ANGLE

Included angle is the name given to that angle which includes kingpin inclination and camber. It is the relationship between the centerline of the wheel and the centerline of the kingpin (or the knuckle support pivots). This angle is built into the knuckle (spindle) forging and will remain constant throughout the life of the vehicle, unless the spindle itself is damaged.

When checking a vehicle on the front end alignment machine, always measure kingpin inclination as well as camber unless some provision is made on the stand for checking condition of the spindle. Where no such provision is made, add the kingpin inclination inclination to the camber for each side of the vehicle. These totals should be exactly the same, regardless of how far from the norm the readings may be.

Since the most common cause of a bent spindle is striking the curb when parking, which causes the spindle to bend upward, the side having the greater included angle usually has the bent spindle. It will be found impossible to achieve good alignment and minimum tire wear unless the bent spindle is replaced.

TOE-IN

Toe-in is the amount that the front wheels are closer together at the front than they are at the back. This dimension is usually measured, in inches or fractions of an inch.

Generally speaking, the wheels are toed-in because they are cambered. When a vehicle operates with zero degrees camber it will be found to operate with zero toe-in. As the required camber increases, so does the toe-in. The reason for this is that the cambered wheel tends to steer in the direction in which it is cambered. Therefore it is necessary to overcome this tendency by compensating very slightly in the direction opposite to that in which it tends to roll. Caster and camber both have an effect on toe-in. Therefore toe-in is the last component on the front end which should be corrected.

TOE-OUT

When a vehicle is steered into a turn, the outside wheel of the vehicle scribes a much larger circle than the inside wheel. Therefore, the outside wheel must be steered to a somewhat less angle than the inside wheel. This difference in the angle is often called toe-out.

The change in angle from toe-in in the straight ahead position to toe-out in the turn is caused by the relative position of the steering arms to the king pin and to each other.

If a line were drawn from the center of the king pin through the center of the steering arm tie rod attaching hole at each wheel the lines would be found to cross almost exactly in the center of the rear axle.

If the front end angles, including toe-in, are set correctly, and the toe-out is found to be incorrect, one or both of the steering arms are bent.

TRACKING

While tracking is more a function of the rear axle and frame, it is difficult to align the front suspension when the vehicle does not track straight. Tracking means that the centerline of the rear axle follows exactly the path of the centerline of the front axle when the vehicle is moving in a straight line.

On vehicles that have equal tread, front and rear, the rear tires will follow in exactly the thread of the front tires, when moving in a straight line. However, there are many vehicles whose rear tread is wider than the front tread. On such vehicles, the rear axle tread will straddle the front axle tread an equal amount on both sides, when moving in a straight line.

Perhaps the easiest way to check a

vehicle for tracking is to stand directly in back of it and watch it more in a straight line down the street. If the observer will stand as near to the center of the vehicle as possible, he can readily observe, even with the difference in perspective between the front and rear wheels, whether or not they are tracking properly. If the vehicle is found to track incorrectly, the difficulty will be found in either the frame or in the rear axle alignment.

Another more accurate method to check tracking is to park the vehicle on a level floor and drop a plumb line from the extreme outer edge of the front suspension lower A-frame. Use the same drop point on each side of the vehicle. Make a chalk lie where the plumb line strikes the floor. Do the same with the rear axle, selecting a point on the rear axle housing for the plumb line.

Measure diagonally from the left rear mark to the right front mark and from the right rear mark to the left front mark. These diagonal measurements should be the same but a ¼ in. variation is acceptable.

If the diagonal measurements taken are different, measure from the right rear mark to the right front mark and from the left rear to the left front. These measurements should also be the same within ¼ in.

If the diagonal measurements are different, but the longitudinal measurements are the same, the frame is swayed (diamond shaped).

However, in the event that the diagonal measurements are unequal and the longitudinal measurements are also unequal, and the vehicle is tracking incorrectly, the rear axle is misaligned.

If the diagonal and longitudinal measurements are both unequal, but the vehicle appears to track correctly on the street, a kneeback is indicated.

NOTE: A kneeback means that a complete side of the front suspension is bent back. This is often caused by crimping the front wheels against the curb when parking the vehicle, then starting up without straightening the wheels out.

Tire and Wheel Service

TIRE AND WHEEL BALANCE

There are 2 types of tire and wheel balancing procedures. They are the dynamic balance and the static balance.

The dynamic balance is the equal distribution of weight on each side of the centerline, so that when the tire and wheel assembly spins there is no tendency for the assembly to move from side to side. Tire and wheel assemblies that are dynamically unbalanced may cause wheel shimmy.

The static balance is the equal distribution of weight around the wheel. Tire and wheel assemblies that are statically unbalanced cause a bouncing action called wheel tramp. This condition will eventually cause uneven tire wear.

Before the tire and wheel assembly can be properly balanced all deposits of mud, etc. must be removed from the inside of the rim area. Stones and other foreign matter should be removed from the tire tread area. The tire and wheel assembly should be inspected for any signs of external damage. Once these conditions have been met the tire and wheel assembly is ready to be balanced according to manufacturers instructions.

TIRE ROTATION

To ensure that all tires wear evenly tire rotation should be done every 8000 miles. If a tire shows excessive wear the wear problem should be corrected before rotating the tires. If the vehicle is equipped with a temporary spare tire, do not include it in the tire rotation procedure.

TIRE REPLACEMENT

Specialized tools and equipment have been designed for use in the replacement of a tire on multipiece rims. The manufacturers instructions should be followed in the use of the machines in the mounting and dismounting of tires to avoid personal injury.

CHRYSLER CORPORATION

Wheel Alignment

The front wheel alignment positions must be retained within the specified limits to prevent abnormal tire tread wear and to ensure steering ease.

Toe-In Adjustment

1. Secure the steering wheel with the front wheels in a straight ahead position.
2. The engine should be operating during the toe-in adjustment procedure on vehicles equipped with power steering.
3. Loosen the tie rod adjustment sleeve lock bolts or the tie rod locknuts, if equipped.
4. Rotate the tie rod adjustment sleeve or the tie rods, if equipped, to adjust the wheel toe position.
5. Tighten the adjustment sleeve lock bolts or the tie rod locknuts, if equipped.

Camber and Caster Adjustment
FRONT WHEEL DRIVE VAN/WAGON
1. Loosen the cam adjusting bolts
2. Rotate the cam bolt to move the

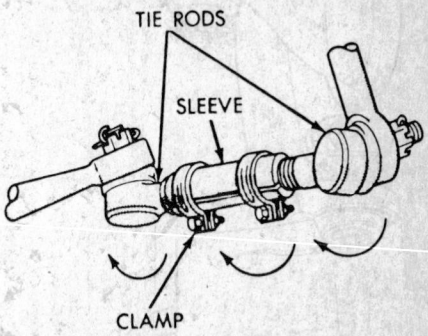

Tie rod adjustment—Except Front wheel drive Van/Wagon

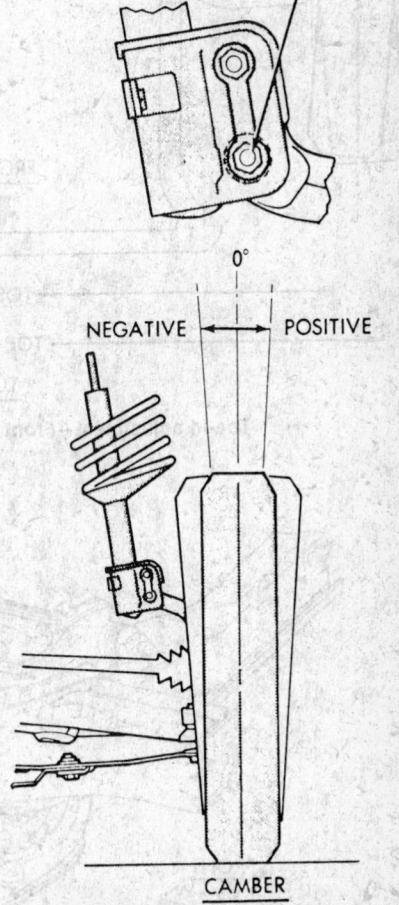

Camber adjustment—Except Front wheel drive Van/Wagon

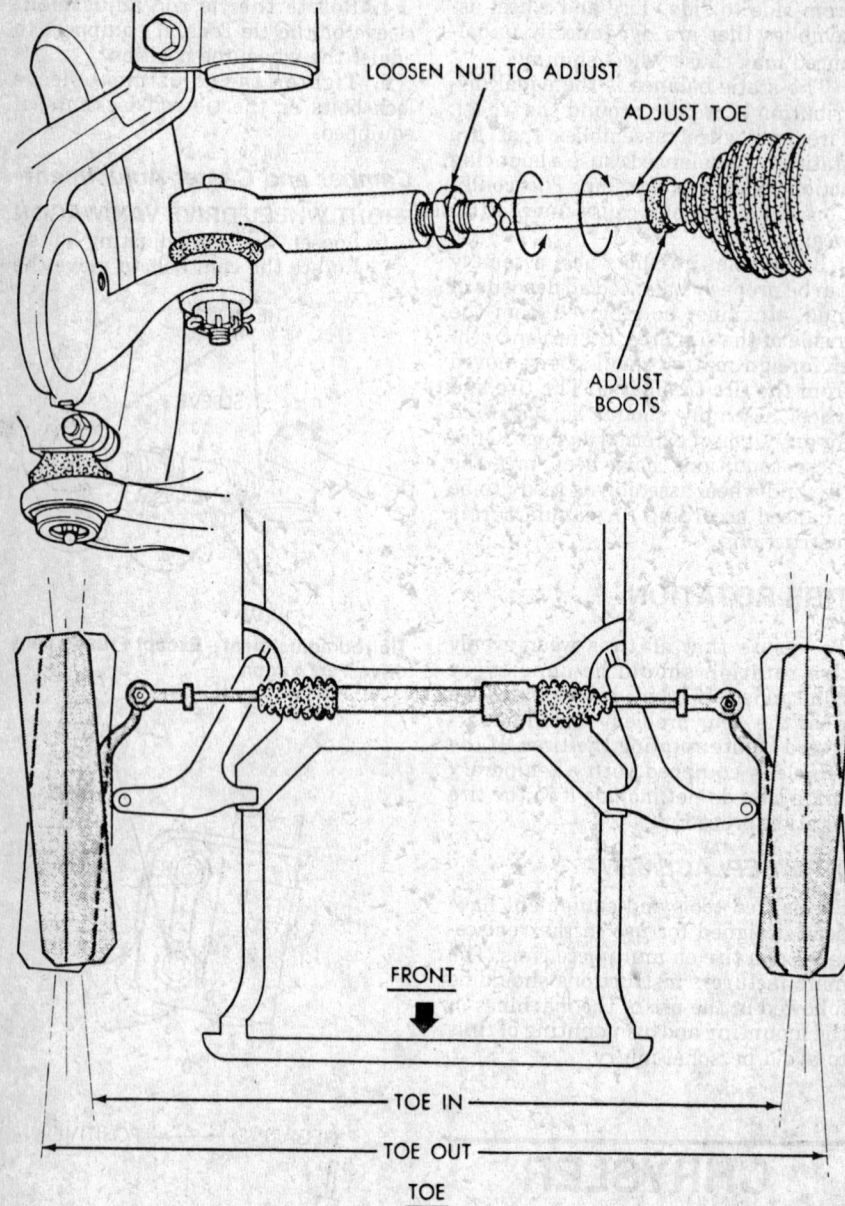

Toe-in adjustment – Front wheel drive Van/Wagon

top of the wheel in or out to reach the specified camber, using the proper guage.

3. Tighten the cam bolts to the specified torgue plus ¼ turn beyond.

4. There is no caster adjustment on the these vehicles.

EXCEPT FRONT WHEEL DRIVE VAN/WAGON

1989–1990

1. Adjustments are made by loosening the retaining nuts and changing the position of the eccentric (cam) bolts, using the proper equipment.

2. Remove all the foreign material from the eccentric (cam) bolt threads.

3. Record the camber and caster measurements before making any adjustments.

4. The camber angle should be adjusted as near as possible to the preferred angle.

5. The camber angle is factory preset and cannot be adjusted on 4WD vehicles.

6. The caster can be adjusted by installing the proper size tapered shims between the front axle pads and the spring brackets.

7. The caster angle should be adjusted as near as possible to the preferred angle.

8. The caster should be the same on both sides of the vehicle.

1986–1988

1. Move both ends of the pivot bar in or out exactly equal amounts to adjust the camber.

2. The camber setting should be

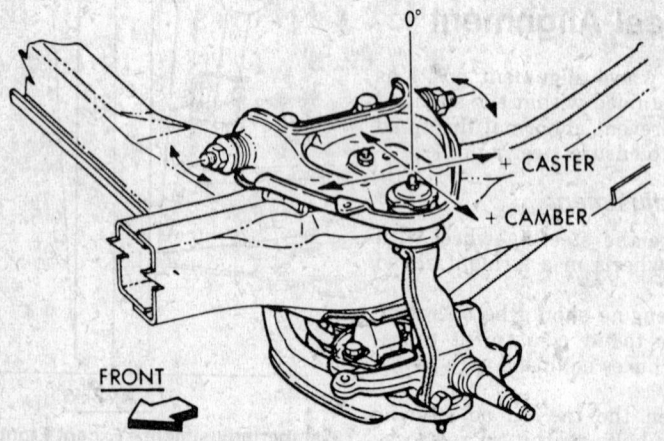

Camber and caster adjustments – 2WD Ram truck and Ramcharger

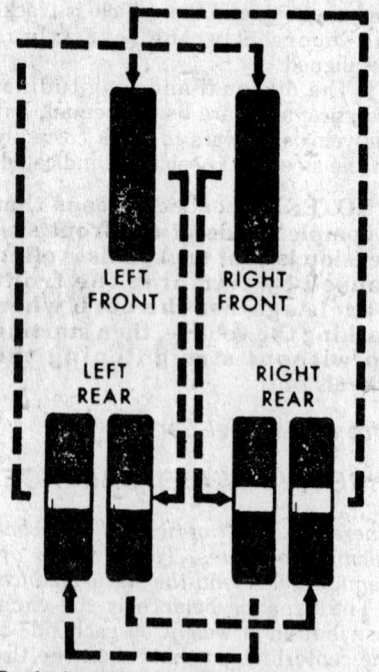

Dual wheel tire rotation sequence – Ram Van/Wagon

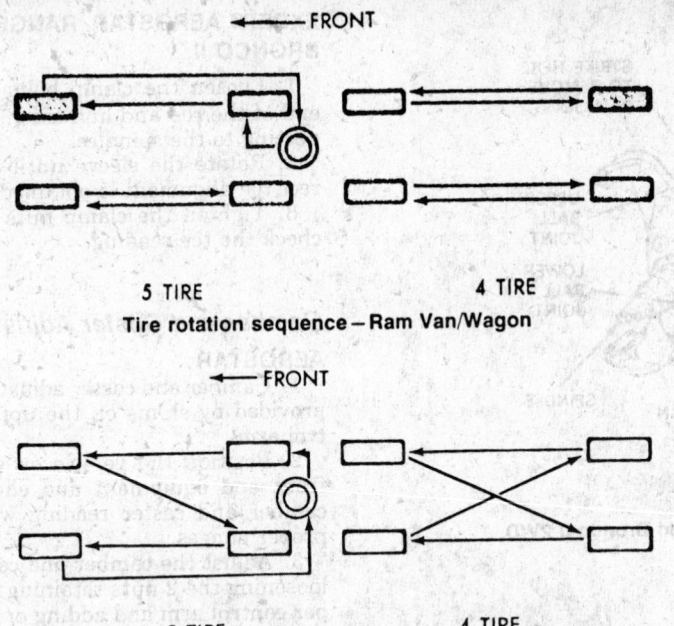

5 TIRE   **4 TIRE**

Tire rotation sequence—Ram Van/Wagon

5 TIRE   **4 TIRE**

Tire rotation sequence—Ram truck and Ramcharger

held as close as possible to the preferred setting.

3. Rotate the pivot bar to adjust the caster. Move each end of the pivot bar exactly the same amount in opposite directions.

4. The caster should be the same on both wheels.

Steering Angle

The steering angle is measured in degrees. It has a fixed relationship with the camber angle and will not change. The angle is not adjustable.

Wheel Service

Tire Rotation

Tire rotation is recommended at 7500 miles and then at intervals of 15000 miles. More frequent rotation is permissable, if desired. When types and sizes of tires differ from front to back, tire rotation is not possible.

FORD MOTOR COMPANY

Wheel Alignment

Toe-In Adjustment

AEROSTAR

1. Start the engine and move the steering wheel back and forth several

times to place the wheel in the center position.

2. Turn the engine off and lock the steering wheel in place. Check the toe reading using the proper tools.

3. To adjust, remove the bellows seal clamp, if one was used. Free the boot from the rod and do not allow the boot to twist when the rod is turned.

4. Loosen the jam nuts and adjust the toe-in to the preferred setting with the proper tools.

5. Tighten the jam nut and replace the bellows seal clamp, if one was used.

RANGER AND BRONCO II

1. Loosen the clamp bolts at each end of the spindle connecting rod tube.

2. Rotate the sleeve until the correct toe-in reading is obtained.

3. With the clamps centered between the adjustment sleeve lockring nibbs, position the bolts horizontally in the proper direction.

4. Tighten the clamp nuts and recheck the toe-in.

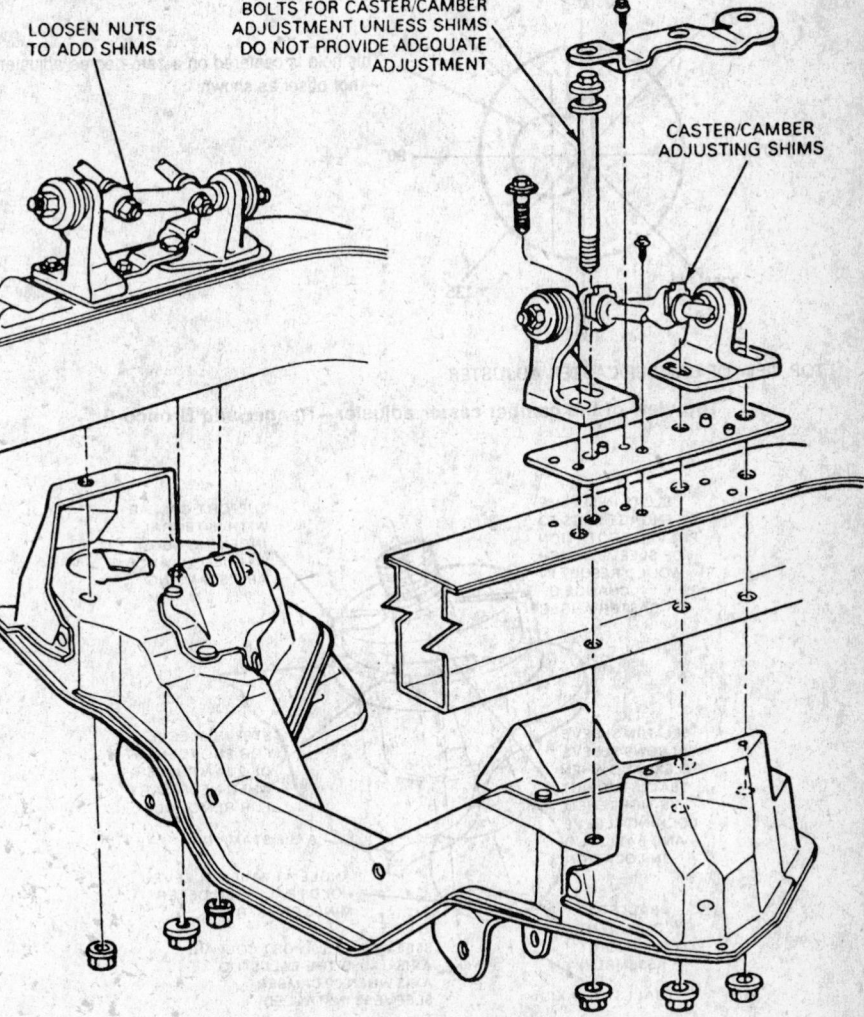

Camber and caster adjustment—Aerostar

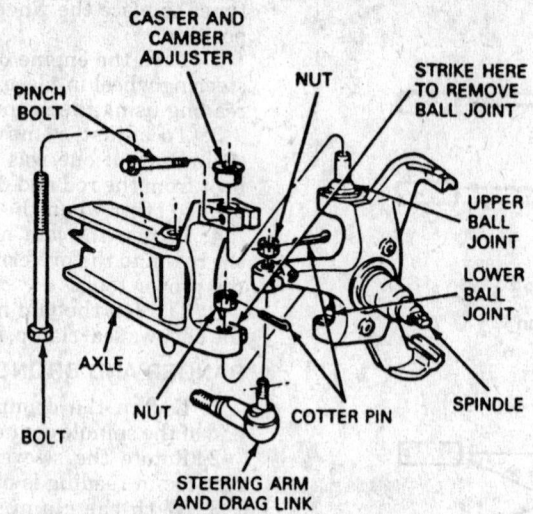

Camber and caster adjustment—Ranger and Bronco II 2WD

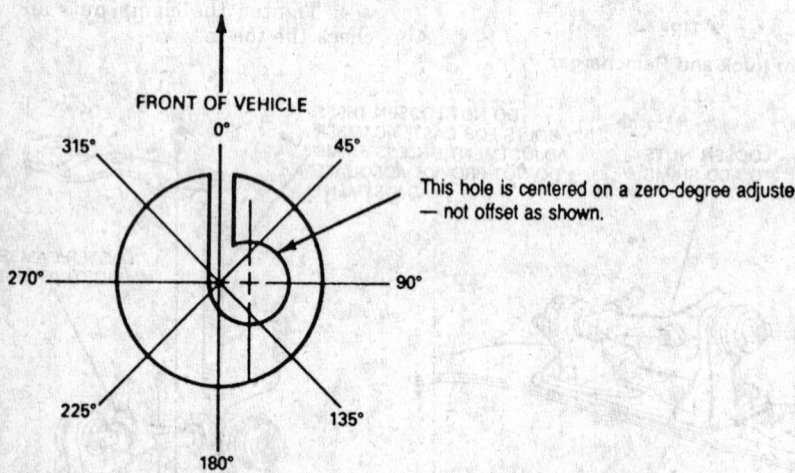

This hole is centered on a zero-degree adjuster — not offset as shown.

TOP VIEW OF CAMBER/CASTER ADJUSTER

Top view of the camber/caster adjuster—Ranger and Bronco II

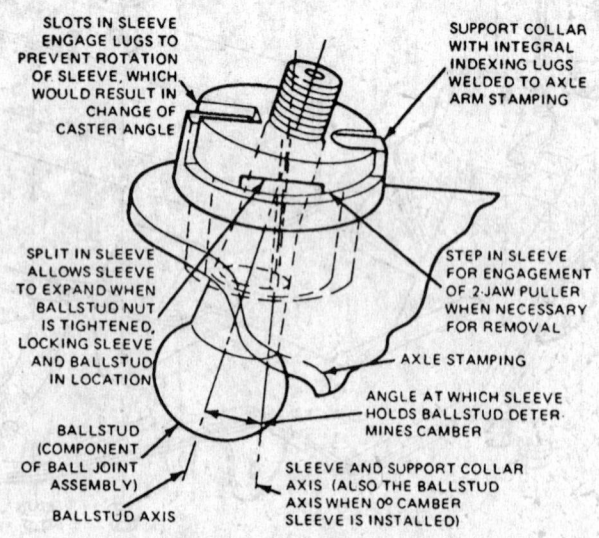

Camber adjustment—Ranger and Bronco II

EXCEPT AEROSTAR, RANGER AND BRONCO II

1. Loosen the clamp bolts at each end of the rod and link assembly connecting to the spindles.
2. Rotate the sleeve until the correct toe alignment is obtained.
3. Tighten the clamp nuts and recheck the toe reading.

Camber and Caster Adjustment
AEROSTAR

1. Camber and caster adjustment is provided by shims on the upper control arm.
2. Position the vehicle on suitable front end equipment and check the camber and caster reading with the proper guages.
3. Adjust the camber and caster by loosening the 2 nuts retaining the upper control arm and adding or removing the necessary amount of shims.
4. Retighten the nuts on the upper control arm to the proper torque to prevent the loss of shims or the movement of the upper ball joint.

RANGER AND BRONCO II—2 WD

1. Camber and caster adjustment is possible using service adjusters, the adjusters are available in various degree increments. One adjuster is used to adjust both camber and caster.
2. Raise and safely support the vehicle. Remove the front wheel.
3. Remove the pinchbnolt at the upper ball joint and pry the adjuster out of the axle with the proper tool.
4. Install the new service adjuster by rotating into position with the proper tool.
5. Tighten the pinchbolt, lower the vehicle and install the front wheel.

RANGER AND BRONCO II—4WD

1. The caster is not adjustable on 4WD vehicles.
2. To adjust the camber, raise and safely support the vehicle. Remove the front wheels.
3. Remove the upper ball joint cotter pin and nut.
4. Remove the camber adjuster sleeve and the adjuster from the spindle with the proper tool.
5. Install the camber adjuster on the top ball joint stud with the arrow pointing outward for positive camber and inward for negative camber. Zero camber bushings will not have an arrow and may be rotated in either direction as long as the lugs on the yoke engage the slots in the bushing.
6. Install a new nut on the lower ball joint stud and tighten.

NOTE: The camber adjuster will seat itself in the spindle at a predetermined position during the tightening sequence. Do not attempt to adjust this position.

7. Install the wheels and lower the vehicle.

E150–350

1. Camber is not adjustable on the Econoline.

2. The caster angle can be adjusted with use of a service kit or equivalent.

3. The kit includes instructions for increasing caster in ½ degree increments by installing an adjustment cam on the bottom of the front radius arm.

BRONCO AND F150–350–2WD

1. Camber and caster adjustment is possible using service adjusters, the adjusters are available in various degree increments. One adjuster is used to adjust both camber and caster.

2. Raise and safely support the vehicle. Remove the front wheel.

3. Remove the pinchbolt at the upper ball joint and pry the adjuster out of the axle with the proper tool.

4. Install the new service adjuster by rotating into position with the proper tool.

5. Tighten the pinchbolt, lower the vehicle and install the front wheel.

BRONCO AND F150–250–4WD

1. The camber adjustment is provided by means of a series of interchangeable mounting sleeves (camber adjusters) for the upper ball joint stud.

2. To adjust, raise and safely support the vehicle. Remove the front wheels.

3. Remove the upper ball joint nut and loosen the lower ball joint nut at the end of the stud.

4. Break the spindle loose from the ball joint studs with the proper tool.

5. Remove the camber adjuster

sleeve and remove the adjuster out of the spindle with the proper tool.

6. Install the camber adjuster on the top ball joint stud with the arrow pointing outward for positive camber and inward for negative camber.

7. Install the nut on the upper ball joint and the lower ball joint stud and tighten.

8. Reinstall the wheel and lower the vehicle.

9. The caster is not adjustable on the Bronco and the F150–4WD vehicles.

10. The caster on the F250–4WD can be adjusted by inserting a shim between the spring and the axle.

11. To adjust the caster, raise and safely support the vehicle. Support the front axle independently.

NOTE: The caster should always be done on the right front axle to avoid changing the front driveshaft alignment.

12. Loosen the U-bolt nuts and separate the spring from the axle.

13. Install caster shims between the spring and the axle.

14. Position the thin edge of the shim toward the front of the vehicle to increase caster and to the rear of the vehicle to decrease caster.

15. Tighten the U-bolt nuts.

F350–4WD AND F-SUPER DUTY VEHICLES

The camber and caster is not adjustable on these vehicles.

Wheel Service

Tire Rotation

To equalize tire wear the tires and wheels should be rotated. Tire rotation is recommended at 5000–10000 mile intervals.

Wheel Alignment

Toe-In Adjustment
LUMINA APV

1. Loosen the locknut on the tie rod ends.

2. Place the steering wheel in a straight ahead position and loosen the jam nuts on the tie rod ends. Make sure that the seals (boots) do not twist.

3. Rotate the inner tie rod to obtain the proper toe angle. Check to see that the number of threads showing on each tie rod are equal.

4. Make sure that the tie rods are square then tighten the locknuts.

EXCEPT LUMINA APV

1. Determine the proper toe-in setting using suitable alignment equipment.

2. To adjust the toe-in, loosen the locknuts on the adjusting sleeve and turn the tie rod to increase or decrease the length of the tie rods.

3. When the tie rods are mounted ahead of the steering knuckle they must be decreased in length in order to increase toe-in. When the tie rod ends are mounted behind the steering knuckle they must be lengthened to increase the toe-in.

4. Tighten the locknuts on the adjusting sleeve.

Camber and Caster
LUMINA APV

1. Loosen both strut to knuckle nuts, just enough to allow movement.

2. Move the top of the wheel in or out to adjust the camber.

3. Tighten the strut to knuckle nuts.

4. The caster is not adjustable.

EXCEPT LUMINA APV AND T SERIES

1. Place the vehicle on a smooth flat surface.

2. To adjust the camber and caster, loosen the upper control arm shaft-to-frame nuts.

3. Add or subtact shims as required, tighten the nuts.

4. To adjust the camber, change shims at both the front and the rear of the shaft.

5. To adjust the caster, transfer the shims from the front to the rear or the rear to the front.

6. Tighten the nuts after selecting the proper position.

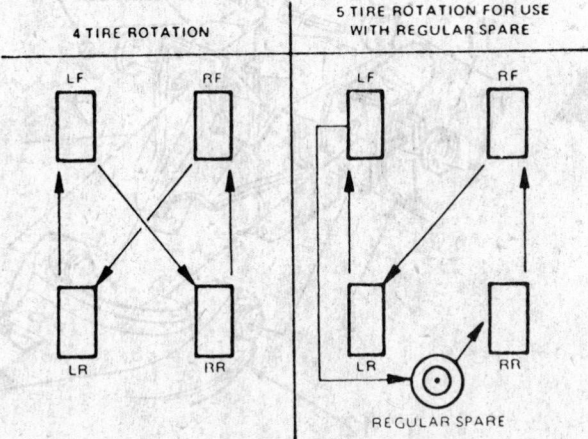

Tire rotation sequence – Ford trucks

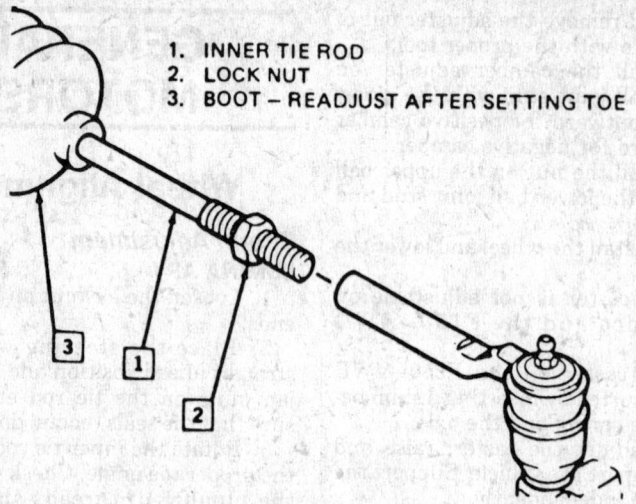

1. INNER TIE ROD
2. LOCK NUT
3. BOOT – READJUST AFTER SETTING TOE

Tie rod adjustment – Lumina APV

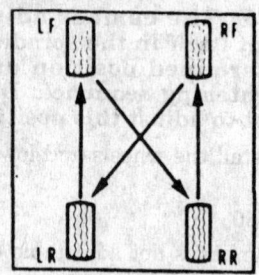

DO NOT INCLUDE
TEMPORARY USE ONLY
SPARE TIRE IN ROTATION

Tire rotation sequence – Chevy trucks

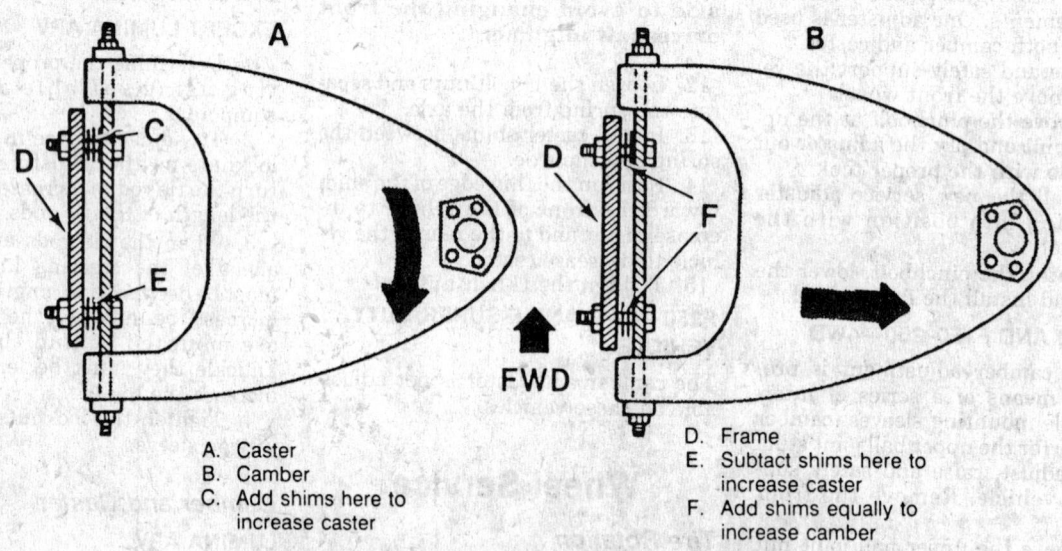

A. Caster
B. Camber
C. Add shims here to increase caster

D. Frame
E. Subtact shims here to increase caster
F. Add shims equally to increase camber

FWD

Camber and caster adjustment – Chevy truck T model

9. Frame
10. Cam nut
11. Cam
12. Upper control arm

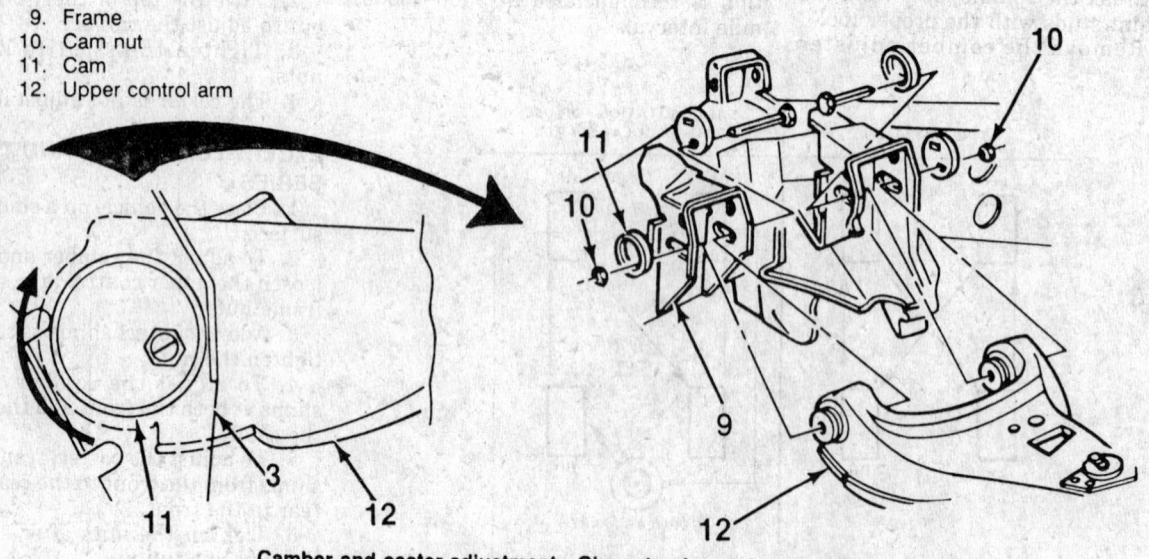

Camber and caster adjustment – Chevy truck exc. Lumina APV

10 SERIES

1. The camber and caster adjustments are made by moving the cams on the upper control arm frame mounting bolts.

2. To adjust the camber and caster, loosen the upper control arm shaft-to-frame nuts.

3. To increase positive caster, move the front cam lobe inboard and the rear cam lobe outboard.

4. To increase positive camber, move both the front and the rear cam lobes inboard.

5. Tighten the nuts after selecting the proper cam position.

Wheel Service

Tire Rotation

To equalize wear, rotate the tire and wheel assemblies every 7500 miles. Radial tires in a non-drive location may develop an irregular wear pattern that can cause increase tire noise if not rotated.

JEEP

Wheel Alignment

Toe-In Adjustment

1. Center the steering wheel and lock in position.

2. To adjust the right wheel toe, Loosen the drag link adjustment sleeve clamp bolts.

3. Turn the drag link with the proper tool until the desired setting is obtained.

4. Tighten the drag link adjustment sleeve clamp bolts.

5. To adjust the left wheel toe, loosen the clamp bolts at both ends of the tie rod.

6. Turn the tie rod with the proper tool until the desired setting is obtained.

7. Tighten the tie rod clamp bolts.

Camber and Caster

The correct wheel camber angle is preset at the factory and cannot be altered by adjustment. It is important that the camber angle be the same for both wheels. If the camber angle is incorrect, the component causing the incorrect camber angle should be replaced.

1. The caster angle should be checked using suitable wheel alignment equipment.

2. Adjust the camber angle by adding or removing shims at the rear of the lower suspension arms, except on 15 and 81 model vehicles.

3. On models 15 and 81, adjust the caster angle by installing tapered

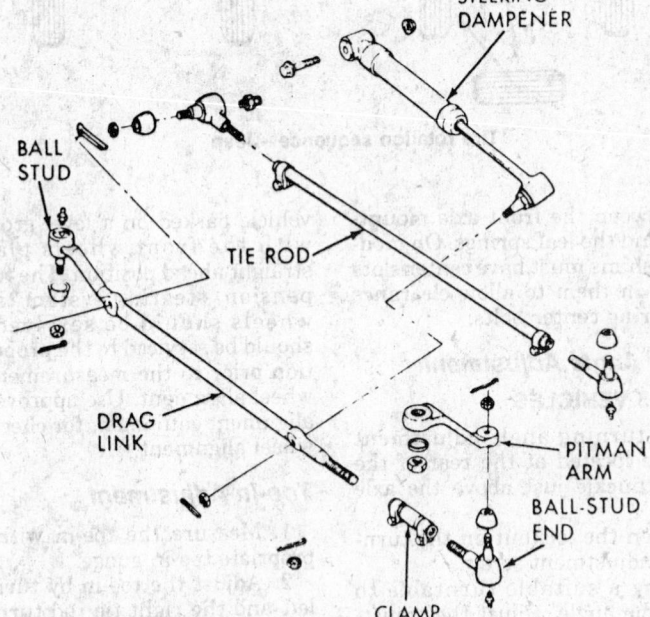

Toe-In adjustment—Jeep

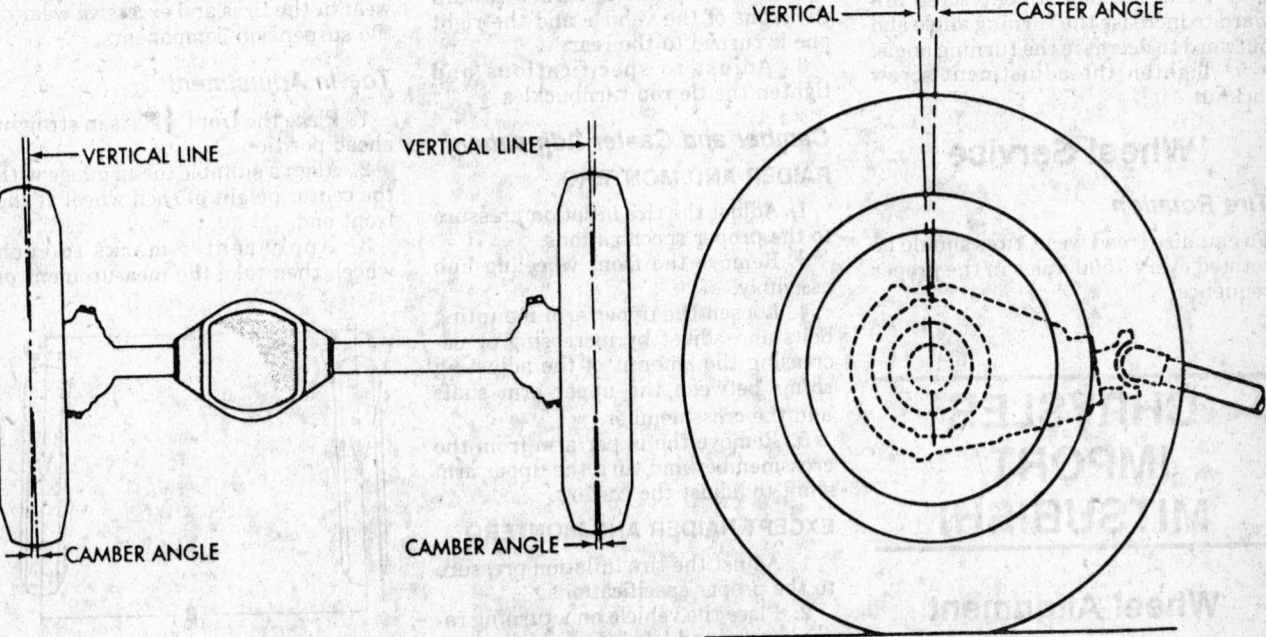

Camber and caster angles—Jeep

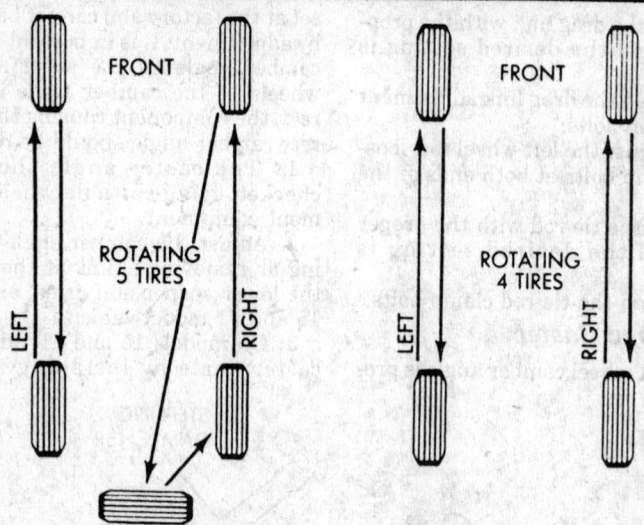

Tire rotation sequence—Jeep

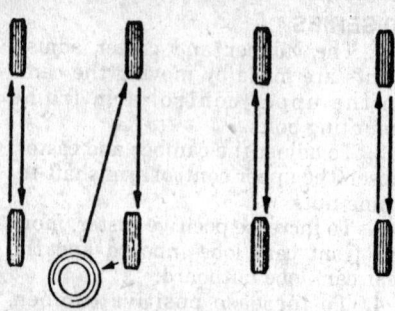

Tire rotation sequence—Chrysler Import/Mitsubishi

shims between the front axle mounting pads and the leaf springs. On model 81, the shims must have center slots machined in them to allow clearance for the spring center bolts.

Steering Angle Adjustment

MODEL 15 VEHICLES

1. The turning angle adjustment screws are located at the rear of the steering knuckle just above the axle centerline.
2. Loosen the locknut on the turning angle adjustment screw.
3. Using a suitable turntable to measure the angle, adjust the adjustment screw to obtain the proper turning angle.
4. Turn the adjustment screw inward to increase the turning angle and outward to decrease the turning angle.
5. Tighten the adjustment screw locknut.

Wheel Service

Tire Rotation

To equalize tread wear, tires should be rotated every 7500 miles in the proper sequence.

CHRYSLER IMPORT/ MITSUBISHI

Wheel Alignment

Measure the wheel alignment with the vehicle parked on a level ground and with the front wheels place in a straight ahead position. The front suspension, steering system tires and wheels should be serviced to the should be serviced to the proper condition prior to the measurement of the wheel alignment. Use approved wheel alignment equipment for checking the wheel alignment.

Toe-In Adjustment

1. Measure the toe-in with an appropriate toe-in guage.
2. Adjust the toe-in by turning the left and the right tie rod turnbuckles the same amount in opposite directions. The toe-in value will decrease if the left turnbuckle is turned toward the front of the vehicle and the right one is turned to the rear.
3. Adjust to specifications and tighten the tie rod turnbuckles.

Camber and Caster Adjustment

RAIDER AND MONTERO

1. Adjust the tire inflation pressure to the proper specifications.
3. Remove the front wheeling hub assembly.
4. Loosen the upper arm mounting bolts and adjust by increasing or decreasing the amount of the adjusting shims between the upper arm shaft and the crossmember.
5. Remove the upper arm from the crossmember and turn the upper arm shaft to adjust the caster.

EXCEPT RAIDER AND MONTERO

1. Adjust the tire inflation pressure to the proper specifications.
2. Place the vehicle on a turning radius guage and level the vehicle.
3. Adjust the camber by rotating

the lower arm's shaft assembly on the Van/Wagon.
4. Loosen the upper mounting bolts and increase or decrease the amount of shims between the upper arm shaft and the crossmember to adjust the camber and caster except on the Van/wagon.

Wheel Service

Tire Rotation

Rotate the tires periodically to ensure uniform wear of the tires. If the spare wheel is of a different type from the other 4 wheels, the 4 wheel rotation method should be used.

ISUZU

Wheel Alignment

The wheel alignment should be checked periodically to ensure proper wear of the tires and excessive wear of the suspension components.

Toe-In Adjustment

1. Place the front wheels in straight ahead position.
2. Align a suitable toe-in guage with the center height of each wheel at the front end.
3. Apply center marks to each wheel, then take the measurement of

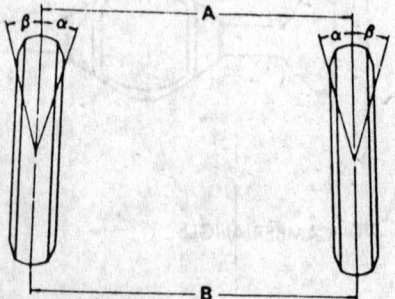

Measuring distance for toe-in adjustment—Isuzu

distance A between the center marks on each wheel.

4. Slowly, move the vehicle rearward until the center marks reach the rear end position.

5. Take the measurement of distance B between the center marks at the rear end.

6. The toe-in equals B - A.

7. To adjust the toe-in, loosen the locknut on the outer tie rod end and turn the outer track rod the same distance right and left.

Camber and Caster Adjustment

1. The camber angle can be adjusted by the means of camber pins installed in position between the chassis frame and the fulcrum pins. Adding shims decreases the camber angle and removing shims increases the camber angle.

2. The caster angle on the 2WD vehicles can be adjusted by varying the length of the strut bar. Adjust with the locknut. Shims should not be used for this adjustment.

3. The caster angle on the 4WD and Amigo can be adjusted by means of the caster shims installed in position between the frame and fulcrum pins. Adding shims to the front side decreases the caster angle while removing shims increases the angle. Adding shims to the rear side increases the caster angle while removing shims increases the angle.

Steering Angle Adjustment

1. The maximum steering angle on the front wheels can be adjusted with the stopper bolts under the frame side members.

2. Position each front wheel on a suitable turning radius guage in a straight ahead position.

3. Adjust the inside angle of each side with the stopper bolts.

NOTE: The maximum steering angles should be set after adjusting the front wheel alignment.

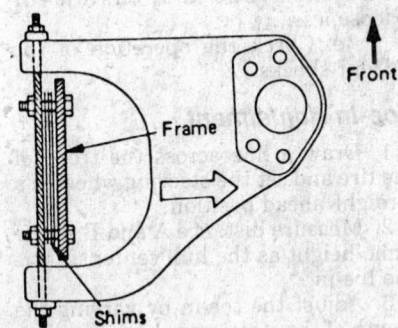

Caster angle adjustment—Isuzu 4WD and Amigo

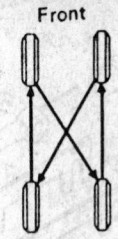

Bias ply tire
4 Wheel rotation

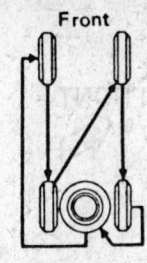

Bias ply tire
5 Wheel rotation

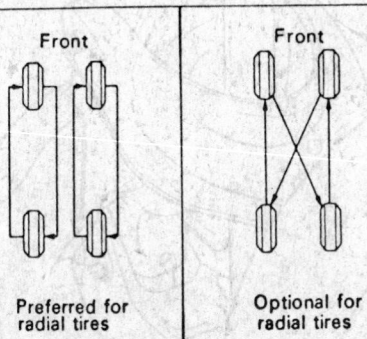

Preferred for radial tires

Optional for radial tires

Tire rotation sequence—Isuzu

Wheel Service

Tire Rotation

Rotating the tires evens out tire wear and prolongs the tire life. Tire rotation should be performed every 7500 miles for maximum performance.

MAZDA

Wheel Alignment

1. Measure the wheel alignment with the vehicle parked on a level ground and with the front wheels place in a straight ahead position.

2. The front suspension, steering system tires and wheels should be serviced to the should be serviced to the proper condition prior to the measurement of the wheel alignment. Use approved wheel alignment equipment for checking the wheel alignment.

Toe-In Adjustment

1. Raise and safely support the front of the vehicle.

2. Turn the wheels by hand, mark a line in the center of each tire tread using a proper scribing tool.

3. Place the front wheels in a straight ahead position and lower the vehicle.

4. Measure the distance between

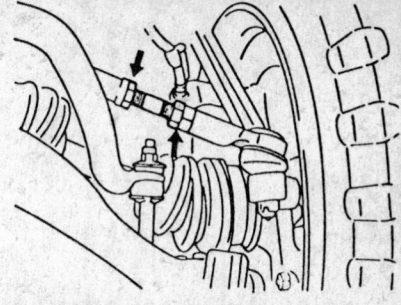

Toe-in adjustment on the vehicle— Mazda B2200 and B2600

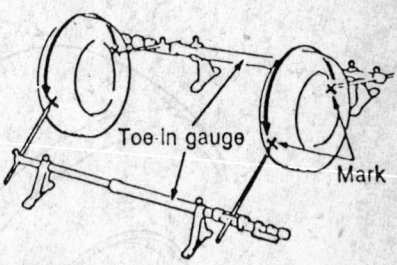

Checking toe-in with a suitable guage— Mazda

the marked lines at the front and the rear of the wheels.

5. To adjust, loosen the left and the right tie rod locknuts and turn the tie rods by the same degree.

6. Adjust the toe-in after adjusting the steering angle except on the MPV.

Camber and Caster Adjustment

EXCEPT MPV

1. Position the front wheels on a turning radius guage.

2. Remove the front wheel hub on 4WD vehicles.

3. Remove the wheel cap and the wheel hub nut on 2WD vehicles. Attach a suitable guage adapter to the hub.

4. Attach a suitable camber and caster guage to the hub or the adapter and measure the camber and the caster.

5. The camber and the caster are adjusted by adding or subtracting the amount of the shims.

MPV

1. Raise and safely support the front of the vehicle.

2. Remove the blocks that hold the mounting block to the fender.

3. Push the mounting block down and turn it to the desired position.

4. Adjust the camber and the caster by changing the position of the mounting block. Retighten the nuts to the specified torque.

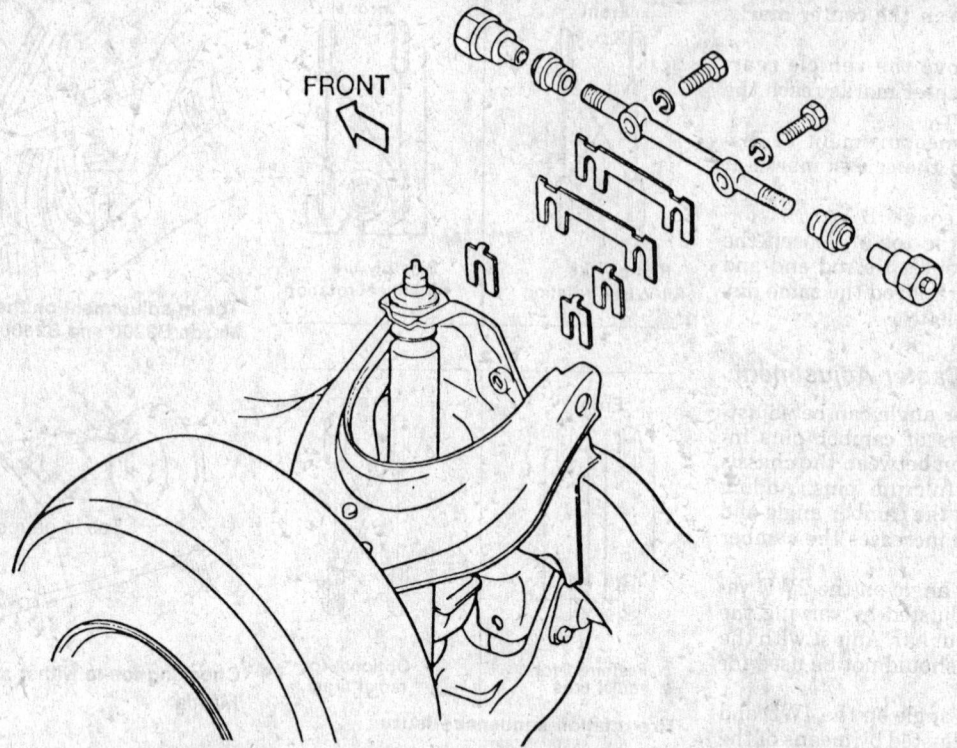

FRONT

Location of the adjusting shims for the camber and caster—Mazda

4WD

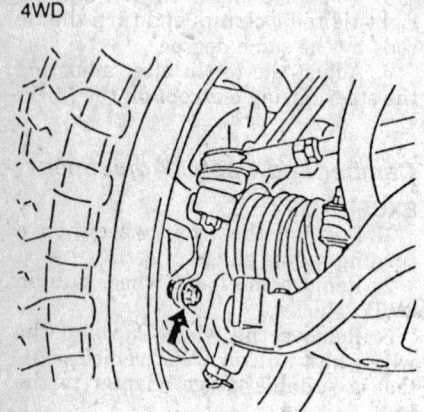

2WD

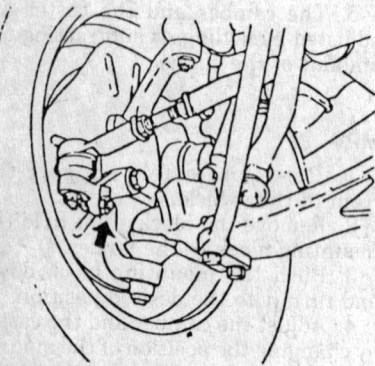

Checking the steering angle—Mazda
2WD and 4WD vehicles

4-wheel rotation

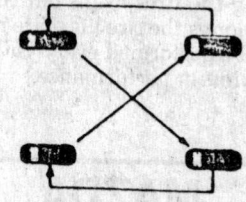

5-wheel rotation

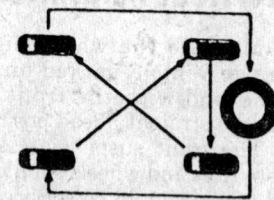

Tire rotation sequence—Mazda

Steering Angle

The steering angle is measured by placing the front wheels on a turning radius guage. Adjust the steering angle by adjusting the bolt fitted on the steering knuckle arm except on the MPV.

Wheel Service

Tire Rotation

In order to prolong tire life and assure uniform wear, the tires should be rotated every 3,750 miles. Do not include the temporary spare in the rotation.

NISSAN

Wheel Alignment

Prior to any adjustments to the front wheel alignment, perform a preliminary inspection of the following:

    a. Check the tires for wear and proper inflation.

    b. Check the wheel runout.

    c. Check the front wheel bearings for looseness.

    d. Check the steering linkage and the front suspension for looseness.

    e. Check the operation of the front shocks.

Toe-In Adjustment

1. Draw a line across the tread of the tire and set the steering wheel in a straight ahead position.

2. Measure distance A and B at the same height as the hub center, this is the toe-in.

3. Adjust the toe-in by varying the length of the steering rods and turning the tie-rod tubes clockwise and counterclockwise an equal amount.

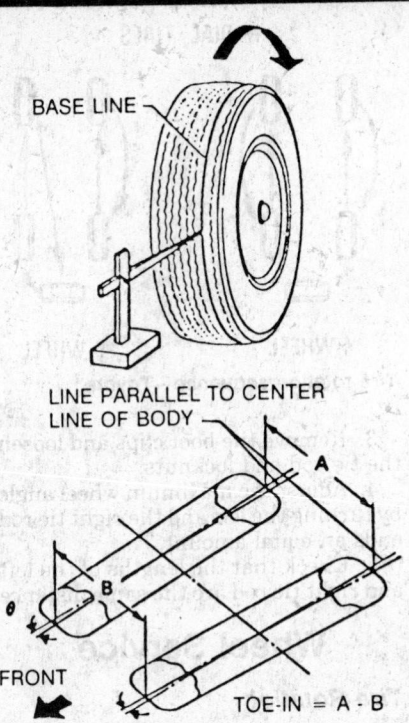

Toe-in adjustment — Nissan

Camber and Caster Adjustment

1. To adjust the camber and caster, increase or decrease the number of adjusting shims inserted between the upper link spindle and the frame.

NOTE: The camber, caster and the king pin inclination are preset at the factory and cannot be adjusted on the Axxess.

2. Do not install 3 or more shims at one place.

3. To adjust the camber, equalize the thickness of the front and rear shims by adding or subtracting shim(s)

4. To adjust the caster, make a difference between the front and rear shims.

5. When the caster is adjusted the camber angle changes and must be readjusted.

Steering Angle Adjustment

1. Set the wheels in a straight ahead position.

2. Move the vehicle forward until the front wheels rest on the turning radius guage properly.

3. Rotate the steering wheel all the way to the right and the left and check the turning angle.

Wheel Service

Tire Rotation

Do not include the T type spare and small type spares in the rotation.

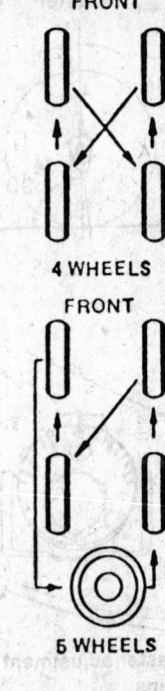

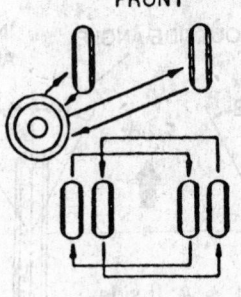

Tire rotation sequence — Nissan

Make sure that the wheel nuts are tightened to specifications when completed.

SUZUKI/GEO

Wheel Alignment

The wheel alignment refers to the angular relationship between the front wheels, the front suspension mounting parts and the ground. Generally the only adjustment required is toe-in.

Toe-In Adjustment

The toe is adjusted by changing the tie rod length. Adjust by loosening the locknuts first the rotate the left and the right tie-rod by the same amount. In this adjustment the left and the right tie rod should become equal in length.

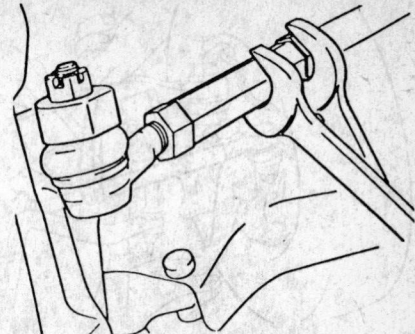

Toe-in adjustment — Suzuki/Geo

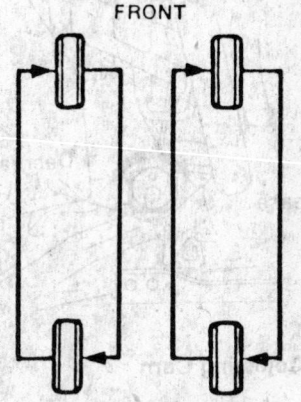

4 WHEEL ROTATION
[Radial Tires]
Tire rotation sequence — Suzuki/Geo

Camber and Caster Adjustment

The camber and caster cannot be adjusted for this vehicle. To prevent any possible incorrect reading of the camber or caster, move the front of the vehicle up and down a few times before inspection.

Wheel Service

Tire Rotation

To equalize tire wear, rotate the tire and wheels at 7500 miles and every 15000 miles thereafter. The tire and wheel assemblies should be rotated whenever uneven tire wear is noticed.

TOYOTA

Wheel Alignment

Follow the specific instructions of the equipment manufacturer the wheel alignment.

Toe-In Adjustment

1. Rock the vehicle up and down to stabilize the suspension.

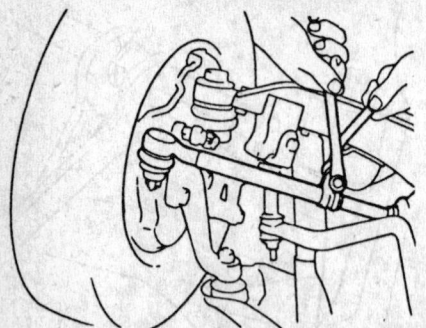

Toe-in adjustment—Toyota

Front Adjusting Cam

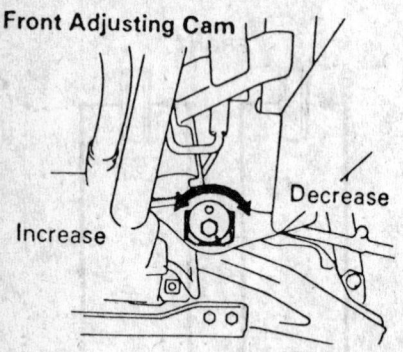

Increase   Decrease

Rear Adjusting Cam

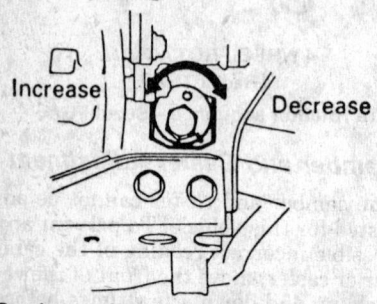

Increase   Decrease

Camber and caster adjustment—Toyota 4WD vehicles

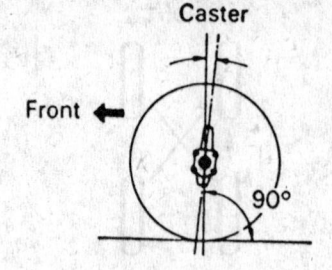

Caster

Front ←

90°

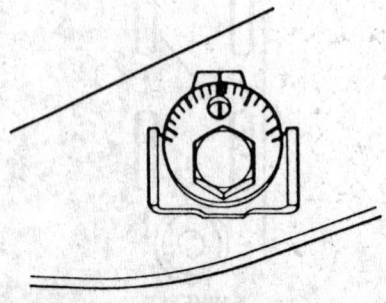

Camber and caster adjustment—Toyota 2WD vans

OUTSIDE ANGLE   INSIDE ANGLE

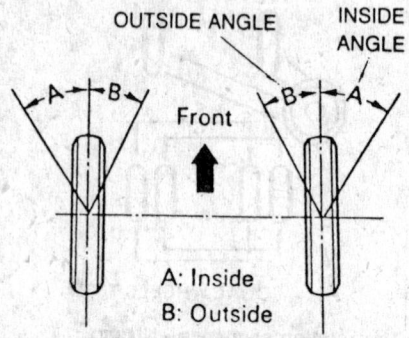

A─B   B─A

Front

A: Inside
B: Outside

Steering wheel angle—Toyota

RADIAL TIRES

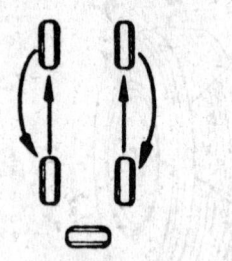
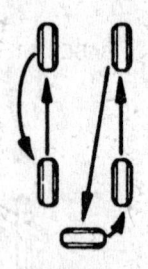

4-WHEEL   5-WHEEL

Tire rotation sequence—Toyota

2. Mark the center of each rear tread at spindle height and measure the distance between the marks on the right and left tires.

3. Advance the vehicle until the marks on the rear side of the tires come to the front.

NOTE: The toe-in should be measured at the same point on the tire and at the same level.

4. Measure the distance between the marks on the front side of the tires.

5. Adjust the toe-in by turning the left and the right tie rod tubes an equal amount. Make sure that the tie rods are the same length.

Camber and Caster Adjustment

1. For 2WD trucks, adjust the camber and the caster by adding or removing shims on the upper arm. If the steering axis inclination is not as speci-

fied after the camber and caster have been properly adjusted, recheck the steering knuckle for bending or looseness.

2. For 2WD vans, adjust the camber and caster by turning the adjusting cam or the strut bar nut. The adjusting cam should not be turned more than 4.5 graduations from the neutral position. Do not turn the strut bar more than 3 threads from the original position.

3. For 4WD vehicles, adjust the camber and the caster by adjusting the front and rear adjusting cams. Refer to the service specifications for the adjustment standards.

Steering Angle Adjustment

1. Remove the caps of the steering knuckle stopper bolts.

2. Loosen the knuckle stopper bolt locknuts and tighten the stopper bolts by hand.

3. Remove the boot clips and loosen the tie rod end locknuts.

4. Adjust the maximum wheel angle by turning the left and the right tie rod ends an equal amount.

5. Check that the lengths of the left and right tie rod are the same distance

Wheel Service

Tire Rotation

It is recommended that the tires and wheels be rotated every 7500 miles to ensure uniform wear.

VOLKSWAGEN

Wheel Alignment

The vehicle should have been run from 625–1250 miles to let the coil springs settle before performing an alignment. The vehicle should be empty.

Toe-In Adjustment
FRONT WHEELS

1. Place the steering gear in the center position by turning the steering wheel from lock to lock and count the number of turns. Turn the wheel back 1/2 the number of turns.

2. Align the lug on the rubber washer on the pinion shaft with the notch in the steering housing.

3. Loosen the locknuts and turn both tie rods until the desired setting is reached. Retighten the locknuts.

4. Check that the bellows are not twisted after turning the tie rods.

5. The specification of the total toe at the front axle is determined by the ride height of the vehicle.

Camber and Caster Adjustment
FRONT WHEELS

1. Adjust the caster by adjusting the length of the radius bar A at location B.

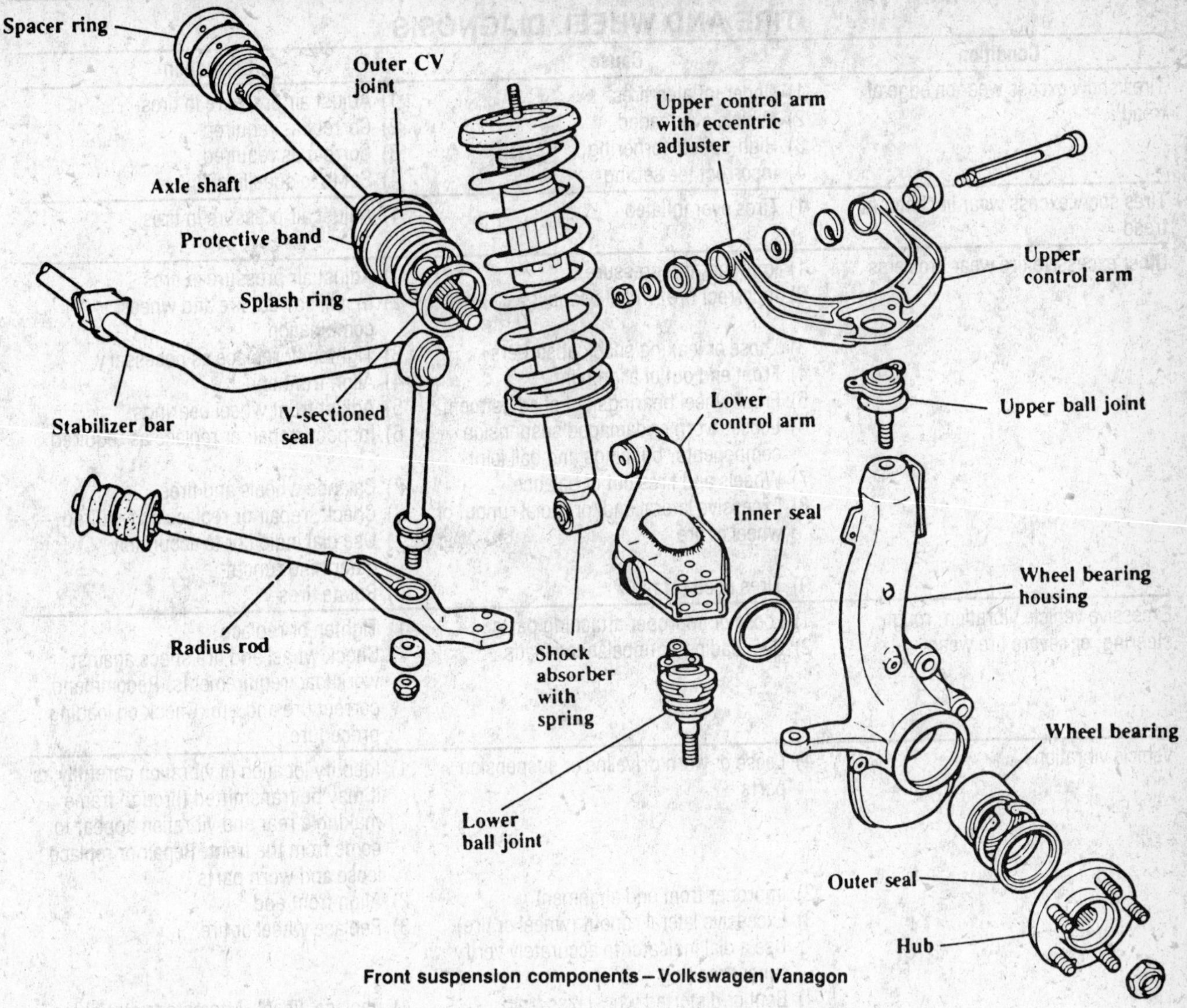

Spacer ring

Outer CV joint

Axle shaft

Protective band

Splash ring

Stabilizer bar

V-sectioned seal

Radius rod

Upper control arm with eccentric adjuster

Upper control arm

Upper ball joint

Lower control arm

Inner seal

Shock absorber with spring

Lower ball joint

Wheel bearing housing

Wheel bearing

Outer seal

Hub

Front suspension components—Volkswagen Vanagon

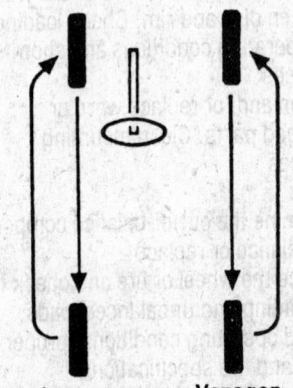

Tire rotation sequence—Vanagon

2. To adjust the camber, loosen the nut on the upper control arm shaft. Turn the upper control arm shaft to obtain the desired setting. Tighten the nut.

Toe-In Adjustment
REAR WHEELS

1. Loosen the inner bolt on the trailing arm.
2. Adjust the toe by moving the trailing arm to the front or rear, using the proper tool.
3. Tighten the bolt.

Camber Adjustment
REAR WHEELS

1. Loosen the outer bolt on the trailing arm.
2. Adjust the camber by moving the trailing arm up or down with the proper tool.
3. Tighten the bolt.

Wheel Service

Tire Rotation

It is recommended that the tires and wheels be rotated every 7500 miles to ensure uniform wear. After the rotation torque the wheel nuts diagonally.

TIRE AND WHEEL DIAGNOSIS

Condition	Cause	Correction
Tires show excess wear on edge of tread	1) Under inflated tires 2) Vehicle overloaded 3) High-speed cornering 4) Incorrect toe setting	1) Adjust air pressure in tires 2) Correct as required 3) Correct as required 4) Set to to specification
Tires show excess wear in center of tread	1) Tires over inflated	1) Adjust air pressure in tires
Other excessive tire wear problems	1) Improper tire pressure 2) Incorrect tire/wheel usage 3) Loose or leaking shock absorbers 4) Front end out of alignment 5) Front wheel bearings out of adjustment 6) Loose, worn or damaged suspension components, bushings and ball joints 7) Wheels and tires out of balance 8) Excessive lateral and/or radial runout of wheel or tire 9) Tires need rotating	1) Adjust air pressure in tires 2) Install correct tire and wheel combination 3) Tighten or replace as necessary 4) Align front end 5) Adjust front wheel bearings 6) Inspect, repair or replace as required 7) Balance wheels and tires 8) Check, repair or replace as required. Use dial indicator to accurately determine runout 9) Rotate tires
Excessive vehicle vibration, rough steering, or severe tire wear	1) Loose or improper attaching parts 2) Overloading or unbalanced loads	1) Tighten or replace 2) Check wheel and tire specs against work load requirements. Recommend correct tire and rim. Check on loading procedure
Vehicle vibrations	1) Loose or worn driveline or suspension parts 2) Improper front end alignment 3) Excessive lateral runout (wheel or tire). Use a dial indicator to accurately verify runout reading 4) Bent or distorted wheel disc from overloading, road impact hazards or improper handling 5) Loose mountings—damaged studs, cap nuts, enlarged stud holes, worn or broken hub face or foreign material on mounting surfaces 6) Out-of-balance wheel and/or tire or hub and drum assembly 7) Out-of-round wheel or tire (excessive radial runout). Use a dial indicator to accurately verify runout reading 8) Wheel stud runout 9) Water in tires	1) Identify location of vibration carefully as it may be transmitted through frame making a rear end vibration appear to come from the front. Repair or replace loose and worn parts 2) Align front end 3) Replace wheel or tire 4) Replace wheel. Attempts to straighten wheel can result in fractures in the steel and weakening of the disc or the weld between disc and rim. Check loading and operating conditions and shop practices 5) Tighten and/or replace worn or damaged parts. Clean mounting surfaces 6) Determine the out-of-balance component and balance or replace 7) Replace the wheel or tire and check for overloading and unbalanced loads, rugged operating conditions, proper wheel and tire specifications 8) Replace hub or axle shaft 9) Remove water

TIRE AND WHEEL DIAGNOSIS

Condition	Cause	Correction
Wheel mounting is difficult	1) Improper application or mismatched parts, including studs and nuts 2) Corroded or worn parts	1) Follow manufacturers' specifications 2) Clean or replace
Wheel-rust or corrosion	1) Poor maintenance	1) Keep clean and protect with paint
Cracked or broken wheel discs. Cracks develop in the wheel disc from hand hole to hand hole, from hand hole to rim, or from hand hole to stud. Stud holes become worn, elongated or deformed. Metal builds up around stud hole edges, cracks develop from stud hole to stud hole. Related driver complaints: unusual operating noise or vibration and on the road failures	1) Metal fatigue resulting from abusive handling 2) Vehicle operated with loose wheel mounting	1) Replace wheel. Check position of wheel on vehicle for working load specifications 2) Replace wheel and check for: — Installation or correct studs and nuts, and recommend exact specifications — Cracked or broken studs, and replace — Worn hub face. Machine if not excessive, or replace if severe — Broken or cracked hub barrel, replace — Worn stud grooves, replace or install recommended serrated bolts — Clean mounting surfaces and re-torque cap nut periodically NOTE: Rust streaks fanning out from stud holes are a sure indication that the cap nuts are or have been loose
Cracks develop in rim base back flange (rim bead seat) or the gutter area (drop well radii)	1) Overloading or abuse 2) Improper use of tools	1) Replace rim or wheel. Check loading and operating conditions. Avoid over inflation of tires. Check specs for rim load capacity, working loads, tire size, ply rating and tire construction 2) Check mounting, demounting, and maintenance procedures
Damaged stud threads	1) Sliding wheel across studs during assembly	1) Replace studs. Follow proper wheel installation procedure
Loose drum	1) Stud too long	1) Replace stud with proper length stud
Broken studs	1) Loose lug nuts 2) Overloading	1) Replace studs. Follow proper torque procedure 2) Replace studs. Compare actual load against vehicle load ratings
Stripping threads	1) Excessive torque	1) Replace studs. Follow proper torque procedure
Rust streaks from stud holes	1) Loose lug nuts	1) Check complete assembly. Replace damaged parts. Follow proper torque procedure
Damaged lug nuts	1) Loose wheel assembly	1) Replace lug nuts. Check for proper stud standout. Follow proper torque procedure

TIRE AND WHEEL DIAGNOSIS

Condition	Cause	Correction
Frozen lug nuts	1) Corrosion or galling	1) If corrosion is slight, wire brush away corrosion If corrosion is excessive, replace studs and nuts If condition persists, lubricate first three threads of each stud with a graphite-based lubricant NOTE: Do not permit lubricant to get on cone seats of stud holes or on cone face of lug nuts

Drive Axles 22

DOMESTIC DRIVE AXLES

Drive Axle Types

FULL FLOATING AXLES

Support of the vehicle and the payload weight is by the axle housing. The wheels are driven by splined shafts which float within the axle housing.

SEMI-FLOATING AXLE

This axle design provides for the support of the payload and vehicle weight to be carried by the axle shaft through the wheel bearings to the axle housing.

SINGLE REDUCTION AXLE

Final drive ratio is obtained by the use of a single ring gear and pinion set.

LOCK-UP TYPE DIFFERENTIALS

Unlike the standard differential, the locking differential equally divides the torque load between the driving wheels. The vehicle equipped with a locking differential can be operated on any surface (sand, snow, etc.) with a minimum of slippage through one wheel and provides the greatest power to the wheel getting traction. The vehicle with the standard differential provides power to the wheel that's easiest to turn; that is, the one experiencing the poorest traction while the other wheel may be gripping well.

When negotiating a turn, the locking differential allows the outer wheel to turn faster than the inner. When traveling in a straight direction, and the vehicle loses traction over a rough or slippery road, the clutches will lockup and neither wheel will spin. A speci-

Checking the drive gear run-out

fied lubricant must be used for locking differentials.

The overhaul procedures are basically the same as for the conventional rear axle assemblies. The noted differences are within the differential carrier case where the lock-up mechanism is located. In some instances where wear is noted within the lock-up mechanism, the lock-up assembly should be replaced as a unit.

Drive Axle Service

COMPONENT INSPECTION AND REPLACING

Cleaning Bearings

Proper bearing cleaning is important. Bearings should always be cleaned separately from other rear axle parts.

1. Soak all bearings in clean kerosene or diesel fuel oil.

NOTE: Ordinary gasoline should not be used nor should bearings should not be cleaned in hot solution tank.

2. Slush bearings in cleaning solution until all oil lubricant is loosened. Brush bearings with soft bristled brush until all dirt has been removed. Remove loose particles of dirt by striking flat against a wood block.

3. Rinse bearings in clean fluid. While holding races to prevent rotation, blow dry with compressed air.

NOTE: Do not spin bearings while drying.

4. After bearings have been inspected, lubricate thoroughly with regular axle lubricant; then wrap each bearing in clean cloth until ready to use.

Cleaning Parts

Immerse all parts in cleaning fluid and clean thoroughly. Use a stiff bristle brush as required to remove foreign deposits. Clean all lubricant passages or channels in pinion cage, carrier, caps and retainers. Make certain the interior of housing is thoroughly cleaned. Clean vent plugs and breathers.

Small parts such as cap screws, bolts, studs, nuts and etc., should be cleaned thoroughly.

Inspection

Magna flux all steel parts, except ball and roller bearings, to detect presence of wear and cracks.

Bearings

Rotate each bearing and check to see if the rollers are worn, chipped, rough or

in any other way damaged. Check the cage to see if it is in any way damaged. If either the bearing rollers or the cage are damaged the bearing must be replaced.

Gears

Examine drive gear and drive pinion, differential pinions and differential side gears carefully, for damaged teeth, worn spots in surface hardening, distortion and where drive gear is attached to differential case with rivets, inspect rivets for looseness, replace loose rivets. Check radial clearances between differential side gears and differential case. Check fit of differential pinions on spider.

Differential Case

Inspect case for cracks, distortion or damage, if in good condition, thoroughly clean case and cover; then assemble case with bolts and mount in lathe centers of V-block stand. If lathe is not available, install differential side bearings and mount case in differential carrier. Install dial indicator and check differential case run out.

Differential case with drive gear installed is checked in the same manner, except the dial indicator reading must be taken at the gear instead of the case flange.

Axle Shafts

Examine splined end of axle shaft for twisted or cracked splines, twisted shaft and worn dowel holes in flange; install new shafts, if necessary.

Install axle shaft assembly in lathe centers and check shaft run-out with dial indicator so the indicator shaft end contacts inner surface of flange near outer edge of flange and check flange run-out.

Shims

Carefully inspect shims for uniform thickness. Where various thickness of shims are used in a pack, it is recommended the thickest shims be used between the thin shims.

Thrust Washers

Replace all thrust washers.

Spider Arms

Carefully inspect spider arms for wear or defects.

Differential Pinion Bushings

Examine bushings (when used) for excessive wear, looseness or damage. Check fit or gears on spider for excessive clearance.

Axle Housing Sleeves

Sleeves showing damaged threads, wear or other damage should be replaced, if a hydraulic press is available, otherwise replace housing.

Housing Check
BEFORE REMOVAL

A check for bent axle housing can be made with unit in vehicle; however, conventional alignment instruments can be used, if available.

1. Raise and safely support rear axle. Block the axle under each spring seat.

2. Check wheel bearing adjustment and adjust, if necessary; then, check wheels for looseness and tighten wheel nuts, if necessary.

3. Place a chalk mark on outer side wall of tires at bottom. Measure across tires at chalk marks with a toe-in gauge.

4. Turn wheels half-way around so the chalk marks are positioned at top of wheel. Measure across tires again. If measurement at top is ⅓ in. or more, smaller than measurement at bottom of wheels, axle housing has sagged and is bent. If measurement at top exceeds bottom dimension by ⅓ in. or more, axle housing is bent at ends.

5. Turn chalk marks on both wheels so the marks are level with axle and at rear of vehicle. Take measurement with toe-in gauge at chalk marks; then, turn both chalk marks to front and level with axle and take another measurement. If measurement at front exceeds rear dimension by ⅓ in. or more, axle is bent to the rear. If the measurement condition is the reverse, the axle is bent forward.

AFTER REMOVAL

Place two straightedges across the housing flanges and measure the distance between the ends of the straightedges at a point 11 inches from the tube center. Relocate the straightedge 180 degrees and remeasure. If the straightedges are parallel in both measurements within ³/₃₂ in., the housing is serviceable.

Oil Seal

Surface of parts, contacted by oil seals must be free of corrosion, pits and grooves. When abrasive cleaning fails to clean up the seal contact surface and restore smooth finish, a new part must be installed.

Oil seals can be removed with a drift pin. When removing a seal, be careful it does not become cocked and result in damage to the retainer. Clean surface of retainer carefully, so the seal will seat properly in retainer. Coat outer surface of seal retainer with a light coat of sealer, to prevent lubricant leaks. Carefully start seal in retainer. Cutting, scratching or curling of lip of seal seriously impairs its efficiency and usually results in premature replacement. Lip of seal should be coated with a high temperature grease containing zinc oxide to help prevent scoring and damage to parts during installation. Seals must always be installed so the seal lip is toward the lubricant.

Pinion Bearing (Preload)

Pinion bearing must be adjusted for preload before assembly is installed in carrier.

1. With pinion bearings and adjusting spacers (or shims) installed in cage, check bearing contact by rotating cage.

2. Using a press, apply pressure (approx. 20,000 lbs.) to outer bearing.

3. Wrap soft wire around cage and pull on horizontal line with spring scale. Rotating (not starting) torque should be within limits recommended by manufacturer.

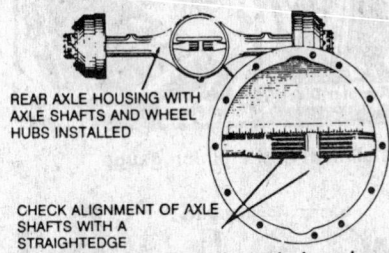

REAR AXLE HOUSING WITH
AXLE SHAFTS AND WHEEL
HUBS INSTALLED

CHECK ALIGNMENT OF AXLE
SHAFTS WITH A
STRAIGHTEDGE

Method of checking the axle housing alignment with full floating axles

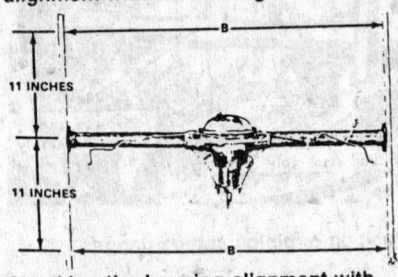

Checking the housing alignment with straight edge bars

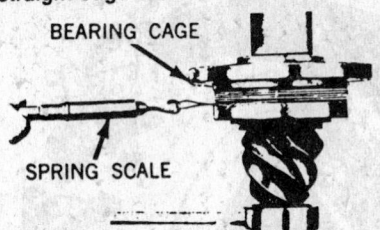

BEARING CAGE

SPRING SCALE

Checking the pinion pre-load

NOTE: Method of determining inch-pounds torque with scale is to determine radius of cage. Multiply radius in inches by pounds pull required to rotate cage to determine inch-pounds torque. Example: An 8 in. diameter divided by 2 equals 4 in. radius. Multiply 4 in. (radius) by 5 pounds (pull) equals 20 in. pounds torque.

4. If press is not available, check preload torque by installing propeller shaft yoke, washer, and nut and torque to specifications; then check as previously explained. Remove yoke after correct adjustment is obtained.

Bevel Gear Shaft Bearing Adjustment

Bevel gear shaft bearings must be adjusted for preload before pinion and cage assembly and differential assembly are installed in carrier.

1. Wrap several turns of soft wire around gear teeth on cross shaft and pull on a horizontal line with spring scale. Rotating (not starting) torque should be used.

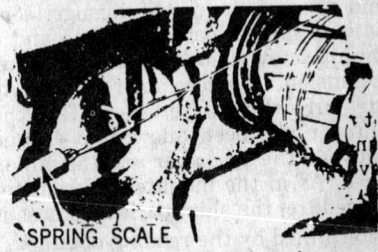

SPRING SCALE

Checking the pre-load on bevel gear cross shaft

NOTE: Method of determining inch lbs. torque with scale is to determine radius. Multiply radius in inches by pounds pull required to rotate shaft to determine inch-pounds torque. Example: An 8 in. diameter divided by 2 equals 4 in. radius times 5 pounds (pull) equals 20 inch lbs. torque.

2. Remove or add shims from under cage or cap opposite bevel gear to obtain specified bearing preload.

3. When making bevel gear and pinion tooth contact or backlash adjustments it is sometimes necessary to remove or add shims from one side.

NOTE: Always remove or add an equal thickness to the opposite side so to maintain correct preload.

Gear Tooth Contact and Backlash

PINION DEPTH MEASUREMENT METHOD

Methods of adjusting pinions to obtain the proper depths will vary with the axle type and the manufacturers recommendations. Pinion depth settings and gear teeth contact may be determined by the use of pinion setting gauges or by the use of marking dye on the gear teeth.

When using the gauge method, backlash is established after the pinion has been properly set. With the dye method, backlash is obtained first, then, the proper pinion tooth contact is established.

The pinion gauge method can be a direct reading micrometer, mounted on or through an arbor bar, set in adapter discs and located in the side carrier bearing cup locations on the differential housing and held in place by the bearing cup caps. The arbor bar coincides and represents the center line of the axle shafts. A reading is taken by the mounted micrometer, from the arbor bar to the head of the pinion to determine the need to add to or remove shims from the shim pack total, to adjust the pinion to the proper nominal assembly dimension or standard pinion depth.

Another method using the arbor bar and discs, is the use of a gauge block with a spring loaded plunger and a thumb screw to lock the plunger upon expansion. A micrometer is used to measure the gauge block after the plunger has been allowed to expand between the arbor bar and the pinion head. As in the mounted micrometer procedure, the shim pack thickness is determined by the reading obtained.

A third method is the use of a gauge block tool, installed in the housing in place of the pinion gear, and a large arbor bar placed in the axle housing differential bearing seats and tightened securely. A measurement is taken between the arbor bar and the pinion tool by either a feeler gauge or the use of individual shims from the shim pack. This measurement represents the shim pack needed for a zero marked pinion.

SETTING NEW PINION (WITHOUT GAUGE)

Whenever a pinion setting gauge is not available, the approximate thickness of the pinion shim pack at the rear pinion bearing cup, change the sign of the marking (individual variation distance) on the new pinion (plus to minus or minus to plus), then add the variation of the old pinion (sign unchanged) which will determine the amount the original shim pack must be changed when installing a new pinion.

On those types of axles where the shims are located between the pinion cage and differential carrier, change the sign of the marking (individual variation distance) on the old pinion (plus to minus or minus to plus), then, add variation of the new pinion (sign unchanged) which will determine how much the original shim pack must be altered when installing a new pinion.

When the approximate thickness of shim pack has been determined, final check of gear tooth contact must be made using dye method.

GEAR TOOTH CONTACT (DYE)

Gear tooth contact cannot be successfully accomplished until pinion and bevel gear bearings are in proper ad-

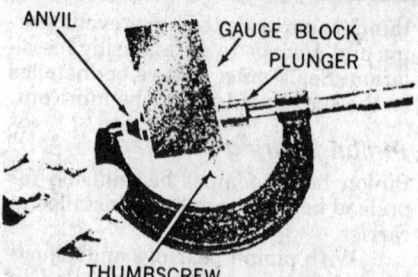

ANVIL · GAUGE BLOCK · PLUNGER · THUMBSCREW

Method of measurement of the gauge block

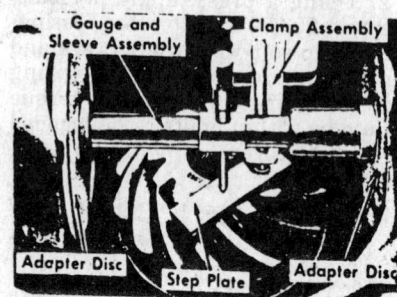

Gauge and Sleeve Assembly · Clamp Assembly · Adapter Disc · Step Plate · Adapter Disc

Installment of the pinion gauge

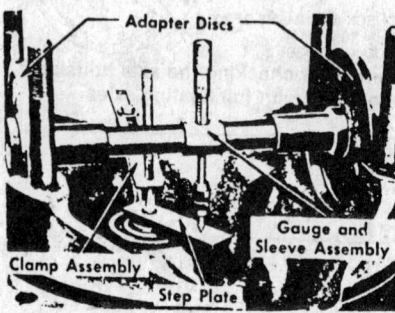

Adapter Discs · Gauge and Sleeve Assembly · Clamp Assembly · Step Plate

Position of pinion setting gauge

Checking the gear backlash-bevel gear

justment and gear backlash is within specified limits.

Check for proper tooth contact by painting a few teeth of bevel gear with marking dye. Turn pinion in direction of normal rotation, then check tooth impression on bevel gear.

GEAR BACKLASH

Gears used in extended service, form running contacts due to wear of teeth; therefore, the original shim pack (be-

tween pinion cage and carrier) should be maintained when checking backlash. If backlash exceeds maximum tolerance, reduce backlash only in the amount that will avoid overlap of worn tooth section. Smoothness and roughness can be noted by rotating bevel gear.

If a slight overlap is present at worn tooth section, rotation will be rough.

If new gears are installed, check backlash with dial indicator.

Backlash is increased by moving bevel gear away from pinion and may be decreased by moving bevel gear toward pinion.

When the drive gear is attached to the differential, backlash is accomplished is differential bearing adjusting rings. It should be remembered, when one ring is tightened, the opposite ring must be loosened an equal amount to maintain previously established bearing adjustment.

On axles where the bevel gear is supported by cross shaft, backlash is accomplished by adding or removing shims under bearing cages.

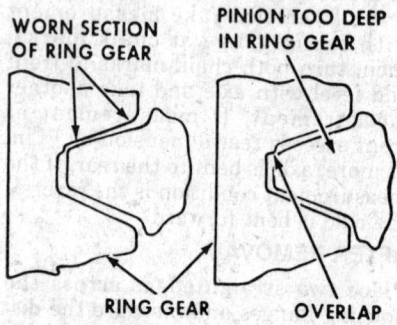

WORN SECTION OF RING GEAR · PINION TOO DEEP IN RING GEAR · RING GEAR · OVERLAP · CORRECT · INCORRECT

Avoid overlap of worn section of gear teeth during backlash adjustments when using original gears

TERMS USED

Certain dimensions must be determined when using the pinion setting gauge:

1. **Nominal Assembly Dimension:** (standard pinion depth) This dimension (varying with axle model) is the distance between the center line of the drive gear (or differential carrier bore) and the end of the drive pinion. This dimension may be marked on the pinion or listed on the Nominal Assembly Dimension and Adapter Disc chart.

2. **Individual Variation Distance:** (pinion depth variance) This dimension is a plus or minus variation of the **Nominal Assembly Dimension** on each individual pinion which may be caused by manufacturing variations.

3. **Corrected Nominal Dimension:** (desired pinion depth) This dimension is the **Nominal Assembly Dimension** plus or minus the **Individual Variation Distance.**

4. **Corrected Micrometer Distance is the Corrected Nominal Dimension** less the thickness of the gauge set step plate (0.400 in.) mounted on end of pinion.

5. **Initial Micrometer Reading** is the dimension taken by micrometer to the gauge step plate.

6. **Shim Pack Correction** is determined by the difference between the **Corrected Micrometer Distance** and the **Initial Micrometer Reading** and represents the amount of shim pack to be added or removed as later explained.

7. **Measured Pinion Depth.** This measurement is the distance between the axle center line and the top of the pinion gear. If a step plate or other type gauge tool is used, this measurement is included in the total.

Pinion and Drive Gear Identification

Drive gears and pinions are tested at the time of manufacture to detect machining variances and to obtain desirable tooth contact and quietness. When the correct setting is achieved, the gears are considered matched and a set of numbers, along with other identifying marks are etched on the gear set.

A plus (+) or minus (−) sign is used, followed by a digit to represent the factory setting where the tooth contact and quietness were the best. This is called the pinion depth variance or individual variation distance.

If the pinion is marked +5 for example, this means the distance from the pinion gear rear face to the axle shaft center line is 0.005 in. more than the standard setting or if the pinion gear is marked −5, this means the distance is 0.005 in. less than the standard setting. To move the pinion to the standard setting, compensating for the variation, shims must be either added to subtracted from the total shim pack, located under the rear pinion bearing cup, between the pinion cage and the differential carrier or under the rear pinion bearing, depending upon the differential model being serviced.

The procedures to follow in the adjustment of the pinion and drive gears are outlined in the respective differential model disassembly and assembly chapters.

As a rule of thumb on the addition or removal of shims for the pinion depth adjustment, draw a diagram as shown and determine which way the

pinion must be moved to obtain the desired pinion depth.

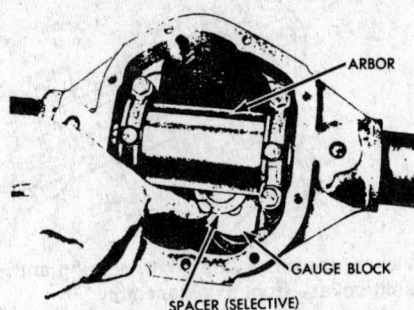

Determining proper shim pack thickness for drive pinion depth of mesh

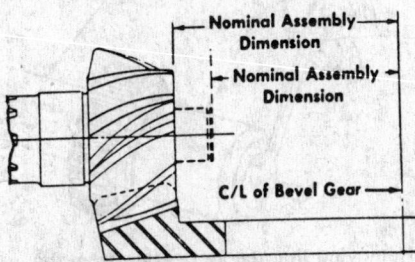

Nominal assembly dimension

Pinion Markings Old Pinion	New Pinion	Difference Between Markings
+5	+8	−3 Remove .003" Shim
+8	+5	+3 Add .003" Shim
−3	−5	+2 Add .002" Shim
−5	−3	−2 Remove .002" Shim
−3	+4	−7 Remove .007" Shim
+2	+6	+6 Add .006" Shim

The sign of the new pinion is changed and then added algebraically to the old pinion sign.

Determining pinion shim pack thickness, if the shim pack is located at the rear pinion bearing cup

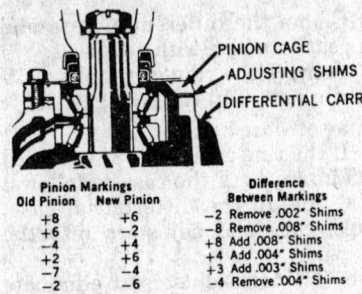

Pinion Markings Old Pinion	New Pinion	Difference Between Markings
+8	+6	−2 Remove .002" Shims
+6	−2	−8 Remove .008" Shims
−4	+4	+8 Add .008" Shims
+2	+6	+4 Add .004" Shims
−7	−4	+3 Add .003" Shims
−2	−6	−4 Remove .004" Shims

The sign of the old pinion is changed and then added algebraically to the new pinion sign.

Determining pinion shim pack thickness, if the shim pack is located between the pinion cage and differential

Standard Torque Specifications and Capscrew Markings

Because of the varied bolt sizes used in the many models of differentials, the torque specifications are not always available for a specific bolt. By deter-

mining the grade of bolt, size and thread, the proper torque limit can be determined.

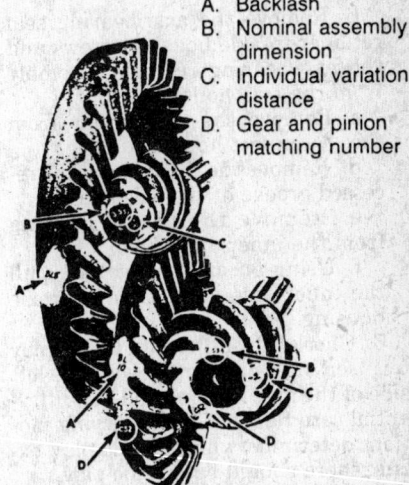

A. Backlash
B. Nominal assembly dimension
C. Individual variation distance
D. Gear and pinion matching number

Typical gear set marking code

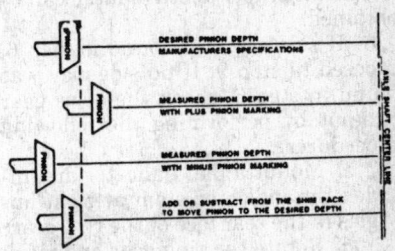

Movement of pinion to obtain desired pinion depth

CHRYSLER CORPORATION

Front Axle Assembly

7¼ INCH RING GEAR AXLE

Disassembly

DIFFERENTIAL

1. Remove the axle housing cover.
2. Clean the inside of the differential case with solvent and blow dry with compressed air.
3. Turn the differential case to make the differential pinion shaft lock screw accessible and remove the lock screw and pinion shaft.
4. Remove the left axle shaft by performing the following procedures:
   a. Remove the C-lock from the recessed groove of the axle shaft.
   b. Remove the axle shaft from the housing.
5. Remove the right axle shaft and inner axle shaft by performing the following procedures:

a. Remove the disconnect housing cover assembly-to-axle housing screws and the assembly and the gasket.

b. Remove the axle bearing seal retainer-to-axle housing screws and the axle/bearing retainer assembly from the axle housing.

c. Remove the shift collar from the disconnect housing.

d. Remove the C-lock from the recessed groove of the axle shaft.

e. Remove the needle bearing from the inner axle shaft.

f. Using an axle puller tool, pull the inner axle shaft from the axle housing.

6. Check for differential side-play by inserting a pry-bar between the left side of the axle housing and the differential case flange. Using a prying motion, determine whether side-play exists; there should be no side-play.

7. Paint the ring gear teeth and make a gear tooth contact pattern. Determine if proper depth of mesh can be obtained.

8. If side-play was found in step 6, proceed to step 9. If no side-play was found in step 6, check the drive gear run-out by performing the following procedures:

a. Mount a dial indicator and index the indicator stem at right angles in the rear face of the ring gear.

b. Rotate the ring gear and mark the ring gear and case at the point of greatest run-out.

c. Total indicator reading should not exceed 0.005 in.; if it does, the possibility exists that the case must be replaced.

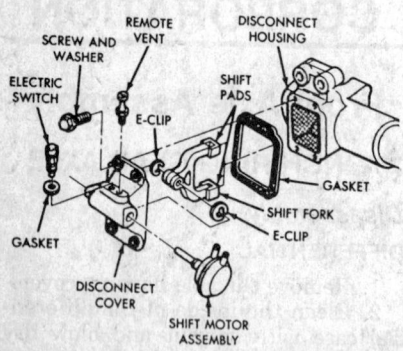

Exploded view of the disconnect housing cover assembly—front axle assembly

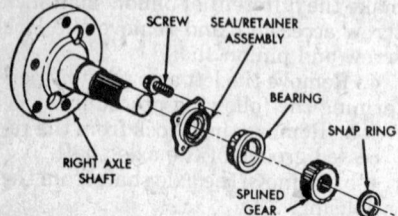

Exploded view of the right axle shaft—front axle assembly

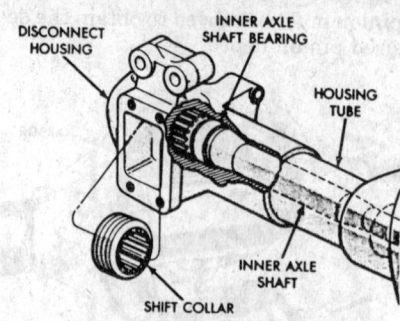

View of the inner axle shaft, bearing and shift collar—front axle assembly

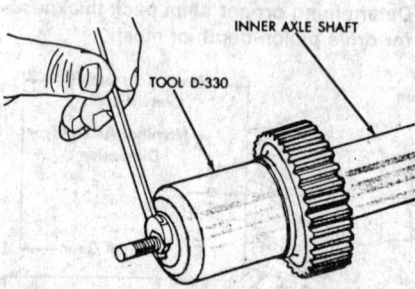

Removing the needle bearing from the inner axle shaft—front axle assembly

9. Using an inch lb. torque wrench, measure and record the pinion bearing preload.

10. Matchmark the axle housing and the differential bearing caps.

11. Remove the threaded adjusters and the differential bearing caps; there is a special wrench to do this through the axle tube.

12. Remove the differential case from the housing.

NOTE: The differential bearing cups and threaded adjuster must be kept together so they can be installed in their original position.

13. Clamp the differential case and ring gear in a vise with soft jaws.

14. Remove the ring gear bolts (left-hand thread). Tap the ring gear loose with a soft-faced mallet.

15. If the ring gear run-out exceeded 0.005 in., recheck the case as follows:

a. Install the differential case, cups, caps and adjusters into the housing.

b. Turn the adjusters to eliminate all side-play and tighten the differential cap bolts snugly.

c. Measure the run-out at the ring gear flange face; total indicator reading should not exceed 0.003 in.

NOTE: It is often possible to reduce run-out by removing the ring gear and reinstalling it 180 degrees from its original position.

d. Remove the differential case from the housing.

16. Remove the pinion shaft lockscrew and remove the pinion shaft.

17. Rotate the differential side gears until the differential pinion shafts can be removed through the opening in the case.

18. Remove the differential side gears and thrust washers.

19. Using a bearing puller, press side bearing from the differential.

PINION GEAR

1. Remove the pinion nut, washer and pinion flange.

2. Remove and discard the pinion oil seal.

3. Using a brass hammer, drive the pinion rearward out of the bearing. This will result in damage to the bearing and cup. The bearing cone and cup must be replaced with new parts. Discard the collapsible spacer.

4. Using a brass drift and a hammer, drive the front and rear bearing cups from the housing.

5. Remove the shim from behind the rear bearing cup and record the thickness.

6. Using a bearing puller, press the rear bearing cone from the pinion stem.

Inspection

1. Clean the differential components in solvent and use compressed air to dry them; do not use compressed air on the bearings, only shop towels.

2. Check the components for wear or damage; replace them, if necessary.

3. Inspect the bearings and bearing cups for wear, cracks or scoring; replace them, if necessary.

4. Inspect the differential side and pinion gears for wear, cracks or chips; replace them, if necessary.

5. Inspect the ring and pinion gears for wear and/or damage; replace them, if necessary.

6. Inspect the differential case for cracks or damage; replace it, if necessary.

Assembly

PINION GEAR

1. The proper pinion setting (relative to the ring gear) is determined by a shim which has been selected before the pinion is to be installed in the carrier. Pinion bearing shims are available in 0.001 in. increments.

2. The head of the pinion is marked with a plus $(+)$ or a minus $(-)$ mark that is followed by a number ranging from 0-4. If the old and new pinions have the same marking and the old bearing is being installed, use a shim of the original thickness. If the old pinion is marked zero (0), however and the new pinion is marked +2, try a

shim that is 0.002 in. thinner. If the new pinion is marked axle housing cup bore and install − 2, try a shim that is 0.002 in. thicker.

3. Position the selected shim in the bore of the rear bearing cup. Install the cup.

NOTE: Special pinion depth measuring tools are available. When using the special tools, follow the manufacturer's recommended procedures. Without the special tools, complete the following procedure and check the pinion depth by examining the pinion to ring gear tooth contact pattern. Correct as required by adding or subtracting shims controlling the pinion depth.

4. Place the rear pinion bearing cone on the pinion stem (small side away from pinion head).

5. Lubricate the front and rear bearing cones and install the rear pinion bearing cone onto the pinion stem with an arbor press.

6. Insert the pinion bearing and collapsible spacer assembly through the carrier and install the front bearing cone. Install the companion flange.

NOTE: During installation of the pinion bearing do not collapse the spacer.

7. Install the drive pinion oil seal into the carrier; be sure to properly seat the seal.

8. Support the pinion in the carrier.

9. Install the belleville washer (convex side up) and pinion nut.

10. Hold the companion flange and tighten the pinion nut to remove all endplay, while rotating the pinion to ensure proper bearing seating. Remove the tools and rotate the pinion several revolutions.

11. Torque the pinion nut to 210 ft. lbs. (285 Nm). Using an inch lbs. torque wrench, measure the pinion bearing preload; the torque is 20–35 inch lbs. for new bearings or 10 inch lbs. over the original figure for the old pinion bearing.

NOTE: The correct preload reading can only be obtained with the carrier nose upright. The final assembly is incorrect if the final pinion nut torque is below 210 ft. lbs. (285 Nm) or if the pinion bearing preload is not within specifications. Under no circumstances should the pinion nut be backed off to reduce the pinion bearing preload; if this is done, a new collapsible spacer will have to be installed and the unit adjusted again until proper preload is obtained.

DIFFERENTIAL

1. Lubricate all parts, before assembly, with rear axle lubricant.

2. Install the thrust washers on the differential side gears and install the side gears into the case.

3. Place thrust washers on both differential pinions and, working through the opening in the case, mesh the pinion gears with the side gears. The pinions should be exactly 180 degrees apart.

4. Rotate the side gears 90 degrees to align the pinions and thrust washers with the pinion shaft holes.

5. From the pinion shaft lockpin hole side of the case, insert the slotted end of the pinion shaft through the case and conical thrust washer; install the pinion shaft through one of the pinion gears.

6. Install a thrust block through the side gear hub, so the slot is centered between the side gears.

7. Hold all of the parts in alignment and align the lockpin holes in the pinion shaft and case. Install the lockpin from the pinion shaft side of the ring gear flange, temporarily.

8. With a stone, relieve the edge of the chamfer on the inside diameter of the ring gear.

9. Heat the ring gear (fluid bath or heat lamp) to a temperature not exceeding 300°F; do not heat ring gear with a torch.

10. Align the ring gear with the case. Insert the ring gear screws through the case flange and into the ring gear.

11. Alternately, tighten each cap screw to 80 ft. lbs. (108 Nm).

12. Position each differential bearing cone on the hub of the differential case (taper away from ring gear) and install the bearing cones; a shop press may be helpful.

13. Install the differential into the axle housing.

14. Install the right and inner axle shaft by performing the following procedures:

   a. Using the axle shaft removal/installation tool, install the inner axle shaft into the housing.

   b. Install the C-clip on the other end of the shaft to retain it.

   c. Install the needle bearing into the end of the inner axle shaft.

   d. Install the shift collar onto the splined end of the inner axle shaft.

   e. Install the axle shaft/bearing retainer assembly into the disconnect housing and torque the bearing retainer-to-disconnect housing screws to 200 inch lbs. (23 Nm).

   f. Using a new gasket, install the disconnect housing cover assembly onto the disconnect housing and torque the bolts to 10 ft. lbs. (14 Nm).

15. Install the left axle shaft into the axle housing and secure it with the C-clip.

16. To complete the installation, reverse the removal procedures.

Adjustment

DIFFERENTIAL BEARING PRELOAD AND RING GEAR-TO-PINION BACKLASH

The threaded adjuster uses a hex drive hole and requires a special tool C–4164 to adjust the side bearing preload through the axle tube. An adjuster lock with 2 pointed teeth which engage in the exposed adjuster thread when the lock is tightened is provided. The shims will range from 0.020–0.038 in. and will be equipped with internal centering tabs. The shims, marked with a number which represents its thickness in thousandths of an in., can be installed with either side against the pinion head.

1. Index the gears so the same gear teeth are in contact throughout the adjustment.

2. The differential bearing cups will not always move with the adjusters. It is important to seat the bearings by rotating them 5–10 times in each direction, each time the adjusters are moved.

3. With the pinion bearings installed and the preload set, install the differential with adjusters, caps and bearings. Lubricate the bearings and adjuster threads. Tighten the top cap screws on the right and left to 10 ft. lbs. Tighten the bottom cap screws finger tight until the head is just seated on the bearing cap.

4. Using the tool, make sure the adjuster rotates freely. Turn both adjusters in until bearing play is eliminated with some drive gear backlash (0.010 in.). Seat the bearing rollers.

5. Install and register a dial indicator against the drive side of a gear tooth. Check the backlash at 4 positions to find the point of minimum backlash. Rotate the gear to the position of least backlash and mark the tooth so the readings will be taken at the same point.

6. Loosen the right adjuster and turn it until the backlash is 0.003–0.004 in. with each adjuster tightened to 10 ft. lbs. Seat the bearings rollers.

7. Tighten the differential bearing cap screws to 45 ft. lbs. (61 Nm).

8. Tighten the right adjuster to 70 ft. lbs. and seat the rollers, until the torque remains constant at 70 ft. lbs. Measure the backlash, if the backlash is not 0.006–0.008 in. increase the torque on the right adjusters and seat the rollers until the correct backlash is obtained.

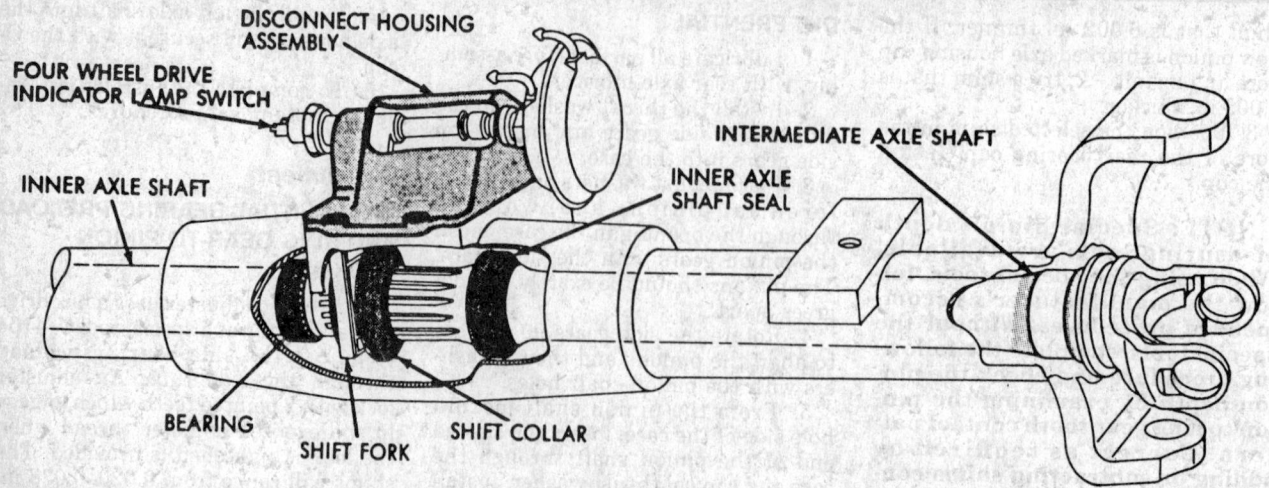

FOUR WHEEL DRIVE INDICATOR LAMP SWITCH — **DISCONNECT HOUSING ASSEMBLY** — **INNER AXLE SHAFT** — **INTERMEDIATE AXLE SHAFT** — **INNER AXLE SHAFT SEAL** — **BEARING** — **SHIFT FORK** — **SHIFT COLLAR**

Sectional view of the of the left disconnect axle assembly—model 44 front axle assembly

9. Tighten the left adjuster to 70 ft. lbs. and seat the bearings until the torque remains constant.

10. If the assembly is properly done, the initial reading on the left adjuster will be approximately 70 ft. lbs. If it is substantially less, the entire procedure should be repeated.

11. After the adjustments are complete, install the adjuster locks; be sure the teeth are engaged in the adjuster threads. Torque the lockscrews to 90 inch lbs.

SPICER MODEL 44 AXLE

Disassembly
DIFFERENTIAL

1. Remove the housing cover and drain the lubricant.

2. Remove the left axle shaft and the inner axle shaft by performing the following procedures:

   a. Using an axle puller, press the left axle shaft from the axle housing.

   b. Remove the disconnect housing assembly-to-axle housing screws, the housing and the gasket.

   c. Remove the shift collar from the inner axle shaft.

   d. Using an inner axle shaft removal/installation tool, attach it to the inner axle shaft and carefully, slide the inner axle shaft from the axle housing.

3. Using an axle spreader tool, mount it onto the axle housing and spread the housing enough to remove the differential.

4. Using a dial indicator, measure the amount the opening is being spread; do not spread the housing more than 0.015 in. (0.38mm), for damage to the housing may occur.

5. Mark the differential bearing caps for reassembly purposes.

6. Loosen the bearing caps until 2–3 threads are engaged.

7. Using a prybar, pry the differential loose.

8. Remove the bearing caps and the differential.

9. Mount the differential into a vise.

10. Remove and discard the ring gear bolts; they are not reusable.

11. Using a brass drift and a hammer, tap the ring gear from the differential.

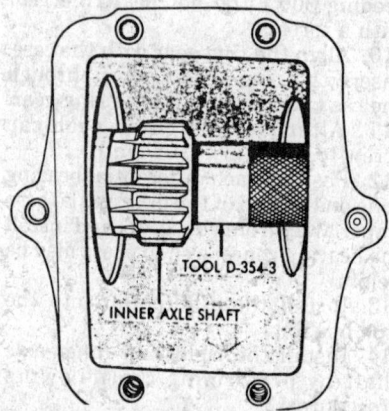

TOOL D-354-3 — **INNER AXLE SHAFT**

Removing the inner axle shaft— model 44 front axle assembly

12. Using a differential bearing puller, press the differential bearings from the differential.

13. Remove the differential bearing shims.

14. Remove the pinion shaft lockpin and pinion shaft.

15. Rotate the pinions to remove them through the case opening.

16. Remove the side gears and thrust washers.

PINION GEAR

1. Remove the differential case from the axle housing.

2. Using a pinion yoke holding tool, remove the pinion gear nut.

3. Remove the pinion washer. Using a pinion yoke holder tool and a pinion puller tool, press the yoke from the pinion gear.

4. Using an oil seal remover tool, remove the pinion gear oil seal and discard it. Remove the slinger, the gasket, the upper pinion bearing cone and preload shim pack; measure and record the shim thicknesses.

5. Press drive the pinion gear and inner bearing cone assembly from the axle housing.

6. Using a shop press, press the inner bearing cone from the pinion gear.

7. Remove and record the shim thicknesses from behind the inner pinion bearing cup.

Inspection

1. Clean the differential components in solvent and use compressed air to dry them; do not use compressed air on the bearings, only shop towels.

2. Check the components for wear or damage; replace them, if necessary.

3. Inspect the bearings and bearing cups for wear, cracks or scoring; replace them, if necessary.

4. Inspect the differential side and pinion gears for wear, cracks or chips; replace them, if necessary.

5. Inspect the ring and pinion gears for wear and/or damage; replace them, if necessary.

6. Inspect the differential case for cracks or damage; replace it, if necessary.

Assembly
PINION GEAR

1. Make sure the ring and pinion are a matched set.

2. Measure the thickness of the

original pinion shim and note the variance on the pinion gear.

3. Perform the following procedure to determine the pinion starter shim thickness:

   a. If the original ring and pinion are being installed, use the original shim.

   b. If a replacement gear set is being installed, determine the best starter shim thickness.

   c. Refer to the pinion variance chart and observe where the old and new pinion marking column intersect.

   d. If the old pinion is +2 and the new pinion is −2, the intersecting figure is +0.004 in. (0.10mm); add this amount to the original shim. Or, if the old pinion is −3 and the new pinion is −2, the intersecting figure is −0.001 in. (−0.025mm); subtract this amount from the original shim.

4. Install the starter shim in the pinion rear bearing cup bore; if the shim is chamfer on one side, position the chamfered side so it faces the bottom of the bore.

5. Using a driver tool, install the pinion rear bearing cup into the axle housing.

6. Install the front bearing cup.

7. Using a shop press, press the rear bearing onto the pinion gear.

8. Install the pinion gear into the axle housing and the front bearing onto the pinion; do not install the slinger or the seal.

9. Install the yoke, the pinion nut washer and the old pinion nut; tighten the nut enough to remove the endplay.

10. Adjust the backlash and gear tooth contact.

DIFFERENTIAL

1. Install the side gears, the thrust washers and the pinion gears into the differential case.

NOTE: If new gears and washers are used, it will not be necessary to check the gear backlash. Correct fit is provided due to the close manufacturing tolerances.

2. Install the pinion shaft and lockpin into the case.

3. Assemble the original differential bearing shim packs, then, remove approximately 0.20 in. (0.50mm) shim thickness from each pack; the remaining shims will serve as a starter shim pack.

4. Install the starter shim packs and bearing onto the case.

5. Align and install the ring gear. Using new bolts, torque the ring gear-to-differential case bolts to 45–60 ft. lbs. (61–81 Nm).

6. Install the spreader tool onto the axle housing. Using a dial indicator, spread the axle housing no more than 0.015 in. (0.38mm).

NOTE: Do not spread the housing more than 0.015 in. (0.038mm), for damage may occur to the case.

7. Install the differential; it may be necessary to tap the differential bearing cups, with a soft hammer, to seat them.

8. Install the differential bearing caps and torque the bolts to 70–90 ft. lbs. (95–122 Nm).

9. Remove the spreader tool and dial indicator.

10. Adjust the ring and pinion gear.

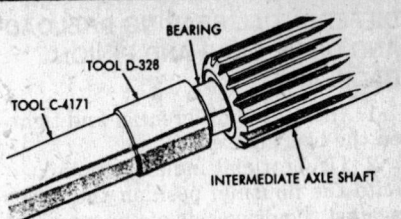

Installing the inner axle shaft bearing—model 44 front axle assembly

11. Using silicone sealant, apply a bead of it to the axle housing. Install the cover and torque the bolts to 35 ft. lbs. (47 Nm).

Adjustment
PINION BEARING PRELOAD

1. Using a yoke holding tool, torque the pinion nut to 200–220 ft. lbs. (271–298 Nm). Rotate the drive pinion several complete revolutions to seat the bearing rollers.

2. Using an inch lb. torque wrench, measure the torque necessary to rotate the pinion; the torque should be 20–40 inch lbs. (2–5 Nm).

NOTE: If the torque is not correct, add a shim to decrease the preload or subtract a shim to increase the preload.

3. Remove the pinion nut, the washer and the yoke after setting the preload; be sure to discard the old nut.

4. Lubricate the new pinion oil seal lip and install the seal into the axle housing.

5. Install the yoke, the washer and a new pinion nut; torque the nut to 200–220 ft. lbs. (271–298 Nm).

Original Pinion Gear Depth Variance	Replacement Pinion Gear Depth Variance								
	−4	−3	−2	−1	0	+1	+2	+3	+4
+4	+0.008	+0.007	+0.006	+0.005	+0.004	+0.003	+0.002	+0.001	0
+3	+0.007	+0.006	+0.005	+0.004	+0.003	+0.002	+0.001	0	−0.001
+2	+0.006	+0.005	+0.004	+0.003	+0.002	+0.001	0	−0.001	−0.002
+1	+0.005	+0.004	+0.003	+0.002	+0.001	0	−0.001	−0.002	−0.003
0	+0.004	+0.003	+0.002	+0.001	0	−0.001	−0.002	−0.003	−0.004
−1	+0.003	+0.002	+0.001	0	−0.001	−0.002	−0.003	−0.004	−0.005
−2	+0.002	+0.001	0	−0.001	−0.002	−0.003	−0.004	−0.005	−0.006
−3	+0.001	0	−0.001	−0.002	−0.003	−0.004	−0.005	−0.006	−0.007
−4	0	−0.001	−0.002	−0.003	−0.004	−0.005	−0.006	−0.007	−0.008

DIFFERENTIAL BEARING PRELOAD AND DRIVE GEAR AND PINION BACKLASH

1. Install the differential and tighten the bearing caps.

2. Using a dial indicator, mount it onto the housing, position the stylus against the drive side of one ring gear tooth; be sure the stylus is at a right angle to the tooth.

3. Move the ring gear toward the dial indicator and zero it.

4. Move the ring gear away from the pinion until the backlash and note the reading on the dial indicator.

NOTE: The reading represents the thickness of the shim pack necessary to take up the clearance between the bearing cup and the case on the ring gear side of the differential assembly.

5. Subtract this reading from the previously recorded total reading to obtain the amount of shims necessary to take up the clearance between the bearing cup and the case at the pinion side of the differential.

6. Remove the differential and ring gear assembly from the carrier.

7. Remove the differential bearing covers. Install the correct thickness shim pack between the bearing cone and differential case hub shoulder. Add an additional 0.015 in. (0.38mm) shim to the drive gear side of the differential and install the differential bearing cones.

8. Install the spreader tool to the axle housing. Using a dial indicator, spread the housing to 0.015–0.020 (0.38–0.51mm). Install the differential assembly into the axle housing.

9. Install the bearing caps, remove the spreader tool and torque the bearing caps snugly.

10. Using a soft hammer, tap the drive gear to seat the differential bearing and cups.

NOTE: When seating the bearing and cups, be careful not to nick the ring gear or drive pinion teeth.

11. Torque the bearing cap bolts to 70–90 ft. lbs. (95–122 Nm).

12. Attach a dial indicator to the carrier, with the stylus contacting a ring gear tooth, and measure the backlash between the ring gear and the drive pinion.

13. Check the backlash at 4 equally spaced points around the circumference of the ring gear; the backlash

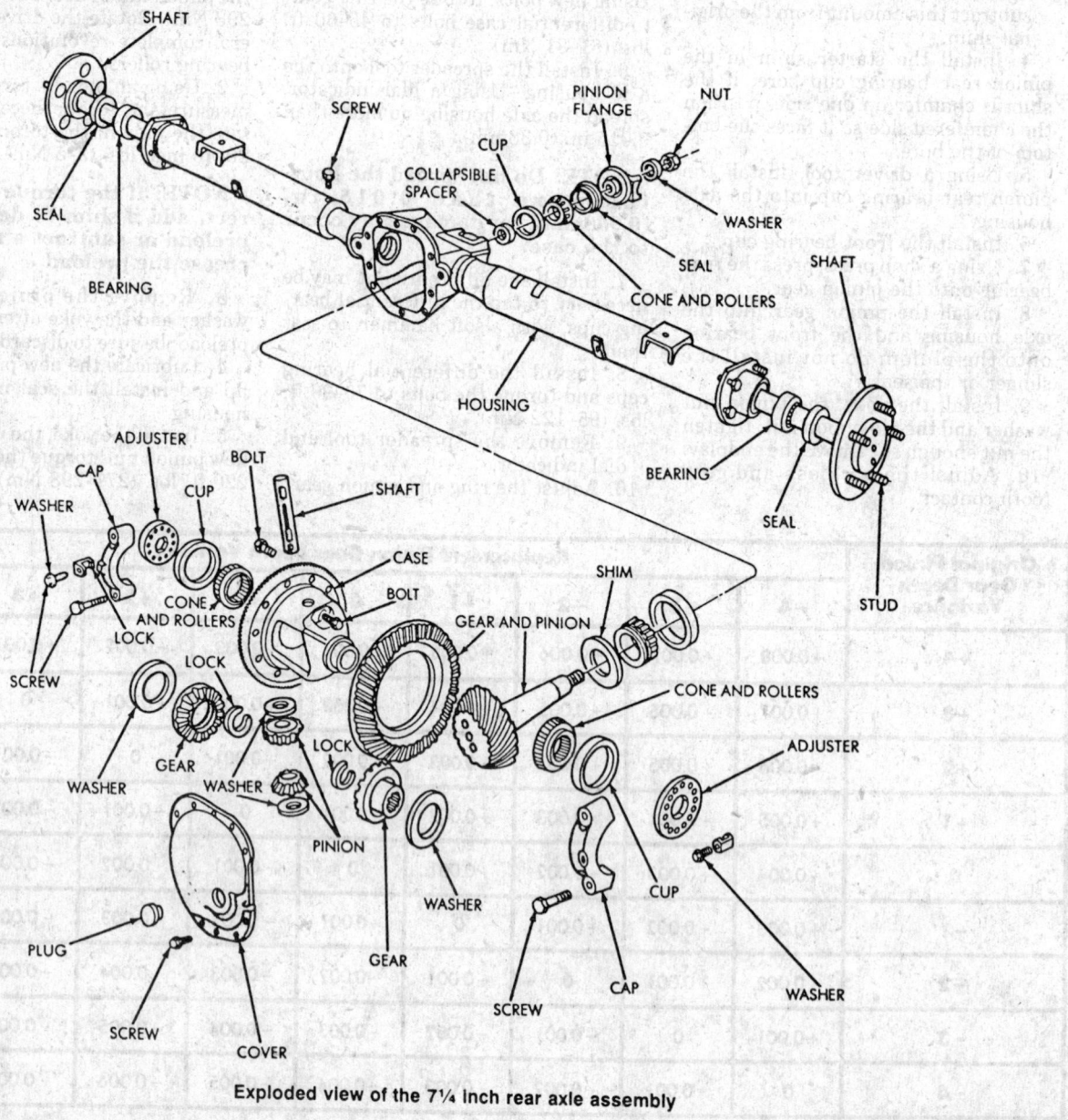

Exploded view of the 7¼ inch rear axle assembly

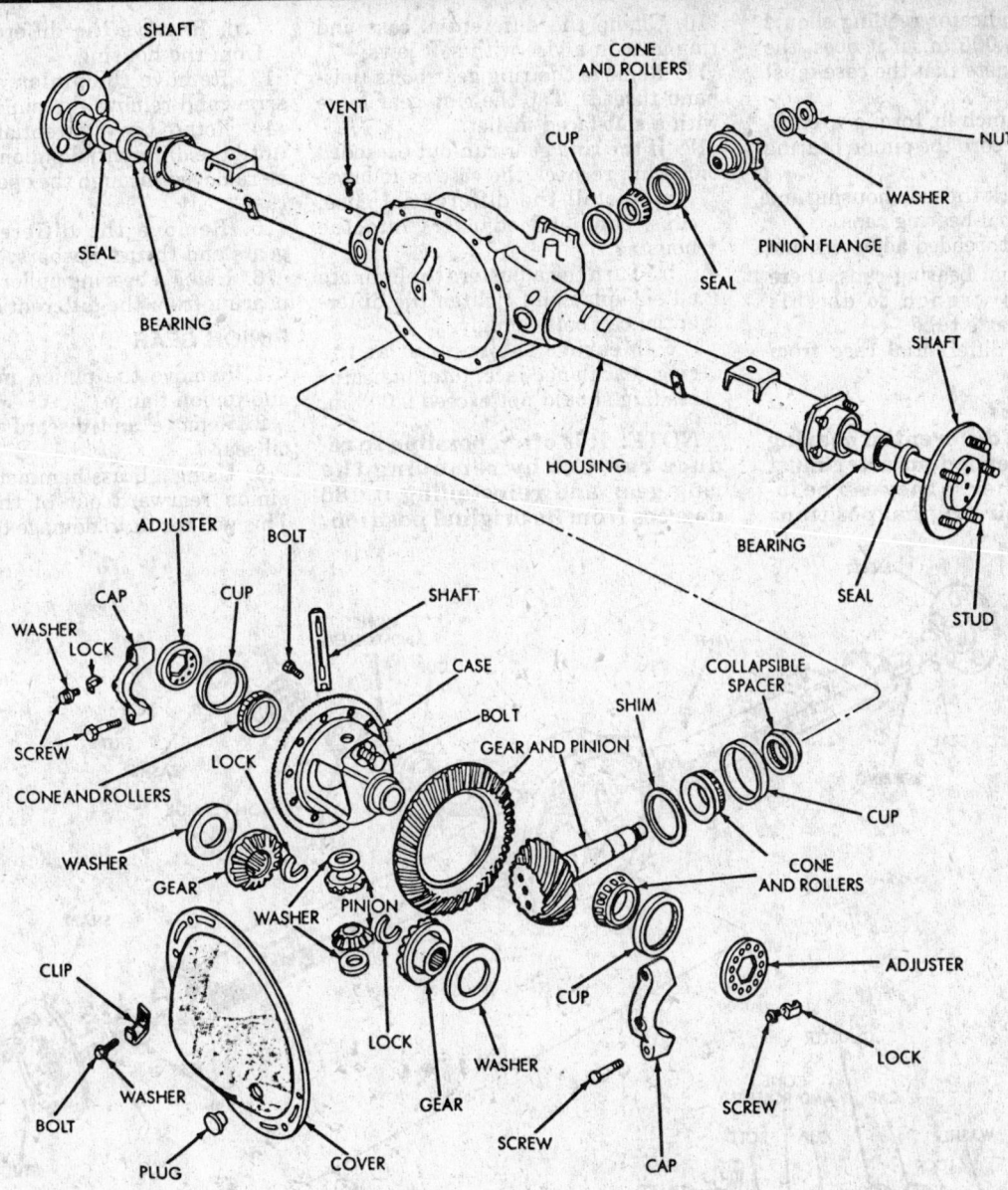

Exploded view of the 8¼ inch rear axle assembly

must be between 0.04–0.009 in. (0.102–0.229mm) and cannot vary more than 0.002 in. (0.508mm) between the 4 check positions.

NOTE: If the backlash does not fall within these specifications, change the shim pack thickness on both the differential bearing hubs to maintain the proper bearing preload and backlash.

SPICER—MODEL 60 AXLE

For overhaul information, refer to the Spicer 60 axle, in the rear axle assembly section.

Rear Axle Assembly

While some Dodge/Plymouth trucks use a 7¼, 8¼, 8⅜ or 9¼ in. semi-floating axle others use a Spicer 60, 60 HD (heavy duty) or 70 full-floating axle.

7¼, 8¼, 8⅜ AND 9¼ INCH RING GEAR AXLE

Disassembly
DIFFERENTIAL
1. Remove the axle housing cover.
2. Clean the inside of the differential case with solvent and blow dry with compressed air.
3. Check for differential side-play by inserting a pry-bar between the left side of the axle housing and the differential case flange. Using a prying motion, determine whether side-play exists; there should be no side-play.
4. Paint the ring gear teeth and make a gear tooth contact pattern. Determine if proper depth of mesh can be obtained.
5. If side-play was found in step 3, proceed to step 6. If no side-play was found in step 3, check the drive gear run-out by performing the following procedures:
  a. Mount a dial indicator and index the indicator stem at right angles in the rear face of the ring gear.
  b. Rotate the ring gear and mark the ring gear and case at the point of greatest run-out.

c. Total indicator reading should not exceed 0.005 in.; if it does, the possibility exists that the case must be replaced.

6. Using an inch lb. torque wrench, measure and record the pinion bearing preload.

7. Matchmark the axle housing and the differential bearing caps.

8. Remove the threaded adjusters and the differential bearing caps; there is a special wrench to do this through the axle tube.

9. Remove the differential case from the housing.

NOTE: The differential bearing cups and threaded adjuster must be kept together so they can be installed in their original position.

10. Clamp the differential case and ring gear in a vise with soft jaws.

11. Remove the ring gear bolts (left-hand thread). Tap the ring gear loose with a soft-faced mallet.

12. If the ring gear run-out exceeded 0.005 in., recheck the case as follows:

a. Install the differential case, cups, caps and adjusters into the housing.

b. Turn the adjusters to eliminate all side-play and tighten the differential cap bolts snugly.

c. Measure the run-out at the ring gear flange face; total indicator reading should not exceed 0.003 in.

NOTE: It is often possible to reduce run-out by removing the ring gear and reinstalling it 180 degrees from its original position.

d. Remove the differential case from the housing.

13. Remove the pinion shaft lock-screw and remove the pinion shaft.

14. Rotate the differential side gears until the differential pinion shafts can be removed through the opening in the case.

15. Remove the differential side gears and thrust washers.

16. Using a bearing puller, press side bearing from the differential.

PINION GEAR

1. Remove the pinion nut, washer and pinion flange.

2. Remove and discard the pinion oil seal.

3. Using a brass hammer, drive the pinion rearward out of the bearing. This will result in damage to the bear-

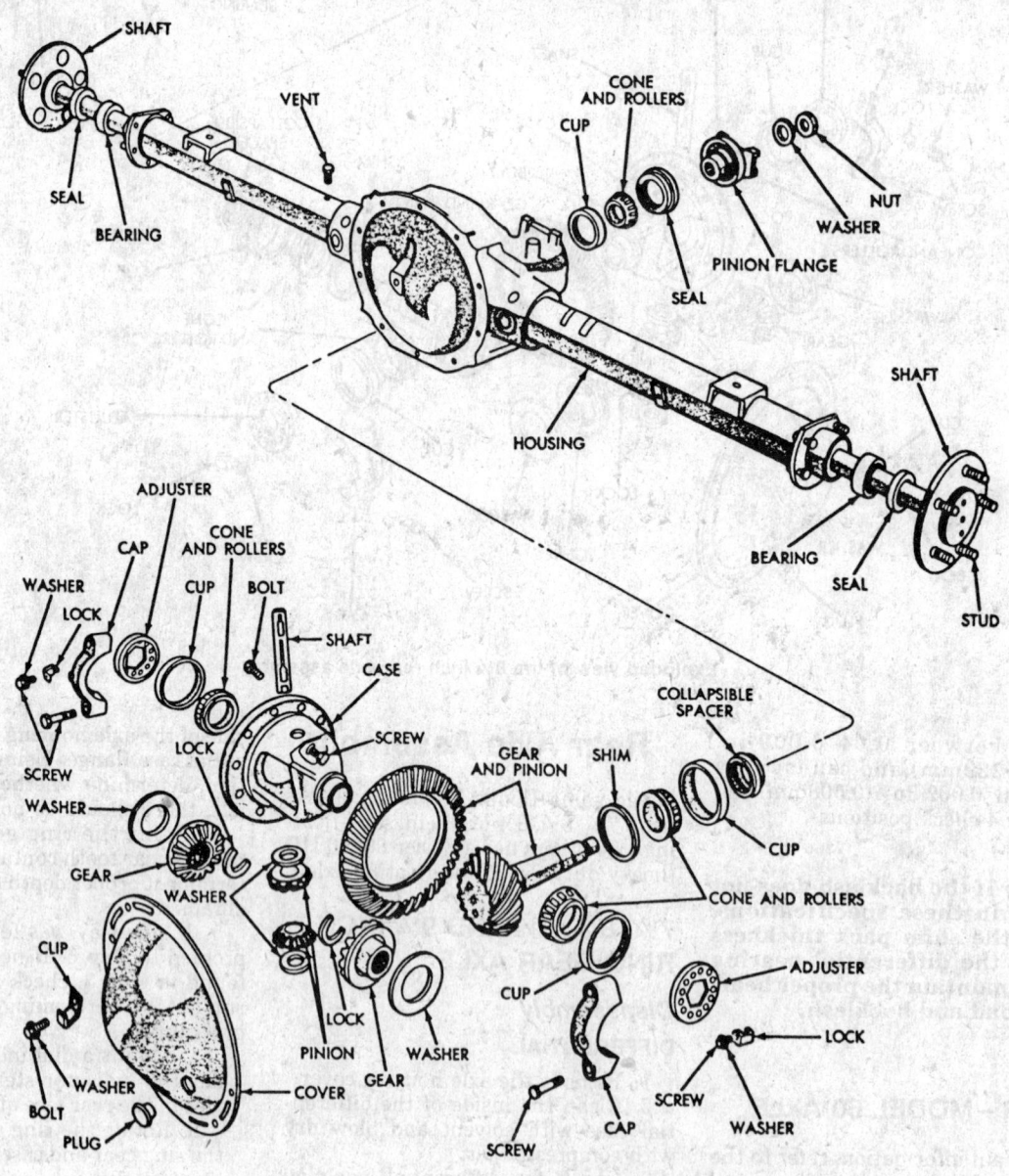

Exploded view of the 8⅜ inch rear axle assembly

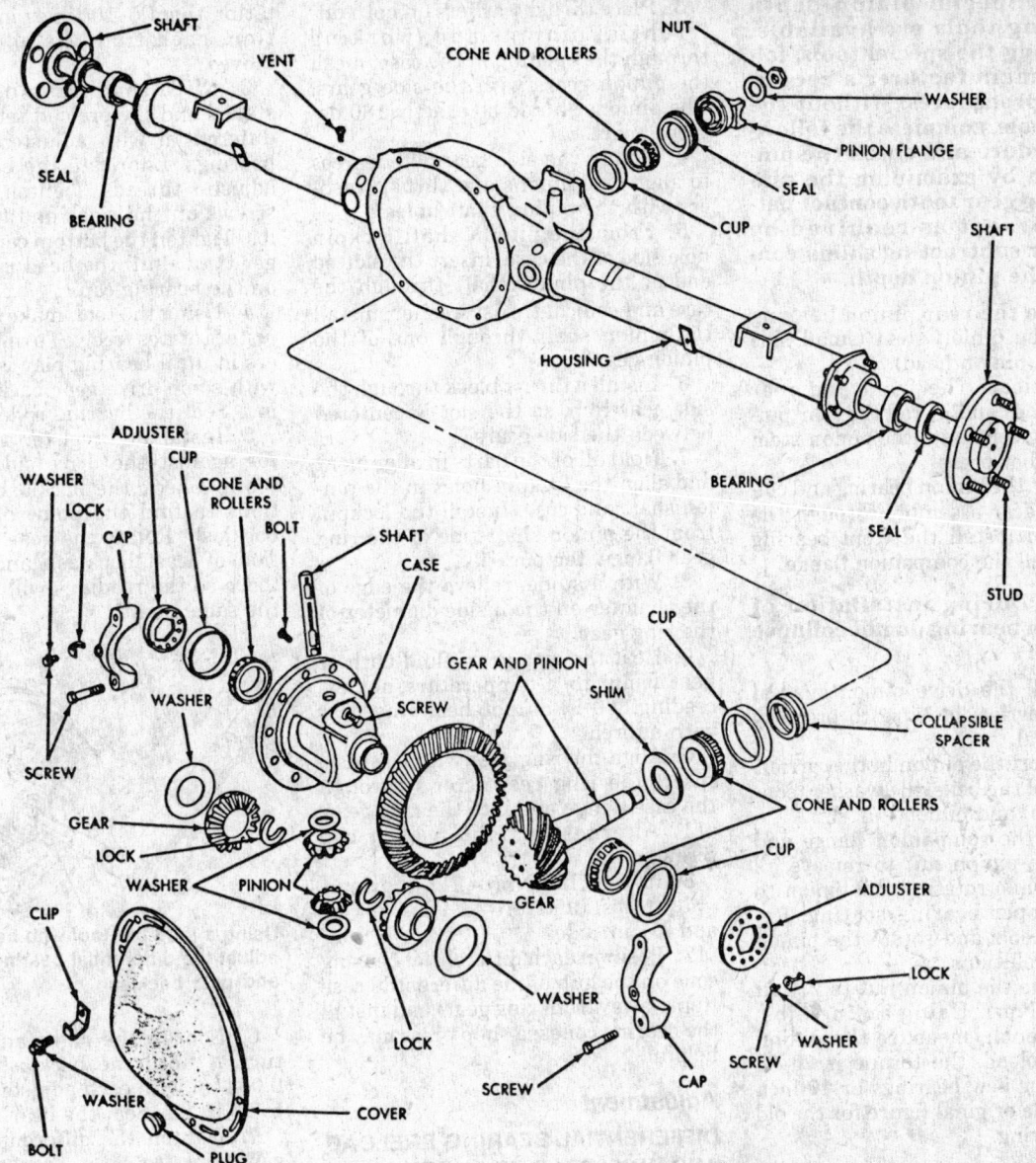

Exploded view of the 9¼ inch rear axle assembly

ing and cup. The bearing cone and cup must be replaced with new parts. Discard the collapsible spacer.

4. Using a brass drift and a hammer, drive the front and rear bearing cups from the housing.

5. Remove the shim from behind the rear bearing cup and record the thickness.

6. Using a bearing puller, press the rear bearing cone from the pinion stem.

Inspection

1. Clean the differential components in solvent and use compressed air to dry them; do not use compressed air on the bearings, only shop towels.

2. Check the components for wear or damage; replace them, if necessary.

3. Inspect the bearings and bearing cups for wear, cracks or scoring; replace them, if necessary.

4. Inspect the differential side and pinion gears for wear, cracks or chips; replace them, if necessary.

5. Inspect the ring and pinion gears for wear and/or damage; replace them, if necessary.

6. Inspect the differential case for cracks or damage; replace it, if necessary.

Assembly

PINION GEAR

1. The proper pinion setting (relative to the ring gear) is determined by a shim which has been selected before the pinion is to be installed in the car-

rier. Pinion bearing shims are available in 0.001 in. increments.

2. The head of the pinion is marked with a plus (+) or a minus (−) mark that is followed by a number ranging from 0–4. If the old and new pinions have the same marking and the old bearing is being installed, use a shim of the original thickness. If the old pinion is marked zero (0), however and the new pinion is marked +2, try a shim that is 0.002 in. thinner. If the new pinion is marked axle housing cup bore and install −2, try a shim that is 0.002 in. thicker.

3. Position the selected shim in the bore of the rear bearing cup. Install the cup.

NOTE: Special pinion depth measuring tools are available. When using the special tools, follow the manufacturer's recommended procedures. Without the special tools, complete the following procedure and check the pinion depth by examining the pinion to ring gear tooth contact pattern. Correct as required by adding or subtracting shims controlling the pinion depth.

4. Place the rear pinion bearing cone on the pinion stem (small side away from pinion head).

5. Lubricate the front and rear bearing cones and install the rear pinion bearing cone onto the pinion stem with an arbor press.

6. Insert the pinion bearing and collapsible spacer assembly through the carrier and install the front bearing cone. Install the companion flange.

NOTE: During installation of the pinion bearing do not collapse the spacer.

7. Install the drive pinion oil seal into the carrier; be sure to properly seat the seal.

8. Support the pinion in the carrier.

9. Install the belleville washer (convex side up) and pinion nut.

10. Hold the companion flange and tighten the pinion nut to remove all endplay, while rotating the pinion to ensure proper bearing seating. Remove the tools and rotate the pinion several revolutions.

11. Torque the pinion nut to 210 ft. lbs. (285 Nm). Using an inch lbs. torque wrench, measure the pinion bearing preload; the torque is 20–35 inch lbs. for new bearings or 10 inch lbs. over the original figure for the old pinion bearing.

NOTE: The correct preload reading can only be obtained with the carrier nose upright. The final assembly is incorrect if the final pinion nut torque is below 210 ft. lbs. (285 Nm) or if the pinion bearing preload is not within specifications. Under no circumstances should the pinion nut be backed off to reduce the pinion bearing preload; if this is done, a new collapsible spacer will have to be installed and the unit adjusted again until proper preload is obtained.

DIFFERENTIAL

1. Lubricate all parts, before assembly, with rear axle lubricant.

2. Install the thrust washers on the differential side gears and install the side gears into the case.

3. Place thrust washers on both differential pinions and, working through the opening in the case, mesh the pinion gears with the side gears. The pinions should be exactly 180 degrees apart.

4. Rotate the side gears 90 degrees to align the pinions and thrust washers with the pinion shaft holes.

5. From the pinion shaft lockpin hole side of the case, insert the slotted end of the pinion shaft through the case and conical thrust washer; install the pinion shaft through one of the pinion gears.

6. Install a thrust block through the side gear hub, so the slot is centered between the side gears.

7. Hold all of the parts in alignment and align the lockpin holes in the pinion shaft and case. Install the lockpin from the pinion shaft side of the ring gear flange, temporarily.

8. With a stone, relieve the edge of the chamfer on the inside diameter of the ring gear.

9. Heat the ring gear (fluid bath or heat lamp) to a temperature not exceeding 300°F; do not heat ring gear with a torch.

10. Align the ring gear with the case. Insert the ring gear screws through the case flange and into the ring gear.

11. Alternately tighten each cap screw to:

80 ft. lbs. (108 Nm) – 7¼ in axle
70 ft. lbs. (95 Nm) – 8¼ in., 8⅜ in. and 9¼ in. axles

12. Position each differential bearing cone on the hub of the differential case (taper away from ring gear) and install the bearing cones; a shop press may be helpful.

Adjustment

DIFFERENTIAL BEARING PRELOAD AND RING GEAR-TO-PINION BACKLASH

The threaded adjuster uses a hex drive hole and requires a special tool C–4164 to adjust the side bearing preload through the axle tube. An adjuster lock with 2 pointed teeth which engage in the exposed adjuster thread when the lock is tightened is provided. The shims will range from 0.020–0.038 in. and will be equipped with internal centering tabs. The shims, marked with a number which represents its thickness in thousandths of an in., can be installed with either side against the pinion head.

1. Index the gears so the same gear teeth are in contact throughout the adjustment.

2. The differential bearing cups will not always move with the adjusters. It is important to seat the bearings by rotating them 5–10 times in each direction, each time the adjusters are moved.

3. With the pinion bearings installed and the preload set, install the differential with adjusters, caps and bearings. Lubricate the bearings and adjuster threads. Tighten the top cap screws on the right and left to 10 ft. lbs. Tighten the bottom cap screws finger tight until the head is just seated on the bearing cap.

4. Using the tool, make sure the adjuster rotates freely. Turn both adjusters in until bearing play is eliminated with some drive gear backlash (0.010 in.). Seat the bearing rollers.

5. Install and register a dial indicator against the drive side of a gear tooth. Check the backlash at 4 positions to find the point of minimum backlash. Rotate the gear to the position of least backlash and mark the tooth so the readings will be taken at the same point.

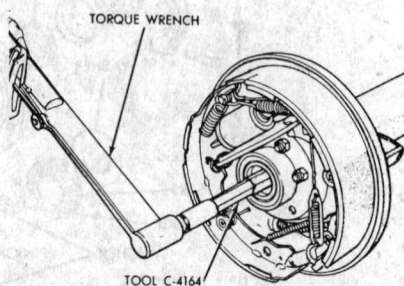

Using a long bar tool with hex end to adjust the differential bearing preload and gear backlash

6. Loosen the right adjuster and turn it until the backlash is 0.003–0.004 in. with each adjuster tightened to 10 ft. lbs. Seat the bearings rollers.

7. Tighten the differential bearing cap screws to:
100 ft. lbs. (136 Nm) – 8¼ in. and 9¼ in axles
45 ft. lbs. (61 Nm) – 7¼ in. axle
70 ft. lbs. (95 Nm) – 8⅜ in axle

8. Tighten the right adjuster to 70 ft. lbs. and seat the rollers, until the torque remains constant at 70 ft. lbs. Measure the backlash, if the backlash is not 0.006–0.008 in. increase the torque on the right adjusters and seat the rollers until the correct backlash is obtained.

9. Tighten the left adjuster to 70 ft. lbs. and seat the bearings until the torque remains constant.

10. If the assembly is properly done, the initial reading on the left adjuster will be approximately 70 ft. lbs. If it is substantially less, the entire procedure should be repeated.

11. After the adjustments are complete, install the adjuster locks; be

sure the teeth are engaged in the adjuster threads. Torque the lockscrews to 90 inch lbs.

SPICER—MODELS 60, 60 HD AND 70 RING GEAR AXLES

Disassembly

DIFFERENTIAL

1. Remove the housing cover and drain the lubricant.

2. Using an axle spreader tool, mount it onto the axle housing and spread the housing enough to remove the differential.

3. Using a dial indicator, measure the amount the opening is being spread; do not spread the housing more than 0.015 in. (0.38mm), for damage to the housing may occur.

4. Mark the differential bearing caps for reassembly purposes.

5. Loosen the bearing caps until 2-3 threads are engaged.

6. Using a prybar, pry the differential loose.

7. Remove the bearing caps and the differential.

8. Mount the differential into a vise.

9. Remove and discard the ring gear bolts; they are not reusable.

10. Using a brass drift and a hammer, tap the ring gear from the differential.

11. Using a differential bearing puller, press the differential bearings from the differential.

12. Remove the differential bearing shims.

13. Remove the pinion shaft lockpin and pinion shaft.

14. Rotate the pinions to remove them through the case opening.

15. Remove the side gears and thrust washers.

PINION GEAR

1. Remove the differential case from the axle housing.

2. Using a pinion yoke holding tool, remove the pinion gear nut.

3. Remove the pinion washer. Using a pinion yoke holder tool and a pinion puller tool, press the yoke from the pinion gear.

4. Using an oil seal remover tool, remove the pinion gear oil seal and discard it. Remove the slinger, the gasket, the upper pinion bearing cone and pre-

load shim pack; measure and record the shim thicknesses.

5. Press drive the pinion gear and inner bearing cone assembly from the axle housing.

6. Using a shop press, press the inner bearing cone from the pinion gear.

7. Remove and record the shim thicknesses from behind the inner pinion bearing cup.

Inspection

1. Clean the differential components in solvent and use compressed air to dry them; do not use compressed air on the bearings, only shop towels.

2. Check the components for wear or damage; replace them, if necessary.

3. Inspect the bearings and bearing cups for wear, cracks or scoring; replace them, if necessary.

4. Inspect the differential side and pinion gears for wear, cracks or chips; replace them, if necessary.

5. Inspect the ring and pinion gears for wear and/or damage; replace them, if necessary.

6. Inspect the differential case for cracks or damage; replace it, if necessary.

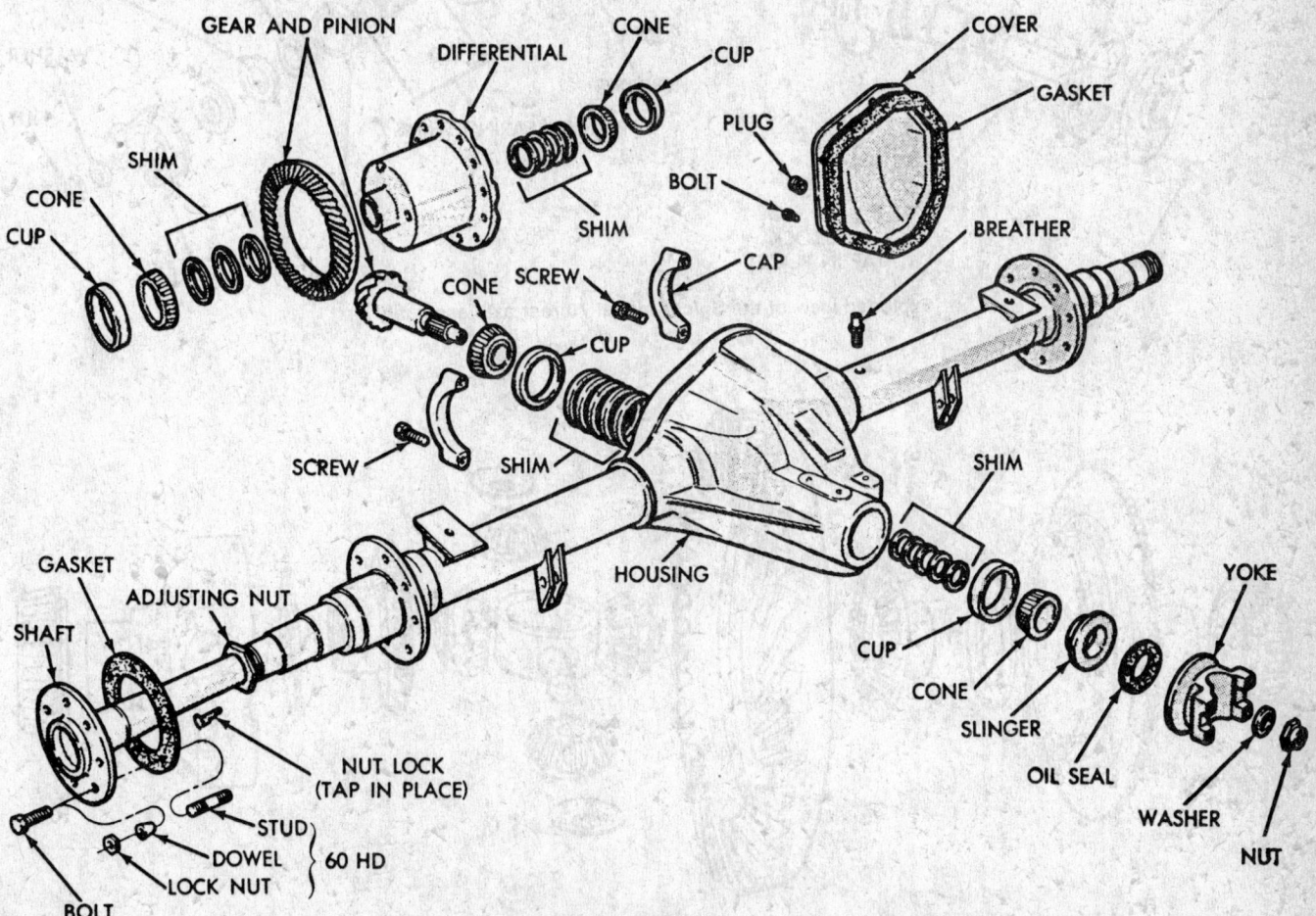

Exploded view of the Spicer 60 and 60 HD rear axle assembly

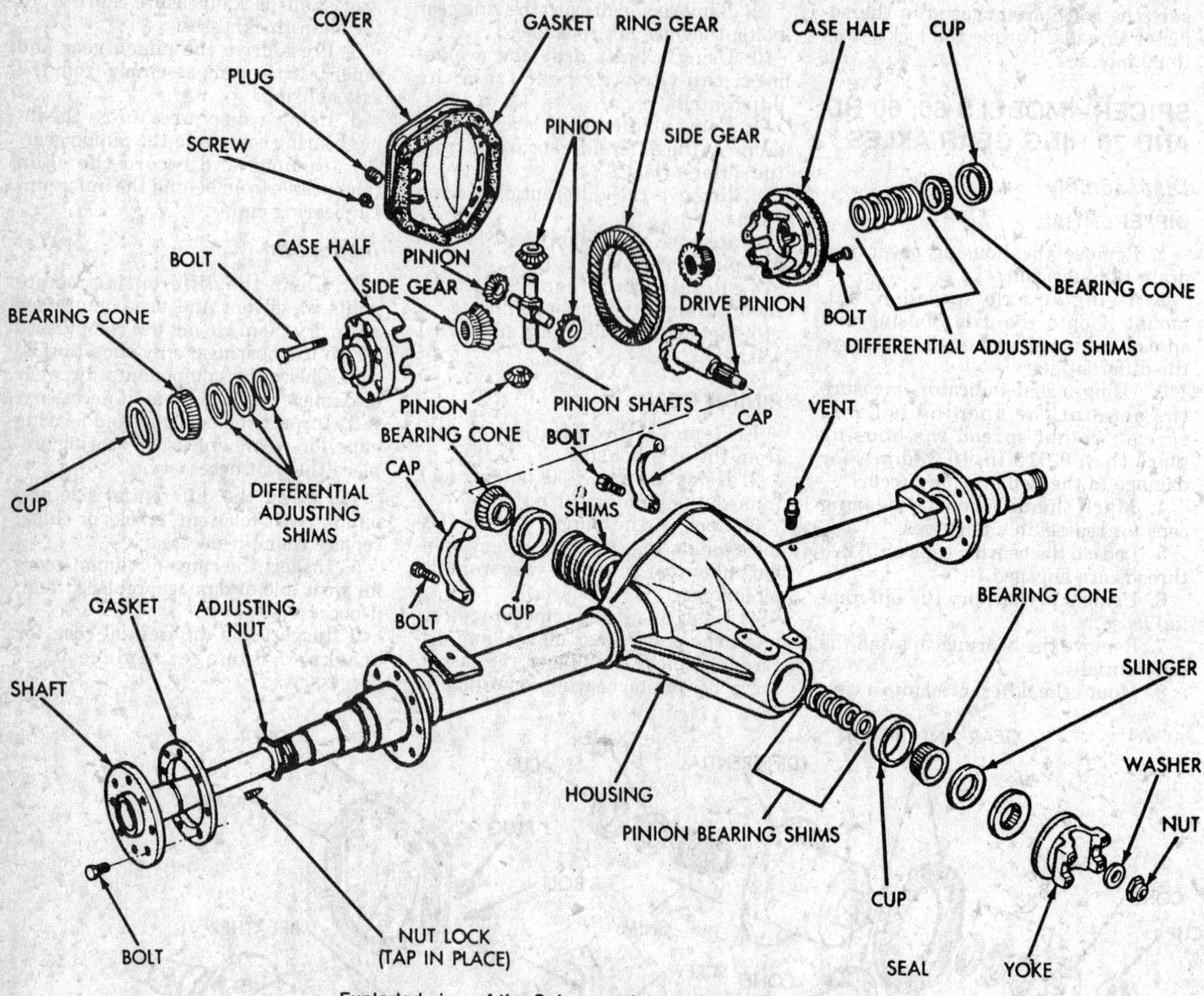

COVER GASKET RING GEAR CASE HALF CUP

PLUG

SCREW PINION SIDE GEAR

BOLT CASE HALF PINION DRIVE PINION

BEARING CONE SIDE GEAR BOLT

DIFFERENTIAL ADJUSTING SHIMS

BEARING CONE

CUP PINION PINION SHAFTS CAP VENT

DIFFERENTIAL ADJUSTING SHIMS

BEARING CONE CAP BOLT CAP

CAP BOLT SHIMS

CUP

BEARING CONE

GASKET ADJUSTING NUT BOLT

SHAFT SLINGER

WASHER

NUT

HOUSING

PINION BEARING SHIMS

NUT LOCK (TAP IN PLACE) CUP

BOLT SEAL YOKE

Exploded view of the Spicer model 70 rear axle assembly

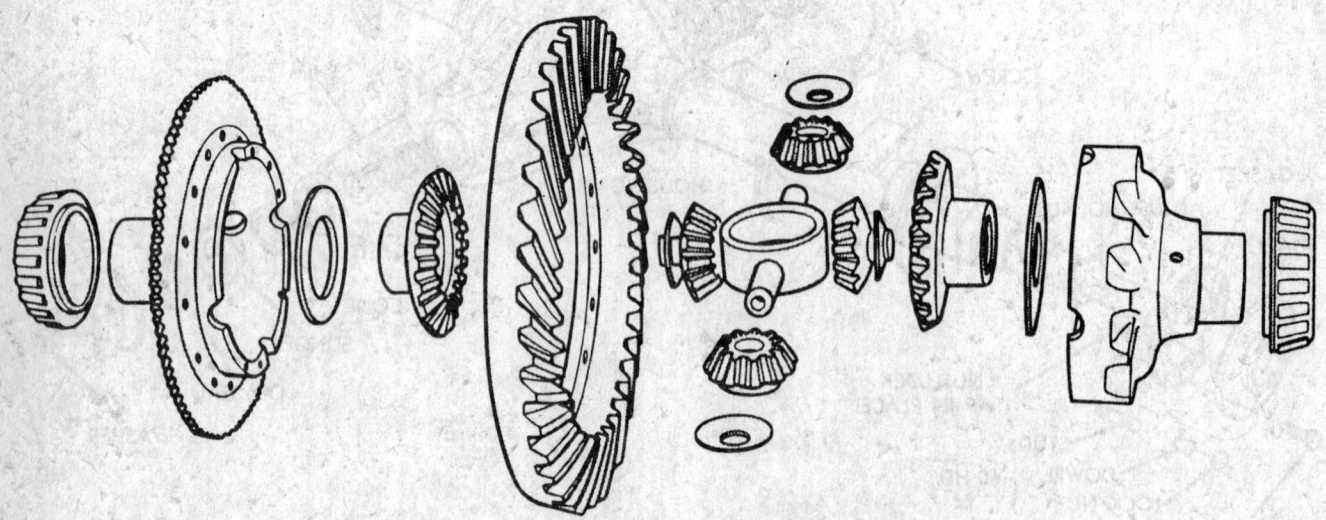

Exploded view of the Spicer model 70 differential

Assembly

PINION GEAR

1. Make sure the ring and pinion are a matched set.

2. Measure the thickness of the original pinion shim and note the variance on the pinion gear.

3. Perform the following procedure to determine the pinion starter shim thickness:

    a. If the original ring and pinion are being installed, use the original shim.

    b. If a replacement gear set is being installed, determine the best starter shim thickness.

    c. Refer to the pinion variance chart and observe where the old and new pinion marking column intersect.

    d. If the old pinion is +2 and the new pinion is −2, the intersecting figure is +0.004 in. (0.10mm); add this amount to the original shim. Or, if the old pinion is −3 and the new pinion is −2, the intersecting figure is −0.001 in. (−0.025mm); subtract this amount from the original shim.

4. Install the starter shim in the pinion rear bearing cup bore; if the shim is chamfer on one side, position the chamfered side so it faces the bottom of the bore.

5. Using a driver tool, install the pinion rear bearing cup into the axle housing.

6. Install the front bearing cup.

7. Using a shop press, press the rear bearing onto the pinion gear.

8. Install the pinion gear into the axle housing and the front bearing onto the pinion; do not install the slinger or the seal.

9. Install the yoke, the pinion nut washer and the old pinion nut; tighten the nut enough to remove the endplay.

10. Adjust the backlash and gear tooth contact.

DIFFERENTIAL

1. Install the side gears, the thrust washers and the pinion gears into the differential case.

NOTE: If new gears and washers are used, it will not be necessary to check the gear backlash. Correct fit is provided due to the close manufacturing tolerances.

2. Install the pinion shaft and lockpin into the case.

3. Assemble the original differential bearing shim packs, then, remove approximately 0.20 in. (0.50mm) shim thickness from each pack; the remaining shims will serve as a starter shim pack.

4. Install the starter shim packs and bearing onto the case.

5. Align and install the ring gear. Using new bolts, torque the ring gear-to-differential case bolts to 100–120 ft. lbs. (136–163 Nm).

6. Install the spreader tool onto the axle housing. Using a dial indicator, spread the axle housing no more than 0.015 in. (0.38mm).

NOTE: Do not spread the housing more than 0.015 in. (0.038mm), for damage may occur to the case.

7. Install the differential; it may be necessary to tap the differential bearing cups, with a soft hammer, to seat them.

8. Install the differential bearing caps and torque the bolts to 70–90 ft. lbs. (95–122 Nm).

9. Remove the spreader tool and dial indicator.

10. Adjust the ring and pinion gear.

11. Using silicone sealant, apply a bead of it to the axle housing. Install the cover and torque the bolts to 35 ft. lbs. (47 Nm).

Adjustment

PINION BEARING PRELOAD

1. Using a yoke holding tool, torque the pinion nut to 250–270 ft. lbs. (339–366 Nm). Rotate the drive pinion several complete revolutions to seat the bearing rollers.

2. Using an inch lb. torque wrench, measure the torque necessary to ro-

tate the pinion; the torque should be 10–20 inch lbs. (1–3 Nm).

NOTE: If the torque is not correct, add a shim to decrease the preload or subtract a shim to increase the preload.

3. Remove the pinion nut, the washer and the yoke after setting the preload; be sure to discard the old nut.

4. Lubricate the new pinion oil seal lip and install the seal into the axle housing.

5. Install the yoke, the washer and a new pinion nut; torque the nut to 250–270 ft. lbs. (339–366 Nm).

DIFFERENTIAL BEARING PRELOAD AND DRIVE GEAR AND PINION BACKLASH

1. Install the differential and tighten the bearing caps.

2. Using a dial indicator, mount it onto the housing, position the stylus against the drive side of one ring gear tooth; be sure the stylus is at a right angle to the tooth.

3. Move the ring gear toward the dial indicator and zero it.

4. Move the ring gear away from the pinion until the backlash and note the reading on the dial indicator.

NOTE: The reading represents the thickness of the shim pack necessary to take up the clearance between the bearing cup and the case on the ring gear side of the differential assembly.

5. Subtract this reading from the previously recorded total reading to obtain the amount of shims necessary to take up the clearance between the bearing cup and the case at the pinion side of the differential.

6. Remove the differential and ring gear assembly from the carrier.

7. Remove the differential bearing covers. Install the correct thickness shim pack between the bearing cone and differential case hub shoulder. Add an additional 0.015 in. (0.38mm) shim to the drive gear side of the dif-

Old Pinion Marking	New Pinion Marking (U.S. Standards)								
	−4	−3	−2	−1	0	+1	+2	+3	+4
+4	+0.008	+0.007	+0.006	+0.005	+0.004	+0.003	+0.002	+0.001	0
+3	+0.007	+0.006	+0.005	+0.004	+0.003	+0.002	+0.001	0	−0.001
+2	+0.006	+0.005	+0.004	+0.003	+0.002	+0.001	0	−0.001	−0.002
+1	+0.005	+0.004	+0.003	+0.002	+0.001	0	−0.001	−0.002	−0.003
0	+0.004	+0.003	+0.002	+0.001	0	−0.001	−0.002	−0.003	−0.004
−1	+0.003	+0.002	+0.001	0	−0.001	−0.002	−0.003	−0.004	−0.005
−2	+0.002	+0.001	0	−0.001	−0.002	−0.003	−0.004	−0.005	−0.006
−3	+0.001	0	−0.001	−0.002	−0.003	−0.004	−0.005	−0.006	−0.007
−4	0	−0.001	−0.002	−0.003	−0.004	−0.005	−0.006	−0.007	−0.008

ferential and install the differential bearing cones.

8. Install the spreader tool to the axle housing. Using a dial indicator, spread the housing to 0.015-0.020 (0.38-0.51mm). Install the differential assembly into the axle housing.

9. Install the bearing caps, remove the spreader tool and torque the bearing caps snugly.

10. Using a soft hammer, tap the drive gear to seat the differential bearing and cups.

NOTE: When seating the bearing and cups, be careful not to nick the ring gear or drive pinion teeth.

11. Torque the bearing cap bolts to 70-90 ft. lbs. (95-122 Nm).

12. Attach a dial indicator to the carrier, with the stylus contacting a ring gear tooth, and measure the backlash between the ring gear and the drive pinion.

13. Check the backlash at 4 equally spaced points around the circumference of the ring gear; the backlash must be between 0.04-0.009 in. (0.102-0.229mm) and cannot vary more than 0.002 in. (0.508mm) between the 4 check positions.

NOTE: If the backlash does not fall within these specifications, change the shim pack thickness on both the differential bearing hubs to maintain the proper bearing preload and backlash.

TRAC-LOK DIFFERENTIAL

Operational Test

If a noisy or rough operation such as a chatter occurs when turning corners, the most probable cause of this chatter or noise is incorrect or contaminated lubricant. Before removing the Trac-Lok unit for repair, drain, flush and refill the axle with the specified lubricant. A complete lubricant drain and

refill with the specified fluid will usually correct the chatter problem. A quick operational test of the Trac-Lok differential can be done easily by performing the following steps.

1. Position one wheel on solid dry pavement and the opposite wheel on ice, mud grease or a similar low traction surface.

2. Gradually, increase the engine rpm to obtain the maximum traction prior to a breakaway. The ability to move the vehicle effectively will demonstrate the proper performance.

NOTE: If the test is performed on extremely slick surfaces such as ice or grease coated surfaces, some question may exist as to proper performance. In these extreme cases, a properly performing Trac-Lok will provide greater pulling power by lightly applying the parking brake.

Disassembly
DIFFERENTIAL

1. Remove the housing cover and drain the lubricant.

2. Using an axle spreader tool, mount it onto the axle housing and spread the housing enough to remove the differential.

3. Using a dial indicator, measure the amount the opening is being spread; do not spread the housing more than 0.015 in. (0.38mm), for damage to the housing may occur.

4. Mark the differential bearing caps for reassembly purposes.

5. Loosen the bearing caps until 2-3 threads are engaged.

6. Using a prybar, pry the differential loose.

7. Remove the bearing caps and the differential.

8. Mount the differential into a vise.

9. Remove and discard the ring gear bolts; they are not reusable.

10. Using a brass drift and a hammer, tap the ring gear from the differential.

11. Using a differential bearing puller, press the differential bearings from the differential.

12. Remove the differential bearing shims.

13. Remove the pinion shaft lockpin and pinion shaft.

14. Rotate the pinions to remove them through the case opening.

15. Remove the side gears and thrust washers.

Inspection

If any member of either clutch pack shows evidence of excessive wear or scoring, the complete clutch pack must be replaced on both sides.

1. Thoroughly, clean each part in solvent.

2. Towel dry bearings or allow them to air dry, do not use compressed air to dry bearings as damage might result. Dry all other parts with compressed air or shop towels. If the parts are not to be assembled immediately, cover them to prevent dust or dirt contamination.

3. Inspect the housing for cracks and sand holes. Replace the housing if it is cracked or porous. Check for burrs and deep scratches or nicks on the gasket and oil seal surfaces. An oil stone or fine tooth file may be used to remove nicks or burrs.

4. Inspect the bearing cup bores for nicks or burrs that may have been created during bearing cup removal.

5. Inspect and clean the axle tubes. Inspect the vent to be sure it is not obstructed.

6. Check housing for bent or loose tubes or other physical damage.

7. Inspect the side gears for worn, cracked or chipped teeth. The gears should fit snugly on the axle shaft splines. Also, inspect the fit of the gears in the differential case bore.

Assembly
DIFFERENTIAL

1. Install the side gears, the thrust washers and the pinion gears into the differential case.

NOTE: If new gears and washers are used, it will not be necessary to check the gear backlash. Correct fit is provided due to the close manufacturing tolerances.

2. Install the pinion shaft and lockpin into the case.

3. Assemble the original differential bearing shim packs, then, remove approximately 0.20 in. (0.50mm) shim thickness from each pack; the remaining shims will serve as a starter shim pack.

4. Install the starter shim packs and bearing onto the case.

5. Align and install the ring gear. Using new bolts, torque the ring gear-to-differential case bolts to 100-120 ft. lbs. (136-163 Nm).

6. Install the spreader tool onto the axle housing. Using a dial indicator, spread the axle housing no more than 0.015 in. (0.38mm).

NOTE: Do not spread the housing more than 0.015 in. (0.038mm), for damage may occur to the case.

7. Install the differential; it may be necessary to tap the differential bearing cups, with a soft hammer, to seat them.

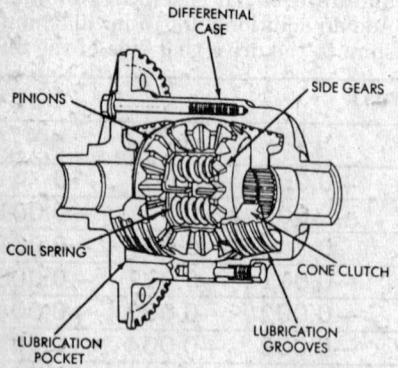

Sectional view of the Spicer model 70 Trak-Lok differential

8. Install the differential bearing caps and torque the bolts to 70–90 ft. lbs. (95–122 Nm).

9. Remove the spreader tool and dial indicator.

10. Adjust the ring and pinion gear.

11. Using silicone sealant, apply a bead of it to the axle housing. Install the cover and torque the bolts to 35 ft. lbs. (47 Nm).

FORD MOTOR COMPANY

Front Axle Assembly

DANA—MODEL 28, 44 AND 50 AXLES

Disassembly

DIFFERENTIAL

1. Remove the left axle arm-to-carrier case bolts, the left arm and drain the lubricant.

2. Remove the right axle stub shaft by performing the following procedures:

    a. Rotate the stub shaft so the open end of the snapring is exposed.

    b. Using 2 prybars, force the snapring from the stub shaft.

    c. Remove the stub shaft from the carrier.

3. Using an axle spreader tool, mount it onto the axle housing and spread the housing enough to remove the differential.

4. Using a dial indicator, measure the amount the opening is being spread; do not spread the housing more than 0.010 in. (0.25mm), for damage to the housing may occur.

5. Mark the differential bearing caps for reassembly purposes.

6. Loosen the bearing caps until 2–3 threads are engaged.

7. Using a prybar, pry the differential loose.

8. Remove the bearing caps and the differential.

9. Mount the differential into a vise.

10. Remove and discard the ring gear bolts; they are not reusable.

11. Using a brass drift and a hammer, tap the ring gear from the differential.

12. Using a differential bearing puller, press the differential bearings from the differential.

13. Remove the differential bearing shims.

14. Remove the pinion shaft lockpin and pinion shaft.

15. Rotate the pinions to remove them through the case opening.

16. Remove the side gears and thrust washers.

PINION GEAR

1. Remove the differential case from the axle carrier housing.

2. Using a pinion yoke holding tool, remove the pinion gear nut.

3. Remove the pinion washer. Using a pinion yoke holder tool and a pinion puller tool, press the yoke from the pinion gear.

4. Using a soft hammer, drive the pinion gear assembly from the axle carrier housing.

5. Using a bearing cup puller tool and a slide hammer, remove the pinion gear oil seal and discard it. Remove the outer pinion bearing and the oil slinger from the carrier input bore.

6. Remove the pinion bearing preload shims.

7. Using a shop press, press the inner pinion bearing cup and baffle from the bore.

8. Rotate the carrier housing. Using a shop press, press the outer pinion bearing cup from the bore.

9. Using a universal bearing remover tool, press the bearing and the oil slinger from the pinion gear.

Inspection

1. Clean the differential components in solvent and use compressed air to dry them; do not use compressed air on the bearings, only shop towels.

2. Check the components for wear or damage; replace them, if necessary.

3. Inspect the bearings and bearing cups for wear, cracks or scoring; replace them, if necessary.

4. Inspect the differential side and pinion gears for wear, cracks or chips; replace them, if necessary.

5. Inspect the ring and pinion gears for wear and/or damage; replace them, if necessary.

6. Inspect the differential case for cracks or damage; replace it, if necessary.

Assembly

PINION GEAR

1. Make sure the ring and pinion are a matched set.

2. Perform the depth gauge check, performing the following procedures:

    a. Using pinion bearing cup replacer tools and the forcing screw from the depth gauge tool set, install the inner and outer pinion cups.

    b. Place a new rear pinion bearing over the aligning adapter and insert it into the pinion bearing retainer assembly. Place the front pinion bearing into the bearing cup in

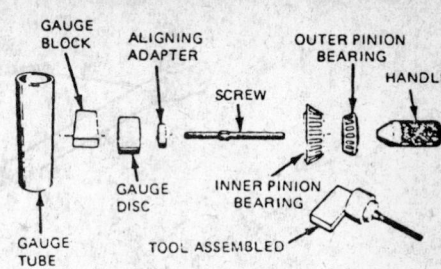

View of the depth gauge tool set

the carrier and assemble the handle onto the screw and hand tighten.

NOTE: The ³⁄₈ in. drive in the handle is to be used for obtaining the proper pinion bearing preload.

    c. Center the gauge tube into the differential bearing bore. Install the bearing caps and tighten the handle until the bearing preload is 20–40 inch lbs. (2.3–4.5 Nm).

    d. Using a feeler gauge or shims, select the thickest feeler shim that will fit between the gauge tube and the gauge block. Insert the feeler gauge or shims directly along the gauge block to insure a correct reading. The feeler gauge, installed between the gauge tube and the gauge block, should have a slight drag feeling.

    e. After the correct shims or feeler gauge thickness is obtained, check the reading, this is the thickness of shim(s) required, provided that upon inspection of the service pinion gear, the button is etched **0**.

NOTE: If the service pinion gear is marked with a (+) plus reading, this amount must be subtracted from the thickness dimension obtained in step d; example: +2 (−0.002 in.). If the service pinion gear is marked with a (−) minus reading, this amount must be added to the thickness dimension obtained in step d; example: −2 (+0.002 in.). Be sure to use the exact same new rear pinion bearing that was used in the previous steps.

    f. Using a micrometer, measure the shims to verify the sizes.

3. Place the oil slinger (if used) onto the pinion. Using a pinion bearing cone installer tool, press the bearing onto the pinion gear.

4. Remove the bearing cup and install the oil baffle (1st) and the required amount of shims into the inner pinion bearing bore.

5. Using the inner pinion bearing cup replacer tool and the forcing screw tool, press the inner pinion bearing

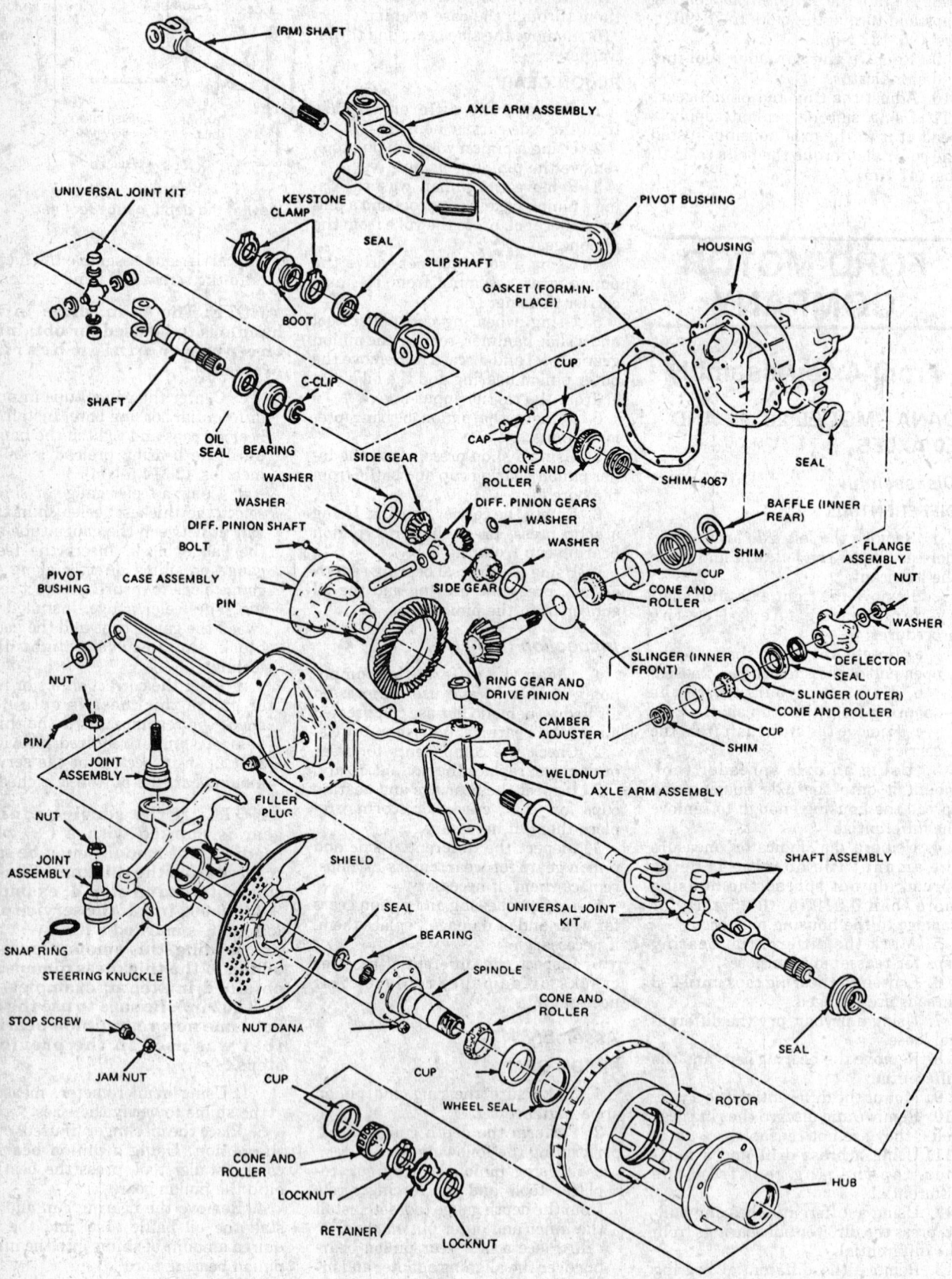

Exploded view of the Dana model 44 front axle assembly—model 50 is similar

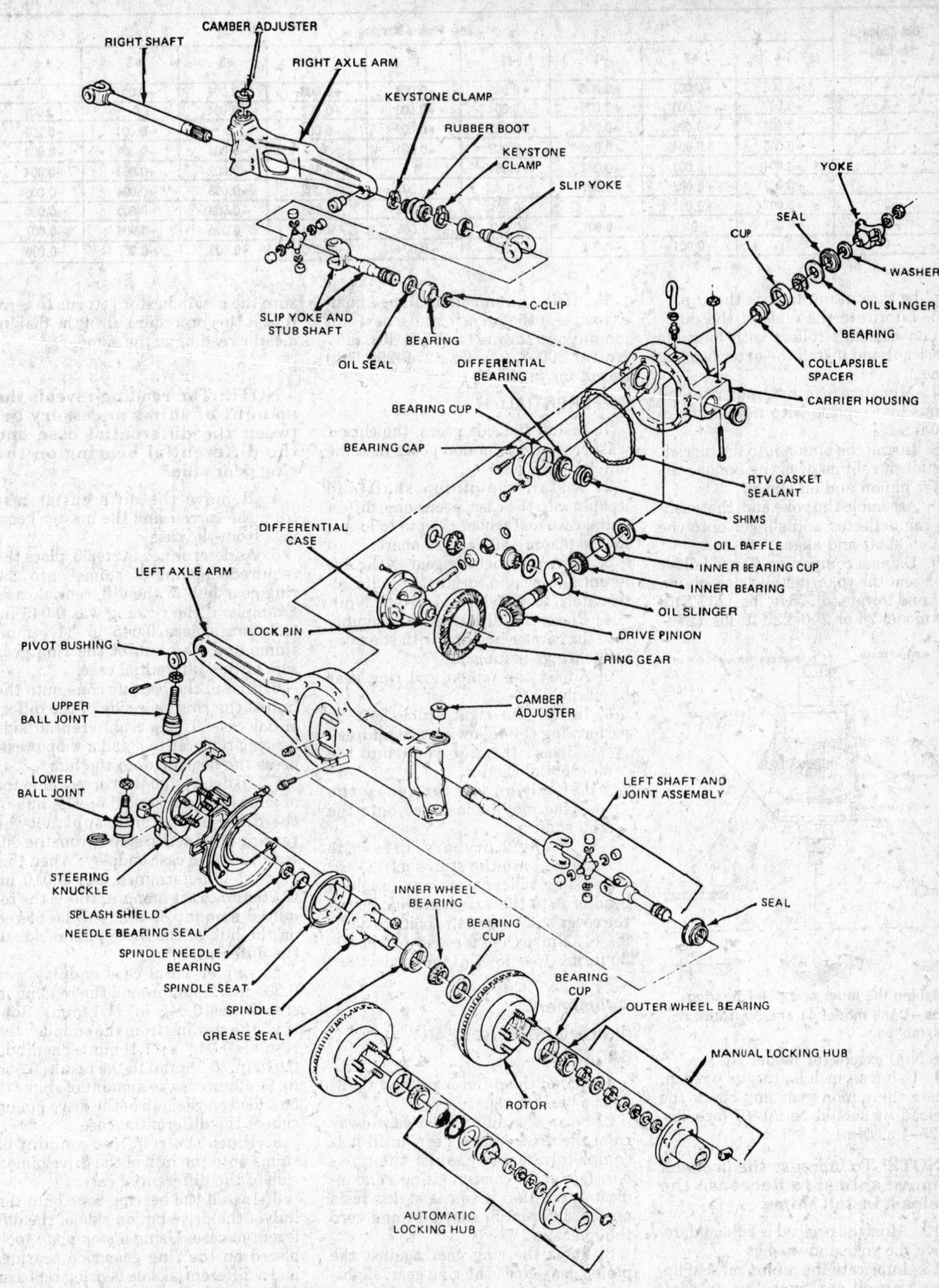

Exploded view of the Dana model 28 front axle assembly

RIGHT SHAFT
CAMBER ADJUSTER
RIGHT AXLE ARM
KEYSTONE CLAMP
RUBBER BOOT
KEYSTONE CLAMP
SLIP YOKE
YOKE
SEAL
CUP
WASHER
OIL SLINGER
BEARING
COLLAPSIBLE SPACER
CARRIER HOUSING
SLIP YOKE AND STUB SHAFT
BEARING
OIL SEAL
C-CLIP
DIFFERENTIAL BEARING
BEARING CUP
BEARING CAP
RTV GASKET SEALANT
SHIMS
OIL BAFFLE
INNER BEARING CUP
INNER BEARING
OIL SLINGER
DRIVE PINION
RING GEAR
DIFFERENTIAL CASE
LEFT AXLE ARM
LOCK PIN
PIVOT BUSHING
UPPER BALL JOINT
CAMBER ADJUSTER
LEFT SHAFT AND JOINT ASSEMBLY
LOWER BALL JOINT
STEERING KNUCKLE
SPLASH SHIELD
NEEDLE BEARING SEAL
SPINDLE NEEDLE BEARING
SPINDLE SEAT
SPINDLE
GREASE SEAL
INNER WHEEL BEARING
BEARING CUP
SEAL
BEARING CUP
OUTER WHEEL BEARING
MANUAL LOCKING HUB
ROTOR
AUTOMATIC LOCKING HUB

Old Pinion Marking	New Pinion Marking								
	−4	−3	−2	−1	0	+1	+2	+3	+4
+4	+0.008	+0.007	+0.006	+0.005	+0.004	+0.003	+0.002	+0.001	0
+3	+0.007	+0.006	+0.005	+0.004	+0.003	+0.002	+0.001	0	−0.001
+2	+0.006	+0.005	+0.004	+0.003	+0.002	+0.001	0	−0.001	−0.002
+1	+0.005	+0.004	+0.003	+0.002	+0.001	0	−0.001	−0.002	−0.003
0	+0.004	+0.003	+0.002	+0.001	0	−0.001	−0.002	−0.003	−0.004
−1	+0.003	+0.002	+0.001	0	−0.001	−0.002	−0.003	−0.004	−0.005
−2	+0.002	+0.001	0	−0.001	−0.002	−0.003	−0.004	−0.005	−0.006
−3	+0.001	0	−0.001	−0.002	−0.003	−0.004	−0.005	−0.006	−0.007
−4	0	−0.001	−0.002	−0.003	−0.004	−0.005	−0.006	−0.007	−0.008

cup; be careful not to cock the cup.

6. Lubricate the ends of the outer pinion bearings rollers with long life lubricant and install the outer bearing cone.

7. Measure the original preload shims and replace with new shims of equal size.

8. Install the pinion into the carrier. Install the shims over the pinion, the outer pinion and oil slinger.

9. Assemble the yoke end, the washer, the deflector and slinger onto the pinion shaft and align.

10. Using a companion flange holder tool, seat the yoke, install a new pinion nut and torque to 175 ft. lbs. (237 Nm) for model 28 or 200–220 ft. lbs. (271–

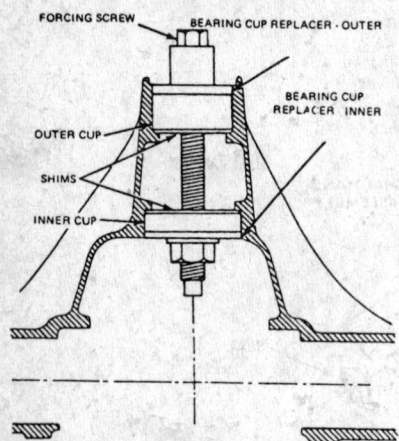

Installing the inner and outer bearing cups — Dana model 44 and 50 front axle assemblies

298 Nm) except for model 28.

11. Using an inch lb. torque wrench, rotate the pinion gear and check the preload; it should be 20–40 inch lbs. (2.25–4.52 Nm).

NOTE: To increase the preload, remove shims; to decrease the preload, install shims.

12. After the preload is adjusted, remove the yoke and washer.

13. Lubricate the pinion oil seal lip. Using an oil seal installation tool, drive a new oil seal into the carrier housing.

14. Using a companion flange holder tool, seat the yoke, install a new pinion nut and torque to 175 ft. lbs. (237 Nm) or 200–220 ft. lbs. (271–298 Nm) except for model 28.

DIFFERENTIAL

1. Install the side gears, the thrust washers and the pinion gears into the differential case.

2. Install the pinion shaft and lockpin into the case; peen some differential case metal over the pin to lock it in two places 180 degress apart.

3. Assemble the ring gear to the differential case and torque the bolts, alternately, to 45–60 ft. lbs. (61–81 Nm).

4. Place the differential assembly into the carrier housing with the master bearings installed.

5. Adjust the pinion and ring gear backlash.

6. Install the right stub shaft by performing the following procedures:

  a. Insert the stub shaft into the differential carrier.

  b. Position the carrier so the snapring may be installed onto the stub shaft.

  c. Using 2 prybars, press the snapring onto the stub shaft.

7. Using silicone sealant, apply a bead of it to the axle housing. Install the cover and torque the bolts to 35 ft. lbs. (47 Nm) except for model 28 or 40–50 ft. lbs. (54–68 Nm) for model 28.

Adjustment

PINION AND RING GEAR BACKLASH

1. Install the differential and tighten the bearing caps.

2. Force the differential case away from the drive pinion gear, until it is completely seated against the cross bore face of the carrier. Using a dial indicator, position it so the stylus rests on the differential case bolt and zero the indicator.

3. Force the ring gear against the pinion gear. Rock the ring gear, slightly, to make sure the gear teeth are in contact. Then, force the ring gear away from the pinion gear, making sure the dial indicator returns to zero; repeat this procedure until the dial indicator reading is the same.

NOTE: The reading reveals the amount of shims necessary between the differential case and the differential bearing on the ring gear side.

4. Remove the differential case from the carrier and the master bearings from the case.

5. As determined in step 3, place the required amount of shims onto the ring gear hub of the differential case. Example: if the reading was 0.045 in. (1.14mm), place 0.045 in. (1.14mm) shims onto the hub of the ring gear side of the differential case.

6. Install the bearing cone onto the hub of the ring gear side of the differential case. Using a differential side bearing replacer tool and a shop press, press the bearing onto the hub.

7. To determine the correct amount of shims to be installed on the hub of the drive pinion side, subtract the reading obtained in step 3 from the differential total case endplay. When this amount is determined, add 0.010 in. (0.26mm) to the amount; this is the required amount of shims to be placed on the hub of the drive pinion side of the differential.

Example: Total case endplay was 0.091 in. (2.30mm) and the reading in step 3 was 0.045 in. (1.14mm). Subtract the reading from the endplay; the result is 0.046 in. (1.16mm). Then, add 0.010 in. (0.26mm) to the result; 0.056 in. (1.42mm) is the amount of shims to be added on the hub of the drive pinion side of the differential case.

8. Place the required amount of shims onto the hub of the drive pinion side of the differential case.

9. Install the bearing cone onto the hub of the drive pinion side of the differential case. Using a step plate tool, placed on the ring gear side bearing, and a differential side bearing replacer tool, drive the bearing onto the hub of the drive pinion side of the differential case.

10. Using a differential bearing replacer tool, install the bearing cone onto the pinion side of the differential case; be sure to position a pinion bearing cone replacer tool on the ring gear bearing to prevent damage to it.

11. Install the differential bearing cups onto the bearing cones.

12. Using a case spreader tool and a dial indicator, spread the carrier housing to 0.015 in. (0.25mm) max.

13. Install the differential case into the carrier; it may be necessary to use a soft hammer to seat the differential case into the carrier case cross bore.

14. Install the bearing caps; make sure the letters or numbers stamped on the caps correspond in both direction and position with the numbers stamped in the carrier. Torque the bolts to 35–40 ft. lbs. (48–54 Nm) for model 28 or 80–90 ft. lbs. (108–122 Nm) except for model 28.

15. Install a dial indicator to the case and check the ring gear backlash at 3 equally spaced points; the backlash should be 0.005–0.009 in. (0.13–0.23mm) and should not vary more than 0.003 in. (0.08mm) between the points.

NOTE: If the backlash is high, the ring gear must be moved closer to the pinion, by moving the shims to the ring gear side from the opposite side. If the backlash is low, the ring gear must be moved away from the pinion by moving the shims from the ring gear side to the opposite side.

16. Check and/or adjust the gear tooth pattern.

17. Install the right stub shaft by performing the following procedures:

a. Insert the stub shaft into the differential carrier.

b. Position the carrier so the snapring may be installed onto the stub shaft.

c. Using 2 prybars, press the snapring onto the stub shaft.

18. Using silicone sealant, apply it to the mating surface of the carrier support arm.

NOTE: Allow silicone sealant 1 hour to cure.

19. Using 2 guide pins, assemble the carrier housing to the carrier support arm and torque the new bolts to 30–40 ft. lbs. (41–54 Nm) except for model 28 or 40–50 ft. lbs. (54–68 Nm) for model 28.

20. Except for model 28, install the support arm tab bolt to the side of the carrier and torque the bolts to 85 ft. lbs. (115–136 Nm).

21. For model 28, assemble the carrier shear bolt and nut, then, torque to 75–95 ft. lbs. (102–129 Nm).

DANA 60 MONOBEAM AXLE

The differential side bearing shims are located between the side bearing cup assembly and the differential case. The axle use inner and outer shims on the pinion gear. The inner shims are used to control the pinion depth in the housing, while the outer shims are used to preload the pinion bearings. The axle uses a solid differential carrier with a removable side and pinion gear shaft.

Disassembly
DIFFERENTIAL CARRIER

1. The axle assembly can be overhauled either in or out of the vehicle. Either way, the free-floating axles must be removed.

2. Drain the lubricant and remove the rear cover and gasket.

3. Matchmark the bearing caps and the housing for reassembly in the same position. Remove the bearing caps and bolts.

4. Using a spreader tool mounted to the carrier housing, spread the housing a maximum of 0.015 in.

NOTE: Do not exceed this measurement. The housing could be permanently damaged. The use of a dial indicator is recommended to prevent over-stretching the housing.

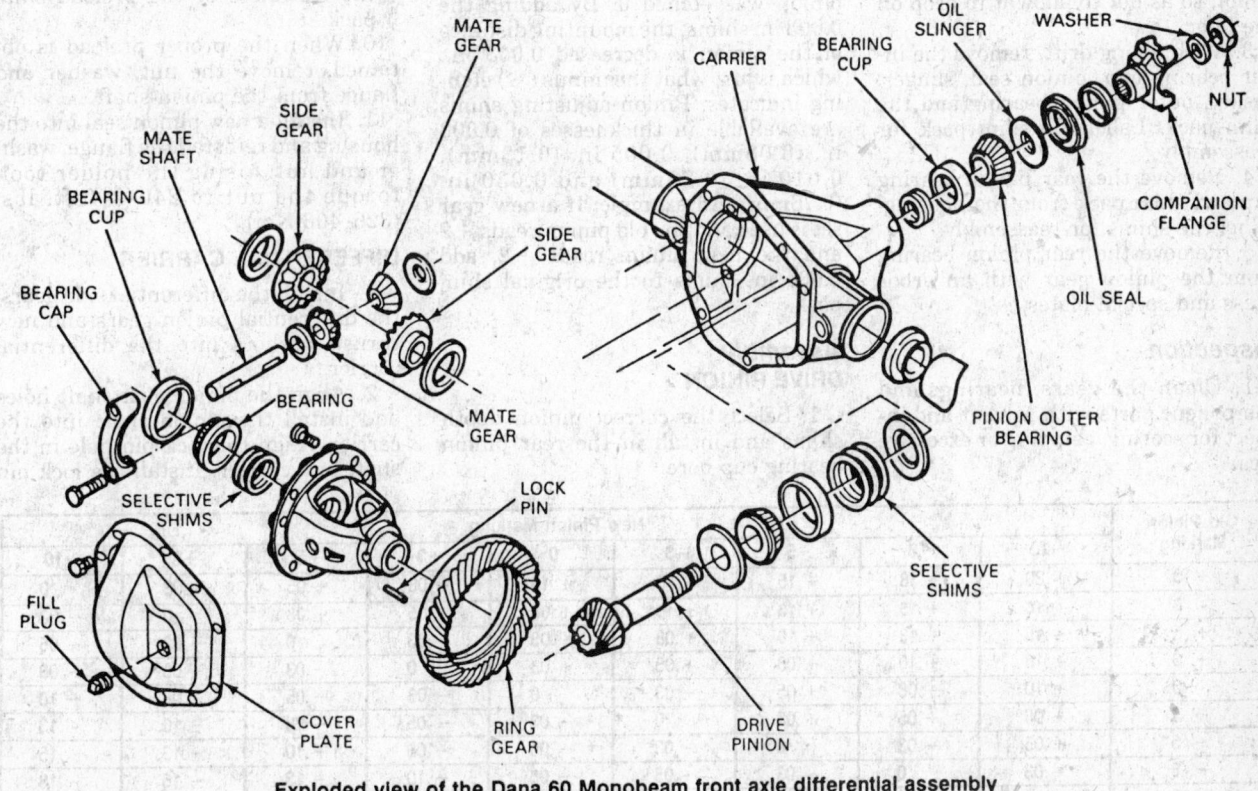

Exploded view of the Dana 60 Monobeam front axle differential assembly

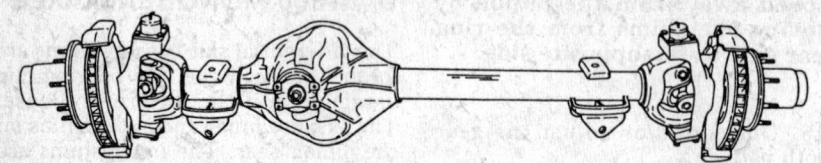

Assembled view of the Dana 60 Monobeam front axle assembly

5. Using a pry bar, remove the differential case from the housing. Remove the spreader tool from the housing.

6. Remove the differential side bearing cups and tag to identify the side, if they are to be used again.

7. Remove the differential gear pinion shaft lock pin and remove the shaft. Rotate the side and pinion gears to remove them from the carrier. Remove the thrust bearings.

8. Remove the bearing cones and rollers from the carrier, marking and noting the shim locations.

9. Remove the ring gear bolts and tap the ring gear from the carrier housing.

10. Inspect the components.

DRIVE PINION

1. Remove the pinion nut and flange from the pinion gear.

2. Remove the pinion gear assembly from the housing. It may be necessary to tap the pinion from the housing with a soft faced hammer. Catch the pinion so as not to allow it to drop on the floor.

3. With a long drift, remove the inner bearing cup, pinion seal, slinger, gasket, outer pinion bearing and the shim pack. Label the shim pack for reassembly.

4. Remove the rear pinion bearing cup and shim pack from the housing. Label the shims for reassembly.

5. Remove the rear pinion bearing from the pinion gear with an arbor press and special plates.

Inspection

1. Clean the gears, bearings and component parts with solvent and inspect for scoring, chipping or excessive wear.

2. Inspect the flanges and splines for excessive wear.

3. Replace the necessary parts as required.

PINION SHIM SELECTION

Ring gears and pinions are supplied in matched sets only. The matched numbers are etched on both gears for verification. On the rear face of the pinion, a plus (+) or a minus (–) number will be etched, indicating the best running position for each particular gear set. This dimension is controlled by the shimming behind the inner bearing cup. Whenever baffles or oil slingers are used, they become part of the adjusting shim pack. An example: If a pinion is etched + 3, this pinion would require 0.003 in. less shims than a pinion etched 0. This means by removing shims, the mounting distance of the pinion is increased by 0.003 in., which is just what a plus (+) etching indicates. If a pinion is etched – 3, it would be necessary to add 0.003 in. more shims than would be required if the pinion was etched 0. By adding the 0.003 in. shims, the mounting distance of the pinion is decreased 0.003 in., which is just what the minus (–) etching indicates. Pinion adjusting shims are available in thicknesses of 0.003 in. (0.08mm), 0.005 in. (0.13mm), 0.010 in. (0.25mm) and 0.030 in. (0.76mm). An example: If a new gear set is used and the old pinion reads + 2 and the new pinion reads – 2, add 0.004 in. shims to the original shim pack.

Assembly
DRIVE PINION

1. Select the correct pinion depth shims and install in the rear pinion bearing cup bore.

2. Install the rear bearing cup in the axle housing.

3. Add or subtract an equal amount of shim thickness to or from the preload or outer shim pack, as was added or subtracted from the inner shim pack.

4. Install the front pinion bearing cup into its bore in the axle housing.

5. Press the rear pinion bearing onto the pinion gear shaft and install the pinion gear with bearing into the axle housing.

6. Install the preload shims and the front pinion bearing; do not install the oil seal at this time.

7. Install the flange with the holding bar tool attached, the washer and the nut on the pinion shaft end. Torque the nut to 240–300 ft. lbs. (326–406 Nm).

8. Remove the holding bar from the flange and with an inch lb. torque wrench, measure the rotating torque of the pinion gear. The rotating torque should be 10–20 inch lbs. with the original bearings or 20–40 inch lbs. with new bearings. Disregard the torque reading necessary to start the shaft to turn.

9. If the preload torque is not in specifications, adjust the shim pack as required.

   a. To increase preload, decrease the thickness of the preload shim pack.

   b. To decrease preload, increase the thickness of the preload shim pack.

10. When the proper preload is obtained, remove the nut, washer and flange from the pinion shaft.

11. Install a new pinion seal into the housing and reinstall the flange, washer and nut. Using the holder tool, torque the nut to 240–300 ft. lbs. (326–406 Nm).

DIFFERENTIAL CARRIER

1. Install the differential side gears, the differential pinion gears and new thrust washers into the differential carrier.

2. Align the pinion gear shaft holes and install the pinion shaft into the carrier. Align the lock pin hole in the shaft and carrier. Install the lock pin

Old Pinion Marking	New Pinion Marking								
	– 10	– 8	– 5	– 3	0	+ 3	+ 5	+ 8	+ 10
+ 10	+ .20	+ .18	+ .15	+ .13	+ .10	+ .08	+ .05	+ .03	0
+ 8	+ .18	+ .15	+ .13	+ .10	+ .08	+ .05	+ .03	0	– .03
+ 5	+ .15	+ .13	+ .10	+ .08	+ .05	+ .03	0	– .03	– .05
+ 3	+ .13	+ .10	+ .08	+ .05	+ .03	0	– .03	– .05	– .08
0	+ .10	+ .08	+ .05	+ .03	0	– .03	– .05	– .08	– .10
– 3	+ .08	+ .05	+ .03	0	– .03	– .05	– .08	– .10	– .13
– 5	+ .05	+ .03	0	– .03	– .05	– .08	– .10	– .13	– .15
– 8	+ .03	0	– .03	– .05	– .08	– .10	– .13	– .15	– .18
– 10	0	– .03	– .05	– .08	– .10	– .13	– .15	– .18	– .20

and peen the hole to avoid having the pin drop from the carrier.

3. Install the differential case side bearings with the proper installation tools. Do not install the shims at this time.

4. Place the carrier assembly into the axle housing with the bearing cups on the bearing cones. Install the bearing caps in their original position and tighten the bearing cap bolts enough to keep the bearing caps in place.

5. Install a dial indicator on the housing so the indicator button contacts the carrier flange. Press the differential carrier to prevent sideplay and center the dial indicator. Rotate the carrier and check the flange for run-out. If the run-out is greater than 0.002 in., the defect is probably due to the bearings or to the carrier and should be corrected.

6. Remove the assembly and install the ring gear. Torque the retaining bolts and reinstall the assembly into the housing. Install the bearing caps in their original position and tighten the cap bolts to keep the bearings caps in place.

7. Install the dial indicator and position the indicator button to contact the ring gear back surface. Rotate the assembly and the run-out should be less than 0.002 in. If over 0.002 in., remove the assembly and relocate the ring gear 180 degrees. Reinstall the assembly and recheck. If the run-out remains over the 0.002 in. tolerance, the ring gear is defective. If the measurement is within tolerances, continue on with the assembly.

8. Position 2 pry bars between the bearing cap and the housing on the side opposite the ring gear. Pull on the pry bars and force the differential carrier as far as possible towards the dial indicator. Rock the assembly to seat the bearings and reset the dial indicator to zero.

9. Reposition the prybars to the opposite side of the carrier and force the carrier assembly as far towards the center of the housing. Read the dial indicator scale. This will be the total amount of shims required for setting the backlash during the reassembly, less the bearing preload. Record the measurement.

10. With the pinion gear installed and properly set, position the differential carrier assembly into the axle housing and install the bearing caps in their proper positions. Tighten the cap bolts just to hold the bearing cups in place.

11. Install a dial indicator on the axle housing with the indicator button contacting the back of the ring gear.

12. Position 2 prybars between the

bearing cup and the axle housing on the ring gear side of the case and pry the ring gear into mesh with the pinion gear teeth, as far as possible. Rock the ring gear to allow the teeth to mesh and the bearings to seat. With the pressure still applied by the prybars, set the dial indicator to zero.

13. Reposition the prybars on the opposite side of ring gear and pry the gear as far as it will go. Take the dial indicator reading. Repeat this procedure until the same reading is obtained each time. This reading represents the necessary amount of shims between the differential carrier and the bearing on the ring gear side.

14. Remove the bearing from the differential carrier on the ring gear side and install the proper amount of shims. Reinstall the bearing.

15. Remove the differential carrier bearing from the opposite side of the ring gear. To determine the amount of shims needed, use the following method.

   a. Subtract the size of the shim pack just installed on the ring gear side of the carrier from the reading obtained and recorded when measurement was taken without the pinion gear in place.

   b. To this figure, add an additional 0.015 in. to compensate for preload and backlash. An example: If the first reading was 0.085 in. and the shims installed on the ring gear side of the carrier were 0.055 in., the correct amount of shims would be $0.085 - 0.055 + 0.015 = 0.045$ in.

16. Install the required shims as determined under step 15 and install the differential side bearing. The installation of the shims should give the proper preload to the bearings and the proper backlash to the ring and pinion gears.

17. Spread the axle housing with the spreader tool no more than 0.015 in. Install the differential bearing outer cups in their correct locations and install the cups in their respective locations.

18. Install the bolts and tighten finger-tight. Rotate the differential carrier and ring gear and tap with a soft-faced hammer to insure proper seating of the assembly in the axle housing.

19. Remove the spreader tool and torque the cap bolts to 80–90 ft. lbs. (108–122 Nm).

20. Install a dial indicator and check the ring gear backlash at 4 equally spaced points of the ring gear circle. The backlash must be within a range of 0.004–0.009 in. and must not vary more than 0.003 in. between the points checked.

21. If the backlash is not within

specifications, the shim packs must be corrected to bring the backlash within limits.

22. Check the tooth contact pattern and verify.

23. Install the cover and torque the bolts to 30–40 ft. lbs. (41–54 Nm). Refill to proper level with lubricant and operate to verify proper assembly.

Rear Axle Assembly

FORD – 10¼ AND 8.8 INCH RING GEAR AXLES

Disassembly
DIFFERENTIAL CARRIER

1. Remove the cover and clean the lubricant from the internal parts.

2. Using a dial indicator, measure and record the ring gear backlash and the runout; the backlash should be 0.008–0.015 in. and the runout should be less than 0.004 in.

3. Mark 1 differential bearing cap to ensure it is installed its original position.

4. Loosen the differential bearing cap bolts.

5. Using a prybar, pry the differential carrier until the bearing cups and shims are loose in the bearing caps.

6. Remove the bearing caps and the differential assembly.

7. If necessary, remove ring gear-to-differential case bolts. Using a hammer and a punch, strike the alternate bolt holes around the ring gear to dislodge it from the differential.

8. If necessary, remove the excitor ring by striking it with a soft hammer.

9. Remove the pinion shaft lock bolt from the differential case. Remove the differential pinion shaft, the pinion gears and the thrust washers.

10. Remove the side gears and thrust washers.

11. Using a bearing puller tool, press the bearings from the differential carrier.

PINION GEAR

1. Remove the differential carrier assembly.

2. Using a companion flange holding tool, remove the companion flange nut.

3. Using a puller tool, press the companion flange from the pinion gear.

4. Using a soft hammer, drive the pinion gear from the housing.

5. Using a prybar, remove the pinion gear oil seal from the housing.

6. Remove the oil slinger and the front pinion bearing.

7. Using a shop press, press the

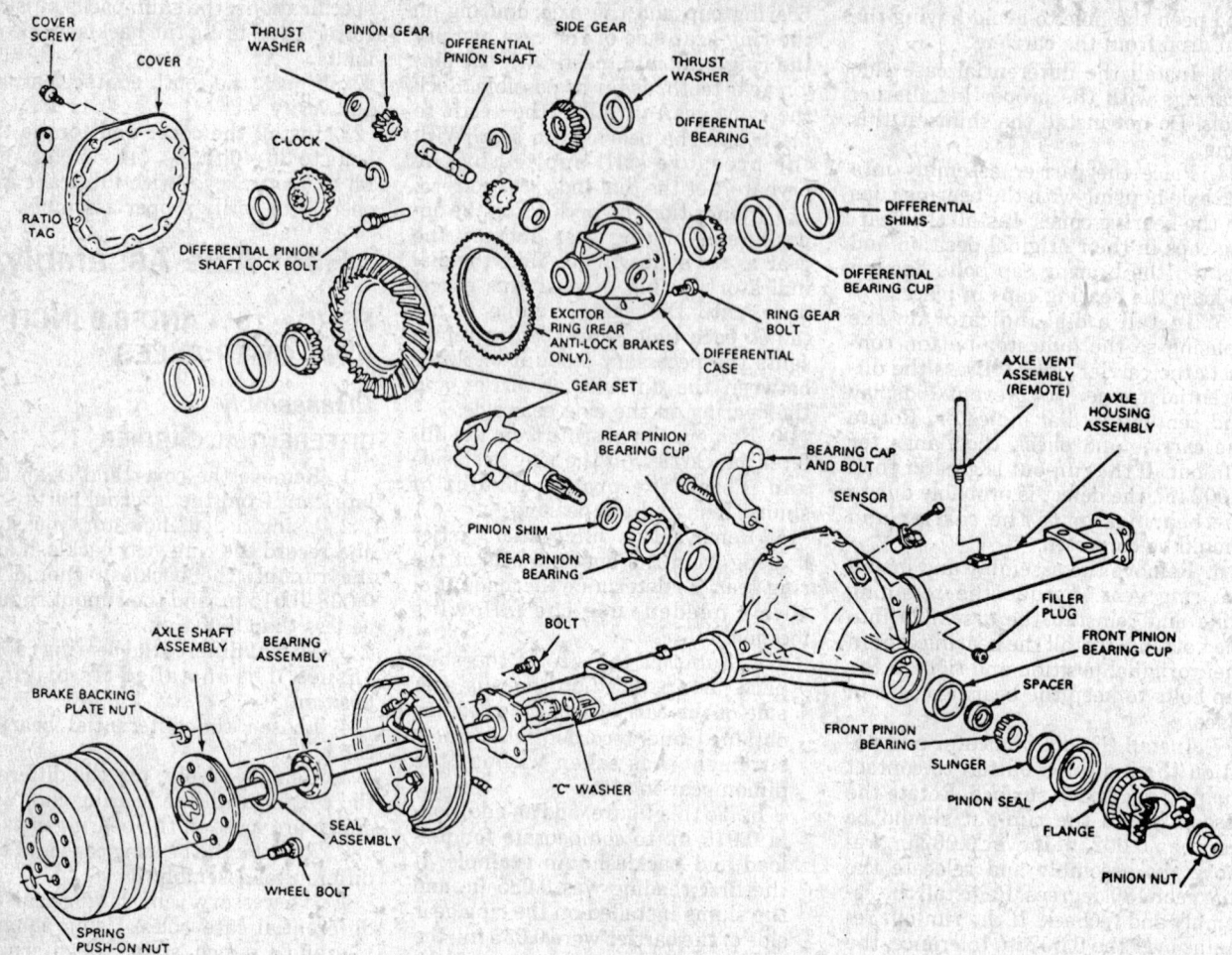

Exploded view of the 8.8 inch ring gear axle—10.25 inch is similar

bearing cone from from the pinion gear.

8. Remove and record the shim from the pinion gear.

Inspection

1. Clean the differential components in solvent and use compressed air to dry them; do not use compressed air on the bearings, only shop towels.

2. Check the components for wear or damage; replace them, if necessary.

3. Inspect the bearings and bearing cups for wear, cracks or scoring; replace them, if necessary.

4. Inspect the differential side and pinion gears for wear, cracks or chips; replace them, if necessary.

5. Inspect the ring and pinion gears for wear and/or damage; replace them, if necessary.

6. Inspect the differential case for cracks or damage; replace it, if necessary.

Assembly

PINION GEAR

NOTE: When replacing the ring and pinion gear, the correct shim thickness for the new gear set to be installed is determined by following procedure using a pinion depth gauge tool set.

1. Assemble the appropriate aligning adapter, the gauge disc and gauge block to the screw.

2. Place the rear pinion bearing over the aligning tool and insert it into the rear portion of the bearing cup of the carrier. Place the front bearing into the front bearing cup and assemble the tool handle into the screw. Roll the assembly back and forth a few times to seat the bearings while tightening the tool handle, by hand, to 20 ft. lbs. (27 Nm).

NOTE: The gauge block must be offset 45 degrees to obtain an accurate reading.

3. Center the gauge tube into the differential bearing bore. Install the bearing caps and tighten the bolts to 70–85 ft. lbs. (96–115 Nm); be sure to install the caps with the triangles pointing outward.

4. Place the selected shim(s) on the pinion and press the pinion bearing cone and roller assembly until it is firmly seated on the shaft, using the pinion bearing cone replacer and the axle bearing/seal plate.

5. Place the collapsible spacer on the pinion stem against the pinion stem shoulder.

6. Install the front pinion bearing and oil slinger in the housing bore and install the pinion seal on the pinion seal replacer. Using a hammer, install the seal until it seats.

7. From the rear of the axle housing, install the drive pinion assembly into the housing pinion shaft bore.

8. Lubricate the pinion shaft splines and install the companion flange.

9. Using a companion flange holder tool, torque the pinion nut to 160 ft. lbs. (217 Nm); rotate the pinion gear, occasionally, to ensure proper bearing seating.

10. Using an inch pound torque wrench, frequently, measure the pinion bearing preload; it should be 9–14 inch lbs. for used bearings or 16–29 inch lbs. for new bearings.

NOTE: If the preload is higher than the specification, tighten to the original reading as recorded; never back off the pinion nut.

DIFFERENTIAL CARRIER

1. If the bearings were removed, use a shop press to press them onto the differential case.

2. Install the side gears and thrust washers. Install the pinion gears, the thrust washer, the pinion shaft and the pinion shaft lock bolt.

3. Using a shop press, align the excitor ring tab with the differential case slot and press the ring gear and excitor ring onto the differential case. Install the ring gear-to-differential case bolts and torque them to 100–120 ft. lbs. (135–162 Nm).

4. Place the differential case with the bearing cups into the housing.

5. On the left side, install a 0.265 in shim. Install the bearing cap and tighten the bolts finger tight.

6. On the right side, install progressively larger shims until the largest can be installed by hand. Install the bearing cap.

7. Torque the bearing cap-to-housing bolts to 70–85 ft. lbs. (95–115 Nm).

8. Rotate the assembly to ensure free rotation.

9. Adjust the ring gear backlash.

Adjustment
RING GEAR AND PINION BACKLASH

1. Using a dial indicator, measure the ring gear and pinion backlash; it should be 0.008–0.015 in. If the backlash is 0.001–0.007 in. or greater than 0.015 in. proceed to step 3. If the backlash is zero, proceed to step 2.

Backlash Change Required	Thickness Change Required	Backlash Change Required	Thickness Change Required
.001	.002	.009	.012
.002	.002	.010	.014
.003	.004	.011	.014
.004	.006	.012	.016
.005	.006	.013	.018
.006	.008	.014	.018
.007	.010	.015	.020
.008	.010		

2. If the backlash is zero, add 0.020 in. shim(s) to the right side and subtract a 0.020 in. shim(s) from the left side.

3. If the backlash is within specification, go to step 7. If the backlash is 0.001–0.007 in. or greater than 0.015 in., increase the thickness of a shim on 1 side and decrease the same thickness of another shim on the other side, until the backlash comes within range.

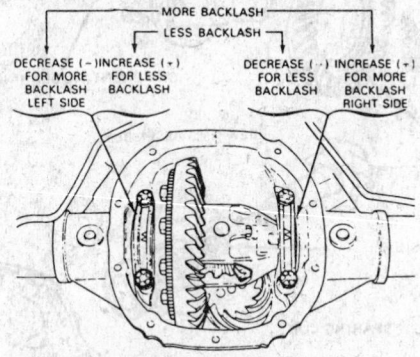

New of te backlash adjustment by changing shims

4. Install and torque the bearing cap bolts to 80–95 ft. lbs. (109–128 Nm).

5. Rotate the assembly several times to ensure proper seating.

6. Recheck the backlash, if it is not within specification, go to step 7.

7. Remove the bearing caps. Increase the shim sizes, on both sides by 0.006 in.; make sure the shims are fully seated and the assembly turns freely. Use a shim driver to install the shims.

8. Install the bearing caps and torque the bearing caps to 80–95 ft. lbs. (109–128 Nm). Recheck the backlash; if not to specification, repeat this entire procedure.

FORD—7½ INCH RING GEAR AXLE

Disassembly
DIFFERENTIAL CASE

1. Check and record the ring gear runout and backlash.

2. Mark on differential bearing cap to help position the caps properly during assembly.

3. Loosen the differential bearing cap bolts and bearing caps.

NOTE: The direction of arrows on bearing caps must be noted. When reassembled, the arrows must be pointing in the same direction as before removal.

4. Pry the differential case, bearing cups and shims out until they are loose in the bearing caps. Remove the bearing caps and the differential assembly from the carrier.

If the ring is removed, discard the bolts. Install new bolts, coated with Loctite® or equivalent. Tighten to 70–85 ft. lbs. (95–115 Nm).

DRIVE PINION

1. Hold the rear axle companion flange with the proper tool and remove the pinion nut.

2. Remove the companion flange. With a soft-faced hammer, drive the pinion out of the front bearing cone and remove it through the rear of the carrier casting.

3. Remove the drive pinion oil seal with tool 1125–AC and T50T–100–A or their equivalent. Remove the front pinion bearing cone, the roller and slinger from the housing.

4. To remove the pinion rear bearing cone, use tool T71P–4621–B or equivalent. Measure the shim found under the bearing cone with a micrometer. Record the thickness of the shim.

NOTE: Before assembling the rear bearing cone to the pinion, it will be necessary to adjust pinion depth.

Inspection

1. Clean the differential components in solvent and use compressed air to dry them; do not use compressed air on the bearings, only shop towels.

2. Check the components for wear or damage; replace them, if necessary.

3. Inspect the bearings and bearing cups for wear, cracks or scoring; replace them, if necessary.

4. Inspect the differential side and pinion gears for wear, cracks or chips; replace them, if necessary.

5. Inspect the ring and pinion gears for wear and/or damage; replace them, if necessary.

6. Inspect the differential case for cracks or damage; replace it, if necessary.

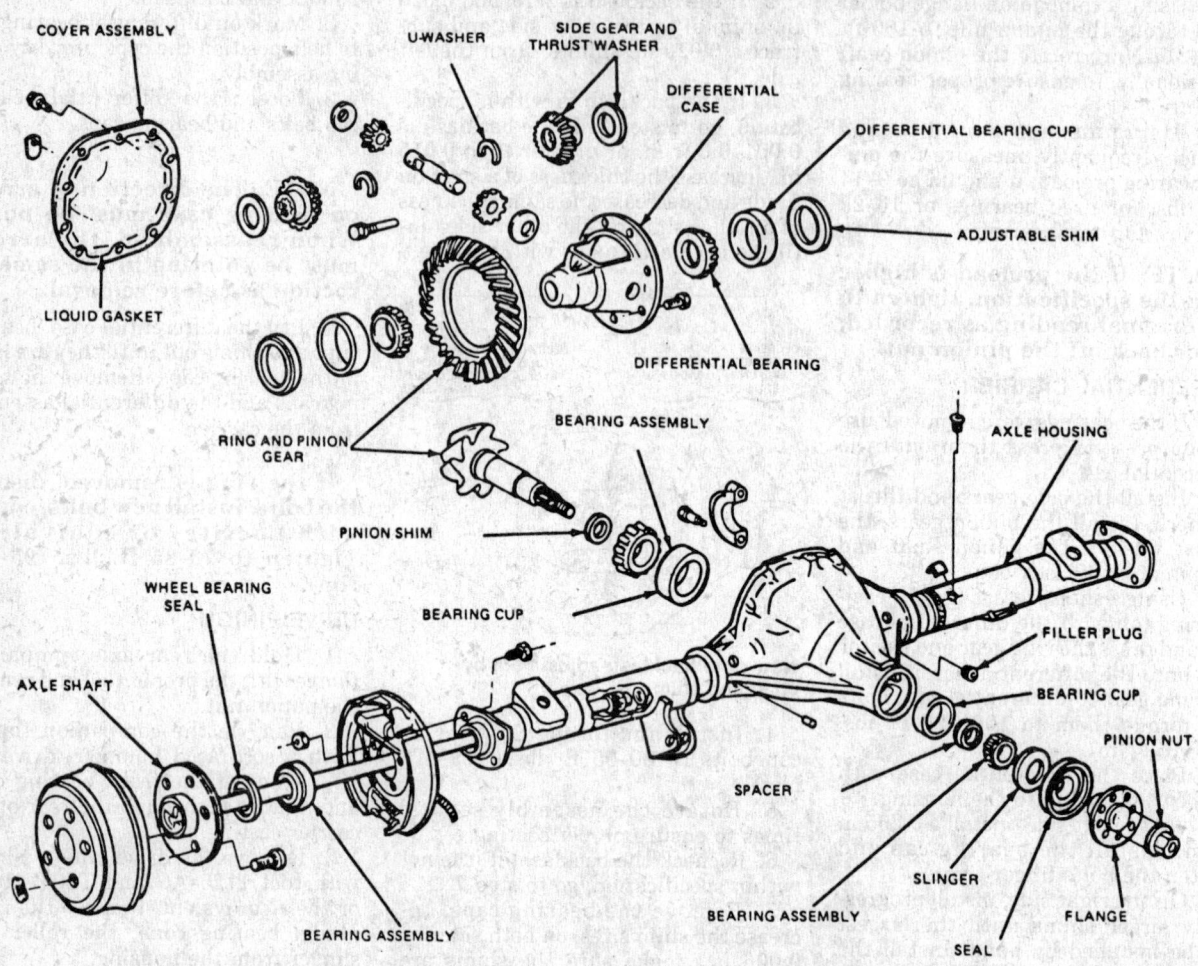

Exploded view of the 7½ inch ring gear axle assembly—Ford Motor Company

Assembly

DRIVE PINION

1. Place the selected shim(s) on the pinion shaft an depress the pinion bearing until firmly seated on the shaft.

2. Place the rear pinion bearing (new or used if in good condition) over the aligning disc and insert it into the pinion bearing cup of the carrier. Place the front bearing into the front bearing cup and assemble the tool handle into the screw and tighten to 20 ft. lbs. (27 Nm).

NOTE: The gauge block must be offset to obtain an accurate reading.

3. Center the gauge tube into the differential bearing bore. Install the bearing caps and torque the bolts to specification; the caps must be installed with the arrows point outboard.

4. Make sure the gauge handle adapter screw, aligning adapter, gauge disc and gauge block assembly are se-curely mounted between front and rear bearing. Recheck tool handle torque prior to gauging to ensure the bearings are properly seated. This can affect final shim selection when im-properly assembled. Clean bearing cups and differential pedestal surfaces thoroughly. Apply only light oil film on bearing assemblies prior to gauging.

5. Gauge block should then be ro-tated several ½ turns to ensure rollers are properly seated in bearing cups. Rotational torque on the gauge assem-bly should be 20 inch lbs. with new bearings. Final position should be ap-proximately 45 degrees across the gauge tube to ensure the gauge block is aligned with gauge tube high point. This area should be utilized for pinion shim selection. Selection of pinion shim with gauge block not aligned with tube high point will cause im-proper shim selection and may result in axle noise.

6. Utilize pinion shims as the gauge for shim selection. This will minimize errors in attempting to stack feeler gauge stock together or simple addi-tion errors in calculating correct shim thickness.

NOTE: Shims must be flat. Do not use dirty, bent, nicked or mu-tilated shims as a gauge.

7. It is important to utilize a light drag on the shim for the correct selec-tion. Do not attempt to force the shim between the gauge block and gauge tube. This will minimize selection of a shim thicker than required which re-sults in a deep tooth contact in final as-sembly for integral axles.

8. If the pinion has a plug (+) marking, subtract his amount from the feeler gauge measurement. If the pinion has a minimum (−) marking, add this amount to the feeler gauge measurement.

DIFFERENTIAL CASE

For shim selection after a complete re-placement of the rear axle housing, the differential assembly or differential side bearings use the following in-structions. For a ring and pinion re-placement only or a backlash adjustment, follow Steps 9 through 13 and Step 15, using the side bearing shims that were originally in the axle.

1. With pinion depth set and pinion installed, place differential case gear assembly with bearings and cups into the carrier.

2. Install a 0.265 in. (6.73mm) shim on left side.

3. Install left bearing cap and tighten bolts finger tight.

4. Install progressively larger shims on the right side until the largest shim selected can be assembled with a slight drag feel.

NOTE: Apply pressure towards the left side to ensure bearing cup is seated.

5. Install right side bearing cap and tighten bearing cup bolts to 70–85 ft. lbs. (95–115 Nm).

6. Rotate assembly to ensure free rotation.

7. Check the ring gear and pinion backlash. If the backlash is 0.008–0.015 in. (0.20–0.38mm) with 0.012–0.015 in. (0.304–0.381mm) preferred, proceed to Step 14. If backlash is not within specifications, go to Step 10, unless zero backlash is measured, then, go to Step 8.

8. If a zero backlash conditions occurs, add 0.020 in. to the right side and subtract 0.020 in. from the left side to allow backlash indication.

9. Recheck backlash.

10. If backlash is not to specification, correct backlash by increasing thickness of one shim and decreasing thickness on the other shim the same amount.

11. Install shim and bearing caps. Tighten cap bolts to 70–85 ft. lbs. (95–115 Nm).

12. Rotate assembly several times.

13. Recheck backlash. If backlash is within specification, go to Step 14. If backlash is not within specification, repeat Step 10. Backlash specification is 0.008–0.015 in. (0.20–0.38mm). Preferred range is 0.012–0.015 in. (0.304–0.381mm).

14. Increase both left and right shim sizes by 0.006 in. and install for correct differential bearing preload; make sure shims are fully seated and assembly turns freely.

15. Utilize white marking compound to obtain a tooth mesh contact pattern in the assembly.

NOTE: Pattern inspection is intended to detect gross errors in set up prior to complete reassembly. Pattern contact should be within the primary area of the ring gear tooth surface avoiding any narrow or hard contact with outer perimeter of tooth (top to root, toe to heel). Pattern inspec-

SHIM CODE CHART

NUMBER OF STRIPES AND COLOR CODE	DIM.
2 – C-COAL	.3070-.3075
1 – C-COAL	.3050-.3055
5 – BLU	.3030-.3035
4 – BLU	.3010-.3015
3 – BLU	.2990-.2995
2 – BLU	.2970-.2975
5 – PINK	.2930-.2935
4 – PINK	.2910-.2915
3 – PINK	.2890-.2895
2 – PINK	.2870-.2875
1 – PINK	.2850-.2855
5 – GRN	.2830-.2835
4 – GRN	.2810-.2815
3 – GRN	.2790-.2795
2 – GRN	.2770-.2775
1 – GRN	.2750-.2755
5 – WH	.2730-.2735
4 – WH	.2710-.2715
3 – WH	.2690-.2695
2 – WH	.2670-.2675
1 – WH	.2650-.2655
5 – YEL	.2630-.2635
4 – YEL	.2610-.2615
3 – YEL	.2590-.2595
2 – YEL	.2570-.2575
1 – YEL	.2550-.2555
5 – ORNG	.2530-.2535
4 – ORNG	.2510-.2515
3 – ORNG	.2490-.2495
2 – ORNG	.2470-.2475
1 – ORNG	.2450-.2455
2 – RED	.2430-.2435
1 – RED	.2410-.2415

BACKLASH ADJUSTMENTS

BACKLASH SPECIFICATIONS

BACKLASH CHANGE REQUIRED (INCH)	THICKNESS CHANGE REQUIRED (INCH)	BACKLASH CHANGE REQUIRED (INCH)	THICKNESS CHANGE REQUIRED (INCH)
.001	.002	.009	.012
.002	.002	.010	.014
.003	.004	.011	.014
.004	.006	.012	.016
.005	.006	.013	.018
.006	.008	.014	.018
.007	.010	.015	.020
.008	.010		

Shim changes for ring gear and pinion backlash

tion should be on the drive (pull) side. Correct assembly of drive pattern will result in satisfactory coast performance. If gross pattern error is detected, the preferred backlash of 0.012–0.015 in. (0.30–0.38mm), recheck pinion shim selection.

16. Install bearing caps and tighten cap bolts to 70–85 ft. lbs. (95–115 Nm).

17. Install the axle shafts.

18. Remove the oil seal replacer from the transmission extension housing. Install the driveshaft in the extension housing. align the scribe marks on the flange and driveshaft and connect the driveshaft at the drive pinion flange. Apply Loctite® to the threads of the attaching bolts and torque to 70–95 ft. lbs. (90–128 Nm).

19. Install the brake drum and attaching shakeproof retainers. Install the wheel and tire on the brake drum. Install the wheel covers.

20. Clean the gasket mating surface of the rear axle housing and cover. Apply a new continuous bead of silicone rubber sealant to the carrier casting face.

NOTE: Make sure the machined surfaces on both cover and carrier are clean before installing the new silicone sealant. Inside of axle must be covered when cleaning the machined surface to pre-

vent axle contamination.

21. Install the cover and torque the bolts to 25–35 ft. lbs. (34–47 Nm), except the ratio tag bolt, which is tightened to 15–25 ft. lbs. (20–34 Nm).

NOTE: Cover assembly must be installed within 15 minutes of application of the silicone or new sealant must be reapplied.

22. Add EOAZ-19580-A (ESP-MC2154-A) or equivalent, through the filler hole until the lubricant level is ⅜ in. (9.5mm) below the filler hole with the axle in the running position.

23. Lower vehicle and road test.

Adjustment

DRIVE PINION AND DRIVE PINION BEARING PRELOAD

1. Install the pinion front bearing and slinger.

2. Apply grease, C1AZ-19590-B or equivalent, between the lips of the pinion seal and install the pinion seal.

3. Insert the companion flange into the seal and hold it firmly against the pinion front bearing cone. From the rear of the carrier casting, insert the pinion shaft, with a new spacer, into the flange.

4. Start new pinion nut. Hold the flange with special tool T78P-4851-A or equivalent, and tighten the pinion nut. As the nut is tightened, the pinion

shaft is pulled into the front bearing cone and into the flange.

5. As the pinion shaft is pulled into the front bearing cone, pinion shaft endplay is reduced. While there is still endplay in the pinion shaft, the flange and bearing cone will be felt to bottom on the collapsible spacer.

6. From this point, a much greater torque must be applied to turn the pinion nut, since the spacer must be collapsed. Very slowly, tighten the nut but check the pinion shaft endplay often to see the pinion bearing preload does not exceed the limits.

7. If the pinion nut is tightened to the point that pinion bearing preload exceeds the limits, the pinion shaft must be removed and a new collapsible spacer installed.

NOTE: Do not decrease the preload by loosening the pinion nut. This will remove the compression between the pinion front and rear bearing cones and the collapsible spacer and may permit the front bearing cone to turn on the pinion shaft.

8. As soon as there is a preload on the bearings, turn the pinion shaft in both directions several times to set the bearing rollers.

9. Adjust the bearing preload to specification. Measure the preload.

FORD—9.0 INCH RING GEAR REAR AXLE WITH REMOVABLE CARRIER
Disassembly
DIFFERENTIAL CARRIER

1. Remove the carrier-to-axle housing bolts, the carrier and drain the gear lube.

2. Place the carrier assembly into a holding fixture, check and record the ring gear runout and the ring gear backlash.

3. Mark one of the differential bearing caps and the mating bearing support with punch marks to help position the parts during reassembly of the carrier. Remove the adjusting nut locks, bearing caps, (remove the bearing caps with a soft mallet) and adjusting nuts. Lift the differential case assembly from the carrier.

4. Remove the differential side bearings with special tools T57L–4220–A or T66P–4220–A. Mark the differential case, differential cover and ring gear for assembly in the original position.

5. Separate the differential cover from the differential case. Remove the side gear thrust washer and the side gear. Using a drift, drive out the 3 differential pinion shaft lock pins. Drive out the long differential pinion shaft

with a brass drift.

6. Using a brass drift carefully positioned inside the case, drive out the 2 short differential pinion shafts. Remove the positioning block, differential pinions and the thrust washers from the differential case.

7. Remove the side gear and side gear thrust washer from the case.

8. Position the carrier assembly in a manner that will permit the removal of the drive pinion shaft nut. Remove the companion flange from the drive pinion shaft and remove the pinion seal.

9. Remove the pinion and retainer assembly from the carrier housing. If a new pinion bearing and or gear set is installed, a new shim will have to installed. Be very careful not to damage the mounting surfaces of the retainer and the carrier.

10. Place a protective sleeve on the pinion pilot bearing surface. Press the pinion shaft out of the pinion retainer. Press the pinion shaft out of the pinion rear bearing cone, using tool T71P–4621–B.

11. Remove and install a new pinion shaft pilot bearing. Remove the old pinion shaft pilot bearing by pressing it off and install the new one by pressing it on. Install a new pinion shaft pilot bearing retainer, concave side up, on the same press.

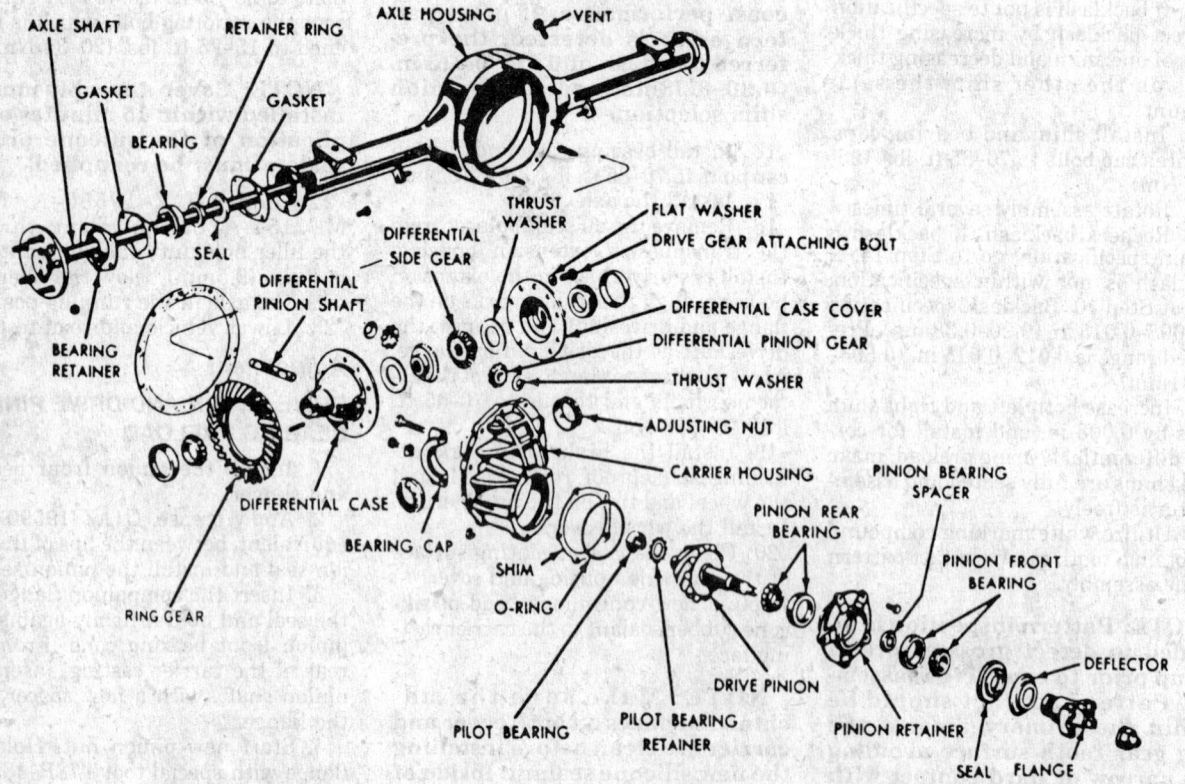

Exploded view of the 9 in. ring gear axle with a removable carrier—Ford Motor Company

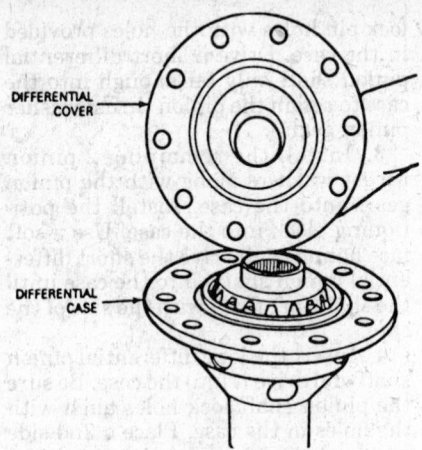

Removing the differential cover

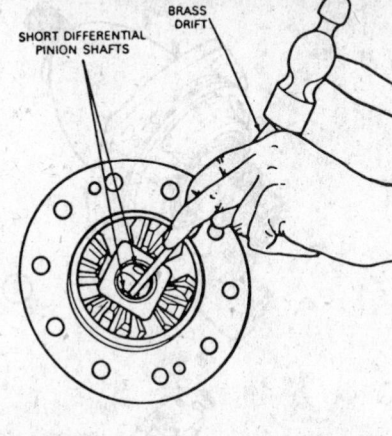

Removing the side gear and thrust washer

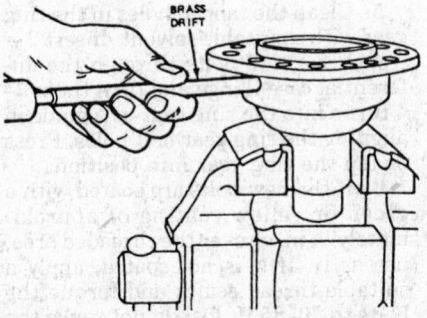

Removing the long differential pinion shaft

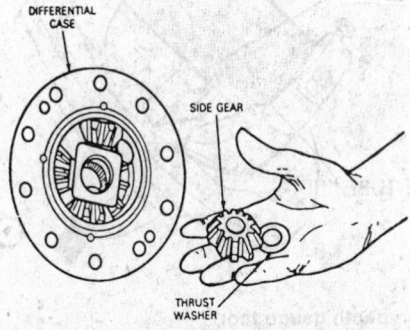

Removing the short differential pinion shaft

NOTE: Do not remove the drive pinion bearing cups from the retainer unless the cups are worn or damaged or if the cone and roller assemblies are damaged.

12. If the cups are worn or damaged, remove them with the use of bearing cup puller T77F-1102-A or T78P-1225-B or equivalent. Install the new bearing cups by pressing them into the retainer with pinion bearing cup replacer T71P-4616-A or equivalent.

13. After the new cups have been installed, make sure they are seated in the retainer by trying to insert a 0.0015 in. feeler gauge between the cup and the bottom of the bore. Whenever the cups are replaced, the cone and roller assemblies should also be replaced.

DRIVE PINION AND BEARING RETAINER

1. Install a holding fixture onto the flange, then, remove the pinion nut and washer. Leave the holding fixture on the flange and using a puller, remove the flange from the pinion shaft.

2. Using a seal puller, remove the pinion seal from the retainer assembly.

3. Remove the retainer assembly-to-carrier bolts and the carrier. Measure and record the thickness of the shim that was between the retainer and the carrier assembly.

4. Install a piece of hose on the pinion pilot bearing surface in front of the pinion gear. Mount the retainer assembly in a press and press the pinion gear out of the retainer.

5. Mount the pinion shaft in a press and press the rear bearing from the pinion shaft.

6. Mount the retainer assembly into a shop press and press the front and rear bearing cups from the assembly.

7. Using a bearing driver, drive the pilot bearing and retainer out of the carrier assembly.

Inspection

1. Clean the differential components in solvent and use compressed air to dry them; do not use compressed air on the bearings, only shop towels.

2. Check the components for wear or damage; replace them, if necessary.

3. Inspect the bearings and bearing cups for wear, cracks or scoring; replace them, if necessary.

4. Inspect the differential side and pinion gears for wear, cracks or chips; replace them, if necessary.

5. Inspect the ring and pinion gears for wear and/or damage; replace them, if necessary.

6. Inspect the differential case for cracks or damage; replace it, if necessary.

Assembly

DRIVE PINION AND RING GEAR SET

NOTE: When replacing a ring gear and a drive pinion or pinion bearings, select the proper pinion shim thickness by using the following procedure and tool T79P-4020-A or equivalent.

1. Select the proper rear pinion bearing aligning adapter and gauge disc to correspond to the axle size. Slide these adapters over the screw or threaded shaft and install the gauge block on the threaded shaft and tighten it securely.

2. Place this assembly, along with the rear drive pinion bearing, into the pinion bearing retainer assembly. Install the front pinion bearing (new or used, if in good condition) and screw the handle onto the threaded shaft, with the tapered end into the front pinion bearing.

3. The flat end of the handle has a ³⁄₈ in. square hole broached in it. This is designed so an inch pound torque wrench may be used to obtain the proper pinion bearing preload.

4. Install the pinion bearing retainer assembly into the carrier (without a pinion shim) and tighten the attaching bolts to 30-45 ft. lbs. Rotate the gauge block so it rests against the pilot boss.

5. Place the differential gauge tube into the differential bearing bore and tighten the bearing caps to the specified torque. Using a feeler gauge, gauge the space between the differential bearing gauge block and gauge tube. Insert a feeler blade directly along the gauge block top to ensure a correct reading. The fit should be a slight drag-type feeling.

6. After a correct feeler gauge is obtained, use the conversion chart provided to find the correct shim thickness needed according to the feeler gauge reading.

7. After determining the correct shim thickness as just outlined in this procedure, assemble the pinion bearing retainer as follows.

NOTE: A new ring gear and drive pinion should always be installed in an axle as a matched set, never separately. Be sure the same matching number appears

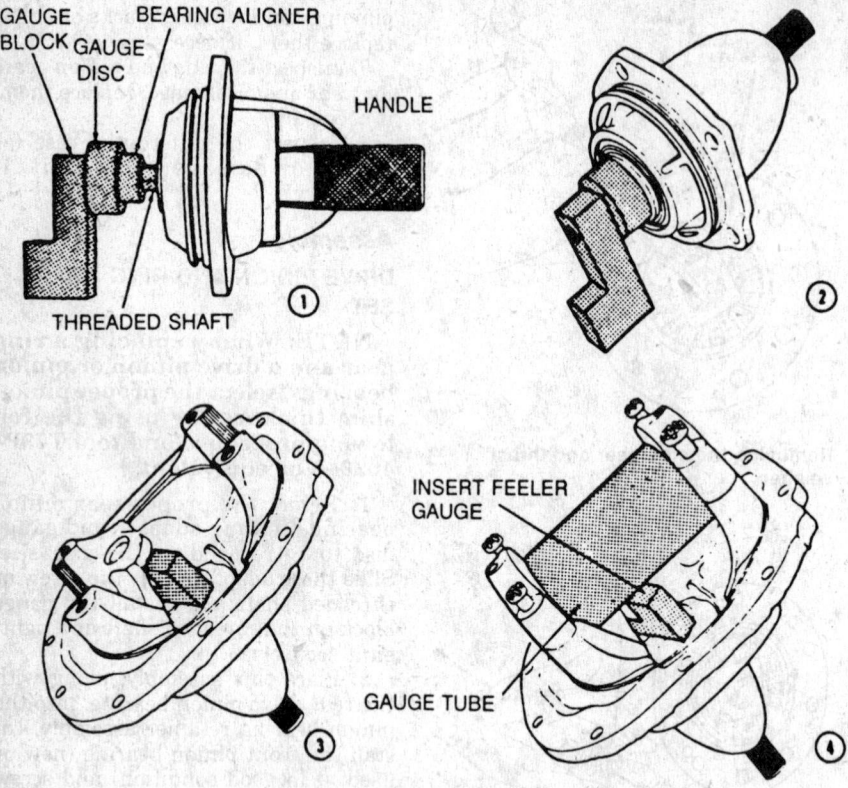

Proper use of the pinion depth gauge tool

on the ring gear and on the head of the drive pinion.

8. Install the pinion retainer attaching bolts and torque them to 30–45 ft. lbs. Install the oil slinger, if equipped.

9. Install a new pinion oil seal in the bearing retainer, install the companion flange. Start a new pinion nut on the drive pinion shaft and apply a small amount of thread lubricant to the flange side of the nut.

10. Hold the flange with tool T57T-4851–B and tighten the pinion nut: do not use impact tools. Check the pinion bearing preload, the correct preload

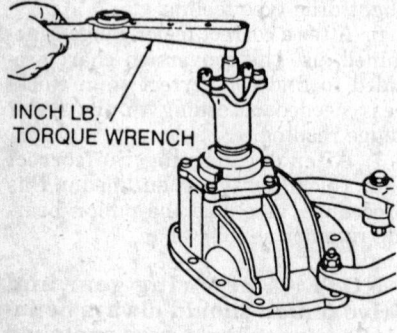

Checking the pinion bearing preload

will be obtained when the torque required to rotate the pinion in the retainer is as specified in the specifications.

11. If the torque required to rotate the pinion is less than specified, tighten the pinion shaft nut a little at a time until the proper preload is established.

NOTE: Do not over-tighten the pinion nut. If excessive preload is obtained as a result of over tightening, replace the collapsible bearing spacer. Do not back off the pinion shaft nut to establish pinion bearing preload.

DIFFERENTIAL CARRIER

NOTE: Lubricate all the differential components liberally with hypoid gear lubricant, EOAZ-19580–A or equivalent, during assembly.

1. Place a side gear thrust washer and side gear in the differential case bore. With a soft faced hammer, drive a short differential pinion shaft into the case far enough to retain a pinion thrust washer and pinion gear.

2. Carefully, align the pinion shaft

lock pin holes with the holes provided in the case. Drive a short differential pinion shaft only far enough into the case to retain the pinion thrust washer pinion gear.

3. Install the remaining 2 pinion thrust washers along with the pinion gears into the case. Install the positioning block into the case. Use a soft face hammer, to drive the short differential pinion shafts into the case until the shafts are flush with the side of the case.

4. Insert the long differential pinion shaft and drive it into the case. Be sure the pinion shaft lock holes align with the holes in the case. Place a 2nd side gear and thrust washer into position. Install the 3 pinion shaft lock pins. Press the differential cover on the case.

5. Clean the tapped holes in the ring gear with a suitable solvent. Insert 2 – $^7/_{16}$ (N.F.) × 2 bolts through the differential case flange and turn them 3–4 turns into the ring gear as a guide in aligning the ring gear bolt holes. Press or tap the ring gear into position.

6. If the new bolts are coated with a green or yellow coating of approximately ½ in. or over the threaded area, use as is. If it is not coated, apply a suitable thread sealer and torque the bolts to 70–85 ft. lbs; do not reuse the old bolts.

7. If the differential bearings have been removed, press them in using tool T57L-4221–A2 or equivalent. Wipe a thin coating of axle lubricant on the differential bearing bores so the differential bearing cups will move easily.

8. Place the cups on the bearings and set the differential case assembly in the carrier. Assemble the differential case and ring gear assembly in the carrier so the marked tooth on the drive pinion indexes between the marked teeth on the ring gear; be sure to match the marked gears as indicated. When assembled out of time, the result is noise and improper mating.

9. Slide the assembly along the bores until a slight backlash is felt between the gear teeth. Set the adjusting nuts in the bores using differential bearing nut wrench T70P-4067–A so they just contact the bearing cups.

10. The nuts should be engaged about the same number of threads (turns) on each side. Carefully position the differential bearing caps on the carrier. Match the marks made when the caps were removed. Before tightening the bearing cap bolts, be sure the adjuster nuts are properly threaded in the cap and carrier and turn freely.

11. Install the bearing cap bolts and

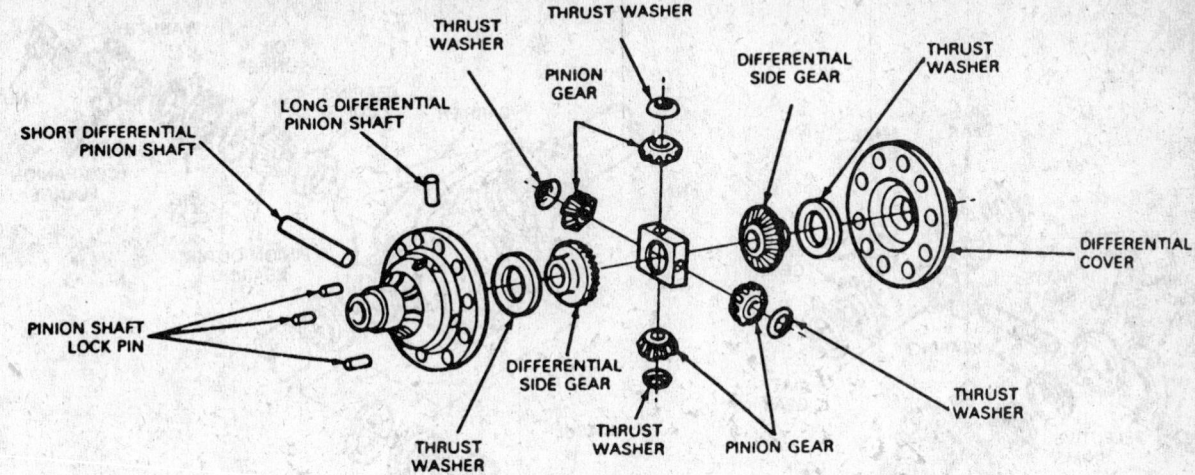

Exploded view of the differential case

alternately torque them to 70–85 ft. lbs. If the adjusting nuts do not turn freely as the cap bolts are tightened, remove the differential bearing caps and again inspect for damaged threads or incorrectly positioned caps.

12. Tightening the bolts to the specified torque is done to be sure the cups and adjusting nuts are seated. Loosen the cap bolts and tighten them to only 25 ft. lbs. before making adjustments.

13. Adjust the backlash between the ring gear and pinion and the differential bearing preload as described in the following section.

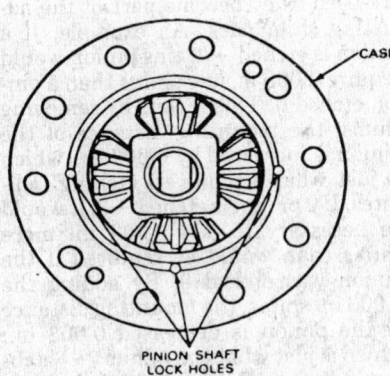

Location of the pinion shaft lock holes

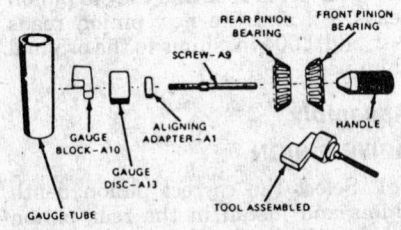

Typical rear axle pinion depth gauge tool

14. Using a new gasket and silicone sealant, position the differential carrier on the studs in the axle housing. Install the carrier-to-housing nuts and washers; torque them to 25–40 ft. lbs.

Adjustment
BACKLASH AND DIFFERENTIAL BEARING PRELOAD

1. Remove the adjusting nut locks, loosen the differential bearing cap bolts and torque the bolts to 15–20 ft. lbs. before making adjustments.

NOTE: The left adjusting nut is on the ring gear side of the carrier and the right nut is on the pinion side.

2. Loosen the right nut until it is away from the cup. Tighten the left nut until the ring gear is just forced into the pinion with zero backlash, then, rotate the pinion several revolutions to be sure there is no binding. Recheck the right nut at this time to make sure it is still loose.

3. Install a dial indicator and tighten the right nut until it first contacts the bearing cups. Set the dial indicator to zero and apply pressure to the bearing by tightening the right nut until the indicator reading shows 0.008–0.012 in. case spread.

4. Turn the pinion gear several times in each direction to seat the bearings in the cups and be sure no bind is evident (this step is important). Tighten the bearing cap bolts to 70–85 ft. lbs.

5. Measure the backlash on several teeth around the ring gear. If the backlash is out of specification, loosen 1 adjusting nut and tighten the opposite nut an equal amount, to move the ring gear away from or toward the pinion.

6. Tightening the left nut moves the ring gear into the pinion to decrease the backlash and tightening the right nut moves the ring gear away.

NOTE: When moving the adjusting nuts, the final movement should always be made in a tightening direction. An example of this is, if the left nut had to be loosened 1 notch, loosen the nut 2 turns and tighten it 1; this ensures the nut is contacting the bearing cup and the cup can not shift after being put in service. After all such adjustments, check to be sure the case spread remains as specified for the new or original bearings used.

7. Use a white marking compound to obtain a tooth mesh contact pattern in the assembly. Pattern contact should be within the primary area of the ring gear tooth surface avoiding any narrow or hard contact with the outer perimeter of tooth (top to root, toe to heel).

8. The pattern inspection should be on the drive (pull) side. The correct assembly of the drive pattern will result in a satisfactory coast performance. If gross pattern error is detected with preferred backlash of 0.012–0.015 in. recheck the pinion shim selection.

DANA—MODEL 30 RING GEAR AXLE

Disassembly
DIFFERENTIAL CARRIER

1. The axle assembly can be overhauled either in or out of the vehicle. Either way, the free-floating axles must be removed.

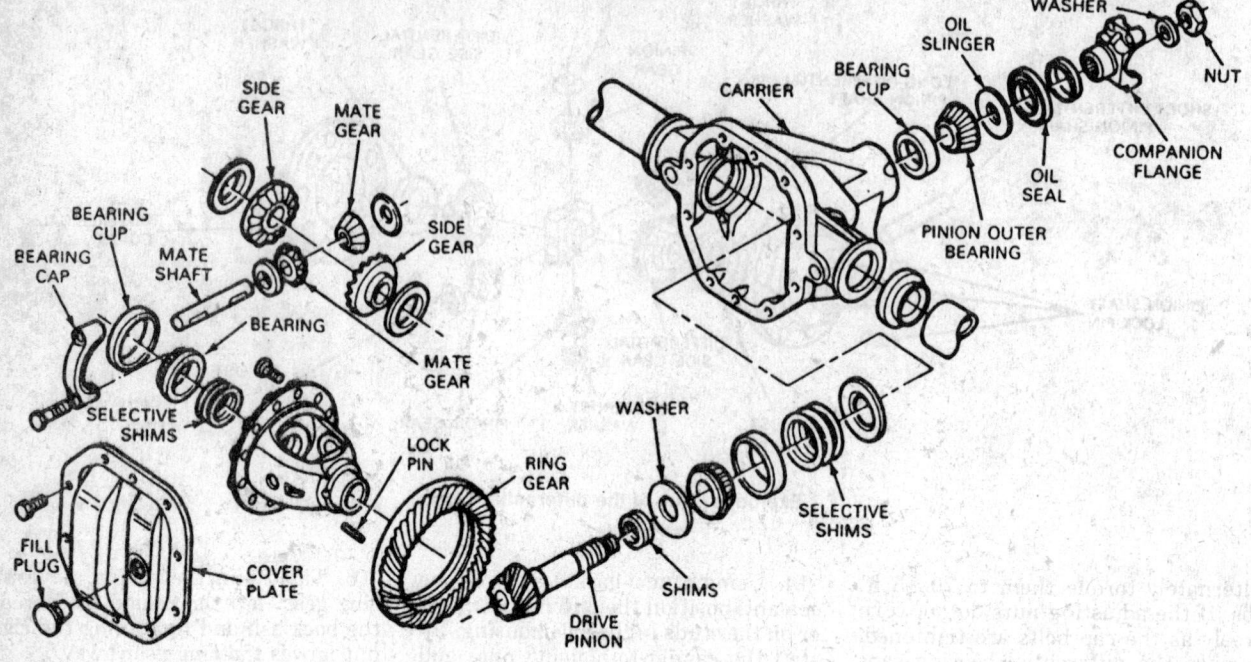

Exploded view of the Dana model 30 rear axle assembly—Ford Motor Company

2. Drain the lubricant and remove the rear cover and gasket.

3. Matchmark the bearing caps and the housing for reassembly in the same position. Remove the bearing caps and bolts.

4. Using a spreader tool mounted to the carrier housing, spread the housing a maximum of 0.015 in.

NOTE: Do not exceed this measurement. The housing could be permanently damaged. The use of a dial indicator is recommended to prevent over-stretching the housing.

5. Using a pry bar, remove the differential case from the housing. Separate the shims and record the dimensions. Remove the spreader tool from the housing.

6. Remove the differential side bearing cups and tag to identify the side, if they are to be used again.

7. Remove the differential gear pinion shaft lock pin and remove the shaft. Rotate the side and pinion gears to remove them from the carrier. Remove the thrust bearings.

8. Remove the bearing cones and rollers from the carrier, marking and noting the shim locations.

9. Remove the ring gear bolts and tap the ring gear from the carrier housing.

10. Inspect the components.

DRIVE PINION

1. Remove the pinion nut and flange from the pinion gear.

2. Remove the pinion gear assembly from the housing. It may be necessary to tap the pinion from the housing with a soft faced hammer. Catch the pinion so as not to allow it to drop on the floor.

3. With a long drift, remove the inner bearing cup, pinion seal, slinger, gasket, outer pinion bearing and the shim pack. Label the shim pack for reassembly.

4. Remove the rear pinion bearing cup and shim pack from the housing. Label the shims for reassembly.

5. Remove the rear pinion bearing from the pinion gear with an arbor press and special plates.

Inspection

1. Clean the gears, bearings and component parts with solvent and inspect for scoring, chipping or excessive wear.

2. Inspect the flanges and splines for excessive wear.

3. Replace the necessary parts as required.

PINION SHIM SELECTION

Ring gears and pinions are supplied in matched sets only. The matched numbers are etched on both gears for veri-

fication. On the rear face of the pinion, a plus (+) or a minus (−) number will be etched, indicating the best running position for each particular gear set. This dimension is controlled by the shimming behind the inner bearing cup. Whenever baffles or oil slingers are used, they become part of the adjusting shim pack. An example: If a pinion is etched +3, this pinion would require 0.003 in. less shims than a pinion etched 0. This means by removing shims, the mounting distance of the pinion is increased by 0.003 in., which is just what a plus (+) etching indicates. If a pinion is etched −3, it would be necessary to add 0.003 in. more shims than would be required if the pinion was etched 0. By adding the 0.003 in. shims, the mounting distance of the pinion is decreased 0.003 in., which is just what the minus (−) etching indicates. Pinion adjusting shims are available in thicknesses of 0.003, 0.005 and 0.010 in. An example: If a new gear set is used and the old pinion reads +2 and the new pinion reads −2, add 0.004 in. shims to the original shim pack.

Assembly
DRIVE PINION

1. Select the correct pinion depth shims and install in the rear pinion bearing cup bore.

2. Install the rear bearing cup in the axle housing.

3. Add or subtract an equal amount of shim thickness to or from the preload or outer shim pack, as was added or subtracted from the inner shim pack.

4. Install the front pinion bearing cup into its bore in the axle housing.

5. Press the rear pinion bearing onto the pinion gear shaft and install the pinion gear with bearing into the axle housing.

6. Install the preload shims and the front pinion bearing; do not install the oil seal at this time.

7. Install the flange with the holding bar tool attached, the washer and the nut on the pinion shaft end. Torque the nut to 200–220 ft. lbs. (271–298 Nm).

8. Remove the holding bar from the flange and with an inch lb. torque wrench, measure the rotating torque of the pinion gear. The rotating torque should be 10–20 inch lbs. with the original bearings or 20–40 inch lbs. with new bearings. Disregard the torque reading necessary to start the shaft to turn.

9. If the preload torque is not in specifications, adjust the shim pack as required.

   a. To increase preload, decrease the thickness of the preload shim pack.

   b. To decrease preload, increase the thickness of the preload shim pack.

10. When the proper preload is obtained, remove the nut, washer and flange from the pinion shaft.

11. Install a new pinion seal into the housing and reinstall the flange, washer and nut. Using the holder tool, torque the nut to 200–220 ft. lbs. (271–298 Nm).

DIFFERENTIAL CARRIER

1. Install the differential side gears, the differential pinion gears and new thrust washers into the differential carrier.

2. Align the pinion gear shaft holes and install the pinion shaft into the carrier. Align the lock pin hole in the shaft and carrier. Install the lock pin and peen the hole to avoid having the pin drop from the carrier.

3. Install the differential case side bearings with the proper installation tools. Do not install the shims at this time.

4. Place the carrier assembly into the axle housing with the bearing cups on the bearing cones. Install the bearing caps in their original position and tighten the bearing cap bolts enough to keep the bearing caps in place.

5. Install a dial indicator on the housing so the indicator button contacts the carrier flange. Press the differential carrier to prevent sideplay and center the dial indicator. Rotate the carrier and check the flange for run-out. If the run-out is greater than 0.002 in., the defect is probably due to the bearings or to the carrier and should be corrected.

6. Remove the assembly and install the ring gear. Torque the retaining bolts and reinstall the assembly into the housing. Install the bearing caps in their original position and tighten the cap bolts to keep the bearings caps in place.

7. Install the dial indicator and position the indicator button to contact the ring gear back surface. Rotate the assembly and the run-out should be less than 0.002 in. If over 0.002 in., remove the assembly and relocate the ring gear 180 degrees. Reinstall the assembly and recheck. If the run-out remains over the 0.002 in. tolerance, the ring gear is defective. If the measurement is within tolerances, continue on with the assembly.

8. Position 2 pry bars between the bearing cap and the housing on the side opposite the ring gear. Pull on the pry bars and force the differential carrier as far as possible towards the dial indicator. Rock the assembly to seat the bearings and reset the dial indicator to zero.

9. Reposition the prybars to the opposite side of the carrier and force the carrier assembly as far towards the center of the housing. Read the dial indicator scale. This will be the total amount of shims required for setting the backlash during the reassembly, less the bearing preload. Record the measurement.

10. With the pinion gear installed and properly set, position the differential carrier assembly into the axle housing and install the bearing caps in their proper positions. Tighten the cap bolts just to hold the bearing cups in place.

11. Install a dial indicator on the axle housing with the indicator button contacting the back of the ring gear.

12. Position 2 prybars between the bearing cup and the axle housing on the ring gear side of the case and pry the ring gear into mesh with the pinion gear teeth, as far as possible. Rock the ring gear to allow the teeth to mesh and the bearings to seat. With the pressure still applied by the prybars, set the dial indicator to zero.

13. Reposition the prybars on the opposite side of ring gear and pry the gear as far as it will go. Take the dial indicator reading. Repeat this procedure until the same reading is obtained each time. This reading represents the necessary amount of shims between the differential carrier and the bearing on the ring gear side.

14. Remove the bearing from the differential carrier on the ring gear side and install the proper amount of shims. Reinstall the bearing.

15. Remove the differential carrier bearing from the opposite side of the ring gear. To determine the amount of shims needed, use the following method.

   a. Subtract the size of the shim pack just installed on the ring gear side of the carrier from the reading obtained and recorded when measurement was taken without the pinion gear in place.

   b. To this figure, add an additional 0.015 in. to compensate for preload and backlash. An example: If the first reading was 0.085 in. and the shims installed on the ring gear side of the carrier were 0.055 in., the correct amount of shims would be 0.085 − 0.055 + 0.015 = 0.045 in.

16. Install the required shims as determined under step 15 and install the differential side bearing. The installation of the shims should give the proper preload to the bearings and the proper backlash to the ring and pinion gears.

17. Spread the axle housing with the spreader tool no more than 0.015 in. Install the differential bearing outer cups in their correct locations and install the cups in their respective locations.

18. Install the bolts and tighten finger-tight. Rotate the differential carrier and ring gear and tap with a soft-faced hammer to insure proper seating of the assembly in the axle housing.

19. Remove the spreader tool and torque the cap bolts to specifications.

20. Install a dial indicator and check the ring gear backlash at 4 equally spaced points of the ring gear circle. The backlash must be within a range of 0.004–0.009 in. and must not vary more than 0.002 in. between the points checked.

21. If the backlash is not within specifications, the shim packs must be corrected to bring the backlash within limits.

22. Check the tooth contact pattern and verify.

23. Complete the assembly, refill to proper level with lubricant and operate to verify proper assembly.

GENERAL MOTORS CORPORATION

Front Axle Assembly

GMC – 8½ INCH RING GEAR AXLE

For overhaul information, refer to the General Motors Corporation 8½ in. ring gear axle, in the rear axle assembly section.

DANA – 9¾ INCH RING GEAR AXLE

For overhaul information, refer to the Dana 9¾ in. ring gear axle, in the rear axle assembly section.

GMC – 8¼ AND 9¼ INCH RING GEAR AXLE

Disassembly

1. Remove the drain plug and drain the fluid from the axle. Remove the axle from the truck.
2. Measure and record the backlash for it may help to determine the cause of an axle problem.
3. Remove the solenoid and the indicator switch.
4. Remove the right axle tube-to-carrier case bolts and the tube with the shaft.
5. Remove the right output shaft sleeve, the shift shaft, damper spring, shift fork and clip assembly.
6. Remove the inner spring and shim.
7. Clamp the axle shaft tube in a vise, using the mounting flange. Strike the inside of the shaft flange to dislodge the carrier connector. Remove the carrier connector and the retaining ring.
8. On K35 Series, remove the snapring, the washer and the thrust washer from the right axle tube.
9. Remove the right side axle shaft with the deflector.
10. Using pullers J-29362 for 15/25 series or J-J-29369-2 for 35 series, along with J-29307, or their equivalents, remove the seal and bearing from the axle shaft tube.
11. Remove the output shaft.
12. Using a tool J-34011 or equivalent, press the differential pilot bearing.

13. Using a prybar and a soft faced hammer, remove the left side output shaft and delector.
14. Using a prybar, remove the seal from the axle tube.
15. Using pullers J-29362 for 15/25 series or J-J-29369-2 for 35 series, along with J-29307, or their equivalents, remove the left output shaft bearings from the case.
16. Remove the carrier case bolts, tap on the carrier case lugs and separate the right and left side carrier case halves.
17. Remove the differential assembly. On the right side of K35 series, you'll have to pry up the locks first.
18. On the K35 series, remove the bolt, the lock, the sleeves and the side bearing cups; turn the sleeves to push the cups from the bores.
19. On the K35 series, remove the adjuster plug with the side bearing cup and the O-ring, if equipped, using tool J-36615.
20. Using a holding tool such as J-8614-01, remove the pinion flange nut and washer.
21. Using a pinion flange removal tool, press the pinion flange from the differential.
22. Remove the pinion with the shim, the bearing cone and the spacer, using tool J-36598.
23. Press the spacer from the pinion.
24. Using a bearing removal tool such as J-36598, press the bearing from the pinion.
25. Remove the shim, the seal, the bearing cup and the cone.
26. Using the puller tool, press the bearing cup from the pinion.
27. Remove the side bearings, using tools J-22888-D and J-8107-2 (15/25 series) or J-36597 (35 series).

NOTE: The ring gear bolts are equipped with left handed threads.

28. Remove the ring gear bolts. Using a brass drift, drive the ring gear from the differential case.
29. Using a drift and a hammer, drive the roll pin from the differential case.
30. Remove the retaining bolt (35 series only) and pinion gear shaft.
31. Roll the pinion gears out of the case with their thrust washers.
32. Remove the side gears and their thrust washers.
33. Place identifying marks on the gears and differential case halves.
34. On K35 series, remove the spacer from the carrier.
35. Using a 6-point deep-well socket, remove the vent plug.
36. Drive the carrier bushings out using tool J-36616.

Assembly

PINION BEARING CUP

1. Mount the left case half in a holding fixture such as J-36598. For 15/25 series, use adapter J-36598-6. Tighten the fixture bolts securely.
2. Using J-36598-3 (15/25 series) or J-36598-4 (35 series), install the outer bearing cup.
3. Remove J-36598-3 (15/25 series) or J-36598-4 (35 series), and place J-36598-15 in the pinion seal bore. Extend the forcing screw through the bore.
4. Install J-36598-3 (15/25 series) or J-36598-4 (35 series) on the forcing screw. Turn the forcing screw until the installer is snug against the bearing cup. Rotate the tool several times to make sure it's not cocked in the bore.
5. Pull the inner bearing cup into place in the bore.

PINION DEPTH MEASUREMENT

1. Coat the pinion bearings with axle fluid.
2. Assemble dial indicator J-29763 on gauging tool J-36601-4 (15/25 series) or J-36601-3 (35 series).
3. Install the pinion bearings and hold them in place.
4. Insert the threaded rod of the gauging tool through the bearings.
5. Install the pilot, flat washer and nut on the tool.
6. Tighten the nut while holding the threaded rod. Adjust the nut to obtain a preload of 10-15 inch lbs. Rotate the shaft several times to seat the bearings, then re-measure the torque and adjust as necessary.
7. Push the dial indicator downward until the needle make 3 revolutions and tighten the tool at this position.
8. Place the button of the gauging tool on the differential bearing bore.
9. Rotate the tool, slowly, back and forth, until the dial inidactor reads the lowest point in the bore. Set the dial indicator to 0. Repeat, to verify the reading.
10. Hold the gauging arm by its flats and move the tool button out of the bearing bore. Record the dial indicator reading. The reading is equal to the required shim size.

    Example: if the reading is 0.508mm, a 0.508 shim is required.
11. Remove the tools and bearing cones.

Pinion Installation

1. Place the proper shim, determined earlier, on the pinion gear.
2. Install the bearing on the pinion gear using j-35512 (15/25 series) or J-36614 (35 series).

3. Place a new spacer on the pinion gear.

4. Place a new seal in the case using J-36266.

5. Place the pinion gear assembly in the case.

6. Install the deflector, flange, washer and nut. Apply sealant, such as PST Sealant® on the pinion gear threads and both sides of the washer.

7. Hold the flange with J-8614-01 and tighten the nut until no endplay is detected. No further tightening should be attempted at this point!

8. Place an inch lbs. torque wrench on the nut and rotate the pinion, observing the reading. The correct preload reading is 15-25 inch lbs.

9. If the reading is low, tighten the nut in very small increments, checking the preload after each tightening. **Do not exceed the recommended preload!** If the recommended preload is exceeded, the pinion will have to be disassembled and a new spacer installed.

10. When the proper preload is set, rotate the pinion several times to make sure the bearings are seated and recheck the preload. Reset, if necessary.

Differential Case

1. On 35 series, install the side gear spacer.

2. Install the thrust washers and side gears in the differential case. On the 35 series, the side gear spacer goes on the left shaft. If used gears and washers are being installed, they must be installed in the same places from which they were removed. Check your identifying marks.

3. Position one pinion gear between the side gears and rotate the gears until the pinion gear is directly opposite the opening in the case.

4. Place the other pinion gear between the side gears, making sure that the holes in the gears line up.

5. Install the thrust washers. Rotate the pinion gears as needed.

6. Install the shaft and pin (15/25 series) or bolt (35 series).

7. Install the ring gear using **new** bolts. Never reuse old bolts! Using an alternating sequence, torque the bolts in 3 progressive steps, to a maximum torque of 88 ft. lbs.

8. Using tools J-22761 (15/25 series) or J-29710 (35 series), along with J-8092, install the side bearings.

9. Using tools J-8092 and J-36612 (15/25 series) or J-36613 (35 series) install the carrier bearings on the sleeves or adjuster plug (35 series).

10. On the 35 series, install a new O-ring on the adjuster plug.

11. On 15/25 series, install the sleeves in the carrier case using J-36599.

12. On 35 series, install the right sleeve using J-36599; install the adjuster plug using J-36615.

13. Using J-36602 and J-8092, install the side bearing cups.

14. Place the differential assembly in the carrier case half containing the pinion gear assembly.

Turn the left sleeve (15/25 series) or adjuster plug (35 series) inward until backlash is felt between the ring and pinion.

15. Remove the case from the holding fixture.

16. Join the case halves **without using sealer at this time**. If the case halves do not mate completely, back out the right sleeve until they do. Tighten the bolts to 35 ft. lbs.

BACKLASH ADJUSTMENT

1. Tighten the right sleeve to 100 ft. lbs.

2. Tighten the left sleeve (15/25 series) or adjuster plug (35 series) to 100 ft. lbs.

3. Matchmark the sleeves and/or plug and the case so that the notches can be counted when turned.

4. Turn the right sleeve **out** 2 notches.

5. Turn the left sleeve or adjuster plug **in** 1 notch.

6. Rotate the pinion several times to seat the bearings.

7. Install indicator and fixture J-34047-3, J-34047-1 and J-8001-1 in the filler plug hole so the stem of the indicator is at the heel end of a tooth.

8. Check and record the backlash at 3 or 4 points around the gear. The pinion must be held stationary while checking backlash. The backlash should be the same at each point ± 0.05mm. If the backlash varies more than ± 0.05mm, check for burrs, a distorted case flange, uneven bolt torques or foreign matter.

Gear backlash should be 0.08-0.25mm with a preferred setting of 0.13-0.18mm.

9. If backlash is incorrect, adjust the sleeves and/or plug as necessary. Follow this example: If it is necessary to turn the right sleeve **in** 1 notch, turn the left sleeve or plug **out** 1 notch.

To increase backlash, turn the left sleeve or plug **in** and the right sleeve **out** an identical amount.

To decrease backlash turn the right sleeve **in** and the left sleeve or plug **out** an identical amount.

A 1 notch change will cause an 0.08mm change in backlash.

GEAR PATTERN CHECK

Before final assembly of the differential, a pattern check of the gear teeth must be made. This determines if the teeth of the ring and pinion gears are meshing properly, for low noise level and long life of the gear teeth. The most important thing to note is if the pattern is located centrally up and down on the face of the ring gear.

1. Wipe any oil from the carrier and all dirt and oil from the teeth of the ring gear.

2. Coat the teeth of the ring gear with a gear marking compound.

3. Apply a load until it takes 40-50 ft. lbs. of torque to turn the pinion gear.

4. Turn the companion flange so the ring gear makes a full rotation in 1 direction, then, turn it a full rotation in the opposite direction.

5. Check the pattern on the teeth and refer to the chart for any adjustments necessary. Contact pattern can be adjusted by either a change in the pinion depth or a change in backlash. Adjust the pattern as necessary.

Carrier Case

1. Bend the locks over the sleeves.

2. Install the bolt and lock over the adjuster plug on the 35 series.

3. Remove the bolts and lift off the right case half.

4. Thoroughly clean the case half mating surfaces with a solvent such as carburetor cleaner.

5. Apply a bead of Loctite®518, or equivalent, on the mating surface of either carrier half. Install the right half and torque the bolts to 35 ft. lbs.

6. Install the left seal, driving it into place with a seal driver and hammer.

7. Drive the shaft and deflector into place.

8. Using J-33842, install the bearing on the output shaft.

9. Install the output shaft in the carrier.

10. Install the vent plug, using Loctite®518 on the threads.

11. Using J-36616, install the carrier bushings.

Axle Tube

1. Drive the bearing into the tube.

2. Install a new seal.

3. Install the shaft and deflector in the tube.

4. Install the washer, aligning the tabs with the slots in the tube.

5. Install the connector gear and retaining ring into place with a plastic mallet.

6. Install the washer and a new snapring (35 series). Make sure that the snapring is fully seated.

7. If any of the following components were replaced, it will be necessary to select the proper size output shaft shim:

- Output shaft
- Tube

- Outer shaft
- Carrier case
- Ring and pinion
- Differential case
- Side bearings
- Connector

a. Push outward on the inner end of the outer shaft and move the shaft outward as far as it will go. The shaft must remain in this position for all measurements.

b. On 15/25 series, measure between the tube flange machined end and the inner surface of the connector. Record the distance.

On 35 series, measure between the tube flange machined surface and the inner surface of the axle shaft shoulder. Record the distance.

This figure is distance "A".

d. On all models, measure the distance between the carrier machined surface and the outer surface of the output shaft. Record the distance. This figure is distance "B".

e. Subtract distance "A" from distance "B". The correct shim size will be one size smaller than the resultant figure. For example: If the resultant figure was 3.53mm, the necessary shim will be 3.30mm.

If the resultant figure was 3.30mm, the necessary shim will be 2.70mm (15/25 series) or 2.80mm (35 series).

Shims are available in the following sizes:

- 15/25 series — 1.27mm, 1.78mm, 2.29mm, 2.70mm, 3.30mm, 3.81mm
- 35 series — 1.80mm, 2.30mm, 2.80mm, 3.30mm, 3.80mm, 4.30mm, 4.80mm

Final Assembly

1. Install the shim selected above on the output shaft. Use grease to hold it in place.

2. Install the sleeve.

3. Install the spring.

4. Install the shift shaft, spring, shift fork and clip in the case. The damper spring fits into the fork indentation. Make sure the clip is seated in the groove on the shift shaft.

5. Apply a bead of Loctite®518 on the tube mating surface and assemble the tube and carrier. Torque the bolts to 30 ft. lbs.

6. Using a brass drift, reach through the actuator hole and manually operate the shift mechanism. The mechanism should move smoothly, without binding.

7. Apply Loctite®518 on the solenoid threads. Tighten the solenoid to 16 ft. lbs.

8. Apply Loctite®518 on the switch threads. Tighten the switch to 45 inch lbs.

9. Refill the axle and install the filler plug and washer. Torque the plug to 24 ft. lbs.

Rear Axle Assembly

The rear axles are categorized by the ring gear diameter and are identified as follows: 7½, 8½, 9½ in. (GMC) semi-floating axles; 9¾, 10½ in. (Dana) full floating axles; 10½ in. (GMC) full floating axles and 12 in. (Rockwell) full floating axles.

GMC — 7½ INCH RING GEAR AXLE

Disassembly

REAR AXLE CASE

1. Before removing the rear axle case from the housing, check the ring gear-to-drive pinion backlash. This will indicate gear or bearing wear or an error in backlash or preload setting which will help in determining cause of axle noise.

2. Mark the bearing caps **R** and **L** to make sure they will be reassembled in their original location. Remove rear axle bearing cap bolts.

3. Remove rear axle case. Exercise caution in prying on carrier so the gasket sealing surface is not damaged. Place right and left bearing outer races and shims in sets with marked bearing caps so they can be reinstalled in their original positions.

4. If rear axle side bearings are to be replaced, use a puller to remove them.

5. Remove rear axle pinions, side gears and thrust washers from case. Mark the side gear and case after removing bolts. Using a brass drift and hammer, drive it off; do not pry between ring gear and case.

DRIVE PINION, BEARING AND RACES

1. Check drive pinion bearing preload. If there is no preload reading, check for looseness of pinion assembly by shaking. Looseness could be caused by defective bearings or worn pinion flange. If rear axle was operated for an extended period with very loose bear-

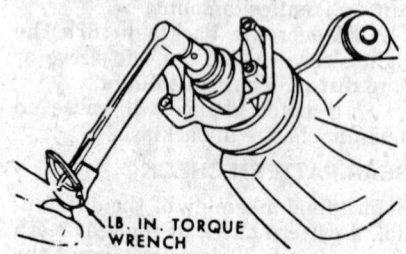

Checking the pinion pre-load

ings, the ring gear and drive pinion will also require replacement.

2. Remove pinion flange nut and washer.

3. Remove pinion flange.

4. Install a drive pinion remover tool and drive out pinion. Apply heavy hand pressure on the pinion remover toward rear axle housing to keep front bearing seated to avoid damage to outer race.

NOTE: The rear pinion bearing must be removed when it becomes necessary to change the pinion depth adjustment.

5. With drive pinion removed from carrier, press bearing from the pinion gear.

6. Drive pinion oil seal from carrier and remove front pinion bearing. If this bearing is to be replaced, remove outer race from carrier.

7. If rear pinion bearing is to be replaced, remove outer race from carrier using a punch in slots provided for this purpose.

Cleaning and Inspection

1. Clean all rear axle bearings thoroughly in clean solvent (do not use a brush). Examine bearings visually and by feel. All bearings should feel smooth when oiled and rotated while applying as much hand pressure as possible. Minute scratches and pits appear on rollers and races at low mileage are due to the initial preload and bearings having these marks should not be rejected.

2. Examine sealing surface of pinion flange for nicks, burrs or rough tool marks which would cause damage to the seal and result in an oil leak. Replace if damaged.

3. Examine carrier bore and remove any burrs that might cause leaks around the O.D. of the pinion seal.

4. Examine the ring gear and drive pinion teeth for excessive wear and scoring. If any of these conditions exist, replacement of the gear set will be required.

5. Inspect the pinion gear shaft for unusual wear; also check the pinion and side gears and thrust washers.

6. Check the press fit of the side bearing inner race on the rear axle case hub by prying against the shoulder at the puller recess in the case. Side bearings must be a tight press fit on the hub.

7. Diagnosis of a rear axle failure such as: chipped bearings, loose (lapped-in) bearings, chipped gears, etc., is a warning that some foreign material is present; therefore, the axle housing must be cleaned.

Assembly

DRIVE PINION

1. If a new rear pinion bearing is to be installed, install new outer races.

2. If a new front pinion bearing is to be installed, install new outer race.

NOTE: Pinion depth is set with pinion setting gauge. The pinion setting gauge provides in effect, a normal or zero pinion as a gauging reference. Instructions are included in gauge set.

3. Make certain all of the gauge parts are clean.

4. Lubricate front and rear pinion bearings liberally with rear axle lubricant.

5. While holding bearings in position, install depth setting gauge assembly.

6. Hold stud stationary with a wrench positioned over the flats on the ends of stud and tighten nut to 20 inch lbs. (2.2 Nm) torque. Rotate gauge plate assembly several complete revolutions to seat the bearings. Tighten nut until a torque between 15–25 inch lbs. (1.6–2.2 Nm) is obtained to keep the gauge plate in rotation.

7. Rotate the gauge plate tool until the gauging areas are parallel with the discs.

8. Make certain rear axle side bearing support bores are clean and free of burrs.

9. Install the correct discs on the gauge shaft.

10. Position the gauge shaft assembly in the carrier so the dial indicator rod is centered on the gauging area of the gauge block and the discs seated fully in the side bearing bores. Install side bearing caps and torque bolts to 55 ft. lbs. (75 Nm). Use a dial indicator reading from 0.00–100.0 inch (0.0–2.5mm).

11. Set dial indictor at zero. Then position on mounting post of the gauge shaft with the contact button touching the indicator pad. Push dial indicator downward until the needle rotates approximately ¾ turn clockwise. Tighten the dial indicator in this position and recheck.

12. Rotate gauge shaft slowly back and forth until the dial indicator reads the greatest deflection. At the point of greatest deflection, set the dial indicator to zero. Repeat rocking action of gauge shaft to verify the zero setting.

13. After the zero setting is obtained, rotate gauge shaft until the dial indicator rod does not touch the gauge block.

14. Record dial reading at pointer position. Example: If pointer moved counterclockwise 0.067 in. (1.70mm) to a dial reading of 0.033 in. (0.84mm) except as follows: dial indicator read-

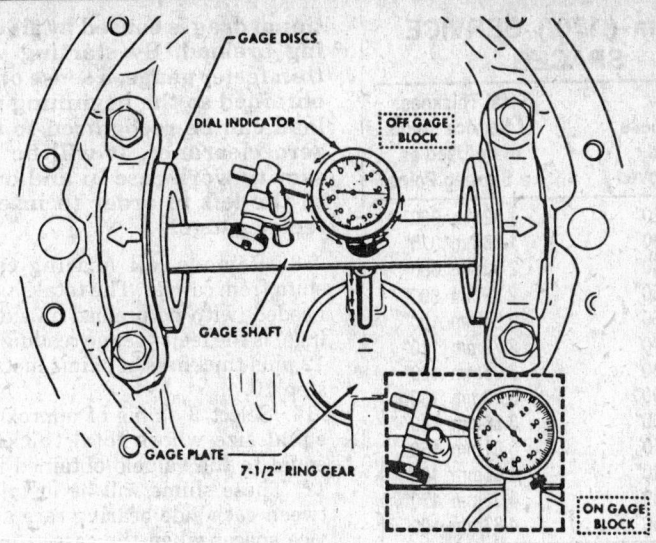

Checking the pinion depth

ing should be within the range of 0.50–0.020–0.050 in. (1.27mm).

15. Loosen the stud tool and remove gauge plate, washer and both bearings from carrier.

16. Position correct shim on drive pinion and install the drive pinion rear bearings.

REAR AXLE CASE

Before assembling the rear axle case, lubricate all parts with rear axle lubricant.

1. Place side gear thrust washer over side gear hubs and install side gears in case. If same parts are reused, install in original sides.

2. Position one pinion, without washer, between the side gears and rotate gears until pinion is directly opposite from loading opening in case. Place other pinion between side gears so the pinion shaft holes are in line; then, rotate gears to make sure holes in pinions align with the holes in case.

3. If the holes align, rotate pinions back toward loading opening just enough to permit sliding in pinion thrust washers.

4. After making certain the mating surfaces of case and ring gear are clean and free of burrs, thread 2 bolts into opposite sides of ring gear; then, install ring gear on case. Install new ring gear attaching bolts snug; never reuse old bolts. Torque bolts alternately in progressive stages to 90 ft. lbs. (120 Nm).

5. If case side bearings were removed, reinstall bearings.

NOTE: The side bearing preload adjustment is to be made before installing the pinion.

6. If the pinion is installed, remove ring gear. Case side bearing preload is adjusted by changing the thickness of

both the right and left shims by an equal amount. By changing the thickness of both shims equally, the original backlash will be maintained. Pro-

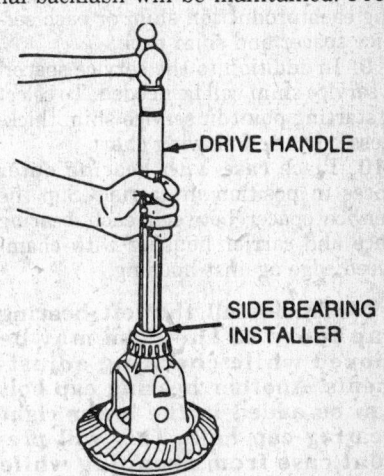

Installing the case side bearings

duction shims are cast iron and vary in thickness from 0.210–0.272 in. (5.33–6.91mm) in increments of 0.002 in. (0.05mm). Standard service spacers are 0.170 in. (4.32mm) thick and steel service shims are available from 0.040–0.082 in. (1.02–2.08mm) in increments of 0.002 in. (0.05mm).

NOTE: Do not attempt to reinstall the production shims as they may break when tapped into place. If service shims were previously installed, they can be reused but, whether using new or old bearings, adhere to the following procedure in all cases.

7. Before installing of the case assembly, make sure the side bearing surfaces in the carrier are clean and free of burrs. If the same bearings are

4.32mm (.170″) SERVICE SPACER

Total Thickness of Both Prod. Shims Removed	Total Thickness of Service Shims to be Used as a Starting Point
10.57mm .420″	1.52mm .060″
10.92mm .430″	1.78mm .070″
11.18mm .440″	2.03mm .080″
11.43mm .450″	2.29mm .090″
11.68mm .460″	2.54mm .100″
11.94mm .470″	2.79mm .110″
12.19mm .480″	3.05mm .120″
12.45mm .490″	3.30mm .130″
12.70mm .500″	3.56mm .140″
12.95mm .510″	3.81mm .150″
13.21mm .520″	4.06mm .160″
13.46mm .530″	4.32mm .170″
13.97mm .550″	4.83mm .190″

being reused, they must have the original outer races in place.

8. Determine the approximate thickness of shims needed by measuring each production shim or each service spacer and shim pack.

9. In addition to the service spacer, a service shim will be needed. To select a starting point in service shim thickness, use the following chart.

10. Place case with bearing outer races in position in carrier. Slip the service spacer between each bearing race and carrier housing with chamfered edge against housing.

NOTE: Install the left bearing cap loose so the case may be moved while checking adjustments. Another bearing cap bolt can be added in the lower right bearing cap hole. This will prevent case from dropping while making shim adjustments.

11. Select 1 or 2 shims totaling the amount shown in the right-hand column and position between the right bearing race and the service spacer. Be sure left bearing race and spacer are against left side of housing.

12. Insert progressively larger feeler gauge sizes 0.010 in. (0.25mm), 0.012 in. (0.30mm), 0.014 in. (0.36mm) or etc. between the right shim and service spacer until there is noticeable increased drag. Push the feeler gauge downward until the end of the gauge makes contact with the carrier bore so as to obtain a correct reading. The point just before additional drag begins is correct feeler gauge thickness. Rotate case while using feeler gauge to assure an even reading.

NOTE: The original light drag is caused by weight of the case against the carrier while addi-

tional drag is caused by side bearing preload. By starting with a thin feeler gauge, a sense of feel is obtained so the beginning of preload can be recognized to obtain zero clearance. It will be necessary to work case in and out and to the left in order to insert the feeler gauge.

13. Remove left bearing cap and shim from carrier. The total shim pack needed (with no preload on side bearings) is the feeler gauge reading in step 12 plus thickness of shims installed in step 10.

14. Select 2 shims of approximately equal size whose total thickness is equal to the valuelo obtained in Step 12. These shims will be installed between each side bearing race and service spacer when the case is installed in the carrier.

NOTE: The objective is to obtain the equivalent of a slip fit of the case in the carrier. For convenience in setting backlash, the preload will not be added until the final step.

15. If the pinion is in position, install the ring gear and adjust the rear axle backlash.

DRIVE PINION, BEARING AND RACES

1. Install a new collapsible spacer on pinion and position assembly in carrier. Lubricate the pinion bearings with rear axle lubricant before installing pinion.

2. Hold the pinion forward in the case assembly.

3. Install the front bearing onto pinion and the drive bearing onto pinion shaft until sealed in the race.

4. Position and install the pinion oil seal in the carrier.

5. Lubricate the pinion oil seal lips and seal surface of pinion flange. Install pinion flange on pinion by tapping with a soft hammer until a few pinion threads project through the flange.

6. Install the pinion washer and nut. While holding the pinion flange, intermittently, rotate the pinion to seat pinion bearings. Tighten the pinion flange nut until the endplay begins to disappear. When no further endplay is detectable and the holder will no longer pivot freely as the pinion is rotated, the preload specifications are being approached; no further tightening should be attempted until the preload has been checked.

7. Check the preload by using an inch lb. torque wrench. After the preload has been checked, final tightening should be done very carefully.

NOTE: If when checking, preload was found to be 5 inch lbs. (0.6 Nm), any additional tightening of the pinion nut can add many additional inch lbs. of torque. Therefore, the pinion nut should be further tightened only a little at a time, for the preload specifications will compress the collapsible spacer too far and require the installation of a new collapsible spacer.

8. While observing the preceding note, carefully set preload at 24–32 inch lbs. (2.7–3.6 Nm) on new bearings or 8–12 inch lbs. (1.0–1.4 Nm) for used bearings.

9. Rotate pinion several times to assure the bearings have been seated. Check the preload again, if preload has been reduced by rotating pinion, reset preload to specifications.

Adjustment

REAR AXLE BACKLASH

1. Install rear axle case into carrier, using shims as determined by the side bearing preload adjustment.

2. Rotate the rear axle case several times to seat bearings, then, mount a dial indicator. Use a small button on the indicator stem so contact can be made near heel end of tooth. Set the dial indicator so the stem is aligned as nearly as possible with gear rotation and perpendicular to tooth angle for accurate backlash reading.

3. Check the backlash at 3 or 4 points around ring gear. Lash must not vary more than 0.002 in. (0.05mm) around ring gear. The pinion must be held stationary when checking backlash; if variation is greater than 0.002 in. (0.05mm) check for burrs, uneven bolting conditions or distorted case flange and make corrections, as necessary.

4. Backlash at the point of minimum lash should be between 0.005–0.009 in. (0.13–0.23mm) for all new gears.

5. If backlash is not within specifications, correct by increasing thickness of one shim and decreasing thickness of other shim the same amount. This will maintain correct rear axle side bearing preload.

   a. For each 0.001 in. (0.03mm) change in backlash desired, transfer 0.002 in. (0.05mm) in shim thickness.

   b. To decrease backlash 0.001 in. (0.03mm), decrease thickness of right shim 0.002 in. (0.05mm) and increase thickness of left shim 0.002 in. (0.05mm).

   c. To increase backlash 0.002 in. (0.05mm), increase thickness of right shim 0.004 in. (0.10mm) and decrease thickness of left shim 0.004 in. (0.10mm).

6. When the backlash is correctly adjusted, remove both bearing caps and both shim packs. Keep the packs in their respective position, right or left side. Select a shim 0.004 in. (0.10mm) thicker than the one removed from left side, then, insert left side shim pack between the spacer and the left bearing race. Loosely install bearing cap.

7. Select a shim 0.004 in. (0.10mm) thicker than the one removed from right side and insert between the spacer and the right bearing race; it will be necessary to drive the right shim into position.

8. Torque to 55 ft. lbs. (75 Nm).

9. Recheck backlash and correct if necessary.

10. Install the axles.

11. Install a new cover gasket, the cover and torque the bolts to 20 ft. lbs. (27 Nm).

12. Refill the rear axle to the proper level.

GMC — 8½ AND 9½ INCH RING GEAR AXLE

This axle assembly is the semi-floating type with a hypoid type drive pinion and ring gears. The drive pinion gear is supported by 2 bearings. The differential case contains 2 pinion gears. The carrier assembly is not removable since it is part of the axle assembly but the design allows for the differential assembly to be serviced while the axle is still in the vehicle. The ring gear bolted to a one piece differential case is supported by 2 preloaded roller bearings.

Disassembly

DIFFERENTIAL CASE

1. Remove the inspection cover from the axle housing and drain the gear lubricant into a pan.

2. Remove the screw or pin that holds the pinion shaft in place and remove the shaft.

3. Push the axle shaft(s) in a little and remove the C-locks from the ends of the shafts. Remove the axle shafts from the housing.

4. Measure and record the backlash; this will allow the old gears to be reassembled at the same amount of lash to avoid changing the gear tooth pattern. It also helps to indicate if there is gear or bearing wear and if there is any error in the original backlash setting.

5. Remove the differential pinions, the side gears and thrust washers from the case; be sure to mark the pinions and side gears so they may be reassembled in their original position.

6. Mark the bearing caps and housing and loosen the retaining bolts. Tap the caps lightly to loosen them. When the caps are loose, remove the bolts and reinstall them, just a few turns; this will keep the case from falling out of the housing when it is pried loose.

7. Using a pry bar, carefully, pry the case assembly loose; be careful not to damage the gasket surface on the housing when prying. The case assembly may suddenly come free if the bearings were preloaded, so pry very slowly.

8. When the case assembly is loose, remove the bearing cap bolts and the caps. Place the caps so they may be reinstalled in the same position. Place any shims that were removed with the cap.

9. Using a bearing puller, pull the differential bearing from the case.

10. To remove the drive pinion bearing, perform the following procedures:

   a. Depending on the bearing that is being replaced, remove the front or rear bearing cup from the carrier assembly.

   b. With the pinion gear mounted in a press, press the rear bearing from the pinion shaft. Be sure to record the thicknesses of the shims that were removed from between the bearing and the gear.

DRIVE PINION

1. With the differential removed, check the pinion preload. Do this by checking the amount of torque needed to turn the pinion gear. For a new bearing, it should be 20–25 inch lbs. and for a used bearing it should be 10–15 inch lbs. If there is no preload reading check the pinion for looseness. If there is any looseness, replace the bearing.

2. Using a holder assembly to secure the flange, remove the flange nut and washer.

3. Using a puller, press the flange from the pinion splines.

4. Thread the pinion nut, a few turns, onto the pinion shaft. Using a brass drift and hammer, lightly tap the end of the pinion shaft to remove the pinion from the carrier; be careful not to allow the pinion to fall out of the carrier.

5. With the pinion removed from the carrier, discard the old seal pinion nut and collapsible spacer; install new ones when reassembling.

RING GEAR

1. Remove the ring gear-to-differential case bolts and tap the ring gear from the case with a soft hammer.

1. Companion Flange		15. Ring gear
2. Deflector		16. Side gear
3. Pinion oil seal		17. Bearing cap
4. Pinion front bearing		18. Axle shaft
5. Pinion bearing spacer		19. Thrust washer
6. Differential carrier		20. Differential pinion
7. Differential case		21. Shim
8. Shim (A) with service shim		22. Pinion rear bearing
9. Gasket		23. Drive pinion
10. Differential bearing		
11. C-lock		
12. Pinion shaft lock bolt		
13. Cover		
14. Pinion shaft		

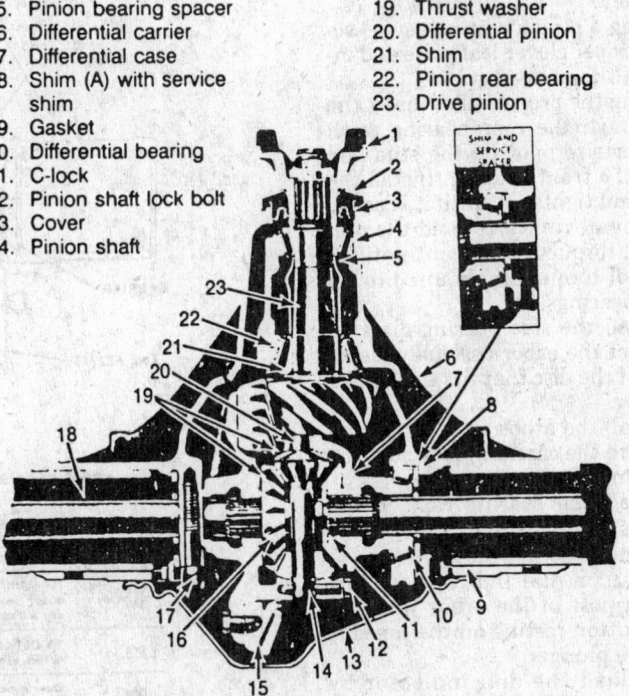

Cross-sectional view of the General Motors 8½ and 9½ inch rear axle assembly

NOTE: Do not try to pry the ring gear off the case. This will damage the machined surfaces.

2. Clean all dirt from the case assembly and lubricate the case with gear lube.

Cleaning and Inspection

1. Clean all parts in solvent and blow dry.

2. Check all of the parts for any signs of wear, chips, cracks or distortion; replace any parts that are defective.

3. Check the fit of the differential side gears in the case and the fit of the side gear and axle shaft splines.

Assembly

DRIVE PINION BEARING

1. Using a bearing driver, install a new bearing cup for each one that was removed; make sure the cups are seated fully against the shoulder in the housing.

2. The pinion depth must be checked to determine the nominal setting. This allows for machining variations in the housing and enables selection of the proper shim so the pinion depth can be set for the best bear tooth contact.

3. Clean the housing and carrier assemblies to insure accurate measurement of the pinion depth.

4. Lubricate the front and rear pinion bearings with gear lubricant and install them in their races in the carrier assembly.

5. Using a pinion setting gauge, select the proper clover leaf plate and install it on the preload stud.

6. Using the proper pilot, insert the stud through the rear bearing, with the proper size pilot on the stud and through the front bearing. Install the hex nut and tighten it until it is snug.

7. Using a wrench to hold the preload stud, torque the hex nut until 20 inch lbs. of torque are required to rotate the bearings.

8. Install the side bearing discs on the ends of the arbor assembly, using the step of the disc that fits the bore of the carrier.

9. Install the arbor and plunger assembly into the carrier; make sure the side bearing discs fit properly.

10. Install the bearing caps in the carrier assembly finger tight; make sure the discs do not move.

11. Mount a dial indicator on the mounting post of the arbor with the contact button resting on the top surface of the plunger.

13. Preload the dial indicator by turning it ½ revolution and tightening it in this position.

14. Use the button on the gauge plate that corresponds to the ring gear size and turn the plate so the plunger rests on top of it.

15. Rock the plunger rod back and forth across the top of the button until the dial indicator reads the greatest amount of variation. Set the dial indicator to zero at the point of most variation. Repeat the rocking of the plunger several times to check the setting.

16. Turn the plunger until it is removed from the gauge plate button. The dial indicator will now read the pinion shim thickness required to set the nominal pinion depth; record the reading.

17. Check for the pinion code number, located on the rear face of the pinion gear; the number will indicate the necessary change to the pinion shim thickness.

NOTE: If the pinion is marked with a plus (+) and a number, add that much to the reading of the dial indicator. If the pinion has no mark, use the reading from the dial indicator as the correct shim thickness. If the pinion is marked with a minus (−) and a number, subtract that amount from the reading on the dial indicator.

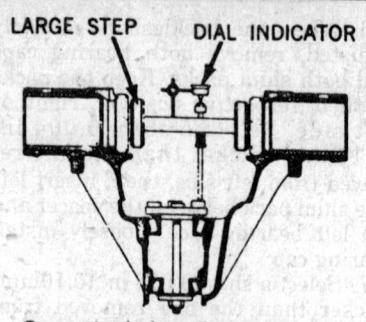

Gauge tools installed in the carrier

18. Remove the depth gauge tools from the carrier assembly and install the proper size shim on the pinion gear.

19. Lubricate the bearing with gear lubricant and use a shop press to press the bearing onto the pinion shaft.

DRIVE PINION

1. Lubricate the front bearing and install it into the front cup.

2. Using a seal driver and a gauge plate, drive the pinion seal into the bore until the gauge plate is flush with the shoulder of the carrier.

3. Lubricate the seal lips and install a new bearing spacer on the pinion gear.

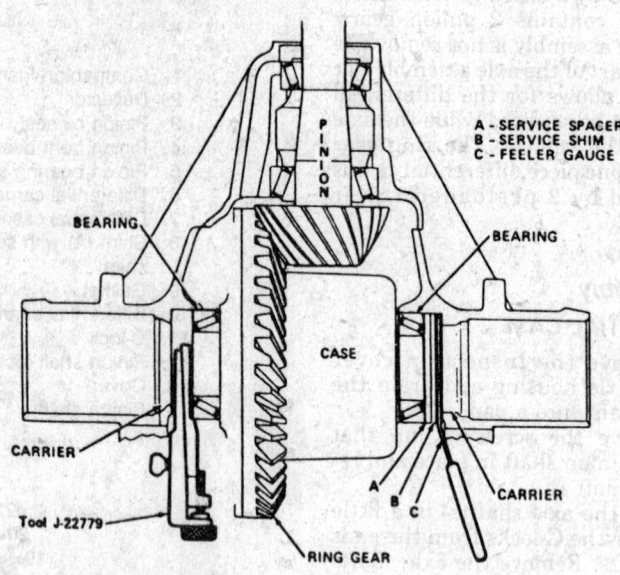

A - SERVICE SPACER
B - SERVICE SHIM
C - FEELER GAUGE

EXAMPLE

	RING GEAR SIDE		OPPOSITE SIDE	
.250"	Thickness of Tool J-22779 required to force ring gear into contact with pinion		Combined total of: Service Spacer (A) Service Shim (B) Feeler Gauge (C)	.265"
− .010" .240"	TO MAINTAIN PROPER BACKLASH (.005" - .008"), ring gear is moved away from pinion by subtracting .010" shims from ring gear side and adding .010" shims to other side			+ .010" .275"
+ .004"	TO OBTAIN PROPER PRELOAD on side bearings, add .004" shims to each side			+ .004"
.244"	Shim dimension required for ring gear side		Shim dimension required for opposite side	.279"

Shim pack selection chart

4. Install the pinion gear into the carrier assembly. Using a large washer and nut, draw the pinion gear through the front bearing, far enough to install the companion flange.

5. With the companion flange installed on the pinion shaft, use a holder assembly and tighten the pinion nut until all of the endplay is removed from the drive pinion.

6. When no more endplay exists, check the preload; the preload should be 20-25 inch lbs. on new bearings or 10-15 inch lbs. on used bearings. Tighten the pinion nut until these figures are reached; do not over tighten the pinion, for this will collapse the spacer too much and make it necessary to replace it.

7. Turn the pinion gear several times to make sure the bearings are seated and recheck the preload.

RING GEAR

1. Align the ring gear bolt holes with the carrier holes and lightly press the ring gear onto the case assembly.

2. Install the bolts and tighten them all evenly, using a criss-cross pattern to avoid cocking the ring gear.

3. When the ring gear is firmly seated against the case, torque the bolts to 60 ft. lbs.

DIFFERENTIAL CASE ASSEMBLY

1. Place the new differential bearing onto the case hub with the thick side of the inner race toward the case. Using a bearing driver, drive the bearing onto the case until it seats against the shoulder on the case.

2. Install the thrust washers and side gears into the case assembly. If the original parts are being used, be sure to place them in their original position.

3. Place the pinions in the case so they are 180 degrees apart as they engage the side gears.

4. Turn the pinion gears so the case holes align with the gear holes. When the holes are aligned, install the pinion shaft and lock screw; do not tighten the lock screw too tightly at this time.

5. Check the bearings, bearing cups, cup seat and carrier caps to make sure they are in good condition.

6. Lubricate the bearings with gear lube. Install the cups on the bearings and the differential assembly into the carrier. Support the carrier assembly to keep it from falling.

7. Install a support strap on the left side bearing and tighten the bearing bolts to an even, snug fit.

8. With the ring gear tight against the pinion gear, insert a gauge tool between the left side bearing cup and carrier housing.

9. While lightly shaking the tool back and forth, turn the adjusting wheel until a slight drag is felt and tighten the locknut.

10. Between the right side bearing and carrier, install a 0.170 in. thick service spacer, a service shim and a feeler gauge. The feeler gauge must be thick enough so a light drag is felt when it is moved between the carrier and the shim.

11. Add the total of the service spacer, service shim and the feeler gauge. Remove the gauge tool from the left side of the carrier, then, using a micrometer, measure the thickness in at least 3 places. Average the readings and record the result.

12. Refer to the chart to determine the proper thickness of the shim packs.

13. Install the left side shim first, then, the right side shim between the bearing cup and spacer. Position the shim so the chamfered side is facing outward or next to the spacer.

NOTE: If there is not enough chamfer around the outside of the shim, file or grind the chamfer a little to allow for easy installation.

14. If there is difficulty in installing the shim, partially remove the case from the carrier and slide both the shim and case back into place.

15. Install the bearing caps and torque them to 60 ft. lbs. Tighten the pinion shaft lock screw.

NOTE: The differential side bearings are now preloaded. If any adjustments are made in later procedures, make sure not to change the preload. Do not change the total thickness of the shim packs.

16. Mount a dial indicator on the carrier assembly with the indicator button perpendicular to the tooth angle and aligned with the gear rotation.

17. Measure the amount of backlash between the ring and pinion gears; it should be between 0.005-0.008 in. Take readings at 4 different spots on the gear; there should not be variations greater than 0.002 in.

18. If there are variations greater than 0.002 in. between the readings, check the runout between the case and ring gear; the gear runout should not be greater than 0.003 in. If the runout exceeds 0.003 in., check the case and ring gear for the deformation or dirt between the case and gear.

19. If the gear backlash exceeds 0.008 in., increase the thickness of the shims on the ring gear side and decrease the thickness of the shims on the opposite side, an equal amount.

20. If the backlash is less than 0.005 in., decrease the shim thickness on the ring gear side and increase the shim thickness on the opposite side an equal amount.

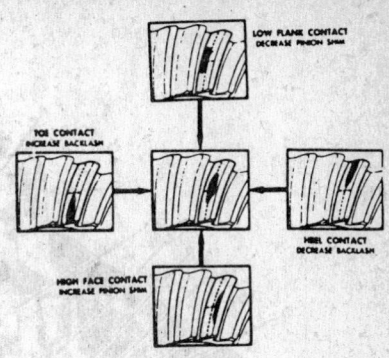

Checking the gear tooth contact

GEAR PATTERN CHECK

Before final assembly of the differential, a pattern check of the gear teeth must be made. This determines if the teeth of the ring and pinion gears are meshing properly, for low noise level and long life of the gear teeth. The most important thing to note is if the pattern is located centrally up and down on the face of the ring gear.

1. Wipe any oil from the carrier and all dirt and oil from the teeth of the ring gear.

2. Coat the teeth of the ring gear with a gear marking compound.

3. With the bearing caps torqued to 55 ft. lbs., expand the brake shoes until it takes 20-30 ft. lbs. of torque to turn the pinion gear.

4. Turn the companion flange so the ring gear makes a full rotation in 1 direction, then, turn it a full rotation in the opposite direction.

5. Check the pattern on the teeth and refer to the chart for any adjustments necessary.

6. With the gear tooth pattern checked and properly adjusted, install the axle housing cover gasket and cover and tighten securely. Refill the axle with gear lube to the correct level.

7. Road test the vehicle to check for any noise and proper operation of the rear.

GMC—10½ INCH RING GEAR AXLE

This axle is a full floating type that uses special hypoid type drive and pinion gears. The pinion gear is supported by 3 bearings, 2 in front of the pinion gear and 1 behind. The differential assembly has either 2 or 4 pinions depending on the application of the axle. This axle assembly must be removed from the vehicle to remove and service the differential.

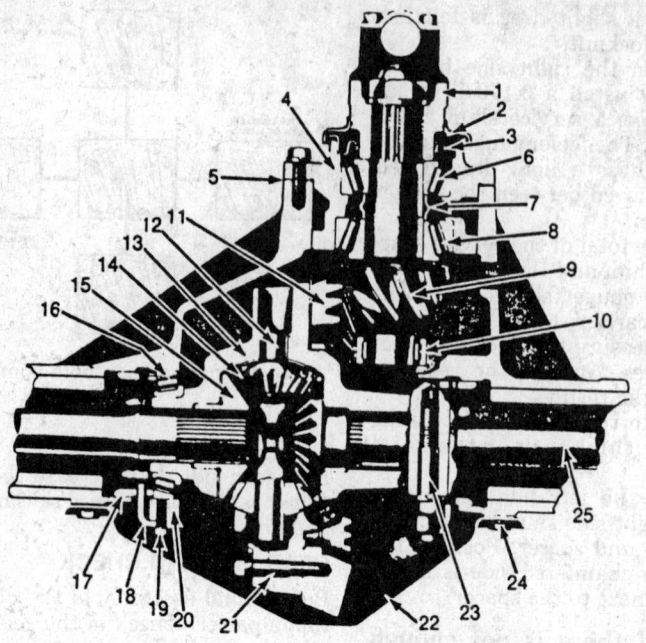

1. Companion flange
2. Oil deflector
3. Oil seal
4. Bearing retainer
5. Shim
6. Pinion front bearing
7. Collapsible spacer
8. Pinion rear bearing
9. Drive pinion
10. Straddle bearing
11. Ring gear
12. Differential spider
13. Differential case
14. Differential pinion
15. Differential side gear
16. Side bearing
17. Side bearing adjusting nut
18. Adjusting nut retainer
19. Retainer screw
20. Bearing cap
21. Case-to-ring gear bolt
22. Differential cover
23. Bearing cap bolt
24. Cover screw
25. Axle shaft

Cross-sectional view of the General Motors 10½ inch rear axle

Disassembly

DIFFERENTIAL

1. Place the axle assembly in a vise or holding fixture.
2. Remove the cover bolts, the cover and allow the gear lubricant to drain into a pan.
3. Remove the axle shafts from the axle assembly.

NOTE: Measure and record the pinion backlash so if the same gears are reused they may be installed at the same backlash to avoid changing the gear tooth pattern.

4. From the bearing caps, remove the adjusting nut lock retainers.
5. Mark the bearing caps so they may be reinstalled in the same position and remove the bearing caps.
6. Loosen the side bearing adjusting nut and remove the differential carrier from the axle housing.

PINION

1. Remove the differential assembly from the axle.

2. Check the pinion bearing for the proper preload. The force required to turn the pinion should be 25–35 inch lbs. for used bearings. If there is no reading, shake the companion flange to check for any looseness in the bearing. If there is any looseness present, replace the bearing.
3. Remove the pinion bearing retainer-to-axle housing bolts.
4. Remove the bearing retainer and pinion assembly from the axle housing. It may be necessary to tap the pilot end of the pinion shaft to help remove the pinion assembly from the carrier.
5. Record the thickness of the shims that are removed from between the carrier assembly and the bearing retainer assembly.

DRIVE PINION

1. With the pinion assembly clamped in a vise, install a holder assembly on the flange.
2. Using the proper size socket, remove the pinion nut and washer from the pinion.

NOTE: When reassembling the pinion use a new nut and washer assembly.

3. With the holder assembly still in place, use a puller to remove the flange from the pinion.
4. With the bearing retainer supported in a shop press, press the pinion out of the retainer assembly; be careful not to allow the pinion gear to fall onto the floor because this can damage the gear.
5. Separate the pinion flange, the oil seal, the front bearing and the bearing retainer; if the oil seal needs to be replaced it may have to be driven from the retainer.
6. Using a drift, drive the front and rear bearing cups from the bearing retainer.
7. Support the pinion assembly in a press, with the bearing supported. Press the bearing from the pinion gear.
8. Using a drift, drive the straddle bearing from the carrier assembly.

DIFFERENTIAL CASE

1. Scribe a line across both halves of the differential case so they may be reassembled in the same position and with the ring gear removed, separate the halves.
2. Remove the ring gear bolts and washers; using a soft hammer tap the ring gear from the case.
3. Remove the internal parts from the case and set them aside in order so they may be reassembled in the same position.
4. If removing the side bearing, perform the following procedures:
   a. Install a bearing puller on the bearing and press the bearing assembly from the differential case.
   b. Check the bearings for any signs of wear on distortion.

Cleaning and Inspection

1. Clean all of the parts in solvent and blow dry.
2. Check the differential gears, pinions, thrust washers and spider for any signs of unusual wear, chips, cracks or pitting.
3. Check the pinion gear for signs of wear, chips, cracks or any other imperfections. Check the splines for signs of wear or distortion.
4. Check all mating surfaces for signs of wear.
5. Check the bearings for signs of wear or pitting on the rollers and races and check the bearing cage for dents and bends. Check the bearing retainer for any cracks, pits, grooves or corrosion.
6. Check the pinion flange splines for any signs of wear or distortion.

7. Replace parts that show any of the signs mentioned above.

Assembly

DIFFERENTIAL CASE

1. If the side bearing was removed, perform the following procedures:

   a. Install a new bearing on the differential case.

   b. Using a bearing driver, drive the bearing onto the case assembly until it seats against the shoulder on the case.

2. Using a good quality gear lubricant coat all of the parts.

3. Assemble the differential pinions and thrust washers onto the spider and install the assembly into the differential case.

4. Align the scribe marks on both halves of the differential case and install the ring gear.

5. Install the ring gear washers and bolts and torque the bolts to approx. 10 ft. lbs.

DRIVE PINION

1. Coat all of the parts with a good quality gear lubricant.

2. Position the pinion gear into a shop press and press the rear bearings onto the pinion assembly.

3. Using a bearing driver, install the front and rear bearing cups into the bearing retainer.

4. Using a bearing driver, install the straddle bearing assembly in the axle housing.

5. Install the bearing retainer, with the bearing cups, onto the pinion gear and install a new collapsible spacer.

6. Using a shop press, press the front bearing onto the pinion gear.

7. Lubricate the oil seal with a good quality high pressure grease and install the seal into the retainer bore; be sure to press the seal until it rests against the internal shoulder.

8. Install the pinion flange and oil deflector onto the pinion gear splines, then, install a new lock washer and pinion nut.

9. With the pinion flange clamped in a vise and a holder assembly installed on the flange, tighten the nut to obtain the proper preload; 25–35 inch lbs. for a new bearing or 5–15 inch lbs. for a used bearing. To preload the bearing, tighten the pinion nut to approx. 350 ft. lbs. and take a reading of the torque required to turn the pinion. Continue tightening the nut until the proper preload is obtained.

   NOTE: Do not tighten the nut too tightly because it will collapse the spacer too much. This will make replacement necessary.

DRIVE PINION

1. If installing a new pinion gear, check the top of the new gear for the depth code number.

2. Compare the new number with the old number on top of the old pinion and check the pinion depth chart for preliminary setting of the pinion depth.

3. Check the thickness of the original shims removed from the pinion and either add or subtract from the shims according to the chart.

4. Place the shim on the carrier assembly and align the holes with those in the axle housing; make sure the surfaces are clean of all dirt and grease.

5. Install the retainer and pinion assembly in the housing making sure the holes align. Install the retaining bolts and torque to approx. 45 ft. lbs.

DIFFERENTIAL CASE

1. Place the bearing cups over the side bearings on the differential assembly and place the unit into the carrier in the axle housing.

2. Align the marks and install the bearing caps and the bolts. Tighten the bearing retaining bolts.

3. Loosen the right side nut and tighten the left side nut until the ring gear comes in contact with the pinion gear; do not force the gears together. This brings the gears to zero lash.

4. Back off the left side adjusting nut about 2 slots and install the lock fingers into the nut.

5. In this order, tighten the right side adjusting nut firmly to force the case assembly into tight contact with the left side adjusting nut, then, loosen the right side nut until it is free from the bearing.

6. Again, retighten the right side adjusting nut until it comes in contact with the bearing. Tighten the right adjusting nut about 2 slots for an old bearing or 3 slots for a new bearing.

7. Install the lock retainers into the slots and torque the bearing cap bolts to 100 ft. lbs.; this procedure now insures the bearings are preloaded properly. If more adjustments are made, make sure the preload stays the same.

		CODE NUMBER ON ORIGINAL PINION			
		+2	+1	0	-1	-2
CODE NUMBER ON SERVICE PINION	+2	–	ADD .001	ADD .002	ADD .001	ADD .004
	+1	SUBT. .001	–	ADD .001	ADD .002	ADD .003
	0	SUBT. .002	SUBT. .001	–	ADD .001	ADD .002
	-1	SUBT. .003	SUBT. .002	SUBT. .001	–	ADD .001
	-2	SUBT. .004	SUBT. .003	SUBT. .002	SUBT. .001	–

Pinion depth codes and corresponding shim thicknesses

To do this, 1 adjusting nut must be loosened the same amount the other nuts is tightened.

8. Install a dial indicator on the housing and measure the amount of backlash between the ring and pinion gear. The backlash should measure between 0.003–0.012 in. with the best figure being between 0.005–0.008 in.

9. If the backlash is more than 0.012 in., loosen the right side adjusting nut 1 slot and tighten the left side 1 slot. If the backlash is less than 0.003 in., loosen the left side nut 1 slot and tighten the right side 1 slot. These adjustments should bring the backlash measurement into an acceptable range.

PATTERN CHECK

1. Clean all the oil from the ring gear. Using a gear marking compound, coat the teeth of the ring gear.

2. Make sure the bearing caps are torqued to 110 ft. lbs. and apply load to the gears while rotating the pinion. Rotate the ring gear a full turn in both directions.

   NOTE: Load must be applied to the assembly while rotating or the pattern will not show completely.

3. Check the pattern on the ring gear, adjust the assembly to get the contact pattern located centrally on the face of the ring gear teeth.

DANA – 9¾ AND 10½ INCH RING GEAR AXLES

The Dana Corporation's 9¾ and 10½ in. ring gear axle assemblies are basically the same but with certain exceptions. The differential side bearing shims are located between the side bearing cup assembly and the differential case on the 9¾ in. ring gear axle assembly, while on the 10½ in. ring gear axle assembly, the side bearing shims are located between the side bearing cup and the axle housing. Both axles use inner and outer shims on the pinion gear. The inner shims are used to control the pinion depth in the housing, while the outer shims are used to preload the pinion bearings. The 9¾ in. ring gear axle uses a solid differential carrier with a removable side and pinion gear shaft. The 10½ in. ring gear axle uses a split differential carrier with the side and pinion gears mounted on a cross shaft.

Disassembly

DIFFERENTIAL CARRIER

9¾ Inch

1. The axle assembly can be overhauled either in or out of the vehicle.

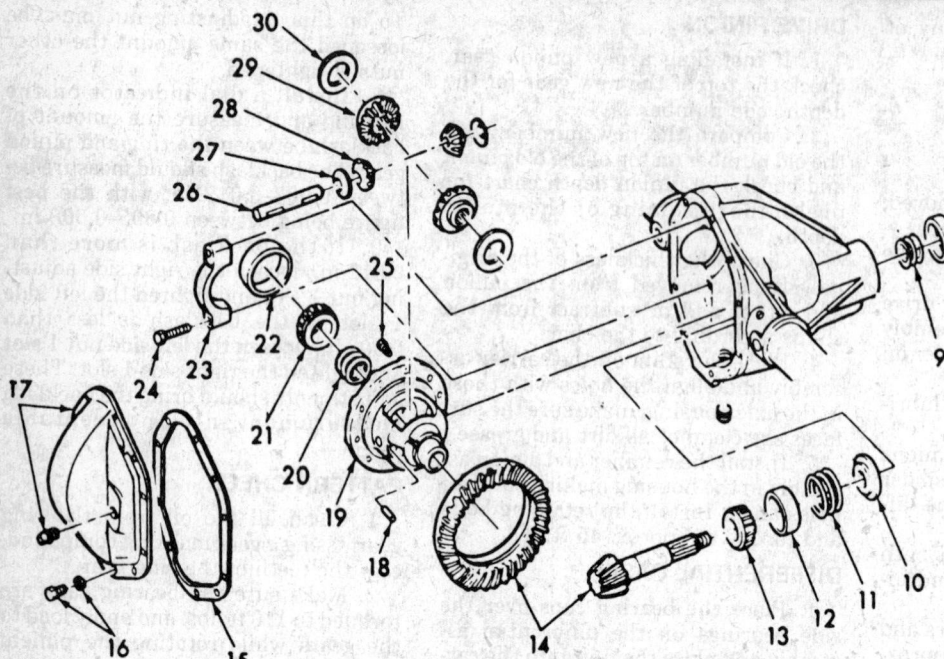

15. Gasket (housing cover)
16. Screw and washer (cover)
17. Cover and plug
18. Lock pin (pinion shaft)
19. Differential case
20. Shims (differential bearing)
21. Cone and roller (differential bearings)
22. Cup (differential bearing)
23. Cap (differential bearing)
24. Bolt (differential bearing cap)
25. Bolt (ring gear)
26. Pinion shaft
27. Thrust washer (pinion)
28. Pinion
29. Side gear
30. Thrust washer (side gear)

1. Nut
2. Washer
3. Companion flange
4. Pinion oil seal
5. Gasket
6. Outer pinion oil slinger

7. Cone and roller (outer pinion bearing)
8. Cone and roller (outer pinion bearing)
9. Shims (outer pinion bearing)

10. Inner pinion oil slinger
11. Shims (inner pinion bearing)
12. Cup (inner pinion bearing)
13. Cone and roller (inner pinion)
14. Ring and pinion

Exploded view of the Dana 9¾ inch differential assembly

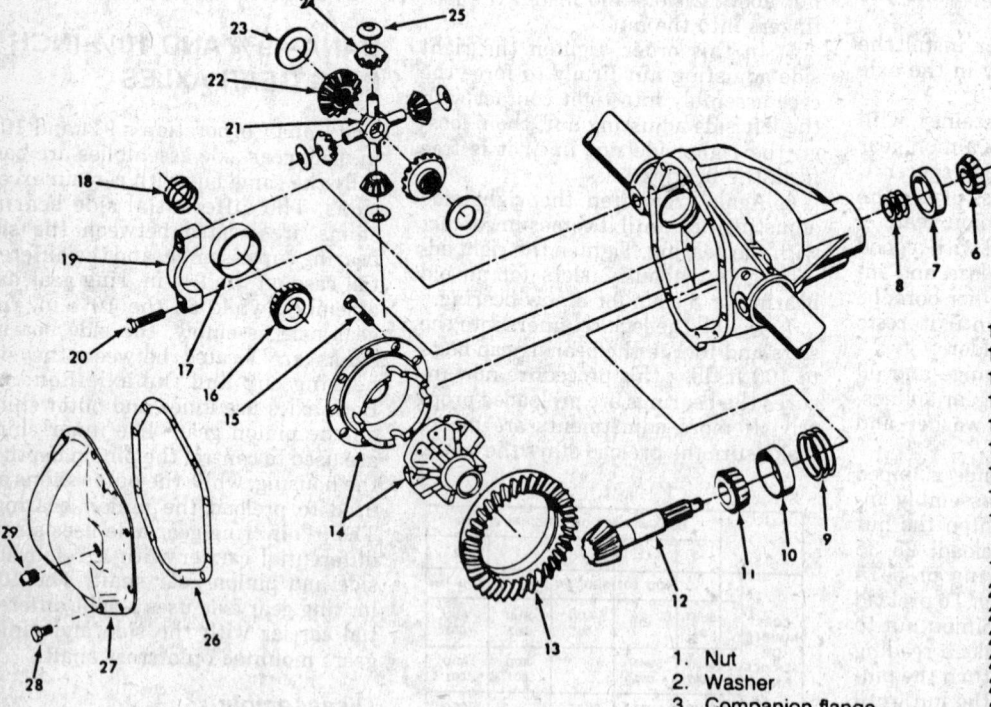

8. Preload shim pack
9. Pinion depth shim pack
10. Rear bearing cup
11. Pinion rear bearing
12. Drive pinion
13. Ring gear
14. Differential case
15. Ring gear bolt
16. Differential side bearing
17. Side bearing cup
18. Side bearing adjusting shims
19. Bearing cap
20. Bearing cap bolt
21. Differential spider
22. Differential side gear
23. Washer
24. Pinion gear
25. Washer
26. Gasket
27. Cover
28. Cover screw
29. Drain plug

1. Nut
2. Washer
3. Companion flange
4. Oil seal
5. Oil slinger
6. Pinion front bearing
7. Front bearing cup

Exploded view of the Dana 10½ inch differential assembly

Either way, the free-floating axles must be removed.

2. Drain the lubricant and remove the rear cover and gasket.

3. Matchmark the bearing caps and the housing for reassembly in the same position. Remove the bearing caps and bolts.

4. Using a spreader tool mounted to the carrier housing, spread the housing a maximum of 0.015 in.

NOTE: Do not exceed this measurement. The housing could be permanently damaged. The use of a dial indicator is recommended to prevent over-stretching the housing.

5. Using a pry bar, remove the differential case from the housing. Separate the shims and record the dimensions and location on the 10½ in. ring gear axle. Remove the spreader tool from the housing.

6. Remove the differential side bearing cups and tag to identify the side, if they are to be used again.

7. Remove the differential gear pinion shaft lock pin and remove the shaft. Rotate the side and pinion gears to remove them from the carrier. Remove the thrust bearings.

8. Remove the bearing cones and rollers from the carrier, marking and noting the shim locations.

9. Remove the ring gear bolts and tap the ring gear from the carrier housing.

10. Inspect the components.

10½ Inch

1. The axle assembly can be overhauled either in or out of the vehicle. Either way, the free-floating axles must be removed.

2. Drain the lubricant and remove the rear cover and gasket.

3. Matchmark the bearing caps and the housing for reassembly in the same position. Remove the bearing caps and bolts.

4. Using a spreader tool mounted to the carrier housing, spread the housing a maximum of 0.015 in.

NOTE: Do not exceed this measurement. The housing could be permanently damaged. The use of a dial indicator is recommended to prevent over-stretching the housing.

5. Using a pry bar, remove the differential case from the housing. Separate the shims and record the dimensions and location on the 10½ in. ring gear axle. Remove the spreader tool from the housing.

6. Using puller tools, press the differential side bearings from the case.

7. Remove the ring gear bolts and tap the ring gear from the case with a soft-faced hammer.

8. Matchmark the case halves for reassembly and remove the retaining bolts.

9. Tap the upper case ½ to separate it from the bottom ½. Remove the internal gears, washers and cross.

DRIVE PINION

1. Remove the pinion nut and flange from the pinion gear.

2. Remove the pinion gear assembly from the housing. It may be necessary to tap the pinion from the housing with a soft faced hammer. Catch the pinion so as not to allow it to drop on the floor.

3. With a long drift, remove the inner bearing cup, pinion seal, slinger, gasket, outer pinion bearing and the shim pack. Label the shim pack for reassembly.

4. Remove the rear pinion bearing cup and shim pack from the housing. Label the shims for reassembly.

5. Remove the rear pinion bearing from the pinion gear with an arbor press and special plates.

Inspection

1. Clean the gears, bearings and component parts with solvent and inspect for scoring, chipping or excessive wear.

2. Inspect the flanges and splines for excessive wear.

3. Replace the necessary parts as required.

SIDE BEARING SHIM SELECTION

10½ Inch

1. With the pinion gear not in the axle housing, place the bearing cups over the side bearings and install the differential carrier into the axle housing.

2. Place the shim that was originally installed on the ring gear side into its original position.

3. Install the bearing caps in their proper positions and tighten the bolts to keep the bearings in place.

4. Mount a dial indicator on the axle housing with the indicator button contacting the back of the ring gear.

5. Position 2 prybars between the bearing shim and the housing on the ring gear side of the differential carrier. Force the differential carrier away from the dial indicator and set the indicator to zero.

6. Reposition the prybars to the opposite side of the differential carrier

Old Pinion Marking	New Pinion Marking								
	− 4	− 3	− 2	− 1	0	+ 1	+ 2	+ 3	+ 4
+ 4	+ 0.008	+ 0.007	+ 0.006	+ 0.005	+ 0.004	+ 0.003	+ 0.002	+ 0.001	0
+ 3	+ 0.007	+ 0.006	+ 0.005	+ 0.004	+ 0.003	+ 0.002	+ 0.001	0	− 0.001
+ 2	+ 0.006	+ 0.005	+ 0.004	+ 0.003	+ 0.002	+ 0.001	0	− 0.001	− 0.002
+ 1	+ 0.005	+ 0.004	+ 0.003	+ 0.002	+ 0.001	0	− 0.001	− 0.002	− 0.003
0	+ 0.004	+ 0.003	+ 0.002	+ 0.001	0	− 0.001	− 0.002	− 0.003	− 0.004
− 1	+ 0.003	+ 0.002	+ 0.001	0	− 0.001	− 0.002	− 0.003	− 0.004	− 0.005
− 2	+ 0.002	+ 0.001	0	− 0.001	− 0.002	− 0.003	− 0.004	− 0.005	− 0.006
− 3	+ 0.001	0	− 0.001	− 0.002	− 0.003	− 0.004	− 0.005	− 0.006	− 0.007
− 4	0	− 0.001	− 0.002	0.003	− 0.004	− 0.005	− 0.006	− 0.007	− 0.008

and force the carrier back towards the dial indicator. Repeat several times until the same reading is obtained each time.

7. To the dial indicator reading, add the thickness of the shim and record the results to be used later in the assembly.

PINION SHIM SELECTION

Ring gears and pinions are supplied in matched sets only. The matched numbers are etched on both gears for verification. On the rear face of the pinion, a plus (+) or a minus (−) number will be etched, indicating the best running position for each particular gear set. This dimension is controlled by the shimming behind the inner bearing cup. Whenever baffles or oil slingers are used, they become part of the adjusting shim pack. An example: If a pinion is etched +3, this pinion would require 0.003 in. less shims than a pinion etched 0. This means by removing shims, the mounting distance of the pinion is increased by 0.003 in., which is just what a plus (+) etching indicates. If a pinion is etched −3, it would be necessary to add 0.003 in. more shims than would be required if the pinion was etched 0. By adding the 0.003 in. shims, the mounting distance of the pinion is decreased 0.003 in., which is just what the minus (−) etching indicates. Pinion adjusting shims are available in thicknesses of 0.003, 0.005 and 0.010 in. An example: If a new gear set is used and the old pinion reads +2 and the new pinion reads −2, add 0.004 in. shims to the original shim pack.

Assembly
DRIVE PINION

1. Select the correct pinion depth shims and install in the rear pinion bearing cup bore.
2. Install the rear bearing cup in the axle housing.
3. Add or subtract an equal amount of shim thickness to or from the preload or outer shim pack, as was added or subtracted from the inner shim pack.
4. Install the front pinion bearing cup into its bore in the axle housing.
5. Press the rear pinion bearing onto the pinion gear shaft and install the pinion gear with bearing into the axle housing.
6. Install the preload shims and the front pinion bearing; do not install the oil seal at this time.
7. Install the flange with the holding bar tool attached, the washer and the nut on the pinion shaft end. Torque the nut to 250 ft. lbs. for the 10½ in. and 255 ft. lbs. for the 9¾ in.

8. Remove the holding bar from the flange and with an inch lb. torque wrench, measure the rotating torque of the pinion gear. The rotating torque should be 10–20 inch lbs. with the original bearings or 20–40 inch lbs. with new bearings. Disregard the torque reading necessary to start the shaft to turn.
9. If the preload torque is not in specifications, adjust the shim pack as required.
   a. To increase preload, decrease the thickness of the preload shim pack.
   b. To decrease preload, increase the thickness of the preload shim pack.
10. When the proper preload is obtained, remove the nut, washer and flange from the pinion shaft.
11. Install a new pinion seal into the housing and reinstall the flange, washer and nut. Using the holder tool, torque the nut to 250 ft. lbs. for the 10½ in. and 255 ft. lbs. for the 9¾ in.

DIFFERENTIAL CARRIER
9¾ Inch

1. Install the differential side gears, the differential pinion gears and new thrust washers into the differential carrier.
2. Align the pinion gear shaft holes and install the pinion shaft into the carrier. Align the lock pin hole in the shaft and carrier. Install the lock pin and peen the hole to avoid having the pin drop from the carrier.
3. Install the differential case side bearings with the proper installation tools. Do not install the shims at this time.
4. Place the carrier assembly into the axle housing with the bearing cups on the bearing cones. Install the bearing caps in their original position and tighten the bearing cap bolts enough to keep the bearing caps in place.
5. Install a dial indicator on the housing so the indicator button contacts the carrier flange. Press the differential carrier to prevent sideplay and center the dial indicator. Rotate the carrier and check the flange for run-out. If the run-out is greater than 0.002 in., the defect is probably due to the bearings or to the carrier and should be corrected.
6. Remove the assembly and install the ring gear. Torque the retaining bolts and reinstall the assembly into the housing. Install the bearing caps in their original position and tighten the cap bolts to keep the bearings caps in place.
7. Install the dial indicator and position the indicator button to contact the ring gear back surface. Rotate the assembly and the run-out should be

less than 0.002 in. If over 0.002 in., remove the assembly and relocate the ring gear 180 degrees. Reinstall the assembly and recheck. If the run-out remains over the 0.002 in. tolerance, the ring gear is defective. If the measurement is within tolerances, continue on with the assembly.

8. Position 2 pry bars between the bearing cap and the housing on the side opposite the ring gear. Pull on the pry bars and force the differential carrier as far as possible towards the dial indicator. Rock the assembly to seat the bearings and reset the dial indicator to zero.
9. Reposition the prybars to the opposite side of the carrier and force the carrier assembly as far towards the center of the housing. Read the dial indicator scale. This will be the total amount of shims required for setting the backlash during the reassembly, less the bearing preload. Record the measurement.
10. With the pinion gear installed and properly set, position the differential carrier assembly into the axle housing and install the bearing caps in their proper positions. Tighten the cap bolts just to hold the bearing cups in place.
11. Install a dial indicator on the axle housing with the indicator button contacting the back of the ring gear.
12. Position 2 prybars between the bearing cup and the axle housing on the ring gear side of the case and pry the ring gear into mesh with the pinion gear teeth, as far as possible. Rock the ring gear to allow the teeth to mesh and the bearings to seat. With the pressure still applied by the prybars, set the dial indicator to zero.
13. Reposition the prybars on the opposite side of ring gear and pry the gear as far as it will go. Take the dial indicator reading. Repeat this procedure until the same reading is obtained each time. This reading represents the necessary amount of shims between the differential carrier and the bearing on the ring gear side.
14. Remove the bearing from the differential carrier on the ring gear side and install the proper amount of shims. Reinstall the bearing.
15. Remove the differential carrier bearing from the opposite side of the ring gear. To determine the amount of shims needed, use the following method.

   a. Subtract the size of the shim pack just installed on the ring gear side of the carrier from the reading obtained and recorded when measurement was taken without the pinion gear in place.
   b. To this figure, add an additional 0.015 in. to compensate for pre-

load and backlash. An example: If the first reading was 0.085 in. and the shims installed on the ring gear side of the carrier were 0.055 in., the correct amount of shims would be $0.085 - 0.055 + 0.015 = 0.045$ in.

16. Install the required shims as determined under step 15 and install the differential side bearing. The installation of the shims should give the proper preload to the bearings and the proper backlash to the ring and pinion gears.

17. Spread the axle housing with the spreader tool no more than 0.015 in. Install the differential bearing outer cups in their correct locations and install the cups in their respective locations.

18. Install the bolts and tighten finger-tight. Rotate the differential carrier and ring gear and tap with a soft-faced hammer to insure proper seating of the assembly in the axle housing.

19. Remove the spreader tool and torque the cap bolts to specifications.

20. Install a dial indicator and check the ring gear backlash at 4 equally spaced points of the ring gear circle. The backlash must be within a range of 0.004–0.009 in. and must not vary more than 0.002 in. between the points checked.

21. If the backlash is not within specifications, the shim packs must be corrected to bring the backlash within limits.

22. Check the tooth contact pattern and verify.

23. Complete the assembly, refill to proper level with lubricant and operate to verify proper assembly.

10½ Inch

1. Install new thrust washers to the side gears and lubricate the contact surfaces.

2. Assemble the side gears, pinion bears, washers and cross shaft into the flanged case ½.

3. Install the upper case ½ to the bottom ½, making sure the scribe marks are aligned.

4. Install the retaining bolts finger tight, then, torque the bolts alternately.

5. If a new ring gear is to be installed or the old one used, install it to the differential case and align the bolt holes and torque the bolts alternately.

6. Install the side carrier bearings.

7. Install the differential carrier, with the side bearings and cups installed, in place in the axle housing.

8. Select the smallest of the original shims as a gauge shim and place it between the bearing cup and the housing on the ring gear side.

9. Install the bearing caps and tighten the bolts to hold the cups in place.

10. Mount a dial indicator on the ring gear side of the axle housing and position the indicator button on the rear side of the ring gear.

11. Position 2 prybars between the bearing cup and the housing on the side opposite the ring gear. With the prybars, force the differential carrier towards the dial indicator and set the indicator dial to zero.

12. Reposition the prybars on the ring gear side of the carrier and force the ring gear into mesh with the pinion gear while observing the dial indicator reading. Repeat this operation until the same reading is obtained each time.

13. Add this indicator reading to the gauging shim thickness to determine the correct shim dimension for installation on the ring gear side of the differential carrier.

14. An example: If the gauging shim was 0.115 in. and the indicator reading was 0.017 in., the correct shim would be $0.115 + 0.017 = 0.172$ in.

15. Remove the gauge shim and install the correct shim into position between the bearing cup and the axle housing on the ring gear side of the housing.

16. To determine the correct dimension for the remaining shim, refer to the side bearing shim selection for the 10½ in. and obtain the recorded shim size. From that figure, subtract the size of the shim installed in step 14 and add 0.006 in. for the bearing preload and backlash.

17. An example: If the reading of the shim just installed on the ring gear side of the carrier was 0.172 in. and the reading obtained during the checking of clearance without the pinion installed was 0.329, the correct shim dimension would be as follows: $0.329 - 0.172 = 0.157 + 0.006 = 0.163$ in.

18. Spread the axle housing with a spreader tool, no more than 0.015 in. The carrier assembly is in place in the housing.

19. Assemble the shim, as determined previously, into place between the bearing cup and the housing. Remove the spreader tool.

20. Install the bearing caps in their marked positions and torque the bolts to specifications.

21. Install a dial indicator and check the ring gear backlash at 4 equally spaced points around the ring gear.

22. The backlash must be within 0.004–0.009 in. and must not vary more than 0.002 in. between the positions checked.

23. Whenever the backlash is not within the allowable limits, it must be corrected. Changing of the shim packs is required.

   a. Low backlash is corrected by decreasing the shim on the ring gear

side and increasing the opposite side shim an equal amount.

   b. High backlash is corrected by increasing the shim on the ring gear side and decreasing the opposite side shim an equal amount.

24. Check the tooth contact pattern and correct as required.

25. Complete the assembly, refill to the correct level and operate to verify correct repairs.

ROCKWELL – 12 INCH RING GEAR AXLE

Disassembly
DIFFERENTIAL

1. Remove locknut, adjusting screw and thrust block.

2. Remove 2 adjuster lock cap screws and locks.

3. Punch-mark bearing caps and carrier to help in locating caps for assembly. Remove bearing adjusters and bearing caps.

NOTE: Do not pry caps free with a prybar or distort locating dowels.

4. Carefully remove differential assembly from carrier.

5. Use a differential side bearing remover to pull bearing cones off each side of case.

6. Make sure the differential case halves are punch-marked so they can be reassembled in same position.

7. Remove drive gear and separate case halves.

8. Remove 2 side gears, the differential spider and the 4 differential pinions.

9. Remove pinion and side gear thrust washers.

DRIVE PINION

1. Remove seal retainer and gasket from carrier.

2. Using brass drift against inner end of pinion, drive out the pinion and bearing assembly.

3. Remove shim pack from carrier from those models having tapered roller outer bearings.

4. It may be necessary to use a drift to remove the pinion rear bearing.

5. Clamp yoke in soft-jawed vise. Remove yoke nut and washer and separate drive pinion from yoke.

6. Separate yoke from oil seal retainer.

7. Place retainer in a soft-jawed vise and, using a hammer and chisel, remove oil seal and then the felt oil seal.

8. If equipped with a tapered roller outer bearing, remove bearing cup, outer tapered bearing cone and bearing spacer from drive pinion.

9. Using a bearing remover press plate with shop press, press the bearing cone or roller bearing from the drive pinion.

10. If equipped, remove the bearing lock ring and use press plates with shop press to press the roller bearing from inner end of drive pinion.

Inspection

1. Clean the gears, bearings and component parts with solvent and inspect for scoring, chipping or excessive wear.

2. Inspect the flanges and splines for excessive wear.

3. Replace the necessary parts as required.

Assembly

NOTE: Thoroughly, clean and lubricate all components with axle lubricant before reassembling.

DRIVE PINION

1. Clean counterbore of oil seal retainer. Saturate the felt seal in oil and install evenly in retainer. Soak oil seal in light engine oil for about 1 hour before installing. Coat outer surface of seal lightly with sealing compound to prevent oil leaks between seal and retainer.

2. Install oil seal into retainer with lip of seal toward inner side of retainer. Using a seal installer, press oil seal into retainer with face of seal flush with retainer face.

3. Retainer surface must be clean and smooth to prevent oil leaks between retainer and carrier.

4. Using a shop press, press bearing into place into carrier bore or the roller bearing into position on drive pinion with chamfered side of inner race facing toward pinion shoulder. Position bearing lock ring to secure bearing on drive pinion.

NOTE: Opposed tapered roller bearing cones, 2 bearing cups and spacer are serviced and replaced as a unit. The spacer is a preselected one to provide proper bearing adjustment.

5. Models with tapered roller bearings:

a. Press inner bearing cone into place with largest side of cone facing the pinion gear end.

b. Install original shim pack in carrier. If original ring gear and pinion are reinstalled, use shims that were removed. Shims are available in 5 thicknesses: 0.012, 0.015, 0.018, 0.021 and 0.024 in. When using new

gears, start with a 0.021 in. shim and check the pinion depth.

c. Insert the pinion assembly into carrier, align roller bearing with carrier boss. Install bearing spacer, bearing cup and bearing cone with the wide side facing pinion splines.

6. If equipped with double-row ball bearing: Using a 2 in. pipe or tubing, drive bearing unit into proper seating position.

7. With pinion assembly properly positioned in carrier, install new gasket. Install seal retainer onto yoke and assemble yoke and retainer assembly onto splined end of drive pinion.

8. Secure the retainer to carrier with lock washers and cap screws and torque.

9. Secure the pinion assembly with yoke washer and nut and torque to 220 ft. lbs.

DIFFERENTIAL

1. To facilitate installation of drive gear, install 2 guide pins ($\frac{1}{2} \times 20 \times 2$ in. bolts) in gear. Start the guide pins through the case flange holes and tap drive gear onto case. If a differential gear is bad, the complete set should be replaced.

2. Lubricate the differential case inner walls and all component parts with axle lubricant. Place the differential pinions and thrust washers onto the spider.

3. Assemble the side gears, pinions, side gear and pinion thrust washers onto the left half of differential.

4. Assemble the drive gear half, right half of differential, being sure to align marks on both halves.

5. Install differential-to-drive gear cap screw and lock washers and tighten evenly until drive gear is flush with case flange. Remove guide pins and install cap screws and torque.

6. Using an installer tool, install the differential side bearing cones.

7. Install the bearing cap locating dowels into the caps. Lubricate the side bearings and place the bearing cups on bearings.

8. Install the differential assembly into the carrier. Carefully install the bearing adjusters into the carrier.

9. Install the bearing caps by aligning the punch marks previously made. Be sure the bearing adjuster threads are engaged with carrier and caps. Tighten the adjusters alternately and evenly. Tighten the bearing cap screws until the lock washers are flat.

Adjustment

DRIVE GEAR AND PINION

1. Loosen the bearing cap screws, enough, to loosen the right-hand bearing adjuster (pinion side) and tighten

the left-hand bearing adjuster (opposite pinion side.) Using the adjuster, remove all backlash between the drive gear and pinion.

2. Back off the left-hand bearing adjuster about 2 notches to point where notch in adjuster is aligned with lock. Tighten the right-hand bearing adjuster solidly to seat the bearing. Loosen right-hand adjuster, enough, to free the bearing; then, retighten against the bearing. Draw up right-hand adjuster 1–2 more notches until adjuster notch aligns with the lock.

3. Using a dial indicator on carrier adjuster, slowly oscillate the drive gear and measure the backlash; it should be 0.005–0.008 in.

4. If backlash exceeds 0.008 in., loosen the right-hand adjuster 1 notch; then, tighten the left-hand adjuster 1 notch. If less than 0.005 in., loosen the left-hand adjuster 1 notch and tighten the right-hand adjuster 1 notch.

5. After the backlash has been adjusted, tighten the bearing cap screws until their respective lock washers flatten out.

6. Check the drive gear run-out.

7. Install the side bearing adjusting locknut and secure with cap screws and lock washers.

CHECKING PINION DEPTH

NOTE: This procedure is performed if the pinion is equipped with tapered roller bearings.

1. Coat the drive gear with red lead.

2. Turn the pinion shaft several revolutions in both directions while applying considerable drag on drive gear.

3. Pinion depth is determined by shim pack selection. Shim packs are available in thicknesses of: 0.012, 0.015, 0.018, 0.021 and 0.024 in.

4. Changing the pinion depth will again require adjusting backlash. After pinion depth and backlash have been adjusted, torque bearing caps.

THRUST BLOCK

1. Install the thrust block and locknut to the adjusting screw.

2. Thread the screw and block into the carrier until the block contacts the drive gear.

3. Rotate the gear and note the drag change.

4. Adjust these parts until point of greatest drag is reached. Back off the screw about a 30 degrees to provide 0.005–0.007 in. clearance between the block and gear. Make certain screw does not turn at all when torquing the locknut to 135 ft. lbs.

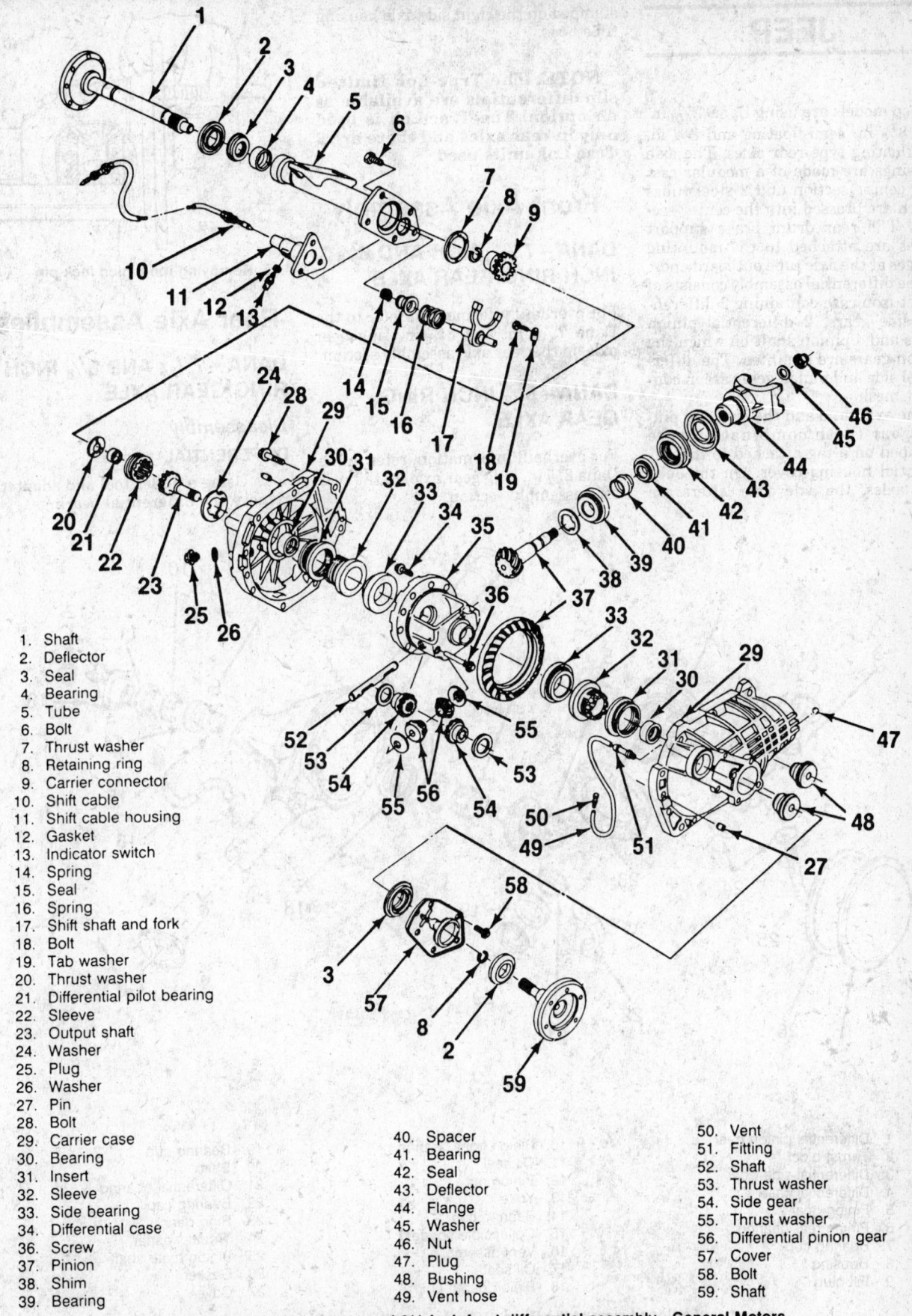

1. Shaft
2. Deflector
3. Seal
4. Bearing
5. Tube
6. Bolt
7. Thrust washer
8. Retaining ring
9. Carrier connector
10. Shift cable
11. Shift cable housing
12. Gasket
13. Indicator switch
14. Spring
15. Seal
16. Spring
17. Shift shaft and fork
18. Bolt
19. Tab washer
20. Thrust washer
21. Differential pilot bearing
22. Sleeve
23. Output shaft
24. Washer
25. Plug
26. Washer
27. Pin
28. Bolt
29. Carrier case
30. Bearing
31. Insert
32. Sleeve
33. Side bearing
34. Differential case
36. Screw
37. Pinion
38. Shim
39. Bearing

40. Spacer
41. Bearing
42. Seal
43. Deflector
44. Flange
45. Washer
46. Nut
47. Plug
48. Bushing
49. Vent hose

50. Vent
51. Fitting
52. Shaft
53. Thrust washer
54. Side gear
55. Thrust washer
56. Differential pinion gear
57. Cover
58. Bolt
59. Shaft

Exploded view of the 8¼ and 9¼ inch front differential assembly—General Motors

JEEP

Jeep models are using Dana $7\frac{9}{16}$ in. and $8\frac{7}{8}$ in. semi-floating and $8\frac{1}{2}$ in. full-floating type rear axles. The axle housings are made of a modular cast iron center section and 2 steel tubes which are pressed into the center section. The rear drum brake support plates are attached to the mounting flanges at the axle tube outboard ends.

The differential assembly consists of a cast iron case containing 2 differential side gears, 2 differential pinion gears and a pinion shaft on which the pinion gears are mounted. The differential side and pinion gears are in constant mesh.

The axle ratio and the ring and pinion gear tooth combinations are stamped on a tag attached to the differential housing cover. On the Jeep rear axles, the axle code letters are stamped on the right side axle housing tube boss.

NOTE: The Trac-Lok limited slip differentials are available as an option. The Trac-Lok is used only in rear axles and there are 2 Trac-Lok units used.

Front Axle Assembly

DANA – $7\frac{9}{16}$ INCH AND $8\frac{7}{8}$ INCH RING GEAR AXLE

For overhaul information, refer to the Dana $7\frac{9}{16}$ in. and $8\frac{7}{8}$ in. ring gear axle, in the rear axle assembly section.

DANA – $8\frac{1}{2}$ INCH RING GEAR AXLE

For overhaul information, refer to the Dana $8\frac{1}{2}$ in. ring gear axle, in the rear axle assembly section.

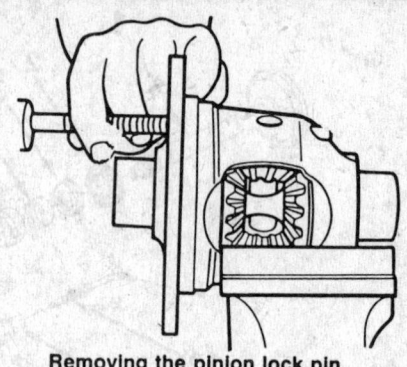

Removing the pinion lock pin

Rear Axle Assemblies

DANA – $7\frac{9}{16}$ AND $8\frac{7}{8}$ INCH RING GEAR AXLE

Disassembly
DIFFERENTIAL

1. Using a puller tool and adapters, remove the differential bearings.

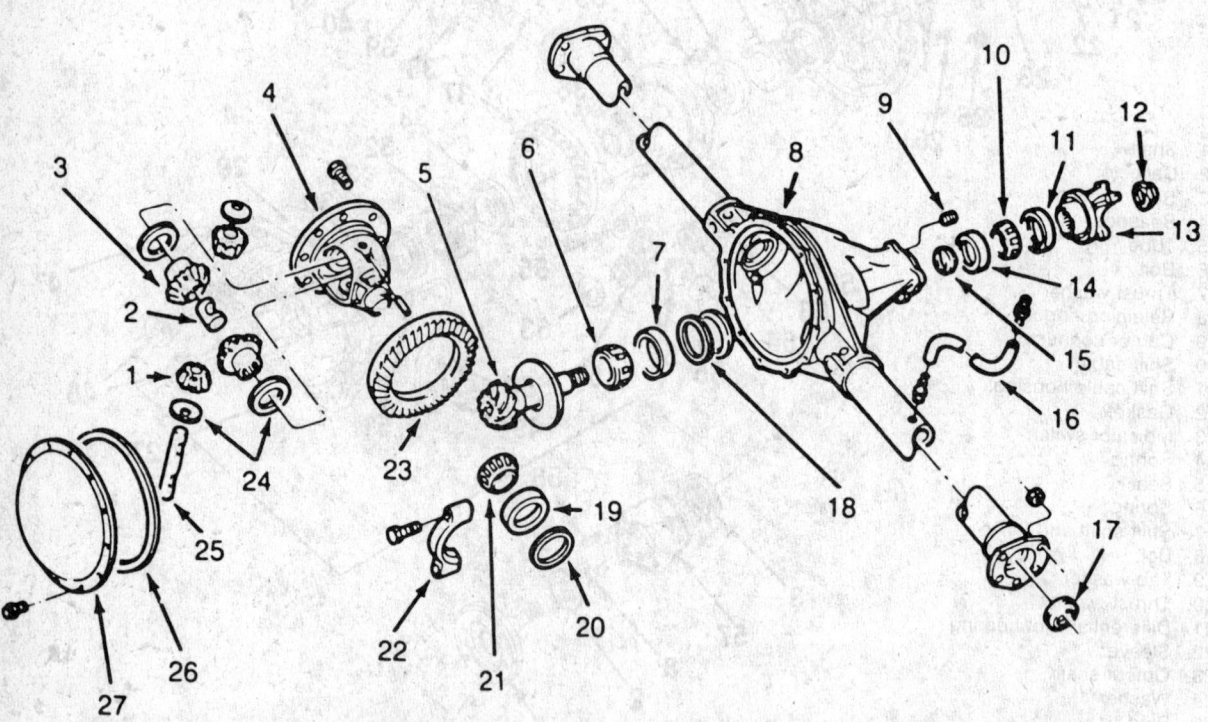

1. Differential pinion gear	10. Pinion front bearing
2. Thrust block	11. Oil seal
3. Differential side gear	12. Pinion nut
4. Differential case	13. Yoke
5. Pinion gear	14. Bearing cup
6. Pinion rear bearing	15. Collapsible spacer
7. Bearing cup	16. Vent assembly
8. Housing	17. Oil seal
9. Fill plug	18. Pinion depth shim

19. Bearing cup	
20. Shim	
21. Differential bearing	
22. Bearing cap	
23. Ring gear	
24. Thrust washer	
25. Pinion mate shaft	
26. Gasket	
27. Cover	

Exploded view of the Dana $8\frac{7}{8}$ inch rear differential assembly

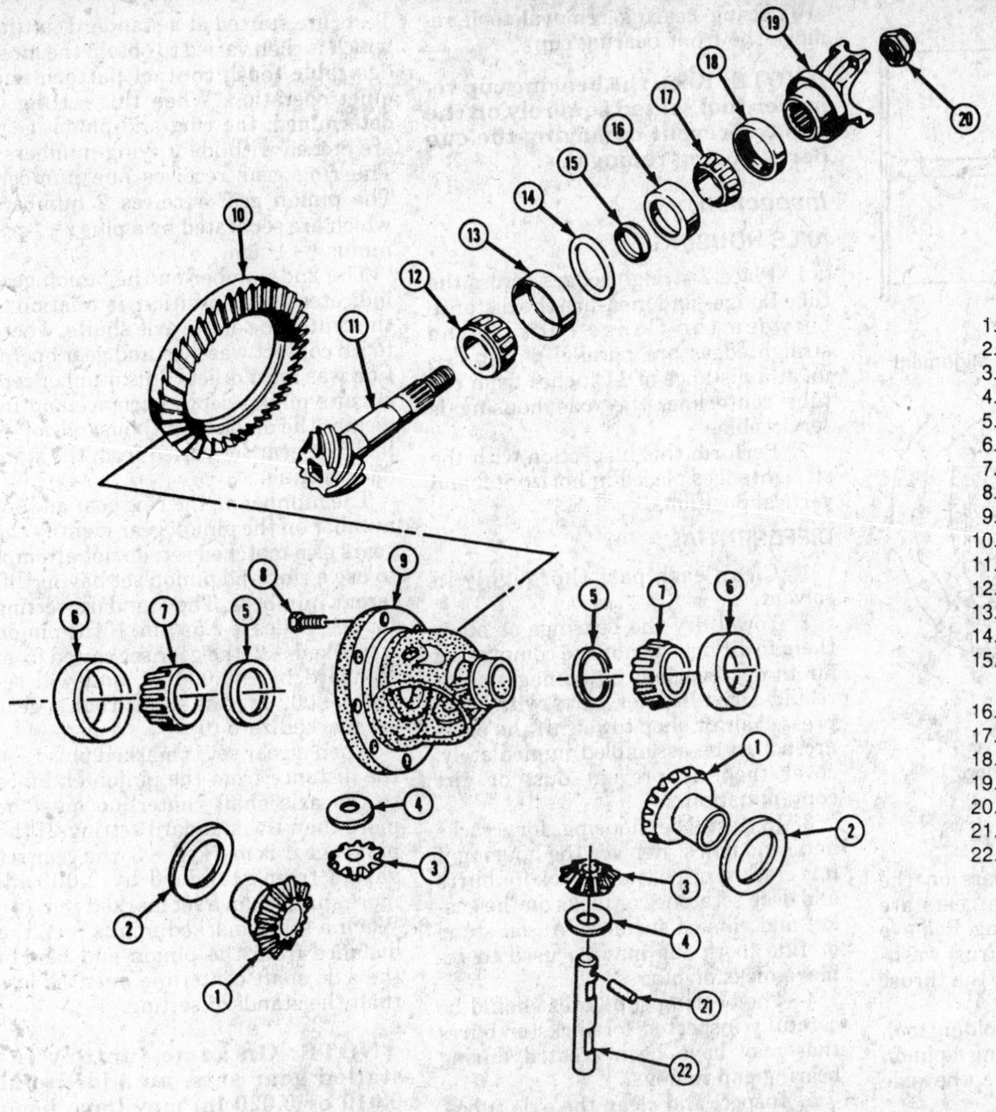

1. Differential side gear
2. Side gear thrust washer
3. Differential pinion
4. Pinion thrust washer
5. Differential bearing shim
6. Differential bearing cup
7. Differential bearing
8. Rear gear bolt
9. Differential case
10. Ring gear
11. Pinion gear
12. Pinion gear rear bearing
13. Rear bearing cup
14. Pinion depth shim
15. Pinion bearing preload spacer
16. Front bearing cup
17. Pinion gear front bearing
18. Pinion seal
19. Pinion yoke
20. Pinion nut
21. Pinion shaft lock pin
22. Differential pinion shaft

Exploded view of the Dana 7⁹/₁₆ inch rear differential assembly

NOTE: When using this tool, be sure the differential case is secure. When the bearing is removed the differential case can drop if not supported.

2. Remove the ring gear-to-differential case bolts.

NOTE: Do not chisel or wedge the gear from the case.

3. Remove the ring gear from the case. Using a brass drift and hammer, tap the ring gear from the case; do not nick the ring gear face of the differential case or drop the gear.

4. Using a drift, remove the pinion mate shaft lockpin. Remove the pinion mate shaft and the thrust block.

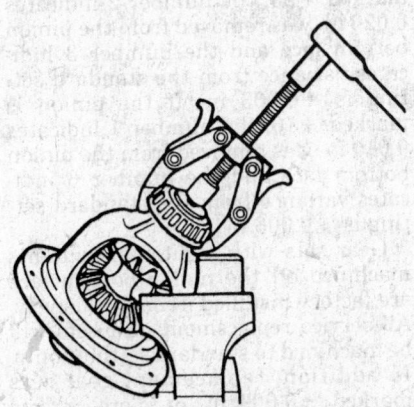

Removing the differential bearing

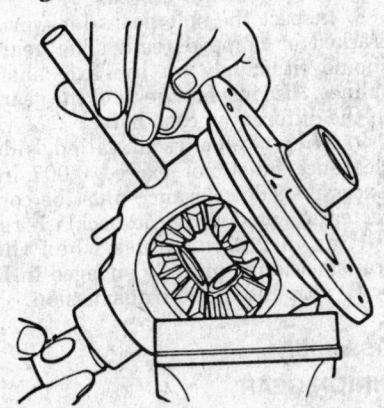

Removing the pinion shaft and thrust block

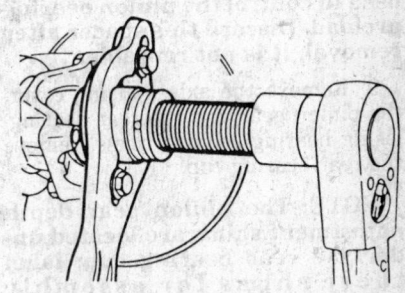

Removing the axle yoke

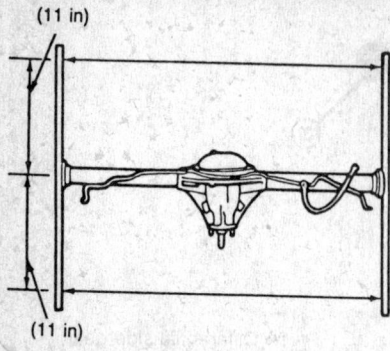

(11 in)

(11 in)

Checking the axle housing alignment

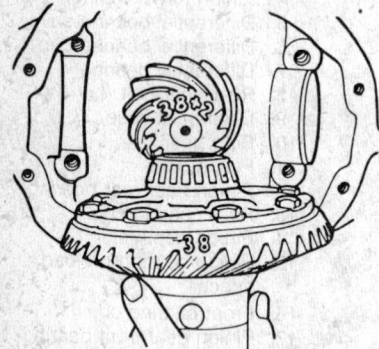

Pinion and ring gear identifying numbers

5. Rotate the pinion gears on the side gears until the pinion gears are aligned with the case opening. Remove the pinion gears with the thrust washers and the side gears with the thrust washers.

6. Using a pinion yoke holding tool, remove the pinion nut. Using a pinion yoke removal tool, remove the axle yoke.

7. Install the axle housing cover to prevent the pinion gear from falling out when the gear is driven out of the bearings and housing. Loosely attach the cover using 2 bolts.

8. Using a seal removal tool, remove the pinion seal. Tap the end of the pinion gear with a soft face mallet to drive the pinion gear out of the front bearing. Remove the front bearing and collapsible spacer. Discard the spacer.

NOTE: The collapsible spacer is used to control the pinion bearing preload. Discard this spacer after removal, it is not reusable.

9. Remove the axle housing cover, the pinion gear and the rear bearing. Using bearing removal tools, remove the rear bearing cup.

NOTE: The pinion gear depth adjustment shims are located under the rear bearing cup; label these shims for assembly reference.

10. Using bearing removal tool, remove the front bearing cup.

NOTE: Keep the bearing cup remover tool seated squarely on the cup to prevent damaging the cup bores during removal.

Inspection

AXLE HOUSING

1. Place 2 straight-edges across the tube flanges and measure the distance between the flange ends. If the straightedges are parallel within $^3/_{32}$ in. at a distance of 11 inches from the tube centerline, the axle housing is serviceable.

2. Perform this inspection with the straightedges placed in horizontal and vertical positions.

DIFFERENTIAL

1. Clean each part thoroughly in solvent.

2. Towel dry the bearings or allow them to air dry, do not use compressed air to dry bearings as damage might result. Dry all other parts with compressed air or shop towels. If the parts are not to be assembled immediately, cover them to prevent dust or dirt contamination.

3. Inspect the housing for cracks and sand holes. Replace the housing if it is cracked or porous. Check for burrs and deep scratches or nicks on the gasket and oil seal surfaces. An oil stone or fine tooth file may be used to remove nicks or burrs.

4. The bearing cup bores should be carefully inspected for nicks or burrs that may have been created during bearing cup removal.

5. Inspect and clean the axle tubes. Inspect the vent to be sure it is not obstructed.

6. Check housing for bent or loose tubes or other physical damage.

7. Inspect the shaft for scoring and wear. The shaft should be a press fit of 0.000–0.010 in. in the case. Replace the shaft if worn or scored.

8. Inspect the side gears for worn, cracked or chipped teeth. The gears should fit snugly on the axle shaft splines. Also inspect the fit of the gears in the differential case bore.

9. With the gears installed, side clearance must not exceed 0.007 in. Excessive side clearance must be corrected to avoid driveline backlash resulting in a clunk noise when the transmission is initially engaged in **D** or **R** with automatic transmission.

Assembly

PINION GEAR

Ring and pinion gear sets are factory tested to detect machining variances.

Tests are started at a standard setting which is then varied to obtain the most desirable tooth contact pattern and quiet operation. When this setting is determined, the ring and pinion gear are etched with identifying numbers. The ring gear receives one number. The pinion gear receives 2 numbers which are separated by a plus (+) or a minus (−) sign.

The 2nd number on the pinion gear indicates pinion position, in relation to the centerline of the axle shafts, where tooth contact was best and gear operation was most quiet. This number represents pinion depth variance and indicates the amount in thousands of an inch the gear set varied from the standard setting.

The number on the ring gear and 1st number on the pinion gear identify the gears as a matched set; do not attempt to use a ring and pinion set having different numbers. The standard setting for Jeep axles is 2.547 in. If the pinion is marked +2, the gear set varied from standard by +0.002 in. and will require 0.002 in. less shims than a gear set marked zero (0).

When a gear set is marked plus (+), the distance from the pinion end face to the axle shaft centerline must be more than the standard setting. If the pinion gear is marked −3, the gear set varied from standard by 0.003 in. more shims than a set marked zero (0). When a set is marked minus (−), the distance from the pinion end face to the axle shaft centerline must be less than the standard setting.

NOTE: On some factory installed gear sets, an additional 0.010 or 0.020 in. may have been machined off the pinion gear bottom face. This does not affect the gear operation but does affect the pinion gear marking and depth measurement.

Pinion gears machined in this fashion have different identifying numbers. For example, if the pinion is marked +23, the number 2 indicates 0.020 in. was removed from the pinion bottom face and the number 3 indicates variance from the standard setting is +0.003 in. If the pinion is marked +16, the number 1 indicates 0.010 in. was removed from the pinion bottom face and the number 6 indicates variance from the standard setting is +0.006 in.

Gear sets with additional amounts machined off the pinion bottom face are factory installed items exclusively. All service replacement gear sets will be machined to standard settings only. In addition, replacement gear sets marked ±0.009 in. or more or sets with mismatched identifying numbers

Old Pinion Marking	New Pinion Marking								
	- 4	- 3	- 2	- 1	0	+ 1	+ 2	+ 3	+ 4
+ 4	+ 0.008	+ 0.007	+ 0.006	+ 0.005	+ 0.004	+ 0.003	+ 0.002	+ 0.001	0
+ 3	+ 0.007	+ 0.006	+ 0.005	+ 0.004	+ 0.003	+ 0.002	+ 0.001	0	- 0.001
+ 2	+ 0.006	+ 0.005	+ 0.004	+ 0.003	+ 0.002	+ 0.001	0	- 0.001	- 0.002
+ 1	+ 0.005	+ 0.004	+ 0.003	+ 0.002	+ 0.001	0	- 0.001	- 0.002	- 0.003
0	+ 0.004	+ 0.003	+ 0.002	+ 0.001	0	- 0.001	- 0.002	- 0.003	- 0.004
- 1	+ 0.003	+ 0.002	+ 0.001	0	- 0.001	- 0.002	- 0.003	- 0.004	- 0.005
- 2	+ 0.002	+ 0.001	0	- 0.001	- 0.002	- 0.003	- 0.004	- 0.005	- 0.006
- 3	+ 0.001	0	- 0.001	- 0.002	- 0.003	- 0.004	- 0.005	- 0.006	- 0.007
- 4	0	- 0.001	- 0.002	0.003	- 0.004	- 0.005	- 0.006	- 0.007	- 0.008

must be returned to the parts distributor center; do not attempt to install these gear sets.

The chart provided in this section will help to determine the approximate starter shim thickness needed for the initial pinion depth measurement. However, the chart will not provide the exact shim thickness required for final adjustment and must not be used as a substitute for an actual pinion depth measurement. The chart should be used as follows.

1. Measure the thickness of the original pinion depth shim. Note the pinion depth variance numbers marked on the old and new pinion gears.

2. Now use the chart to determine the starter shim thickness. An example of this is as follows:

If the old pinion is marked −3 and the new pinion is marked +2, the chart procedure would be as follows. Go to the old pinion column and locate the − 3, then go across the chart until the + 2 figure is reached in the new pinion column. The box where the 2 columns intersect will indicate the amount of starter shim thickness required.

DIFFERENTIAL

1. Install the differential bearings on the case using tools J–21784 and J–8092 or equivalent.

2. Install the thrust washers on the differential side gears and install the gears in the differential case. Install the differential pinion gears in the case. Install the thrust washers behind the pinion gears and align the pinion gear bores.

3. Rotate the differential side and pinion gears until the pinion mate shaft bores in the pinion gears are aligned with the shaft bores in the case.

4. Install the thrust block in the case. Insert the block through the side gear bore. Align the bore in the block with the pinion mate shaft bores in the pinion gears and case.

5. Install the pinion mate shaft. Align the lockpin bore in the shaft with the bore in the case and install the shaft lockpin.

RING GEAR

1. Position the ring gear on the differential case. Install the 2 ring gear bolts in the opposite holes and tighten the bolts to pull the gear into position.

2. Install the remaining ring gear bolts and tighten to within 105 ft. lbs. torque.

3. Position the shims previously selected to remove the differential bearing sideplay on the bearing cups and install the differential assembly in the axle housing. Install the bearing cap bolts and tighten the bolts with 85 ft. lbs. torque.

4. Attach the dial indicator to the housing. Position the indicator so the indicator stylus contacts the drive side of a ring gear tooth and at a right angle to the tooth. Move the ring gear back and forth and note the movement registered on the dial indicator. The ring gear backlash should be 0.005–0.009 in., with 0.008 in. desired.

5. Adjust the backlash as follows: to increase the backlash, install the thinner shim on the ring gear side and the thicker shim on the opposite side. To decrease the backlash, reverse the procedure; however, do not change the total thickness of the shims.

NOTE: The following is an example on how to decrease backlash. The sideplay was removed using 0.090 in. shims on each side totaling 0.180 in. Backlash is checked and found to be 0.011 in. To correct the backlash, add 0.004 in. the the shim on the ring gear side and subtract 0.004 in. from the shim on the opposite side. This will result in 0.094 in. shim on the ring gear side and 0.086 in. shim on the other side. The backlash will be approximately 0.007–0.008 in. The total shim thickness remains 0.180 in.

Adjustment

PINION DEPTH MEASUREMENT

1. Measure the thickness of the original pinion depth shim. Note the pinion depth variance numbers marked on the old and new pinion gears.

2. Determine the starter shim thickness. Using the chart, determine the amount to be added or subtracted

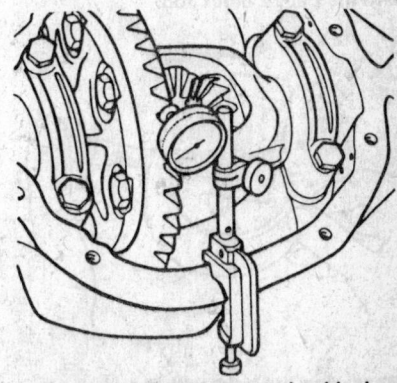

Measuring the ring gear backlash

from the original shim thickness for starter shim thickness.

NOTE: The starter shim thickness must not be used as a final shim setting. An actual pinion depth measurement must be performed and the final shim thickness adjusted, as necessary.

3. Install the ring bearing on the pinion gear with the large diameter of the bearing cage facing the gear end of the pinion. Press the bearing against the rear face of the gear.

4. Clean the pinion bearing bores in the axle housing thoroughly. This is important in obtaining the correct pinion gear depth adjustment. Install the starter pinion depth shim in the housing rear bearing cup bore. Be sure the shim is centered in the bearing cup bore.

NOTE: If the shim is chamfered, be sure the chamfered side faces the bottom of the bearing cup bore.

5. Install the ring bearing cup using tools J–8092 and J–8608 or equivalent. Install the front bearing cup using tools J–8092 and J–8611–01 or equivalent. Install the pinion gear in the rear bearing cup.

6. Install the front bearing, rear universal joint yoke and original pinion nut on the pinion gear. Tighten the pinion nut only enough to remove the bearing endplay.

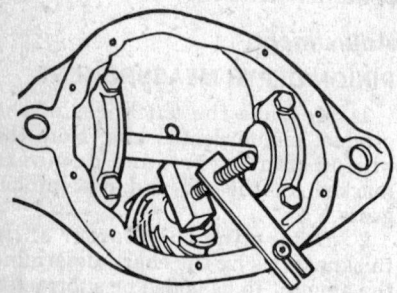

Installing the gauge arbor tool, discs and the gauge block tool

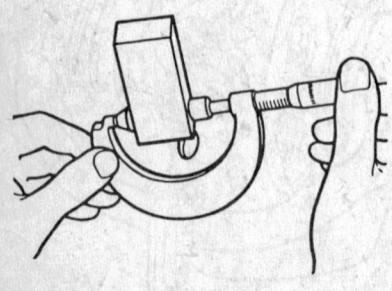

Measuring the anvil with a micrometer

NOTE: Do not install a replacement pinion nut and collapsible spacer at this time as the pinion gear will be removed after depth measurement.

7. Note the pinion depth variance marked on the pinion gear. If the number is preceded by a plus (+) sign, add that amount (in thousandths) to the standard setting for the axle model being overhauled. If the number is preceded by a minus (−) sign, subtract that amount (in thousandths) from the standard setting. The result of this addition or subtraction is the desired pinion depth. Record this figure for future reference.

8. Assemble an arbor tool J–5223–4 and discs J–5223–23 or equivalent, install the assembled tools in the differential bearing cup bores; be sure the discs are completely seated in bearing cup bores.

9. Install the bearing caps over the discs and install the bearing cap bolts. Tighten the bearing cap bolts securely but not with the specified torque.

10. Position a gauge block tool J–5223–20 or equivalent, on the end face of the pinion gear with the anvil end of the gauge block seated on the gear and the gauge plunger under the arbor tool J–5223–4 or equivalent.

11. Assemble and mount the clamp tool J–5223–24 and bolt J–5223–29 or equivalent, on the axle housing. Use the axle housing cover bolt to attach the clamp to the housing.

12. Extend the clamp bolt until it presses against the gauge block with enough force to prevent the gauge block from moving. Loosen the gauge block thumb screw to release the gauge block plunger. When the plunger contacts the arbor tool, tighten the thumbscrew to lock the plunger in position; do not disturb the plunger position.

13. Remove the clamp and bolt assembly from the axle housing. Remove the gauge block and measure the distance from the end of the anvil to the end of the plunger using a 2–3 in. micrometer. This dimension represents the measured pinion depth. Record this dimension for assembly reference.

14. Remove the bearing caps, the arbor tool and discs from the axle housing. Remove the pinion gear, the rear bearing cup and pinion depth shim from the axle housing.

15. Measure the thickness or the depth shim. Add this dimension to the measured pinion depth. From this total, subtract the desired pinion depth. The result represents the correct shim thickness required.

NOTE: The desired pinion depth is the standard setting plus

or minus the pinion depth variance.

PINION GEAR BEARING PRELOAD

1. Install the correct thickness pinion depth shim(s) in the axle housing bearing cup bore. Install the rear bearing cup and pinion gear.

NOTE: The collapsible spacer controls the pinion bearing preload; do not reuse the old spacer, use a replacement spacer only.

2. Install the replacement collapsible spacer and front bearing on the pinion gear. Install the pinion oil seal using tool J–22661 or equivalent.

3. Install the pinion yoke, replace the pinion nut and tighten the pinion nut finger-tight. Tighten the pinion nut enough to remove endplay and seat the pinion bearings. Using tool J–22575 or equivalent, tighten the nut and use tool J–86141–01 or equivalent, to hold the yoke while tightening the nut.

4. Rotate the pinion, while tightening the nut, to seat the bearings evenly. Remove the tools.

NOTE: Do not exceed the specified preload torque or loosen the nut to reduce the preload torque, if the specified torque is exceeded.

5. Using an inch lb. torque wrench and adapter tool J–22575 or equivalent, measure the torque required to turn the pinion gear; the correct pinion bearing preload torque is 17–25 inch lbs. Continue tightening the pinion nut until the required preload torque is obtained.

6. If the pinion bearing preload torque is exceeded, remove the pinion gear, replace the collapsible spacer and pinion nut and adjust the preload again.

DIFFERENTIAL BEARING

1. Place the bearing cup over each differential bearing and install the dif-

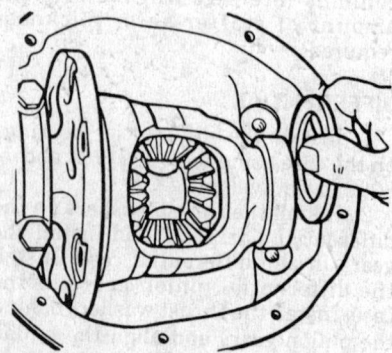

Installing the shim(s) on the side of the differential bearing cup

ferential case assembly in the axle housing.

2. Install the shim on each side between the bearing cup and the housing; use 0.080 in. shims as the starting point.

3. Install the bearing caps and tighten the bolts finger-tight. Mount the dial indicator on the housing. Using a prybar, pry between the shims and housing. Pry the assembly to one side and zero the indicator, then, pry the assembly to the opposite side and read the indicator.

NOTE: Do not zero or read the indicator while prying.

4. The amount read on the indicator is the shim thickness that should be added to arrive at the zero preload and zero endplay. Repeat the procedure to ensure accuracy and adjust if necessary; shims are available in thicknesses from 0.080–0.110 in. in increments of 0.002 in.

5. When the sideplay is eliminated, a slight bearing drag will be noticed. Install the bearing caps and tighten the bearing cap bolts to with 85 ft. lbs. torque.

6. Attach the dial indicator to the axle housing and check the ring gear mounting face of the differential case for runout; runout should not exceed 0.002 in.

7. Remove the case from the housing. Retain the shims used to adjust the sideplay.

DIFFERENTIAL AND BEARING PRELOAD

NOTE: The differential bearings must be preloaded to compensate for heat and loads during operation. The differential bearings are preloaded by increasing the shim pack thickness at each side of the differential by 0.004 in. for a total of 0.008 in.

1. Remove the differential assembly from the housing. Be sure to keep the differential bearing shim packs together for the proper assembly. Do not distort the shims in the axle housing bearing bores.

2. Install the differential bearing cups on the differential bearings. The cups should cover the differential bearing rollers completely. Position the differential assembly in the housing so the bearings just start into the housing bearing bores.

NOTE: Slightly tipping the bearing cups will ease starting them into the bores. Also keep the differential assembly square in the housing during installation and push it in as far as possible.

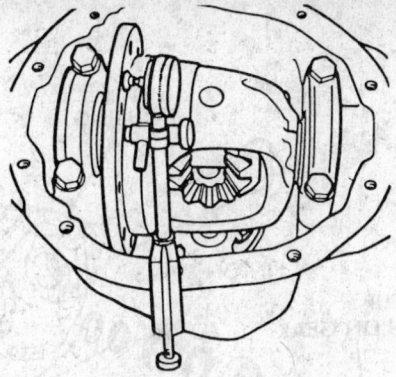

Installation of the dial indicator

3. Tap the outer edge of the bearing cups until the differential is seated in the housing.

4. Install the differential bearing caps. Position the caps accordingly to the alignment punch marks made at disassembly. Tighten the bearing cap bolts with 85 ft. lbs. torque. Preloading the differential bearings may change the backlash setting. Check and correct the backlash, if necessary.

5. Install the propeller shaft, aligning the index marks made at disassembly. Install the axle shafts, bearings, seals and brake support plates. Fill the rear axle with the specified axle lubricant.

6. Check and adjust the axle shaft endplay, if necessary. Adjust the endplay at the left side of the axle shaft only. Install the hubs, drums and wheels.

Lower the vehicle and road test the vehicle to check the rear axle assembly for proper operation.

DANA — 8½ INCH RING GEAR AXLE

Disassembly

DIFFERENTIAL

1. Remove the housing cover and drain the lubricant.

2. Using an axle spreader tool, mount it onto the axle housing and spread the housing enough to remove the differential.

3. Using a dial indicator, measure the amount the opening is being spread; do not spread the housing more than 0.015 in. (0.38mm), for damage to the housing may occur.

4. Mark the differential bearing caps for reassembly purposes.

5. Loosen the bearing caps until 2–3 threads are engaged.

6. Using a prybar, pry the differential loose.

7. Remove the bearing caps and the differential.

8. Mount the differential into a vise.

9. Remove and discard the ring gear bolts; they are not reusable.

10. Using a brass drift and a hammer, tap the ring gear from the differential.

11. Using a differential bearing puller, press the differential bearings from the differential.

12. Remove the differential bearing shims.

13. Using 2 sets of feeler gauges, insert them between each side of the side gear thrust washer and differential case and measure the side gear clearance; the clearance should not exceed 0.007 in. (0.18mm). Replace both thrust washers, if the clearance exceeds the tolerance.

14. Remove the pinion shaft lockpin and pinion shaft.

15. Rotate the pinions to remove them through the case opening.

16. Remove the side gears and thrust washers.

PINION GEAR

1. Remove the differential case from the axle housing.

2. Using a pinion yoke holding tool, remove the pinion gear nut.

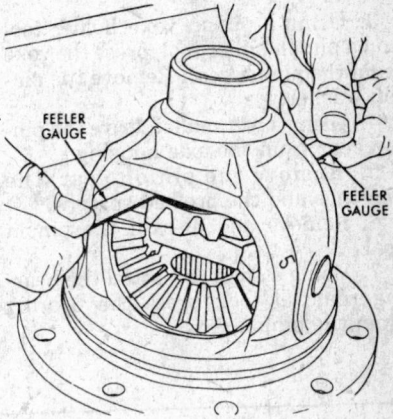

Measuring the side gear clearance

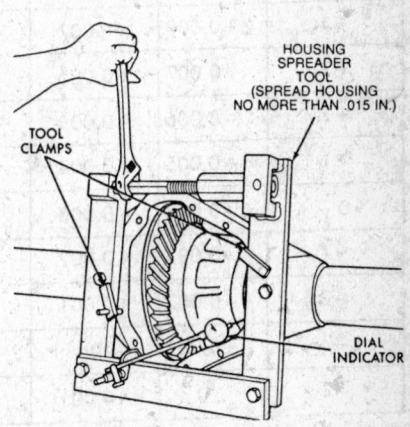

Spreading the axle housing

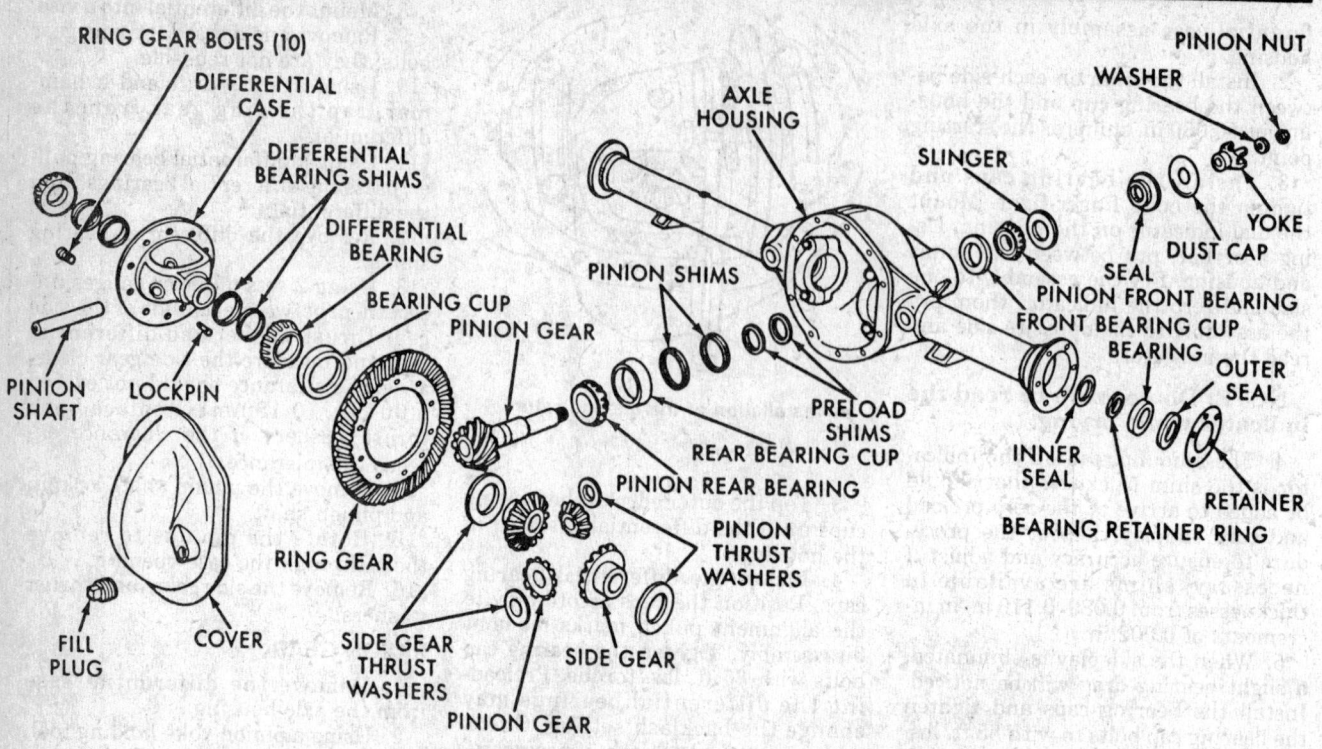

Exploded view of the Dana 8½ inch rear differential assembly

3. Using a pinion yoke holder tool and a pinion puller tool, press the yoke from the pinion gear. Remove the pinion washer.

4. Using a soft mallet, drive the pinion gear from the axle housing.

5. Remove the pinion gear, the bearings and the preload spacers.

6. Remove and discard the pinion seal.

7. Using a shop press and the bearing removal tool, press the bearing from the pinion gear.

Inspection

1. Clean the differential components in solvent and use compressed air to dry them; do not use compressed air on the bearings, only shop towels.

2. Check the components for wear or damage; replace them, if necessary.

3. Inspect the bearings and bearing cups for wear, cracks or scoring; replace them, if necessary.

4. Inspect the differential side and pinion gears for wear, cracks or chips; replace them, if necessary.

5. Inspect the ring and pinion gears for wear and/or damage; replace them, if necessary.

6. Inspect the differential case for cracks or damage; replace it, if necessary.

Assembly

PINION GEAR

1. Make sure the ring and pinion are a matched set.

2. Measure the thickness of the

OLD PINION MARKING	NEW PINION MARKING								
	−4	−3	−2	−1	+0	+1	+2	+3	+4
+4	+0.008	+0.007	+0.006	+0.005	+0.004	+0.003	+0.002	+0.001	0
+3	+0.007	+0.006	+0.005	+0.004	+0.003	+0.002	+0.001	0	−0.001
+2	+0.006	+0.005	+0.004	+0.003	+0.002	+0.001	0	−0.001	−0.002
+1	+0.005	+0.004	+0.003	+0.002	+0.001	0	−0.001	−0.002	−0.003
0	+0.004	+0.003	+0.002	+0.001	0	−0.001	−0.002	−0.003	−0.004
−1	+0.003	+0.002	+0.001	0	−0.001	−0.002	−0.003	−0.004	−0.005
−2	+0.002	+0.001	0	−0.001	−0.002	−0.003	−0.004	−0.005	−0.006
−3	+0.001	0	−0.001	−0.002	−0.003	−0.004	−0.005	−0.006	−0.007
−4	0	−0.001	−0.002	−0.003	−0.004	−0.005	−0.006	−0.007	−0.008

original pinion shim and note the variance on the pinion gear.

3. Perform the following procedure to determine the pinion starter shim thickness:

a. If the original ring and pinion are being installed, use the original shim.

b. If a replacement gear set is being installed, determine the best starter shim thickness.

c. Refer to the pinion variance chart and observe where the old and new pinion marking column intersect.

d. If the old pinion is +1 and the new pinion is −3, the intersecting figure is +0.004 in. (0.10mm); add this amount to the original shim. Or, if the old pinion is −3 and the new pinion is −2, the intersecting figure is −0.001 in. (−0.025mm); subtract this amount from the original shim.

4. Install the starter shim in the pinion rear bearing cup bore; if the shim is chamfer on one side, position the chamfered side so it faces the bottom of the bore.

5. Using a driver tool, install the pinion rear bearing cup into the axle housing.

6. Install the front bearing cup.

7. Using a shop press, press the rear bearing onto the pinion gear.

8. Install the pinion gear into the axle housing and the front bearing onto the pinion; do not install the slinger or the seal.

9. Install the yoke, the pinion nut washer and the old pinion nut; tighten the nut enough to remove the endplay.

10. Adjust the backlash and gear tooth contact.

DIFFERENTIAL

1. Install the side gears, the thrust washers and the pinion gears into the differential case.

NOTE: Be sure to install new side gear thrust washers, if the clearance measured at disassembly exceeded 0.007 in. (0.18mm).

2. Using 2 sets of feeler gauges, insert them between each side of the side gear thrust washer and differential case and measure the side gear clearance; the clearance should not exceed 0.007 in. (0.18mm). Replace both thrust washers, if the clearance exceeds the tolerance.

3. Install the pinion shaft and lockpin into the case.

4. Assemble the original differential bearing shim packs, then, remove approximately 0.20 in. (0.50mm) shim thickness from each pack; the remaining shims will serve as a starter shim pack.

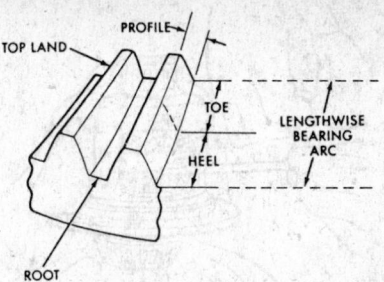

View of the gear tooth nomenclature

5. Install the starter shim packs and bearing onto the case.

6. Align and install the ring gear. Using new bolts, torque the ring gear-to-differential case bolts to 55 ft. lbs. (75 Nm).

7. Install the spreader tool onto the axle housing. Using a dial indicator, spread the axle housing no more than 0.015 in. (0.38mm).

NOTE: Do not spread the housing more than 0.015 in. (0.038mm), for damage may occur to the case.

8. Install the differential; it may be necessary to tap the differential bearing cups, with a soft hammer, to seat them.

9. Install the differential bearing caps and torque the bolts to 57 ft. lbs. (77 Nm).

10. Remove the spreader tool and dial indicator.

11. Adjust the ring and pinion gear.

12. Using silicone sealant, apply a bead of it to the axle housing. Install the cover and torque the bolts to 35 ft. lbs. (47 Nm).

Adjustment

RING AND PINION

1. Using yellow ferrous oxide compound, coat the drive and coast sides of the ring gear teeth.

2. Install the differential and tighten the bearing caps.

3. Using a dial indicator, mount it onto the housing, position the stylus against the drive side of one ring gear tooth; be sure the stylus is at a right angle to the tooth.

4. Move the ring gear toward the dial indicator and zero it.

5. Move the ring gear away from the pinion until the backlash is 0.005–0.009 in. (0.1 3–0.23mm). Insert shims or feeler gauges between the differential bearings and housing to maintain the backlash during the remainder of the adjustment.

NOTE: To increase the backlash, subtract shims from the ring gear side of the case. To decrease

the backlash, add shims to the ring gear side of the case.

6. Move the dial indicator aside and recoat any necessary ring gear teeth with the ferrous oxide compound.

7. Rotate the ring gear a complete revolution in both directions and note the tooth contact pattern imprinted in the compound.

8. Adjustments can be made as follows:

a. Decreasing the backlash moves the drive and coast side pattern slightly lower and toward the toe.

b. Increasing the backlash move the drive and coast side pattern slightly higher and toward the heel.

c. A thicker pinion shim moves the pinion closer to the ring gear. The drive pattern moves deeper on the tooth and slightly towards the heel.

d. A thinner pinion shim moves the pinion away from the ring gear. Drive pattern moves toward the top of the tooth and toward the heel. Coast pattern moves toward the top of the tooth and slightly toward the toe.

9. Remove the differential and pinion gear.

10. Change the pinion and differential shims, as necessary.

11. Install the pinion and differential.

12. Recheck the tooth contact pattern and backlash; adjust, as necessary.

PINION BEARING PRELOAD

1. Using a yoke holding tool, torque the pinion nut to a minimum of 210 ft. lbs. (285 Nm).

2. Using an inch lb. torque wrench, measure the torque necessary to rotate the pinion; the torque should be 20–40 inch lbs. (2–5 Nm).

NOTE: If the torque is not correct, add shims to increase the preload or subtract shims to decrease the preload.

3. Remove the pinion nut, the washer and the yoke after setting the preload; be sure to discard the old nut.

4. Lubricate the new pinion oil seal lip and install the seal into the axle housing.

5. Install the yoke, the washer and a new pinion nut; torque the nut to a minimum of 210 ft. lbs. (285 Nm).

DIFFERENTIAL BEARING PRELOAD

1. Remove the differential bearing from the side of the case opposite the ring gear.

2. Add a 0.015 in (0.38mm) shim the shim pack on the side of the case.

3. Reistall the bearing onto the case.

4. Clean the ferrous oxide compound from the ring and pinion gears, then, lubricate the differential gears and bearings with axle lubricant.

5. Install the spreader tool onto the axle housing. Using a dial indicator, spread the axle housing no more than 0.015 in. (0.38mm).

NOTE: Do not spread the housing more than 0.15 in. (0.38mm), for damage may occur to the case.

6. Install the differential; it may be necessary to tap the differential bearing cups, with a soft hammer, to seat them.

7. Install the differential bearing caps and torque the bolts to 57 ft. lbs. (77 Nm).

8. Recheck the backlash; it should be 0.005–0.009 in. (0.13–0.23mm). If necessary, reset the backlash.

TRAC-LOK DIFFERENTIAL

Operational Test

If a noisy or rough operation such as a chatter occurs when turning corners, the most probable cause of this chatter or noise is incorrect or contaminated lubricant. Before removing the Trac-Lok unit for repair, drain, flush and refill the axle with the specified lubricant. A complete lubricant drain and refill with the specified fluid will usually correct the chatter problem. A quick operational test of the Trac-Lok differential can be done easily by performing the following steps.

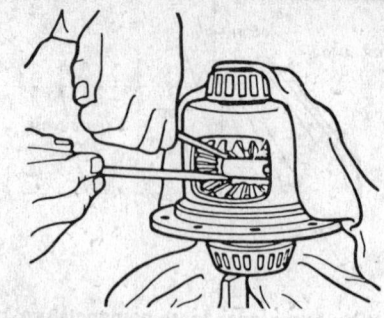

Removing the snapring from the pinion mate shaft

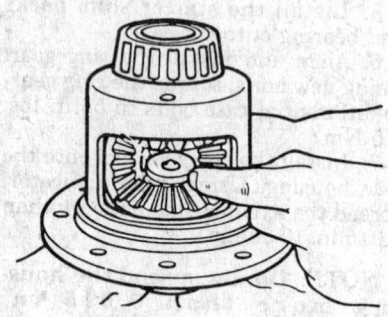

Install the step plate tool

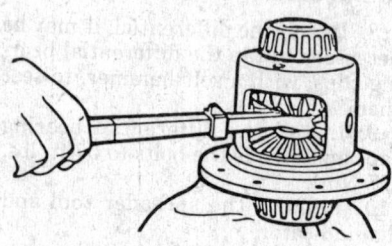

Install the gear rotating tool

1. Position one wheel on solid dry pavement and the opposite wheel on ice, mud grease or a similar low traction surface.

2. Gradually, increase the engine rpm to obtain the maximum traction prior to a breakaway. The ability to move the vehicle effectively will demonstrate the proper performance.

NOTE: If the test is performed on extremely slick surfaces such as ice or grease coated surfaces, some question may exist as to proper performance. In these extreme cases, a properly performing Trac-Lok will provide greater pulling power by lightly applying the parking brake.

Disassembly

DIFFERENTIAL

1. Remove the differential from the axle housing as previously outlined in this section. Install one axle shaft in the vise with the spline end facing upward and tighten the vise.

2. Do not allow more than 2¾ in. of the shaft to extend above the top of the vise. This prevents the shaft from fully entering the side gear, causing interference with the step plate tool used to remove the differential gears.

3. Mount the differential case on the axle shaft with the ring gear bolt heads facing upward. Place some shop towels under the ring gear to protect the gear when it is removed from the case.

4. Remove and discard the ring gear bolts. Remove the ring gear from the

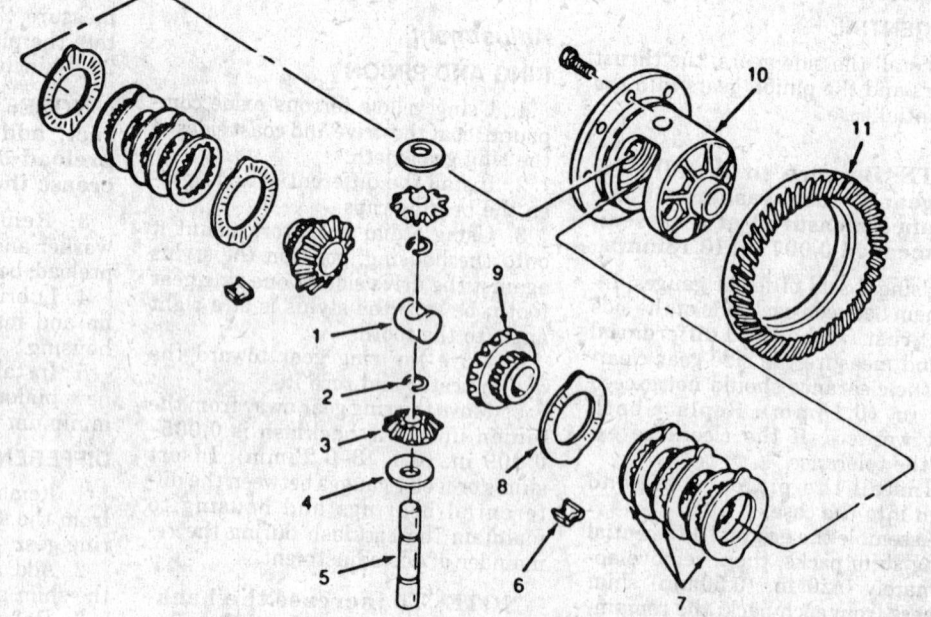

1. Thrust block
2. Snapring
3. Pinion gear
4. Thrust washer
5. Pinion shaft
6. Retainer clip
7. Clutch pack
8. Belleville spring
9. Side gear
10. Case
11. Ring gear

Exploded view of the Trac-Loc differential

case, using a rawhide hammer. Remove the differential case from the axle shaft and remove the ring gear and remount the differential case on the axle shaft.

5. Use suitable tools to disengage the snaprings from the pinion mate shaft. Place a shop towel on the opposite opening of the case to prevent the snaprings from flying out of the case. Remove the pinion mate shaft using a hammer and brass drift.

NOTE: A special gear rotating tool J-23781-3 or equivalent, is required to perform the following steps. The tool consists of 3 parts; the gear rotating tool, forcing screw and step plate.

6. Install step plate tool into the lower differential side gear. Position the pawl end of the gear rotating tool onto the step plate.

7. Insert the forcing screw tool through the top of the case and thread it into the gear rotating tool. Before using the forcing screw tool, apply a small amount of grease to the centering hole in the step plate and oil the threads of the forcing screw.

8. Center the forcing screw in the step plate and tighten the screw to move the differential side gears away from the differential pinion gears. Remove the differential pinion gear thrust washers using a feeler gauge or a shim stock of 0.030 in. thickness. Insert the feeler gauge or shim stock between the washer and the case and withdraw the shim stock with the thrust washer.

9. Tighten the forcing screw until a slight movement of the differential pinion gear is observed. Insert the pawl end of the gear rotating tool between the teeth of one differential side gear.

10. Pull the handle of the tool to rotate the side gears and pinion gears. Remove the pinion gears as they appear in the case opening. It could be necessary to adjust the tension applied on the belleville springs by the forcing screw before the gears can be rotated in the case.

11. Retain the upper side gear and clutch pack in the case by placing a hand on the bottom of the rotating tool while removing the forcing screw. Remove the rotating tool, upper side gear and clutch pack.

12. Remove the differential case from the axle shaft. Invert the case with the flange or ring gear side up and remove the step plate tool, lower the side gear and clutch pack from the case. Remove the retainer clips from both the clutch packs to allow separation of the plates and discs.

Inspection

If any member of either clutch pack shows evidence of excessive wear or scoring, the complete clutch pack must be replaced on both sides.

1. Thoroughly, clean each part in solvent.

2. Towel dry bearings or allow them to air dry, do not use compressed air to dry bearings as damage might result. Dry all other parts with compressed air or shop towels. If the parts are not to be assembled immediately, cover them to prevent dust or dirt contamination.

3. Inspect the housing for cracks and sand holes. Replace the housing if it is cracked or porous. Check for burrs and deep scratches or nicks on the gasket and oil seal surfaces. An oil stone or fine tooth file may be used to remove nicks or burrs.

4. Inspect the bearing cup bores for nicks or burrs that may have been created during bearing cup removal.

5. Inspect and clean the axle tubes. Inspect the vent to be sure it is not obstructed.

6. Check housing for bent or loose tubes or other physical damage.

7. Inspect the side gears for worn, cracked or chipped teeth. The gears should fit snugly on the axle shaft splines. Also, inspect the fit of the gears in the differential case bore.

Assembly

DIFFERENTIAL

1. Lubricate all the differential components with the specified gear lubricant. Assemble the clutch packs. Install the plates and discs in the same position as when removed regardless of whether they are replacement or original parts.

2. Install the clutch retainer clips on the ears of the clutch plates. Be sure the clutch packs are completely assembled and seated on the ears of the plates. Install the clutch packs on the differential side gears and install the assembly in the case.

3. Make sure the clutch pack stays assembled on the side gear splines and

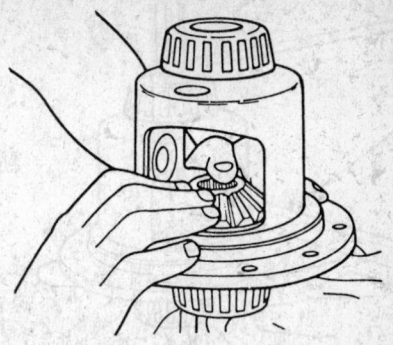

Installing the clutch packs

the retainer clips are completely seated in the case pockets. To prevent the pack from falling out of the case, it will be necessary to hold it in place by hand while mounting the case on the axle shaft.

NOTE: When installing the differential case on the axle shaft, make sure the splines of the side gears are aligned with those of the axle shaft. Make sure the clutch pack is still properly assembled in the case after installing the case on the axle shaft.

4. Mount the case assembly on the axle shaft. Install the step plate tool in the side gear and apply a small amount of grease in the centering hole of the step plate.

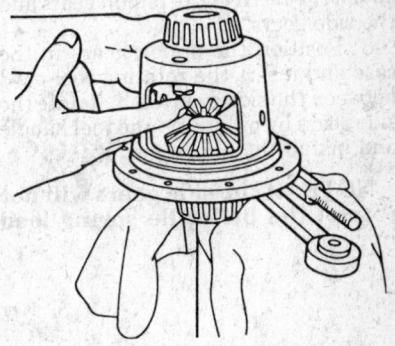

Keeping the side gear rotating tool in position

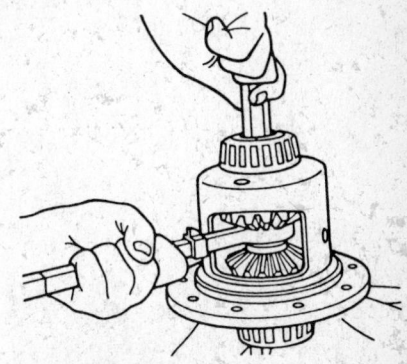

Installing the forcing screw into the rotating tool

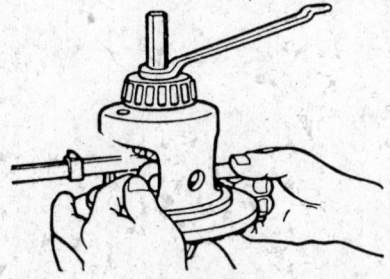

Removing the pinion gear thrust washers

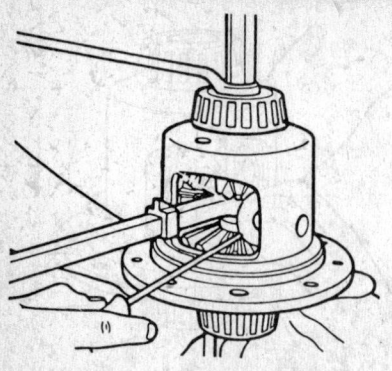

Installing the thrust washers

Component	Service Set-To Torque	Service Recheck Torque
Wheel Lug Nuts	102 N·m (75 ft-lbs)	81-122 N·m (60-90 ft-lbs)
Brake Support Plate Nuts	43 N·m (32 ft-lbs)	34-54 N·m (25-40 ft-lbs)
U-Joint Strap Bolts	19 N·m (170 in-lbs)	15-23 N·m (140-200 in-lbs)
Differential Bearing Cap Bolts	77 N·m (57 ft-lbs)	64-91 N·m (47-67 ft-lbs)
Ring Gear-to-Case Bolts	70 N·m (52 ft-lbs)	57-88 N·m (42-65 ft-lbs)
Rear Axle Cover Screws	19 N·m (170 in-lbs)	17-21 N·m (150-190 in-lbs)
Rear Axle Filler Plug	34 N·m (25 ft-lbs)	27-41 N·m (20-30 ft-lbs)

Component	Service Set-To Torque	Service Recheck Torque
Axle Housing Cover Bolts	19 N·m (170 in-lbs)	17-21 N·m (150-190 in-lbs)
Brake Tube-to-Rear Wheel Cylinder	11 N·m (97 in-lbs)	10-12 N·m (90-105 in-lbs)
Differential Bearing Cap Bolts	115 N·m (85 ft-lbs)	102-129 N·m (75-95 ft-lbs)
Ring Gear-to-Case Bolt	142 N·m (105 ft-lbs)	135-149 N·m (95-115 ft-lbs)
Rear Brake Support Plate Bolts	43 N·m (32 ft-lbs)	34-54 N·m (25-40 ft-lbs)
Universal Joint Strap Bolts	19 N·m (170 in-lbs)	16-22 N·m (140-200 in-lbs)

Torque specifications for the Trac-Lock differential

5. Install the remaining clutch pack and side gear. Make sure the clutch pack stays assembled on the side gear splines and the retainer clips are completely seated in the pockets of the case.

6. Position the gear rotating tool in the upper side of the gear. Keep the side gear and rotating tool in position by holding them with your hand. Insert the forcing screw through the top of the case and thread it into the rotating tool.

7. Install both of the differential pinion gears in the case; be sure the bores of the gears are aligned. Hold the gears in place by hand. Tighten the forcing screw to compress the belleville springs and provide clearance between the teeth of the pinion gears and the side gears.

8. Position the pinion gears in the case and insert the rotating tool pawl between the side gear teeth. Rotate the side gears by pulling on the tool handle and install the pinion gears.

NOTE: If the side gears will not rotate, the belleville spring load will have to be adjusted. If adjustment is necessary, loosen or tighten the forcing screw slightly until the gears will rotate.

9. Rotate the side gears, using the rotating tool handle, until the shaft bores in both the pinion gears are aligned with the case bore. Lubricate both sides of the pinion gear thrust washers.

10. Tighten or loosen the forcing screw to permit the thrust washer installation. Install the thrust washers and using a suitable tool, guide the washers into position. Make sure the shaft bores in the washers and gears are aligned with the case bores.

11. Remove the forcing screw, rotating tool and step plate. Lubricate the pinion mate shaft and seat the shaft in the case. Be sure the snapring grooves in the shaft are exposed to allow the snapring installation.

12. Install the pinion mate shaft snaprings, remove the case from the axle shaft and install the ring gear on the case. Be sure to use replacement ring bolts only; do not reuse the original bolts.

13. Align the ring gear and case bolt holes and install the ring gear bolts finger tight only. Remove the case on the axle shaft and tighten the bolts evenly to the proper torque specifications.

14. Install the differential assembly into the axle housing.

IMPORT DRIVE AXLES

General Information

DIFFERENTIAL OPERATION

The differential is an arrangement of gears that permits the wheels to turn at different speeds when cornering and divides the torque between the axle shafts. The differential gears are mounted on a pinion shaft and the gears are free to rotate on this shaft. The pinion shaft is fitted in a bore in the differential case and is at right angles to the axle shafts.

Power flow through the differential is as follows. The drive pinion, which is turned by the driveshaft, turns the ring gear. The ring gear, which is bolted to the differential case, rotates the case. The differential pinion forces the pinion gears against the side gears. In cases where both wheels have equal traction, the pinion gears do not rotate on the pinion shaft, because the input force of the pinion gear is divided equally between the 2 side gears. Consequently the pinion gears revolve with the pinion shaft, although they do not revolve on the pinion shaft itself. The side gears, which are splined to the axle shafts and meshed with the pinion gears, rotate the axle shafts.

GEAR RATIOS

The drive axle of a vehicle is said to have a certain axle ratio. This number (usually a whole number and a decimal fraction) is actually a comparison of the number of gear teeth on the ring gear and the pinion gear. For example, a 4:11 rear means that theoretically, there are 4:11 teeth on the ring gear and one tooth on the pinion. Actually, on a 4:11 rear, there are 37 teeth on the ring gear and nine teeth on the pinion gear. By dividing the number of teeth on the pinion gear into the number of teeth on the ring gear, the numerical axle ratio (4:11) is obtained.

Differential Diagnosis

The most essential part of rear axle service is proper diagnosis of the problem. Any gear driven unit will produce a certain amount of noise. Acceptable or normal noise can be classified as a slight noise heard only at certain speeds or under unusual conditions. This noise tends to reach a peak at 40–60 mph depending on the road condition, load, gear ratio and tire size.

Frequently, other noises are mistakenly diagnosed as coming from the axle assembly. Vehicle noises from tires, transmission, driveshaft, U-joints and front and rear wheel bearings will often be mistaken as emanating from the front or rear axle assemblies.

EXTERNAL NOISE ELIMINATION

It is advisable to make a thorough road test to determine whether the noise originates in the axle assembly or whether it originates from the tires, engine transmission, wheel bearings or road surface. Noise originating from other places cannot be corrected by overhauling the axle assemblies.

ROAD NOISE

Brick roads or rough surfaced concrete, may cause a noise which can be mistaken as coming from the axle assembly. Driving on a different type of road (smooth asphalt or dirt) will determine whether the road is the cause of the noise. Road noise is usually the same on drive or coast conditions.

TIRE NOISE

Tire noise can be mistaken as axle assembly noises, even though the tires are at fault. Snow tread and mud tread tires or tires worn unevenly will frequently cause vibrations which seem to originate elsewhere; temporarily and for test purposes only, inflate the tires to 40–50 lbs. This will significantly alter the noise produced by the tires, but will not alter noise from the axle assembly. Noises from the axle assembly will normally cease at speeds below 30 mph on coast, while tire noise will continue at lower tone as vehicle speed is decreased. Th axle noise will usually change from drive conditions to coast conditions, while tire noise will not. Do not forget to lower the tire pressure to normal after the test is complete.

ENGINE AND TRANSMISSION NOISE

Engine and transmission noises also seem to originate in the axle assemblies. Road test the vehicle and determine at which speeds the noise is most pronounced. Stop the vehicle in a quiet place to avoid interfering noises. With the transmission in **N** position, run the engine slowly through the engine speeds corresponding to the vehicle speed at which the noise was most noticeable. If a similar noise was produced with the vehicle standing still,

the noise is not in the axle assemblies, but somewhere in the engine or transmission.

FRONT WHEEL BEARING NOISE

Front wheel bearing noises, sometimes confused with axle noises, will not change when comparing drive and coast conditions. While holding the vehicle speed steady, lightly apply the footbrake. This will often cause wheel bearing noise to lessen, as some of the weight is taken off the bearing. Front wheel bearings are easily checked by jacking up the wheels and spinning the wheels. Shaking the wheels will also determine if the wheel bearings are excessively loose.

AXLE ASSEMBLY NOISE

If a logical test of the vehicle shows that the noise is not caused by external items, it can be assumed that the noise originates from the axle assembly. The axle assembly should be tested on a smooth level road to avoid road noise. It is not advisable to test the axle by jacking up the wheels and running the vehicle.

True axle noises generally fall into 2 classes; gear noise and bearing noises and can be caused by a faulty driveshaft, faulty wheel bearings, worn differential or pinion shaft bearings, U-joint misalignment, worn differential side gears and pinions, or mismatched, improperly adjusted, or scored ring and pinion gears.

REAR WHEEL BEARING NOISE

A rough rear wheel bearing causes a vibration or growl which will continue with the vehicle coasting or in neutral. A brinelled rear wheel bearing will also cause a knock or click noise. Jack up the rear wheels and spin the wheel slowly, listening for signs of a rough or brinelled wheel bearing.

DIFFERENTIAL SIDE GEAR AND PINION NOISE

Differential side gears and pinions seldom cause noise since their movement is relatively slight on straight ahead driving. Noise produced by these gears will be more noticeable on turns.

PINION BEARING NOISE

Pinion bearing failures can be distinguished by their speed of rotation, which is higher than side bearings or axle bearings. Rough or brinelled pinion bearings cause a continuous low pitch whirring or scraping noise beginning at low speeds.

SIDE BEARING NOISE

Side bearings produce a constant rough noise, which is slower than the pinion bearing noise.

Bearing Diagnosis

This section will help in the diagnosis of bearing failure and the causes. Bearing diagnosis can be very helpful in determining the cause of axle assembly failure.

When disassembling a axle assembly, the general condition of all bearings should be noted and classified where possible. Proper recognition of the cause will help in correcting the problem and avoiding a repetition of the failure. Some of the common causes of bearing failure are:

1. Abuse during assembly or disassembly.
2. Improper assembly methods.
3. Improper or inadequate lubrication.
4. Bearing contact with dirt or water.
5. Wear caused by dirt or metal chips.
6. Corrosion or rust.
7. Seizing due to overloading.
8. Overheating.
9. Frettage of the bearing seats.
10. Brinelling from impact or shock loading.
11. Manufacturing defects.
12. Pitting due to fatigue.

To avoid damage to the bearing from improper handling, it is best to treat a used bearing the same as a new bearing. Always work in a clean area with clean tools. Remove all outside dirt from the housing before exposing a bearing and clean all bearing seats before installing a bearing. Never spin a bearing, either by hand or with compressed air.

General Service and Inspection

CLEANING BEARINGS

Proper bearing cleaning is important. Bearings should always be cleaned separately from other rear axle parts.

1. Soak all bearings in clean kerosene or diesel fuel oil. Ordinary gasoline should not be used. Bearings should not be cleaned in hot solution tank.
2. Slush bearings in cleaning solution until all oil lubricant is loosened.

Brush bearings with soft bristled brush until all dirt has been removed. Remove loose particles of dirt by striking flat against a wood block.

3. Rinse bearings in clean fluid. While holding races to prevent rotation, blow dry with compressed air. Do not spin bearings while drying.

4. After bearings have been inspected, lubricate thoroughly with regular axle lubricant; then wrap each bearing in clean cloth until ready to use.

CLEANING PARTS

Immerse all parts in suitable cleaning fluid and clean thoroughly. Use a stiff bristle brush as required to remove foreign deposits. Clean all lubricant passages or channels in pinion cage, carrier, caps and retainers. Make certain that interior of housing is thoroughly cleaned. Clean vent plugs and breathers. Small parts such as cap screws, bolts, studs, nuts etc., should be cleaned thoroughly.

INSPECTION

Magna Flux or equivalent test all steel parts, except ball and roller bearings, to detect presence of wear and cracks.

BEARINGS

Rotate each bearing and check to see if the rollers are worn, chipped, rough or in any other way damaged. Check the cage to see if it is in any way damaged. If either the bearing rollers or the cage are damaged the bearing must be replaced.

GEARS

Examine drive gear and drive pinion, differential pinions and differential side gears carefully, for damaged teeth, worn spots in surface hardening, distortion and where drive gear is attached to differential case with rivets, inspect rivets for looseness, replace loose rivets. Check radial clearances between differential side gears and differential case. Check fit of differential pinions on spider.

DIFFERENTIAL CASE

Inspect case for cracks, distortion or damage, if in good condition, thoroughly clean case and cover; then assemble case with bolts and mount in lathe centers of block stand. If lathe is not available, install differential side bearings and mount case in differential carrier. Install dial indicator and check differential case run out.

Differential case with drive gear installed is checked in the same manner, except that dial indicator reading must be taken at gear instead of at case flange.

AXLE SHAFTS

Examine splined end of axle shaft for twisted or cracked splines, twisted shaft and worn dowel holes in flange. Install new shafts if necessary.

Install axle shaft assembly in lathe centers and check shaft run out with dial indicator so that indicator shaft end contacts inner surface of flange near outer edge of flange and check flange run out.

SHIMS

Carefully inspect shims for uniform thickness. Where various thickness of shims are used in a pack, it is recommended that the thickest shims be used between the thin shims.

THRUST WASHERS

Replace all thrust washers upon installation.

SPIDER ASSEMBLY

Carefully inspect spider componets for wear or defects.

PINION BUSHINGS

Examine bushings (when used) for excessive wear, looseness, or damage. Check fit or gears on spider for excessive clearance.

AXLE HOUSING SLEEVES

Sleeves showing damaged threads, wear, or other damage should be replaced as necessary.

CHRYSLER IMPORT/ MITSUBISHI

Front Drive Axle

Inspection Before Disassembly

1. Position the removed differential carrier assembly in a suitable holding fixture. Remove the differential cover.
2. With the drive pinion locked in place, measure the final drive gear

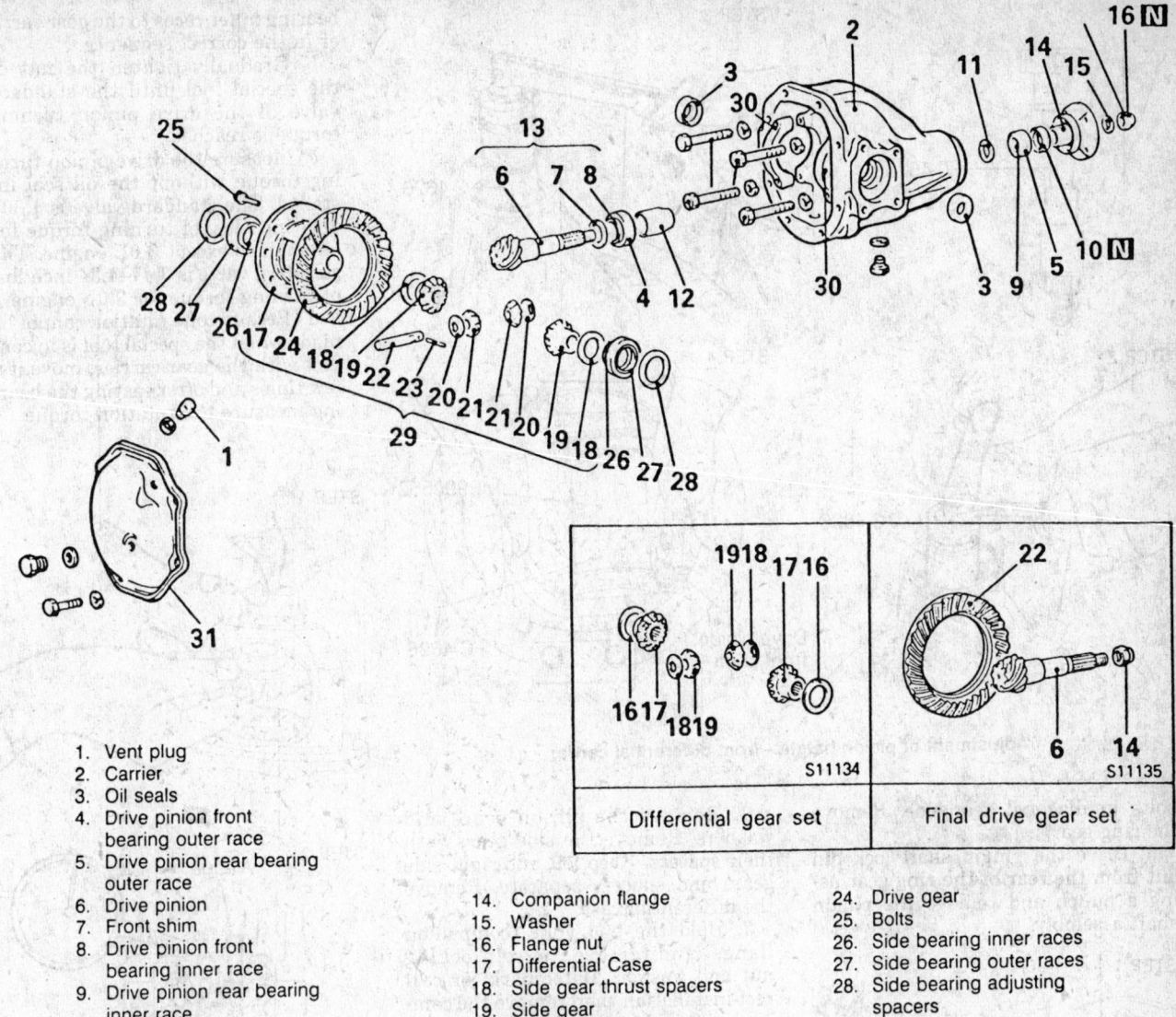

1. Vent plug
2. Carrier
3. Oil seals
4. Drive pinion front bearing outer race
5. Drive pinion rear bearing outer race
6. Drive pinion
7. Front shim
8. Drive pinion front bearing inner race
9. Drive pinion rear bearing inner race
10. Oil seal
11. Rear shim
12. Drive pinion spacer
13. Drive pinion assembly
14. Companion flange
15. Washer
16. Flange nut
17. Differential Case
18. Side gear thrust spacers
19. Side gear
20. Pinion washers
21. Pinion gears
22. Pinion shaft
23. Lock pin
24. Drive gear
25. Bolts
26. Side bearing inner races
27. Side bearing outer races
28. Side bearing adjusting spacers
29. Differential case assembly
30. Bearing caps
31. Cover

Differential gear set

Final drive gear set

Exploded view of front differential assembly – Chrysler Import/Mitsubishi

(ring gear) backlash with a dial indicator on the drive gear. Measure at 4 points or more on the drive gear. If the backlash is not within the standard valve, adjust it by using the correct side bearing spacer.

3. The final drive gear backlash standard valve is 0.0043–0.0063 in. on all engines.

4. Measure the drive gear runout at the shoulder on the reverse side of the drive gear the limit is 0.0020 in. on all engines.

5. While locking the side gear with a wedge or equivalent, measure the differential gear backlash with a dial indicator on the pinion gear. The standard valve is 0–0.0030 in. The service limit is 0.008 in. on all engines.

6. Check the final drive gear tooth contact by the following steps:
   a. Apply a thin uniform coat of machine blue or equivalent to both surfaces of the drive gear teeth.
   b. Insert a tool between the differential carrier and the differential case and then rotate the companion flange by hand (once in normal direction and once reverse direction) while applying a load to the drive gear so that the revolution torque (about 2.5 ft. lbs.) is applied to the drive pinion.
   c. If the drive gear is rotated too much, contact pattern will become unclear and difficult to read. If a correct pattern cannot be obtained (drive gear and pinion

worn beyond allowable limits), replace the drive gear and the drive pinion as a set.

Disassembly and Assembly

1. Position the removed differential carrier assembly in a suitable holding fixture. Remove the differential cover.

2. Matchmark and remove the bearing (carrier) caps and pry out the differential assembly.

3. Remove the differential side bearings. Be sure to keep the right and left bearings and shims separated. Remove side bearings inner and outer races.

4. Matchmark ring gear assembly and loosen the ring gear mounting

STEP 1

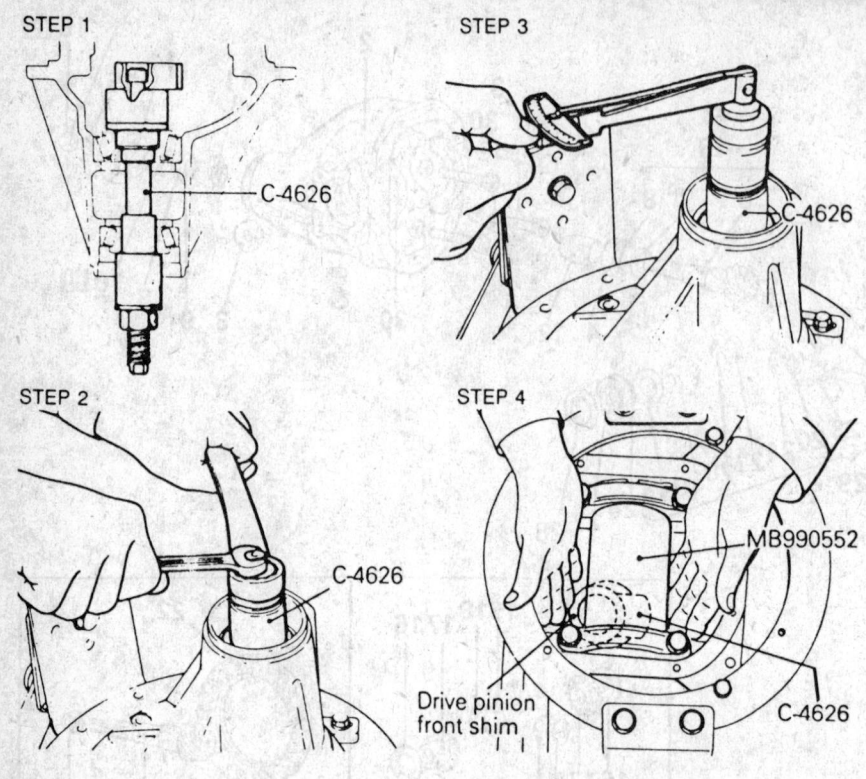

C-4626

STEP 2

C-4626

STEP 3

C-4626

STEP 4

MB990552

Drive pinion
front shim

C-4626

Adjustment of pinion height—front differential carrier

bearing inner races to the gear carrier in the correct sequence.

b. Gradually tighten the nut of the special tool until the standard valve of the drive pinion turning torque is reached.

c. Measure the drive pinion turning torque without the oil seal installed. The standard valve is 1.30–2.17 inch lbs. of turning torque for all engines except 3.0L engine. The standard valve is 3.47–4.34 inch lbs. of turning torque for 3.0L engine.

d. Because one rotation cannot be made when the special tool is in contact with the gear carrier, move it a few times and after seating the bearing measure the rotation torque.

STEP 1

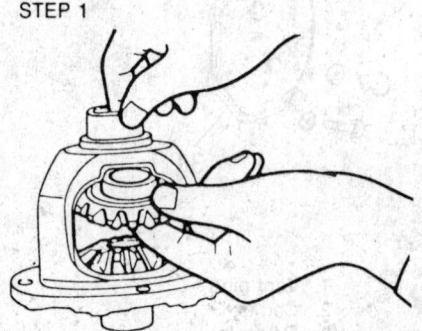

bolts in diagonal sequence. Remove the ring gear.

5. Drive the pinion shaft lock pin out from the rear of the ring gear using a punch and remove the pinion shaft assembly.

STEP 1

C-3281

STEP 2

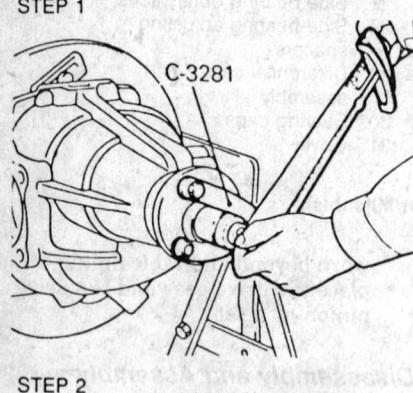

Adjustment of drive pinion preload—front differential carrier

6. Remove the pinion gears and washers. Remove the side gears with their spacers. Keep left and right side gears and spacers separate. Remove the differential case.

7. Hold the end yoke (companion flange) and remove the self-locking nut and washer. Matchmark for correct installation then remove the companion flange.

8. Tap the end of the drive pinion shaft (matchmark shaft assembly for correct installation) with a plastic hammer and force out the drive pinion along with its adjusting shim, the rear inner race, the drive pinion spacer and the preload adjusting shim. The rear bearing inner race can be pressed off the pinion shaft.

9. Remove the front and rear pinion bearing outer races. Remove the oil seals. Remove the vent plug if necessary and gear carrier case.

10. Install the vent plug to the gear carrier case if necessary. Install oil seals.

11. Press the drive pinion front and rear bearing outer races into the differential carrier case. Make sure that the races do not tilt and that they sit fully in the case.

12. At this point of the reassembly, adjust the drive pinion height by following this procedure:

a. Install special service tool C-4626 and drive pinion front and rear

STEP 2

Wedge

STEP 3

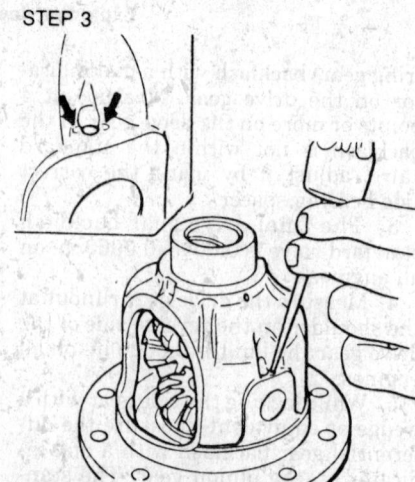

Adjustment of differential gear backlash—front differential carrier

e. Mount the special service tool MB990552 in the side bearing seat of the gear carrier and then select a drive pinion front height adjusting shim (keep front shims to a minimum) of a thickness which corresponds to the gap between the special tools. Be sure to clean the side bearing seat thoroughly. When mounting the special tool be sure that the cut-out sections are in the correct position and also confirm that the tool is in close contact with the side bearing seat.

f. Install the selected drive pinion front shim(s) to the drive pinion. Using special tool MB990802, press fit the drive pinion front bearing inner race.

13. At this point of the reassembly, adjust the drive pinion preload (turning torque) by following this procedure:

a. Without the oil seal installed insert the drive pinion into the gear carrier and then install from the front side of the carrier, the driver pinion spacer, drive pinion rear shim, drive pinion rear bearing inner race and the companion flange in that order.

b. Tighten the companion flange to 116–159 ft. lbs. on all engines except the 3.0L engine. On the 3.0L engine torque the companion flange to 137–181 ft. lbs. Measure the drive pinion turning torque without the

oil seal installed. The standard valve is 1.30–2.17 inch lbs. turning torque for all engines except the 3.0L engine. The standard valve is 3.47–4.34 inch lbs. turning torque for the 3.0L engine.

c. Adjust the turning torque by replacing the drive pinion rear shim(s) or the drive pinion spacer. Reduce number of drive pinion rear shims by replacing drive pinion spacer(s) as necessary.

d. Remove the companion flange and drive pinion once again.

14. Install drive pinion rear bearing inner race. Install oil seal into the gear carrier using suitable tools.

15. Install the drive pinion assembly and companion flange in the correct position. Tighten the companion flange to 116–159 ft. lbs. on all engines except the 3.0L engine. On the 3.0L engine torque the companion flange to 137–181 ft. lbs.

16. Measure the drive pinion turning torque with oil seal installed. The standard valve is 3.04–3.91 inch lbs. of turning torque for all engines except the 3.0L engine. The standard valve for the 3.0L engine is 5.21–6.08 inch lbs. of turning torque.

17. If not within the standard valve, check for faulty installation of the oil seal or wrong torque on the companion flange locking nut.

18. At this point of the reassembly,

adjust the differential gear backlash by following this procedure:

a. Install the side gears, side gear thrust spacers, pinion gear and pinion washers into the differential case. Temporarily install the pinion shaft. Do not drive the lock pin in place yet.

b. While locking the side gear with a wedge or equivalent, measure the differential gear backlash with a dial indicator on the pinion gear. The standard valve is 0–0.0030 in. and the service limit is 0.008 in. on all engines.

c. If the differential gear backlash exceeds the limit adjust the backlash by installing thicker side gear thrust spacers. If adjustment is not possible replace the side gears and pinion gears as a set.

19. Align the pinion shaft lock pin hole with the differential case lock pin hole.

20. Drive the lock pin in and stake the pin with a suitable tool in 2 different locations.

21. Install the drive gear onto the differential case with matchmarks in the correct position. Torque (with locking adhesive or equivalent on threads) drive gear attaching bolts to 58–65 ft. lbs. in a diagonal sequence for all engines.

22. Press fit the side bearing inner races to the differential case using suitable tools.

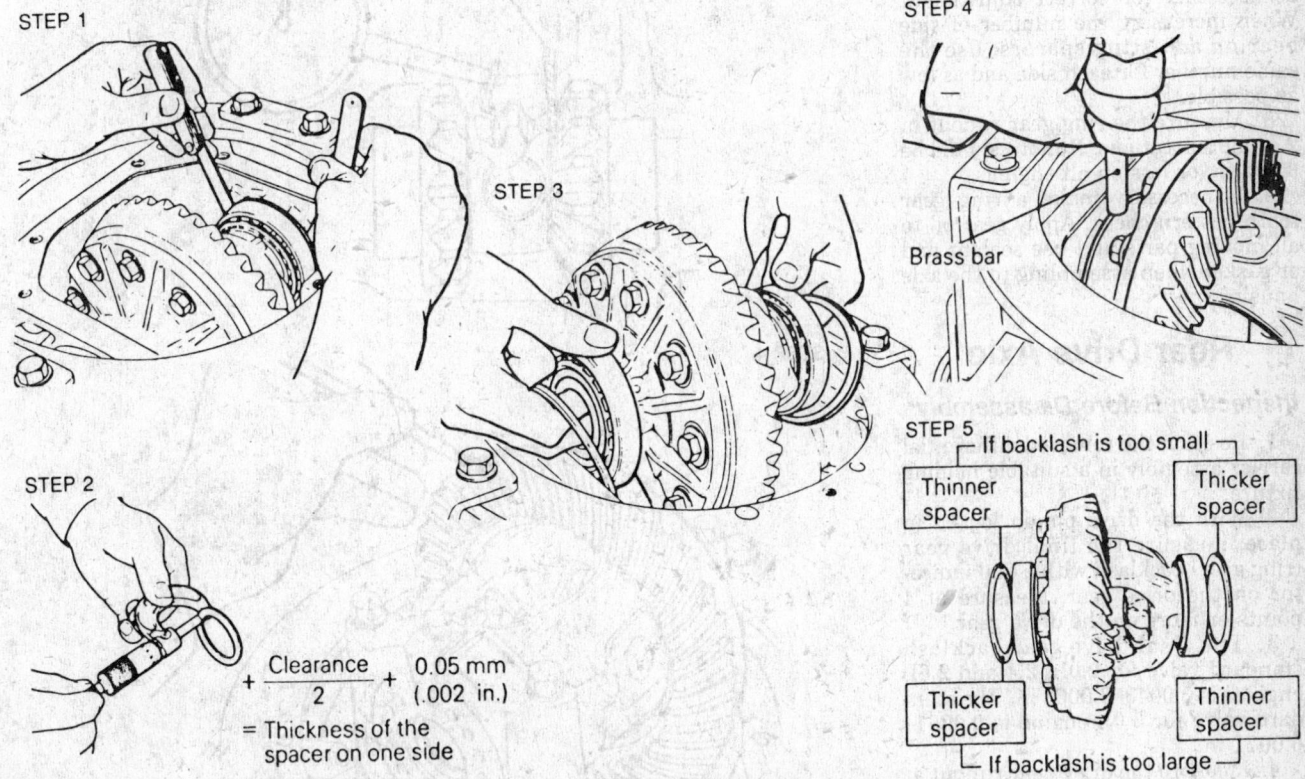

STEP 1

STEP 2

STEP 3

STEP 4

Brass bar

STEP 5

$$+ \frac{Clearance}{2} + \begin{matrix} 0.05 \text{ mm} \\ (.002 \text{ in.}) \end{matrix}$$

= Thickness of the spacer on one side

If backlash is too small
Thinner spacer   Thicker spacer
Thicker spacer   Thinner spacer
If backlash is too large

Adjustment of final drive backlash—front differential carrier

23. At this point of the reassembly, adjust the final drive gear backlash as follows:

 a. Install the side bearing adjusting spacers (both sides same size-thin spacers as possible) to the side bearing outer races, then mount ther differential case assembly into the gear carrier.

 b. Push the differential case assembly to one side. Measure the clearance between the gear carrier and the side bearing adjusting spacer with a feeler gauge.

 c. Measure the thickness of the side bearing adjusting spacers on a side. Select 2 pairs of spacers which correspond to that thickness plus ½ of the clearance plus 0.002 in. and then install one pair each to the drive pinion side and the drive gear side.

24. Install the side bearing adjusting spacers and differential case assembly in the gear carrier.

25. Tap the side bearing adjusting spacers with a brass tool to position them to the side bearing outer races.

26. Align the mating marks on the gear carrier and the bearing caps and tighten the bearing caps to 40–47 ft. lbs. on all engines.

27. Attach a dial indicator to the ring gear teeth and measure the final drive gear backlash at 4 different locations. Final drive gear backlash standard valve is 0.0043–0.0063 in. for all engines. Change the side bearing spacers as necessary for correct adjustment. When increasing the number of side bearing adjusting spacers, use the same number for each side and as few as possible.

28. Measure the ring gear runout in 2 or more locations. Runout should be 0.002 in. or less on all engines.

29. If necessary make a ring gear tooth pattern check. Apply gear oil to all moving parts and use sealant and or gasket when assembling to the axle housing.

Rear Drive Axle

Inspection Before Disassembly

1. Position the removed differential carrier assembly in a suitable holding fixture.

2. With the drive pinion locked in place, measure the final drive gear (ring gear) backlash with a dial indicator on the drive gear. Measure at 4 points or more on the drive gear.

3. The final drive gear backlash standard valve for 2.0L, 2.4 and 2.6L engines is 0.0043–0.0063 in. The standard valve for 3.0L engine is 0.0051–0.0071 in.

4. Measure the drive gear runout at the shoulder on the reverse side of the

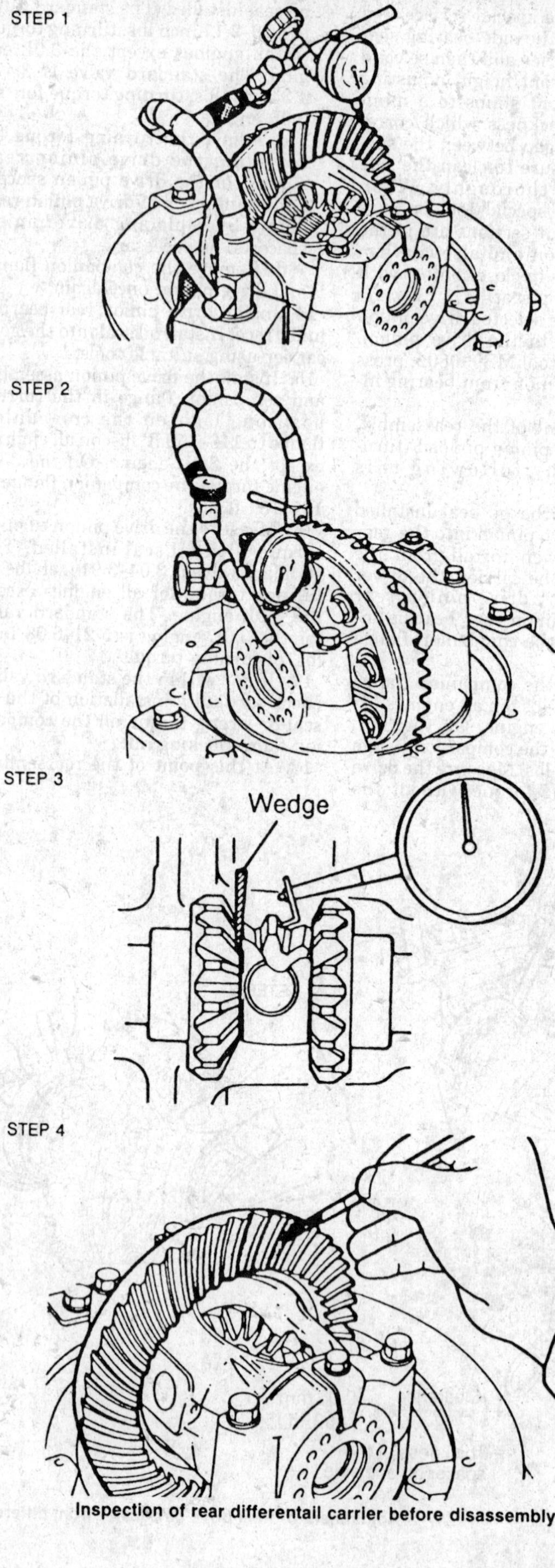

STEP 1

STEP 2

STEP 3

Wedge

STEP 4

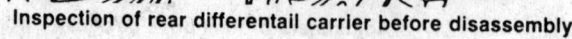

Inspection of rear differentail carrier before disassembly

Standard tooth contact pattern

1 Toe
2 Drive-side
3 Heel
4 Coast-side

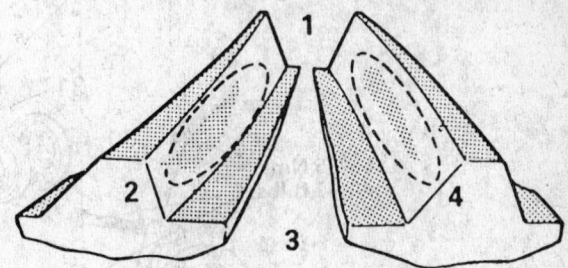

Problem	Solution

Tooth contact pattern resulting from excessive pinion height

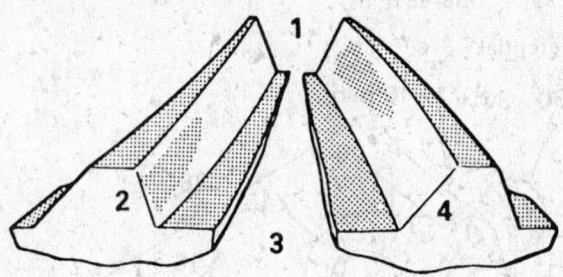

The drive pinion is positioned too far from the center of the drive gear.

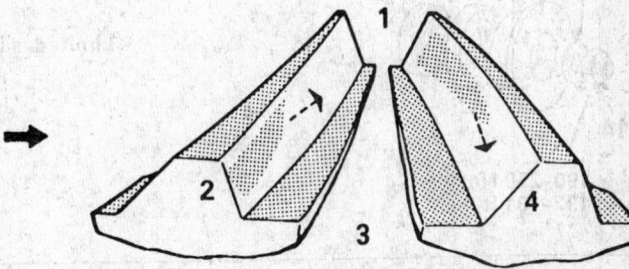

Increase the thickness of the pinion height adjusting shim, and position the drive pinion closer to the center of the drive gear.
Also, for backlash adjustment, position the drive gear farther from the drive pinion.

Tooth contact pattern resulting from insufficient pinion height

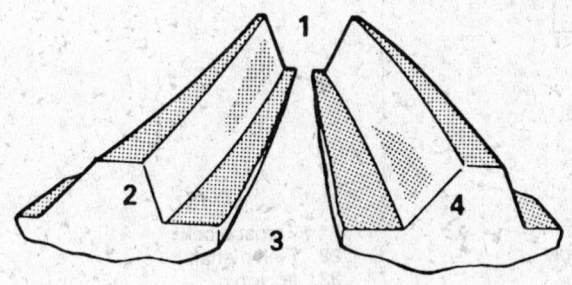

The drive pinion is positioned too close to the center of the drive gear.

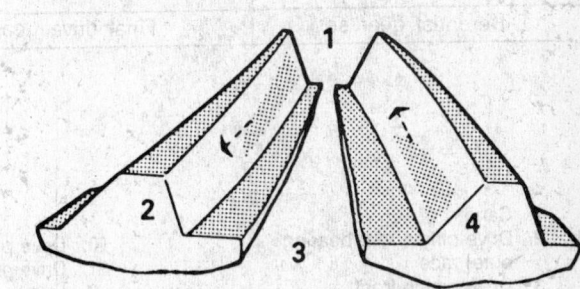

Decrease the thickness of the pinion height adjusting shim, and position the drive pinion farther from the center of the drive gear.
Also, for backlash adjustment, position the drive gear closer to the drive pinion.

Differential tooth contact pattern—Chrysler Import/Mitsubishi

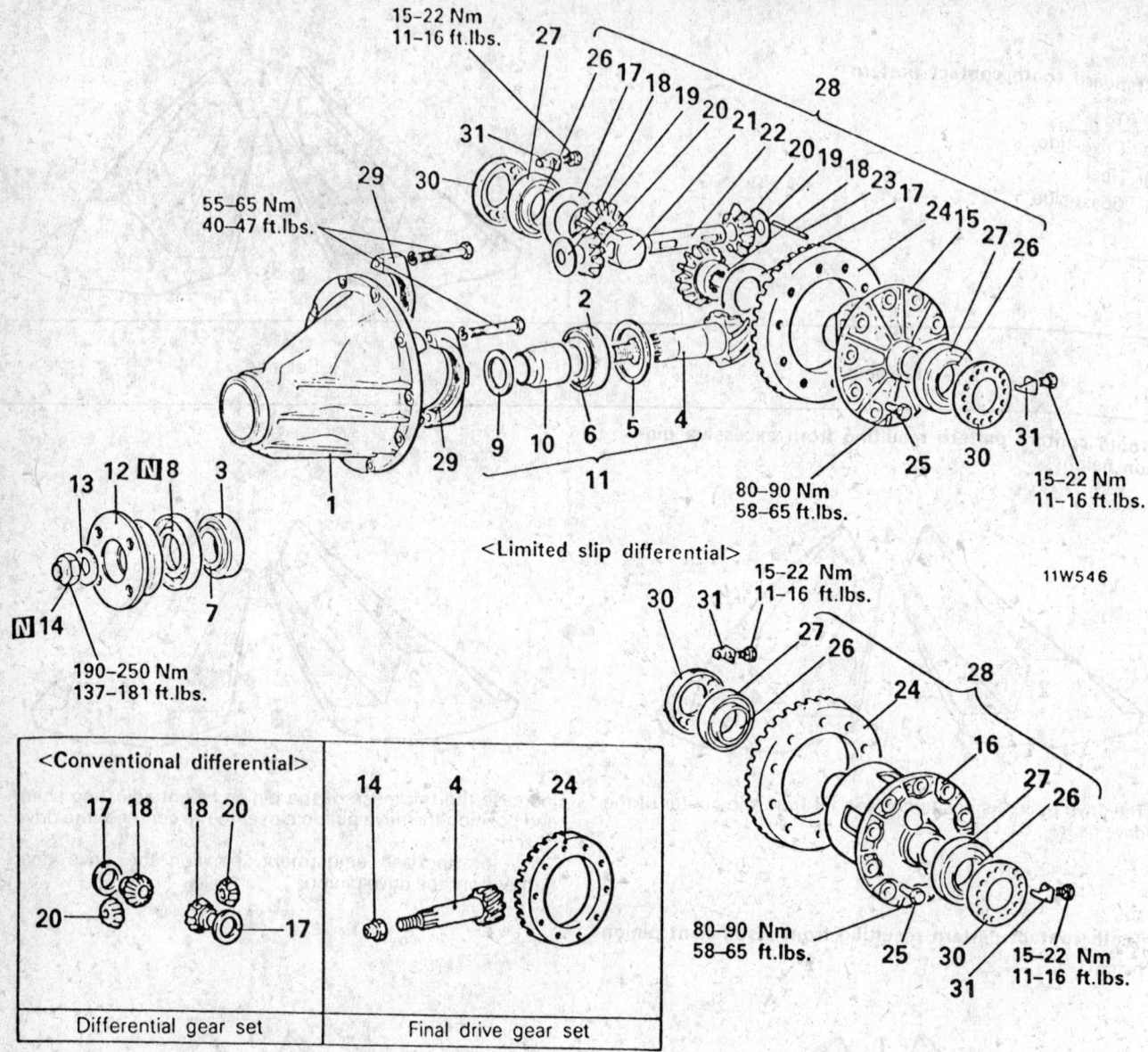

15–22 Nm
11–16 ft.lbs.

55–65 Nm
40–47 ft.lbs.

190–250 Nm
137–181 ft.lbs.

80–90 Nm
58–65 ft.lbs.

15–22 Nm
11–16 ft.lbs.

<Limited slip differential>

11W546

15–22 Nm
11–16 ft.lbs.

80–90 Nm
58–65 ft.lbs.

15–22 Nm
11–16 ft.lbs.

<Conventional differential>

Differential gear set | Final drive gear set

1. Carrier
2. Drive pinion rear bearing outer race
3. Drive pinion front bearing outer race
4. Drive pinion
5. Rear shim
6. Drive pinion rear bearing inner race
7. Drive pinion front bearing inner race
8. Oil seal
9. Front shim

10. Drive pinion spacer
11. Drive pinion assembly
12. Companion flange
13. Washer
14. Flange nut
15. Differential Case
16. Limited slip differential case assembly
17. Side gear thrust spacers
18. Side gear
19. Pinion washers
20. Pinion gears

21. Thrust block
22. Pinion shaft
23. Lock pin
24. Drive gear
25. Bolts
26. Side bearing inner races
27. Side bearing outer races
28. Differential case assembly
29. Bearing caps
30. Side bearing nuts
31. Lock plates

Exploded view of rear differential assembly – Chrysler Import/Mitsubishi

drive gear the limit is 0.0020 in. on all engines.

5. On a conventional differential asssembly the differential gear backlash must be checked. While locking the side gear with a wedge or equivalent, measure the differential gear backlash with a dial indicator on the pinion gear. The differential gear backlash (conventional) for the 2.0L, 2.4L and 2.6L engines standard valve is 0.0004–0.0030 in. The standard valve (conventional) for the 3.0L engine is 0–0.0030 in. The service limit is 0.008 in. on all engines.

6. Check the final drive gear tooth contact by the following steps:

a. Apply a thin uniform coat of machine blue or equivalent to both surfaces of the drive gear teeth.

b. Insert a tool between the differential carrier and the differential case and then rotate the companion flange by hand (once in normal direction and once reverse direction) while applying a load to the drive gear so that the revolution torque (about 2.5 ft. lbs.) is applied to the drive pinion.

c. If the drive gear is rotated too much, contact pattern will become unclear and difficult to read. If a correct pattern cannot be obtained (drive gear and pinion worn beyond allowable limits), replace the drive gear and the drive pinion as a set.

Disassembly and Assembly

1. Position the differential assembly in a suitable holding fixture. Remove the lock bolts and plates holding the side bearing nut in place.

2. Remove the side bearing nuts with the special adjusting tool spanner wrench No. MB990201 or equivalent.

3. Matchmark and remove the carrier caps and pry out the differential.

4. Pull off the differential side bearings. Be sure to keep the right and left bearings and shims separated. Remove side bearings inner and outer races.

5. Matchmark ring gear assembly and loosen the ring gear mounting bolts in diagonal sequence. Remove the ring gear.

6. Drive the pinion shaft lock pin out from the rear of the ring gear using a punch and remove the pinion shaft assembly.

7. Remove the thrust block, pinion gears and washers. Remove the side gears with their spacers. Keep left and right side gears and spacers separate. Remove the differential case.

8. On limited slip differential vehicles, remove limited slip differential case assembly.

9. Hold the end yoke (companion flange) and remove the self-locking

nut and washer. Matchmark for correct installation then remove the companion flange.

10. Tap the end of the drive pinion shaft (matchmark shaft assembly for correct installation) with a plastic hammer and force out the drive pinion

11. Remove the front and rear pinion bearing outer races. The front inner

along with its adjusting shim, the rear inner race, the drive pinion spacer and the preload adjusting shim. The rear bearing inner race can be pressed off the pinion shaft.

11. Remove the front and rear pinion bearing outer races. The front inner

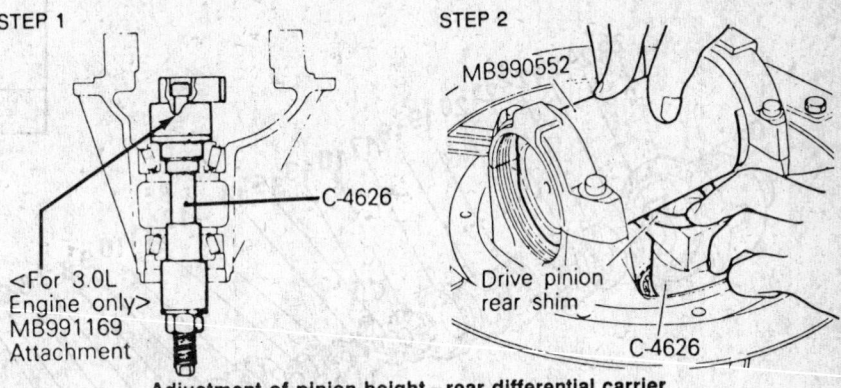

Adjustment of pinion height – rear differential carrier

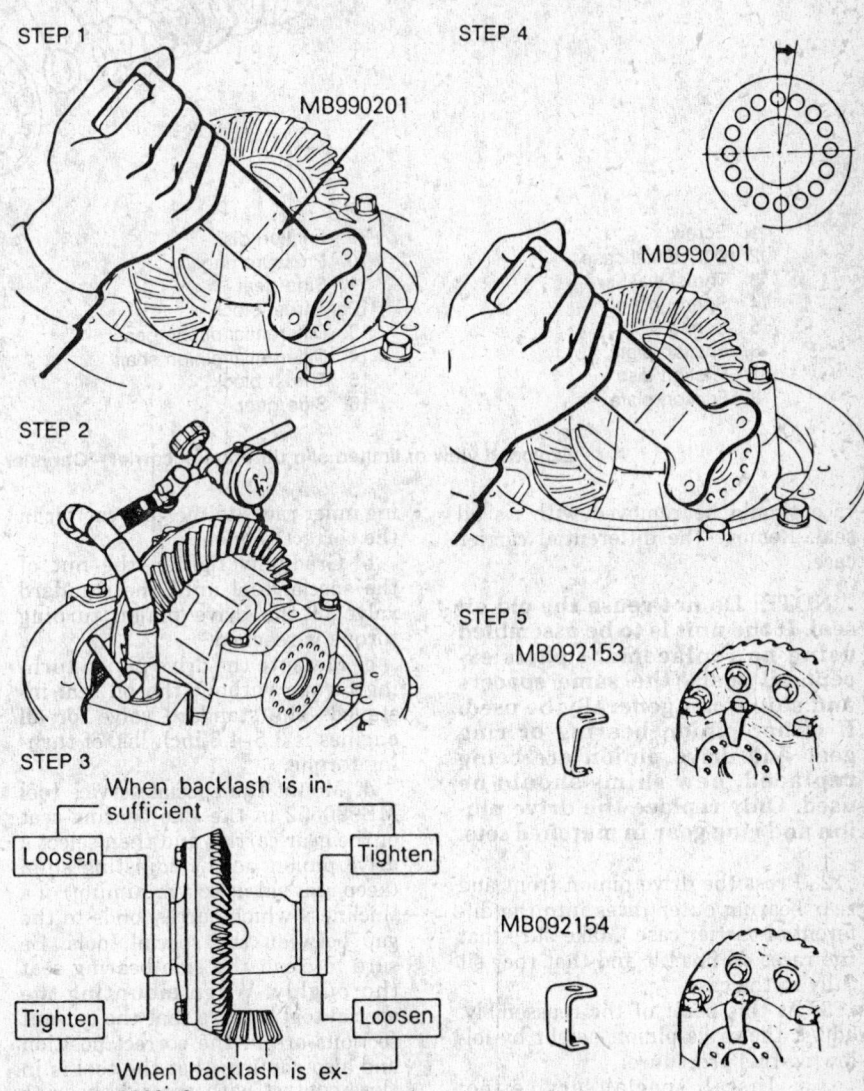

Adjustment of final drive gear backlash – rear differential carrier

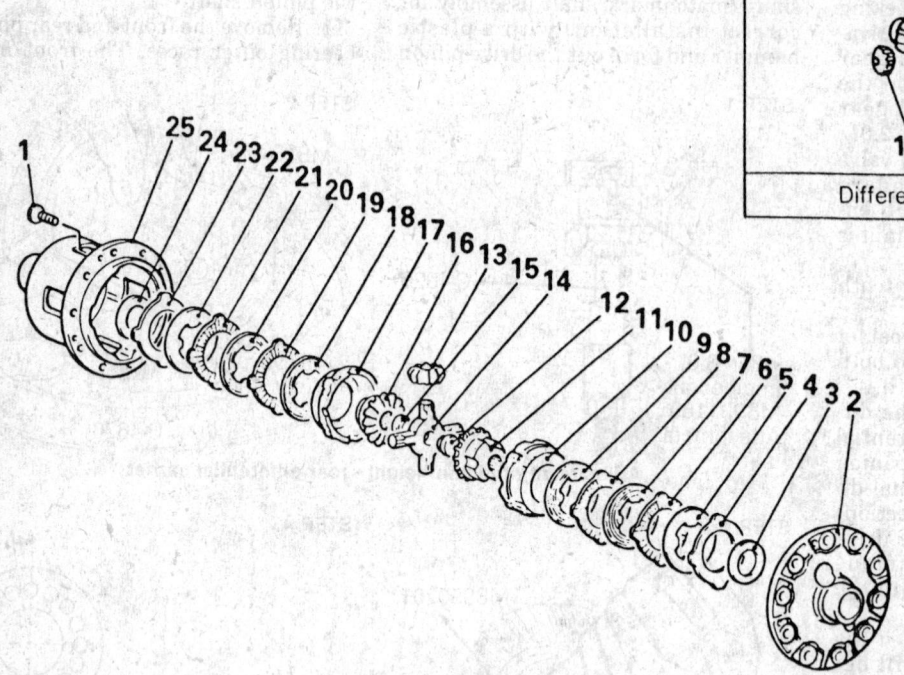

Differential gear set

1. Screw
2. Differential case
3. Thrust washer
4. Spring plate
5. Spring disc
6. Friction plate
7. Friction disc
8. Friction plate

9. Friction disc
10. Pressure ring
11. Side gear
12. Thrust block
13. Differential pinion gear
14. Differential pinion shaft
15. Thrust block
16. Side gear

17. Pressure ring
18. Friction disc
19. Friction plate
20. Friction disc
21. Friction plate
22. Spring disc
23. Spring plate
24. Thrust washer
25. Differential case

Exploded view of limited slip differential carrier — Chrysler Import/Mitsubishi

race should be removed with its oil seal. Remove the differential carrier case.

NOTE: Do not reuse the old oil seal. If the unit is to be assembled using no replacement parts except oil seals, the same spacers and shims can generally be used. If either pinion bearing or ring gear and drive pinion are being replaced, new shims should be used. Only replace the drive pinion and ring gear in matched sets.

12. Press the drive pinion front and rear bearing outer races into the differential carrier case. Make sure that the races do not tilt and that they sit fully in the case.

13. At this point of the reassembly, adjust the drive pinion height by following this procedure:

a. Install special service tool MB991169 (use special adapter head tool C–4626 for 3.0L engine only) and drive pinion front and rear bear-

ing inner races to the gear carrier in the correct sequence.

b. Gradually tighten the nut of the special tool until the standard valve of the drive pinion turning torque is reached.

c. Measure the drive pinion turning torque without the oil seal installed. The standard valve for all engines is 3.5–4.3 inch lbs. of turning torque.

d. Mount the special service tool MB990552 in the side bearing seat of the gear carrier and then select a drive pinion height adjusting shim (keep rear shims to a minimum) of a thickness which corresponds to the gap between the special tools. Be sure to clean the side bearing seat thoroughly. When mounting the special tool be sure that the cut-out sections are in the correct position and also confirm that the tool is in close contact with the side bearing seat.

e. Install the selected drive pinion rear shim(s) to the drive pinion. Us-

ing special tool MB990802, press fit the drive pinion rear bearing inner race.

14. At this point of the reassembly, adjust the drive pinion preload (turning torque) by following this procedure:

a. Without the oil seal installed fit the drive pinion front shims between the drive pinion spacer and the drive pinion front bearing inner race.

b. Tighten the companion flange to 137–181 ft. lbs. on all engines. Measure the drive pinion turning torque without the oil seal installed. The standard valve is 3.5–4.3 inch lbs. turning torque for all engines.

c. Adjust the turning torque by replacing the drive pinion front shim(s) or the drive pinion spacer. Reduce number of drive pinion front shims by replacing drive pinion spacer as necessary.

d. Remove the companion flange and drive pinion once again.

15. Install drive pinion front bearing

inner race. Install oil seal into the gear carrier using suitable tools.

16. Install the drive pinion assembly and companion flange in the correct position. Torque the companion flange locking nut to 137–181 ft. lbs. on all engines.

17. Measure the drive pinion turning torque with oil seal installed. The standard valve is 5.6–6.5 inch lbs. of turning torque on all engines.

18. On limited slip differential vehicles, install limited slip differential case assembly.

19. At this point of the reassembly (conventional differential), adjust the differential gear backlash by following this procedure:

   a. Install the side gears, side gear thrust spacers, pinion gear and pinion washers into the differential case. Temporarily install the pinion shaft. Do not drive the lock pin in place yet.

   b. While locking the side gear with a wedge or equivalent, measure the differential gear backlash with a dial indicator on the pinion gear. The differential gear backlash for the 2.0L, 2.4L and 2.6L engines standard valve is 0.0004–0.0030 in. The standard valve for the 3.0L engine is 0–0.0030 in. The service limit is 0.008 in. on all engines.

   c. If the differential gear backlash exceeds the limit adjust the backlash by installing thicher side gear thrust spacers. If adjustment is not possible replace the side gears and pinion gears as a set.

20. Install the thrust block. Align the pinion shaft lock pin hole with the differential case lock pin hole.

21. Drive the lock pin in and stake the pin with a suitable tool in 2 different locations.

22. Install the drive gear onto the differential case with matchmarks in the correct position. Torque (with locking adhesive or equivalent on threads) drive gear attaching bolts to 58–65 ft. lbs. on conventional and limited slip differentials in a diagonal sequence.

23. Press fit the side bearing inner races to the differential case using suitable tools.

24. Install the differential case assembly.

25. Install the carrier (bearing) caps with their mating marks in line with the marks on the carriers and finger tighten the 4 set bolts.

26. Install the side bearing nuts and tighten the carrier cap bolts to 40–47 ft. lbs. on all engines.

27. Screw in the side bearing nuts to adjust the final drive gear backlash. Each nut should be tightened to the state just before preloading of the side bearing.

28. Attach a dial indicator to the ring gear teeth and measure the final drive gear backlash in 4 different locations. The final drive gear backlash standard valve for 2.0L, 2.4 and 2.6L engines is 0.0043–0.0063 in. The standard valve for 3.0L engine is 0.0051–0.0071 in.

NOTE: If the backlash has to be adjusted, loosen the bearing nut on the back side of the ring gear and tighten the bearing nut on the teeth side by the same amount.

29. After adjusting backlash, tighten the bearing nuts $1/2$ pitch. One pitch is the space between 2 adjacent holes on the side of the bearing nut.

30. Again measure the backlash and install a 1 or 2 pronged lock plate whichever lines up with the bearing nut holes. Tighten the lock plate bolts to 11–16 ft.lb.

31. Measure the ring gear runout in 2 or more locations. Runout should be 0.002 in. or less on all engines.

32. If necessary make a ring gear tooth pattern check. Apply gear oil to all moving parts and use sealant and or gasket when assembling to the axle housing.

ISUZU

Front Drive Axle

Disassembly and Assembly

1. Position the removed differential carrier assembly in a suitable holding fixture.

2. Mark and remove the side bearing caps.

3. Remove the differential cage assembly. Keep the right and left side bearing races and shims if so equiped in separate groups for reinstallation in same positions.

4. Remove the differential side bearings from the case by using puller J–22888 and plug J–8107–2 or equivalent. Carefully record the thickness of each side bearing and each shim(s) removed for later use in reassembly and keep separated. Puller arms must not pull against roller cage. Use care to position legs against inner race. As bearing is being removed, check for free rotation of bearing. If bearing does not rotate freely, check position of puller legs.

5. Remove the companion flange nut while holding flange with suitable tools. Remove the flange assembly with puller if necessary.

6. Drive the pinion assembly from the carrier using suitable tools and brass punch. The outer (front) bearing will fall loose in the carrier, while the inner (rear) bearing will remain pressed on the drive pinion. Both races will remain in the carrier bores.

7. Remove rear bearing from the drive pinion by use of a press and tool J–22912–01 or equivalent.

8. Remove the oil seal and then drive the 2 outer races from the carrier by use of a brass drift and suitable tools.

9. Remove the ring gear bolts and separate the ring gear from the differential case. Use care when removing ring gear to prevent damage to differential case or ring gear. Do not force a chisel or other tools between the joining faces.

10. Drive out the pinion shaft lock pin with a long drift. It may be necessary to first break the stake on lock pin using a 0.020 diameter drill or equivalent.

11. Remove the pinion shaft with a drift pin and take out the pinion gears, side gears and thrust washers from the differential case.

NOTE: It is important to clean and assemble parts with care and to follow adjustment procedures. Axles which are contaminated with dirt or other foreign material, or which are incorrectly adjusted may be noisy and have short life. Be sure to use new seals, gaskets and flange nut when reassembling axle.

12. Wash the bearings in a suitable solvent. Then examine bearings carefully for wear, separation, cracks, seizure and other abnormal conditions. Replace bearings as necessary.

13. Check the ring gear and drive pinion teeth for wear, chipping, cracks, pitting and abnormal contact. Check the drive pinion splines for cracks, distortion and step wear. Replace parts if needed. Ring gear (torque for ring gear replacement is 73–87 ft. lbs. in diagonal sequence) and drive pinion come only in matched sets. If either item is defective, both parts must be replaced.

14. Check the pinion gears and side gears for wear, chipped teeth and separation and replace if needed.

15. Check and replace the thrust blocks if so equipped and thrust washers, if worn or damaged.

16. Check the lock pin for bending, dents and other abnormal conditions. Replace if necessary.

17. Check the side gear-to-axle shaft fit. Also check the fit of the pinion shaft to pinion gears. Examine the contact surfaces between side gears

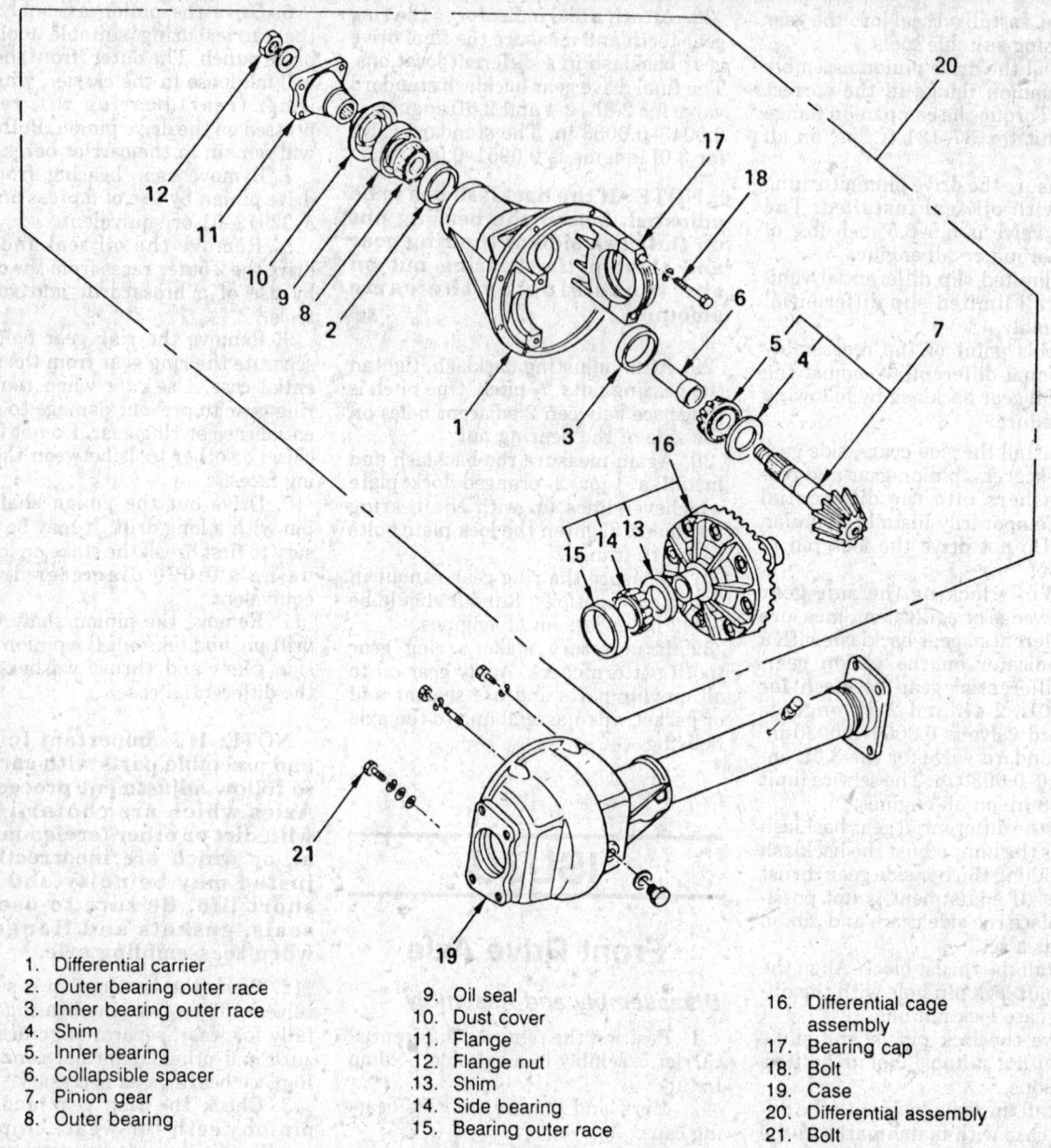

1. Differential carrier
2. Outer bearing outer race
3. Inner bearing outer race
4. Shim
5. Inner bearing
6. Collapsible spacer
7. Pinion gear
8. Outer bearing
9. Oil seal
10. Dust cover
11. Flange
12. Flange nut
13. Shim
14. Side bearing
15. Bearing outer race
16. Differential cage assembly
17. Bearing cap
18. Bolt
19. Case
20. Differential assembly
21. Bolt

Exploded view of front differential assembly—Isuzu

and differential case and between ring gear and case.

18. The standard valve for the clearance between the differential pinion and the cross pin is 0.002–0.005 in. and the service limit is 0.008 in.

19. The standard valve for the clearance between the side gear and the differential assembly is 0.001–0.004 in. and the service limit is 0.006 in.

20. The standard valve for play (backlash) in splines betwwen the side gear and the axle shaft is 0.001–0.006 in. and the service limit is 0.010 in.

21. Install the side gears and thrust washers in the differential case.

22. Position the pinion gears 180 degrees apart. Roll gears into position,

making sure they are in alignment, to allow installation of the pinion shaft.

23. Place the thrust block if equipped between the pinion gears and drive the pinion shaft into position. Make sure that the lock pin hole in cross shaft aligns with the hole in the case.

24. Measure the amount of backlash between the differential gears and the pinion gears. If the backlash is greater than 0.001–0.003 in., make the necessary adjustment with the thrust washers.

25. Install lock pin into cross shaft and stake the cage to prevent loosening of the pin.

26. Apply Loctite or equivalent to the threaded portion of the bolts. Install

the ring gear in position on the differential case. Tighten the bolts (always use new retaining bolts) in diagonal sequence to 73–87 ft. lbs.

27. To set pinion depth install the drive pinion front and rear bearing outer races into carrier bores. Use drive handle J–8092 with J–24256 for front bearing race and J–24252 or equivalent for rear bearing race.

28. Lubricate and position the front and rear bearings to be used for final assembly into their respective races.

29. Install gauging plate J–23597-7 and preload stud and pilot J–23597-9 or their equivalent through front and rear bearings and tighten nut snugly.

30. Rotate the bearings to insure

Pinion marking → Dial indicator reading (inches) ↓	+10	+8	+6	+4	+2	0	−2	−4	−6	−8	−10
0 081											2.18(0.0858)
0 082										2.18(0.0858)	2.20(0.0866)
0 083									2.18(0.0858)	2.20(0.0866)	2.24(0.0882)
0 084								2.18(0.0858)	2.20(0.0866)	2.24(0.0882)	2.26(0.0890)
0 085							2.18(0.0858)	2.20(0.0866)	2.24(0.0882)	2.26(0.0890)	2.28(0.0898)
0 086						2.18(0.0858)	2.20(0.0866)	2.24(0.0882)	2.26(0.0890)	2.28(0.0898)	2.32(0.0914)
0 087					2.18(0.0858)	2.20(0.0866)	2.24(0.0882)	2.26(0.0890)	2.28(0.0898)	2.32(0.0914)	2.34(0.0921)
0 088				2.18(0.0858)	2.20(0.0866)	2.24(0.0882)	2.26(0.0890)	2.28(0.0898)	2.32(0.0914)	2.34(0.0921)	2.36(0.0929)
0 089			2.18(0.0858)	2.20(0.0866)	2.24(0.0882)	2.26(0.0890)	2.28(0.0898)	2.32(0.0914)	2.34(0.0921)	2.36(0.0929)	2.38(0.0937)
0 090		2.18(0.0858)	2.20(0.0866)	2.24(0.0882)	2.26(0.0890)	2.28(0.0898)	2.32(0.0914)	2.34(0.0921)	2.36(0.0929)	2.38(0.0937)	2.42(0.0953)
0 091	2.18(0.0858)	2.20(0.0866)	2.24(0.0882)	2.26(0.0890)	2.28(0.0898)	2.32(0.0914)	2.34(0.0921)	2.36(0.0929)	2.38(0.0937)	2.42(0.0953)	2.44(0.0961)
0 092	2.20(0.0866)	2.24(0.0882)	2.26(0.0890)	2.28(0.0898)	2.32(0.0914)	2.34(0.0921)	2.36(0.0929)	2.38(0.0937)	2.42(0.0953)	2.44(0.0961)	2.46(0.0969)
0 093	2.24(0.0882)	2.26(0.0890)	2.28(0.0898)	2.32(0.0914)	2.34(0.0921)	2.36(0.0929)	2.38(0.0937)	2.42(0.0953)	2.44(0.0961)	2.46(0.0969)	2.48(0.0977)
0 094	2.26(0.0890)	2.28(0.0898)	2.32(0.0914)	2.34(0.0921)	2.36(0.0929)	2.38(0.0937)	2.42(0.0953)	2.44(0.0961)	2.46(0.0969)	2.48(0.0977)	2.52(0.0992)
0 095	2.28(0.0898)	2.32(0.0914)	2.34(0.0921)	2.36(0.0929)	2.38(0.0937)	2.42(0.0953)	2.44(0.0961)	2.46(0.0969)	2.48(0.0977)	2.52(0.0992)	2.54(0.1000)
0 096	2.32(0.0914)	2.34(0.0921)	2.36(0.0929)	2.38(0.0937)	2.42(0.0953)	2.44(0.0961)	2.46(0.0969)	2.48(0.0977)	2.52(0.0992)	2.54(0.1000)	2.56(0.1008)
0 097	2.34(0.0921)	2.36(0.0929)	2.38(0.0937)	2.42(0.0953)	2.44(0.0961)	2.46(0.0969)	2.48(0.0977)	2.52(0.0992)	2.54(0.1000)	2.56(0.1008)	
0 098	2.36(0.0929)	2.38(0.0937)	2.42(0.0953)	2.44(0.0961)	2.46(0.0969)	2.48(0.0977)	2.52(0.0992)	2.54(0.1000)	2.56(0.1008)		
0 099	2.38(0.0937)	2.42(0.0953)	2.44(0.0961)	2.46(0.0969)	2.48(0.0977)	2.52(0.0992)	2.54(0.1000)	2.56(0.1008)			
0 000	2.42(0.0953)	2.44(0.0961)	2.46(0.0969)	2.48(0.0977)	2.52(0.0992)	2.54(0.1000)	2.56(0.1008)				
0 001	2.44(0.0961)	2.46(0.0969)	2.48(0.0977)	2.52(0.0992)	2.54(0.1000)	2.56(0.1008)					
0 002	2.46(0.0969)	2.48(0.0977)	2.52(0.0992)	2.54(0.1000)	2.56(0.1008)						
0 003	2.48(0.0977)	2.52(0.0992)	2.54(0.1000)	2.56(0.1008)							
0 004	2.52(0.0992)	2.54(0.1000)	2.56(0.1008)								
0 005	2.54(0.1000)	2.56(0.1008)									
0 006	2.56(0.1008)										

Shim selection chart — Isuzu front and rear conventional drive axles

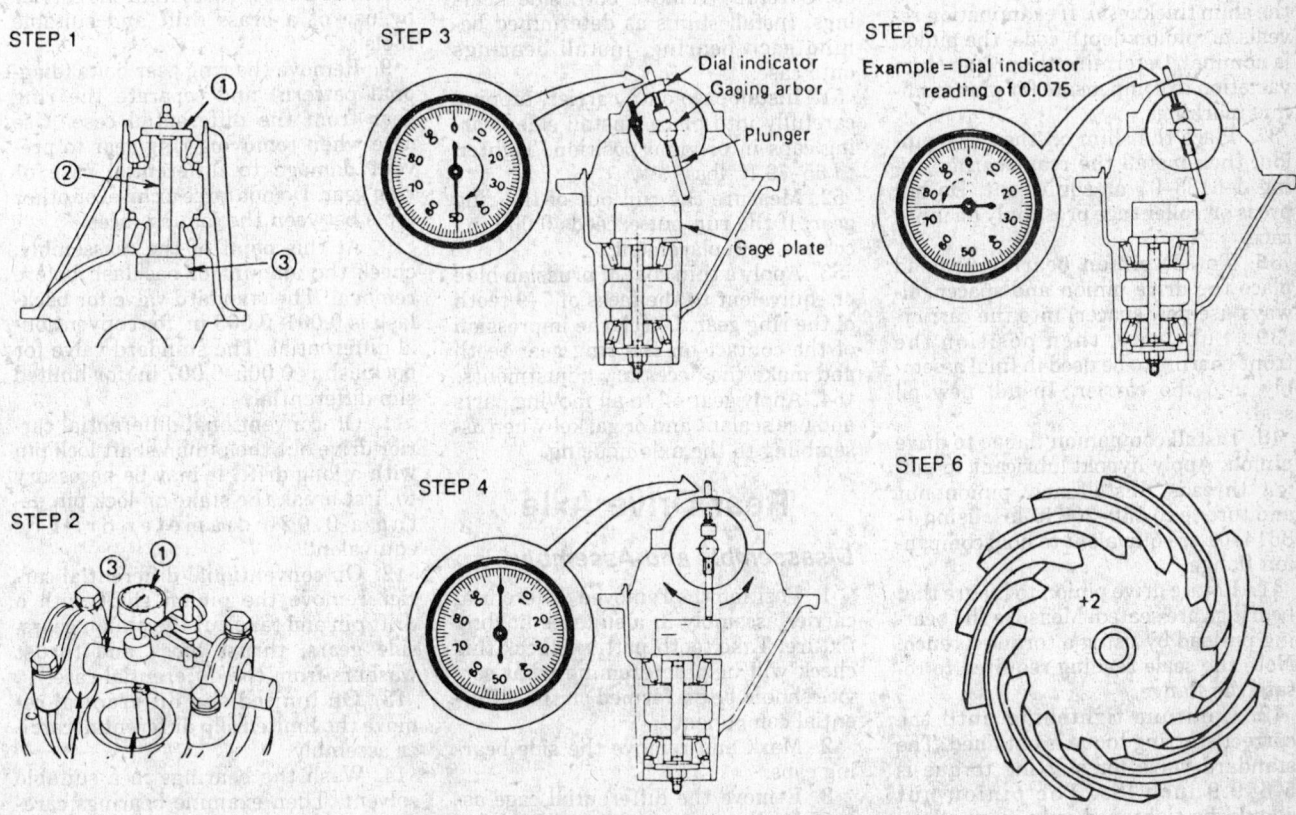

STEP 1

STEP 2

STEP 3 — Dial indicator / Gaging arbor / Plunger / Gage plate

STEP 4

STEP 5 — Example = Dial indicator reading of 0.075

STEP 6 — +2

Adjustment of drive pinion mounting distance

proper seating and tighten locknut until 20 inch lbs. of torque are required to rotate new bearings; 8–10 inch lbs. for used bearings.

31. Place discs J-23597-8 onto arbor J-23597-1 or equivalent and place tool into position in side bearing bores.

32. Install bearing caps and torque to 65–72 ft. lbs.

33. Mount dial indicator J-8001 or equivalent on arbor post and preload dial ½ revolution. Tighten indicator in this position.

34. Position the indicator plunger on the gauge plate and slowly swing across until the highest reading is obtained. Zero the indicator on the highest reading of the gauge plate.

35. Carefully swing the plunger off the gauge plate. Note the indicator reading. Recheck to verify the reading. Record the number the dial indicator needle points to. The reading on the dial is the correct dimension for the rear pinion depth shim. Convert the dial indicator reading (inches) to pinion marking (depth code) specification.

36. Record pinion depth code on the head of the drive pinion. The number indicates a necessary change in the pinion mounting distance. A plus number indicates the need for a greater mounting distance (which can be achieved by decreasing the shim thickness). A minus number indicates the need for a smaller mounting distance (which can be achieved by increasing the shim thickness). If examination reveals no pinion depth code, the pinion is nominal. Determine the proper shim variation to compensate for plus or minus markings.

37. Place the shim on the drive pinion, then install the rear bearing, using J-6133-01 or equivalent. Do not press on roller cage press only on inner race.

38. To set pinion bearing preload place the drive pinion and spacer (always use new spacer) into the carrier.

39. Lubricate, then position the front bearing to be used in final assembly into the carrier. Install new oil seal.

40. Install companion flange to drive pinion. Apply hypoid lubricant to pinion threads. Install new pinion nut and torque to 130–202 ft. lbs. using J-8614-01 or equivalent to hold companion flange.

41. Rotate drive pinion to insure that bearings are seated. Measure the bearing preload by using a torque wrench. Note the scale reading required to rotate the flange.

42. Continue tightening until the correct starting toque is obtained. The standard valve for starting torque is 5.6–9.9 inch lbs. The pinion nut should be tightened only in small in-

crements and the scale should be checked after each small amount of tightening. Exceeding preload specifications may compress the collapsible spacer too far and require its replacement.

43. Install the side bearings to be used in final assembly onto the differential case. Do not install shims at this time. Use J-24244 or equivalent for the first bearing installation.

44. Support case on plug J-8107-2 or equivalent for opposite bearing installation.

45. Install the differential cage assembly into carrier bores.

46. Using 2 sets of feeler gauges, insert a feeler gauge of sufficient thickness between each bearing outer race and the carrier to remover all endplay. Make certain the feeler gauge is pushed to the bottom of the bearing bores.

47. Install a dial indicator on the carrier so that the stem of the tool is at a right angle to a tooth on the ring gear.

48. Adjust feeler gauge thickness from side to side until the gear backlash is 0.005–0.007 in.

49. With zero endplay and correct backlash established, remove the feeler gauge. Determine the thickness of the shims (always use new shims) required and add 0.002 in. to each shim pack to provide side bearing preload.

50. Remove the case from the carrier. Carefully remove both side bearings. Install shims as determined behind each bearing, install bearings onto case.

51. Install case onto carrier, tapping carefully into place. Install side bearing caps in original position. Tighten to 65–79 ft. lbs.

52. Measure the run out of the ring gear. If the run out exceeds 0.002 in., correct by replacement.

53. Apply a thin coat of prussian blue or equivalent to the faces of 7–9 teeth of the ring gear. Check the impression of the contact on the ring gear teeth and make the necessary adjustments.

54. Apply gear oil to all moving parts and use sealant and or gasket when assembling to the axle housing.

Rear Drive Axle

Disassembly and Assembly

1. Position the removed differential carrier assembly in a suitable holding fixture. Take tooth pattern check this check will help deteremine what service should be performed on the differential components.

2. Mark and remove the side bearing caps.

3. Remove the differential cage assembly. Keep the right and left side

bearing races and shims if so equiped in separate groups for reinstallation in same positions.

4. Remove the differential side bearings from the case by using puller J-22888 and plug J-8107-2 or equivalent (some special tools may be different on the limited slip differential carrier). Carefully record the thickness of each side bearing and each shim(s) removed for later use in reassembly and keep separated. Puller arms must not pull against roller cage. Use care to position legs against inner race. As bearing is being removed, check for free rotation of bearing. If bearing does not rotate freely, check position of puller legs.

5. Remove the companion flange nut while holding flange with suitable tools. Remove the flange assembly with puller if necessary.

6. Drive the pinion shaft assembly from the carrier using suitable tools and brass punch. The outer (front) bearing will fall loose in the carrier, while the inner (rear) bearing will remain pressed on the drive pinion. Both races will remain in the carrier bores.

7. Remove rear bearing from the drive pinion by use of a press and tool J-22912-01 or equivalent on conventional differential carrier. On limited slip differential carrier use special tool J-37452 and a press.

8. Remove the oil seal and then drive the 2 outer races from the carrier by use of a brass drift and suitable tools.

9. Remove the ring gear bolts (diagonal pattern) and separate the ring gear from the differential case. Use care when removing ring gear to prevent damage to differential case or ring gear. Do not force a chisel or other tools between the joining faces.

10. At this point of the reassembly, check the amount of backlash before removal. The standard valve for backlash is 0.001–0.003 in. for conventional differential. The standard valve for backlash is 0.005–0.007 in. for limited slip differential.

11. On conventional differential carrier drive out the pinion shaft lock pin with a long drift. It may be necessary to first break the stake on lock pin using a 0.020 diameter drill or equivalent.

12. On conventional differential carrier remove the pinion shaft with a drift pin and take out the pinion gears, side gears, thrust block and thrust washers from the differential case.

13. On limited slip differential remove the limited slip differential carrier assembly.

14. Wash the bearings in a suitable solvent. Then examine bearings carefully for wear, separation, cracks, sei-

zure and other abnormal conditions. Replace bearings as necessary.

15. Check the ring gear and drive pinion teeth for wear, chipping, cracks, pitting and abnormal contact. Check the drive pinion splines for cracks, distortion and step wear. Replace parts if needed. Ring gear and drive pinion come only in matched sets. If either item is defective, both parts must be replaced.

16. Check the pinion gears and side gears for wear, chipped teeth and separation and replace if needed.

17. Check and replace the thrust blocks on conventional differential and thrust washers, if worn or damaged.

18. On conventional differential check the lock pin for bending, dents and other abnormal conditions. Replace if necessary.

19. Check the clearance between pinion gear and cross pin. The standard valve for the clearance between the differential pinion and the cross pin is 0.002–0.005 in. and the service limit is 0.008 in. on both differential carriers.

20. Check the clearance between the side gear and differential box. The standard valve for the clearance between the side gear and the differential assembly is 0.001–0.004 in. and the service limit is 0.006 in. on conven-

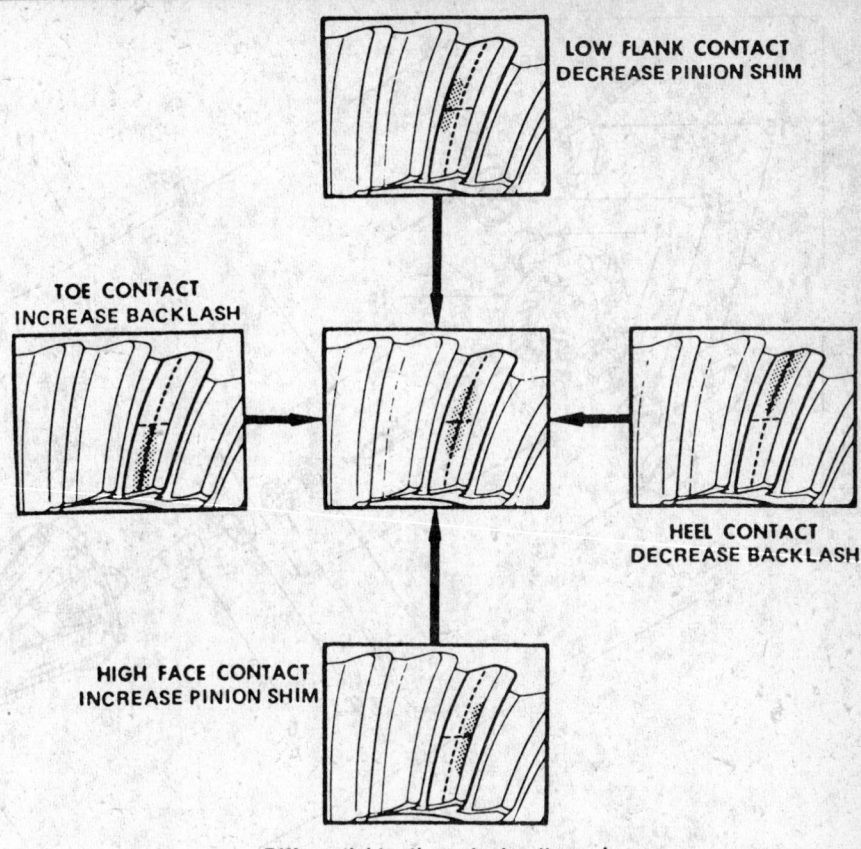

Differential tooth contact pattern – Isuzu

Pinion marking / Dial indicator reading (inches)	+10	+8	+6	+4	+2	0	−2	−4	−6	−8	−10
0 073										1.94(0.0764)	1.96(0.0772)
0 074									1.94(0.0764)	1.96(0.0772)	1.98(0.0779)
0 075								1.94(0.0764)	1.96(0.0772)	1.98(0.0779)	2.00(0.0787)
0 076						1.94(0.0764)	1.96(0.0772)	1.98(0.0779)	2.00(0.0787)	2.02(0.0795)	2.04(0.0803)
0 077						1.96(0.0772)	1.98(0.0779)	2.00(0.0787)	2.02(0.0795)	2.04(0.0803)	2.06(0.0811)
0 078				1.94(0.0764)	1.96(0.0772)	1.98(0.0779)	2.00(0.0787)	2.02(0.0795)	2.04(0.0803)	2.06(0.0811)	2.08(0.0819)
0 079			1.94(0.0764)	1.96(0.0772)	1.98(0.0779)	2.00(0.0787)	2.02(0.0795)	2.04(0.0803)	2.06(0.0811)	2.08(0.0819)	2.10(0.0827)
0 080	1.94(0.0764)	1.96(0.0772)	1.98(0.0779)	2.00(0.0787)	2.02(0.0795)	2.04(0.0803)	2.06(0.0811)	2.08(0.0819)	2.10(0.0827)	2.12(0.0835)	2.14(0.0842)
0 081	1.96(0.0772)	1.98(0.0779)	2.00(0.0787)	2.02(0.0795)	2.04(0.0803)	2.06(0.0811)	2.08(0.0819)	2.10(0.0827)	2.12(0.0835)	2.14(0.0842)	2.16(0.0850)
0 082	1.98(0.0779)	2.00(0.0787)	2.02(0.0795)	2.04(0.0803)	2.06(0.0811)	2.08(0.0819)	2.10(0.0827)	2.12(0.0835)	2.14(0.0842)	2.16(0.0850)	2.18(0.0858)
0 083	2.00(0.0787)	2.02(0.0795)	2.04(0.0803)	2.06(0.0811)	2.08(0.0819)	2.10(0.0827)	2.12(0.0835)	2.14(0.0842)	2.16(0.0850)	2.18(0.0858)	2.20(0.0866)
0 084	2.04(0.0803)	2.06(0.0811)	2.08(0.0819)	2.10(0.0827)	2.12(0.0835)	2.14(0.0842)	2.16(0.0850)	2.18(0.0858)	2.20(0.0866)	2.22(0.0874)	2.24(0.0882)
0 085	2.06(0.0811)	2.08(0.0819)	2.10(0.0827)	2.12(0.0835)	2.14(0.0842)	2.16(0.0850)	2.18(0.0858)	2.20(0.0866)	2.22(0.0874)	2.24(0.0882)	2.26(0.0890)
0 086	2.08(0.0819)	2.10(0.0827)	2.12(0.0835)	2.14(0.0842)	2.16(0.0850)	2.18(0.0858)	2.20(0.0866)	2.22(0.0874)	2.24(0.0882)	2.26(0.0890)	2.28(0.0898)
0 087	2.12(0.0835)	2.14(0.0842)	2.16(0.0850)	2.18(0.0858)	2.20(0.0866)	2.22(0.0874)	2.24(0.0882)	2.26(0.0890)	2.28(0.0898)	2.30(0.0906)	2.32(0.0913)
0 088	2.14(0.0842)	2.16(0.0850)	2.18(0.0858)	2.20(0.0866)	2.22(0.0874)	2.24(0.0882)	2.26(0.0890)	2.28(0.0898)	2.30(0.0906)	2.32(0.0913)	2.34(0.0921)
0 089	2.16(0.0850)	2.18(0.0858)	2.20(0.0866)	2.22(0.0874)	2.24(0.0882)	2.26(0.0890)	2.28(0.0898)	2.30(0.0906)	2.32(0.0913)	2.34(0.0921)	2.36(0.0929)
0 090	2.18(0.0858)	2.20(0.0866)	2.22(0.0874)	2.24(0.0882)	2.26(0.0890)	2.28(0.0898)	2.30(0.0906)	2.32(0.0913)	2.34(0.0921)	2.36(0.0929)	
0 091	2.22(0.0874)	2.24(0.0882)	2.26(0.0890)	2.28(0.0898)	2.30(0.0906)	2.32(0.0913)	2.34(0.0921)	2.36(0.0929)			
0 092	2.24(0.0882)	2.26(0.0890)	2.28(0.0898)	2.30(0.0906)	2.32(0.0913)	2.34(0.0921)	2.36(0.0929)				
0 093	2.26(0.0890)	2.28(0.0898)	2.30(0.0906)	2.32(0.0913)	2.34(0.0921)	2.36(0.0929)					
0 094	2.28(0.0898)	2.30(0.0906)	2.32(0.0913)	2.34(0.0921)	2.36(0.0929)						
0 095	2.32(0.0913)	2.34(0.0921)	2.36(0.0929)								
0 096	2.34(0.0921)	2.36(0.0929)									
0 097	2.36(0.0929)										

Shim selection chart – Isuzu rear limited slip differential drive axle

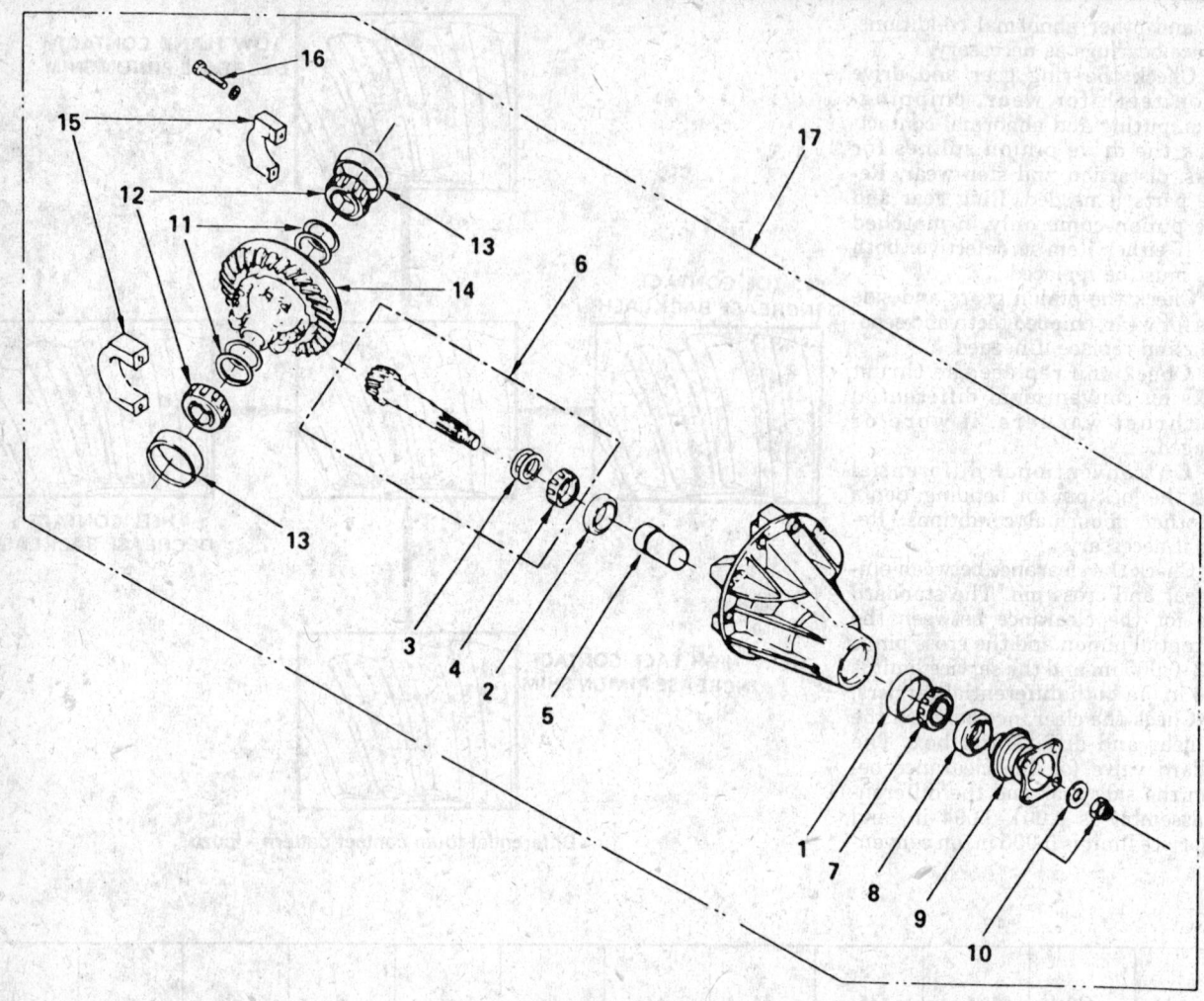

1. Outer bearing outer race
2. Inner bearing outer race
3. Shim
4. Inner bearing
5. Collapsible spacer
6. Drive pinion assembly
7. Outer bearing
8. Oil seal
9. Flange assembly
10. Flange nut
11. Shim
12. Side bearing
13. Side bearing outer race
14. Differential cage
   assembly and ring gear
15. Bearing cap
16. Bolt
17. Differential assembly

Exploded view rear differential assembly—Isuzu

tional carrier. The standard valve for the clearance between the side gear and the differential assembly is 0.002–0.004 in. and the service limit is 0.006 in. on limited slip differential carrier.

21. Check the play in splines betwwen the side gear and axle shafts. The standard valve for play (backlash) in splines betwwen the side gear and the axle shaft is 0.001–0.006 in. and the service limit is 0.010 in. on the conventional differential. The standard valve for play in splines betwwen the side gear and the axle shaft is 0.03–0.15 in. and the service limit is 0.010 in. on the limited slip differential.

22. Install the side gears and thrust washers in the differential case. Position the pinion gears 180 degrees apart. Roll gears into position, making

sure they are in alignment, to allow installation of the pinion shaft.

23. Place the thrust block on conventional differential between the pinion gears and drive the pinion shaft into position. Make sure that the lock pin hole in cross shaft aligns with the hole in the case.

24. Measure the amount of backlash between the differential gears and the pinion gears. The standard valve is 0.001–0.003 in. on conventional differential and 0.005–0.007 in. on limited slip differential. Make the necessary adjustment with the thrust washers.

25. Install lock pin into cross shaft and stake the cage to prevent loosening of the pin.

26. Apply Loctite or equivalent to the threaded portion of the bolts. Install

the ring gear in position on the differential case. Tighten the bolts (always use new retaining bolts) in diagonal sequence to 73–87 ft. lbs. on conventional and limited slip differentials.

27. To set pinion depth install the drive pinion front and rear bearing outer races into carrier bores. Use drive handle J-8092 with J-24256 (conventional carrier) or J-37262 for limited slip differential carrier for front bearing race and J-24252 (conventional carrier) and J-24252 for limited slip differential carrier for rear bearing race.

28. Lubricate and position the front and rear bearings to be used for final assembly into their respective races.

29. Install gauging plate J-23597-7 and preload stud and pilot J-23597-9

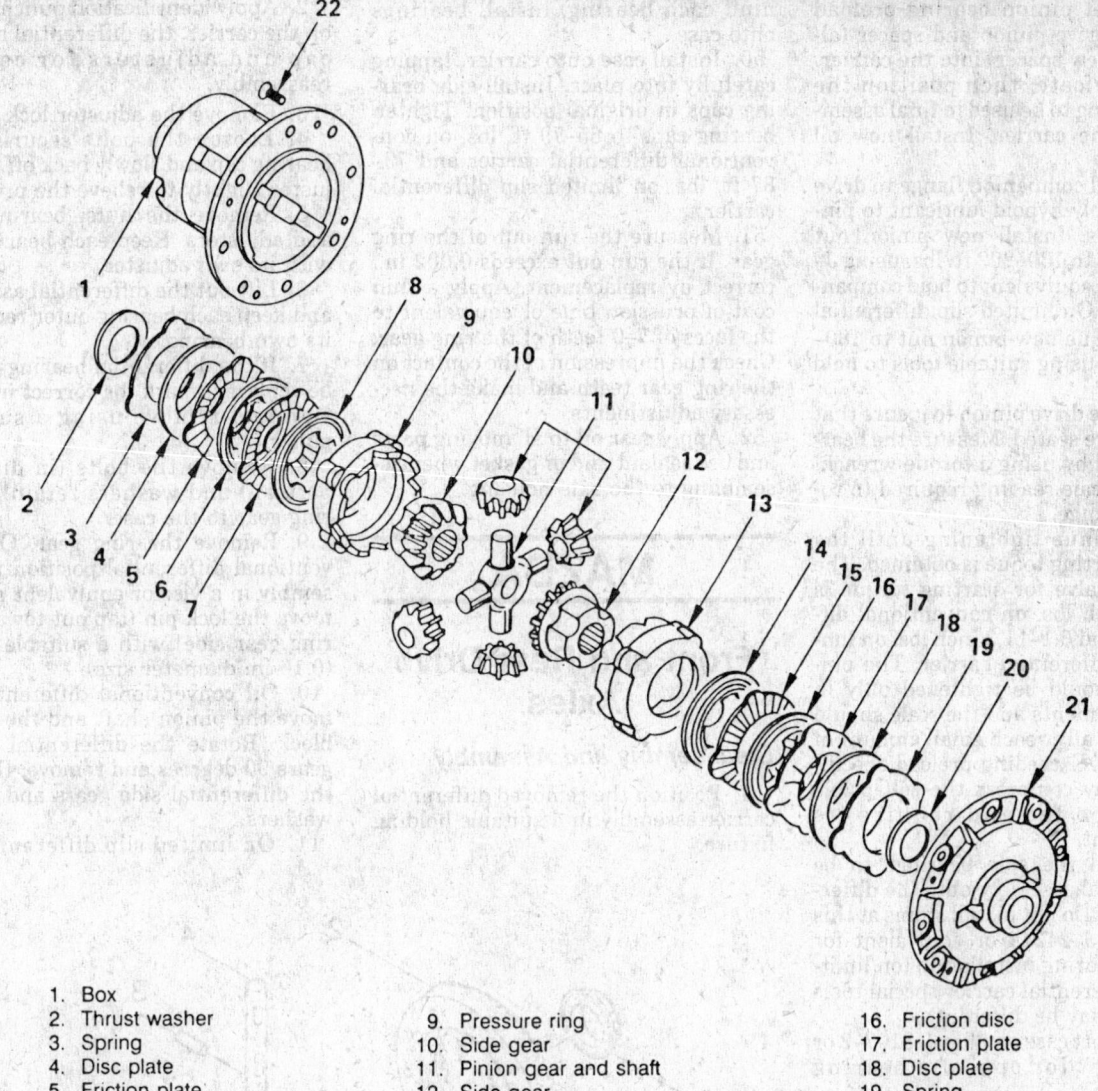

1. Box
2. Thrust washer
3. Spring
4. Disc plate
5. Friction plate
6. Friction disc
7. Friction plate
8. Friction disc
9. Pressure ring
10. Side gear
11. Pinion gear and shaft
12. Side gear
13. Pressure ring
14. Friction disc
15. Friction plate
16. Friction disc
17. Friction plate
18. Disc plate
19. Spring
20. Thrust washer
21. Box cover
22. Bolt

Exploded view limited slip differentail carrier—Isuzu

or their equivalent through front and rear bearings and tighten nut snugly. On limited slip differential carriers special tool numbers may change.

30. Rotate the bearings to insure proper seating and tighten locknut until 20 inch lbs. of torque are required to rotate new bearings; 8–10 inch lbs. for used bearings.

31. Place discs J–23597–8 onto arbor J–23597–1 or equivalent (special tool numbers may change on limited slip differential carrier) and place tool into position in side bearing bores. Install bearing caps and torque to 65–72 ft. lbs. on conventional carrier. Torque the bearing caps 73–87 ft. lbs. on limited slip differential carrier.

32. Mount dial indicator J–8001 or equivalent on arbor post and preload

dial ½ revolution. Tighten indicator in this position.

33. Position the indicator plunger on the gauge plate and slowly swing across until the highest reading is obtained. Zero the indicator on the highest reading of the gauge plate. Repeat procedure if necessary for correct readings.

34. After zero setting is obtained carefully swing the plunger off the gauge plate. Note the indicator reading. Recheck to verify the reading. Record the number the dial indicator needle points to. The reading on the dial is the correct dimension for the rear pinion depth shim. Convert the dial indicator reading (inches) to the correct pinion marking (depth code) specification.

35. Record pinion depth code on the head of the drive pinion. The number indicates a necessary change in the pinion mounting distance. A plus number indicates the need for a greater mounting distance (which can be achieved by decreasing the shim thickness). A minus number indicates the need for a smaller mounting distance (which can be achieved by increasing the shim thickness). If examination reveals no pinion depth code, the pinion is nominal. Determine the proper shim variation to compensate for plus or minus markings.

36. Place the shim on the drive pinion with chamfered side toward the pinion head then install the rear bearing, using J–6133–01 or equivalent. Do not press on roller cage press only on inner race.

37. To set pinion bearing preload place the drive pinion and spacer (always use new spacer) into the carrier.

38. Lubricate, then position the front bearing to be used in final assembly into the carrier. Install new oil seal.

39. Install companion flange to drive pinion. Apply hypoid lubricant to pinion threads. Install new pinion nut and torque to 130–202 ft. lbs. using J-8614-01 or equivalent to hold companion flange. On limited slip differential carrier torque new pinion nut to 180–217 ft. lbs. using suitable tools to hold flange.

40. Rotate drive pinion to insure that bearings are seated. Measure the bearing preload by using a torque wrench. Note the scale reading required to rotate the flange.

41. Continue tightening until the correct starting toque is obtained. The standard valve for starting torque is 5.6–9.9 inch lbs. on conventional differential and 6.1–11.3 inch lbs. on limited slip differential carrier. The pinion nut should be tightened only in small increments and the scale should be checked after each small amount of tightening. Exceeding preload specifications may compress the collapsible spacer too far and require its replacement.

42. Install the side bearings to be used in final assembly onto the differential case. Do not install shims at this time. Use J-24244 or equivalent for the first bearing installation (on limited slip differential carrier special tools numbers may be different).

43. Support case on plug J-8107-2 or equivalent for opposite bearing installation.

44. Install the differential cage assembly into carrier bores.

45. Using 2 sets of feeler gauges, insert feeler gauge of sufficient thickness between each bearing outer race and the carrier to remover all endplay. Make certain the feeler gauge is pushed to the bottom of the bearing bores.

46. Install a dial indicator on the carrier so that the stem of the tool is at a right angle to a tooth on the ring gear.

47. Adjust feeler gauge thickness from side to side until the gear backlash is 0.005–0.007 in. on conventional differential and 0.006–0.008 in. on limited slip differential.

48. With zero endplay and correct backlash established, remove the feeler gauge. Determine the thickness of the shims (always use new shims) required and add 0.002 in. to each shim pack to provide side bearing preload.

49. Remove the case from the carrier. Carefully remove both side bearings. Install shims as determined behind each bearing, install bearings onto case.

50. Install case onto carrier, tapping carefully into place. Install side bearing caps in original position. Tighten bearing caps to 65–79 ft. lbs. on conventional differential carrier and 73–87 ft. lbs. on limited slip differential carrier.

51. Measure the run out of the ring gear. If the run out exceeds 0.002 in., correct by replacement. Apply a thin coat of prussian blue or equivalent to the faces of 7–9 teeth of the ring gear. Check the impression of the contact on the ring gear teeth and make the necessary adjustments.

52. Apply gear oil to all moving parts and use sealant and or gasket when assembling to the axle housing.

MAZDA

Front and Rear Drive Axles

Disassembly and Assembly

1. Position the removed differential carrier assembly in a suitable holding fixture.

2. Apply identification punch marks on the carrier, the differential bearing cap and adjusters for correct reassembly.

3. Remove the adjuster lock plates.

4. Loosen the bolts securing the bearing cap and slowly back off the adjuster slightly to relieve the preload.

5. Remove the nuts, bearing caps and adjusters. Keep each bearing cap with its own adjuster.

6. Lift out the differential assembly and keep each bearing outer race with its own bearing.

7. If the differential bearings are to be replaced, mark for correct installation and remove using a suitable puller.

8. Remove the bolts (in diagonal pattern) and washers retaining the ring gear to the case.

9. Remove the ring gear. On conventional differential position the assembly in a vise or equivalent and remove the lock pin (tap out toward the ring gear side) with a suitable punch (0.16 in. diameter size).

10. On conventional differential remove the pinion shaft and the thrust block. Rotate the differential pinion gears 90 degrees and remove. Remove the differential side gears and thrust washers.

11. On limited slip differential re-

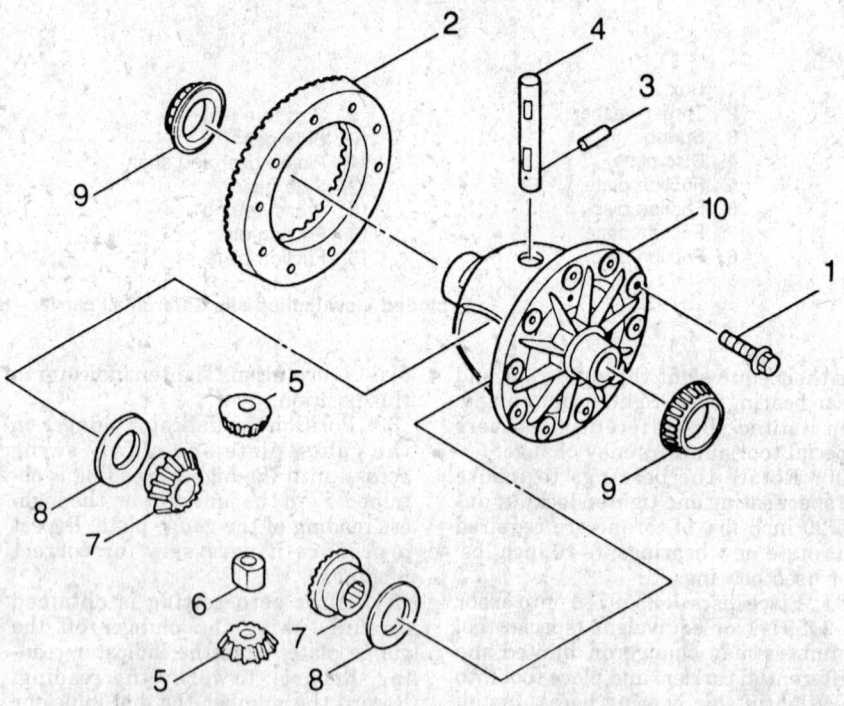

1. Bolt
2. Ring gear
3. Pin
4. Pinion shaft
5. Pinion gears
6. Thrust block
7. Side gears
8. Thrust washer
9. Bearings
10. Gear case

Exploded view standard differential assembly – Mazda

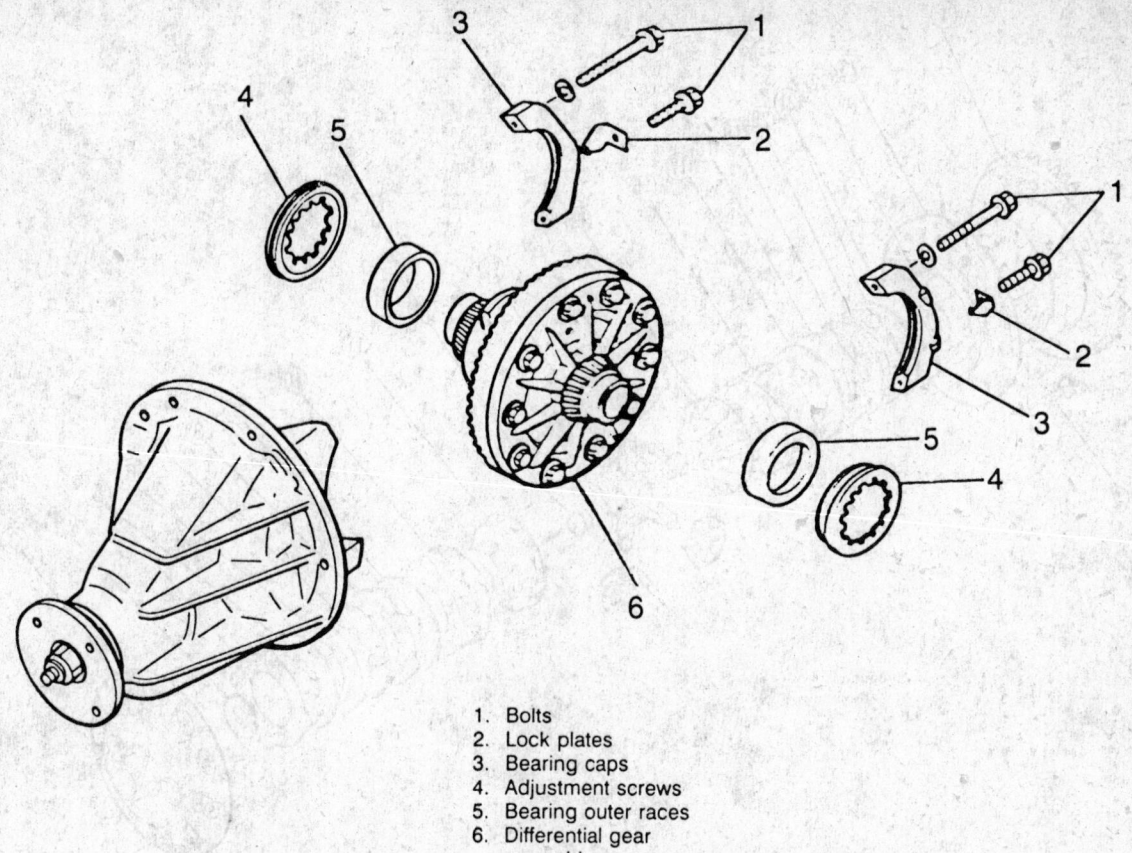

1. Bolts
2. Lock plates
3. Bearing caps
4. Adjustment screws
5. Bearing outer races
6. Differential gear
   assembly

Exploded view of differential carrier – Mazda

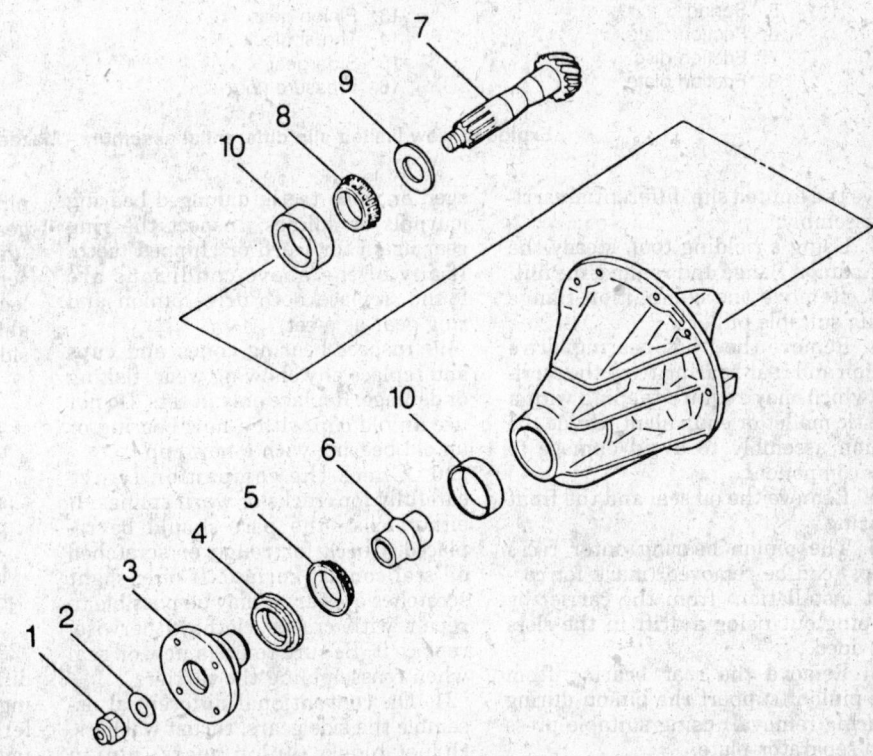

1. Locknut
2. Washer
3. Flange
4. Oil seal
5. Front bearing
6. Collapsible spacer
7. Drive pinion
8. Rear bearing
9. Spacer
10. Bearing outer races

Exploded view of drive pinion assembly – Mazda

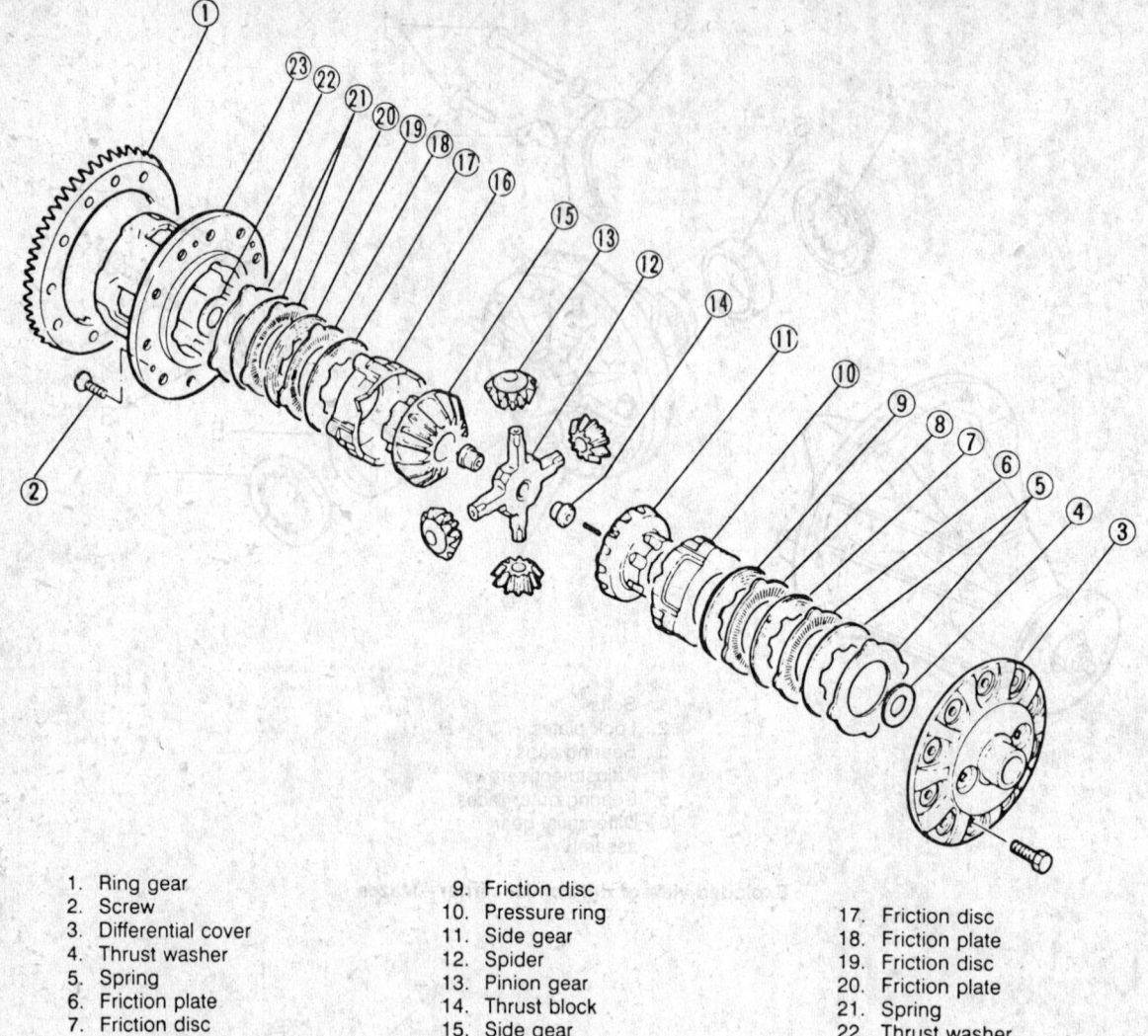

1. Ring gear
2. Screw
3. Differential cover
4. Thrust washer
5. Spring
6. Friction plate
7. Friction disc
8. Friction plate

9. Friction disc
10. Pressure ring
11. Side gear
12. Spider
13. Pinion gear
14. Thrust block
15. Side gear
16. Pressure ring

17. Friction disc
18. Friction plate
19. Friction disc
20. Friction plate
21. Spring
22. Thrust washer
23. Differential case

Exploded view limited slip differential assembly—Mazda

move the limited slip differential carrier assembly.

12. Using a holding tool, steady the companion flange and remove the nut.

13. Remove the companion flange using suitable puller.

14. Remove the front bearing, drive pinion and rear bearing from the carrier, which may require tapping with a plastic mallet or equivalent. Guide the pinion assembly to avoid damage to any component.

15. Remove the oil seal and the front bearing.

16. The pinion bearing outer races (cups) can be removed (mark for correct installation) from the carrier by tapping out using a drift in the slots provided.

17. Remove the rear bearing from the pinion (support the pinion during bearing removal) using suitable press and separator plate.

18. Check the drive pinion for dam-aged or worn teeth, damaged bearing journals or splines. Inspect the ring gear again for worn or chipped teeth. If any of the above conditions are found, replace both drive pinion and ring gear as a set.

19. Inspect bearing cones and cups and replace any showing wear, flaking or damage. Replace only in sets. Do not use an old cup with a new bearing or an old bearing with a new cup.

20. Check the companion flange carefully for cracks or worn splines. If either exist, the part should be replaced. Check for rough or scratched oil seal contact surface. If only slight scratches appear, it may be possible to repair with crocus cloth. Otherwise, replace it. Be sure to use a new oil seal when reassembling the carrier.

21. On conventional differential assemble the side gears, thrust washers, thrust block, pinion gears, pinion shaft and lock pin. After installing lock pin stake pin so pin cannot come out of gear case.

22. On conventional differential assemble press the side bearings (correct location) onto the gear case using suitable tools. Adjust the backlash of the side gears and pinion gear as follows:

   a. Position a dial indicator against the pinion gear. Secure one of the side gears.

   b. Move the pinion gear and measure the backlash at the end of the pinion gear.

   c. The standard valve for backlash is 0–0.004 in. on all vehicles. If backlash exceeds the standard valve use suitable thrust washer to adjust.

23. On conventional and limited slip differetial carriers apply thread locking compound install ring gear to differential assembly. Torque the retaining bolts to 51–61 ft. lbs. all vehicles in a diagonal pattern.

24. Press fit the companion flange side bearing outer races using bearing installer set.

25. Press fit the ring gear side bearing outer races using bearing installer set.

NOTE: Special tools are required to check and adjust the pinion height, these include a drive pinion model tool 498531565, a pinion height adjustment gauge body tool 490727570 and a gauge block tool 400305555.

26. At this point of the reassembly, adjust the pinion height as follows:

a. Fit the spacer, rear bearing and collar onto the drive pinion model. Secure the collar with an O-ring and install in the assembly in the carrier.

b. Attach the front bearing, collar, companion flange, washer and nut to the drive pinion model. Use the same spacer and nut which were removed at disassembly. Be careful to install collars in their correct position facing in the correct direction.

c. Tighten the nut until the drive pinion model can be turned by hand without any apparant play.

d. Install a dial indicator on the pinion height adjustment gauge body. Place the gauge block on top of the drive pinion model and then set the pinion height adjustment gauge body on top of the gauge block.

e. Place the measuring probe of the dial indicator so that it contacts the location where the side bearing is installed in the carrier. Zero and set up the indicator to measure the lowest point. Measure both the left and right sides. Add the 2 values (right and left side readings) and divide the total by 2. The standard valve of this specification is 0 in. on all vehicles. Install correct spacer as necessary to adjust pinion height.

27. Install the correct spacer(s) on the pinion shaft (facing in the proper direction) and press the bearing on the pinion shaft.

28. Install the drive pinion, spacer, front bearing, collapsible spacer and companion flange in the carrier and temporarily tighten the locknut. Do Not install the pinion oil seal at this time.

29. At this point of the reassembly, adjust preload of the drive pinion as follows:

a. Rotate the companion flange by hand to seat the bearing.

b. Use a torque wrench to tighten the locknut. Tighten slowly until the required preload drag of 7.8–12.2 inch lbs. is reached with a locknut torque of 94–130 ft. lbs. for all applications except MPV vehicle.

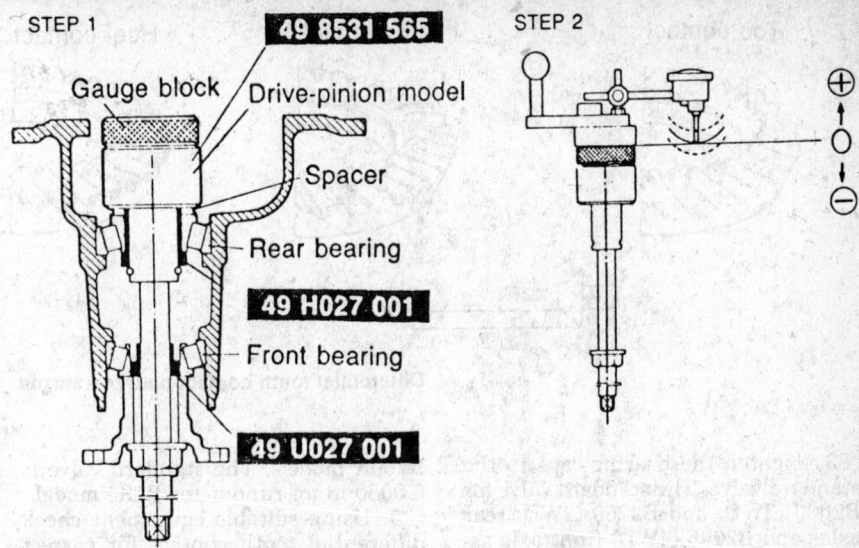

STEP 1

49 8531 565

Gauge block / Drive-pinion model

Spacer

Rear bearing

49 H027 001

Front bearing

49 U027 001

STEP 2

Adjustment of drive pinion height

c. On MPV vehicle, slowly tighten until the preload drag of 11.3–15.6 inch lbs. is reached with a locknut torque of 94–210 ft. lbs.

d. If the specified preload cannot be maintained within the locknut tightening range, install a new collapsible spacer and repeat the process if necessary.

e. On all applications except MPV vehicle remove the locknut and flange. Install the pinion seal, flange and new locknut. Tighten the locknut to the ft. lbs. specification torque giving the correct preload.

f. On MPV vehicle, remove the locknut and flange. Install the pinion seal, flange and new locknut. Retighten the locknut to the preload drag of 13.9–18.2 inch lbs. is reached with a locknut torque of 94–210 ft. lbs.

30. Install the differential gear assembly in the carrier. Note the identification marks on the adjusters and install each to its respective side. Install the bearing caps making sure that the identification marks on the caps correspond with those on the carrier and install the bolts.

31. Loosely tighten the bearing cap mounting bolts and completely tighten the adjustment screws by hand. Then, while turning the ring gear, alternately tighten the left and right adjustment screws using a suitable tool.

32. At this point of the reassembly, adjust the drive pinion and ring gear backlash and side bearing preload as follows:

a. Mark the ring gear at 4 points at approximately 90 degree intervals and mount a dial indicator to the carrier flange so that the feeler comes in contact at right angles with one of the ring gear teeth.

b. Turn both bearing adjusters equally until backlash becomes (standard valve for all vehicles) 0.0035–0.0043 in.

c. Check the backlash at the other 3 marked points and make sure that minimum backlash is more than 0.002 in. and the difference in the valve of the maximum and minimum backlash is less than 0.0028 in.

d. After adjusting the backlash, tighten the adjustment screws equally until the distance between both pilot sections on the bearing caps becomes the standard distance. The standard distance is 7.3004–7.3031 in. for B2000 (2WD) and B2200 (2WD) rear axle assemblies and B2600 (4WD) front axle assembly. The standard distance is 8.0484 in. plus or minus 0.028 in. for B2600 (2WD and 4WD) rear axle assemblies and the MPV rear axle assembly. When adjusting the differential bearing preload, care must be taken not to affect the backlash of the drive pinion gear and ring gear.

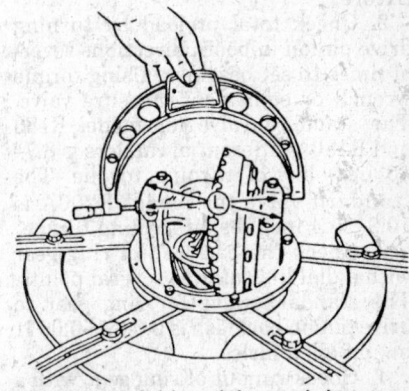

Adjustment of drive pinion and ring gear backlash

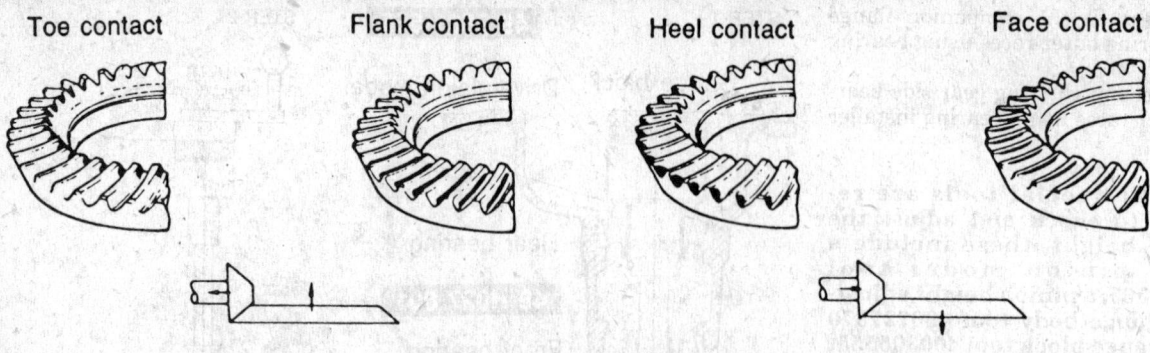

Toe contact   Flank contact   Heel contact   Face contact

Differential tooth contact pattern—Mazda

33. Tighten the bearing caps to the standard valve. The standard valve for B2000 (2WD) and B2200 (2WD) rear axles and B2600 (4WD) front axle assembly is 27–38 ft. lbs. The standard valve for B2600 (2WD and 4WD) rear axle assemblies is 41–59 ft. lbs. The standard valve for the MPV vehicle is 51–61 ft. lbs.

34. Install the adjuster lock plates.

35. Coat both surfaces of 6–8 teeth of the ring gear uniformly with a thin coat of red lead or equivalent.

36. While moving the ring gear back and forth by hand, rotate the drive pinion several times and check the tooth contact.

37. If pattern is not correct readjust the pinion height then the backlash.

38. Apply gear oil to all moving parts and use sealant and or gasket when assembling to the axle housing.

NISSAN

Front Drive Axle

Inspection Before Disassembly

1. Position the removed differential carrier assembly in a suitable holding fixture.

2. Check total preload by turning drive pinion in booth directions several times to set bearings. Using torque wrench or equivalent measure valve. The standard valve for model R180 and R180A differential carriers is 8.7–20 inch lbs. of turning torque. The standard valve for model R200A is 10.9–20.4 inch lbs. of turning torque.

3. Check the backlash of ring gear with a dial indicator at several points. The standard valve for ring gear to drive pinion backlash is 0.0051–0.0071 in., on all vehicles.

4. Check runout of ring gear with a dial indicator. The standard valve is 0.0020 in. of runout for R180A and R200A models. The standard valve is 0.0030 in. of runout for R180 model.

5. Using suitable equipment check differential tooth contact for correct pattern as follows:

  a. Thoroughly clean ring gear and drive pinion teeth.

  b. Sparingly apply red lead or equivalent to 3 or 4 teeth of ring gear drive side.

  c. Hold companion flange steady by hand and rotate the ring gear in both directions.

  d. Interrupt contact marks left on the ring gear assembly. Gear tooth contact pattern check is necessary to verify correct relationship between ring gear and drive pinion.

6. On R180 model, check backlash of side gear. Using a thickness gauge, measure clearance between side gear and differential case. The standard valve is 0.006 in. of side gear clearance.

Disassembly and assembly

1. Position the removed differential carrier assembly in a suitable holding fixture.

2. On R180A and R200A models remove the extension tube and differential side shaft assembly. Remove differential side flanges.

3. Matchmark side retainers and shims (some have retaining bolts) with paint to ensure that they are replaced in proper position during reassembly.

4. Remove the differential case from the final drive housing.

5. Remove the side bearing outer races. Mark all parts for correct installation.

6. Remove the side bearing oil seals using suitable puller.

7. Attach a flange holding wrench or equivalent to hold flange from turning. Loosen the drive pinion nut.

8. Remove the companion flange using suitable puller.

9. Remove the drive pinion together with pinion rear bearing inner cone, drive pinion bearing spacer and pinion bearing adjusting washer.

10. Remove front oil seal and pinion front bearing inner cone.

11. Remove the pinion front and rear bearing outer races with brass drift from housing.

12. Remove the pinion rear bearing inner cone and drive pinion adjusting washer. Remove rear bearing with suitable press and separator plate.

13. Remove side bearing inner race assembly with a puller from the differential carrier. Mark all parts for correct installation.

14. Remove ring gear by spreading out lock straps if so equipped and loosening ring gear bolts in a criss-cross pattern. Remove all retaining bolts.

15. Tap ring gear off gear case using a soft hammer or equivalent. Tap evenly all around to keep ring gear from binding.

16. On R180A model, separate the left and right differential case halves.

17. On R180 and R200A models, drive out pinion mate shaft lock pin with drift pin from ring gear side.

18. Draw out pinion mate shaft and thrust block if so equipped. Rotate pinion mate gears out of case and remove side gears and thrust washers. Put marks on gears and thrust washers so that they can be reinstalled in their original positions from which they were removed.

19. Clean disassembled parts completely. Repair or replace any damaged or faulty parts. If replacing drive pinion or ring gear, replace with a new hypoid gear as a set.

20. On R180A model, assemble the differential carrier as follows:

  a. Measure the clearance between the side gear thrust washer and differential case.

  b. The standard valve for the clearance between side gear thrust washer and differential case is 0.0039–0.0079 in. The clearance can be adjusted with side gear thrust washer.

  c. Apply gear oil to gear tooth surfaces and thrust surfaces and check to see they turn properly.

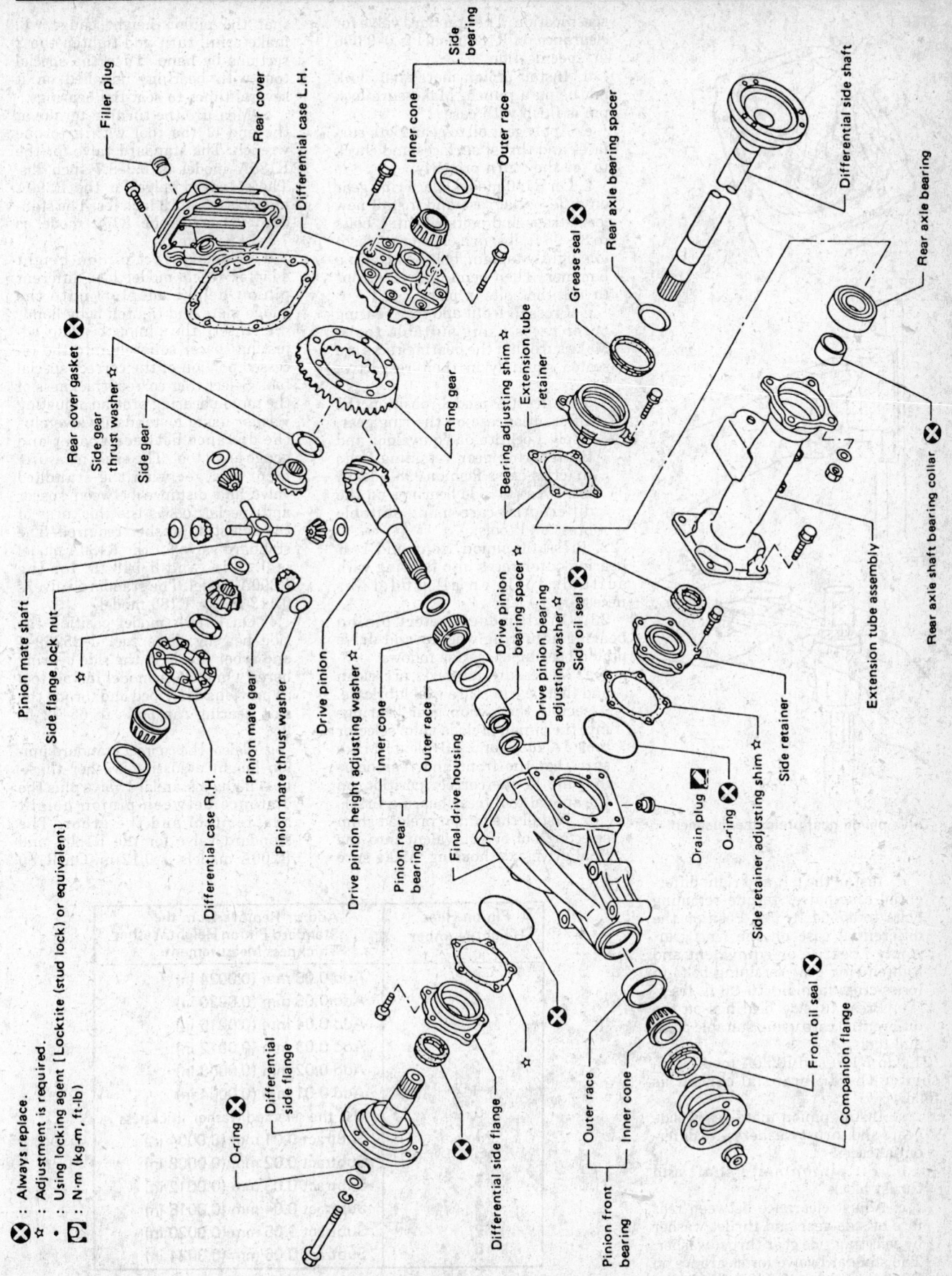

Filler plug

Rear cover

Differential case L.H.

Inner cone

Side bearing

Outer race

Rear axle bearing spacer

Grease seal

Differential side shaft

Rear axle bearing

Rear cover gasket

Side gear thrust washer ☆

Side gear

Ring gear

Bearing adjusting shim ☆

Extension tube retainer

Rear axle shaft bearing collar

Pinion mate shaft

Side flange lock nut ☆

Differential case R.H.

Pinion mate gear

Pinion mate thrust washer

Drive pinion

Pinion rear bearing

Inner cone

Outer race

Drive pinion height adjusting washer ☆

Final drive housing

Drive pinion bearing spacer ☆

Drive pinion bearing adjusting washer ☆

Side oil seal

Drain plug

O-ring

Side retainer adjusting shim ☆

Side retainer

Extension tube assembly

O-ring

Differential side flange

Differential side flange

Outer race

Inner cone

Pinion front bearing

Front oil seal

Companion flange

⊗ : Always replace.
☆ : Adjustment is required.
• : Using locking agent [Locktite (stud lock) or equivalent]
⊡ : N·m (kg-m, ft-lb)

Exploded view of front differential assembly–Nissan

STEP 1

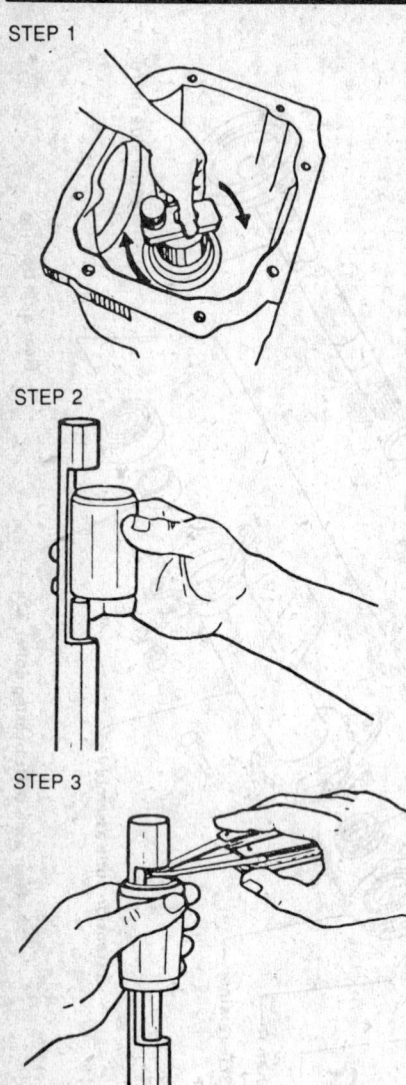

STEP 2

STEP 3

Drive pinion gear preload adjustment

d. Install the left and right differential case halves torque retaining bolts to 47–54 ft. lbs. Position the differential case on the ring gear. Apply Locktite or equivalent and tighten ring gear retaining bolts to (criss-cross fashion) 76–90 ft. lbs.

e. Press fit side bearings on the differential case using suitable press and tools.

21. On R180 and R220A models, assemble the differential carrier as follows:

a. Install pinion mate gears, side gears and thrust washers into differential case.

b. Fit pinion mate shaft and thrust block.

c. Adjust clearance between rear face of side gear and thrust washer by selecting side gear thrust washer. The standard valve for clearance on R200A model is 0.0039–0.0079 in.

specification. The standard valve for clearance on R180 model is 0–0.006 in. specification.

d. Install pinion mate shaft lock pin using a punch. Make sure lock pin is flush with case.

e. Apply gear oil to gear tooth surfaces and thrust surfaces and check to see they turn properly.

f. On R180 model, place ring gear on differential case and install new lock straps and bolts. Tighten bolts to 88–98 ft. lbs. in a criss-cross fashion, lightly tapping bolt head with a hammer. Then bend up lock straps to lock the bolts in place.

g. Press fit front and rear bearing outer races using suitable tools. Make sure that the bearing races are seated squarely in their respective bores.

h. On R200A model, position the differential case on the ring gear. Apply Locktite or equivalent and tighten ring gear retaining bolts to (criss-cross fashion) 98–112 ft. lbs. Press fit side bearings on the differential case using suitable press and tools.

22. Press fit pinion front and rear bearing outer races into housing with suitable tools on all models as necessary.

23. On all models select pinion bearing adjusting washer and drive pinion bearing spacer as follows:

a. Make sure all parts are clean and that bearings are well lubricate. Assemble the pinion gear bearings into the pinion preload shim selector tool J–34309 or equivalent. Make sure that the front and rear bearings are in the correct position on the special tool (rear bearing first).

b. Install the pinion preload shim selector tool or equivalent into the final drive axle housing. Make sure

that the pinion height gauge will make a full turn and tighten the 2 sections by hand. Turn the special tool with bearings installed on it several times to seat the bearings.

c. Measure the turning torque at the end of the tool with a torque wrench. The standard valve for the R180A model is 5.2–8.7 inch lbs. The standard valve for the R200A model is 8.7–11.3 inch lbs. The standard valve for the R180 model is 7.8–14.8. inch lbs.

d. Place correct pinion height adapter (each model has different pinion height adapter) onto the gauge plate and tighten it by hand.

e. Insert the pinion bearing adjusting spacer squarely into the recessed portion of the correct special tool. Select ther correct thickness of the pinion bearing preload adjusting washer using a standard valve plus the distance between spacer and gauge anvil tool. The exact measurement you get with the standard valve and distance between spacer and special tool is the thickness of the adjusting washer required. The standard valve for the R180A model is 0.24 in. and 0.138 in. for the R200A model. The standard valve is 0.118 in. for R180 model.

f. On R180A model position the side bearing discs tool J–25269–4 and arbor firmly in the side bearing bores. On R200A model install tool as previous described and torque the side bearing cap bolts to 65–72 ft. lbs.

g. Select the correct standard pinion height adjusting washer thickness using a standard valve plus the distance between pinion height adapter tool and the arbor. The standard valve for the R180A and R200A models is 0.12 in. On R180

Pinion Head Height Number	Add or Remove from the Standard Pinion Height Washer Thickness Measurement
−6	Add 0.06 mm (0.0024 in)
−5	Add 0.05 mm (0.0020 in)
−4	Add 0.04 mm (0.0016 in)
−3	Add 0.03 mm (0.0012 in)
−2	Add 0.02 mm (0.0008 in)
−1	Add 0.01 mm (0.0004 in)
0	Use the selected washer thickness
+1	Subtract 0.01 mm (0.0004 in)
+2	Subtract 0.02 mm (0.0008 in)
+3	Subtract 0.03 mm (0.0012 in)
+4	Subtract 0.04 mm (0.0016 in)
+5	Subtract 0.05 mm (0.0020 in)
+6	Subtract 0.06 mm (0.0024 in)

Drive pinion gear height chart—Nissan

model use the appropriate service tool for this operation and measure the thickness of the lead washer.

h. Correct the pinion height washer size by referring to the pinion head number. There are 2 numbers on the pinion gear. The first one refers to the pinion and ring gear as a matched set and the number should be the same on the ring gear. The second number is the pinion head height number and it refers to the ideal pinion height for proper operation. Use this number to determine the correct pinion height washer. Select the correct pinion height washer and remove the special tools with pinion bearings from housing.

24. Install pinion height adjusting washer in drive pinion, bevel side toward gear and press fit rear bearing inner race in it, using press and special tools.

25. Lubricate front bearing with gear oil and place it in gear housing.

26. Carefully fit a new oil seal into carrier. Make sure oil seal is flush with end of carrier and apply multi-purpose grease into cavity between lips.

27. Install the drive pinion bearing spacer, pinion bearing adjusting washer and drive pinion in housing.

28. Install companion flange into drive pinion by tapping with a soft hammer or equivalent.

29. Hold companion flange and temporarily tighten pinion nut, until there is no axial play. Ascertain that threaded portion of drive pinion and pinion nut are free from oil or grease. Tighten pinion nut by degrees to the specified preload while checking the preload with torque wrench. Preload with oil seal is 7.8–14.8 inch lbs. for R180 and R180A models and 10.0–15.2 inch lbs. for the R200A model.

30. When checking preload, turn drive pinion in both directions several times set bearing rollers. After preload is reached torque pinion nut to 123–145 ft. lbs. on R180 and R180A models and 137–217 ft. lbs. on R200A model. Preload and final pinion nut torque **MUST** both be attained for correct operation of axle unit.

31. On the R180 and R180A models select the side retaining washer as follows:

a. Make sure all parts are cleaned and well lubricated with auotmatic transmission fluid.

b. Install the differentrial carrier and side bearing assembly into the housing.

c. Install all old bearing preload shims onto side bearing retainer. Install both bearing retainers onto housing and torque retaining bolts to 6.5–7.2 ft. lbs.

d. Turn carrier several times to seat bearings. Measure the carrier turning torque with a tool at the ring gear retaining bolt. The standard valve of pulling force at the ring gear bolt is 7.7–8.8 ft. lbs. Add or subtract shims as necessary. Increase shim thickness to decrease turning torque and decrease shim thickness to increase turning torque. Record correct shim(s) thickness and remove carrier and bearings.

32. On the R220A model select the side retaining washer as follows:

a. Make sure all parts are cleaned and well lubricated with auotmatic transmission fluid.

b. Install the differential carrier with side bearings and bearing races installed into gear housing.

c. Install side bearing spacer on the ring gear end of the carrier.

d. Install old preload shims using suitable tools on the carrier end opposite the ring gear.

e. Install side bearing caps and torque retaining bolts to 65–72 ft. lbs.

f. Turn carrier several times to seat bearings. Measure the carrier turning torque with a tool at the ring gear retaining bolt. The standard valve of pulling force at the ring gear bolt is 7.7–8.8 ft. lbs. Add or subtract shims as necessary. Increase shim thickness to decrease turning torque and decrease shim thickness to increase turning torque. Record correct shim(s) thickness and remove carrier and bearings.

33. On R180 and R180A model, press fit side bearing outer race into side retainer using suitable tools.

34. Install oil seals. Install differential case assembly. Place side retainer (correct marked position) adjusting shims and O-ring on side retainer and install on housing.

35. On R200A model install differential case assembly with side bearings outer races into housing. Insert the correct bearing adjusting washers between side bearings and housing. Drive in side bearing spacer with suitable tools.

36. Align mark on bearing caps and install caps torque to 65–72 ft. lbs.

37. Measure the ring gear to drive pinion backlash witrh a dial indicator. The standard valve is 0.0051–0.0071 in. If backlash is too small, decrease thickness of right shim and increase thickness of left shim by the same amount. If backlash is to great reverse the above procedure. Never change the toatal amount of shims as it will change the bearing preload.

38. Check total preload with torque wrench, turn drive pinion several times in both directions to set bearings. The standard valve for total preload is 8.7–20.0 inch lbs. for R180 and R180A models. The standard valve for total preload is 10.9–20.4 inch lbs. for R200A model.

39. On R180 and R180A models if preload is too great, add the same amount of shims to each side. If preload is too small, remove the same amount of shims from each side. Never add or remove a different number of shims for each side as it will change ring gear to drive pinion backlash.

40. On R200A model if preload is too great, remove the same amount of shims from each side. If preload is too small, add the same amount of shims to each side. Never add or remove a different number of shims for each side as it will change ring gear to drive pinion backlash.

41. Recheck ring gear to drive pinion backlash. Check runout of ring gear the standard valve for R180 and R180A models is 0.0031 in. and 0.0020 in. for the R200A model.

42. Check tooth contact pattern as necessary.

43. Install rear cover and gasket on all models. Install extension tube and differential side shaft assembly on R180A and R200A models. Install the side flange in correct position on R180 model.

Rear Drive Axle

Inspection Before Disassembly

1. Position the removed differential carrier assembly in a suitable holding fixture.

2. Check total preload by turning drive pinion in booth directions several times to set bearings. Using torque wrench or equivalent measure valve. The standard valve for model H190A and C220 models is 10–19 inch lbs. of turning torque. The standard valve for H233B model is 8.7–17.4 inch lbs. of turning torque.

3. Check backlash of ring gear with a dial indicator at several points. The standard valve for models H190A and C220 ring gear to drive pinion backlash is 0.0051–0.0071 in. and 0.0059–0.0079 in. for H233B model.

4. Check runout of ring gear with a dial indicator. The standard valve is 0.0031 in. of runout for all models.

5. Using suitable equipment check differential tooth contact for correct pattern as follows:

a. Thoroughly clean ring gear and drive pinion teeth.

b. Sparingly apply red lead or equivalent to 3 or 4 teeth of ring gear drive side.

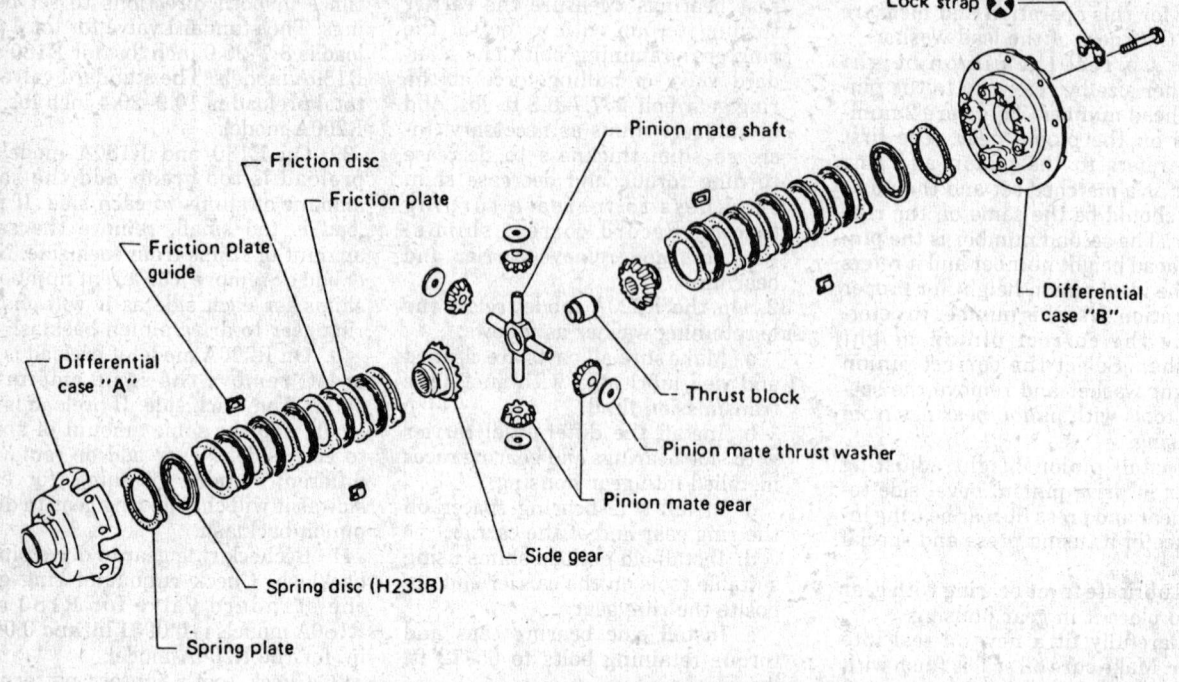

Exploded view of limited slip differential assembly—Nissan

c. Hold companion flange steady by hand and rotate the ring gear in both directions.

d. Interrupt contact marks left on the ring gear assembly. Gear tooth contact pattern check is necessary to verify correct relationship between ring gear and drive pinion.

6. On all models, check backlash of side gear. Using a thickness gauge, measure clearance between side gear thrust washer and differential case. The standard valve is 0.039–0.0079 in. of side gear clearance.

Disassembly and assembly

1. Position the removed differential carrier assembly in a suitable holding fixture.

2. Matchmark side bearing caps and shims if so equipped to ensure that they are replaced in proper position during reassembly.

3. On H233B model, remove the side lock fingers and side bearing caps. Remove side bearings with suitable tools.

4. Remove the differential case from the final drive housing. Mark all parts for correct installation.

5. Attach a flange holding wrench or equivalent to hold flange from turning. Remove the drive pinion nut.

6. Remove the companion flange using suitable puller.

7. Remove the drive pinion together with pinion rear bearing inner cone, drive pinion bearing spacer and pinion bearing adjusting washer.

8. Remove front oil seal and pinion front bearing inner cone.

9. Remove the side oil seals as required.

10. Remove the pinion front and rear bearing outer races with brass drift from housing.

11. Remove the pinion rear bearing inner cone and drive pinion adjusting washer. Remove rear bearing with suitable press and separator plate.

12. Remove side bearing inner race assembly with a puller from the differential carrier. Mark all parts for correct installation.

13. Remove ring gear by spreading out lock straps if so equipped and loosening ring gear bolts in a criss-cross pattern. Remove all retaining bolts.

14. Tap ring gear off gear case using a soft hammer or equivalent. Tap evenly all around to keep ring gear from binding.

15. On C200 and H233B models, separate the left and right differential case halves.

16. On R190A model, drive out pinion mate shaft lock pin with drift pin from ring gear side.

17. Draw out pinion mate shaft and thrust block if so equipped. Rotate pinion mate gears out of case and remove side gears and thrust washers. Put marks on gears and thrust washers so that they can be reinstalled in their original positions from which they were removed.

18. Clean disassembled parts completely. Repair or replace any damaged or faulty parts. If replacing drive pinion or ring gear, replace with a new hypoid gear as a set.

20. On C200 and H233B models, assemble the differential carrier as follows:

a. Measure the clearance between the side gear thrust washer and differential case.

b. The standard valve for the clearance between side gear thrust washer and differential case is 0.0039–0.0079 in. The clearance can be adjusted with side gear thrust washer.

c. Apply gear oil to gear tooth surfaces and thrust surfaces and check to see they turn properly.

d. Install the left and right differential case halves torque retaining bolts to 40–47 ft. lbs. Position the differential case on the ring gear. Apply Locktite or equivalent and tighten ring gear retaining bolts to (criss-cross fashion) 98–112 ft. lbs. bend up lock straps if so equipped.

e. Press fit side bearings on the differential case using suitable press and tools.

21. On R190A model, assemble the differential carrier as follows:

a. Install pinion mate gears, side gears and thrust washers into differential case.

b. Fit pinion mate shaft and thrust block.

c. Adjust clearance between rear face of side gear and thrust washer by selecting side gear thrust washer.

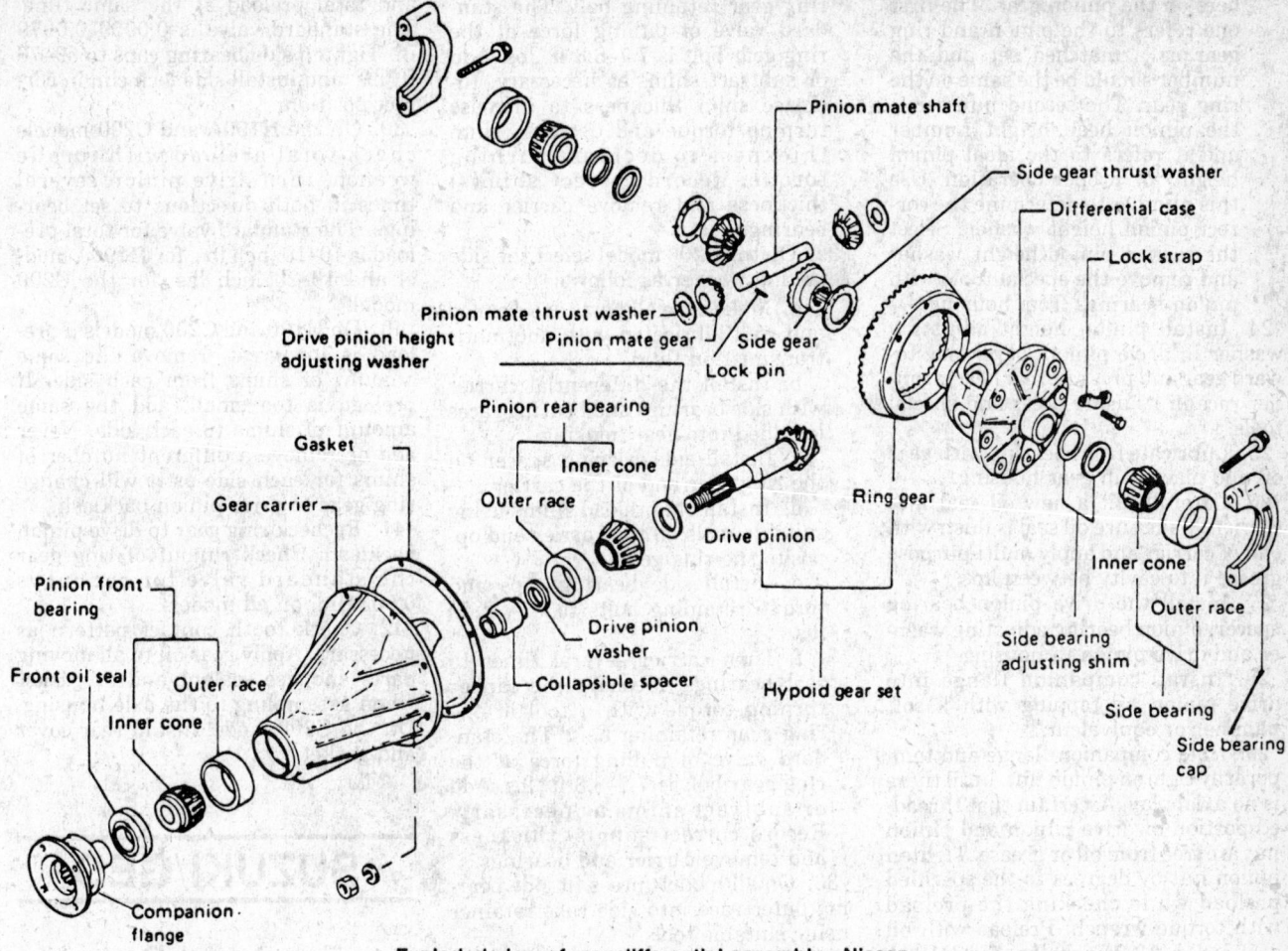

Exploded view of rear differential assembly—Nissan

The standard valve for clearance on R190A model is 0.0039–0.0079 in. specification.

d. Install pinion mate shaft lock pin using a punch. Make sure lock pin is flush with case.

e. Apply gear oil to gear tooth surfaces and thrust surfaces and check to see they turn properly.

f. Place ring gear on differential case and install new lock straps and bolts. Tighten bolts to 98–112 ft. lbs. in a criss-cross fashion, lightly tapping bolt head with a hammer. Then bend up lock straps to lock the bolts in place.

g. Press fit front and rear bearing outer races using suitable tools. Make sure that the bearing races are seated squarely in their respective bores.

22. Press fit pinion front and rear bearing outer races into housing with suitable tools on all models as necessary.

23. On all models select pinion bearing adjusting washer and drive pinion bearing spacer as follows:

a. Make sure all parts are clean

and that bearings are well lubricate. Assemble the pinion gear bearings into the pinion preload shim selector tool J–34309 or equivalent. Make sure that the front and rear bearings are in the correct position on the special tool (rear bearing first).

b. Install the pinion preload shim selector tool or equivalent into the final drive axle housing. Make sure that the pinion height gauge will make a full turn and tighten the 2 sections by hand. Turn the special tool with bearings installed on it several times to seat the bearings.

c. Measure the turning torque at the end of the tool with a torque wrench. The standard valve for the H190A and C200 models is 8.7–11.3 inch lbs. The standard valve for the R233B model is 3.5–7.8 inch lbs. The standard valve for the R180 model is 7.8–14.8. inch lbs.

d. Place correct pinion height adapter (each model has different pinion height adapter) onto the gauge plate and tighten it by hand.

e. Insert the pinion bearing adjusting spacer squarely into the re-

cessed portion of the correct special tool. Select ther correct thickness of the pinion bearing preload adjusting washer using the distance between spacer and gauge anvil tool. The exact measurement you get with the distance between spacer and special tool is the thickness of the adjusting washer required.

f. Position the side bearing discs tool J–25269–4 and arbor firmly in the side bearing bores and torque the side bearing cap bolts to 36–43 ft. lbs. on the H190A model, 65–72 ft. lbs. on the C200 model and 69–76 ft. lbs. on the H233B model.

g. Select the correct standard pinion height adjusting washer thickness using a standard valve plus the distance between pinion height adapter tool and the arbor. The standard valve for the C200 is 0.138 in. The standard valve for the H190A is 0. The standard valve for the H233B is either 0.098, 0.118 or 0.138 in. specification.

h. Correct the pinion height washer size by referring to the pinion head number. There are 2 numbers on the pinion gear. The first one refers to the pinion and ring gear as a matched set and the number should be the same on the ring gear. The second number is the pinion head height number and it refers to the ideal pinion height for proper operation. Use this number to determine the correct pinion height washer. Select the correct pinion height washer and remove the special tools with pinion bearings from housing.

24. Install pinion height adjusting washer in drive pinion, bevel side toward gear and press fit rear bearing inner race in it, using press and special tools.

25. Lubricate front bearing with gear oil and place it in gear housing.

26. Carefully fit a new oil seal into carrier. Make sure oil seal is flush with end of carrier and apply multi-purpose grease into cavity between lips.

27. Install the drive pinion bearing spacer, pinion bearing adjusting washer and drive pinion in housing.

28. Install companion flange into drive pinion by tapping with a soft hammer or equivalent.

29. Hold companion flange and temporarily tighten pinion nut, until there is no axial play. Ascertain that threaded portion of drive pinion and pinion nut are free from oil or grease. Tighten pinion nut by degrees to the specified preload while checking the preload with torque wrench. Preload with oil seal is 9.5–13.9 inch lbs. for H190A model and 9.5–14.8 inch lbs. for the C200 model. The preload with oil seal is 4.3–8.7 inch lbs. for H233B model.

30. When checking preload, turn drive pinion in both directions several times set bearing rollers. After preload is reached torque pinion nut t0 94–217 ft. lbs. for H190A and C200 models. Torque the pinion nut to 145–181 ft. lbs. on H233B model. Preload and final pinion nut torque must both be attained for correct operation of axle unit.

31. On the H190A model select the side retaining washer as follows:

a. Make sure all parts are cleaned and well lubricated with auotmatic transmission fluid.

b. Remove the side bearings with suitable tools. Install side bearing old shims on the carrier side away from the ring gear.

c. Reinstall carrier side bearings using suitable press.

d. Install the assembly into the housing. Install side bearing caps and torque retaining bolts to 36–43 ft. lbs.

e. Turn carrier several times to seat bearings. Measure the carrier turning torque with a tool at the ring gear retaining bolt. The standard valve of pulling force at the ring gear bolt is 7.7–8.8 ft. lbs. Add or subtract shims as necessary. Increase shim thickness to increase turning torque and decrease shim thickness to decrease turning torque. Record correct shim(s) thickness and remove carrier and bearings.

32. On the C200 model select the side retaining washer as follows:

a. Make sure all parts are cleaned and well lubricated with auotmatic transmission fluid.

b. Install the differential carrier with side bearings and bearing races installed into gear housing.

c. Install side bearing spacer on the ring gear end of the carrier.

d. Install old preload shims using suitable tools on the carrier end opposite the ring gear.

e. Install side bearing caps and torque retaining bolts to 65–72 ft. lbs.

f. Turn carrier several times to seat bearings. Measure the carrier turning torque with a tool at the ring gear retaining bolt. The standard valve of pulling force at the ring gear bolt is 7.7–8.8 ft. lbs. Add or subtract shims as necessary. Record correct shim(s) thickness and remove carrier and bearings.

33. On all models press fit side bearing outer race into side case retainer using suitable tools.

34. On H190A model install differential case assembly with side bearings outer races into housing. On C200 model insert left and right side bearing adjusting washers between side bearing and carrier. Install differential assembly and side bearing spacer using suitable tools.

35. On H233B model install the side bearing adjusters on gear carrier, screw adjusters lightly at this stage.

36. Align mark on bearing caps and install bearing caps torque to 36–43 ft. lbs. on H190A model and 65–72 ft. lbs. on C200 model. On the H223 model do not fully tighten at this point to allow further tightening of side bearing adjusters.

37. On the H190A and C200 models measure the ring gear to drive pinion backlash with a dial indicator. The standard valve is 0.0051–0.0071 in. If backlash is too small, decrease thickness of right shim and increase thickness of left shim by the same amount. If backlash is to great reverse the above procedure. Never change the toatal amount of shims as it will change the bearing preload.

38. On the H233B model tighten both right and left bearing adjusters alternately and measure the backlash and total preload at the same time. The standard valve is 0.0059–0.0079 in. Tighten side bearing caps to 69–76 ft. lbs. and install side lock clip in correct position.

39. On the H190A and C200 models check total preload with torque wrench, turn drive pinion several times in both directions to set bearings. The standard valve for total preload is 10–19 inch lbs. for H190A model and 10–20 inch lbs. for the C200 model.

40. On H190 and C200 models if preload is too great, remove the same amount of shims from each side. If preload is too small, add the same amount of shims to each side. Never add or remove a different number of shims for each side as it will change ring gear to drive pinion backlash.

41. Recheck ring gear to drive pinion backlash. Check runout of ring gear the standard valve for runout is 0.0031 in. on all models.

42. Check tooth contact pattern as necessary. Apply gear oil to all moving parts and use sealant and or gasket when assembling to the axle housing. On the C200 model install rear cover and gasket.

SUZUKI/GEO

Front and Rear Drive Axles

Disassembly and Assembly

1. Position the removed differential carrier assembly in a suitable holding fixture.

2. Mark side differential side bearing caps for correct installation.

3. Remove the bearing side lock plates and bearing caps. Remove bearing adjusters and outer races. Remove ring gear complete assembly.

4. Remove the flange nut using suitable tools to hold flange.

5. Remove the pinion shaft assembly from the carrier. Remove the pinion shaft oil seal using suitable tools.

6. Remove the outer pinion bearing from the differential carrier.

7. Remove the outer and inner pinion bearing races from the differential carrier.

8. Remove the ring gear bolts and remove the ring gear from the differential assembly.

9. Drive out spring pin (3 spring pins are used on 4 pinion type rear assembly) and remove the differential side gears, selective shims, pinion gears, thrust washers and cross shafts.

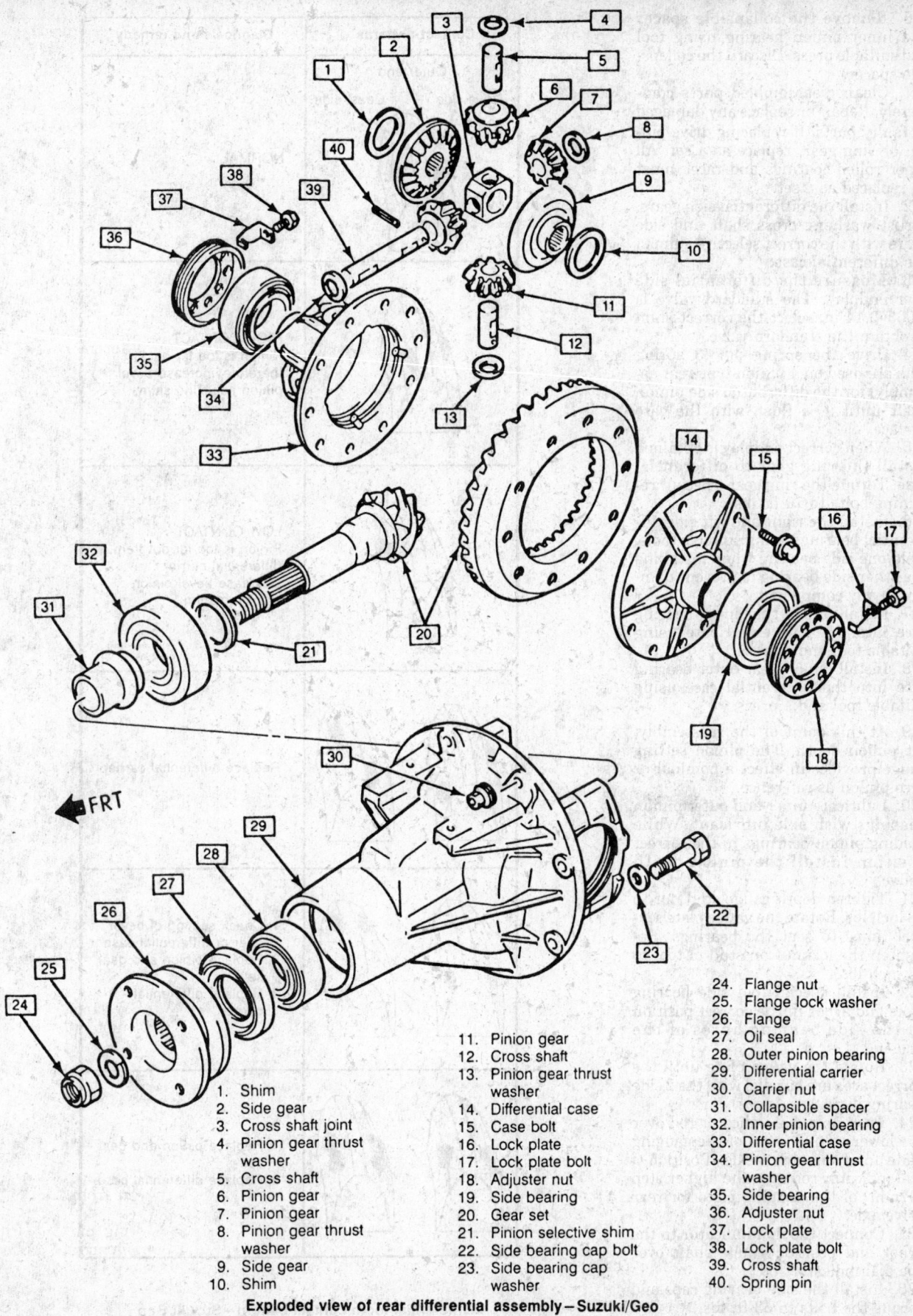

1. Shim
2. Side gear
3. Cross shaft joint
4. Pinion gear thrust washer
5. Cross shaft
6. Pinion gear
7. Pinion gear
8. Pinion gear thrust washer
9. Side gear
10. Shim
11. Pinion gear
12. Cross shaft
13. Pinion gear thrust washer
14. Differential case
15. Case bolt
16. Lock plate
17. Lock plate bolt
18. Adjuster nut
19. Side bearing
20. Gear set
21. Pinion selective shim
22. Side bearing cap bolt
23. Side bearing cap washer
24. Flange nut
25. Flange lock washer
26. Flange
27. Oil seal
28. Outer pinion bearing
29. Differential carrier
30. Carrier nut
31. Collapsible spacer
32. Inner pinion bearing
33. Differential case
34. Pinion gear thrust washer
35. Side bearing
36. Adjuster nut
37. Lock plate
38. Lock plate bolt
39. Cross shaft
40. Spring pin

Exploded view of rear differential assembly—Suzuki/Geo

10. Remove the collapsible spacer and inner pinion bearing using tool and suitable press. Discard the collapsible spacer.

11. Clean disassembled parts completely. Repair or replace any damaged or faulty parts. If replacing drive pinion or ring gear, replace as a set. All taper roller bearings and races must be replaced as a set.

12. Install the differential side gears, thrust washers, cross shaft and side gears with the correct selective shim in the differential case.

13. Measure the differential side gear endplay. The standard valve is 0.005–0.14 in. select the correct shim to obtain the standard valve.

14. Drive the spring pin (3 spring pins are used on 4 pinion type rear assembly) for the differential side pinion shaft until it is flush with the case surface.

15. When correct endplay is obtained install the ring gear to differential case. Torque the ring gear (special) retaining bolts to 70 ft. lbs.

16. Install the right and left side differential bearings using suitable tools. Hold one side bearing when installing the other side bearing to prevent damage to any component.

17. Install pinion gear inner bearing race into the differential case using suitable tool and a press.

18. Install pinion gear outer bearing race into the differential case using suitable tool and a press.

19. At this point of the reassembly, set pinion depth. The pinion setting gauge provides in effect a nominal or zero pinion as reference.

20. Lubricate inner and outer pinion bearings with axle lubricant. While holding pinion bearings in the correct position install the pinion depth gauge.

21. Tighten depth gauge stud nut to 18 inch lbs. Rotate the gauge plate several times to seat the bearings. Retighten the locknut on stud of tool to 18 inch lbs.

22. Install special tool side bearing discs and arbor in the correct position in the side bearing bores of the housing.

23. Rotate the gauge plate until the correct area are parallel with the 2 side bearing discs.

24. Position a dial indicator rod over the lower step (94mm) of the gauging plate for front drive axle. Position a dial indicator rod over the higher step (97mm) of the gauging plate for rear drive axle.

25. Connect the dial indicator to the arbor and position gauge shaft over the dial indicator rod.

26. Install the side bearing caps and torque the bolts to 63 ft. lbs.

Contact patterns	Diagnosis and remedy
Outer end — Drive side — Coast side — Inner end	NORMAL
	HIGH CONTACT Pinion is too far back, therefore, increase bevel pinion adjusting shim.
	LOW CONTACT Pinion is too far out from differential carrier. Decrease bevel pinion adjusting shim.
	Replace differential carrier.
	1. Check seating of bevel gear or differential case. 2. Replace pinion and gear set. 3. Replace differential carrier.
	1. Replace pinion and gear set. 2. Replace differential case.

Differential tooth contact pattern—Suzuki/Geo

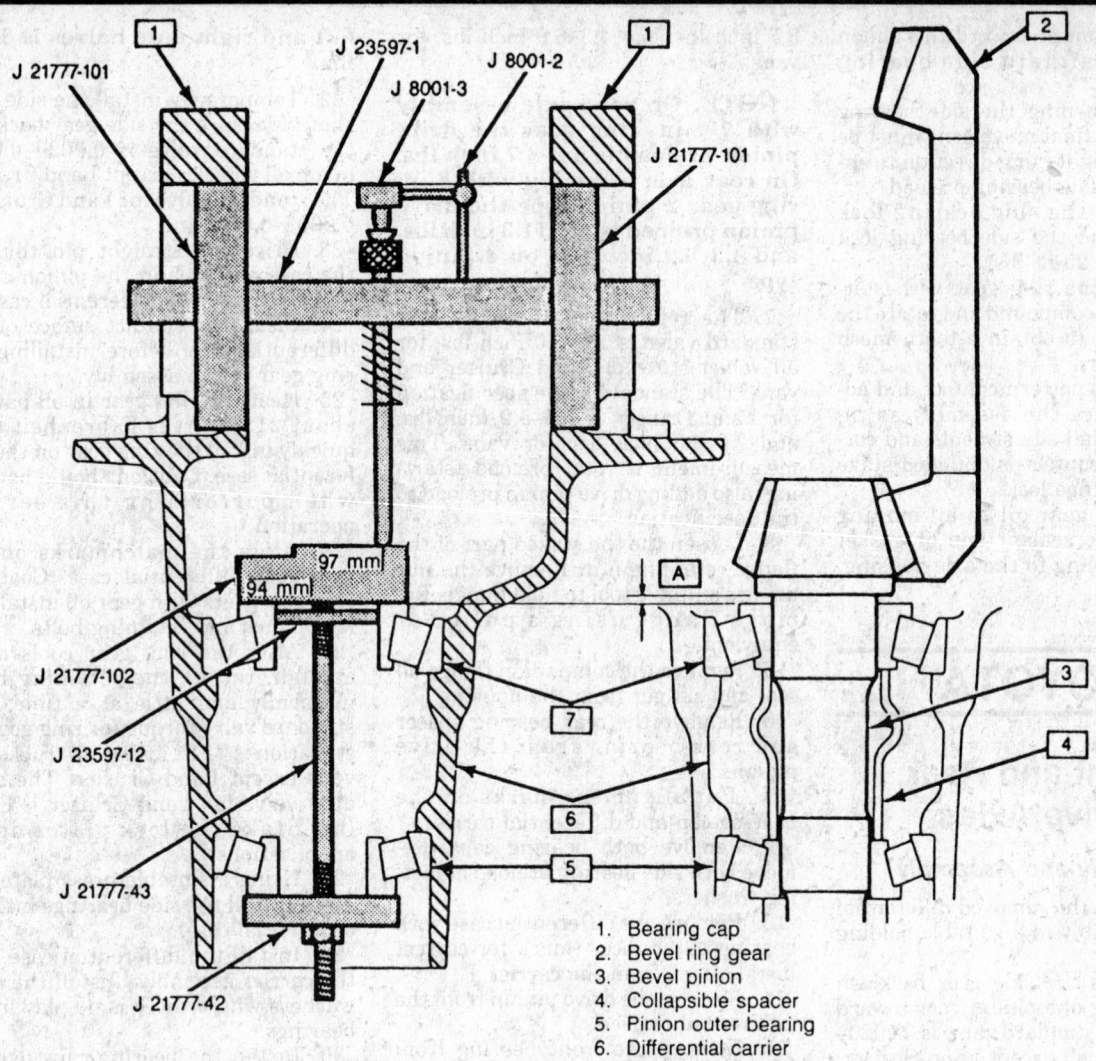

1. Bearing cap
2. Bevel ring gear
3. Bevel pinion
4. Collapsible spacer
5. Pinion outer bearing
6. Differential carrier
7. Pinion inner bearing

Pinion depth measurement

27. Position the dial indicator to the zero setting. Slowly rotate the arbor on the lower step (94mm) front drive axle or higher step (97mm) rear driver axle of the gauge plate to determine the point of the greatest deflection. At this point reset the dial indicator to zero setting.

28. Slowly rotate the arbor until the dial indicator is no longer on the gauge plate.

29. Record the dial indicator reading. This reading indicates the selective shim required for the correct pinion depth. The selective shims range from 0.012–0.050 in.

30. Install the correct pinion shim on the pinion gear shaft.

31. Assembly the inner pinion bearing on the pinion gear shaft using suitable press plate and press.

32. Install new collapsible spacer on the pinion gear shaft.

33. Install the outer pinion bearing in the differential carrier using suitable tools.

34. Apply lubricant to the oil seal lip. Install the new pinion oil seal using a suitable seal installer or equivalent.

35. Install pinion flange, pinion washer and nut. Hold the pinion flange (suitable flange holding tool) while frequently rotating pinion to seal pinion bearings. Tighten pinion flange nut (little at a time) until endplay is taken up. Measure preload with torque wrench the standard valve for all applications is 11 inch lbs. of starting torque not rotating torque. Exceeding preload specification will compress the collapsible spacer.

36. After final preload has be reached rotate the pinion several times to make sure the bearings have been seated. Recheck the preload specification.

37. Install the ring gear and differential carrier assembly, side bearing outer races, bearing adjusters and bearing caps to housing. Align all identification marks.

38. Rotate pinion gear to ensure the side bearings seat. Tighten bearing caps 15 ft. lbs. on the front and rear drive axle assembly.

39. Using suitable tool turn the bearing adjusters to push side bearings lightly from outside so that outer races are in contact with inner races. Apply a small amount of lubricant to bearings. With dial indicator installed to differential carrier and carrier at 0.0 backlash (ring gear fully engaged into pinion gear) set or check preload. The differential side bearing preload is 0.002–0.006 in.

40. Rotate pinion gear to ensure the side bearings seat. Tighten bearing caps to 44 ft. lbs. on the front drive axle assembly and 63 ft. lbs. on the rear drive axle assembly.

41. Check backlash using a dial indicator installed to the differential ring gear. The standard valve for all applications is 0.008–0.015 in. If adjustment is necessary loosen one adjuster nut and tighten the opposite nut an equal amount. This will move the ring

gear away from or toward the pinion gear and maintain side bearing preload.

42. When turning the side bearing adjusters the final movement must be made in the tightening direction to ensure correct side bearing preload.

43. Install the side bearing lock plates. Torque the side bearing lock plate bolts to 25 ft. lbs.

44. Paint the ring gear with suitable marking compound and rotate the pinion flange to obtain a tooth mesh contact pattern.

45. Inspect pattern contact and adjust or service the assembly as required. After all adjustments and correct pattern contact is obtained stake the pinion flange locknut.

46. Apply gear oil to all moving parts and use sealant and or gasket when assembling to the axle housing.

TOYOTA

Front and Rear Drive Axles

Disassembly and Assembly

1. Position the removed differential carrier assembly in a suitable holding fixture.

2. Measure the side gear backlash while holding one pinion gear toward the case. The standard valve is 0.0020–0.0079 in. on all except Land Cruiser. The standard valve for Land Cruiser is 0.0008–0.0079 in.

3. Check ring gear runout. Maximum runout is 0.0028 in. on all except Land Criuser. The standard valve for Land Cruiser is 0.0039 in.

NOTE: On rear axle assembly with 7½ in. ring gear the runout is 0.0028 in. On rear axle assembly with 8 in. ring gear the runout is 0.0039 in.

4. Hold drive pinion flange check ring gear backlash (several different locations). The standard valve is 0.0051–0.0071 in. on all except Land Criuser. The standard valve for Land Cruiser is 0.0059–0.0079 in.

5. Inspect tooth contact between ring gear and drive pinion using suitable marking compound.

6. Using a torque wrench measure the (starting) preload of the backlash between the drive pinion and ring gear. The standard valve is 5.2–8.7 inch lbs. for all vehicles except Land Cruiser and vans. The standard valve specification for Land Cruiser is 6.1–

8.7 inch lbs. and 4.3–6.9 inch lbs. for vans.

NOTE: On rear axle assembly with 7½ in. ring gear the drive pinion preload is 5.2–8.7 inch lbs. On rear axle assembly with 8 in. ring gear 2 pinion type the drive pinion preload is 7.8–11.3 inch lbs. and 4.3–6.9 inch lbs. on 4 pinion type.

7. The total preload specification standard valve is 1.6–3.0 inch lbs. for all vehicles except Land Cruiser and vans. The standard valve specification for Land Cruiser is 3.5–5.2 inch lbs. and 2.6–4.3 inch lbs. for vans. This measurement is total preload (starting) also adding drive pinion preload to the specification.

8. Loosen the the staked part of the flange retaining nut. Remove the nut using a suitable tool to hold flange. Remove flange using a puller or equivlent.

9. Remove the companion flange oil seal and slinger from the housing.

10. Remove the rear bearing spacer and rear bearing from the drive pinion.

11. Put alignment marks on the bearing cap and differential carrier.

12. Remove both bearing caps. Remove both side bearing preload adjusting washers.

13. Remove the differential case with bearing outer races (mark for correct installation) from the carrier.

14. Remove the drive pinion from the differential carrier.

15. Remove the front bearing from the drive pinion using a suitable tool and press.

16. Remove the drive pinion front and rear bearing outer races from the housing.

17. Remove the side bearing from the differential case using suitable puller. Puller jaws must align with notches in the differential case.

18. Remove the ring gear retaining bolts and lock plates.

19. Place matchmarks on the ring gear then using a plastic tool or equivalent separate it from the differential case.

20. Drive out the straight pin and remove the pinion shaft, pinion gears, side gears and thrust washers.

21. Assemble the differential case by installing the correct (check backlash with dial indicator) thrust washer to side gears. Try to select washers of the same thickness for both sides. Install the thrust washer and side gears in the differrential case.

NOTE: On some rear axle assemblies the differential case splits. The retaining torque for

left and right case halves is 35 ft. lbs.

22. Temporarily install the side gear shaft. Measure the side gear backlash the standard valve is 0.0020–0.0079 in. on all vehicles except Land Cruiser. The standard valve for Land Cruiser is 0.0008–0.0079 in.

23. Drive the straight pin through the case and hole in the pinion shaft. Stake the pin and differential case.

24. Clean the contact suface of the differential case before installing the ring gear to the assembly.

25. Heat the ring gear in oil bath to about 212 degrees Fahrenheit then quickly install the ring gear on the differentiai case. Caution should be used when performing this service operation.

26. Align the matchmarks on the ring gear. differential case. Coat the ring gear bolts with gear oil install the lock plates and retaining bolts.

27. After the ring gear cools down enough, torque the retaining bolts uniformly and little at a time. The standard valve torque for ring gear installation is 71 ft. lbs. on all trucks and vans except Land Crusier. The standard valve for Land Cruiser is 81 ft. lbs. Stake the lock plates on all appplications.

28. Using a suitable press plate and press install the side bearings into the differential case.

29. Install the differential case onto the carrier assembly. Install the plate washers where there is no play in the bearings.

30. Install the bearing caps using a dial indicator measure the runout of the ring gear. The standard valve for runout is 0.0028 in. on all except Land Criuser. The standard valve for Land Criuser is 0.0039 in. after runout inspection is performed remove the assembly from the carrier.

NOTE: On rear axle assembly with 7½ in. ring gear the runout is 0.0028 in. On rear axle assembly with 8 in. ring gear the runout is 0.0039 in.

31. Install the front and rear pinion bearing outer races to the carrier.

32. Install the drive pinion assembly (rear bearing) and front bearing in housing. Install the companion flange with a suitable tool to hold flange in place.

33. Rotate the flange several times to seat the bearing. Adjust the drive pinion preload by tightening the flange nut. Use a torque wrench to measure the preload. The standard valve for preload (starting torque) is 10.4–16.5 inch lbs. on trucks. The standard valve for preload (starting torque) is 8.7–13.9 inch lbs. on vans. The standard

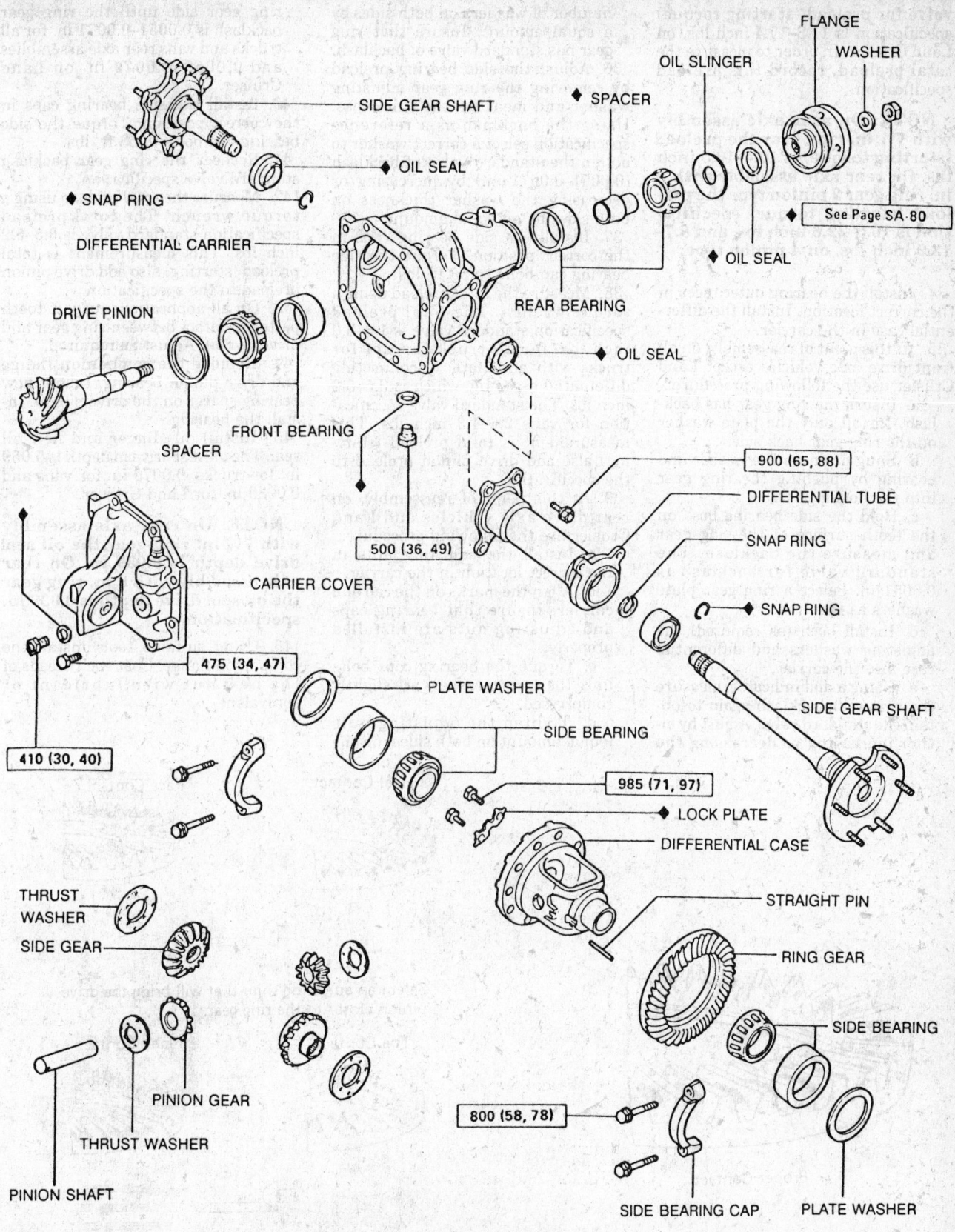

SIDE GEAR SHAFT

◆ OIL SEAL

◆ SNAP RING

DIFFERENTIAL CARRIER

DRIVE PINION

FRONT BEARING

SPACER

CARRIER COVER

475 (34, 47)

410 (30, 40)

PLATE WASHER

SIDE BEARING

THRUST WASHER

SIDE GEAR

PINION GEAR

THRUST WASHER

PINION SHAFT

SPACER

OIL SLINGER

FLANGE

WASHER

◆ See Page SA-80

◆ OIL SEAL

REAR BEARING

◆ OIL SEAL

500 (36, 49)

900 (65, 88)

DIFFERENTIAL TUBE

SNAP RING

◆ SNAP RING

SIDE GEAR SHAFT

985 (71, 97)

◆ LOCK PLATE

DIFFERENTIAL CASE

STRAIGHT PIN

RING GEAR

SIDE BEARING

800 (58, 78)

SIDE BEARING CAP

PLATE WASHER

Exploded view of front differential assembly—Toyota

valve for preload (starting torque) specification is 11.3–17.4 inch lbs. on Land Crusier. In order to measure the total preload, record this preload specification.

NOTE: On rear axle assembly with 7½ in. ring gear the preload (starting torque) is 10.4–16.5 inch lbs. On rear axle assembly with 8 in. ring gear 2 pinion type the preload (starting torque) specification is 16.5–22.6 inch lbs. and 8.7–13.9 inch lbs. on 4 pinion type.

34. Install the bearing outer races in the correct location. Install the differential case in the carrier.

35. At this point of reassembly, on all front drive axle vehicles except Land Crusier use the following procedure:

   a. Insure the ring gear has backlash. Install only the plate washer on the ring gear back side.

   b. Snug down on the washer and bearing by pushing the ring gear into the housing.

   c. Hold the side bearing boss on the teeth surface of the ring gear and measure the backlash. The standard valve for backlash is 0.0051 in. Select a ring gear plate washers as required.

   d. Install both (as required) the adjusting washers and differential case into the carrier.

   e. Using a dial indicator, measure the ring gear backlash again to obtain the standard valve. Adjust by either increasing or decreasing the number of washers on both sides by a equal amount. Insure that ring gear has standard valve of backlash.

36. Adjust the side bearing preload by removing the ring gear adjusting washer and measure the thickness. Using the backlash as a reference specification select a correct washer to obtain the standard valve of backlash (0.0051–0.0071 in.) by increasing or decreasing the washer thickness on both sides by an equal amount.

37. Install the side bearing caps in the correct position. Torque the side bearing cap bolts to 58 ft. lbs.

38. Measure the total preload using a torque wrench. The total preload specification standard valve is 1.6–3.0 inch lbs. for all trucks except for trucks with automatic disconnecting differential assembly which is 3.5–5.2 inch lbs. The standard valve specification for vans 2.6–4.3 inch lbs. This measurement is total preload (starting) also add drive pinion preload to the specification.

39. At this point of reassembly, on rear drive axle vehicles and Land Crusier use the following procedure:

   a. Install the adjusting nuts in the correct location in the carrier.

   b. Align the marks on the cap and carrier. Insure that bearing caps and adjusting nuts are installed properly.

   c. Torque the bearing caps bolts until the spring washers are slightly compressed.

   d. Tighten the adjusting nuts (equal amount on both sides) on the ring gear side until the ring gear backlash is 0.0051–0.0071 in. for all trucks and vans rear axle assemblies and 0.0059–0.0079 in. on Land Cruiser.

40. Install the side bearing caps in the correct position. Torque the side bearing cap bolts to 58 ft. lbs.

41. Recheck the ring gear backlash standard valve specification.

42. Measure the total preload using a torque wrench. The total preload specification standard valve is 3.5–5.2 inch lbs. This measurement is total preload (starting) also add drive pinion preload to the specification.

43. On all applications inspect tooth pattern contact between ring gear and drive pinion. Adjust as required.

44. Remove the companion flange and drive pinion bearing. Install new bearing spacer on the drive pinion. Install the bearing.

45. Install oil slinger and new oil seal. The oil seal drive in depth is 0.059 in. for trucks, 0.0079 in. for vans and 0.0039 in. for Land Cruiser.

NOTE: On rear axle assembly with 7½ in. ring gear the oil seal drive depth is 0.059 in. On rear axle assembly with 8 in. ring gear the oil seal drive depth is 0.039 in. specification.

46. Using suitable tools install the companion flange. Coat the threads of the new nut with lubricant or equivalent.

Heel Contact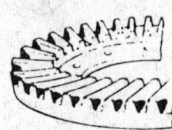

Face Contact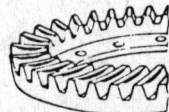

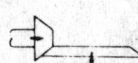

Select an adjusting shim that will bring the drive pinion closer to the ring gear.

Toe Contact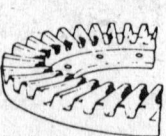

Flank Contact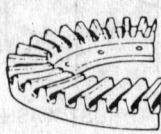

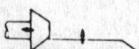

Select an adjusting shim that will shift the drive pinion away from the ring gear.

Proper Contact

Differential tooth contact pattern—Toyota